A MASTER LIST

OF NONSTELLAR OPTICAL ASTRONOMICAL OBJECTS

A Master List Of Nonstellar Optical Astronomical Objects

COMPILED BY ROBERT S. DIXON
AND GEORGE SONNEBORN

Ohio State University Press

Publication of this *Master List of Nonstellar Optical Astronomical Objects* has been supported by the National Science Foundation under Grant No. AST 7824455.

Any opinions, findings, and conclusions or recommendations expressed in this publication are those of the authors and do not necessarily reflect the views of the National Science Foundation.

Library of Congress Cataloguing in Publication Data

Dixon, Robert S
A master list of nonstellar optical astronomical objects.

Bibliography: p.
1. Astronomy—Catalogs. I. Sonneborn, George, joint author. II. Title.
QB65.D56 523 79-27627
ISBN 0-8142-0250-0

Contents

Preface

This work is a compendium of all known catalogues of nonstellar objects. It was constructed by entering all the catalogues onto punched cards, converting them to a standard format, and combining them in right ascension order. Every object from every catalogue is included, so that multiple listings of the same object may be intercompared. The data given for each object includes 1950.0 position, angular diameter, magnitude, and description.

Approximately 185,000 listings from 270 catalogues are included. The characteristics of each catalogue are described in table 3 (see page 27), and complete bibliographical references are provided at the end. The compendium itself follows immediately after this explanatory text. The acronym MOL is used when referring to this work.

Acknowledgments

This work was supported principally by grants from the National Science Foundation, with additional financial support being gratefully received from Professor John D. Kraus. The work was greatly facilitated by the excellent Astronomy Library located at Perkins Observatory. A special acknowledgement goes to Mirjana Gearhart, who very capably and diligently assisted with many of the foreign-language papers, library searches, and preparation of the bibliographical material. Patricia Pitts was also very helpful during the library search phase of this work. To Peggy Shaner, whose dedication to detail and thoroughness in carrying this project through to completion sets an example that will not soon be equaled, we offer our most heartfelt gratitude. To Sanjoy Ghosh, who maintained his composure while creating and filing the mountains of photocopies and computer cards used for this project, we offer our congratulations. Judith Dixon was most helpful in typing the manuscript.

Those who assisted in the keypunching include Diane Cole, Virginia Howlett, and the staff and trainees of the Lebanon Correctional Institute. Clifton Cavanaugh gathered and prepared the material used in Appendix A.

We thank the following for sending us data in computer-readable form: M. J. E. Peebles, M. G. Hauser, C. T. Kowal, W. L.W. Sargent, and F. Ghigo. Finally, we acknowledge the assistance of the Instruction and Research Computer Center of the Ohio State University for providing reliable and accessible computer facilities.

Introduction

The field of cataloguing is unglamorous, generally thankless, and has been described as "scientifically unexciting," yet catalogues are probably referred to more frequently than any other portion of the scientific literature. The large astronomical catalogues are seldom included in the bibliographies of published papers, yet many of those same papers refer to astronomical objects by their catalogue name. Thus the criterion of citation frequency does not accurately measure the significance of catalogues.

A catalogue is a means of disseminating knowledge in a systematic manner. A catalogue is synergistic because as a whole it is more valuable than the individual entities it contains. As a whole, it is amenable to statistical investigations that cannot be carried out on individual members.

The MOL is one step beyond a catalogue; it is a collection of catalogues. It too is a means of disseminating knowledge. In fact, it has been designed with the express purpose of disseminating as widely as possible the knowledge contained in some 270 catalogues. It too is synergistic. By having all of these catalogues together in one place, with common notation and units, they may be intercompared much more easily than in their original separate forms.

It is often a source of amazement to persons outside astronomy to learn how disorganized the field is in some respects. Let such a person point to a fuzzy object on an astronomical photograph and ask the seemingly simple question "What object is that?" Then try to answer him in a definitive way, or try to refer him to a reference work that will answer his question. This simple example is symptomatic of a larger problem—namely, that astronomers place greater emphasis on acquiring new knowledge than on organizing and disseminating that knowledge. This emphasis works to the detriment of non-astronomers who need astronomical data, as well as to those astronomers who continue to rediscover the same objects (many examples can be found in this compendium). It is hoped that the MOL will help rebalance the emphasis scale, and if nothing else will provide an answer to the question "What object is that?"

During the preparation of this work many decisions had to be made regarding its form and content. Often these decisions pitted the idealism of including all possible data for all known objects against the practicalities of cost, time, and ease of use. We struck our balance by including all known objects, but limiting the data about each one to those quantities that are known for a large fraction of the objects. These quantities are position, angular diameter, and magnitude. Other quantities, such as velocity, distance, spectrum, and so on, are known for only a tiny fraction of the objects, and we could not justify providing for this data additional columns that would be blank most of the time. Nor could we justify decreasing the readability and ease of use of this work by compressing the primary data to make room for a general-purpose "miscellaneous information" column to contain the additional data. Instead, we have provided extensive references to the original catalogues, from which the reader may extract the additional data himself. In addition, we did include much of the additional data in our original computer card preparation (since it was little extra work to do so), and this data is available separately from the authors.

In many cases, the same physical object appears in several different catalogues. For several reasons, we have not attempted to combine these multiple listings into a single listing containing "preferred" or "averaged" data. Chief among these reasons is the fact that it is impossible to do so in many cases. The same object may be listed as a single entity in one catalogue, and as a number of independent components in another catalogue. Since the angular diameters of these objects are appreciable, there is often unresolvable ambiguity as to which object in one catalogue is the same as which object(s) in another catalogue. To study each case in detail would require inspection of all 185,000 listings on photographs such as the Palomar Sky Survey. Although such a study would be useful, it is beyond the means available at the present time. The preparation of the MOL has spanned nearly six years, and to delay its publication until such a study could be done is not warranted.

With regard to deriving preferred or averaged data values, we believe that such work should be carried out by experts in the study of the various types of objects included in the MOL. Since we cannot be specialists in all the astronomical subfields spanned by the MOL, it would be presumptuous for us to do so.

The positive reason for not combining the multiple listings is that the redundancy is useful. by intercomparing the different listings, the reader can gain a better understanding of the accuracy of each quantity. He is free to select and combine data as his own judgment dictates, knowing that all the originally catalogued objects have been included.

The MOL is of course heterogeneous, since the catalogues it contains have a wide range of characteristics. The MOL is not a complete statistical sample of anything. It is, however, a summing up of man's knowledge of the existence of nonstellar objects at this point in time. Since man's knowledge will be always heterogeneous and incomplete, so must be any compendium of that knowledge.

Historical Background

The first widely known catalogue of nonstellar astronomical objects was published by Messier in 1784. It contains the 103 "nebulous objects" known at that time, north of declination −35°. Messier's numbers remain in common usage today; for example, M31 is the large galaxy in Andromeda.

In 1864 Sir John Herschel published his *General Catalog of Nebulae,* containing 5,079 objects, based primarily on his own observations and those of his father, Sir William Herschel. The GC Catalog, as it was known, was later supplanted by the *New General Catalogue of Nebulae and Clusters of Stars,* published by Dreyer in 1888. Dreyer's catalogue contains 7,840 objects, and is a compendium of observations and discoveries by some 37 earlier authors. Dreyer later supplemented his original catalogue with two Index Catalogues published in 1895 and 1908, containing an additional 5,386 objects. The NGC and IC numbers assigned in these catalogues remain today the most commonly used designations for many objects.

Since that time, no general catalogues of nonstellar objects have been made. It should be noted that Lundmark announced his intention to do so in 1930, but apparently never completed the work. Nilson (1973) has given a detailed historical summary of work in this area. More recently, Sulentic and Tifft (1973) have published a *Revised New General Catalog* (RNGC), which provides modern data, but only for the original NGC objects. Corwin (1975) has announced his intention to produce a similar Revised Index Catalog.

The Catalogues of Today

The number of known nonstellar objects has increased by an order of magnitude since Dreyer's time, and their data are scattered throughout the astronomical literature, generally being catalogued in specialized lists of physically similar objects.

For some types of nonstellar objects there exist single, authoritative catalogues, which are themselves compendia of all previous catalogues of that type of object. An example of such a catalogue is the *Catalogue of Star Clusters and Associations* by Alter et al. (1970). Even in these cases, however, a catalogue may be limited in some way, such as to those objects in our own galaxy, as in the *Catalog of Galactic Planetary Nebulae,* by Perek and Kohoutek. For other types of nonstellar objects, no such compendium exists. For example, galaxies have been catalogued in various parts of the sky, by various authors, to various limiting magnitudes, or for certain of their characteristics.

The specialized catalogues have been published in a wide variety of journals (occasionally little-known ones), as books, and in various other forms. They contain in many cases all that is known about a specific object, in a compact concise arrangement. Each has its own distinctive symbolism, content, and format. To use such a catalogue intelligently usually requires a considerable preparatory period of familiarization, until all the symbols, quantities, and subsidiary tables are clearly understood. This is no burden to a specialist in the field, but it tends to discourage the nonspecialist from using such a catalogue as an occasional reference for needed facts. Due to the diverse locations and forms of the catalogues, the nonspecialist may well be unaware of the existence or location of the data he needs. Only rarely are the specialized catalogues available in a computer-readable form.

For these reasons, and since it has been nearly seventy years since a general catalogue of nonstellar objects has been prepared, the authors decided to prepare a compendium of all such objects that would be uniform, self-explanatory, easily used, and widely available.

Purposes of This Work

The intended purposes of this work are enumerated below.

Purpose 1. To serve as a pointer to the astronomical literature. This work is not intended to replace any of the original catalogues. Rather, it is intended to call attention to the existence of those catalogues, and to the data contained therein.

Purpose 2. To enable cross-referencing among different catalogues of the same kind of objects. The day has not yet (and may never) come when all objects are referred to by a single, universally used name. For objects with appreciable angular diameters (i.e., most nonstellar objects), the problem of cross-referencing is complex because a single object in one catalogue may be listed as several in another catalogue. Thus there is no unique one-to-one correspondence. (Marsalkova [1974] has dealt with this problem in a compilation of H II region data.) Each object is listed as many times in this work as it has been independently catalogued in the literature. For example, many galaxies are listed three times, as catalogued independently by Nilson, by Vorontsov-Velyaminov, and by Zwicky. This multiple listing allows easy intercomparison of the data from each of the original catalogues, and enables statistical studies of the systematic differences among each of the catalogues as a whole. These multiple listings are by no means completely duplicative, since no catalogue is in complete agreement with any other either as to numerical data values or as to the objects included.

By way of example, there is a faint blue object near $12^h20^m19^s$, $+13°46'$ that appears to have been catalogued independently by Rubin et al. (RMB 099), by Luyten (LB 04392), and at Asiago Observatory (A3 057). Similarly there is a galaxy near $10^h50^m00^s$, $+73°57'$ that has been catalogued twice by Zwicky (ZWG 334.004 and 333.062), by Nilson (UGC 05997), by Vorontsov-Velyaminov (MCG +12−10−089), by the Astronomer Royal (AR 25) who called it a nebula, which was the terminology used before galaxies were recognized as such, and it is also listed as RNGC 3403.

Purpose 3. To enable studies of coincidences between objects of nominally different types to be made, such as between blue stellar objects and galaxies. A few illustrative examples follow:

The galaxy near $10^h50^m12^s$, + $50°34^m$ is catalogued not only in the usual galaxy catalogues (MCG +80−20−037, ZWG 267.023, UGC 05998), but also has been singled out for its unusual spectral characteristics as a Markarian galaxy (MRK 156).

The double galaxy near $12^h19^m00^s$, + 14°53′ has been catalogued for both of its components in all the usual galaxy catalogues, as a double galaxy by Holmberg (HOLM 337 A and B) and by Karachentseva (KARA.72 332 A and B), as a peculiar galaxy by Schanberg (SCH 34), and as a faint blue object by Rubin et al (RMB 054).

It is suggested that the reader look up these various examples in the MOL as a means of becoming familiar with the search technique suggested below.

Purpose 4. To call attention to the known existence of each nonstellar object, and to present a moderate amount of information about each one. For many well-known objects data of far greater scope and precision exists elsewhere in the literature. Most of this information is accessible through the references given in this work, through those references given in the original catalogues, or through the *Science Citation Index* (published annually). This latter publication is very useful in that it enables one to proceed *forward* in time from an original reference, rather than backward only as is the usual case with bibliographical references. Another very useful general reference work is the bibliographical guide to astronomy and astrophysics by Kemp (1970).

It is impossible to include all known information about each object in any single reference work such as this. It is possible, however, to include a reasonable amount about every object. The information contained here is generally sufficient to decide whether further reference searching is needed.

Purpose 5. To make astronomical data dealing with nonstellar objects available to as wide an audience as possible in an easily understood manner. Persons other than professional astronomers, such as physicists, engineers, amateur astronomers, and others, may have needs for data on nonstellar objects but have difficulty obtaining it. Such persons generally are not familiar with the astronomical literature and do not wish to invest the time required to become so. Much of the literature may in fact be physically inaccessible to them. To this end considerable attention has been given to the choice of data to be included and the manner of presentation, as will be explained below.

Purpose 6. To make all catalogues of nonstellar objects available in a computer readable form. Most of the catalogues included in this work were not previously available in that form. Now that they are, various types of automated data correlation, retrieval, and selection projects can be done. Employing the computer-readable form of this work, users can, for example, process the data to select out only that which interests them at the moment (such as specific kinds of objects, objects in certain areas of the sky, objects with certain angular diameters, etc.). Users interested in only one specific catalogue could select it out of all the others. In essence, a user may use this catalogue to construct a new catalogue based upon his own individual needs, criteria, and tastes.

Examples of automated correlations that might be done are listed above under purposes 2 and 3.

Now that the initial computer data file required to create the present work has been established, it should be a relatively simple matter to update it periodically in the future, and we hope to do so indefinitely.

Purpose 7. To make all catalogues available in a common epoch (specifically 1950.0) and with consistent notation and units.

Human Engineering Aspects

The intended audience for this work, in addition to professional astronomers, is all persons with at least an elementary knowledge of astronomy, including the concepts of nonstellar object, right ascension, declination, magnitude, and angular diameter. It is hoped that a person of such qualifications could open this compendium to a random page and understand almost at sight the information presented there.

The following human-engineering techniques were used toward the goal of achieving this end. Only the reader can judge how well these goals have been met for his particular case. There are of course exceptions, but these are the general techniques that have been used.

1. The data are presented in a way that is hopefully self-explanatory to the intended audience.
2. Abbreviations and special symbols are avoided wherever possible.
3. Where used, abbreviations and special symbols are mnemonic in nature and have obvious meanings.
4. No subsidiary tables are necessary to explain the main body of the MOL. The related small tables contain *supplementary* information, rather than *prerequisite* information.
5. Data are presented consistently, always in the same units. Right ascensions are always hours, minutes, and seconds of time, and declinations are always degrees, minutes, and seconds of arc, as opposed to mixtures such as tenths of minutes. Epochs are always 1950.0. Angular diameters are always arc seconds, since they can be more readily compared than if they were a mixture of degrees, minutes, and seconds of arc. The only exception to the use of consistent units is in the magnitude column. The magnitudes are whatever type was used in the original catalogue. To attempt conversion to some standard system would only lead to confusion and inaccuracy. However, the various magnitude types do not vary greatly from one another, so those given are still useful in an approximate sense, regardless of what type they are. The magnitude type used in each catalogue is listed in table 3, beginning on page 27.
6. Data are ordered by a single index (right ascension), for ease in proceeding directly to a specific area. A useful search technique is to simply read down the declination degrees column on the pages preceding, containing, and following the desired right ascension (a total of 3 pages). The objects near a desired position can generally be found using this technique, in less than a minute. The data near 00^h have been duplicated after 24^h to facilitate this process.
7. Nothing has been omitted to be assumed by the reader. Positive declination is indicated by a plus sign, rather than a blank. Leading zeros in the positions have been included wherever needed. Trailing zeros have been included only where required to establish proper units and column positions. Blank columns indicate that the data for those columns were not contained in the original catalogue.
8. We have attempted to minimize the production costs and hence the purchase price of this work by printing the catalogue directly from computer listings. In addition the computer pages have been printed four per MOL page, a reduction of 40 percent. This has accordingly reduced costs by a factor of four with a negligible loss of readability.

Scope

This work is intended to contain all catalogued nonstellar objects, excluding solar system objects. There is no precise definition of the term *nonstellar object,* nor does there need to be one. Those objects that most astronomers would agree are not stars are included. Those classes of objects that include some nonstars are included in this work. The adjective *stellar* has a semantic ambiguity, in that it is used commonly to refer both to actual stars and to nonstars that look like stars on photographs.

Thus this compendium of "nonstellar" objects quite properly contains some "stellar" objects (examples are the blue stellar objects and supernovae). For those cases where there is doubt as to the "stellar" character of an object, they have been specifically *included* in this work. They constitute only a small portion of the entire work, and their presence may aid many while harming none. Star clusters are generally considered to be nonstellar objects, and hence are included in this work. However, binary and multiple star systems are not generally so considered, and hence are omitted. The dividing line between these two kinds of star groups is generally clear, and should cause no confusion among users of this work.

This work is by no means a complete catalogue of all existing objects down to some limiting magnitude. It is only as complete as its constituent catalogues. In a sense, it merely reflects the current status of man's exploration of the universe in search of these objects.

For certain classes of objects, there already exist compendia of all previous catalogues (such as Alter et al., *Catalogue of Star Clusters and Associations*). In such cases, we incorporated these compendia and then ignored all previous work dealing with that class of objects. This explains, for example, why Messier's famous objects are not included herein, since they have been superseded by many subsequent catalogues (such as the NGC and IC.) Discoveries of new objects published subsequent to such compendia were, however, included in this work.

Many papers have been published that appear to be catalogues, but do not fit our working definition of a catalogue. If a publication contains a list of (apparently) independently discovered and catalogued objects, presumably resulting from some survey efforts, we have included it regardless of whether it is commonly referred to as a catalogue. If, instead, a publication only contains more accurate or complete information about previously catalogued objects, we have excluded it, even though it may be commonly referred to as a "catalogue." An example of the latter case is de Vaucouleurs's *Reference Catalogue of Bright Galaxies* (1964). These general policies are in keeping with one of the stated purposes of including all objects, as opposed to all information about all objects.

Object Notation

In keeping with one of the stated purposes, that this work not replace any of its constituent catalogues, no new designations have been assigned to the objects, except where the original author of a constituent catalogue did not assign specific designations himself. In these cases we assigned designations in such a way as to enable that object to be found most readily in the original catalogue (e.g., sequential numbering). This is in keeping with another stated purpose, that this work is a "pointer" to the original catalogues. Our assigned designations should be regarded as "pointers" rather than "names."

It is of interest to note that Dreyer, in his famous NGC compendium, stated, "It was with much regret that I found it necessary to introduce new numbers, and it is greatly to be hoped that these will be *quoted* as little as possible, but that old nebulae, as hitherto, will be chiefly mentioned by their (original) number." As we all know, however, human nature prevented Dreyer's hope from being realized. Today his numbers are used everywhere, and the original designations are virtually forgotten.

We believe we have prevented a recurrence of this problem by not creating any new over-all object-numbering system.

The prefix designations assigned to each originally undesignated catalogue generally were chosen from some mnemonic combination of the author's initials. With multiple authors, last-name initials were used, in the order of appearance in the original catalogue. This procedure was modified occasionally to avoid duplication with other named catalogues, or with other authors having the same initials. In cases where the same authors published different catalogues, they were generally distinguished by appending two digits representing the year of publication. In a few cases of multiple catalogues published in the same year, a further letter was appended to distinguish them in a mnemonic way. For example, VDB.66G prefixes Van den Bergh's catalogue of dwarf galaxies (Van den Bergh, 1966a) as opposed to his VDB.66B catalogue of blue objects (Van den Bergh, 1966b) and his VDB.66N catalogue of nebulae (Van den Bergh, 1966c).

The various catalogues of galaxies authored by Zwicky et al. pose a particularly difficult notational problem. So as to avoid confusion within our work, the designations used for the Zwicky catalogues are explained here.

nZW (n=1 thru 7) prefixes the seven lists of "Compact Galaxies, Compact parts of Galaxies, Eruptive Galaxies, and Post-Eruptive Galaxies," circulated privately by Zwicky (1961–68).

8ZW prefixes the posthumous list of compact galaxies published with Sargent and Kowal (1975). Since the objects were not numbered in the original catalogue, we assigned positional designations.

ZCG ("Zwicky Compact Galaxy") prefixes the additional galaxies listed in the red-covered single volume *Catalogue of Selected Compact Galaxies and of Post-Eruptive Galaxies* (1971) that were not included in the original seven privately circulated lists. Since the objects were not numbered in the red catalogue, we assigned positional designations to most of them. To those that Zwicky placed in special tables, apart from his main catalogue, we assigned designations consisting of the original page number followed by a running sequence number.

ZWG ("Zwicky Galaxy" prefixes the galaxies listed in the blue-covered six-volume *Catalogue of Galaxies and of Clusters of Galaxies* (1960–68). Each galaxy is designated first by its Zwicky field number and then by a running sequence number within that field. This expedites finding the galaxy in the original catalogue.

ZC ("Zwicky Cluster") prefixes the clusters of galaxies in the above six-volume catalgue. Each cluster is designated by a positional notation, as used by Zwicky himself (Rudnicki and Zwicky, 1967). It was not possible to use the field numbers as we did with the galaxies because we obtained these data from a magnetic tape prepared elsewhere that omitted the field numbers.

ZH and ZL are the remaining Zwicky prefixes, but these are unrelated to any of the above galaxy catalogues, and so should cause no confusion.

Accuracy

Objects from catalogues and lists with equatorial coordinates referred to an equinox other than 1950.0 were precessed by the use of Newcomb's constant and rigorous formulas. A complete discussion of the methods used may be found in the *Explanatory Supplement to the Nautical Almanac* (Her Majesty's Stationery Office, 1961). Our precession calculations are believed to be accurate to at least 1″.0. Wherever possible, the precision of the precessed coordinates used in this work is the same as that in the original catalogue. When an original catalogue employed a mixed system of units, such as tenths of minutes, unit consistency required that this be converted to seconds, resulting in an apparent precision increase by a factor of six. For this reason the precision of position and angular diameter used in each original catalogue is tabulated in the main index (table 5).

With the exception of a few catalogues, which were already available on magnetic tape, all data were keypunched and subsequently proofread at least once by us. Nevertheless, it is unlikely that a body of data as large as this work is completely error-free. The authors would be pleased to learn of any errors or apparent discrepancies found so that the magnetic-tape version can be corrected and appropriate errata made available to users of the printed version.

Completeness

A systematic search of the astronomical literature through 1975 was made in an attempt to make this work as complete as possible. The literature was searched primarily by article title, which eliminated a majority of articles from further investigation. If a title appeared promising, the article was examined further for catalogues or listings of observations of new objects. We may have missed articles where the title was not sufficiently indicative of its content. The journals and other publications that were systematically searched are given in table 1, along with the years of publication searched. Many articles were omitted from further consideration because their catalogues contained only new observations of previously catalogued objects. Unfortunately, in several papers, authors presented new observations on previously uncatalogued objects but included no positions. Since our compendium is organized by equatorial coordinates, we were unable to include such catalogues.

An example of this is the work of Hodge (1966), where H II regions in several external galaxies were catalogued. In order to provide data for future observations, X-Y arc-second positions of these H II regions were given with respect to the "center" of their respective galaxies. Since the equatorial coordinates of the assumed center of the galaxies used in this study were not given, it was not possible to compute equatorial positions for the H II regions.

An area of possible incompleteness is illustrated by the following examples. A fictitious astronomer named Ajax catalogues a number of physically real objects that he calls planetary nebulae. Later, while compiling a comprehensive catalogue, Perek and Kohoutek conclude that some of these objects are in fact not planetary nebulae and therefore omit them from their compendium. Since the Perek and Kohoutek catalogue is published after Ajax's list and is a compendium of all earlier catalogues of planetary nebulae, we take Perek and Kohoutek to be complete and ignore all earlier planetary nebulae catalogues. We therefore would have unintentionally omitted those objects from Ajax's catalogue that Perek and Kohoutek omitted, even though they are physically real, most likely being nebulae of other types.

Although several installments of the ESO survey have been published, they have been omitted from this compendium at the request of their authors. They plan to revise the ESO survey after it is completed, and therefore do not wish the currently published preliminary results disseminated further.

The Matter of Nomenclature

There are two methods widely used by cataloguers for naming their objects. One method simply numbers the objects serially from beginning to end, usually in right ascension order (e.g., UGC 12195); the other method employs a positional notation that gives the approximate position of the object (e.g., XYZ 2016-18). This latter method is commonly known as the Parkes naming system, in reference to its use in the Parkes catalogue of radio sources (Bolton, Gardner, and Mackay, 1964). Actually, the positional notational method was first used by Shapley in 1938, but it apparently was an idea ahead of its time. Recently, the International Astronomical Union has recommended that the Parkes-type nomenclature system be used for all newly published catalogues of nonstellar objects.

Any system has its drawbacks, however. Those of the Parkes system include the fact that the same object may have several different numerical designations due to slight positional errors (e.g., XYZ 0716-25 may be the same object as ABC 0715-24). In addition, if the positional designations are assigned for one epoch, they are incorrect for all other epochs. Such a case exists in the MOL where Shapley's names were originally assigned for epoch 2000.0, but are given herein with epoch 1950.0 positions.

There is now an unfortunate tendency on the part of various authors to retroactively invent positional names for objects that did not originally have them, and to omit the identifying prefix (e.g.,

TABLE 1

Journals Searched Systematically

Title	Years
Acta Astronomica vols. 1–21	(1926–71)
Acta Cosmologica vols. 1	
Annales d'Astrophysique vols. 1–38	(1938–68)
Arkiv für Astronomi vols. 1–5	(1959–67)
Astrofizika vols. 1–10	(1966–74)
Astronomical Journal vols. 26–80	(1908–75)
Astronomical Society of the Pacific, Publications vols. 20–87	(1908–75)
Astronomicheskii Tsirkuliar Akademiia Nauk SSSR nos. 1–1000	(1940–75)
Astronomicheskii Zhurnal Akademiia Nauk SSSR vols. 1–52	(1924–75)
Astronomische Nachrichten vols. 178–296	(1908–75)
Astronomy and Astrophysics vols. 1–44	(1969–75)
Astronomy and Astrophysics Supplement Series vols. 1–22	(1970–75)
Astrophysical Letters vols. 1–16	(1967–75)
Astrophysical Journal vols. 27–202	(1908–75)
Astrophysical Journal Supplement Series vols. 1–29	(1954–75)
Astrophysics and Space Science vols. 1–38	(1968–75)
Bulleten Ambastumani Astrofizicheskaia Observatoriia vols. 1–39	(1937–70)
Bulletin of the Astronomical Institute of Czechoslovakia vols. 1–26	(1947–75)
Bulletin of the Astronomical Institute of the Netherlands vols. 1–20	(1921–69)
Bulletin of the Astronomical Institute of the Netherlands. Supplements vols. 1–3	(1966–69)
Comptes rendus de l'Académie des Sciences, Paris vols. 152–281	(1911–75)
Harvard College Observatory Annals vols. 60–118	(1908–51)
Harvard College Observatory Bulletins nos. 501–921	(1921–52)
Harvard College Observatory Circulars nos. 101–450	(1904–47)
Heidelberg Astronomisches Rechen-Institute Mitteilungen	
Serie A nos. 1–95	(1954–75)
Serie B nos. 1–55	(1954–75)
Heidelberg Sternwarte Veroffentlichungen vols. 1–17	(1900–60)
Irish Astronomical Journal vols. 1–9	(1924–70)
Liège Institut d'Astrophysique Mémoirs vols. 1–29	(1924–73)
National Academy of Sciences Proceedings vols. 1–71	(1915–74)
Nature vols. 78–258	(1908–75)
Observatory vols. 31–95	(1908–75)
Pulkova Glavnaia Astronomicheskaia Observatoriia Trudy (Publications) vols. 19–69	(1911–52)
Royal Astronomical Society, Memoirs vols. 58–79	(1908–75)
Royal Astronomical Society, Monthly Notices vols. 69–173	(1908–75)
Royal Astronomical Society, Quarterly Journal vols. 1–16	(1960–75)
Royal Astronomical Society of Canada, Journal vols. 1–69	(1907–75)
Uppsala Astronomiska Observatorium Meddelanden nos. 10–150	(1926–65)

1024-17). This practice may be the result of misunderstanding the IAU recommendation (which applies only to *new* lists), or of poorer reasons, but in any case it should be avoided. If everyone goes about renaming objects willy-nilly, only hopeless confusion will result. Different authors may invent different positional names to refer to what was originally a single object, since they may measure slightly different positions. Authors writing in different specialties (optical, radio, X ray, etc.) may well invent the same positional notation for completely different objects. The prefixes used by an original cataloguer convey important information, such as in what region of the electromagnetic spectrum the object was catalogued, by whom, and for what reason. By omitting such prefixes, only blank anonymity remains.

At the 1976 IAU General Assembly, a working group of Commission 5 was formed to study the question of a universal nomenclature for all astronomical objects. One of the present authors (RSD) is a member of that group. Regardless of whatever nomenclature may be adopted, it will be futile unless it is universally used. Probably the only practical way to ensure its use is through journal editorial standards. This need has been recognized by the IAU working group, and is being incorporated into the study from the beginning. If such a system can ultimately be adopted, it will be a milestone not only in terms of over-all astronomical organization, but also of cooperation among the many diverse subfields of astronomy.

For many purposes, and particularly when using the MOL, if one simply regards the designation as a pointer to the object in its original catalogue (as discussed elsewhere in this work), and uses whatever prefix designation the author has used, it makes little difference which notation system is employed.

The Matter of Epoch

It is traditional to change the standard epoch every 25 or 50 years, to more closely approximate the current year at any given time. This practice is partially an outgrowth of the desire to facilitate manual pointing of optical telescopes, using setting circles that are correct only for the current year. With the explosive growth of computers in our society, however, this need may well become of minor significance. Large optical telescopes, practically all nonoptical telescopes (radio, etc.), and many small optical telescopes are already controlled by computers, either directly or indirectly. Such computers can make precession and other corrections virtually instantaneously, making the choice of standard epoch arbitrary. With the growing use of the ubiquitous programmable pocket calculator, even the smallest of telescopes can be positioned in any epoch. The choice of standard epoch becomes more arbitrary when it is realized that there is virtually as much work involved in accurately precessing over a 1-year period as there is over a 100-year period.

The practice of changing the standard epoch from time to time has several undesirable effects. An obvious one is that position cannot be directly compared between papers published in widely separated years. In a sense, a person who publishes a position using the epoch in vogue at that time has not removed all his "observational" effects. In this case, the observational effect is the quarter- or half-century in which his paper happens to be published. It would seem desirable to have some method of removing this effect. One could argue that a reader could do the precession himself (perhaps using his pocket calculator), but the fact is that there may be hundreds or thousands of readers but there is only one author. A little effort on the part of the author would save manyfold that effort thereafter.

A less obvious effect of epoch-changing is that when positional-type names are assigned to objects (i.e., XYZ 2024-15), they are correct only for the epoch in which they are published. If their positions are changed to another epoch, the names become misleading. If the names themselves were changed to the new epoch, only confusion would result.

We suggest that the time has come to abandon the traditional practice of epoch-changing and adopt some standard epoch to be used forever. The standard epoch should be regarded as being in the same philosophical category as the origin of the Julian Date, which is fixed at some specific value and is then forever left constant and universally used.

For ultimate accuracy in astronomic applications, however, it is necessary to redefine the fundamental astronomical constants occasionally. Since the precessional constants are known only to limited accuracy, one cannot precess over arbitrarily large time spans if one requires the ultimate in position accuracy. For all other astronomical purposes, however, use of an ''eternal'' epoch has no apparent disadvantages and has the advantages given above.

We further suggest that this eternal epoch be chosen as 1950.0, *not* because that is the one commonly used now, but because:

1. More published positions exist for epoch 1950.0 than for any other epoch. Thus more published data would already be in the ''eternal''epoch than if any other epoch were chosen.
2. Most positional designations assigned to objects are for epoch 1950.0, since this naming practice has become popular only in recent years. Thus these names are meaningful only if 1950.0 is chosen as the ''eternal'' epoch.

It is for these reasons that positional coordinates in the MOL are given for epoch 1950.0. There are, however, workers who already use epoch 1975.0 (this has been done consistently in catalogues of objects in the Megallanic Clouds), and Harlow Shapley and coworkers at Harvard College Observatory used epoch 2000.0 in their galaxy catalogue published in 1938.

It is hoped that those who would adopt new epochs would give consideration to conditions as they exist today, rather than proceeding primarily out of tradition from the past.

Explanatory Notes on Certain Catalogues and Types of Objects

In this section we have included notes, comments, and discussion of the manner in which certain types of objects were treated, along with problems and peculiarities of individual catalogues. Table 2 contains an outline of all catalogues used in this work, organized by type of object. The nonstellar objects have been arbitrarily divided into two categories: extragalactic and galactic objects. Under the latter heading we have included star clusters, H II regions, reflection nebulae, supernova remnants, and so on, located in our galaxy and its environs, including the Magellanic Clouds. In the ''extragalactic'' category we include galaxies, ''nebulae'' and extragalactic nebulae, cluster of galaxies, quasi-stellar objects, and so on. We have reserved the term *nebula* to refer only to those objects catalogued before the true extragalactic nature of most of these objects was recognized. The ''nebulae'' are included under this category because the majority are probably extragalactic, even though there are undoubtedly a few galactic objects in the group.

Index Catalogue. The *Index Catalogue* (Dreyer, 1908) presented a special problem, since there has been no systematic revision or updating of the IC as a whole since its publication. The information needed for revision of coordinates, type of object, magnitude, and so on, for the IC objects is scattered throughout the astronomical literature from 1908 to the present.

We have therefore attempted to extract modern data on the IC objects from the astronomical literature. In the process of systematically searching the literature for catalogues, special note was made of articles containing information on any IC objects. A record was made, object by object, of any pertinent data that could be used to update the original IC.

Several hundred articles containing data about the IC objects were found. The data and references were entered into a handwritten card file. This information is too extensive to reproduce or index here, but it may be consulted by contacting the authors.

These efforts resulted in modern data of various kinds for many, but not all, IC objects. In general it was found that if there were data available for an object, they usually came from more than one source. The following guidelines were used when deciding upon which data to use to update the IC.

TABLE 2
CATALOGUES INCLUDED IN THE MOL
ARRANGED BY TYPE OF OBJECT

I. Extragalactic Objects

A. Galaxies

AGU	CR	HOD.60	MRK	SEY	T
AND	DV.56	HOLM	RB	SHAH	UGC
ARP	DVDV	HPW	RDS	SHAP	VDB.66G
ARP.65	FAI	HSN	REA	SHER	VV
BAA2	FIT.G	KARA.68	REA.66	SHP	VVI
BAK2	GRA	KARA.73	REIZ	SHRV.73B	ZCG
BB	HARO	KARA.73B	SA	SM1	ZH
BFGS	HAW	KEEN	SB	SPA	ZWG
BEM	HMS	KW	SCH	SPC	1ZW thru 8ZW
BON	HO	MAI	SEA	SVEN	
CHR	HOAG	MCG			

B. Nebulae and Extragalactic Nebulae

AMES	BIGO	FATH	KEEL	REIN1 thru REIN7
AR	CAR	HELW	KN10 thru KN16	SC
BAK1	DUN	HN	MEL	WK

C. Clusters of Galaxies

ARC	HMS	KLEM	MKW	SNO	ZC

D. Quasi-stellar Objects

BC	SHB
NAB	

E. Seyfert Galaxies

KW
VVI

F. Blue Objects and UV Concentrations

A1	BOR	HZ	RLWT	SM2	VDB.66B
A2	BRON	LB	RMB	STOCK	WEED
A3	BSO	PHL	RRS	SV	WEI
BFG	BV	PRA	RS	TON-N	ZL
BFL	FEIG	RIC	SAAK	TON-S	

G. Supernovae

SN

H. Miscellaneous (catalogues of more than one type of object, mixed together)

KON
SER
YC

II. Galactic Objects

A. Groupings of Stars

1. Globular Clusters

a. Galactic

GCL

b. Extragalactic

BAA3 (M31)
HOD.61 (Fornax)
HODG (LMC)
SHRV.73 (M31)

TABLE 2 (*continued*)

2. Open Clusters
   a. Galactic
      OCL
      PIS
      TER
      VHA
   b. Extragalactic
      HOW (LMC)
      HSE (LMC)
      LIN.CL (SMC)
      LW (LMC)
      MOHR (SMC)
      SL (LMC)
3. OB Associations
   a. Galactic
      ASS
      LYNG
   b. Extragalactic
      LH (LMC)
4. Stellar Groups
   a. Galactic
      BA
      SAP
5. Star Chains, Rings, etc.
   a. Galactic
      GO
      ISS
      URA
      VB

B. Nebulae
1. H II Regions and $H\alpha$ Emission Nebulae
   a. Galactic

COU	MRSL
CS	SG1 thru SG3
DV.55	SIV
EM	ST
LBN	WIL
	WRAY

   b. Extragalactic
      BAA1 (M31)
      BCL
      CM
      LH115 (LMC)
      LH120 (LMC)
      LIN (SMC)
2. Planetary Nebulae
   a. Galactic
      ACK
      ALL
      KAZ
      PK

TABLE 2 (*continued*)

b. Extragalactic
   WS (LMC)

3. Diffuse Objects
   a. Galactic

BOH	MIN.48
CED	SMI
HUB	SS
MIN.46	YM
MIN.47	

   b. Extragalactic
      BAA4 (M31)
   c. Miscellaneous
      1. Reflection Nebulae
         a. Galactic
            DDKS
            DG
            VDB.66N
            VHE
      2. Dark Nebulae
         a. Galactic
            B
            FIT.N
            HOD.72
            HOFF
            KHAV
            LDN
            SCHO

3. Herbig-Haro and IR Objects
   a. Galactic
      HH
      MAFFEI

4. Supernova Remnants
   a. Galactic
      MIL
      VMT.
   b. Extragalactic
      MACL (SMC)
      MHW (LMC)

---

1. Equatorial coordinates with the greatest precision were used.
2. The largest angular diameter was used.
3. Photographic apparent magnitudes were chosen over apparent magnitudes of other types (e.g., the Harvard magnitude system).
4. Type of object: for galaxies, the classification was included where available. Unfortunately, the classifications were not always made on the same system.

In general, more recent data were taken over older data. Of the 5,386 objects in the IC, 1,031 of them were updated to some extent. The fraction of objects for which updated data exists is surprisingly small considering the extent of the literature search involved. The conclusion to be made is that current knowledge of the objects in the IC, as a whole, is not much better than it was in 1908; the majority of IC objects have not been studied in detail.

MCG Catalog. The *Morphological Catalog of Galaxies* (MCG) (Vorontsov-Velyaminov et al., 1962–74) is one of the longest and most comprehensive works in this area. Unfortunately, we have been unable to include the classifications for the galaxies in the present version of this work. This is a result of (a) the lack of straightforward and practical means of expressing the MCG classifications in a standard computer character set such as BCD or ASCII, and (b) the size of the classification field (i.e., the number of characters) varies greatly and often exceeds the space available in the Type-of-Object field used at present. Since the primary purpose of the MCG was to classify as many galaxies as possible on one system, it is most unfortunate that we were unable to include these classifications. However, we hope to obtain a magnetic tape version of the MCG from the Soviet Union, and it will enable the classifications to be added.

HOLM list. There have been questions raised (Zonn, 1963; Reaves, 1975) about the reliability of the identifications in the catalogue of double and multiple galaxies (Holmberg, 1937). Zonn (1963) found that nearly 30 percent of Holmberg's double or multiple systems do not appear to be galaxies, and may in fact be stars. Nonetheless, we have included Holmberg's catalogue in this work because most objects in it are not misidentifications.

Quasi-stellar objects. A catalogue of quasi-stellar objects is maintained by C. Barbieri and M. Capaccioli at the Univeristy of Padova Observatory. We received the January 1975 version of the catalogue and have used this in our compendium. For notation, Barbieri and Capaccioli have used the name by which each QSO was originally catalogued. In most cases this is a radio source designation to which we have added the prefix BC to indicate the source of the data. In addition to the BC catalogue, we have included the NAB catalogue of Bahcall et al. (1973), and the SHB catalogue of Smith-Haenni (1977).

Supernovae. Although supernovae are stellar phenomena, they have been included in this work because of their importance in extragalactic research. The original supernova catalogue of Kowal and Sargent (1971) has been kept up to date by them, and we have used a recent version (Kowal and Sargent, 1975). The catalogue at present contains supernovae up through SN 1974C, a total of 414 objects in all.

Blue Objects. There has been considerable effort over the last two decades to catalogue and determine the nature of the numerous faint blue objects located at high galactic latitudes. Although stellar in appearance, a significant fraction of these objects are nonstellar and extragalactic in nature. Sandage (1965) first estimated the majority of these blue objects to be "superluminous galaxies with very large redshifts" on the basis of number counts and photoelectric and spectrographic observations. Subsequent investigation by Sandage and Luyten (1967, 1969), Schmidt (1974), and others have found that though the blue objects identified as extragalactic (e.g., as galactic nuclei or quasi-stellar objects) are not a majority of the objects sampled, they are still quite numerous. The majority of the blue objects have in fact been identified as either white dwarfs or F and G subdwarfs. Schmidt's (1974) investigation of 120 faint blue objects in the PHL and LB catalogues found 58 percent to be galactic stars and 37 percent to be quasi-stellar objects. Due to the cosmological importance (Sandage, 1965) and extragalactic nature of many faint blue objects, they have been included in this work. Unfortunately, since it is not presently possible to separate stellar and nonstellar faint blue objects, we have included the complete catalogues of these objects. However, catalogues of blue objects specifically identified by their authors as stars have been excluded.

Star Clusters. All catalogued globular and open clusters and OB associations located in the galaxy are, with a few recent exceptions, contained in the compendium of Alter et al. (1970). It is therefore more efficient to use this compendium than the constituent catalogues. We have used their notation GCL, OCL, and ASS for globular clusters, open clusters, and OB associations, respectively. For clusters and associations where multiple data were present, we chose the faintest apparent magnitude and largest angular diameter with the idea of representing, respectively, lower and upper limits to these quantities. The limits are believed to be those in best keeping with the purposes of the MOL.

Most other catalogues of star clusters or associations included herein contain objects in other galaxies, or in the Magellanic clouds.

Star Chains and Rings. The astronomical significance and the physical reality of star rings and star chains continues to be a debated subject. Since this matter has not yet been settled and since they are easily seen and probably often rediscovered on photographs, we have included the catalogues of star chains and star rings (SO, ISS, URA, VB).

H II Regions. Marsalkova (1974) has investigated the apparent distribution of H II regions along the entire galactic equator. Data were obtained by compiling a compendium catalogue based on 13 source cataogues, including among others the works of Gum (1955), Rogers et al. (1960), and Sharpless (1969). We have used Marsalkova's synthesis of these 13 catalogues in lieu of incorporating the source catalogues individually.

Planetary Nebulae. The Perek and Kohoutek (1967) catalogue was used, with a number of additions and corrections given by Kohoutek (1968, 1969) and Acker (1975). In addition we also include objects from two more recent catalogues (ALL, KAZ).

TABLE 3

CHARACTERISTICS OF THE CATALOGUES USED IN THIS COMPENDIUM ARRANGED IN ALPHABETICAL ORDER BY POINTER

List Pointer	Type of Object	Right Ascension Range	Declination Range	Original Right Ascension Precision	Original Declination Precision	Number of Objects	Original Angular Diameter Precision	Faintest Magnitude	Type of Magnitude	Reference
ACK	Planetary nebula	00^h to 24^h	$-90°$ to $+90°$	$0^m.1$	$1'$	148	–	—	—	Acker (1975)
AGU	Interesting galaxy	00^h to 24^h	$-45°$ to $-38°$	$0^m.1$	$0'.1$	48	$0'.1$	18.	pg	Aguero (1971)
ALL	Planetary nebula	$18^h\ 31^m$	$-27°\ 08'$	$0^m.01$	$0'.1$	1	$0'.5$	12.	—	Allen (1973)
AMES	Nebula	12^h02^m to 12^h48^m	$+5°$ to $+19°$	1^s	$0'.1$	2161	$1''$	18.2	pg	Ames (1932)
AND	Dwarf spheroidal galaxy	00^h10^m to 01^h20^m	$+32°$ to $+47°$	$0^m.1$	$1'$	4	$0'.1$	—	—	Van den Bergh (1972)
AR	Nebula	09^h30^m to 12^h	$+74°$ to $+78°$	$0^s.01$	$0''.1$	43	—	—	—	Astronomer Royal (1910)
ARC	Rich galaxy cluster	00^h to 24^h	$-27°$ to $+90°$	$0^m.1$	$1'$	2712	—	18.4	pg	Abell (1958)
ARP	Peculiar galaxy	00^h to 24^h	$-43°$ to $+90°$	$0^m.1$	$1'$	338	—	—	—	Arp (1966)
ARP.65	Compact dwarf galaxy	$11^h\ 17^m$	$+51°\ 46'$	1^s	$1'$	4	$0''.1$	17.9	pg	Arp (1965)
ASS	OB association	00^h to 24^h	$-90°$ to $+90°$	$0^m.1$	$1'$	69	$5'$	—	—	Alter et al. (1970)
A1	Faint blue object	11^h17^m to 11^h40^m	$+12°$ to $+17°$	$0^s.01$	$0''.1$	129	—	17.5	pg	Barbieri et al. (1968)
A2	Faint blue object	12^h40^m to 13^h06^m	$+26°$ to $+31°$	$0^s.1$	$1''$	255	—	18.7	B	Barbieri & Rosino (1972)
A3	Faint blue object	12^h15^m to $+12^h36^m$	$+12°30'$ to $+17°30'$	$0^s.1$	$1''$	325	—	18.0	B	Barbieri & Benvenuti (1974)
	UV concentration					19	—		B	
B	Dark object	Selected galactic plane regions		1^s	$1'$	349	$1'$	—	—	Barnard (1927)
BA	Stellar group	Various		$0^m.1$	$1'$	6	$1'$	—	—	Kiral (1969a, 1969b), Wagner (1971), Wooden (1971), Grubissich (1973)
BAA 1	Hα emission nebula in M31	00^h30^m to 00^h44^m	$+40°$ to $+42°$	$0^s.01$	$0''.1$	688	—	—	—	Baade & Arp (1964)
BAA 2	Galaxy near M31	00^h30^m to 00^h44^m	$+40°$ to $+42°$	$0^s.01$	$0''.1$	68	—	—	—	Baade & Arp (1964)
BAA 3	Globular cluster in M31	00^h30^m to 00^h44^m	$+40°$ to $+42°$	$0^s.01$	$0''.1$	30	–	–	–	Baade & Arp (1964)
BAA 4	Diffuse object in M31	00^h30^m to 00^h44^m	$+40°$ to $+42°$	$0^s.01$	$0''.1$	40	—	—	—	Baade & Arp (1964)

TABLE 3 (*continued*)

List Pointer	Type of Object	Right Ascension Range	Declination Range	Original Right Ascension Precision	Original Declination Precision	Number of Objects	Original Angular Diameter Precision	Faintest Magnitude	Type of Magnitude	Reference
BAK 1	Extragalactic nebula	02^h50^m to 03^h50^m	−20° to −33°	1^s	0′.1	918	1″	19.	H	Baker (1933)
BAK 2	Galaxy	03^h30^m to 04^h10^m	−16° to −28°	1^s	0′.1	1082	1″	19.	H	Baker (1937)
BB	Galaxy near QSO	Various		$0^s.05$	0″.5	88	—	—	—	Bahcall & Bahcall (1970)
BC	Quasi-stellar object	00^h to 24^h	−90° to +90°	$0^s.1$	1″	412	—	21.2	V	Barbieri & Capaccioli (1974)
BCL	H II region in M33	01^h31^m	+30°24′	$0^m.01$	0′.1	369	1″	—	—	Boulesteix et al. (1974)
BEM	Galaxy	09^h42^m	+68°	$0^m.1$	1′.	2	0′.1	14.7	B	Bertola & Maffei (1974)
BFG	Ultraviolet excess object	12^h45^m to 13^h15^m	+33° to +39°	$0^s.01$	0″.1	175	—	19.5	V	Bracessi et al. (1970, 1973)
BFGS	Galaxy with UV knots	09^h to 13^h	+28° to +32°	$0^m.1$	1′	2	1″	16.5	pg	Bracessi et al. (1972)
BFL	Peculiar blue object	00^h32^m to 00^h41^m	+39° to +40°	1^s	0′.1	5	—	19.47	V	Börngen et al. (1970)
BIGO	Nebula	$00^h$to 24^h	−20° to +80°	1^s	1′	104	—	—	—	Bigourdan (1912)
BEM	Galaxy	09^h42^m	+69°	$0^m.1$	1′	2	0′.1	14.7	B	Bertola & Maffei (1974)
BOH	Nebulous object	04^h14^m	+28°	1^s	1′	1	—	—	—	Bohlin (1921)
BON	Variable galaxy	Various		1^s	0′.1	3	1″	17.7	pg	Bond (1972, 1973)
BOR	Blue object	00^h36^m	+39°30′	1^s	0′.1	7	—	20.0	V	Börngen (1966)
BRON	Faint blue object	North Galactic Pole region		$0^s.1$	1″	54	—	20.1	V	Bronkalla (1971)
BSO	Blue stellar object	12^h45^m to 15^h15^m	+33° to +46°	1^s	1″	31	—	19.1	V	Sandage & Veron (1965)
BV	Faint blue variable	03^h to 24^h	−30° to 0°	1^s	0′.1	11	—	19.1	pg	Luyten & Haro (1959), Haro & Luyten (1960)
CAR	Nebula	08^h11^m	+21°	$0^s.1$	1″	52	1″	16.7	pg	Carpenter (1931)
CED	Diffuse galactic nebula	Galactic plane		$0^m.1$	1′	330	0′.5	13.7	pg	Cederblad (1946)

CHR	Galaxy in Leo cluster	10^h22^m	+11°	$0^m.1$	0′.1	6	—	18.1	pg	Christie (1931)
CM	H II region in LMC	04^h53^m to 05^h37^m	−66° to −70°	$0^m.1$	1′	41	—	—	—	Cheriguene & Monnet (1972)
COU	Hα emission nebula	Galactic plane		1^m	1′	106	1′	—	—	Courtès (1951)
CR	Galaxy in Perseus cluster	03^h07^m to 03^h18^m	+40° to +42°	$0^m.1$	1′	50	—	15.7	pg	Chincarini & Rood (1971)
CS	Hα emission region	Various		1^m	1′	2	30′	—	—	Courtès & Sivan (1972)
DDKS	Reflection nebula	17^h15^m	−78°40′	1^m	1′	1	1°	—	—	Danziger et al. (1974)
DG	Reflection nebula	00^h to 24^h	−30° to +90°	$0^m.1$	1′	192	1′	—	—	Dorschner & Gürtler (1964)
DUN	Small nebula	12^h56^m	+28°	$0^m.1$	1′	10	1″	—	—	Duncan (1923)
DV.55	Hα emission nebulosity	$07^h$to 08^h	−79°	1^h	1°	2	0°.5	—	—	de Vaucouleurs (1955)
DV.56	Galaxy	00^h to 24^h	−35° to −90°	$0^m.1$	1′	127	0′.1	—	—	de Vaucouleurs (1956)
DVDV	Galaxy	00^h to 24^h	−35° to 90°	$0^m.1$	1′	6	0′.1	—	—	de Vaucouleurs & de Vaucouleurs (1961)
EM	Emission object	18^h30^m	−30°	$0^m.1$	1′	7	—	13.1	pg	McCluskey (1961)
FAI	Compact galaxy	13^h to 17^h	+80°to +20°	$0^m.1$	1′	168	–	—	—	Fairall (1968, 1970)
FATH	Nebula	Various northern areas		1^s	1′	1032	1″	—	—	Fath (1914)
FEIG	Faint blue star	Galactic Pole regions		$0^m.1$	1′	114	—	15.0	pg	Feige (1958)
FIT.G	Galaxy	07^h42^m to 08^h04^m	−25° to −27°	$0^m.1$	1′	18	—	19.	pg	Fitzgerald (1974a)
FIT.N	Dark nebula	12^h38^m	−78°	$0^m.1$	1′	1	0′.1	21.	pg	Fitzgerald (1974b)
GCL	Globular star cluster	00^h to 24^h	−90° to +90°	$0^m.1$	1′	125	0′.1	14.4	Var	Alter et al. (1970)
GO	Star chain	10^h40^m	−58°	$0^m.1$	1′	75	0′.1	19.9	pg	Grossie & Öpik (1968)
GRA	Peculiar ring galaxy	06^h43^m	−74°	$0^m.1$	1′	1	1″	—	—	Graham (1974)
HARO	Blue emission line galaxy	Various Areas		$0^m.1$	1′	44	—	—	—	Haro (1956)
HAW	Galaxy	Various		$0^m.1$	1′	2	1′	20.0	pg	Harrington & Wilson (1950)
HELW	Nebula	Various		$0^m.1$	1′	508	—	—	—	Knox-Shaw (1912), 1915), Gregory (1921a,b), and Madwar (1932)
HH	Herbig-Haro object	Various areas		$0^s.1$	1″	77	1″	—	—	Herbig (1974)

TABLE 3 (*continued*)

List Pointer	Type of Object	Right Ascension Range	Declination Range	Original Right Ascension Precision	Original Declination Precision	Number of Objects	Original Angular Diameter Precision	Faintest Magnitude	Type of Magnitude	Reference
HMS	Galaxy	00^h to 24^h	−90° to +90°	$0^m.1$	1′	35	—	—	—	Humason et al. (1956)
	Galaxy cluster	00^h to 24^h	−90° to +90°	$0^m.1$	1′	21	—	—	—	Humason et al. (1956)
HN	Nebula	Various		$0^m.1$	1′	1248	—	—	—	Harvard Annals (1908)
	Nebula	Various		$0^s.1$	0′.1	1650	0′.1	—	—	Bailey (1913)
HO	Galaxy	Various		$0^m.1$	1′	5	1′	13.2	pg	Holmberg (1950)
HOAG	Compact galaxy with ring	15^h15^m	+21°46′	$0^m.1$	1′	1	1″	15.	pg	O'Connell et al. (1974), Hoag (1950)
HOD.60	Galaxy in Fornax cluster	03^h32^m to 03^h42^m	−33° to −36°	$0^m.1$	1′	17	0′.1	17.5	pg	Hodge (1959, 1960b)
HOD.61	Globular cluster in Fornax dwarf galaxy	02^h35^m to 02^h40^m	−34° to —35°	$0^m.1$	1′	5	0′.1	—	—	Hodge (1961)
HOD.72	Dark nebula in LMC	05^h to 06^h	−67° to −71°	$0^m.1$	1′	68	1′	—	—	Hodge (1972)
HODG	Red globular cluster in LMC	05^h to 06^h	−67° to −74°	$0^m.1$	1′	12	0′.1	—	—	Hodge (1960a)
HOFF	Dark hole	10^h20^m to 11^h30^m	−60°	1^m	1′	10	0°.01	—	—	Hoffleit (1953)
HOLM	Multiple galaxy	00^h to 24^h	−20° to +90°	$0^m.1$	1′	1854*	0′.1	16.1	pg	Holmberg (1937)
HOW	Star cluster in SMC	00^h16^m to 00^h42^m	−71° to 75°	$0^m.1$	0′.1	86	0′.1	—	—	Hodge & Wright (1974)
HPW	Dwarf galaxy in Fornax cluster	03^h13^m to 03^h41^m	−33° to −41°	$0^m.1$	1′	50	—	—	—	Hodge et al. (1964)
HSE	Star cluster in LMC	04^h15^m to 06^h23^m	−64° to −74°	$0^m.1$	1′	458	—	—	—	Hodge & Sexton (1966)
HSN	Bright galaxy behind SMC	00^h36^m to 01^h18^m	−70° to −75°	$0^m.1$	1′	30	—	—	—	Hodge & Snow (1975)
HUB	Diffuse nebula	Galactic plane		4′	0°.1	62	—	—	—	Hubble (1922)
HZ	Faint blue star	Hyades; North Galactic Pole		$0^m.1$	1′	48	—	15.1	pg	Humason & Zwicky (1947)

IC	Nonstellar object	00^h to 24^h	−90° to +90°	Var	Var	5398	Var	—	—	Dreyer (1895, 1908) and others (see text)
ISS	Stellar ring	00^h to 24^h	−33° to +67°	$0^m.1$	1′	1002	0′.01	—	—	Isserstedt (1968)
	Same	Same	−45° to −33°	$0^m.1$	1′	65	0′.1	—	—	Isserstedt (1970)
KARA.68	Dwarf galaxy	00^h to 24^h	−27° to +90°	$0^m.1$	1′	241	6″.7	—	—	Karachentseva (1968)
KARA.72	Isolated pair of galaxies	00^h to 24^h	−30° to +90°	$0^m.1$	1′	602	0′.1	16.	pg	Karachentseva (1972)
KARA.73	Dwarf galaxy	00^h to 24^h	−45° to −33°	$0^m.1$	1′	59	6″.7	—	—	Karachentseva (1973a)
KARA.73B	Isolated galaxy	00^h to 24^h	−90° to +90°	$0^m.1$	1′	1051	0′.1	16.	pg	Karachentseva (1973b)
KAZ	Planetary nebula	Various		1^m	1′	2	1″	12.98	pv	Kazaryan (1966)
KEEL	Nebula	Various areas		$0^s.1$	1″	744	—	18.	pg	Keeler (1900)
KEEN	Galaxy	04^h to 18^h	+60° to +80°	$0^m.1$	1′	32	0′.1	14.	pg	Keenan (1935)
KHAV	Dark nebula	Galactic Plane		1^m	0°.1	797	0′.1	—	—	Khavtassi (1955)
KLEM	Galaxy group	00^h to 24^h	−25° to −50°	$0^m.1$	1′	44	1′	16.	pg	Klemola (1969)
KN 10	Nebula	00^h45^m to 01^h07^m	+05° to +09°	$0^s.1$	1″	62	—	—	—	Wolf (1909a)
KN 11	Nebula	02^h45^m to 03^h05^m	+01° to +06°	$0^s.1$	1″	94	—	—	—	Wolf (1909b)
KN 12	Nebula	11^h25^m to 12^h	+45° to +51°	$0^s.1$	1″	279	—	—	—	Wolf (1911)
KN 13	Nebula	23^h45^m to 00^h15^m	+10° to +15°	$0^s.1$	1″	113	—	—	—	Wolf (1912)
KN 14	Nebula	01^h10^m to 01^h40^m	+27° to +33°	$0^s.1$	1″	516	—	—	—	Wolf & Ernst (1913)
KN 15	Nebula	09^h45^m to 10^h15^m	+33° to +39°	$0^s.1$	1″	189	—	—	—	Wolf (1916)
KN 16	Nebula	12^h25^m to 12^h50^m	+27° to +30°	$0^s.1$	0′.1	702	—	—	—	Wolf (1928)
KON	Extragalactic object	08^h34^m	−26°	$0^m.1$	1′	1	—	13.	pg	Kondratjeva (1972)
KW	Seyfert galaxy	00^h to 24^h	−90° to +90°	$0^m.1$	1′	71	1″	—	—	Khachikian & Weedman (1974)
LB	Faint blue star	Various areas (see appendix 1)		Var. (see appendix 1)		11444	—	20.8	pg	Luyten et al. (1967a, b) (50 papers)
LBN	Bright nebula	00^h to 24^h	−33° to +90°	$0^m.1$	1′	1125	1′	—	—	Lynds (1965)
LDN	Dark nebula	00^h to 24^h	−33° to +90°	$0^m.1$	1′	1800	20″	—	—	Lynds (1962)
LH	Stellar association in LMC	04^h50^m to 05^h55^m	−66° to −71°31′	$0^m.1$	1′	122	0′.1	—	—	Lucke & Hodge (1970)
LH 115	Hα emission nebula in SMC	00^h20^m to 01^h30^m	−70° to −74°	$0^m.1$	1′	117	1″	—	—	Henize (1956)
LH 120	Hα emission nebula in LMC	04^h40^m to 06^h20^m	−66° to −70°	$0^m.1$	1′	411	1″	—	—	Henize (1956)
LIN	SMC Hα emission line object	00^h15^m to 02^h15^m	−72°30′ to −76°30′	$0^m.1$	1′	596	—	17.8	pg	Lindsay (1961)
LIN.CL	Star cluster in SMC	00^h to 02^h	−70° to −78°	$0^m.1$	1′	116	0′.1	16.4	pg	Lindsay (1958)

TABLE 3 (*continued*)

List Pointer	Type of Object	Right Ascension Range	Declination Range	Original Right Ascension Precision	Original Declination Precision	Number of Objects	Original Angular Diameter Precision	Faintest Magnitude	Type of Magnitude	Reference
LW	Star cluster in LMC	03^h45^m to 06^h35^m	−62° to −76°	$0^m.1$	1′	484	—	—	—	Lynga & Westerlund (1963)
LYNG	OB concentration	10^h to 16^h	−50° to −60°	1^s	0′.1	12	0°.1	—	—	Lynga (1970)
MACL	Supernova remnant in SMC	00^h45^m	−73°	1^s	1″	1	3″	—	—	Mathewson & Clarke (1972)
MAFFEI	Infrared object	02^h36^m	+59°24′	$0^m.1$	0′.1	2	1″	—	—	Maffei (1968)
MAI	Dwarf spheroidal galaxy	00^h to 24^h	+56° to +90°	1^s	1″	104	6″.7	—	—	Mailyan (1973)
MCG (pt. I)	Galaxy	00^h to 24^h	+45° to +90°	$0^m.1$	1′	~5400	0′.1	20.	pg	Vorontsov-Velyaminov & Krasnogorskaja (1962)
(pt. II)	Galaxy	00^h to 24^h	+15° to +45°	$0^m.1$	1′	~9500	0′.1	20.	pg	Vorontsov-Velyaminov & Arhipova (1964)
(pt. III)	Galaxy	00^h to 24^h	−09° to +15°	$0^m.1$	1′	~7100	0′.1	20.	pg	Vorontsov-Velyaminov & Arhipova (1963)
(pt. IV)	Galaxy	00^h to 24^h	−33° to −09°	$0^m.1$	1′	~6800	0′.1	20.	pg	Vorontsov-Velyaminov & Arhipova (1968)
(pt. V)	Galaxy	00^h to 24^h	−33° to −45°	$0^m.1$	1′	1200	0′.1	20.	pg	Vorontsov-Velyaminov & Arhipova (1974)
					(total = 29,981)					
MEL	Nebula	Various		$0^m.1$	1′	4	—	—	—	Melotte (1926)
MHW	Supernova remnant in LMC	05^h26^m	−66°	$0^m.1$	1′	1	1′	—	—	Mathewson et al. (1963)
MIL	Supernova remnant	00^h to 24^h	−90° to +90°	1^s	0′.1	97	0′.1	—	—	Milne (1970)
MIN.46	Diffuse nebula	Galactic plane		$0^m.1$	1′	23	—	—	—	Minkowski (1946)
MIN.47	Diffuse nebula	Galactic plane		$0^m.1$	1′	18	—	—	—	Minkowski (1947)

MIN.48	Diffuse nebula	Galactic plane		$0^{m}.1$	1′	7	—	—	—	Minkowski (1948)
MKW	Poor galaxy cluster	09^{h} to 16^{h}	−04° to +29°	$0^{m}.1$	1′	16	—	15.5	—	Morgan et al. (1975)
MOHR	Star cluster in SMC	00^{h} to 02^{h}	−70° to −75°	$0^{m}.1$	0′.1	6	—	16.5	H	Mohr (1935)
MRK	Galaxy with UV continuum	00^{h} to 24^{h}	−11° to +90°	$0^{m}.1$	1′	700	1″	17.5	pg	Markarian (1967, 1969a,b, 1971), Markarian and Lipovetskii (1972, 1973a, 1974)
MRSL	H II region	00^{h} to 24^{h}	−90° to +90°	$0^{m}.1$	1′	697	0′.1	—	—	Marsalkova (1974)
NAB	Quasi-stellar object	Various		1^{s}	1″	9	—	17.6	B	Bahcall et al. (1973)
OCL	Open star cluster	00^{h} to 24^{h}	−90° to +90°	$0^{m}.1$	1′	1039	0′.1	18.	var	Alter et al. (1970)
PHL	Blue stellar object	21^{h} to 04^{h}	−33° to +17°	$0^{m}.1$	1′	8745	—	19.0	pg	Haro & Luyten (1962)
PIS	Star cluster	06^{h}	+20°	$0^{m}.1$	0′.1	1	0′.1	—	—	Pismis (1970)
PK	Planetary nebula	00^{h} to 24^{h}	−90° to +90°	$0^{s}.1$	1″	1037	1″	18.9	pg	Perek & Kohoutek (1967; Errata: 1968, 1969), Update: Acker (1975)
PRA	Blue nonstellar object	Various		$0^{s}.4$	0′.1	6	—	17.	pg	Prata (1966)
RB	Galaxy in Coma cluster	$12^{h}57^{m}$	+28°	$0^{m}.1$	1′	271	0″.1	—	—	Rood & Baum (1967)
RDS	Galaxy in ARC 2199	$16^{h}27^{m}$	+39°	$0^{m}.1$	0′.1	170	1″	—	—	Rood & Sastry (1972)
REA	Dwarf galaxy	12^{h} to 13^{h}	+03° to +20°	$0^{m}.1$	1′	90	1″	—	—	Reaves (1952, 1962)
REA.66	Dwarf galaxy	$12^{h}57^{m}$	+28°	1^{s}	0.1	4	—	—	—	Reaves (1966)
REIN 1	Nebula	Various		$0^{s}.01$	0″.1	141	—	—	—	Reinmuth (1916)
REIN 2	Nebula	Various		$0^{s}.01$	0″.1	336	—	—	—	Reinmuth (1927)
REIN 3	Nebula	$12^{h}00^{m}$ to $12^{h}45^{m}$	+08° to +14°	$0^{s}.01$	0″.1	368	—	—	—	Reinmuth (1927)
REIN 4	Nebula	Various		$0^{s}.01$	0″.1	326	—	—	—	Reinmuth (1928)
REIN 5	Nebula	$13^{h}15^{m}$ to $13^{h}30^{m}$	+10° to +13°	$0^{s}.01$	0″.1	288	—	—	—	Reinmuth (1929)
REIN 6	Nebula	Various		$0^{s}.01$	0″.1	263	—	—	—	Reinmuth (1932)
REIN 7	Nebula	Various		$0^{s}.01$	0″.1	240	—	—	—	Reinmuth (1940)
REIZ	Galaxy	10^{h} to 16^{h}	−13° to +70°	$0^{m}.1$	1′	4666	0′.1	16.5	pg	Reiz (1941)
RIC	Blue object	$00^{h}30^{m}$ to $00^{h}50^{m}$	+39° to +42°	$0^{s}.01$	0″.1	998	—	21.3	V	Richter (1974)
RLWT	Faint very blue star	Various		$0^{m}.1$	1′	152	—	20.	V	Rubin & Losee (1971), Rubin et al. (1974)
RMB	Faint blue object	$12^{h}15^{m}$ to $12^{h}40^{m}$	+10° to +16°	$0^{m}.1$	1′	228	—	—	—	Rubin et al. (1967)

TABLE 3 (*continued*)

List Pointer	Type of Object	Right Ascension Range	Declination Range	Original Right Ascension Precision	Original Declination Precision	Number of Objects	Original Angular Diameter Precision	Faintest Magnitude	Type of Magnitude	Reference
RNGC	Various	00^h to 24^h	−90° to +90°	$0^m.1$	1′	8164	—	17.0	var	Sulentic & Tifft (1973)
RRS	Blue object	13^h25^m to 13^h50^m	+28° to +32°	1^s	1′	304	—	19.76	V	Richter et al. (1968)
RS	Blue stellar object	13^h25^m to 13^h40^m	+25° to +29°	$0^s.1$	0′.1	53	—	18.7	V	Richter & Sahakjan (1965)
SA	Galaxy	Various		$0^m.1$	1′	6	0′.1	12.9	H	Shapley & Ames (1932)
SAAK	Extremely blue object	17^h24^m	+42°	$0^m.1$	1′	3	—	18.2	pg	Saakyan (1965)
SAP	Stellar group	05^h18^m	−68°	1^m	0°.5	1	1′	—	—	Sanduleak & Philip (1963)
SB	Galaxy with emission lines	00^h to 01^h	−17° to −40°	$0^m.1$	1′	3	—	16.	pg	Slettebak & Brundage (1971)
SC	Nebula	00^h to 24^h	−90° to +90°	1^s	0′.1	2621	—	—	—	Shapley et al. (1924)
SCH	Peculiar galaxy	Various		$0^m.1$	1′	91	—	15.4	pg	Schanberg (1973)
SCHO	Dark cloud	00^h to 24^h	−36° to +90°	$0^m.1$	0′.1	1456	1′	—	—	Schoenberg (1964)
SEA	Galaxy chain	12^h40^m	−40°	$0^m.1$	0′.1	9	0′.1	16.6	pg	Sersic & Aguero (1972)
SER	Peculiar object	00^h to 24^h	−90° to −43°	$0^m.1$	1′	184	10″	18.5	pg	Sersic (1974)
SEY	Faint galaxy	03^h to 12^h	+30° to +75°	1^s	0′.1	193	0′.1	15.4	H	Seyfert (1937)
SG 1	Diffuse emission nebula	Galactic plane		$0^m.1$	1′	36	1′	—	—	Shain & Gaze (1950)
SG 2	Diffuse emission nebula	Galactic plane		$0^m.1$	1′	101	1′	—	—	Gaze & Shain (1951)
SG 3	Diffuse emission nebula	Galactic plane		$0^m.1$	1′	250	1′	—	—	Gaze & Shain (1952)
SHAH	Group of compact galaxies	Various areas		$0^m.1$	1′	175	0′.1	18.7	red	Shahbazian (1973), Shahbazian & Petrozian (1974), Baier et al. (1974)
SHAP	Galaxy	02^h45^m to 04^h45^m	−60° to −45°	1^s	0′.1	7893	1″	18.0	H	Shapley (1935)
SHB	QSO	00^h to 24^h	−50° to +90°	Various		403	—	20.	V	Smith-Haenni (1977)

SHER	Dwarf galaxy	05^h10^m	$-33°$	1^m	$0°.1$	1	$10''$	—	—	Sher (1963)
SHP	Sculptor dwarf galaxy	00^h55^m	$-34°$	$0^m.1$	$1'$	1	$1'$	10.	H	Shapley (1938)
SHRV.73A	Globular cluster in M31	00^h30^m to 00^h50^m	$+39°$ to $+42°$	$0^m.1$	$1'$	25	$0''.1$	18.2	V	Sharov (1973a)
SHRV.73B	Compact galaxy near M31	00^h25^m to 00^h55^m	$+39°$ to $+43°$	$0^m.1$	$1'$	50	$0''.1$	17.2	V	Sharov (1973b)
SIV	Faint H emission region	Various		$10'$	$10'$	11	$10'$	—	—	Sivan (1974)
SL	Star cluster in LMC	04^h30^m to 06^h30^m	$-64°$ to $-75°$	$0^m.1$	$1'$	898	$1''$	—	—	Shapley & Lindsay (1963), Lindsay (1974)
SM 1	Blue galaxy	17^h17^m	$+44°$	$0^m.1$	$1'$	2	—	18.7	V	Saakyan & Mnatsakanyan (1965a)
SM 2	Faint blue object	17^h18^m	$+43°$	$0^m.1$	$1'$	28	—	19.0	B	Saakyan & Mnatsakanyan (1965b)
SMI	Faint nebulosity	13^h to 16^h	$-51°$ to $-63°$	$0^m.1$	$1'$	14	$0°.05$	—	—	Smith (1972)
SN	Supernova	00^h to 24^h	$-90°$ to $+90°$	$0^m.1$	$1'$	414	—	20.5	pg	Kowal & Sargent (1971), Kowal (1975)
SNO	Group of galaxies	00^h to 24^h	$-22°$ to $-46°$	$0^m.1$	$0'.1$	34	$1'$	19.	B	Snow (1970)
SPA	Galaxy	12^h45^m	$+14°$	$0^m.1$	$1'$	7	$0'.1$	16.	H	Shapley & Paraskevopoulos (1940)
SPC	Galaxy trio	00^h04^m	$-42°$	$0^m.1$	$1'$	3	$0'.1$	12.	pg	Sersic et al. (1968)
SS	Diffuse galactic nebula	Various		$0^m.1$	$1'$	74	—	—	—	Struve & Straka (1962)
ST	H emission nebula	05^h to 07^h30^m	$-20°$ to $+22°$	$0^m.1$	$1'$	42	$1'$	—	—	Strohmeier (1950)
STOCK	Blue knot near galaxy	01^h to 14^h	$-03°$ to $+39°$	$0^m.1$	$1'$	49	—	—	—	Stockton (1968)
SV	Blue stellar object	Various		$0^s.01$	$0''.1$	4	—	19.5	V	Sandage & Veron (1965)
SVEN	Galaxy	00^h to 24^h	$-40°$ to $+15°$	$0^m.1$	$1'$	461	$0'.1$	6.8	—	Svenonius (1937)
T	Galaxy in Cancer cluster	08^h17^m	$+21°$	$0^m.1$	$0'.1$	8	—	16.5	pg	Tifft et al. (1973)
TER	Star cluster	17^h30^m to 18^h10^m	$-35°$ to $-22°$	$0^s.01$	$0''.1$	11	$5''$	—	—	Terzan (1971)
TON-N	Blue star	North Galactic Pole		$0^m.1$	$1'$	1588	—	17.1	pg	Iriarte & Chavira (1957), Chavira (1959)
TON-S	Blue star	South Galactic Pole		$0^m.1$	$1'$	419	—	16.2	pg	Chavira (1958)
UGC	Galaxy	00^h to 24^h	$-02°30'$ to $+90°$	$0^m.1$	$1'$	12940	$0'.1$	19.	pg	Nilson (1973)

TABLE 3 (*continued*)

List Pointer	Type of Object	Right Ascension Range	Declination Range	Original Right Ascension Precision	Original Declination Precision	Number of Objects	Original Angular Diameter Precision	Faintest Magnitude	Type of Magnitude	Reference
URA	Stellar ring	Various		1^s	1′	89	0.′1	—	—	Uranova (1973a,b)
VB	Stellar ring	00^h to 04^h	+50° to +70°	$0^m.1$	1′	257	7″	—	⅛	Vidal & Bern (1973)
VDB.66B	Blue object	00^h36^m	+39°	1^s	0.′1	33	—	20.5	B	van den Bergh (1966a)
VDB.66G	Dwarf galaxy	00^h to 24^h	−33° to +90°	1^m	1′	243	35″	—	—	van den Bergh (1966c)
VDB.66N	Reflection nebula	00^h to 24^h	−33 to +90°	0.°1	0.°1	158	—	10.12	V	van den Bergh (1966b)
VHA	Star cluster	07^h to 18^h	−70° to −27°	$0^m.1$	1′	262	0.′5	—	—	van den Bergh & Hagen (1975)
VHE	Reflection nebula	07^h to 18^h	−70° to -27^h	$0^m.1$	1′	136	0.′1	—	—	van den Bergh & Herbst (1975)
VMT	Supernova remnant	00^h to 24^h	−90° to +90°	1^s	0.′1	24	0.′1	—	—	van den Bergh et al. (1973)
VV	Interacting galaxy	00^h to 24^h	−90° to +90°	$0^m.1$	1′	922	0.′1	20.	pg	Vorontsov-Velyaminov (1959)
VVI	Seyfert galaxy	00^h to 24^h	−90° to +90°	$0^m.1$	1′	95	1″	19.	pg	Vorontsov-Velyaminov & Ivanisevic (1974)
WEED	Very blue stellar object	Various		$0^m.1$	1′	9	—	19.0	B	Weedman (1971)

WEI	Faint blue object	13^h05^m	+29°	$0^m.1$	1′	11	–	18.0	V	Weistrop (1973)
WIL	Emission line object	11^h44^m	−03°30′	1^s	1″	1	—	15.	pg	Wild (1953), Zwicky & Humason (1961)
WK	Nebula	03^h12^m	+41°	$0^s.01$	0″.1	124	—	—	—	Wolf & Kaiser (1913)
WRAY	H_α emission object	08^h to 18^h	−64° to −23°	$0^s.1$	1″	55	—	10.	V	Wray (1966)
WS	Planetary nebula in LMC	04^h10^m to 06^h20^m	−74° to −67°	1^s	0′.1	42	–	18.1	B	Westerlund & Smith (1963)
YC	Unusual object	00^h to 24^h	−58° to −34°	1^s	0′.1	40	—	—	—	Lü (1971)
YM	Symmetric galactic nebula	Various		0°.1	0°.1	47	0′.1	—	—	Johnson (1955, 1956)
ZC	Cluster of galaxies	00^h to 24^h	—30° to +90°	$0^m.1$	1′	9133	1′	—	—	Zwicky et al. (1961), Zwicky & Herzog (1963, 1966, 1968), Zwicky et al. (1968), Zwicky & Kowal (1968)
ZCG	Compact galaxy	Various		$0^m.1$	1	280	—	19.3	pg	Zwicky (1971)
ZH	Galaxy	01^h to 24^h	−01°	$0^m.1$	1′	85	—	16.7	pg	Zwicky & Humason (1964)
ZL	Ultrafaint blue star	Various		1^s	0′.1	218	—	22.8	pg	Zwicky & Lyuten (1967)
ZWG	Galaxy	00^h to 24^h	−30° to +90°	$0^m.1$	1′	29378	—	15.9	pg	Zwicky et al. (1961), Zwicky & Herzog (1963, 1966, 1968), Zwicky et al. (1968), Zwicky & Kowal (1968)
1ZW to 7ZW	Compact galaxy	Various		$0^m.1$	1′	2310	—	—	—	Zwicky (1961–68)
8ZW	Compact galaxy	Various		$0^m.1$	1′	504	—	20.3	pg	Zwicky et al. (1975)

*(Components)

Bibliography

The letters preceeding the references are the catalogue pointers used in this work. The same pointer may occur with several different references, so as to include all parts of multipartite catalogues, errata, explanatory papers, and so on. Journal names have either been spelled out entirely, or are abbreviated as by Kemp (1970).

ARC Abell, G. O. 1958. The distribution of rich clusters of galaxies. *Astrophys. J. Suppl.* 3:211.

ACK Acker, A. 1975. Private communication.

AGU Aguero, E. L. 1971. A list of interesting southern galaxies. *Publ. Astr. Soc. Pacific* 83:310.

ALL Allen, D. A. 1973. A new planetary nebula. *Observatory* 93:85.

ASS OCL GCL Alter, G., B. Balasz, and J. Ruprecht. 1970. *Catalogue of star clusters and associations.* Budapest: Akademia Kiado.

AMES Ames, A. 1932. A catalogue of 2778 nebulae, including the Comma-Virgo group. *Harvard Annals* 88:1.

ARP.65 Arp, H. 1965. A very small condensed galaxy. Astrophys. J. 142:402.

ARP Arp, H. 1966. Atlas of peculiar galaxies. *Astrophys. J. Suppl.* 14:1.

AR Astronomer Royal (Communicated by). 1910. Observations of nebulae made at the Royal Observatory, Greenwich. *Mon. Not. Roy. Astr. Soc.* 71:509.

BAA Baade, W., and H. Arp. 1964. Positions of emission nebulae in M31. *Astrophys. J.* 139:1027.

BB Bahcall, N. A., and J. N. Bahcall. 1970. Galaxies in the direction of QSO's with small redshifts. *Publ. Astr. Soc. Pacific* 82:1276.

NAB Bahcall, N. A., J. N. Bahcall, and M. Schmidt. 1973. Observations of QSO's in the direction of clusters of galaxies. *Astrophys. J.* 183:777.

SHAH Baier, F. W., M. B. Petrosian, H. Tiersch, and R. K. Shabhazian. 1974. Compact groups of compct galaxies III. *Astrofizika* 10:327.

HN Bailey, S. I. 1913. 1649 new nebulae. *Harvard Annals* 72:17.

BAK1 Baker, R. H. 1933. A catalogue of 985 galaxies in Fornax and Eridanus. *Harvard Annals* 88:77.

BAK2 Baker, R. H. 1937. A catalogue of 1113 galaxies in a region of Fornax and Eridanus. *Harvard Annals* 88:163.

A3 Barbieri, C., and P. Benvenuti. 1974. Studies of blue objects at high galactic latitudes III. Faint blue objects in the field of BD+15°2469. *Ast. & Astrophys. Suppl.* 13:269.

BC Barbieri, C., and M. Capacciolo. 1974. Catalogue of quasistellar objects. *Publ. Oss. Astr. Padov.* Edited July 1, 1974.

AL Barbieri, C., L. Erculiani, and J. Rosino. 1968. Faint blue objects in the field of 88 Leonis. *Publ. Oss. Astr. Padova,* No. 143.

A2 Barbieri, C., and L. Rosino. 1972. Studies of blue objects at high galactic latitudes. *Astrophys. and Sp. Sci.* 16:324.

B Barnard, E. E. 1927. A photographic atlas of selected regions of the Milky Way. Carnegie Institute of Washington, No. 247.

BEM Bertola, F., and P. Maffei. 1974. Two faint companions to M81. *Ast. & Astrophys.* 32:117.

BIGO Bigourdan, G. 1912. Cinquième liste de nébuleuses découverts à l'Observatoire de Paris. *Comptes rendus* 155:1049.

BOH Bohlin, K. 1922. Ein Nebelobjekt im Taurus. *Ast. Nach* 216:31.

Bolton, J. G., F. F. Gardner, and M. B. Mackey. 1964. The Parkes catalogue of radio sources declination zone −20° to −60°. *Aust. J. Phys.* 17:340.

BON Bond, H. E. 1972. The optically variable galaxies: V 395 Herculis. *Astrophys. J.* 174:L163.

BON — Bond, H. E. 1973. A search for extragalactic objects in the General catalogue of variable stars. *Astrophys J.* 181:L23.

BOR — Börngen, F. 1966. Blaue Objekte in einem Feld bei −22° Galakteischer Breite. *Mitt. Karl Schwarzchild Obs. Tautenberg,* Nr. 28.

BFL — Börngen, F., G. Friedrich, G. Lenk, R. Richter, and N. B. Richter. 1970. OB-stars on the outmost borders of M31. *Mitt. Karl Schwarzchild Obs. Tautenberg,* Nr. 50.

BCL — Boulesteix, J. G. Courtes, et al. 1974. An optical study of M33: I. Morphology of the gas. *Ast. & Astrophys.* 37:33.

BFG — Bracessi, A., L. Formiggini, and E. Gandolfi. 1970. Magnitudes, colours, and coordinates of 175 ultraviolet-excess objects in the field 13^h, + 36°. *Ast. & Astrophys.* 5:264.

BFG — Bracessi, A., L. Formiggini, and E. Gandolfi. 1973. Erratum: Correction of an error in the paper: Magnitudes, colours and coordinates of 175 UV-excess objects in the field 13^h, +36°. *Ast. & Astrophys.* 23:159.

BFGS — Bracessi, A., L. Formiggini, and I. Gioia. 1972. Two galaxies dominates by bright ultraviolet knots. *Publ. Astr. Soc. Pacific* 84:592.

BRON — Bronkalla, W. 1971. Schwache blaue Objekte in der Nahe des galaktischen Nordpols (Umgebung von M3). *Ast. Nach* 292:263.

CAR — Carpenter, E. F. 1931. A cluster of extragalactic nebulae in Cancer. *Publ. Astr. Soc. Pacific* 43:247.

CED — Cederblad, S. 1946. Catalogue of bright diffuse galactic nebulae. *Lund Annals,* Series 2, No. 119.

TON-S — Chavira, E. 1958. Estrellas azules en el casquete galactico, sur. *Bol. Obs. Ton. y Tacubaya* Vol. 2, No. 17, 15.

TON-N — Chavira, E. 1959. Estrellas Azules en el casquete galactico norte II. *Bol. Obs. Ton. y Tacubaya* Vol. 2, No. 18, 3.

CM — Cheriguene, M. G., and Monnet, G. 1972. Etude cinématique de l'hydrogen ionise dans le Grand Nuage de Magellan. *Ast. & Astrophys.* 16:28.

CR — Chincarini, G., and Rood, H. J. 1971. Dynamics of the Perseus cluster of galaxies. *Astrophys. J.* 168:321.

CHR — Christie, W. H. 1931. Note on the cluster of extragalactic nebulae in Leo. *Publ. Astr. Soc. Pacific* 43:350.

IC — Corwin, G. 1974. Private communications.

COU — Courtès, G. 1951. Etude de la voie Lactée Enluminière Monochromatique H de 320° à 350° et de 25° à 80° de longitude galactique. *Comptes rendus* 232:795.

CS — Courtès, G., and Sivan, J. P. 1972, Two new large-diameter galactic regions of faint H-alpha emission. *Astrophys. Lett.* 11:159.

DDKS — Danziger, I. J., Dennefeld, H., Kunth, D., and Schuster, H. E. 1974. A large southern reflection nebula at high galactic latitude. *Ast. & Astrophys.* 37:419.

DV.55 — de Vaucouleurs, G. 1955. Emission nebulosities near the South Pole. *Observatory* 75:129.

DV.56 — de Vaucouleurs, G. 1956. Survey of brighter galaxies south of $\delta = -35°$. *Mem. of the Commonwealth Obs.* Vol. 3, No. 3.

DVDV — de Vaucouleurs, G., and de Vaucouleurs, A. 1961. Classification and radial velocities of bright southern galaxies. *Mem. Roy. Astr. Soc.* 68:69.

de Vaucouleurs, G., and de Vaucouleurs, A. 1974. *Reference catalogue of bright galaxies.* University of Texas Press, Austin.

DG — Dorschner, V. J., and Gürtler, J. 1964. Untersuchungen über Reflexionsnebel am Palomar Sky Survey I. Catalog. *Ast. Nach* 287:257.

IC — Dreyer, J. L. E. 1962. *New general catalogue (1888), Index catalogue (1895), and Second index catalogue (1908).* Royal Astronomical Society.

DUN — Duncan, J. C. 1923. Photographic studies of nebulae. *Astrophys. J.* 57:137.

FAI — Fairall, A. P. 1968. A list of compact and bright-nucleus galaxies. *Mon. Not. Astr. Soc. S. Afr.* 27:67.

FAI — Fairall, A. P. 1970. A second list of compact and bright-nucleus galaxies. *Mon. Not. Astr. Soc. S. Afr.* 29:48.

FATH — Fath, E. A. 1914. A study of nebulae. *Astr. J.* 28:75.

FEIG — Feige, J. 1958. A search for under-luminous hot stars. *Astr. J.* 128:267.

FIT.G — Fitzgerald, M. P. 1974a. Eighteen possible galaxies in Puppis. *Ast. & Astrophys.* 31:467.

FIT.N — Fitzgerald, M. P. 1974b. A remarkable "bright dark nebula" in Chamaeleon. *Ast. & Astrophys.* 32:465.

SG2 — Gaze, V. F., and Shain, G. A. 1951. A second list of diffuse nebulae. *Simeiz Crimean Astrophys. Obs. Contr.* 7:93.

SG3 — Gaze, V. F. and Shain, G. A. 1952. A third list of diffuse emission nebulae. *Simeiz Crimean Astrophys. Obs. Contr.* 9:52.

GRA — Graham, J. A. 1974. A peculiar southern ring galaxy. *Observatory* 94:290.

HELW — Gregory, C. C. L. 1921a. Third list of nebulae photographed with the Reynolds reflector. *Helwan Obs.* 21:201.

HELW — Gregory, C. C. L. 1921b. Fourth list of nebulae photographed with the Reynolds reflector. *Helwan Obs.* 22:219.

GO — Grossie, H. H. R., and Öpik, E. J. 1968. Star chains around Eta Carinae. *Irish Astr. J.* 8:249.

BA 11 — Grubissich, C. 1973. The stellar group Basel 11 in Taurus. *Ast. and Astrophys. Suppl.* 11, 283.

Gum, C. S. 1955. A survey of southern H II regions. *Mem. Roy. Astr. Soc.* 67:155.

HARO — Haro, G. 1956. Nota preliminar sombre galaxias azules con lineas de emision. *Bol. Obs. Ton. y Tacubaya* Vol. 2, No. 14, 8.

BV — Haro G., and Luyten, W. J. 1960. Note on some further faint blue variables. *Bol. Obs. Ton. y Tacubaya* Vol. 2, No. 19, 17.

PHL — Haro, G., and Luyten, W. J. 1962. Faint blue stars in the region near the South Galactic Pole. *Bol. Obs. Ton. y Tacubaya* Vol. 3, No. 22, 37.

HAW — Harrington, R. G., and Wilson, A. G. 1950. Two new stellar systems in Leo. *Publ. Astr. Soc. Pacific* 62:118.

HN — *Harvard Annals* (no author given). 1908. Nebulae discovered at the Harvard College Observatory. *Harvard Annals* 60:147.

LH115
LH120 — Henize, K. G. 1956. Catalogues of H-alpha emission stars and nebulae in the Large and Small Magellanic Clouds. *Astrophys. J. Suppl.* 2:315.

Her Majesty's Stationery Office. 1961. *Explanatory supplement to the Astronomical Ephemeris and the American Ephemeris and Nautical Almanac.*

HH — Herbig, G. H. 1974, Draft catalog of Herbig-Haro Objects. *Lick Obs. Bull.* No. 658, February.

Herschel, J. 1864, The general catalogue of nebulae. *Philosophical Transactions.*

HOAG — Hoag, A. A. 1950. A peculiar object in Serpens. *Astr. J.* 55:170.

HOD.60 — Hodge, P. W. 1959. Dwarf members of a southern cluster I. *Publ. Astr. Soc. Pacific* 71:28.

HODG — Hodge, P. W. 1960a. Studies of the Large Magellanic Cloud I. The red globular clusters. *Astrophys, J.* 131:351.

HOD.60 — Hodge, P. W. 1960b. Dwarf members of a southern cluster II. *Publ. Astr. Soc. Pacific* 72:188.

HOD.61 — Hodge, P. W. 1961. The Fornax dwarf galaxy. I. The globular clusters. *Astr. J.* 66:83.

Hodge, P. W. 1966. *An atlas and catalogue of H II regions in Galaxies.* University of Washington, Seattle.

HOD.72 — Hodge, P. W. 1972. Dark nebulae in the Large Magellanic Cloud. *Publ. Astr. Soc. Pacific.* 84:365.

HPW — Hodge, P. W., Pyper, D. M., and Webb, C. J. 1965. Dwarf galaxies in the Fornax Cluster. *Astr. J.* 70:559.

HSE — Hodge, P. W., and Sexton, J. A. 1966. 457 new star clusters in the Large Magellanic Cloud. *Astr.J.* 71:363.

HSN — Hodge, P. W., and Snow, T. P. 1975. Finding list of bright galaxies behind the Small Magellanic Cloud. *Astr. J.* 80:9.

HOW — Hodge, P. W., and Wright, F. W. 1974. Catalog of 86 new star clusters in the Small Magellanic Cloud. *Astr. J.* 79:858.

HOFF — Hoffleit, D. 1953. A preliminary survey of nebulosities and associated B-stars in Carina. *Harvard Annals* 119:37.

HOLM — Holmberg, E. 1937. A study of double and multiple galaxies together with inquiries into some general metagalactic problems. *Lund Annals* 6.

HO — Holmberg, E. 1950. A photometric study of nearby galaxies. *Lund Annals* Series 2, Vol. 6, No. 5.

HUB — Hubble, E. 1922. A general study of diffuse galactic nebulae. *Astrophys. J.* 56:162.

HMS — Humason, M. L., Mayall, N. U., and Sandage, A. R. 1956. Redshifts and magnitudes of extragalactic nebulae. *Astr. J.* 61:97.

HZ — Humason, M. L., and Zwicky, F. 1947. A search for faint blue stars. *Astrophys. J.* 105:85.

Institute for Scientific Information. 1961 to present. *Science citation index*. Philadelphia.

TON-N — Iriarte, B., and Chavira, E. 1957. Estrellas azules en el Casquete Galactico Norte. *Bol. Obs. Ton. y Tacubaya* 2, No. 16, 3.

ISS — Isserstedt, J. 1968. Sternringe, eine Neue Art von sternaggregaten und ihre Berzichung zur galaktischen Strutur. *Veroff. des Astronomischen Instituts der Ruhr-Universität Bochum* 1:1.

ISS — Isserstedt, J. 1970. Optical spiral tracers and galactic structure. *Ast. & Astrophys.* 9:70.

YM — Johnson, H. M. 1955. Symmetraic galactic nebulae. *Astrophys. J.* 121:604.

YM — Johnson, H. M. 1956. Symmetric galactic nebulae II. *Astrophys. J.* 124:90.

KARA.72 — Karachentsev, I. D. 1972. A catalog of isolated pairs of galaxies in the Northern Hemisphere. *Astrofiz. Issled. Izu, Spets. Astrofiz.* 7:3.

KARA.68 — Karachentseva, V. E. 1968. The distribution of sculptor-type dwarf galaxies. *Publ. Byurakan Obs.* 39:61.

KARA.73 — Karachentseva, V. E. 1973a. Dwarf galaxies of the sculptor type in the zones −36° to −42° of the Palomar Sky Atlas. *Astrofiz. Issled. Izv. Spets. Astrofiz.* 5:10.

KARA.73B — Karachentseva, V. E. 1973b. A catalog of isolated galaxies. *Astrofiz. Issled. Izv. Spets. Astrofiz.* 8, 3.

KAZ — Kazaryan, M. A. 1966. Two new planetary nebulae. *Astrofizika* 2:371.

KEEL — Keeler, J. E. 1900. Photographs of nebulae and clusters made with the Crossley Reflector. *Astrophys. J.* 11:325.

KEEN — Keenan, P. C. 1935. Studies of extragalactic nebulae. *Astrophys. J.* 83:62.

Kemp, D. A. 1970. *Astronomy and astrophysics: a bibliographical guide*. Archon Books, Hamden, Conn.

KW — Khachikian, E. Y., and Weedman, D. W. 1974, An atlas of Seyfert Galaxies. *Astrophys. J.* 192:581.

KHAV — Khavtassi, J. 1955. A statistical study of dark nebulae. *Buyll. Abastuman. Astrofiz. Obs.* No. 18, *29*.

BA6 — Kiral, A. 1969a. Die Sterngruppe Ba 6. *Ast. & Astrophys.* 2:22.

BA7 — Kiral, A. 1969b. Die Sterngruppe Ba 7. *Ast. & Astrophys.* 3:327.

KLEM — Klemola, A. R. 1969. Groups and clusters of southern galaxies. *Astr. J.* 74:804.

HELW — Knox-Shaw, H. 1912. Observations of nebulae made during 1909–1911. *Helwan Obs.* 9:69.

HELW — Knox-Shaw, H. 1915. Observations of nebulae made during 1912–1914. *Helwan Obs.* 15:129.

KON — Kondratjeva, L. N. 1972. He 2-10: an extragalactic object. *Asst. Tsirk.* No. 683, 7.

SN — Kowal, C. T., and Sargent, W. L. W. 1971. Supernovae discovered since 1885. *Astr. J.* 76:756.

SN — Kowal, C. T. 1975. Private communication.

LIN.CL — Lindsay, E. M. 1958. Cluster system of the Small Magellanic Cloud, *Mon. Not. Roy. Astr. Soc.* 118:172.

LIN — Lindsay, E. M. 1961. A new catalogue of emission-line stars and planetary nebulae in the Small Magellanic Cloud. *Astr. J.* 66:169.

SL — Lindsay, E. M. 1974. Remarks on some Lynga-Westerlund objects in the Large Magellanic Cloud *Irish Astr. J.* 6:233.

YC Lü, P. K. 1971. Some unusual Southern Hemisphere objects. *Astr. J.* 76:775.

LH Lucke, P. B., and Hodge, P. W. 1970. Catalogue of stellar associations in the Large Magellanic Cloud. *Astr. J.* 75:171.

Lundmark, K. 1930. A new general catalogue of nebulae. *Publ. Astr. Soc. Pacific* 423:31.

LB Luyten, W. 1967a. A search for faint blue stars. Papers 1–30. The Observatory, U. of Minn.

LB Luyten, W. 1967b. A search for faint blue stars. Papers 31–50. The Observatory, U. of Minn.

BV Luyten, W., and Haro, G. 1959. Note on some faint blue variables. *Publ. Astr. Soc. Pacific* 71:469.

LDN Lynds, B. T. 1962. Catalogue of dark nebulae. *Astrophys. J. Suppl.* 7:1.

LBN Lynds, B. T. 1965. Catalogue of bright nebulae. *Astrophys. J. Suppl.* 7:163.

LYNG Lynga, G. 1970. On the distribution of OB stars in the Southern Milky Way. *Ast. & Astrophys.* 8:41.

LW Lynga, G., and Westerlund, B. E. 1963. Catalogue of clusters in the outer parts of the Large Magellanic Cloud. *Mon. Not. Roy. Astr. Soc.* 127:31.

HELW Madwar, M. R. 1932. Sixth list of nebulae photographed with the Reynolds Reflector. *Helwan Obs.* 38:1.

MAFFEI Maffei, P. 1968. Infrared object in the region of IC 1805. *Publ. Astr. Soc. Pacific* 80:618.

MAI Mailyan, N. Sh. 1973. Search for dwarf spheroidal galaxies. *Astrofizika* 9:63.

MRK Markarian, B. E. 1967. Galaxies with an ultraviolet continuum I. *Astrofizika* 3:55.

MRK Markarian, B. E. 1969a. Galaxies with an ultraviolet continuum II. *Astrofizika* 5:443.

MRK Markarian, B. E. 1969b. Galaxies with an ultraviolet continuum III. *Astrofizika* 5:581.

MRK Markarian, B. E. 1971. Galaxies with an ultraviolet continuum IV. *Astrofizika* 7:229.

MRK Markarian, B. E., and Lipovetskii, V. A. 1972. Galaxies with an ultraviolet continuum V, *Astrofizika* 8:89.

MRK Markarian, B. E., and Lipovetskii, V. A. 1973. Galaxies with an ultraviolet continuum VI. *Astrofizika* 9:487.

MRK Markarian, B. E., and Lipovetskii, V. A. 1974. Galaxies with an ultraviolet continuum VII. *Astrofizika* 10:307.

MRSL Marsalkova, P. 1974. A comparison catalogue of H II Regions. *Astrophys. & Sp. Sci.* 27:3.

MACL Mathewson, D. S., and Clark, J. N. 1972. A supernova remnant in the Small Magellanic Cloud. *Astrophys. J.* 178:L105.

MHW Mathewson, D. S., Healey, J. R., and Westerlund, B. E. 1963. A supernova remnant in the Large Magellanic Cloud. *Nature* 199:681.

EM McCluskey, S. W. 1961. Emission objects near Selected Area 158. *Publ. Astr. Soc. Pacific* 73:264.

MEL Melotte, P. J. 1926. New nebulae shown on Franklin-Adams chart plates. *Mon. Not. Roy. Astr. Soc.* 86:636.

Messier, C. 1784. List of nebulous objects. Originally published in *Connaissance des temps* 1784; more recently published by H. Shapley and H. Davis: Catalogue of nebulae and clusters. *Observatory* 41:318–21, August, 1918.

MIL Milne, D. K. 1970. Non-thermal galactic radio sources. *Austral. J. Phys.* 23:425.

MIN.46 Minkowski, R. 1946. New emission nebulae. *Publ. Astr. Soc. Pacific* 53:305.

MIN.47 Minkowski, R. 1947. New emission nebulae. *Publ. Astr. Soc. Pacific* 59:257.

MIN.48 Minkowski, R. 1948. New emission nebulae. *Publ. Astr. Soc. Pacific* 60:386.

MOHR Mohr, J. 1935. Outlying star clusters of the Small Magellanic Cloud. *Harvard Bulletin* No. 899, 15.

MKW Morgan, W. W., Kayser, S., and White, R. A. 1975. CD galaxies in poor clusters. *Astrophys. J.* 199:545.

UGC Nilson, P. 1973. *Uppsala general catalogue of galaxies*. Nova Acta Regiae Societatis Scientiarum Ser. V, Vol. 1, Uppsala.

HOAG	O'Connell, R. W., Scargle, J. D., and Sargent, W. L. W. 1974. The nature of Hoag's Object. *Astrophys. J.* 191:61.
PK	Perek, L., and Kohoutek, L. 1967. *Catalogue of galactic planetary nebulae.* Academia Publishing House of the Czechoslovakian Academy of Sciences, Prague.
PK	Perek, L., and Kohoutek, L. 1968. Errata-Catalog of galactic planetary nebulae. *Bull. Astr. Soc. Czech.* 18:252.
PK	Perek, L., and Kohoutek, L. 1969. Errata-Catalog of galactic planetary nebulae. *Bull. Astr. Soc.Czech.* 20:381.
SHAH	Petrosian, M. B. 1974. Compact groups of compact galaxies IV. *Astrofizika* 10:471.
PIS	Pismis, P. 1970. Studies on star clusters: a new small cluster near NGC 2175 and some remarks on the latter. *Bol. Ton. y Tacubaya* Vol. 5, No. 34, 219.
PRA	Prata, S. W. 1966. Non-stellar objects in the Tonantzintla catalogue of blue stars. *Publ. Astr. Soc. Pacific* 78:61.
REA	Reaves, G. 1952. Dwarf galaxies in the Virgo Cluster. Ph.D. Dissertation, University of California, Berkeley.
REA	Reaves, G. 1962. Dwarf galaxies in the Virgo Cluster. *Publ. Astr. Soc. Pacific* 74:392.
REA.66	Reaves, G. 1966. Dwarf galaxies in the Coma Cluster. *Publ. Astr. Soc. Pacific* 78:407.
	Reaves, G. 1975. Private communication.
REIN1	Reinmuth, K. 1916. 141 photographische Nebel Positionen. *Veroff. der Sternwarte zu Heidelberg* 7:175.
REIN2	Reinmuth, K. 1927. Photographische Positionsbestimmung von Nebelflecken. *Veroff. der Sternwarte zu Heidelberg* 8:25.
REIN3	Reinmuth, K. 1927. Photographische Positionsbestimmung von Nebelflecken in Virgo. *Veroff. der Sternwarte zu Heidelberg* 8:69.
REIN4	Reinmuth, K. 1928. Photographische Positionsbestimmung von 317 Nebelflecken. *Veroff. der Sternwarte zu Heidelberg* 8:133.
REIN5	Reinmuth, K. 1929. Photographische Positionsbestimmung von 351 Nebelflecken und schwacharen Sternen. *Veroff. der Sternwarte zu Heidelberg* 8:167.
REIN6	Reinmuth, K. 1932. Photographische Positionsbestimmung von 234 Nebelflecken. *Veroff. der Sternwarte zu Heidelberg* 8:191.
REIN7	Reinmuth, K. 1940. Photographische Positionsbestimmung von 207 Nebelflecken. *Veroff. der Sternwarte zu Heidelberg* 12:27.
REIZ	Reiz, A. 1941. A study of external galaxies with special regard to the distribution problem. *Annals Obs. Lund.* 9.
RIC	Richter, G. A. 1974. Katalog blauer Objekte in der weiteren Ungebung von M31. *Veroff. Sternwarte Sonneberg* 8:75.
RRS	Richter, L., Richter, N., and Schnell, A. 1968. Statistik blauer Objekte in der Nahe des galaktisches Nordpoles: Teil II. *Mitt. Karl Schwarzchild Obs.* No. 38.
RS	Richter, N., and Sahakjan, K. 1965. Statistik blauer Objekte in der Nahe des galaktisches Nordpoles: Teil I. *Mitt. Karl-Schwarzchild-Obs.* No. 24.
	Rodgers, A. W., Campbell, C. T., and Whiteoak, J. B. 1960. A catalogue of H alpha emission regions in the Southern Milky Way. *Mon. Not. Roy. Astr. Soc.* 121:103.
RB	Rood, H. J., and Baum, W. A. 1967. Photographic brightness profiles of Coma Cluster galaxies: Catalog of program galaxies. *Astr. J.* 72:398.
RDS	Rood, H. J., and Sastry, G. N. 1972. Static properties of galaxies in the Abell Cluster 2199. *Astr. J.* 77:451.
RLWT	Rubin, V. C., and Losee, J. 1972. A finding list of faint blue stars in the anticenter region of the galaxy. *Astr. J.* 76:1099.
RMB	Rubin, V. C., Moore, S., and Bertiau, F. C. 1967. Faint blue objects in the Virgo Cluster Region. *Astr. J.* 72:59.

RLWT Rubin, V. C., Westpfahl, D., and Tuve, M. 1974. Second finding list of faint blue stars in the anticenter region of the galaxy. *Astr. J.* 79:1406.

Rudnicki, K., and Zwicky, F. 1967. A supernova on the intergalactic bridge. *A. J.* 72:407.

SAAK Saakyan, K. A. 1965. Two extremely blue objects near a D galaxy. *Astrofizika* 1:126.

SM1 Saakyan, K. A., and Mnatsakanyan, R. G. 1965a. On two blue galaxies. *Astrofizika* 1:125.

SM2 Saakyan, K. A., and Mnatsakanyan, R. G. 1965b. Faint blue stars in the region $\alpha = 17^h 18^m$, $\delta = 43°30'$ (1950.0). *Astrofizika* 1:229.

Sandage, A. 1965. The existence of a major new constituent of the universe: The quasi-stellar galaxies. *Ap. J.* 141:1560.

Sandage, A., and Luyten, W. J. 1967. On the nature of faint blue objects in high galactic latitudes I. Photometry, proper motions and spectra in PHL Field 1:36 +6° and Richter Field M3, II. *Astrophys. J.* 141:1560.

Sandage, A., and Luyten, W. J. 1967. On the nature of faint blue objects in high galactic latitudes. II. Summary of photometric results for 301 objects in seven survey fields. *Astrophys. J.* 155:913.

SV
BSO Sandage, A. R., and Vernon, P. 1965. Photometric results of a special survey of interlopers. *Astrophys. J.* 142:412.

SAP Sanduleak, N., and Philip, A. G. D. 1968. A stellar group in line of sight with the Large Magellanic Cloud. *Astrophys. J.* 73:113.

SCH Schanberg, B. C. 1973. An isophotometric and photographic atlas of peculiar galaxies. *Astrophys J. Suppl.* 26:115.

Schmidt, M. 1974. On the nature of faint blue objects in high galactic latitudes. III. A spectroscopic search for quasars in four survey fields. *Astrophys. J.* 193:509.

SCHO Schoenberg, E. 1964. Katalog von 1456 Dunkelwolken der nordlichen Milchstrasse bis zu sudlichen Deklination = −36°, *Bayerische Akad. Wissen.* Vol. 5, No. 26, 5.

SER Sersic, J. L. 1974. A list of peculiar galaxies, interacting pairs, groups, and clusters south of declination −43°. *Astrophys. & Sp. Sci.* 28:365.

SEA Sersic, J. L., and Aguero, E. L. 1972. Chain of galaxies in Centaurus. *Astrophys. & Sp. Sci.* 19:387.

SPC Sersic, J. L., Pastoriza, M. G., and Carranza, G. J. 1968. A trio of southern galaxies. *Astrophys. Lett.* 2:45.

SEY Seyfert, C. K. 1937. A study of faint northern galaxies. *Harvard Annals* 105:219.

SHAH Shahbazian, R. K. 1973. Compact groups of compact galaxies. I. *Astrofizika* 9:495.

SHAH Shdahbazian, R. K., and Petrozian, M. B. 1974. Compact groups of compact galaxies II. *Astrofizika* 10:13.

SG1 Shain, G. A., and Gaze, V. F. 1950. Certain results from the investigation of bright galactic nebulae. *Simeiz Crimean Astrophys. Obs. Contr.* 6:3.

SHAP Shapley, H. 1935. A catalog of 7889 external galaxies in Horogium and surrounding regions. *Harvard Annals* 88:107.

SHP Shapley, H. 1938. A stellar system of a new type. *Harvard Bulletin* No. 908, 1.

SA Shapley, H., and Ames, A. 1932. A survey of the external galaxies brighter than $13^m\!.0$. *Harvard Annals* 88:43.

SL Shapley, H., and Lindsay, E. M. 1963. A catalogue of clusters in the Large Magellanic Cloud. *Irish Astr. J.*6:74.

SPA Shapley, H., and Paraskevopoulos, J. 1940. Southern clusters and galaxies. *Harvard Bulletin* No. 914, 1.

SC Shapley, H., Menzel, D., and Campbell, L. 1924. Descriptions and positions of 2829 new nebulae. *Harvard Annals* 85:113.

SHRV.73A Sharov, A. S. 1973a. Compact galaxies in the region of Andromeda Nebula. *Astr. Zh.* 50:1023.

SHRV.73B Sharov, A. S. 1973b. Compact galaxies in the neighborhood of the Andromeda Nebula. *Astr. Zh.* 50:1023.

Sharpless, S. 1959. A catalog of H II regions. *Astrophys. J. Suppl.* 4:257.

SHER — Sher, D. 1963. An extended object in the southern sky. *Observatory* 83:256.

SIV — Sivan, J. P. 1974. Interstellar hydrogen wide field photographic H-alpha survey. *Ast. & Astrophys. Suppl.* 16:163.

SB — Slettebak, A., and Brundage, R. 1971. A finding list of early-type stars near the South Galactic Pole. *Astr. J.* 76:338.

SMI — Smith, B. M. 1972. Catalogue of nebulae in Crux, Centaurus, Ciccinus, and Norma. *U. of Arizona, Steward Obs.* September.

SHB — Smith-Haenni, A. L. 1977. A comprehensive catalogue of quasi-stellar objects. *Astr. & Astrophys. Suppl.* 27:205.

SNO — Snow. T. P. 1970. Groups and clusters of galaxies. *Astr. J.* 75:237.

STOCK — Stockton, A. N. 1968. Blue condensations associated with galaxies. Ph.D. diss., Univ. of Arizona.

ST — Strohmeier, W. 1950. Rote-Nebel in der Wintermilchstrasse. *Zs. f. Astrophys.* 27:49.

SS — Struve, O., and Straka, W. C. 1962. Notes on diffuse galactic nebulae. *Publ. Astr. Soc. Pacific* 74:474.

RNGC — Sulentic, J. W., and Tifft, W. F. 1973, *The revised new general catalog of nonstellar astronomical objects.* U. of Ariz. Press, Tucson.

SVEN — Svenonius, B. 1937. Catalog of anagalactic objects photographed at Helwan for which resolution, dimensions and total magnitude has been determined. *Annals Obs. Lund* 6.

TER — Terzan, A. 1971. Quatre nouveaux amas stellaires dans la direction de la region centrale de la galaxie. *Ast. & Astrophys.* 12:477.

T — Tifft, W. F., Jewsbury, C. P., and Sargent, T. A. 1973. Investigation of the Cancer Cluster of galaxies. *Astrophys. J.* 185:115.

URA — Uranova, T. A. 1973a. Supplement list of stellar rings. I. *Ast. Tsirk.* No. 772, 4.

URA — Uranova, T. A. 1973b. Supplement list of stellar rings II. *Ast. Tsirk.* No. 773, 3.

VDB.66B — van den Bergh, S. 1966a, Observation of faint blue objects. *Astrophys. J.* 144:866.

VDB.66N — van den Bergh, S. 1966b. A study of reflection nebulae. *Astr. J.* 71:990.

VDB.66G — van den Bergh, S. 1966c. Luminosity classifications of dwarf galaxies. *Astr. J.* 71:922.

AND — van den Bergh, S. 1972. Search for faint companions to M31. *Astrophys. J.* 171:231.

VHA — van den Bergh, S., and Hagen, G. L. 1975. Uniform survey of clusters in the Southern Milky Way. *Astr. J.* 80:11.

VHE — van den Bergh, S., and Herbst, W. 1975. Catalogue of southern stars imbedded in nebulosity. *Astr. J.* 80:208.

VMT — van den Bergh, S., Marschner, A. P., and Terzian, Y. 1973. An optical atlas of galactic supernovae remnants. *Astrophys. J. Suppl.* 26:19.

VB — Vidal, N. V., and Bern, K. A. 1973. Stellar rings and the structure of the galaxy. *Astr. & Astrophys.* 29:277.

VV — Vorontsov-Velyaminov, B. A. 1959. *Atlas and catalog of interacting galaxies*. Sternberg, Inst., Moscow State University.

MCG — Vorontsov-Velyaminov, B. A., and Arhipova, V. P. 1963. *Morphological catalog of galaxies*, Part III. Moscow State University.

MCG — Vorontsov-Velyaminov, B. A., and Arhipova, V. P. 1964. *Morphological catalog of galaxies*, Part II. Moscow State University.

MCG — Vorontsov-Velyaminov, B. A., and Arhipova, V. P. 1968. *Morphological catalog of galaxies*, Part IV. Moscow State University.

MCG — Vorontsov-Velyaminov, B. A., and Arhipova, V. P. 1974. *Morphological catalog of galaxies*. Part V. Moscow State University.

VVI — Vorontsov-Velyaminov, B. A., and Ivanisevic, G. 1974. Survey of galaxies with Seyfert and Seyfert-like spectra. *Astron. Zh.* 51:300.

MCG — Vorontsov-Velyaminov, B. A., and Krasnogorskaja, A. A. 1962. *Morphological catalog of galaxies*, Part I. Moscow State University.

BA8 BA9	Wagner, R. 1971. The stellar groups Ba 8, Ba 9, and Be 68. *Ast. & Astrophys.* 14:283.
WEED	Weedman, D. W. 1971. Very blue stellar objects near galaxies. *Astrophys. Lett.* 9:49.
WEI	Weistrop, D. 1973. A search for faint blue objects near the North Galactic Pole. *Ast. & Astrophys.* 23:215.
WS	Westerlund, B. E., and Smith, L. F. 1963. Planetary nebulae in the Large Magellanic Cloud. *Mon Not. Roy. Astr. Soc.* 127:449.
WIL	Wild, P. 1953. An interesting group of galaxies. *Publ. Astr. Soc. Pacific* 65:202.
KN	Wolf, M. 1909a. Konigstuhl Nebel-Liste 10. *Veroff. der Sternwarte zu Heidelberg* 6:1.
KN	Wolf, M. 1909b. Konigstuhl Nebel-Liste 11. *Veroff. der Sternwarte zu Heidelberg* 6:5.
KN	Wolf, M. 1911. Konigstuhl Nebel-Liste 12. *Veroff. der Sternwarte zu Heidelberg* 6:9.
KN	Wolf, M. 1912. Konigstuhl Nebel-Liste 13, *Veroff. der Sternwarte zu Heidelberg* 6:85.
KN	Wolf, M., and Ernst E. 1913. Konigstuhl Nebel-Liste 14. *Veroff. der Sternwarte zu Heidelberg* 6:115.
KN	Wolf, M. 1916. Konigstuhl Nebel-Liste 15. *Veroff. der Sternwarte zu Heidelberg* 7:169.
KN	Wolf, M. 1928. Konigstuhl Nebel-Liste 16. *Veroff. der Sternwarte zu Heidelberg* 8:113.
WK	Wolf, M., and Kaiser, F. 1913. Positionsbestimmungen von 124 Nebelflecken in Perseue-Nebelhaufen. *Veroff. der Sternwarte zu Heidelberg* 6:131.
BA	Wooden, W. H. 1971. RGU photographic photometry of the galactic clusters K4 and Ba 10. *Astr. & Astrophys.* 13:218.
WRA	Wray, J. D. 1966. A study of H emission objects in the Southern Milky Way. Ph.D. Dissertation, Northwestern Univ.
	Zonn, W. 1963. Identification of double and multiple galaxies. *Publ. Astr. Soc. Pacific* 75:184.
1ZW to 7ZW	Zwicky, F. 1961–68. Seven privately circulated lists.
ZCG	Zwicky, F. 1971. *Catalogue of selected compact galaxies and post-eruptive galaxies.* Offsetdruk L. Speich, Zurich.
ZC ZWG	Zwicky, F., and Herzog, E. 1963. *Catalogue of galaxies and clusters of galaxieis,* Volume II. Cal. Tech. Press, Pasadena.
ZC ZWG	Zwicky, F., and Herzog, E. 1966. *Catalogue of galaxies and clusters of galaxies,* Volume III. Cal. Tech. Press, Pasadena.
ZC ZWG	Zwicky, F., and Herzog, E. 1968. *Catalogue of galaxies and clusters of galaxies.* Volume IV. Cal. Tech. Press, Pasadena.
ZC ZWG	Zwicky, F., Herzog, E., and Wild, P. 1961 *Catalogue of galaxies and clusters of galaxies,* Volume I. Cal. Tech. Press, Pasadena.
WIL	Zwicky, F., and Humason, M. L. 1961. Spectra and other characteristics of interconnected galaxies and of galaxies in groups and in clusters II. *Astrophys. J.* 133:794.
ZH	Zwicky, F., and Humason, M. L. 1964. Spectra and other characteristics of interconnected galaxies and of galaxies in groups and in clusters III. *Astrophys. J.* 139:269.
ZC ZWG	Zwicky, F., Karpowicz, M., and Kowal, C. T. 1968. *Catalogue of galaxies and clusters of galaxies,* Volume V. Cal. Tech. Press, Pasadena.
ZC ZWG	Zwicky, F., and Kowal, C. T. 1968. *Catalogue of galaxies and clusters of galaxies,* Volume VI. Cal. Tech. Press, Pasadena.
ZL	Zwicky, F., and Luyten, W. 1967. A search for faint blue stars XLV. Ultra faint blue stars observed with the 200″ Telescope. *The Observatory, U. of Minn.*
8ZW	Zwicky, F., Sargent, W. L. W., and Kowal, C. T. 1975. Eighth list of compact galaxies. *Astr. J.* 80:545.

A MASTER LIST
OF NONSTELLAR OPTICAL ASTRONOMICAL OBJECTS

OBJECT NAME	RIGHT ASCEN.	DECLINATION	DIAM.	MAGN.	TYPE OF OBJECT
VDB.66G 222	00 00	+ 15 02	100		DWARF GALAXY
LBN 0583	00 00	+ 67 00	6300		BRIGHT NEBULA
LBN 0582	00 00	+ 67 00	9900		BRIGHT NEBULA
LBN 0587	00 00	+ 68 20	3420		BRIGHT NEBULA
PHL 6201	00 00 00.	+ 04 00		18.2	BLUE STELLAR OBJECT
PHL 6200	00 00 00.	+ 05 44		18.4	BLUE STELLAR OBJECT
ZWG 456.018	00 00 00.	+ 16 22		14.9	GALAXY
UGC 00001	00 00 00.	+ 16 22	90	14.9	GALAXY DBL SYS
ZWG 456.019	00 00 00.	+ 21 19		15.7	GALAXY
MCG+03-01-014	00 00 00.	+ 21 20	27	17.	GALAXY
UGC 00002	00 00 00.	+ 44 39	66	17.	GALAXY
LDN 1272	00 00 00.	+ 67 00	11940		DARK NEBULA
OCL 0286	00 00 00.	+ 67 06	600	11.	OPEN STAR CLUSTER
LDN 1273	00 00 00.	+ 68 15	1800		DARK NEBULA
MCG+13-01-003	00 00 00.	+ 77 00	57	15.	GALAXY
SC 2357-0416.0	00 00 00.	- 03 59 18.	48		NEBULA
PHL 6199	00 00 00.	- 08 30		16.5	BLUE STELLAR OBJECT
PHL 2564	00 00 00.	- 18 38		17.8	BLUE STELLAR OBJECT
PHL 6198	00 00 00.	- 18 49		17.5	BLUE STELLAR OBJECT
PHL 6202	00 00 00.	- 28 54		18.6	BLUE STELLAR OBJECT
LB 04921	00 00 00.	- 29 51		18.5	FAINT BLUE STAR
LB 04922	00 00 00.	- 30 41		18.9	FAINT BLUE STAR
MCG-05-01-026	00 00 00.	- 30 55	30	15.	GALAXY
PHL 0633	00 00 00.	- 32 28		16.6	BLUE STELLAR OBJECT
RNGC 1638	00 00 01.	- 00 00		13.0	GALAXY
RNGC 1072	00 00 01.	- 00 00		14.5	GALAXY
KN 13.047	00 00 03.9	+ 16 22 01.			NEBULA
IC 5378	00 00 04.	+ 16 20 40.			NONSTELLAR OBJECT
SC 2357-0411.4	00 00 05.	- 03 54 42.	12		NEBULA
ZC 0000.1+0806	00 00 06.	+ 08 06	2420		CLUSTER OF GALAXIES
ZWG 408.013	00 00 06.	+ 08 27		15.7	GALAXY
MCG+01-01-013	00 00 06.	+ 08 27 30.	36	15.	GALAXY
ZWG 456.020	00 00 06.	+ 16 18		15.7	GALAXY
MCG+03-01-016	00 00 06.	+ 16 21	48	15.	GALAXY
MCG+03-01-015	00 00 06.	+ 16 21	30	14.5	GALAXY
VV 263B	00 00 06.	+ 16 22	48	16.	INTERACTING GALAXY
VV 263A	00 00 06.	+ 16 22	42	15.	INTERACTING GALAXY
VV 263	00 00 06.	+ 16 22	72		INTERACTING GALAXY
PHL 2565	00 00 06.	- 02 37		18.9	BLUE STELLAR OBJECT
MCG-01-01-024	00 00 06.	- 03 57 30.	60	14.	GALAXY
PHL 6203	00 00 06.	- 09 37		18.0	BLUE STELLAR OBJECT
PHL 6204	00 00 06.	- 12 18		18.0	BLUE STELLAR OBJECT
LB 04923	00 00 06.	- 30 50		18.0	FAINT BLUE STAR
KN 13.048	00 00 06.8	+ 16 19 26.			NEBULA
IC 5379	00 00 07.	+ 16 18 16.			NONSTELLAR OBJECT
ARC 2695	00 00 10.	+ 18 29		17.7	RICH CLUSTER OF GALAXIES
SC 2357-0409.9	00 00 10.	- 03 53 12.	60		NEBULA
HN 1237	00 00 10.	- 66 28			NEBULA
IC 5380	00 00 10.	- 66 28			NONSTELLAR OBJECT
SC 2357-6805.4	00 00 10.	- 67 48 42.	6		NEBULA
PHL 6207	00 00 12.	+ 04 11		18.3	BLUE STELLAR OBJECT
PHL 0634	00 00 12.	+ 07 22		17.8	BLUE STELLAR OBJECT
MCG+03-01-017	00 00 12.	+ 16 18	24	17.5	GALAXY
ZWG 456.021	00 00 12.	+ 18 37		14.8	GALAXY
UGC 00003	00 00 12.	+ 18 37	120	14.8	GALAXY SBa
SCHO 0001	00 00 12.	+ 61 30 30.	320		ISOLATED DARK CLOUD
PHL 2566	00 00 12.	- 03 18		18.8	BLUE STELLAR OBJECT
8ZW 0000-03.8	00 00 12.	- 03 50		15.5	COMPACT GALAXY
MCG-01-01-025	00 00 12.	- 03 52	36	16.	GALAXY
PHL 6205	00 00 12.	- 05 06		18.0	BLUE STELLAR OBJECT
PHL 6206	00 00 12.	- 05 12		17.8	BLUE STELLAR OBJECT
PHL 6208	00 00 12.	- 18 08		18.6	BLUE STELLAR OBJECT
LB 04924	00 00 12.	- 29 48		18.3	FAINT BLUE STAR
LB 04925	00 00 12.	- 30 54		17.6	FAINT BLUE STAR
MCG+03-01-018	00 00 15.	+ 18 38	66	14.5	GALAXY
SC 2357-0406.9	00 00 15.	- 03 50 12.	12		NEBULA
ZWG 382.020	00 00 18.	+ 03 13		15.6	GALAXY
ZWG 499.039	00 00 18.	+ 31 12		15.6	GALAXY
ZWG 498.067	00 00 18.	+ 31 12		15.6	GALAXY
MCG+05-01-027	00 00 18.	+ 31 12 30.	36	15.	GALAXY
MRSL 118+06/1	00 00 18.	+ 68 13	6600		HII REGION
PHL 0635	00 00 18.	- 08 14		18.0	BLUE STELLAR OBJECT
LB 04926	00 00 18.	- 29 46		18.4	FAINT BLUE STAR
LB 04927	00 00 18.	- 30 30		18.7	FAINT BLUE STAR
MCG-05-01-027	00 00 18.	- 32 30	12	16.	GALAXY
RNGC 7812	00 00 19.	- 34 31			GALAXY
KN 13.049	00 00 19.6	+ 13 08 31.			NEBULA
SCHO 0002	00 00 20.	+ 61 57 42.	250		ISOLATED DARK CLOUD
MCG-01-01-026	00 00 21.	- 03 50 30.	60	15.	GALAXY
ARP 130	00 00 23.	+ 16 22			PECULIAR GALAXY
MCG+01-01-014	00 00 24.	+ 03 54	36	14.5	GALAXY
PHL 2568	00 00 24.	+ 14 17		18.6	BLUE STELLAR OBJECT
ZC 0000.4+1931	00 00 24.	+ 19 31	4300		CLUSTER OF GALAXIES
PHL 6209	00 00 24.	- 12 16		18.1	BLUE STELLAR OBJECT
PHL 0636	00 00 24.	- 26 21		17.1	BLUE STELLAR OBJECT
PHL 2567	00 00 24.	- 28 10		18.5	BLUE STELLAR OBJECT
LB 04928	00 00 24.	- 30 08		18.7	FAINT BLUE STAR
LB 04929	00 00 24.	- 30 26		19.5	FAINT BLUE STAR
MCG-06-01-016	00 00 24.	- 34 31	36	13.5	GALAXY
HOLM 827B	00 00 25.	- 02 14	30	14.6	PART OF MULTIPLE GALAXY
ZWG 408.014	00 00 30.	+ 03 56		15.5	GALAXY
UGC 00004	00 00 30.	+ 03 56	78	15.5	GALAXY Sb-c
ZWG 456.022	00 00 30.	+ 18 36		15.5	GALAXY
MCG+00-01-021	00 00 30.	- 02 12	72	12.8	GALAXY
PHL 2569	00 00 30.	- 06 44		18.4	BLUE STELLAR OBJECT
MCG-02-01-010	00 00 30.	- 12 40	60	15.	GALAXY
PHL 6210	00 00 30.	- 29 12		18.4	BLUE STELLAR OBJECT
LB 04930	00 00 30.	- 29 51		18.5	FAINT BLUE STAR
SC 2357-0228.9	00 00 32.	- 02 12 12.	60		NEBULA
HOLM 827A	00 00 34.	- 02 11	66	12.8	PART OF MULTIPLE GALAXY
KN 13.050	00 00 34.9	+ 15 50 32.			NEBULA
MCG+00-01-022	00 00 36.	+ 03 20 30.	42	13.5	GALAXY
PHL 0637	00 00 36.	+ 07 00		18.2	BLUE STELLAR OBJECT
4ZW 001	00 00 36.	+ 21 42			COMPACT GALAXY
ZWG 478.015	00 00 36.	+ 21 42		14.4	GALAXY
ZWG 477.043	00 00 36.	+ 21 42		14.4	GALAXY
MRK 334	00 00 36.	+ 21 42	18	15.	GALAXY WITH UV CONTINUUM
UGC 00006	00 00 36.	+ 21 42	60	14.4	GALAXY PECULR
4ZW 002	00 00 36.	+ 24 54			COMPACT GALAXY
ZWG 478.016	00 00 36.	+ 24 54		15.5	GALAXY
ZWG 477.044	00 00 36.	+ 24 54		15.5	GALAXY
ZWG 499.040	00 00 36.	+ 30 45		15.0	GALAXY
ZWG 498.068	00 00 36.	+ 30 45		15.0	GALAXY
ZWG 382.021	00 00 36.	- 02 11		14.3	GALAXY
UGC 00005	00 00 36.	- 02 11	84	14.3	GALAXY SBb/Sc
KARA.73B 0001	00 00 36.	- 02 11	144	14.3	ISOLATED GALAXY S
PHL 6212	00 00 36.	- 03 09		17.0	BLUE STELLAR OBJECT
PHL 2570	00 00 36.	- 04 09		18.1	BLUE STELLAR OBJECT

OBJECT NAME	RIGHT ASCEN.	DECLINATION	DIAM.	MAGN.	TYPE OF OBJECT
PHL 6211	00 00 36.	- 12 58		15.8	BLUE STELLAR OBJECT
PHL 2571	00 00 36.	- 16 00		18.3	BLUE STELLAR OBJECT
PHL 2572	00 00 36.	- 28 08		18.7	BLUE STELLAR OBJECT
LB 01539	00 00 36.	- 48 03		14.2	FAINT BLUE STAR
KN 13.051	00 00 37.6	+ 15 41 12.			NEBULA
IC 5381	00 00 37.6	+ 15 41 14.			GALAXY
KEEL 740	00 00 37.6	+ 15 41 16.			NEBULA
MCG+03-01-019	00 00 39.	+ 15 41	60	15.	GALAXY
SC 2358-0404.3	00 00 40.	- 03 47 36.	12		NEBULA
KN 13.052	00 00 40.7	+ 15 51 52.			NEBULA
SC 2358-6555.4	00 00 41.	- 65 38 42.	66		NEBULA
ZWG 382.022	00 00 42.	+ 03 20		15.3	GALAXY
PHL 2574	00 00 42.	+ 05 42		18.4	BLUE STELLAR OBJECT
ZWG 456.023	00 00 42.	+ 15 42		14.9	GALAXY
UGC 00007	00 00 42.	+ 15 42	90	14.9	GALAXY PECULR
ZWG 456.024	00 00 42.	+ 15 52		12.0	GALAXY
UGC 00008	00 00 42.	+ 15 52	390	12.0	GALAXY Sa-b
MCG+03-01-020	00 00 42.	+ 15 52	300	11.7	GALAXY
MCG+04-01-013	00 00 42.	+ 21 40	60	14.5	GALAXY
MCG+05-01-028	00 00 42.	+ 29 31	30	15.	GALAXY
8ZW 0000-07.5	00 00 42.	- 07 28		17.5	COMPACT GALAXY
PHL 6214	00 00 42.	- 08 36		18.7	BLUE STELLAR OBJECT
MCG-02-01-011	00 00 42.	- 10 15	48	15.5	GALAXY
PHL 0638	00 00 42.	- 14 36		18.3	BLUE STELLAR OBJECT
PHL 2575	00 00 42.	- 20 24		18.5	BLUE STELLAR OBJECT
PHL 6213	00 00 42.	- 20 27		18.2	BLUE STELLAR OBJECT
TON-S 0135	00 00 42.	- 23 57		12.5	BLUE STAR
PHL 2573	00 00 42.	- 24 56		17.9	BLUE STELLAR OBJECT
LB 04931	00 00 42.	- 30 15		18.3	FAINT BLUE STAR
LB 04932	00 00 42.	- 30 46		17.5	FAINT BLUE STAR
MCG-06-01-017	00 00 42.	- 36 12	36	15.	GALAXY
MCG+01-01-015	00 00 45.	+ 08 20 30.	60	14.5	GALAXY
MCG-02-01-012	00 00 45.	- 11 02	48	14.5	GALAXY
ARC 2696	00 00 46.	+ 00 38		16.9	RICH CLUSTER OF GALAXIES
ARC 2697	00 00 46.	- 06 23		17.2	RICH CLUSTER OF GALAXIES
UGC 00009	00 00 48.	+ 04 21	78	16.5	GALAXY
ZC 0000.8+0452	00 00 48.	+ 04 52	8130		CLUSTER OF GALAXIES
MCG+01-01-016	00 00 48.	+ 05 25	36	14.	GALAXY
ZWG 408.015	00 00 48.	+ 08 20		15.4	GALAXY
UGC 00010	00 00 48.	+ 08 20	120	15.4	GALAXY Sc
MCG+04-01-014	00 00 48.	+ 21 48 30.	48	15.5	GALAXY
UGC 00011	00 00 48.	+ 21 50	66	16.0	GALAXY
ZWG 499.041	00 00 48.	+ 29 31		15.7	GALAXY
ZWG 498.069	00 00 48.	+ 29 31		15.7	GALAXY
UGC 00012	00 00 48.	+ 29 31	66	15.7	GALAXY Sc
KARA.73B 0002	00 00 48.	+ 29 31	42	15.7	ISOLATED GALAXY S
ZWG 499.042	00 00 48.	+ 30 30		15.7	GALAXY
ZWG 498.070	00 00 48.	+ 30 30		15.7	GALAXY
KARA.73B 0003	00 00 48.	+ 30 30	18	15.7	ISOLATED GALAXY S
ZCG 0000+34	00 00 48.	+ 34 29		17.6	COMPACT GALAXY
4ZW 003	00 00 48.	+ 34 37			COMPACT GALAXY
PHL 6215	00 00 48.	- 02 48		17.8	BLUE STELLAR OBJECT
PHL 6216	00 00 48.	- 04 20		18.1	BLUE STELLAR OBJECT
PHL 2576	00 00 48.	- 04 51		18.4	BLUE STELLAR OBJECT
PHL 2577	00 00 48.	- 05 17		18.3	BLUE STELLAR OBJECT
PHL 2578	00 00 48.	- 09 19		18.4	BLUE STELLAR OBJECT
PHL 2579	00 00 48.	- 09 41		13.3	BLUE STELLAR OBJECT
PHL 6217	00 00 48.	- 12 24		18.4	BLUE STELLAR OBJECT
PHL 6218	00 00 48.	- 14 29		18.0	BLUE STELLAR OBJECT
PHL 2580	00 00 48.	- 23 57		13.2	BLUE STELLAR OBJECT
PHL 6219	00 00 48.	- 25 26		18.2	BLUE STELLAR OBJECT
PHL 2581	00 00 48.	- 28 57		18.2	BLUE STELLAR OBJECT
RNGC 7815	00 00 49.	+ 20 25			NON-EXISTENT OBJECT
KEEL 741	00 00 51.7	+ 21 03 50.			NEBULA
ARC 2698	00 00 52.	+ 04 22		17.0	RICH CLUSTER OF GALAXIES
RN 1238	00 00 52.	- 65 28			NEBULA
IC 5382	00 00 52.	- 65 28			NONSTELLAR OBJECT
SCHO 0003	00 00 53.	+ 59 46 18.	270		ISOLATED DARK CLOUD
ZC 0000.9+0417	00 00 54.	+ 04 17	2890		CLUSTER OF GALAXIES
ZWG 408.016	00 00 54.	+ 05 25		15.2	GALAXY
PHL 6220	00 00 54.	+ 06 56		18.1	BLUE STELLAR OBJECT
ZWG 456.025	00 00 54.	+ 16 28		15.5	GALAXY
MCG+04-01-015	00 00 54.	+ 27 04	66	15.	GALAXY
ZWG 478.017	00 00 54.	+ 27 05		15.0	GALAXY
ZWG 477.045	00 00 54.	+ 27 05		15.0	GALAXY
UGC 00013	00 00 54.	+ 27 05	66	15.0	GALAXY SB0/a
OCL 0282	00 00 54.	+ 63 19	240	16.	OPEN STAR CLUSTER
PHL 2582	00 00 54.	- 04 02		17.8	BLUE STELLAR OBJECT
PHL 6223	00 00 54.	- 05 20		16.9	BLUE STELLAR OBJECT
PHL 6222	00 00 54.	- 09 36		13.9	BLUE STELLAR OBJECT
PHL 6221	00 00 54.	- 09 40		18.2	BLUE STELLAR OBJECT
MCG-02-01-013	00 00 54.	- 11 01	78	14.	GALAXY
PHL 0639	00 00 54.	- 17 00		14.8	BLUE STELLAR OBJECT
PHL 2583	00 00 54.	- 19 51		18.3	BLUE STELLAR OBJECT
LB 04933	00 00 54.	- 29 48		19.2	FAINT BLUE STAR
LB 04934	00 00 54.	- 30 25		18.4	FAINT BLUE STAR
RNGC 7808	00 00 55.	- 11 01		14.0	GALAXY
MCG+04-01-016	00 00 57.	+ 24 49	30	16.	GALAXY
SG 3.249	00 00 57.	+ 66 57	3000		DIFFUSE EMISSION NEBULA
LBN 0591	00 01	+ 65 20	120		BRIGHT NEBULA
LBN 0584	00 01	+ 66 40	2100		BRIGHT NEBULA
LBN 0586	00 01	+ 67 10	1320		BRIGHT NEBULA
LBN 0589	00 01	+ 68 20	720		BRIGHT NEBULA
PHL 2584	00 01 00.	+ 06 36		17.9	BLUE STELLAR OBJECT
MCG+02-01-016	00 01 00.	+ 10 20	36	15.	GALAXY
ZWG 433.019	00 01 00.	+ 11 13		15.6	GALAXY
ZWG 456.026	00 01 00.	+ 16 44		15.7	GALAXY
PHL 6225	00 01 00.	+ 17 55		18.2	BLUE STELLAR OBJECT
ZWG 478.018	00 01 00.	+ 21 49		15.7	GALAXY
ZWG 477.046	00 01 00.	+ 21 49		15.7	GALAXY
MCG+04-01-017	00 01 00.	+ 22 55	108	14.	GALAXY
ZWG 478.019	00 01 00.	+ 22 56		14.0	GALAXY
ZWG 477.047	00 01 00.	+ 22 56		14.0	GALAXY
UGC 00014	00 01 00.	+ 22 56	120	14.0	GALAXY Sc
MCG+06-01-009	00 01 00.	+ 37 04	30	15.	GALAXY
PHL 6226	00 01 00.	- 05 49		17.9	BLUE STELLAR OBJECT
MCG-01-01-027	00 01 00.	- 07 58	36	14.5	GALAXY
PHL 6224	00 01 00.	- 21 23		16.6	BLUE STELLAR OBJECT
LB 04935	00 01 00.	- 29 55		17.1	FAINT BLUE STAR
LB 04936	00 01 00.	- 30 40		18.6	FAINT BLUE STAR
RNGC 7822	00 01 01.	+ 68 20			DIFFUSE NEBULA
KN 13.053	00 01 02.2	+ 10 19 32.			NEBULA
KN 13.054	00 01 03.0	+ 13 57 46.			NEBULA
KN 13.055	00 01 05.0	+ 11 13 08.			NEBULA
ZC 0001.1+0030	00 01 06.	+ 00 30	3760		CLUSTER OF GALAXIES
MCG+01-01-017	00 01 06.	+ 03 59	60	15.	GALAXY
PHL 6230	00 01 06.	+ 04 50		18.4	BLUE STELLAR OBJECT
PHL 6229	00 01 06.	+ 05 01		18.3	BLUE STELLAR OBJECT

OBJECT NAME	RIGHT ASCEN.	DECLINATION	DIAM.	MAGN.	TYPE OF OBJECT
ZWG 408.017	00 01 06.	+ 08 21		15.6	GALAXY
MCG+02-01-017	00 01 06.	+ 14 57	108	16.	GALAXY
PHL 6227	00 01 06.	+ 18 38		18.4	BLUE STELLAR OBJECT
4ZW 004	00 01 06.	+ 24 20			COMPACT GALAXY
4ZW 005	00 01 06.	+ 25 41			COMPACT GALAXY
PHL 6231	00 01 06.	- 09 38		18.1	BLUE STELLAR OBJECT
PHL 2585	00 01 06.	- 12 08		18.3	BLUE STELLAR OBJECT
PHL 6228	00 01 06.	- 28 42		8.0	BLUE STELLAR OBJECT
LB 04937	00 01 06.	- 29 55		17.8	FAINT BLUE STAR
LB 01540	00 01 06.	- 57 49		14.2	FAINT BLUE STAR
KN 13.056	00 01 09.0	+ 14 56 13.			NEBULA
ARC 2699	00 01 10.	- 05 33		17.6	RICH CLUSTER OF GALAXIES
ZC 0001.2+0140	00 01 12.	+ 01 40	3430		CLUSTER OF GALAXIES
UGC 00015	00 01 12.	+ 04 00	90	16.0	GALAXY Sa-b
PHL 6233	00 01 12.	+ 04 40		15.2	BLUE STELLAR OBJECT
PHL 6235	00 01 12.	+ 07 04		17.1	BLUE STELLAR OBJECT
ZWG 408.018	00 01 12.	+ 07 11		14.0	GALAXY
UGC 00016	00 01 12.	+ 07 11	126	14.0	GALAXY Sb/Sc
MCG+01-01-018	00 01 12.	+ 07 11	84	13.5	GALAXY
PHL 2587	00 01 12.	+ 08 26		18.5	BLUE STELLAR OBJECT
ZWG 456.027	00 01 12.	+ 15 55		15.4	GALAXY
MCG+03-01-025	00 01 12.	+ 19 03	42	15.5	GALAXY
ZCG 0001+24.3	00 01 12.	+ 24 18		19.3	COMPACT GALAXY
PHL 2586	00 01 12.	- 05 54		17.1	BLUE STELLAR OBJECT
PHL 0640	00 01 12.	- 10 14		16.9	BLUE STELLAR OBJECT
PHL 6232	00 01 12.	- 17 37		4.6	BLUE STELLAR OBJECT
PHL 6234	00 01 12.	- 30 18		18.0	BLUE STELLAR OBJECT
LB 04938	00 01 12.	- 30 35		19.6	FAINT BLUE STAR
LB 04939	00 01 12.	- 30 51		19.5	FAINT BLUE STAR
SC 2358-6813.8	00 01 12.	- 67 57 06.	18		NEBULA
RNGC 7816	00 01 13.	+ 07 11		14.0	GALAXY
IC 5383	00 01 13.	+ 15 43 28.			NONSTELLAR OBJECT
ARC 2700	00 01 16.	+ 01 48		16.0	RICH CLUSTER OF GALAXIES
HN 2891	00 01 16.4	- 43 53 48.	24		NEBULA
UGC 00017	00 01 18.	+ 14 56	180	17.	GALAXY DWARF
ZC 0001.3+3230	00 01 18.	+ 32 30	610		CLUSTER OF GALAXIES
5ZW 001	00 01 18.	+ 51 28			COMPACT GALAXY
PHL 2590	00 01 18.	- 04 24		18.1	BLUE STELLAR OBJECT
PHL 2591	00 01 18.	- 05 00		18.3	BLUE STELLAR OBJECT
PHL 2588	00 01 18.	- 06 23		16.9	BLUE STELLAR OBJECT
PHL 0641	00 01 18.	- 14 39		17.2	BLUE STELLAR OBJECT
MCG-03-01-016	00 01 18.	- 14 58	36	14.5	GALAXY
PHL 2589	00 01 18.	- 22 58		17.8	BLUE STELLAR OBJECT
LB 04940	00 01 18.	- 30 08		18.2	FAINT BLUE STAR
KN 13.057	00 01 19.9	+ 12 29 15.			NEBULA
KN 13.058	00 01 22.5	+ 10 35 53.			NEBULA
PHL 0642	00 01 24.	+ 09 20		18.0	BLUE STELLAR OBJECT
ZWG 433.020	00 01 24.	+ 10 35		15.7	GALAXY
UGC 00018	00 01 24.	+ 10 35	66	15.7	GALAXY COMPACT
PHL 6238	00 01 24.	+ 17 58		18.0	BLUE STELLAR OBJECT
ZWG 456.028	00 01 24.	+ 20 28		12.7	GALAXY
UGC 00019	00 01 24.	+ 20 28	240	12.7	GALAXY Sb/Sc
KARA.73B 0004	00 01 24.	+ 20 28	216	12.7	ISOLATED GALAXY S
MCG+03-01-021	00 01 24.	+ 20 30	180	12.	GALAXY
MCG+13-01-004	00 01 24.	+ 80 01	39	16.	GALAXY
PHL 2593	00 01 24.	- 05 20		18.3	BLUE STELLAR OBJECT
PHL 6236	00 01 24.	- 06 12		18.5	BLUE STELLAR OBJECT
MCG-02-01-015	00 01 24.	- 11 27 30.	12	15.	GALAXY
MCG-02-01-014	00 01 24.	- 11 27 30.	60	14.5	GALAXY
PHL 6237	00 01 24.	- 17 42		18.9	BLUE STELLAR OBJECT
PHL 2592	00 01 24.	- 19 04		18.0	BLUE STELLAR OBJECT
PHL 2594	00 01 24.	- 22 39		18.5	BLUE STELLAR OBJECT
SNO 34	00 01 24.	- 27 32 03.	600	17.	GROUP OF 8 GALAXIES
LB 04941	00 01 24.	- 29 55		20.2	FAINT BLUE STAR
KN 13.059	00 01 24.9	+ 13 26 43.			NEBULA
RNGC 7817	00 01 25.	+ 20 28		12.5	GALAXY
HN 2892	00 01 25.6	- 46 28 30.	18		NEBULA
KN 13.060	00 01 26.3	+ 14 25 30.			NEBULA
KN 13.061	00 01 26.8	+ 15 44 30.			NEBULA
LIN.CL 001	00 01 27.	- 73 45 12.	246	11.7	STAR CLUSTER IN SMC
KEEL 742	00 01 27.0	+ 15 44 30.			NEBULA
MCG+01-01-019	00 01 30.	+ 07 05	60	14.	GALAXY
UGC 00020	00 01 30.	+ 80 01	132	16.0	GALAXY
MCG+00-01-023	00 01 30.	- 01 10 30.	42	14.5	GALAXY
PHL 6239	00 01 30.	- 13 25		16.7	BLUE STELLAR OBJECT
PHL 6240	00 01 30.	- 21 33		18.4	BLUE STELLAR OBJECT
TON-S 0136	00 01 30.	- 26 59		15.7	BLUE STAR
LB 04942	00 01 30.	- 29 54		16.7	FAINT BLUE STAR
LB 04943	00 01 30.	- 30 03		19.8	FAINT BLUE STAR
SC 2358-6820.4	00 01 30.	- 68 03 42.	12		NEBULA
KN 13.062	00 01 31.3	+ 10 01 09.			NEBULA
SCHO 0005	00 01 34.	+ 58 19 54.	250		ISOLATED DARK CLOUD
SCHO 0004	00 01 34.	+ 62 42 06.	300		ISOLATED DARK CLOUD
PHL 2597	00 01 36.	+ 01 55		18.6	BLUE STELLAR OBJECT
ZWG 408.019	00 01 36.	+ 07 05		15.1	GALAXY
UGC 00021	00 01 36.	+ 07 05	66	15.1	GALAXY Sc
MCG+02-01-018	00 01 36.	+ 10 00	72	15.	GALAXY
PHL 6241	00 01 36.	+ 16 12		17.0	BLUE STELLAR OBJECT
PHL 6242	00 01 36.	+ 17 36		18.7	BLUE STELLAR OBJECT
ZWG 382.023	00 01 36.	- 01 08		15.6	GALAXY
PHL 2595	00 01 36.	- 03 56		17.0	BLUE STELLAR OBJECT
PHL 2596	00 01 36.	- 11 02		18.5	BLUE STELLAR OBJECT
MCG-02-01-016	00 01 36.	- 12 15	48	15.	GALAXY
PHL 6244	00 01 36.	- 16 10		18.5	BLUE STELLAR OBJECT
PHL 6243	00 01 36.	- 16 12		18.3	BLUE STELLAR OBJECT
RNGC 7818	00 01 37.	+ 07 05		15.0	GALAXY
RNGC 7813	00 01 37.	- 12 15		15.0	GALAXY
IC 5384	00 01 37.	- 12 15 44.			NONSTELLAR OBJECT
KN 13.063	00 01 39.1	+ 10 30 54.			NEBULA
ARC 2701	00 01 40.	- 09 52		17.8	RICH CLUSTER OF GALAXIES
UGC 00022	00 01 42.	+ 10 02	78	16.0	GALAXY S0
ZWG 433.021	00 01 42.	+ 10 30		15.2	GALAXY
UGC 00023	00 01 42.	+ 10 30	78	15.2	GALAXY SBb
MCG+02-01-019	00 01 42.	+ 10 30	84	14.	GALAXY
PHL 0643	00 01 42.	+ 15 13		16.9	BLUE STELLAR OBJECT
MCG+04-01-018	00 01 42.	+ 22 18 30.	48	15.5	GALAXY
ZWG 478.020	00 01 42.	+ 22 19		15.4	GALAXY
ZWG 477.048	00 01 42.	+ 22 19		15.4	GALAXY
UGC 00024	00 01 42.	+ 22 19	72	15.4	GALAXY SBc
ZCG 0001+24.4	00 01 42.	+ 24 23		17.0	COMPACT GALAXY
ZC 0001.7+2746	00 01 42.	+ 27 46	1340		CLUSTER OF GALAXIES
ZWG 499.043	00 01 42.	+ 31 47		15.7	GALAXY
ZWG 498.071	00 01 42.	+ 31 47		15.7	GALAXY
SC 2359-0225.0	00 01 42.	- 02 08 18.	12		NEBULA
SC 2359-0228.4	00 01 42.	- 02 11 42.			NEBULA
MCG-02-01-017	00 01 42.	- 08 18	30	15.	GALAXY
PHL 6245	00 01 42.	- 08 34		11.4	BLUE STELLAR OBJECT
MCG-03-01-017	00 01 42.	- 14 47	54	15.	GALAXY

OBJECT NAME	RIGHT ASCEN.	DECLINATION	DIAM.	MAGN.	TYPE OF OBJECT
PHL 6246	00 01 42.	- 16 16		17.9	BLUE STELLAR OBJECT
PHL 6248	00 01 42.	- 29 19		18.2	BLUE STELLAR OBJECT
LB 04944	00 01 42.	- 29 39		19.6	FAINT BLUE STAR
OCL 0043	00 01 42.	- 30 13	5400	8.	OPEN STAR CLUSTER
PHL 6247	00 01 42.	- 30 25		7.0	BLUE STELLAR OBJECT
LB 04945	00 01 42.	- 30 30		19.5	FAINT BLUE STAR
MOHR 1	00 01 42.	- 73 38 48.		14.0	STAR CLUSTER IN SMC
UGC 00025	00 01 48.	+ 05 53	66	16.0	GALAXY Sc
MCG+01-01-020	00 01 48.	+ 05 53	60	15.	GALAXY
PHL 6253	00 01 48.	+ 17 12		18.6	BLUE STELLAR OBJECT
ZWG 499.044	00 01 48.	+ 31 12		14.3	GALAXY
ZWG 498.072	00 01 48.	+ 31 12		14.3	GALAXY
UGC 00026	00 01 48.	+ 31 12	120	14.3	GALAXY SBb
MCG+05-01-029	00 01 48.	+ 31 12	90	14.	GALAXY
MCG+06-01-010	00 01 48.	+ 37 43	36	16.	GALAXY
MCG+08-01-020	00 01 48.	+ 47 12	72	15.	GALAXY
OCL 0274	00 01 48.	+ 55 45	180	8.	OPEN STAR CLUSTER
PHL 6250	00 01 48.	- 02 28		17.4	BLUE STELLAR OBJECT
PHL 6249	00 01 48.	- 03 18		16.5	BLUE STELLAR OBJECT
PHL 6251	00 01 48.	- 03 50		18.7	BLUE STELLAR OBJECT
PHL 2598	00 01 48.	- 05 57		18.5	BLUE STELLAR OBJECT
PHL 0644	00 01 48.	- 09 45		17.3	BLUE STELLAR OBJECT
PHL 6252	00 01 48.	- 11 56		18.0	BLUE STELLAR OBJECT
LB 04946	00 01 48.	- 29 52		19.0	FAINT BLUE STAR
LB 04947	00 01 48.	- 30 14		18.2	FAINT BLUE STAR
LB 04949	00 01 48.	- 30 20		18.4	FAINT BLUE STAR
LB 04948	00 01 48.	- 30 20		18.2	FAINT BLUE STAR
LB 04950	00 01 48.	- 30 32		19.2	FAINT BLUE STAR
RNGC 7819	00 01 49.	+ 31 12		14.5	GALAXY
MCG+01-01-021	00 01 54.	+ 05 32 30.	72	14.	GALAXY
ZWG 408.020	00 01 54.	+ 05 34		15.3	GALAXY
UGC 00027	00 01 54.	+ 05 34	132	15.3	GALAXY Sc
PHL 6255	00 01 54.	+ 16 55		18.4	BLUE STELLAR OBJECT
ZC 0001.9+2351	00 01 54.	+ 23 51	740		CLUSTER OF GALAXIES
ZWG 499.045	00 01 54.	+ 27 42		15.7	GALAXY
ZWG 498.073	00 01 54.	+ 27 42		15.7	GALAXY
ZWG 549.015	00 01 54.	+ 47 13		15.1	GALAXY
ZWG 548.024	00 01 54.	+ 47 13		15.1	GALAXY
PHL 6256	00 01 54.	- 04 26		18.1	BLUE STELLAR OBJECT
PHL 6254	00 01 54.	- 13 20		15.9	BLUE STELLAR OBJECT
PHL 2599	00 01 54.	- 15 00		19.0	BLUE STELLAR OBJECT
LB 04951	00 01 54.	- 30 42		18.3	FAINT BLUE STAR
LB 04952	00 01 54.	- 30 55		18.8	FAINT BLUE STAR
KEEL 743	00 01 54.5	+ 15 42 14.			NEBULA
KEEL 744	00 01 56.9	+ 20 26 22.			NEBULA
MCG+01-01-022	00 01 57.	+ 04 54 30.	60	14.	GALAXY
KN 13.065	00 01 57.	+ 15 35			NEBULA
KN 13.064	00 01 57.3	+ 11 25 56.			NEBULA
KN 13.066	00 01 59.3	+ 11 25 07.			NEBULA
ZWG 408.021	00 02 00.	+ 04 55		13.9	GALAXY
UGC 00028	00 02 00.	+ 04 55	96	13.9	GALAXY S0-a
ZC 0002.0+1820	00 02 00.	+ 18 20	2490		CLUSTER OF GALAXIES
MCG+05-01-030	00 02 00.	+ 28 00 30.	21	15.	GALAXY
ZWG 499.046	00 02 00.	+ 28 02		15.2	GALAXY
ZWG 498.074	00 02 00.	+ 28 02		15.2	GALAXY
UGC 00029	00 02 00.	+ 28 02	78	15.2	GALAXY E
UGC 00030	00 02 00.	+ 33 17	72	16.0	GALAXY S
MRSL 118+04/1	00 02 00.	+ 66 53	10800		HII REGION
PHL 2600	00 02 00.	- 07 14		15.6	BLUE STELLAR OBJECT
MCG-02-01-018	00 02 00.	- 09 18	30	15.5	GALAXY
PHL 2601	00 02 00.	- 10 22		18.7	BLUE STELLAR OBJECT
PHL 6257	00 02 00.	- 16 24		18.2	BLUE STELLAR OBJECT
PHL 0645	00 02 00.	- 24 41		13.3	BLUE STELLAR OBJECT
TON-S 0137	00 02 00.	- 24 41		13.3	BLUE STAR
PHL 2602	00 02 00.	- 25 24		18.3	BLUE STELLAR OBJECT
PHL 2603	00 02 00.	- 26 51		18.3	BLUE STELLAR OBJECT
RNGC 7820	00 02 01.	+ 04 55		14.0	GALAXY
HN 2893	00 02 02.5	- 45 45 24.	12		NEBULA
MCG-01-01-028	00 02 03.	- 07 21 30.	138	13.	GALAXY
MCG-01-01-029	00 02 03.	- 08 22	30	16.	GALAXY
MCG-01-01-030	00 02 03.	- 08 23	24	13.5	GALAXY
CED 214B	00 02 04.	+ 66 53	1500		DIFFUSE GALACTIC NEBULA
SC 2359-0232.1	00 02 05.	- 02 15 24.	12		NEBULA
DG 001	00 02 06.	+ 66 53	3900		REFLECTION NEBULA
PHL 6258	00 02 06.	- 04 39		18.0	BLUE STELLAR OBJECT
PHL 2605	00 02 06.	- 06 23		18.8	BLUE STELLAR OBJECT
PHL 2604	00 02 06.	- 09 02		15.9	BLUE STELLAR OBJECT
PHL 2606	00 02 06.	- 21 23		18.3	BLUE STELLAR OBJECT
MCG-05-01-028	00 02 06.	- 30 46	48	14.5	GALAXY
HN 2894	00 02 06.8	- 45 54 24.	18		NEBULA
HN 2895	00 02 07.7	- 45 45 42.	12		NEBULA
SCHO 0006	00 02 08.	+ 58 50 48.	280		ISOLATED DARK CLOUD
CED 215	00 02 10.	+ 68 17	3600		DIFFUSE GALACTIC NEBULA
ZWG 478.021	00 02 12.	+ 26 33		14.6	GALAXY
ZWG 477.049	00 02 12.	+ 26 33		14.6	GALAXY
ZWG 499.047	00 02 12.	+ 32 00		15.6	GALAXY
ZWG 498.075	00 02 12.	+ 32 00		15.6	GALAXY
MCG+00-01-024	00 02 12.	- 01 48	24	14.	GALAXY
MCG+00-01-025	00 02 12.	- 01 51	36	14.	GALAXY
PHL 0646	00 02 12.	- 10 24		13.8	BLUE STELLAR OBJECT
PHL 2607	00 02 12.	- 14 19		13.9	BLUE STELLAR OBJECT
PHL 6259	00 02 12.	- 16 42		18.7	BLUE STELLAR OBJECT
RNGC 7823	00 02 14.	- 62 21			UNVERIFIED SOUTHERN OBJECT
MCG+01-01-023	00 02 15.	+ 05 25 30.	36	15.5	GALAXY
LB 00401	00 02 16.	+ 29 23 06.		17.4	FAINT BLUE STAR
ARC 2702	00 02 16.	+ 31 08		17.1	RICH CLUSTER OF GALAXIES
ZWG 408.022	00 02 18.	+ 05 27		15.7	GALAXY
PHL 6260	00 02 18.	+ 06 59		18.1	BLUE STELLAR OBJECT
PHL 0647	00 02 18.	+ 16 35		15.3	BLUE STELLAR OBJECT
ZWG 456.029	00 02 18.	+ 16 55		15.3	GALAXY
UGC 00031	00 02 18.	+ 16 55	84	15.3	GALAXY IRR
MCG+03-01-022	00 02 18.	+ 16 55	48	14.5	GALAXY
MCG+07-01-004	00 02 18.	+ 41 27	60	17.	GALAXY
ZWG 382.025	00 02 18.	- 01 46		15.0	GALAXY
MRK 544	00 02 18.	- 01 46	19	15.5	GALAXY WITH UV CONTINUUM
ZWG 382.024	00 02 18.	- 01 51		15.1	GALAXY
MCG-03-01-018	00 02 18.	- 16 19	78	14.	GALAXY
LB 04953	00 02 18.	- 29 48		18.8	FAINT BLUE STAR
LB 04954	00 02 18.	- 30 30		20.5	FAINT BLUE STAR
HN 2896	00 02 18.9	- 45 45 54.			NEBULA
HN 2897	00 02 19.2	- 45 45 54.	12		NEBULA
KN 13.067	00 02 20.3	+ 13 19 20.			NEBULA
SC 2359-0204.2	00 02 21.	- 01 47 30.	18		NEBULA
SC 2359-0425.8	00 02 23.	- 04 09 06.	36		NEBULA
MCG+01-01-024	00 02 24.	+ 04 49	48	14.5	GALAXY
PHL 2608	00 02 24.	+ 06 16		18.2	BLUE STELLAR OBJECT
PHL 6262	00 02 24.	+ 07 10		18.3	BLUE STELLAR OBJECT
ZC 0002.4+0744	00 02 24.	+ 07 44	9410		CLUSTER OF GALAXIES

OBJECT NAME	RIGHT ASCEN.	DECLINATION	DIAM.	MAGN.	TYPE OF OBJECT
PHL 2612	00 02 24.	+ 07 54		18.6	BLUE STELLAR OBJECT
PHL 6261	00 02 24.	+ 08 10		17.1	BLUE STELLAR OBJECT
PHL 6263	00 02 24.	+ 08 31		18.0	BLUE STELLAR OBJECT
UGC 00032	00 02 24.	+ 11 26	78	17.	GALAXY
ZWG 478.022	00 02 24.	+ 21 52		15.6	GALAXY
ZWG 477.050	00 02 24.	+ 21 52		15.6	GALAXY
PHL 2609	00 02 24.	- 02 42		18.6	BLUE STELLAR OBJECT
PHL 2610	00 02 24.	- 15 58		18.4	BLUE STELLAR OBJECT
PHL 2611	00 02 24.	- 29 58		17.3	BLUE STELLAR OBJECT
LB 04955	00 02 24.	- 30 07		17.5	FAINT BLUE STAR
MCG-06-01-018	00 02 24.	- 35 58	36	16.	GALAXY
KN 13.068	00 02 24.7	+ 11 25 22.			NEBULA
MCG+04-01-019	00 02 27.	+ 21 50	42	15.5	GALAXY
MCG-05-01-029	00 02 27.	- 28 00	24	14.5	GALAXY
MCG-05-01-030	00 02 27.	- 30 47	30	15.	GALAXY
SC 2359-0207.9	00 02 29.	- 01 51 12.	18		NEBULA
ZWG 408.023	00 02 30.	+ 04 51		15.4	GALAXY
UGC 00033	00 02 30.	+ 04 51	78	15.4	GALAXY SB0-a
MCG+01-01-025	00 02 30.	+ 06 37	72	14.	GALAXY
ZWG 478.023	00 02 30.	+ 21 55		15.7	GALAXY
ZWG 477.051	00 02 30.	+ 21 55		15.7	GALAXY
ZWG 382.026	00 02 30.	- 01 57		15.7	GALAXY
8ZW 0002-07.4	00 02 30.	- 07 25		15.7	COMPACT GALAXY
LB 04956	00 02 30.	- 29 40		17.7	FAINT BLUE STAR
LB 04957	00 02 30.	- 30 20		16.7	FAINT BLUE STAR
MCG-06-01-019	00 02 30.	- 35 58 30.	30	15.	GALAXY
IC 1528	00 02 31.	- 03 23 44.			NONSTELLAR OBJECT
SC 2359-0331.1	00 02 33.	- 03 14 24.			NEBULA
MIL 01	00 02 35.	+ 72 20	7500		SUPERNOVA REMNANT
SC 0000-0422.1	00 02 35.	- 04 05 24.	36		NEBULA
ZWG 408.024	00 02 36.	+ 04 56		15.5	GALAXY
MCG+01-01-026	00 02 36.	+ 06 28	42	14.5	GALAXY
ZWG 408.025	00 02 36.	+ 06 38		14.5	GALAXY
UGC 00034	00 02 36.	+ 06 38	120	14.5	GALAXY Sa/Sb
SCHO 0209	00 02 36.	+ 09 20 00.	510		ISOLATED DARK CLOUD
PHL 6266	00 02 36.	+ 09 29		18.3	BLUE STELLAR OBJECT
PHL 6268	00 02 36.	+ 17 44		19.0	BLUE STELLAR OBJECT
PHL 0648	00 02 36.	- 04 20		18.2	BLUE STELLAR OBJECT
PHL 2613	00 02 36.	- 06 58		18.3	BLUE STELLAR OBJECT
PHL 6265	00 02 36.	- 09 19		16.3	BLUE STELLAR OBJECT
PHL 6264	00 02 36.	- 13 12		14.1	BLUE STELLAR OBJECT
PHL 2614	00 02 36.	- 16 58		18.4	BLUE STELLAR OBJECT
PHL 6267	00 02 36.	- 20 05		18.9	BLUE STELLAR OBJECT
TON-S 0138	00 02 36.	- 26 49		15.6	BLUE STAR
PHL 0649	00 02 36.	- 26 50		15.8	BLUE STELLAR OBJECT
LB 04958	00 02 36.	- 29 56		18.6	FAINT BLUE STAR
LB 04959	00 02 36.	- 30 23		18.6	FAINT BLUE STAR
LB 01541	00 02 36.	- 57 13		14.0	FAINT BLUE STAR
LB 03127	00 02 36.	- 62 54		12.9	FAINT BLUE STAR
RNGC 7825	00 02 37.	+ 04 56		15.5	GALAXY
RNGC 7824	00 02 37.	+ 06 38		14.5	GALAXY
RNGC 7826	00 02 37.	- 21 00			NON-EXISTENT OBJECT
SCHO 0007	00 02 38.	+ 57 17 36.	230		ISOLATED DARK CLOUD
IC 1529	00 02 38.	- 11 47 08.			NONSTELLAR OBJECT
UGC 00035	00 02 42.	+ 05 59	102	17.	GALAXY DWRF SP
ZWG 408.026	00 02 42.	+ 06 30		14.7	GALAXY
UGC 00036	00 02 42.	+ 06 30	102	14.7	GALAXY Sa
PHL 2617	00 02 42.	+ 06 56		18.1	BLUE STELLAR OBJECT
ZWG 499.048	00 02 42.	+ 32 42		15.4	GALAXY
ZWG 498.076	00 02 42.	+ 32 42		15.4	GALAXY
PHL 2616	00 02 42.	- 03 11		18.6	BLUE STELLAR OBJECT
MCG-01-01-031	00 02 42.	- 04 04	48	15.	GALAXY
MCG-02-01-019	00 02 42.	- 11 45	90	14.	GALAXY
MCG-03-01-019	00 02 42.	- 16 45 30.	72	14.	GALAXY
PHL 2615	00 02 42.	- 17 04		17.9	BLUE STELLAR OBJECT
LB 04960	00 02 42.	- 30 16		17.5	FAINT BLUE STAR
RNGC 7814	00 02 43.	+ 15 52		12.0	GALAXY
RNGC 7821	00 02 43.	- 16 45		14.0	GALAXY
SCHO 0008	00 02 44.	+ 56 12 06.	260		ISOLATED DARK CLOUD
ARC 2703	00 02 46.	+ 15 50		17.1	RICH CLUSTER OF GALAXIES
SCHO 0009	00 02 47.	+ 58 56 12.	200		ISOLATED DARK CLOUD
PHL 0651	00 02 48.	+ 03 20		17.9	BLUE STELLAR OBJECT
MCG+01-01-028	00 02 48.	+ 04 53	48	15.	GALAXY
MCG+01-01-027	00 02 48.	+ 04 55 30.	48	14.	GALAXY
PHL 0650	00 02 48.	+ 05 06		16.3	BLUE STELLAR OBJECT
PHL 2619	00 02 48.	+ 14 28		18.3	BLUE STELLAR OBJECT
MCG+05-01-031	00 02 48.	+ 32 14 30.	42	15.	GALAXY
ZWG 499.049	00 02 48.	+ 32 15		15.0	GALAXY
ZWG 498.077	00 02 48.	+ 32 15		15.0	GALAXY
PHL 2618	00 02 48.	- 13 50		18.1	BLUE STELLAR OBJECT
PHL 0652	00 02 48.	- 15 19		17.9	BLUE STELLAR OBJECT
KN 13.069	00 02 50.1	+ 15 46 26.			NEBULA
SC 0000-6815.2	00 02 51.	- 67 58 29.	6		NEBULA
KN 13.070	00 02 53.1	+ 13 32 04.			NEBULA
ZWG 408.027	00 02 54.	+ 04 54		15.5	GALAXY
UGC 00037	00 02 54.	+ 04 54	78	15.5	GALAXY SBb
ZWG 408.028	00 02 54.	+ 04 57		14.6	GALAXY
UGC 00038	00 02 54.	+ 04 57	84	14.6	GALAXY SB0
ZWG 549.016	00 02 54.	+ 46 16		15.3	GALAXY
ZWG 548.025	00 02 54.	+ 46 16		15.3	GALAXY
PHL 6269	00 02 54.	- 05 10		18.0	BLUE STELLAR OBJECT
MCG-05-01-031	00 02 54.	- 28 22	24	16.	GALAXY
LB 04961	00 02 54.	- 30 18		17.6	FAINT BLUE STAR
MCG-06-01-020	00 02 54.	- 36 12	60	16.	GALAXY
RNGC 7827	00 02 55.	+ 04 57		14.5	GALAXY
ARC 2704	00 02 58.	- 12 09		17.7	RICH CLUSTER OF GALAXIES
SG 3.001	00 02 59.	+ 65 47	360		DIFFUSE EMISSION NEBULA
LBN 0588	00 03	+ 66 50	1320		BRIGHT NEBULA
PHL 0654	00 03 00.	+ 15 08		18.1	BLUE STELLAR OBJECT
PHL 6270	00 03 00.	+ 17 27		18.5	BLUE STELLAR OBJECT
ZC 0003.0+2008	00 03 00.	+ 20 08	1340		CLUSTER OF GALAXIES
UGC 00039	00 03 00.	+ 53 22	78	17.	GALAXY Sc
MCG+15-01-001	00 03 00.	+ 88 04	96	16.	GALAXY
PHL 2620	00 03 00.	- 06 00		16.8	BLUE STELLAR OBJECT
PHL 0653	00 03 00.	- 21 56		17.2	BLUE STELLAR OBJECT
LB 04962	00 03 00.	- 30 46		18.2	FAINT BLUE STAR
MCG-05-01-032	00 03 00.	- 30 52	30	16.	GALAXY
KEEL 001	00 03 01.2	+ 20 51 39.			NEBULA
SCHO 0010	00 03 02.	+ 57 37 00.	180		ISOLATED DARK CLOUD
MCG-02-01-020	00 03 04.	- 13 53	9	15.	GALAXY
ZC 0003.1+1114	00 03 06.	+ 11 14	2960		CLUSTER OF GALAXIES
PHL 2621	00 03 06.	- 04 02		18.2	BLUE STELLAR OBJECT
MCG-02-01-021	00 03 06.	- 13 52	60	16.5	GALAXY
LB 04963	00 03 06.	- 29 44		17.9	FAINT BLUE STAR
LB 04964	00 03 06.	- 29 54		18.5	FAINT BLUE STAR
MCG-06-01-021	00 03 06.	- 36 13	30	15.5	GALAXY
ARP 051	00 03 11.	- 13 43			PECULIAR GALAXY
PHL 0655	00 03 12.	+ 06 33		16.7	BLUE STELLAR OBJECT

OBJECT NAME	RIGHT ASCEN.	DECLINATION	DIAM.	MAGN.	TYPE OF OBJECT
ZC 0003.2+2410	00 03 12.	+ 24 10	740		CLUSTER OF GALAXIES
ZWG 478.024	00 03 12.	+ 27 10		15.2	GALAXY
ZWG 477.052	00 03 12.	+ 27 10		15.2	GALAXY
UGC 00040	00 03 12.	+ 27 10	72	15.2	GALAXY SB
PHL 2622	00 03 12.	- 05 21		18.6	BLUE STELLAR OBJECT
8ZW 0003-06.7	00 03 12.	- 06 44		15.1	COMPACT GALAXY
PHL 6271	00 03 12.	- 07 42		17.8	BLUE STELLAR OBJECT
MCG-01-01-032	00 03 12.	- 07 55	30	15.	GALAXY
MCG+04-01-020	00 03 15.	+ 22 13	36	15.5	GALAXY
MCG+04-01-021	00 03 15.	+ 27 10	66	14.5	GALAXY
PHL 2629	00 03 18.	- 04 47		17.9	BLUE STELLAR OBJECT
PHL 0656	00 03 18.	- 07 26		18.0	BLUE STELLAR OBJECT
PHL 0657	00 03 18.	- 10 19		18.2	BLUE STELLAR OBJECT
MCG-02-01-022	00 03 18.	- 14 16	60	15.	GALAXY
MCG-02-01-023	00 03 18.	- 14 17	60	15.	GALAXY
LB 04965	00 03 18.	- 29 43		19.0	FAINT BLUE STAR
LB 04966	00 03 18.	- 30 05		18.9	FAINT BLUE STAR
MCG+04-01-022	00 03 21.	+ 22 11	60	15.5	GALAXY
SC 0000-6828.4	00 03 21.	- 68 11 41.	6		NEBULA
ARC 2705	00 03 22.	+ 15 32		17.1	RICH CLUSTER OF GALAXIES
ARP 144	00 03 23.	- 13 41			PECULIAR GALAXY
PHL 6322	00 03 24.	+ 03 32		18.3	BLUE STELLAR OBJECT
ZC 0003.4+1035	00 03 24.	+ 10 35	940		CLUSTER OF GALAXIES
PHL 0658	00 03 24.	+ 15 51		16.2	BLUE STELLAR OBJECT
UGC 00041	00 03 24.	+ 22 12	72	16.0	GALAXY SBc
5ZW 002	00 03 24.	+ 47 07			COMPACT GALAXY
PHL 6272	00 03 24.	- 03 28		17.2	BLUE STELLAR OBJECT
PHL 2625	00 03 24.	- 06 39		18.4	BLUE STELLAR OBJECT
PHL 2623	00 03 24.	- 12 58		18.7	BLUE STELLAR OBJECT
PHL 6274	00 03 24.	- 25 28		18.2	BLUE STELLAR OBJECT
PHL 6273	00 03 24.	- 28 16		17.7	BLUE STELLAR OBJECT
PHL 2624	00 03 24.	- 30 39		16.6	BLUE STELLAR OBJECT
MCG-05-01-033	00 03 24.	- 31 24	36	15.	GALAXY
MCG-06-01-022	00 03 24.	- 36 23	9	15.	GALAXY
SER 164.01	00 03 24.	- 54 09	1500	16.	CLUSTER OF GALAXIES
SC 0000-0417.2	00 03 25.	- 04 00 30.	12		NEBULA
BC 4C15.01	00 03 25.01	+ 15 53 06C6		16.40	QUASI-STELLAR OBJECT
SHB 001	00 03 26.0	+ 15 53 00.		16.4	QUASI-STELLAR OBJECT
ARC 2706	00 03 28.	+ 10 52		17.2	RICH CLUSTER OF GALAXIES
KN 13.071	00 03 28.1	+ 14 08 14.			NEBULA
KN 13.072	00 03 28.2	+ 15 42 43.			NEBULA
UGC 00042	00 03 30.	+ 12 51	60	17.	GALAXY DISTRBD
UGC 00043	00 03 30.	+ 14 09	72	16.5	GALAXY
MCG+04-01-023	00 03 30.	+ 27 04	42	17.	GALAXY
PHL 0659	00 03 30.	- 02 26		17.8	BLUE STELLAR OBJECT
PHL 2626	00 03 30.	- 04 46		18.4	BLUE STELLAR OBJECT
LB 04967	00 03 30.	- 29 54		18.7	FAINT BLUE STAR
LB 04968	00 03 30.	- 30 09		18.5	FAINT BLUE STAR
LB 04969	00 03 30.	- 30 19		16.8	FAINT BLUE STAR
MCG-05-01-034	00 03 30.	- 30 55	30	15.	GALAXY
KN 13.073	00 03 35.3	+ 10 03 27.			NEBULA
PHL 2630	00 03 36.	+ 07 38		18.0	BLUE STELLAR OBJECT
MCG+01-01-029	00 03 36.	+ 08 37	60	14.5	GALAXY
PHL 2627	00 03 36.	+ 08 41		17.0	BLUE STELLAR OBJECT
ZC 0003.6+1331	00 03 36.	+ 13 31	1550		CLUSTER OF GALAXIES
UGC 00044	00 03 36.	+ 19 28	60	17.	GALAXY
ZC 0003.6+3250	00 03 36.	+ 32 50	4370		CLUSTER OF GALAXIES
ZWG 556.001	00 03 36.	+ 54 16		15.5	GALAXY
UGC 00045	00 03 36.	+ 54 16	60	15.5	GALAXY Sc-IRR
PHL 2628	00 03 36.	- 03 36		17.8	BLUE STELLAR OBJECT
PHL 0660	00 03 36.	- 05 01		18.1	BLUE STELLAR OBJECT
VV 272B	00 03 36.	- 13 41	18	16.	INTERACTING GALAXY
VV 272A	00 03 36.	- 13 41	48	14.	INTERACTING GALAXY
VV 272	00 03 36.	- 13 41	90		INTERACTING GALAXY
MCG-02-01-024	00 03 36.	- 13 42	30	15.	GALAXY
PHL 0661	00 03 36.	- 22 18		18.4	BLUE STELLAR OBJECT
LB 04970	00 03 36.	- 29 48		18.2	FAINT BLUE STAR
LB 04971	00 03 36.	- 30 24		18.6	FAINT BLUE STAR
RNGC 7830	00 03 37.	+ 08 06			NON-EXISTENT OBJECT
KN 13.074	00 03 38.6	+ 16 12 15.			NEBULA
PHL 2631	00 03 42.	+ 06 45		18.7	BLUE STELLAR OBJECT
ZWG 408.029	00 03 42.	+ 08 36		15.7	GALAXY
ZC 0003.7+2608	00 03 42.	+ 26 08	1080		CLUSTER OF GALAXIES
MCG-04-01-014	00 03 42.	- 23 09	24	15.	GALAXY
PHL 2632	00 03 42.	- 24 00		18.5	BLUE STELLAR OBJECT
PHL 0662	00 03 42.	- 29 29		18.1	BLUE STELLAR OBJECT
LB 04973	00 03 42.	- 30 24		18.4	FAINT BLUE STAR
LB 04972	00 03 42.	- 30 50		18.1	FAINT BLUE STAR
ARC 2707	00 03 46.	- 10 41		17.6	RICH CLUSTER OF GALAXIES
PHL 2633	00 03 48.	+ 03 24		18.6	BLUE STELLAR OBJECT
PHL 2636	00 03 48.	+ 12 20		17.5	BLUE STELLAR OBJECT
ZWG 456.030	00 03 48.	+ 17 09		14.5	GALAXY
UGC 00046	00 03 48.	+ 17 09	48	14.5	GALAXY DBL SYS
KARA.72 001A	00 03 48.	+ 17 09	36	14.5	PART OF DOUBLE GALAXY
MCG+03-01-024	00 03 48.	+ 17 09 30.	36	16.	GALAXY
KARA.72 001B	00 03 48.	+ 17 10	30		PART OF DOUBLE GALAXY
MCG+03-01-023	00 03 48.	+ 17 10	24	16.	GALAXY
PHL 2635	00 03 48.	+ 17 43		17.4	BLUE STELLAR OBJECT
MRK 335	00 03 48.	+ 19 55	15	14.	GALAXY WITH UV CONTINUUM
KW 31	00 03 48.	+ 19 55	16		SEYFERT GALAXY
VVI 01	00 03 48.	+ 19 55	15	14.18	SEYFERT GALAXY
PHL 6275	00 03 48.	- 06 05		17.6	BLUE STELLAR OBJECT
PHL 6278	00 03 48.	- 07 22		18.6	BLUE STELLAR OBJECT
PHL 2634	00 03 48.	- 07 22		18.2	BLUE STELLAR OBJECT
PHL 6276	00 03 48.	- 13 04		11.0	BLUE STELLAR OBJECT
MCG-02-01-025	00 03 48.	- 13 41 30.	48	14.	GALAXY
PHL 2637	00 03 48.	- 15 00		18.3	BLUE STELLAR OBJECT
PHL 2638	00 03 48.	- 15 46		18.3	BLUE STELLAR OBJECT
PHL 0663	00 03 48.	- 16 12		16.8	BLUE STELLAR OBJECT
PHL 2639	00 03 48.	- 18 48		18.4	BLUE STELLAR OBJECT
PHL 6277	00 03 48.	- 22 08		17.1	BLUE STELLAR OBJECT
TON-S 0139	00 03 48.	- 22 09		15.1	BLUE STAR
LB 04974	00 03 48.	- 29 35		17.4	FAINT BLUE STAR
LB 04975	00 03 48.	- 29 43		18.9	FAINT BLUE STAR
LB 04976	00 03 48.	- 30 35		17.4	FAINT BLUE STAR
AGU 01	00 03 48.	- 41 45 00.		12.5	2 INTERACTING GALAXIES
LB 01542	00 03 48.	- 48 03		14.3	FAINT BLUE STAR
SHB 002	00 03 48.7	- 00 21 06.		19.4	QUASI-STELLAR OBJECT
BC 3CR2	00 03 48.70	- 00 21 06.6		19.35	QUASI-STELLAR OBJECT
RNGC 7829	00 03 49.	- 13 41		14.0	GALAXY
RNGC 7828	00 03 49.	- 13 41		14.0	GALAXY
IC 5385	00 03 50.	- 00 21			NONSTELLAR OBJECT
PHL 6279	00 03 54.	- 06 14		18.1	BLUE STELLAR OBJECT
LB 04977	00 03 54.	- 30 23		19.2	FAINT BLUE STAR
LB 01543	00 03 54.	- 46 31		15.1	FAINT BLUE STAR
RNGC 7833	00 03 55.	+ 27 22			NON-EXISTENT OBJECT
IC 5386	00 03 55.	- 03 59 44.			SAME AS NGC 7832
MCG+08-01-021	00 03 57.	+ 47 35 30.	72	14.	GALAXY

OBJECT NAME	RIGHT ASCEN.	DECLINATION	DIAM.	MAGN.	TYPE OF OBJECT
MCG-05-01-035	00 03 57.	- 32 15	60	15.	GALAXY
HOLM 001B	00 03 58.	+ 04 51	12	14.8	PART OF MULTIPLE GALAXY
LB 00402	00 03 58.	+ 29 56 06.		17.7	FAINT BLUE STAR
ARC 2708	00 03 58.	- 17 12		17.4	RICH CLUSTER OF GALAXIES
ARP 146	00 03 59.	- 07 01			PECULIAR GALAXY
LBN 0585	00 04	+ 65 21	60		BRIGHT NEBULA
PHL 0664	00 04 00.	+ 07 56		17.1	BLUE STELLAR OBJECT
MCG+01-01-030	00 04 00.	+ 08 05 30.	60	15.	GALAXY
UGC 00047	00 04 00.	+ 17 00	84	17.	GALAXY
ZC 0004.0+2446	00 04 00.	+ 24 46	1810		CLUSTER OF GALAXIES
UGC 00048	00 04 00.	+ 47 36	126	16.0	GALAXY SB c
ISS 0001	00 04 00.	+ 63 31	136		STELLAR RING
LDN 1275	00 04 00.	+ 67 10	540		DARK NEBULA
MCG-01-01-033	00 04 00.	- 03 58	48	13.	GALAXY
PHL 2642	00 04 00.	- 04 18		18.3	BLUE STELLAR OBJECT
PHL 6280	00 04 00.	- 06 10		17.9	BLUE STELLAR OBJECT
PHL 2640	00 04 00.	- 09 54		18.6	BLUE STELLAR OBJECT
PHL 0665	00 04 00.	- 14 52		17.3	BLUE STELLAR OBJECT
SC 0001-1711.5	00 04 00.	- 16 54 47.	12		NEBULA
MCG-03-01-020	00 04 00.	- 18 20	30	15.	GALAXY
PHL 2641	00 04 00.	- 22 50		18.6	BLUE STELLAR OBJECT
PHL 2643	00 04 00.	- 27 26		18.5	BLUE STELLAR OBJECT
LB 04978	00 04 00.	- 29 48		19.8	FAINT BLUE STAR
LB 04979	00 04 00.	- 30 12		19.0	FAINT BLUE STAR
LB 04980	00 04 00.	- 30 24		19.5	FAINT BLUE STAR
LB 04981	00 04 00.	- 30 29		18.2	FAINT BLUE STAR
RNGC 7832	00 04 01.	- 03 58		13.5	GALAXY
HOLM 001A	00 04 04.	+ 04 51	30	14.0	PART OF MULTIPLE GALAXY
ARC 2709	00 04 04.	- 10 15		17.2	RICH CLUSTER OF GALAXIES
ARC 2710	00 04 04.	- 15 39		17.2	RICH CLUSTER OF GALAXIES
PHL 6281	00 04 06.	+ 07 16		18.3	BLUE STELLAR OBJECT
ZWG 408.030	00 04 06.	+ 08 05		15.4	GALAXY
UGC 00049	00 04 06.	+ 08 05	90	15.4	GALAXY Sc
MCG+04-01-024	00 04 06.	+ 25 51 30.	54	15.	GALAXY
ZWG 478.025	00 04 06.	+ 25 53		14.9	GALAXY
ZWG 477.053	00 04 06.	+ 25 53		14.9	GALAXY
UGC 00050	00 04 06.	+ 25 53	60	14.9	GALAXY Sa-b
PHL 2644	00 04 06.	- 05 12		18.8	BLUE STELLAR OBJECT
RNGC 7834	00 04 07.	+ 08 05		15.5	GALAXY
MCG+01-01-032	00 04 09.	+ 04 48 30.	42	14.5	GALAXY
MCG+01-01-031	00 04 09.	+ 08 09	24	15.5	GALAXY
SS 01	00 04 11.	+ 65 21			DIFFUSE GALACTIC NEBULA
PHL 6284	00 04 12.	+ 04 34		18.2	BLUE STELLAR OBJECT
ZWG 408.031	00 04 12.	+ 04 50		15.3	GALAXY
UGC 00051	00 04 12.	+ 04 50	66	15.3	GALAXY SBb-c
PHL 6282	00 04 12.	+ 07 01		17.7	BLUE STELLAR OBJECT
ZWG 408.032	00 04 12.	+ 08 21		15.0	GALAXY
UGC 00052	00 04 12.	+ 08 21	108	15.0	GALAXY Sc
ZC 0004.2+1516	00 04 12.	+ 15 16	4840		CLUSTER OF GALAXIES
PHL 6283	00 04 12.	+ 17 02		16.9	BLUE STELLAR OBJECT
DG 002	00 04 12.	+ 65 21	60		REFLECTION NEBULA
PHL 0666	00 04 12.	- 27 38		13.4	BLUE STELLAR OBJECT
LB 04982	00 04 12.	- 30 04		17.3	FAINT BLUE STAR
RNGC 7835	00 04 13.	+ 08 09		15.5	GALAXY
PHL 2646	00 04 18.	+ 04 04		18.1	BLUE STELLAR OBJECT
MCG+01-01-033	00 04 18.	+ 04 38	24	15.	GALAXY
ZWG 408.033	00 04 18.	+ 04 40		15.6	GALAXY
MCG+01-01-034	00 04 18.	+ 08 21 30.	18	14.	GALAXY
ZWG 456.031	00 04 18.	+ 19 02		15.5	GALAXY
UGC 00053	00 04 18.	+ 19 02	84	15.5	GALAXY Sb/SBc
UGC 00054	00 04 18.	+ 41 28	72	17.	GALAXY Sc
VMT 01	00 04 18.	+ 72 04 30.	5400		SUPERNOVA REMNANT
SC 0001-0417.0	00 04 18.	- 04 00 18.	12		NEBULA
PHL 2645	00 04 18.	- 05 25		16.8	BLUE STELLAR OBJECT
PHL 2647	00 04 18.	- 05 35		18.7	BLUE STELLAR OBJECT
PHL 0667	00 04 18.	- 08 04		18.5	BLUE STELLAR OBJECT
PHL 6286	00 04 18.	- 11 15		18.7	BLUE STELLAR OBJECT
PHL 2648	00 04 18.	- 11 54		18.6	BLUE STELLAR OBJECT
PHL 6285	00 04 18.	- 13 07		14.0	BLUE STELLAR OBJECT
PHL 0668	00 04 18.	- 15 48		18.2	BLUE STELLAR OBJECT
MCG-04-01-015	00 04 18.	- 25 12	36	17.	GALAXY
TON-S 0140	00 04 18.	- 27 38		13.7	BLUE STAR
LB 04983	00 04 18.	- 30 40		18.8	FAINT BLUE STAR
RNGC 7837	00 04 19.	+ 08 04		16.0	GALAXY
MCG+00-01-026	00 04 21.	+ 03 11 30.	9	15.5	GALAXY
MCG+01-01-036	00 04 21.	+ 08 04	36	16.	GALAXY
MCG+01-01-035	00 04 21.	+ 08 04	24	16.	GALAXY
MCG-01-01-034	00 04 21.	- 07 13 30.	60	16.	GALAXY
ARP 246	00 04 22.	+ 08 05			PECULIAR GALAXY
ARC 2711	00 04 22.	+ 24 50		17.5	RICH CLUSTER OF GALAXIES
ARC 2712	00 04 22.	- 18 21		17.5	RICH CLUSTER OF GALAXIES
SC 0001-6816.4	00 04 22.	- 67 59 41.	12		NEBULA
ZWG 382.027	00 04 24.	+ 02 50		15.5	GALAXY
PHL 2650	00 04 24.	+ 03 59		18.4	BLUE STELLAR OBJECT
PHL 2649	00 04 24.	+ 04 14		18.6	BLUE STELLAR OBJECT
PHL 6287	00 04 24.	+ 04 47		18.6	BLUE STELLAR OBJECT
PHL 0670	00 04 24.	+ 06 09		16.2	BLUE STELLAR OBJECT
PHL 0669	00 04 24.	+ 06 28		18.7	BLUE STELLAR OBJECT
ZWG 408.034	00 04 24.	+ 08 04		15.3	GALAXY
UGC 00055	00 04 24.	+ 46 23	72	17.	GALAXY
MCG+08-01-022	00 04 24.	+ 46 23	60	16.	GALAXY
PHL 6288	00 04 24.	- 02 36		18.2	BLUE STELLAR OBJECT
MCG-01-01-035	00 04 24.	- 03 58	21	16.	GALAXY
PHL 2651	00 04 24.	- 16 44		19.0	BLUE STELLAR OBJECT
LB 04984	00 04 24.	- 30 44		18.7	FAINT BLUE STAR
LB 01544	00 04 24.	- 46 07		14.5	FAINT BLUE STAR
RNGC 7838	00 04 25.	+ 08 04		15.5	GALAXY
RNGC 7839	00 04 25.	+ 27 23			GALAXY
KN 13.075	00 04 27.5	+ 13 49 33.			NEBULA
PHL 0672	00 04 30.	+ 13 10		18.9	BLUE STELLAR OBJECT
ZWG 433.022	00 04 30.	+ 13 48		15.6	GALAXY
UGC 00056	00 04 30.	+ 13 48	60	15.6	GALAXY Sc
ZWG 518.011	00 04 30.	+ 35 32		15.6	GALAXY
ZWG 517.016	00 04 30.	+ 35 32		15.6	GALAXY
MCG+06-01-012	00 04 30.	+ 35 33	45	14.5	GALAXY
MCG+06-01-011	00 04 30.	+ 37 55 30.	24	16.	GALAXY
PHL 2652	00 04 30.	- 04 48		17.8	BLUE STELLAR OBJECT
LB 04985	00 04 30.	- 29 36		18.3	FAINT BLUE STAR
LB 04986	00 04 30.	- 30 42		18.2	FAINT BLUE STAR
PHL 0671	00 04 30.	- 31 30		16.7	BLUE STELLAR OBJECT
TON-S 0141	00 04 30.	- 31 31		15.7	BLUE STAR
TON-S 0142	00 04 30.	- 33 45		14.2	BLUE STAR
SPC 3	00 04 30.	- 41 44	30		MEMBER OF GALAXY TRIPLET
SPC 2	00 04 30.	- 41 44	36		MEMBER OF GALAXY TRIPLET
SPC 1	00 04 30.	- 41 44	168	12.	MEMBER OF GALAXY TRIPLET
RNGC 7840	00 04 31.	+ 08 09			NON-EXISTENT OBJECT
PHL 6290	00 04 36.	+ 04 39		16.9	BLUE STELLAR OBJECT
PHL 2653	00 04 36.	+ 04 43		17.1	BLUE STELLAR OBJECT

OBJECT NAME	RIGHT ASCEN.	DECLINATION	DIAM.	MAGN.	TYPE OF OBJECT
PHL 6289	00 04 36.	+ 04 50		18.1	BLUE STELLAR OBJECT
ZWG 433.023	00 04 36.	+ 13 29		15.7	GALAXY
ZWG 478.026	00 04 36.	+ 27 26		13.4	GALAXY
ZWG 477.054	00 04 36.	+ 27 26		13.4	GALAXY
UGC 00057	00 04 36.	+ 27 26	108	13.4	GALAXY Sb
KARA.72 002A	00 04 36.	+ 27 26	108	13.4	PART OF DOUBLE GALAXY
MCG+00-01-027	00 04 36.	- 00 42 30.	9	15.	GALAXY
PHL 0673	00 04 36.	- 06 25		18.0	BLUE STELLAR OBJECT
LB 04987	00 04 36.	- 30 40		18.0	FAINT BLUE STAR
LB 01545	00 04 36.	- 58 13		14.9	FAINT BLUE STAR
KN 13.076	00 04 37.5	+ 13 29 24.			NEBULA
HN 1239	00 04 38.3	- 44 38 48.	18		NEBULA
MCG+01-01-037	00 04 39.	+ 08 01	30	14.5	GALAXY
MCG-04-01-016	00 04 39.	- 22 04	84	15.	GALAXY
HOLM 002A	00 04 40.	+ 27 26	18	14.3	PART OF MULTIPLE GALAXY
HOLM 002B	00 04 41.	+ 27 24	36	14.6	PART OF MULTIPLE GALAXY
SC 0002-0421.4	00 04 41.	- 04 04 42.	18		NEBULA
ZWG 408.035	00 04 42.	+ 08 02		14.6	GALAXY
UGC 00058	00 04 42.	+ 08 02	72	14.6	GALAXY S0?
ZWG 478.027	00 04 42.	+ 27 24		14.8	GALAXY
ZWG 477.055	00 04 42.	+ 27 24		14.8	GALAXY
UGC 00059	00 04 42.	+ 27 24	66	14.8	GALAXY Sa-b
KARA.72 002B	00 04 42.	+ 27 24	72	14.8	PART OF DOUBLE GALAXY
MCG+04-01-026	00 04 42.	+ 27 24	72	15.	GALAXY
MCG+04-01-025	00 04 42.	+ 27 26	102	14.	GALAXY
4ZW 006	00 04 42.	+ 28 05			COMPACT GALAXY
ZWG 499.050	00 04 42.	+ 32 20		13.4	GALAXY
ZWG 498.078	00 04 42.	+ 32 20		13.4	GALAXY
UGC 00060	00 04 42.	+ 32 20	108	13.4	GALAXY Sb
ZC 0004.7-0216	00 04 42.	- 02 16	4700		CLUSTER OF GALAXIES
PHL 2654	00 04 42.	- 06 21		18.4	BLUE STELLAR OBJECT
PHL 6291	00 04 42.	- 18 29		17.7	BLUE STELLAR OBJECT
PHL 6292	00 04 42.	- 28 54		18.3	BLUE STELLAR OBJECT
LB 04988	00 04 42.	- 29 45		16.6	FAINT BLUE STAR
LB 04989	00 04 42.	- 30 07		19.2	FAINT BLUE STAR
AGU 02	00 04 42.	- 41 37 30.			COMPACT GALAXY
RNGC 0003	00 04 43.	+ 08 01		14.5	GALAXY
RNGC 0002	00 04 43.	+ 27 24		15.0	GALAXY
RNGC 0001	00 04 43.	+ 27 26		13.5	GALAXY
RNGC 7831	00 04 43.	+ 32 19		14.0	GALAXY
IC 1530	00 04 43.	+ 32 20			NONSTELLAR OBJECT
MCG+05-01-032	00 04 45.	+ 32 20	72	14.	GALAXY
MCG+08-01-023	00 04 45.	+ 46 42	48	14.	GALAXY
MCG+08-01-024	00 04 45.	+ 46 44 30.	36	14.	GALAXY
PHL 2656	00 04 48.	+ 04 02		18.4	BLUE STELLAR OBJECT
PHL 0674	00 04 48.	+ 04 30		18.3	BLUE STELLAR OBJECT
ZC 0004.8+1609	00 04 48.	+ 16 09	2350		CLUSTER OF GALAXIES
ZC 0004.8+2304	00 04 48.	+ 23 04	1210		CLUSTER OF GALAXIES
MCG+08-01-025	00 04 48.	+ 46 45	30	15.	GALAXY
5ZW 003	00 04 48.	+ 46 46			COMPACT GALAXY
PHL 2657	00 04 48.	- 02 50		18.2	BLUE STELLAR OBJECT
PHL 6294	00 04 48.	- 03 44		17.9	BLUE STELLAR OBJECT
PHL 6295	00 04 48.	- 11 10		18.3	BLUE STELLAR OBJECT
MCG-02-01-026	00 04 48.	- 12 05	42	15.	GALAXY
PHL 6293	00 04 48.	- 12 26		17.8	BLUE STELLAR OBJECT
PHL 2655	00 04 48.	- 18 46		18.1	BLUE STELLAR OBJECT
MCG-05-01-036	00 04 48.	- 28 24	24	16.	GALAXY
LB 04990	00 04 48.	- 30 47		18.0	FAINT BLUE STAR
LB 04991	00 04 48.	- 30 51		18.9	FAINT BLUE STAR
MCG-06-01-023	00 04 48.	- 37 44	42	16.	GALAXY
LB 01546	00 04 48.	- 49 36		14.4	FAINT BLUE STAR
SC 0002-0328.8	00 04 52.	- 03 12 06.	6		NEBULA
PHL 0675	00 04 54.	+ 06 02		18.2	BLUE STELLAR OBJECT
PHL 0676	00 04 54.	+ 14 12		17.8	BLUE STELLAR OBJECT
MCG+05-01-033	00 04 54.	+ 28 23	30	16.	GALAXY
ZWG 549.017	00 04 54.	+ 46 43		15.3	GALAXY
ZWG 548.026	00 04 54.	+ 46 43		15.3	GALAXY
ZWG 549.018	00 04 54.	+ 46 46		14.3	GALAXY
ZWG 548.027	00 04 54.	+ 46 46		14.3	GALAXY
UGC 00061	00 04 54.	+ 46 46	96	14.3	GALAXY S0
PHL 2658	00 04 54.	- 11 42		17.9	BLUE STELLAR OBJECT
LB 04993	00 04 54.	- 30 20		20.3	FAINT BLUE STAR
LB 04994	00 04 54.	- 30 41		17.2	FAINT BLUE STAR
LB 04992	00 04 54.	- 30 47		19.5	FAINT BLUE STAR
PK118+02.1	00 04 55.	+ 64 41	154		PLANETARY NEBULA
SC 0002-0212.0	00 04 56.	- 01 55 18.			NEBULA
ARC 0001	00 04 58.	+ 16 15		17.1	RICH CLUSTER OF GALAXIES
ZC 0005.0+0158	00 05 00.	+ 01 58	1080		CLUSTER OF GALAXIES
PHL 2659	00 05 00.	+ 05 20		18.5	BLUE STELLAR OBJECT
ZC 0005.0+1003	00 05 00.	+ 10 03	2690		CLUSTER OF GALAXIES
PHL 0678	00 05 00.	+ 13 19		12.0	BLUE STELLAR OBJECT
ZC 0005.0+1422	00 05 00.	+ 14 22	1550		CLUSTER OF GALAXIES
LB 00403	00 05 00.	+ 14 47 54.		15.9	FAINT BLUE STAR
PHL 2662	00 05 00.	+ 17 56		18.1	BLUE STELLAR OBJECT
ZWG 549.019	00 05 00.	+ 46 46		15.6	GALAXY
ZWG 548.028	00 05 00.	+ 46 46		15.6	GALAXY
PHL 6297	00 05 00.	- 10 01		18.3	BLUE STELLAR OBJECT
PHL 0677	00 05 00.	- 16 22		16.6	BLUE STELLAR OBJECT
PHL 2660	00 05 00.	- 19 44		18.2	BLUE STELLAR OBJECT
PHL 6296	00 05 00.	- 28 29		17.0	BLUE STELLAR OBJECT
LB 04995	00 05 00.	- 29 52		18.2	FAINT BLUE STAR
PHL 2661	00 05 00.	- 29 58		18.2	BLUE STELLAR OBJECT
LB 04996	00 05 00.	- 30 20		19.5	FAINT BLUE STAR
MCG-02-01-027	00 05 03.	- 12 07	72	15.	GALAXY
PHL 6298	00 05 06.	+ 05 12		18.5	BLUE STELLAR OBJECT
PHL 0679	00 05 06.	+ 13 22		17.5	BLUE STELLAR OBJECT
ZWG 478.028	00 05 06.	+ 25 18		15.5	GALAXY
ZWG 477.056	00 05 06.	+ 25 18		15.5	GALAXY
MCG+07-01-005	00 05 06.	+ 40 35 30.	48	15.	GALAXY
OCL 0283	00 05 06.	+ 61 08	540		OPEN STAR CLUSTER
8ZW 0005-06.9	00 05 06.	- 06 57		17.9	COMPACT GALAXY
PHL 2663	00 05 06.	- 31 20		17.4	BLUE STELLAR OBJECT
PHL 2664	00 05 12.	+ 05 34		18.5	BLUE STELLAR OBJECT
PHL 0680	00 05 12.	+ 08 05		17.1	BLUE STELLAR OBJECT
PHL 6299	00 05 12.	+ 15 14		18.7	BLUE STELLAR OBJECT
PHL 6301	00 05 12.	+ 15 58		18.1	BLUE STELLAR OBJECT
4ZW 007	00 05 12.	+ 35 05			COMPACT GALAXY
ZWG 518.012	00 05 12.	+ 35 05		14.6	GALAXY
ZWG 517.017	00 05 12.	+ 35 05		14.6	GALAXY
UGC 00062	00 05 12.	+ 35 05	72	14.6	GALAXY (E)
MCG+06-01-013	00 05 12.	+ 35 06	21	14.	GALAXY
ZWG 518.013	00 05 12.	+ 35 41		15.6	GALAXY
ZWG 517.018	00 05 12.	+ 35 41		15.6	GALAXY
UGC 00063	00 05 12.	+ 35 41	60	15.6	GALAXY IRR
ZWG 534.006	00 05 12.	+ 40 36		15.5	GALAXY
UGC 00064	00 05 12.	+ 40 36	84	15.5	GALAXY KNOTS
SC 0002-0230.6	00 05 12.	- 02 13 54.	18		NEBULA
PHL 0681	00 05 12.	- 02 48		6.3	BLUE STELLAR OBJECT

OBJECT NAME	RIGHT ASCEN.	DECLINATION	DIAM.	MAGN.	TYPE OF OBJECT
PHL 0682	00 05 12.	- 04 10		18.1	BLUE STELLAR OBJECT
PHL 6300	00 05 12.	- 23 18		18.5	BLUE STELLAR OBJECT
RNGC 0005	00 05 13.	+ 35 05		14.5	GALAXY
PHL 0683	00 05 18.	+ 06 27		17.8	BLUE STELLAR OBJECT
ZWG 433.024	00 05 18.	+ 11 43		15.6	GALAXY
ZC 0005.3+1307	00 05 18.	+ 13 07	2150		CLUSTER OF GALAXIES
ZWG 456.032	00 05 18.	+ 20 08		15.5	GALAXY
KARA.73B 0005	00 05 18.	+ 20 08	36	15.5	ISOLATED GALAXY S
MCG+06-01-014	00 05 18.	+ 35 42	48	15.	GALAXY
ZWG 518.014	00 05 18.	+ 39 27		15.4	GALAXY
ZWG 517.019	00 05 18.	+ 39 27		15.4	GALAXY
PHL 6302	00 05 18.	- 09 37		18.2	BLUE STELLAR OBJECT
PHL 2665	00 05 18.	- 15 10		18.5	BLUE STELLAR OBJECT
MCG-04-01-017	00 05 18.	- 20 58	48	16.	GALAXY
KN 13.077	00 05 19.0	+ 11 34 56.			NEBULA
KN 13.078	00 05 21.6	+ 11 22 44.			NEBULA
SC 0002-0233.1	00 05 22.	- 02 16 24.	18		NEBULA
SC 0002-6816.2	00 05 23.	- 67 59 29.	6		NEBULA
PHL 0684	00 05 24.	+ 05 10		17.0	BLUE STELLAR OBJECT
MCG+02-01-020	00 05 24.	+ 11 35	36	15.5	GALAXY
ZWG 499.051	00 05 24.	+ 32 48		13.8	GALAXY
ZWG 498.079	00 05 24.	+ 32 48		13.8	GALAXY
MRK 336	00 05 24.	+ 32 48	25	16.	GALAXY WITH UV CONTINUUM
UGC 00065	00 05 24.	+ 32 48	66	13.8	GALAXY
ZWG 534.007	00 05 24.	+ 41 04		15.6	GALAXY
PHL 2668	00 05 24.	- 04 26		18.7	BLUE STELLAR OBJECT
PHL 2667	00 05 24.	- 04 26		18.5	BLUE STELLAR OBJECT
PHL 6305	00 05 24.	- 06 20		18.5	BLUE STELLAR OBJECT
PHL 6306	00 05 24.	- 07 55		18.4	BLUE STELLAR OBJECT
PHL 6303	00 05 24.	- 10 11		14.8	BLUE STELLAR OBJECT
PHL 2666	00 05 24.	- 15 20		18.4	BLUE STELLAR OBJECT
PHL 6304	00 05 24.	- 23 57		16.8	BLUE STELLAR OBJECT
TON-S 0143	00 05 24.	- 23 57		15.7	BLUE STAR
RNGC 7836	00 05 25.	+ 32 48			GALAXY
SC 0002-1629.1	00 05 27.	- 16 12 23.	6		NEBULA
ZWG 408.036	00 05 30.	+ 09 26		14.9	GALAXY
UGC 00066	00 05 30.	+ 09 26	90	14.9	GALAXY Sb/SBb
MCG+01-01-038	00 05 30.	+ 09 27	72	14.5	GALAXY
ZWG 433.025	00 05 30.	+ 14 33		15.7	GALAXY
LB 00404	00 05 31.	+ 30 57 06.		17.6	FAINT BLUE STAR
LB 00405	00 05 34.	+ 16 03 36.		16.9	FAINT BLUE STAR
MCG+01-01-039	00 05 36.	+ 07 29	60	14.5	GALAXY
ZWG 408.037	00 05 36.	+ 07 30		15.4	GALAXY
UGC 00067	00 05 36.	+ 07 30	90	15.4	GALAXY Sa-b
PHL 0685	00 05 36.	+ 08 02		16.6	BLUE STELLAR OBJECT
4ZW 008	00 05 36.	+ 25 07			COMPACT GALAXY
ZWG 478.029	00 05 36.	+ 26 44		14.4	GALAXY
ZWG 477.057	00 05 36.	+ 26 44		14.4	GALAXY
UGC 00068	00 05 36.	+ 26 44	60	14.4	GALAXY SB
ZWG 478.030	00 05 36.	+ 27 15		14.7	GALAXY
ZWG 477.058	00 05 36.	+ 27 15		14.7	GALAXY
UGC 00069	00 05 36.	+ 27 15	78	14.7	GALAXY Sc
MCG+04-01-027	00 05 36.	+ 27 15	66	14.	GALAXY
PHL 6307	00 05 36.	- 08 46		13.5	BLUE STELLAR OBJECT
PHL 0686	00 05 36.	- 13 46		18.1	BLUE STELLAR OBJECT
RNGC 0004	00 05 37.	+ 07 53			GALAXY
MCG+04-01-028	00 05 39.	+ 26 44	51	15.	GALAXY
SC 0003-6715.2	00 05 41.	- 66 58 29.	18		NEBULA
PHL 2669	00 05 42.	+ 08 18		18.3	BLUE STELLAR OBJECT
PHL 6308	00 05 42.	+ 09 21		18.4	BLUE STELLAR OBJECT
PHL 6309	00 05 42.	+ 09 22		18.2	BLUE STELLAR OBJECT
PHL 2670	00 05 42.	+ 18 30		18.7	BLUE STELLAR OBJECT
UGC 00070	00 05 42.	+ 49 30	60	16.5	GALAXY
ZWG 549.020	00 05 42.	+ 49 59		15.7	GALAXY
ZWG 548.029	00 05 42.	+ 49 59		15.7	GALAXY
PHL 2671	00 05 42.	- 22 30		18.7	BLUE STELLAR OBJECT
RNGC 0006	00 05 43.	+ 32 14			NON-EXISTENT OBJECT
PHL 6314	00 05 48.	+ 06 04		18.5	BLUE STELLAR OBJECT
PHL 0687	00 05 48.	+ 17 51		13.7	BLUE STELLAR OBJECT
PHL 6310	00 05 48.	+ 17 59		17.8	BLUE STELLAR OBJECT
ZWG 499.052	00 05 48.	+ 30 50		15.7	GALAXY
ZWG 498.080	00 05 48.	+ 30 50		15.7	GALAXY
PHL 2672	00 05 48.	- 04 16		16.9	BLUE STELLAR OBJECT
PHL 6311	00 05 48.	- 08 10		18.2	BLUE STELLAR OBJECT
PHL 2673	00 05 48.	- 09 30		13.5	BLUE STELLAR OBJECT
PHL 6312	00 05 48.	- 16 14		18.0	BLUE STELLAR OBJECT
PHL 6315	00 05 48.	- 19 06		18.5	BLUE STELLAR OBJECT
PHL 6313	00 05 48.	- 20 50		18.2	BLUE STELLAR OBJECT
MCG-05-01-037	00 05 48.	- 30 13	120	13.5	GALAXY
RNGC 0007	00 05 50.	- 30 11		13.0	GALAXY
ARC 0002	00 05 51.	- 19 55		17.3	RICH CLUSTER OF GALAXIES
IC 0001	00 05 52.	+ 27 26 06.			TWO STARS
PHL 0688	00 05 54.	+ 04 55		17.9	BLUE STELLAR OBJECT
ZWG 433.026	00 05 54.	+ 09 56		15.4	GALAXY
PHL 6316	00 05 54.	+ 14 36		18.1	BLUE STELLAR OBJECT
ZWG 456.033	00 05 54.	+ 17 52		15.4	GALAXY
MCG-02-01-028	00 05 54.	- 11 13	66	14.	GALAXY
LBN 0525	00 06	+ 20 50	1440		BRIGHT NEBULA
PHL 6318	00 06 00.	+ 05 23		18.1	BLUE STELLAR OBJECT
PHL 0689	00 06 00.	+ 07 36		18.4	BLUE STELLAR OBJECT
ZWG 534.008	00 06 00.	+ 43 50		15.7	GALAXY
UGC 00071	00 06 00.	+ 43 50	72	15.7	GALAXY Sb-c
PHL 2674	00 06 00.	- 04 05		18.1	BLUE STELLAR OBJECT
PHL 6319	00 06 00.	- 06 28		16.7	BLUE STELLAR OBJECT
8ZW 0006-07.8	00 06 00.	- 07 49		18.0	COMPACT GALAXY
PHL 0690	00 06 00.	- 12 38		18.2	BLUE STELLAR OBJECT
PHL 6317	00 06 00.	- 22 29		17.8	BLUE STELLAR OBJECT
SCHO 0011	00 06 02.	+ 59 21 06.	210		ISOLATED DARK CLOUD
MCG-01-01-036	00 06 03.	- 05 29	66	15.	GALAXY
PHL 0691	00 06 06.	+ 13 56		16.5	BLUE STELLAR OBJECT
MCG+02-01-021	00 06 06.	+ 15 32 30.	180	12.5	GALAXY
HOLM 003B	00 06 06.	+ 23 34	24	14.9	PART OF MULTIPLE GALAXY
ZWG 518.015	00 06 06.	+ 37 10		14.5	GALAXY
ZWG 517.020	00 06 06.	+ 37 10		14.5	GALAXY
RNGC 0011	00 06 06.	+ 37 10		14.5	GALAXY
UGC 00073	00 06 06.	+ 37 10	102	14.5	GALAXY Sa
ZWG 382.028	00 06 06.	- 01 00		15.5	GALAXY
UGC 00072	00 06 06.	- 01 00	66	15.5	GALAXY Sa
MCG+00-01-028	00 06 06.	- 01 01	54	14.5	GALAXY
MCG-01-01-037	00 06 06.	- 07 16	48	14.	GALAXY
MCG-06-01-024	00 06 06.	- 34 08	102	13.	GALAXY
RNGC 0014	00 06 07.	+ 15 32		13.5	GALAXY
RNGC 0010	00 06 08.	- 34 09			GALAXY
ARP 235	00 06 10.	+ 15 32			PECULIAR GALAXY
ZWG 408.038	00 06 12.	+ 04 20		14.5	GALAXY
UGC 00074	00 06 12.	+ 04 20	126	14.5	GALAXY Sc
PHL 2675	00 06 12.	+ 04 22		18.4	BLUE STELLAR OBJECT
MCG+03-01-026	00 06 12.	+ 15 31	156	13.5	GALAXY
ZWG 456.034	00 06 12.	+ 15 32		13.3	GALAXY
UGC 00075	00 06 12.	+ 15 32	180	13.3	GALAXY IRR
VV 080	00 06 12.	+ 15 32	90	13.5	INTERACTING GALAXY
UGC 00076	00 06 12.	+ 24 15	66	16.0	GALAXY
ZWG 499.053	00 06 12.	+ 33 10		14.2	GALAXY
ZWG 498.081	00 06 12.	+ 33 10		14.2	GALAXY
UGC 00077	00 06 12.	+ 33 10	150	14.2	GALAXY Sb
MCG+05-01-034	00 06 12.	+ 33 10	132	14.5	GALAXY
MCG+06-01-015	00 06 12.	+ 37 10	84	13.	GALAXY
PHL 0692	00 06 12.	- 05 28		17.9	BLUE STELLAR OBJECT
MCG-02-01-029	00 06 12.	- 13 16 30.	36	15.	GALAXY
PHL 0693	00 06 12.	- 23 21		16.9	BLUE STELLAR OBJECT
KN 13.079	00 06 12.0	+ 15 32 14.			NEBULA
KN 13.080	00 06 12.1	+ 15 34 58.			NEBULA
RNGC 0012	00 06 13.	+ 04 20		14.5	GALAXY
RNGC 0008	00 06 13.	+ 23 34		15.0	GALAXY
RNGC 0013	00 06 13.	+ 33 09		14.0	GALAXY
SC 0003-0259.8	00 06 13.	- 02 43 06.	54		NEBULA
MCG+01-01-040	00 06 15.	+ 04 18 30.	72	13.5	GALAXY
MCG+04-01-029	00 06 15.	+ 24 15	66	16.	GALAXY
HOLM 003A	00 06 16.	+ 23 33	54	13.8	PART OF MULTIPLE GALAXY
SCHO 0012	00 06 16.	+ 59 23 24.	210		ISOLATED DARK CLOUD
PHL 6320	00 06 18.	+ 18 14		16.6	BLUE STELLAR OBJECT
PHL 2677	00 06 18.	+ 18 16		17.5	BLUE STELLAR OBJECT
MCG+04-01-030	00 06 18.	+ 23 32	78	15.	GALAXY
ZWG 478.031	00 06 18.	+ 23 33		14.5	GALAXY
ZWG 477.059	00 06 18.	+ 23 33		14.5	GALAXY
UGC 00078	00 06 18.	+ 23 33	78	14.5	GALAXY PECULR
KARA.73B 0006	00 06 18.	+ 23 33	48	14.5	ISOLATED GALAXY S
PHL 0694	00 06 18.	- 03 01		14.0	BLUE STELLAR OBJECT
PHL 2676	00 06 18.	- 07 02		18.3	BLUE STELLAR OBJECT
MCG-06-01-025	00 06 18.	- 37 37	18	15.5	GALAXY
RNGC 0009	00 06 19.	+ 23 32		14.5	GALAXY
HELW 311	00 06 23.	- 30 19 30.			NEBULA
PHL 6321	00 06 24.	+ 07 27		18.3	BLUE STELLAR OBJECT
PHL 6324	00 06 24.	+ 15 32		18.4	BLUE STELLAR OBJECT
ZWG 478.032	00 06 24.	+ 25 21		15.7	GALAXY
ZWG 477.060	00 06 24.	+ 25 21		15.7	GALAXY
UGC 00079	00 06 24.	+ 25 21	102	15.7	GALAXY Sc
ZWG 478.033	00 06 24.	+ 27 27		12.5	GALAXY
ZWG 477.061	00 06 24.	+ 27 27		12.5	GALAXY
UGC 00080	00 06 24.	+ 27 27	108	12.5	GALAXY S0
MCG+05-01-035	00 06 24.	+ 27 30	24	18.	GALAXY
ZC 0006.4+4050	00 06 24.	+ 40 50	2220		CLUSTER OF GALAXIES
MCG-01-01-038	00 06 24.	- 02 39	66	15.	GALAXY
PHL 2678	00 06 24.	- 12 30		18.6	BLUE STELLAR OBJECT
PHL 6323	00 06 24.	- 24 10		16.9	BLUE STELLAR OBJECT
SC 0003-0257.0	00 06 25.	- 02 40 18.	18		NEBULA
MCG+03-01-027	00 06 27.	+ 21 20	45	14.	GALAXY
SC 0003-1308.6	00 06 27.	- 12 51 53.	6		NEBULA
LB 00406	00 06 29.	+ 16 54 18.		17.0	FAINT BLUE STAR
ZWG 433.027	00 06 30.	+ 10 17		15.6	GALAXY
ZWG 433.028	00 06 30.	+ 10 37		15.7	GALAXY
UGC 00081	00 06 30.	+ 10 37	90	15.7	GALAXY Sb
MCG+02-01-022	00 06 30.	+ 10 37	84	15.	GALAXY
ZWG 456.035	00 06 30.	+ 21 21		14.9	GALAXY
UGC 00082	00 06 30.	+ 21 21	72	14.9	GALAXY Sa
MCG+04-01-031	00 06 30.	+ 25 20	90	15.	GALAXY
MCG+04-01-032	00 06 30.	+ 27 27 30.	66	13.5	GALAXY
PHL 2680	00 06 30.	- 10 52		18.6	BLUE STELLAR OBJECT
PHL 2679	00 06 30.	- 11 00		18.4	BLUE STELLAR OBJECT
SC 0003-1557.4	00 06 30.	- 15 40 42.	6		NEBULA
MCG-06-01-026	00 06 30.	- 37 38	48	15.	GALAXY
KN 13.081	00 06 30.3	+ 10 38 29.			NEBULA
KN 13.082	00 06 30.5	+ 10 17 16.			NEBULA
RNGC 0015	00 06 31.	+ 21 20		15.0	GALAXY
RNGC 0016	00 06 31.	+ 27 27		13.0	GALAXY
RNGC 0017	00 06 31.	- 12 24			NON-EXISTENT OBJECT
PHL 6329	00 06 36.	+ 05 11		18.2	BLUE STELLAR OBJECT
PHL 6327	00 06 36.	+ 09 20		18.3	BLUE STELLAR OBJECT
PHL 6326	00 06 36.	+ 13 08		16.7	BLUE STELLAR OBJECT
ZWG 518.016	00 06 36.	+ 38 49		15.7	GALAXY
ZWG 517.021	00 06 36.	+ 38 49		15.7	GALAXY
MCG+00-01-029	00 06 36.	- 01 36 30.	54	15.5	GALAXY
MCG-01-01-039	00 06 36.	- 02 38	36	14.	GALAXY
MCG-01-01-040	00 06 36.	- 07 12	48	15.	GALAXY
PHL 6328	00 06 36.	- 10 44		16.6	BLUE STELLAR OBJECT
PHL 6325	00 06 36.	- 23 50		16.4	BLUE STELLAR OBJECT
LB 04997	00 06 36.	- 27 32		17.8	FAINT BLUE STAR
LB 04998	00 06 36.	- 27 39		17.2	FAINT BLUE STAR
LB 04999	00 06 36.	- 27 42		18.2	FAINT BLUE STAR
LB 05000	00 06 36.	- 27 54		18.8	FAINT BLUE STAR
MCG-06-01-027	00 06 36.	- 33 12	24	16.	GALAXY
ARC 0003	00 06 40.	+ 03 46		17.0	RICH CLUSTER OF GALAXIES
PHL 6330	00 06 42.	+ 05 48		18.6	BLUE STELLAR OBJECT
4ZW 009	00 06 42.	+ 39 19			COMPACT GALAXY
SC 0004-1632.2	00 06 42.	- 16 15 30.	18		NEBULA
LB 05001	00 06 42.	- 28 14		20.0	FAINT BLUE STAR
IC 1531	00 06 43.	- 32 33 14.			NONSTELLAR OBJECT
ARC 0004	00 06 46.	+ 06 31		17.8	RICH CLUSTER OF GALAXIES
UGC 00083	00 06 48.	+ 00 20	66	16.5	GALAXY Sb
ZC 0006.8+0350	00 06 48.	+ 03 50	2960		CLUSTER OF GALAXIES
PHL 6332	00 06 48.	+ 07 52		17.7	BLUE STELLAR OBJECT
ZC 0006.8+2357	00 06 48.	+ 23 57	670		CLUSTER OF GALAXIES
RNGC 0019	00 06 48.	+ 32 34			NON-EXISTENT OBJECT
MCG+05-01-036	00 06 48.	+ 33 03	78	14.	GALAXY
ZC 0006.8+3401	00 06 48.	+ 34 01	870		CLUSTER OF GALAXIES
MCG+08-01-026	00 06 48.	+ 47 03	84	14.	GALAXY
PHL 6331	00 06 48.	- 04 34		18.7	BLUE STELLAR OBJECT
PHL 0695	00 06 48.	- 06 24		17.8	BLUE STELLAR OBJECT
PHL 6333	00 06 48.	- 08 13		18.0	BLUE STELLAR OBJECT
PHL 6334	00 06 48.	- 17 38		18.4	BLUE STELLAR OBJECT
PHL 2681	00 06 48.	- 18 50		18.8	BLUE STELLAR OBJECT
LB 05002	00 06 48.	- 27 56		19.9	FAINT BLUE STAR
LB 05004	00 06 48.	- 27 57		19.7	FAINT BLUE STAR
LB 05003	00 06 48.	- 27 58		19.1	FAINT BLUE STAR
LB 05005	00 06 48.	- 28 02		15.1	FAINT BLUE STAR
LB 05006	00 06 48.	- 28 34		17.9	FAINT BLUE STAR
RNGC 0018	00 06 49.	+ 27 27			NON-EXISTENT OBJECT
ZC 0006.9+0931	00 06 54.	+ 09 31	1610		CLUSTER OF GALAXIES
ZWG 499.054	00 06 54.	+ 33 03		14.5	GALAXY
ZWG 498.082	00 06 54.	+ 33 03		14.5	GALAXY
UGC 00084	00 06 54.	+ 33 03	120	14.5	GALAXY E-S0
ZWG 549.021	00 06 54.	+ 47 05		14.5	GALAXY
ZWG 548.030	00 06 54.	+ 47 05		14.5	GALAXY
UGC 00085	00 06 54.	+ 47 05	60	14.5	GALAXY Sb-c
MCG+08-01-027	00 06 54.	+ 47 50 30.	60	14.	GALAXY
5ZW 004	00 06 54.	+ 47 51			COMPACT GALAXY

OBJECT NAME	RIGHT ASCEN.	DECLINATION	DIAM.	MAGN.	TYPE OF OBJECT
PHL 2682	00 06 54.	- 03 10		18.0	BLUE STELLAR OBJECT
PHL 6335	00 06 54.	- 03 28		17.9	BLUE STELLAR OBJECT
LB 05008	00 06 54.	- 27 31		19.5	FAINT BLUE STAR
LB 05009	00 06 54.	- 27 49		19.3	FAINT BLUE STAR
LB 05007	00 06 54.	- 28 36		19.0	FAINT BLUE STAR
LB 05010	00 06 54.	- 28 38		18.0	FAINT BLUE STAR
MCG-06-01-028	00 06 54.	- 37 50	42	15.	GALAXY
MCG-06-01-029	00 06 54.	- 37 53	30	17.	GALAXY
MCG+00-01-030	00 06 57.	+ 00 36	21	15.5	GALAXY
KHAV 001	00 07	+ 61 41	5310		DARK NEBULA
PHL 2683	00 07 00.	+ 17 46		17.5	BLUE STELLAR OBJECT
RNGC 0020	00 07 00.	+ 33 02		14.5	GALAXY
ZWG 549.022	00 07 00.	+ 47 51		15.0	GALAXY
ZWG 548.031	00 07 00.	+ 47 51		15.0	GALAXY
LDN 1265	00 07 00.	+ 58 35	180		DARK NEBULA
OCL 0284	00 07 00.	+ 60 09	360		OPEN STAR CLUSTER
PHL 6338	00 07 00.	- 04 44		18.0	BLUE STELLAR OBJECT
PHL 6336	00 07 00.	- 06 40		18.1	BLUE STELLAR OBJECT
PHL 6339	00 07 00.	- 07 54		18.0	BLUE STELLAR OBJECT
PHL 6337	00 07 00.	- 08 09		18.5	BLUE STELLAR OBJECT
PHL 6340	00 07 00.	- 13 34		18.2	BLUE STELLAR OBJECT
MCG-06-01-030	00 07 00.	- 37 57	30	15.	GALAXY
HELW 312	00 07 03.	- 30 06 54.			NEBULA
HN 1240	00 07 04.6	- 46 28 06.	18		NEBULA
ZC 0007.1+1435	00 07 06.	+ 14 35	1880		CLUSTER OF GALAXIES
ZWG 549.023	00 07 06.	+ 49 34		15.2	GALAXY
ZWG 548.032	00 07 06.	+ 49 34		15.2	GALAXY
PHL 0696	00 07 06.	- 09 25		18.0	BLUE STELLAR OBJECT
PHL 6341	00 07 06.	- 12 21		18.6	BLUE STELLAR OBJECT
MCG-05-01-038	00 07 06.	- 32 34	90	14.	GALAXY
LB 00407	00 07 11.	+ 14 26 48.		16.4	FAINT BLUE STAR
PHL 0697	00 07 12.	+ 04 27		18.8	BLUE STELLAR OBJECT
ZC 0007.2+1217	00 07 12.	+ 12 17	3230		CLUSTER OF GALAXIES
PHL 0701	00 07 12.	+ 13 42		17.9	BLUE STELLAR OBJECT
PHL 0700	00 07 12.	+ 14 26		17.4	BLUE STELLAR OBJECT
MCG+05-01-039	00 07 12.	+ 27 32	24	14.	GALAXY
ZWG 499.055	00 07 12.	+ 27 34		14.9	GALAXY
UGC 00086	00 07 12.	+ 27 34	108	14.9	GALAXY Sb
MCG+05-01-038	00 07 12.	+ 28 02 30.	48	14.5	GALAXY
ZWG 499.056	00 07 12.	+ 28 05		15.1	GALAXY
UGC 00087	00 07 12.	+ 28 05	84	15.1	GALAXY E?
ZWG 499.057	00 07 12.	+ 28 12		15.7	GALAXY
MCG+05-01-037	00 07 12.	+ 28 23	24	16.	GALAXY
PHL 6343	00 07 12.	- 06 47		18.6	BLUE STELLAR OBJECT
PHL 2684	00 07 12.	- 11 15		18.6	BLUE STELLAR OBJECT
PHL 6342	00 07 12.	- 16 38		18.3	BLUE STELLAR OBJECT
PHL 0698	00 07 12.	- 16 43		17.7	BLUE STELLAR OBJECT
PHL 0699	00 07 12.	- 22 12		14.7	BLUE STELLAR OBJECT
LB 05011	00 07 12.	- 27 34		15.5	FAINT BLUE STAR
LB 05012	00 07 12.	- 27 58		16.3	FAINT BLUE STAR
RNGC 0022	00 07 13.	+ 27 34		15.0	GALAXY
ZC 0007.3+0156	00 07 18.	+ 01 56	1950		CLUSTER OF GALAXIES
ZWG 408.039	00 07 18.	+ 04 25		15.3	GALAXY
UGC 00088	00 07 18.	+ 04 25	72	15.3	GALAXY S0-a
ZC 0007.3+0608	00 07 18.	+ 06 08	400		CLUSTER OF GALAXIES
ZWG 433.029	00 07 18.	+ 15 27		15.7	GALAXY
MRK 545	00 07 18.	+ 25 38	60	14.5	GALAXY WITH UV CONTINUUM
MCG+04-01-033	00 07 18.	+ 25 38	144	13.5	GALAXY
ZWG 478.034	00 07 18.	+ 25 39		12.5	GALAXY
ZWG 477.062	00 07 18.	+ 25 39		12.5	GALAXY
UGC 00089	00 07 18.	+ 25 39	132	12.5	GALAXY SBa
ZWG 478.035	00 07 18.	+ 26 31		15.5	GALAXY
ZWG 477.063	00 07 18.	+ 26 31		15.5	GALAXY
MCG+05-01-040	00 07 18.	+ 27 59	36	16.	GALAXY
MCG-01-01-041	00 07 18.	- 08 10	42	14.5	GALAXY
PHL 0702	00 07 18.	- 08 16		18.4	BLUE STELLAR OBJECT
LB 05013	00 07 18.	- 28 02		20.0	FAINT BLUE STAR
LB 05014	00 07 18.	- 28 16		17.3	FAINT BLUE STAR
LB 05015	00 07 18.	- 28 21		18.8	FAINT BLUE STAR
LB 05016	00 07 18.	- 29 33		20.5	FAINT BLUE STAR
LB 05017	00 07 18.	- 28 38		19.4	FAINT BLUE STAR
LB 05018	00 07 18.	- 28 40		18.3	FAINT BLUE STAR
MCG-06-01-031	00 07 18.	- 35 33	18	15.	GALAXY
SER 164.04	00 07 18.	- 57 19	2100	15.5	CLUSTER OF 50 GALAXIES
LB 03128	00 07 18.	- 71 37		14.0	FAINT BLUE STAR
RNGC 0023	00 07 19.	+ 25 39		13.0	GALAXY
SN 1955C	00 07 19.	+ 25 39		16.0	SUPERNOVA
MCG+01-01-041	00 07 21.	+ 04 22 30.	17	15.	GALAXY
MCG+05-01-041	00 07 21.	+ 27 55	48	14.5	GALAXY
HELW 313	00 07 22.	- 30 44 36.			NEBULA
PHL 2687	00 07 24.	+ 13 27		18.2	BLUE STELLAR OBJECT
UGC 00090	00 07 24.	+ 16 40	66	18.	GALAXY DWARF
ZWG 499.058	00 07 24.	+ 27 53		15.7	GALAXY
UGC 00091	00 07 24.	+ 27 53	84	15.7	GALAXY S
ZWG 499.059	00 07 24.	+ 27 56		14.7	GALAXY
UGC 00092	00 07 24.	+ 27 56	96	14.7	GALAXY Sb
ZC 0007.4+3254	00 07 24.	+ 32 54	2550		CLUSTER OF GALAXIES
PHL 2686	00 07 24.	- 04 24		18.8	BLUE STELLAR OBJECT
PHL 2685	00 07 24.	- 08 23		17.3	BLUE STELLAR OBJECT
PHL 6344	00 07 24.	- 21 26		13.9	BLUE STELLAR OBJECT
LB 05019	00 07 24.	- 27 46		19.3	FAINT BLUE STAR
LB 05021	00 07 24.	- 28 03		17.2	FAINT BLUE STAR
LB 05020	00 07 24.	- 28 03		16.4	FAINT BLUE STAR
HN 0786	00 07 24.	- 64 38	120		NEBULA
IC 1532	00 07 24.	- 64 38			NONSTELLAR OBJECT
RNGC 0024	00 07 26.	- 25 15		12.0	GALAXY
MCG-04-01-018	00 07 27.	- 25 15	390	12.	GALAXY
RNGC 0025	00 07 27.	- 57 19			GALAXY
ZCG 0007+47	00 07 30.	+ 47 48		15.8	COMPACT GALAXY
OCL 0285	00 07 30.	+ 60 56	480	12.	OPEN STAR CLUSTER
MCG-01-01-043	00 07 30.	- 04 59	36	14.	GALAXY
MCG-01-01-042	00 07 30.	- 06 35 30.	72	15.	GALAXY
MCG-06-01-032	00 07 30.	- 37 03	15	14.5	GALAXY
SC 0004-1556.5	00 07 31.	- 15 39 48.	18		NEBULA
HOLM 004A	00 07 33.	- 04 59	36	13.8	PART OF MULTIPLE GALAXY
MEL 4	00 07 33.	- 15 42	600		NEBULA
HELF 314A	00 07 33.	- 30 10 18.			NEBULA
MCG+00-01-031	00 07 36.	+ 00 03	30	16.	GALAXY
PHL 2688	00 07 36.	+ 06 00		18.6	BLUE STELLAR OBJECT
PHL 6345	00 07 36.	+ 09 31		17.9	BLUE STELLAR OBJECT
MCG+05-01-042	00 07 36.	+ 30 33 30.	108	14.	GALAXY
PHL 6348	00 07 36.	- 04 00		18.4	BLUE STELLAR OBJECT
PHL 6346	00 07 36.	- 10 48		17.0	BLUE STELLAR OBJECT
PHL 6347	00 07 36.	- 15 54		17.1	BLUE STELLAR OBJECT
PHL 0703	00 07 36.	- 26 30		13.0	BLUE STELLAR OBJECT
LB 05022	00 07 36.	- 28 02		18.5	FAINT BLUE STAR
LB 05023	00 07 36.	- 28 16		17.5	FAINT BLUE STAR
HOLM 004B	00 07 37.	- 05 00	12	15.2	PART OF MULTIPLE GALAXY

OBJECT NAME	RIGHT ASCEN.	DECLINATION	DIAM.	MAGN.	TYPE OF OBJECT
LB 02697	00 07 38.	+ 16 09 06.		16.7	FAINT BLUE STAR
ZC 0007.7+1056	00 07 42.	+ 10 56	4970		CLUSTER OF GALAXIES
ZC 0007.7+1725	00 07 42.	+ 17 25	1810		CLUSTER OF GALAXIES
PHL 2690	00 07 42.	+ 18 23		17.0	BLUE STELLAR OBJECT
ZWG 499.060	00 07 42.	+ 30 35		15.6	GALAXY
UGC 00093	00 07 42.	+ 30 35	156	15.6	GALAXY S IV
ZWG 499.061	00 07 42.	+ 31 50		15.2	GALAXY
ABC 0005	00 07 42.	+ 32 50		17.1	RICH CLUSTER OF GALAXIES
PHL 2689	00 07 42.	- 04 40		18.3	BLUE STELLAR OBJECT
PHL 6350	00 07 42.	- 05 54		18.2	BLUE STELLAR OBJECT
PHL 6349	00 07 42.	- 06 58		18.6	BLUE STELLAR OBJECT
PHL 0704	00 07 42.	- 07 01		18.3	BLUE STELLAR OBJECT
PHL 2691	00 07 42.	- 17 40		18.2	BLUE STELLAR OBJECT
MCG-03-01-021	00 07 42.	- 18 34	78	15.	GALAXY
TON-S 0144	00 07 42.	- 26 29		13.3	BLUE STAR
LB 05024	00 07 42.	- 27 26		18.4	FAINT BLUE STAR
LB 05025	00 07 42.	- 28 39		18.6	FAINT BLUE STAR
SCHO 0013	00 07 45.	+ 63 59 06.	330		ISOLATED DARK CLOUD
MCG-01-01-044	00 07 45.	- 07 22	30	15.	GALAXY
HELW 314B	00 07 46.	- 30 36 00.			NEBULA
KN 13.083	00 07 47.6	+ 11 51 42.			NEBULA
ZWG 433.030	00 07 48.	+ 11 52		15.7	GALAXY
3ZW 001	00 07 48.	+ 21 16			COMPACT GALAXY
ZWG 478.036	00 07 48.	+ 25 33		13.9	GALAXY
ZWG 477.064	00 07 48.	+ 25 33		13.9	GALAXY
UGC 00094	00 07 48.	+ 25 33	120	13.9	GALAXY Sb
MCG+04-01-034	00 07 48.	+ 25 33	138	13.5	GALAXY
MCG+05-01-043	00 07 48.	+ 28 42	114	16.	GALAXY
ZWG 499.062	00 07 48.	+ 28 43		15.7	GALAXY
UGC 00095	00 07 48.	+ 28 43	114	15.7	GALAXY Sc
KARA.72 003A	00 07 48.	+ 28 43	102	15.7	PART OF DOUBLE GALAXY
ZC 0007.8+3348	00 07 48.	+ 33 48	3160		CLUSTER OF GALAXIES
LB 05026	00 07 48.	- 27 35		18.8	FAINT BLUE STAR
LB 05027	00 07 48.	- 28 36		19.0	FAINT BLUE STAR
HN 1241	00 07 48.4	- 46 41 42.	78		NEBULA
RNGC 0026	00 07 49.	+ 25 33		14.0	GALAXY
RNGC 0028	00 07 51.	- 57 17			GALAXY
KN 13.084	00 07 51.0	+ 14 00 34.			NEBULA
ABC 0006	00 07 53.	+ 17 26		17.5	RICH CLUSTER OF GALAXIES
HN 1242	00 07 53.6	- 46 46 06.	78		NEBULA
ZC 0007.9+0859	00 07 54.	+ 08 59	1680		CLUSTER OF GALAXIES
PHL 2693	00 07 54.	+ 15 15		18.1	BLUE STELLAR OBJECT
MCG+05-01-044	00 07 54.	+ 28 42	54	14.5	GALAXY
ZWG 499.063	00 07 54.	+ 28 44		14.5	GALAXY
RNGC 0027	00 07 54.	+ 28 44		14.5	GALAXY
UGC 00096	00 07 54.	+ 28 44	90	14.5	GALAXY S
KARA.72 003B	00 07 54.	+ 28 44	90	14.5	PART OF DOUBLE GALAXY
MCG+00-01-032	00 07 54.	- 01 52	18	15.5	GALAXY
PHL 2692	00 07 54.	- 27 47		18.1	BLUE STELLAR OBJECT
LB 02698	00 07 57.	+ 15 55 48.		16.2	FAINT BLUE STAR
MCG+05-01-045	00 07 57.	+ 28 22 30.	42	15.	GALAXY
LBN 0530	00 08	+ 20 20	1080		BRIGHT NEBULA
LBN 0578	00 08	+ 58 30	300		BRIGHT NEBULA
PHL 0706	00 08 00.	+ 09 10		17.8	BLUE STELLAR OBJECT
PHL 0705	00 08 00.	+ 09 29		16.4	BLUE STELLAR OBJECT
3ZW 002	00 08 00.	+ 10 42			COMPACT GALAXY
KW 65	00 08 00.	+ 10 42	12		SEYFERT GALAXY
VVI 02	00 08 00.	+ 10 42		15.96	SEYFERT GALAXY
ZWG 456.036	00 08 00.	+ 16 19		15.2	GALAXY
PHL 6352	00 08 00.	+ 17 07		18.3	BLUE STELLAR OBJECT
ZWG 499.064	00 08 00.	+ 28 24		15.6	GALAXY
UGC 00097	00 08 00.	+ 28 24	78	15.6	GALAXY S0
ZC 0008.0+3113	00 08 00.	+ 31 13	1480		CLUSTER OF GALAXIES
MCG+05-01-046	00 08 00.	+ 32 42	72	14.	GALAXY
ZWG 499.065	00 08 00.	+ 32 43		13.9	GALAXY
UGC 00098	00 08 00.	+ 32 43	72	13.9	GALAXY SBb
ZC 0008.0+3351	00 08 00.	+ 33 51	940		CLUSTER OF GALAXIES
PHL 0707	00 08 00.	- 05 18		18.1	BLUE STELLAR OBJECT
SN 1951G	00 08 00.	- 06 34		18.8	SUPERNOVA
MCG-01-01-045	00 08 00.	- 06 55	48	15.	GALAXY
PHL 0708	00 08 00.	- 07 20		16.7	BLUE STELLAR OBJECT
PHL 6351	00 08 00.	- 13 12		16.6	BLUE STELLAR OBJECT
PHL 0709	00 08 00.	- 16 14		18.6	BLUE STELLAR OBJECT
PHL 0710	00 08 00.	- 17 02		18.5	BLUE STELLAR OBJECT
LB 05028	00 08 00.	- 27 35		19.5	FAINT BLUE STAR
LB 05029	00 08 00.	- 28 00		15.3	FAINT BLUE STAR
IC 1533	00 08 03.	- 07 41 38.			NONSTELLAR OBJECT
RNGC 0031	00 08 03.	- 57 16			GALAXY
KN 13.085	00 08 03.2	+ 11 13 13.			NEBULA
HELW 080	00 08 04.	- 24 58 06.			NEBULA
MCG+00-01-033	00 08 06.	+ 01 15 30.	27	15.5	GALAXY
PHL 6353	00 08 06.	+ 05 54		18.2	BLUE STELLAR OBJECT
MCG+02-01-023	00 08 06.	+ 13 26	120	14.5	GALAXY
UGC 00099	00 08 06.	+ 13 27	180	16.0	GALAXY DWRF SP
PHL 0711	00 08 06.	+ 17 22		18.0	BLUE STELLAR OBJECT
MCG+05-01-047	00 08 06.	+ 28 15 30.	30	15.	GALAXY
RNGC 0021	00 08 06.	+ 32 42		14.0	GALAXY
ZWG 499.066	00 08 06.	+ 33 05		13.5	GALAXY
UGC 00100	00 08 06.	+ 33 05	96	13.5	GALAXY Sb-c
ZWG 518.017	00 08 06.	+ 35 35		15.5	GALAXY
MCG+06-01-016	00 08 06.	+ 35 35	42	14.	GALAXY
ZWG 382.029	00 08 06.	- 00 19		15.3	GALAXY
MCG+00-01-034	00 08 06.	- 00 21 30.	36	14.	GALAXY
MCG-01-01-046	00 08 06.	- 05 07	30	14.	GALAXY
PHL 2694	00 08 06.	- 13 25		18.5	BLUE STELLAR OBJECT
LB 05030	00 08 06.	- 27 41		16.2	FAINT BLUE STAR
LB 05031	00 08 06.	- 28 16		18.2	FAINT BLUE STAR
MCG-05-01-039	00 08 06.	- 29 32	30	15.	GALAXY
KN 13.086	00 08 06.0	+ 13 25 51.			NEBULA
KN 13.087	00 08 08.9	+ 11 16 35.			NEBULA
HELW 081	00 08 10.	- 25 22 06.			NEBULA
SC 0005-6803.9	00 08 10.	- 67 47 12.	6		NEBULA
ZWG 433.031	00 08 12.	+ 14 33		15.7	GALAXY
ZWG 499.067	00 08 12.	+ 27 36		15.7	GALAXY
ZWG 499.068	00 08 12.	+ 28 18		15.0	GALAXY
MCG+05-01-048	00 08 12.	+ 33 04	90	14.	GALAXY
RNGC 0029	00 08 12.	+ 33 05		13.5	GALAXY
PHL 2695	00 08 12.	- 08 34		18.5	BLUE STELLAR OBJECT
PHL 2697	00 08 12.	- 09 12		18.4	BLUE STELLAR OBJECT
PHL 6354	00 08 12.	- 10 22		18.3	BLUE STELLAR OBJECT
PHL 6355	00 08 12.	- 12 54		15.9	BLUE STELLAR OBJECT
PHL 2696	00 08 12.	- 17 14		18.6	BLUE STELLAR OBJECT
PHL 0712	00 08 12.	- 20 54		17.1	BLUE STELLAR OBJECT
LB 05032	00 08 12.	- 28 04		20.0	FAINT BLUE STAR
RNGC 0030	00 08 13.	+ 21 41			NON-EXISTENT OBJECT
KN 13.088	00 08 13.4	+ 14 34 14.			NEBULA
HELW 082	00 08 15.	- 25 21 54.			NEBULA
HELW 315	00 08 15.	- 30 15 00.			NEBULA

OBJECT NAME	RIGHT ASCEN.	DECLINATION	DIAM.	MAGN.	TYPE OF OBJECT
HN 1243	00 08 15.2	- 47 21 06.			NEBULA
SC 0005-6913.4	00 08 17.	- 68 56 42.	6		NEBULA
PHL 6356	00 08 18.	+ 06 18		18.3	BLUE STELLAR OBJECT
PHL 2698	00 08 18.	+ 16 00		18.5	BLUE STELLAR OBJECT
ZWG 478.037	00 08 18.	+ 25 17		15.4	GALAXY
UGC 00101	00 08 18.	+ 25 17	84	15.4	GALAXY SBa
PHL 6357	00 08 18.	- 08 14		18.0	BLUE STELLAR OBJECT
MCG-02-01-030	00 08 18.	- 09 56	30	15.	GALAXY
LB 05033	00 08 18.	- 28 14		17.5	FAINT BLUE STAR
LB 05034	00 08 18.	- 28 19		18.9	FAINT BLUE STAR
RNGC 0033	00 08 19.	+ 03 24			NON-EXISTENT OBJECT
RNGC 0032	00 08 19.	+ 18 31			NON-EXISTENT OBJECT
HELW 083	00 08 20.	- 25 33 42.			NEBULA
VDB.66N 001	00 08 21.	+ 58 29	516		REFLECTION NEBULA
PHL 6358	00 08 24.	+ 03 55		18.3	BLUE STELLAR OBJECT
PHL 0713	00 08 24.	+ 04 20		18.0	BLUE STELLAR OBJECT
PHL 6361	00 08 24.	+ 04 35		18.2	BLUE STELLAR OBJECT
ZC 0008.4+0608	00 08 24.	+ 06 08	810		CLUSTER OF GALAXIES
PHL 6359	00 08 24.	+ 17 54		18.6	BLUE STELLAR OBJECT
MCG+04-01-035	00 08 24.	+ 25 16	60	15.	GALAXY
ZC 0008.4+2851	00 08 24.	+ 28 51	1340		CLUSTER OF GALAXIES
MCG+05-01-049	00 08 24.	+ 29 46	84	14.5	GALAXY
ZWG 499.069	00 08 24.	+ 29 47		14.8	GALAXY
UGC 00102	00 08 24.	+ 29 47	96	14.8	GALAXY SBa/Sa
PHL 0714	00 08 24.	- 04 09		17.1	BLUE STELLAR OBJECT
PHL 2699	00 08 24.	- 06 16		18.4	BLUE STELLAR OBJECT
PHL 0715	00 08 24.	- 07 56		18.2	BLUE STELLAR OBJECT
PHL 6360	00 08 24.	- 12 00		17.3	BLUE STELLAR OBJECT
MCG-02-01-031	00 08 24.	- 13 07 30.	42	15.	GALAXY
MCG-04-01-019	00 08 24.	- 21 21 30.	36	15.	GALAXY
LB 05035	00 08 24.	- 27 35		19.0	FAINT BLUE STAR
LB 05036	00 08 24.	- 27 46		17.4	FAINT BLUE STAR
LB 05037	00 08 24.	- 28 00		19.0	FAINT BLUE STAR
LB 05038	00 08 24.	- 28 05		18.8	FAINT BLUE STAR
MCG-06-01-033	00 08 24.	- 35 30	60	15.5	GALAXY
IC 0002	00 08 27.6	- 13 05 59.			GALAXY SA(r)
ZWG 382.030	00 08 30.	+ 02 24		15.6	GALAXY
KARA.73B 0007	00 08 30.	+ 02 24	36	15.6	ISOLATED GALAXY S
UGC 00103	00 08 30.	+ 04 00	66	17.	GALAXY S
PHL 2701	00 08 30.	+ 06 00		18.6	BLUE STELLAR OBJECT
PHL 6362	00 08 30.	+ 06 30		18.4	BLUE STELLAR OBJECT
PHL 2700	00 08 30.	- 03 58		18.2	BLUE STELLAR OBJECT
MCG-02-01-032	00 08 30.	- 12 22	120	13.5	GALAXY
LB 05039	00 08 30.	- 27 59		18.5	FAINT BLUE STAR
RNGC 0034	00 08 31.	- 12 22		13.0	GALAXY
RNGC 0034	00 08 31.	- 12 22		13.0	GALAXY
PHL 2703	00 08 36.	+ 06 06		18.5	BLUE STELLAR OBJECT
UGC 00104	00 08 36.	+ 07 43	84	16.0	GALAXY SBa
MCG+01-01-042	00 08 36.	+ 07 43	36	15.	GALAXY
PHL 6364	00 08 36.	+ 08 27		18.4	BLUE STELLAR OBJECT
ZC 0008.6+1458	00 08 36.	+ 14 58	1010		CLUSTER OF GALAXIES
MCG+05-01-050	00 08 36.	+ 28 37 30.	66	15.	GALAXY
ZWG 499.070	00 08 36.	+ 28 50		15.7	GALAXY
PHL 2702	00 08 36.	- 03 08		16.6	BLUE STELLAR OBJECT
MCG-02-01-033	00 08 36.	- 12 16	30	14.	GALAXY
PHL 6363	00 08 36.	- 15 11		18.5	BLUE STELLAR OBJECT
RNGC 0035	00 08 37.	- 12 16		14.0	GALAXY
MCG+00-01-035	00 08 42.	+ 02 24	36	13.5	GALAXY
PHL 0716	00 08 42.	+ 05 01		18.1	BLUE STELLAR OBJECT
MCG+01-01-043	00 08 42.	+ 06 03	96	13.	GALAXY
PHL 6365	00 08 42.	+ 07 03		17.1	BLUE STELLAR OBJECT
ZWG 499.071	00 08 42.	+ 28 38		15.1	GALAXY
UGC 00105	00 08 42.	+ 28 38	72	15.1	GALAXY SBO
PHL 6367	00 08 42.	- 02 42		18.0	BLUE STELLAR OBJECT
PHL 2704	00 08 42.	- 04 36		17.5	BLUE STELLAR OBJECT
PHL 6368	00 08 42.	- 07 02		18.1	BLUE STELLAR OBJECT
PHL 2705	00 08 42.	- 16 00		18.3	BLUE STELLAR OBJECT
PHL 6366	00 08 42.	- 27 42		18.2	BLUE STELLAR OBJECT
LB 05040	00 08 42.	- 27 48		20.0	FAINT BLUE STAR
LB 05041	00 08 42.	- 28 33		18.2	FAINT BLUE STAR
LB 01547	00 08 42.	- 45 48		14.5	FAINT BLUE STAR
RNGC 0036	00 08 43.	+ 06 08		14.5	GALAXY
LB 00408	00 08 46.	+ 16 19 30.		17.2	FAINT BLUE STAR
LB 02699	00 08 47.	+ 15 26 24.		16.8	FAINT BLUE STAR
ZWG 408.040	00 08 48.	+ 06 06		14.5	GALAXY
UGC 00106	00 08 48.	+ 06 06	162	14.5	GALAXY Sb/SBb
PHL 2707	00 08 48.	+ 18 38		18.6	BLUE STELLAR OBJECT
PHL 2706	00 08 48.	- 08 20		18.9	BLUE STELLAR OBJECT
PHL 0717	00 08 48.	- 12 46		14.0	BLUE STELLAR OBJECT
LB 05042	00 08 48.	- 27 38		18.9	FAINT BLUE STAR
LB 05043	00 08 48.	- 27 56		17.7	FAINT BLUE STAR
LB 05044	00 08 48.	- 28 15		19.2	FAINT BLUE STAR
LB 05045	00 08 48.	- 28 19		19.3	FAINT BLUE STAR
LB 05046	00 08 48.	- 28 26		18.5	FAINT BLUE STAR
MCG-05-01-040	00 08 48.	- 29 09	60	15.5	GALAXY
MCG-06-01-034	00 08 48.	- 33 52	90	14.5	GALAXY
MCG+01-01-044	00 08 51.	+ 06 05 30.	48	16.	GALAXY
RNGC 0037	00 08 51.	- 57 14			GALAXY
PHL 6370	00 08 54.	+ 05 28		18.8	BLUE STELLAR OBJECT
ZC 0008.9+3207	00 08 54.	+ 32 07	2490		CLUSTER OF GALAXIES
PHL 6371	00 08 54.	- 05 08		18.5	BLUE STELLAR OBJECT
PHL 6369	00 08 54.	- 05 38		17.8	BLUE STELLAR OBJECT
PHL 0718	00 08 54.	- 11 46		14.7	BLUE STELLAR OBJECT
KN 13.089	00 08 56.3	+ 10 59 04.			NEBULA
MCG+04-01-036	00 08 57.	+ 21 44	30	16.	GALAXY
KN 13.090	00 08 57.2	+ 13 13 59.			NEBULA
KN 13.091	00 08 59.1	+ 13 14 03.			NEBULA
KHAV 002	00 09	+ 70 59	8540		DARK NEBULA
PHL 2710	00 09 00.	+ 04 58		17.7	BLUE STELLAR OBJECT
ZWG 433.032	00 09 00.	+ 13 14		15.7	GALAXY
PHL 0719	00 09 00.	+ 18 33		16.6	BLUE STELLAR OBJECT
UGC 00107	00 09 00.	+ 27 40	66	16.0	GALAXY S
MCG+14-01-006	00 09 00.	+ 85 24	24	17.	GALAXY
PHL 6372	00 09 00.	- 05 06		17.8	BLUE STELLAR OBJECT
PHL 6373	00 09 00.	- 06 22		18.4	BLUE STELLAR OBJECT
PHL 2711	00 09 00.	- 07 44		18.6	BLUE STELLAR OBJECT
PHL 2712	00 09 00.	- 08 45		18.6	BLUE STELLAR OBJECT
PHL 6374	00 09 00.	- 10 14		18.1	BLUE STELLAR OBJECT
PHL 2709	00 09 00.	- 14 42		19.0	BLUE STELLAR OBJECT
LB 05047	00 09 00.	- 27 38		20.0	FAINT BLUE STAR
LB 05048	00 09 00.	- 27 57		18.3	FAINT BLUE STAR
PHL 2708	00 09 00.	- 28 36		18.6	BLUE STELLAR OBJECT
KN 13.092	00 09 01.5	+ 13 31 36.			NEBULA
SCHO 0014	00 09 04.	+ 57 15 54.	240		ISOLATED DARK CLOUD
PHL 6377	00 09 06.	+ 06 30		18.4	BLUE STELLAR OBJECT
ZWG 456.037	00 09 06.	+ 20 42		15.0	GALAXY
MRK 337	00 09 06.	+ 20 42	20	15.5	GALAXY WITH UV CONTINUUM
MCG+05-01-051	00 09 06.	+ 28 12 30.	48	15.	GALAXY

OBJECT NAME	RIGHT ASCEN.	DECLINATION	DIAM.	MAGN.	TYPE OF OBJECT
ARC 0007	00 09 06.	+ 32 09		17.1	RICH CLUSTER OF GALAXIES
ZC 0009.1+5040	00 09 06.	+ 50 40	2290		CLUSTER OF GALAXIES
ZWG 382.031	00 09 06.	- 00 44		15.6	GALAXY
PHL 6375	00 09 06.	- 05 14		16.6	BLUE STELLAR OBJECT
PHL 2713	00 09 06.	- 09 34		18.2	BLUE STELLAR OBJECT
PHL 6376	00 09 06.	- 11 18		17.8	BLUE STELLAR OBJECT
PHL 6378	00 09 06.	- 25 38		18.4	BLUE STELLAR OBJECT
LB 05049	00 09 06.	- 28 00		18.3	FAINT BLUE STAR
MCG+00-01-036	00 09 09.	- 00 45 30.	24	15.	GALAXY
PHL 0720	00 09 12.	+ 05 26		18.3	BLUE STELLAR OBJECT
PHL 6381	00 09 12.	+ 09 18		17.0	BLUE STELLAR OBJECT
PHL 2715	00 09 12.	+ 12 24		18.1	BLUE STELLAR OBJECT
LB 00409	00 09 12.	+ 15 00 48.		17.7	FAINT BLUE STAR
ZWG 499.072	00 09 12.	+ 28 14		14.8	GALAXY
UGC 00108	00 09 12.	+ 28 14	66	14.8	GALAXY SB:b
ZWG 499.073	00 09 12.	+ 32 50		15.4	GALAXY
PHL 6382	00 09 12.	- 02 42		18.5	BLUE STELLAR OBJECT
PHL 0721	00 09 12.	- 03 22		18.6	BLUE STELLAR OBJECT
PHL 2714	00 09 12.	- 04 05		17.8	BLUE STELLAR OBJECT
PHL 0722	00 09 12.	- 10 56		15.7	BLUE STELLAR OBJECT
PHL 6379	00 09 12.	- 11 29		13.8	BLUE STELLAR OBJECT
PHL 6380	00 09 12.	- 12 45		17.8	BLUE STELLAR OBJECT
PHL 0723	00 09 12.	- 17 58		18.1	BLUE STELLAR OBJECT
LB 05050	00 09 12.	- 27 58		18.4	FAINT BLUE STAR
LB 05051	00 09 12.	- 28 37		18.5	FAINT BLUE STAR
MCG-05-01-041	00 09 12.	- 30 26	60	15.	GALAXY
SC 0006-6835.7	00 09 13.	- 68 19 00.			NEBULA
MCG+00-01-037	00 09 15.	- 01 23	36	15.5	GALAXY
MCG-01-01-047	00 09 15.	- 05 52	60	13.5	GALAXY
HELW 316	00 09 16.	- 30 25 24.			NEBULA
KN 13.093	00 09 17.2	+ 13 29 48.			NEBULA
ZC 0009.3+0253	00 09 18.	+ 02 53	1010		CLUSTER OF GALAXIES
PHL 2716	00 09 18.	+ 04 46		18.3	BLUE STELLAR OBJECT
ZWG 478.038	00 09 18.	+ 22 45		15.7	GALAXY
UGC 00109	00 09 18.	- 01 22	66	17.	GALAXY S-IRR
PHL 2717	00 09 18.	- 09 38		18.5	BLUE STELLAR OBJECT
PHL 6383	00 09 18.	- 10 43		17.7	BLUE STELLAR OBJECT
LB 05052	00 09 18.	- 27 30		18.6	FAINT BLUE STAR
LB 00410	00 09 19.	+ 14 30 42.		13.0	FAINT BLUE STAR
RNGC 0038	00 09 19.	- 05 52		13.5	GALAXY
FEIG 001	00 09 24.	+ 11 49		12.7	FAINT BLUE STAR
ZWG 478.039	00 09 24.	+ 26 07		15.3	GALAXY
UGC 00110	00 09 24.	+ 26 07	78	15.3	GALAXY Sc
PHL 2719	00 09 24.	- 05 26		18.1	BLUE STELLAR OBJECT
PHL 6385	00 09 24.	- 12 14		18.1	BLUE STELLAR OBJECT
PHL 6386	00 09 24.	- 13 58		18.5	BLUE STELLAR OBJECT
PHL 2718	00 09 24.	- 20 36		18.9	BLUE STELLAR OBJECT
PHL 6384	00 09 24.	- 24 21		17.6	BLUE STELLAR OBJECT
LB 03129	00 09 24.	- 63 51		13.0	FAINT BLUE STAR
PHL 2721	00 09 30.	+ 09 18		18.5	BLUE STELLAR OBJECT
ZC 0009.5+2349	00 09 30.	+ 23 49	740		CLUSTER OF GALAXIES
ZWG 478.040	00 09 30.	+ 24 43		15.7	GALAXY
PHL 2720	00 09 30.	- 25 22		17.9	BLUE STELLAR OBJECT
IC 0003	00 09 32.2	- 00 41 33.			GALAXY E3
ARC 0008	00 09 33.	- 11 28		17.2	RICH CLUSTER OF GALAXIES
KN 13.094	00 09 33.2	+ 11 44 27.			NEBULA
KN 13.095	00 09 35.0	+ 11 46 03.			NEBULA
ZWG 382.033	00 09 36.	+ 01 30		15.7	GALAXY
ZC 0009.6+0910	00 09 36.	+ 09 10	1480		CLUSTER OF GALAXIES
ZWG 433.033	00 09 36.	+ 11 46		15.4	GALAXY
UGC 00111	00 09 36.	+ 11 46	78	15.4	GALAXY Sc
MCG+02-01-024	00 09 36.	+ 11 46	72	15.	GALAXY
KARA.73B 0008	00 09 36.	+ 11 46	78	15.4	ISOLATED GALAXY S
PHL 0725	00 09 36.	+ 13 26		18.5	BLUE STELLAR OBJECT
ZWG 499.074	00 09 36.	+ 27 38		15.3	GALAXY
ZWG 499.075	00 09 36.	+ 29 03		15.6	GALAXY
UGC 00112	00 09 36.	+ 41 28	120	17.	GALAXY DWRF SP
ZWG 382.032	00 09 36.	- 00 40		15.1	GALAXY
MCG+00-01-038	00 09 36.	- 00 42	24	13.	GALAXY
PHL 6387	00 09 36.	- 04 08		18.3	BLUE STELLAR OBJECT
PHL 6388	00 09 36.	- 08 05		18.3	BLUE STELLAR OBJECT
PHL 2722	00 09 36.	- 18 16		18.6	BLUE STELLAR OBJECT
PHL 0724	00 09 36.	- 21 58		18.2	BLUE STELLAR OBJECT
LB 05053	00 09 36.	- 27 03		16.8	FAINT BLUE STAR
MCG-05-01-042	00 09 36.	- 27 29	36	15.	GALAXY
MCG+04-01-037	00 09 39.	+ 22 02 30.	66	15.	GALAXY
ZWG 478.041	00 09 42.	+ 22 03		14.9	GALAXY
UGC 00113	00 09 42.	+ 22 03	72	14.9	GALAXY Sa
SN 1971H	00 09 42.	+ 28 22		17.0	SUPERNOVA
ZWG 499.076	00 09 42.	+ 30 47		14.4	GALAXY
RNGC 0039	00 09 42.	+ 30 47		14.5	GALAXY
UGC 00114	00 09 42.	+ 30 47	78	14.4	GALAXY Sc
MCG+05-01-052	00 09 42.	+ 30 47	66	14.	GALAXY
PHL 6389	00 09 42.	- 25 18		18.4	BLUE STELLAR OBJECT
LB 05054	00 09 42.	- 27 53		18.9	FAINT BLUE STAR
LB 05055	00 09 42.	- 28 02		18.8	FAINT BLUE STAR
LB 05056	00 09 42.	- 28 08		19.0	FAINT BLUE STAR
LB 00411	00 09 45.	+ 16 21 48.		17.2	FAINT BLUE STAR
MCG+04-01-038	00 09 45.	+ 22 02 30.	30	17.	GALAXY
PHL 2723	00 09 45.	- 29 28		17.8	BLUE STELLAR OBJECT
SG 3.002	00 09 46.	+ 67 56	3000		DIFFUSE EMISSION NEBULA
ARC 0009	00 09 47.	+ 09 12		18.0	RICH CLUSTER OF GALAXIES
PHL 2726	00 09 48.	+ 03 38		13.6	BLUE STELLAR OBJECT
PHL 6390	00 09 48.	+ 04 04		18.5	BLUE STELLAR OBJECT
PHL 2724	00 09 48.	+ 09 36		18.2	BLUE STELLAR OBJECT
FEIG 002	00 09 48.	+ 15 06		13.1	FAINT BLUE STAR
PHL 0727	00 09 48.	+ 16 20		17.2	BLUE STELLAR OBJECT
PHL 6391	00 09 48.	+ 17 03		18.5	BLUE STELLAR OBJECT
ZC 0009.8+1913	00 09 48.	+ 19 13	2150		CLUSTER OF GALAXIES
ZWG 499.077	00 09 48.	+ 32 29		15.7	GALAXY
PHL 6392	00 09 48.	- 05 18		17.2	BLUE STELLAR OBJECT
PHL 0726	00 09 48.	- 06 23		16.6	BLUE STELLAR OBJECT
PHL 6393	00 09 48.	- 11 28		18.2	BLUE STELLAR OBJECT
LB 05057	00 09 48.	- 27 28		16.4	FAINT BLUE STAR
PHL 2725	00 09 48.	- 28 56		18.3	BLUE STELLAR OBJECT
LB 02700	00 09 52.	+ 16 55 12.		17.0	FAINT BLUE STAR
KN 13.096	00 09 53.9	+ 13 53 58.			NEBULA
PHL 0728	00 09 54.	+ 04 20		18.0	BLUE STELLAR OBJECT
PK119+06.1	00 09 54.	+ 68 54	47	18.3	PLANETARY NEBULA
MCG+00-01-039	00 09 54.	- 00 21	48	14.5	GALAXY
PHL 6395	00 09 54.	- 04 12		18.3	BLUE STELLAR OBJECT
PHL 6394	00 09 54.	- 19 50		17.9	BLUE STELLAR OBJECT
LB 05058	00 09 54.	- 28 38		18.6	FAINT BLUE STAR
LB 02701	00 09 59.	+ 15 44 06.		16.3	FAINT BLUE STAR
5ZW 005	00 10 00.	+ 52 21			COMPACT GALAXY
LDN 1276	00 10 00.	+ 63 45	240		DARK NEBULA
UGC 00115	00 10 00.	+ 88 06	138	16.5	GALAXY DWRF SP
ZC 0010.0-0154	00 10 00.	- 01 54	1340		CLUSTER OF GALAXIES

OBJECT NAME	RIGHT ASCEN.	DECLINATION	DIAM.	MAGN.	TYPE OF OBJECT
PHL 2728	00 10 00.	- 02 50		16.7	BLUE STELLAR OBJECT
MCG-01-01-048	00 10 00.	- 03 33	36	15.	GALAXY
PHL 0729	00 10 00.	- 05 50		16.4	BLUE STELLAR OBJECT
PHL 6396	00 10 00.	- 11 42		15.5	BLUE STELLAR OBJECT
PHL 2727	00 10 00.	- 13 04		17.9	BLUE STELLAR OBJECT
PHL 6408	00 10 00.	- 16 01		17.6	BLUE STELLAR OBJECT
PHL 2729	00 10 00.	- 18 30		19.0	BLUE STELLAR OBJECT
PHL 0730	00 10 00.	- 20 00		17.7	BLUE STELLAR OBJECT
LB 05059	00 10 00.	- 27 38		17.4	FAINT BLUE STAR
LB 05060	00 10 00.	- 27 58		19.7	FAINT BLUE STAR
LB 05061	00 10 00.	- 28 33		18.8	FAINT BLUE STAR
MCG+01-01-045	00 10 03.	+ 05 12	42	15.	GALAXY
MCG+00-01-040	00 10 03.	- 00 14	48	16.	GALAXY
ZWG 408.041	00 10 06.	+ 05 14		15.4	GALAXY
UGC 00116	00 10 06.	+ 05 14	72	15.4	GALAXY S
KARA.73B 0009	00 10 06.	+ 05 14	48	15.4	ISOLATED GALAXY S
PHL 2730	00 10 06.	- 06 30		17.1	BLUE STELLAR OBJECT
LB 05062	00 10 06.	- 28 17		19.5	FAINT BLUE STAR
PHL 6397	00 10 06.	- 30 52		18.5	BLUE STELLAR OBJECT
ZC 0010.2+0002	00 10 12.	+ 00 02	1340		CLUSTER OF GALAXIES
PHL 2733	00 10 12.	+ 05 58		18.4	BLUE STELLAR OBJECT
PHL 2732	00 10 12.	+ 06 24		17.5	BLUE STELLAR OBJECT
PHL 6398	00 10 12.	+ 07 00		18.4	BLUE STELLAR OBJECT
PHL 2737	00 10 12.	+ 14 22		18.6	BLUE STELLAR OBJECT
LB 02702	00 10 12.	+ 14 45 48.		16.9	FAINT BLUE STAR
ZWG 478.042	00 10 12.	+ 21 46		14.6	GALAXY
ZWG 499.078	00 10 12.	+ 33 06		15.2	GALAXY
UGC 00117	00 10 12.	+ 33 06	102	15.2	GALAXY Sc
PHL 2731	00 10 12.	- 05 13		16.6	BLUE STELLAR OBJECT
PHL 2734	00 10 12.	- 07 54		18.4	BLUE STELLAR OBJECT
PHL 0731	00 10 12.	- 09 03		17.2	BLUE STELLAR OBJECT
PHL 2735	00 10 12.	- 14 49		18.2	BLUE STELLAR OBJECT
PHL 2736	00 10 12.	- 15 28		18.4	BLUE STELLAR OBJECT
PHL 6399	00 10 12.	- 22 02		17.5	BLUE STELLAR OBJECT
LB 05063	00 10 12.	- 28 31		20.2	FAINT BLUE STAR
LB 05064	00 10 12.	- 28 32		17.6	FAINT BLUE STAR
LB 05065	00 10 12.	- 28 40		18.0	FAINT BLUE STAR
RNGC 0041	00 10 13.	+ 21 46		14.5	GALAXY
MCG+04-01-039	00 10 15.	+ 21 44	48	14.5	GALAXY
ARC 0010	00 10 15.	- 06 16		17.2	RICH CLUSTER OF GALAXIES
ARC 0011	00 10 15.	- 16 43		17.2	RICH CLUSTER OF GALAXIES
KN 13.097	00 10 15.1	+ 13 19 47.			NEBULA
PK120+09.1	00 10 16.79	+ 72 14 39.5	60	10.7	PLANETARY NEBULA
ZWG 478.043	00 10 18.	+ 21 50		15.0	GALAXY
UGC 00118	00 10 18.	+ 21 50	66	15.0	GALAXY E-S0
LB 00412	00 10 18.	+ 29 27 24.		17.5	FAINT BLUE STAR
MCG+05-01-053	00 10 18.	+ 33 05	78	15.	GALAXY
LB 05066	00 10 18.	- 27 50		19.4	FAINT BLUE STAR
LB 05067	00 10 18.	- 28 14		17.3	FAINT BLUE STAR
LB 03130	00 10 18.	- 72 12		12.9	FAINT BLUE STAR
RNGC 0042	00 10 19.	+ 21 50		15.0	GALAXY
RNGC 0040	00 10 20.	+ 72 15		10.0	PLANETARY NEBULA
MCG+04-01-041	00 10 21.	+ 21 48	48	15.	GALAXY
MCG+04-01-040	00 10 21.	+ 23 26	42	15.5	GALAXY
SC 0007-6813.8	00 10 23.	- 67 57 06.	36		NEBULA
PHL 0732	00 10 24.	+ 06 09		18.2	BLUE STELLAR OBJECT
ZWG 433.034	00 10 24.	+ 14 08		14.2	GALAXY
UGC 00119	00 10 24.	+ 14 08	66	14.2	GALAXY PECULR
RNGC 0043	00 10 24.	+ 30 38		14.0	GALAXY
MCG+05-01-054	00 10 24.	+ 30 38	90	14.	GALAXY
ZWG 499.079	00 10 24.	+ 30 39		13.9	GALAXY
UGC 00120	00 10 24.	+ 30 39	90	13.9	GALAXY SB0
ZWG 518.018	00 10 24.	+ 38 58		15.4	GALAXY
UGC 00121	00 10 24.	+ 38 58	72	15.4	GALAXY SBc
KARA.73B 0010	00 10 24.	+ 38 58	72	15.4	ISOLATED GALAXY S
PHL 2739	00 10 24.	- 02 54		18.4	BLUE STELLAR OBJECT
PHL 6400	00 10 24.	- 04 45		18.7	BLUE STELLAR OBJECT
MCG-01-01-049	00 10 24.	- 05 48	60	14.5	GALAXY
PHL 2740	00 10 24.	- 12 40		18.6	BLUE STELLAR OBJECT
MCG-02-01-034	00 10 24.	- 13 21	30	15.5	GALAXY
LB 05068	00 10 24.	- 28 07		18.2	FAINT BLUE STAR
LB 05069	00 10 24.	- 28 28		19.4	FAINT BLUE STAR
PHL 2738	00 10 24.	- 31 29		17.9	BLUE STELLAR OBJECT
MCG+06-01-017	00 10 27.	+ 38 58	66	14.	GALAXY
KN 13.098	00 10 28.2	+ 14 07 46.			NEBULA
MCG+02-01-025	00 10 30.	+ 14 08	60	14.5	GALAXY
PHL 2742	00 10 30.	+ 17 51		17.8	BLUE STELLAR OBJECT
ZWG 456.038	00 10 30.	+ 18 14		15.5	GALAXY
RNGC 0044	00 10 30.	+ 31 01			NON-EXISTENT OBJECT
PHL 2741	00 10 30.	- 05 12		18.7	BLUE STELLAR OBJECT
MCG-01-01-050	00 10 30.	- 06 00	30	15.5	GALAXY
LB 05070	00 10 30.	- 27 26		18.9	FAINT BLUE STAR
MCG-05-01-043	00 10 30.	- 31 01	42	16.	GALAXY
VDB.66N 002	00 10 35.	+ 65 18	204		REFLECTION NEBULA
PHL 2745	00 10 36.	+ 15 02		18.1	BLUE STELLAR OBJECT
MCG+03-01-028	00 10 36.	+ 16 45	114	14.	GALAXY
DG 003	00 10 36.	+ 65 20	180		REFLECTION NEBULA
PHL 2744	00 10 36.	- 02 30		18.8	BLUE STELLAR OBJECT
SN 1954V	00 10 36.	- 07 52		20.0	SUPERNOVA
MCG-02-01-035	00 10 36.	- 12 50 30.	72	15.	GALAXY
PHL 0733	00 10 36.	- 17 14		18.7	BLUE STELLAR OBJECT
PHL 2743	00 10 36.	- 23 18		17.5	BLUE STELLAR OBJECT
MCG-04-01-020	00 10 36.	- 24 30 30.	72	13.5	GALAXY
LB 05071	00 10 36.	- 27 44		18.5	FAINT BLUE STAR
LB 05072	00 10 36.	- 28 12		17.3	FAINT BLUE STAR
LIN.CL 002	00 10 36.	- 73 46 00.	270	16.0	STAR CLUSTER IN SMC
LB 02703	00 10 37.	+ 15 55 48.		16.9	FAINT BLUE STAR
CED 001	00 10 40.	+ 65 20	180		DIFFUSE GALACTIC NEBULA
PHL 0734	00 10 42.	+ 14 54		2.7	BLUE STELLAR OBJECT
ZWG 456.039	00 10 42.	+ 16 45		15.2	GALAXY
UGC 00122	00 10 42.	+ 16 45	156	15.2	GALAXY IRR
PHL 2746	00 10 42.	- 13 32		16.7	BLUE STELLAR OBJECT
LB 05073	00 10 42.	- 28 07		19.5	FAINT BLUE STAR
MCG-06-01-035	00 10 42.	- 37 48	30	15.	GALAXY
LB 03131	00 10 42.	- 62 09		12.4	FAINT BLUE STAR
ARC 0012	00 10 45.	- 07 53		17.2	RICH CLUSTER OF GALAXIES
ZC 0010.8+0231	00 10 48.	+ 02 31	400		CLUSTER OF GALAXIES
ZWG 456.040	00 10 48.	+ 17 13		14.2	GALAXY
UGC 00123	00 10 48.	+ 17 13	78	14.2	GALAXY S
ZWG 499.080	00 10 48.	+ 28 06		15.5	GALAXY
UGC 00124	00 10 48.	+ 28 06	72	15.5	GALAXY Sa
PHL 6401	00 10 48.	- 07 40		17.8	BLUE STELLAR OBJECT
PHL 2747	00 10 48.	- 17 17		17.6	BLUE STELLAR OBJECT
LB 05074	00 10 48.	- 27 32		18.8	FAINT BLUE STAR
LB 05075	00 10 48.	- 28 04		18.9	FAINT BLUE STAR
PHL 6402	00 10 49.	- 31 08		18.2	BLUE STELLAR OBJECT
LB 03132	00 10 48.	- 60 47		12.7	FAINT BLUE STAR
MCG-01-01-051	00 10 51.	- 07 05	18	15.	GALAXY

OBJECT NAME	RIGHT ASCEN.	DECLINATION	DIAM.	MAGN.	TYPE OF OBJECT
IC 0004	00 10 52.1	+ 17 12 33.			GALAXY SB(s)
SC 0008-6805.6	00 10 53.	- 67 48 54.	12		NEBULA
PHL 2749	00 10 54.	+ 07 32		18.3	BLUE STELLAR OBJECT
MCG+03-01-029	00 10 54.	+ 17 12	66	14.	GALAXY
PHL 6403	00 10 54.	+ 18 30		17.4	BLUE STELLAR OBJECT
ZC 0010.9+3457	00 10 54.	+ 34 57	1210		CLUSTER OF GALAXIES
PHL 2748	00 10 54.	- 03 31		17.1	BLUE STELLAR OBJECT
PHL 6404	00 10 54.	- 04 02		17.9	BLUE STELLAR OBJECT
8ZW 0010-06.4	00 10 54.	- 06 22		17.9	COMPACT GALAXY
PHL 6405	00 10 54.	- 20 00		18.6	BLUE STELLAR OBJECT
LB 05076	00 10 54.	- 28 39		17.2	FAINT BLUE STAR
VDB.66G 223	00 11	- 23 25	470		DWARF GALAXY
PHL 6409	00 11 00.	+ 03 49		18.2	BLUE STELLAR OBJECT
PHL 6407	00 11 00.	+ 03 54		17.9	BLUE STELLAR OBJECT
MCG+08-01-028	00 11 00.	+ 47 52	42	15.	GALAXY
MCG+08-01-029	00 11 00.	+ 48 24	42	15.	GALAXY
ZWG 549.024	00 11 00.	+ 48 25		15.1	GALAXY
LDN 1277	00 11 00.	+ 63 45	300		DARK NEBULA
PHL 0735	00 11 00.	- 03 30		18.2	BLUE STELLAR OBJECT
MCG-02-01-036	00 11 00.	- 09 02 30.	42	14.	GALAXY
PHL 6410	00 11 00.	- 10 46		18.6	BLUE STELLAR OBJECT
PHL 6406	00 11 00.	- 13 06		16.9	BLUE STELLAR OBJECT
PHL 2750	00 11 00.	- 23 18		18.3	BLUE STELLAR OBJECT
LB 05077	00 11 00.	- 27 44		17.9	FAINT BLUE STAR
MCG-01-01-052	00 11 03.	- 05 22	66	13.	GALAXY
ZWG 499.081	00 11 06.	+ 29 55		14.7	GALAXY
MCG+05-01-055	00 11 06.	+ 29 55	48	15.	GALAXY
MCG+06-01-018	00 11 06.	+ 36 20 30.	12	15.	GALAXY
5ZW 006	00 11 06.	+ 47 52			COMPACT GALAXY
ZWG 549.025	00 11 06.	+ 47 53		15.2	GALAXY
UGC 00125	00 11 06.	+ 47 53	72	15.2	GALAXY S0
PHL 2751	00 11 06.	- 02 28		18.4	BLUE STELLAR OBJECT
PHL 2752	00 11 06.	- 03 02		18.2	BLUE STELLAR OBJECT
PHL 2753	00 11 06.	- 04 08		18.5	BLUE STELLAR OBJECT
LB 05078	00 11 06.	- 28 40		18.6	FAINT BLUE STAR
IC 1534	00 11 08.	+ 47 52 15.			NONSTELLAR OBJECT
ARC 0013	00 11 08.	- 19 47		16.6	RICH CLUSTER OF GALAXIES
KN 13.099	00 11 08.4	+ 12 54 59.			NEBULA
KN 13.100	00 11 09.8	+ 14 19 34.			NEBULA
PHL 6411	00 11 12.	+ 13 09		17.9	BLUE STELLAR OBJECT
ZWG 433.035	00 11 12.	+ 14 20		15.7	GALAXY
UGC 00126	00 11 12.	+ 14 20	66	15.7	GALAXY S
UGC 00127	00 11 12.	+ 26 42	108	16.5	GALAXY SB:c
UGC 00128	00 11 12.	+ 35 43	138	16.5	GALAXY S IV-V
MCG+06-01-019	00 11 12.	+ 35 43	114	14.	GALAXY
MCG+08-01-030	00 11 12.	+ 47 52	90	15.	GALAXY
PHL 2754	00 11 12.	- 02 22		17.2	BLUE STELLAR OBJECT
LB 05079	00 11 12.	- 27 41		17.0	FAINT BLUE STAR
LB 05080	00 11 12.	- 28 22		19.4	FAINT BLUE STAR
SVEN 001	00 11 14.	- 23 41	18	16.8	GALAXY
MCG+05-01-056	00 11 15.	+ 28 28	45	15.	GALAXY
ZWG 382.034	00 11 18.	+ 03 26		15.5	GALAXY
UGC 00129	00 11 18.	+ 03 26	60	15.5	GALAXY SBb
MCG+00-01-041	00 11 18.	+ 03 26 30.	54	14.	GALAXY
PHL 0736	00 11 18.	+ 09 17		17.6	BLUE STELLAR OBJECT
PHL 6414	00 11 18.	+ 17 04		18.5	BLUE STELLAR OBJECT
PHL 6413	00 11 18.	+ 17 15		19.0	BLUE STELLAR OBJECT
ZWG 456.041	00 11 18.	+ 17 42		15.7	GALAXY
MCG+04-01-042	00 11 18.	+ 26 42 30.	72	15.	GALAXY
ZWG 499.082	00 11 18.	+ 28 30		15.7	GALAXY
ZWG 499.083	00 11 18.	+ 30 37		14.2	GALAXY
UGC 00130	00 11 18.	+ 30 37	27	14.2	GALAXY COMPACT
ZWG 549.026	00 11 18.	+ 47 53		15.2	GALAXY
UGC 00131	00 11 18.	+ 47 53	66	15.2	GALAXY S
MCG+08-01-031	00 11 18.	+ 47 58	90	13.	GALAXY
MCG-01-01-053	00 11 18.	- 04 45	66	15.	GALAXY
PHL 6412	00 11 18.	- 13 40		16.6	BLUE STELLAR OBJECT
LB 05081	00 11 18.	- 27 58		18.6	FAINT BLUE STAR
KN 13.101	00 11 19.	+ 12 49			NEBULA
IC 1535	00 11 20.	+ 47 52 39.			NONSTELLAR OBJECT
SVEN 002	00 11 20.	- 23 27	480	12.3	GALAXY
KN 13.102	00 11 20.1	+ 13 47 54.			NEBULA
RNGC 0048	00 11 23.	+ 47 57		15.0	GALAXY
PHL 6415	00 11 24.	+ 05 12		17.9	BLUE STELLAR OBJECT
PHL 6416	00 11 24.	+ 06 18		18.5	BLUE STELLAR OBJECT
MCG+02-01-026	00 11 24.	+ 12 41	96	14.5	GALAXY
ZWG 433.036	00 11 24.	+ 12 42		15.3	GALAXY
UGC 00132	00 11 24.	+ 12 42	108	15.3	GALAXY
PHL 2757	00 11 24.	+ 16 14		18.4	BLUE STELLAR OBJECT
PHL 6418	00 11 24.	+ 17 34		18.2	BLUE STELLAR OBJECT
ZWG 549.027	00 11 24.	+ 47 58		15.0	GALAXY
UGC 00133	00 11 24.	+ 47 58	102	15.0	GALAXY Sb
PHL 2758	00 11 24.	- 04 45		17.9	BLUE STELLAR OBJECT
PHL 2755	00 11 24.	- 04 45		18.3	BLUE STELLAR OBJECT
PHL 2756	00 11 24.	- 05 59		18.1	BLUE STELLAR OBJECT
PHL 6417	00 11 24.	- 08 33		18.2	BLUE STELLAR OBJECT
PHL 6420	00 11 24.	- 15 14		18.9	BLUE STELLAR OBJECT
PHL 6419	00 11 24.	- 17 49		7.1	BLUE STELLAR OBJECT
PHL 0737	00 11 24.	- 21 37		17.0	BLUE STELLAR OBJECT
PHL 2760	00 11 24.	- 22 40		18.2	BLUE STELLAR OBJECT
LB 05082	00 11 24.	- 27 44		19.7	FAINT BLUE STAR
MCG-05-01-044	00 11 24.	- 31 00	48	15.	GALAXY
KN 13.103	00 11 25.8	+ 12 41 04.			NEBULA
RNGC 0045	00 11 26.	- 23 27		11.0	GALAXY
LB 02704	00 11 27.	+ 15 34 18.		16.4	FAINT BLUE STAR
KN 13.104	00 11 29.0	+ 12 36 53.			NEBULA
UGC 00134	00 11 30.	+ 12 38	60	17.	GALAXY S IV-V
PHL 2761	00 11 30.	+ 17 15		18.5	BLUE STELLAR OBJECT
PHL 6421	00 11 30.	- 13 27		15.3	BLUE STELLAR OBJECT
MCG-03-01-022	00 11 30.	- 19 57	36	14.5	GALAXY
MCG-04-01-021	00 11 30.	- 23 28	480	11.	GALAXY
LB 05083	00 11 30.	- 28 04		19.8	FAINT BLUE STAR
RNGC 0046	00 11 31.	+ 05 43			NON-EXISTENT OBJECT
SVEN 003	00 11 32.	- 23 36 49.	30	16.1	GALAXY
SS 02	00 11 34.	+ 65 26			DIFFUSE GALACTIC NEBULA
PHL 0738	00 11 36.	+ 03 58		18.2	BLUE STELLAR OBJECT
PHL 6422	00 11 36.	+ 06 25		17.9	BLUE STELLAR OBJECT
UGC 00135	00 11 36.	+ 07 09	72	17.	GALAXY
ZC 0011.6+1017	00 11 36.	+ 10 17	2490		CLUSTER OF GALAXIES
PHL 6426	00 11 36.	+ 13 28		18.6	BLUE STELLAR OBJECT
PHL 2762	00 11 36.	+ 14 34		18.6	BLUE STELLAR OBJECT
MCG+04-01-043A	00 11 36.	+ 22 26 30.	18	15.	GALAXY
MCG+04-01-043	00 11 36.	+ 22 28 30.	42	15.	GALAXY
ZWG 478.044	00 11 36.	+ 22 30		15.6	GALAXY
ZWG 499.084	00 11 36.	+ 28 16		15.3	GALAXY
ZC 0011.6+2828	00 11 36.	+ 28 28	1750		CLUSTER OF GALAXIES
ZC 0011.6+3512	00 11 36.	+ 35 12	610		CLUSTER OF GALAXIES
MCG+08-01-032	00 11 36.	+ 47 51	42	15.	GALAXY

OBJECT NAME	RIGHT ASCEN.	DECLINATION	DIAM.	MAGN.	TYPE OF OBJECT
PHL 6425	00 11 36.	- 03 14		17.8	BLUE STELLAR OBJECT
PHL 6424	00 11 36.	- 08 39		18.2	BLUE STELLAR OBJECT
PHL 2763	00 11 36.	- 12 30		17.3	BLUE STELLAR OBJECT
PHL 6423	00 11 36.	- 18 49		17.9	BLUE STELLAR OBJECT
PHL 2764	00 11 36.	- 24 42		18.3	BLUE STELLAR OBJECT
IC 1536	00 11 41.	+ 47 50 33.			NONSTELLAR OBJECT
RNGC 0049	00 11 41.	+ 47 58		15.5	GALAXY
ZC 0011.7+1435	00 11 42.	+ 14 35	4570		CLUSTER OF GALAXIES
PHL 2765	00 11 42.	+ 18 11		18.8	BLUE STELLAR OBJECT
ZWG 549.028	00 11 42.	+ 47 52		15.7	GALAXY
ZWG 549.029	00 11 42.	+ 47 58		15.3	GALAXY
UGC 00136	00 11 42.	+ 47 58	66	15.3	GALAXY S0?
MCG+08-01-033	00 11 42.	+ 47 58	60	15.	GALAXY
8ZW 0011-05.6	00 11 42.	- 05 35		17.8	COMPACT GALAXY
PHL 0739	00 11 42.	- 06 04		18.2	BLUE STELLAR OBJECT
PHL 6427	00 11 42.	- 06 27		17.1	BLUE STELLAR OBJECT
8ZW 0011-07.2	00 11 42.	- 07 14		14.3	COMPACT GALAXY
PHL 0740	00 11 42.	- 07 36		17.2	BLUE STELLAR OBJECT
MCG-01-01-054	00 11 45.	- 07 44	18	16.	GALAXY
ZWG 499.085	00 11 48.	+ 28 14		15.5	GALAXY
MCG+08-01-034	00 11 48.	+ 48 22	72	16.	GALAXY
ZWG 549.030	00 11 48.	+ 48 24		15.7	GALAXY
UGC 00137	00 11 48.	+ 48 24	72	15.7	GALAXY Sb-c
PHL 6428	00 11 48.	- 13 36		10.5	BLUE STELLAR OBJECT
MCG-02-01-037	00 11 48.	- 14 02	30	15.5	GALAXY
MCG-03-01-023	00 11 48.	- 19 59	54	15.	GALAXY
LB 05084	00 11 48.	- 27 40		17.4	FAINT BLUE STAR
MCG-01-01-055	00 11 51.	- 07 27 30.	138	13.	GALAXY
RNGC 0051	00 11 53.	+ 47 59		14.5	GALAXY
ZWG 549.031	00 11 54.	+ 47 59		14.6	GALAXY
UGC 00138	00 11 54.	+ 47 59	102	14.6	GALAXY (S0-a)
MCG+08-01-035	00 11 54.	+ 47 59	84	13.	GALAXY
MCG+00-01-042	00 11 54.	- 00 17	18	15.	GALAXY
MCG+00-01-043	00 11 54.	- 01 02	120	13.	GALAXY
8ZW 0011-03.3	00 11 54.	- 03 21		18.0	COMPACT GALAXY
PHL 2766	00 11 54.	- 13 39		14.8	BLUE STELLAR OBJECT
PHL 2767	00 11 54.	- 14 52		18.2	BLUE STELLAR OBJECT
LB 05085	00 11 54.	- 27 36		19.6	FAINT BLUE STAR
LB 01548	00 11 54.	- 57 55		13.5	FAINT BLUE STAR
RNGC 0047	00 11 55.	- 07 27		13.0	GALAXY
SCHO 0016	00 11 56.	+ 56 50 36.	240		ISOLATED DARK CLOUD
MCG+00-01-044	00 11 57.	- 00 19	12	15.	GALAXY
SIV 02	00 12	+ 50 30	19800		FAINT H EMISSION REGION
CS 1	00 12	+ 50 30	19800		FAINT H-ALPHA REGION
KHAV 003	00 12	+ 63 53			DARK NEBULA
LB 09770	00 12	- 82 37		11.5	FAINT BLUE STAR
PHL 6429	00 12 00.	+ 06 09		18.2	BLUE STELLAR OBJECT
PHL 0741	00 12 00.	+ 07 48		16.8	BLUE STELLAR OBJECT
ZWG 433.037	00 12 00.	+ 10 30		15.7	GALAXY
ZWG 456.042	00 12 00.	+ 18 17		14.6	GALAXY
UGC 00140	00 12 00.	+ 18 17	156	14.6	GALAXY S
ZWG 499.086	00 12 00.	+ 28 10		15.7	GALAXY
UGC 00141	00 12 00.	+ 28 10	72	15.7	GALAXY SB0-a
ZWG 382.035	00 12 00.	- 01 00		14.8	GALAXY
UGC 00139	00 12 00.	- 01 00	144	14.8	GALAXY Sc
KARA.73B 0011	00 12 00.	- 01 00	132	14.8	ISOLATED GALAXY S
MRK 546	00 12 00.	- 01 50	9	16.5	GALAXY WITH UV CONTINUUM
MCG-01-01-056	00 12 00.	- 07 33	15	16.	GALAXY
MCG-01-01-057	00 12 00.	- 07 34 30.	36	14.5	GALAXY
PHL 0742	00 12 00.	- 10 44		17.7	BLUE STELLAR OBJECT
PHL 2768	00 12 00.	- 13 45		11.5	BLUE STELLAR OBJECT
PHL 6430	00 12 00.	- 14 20		18.1	BLUE STELLAR OBJECT
PHL 2769	00 12 00.	- 26 31		18.4	BLUE STELLAR OBJECT
LB 05086	00 12 00.	- 28 25		18.0	FAINT BLUE STAR
MCG-06-01-036	00 12 00.	- 37 57	60	15.	GALAXY
RNGC 0052	00 12 01.	+ 18 17		14.5	GALAXY
MCG+03-01-030	00 12 03.	+ 18 20	114	13.5	GALAXY
PHL 6431	00 12 06.	- 02 44		16.6	BLUE STELLAR OBJECT
MCG-01-01-058	00 12 06.	- 07 37 30.	60	12.	GALAXY
LB 05087	00 12 06.	- 27 54		19.3	FAINT BLUE STAR
RNGC 0050	00 12 07.	- 07 37		12.0	GALAXY
PHL 6434	00 12 12.	+ 04 34		17.6	BLUE STELLAR OBJECT
ZWG 408.042	00 12 12.	+ 04 57		15.6	GALAXY
UGC 00142	00 12 12.	+ 04 57	78	15.6	GALAXY SBb
MCG+00-01-045	00 12 12.	- 00 20	12	16.	GALAXY
PHL 6432	00 12 12.	- 02 25		15.9	BLUE STELLAR OBJECT
PHL 2770	00 12 12.	- 13 14		18.4	BLUE STELLAR OBJECT
PHL 0743	00 12 12.	- 21 46		17.2	BLUE STELLAR OBJECT
PHL 6433	00 12 12.	- 22 52		16.6	BLUE STELLAR OBJECT
SVEN 004	00 12 13.	- 39 53	18	14.3	GALAXY
HELW 001	00 12 13.	- 39 53			NEBULA
RNGC 0053	00 12 16.	- 60 36			UNVERIFIED SOUTHERN OBJECT
ZWG 478.045	00 12 18.	+ 26 03		14.8	GALAXY
MCG+04-01-044	00 12 18.	+ 26 04	36	15.	GALAXY
ZWG 518.019	00 12 18.	+ 38 24		15.7	GALAXY
MCG+00-01-046	00 12 18.	- 00 02 30.	21	14.	GALAXY
PHL 0744	00 12 18.	- 03 36		18.1	BLUE STELLAR OBJECT
PHL 6435	00 12 18.	- 26 13		18.3	BLUE STELLAR OBJECT
LB 05088	00 12 18.	- 27 42		19.6	FAINT BLUE STAR
SC 0009-6926.1	00 12 19.	- 69 09 25.	12		NEBULA
MCG-01-01-059	00 12 21.	- 07 33	42	16.	GALAXY
PHL 0745	00 12 24.	+ 05 16		18.1	BLUE STELLAR OBJECT
ZWG 456.043	00 12 24.	+ 16 33		15.4	GALAXY
5ZW 007	00 12 24.	+ 51 05			COMPACT GALAXY
ZWG 382.036	00 12 24.	- 00 01		15.6	GALAXY
PHL 2772	00 12 24.	- 05 56		18.4	BLUE STELLAR OBJECT
PHL 6436	00 12 24.	- 06 24		17.9	BLUE STELLAR OBJECT
PHL 2774	00 12 24.	- 09 50		5.7	BLUE STELLAR OBJECT
PHL 6437	00 12 24.	- 12 37		18.3	BLUE STELLAR OBJECT
PHL 2771	00 12 24.	- 15 40		17.8	BLUE STELLAR OBJECT
PHL 0746	00 12 24.	- 18 08		18.8	BLUE STELLAR OBJECT
PHL 2773	00 12 24.	- 20 35		18.6	BLUE STELLAR OBJECT
MCG-04-01-022	00 12 24.	- 24 22 30.	36	15.	GALAXY
PHL 6438	00 12 24.	- 26 54		18.5	BLUE STELLAR OBJECT
LB 05089	00 12 24.	- 27 36		16.2	FAINT BLUE STAR
MCG-02-01-038	00 12 27.	- 12 03	30	15.	GALAXY
ZWG 433.038	00 12 30.	+ 14 48		15.7	GALAXY
LB 02705	00 12 30.	+ 17 06 12.		16.5	FAINT BLUE STAR
PHL 2775	00 12 30.	- 03 40		18.4	BLUE STELLAR OBJECT
8ZW 0012-06.4	00 12 30.	- 06 23		17.7	COMPACT GALAXY
MCG-01-01-060	00 12 30.	- 07 24	72	14.	GALAXY
MCG-05-01-045	00 12 30.	- 28 51	36	15.	GALAXY
SB 968	00 12 30.	- 39 30		8.	PEC. EMISSION-LINE GALAXY
LB 02706	00 12 31.	+ 17 07 36.		15.8	FAINT BLUE STAR
RNGC 0054	00 12 31.	- 07 24		14.0	GALAXY
SVEN 006	00 12 31.	- 39 29	1680	9.2	GALAXY
SVEN 005	00 12 32.	- 23 26 49.	18	16.8	GALAXY
RNGC 0055	00 12 33.	- 39 30		8.0	GALAXY

OBJECT NAME	RIGHT ASCEN.	DECLINATION	DIAM.	MAGN.	TYPE OF OBJECT
LB 00413	00 12 35.	+ 15 04 12.		17.7	FAINT BLUE STAR
PHL 2780	00 12 36.	+ 03 50		18.5	BLUE STELLAR OBJECT
PHL 2777	00 12 36.	+ 04 32		18.2	BLUE STELLAR OBJECT
PHL 0747	00 12 36.	+ 07 54		16.6	BLUE STELLAR OBJECT
PHL 6439	00 12 36.	+ 15 58		17.2	BLUE STELLAR OBJECT
ZWG 499.087	00 12 36.	+ 31 13		15.6	GALAXY
PHL 0748	00 12 36.	- 02 48		17.9	BLUE STELLAR OBJECT
PHL 2776	00 12 36.	- 03 20		16.6	BLUE STELLAR OBJECT
PHL 6440	00 12 36.	- 04 50		17.9	BLUE STELLAR OBJECT
PHL 6441	00 12 36.	- 05 32		18.5	BLUE STELLAR OBJECT
PHL 2778	00 12 36.	- 15 00		18.4	BLUE STELLAR OBJECT
PHL 2779	00 12 36.	- 18 50		18.7	BLUE STELLAR OBJECT
MCG-04-01-023	00 12 36.	- 24 59	42	15.	GALAXY
KN 13.105	00 12 37.0	+ 13 44 11.			NEBULA
MCG-01-01-061	00 12 39.	- 07 50	48	15.5	GALAXY
MCG-04-01-025	00 12 39.	- 24 20	42	15.	GALAXY
MCG-04-01-024	00 12 39.	- 24 23	36	15.5	GALAXY
ZC 0012.7+0408	00 12 42.	+ 04 08	1750		CLUSTER OF GALAXIES
MCG+01-01-046	00 12 42.	+ 05 36	36	15.	GALAXY
UGC 00143	00 12 42.	+ 05 37	84	16.0	GALAXY SBa
PHL 0750	00 12 42.	+ 13 37		17.6	BLUE STELLAR OBJECT
ZWG 499.088	00 12 42.	+ 29 05		15.7	GALAXY
PHL 0749	00 12 42.	- 03 56		18.3	BLUE STELLAR OBJECT
8ZW 0012-06.6	00 12 42.	- 06 36		13.5	COMPACT GALAXY
LB 05090	00 12 42.	- 27 49		18.6	FAINT BLUE STAR
LB 05091	00 12 42.	- 28 19		17.6	FAINT BLUE STAR
LB 05093	00 12 42.	- 28 37		20.4	FAINT BLUE STAR
PHL 2781	00 12 42.	- 29 12		17.5	BLUE STELLAR OBJECT
MCG-05-01-046	00 12 42.	- 29 30	60	15.	GALAXY
ARC 0014	00 12 43.	- 24 10		15.2	RICH CLUSTER OF GALAXIES
ARC 0015	00 12 43.	- 26 18		17.4	RICH CLUSTER OF GALAXIES
MCG+00-01-047	00 12 45.	+ 02 16 30.	36	15.	GALAXY
LB 00414	00 12 48.	+ 15 13 12.		16.8	FAINT BLUE STAR
ZWG 456.044	00 12 48.	+ 15 57		15.7	GALAXY
UGC 00144	00 12 48.	+ 15 57	60	15.7	GALAXY Sb-c
MCG+03-01-032	00 12 48.	+ 15 57 30.	42	16.	GALAXY
MCG+03-01-031	00 12 48.	+ 17 02 30.	90	13.	GALAXY
ZWG 456.045	00 12 48.	+ 18 30		15.6	GALAXY
ZC 0012.8+1850	00 12 48.	+ 18 50	1610		CLUSTER OF GALAXIES
3ZW 003	00 12 48.	+ 24 12			COMPACT GALAXY
ZC 0012.8+2446	00 12 48.	+ 24 46	870		CLUSTER OF GALAXIES
MRSL 118-01/1	00 12 48.	+ 60 59	60		HII REGION
PHL 2782	00 12 48.	- 04 48		18.8	BLUE STELLAR OBJECT
PHL 2783	00 12 48.	- 09 52		18.6	BLUE STELLAR OBJECT
PHL 0751	00 12 48.	- 15 29		18.0	BLUE STELLAR OBJECT
PHL 0752	00 12 48.	- 17 20		16.7	BLUE STELLAR OBJECT
LB 05092	00 12 48.	- 27 36		19.3	FAINT BLUE STAR
RNGC 0056	00 12 49.	+ 12 10			NON-EXISTENT OBJECT
RNGC 0057	00 12 49.	+ 17 02		13.5	GALAXY
KN 13.106	00 12 49.2	+ 12 40 20.			NEBULA
MCG-04-01-026	00 12 51.	- 21 43	36	13.	GALAXY
PHL 6446	00 12 54.	+ 05 25		18.5	BLUE STELLAR OBJECT
PHL 6443	00 12 54.	+ 06 40		18.3	BLUE STELLAR OBJECT
ZWG 456.046	00 12 54.	+ 17 03		13.7	GALAXY
UGC 00145	00 12 54.	+ 17 03	168	13.7	GALAXY E
PHL 6444	00 12 54.	+ 18 10		18.1	BLUE STELLAR OBJECT
PHL 2785	00 12 54.	- 03 56		9.2	BLUE STELLAR OBJECT
PHL 2784	00 12 54.	- 03 56		9.2	BLUE STELLAR OBJECT
PHL 2786	00 12 54.	- 07 40		18.0	BLUE STELLAR OBJECT
PHL 6445	00 12 54.	- 08 28		17.7	BLUE STELLAR OBJECT
PHL 6442	00 12 54.	- 08 33		17.1	BLUE STELLAR OBJECT
MCG-04-01-027	00 12 54.	- 24 21	24	15.5	GALAXY
TON-S 0145	00 12 54.	- 26 00		16.0	BLUE STAR
LB 05094	00 12 54.	- 27 37		18.8	FAINT BLUE STAR
REIN 2.001	00 12 55.91	+ 17 03 01.5			NEBULA
RNGC 0059	00 12 56.	- 21 43		13.0	GALAXY
LB 09771	00 13	- 81 57		15.5	FAINT BLUE STAR
PHL 2788	00 13 00.	+ 05 54		18.5	BLUE STELLAR OBJECT
PHL 6447	00 13 00.	+ 07 49		17.7	BLUE STELLAR OBJECT
MCG+03-01-033	00 13 00.	+ 19 30	24	15.5	GALAXY
LDN 1278	00 13 00.	+ 63 35	480		DARK NEBULA
PHL 0753	00 13 00.	- 17 00		18.8	BLUE STELLAR OBJECT
LB 07493	00 13 00.	- 22 34		16.9	FAINT BLUE STAR
PHL 2787	00 13 00.	- 28 45		17.8	BLUE STELLAR OBJECT
LB 03133	00 13 00.	- 62 56		12.8	FAINT BLUE STAR
SVEN 007	00 13 01.	- 39 43 48.	30	14.6	GALAXY
LB 00415	00 13 05.	+ 13 25 24.		16.6	FAINT BLUE STAR
PHL 6448	00 13 06.	+ 05 14		18.1	BLUE STELLAR OBJECT
PHL 0754	00 13 06.	+ 13 25		17.1	BLUE STELLAR OBJECT
LB 00416	00 13 06.	+ 17 15 42.		16.9	FAINT BLUE STAR
ZWG 456.047	00 13 06.	+ 19 39		15.7	GALAXY
ZWG 478.046	00 13 06.	+ 24 51		15.7	GALAXY
ZWG 478.047	00 13 06.	+ 27 10		15.4	GALAXY
UGC 00146	00 13 06.	+ 27 10	72	15.4	GALAXY S0
MCG+04-01-045	00 13 06.	+ 27 11	63	15.	GALAXY
MCG+05-01-057	00 13 06.	+ 29 21	90	15.5	GALAXY
ZWG 499.089	00 13 06.	+ 29 23		14.6	GALAXY
UGC 00147	00 13 06.	+ 29 23	102	14.6	GALAXY S0-a
MCG-02-01-039	00 13 06.	- 08 34 30.	18	15.	GALAXY
SNO 01	00 13 06.	- 24 10	1500	19.	CLUSTER OF 50 GALAXIES
SC 0010-6807.3	00 13 06.	- 67 50 37.	12		NEBULA
RNGC 0058	00 13 07.	- 07 27			NON-EXISTENT OBJECT
SVEN 008	00 13 07.	- 39 44	18	14.8	GALAXY
LB 02707	00 13 08.	+ 15 16 36.		16.6	FAINT BLUE STAR
SCHO 0017	00 13 10.	+ 60 05 18.	260		ISOLATED DARK CLOUD
ZWG 456.048	00 13 12.	+ 15 48		14.0	GALAXY
UGC 00148	00 13 12.	+ 15 48	90	14.0	GALAXY S
MCG+03-01-034	00 13 12.	+ 15 49	63	14.	GALAXY
PHL 2790	00 13 12.	+ 17 14		18.4	BLUE STELLAR OBJECT
ZWG 456.049	00 13 12.	+ 19 30		15.0	GALAXY
ZWG 456.050	00 13 12.	+ 19 44		15.4	GALAXY
SCHO 0018	00 13 12.	+ 60 45 12.	260		ISOLATED DARK CLOUD
PHL 0755	00 13 12.	- 11 42		18.4	BLUE STELLAR OBJECT
PHL 2789	00 13 12.	- 13 14		18.6	BLUE STELLAR OBJECT
PHL 0756	00 13 12.	- 26 02		16.6	BLUE STELLAR OBJECT
LB 05095	00 13 12.	- 27 36		19.5	FAINT BLUE STAR
AGU 03	00 13 12.	- 42 10 00.	30	16.	PECULIAR GALAXY
SVEN 009	00 13 13.	- 39 41 49.	12	15.2	GALAXY
KN 13.107	00 13 15.4	+ 13 58 24.			NEBULA
KN 13.108	00 13 15.9	+ 15 48 37.			NEBULA
FATE 1.001	00 13 16.	+ 15 48	81		NEBULA
PHL 6449	00 13 18.	+ 03 36		17.9	BLUE STELLAR OBJECT
PHL 6450	00 13 18.	+ 06 34		18.2	BLUE STELLAR OBJECT
PHL 2791	00 13 18.	+ 13 16		18.3	BLUE STELLAR OBJECT
ZWG 433.039	00 13 18.	+ 13 47		15.6	GALAXY
UGC 00149	00 13 18.	+ 13 47	78	15.6	GALAXY S
KARA.73B 0012	00 13 18.	+ 13 47	90	15.6	ISOLATED GALAXY S
ZWG 499.090	00 13 18.	+ 29 48		15.7	GALAXY

OBJECT NAME	RIGHT ASCEN.	DECLINATION	DIAM.	MAGN.	TYPE OF OBJECT
LB 05096	00 13 18.	- 28 14		20.2	FAINT BLUE STAR
MCG-06-01-037	00 13 18.	- 33 18	54	14.5	GALAXY
SCHO 0019	00 13 20.	+ 59 15 06.	240		ISOLATED DARK CLOUD
KN 13.109	00 13 22.3	+ 13 47 52.			NEBULA
LB 00417	00 13 23.	+ 16 07 12.		17.8	FAINT BLUE STAR
ZC 0013.4+0126	00 13 24.	+ 01 26	1410		CLUSTER OF GALAXIES
PHL 2794	00 13 24.	+ 03 43		18.2	BLUE STELLAR OBJECT
PHL 2795	00 13 24.	+ 04 50		18.2	BLUE STELLAR OBJECT
MCG+02-01-027	00 13 24.	+ 13 48	72	15.	GALAXY
PHL 2793	00 13 24.	+ 17 45		18.9	BLUE STELLAR OBJECT
ZC 0013.4+1805	00 13 24.	+ 18 05	8530		CLUSTER OF GALAXIES
ZWG 382.037	00 13 24.	- 00 34		15.4	GALAXY
UGC 00150	00 13 24.	- 00 34	96	15.4	GALAXY S
MCG+00-01-048	00 13 24.	- 00 37	66	13.	GALAXY
PHL 6452	00 13 24.	- 03 42		18.5	BLUE STELLAR OBJECT
PHL 6451	00 13 24.	- 04 00		18.4	BLUE STELLAR OBJECT
PHL 2792	00 13 24.	- 04 10		18.3	BLUE STELLAR OBJECT
PHL 0757	00 13 24.	- 17 22		17.9	BLUE STELLAR OBJECT
LB 05097	00 13 24.	- 27 25		19.8	FAINT BLUE STAR
LB 05098	00 13 24.	- 27 56		18.7	FAINT BLUE STAR
SC 0011-6933.4	00 13 24.	- 69 16 43.	6		NEBULA
RNGC 0060	00 13 25.	- 00 34		15.5	GALAXY
KN 13.110	00 13 25.5	+ 13 48 13.			NEBULA
LB 00418	00 13 26.	+ 13 26 18.		15.8	FAINT BLUE STAR
SNO 02	00 13 27.	- 24 24 53.	600	15.	LOOSE GRP OF 4 GALAXIES
5ZW 008	00 13 30.	+ 47 20			COMPACT GALAXY
8ZW 0013-06.9	00 13 30.	- 06 55		17.8	COMPACT GALAXY
TON-S 0146	00 13 30.	- 26 05		15.2	BLUE STAR
LB 05099	00 13 30.	- 28 07		18.7	FAINT BLUE STAR
PHL 2796	00 13 30.	- 30 06		18.2	BLUE STELLAR OBJECT
LB 03134	00 13 30.	- 62 25		12.5	FAINT BLUE STAR
SCHO 0020	00 13 31.	+ 62 13 24.	280		ISOLATED DARK CLOUD
IC 1537	00 13 31.	- 39 35 28.			NONSTELLAR OBJECT
MCG+00-01-049	00 13 33.	+ 02 09 30.	27	15.	GALAXY
SC 0011-6546.9	00 13 34.	- 65 30 13.	12		NEBULA
PHL 0758	00 13 36.	+ 03 44		18.9	BLUE STELLAR OBJECT
ZWG 408.043	00 13 36.	+ 04 36		15.6	GALAXY
ZWG 433.040	00 13 36.	+ 10 03		14.7	GALAXY
UGC 00151	00 13 36.	+ 10 03	90	14.7	GALAXY S0?
KARA.73B 0013	00 13 36.	+ 10 03	72	14.7	ISOLATED GALAXY S
LB 02708	00 13 36.	+ 15 47 30.		16.7	FAINT BLUE STAR
PHL 0760	00 13 36.	+ 16 06		18.5	BLUE STELLAR OBJECT
PHL 2798	00 13 36.	+ 17 30		18.2	BLUE STELLAR OBJECT
ZC 0013.6+2927	00 13 36.	+ 29 27	6380		CLUSTER OF GALAXIES
PHL 2797	00 13 36.	- 03 10		18.0	BLUE STELLAR OBJECT
PHL 6453	00 13 36.	- 07 37		18.0	BLUE STELLAR OBJECT
PHL 2799	00 13 36.	- 07 37		18.5	BLUE STELLAR OBJECT
MCG-02-01-040	00 13 36.	- 09 23	36	16.	GALAXY
PHL 0759	00 13 36.	- 14 29		18.4	BLUE STELLAR OBJECT
PHL 6454	00 13 36.	- 25 06		18.1	BLUE STELLAR OBJECT
LB 00419	00 13 38.	+ 14 37 24.		15.8	FAINT BLUE STAR
MCG+00-01-050	00 13 39.	+ 02 10	24	16.	GALAXY
MCG+00-01-051	00 13 42.	+ 01 15	10	15.5	GALAXY
MCG+02-01-028	00 13 42.	+ 10 03	54	14.5	GALAXY
PHL 2800	00 13 42.	- 07 52		18.2	BLUE STELLAR OBJECT
PHL 2801	00 13 42.	- 14 44		17.0	BLUE STELLAR OBJECT
PHL 0761	00 13 42.	- 24 06		15.5	BLUE STELLAR OBJECT
TON-S 0147	00 13 42.	- 24 07		14.8	BLUE STAR
KN 13.111	00 13 42.4	+ 12 08 17.			NEBULA
BC PKS0013-14	00 13 46.5	- 14 46 45.		17.	QUASI-STELLAR OBJECT
PHL 2803	00 13 48.	+ 05 21		18.3	BLUE STELLAR OBJECT
PHL 6456	00 13 48.	+ 05 23		18.6	BLUE STELLAR OBJECT
PHL 0762	00 13 48.	+ 08 04		18.2	BLUE STELLAR OBJECT
ZWG 499.091	00 13 48.	+ 28 42		15.2	GALAXY
MCG+05-01-058	00 13 48.	+ 30 03	48	15.5	GALAXY
ZWG 499.092	00 13 48.	+ 30 06		14.8	GALAXY
PHL 6455	00 13 48.	- 04 00		17.9	BLUE STELLAR OBJECT
PHL 2804	00 13 48.	- 06 35		18.2	BLUE STELLAR OBJECT
PHL 2802	00 13 48.	- 19 40		17.8	BLUE STELLAR OBJECT
PHL 2805	00 13 48.	- 21 34		18.5	BLUE STELLAR OBJECT
PHL 6457	00 13 48.	- 26 07		15.8	BLUE STELLAR OBJECT
LB 00420	00 13 49.	+ 15 34 12.		18.6	FAINT BLUE STAR
MCG+05-01-059	00 13 51.	+ 29 39	90	15.	GALAXY
MCG-01-01-063	00 13 51.	- 06 30 30.	48	15.5	GALAXY
MCG-01-01-062	00 13 51.	- 06 30 30.	60	15.	GALAXY
LB 00422	00 13 53.	+ 31 40 24.		16.5	FAINT BLUE STAR
LB 00421	00 13 53.	+ 32 04 24.		16.3	FAINT BLUE STAR
MCG+00-01-052	00 13 54.	+ 02 02	24	15.5	GALAXY
PHL 6462	00 13 54.	+ 13 10		18.6	BLUE STELLAR OBJECT
PHL 6458	00 13 54.	+ 15 13		18.6	BLUE STELLAR OBJECT
ZWG 456.051	00 13 54.	+ 15 55		15.7	GALAXY
LB 00423	00 13 54.	+ 17 34 18.		15.0	FAINT BLUE STAR
ZWG 499.093	00 13 54.	+ 29 39		15.7	GALAXY
UGC 00152	00 13 54.	+ 29 39	90	15.7	GALAXY Sc
ZWG 499.094	00 13 54.	+ 30 05		15.6	GALAXY
ZC 0013.9-0121	00 13 54.	- 01 21	1210		CLUSTER OF GALAXIES
PHL 0763	00 13 54.	- 03 06		18.4	BLUE STELLAR OBJECT
PHL 6460	00 13 54.	- 04 38		18.1	BLUE STELLAR OBJECT
8ZW 0013-05.1	00 13 54.	- 05 06		16.2	COMPACT GALAXY
PHL 6459	00 13 54.	- 05 12		16.7	BLUE STELLAR OBJECT
PHL 6461	00 13 54.	- 05 54		18.6	BLUE STELLAR OBJECT
PHL 2806	00 13 54.	- 05 54		18.2	BLUE STELLAR OBJECT
PHL 2807	00 13 54.	- 09 56		18.7	BLUE STELLAR OBJECT
RNGC 0061B	00 13 55.	- 06 30		15.5	GALAXY
RNGC 0061A	00 13 55.	- 06 30		15.0	GALAXY
FATH 1.002	00 13 57.	+ 15 54	11		NEBULA
MCG+06-01-020	00 13 57.	+ 35 55	48	14.	GALAXY
SNO 03	00 13 58.	- 24 09 11.	1200	19.	CLUSTER OF 15 GALAXIES
MCG+00-01-053	00 14 00.	+ 01 14 30.	30	14.5	GALAXY
ZWG 382.038	00 14 00.	+ 01 16		15.7	GALAXY
ZWG 382.039	00 14 00.	+ 01 20		14.9	GALAXY
PHL 6464	00 14 00.	+ 03 48		18.4	BLUE STELLAR OBJECT
ZC 0014.0+0826	00 14 00.	+ 08 26	3700		CLUSTER OF GALAXIES
ZWG 518.020	00 14 00.	+ 35 55		15.5	GALAXY
UGC 00153	00 14 00.	+ 47 40	72	17.	GALAXY
PHL 6465	00 14 00.	- 02 30		18.8	BLUE STELLAR OBJECT
PHL 2809	00 14 00.	- 03 24		18.4	BLUE STELLAR OBJECT
PHL 2810	00 14 00.	- 05 42		18.9	BLUE STELLAR OBJECT
PHL 2808	00 14 00.	- 06 00		17.9	BLUE STELLAR OBJECT
MCG-02-01-041	00 14 00.	- 10 48	24	15.	GALAXY
PHL 0764	00 14 00.	- 14 23		18.6	BLUE STELLAR OBJECT
PHL 0765	00 14 00.	- 25 02		16.6	BLUE STELLAR OBJECT
LB 05100	00 14 00.	- 28 00		18.5	FAINT BLUE STAR
LB 00424	00 14 01.	+ 32 11 54.		15.7	FAINT BLUE STAR
MCG+00-01-054	00 14 03.	+ 01 18 30.	30	15.5	GALAXY
SCHO 0015	00 14 05.	+ 56 40 24.	240		ISOLATED DARK CLOUD
8ZW 0014-02.4	00 14 06.	- 02 22		17.1	COMPACT GALAXY
PHL 6466	00 14 06.	- 21 39		15.7	BLUE STELLAR OBJECT

OBJECT NAME	RIGHT ASCEN.	DECLINATION	DIAM.	MAGN.	TYPE OF OBJECT
LB 02709	00 14 09.	+ 14 40 54.		17.1	FAINT BLUE STAR
ARC 0016	00 14 11.	+ 06 29		17.0	RICH CLUSTER OF GALAXIES
ZWG 382.040	00 14 12.	+ 00 26		15.5	GALAXY
MCG+00-01-055	00 14 12.	+ 01 19	60	14.5	GALAXY
UGC 00154	00 14 12.	+ 02 11	72	18.	GALAXY DWARF
PHL 2812	00 14 12.	+ 05 32		17.9	BLUE STELLAR OBJECT
ZWG 408.044	00 14 12.	+ 06 47		14.6	GALAXY
UGC 00155	00 14 12.	+ 06 47	96	14.6	GALAXY S
MCG+01-01-047	00 14 12.	+ 06 47	84	14.	GALAXY
PHL 0766	00 14 12.	+ 06 48		16.3	BLUE STELLAR OBJECT
ZWG 433.041	00 14 12.	+ 12 04		15.3	GALAXY
UGC 00156	00 14 12.	+ 12 04	180	15.3	GALAXY DWRF IR
ZWG 499.095	00 14 12.	+ 29 20		14.8	GALAXY
PHL 0767	00 14 12.	- 04 20		17.1	BLUE STELLAR OBJECT
PHL 6467	00 14 12.	- 19 15		17.2	BLUE STELLAR OBJECT
PHL 2811	00 14 12.	- 19 38		18.1	BLUE STELLAR OBJECT
PHL 0768	00 14 12.	- 22 08		18.2	BLUE STELLAR OBJECT
PHL 2813	00 14 12.	- 31 44		17.7	BLUE STELLAR OBJECT
MCG-06-01-038	00 14 12.	- 36 33	15	15.	GALAXY
KN 13.112	00 14 13.2	+ 12 04 19.			NEBULA
FATH 1.003	00 14 14.	- 14 08	27		NEBULA
SC 0011-6444.7	00 14 14.	- 64 28 01.	18		NEBULA
MCG+02-01-029	00 14 15.	+ 12 03	120	14.	GALAXY
LB 00425	00 14 15.	+ 16 42 48.		15.8	FAINT BLUE STAR
ZC 0014.3+0524	00 14 18.	+ 05 24	2080		CLUSTER OF GALAXIES
ZC 0014.3+0627	00 14 18.	+ 06 27	4100		CLUSTER OF GALAXIES
ZWG 478.048	00 14 18.	+ 22 15		15.1	GALAXY
MCG+05-01-060	00 14 18.	+ 27 33 30.	48	15.	GALAXY
ZWG 499.096	00 14 18.	+ 27 35		15.1	GALAXY
UGC 00157	00 14 18.	+ 27 35	66	15.1	GALAXY SBb
ZWG 499.097	00 14 18.	+ 27 37		14.7	GALAXY
ZC 0014.3+4427	00 14 18.	+ 44 27	2350		CLUSTER OF GALAXIES
MCG+00-01-056	00 14 18.	- 00 23	36	15.	GALAXY
8ZW 0014-03.0	00 14 18.	- 03 03		17.8	COMPACT GALAXY
MCG-01-01-064	00 14 18.	- 05 33	72	14.	GALAXY
PHL 6470	00 14 18.	- 07 44		18.2	BLUE STELLAR OBJECT
PHL 6468	00 14 18.	- 12 00		14.4	BLUE STELLAR OBJECT
PHL 6469	00 14 18.	- 20 29		6.4	BLUE STELLAR OBJECT
SC 0011-6947.4	00 14 21.	- 69 30 43.	6		NEBULA
ARC 0017	00 14 23.	+ 08 32		17.6	RICH CLUSTER OF GALAXIES
LB 02710	00 14 23.	+ 17 04 54.		17.3	FAINT BLUE STAR
PHL 2814	00 14 24.	+ 07 48		17.8	BLUE STELLAR OBJECT
PHL 2815	00 14 24.	+ 08 48		18.5	BLUE STELLAR OBJECT
PHL 6472	00 14 24.	+ 15 36		18.3	BLUE STELLAR OBJECT
MCG+04-01-046	00 14 24.	+ 22 15	24	15.	GALAXY
ZWG 478.049	00 14 24.	+ 23 28		15.7	GALAXY
ZWG 499.098	00 14 24.	+ 29 40		14.8	GALAXY
ZWG 534.009	00 14 24.	+ 41 53		15.5	GALAXY
UGC 00158	00 14 24.	+ 41 53	114	15.5	GALAXY Sb
MCG+07-01-006	00 14 24.	+ 41 53	72	15.	GALAXY
ZC 0014.4-0116	00 14 24.	- 01 16	5170		CLUSTER OF GALAXIES
PHL 6473	00 14 24.	- 02 56		17.7	BLUE STELLAR OBJECT
PHL 6474	00 14 24.	- 03 18		18.0	BLUE STELLAR OBJECT
PHL 2816	00 14 24.	- 04 17		18.4	BLUE STELLAR OBJECT
PHL 6475	00 14 24.	- 05 40		18.6	BLUE STELLAR OBJECT
PHL 0769	00 14 24.	- 07 10		18.5	BLUE STELLAR OBJECT
MCG-02-01-042	00 14 24.	- 09 55	24	15.5	GALAXY
PHL 6476	00 14 24.	- 15 28		18.2	BLUE STELLAR OBJECT
PHL 6471	00 14 24.	- 22 38		18.1	BLUE STELLAR OBJECT
SNO 04	00 14 25.	- 24 21 47.	420	18.	GROUP OF 6 GALAXIES
MCG+06-01-021	00 14 27.	+ 34 13	66	14.5	GALAXY
ARC 0018	00 14 28.	- 03 04		17.1	RICH CLUSTER OF GALAXIES
SNO 05	00 14 29.	- 24 04 23.	600	18.	GROUP OF 6 GALAXIES
ZC 0014.5+1152	00 14 30.	+ 11 52	870		CLUSTER OF GALAXIES
UGC 00159	00 14 30.	+ 17 15	84	18.	GALAXY DWARF
ZC 0014.5+2315	00 14 30.	+ 23 15	15860		CLUSTER OF GALAXIES
UGC 00160	00 14 30.	+ 34 12	96	16.5	GALAXY S IV-V
MCG+06-01-022	00 14 30.	+ 34 59 30.	24	15.	GALAXY
ZWG 518.021	00 14 30.	+ 35 00		15.7	GALAXY
MCG-02-01-043	00 14 30.	- 13 44	30	14.	GALAXY
RNGC 0062	00 14 32.	- 13 44		14.0	GALAXY
MCG+00-01-057	00 14 33.	- 01 16	30	15.	GALAXY
MCG-01-01-065	00 14 33.	- 06 39 30.	60	13.5	GALAXY
HOLM 005A	00 14 33.	- 13 46	48	14.5	PART OF MULTIPLE GALAXY
PHL 6477	00 14 36.	+ 05 16		17.9	BLUE STELLAR OBJECT
UGC 00161	00 14 36.	+ 06 27	66	18.	GALAXY DWARF
PHL 2818	00 14 36.	+ 13 40		18.5	BLUE STELLAR OBJECT
ZWG 456.052	00 14 36.	+ 17 09		15.6	GALAXY
UGC 00162	00 14 36.	+ 17 09	78	15.6	GALAXY S
MCG+03-01-035	00 14 36.	+ 17 09	36	15.5	GALAXY
PHL 2817	00 14 36.	+ 17 17		18.2	BLUE STELLAR OBJECT
MCG+05-01-061	00 14 36.	+ 29 55	30	15.	GALAXY
ZWG 518.022	00 14 36.	+ 35 36		15.7	GALAXY
MCG+06-01-023	00 14 36.	+ 35 37	36	14.5	GALAXY
PHL 0770	00 14 36.	- 07 45		18.1	BLUE STELLAR OBJECT
MCG-03-01-024	00 14 36.	- 19 36	78	13.	GALAXY
MCG-04-01-028	00 14 36.	- 24 37 30.	42	15.	GALAXY
LB 02711	00 14 38.	+ 15 42 42.		15.9	FAINT BLUE STAR
MCG-01-01-066	00 14 39.	- 04 53 30.	78	14.5	GALAXY
PHL 2820	00 14 42.	+ 05 05		16.7	BLUE STELLAR OBJECT
ZWG 408.045	00 14 42.	+ 06 27		15.4	GALAXY
UGC 00163	00 14 42.	+ 06 27	66	15.4	GALAXY SBc
PHL 0771	00 14 42.	+ 17 42		17.0	BLUE STELLAR OBJECT
ZWG 499.099	00 14 42.	+ 29 56		15.3	GALAXY
MCG-01-01-067	00 14 42.	- 03 59 30.	36	14.5	GALAXY
PHL 2819	00 14 42.	- 04 08		18.7	BLUE STELLAR OBJECT
MCG-02-01-045	00 14 42.	- 09 47	12	15.	GALAXY
MCG-02-01-044	00 14 42.	- 12 30	24	15.	GALAXY
PHL 6478	00 14 42.	- 14 24		9.8	BLUE STELLAR OBJECT
HOLM 005B	00 14 44.	- 13 45	36	15.0	PART OF MULTIPLE GALAXY
MCG+01-01-048	00 14 45.	+ 06 27	48	15.	GALAXY
MCG+03-01-036	00 14 45.	+ 17 48	66	14.	GALAXY
SCHO 0021	00 14 45.	+ 60 38 30.	240		ISOLATED DARK CLOUD
LB 02712	00 14 47.	+ 14 32 48.		16.7	FAINT BLUE STAR
ZC 0014.8+0124	00 14 48.	+ 01 24	3830		CLUSTER OF GALAXIES
PHL 2823	00 14 48.	+ 03 35		18.2	BLUE STELLAR OBJECT
PHL 6479	00 14 48.	+ 05 01		18.5	BLUE STELLAR OBJECT
PHL 0772	00 14 48.	+ 13 54		17.5	BLUE STELLAR OBJECT
ZWG 456.053	00 14 48.	+ 17 47		15.2	GALAXY
UGC 00164	00 14 48.	+ 17 47	108	15.2	GALAXY SBb/SBc
MCG+00-01-058	00 14 48.	- 01 17 30.	48	14.	GALAXY
MCG-02-01-046	00 14 48.	- 09 49	36	15.	GALAXY
PHL 2821	00 14 48.	- 19 34		17.3	BLUE STELLAR OBJECT
PHL 2870	00 14 48.	- 23 54		17.9	BLUE STELLAR OBJECT
PHL 2822	00 14 48.	- 30 18		18.5	BLUE STELLAR OBJECT
MCG-02-01-047	00 14 51.	- 09 50	48	14.5	GALAXY
PHL 2825	00 14 54.	+ 17 20		18.7	BLUE STELLAR OBJECT
ZWG 478.050	00 14 54.	+ 25 59		15.4	GALAXY

OBJECT NAME	RIGHT ASCEN.	DECLINATION	DIAM.	MAGN.	TYPE OF OBJECT
MCG+05-01-062	00 14 54.	+ 29 55	66	14.5	GALAXY
5ZW 009	00 14 54.	+ 48 02			COMPACT GALAXY
ZWG 382.041	00 14 54.	- 01 14		15.3	GALAXY
PHL 2824	00 14 54.	- 02 32		17.3	BLUE STELLAR OBJECT
MCG-01-01-068	00 14 54.	- 07 07	84	14.	GALAXY
RNGC 0064	00 14 55.	- 07 07		14.0	GALAXY
LB 00426	00 14 58.	+ 14 02 18.		17.3	FAINT BLUE STAR
VDB.66G 001	00 15	- 19 18	70		DWARF GALAXY
PHL 6480	00 15 00.	+ 04 34		18.4	BLUE STELLAR OBJECT
PHL 2827	00 15 00.	+ 07 44		18.4	BLUE STELLAR OBJECT
ZWG 478.051	00 15 00.	+ 24 23		15.4	GALAXY
UGC 00165	00 15 00.	+ 24 23	60	15.4	GALAXY Sa-b
MCG+04-01-047	00 15 00.	+ 24 24	60	15.	GALAXY
ZWG 499.100	00 15 00.	+ 29 56		15.5	GALAXY
UGC 00166	00 15 00.	+ 29 56	144	15.5	GALAXY Sc
ZC 0015.0+4123	00 15 00.	+ 41 23	2220		CLUSTER OF GALAXIES
OCL 0288	00 15 00.	+ 60 41	240	14.	OPEN STAR CLUSTER
LDN 1279	00 15 00.	+ 63 30	600		DARK NEBULA
MCG-01-01-069	00 15 00.	- 03 30	72	15.5	GALAXY
PHL 6481	00 15 00.	- 04 48		18.4	BLUE STELLAR OBJECT
PHL 2828	00 15 00.	- 05 30		18.3	BLUE STELLAR OBJECT
PHL 2826	00 15 00.	- 07 22		17.9	BLUE STELLAR OBJECT
MCG-04-01-029	00 15 00.	- 24 30	6	16.	GALAXY
IC 0005	00 15 02.	- 09 49 05.			NONSTELLAR OBJECT
BC PKS0015+17	00 15 02.5	+ 17 52 43.		18.	QUASI-STELLAR OBJECT
MCG-01-01-070	00 15 03.	- 04 54 30.	48	14.5	GALAXY
LB 00427	00 15 05.	+ 14 34 42.		13.6	FAINT BLUE STAR
SHO 06	00 15 05.	- 24 34 17.	300	19.	CLUSTER OF 15 GALAXIES
ZWG 433.042	00 15 06.	+ 11 10		12.6	GALAXY
UGC 00167	00 15 06.	+ 11 10	120	12.6	GALAXY S
ZC 0015.1+2531	00 15 06.	+ 25 31	1080		CLUSTER OF GALAXIES
ZC 0015.1-0108	00 15 06.	- 01 08	540		CLUSTER OF GALAXIES
PHL 0773	00 15 06.	- 19 14		17.9	BLUE STELLAR OBJECT
LB 07416	00 15 06.	- 26 06		19.8	FAINT BLUE STAR
LB 02713	00 15 08.	+ 16 15 42.		16.5	FAINT BLUE STAR
SCHO 0022	00 15 08.	+ 58 11 12.	190		ISOLATED DARK CLOUD
KN 13.113	00 15 10.5	+ 11 10 25.			NEBULA
PHL 6482	00 15 12.	+ 04 54		18.3	BLUE STELLAR OBJECT
MCG+02-01-030	00 15 12.	+ 11 10	108	13.	GALAXY
PHL 6483	00 15 12.	+ 17 42		18.4	BLUE STELLAR OBJECT
ZWG 499.101	00 15 12.	+ 29 53		15.7	GALAXY
ZC 0015.2+4238	00 15 12.	+ 42 38	1480		CLUSTER OF GALAXIES
PHL 0774	00 15 12.	- 04 46		17.6	BLUE STELLAR OBJECT
PHL 2829	00 15 12.	- 14 38		18.0	BLUE STELLAR OBJECT
PHL 2830	00 15 12.	- 14 54		18.6	BLUE STELLAR OBJECT
MCG-03-01-025	00 15 12.	- 18 19	54	14.	GALAXY
PHL 6484	00 15 12.	- 19 08		18.1	BLUE STELLAR OBJECT
LB 07417	00 15 12.	- 24 56		16.4	FAINT BLUE STAR
LB 07418	00 15 12.	- 26 33		18.8	FAINT BLUE STAR
RNGC 0063	00 15 13.	+ 11 10		12.5	GALAXY
HOW 01	00 15 15.	- 73 40 56.	18		STAR CLUSTER IN SMC
ZWG 382.042	00 15 18.	+ 00 03		15.5	GALAXY
PHL 2831	00 15 18.	+ 06 36		18.2	BLUE STELLAR OBJECT
PHL 6485	00 15 18.	+ 06 54		18.1	BLUE STELLAR OBJECT
PHL 2832	00 15 18.	+ 16 56		18.4	BLUE STELLAR OBJECT
ZWG 478.052	00 15 18.	+ 24 17		14.8	GALAXY
5ZW 010	00 15 18.	+ 45 56			COMPACT GALAXY
MCG-02-01-048	00 15 18.	- 09 33	30	15.	GALAXY
LB 07419	00 15 18.	- 26 05		18.0	FAINT BLUE STAR
LB 02714	00 15 22.	+ 16 00 48.		16.2	FAINT BLUE STAR
MCG+00-01-059	00 15 24.	+ 00 02	54	14.	GALAXY
PHL 2834	00 15 24.	+ 04 54		17.6	BLUE STELLAR OBJECT
PHL 2835	00 15 24.	+ 13 26		18.4	BLUE STELLAR OBJECT
PHL 6488	00 15 24.	+ 18 16		18.8	BLUE STELLAR OBJECT
MCG+04-01-048	00 15 24.	+ 24 17 30.	48	14.5	GALAXY
ZWG 499.102	00 15 24.	+ 29 11		14.8	GALAXY
MCG+05-01-063	00 15 24.	+ 29 11	36	15.	GALAXY
PHL 2833	00 15 24.	- 05 34		13.4	BLUE STELLAR OBJECT
PHL 6487	00 15 24.	- 06 06		16.9	BLUE STELLAR OBJECT
IC 1538	00 15 26.	+ 29 47			MAY NOT EXIST
SC 0013-6909.3	00 15 27.	- 68 52 37.	12		NEBULA
SC 0013-6854.6	00 15 29.	- 68 37 55.	48		NEBULA
UGC 00168	00 15 30.	+ 18 01	78	16.0	GALAXY Sa
MCG+03-01-037	00 15 30.	+ 18 01	66	14.5	GALAXY
LB 07420	00 15 30.	- 22 54		17.6	FAINT BLUE STAR
PHL 2838	00 15 36.	+ 12 28		18.0	BLUE STELLAR OBJECT
MCG+02-01-031	00 15 36.	+ 12 54	48	15.	GALAXY
ZWG 499.103	00 15 36.	+ 29 40		15.6	GALAXY
ZWG 499.104	00 15 36.	+ 29 47		15.7	GALAXY
RNGC 0067	00 15 36.	+ 29 47		15.5	GALAXY
HOLM 006E	00 15 36.	+ 29 47	18	15.3	PART OF MULTIPLE GALAXY
4ZW 011	00 15 36.	+ 37 57			COMPACT GALAXY
PHL 2836	00 15 36.	- 03 34		17.4	BLUE STELLAR OBJECT
PHL 2837	00 15 36.	- 15 10		18.4	BLUE STELLAR OBJECT
LB 07421	00 15 36.	- 21 18		19.3	FAINT BLUE STAR
LB 07422	00 15 36.	- 22 49		17.7	FAINT BLUE STAR
LB 07423	00 15 36.	- 23 22		18.3	FAINT BLUE STAR
LB 07424	00 15 36.	- 25 41		18.1	FAINT BLUE STAR
LB 07425	00 15 36.	- 25 56		20.0	FAINT BLUE STAR
PHL 0775	00 15 36.	- 26 54		17.1	BLUE STELLAR OBJECT
MCG-06-01-039	00 15 36.	- 34 10	30	15.	GALAXY
SC 0013-6915.1	00 15 37.	- 68 58 25.	18		NEBULA
HOLM 006G	00 15 38.	+ 29 48	18	15.5	PART OF MULTIPLE GALAXY
MCG+05-01-064	00 15 39.	+ 29 46	18	17.5	GALAXY
SC 0013-6934.1	00 15 40.	- 69 17 25.	6		NEBULA
SC 0013-6942.9	00 15 40.	- 69 26 13.	6		NEBULA
PHL 6489	00 15 42.	+ 06 18		17.1	BLUE STELLAR OBJECT
PHL 0776	00 15 42.	+ 17 49		18.6	BLUE STELLAR OBJECT
ZWG 456.054	00 15 42.	+ 19 06		14.8	GALAXY
UGC 00169	00 15 42.	+ 19 06	66	14.8	GALAXY SB?c
MCG+05-01-066	00 15 42.	+ 29 45	15	17.	GALAXY
ZWG 499.105	00 15 42.	+ 29 46		15.7	GALAXY
RNGC 0069	00 15 42.	+ 29 46		15.5	GALAXY
MCG+05-01-065	00 15 42.	+ 29 47	24	15.	GALAXY
ZWG 499.106	00 15 42.	+ 29 48		14.5	GALAXY
RNGC 0068	00 15 42.	+ 29 48		14.5	GALAXY
HOLM 006A	00 15 42.	+ 29 48	18	14.6	PART OF MULTIPLE GALAXY
UGC 00170	00 15 42.	+ 29 48	90	14.5	GALAXY (S0)
MCG+08-01-036	00 15 42.	+ 48 26	78	14.	GALAXY
ZWG 549.032	00 15 42.	+ 48 27		14.4	GALAXY
UGC 00171	00 15 42.	+ 48 27	60	14.4	GALAXY IRR
LB 07426	00 15 42.	- 21 50		18.0	FAINT BLUE STAR
LB 07427	00 15 42.	- 25 16		19.7	FAINT BLUE STAR
LB 00428	00 15 44.	+ 16 21 18.		16.6	FAINT BLUE STAR
HOLM 006F	00 15 44.	+ 29 46	18	15.3	PART OF MULTIPLE GALAXY
VV 166C	00 15 45.	+ 29 47 30.	42	15.	INTERACTING GALAXY
VV 166B	00 15 45.	+ 29 47 30.	9	19.	INTERACTING GALAXY
VV 166A	00 15 45.	+ 29 47 30.	72	14.9	INTERACTING GALAXY
MCG+05-01-067	00 15 45.	+ 29 48	84	14.	GALAXY
MCG+06-01-024	00 15 45.	+ 37 57	36	16.	GALAXY
HOLM 006C	00 15 46.	+ 29 49	18	14.9	PART OF MULTIPLE GALAXY
HOLM 006B	00 15 47.	+ 29 48	18	14.8	PART OF MULTIPLE GALAXY
PHL 6492	00 15 48.	+ 05 16		18.5	BLUE STELLAR OBJECT
PHL 6490	00 15 48.	+ 17 20		17.9	BLUE STELLAR OBJECT
UGC 00172	00 15 48.	+ 17 34	60	17.	GALAXY Sc
ZWG 456.055	00 15 48.	+ 18 14		15.3	GALAXY
ZWG 499.107	00 15 48.	+ 29 47		14.8	GALAXY
RNGC 0071	00 15 48.	+ 29 47		15.0	GALAXY
UGC 00173	00 15 48.	+ 29 47	90	14.8	GALAXY (E)
MCG+05-01-068	00 15 48.	+ 29 47	21	14.	GALAXY
RNGC 0070	00 15 48.	+ 29 48		14.5	GALAXY
ZWG 499.108	00 15 48.	+ 29 49		14.5	GALAXY
UGC 00174	00 15 48.	+ 29 49	120	14.5	GALAXY Sb
ZWG 549.033	00 15 48.	+ 49 43		15.5	GALAXY
UGC 00175	00 15 48.	+ 49 43	78	15.5	GALAXY Sc/SBc
PHL 6491	00 15 48.	- 06 24		18.4	BLUE STELLAR OBJECT
PHL 2839	00 15 48.	- 17 18		18.2	BLUE STELLAR OBJECT
LB 07428	00 15 48.	- 25 49		19.7	FAINT BLUE STAR
IC 1539	00 15 49.	+ 29 49			SAME AS NGC 70
MCG+05-01-069	00 15 51.	+ 29 45	60	15.	GALAXY
ARP 113	00 15 51.	+ 29 48			PECULIAR GALAXY
HOLM 006D	00 15 52.	+ 29 46	18	15.0	PART OF MULTIPLE GALAXY
LB 02715	00 15 54.	+ 17 54 12.		16.8	FAINT BLUE STAR
ZWG 499.109	00 15 54.	+ 29 46		15.0	GALAXY
RNGC 0072	00 15 54.	+ 29 46		15.0	GALAXY
UGC 00176	00 15 54.	+ 29 46	78	15.0	GALAXY SBa
MCG-01-01-071	00 15 54.	- 07 49	36	13.5	GALAXY
LB 07429	00 15 54.	- 21 10		17.7	FAINT BLUE STAR
LB 07430	00 15 54.	- 21 50		17.4	FAINT BLUE STAR
LB 07431	00 15 54.	- 22 38		19.0	FAINT BLUE STAR
LB 07432	00 15 54.	- 25 17		18.8	FAINT BLUE STAR
LB 07433	00 15 54.	- 25 22		17.1	FAINT BLUE STAR
SC 0013-6911.6	00 15 54.	- 68 54 55.	18		NEBULA
LB 02716	00 15 56.	+ 15 39 06.		15.4	FAINT BLUE STAR
MCG-01-01-072	00 15 57.	- 06 34	30	16.	GALAXY
VDB.66G 002	00 16	+ 10 37	70		DWARF GALAXY
PHL 2842	00 16 00.	+ 05 25		18.5	BLUE STELLAR OBJECT
PHL 2841	00 16 00.	+ 17 12		17.5	BLUE STELLAR OBJECT
ZWG 499.110	00 16 00.	+ 29 45		15.1	GALAXY
HMS 1.01	00 16 00.	+ 29 46			E4 GALAXY
RNGC 0072A	00 16 00.	+ 29 46			GALAXY
MCG+05-01-070	00 16 00.	+ 29 46	15	16.	GALAXY
LDN 1281	00 16 00.	+ 63 40	300		DARK NEBULA
ZWG 344.008	00 16 00.	+ 79 26		15.4	GALAXY
UGC 00177	00 16 00.	+ 79 26	78	15.4	GALAXY S
ZC 0016.0-0030	00 16 00.	- 00 30	1480		CLUSTER OF GALAXIES
8ZW 0016-04.9	00 16 00.	- 04 52		16.2	COMPACT GALAXY
MCG-01-01-073	00 16 00.	- 06 36	30	15.5	GALAXY
PHL 0777	00 16 00.	- 06 46		18.2	BLUE STELLAR OBJECT
PHL 6493	00 16 00.	- 08 23		17.8	BLUE STELLAR OBJECT
PHL 6495	00 16 00.	- 09 48		18.4	BLUE STELLAR OBJECT
PHL 0778	00 16 00.	- 10 12		15.4	BLUE STELLAR OBJECT
PHL 2840	00 16 00.	- 11 40		18.5	BLUE STELLAR OBJECT
LB 07434	00 16 00.	- 22 58		17.8	FAINT BLUE STAR
PHL 6494	00 16 00.	- 24 32		17.0	BLUE STELLAR OBJECT
LB 07435	00 16 00.	- 25 18		19.6	FAINT BLUE STAR
LB 07436	00 16 00.	- 25 34		16.3	FAINT BLUE STAR
LB 07437	00 16 00.	- 25 55		18.1	FAINT BLUE STAR
PK118-08.1	00 16 01.5	+ 53 35 41.			PLANETARY NEBULA
LB 00429	00 16 02.	+ 15 05 06.		16.0	FAINT BLUE STAR
LB 00430	00 16 03.	+ 14 21 24.		15.4	FAINT BLUE STAR
PHL 6496	00 16 06.	+ 05 17		16.4	BLUE STELLAR OBJECT
PHL 6498	00 16 06.	+ 07 05		18.5	BLUE STELLAR OBJECT
MCG+01-01-049	00 16 06.	+ 07 12 30.	30	15.	GALAXY
ZWG 408.046	00 16 06.	+ 07 14		15.7	GALAXY
ZWG 549.034	00 16 06.	+ 47 44		14.5	GALAXY
UGC 00178	00 16 06.	+ 47 44	72	14.5	GALAXY S
PHL 6499	00 16 06.	- 07 10		18.2	BLUE STELLAR OBJECT
PHL 6500	00 16 06.	- 07 56		18.1	BLUE STELLAR OBJECT
PHL 6497	00 16 06.	- 13 15		17.9	BLUE STELLAR OBJECT
MCG-02-01-049	00 16 06.	- 14 05	24	16.	GALAXY
PHL 2843	00 16 06.	- 22 16		18.3	BLUE STELLAR OBJECT
LB 07438	00 16 06.	- 22 24		18.8	FAINT BLUE STAR
LB 07439	00 16 06.	- 23 36		18.7	FAINT BLUE STAR
LB 07440	00 16 06.	- 23 37		18.7	FAINT BLUE STAR
LB 07441	00 16 06.	- 24 28		18.7	FAINT BLUE STAR
LB 07442	00 16 06.	- 24 54		17.5	FAINT BLUE STAR
LB 07443	00 16 06.	- 26 06		18.0	FAINT BLUE STAR
PHL 6502	00 16 06.	- 27 16		18.3	BLUE STELLAR OBJECT
LB 03135	00 16 06.	- 74 23		13.3	FAINT BLUE STAR
LIN.CL 003	00 16 06.	- 74 35 44.	342	15.5	STAR CLUSTER IN SMC
PHL 6501	00 16 09.	- 12 25		18.5	BLUE STELLAR OBJECT
ARP 256	00 16 11.	- 10 39			PECULIAR GALAXY
PHL 6504	00 16 12.	+ 06 02		18.0	BLUE STELLAR OBJECT
PHL 6503	00 16 12.	+ 07 12		15.7	BLUE STELLAR OBJECT
ZWG 433.043	00 16 12.	+ 10 20		14.7	GALAXY
KARA.73B 0014	00 16 12.	+ 10 20	30	14.7	ISOLATED GALAXY F
PHL 0780	00 16 12.	+ 15 04		16.5	BLUE STELLAR OBJECT
MCG+08-01-037	00 16 12.	+ 47 44	72	15.	GALAXY
PHL 6505	00 16 12.	- 04 34		17.1	BLUE STELLAR OBJECT
PHL 6506	00 16 12.	- 05 23		18.3	BLUE STELLAR OBJECT
MCG-02-01-050	00 16 12.	- 08 42	54	14.5	GALAXY
MCG-03-01-026	00 16 12.	- 15 37	72	13.5	GALAXY
PHL 0779	00 16 12.	- 18 58		18.1	BLUE STELLAR OBJECT
LB 07444	00 16 12.	- 25 03		16.8	FAINT BLUE STAR
LB 07445	00 16 12.	- 25 17		19.9	FAINT BLUE STAR
LB 07446	00 16 12.	- 25 22		16.6	FAINT BLUE STAR
LB 07447	00 16 12.	- 25 58		18.8	FAINT BLUE STAR
LB 02717	00 16 14.	+ 14 43 54.		15.8	FAINT BLUE STAR
RNGC 0073	00 16 14.	- 15 37		13.0	GALAXY
MCG+05-01-071	00 16 15.	+ 29 47	42	16.	GALAXY
MCG-01-01-074	00 16 15.	- 03 41	60	15.	GALAXY
MCG-03-01-027	00 16 15.	- 19 18	78	15.	GALAXY
PHL 6507	00 16 18.	+ 16 00		17.7	BLUE STELLAR OBJECT
MCG+03-01-038	00 16 18.	+ 19 07	54	15.5	GALAXY
MCG+04-01-049	00 16 18.	+ 23 12	66	14.5	GALAXY
RNGC 0074	00 16 18.	+ 29 47		16.0	GALAXY
PHL 6509	00 16 18.	- 03 48		18.0	BLUE STELLAR OBJECT
PHL 2844	00 16 18.	- 08 01		18.4	BLUE STELLAR OBJECT
8ZW 0016-10.6	00 16 19.	- 10 37		14.5	COMPACT GALAXY
VV 352A	00 16 18.	- 10 38	90	14.	INTERACTING GALAXY
VV 352B	00 16 18.	- 10 39	72	14.	INTERACTING GALAXY
MCG-02-01-052	00 16 18.	- 10 39	60	13.5	GALAXY
MCG-02-01-051	00 16 18.	- 10 39	60	14.	GALAXY
PHL 0781	00 16 18.	- 15 03		19.0	BLUE STELLAR OBJECT
PHL 6508	00 16 18.	- 16 08		18.5	BLUE STELLAR OBJECT

OBJECT NAME	RIGHT ASCEN.	DECLINATION	DIAM.	MAGN.	TYPE OF OBJECT
PHL 2845	00 16 18.	- 25 36		18.1	BLUE STELLAR OBJECT
LIN 001	00 16 18.	- 74 04 50.		17.00	PLANETARY NEBULA IN SMC
HOLM 007B	00 16 20.	- 10 38	30	14.4	PART OF MULTIPLE GALAXY
ABC 0019	00 16 21.	- 06 28		17.8	RICH CLUSTER OF GALAXIES
HOLM 007A	00 16 21.	- 10 39	24	14.2	PART OF MULTIPLE GALAXY
MCG-02-01-053	00 16 21.	- 12 41	30	16.	GALAXY
IC 0006	00 16 22.	- 03 33 11.			NONSTELLAR OBJECT
PHL 0782	00 16 24.	+ 04 34		18.5	BLUE STELLAR OBJECT
LB 02718	00 16 24.	+ 14 17 54.		15.9	FAINT BLUE STAR
ZWG 478.053	00 16 24.	+ 23 12		15.3	GALAXY
UGC 00179	00 16 24.	+ 23 12	78	15.3	GALAXY Sc
5ZW 011	00 16 24.	+ 46 03			COMPACT GALAXY
MCG+08-01-038	00 16 24.	+ 49 42	90	15.	GALAXY
MCG-01-01-075	00 16 24.	- 03 33	30	14.5	GALAXY
PHL 0783	00 16 24.	- 06 45		16.6	BLUE STELLAR OBJECT
PHL 2846	00 16 24.	- 08 42		18.7	BLUE STELLAR OBJECT
8ZW 0016-11.5	00 16 24.	- 11 28		14.7	COMPACT GALAXY
PHL 2847	00 16 24.	- 16 58		18.4	BLUE STELLAR OBJECT
PHL 0784	00 16 24.	- 20 01		19.0	BLUE STELLAR OBJECT
LB 07448	00 16 24.	- 21 50		17.8	FAINT BLUE STAR
PHL 0785	00 16 24.	- 22 09		18.2	BLUE STELLAR OBJECT
LB 07449	00 16 24.	- 22 44		17.4	FAINT BLUE STAR
LB 07450	00 16 24.	- 26 22		18.0	FAINT BLUE STAR
PHL 0786	00 16 24.	- 32 12		14.1	BLUE STELLAR OBJECT
TON-S 0148	00 16 24.	- 32 12		13.9	BLUE STAR
LB 00431	00 16 25.	+ 13 55 36.		17.3	FAINT BLUE STAR
IC 0007	00 16 29.	+ 10 16 13.			NONSTELLAR OBJECT
ZC 0016.5+0715	00 16 30.	+ 07 15	1680		CLUSTER OF GALAXIES
ZWG 433.044	00 16 30.	+ 15 28		15.3	GALAXY
UGC 00180	00 16 30.	+ 15 28	72	15.3	GALAXY Sa-b
PHL 0787	00 16 30.	+ 17 26		18.1	BLUE STELLAR OBJECT
ZWG 457.001	00 16 30.	+ 19 07		15.4	GALAXY
ZWG 456.056	00 16 30.	+ 19 07		15.4	GALAXY
UGC 00181	00 16 30.	+ 19 07	78	15.4	GALAXY Sa
IC 0008	00 16 30.	- 03 29 53.			NONSTELLAR OBJECT
PHL 6510	00 16 30.	- 11 14		18.4	BLUE STELLAR OBJECT
MCG-04-02-001	00 16 30.	- 23 10	24	15.	GALAXY
LB 07451	00 16 30.	- 26 25		18.9	FAINT BLUE STAR
PHL 6511	00 16 30.	- 28 09		18.1	BLUE STELLAR OBJECT
RNGC 0065	00 16 32.	- 23 10		15.0	GALAXY
MCG-01-01-076	00 16 33.	- 03 30	42	15.	GALAXY
ZC 0016.6+0746	00 16 36.	+ 07 46	1810		CLUSTER OF GALAXIES
PHL 0788	00 16 36.	+ 13 58		18.4	BLUE STELLAR OBJECT
PHL 6512	00 16 36.	+ 14 34		18.8	BLUE STELLAR OBJECT
MCG+02-01-032	00 16 36.	+ 15 29	60	15.	GALAXY
PHL 6513	00 16 36.	- 03 10		18.1	BLUE STELLAR OBJECT
PHL 2848	00 16 36.	- 06 25		18.0	BLUE STELLAR OBJECT
MCG-03-01-028	00 16 36.	- 14 43	54	15.	GALAXY
PHL 6514	00 16 36.	- 17 58		18.6	BLUE STELLAR OBJECT
MCG-04-02-002	00 16 36.	- 23 13 30.	72	14.	GALAXY
LB 07452	00 16 36.	- 23 37		18.7	FAINT BLUE STAR
LB 07453	00 16 36.	- 25 00		18.3	FAINT BLUE STAR
LB 07454	00 16 36.	- 25 25		19.7	FAINT BLUE STAR
SC 0014-6848.5	00 16 36.	- 68 31 50.	18		NEBULA
RNGC 0077	00 16 38.	- 22 58		15.0	GALAXY
RNGC 0066	00 16 38.	- 23 13		14.0	GALAXY
MCG-04-02-003	00 16 39.	- 22 58	24	15.	GALAXY
SC 0014-6653.2	00 16 41.	- 66 36 32.			NEBULA
ZC 0016.7+0257	00 16 42.	+ 02 57	1010		CLUSTER OF GALAXIES
ZWG 408.047	00 16 42.	+ 03 51		15.3	GALAXY
3ZW 004	00 16 42.	+ 03 52			COMPACT GALAXY
PHL 2850	00 16 42.	+ 15 38		18.2	BLUE STELLAR OBJECT
ZWG 534.010	00 16 42.	+ 39 33		15.7	GALAXY
PHL 6515	00 16 42.	- 06 41		17.4	BLUE STELLAR OBJECT
PHL 2849	00 16 42.	- 07 40		17.9	BLUE STELLAR OBJECT
MCG-02-01-054	00 16 42.	- 11 11	42	14.5	GALAXY
FATH 1.004	00 16 42.	- 14 42	16		NEBULA
MCG-04-02-004	00 16 42.	- 23 03 30.	36	15.	GALAXY
TON-S 0149	00 16 42.	- 25 29		13.2	BLUE STAR
LB 07455	00 16 42.	- 26 36		18.4	FAINT BLUE STAR
MCG-06-01-040	00 16 42.	- 36 53	24	16.5	GALAXY
PHL 2851	00 16 48.	+ 04 56		18.2	BLUE STELLAR OBJECT
ZWG 408.048	00 16 48.	+ 06 10		14.8	GALAXY
UGC 00182	00 16 48.	+ 06 10	114	14.8	GALAXY S0
MCG+01-01-050	00 16 48.	+ 06 20	18	15.5	GALAXY
ZC 0016.8+1157	00 16 48.	+ 11 57	2820		CLUSTER OF GALAXIES
ZC 0016.8+1244	00 16 48.	+ 12 44	1880		CLUSTER OF GALAXIES
ZWG 457.002	00 16 48.	+ 16 35		15.6	GALAXY
ZWG 456.057	00 16 48.	+ 16 35		15.6	GALAXY
MCG+08-01-039	00 16 48.	+ 46 58	96	13.	GALAXY
ZC 0016.8-0215	00 16 48.	- 02 15	1010		CLUSTER OF GALAXIES
PHL 2852	00 16 48.	- 06 02		18.6	BLUE STELLAR OBJECT
PHL 2855	00 16 48.	- 08 57		18.1	BLUE STELLAR OBJECT
LB 07456	00 16 48.	- 21 33		17.5	FAINT BLUE STAR
PHL 2853	00 16 48.	- 22 24		18.3	BLUE STELLAR OBJECT
MCG-04-02-005	00 16 48.	- 24 48 30.	48	14.	GALAXY
MCG-05-02-001	00 16 48.	- 26 59	30	15.	GALAXY
PHL 6516	00 16 48.	- 30 00		18.3	BLUE STELLAR OBJECT
PHL 2854	00 16 48.	- 30 10		18.2	BLUE STELLAR OBJECT
TON-S 0150	00 16 48.	- 30 20		14.2	BLUE STAR
PHL 0789	00 16 48.	- 32 16		15.7	BLUE STELLAR OBJECT
MCG+01-01-051	00 16 54.	+ 06 09 30.	48	14.	GALAXY
PHL 6517	00 16 54.	+ 12 36		16.6	BLUE STELLAR OBJECT
MCG+05-01-072	00 16 54.	+ 29 39	48	15.	GALAXY
ZC 0016.9+3148	00 16 54.	+ 31 48	1340		CLUSTER OF GALAXIES
ZWG 549.035	00 16 54.	+ 46 58		13.8	GALAXY
UGC 00183	00 16 54.	+ 46 58	96	13.8	GALAXY Sb
MCG-01-01-077	00 16 54.	- 04 22	60	14.	GALAXY
LB 07457	00 16 54.	- 20 43		17.0	FAINT BLUE STAR
LB 07458	00 16 54.	- 21 58		18.1	FAINT BLUE STAR
PHL 2856	00 16 54.	- 22 06		15.6	BLUE STELLAR OBJECT
LB 07459	00 16 54.	- 23 24		18.6	FAINT BLUE STAR
LB 07460	00 16 54.	- 25 54		17.8	FAINT BLUE STAR
MCG-05-02-002	00 16 54.	- 26 58	9	15.	GALAXY
PHL 6518	00 16 54.	- 27 59		17.5	BLUE STELLAR OBJECT
RNGC 0075	00 16 55.	+ 06 09		15.0	GALAXY
SCHO 0023	00 16 55.	+ 57 51 18.	320		ISOLATED DARK CLOUD
FATH 1.005	00 16 55.	- 14 01	27		NEBULA
FATH 1.006	00 16 55.	- 14 03			NEBULA
MCG-02-01-055	00 16 57.	- 09 55	30	16.	GALAXY
CED 002	00 17	+ 56 19			DIFFUSE GALACTIC NEBULA
ZC 0017.0+0320	00 17 00.	+ 03 20	470		CLUSTER OF GALAXIES
PHL 0790	00 17 00.	+ 06 08		15.7	BLUE STELLAR OBJECT
UGC 00184	00 17 00.	+ 12 55	60	19.	GALAXY DWARF
PHL 6463	00 17 00.	+ 15 32		18.4	BLUE STELLAR OBJECT
MCG+05-01-073	00 17 00.	+ 29 39	18	16.	GALAXY
ZWG 499.111	00 17 00.	+ 29 40		14.0	GALAXY
RNGC 0076	00 17 00.	+ 29 40		14.0	GALAXY
UGC 00185	00 17 00.	+ 29 40	84	14.0	GALAXY COMPACT
LDN 1280	00 17 00.	+ 62 23	540		DARK NEBULA
PHL 2857	00 17 00.	- 15 22		18.5	BLUE STELLAR OBJECT
TON-S 0152	00 17 00.	- 22 05		14.9	BLUE STAR
PHL 0791	00 17 00.	- 23 10		18.1	BLUE STELLAR OBJECT
LB 07461	00 17 00.	- 24 48		19.6	FAINT BLUE STAR
TON-S 0151	00 17 00.	- 32 15		16.0	BLUE STAR
LB 01549	00 17 00.	- 49 39		14.6	FAINT BLUE STAR
LB 00432	00 17 05.	+ 17 24 36.		16.5	FAINT BLUE STAR
MCG+00-02-001	00 17 06.	+ 00 05	48	15.	GALAXY
HOLM 008A	00 17 06.	+ 29 40	24	14.4	PART OF MULTIPLE GALAXY
PHL 2858	00 17 06.	- 05 27		18.4	BLUE STELLAR OBJECT
ABC 0020	00 17 06.	- 22 54		17.1	RICH CLUSTER OF GALAXIES
LB 07462	00 17 06.	- 22 56		17.6	FAINT BLUE STAR
LB 07463	00 17 06.	- 25 36		18.5	FAINT BLUE STAR
PHL 2859	00 17 06.	- 27 08		17.3	BLUE STELLAR OBJECT
MCG+04-01-050	00 17 09.	+ 23 30	48	15.	GALAXY
IC 0009	00 17 11.	- 14 24 06.			NONSTELLAR OBJECT
MCG+00-02-002	00 17 12.	+ 00 18	12	15.	GALAXY
ZC 0017.2+1008	00 17 12.	+ 10 08	3760		CLUSTER OF GALAXIES
ZWG 457.003	00 17 12.	+ 18 05		15.6	GALAXY
ZWG 456.058	00 17 12.	+ 18 05		15.6	GALAXY
ZWG 478.054	00 17 12.	+ 23 30		14.9	GALAXY
UGC 00186	00 17 12.	+ 23 30	72	14.9	GALAXY SBb
HOLM 008B	00 17 12.	+ 29 40	24	15.3	PART OF MULTIPLE GALAXY
5ZW 012	00 17 12.	+ 44 58			COMPACT GALAXY
ZC 0017.2+4501	00 17 12.	+ 45 01	1410		CLUSTER OF GALAXIES
PHL 6519	00 17 12.	- 09 34		15.8	BLUE STELLAR OBJECT
FEIG 003	00 17 12.	- 09 34		15.0	FAINT BLUE STAR
PHL 2861	00 17 12.	- 11 30		18.6	BLUE STELLAR OBJECT
MCG-02-02-001	00 17 12.	- 14 24	24	15.	GALAXY
PHL 2860	00 17 12.	- 18 21		19.0	BLUE STELLAR OBJECT
LB 07464	00 17 12.	- 20 59		16.0	FAINT BLUE STAR
LB 07465	00 17 12.	- 25 27		20.0	FAINT BLUE STAR
IC 1540	00 17 14.	+ 23 26 12.			NONSTELLAR OBJECT
SCHO 0024	00 17 14.	+ 58 27 36.	300		ISOLATED DARK CLOUD
FATH 2.001	00 17 14.	- 14 25			NEBULA
PK119+00.1	00 17 14.5	+ 62 42 22.			PLANETARY NEBULA
ZC 0017.3+0439	00 17 18.	+ 04 39	1280		CLUSTER OF GALAXIES
ZC 0017.3+1711	00 17 18.	+ 17 11	1080		CLUSTER OF GALAXIES
ZWG 457.004	00 17 18.	+ 19 42		15.5	GALAXY
ZWG 456.059	00 17 18.	+ 19 42		15.5	GALAXY
KARA.72 004A	00 17 18.	+ 19 42	36	15.5	PART OF DOUBLE GALAXY
4ZW 012	00 17 18.	+ 31 51			COMPACT GALAXY
5ZW 013	00 17 18.	+ 43 58			COMPACT GALAXY
PHL 0792	00 17 18.	- 04 14		18.9	BLUE STELLAR OBJECT
PHL 2862	00 17 18.	- 13 20		18.2	BLUE STELLAR OBJECT
LB 07466	00 17 18.	- 22 53		16.7	FAINT BLUE STAR
LB 07467	00 17 18.	- 25 16		19.0	FAINT BLUE STAR
MCG-06-01-041	00 17 18.	- 34 46	48	15.	GALAXY
SC 0014-6648.5	00 17 20.	- 66 31 50.	42		NEBULA
ZC 0017.4+0550	00 17 24.	+ 05 50	1410		CLUSTER OF GALAXIES
UGC 00187	00 17 24.	+ 08 52	66	16.5	GALAXY S
PHL 0795	00 17 24.	+ 13 36		14.2	BLUE STELLAR OBJECT
FEIG 004	00 17 24.	+ 13 36		14.0	FAINT BLUE STAR
ZWG 457.005	00 17 24.	+ 19 07		15.6	GALAXY
ZWG 456.060	00 17 24.	+ 19 07		15.6	GALAXY
ZWG 457.006	00 17 24.	+ 19 43		15.7	GALAXY
ZWG 456.061	00 17 24.	+ 19 43		15.7	GALAXY
UGC 00188	00 17 24.	+ 19 43	84	15.7	GALAXY SB
KARA.72 004B	00 17 24.	+ 19 43	66	15.7	PART OF DOUBLE GALAXY
ZWG 478.055	00 17 24.	+ 21 44		15.5	GALAXY
ZCG 0017+31	00 17 24.	+ 31 51		18.5	COMPACT GALAXY
PHL 0793	00 17 24.	- 06 54		17.5	BLUE STELLAR OBJECT
PHL 0794	00 17 24.	- 07 06		18.4	BLUE STELLAR OBJECT
PHL 2863	00 17 24.	- 21 52		16.6	BLUE STELLAR OBJECT
LB 07468	00 17 24.	- 21 57		18.8	FAINT BLUE STAR
LB 07469	00 17 24.	- 23 37		19.5	FAINT BLUE STAR
LB 07470	00 17 24.	- 24 04		18.0	FAINT BLUE STAR
LB 07471	00 17 24.	- 24 16		19.5	FAINT BLUE STAR
LB 07472	00 17 24.	- 24 26		18.7	FAINT BLUE STAR
LB 07473	00 17 24.	- 24 39		17.2	FAINT BLUE STAR
LB 07474	00 17 24.	- 25 42		18.3	FAINT BLUE STAR
LB 07475	00 17 24.	- 26 04		19.6	FAINT BLUE STAR
LB 07476	00 17 24.	- 26 12		19.4	FAINT BLUE STAR
LB 03136	00 17 24.	- 63 26		13.4	FAINT BLUE STAR
IC 1541	00 17 25.	+ 21 43 24.			NONSTELLAR OBJECT
LB 00433	00 17 26.	+ 13 35 42.		14.3	FAINT BLUE STAR
MCG+01-02-002	00 17 27.	+ 06 10	36	15.5	GALAXY
MCG+01-02-001	00 17 27.	+ 06 10	24	15.5	GALAXY
LB 02719	00 17 27.	+ 16 11 06.		16.8	FAINT BLUE STAR
MCG-01-02-001	00 17 27.	- 06 37	84	14.	GALAXY
ZWG 409.001	00 17 30.	+ 06 12		14.8	GALAXY
KARA.72 005B	00 17 30.	+ 06 12	30		PART OF DOUBLE GALAXY
KARA.72 005A	00 17 30.	+ 06 12	24	14.8	PART OF DOUBLE GALAXY
MCG+02-02-001	00 17 30.	+ 10 35	84	14.	GALAXY
MCG+02-01-033	00 17 30.	+ 10 36	72	14.	GALAXY
UGC 00189	00 17 30.	+ 14 50	72	17.	GALAXY
MCG+03-02-001	00 17 30.	+ 19 08	45	15.	GALAXY
MCG+03-02-002	00 17 30.	+ 19 42	18	16.	GALAXY
MCG+03-02-003	00 17 30.	+ 19 44	66	16.	GALAXY
UGC 00190	00 17 30.	+ 42 17	84	16.5	GALAXY
MCG+07-01-007	00 17 30.	+ 42 17	36	16.	GALAXY
MCG+10-01-001	00 17 30.	+ 59 02	300	13.6	GALAXY
IC 0010	00 17 30.	+ 59 02	300		GALAXY Sc
SCHO 0025	00 17 30.	+ 61 37 48.	240		ISOLATED DARK CLOUD
PHL 2864	00 17 30.	- 02 46		17.5	BLUE STELLAR OBJECT
MCG-01-02-002	00 17 30.	- 04 20	48	15.	GALAXY
YC 0117-40	00 17 30.	- 40 30			UNUSUAL SOUTHERN OBJECT
ZWG 434.001	00 17 36.	+ 10 36		15.6	GALAXY
UGC 00191	00 17 36.	+ 10 36	156	15.6	GALAXY DWRF SP
PHL 0796	00 17 36.	+ 17 13		18.3	BLUE STELLAR OBJECT
ZC 0017.6+3327	00 17 36.	+ 33 27	810		CLUSTER OF GALAXIES
UGC 00192	00 17 36.	+ 59 02	600		GALAXY DWRF IR
MCG+00-02-003	00 17 36.	- 00 18	48	16.	GALAXY
PHL 2865	00 17 36.	- 03 14		14.5	BLUE STELLAR OBJECT
PHL 2866	00 17 36.	- 04 24		15.3	BLUE STELLAR OBJECT
PHL 6520	00 17 36.	- 09 54		13.9	BLUE STELLAR OBJECT
PHL 2867	00 17 36.	- 14 51		18.5	BLUE STELLAR OBJECT
LB 07477	00 17 36.	- 23 20		19.5	FAINT BLUE STAR
PHL 2868	00 17 36.	- 27 26		18.3	BLUE STELLAR OBJECT
SC 0015-6716.6	00 17 40.	- 66 59 56.			NEBULA
LB 02720	00 17 41.	+ 14 38 36.		16.0	FAINT BLUE STAR
MCG+01-02-003	00 17 42.	+ 07 24	84	14.	GALAXY
ZC 0017.7+2333	00 17 42.	+ 23 33	1480		CLUSTER OF GALAXIES
ZC 0017.7+3434	00 17 42.	+ 34 34	940		CLUSTER OF GALAXIES
MCG-01-02-003	00 17 42.	- 02 56	42	16.	GALAXY
IC 0012	00 17 42.	- 02 56 12.			NONSTELLAR OBJECT

OBJECT NAME	RIGHT ASCEN.	DECLINATION	DIAM.	MAGN.	TYPE OF OBJECT
FEIG 005	00 17 42.	- 03 15		13.5	FAINT BLUE STAR
LB 07478	00 17 42.	- 23 04		18.8	FAINT BLUE STAR
LB 07479	00 17 42.	- 23 56		18.2	FAINT BLUE STAR
LB 07480	00 17 42.	- 24 38		19.7	FAINT BLUE STAR
LB 07481	00 17 42.	- 25 40		18.0	FAINT BLUE STAR
LB 07482	00 17 42.	- 25 49		18.5	FAINT BLUE STAR
BIGO 456	00 17 43.	+ 00 34			NEBULA
IC 0013	00 17 46.	+ 07 25 06.			NONSTELLAR OBJECT
ZWG 383.001	00 17 48.	+ 00 33		14.5	GALAXY
MRK 547	00 17 48.	+ 00 33	20	15.5	GALAXY WITH UV CONTINUUM
UGC 00194	00 17 48.	+ 00 33	78	14.5	GALAXY S0
UGC 00193	00 17 48.	+ 00 33	78	14.5	GALAXY SB0-a
KARA.72 006A	00 17 48.	+ 00 33	60	14.5	PART OF DOUBLE GALAXY
ZWG 409.002	00 17 48.	+ 07 25		15.0	GALAXY
UGC 00195	00 17 48.	+ 07 25	96	15.0	GALAXY Sb-c
MCG+02-02-002	00 17 48.	+ 09 57	36	15.5	GALAXY
MCG+06-01-025	00 17 48.	+ 37 32 30.	42	16.	GALAXY
ZWG 519.001	00 17 48.	+ 38 17		15.7	GALAXY
ZWG 518.023	00 17 48.	+ 38 17		15.7	GALAXY
MCG+08-01-040	00 17 49.	+ 47 09	90	13.	GALAXY
MCG-01-02-004	00 17 48.	- 02 56	36	16.	GALAXY
PHL 2869	00 17 48.	- 03 28		17.0	BLUE STELLAR OBJECT
MCG-02-02-002	00 17 48.	- 14 17 30.	36	16.	GALAXY
LB 07483	00 17 48.	- 20 46		18.7	FAINT BLUE STAR
LB 07484	00 17 48.	- 21 22		17.3	FAINT BLUE STAR
LB 07485	00 17 48.	- 24 54		19.4	FAINT BLUE STAR
LB 07486	00 17 48.	- 26 34		18.5	FAINT BLUE STAR
RNGC 0078B	00 17 49.	+ 00 33			GALAXY
RNGC 0078A	00 17 49.	+ 00 33		14.5	GALAXY
IC 0011	00 17 49.	+ 56 19			NONSTELLAR OBJECT
SHB 003	00 17 49.8	+ 15 24 16.		18.2	QUASI-STELLAR OBJECT
BC 3CR9	00 17 49.94	+ 15 24 16.2		18.21	QUASI-STELLAR OBJECT
SC 0015-6713.6	00 17 50.	- 66 56 56.	12		NEBULA
MCG+00-02-004	00 17 51.	+ 00 33	60	14.	GALAXY
KARA.72 006B	00 17 54.	+ 00 33	66		PART OF DOUBLE GALAXY
MCG+00-02-005	00 17 54.	+ 00 33	24	14.	GALAXY
ZC 0017.9+3641	00 17 54.	+ 36 41	1010		CLUSTER OF GALAXIES
ZWG 549.036	00 17 54.	+ 47 10		14.2	GALAXY
UGC 00196	00 17 54.	+ 47 10	90	14.2	GALAXY SBc
MCG+00-02-006	00 17 54.	- 00 50	18	15.5	GALAXY
PHL 0797	00 17 54.	- 04 00		13.0	BLUE STELLAR OBJECT
LB 07487	00 17 54.	- 20 58		16.8	FAINT BLUE STAR
LB 07488	00 17 54.	- 21 03		19.3	FAINT BLUE STAR
LB 07489	00 17 54.	- 21 35		20.1	FAINT BLUE STAR
LB 07490	00 17 54.	- 22 28		19.7	FAINT BLUE STAR
LB 07491	00 17 54.	- 25 02		19.0	FAINT BLUE STAR
LB 07492	00 17 54.	- 26 24		15.5	FAINT BLUE STAR
ABC 0021	00 17 57.	+ 28 22		16.2	RICH CLUSTER OF GALAXIES
PHL 2871	00 17 54.	+ 15 23		18.6	BLUE STELLAR OBJECT
LBN 0590	00 18	+ 59 00	300		BRIGHT NEBULA
LBN 0592	00 18	+ 61 30	720		BRIGHT NEBULA
KHAV 004	00 18	+ 65 17			DARK NEBULA
ZWG 434.002	00 18 00.	+ 15 04		15.7	GALAXY
MCG+04-02-001	00 18 00.	+ 22 20	39	15.	GALAXY
4ZW 013	00 18 00.	+ 31 14			COMPACT GALAXY
SG 2.001	00 18 00.	+ 59 02	240		DIFFUSE EMISSION NEBULA
MCG+00-02-007	00 18 00.	- 00 45	60	15.5	GALAXY
PHL 0798	00 18 00.	- 16 35		16.7	BLUE STELLAR OBJECT
LB 07494	00 18 00.	- 24 22		19.9	FAINT BLUE STAR
MCG-06-02-003	00 18 00.	- 33 47	18	15.5	GALAXY
MCG-06-02-001	00 18 00.	- 34 47	24	16.	GALAXY
MCG-06-02-002	00 18 01.	- 34 47	24	16.	GALAXY
MCG-02-02-003	00 18 03.	- 13 37	54	15.5	GALAXY
IC 1542	00 18 06.	+ 22 18 54.			NONSTELLAR OBJECT
ZWG 479.001	00 18 06.	+ 22 20		15.0	GALAXY
ZC 0018.1+3116	00 18 06.	+ 31 16	740		CLUSTER OF GALAXIES
MCG-01-02-005	00 18 06.	- 04 43	60	14.5	GALAXY
LB 07495	00 18 06.	- 23 39		19.9	FAINT BLUE STAR
LB 07496	00 18 06.	- 26 29		17.8	FAINT BLUE STAR
ZC 0018.2+0117	00 18 12.	+ 01 17	810		CLUSTER OF GALAXIES
MCG+00-02-008	00 18 12.	+ 01 41 30.	12	15.5	GALAXY
PHL 6521	00 18 12.	+ 16 24		18.2	BLUE STELLAR OBJECT
ZWG 457.007	00 18 12.	+ 19 45		15.1	GALAXY
UGC 00197	00 18 12.	+ 19 45	84	15.1	GALAXY Sa
ZWG 457.008	00 18 12.	+ 21 15		15.3	GALAXY
PHL 2872	00 18 12.	- 20 21		18.3	BLUE STELLAR OBJECT
LB 07497	00 18 12.	- 20 41		18.6	FAINT BLUE STAR
LB 07498	00 18 12.	- 22 15		18.7	FAINT BLUE STAR
LB 07499	00 18 12.	- 23 04		19.8	FAINT BLUE STAR
LB 07500	00 18 12.	- 23 21		19.7	FAINT BLUE STAR
LB 07501	00 18 12.	- 24 30		18.0	FAINT BLUE STAR
LB 07502	00 18 12.	- 25 58		20.0	FAINT BLUE STAR
ABC 0022	00 18 12.	- 25 59		17.5	RICH CLUSTER OF GALAXIES
LB 00434	00 18 14.	+ 14 43 06.		17.3	FAINT BLUE STAR
MCG-04-02-006	00 18 15.	- 23 33	96	14.	GALAXY
ZWG 409.003	00 18 18.	+ 06 32		15.7	GALAXY
MCG+03-02-004	00 18 18.	+ 19 46	48	15.	GALAXY
MCG+04-02-002	00 18 18.	+ 21 35	42	14.5	GALAXY
ZWG 479.002	00 18 18.	+ 21 36		14.2	GALAXY
UGC 00198	00 18 18.	+ 21 36	48	14.2	GALAXY S
ZC 0018.3-0009	00 18 18.	- 00 09	1210		CLUSTER OF GALAXIES
LB 07503	00 18 18.	- 21 52		18.8	FAINT BLUE STAR
LB 07504	00 18 18.	- 23 46		19.0	FAINT BLUE STAR
LB 07505	00 18 18.	- 23 56		18.4	FAINT BLUE STAR
LB 07506	00 18 18.	- 25 28		16.5	FAINT BLUE STAR
SCHO 0026	00 18 19.	+ 61 01 30.	210		ISOLATED DARK CLOUD
IC 1543	00 18 20.	+ 21 35 30.			NONSTELLAR OBJECT
MCG+00-02-009	00 18 21.	- 02 06	42	15.5	GALAXY
LB 00435	00 18 23.	+ 16 22 54.		16.5	FAINT BLUE STAR
UGC 00199	00 18 24.	+ 12 35	60	19.	GALAXY DWARF
UGC 00200	00 18 24.	+ 17 25	66	18.	GALAXY DWRF IR
MCG+03-02-005	00 18 24.	+ 21 16	42	15.	GALAXY
ZWG 479.003	00 18 24.	+ 22 18		14.9	GALAXY
RNGC 0079	00 18 24.	+ 22 18		15.0	GALAXY
MCG+04-02-003	00 18 24.	+ 22 18	18	15.	GALAXY
MCG+06-01-026	00 18 24.	+ 34 13 30.	36	14.5	GALAXY
LB 07507	00 18 24.	- 21 44		17.1	FAINT BLUE STAR
LB 07508	00 18 24.	- 22 05		18.8	FAINT BLUE STAR
LB 07509	00 18 24.	- 26 18		18.8	FAINT BLUE STAR
LB 07510	00 18 24.	- 26 39		17.4	FAINT BLUE STAR
SCHO 0027	00 18 25.	+ 61 07 36.	240		ISOLATED DARK CLOUD
MCG+00-02-010	00 18 27.	- 02 05	36	15.5	GALAXY
SCHO 0028	00 18 28.	+ 58 28 36.	260		ISOLATED DARK CLOUD
ZWG 409.004	00 18 30.	+ 06 10		15.7	GALAXY
MRK 548	00 18 30.	+ 06 10	18	16.	GALAXY WITH UV CONTINUUM
UGC 00201	00 18 30.	+ 07 20	72	16.5	GALAXY Sc
ZWG 479.004	00 18 30.	+ 26 14		15.5	GALAXY
SN 1954F	00 18 30.	+ 26 14		18.8	SUPERNOVA
KARA.73B 0015	00 18 30.	+ 26 14	36	15.5	ISOLATED GALAXY S
ZWG 479.005	00 18 30.	+ 26 57		14.5	GALAXY
UGC 00202	00 18 30.	+ 26 57	84	14.5	GALAXY SB0
PHL 0799	00 18 30.	- 19 19		17.8	BLUE STELLAR OBJECT
LB 07511	00 18 30.	- 22 00		18.0	FAINT BLUE STAR
LB 07512	00 18 30.	- 24 11		18.8	FAINT BLUE STAR
LB 07513	00 18 30.	- 25 08		16.1	FAINT BLUE STAR
LB 02721	00 18 32.	+ 13 38 54.		16.3	FAINT BLUE STAR
SC 0016-6851.6	00 18 35.	- 68 34 56.	18		NEBULA
PHL 2873	00 18 36.	+ 16 34		18.3	BLUE STELLAR OBJECT
ZWG 479.006	00 18 36.	+ 22 05		13.7	GALAXY
RNGC 0080	00 18 36.	+ 22 05		13.5	GALAXY
UGC 00203	00 18 36.	+ 22 05	138	13.7	GALAXY S0
MCG+04-02-004	00 18 36.	+ 22 05	27	13.5	GALAXY
RNGC 0081	00 18 36.	+ 22 06			GALAXY
ZWG 500.001	00 18 36.	+ 29 22		15.3	GALAXY
ZWG 499.112	00 18 36.	+ 29 22		15.3	GALAXY
ZWG 500.002	00 18 36.	+ 30 11		15.2	GALAXY
ZWG 499.113	00 18 36.	+ 30 11		15.2	GALAXY
KARA.72 007A	00 18 36.	+ 30 11	36	15.2	PART OF DOUBLE GALAXY
MCG+05-02-001	00 18 36.	+ 30 12 30.	30	15.5	GALAXY
4ZW 014	00 18 36.	+ 34 05			COMPACT GALAXY
LB 07514	00 18 36.	- 21 14		19.5	FAINT BLUE STAR
LB 07515	00 18 36.	- 21 19		19.1	FAINT BLUE STAR
LB 07516	00 18 36.	- 21 43		17.9	FAINT BLUE STAR
LB 07517	00 18 36.	- 24 40		17.3	FAINT BLUE STAR
MCG+05-02-002	00 18 39.	+ 30 14	48	15.	GALAXY
ZC 0018.7+0256	00 18 42.	+ 02 56	540		CLUSTER OF GALAXIES
MCG+04-02-005	00 18 42.	+ 22 10	24	13.5	GALAXY
IC 1544	00 18 42.	+ 22 45 36.			NONSTELLAR OBJECT
ZWG 479.007	00 18 42.	+ 22 50		14.6	GALAXY
UGC 00204	00 18 42.	+ 22 50	84	14.6	GALAXY Sc
MCG+04-02-006	00 18 42.	+ 22 50	54	14.5	GALAXY
MCG+05-02-003	00 18 42.	+ 29 21	30	15.5	GALAXY
ZWG 500.003	00 18 42.	+ 29 35		15.6	GALAXY
ZWG 499.114	00 18 42.	+ 29 35		15.6	GALAXY
ZWG 500.004	00 18 42.	+ 30 14		15.3	GALAXY
ZWG 499.115	00 18 42.	+ 30 14		15.3	GALAXY
KARA.72 007B	00 18 42.	+ 30 14	42	15.3	PART OF DOUBLE GALAXY
ZWG 519.002	00 18 42.	+ 36 55		15.7	GALAXY
ZWG 518.024	00 18 42.	+ 36 55		15.7	GALAXY
ASS 35	00 18 42.	+ 62 17			OB ASSOCIATION
8ZW 0018-09.3	00 18 42.	- 09 15		16.8	COMPACT GALAXY
LB 07518	00 18 42.	- 21 06		19.0	FAINT BLUE STAR
LB 07519	00 18 42.	- 23 18		16.6	FAINT BLUE STAR
LB 07520	00 18 42.	- 24 22		19.7	FAINT BLUE STAR
LB 07521	00 18 42.	- 25 49		17.9	FAINT BLUE STAR
LB 01550	00 18 42.	- 52 18		14.2	FAINT BLUE STAR
SCHO 0029	00 18 44.	+ 61 13 00.	240		ISOLATED DARK CLOUD
IC 1545	00 18 45.	+ 21 42 24.			NONSTELLAR OBJECT
MCG-02-02-004	00 18 45.	- 14 32	15	15.5	GALAXY
RNGC 0087	00 18 46.	- 48 54			GALAXY
LB 00436	00 18 47.	+ 14 00 36.		16.2	FAINT BLUE STAR
ZWG 457.009	00 18 48.	+ 17 03		15.6	GALAXY
UGC 00205	00 18 48.	+ 17 03	84	15.6	GALAXY SB
ZWG 479.008	00 18 48.	+ 22 09		14.3	GALAXY
RNGC 0083	00 18 48.	+ 22 09		14.5	GALAXY
UGC 00206	00 18 48.	+ 22 09	78	14.3	GALAXY E
RNGC 0082	00 18 48.	+ 22 11			NON-EXISTENT OBJECT
ZWG 479.009	00 18 48.	+ 22 14		15.7	GALAXY
RNGC 0085B	00 18 48.	+ 22 14		15.5	GALAXY
RNGC 0085A	00 18 48.	+ 22 14		15.5	GALAXY
MCG+04-02-007	00 18 48.	+ 22 14	27	15.5	GALAXY
ZWG 500.005	00 18 48.	+ 29 21		15.1	GALAXY
MCG+05-02-004	00 18 48.	+ 29 24	54	15.5	GALAXY
MCG+06-01-027	00 18 48.	+ 36 52	24	16.	GALAXY
MCG+06-01-028	00 18 48.	+ 37 48	24	15.	GALAXY
ZC 0018.8+4127	00 18 48.	+ 41 27	1280		CLUSTER OF GALAXIES
LB 07522	00 18 48.	- 21 00		18.3	FAINT BLUE STAR
LB 07523	00 18 48.	- 22 55		17.5	FAINT BLUE STAR
LB 07524	00 18 48.	- 23 16		19.9	FAINT BLUE STAR
LB 07525	00 18 48.	- 23 33		17.7	FAINT BLUE STAR
LB 07526	00 18 48.	- 24 54		18.2	FAINT BLUE STAR
LB 07527	00 18 48.	- 25 04		19.6	FAINT BLUE STAR
MCG+03-02-006	00 18 51.	+ 17 04	42	15.	GALAXY
RNGC 0088	00 18 52.	- 48 55			GALAXY
RNGC 0089	00 18 52.	- 48 57			GALAXY
ZWG 383.002	00 18 54.	+ 00 53		15.6	GALAXY
KARA.73B 0016	00 18 54.	+ 00 53	30	15.6	ISOLATED GALAXY S
IC 1546	00 18 54.	+ 22 13 41.			NONSTELLAR OBJECT
ZWG 479.010	00 18 54.	+ 22 14		15.6	GALAXY
MCG+04-02-008	00 18 54.	+ 22 14	48	15.5	GALAXY
RNGC 0086	00 18 54.	+ 22 17		15.0	GALAXY
MCG+04-02-009	00 18 54.	+ 22 17 30.	33	16.	GALAXY
ZWG 479.011	00 18 54.	+ 22 18		14.9	GALAXY
RNGC 0084	00 18 54.	+ 22 20		17.0	GALAXY
MCG+04-02-010	00 18 54.	+ 22 20	45	17.	GALAXY
4ZW 015	00 18 54.	+ 27 47			COMPACT GALAXY
ZC 0018.9+2818	00 18 54.	+ 28 18	4700		CLUSTER OF GALAXIES
ZWG 500.006	00 18 54.	+ 29 23		15.6	GALAXY
UGC 00207	00 18 54.	+ 29 23	60	15.6	GALAXY Sb
MCG+06-01-029	00 18 54.	+ 37 48	24	15.	GALAXY
ZWG 519.003	00 18 54.	+ 37 49		15.3	GALAXY
ZWG 518.025	00 18 54.	+ 37 49		15.3	GALAXY
MCG+06-01-030	00 18 54.	+ 37 49 30.	42	14.5	GALAXY
LB 07528	00 18 54.	- 21 16		18.1	FAINT BLUE STAR
LB 07529	00 18 54.	- 21 42		19.1	FAINT BLUE STAR
LB 07530	00 18 54.	- 22 44		16.8	FAINT BLUE STAR
LB 07531	00 18 54.	- 23 00		17.6	FAINT BLUE STAR
LB 07532	00 18 54.	- 25 06		19.0	FAINT BLUE STAR
LB 07533	00 18 54.	- 26 24		19.9	FAINT BLUE STAR
LB 07534	00 18 54.	- 26 36		19.2	FAINT BLUE STAR
MCG+00-02-011	00 18 57.	+ 03 19 30.	48	15.	GALAXY
LBN 0556	00 19	+ 19 40	2100		BRIGHT NEBULA
PHL 2874	00 19 00.	+ 16 36		17.5	BLUE STELLAR OBJECT
ZCG 0019+25.3	00 19 00.	+ 25 14		16.5	COMPACT GALAXY
ZWG 535.001	00 19 00.	+ 43 00		15.7	GALAXY
ZWG 534.011	00 19 00.	+ 43 00		15.7	GALAXY
LDN 1282	00 19 00.	+ 61 30	480		DARK NEBULA
LDN 1284	00 19 00.	+ 62 55	360		DARK NEBULA
ZC 0019.0-0112	00 19 00.	- 01 12	2290		CLUSTER OF GALAXIES
LB 07535	00 19 00.	- 20 56		17.0	FAINT BLUE STAR
LB 07536	00 19 00.	- 22 12		18.8	FAINT BLUE STAR
LB 07537	00 19 00.	- 25 54		18.4	FAINT BLUE STAR
IC 1547	00 19 01.	+ 22 14			MAY NOT EXIST
RNGC 0092	00 19 04.	- 48 54			GALAXY
ZWG 479.012	00 19 06.	+ 22 02		15.7	GALAXY
5ZW 014	00 19 06.	+ 43 33			COMPACT GALAXY

OBJECT NAME	RIGHT ASCEN.	DECLINATION	DIAM.	MAGN.	TYPE OF OBJECT
ZWG 549.037	00 19 06.	+ 51 17		15.7	GALAXY
MRSL 119-00/1	00 19 06.	+ 61 28	1800		HII REGION
LB 07538	00 19 06.	- 20 43		18.0	FAINT BLUE STAR
LB 07539	00 19 06.	- 22 26		18.8	FAINT BLUE STAR
TON-S 0153	00 19 06.	- 22 51		14.7	BLUE STAR
LB 07540	00 19 06.	- 24 46		18.3	FAINT BLUE STAR
SG 2.002	00 19 07.	+ 61 27	1800		DIFFUSE EMISSION NEBULA
SC 0016-6947.7	00 19 09.	- 69 31 02.	30		NEBULA
ARC 0023	00 19 10.	- 01 10		17.0	RICH CLUSTER OF GALAXIES
ZWG 479.013	00 19 12.	+ 22 08		14.5	GALAXY
RNGC 0091	00 19 12.	+ 22 08		14.5	GALAXY
UGC 00208	00 19 12.	+ 22 08	144	14.5	GALAXY S
MCG+04-02-011	00 19 12.	+ 22 08	132	14.5	GALAXY
ZCG 0019+25.2	00 19 12.	+ 25 13		18.3	COMPACT GALAXY
ZWG 519.004	00 19 12.	+ 39 18		15.7	GALAXY
ZWG 518.026	00 19 12.	+ 39 18		15.7	GALAXY
OCL 0290	00 19 12.	+ 64 07	450	13.	OPEN STAR CLUSTER
LB 07541	00 19 12.	- 23 24		18.4	FAINT BLUE STAR
LB 07542	00 19 12.	- 25 09		17.7	FAINT BLUE STAR
LB 07543	00 19 12.	- 25 28		20.1	FAINT BLUE STAR
LB 07544	00 19 12.	- 26 15		18.3	FAINT BLUE STAR
LB 07545	00 19 12.	- 26 34		13.3	FAINT BLUE STAR
MCG+07-01-008	00 19 15.	+ 41 31 30.	36	15.	GALAXY
ARP 065	00 19 16.	+ 22 06			PECULIAR GALAXY
PHL 2875	00 19 18.	+ 15 45		18.5	BLUE STELLAR OBJECT
ZWG 479.014	00 19 18.	+ 21 44		15.6	GALAXY
RNGC 0090	00 19 18.	+ 22 09			NON-EXISTENT OBJECT
MCG+06-02-001	00 19 18.	+ 39 18	30	15.	GALAXY
ZWG 535.002	00 19 18.	+ 41 32		15.6	GALAXY
ZWG 534.012	00 19 18.	+ 41 32		15.6	GALAXY
5ZW 015	00 19 18.	+ 41 33			COMPACT GALAXY
ZWG 549.038	00 19 18.	+ 48 51		14.8	GALAXY
MCG+08-01-041	00 19 18.	+ 48 51	72	14.	GALAXY
MRK 549	00 19 18.	- 01 53	10	15.5	GALAXY WITH UV CONTINUUM
LB 07546	00 19 18.	- 22 06		19.3	FAINT BLUE STAR
LB 07547	00 19 18.	- 22 52		19.4	FAINT BLUE STAR
LB 07548	00 19 18.	- 25 04		19.5	FAINT BLUE STAR
LB 07549	00 19 18.	- 25 33		18.9	FAINT BLUE STAR
LB 07550	00 19 18.	- 25 43		19.2	FAINT BLUE STAR
LB 07551	00 19 18.	- 25 48		18.8	FAINT BLUE STAR
IC 1548	00 19 19.	+ 21 43 47.			NONSTELLAR OBJECT
ZWG 479.015	00 19 24.	+ 22 08		14.7	GALAXY
UGC 00209	00 19 24.	+ 22 08	96	14.7	GALAXY S
RNGC 0093	00 19 24.	+ 22 09		14.5	GALAXY
MCG+04-02-012	00 19 24.	+ 22 09	78	14.5	GALAXY
MCG+04-02-013	00 19 24.	+ 23 27	60	15.	GALAXY
ZWG 479.016	00 19 24.	+ 23 28		15.0	GALAXY
UGC 00210	00 19 24.	+ 23 28	78	15.0	GALAXY Sb
MCG+07-01-009	00 19 24.	+ 41 31	42	15.	GALAXY
ZWG 535.003	00 19 24.	+ 41 32		15.6	GALAXY
ZWG 534.013	00 19 24.	+ 41 32		15.6	GALAXY
ISS 0027	00 19 24.	+ 60 50	345		STELLAR RING
MCG-02-02-005	00 19 24.	- 09 46	30	15.	GALAXY
LB 07552	00 19 24.	- 21 07		18.9	FAINT BLUE STAR
LB 07553	00 19 24.	- 24 47		18.3	FAINT BLUE STAR
LB 07554	00 19 24.	- 25 59		17.3	FAINT BLUE STAR
LB 07555	00 19 24.	- 26 31		18.4	FAINT BLUE STAR
LB 00437	00 19 26.	+ 15 12 06.		18.0	FAINT BLUE STAR
MCG-03-02-001	00 19 27.	- 16 16	24	14.5	GALAXY
PHL 2876	00 19 30.	+ 18 15		18.9	BLUE STELLAR OBJECT
MCG-03-02-002	00 19 30.	- 19 15	36	15.	GALAXY
LB 07556	00 19 30.	- 21 37		20.0	FAINT BLUE STAR
LB 07557	00 19 30.	- 22 35		17.8	FAINT BLUE STAR
LB 07558	00 19 30.	- 22 41		17.8	FAINT BLUE STAR
LB 07559	00 19 30.	- 25 19		18.8	FAINT BLUE STAR
SCHO 0030	00 19 32.	+ 60 30 18.	260		ISOLATED DARK CLOUD
MCG+00-02-012	00 19 33.	- 01 43	54	15.5	GALAXY
ZWG 409.005	00 19 36.	+ 06 59		15.6	GALAXY
MCG+02-02-003	00 19 36.	+ 10 13	108	13.	GALAXY
PHL 0801	00 19 36.	+ 13 28		18.3	BLUE STELLAR OBJECT
PHL 0800	00 19 36.	+ 15 12		18.5	BLUE STELLAR OBJECT
UGC 00211	00 19 36.	+ 20 20	66	17.	GALAXY
ZWG 479.017	00 19 36.	+ 22 12		15.6	GALAXY
RNGC 0094	00 19 36.	+ 22 12		15.5	GALAXY
ZC 0019.6+2258	00 19 36.	+ 22 58	1480		CLUSTER OF GALAXIES
ZCG 0019+25.1	00 19 36.	+ 25 11		17.7	COMPACT GALAXY
MCG+05-02-005	00 19 36.	+ 30 00	30	15.	GALAXY
LB 07560	00 19 36.	- 20 56		17.5	FAINT BLUE STAR
LB 07561	00 19 36.	- 21 09		18.7	FAINT BLUE STAR
LB 07562	00 19 36.	- 23 07		17.4	FAINT BLUE STAR
LB 07563	00 19 36.	- 23 23		18.0	FAINT BLUE STAR
LB 07564	00 19 36.	- 23 39		19.5	FAINT BLUE STAR
LB 07565	00 19 36.	- 24 37		18.8	FAINT BLUE STAR
TON-S 0154	00 19 36.	- 24 43		13.9	BLUE STAR
RNGC 0095	00 19 37.	+ 10 13		13.5	GALAXY
MCG-01-02-006	00 19 39.	- 04 37	60	14.	GALAXY
ARP 035	00 19 41.	- 01 41			PECULIAR GALAXY
RNGC 0096	00 19 42.	+ 22 17		17.0	GALAXY
MCG+04-02-014	00 19 42.	+ 22 17	30	17.	GALAXY
ZWG 500.007	00 19 42.	+ 29 58		15.2	GALAXY
ZWG 383.003	00 19 42.	- 01 12		15.7	GALAXY
LB 07566	00 19 42.	- 23 24		19.8	FAINT BLUE STAR
MCG+05-02-006	00 19 45.	+ 29 14	66	14.5	GALAXY
VV 257B	00 19 45.	- 01 34 30.	15	16.	INTERACTING GALAXY
VV 257B	00 19 45.	- 01 34 30.	18	16.	INTERACTING GALAXY
VV 257	00 19 45.	- 01 34 30.	108	15.	INTERACTING GALAXY
LB 00438	00 19 47.	+ 15 53 12.		16.1	FAINT BLUE STAR
RNGC 0097	00 19 47.	+ 29 28		13.5	GALAXY
ZWG 409.006	00 19 48.	+ 06 10		15.7	GALAXY
UGC 00213	00 19 48.	+ 06 10	78	15.7	GALAXY (S0)
ZWG 434.003	00 19 46.	+ 10 13		13.4	GALAXY
UGC 00214	00 19 48.	+ 10 13	108	13.4	GALAXY Sc
3ZW 005	00 19 48.	+ 25 15			COMPACT GALAXY
ZWG 500.008	00 19 48.	+ 29 13		14.9	GALAXY
UGC 00215	00 19 48.	+ 29 13	84	14.9	GALAXY SBa
MCG+05-02-007	00 19 48.	+ 29 30	30	14.	GALAXY
MCG+00-02-013	00 19 48.	- 01 13 30.	36	14.5	GALAXY
ZWG 383.004	00 19 48.	- 01 34		15.0	GALAXY
UGC 00212	00 19 48.	- 01 34	102	15.0	GALAXY SB
MCG+00-02-015	00 19 48.	- 01 35	15	15.	GALAXY
MCG+00-02-014	00 19 48.	- 01 35	90	15.	GALAXY
MCG-03-02-003	00 19 48.	- 14 52 30.	48	15.	GALAXY
MCG-03-02-004	00 19 48.	- 20 05 30.	48	15.	GALAXY
LB 07567	00 19 48.	- 24 21		19.3	FAINT BLUE STAR
LB 07568	00 19 48.	- 25 10		17.9	FAINT BLUE STAR
LB 07569	00 19 48.	- 26 04		18.8	FAINT BLUE STAR
LB 07570	00 19 48.	- 26 36		19.7	FAINT BLUE STAR
SC 0017-6554.4	00 19 49.	- 65 37 44.	6		NEBULA

OBJECT NAME	RIGHT ASCEN.	DECLINATION	DIAM.	MAGN.	TYPE OF OBJECT
LB 00439	00 19 50.	+ 15 02 18.		16.6	FAINT BLUE STAR
MCG-02-02-006	00 19 51.	- 13 32	30	15.	GALAXY
ZWG 500.009	00 19 54.	+ 29 28		13.5	GALAXY
UGC 00216	00 19 54.	+ 29 28	90	13.5	GALAXY (E)
LB 07571	00 19 54.	- 21 21		18.4	FAINT BLUE STAR
LB 07572	00 19 54.	- 23 48		18.4	FAINT BLUE STAR
LB 07573	00 19 54.	- 24 17		18.9	FAINT BLUE STAR
LB 07574	00 19 54.	- 26 06		16.5	FAINT BLUE STAR
ARC 0024	00 19 56.	+ 23 02		17.5	RICH CLUSTER OF GALAXIES
LBN 0593	00 20	+ 61 30	1320		BRIGHT NEBULA
YC 0020-41	00 20	- 41 51			UNUSUAL SOUTHERN GALAXY
IC 0014	00 20 00.	+ 10 12			NONSTELLAR OBJECT
PHL 0802	00 20 00.	+ 15 00		17.2	BLUE STELLAR OBJECT
ZC 0020.0+2324	00 20 00.	+ 23 24	2080		CLUSTER OF GALAXIES
UGC 00217	00 20 00.	+ 27 50	72	17.	GALAXY DWARF
MCG+05-02-008	00 20 00.	+ 29 40	36	15.5	GALAXY
LDN 1283	00 20 00.	+ 61 26	360		DARK NEBULA
MCG-03-02-005	00 20 00.	- 17 21	60	15.	GALAXY
MCG-03-02-006	00 20 00.	- 19 57	48	15.5	GALAXY
LB 07575	00 20 00.	- 20 58		17.1	FAINT BLUE STAR
LB 07576	00 20 00.	- 23 50		18.2	FAINT BLUE STAR
LB 07577	00 20 00.	- 24 31		19.0	FAINT BLUE STAR
LB 07578	00 20 00.	- 24 50		18.8	FAINT BLUE STAR
MCG-03-02-007	00 20 03.	- 19 57 30.	18	15.5	GALAXY
MCG-04-02-007	00 20 03.	- 22 34	36	15.5	GALAXY
4ZW 016	00 20 06.	+ 33 30			COMPACT GALAXY
ZC 0020.1-0023	00 20 06.	- 00 23	2020		CLUSTER OF GALAXIES
LB 07579	00 20 06.	- 22 10		17.1	FAINT BLUE STAR
LB 07580	00 20 06.	- 22 19		18.5	FAINT BLUE STAR
LB 07581	00 20 06.	- 23 12		18.7	FAINT BLUE STAR
LB 07582	00 20 06.	- 25 25		19.0	FAINT BLUE STAR
ZWG 479.018	00 20 12.	+ 23 53		15.4	GALAXY
ZWG 500.010	00 20 12.	+ 29 40		15.5	GALAXY
MCG-01-02-007	00 20 12.	- 06 58 30.	48	16.	GALAXY
MCG-02-02-007	00 20 12.	- 08 44	90	15.	GALAXY
LB 07583	00 20 12.	- 20 40		17.4	FAINT BLUE STAR
LB 07584	00 20 12.	- 20 43		20.4	FAINT BLUE STAR
LB 07585	00 20 12.	- 20 51		17.8	FAINT BLUE STAR
LB 07586	00 20 12.	- 23 20		19.6	FAINT BLUE STAR
LB 07587	00 20 12.	- 24 15		19.1	FAINT BLUE STAR
LB 07588	00 20 12.	- 25 33		18.6	FAINT BLUE STAR
LB 07589	00 20 12.	- 25 52		16.6	FAINT BLUE STAR
LB 07590	00 20 12.	- 26 20		16.2	FAINT BLUE STAR
MCG+01-02-004	00 20 15.	+ 06 31	60	15.	GALAXY
MCG-04-02-009	00 20 15.	- 24 24 30.	60	15.	GALAXY
MCG-04-02-008	00 20 15.	- 24 24 30.	12	16.	GALAXY
ZWG 409.007	00 20 18.	+ 06 33		15.3	GALAXY
MCG+01-02-005	00 20 18.	+ 06 40	24	14.	GALAXY
ZWG 409.008	00 20 18.	+ 06 42		14.9	GALAXY
UGC 00218	00 20 18.	+ 06 42	96	14.9	GALAXY S
ZWG 479.019	00 20 18.	+ 24 57		15.6	GALAXY
3ZW 006	00 20 18.	+ 25 13			COMPACT GALAXY
ZWG 519.005	00 20 18.	+ 37 56		15.6	GALAXY
5ZW 016	00 20 18.	+ 49 15			COMPACT GALAXY
LB 07591	00 20 18.	- 22 27		15.7	FAINT BLUE STAR
LB 07592	00 20 18.	- 22 41		19.1	FAINT BLUE STAR
LB 07593	00 20 18.	- 23 04		16.8	FAINT BLUE STAR
LB 07594	00 20 18.	- 24 30		19.6	FAINT BLUE STAR
LB 07595	00 20 18.	- 25 35		18.5	FAINT BLUE STAR
RNGC 0098	00 20 22.	- 45 33			UNVERIFIED SOUTHERN OBJECT
ZWG 457.010	00 20 24.	+ 15 58		15.7	GALAXY
UGC 00219	00 20 24.	+ 15 58	72	15.7	GALAXY
MCG+03-02-007	00 20 24.	+ 15 58	42	16.	GALAXY
PHL 2877	00 20 24.	+ 16 10		18.3	BLUE STELLAR OBJECT
ZC 0020.4+3631	00 20 24.	+ 36 31	1810		CLUSTER OF GALAXIES
8ZW 0020-09.5	00 20 24.	- 09 32		16.4	COMPACT GALAXY
MCG-03-02-009	00 20 24.	- 14 48	36	14.5	GALAXY
MCG-03-02-008	00 20 24.	- 19 20	36	15.	GALAXY
LB 07596	00 20 24.	- 22 06		17.0	FAINT BLUE STAR
LB 07597	00 20 24.	- 23 34		19.6	FAINT BLUE STAR
LB 07598	00 20 24.	- 25 02		19.0	FAINT BLUE STAR
LB 07599	00 20 24.	- 25 41		17.9	FAINT BLUE STAR
LB 07600	00 20 24.	- 26 06		17.7	FAINT BLUE STAR
LB 07601	00 20 24.	- 26 40		18.7	FAINT BLUE STAR
LIN.CL 004	00 20 26.	- 74 01 39.	816	13.9	STAR CLUSTER IN SMC
MCG-03-02-010	00 20 27.	- 16 16	60	15.	GALAXY
ARC 0025	00 20 28.	- 00 26		17.8	RICH CLUSTER OF GALAXIES
UGC 00220	00 20 30.	+ 02 10	102	16.0	GALAXY S0
MCG+02-02-004	00 20 30.	+ 13 25	36	15.5	GALAXY
4ZW 017	00 20 30.	+ 23 55			COMPACT GALAXY
ZC 0020.5+4429	00 20 30.	+ 44 29	870		CLUSTER OF GALAXIES
ZWG 549.039	00 20 30.	+ 46 26		15.7	GALAXY
LB 07602	00 20 30.	- 21 30		18.0	FAINT BLUE STAR
LB 07603	00 20 30.	- 22 01		17.9	FAINT BLUE STAR
LB 07604	00 20 30.	- 22 53		18.7	FAINT BLUE STAR
LB 07605	00 20 30.	- 24 28		18.8	FAINT BLUE STAR
LB 07606	00 20 30.	- 25 20		17.1	FAINT BLUE STAR
LB 07607	00 20 30.	- 25 45		16.5	FAINT BLUE STAR
MCG-04-02-010	00 20 33.	- 22 23 30.	48	15.	GALAXY
PHL 6522	00 20 36.	+ 13 24		18.3	BLUE STELLAR OBJECT
PHL 0803	00 20 36.	+ 16 04		18.8	BLUE STELLAR OBJECT
ZWG 479.020	00 20 36.	+ 27 10		14.5	GALAXY
UGC 00221	00 20 36.	+ 27 10	42	14.5	GALAXY PECULR
ZWG 500.011	00 20 36.	+ 29 33		15.4	GALAXY
LB 07608	00 20 36.	- 21 11		17.8	FAINT BLUE STAR
LB 07609	00 20 36.	- 21 23		17.7	FAINT BLUE STAR
LB 07610	00 20 36.	- 21 54		18.5	FAINT BLUE STAR
LB 07611	00 20 36.	- 22 17		17.1	FAINT BLUE STAR
LB 07612	00 20 36.	- 23 37		17.4	FAINT BLUE STAR
LB 07613	00 20 36.	- 24 54		17.0	FAINT BLUE STAR
LB 07614	00 20 36.	- 26 25		19.5	FAINT BLUE STAR
LIN.CL 005	00 20 36.	- 75 20 45.	270	15.7	STAR CLUSTER IN SMC
IC 1549	00 20 38.	+ 06 42 04.			NONSTELLAR OBJECT
SCHO 0031	00 20 40.	+ 62 20 12.	190		ISOLATED DARK CLOUD
ARC 0026	00 20 41.	+ 36 35		17.4	RICH CLUSTER OF GALAXIES
ZWG 383.005	00 20 42.	+ 01 33		15.7	GALAXY
KARA.73B 0017	00 20 42.	+ 01 33	36	15.7	ISOLATED GALAXY S
ZWG 409.009	00 20 42.	+ 06 17		15.7	GALAXY
ZWG 434.004	00 20 42.	+ 13 24		15.7	GALAXY
ZWG 479.021	00 20 42.	+ 24 40		15.5	GALAXY
UGC 00222	00 20 42.	+ 34 52	72	18.	GALAXY
LB 07615	00 20 42.	- 23 56		17.0	FAINT BLUE STAR
LB 07616	00 20 42.	- 25 03		17.7	FAINT BLUE STAR
LB 07617	00 20 42.	- 25 11		17.7	FAINT BLUE STAR
LB 07618	00 20 42.	- 26 19		17.8	FAINT BLUE STAR
LB 03137	00 20 42.	- 61 18		13.2	FAINT BLUE STAR
LIN.CL 006	00 20 44.	- 73 56 45.	408	15.3	STAR CLUSTER IN SMC
MCG+00-02-016	00 20 45.	+ 01 34 30.	48	15.	GALAXY

OBJECT NAME	RIGHT ASCEN.	DECLINATION	DIAM.	MAGN.	TYPE OF OBJECT
SNO 07	00 20 46.	- 23 41 14.	300	19.	GROUP OF 6 GALAXIES
LB 07619	00 20 48.	- 22 56		18.7	FAINT BLUE STAR
LB 07620	00 20 48.	- 25 02		18.0	FAINT BLUE STAR
LB 07621	00 20 48.	- 25 56		18.7	FAINT BLUE STAR
MCG-02-02-008	00 20 51.	- 08 45	48	16.	GALAXY
ZC 0020.9+0121	00 20 54.	+ 01 21	940		CLUSTER OF GALAXIES
UGC 00223	00 20 54.	+ 19 58	96	16.5	GALAXY (DWARF)
LB 07622	00 20 54.	- 21 29		18.6	FAINT BLUE STAR
LB 07623	00 20 54.	- 21 39		19.5	FAINT BLUE STAR
LB 07624	00 20 54.	- 22 27		18.0	FAINT BLUE STAR
LB 07625	00 20 54.	- 23 20		17.7	FAINT BLUE STAR
LB 07626	00 20 54.	- 25 48		18.8	FAINT BLUE STAR
MCG+02-02-005	00 21 00.	+ 14 24	60	15.	GALAXY
PHL 0804	00 21 00.	+ 15 30		18.1	BLUE STELLAR OBJECT
UGC 00225	00 21 00.	+ 20 34	72	16.0	GALAXY S0
ZWG 500.012	00 21 00.	+ 30 00		15.6	GALAXY
MCG+00-02-017	00 21 00.	- 00 23 30.	24	15.5	GALAXY
MCG+00-02-018	00 21 00.	- 00 46	24	16.	GALAXY
ZWG 383.006	00 21 00.	- 00 47		15.4	GALAXY
UGC 00224	00 21 00.	- 00 47	78	15.4	GALAXY DISRPTD
MCG+00-02-019	00 21 00.	- 00 47	24	16.	GALAXY
ZC 0021.0-0137	00 21 00.	- 01 37	4910		CLUSTER OF GALAXIES
LB 07627	00 21 00.	- 21 08		19.6	FAINT BLUE STAR
LB 07628	00 21 00.	- 21 16		19.3	FAINT BLUE STAR
LB 07629	00 21 00.	- 22 15		19.3	FAINT BLUE STAR
LB 07630	00 21 00.	- 25 32		18.7	FAINT BLUE STAR
LB 07631	00 21 00.	- 26 40		19.8	FAINT BLUE STAR
VV 038B	00 21 03.	- 00 46	15	16.5	INTERACTING GALAXY
VV 038A	00 21 03.	- 00 46 30.	18	16.	INTERACTING GALAXY
SC 0018-6658.9	00 21 03.	- 66 42 15.	12		NEBULA
LB 00440	00 21 05.	+ 17 12 06.		16.1	FAINT BLUE STAR
ARP 201	00 21 05.	- 00 47			PECULIAR GALAXY
PHL 2878	00 21 06.	+ 12 32		18.2	BLUE STELLAR OBJECT
MCG+03-02-008	00 21 06.	+ 20 36	24	16.	GALAXY
ZWG 500.013	00 21 06.	+ 30 17		15.1	GALAXY
LB 07632	00 21 06.	- 23 12		17.1	FAINT BLUE STAR
LB 07633	00 21 06.	- 23 39		18.7	FAINT BLUE STAR
LB 07634	00 21 06.	- 23 50		19.2	FAINT BLUE STAR
LB 07635	00 21 06.	- 26 21		18.9	FAINT BLUE STAR
LB 01551	00 21 06.	- 45 32		13.2	FAINT BLUE STAR
ZWG 434.005	00 21 12.	+ 14 24		15.1	GALAXY
MRK 338	00 21 12.	+ 14 24	26	15.5	GALAXY WITH UV CONTINUUM
UGC 00226	00 21 12.	+ 14 24	72	15.1	GALAXY
RNGC 0099	00 21 12.	+ 15 29		14.0	GALAXY
MCG+02-02-006	00 21 12.	+ 15 29	90	14.	GALAXY
ZC 0021.2+2504	00 21 12.	+ 25 04	2960		CLUSTER OF GALAXIES
ZWG 479.022	00 21 12.	+ 26 40		14.7	GALAXY
UGC 00227	00 21 12.	+ 26 40	90	14.7	GALAXY COMPACT
ZWG 519.006	00 21 12.	+ 39 16		15.7	GALAXY
LB 07636	00 21 12.	- 20 55		16.8	FAINT BLUE STAR
LB 07637	00 21 12.	- 21 57		18.8	FAINT BLUE STAR
LB 07638	00 21 12.	- 22 28		18.7	FAINT BLUE STAR
LB 07639	00 21 12.	- 26 32		18.2	FAINT BLUE STAR
ZWG 479.023	00 21 18.	+ 24 02		14.8	GALAXY
UGC 00228	00 21 18.	+ 24 02	78	14.8	GALAXY Sb-c
MCG+04-02-015	00 21 18.	+ 24 02	54	14.5	GALAXY
ZWG 500.014	00 21 18.	+ 28 03		15.3	GALAXY
UGC 00229	00 21 18.	+ 28 03	60	15.3	GALAXY Sb
KARA.73B 0018	00 21 18.	+ 28 03	54	15.3	ISOLATED GALAXY S
SCHO 0032	00 21 18.	+ 62 20 54.	180		ISOLATED DARK CLOUD
ZWG 383.007	00 21 18.	- 01 00		15.7	GALAXY
MCG-01-02-008	00 21 18.	- 03 20	24	15.5	GALAXY
LB 07640	00 21 18.	- 22 45		18.7	FAINT BLUE STAR
HOLM 009A	00 21 23.	+ 16 13	150	13.8	PART OF MULTIPLE GALAXY
PHL 2880	00 21 24.	+ 13 25		18.3	BLUE STELLAR OBJECT
ZWG 457.011	00 21 24.	+ 15 30		14.0	GALAXY
ZWG 434.006	00 21 24.	+ 15 30		14.0	GALAXY
UGC 00230	00 21 24.	+ 15 30	90	14.0	GALAXY Sc
RNGC 0100	00 21 24.	+ 16 13		14.5	GALAXY
PHL 2879	00 21 24.	+ 17 56		17.4	BLUE STELLAR OBJECT
ZWG 479.024	00 21 24.	+ 25 23		15.2	GALAXY
MCG+04-02-016	00 21 24.	+ 25 23	30	16.	GALAXY
LB 07641	00 21 24.	- 20 56		19.4	FAINT BLUE STAR
LB 07642	00 21 24.	- 21 30		18.0	FAINT BLUE STAR
LB 07643	00 21 24.	- 24 19		19.3	FAINT BLUE STAR
LB 07644	00 21 24.	- 26 19		18.7	FAINT BLUE STAR
RNGC 0101	00 21 27.	- 32 50		13.0	GALAXY
ZWG 457.012	00 21 30.	+ 16 12		14.6	GALAXY
UGC 00231	00 21 30.	+ 16 12	348	14.6	GALAXY Sc
MCG+03-02-009	00 21 30.	+ 16 12	420	13.	GALAXY
ZC 0021.5+3247	00 21 30.	+ 32 47	1140		CLUSTER OF GALAXIES
LB 07645	00 21 30.	- 21 14		19.3	FAINT BLUE STAR
LB 07646	00 21 30.	- 22 08		18.2	FAINT BLUE STAR
LB 07647	00 21 30.	- 22 23		17.8	FAINT BLUE STAR
LB 07648	00 21 30.	- 22 26		18.1	FAINT BLUE STAR
LB 07649	00 21 30.	- 22 33		18.2	FAINT BLUE STAR
LB 07650	00 21 30.	- 22 59		20.1	FAINT BLUE STAR
TON-S 0155	00 21 30.	- 23 28		14.9	BLUE STAR
TON-S 0156	00 21 30.	- 28 09		15.2	BLUE STAR
MCG-05-02-003	00 21 30.	- 32 50	48	13.	GALAXY
HOLM 009B	00 21 32.	+ 16 16	42	14.9	PART OF MULTIPLE GALAXY
SC 0019-6654.9	00 21 32.	- 66 38 15.	12		NEBULA
ZWG 434.007	00 21 36.	+ 13 58		15.4	GALAXY
KARA.73B 0019	00 21 36.	+ 13 58	54	15.4	ISOLATED GALAXY S
PHL 2881	00 21 36.	+ 14 24		16.5	BLUE STELLAR OBJECT
PHL 6523	00 21 36.	+ 15 18		18.2	BLUE STELLAR OBJECT
ZWG 479.025	00 21 36.	+ 23 38		15.7	GALAXY
ZWG 519.007	00 21 36.	+ 36 18		15.7	GALAXY
4ZW 018	00 21 36.	+ 36 19			COMPACT GALAXY
MCG-02-02-010	00 21 36.	- 14 30	18	14.5	GALAXY
LB 07651	00 21 36.	- 24 24		18.0	FAINT BLUE STAR
LB 07652	00 21 36.	- 24 30		17.4	FAINT BLUE STAR
MCG+02-02-007	00 21 42.	+ 09 49	24	15.5	GALAXY
ZC 0021.7+4444	00 21 42.	+ 44 44	1480		CLUSTER OF GALAXIES
ZWG 383.008	00 21 42.	- 01 55		15.6	GALAXY
LB 07653	00 21 42.	- 20 42		19.8	FAINT BLUE STAR
LB 07654	00 21 42.	- 23 55		18.3	FAINT BLUE STAR
LB 07655	00 21 42.	- 25 34		19.5	FAINT BLUE STAR
LIN 002	00 21 43.	- 73 54 40.		16.94	PLANETARY NEBULA IN SMC
MCG+00-02-020	00 21 45.	- 01 55	36	15.	GALAXY
ZWG 519.008	00 21 48.	+ 37 55		15.1	GALAXY
MCG+06-02-002	00 21 48.	+ 37 55	33	15.	GALAXY
ZC 0021.8+4309	00 21 48.	+ 43 09	1550		CLUSTER OF GALAXIES
MCG-02-02-011	00 21 48.	- 14 13	36	14.	GALAXY
LB 07656	00 21 48.	- 22 36		19.7	FAINT BLUE STAR
LB 07657	00 21 48.	- 23 38		17.6	FAINT BLUE STAR
LB 07658	00 21 48.	- 23 40		18.2	FAINT BLUE STAR
IC 1550	00 21 49.	+ 37 54 52.			NONSTELLAR OBJECT

OBJECT NAME	RIGHT ASCEN.	DECLINATION	DIAM.	MAGN.	TYPE OF OBJECT
RNGC 0102	00 21 50.	- 14 13		14.0	GALAXY
MCG-01-02-009	00 21 51.	- 04 08	27	15.	GALAXY
RNGC 0104	00 21 53.	- 72 21		4.5	GLOBULAR CLUSTER
ZWG 383.009	00 21 54.	+ 03 03		15.7	GALAXY
ZC 0021.9+1214	00 21 54.	+ 12 14	470		CLUSTER OF GALAXIES
ZC 0021.9+3648	00 21 54.	+ 36 48	1750		CLUSTER OF GALAXIES
MCG+06-02-003	00 21 54.	+ 37 52	42	16.5	GALAXY
LB 07659	00 21 54.	- 21 42		19.7	FAINT BLUE STAR
LB 07660	00 21 54.	- 22 01		18.6	FAINT BLUE STAR
LB 07661	00 21 54.	- 23 31		19.0	FAINT BLUE STAR
LB 07662	00 21 54.	- 25 15		18.2	FAINT BLUE STAR
LB 07663	00 21 54.	- 25 45		17.2	FAINT BLUE STAR
LB 07664	00 21 54.	- 26 10		16.7	FAINT BLUE STAR
LB 03138	00 21 54.	- 61 58		13.0	FAINT BLUE STAR
GCL 001	00 21 54.	- 72 21	3360	4.68	GLOBULAR STAR CLUSTER
LH115-N001	00 21 54.	- 73 54			EMISSION NEBULA IN SMC
ZWG 383.010	00 22 00.	+ 03 01		15.4	GALAXY
MCG+00-02-021	00 22 00.	+ 03 03 30.	36	15.5	GALAXY
ZWG 479.026	00 22 00.	+ 24 44		15.7	GALAXY
ZWG 500.015	00 22 00.	+ 29 00		14.8	GALAXY
SN 1968O	00 22 00.	+ 29 00		15.8	SUPERNOVA
ZWG 500.016	00 22 00.	+ 32 59		14.7	GALAXY
UGC 00232	00 22 00.	+ 32 59	90	14.7	GALAXY SBa
MCG+05-02-009	00 22 00.	+ 33 00	48	14.	GALAXY
LDN 1285	00 22 00.	+ 62 55	420		DARK NEBULA
LB 07665	00 22 00.	- 21 52		19.4	FAINT BLUE STAR
LB 07666	00 22 00.	- 23 07		19.7	FAINT BLUE STAR
LB 07667	00 22 00.	- 24 59		18.3	FAINT BLUE STAR
LB 07668	00 22 00.	- 26 28		18.4	FAINT BLUE STAR
MCG-02-02-012	00 22 03.	- 11 46	90	16.	GALAXY
SCHO 0033	00 22 05.	+ 62 21 42.	190		ISOLATED DARK CLOUD
MCG+00-02-022	00 22 06.	+ 03 01	14	15.5	GALAXY
ZC 0022.1+0810	00 22 06.	+ 08 10	470		CLUSTER OF GALAXIES
ZWG 434.008	00 22 06.	+ 14 33		14.6	GALAXY
MRK 339	00 22 06.	+ 14 33	20	15.	GALAXY WITH UV CONTINUUM
UGC 00233	00 22 06.	+ 14 33	60	14.6	GALAXY PECULR
PHL 6524	00 22 06.	+ 17 52		18.9	BLUE STELLAR OBJECT
OCL 0292	00 22 06.	+ 62 22	60	13.	OPEN STAR CLUSTER
MCG-01-02-010	00 22 06.	- 06 19	84	14.	GALAXY
LB 07669	00 22 06.	- 20 43		18.2	FAINT BLUE STAR
LB 07670	00 22 06.	- 22 34		18.8	FAINT BLUE STAR
LB 07671	00 22 06.	- 22 50		16.7	FAINT BLUE STAR
LB 07672	00 22 06.	- 24 47		15.8	FAINT BLUE STAR
LB 07673	00 22 06.	- 26 35		17.0	FAINT BLUE STAR
3ZW 007	00 22 12.	+ 03 01			COMPACT GALAXY
ZWG 457.013	00 22 12.	+ 20 48		15.5	GALAXY
ZWG 500.017	00 22 12.	+ 29 17		15.6	GALAXY
UGC 00234	00 22 12.	+ 29 17	72	15.6	GALAXY Sc
UGC 00235	00 22 12.	+ 44 53	66	16.0	GALAXY E-S
LB 07674	00 22 12.	- 21 39		18.8	FAINT BLUE STAR
LB 07675	00 22 12.	- 25 02		16.9	FAINT BLUE STAR
LB 07676	00 22 12.	- 26 01		16.8	FAINT BLUE STAR
RNGC 0106	00 22 13.	- 05 26			GALAXY
ZWG 383.011	00 22 18.	+ 00 15		15.7	GALAXY
MCG+00-02-023	00 22 18.	+ 00 16	30	15.	GALAXY
MCG+01-02-006	00 22 18.	+ 06 22 30.	60	15.	GALAXY
PHL 2882	00 22 18.	+ 15 55		18.1	BLUE STELLAR OBJECT
ZC 0022.3+3424	00 22 18.	+ 34 24	870		CLUSTER OF GALAXIES
ZWG 535.004	00 22 18.	+ 43 23		15.0	GALAXY
UGC 00236	00 22 18.	+ 43 23	66	15.0	GALAXY E
KARA.73B 0020	00 22 18.	+ 43 23	36	15.0	ISOLATED GALAXY S
MCG+00-02-024	00 22 18.	- 01 41 30.	30	15.	GALAXY
MCG-03-02-011	00 22 18.	- 20 16	24	15.5	GALAXY
LB 07677	00 22 18.	- 20 58		16.5	FAINT BLUE STAR
ARC 0027	00 22 18.	- 20 59		16.5	RICH CLUSTER OF GALAXIES
LB 07679	00 22 18.	- 21 17		19.2	FAINT BLUE STAR
LB 07678	00 22 18.	- 21 17		18.5	FAINT BLUE STAR
LB 07680	00 22 18.	- 21 21		17.9	FAINT BLUE STAR
LB 07681	00 22 18.	- 22 18		15.8	FAINT BLUE STAR
LB 07682	00 22 18.	- 22 34		19.0	FAINT BLUE STAR
LB 07683	00 22 18.	- 22 40		18.2	FAINT BLUE STAR
LB 07684	00 22 18.	- 24 18		18.7	FAINT BLUE STAR
LB 07685	00 22 18.	- 26 06		19.4	FAINT BLUE STAR
LIN.CL 007	00 22 18.	- 74 02 04.	678	13.2	STAR CLUSTER IN SMC
SNO 08	00 22 19.	- 23 58 09.	24	19.	GROUP OF 5 GALAXIES
ZWG 409.010	00 22 24.	+ 06 23		15.4	GALAXY
UGC 00237	00 22 24.	+ 06 23	60	15.4	GALAXY SBa
ZWG 500.018	00 22 24.	+ 31 05		14.4	GALAXY
UGC 00238	00 22 24.	+ 31 05	120	14.4	GALAXY S
MCG+05-02-010	00 22 24.	+ 31 05	108	14.5	GALAXY
5ZW 017	00 22 24.	+ 39 22			COMPACT GALAXY
LB 07686	00 22 24.	- 20 59		18.4	FAINT BLUE STAR
MCG-04-02-011	00 22 24.	- 20 59 30.	6	16.	GALAXY
LB 07687	00 22 24.	- 22 58		20.0	FAINT BLUE STAR
LB 07688	00 22 24.	- 25 21		18.8	FAINT BLUE STAR
MCG-04-02-012	00 22 27.	- 20 59	6	16.	GALAXY
MIL 02	00 22 28.	+ 63 51 54.	420		SUPERNOVA REMNANT
MCG+01-02-007	00 22 30.	+ 06 11 30.	180	14.	GALAXY
PHL 0805	00 22 30.	+ 18 31		18.2	BLUE STELLAR OBJECT
UGC 00239	00 22 30.	+ 41 47	66	16.5	GALAXY IRR
OCL 0289	00 22 30.	+ 60 07	240	15.	OPEN STAR CLUSTER
OCL 0291	00 22 30.	+ 61 04	780	10.8	OPEN STAR CLUSTER
MCG-04-02-013	00 22 30.	- 20 59 30.	6	16.	GALAXY
LB 07689	00 22 30.	- 21 39		19.3	FAINT BLUE STAR
LB 07690	00 22 30.	- 23 01		17.6	FAINT BLUE STAR
LB 07691	00 22 30.	- 23 02		19.0	FAINT BLUE STAR
LB 07692	00 22 30.	- 24 58		18.9	FAINT BLUE STAR
LB 07693	00 22 30.	- 25 24		20.0	FAINT BLUE STAR
LB 07694	00 22 30.	- 25 36		19.0	FAINT BLUE STAR
LB 07695	00 22 30.	- 26 06		18.4	FAINT BLUE STAR
LB 01552	00 22 30.	- 46 09		14.6	FAINT BLUE STAR
RNGC 0103	00 22 31.	+ 61 04		11.0	OPEN CLUSTER
LIN.CL 008	00 22 31.	- 73 04 28.	1224	11.6	STAR CLUSTER IN SMC
VMT 02	00 22 33.	+ 63 51 48.	480		SUPERNOVA REMNANT
ARC 0028	00 22 35.	+ 07 52		17.6	RICH CLUSTER OF GALAXIES
ZC 0022.6+0749	00 22 36.	+ 07 49	740		CLUSTER OF GALAXIES
MCG+02-02-008	00 22 36.	+ 12 36	78	13.5	GALAXY
PHL 0806	00 22 36.	+ 18 17		18.0	BLUE STELLAR OBJECT
ZWG 500.019	00 22 36.	+ 29 46		15.5	GALAXY
MCG+05-02-011	00 22 36.	+ 29 47	48	15.	GALAXY
5ZW 018	00 22 36.	+ 39 40			COMPACT GALAXY
ZWG 383.012	00 22 36.	- 01 23		15.3	GALAXY
LB 07696	00 22 36.	- 22 44		17.8	FAINT BLUE STAR
LB 07697	00 22 36.	- 23 06		17.8	FAINT BLUE STAR
LB 07698	00 22 36.	- 23 30		19.5	FAINT BLUE STAR
LB 07699	00 22 36.	- 24 31		17.8	FAINT BLUE STAR
LB 07700	00 22 36.	- 24 42		19.5	FAINT BLUE STAR
LB 07701	00 22 36.	- 24 46		18.0	FAINT BLUE STAR

OBJECT NAME	RIGHT ASCEN.	DECLINATION	DIAM.	MAGN.	TYPE OF OBJECT
SN 1968N	00 22 38.	+ 29 46		19.0	SUPERNOVA
ZWG 409.011	00 22 42.	+ 06 13		14.9	GALAXY
UGC 00240	00 22 42.	+ 06 13	66	14.9	GALAXY Sb/SBb
MCG+01-02-008	00 22 42.	+ 06 25	24	15.	GALAXY
RNGC 0105	00 22 42.	+ 12 36		14.0	GALAXY
ZWG 550.001	00 22 42.	+ 48 13		15.5	GALAXY
ZWG 549.040	00 22 42.	+ 48 13		15.5	GALAXY
MCG+00-02-025	00 22 42.	- 01 23 30.	60	14.5	GALAXY
8ZW 0022-13.5	00 22 42.	- 13 33		14.0	COMPACT GALAXY
LB 07702	00 22 42.	- 20 44		17.5	FAINT BLUE STAR
LB 07703	00 22 42.	- 22 22		19.6	FAINT BLUE STAR
LB 07704	00 22 42.	- 22 45		16.7	FAINT BLUE STAR
LB 07705	00 22 42.	- 22 47		17.4	FAINT BLUE STAR
LB 07706	00 22 42.	- 23 44		19.7	FAINT BLUE STAR
ZWG 434.009	00 22 48.	+ 12 37		14.1	GALAXY
UGC 00241	00 22 48.	+ 12 37	60	14.1	GALAXY Sa
PHL 2883	00 22 48.	+ 16 34		18.1	BLUE STELLAR OBJECT
ZWG 457.014	00 22 48.	+ 19 56		14.3	GALAXY
UGC 00242	00 22 48.	+ 19 56	126	14.3	GALAXY SBc
MCG+04-02-017	00 22 48.	+ 24 32	54	14.5	GALAXY
ZWG 550.002	00 22 48.	+ 45 39		14.9	GALAXY
UGC 00243	00 22 48.	+ 45 39	114	14.9	GALAXY Sb
LB 07707	00 22 48.	- 21 39		18.9	FAINT BLUE STAR
LB 07708	00 22 48.	- 22 00		18.8	FAINT BLUE STAR
LB 07709	00 22 48.	- 23 59		17.5	FAINT BLUE STAR
MCG+08-02-001	00 22 51.	+ 45 38	108	14.	GALAXY
SC 0020-6654.1	00 22 52.	- 66 37 28.	12		NEBULA
ZWG 409.012	00 22 54.	+ 06 26		15.7	GALAXY
MCG+01-02-009	00 22 54.	+ 07 11	42	15.	GALAXY
ZWG 479.027	00 22 54.	+ 24 33		14.9	GALAXY
UGC 00244	00 22 54.	+ 24 33	66	14.9	GALAXY SB?b-c
MCG-03-02-012	00 22 54.	- 14 50	60	15.	GALAXY
LB 07710	00 22 54.	- 21 34		18.5	FAINT BLUE STAR
LB 07711	00 22 54.	- 22 03		18.6	FAINT BLUE STAR
LB 07712	00 22 54.	- 22 32		18.8	FAINT BLUE STAR
LB 07713	00 22 54.	- 22 58		19.0	FAINT BLUE STAR
LB 07714	00 22 54.	- 26 34		18.4	FAINT BLUE STAR
LB 01553	00 22 54.	- 50 18		13.3	FAINT BLUE STAR
ZC 0023.0+1101	00 23 00.	+ 11 01	1080		CLUSTER OF GALAXIES
PHL 2884	00 23 00.	+ 14 20		17.8	BLUE STELLAR OBJECT
PHL 6525	00 23 00.	+ 15 20		16.2	BLUE STELLAR OBJECT
PHL 6526	00 23 00.	+ 17 49		18.6	BLUE STELLAR OBJECT
MCG+03-02-010	00 23 00.	+ 19 58	90	13.5	GALAXY
MCG+07-02-001	00 23 00.	+ 40 42	18	18.	GALAXY
MCG-02-02-013	00 23 00.	- 08 40	36	14.5	GALAXY
LB 07715	00 23 00.	- 22 30		18.9	FAINT BLUE STAR
LB 07716	00 23 00.	- 23 35		20.0	FAINT BLUE STAR
LB 07717	00 23 00.	- 25 30		18.9	FAINT BLUE STAR
SCHO 0034	00 23 01.	+ 62 59 42.	330		ISOLATED DARK CLOUD
SC 0020-6921.1	00 23 01.	- 69 04 28.	12		NEBULA
LB 02722	00 23 05.	+ 15 10 54		16.1	FAINT BLUE STAR
ZC 0023.1+0029	00 23 06.	+ 00 29	1010		CLUSTER OF GALAXIES
PHL 2885	00 23 06.	+ 13 24		18.6	BLUE STELLAR OBJECT
PHL 0807	00 23 06.	+ 15 26		17.6	BLUE STELLAR OBJECT
ZWG 457.015	00 23 06.	+ 17 06		15.7	GALAXY
KARA.73B 0021	00 23 06.	+ 17 06	36	15.7	ISOLATED GALAXY S
8ZW 0023-08.5	00 23 06.	- 08 33		17.5	COMPACT GALAXY
MCG-03-02-013	00 23 06.	- 14 50	90	15.	GALAXY
LB 07718	00 23 06.	- 20 58		16.8	FAINT BLUE STAR
LB 07719	00 23 06.	- 21 44		19.4	FAINT BLUE STAR
LB 07720	00 23 06.	- 21 54		08.0	FAINT BLUE STAR
RNGC 0107	00 23 08.	- 08 33			GALAXY
SC 0020-6651.1	00 23 09.	- 66 34 28.			NEBULA
MCG+01-02-010	00 23 12.	+ 05 07	72	15.	GALAXY
ZC 0023.2+0802	00 23 12.	+ 08 02	1080		CLUSTER OF GALAXIES
ZWG 409.013	00 23 12.	+ 08 22		15.7	GALAXY
MCG+07-02-002	00 23 12.	+ 40 41	12		GALAXY
LB 07721	00 23 12.	- 21 12		18.5	FAINT BLUE STAR
LB 07722	00 23 12.	- 22 10		17.7	FAINT BLUE STAR
LB 07723	00 23 12.	- 22 53		19.6	FAINT BLUE STAR
LB 07724	00 23 12.	- 24 54		19.5	FAINT BLUE STAR
LB 07725	00 23 12.	- 26 05		18.6	FAINT BLUE STAR
LB 07726	00 23 12.	- 26 19		18.8	FAINT BLUE STAR
ZL 001	00 23 14.	+ 16 55 42.		17.1	ULTRAFAINT BLUE STAR
MCG-01-02-011	00 23 15.	- 02 33 30.	132	14.	GALAXY
MCG-02-02-014	00 23 15.	- 08 32	36	15.	GALAXY
LIN.CL 009	00 23 16.	- 74 21 10.	270	14.7	STAR CLUSTER IN SMC
RNGC 0108	00 23 17.	+ 28 57		13.5	GALAXY
ZWG 409.014	00 23 18.	+ 05 08		15.7	GALAXY
UGC 00245	00 23 18.	+ 05 08	90	15.7	GALAXY SBa-b
PHL 2886	00 23 18.	+ 15 28		18.8	BLUE STELLAR OBJECT
ZWG 500.020	00 23 18.	+ 28 56		13.3	GALAXY
UGC 00246	00 23 18.	+ 28 56	138	13.3	GALAXY SB0
MCG+05-02-012	00 23 18.	+ 28 57	84	13.5	GALAXY
MCG+07-02-003	00 23 18.	+ 40 41	15	17.	GALAXY
LB 07727	00 23 18.	- 20 50		19.4	FAINT BLUE STAR
LB 07728	00 23 18.	- 21 53		18.0	FAINT BLUE STAR
LB 07729	00 23 18.	- 22 24		17.5	FAINT BLUE STAR
LB 07730	00 23 18.	- 23 00		18.2	FAINT BLUE STAR
LB 07731	00 23 18.	- 23 07		15.3	FAINT BLUE STAR
LB 07732	00 23 18.	- 23 15		18.0	FAINT BLUE STAR
LB 07733	00 23 18.	- 24 16		18.2	FAINT BLUE STAR
LB 07734	00 23 18.	- 25 09		17.6	FAINT BLUE STAR
MCG-06-02-004	00 23 18.	- 33 22	12	16.	GALAXY
ZL 002	00 23 20.	+ 16 52 06.		20.7	ULTRAFAINT BLUE STAR
MCG-01-02-012	00 23 21.	- 03 42	24	16.	GALAXY
MCG+02-02-009	00 23 24.	+ 13 22	72	14.5	GALAXY
UGC 00247	00 23 24.	+ 14 04	66	16.5	GALAXY Sc
ZWG 479.028	00 23 24.	+ 21 31		15.7	GALAXY
KARA.72 008A	00 23 24.	+ 21 31	48	15.7	PART OF DOUBLE GALAXY
MCG+04-02-019	00 23 24.	+ 21 32	36	15.	GALAXY
ZWG 479.029	00 23 24.	+ 25 27		14.9	GALAXY
UGC 00248	00 23 24.	+ 25 27	114	14.9	GALAXY DISRPTD
MCG+04-02-018	00 23 24.	+ 25 27	90	14.5	GALAXY
ZC 0023.4+3051	00 23 24.	+ 30 51	1280		CLUSTER OF GALAXIES
MCG-02-02-009	00 23 24.	- 09 55 30.	48	15.	GALAXY
LB 07735	00 23 24.	- 20 53		19.0	FAINT BLUE STAR
LB 07736	00 23 24.	- 21 14		13.7	FAINT BLUE STAR
LB 07737	00 23 24.	- 21 36		19.9	FAINT BLUE STAR
LB 07738	00 23 24.	- 22 24		19.8	FAINT BLUE STAR
LB 07739	00 23 24.	- 23 44		16.8	FAINT BLUE STAR
LB 07740	00 23 24.	- 24 14		18.4	FAINT BLUE STAR
LB 07741	00 23 24.	- 25 45		17.5	FAINT BLUE STAR
MCG-01-02-013	00 23 27.	- 03 12	48	17.	GALAXY
PHL 2887	00 23 30.	+ 12 00		18.8	BLUE STELLAR OBJECT
KARA.72 009A	00 23 30.	+ 13 22	30	14.8	PART OF DOUBLE GALAXY
ZWG 479.030	00 23 30.	+ 25 23		15.7	GALAXY
LB 07742	00 23 30.	- 20 53		19.5	FAINT BLUE STAR

OBJECT NAME	RIGHT ASCEN.	DECLINATION	DIAM.	MAGN.	TYPE OF OBJECT
LB 07743	00 23 30.	- 21 35		19.8	FAINT BLUE STAR
LB 07744	00 23 30.	- 22 17		19.2	FAINT BLUE STAR
LB 07745	00 23 30.	- 23 15		18.8	FAINT BLUE STAR
LB 07746	00 23 30.	- 23 23		19.6	FAINT BLUE STAR
LB 07747	00 23 30.	- 23 25		18.8	FAINT BLUE STAR
LB 07748	00 23 30.	- 23 36		18.5	FAINT BLUE STAR
LB 07749	00 23 30.	- 25 02		18.4	FAINT BLUE STAR
LB 07750	00 23 30.	- 25 27		19.6	FAINT BLUE STAR
MCG+03-02-011	00 23 33.	+ 16 10	66	15.	GALAXY
ZWG 434.010	00 23 36.	+ 12 31		15.2	GALAXY
ZWG 434.011	00 23 36.	+ 13 22		14.8	GALAXY
UGC 00249	00 23 36.	+ 13 22	102	14.8	GALAXY S-IRR
KARA.72 009B	00 23 36.	+ 13 22	54		PART OF DOUBLE GALAXY
ZWG 457.016	00 23 36.	+ 16 08		15.6	GALAXY
UGC 00250	00 23 36.	+ 16 08	84	15.6	GALAXY SB:c
ZL 003	00 23 36.	+ 16 58 42.		19.9	ULTRAFAINT BLUE STAR
ZWG 479.031	00 23 36.	+ 21 32		15.0	GALAXY
UGC 00251	00 23 36.	+ 21 32	78	15.0	GALAXY SBa
KARA.72 008B	00 23 36.	+ 21 32	66	15.0	PART OF DOUBLE GALAXY
RNGC 0109	00 23 36.	+ 21 33		15.0	GALAXY
MCG+04-02-020	00 23 36.	+ 21 33	60	15.	GALAXY
ZWG 383.013	00 23 36.	- 02 18		15.7	GALAXY
LB 07751	00 23 36.	- 22 20		20.0	FAINT BLUE STAR
LB 07752	00 23 36.	- 23 06		18.8	FAINT BLUE STAR
LB 07753	00 23 36.	- 23 23		19.8	FAINT BLUE STAR
ZL 004	00 23 40.	+ 16 55 48.		21.2	ULTRAFAINT BLUE STAR
SC 0021-6836.6	00 23 40.	- 68 19 58.	60		NEBULA
MCG+01-02-011	00 23 42.	+ 05 59 30.	60	15.	GALAXY
PHL 0808	00 23 42.	+ 18 30		18.7	BLUE STELLAR OBJECT
LB 07754	00 23 42.	- 21 24		17.5	FAINT BLUE STAR
LB 07755	00 23 42.	- 26 20		19.2	FAINT BLUE STAR
LB 07756	00 23 42.	- 26 34		17.8	FAINT BLUE STAR
SC 0021-6845.0	00 23 44.	- 68 28 22.	12		NEBULA
MCG-01-02-014	00 23 45.	- 04 46 30.	72	14.	GALAXY
MCG+00-02-026	00 23 48.	+ 00 44 30.	39	15.5	GALAXY
UGC 00252	00 23 48.	+ 00 45	66	16.5	GALAXY Sc
PHL 2888	00 23 48.	+ 14 28		17.7	BLUE STELLAR OBJECT
ZC 0023.8+2631	00 23 48.	+ 26 31	1680		CLUSTER OF GALAXIES
LB 07757	00 23 48.	- 21 52		17.0	FAINT BLUE STAR
LB 07758	00 23 48.	- 22 02		17.7	FAINT BLUE STAR
LB 07759	00 23 48.	- 22 47		17.9	FAINT BLUE STAR
LB 07760	00 23 48.	- 23 14		19.9	FAINT BLUE STAR
TON-S 0157	00 23 48.	- 23 19		14.5	BLUE STAR
LB 07761	00 23 48.	- 24 11		16.7	FAINT BLUE STAR
LB 01554	00 23 48.	- 58 15		14.4	FAINT BLUE STAR
ZL 006	00 23 50.	+ 16 55 42.		22.8	ULTRAFAINT BLUE STAR
ZL 005	00 23 50.	+ 17 02 06.		22.7	ULTRAFAINT BLUE STAR
MCG+01-02-012	00 23 51.	+ 04 15	36	15.	GALAXY
ZL 007	00 23 51.	+ 16 55 00.		19.8	ULTRAFAINT BLUE STAR
SC 0021-6604.5	00 23 53.	- 65 47 52.	12		NEBULA
ZWG 409.015	00 23 54.	+ 06 01		15.3	GALAXY
UGC 00253	00 23 54.	+ 06 01	84	15.3	GALAXY Sb/SB
3ZW 008	00 23 54.	+ 21 29			COMPACT GALAXY
UGC 00254	00 23 54.	+ 39 13	66	16.5	GALAXY S
LB 07762	00 23 54.	- 22 28		20.0	FAINT BLUE STAR
LB 07763	00 23 54.	- 23 30		18.7	FAINT BLUE STAR
LB 07764	00 23 54.	- 23 37		18.6	FAINT BLUE STAR
LB 07765	00 23 54.	- 25 29		18.4	FAINT BLUE STAR
LB 07766	00 23 54.	- 25 40		20.1	FAINT BLUE STAR
MCG-06-02-005	00 23 54.	- 33 41	18	15.5	GALAXY
ZL 008	00 23 58.	+ 16 48 48.		20.8	ULTRAFAINT BLUE STAR
SS 03	00 23 59.	+ 64 25			DIFFUSE GALACTIC NEBULA
LBN 0596	00 24	+ 64 20	60		BRIGHT NEBULA
LB 02723	00 24 00.	+ 15 51 12.		16.2	FAINT BLUE STAR
ZC 0024.0+1652	00 24 00.	+ 16 52	340		CLUSTER OF GALAXIES
ZC 0024.0+1922	00 24 00.	+ 19 22	2690		CLUSTER OF GALAXIES
MCG+08-02-002	00 24 00.	+ 48 50	24	15.	GALAXY
LB 07767	00 24 00.	- 21 32		17.7	FAINT BLUE STAR
ZL 009	00 24 02.	+ 16 43 12.		18.8	ULTRAFAINT BLUE STAR
SCHO 0035	00 24 03.	+ 57 35 18.	270		ISOLATED DARK CLOUD
RNGC 0112	00 24 05.	+ 31 27		14.5	GALAXY
MCG+05-02-013	00 24 06.	+ 31 27	48	15.	GALAXY
4ZW 019	00 24 06.	+ 37 20			COMPACT GALAXY
5ZW 019	00 24 06.	+ 49 01			COMPACT GALAXY
LB 07768	00 24 06.	- 21 45		16.6	FAINT BLUE STAR
LB 07769	00 24 06.	- 23 40		19.5	FAINT BLUE STAR
RNGC 0111	00 24 07.	- 02 54			NON-EXISTENT OBJECT
LB 02724	00 24 08.	+ 15 06 00.		15.8	FAINT BLUE STAR
ZL 010	00 24 09.	+ 16 45 54.		20.8	ULTRAFAINT BLUE STAR
MCG+06-02-004	00 24 09.	+ 36 40	42	16.	GALAXY
ZL 011	00 24 10.	+ 16 47 12.		20.7	ULTRAFAINT BLUE STAR
MCG+02-02-010	00 24 12.	+ 11 17	48	15.	GALAXY
PHL 6527	00 24 12.	+ 17 22		18.9	BLUE STELLAR OBJECT
ZWG 500.021	00 24 12.	+ 31 25		14.5	GALAXY
UGC 00255	00 24 12.	+ 31 25	78	14.5	GALAXY S
ZWG 550.003	00 24 12.	+ 48 50		15.0	GALAXY
5ZW 020	00 24 12.	+ 48 51			COMPACT GALAXY
ZWG 550.004	00 24 12.	+ 49 45		15.6	GALAXY
UGC 00256	00 24 12.	+ 49 45	78	15.6	GALAXY Sb-c
MCG+08-02-003	00 24 12.	+ 49 45	108	15.	GALAXY
MCG-01-02-015	00 24 12.	- 03 52	51	16.	GALAXY
LB 07770	00 24 12.	- 21 23		18.2	FAINT BLUE STAR
LB 07771	00 24 12.	- 22 00		19.5	FAINT BLUE STAR
LB 07772	00 24 12.	- 26 03		18.0	FAINT BLUE STAR
MCG-05-02-004	00 24 12.	- 30 51	30	15.	GALAXY
ARC 0030	00 24 13.	- 12 28		17.8	RICH CLUSTER OF GALAXIES
UGC 00257	00 24 18.	+ 01 12	84	16.0	GALAXY
ZWG 434.012	00 24 18.	+ 11 18		15.7	GALAXY
KARA.72 010A	00 24 18.	+ 11 18	42	15.7	PART OF DOUBLE GALAXY
LB 07773	00 24 18.	- 22 26		18.1	FAINT BLUE STAR
LB 07774	00 24 18.	- 24 21		18.4	FAINT BLUE STAR
MCG-06-02-006	00 24 18.	- 33 57	180	14.	GALAXY
LB 03139	00 24 18.	- 73 33		14.0	FAINT BLUE STAR
ZL 012	00 24 19.	+ 16 47 54.		20.8	ULTRAFAINT BLUE STAR
ARC 0029	00 24 19.	+ 38 18		17.5	RICH CLUSTER OF GALAXIES
ZL 013	00 24 20.	+ 16 52 24.		20.7	ULTRAFAINT BLUE STAR
MCG-01-02-016	00 24 21.	- 02 46 30.	24	13.5	GALAXY
RNGC 0115	00 24 22.	- 33 57			GALAXY
ZC 0024.4+0513	00 24 24.	+ 05 13	1750		CLUSTER OF GALAXIES
MCG+02-02-011	00 24 24.	+ 11 18	180	13.	GALAXY
PHL 0809	00 24 24.	+ 13 16		15.8	BLUE STELLAR OBJECT
PHL 2889	00 24 24.	+ 13 53		14.3	BLUE STELLAR OBJECT
ZL 014	00 24 24.	+ 16 47 42.		22.5	ULTRAFAINT BLUE STAR
PHL 6532	00 24 24.	+ 16 55		16.7	BLUE STELLAR OBJECT
ZC 0024.4+3014	00 24 24.	+ 30 14	10950		CLUSTER OF GALAXIES
MCG+06-02-005	00 24 24.	+ 36 40	33	16.	GALAXY
ZC 0024.4+3818	00 24 24.	+ 38 18	1480		CLUSTER OF GALAXIES
UGC 00258	00 24 24.	+ 50 49	66	16.0	GALAXY Sa-b

OBJECT NAME	RIGHT ASCEN.	DECLINATION	DIAM.	MAGN.	TYPE OF OBJECT
OCL 0300	00 24 24.	+ 71 07			OPEN STAR CLUSTER
MCG+00-02-027	00 24 24.	- 02 03 30.	48	14.	GALAXY
LB 07775	00 24 24.	- 23 01		17.4	FAINT BLUE STAR
LB 07776	00 24 24.	- 23 50		19.0	FAINT BLUE STAR
LB 07777	00 24 24.	- 25 49		17.8	FAINT BLUE STAR
LB 07778	00 24 24.	- 26 09		19.5	FAINT BLUE STAR
RNGC 0113	00 24 25.	- 02 46		13.5	GALAXY
RNGC 0110	00 24 26.	+ 71 07			OPEN CLUSTER
ARC 0032	00 24 26.	- 09 23		18.0	RICH CLUSTER OF GALAXIES
ZL 015	00 24 28.	+ 16 50 54.		18.7	ULTRAFAINT BLUE STAR
MCG+00-02-028	00 24 30.	+ 00 11 30.	36	15.5	GALAXY
ZWG 434.013	00 24 30.	+ 11 18		14.0	GALAXY
UGC 00260	00 24 30.	+ 11 18	186	14.0	GALAXY Sc
KARA.72 010B	00 24 30.	+ 11 18	168	14.0	PART OF DOUBLE GALAXY
PHL 6528	00 24 30.	+ 13 55		16.8	BLUE STELLAR OBJECT
ZL 016	00 24 30.	+ 16 53 36.		21.6	ULTRAFAINT BLUE STAR
MRSL 120+02/1	00 24 30.	+ 64 26	120		HII REGION
ZWG 383.014	00 24 30.	- 02 03		15.0	GALAXY
UGC 00259	00 24 30.	- 02 03	78	15.0	GALAXY SBO
LB 07779	00 24 30.	- 22 34		18.5	FAINT BLUE STAR
LB 07780	00 24 30.	- 24 02		18.4	FAINT BLUE STAR
LB 07781	00 24 30.	- 25 01		18.0	FAINT BLUE STAR
LB 07782	00 24 30.	- 25 40		17.8	FAINT BLUE STAR
LB 07783	00 24 30.	- 26 15		17.7	FAINT BLUE STAR
RNGC 0114	00 24 31.	- 02 03		15.0	GALAXY
ZL 017	00 24 32.	+ 16 49 48.		19.9	ULTRAFAINT BLUE STAR
MCG+00-02-029	00 24 33.	+ 01 04 30.	48	15.	GALAXY
ARC 0031	00 24 33.	+ 22 22		17.7	RICH CLUSTER OF GALAXIES
MCG+00-02-031	00 24 33.	- 01 08	24	17.	GALAXY
MCG+00-02-030	00 24 33.	- 01 08	24	17.	GALAXY
SC 0022-6720.5	00 24 33.	- 67 04 16.	6		NEBULA
ARC 0033	00 24 35.	- 19 47		17.9	RICH CLUSTER OF GALAXIES
ZWG 383.015	00 24 36.	+ 01 03		15.5	GALAXY
ZC 0024.6+0210	00 24 36.	+ 02 10	1280		CLUSTER OF GALAXIES
ZC 0024.6+2222	00 24 36.	+ 22 22	1480		CLUSTER OF GALAXIES
ZWG 479.032	00 24 36.	+ 23 54		14.8	GALAXY
UGC 00261	00 24 36.	+ 23 54	78	14.8	GALAXY PECULR
ZWG 535.005	00 24 36.	+ 39 31		15.1	GALAXY
UGC 00262	00 24 36.	+ 39 31	72	15.1	GALAXY PECULR
UGC 00263	00 24 36.	+ 43 46	102	16.5	GALAXY Sc-IRR
MCG-01-02-017	00 24 36.	- 07 58	36	14.5	GALAXY
LB 07784	00 24 36.	- 21 48		17.9	FAINT BLUE STAR
LB 07785	00 24 36.	- 22 06		16.0	FAINT BLUE STAR
LB 07786	00 24 36.	- 24 49		18.0	FAINT BLUE STAR
LB 07787	00 24 36.	- 24 55		18.3	FAINT BLUE STAR
RNGC 0121	00 24 36.	- 71 49		11.0	GLOBULAR CLUSTER IN SMC
RNGC 0117	00 24 37.	+ 01 03		15.5	GALAXY
LIN.CL 010	00 24 37.	- 71 48 53.	408	11.1	STAR CLUSTER IN SMC
NAB 0024+22	00 24 38.	+ 22 25 00.		16.6	QUASI-STELLAR OBJECT
SHB 004	00 24 38.	+ 22 25 00.		16.6	QUASI-STELLAR OBJECT
RNGC 0116	00 24 38.	- 07 58		14.5	GALAXY
LIN 003	00 24 39.	- 74 28 47.		16.34	SMC EMISSION LINE OBJECT
ZWG 457.017	00 24 42.	+ 19 47		15.2	GALAXY
UGC 00265	00 24 42.	+ 19 47	78	15.2	GALAXY S
MCG+03-02-012	00 24 42.	+ 19 48	42	15.	GALAXY
MCG+04-02-021	00 24 42.	+ 22 23	18	18.	GALAXY
MCG+04-02-022	00 24 42.	+ 23 54	66	15.	GALAXY
4ZW 020	00 24 42.	+ 39 31			COMPACT GALAXY
MCG+07-02-004	00 24 42.	+ 43 47	90	17.	GALAXY
3ZW 009	00 24 42.	- 02 03			COMPACT GALAXY
ZWG 383.016	00 24 42.	- 02 03		14.9	GALAXY
UGC 00264	00 24 42.	- 02 03	78	14.9	GALAXY COMPACT
MCG+00-02-032	00 24 42.	- 02 04	36	14.	GALAXY
MCG-01-02-018	00 24 42.	- 08 05	24	15.	GALAXY
LB 07788	00 24 42.	- 22 02		16.6	FAINT BLUE STAR
LB 07789	00 24 42.	- 22 14		18.8	FAINT BLUE STAR
LB 07790	00 24 42.	- 22 34		18.9	FAINT BLUE STAR
LB 07791	00 24 42.	- 25 34		18.6	FAINT BLUE STAR
LB 03140	00 24 42.	- 61 55		13.8	FAINT BLUE STAR
SC 0022-6656.5	00 24 42.	- 66 39 53.	6		NEBULA
RNGC 0118	00 24 43.	- 02 03		15.0	GALAXY
RNGC 0119	00 24 43.	- 57 15			GALAXY
ARC 0034	00 24 44.	- 09 05		18.0	RICH CLUSTER OF GALAXIES
MRK 550	00 24 48.	+ 01 10	8	16.5	GALAXY WITH UV CONTINUUM
UGC 00266	00 24 48.	+ 10 33	72	16.0	GALAXY S?
ZWG 535.006	00 24 48.	+ 39 37		15.7	GALAXY
MCG-01-02-019	00 24 48.	- 05 10 30.	48	16.	GALAXY
LB 07792	00 24 48.	- 20 40		20.6	FAINT BLUE STAR
LB 07793	00 24 48.	- 22 30		18.9	FAINT BLUE STAR
LB 07794	00 24 48.	- 22 46		20.0	FAINT BLUE STAR
LB 07795	00 24 48.	- 22 58		20.1	FAINT BLUE STAR
LB 07796	00 24 48.	- 23 16		16.8	FAINT BLUE STAR
LB 07797	00 24 48.	- 24 48		18.3	FAINT BLUE STAR
LB 07798	00 24 49.	- 24 53		19.1	FAINT BLUE STAR
LB 07799	00 24 48.	- 25 20		17.5	FAINT BLUE STAR
ARC 0035	00 24 53.	- 21 57		17.1	RICH CLUSTER OF GALAXIES
MCG+04-02-023	00 24 54.	+ 22 23	3	20.	GALAXY
ZWG 500.022	00 24 54.	+ 30 43		15.6	GALAXY
ZWG 535.007	00 24 54.	+ 45 19		15.7	GALAXY
ZWG 383.017	00 24 54.	- 01 47		14.8	GALAXY
UGC 00267	00 24 54.	- 01 47	96	14.8	GALAXY SO
MCG+00-02-033	00 24 54.	- 01 47	60	13.5	GALAXY
LB 07800	00 24 54.	- 25 20		18.6	FAINT BLUE STAR
LB 01555	00 24 54.	- 47 46		13.4	FAINT BLUE STAR
FATH 1.007	00 24 55.	+ 30 43	14		NEBULA
RNGC 0120	00 24 55.	- 01 47		15.0	GALAXY
IC 1551	00 24 59.	+ 08 35 43.			NONSTELLAR OBJECT
HMS 0025+2223	00 25	+ 22 23			CLUSTER OF GALAXIES
YC 0025-41	00 25	- 41 45			UNUSUAL SOUTHERN NEBULA
ZC 0025.0+0434	00 25 00.	+ 04 34	740		CLUSTER OF GALAXIES
ZWG 409.016	00 25 00.	+ 08 36		15.0	GALAXY
UGC 00268	00 25 00.	+ 08 36	162	15.0	GALAXY S
MCG+01-02-013	00 25 00.	+ 08 36	144	15.	GALAXY
KARA.73B 0022	00 25 00.	+ 08 36	156	15.0	ISOLATED GALAXY S
PHL 6529	00 25 00.	+ 14 58		18.5	BLUE STELLAR OBJECT
PHL 6530	00 25 00.	+ 15 00		18.5	BLUE STELLAR OBJECT
MCG-02-02-015	00 25 00.	- 11 47	30	15.	GALAXY
8ZW 0025-13.6	00 25 00.	- 13 38		15.5	COMPACT GALAXY
LB 07801	00 25 00.	- 22 23		18.0	FAINT BLUE STAR
LB 07802	00 25 00.	- 22 49		18.6	FAINT BLUE STAR
LB 07803	00 25 00.	- 26 41		17.9	FAINT BLUE STAR
MCG-06-02-007	00 25 00.	- 32 55	18	15.	GALAXY
MCG-06-02-008	00 25 00.	- 34 22	60	14.5	GALAXY
SC 0022-6657.6	00 25 00.	- 66 40 59.	6		NEBULA
MCG+00-02-034	00 25 06.	+ 00 15	36	15.	GALAXY
UGC 00269	00 25 06.	+ 03 58	66	16.0	GALAXY S
ZC 0025.1+1653	00 25 06.	+ 16 53	610		CLUSTER OF GALAXIES
MCG+05-02-014	00 25 06.	+ 30 20	30	15.5	GALAXY

OBJECT NAME	RIGHT ASCEN.	DECLINATION	DIAM.	MAGN.	TYPE OF OBJECT
ZC 0025.1+3310	00 25 06.	+ 33 10	2350		CLUSTER OF GALAXIES
LB 07804	00 25 06.	- 20 42		20.4	FAINT BLUE STAR
LB 07805	00 25 06.	- 23 20		20.0	FAINT BLUE STAR
LB 07806	00 25 06.	- 24 16		19.0	FAINT BLUE STAR
LB 07807	00 25 06.	- 25 27		17.5	FAINT BLUE STAR
LB 07808	00 25 06.	- 26 37		19.0	FAINT BLUE STAR
MCG-06-02-009	00 25 06.	- 34 30	90	15.	GALAXY
RNGC 0122	00 25 07.	- 01 54			NON-EXISTENT OBJECT
ARC 0036	00 25 07.	- 13 04		17.6	RICH CLUSTER OF GALAXIES
MCG+00-02-035	00 25 12.	+ 03 12 30.	30	15.5	GALAXY
PHL 6531	00 25 12.	+ 13 34		16.0	BLUE STELLAR OBJECT
HMS 1.02	00 25 12.	+ 22 25			Sb GALAXY
MCG+04-02-024	00 25 12.	+ 22 25	9	18.5	GALAXY
UGC 00270	00 25 12.	+ 32 30	72	17.	GALAXY
MCG+00-02-036	00 25 12.	- 01 29 30.	72	14.	GALAXY
LB 07809	00 25 12.	- 20 49		18.1	FAINT BLUE STAR
LB 07811	00 25 12.	- 20 56		17.6	FAINT BLUE STAR
LB 07810	00 25 12.	- 20 56		16.8	FAINT BLUE STAR
LB 07812	00 25 12.	- 22 18		17.9	FAINT BLUE STAR
LB 07813	00 25 12.	- 22 25		18.4	FAINT BLUE STAR
LB 07814	00 25 12.	- 22 27		18.2	FAINT BLUE STAR
LB 07815	00 25 12.	- 23 18		16.9	FAINT BLUE STAR
LB 07816	00 25 12.	- 23 49		19.0	FAINT BLUE STAR
LB 07817	00 25 12.	- 26 27		18.2	FAINT BLUE STAR
LB 03141	00 25 12.	- 70 45		13.8	FAINT BLUE STAR
RNGC 0123	00 25 13.	- 01 52			NON-EXISTENT OBJECT
FATH 1.008	00 25 15.	+ 30 19	5		NEBULA
MCG+00-02-037	00 25 18.	+ 02 15	120	13.5	GALAXY
ZWG 383.020	00 25 18.	+ 03 12		15.5	GALAXY
UGC 00273	00 25 18.	+ 25 44	72	17.	GALAXY S?IV-V
ZWG 500.023	00 25 18.	+ 30 20		15.6	GALAXY
UGC 00274	00 25 18.	+ 30 20	72	15.6	GALAXY SBO-a
MCG+05-02-015	00 25 18.	+ 30 30	48	16.	GALAXY
ZWG 383.019	00 25 18.	- 01 28		15.5	GALAXY
UGC 00272	00 25 18.	- 01 28	108	15.5	GALAXY Sc
ZWG 383.018	00 25 18.	- 02 05		13.8	GALAXY
UGC 00271	00 25 18.	- 02 05	96	13.8	GALAXY Sc
MCG+00-02-038	00 25 18.	- 02 05 30.	84	13.	GALAXY
ZWG 383.021	00 25 24.	+ 02 14		14.9	GALAXY
UGC 00275	00 25 24.	+ 02 14	114	14.9	GALAXY SB
ZC 0025.4+0540	00 25 24.	+ 05 40	1080		CLUSTER OF GALAXIES
ZC 0025.4+1342	00 25 24.	+ 13 42	1950		CLUSTER OF GALAXIES
ZWG 500.024	00 25 24.	+ 30 29		15.6	GALAXY
UGC 00276	00 25 24.	+ 30 29	72	15.6	GALAXY Sa-b
MCG+06-02-006	00 25 24.	+ 37 46	36	17.	GALAXY
IC 0015	00 25 24.	- 00 20 47.			NONSTELLAR OBJECT
MCG-02-02-016	00 25 24.	- 11 49 30.	9	15.	GALAXY
MCG-02-02-017	00 25 24.	- 13 22	30	15.	GALAXY
MCG-02-02-018	00 25 24.	- 14 10	30	15.	GALAXY
LB 07818	00 25 24.	- 21 46		16.0	FAINT BLUE STAR
LB 07819	00 25 24.	- 22 34		17.8	FAINT BLUE STAR
LB 07820	00 25 24.	- 22 41		19.8	FAINT BLUE STAR
LB 07821	00 25 24.	- 22 50		16.5	FAINT BLUE STAR
LB 07822	00 25 24.	- 24 48		19.0	FAINT BLUE STAR
RNGC 0124	00 25 25.	- 02 05		14.0	GALAXY
ARC 0037	00 25 25.	- 10 48		18.0	RICH CLUSTER OF GALAXIES
LIN.CL 011	00 25 28.	- 73 04 05.	612	14.0	STAR CLUSTER IN SMC
4ZW 021	00 25 30.	+ 32 53			COMPACT GALAXY
ZWG 550.005	00 25 30.	+ 48 49		15.4	GALAXY
MCG+14-01-007	00 25 30.	+ 85 44	36	17.	GALAXY
MCG-01-02-020	00 25 30.	- 08 25 30.	48	15.	GALAXY
LB 07823	00 25 30.	- 20 58		18.7	FAINT BLUE STAR
LB 07824	00 25 30.	- 21 08		18.3	FAINT BLUE STAR
LB 07825	00 25 30.	- 21 14		18.8	FAINT BLUE STAR
LB 07826	00 25 30.	- 22 03		18.0	FAINT BLUE STAR
LB 07827	00 25 30.	- 22 06		19.7	FAINT BLUE STAR
LB 07828	00 25 30.	- 23 49		19.1	FAINT BLUE STAR
LB 07829	00 25 30.	- 24 01		17.7	FAINT BLUE STAR
LB 07830	00 25 30.	- 25 00		19.8	FAINT BLUE STAR
PK119-06.1	00 25 30.2	+ 55 41 20.	5	13.3	PLANETARY NEBULA
MCG-02-02-019	00 25 33.	- 13 48	60	15.	GALAXY
MCG+00-02-039	00 25 36.	+ 02 58 30.	60	14.5	GALAXY
MCG+00-02-040	00 25 36.	+ 03 07	90	14.5	GALAXY
ZWG 479.033	00 25 36.	+ 23 10		15.7	GALAXY
MCG+05-02-016	00 25 36.	+ 30 32	96	14.5	GALAXY
ZWG 500.025	00 25 36.	+ 33 00		15.6	GALAXY
ASS 36	00 25 36.	+ 62 25			OB ASSOCIATION CAS OB4
IC 0016	00 25 36.	- 13 22 41.			NONSTELLAR OBJECT
MCG-03-02-014	00 25 36.	- 15 13	36	15.	GALAXY
LB 07831	00 25 36.	- 21 05		18.7	FAINT BLUE STAR
LB 07832	00 25 36.	- 22 30		18.5	FAINT BLUE STAR
LB 07833	00 25 36.	- 22 54		18.6	FAINT BLUE STAR
LB 07834	00 25 36.	- 25 16		19.9	FAINT BLUE STAR
LB 01556	00 25 36.	- 57 55		15.1	FAINT BLUE STAR
FATH 1.009	00 25 38.	+ 30 32	27		NEBULA
SN 1972N	00 25 40.	+ 30 32		17.0	SUPERNOVA
FATH 1.010	00 25 41.	+ 30 55	14		NEBULA
MCG+00-02-042	00 25 42.	+ 02 16	60	14.	GALAXY
ZWG 383.022	00 25 42.	+ 02 57		15.7	GALAXY
UGC 00277	00 25 42.	+ 02 57	102	15.7	GALAXY Sc:
MCG+00-02-041	00 25 42.	+ 03 07	60	14.	GALAXY
ZWG 409.017	00 25 42.	+ 06 51		15.7	GALAXY
MCG+04-02-025	00 25 42.	+ 27 06	90	15.5	GALAXY
ZWG 479.034	00 25 42.	+ 27 07		15.3	GALAXY
UGC 00278	00 25 42.	+ 27 07	90	15.3	GALAXY SBb-c
ZWG 500.026	00 25 42.	+ 30 32		14.3	GALAXY
UGC 00279	00 25 42.	+ 30 32	108	14.3	GALAXY SB?
ZWG 500.027	00 25 42.	+ 30 55		15.1	GALAXY
LB 07835	00 25 42.	- 24 51		19.8	FAINT BLUE STAR
MCG+01-02-014	00 25 45.	+ 04 42 30.	36	14.5	GALAXY
ZWG 383.023	00 25 48.	+ 02 15		15.7	GALAXY
UGC 00281	00 25 48.	+ 02 15	84	15.7	GALAXY Sc
ZWG 383.024	00 25 48.	+ 03 07		15.0	GALAXY
UGC 00282	00 25 48.	+ 03 07	120		GALAXY
UGC 00283	00 25 48.	+ 07 07	84	15.0	GALAXY Sc
MCG+04-02-026	00 25 48.	+ 23 12 30.	36	15.	GALAXY
ZWG 479.035	00 25 48.	+ 23 14		15.6	GALAXY
UGC 00280	00 25 48.	- 00 30	84	16.0	GALAXY SBb
MCG+00-02-043	00 25 48.	- 00 30	60	15.	GALAXY
MCG-02-02-020	00 25 48.	- 09 45	36	15.	GALAXY
LB 07836	00 25 48.	- 21 00		19.3	FAINT BLUE STAR
LB 07837	00 25 48.	- 22 40		19.2	FAINT BLUE STAR
LB 07838	00 25 48.	- 23 13		20.2	FAINT BLUE STAR
LB 07839	00 25 48.	- 23 34		19.5	FAINT BLUE STAR
LB 07840	00 25 48.	- 24 32		19.5	FAINT BLUE STAR
MCG-05-02-005	00 25 48.	- 28 15	84	16.	GALAXY
ARC 0038	00 25 49.	+ 13 40		17.6	RICH CLUSTER OF GALAXIES
ARC 0039	00 25 49.	- 11 40		18.0	RICH CLUSTER OF GALAXIES

OBJECT NAME	RIGHT ASCEN.	DECLINATION	DIAM.	MAGN.	TYPE OF OBJECT
MCG+00-02-044	00 25 51.	+ 02 21	24	15.5	GALAXY
LIN.CL 012	00 25 51.	- 73 35 11.	408	15.0	STAR CLUSTER IN SMC
MCG+00-02-046	00 25 54.	+ 00 30	24	15.5	GALAXY
ZWG 383.025	00 25 54.	+ 02 22		14.8	GALAXY
MCG+00-02-045	00 25 54.	+ 02 40 30.	54	14.5	GALAXY
ZWG 409.018	00 25 54.	+ 04 44		15.3	GALAXY
PHL 2890	00 25 54.	+ 18 39		18.6	BLUE STELLAR OBJECT
ZWG 500.028	00 25 54.	+ 30 53		15.3	GALAXY
MRK 340	00 25 54.	+ 30 53	14	16.5	GALAXY WITH UV CONTINUUM
ZWG 500.029	00 25 54.	+ 33 00		15.1	GALAXY
UGC 00284	00 25 54.	+ 33 00	96	15.1	GALAXY SBc/Sc
MCG+05-02-017	00 25 54.	+ 33 01	66	14.5	GALAXY
ZC 0025.9-0021	00 25 54.	- 00 21	610		CLUSTER OF GALAXIES
MCG-01-02-021	00 25 54.	- 07 40 30.	54	15.5	GALAXY
LB 07841	00 25 54.	- 20 44		18.8	FAINT BLUE STAR
LB 07842	00 25 54.	- 23 37		18.6	FAINT BLUE STAR
LB 07843	00 25 54.	- 26 38		17.7	FAINT BLUE STAR
IC 0017	00 25 55.	+ 02 22 07.			NONSTELLAR OBJECT
LIN 004	00 25 55.	- 74 40 12.		14.12	SMC EMISSION LINE OBJECT
PATH 1.011	00 25 56.	+ 30 53	5		NEBULA
LB 03838	00 25 58.	+ 59 36 48.		18.0	FAINT BLUE STAR
ARP 100	00 25 59.	- 11 52			PECULIAR GALAXY
LBN 0594	00 26	+ 56 25	120		BRIGHT NEBULA
ZC 0026.0+0600	00 26 00.	+ 06 00	1480		CLUSTER OF GALAXIES
ZC 0026.0+0732	00 26 00.	+ 07 32	940		CLUSTER OF GALAXIES
ZWG 535.008	00 26 00.	+ 43 46		15.5	GALAXY
ASS 37	00 26 00.	+ 63 05			OB ASSOCIATION CAS OB14
8ZW 0026-11.8	00 26 00.	- 11 51		14.6	COMPACT GALAXY
MCG-02-02-022	00 26 00.	- 12 55	36	15.5	GALAXY
MCG-02-02-021	00 26 00.	- 13 17	36	14.	GALAXY
LB 07845	00 26 00.	- 21 10		19.5	FAINT BLUE STAR
LB 07844	00 26 00.	- 21 10		17.7	FAINT BLUE STAR
LB 07846	00 26 00.	- 21 14		16.8	FAINT BLUE STAR
LB 07847	00 26 00.	- 22 06		19.7	FAINT BLUE STAR
LB 07848	00 26 00.	- 22 26		19.6	FAINT BLUE STAR
LB 07849	00 26 00.	- 24 40		18.0	FAINT BLUE STAR
LB 07850	00 26 00.	- 25 14		19.1	FAINT BLUE STAR
LB 07851	00 26 00.	- 25 20		19.7	FAINT BLUE STAR
LB 07852	00 26 00.	- 26 13		17.4	FAINT BLUE STAR
ARC 0040	00 26 02.	+ 16 08		17.5	RICH CLUSTER OF GALAXIES
PATH 1.012	00 26 03.	+ 30 26	14		NEBULA
IC 0018	00 26 03.	- 11 51 42.			NONSTELLAR OBJECT
ARC 0042	00 26 04.	- 23 55		17.1	RICH CLUSTER OF GALAXIES
MCG+00-02-048	00 26 06.	+ 02 34 30.	114	13.	GALAXY
BMS 1.03	00 26 06.	+ 02 40			Sa GALAXY
MCG+00-02-047	00 26 06.	+ 03 10	24	15.	GALAXY
ZC 0026.1+1148	00 26 06.	+ 11 48	3430		CLUSTER OF GALAXIES
MCG+05-02-018	00 26 06.	+ 28 40	60	15.	GALAXY
ZWG 500.030	00 26 06.	+ 30 25		15.4	GALAXY
VV 234	00 26 06.	- 11 51	42	15.	INTERACTING GALAXY
MCG-02-02-023	00 26 06.	- 11 51	36	15.	GALAXY
LB 07853	00 26 06.	- 20 59		18.0	FAINT BLUE STAR
LB 07854	00 26 06.	- 21 41		19.2	FAINT BLUE STAR
LB 07855	00 26 06.	- 23 48		18.2	FAINT BLUE STAR
SC 0023-6934.4	00 26 06.	- 69 17 47.	12		NEBULA
IC 0019	00 26 07.	- 11 55 00.			NONSTELLAR OBJECT
IC 0020	00 26 09.	- 13 17 36.			NONSTELLAR OBJECT
HOW 02	00 26 09.	- 74 16 30.	18		STAR CLUSTER IN SMC
ZWG 383.026	00 26 12.	+ 03 09		15.0	GALAXY
ARC 0041	00 26 12.	+ 07 35		17.6	RICH CLUSTER OF GALAXIES
ZWG 434.014	00 26 12.	+ 11 22		15.5	GALAXY
PHL 0810	00 26 12.	+ 13 38		15.0	BLUE STELLAR OBJECT
ZWG 500.031	00 26 12.	+ 28 40		15.0	GALAXY
UGC 00285	00 26 12.	+ 28 40	60	15.0	GALAXY Sa
MCG-02-02-024	00 26 12.	- 11 54 30.	36	15.	GALAXY
LB 07856	00 26 12.	- 23 01		18.3	FAINT BLUE STAR
LB 07857	00 26 12.	- 23 16		18.7	FAINT BLUE STAR
LB 07858	00 26 12.	- 24 06		17.6	FAINT BLUE STAR
LB 07859	00 26 12.	- 24 41		17.8	FAINT BLUE STAR
LB 07860	00 26 12.	- 25 29		18.4	FAINT BLUE STAR
LB 07861	00 26 12.	- 25 48		19.7	FAINT BLUE STAR
LB 07862	00 26 12.	- 26 10		19.8	FAINT BLUE STAR
ZWG 383.027	00 26 18.	+ 02 33		14.2	GALAXY
UGC 00286	00 26 18.	+ 02 33	102	14.2	GALAXY S0
ZWG 434.015	00 26 18.	+ 09 52		15.6	GALAXY
UGC 00287	00 26 18.	+ 09 52	78	15.6	GALAXY Sc
MCG+02-02-012	00 26 18.	+ 09 53	36	14.5	GALAXY
PHL 2891	00 26 18.	+ 13 20		17.7	BLUE STELLAR OBJECT
ZC 0026.3+1616	00 26 18.	+ 16 16	2690		CLUSTER OF GALAXIES
ZC 0026.3+1719	00 26 18.	+ 17 19	3090		CLUSTER OF GALAXIES
UGC 00288	00 26 18.	+ 43 09	84	16.0	GALAXY PECULR
MCG-02-02-025	00 26 18.	- 08 34	30	15.	GALAXY
LB 07863	00 26 18.	- 21 23		18.8	FAINT BLUE STAR
LB 07864	00 26 18.	- 21 37		18.4	FAINT BLUE STAR
LB 07865	00 26 18.	- 21 54		19.6	FAINT BLUE STAR
LB 07866	00 26 19.	- 22 12		20.0	FAINT BLUE STAR
LB 07867	00 26 18.	- 24 14		18.7	FAINT BLUE STAR
LB 07868	00 26 18.	- 24 24		17.9	FAINT BLUE STAR
LB 07869	00 26 18.	- 25 35		19.1	FAINT BLUE STAR
LB 07870	00 26 18.	- 26 24		16.0	FAINT BLUE STAR
RNGC 0125	00 26 19.	+ 02 34		14.0	GALAXY
ARC 0043	00 26 20.	+ 17 19		15.9	RICH CLUSTER OF GALAXIES
UGC 00289	00 26 24.	+ 15 41	66	17.	GALAXY S
LB 07871	00 26 24.	- 20 54		18.0	FAINT BLUE STAR
LB 07872	00 26 24.	- 22 40		19.5	FAINT BLUE STAR
LB 07873	00 26 24.	- 23 18		17.3	FAINT BLUE STAR
LB 07874	00 26 24.	- 23 38		17.5	FAINT BLUE STAR
LB 07875	00 26 24.	- 25 46		19.2	FAINT BLUE STAR
LB 07876	00 26 24.	- 26 41		17.3	FAINT BLUE STAR
MCG-05-02-006	00 26 24.	- 32 42	60	14.	GALAXY
LIN 005	00 26 25.	- 74 28 48.		13.77	SMC EMISSION LINE OBJECT
MCG+00-02-049	00 26 27.	+ 02 32	48	14.5	GALAXY
PATH 1.013	00 26 27.	+ 30 44	19		NEBULA
MCG+00-02-050	00 26 30.	+ 02 36	27	15.	GALAXY
UGC 00290	00 26 30.	+ 15 37	126	18.	GALAXY DWARF
ZWG 500.032	00 26 30.	+ 30 43		15.0	GALAXY
UGC 00291	00 26 30.	+ 32 51	60	17.	GALAXY (DWARF)
MCG+06-02-007	00 26 30.	+ 39 12	12	16.5	GALAXY
4ZW 022	00 26 30.	+ 39 13			COMPACT GALAXY
LB 07877	00 26 30.	- 21 20		19.5	FAINT BLUE STAR
LB 07878	00 26 30.	- 22 54		17.2	FAINT BLUE STAR
LB 07879	00 26 30.	- 26 03		19.8	FAINT BLUE STAR
LB 07880	00 26 30.	- 26 19		18.9	FAINT BLUE STAR
LB 01557	00 26 30.	- 48 47		13.7	FAINT BLUE STAR
MCG+00-02-051	00 26 33.	+ 02 36	144	12.7	GALAXY
ZWG 383.028	00 26 36.	+ 02 32		15.5	GALAXY
MCG+00-02-052	00 26 36.	+ 02 36	30	15.	GALAXY
VB 089	00 26 36.	+ 61 13	316		STELLAR RING
MCG+00-02-053	00 26 36.	- 00 26 30.	24	14.5	GALAXY
LB 07881	00 26 36.	- 20 56		18.4	FAINT BLUE STAR
LB 07882	00 26 36.	- 21 01		16.9	FAINT BLUE STAR
LB 07883	00 26 36.	- 22 18		17.3	FAINT BLUE STAR
LB 07884	00 26 36.	- 22 27		17.8	FAINT BLUE STAR
LB 07885	00 26 36.	- 22 36		19.5	FAINT BLUE STAR
LB 07886	00 26 36.	- 23 22		19.4	FAINT BLUE STAR
LB 07887	00 26 36.	- 24 29		20.2	FAINT BLUE STAR
MCG-05-02-008	00 26 36.	- 31 06 30.	9	16.	GALAXY
MCG-05-02-007	00 26 36.	- 31 07	48	15.	GALAXY
RNGC 0126	00 26 37.	+ 02 32		15.5	GALAXY
RNGC 0127	00 26 37.	+ 02 36		15.0	GALAXY
IC 0021	00 26 37.	- 00 26 24.			NONSTELLAR OBJECT
LB 03839	00 26 39.	+ 59 44 18.		17.3	FAINT BLUE STAR
LB 03840	00 26 40.	+ 60 17 00.		19.0	FAINT BLUE STAR
ZWG 383.029	00 26 42.	+ 02 35		13.2	GALAXY
UGC 00292	00 26 42.	+ 02 35	192	13.2	GALAXY S0
ZC 0026.7+1025	00 26 42.	+ 10 25	6120		CLUSTER OF GALAXIES
UGC 00293	00 26 42.	+ 26 08	66	17.	GALAXY
4ZW 023	00 26 42.	+ 30 17			COMPACT GALAXY
UGC 00294	00 26 42.	+ 31 06	72	17.	GALAXY Sc
MCG-03-02-015	00 26 42.	- 16 09	54	13.	GALAXY
LB 07888	00 26 42.	- 21 21		15.7	FAINT BLUE STAR
LB 07889	00 26 42.	- 21 28		19.5	FAINT BLUE STAR
LB 07890	00 26 42.	- 22 09		18.6	FAINT BLUE STAR
LB 07891	00 26 42.	- 23 08		18.5	FAINT BLUE STAR
TON-S 0158	00 26 42.	- 23 58		14.6	BLUE STAR
LB 07892	00 26 42.	- 25 02		17.8	FAINT BLUE STAR
LB 07893	00 26 42.	- 25 05		17.9	FAINT BLUE STAR
TON-S 0159	00 26 42.	- 30 39		14.4	BLUE STAR
RNGC 0130	00 26 43.	+ 02 35		15.0	GALAXY
RNGC 0128	00 26 43.	+ 02 35		13.0	GALAXY
SCHO 0036	00 26 43.	+ 58 57 06.	390		ISOLATED DARK CLOUD
PATH 1.014	00 26 47.	+ 30 17	8		NEBULA
ZWG 500.033	00 26 48.	+ 30 17		14.8	GALAXY
MRK 551	00 26 48.	+ 30 17	14	16.5	GALAXY WITH UV CONTINUUM
KARA.72 011B	00 26 48.	+ 30 17	18		PART OF DOUBLE GALAXY
KARA.72 011A	00 26 48.	+ 30 17	24	14.8	PART OF DOUBLE GALAXY
5ZW 021	00 26 48.	+ 51 35			COMPACT GALAXY
LB 07894	00 26 48.	- 20 41		17.7	FAINT BLUE STAR
LB 07895	00 26 48.	- 21 15		19.8	FAINT BLUE STAR
LB 07896	00 26 48.	- 22 38		18.7	FAINT BLUE STAR
LB 07897	00 26 48.	- 25 30		19.0	FAINT BLUE STAR
LIN 007	00 26 50.	- 73 30 54.		17.3	SMC EMISSION LINE OBJECT
LIN 006	00 26 51.	- 72 52 42.		17.8	SMC EMISSION LINE OBJECT
MCG+02-02-013	00 26 54.	+ 11 18	36	15.	GALAXY
MCG+03-02-014	00 26 54.	+ 17 13	9	18.	GALAXY
MCG+03-02-013	00 26 54.	+ 17 13	27	17.	GALAXY
MCG+00-02-054	00 26 54.	- 00 30	30	15.	GALAXY
UGC 00295	00 26 54.	- 01 22	66	16.0	GALAXY SBb
LB 07898	00 26 54.	- 21 08		19.5	FAINT BLUE STAR
LB 07899	00 26 54.	- 21 37		18.2	FAINT BLUE STAR
MCG+00-02-055	00 26 57.	- 00 31 30.	18	15.5	GALAXY
LBN 0570	00 27	+ 18 10	3000		BRIGHT NEBULA
LBN 0595	00 27	+ 56 28	120		BRIGHT NEBULA
ZWG 409.019	00 27 00.	+ 04 51		15.7	GALAXY
UGC 00296	00 27 00.	+ 11 46	72	16.5	GALAXY TRP SYS
ZWG 457.018	00 27 00.	+ 21 12		15.4	GALAXY
UGC 00297	00 27 00.	+ 21 12	66	15.4	GALAXY S
KARA.73B 0023	00 27 00.	+ 21 12	60	15.4	ISOLATED GALAXY S
ZC 0027.0-0036	00 27 00.	- 00 36	5040		CLUSTER OF GALAXIES
MCG+00-02-056	00 27 00.	- 01 22 30.	60	14.5	GALAXY
MCG-02-02-026	00 27 00.	- 14 16	30	15.	GALAXY
LB 07900	00 27 00.	- 22 03		18.7	FAINT BLUE STAR
LB 07901	00 27 00.	- 22 17		18.1	FAINT BLUE STAR
LB 07902	00 27 00.	- 23 10		16.6	FAINT BLUE STAR
LB 07903	00 27 00.	- 24 07		18.0	FAINT BLUE STAR
ARC 0044	00 27 01.	+ 11 46		17.0	RICH CLUSTER OF GALAXIES
IC 0022	00 27 01.	- 09 21 36.			NONSTELLAR OBJECT
BC 3C13 7	00 27 03.	+ 39 32 12.			QUASI-STELLAR OBJECT
SHB 005	00 27 03.	+ 39 32 12.		19.	QUASI-STELLAR OBJECT
RNGC 0131	00 27 04.	- 33 33			GALAXY
ZWG 383.030	00 27 06.	+ 00 08		15.7	GALAXY
MCG+00-02-057	00 27 06.	+ 00 08 30.	42	14.5	GALAXY
ZWG 434.016	00 27 06.	+ 11 20		15.5	GALAXY
ZWG 500.034	00 27 06.	+ 29 47		15.4	GALAXY
5ZW 022	00 27 06.	+ 39 56			COMPACT GALAXY
URA 38	00 27 06.	+ 59 53	102		STELLAR RING
OCL 0294	00 27 06.	+ 59 57	3000	10.0	OPEN STAR CLUSTER
RNGC 0129	00 27 06.	+ 59 57		10.0	OPEN CLUSTER
MCG-02-02-027	00 27 06.	- 09 20	36	15.	GALAXY
MCG-03-02-016	00 27 06.	- 15 22	18	15.	GALAXY
MCG-03-02-017	00 27 06.	- 15 23	48	15.	GALAXY
LB 07904	00 27 06.	- 21 36		19.1	FAINT BLUE STAR
LB 07905	00 27 06.	- 22 28		16.7	FAINT BLUE STAR
LB 07906	00 27 06.	- 22 34		19.4	FAINT BLUE STAR
LB 07907	00 27 06.	- 23 21		18.6	FAINT BLUE STAR
LB 07908	00 27 06.	- 25 55		19.8	FAINT BLUE STAR
IC 1552	00 27 07.	+ 21 11 30.			NONSTELLAR OBJECT
SC 0024-6709.3	00 27 07.	- 66 52 42.	6		NEBULA
MCG+00-02-058	00 27 09.	+ 01 35	48	15.	GALAXY
MCG+00-02-059	00 27 12.	+ 02 40	54	14.5	GALAXY
UGC 00298	00 27 12.	+ 02 42	60	16.0	GALAXY Sb
MCG+03-02-015	00 27 12.	+ 21 12	48	15.	GALAXY
ZWG 500.035	00 27 12.	+ 31 07		15.3	GALAXY
UGC 00299	00 27 12.	+ 31 07	96	15.3	GALAXY Sc
MCG+05-02-019	00 27 12.	+ 31 08	48	15.	GALAXY
MCG+00-02-060	00 27 12.	- 00 07	36	14.5	GALAXY
MCG+00-02-061	00 27 12.	- 00 31 30.	18	15.5	GALAXY
LB 07909	00 27 12.	- 20 50		18.3	FAINT BLUE STAR
LB 07910	00 27 12.	- 21 55		19.5	FAINT BLUE STAR
LB 07911	00 27 12.	- 22 32		18.3	FAINT BLUE STAR
LB 07912	00 27 12.	- 22 50		19.7	FAINT BLUE STAR
LB 07913	00 27 12.	- 23 22		20.0	FAINT BLUE STAR
LB 07914	00 27 12.	- 25 19		18.6	FAINT BLUE STAR
LB 07915	00 27 12.	- 26 22		18.8	FAINT BLUE STAR
MCG-06-02-010	00 27 12.	- 33 33	90	13.5	GALAXY
MCG-01-02-022	00 27 15.	- 05 24	48	15.	GALAXY
ZWG 500.036	00 27 18.	+ 32 37		15.7	GALAXY
OCL 0293	00 27 18.	+ 57 42	300	12.	OPEN STAR CLUSTER
ZWG 383.031	00 27 18.	- 00 31		15.6	GALAXY
LB 07916	00 27 18.	- 23 07		20.1	FAINT BLUE STAR
LB 07917	00 27 18.	- 23 24		17.5	FAINT BLUE STAR
LB 07918	00 27 18.	- 23 40		19.6	FAINT BLUE STAR
LB 07919	00 27 18.	- 26 13		17.4	FAINT BLUE STAR
ARC 0045	00 27 19.	- 12 34		17.8	RICH CLUSTER OF GALAXIES
MCG+00-02-062	00 27 24.	+ 03 15	60	15.5	GALAXY
ZC 0027.4+2941	00 27 24.	+ 29 41	1410		CLUSTER OF GALAXIES

OBJECT NAME	RIGHT ASCEN.	DECLINATION	DIAM.	MAGN.	TYPE OF OBJECT
ZC 0027.4+3107	00 27 24.	+ 31 07	1280		CLUSTER OF GALAXIES
ZWG 519.009	00 27 24.	+ 38 03		15.5	GALAXY
URA 36	00 27 24.	+ 59 54	216		STELLAR RING
LB 07920	00 27 24.	- 25 39		18.6	FAINT BLUE STAR
LB 07921	00 27 24.	- 26 26		18.0	FAINT BLUE STAR
LB 07922	00 27 24.	- 26 34		17.9	FAINT BLUE STAR
MCG-05-02-009	00 27 24.	- 27 45	60	15.	GALAXY
HOLM 010B	00 27 26.	- 11 22	30	14.8	PART OF MULTIPLE GALAXY
LIN 008	00 27 26.	- 73 24 54.		16.36	SMC EMISSION LINE OBJECT
SC 0025-6543.7	00 27 29.	- 65 27 06.	12		NEBULA
UGC 00300	00 27 30.	+ 03 15	84	18.	GALAXY DWARF
MCG-02-02-029	00 27 30.	- 11 23	36	15.5	GALAXY
MCG-02-02-028	00 27 30.	- 11 34	60	15.5	GALAXY
LB 07923	00 27 30.	- 20 46		16.8	FAINT BLUE STAR
LB 07924	00 27 30.	- 21 46		19.0	FAINT BLUE STAR
LB 07925	00 27 30.	- 22 15		17.8	FAINT BLUE STAR
LB 07926	00 27 30.	- 22 58		17.0	FAINT BLUE STAR
LB 07927	00 27 30.	- 24 48		16.9	FAINT BLUE STAR
LB 07928	00 27 30.	- 24 54		17.9	FAINT BLUE STAR
SC 0025-6539.5	00 27 32.	- 65 22 54.	18		NEBULA
MCG+00-02-063	00 27 33.	+ 01 49 30.	108	12.5	GALAXY
FATH 1.015	00 27 33.	+ 30 25	8		NEBULA
ZWG 383.032	00 27 36.	+ 01 49		13.8	GALAXY
UGC 00301	00 27 36.	+ 01 49	114	13.8	GALAXY SBb/Sc
FATH 1.016	00 27 36.	+ 30 19	5		NEBULA
MCG-02-02-030	00 27 36.	- 11 23	120	13.	GALAXY
LB 07930	00 27 36.	- 20 43		19.3	FAINT BLUE STAR
LB 07929	00 27 36.	- 20 43		19.2	FAINT BLUE STAR
LB 07931	00 27 36.	- 21 19		19.8	FAINT BLUE STAR
LB 07932	00 27 36.	- 24 14		19.0	FAINT BLUE STAR
LB 07933	00 27 36.	- 25 22		17.2	FAINT BLUE STAR
RNGC 0132	00 27 37.	+ 01 49		14.0	GALAXY
HOLM 010A	00 27 38.	- 11 22	66	13.1	PART OF MULTIPLE GALAXY
HOW 03	00 27 39.	- 73 58 47.	24		STAR CLUSTER IN SMC
LB 03841	00 27 40.	+ 59 51 24.		17.8	FAINT BLUE STAR
ZC 0027.7+1822	00 27 42.	+ 18 22	2550		CLUSTER OF GALAXIES
UGC 00302	00 27 42.	+ 24 49	60	17.	GALAXY DWRF SP
URA 37	00 27 42.	+ 59 56	66		STELLAR RING
LB 07934	00 27 42.	- 20 42		16.5	FAINT BLUE STAR
LB 07935	00 27 42.	- 21 22		18.8	FAINT BLUE STAR
LB 07936	00 27 42.	- 21 54		19.0	FAINT BLUE STAR
LB 07937	00 27 42.	- 23 44		17.5	FAINT BLUE STAR
LB 07938	00 27 42.	- 26 08		18.6	FAINT BLUE STAR
SC 0025-6940.5	00 27 42.	- 69 23 54.	12		NEBULA
ABC 0046	00 27 43.	- 13 09		17.6	RICH CLUSTER OF GALAXIES
MCG+01-02-015	00 27 48.	+ 05 34	36	15.	GALAXY
ZWG 409.020	00 27 48.	+ 06 41		15.5	GALAXY
MCG+02-02-014	00 27 48.	+ 13 05	72	15.	GALAXY
4ZW 024	00 27 48.	+ 32 27			COMPACT GALAXY
SHRV.73B 01	00 27 48.	+ 41 20	10	15.91	COMPACT GALAXY
UGC 00303	00 27 48.	+ 41 49	126	16.5	GALAXY IRR
LB 07939	00 27 48.	- 21 31		19.6	FAINT BLUE STAR
LB 07940	00 27 48.	- 22 20		15.8	FAINT BLUE STAR
LB 07941	00 27 48.	- 22 24		17.8	FAINT BLUE STAR
LB 07942	00 27 48.	- 22 43		17.0	FAINT BLUE STAR
LB 07943	00 27 48.	- 23 03		16.9	FAINT BLUE STAR
LB 07944	00 27 48.	- 23 34		18.3	FAINT BLUE STAR
LB 07945	00 27 48.	- 24 00		18.7	FAINT BLUE STAR
LB 07946	00 27 48.	- 24 09		18.9	FAINT BLUE STAR
LB 07947	00 27 48.	- 26 34		19.8	FAINT BLUE STAR
MCG-06-02-011	00 27 48.	- 37 11	60	14.5	GALAXY
SNO 09	00 27 52.	- 23 14 48.	600	16.	LOOSE CLUSTER OF GALAXIES
RNGC 0134	00 27 52.	- 33 32		11.0	GALAXY
ZWG 409.021	00 27 54.	+ 05 36		15.2	GALAXY
MCG+02-02-015	00 27 54.	+ 12 23	36	15.5	GALAXY
ZWG 479.036	00 27 54.	+ 22 51		15.7	GALAXY
MCG+07-02-005	00 27 54.	+ 41 49	90	15.	GALAXY
MCG-03-02-018	00 27 54.	- 16 59	48	14.5	GALAXY
LB 07948	00 27 54.	- 21 13		19.2	FAINT BLUE STAR
LB 07949	00 27 54.	- 22 32		18.9	FAINT BLUE STAR
LB 07950	00 27 54.	- 24 06		17.7	FAINT BLUE STAR
LB 07951	00 27 54.	- 26 29		16.8	FAINT BLUE STAR
MCG-06-02-012	00 27 54.	- 33 32	420	11.4	GALAXY
LB 01558	00 27 54.	- 46 49		13.6	FAINT BLUE STAR
SC 0025-6537.7	00 27 54.	- 65 21 06.	12		NEBULA
YC 0127-42	00 27 55.	- 42 56 42.			UNUSUAL SOUTHERN NEBULA
LIN.CL 013	00 27 55.	- 73 40 01.	474	15.3	STAR CLUSTER IN SMC
LB 09772	00 28	- 82 15		14.2	FAINT BLUE STAR
ZWG 434.017	00 28 00.	+ 12 24		15.7	GALAXY
UGC 00304	00 28 00.	+ 12 24	72	15.7	GALAXY S0-a
ZWG 434.018	00 28 00.	+ 13 05		15.3	GALAXY
UGC 00305	00 28 00.	+ 13 05	96	15.3	GALAXY Sc
LB 07952	00 28 00.	- 21 05		19.2	FAINT BLUE STAR
LB 07953	00 28 00.	- 22 26		20.1	FAINT BLUE STAR
LB 07954	00 28 00.	- 23 19		18.2	FAINT BLUE STAR
LB 07955	00 28 00.	- 24 30		18.8	FAINT BLUE STAR
LIN 009	00 28 00.	- 73 53 31.		16.87	SMC EMISSION LINE OBJECT
ABC 0047	00 28 03.	- 24 26		17.5	RICH CLUSTER OF GALAXIES
MCG-02-02-031	00 28 06.	- 08 40	30	15.	GALAXY
MCG-02-02-033	00 28 06.	- 09 02	66	14.	GALAXY
MCG-02-02-034	00 28 06.	- 11 40	60	15.	GALAXY
MCG-02-02-032	00 28 06.	- 12 59	36	14.5	GALAXY
LB 07956	00 28 06.	- 20 41		17.7	FAINT BLUE STAR
LB 07957	00 28 06.	- 21 01		16.3	FAINT BLUE STAR
LB 07958	00 28 06.	- 21 14		19.5	FAINT BLUE STAR
LB 07959	00 28 06.	- 21 36		18.0	FAINT BLUE STAR
LB 07960	00 28 06.	- 21 47		16.7	FAINT BLUE STAR
LB 07961	00 28 06.	- 25 16		17.8	FAINT BLUE STAR
MCG-06-02-013	00 28 06.	- 37 42	48	15.	GALAXY
MCG+02-02-016	00 28 12.	+ 10 22	36	15.5	GALAXY
ZWG 535.009	00 28 12.	+ 41 55		15.7	GALAXY
UGC 00306	00 28 12.	+ 41 55	60	15.7	GALAXY Sc
LB 07962	00 28 12.	- 21 55		17.7	FAINT BLUE STAR
LB 07963	00 28 12.	- 23 24		18.3	FAINT BLUE STAR
LB 07964	00 28 12.	- 23 32		18.6	FAINT BLUE STAR
LB 07965	00 28 12.	- 25 57		16.4	FAINT BLUE STAR
MCG-05-02-011	00 28 12.	- 29 00	48	15.	GALAXY
MCG-05-02-010	00 28 12.	- 29 49	24	16.	GALAXY
RNGC 0133	00 28 17.	+ 63 05		9.5	OPEN CLUSTER
MCG+02-02-017	00 28 18.	+ 09 56	72	14.	GALAXY
UGC 00307	00 28 18.	+ 10 23	78	16.0	GALAXY S0
ZC 0028.3+1423	00 28 18.	+ 14 23	3230		CLUSTER OF GALAXIES
MCG+07-02-006	00 28 18.	+ 41 54	60	15.5	GALAXY
LB 07966	00 28 18.	- 23 00		19.2	FAINT BLUE STAR
MCG-05-02-012	00 28 18.	- 29 38	48	15.	GALAXY
IC 0023	00 28 20.	- 13 00 32.			NONSTELLAR OBJECT
HOW 04	00 28 22.	- 74 05 11.	18		STAR CLUSTER IN SMC
ZWG 409.022	00 28 24.	+ 04 49		15.5	GALAXY

OBJECT NAME	RIGHT ASCEN.	DECLINATION	DIAM.	MAGN.	TYPE OF OBJECT
MCG+01-02-016	00 28 24.	+ 04 52	72	14.	GALAXY
ZWG 409.023	00 28 24.	+ 04 54		14.8	GALAXY
UGC 00308	00 28 24.	+ 04 54	84	14.8	GALAXY Sa
ZWG 409.024	00 28 24.	+ 05 21		15.6	GALAXY
KARA.73B 0024	00 28 24.	+ 05 21	24	15.6	ISOLATED GALAXY S0
4ZW 025	00 28 24.	+ 25 47			COMPACT GALAXY
OCL 0296	00 28 24.	+ 63 05	450	9.6	OPEN STAR CLUSTER
MCG-02-02-035	00 28 24.	- 09 27	48	15.	GALAXY
LB 07967	00 28 24.	- 22 50		19.5	FAINT BLUE STAR
LB 07968	00 28 24.	- 23 49		18.5	FAINT BLUE STAR
LB 07969	00 28 24.	- 24 59		17.4	FAINT BLUE STAR
LB 07970	00 28 24.	- 26 00		18.5	FAINT BLUE STAR
LB 07971	00 28 24.	- 26 43		16.0	FAINT BLUE STAR
RNGC 0139	00 28 25.	+ 04 49		15.5	GALAXY
RNGC 0138	00 28 25.	+ 04 54		15.0	GALAXY
MCG-01-02-023	00 28 27.	- 07 30 30.	60	15.	GALAXY
MCG+01-02-017	00 28 30.	+ 04 16	36	15.	GALAXY
ZWG 434.019	00 28 30.	+ 09 55		14.2	GALAXY
RNGC 0137	00 28 30.	+ 09 55		14.0	GALAXY
UGC 00309	00 28 30.	+ 09 55	120	14.2	GALAXY S0
KARA.73B 0025	00 28 30.	+ 09 55	60	14.2	ISOLATED GALAXY F
MCG-02-02-036	00 28 30.	- 09 40	48	15.	GALAXY
LB 07972	00 28 30.	- 24 57		16.7	FAINT BLUE STAR
SVEN 010	00 28 32.	- 10 05	18	15.5	GALAXY
MCG-01-02-024	00 28 33.	- 03 20	48	15.	GALAXY
LB 03842	00 28 35.	+ 60 15 54.		17.9	FAINT BLUE STAR
ZWG 409.025	00 28 36.	+ 04 14		15.6	GALAXY
ZC 0028.6+0642	00 28 36.	+ 06 42	340		CLUSTER OF GALAXIES
MCG+01-02-018	00 28 36.	+ 08 12	36	15.	GALAXY
MCG+05-02-020	00 28 36.	+ 28 42 30.	66	15.	GALAXY
ZWG 500.037	00 28 36.	+ 28 43		14.8	GALAXY
UGC 00310	00 28 36.	+ 28 43	78	14.8	GALAXY Sc
IC 0024	00 28 36.	+ 30 34			OPEN CLUSTER
ZC 0028.6+4013	00 28 36.	+ 40 13	1210		CLUSTER OF GALAXIES
LB 07973	00 28 36.	- 22 46		18.7	FAINT BLUE STAR
LB 07974	00 28 36.	- 23 59		18.9	FAINT BLUE STAR
LB 07975	00 28 36.	- 24 54		18.8	FAINT BLUE STAR
LB 07976	00 28 36.	- 25 25		18.8	FAINT BLUE STAR
LB 07977	00 28 36.	- 26 15		18.7	FAINT BLUE STAR
ABC 0048	00 28 37.	+ 12 17		17.6	RICH CLUSTER OF GALAXIES
IC 0025	00 28 39.	- 00 40 14.			NONSTELLAR OBJECT
MCG+00-02-064	00 28 39.	- 00 41	48	15.	GALAXY
RNGC 0140	00 28 41.	+ 30 31		14.0	GALAXY
RNGC 0136	00 28 41.	+ 61 15		11.5	OPEN CLUSTER
MCG+01-02-019	00 28 42.	+ 08 11 30.	72	13.5	GALAXY
ZWG 409.026	00 28 42.	+ 08 12		15.0	GALAXY
MRK 552	00 28 42.	+ 08 12	20	15.	GALAXY WITH UV CONTINUUM
HOLM 011B	00 28 42.	+ 08 13	30	14.0	PART OF MULTIPLE GALAXY
ZWG 500.038	00 28 42.	+ 30 31		14.2	GALAXY
UGC 00311	00 28 42.	+ 30 31	90	14.2	GALAXY Sc
MCG+05-02-021	00 28 42.	+ 30 31	90	14.	GALAXY
FATH 2.002	00 28 42.	+ 30 31	5		NEBULA
OCL 0295	00 28 42.	+ 61 15	360	14.	OPEN STAR CLUSTER
MRSL 120-00/1	00 28 42.	+ 62 12	2400		HII REGION
ZWG 383.033	00 28 42.	- 00 40		15.6	GALAXY
LB 07978	00 28 42.	- 22 09		17.4	FAINT BLUE STAR
MCG-04-02-014	00 28 42.	- 22 53	60	14.	GALAXY
LB 07979	00 28 42.	- 24 09		18.8	FAINT BLUE STAR
TON-S 0160	00 28 42.	- 25 52		15.2	BLUE STAR
TON-S 0161	00 28 42.	- 27 29		15.2	BLUE STAR
MCG-06-02-014	00 28 42.	- 37 12	24	15.	GALAXY
MCG+01-02-021	00 28 45.	+ 05 55 30.	60	14.5	GALAXY
MCG+01-02-020	00 28 45.	+ 08 08 30.	60	14.	GALAXY
MCG-01-02-025	00 28 45.	- 08 09	72	14.5	GALAXY
MCG-02-02-038	00 28 45.	- 10 45	78	13.5	GALAXY
MCG-02-02-037	00 28 45.	- 11 08	48	15.	GALAXY
RNGC 0142	00 28 45.	- 22 53		14.0	GALAXY
HOLM 011A	00 28 47.	+ 08 12	108	13.5	PART OF MULTIPLE GALAXY
ZWG 409.027	00 28 48.	+ 04 55		15.4	GALAXY
ZC 0028.8+0613	00 28 48.	+ 06 13	2550		CLUSTER OF GALAXIES
ZWG 409.028	00 28 48.	+ 08 11		14.6	GALAXY
UGC 00312	00 28 48.	+ 08 11	102	14.6	GALAXY SB
SHRV.73B 02	00 28 48.	+ 40 37	10	15.94	COMPACT GALAXY
MCG-02-02-039	00 28 48.	- 12 06 30.	48	16.	GALAXY
LB 07980	00 28 48.	- 21 42		18.8	FAINT BLUE STAR
LB 07981	00 28 48.	- 22 01		18.5	FAINT BLUE STAR
MCG-04-02-015	00 28 48.	- 22 50	48	15.	GALAXY
LB 07982	00 28 48.	- 24 46		18.2	FAINT BLUE STAR
MCG-05-02-013	00 28 48.	- 28 41	30	15.5	GALAXY
MCG-06-02-015	00 28 48.	- 37 10	48	14.	GALAXY
RNGC 0141	00 28 49.	+ 04 55		15.5	GALAXY
SVEN 011	00 28 50.	- 09 46 56.	36	15.7	GALAXY
RNGC 0143	00 28 51.	- 22 50		15.0	GALAXY
ABC 0050	00 28 52.	- 22 30		17.4	RICH CLUSTER OF GALAXIES
ZWG 409.029	00 28 54.	+ 05 56		14.3	GALAXY
UGC 00313	00 28 54.	+ 05 56	84	14.3	GALAXY S
ZWG 409.030	00 28 54.	+ 08 07		15.4	GALAXY
UGC 00314	00 28 54.	+ 08 07	78	15.4	GALAXY S
MRSL 120-05/1	00 28 54.	+ 57 01	600		HII REGION
MCG-03-02-019	00 28 54.	- 20 03	108	14.5	GALAXY
LB 07983	00 28 54.	- 21 13		18.8	FAINT BLUE STAR
LB 07984	00 28 54.	- 21 38		19.7	FAINT BLUE STAR
LB 07985	00 28 54.	- 22 30		17.9	FAINT BLUE STAR
MCG-04-02-016	00 28 54.	- 22 55	36	14.	GALAXY
LB 07986	00 28 54.	- 24 00		20.0	FAINT BLUE STAR
TON-S 0162	00 28 54.	- 24 15		14.6	BLUE STAR
LB 07987	00 28 54.	- 25 23		18.2	FAINT BLUE STAR
LB 07988	00 28 54.	- 25 36		17.8	FAINT BLUE STAR
LB 07989	00 28 54.	- 26 00		18.6	FAINT BLUE STAR
LIN 010	00 28 54.	- 73 31 31.		14.35	SMC EMISSION LINE OBJECT
ABC 0049	00 28 55.	- 11 42		18.0	RICH CLUSTER OF GALAXIES
RNGC 0144	00 28 57.	- 22 55		14.0	GALAXY
HOW 05	00 28 57.	- 72 36 41.	12		STAR CLUSTER IN SMC
LBN 0574	00 29	+ 17 40	1500		BRIGHT NEBULA
LBN 0597	00 29	+ 57 00	480		BRIGHT NEBULA
UGC 00315	00 29 00.	+ 08 07	60	16.5	GALAXY Sb-c
ZC 0029.0+1154	00 29 00.	+ 11 54	1140		CLUSTER OF GALAXIES
ZC 0029.0+1204	00 29 00.	+ 12 04	3290		CLUSTER OF GALAXIES
ZWG 434.020	00 29 00.	+ 14 20		15.7	GALAXY
UGC 00316	00 29 00.	+ 14 20	84	15.7	GALAXY Sc
OCL 0297	00 29 00.	+ 62 53	420	10.	OPEN STAR CLUSTER
MCG-01-02-026	00 29 00.	- 06 23 30.	60	14.5	GALAXY
LB 07990	00 29 00.	- 22 18		16.9	FAINT BLUE STAR
LB 07991	00 29 00.	- 22 20		16.9	FAINT BLUE STAR
LB 07992	00 29 00.	- 22 36		18.7	FAINT BLUE STAR
LB 07993	00 29 00.	- 22 59		16.8	FAINT BLUE STAR
LB 07994	00 29 00.	- 23 09		18.0	FAINT BLUE STAR
LB 07995	00 29 00.	- 24 57		19.4	FAINT BLUE STAR

OBJECT NAME	RIGHT ASCEN.	DECLINATION	DIAM.	MAGN.	TYPE OF OBJECT
BC PKS0029-414	00 29 01.3	- 41 24 39.		17.5	QUASI-STELLAR OBJECT
ARC 0051	00 29 03.	- 23 56		17.4	RICH CLUSTER OF GALAXIES
UGC 00317	00 29 06.	+ 00 39	60	17.	GALAXY
MCG-02-02-040	00 29 06.	- 10 46	78	13.5	GALAXY
LB 07996	00 29 06.	- 21 15		17.0	FAINT BLUE STAR
LB 07997	00 29 06.	- 21 51		18.8	FAINT BLUE STAR
LB 07998	00 29 06.	- 23 29		18.9	FAINT BLUE STAR
LB 07999	00 29 06.	- 25 52		17.5	FAINT BLUE STAR
LB 08000	00 29 06.	- 26 16		18.3	FAINT BLUE STAR
LB 03142	00 29 06.	- 61 33		13.0	FAINT BLUE STAR
ARP 019	00 29 11.	- 05 26			PECULIAR GALAXY
ZWG 519.010	00 29 12.	+ 37 24		15.7	GALAXY
KARA.73B 0026	00 29 12.	+ 37 24	36	15.7	ISOLATED GALAXY S
MCG+06-02-008	00 29 12.	+ 37 25	30	15.	GALAXY
SHRV.73B 03	00 29 12.	+ 42 36	8	17.25	COMPACT GALAXY
MCG-01-02-027	00 29 12.	- 05 26 30.	96	12.	GALAXY
MCG-02-02-041	00 29 12.	- 11 16	24	15.	GALAXY
MCG-02-02-042	00 29 12.	- 13 14	30	15.	GALAXY
LB 08001	00 29 12.	- 21 20		20.1	FAINT BLUE STAR
LB 08002	00 29 12.	- 23 40		19.8	FAINT BLUE STAR
LB 08003	00 29 12.	- 24 01		18.8	FAINT BLUE STAR
LB 08004	00 29 12.	- 25 14		18.5	FAINT BLUE STAR
LB 08005	00 29 12.	- 26 22		19.2	FAINT BLUE STAR
RNGC 0145	00 29 14.	- 05 26		12.0	GALAXY
IC 0026	00 29 14.	- 13 36 56.			NONSTELLAR OBJECT
MCG-01-02-028	00 29 15.	- 02 47 30.	60	16.	GALAXY
SNO 10	00 29 17.	- 23 57 07.	300	19.	GROUP OF 10 GALAXIES
UGC 00318	00 29 18.	+ 37 24	66	16.5	GALAXY SBc
LB 08006	00 29 18.	- 20 43		19.4	FAINT BLUE STAR
LB 08007	00 29 18.	- 25 42		18.7	FAINT BLUE STAR
MCG-05-02-014	00 29 18.	- 26 59	162	14.	GALAXY
AGU 04	00 29 18.	- 44 26 00.	36	15.	PECULIAR GALAXY
LB 01559	00 29 18.	- 47 41		10.9	FAINT BLUE STAR
LIN 011	00 29 18.	- 73 33 55.		15.47	SMC EMISSION LINE OBJECT
ARC 0052	00 29 19.	- 12 28		18.0	RICH CLUSTER OF GALAXIES
ZWG 383.034	00 29 24.	+ 02 21		15.7	GALAXY
ZWG 500.039	00 29 24.	+ 30 51		15.7	GALAXY
LB 08008	00 29 24.	- 21 36		18.3	FAINT BLUE STAR
LB 08009	00 29 24.	- 22 16		16.9	FAINT BLUE STAR
LB 08010	00 29 24.	- 26 14		17.6	FAINT BLUE STAR
SN 1954D	00 29 29.	+ 31 24		17.4	SUPERNOVA
ZC 0029.5+1750	00 29 30.	+ 17 50	470		CLUSTER OF GALAXIES
ZWG 500.040	00 29 30.	+ 31 08		15.7	GALAXY
ZWG 500.041	00 29 30.	+ 31 24		15.1	GALAXY
UGC 00319	00 29 30.	+ 31 24	96	15.1	GALAXY Sb-c
MCG+05-02-022	00 29 30.	+ 31 25	72	14.5	GALAXY
SHRV.73B 04	00 29 30.	+ 40 10	5	17.01	COMPACT GALAXY
MCG+08-02-004	00 29 30.	+ 50 30	24	16.	GALAXY
MCG-02-02-043	00 29 30.	- 08 36	72	15.	GALAXY
LB 08011	00 29 30.	- 21 44		16.8	FAINT BLUE STAR
LB 08012	00 29 30.	- 23 50		18.7	FAINT BLUE STAR
LB 08013	00 29 30.	- 26 24		19.5	FAINT BLUE STAR
LB 08014	00 29 30.	- 26 34		18.4	FAINT BLUE STAR
ZWG 535.010	00 29 36.	+ 39 40		15.7	GALAXY
LB 08015	00 29 36.	- 22 37		17.8	FAINT BLUE STAR
LB 08016	00 29 36.	- 23 04		18.2	FAINT BLUE STAR
LB 08017	00 29 36.	- 24 09		19.4	FAINT BLUE STAR
LB 08018	00 29 36.	- 24 22		19.3	FAINT BLUE STAR
SNO 11	00 29 39.	- 22 53 50.	480	12.	GROUP OF 3 GALAXIES
LB 03843	00 29 41.	+ 60 16 42.		17.8	FAINT BLUE STAR
MCG+07-02-007	00 29 42.	+ 39 40	30	17.	GALAXY
LB 03845	00 29 42.	+ 59 49 54.		17.5	FAINT BLUE STAR
LB 03844	00 29 42.	+ 59 49 54.		16.7	FAINT BLUE STAR
MCG-02-02-044	00 29 42.	- 12 12	42	15.5	GALAXY
LB 08019	00 29 42.	- 25 52		18.2	FAINT BLUE STAR
LH115-N003	00 29 42.	- 74 04	48		EMISSION NEBULA IN SMC
SNO 12	00 29 44.	- 23 17 44.	420	16.	GROUP OF 3 GALAXIES
MCG-01-02-029	00 29 45.	- 06 23 30.	18	15.	GALAXY
LIN 012	00 29 46.	- 74 03 50.		14.32	SMC EMISSION LINE OBJECT
VB 001	00 29 48.	+ 68 37	873		STELLAR RING
MCG+00-02-065	00 29 48.	- 02 02	30	15.	GALAXY
MCG-02-02-045	00 29 48.	- 12 11	24	16.	GALAXY
MCG-02-02-046	00 29 48.	- 12 20	30	14.5	GALAXY
LB 08020	00 29 48.	- 22 18		16.1	FAINT BLUE STAR
LB 08021	00 29 48.	- 22 49		16.6	FAINT BLUE STAR
LB 08022	00 29 48.	- 23 40		18.4	FAINT BLUE STAR
LB 08023	00 29 48.	- 23 43		17.7	FAINT BLUE STAR
LB 08024	00 29 48.	- 25 29		17.8	FAINT BLUE STAR
LB 08025	00 29 48.	- 26 04		20.0	FAINT BLUE STAR
SC 0027-5956.7	00 29 49.	- 59 40 08.	72		NEBULA
LIN 013	00 29 52.	- 74 03 56.		15.94	SMC EMISSION LINE OBJECT
UGC 00320	00 29 54.	+ 02 18	78	16.5	GALAXY Sc
MCG+00-02-066	00 29 54.	+ 02 18 30.	72	14.	GALAXY
MCG+04-02-027	00 29 54.	+ 23 07	66	15.	GALAXY
ZWG 479.037	00 29 54.	+ 23 08		15.4	GALAXY
UGC 00321	00 29 54.	+ 23 08	96	15.4	GALAXY SB?c
MCG-02-02-047	00 29 54.	- 10 43	24	15.5	GALAXY
LB 08026	00 29 54.	- 21 42		16.2	FAINT BLUE STAR
LB 08027	00 29 54.	- 23 29		18.3	FAINT BLUE STAR
LB 08028	00 29 54.	- 24 28		18.6	FAINT BLUE STAR
LB 08029	00 29 54.	- 26 38		18.9	FAINT BLUE STAR
SC 0027-5948.4	00 29 55.	- 59 31 50.	72		NEBULA
VDB.66G 003	00 30	+ 48 12	400		DWARF GALAXY
UGC 00322	00 30 00.	+ 01 13	78	16.5	GALAXY Sc
ZC 0030.0+0448	00 30 00.	+ 04 48	1340		CLUSTER OF GALAXIES
ZWG 500.042	00 30 00.	+ 30 28		15.2	GALAXY
MCG+05-02-023	00 30 00.	+ 30 29	66	15.	GALAXY
SHRV.73A 01	00 30 00.	+ 39 45	10	15.62	GLOBULAR CLUSTER IN M31
ZWG 557.001	00 30 00.	+ 53 00		15.6	GALAXY
MCG-02-02-048	00 30 00.	- 10 17	48	14.5	GALAXY
LB 08030	00 30 00.	- 22 25		18.8	FAINT BLUE STAR
LB 08031	00 30 00.	- 23 10		17.2	FAINT BLUE STAR
LB 08032	00 30 00.	- 24 03		16.3	FAINT BLUE STAR
LB 08033	00 30 00.	- 25 33		19.1	FAINT BLUE STAR
LB 08034	00 30 00.	- 26 08		18.7	FAINT BLUE STAR
LB 03143	00 30 00.	- 71 26		13.4	FAINT BLUE STAR
IC 1553	00 30 05.	- 25 52			NONSTELLAR OBJECT
MCG+00-02-067	00 30 06.	+ 01 11 30.	60	15.	GALAXY
ZC 0030.1+0344	00 30 06.	+ 03 44	1080		CLUSTER OF GALAXIES
ZWG 479.038	00 30 06.	+ 25 51		15.7	GALAXY
OCL 0298	00 30 06.	+ 61 35	90	18.	OPEN STAR CLUSTER
LB 08035	00 30 06.	- 20 56		16.0	FAINT BLUE STAR
LB 08036	00 30 06.	- 21 28		20.0	FAINT BLUE STAR
LB 08037	00 30 06.	- 22 56		19.5	FAINT BLUE STAR
LB 08038	00 30 06.	- 23 48		19.2	FAINT BLUE STAR
LB 08039	00 30 06.	- 24 58		18.7	FAINT BLUE STAR
LB 08040	00 30 06.	- 25 20		18.9	FAINT BLUE STAR
HN 0109	00 30 06.	- 25 52			NEBULA

OBJECT NAME	RIGHT ASCEN.	DECLINATION	DIAM.	MAGN.	TYPE OF OBJECT
LB 08041	00 30 06.	- 25 56		18.0	FAINT BLUE STAR
LB 08042	00 30 06.	- 26 00		19.5	FAINT BLUE STAR
ARC 0053	00 30 08.	- 07 50		17.2	RICH CLUSTER OF GALAXIES
MCG+00-02-068	00 30 09.	- 02 05 30.	42	15.	GALAXY
RNGC 0146	00 30 10.	+ 63 00		9.5	OPEN CLUSTER
MCG+02-02-018	00 30 12.	+ 11 27	48	14.5	GALAXY
ZWG 500.043	00 30 12.	+ 30 22		15.1	GALAXY
OCL 0299	00 30 12.	+ 63 01	480	9.8	OPEN STAR CLUSTER
MCG-03-02-020	00 30 12.	- 16 55	18	16.	GALAXY
LB 08043	00 30 12.	- 21 41		19.7	FAINT BLUE STAR
LB 08044	00 30 12.	- 21 46		18.7	FAINT BLUE STAR
LB 08045	00 30 12.	- 25 51		17.2	FAINT BLUE STAR
MCG-04-02-017	00 30 12.	- 25 51 30.	60	14.	GALAXY
LB 08046	00 30 12.	- 26 13		18.9	FAINT BLUE STAR
MCG-06-02-016	00 30 12.	- 33 27	24	16.5	GALAXY
MCG-02-02-049	00 30 15.	- 11 35	108	13.5	GALAXY
HELW 002	00 30 16.	- 33 27			NEBULA
MCG+01-02-022	00 30 18.	+ 05 00 30.	30	16.	GALAXY
ZWG 434.021	00 30 18.	+ 11 27		15.3	GALAXY
UGC 00323	00 30 18.	+ 11 27	60	15.3	GALAXY S
MCG+08-02-005	00 30 18.	+ 48 12 30.	300	10.5	GALAXY
MCG-03-02-021	00 30 18.	- 18 54	36	15.	GALAXY
LB 08047	00 30 18.	- 23 32		19.6	FAINT BLUE STAR
LB 08048	00 30 18.	- 23 48		19.2	FAINT BLUE STAR
LB 08049	00 30 18.	- 24 41		18.7	FAINT BLUE STAR
LB 08050	00 30 18.	- 25 51		17.0	FAINT BLUE STAR
LB 08051	00 30 18.	- 26 23		17.4	FAINT BLUE STAR
LB 01560	00 30 18.	- 56 13		13.6	FAINT BLUE STAR
SVEN 013	00 30 20.	- 09 36	42	15.5	GALAXY
SVEN 012	00 30 20.	- 10 16 56.	36	14.6	GALAXY
HOLM 012B	00 30 20.	- 17 03	42	14.4	PART OF MULTIPLE GALAXY
HN 0110	00 30 22.	- 32 18			NEBULA
UGC 00324	00 30 24.	+ 02 54	60	16.5	GALAXY Sb
ZWG 409.031	00 30 24.	+ 04 59		15.7	GALAXY
ZWG 409.032	00 30 24.	+ 05 00		15.7	GALAXY
MCG+01-02-023	00 30 24.	+ 05 00	36	14.5	GALAXY
MCG+04-02-028	00 30 24.	+ 21 54	42	16.	GALAXY
UGC 00325	00 30 24.	+ 21 57	60	16.0	GALAXY
ZWG 550.006	00 30 24.	+ 48 14		12.0	GALAXY
UGC 00326	00 30 24.	+ 48 14	900	12.0	GALAXY DWRF EL
LB 08052	00 30 24.	- 22 00		18.8	FAINT BLUE STAR
LB 08053	00 30 24.	- 22 09		19.8	FAINT BLUE STAR
LB 08054	00 30 24.	- 23 05		17.0	FAINT BLUE STAR
LB 08055	00 30 24.	- 23 40		18.4	FAINT BLUE STAR
LB 08056	00 30 24.	- 24 17		19.4	FAINT BLUE STAR
LB 08057	00 30 24.	- 24 24		18.2	FAINT BLUE STAR
LB 08058	00 30 24.	- 25 53		17.1	FAINT BLUE STAR
HOLM 012A	00 30 25.	- 17 03	72	12.2	PART OF MULTIPLE GALAXY
MCG+01-02-024	00 30 27.	+ 04 59	36	15.	GALAXY
SC 0028-6830.9	00 30 28.	- 68 14 20.	12		NEBULA
MCG-02-02-050	00 30 30.	- 10 10	24	15.	GALAXY
LB 08059	00 30 30.	- 23 16		17.5	FAINT BLUE STAR
LB 08060	00 30 30.	- 24 57		18.8	FAINT BLUE STAR
LB 08061	00 30 30.	- 26 11		18.6	FAINT BLUE STAR
LH115-N002	00 30 30.	- 71 58			EMISSION NEBULA IN SMC
LIN.CL 014	00 30 30.	- 72 52 56.	342	15.3	STAR CLUSTER IN SMC
RNGC 0147	00 30 32.	+ 48 14		11.5	GALAXY
MCG-01-02-030	00 30 33.	- 08 25	60	15.	GALAXY
IC 0027	00 30 35.	- 13 38 58.			NONSTELLAR OBJECT
MCG+01-02-025	00 30 36.	+ 07 37	60	15.	GALAXY
ZWG 409.033	00 30 36.	+ 07 38		15.7	GALAXY
UGC 00327	00 30 36.	+ 07 38	72	15.7	GALAXY Sb-c
KARA.72 012A	00 30 36.	+ 07 38	78	15.7	PART OF DOUBLE GALAXY
ZC 0030.6+1722	00 30 36.	+ 17 22	1950		CLUSTER OF GALAXIES
LB 08062	00 30 36.	- 22 08		19.5	FAINT BLUE STAR
LB 08063	00 30 36.	- 23 22		18.0	FAINT BLUE STAR
LB 08064	00 30 36.	- 23 23		18.7	FAINT BLUE STAR
LB 08065	00 30 36.	- 23 42		18.7	FAINT BLUE STAR
MCG-05-02-015	00 30 36.	- 32 33	72	14.	GALAXY
IC 0028	00 30 37.	- 13 43 52.			NONSTELLAR OBJECT
SVEN 015	00 30 38.	- 09 39 57.	60	13.5	GALAXY
SVEN 014	00 30 38.	- 09 41	90	13.7	GALAXY
LIN 015	00 30 39.	- 73 56 08.		15.93	SMC EMISSION LINE OBJECT
IC 1554	00 30 39.5	- 32 32 04.			GALAXY
MCG+00-02-069	00 30 42.	- 00 04	36	15.	GALAXY
MCG-02-02-051	00 30 42.	- 13 25	60	14.	GALAXY
LB 08067	00 30 42.	- 21 19		20.3	FAINT BLUE STAR
LB 08066	00 30 42.	- 21 19		19.1	FAINT BLUE STAR
LB 08068	00 30 42.	- 22 59		18.1	FAINT BLUE STAR
LB 08069	00 30 42.	- 23 30		19.1	FAINT BLUE STAR
LB 08070	00 30 42.	- 26 02		19.8	FAINT BLUE STAR
LB 08071	00 30 42.	- 26 08		19.9	FAINT BLUE STAR
LIN 014	00 30 43.	- 71 59 08.		16.39	PLANETARY NEBULA IN SMC
RNGC 0135	00 30 44.	- 13 38			GALAXY
MCG+01-02-026	00 30 45.	+ 07 34 30.	48	15.	GALAXY
RNGC 0152	00 30 46.	- 73 25		12.0	GLOBULAR CLUSTER IN SMC
ZC 0030.8+0712	00 30 48.	+ 07 12	340		CLUSTER OF GALAXIES
ZWG 409.034	00 30 48.	+ 07 35		15.6	GALAXY
KARA.72 012B	00 30 48.	+ 07 35	48	15.6	PART OF DOUBLE GALAXY
ZC 0030.8+1806	00 30 48.	+ 18 06	2690		CLUSTER OF GALAXIES
MCG+04-02-029	00 30 48.	+ 22 37	36	16.	GALAXY
4ZW 026	00 30 48.	+ 30 35			COMPACT GALAXY
SHRV.73B 05	00 30 48.	+ 39 25	8	16.50	COMPACT GALAXY
UGC 00328	00 30 48.	- 01 23	96	16.0	GALAXY
MCG+00-02-070	00 30 48.	- 01 23 30.	36	15.	GALAXY
LB 08072	00 30 48.	- 22 28		17.8	FAINT BLUE STAR
LB 08073	00 30 48.	- 22 48		18.2	FAINT BLUE STAR
LB 08074	00 30 48.	- 25 02		19.1	FAINT BLUE STAR
LB 08075	00 30 48.	- 26 20		17.5	FAINT BLUE STAR
SVEN 016	00 30 50.	- 10 09 57.	12	15.4	GALAXY
MCG-02-02-052	00 30 51.	- 14 17	48	15.	GALAXY
SC 0028-6801.5	00 30 51.	- 67 44 56.	12		NEBULA
LIN.CL 015	00 30 52.	- 73 25 14.	816	11.9	STAR CLUSTER IN SMC
ZWG 457.019	00 30 54.	+ 19 40		15.5	GALAXY
ZWG 479.039	00 30 54.	+ 22 38		15.7	GALAXY
KARA.73B 0027	00 30 54.	+ 22 38	54	15.7	ISOLATED GALAXY S
MCG+04-02-030	00 30 54.	+ 23 07	48	15.5	GALAXY
MCG+06-02-009	00 30 54.	+ 39 18	66	15.	GALAXY
ZC 0030.9+4255	00 30 54.	+ 42 55	4700		CLUSTER OF GALAXIES
DG 004	00 30 54.	+ 55 32	240		REFLECTION NEBULA
URA 41	00 30 54.	+ 57 49	714		STELLAR RING
LB 08076	00 30 54.	- 20 41		17.0	FAINT BLUE STAR
LB 08077	00 30 54.	- 22 13		18.7	FAINT BLUE STAR
LB 08078	00 30 54.	- 23 09		17.0	FAINT BLUE STAR
LB 08079	00 30 54.	- 24 47		19.4	FAINT BLUE STAR
LB 08080	00 30 54.	- 25 11		17.6	FAINT BLUE STAR
LB 08081	00 30 54.	- 25 14		19.7	FAINT BLUE STAR
LB 08082	00 30 54.	- 25 34		18.7	FAINT BLUE STAR

OBJECT NAME	RIGHT ASCEN.	DECLINATION	DIAM.	MAGN.	TYPE OF OBJECT
LB 08083	00 30 54.	- 26 29		19.4	FAINT BLUE STAR
LB 08084	00 30 54.	- 26 44		16.0	FAINT BLUE STAR
SVEN 017	00 30 56.	- 09 49	36	16.2	GALAXY
HOW 06	00 30 58.	- 72 55 28.	18		STAR CLUSTER IN SMC
VDB.66G 004	00 31	+ 31 11	70		DWARF GALAXY
LBN 0599	00 31	+ 56 00	2220		BRIGHT NEBULA
KHAV 005	00 31	+ 60 53	8080		DARK NEBULA
VDB.66G 224	00 31	- 31 05	100		DWARF GALAXY
ZWG 383.035	00 31 00.	+ 02 24		15.7	GALAXY
UGC 00329	00 31 00.	+ 02 24	126	15.7	GALAXY Sc
MCG+00-02-071	00 31 00.	+ 02 25	120	14.	GALAXY
ZWG 479.040	00 31 00.	+ 23 08		15.4	GALAXY
ZWG 519.011	00 31 00.	+ 39 17		15.0	GALAXY
UGC 00330	00 31 00.	+ 39 17	84	15.0	GALAXY S0
LB 08085	00 31 00.	- 22 20		17.0	FAINT BLUE STAR
LB 08086	00 31 00.	- 22 58		18.2	FAINT BLUE STAR
LB 08087	00 31 00.	- 23 31		19.4	FAINT BLUE STAR
LB 08088	00 31 00.	- 24 21		18.9	FAINT BLUE STAR
LB 08089	00 31 00.	- 24 23		18.7	FAINT BLUE STAR
LB 08090	00 31 00.	- 24 30		18.6	FAINT BLUE STAR
LB 08091	00 31 00.	- 25 26		17.8	FAINT BLUE STAR
LB 03144	00 31 00.	- 76 44		12.5	FAINT BLUE STAR
RNGC 0149	00 31 04.	+ 30 27		15.0	GALAXY
ARC 0054	00 31 06.	+ 26 29		17.7	RICH CLUSTER OF GALAXIES
MCG+05-02-024	00 31 06.	+ 30 27	48	15.	GALAXY
ZC 0031.1+3141	00 31 06.	+ 31 41	1080		CLUSTER OF GALAXIES
VDB.66N 003	00 31 06.	+ 69 09	108		REFLECTION NEBULA
LB 08092	00 31 06.	- 21 54		17.5	FAINT BLUE STAR
LB 08093	00 31 06.	- 22 00		19.5	FAINT BLUE STAR
LB 08094	00 31 06.	- 26 26		18.6	FAINT BLUE STAR
LB 08095	00 31 06.	- 26 35		16.1	FAINT BLUE STAR
MCG+01-02-027	00 31 09.	+ 06 58	48	15.	GALAXY
ZC 0031.2+0602	00 31 12.	+ 06 02	870		CLUSTER OF GALAXIES
ZWG 409.035	00 31 12.	+ 06 58		15.2	GALAXY
UGC 00331	00 31 12.	+ 06 58	66	15.2	GALAXY SBa
ZC 0031.2+2629	00 31 12.	+ 26 29	1880		CLUSTER OF GALAXIES
ZWG 500.044	00 31 12.	+ 30 27		15.0	GALAXY
UGC 00332	00 31 12.	+ 30 27	90	15.0	GALAXY S0
MCG+05-02-025	00 31 12.	+ 31 10 30.	60	15.	GALAXY
ZWG 550.007	00 31 12.	+ 48 40		15.7	GALAXY
UGC 00333	00 31 12.	+ 48 40	78	15.7	GALAXY S
MCG+08-02-006	00 31 12.	+ 48 40	60	15.	GALAXY
LB 08096	00 31 12.	- 20 48		18.1	FAINT BLUE STAR
LB 08097	00 31 12.	- 20 54		18.6	FAINT BLUE STAR
LB 08098	00 31 12.	- 21 22		19.6	FAINT BLUE STAR
LB 08099	00 31 12.	- 24 23		17.0	FAINT BLUE STAR
LB 08100	00 31 12.	- 26 00		18.1	FAINT BLUE STAR
ZC 0031.3+0101	00 31 18.	+ 01 01	1340		CLUSTER OF GALAXIES
MCG+01-02-028	00 31 18.	+ 07 00	36	15.5	GALAXY
ZWG 500.045	00 31 18.	+ 31 10		15.7	GALAXY
UGC 00334	00 31 18.	+ 31 10	120	15.7	GALAXY DWRF SP
LB 08101	00 31 18.	- 26 44		18.4	FAINT BLUE STAR
REIN 4.003	00 31 18.76	- 09 52 21.3			NEBULA
ARC 0056	00 31 20.	- 08 05		18.0	RICH CLUSTER OF GALAXIES
MCG+01-02-029	00 31 21.	+ 06 59 30.	48	15.	GALAXY
REIN 4.004	00 31 21.56	- 09 55 23.7			NEBULA
RIC 001	00 31 23.80	+ 39 17 06.5		19.65	BLUE OBJECT
ARC 0055	00 31 24.	+ 06 03		17.2	RICH CLUSTER OF GALAXIES
ZWG 409.036	00 31 24.	+ 06 59		14.2	GALAXY
UGC 00335	00 31 24.	+ 06 59	150	14.2	GALAXY E+COMP
ZWG 535.011	00 31 24.	+ 43 52		15.6	GALAXY
UGC 00336	00 31 24.	+ 43 52	102	15.6	GALAXY Sc
KARA.73B 0028	00 31 24.	+ 43 52	66	15.6	ISOLATED GALAXY S
LB 08102	00 31 24.	- 23 14		17.2	FAINT BLUE STAR
LP 08103	00 31 24.	- 23 22		19.1	FAINT BLUE STAR
LB 08104	00 31 24.	- 23 31		18.7	FAINT BLUE STAR
LB 08105	00 31 24.	- 25 28		17.3	FAINT BLUE STAR
LB 08106	00 31 24.	- 26 13		19.8	FAINT BLUE STAR
RIC 002	00 31 24.06	+ 39 19 18.9		17.98	BLUE OBJECT
REIN 4.005	00 31 24.51	- 09 54 49.4			NEBULA
ARC 0057	00 31 25.	- 09 12		17.6	RICH CLUSTER OF GALAXIES
REIN 4.006	00 31 28.87	- 09 53 56.6			NEBULA
MCG+02-02-019	00 31 30.	+ 09 48	36	16.	GALAXY
MCG+07-02-008	00 31 30.	+ 43 52	60	14.5	GALAXY
MCG-02-02-053	00 31 30.	- 12 55	21	14.	GALAXY
LB 08107	00 31 30.	- 23 47		19.2	FAINT BLUE STAR
TON-S 0163	00 31 30.	- 27 23		14.2	BLUE STAR
REIN 2.002	00 31 30.75	- 09 58 52.1			NEBULA
SS 04	00 31 32.	+ 69 10			DIFFUSE GALACTIC NEBULA
ARC 0058	00 31 32.	- 07 03		17.2	RICH CLUSTER OF GALAXIES
SVEN 018	00 31 32.	- 09 58	180	12.0	GALAXY
RNGC 0154	00 31 32.	- 12 55		14.0	GALAXY
MCG-01-02-031	00 31 33.	- 03 26	60	14.5	GALAXY
HOW 07	00 31 34.	- 72 39 04.	12		STAR CLUSTER IN SMC
SCHO 0037	00 31 35.	+ 61 55 18.	290		ISOLATED DARK CLOUD
MCG+02-02-020	00 31 36.	+ 09 55	48	15.	GALAXY
UGC 00337	00 31 36.	+ 24 22	72	16.5	GALAXY Sc
ZC 0031.6+2859	00 31 36.	+ 28 59	610		CLUSTER OF GALAXIES
MCG+06-02-010	00 31 36.	+ 39 21	48	15.	GALAXY
SHRV.73B 06	00 31 36.	+ 41 12	7	16.95	COMPACT GALAXY
DG 005	00 31 36.	+ 69 10	120		REFLECTION NEBULA
ZWG 383.036	00 31 36.	- 02 27		15.6	GALAXY
MCG-02-02-054	00 31 36.	- 09 58	216	11.	GALAXY
LB 08108	00 31 36.	- 21 16		18.8	FAINT BLUE STAR
TON-S 0164	00 31 36.	- 23 16		15.7	BLUE STAR
LB 08109	00 31 36.	- 24 06		19.7	FAINT BLUE STAR
LB 08110	00 31 36.	- 25 29		18.7	FAINT BLUE STAR
MCG-06-02-017	00 31 36.	- 33 17	30	15.	GALAXY
MCG-06-02-019	00 31 36.	- 34 34	42	15.	GALAXY
MCG-06-02-018	00 31 36.	- 34 34	42	15.	GALAXY
IC 0029	00 31 37.	- 02 27 23.			NONSTELLAR OBJECT
RNGC 0151	00 31 38.	- 09 58		12.5	GALAXY
MCG+00-02-072	00 31 39.	- 02 27 30.	36	15.	GALAXY
IC 0030	00 31 41.	- 02 21 41.			NONSTELLAR OBJECT
ZC 0031.7+0034	00 31 42.	+ 00 34	670		CLUSTER OF GALAXIES
MCG+00-02-073	00 31 42.	+ 02 11 30.	18	15.5	GALAXY
ZWG 409.037	00 31 42.	+ 05 16		15.1	GALAXY
ZWG 409.038	00 31 42.	+ 05 40		15.7	GALAXY
MCG+02-02-021	00 31 42.	+ 11 59	108	15.	GALAXY
ZC 0031.7+2916	00 31 42.	+ 29 16	1010		CLUSTER OF GALAXIES
ZC 0031.7+3453	00 31 42.	+ 34 53	1610		CLUSTER OF GALAXIES
ZWG 519.012	00 31 42.	+ 39 20		15.5	GALAXY
UGC 00338	00 31 42.	+ 39 20	60	15.5	GALAXY SB?b-c
SHRV.73B 07	00 31 42.	+ 42 49	3	16.25	COMPACT GALAXY
5ZW 023	00 31 42.	+ 42 50			COMPACT GALAXY
MCG+00-02-074	00 31 42.	- 02 21 30.	24	15.	GALAXY
MCG-01-02-032	00 31 42.	- 07 52	36	15.5	GALAXY
LB 08111	00 31 42.	- 21 30		18.5	FAINT BLUE STAR
LB 08112	00 31 42.	- 23 00		17.1	FAINT BLUE STAR
LB 08113	00 31 42.	- 23 24		16.9	FAINT BLUE STAR
LB 08114	00 31 42.	- 24 24		18.0	FAINT BLUE STAR
LB 08115	00 31 42.	- 25 38		19.8	FAINT BLUE STAR
MCG-05-02-016	00 31 42.	- 31 03	120	14.	GALAXY
RIC 003	00 31 43.69	+ 39 27 30.		18.04	BLUE OBJECT
RNGC 0153	00 31 44.	- 09 59			NON-EXISTENT OBJECT
RIC 004	00 31 44.92	+ 41 10 48.1		19.16	BLUE OBJECT
MCG+04-02-031	00 31 45.	+ 24 18	60	17.	GALAXY
MCG-01-02-033	00 31 45.	- 07 35	48	15.5	GALAXY
MCG-04-02-018	00 31 45.	- 21 41 30.	78	13.5	GALAXY
RIC 005	00 31 45.31	+ 41 42 27.9		18.36	BLUE OBJECT
RNGC 0150	00 31 46.	- 28 05		12.5	GALAXY
RNGC 0148	00 31 46.	- 32 04		13.0	GALAXY
HOW 08	00 31 47.	- 73 54 22.	18		STAR CLUSTER IN SMC
RIC 006	00 31 47.74	+ 39 33 25.9		19.66	BLUE OBJECT
ZWG 383.037	00 31 48.	+ 02 10		15.7	GALAXY
ZWG 434.022	00 31 48.	+ 09 55		15.6	GALAXY
UGC 00339	00 31 48.	+ 09 55	78	15.6	GALAXY DBL SYS
ZWG 434.023	00 31 48.	+ 12 00		15.5	GALAXY
UGC 00340	00 31 48.	+ 12 00	84	15.5	GALAXY Sa-b
ZC 0031.8+3004	00 31 48.	+ 30 04	940		CLUSTER OF GALAXIES
SN 1966I	00 31 48.	+ 30 07		16.0	SUPERNOVA
LB 08116	00 31 48.	- 21 01		17.9	FAINT BLUE STAR
MCG-04-02-019	00 31 48.	- 21 44	36	15.	GALAXY
LB 08117	00 31 48.	- 23 38		19.6	FAINT BLUE STAR
LB 08118	00 31 48.	- 25 46		18.2	FAINT BLUE STAR
MCG-05-02-018	00 31 48.	- 28 04	270	12.	GALAXY
MCG-05-02-017	00 31 48.	- 32 05	108	13.5	GALAXY
IC 0031	00 31 49.	+ 11 59 43.			NONSTELLAR OBJECT
ARC 0059	00 31 49.	+ 30 00		17.5	RICH CLUSTER OF GALAXIES
RIC 007	00 31 50.03	+ 41 17 08.4		18.54	BLUE OBJECT
RIC 008	00 31 50.95	+ 39 02 04.7		17.63	BLUE OBJECT
RIC 009	00 31 51.24	+ 40 15 11.1		17.63	BLUE OBJECT
MCG+00-02-076	00 31 54.	+ 01 22 30.	24	16.	GALAXY
UGC 00341	00 31 54.	+ 01 24	72	16.0	GALAXY S
ZWG 383.038	00 31 54.	+ 03 18		15.7	GALAXY
UGC 00342	00 31 54.	+ 03 18	66	15.7	GALAXY Sc
MCG+00-02-075	00 31 54.	+ 03 20	60	15.	GALAXY
ZWG 409.039	00 31 54.	+ 08 10		15.7	GALAXY
SHRV.73B 08	00 31 54.	+ 39 18	10	15.36	COMPACT GALAXY
LB 08119	00 31 54.	- 21 12		19.6	FAINT BLUE STAR
LB 08120	00 31 54.	- 21 36		16.9	FAINT BLUE STAR
LB 08121	00 31 54.	- 22 40		18.2	FAINT BLUE STAR
LB 08122	00 31 54.	- 23 12		19.0	FAINT BLUE STAR
LB 08123	00 31 54.	- 26 24		19.2	FAINT BLUE STAR
LB 08124	00 31 54.	- 26 33		17.7	FAINT BLUE STAR
MCG+00-02-077	00 31 57.	- 02 30	36	17.	GALAXY
LBN 0602	00 32	+ 69 10	60		BRIGHT NEBULA
MCG+00-02-078	00 32 00.	+ 02 10	36	15.	GALAXY
MCG+06-02-011	00 32 00.	+ 39 18	54	15.	GALAXY
UGC 00343	00 32 00.	+ 53 10	84	16.0	GALAXY Sc-IRR
LDN 1286	00 32 00.	+ 58 40	960		DARK NEBULA
MCG+11-01-001	00 32 00.	+ 65 11	2	16.	GALAXY
MCG+11-01-001	00 32 00.	+ 65 11	2	16.	GALAXY
LB 08125	00 32 00.	- 21 48		18.7	FAINT BLUE STAR
LB 08126	00 32 00.	- 22 12		19.1	FAINT BLUE STAR
LB 08127	00 32 00.	- 22 18		18.0	FAINT BLUE STAR
LB 08128	00 32 00.	- 23 50		18.8	FAINT BLUE STAR
LB 08129	00 32 00.	- 24 34		19.2	FAINT BLUE STAR
LB 08130	00 32 00.	- 25 22		17.7	FAINT BLUE STAR
LB 08131	00 32 00.	- 25 52		18.4	FAINT BLUE STAR
RNGC 0156	00 32 02.	- 08 37			NON-EXISTENT OBJECT
RIC 010	00 32 05.04	+ 40 20 52.0		18.64	BLUE OBJECT
MCG+00-02-079	00 32 06.	+ 01 33	24	16.	GALAXY
ZWG 519.013	00 32 06.	+ 39 17		15.4	GALAXY
UGC 00344	00 32 06.	+ 39 17	72	15.4	GALAXY Sc/SBc
MCG-02-02-055	00 32 06.	- 11 02	30	13.5	GALAXY
LB 08132	00 32 06.	- 20 40		16.6	FAINT BLUE STAR
LB 08133	00 32 06.	- 24 51		19.8	FAINT BLUE STAR
LB 08134	00 32 06.	- 24 58		18.5	FAINT BLUE STAR
MCG-05-02-019	00 32 06.	- 30 18	72	14.	GALAXY
RNGC 0155	00 32 08.	- 11 02		13.0	GALAXY
IC 1555	00 32 08.	- 30 16 23.			NONSTELLAR OBJECT
RIC 011	00 32 08.84	+ 40 17 01.8		18.73	BLUE OBJECT
MCG-01-02-034	00 32 09.	- 08 12 30.	48	14.	GALAXY
ZWG 409.040	00 32 12.	+ 07 10		15.6	GALAXY
ZWG 500.046	00 32 12.	+ 28 08		15.6	GALAXY
UGC 00345	00 32 12.	+ 28 08	66	15.6	GALAXY SB?c
KARA.73B 0029	00 32 12.	+ 28 08	66	15.6	ISOLATED GALAXY S
MCG+05-02-026	00 32 12.	+ 31 40	48	14.5	GALAXY
8ZW 0032-13.6	00 32 12.	- 13 38		15.9	COMPACT GALAXY
LB 08135	00 32 12.	- 21 52		17.8	FAINT BLUE STAR
LB 08136	00 32 12.	- 22 01		19.6	FAINT BLUE STAR
LB 08137	00 32 12.	- 22 33		17.9	FAINT BLUE STAR
LB 08138	00 32 12.	- 22 36		17.4	FAINT BLUE STAR
LB 08139	00 32 12.	- 23 48		18.3	FAINT BLUE STAR
LB 08140	00 32 12.	- 26 36		18.4	FAINT BLUE STAR
TON-S 0165	00 32 12.	- 27 44		15.8	BLUE STAR
BAA 3.21	00 32 12.21	+ 39 33 09.8			GLOBULAR CLUSTER IN M31
RIC 012	00 32 13.61	+ 40 25 59.0		21.01	BLUE OBJECT
RNGC 0159	00 32 14.	- 56 03			GALAXY
RIC 013	00 32 14.27	+ 39 25 06.1		18.46	BLUE OBJECT
REIN 2.003	00 32 14.40	- 08 40 19.6			NEBULA
MCG-02-02-056	00 32 15.	- 08 40	270	10.	GALAXY
RIC 014	00 32 15.40	+ 41 12 34.2		19.77	BLUE OBJECT
RIC 015	00 32 16.71	+ 41 54 14.2		19.01	BLUE OBJECT
RIC 016	00 32 16.74	+ 41 27 48.4		18.37	BLUE OBJECT
RIC 017	00 32 17.78	+ 40 40 29.8		17.64	BLUE OBJECT
RIC 018	00 32 17.83	+ 41 47 28.8		18.33	BLUE OBJECT
ZWG 409.041	00 32 18.	+ 05 22		15.7	GALAXY
ZC 0032.3+2947	00 32 18.	+ 29 47	1340		CLUSTER OF GALAXIES
ZWG 500.047	00 32 18.	+ 31 40		14.8	GALAXY
UGC 00346	00 32 18.	+ 31 40	66	14.8	GALAXY Sb-c
SHRV.73B 09	00 32 18.	+ 42 12	8	16.52	COMPACT GALAXY
SHRV.73B 10	00 32 18.	+ 42 19	8	16.44	COMPACT GALAXY
UGC 00347	00 32 18.	+ 45 15	60	16.5	GALAXY Sc
MCG-01-02-035	00 32 18.	- 07 56 30.	60	15.5	GALAXY
MCG-02-02-057	00 32 18.	- 09 37	48	14.5	GALAXY
LB 08141	00 32 18.	- 22 42		17.8	FAINT BLUE STAR
LB 08142	00 32 18.	- 22 53		18.8	FAINT BLUE STAR
LB 08143	00 32 18.	- 22 58		16.0	FAINT BLUE STAR
LB 01561	00 32 18.	- 58 40		14.2	FAINT BLUE STAR
REIN 4.007	00 32 18.20	- 08 39 22.5			NEBULA
ARC 0060	00 32 19.	+ 29 45		17.9	RICH CLUSTER OF GALAXIES
RIC 019	00 32 19.70	+ 41 41 22.0		19.56	BLUE OBJECT
RNGC 0158	00 32 20.	- 08 35			NON-EXISTENT OBJECT
RNGC 0157	00 32 20.	- 08 40		11.5	GALAXY

OBJECT NAME	RIGHT ASCEN.	DECLINATION	DIAM.	MAGN.	TYPE OF OBJECT
ABC 0061	00 32 21.	- 23 29		17.7	RICH CLUSTER OF GALAXIES
LIN 016	00 32 21.	- 73 28 45.		17.06	PLANETARY NEBULA IN SMC
RIC 020	00 32 21.30	+ 40 15 35.2		18.37	BLUE OBJECT
RIC 021	00 32 22.13	+ 41 43 54.7		17.96	BLUE OBJECT
SHB 006	00 32 23.	+ 42 21 24.		18.3	QUASI-STELLAR OBJECT
BC 4C42.01	00 32 23.22	+ 42 21 48.3			QUASI-STELLAR OBJECT
RIC 022	00 32 23.24	+ 41 28 33.4		18.26	BLUE OBJECT
VB 002	00 32 24.	+ 65 36	309		STELLAR RING
MCG-02-02-058	00 32 24.	- 12 46	72	14.	GALAXY
LB 08144	00 32 24.	- 21 38		20.1	FAINT BLUE STAR
MCG-04-02-020	00 32 24.	- 22 09 30.	24	16.	GALAXY
LB 08145	00 32 24.	- 23 39		18.0	FAINT BLUE STAR
LB 08146	00 32 24.	- 25 37		19.2	FAINT BLUE STAR
LH115-N004	00 32 24.	- 73 30			EMISSION NEBULA IN SMC
RIC 023	00 32 25.31	+ 41 03 16.0		19.23	BLUE OBJECT
RIC 024	00 32 25.98	+ 41 58 47.2		18.26	BLUE OBJECT
SVEF 019	00 32 26.	- 09 36	12	14.5	GALAXY
RIC 025	00 32 26.06	+ 40 41 11.4		18.89	BLUE OBJECT
REIN 4.008	00 32 26.76	- 08 43 25.0			NEBULA
MCG+00-02-080	00 32 27.	- 02 25	24	15.	GALAXY
RIC 026	00 32 27.11	+ 41 45 38.2		18.19	BLUE OBJECT
RIC 027	00 32 27.81	+ 41 06 01.2		18.01	BLUE OBJECT
IC 0032	00 32 28.	- 02 25 24.			NONSTELLAR OBJECT
RIC 028	00 32 28.54	+ 41 58 09.8		17.72	BLUE OBJECT
MCG+01-02-031	00 32 30.	+ 05 15	36	15.5	GALAXY
MCG+01-02-030	00 32 30.	+ 05 15	18	16.	GALAXY
ZC 0032.5+0551	00 32 30.	+ 05 51	2020		CLUSTER OF GALAXIES
ZWG 434.024	00 32 30.	+ 14 05		15.3	GALAXY
ZWG 479.041	00 32 30.	+ 23 03		15.1	GALAXY
SHRV.73B 11	00 32 30.	+ 42 18	8	16.32	COMPACT GALAXY
MCG-02-02-059	00 32 30.	- 09 36	48	14.5	GALAXY
LB 08147	00 32 30.	- 24 55		19.2	FAINT BLUE STAR
MCG-05-02-020	00 32 30.	- 26 42	24	15.	GALAXY
BAA 3.22	00 32 31.92	+ 39 29 06.6			GLOBULAR CLUSTER IN M31
IC 0033	00 32 32.	- 02 25 00.			NONSTELLAR OBJECT
SVEN 020	00 32 32.	- 09 37	36	14.7	GALAXY
IC 1556	00 32 32.	- 09 50 30.			MAY NOT EXIST
SVEN 021	00 32 32.	- 10 14 58.	6	16.7	GALAXY
SC 0030-6535.4	00 32 32.	- 65 18 51.	12		NEBULA
RIC 029	00 32 32.24	+ 39 57 44.2		18.79	BLUE OBJECT
MCG+00-02-081	00 32 33.	- 00 24	30	15.5	GALAXY
MCG+00-02-082	00 32 33.	- 02 24 30.	30	15.	GALAXY
RIC 030	00 32 33.70	+ 40 51 00.6		18.82	BLUE OBJECT
MCG+00-02-084	00 32 36.	+ 00 26	45	15.5	GALAXY
ZC 0032.6+0113	00 32 36.	+ 01 13	4230		CLUSTER OF GALAXIES
ZC 0032.6+0207	00 32 36.	+ 02 07	10750		CLUSTER OF GALAXIES
MCG+00-02-083	00 32 36.	+ 03 09	18	16.	GALAXY
ZC 0032.6+2536	00 32 36.	+ 25 36	940		CLUSTER OF GALAXIES
ZWG 500.048	00 32 36.	+ 29 32		15.0	GALAXY
AND 3	00 32 36.	+ 36 14	270		DWARF SPHEROIDAL GALAXY
SHRV.73B 12	00 32 36.	+ 42 58	10	16.34	COMPACT GALAXY
LB 08148	00 32 36.	- 21 04		18.5	FAINT BLUE STAR
LB 08149	00 32 36.	- 22 03		17.8	FAINT BLUE STAR
LB 08150	00 32 36.	- 22 35		18.2	FAINT BLUE STAR
LB 03145	00 32 36.	- 62 45		13.6	FAINT BLUE STAR
RIC 031	00 32 37.15	+ 41 45 25.0		18.87	BLUE OBJECT
ABC 0062	00 32 40.	+ 20 07		17.1	RICH CLUSTER OF GALAXIES
RIC 032	00 32 40.83	+ 39 17 22.8		19.32	BLUE OBJECT
RIC 033	00 32 41.04	+ 42 16 28.2		18.25	BLUE OBJECT
RIC 034	00 32 41.93	+ 41 46 21.6		19.54	BLUE OBJECT
ZC 0032.7+0317	00 32 42.	+ 03 17	1340		CLUSTER OF GALAXIES
LB 08151	00 32 42.	- 20 46		19.2	FAINT BLUE STAR
LB 08152	00 32 42.	- 22 24		18.2	FAINT BLUE STAR
LB 08153	00 32 42.	- 23 32		18.4	FAINT BLUE STAR
LB 08154	00 32 42.	- 23 39		17.9	FAINT BLUE STAR
LB 08155	00 32 42.	- 25 14		17.4	FAINT BLUE STAR
LB 08156	00 32 42.	- 25 26		17.0	FAINT BLUE STAR
MCG-04-02-021	00 32 42.	- 26 21 30.	36	15.	GALAXY
RIC 035	00 32 43.03	+ 40 55 35.6		18.67	BLUE OBJECT
RIC 036	00 32 43.90	+ 39 20 27.6		19.28	BLUE OBJECT
MCG+00-02-085	00 32 45.	+ 03 08 30.	42	15.	GALAXY
SC 0030-6845.5	00 32 46.	- 68 28 57.	6		NEBULA
RIC 037	00 32 46.27	+ 41 54 28.3		19.24	BLUE OBJECT
RIC 038	00 32 47.27	+ 39 40 07.0		18.31	BLUE OBJECT
ZWG 383.039	00 32 48.	+ 02 39		15.7	GALAXY
UGC 00348	00 32 48.	+ 02 39	72	15.7	GALAXY
ZWG 383.040	00 32 48.	+ 03 07		15.6	GALAXY
UGC 00349	00 32 48.	+ 03 07	90	15.6	GALAXY SBb
MCG+00-02-086	00 32 48.	+ 03 10	48	16.	GALAXY
ZWG 409.042	00 32 48.	+ 04 36		15.7	GALAXY
UGC 00350	00 32 48.	+ 04 36	78	15.7	GALAXY S0-a
SHRV.73B 13	00 32 48.	+ 41 58	8	15.70	COMPACT GALAXY
MCG-02-02-060	00 32 48.	- 12 33	24	14.5	GALAXY
LB 08157	00 32 48.	- 22 19		17.7	FAINT BLUE STAR
LB 00151	00 32 48.	- 22 30		16.5	FAINT BLUE STAR
LB 08158	00 32 48.	- 24 15		19.7	FAINT BLUE STAR
LB 08159	00 32 48.	- 24 32		17.1	FAINT BLUE STAR
LB 08160	00 32 48.	- 24 53		18.9	FAINT BLUE STAR
LB 08161	00 32 48.	- 25 50		19.2	FAINT BLUE STAR
LB 08162	00 32 48.	- 26 06		19.8	FAINT BLUE STAR
MCG-05-02-021	00 32 48.	- 30 17	36	15.5	GALAXY
LB 01562	00 32 48.	- 59 15		14.3	FAINT BLUE STAR
RIC 039	00 32 48.02	+ 39 19 35.9		18.90	BLUE OBJECT
RIC 040	00 32 48.12	+ 40 27 54.1		18.26	BLUE OBJECT
BAA 2.55	00 32 48.79	+ 39 35 52.0			GALAXY
BPL 14	00 32 49.	+ 39 30 12.		19.03	PEC. BLUE OBJECT NEAR M31
RIC 041	00 32 49.88	+ 41 36 38.3		18.11	BLUE OBJECT
RNGC 0167	00 32 51.	- 23 38		13.0	GALAXY
RIC 042	00 32 52.92	+ 39 56 01.8		18.52	BLUE OBJECT
MCG+00-02-087	00 32 54.	+ 02 40	36	14.5	GALAXY
MCG+01-02-032	00 32 54.	+ 08 50 30.	144	13.5	GALAXY
ZWG 409.043	00 32 54.	+ 09 27		15.7	GALAXY
ZC 0032.9+2008	00 32 54.	+ 20 08	1550		CLUSTER OF GALAXIES
LB 08163	00 32 54.	- 21 09		19.6	FAINT BLUE STAR
LB 08164	00 32 54.	- 21 24		17.7	FAINT BLUE STAR
LB 08165	00 32 54.	- 22 16		17.6	FAINT BLUE STAR
LB 08166	00 32 54.	- 23 12		17.9	FAINT BLUE STAR
LB 08167	00 32 54.	- 23 30		19.1	FAINT BLUE STAR
MCG-04-02-022	00 32 54.	- 23 38 30.	48	13.5	GALAXY
MCG-06-02-020	00 32 54.	- 37 53	42	14.5	GALAXY
RIC 043	00 32 54.74	+ 39 52 40.8		19.38	BLUE OBJECT
RIC 044	00 32 54.91	+ 41 24 01.6		19.45	BLUE OBJECT
RNGC 0161	00 32 55.	- 03 06		15.0	GALAXY
RIC 045	00 32 55.22	+ 39 19 21.7		17.52	BLUE OBJECT
RIC 046	00 32 55.27	+ 42 05 49.8		19.23	BLUE OBJECT
RIC 074	00 32 55.86	+ 39 29 56.8		18.10	BLUE OBJECT
SVEF 022	00 32 56.	- 10 01	24	15.5	GALAXY
RIC 048	00 32 56.01	+ 39 02 15.2		20.49	BLUE OBJECT

OBJECT NAME	RIGHT ASCEN.	DECLINATION	DIAM.	MAGN.	TYPE OF OBJECT
RIC 049	00 32 56.49	+ 40 31 06.6		17.78	BLUE OBJECT
RIC 050	00 32 56.69	+ 41 48 00.2		20.26	BLUE OBJECT
MCG-01-02-036	00 32 57.	- 03 06 30.	23	15.	GALAXY
RIC 051	00 32 57.58	+ 39 28 42.6		18.50	BLUE OBJECT
RIC 052	00 32 57.59	+ 40 00 36.0		18.06	BLUE OBJECT
RIC 053	00 32 57.94	+ 41 49 14.1		18.56	BLUE OBJECT
MCG-01-02-037	00 32 58.	- 03 08	36	16.	GALAXY
RIC 054	00 32 59.14	+ 40 11 05.8		18.61	BLUE OBJECT
RIC 055	00 32 59.95	+ 41 46 07.3		18.22	BLUE OBJECT
VDB.66G 225	00 33	- 25 35	200		DWARF GALAXY
ZC 0033.0+0212	00 33 00.	+ 02 12	1080		CLUSTER OF GALAXIES
MCG+01-02-033	00 33 00.	+ 08 43	60	15.	GALAXY
ZWG 409.044	00 33 00.	+ 08 51		13.9	GALAXY
UGC 00351	00 33 00.	+ 08 51	204	13.9	GALAXY SBa
ZC 0033.0+1209	00 33 00.	+ 12 09	3230		CLUSTER OF GALAXIES
UGC 00352	00 33 00.	+ 24 00	84	16.5	GALAXY
ZWG 479.042	00 33 00.	+ 25 24		15.7	GALAXY
LDN 1287	00 33 00.	+ 63 10	540		DARK NEBULA
8ZW 0033-11.5	00 33 00.	- 11 33		17.2	COMPACT GALAXY
MCG-02-02-061	00 33 00.	- 11 43	30	15.	GALAXY
MCG-03-02-022	00 33 00.	- 20 26	138	13.5	GALAXY
LB 00152	00 33 00.	- 21 14		16.6	FAINT BLUE STAR
LB 08168	00 33 00.	- 23 39		18.4	FAINT BLUE STAR
LB 08169	00 33 00.	- 23 56		17.4	FAINT BLUE STAR
LB 08170	00 33 00.	- 23 57		19.5	FAINT BLUE STAR
BAA 2.54	00 33 00.19	+ 39 41 57.8			GALAXY
IC 0034	00 33 01.	+ 08 51 30.			NONSTELLAR OBJECT
LIN 017	00 33 01.	- 73 58 10.		15.50	SMC EMISSION LINE OBJECT
IC 1557	00 33 02.	- 03 09 12.			NONSTELLAR OBJECT
RIC 056	00 33 02.08	+ 40 40 15.6		18.49	BLUE OBJECT
RIC 057	00 33 02.62	+ 39 47 55.7		18.48	BLUE OBJECT
RIC 058	00 33 02.64	+ 41 21 21.2		17.94	BLUE OBJECT
MCG-01-02-038	00 33 03.	- 05 01	60	14.	GALAXY
RIC 059	00 33 03.12	+ 40 29 03.5		18.77	BLUE OBJECT
RIC 060	00 33 03.91	+ 41 17 27.3		19.35	BLUE OBJECT
RIC 061	00 33 04.07	+ 40 29 49.4		18.61	BLUE OBJECT
ABC 0063	00 33 05.	+ 49 25		16.1	RICH CLUSTER OF GALAXIES
RIC 062	00 33 05.46	+ 42 03 19.4		18.98	BLUE OBJECT
ZWG 409.045	00 33 06.	+ 08 43		15.6	GALAXY
UGC 00353	00 33 06.	+ 08 43	90	15.6	GALAXY Sc
ZC 0033.1+2925	00 33 06.	+ 29 25	1280		CLUSTER OF GALAXIES
ZC 0033.1+4926	00 33 06.	+ 49 26	2220		CLUSTER OF GALAXIES
LB 08171	00 33 06.	- 20 50		16.8	FAINT BLUE STAR
LB 08172	00 33 06.	- 21 59		18.0	FAINT BLUE STAR
LB 08173	00 33 06.	- 23 41		18.3	FAINT BLUE STAR
LB 08174	00 33 06.	- 25 59		18.8	FAINT BLUE STAR
RIC 063	00 33 06.56	+ 39 08 10.5		17.86	BLUE OBJECT
LIN 018	00 33 07.	- 73 53 46.		15.01	SMC EMISSION LINE OBJECT
RIC 064	00 33 07.36	+ 41 10 40.2		17.95	BLUE OBJECT
RIC 065	00 33 07.55	+ 40 37 25.6		17.32	BLUE OBJECT
BAA 2.52	00 33 07.84	+ 39 47 32.0			GALAXY
MCG-06-02-024	00 33 09.	- 33 15	48	15.5	GALAXY
RIC 066	00 33 09.48	+ 41 27 06.6		18.92	BLUE OBJECT
RIC 067	00 33 09.72	+ 39 18 29.0		18.34	BLUE OBJECT
BAA 3.20	00 33 10.51	+ 39 40 37.7			GLOBULAR CLUSTER IN M31
RIC 068	00 33 10.64	+ 41 35 51.0		17.98	BLUE OBJECT
MCG+05-02-027	00 33 12.	+ 31 36	60	14.5	GALAXY
ZC 0033.2-0312	00 33 12.	- 03 12	7260		CLUSTER OF GALAXIES
SN 1954S	00 33 12.	- 08 47		17.4	SUPERNOVA
MCG-02-02-062	00 33 12.	- 08 50 30.	30	15.	GALAXY
MCG-02-02-063	00 33 12.	- 13 53	54	15.	GALAXY
LB 08175	00 33 12.	- 21 14		19.5	FAINT BLUE STAR
LB 08176	00 33 12.	- 21 23		19.4	FAINT BLUE STAR
LB 08177	00 33 12.	- 21 54		19.3	FAINT BLUE STAR
LB 08178	00 33 12.	- 23 12		18.5	FAINT BLUE STAR
LB 08179	00 33 12.	- 23 48		19.0	FAINT BLUE STAR
LB 08180	00 33 12.	- 23 50		18.8	FAINT BLUE STAR
LB 08181	00 33 12.	- 23 51		18.6	FAINT BLUE STAR
MCG-04-02-023	00 33 12.	- 24 39 30.	60	15.5	GALAXY
LB 08182	00 33 12.	- 26 07		18.4	FAINT BLUE STAR
RIC 069	00 33 13.62	+ 41 34 56.2		19.46	BLUE OBJECT
RNGC 0166	00 33 14.	- 13 53		15.0	GALAXY
RIC 070	00 33 15.18	+ 40 55 36.4		18.91	BLUE OBJECT
RIC 071	00 33 15.36	+ 39 53 34.8		18.13	BLUE OBJECT
RIC 072	00 33 15.62	+ 40 55 50.2		19.60	BLUE OBJECT
RIC 073	00 33 17.55	+ 39 22 52.4		18.38	BLUE OBJECT
MCG+01-02-034	00 33 18.	+ 08 26	30	15.	GALAXY
ZC 0033.3+2028	00 33 18.	+ 20 28	740		CLUSTER OF GALAXIES
UGC 00354	00 33 18.	+ 23 42	60	16.0	GALAXY Sb-c
MCG+04-02-032	00 33 18.	+ 23 45	48	16.	GALAXY
ZWG 500.049	00 33 18.	+ 31 36		14.5	GALAXY
UGC 00355	00 33 18.	+ 31 36	72	14.5	GALAXY SBa-b
LB 08183	00 33 18.	- 22 04		17.3	FAINT BLUE STAR
LB 08184	00 33 18.	- 22 42		19.6	FAINT BLUE STAR
LB 08185	00 33 18.	- 23 32		19.8	FAINT BLUE STAR
LB 08186	00 33 18.	- 24 16		17.9	FAINT BLUE STAR
LB 08187	00 33 18.	- 26 34		16.7	FAINT BLUE STAR
MCG-05-02-022	00 33 18.	- 28 45	30	15.	GALAXY
RIC 074	00 33 18.34	+ 40 34 59.2		18.39	BLUE OBJECT
RIC 075	00 33 18.69	+ 40 56 46.0		18.91	BLUE OBJECT
SVEN 023	00 33 20.	- 10 20	12	15.9	GALAXY
RIC 076	00 33 20.19	+ 40 47 15.0		18.04	BLUE OBJECT
RIC 077	00 33 20.78	+ 40 26 05.0		17.77	BLUE OBJECT
RIC 078	00 33 20.88	+ 41 23 40.4		18.40	BLUE OBJECT
RIC 079	00 33 21.03	+ 39 05 30.9		19.69	BLUE OBJECT
RIC 080	00 33 21.68	+ 41 32 56.2		18.69	BLUE OBJECT
RIC 081	00 33 21.93	+ 38 53 49.2		20.04	BLUE OBJECT
SC 0031-6547.5	00 33 22.	- 65 31 21.	12		NEBULA
RIC 082	00 33 22.09	+ 39 28 24.2		18.69	BLUE OBJECT
RNGC 0160	00 33 23.	+ 23 41		13.5	GALAXY
HN 0111	00 33 23.	- 25 39			NEBULA
IC 1558	00 33 23.	- 25 39			NONSTELLAR OBJECT
RIC 083	00 33 23.88	+ 39 55 42.4		18.52	BLUE OBJECT
MCG+00-02-088	00 33 24.	+ 01 36 30.	60	14.	GALAXY
ZWG 409.046	00 33 24.	+ 08 27		15.0	GALAXY
MCG+02-02-022	00 33 24.	+ 12 21	36	15.	GALAXY
MCG+04-02-033	00 33 24.	+ 23 40	120	13.	GALAXY
ZWG 479.043	00 33 24.	+ 23 41		13.7	GALAXY
UGC 00356	00 33 24.	+ 23 41	192	13.7	GALAXY Sa
ZC 0033.4-0230	00 33 24.	- 02 30	1080		CLUSTER OF GALAXIES
LB 08188	00 33 24.	- 21 17		19.7	FAINT BLUE STAR
LB 08189	00 33 24.	- 21 26		19.1	FAINT BLUE STAR
LB 08190	00 33 24.	- 21 30		17.3	FAINT BLUE STAR
LB 08191	00 33 24.	- 22 19		19.7	FAINT BLUE STAR
MCG-04-02-024	00 33 24.	- 25 38 30.	192	13.	GALAXY
LB 08192	00 33 24.	- 26 32		19.4	FAINT BLUE STAR
RIC 084	00 33 24.75	+ 41 51 08.6		18.87	BLUE OBJECT
RIC 085	00 33 25.22	+ 41 28 56.2		18.12	BLUE OBJECT

OBJECT NAME	RIGHT ASCEN.	DECLINATION	DIAM.	MAGN.	TYPE OF OBJECT
RIC 086	00 33 25.32	+ 41 19 44.4		18.86	BLUE OBJECT
RIC 087	00 33 25.68	+ 42 06 08.4		19.25	BLUE OBJECT
SVEN 024	00 33 26.	- 10 22	12	15.5	GALAXY
RIC 088	00 33 27.14	+ 40 19 14.9		18.61	BLUE OBJECT
RIC 089	00 33 27.44	+ 39 49 13.4		20.12	BLUE OBJECT
RIC 090	00 33 27.61	+ 41 58 17.6		18.00	BLUE OBJECT
RIC 091	00 33 28.98	+ 39 06 35.2		19.55	BLUE OBJECT
RNGC 0162	00 33 29.	+ 23 42			GALAXY
RIC 092	00 33 29.02	+ 39 05 54.0		19.84	BLUE OBJECT
RIC 093	00 33 29.23	+ 41 52 30.8		18.19	BLUE OBJECT
RIC 094	00 33 29.73	+ 41 04 04.4		19.67	BLUE OBJECT
RIC 095	00 33 29.95	+ 39 06 25.8		17.98	BLUE OBJECT
ZWG 457.020	00 33 30.	+ 21 12		15.6	GALAXY
UGC 00357	00 33 30.	+ 21 12	78	15.6	GALAXY Sb-c
VB 003	00 33 30.	+ 65 02	517		STELLAR RING
MCG+14-01-008	00 33 30.	+ 81 27 30.	36	17.	GALAXY
MCG-02-02-064	00 33 30.	- 10 10	60	14.	GALAXY
MCG-02-02-066	00 33 30.	- 10 23	24	13.	GALAXY
MCG-02-02-065	00 33 30.	- 10 24	18	16.	GALAXY
LB 08193	00 33 30.	- 21 34		18.9	FAINT BLUE STAR
LB 08194	00 33 30.	- 24 30		19.5	FAINT BLUE STAR
LB 08195	00 33 30.	- 26 07		19.5	FAINT BLUE STAR
RIC 096	00 33 30.32	+ 40 30 01.0		18.18	BLUE OBJECT
BAA 2.53	00 33 30.65	+ 39 39 28.3			GALAXY
RNGC 0163	00 33 32.	- 10 24		13.0	GALAXY
RIC 097	00 33 32.12	+ 40 37 07.3		17.75	BLUE OBJECT
RIC 098	00 33 32.30	+ 41 29 58.9		19.75	BLUE OBJECT
RIC 099	00 33 32.96	+ 40 16 12.6		18.35	BLUE OBJECT
MCG+01-02-035	00 33 33.	+ 04 20 30.	48	15.5	GALAXY
RIC 100	00 33 33.16	+ 41 29 45.5		19.04	BLUE OBJECT
RIC 101	00 33 33.89	+ 40 07 03.4		17.70	BLUE OBJECT
BPL 15	00 33 34.	+ 39 06 18.		17.71	PEC. BLUE OBJECT NEAR M31
SC 0031-6548.2	00 33 34.	- 65 31 39.	12		NEBULA
RIC 102	00 33 34.30	+ 40 36 25.6		18.32	BLUE OBJECT
RIC 103	00 33 35.41	+ 39 07 53.6		20.57	BLUE OBJECT
ZWG 383.041	00 33 36.	+ 01 26		15.3	GALAXY
UGC 00358	00 33 36.	+ 01 26	78	15.3	GALAXY S
ZC 0033.6+0417	00 33 36.	+ 04 17	740		CLUSTER OF GALAXIES
ZWG 409.047	00 33 36.	+ 04 22		15.5	GALAXY
UGC 00359	00 33 36.	+ 04 22	66	15.5	GALAXY S
ZWG 434.025	00 33 36.	+ 12 22		15.3	GALAXY
UGC 00360	00 33 36.	+ 25 32	102	16.0	GALAXY
MCG+05-02-028	00 33 36.	+ 32 38	45	15.	GALAXY
MCG-01-02-040	00 33 36.	- 03 47	36	15.	GALAXY
MCG-01-02-039	00 33 36.	- 07 50	72	14.5	GALAXY
SN 1964J	00 33 36.	- 10 10		17.0	SUPERNOVA
MCG-02-02-067	00 33 36.	- 12 02	42	15.5	GALAXY
LB 08196	00 33 36.	- 22 20		17.3	FAINT BLUE STAR
LB 08197	00 33 36.	- 22 24		16.7	FAINT BLUE STAR
LB 08198	00 33 36.	- 22 32		17.0	FAINT BLUE STAR
LB 08199	00 33 36.	- 23 18		18.7	FAINT BLUE STAR
LB 08200	00 33 36.	- 25 15		17.8	FAINT BLUE STAR
RIC 104	00 33 36.74	+ 41 01 16.0		19.38	BLUE OBJECT
RIC 105	00 33 37.60	+ 39 53 30.8		17.94	BLUE OBJECT
RIC 106	00 33 37.83	+ 40 52 05.6		18.17	BLUE OBJECT
SVEN 025	00 33 38.	- 10 10	60	14.2	GALAXY
RIC 107	00 33 38.90	+ 40 20 57.2		18.34	BLUE OBJECT
MCG+01-02-036	00 33 39.	+ 07 32 30.	60	15.	GALAXY
RIC 108	00 33 39.37	+ 41 42 22.3		19.49	BLUE OBJECT
RIC 109	00 33 39.93	+ 40 26 04.6		17.76	BLUE OBJECT
RIC 110	00 33 40.69	+ 41 41 02.8		19.61	BLUE OBJECT
BAA 2.33	00 33 41.51	+ 39 22 07.4			GALAXY
RIC 111	00 33 41.65	+ 41 47 12.2		18.39	BLUE OBJECT
ZWG 500.050	00 33 42.	+ 32 38		15.2	GALAXY
SHRV.73B 14	00 33 42.	+ 41 41	7	16.79	COMPACT GALAXY
SHRV.73B 15	00 33 42.	+ 42 47	10	16.40	COMPACT GALAXY
ZWG 535.012	00 33 42.	+ 45 24		15.3	GALAXY
ZCG 0033+45	00 33 42.	+ 45 24		15.3	COMPACT GALAXY
ZWG 550.008	00 33 42.	+ 48 44		15.7	GALAXY
UGC 00361	00 33 42.	+ 48 44	72	15.7	GALAXY COMPACT
MCG-02-02-068	00 33 42.	- 12 32	72	14.	GALAXY
LB 08201	00 33 42.	- 22 00		17.7	FAINT BLUE STAR
LB 08202	00 33 42.	- 23 25		17.3	FAINT BLUE STAR
MCG-06-02-021	00 33 42.	- 32 53	24	14.5	GALAXY
RIC 112	00 33 42.15	+ 42 05 59.9		20.42	BLUE OBJECT
RIC 113	00 33 42.87	+ 41 43 31.8		19.10	BLUE OBJECT
RIC 114	00 33 43.09	+ 42 03 56.3		19.04	BLUE OBJECT
RIC 115	00 33 44.06	+ 41 13 23.5		18.66	BLUE OBJECT
MCG+03-02-016	00 33 45.	+ 18 25	24	17.	GALAXY
MCG-06-02-022	00 33 45.	- 32 52	36	14.5	GALAXY
RIC 116	00 33 45.56	+ 40 04 11.6		17.62	BLUE OBJECT
RIC 117	00 33 47.77	+ 38 55 22.7		19.87	BLUE OBJECT
MRK 553	00 33 48.	+ 02 54	13	16.5	GALAXY WITH UV CONTINUUM
ZC 0033.8+0538	00 33 48.	+ 05 38	9880		CLUSTER OF GALAXIES
ZC 0033.8+0615	00 33 48.	+ 06 15	1080		CLUSTER OF GALAXIES
LB 08203	00 33 48.	- 21 50		17.0	FAINT BLUE STAR
LB 08204	00 33 48.	- 23 29		18.6	FAINT BLUE STAR
LB 08205	00 33 48.	- 23 36		17.7	FAINT BLUE STAR
LB 08206	00 33 48.	- 25 42		19.8	FAINT BLUE STAR
RNGC 0176	00 33 48.	- 73 27		12.0	OPEN CLUSTER IN SMC
LIN.CL 017	00 33 48.	- 73 53 04.	816		STAR CLUSTER IN SMC
LIN.CL 016	00 33 49.	- 73 27 04.	678	11.9	STAR CLUSTER IN SMC
LIN 019	00 33 49.	- 73 38 22.		16.42	SMC EMISSION LINE OBJECT
RIC 118	00 33 49.67	+ 41 40 47.4		18.84	BLUE OBJECT
RIC 119	00 33 49.80	+ 41 17 26.4		18.13	BLUE OBJECT
SVEN 026	00 33 50.	- 09 39	6	16.4	GALAXY
RIC 120	00 33 50.39	+ 41 32 44.0		18.18	BLUE OBJECT
RIC 121	00 33 50.67	+ 41 31 47.0		18.42	BLUE OBJECT
MCG-01-02-041	00 33 51.	- 06 56 30.	30	15.	GALAXY
ARC 0064	00 33 52.	+ 18 38		17.5	RICH CLUSTER OF GALAXIES
RIC 122	00 33 52.87	+ 40 57 23.2		17.65	BLUE OBJECT
RIC 123	00 33 53.39	+ 41 51 22.9		18.30	BLUE OBJECT
RIC 124	00 33 53.52	+ 40 21 10.6		19.20	BLUE OBJECT
MCG+00-02-090	00 33 54.	+ 02 10	66	15.	GALAXY
MCG+00-02-089	00 33 54.	+ 02 29	24	16.	GALAXY
UGC 00362	00 33 54.	+ 32 29	72	17.	GALAXY DWRF SP
UGC 00363	00 33 54.	+ 43 32	60	16.5	GALAXY S
MCG-01-02-042	00 33 54.	- 06 36	60	15.5	GALAXY
LB 08207	00 33 54.	- 21 52		19.6	FAINT BLUE STAR
MCG-04-02-025	00 33 54.	- 23 17	60	15.	GALAXY
LB 08208	00 33 54.	- 23 22		19.7	FAINT BLUE STAR
LB 08209	00 33 54.	- 23 46		18.2	FAINT BLUE STAR
LB 08210	00 33 54.	- 23 58		19.5	FAINT BLUE STAR
LB 08211	00 33 54.	- 24 29		18.6	FAINT BLUE STAR
LB 08212	00 33 54.	- 25 40		20.0	FAINT BLUE STAR
MCG-05-02-023	00 33 54.	- 28 03	48	14.5	GALAXY
RIC 125	00 33 54.26	+ 40 35 48.4		18.48	BLUE OBJECT
RIC 126	00 33 54.32	+ 40 43 50.6		18.16	BLUE OBJECT
RIC 127	00 33 54.40	+ 41 24 02.5		18.40	BLUE OBJECT
RNGC 0164	00 33 55.	+ 02 29		16.0	GALAXY
RIC 128	00 33 55.67	+ 40 52 49.1		18.57	BLUE OBJECT
SVEN 027	00 33 56.	- 10 23	60	14.0	GALAXY
RNGC 0165	00 33 56.	- 10 24		13.0	GALAXY
RIC 129	00 33 56.02	+ 40 16 07.4		18.38	BLUE OBJECT
BAA 1.502	00 33 57.37	+ 39 29 38.5			EMISSION NEBULA IN M31
RIC 130	00 33 57.65	+ 40 32 26.4		18.48	BLUE OBJECT
KHAV 007	00 34	+ 65 35	5800		DARK NEBULA
KHAV 006	00 34	+ 69 23	6990		DARK NEBULA
ZWG 457.021	00 34 00.	+ 21 20		15.6	GALAXY
SHRV.73A 02	00 34 00.	+ 40 03	8	16.59	GLOBULAR CLUSTER IN M31
SHRV.73A 03	00 34 00.	+ 40 51	10	16.62	GLOBULAR CLUSTER IN M31
MCG-02-02-069	00 34 00.	- 10 22 30.	72	13.	GALAXY
LB 08213	00 34 00.	- 21 33		19.6	FAINT BLUE STAR
LB 08214	00 34 00.	- 24 07		19.8	FAINT BLUE STAR
LB 08215	00 34 00.	- 24 50		17.8	FAINT BLUE STAR
LB 08216	00 34 00.	- 25 06		17.0	FAINT BLUE STAR
LB 08217	00 34 00.	- 25 52		16.8	FAINT BLUE STAR
RIC 131	00 34 01.04	+ 42 01 04.8		18.48	BLUE OBJECT
RIC 132	00 34 01.67	+ 40 07 37.8		18.57	BLUE OBJECT
RIC 133	00 34 01.80	+ 41 42 32.4		18.79	BLUE OBJECT
RIC 134	00 34 01.82	+ 40 51 56.8		18.39	BLUE OBJECT
SVEN 028	00 34 02.	- 10 18 59.	24	15.2	GALAXY
RIC 135	00 34 02.26	+ 41 42 32.5		19.11	BLUE OBJECT
RIC 136	00 34 02.40	+ 42 11 10.9		19.84	BLUE OBJECT
RIC 137	00 34 02.60	+ 40 59 53.7		18.92	BLUE OBJECT
BAA 3.19	00 34 04.42	+ 40 03 12.3			GLOBULAR CLUSTER IN M31
RIC 138	00 34 05.90	+ 41 45 50.0		18.14	BLUE OBJECT
ZWG 383.042	00 34 06.	+ 01 35		15.4	GALAXY
ZWG 457.022	00 34 06.	+ 21 17		14.5	GALAXY
UGC 00364	00 34 06.	+ 21 17	72	14.5	GALAXY S
ZWG 500.051	00 34 06.	+ 30 34		15.6	GALAXY
MCG+08-02-007	00 34 06.	+ 50 11	36	16.	GALAXY
BZW 0034-11.1	00 34 06.	- 11 09		17.6	COMPACT GALAXY
BZW 0034-11.6	00 34 06.	- 11 39		18.3	COMPACT GALAXY
LB 08218	00 34 06.	- 21 15		17.5	FAINT BLUE STAR
LB 08219	00 34 06.	- 22 07		18.0	FAINT BLUE STAR
LB 08220	00 34 06.	- 23 09		19.8	FAINT BLUE STAR
LB 08221	00 34 06.	- 25 19		18.4	FAINT BLUE STAR
MCG-05-02-024	00 34 06.	- 28 03	60	14.5	GALAXY
MCG-05-02-025	00 34 06.	- 28 34	36	15.5	GALAXY
RIC 139	00 34 06.18	+ 40 58 50.2		19.10	BLUE OBJECT
RIC 140	00 34 06.69	+ 41 40 22.3		18.62	BLUE OBJECT
RNGC 0170	00 34 07.	+ 01 35		15.5	GALAXY
RIC 141	00 34 07.90	+ 40 20 33.8		18.82	BLUE OBJECT
ARC 0066	00 34 08.	- 05 28		17.8	RICH CLUSTER OF GALAXIES
RIC 142	00 34 08.49	+ 41 45 11.0		18.86	BLUE OBJECT
MCG+00-02-091	00 34 09.	+ 01 37	24	15.5	GALAXY
BAA 3.16	00 34 09.39	+ 39 22 09.0			GLOBULAR CLUSTER IN M31
RIC 143	00 34 09.73	+ 38 59 14.4		19.08	BLUE OBJECT
RIC 144	00 34 10.22	+ 41 21 21.7		18.38	BLUE OBJECT
RNGC 0169	00 34 11.	+ 23 43		13.5	GALAXY
RIC 145	00 34 11.51	+ 39 06 47.4		17.75	BLUE OBJECT
RIC 146	00 34 11.71	+ 40 18 39.8		19.02	BLUE OBJECT
ZC 0034.2+0223	00 34 12.	+ 02 23	740		CLUSTER OF GALAXIES
MCG+03-02-017	00 34 12.	+ 21 18	45	14.	GALAXY
MRK 341	00 34 12.	+ 23 42	20	15.	GALAXY WITH UV CONTINUUM
KARA.72 013B	00 34 12.	+ 23 42	42		PART OF DOUBLE GALAXY
MCG+04-02-035	00 34 12.	+ 23 42	132	14.	GALAXY
MCG+04-02-034	00 34 12.	+ 23 42	42	14.	GALAXY
ZWG 479.044	00 34 12.	+ 23 43		13.3	GALAXY
UGC 00365	00 34 12.	+ 23 43	210	13.3	GALAXY Sb
KARA.72 013A	00 34 12.	+ 23 43	138	13.3	PART OF DOUBLE GALAXY
UGC 00366	00 34 12.	+ 28 38	72	16.0	GALAXY COMPACT
LB 08222	00 34 12.	- 22 11		17.5	FAINT BLUE STAR
LB 08223	00 34 12.	- 23 44		18.7	FAINT BLUE STAR
LB 08224	00 34 12.	- 24 02		19.5	FAINT BLUE STAR
LB 08225	00 34 12.	- 24 04		19.3	FAINT BLUE STAR
LB 08226	00 34 12.	- 24 55		18.2	FAINT BLUE STAR
LB 08227	00 34 12.	- 25 09		17.7	FAINT BLUE STAR
MCG-05-02-026	00 34 12.	- 28 05	60	15.	GALAXY
RIC 147	00 34 12.30	+ 41 07 01.6		19.10	BLUE OBJECT
RIC 148	00 34 12.58	+ 41 47 29.1		18.28	BLUE OBJECT
RIC 149	00 34 13.62	+ 41 46 43.5		18.63	BLUE OBJECT
IC 1559	00 34 13.9	+ 23 42 34.			GALAXY
MCG+03-02-018	00 34 15.	+ 18 30	36	16.	GALAXY
ARP 282	00 34 15.	+ 23 43			PECULIAR GALAXY
RNGC 0171	00 34 15.	- 18 46			NON-EXISTENT OBJECT
RNGC 0168	00 34 15.	- 22 52		14.0	GALAXY
MCG-04-02-026	00 34 15.	- 22 52 30.	48	14.5	GALAXY
RIC 150	00 34 15.16	+ 40 53 04.3		19.10	BLUE OBJECT
ARC 0065	00 34 16.	+ 18 48		17.5	RICH CLUSTER OF GALAXIES
RNGC 0169A	00 34 17.	+ 23 43			GALAXY
RIC 151	00 34 17.46	+ 40 23 17.4		18.83	BLUE OBJECT
SZW 024	00 34 18.	+ 50 06			COMPACT GALAXY
MCG-02-02-070	00 34 18.	- 12 57	48	15.5	GALAXY
LB 08228	00 34 18.	- 21 25		17.7	FAINT BLUE STAR
LB 08229	00 34 18.	- 23 55		18.8	FAINT BLUE STAR
LB 08230	00 34 18.	- 26 06		18.4	FAINT BLUE STAR
SVEN 029	00 34 20.	- 10 07	42	16.0	GALAXY
ARC 0067	00 34 22.	+ 18 59		17.5	RICH CLUSTER OF GALAXIES
RIC 152	00 34 22.33	+ 41 44 45.1		18.45	BLUE OBJECT
RIC 153	00 34 23.56	+ 41 47 08.5		18.52	BLUE OBJECT
ZC 0034.4+0851	00 34 24.	+ 08 51	740		CLUSTER OF GALAXIES
MCG+02-02-023	00 34 24.	+ 12 46	36	16.	GALAXY
ZC 0034.4+2117	00 34 24.	+ 21 17	610		CLUSTER OF GALAXIES
ZC 0034.4+2456	00 34 24.	+ 24 56	1340		CLUSTER OF GALAXIES
MCG+04-02-037	00 34 24.	+ 25 24	12	16.	GALAXY
MCG+04-02-036	00 34 24.	+ 25 24	18	15.	GALAXY
ZWG 479.045	00 34 24.	+ 25 25		14.8	GALAXY
UGC 00367	00 34 24.	+ 25 25	102	14.8	GALAXY COMPACT
ZC 0034.4+2532	00 34 24.	+ 25 32	4030		CLUSTER OF GALAXIES
ZWG 500.052	00 34 24.	+ 28 34		15.0	GALAXY
UGC 00368	00 34 24.	+ 38 50	60	16.5	GALAXY S
SHRV.73B 16	00 34 24.	+ 43 13	8	15.84	COMPACT GALAXY
LB 08231	00 34 24.	- 21 50		19.6	FAINT BLUE STAR
LB 08232	00 34 24.	- 22 32		19.0	FAINT BLUE STAR
LB 08233	00 34 24.	- 22 52		17.3	FAINT BLUE STAR
LB 08234	00 34 24.	- 25 46		18.5	FAINT BLUE STAR
LB 08235	00 34 24.	- 25 58		19.2	FAINT BLUE STAR
MCG-05-02-027	00 34 24.	- 28 39	120	14.5	GALAXY
SE 969	00 34 24.	- 33 50		16.	PEC. EMISSION-LINE GALAXY
HOW 09	00 34 24.	- 73 15 27.	18		STAR CLUSTER IN SMC
RIC 154	00 34 24.55	+ 41 03 32.0		18.07	BLUE OBJECT
ARC 0068	00 34 25.	+ 08 52		18.0	RICH CLUSTER OF GALAXIES
RIC 155	00 34 25.32	+ 41 42 11.2		18.56	BLUE OBJECT
RIC 156	00 34 26.75	+ 40 53 20.2		18.38	BLUE OBJECT

OBJECT NAME	RIGHT ASCEN.	DECLINATION	DIAM.	MAGN.	TYPE OF OBJECT
RNGC 0174	00 34 28.	- 29 45		14.0	GALAXY
RIC 157	00 34 28.86	+ 41 17 06.5		19.40	BLUE OBJECT
ZC 0034.5+1903	00 34 30.	+ 19 03	3970		CLUSTER OF GALAXIES
ZWG 479.046	00 34 30.	+ 25 34		15.3	GALAXY
ZWG 519.014	00 34 30.	+ 35 38		15.5	GALAXY
LB 08236	00 34 30.	- 21 02		17.0	FAINT BLUE STAR
LB 08237	00 34 30.	- 22 38		17.5	FAINT BLUE STAR
LB 08238	00 34 30.	- 24 36		18.6	FAINT BLUE STAR
MCG-05-02-028	00 34 30.	- 29 45	60	14.	GALAXY
MCG-05-02-029	00 34 30.	- 29 50	12	15.5	GALAXY
HOW 10	00 34 30.	- 73 15 09.	24		STAR CLUSTER IN SMC
RIC 158	00 34 30.64	+ 41 18 43.3		19.38	BLUE OBJECT
RIC 159	00 34 30.69	+ 40 20 59.2		18.68	BLUE OBJECT
RIC 160	00 34 30.82	+ 41 44 30.7		19.89	BLUE OBJECT
RIC 161	00 34 32.00	+ 41 02 05.8		17.61	BLUE OBJECT
RIC 162	00 34 32.54	+ 41 02 10.4		19.02	BLUE OBJECT
RIC 163	00 34 34.68	+ 41 16 33.3		18.44	BLUE OBJECT
RIC 164	00 34 35.84	+ 39 05 20.5		19.61	BLUE OBJECT
ZWG 383.043	00 34 36.	+ 01 40		14.5	GALAXY
UGC 00369	00 34 36.	+ 01 40	240	14.5	GALAXY Sb/Sc
MCG+00-02-092	00 34 36.	+ 01 40 30.	162	13.	GALAXY
UGC 00370	00 34 36.	+ 12 47	78	16.0	GALAXY S IV-V
MCG+05-02-029	00 34 36.	+ 28 52	90	15.	GALAXY
MCG-03-02-023	00 34 36.	- 18 16	60	15.	GALAXY
LB 08239	00 34 36.	- 22 48		17.5	FAINT BLUE STAR
LIN 020	00 34 36.	- 73 29 29.		14.36	SMC EMISSION LINE OBJECT
LB 03146	00 34 36.	- 73 34		13.7	FAINT BLUE STAR
BAA 4.38	00 34 36.88	+ 40 05 51.9			DIFFUSE OBJECT IN M31
RNGC 0173	00 34 37.	+ 01 40		14.5	GALAXY
BAA 2.32	00 34 37.53	+ 39 20 56.8			GALAXY
RIC 165	00 34 37.53	+ 40 59 21.9		18.97	BLUE OBJECT
RIC 166	00 34 37.83	+ 42 12 40.7		19.16	BLUE OBJECT
RIC 167	00 34 39.34	+ 41 39 37.2		18.44	BLUE OBJECT
HAPO 11	00 34 40.	- 33 51			BLUE EMISSION-LINE GALAXY
RIC 168	00 34 41.16	+ 41 31 30.2		18.15	BLUE OBJECT
RIC 169	00 34 41.58	+ 39 02 10.7		20.13	BLUE OBJECT
ZWG 500.053	00 34 42.	+ 28 53		15.0	GALAXY
UGC 00371	00 34 42.	+ 28 53	108	15.0	GALAXY Sc
ZC 0034.7+4536	00 34 42.	+ 45 36	2760		CLUSTER OF GALAXIES
5ZW 025	00 34 42.	+ 45 43			COMPACT GALAXY
8ZW 0034-09.7	00 34 42.	- 09 43		17.8	COMPACT GALAXY
MCG-02-02-071	00 34 42.	- 14 29	42	15.	GALAXY
LB 08240	00 34 42.	- 20 53		20.1	FAINT BLUE STAR
LB 08241	00 34 42.	- 24 06		18.0	FAINT BLUE STAR
LB 08242	00 34 42.	- 26 33		20.2	FAINT BLUE STAR
LB 01563	00 34 42.	- 46 08		12.7	FAINT BLUE STAR
RIC 170	00 34 42.18	+ 41 09 45.4		18.56	BLUE OBJECT
RIC 171	00 34 42.81	+ 39 07 44.8		17.92	BLUE OBJECT
RIC 172	00 34 42.84	+ 41 00 35.1		18.52	BLUE OBJECT
RIC 173	00 34 43.42	+ 41 02 33.2		18.11	BLUE OBJECT
MCG-03-02-024	00 34 45.	- 20 14	132	12.5	GALAXY
RNGC 0172	00 34 45.	- 22 52		14.0	GALAXY
RIC 174	00 34 46.52	+ 41 51 00.9		18.91	BLUE OBJECT
BAA 1.493	00 34 46.99	+ 40 06 28.6			EMISSION NEBULA IN M31
LIN 021	00 34 47.	- 73 50 17.		15.09	SMC EMISSION LINE OBJECT
BAA 1.494	00 34 47.13	+ 40 06 22.2			EMISSION NEBULA IN M31
ZC 0034.8+2329	00 34 48.	+ 23 29	5240		CLUSTER OF GALAXIES
ZC 0034.8+3059	00 34 48.	+ 30 59	1950		CLUSTER OF GALAXIES
MCG+07-02-009	00 34 48.	+ 42 37	60	15.5	GALAXY
UGC 00372	00 34 48.	+ 42 38	138	17.	GALAXY S IV-V
LB 08243	00 34 48.	- 22 12		18.7	FAINT BLUE STAR
LB 08244	00 34 48.	- 22 36		18.1	FAINT BLUE STAR
MCG-04-02-027	00 34 48.	- 22 52	120	14.	GALAXY
LB 08245	00 34 48.	- 23 07		19.7	FAINT BLUE STAR
LB 08246	00 34 48.	- 23 09		19.1	FAINT BLUE STAR
LB 08247	00 34 48.	- 23 50		19.5	FAINT BLUE STAR
LB 08248	00 34 48.	- 24 01		16.9	FAINT BLUE STAR
ARC 0070	00 34 49.	- 07 43		17.5	RICH CLUSTER OF GALAXIES
BAA 1.501	00 34 49.35	+ 39 43 51.7			EMISSION NEBULA IN M31
RIC 175	00 34 49.80	+ 41 59 55.1		19.72	BLUE OBJECT
RIC 176	00 34 49.95	+ 40 51 01.6		18.16	BLUE OBJECT
RIC 177	00 34 50.06	+ 38 52 55.2		17.19	BLUE OBJECT
BAA 1.498	00 34 50.39	+ 39 44 36.7			EMISSION NEBULA IN M31
BAA 1.499	00 34 50.56	+ 39 44 32.1			EMISSION NEBULA IN M31
BAA 1.497	00 34 50.99	+ 39 44 54.0			EMISSION NEBULA IN M31
BAA 1.500	00 34 51.70	+ 39 44 01.7			EMISSION NEBULA IN M31
KEEL 002	00 34 52.2	+ 48 12 01.			NEBULA
RIC 178	00 34 52.55	+ 41 37 14.1		19.17	BLUE OBJECT
KEEL 003	00 34 52.6	+ 48 17 54.			NEBULA
KEEL 004	00 34 53.7	+ 47 53 56.			NEBULA
MCG+02-02-024	00 34 54.	+ 10 04	60	14.	GALAXY
ZC 0034.9+1802	00 34 54.	+ 18 02	2760		CLUSTER OF GALAXIES
ZC 0034.9+3916	00 34 54.	+ 39 16	1410		CLUSTER OF GALAXIES
MCG+07-02-010	00 34 54.	+ 39 34	9	17.	GALAXY
MCG+00-02-093	00 34 54.	- 01 58	48	16.	GALAXY
8ZW 0034-12.7	00 34 54.	- 12 42		17.5	COMPACT GALAXY
LB 08249	00 34 54.	- 21 22		18.7	FAINT BLUE STAR
LB 08250	00 34 54.	- 25 05		18.7	FAINT BLUE STAR
RIC 179	00 34 54.26	+ 39 21 42.0		17.87	BLUE OBJECT
BC 5C3.44	00 34 54.4	+ 39 21 42.		17.95	QUASI-STELLAR OBJECT
RIC 180	00 34 54.44	+ 40 57 49.6		19.07	BLUE OBJECT
SHB 007	00 34 54.5	+ 39 21 41.		18.	QUASI-STELLAR OBJECT
BAA 2.31	00 34 54.56	+ 39 33 41.9			GALAXY
RIC 181	00 34 55.06	+ 41 38 40.9		18.77	BLUE OBJECT
RIC 182	00 34 55.40	+ 40 55 52.8		19.78	BLUE OBJECT
BAA 1.495	00 34 55.58	+ 40 04 52.5			EMISSION NEBULA IN M31
BAA 1.496	00 34 55.74	+ 40 04 16.5			EMISSION NEBULA IN M31
RIC 183	00 34 56.80	+ 41 24 22.4		18.75	BLUE OBJECT
BFL 16	00 34 57.	+ 39 19 00.		17.84	PEC. BLUE OBJECT NEAR M31
RNGC 0175	00 34 57.	- 20 12		13.0	GALAXY
RIC 184	00 34 57.33	+ 41 44 42.3		19.02	BLUE OBJECT
ARC 0069	00 34 58.	+ 18 04		17.1	RICH CLUSTER OF GALAXIES
BAA 1.489	00 34 58.73	+ 40 08 34.3			EMISSION NEBULA IN M31
RIC 185	00 34 59.28	+ 40 55 51.6		19.39	BLUE OBJECT
RIC 186	00 34 59.34	+ 41 41 33.6		18.74	BLUE OBJECT
KHAV 008	00 35	+ 58 35			DARK NEBULA
KHAV 009	00 35	+ 59 23	9800		DARK NEBULA
YC 0035-34	00 35	- 34 01			UNUSUAL SOUTHERN NEBULA
MCG+00-02-094	00 35 00.	+ 00 00	36	15.	GALAXY
UGC 00373	00 35 00.	+ 09 17	72	16.0	GALAXY S?
MCG+04-02-038	00 35 00.	+ 25 20 30.	36	15.	GALAXY
LDN 1288	00 35 00.	+ 65 50	2340		DARK NEBULA
MCG-03-02-025	00 35 00.	- 18 15	36	15.	GALAXY
LB 08251	00 35 00.	- 20 40		18.5	FAINT BLUE STAR
LB 08252	00 35 00.	- 21 22		20.0	FAINT BLUE STAR
LB 08253	00 35 00.	- 21 49		18.0	FAINT BLUE STAR
LB 08254	00 35 00.	- 22 39		19.4	FAINT BLUE STAR
LB 08255	00 35 00.	- 23 58		19.7	FAINT BLUE STAR
LB 08256	00 35 00.	- 24 39		19.5	FAINT BLUE STAR
LB 08257	00 35 00.	- 24 46		19.0	FAINT BLUE STAR
LB 08258	00 35 00.	- 25 03		18.1	FAINT BLUE STAR
LB 08259	00 35 00.	- 25 08		18.0	FAINT BLUE STAR
LB 08260	00 35 00.	- 25 12		19.5	FAINT BLUE STAR
LB 08261	00 35 00.	- 25 56		16.7	FAINT BLUE STAR
MCG-06-02-022A	00 35 00.	- 34 01	78	13.	GALAXY
BAA 1.492	00 35 01.68	+ 40 07 13.4			EMISSION NEBULA IN M31
RIC 187	00 35 02.07	+ 39 15 10.8		18.77	BLUE OBJECT
BAA 1.487	00 35 02.89	+ 40 08 45.5			EMISSION NEBULA IN M31
RNGC 0177	00 35 03.	- 22 50		13.0	GALAXY
RIC 188	00 35 03.44	+ 39 13 22.7		17.80	BLUE OBJECT
BAA 1.488	00 35 03.76	+ 40 08 36.8			EMISSION NEBULA IN M31
BAA 1.490	00 35 04.07	+ 40 07 01.4			EMISSION NEBULA IN M31
BAA 1.491	00 35 04.41	+ 40 06 55.7			EMISSION NEBULA IN M31
RIC 189	00 35 04.74	+ 40 40 28.3		18.53	BLUE OBJECT
IC 1560	00 35 05.	+ 02 24			NONSTELLAR OBJECT
RIC 190	00 35 05.38	+ 40 20 43.6		18.51	BLUE OBJECT
RIC 191	00 35 05.62	+ 41 54 33.0		18.21	BLUE OBJECT
RIC 192	00 35 05.67	+ 38 52 28.9		18.39	BLUE OBJECT
ZWG 383.044	00 35 06.	+ 00 00		15.4	GALAXY
IC 0035	00 35 06.	+ 10 04 58.			NONSTELLAR OBJECT
ZWG 434.026	00 35 06.	+ 10 05		15.0	GALAXY
UGC 00374	00 35 06.	+ 10 05	78	15.0	GALAXY Sc
KARA.73B 0030	00 35 06.	+ 10 05	54	15.0	ISOLATED GALAXY S
ZWG 479.047	00 35 06.	+ 25 22		15.0	GALAXY
UGC 00375	00 35 06.	+ 25 22	72	15.0	GALAXY COMPACT
UGC 00376	00 35 06.	+ 32 27	84	16.0	GALAXY Sc
ZC 0035.1-0007	00 35 06.	- 00 07	2080		CLUSTER OF GALAXIES
LB 08262	00 35 06.	- 20 41		18.7	FAINT BLUE STAR
MCG-04-02-028	00 35 06.	- 22 50	144	13.5	GALAXY
LB 00153	00 35 06.	- 23 39		16.3	FAINT BLUE STAR
LB 08263	00 35 06.	- 23 56		19.3	FAINT BLUE STAR
LB 08264	00 35 06.	- 24 41		19.4	FAINT BLUE STAR
LB 08265	00 35 06.	- 25 09		18.5	FAINT BLUE STAR
LB 08266	00 35 06.	- 26 20		17.8	FAINT BLUE STAR
LIN.CL 018	00 35 06.	- 73 16 47.	1566		STAR CLUSTER IN SMC
RIC 193	00 35 06.53	+ 41 24 30.8		19.01	BLUE OBJECT
RIC 194	00 35 06.88	+ 40 21 07.7		18.13	BLUE OBJECT
RIC 195	00 35 07.19	+ 40 46 11.6		19.00	BLUE OBJECT
ARC 0071	00 35 08.	+ 29 19		15.6	RICH CLUSTER OF GALAXIES
BAA 2.50	00 35 08.02	+ 39 54 11.0			GALAXY
RIC 196	00 35 10.09	+ 40 43 27.2		18.44	BLUE OBJECT
BAA 2.51	00 35 10.47	+ 39 49 07.0			GALAXY
BAA 2.30	00 35 10.61	+ 39 38 05.0			GALAXY
UGC 00377	00 35 12.	+ 03 45	78	17.	GALAXY
ZWG 409.048	00 35 12.	+ 05 56		15.5	GALAXY
ZWG 457.023	00 35 12.	+ 17 06		15.6	GALAXY
MCG+04-02-039	00 35 12.	+ 26 02	39	15.5	GALAXY
SHRV.73B 17	00 35 12.	+ 40 25	7	17.01	COMPACT GALAXY
5ZW 026	00 35 12.	+ 45 20			COMPACT GALAXY
UGC 00378	00 35 12.	+ 47 57	60	16.5	GALAXY PECULR
MCG-03-02-026	00 35 12.	- 18 09	42	14.	GALAXY
LB 08267	00 35 12.	- 20 50		17.5	FAINT BLUE STAR
LB 08268	00 35 12.	- 21 54		18.8	FAINT BLUE STAR
LB 08269	00 35 12.	- 22 24		18.5	FAINT BLUE STAR
LB 08270	00 35 12.	- 23 07		18.8	FAINT BLUE STAR
LB 08271	00 35 12.	- 23 36		20.0	FAINT BLUE STAR
LB 08272	00 35 12.	- 25 03		18.5	FAINT BLUE STAR
LB 08273	00 35 12.	- 25 05		19.0	FAINT BLUE STAR
LB 08274	00 35 12.	- 25 30		18.8	FAINT BLUE STAR
RIC 197	00 35 12.00	+ 40 46 30.6		19.64	BLUE OBJECT
KEEL 005	00 35 13.3	+ 47 55 36.			NEBULA
RIC 198	00 35 13.47	+ 40 48 20.6		19.45	BLUE OBJECT
MCG+03-02-019	00 35 15.	+ 17 06	30	15.	GALAXY
MCG+07-02-012	00 35 15.	+ 39 37	24	18.	GALAXY
MCG+07-02-011	00 35 15.	+ 39 38	9	17.	GALAXY
MCG+08-02-008	00 35 15.	+ 47 55	48	16.	GALAXY
RNGC 0179	00 35 15.	- 18 09		14.0	GALAXY
RIC 199	00 35 15.13	+ 39 02 49.9		18.92	BLUE OBJECT
RIC 200	00 35 15.88	+ 39 09 36.8		18.98	BLUE OBJECT
RIC 201	00 35 16.00	+ 41 14 27.0		18.77	BLUE OBJECT
RIC 202	00 35 16.15	+ 41 40 49.0		18.01	BLUE OBJECT
RIC 203	00 35 17.73	+ 41 22 59.1		18.69	BLUE OBJECT
ZWG 409.049	00 35 18.	+ 04 37		15.4	GALAXY
MCG+01-02-038	00 35 18.	+ 04 38	48	14.5	GALAXY
UGC 00379	00 35 18.	+ 04 52	78	16.0	GALAXY Sc
MCG+01-02-037	00 35 18.	+ 04 52	84	15.	GALAXY
VB 090	00 35 18.	+ 60 02	275		STELLAR RING
MCG-02-02-072	00 35 18.	- 14 08	60	15.5	GALAXY
LB 08275	00 35 18.	- 23 29		19.0	FAINT BLUE STAR
LB 08276	00 35 18.	- 25 04		19.6	FAINT BLUE STAR
MCG-05-02-030	00 35 18.	- 26 54	48	15.	GALAXY
RIC 204	00 35 18.24	+ 41 33 55.9		18.54	BLUE OBJECT
RIC 205	00 35 18.81	+ 41 16 02.6		19.18	BLUE OBJECT
IC 0036	00 35 19.	- 15 42 51.			NONSTELLAR OBJECT
SC 0033-6752.6	00 35 19.	- 67 36 04.	6		NEBULA
SHB 008	00 35 20.3	+ 23 50 35.		19.0	QUASI-STELLAR OBJECT
RIC 206	00 35 20.38	+ 39 03 02.4		18.02	BLUE OBJECT
MCG+01-02-039	00 35 21.	+ 08 21	120	13.	GALAXY
RIC 207	00 35 21.68	+ 38 55 15.1		18.80	BLUE OBJECT
HSN 01	00 35 22.	- 70 50			BRIGHT GALAXY BEHIND SMC
RIC 208	00 35 22.53	+ 40 36 39.4		18.95	BLUE OBJECT
RIC 209	00 35 23.85	+ 40 52 03.4		19.06	BLUE OBJECT
ZWG 409.050	00 35 24.	+ 08 22		14.3	GALAXY
RNGC 0180	00 35 24.	+ 08 22		14.5	GALAXY
UGC 00380	00 35 24.	+ 08 22	168	14.3	GALAXY SBc
ZC 0035.4+2015	00 35 24.	+ 20 15	2960		CLUSTER OF GALAXIES
ZWG 479.048	00 35 24.	+ 22 45		15.7	GALAXY
MCG-02-02-073	00 35 24.	- 09 31	84	13.5	GALAXY
LB 08277	00 35 24.	- 23 26		18.4	FAINT BLUE STAR
LB 08278	00 35 24.	- 24 30		19.9	FAINT BLUE STAR
LB 08279	00 35 24.	- 25 20		17.8	FAINT BLUE STAR
LB 08280	00 35 24.	- 25 50		18.6	FAINT BLUE STAR
LB 08281	00 35 24.	- 25 53		18.1	FAINT BLUE STAR
LB 08282	00 35 24.	- 26 44		17.5	FAINT BLUE STAR
MCG-05-02-030A	00 35 24.	- 29 11	42	15.	GALAXY
RIC 210	00 35 24.25	+ 41 34 48.5		18.25	BLUE OBJECT
RIC 211	00 35 24.47	+ 42 08 18.4		19.54	BLUE OBJECT
HELW 317	00 35 26.	- 09 32 41.			NEBULA
RIC 212	00 35 26.84	+ 40 44 33.4		18.54	BLUE OBJECT
RIC 213	00 35 27.78	+ 41 13 48.0		18.46	BLUE OBJECT
RIC 214	00 35 27.98	+ 41 41 45.9		18.96	BLUE OBJECT
RIC 215	00 35 28.13	+ 41 47 04.0		18.81	BLUE OBJECT
RIC 216	00 35 29.18	+ 40 50 47.9		19.63	BLUE OBJECT
MCG+00-02-095	00 35 30.	+ 02 28	96	13.4	GALAXY
MCG+05-02-030	00 35 30.	+ 30 36	30	15.	GALAXY
ZWG 500.054	00 35 30.	+ 30 37		15.3	GALAXY

OBJECT NAME	RIGHT ASCEN.	DECLINATION	DIAM.	MAGN.	TYPE OF OBJECT
UGC 00381	00 35 30.	+ 30 37	96	15.3	GALAXY VY CMPT
LB 00154	00 35 30.	- 22 53		16.5	FAINT BLUE STAR
LB 08283	00 35 30.	- 24 01		17.7	FAINT BLUE STAR
LB 08284	00 35 30.	- 24 02		16.6	FAINT BLUE STAR
LB 08285	00 35 30.	- 25 10		18.3	FAINT BLUE STAR
RNGC 0181	00 35 34.	+ 29 12		15.5	GALAXY
MCG+05-02-032	00 35 36.	+ 29 12	30	15.5	GALAXY
MCG+05-02-031	00 35 36.	+ 29 21	48	15.5	GALAXY
SHRV.73A 04	00 35 36.	+ 39 47	7	16.95	GLOBULAR CLUSTER IN M31
LB 08286	00 35 36.	- 23 16		18.0	FAINT BLUE STAR
LB 08287	00 35 36.	- 24 00		18.2	FAINT BLUE STAR
LB 08288	00 35 36.	- 24 19		16.5	FAINT BLUE STAR
LB 08289	00 35 36.	- 24 40		17.1	FAINT BLUE STAR
RIC 217	00 35 36.08	+ 41 04 02.0		18.71	BLUE OBJECT
RIC 218	00 35 36.30	+ 41 46 49.6		18.50	BLUE OBJECT
RIC 219	00 35 36.40	+ 41 35 52.6		18.03	BLUE OBJECT
RNGC 0182	00 35 37.	+ 02 27		14.0	GALAXY
HOW 11	00 35 37.	- 73 53 15.	36		STAR CLUSTER IN SMC
RIC 220	00 35 37.41	+ 39 15 25.8		17.79	BLUE OBJECT
BAA 3.15	00 35 38.96	+ 39 47 08.0			GLOBULAR CLUSTER IN M31
MCG+05-02-033	00 35 39.	+ 32 22	60	14.5	GALAXY
RNGC 0183	00 35 40.	+ 29 14		14.0	GALAXY
HSN 02	00 35 40.	- 71 02			BRIGHT GALAXY BEHIND SMC
RIC 221	00 35 40.74	+ 39 11 46.8		18.79	BLUE OBJECT
RIC 222	00 35 41.28	+ 41 20 38.3		20.61	BLUE OBJECT
ZWG 383.045	00 35 42.	+ 02 27		13.8	GALAXY
UGC 00382	00 35 42.	+ 02 27	138	13.8	GALAXY Sa
UGC 00383	00 35 42.	+ 08 31	78	17.	GALAXY
MRK 342	00 35 42.	+ 13 15	10	16.5	GALAXY WITH UV CONTINUUM
MCG+02-02-025	00 35 42.	+ 14 46	60	15.	GALAXY
ZWG 500.055	00 35 42.	+ 29 12		15.4	GALAXY
MCG+05-02-035	00 35 42.	+ 29 14	27	14.	GALAXY
ZWG 500.056	00 35 42.	+ 32 22		14.9	GALAXY
UGC 00384	00 35 42.	+ 32 22	114	14.9	GALAXY Sc
LB 08290	00 35 42.	- 22 20		17.4	FAINT BLUE STAR
LB 08291	00 35 42.	- 22 36		16.8	FAINT BLUE STAR
LB 00155	00 35 42.	- 22 42		16.8	FAINT BLUE STAR
LB 08292	00 35 42.	- 22 56		19.7	FAINT BLUE STAR
LB 08293	00 35 42.	- 25 32		18.4	FAINT BLUE STAR
LB 08294	00 35 42.	- 25 42		18.6	FAINT BLUE STAR
ARC 0073	00 35 43.	+ 09 09		18.0	RICH CLUSTER OF GALAXIES
SC 0033-6621.3	00 35 43.	- 66 04 47.	18		NEBULA
RIC 223	00 35 44.38	+ 40 50 23.1		18.37	BLUE OBJECT
BAA 4.37	00 35 44.50	+ 40 50 23.6			DIFFUSE OBJECT IN M31
RIC 224	00 35 44.66	+ 40 49 32.4		18.69	BLUE OBJECT
MCG+00-02-096	00 35 45.	- 00 10 30.	24	16.	GALAXY
MCG+00-02-097	00 35 45.	- 01 13 30.	36	16.	GALAXY
MCG-03-02-027	00 35 45.	- 15 08	36	15.	GALAXY
MCG-03-02-028	00 35 45.	- 19 10	18	16.	GALAXY
RIC 225	00 35 45.26	+ 40 48 26.4		18.57	BLUE OBJECT
RIC 226	00 35 45.44	+ 41 39 57.4		19.38	BLUE OBJECT
ARC 0072	00 35 46.	+ 45 27		16.3	RICH CLUSTER OF GALAXIES
LIN 022	00 35 46.	- 74 03 05.		14.57	SMC EMISSION LINE OBJECT
RIC 227	00 35 47.43	+ 42 03 24.0		17.84	BLUE OBJECT
RIC 228	00 35 47.86	+ 41 36 11.2		19.48	BLUE OBJECT
MCG+00-02-099	00 35 48.	+ 01 42 30.	30	15.	GALAXY
MCG+00-02-098	00 35 48.	+ 02 55	60	14.	GALAXY
ZC 0035.8+0913	00 35 48.	+ 09 13	740		CLUSTER OF GALAXIES
ZWG 434.027	00 35 48.	+ 13 12		15.7	GALAXY
UGC 00385	00 35 48.	+ 13 12	138	15.7	GALAXY PECULR
ZWG 434.028	00 35 48.	+ 14 46		15.2	GALAXY
MRK 343	00 35 48.	+ 14 46	12	16.	GALAXY WITH UV CONTINUUM
UGC 00386	00 35 48.	+ 14 46	102	15.2	GALAXY S0
ZWG 500.057	00 35 48.	+ 29 15		13.8	GALAXY
UGC 00387	00 35 48.	+ 29 15	126	13.8	GALAXY (E)
ZWG 500.058	00 35 48.	+ 29 21		15.2	GALAXY
MCG+05-02-034	00 35 48.	+ 30 00 30.	54	15.	GALAXY
UGC 00388	00 35 48.	+ 30 01	102	16.0	GALAXY Sc
PK121-02.1	00 35 48.	+ 60 00	39		PLANETARY NEBULA
LB 08295	00 35 48.	- 21 15		17.0	FAINT BLUE STAR
LB 08296	00 35 48.	- 22 37		17.9	FAINT BLUE STAR
LB 08297	00 35 48.	- 23 22		20.3	FAINT BLUE STAR
RIC 229	00 35 49.83	+ 41 45 47.8		18.17	BLUE OBJECT
RIC 230	00 35 49.86	+ 41 12 21.7		16.33	BLUE OBJECT
RIC 231	00 35 50.63	+ 41 07 11.0		20.24	BLUE OBJECT
BAA 2.29	00 35 50.70	+ 39 46 06.5			GALAXY
MCG+00-02-101	00 35 51.	- 00 17 30.	42	15.5	GALAXY
MCG+00-02-100	00 35 51.	- 01 09	36	15.	GALAXY
RIC 232	00 35 51.08	+ 40 48 08.0		18.46	BLUE OBJECT
RIC 233	00 35 51.29	+ 41 43 19.9		18.54	BLUE OBJECT
RIC 234	00 35 51.32	+ 40 33 17.8		18.52	BLUE OBJECT
RNGC 0184	00 35 52.	+ 29 10		15.5	GALAXY
RNGC 0189	00 35 52.	+ 60 47		11.0	OPEN CLUSTER
LIN.CL 019	00 35 52.	- 74 11 11.	678		STAR CLUSTER IN SMC
RIC 235	00 35 52.37	+ 41 45 44.3		18.54	BLUE OBJECT
RIC 236	00 35 53.62	+ 40 37 16.4		19.38	BLUE OBJECT
BAA 1.380	00 35 53.76	+ 39 36 28.2			EMISSION NEBULA IN M31
RIC 237	00 35 53.92	+ 41 11 58.8		17.75	BLUE OBJECT
ZWG 383.046	00 35 54.	+ 01 42		15.4	GALAXY
UGC 00389	00 35 54.	+ 01 42	60	15.4	GALAXY
ZWG 383.047	00 35 54.	+ 02 53		14.8	GALAXY
UGC 00390	00 35 54.	+ 02 53	102	14.8	GALAXY SB0-a
MCG+01-02-040	00 35 54.	+ 08 28	60	16.	GALAXY
ZWG 479.049	00 35 54.	+ 23 20		15.5	GALAXY
MRK 344	00 35 54.	+ 23 20	20	16.5	GALAXY WITH UV CONTINUUM
ZWG 500.059	00 35 54.	+ 29 10		15.5	GALAXY
MCG+08-02-009	00 35 54.	+ 49 44	54	16.	GALAXY
UGC 00391	00 35 54.	+ 49 45	66	16.0	GALAXY SBb
LB 08298	00 35 54.	- 24 20		16.9	FAINT BLUE STAR
LB 08299	00 35 54.	- 24 53		19.7	FAINT BLUE STAR
LB 08300	00 35 54.	- 25 24		20.0	FAINT BLUE STAR
LB 03147	00 35 54.	- 61 21		12.2	FAINT BLUE STAR
RIC 238	00 35 54.06	+ 40 46 08.6		18.32	BLUE OBJECT
BAA 1.381	00 35 54.09	+ 39 36 25.0			EMISSION NEBULA IN M31
RIC 239	00 35 54.82	+ 38 56 31.0		18.07	BLUE OBJECT
RIC 240	00 35 54.83	+ 40 35 49.9		18.68	BLUE OBJECT
RNGC 0186	00 35 55.	+ 02 53		15.0	GALAXY
RIC 241	00 35 55.76	+ 41 20 07.6		19.06	BLUE OBJECT
RIC 242	00 35 57.00	+ 42 01 17.5		19.21	BLUE OBJECT
RIC 243	00 35 57.10	+ 41 30 16.2		19.69	BLUE OBJECT
RIC 244	00 35 57.66	+ 41 00 20.6		18.36	BLUE OBJECT
RIC 245	00 35 57.77	+ 41 18 51.2		19.88	BLUE OBJECT
RIC 246	00 35 58.64	+ 39 11 50.6		19.68	BLUE OBJECT
HN 0112	00 35 59.	- 24 36			NEBULA
IC 1561	00 35 59.	- 24 36			NONSTELLAR OBJECT
LBN 0600	00 36	+ 50 00	360		BRIGHT NEBULA
ZWG 383.049	00 36 00.	+ 01 25		15.5	GALAXY
ZC 0036.0+0659	00 36 00.	+ 06 59	1480		CLUSTER OF GALAXIES

OBJECT NAME	RIGHT ASCEN.	DECLINATION	DIAM.	MAGN.	TYPE OF OBJECT
MCG+03-02-020	00 36 00.	+ 17 07	120	16.	GALAXY
UGC 00393	00 36 00.	+ 17 08	180	16.5	GALAXY S-IRR
MCG+07-02-013	00 36 00.	+ 41 42 30.	90	15.	GALAXY
ZWG 535.013	00 36 00.	+ 41 43		15.7	GALAXY
UGC 00394	00 36 00.	+ 41 43	132	15.7	GALAXY SB IV
KARA.73B 0031	00 36 00.	+ 41 43	132	15.7	ISOLATED GALAXY S
ZWG 360.003	00 36 00.	+ 82 57		14.8	GALAXY
UGC 00392	00 36 00.	+ 82 57	108	14.8	GALAXY S
MCG+14-01-009	00 36 00.	+ 82 57 30.	90	15.	GALAXY
ZWG 383.048	00 36 00.	- 00 08		15.5	GALAXY
8ZW 0036-09.4	00 36 00.	- 09 23		14.0	COMPACT GALAXY
MCG-02-02-074	00 36 00.	- 11 19	48	15.5	GALAXY
MCG-03-02-029	00 36 00.	- 15 39	42	15.	GALAXY
LB 08301	00 36 00.	- 21 48		18.3	FAINT BLUE STAR
LB 08302	00 36 00.	- 21 51		16.4	FAINT BLUE STAR
LB 08303	00 36 00.	- 22 40		19.5	FAINT BLUE STAR
LB 08304	00 36 00.	- 24 52		19.5	FAINT BLUE STAR
LB 08305	00 36 00.	- 25 44		19.0	FAINT BLUE STAR
RIC 247	00 36 01.03	+ 42 10 27.4		20.95	BLUE OBJECT
BAA 4.36	00 36 02.71	+ 40 02 38.7			DIFFUSE OBJECT IN M31
MCG-04-02-029	00 36 03.	- 24 36 30.	78	14.	GALAXY
RIC 248	00 36 03.97	+ 40 45 41.8		18.43	BLUE OBJECT
IC 0037	00 36 04.	- 15 38 21.			NONSTELLAR OBJECT
HN 0113	00 36 05.	- 24 32			NEBULA
IC 1562	00 36 05.	- 24 32			NONSTELLAR OBJECT
RIC 249	00 36 05.50	+ 42 11 13.3		21.20	BLUE OBJECT
RIC 250	00 36 05.59	+ 41 30 41.2		18.64	BLUE OBJECT
RIC 251	00 36 05.79	+ 39 11 17.4		18.71	BLUE OBJECT
ZC 0036.1+2035	00 36 06.	+ 20 35	870		CLUSTER OF GALAXIES
ZWG 479.050	00 36 06.	+ 25 26		15.6	GALAXY
MCG+05-02-036	00 36 06.	+ 31 16	24	16.	GALAXY
ZWG 500.060	00 36 06.	+ 32 08		15.3	GALAXY
5ZW 027	00 36 06.	+ 40 18			COMPACT GALAXY
MCG+08-02-010	00 36 06.	+ 48 02	180	10.2	GALAXY
MCG-03-02-030	00 36 06.	- 15 42	36	14.5	GALAXY
LB 08306	00 36 06.	- 23 20		20.2	FAINT BLUE STAR
MCG-04-02-030	00 36 06.	- 24 32 30.	78	13.	GALAXY
LB 08307	00 36 06.	- 26 07		16.9	FAINT BLUE STAR
RIC 252	00 36 06.29	+ 39 15 25.7		20.71	BLUE OBJECT
RIC 253	00 36 07.01	+ 41 35 45.0		18.52	BLUE OBJECT
RIC 254	00 36 07.86	+ 39 06 48.0		17.92	BLUE OBJECT
IC 0038	00 36 08.	- 15 42 04.			NONSTELLAR OBJECT
KEEL 006	00 36 08.8	+ 48 11 36.			NEBULA
RIC 255	00 36 09.36	+ 40 54 05.8		18.18	BLUE OBJECT
LIN 023	00 36 10.	- 73 55 18.		14.44	SMC EMISSION LINE OBJECT
LIN 024	00 36 10.	- 74 02 12.		15.04	SMC EMISSION LINE OBJECT
ZWG 500.061	00 36 12.	+ 29 15		15.7	GALAXY
ZWG 500.062	00 36 12.	+ 31 16		15.4	GALAXY
UGC 00395	00 36 12.	+ 31 16	72	15.4	GALAXY COMPACT
SHRV.73A 05	00 36 12.	+ 40 54	7	17.24	GLOBULAR CLUSTER IN M31
ZWG 550.009	00 36 12.	+ 48 03		11.0	GALAXY
UGC 00396	00 36 12.	+ 48 03	864	11.0	GALAXY DWRF EL
MCG+10-02-001	00 36 12.	+ 58 19	42	15.	GALAXY
LB 08308	00 36 12.	- 20 44		19.0	FAINT BLUE STAR
LB 08309	00 36 12.	- 23 51		19.4	FAINT BLUE STAR
LB 08310	00 36 12.	- 23 56		16.8	FAINT BLUE STAR
LB 08311	00 36 12.	- 24 08		19.6	FAINT BLUE STAR
LB 08312	00 36 12.	- 24 28		19.1	FAINT BLUE STAR
LB 08313	00 36 12.	- 24 49		17.9	FAINT BLUE STAR
LB 03148	00 36 12.	- 66 29		14.0	FAINT BLUE STAR
LIN.CL 020	00 36 12.	- 72 45 35.	342	14.8	STAR CLUSTER IN SMC
RIC 256	00 36 12.09	+ 40 49 41.4		18.88	BLUE OBJECT
RIC 257	00 36 12.26	+ 41 53 40.2		18.52	BLUE OBJECT
BAA 4.35	00 36 12.44	+ 39 41 00.3			DIFFUSE OBJECT IN M31
RNGC 0185	00 36 13.	+ 48 04		11.0	GALAXY
RIC 258	00 36 14.97	+ 41 17 29.2		19.65	BLUE OBJECT
MCG+01-02-042	00 36 15.	+ 06 46	12	15.	GALAXY
MCG+01-02-041	00 36 15.	+ 06 47	42	15.	GALAXY
MCG+04-02-040	00 36 15.	+ 25 21 30.	66	14.	GALAXY
3ZW 010	00 36 18.	+ 06 46			COMPACT GALAXY
ZWG 409.051	00 36 18.	+ 06 46		14.4	GALAXY
UGC 00397	00 36 18.	+ 06 46	72	14.4	GALAXY Sa-b
ZWG 479.051	00 36 18.	+ 25 22		14.6	GALAXY
UGC 00398	00 36 18.	+ 25 22	90	14.6	GALAXY Sa
ZWG 479.052	00 36 18.	+ 26 37		15.3	GALAXY
ZC 0036.3+2914	00 36 18.	+ 29 14	5710		CLUSTER OF GALAXIES
5ZW 028	00 36 18.	+ 45 36			COMPACT GALAXY
MCG+00-02-102	00 36 18.	- 01 04	30	16.	GALAXY
LB 08314	00 36 18.	- 22 32		17.6	FAINT BLUE STAR
LB 08315	00 36 18.	- 24 18		19.4	FAINT BLUE STAR
LB 08316	00 36 18.	- 25 28		19.0	FAINT BLUE STAR
RIC 259	00 36 18.48	+ 41 02 00.3		17.14	BLUE OBJECT
RNGC 0190	00 36 19.	+ 06 46		14.5	GALAXY
RIC 260	00 36 21.80	+ 41 13 19.2		18.48	BLUE OBJECT
RIC 261	00 36 22.18	+ 41 02 20.6		18.15	BLUE OBJECT
RIC 262	00 36 22.59	+ 41 59 16.8		18.47	BLUE OBJECT
ARP 127	00 36 23.	- 09 17			PECULIAR GALAXY
SC 0034-6621.1	00 36 23.	- 66 04 35.	6		NEBULA
RIC 263	00 36 23.52	+ 41 40 45.4		19.12	BLUE OBJECT
RIC 264	00 36 23.77	+ 41 40 02.3		18.56	BLUE OBJECT
BAA 1.482	00 36 23.90	+ 40 20 41.3			EMISSION NEBULA IN M31
ZWG 409.052	00 36 24.	+ 06 32		15.4	GALAXY
MCG+01-02-043	00 36 24.	+ 06 32	30	15.5	GALAXY
ZC 0036.4+2055	00 36 24.	+ 20 55	2350		CLUSTER OF GALAXIES
4ZW 027	00 36 24.	+ 25 17			COMPACT GALAXY
5ZW 029	00 36 24.	+ 40 18			COMPACT GALAXY
MCG-03-02-031	00 36 24.	- 18 54	48	14.5	GALAXY
LB 08317	00 36 24.	- 21 22		18.6	FAINT BLUE STAR
LB 08318	00 36 24.	- 21 37		18.5	FAINT BLUE STAR
LB 08319	00 36 24.	- 22 07		17.5	FAINT BLUE STAR
LB 08320	00 36 24.	- 22 47		16.3	FAINT BLUE STAR
LB 08321	00 36 24.	- 23 22		18.1	FAINT BLUE STAR
MCG-04-02-031	00 36 24.	- 24 30	30	15.	GALAXY
LB 08322	00 36 24.	- 25 04		19.3	FAINT BLUE STAR
LB 08323	00 36 24.	- 25 09		18.8	FAINT BLUE STAR
LB 01564	00 36 24.	- 59 14		13.5	FAINT BLUE STAR
RIC 265	00 36 24.55	+ 41 39 00.3		18.68	BLUE OBJECT
RIC 266	00 36 24.64	+ 39 28 06.3		19.23	BLUE OBJECT
RIC 267	00 36 24.82	+ 41 30 46.3		18.99	BLUE OBJECT
BAA 1.483	00 36 24.88	+ 40 18 33.6			EMISSION NEBULA IN M31
RIC 268	00 36 24.98	+ 41 49 53.8		20.28	BLUE OBJECT
RIC 269	00 36 25.44	+ 38 52 00.7		19.31	BLUE OBJECT
RIC 270	00 36 25.62	+ 39 03 02.0		19.20	BLUE OBJECT
BAA 3.18	00 36 25.74	+ 40 17 52.2			GLOBULAR CLUSTER IN M31
HOLM 013A	00 36 26.	- 09 16	54	14.2	PART OF MULTIPLE GALAXY
RNGC 0191	00 36 26.	- 09 17		12.0	GALAXY
HARO 12	00 36 26.	- 21 23			BLUE EMISSION-LINE GALAXY
ARC 0074	00 36 26.	- 22 36		15.9	RICH CLUSTER OF GALAXIES

OBJECT NAME	RIGHT ASCEN.	DECLINATION	DIAM.	MAGN.	TYPE OF OBJECT
BAA 1.486	00 36 26.70	+ 40 10 15.4			EMISSION NEBULA IN M31
BAA 1.484	00 36 26.90	+ 40 10 35.1			EMISSION NEBULA IN M31
BAA 1.481	00 36 26.96	+ 40 20 46.2			EMISSION NEBULA IN M31
MCG+01-02-044	00 36 27.	+ 05 44 30.	42	14.	GALAXY
HOLM 013B	00 36 27.	- 09 17	36	14.2	PART OF MULTIPLE GALAXY
BAA 1.480	00 36 27.05	+ 40 20 51.6			EMISSION NEBULA IN M31
BAA 1.478	00 36 27.10	+ 40 20 59.2			EMISSION NEBULA IN M31
BAA 1.479	00 36 27.20	+ 40 20 54.0			EMISSION NEBULA IN M31
RIC 271	00 36 27.83	+ 39 24 18.4		17.86	BLUE OBJECT
IC 1563	00 36 28.1	- 09 17 21.			GALAXY
BAA 1.485	00 36 28.58	+ 40 10 12.5			EMISSION NEBULA IN M31
RIC 272	00 36 29.12	+ 41 34 43.5		18.53	BLUE OBJECT
BAA 1.372	00 36 29.25	+ 40 07 14.1			EMISSION NEBULA IN M31
RIC 273	00 36 29.42	+ 40 56 52.2		18.79	BLUE OBJECT
BAA 1.477	00 36 29.75	+ 40 24 08.2			EMISSION NEBULA IN M31
ZWG 383.050	00 36 30.	+ 01 36		15.6	GALAXY
ZWG 409.053	00 36 30.	+ 05 45		14.8	GALAXY
UGC 00399	00 36 30.	+ 05 45	72	14.8	GALAXY Sb/SBc
ZWG 409.054	00 36 30.	+ 07 12		15.7	GALAXY
MCG+05-02-037	00 36 30.	+ 29 22 30.	30	15.	GALAXY
MCG-02-02-077	00 36 30.	- 09 17 30.	72	12.5	GALAXY
MCG-02-02-076	00 36 30.	- 09 17 30.	30	15.	GALAXY
8ZW 0036-10.4	00 36 30.	- 10 23		15.4	COMPACT GALAXY
MCG-02-02-075	00 36 30.	- 10 47	48	15.	GALAXY
LB 08324	00 36 30.	- 23 16		18.8	FAINT BLUE STAR
LB 08325	00 36 30.	- 23 17		18.6	FAINT BLUE STAR
LB 08326	00 36 30.	- 23 57		17.9	FAINT BLUE STAR
BAA 1.369	00 36 30.04	+ 40 07 24.2			EMISSION NEBULA IN M31
BAA 1.469	00 36 30.52	+ 40 31 13.1			EMISSION NEBULA IN M31
BAA 1.475	00 36 31.54	+ 40 25 45.2			EMISSION NEBULA IN M31
RIC 274	00 36 32.44	+ 41 30 26.9		20.11	BLUE OBJECT
BAA 2.45	00 36 32.57	+ 40 20 16.1			GALAXY
RNGC 0178	00 36 33.	- 14 27		13.0	GALAXY
BAA 2.46	00 36 33.03	+ 40 20 00.2			GALAXY
BAA 1.476	00 36 33.42	+ 40 24 36.6			EMISSION NEBULA IN M31
IC 1564	00 36 34.	+ 05 42			NONSTELLAR OBJECT
LIN 025	00 36 34.	- 73 39 42.		14.27	SMC EMISSION LINE OBJECT
BAA 1.375	00 36 35.17	+ 40 06 00.1			EMISSION NEBULA IN M31
BAA 2.47	00 36 35.46	+ 40 16 29.2			GALAXY
BAA 1.470	00 36 35.87	+ 40 29 46.5			EMISSION NEBULA IN M31
BAA 2.48	00 36 35.91	+ 40 13 04.8			GALAXY
BAA 1.379	00 36 35.94	+ 40 05 30.4			EMISSION NEBULA IN M31
ZC 0036.6+1055	00 36 36.	+ 10 55	1810		CLUSTER OF GALAXIES
MCG+02-02-026	00 36 36.	+ 12 49	48	16.	GALAXY
UGC 02809	00 36 36.	+ 19 34	108	16.5	GALAXY DWARF
ZC 0036.6+2129	00 36 36.	+ 21 29	1610		CLUSTER OF GALAXIES
MCG+04-02-041	00 36 36.	+ 24 27	36	16.	GALAXY
ZWG 500.063	00 36 36.	+ 29 23		15.2	GALAXY
UGC 00400	00 36 36.	+ 29 23	72	15.2	GALAXY
SHRV.73A 06	00 36 36.	+ 40 06	7	16.67	GLOBULAR CLUSTER IN M31
8ZW 0036-14.4	00 36 36.	- 14 25		13.0	COMPACT GALAXY
MCG-02-02-078	00 36 36.	- 14 28	108	13.	GALAXY
LB 08327	00 36 36.	- 22 35		20.1	FAINT BLUE STAR
LB 08328	00 36 36.	- 22 57		19.7	FAINT BLUE STAR
LB 08329	00 36 36.	- 23 12		18.7	FAINT BLUE STAR
LB 08330	00 36 36.	- 24 30		18.8	FAINT BLUE STAR
LB 01565	00 36 36.	- 53 09		13.7	FAINT BLUE STAR
RIC 275	00 36 36.02	+ 40 47 27.2		19.20	BLUE OBJECT
IC 0039	00 36 37.	- 14 26 40.			SAME AS NGC 178
BAA 2.28	00 36 37.46	+ 40 03 05.2			GALAXY
BAA 2.27	00 36 37.82	+ 40 03 36.0			GALAXY
BAA 2.49	00 36 37.98	+ 40 12 09.2			GALAXY
HARO 13	00 36 38.	- 21 20			BLUE EMISSION-LINE GALAXY
RIC 276	00 36 38.06	+ 41 35 28.2		19.74	BLUE OBJECT
BAA 1.378	00 36 38.58	+ 40 05 13.7			EMISSION NEBULA IN M31
BAA 2.26	00 36 38.77	+ 40 03 59.8			GALAXY
BAA 2.25	00 36 38.92	+ 40 06 56.8			GALAXY
MCG+00-02-104	00 36 39.	+ 00 35 30.	102	13.5	GALAXY
MCG+00-02-103	00 36 39.	+ 03 05	24	13.	GALAXY
MCG+01-02-046	00 36 39.	+ 03 40	48	14.	GALAXY
MCG+01-02-045	00 36 39.	+ 04 17 30.	48	14.	GALAXY
RNGC 0209	00 36 39.	- 18 54		14.0	GALAXY
RIC 277	00 36 39.02	+ 39 31 53.3		20.07	BLUE OBJECT
BAA 1.471	00 36 39.19	+ 40 28 49.4			EMISSION NEBULA IN M31
RIC 278	00 36 39.38	+ 41 37 23.5		19.54	BLUE OBJECT
LIN 026	00 36 40.	- 73 29 06.		14.28	SMC EMISSION LINE OBJECT
BAA 1.474	00 36 40.21	+ 40 27 50.9			EMISSION NEBULA IN M31
SHB 009	00 36 41.	+ 03 01		18.	QUASI-STELLAR OBJECT
ZWG 383.051	00 36 42.	+ 00 35		13.9	GALAXY
UGC 00401	00 36 42.	+ 00 35	144	13.9	GALAXY SBa
MCG+00-02-107	00 36 42.	+ 00 38 30.	72	14.	GALAXY
MCG+00-02-106	00 36 42.	+ 02 11 30.	54	15.	GALAXY
KARA.68 001	00 36 42.	+ 02 43	27		DWARF GALAXY
MCG+00-02-105	00 36 42.	+ 02 46	66	13.3	GALAXY
ZWG 409.055	00 36 42.	+ 03 41		14.8	GALAXY
MRK 554	00 36 42.	+ 03 41	15	15.5	GALAXY WITH UV CONTINUUM
UGC 00402	00 36 42.	+ 03 41	96	14.8	GALAXY S0
ZWG 434.029	00 36 42.	+ 12 50		15.3	GALAXY
UGC 00404	00 36 42.	+ 12 50	78	15.3	GALAXY Sc-IRR
UGC 00403	00 36 42.	+ 12 50	72		GALAXY S
MCG+02-02-027	00 36 42.	+ 12 50	60	16.	GALAXY
MCG+06-02-012	00 36 42.	+ 36 04	48	15.	GALAXY
OCL 0301	00 36 42.	+ 60 48	240	11.1	OPEN STAR CLUSTER
MCG+00-02-108	00 36 42.	- 02 25	48	15.	GALAXY
LB 08331	00 36 42.	- 24 32		18.2	FAINT BLUE STAR
RIC 279	00 36 42.20	+ 39 31 02.4		18.00	BLUE OBJECT
RIC 280	00 36 42.85	+ 41 55 01.6		19.07	BLUE OBJECT
RNGC 0192	00 36 43.	+ 00 35		14.0	GALAXY
RNGC 0196	00 36 43.	+ 00 38		14.0	GALAXY
RNGC 0194	00 36 43.	+ 02 46		14.0	GALAXY
RIC 281	00 36 43.94	+ 41 01 07.2		19.36	BLUE OBJECT
RIC 282	00 36 44.62	+ 41 02 45.6		20.55	BLUE OBJECT
MCG+00-02-110	00 36 45.	+ 00 37	78	15.	GALAXY
MCG+00-02-109	00 36 45.	+ 02 31 30.	54	13.	GALAXY
MCG+01-02-047	00 36 45.	+ 06 27	60	14.5	GALAXY
MCG-03-02-032	00 36 45.	- 19 12 30.	66	14.	GALAXY
RIC 283	00 36 45.20	+ 40 56 26.9		18.97	BLUE OBJECT
RIC 284	00 36 46.33	+ 40 43 40.6		19.15	BLUE OBJECT
RIC 285	00 36 46.71	+ 40 57 16.4		19.66	BLUE OBJECT
BAA 1.373	00 36 47.10	+ 40 04 38.2			EMISSION NEBULA IN M31
BAA 1.370	00 36 47.80	+ 40 04 44.9			EMISSION NEBULA IN M31
ZWG 383.053	00 36 48.	+ 00 38		14.2	GALAXY
UGC 00406	00 36 48.	+ 00 38	84	14.2	GALAXY S?
UGC 00405	00 36 48.	+ 00 38	78		GALAXY SB0
ZWG 383.054	00 36 48.	+ 02 45		13.9	GALAXY
UGC 00407	00 36 48.	+ 02 45	120	13.9	GALAXY COMPACT
ZWG 383.055	00 36 48.	+ 03 03		14.3	GALAXY
UGC 00408	00 36 48.	+ 03 03	120	14.3	GALAXY COMPACT
ZWG 409.056	00 36 48.	+ 04 17		15.6	GALAXY
UGC 00409	00 36 48.	+ 04 17	72	15.6	GALAXY S-IRR
ZWG 409.057	00 36 48.	+ 06 27		14.9	GALAXY
UGC 00410	00 36 48.	+ 06 27	126	14.9	GALAXY COMPACT
ZWG 479.053	00 36 48.	+ 22 48		15.7	GALAXY
ZWG 479.054	00 36 48.	+ 24 59		15.0	GALAXY
ZWG 479.055	00 36 48.	+ 25 22		14.5	GALAXY
UGC 00411	00 36 48.	+ 25 22	72	14.5	GALAXY COMPACT
MCG+04-02-042	00 36 48.	+ 25 22	21	14.5	GALAXY
ZWG 500.064	00 36 48.	+ 29 29		15.5	GALAXY
UGC 00412	00 36 48.	+ 29 29	60	15.5	GALAXY S0-a
ZC 0036.8+3056	00 36 48.	+ 30 56	810		CLUSTER OF GALAXIES
ZWG 519.015	00 36 48.	+ 36 05		14.9	GALAXY
OCL 0302	00 36 48.	+ 61 41	240	13.	OPEN STAR CLUSTER
ZWG 383.052	00 36 48.	- 02 24		15.6	GALAXY
LB 08332	00 36 48.	- 20 41		18.5	FAINT BLUE STAR
LB 08333	00 36 48.	- 23 48		17.8	FAINT BLUE STAR
MCG-05-02-031	00 36 48.	- 27 37	66	14.5	GALAXY
MCG-05-02-032	00 36 48.	- 30 13	30	14.5	GALAXY
RIC 286	00 36 48.00	+ 42 15 12.2		19.48	BLUE OBJECT
BAA 1.472	00 36 48.21	+ 40 27 27.0			EMISSION NEBULA IN M31
BAA 1.368	00 36 48.55	+ 40 04 47.9			EMISSION NEBULA IN M31
RIC 287	00 36 48.80	+ 39 17 38.4		18.82	BLUE OBJECT
RNGC 0197	00 36 49.	+ 00 37		15.0	GALAXY
IC 0040	00 36 49.	+ 02 11 08.			NONSTELLAR OBJECT
RNGC 0198	00 36 49.	+ 02 31		14.0	GALAXY
RNGC 0193	00 36 49.	+ 03 03		14.5	GALAXY
IC 1565	00 36 49.	+ 06 27 14.			NONSTELLAR OBJECT
BAA 3.17	00 36 50.69	+ 40 14 46.5			GLOBULAR CLUSTER IN M31
BAA 1.374	00 36 50.91	+ 40 03 57.1			EMISSION NEBULA IN M31
MCG+01-02-048	00 36 51.	+ 06 31 30.	36	15.	GALAXY
RIC 288	00 36 51.20	+ 40 51 11.0		19.70	BLUE OBJECT
BAA 2.43	00 36 52.03	+ 40 31 47.6			GALAXY
RIC 289	00 36 52.03	+ 40 48 37.0		19.47	BLUE OBJECT
BAA 1.348	00 36 52.34	+ 40 11 03.1			EMISSION NEBULA IN M31
BAA 1.377	00 36 52.38	+ 40 03 15.4			EMISSION NEBULA IN M31
BAA 1.347	00 36 52.64	+ 40 11 43.3			EMISSION NEBULA IN M31
RIC 290	00 36 53.17	+ 39 03 45.3		18.84	BLUE OBJECT
RIC 291	00 36 53.29	+ 41 08 41.1		18.75	BLUE OBJECT
ZWG 383.056	00 36 54.	+ 02 10		15.1	GALAXY
UGC 00413	00 36 54.	+ 02 10	72	15.1	GALAXY S
ZWG 383.057	00 36 54.	+ 02 31		14.1	GALAXY
UGC 00414	00 36 54.	+ 02 31	78	14.1	GALAXY Sc
MCG+00-02-112	00 36 54.	+ 02 37 30.	90	12.	GALAXY
MCG+00-02-111	00 36 54.	+ 02 53	60	14.	GALAXY
MCG+01-02-049	00 36 54.	+ 03 39	6	15.5	GALAXY
ZWG 409.058	00 36 54.	+ 06 32		15.4	GALAXY
SHRV.73A 07	00 36 54.	+ 41 53	10	16.84	GLOBULAR CLUSTER IN M31
HELW 318	00 36 54.	- 09 18 42.			NEBULA
LB 08334	00 36 54.	- 20 43		19.4	FAINT BLUE STAR
LB 08335	00 36 54.	- 20 46		17.2	FAINT BLUE STAR
LB 08336	00 36 54.	- 21 02		16.6	FAINT BLUE STAR
LB 08337	00 36 54.	- 22 58		17.4	FAINT BLUE STAR
LB 08338	00 36 54.	- 26 20		16.7	FAINT BLUE STAR
BAA 1.371	00 36 54.41	+ 40 03 44.0			EMISSION NEBULA IN M31
RIC 292	00 36 55.06	+ 39 26 04.4		19.88	BLUE OBJECT
RIC 293	00 36 55.71	+ 40 49 42.0		18.09	BLUE OBJECT
BAA 1.466	00 36 55.90	+ 40 33 50.1			EMISSION NEBULA IN M31
RIC 294	00 36 55.94	+ 40 40 12.7		20.16	BLUE OBJECT
BAA 1.376	00 36 55.99	+ 40 02 54.2			EMISSION NEBULA IN M31
HOW 12	00 36 56.	- 73 39 02.	24		STAR CLUSTER IN SMC
BAA 1.346	00 36 56.01	+ 40 11 55.1			EMISSION NEBULA IN M31
BAA 1.344	00 36 56.75	+ 40 12 28.1			EMISSION NEBULA IN M31
BAA 1.366	00 36 56.83	+ 40 04 02.2			EMISSION NEBULA IN M31
BPL 17	00 36 57.	+ 38 51 42.		17.91	PEC. BLUE OBJECT NEAR M31
MCG+01-02-050	00 36 57.	+ 08 41	132	14.	GALAXY
RIC 295	00 36 57.21	+ 41 36 14.1		18.99	BLUE OBJECT
RIC 296	00 36 57.24	+ 41 04 10.1		19.80	BLUE OBJECT
RIC 297	00 36 57.31	+ 41 34 45.2		19.41	BLUE OBJECT
RIC 298	00 36 57.37	+ 39 32 38.4		18.90	BLUE OBJECT
BAA 1.364	00 36 57.51	+ 40 04 13.2			EMISSION NEBULA IN M31
IC 1566	00 36 58.	+ 06 32 26.			NONSTELLAR OBJECT
RIC 299	00 36 58.18	+ 39 07 14.2		18.61	BLUE OBJECT
BAA 1.358	00 36 58.48	+ 40 05 05.8			EMISSION NEBULA IN M31
RIC 300	00 36 58.48	+ 40 44 35.2		19.78	BLUE OBJECT
BAA 1.363	00 36 58.90	+ 40 04 18.0			EMISSION NEBULA IN M31
IC 1567	00 36 59.	+ 06 21			NONSTELLAR OBJECT
RIC 301	00 36 59.46	+ 40 44 56.8		18.62	BLUE OBJECT
RIC 302	00 36 59.58	+ 40 53 42.4		18.88	BLUE OBJECT
BAA 1.362	00 36 59.88	+ 40 04 10.2			EMISSION NEBULA IN M31
KHAV 010	00 37	+ 54 52	3610		DARK NEBULA
MCG+00-02-115	00 37 00.	+ 00 35 30.	102	12.	GALAXY
ZWG 383.058	00 37 00.	+ 02 51		15.0	GALAXY
UGC 00415	00 37 00.	+ 02 51	84	15.0	GALAXY S0
MCG+00-02-114	00 37 00.	+ 03 11	48	14.5	GALAXY
MCG+00-02-113	00 37 00.	+ 03 17	48	15.	GALAXY
UGC 00416	00 37 00.	+ 03 40	84	16.5	GALAXY
UGC 00417	00 37 00.	+ 03 55	60	18.	GALAXY DWARF
ZWG 409.059	00 37 00.	+ 08 41		15.2	GALAXY
UGC 00418	00 37 00.	+ 08 41	126	15.2	GALAXY Sb
KARA.72 014A	00 37 00.	+ 08 41	114	15.2	PART OF DOUBLE GALAXY
MCG+01-02-051	00 37 00.	+ 08 43	60	14.	GALAXY
MCG+03-02-021	00 37 00.	+ 20 58	24	16.	GALAXY
ZWG 500.065	00 37 00.	+ 32 53		14.9	GALAXY
ZWG 519.016	00 37 00.	+ 35 35		15.5	GALAXY
VB 004	00 37 00.	+ 69 20	376		STELLAR RING
MCG-02-02-079	00 37 00.	- 09 28	60	14.5	GALAXY
MCG-03-02-034	00 37 00.	- 14 57	78	13.5	GALAXY
MCG-03-02-033	00 37 00.	- 19 07	18	16.	GALAXY
LB 08339	00 37 00.	- 22 10		18.6	FAINT BLUE STAR
LB 08340	00 37 00.	- 22 40		19.1	FAINT BLUE STAR
LB 08341	00 37 00.	- 23 22		19.0	FAINT BLUE STAR
LB 08342	00 37 00.	- 23 30		17.9	FAINT BLUE STAR
LB 08343	00 37 00.	- 24 22		16.3	FAINT BLUE STAR
LB 08344	00 37 00.	- 26 09		20.1	FAINT BLUE STAR
LB 08345	00 37 00.	- 26 24		18.0	FAINT BLUE STAR
LB 08346	00 37 00.	- 26 25		19.7	FAINT BLUE STAR
RIC 303	00 37 00.17	+ 41 20 54.2		18.50	BLUE OBJECT
RNGC 0200	00 37 01.	+ 02 37		14.0	GALAXY
RNGC 0199	00 37 01.	+ 02 51		15.0	GALAXY
RIC 304	00 37 01.39	+ 41 52 18.4		18.86	BLUE OBJECT
BAA 1.361	00 37 01.40	+ 40 04 02.5			EMISSION NEBULA IN M31
BAA 1.367	00 37 01.70	+ 40 03 16.7			EMISSION NEBULA IN M31
RIC 305	00 37 01.70	+ 42 02 09.8		19.40	BLUE OBJECT
BAA 1.359	00 37 01.84	+ 40 04 30.1			EMISSION NEBULA IN M31
RIC 306	00 37 01.87	+ 41 37 21.6		18.10	BLUE OBJECT
BAA 1.360	00 37 01.92	+ 40 04 15.4			EMISSION NEBULA IN M31
BAA 1.338	00 37 01.96	+ 40 14 02.0			EMISSION NEBULA IN M31

OBJECT NAME	RIGHT ASCEN.	DECLINATION	DIAM.	MAGN.	TYPE OF OBJECT
BAA 1.339	00 37 01.98	+ 40 13 22.9			EMISSION NEBULA IN M31
RIC 307	00 37 02.08	+ 38 51 48.8		18.24	BLUE OBJECT
RIC 308	00 37 02.27	+ 41 43 33.0		19.03	BLUE OBJECT
BAA 1.365	00 37 02.48	+ 40 03 20.6			EMISSION NEBULA IN M31
RNGC 0187	00 37 03.	- 14 57		13.0	GALAXY
MCG-04-02-032	00 37 03.	- 20 39 30.	66	15.	GALAXY
BAA 1.351	00 37 03.24	+ 40 06 48.5			EMISSION NEBULA IN M31
BAA 2.44	00 37 03.37	+ 40 29 57.3			GALAXY
RIC 309	00 37 03.85	+ 41 15 19.6		19.89	BLUE OBJECT
BAA 1.357	00 37 04.95	+ 40 04 39.0			EMISSION NEBULA IN M31
VDB.66B 25	00 37 05.	+ 39 10 23.		19.20	BLUE OBJECT
RIC 310	00 37 05.18	+ 40 53 40.6		18.07	BLUE OBJECT
BAA 3.14	00 37 05.41	+ 40 15 02.7			GLOBULAR CLUSTER IN M31
RIC 311	00 37 05.68	+ 39 35 03.3		18.67	BLUE OBJECT
RIC 312	00 37 05.82	+ 39 10 58.6		18.20	BLUE OBJECT
ZWG 383.059	00 37 06.	+ 00 35		14.7	GALAXY
UGC 00419	00 37 06.	+ 00 35	138	14.7	GALAXY Sc/SBc
ZWG 383.060	00 37 06.	+ 02 37		14.0	GALAXY
UGC 00420	00 37 06.	+ 02 37	120	14.0	GALAXY SBc
MCG+00-02-116	00 37 06.	+ 03 02 30.	48	14.	GALAXY
ZWG 383.061	00 37 06.	+ 03 10		15.0	GALAXY
ZWG 383.062	00 37 06.	+ 03 15		15.5	GALAXY
UGC 00421	00 37 06.	+ 03 15	60	15.5	GALAXY S
ZWG 409.060	00 37 06.	+ 08 43		15.0	GALAXY
UGC 00422	00 37 06.	+ 08 43	84	15.0	GALAXY SB
KARA.72 014B	00 37 06.	+ 08 43	66	15.0	PART OF DOUBLE GALAXY
LB 08347	00 37 06.	- 23 04		17.2	FAINT BLUE STAR
LB 08348	00 37 06.	- 23 06		16.9	FAINT BLUE STAR
LB 08349	00 37 06.	- 23 34		18.8	FAINT BLUE STAR
MCG-05-02-033	00 37 06.	- 30 38	54	15.5	GALAXY
LB 03149	00 37 06.	- 68 13		13.3	FAINT BLUE STAR
BAA 1.355	00 37 06.12	+ 40 04 57.7			EMISSION NEBULA IN M31
RNGC 0201	00 37 07.	+ 00 35		14.5	GALAXY
RNGC 0203	00 37 07.	+ 03 10		15.0	GALAXY
RNGC 0202	00 37 07.	+ 03 15		15.5	GALAXY
BAA 1.356	00 37 07.28	+ 40 04 31.6			EMISSION NEBULA IN M31
RIC 314	00 37 07.54	+ 39 23 04.5		18.21	BLUE OBJECT
BAA 1.462	00 37 07.55	+ 40 37 04.8			EMISSION NEBULA IN M31
RIC 313	00 37 07.56	+ 39 26 05.0		20.29	BLUE OBJECT
RIC 315	00 37 07.57	+ 39 43 20.8		17.99	BLUE OBJECT
RIC 316	00 37 07.74	+ 41 09 16.2		19.79	BLUE OBJECT
VDB.66B 28	00 37 08.	+ 39 22 47.		18.72	BLUE OBJECT
RNGC 0195	00 37 08.	- 09 28		14.0	GALAXY
BAA 1.334	00 37 08.94	+ 40 16 39.7			EMISSION NEBULA IN M31
VDB.66B 29	00 37 09.	+ 39 23 11.		20.27	BLUE OBJECT
SC 0034-6622.7	00 37 09.	- 66 06 12.	6		NEBULA
BAA 1.337	00 37 09.40	+ 40 13 16.0			EMISSION NEBULA IN M31
RIC 317	00 37 09.70	+ 41 00 00.2		17.70	BLUE OBJECT
BAA 1.353	00 37 09.90	+ 40 04 41.2			EMISSION NEBULA IN M31
RIC 319	00 37 09.93	+ 41 28 55.0		18.43	BLUE OBJECT
IC 0041	00 37 10.	- 14 26 53.			NONSTELLAR OBJECT
RIC 319	00 37 10.23	+ 39 07 18.5		17.37	BLUE OBJECT
ABC 0075	00 37 11.	+ 20 59		16.5	RICH CLUSTER OF GALAXIES
RIC 320	00 37 11.14	+ 39 39 25.0		18.73	BLUE OBJECT
BAA 1.343	00 37 11.33	+ 40 11 09.5			EMISSION NEBULA IN M31
ZWG 383.063	00 37 12.	+ 03 01		14.6	GALAXY
UGC 00423	00 37 12.	+ 03 01	102	14.6	GALAXY COMPACT
ABC 0076	00 37 12.	+ 06 30		15.0	RICH CLUSTER OF GALAXIES
ZWG 457.024	00 37 12.	+ 20 56		15.3	GALAXY
MCG-03-02-035	00 37 12.	- 14 31	36	14.	GALAXY
LB 00072	00 37 12.	- 21 40		16.8	FAINT BLUE STAR
LB 08350	00 37 12.	- 22 01		19.4	FAINT BLUE STAR
LB 08351	00 37 12.	- 22 18		19.1	FAINT BLUE STAR
LB 08352	00 37 12.	- 22 33		19.8	FAINT BLUE STAR
LB 00156	00 37 12.	- 23 59		16.2	FAINT BLUE STAR
LB 08353	00 37 12.	- 25 53		20.0	FAINT BLUE STAR
LB 08354	00 37 12.	- 26 18		17.4	FAINT BLUE STAR
LB 08355	00 37 12.	- 26 40		18.8	FAINT BLUE STAR
BAA 1.459	00 37 12.06	+ 40 39 22.3			EMISSION NEBULA IN M31
RIC 321	00 37 12.22	+ 39 13 03.8		17.32	BLUE OBJECT
BAA 1.336	00 37 12.30	+ 40 13 29.9			EMISSION NEBULA IN M31
RIC 322	00 37 12.52	+ 39 07 33.0		17.90	BLUE OBJECT
RNGC 0204	00 37 13.	+ 03 01		14.5	GALAXY
RIC 323	00 37 13.22	+ 41 14 46.6		19.99	BLUE OBJECT
BAA 2.24	00 37 14.03	+ 40 13 10.9			GALAXY
BAA 1.354	00 37 14.45	+ 40 04 01.1			EMISSION NEBULA IN M31
BAA 1.352	00 37 14.90	+ 40 04 47.9			EMISSION NEBULA IN M31
MCG+03-02-022	00 37 15.	+ 20 56	27	15.	GALAXY
RIC 324	00 37 15.56	+ 39 07 45.8		18.62	BLUE OBJECT
RIC 325	00 37 16.26	+ 41 13 26.8		19.74	BLUE OBJECT
BAA 1.341	00 37 16.96	+ 40 10 49.3			EMISSION NEBULA IN M31
HELW 319	00 37 17.	- 09 25 06.			NEBULA
LIN.CL 021	00 37 17.	- 72 59 36.	342	13.5	STAR CLUSTER IN SMC
RIC 326	00 37 17.00	+ 39 19 19.0		18.25	BLUE OBJECT
BAA 1.473	00 37 17.36	+ 40 22 44.6			EMISSION NEBULA IN M31
BAA 1.349	00 37 17.45	+ 40 05 50.4			EMISSION NEBULA IN M31
RIC 327	00 37 17.58	+ 39 30 59.4		18.95	BLUE OBJECT
RIC 328	00 37 17.82	+ 41 54 03.4		18.18	BLUE OBJECT
ZWG 409.061	00 37 18.	+ 06 34		15.6	GALAXY
MCG+01-02-052	00 37 18.	+ 06 34	48	15.	GALAXY
UGC 00424	00 37 18.	+ 20 16	78	18.	GALAXY
4ZW 028	00 37 18.	+ 26 54			COMPACT GALAXY
LB 08356	00 37 18.	- 22 03		15.5	FAINT BLUE STAR
RIC 329	00 37 18.24	+ 39 37 16.8		19.16	BLUE OBJECT
RIC 330	00 37 20.03	+ 38 52 26.2		20.54	BLUE OBJECT
BAA 1.452	00 37 20.60	+ 40 42 27.3			EMISSION NEBULA IN M31
VDB.66B 22	00 37 21.	+ 38 50 35.		17.42	BLUE OBJECT
MCG-04-02-033	00 37 21.	- 22 21 30.	36	16.	GALAXY
IC 1568	00 37 22.	+ 06 34 31.			NONSTELLAR OBJECT
BAA 2.23	00 37 22.26	+ 40 14 11.0			GALAXY
BAA 2.42	00 37 22.29	+ 40 33 06.8			GALAXY
RIC 331	00 37 23.70	+ 38 55 21.3		17.99	BLUE OBJECT
BAA 1.451	00 37 23.80	+ 40 42 14.8			EMISSION NEBULA IN M31
5ZW 030	00 37 24.	+ 48 17			COMPACT GALAXY
ZC 0037.4-0113	00 37 24.	- 01 13	1280		CLUSTER OF GALAXIES
MCG+00-02-117	00 37 24.	- 02 00	42	15.	GALAXY
LB 08357	00 37 24.	- 20 46		18.0	FAINT BLUE STAR
LB 08358	00 37 24.	- 22 12		19.3	FAINT BLUE STAR
LB 08359	00 37 24.	- 22 14		17.7	FAINT BLUE STAR
LB 08360	00 37 24.	- 24 06		18.1	FAINT BLUE STAR
LB 08361	00 37 24.	- 24 31		19.0	FAINT BLUE STAR
LB 08362	00 37 24.	- 24 51		18.6	FAINT BLUE STAR
LB 08363	00 37 24.	- 26 34		17.4	FAINT BLUE STAR
RIC 332	00 37 24.52	+ 42 04 36.7		20.63	BLUE OBJECT
RIC 333	00 37 25.42	+ 41 16 14.6		17.99	BLUE OBJECT
RIC 334	00 37 25.60	+ 38 57 37.9		17.84	BLUE OBJECT
PK121+00.1	00 37 25.7	+ 62 35 02.			PLANETARY NEBULA
RIC 335	00 37 25.79	+ 39 49 50.8		20.53	BLUE OBJECT
RIC 336	00 37 26.09	+ 39 48 36.2		20.58	BLUE OBJECT
RIC 337	00 37 26.85	+ 39 49 32.4		20.27	BLUE OBJECT
BAA 1.446	00 37 26.88	+ 40 44 58.3			EMISSION NEBULA IN M31
BAA 1.465	00 37 26.91	+ 40 29 29.3			EMISSION NEBULA IN M31
RNGC 0207	00 37 27.	- 15 26			NON-EXISTENT OBJECT
RIC 338	00 37 27.27	+ 39 49 37.6		21.35	BLUE OBJECT
BAA 1.468	00 37 28.10	+ 40 25 18.8			EMISSION NEBULA IN M31
BAA 1.345	00 37 28.14	+ 40 07 44.3			EMISSION NEBULA IN M31
RIC 339	00 37 28.86	+ 42 13 50.9		18.67	BLUE OBJECT
RIC 340	00 37 28.95	+ 41 52 32.9		18.55	BLUE OBJECT
BAA 1.340	00 37 29.26	+ 40 09 21.6			EMISSION NEBULA IN M31
RIC 341	00 37 29.26	+ 42 17 31.3		17.90	BLUE OBJECT
ZWG 479.056	00 37 30.	+ 22 26		14.5	GALAXY
UGC 00425	00 37 30.	+ 22 26	42	14.5	GALAXY SB
SHRV.73B 18	00 37 30.	+ 43 01	8	16.12	COMPACT GALAXY
5ZW 031	00 37 30.	+ 53 16			COMPACT GALAXY
LB 08364	00 37 30.	- 20 42		19.5	FAINT BLUE STAR
LB 08365	00 37 30.	- 22 23		18.3	FAINT BLUE STAR
LB 08366	00 37 30.	- 22 42		17.8	FAINT BLUE STAR
LB 08367	00 37 30.	- 26 06		20.1	FAINT BLUE STAR
BAA 1.453	00 37 30.64	+ 40 38 49.6			EMISSION NEBULA IN M31
RIC 342	00 37 30.77	+ 39 58 12.4		18.36	BLUE OBJECT
RIC 343	00 37 31.16	+ 41 39 05.7		19.41	BLUE OBJECT
RIC 344	00 37 31.18	+ 39 13 28.4		18.64	BLUE OBJECT
HOW 13	00 37 32.	- 73 41 44.	18		STAR CLUSTER IN SMC
BAA 1.445	00 37 32.02	+ 40 44 35.9			EMISSION NEBULA IN M31
RIC 345	00 37 34.02	+ 39 41 29.3		18.44	BLUE OBJECT
RIC 346	00 37 34.50	+ 41 39 01.4		19.96	BLUE OBJECT
RIC 347	00 37 34.94	+ 41 38 47.2		19.42	BLUE OBJECT
RIC 348	00 37 35.32	+ 39 37 28.7		19.21	BLUE OBJECT
BAA 1.461	00 37 35.58	+ 40 33 54.6			EMISSION NEBULA IN M31
ZWG 457.025	00 37 36.	+ 20 48		15.7	GALAXY
VDB.66B 32	00 37 36.	+ 39 38 10.		19.20	BLUE OBJECT
ZWG 535.014	00 37 36.	+ 41 25		9.4	GALAXY
UGC 00426	00 37 36.	+ 41 25	1200	9.4	GALAXY E
ZWG 550.010	00 37 36.	+ 50 04		15.7	GALAXY
UGC 00427	00 37 36.	+ 50 04	90	15.7	GALAXY Sc
8ZW 0037-10.7	00 37 36.	- 10 42		14.0	COMPACT GALAXY
LB 08368	00 37 36.	- 23 17		17.6	FAINT BLUE STAR
LB 08369	00 37 36.	- 23 20		18.6	FAINT BLUE STAR
LB 08370	00 37 36.	- 24 45		19.1	FAINT BLUE STAR
LB 08371	00 37 36.	- 26 42		17.3	FAINT BLUE STAR
MCG-05-02-034	00 37 36.	- 30 53	36	15.	GALAXY
BAA 1.458	00 37 36.39	+ 40 35 55.7			EMISSION NEBULA IN M31
RIC 349	00 37 36.42	+ 39 56 08.0		18.96	BLUE OBJECT
RIC 350	00 37 36.54	+ 39 34 49.8		18.04	BLUE OBJECT
RIC 351	00 37 36.78	+ 39 47 47.0		20.74	BLUE OBJECT
RIC 352	00 37 37.06	+ 39 28 05.6		18.56	BLUE OBJECT
RIC 353	00 37 37.37	+ 39 48 14.0		21.30	BLUE OBJECT
RIC 354	00 37 37.66	+ 39 28 33.0		17.45	BLUE OBJECT
RIC 355	00 37 37.98	+ 39 28 12.2		19.63	BLUE OBJECT
RNGC 0205	00 37 38.	+ 41 25		10.0	GALAXY
BAA 1.457	00 37 38.08	+ 40 36 05.3			EMISSION NEBULA IN M31
RIC 356	00 37 38.22	+ 41 33 33.9		20.42	BLUE OBJECT
RIC 357	00 37 38.55	+ 39 47 44.0		19.86	BLUE OBJECT
RIC 358	00 37 39.07	+ 39 48 02.0		19.44	BLUE OBJECT
BAA 1.456	00 37 39.10	+ 40 37 25.7			EMISSION NEBULA IN M31
RIC 359	00 37 39.32	+ 39 47 28.8		19.84	BLUE OBJECT
BAA 4.34	00 37 39.93	+ 40 14 20.2			DIFFUSE OBJECT IN M31
BAA 1.332	00 37 40.16	+ 40 17 46.8			EMISSION NEBULA IN M31
BAA 1.467	00 37 41.39	+ 40 26 25.7			EMISSION NEBULA IN M31
BAA 2.22	00 37 41.41	+ 40 21 23.0			GALAXY
RIC 360	00 37 41.41	+ 42 03 54.8		19.26	BLUE OBJECT
BAA 1.460	00 37 41.78	+ 40 33 06.3			EMISSION NEBULA IN M31
MCG+00-02-118	00 37 42.	+ 02 29 30.	48	15.	GALAXY
ZWG 479.057	00 37 42.	+ 24 10		15.3	GALAXY
MCG+04-02-043	00 37 42.	+ 24 10	15	15.	GALAXY
UGC 00428	00 37 42.	+ 29 17	90	16.0	GALAXY
HOLM 017C	00 37 42.	+ 41 24	660	10.1	PART OF MULTIPLE GALAXY
MCG+07-02-014	00 37 42.	+ 41 24	300		GALAXY
MCG+08-02-011	00 37 42.	+ 50 02	78	15.	GALAXY
LB 08372	00 37 42.	- 21 44		19.8	FAINT BLUE STAR
LB 08373	00 37 42.	- 25 51		17.2	FAINT BLUE STAR
BAA 2.21	00 37 42.61	+ 40 21 14.2			GALAXY
RIC 361	00 37 43.24	+ 39 56 49.3		18.24	BLUE OBJECT
BAA 1.464	00 37 44.12	+ 40 27 36.0			EMISSION NEBULA IN M31
BAA 1.335	00 37 44.39	+ 40 11 26.1			EMISSION NEBULA IN M31
BAA 1.333	00 37 44.39	+ 40 14 08.4			EMISSION NEBULA IN M31
BAA 4.33	00 37 46.07	+ 40 19 47.9			DIFFUSE OBJECT IN M31
RIC 362	00 37 46.60	+ 41 51 14.2		18.99	BLUE OBJECT
BAA 1.315	00 37 47.17	+ 40 24 23.6			EMISSION NEBULA IN M31
BAA 1.310	00 37 47.50	+ 40 26 03.4			EMISSION NEBULA IN M31
RIC 363	00 37 47.80	+ 39 43 37.4		18.30	BLUE OBJECT
ZWG 383.064	00 37 48.	+ 02 28		15.5	GALAXY
MCG+01-02-053	00 37 48.	+ 06 26	9	15.5	GALAXY
ZC 0037.8+1802	00 37 48.	+ 18 02	2960		CLUSTER OF GALAXIES
ZC 0037.8+2434	00 37 48.	+ 24 34	1610		CLUSTER OF GALAXIES
ZWG 479.058	00 37 48.	+ 24 45		15.1	GALAXY
MRK 345	00 37 48.	+ 24 45	12	15.5	GALAXY WITH UV CONTINUUM
SHRV.73A 08	00 27 48.	+ 40 10	9	18.2	GLOBULAR CLUSTER IN M31
MCG-02-02-080	00 37 48.	- 10 34	36	15.	GALAXY
LB 08374	00 37 48.	- 22 23		18.0	FAINT BLUE STAR
LB 08375	00 37 48.	- 25 19		18.5	FAINT BLUE STAR
LB 08376	00 37 48.	- 25 44		17.8	FAINT BLUE STAR
LB 08377	00 37 48.	- 25 49		17.9	FAINT BLUE STAR
LB 08378	00 37 48.	- 26 04		20.0	FAINT BLUE STAR
BAA 1.312	00 37 48.21	+ 40 24 56.1			EMISSION NEBULA IN M31
RIC 364	00 37 48.50	+ 41 11 23.4		20.89	BLUE OBJECT
RIC 365	00 37 48.72	+ 41 56 24.3		20.20	BLUE OBJECT
RNGC 0208	00 37 49.	+ 02 28		15.5	GALAXY
BAA 1.455	00 37 49.14	+ 40 35 00.8			EMISSION NEBULA IN M31
BAA 1.450	00 37 49.15	+ 40 39 02.7			EMISSION NEBULA IN M31
BAA 1.318	00 37 49.71	+ 40 22 51.8			EMISSION NEBULA IN M31
BAA 1.319	00 37 49.80	+ 40 22 31.5			EMISSION NEBULA IN M31
RNGC 0206	00 37 50.	+ 40 28			CLST/NEB IN EXTRNL GALAXY
SC 0035-6617.6	00 37 50.	- 66 01 06.	18		NEBULA
BAA 1.448	00 37 50.26	+ 40 40 50.7			EMISSION NEBULA IN M31
BAA 1.331	00 37 50.54	+ 40 16 18.6			EMISSION NEBULA IN M31
MCG+00-02-119	00 37 51.	+ 02 59 30.	60	15.	GALAXY
BAA 1.326	00 37 51.32	+ 40 19 12.9			EMISSION NEBULA IN M31
BAA 1.324	00 37 51.34	+ 40 20 20.5			EMISSION NEBULA IN M31
RIC 366	00 37 51.76	+ 39 47 17.4		19.65	BLUE OBJECT
BAA 1.328	00 37 51.76	+ 40 18 44.4			EMISSION NEBULA IN M31
BAA 1.323	00 37 51.79	+ 40 20 34.1			EMISSION NEBULA IN M31
RNGC 0212	00 37 52.	- 56 26			GALAXY
BAA 1.442	00 37 52.37	+ 40 44 47.7			EMISSION NEBULA IN M31
BAA 1.322	00 37 52.40	+ 40 20 56.9			EMISSION NEBULA IN M31
BAA 1.441	00 37 52.74	+ 40 44 52.5			EMISSION NEBULA IN M31

OBJECT NAME	RIGHT ASCEN.	DECLINATION	DIAM.	MAGN.	TYPE OF OBJECT
RIC 367	00 37 53.12	+ 41 18 04.6		18.69	BLUE OBJECT
BAA 1.417	00 37 53.47	+ 40 58 36.9			EMISSION NEBULA IN M31
RIC 368	00 37 53.51	+ 42 13 07.0		18.45	BLUE OBJECT
IC 1569	00 37 54.	+ 06 26 43.			NONSTELLAR OBJECT
ZWG 409.062	00 37 54.	+ 06 38		15.6	GALAXY
UGC 00429	00 37 54.	+ 06 38	96	15.6	GALAXY S0
ZWG 535.015	00 37 54.	+ 44 42		15.6	GALAXY
OCL 0303	00 37 54.	+ 60 35	1080		OPEN STAR CLUSTER
LB 08379	00 37 54.	- 20 50		17.6	FAINT BLUE STAR
LB 08380	00 37 54.	- 21 53		18.6	FAINT BLUE STAR
LB 08381	00 37 54.	- 23 41		17.7	FAINT BLUE STAR
LB 08382	00 37 54.	- 24 00		18.6	FAINT BLUE STAR
LB 08383	00 37 54.	- 24 14		16.2	FAINT BLUE STAR
RIC 369	00 37 54.36	+ 41 33 13.8		18.57	BLUE OBJECT
RIC 370	00 37 54.38	+ 41 24 32.9		20.03	BLUE OBJECT
RIC 371	00 37 54.76	+ 42 16 37.9		18.88	BLUE OBJECT
RIC 372	00 37 54.86	+ 41 35 29.6		18.24	BLUE OBJECT
BAA 1.440	00 37 54.93	+ 40 46 31.5			EMISSION NEBULA IN M31
ARC 0080	00 37 55.	- 24 58		17.1	RICH CLUSTER OF GALAXIES
HSN 03	00 37 55.	- 72 15			BRIGHT GALAXY BEHIND SMC
BAA 1.309	00 37 55.13	+ 40 24 57.8			EMISSION NEBULA IN M31
RIC 373	00 37 55.18	+ 41 54 55.0		18.04	BLUE OBJECT
BAA 1.329	00 37 55.44	+ 40 18 04.3			EMISSION NEBULA IN M31
RIC 374	00 37 55.68	+ 41 39 19.7		18.94	BLUE OBJECT
BAA 1.321	00 37 55.94	+ 40 21 06.1			EMISSION NEBULA IN M31
ARC 0078	00 37 56.	+ 26 28		17.9	RICH CLUSTER OF GALAXIES
BAA 1.463	00 37 56.04	+ 40 29 27.3			EMISSION NEBULA IN M31
BAA 1.327	00 37 56.57	+ 40 18 19.9			EMISSION NEBULA IN M31
BAA 1.307	00 37 56.82	+ 40 25 55.0			EMISSION NEBULA IN M31
ARC 0077	00 37 57.	+ 29 15		16.5	RICH CLUSTER OF GALAXIES
RNGC 0210	00 37 57.	- 14 09		12.0	GALAXY
BAA 1.350	00 37 57.84	+ 39 59 47.0			EMISSION NEBULA IN M31
BAA 1.342	00 37 57.85	+ 40 02 35.6			EMISSION NEBULA IN M31
BAA 1.454	00 37 57.98	+ 40 34 37.5			EMISSION NEBULA IN M31
ARC 0079	00 37 58.	+ 17 52		17.1	RICH CLUSTER OF GALAXIES
NAB 0037+18	00 37 58.	+ 18 00 15.		17.9	QUASI-STELLAR OBJECT
SHB 010	00 37 58.	+ 18 00 15.		17.9	QUASI-STELLAR OBJECT
BAA 1.425	00 37 59.50	+ 40 52 19.4			EMISSION NEBULA IN M31
RIC 375	00 37 59.60	+ 42 16 13.6		18.43	BLUE OBJECT
BAA 1.306	00 37 59.98	+ 40 27 07.6			EMISSION NEBULA IN M31
MCG+00-02-120	00 38 00.	+ 01 24 30.	24	15.	GALAXY
ZWG 383.065	00 38 00.	+ 01 25		15.2	GALAXY
UGC 00430	00 38 00.	+ 02 58	78	16.0	GALAXY S0
IC 1570	00 38 00.	+ 09 28 48.			NONSTELLAR OBJECT
ZC 0038.0+2914	00 38 00.	+ 29 14	2690		CLUSTER OF GALAXIES
ZWG 500.066	00 38 00.	+ 29 53		15.1	GALAXY
UGC 00431	00 38 00.	+ 29 53	72	15.1	GALAXY VY CMPT
MCG+00-02-121	00 38 00.	- 00 35 30.	60	13.	GALAXY
MCG-02-02-081	00 38 00.	- 14 10	270	11.	GALAXY
LB 08384	00 38 00.	- 21 38		18.3	FAINT BLUE STAR
LB 08385	00 38 00.	- 23 24		17.7	FAINT BLUE STAR
LB 08386	00 38 00.	- 23 34		17.7	FAINT BLUE STAR
LB 08387	00 38 00.	- 24 09		19.9	FAINT BLUE STAR
LB 08388	00 38 00.	- 26 14		18.7	FAINT BLUE STAR
LB 08389	00 38 00.	- 26 31		17.8	FAINT BLUE STAR
BAA 1.444	00 38 00.07	+ 40 41 06.4			EMISSION NEBULA IN M31
RIC 376	00 38 00.13	+ 38 53 39.9		19.11	BLUE OBJECT
RIC 377	00 38 00.14	+ 39 42 37.0		19.48	BLUE OBJECT
BAA 1.325	00 38 00.40	+ 40 18 48.3			EMISSION NEBULA IN M31
BAA 2.19	00 38 00.81	+ 40 28 19.8			GALAXY
BAA 1.303	00 38 00.83	+ 40 28 47.9			EMISSION NEBULA IN M31
HSN 04	00 38 01.	- 72 16			BRIGHT GALAXY BEHIND SMC
RIC 378	00 38 01.50	+ 39 09 29.6		17.72	BLUE OBJECT
RIC 379	00 38 02.12	+ 40 04 07.2		18.68	BLUE OBJECT
RIC 380	00 38 02.48	+ 39 24 08.6		20.33	BLUE OBJECT
RIC 381	00 38 02.62	+ 39 20 21.7		18.11	BLUE OBJECT
BAA 1.447	00 38 03.57	+ 40 38 58.6			EMISSION NEBULA IN M31
IC 1571	00 38 04.	- 00 36 42.			NONSTELLAR OBJECT
BAA 1.443	00 38 04.26	+ 40 42 31.7			EMISSION NEBULA IN M31
RIC 382	00 38 04.98	+ 39 31 23.8		18.08	BLUE OBJECT
RIC 383	00 38 05.14	+ 39 42 06.1		18.60	BLUE OBJECT
BAA 1.449	00 38 05.15	+ 40 38 22.9			EMISSION NEBULA IN M31
RIC 384	00 38 05.97	+ 41 49 54.4		18.36	BLUE OBJECT
ZWG 383.066	00 38 06.	- 00 35		14.8	GALAXY
UGC 00432	00 38 06.	- 00 35	78	14.8	GALAXY Sb
SN 1954R	00 38 06.	- 14 10		15.9	SUPERNOVA
LB 08390	00 38 06.	- 21 15		16.6	FAINT BLUE STAR
LB 08391	00 38 06.	- 23 39		19.9	FAINT BLUE STAR
LB 01566	00 38 06.	- 55 18		12.7	FAINT BLUE STAR
SC 0035-6533.9	00 38 06.	- 65 17 24.	12		NEBULA
RIC 385	00 38 06.57	+ 42 02 52.9		18.71	BLUE OBJECT
RIC 386	00 38 06.83	+ 41 46 49.2		19.32	BLUE OBJECT
RIC 387	00 38 06.97	+ 41 47 12.8		18.35	BLUE OBJECT
BAA 1.432	00 38 07.47	+ 40 48 22.0			EMISSION NEBULA IN M31
BAA 1.320	00 38 08.50	+ 40 19 37.4			EMISSION NEBULA IN M31
RIC 388	00 38 08.66	+ 39 43 34.5		19.59	BLUE OBJECT
BAA 1.388	00 38 08.73	+ 41 09 12.6			EMISSION NEBULA IN M31
RIC 389	00 38 08.77	+ 41 16 01.2		19.56	BLUE OBJECT
RIC 390	00 38 09.22	+ 40 04 35.6		19.40	BLUE OBJECT
BAA 1.330	00 38 09.64	+ 40 15 50.6			EMISSION NEBULA IN M31
BAA 1.408	00 38 09.73	+ 41 00 11.8			EMISSION NEBULA IN M31
RIC 391	00 38 11.68	+ 41 15 51.5		19.17	BLUE OBJECT
BAA 1.316	00 38 11.69	+ 40 20 34.7			EMISSION NEBULA IN M31
MCG+00-02-122	00 38 12.	+ 02 35	48	15.5	GALAXY
ZC 0038.2+2159	00 38 12.	+ 21 59	1210		CLUSTER OF GALAXIES
MCG-01-02-043	00 38 12.	- 02 48	36	16.	GALAXY
LB 08392	00 38 12.	- 20 51		18.6	FAINT BLUE STAR
LB 08393	00 38 12.	- 21 59		18.1	FAINT BLUE STAR
LB 08394	00 38 12.	- 23 58		19.4	FAINT BLUE STAR
LB 08395	00 38 12.	- 24 50		16.9	FAINT BLUE STAR
LB 08396	00 38 12.	- 26 14		18.9	FAINT BLUE STAR
RIC 392	00 38 12.85	+ 39 25 51.1		19.24	BLUE OBJECT
BAA 1.317	00 38 13.87	+ 40 19 41.1			EMISSION NEBULA IN M31
BAA 1.439	00 38 13.96	+ 40 46 08.1			EMISSION NEBULA IN M31
BAA 1.437	00 38 14.53	+ 40 46 38.0			EMISSION NEBULA IN M31
MCG+00-02-123	00 38 15.	- 00 32 30.	24	17.	GALAXY
HOW 14	00 38 15.	- 74 09 02.	30		STAR CLUSTER IN SMC
BAA 1.431	00 38 15.02	+ 40 47 23.1			EMISSION NEBULA IN M31
BAA 1.434	00 38 15.04	+ 40 47 05.2			EMISSION NEBULA IN M31
BAA 1.300	00 38 15.11	+ 40 29 30.0			EMISSION NEBULA IN M31
BAA 1.438	00 38 15.24	+ 40 46 07.5			EMISSION NEBULA IN M31
RIC 393	00 38 15.49	+ 39 23 08.2		19.26	BLUE OBJECT
RIC 394	00 38 15.95	+ 39 59 44.9		18.42	BLUE OBJECT
BAA 1.433	00 38 16.46	+ 40 46 53.6			EMISSION NEBULA IN M31
BAA 1.313	00 38 16.50	+ 40 20 27.0			EMISSION NEBULA IN M31
BAA 1.430	00 38 16.90	+ 40 47 14.8			EMISSION NEBULA IN M31
RIC 395	00 38 17.07	+ 42 00 12.3		18.07	BLUE OBJECT
BAA 1.314	00 38 17.15	+ 40 20 08.7			EMISSION NEBULA IN M31

OBJECT NAME	RIGHT ASCEN.	DECLINATION	DIAM.	MAGN.	TYPE OF OBJECT
BAA 1.436	00 38 17.15	+ 40 46 19.9			EMISSION NEBULA IN M31
BAA 1.429	00 38 17.63	+ 40 47 42.0			EMISSION NEBULA IN M31
BAA 1.398	00 38 17.74	+ 41 04 26.7			EMISSION NEBULA IN M31
ZWG 383.067	00 38 18.	+ 02 33		15.7	GALAXY
ZWG 500.067	00 38 18.	+ 31 27		14.6	GALAXY
UGC 00433	00 38 18.	+ 31 27	114	14.6	GALAXY Sc
VDB.66B 24	00 38 18.	+ 38 57 09.		19.24	BLUE OBJECT
UGC 00434	00 38 18.	+ 50 28	72	16.5	GALAXY Sb-c
LB 08397	00 38 18.	- 21 12		19.2	FAINT BLUE STAR
LB 08398	00 38 18.	- 25 35		18.7	FAINT BLUE STAR
LB 08399	00 38 18.	- 25 39		17.8	FAINT BLUE STAR
BAA 1.427	00 38 18.25	+ 40 48 19.7			EMISSION NEBULA IN M31
BAA 1.435	00 38 18.38	+ 40 46 28.8			EMISSION NEBULA IN M31
BAA 1.428	00 38 18.53	+ 40 48 03.8			EMISSION NEBULA IN M31
BAA 2.41	00 38 19.55	+ 40 44 29.4			GALAXY
HOLM 014B	00 38 20.	- 14 03	30	15.1	PART OF MULTIPLE GALAXY
SC 0036-6804.3	00 38 20.	- 67 47 49.	12		NEBULA
BAA 1.421	00 38 20.25	+ 40 52 37.9			EMISSION NEBULA IN M31
RIC 396	00 38 20.28	+ 41 19 56.8		17.98	BLUE OBJECT
RIC 398	00 38 20.55	+ 38 56 53.0		19.97	BLUE OBJECT
BAA 1.412	00 38 20.82	+ 40 58 28.2			EMISSION NEBULA IN M31
MCG-01-02-044	00 38 21.	- 04 43 30.	36	15.	GALAXY
RIC 397	00 38 22.38	+ 39 49 11.0		18.25	BLUE OBJECT
BAA 1.311	00 38 22.60	+ 40 20 54.2			EMISSION NEBULA IN M31
BAA 1.424	00 38 22.81	+ 40 50 21.2			EMISSION NEBULA IN M31
BAA 1.426	00 38 23.65	+ 40 47 46.3			EMISSION NEBULA IN M31
BC PKS0038-020	00 38 23.8	- 02 02 54.		18.	QUASI-STELLAR OBJECT
SHB 011	00 38 23.8	- 02 03 08.		18.5	QUASI-STELLAR OBJECT
ARC 0081	00 38 24.	+ 22 00		17.7	RICH CLUSTER OF GALAXIES
ZC 0038.4+2515	00 38 24.	+ 25 15	3360		CLUSTER OF GALAXIES
MCG+05-02-038	00 38 24.	+ 31 28	102	14.	GALAXY
ZWG 345.001	00 38 24.	+ 79 00		15.7	GALAXY
ZWG 344.009	00 38 24.	+ 79 00		15.7	GALAXY
MCG+00-02-124	00 38 24.	- 01 54	48	14.5	GALAXY
MCG-02-02-082	00 38 24.	- 14 04	48	14.5	GALAXY
LB 08400	00 38 24.	- 20 49		19.0	FAINT BLUE STAR
LB 08401	00 38 24.	- 22 19		16.8	FAINT BLUE STAR
LB 08402	00 38 24.	- 22 29		18.6	FAINT BLUE STAR
LB 08403	00 38 24.	- 26 27		18.9	FAINT BLUE STAR
MCG-05-02-035	00 38 24.	- 30 45	30	15.5	GALAXY
LB 01567	00 38 24.	- 52 29		14.6	FAINT BLUE STAR
BAA 1.298	00 38 24.10	+ 40 31 46.4			EMISSION NEBULA IN M31
BAA 1.296	00 38 24.20	+ 40 32 05.4			EMISSION NEBULA IN M31
RNGC 0211	00 38 25.	+ 03 10			NON-EXISTENT OBJECT
HOLM 014A	00 38 25.	- 14 04	36	14.3	PART OF MULTIPLE GALAXY
RIC 399	00 38 27.35	+ 39 57 22.4		18.95	BLUE OBJECT
BAA 1.295	00 38 27.36	+ 40 33 32.3			EMISSION NEBULA IN M31
BAA 1.401	00 38 27.78	+ 41 00 29.8			EMISSION NEBULA IN M31
RIC 400	00 38 28.04	+ 41 40 48.3		17.01	BLUE OBJECT
RIC 401	00 38 28.18	+ 41 33 34.4		20.30	BLUE OBJECT
KEEL 007	00 38 28.7	+ 48 02 47.			NEBULA
BAA 2.20	00 38 28.96	+ 40 24 08.9			GALAXY
RIC 402	00 38 29.40	+ 39 59 03.5		18.25	BLUE OBJECT
RIC 403	00 38 29.66	+ 41 24 47.8		18.55	BLUE OBJECT
RIC 404	00 38 29.72	+ 40 02 51.6		18.62	BLUE OBJECT
VB 005	00 38 30.	+ 64 48	410		STELLAR RING
ZWG 383.068	00 38 30.	- 01 54		15.6	GALAXY
UGC 00435	00 38 30.	- 01 54	72	15.6	GALAXY
LB 08404	00 38 30.	- 24 16		19.0	FAINT BLUE STAR
LB 08405	00 38 30.	- 25 19		17.9	FAINT BLUE STAR
LB 08406	00 38 30.	- 26 00		18.5	FAINT BLUE STAR
BAA 1.422	00 38 30.12	+ 40 51 09.6			EMISSION NEBULA IN M31
BAA 1.305	00 38 30.18	+ 40 22 51.8			EMISSION NEBULA IN M31
BAA 1.420	00 38 30.45	+ 40 51 54.9			EMISSION NEBULA IN M31
BAA 1.418	00 38 31.08	+ 40 52 57.5			EMISSION NEBULA IN M31
RIC 405	00 38 31.23	+ 39 46 20.8		18.38	BLUE OBJECT
LIN.CL 022	00 38 32.	- 73 39 49.	678	11.7	STAR CLUSTER IN SMC
RNGC 0220	00 38 32.	- 73 40		11.5	OPEN CLUSTER IN SMC
LIN 028	00 38 32.	- 73 49 31.		14.43	SMC EMISSION LINE OBJECT
RIC 406	00 38 32.36	+ 41 33 58.8		18.62	BLUE OBJECT
MCG+03-02-023	00 38 33.	+ 16 11	90	14.	GALAXY
MCG-04-02-034	00 38 33.	- 22 21 30.	48	15.	GALAXY
BAA 1.396	00 38 33.35	+ 41 03 25.4			EMISSION NEBULA IN M31
IC 0042	00 38 34.	- 15 42 07.			NONSTELLAR OBJECT
SC 0036-6802.7	00 38 34.	- 67 46 13.	18		NEBULA
BAA 2.37	00 38 34.51	+ 41 00 19.4			GALAXY
LIN 027	00 38 35.	- 72 14 37.		14.78	SMC EMISSION LINE OBJECT
BAA 1.394	00 38 35.44	+ 41 03 22.0			EMISSION NEBULA IN M31
BAA 1.294	00 38 35.59	+ 40 32 29.4			EMISSION NEBULA IN M31
BAA 1.393	00 38 35.68	+ 41 03 22.1			EMISSION NEBULA IN M31
BAA 1.395	00 38 35.75	+ 41 03 10.9			EMISSION NEBULA IN M31
RIC 407	00 38 35.84	+ 41 34 52.2		18.84	BLUE OBJECT
ZC 0038.6+0217	00 38 36.	+ 02 17	3360		CLUSTER OF GALAXIES
ZWG 434.030	00 38 36.	+ 14 55		15.7	GALAXY
ZWG 457.026	00 38 36.	+ 16 12		14.8	GALAXY
RNGC 0213	00 38 36.	+ 16 12		15.0	GALAXY
UGC 00436	00 38 36.	+ 16 12	114	14.8	GALAXY SBa
MCG-03-02-036	00 38 36.	- 15 42 30.	36	15.	GALAXY
LB 08407	00 38 36.	- 22 40		17.6	FAINT BLUE STAR
LB 08408	00 38 36.	- 24 14		19.5	FAINT BLUE STAR
LB 08409	00 38 36.	- 24 18		18.8	FAINT BLUE STAR
LB 08410	00 38 36.	- 24 45		16.8	FAINT BLUE STAR
LB 08411	00 38 36.	- 26 26		19.2	FAINT BLUE STAR
BAA 1.419	00 38 36.19	+ 40 51 54.3			EMISSION NEBULA IN M31
BAA 1.409	00 38 36.78	+ 40 56 16.1			EMISSION NEBULA IN M31
IC 1872	00 38 37.	+ 16 00			NONSTELLAR OBJECT
HSN 05	00 38 37.	- 72 19			BRIGHT GALAXY BEHIND SMC
RIC 408	00 38 37.02	+ 41 33 23.7		18.81	BLUE OBJECT
BAA 1.392	00 38 37.83	+ 41 03 21.8			EMISSION NEBULA IN M31
BAA 1.390	00 38 37.88	+ 41 04 15.9			EMISSION NEBULA IN M31
RNGC 0222	00 38 38.	- 73 40		11.5	OPEN CLUSTER IN SMC
LIN.CL 024	00 38 38.	- 73 40 07.	342	11.7	STAR CLUSTER IN SMC
BAA 1.292	00 38 38.01	+ 40 33 16.5			EMISSION NEBULA IN M31
RNGC 0215	00 38 40.	- 56 29			GALAXY
LIN.CL 023	00 38 40.	- 73 01 55.	204	15.4	STAR CLUSTER IN SMC
BAA 1.410	00 38 40.54	+ 40 55 40.7			EMISSION NEBULA IN M31
RIC 409	00 38 40.84	+ 42 13 22.7		18.79	BLUE OBJECT
BAA 1.411	00 38 41.67	+ 40 55 28.0			EMISSION NEBULA IN M31
BAA 1.301	00 38 41.79	+ 40 25 33.8			EMISSION NEBULA IN M31
RIC 410	00 38 41.90	+ 41 45 27.5		18.36	BLUE OBJECT
ZC 0038.7+2107	00 38 42.	+ 21 07	2420		CLUSTER OF GALAXIES
MCG+04-02-044	00 38 42.	+ 25 13	114	13.	GALAXY
ZC 0038.7+3756	00 38 42.	+ 37 56	1550		CLUSTER OF GALAXIES
SHRV.73B 19	00 38 42.	+ 40 40	8	16.17	COMPACT GALAXY
SHRV.73A 09	00 38 42.	+ 42 02	8	17.09	GLOBULAR CLUSTER IN M31
MCG-02-02-083	00 38 42.	- 13 50	72	14.	GALAXY
MCG-03-02-037	00 38 42.	- 15 13	36	15.	GALAXY
LB 08412	00 38 42.	- 21 36		17.8	FAINT BLUE STAR

OBJECT NAME	RIGHT ASCEN.	DECLINATION	DIAM.	MAGN.	TYPE OF OBJECT
LB 08413	00 38 42.	- 23 29		19.5	FAINT BLUE STAR
LB 08414	00 38 42.	- 23 46		18.7	FAINT BLUE STAR
LB 08415	00 38 42.	- 25 02		19.4	FAINT BLUE STAR
LB 08416	00 38 42.	- 25 09		18.8	FAINT BLUE STAR
RIC 411	00 38 42.11	+ 41 38 41.6		19.40	BLUE OBJECT
RIC 412	00 38 42.38	+ 39 57 35.0		19.61	BLUE OBJECT
BAA 1.407	00 38 42.75	+ 40 55 39.5			EMISSION NEBULA IN M31
ARC 0082	00 38 43.	+ 25 08		17.1	RICH CLUSTER OF GALAXIES
SC 0036-6551.6	00 38 43.	- 65 35 07.	18		NEBULA
BAA 1.406	00 38 43.05	+ 40 55 59.1			EMISSION NEBULA IN M31
BAA 1.308	00 38 43.11	+ 40 18 03.9			EMISSION NEBULA IN M31
BAA 1.405	00 38 44.25	+ 40 55 56.5			EMISSION NEBULA IN M31
BAA 4.32	00 38 44.85	+ 40 36 47.4			DIFFUSE OBJECT IN M31
BAA 1.403	00 38 44.90	+ 40 56 46.8			EMISSION NEBULA IN M31
MCG-01-02-045	00 38 45.	- 05 34	48	15.	GALAXY
BAA 1.304	00 38 45.25	+ 40 22 15.6			EMISSION NEBULA IN M31
BAA 2.40	00 38 45.25	+ 40 45 45.4			GALAXY
BAA 1.404	00 38 45.50	+ 40 56 14.3			EMISSION NEBULA IN M31
BAA 1.414	00 38 45.80	+ 40 53 41.3			EMISSION NEBULA IN M31
RNGC 0214	00 38 46.	+ 25 14		13.0	GALAXY
BAA 1.289	00 38 46.20	+ 40 34 32.1			EMISSION NEBULA IN M31
BAA 4.31	00 38 46.33	+ 40 39 21.7			DIFFUSE OBJECT IN M31
BAA 1.423	00 38 46.50	+ 40 48 35.9			EMISSION NEBULA IN M31
RIC 413	00 38 47.18	+ 41 42 36.9		19.71	BLUE OBJECT
UGC 00437	00 38 48.	+ 03 13	66	18.	GALAXY DWRF SP
ZC 0038.8+1346	00 38 48.	+ 13 46	1610		CLUSTER OF GALAXIES
ZC 0038.8+1950	00 38 48.	+ 19 50	2760		CLUSTER OF GALAXIES
ZWG 479.059	00 38 48.	+ 25 13		13.0	GALAXY
UGC 00438	00 38 48.	+ 25 13	132	13.0	GALAXY Sc
SHRV.73B 20	00 38 48.	+ 42 04	8	16.69	COMPACT GALAXY
MCG+00-02-125	00 38 48.	- 02 00	54	13.	GALAXY
PHL 2894	00 38 48.	- 03 34		18.1	BLUE STELLAR OBJECT
LB 08417	00 38 48.	- 20 41		19.9	FAINT BLUE STAR
LB 08418	00 38 48.	- 21 08		16.7	FAINT BLUE STAR
LB 08419	00 38 48.	- 21 50		17.0	FAINT BLUE STAR
LB 08420	00 38 48.	- 24 44		19.1	FAINT BLUE STAR
LB 08421	00 38 48.	- 25 08		18.0	FAINT BLUE STAR
LB 08422	00 38 48.	- 25 20		16.6	FAINT BLUE STAR
LB 08423	00 38 48.	- 26 08		18.8	FAINT BLUE STAR
BAA 1.402	00 38 48.11	+ 40 57 06.1			EMISSION NEBULA IN M31
RIC 414	00 38 48.34	+ 39 56 12.4		18.49	BLUE OBJECT
RIC 415	00 38 48.78	+ 39 56 05.2		18.00	BLUE OBJECT
SHB 012	00 38 49.2	- 01 59 18.		15.0	QUASI-STELLAR OBJECT
BAA 1.400	00 38 49.64	+ 40 57 38.0			EMISSION NEBULA IN M31
BAA 1.399	00 38 49.92	+ 40 57 45.1			EMISSION NEBULA IN M31
BAA 4.30	00 38 50.11	+ 40 44 20.8			DIFFUSE OBJECT IN M31
RIC 416	00 38 50.22	+ 39 10 57.1		18.14	BLUE OBJECT
RIC 417	00 38 50.28	+ 41 45 08.5		18.15	BLUE OBJECT
BAA 1.413	00 38 50.79	+ 40 53 26.4			EMISSION NEBULA IN M31
BAA 2.39	00 38 50.85	+ 40 46 21.8			GALAXY
ARC 0083	00 38 51.	+ 13 46		17.8	RICH CLUSTER OF GALAXIES
BAA 2.18	00 38 51.49	+ 40 29 22.9			GALAXY
BAA 1.416	00 38 51.81	+ 40 50 27.4			EMISSION NEBULA IN M31
BAA 2.16	00 38 53.18	+ 40 30 58.5			GALAXY
ZWG 500.068	00 38 54.	+ 29 15		15.4	GALAXY
ZC 0038.9+3532	00 38 54.	+ 35 32	2220		CLUSTER OF GALAXIES
ZWG 383.069	00 38 54.	- 01 59		14.4	GALAXY
UGC 00439	00 38 54.	- 01 59	72	14.4	GALAXY Sa
MCG-03-02-038	00 38 54.	- 15 51	36	15.	GALAXY
LB 08424	00 38 54.	- 21 41		18.3	FAINT BLUE STAR
LB 08425	00 38 54.	- 24 38		17.7	FAINT BLUE STAR
LB 08426	00 38 54.	- 24 46		18.4	FAINT BLUE STAR
LB 08427	00 38 54.	- 24 49		20.2	FAINT BLUE STAR
LB 08428	00 38 54.	- 24 53		18.7	FAINT BLUE STAR
LB 08429	00 38 54.	- 25 02		17.7	FAINT BLUE STAR
RIC 418	00 38 54.56	+ 41 52 51.3		18.18	BLUE OBJECT
BAA 1.297	00 38 55.12	+ 40 27 34.0			EMISSION NEBULA IN M31
BAA 1.302	00 38 55.51	+ 40 21 29.4			EMISSION NEBULA IN M31
RIC 419	00 38 55.93	+ 42 14 28.5		18.88	BLUE OBJECT
BAA 2.35	00 38 56.95	+ 40 58 18.7			GALAXY
MCG+06-02-013	00 38 57.	+ 36 05 30.	60	15.	GALAXY
RNGC 0216	00 38 57.	- 21 19		13.0	GALAXY
HOW 15	00 38 57.	- 74 16 56.	18		STAR CLUSTER IN SMC
BAA 2.36	00 38 57.66	+ 40 57 46.1			GALAXY
RIC 420	00 38 58.37	+ 39 07 14.6		19.96	BLUE OBJECT
HELW 320	00 38 59.	- 09 32 31.			NEBULA
LIN 029	00 38 59.	- 72 18 32.		16.08	SMC EMISSION LINE OBJECT
BAA 1.382	00 38 59.12	+ 41 04 15.8			EMISSION NEBULA IN M31
BAA 2.62	00 38 59.22	+ 41 13 29.0			GALAXY
BAA 1.397	00 38 59.25	+ 40 59 25.6			EMISSION NEBULA IN M31
BAA 1.387	00 38 59.65	+ 41 01 53.3			EMISSION NEBULA IN M31
BAA 4.29	00 38 59.70	+ 40 39 59.5			DIFFUSE OBJECT IN M31
BAA 2.58	00 38 59.93	+ 41 10 35.7			GALAXY
ZC 0039.0+0822	00 39 00.	+ 08 22	1480		CLUSTER OF GALAXIES
ZWG 500.069	00 39 00.	+ 28 55		15.4	GALAXY
ZWG 519.017	00 39 00.	+ 36 05		15.5	GALAXY
UGC 00440	00 39 00.	+ 36 05	78	15.5	GALAXY S
5ZW 032	00 39 00.	+ 40 01			COMPACT GALAXY
LDN 1290	00 39 00.	+ 60 50	420		DARK NEBULA
LDN 1293	00 39 00.	+ 62 50	420		DARK NEBULA
MCG-04-02-035	00 39 00.	- 21 19 30.	72	13.	GALAXY
LB 08430	00 39 00.	- 21 28		18.7	FAINT BLUE STAR
LB 08431	00 39 00.	- 21 59		19.5	FAINT BLUE STAR
LB 08432	00 39 00.	- 22 02		18.4	FAINT BLUE STAR
LB 08433	00 39 00.	- 22 16		17.8	FAINT BLUE STAR
LB 08434	00 39 00.	- 23 52		18.7	FAINT BLUE STAR
LB 08435	00 39 00.	- 24 58		19.4	FAINT BLUE STAR
MCG-06-02-023	00 39 00.	- 36 06	48	15.	GALAXY
SC 0036-6626.4	00 39 00.	- 66 09 55.	6		NEBULA
BAA 1.293	00 39 00.86	+ 40 28 51.6			EMISSION NEBULA IN M31
LIN 030	00 39 01.	- 73 47 32.		15.62	SMC EMISSION LINE OBJECT
BAA 1.391	00 39 01.92	+ 41 00 36.5			EMISSION NEBULA IN M31
RNGC 0217	00 39 02.	- 10 18		13.0	GALAXY
LIN.CL 025	00 39 02.	- 73 38 13.	546	11.8	STAR CLUSTER IN SMC
BAA 1.386	00 39 02.83	+ 41 02 20.1			EMISSION NEBULA IN M31
RNGC 0218	00 39 03.	+ 36 05		15.5	GALAXY
MCG-02-02-084	00 39 03.	- 09 42	12	16.	GALAXY
MCG-02-02-085	00 39 03.	- 10 18	156	13.	GALAXY
RNGC 0231	00 39 03.	- 73 38		12.0	OPEN CLUSTER IN SMC
BAA 1.389	00 39 03.07	+ 41 00 39.7			EMISSION NEBULA IN M31
BAA 1.415	00 39 03.27	+ 40 50 17.3			EMISSION NEBULA IN M31
BAA 1.299	00 39 03.39	+ 40 24 23.5			EMISSION NEBULA IN M31
RIC 421	00 39 03.79	+ 39 47 42.9		18.86	BLUE OBJECT
BAA 1.385	00 39 04.19	+ 41 02 31.9			EMISSION NEBULA IN M31
BAA 1.384	00 39 04.23	+ 41 02 37.9			EMISSION NEBULA IN M31
BAA 1.383	00 39 04.27	+ 41 02 53.8			EMISSION NEBULA IN M31
RIC 422	00 39 04.42	+ 38 53 26.6		18.81	BLUE OBJECT
BAA 2.61	00 39 04.45	+ 41 12 38.7			GALAXY
RIC 423	00 39 05.21	+ 41 43 18.7		17.77	BLUE OBJECT
BAA 2.17	00 39 05.86	+ 40 27 35.6			GALAXY
MCG+00-02-126	00 39 06.	+ 00 54	36	15.	GALAXY
UGC 00441	00 39 06.	+ 10 33	60	16.5	GALAXY S
ZC 0039.1+2227	00 39 06.	+ 22 27	2620		CLUSTER OF GALAXIES
MCG+06-02-014	00 39 06.	+ 36 50	36	15.	GALAXY
5ZW 033	00 39 06.	+ 46 23			COMPACT GALAXY
ARC 0085	00 39 06.	- 09 38		15.7	RICH CLUSTER OF GALAXIES
LB 08436	00 39 06.	- 20 43		18.8	FAINT BLUE STAR
LB 08437	00 39 06.	- 23 04		16.9	FAINT BLUE STAR
LB 08438	00 39 06.	- 23 09		19.8	FAINT BLUE STAR
LB 08439	00 39 06.	- 23 20		19.5	FAINT BLUE STAR
LB 03150	00 39 06.	- 62 21		14.2	FAINT BLUE STAR
RIC 424	00 39 06.41	+ 42 06 43.8		18.61	BLUE OBJECT
RIC 425	00 39 07.22	+ 40 04 52.0		18.91	BLUE OBJECT
RIC 426	00 39 07.37	+ 39 20 40.4		17.98	BLUE OBJECT
BAA 2.57	00 39 07.48	+ 41 08 32.7			GALAXY
BAA 2.56	00 39 07.88	+ 41 08 15.7			GALAXY
VDB.66B 27	00 39 08.	+ 39 21 08.		19.08	BLUE OBJECT
BAA 1.291	00 39 08.02	+ 40 29 52.8			EMISSION NEBULA IN M31
BAA 2.13	00 39 09.09	+ 40 32 26.2			GALAXY
LIN 031	00 39 10.	- 72 37 08.		16.18	SMC EMISSION LINE OBJECT
RIC 427	00 39 10.14	+ 39 34 15.8		19.14	BLUE OBJECT
RIC 428	00 39 10.49	+ 39 49 44.1		18.49	BLUE OBJECT
BAA 1.290	00 39 11.04	+ 40 30 48.5			EMISSION NEBULA IN M31
BAA 2.38	00 39 11.74	+ 40 49 12.8			GALAXY
ZWG 383.070	00 39 12.	+ 00 53		15.7	GALAXY
ZC 0039.2+0430	00 39 12.	+ 04 30	610		CLUSTER OF GALAXIES
ZWG 409.063	00 39 12.	+ 08 06		15.6	GALAXY
MCG+01-02-054	00 39 12.	+ 08 08	12	15.	GALAXY
ARC 0084	00 39 12.	+ 21 08		16.8	RICH CLUSTER OF GALAXIES
ZWG 500.070	00 39 12.	+ 32 43		15.4	GALAXY
UGC 00442	00 39 12.	+ 32 43	108	15.4	GALAXY Sc
ZWG 519.018	00 39 12.	+ 36 50		15.7	GALAXY
ZCG 0039+40.1	00 39 12.	+ 40 05		15.7	COMPACT GALAXY
6ZW 001	00 39 12.	+ 40 07			COMPACT GALAXY
SHRV.73A 10	00 39 12.	+ 40 58		16.6	GLOBULAR CLUSTER IN M31
MCG+00-02-127	00 39 12.	- 00 41	33	17.	GALAXY
PHL 2892	00 39 12.	- 02 43		17.5	BLUE STELLAR OBJECT
MCG-03-02-039	00 39 12.	- 17 08 30.	18	15.5	GALAXY
LB 00157	00 39 12.	- 20 03		17.5	FAINT BLUE STAR
LB 08440	00 39 12.	- 22 47		18.2	FAINT BLUE STAR
LB 08441	00 39 12.	- 23 00		17.9	FAINT BLUE STAR
LB 08442	00 39 12.	- 23 49		18.5	FAINT BLUE STAR
LB 08443	00 39 12.	- 24 10		19.3	FAINT BLUE STAR
RIC 429	00 39 12.84	+ 39 41 46.6		19.66	BLUE OBJECT
RIC 430	00 39 14.64	+ 39 19 54.4		19.95	BLUE OBJECT
MCG-03-02-040	00 39 15.	- 17 09	90	14.	GALAXY
RIC 431	00 39 15.79	+ 41 40 46.0		20.13	BLUE OBJECT
BAA 1.288	00 39 16.71	+ 40 31 20.7			EMISSION NEBULA IN M31
HOLM 015B	00 39 17.	- 09 34	24	14.6	PART OF MULTIPLE GALAXY
RIC 432	00 39 17.29	+ 39 53 42.2		19.18	BLUE OBJECT
BAA 2.60	00 39 17.68	+ 41 08 36.5			GALAXY
ZWG 409.064	00 39 18.	+ 08 05		15.6	GALAXY
ZWG 479.060	00 39 18.	+ 25 18		15.4	GALAXY
MCG+06-02-015	00 39 18.	+ 36 31	66	14.	GALAXY
LDN 1291	00 39 18.	+ 61 45	240		DARK NEBULA
ZWG 383.072	00 39 18.	- 02 00		15.3	GALAXY
ZWG 383.071	00 39 18.	- 02 11		15.4	GALAXY
MCG-02-02-086	00 39 18.	- 09 35	48	14.5	GALAXY
MCG-02-02-087	00 39 18.	- 09 45	48	16.	GALAXY
MCG-02-02-088	00 39 18.	- 09 45 30.	18	16.	GALAXY
LB 08444	00 39 18.	- 23 33		18.6	FAINT BLUE STAR
LB 08445	00 39 18.	- 25 04		17.9	FAINT BLUE STAR
LB 08446	00 39 18.	- 25 37		19.1	FAINT BLUE STAR
LB 08447	00 39 18.	- 25 44		18.1	FAINT BLUE STAR
LB 08448	00 39 18.	- 26 36		19.7	FAINT BLUE STAR
BAA 2.59	00 39 18.51	+ 41 08 08.6			GALAXY
BAA 2.14	00 39 18.93	+ 40 30 04.6			GALAXY
HELW 321	00 39 19.	- 09 34 38.			NEBULA
HOLM 016B	00 39 19.	- 09 48	48	15.2	PART OF MULTIPLE GALAXY
BAA 1.504	00 39 19.08	+ 41 06 04.3			EMISSION NEBULA IN M31
RIC 433	00 39 19.78	+ 41 30 28.2		19.67	BLUE OBJECT
RIC 434	00 39 19.80	+ 39 57 38.8		19.18	BLUE OBJECT
BAA 1.282	00 39 19.81	+ 40 41 22.3			EMISSION NEBULA IN M31
RIC 435	00 39 19.90	+ 42 07 46.6		19.75	BLUE OBJECT
HOLM 015A	00 39 20.	- 09 35	48	13.7	PART OF MULTIPLE GALAXY
HOLM 016A	00 39 20.	- 09 47	48	15.1	PART OF MULTIPLE GALAXY
RIC 436	00 39 20.12	+ 40 12 23.7		20.50	BLUE OBJECT
BAA 1.287	00 39 20.14	+ 40 32 41.9			EMISSION NEBULA IN M31
BAA 1.509	00 39 20.53	+ 41 09 28.1			EMISSION NEBULA IN M31
BAA 1.503	00 39 20.54	+ 41 05 39.8			EMISSION NEBULA IN M31
RIC 437	00 39 20.75	+ 40 07 36.0		20.42	BLUE OBJECT
MCG+01-02-055	00 39 21.	+ 08 06	12	15.5	GALAXY
MCG-02-02-089	00 39 21.	- 14 24	36	15.	GALAXY
LIN.CL 026	00 39 21.	- 72 51 44.	408	12.7	STAR CLUSTER IN SMC
BAA 2.12	00 39 21.22	+ 40 34 27.3			GALAXY
BAA 1.286	00 39 21.24	+ 40 33 11.8			EMISSION NEBULA IN M31
RIC 438	00 39 21.76	+ 41 46 36.7		18.89	BLUE OBJECT
RIC 439	00 39 22.22	+ 41 51 46.3		18.52	BLUE OBJECT
BAA 2.09	00 39 22.65	+ 40 36 51.6			GALAXY
BAA 2.34	00 39 22.69	+ 41 01 40.0			GALAXY
BAA 2.15	00 39 22.96	+ 40 29 00.5			GALAXY
BAA 2.08	00 39 23.14	+ 40 46 47.6			GALAXY
BAA 1.280	00 39 23.33	+ 40 41 37.2			EMISSION NEBULA IN M31
RIC 440	00 39 23.81	+ 42 12 47.0		19.15	BLUE OBJECT
ZC 0039.4+3117	00 39 24.	+ 31 17	2820		CLUSTER OF GALAXIES
ZWG 500.071	00 39 24.	+ 33 08		15.5	GALAXY
UGC 00443	00 39 24.	+ 33 08	66	15.5	GALAXY Sc
ZWG 519.019	00 39 24.	+ 36 32		14.0	GALAXY
UGC 00444	00 39 24.	+ 36 32	72	14.0	GALAXY S?
LDN 1294	00 39 24.	+ 62 00	180		DARK NEBULA
OCL 0309	00 39 24.	+ 85 04	4020	9.3	OPEN STAR CLUSTER
LB 08449	00 39 24.	- 20 59		18.0	FAINT BLUE STAR
LB 08450	00 39 24.	- 21 06		18.9	FAINT BLUE STAR
LB 08451	00 39 24.	- 21 44		18.7	FAINT BLUE STAR
MCG-06-02-025	00 39 24.	- 33 15	42	15.	GALAXY
LB 03151	00 39 24.	- 69 01		14.2	FAINT BLUE STAR
LH115-N005	00 39 24.	- 73 02			EMISSION NEBULA IN SMC
RIC 441	00 39 24.44	+ 40 13 02.0		19.39	BLUE OBJECT
BAA 1.278	00 39 24.62	+ 40 42 19.3			EMISSION NEBULA IN M31
BAA 1.506	00 39 24.70	+ 41 07 43.6			EMISSION NEBULA IN M31
BAA 2.63	00 39 24.75	+ 41 10 13.9			GALAXY
BAA 2.10	00 39 25.05	+ 40 34 27.1			GALAXY
BAA 1.513	00 39 25.40	+ 41 12 06.2			EMISSION NEBULA IN M31
BAA 2.65	00 39 25.43	+ 41 10 41.9			GALAXY
BAA 1.507	00 39 25.75	+ 41 07 45.8			EMISSION NEBULA IN M31
BAA 2.64	00 39 26.12	+ 41 10 32.8			GALAXY

OBJECT NAME	RIGHT ASCEN.	DECLINATION	DIAM.	MAGN.	TYPE OF OBJECT
BAA 1.505	00 39 26.37	+ 41 06 55.2			EMISSION NEBULA IN M31
LIN 032	00 39 27.	- 73 02 26.		16.35	PLANETARY NEBULA IN SMC
BAA 1.281	00 39 27.15	+ 40 40 59.4			EMISSION NEBULA IN M31
BAA 1.510	00 39 27.96	+ 41 09 11.7			EMISSION NEBULA IN M31
RNGC 0188	00 39 28.	+ 85 04		9.5	OPEN CLUSTER
BAA 1.508	00 39 28.18	+ 41 07 44.0			EMISSION NEBULA IN M31
BAA 1.526	00 39 28.62	+ 41 18 55.0			EMISSION NEBULA IN M31
BAA 1.279	00 39 29.64	+ 40 40 59.5			EMISSION NEBULA IN M31
RIC 442	00 39 29.70	+ 38 59 29.3		17.95	BLUE OBJECT
RIC 443	00 39 29.98	+ 39 06 15.4		17.96	BLUE OBJECT
ZWG 434.031	00 39 30.	+ 11 44		15.7	GALAXY
ZCG 0039+40.0	00 39 30.	+ 40 01		17.2	COMPACT GALAXY
4ZW 029	00 39 30.	+ 40 03			COMPACT GALAXY
KW 61	00 39 30.	+ 40 03	8		SEYFERT GALAXY
VVI 03	00 39 30.	+ 40 03	3	16.18	SEYFERT GALAXY
MCG-02-02-091	00 39 30.	- 09 49	36	16.	GALAXY
MCG-02-02-090	00 39 30.	- 11 05	12	16.	GALAXY
LB 08452	00 39 30.	- 23 38		18.2	FAINT BLUE STAR
BAA 1.284	00 39 30.02	+ 40 34 51.9			EMISSION NEBULA IN M31
RIC 444	00 39 30.58	+ 41 40 05.1		18.13	BLUE OBJECT
PK118-74.1	00 39 31.	- 12 41 41.	273	8.	PLANETARY NEBULA
RIC 445	00 39 32.51	+ 40 12 02.0		20.10	BLUE OBJECT
RIC 446	00 39 33.60	+ 39 37 45.4		18.75	BLUE OBJECT
VDB.66B 31	00 39 34.	+ 39 37 44.		19.65	BLUE OBJECT
BAA 1.515	00 39 34.59	+ 41 11 28.1			EMISSION NEBULA IN M31
SC 0037-6757.3	00 39 35.	- 67 40 49.	18		NEBULA
BAA 1.512	00 39 35.41	+ 41 09 52.2			EMISSION NEBULA IN M31
BAA 1.516	00 39 35.76	+ 41 11 23.9			EMISSION NEBULA IN M31
MCG+00-02-128	00 39 36.	+ 00 38 30.	24	16.	GALAXY
ZC 0039.6+1458	00 39 36.	+ 14 58	4100		CLUSTER OF GALAXIES
MCG+05-02-040	00 39 36.	+ 29 22	90	14.	GALAXY
MCG+05-02-039	00 39 36.	+ 29 25	54	15.	GALAXY
UGC 00445	00 39 36.	+ 30 02	60	16.0	GALAXY SBb
UGC 00446	00 39 36.	+ 32 56	72	17.	GALAXY DWRF IR
LB 08453	00 39 36.	- 21 16		19.0	FAINT BLUE STAR
LB 08454	00 39 36.	- 22 52		17.3	FAINT BLUE STAR
LB 08455	00 39 36.	- 23 15		20.0	FAINT BLUE STAR
LB 08456	00 39 36.	- 26 01		18.0	FAINT BLUE STAR
LB 08457	00 39 36.	- 26 28		18.8	FAINT BLUE STAR
LB 08458	00 39 36.	- 26 30		15.8	FAINT BLUE STAR
LH115-N006	00 39 36.	- 74 03			EMISSION NEBULA IN SMC
LIN 033	00 39 36.	- 74 03 56.		16.46	PLANETARY NEBULA IN SMC
BAA 1.285	00 39 36.80	+ 40 31 32.5			EMISSION NEBULA IN M31
BAA 2.11	00 39 36.92	+ 40 32 17.5			GALAXY
BAA 1.514	00 39 37.11	+ 41 10 46.1			EMISSION NEBULA IN M31
RIC 447	00 39 37.22	+ 39 54 38.4		19.49	BLUE OBJECT
BAA 1.517	00 39 37.61	+ 41 12 01.2			EMISSION NEBULA IN M31
BAA 1.275	00 39 37.84	+ 40 43 47.7			EMISSION NEBULA IN M31
MCG+00-02-129	00 39 39.	+ 00 35	30	14.	GALAXY
IC 0043	00 39 39.	+ 29 23			NONSTELLAR OBJECT
BAA 4.28	00 39 39.34	+ 40 42 47.4			DIFFUSE OBJECT IN M31
RIC 448	00 39 39.46	+ 39 19 29.8		18.65	BLUE OBJECT
RIC 449	00 39 39.87	+ 39 18 35.8		18.54	BLUE OBJECT
HN 0028	00 39 40.	+ 00 37			NEBULA
HN 0007	00 39 40.	+ 00 37			NEBULA
SN 1973U	00 39 41.	+ 29 22		16.5	SUPERNOVA
IC 1573	00 39 41.	- 23 48			NONSTELLAR OBJECT
HN 0114	00 39 41.	- 23 49			NEBULA
BAA 1.283	00 39 41.38	+ 40 33 20.1			EMISSION NEBULA IN M31
ZWG 383.073	00 39 42.	+ 00 38		15.6	GALAXY
UGC 00447	00 39 42.	+ 10 28	60	16.5	GALAXY S
MCG+04-02-045	00 39 42.	+ 25 17	30	15.5	GALAXY
ZWG 500.072	00 39 42.	+ 29 22		14.4	GALAXY
UGC 00448	00 39 42.	+ 29 22	132	14.4	GALAXY Sc
ZWG 500.073	00 39 42.	+ 29 25		15.3	GALAXY
UGC 00449	00 39 42.	+ 29 25	60	15.3	GALAXY PECULR
SHRV.73A 11	00 39 42.	+ 40 09	8	16.57	GLOBULAR CLUSTER IN M31
PHL 2893	00 39 42.	- 20 10		17.9	BLUE STELLAR OBJECT
LB 08459	00 39 42.	- 22 18		17.1	FAINT BLUE STAR
LB 08460	00 39 42.	- 22 44		18.7	FAINT BLUE STAR
LB 08461	00 39 42.	- 22 48		20.0	FAINT BLUE STAR
LB 08462	00 39 42.	- 22 56		18.5	FAINT BLUE STAR
MCG-04-02-036	00 39 42.	- 23 52 30.	60	15.5	GALAXY
LB 08463	00 39 42.	- 24 44		19.5	FAINT BLUE STAR
LB 08464	00 39 42.	- 24 54		18.8	FAINT BLUE STAR
LB 08465	00 39 42.	- 25 22		17.5	FAINT BLUE STAR
LB 08466	00 39 42.	- 26 12		19.4	FAINT BLUE STAR
BAA 1.518	00 39 42.57	+ 41 11 42.3			EMISSION NEBULA IN M31
RNGC 0219	00 39 43.	+ 00 38		15.5	GALAXY
RIC 450	00 39 43.62	+ 39 11 19.7		20.83	BLUE OBJECT
VDB.66N 004	00 39 44.	+ 61 35	504		REFLECTION NEBULA
IC 0044	00 39 45.	+ 00 36 10.			NONSTELLAR OBJECT
MCG+04-02-046	00 39 45.	+ 25 20 30.	36	15.5	GALAXY
MCG-02-02-092	00 39 45.	- 11 45	60	14.5	GALAXY
MCG-03-02-041	00 39 45.	- 18 27 30.	180	13.	GALAXY
BAA 1.522	00 39 46.09	+ 41 13 09.3			EMISSION NEBULA IN M31
ZWG 383.074	00 39 48.	+ 00 34		14.5	GALAXY
UGC 00450	00 39 48.	+ 00 34	90	14.5	GALAXY PECULR
MCG+03-02-024	00 39 48.	+ 18 32	18	16.	GALAXY
UGC 00451	00 39 48.	+ 18 33	114	16.5	GALAXY
ZC 0039.8+2425	00 39 48.	+ 24 25	740		CLUSTER OF GALAXIES
AND 4	00 39 48.	+ 40 18	60		DWARF SPHEROIDAL GALAXY
SHRV.73A 12	00 39 48.	+ 42 46	8	17.14	GLOBULAR CLUSTER IN M31
PHL 6533	00 39 48.	- 06 14		18.4	BLUE STELLAR OBJECT
LB 08467	00 39 48.	- 21 58		18.2	FAINT BLUE STAR
LB 08468	00 39 48.	- 22 04		18.7	FAINT BLUE STAR
LB 00158	00 39 48.	- 22 20		16.6	FAINT BLUE STAR
LB 08469	00 39 48.	- 22 31		16.8	FAINT BLUE STAR
LB 08470	00 39 48.	- 23 38		19.3	FAINT BLUE STAR
LB 08471	00 39 48.	- 25 28		16.0	FAINT BLUE STAR
RIC 451	00 39 48.11	+ 42 02 18.4		19.25	BLUE OBJECT
BAA 1.523	00 39 48.89	+ 41 12 51.7			EMISSION NEBULA IN M31
HSN 06	00 39 49.	- 71 54			BRIGHT GALAXY BEHIND SMC
RIC 452	00 39 49.31	+ 39 56 54.8		18.26	BLUE OBJECT
RIC 453	00 39 49.44	+ 39 31 01.3		19.48	BLUE OBJECT
RIC 454	00 39 49.74	+ 39 09 01.6		18.51	BLUE OBJECT
BAA 1.524	00 39 50.14	+ 41 13 37.2			EMISSION NEBULA IN M31
BAA 1.525	00 39 51.54	+ 41 15 17.8			EMISSION NEBULA IN M31
BAA 1.532	00 39 52.31	+ 41 16 43.9			EMISSION NEBULA IN M31
BAA 1.527	00 39 53.33	+ 41 15 30.7			EMISSION NEBULA IN M31
RIC 455	00 39 53.40	+ 39 21 56.8		19.09	BLUE OBJECT
BAA 1.529	00 39 53.75	+ 41 15 39.3			EMISSION NEBULA IN M31
BAA 4.27	00 39 53.85	+ 41 55 05.3			DIFFUSE OBJECT IN M31
ZWG 535.016	00 39 54.	+ 40 36		9.2	GALAXY
UGC 00452	00 39 54.	+ 40 36	660	9.2	GALAXY E
LB 08472	00 39 54.	- 22 26		19.8	FAINT BLUE STAR
LB 08473	00 39 54.	- 26 14		16.5	FAINT BLUE STAR
TON-S 0166	00 39 54.	- 26 33		15.3	BLUE STAR
LB 08474	00 39 54.	- 26 42		19.3	FAINT BLUE STAR
BAA 1.528	00 39 54.37	+ 41 15 23.6			EMISSION NEBULA IN M31
RIC 456	00 39 55.04	+ 40 02 38.6		18.47	BLUE OBJECT
RIC 457	00 39 55.76	+ 39 19 22.5		19.14	BLUE OBJECT
BAA 1.269	00 39 55.84	+ 40 45 57.0			EMISSION NEBULA IN M31
BAA 1.267	00 39 56.52	+ 40 46 17.2			EMISSION NEBULA IN M31
BAA 1.274	00 39 56.87	+ 40 43 18.8			EMISSION NEBULA IN M31
BAA 1.276	00 39 56.99	+ 40 42 45.1			EMISSION NEBULA IN M31
BAA 1.511	00 39 57.07	+ 41 06 21.0			EMISSION NEBULA IN M31
RIC 458	00 39 57.79	+ 39 18 18.9		20.56	BLUE OBJECT
RNGC 0230	00 39 58.	- 23 54		15.0	GALAXY
BAA 1.530	00 39 58.16	+ 41 15 27.7			EMISSION NEBULA IN M31
BAA 1.265	00 39 58.17	+ 40 46 37.7			EMISSION NEBULA IN M31
BAA 1.531	00 39 58.76	+ 41 15 34.3			EMISSION NEBULA IN M31
IC 0045	00 39 59.	+ 29 24			NONSTELLAR OBJECT
SN 1885A	00 39 59.	+ 41 00		5.8	SUPERNOVA
RIC 459	00 39 59.56	+ 39 20 39.4		19.04	BLUE OBJECT
BAA 1.533	00 39 59.87	+ 41 16 31.3			EMISSION NEBULA IN M31
LBN 0603	00 40	+ 52 00	4020		BRIGHT NEBULA
VDB.66G 226	00 40	- 22 35	70		DWARF GALAXY
MCG+00-02-130	00 40 00.	+ 02 29	60	15.	GALAXY
UGC 00453	00 40 00.	+ 29 37	78	16.5	GALAXY Sc
MCG+07-02-015	00 40 00.	+ 40 35	900		GALAXY
ZWG 535.017	00 40 00.	+ 41 00		4.3	GALAXY
UGC 00454	00 40 00.	+ 41 00	12000	4.3	GALAXY Sb
MCG+07-02-016	00 40 00.	+ 41 00	9600		GALAXY
LDN 1297	00 40 00.	+ 62 30	960		DARK NEBULA
ISS 0003	00 40 00.	+ 64 01	320		STELLAR RING
VB 006	00 40 00.	+ 64 40	671		STELLAR RING
LDN 1303	00 40 00.	+ 65 00	3360		DARK NEBULA
LDN 1304	00 40 00.	+ 69 30	8100		DARK NEBULA
MCG+00-02-131	00 40 00.	- 02 16 30.	66	14.	GALAXY
MCG+00-02-132	00 40 00.	- 02 17 30.	18	14.	GALAXY
MCG-02-02-093	00 40 00.	- 12 22	48	16.	GALAXY
PHL 6534	00 40 00.	- 17 26		18.3	BLUE STELLAR OBJECT
LB 08475	00 40 00.	- 21 52		18.8	FAINT BLUE STAR
MCG-04-02-037	00 40 00.	- 23 54 30.	54	15.	GALAXY
LB 08476	00 40 00.	- 24 06		19.1	FAINT BLUE STAR
LB 08477	00 40 00.	- 26 14		19.4	FAINT BLUE STAR
RIC 460	00 40 00.84	+ 39 57 26.3		18.41	BLUE OBJECT
HOLM 017B	00 40 01.	+ 40 35	162	9.6	PART OF MULTIPLE GALAXY
ARP 168	00 40 01.	+ 40 35			PECULIAR GALAXY
HOLM 017A	00 40 01.	+ 40 59			PART OF MULTIPLE GALAXY
ARC 0086	00 40 01.	- 22 05		15.9	RICH CLUSTER OF GALAXIES
RIC 461	00 40 01.00	+ 40 15 02.2		17.91	BLUE OBJECT
BAA 1.277	00 40 01.30	+ 40 38 26.1			EMISSION NEBULA IN M31
BAA 1.535	00 40 01.95	+ 41 17 35.5			EMISSION NEBULA IN M31
RNGC 0221	00 40 02.	+ 40 36		10.0	GALAXY
RNGC 0224	00 40 02.	+ 41 00		4.5	GALAXY
BAA 1.534	00 40 02.53	+ 41 17 10.9			EMISSION NEBULA IN M31
RIC 462	00 40 02.54	+ 39 17 04.8		18.71	BLUE OBJECT
BAA 1.520	00 40 02.68	+ 41 09 04.1			EMISSION NEBULA IN M31
RIC 463	00 40 03.44	+ 40 16 05.4		19.17	BLUE OBJECT
BAA 1.271	00 40 03.48	+ 40 44 43.1			EMISSION NEBULA IN M31
BAA 1.266	00 40 03.54	+ 40 45 22.4			EMISSION NEBULA IN M31
BAA 1.537	00 40 03.75	+ 41 17 38.8			EMISSION NEBULA IN M31
RIC 464	00 40 04.20	+ 40 03 08.9		19.60	BLUE OBJECT
RIC 465	00 40 04.41	+ 40 04 03.9		19.42	BLUE OBJECT
BAA 1.536	00 40 04.60	+ 41 17 26.0			EMISSION NEBULA IN M31
BAA 1.272	00 40 04.63	+ 40 44 14.2			EMISSION NEBULA IN M31
BAA 1.519	00 40 05.15	+ 41 08 42.4			EMISSION NEBULA IN M31
UGC 00455	00 40 06.	- 02 16	96	15.0	GALAXY COMPACT
MCG+00-02-134	00 40 06.	+ 00 00 30.	42	15.	GALAXY
ZC 0040.1+0101	00 40 06.	+ 01 01	340		CLUSTER OF GALAXIES
ZWG 383.077	00 40 06.	+ 02 27		15.6	GALAXY
MCG+00-02-133	00 40 06.	+ 02 28 30.	60	15.	GALAXY
MCG+04-02-047	00 40 06.	+ 23 11 30.	36	15.5	GALAXY
ZWG 479.061	00 40 06.	+ 23 13		15.7	GALAXY
ZWG 500.074	00 40 06.	+ 29 39		14.7	GALAXY
ZC 0040.1+3019	00 40 06.	+ 30 19	1280		CLUSTER OF GALAXIES
ZWG 500.075	00 40 06.	+ 33 15		15.5	GALAXY
UGC 00457	00 40 06.	+ 33 15	84	15.5	GALAXY Sb-c
ZWG 383.076	00 40 06.	- 01 48		13.7	GALAXY
UGC 00456	00 40 06.	- 01 48	102	13.4	GALAXY E
MCG+00-02-135	00 40 06.	- 01 48	78	13.5	GALAXY
ZWG 383.075	00 40 06.	- 02 16		15.0	GALAXY
8ZW 0040-11.1	00 40 06.	- 11 05		14.6	COMPACT GALAXY
PHL 0811	00 40 06.	- 16 16		17.4	BLUE STELLAR OBJECT
LB 08478	00 40 06.	- 24 01		18.9	FAINT BLUE STAR
LB 08479	00 40 06.	- 24 19		19.0	FAINT BLUE STAR
RNGC 0223	00 40 07.	+ 00 35		14.5	GALAXY
RNGC 0227	00 40 07.	- 01 48		13.0	GALAXY
LIN 034	00 40 08.	- 75 16 56.		14.65	SMC EMISSION LINE OBJECT
BAA 1.540	00 40 08.51	+ 41 19 02.2			EMISSION NEBULA IN M31
BAA 1.539	00 40 08.59	+ 41 18 58.6			EMISSION NEBULA IN M31
RNGC 0226	00 40 09.	+ 32 18		14.5	GALAXY
BAA 1.538	00 40 09.11	+ 41 18 51.1			EMISSION NEBULA IN M31
BAA 1.521	00 40 09.39	+ 41 09 24.5			EMISSION NEBULA IN M31
BAA 1.541	00 40 09.84	+ 41 19 12.7			EMISSION NEBULA IN M31
RIC 466	00 40 09.85	+ 39 07 00.9		17.85	BLUE OBJECT
BAA 1.268	00 40 09.86	+ 40 44 13.2			EMISSION NEBULA IN M31
BAA 1.273	00 40 10.37	+ 40 42 06.4			EMISSION NEBULA IN M31
BAA 1.264	00 40 10.52	+ 40 46 00.2			EMISSION NEBULA IN M31
BAA 1.560	00 40 10.73	+ 41 29 33.4			EMISSION NEBULA IN M31
RNGC 0228	00 40 11.	+ 23 14		15.0	GALAXY
BAA 1.542	00 40 11.64	+ 41 19 18.9			EMISSION NEBULA IN M31
ZWG 383.078	00 40 12.	+ 00 00		15.7	GALAXY
PHL 2895	00 40 12.	+ 01 46		18.5	BLUE STELLAR OBJECT
MCG+04-02-048	00 40 12.	+ 23 12 30.	72	14.5	GALAXY
ZWG 479.062	00 40 12.	+ 23 14		14.9	GALAXY
UGC 00458	00 40 12.	+ 23 14	72	14.9	GALAXY SBa
ZC 0040.2+2714	00 40 12.	+ 27 14	1340		CLUSTER OF GALAXIES
ZWG 500.076	00 40 12.	+ 32 18		14.4	GALAXY
UGC 00459	00 40 12.	+ 32 18	60	14.4	GALAXY PECULR
ZC 0040.2-0004	00 40 12.	- 00 04	270		CLUSTER OF GALAXIES
PHL 6537	00 40 12.	- 10 49		18.3	BLUE STELLAR OBJECT
PHL 6535	00 40 12.	- 12 35		17.7	BLUE STELLAR OBJECT
MCG-04-02-038	00 40 12.	- 22 05 30.	12	15.5	GALAXY
LB 08480	00 40 12.	- 22 41		18.8	FAINT BLUE STAR
LB 08481	00 40 12.	- 24 11		16.8	FAINT BLUE STAR
LB 08482	00 40 12.	- 24 29		18.0	FAINT BLUE STAR
LB 08483	00 40 12.	- 26 15		19.4	FAINT BLUE STAR
LB 08484	00 40 12.	- 26 31		18.4	FAINT BLUE STAR
LB 01568	00 40 12.	- 48 39		15.1	FAINT BLUE STAR
RIC 467	00 40 12.96	+ 39 31 22.4		20.29	BLUE OBJECT
LIN.CL 027	00 40 14.	- 73 10 38.	816	11.9	STAR CLUSTER IN SMC
BAA 1.545	00 40 14.23	+ 41 20 46.4			EMISSION NEBULA IN M31
BAA 1.270	00 40 14.31	+ 40 43 09.2			EMISSION NEBULA IN M31

OBJECT NAME	RIGHT ASCEN.	DECLINATION	DIAM.	MAGN.	TYPE OF OBJECT
BAA 1.561	00 40 14.36	+ 41 30 10.5			EMISSION NEBULA IN M31
RIC 468	00 40 14.46	+ 39 12 39.9		19.98	BLUE OBJECT
MCG-01-02-046	00 40 15.	- 04 06 30.	48	15.	GALAXY
BAA 1.543	00 40 15.04	+ 41 19 15.5			EMISSION NEBULA IN M31
BAA 1.544	00 40 15.21	+ 41 20 19.4			EMISSION NEBULA IN M31
BAA 1.557	00 40 15.32	+ 41 27 11.1			EMISSION NEBULA IN M31
RIC 469	00 40 15.47	+ 41 57 36.6		18.44	BLUE OBJECT
RNGC 0232	00 40 16.	- 23 50		14.0	GALAXY
HOW 16	00 40 16.	- 74 01 07.	24		STAR CLUSTER IN SMC
RIC 470	00 40 16.47	+ 39 03 51.8		19.63	BLUE OBJECT
BAA 1.546	00 40 17.47	+ 41 20 31.0			EMISSION NEBULA IN M31
BAA 1.549	00 40 17.90	+ 41 21 23.7			EMISSION NEBULA IN M31
IC 0046	00 40 18.	+ 26 58 40.			NONSTELLAR OBJECT
ZWG 479.063	00 40 18.	+ 26 59		14.7	GALAXY
KARA.73B 0032	00 40 18.	+ 26 59	36	14.7	ISOLATED GALAXY S0
VB 007	00 40 18.	+ 65 25	618		STELLAR RING
VB 008	00 40 18.	+ 65 28	369		STELLAR RING
PHL 2896	00 40 18.	- 01 13		18.6	BLUE STELLAR OBJECT
MCG-02-02-094	00 40 18.	- 11 23	24	16.	GALAXY
PHL 6538	00 40 18.	- 11 45		18.5	BLUE STELLAR OBJECT
MCG-04-02-039	00 40 18.	- 22 02 30.	30	15.5	GALAXY
LB 08485	00 40 18.	- 23 12		18.1	FAINT BLUE STAR
LB 08486	00 40 18.	- 23 18		17.5	FAINT BLUE STAR
MCG-04-02-040	00 40 18.	- 23 50 30.	48	14.	GALAXY
LB 08487	00 40 18.	- 26 18		17.5	FAINT BLUE STAR
LB 01569	00 40 18.	- 49 27		14.6	FAINT BLUE STAR
BAA 1.567	00 40 18.42	+ 41 31 43.8			EMISSION NEBULA IN M31
BAA 4.39	00 40 18.54	+ 40 25 07.5			DIFFUSE OBJECT IN M31
BAA 1.259	00 40 18.54	+ 40 46 50.2			EMISSION NEBULA IN M31
BAA 1.548	00 40 18.81	+ 41 20 54.8			EMISSION NEBULA IN M31
BAA 2.66	00 40 19.74	+ 41 17 41.7			GALAXY
BAA 1.553	00 40 19.74	+ 41 22 04.6			EMISSION NEBULA IN M31
BAA 1.550	00 40 20.01	+ 41 21 18.1			EMISSION NEBULA IN M31
BAA 1.552	00 40 20.99	+ 41 21 49.9			EMISSION NEBULA IN M31
BAA 1.559	00 40 21.33	+ 41 26 41.4			EMISSION NEBULA IN M31
RNGC 0235B	00 40 22.	- 23 49		14.0	GALAXY
RNGC 0235A	00 40 22.	- 23 49		14.0	GALAXY
BAA 1.263	00 40 22.64	+ 40 45 48.0			EMISSION NEBULA IN M31
RIC 471	00 40 22.81	+ 39 08 17.3		19.29	BLUE OBJECT
RNGC 0229	00 40 23.	+ 23 13		14.5	GALAXY
ARC 0088	00 40 23.	- 26 20		15.6	RICH CLUSTER OF GALAXIES
LIN 035	00 40 23.	- 74 13 39.		16.08	SMC EMISSION LINE OBJECT
BAA 1.551	00 40 23.90	+ 41 21 12.3			EMISSION NEBULA IN M31
MCG+04-02-049	00 40 24.	+ 23 13	48	15.	GALAXY
ZWG 479.064	00 40 24.	+ 23 14		14.7	GALAXY
ISS 0004	00 40 24.	+ 65 31	352		STELLAR RING
PHL 6539	00 40 24.	- 01 12		12.0	BLUE STELLAR OBJECT
PHL 6541	00 40 24.	- 08 46		18.1	BLUE STELLAR OBJECT
MCG-02-02-095	00 40 24.	- 11 52	18	16.	GALAXY
IC 0047	00 40 24.	- 14 01 09.			MAY NOT EXIST
LB 08488	00 40 24.	- 20 54		16.1	FAINT BLUE STAR
LB 08489	00 40 24.	- 23 05		18.5	FAINT BLUE STAR
LB 08490	00 40 24.	- 23 34		18.8	FAINT BLUE STAR
LB 08491	00 40 24.	- 23 57		18.5	FAINT BLUE STAR
BAA 1.260	00 40 24.20	+ 40 45 36.8			EMISSION NEBULA IN M31
RIC 472	00 40 25.10	+ 39 19 34.9		19.31	BLUE OBJECT
BAA 1.556	00 40 25.17	+ 41 22 24.3			EMISSION NEBULA IN M31
BAA 1.258	00 40 25.41	+ 40 48 01.8			EMISSION NEBULA IN M31
BAA 1.555	00 40 25.47	+ 41 21 26.0			EMISSION NEBULA IN M31
HOW 17	00 40 26.	- 72 43 37.	12		STAR CLUSTER IN SMC
BAA 1.257	00 40 26.15	+ 40 48 31.5			EMISSION NEBULA IN M31
BAA 1.554	00 40 26.52	+ 41 21 06.8			EMISSION NEBULA IN M31
BAA 1.262	00 40 26.53	+ 40 45 10.5			EMISSION NEBULA IN M31
BAA 1.254	00 40 26.83	+ 40 48 37.5			EMISSION NEBULA IN M31
BAA 1.256	00 40 26.94	+ 40 48 31.8			EMISSION NEBULA IN M31
MCG-04-02-042	00 40 27.	- 23 49 30.	12	14.5	GALAXY
MCG-04-02-041	00 40 27.	- 23 49 30.	21	14.5	GALAXY
BAA 1.255	00 40 27.12	+ 40 48 34.8			EMISSION NEBULA IN M31
RIC 473	00 40 27.18	+ 41 49 52.8		19.79	BLUE OBJECT
BAA 1.261	00 40 28.06	+ 40 45 01.1			EMISSION NEBULA IN M31
BFL 18	00 40 29.	+ 40 08 36.		19.47	PEC. BLUE OBJECT NEAR M31
BAA 1.253	00 40 29.51	+ 40 50 08.4			EMISSION NEBULA IN M31
PHL 6542	00 40 30.	+ 00 56		18.6	BLUE STELLAR OBJECT
ZC 0040.5+3200	00 40 30.	+ 32 00	1080		CLUSTER OF GALAXIES
SHEV.73A 13	00 40 30.	+ 38 55	9	16.26	GLOBULAR CLUSTER IN M31
LDN 1298	00 40 30.	+ 60 50	480		DARK NEBULA
OCL 0305	00 40 30.	+ 61 31	1500	9.1	OPEN STAR CLUSTER
VB 009	00 40 30.	+ 65 01	329		STELLAR RING
PHL 2900	00 40 30.	- 06 09		18.4	BLUE STELLAR OBJECT
FEIG 006	00 40 30.	- 09 02		14.7	FAINT BLUE STAR
PHL 2897	00 40 30.	- 09 28		18.6	BLUE STELLAR OBJECT
ARC 0087	00 40 30.	- 10 05		16.6	RICH CLUSTER OF GALAXIES
MCG-02-02-096	00 40 30.	- 12 28	48	16.	GALAXY
PHL 2898	00 40 30.	- 15 02		18.7	BLUE STELLAR OBJECT
PHL 2899	00 40 30.	- 15 19		18.4	BLUE STELLAR OBJECT
LB 08492	00 40 30.	- 20 45		16.0	FAINT BLUE STAR
LB 08493	00 40 30.	- 22 31		18.5	FAINT BLUE STAR
LB 08494	00 40 30.	- 24 05		19.7	FAINT BLUE STAR
LB 08495	00 40 30.	- 24 30		19.3	FAINT BLUE STAR
MCG-06-02-026	00 40 30.	- 37 10	30	14.5	GALAXY
LB 01570	00 40 30.	- 54 13		14.5	FAINT BLUE STAR
BAA 4.40	00 40 30.34	+ 40 37 58.1			DIFFUSE OBJECT IN M31
RIC 474	00 40 30.67	+ 39 18 24.4		18.62	BLUE OBJECT
RIC 475	00 40 31.57	+ 39 36 27.0		19.11	BLUE OBJECT
BAA 1.565	00 40 31.59	+ 41 28 47.1			EMISSION NEBULA IN M31
RNGC 0225	00 40 32.	+ 61 31		9.0	OPEN CLUSTER
LIN 036	00 40 34.	- 72 12 57.		15.33	SMC EMISSION LINE OBJECT
RIC 476	00 40 34.28	+ 40 08 27.7		19.85	BLUE OBJECT
RIC 477	00 40 34.57	+ 39 21 11.1		19.55	BLUE OBJECT
HN 0115	00 40 35.	- 22 31			NEBULA
IC 1574	00 40 35.0	- 22 31 21.			GALAXY
BAA 4.01	00 40 35.00	+ 41 53 53.0			DIFFUSE OBJECT IN M31
PHL 2901	00 40 36.	+ 00 58		18.4	BLUE STELLAR OBJECT
ZC 0040.6+0715	00 40 36.	+ 07 15	3630		CLUSTER OF GALAXIES
4ZW 030	00 40 36.	+ 39 33			COMPACT GALAXY
ZWG 550.011	00 40 36.	+ 50 24		14.6	GALAXY
UGC 00460	00 40 36.	+ 50 24	72	14.6	GALAXY Sc
PHL 2902	00 40 36.	- 01 12		18.4	BLUE STELLAR OBJECT
PHL 2903	00 40 36.	- 08 38		18.0	BLUE STELLAR OBJECT
PHL 0812	00 40 36.	- 09 03		13.9	BLUE STELLAR OBJECT
ARC 0089	00 40 36.	- 09 43		16.6	RICH CLUSTER OF GALAXIES
MCG-02-02-097	00 40 36.	- 11 31	12	16.	GALAXY
PHL 6544	00 40 36.	- 11 58		18.1	BLUE STELLAR OBJECT
PHL 6545	00 40 36.	- 12 01		18.5	BLUE STELLAR OBJECT
PHL 2904	00 40 36.	- 13 36		18.6	BLUE STELLAR OBJECT
PHL 6543	00 40 36.	- 18 56		17.4	BLUE STELLAR OBJECT
LB 08496	00 40 36.	- 22 02		16.5	FAINT BLUE STAR
MCG-04-02-043	00 40 36.	- 22 32	96	14.	GALAXY

OBJECT NAME	RIGHT ASCEN.	DECLINATION	DIAM.	MAGN.	TYPE OF OBJECT
LB 08497	00 40 36.	- 23 34		18.5	FAINT BLUE STAR
LB 08498	00 40 36.	- 23 38		17.7	FAINT BLUE STAR
LB 08499	00 40 36.	- 24 52		16.8	FAINT BLUE STAR
LB 08500	00 40 36.	- 25 30		19.8	FAINT BLUE STAR
LB 08501	00 40 36.	- 26 13		17.0	FAINT BLUE STAR
TON-S 0167	00 40 36.	- 28 39		16.0	BLUE STAR
BAA 1.558	00 40 36.42	+ 41 24 15.2			EMISSION NEBULA IN M31
LIN 037	00 40 37.	- 73 24 21.		15.86	SMC EMISSION LINE OBJECT
HELW 084	00 40 38.	- 08 40 33.			NEBULA
RIC 478	00 40 38.38	+ 39 20 55.4		20.27	BLUE OBJECT
RIC 479	00 40 38.96	+ 39 24 33.6		19.32	BLUE OBJECT
SC 0038-6809.2	00 40 39.	- 67 52 44.	6		NEBULA
BAA 1.251	00 40 40.25	+ 40 51 37.3			EMISSION NEBULA IN M31
BAA 1.562	00 40 41.22	+ 41 26 29.1			EMISSION NEBULA IN M31
PHL 6546	00 40 42.	+ 02 54		18.2	BLUE STELLAR OBJECT
OCL 0306	00 40 42.	+ 63 55	240	11.	OPEN STAR CLUSTER
ARC 0091	00 40 42.	- 10 55		17.6	RICH CLUSTER OF GALAXIES
PHL 6547	00 40 42.	- 11 46		18.4	BLUE STELLAR OBJECT
LB 08502	00 40 42.	- 20 41		17.9	FAINT BLUE STAR
LB 08503	00 40 42.	- 22 24		20.4	FAINT BLUE STAR
LB 08504	00 40 42.	- 23 54		19.4	FAINT BLUE STAR
LB 08505	00 40 42.	- 25 08		16.6	FAINT BLUE STAR
LB 08506	00 40 42.	- 25 23		17.7	FAINT BLUE STAR
LB 08507	00 40 42.	- 26 20		18.2	FAINT BLUE STAR
MCG-05-02-036	00 40 42.	- 27 45	24	15.	GALAXY
BAA 1.563	00 40 42.05	+ 41 26 37.3			EMISSION NEBULA IN M31
BAA 1.252	00 40 42.11	+ 40 48 44.8			EMISSION NEBULA IN M31
BAA 1.023	00 40 43.43	+ 41 02 07.0			EMISSION NEBULA IN M31
RIC 480	00 40 43.71	+ 39 19 41.7		18.71	BLUE OBJECT
RIC 481	00 40 44.81	+ 42 05 44.3		18.63	BLUE OBJECT
MCG-01-02-047	00 40 45.	- 06 55 30.	36	14.	GALAXY
MCG-01-02-048	00 40 45.	- 06 57	48	15.	GALAXY
SC 0038-6755.8	00 40 45.	- 67 39 20.	24		NEBULA
BAA 1.024	00 40 45.01	+ 41 02 09.8			EMISSION NEBULA IN M31
RIC 482	00 40 45.72	+ 39 30 01.8		18.61	BLUE OBJECT
RIC 483	00 40 45.98	+ 40 00 42.8		19.07	BLUE OBJECT
RNGC 0233	00 40 46.	+ 30 19		14.0	GALAXY
RIC 484	00 40 46.63	+ 39 37 19.4		18.79	BLUE OBJECT
RIC 485	00 40 46.92	+ 39 40 20.4		20.31	BLUE OBJECT
ARC 0090	00 40 47.	+ 01 54		17.2	RICH CLUSTER OF GALAXIES
LIN 039	00 40 47.	- 74 04 27.		13.92	SMC EMISSION LINE OBJECT
RIC 486	00 40 47.33	+ 42 05 57.8		18.53	BLUE OBJECT
BAA 1.566	00 40 47.52	+ 41 26 30.5			EMISSION NEBULA IN M31
BAA 1.006	00 40 47.57	+ 40 55 33.4			EMISSION NEBULA IN M31
BAA 1.564	00 40 47.98	+ 41 26 19.3			EMISSION NEBULA IN M31
RNGC 0234	00 40 48.	+ 14 04		13.5	GALAXY
MCG+02-02-028	00 40 48.	+ 14 04	84	13.	GALAXY
ZC 0040.8+2342	00 40 48.	+ 23 42	870		CLUSTER OF GALAXIES
ZC 0040.8+2404	00 40 48.	+ 24 04	2490		CLUSTER OF GALAXIES
ZWG 500.077	00 40 48.	+ 29 47		15.7	GALAXY
MCG+05-02-041	00 40 48.	+ 30 19	66	14.	GALAXY
VDB.66B 33	00 40 48.	+ 39 40 18.		20.34	BLUE OBJECT
OCL 0304	00 40 48.	+ 59 53	600		OPEN STAR CLUSTER
LDN 1301	00 40 48.	+ 62 11	180		DARK NEBULA
PHL 6551	00 40 48.	- 10 53		16.6	BLUE STELLAR OBJECT
PHL 6548	00 40 48.	- 10 54		18.1	BLUE STELLAR OBJECT
PHL 2905	00 40 48.	- 12 39		18.4	BLUE STELLAR OBJECT
PHL 6549	00 40 48.	- 14 30		18.3	BLUE STELLAR OBJECT
PHL 6550	00 40 48.	- 14 57		18.7	BLUE STELLAR OBJECT
MCG-03-02-042	00 40 48.	- 17 54	24	15.	GALAXY
LB 08508	00 40 48.	- 20 47		19.7	FAINT BLUE STAR
LB 08509	00 40 48.	- 21 38		18.5	FAINT BLUE STAR
LB 08510	00 40 48.	- 25 46		18.4	FAINT BLUE STAR
LIN 038	00 40 48.	- 73 46 39.		14.59	SMC EMISSION LINE OBJECT
RIC 487	00 40 48.10	+ 39 32 00.4		18.99	BLUE OBJECT
BAA 1.007	00 40 48.85	+ 40 55 45.8			EMISSION NEBULA IN M31
BAA 1.001	00 40 49.35	+ 40 53 27.4			EMISSION NEBULA IN M31
RIC 488	00 40 49.83	+ 39 22 03.2		19.07	BLUE OBJECT
VDB.66B 30	00 40 50.	+ 39 33 06.		19.24	BLUE OBJECT
RIC 489	00 40 50.28	+ 39 41 27.0		19.96	BLUE OBJECT
BAA 1.002	00 40 50.58	+ 40 53 29.7			EMISSION NEBULA IN M31
RIC 490	00 40 50.69	+ 40 07 12.3		18.22	BLUE OBJECT
RIC 491	00 40 50.78	+ 39 55 33.3		19.37	BLUE OBJECT
RIC 492	00 40 50.91	+ 38 50 21.0		17.64	BLUE OBJECT
BAA 1.003	00 40 51.10	+ 40 54 12.0			EMISSION NEBULA IN M31
BAA 1.547	00 40 53.46	+ 41 15 36.8			EMISSION NEBULA IN M31
BAA 1.034	00 40 53.62	+ 41 06 23.7			EMISSION NEBULA IN M31
BAA 1.568	00 40 53.96	+ 41 28 58.2			EMISSION NEBULA IN M31
PHL 6540	00 40 54.	+ 01 26		18.3	BLUE STELLAR OBJECT
ZWG 383.080	00 40 54.	+ 02 40		14.5	GALAXY
UGC 00462	00 40 54.	+ 02 40	84	14.5	GALAXY Sc
PHL 2906	00 40 54.	+ 10 18		17.4	BLUE STELLAR OBJECT
PHL 2907	00 40 54.	+ 10 45		18.5	BLUE STELLAR OBJECT
ZWG 435.001	00 40 54.	+ 14 04		13.5	GALAXY
ZWG 434.032	00 40 54.	+ 14 04		13.5	GALAXY
UGC 00463	00 40 54.	+ 14 04	120	13.5	GALAXY Sc
ZC 0040.9+1937	00 40 54.	+ 19 37	1880		CLUSTER OF GALAXIES
ZWG 383.079	00 40 54.	- 00 24		13.6	GALAXY
UGC 00461	00 40 54.	- 00 24	120	13.6	GALAXY Sb-c
MCG+00-02-136	00 40 54.	- 00 24	78	13.2	GALAXY
KARA.73B 0033	00 40 54.	- 00 24	102	13.6	ISOLATED GALAXY S
PHL 6552	00 40 54.	- 02 08		16.8	BLUE STELLAR OBJECT
MCG-01-02-049	00 40 54.	- 06 38	120	14.5	GALAXY
PHL 6553	00 40 54.	- 17 12		17.1	BLUE STELLAR OBJECT
LB 08511	00 40 54.	- 23 05		18.3	FAINT BLUE STAR
LB 08512	00 40 54.	- 24 38		19.2	FAINT BLUE STAR
LB 08513	00 40 54.	- 25 37		19.5	FAINT BLUE STAR
LB 08514	00 40 54.	- 26 24		18.5	FAINT BLUE STAR
LB 08515	00 40 54.	- 26 28		18.6	FAINT BLUE STAR
LB 03152	00 40 54.	- 73 03		12.6	FAINT BLUE STAR
RIC 493	00 40 54.83	+ 40 33 04.4		19.04	BLUE OBJECT
BAA 1.055	00 40 54.93	+ 41 10 28.9			EMISSION NEBULA IN M31
RNGC 0236	00 40 55.	+ 02 40		14.5	GALAXY
RNGC 0237	00 40 55.	- 00 24		13.5	GALAXY
BAA 1.005	00 40 55.04	+ 40 53 55.0			EMISSION NEBULA IN M31
BAA 1.004	00 40 55.12	+ 40 53 44.8			EMISSION NEBULA IN M31
BAA 1.571	00 40 55.24	+ 41 32 49.2			EMISSION NEBULA IN M31
BAA 1.008	00 40 55.76	+ 40 55 40.4			EMISSION NEBULA IN M31
HELW 085	00 40 56.	- 08 33 33.			NEBULA
BAA 1.569	00 40 56.82	+ 41 28 57.3			EMISSION NEBULA IN M31
HELW 086	00 40 57.	- 08 21 51.			NEBULA
BAA 1.570	00 40 57.63	+ 41 31 36.9			EMISSION NEBULA IN M31
BAA 1.039	00 40 57.67	+ 41 06 43.5			EMISSION NEBULA IN M31
BAA 4.02	00 40 57.89	+ 41 06 50.2			DIFFUSE OBJECT IN M31
RIC 494	00 40 58.54	+ 39 20 14.1		19.60	BLUE OBJECT
ARP 231	00 40 59.	- 04 24			PECULIAR GALAXY
BAA 1.014	00 40 59.03	+ 40 55 52.1			EMISSION NEBULA IN M31
BAA 1.009	00 40 59.66	+ 40 55 12.3			EMISSION NEBULA IN M31

OBJECT NAME	RIGHT ASCEN.	DECLINATION	DIAM.	MAGN.	TYPE OF OBJECT
BAA 1.041	00 40 59.99	+ 41 06 55.9			EMISSION NEBULA IN M31
LBN 0604	00 41	+ 61 40	300		BRIGHT NEBULA
KHAV 011	00 41	+ 61 52	6010		DARK NEBULA
MCG+01-03-001	00 41 00.	+ 02 42	66	13.5	GALAXY
PHL 6554	00 41 00.	+ 11 46		15.3	BLUE STELLAR OBJECT
PHL 6555	00 41 00.	+ 13 28		18.3	BLUE STELLAR OBJECT
MCG+04-02-050	00 41 00.	+ 23 11	60	15.	GALAXY
ZWG 479.065	00 41 00.	+ 23 12		15.1	GALAXY
ZWG 500.078	00 41 00.	+ 30 19		13.8	GALAXY
UGC 00464	00 41 00.	+ 30 19	126	13.8	GALAXY E?
VB 091	00 41 00.	+ 59 57	497		STELLAR RING
LDN 1300	00 41 00.	+ 61 10	360		DARK NEBULA
LDN 1302	00 41 00.	+ 61 35	240		DARK NEBULA
MCG-01-03-003	00 41 00.	- 04 23	24	14.	GALAXY
MCG-01-03-002	00 41 00.	- 04 23	48	14.	GALAXY
MCG-01-03-001	00 41 00.	- 08 28 30.	48	14.	GALAXY
LB 08516	00 41 00.	- 22 09		18.3	FAINT BLUE STAR
LB 08517	00 41 00.	- 23 21		18.8	FAINT BLUE STAR
LB 08518	00 41 00.	- 23 44		19.4	FAINT BLUE STAR
LB 08519	00 41 00.	- 23 50		18.4	FAINT BLUE STAR
RIC 495	00 41 00.00	+ 40 12 19.8		18.89	BLUE OBJECT
BAA 1.010	00 41 00.05	+ 40 55 11.0			EMISSION NEBULA IN M31
BAA 1.012	00 41 00.16	+ 40 55 19.7			EMISSION NEBULA IN M31
BAA 1.011	00 41 00.23	+ 40 55 18.2			EMISSION NEBULA IN M31
LB 03916	00 41 01.	+ 59 12 12.		18.7	FAINT BLUE STAR
BAA 1.016	00 41 02.59	+ 40 55 59.4			EMISSION NEBULA IN M31
BAA 1.017	00 41 02.66	+ 40 56 15.5			EMISSION NEBULA IN M31
LB 03917	00 41 03.	+ 59 34 18.		17.3	FAINT BLUE STAR
IC 1575	00 41 03.	- 04 25 28.			NONSTELLAR OBJECT
RIC 496	00 41 03.10	+ 39 14 36.4		19.21	BLUE OBJECT
BAA 1.018	00 41 03.52	+ 40 56 15.9			EMISSION NEBULA IN M31
BAA 1.015	00 41 03.58	+ 40 55 17.5			EMISSION NEBULA IN M31
BAA 1.589	00 41 03.70	+ 41 40 18.3			EMISSION NEBULA IN M31
BAA 1.013	00 41 03.82	+ 40 55 08.5			EMISSION NEBULA IN M31
BAA 1.060	00 41 04.98	+ 41 09 33.7			EMISSION NEBULA IN M31
VDB.66B 26	00 41 05.	+ 39 12 42.		19.20	BLUE OBJECT
IC 0048	00 41 05.	- 08 09 16.			NONSTELLAR OBJECT
BAA 1.019	00 41 05.45	+ 40 56 19.8			EMISSION NEBULA IN M31
ZWG 384.001	00 41 06.	+ 02 07		15.7	GALAXY
ZC 0041.1+1016	00 41 06.	+ 10 16	1810		CLUSTER OF GALAXIES
ABC 0092	00 41 06.	+ 20 23		17.9	RICH CLUSTER OF GALAXIES
ZC 0041.1+2027	00 41 06.	+ 20 27	2690		CLUSTER OF GALAXIES
ZWG 500.079	00 41 06.	+ 32 35		14.5	GALAXY
UGC 00465	00 41 06.	+ 32 35	108	14.5	GALAXY SBa
MCG+05-02-042	00 41 06.	+ 32 35	90	14.5	GALAXY
LB 03918	00 41 06.	+ 59 03 42.		16.8	FAINT BLUE STAR
ZC 0041.1-0008	00 41 06.	- 00 08	1480		CLUSTER OF GALAXIES
LB 08520	00 41 06.	- 21 32		17.0	FAINT BLUE STAR
LB 08521	00 41 06.	- 22 00		18.7	FAINT BLUE STAR
LB 08522	00 41 06.	- 22 40		19.1	FAINT BLUE STAR
LB 08523	00 41 06.	- 24 22		17.5	FAINT BLUE STAR
RIC 497	00 41 07.34	+ 39 48 56.7		19.38	BLUE OBJECT
BAA 2.02	00 41 07.71	+ 41 04 56.2			GALAXY
BAA 2.01	00 41 07.93	+ 42 04 55.0			GALAXY
RNGC 0238	00 41 08.	- 50 27			UNVERIFIED SOUTHERN OBJECT
HOW 18	00 41 08.	- 72 39 49.	12		STAR CLUSTER IN SMC
BAA 1.598	00 41 08.11	+ 41 41 46.3			EMISSION NEBULA IN M31
RIC 498	00 41 08.34	+ 40 00 05.0		18.27	BLUE OBJECT
BAA 1.597	00 41 08.34	+ 41 41 33.5			EMISSION NEBULA IN M31
BAA 1.596	00 41 08.78	+ 41 41 23.9			EMISSION NEBULA IN M31
BAA 1.595	00 41 09.17	+ 41 41 19.6			EMISSION NEBULA IN M31
BAA 1.073	00 41 09.84	+ 41 10 04.0			EMISSION NEBULA IN M31
BAA 1.071	00 41 09.99	+ 41 09 43.9			EMISSION NEBULA IN M31
BAA 1.581	00 41 10.10	+ 41 36 30.4			EMISSION NEBULA IN M31
BAA 1.573	00 41 10.43	+ 41 32 01.6			EMISSION NEBULA IN M31
BAA 1.072	00 41 10.52	+ 41 09 39.9			EMISSION NEBULA IN M31
BAA 1.078	00 41 10.69	+ 41 10 35.9			EMISSION NEBULA IN M31
BAA 1.076	00 41 10.70	+ 41 10 27.2			EMISSION NEBULA IN M31
BAA 1.074	00 41 11.42	+ 41 10 09.7			EMISSION NEBULA IN M31
BAA 1.021	00 41 11.48	+ 40 55 48.0			EMISSION NEBULA IN M31
PHL 2909	00 41 12.	+ 01 20		18.5	BLUE STELLAR OBJECT
PHL 2908	00 41 12.	+ 15 13		12.2	BLUE STELLAR OBJECT
FEIG 007	00 41 12.	- 10 17		15.0	FAINT BLUE STAR
PHL 2910	00 41 12.	- 15 36		18.6	BLUE STELLAR OBJECT
PHL 0813	00 41 12.	- 19 02		18.4	BLUE STELLAR OBJECT
LB 00159	00 41 12.	- 19 59		17.0	FAINT BLUE STAR
PHL 6556	00 41 12.	- 20 02		17.2	BLUE STELLAR OBJECT
LB 08524	00 41 12.	- 21 40		19.5	FAINT BLUE STAR
LB 08525	00 41 12.	- 25 49		19.0	FAINT BLUE STAR
BAA 1.075	00 41 12.07	+ 41 10 04.9			EMISSION NEBULA IN M31
BAA 1.032	00 41 12.12	+ 41 02 59.4			EMISSION NEBULA IN M31
BAA 1.574	00 41 12.22	+ 41 32 06.8			EMISSION NEBULA IN M31
BAA 1.020	00 41 12.80	+ 40 55 33.8			EMISSION NEBULA IN M31
RIC 499	00 41 12.94	+ 38 59 31.5		19.77	BLUE OBJECT
BAA 1.077	00 41 12.96	+ 41 10 13.7			EMISSION NEBULA IN M31
SC 0039-6811.2	00 41 13.	- 67 54 45.	6		NEBULA
LIN 040	00 41 13.	- 72 52 57.		14.81	SMC EMISSION LINE OBJECT
BAA 1.572	00 41 13.68	+ 41 31 04.3			EMISSION NEBULA IN M31
RIC 500	00 41 13.75	+ 41 59 40.7		18.72	BLUE OBJECT
RIC 501	00 41 13.84	+ 41 59 33.6		19.20	BLUE OBJECT
BAA 1.575	00 41 14.01	+ 41 32 13.7			EMISSION NEBULA IN M31
BAA 1.022	00 41 14.13	+ 40 57 06.6			EMISSION NEBULA IN M31
MCG-01-03-004	00 41 15.	- 04 25	42	15.5	GALAXY
MCG-02-03-001	00 41 15.	- 08 26	36	14.	GALAXY
ABC 0093	00 41 15.	- 18 45		16.9	RICH CLUSTER OF GALAXIES
BAA 1.576	00 41 15.48	+ 41 32 34.9			EMISSION NEBULA IN M31
BAA 1.030	00 41 15.64	+ 41 01 36.2			EMISSION NEBULA IN M31
BAA 1.029	00 41 15.80	+ 41 01 32.1			EMISSION NEBULA IN M31
RIC 502	00 41 15.86	+ 39 44 47.6		19.06	BLUE OBJECT
BAA 1.577	00 41 16.94	+ 41 32 44.9			EMISSION NEBULA IN M31
RIC 503	00 41 17.48	+ 39 32 50.8		17.90	BLUE OBJECT
MCG+00-03-002	00 41 18.	+ 00 31	72	14.	GALAXY
ZWG 384.002	00 41 18.	+ 00 33		15.4	GALAXY
UGC 00466	00 41 18.	+ 00 33	90	15.4	GALAXY S
UGC 00467	00 41 18.	+ 28 36	84	17.	GALAXY S IV
MCG-01-03-005	00 41 18.	- 04 31	96	14.5	GALAXY
PHL 0814	00 41 18.	- 10 17		14.6	BLUE STELLAR OBJECT
PHL 2911	00 41 18.	- 11 32		18.3	BLUE STELLAR OBJECT
MCG-03-03-001	00 41 18.	- 18 01	48	15.	GALAXY
LB 08526	00 41 18.	- 22 29		15.6	FAINT BLUE STAR
MCG-04-02-044	00 41 18.	- 24 41	90	15.	GALAXY
LB 08527	00 41 18.	- 24 50		19.1	FAINT BLUE STAR
LB 08528	00 41 18.	- 26 09		18.9	FAINT BLUE STAR
LH115-N008	00 41 18.	- 73 16	21		EMISSION NEBULA IN SMC
BAA 1.580	00 41 18.37	+ 41 33 30.9			EMISSION NEBULA IN M31
RIC 504	00 41 19.13	+ 38 58 16.5		19.48	BLUE OBJECT
BAA 1.027	00 41 19.30	+ 41 00 46.2			EMISSION NEBULA IN M31
BAA 1.028	00 41 19.78	+ 41 00 50.4			EMISSION NEBULA IN M31
BAA 1.578	00 41 19.90	+ 41 32 28.6			EMISSION NEBULA IN M31
IC 0049	00 41 20.	+ 01 35 02.			NONSTELLAR OBJECT
BAA 1.083	00 41 20.67	+ 41 10 53.8			EMISSION NEBULA IN M31
BAA 1.035	00 41 20.83	+ 41 02 21.7			EMISSION NEBULA IN M31
BAA 1.036	00 41 20.86	+ 41 02 24.0			EMISSION NEBULA IN M31
BAA 1.045	00 41 21.10	+ 41 04 14.7			EMISSION NEBULA IN M31
RIC 505	00 41 21.58	+ 39 57 36.8		19.00	BLUE OBJECT
BAA 1.086	00 41 22.38	+ 41 10 35.4			EMISSION NEBULA IN M31
BAA 1.049	00 41 23.18	+ 41 04 44.6			EMISSION NEBULA IN M31
BAA 1.050	00 41 23.32	+ 41 04 50.2			EMISSION NEBULA IN M31
BAA 1.048	00 41 23.40	+ 41 04 35.8			EMISSION NEBULA IN M31
PHL 0815	00 41 24.	+ 01 19		16.8	BLUE STELLAR OBJECT
ZWG 384.003	00 41 24.	+ 01 35		14.5	GALAXY
UGC 00468	00 41 24.	+ 01 35	114	14.5	GALAXY Sc
MCG+00-03-003	00 41 24.	+ 01 35	84	13.5	GALAXY
PHL 2912	00 41 24.	+ 14 56		18.6	BLUE STELLAR OBJECT
ZWG 479.066	00 41 24.	+ 25 55		15.3	GALAXY
UGC 00469	00 41 24.	+ 25 55	96	15.3	GALAXY S
MCG+04-02-051	00 41 24.	+ 25 56	84	15.	GALAXY
ZWG 500.080	00 41 24.	+ 30 05		15.1	GALAXY
PHL 6559	00 41 24.	- 01 45		18.3	BLUE STELLAR OBJECT
PHL 6558	00 41 24.	- 01 54		17.3	BLUE STELLAR OBJECT
ZC 0041.4-0241	00 41 24.	- 02 41	2290		CLUSTER OF GALAXIES
PHL 6557	00 41 24.	- 10 09		16.7	BLUE STELLAR OBJECT
PHL 6560	00 41 24.	- 11 28		18.3	BLUE STELLAR OBJECT
PHL 2914	00 41 24.	- 15 41		18.3	BLUE STELLAR OBJECT
PHL 2913	00 41 24.	- 19 43		18.5	BLUE STELLAR OBJECT
LB 08529	00 41 24.	- 21 39		17.0	FAINT BLUE STAR
LB 08530	00 41 24.	- 25 16		17.6	FAINT BLUE STAR
LB 01571	00 41 24.	- 51 53		14.4	FAINT BLUE STAR
LIN 041	00 41 24.	- 73 16 03.		16.27	PLANETARY NEBULA IN SMC
BAA 1.051	00 41 24.06	+ 41 04 51.3			EMISSION NEBULA IN M31
RIC 506	00 41 24.18	+ 39 30 25.8		18.94	BLUE OBJECT
BAA 1.102	00 41 24.29	+ 41 15 19.2			EMISSION NEBULA IN M31
BAA 1.106	00 41 24.45	+ 41 15 41.0			EMISSION NEBULA IN M31
RIC 507	00 41 24.78	+ 38 56 08.5		19.18	BLUE OBJECT
LB 03919	00 41 25.	+ 59 14 54.		18.2	FAINT BLUE STAR
LIN.CL 028	00 41 25.	- 72 52 03.	207	13.6	STAR CLUSTER IN SMC
BAA 1.591	00 41 25.19	+ 41 38 26.2			EMISSION NEBULA IN M31
RIC 508	00 41 25.23	+ 39 21 53.5		19.31	BLUE OBJECT
BAA 1.044	00 41 25.23	+ 41 03 33.0			EMISSION NEBULA IN M31
BAA 1.623	00 41 25.45	+ 41 46 34.7			EMISSION NEBULA IN M31
BAA 1.114	00 41 25.56	+ 41 16 29.2			EMISSION NEBULA IN M31
BAA 1.608	00 41 25.62	+ 41 42 26.9			EMISSION NEBULA IN M31
BAA 1.116	00 41 25.80	+ 41 16 53.7			EMISSION NEBULA IN M31
VDB.66B 23	00 41 26.	+ 38 57 05.		19.65	BLUE OBJECT
BAA 1.107	00 41 26.00	+ 41 15 35.2			EMISSION NEBULA IN M31
BAA 1.043	00 41 26.15	+ 41 03 24.0			EMISSION NEBULA IN M31
BAA 1.586	00 41 26.39	+ 41 35 07.6			EMISSION NEBULA IN M31
RIC 509	00 41 26.41	+ 39 59 19.0		18.45	BLUE OBJECT
BAA 1.110	00 41 26.42	+ 41 15 42.9			EMISSION NEBULA IN M31
BAA 2.04	00 41 26.76	+ 41 08 59.0			GALAXY
BAA 1.031	00 41 26.94	+ 41 00 05.1			EMISSION NEBULA IN M31
BAA 2.03	00 41 27.02	+ 41 03 09.4			GALAXY
BAA 1.124	00 41 27.30	+ 41 17 34.9			EMISSION NEBULA IN M31
BAA 1.118	00 41 27.31	+ 41 16 56.1			EMISSION NEBULA IN M31
BAA 1.025	00 41 27.36	+ 40 57 12.6			EMISSION NEBULA IN M31
LB 03920	00 41 28.	+ 59 18 54.		18.2	FAINT BLUE STAR
LIN 042	00 41 28.	- 73 58 57.		15.29	SMC EMISSION LINE OBJECT
BAA 1.033	00 41 28.28	+ 41 00 57.0			EMISSION NEBULA IN M31
BAA 1.046	00 41 28.99	+ 41 03 30.4			EMISSION NEBULA IN M31
HOW 19	00 41 29.	- 74 26 25.	18		STAR CLUSTER IN SMC
RIC 510	00 41 29.03	+ 39 04 33.7		19.63	BLUE OBJECT
RIC 511	00 41 29.18	+ 39 34 00.8		19.87	BLUE OBJECT
RIC 512	00 41 29.60	+ 42 10 07.4		20.03	BLUE OBJECT
RIC 513	00 41 29.70	+ 39 46 19.2		19.13	BLUE OBJECT
PHL 2915	00 41 30.	+ 14 58		18.4	BLUE STELLAR OBJECT
ZC 0041.5+1805	00 41 30.	+ 18 05	4100		CLUSTER OF GALAXIES
MCG+04-02-052	00 41 30.	+ 26 34	54	15.	GALAXY
ZWG 479.067	00 41 30.	+ 26 35		14.8	GALAXY
UGC 00470	00 41 30.	+ 26 35	66	14.8	GALAXY IRR
SHRV.73A 14	00 41 30.	+ 41 39	8	16.53	GLOBULAR CLUSTER IN M31
VB 010	00 41 30.	+ 66 38	389		STELLAR RING
MCG-02-03-002	00 41 30.	- 14 03	30	15.	GALAXY
PHL 6561	00 41 30.	- 14 49		18.7	BLUE STELLAR OBJECT
PHL 6562	00 41 30.	- 16 15		18.6	BLUE STELLAR OBJECT
LB 08531	00 41 30.	- 21 12		18.8	FAINT BLUE STAR
LB 08533	00 41 30.	- 22 58		17.4	FAINT BLUE STAR
LB 08532	00 41 30.	- 22 58		16.3	FAINT BLUE STAR
LB 08534	00 41 30.	- 25 06		17.0	FAINT BLUE STAR
LH115-N007	00 41 30.	- 72 55			EMISSION NEBULA IN SMC
BAA 1.621	00 41 30.33	+ 41 45 32.1			EMISSION NEBULA IN M31
BAA 1.054	00 41 30.44	+ 41 05 01.5			EMISSION NEBULA IN M31
BAA 1.026	00 41 31.45	+ 40 57 27.0			EMISSION NEBULA IN M31
RIC 514	00 41 31.47	+ 39 08 32.8		18.84	BLUE OBJECT
RIC 530	00 41 31.54	+ 39 45 21.6		19.90	BLUE OBJECT
BAA 1.037	00 41 31.58	+ 41 00 58.4			EMISSION NEBULA IN M31
RIC 515	00 41 31.64	+ 40 27 33.8		18.13	BLUE OBJECT
BAA 1.038	00 41 32.20	+ 41 01 08.2			EMISSION NEBULA IN M31
BAA 1.053	00 41 32.27	+ 41 04 40.5			EMISSION NEBULA IN M31
BAA 1.047	00 41 32.61	+ 41 03 04.4			EMISSION NEBULA IN M31
BAA 1.579	00 41 32.71	+ 41 31 12.2			EMISSION NEBULA IN M31
RIC 516	00 41 33.19	+ 40 28 10.4		20.30	BLUE OBJECT
LB 03921	00 41 34.	+ 58 59 36.		18.3	FAINT BLUE STAR
RNGC 0241	00 41 34.	- 73 42			NON-EXISTENT OBJECT
BAA 1.130	00 41 34.80	+ 41 17 56.8			EMISSION NEBULA IN M31
RIC 517	00 41 35.45	+ 39 46 04.8		19.72	BLUE OBJECT
PHL 2916	00 41 36.	+ 03 02		18.2	BLUE STELLAR OBJECT
ZWG 435.002	00 41 36.	+ 10 18		15.7	GALAXY
PHL 2917	00 41 36.	+ 12 19		18.0	BLUE STELLAR OBJECT
ZC 0041.6+2748	00 41 36.	+ 27 48	940		CLUSTER OF GALAXIES
MCG-02-03-003	00 41 36.	- 11 34	12	15.5	GALAXY
PHL 6563	00 41 36.	- 15 32		18.7	BLUE STELLAR OBJECT
LB 08535	00 41 36.	- 20 48		19.1	FAINT BLUE STAR
LB 08536	00 41 36.	- 21 02		19.8	FAINT BLUE STAR
LB 08537	00 41 36.	- 22 12		18.0	FAINT BLUE STAR
LB 08538	00 41 36.	- 23 10		17.9	FAINT BLUE STAR
LB 08539	00 41 36.	- 24 02		18.2	FAINT BLUE STAR
LB 08540	00 41 36.	- 24 50		18.5	FAINT BLUE STAR
BAA 1.040	00 41 36.31	+ 41 01 33.2			EMISSION NEBULA IN M31
BAA 1.064	00 41 36.63	+ 41 04 54.8			EMISSION NEBULA IN M31
BAA 1.079	00 41 36.99	+ 41 07 36.6			EMISSION NEBULA IN M31
LIN 043	00 41 37.	- 72 54 46.		16.38	PLANETARY NEBULA IN SMC
RIC 518	00 41 37.20	+ 40 02 25.0		20.54	BLUE OBJECT
BAA 1.052	00 41 37.53	+ 41 03 37.8			EMISSION NEBULA IN M31
BAA 1.042	00 41 37.81	+ 41 01 26.4			EMISSION NEBULA IN M31
BAA 1.058	00 41 38.35	+ 41 04 28.8			EMISSION NEBULA IN M31
RIC 519	00 41 38.46	+ 40 02 29.8		17.13	BLUE OBJECT

OBJECT NAME	RIGHT ASCEN.	DECLINATION	DIAM.	MAGN.	TYPE OF OBJECT
BAA 1.094	00 41 38.60	+ 41 10 43.8			EMISSION NEBULA IN M31
BAA 1.056	00 41 38.99	+ 41 04 20.6			EMISSION NEBULA IN M31
BOR 072	00 41 39.	+ 40 01 53.		17.36	BLUE OBJECT
BAA 1.084	00 41 39.09	+ 41 08 13.8			EMISSION NEBULA IN M31
BAA 1.584	00 41 39.35	+ 41 32 55.4			EMISSION NEBULA IN M31
BAA 1.583	00 41 39.60	+ 41 32 49.7			EMISSION NEBULA IN M31
BAA 1.057	00 41 39.74	+ 41 04 14.0			EMISSION NEBULA IN M31
BAA 1.065	00 41 39.79	+ 41 04 36.3			EMISSION NEBULA IN M31
BAA 1.585	00 41 39.85	+ 41 33 07.9			EMISSION NEBULA IN M31
BAA 1.582	00 41 40.02	+ 41 32 25.6			EMISSION NEBULA IN M31
BAA 1.149	00 41 40.20	+ 41 21 15.5			EMISSION NEBULA IN M31
BAA 1.059	00 41 40.35	+ 41 04 11.2			EMISSION NEBULA IN M31
BAA 1.587	00 41 40.57	+ 41 33 11.5			EMISSION NEBULA IN M31
BAA 1.062	00 41 40.59	+ 41 04 26.1			EMISSION NEBULA IN M31
RIC 520	00 41 40.72	+ 39 54 33.8		18.92	BLUE OBJECT
BAA 1.080	00 41 41.08	+ 41 07 01.1			EMISSION NEBULA IN M31
BAA 1.063	00 41 41.20	+ 41 04 20.9			EMISSION NEBULA IN M31
BAA 1.625	00 41 41.25	+ 41 45 38.2			EMISSION NEBULA IN M31
BAA 1.067	00 41 41.30	+ 41 04 32.5			EMISSION NEBULA IN M31
BAA 1.061	00 41 41.48	+ 41 04 12.6			EMISSION NEBULA IN M31
RIC 521	00 41 41.83	+ 39 54 50.8		19.07	BLUE OBJECT
ZC 0041.7+3859	00 41 42.	+ 38 59	1410		CLUSTER OF GALAXIES
MCG-01-03-006	00 41 42.	- 04 14	60	15.5	GALAXY
HELW 087	00 41 42.	- 08 29 51.			NEBULA
PHL 6564	00 41 42.	- 11 08		18.6	BLUE STELLAR OBJECT
8ZW 0041-13.6	00 41 42.	- 13 34		16.0	COMPACT GALAXY
PHL 6565	00 41 42.	- 16 30		18.5	BLUE STELLAR OBJECT
LB 08541	00 41 42.	- 21 32		17.4	FAINT BLUE STAR
LB 08542	00 41 42.	- 23 50		17.9	FAINT BLUE STAR
LH115-N009	00 41 42.	- 73 19	21		EMISSION NEBULA IN SMC
LIN 044	00 41 42.	- 73 19 46.		17.4	SMC EMISSION LINE OBJECT
RIC 522	00 41 42.32	+ 39 59 09.5		19.75	BLUE OBJECT
BAA 3.01	00 41 42.36	+ 41 12 25.9			GLOBULAR CLUSTER IN M31
RIC 523	00 41 42.72	+ 39 59 29.6		17.86	BLUE OBJECT
BOR 174	00 41 43.	+ 39 58 35.		19.77	BLUE OBJECT
RIC 524	00 41 43.12	+ 39 34 14.0		18.66	BLUE OBJECT
BAA 1.070	00 41 43.32	+ 41 04 37.2			EMISSION NEBULA IN M31
RIC 525	00 41 43.70	+ 39 54 59.8		18.94	BLUE OBJECT
BAA 1.588	00 41 43.81	+ 41 23 32.0			EMISSION NEBULA IN M31
BAA 1.081	00 41 44.31	+ 41 07 11.7			EMISSION NEBULA IN M31
RIC 526	00 41 44.82	+ 39 58 23.9		18.14	BLUE OBJECT
MCG-04-02-045	00 41 45.	- 25 22 30.	36	15.	GALAXY
BAA 1.068	00 41 45.45	+ 41 03 57.6			EMISSION NEBULA IN M31
BAA 1.593	00 41 45.47	+ 41 35 33.5			EMISSION NEBULA IN M31
BAA 1.069	00 41 45.93	+ 41 03 57.1			EMISSION NEBULA IN M31
BAA 1.592	00 41 45.93	+ 41 35 29.1			EMISSION NEBULA IN M31
LB 03922	00 41 46.	+ 59 33 18.		18.4	FAINT BLUE STAR
HN 0116	00 41 46.	- 25 23			NEBULA
RNGC 0242	00 41 46.	- 73 43		12.0	OPEN CLUSTER IN SMC
BAA 1.594	00 41 46.05	+ 41 35 32.9			EMISSION NEBULA IN M31
BAA 1.590	00 41 46.39	+ 41 34 46.1			EMISSION NEBULA IN M31
BAA 1.066	00 41 46.61	+ 41 03 37.7			EMISSION NEBULA IN M31
IC 1576	00 41 47.	- 25 22			NONSTELLAR OBJECT
LIN 046	00 41 47.	- 73 27 04.		15.38	SMC EMISSION LINE OBJECT
LIN.CL 029	00 41 47.	- 73 42 40.	546	12.1	STAR CLUSTER IN SMC
LIN 047	00 41 47.	- 73 44 40.		14.55	SMC EMISSION LINE OBJECT
BAA 1.085	00 41 47.16	+ 41 07 16.7			EMISSION NEBULA IN M31
PHL 6566	00 41 48.	+ 02 05		17.9	BLUE STELLAR OBJECT
3ZW 011	00 41 48.	+ 16 32			COMPACT GALAXY
ZC 0041.8+3014	00 41 48.	+ 30 14	1410		CLUSTER OF GALAXIES
4ZW 031	00 41 48.	+ 37 27			COMPACT GALAXY
SHRV.73A 15	00 41 48.	+ 41 23	10	16.49	GLOBULAR CLUSTER IN M31
SHRV.73A 16	00 41 48.	+ 42 01	8	16.62	GLOBULAR CLUSTER IN M31
MCG-02-03-004	00 41 48.	- 12 52	60	15.	GALAXY
MCG-03-03-002	00 41 48.	- 17 37	72	15.	GALAXY
LB 08543	00 41 48.	- 21 39		17.2	FAINT BLUE STAR
LB 08544	00 41 48.	- 22 08		19.4	FAINT BLUE STAR
LB 08545	00 41 48.	- 22 33		17.8	FAINT BLUE STAR
LB 08546	00 41 48.	- 22 59		17.7	FAINT BLUE STAR
LB 08547	00 41 48.	- 23 18		19.1	FAINT BLUE STAR
LB 08548	00 41 48.	- 23 54		18.3	FAINT BLUE STAR
LB 08549	00 41 48.	- 24 24		18.7	FAINT BLUE STAR
LB 08550	00 41 48.	- 24 48		19.0	FAINT BLUE STAR
LB 08551	00 41 48.	- 25 39		17.6	FAINT BLUE STAR
MCG-05-03-001	00 41 48.	- 28 53	30	15.5	GALAXY
PHL 6568	00 41 48.	- 31 16		18.6	BLUE STELLAR OBJECT
PHL 6567	00 41 48.	- 31 22		18.5	BLUE STELLAR OBJECT
LB 03153	00 41 48.	- 69 01		13.4	FAINT BLUE STAR
LIN 045	00 41 48.	- 73 18 52.		14.03	PLANETARY NEBULA IN SMC
BAA 1.091	00 41 48.06	+ 41 08 56.5			EMISSION NEBULA IN M31
RIC 527	00 41 48.33	+ 40 10 12.1		18.26	BLUE OBJECT
BAA 1.602	00 41 48.73	+ 41 36 24.1			EMISSION NEBULA IN M31
BAA 1.601	00 41 48.99	+ 41 36 17.7			EMISSION NEBULA IN M31
BAA 1.603	00 41 49.04	+ 41 36 27.4			EMISSION NEBULA IN M31
RIC 528	00 41 49.16	+ 40 14 07.9		19.58	BLUE OBJECT
BAA 1.090	00 41 49.18	+ 41 08 40.6			EMISSION NEBULA IN M31
RIC 529	00 41 49.31	+ 40 28 08.9		18.29	BLUE OBJECT
BAA 1.087	00 41 50.22	+ 41 08 00.1			EMISSION NEBULA IN M31
BAA 1.604	00 41 50.82	+ 41 36 30.9			EMISSION NEBULA IN M31
BAA 1.089	00 41 51.67	+ 41 08 17.5			EMISSION NEBULA IN M31
BAA 3.27	00 41 51.67	+ 42 00 56.3			GLOBULAR CLUSTER IN M31
BAA 1.096	00 41 51.69	+ 41 08 55.5			EMISSION NEBULA IN M31
BAA 2.67	00 41 51.86	+ 41 30 24.4			GALAXY
BAA 1.092	00 41 52.09	+ 41 08 22.2			EMISSION NEBULA IN M31
BAA 1.095	00 41 52.68	+ 41 08 39.3			EMISSION NEBULA IN M31
BAA 1.599	00 41 52.73	+ 41 35 28.6			EMISSION NEBULA IN M31
LIN 048	00 41 53.	- 73 43 34.		14.25	SMC EMISSION LINE OBJECT
BAA 1.605	00 41 53.03	+ 41 36 37.4			EMISSION NEBULA IN M31
BAA 1.600	00 41 53.17	+ 41 35 30.1			EMISSION NEBULA IN M31
RIC 531	00 41 53.49	+ 39 55 07.0		19.11	BLUE OBJECT
BAA 1.120	00 41 53.55	+ 41 13 23.3			EMISSION NEBULA IN M31
BAA 1.121	00 41 53.64	+ 41 13 39.2			EMISSION NEBULA IN M31
BAA 1.097	00 41 53.73	+ 41 08 46.9			EMISSION NEBULA IN M31
ZC 0041.9+0052	00 41 54.	+ 00 52	2620		CLUSTER OF GALAXIES
PHL 6569	00 41 54.	+ 15 16		17.8	BLUE STELLAR OBJECT
ZWG 458.001	00 41 54.	+ 16 32		15.3	GALAXY
MCG+07-02-017	00 41 54.	+ 45 08	48	16.	GALAXY
LB 08552	00 41 54.	- 25 55		17.4	FAINT BLUE STAR
BAA 1.098	00 41 54.40	+ 41 09 04.5			EMISSION NEBULA IN M31
BAA 1.609	00 41 55.44	+ 41 38 06.8			EMISSION NEBULA IN M31
BAA 1.103	00 41 56.57	+ 41 10 39.8			EMISSION NEBULA IN M31
BAA 1.606	00 41 56.94	+ 41 36 49.2			EMISSION NEBULA IN M31
RIC 532	00 41 56.99	+ 40 30 19.8		19.44	BLUE OBJECT
BAA 1.082	00 41 57.22	+ 41 05 26.4			EMISSION NEBULA IN M31
BAA 1.104	00 41 57.34	+ 41 10 35.7			EMISSION NEBULA IN M31
RIC 533	00 41 57.47	+ 38 49 42.3		20.37	BLUE OBJECT
BAA 1.108	00 41 57.57	+ 41 11 05.8			EMISSION NEBULA IN M31
BAA 1.607	00 41 57.69	+ 41 37 42.0			EMISSION NEBULA IN M31
LB 03923	00 41 58.	+ 59 26 54.		18.4	FAINT BLUE STAR
HN 0117	00 41 58.	- 25 21			NEBULA
BAA 1.105	00 41 58.36	+ 41 10 34.6			EMISSION NEBULA IN M31
BAA 1.100	00 41 58.83	+ 41 09 52.4			EMISSION NEBULA IN M31
IC 1578	00 41 59.	- 25 20			NONSTELLAR OBJECT
BAA 1.111	00 41 59.29	+ 41 11 21.5			EMISSION NEBULA IN M31
BAA 1.610	00 41 59.30	+ 41 37 35.6			EMISSION NEBULA IN M31
RIC 534	00 41 59.47	+ 39 33 56.7		19.31	BLUE OBJECT
BAA 1.099	00 41 59.69	+ 41 09 35.6			EMISSION NEBULA IN M31
KHAV 012	00 42	+ 52 16	1860		DARK NEBULA
ZWG 410.001	00 42 00.	+ 08 22		15.0	GALAXY
PHL 2918	00 42 00.	+ 09 51		17.5	BLUE STELLAR OBJECT
ZWG 458.002	00 42 00.	+ 16 34		15.7	GALAXY
VDB.66B 02	00 42 00.	+ 38 51 04.		20.27	BLUE OBJECT
VDB.66B 13	00 42 00.	+ 39 33 40.		19.51	BLUE OBJECT
SHRV.73B 21	00 42 00.	+ 42 44	9	15.59	COMPACT GALAXY
ZWG 535.018	00 42 00.	+ 45 09		15.7	GALAXY
UGC 00471	00 42 00.	+ 45 09	120	15.7	GALAXY Sb/Sc
LDN 1295	00 42 00.	+ 52 10	660		DARK NEBULA
LDN 1296	00 42 00.	+ 52 30	720		DARK NEBULA
PHL 6570	00 42 00.	- 03 28		18.1	BLUE STELLAR OBJECT
MCG-01-03-007	00 42 00.	- 04 02	60	14.	GALAXY
PHL 6572	00 42 00.	- 09 00		17.8	BLUE STELLAR OBJECT
PHL 6571	00 42 00.	- 18 52		18.6	BLUE STELLAR OBJECT
PHL 0816	00 42 00.	- 21 15		17.3	BLUE STELLAR OBJECT
LB 08553	00 42 00.	- 21 51		15.9	FAINT BLUE STAR
LB 08554	00 42 00.	- 22 32		17.1	FAINT BLUE STAR
LB 08555	00 42 00.	- 23 06		20.0	FAINT BLUE STAR
LB 08556	00 42 00.	- 23 20		19.8	FAINT BLUE STAR
LB 08557	00 42 00.	- 24 51		18.0	FAINT BLUE STAR
MCG-04-03-001	00 42 00.	- 25 20	36	14.5	GALAXY
LIN 049	00 42 00.	- 73 11 46.		16.33	PLANETARY NEBULA IN SMC
BAA 1.119	00 42 00.10	+ 41 12 17.1			EMISSION NEBULA IN M31
BAA 1.115	00 42 00.22	+ 41 11 27.0			EMISSION NEBULA IN M31
BAA 1.112	00 42 00.87	+ 41 11 12.6			EMISSION NEBULA IN M31
BOR 009	00 42 01.	+ 40 09 04.		19.09	BLUE OBJECT
RIC 535	00 42 01.54	+ 40 09 40.2		18.95	BLUE OBJECT
RIC 536	00 42 01.65	+ 40 08 21.6		18.48	BLUE OBJECT
RIC 537	00 42 01.69	+ 40 03 09.8		20.40	BLUE OBJECT
BAA 3.24	00 42 01.92	+ 41 47 26.9			GLOBULAR CLUSTER IN M31
BOR 159	00 42 02.	+ 40 02 34.		20.00	BLUE OBJECT
BOR 013	00 42 02.	+ 40 07 46.		18.21	BLUE OBJECT
RNGC 0239	00 42 02.	- 04 02		14.0	GALAXY
BAA 2.05	00 42 02.17	+ 41 10 39.3			GALAXY
BAA 1.125	00 42 02.26	+ 41 13 01.5			EMISSION NEBULA IN M31
BAA 3.25	00 42 02.38	+ 41 49 08.4			GLOBULAR CLUSTER IN M31
BAA 1.618	00 42 02.57	+ 41 40 33.9			EMISSION NEBULA IN M31
BAA 1.101	00 42 02.59	+ 41 09 29.9			EMISSION NEBULA IN M31
BAA 1.126	00 42 02.70	+ 41 12 58.5			EMISSION NEBULA IN M31
BAA 1.109	00 42 02.82	+ 41 10 20.6			EMISSION NEBULA IN M31
BAA 1.117	00 42 02.87	+ 41 11 38.4			EMISSION NEBULA IN M31
BAA 1.122	00 42 02.99	+ 41 12 18.3			EMISSION NEBULA IN M31
BAA 1.088	00 42 03.62	+ 41 06 30.8			EMISSION NEBULA IN M31
BAA 4.03	00 42 03.74	+ 41 49 49.9			DIFFUSE OBJECT IN M31
RIC 538	00 42 03.93	+ 39 58 17.0		19.49	BLUE OBJECT
IC 1577	00 42 04.	- 08 24 41.			NONSTELLAR OBJECT
HELW 088	00 42 04.	- 08 36 46.			NEBULA
BAA 1.113	00 42 04.02	+ 41 10 49.3			EMISSION NEBULA IN M31
RIC 539	00 42 04.35	+ 40 03 35.4		19.72	BLUE OBJECT
BAA 1.093	00 42 04.53	+ 41 06 39.0			EMISSION NEBULA IN M31
BAA 1.123	00 42 04.93	+ 41 12 02.0			EMISSION NEBULA IN M31
BOR 195	00 42 05.	+ 39 57 40.		19.79	BLUE OBJECT
BOR 155	00 42 05.	+ 40 02 58.		19.47	BLUE OBJECT
VDB.66B 20	00 42 05.	+ 40 03 04.		19.85	BLUE OBJECT
BAA 1.141	00 42 05.56	+ 41 15 46.1			EMISSION NEBULA IN M31
RIC 540	00 42 05.68	+ 39 31 18.0		18.89	BLUE OBJECT
RIC 541	00 42 05.90	+ 39 19 14.8		19.38	BLUE OBJECT
PHL 2919	00 42 06.	+ 00 02		18.7	BLUE STELLAR OBJECT
SHRV.73B 22	00 42 06.	+ 43 14	8	16.69	COMPACT GALAXY
PHL 6573	00 42 06.	- 02 04		18.2	BLUE STELLAR OBJECT
PHL 6574	00 42 06.	- 10 38		18.1	BLUE STELLAR OBJECT
PHL 6575	00 42 06.	- 16 30		18.6	BLUE STELLAR OBJECT
PHL 0817	00 42 06.	- 21 48		17.5	BLUE STELLAR OBJECT
LB 08558	00 42 06.	- 24 24		18.4	FAINT BLUE STAR
BAA 1.131	00 42 06.25	+ 41 13 28.1			EMISSION NEBULA IN M31
BAA 1.129	00 42 06.36	+ 41 13 01.5			EMISSION NEBULA IN M31
BAA 1.128	00 42 06.45	+ 41 12 42.2			EMISSION NEBULA IN M31
BAA 1.640	00 42 06.59	+ 41 53 10.4			EMISSION NEBULA IN M31
RIC 542	00 42 06.62	+ 42 08 50.2		19.06	BLUE OBJECT
BAA 1.613	00 42 06.99	+ 41 38 02.4			EMISSION NEBULA IN M31
BAA 1.138	00 42 07.45	+ 41 14 02.1			EMISSION NEBULA IN M31
BAA 1.150	00 42 07.88	+ 41 17 24.2			EMISSION NEBULA IN M31
BAA 1.615	00 42 07.90	+ 41 38 36.1			EMISSION NEBULA IN M31
BAA 1.628	00 42 08.59	+ 41 46 10.0			EMISSION NEBULA IN M31
LIN 051	00 42 09.	- 74 03 58.		15.04	SMC EMISSION LINE OBJECT
RIC 543	00 42 09.16	+ 39 41 40.4		18.63	BLUE OBJECT
BAA 1.614	00 42 09.19	+ 41 37 45.9			EMISSION NEBULA IN M31
BAA 1.127	00 42 09.57	+ 41 12 00.2			EMISSION NEBULA IN M31
BAA 1.140	00 42 09.59	+ 41 14 52.9			EMISSION NEBULA IN M31
RIC 544	00 42 09.68	+ 40 58 44.0		18.85	BLUE OBJECT
BAA 1.139	00 42 09.82	+ 41 14 50.3			EMISSION NEBULA IN M31
RIC 545	00 42 10.26	+ 40 09 51.8		17.83	BLUE OBJECT
RIC 546	00 42 10.50	+ 40 39 29.1		19.20	BLUE OBJECT
BAA 4.04	00 42 10.81	+ 41 38 17.1			DIFFUSE OBJECT IN M31
BAA 1.132	00 42 10.93	+ 41 12 56.2			EMISSION NEBULA IN M31
LIN 050	00 42 11.	- 73 29 28.		15.87	SMC EMISSION LINE OBJECT
BAA 4.05	00 42 11.22	+ 41 37 36.8			DIFFUSE OBJECT IN M31
BAA 1.142	00 42 11.24	+ 41 14 59.6			EMISSION NEBULA IN M31
BAA 1.136	00 42 11.50	+ 41 13 01.9			EMISSION NEBULA IN M31
BAA 1.145	00 42 11.75	+ 41 15 15.0			EMISSION NEBULA IN M31
BAA 1.137	00 42 11.95	+ 41 12 58.4			EMISSION NEBULA IN M31
ZWG 410.002	00 42 12.	+ 04 53		15.2	GALAXY
PHL 0818	00 42 12.	+ 09 26		16.2	BLUE STELLAR OBJECT
PHL 6577	00 42 12.	+ 10 16		18.1	BLUE STELLAR OBJECT
MCG+03-03-001	00 42 12.	+ 19 49	24	15.5	GALAXY
MRK 346	00 42 12.	+ 27 12	18	16.5	GALAXY WITH UV CONTINUUM
LDN 1299	00 42 12.	+ 55 28	240		DARK NEBULA
PHL 6536	00 42 12.	- 00 16		18.5	BLUE STELLAR OBJECT
PHL 2921	00 42 12.	- 12 00		18.4	BLUE STELLAR OBJECT
PHL 6576	00 42 12.	- 17 02		18.2	BLUE STELLAR OBJECT
PHL 0819	00 42 12.	- 20 44		16.6	BLUE STELLAR OBJECT
PHL 2920	00 42 12.	- 22 00		17.1	BLUE STELLAR OBJECT
LB 00160	00 42 12.	- 22 31		17.2	FAINT BLUE STAR
LB 08559	00 42 12.	- 23 00		17.5	FAINT BLUE STAR
LB 08560	00 42 12.	- 23 27		18.3	FAINT BLUE STAR
LB 08561	00 42 12.	- 23 52		18.0	FAINT BLUE STAR
LB 08562	00 42 12.	- 25 51		18.4	FAINT BLUE STAR
RIC 547	00 42 12.07	+ 40 01 18.6		17.79	BLUE OBJECT

OBJECT NAME	RIGHT ASCEN.	DECLINATION	DIAM.	MAGN.	TYPE OF OBJECT
BAA 1.143	00 42 12.26	+ 41 15 02.7			EMISSION NEBULA IN M31
BAA 1.617	00 42 12.33	+ 41 39 02.1			EMISSION NEBULA IN M31
BAA 1.616	00 42 12.51	+ 41 38 56.7			EMISSION NEBULA IN M31
BAA 1.147	00 42 12.54	+ 41 15 18.5			EMISSION NEBULA IN M31
BAA 1.144	00 42 12.89	+ 41 15 04.7			EMISSION NEBULA IN M31
BAA 1.620	00 42 13.07	+ 41 39 10.5			EMISSION NEBULA IN M31
BAA 1.170	00 42 13.25	+ 41 24 13.2			EMISSION NEBULA IN M31
BAA 1.622	00 42 13.84	+ 41 39 13.3			EMISSION NEBULA IN M31
HOLM 018B	00 42 14.	- 18 31	60	13.4	PART OF MULTIPLE GALAXY
BAA 1.134	00 42 14.01	+ 41 12 34.5			EMISSION NEBULA IN M31
BAA 1.148	00 42 14.29	+ 41 16 10.8			EMISSION NEBULA IN M31
BAA 4.06	00 42 14.60	+ 41 38 23.9			DIFFUSE OBJECT IN M31
BAA 1.619	00 42 14.68	+ 41 38 48.9			EMISSION NEBULA IN M31
RIC 548	00 42 14.90	+ 39 21 45.2		19.02	BLUE OBJECT
BAA 1.133	00 42 15.84	+ 41 12 13.0			EMISSION NEBULA IN M31
LIN 052	00 42 16.	- 73 42 34.		14.44	SMC EMISSION LINE OBJECT
BAA 1.135	00 42 16.01	+ 41 12 18.5			EMISSION NEBULA IN M31
RIC 549	00 42 16.04	+ 40 20 07.6		18.12	BLUE OBJECT
BAA 1.146	00 42 16.18	+ 41 14 38.6			EMISSION NEBULA IN M31
RIC 550	00 42 16.43	+ 39 47 06.6		19.40	BLUE OBJECT
BAA 1.624	00 42 17.46	+ 41 39 03.8			EMISSION NEBULA IN M31
RIC 551	00 42 17.72	+ 40 15 13.8		18.30	BLUE OBJECT
RIC 552	00 42 17.98	+ 40 39 06.6		20.36	BLUE OBJECT
ZC 0042.3+0729	00 42 18.	+ 07 29	670		CLUSTER OF GALAXIES
UGC 00472	00 42 18.	+ 16 40	90	17.	GALAXY DWARF
ZWG 458.003	00 42 18.	+ 19 49		15.4	GALAXY
ZC 0042.3+3125	00 42 18.	+ 31 25	1480		CLUSTER OF GALAXIES
SHRV.73A 17	00 42 18.	+ 41 07	8	17.22	GLOBULAR CLUSTER IN M31
VB 011	00 42 18.	+ 66 40	356		STELLAR RING
MCG-02-03-005	00 42 18.	- 08 35	48	14.5	GALAXY
PHL 2922	00 42 18.	- 10 46		18.2	BLUE STELLAR OBJECT
PHL 6578	00 42 18.	- 13 54		18.9	BLUE STELLAR OBJECT
HOLM 018A	00 42 18.	- 18 31	48	13.4	PART OF MULTIPLE GALAXY
LB 08563	00 42 18.	- 21 20		18.3	FAINT BLUE STAR
LB 08564	00 42 18.	- 22 34		16.8	FAINT BLUE STAR
LB 08565	00 42 18.	- 22 56		18.1	FAINT BLUE STAR
LB 08566	00 42 18.	- 23 04		18.7	FAINT BLUE STAR
LB 08567	00 42 18.	- 23 34		18.3	FAINT BLUE STAR
LB 08568	00 42 18.	- 24 32		17.6	FAINT BLUE STAR
LB 08569	00 42 18.	- 26 25		18.2	FAINT BLUE STAR
LB 03154	00 42 18.	- 60 56		14.5	FAINT BLUE STAR
RIC 554	00 42 18.75	+ 38 49 42.2		18.26	BLUE OBJECT
RIC 553	00 42 18.84	+ 39 26 56.2		18.20	BLUE OBJECT
RIC 555	00 42 19.32	+ 40 13 36.6		18.48	BLUE OBJECT
RIC 556	00 42 19.54	+ 39 44 23.8		18.19	BLUE OBJECT
RIC 557	00 42 20.28	+ 40 14 08.8		18.84	BLUE OBJECT
RIC 558	00 42 20.46	+ 40 40 40.4		18.53	BLUE OBJECT
BAA 1.611	00 42 20.55	+ 41 34 59.3			EMISSION NEBULA IN M31
BAA 1.171	00 42 20.99	+ 41 23 17.4			EMISSION NEBULA IN M31
BAA 4.07	00 42 20.99	+ 41 42 41.0			DIFFUSE OBJECT IN M31
BAA 1.612	00 42 21.11	+ 41 34 59.7			EMISSION NEBULA IN M31
BAA 1.151	00 42 21.16	+ 41 18 02.6			EMISSION NEBULA IN M31
BAA 1.157	00 42 21.18	+ 41 19 54.1			EMISSION NEBULA IN M31
RIC 559	00 42 21.25	+ 40 19 18.6		18.08	BLUE OBJECT
RIC 560	00 42 21.30	+ 39 20 26.2		18.09	BLUE OBJECT
BAA 1.152	00 42 21.66	+ 41 18 02.3			EMISSION NEBULA IN M31
RIC 561	00 42 22.19	+ 40 02 53.6		18.19	BLUE OBJECT
BAA 1.153	00 42 22.87	+ 41 18 00.5			EMISSION NEBULA IN M31
RIC 562	00 42 23.14	+ 42 03 15.0		20.39	BLUE OBJECT
PHL 6579	00 42 24.	+ 01 28		18.5	BLUE STELLAR OBJECT
PHL 2925	00 42 24.	+ 03 10		18.0	BLUE STELLAR OBJECT
PHL 2924	00 42 24.	+ 13 29		18.0	BLUE STELLAR OBJECT
ZC 0042.4+2437	00 42 24.	+ 24 37	1880		CLUSTER OF GALAXIES
ZWG 479.068	00 42 24.	+ 25 30		15.6	GALAXY
VDB.66B 11	00 42 24.	+ 39 21 04.		17.74	BLUE OBJECT
VB 012	00 42 24.	+ 64 00	443		STELLAR RING
PHL 0820	00 42 24.	- 12 45		15.7	BLUE STELLAR OBJECT
PHL 2923	00 42 24.	- 14 12		17.8	BLUE STELLAR OBJECT
LB 08570	00 42 24.	- 22 04		18.0	FAINT BLUE STAR
LB 08571	00 42 24.	- 23 04		19.3	FAINT BLUE STAR
PHL 2926	00 42 24.	- 24 13		18.8	BLUE STELLAR OBJECT
PHL 6580	00 42 24.	- 28 00		18.2	BLUE STELLAR OBJECT
BAA 1.172	00 42 24.19	+ 41 23 58.9			EMISSION NEBULA IN M31
BAA 1.159	00 42 24.62	+ 41 20 30.9			EMISSION NEBULA IN M31
LIN 053	00 42 25.	- 72 35 04.		14.80	SMC EMISSION LINE OBJECT
BAA 1.630	00 42 25.03	+ 41 45 20.9			EMISSION NEBULA IN M31
BAA 1.158	00 42 25.05	+ 41 19 36.2			EMISSION NEBULA IN M31
BAA 1.156	00 42 25.26	+ 41 18 51.6			EMISSION NEBULA IN M31
BAA 1.637	00 42 25.35	+ 41 46 04.7			EMISSION NEBULA IN M31
BAA 1.154	00 42 25.47	+ 41 18 03.2			EMISSION NEBULA IN M31
BAA 1.629	00 42 25.65	+ 41 43 48.8			EMISSION NEBULA IN M31
BAA 1.155	00 42 26.06	+ 41 18 36.4			EMISSION NEBULA IN M31
BAA 1.161	00 42 26.25	+ 41 20 22.7			EMISSION NEBULA IN M31
BAA 1.164	00 42 26.74	+ 41 20 54.6			EMISSION NEBULA IN M31
BAA 1.169	00 42 26.93	+ 41 22 05.4			EMISSION NEBULA IN M31
BAA 1.162	00 42 26.97	+ 41 20 50.2			EMISSION NEBULA IN M31
RIC 563	00 42 27.14	+ 40 43 40.5		18.81	BLUE OBJECT
BAA 1.163	00 42 27.51	+ 41 20 46.7			EMISSION NEBULA IN M31
RIC 564	00 42 27.52	+ 38 54 19.7		18.19	BLUE OBJECT
BAA 1.165	00 42 27.58	+ 41 20 54.9			EMISSION NEBULA IN M31
LIN 054	00 42 28.	- 73 45 34.		15.73	SMC EMISSION LINE OBJECT
RIC 565	00 42 28.35	+ 39 47 51.5		18.31	BLUE OBJECT
BAA 1.166	00 42 28.84	+ 41 20 45.2			EMISSION NEBULA IN M31
BAA 1.638	00 42 28.85	+ 41 47 54.3			EMISSION NEBULA IN M31
BAA 1.160	00 42 29.06	+ 41 19 53.0			EMISSION NEBULA IN M31
BAA 1.167	00 42 29.41	+ 41 21 02.4			EMISSION NEBULA IN M31
BAA 3.23	00 42 29.46	+ 41 41 17.4			GLOBULAR CLUSTER IN M31
RIC 566	00 42 29.88	+ 40 57 40.4		19.37	BLUE OBJECT
MCG+01-03-001	00 42 30.	+ 05 51	42	13.	GALAXY
ZWG 410.003	00 42 30.	+ 05 52		14.8	GALAXY
UGC 00473	00 42 30.	+ 05 52	96	14.8	GALAXY S0-a
ZWG 410.004	00 42 30.	+ 08 12		15.6	GALAXY
ZC 0042.5+0824	00 42 30.	+ 08 24	1280		CLUSTER OF GALAXIES
PHL 6581	00 42 30.	+ 10 17		13.8	BLUE STELLAR OBJECT
HUB E01	00 42 30.	+ 55 16			DIFFUSE NEBULA
HUB E02	00 42 30.	+ 59 46			DIFFUSE NEBULA
OCL 0307	00 42 30.	+ 64 08	300	18.	OPEN STAR CLUSTER
VB 013	00 42 30.	+ 64 18	316		STELLAR RING
HELW 089	00 42 30.	- 08 36 40.			NEBULA
HOLM 019A	00 42 30.	- 16 47	48	13.6	PART OF MULTIPLE GALAXY
PHL 6582	00 42 30.	- 22 10		18.7	BLUE STELLAR OBJECT
RIC 567	00 42 30.01	+ 39 03 32.9		18.03	BLUE OBJECT
BAA 1.177	00 42 30.03	+ 41 24 07.5			EMISSION NEBULA IN M31
BAA 1.639	00 42 30.24	+ 41 47 56.7			EMISSION NEBULA IN M31
BAA 1.168	00 42 30.53	+ 41 21 31.8			EMISSION NEBULA IN M31
RIC 568	00 42 30.77	+ 39 56 06.6		19.73	BLUE OBJECT
RNGC 0240	00 42 31.	+ 05 52		15.0	GALAXY
BAA 1.189	00 42 31.08	+ 41 33 29.5			EMISSION NEBULA IN M31

OBJECT NAME	RIGHT ASCEN.	DECLINATION	DIAM.	MAGN.	TYPE OF OBJECT
RIC 569	00 42 31.58	+ 39 16 55.4		19.48	BLUE OBJECT
RIC 570	00 42 31.74	+ 40 05 51.0		18.31	BLUE OBJECT
RIC 571	00 42 32.47	+ 39 42 57.0		18.55	BLUE OBJECT
BAA 1.173	00 42 32.50	+ 41 22 51.2			EMISSION NEBULA IN M31
BAA 1.196	00 42 32.93	+ 41 35 46.6			EMISSION NEBULA IN M31
VDB.66B 21	00 42 33.	+ 40 06 04.		18.86	BLUE OBJECT
BAA 1.176	00 42 33.13	+ 41 23 18.2			EMISSION NEBULA IN M31
BAA 1.175	00 42 33.28	+ 41 22 59.3			EMISSION NEBULA IN M31
RIC 572	00 42 33.38	+ 40 38 09.2		19.04	BLUE OBJECT
RIC 573	00 42 33.48	+ 40 06 57.2		18.35	BLUE OBJECT
BAA 1.178	00 42 33.54	+ 41 23 51.4			EMISSION NEBULA IN M31
BAA 2.06	00 42 33.56	+ 41 34 59.9			GALAXY
BAA 4.08	00 42 33.67	+ 41 41 19.1			DIFFUSE OBJECT IN M31
BAA 1.174	00 42 33.89	+ 41 22 44.5			EMISSION NEBULA IN M31
RIC 574	00 42 35.62	+ 40 39 07.2		19.08	BLUE OBJECT
PHL 6584	00 42 36.	+ 05 36		18.0	BLUE STELLAR OBJECT
PHL 2927	00 42 36.	- 02 29		17.2	BLUE STELLAR OBJECT
PHL 6585	00 42 36.	- 06 25		18.4	BLUE STELLAR OBJECT
PHL 6583	00 42 36.	- 12 24		18.3	BLUE STELLAR OBJECT
PHL 0821	00 42 36.	- 15 45		18.5	BLUE STELLAR OBJECT
LB 08572	00 42 36.	- 23 43		18.5	FAINT BLUE STAR
HOLM 019B	00 42 37.	- 16 48	24	14.4	PART OF MULTIPLE GALAXY
BAA 1.183	00 42 37.19	+ 41 27 07.0			EMISSION NEBULA IN M31
BAA 1.179	00 42 37.68	+ 41 24 35.1			EMISSION NEBULA IN M31
RIC 575	00 42 38.71	+ 39 45 04.0		18.61	BLUE OBJECT
BAA 1.180	00 42 38.75	+ 41 24 31.3			EMISSION NEBULA IN M31
BAA 1.181	00 42 39.00	+ 41 24 31.9			EMISSION NEBULA IN M31
BAA 1.626	00 42 39.08	+ 41 37 56.4			EMISSION NEBULA IN M31
RIC 576	00 42 39.26	+ 39 23 39.4		19.76	BLUE OBJECT
LIN 055	00 42 40.	- 73 30 34.		14.98	SMC EMISSION LINE OBJECT
BAA 1.182	00 42 40.15	+ 41 26 25.6			EMISSION NEBULA IN M31
BAA 1.649	00 42 40.39	+ 41 58 01.8			EMISSION NEBULA IN M31
RIC 577	00 42 40.86	+ 40 02 37.6		19.22	BLUE OBJECT
RIC 578	00 42 40.96	+ 40 01 50.4		19.08	BLUE OBJECT
BAA 4.09	00 42 41.06	+ 41 46 39.1			DIFFUSE OBJECT IN M31
BAA 4.10	00 42 41.12	+ 41 46 38.3			DIFFUSE OBJECT IN M31
BAA 4.11	00 42 41.40	+ 41 40 15.2			DIFFUSE OBJECT IN M31
RIC 579	00 42 41.95	+ 39 52 52.8		18.62	BLUE OBJECT
ZC 0042.7+0951	00 42 42.	+ 09 51	940		CLUSTER OF GALAXIES
UGC 00474	00 42 42.	+ 10 15	60	16.0	GALAXY Sc
PK122-04.1	00 42 42.	+ 57 41	36	16.3	PLANETARY NEBULA
PHL 6586	00 42 42.	- 01 02		17.6	BLUE STELLAR OBJECT
PHL 2928	00 42 42.	- 14 17		18.0	BLUE STELLAR OBJECT
PHL 6587	00 42 42.	- 14 42		18.6	BLUE STELLAR OBJECT
PHL 2929	00 42 42.	- 19 58		18.6	BLUE STELLAR OBJECT
LB 08573	00 42 42.	- 20 54		19.2	FAINT BLUE STAR
LB 08574	00 42 42.	- 24 18		17.5	FAINT BLUE STAR
LB 08575	00 42 42.	- 25 22		19.7	FAINT BLUE STAR
LB 08576	00 42 42.	- 26 04		17.6	FAINT BLUE STAR
RIC 580	00 42 42.24	+ 40 53 09.4		19.02	BLUE OBJECT
RIC 581	00 42 42.34	+ 39 12 00.3		18.42	BLUE OBJECT
RIC 582	00 42 42.44	+ 40 33 33.6		18.46	BLUE OBJECT
BAA 4.12	00 42 42.86	+ 41 50 40.5			DIFFUSE OBJECT IN M31
BAA 4.13	00 42 42.96	+ 41 50 39.7			DIFFUSE OBJECT IN M31
BAA 1.635	00 42 43.08	+ 41 43 06.9			EMISSION NEBULA IN M31
BAA 1.194	00 42 43.17	+ 41 33 38.0			EMISSION NEBULA IN M31
BAA 1.187	00 42 43.25	+ 41 30 08.6			EMISSION NEBULA IN M31
BAA 1.641	00 42 43.39	+ 41 50 27.8			EMISSION NEBULA IN M31
BAA 4.14	00 42 43.44	+ 41 33 29.9			DIFFUSE OBJECT IN M31
BAA 1.627	00 42 43.72	+ 41 40 53.6			EMISSION NEBULA IN M31
BAA 1.657	00 42 43.78	+ 42 01 23.9			EMISSION NEBULA IN M31
RIC 583	00 42 44.75	+ 40 13 55.4		18.45	BLUE OBJECT
RIC 584	00 42 44.84	+ 39 01 14.1		19.72	BLUE OBJECT
ARC 0094	00 42 45.	- 02 37		17.2	RICH CLUSTER OF GALAXIES
BAA 1.188	00 42 45.54	+ 41 30 17.5			EMISSION NEBULA IN M31
BAA 1.642	00 42 45.57	+ 41 50 35.5			EMISSION NEBULA IN M31
BAA 1.633	00 42 46.17	+ 41 42 33.1			EMISSION NEBULA IN M31
RIC 585	00 42 46.49	+ 40 04 20.4		18.72	BLUE OBJECT
RIC 586	00 42 46.62	+ 39 10 48.4		20.19	BLUE OBJECT
BAA 1.195	00 42 46.94	+ 41 33 28.3			EMISSION NEBULA IN M31
HELW 003	00 42 47.	- 21 01			NEBULA
BAA 4.15	00 42 47.05	+ 41 33 08.8			DIFFUSE OBJECT IN M31
BAA 1.636	00 42 47.91	+ 41 42 34.9			EMISSION NEBULA IN M31
PHL 2930	00 42 48.	- 03 31		17.8	BLUE STELLAR OBJECT
MCG-01-03-008	00 42 48.	- 03 57 30.	36	15.	GALAXY
MCG-02-03-006	00 42 48.	- 09 53	48	14.	GALAXY
PHL 6588	00 42 48.	- 15 38		16.8	BLUE STELLAR OBJECT
PHL 6590	00 42 48.	- 16 10		18.7	BLUE STELLAR OBJECT
PHL 2931	00 42 48.	- 18 34		18.4	BLUE STELLAR OBJECT
LB 08577	00 42 48.	- 21 11		18.8	FAINT BLUE STAR
LB 08578	00 42 48.	- 21 49		19.7	FAINT BLUE STAR
LB 08579	00 42 48.	- 22 01		17.8	FAINT BLUE STAR
LB 08580	00 42 48.	- 23 02		18.0	FAINT BLUE STAR
PHL 6589	00 42 48.	- 23 34		18.6	BLUE STELLAR OBJECT
LB 08581	00 42 48.	- 24 10		18.4	FAINT BLUE STAR
LB 01572	00 42 48.	- 51 40		13.6	FAINT BLUE STAR
LIN 056	00 42 48.	- 72 54 10.		18.0	PLANETARY NEBULA IN SMC
RIC 587	00 42 48.05	+ 39 07 16.3		19.15	BLUE OBJECT
BAA 3.26	00 42 48.30	+ 41 44 01.5			GLOBULAR CLUSTER IN M31
BAA 1.191	00 42 48.54	+ 41 31 16.2			EMISSION NEBULA IN M31
BAA 1.190	00 42 48.59	+ 41 31 11.6			EMISSION NEBULA IN M31
BAA 1.631	00 42 49.50	+ 41 41 47.9			EMISSION NEBULA IN M31
BAA 1.632	00 42 50.24	+ 41 41 49.6			EMISSION NEBULA IN M31
RIC 588	00 42 50.38	+ 40 53 00.0		19.70	BLUE OBJECT
BAA 1.184	00 42 51.63	+ 41 26 30.3			EMISSION NEBULA IN M31
RIC 589	00 42 51.94	+ 39 10 16.6		18.89	BLUE OBJECT
BAA 1.214	00 42 51.94	+ 41 40 39.2			EMISSION NEBULA IN M31
BAA 1.185	00 42 52.09	+ 41 26 42.0			EMISSION NEBULA IN M31
BAA 1.215	00 42 52.09	+ 41 40 47.9			EMISSION NEBULA IN M31
BAA 1.199	00 42 52.16	+ 41 34 48.0			EMISSION NEBULA IN M31
BAA 1.198	00 42 52.22	+ 41 34 43.3			EMISSION NEBULA IN M31
BAA 1.634	00 42 52.23	+ 41 41 40.5			EMISSION NEBULA IN M31
RIC 590	00 42 52.34	+ 39 24 22.4		19.45	BLUE OBJECT
BAA 1.205	00 42 52.58	+ 41 38 01.8			EMISSION NEBULA IN M31
RIC 591	00 42 52.80	+ 40 14 49.8		18.16	BLUE OBJECT
BAA 1.200	00 42 53.59	+ 41 36 11.2			EMISSION NEBULA IN M31
BAA 1.192	00 42 53.68	+ 41 31 36.1			EMISSION NEBULA IN M31
ZWG 535.019	00 42 54.	+ 44 46		15.7	GALAXY
LB 08582	00 42 54.	- 25 51		20.0	FAINT BLUE STAR
PHL 6592	00 42 54.	- 27 00		18.1	BLUE STELLAR OBJECT
PHL 6591	00 42 54.	- 27 36		17.9	BLUE STELLAR OBJECT
RIC 592	00 42 54.66	+ 40 30 31.8		20.04	BLUE OBJECT
BAA 4.17	00 42 54.77	+ 41 36 24.6			DIFFUSE OBJECT IN M31
BAA 4.16	00 42 54.81	+ 41 42 16.8			DIFFUSE OBJECT IN M31
BAA 1.207	00 42 55.73	+ 41 38 26.0			EMISSION NEBULA IN M31
BAA 4.18	00 42 56.61	+ 41 42 49.8			DIFFUSE OBJECT IN M31
BAA 1.652	00 42 56.85	+ 41 57 22.1			EMISSION NEBULA IN M31
BAA 3.02	00 42 56.87	+ 41 29 11.1			GLOBULAR CLUSTER IN M31

OBJECT NAME	RIGHT ASCEN.	DECLINATION	DIAM.	MAGN.	TYPE OF OBJECT
BAA 1.210	00 42 57.16	+ 41 39 23.4			EMISSION NEBULA IN M31
RIC 593	00 42 57.77	+ 40 03 26.7		18.72	BLUE OBJECT
LIN 057	00 42 58.	- 73 30 41.		14.18	SMC EMISSION LINE OBJECT
BAA 1.204	00 42 58.23	+ 41 36 37.7			EMISSION NEBULA IN M31
RIC 594	00 42 58.32	+ 39 42 50.3		18.27	BLUE OBJECT
RIC 595	00 42 58.47	+ 41 04 07.1		19.91	BLUE OBJECT
BAA 1.212	00 42 58.47	+ 41 39 21.1			EMISSION NEBULA IN M31
BAA 1.186	00 42 58.57	+ 41 26 14.3			EMISSION NEBULA IN M31
BAA 1.201	00 42 58.80	+ 41 35 47.4			EMISSION NEBULA IN M31
RIC 596	00 42 58.89	+ 40 07 23.2		19.23	BLUE OBJECT
BAA 4.19	00 42 58.96	+ 41 38 04.7			DIFFUSE OBJECT IN M31
BAA 4.20	00 42 59.17	+ 41 41 04.9			DIFFUSE OBJECT IN M31
BAA 1.202	00 42 59.18	+ 41 35 44.2			EMISSION NEBULA IN M31
BAA 3.04	00 42 59.60	+ 41 35 36.6			GLOBULAR CLUSTER IN M31
BAA 1.219	00 42 59.60	+ 41 41 04.1			EMISSION NEBULA IN M31
LBN 0605	00 43	+ 55 30	7500		BRIGHT NEBULA
VDB.66G 005	00 43	- 11 49	100		DWARF GALAXY
LB 09773	00 43	- 82 32		13.6	FAINT BLUE STAR
PHL 6597	00 43 00.	+ 03 39		18.2	BLUE STELLAR OBJECT
ZC 0043.0+1543	00 43 00.	+ 15 43	670		CLUSTER OF GALAXIES
ZWG 500.081	00 43 00.	+ 29 33		15.7	GALAXY
AND 1	00 43 00.	+ 37 44	150		DWARF SPHEROIDAL GALAXY
SHRV.73B 23	00 43 00.	+ 39 24	8	15.89	COMPACT GALAXY
MCG+08-02-012	00 43 00.	+ 49 23	42	15.	GALAXY
ZWG 550.012	00 43 00.	+ 49 24		15.1	GALAXY
MCG+00-03-004	00 43 00.	- 01 23	42	15.5	GALAXY
MCG-02-03-007	00 43 00.	- 12 48	18	15.5	GALAXY
PHL 6598	00 43 00.	- 15 06		18.8	BLUE STELLAR OBJECT
PHL 6594	00 43 00.	- 16 49		18.7	BLUE STELLAR OBJECT
PHL 6595	00 43 00.	- 18 16		18.5	BLUE STELLAR OBJECT
PHL 0822	00 43 00.	- 20 44		17.0	BLUE STELLAR OBJECT
LB 08583	00 43 00.	- 23 39		20.0	FAINT BLUE STAR
PHL 6596	00 43 00.	- 23 40		18.6	BLUE STELLAR OBJECT
PHL 0823	00 43 00.	- 23 54		17.2	BLUE STELLAR OBJECT
PHL 6593	00 43 00.	- 32 16		16.6	BLUE STELLAR OBJECT
BAA 3.03	00 43 00.47	+ 41 29 29.5			GLOBULAR CLUSTER IN M31
RIC 597	00 43 00.48	+ 40 53 56.3		19.56	BLUE OBJECT
RIC 598	00 43 00.89	+ 40 54 03.1		20.27	BLUE OBJECT
LIN 059	00 43 04.	- 73 39 41.		13.93	SMC EMISSION LINE OBJECT
BAA 1.644	00 43 04.04	+ 41 52 31.8			EMISSION NEBULA IN M31
BAA 1.203	00 43 04.15	+ 41 35 12.4			EMISSION NEBULA IN M31
RIC 599	00 43 04.18	+ 40 55 38.5		20.47	BLUE OBJECT
RIC 600	00 43 04.20	+ 41 04 30.2		18.78	BLUE OBJECT
HELW 004	00 43 05.	- 20 37			NEBULA
LIN 058	00 43 05.	- 73 17 11.		13.71	SMC EMISSION LINE OBJECT
RIC 601	00 43 05.11	+ 39 21 41.6		18.59	BLUE OBJECT
BAA 1.645	00 43 05.15	+ 41 53 20.5			EMISSION NEBULA IN M31
RIC 602	00 43 05.35	+ 38 53 22.3		18.97	BLUE OBJECT
BAA 1.647	00 43 05.75	+ 41 53 42.2			EMISSION NEBULA IN M31
SHRV.73A 18	00 43 06.	+ 40 14	8	16.89	GLOBULAR CLUSTER IN M31
PHL 6600	00 43 06.	- 16 38		18.6	BLUE STELLAR OBJECT
PHL 6599	00 43 06.	- 26 16		16.9	BLUE STELLAR OBJECT
MCG-05-03-002	00 43 06.	- 26 50	48	15.	GALAXY
LH115-N010	00 43 06.	- 73 27	23		EMISSION NEBULA IN SMC
BAA 1.643	00 43 06.22	+ 41 52 05.0			EMISSION NEBULA IN M31
BAA 1.648	00 43 06.23	+ 41 53 41.1			EMISSION NEBULA IN M31
RIC 603	00 43 06.56	+ 39 46 18.6		17.67	BLUE OBJECT
LB 03924	00 43 07.	+ 59 07 06.		17.8	FAINT BLUE STAR
RIC 604	00 43 07.05	+ 40 06 17.4		17.92	BLUE OBJECT
VDB.66B 04	00 43 08.	+ 38 53 27.		18.79	BLUE OBJECT
BAA 1.646	00 43 08.00	+ 41 53 13.7			EMISSION NEBULA IN M31
RIC 605	00 43 08.28	+ 39 32 36.4		18.52	BLUE OBJECT
BAA 4.21	00 43 08.74	+ 41 45 55.5			DIFFUSE OBJECT IN M31
RIC 606	00 43 08.91	+ 40 05 42.8		19.70	BLUE OBJECT
RIC 607	00 43 09.45	+ 39 05 02.6		19.48	BLUE OBJECT
RIC 608	00 43 09.50	+ 40 05 23.0		17.62	BLUE OBJECT
HOLM 020B	00 43 10.	- 09 38	24	15.3	PART OF MULTIPLE GALAXY
HN 0118	00 43 10.	- 23 51			NEBULA
IC 1579	00 43 10.	- 26 50			NONSTELLAR OBJECT
LIN 061	00 43 10.	- 73 17 47.		14.40	SMC EMISSION LINE OBJECT
LIN 060	00 43 10.	- 73 26 47.		15.19	PLANETARY NEBULA IN SMC
LIN 062	00 43 10.	- 73 33 17.		14.36	PLANETARY NEBULA IN SMC
BAA 4.22	00 43 10.13	+ 41 40 24.1			DIFFUSE OBJECT IN M31
RIC 609	00 43 10.92	+ 40 07 26.8		18.76	BLUE OBJECT
BAA 1.650	00 43 11.25	+ 41 54 48.9			EMISSION NEBULA IN M31
RIC 610	00 43 11.88	+ 39 24 42.0		17.82	BLUE OBJECT
PHL 6601	00 43 12.	+ 10 06		17.1	BLUE STELLAR OBJECT
ZWG 479.069	00 43 12.	+ 24 59		15.4	GALAXY
ZWG 519.020	00 43 12.	+ 35 41		15.7	GALAXY
VDB.66B 06	00 43 12.	+ 39 06 03.		19.51	BLUE OBJECT
SHRV.73A 19	00 43 12.	+ 40 26	8	16.12	GLOBULAR CLUSTER IN M31
ZWG 535.020	00 43 12.	+ 45 12		15.7	GALAXY
PHL 2932	00 43 12.	- 18 32		18.2	BLUE STELLAR OBJECT
PHL 2933	00 43 12.	- 19 38		18.7	BLUE STELLAR OBJECT
PHL 6602	00 43 12.	- 20 38		17.8	BLUE STELLAR OBJECT
PHL 2934	00 43 12.	- 23 01		17.0	BLUE STELLAR OBJECT
TON-S 0169	00 43 12.	- 23 54		15.7	BLUE STAR
LB 08584	00 43 12.	- 24 58		18.6	FAINT BLUE STAR
LB 08585	00 43 12.	- 25 40		18.8	FAINT BLUE STAR
TON-S 0168	00 43 12.	- 26 13		15.9	BLUE STAR
LH115-N011	00 43 12.	- 73 33	24		EMISSION NEBULA IN SMC
HOW 20	00 43 12.	- 74 38 12.	24		STAR CLUSTER IN SMC
BAA 3.05	00 43 12.11	+ 41 37 08.5			GLOBULAR CLUSTER IN M31
RIC 611	00 43 12.20	+ 41 04 44.6		18.42	BLUE OBJECT
RIC 612	00 43 12.51	+ 39 50 09.9		20.50	BLUE OBJECT
RIC 613	00 43 12.91	+ 39 06 35.0		18.57	BLUE OBJECT
BAA 1.659	00 43 13.11	+ 41 57 51.5			EMISSION NEBULA IN M31
HOLM 020A	00 43 14.	- 09 37	30	13.6	PART OF MULTIPLE GALAXY
RIC 614	00 43 14.03	+ 40 13 51.4		19.71	BLUE OBJECT
RIC 615	00 43 14.90	+ 39 19 10.6		18.59	BLUE OBJECT
MCG+04-02-053	00 43 15.	+ 25 00	45	15.	GALAXY
ARC 0095	00 43 15.	- 01 09		17.2	RICH CLUSTER OF GALAXIES
RNGC 0244	00 43 15.	- 15 51		13.0	GALAXY
HARO 14	00 43 15.	- 15 51			BLUE EMISSION-LINE GALAXY
BAA 1.193	00 43 15.41	+ 41 28 25.6			EMISSION NEBULA IN M31
RIC 616	00 43 15.46	+ 39 51 09.6		18.47	BLUE OBJECT
BAA 1.651	00 43 15.54	+ 41 54 13.8			EMISSION NEBULA IN M31
RIC 617	00 43 15.86	+ 39 18 17.6		19.54	BLUE OBJECT
RNGC 0243	00 43 16.	+ 29 40		14.5	GALAXY
LIN 064	00 43 16.	- 73 31 05.		14.41	SMC EMISSION LINE OBJECT
RIC 618	00 43 16.82	+ 39 17 09.6		20.86	BLUE OBJECT
LIN 063	00 43 17.	- 72 58 53.		13.52	SMC EMISSION LINE OBJECT
BAA 1.208	00 43 17.39	+ 41 36 15.2			EMISSION NEBULA IN M31
RIC 619	00 43 17.63	+ 39 16 01.2		19.35	BLUE OBJECT
PHL 6604	00 43 18.	+ 10 12		18.5	BLUE STELLAR OBJECT
PHL 6603	00 43 18.	+ 10 38		18.3	BLUE STELLAR OBJECT
ZWG 479.070	00 43 18.	+ 24 56		15.1	GALAXY
MCG+05-02-043	00 43 18.	+ 29 40	42	15.	GALAXY
ZWG 501.001	00 43 18.	+ 29 41		14.6	GALAXY
ZWG 500.082	00 43 18.	+ 29 41		14.6	GALAXY
SHRV.73B 24	00 43 18.	+ 39 19	8	16.26	COMPACT GALAXY
ZC 0043.3+3920	00 43 18.	+ 39 20	2020		CLUSTER OF GALAXIES
MCG+08-02-013	00 43 18.	+ 49 12	54	16.	GALAXY
MCG+08-02-014	00 43 18.	+ 50 57	60	15.	GALAXY
PHL 6606	00 43 18.	- 02 47		18.4	BLUE STELLAR OBJECT
MCG-01-03-009	00 43 18.	- 04 07	42	15.	GALAXY
PHL 2935	00 43 18.	- 11 06		18.5	BLUE STELLAR OBJECT
PHL 6618	00 43 18.	- 13 36		18.5	BLUE STELLAR OBJECT
MCG-03-03-003	00 43 18.	- 15 53	54	13.	GALAXY
PHL 6605	00 43 18.	- 31 40		12.0	BLUE STELLAR OBJECT
RIC 620	00 43 18.10	+ 39 52 06.4		19.85	BLUE OBJECT
BAA 1.197	00 43 18.13	+ 41 29 54.6			EMISSION NEBULA IN M31
BAA 1.209	00 43 18.19	+ 41 36 15.2			EMISSION NEBULA IN M31
BAA 1.211	00 43 18.92	+ 41 36 14.0			EMISSION NEBULA IN M31
RIC 621	00 43 19.71	+ 41 04 09.6		17.95	BLUE OBJECT
SVEN 030	00 43 20.	- 31 31	36	15.7	GALAXY
KEEL 008	00 43 20.0	- 21 09 23.		18.	NEBULA
BAA 4.23	00 43 20.93	+ 41 49 43.3			DIFFUSE OBJECT IN M31
BAA 3.06	00 43 20.96	+ 41 44 30.5			GLOBULAR CLUSTER IN M31
SVEN 031	00 43 23.	- 20 54	36	13.6	GALAXY
BAA 1.655	00 43 23.13	+ 41 55 10.1			EMISSION NEBULA IN M31
BAA 1.654	00 43 23.27	+ 41 55 02.9			EMISSION NEBULA IN M31
PHL 2936	00 43 24.	+ 01 06		18.2	BLUE STELLAR OBJECT
PHL 6610	00 43 24.	+ 10 52		18.2	BLUE STELLAR OBJECT
PHL 6609	00 43 24.	+ 14 40		18.6	BLUE STELLAR OBJECT
ZWG 550.013	00 43 24.	+ 49 13		15.3	GALAXY
ZWG 550.014	00 43 24.	+ 50 57		14.9	GALAXY
UGC 00475	00 43 24.	+ 50 57	102	14.9	GALAXY Sc
PHL 6608	00 43 24.	- 00 31		18.3	BLUE STELLAR OBJECT
MCG-01-03-010	00 43 24.	- 07 32	78	13.	GALAXY
PHL 6607	00 43 24.	- 08 50		16.6	BLUE STELLAR OBJECT
PHL 2937	00 43 24.	- 13 20		18.1	BLUE STELLAR OBJECT
ARC 0097	00 43 24.	- 23 27		17.1	RICH CLUSTER OF GALAXIES
MCG-05-03-003	00 43 24.	- 31 30	36	15.	GALAXY
RIC 622	00 43 24.96	+ 39 27 21.2		18.01	BLUE OBJECT
BAA 1.656	00 43 25.18	+ 41 55 00.1			EMISSION NEBULA IN M31
BAA 1.660	00 43 25.29	+ 41 56 23.2			EMISSION NEBULA IN M31
LIN 065	00 43 26.	- 71 48 35.		13.84	SMC EMISSION LINE OBJECT
BAA 1.683	00 43 26.02	+ 42 06 49.9			EMISSION NEBULA IN M31
RIC 623	00 43 26.11	+ 39 49 17.6		18.52	BLUE OBJECT
RIC 624	00 43 26.42	+ 39 54 40.2		19.70	BLUE OBJECT
RIC 625	00 43 26.82	+ 38 51 45.2		19.62	BLUE OBJECT
VDB.66B 17	00 43 27.	+ 39 49 14.		18.90	BLUE OBJECT
LIN 066	00 43 27.	- 73 41 05.		17.21	SMC EMISSION LINE OBJECT
BAA 4.24	00 43 27.57	+ 41 52 44.2			DIFFUSE OBJECT IN M31
BAA 1.216	00 43 28.03	+ 41 36 02.7			EMISSION NEBULA IN M31
BAA 1.206	00 43 28.51	+ 41 33 07.4			EMISSION NEBULA IN M31
BAA 1.217	00 43 28.54	+ 41 36 00.2			EMISSION NEBULA IN M31
BAA 1.658	00 43 28.89	+ 41 54 59.3			EMISSION NEBULA IN M31
KEEL 009	00 43 29.0	- 20 52 52.		16.	NEBULA
RIC 626	00 43 29.00	+ 39 34 21.1		19.89	BLUE OBJECT
BAA 1.226	00 43 29.25	+ 41 45 29.8			EMISSION NEBULA IN M31
RIC 627	00 43 29.42	+ 40 33 07.0		20.11	BLUE OBJECT
RIC 628	00 43 29.51	+ 39 55 59.8		18.49	BLUE OBJECT
VDB.66B 01	00 43 30.	+ 38 53 26.		19.51	BLUE OBJECT
ZWG 384.004	00 43 30.	- 01 59		12.9	GALAXY
MRK 555	00 43 30.	- 01 59	10	14.5	GALAXY WITH UV CONTINUUM
UGC 00476	00 43 30.	- 01 59	84	12.9	GALAXY S
MCG+00-03-005	00 43 30.	- 02 00	66	12.9	GALAXY
MCG-02-03-008	00 43 30.	- 09 35	42	14.5	GALAXY
PHL 2938	00 43 30.	- 17 48		18.7	BLUE STELLAR OBJECT
RIC 629	00 43 30.40	+ 40 14 52.0		19.21	BLUE OBJECT
RNGC 0245	00 43 31.	- 02 00		13.0	GALAXY
BAA 1.684	00 43 31.63	+ 42 08 33.0			EMISSION NEBULA IN M31
RIC 630	00 43 32.01	+ 40 30 16.7		19.99	BLUE OBJECT
BAA 1.218	00 43 32.36	+ 41 25 36.3			EMISSION NEBULA IN M31
RIC 631	00 43 32.82	+ 40 55 58.7		18.63	BLUE OBJECT
LIN 067	00 43 33.	- 73 39 41.		13.75	SMC EMISSION LINE OBJECT
RIC 632	00 43 33.02	+ 38 56 34.4		18.46	BLUE OBJECT
RIC 633	00 43 33.46	+ 40 48 54.6		20.89	BLUE OBJECT
RNGC 0249	00 43 34.	- 73 21			DIFFUSE NEBULA IN SMC
BAA 1.213	00 43 34.35	+ 41 34 08.3			EMISSION NEBULA IN M31
RNGC 0248	00 43 35.	- 73 39			DIFFUSE NEBULA IN SMC
BAA 1.221	00 43 35.04	+ 41 39 54.4			EMISSION NEBULA IN M31
RIC 634	00 43 35.20	+ 39 15 19.8		20.51	BLUE OBJECT
RIC 635	00 43 35.30	+ 39 54 56.0		18.08	BLUE OBJECT
RIC 636	00 43 35.42	+ 40 41 15.0		19.75	BLUE OBJECT
RIC 637	00 43 35.54	+ 39 56 49.6		18.08	BLUE OBJECT
BAA 1.662	00 43 35.69	+ 41 56 07.0			EMISSION NEBULA IN M31
PHL 6612	00 43 36.	+ 00 48		18.3	BLUE STELLAR OBJECT
PHL 2939	00 43 36.	+ 13 23		18.0	BLUE STELLAR OBJECT
MCG+03-03-002	00 43 36.	+ 19 12	180	14.5	GALAXY
ZWG 458.004	00 43 36.	+ 19 13		15.4	GALAXY
UGC 00477	00 43 36.	+ 19 13	210	15.4	GALAXY
KARA.72 015A	00 43 36.	+ 19 13	198	15.4	PART OF DOUBLE GALAXY
MCG+05-02-045	00 43 36.	+ 29 20	48	15.	GALAXY
MCG+05-02-044	00 43 36.	+ 29 57	66	15.	GALAXY
ZC 0043.6-0013	00 43 36.	- 00 13	2420		CLUSTER OF GALAXIES
IC 0050	00 43 36.	- 09 46 19.			NONSTELLAR OBJECT
MCG-02-03-009	00 43 36.	- 11 47	120	14.	GALAXY
PHL 2941	00 43 36.	- 17 31		18.1	BLUE STELLAR OBJECT
PHL 6611	00 43 36.	- 17 32		18.3	BLUE STELLAR OBJECT
PHL 2940	00 43 36.	- 18 04		18.6	BLUE STELLAR OBJECT
MCG-04-03-002	00 43 36.	- 21 54	48	15.	GALAXY
PHL 2942	00 43 36.	- 23 25		18.2	BLUE STELLAR OBJECT
LH115-N012B	00 43 36.	- 73 21	124		EMISSION NEBULA IN SMC
LH115-N013B	00 43 36.	- 73 39	59		EMISSION NEBULA IN SMC
LH115-N013A	00 43 36.	- 73 39	35		EMISSION NEBULA IN SMC
BAA 1.653	00 43 36.83	+ 41 52 15.8			EMISSION NEBULA IN M31
RIC 638	00 43 36.84	+ 39 12 43.7		17.84	BLUE OBJECT
VDB.66B 08	00 43 37.	+ 39 16 14.		19.61	BLUE OBJECT
IC 1580	00 43 38.	+ 29 39			NONSTELLAR OBJECT
VDB.66B 07	00 43 38.	+ 39 13 50.		18.72	BLUE OBJECT
LIN 071	00 43 38.	- 73 58 53.		14.62	SMC EMISSION LINE OBJECT
LIN 069	00 43 39.	- 73 29 23.		15.53	SMC EMISSION LINE OBJECT
LIN 070	00 43 39.	- 73 46 59.		14.88	SMC EMISSION LINE OBJECT
RIC 639	00 43 39.04	+ 41 17 59.6		19.53	BLUE OBJECT
ARC 0096	00 43 40.	+ 39 14		17.4	RICH CLUSTER OF GALAXIES
IC 1581	00 43 40.	- 26 10			NONSTELLAR OBJECT
HN 0119	00 43 40.	- 26 11			NEBULA
LIN 068	00 43 40.	- 73 21 11.		13.75	SMC EMISSION LINE OBJECT
BAA 1.661	00 43 40.57	+ 41 54 58.9			EMISSION NEBULA IN M31
RIC 640	00 43 41.00	+ 40 46 39.5		18.82	BLUE OBJECT
BAA 3.08	00 43 41.68	+ 41 45 30.8			GLOBULAR CLUSTER IN M31
PHL 6614	00 43 42.	+ 11 36		18.3	BLUE STELLAR OBJECT
ZWG 501.002	00 43 42.	+ 29 21		15.7	GALAXY

OBJECT NAME	RIGHT ASCEN.	DECLINATION	DIAM.	MAGN.	TYPE OF OBJECT
ZWG 500.083	00 43 42.	+ 29 21		15.7	GALAXY
ZWG 501.003	00 43 42.	+ 29 58		14.5	GALAXY
ZWG 500.084	00 43 42.	+ 29 58		14.5	GALAXY
UGC 00478	00 43 42.	+ 29 58	96	14.5	GALAXY Sa
MCG+05-02-046	00 43 42.	+ 31 32	66	15.	GALAXY
MCG+06-02-016	00 43 42.	+ 36 03	90	13.5	GALAXY
SHRV.73B 25	00 43 42.	+ 40 47	8	16.10	COMPACT GALAXY
SHRV.73A 20	00 43 42.	+ 40 47	8	16.80	GLOBULAR CLUSTER IN M31
MCG-02-03-010	00 43 42.	- 09 46	36	14.5	GALAXY
PHL 6613	00 43 42.	- 13 40		16.7	BLUE STELLAR OBJECT
BAA 1.664	00 43 43.58	+ 41 55 24.8			EMISSION NEBULA IN M31
BAA 1.224	00 43 44.28	+ 41 41 52.2			EMISSION NEBULA IN M31
RIC 641	00 43 44.66	+ 39 14 34.9		18.88	BLUE OBJECT
BAA 1.225	00 43 44.79	+ 41 42 53.1			EMISSION NEBULA IN M31
MCG+05-02-047	00 43 45.	+ 29 27	45	14.5	GALAXY
LIN 072	00 43 45.	- 73 32 47.		15.14	SMC EMISSION LINE OBJECT
KEEL 010	00 43 45.0	- 21 13 18.		18.	NEBULA
RIC 642	00 43 45.39	+ 39 46 02.0		18.24	BLUE OBJECT
KEEL 011	00 43 45.6	- 20 58 37.		18.	NEBULA
BAA 1.231	00 43 45.88	+ 41 44 12.4			EMISSION NEBULA IN M31
HN 0120	00 43 46.	- 24 34			NEBULA
BAA 1.677	00 43 46.01	+ 41 57 23.6			EMISSION NEBULA IN M31
RIC 643	00 43 46.32	+ 39 57 53.9		18.58	BLUE OBJECT
BAA 1.238	00 43 46.59	+ 41 48 18.4			EMISSION NEBULA IN M31
RIC 644	00 43 46.60	+ 39 52 48.1		18.90	BLUE OBJECT
BAA 1.228	00 43 46.89	+ 41 43 22.4			EMISSION NEBULA IN M31
BAA 1.678	00 43 46.91	+ 41 57 16.2			EMISSION NEBULA IN M31
IC 1582	00 43 47.	- 24 33			NONSTELLAR OBJECT
BAA 1.234	00 43 47.04	+ 41 47 20.5			EMISSION NEBULA IN M31
BAA 1.239	00 43 47.20	+ 41 48 14.8			EMISSION NEBULA IN M31
BAA 1.676	00 43 47.42	+ 41 56 14.2			EMISSION NEBULA IN M31
BAA 1.669	00 43 47.76	+ 41 55 16.8			EMISSION NEBULA IN M31
BAA 1.223	00 43 47.79	+ 41 40 41.2			EMISSION NEBULA IN M31
PHL 0824	00 43 48.	+ 01 30		18.6	BLUE STELLAR OBJECT
PHL 6615	00 43 48.	+ 11 38		17.6	BLUE STELLAR OBJECT
ARC 0098	00 43 48.	+ 20 13		16.9	RICH CLUSTER OF GALAXIES
ZWG 480.001	00 43 48.	+ 25 21		15.6	GALAXY
ZWG 479.071	00 43 48.	+ 25 21		15.6	GALAXY
ZWG 501.004	00 43 48.	+ 29 27		15.0	GALAXY
ZWG 500.085	00 43 48.	+ 29 27		15.0	GALAXY
ZWG 501.005	00 43 48.	+ 31 32		15.3	GALAXY
ZWG 500.086	00 43 48.	+ 31 32		15.3	GALAXY
UGC 00479	00 43 48.	+ 31 32	78	15.3	GALAXY SBa
ZWG 519.021	00 43 48.	+ 36 03		13.6	GALAXY
UGC 00480	00 43 48.	+ 36 03	96	13.6	GALAXY S
KARA.72 016A	00 43 48.	+ 36 03	78	13.6	PART OF DOUBLE GALAXY
MCG+06-02-017	00 43 48.	+ 36 03	42	15.	GALAXY
SHRV.73A 21	00 43 48.	+ 39 08	7	16.58	GLOBULAR CLUSTER IN M31
PHL 6617	00 43 48.	- 02 11		16.6	BLUE STELLAR OBJECT
PHL 0825	00 43 48.	- 03 17		16.5	BLUE STELLAR OBJECT
PHL 2943	00 43 48.	- 03 52		18.6	BLUE STELLAR OBJECT
ARP 230	00 43 48.	- 13 44			PECULIAR GALAXY
PHL 6616	00 43 48.	- 24 33		18.2	BLUE STELLAR OBJECT
MCG-04-03-003	00 43 48.	- 24 34	72	15.	GALAXY
HOW 21	00 43 48.	- 74 16 54.	12		STAR CLUSTER IN SMC
BAA 1.227	00 43 48.59	+ 41 42 50.9			EMISSION NEBULA IN M31
BAA 1.667	00 43 48.65	+ 41 55 04.8			EMISSION NEBULA IN M31
BAA 1.230	00 43 48.70	+ 41 43 26.6			EMISSION NEBULA IN M31
BAA 1.665	00 43 48.81	+ 41 55 01.9			EMISSION NEBULA IN M31
BAA 1.663	00 43 48.82	+ 41 54 28.1			EMISSION NEBULA IN M31
BAA 1.671	00 43 48.90	+ 41 55 22.7			EMISSION NEBULA IN M31
BAA 1.222	00 43 48.95	+ 41 40 16.2			EMISSION NEBULA IN M31
BAA 1.668	00 43 49.16	+ 41 55 03.7			EMISSION NEBULA IN M31
BAA 1.666	00 43 49.18	+ 41 55 00.1			EMISSION NEBULA IN M31
BAA 1.670	00 43 49.22	+ 41 55 15.9			EMISSION NEBULA IN M31
BAA 1.675	00 43 49.23	+ 41 55 52.1			EMISSION NEBULA IN M31
RIC 645	00 43 49.26	+ 39 15 14.8		20.78	BLUE OBJECT
BAA 1.672	00 43 49.41	+ 41 55 21.5			EMISSION NEBULA IN M31
BAA 1.673	00 43 49.54	+ 41 55 24.6			EMISSION NEBULA IN M31
RIC 646	00 43 49.59	+ 39 09 41.2		19.43	BLUE OBJECT
BAA 1.674	00 43 49.65	+ 41 55 26.4			EMISSION NEBULA IN M31
BAA 1.679	00 43 49.99	+ 41 57 13.0			EMISSION NEBULA IN M31
RIC 647	00 43 50.52	+ 41 08 19.1		20.25	BLUE OBJECT
RIC 648	00 43 50.63	+ 40 00 26.9		18.89	BLUE OBJECT
RIC 649	00 43 52.54	+ 40 28 03.0		17.73	BLUE OBJECT
IC 0051	00 43 53.	- 13 42 44.			NONSTELLAR OBJECT
BAA 1.243	00 43 53.13	+ 41 54 37.3			EMISSION NEBULA IN M31
RIC 650	00 43 53.48	+ 40 23 27.8		18.35	BLUE OBJECT
RIC 651	00 43 53.54	+ 39 46 40.1		20.77	BLUE OBJECT
PHL 6619	00 43 54.	+ 11 24		15.5	BLUE STELLAR OBJECT
ZWG 501.006	00 43 54.	+ 31 27		15.5	GALAXY
ZWG 500.087	00 43 54.	+ 31 27		15.5	GALAXY
ZWG 519.022	00 43 54.	+ 36 04		15.6	GALAXY
KARA.72 016B	00 43 54.	+ 36 04	48	15.6	PART OF DOUBLE GALAXY
MCG+06-02-018	00 43 54.	+ 37 10	90	14.5	GALAXY
PHL 2944	00 43 54.	- 01 38		18.3	BLUE STELLAR OBJECT
MCG-02-03-012	00 43 54.	- 09 33	30	15.	GALAXY
MCG-02-03-011	00 43 54.	- 13 42	60	13.	GALAXY
PHL 0826	00 43 54.	- 14 36		17.9	BLUE STELLAR OBJECT
PHL 2945	00 43 54.	- 24 58		18.5	BLUE STELLAR OBJECT
BAA 1.237	00 43 54.31	+ 41 46 45.1			EMISSION NEBULA IN M31
RIC 652	00 43 54.34	+ 39 24 28.2		19.69	BLUE OBJECT
BAA 1.681	00 43 54.97	+ 41 59 22.8			EMISSION NEBULA IN M31
RIC 653	00 43 55.56	+ 40 00 10.2		18.03	BLUE OBJECT
RIC 654	00 43 55.94	+ 40 54 24.6		19.10	BLUE OBJECT
SVEN 032	00 43 56.	- 31 08	30	16.0	GALAXY
BAA 1.246	00 43 56.02	+ 41 55 36.3			EMISSION NEBULA IN M31
RIC 655	00 43 56.04	+ 39 59 20.2		18.06	BLUE OBJECT
BAA 1.245	00 43 56.33	+ 41 55 27.5			EMISSION NEBULA IN M31
BAA 1.682	00 43 56.48	+ 41 59 24.7			EMISSION NEBULA IN M31
BAA 1.235	00 43 56.82	+ 41 45 59.7			EMISSION NEBULA IN M31
BAA 1.242	00 43 57.47	+ 41 52 14.2			EMISSION NEBULA IN M31
RIC 656	00 43 57.52	+ 40 55 21.8		18.12	BLUE OBJECT
BAA 1.244	00 43 57.57	+ 41 54 48.3			EMISSION NEBULA IN M31
RIC 657	00 43 57.68	+ 39 51 22.2		18.14	BLUE OBJECT
BAA 1.680	00 43 57.91	+ 41 56 47.4			EMISSION NEBULA IN M31
RIC 658	00 43 57.93	+ 38 58 56.6		19.67	BLUE OBJECT
LB 03925	00 43 58.	+ 59 34 42.		15.8	FAINT BLUE STAR
RIC 659	00 43 58.26	+ 39 50 35.3		18.48	BLUE OBJECT
KEEL 012	00 43 58.6	- 20 46 43.		18.	NEBULA
RIC 660	00 43 58.64	+ 41 19 53.0		20.61	BLUE OBJECT
RIC 661	00 43 59.22	+ 40 47 46.2		20.61	BLUE OBJECT
BAA 1.229	00 43 59.40	+ 41 41 46.5			EMISSION NEBULA IN M31
BAA 1.247	00 43 59.60	+ 41 56 43.9			EMISSION NEBULA IN M31
BAA 1.241	00 43 59.96	+ 41 49 44.2			EMISSION NEBULA IN M31
LBN 0607	00 44	+ 57 40	2700		BRIGHT NEBULA
UGC 00482	00 44 00.	+ 08 12	60	17.	GALAXY Sc
PHL 6620	00 44 00.	+ 13 43		8.8	BLUE STELLAR OBJECT

OBJECT NAME	RIGHT ASCEN.	DECLINATION	DIAM.	MAGN.	TYPE OF OBJECT
5ZW 034	00 44 00.	+ 46 32			COMPACT GALAXY
UGC 00481	00 44 00.	+ 83 29	72	16.0	GALAXY S
ZC 0044.0-0116	00 44 00.	- 01 16	2890		CLUSTER OF GALAXIES
PHL 6621	00 44 00.	- 22 18		18.4	BLUE STELLAR OBJECT
LB 01573	00 44 00.	- 53 50		14.7	FAINT BLUE STAR
LB 01574	00 44 00.	- 58 41		15.0	FAINT BLUE STAR
BAA 1.233	00 44 00.38	+ 41 43 31.6			EMISSION NEBULA IN M31
LIN 073	00 44 01.	- 74 09 54.		15.10	SMC EMISSION LINE OBJECT
RIC 662	00 44 01.26	+ 39 09 17.6		18.49	BLUE OBJECT
RIC 663	00 44 01.81	+ 39 35 06.4		19.90	BLUE OBJECT
LB 03926	00 44 02.	+ 59 25 36.		16.7	FAINT BLUE STAR
RIC 664	00 44 02.02	+ 39 02 52.2		19.70	BLUE OBJECT
BAA 3.29	00 44 03.60	+ 42 28 26.1			GLOBULAR CLUSTER IN M31
BAA 1.685	00 44 04.16	+ 42 11 19.9			EMISSION NEBULA IN M31
RIC 665	00 44 04.86	+ 40 44 50.8		19.21	BLUE OBJECT
RNGC 0256	00 44 05.	- 73 46		12.0	CLUSTER/NEBULOSITY IN SMC
RIC 666	00 44 05.08	+ 41 01 35.6		20.15	BLUE OBJECT
RIC 667	00 44 05.39	+ 39 19 24.2		18.15	BLUE OBJECT
RIC 668	00 44 05.66	+ 39 51 54.2		19.48	BLUE OBJECT
PHL 2946	00 44 06.	+ 13 06		13.8	BLUE STELLAR OBJECT
FEIG 008	00 44 06.	+ 13 06		14.2	FAINT BLUE STAR
PHL 2947	00 44 06.	+ 15 16		18.4	BLUE STELLAR OBJECT
ZWG 501.007	00 44 06.	+ 27 40		15.4	GALAXY
ZWG 500.088	00 44 06.	+ 27 40		15.4	GALAXY
ZWG 501.008	00 44 06.	+ 29 32		15.7	GALAXY
ZWG 500.089	00 44 06.	+ 29 32		15.7	GALAXY
VDB.66B 05	00 44 06.	+ 39 26 25.		19.33	BLUE OBJECT
PHL 2948	00 44 06.	- 03 55		18.0	BLUE STELLAR OBJECT
PHL 2950	00 44 06.	- 11 46		18.5	BLUE STELLAR OBJECT
PHL 2949	00 44 06.	- 31 38		18.0	BLUE STELLAR OBJECT
RIC 669	00 44 06.12	+ 40 53 22.4		18.11	BLUE OBJECT
RIC 670	00 44 06.12	+ 41 02 12.2		18.95	BLUE OBJECT
RIC 671	00 44 06.54	+ 41 12 49.8		19.02	BLUE OBJECT
BAA 1.240	00 44 07.61	+ 41 46 26.9			EMISSION NEBULA IN M31
RIC 672	00 44 07.96	+ 39 22 11.4		20.13	BLUE OBJECT
LIN.CL 030	00 44 08.	- 73 46 17.	270	12.0	STAR CLUSTER IN SMC
BAA 1.236	00 44 09.04	+ 41 44 27.2			EMISSION NEBULA IN M31
RIC 673	00 44 09.06	+ 39 45 58.9		18.14	BLUE OBJECT
RIC 674	00 44 10.16	+ 40 19 37.4		17.89	BLUE OBJECT
RIC 675	00 44 10.78	+ 40 49 45.2		18.98	BLUE OBJECT
RIC 676	00 44 11.87	+ 39 44 48.2		19.52	BLUE OBJECT
RIC 677	00 44 11.90	+ 40 41 38.2		20.58	BLUE OBJECT
PHL 0827	00 44 12.	+ 02 52		18.4	BLUE STELLAR OBJECT
PHL 6623	00 44 12.	+ 12 18		18.2	BLUE STELLAR OBJECT
PHL 2951	00 44 12.	+ 13 11		18.6	BLUE STELLAR OBJECT
PHL 6624	00 44 12.	+ 15 12		18.2	BLUE STELLAR OBJECT
ZWG 480.002	00 44 12.	+ 26 13		15.7	GALAXY
ZWG 479.072	00 44 12.	+ 26 13		15.7	GALAXY
UGC 00483	00 44 12.	+ 26 13	72	15.7	GALAXY IV-V
KARA.73B 0034	00 44 12.	+ 26 13	72	15.7	ISOLATED GALAXY S
ZWG 501.009	00 44 12.	+ 30 00		15.7	GALAXY
ZWG 500.090	00 44 12.	+ 30 00		15.7	GALAXY
MCG+05-03-001	00 44 12.	+ 32 23 30.	114	14.5	GALAXY
ZWG 501.010	00 44 12.	+ 32 24		14.1	GALAXY
ZWG 500.091	00 44 12.	+ 32 24		14.1	GALAXY
UGC 00484	00 44 12.	+ 32 24	168	14.1	GALAXY SBb
KARA.72 017A	00 44 12.	+ 32 24	108	14.1	PART OF DOUBLE GALAXY
SHRV.73A 22	00 44 12.	+ 41 08	7	16.88	GLOBULAR CLUSTER IN M31
PHL 6622	00 44 12.	- 03 38		17.3	BLUE STELLAR OBJECT
MCG-01-03-011	00 44 12.	- 06 48	60	14.5	GALAXY
PHL 2952	00 44 12.	- 14 36		18.7	BLUE STELLAR OBJECT
PHL 2954	00 44 12.	- 21 48		18.8	BLUE STELLAR OBJECT
PHL 2953	00 44 12.	- 21 56		17.6	BLUE STELLAR OBJECT
LIN 074	00 44 12.	- 72 16 30.		14.75	SMC EMISSION LINE OBJECT
LH115-N012	00 44 12.	- 73 22	395		EMISSION NEBULA IN SMC
RIC 678	00 44 13.10	+ 39 33 52.2		19.64	BLUE OBJECT
BAA 1.220	00 44 13.26	+ 41 33 58.7			EMISSION NEBULA IN M31
RIC 679	00 44 14.46	+ 40 38 10.0		18.63	BLUE OBJECT
BAA 1.686	00 44 14.71	+ 42 11 35.4			EMISSION NEBULA IN M31
BAA 3.07	00 44 15.46	+ 41 38 23.0			GLOBULAR CLUSTER IN M31
RIC 680	00 44 15.64	+ 39 22 06.8		18.45	BLUE OBJECT
RIC 681	00 44 15.94	+ 39 01 29.7		20.29	BLUE OBJECT
ZWG 480.003	00 44 18.	+ 24 29		15.5	GALAXY
ZCG 0044+40	00 44 18.	+ 40 57		17.6	COMPACT GALAXY
ZWG 535.021	00 44 18.	+ 44 15		15.7	GALAXY
VB 014	00 44 18.	+ 66 19	262		STELLAR RING
PHL 2955	00 44 18.	- 09 35		17.9	BLUE STELLAR OBJECT
PHL 6625	00 44 18.	- 21 00		18.7	BLUE STELLAR OBJECT
LB C0161	00 44 18.	- 21 54		16.7	FAINT BLUE STAR
RIC 682	00 44 18.68	+ 40 54 41.7		18.69	BLUE OBJECT
RIC 683	00 44 19.83	+ 39 41 18.4		19.93	BLUE OBJECT
RIC 684	00 44 19.97	+ 39 06 58.3		19.47	BLUE OBJECT
ARC 0099	00 44 20.	- 17 56		17.1	RICH CLUSTER OF GALAXIES
LIN 075	00 44 21.	- 73 23 12.		15.19	SMC EMISSION LINE OBJECT
RIC 685	00 44 21.02	+ 39 16 41.4		19.02	BLUE OBJECT
RIC 686	00 44 21.15	+ 39 06 54.1		19.18	BLUE OBJECT
RIC 687	00 44 21.58	+ 40 53 17.0		18.38	BLUE OBJECT
BAA 1.232	00 44 23.80	+ 41 39 51.6			EMISSION NEBULA IN M31
PHL 2957	00 44 24.	+ 01 20		18.2	BLUE STELLAR OBJECT
PHL 2759	00 44 24.	+ 06 50		18.7	BLUE STELLAR OBJECT
PHL 6629	00 44 24.	+ 09 44		17.6	BLUE STELLAR OBJECT
PHL 2956	00 44 24.	+ 10 57		11.5	BLUE STELLAR OBJECT
ZC 0044.4+2012	00 44 24.	+ 20 12	3230		CLUSTER OF GALAXIES
ZWG 501.011	00 44 24.	+ 27 48		15.4	GALAXY
ZWG 500.092	00 44 24.	+ 27 48		15.4	GALAXY
ZWG 501.012	00 44 24.	+ 30 04		14.8	GALAXY
ZWG 500.093	00 44 24.	+ 30 04		14.8	GALAXY
UGC 00485	00 44 24.	+ 30 04	132	14.8	GALAXY Sc
ZWG 501.013	00 44 24.	+ 32 25		15.1	GALAXY
ZWG 500.094	00 44 24.	+ 32 25		15.1	GALAXY
KARA.72 017B	00 44 24.	+ 32 25	36	15.1	PART OF DOUBLE GALAXY
ISS 0005	00 44 24.	+ 66 23	260		STELLAR RING
PHL 6627	00 44 24.	- 10 28		18.4	BLUE STELLAR OBJECT
PHL 6626	00 44 24.	- 11 05		16.7	BLUE STELLAR OBJECT
PHL 6628	00 44 24.	- 15 26		18.7	BLUE STELLAR OBJECT
PHL 2958	00 44 24.	- 19 24		18.7	BLUE STELLAR OBJECT
MCG-06-02-027	00 44 24.	- 36 05	60	16.	GALAXY
LB 03155	00 44 24.	- 73 23		13.8	FAINT BLUE STAR
LH115-N015	00 44 24.	- 73 42	27		EMISSION NEBULA IN SMC
RIC 688	00 44 24.88	+ 39 15 18.0		18.36	BLUE OBJECT
RIC 689	00 44 26.44	+ 40 50 53.2		19.03	BLUE OBJECT
BAA 1.687	00 44 26.74	+ 42 10 35.7			EMISSION NEBULA IN M31
MCG+05-03-002	00 44 27.	+ 30 04 30.	132	15.	GALAXY
RIC 690	00 44 27.66	+ 40 01 36.6		18.07	BLUE OBJECT
BAA 3.28	00 44 28.51	+ 42 05 21.3			GLOBULAR CLUSTER IN M31
PHL 2961	00 44 30.	+ 00 36		18.2	BLUE STELLAR OBJECT
PHL 0828	00 44 30.	+ 03 04		16.2	BLUE STELLAR OBJECT
PHL 2959	00 44 30.	+ 14 42		15.1	BLUE STELLAR OBJECT

OBJECT NAME	RIGHT ASCEN.	DECLINATION	DIAM.	MAGN.	TYPE OF OBJECT
ZWG 480.004	00 44 30.	+ 22 48		15.0	GALAXY
MCG+05-03-003	00 44 30.	+ 32 25	24	15.5	GALAXY
MCG+08-02-015	00 44 30.	+ 50 37	90	14.	GALAXY
ZC 0044.5+7614	00 44 30.	+ 76 14	2890		CLUSTER OF GALAXIES
PHL 0829	00 44 30.	- 12 08		12.2	BLUE STELLAR OBJECT
PHL 2960	00 44 30.	- 18 27		18.6	BLUE STELLAR OBJECT
LH115-N014	00 44 30.	- 73 29	24		EMISSION NEBULA IN SMC
LH115-N016	00 44 30.	- 73 40	64		EMISSION NEBULA IN SMC
RIC 691	00 44 30.24	+ 39 46 49.6		19.00	BLUE OBJECT
RIC 692	00 44 30.30	+ 41 24 26.4		20.09	BLUE OBJECT
RIC 693	00 44 30.76	+ 40 08 05.2		18.92	BLUE OBJECT
RIC 694	00 44 31.62	+ 39 51 43.2		18.07	BLUE OBJECT
RIC 695	00 44 31.63	+ 40 46 49.1		19.17	BLUE OBJECT
KEEL 013	00 44 32.0	- 25 43 43.		17.	NEBULA
IC 1583	00 44 33.	+ 22 48 16.			NONSTELLAR OBJECT
MCG+04-03-001	00 44 33.	+ 22 49	24	15.5	GALAXY
MCG-04-03-004	00 44 33.	- 22 07 30.	48	15.	GALAXY
LIN 077	00 44 33.	- 73 19 00.		13.96	SMC EMISSION LINE OBJECT
RNGC 0247	00 44 34.	- 21 01		10.0	GALAXY
RIC 696	00 44 34.36	+ 41 05 00.9		18.73	BLUE OBJECT
RIC 697	00 44 34.78	+ 41 18 52.1		18.39	BLUE OBJECT
LB 03927	00 44 35.	+ 59 37 00.		18.8	FAINT BLUE STAR
SVEN 033	00 44 35.	- 21 02	1176	9.8	GALAXY
LIN 076	00 44 35.	- 72 39 36.		15.55	SMC EMISSION LINE OBJECT
RIC 698	00 44 35.22	+ 39 41 10.3		19.86	BLUE OBJECT
SHB 013	00 44 35.3	- 07 22 00.		17.7	QUASI-STELLAR OBJECT
RIC 699	00 44 35.36	+ 39 46 52.4		18.06	BLUE OBJECT
PHL 6631	00 44 36.	+ 00 34		18.5	BLUE STELLAR OBJECT
PHL 6637	00 44 36.	+ 11 29		18.7	BLUE STELLAR OBJECT
PHL 6636	00 44 36.	+ 11 35		18.3	BLUE STELLAR OBJECT
PHL 6630	00 44 36.	+ 11 41		17.7	BLUE STELLAR OBJECT
RNGC 0246	00 44 36.	+ 12 09			PLANETARY NEBULA
ZWG 480.005	00 44 36.	+ 22 47		14.9	GALAXY
IC 1584	00 44 36.	+ 27 33			NONSTELLAR OBJECT
5ZW 035	00 44 36.	+ 40 57			COMPACT GALAXY
ZWG 550.015	00 44 36.	+ 50 37		14.8	GALAXY
UGC 00486	00 44 36.	+ 50 37	78	14.8	GALAXY Sb/SBb
MCG-01-03-012	00 44 36.	- 02 59	48	15.	GALAXY
PHL 6638	00 44 36.	- 07 20		18.4	BLUE STELLAR OBJECT
PHL 6639	00 44 36.	- 08 26		18.3	BLUE STELLAR OBJECT
PHL 6633	00 44 36.	- 09 04		17.9	BLUE STELLAR OBJECT
PHL 6640	00 44 36.	- 10 27		18.2	BLUE STELLAR OBJECT
PHL 6632	00 44 36.	- 12 46		18.8	BLUE STELLAR OBJECT
PHL 2962	00 44 36.	- 16 05		17.8	BLUE STELLAR OBJECT
PHL 6634	00 44 36.	- 20 02		17.9	BLUE STELLAR OBJECT
MCG-06-02-028	00 44 36.	- 36 16	36	15.	GALAXY
RIC 700	00 44 36.22	+ 39 44 29.7		18.43	BLUE OBJECT
RIC 701	00 44 36.24	+ 40 34 56.0		18.80	BLUE OBJECT
IC 1585	00 44 37.	+ 22 47 04.			NONSTELLAR OBJECT
LIN 079	00 44 38.	- 73 31 12.		14.27	SMC EMISSION LINE OBJECT
RIC 702	00 44 38.20	+ 40 47 35.4		18.54	BLUE OBJECT
RIC 703	00 44 38.96	+ 40 12 37.3		19.29	BLUE OBJECT
MCG+04-03-002	00 44 39.	+ 22 48	36	15.5	GALAXY
MCG+05-03-004	00 44 39.	+ 27 32	84	15.	GALAXY
LIN 078	00 44 39.	- 73 22 36.		12.96	SMC EMISSION LINE OBJECT
RIC 704	00 44 39.20	+ 39 42 34.2		19.10	BLUE OBJECT
RIC 705	00 44 39.82	+ 40 00 29.4		19.21	BLUE OBJECT
RNGC 0261	00 44 41.	- 73 23			DIFFUSE NEBULA IN SMC
RIC 706	00 44 41.50	+ 41 28 38.2		19.19	BLUE OBJECT
RIC 707	00 44 41.57	+ 39 59 02.7		19.06	BLUE OBJECT
RIC 708	00 44 41.71	+ 39 53 31.5		19.85	BLUE OBJECT
RIC 709	00 44 41.88	+ 41 31 07.4		19.11	BLUE OBJECT
ZWG 410.005	00 44 42.	+ 07 39		14.9	GALAXY
RNGC 0250	00 44 42.	+ 07 39		15.0	GALAXY
UGC 00487	00 44 42.	+ 07 39	84	14.9	GALAXY S0-a
ZWG 435.003	00 44 42.	+ 14 26		15.2	GALAXY
UGC 00488	00 44 42.	+ 14 26	60	15.2	GALAXY Sa-b
MCG+02-03-001	00 44 42.	+ 14 26	48	15.	GALAXY
ZWG 501.014	00 44 42.	+ 27 33		15.0	GALAXY
UGC 00489	00 44 42.	+ 27 33	108	15.0	GALAXY S
PHL 2963	00 44 42.	- 07 04		18.6	BLUE STELLAR OBJECT
PHL 2965	00 44 42.	- 08 58		18.6	BLUE STELLAR OBJECT
PHL 2964	00 44 42.	- 20 57		17.5	BLUE STELLAR OBJECT
MCG-04-03-005	00 44 42.	- 21 02	1140	10.	GALAXY
MCG-05-03-004	00 44 42.	- 26 40	36	16.	GALAXY
LH115-N012A	00 44 42.	- 73 23	113		EMISSION NEBULA IN SMC
RIC 710	00 44 42.57	+ 40 48 48.5		18.86	BLUE OBJECT
RIC 711	00 44 43.04	+ 40 03 01.9		18.01	BLUE OBJECT
RIC 712	00 44 43.94	+ 41 25 52.2		18.44	BLUE OBJECT
LIN 081	00 44 44.	- 73 28 36.		13.56	SMC EMISSION LINE OBJECT
MCG+01-03-002	00 44 45.	+ 07 39	48	14.5	GALAXY
MCG+02-03-002	00 44 45.	+ 15 25	48	15.	GALAXY
MCG+04-03-003	00 44 45.	+ 22 48	33	16.	GALAXY
ARC 0100	00 44 45.	- 02 48		17.7	RICH CLUSTER OF GALAXIES
MCG-01-03-013	00 44 45.	- 07 04	24	15.5	GALAXY
MCG-04-03-006	00 44 45.	- 22 51	72	16.	GALAXY
LIN 080	00 44 45.	- 73 27 18.		14.92	SMC EMISSION LINE OBJECT
RIC 713	00 44 45.96	+ 39 34 25.5		20.40	BLUE OBJECT
RIC 714	00 44 46.18	+ 41 32 34.3		19.42	BLUE OBJECT
RIC 715	00 44 46.30	+ 40 44 57.6		20.16	BLUE OBJECT
VDB.66B 19	00 44 47.	+ 40 01 12.		19.85	BLUE OBJECT
RIC 716	00 44 47.17	+ 39 48 08.6		19.24	BLUE OBJECT
RIC 717	00 44 47.57	+ 41 18 18.3		20.29	BLUE OBJECT
RIC 718	00 44 47.72	+ 40 00 52.2		19.35	BLUE OBJECT
RIC 719	00 44 47.85	+ 39 55 30.2		17.73	BLUE OBJECT
PHL 2969	00 44 48.	+ 01 32		18.1	BLUE STELLAR OBJECT
PHL 2966	00 44 48.	+ 03 20		18.7	BLUE STELLAR OBJECT
ZWG 435.004	00 44 48.	+ 15 25		15.5	GALAXY
ZWG 501.015	00 44 48.	+ 31 23		15.7	GALAXY
ZWG 500.095	00 44 48.	+ 31 23		15.7	GALAXY
ZC 0044.8+3642	00 44 48.	+ 36 42	2220		CLUSTER OF GALAXIES
VB 015	00 44 48.	+ 63 25	389		STELLAR RING
PHL 6642	00 44 48.	- 00 29		18.1	BLUE STELLAR OBJECT
PHL 2967	00 44 48.	- 01 43		18.5	BLUE STELLAR OBJECT
PHL 6643	00 44 48.	- 01 52		18.6	BLUE STELLAR OBJECT
PHL 6645	00 44 48.	- 02 27		18.3	BLUE STELLAR OBJECT
PHL 6646	00 44 48.	- 08 23		18.6	BLUE STELLAR OBJECT
PHL 6644	00 44 48.	- 12 02		18.2	BLUE STELLAR OBJECT
PHL 6641	00 44 48.	- 12 52		17.9	BLUE STELLAR OBJECT
PHL 2968	00 44 48.	- 13 17		18.8	BLUE STELLAR OBJECT
PHL 6647	00 44 48.	- 15 22		18.6	BLUE STELLAR OBJECT
RIC 720	00 44 48.26	+ 40 14 26.3		19.60	BLUE OBJECT
BAA 1.248	00 44 49.12	+ 42 06 36.4			EMISSION NEBULA IN M31
RIC 721	00 44 49.96	+ 39 55 39.9		19.70	BLUE OBJECT
SVEN 034	00 44 50.	- 31 25	18	16.2	GALAXY
RIC 722	00 44 50.95	+ 41 01 38.0		18.70	BLUE OBJECT
MCG+05-03-005	00 44 51.	+ 31 22	48	17.5	GALAXY
HOW 22	00 44 52.	- 72 19 59.	18		STAR CLUSTER IN SMC
RIC 723	00 44 52.80	+ 40 03 16.0		18.45	BLUE OBJECT
RIC 724	00 44 53.18	+ 41 19 29.2		17.77	BLUE OBJECT
RIC 725	00 44 53.24	+ 39 09 10.6		19.57	BLUE OBJECT
RIC 726	00 44 53.32	+ 40 00 31.6		18.30	BLUE OBJECT
PHL 0830	00 44 54.	+ 09 42		9.0	BLUE STELLAR OBJECT
MCG+02-03-003	00 44 54.	+ 13 52 30.	48	15.	GALAXY
PHL 2970	00 44 54.	- 02 36		17.5	BLUE STELLAR OBJECT
PHL 6649	00 44 54.	- 10 48		18.3	BLUE STELLAR OBJECT
PHL 0831	00 44 54.	- 11 28		17.8	BLUE STELLAR OBJECT
PHL 6648	00 44 54.	- 13 54		17.6	BLUE STELLAR OBJECT
PHL 2971	00 44 54.	- 14 06		18.0	BLUE STELLAR OBJECT
PHL 0832	00 44 54.	- 22 01		18.0	BLUE STELLAR OBJECT
PHL 6650	00 44 54.	- 22 11		18.6	BLUE STELLAR OBJECT
PHL 6651	00 44 54.	- 27 38		18.5	BLUE STELLAR OBJECT
LH115-N017	00 44 54.	- 73 48	191		EMISSION NEBULA IN SMC
HOW 23	00 44 54.	- 74 02 41.	24		STAR CLUSTER IN SMC
RIC 727	00 44 55.76	+ 39 50 19.2		18.78	BLUE OBJECT
LIN 082	00 44 56.	- 73 39 24.		14.33	SMC EMISSION LINE OBJECT
ARC 0101	00 44 57.	- 01 12		17.2	RICH CLUSTER OF GALAXIES
RIC 728	00 44 58.21	+ 41 35 51.0		19.03	BLUE OBJECT
SVEN 035	00 44 59.	- 20 41	54	13.2	GALAXY
KEEL 014	00 44 59.5	- 20 40 14.		18.	NEBULA
PHL 6652	00 45 00.	+ 12 40		18.0	BLUE STELLAR OBJECT
ZWG 435.005	00 45 00.	+ 13 52		15.5	GALAXY
PHL 6653	00 45 00.	+ 15 08		18.4	BLUE STELLAR OBJECT
ZWG 519.023	00 45 00.	+ 38 09		15.4	GALAXY
SHRV.73B 27	00 45 00.	+ 39 22	8	16.00	COMPACT GALAXY
SHRV.73B 26	00 45 00.	+ 41 23	7	16.66	COMPACT GALAXY
MCG-01-03-014	00 45 00.	- 05 21 30.	42	15.	GALAXY
MCG-02-03-014	00 45 00.	- 10 07	30	15.	GALAXY
MCG-02-03-013	00 45 00.	- 11 33	36	14.5	GALAXY
PHL 2973	00 45 00.	- 13 25		18.2	BLUE STELLAR OBJECT
PHL 2972	00 45 00.	- 16 00		18.6	BLUE STELLAR OBJECT
MCG-04-03-007	00 45 00.	- 23 17 30.	36	15.5	GALAXY
PHL 0833	00 45 00.	- 25 32		16.9	BLUE STELLAR OBJECT
PHL 2974	00 45 00.	- 30 03		18.3	BLUE STELLAR OBJECT
MCG-05-03-005	00 45 00.	- 31 43	60	13.5	GALAXY
RIC 729	00 45 00.74	+ 41 41 58.2		18.19	BLUE OBJECT
KEEL 015	00 45 01.7	- 25 42 46.		17.	NEBULA
LIN 084	00 45 02.	- 73 26 06.		13.70	SMC EMISSION LINE OBJECT
RIC 730	00 45 02.36	+ 41 35 38.4		18.64	BLUE OBJECT
BAA 3.30	00 45 02.70	+ 42 52 56.4			GLOBULAR CLUSTER IN M31
SN 1940E	00 45 03.	- 25 34		14.0	SUPERNOVA
LIN 083	00 45 03.	- 73 05 18.		16.62	PLANETARY NEBULA IN SMC
RNGC 0253	00 45 04.	- 25 34		7.5	GALAXY
SVEN 038	00 45 04.	- 25 34 06.	1482	9.6	GALAXY
LIN.CL 031	00 45 04.	- 72 59 24.	204	13.9	STAR CLUSTER IN SMC
RIC 731	00 45 04.66	+ 41 25 20.3		17.76	BLUE OBJECT
SVEN 036	00 45 05.	- 20 45	24	14.5	GALAXY
SVEN 037	00 45 05.	- 20 47	36	13.7	GALAXY
MCG+05-03-006	00 45 06.	+ 29 40	24	15.	GALAXY
MCG+06-02-019	00 45 06.	+ 38 10	42	15.	GALAXY
PHL 6654	00 45 06.	- 08 44		16.8	BLUE STELLAR OBJECT
PHL 6656	00 45 06.	- 09 48		18.3	BLUE STELLAR OBJECT
PHL 6655	00 45 06.	- 14 19		18.3	BLUE STELLAR OBJECT
PHL 6658	00 45 06.	- 18 28		18.1	BLUE STELLAR OBJECT
LB 03156	00 45 06.	- 72 07		13.6	FAINT BLUE STAR
RIC 732	00 45 06.34	+ 39 45 59.0		18.71	BLUE OBJECT
KEEL 016	00 45 06.4	- 20 42 04.		14.	NEBULA
LIN 085	00 45 08.	- 73 25 00.		12.53	SMC EMISSION LINE OBJECT
LIN 086	00 45 08.	- 73 30 30.		13.26	SMC EMISSION LINE OBJECT
KEEL 017	00 45 08.5	- 20 45 30.		17.	SPIRAL NEBULA
KEEL 018	00 45 08.7	- 20 43 39.		18.	NEBULA
RIC 733	00 45 08.94	+ 40 02 57.3		17.48	BLUE OBJECT
MCG-04-03-008	00 45 09.	- 23 03 30.	27	15.	GALAXY
KEEL 019	00 45 09.3	- 20 47 31.		15.	SPIRAL NEBULA
KEEL 020	00 45 09.3	- 20 57 30.		18.	NEBULA
RIC 734	00 45 10.27	+ 39 32 36.2		17.50	BLUE OBJECT
RIC 735	00 45 10.28	+ 41 33 12.0		18.07	BLUE OBJECT
BAA 3.10	00 45 10.32	+ 42 12 23.0			GLOBULAR CLUSTER IN M31
RNGC 0251	00 45 11.	+ 19 18		14.5	GALAXY
RNGC 0254	00 45 11.	- 31 42		13.0	GALAXY
RIC 736	00 45 11.74	+ 40 25 46.0		18.67	BLUE OBJECT
RIC 737	00 45 11.76	+ 40 26 16.7		18.75	BLUE OBJECT
PHL 2975	00 45 12.	+ 02 45		18.5	BLUE STELLAR OBJECT
ZWG 435.006	00 45 12.	+ 12 00		15.5	GALAXY
ZWG 458.005	00 45 12.	+ 19 18		14.6	GALAXY
UGC 00490	00 45 12.	+ 19 18	150	14.6	GALAXY Sc
KARA.72 015B	00 45 12.	+ 19 18	150	14.6	PART OF DOUBLE GALAXY
ZWG 501.016	00 45 12.	+ 29 41		15.4	GALAXY
PHL 6659	00 45 12.	- 00 34		18.3	BLUE STELLAR OBJECT
PHL 2976	00 45 12.	- 10 07		18.5	BLUE STELLAR OBJECT
PHL 6661	00 45 12.	- 10 36		18.4	BLUE STELLAR OBJECT
MCG-04-03-009	00 45 12.	- 25 33 30.	1500	7.	GALAXY
PHL 6660	00 45 12.	- 28 20		18.4	BLUE STELLAR OBJECT
PHL 6662	00 45 12.	- 30 46		18.5	BLUE STELLAR OBJECT
LB 01575	00 45 12.	- 47 38		13.3	FAINT BLUE STAR
LH115-N018	00 45 12.	- 73 06			EMISSION NEBULA IN SMC
RIC 738	00 45 12.75	+ 40 00 05.2		19.46	BLUE OBJECT
HOLM 022B	00 45 13.	- 03 04	24	15.3	PART OF MULTIPLE GALAXY
HOLM 021A	00 45 13.	- 10 07	120	14.2	PART OF MULTIPLE GALAXY
HOLM 021B	00 45 13.	- 10 11	102	14.2	PART OF MULTIPLE GALAXY
LIN 087	00 45 13.	- 73 56 36.		14.58	SMC EMISSION LINE OBJECT
RIC 739	00 45 13.48	+ 39 51 06.3		18.36	BLUE OBJECT
RIC 740	00 45 14.68	+ 40 21 58.2		18.47	BLUE OBJECT
VDB.66B 10	00 45 15.	+ 39 22 12.		20.46	BLUE OBJECT
RNGC 0255	00 45 15.	- 11 45		12.5	GALAXY
MCG-04-03-010	00 45 15.	- 20 41 30.	54	14.	GALAXY
MCG-04-03-012	00 45 15.	- 20 43	36	17.	GALAXY
MCG-04-03-011	00 45 15.	- 20 45	36	16.	GALAXY
MCG-04-03-013	00 45 15.	- 20 47	60	14.5	GALAXY
MCG-05-03-006	00 45 15.	- 28 13	48	15.	GALAXY
BAA 4.25	00 45 15.44	+ 41 23 53.5			DIFFUSE OBJECT IN M31
BAA 3.09	00 45 15.52	+ 42 09 12.0			GLOBULAR CLUSTER IN M31
HOLM 023B	00 45 16.	+ 27 21	72	14.1	PART OF MULTIPLE GALAXY
LB 03928	00 45 16.	+ 59 03 48.		19.1	FAINT BLUE STAR
RIC 741	00 45 16.05	+ 41 45 36.4		18.13	BLUE OBJECT
RIC 742	00 45 16.07	+ 41 25 00.2		18.99	BLUE OBJECT
RIC 743	00 45 16.29	+ 39 22 10.2		20.28	BLUE OBJECT
RIC 744	00 45 17.51	+ 41 44 33.0		18.07	BLUE OBJECT
BAA 1.249	00 45 17.60	+ 42 11 09.2			EMISSION NEBULA IN M31
PHL 0834	00 45 18.	+ 02 54		18.7	BLUE STELLAR OBJECT
MCG+03-03-003	00 45 18.	+ 19 19	138	13.5	GALAXY
IC 1586	00 45 18.	+ 22 04 45.			NONSTELLAR OBJECT
3ZW 012	00 45 18.	+ 22 06			COMPACT GALAXY
ZWG 480.006	00 45 18.	+ 22 06		14.9	GALAXY
MRK 347	00 45 18.	+ 22 06	20	15.	GALAXY WITH UV CONTINUUM
ZWG 480.007	00 45 18.	+ 27 21		13.4	GALAXY

OBJECT NAME	RIGHT ASCEN.	DECLINATION	DIAM.	MAGN.	TYPE OF OBJECT
UGC 00491	00 45 18.	+ 27 21	102	13.4	GALAXY S0
SHRV.73A 23	00 45 18.	+ 41 25	7	16.70	GLOBULAR CLUSTER IN M31
ZC 0045.3+4726	00 45 18.	+ 47 26	940		CLUSTER OF GALAXIES
MCG-02-03-015	00 45 18.	- 10 06	96	13.5	GALAXY
MCG-02-03-016	00 45 18.	- 10 10	180	13.5	GALAXY
MCG-02-03-017	00 45 18.	- 11 44	180	11.	GALAXY
PHL 2977	00 45 18.	- 24 16		18.4	BLUE STELLAR OBJECT
LIN 088	00 45 19.	- 73 38 37.		13.28	SMC EMISSION LINE OBJECT
LIN 089	00 45 19.	- 73 43 07.		16.25	SMC EMISSION LINE OBJECT
RIC 745	00 45 19.69	+ 41 05 30.8		19.27	BLUE OBJECT
RIC 746	00 45 19.92	+ 39 12 27.0		18.24	BLUE OBJECT
SVEN 039	00 45 20.	- 31 18	78	13.7	GALAXY
MCG+05-03-007	00 45 21.	+ 29 40	18	15.	GALAXY
HELW 090	00 45 21.	- 12 03 48.			NEBULA
RIC 747	00 45 21.12	+ 41 29 35.7		18.84	BLUE OBJECT
RIC 748	00 45 21.64	+ 40 38 40.6		18.04	BLUE OBJECT
RNGC 0252	00 45 22.	+ 27 21		13.5	GALAXY
RIC 749	00 45 22.08	+ 40 02 12.6		18.35	BLUE OBJECT
BAA 4.26	00 45 22.85	+ 41 24 21.0			DIFFUSE OBJECT IN M31
RIC 750	00 45 23.85	+ 40 42 36.4		17.60	BLUE OBJECT
ZWG 384.005	00 45 24.	+ 00 57		15.5	GALAXY
PHL 6665	00 45 24.	+ 01 00		17.7	BLUE STELLAR OBJECT
PHL 0835	00 45 24.	+ 02 51		18.5	BLUE STELLAR OBJECT
PHL 0836	00 45 24.	+ 02 54		18.8	BLUE STELLAR OBJECT
PHL 6486	00 45 24.	+ 14 14		17.7	BLUE STELLAR OBJECT
MCG+04-03-004	00 45 24.	+ 27 21	60	14.5	GALAXY
ZC 0045.4+2934	00 45 24.	+ 29 34	1280		CLUSTER OF GALAXIES
ZWG 501.017	00 45 24.	+ 29 41		15.6	GALAXY
SHRV.73B 28	00 45 24.	+ 41 52	8	16.09	COMPACT GALAXY
SHRV.73A 24	00 45 24.	+ 42 09	8	17.27	GLOBULAR CLUSTER IN M31
OCL 0308	00 45 24.	+ 66 58	240	14.	OPEN STAR CLUSTER
MCG+00-03-006	00 45 24.	- 01 51	30	14.	GALAXY
MCG-02-03-018	00 45 24.	- 09 38	60	13.	GALAXY
PHL 6666	00 45 24.	- 12 01		18.2	BLUE STELLAR OBJECT
PHL 6664	00 45 24.	- 12 55		16.6	BLUE STELLAR OBJECT
PHL 6663	00 45 24.	- 15 56		17.9	BLUE STELLAR OBJECT
MCG-04-03-014	00 45 24.	- 21 44 30.	33	15.	GALAXY
PHL 2978	00 45 24.	- 21 58		18.3	BLUE STELLAR OBJECT
LIN.CL 032	00 45 24.	- 69 11 54.	204		STAR CLUSTER IN SMC
RIC 751	00 45 24.31	+ 40 16 14.9		19.38	BLUE OBJECT
VDB.66B 16	00 45 25.	+ 39 46 47.		19.51	BLUE OBJECT
LIN 090	00 45 25.	- 73 44 49.		14.47	SMC EMISSION LINE OBJECT
RIC 752	00 45 25.17	+ 40 16 08.0		19.16	BLUE OBJECT
RIC 753	00 45 25.97	+ 41 03 11.9		18.70	BLUE OBJECT
HOLM 022A	00 45 27.	- 03 03	108	14.1	PART OF MULTIPLE GALAXY
RIC 754	00 45 27.66	+ 40 06 05.3		18.58	BLUE OBJECT
RIC 755	00 45 27.70	+ 41 00 49.8		19.37	BLUE OBJECT
RIC 756	00 45 27.80	+ 41 37 46.6		18.45	BLUE OBJECT
HOLM 023D	00 45 28.	+ 27 23	18	15.0	PART OF MULTIPLE GALAXY
RIC 757	00 45 28.22	+ 40 59 35.6		19.95	BLUE OBJECT
RIC 758	00 45 28.88	+ 39 08 18.0		18.66	BLUE OBJECT
VDB.66B 15	00 45 29.	+ 39 47 59.		19.51	BLUE OBJECT
SC 0043-6927.6	00 45 29.	- 69 11 12.	42		NEBULA
MACL 1	00 45 29.	- 73 24 00.	72		SUPERNOVA REMNANT IN SMC
RIC 759	00 45 29.10	+ 39 42 57.0		18.74	BLUE OBJECT
RIC 760	00 45 29.14	+ 41 31 06.4		18.06	BLUE OBJECT
RIC 761	00 45 29.73	+ 39 47 02.1		18.62	BLUE OBJECT
RIC 762	00 45 29.89	+ 41 45 46.6		18.30	BLUE OBJECT
MCG+01-03-003	00 45 30.	+ 08 02	90	13.	GALAXY
ZWG 410.006	00 45 30.	+ 08 03		13.7	GALAXY
RNGC 0257	00 45 30.	+ 08 03		13.5	GALAXY
UGC 00493	00 45 30.	+ 08 03	132	13.7	GALAXY Sc
ZWG 501.018	00 45 30.	+ 28 20		15.0	GALAXY
FATF 1.017	00 45 30.	+ 46 05			NEBULA
VB 016	00 45 30.	+ 65 00	1880		STELLAR RING
ZWG 384.006	00 45 30.	- 01 49		14.6	GALAXY
UGC 00492	00 45 30.	- 01 49	84	14.6	GALAXY E
MCG-01-03-015	00 45 30.	- 03 02	18	12.5	GALAXY
PHL 6669	00 45 30.	- 16 20		18.1	BLUE STELLAR OBJECT
PHL 0837	00 45 30.	- 20 27		17.4	BLUE STELLAR OBJECT
PHL 6668	00 45 30.	- 22 00		5.4	BLUE STELLAR OBJECT
PHL 6667	00 45 30.	- 29 46		18.4	BLUE STELLAR OBJECT
LB 03157	00 45 30.	- 63 02		13.4	FAINT BLUE STAR
RIC 763	00 45 30.48	+ 40 07 21.5		19.83	BLUE OBJECT
RIC 764	00 45 31.14	+ 39 08 36.2		20.33	BLUE OBJECT
RNGC 0259	00 45 32.	- 03 03		12.5	GALAXY
SVEN 040	00 45 32.	- 31 01	18	16.1	GALAXY
SVEN 041	00 45 32.	- 31 22	18	16.0	GALAXY
LIN 091	00 45 32.	- 73 23 13.		12.87	SMC EMISSION LINE OBJECT
RIC 765	00 45 32.11	+ 40 57 05.0		18.95	BLUE OBJECT
RNGC 0258	00 45 34.	+ 27 22		15.0	GALAXY
RIC 766	00 45 34.88	+ 39 11 57.6		18.97	BLUE OBJECT
RIC 767	00 45 35.27	+ 39 08 45.3		19.26	BLUE OBJECT
PHL 0838	00 45 36.	+ 02 18		16.7	BLUE STELLAR OBJECT
ZWG 480.008	00 45 36.	+ 26 12		15.6	GALAXY
MCG+04-03-005	00 45 36.	+ 27 22 30.	30	15.5	GALAXY
ZWG 501.019	00 45 36.	+ 27 43		15.6	GALAXY
PHL 2980	00 45 36.	- 01 42		17.7	BLUE STELLAR OBJECT
PHL 0839	00 45 36.	- 04 06		15.2	BLUE STELLAR OBJECT
PHL 2979	00 45 36.	- 14 14		16.6	BLUE STELLAR OBJECT
PHL 6670	00 45 36.	- 15 40		17.7	BLUE STELLAR OBJECT
LB 03158	00 45 36.	- 65 40		13.5	FAINT BLUE STAR
LIN 093	00 45 37.	- 73 36 19.		14.75	SMC EMISSION LINE OBJECT
RIC 768	00 45 37.00	+ 39 13 37.3		18.11	BLUE OBJECT
KEEL 021	00 45 37.8	- 25 43 13.		18.	NEBULA
RIC 769	00 45 37.93	+ 39 01 04.9		18.68	BLUE OBJECT
LIN 092	00 45 38.	- 73 24 55.		15.74	SMC EMISSION LINE OBJECT
RIC 770	00 45 39.20	+ 39 42 46.8		18.51	BLUE OBJECT
RIC 771	00 45 39.73	+ 39 55 28.0		18.38	BLUE OBJECT
RIC 772	00 45 39.89	+ 39 05 24.8		19.63	BLUE OBJECT
HOLM 023A	00 45 40.	+ 27 24	30	15.7	PART OF MULTIPLE GALAXY
HOW 24	00 45 41.	- 72 44 29.	54		STAR CLUSTER IN SMC
RIC 773	00 45 41.23	+ 40 48 19.0		19.21	BLUE OBJECT
PHL 6671	00 45 42.	+ 12 18		18.0	BLUE STELLAR OBJECT
URA 45	00 45 42.	+ 62 42	234		STELLAR RING
PHL 0840	00 45 42.	- 18 26		17.5	BLUE STELLAR OBJECT
SVEN 042	00 45 44.	- 31 30	18	16.1	GALAXY
RIC 774	00 45 44.26	+ 40 30 44.0		17.72	BLUE OBJECT
RIC 775	00 45 44.48	+ 41 41 13.0		19.06	BLUE OBJECT
KEEL 022	00 45 44.8	- 21 20 54.		18.	NEBULA
RIC 776	00 45 44.84	+ 39 13 27.2		18.49	BLUE OBJECT
RIC 777	00 45 45.35	+ 39 56 44.4		18.37	BLUE OBJECT
RIC 778	00 45 45.64	+ 42 13 32.9		20.35	BLUE OBJECT
RIC 779	00 45 45.86	+ 41 33 36.1		18.09	BLUE OBJECT
BAA 1.250	00 45 46.08	+ 42 13 48.0			EMISSION NEBULA IN M31
RIC 780	00 45 46.21	+ 42 13 48.9		19.69	BLUE OBJECT
IC 0052	00 45 47.	+ 03 48 44.			NONSTELLAR OBJECT
HOLM 023B	00 45 47.	+ 27 26	12	15.2	PART OF MULTIPLE GALAXY
RIC 781	00 45 47.47	+ 42 06 02.3		18.96	BLUE OBJECT
RIC 782	00 45 47.67	+ 41 12 08.4		19.60	BLUE OBJECT
PHL 2982	00 45 48.	+ 03 33		18.2	BLUE STELLAR OBJECT
MCG+01-03-005	00 45 48.	+ 03 48	36	14.	GALAXY
ZWG 410.007	00 45 48.	+ 03 50		15.4	GALAXY
UGC 00494	00 45 48.	+ 03 50	72	15.4	GALAXY S
MCG+01-03-004	00 45 48.	+ 07 03	60	15.	GALAXY
ZWG 410.008	00 45 48.	+ 07 04		15.7	GALAXY
UGC 00495	00 45 48.	+ 07 04	78	15.7	GALAXY Sa
SHRV.73B 29	00 45 48.	+ 39 15	10	16.06	COMPACT GALAXY
MRSL 122+00/1	00 45 48.	+ 62 39	900		HII REGION
PHL 6672	00 45 48.	- 11 36		17.5	BLUE STELLAR OBJECT
PHL 2981	00 45 48.	- 26 06		18.5	BLUE STELLAR OBJECT
PHL 6673	00 45 48.	- 28 19		17.3	BLUE STELLAR OBJECT
RNGC 0265	00 45 48.	- 73 45		12.0	GLOBULAR CLUSTER IN SMC
RIC 783	00 45 48.60	+ 42 14 58.8		18.80	BLUE OBJECT
LIN 094	00 45 49.	- 73 32 49.		15.29	SMC EMISSION LINE OBJECT
LIN.CL 034	00 45 49.	- 73 44 37.	474	11.9	STAR CLUSTER IN SMC
RIC 784	00 45 49.16	+ 41 46 11.4		18.44	BLUE OBJECT
SVEN 044	00 45 50.	- 31 10	12	16.4	GALAXY
SVEN 043	00 45 50.	- 31 50 07.	24	16.6	GALAXY
RIC 785	00 45 50.21	+ 40 07 50.4		17.90	BLUE OBJECT
RIC 786	00 45 50.24	+ 40 04 43.0		19.69	BLUE OBJECT
HOLM 023C	00 45 51.	+ 27 25	36	14.5	PART OF MULTIPLE GALAXY
LIN.CL 033	00 45 51.	- 73 05 13.	204	14.3	STAR CLUSTER IN SMC
RNGC 0260	00 45 52.	+ 27 25		14.5	GALAXY
RIC 787	00 45 52.12	+ 40 51 22.4		18.57	BLUE OBJECT
RIC 788	00 45 52.56	+ 40 27 26.6		19.55	BLUE OBJECT
RIC 789	00 45 52.73	+ 40 59 47.2		18.55	BLUE OBJECT
RIC 790	00 45 53.58	+ 40 16 15.0		19.26	BLUE OBJECT
RIC 791	00 45 53.68	+ 39 53 28.2		19.10	BLUE OBJECT
PHL 0841	00 45 54.	+ 00 02		18.4	BLUE STELLAR OBJECT
MCG+00-03-007	00 45 54.	+ 01 04	24	16.	GALAXY
ZWG 384.007	00 45 54.	+ 01 05		14.6	GALAXY
UGC 00496	00 45 54.	+ 01 05	90	14.6	GALAXY DBL SYS
KARA.72 018B	00 45 54.	+ 01 05	36		PART OF DOUBLE GALAXY
KARA.72 018A	00 45 54.	+ 01 05	30	14.6	PART OF DOUBLE GALAXY
PHL 2983	00 45 54.	+ 15 25		16.4	BLUE STELLAR OBJECT
ZWG 480.009	00 45 54.	+ 27 25		14.3	GALAXY
UGC 00497	00 45 54.	+ 27 25	54	14.3	GALAXY S-IRR
SHRV.73B 30	00 45 54.	+ 39 46	5	17.18	COMPACT GALAXY
PHL 2984	00 45 54.	- 01 10		18.5	BLUE STELLAR OBJECT
HELW 091	00 45 54.	- 12 07 19.			NEBULA
LB 01576	00 45 54.	- 45 48		13.3	FAINT BLUE STAR
LH115-N019	00 45 54.	- 73 24	419		EMISSION NEBULA IN SMC
KEEL 023	00 45 54.6	- 25 23 58.		17.	NEBULA
LIN.CL 035	00 45 55.	- 73 44 55.	204	13.5	STAR CLUSTER IN SMC
RIC 792	00 45 56.10	+ 41 33 55.2		17.50	BLUE OBJECT
RIC 793	00 45 56.22	+ 39 14 14.9		19.04	BLUE OBJECT
RIC 794	00 45 56.32	+ 39 53 08.6		19.11	BLUE OBJECT
KEEL 024	00 45 56.4	- 25 44 34.		18.	NEBULA
RIC 795	00 45 56.65	+ 42 16 38.9		17.67	BLUE OBJECT
LIN 095	00 45 57.	- 72 47 13.		16.80	PLANETARY NEBULA IN SMC
RIC 796	00 45 57.06	+ 39 18 23.5		17.64	BLUE OBJECT
HARO 15	00 45 58.	- 13 03			BLUE EMISSION-LINE GALAXY
RIC 797	00 45 58.54	+ 39 59 05.8		18.08	BLUE OBJECT
BAA 3.11	00 45 59.43	+ 42 07 17.5			GLOBULAR CLUSTER IN M31
LBN 0608	00 46	+ 50 45	1200		BRIGHT NEBULA
LBN 0609	00 46	+ 51 00	360		BRIGHT NEBULA
LBN 0610	00 46	+ 55 20	240		BRIGHT NEBULA
LBN 0611	00 46	+ 57 00	360		BRIGHT NEBULA
ZWG 435.007	00 46 00.	+ 10 04		15.6	GALAXY
PHL 6675	00 46 00.	+ 15 12		16.6	BLUE STELLAR OBJECT
UGC 00498	00 46 00.	+ 16 21	60	16.5	GALAXY Sb-c
MCG+04-03-006	00 46 00.	+ 27 24 30.	48	14.5	GALAXY
VDB.66B 12	00 46 00.	+ 39 26 56.		19.41	BLUE OBJECT
7ZW 001	00 46 00.	+ 85 07			COMPACT GALAXY
MCG-02-03-019	00 46 00.	- 12 59	36	14.	GALAXY
PHL 6674	00 46 00.	- 17 38		18.6	BLUE STELLAR OBJECT
LB 01577	00 46 00.	- 56 13		15.1	FAINT BLUE STAR
LH115-N020	00 46 00.	- 73 32			EMISSION NEBULA IN SMC
LH115-N021	00 46 00.	- 73 34	43		EMISSION NEBULA IN SMC
RIC 798	00 46 00.13	+ 39 27 11.0		20.10	BLUE OBJECT
RIC 799	00 46 00.65	+ 40 13 24.0		19.62	BLUE OBJECT
LIN 096	00 46 01.	- 73 31 49.		13.56	SMC EMISSION LINE OBJECT
BAA 1.688	00 46 01.67	+ 42 35 16.3			EMISSION NEBULA IN M31
RIC 800	00 46 02.18	+ 39 42 40.4		18.25	BLUE OBJECT
RIC 801	00 46 02.26	+ 40 45 21.2		19.53	BLUE OBJECT
RIC 802	00 46 02.42	+ 40 04 36.4		18.62	BLUE OBJECT
RIC 803	00 46 02.44	+ 40 12 41.0		18.81	BLUE OBJECT
ARC 0102	00 46 04.	+ 01 06		15.4	RICH CLUSTER OF GALAXIES
RIC 804	00 46 04.38	+ 39 52 47.3		18.57	BLUE OBJECT
MCG+05-03-008	00 46 06.	+ 31 40 30.	48	14.5	GALAXY
ZWG 501.020	00 46 06.	+ 31 42		15.0	GALAXY
MRK 348	00 46 06.	+ 31 42	24	15.5	GALAXY WITH UV CONTINUUM
KW 32	00 46 06.	+ 31 42	77		SEYFERT GALAXY
VVI 04	00 46 06.	+ 31 42	24	15.29	SEYFERT GALAXY
UGC 00499	00 46 06.	+ 31 42	96	15.0	GALAXY S0
PHL 0842	00 46 06.	- 08 16		16.9	BLUE STELLAR OBJECT
PHL 6676	00 46 06.	- 10 56		16.7	BLUE STELLAR OBJECT
PHL 0843	00 46 06.	- 12 48		18.2	BLUE STELLAR OBJECT
RNGC 0264	00 46 07.	- 38 32			GALAXY
LIN 098	00 46 07.	- 73 25 25.		12.92	SMC EMISSION LINE OBJECT
LIN 099	00 46 07.	- 73 27 25.		12.48	SMC EMISSION LINE OBJECT
LIN 100	00 46 07.	- 73 37 37.		14.57	SMC EMISSION LINE OBJECT
RIC 805	00 46 07.42	+ 39 52 56.6		18.76	BLUE OBJECT
RIC 806	00 46 07.62	+ 39 02 53.2		18.75	BLUE OBJECT
VDB.66B 18	00 46 08.	+ 39 53 22.		18.62	BLUE OBJECT
LIN 097	00 46 08.	- 73 19 31.		14.88	SMC EMISSION LINE OBJECT
RIC 807	00 46 08.58	+ 39 03 51.6		19.49	BLUE OBJECT
RNGC 0262	00 46 09.	+ 31 42		15.0	GALAXY
RIC 808	00 46 09.22	+ 41 33 38.2		18.54	BLUE OBJECT
RIC 809	00 46 09.34	+ 39 56 17.6		17.72	BLUE OBJECT
BAA 2.68	00 46 09.46	+ 42 30 53.8			GALAXY
RIC 810	00 46 10.70	+ 39 26 29.5		20.33	BLUE OBJECT
PHL 2986	00 46 12.	+ 00 50		18.7	BLUE STELLAR OBJECT
ZWG 435.008	00 46 12.	+ 11 06		15.6	GALAXY
UGC 00500	00 46 12.	+ 11 06	78	15.6	GALAXY Sb/Sc
MRSL 122+02/1	00 46 12.	+ 64 56	900		HII REGION
ZWG 384.008	00 46 12.	- 00 49		15.7	GALAXY
PHL 0844	00 46 12.	- 01 32		18.7	BLUE STELLAR OBJECT
ZC 0046.2-0240	00 46 12.	- 02 40	1210		CLUSTER OF GALAXIES
PHL 2987	00 46 12.	- 05 36		18.7	BLUE STELLAR OBJECT
PHL 2985	00 46 12.	- 12 35		17.9	BLUE STELLAR OBJECT
PHL 6677	00 46 12.	- 14 02		16.9	BLUE STELLAR OBJECT
PHL 6681	00 46 12.	- 15 03		18.6	BLUE STELLAR OBJECT
PHL 6678	00 46 12.	- 16 23		18.6	BLUE STELLAR OBJECT
PHL 6679	00 46 12.	- 16 30		18.7	BLUE STELLAR OBJECT

OBJECT NAME	RIGHT ASCEN.	DECLINATION	DIAM.	MAGN.	TYPE OF OBJECT
TON-S 0170	00 46 12.	- 27 19		15.9	BLUE STAR
PHL 6680	00 46 12.	- 28 04		18.2	BLUE STELLAR OBJECT
LH115-N022	00 46 12.	- 73 33	151		EMISSION NEBULA IN SMC
RNGC 0267	00 46 12.	- 73 33			CLUSTER/NEBULOSITY IN SMC
LH115-N023	00 46 12.	- 73 34	56		EMISSION NEBULA IN SMC
RIC 811	00 46 12.18	+ 41 10 19.2		19.07	BLUE OBJECT
HELW 005	00 46 13.	- 12 03			NEBULA
LIN 102	00 46 13.	- 73 23 55.		13.73	SMC EMISSION LINE OBJECT
RIC 812	00 46 13.00	+ 39 48 51.8		18.79	BLUE OBJECT
BAA 3.12	00 46 13.23	+ 42 25 16.6			GLOBULAR CLUSTER IN M31
RIC 813	00 46 13.39	+ 40 38 21.0		18.01	BLUE OBJECT
RIC 814	00 46 14.18	+ 41 43 38.4		19.34	BLUE OBJECT
RIC 815	00 46 14.20	+ 39 59 28.3		19.80	BLUE OBJECT
RIC 816	00 46 14.63	+ 39 57 19.2		18.26	BLUE OBJECT
MCG+02-03-004	00 46 15.	+ 11 05	60	15.	GALAXY
MCG-02-03-020	00 46 15.	- 12 02	72	15.	GALAXY
LIN 101	00 46 15.	- 72 48 37.		14.31	SMC EMISSION LINE OBJECT
RIC 817	00 46 15.58	+ 39 49 31.1		18.20	BLUE OBJECT
RIC 818	00 46 15.91	+ 41 24 27.6		19.57	BLUE OBJECT
IC 1587	00 46 16.	- 23 49			NONSTELLAR OBJECT
HN 0121	00 46 16.	- 23 50			NEBULA
RIC 819	00 46 16.04	+ 39 40 26.9		18.06	BLUE OBJECT
RIC 820	00 46 16.82	+ 41 50 15.6		18.75	BLUE OBJECT
LIN 104	00 46 17.	- 74 03 31.		14.25	SMC EMISSION LINE OBJECT
RIC 821	00 46 17.32	+ 39 06 24.0		19.22	BLUE OBJECT
RIC 822	00 46 17.99	+ 40 14 59.7		19.42	BLUE OBJECT
MRK 556	00 46 18.	+ 04 05	11	16.5	GALAXY WITH UV CONTINUUM
ZWG 410.009	00 46 18.	+ 09 27		15.5	GALAXY
PHL 2988	00 46 18.	+ 13 32		18.3	BLUE STELLAR OBJECT
3ZW 013	00 46 18.	+ 17 47			COMPACT GALAXY
ZC 0046.3+1753	00 46 18.	+ 17 53	3630		CLUSTER OF GALAXIES
ZWG 501.021	00 46 18.	+ 27 57		15.2	GALAXY
UGC 00501	00 46 18.	+ 27 57	108	15.2	GALAXY Sc
UGC 00502	00 46 18.	+ 45 55	84	18.	GALAXY DWRF SP
PHL 2989	00 46 18.	- 06 45		18.6	BLUE STELLAR OBJECT
MCG-02-03-021	00 46 18.	- 13 22	36	14.5	GALAXY
PHL 2990	00 46 18.	- 18 30		18.6	BLUE STELLAR OBJECT
MCG-04-03-015	00 46 18.	- 23 50 30.	48	15.5	GALAXY
PHL 6682	00 46 18.	- 27 22		17.4	BLUE STELLAR OBJECT
PHL 6683	00 46 18.	- 28 36		18.3	BLUE STELLAR OBJECT
RIC 823	00 46 19.37	+ 40 26 32.8		18.58	BLUE OBJECT
RIC 824	00 46 20.52	+ 40 01 09.9		18.27	BLUE OBJECT
RNGC 0263	00 46 21.	- 13 22		14.0	GALAXY
RIC 825	00 46 21.01	+ 40 25 45.3		17.88	BLUE OBJECT
RIC 826	00 46 21.98	+ 41 32 04.0		20.43	BLUE OBJECT
LIN 103	00 46 22.	- 72 11 25.		14.74	SMC EMISSION LINE OBJECT
RIC 827	00 46 22.10	+ 39 52 19.6		17.82	BLUE OBJECT
LIN.CL 036	00 46 23.	- 74 10 13.	270	13.3	STAR CLUSTER IN SMC
RIC 328	00 46 23.49	+ 41 28 13.0		18.97	BLUE OBJECT
RIC 829	00 46 23.72	+ 40 56 12.2		19.10	BLUE OBJECT
PHL 2991	00 46 24.	+ 00 40		18.3	BLUE STELLAR OBJECT
PHL 2992	00 46 24.	+ 01 18		18.5	BLUE STELLAR OBJECT
PHL 0845	00 46 24.	+ 01 22		18.3	BLUE STELLAR OBJECT
PHL 6684	00 46 24.	+ 03 08		18.3	BLUE STELLAR OBJECT
PHL 2993	00 46 24.	- 00 48		18.6	BLUE STELLAR OBJECT
MRK 557	00 46 24.	- 02 38	15	15.5	GALAXY WITH UV CONTINUUM
PHL 6685	00 46 24.	- 17 42		18.7	BLUE STELLAR OBJECT
PHL 0846	00 46 24.	- 21 38		15.6	BLUE STELLAR OBJECT
PHL 6687	00 46 24.	- 22 50		18.5	BLUE STELLAR OBJECT
PHL 6688	00 46 24.	- 23 06		18.8	BLUE STELLAR OBJECT
PHL 6686	00 46 24.	- 28 25		18.1	BLUE STELLAR OBJECT
LH115-N026	00 46 24.	- 73 31			EMISSION NEBULA IN SMC
LH115-N025	00 46 24.	- 73 31	48		EMISSION NEBULA IN SMC
LH115-N024	00 46 24.	- 73 36	71		EMISSION NEBULA IN SMC
LIN 106	00 46 25.	- 73 30 56.		13.51	SMC EMISSION LINE OBJECT
LIN 107	00 46 25.	- 73 31 26.		14.80	PLANETARY NEBULA IN SMC
RIC 830	00 46 25.00	+ 41 42 39.8		20.20	BLUE OBJECT
RIC 831	00 46 25.59	+ 40 50 20.1		18.81	BLUE OBJECT
RIC 832	00 46 25.60	+ 41 47 54.2		18.42	BLUE OBJECT
RIC 833	00 46 27.53	+ 40 52 27.6		18.18	BLUE OBJECT
RIC 834	00 46 27.68	+ 41 10 27.8		18.69	BLUE OBJECT
LIN 105	00 46 28.	- 72 22 38.		15.44	SMC EMISSION LINE OBJECT
RIC 835	00 46 28.16	+ 39 55 09.5		19.13	BLUE OBJECT
RIC 836	00 46 29.35	+ 39 49 19.4		17.39	BLUE OBJECT
RIC 837	00 46 29.68	+ 40 24 56.5		19.09	BLUE OBJECT
UGC 00503	00 46 30.	+ 04 15	60	17.	GALAXY IRR
ZC 0046.5+2300	00 46 30.	+ 23 00	4170		CLUSTER OF GALAXIES
5ZW 036	00 46 30.	+ 49 45			COMPACT GALAXY
PHL 2994	00 46 30.	- 00 05		18.6	BLUE STELLAR OBJECT
PHL 6690	00 46 30.	- 09 02		18.2	BLUE STELLAR OBJECT
PHL 6689	00 46 30.	- 11 44		16.9	BLUE STELLAR OBJECT
RIC 838	00 46 31.32	+ 39 51 06.5		19.28	BLUE OBJECT
SVEN 045	00 46 32.	- 31 45	12	16.7	GALAXY
LIN 107A	00 46 32.	- 73 07 38.		14.19	SMC EMISSION LINE OBJECT
RIC 839	00 46 32.26	+ 41 35 44.8		18.51	BLUE OBJECT
RIC 840	00 46 32.34	+ 39 40 26.6		19.53	BLUE OBJECT
RIC 841	00 46 32.70	+ 41 49 45.8		17.99	BLUE OBJECT
RIC 842	00 46 32.93	+ 41 30 26.1		21.00	BLUE OBJECT
MCG-04-03-016	00 46 33.	- 24 05	15	15.5	GALAXY
RIC 843	00 46 33.41	+ 39 20 55.0		17.83	BLUE OBJECT
RIC 844	00 46 34.36	+ 40 29 46.4		17.65	BLUE OBJECT
RIC 845	00 46 35.82	+ 40 39 28.9		18.52	BLUE OBJECT
RIC 846	00 46 35.86	+ 41 00 55.4		19.75	BLUE OBJECT
ZWG 435.009	00 46 36.	+ 11 52		15.7	GALAXY
PHL 0847	00 46 36.	+ 15 25		18.2	BLUE STELLAR OBJECT
ZC 0046.6+2718	00 46 36.	+ 27 18	810		CLUSTER OF GALAXIES
PHL 2995	00 46 36.	- 05 00		18.3	BLUE STELLAR OBJECT
PHL 6692	00 46 36.	- 10 40		18.4	BLUE STELLAR OBJECT
PHL 6691	00 46 36.	- 13 44		16.6	BLUE STELLAR OBJECT
PHL 2996	00 46 36.	- 20 12		18.5	BLUE STELLAR OBJECT
PHL 6693	00 46 36.	- 27 46		18.5	BLUE STELLAR OBJECT
LB 00083	00 46 36.	- 34 35		15.7	FAINT BLUE STAR
LH115-N027	00 46 36.	- 73 22	51		EMISSION NEBULA IN SMC
RIC 847	00 46 36.18	+ 39 54 41.3		18.64	BLUE OBJECT
LIN 108	00 46 38.	- 73 00 02.		13.45	SMC EMISSION LINE OBJECT
RIC 848	00 46 38.16	+ 40 58 02.6		20.35	BLUE OBJECT
RIC 849	00 46 38.86	+ 40 32 55.6		19.78	BLUE OBJECT
RIC 850	00 46 38.91	+ 41 37 03.6		20.01	BLUE OBJECT
KEEL 025	00 46 39.3	- 21 14 06.		18.	NEBULA
RIC 851	00 46 40.55	+ 40 34 26.6		18.17	BLUE OBJECT
PHL 2998	00 46 42.	+ 11 04		13.8	BLUE STELLAR OBJECT
PHL 2997	00 46 42.	+ 12 31		18.0	BLUE STELLAR OBJECT
SHRV.73B 31	00 46 42.	+ 42 47	7	15.79	COMPACT GALAXY
MCG-02-03-023	00 46 42.	- 12 43	24	15.	GALAXY
MCG-02-03-022	00 46 42.	- 12 44 30.	18	15.	GALAXY
LIN 112	00 46 42.	- 73 31 38.		14.57	SMC EMISSION LINE OBJECT
LH115-N028A	00 46 42.	- 73 32	54		EMISSION NEBULA IN SMC
LIN 113	00 46 42.	- 73 37 26.		15.32	SMC EMISSION LINE OBJECT

OBJECT NAME	RIGHT ASCEN.	DECLINATION	DIAM.	MAGN.	TYPE OF OBJECT
LIN.CL 037	00 46 42.	- 73 47 38.	270	12.2	STAR CLUSTER IN SMC
RIC 852	00 46 42.46	+ 39 16 04.6		18.48	BLUE OBJECT
ARC 0103	00 46 43.	- 06 08		17.2	RICH CLUSTER OF GALAXIES
LIN 110	00 46 43.	- 73 19 14.		14.27	SMC EMISSION LINE OBJECT
LIN 111	00 46 43.	- 73 20 26.		14.09	SMC EMISSION LINE OBJECT
RNGC 0269	00 46 43.	- 73 48		12.0	GLOBULAR CLUSTER IN SMC
LIN 109	00 46 44.	- 72 51 32.		14.11	SMC EMISSION LINE OBJECT
RIC 853	00 46 44.86	+ 41 42 00.1		18.85	BLUE OBJECT
RIC 854	00 46 45.00	+ 39 20 07.3		20.33	BLUE OBJECT
VDB.66B 09	00 46 46.	+ 39 19 09.		19.92	BLUE OBJECT
LIN 117	00 46 47.	- 73 56 20.		15.52	SMC EMISSION LINE OBJECT
RIC 855	00 46 47.47	+ 40 57 44.2		18.65	BLUE OBJECT
RIC 856	00 46 47.74	+ 39 49 52.4		18.43	BLUE OBJECT
PHL 0848	00 46 48.	+ 00 10		18.8	BLUE STELLAR OBJECT
PHL 6696	00 46 48.	+ 11 00		18.3	BLUE STELLAR OBJECT
PHL 6695	00 46 48.	+ 11 20		17.1	BLUE STELLAR OBJECT
UGC 00504	00 46 48.	+ 11 55	78	16.5	GALAXY S
PHL 6694	00 46 48.	+ 15 01		16.5	BLUE STELLAR OBJECT
ZC 0046.8+3924	00 46 48.	+ 39 24	2020		CLUSTER OF GALAXIES
MCG+00-03-008	00 46 48.	- 02 03	60	14.	GALAXY
PHL 0849	00 46 48.	- 13 43		18.2	BLUE STELLAR OBJECT
PHL 2999	00 46 48.	- 14 29		18.4	BLUE STELLAR OBJECT
MCG-04-03-017	00 46 48.	- 24 08	24	15.5	GALAXY
LH115-N029	00 46 48.	- 73 14			EMISSION NEBULA IN SMC
LIN 116	00 46 48.	- 73 30 44.		14.43	SMC EMISSION LINE OBJECT
LH115-N028	00 46 48.	- 73 32	121		EMISSION NEBULA IN SMC
LIN 115	00 46 49.	- 73 14 38.		16.28	PLANETARY NEBULA IN SMC
HSN 07	00 46 49.	- 73 55			BRIGHT GALAXY BEHIND SMC
RIC 857	00 46 49.74	+ 40 01 32.6		17.96	BLUE OBJECT
LIN 114	00 46 51.	- 72 28 20.		13.64	SMC EMISSION LINE OBJECT
RIC 858	00 46 52.67	+ 39 11 32.2		18.18	BLUE OBJECT
SVEN 046	00 46 53.	- 20 42	24	14.5	GALAXY
HELW 006	00 46 53.	- 20 42			NEBULA
RIC 859	00 46 53.10	+ 40 29 22.6		19.27	BLUE OBJECT
RIC 860	00 46 53.28	+ 39 32 43.8		19.26	BLUE OBJECT
RIC 861	00 46 53.88	+ 41 29 22.5		18.89	BLUE OBJECT
PHL 6699	00 46 54.	+ 01 33		17.7	BLUE STELLAR OBJECT
PHL 0850	00 46 54.	+ 11 11		17.8	BLUE STELLAR OBJECT
PHL 6698	00 46 54.	+ 14 04		15.9	BLUE STELLAR OBJECT
4ZW 032	00 46 54.	+ 23 18			COMPACT GALAXY
ZWG 480.010	00 46 54.	+ 23 18		14.7	GALAXY
MCG+04-03-007	00 46 54.	+ 23 20	18	18.	GALAXY
ZWG 520.001	00 46 54.	+ 38 06		15.7	GALAXY
ZWG 519.024	00 46 54.	+ 38 06		15.7	GALAXY
ZWG 384.009	00 46 54.	- 02 01		15.1	GALAXY
UGC 00505	00 46 54.	- 02 01	90	15.1	GALAXY S
KARA.68 002	00 46 54.	- 18 21	47		DWARF GALAXY
PHL 6697	00 46 54.	- 32 24		16.6	BLUE STELLAR OBJECT
LH115-N031	00 46 54.	- 73 43			EMISSION NEBULA IN SMC
LIN 119	00 46 55.	- 73 19 38.		15.17	SMC EMISSION LINE OBJECT
KEEL 026	00 46 55.2	- 20 42 13.		17.	NEBULA
RIC 862	00 46 55.24	+ 41 22 49.7		18.41	BLUE OBJECT
RIC 863	00 46 56.24	+ 41 07 41.9		18.09	BLUE OBJECT
LIN.CL 038	00 46 57.	- 70 09 44.	546	16.2	STAR CLUSTER IN SMC
LIN 118	00 46 57.	- 72 36 26.		15.22	SMC EMISSION LINE OBJECT
RIC 864	00 46 57.95	+ 38 54 29.5		18.55	BLUE OBJECT
LIN 121	00 46 58.	- 74 04 38.		14.11	SMC EMISSION LINE OBJECT
RIC 866	00 46 59.96	+ 38 56 23.5		19.31	BLUE OBJECT
LBN 0612	00 47	+ 50 25	480		BRIGHT NEBULA
LBN 0613	00 47	+ 50 30	120		BRIGHT NEBULA
KHAV 013	00 47	+ 62 10	11530		DARK NEBULA
VDB.66G 006	00 47	- 21 19	70		DWARF GALAXY
MCG+00-03-009	00 47 00.	+ 00 50	114	13.	GALAXY
PHL 3000	00 47 00.	+ 15 17		18.6	BLUE STELLAR OBJECT
ZWG 480.011	00 47 00.	+ 22 40		14.5	GALAXY
UGC 00506	00 47 00.	+ 22 40	102	14.5	GALAXY
MCG+04-03-009	00 47 00.	+ 23 19	12	18.	GALAXY
MCG+04-03-008	00 47 00.	+ 23 19	48	15.5	GALAXY
SHRV.73B 32	00 47 00.	+ 41 23	7	16.36	COMPACT GALAXY
OCL 0310	00 47 00.	+ 63 52	720		OPEN STAR CLUSTER
PHL 6700	00 47 00.	- 12 00		17.0	BLUE STELLAR OBJECT
PHL 6701	00 47 00.	- 16 48		17.9	BLUE STELLAR OBJECT
PHL 6702	00 47 00.	- 16 58		18.4	BLUE STELLAR OBJECT
LB 03159	00 47 00.	- 67 41		14.3	FAINT BLUE STAR
LB 03160	00 47 00.	- 68 43		14.0	FAINT BLUE STAR
LIN 120	00 47 00.	- 73 42 50.		15.46	PLANETARY NEBULA IN SMC
RIC 867	00 47 00.08	+ 40 00 19.3		18.01	BLUE OBJECT
RIC 868	00 47 00.37	+ 41 04 39.2		18.28	BLUE OBJECT
RIC 869	00 47 00.82	+ 40 35 35.6		19.23	BLUE OBJECT
RIC 870	00 47 01.07	+ 40 40 51.1		19.06	BLUE OBJECT
RIC 871	00 47 01.51	+ 41 35 50.0		18.78	BLUE OBJECT
RIC 872	00 47 01.75	+ 39 13 18.2		18.85	BLUE OBJECT
BAA 2.07	00 47 01.88	+ 42 16 57.6			GALAXY
ARC 0105	00 47 02.	- 05 01		17.2	RICH CLUSTER OF GALAXIES
RIC 873	00 47 02.16	+ 39 57 22.5		18.10	BLUE OBJECT
RIC 874	00 47 03.17	+ 39 18 38.0		19.27	BLUE OBJECT
RIC 875	00 47 03.30	+ 39 22 07.0		18.34	BLUE OBJECT
RIC 876	00 47 04.13	+ 39 05 57.6		18.34	BLUE OBJECT
RIC 877	00 47 04.79	+ 39 16 21.3		18.80	BLUE OBJECT
RIC 878	00 47 05.27	+ 42 12 46.3		19.76	BLUE OBJECT
RIC 879	00 47 05.50	+ 39 16 33.0		19.12	BLUE OBJECT
RIC 880	00 47 05.96	+ 39 36 45.0		18.24	BLUE OBJECT
ZC 0047.1+0040	00 47 06.	+ 00 40	1340		CLUSTER OF GALAXIES
ZWG 384.010	00 47 06.	+ 00 51		15.2	GALAXY
UGC 00507	00 47 06.	+ 00 51	138	15.2	GALAXY Sc
MCG+04-03-010	00 47 06.	+ 22 40	78	14.5	GALAXY
ZC 0047.1+2413	00 47 06.	+ 24 13	1410		CLUSTER OF GALAXIES
ZWG 501.022	00 47 06.	+ 32 00		12.6	GALAXY
UGC 00508	00 47 06.	+ 32 00	210	12.6	GALAXY SBa
MCG+05-03-009	00 47 06.	+ 32 00	144	13.	GALAXY
ZC 0047.1+3729	00 47 06.	+ 37 29	2890		CLUSTER OF GALAXIES
VDB.66B 03	00 47 06.	+ 38 55 45.		19.70	BLUE OBJECT
HMS 1.04	00 47 06.	+ 42 19			E2 GALAXY
MCG+07-02-018	00 47 06.	+ 42 19	1	20.	GALAXY
HMS 1.05	00 47 06.	+ 42 20			Sb GALAXY
MCG+07-02-019	00 47 06.	+ 42 20	9	18.	GALAXY
ZC 0047.1+5007	00 47 06.	+ 50 07	1140		CLUSTER OF GALAXIES
PHL 3002	00 47 06.	- 02 09		18.4	BLUE STELLAR OBJECT
PHL 6703	00 47 06.	- 04 24		18.0	BLUE STELLAR OBJECT
PHL 3001	00 47 06.	- 12 11		18.3	BLUE STELLAR OBJECT
PHL 3003	00 47 06.	- 15 22		18.0	BLUE STELLAR OBJECT
LB 00162	00 47 06.	- 21 06		16.2	FAINT BLUE STAR
LB 00084	00 47 06.	- 34 21		15.5	FAINT BLUE STAR
LIN 124	00 47 06.	- 73 32 32.		13.95	SMC EMISSION LINE OBJECT
RIC 881	00 47 06.07	+ 41 26 28.4		18.95	BLUE OBJECT
RIC 882	00 47 06.69	+ 39 20 04.6		18.04	BLUE OBJECT
RIC 883	00 47 07.06	+ 41 45 14.6		18.20	BLUE OBJECT
RIC 884	00 47 07.26	+ 41 07 55.4		18.20	BLUE OBJECT

OBJECT NAME	RIGHT ASCEN.	DECLINATION	DIAM.	MAGN.	TYPE OF OBJECT
RIC 885	00 47 07.58	+ 39 15 11.5		17.42	BLUE OBJECT
HOW 25	00 47 08.	- 74 34 52.	18		STAR CLUSTER IN SMC
RIC 886	00 47 08.46	+ 40 11 53.4		19.35	BLUE OBJECT
RIC 887	00 47 08.53	+ 39 37 59.4		18.70	BLUE OBJECT
ARC 0104	00 47 09.	+ 24 15		15.9	RICH CLUSTER OF GALAXIES
RNGC 0266	00 47 09.	+ 32 00		12.5	GALAXY
LIN 122	00 47 09.	- 72 21 32.		14.32	SMC EMISSION LINE OBJECT
LIN 123	00 47 09.	- 72 36 44.		14.95	SMC EMISSION LINE OBJECT
RIC 888	00 47 09.48	+ 41 34 21.8		18.24	BLUE OBJECT
LIN 128	00 47 11.	- 73 40 32.		14.91	SMC EMISSION LINE OBJECT
LIN 129	00 47 11.	- 73 41 56.		13.97	SMC EMISSION LINE OBJECT
RIC 889	00 47 11.62	+ 40 27 00.5		18.48	BLUE OBJECT
ZWG 410.010	00 47 12.	+ 04 18		15.7	GALAXY
PHL 6707	00 47 12.	+ 11 54		18.4	BLUE STELLAR OBJECT
ZWG 480.012	00 47 12.	+ 25 08		15.7	GALAXY
UGC 00509	00 47 12.	+ 31 20	84	16.5	GALAXY Sc
MRSL 122+01/1	00 47 12.	+ 64 28	120		HII REGION
PHL 3004	00 47 12.	- 13 11		18.3	BLUE STELLAR OBJECT
PHL 6704	00 47 12.	- 18 13		17.4	BLUE STELLAR OBJECT
PHL 3006	00 47 12.	- 21 06		18.1	BLUE STELLAR OBJECT
PHL 3005	00 47 12.	- 23 05		17.8	BLUE STELLAR OBJECT
PHL 6705	00 47 12.	- 28 36		17.9	BLUE STELLAR OBJECT
PHL 6706	00 47 12.	- 30 54		18.3	BLUE STELLAR OBJECT
PHL 6708	00 47 12.	- 31 00		18.3	BLUE STELLAR OBJECT
LB 03161	00 47 12.	- 68 40		14.0	FAINT BLUE STAR
LIN 125	00 47 12.	- 73 24 20.		13.71	SMC EMISSION LINE OBJECT
LH115-N030A	00 47 12.	- 73 26	20		EMISSION NEBULA IN SMC
LIN 126	00 47 12.	- 73 26 20.		14.19	SMC EMISSION LINE OBJECT
LIN 127	00 47 12.	- 73 30 26.		14.06	SMC EMISSION LINE OBJECT
RIC 890	00 47 12.42	+ 41 20 37.5		18.89	BLUE OBJECT
RIC 891	00 47 13.27	+ 41 19 46.3		18.77	BLUE OBJECT
RIC 892	00 47 13.34	+ 40 25 44.8		20.05	BLUE OBJECT
SVEN 047	00 47 14.	- 31 20	12	16.5	GALAXY
RIC 893	00 47 14.25	+ 39 35 51.0		18.92	BLUE OBJECT
RIC 894	00 47 15.62	+ 40 05 48.4		18.30	BLUE OBJECT
RIC 895	00 47 15.70	+ 40 57 08.0		19.31	BLUE OBJECT
RIC 896	00 47 16.09	+ 39 52 44.6		18.29	BLUE OBJECT
RIC 897	00 47 17.10	+ 39 25 47.0		20.67	BLUE OBJECT
RIC 898	00 47 17.58	+ 39 49 53.0		19.12	BLUE OBJECT
RIC 899	00 47 17.69	+ 40 18 49.0		18.85	BLUE OBJECT
MCG+00-03-010	00 47 18.	+ 00 41	33	15.5	GALAXY
PHL 6709	00 47 18.	+ 10 00		11.5	BLUE STELLAR OBJECT
PHL 3007	00 47 18.	+ 13 10		17.3	BLUE STELLAR OBJECT
PHL 6712	00 47 18.	+ 14 08		18.6	BLUE STELLAR OBJECT
PHL 6710	00 47 18.	+ 14 32		16.8	BLUE STELLAR OBJECT
ZWG 458.006	00 47 18.	+ 21 26		15.3	GALAXY
UGC 00510	00 47 18.	+ 21 26	84	15.3	GALAXY Sc
MCG+03-03-004	00 47 18.	+ 21 27	66	15.	GALAXY
SHRV.73B 33	00 47 18.	+ 39 01	7	16.82	COMPACT GALAXY
PHL 6711	00 47 18.	- 16 53		18.3	BLUE STELLAR OBJECT
LB 03162	00 47 18.	- 61 40		12.6	FAINT BLUE STAR
LIN 131	00 47 18.	- 73 21 32.		16.44	SMC EMISSION LINE OBJECT
LIN 132	00 47 18.	- 73 24 20.		15.25	PLANETARY NEBULA IN SMC
LH115-N030	00 47 18.	- 73 25	156		EMISSION NEBULA IN SMC
LIN.CL 039	00 47 18.	- 73 37 44.	1020	11.8	STAR CLUSTER IN SMC
RIC 900	00 47 18.15	+ 41 44 28.7		19.05	BLUE OBJECT
RIC 901	00 47 18.22	+ 41 39 23.8		18.65	BLUE OBJECT
RIC 902	00 47 18.74	+ 40 30 40.9		18.22	BLUE OBJECT
RIC 903	00 47 20.24	+ 41 43 48.4		18.27	BLUE OBJECT
RIC 904	00 47 20.39	+ 41 03 02.9		18.87	BLUE OBJECT
MCG-04-03-018	00 47 21.	- 21 47 30.	24	15.	GALAXY
LIN 130	00 47 21.	- 72 11 26.		14.34	SMC EMISSION LINE OBJECT
VDB.66B 14	00 47 22.	+ 39 40 08.		20.39	BLUE OBJECT
RIC 905	00 47 22.64	+ 41 58 04.5		18.50	BLUE OBJECT
RIC 906	00 47 22.73	+ 39 41 49.8		19.55	BLUE OBJECT
RIC 907	00 47 22.84	+ 41 17 35.3		18.81	BLUE OBJECT
RIC 908	00 47 23.28	+ 41 02 55.2		18.72	BLUE OBJECT
RIC 909	00 47 23.94	+ 39 40 42.9		20.31	BLUE OBJECT
ZWG 384.011	00 47 24.	+ 00 43		15.7	GALAXY
PHL 3008	00 47 24.	+ 10 41		14.2	BLUE STELLAR OBJECT
PHL 6713	00 47 24.	+ 13 01		17.4	BLUE STELLAR OBJECT
ZWG 501.023	00 47 24.	+ 31 28		15.6	GALAXY
UGC 00511	00 47 24.	+ 31 28	108	15.6	GALAXY Sc
MCG+00-03-011	00 47 24.	- 00 24	24	15.5	GALAXY
PHL 6714	00 47 24.	- 10 06		17.3	BLUE STELLAR OBJECT
PHL 6715	00 47 24.	- 10 46		17.7	BLUE STELLAR OBJECT
PHL 6716	00 47 24.	- 11 28		18.0	BLUE STELLAR OBJECT
MCG-02-03-025	00 47 24.	- 12 29	24	14.5	GALAXY
MCG-02-03-024	00 47 24.	- 12 34	48	15.	GALAXY
PHL 6718	00 47 24.	- 17 22		18.3	BLUE STELLAR OBJECT
MCG-04-03-019	00 47 24.	- 21 16 30.	84	16.	GALAXY
RIC 910	00 47 24.85	+ 39 20 17.5		18.38	BLUE OBJECT
RIC 911	00 47 24.93	+ 40 14 53.3		18.52	BLUE OBJECT
RIC 912	00 47 24.94	+ 39 23 43.5		18.62	BLUE OBJECT
RIC 913	00 47 25.24	+ 40 07 24.5		17.78	BLUE OBJECT
HELW 092	00 47 26.	- 12 29 26.			NEBULA
RIC 914	00 47 26.72	+ 41 40 13.8		18.18	BLUE OBJECT
RIC 915	00 47 26.78	+ 39 57 06.8		19.33	BLUE OBJECT
LIN 133	00 47 27.	- 72 23 38.		15.50	SMC EMISSION LINE OBJECT
RIC 916	00 47 27.33	+ 39 27 34.1		20.25	BLUE OBJECT
LIN 134	00 47 28.	- 73 59 50.		16.65	PLANETARY NEBULA IN SMC
LIN 135	00 47 28.	- 74 07 02.		14.87	SMC EMISSION LINE OBJECT
RIC 917	00 47 28.69	+ 39 25 06.3		19.61	BLUE OBJECT
RIC 918	00 47 28.98	+ 42 07 43.6		19.48	BLUE OBJECT
HELW 093	00 47 29.	- 12 34 26.			NEBULA
ZC 0047.5+0051	00 47 30.	+ 00 51	6990		CLUSTER OF GALAXIES
ZWG 410.011	00 47 30.	+ 06 55		15.4	GALAXY
ZWG 410.012	00 47 30.	+ 07 39		15.1	GALAXY
UGC 00512	00 47 30.	+ 07 39	72	15.1	GALAXY S-IRR
PHL 3009	00 47 30.	+ 10 20		18.3	BLUE STELLAR OBJECT
ZWG 480.013	00 47 30.	+ 24 13		15.6	GALAXY
SHRV.73B 34	00 47 30.	+ 39 00	8	16.17	COMPACT GALAXY
MCG-01-03-016	00 47 30.	- 03 49	48	14.5	GALAXY
MCG-03-03-004	00 47 30.	- 14 57	48	15.	GALAXY
TON-S 0171	00 47 30.	- 29 33		15.9	BLUE STAR
RIC 919	00 47 30.74	+ 41 16 45.0		19.90	BLUE OBJECT
RIC 920	00 47 31.02	+ 40 57 02.0		18.02	BLUE OBJECT
RIC 921	00 47 31.18	+ 40 20 05.0		18.40	BLUE OBJECT
BAA 3.13	00 47 31.49	+ 42 19 26.1			GLOBULAR CLUSTER IN M31
RIC 922	00 47 31.54	+ 39 52 59.4		18.51	BLUE OBJECT
RIC 923	00 47 31.68	+ 41 25 30.7		18.19	BLUE OBJECT
RIC 924	00 47 31.73	+ 40 21 53.1		17.55	BLUE OBJECT
RIC 925	00 47 32.63	+ 40 52 52.4		18.11	BLUE OBJECT
LIN 138	00 47 35.	- 73 43 21.		14.71	PLANETARY NEBULA IN SMC
PHL 3012	00 47 36.	+ 01 30		18.7	BLUE STELLAR OBJECT
ZC 0047.6+1040	00 47 36.	+ 10 40	1750		CLUSTER OF GALAXIES
PHL 6722	00 47 36.	+ 14 26		15.7	BLUE STELLAR OBJECT
PHL 0851	00 47 36.	- 00 18		18.4	BLUE STELLAR OBJECT
ZWG 384.012	00 47 36.	- 02 12		15.5	GALAXY
MCG-01-03-017	00 47 36.	- 05 28	78	12.5	GALAXY
PHL 6723	00 47 36.	- 12 14		18.1	BLUE STELLAR OBJECT
PHL 6719	00 47 36.	- 12 26		18.3	BLUE STELLAR OBJECT
MCG-02-03-026	00 47 36.	- 12 27	18	14.5	GALAXY
PHL 0852	00 47 36.	- 14 28		18.5	BLUE STELLAR OBJECT
PHL 6720	00 47 36.	- 14 44		18.7	BLUE STELLAR OBJECT
PHL 3010	00 47 36.	- 16 10		18.1	BLUE STELLAR OBJECT
PHL 3011	00 47 36.	- 19 28		18.8	BLUE STELLAR OBJECT
PHL 6724	00 47 36.	- 20 20		18.2	BLUE STELLAR OBJECT
PHL 6721	00 47 36.	- 28 39		18.4	BLUE STELLAR OBJECT
LB 03163	00 47 36.	- 62 30		14.1	FAINT BLUE STAR
RIC 926	00 47 36.20	+ 42 14 55.8		19.82	BLUE OBJECT
RIC 927	00 47 36.90	+ 42 08 06.9		20.29	BLUE OBJECT
HELW 094	00 47 37.	- 12 27 26.			NEBULA
LIN 136	00 47 37.	- 72 52 09.		14.53	SMC EMISSION LINE OBJECT
LIN 137	00 47 37.	- 73 04 39.		14.97	PLANETARY NEBULA IN SMC
RIC 928	00 47 37.29	+ 39 19 03.6		19.26	BLUE OBJECT
RNGC 0268	00 47 38.	- 05 28		13.5	GALAXY
MCG+04-03-011	00 47 39.	+ 24 13 30.	33	15.5	GALAXY
HOLM 024B	00 47 39.	- 18 36	66	13.3	PART OF MULTIPLE GALAXY
RIC 929	00 47 39.48	+ 42 07 03.7		18.54	BLUE OBJECT
HOLM 024A	00 47 41.	- 18 35	30	12.9	PART OF MULTIPLE GALAXY
RIC 930	00 47 41.40	+ 41 41 11.4		18.65	BLUE OBJECT
RIC 931	00 47 41.58	+ 40 36 29.2		18.97	BLUE OBJECT
RIC 932	00 47 41.85	+ 39 12 27.6		19.21	BLUE OBJECT
ZC 0047.7+2122	00 47 42.	+ 21 22	1210		CLUSTER OF GALAXIES
ZWG 480.014	00 47 42.	+ 24 15		15.0	GALAXY
SHRV.73B 35	00 47 42.	+ 40 57	7	16.51	COMPACT GALAXY
SHRV.73B 36	00 47 42.	+ 41 49	10	16.07	COMPACT GALAXY
PHL 6725	00 47 42.	- 11 29		17.9	BLUE STELLAR OBJECT
LB 01578	00 47 42.	- 54 14		12.8	FAINT BLUE STAR
LH115-N033	00 47 42.	- 73 43			EMISSION NEBULA IN SMC
RIC 933	00 47 42.51	+ 42 03 08.1		18.08	BLUE OBJECT
RIC 934	00 47 43.05	+ 40 15 41.1		19.66	BLUE OBJECT
RIC 935	00 47 43.05	+ 42 11 46.0		19.10	BLUE OBJECT
RIC 936	00 47 43.09	+ 40 37 16.3		20.59	BLUE OBJECT
LIN 140	00 47 44.	- 74 33 57.		13.96	SMC EMISSION LINE OBJECT
RIC 937	00 47 44.14	+ 39 48 01.9		17.93	BLUE OBJECT
RIC 938	00 47 44.44	+ 40 29 32.7		17.92	BLUE OBJECT
RIC 939	00 47 45.34	+ 41 19 47.9		19.18	BLUE OBJECT
RIC 940	00 47 46.32	+ 42 00 17.9		19.94	BLUE OBJECT
RIC 941	00 47 46.80	+ 40 21 27.0		17.70	BLUE OBJECT
LIN 139	00 47 47.	- 73 40 03.		14.44	SMC EMISSION LINE OBJECT
RIC 942	00 47 47.24	+ 41 23 34.0		18.56	BLUE OBJECT
PHL 6726	00 47 48.	+ 00 08		18.7	BLUE STELLAR OBJECT
PHL 3014	00 47 48.	+ 00 36		18.2	BLUE STELLAR OBJECT
ZWG 435.010	00 47 48.	+ 11 27		15.7	GALAXY
UGC 00513	00 47 48.	+ 11 27	66	15.7	GALAXY S
ZC 0047.8+1848	00 47 48.	+ 18 48	1080		CLUSTER OF GALAXIES
PHL 0853	00 47 48.	- 05 07		17.5	BLUE STELLAR OBJECT
MCG-01-03-018	00 47 48.	- 06 07 30.	42	13.	GALAXY
PHL 3015	00 47 48.	- 14 32		18.6	BLUE STELLAR OBJECT
PHL 6727	00 47 48.	- 14 56		18.6	BLUE STELLAR OBJECT
PHL 3013	00 47 48.	- 17 02		17.9	BLUE STELLAR OBJECT
ARC 0107	00 47 49.	- 19 32		17.1	RICH CLUSTER OF GALAXIES
RIC 943	00 47 49.34	+ 41 39 50.0		20.23	BLUE OBJECT
RIC 944	00 47 49.35	+ 39 06 57.8		18.44	BLUE OBJECT
RIC 945	00 47 49.90	+ 42 12 54.4		17.69	BLUE OBJECT
RIC 946	00 47 50.71	+ 40 43 34.1		18.79	BLUE OBJECT
MCG+04-03-012	00 47 51.	+ 24 15	24	15.	GALAXY
IC 0053	00 47 52.	+ 10 20 53.			NONSTELLAR OBJECT
RIC 948	00 47 52.90	+ 40 51 52.1		18.41	BLUE OBJECT
ARC 0106	00 47 53.	- 11 58		17.2	RICH CLUSTER OF GALAXIES
RIC 949	00 47 53.86	+ 42 06 56.9		20.08	BLUE OBJECT
RIC 950	00 47 53.92	+ 39 10 03.6		19.71	BLUE OBJECT
PHL 3017	00 47 54.	- 06 30		18.6	BLUE STELLAR OBJECT
PHL 3018	00 47 54.	- 12 18		18.4	BLUE STELLAR OBJECT
PHL 6728	00 47 54.	- 15 20		18.0	BLUE STELLAR OBJECT
PHL 3016	00 47 54.	- 17 58		17.2	BLUE STELLAR OBJECT
PHL 6729	00 47 54.	- 27 33		18.3	BLUE STELLAR OBJECT
LH115-N032	00 47 54.	- 73 05	33		EMISSION NEBULA IN SMC
LH115-N034	00 47 54.	- 73 27			EMISSION NEBULA IN SMC
LIN 141	00 47 55.	- 72 53 27.		15.02	SMC EMISSION LINE OBJECT
RIC 951	00 47 55.61	+ 41 19 57.6		18.22	BLUE OBJECT
RIC 952	00 47 56.94	+ 40 16 25.2		18.39	BLUE OBJECT
RIC 953	00 47 57.19	+ 40 33 45.3		18.50	BLUE OBJECT
RIC 954	00 47 57.21	+ 39 14 38.7		19.27	BLUE OBJECT
RIC 955	00 47 57.41	+ 41 02 48.6		17.95	BLUE OBJECT
RIC 956	00 47 57.63	+ 40 02 31.1		17.91	BLUE OBJECT
LIN 142	00 47 59.	- 73 26 27.		15.09	PLANETARY NEBULA IN SMC
RIC 957	00 47 59.60	+ 41 07 34.8		18.83	BLUE OBJECT
RIC 958	00 47 59.83	+ 41 31 58.8		19.79	BLUE OBJECT
LBN 0614	00 48	+ 65 00	660		BRIGHT NEBULA
UGC 00514	00 48 00.	+ 00 11	60	16.0	GALAXY
PHL 6732	00 48 00.	+ 10 24		18.5	BLUE STELLAR OBJECT
PHL 6731	00 48 00.	+ 11 46		17.8	BLUE STELLAR OBJECT
UGC 00515	00 48 00.	+ 11 50	66	17.	GALAXY
PHL 6730	00 48 00.	+ 13 12		17.8	BLUE STELLAR OBJECT
MCG+00-03-012	00 48 00.	- 02 12	114	12.	GALAXY
PHL 6733	00 48 00.	- 02 36		18.0	BLUE STELLAR OBJECT
PHL 6734	00 48 00.	- 18 30		18.1	BLUE STELLAR OBJECT
RIC 959	00 48 00.22	+ 42 08 45.1		18.42	BLUE OBJECT
RIC 960	00 48 00.31	+ 39 11 19.0		19.24	BLUE OBJECT
RIC 961	00 48 00.72	+ 42 08 31.6		19.61	BLUE OBJECT
RIC 962	00 48 00.94	+ 41 19 42.7		18.27	BLUE OBJECT
RIC 963	00 48 01.70	+ 39 27 04.0		18.20	BLUE OBJECT
HOW 26	00 48 02.	- 73 59 52.	36		STAR CLUSTER IN SMC
RIC 964	00 48 02.17	+ 40 29 33.9		19.01	BLUE OBJECT
RIC 965	00 48 03.67	+ 41 38 15.2		18.72	BLUE OBJECT
RIC 966	00 48 03.76	+ 42 00 01.8		19.37	BLUE OBJECT
SC 0046-6705.9	00 48 04.	- 66 49 32.	72		NEBULA
LIN 144	00 48 04.	- 74 01 09.		16.01	PLANETARY NEBULA IN SMC
RIC 967	00 48 04.99	+ 40 00 16.7		18.58	BLUE OBJECT
LIN 143	00 48 05.	- 73 39 03.		13.88	SMC EMISSION LINE OBJECT
RIC 968	00 48 05.08	+ 41 45 28.0		18.90	BLUE OBJECT
RIC 969	00 48 05.91	+ 39 55 32.8		20.06	BLUE OBJECT
MCG+02-03-005	00 48 06.	+ 10 19	36	15.	GALAXY
ZWG 435.011	00 48 06.	+ 10 20		15.5	GALAXY
UGC 00516	00 48 06.	+ 10 20	90	15.5	GALAXY E-SO
ZWG 435.012	00 48 06.	+ 11 05		15.7	GALAXY
UGC 00517	00 48 06.	+ 11 05	66	15.7	GALAXY COMPACT
MCG+02-03-006	00 48 06.	+ 11 49	48	16.	GALAXY
PHL 3020	00 48 06.	+ 14 01		18.0	BLUE STELLAR OBJECT
PHL 3019	00 48 06.	+ 14 09		16.6	BLUE STELLAR OBJECT
ZWG 501.024	00 48 06.	+ 28 25		15.6	GALAXY
ZWG 501.025	00 48 06.	+ 29 53		15.2	GALAXY
UGC 00518	00 48 06.	+ 29 53	66	15.2	GALAXY SBc

OBJECT NAME	RIGHT ASCEN.	DECLINATION	DIAM.	MAGN.	TYPE OF OBJECT
MCG+05-03-010	00 48 06.	+ 29 53	48	15.	GALAXY
OCL 0311	00 48 06.	+ 57 55	360	17.	OPEN STAR CLUSTER
MCG-02-03-027	00 48 06.	- 08 55	78	13.	GALAXY
MCG-03-03-005	00 48 06.	- 15 34	84	15.	GALAXY
PHL 3021	00 48 06.	- 16 40		18.6	BLUE STELLAR OBJECT
PHL 3022	00 48 06.	- 17 26		18.1	BLUE STELLAR OBJECT
PHL 6657	00 48 06.	- 18 28		18.1	BLUE STELLAR OBJECT
PHL 0854	00 48 06.	- 23 41		17.2	BLUE STELLAR OBJECT
PHL 6735	00 48 06.	- 28 00		18.2	BLUE STELLAR OBJECT
RIC 970	00 48 06.88	+ 39 54 46.1		18.43	BLUE OBJECT
SVEN 048	00 48 07.	- 31 11	36	16.3	GALAXY
SVEN 049	00 48 07.	- 31 15	60	15.5	GALAXY
RIC 971	00 48 07.05	+ 39 39 47.7		18.67	BLUE OBJECT
RNGC 0270	00 48 08.	- 08 55		13.0	GALAXY
MCG-04-03-020	00 48 09.	- 21 30 30.	48	15.5	GALAXY
RIC 972	00 48 09.28	+ 39 16 40.5		18.55	BLUE OBJECT
RIC 973	00 48 09.64	+ 41 33 35.5		18.75	BLUE OBJECT
RIC 974	00 48 09.81	+ 41 01 22.8		18.74	BLUE OBJECT
RIC 975	00 48 10.49	+ 41 33 36.2		19.60	BLUE OBJECT
PHL 6738	00 48 12.	+ 01 07		17.5	BLUE STELLAR OBJECT
ZC 0048.2+0200	00 48 12.	+ 02 00	1140		CLUSTER OF GALAXIES
PHL 6736	00 48 12.	+ 10 39		18.2	BLUE STELLAR OBJECT
PHL 6737	00 48 12.	+ 10 40		17.9	BLUE STELLAR OBJECT
ZWG 480.015	00 48 12.	+ 25 21		15.7	GALAXY
SHRV.73B 37	00 48 12.	+ 41 34	8	16.61	COMPACT GALAXY
ZC 0048.2+4305	00 48 12.	+ 43 05	4970		CLUSTER OF GALAXIES
ZWG 384.013	00 48 12.	- 02 09		13.2	GALAXY
UGC 00519	00 48 12.	- 02 09	162	13.2	GALAXY SBb
PHL 0855	00 48 12.	- 03 42		17.2	BLUE STELLAR OBJECT
MCG-01-03-019	00 48 12.	- 07 09	108	13.	GALAXY
PHL 6739	00 48 12.	- 09 34		17.9	BLUE STELLAR OBJECT
PHL 0856	00 48 12.	- 09 43		15.0	BLUE STELLAR OBJECT
PHL 6740	00 48 12.	- 10 04		18.4	BLUE STELLAR OBJECT
PHL 3023	00 48 12.	- 11 22		18.5	BLUE STELLAR OBJECT
LH115-N038	00 48 12.	- 74 01			EMISSION NEBULA IN SMC
RIC 976	00 48 12.53	+ 41 06 48.1		17.90	BLUE OBJECT
RNGC 0271	00 48 13.	- 02 09		13.0	GALAXY
IC 0054	00 48 13.	- 02 32 50.			OPEN CLUSTER
LIN 145	00 48 13.	- 72 52 09.		15.02	SMC EMISSION LINE OBJECT
RIC 977	00 48 14.67	+ 41 31 00.3		18.59	BLUE OBJECT
LIN 146	00 48 17.	- 73 37 27.		13.77	SMC EMISSION LINE OBJECT
ZC 0048.3+4119	00 48 18.	+ 41 19	4770		CLUSTER OF GALAXIES
ZWG 536.001	00 48 18.	+ 44 27		15.7	GALAXY
ZWG 535.022	00 48 18.	+ 44 27		15.7	GALAXY
KARA.73B 0035	00 48 18.	+ 44 27	24	15.7	ISOLATED GALAXY S
PHL 6743	00 48 18.	- 02 46		18.5	BLUE STELLAR OBJECT
MCG-01-03-020	00 48 18.	- 07 08	18	18.	GALAXY
PHL 6741	00 48 18.	- 14 46		18.7	BLUE STELLAR OBJECT
PHL 6742	00 48 18.	- 16 22		18.3	BLUE STELLAR OBJECT
MCG-05-03-007	00 48 18.	- 30 42	48	15.5	GALAXY
MCG-05-03-008	00 48 18.	- 31 41	60	15.	GALAXY
RIC 978	00 48 19.32	+ 40 25 55.9		18.39	BLUE OBJECT
RNGC 0273	00 48 20.	- 07 10		13.0	GALAXY
PHL 6755	00 48 22.	- 15 04		18.0	BLUE STELLAR OBJECT
RIC 979	00 48 22.15	+ 39 49 28.7		20.11	BLUE OBJECT
RIC 980	00 48 22.53	+ 41 33 52.6		18.03	BLUE OBJECT
RIC 981	00 48 23.74	+ 42 09 25.3		18.65	BLUE OBJECT
MCG+00-03-013	00 48 24.	+ 01 27	24	15.5	GALAXY
PHL 6775	00 48 24.	+ 10 15		18.1	BLUE STELLAR OBJECT
ZWG 435.013	00 48 24.	+ 10 28		15.5	GALAXY
MCG+02-03-007	00 48 24.	+ 11 15	48	15.	GALAXY
ZWG 501.026	00 48 24.	+ 29 15		15.5	GALAXY
MCG+05-03-011	00 48 24.	+ 29 15	24	16.	GALAXY
SHRV.73B 38	00 48 24.	+ 39 32	7	16.72	COMPACT GALAXY
SHRV.73B 39	00 48 24.	+ 40 55	7	16.88	COMPACT GALAXY
PHL 0857	00 48 24.	- 01 20		18.2	BLUE STELLAR OBJECT
PHL 6744	00 48 24.	- 01 26		18.4	BLUE STELLAR OBJECT
VV 081B	00 48 24.	- 07 21 30.	84	13.	INTERACTING GALAXY
PHL 0858	00 48 24.	- 07 46		16.7	BLUE STELLAR OBJECT
PHL 3025	00 48 24.	- 08 44		18.6	BLUE STELLAR OBJECT
PHL 6746	00 48 24.	- 11 20		18.2	BLUE STELLAR OBJECT
PHL 6747	00 48 24.	- 12 19		18.6	BLUE STELLAR OBJECT
PHL 3024	00 48 24.	- 12 45		18.6	BLUE STELLAR OBJECT
PHL 6745	00 48 24.	- 15 32		18.7	BLUE STELLAR OBJECT
LH115-N035	00 48 24.	- 72 49	49		EMISSION NEBULA IN SMC
RIC 982	00 48 24.82	+ 41 06 04.3		18.42	BLUE OBJECT
RIC 983	00 48 26.12	+ 41 56 48.4		19.01	BLUE OBJECT
LIN 149	00 48 28.	- 73 45 21.		14.89	SMC EMISSION LINE OBJECT
RIC 984	00 48 28.90	+ 41 38 01.8		18.77	BLUE OBJECT
ARP 140	00 48 29.	- 07 20			PECULIAR GALAXY
3ZW 014	00 48 30.	+ 00 52			COMPACT GALAXY
ZWG 384.014	00 48 30.	+ 00 52		15.6	GALAXY
ZC 0048.5+1540	00 48 30.	+ 15 40	3560		CLUSTER OF GALAXIES
UGC 00520	00 48 30.	+ 17 28	60	17.	GALAXY Sb-c
MCG-01-03-021	00 48 30.	- 07 20	60	13.	GALAXY
HOLM 026B	00 48 30.	- 07 21	54	12.7	PART OF MULTIPLE GALAXY
VV 081A	00 48 30.	- 07 21	66	13.	INTERACTING GALAXY
PHL 6748	00 48 30.	- 08 45		18.6	BLUE STELLAR OBJECT
RIC 985	00 48 30.00	+ 39 34 59.5		17.84	BLUE OBJECT
RIC 986	00 48 30.87	+ 41 21 21.4		18.75	BLUE OBJECT
ARC 0108	00 48 31.	- 06 50		17.2	RICH CLUSTER OF GALAXIES
LIN 148	00 48 31.	- 72 40 09.		14.31	SMC EMISSION LINE OBJECT
RIC 987	00 48 31.28	+ 40 25 48.8		19.29	BLUE OBJECT
RNGC 0275	00 48 32.	- 07 20		13.5	GALAXY
RNGC 0274	00 48 32.	- 07 20		13.0	GALAXY
HOLM 026A	00 48 32.	- 07 21	54	12.5	PART OF MULTIPLE GALAXY
LIN 147	00 48 32.	- 72 29 09.		13.10	SMC EMISSION LINE OBJECT
MCG-01-03-022	00 48 33.	- 07 21	66	13.	GALAXY
RIC 988	00 48 33.32	+ 41 24 28.6		19.03	BLUE OBJECT
HN 0122	00 48 34.	- 23 50			NEBULA
IC 1588	00 48 34.	- 23 50			NONSTELLAR OBJECT
RIC 989	00 48 35.2	+ 42 13 42.		19.33	BLUE OBJECT
RIC 990	00 48 35.77	+ 40 01 38.5		18.87	BLUE OBJECT
RIC 991	00 48 35.91	+ 40 36 19.9		19.12	BLUE OBJECT
PHL 0860	00 48 36.	+ 00 26		16.2	BLUE STELLAR OBJECT
PHL 0859	00 48 36.	+ 10 58		18.6	BLUE STELLAR OBJECT
MCG+02-03-008	00 48 36.	+ 11 44	36	14.5	GALAXY
ZWG 435.014	00 48 36.	+ 11 45		15.4	GALAXY
UGC 00521	00 48 36.	+ 11 45	72	15.4	GALAXY IRR
OCL 0312	00 48 36.	+ 35 34			OPEN STAR CLUSTER
KARA.68 003	00 48 36.	- 02 20	27		DWARF GALAXY
PHL 3027	00 48 36.	- 12 21		18.4	BLUE STELLAR OBJECT
PHL 6750	00 48 36.	- 15 30		18.7	BLUE STELLAR OBJECT
PHL 3026	00 48 36.	- 16 30		18.5	BLUE STELLAR OBJECT
PHL 6749	00 48 36.	- 17 34		17.9	BLUE STELLAR OBJECT
PHL 6751	00 48 36.	- 18 34		18.7	BLUE STELLAR OBJECT
PHL 0861	00 48 36.	- 20 18		15.2	BLUE STELLAR OBJECT
LB 00163	00 48 36.	- 20 38		16.3	FAINT BLUE STAR
RNGC 0272	00 48 38.	+ 35 34			OPEN CLUSTER
HOLM 025B	00 48 38.	+ 40 28	24	14.9	PART OF MULTIPLE GALAXY
LIN 150	00 48 38.	- 72 25 09.		15.14	SMC EMISSION LINE OBJECT
RIC 992	00 48 38.6	+ 42 13 33.		19.61	BLUE OBJECT
HOLM 025A	00 48 39.	+ 40 27	60	13.5	PART OF MULTIPLE GALAXY
RIC 993	00 48 39.89	+ 41 30 53.3		19.59	BLUE OBJECT
LIN 152	00 48 40.	- 73 42 16.		14.61	SMC EMISSION LINE OBJECT
LIN 153	00 48 40.	- 73 46 52.		13.46	SMC EMISSION LINE OBJECT
RIC 994	00 48 40.74	+ 40 27 12.6		17.16	BLUE OBJECT
RIC 995	00 48 40.88	+ 41 43 34.7		19.50	BLUE OBJECT
LIN 151	00 48 41.	- 73 15 34.		14.79	SMC EMISSION LINE OBJECT
PHL 0862	00 48 42.	+ 03 23		17.7	BLUE STELLAR OBJECT
ZWG 435.015	00 48 42.	+ 10 32		15.7	GALAXY
MCG+07-02-020	00 48 42.	+ 40 26 30.	60	14.5	GALAXY
ZWG 535.023	00 48 42.	+ 40 27		14.6	GALAXY
UGC 00522	00 48 42.	+ 40 27	66	14.6	GALAXY Sb
KARA.73B 0036	00 48 42.	+ 40 27	54	14.6	ISOLATED GALAXY S
ISS 0032	00 48 42.	+ 62 43	183		STELLAR RING
PHL 0863	00 48 42.	- 06 11		18.5	BLUE STELLAR OBJECT
PHL 3028	00 48 42.	- 12 39		18.7	BLUE STELLAR OBJECT
PHL 6752	00 48 42.	- 17 13		17.8	BLUE STELLAR OBJECT
PHL 6753	00 48 42.	- 20 36		17.1	BLUE STELLAR OBJECT
PHL 0864	00 48 42.	- 29 22		18.1	BLUE STELLAR OBJECT
PHL 6754	00 48 42.	- 30 30		17.0	BLUE STELLAR OBJECT
MCG-05-03-009	00 48 42.	- 32 43	48	15.	GALAXY
LB 01579	00 48 42.	- 58 06		13.9	FAINT BLUE STAR
LH115-N039	00 48 42.	- 73 16			EMISSION NEBULA IN SMC
RIC 996	00 48 42.63	+ 41 06 56.7		18.95	BLUE OBJECT
RIC 997	00 48 42.73	+ 40 37 57.0		19.74	BLUE OBJECT
MCG+00-03-014	00 48 48.	+ 01 06	45	15.5	GALAXY
PHL 6756	00 48 48.	+ 10 35		17.7	BLUE STELLAR OBJECT
ZWG 557.002	00 48 48.	+ 51 50		15.7	GALAXY
UGC 00523	00 48 48.	+ 51 50	72	15.7	GALAXY Sa-b
ISS 0033	00 48 48.	+ 61 58	259		STELLAR RING
VB 092	00 48 48.	+ 62 03	275		STELLAR RING
VB 093	00 48 48.	+ 62 48	342		STELLAR RING
PHL 6757	00 48 48.	- 13 12		18.3	BLUE STELLAR OBJECT
PHL 3030	00 48 48.	- 21 28		18.7	BLUE STELLAR OBJECT
PHL 6758	00 48 48.	- 24 00		18.6	BLUE STELLAR OBJECT
PHL 6790	00 48 48.	- 24 30		18.1	BLUE STELLAR OBJECT
TON-S 0172	00 48 48.	- 28 03		16.0	BLUE STAR
PHL 3029	00 48 48.	- 28 05		18.1	BLUE STELLAR OBJECT
LH115-N036	00 48 48.	- 73 09	263		EMISSION NEBULA IN SMC
LB 03164	00 48 48.	- 74 48		13.0	FAINT BLUE STAR
RIC 998	00 48 49.29	+ 41 45 40.2		19.00	BLUE OBJECT
RIC 999	00 48 50.82	+ 41 13 57.6		19.48	BLUE OBJECT
MCG+05-03-012	00 48 51.	+ 29 25 30.	90	16.	GALAXY
HOW 28	00 48 51.	- 74 23 10.	18		STAR CLUSTER IN SMC
HOW 27	00 48 51.	- 74 53 52.	12		STAR CLUSTER IN SMC
LIN 156	00 48 53.	- 73 09 04.		11.87	SMC EMISSION LINE OBJECT
LIN.CL 040	00 48 53.	- 73 12 46.	750	12.3	STAR CLUSTER IN SMC
MCG+05-03-013	00 48 54.	+ 29 06 30.	54	15.	GALAXY
ZWG 501.027	00 48 54.	+ 29 08		14.5	GALAXY
UGC 00524	00 48 54.	+ 29 08	54	14.5	GALAXY SBb
ZWG 501.028	00 48 54.	+ 29 27		15.5	GALAXY
UGC 00525	00 48 54.	+ 29 27	114	15.5	GALAXY S
MCG+00-03-015	00 48 54.	- 01 32	18	15.5	GALAXY
MCG-01-03-023	00 48 54.	- 03 24	66	14.	GALAXY
PHL 3031	00 48 54.	- 08 18		18.6	BLUE STELLAR OBJECT
MCG-02-03-028	00 48 54.	- 08 51	18	13.5	GALAXY
PHL 6759	00 48 54.	- 17 18		18.0	BLUE STELLAR OBJECT
PHL 6760	00 48 54.	- 19 58		18.1	BLUE STELLAR OBJECT
LB 01580	00 48 54.	- 45 25		12.8	FAINT BLUE STAR
LB 03165	00 48 54.	- 63 21		12.8	FAINT BLUE STAR
LH115-N040	00 48 54.	- 73 59			EMISSION NEBULA IN SMC
LIN 155	00 48 55.	- 72 27 04.		15.66	SMC EMISSION LINE OBJECT
RNGC 0277	00 48 56.	- 08 51		13.0	GALAXY
LIN 154	00 48 56.	- 72 05 40.		15.02	SMC EMISSION LINE OBJECT
LIN 159	00 48 58.	- 73 33 58.		14.02	SMC EMISSION LINE OBJECT
LIN 160	00 48 58.	- 73 43 10.		12.60	SMC EMISSION LINE OBJECT
LIN 158	00 48 59.	- 73 11 22.		14.31	SMC EMISSION LINE OBJECT
LBN 0615	00 49	+ 56 10	960		BRIGHT NEBULA
PHL 0865	00 49 00.	+ 01 25		18.3	BLUE STELLAR OBJECT
PHL 3032	00 49 00.	+ 09 38		17.5	BLUE STELLAR OBJECT
PHL 6763	00 49 00.	+ 12 59		18.0	BLUE STELLAR OBJECT
PHL 6761	00 49 00.	+ 14 50		17.8	BLUE STELLAR OBJECT
ZWG 480.016	00 49 00.	+ 22 15		15.7	GALAXY
PHL 6764	00 49 00.	- 00 10		18.6	BLUE STELLAR OBJECT
PHL 3033	00 49 00.	- 01 24		17.6	BLUE STELLAR OBJECT
ZWG 384.015	00 49 00.	- 01 30		15.4	GALAXY
MCG+00-03-016	00 49 00.	- 02 19	24	16.	GALAXY
MCG-02-03-029	00 49 00.	- 08 47	78	15.	GALAXY
PHL 3034	00 49 00.	- 10 06		18.8	BLUE STELLAR OBJECT
IC 0056	00 49 00.	- 13 06 39.			NONSTELLAR OBJECT
PHL 6762	00 49 00.	- 28 26		18.5	BLUE STELLAR OBJECT
LIN 161	00 49 00.	- 72 50 52.		14.60	SMC EMISSION LINE OBJECT
LIN 157	00 49 00.	- 72 58 22.		12.98	SMC EMISSION LINE OBJECT
LH115-N037	00 49 00.	- 73 03	319		EMISSION NEBULA IN SMC
KARA.73 01	00 49 01.	- 38 00	60		DWARF GALAXY
MCG-02-03-030	00 49 03.	- 13 07	48	14.5	GALAXY
LIN 164	00 49 04.	- 73 23 58.		12.76	SMC EMISSION LINE OBJECT
UGC 00526	00 49 06.	+ 02 55	60	16.5	GALAXY
MCG+01-03-006	00 49 06.	+ 07 27	30	14.5	GALAXY
ZWG 410.013	00 49 06.	+ 07 28		15.0	GALAXY
PHL 6766	00 49 06.	+ 09 44		18.5	BLUE STELLAR OBJECT
MCG+08-02-016	00 49 06.	+ 47 17	114	11.5	GALAXY
PHL 6765	00 49 06.	- 10 04		18.7	BLUE STELLAR OBJECT
PHL 0866	00 49 06.	- 27 00		17.5	BLUE STELLAR OBJECT
LIN 162	00 49 06.	- 72 45 34.		15.48	SMC EMISSION LINE OBJECT
LIN 163	00 49 06.	- 73 01 40.		13.62	SMC EMISSION LINE OBJECT
KN 10.01	00 49 06.9	+ 07 26 53.			NEBULA
IC 0055	00 49 07.	+ 07 26 51.			NONSTELLAR OBJECT
IC 1589	00 49 08.	- 34 44 10.			NONSTELLAR OBJECT
RNGC 0278	00 49 11.	+ 47 17		12.5	GALAXY
LIN 166	00 49 11.	- 73 09 16.		12.92	SMC EMISSION LINE OBJECT
UGC 00527	00 49 12.	+ 02 49	90	16.0	GALAXY
MCG+00-03-017	00 49 12.	+ 02 49	36	15.	GALAXY
PHL 6767	00 49 12.	+ 10 34		18.4	BLUE STELLAR OBJECT
PHL 3035	00 49 12.	+ 11 28		17.2	BLUE STELLAR OBJECT
SHRV.73A 25	00 49 12.	+ 41 19	8	16.84	GLOBULAR CLUSTER IN M31
ZWG 550.016	00 49 12.	+ 47 17		10.5	GALAXY
UGC 00528	00 49 12.	+ 47 17	168	10.5	GALAXY S
VB 094	00 49 12.	+ 59 07	517		STELLAR RING
VB 017	00 49 12.	+ 64 15	188		STELLAR RING
ISS 0006	00 49 12.	+ 64 15	192		STELLAR RING
PHL 6769	00 49 12.	- 00 30		18.5	BLUE STELLAR OBJECT
PHL 6768	00 49 12.	- 00 46		18.0	BLUE STELLAR OBJECT
PHL 6770	00 49 12.	- 10 28		18.1	BLUE STELLAR OBJECT

OBJECT NAME	RIGHT ASCEN.	DECLINATION	DIAM.	MAGN.	TYPE OF OBJECT
PHL 6771	00 49 12.	- 11 06		18.6	BLUE STELLAR OBJECT
PHL 0867	00 49 12.	- 31 00		14.8	BLUE STELLAR OBJECT
MCG-06-03-001	00 49 12.	- 37 56	42	14.5	GALAXY
LIN 165	00 49 12.	- 72 59 34.		14.11	SMC EMISSION LINE OBJECT
LH115-N041	00 49 12.	- 73 10	21		EMISSION NEBULA IN SMC
LIN 169	00 49 15.	- 73 48 40.		14.36	SMC EMISSION LINE OBJECT
LIN 167	00 49 17.	- 72 59 58.		13.79	SMC EMISSION LINE OBJECT
LIN 168	00 49 17.	- 73 15 58.		14.11	SMC EMISSION LINE OBJECT
PHL 0868	00 49 18.	+ 00 44		17.5	BLUE STELLAR OBJECT
PHL 6774	00 49 18.	+ 13 04		18.2	BLUE STELLAR OBJECT
VB 095	00 49 18.	+ 62 20	940		STELLAR RING
PHL 0869	00 49 18.	- 11 16		18.1	BLUE STELLAR OBJECT
PHL 6773	00 49 18.	- 14 44		18.7	BLUE STELLAR OBJECT
PHL 0870	00 49 18.	- 21 31		18.6	BLUE STELLAR OBJECT
PHL 3036	00 49 18.	- 31 12		18.6	BLUE STELLAR OBJECT
PHL 6772	00 49 18.	- 32 10		13.8	BLUE STELLAR OBJECT
MCG+07-02-021	00 49 21.	+ 44 03	60	15.	GALAXY
LIN 172	00 49 22.	- 73 34 16.		14.16	SMC EMISSION LINE OBJECT
LIN.CL 041	00 49 23.	- 72 58 40.	270	13.1	STAR CLUSTER IN SMC
LIN 171	00 49 23.	- 73 03 22.		14.10	SMC EMISSION LINE OBJECT
PHL 3037	00 49 24.	+ 00 46		17.8	BLUE STELLAR OBJECT
PHL 3040	00 49 24.	+ 00 52		18.0	BLUE STELLAR OBJECT
PHL 0871	00 49 24.	+ 09 59		18.0	BLUE STELLAR OBJECT
PHL 3038	00 49 24.	+ 11 12		18.3	BLUE STELLAR OBJECT
ZWG 501.029	00 49 24.	+ 29 24		14.3	GALAXY
UGC 00529	00 49 24.	+ 29 24	60	14.3	GALAXY S0-a
SHRV.73B 40	00 49 24.	+ 42 56	10	15.47	COMPACT GALAXY
ZWG 536.002	00 49 24.	+ 44 04		15.5	GALAXY
ZWG 535.024	00 49 24.	+ 44 04		15.5	GALAXY
UGC 00530	00 49 24.	+ 44 04	66	15.5	GALAXY Sa
PHL 3041	00 49 24.	- 00 12		18.7	BLUE STELLAR OBJECT
MCG+00-03-018	00 49 24.	- 00 46	24	15.	GALAXY
PHL 6777	00 49 24.	- 05 14		16.7	BLUE STELLAR OBJECT
PHL 3039	00 49 24.	- 06 41		18.6	BLUE STELLAR OBJECT
PHL 6776	00 49 24.	- 10 04		18.0	BLUE STELLAR OBJECT
PHL 6778	00 49 24.	- 19 26		18.5	BLUE STELLAR OBJECT
LIN 170	00 49 24.	- 72 42 04.		14.53	SMC EMISSION LINE OBJECT
LIN 174	00 49 25.	- 74 14 04.		15.96	PLANETARY NEBULA IN SMC
HSN 08	00 49 26.	- 73 56			BRIGHT GALAXY BEHIND SMC
LIN.CL 042	00 49 28.	- 73 24 40.	342	11.9	STAR CLUSTER IN SMC
MCG+05-03-014	00 49 30.	+ 29 23 30.	45	15.	GALAXY
ZWG 384.016	00 49 30.	- 00 44		15.1	GALAXY
SN 1955D	00 49 30.	- 16 41		15.5	SUPERNOVA
MCG-03-03-006	00 49 30.	- 16 41	60	14.	GALAXY
PHL 0872	00 49 30.	- 29 29		16.5	BLUE STELLAR OBJECT
LB 03166	00 49 30.	- 72 11		13.3	FAINT BLUE STAR
LH115-N043	00 49 30.	- 74 14			EMISSION NEBULA IN SMC
RNGC 0290	00 49 31.	- 73 25		12.0	OPEN CLUSTER IN SMC
LIN 173	00 49 32.	- 71 57 46.		14.80	SMC EMISSION LINE OBJECT
HSN 09	00 49 32.	- 73 57			BRIGHT GALAXY BEHIND SMC
MCG-01-03-024	00 49 33.	- 05 06	60	16.	GALAXY
LIN 178	00 49 33.	- 73 45 22.		15.19	SMC EMISSION LINE OBJECT
LIN 177	00 49 34.	- 73 21 58.		13.81	SMC EMISSION LINE OBJECT
LIN 176	00 49 35.	- 73 00 22.		14.29	SMC EMISSION LINE OBJECT
PHL 6780	00 49 36.	+ 01 02		18.2	BLUE STELLAR OBJECT
ZWG 384.017	00 49 36.	+ 02 10		15.7	GALAXY
UGC 00531	00 49 36.	+ 02 10	90	15.7	GALAXY SB
MCG+00-03-019	00 49 36.	+ 02 10	33	15.5	GALAXY
PHL 0874	00 49 36.	+ 12 42		18.2	BLUE STELLAR OBJECT
PHL 0873	00 49 36.	+ 13 37		18.1	BLUE STELLAR OBJECT
PHL 3042	00 49 36.	- 04 19		18.4	BLUE STELLAR OBJECT
PHL 6781	00 49 36.	- 11 24		18.8	BLUE STELLAR OBJECT
PHL 6782	00 49 36.	- 21 30		18.4	BLUE STELLAR OBJECT
PHL 6779	00 49 36.	- 24 14		17.2	BLUE STELLAR OBJECT
LIN 175	00 49 36.	- 72 34 04.		14.75	SMC EMISSION LINE OBJECT
RNGC 0276	00 49 40.	- 22 57		14.0	GALAXY
HN C123	00 49 40.	- 22 57			NEBULA
IC 1591	00 49 40.	- 22 57			NONSTELLAR OBJECT
LIN 180	00 49 40.	- 73 13 10.		14.07	SMC EMISSION LINE OBJECT
ZC 0049.7+3533	00 49 42.	+ 35 33	1140		CLUSTER OF GALAXIES
ZWG 384.018	00 49 42.	- 02 28		14.0	GALAXY
MRK 558	00 49 42.	- 02 28	10	14.5	GALAXY WITH UV CONTINUUM
UGC 00532	00 49 42.	- 02 28	96	14.0	GALAXY S0
MCG+00-03-019A	00 49 42.	- 02 28	66	13.	GALAXY
PHL 3044	00 49 42.	- 09 42		16.0	BLUE STELLAR OBJECT
PHL 6783	00 49 42.	- 10 56		12.3	BLUE STELLAR OBJECT
PHL 6784	00 49 42.	- 16 56		18.6	BLUE STELLAR OBJECT
MCG-04-03-021	00 49 42.	- 22 57 30.	54	14.5	GALAXY
PHL 3043	00 49 42.	- 26 55		17.0	BLUE STELLAR OBJECT
LIN 179	00 49 42.	- 72 42 34.		16.32	PLANETARY NEBULA IN SMC
LH115-N042	00 49 42.	- 72 43			EMISSION NEBULA IN SMC
SG 1.01	00 49 43.	+ 56 21	1740		DIFFUSE EMISSION NEBULA
RNGC 0279	00 49 43.	- 02 28		14.0	GALAXY
MCG-01-03-025	00 49 45.	- 05 10	36	16.	GALAXY
ZC 0049.8+0544	00 49 48.	+ 05 44	2490		CLUSTER OF GALAXIES
MCG+02-03-009	00 49 48.	+ 14 14	72	15.	GALAXY
ZWG 435.016	00 49 48.	+ 14 15		15.4	GALAXY
UGC 00533	00 49 48.	+ 14 15	108	15.4	GALAXY Sc
PHL 6789	00 49 48.	+ 14 32		18.3	BLUE STELLAR OBJECT
ZWG 480.017	00 49 48.	+ 24 05		14.6	GALAXY
UGC 00534	00 49 48.	+ 24 05	108	14.6	GALAXY S
ZWG 480.018	00 49 48.	+ 25 52		15.5	GALAXY
UGC 00535	00 49 48.	+ 25 52	60	15.5	GALAXY SBb
VB 018	00 49 48.	+ 68 02	376		STELLAR RING
MCG-01-03-027	00 49 48.	- 04 13	78	13.5	GALAXY
MCG-01-03-026	00 49 48.	- 04 37	48	16.	GALAXY
PHL 6785	00 49 48.	- 11 30		16.7	BLUE STELLAR OBJECT
PHL 6786	00 49 48.	- 15 36		18.7	BLUE STELLAR OBJECT
PHL 6787	00 49 48.	- 20 56		17.7	BLUE STELLAR OBJECT
PHL 6788	00 49 48.	- 23 06		17.5	BLUE STELLAR OBJECT
PHL 3045	00 49 48.	- 26 48		18.1	BLUE STELLAR OBJECT
TON-S 0173	00 49 48.	- 27 18		15.8	BLUE STAR
LIN 181	00 49 48.	- 72 42 53.		16.08	SMC EMISSION LINE OBJECT
ABC 0109	00 49 50.	+ 21 57		17.7	RICH CLUSTER OF GALAXIES
LIN 186	00 49 51.	- 73 45 59.		15.20	SMC EMISSION LINE OBJECT
RNGC 0280	00 49 52.	+ 24 04		14.5	GALAXY
LIN 185	00 49 52.	- 73 21 41.		13.24	SMC EMISSION LINE OBJECT
SVEN 051	00 49 53.	- 38 12 11.	12	16.1	GALAXY
SVEN 050	00 49 53.	- 38 13	24	15.4	GALAXY
LIN 183	00 49 53.	- 73 06 23.		15.06	SMC EMISSION LINE OBJECT
LIN 184	00 49 53.	- 73 06 47.		14.90	PLANETARY NEBULA IN SMC
PHL 3047	00 49 54.	+ 01 32		18.6	BLUE STELLAR OBJECT
PHL 6792	00 49 54.	+ 10 04		18.3	BLUE STELLAR OBJECT
PHL 6791	00 49 54.	+ 12 35		18.3	BLUE STELLAR OBJECT
ZC 0049.9+2201	00 49 54.	+ 22 01	1410		CLUSTER OF GALAXIES
MCG+04-03-013	00 49 54.	+ 24 04	96	14.5	GALAXY
UGC 00536	00 49 54.	+ 28 56	60	16.5	GALAXY
ZWG 536.003	00 49 54.	+ 39 37		15.2	GALAXY

OBJECT NAME	RIGHT ASCEN.	DECLINATION	DIAM.	MAGN.	TYPE OF OBJECT
ZWG 535.025	00 49 54.	+ 39 37		15.2	GALAXY
MRSL 123-06/1	00 49 54.	+ 56 20	2400		HII REGION
OCL 0313	00 49 54.	+ 56 21	960	7.4	OPEN STAR CLUSTER
PHL 3046	00 49 54.	- 00 54		18.7	BLUE STELLAR OBJECT
PHL 0875	00 49 54.	- 13 52		17.1	BLUE STELLAR OBJECT
LIN 182	00 49 54.	- 72 44 29.		14.45	SMC EMISSION LINE OBJECT
HELW 257	00 49 55.	- 26 50 34.			NEBULA
RNGC 0281	00 49 56.	+ 56 21		7.5	OPEN CLUSTER
MCG+04-03-014	00 49 57.	+ 25 50 30.	60	15.	GALAXY
LIN 189	00 49 57.	- 73 29 41.		14.24	SMC EMISSION LINE OBJECT
LIN 190	00 49 57.	- 73 40 53.		14.65	SMC EMISSION LINE OBJECT
LIN 188	00 49 58.	- 73 23 17.		13.77	SMC EMISSION LINE OBJECT
LIN 187	00 49 59.	- 73 01 11.		14.62	SMC EMISSION LINE OBJECT
RIC 865	00 49 59.84	+ 39 59 44.8		19.80	BLUE OBJECT
LBN 0616	00 50	+ 56 20	1920		BRIGHT NEBULA
MCG+00-03-020	00 50 00.	+ 01 31	48	15.	GALAXY
PHL 6793	00 50 00.	+ 02 44		18.4	BLUE STELLAR OBJECT
UGC 00537	00 50 00.	+ 05 45	60	18.	GALAXY DWRF IR
ZWG 435.017	00 50 00.	+ 11 32		15.6	GALAXY
PHL 3048	00 50 00.	+ 11 36		17.8	BLUE STELLAR OBJECT
MCG+05-03-015	00 50 00.	+ 30 21	24	14.5	GALAXY
ZWG 501.030	00 50 00.	+ 30 22		14.7	GALAXY
MCG+07-03-001	00 50 00.	+ 39 35	36	14.5	GALAXY
PHL 0876	00 50 00.	- 01 33		18.3	BLUE STELLAR OBJECT
PHL 3049	00 50 00.	- 11 19		18.6	BLUE STELLAR OBJECT
PHL 6794	00 50 00.	- 15 20		18.3	BLUE STELLAR OBJECT
PHL 3050	00 50 00.	- 23 13		18.8	BLUE STELLAR OBJECT
PHL 3051	00 50 00.	- 26 49		17.5	BLUE STELLAR OBJECT
PHL 3052	00 50 00.	- 27 00		17.7	BLUE STELLAR OBJECT
MCG-06-03-002	00 50 00.	- 35 16	30	14.	GALAXY
LH115-N045	00 50 00.	- 73 30	36		EMISSION NEBULA IN SMC
KARA.73 02	00 50 01.	- 37 51	87		DWARF GALAXY
RNGC 0282	00 50 03.	+ 30 21		14.5	GALAXY
KN 10.02	00 50 04.9	+ 05 46 44.			NEBULA
ZWG 410.014	00 50 06.	+ 05 47		15.7	GALAXY
UGC 00538	00 50 06.	+ 05 47	72	15.7	GALAXY Sc
PHL 6797	00 50 06.	+ 10 39		9.0	BLUE STELLAR OBJECT
PHL 6798	00 50 06.	+ 14 29		16.5	BLUE STELLAR OBJECT
PHL 6795	00 50 06.	+ 14 38		16.1	BLUE STELLAR OBJECT
MCG+07-03-002	00 50 06.	+ 41 40 30.	48	14.	GALAXY
ZWG 536.004	00 50 06.	+ 41 42		14.7	GALAXY
ZWG 535.026	00 50 06.	+ 41 42		14.7	GALAXY
UGC 00539	00 50 06.	+ 41 42	72	14.7	GALAXY Sc
SHRV.73B 41	00 50 06.	+ 42 07	7	16.64	COMPACT GALAXY
PHL 6799	00 50 06.	- 16 24		18.2	BLUE STELLAR OBJECT
PHL 6796	00 50 06.	- 19 20		17.9	BLUE STELLAR OBJECT
LB 01581	00 50 06.	- 59 05		14.8	FAINT BLUE STAR
LH115-N044	00 50 06.	- 71 41			EMISSION NEBULA IN SMC
LIN 192	00 50 06.	- 72 26 05.		15.18	SMC EMISSION LINE OBJECT
LH115-N046	00 50 06.	- 73 07	19		EMISSION NEBULA IN SMC
LIN.CL 043	00 50 07.	- 72 16 11.	612	11.9	STAR CLUSTER IN SMC
LIN 191	00 50 08.	- 71 40 47.		15.98	PLANETARY NEBULA IN SMC
LIN 193	00 50 09.	- 73 26 41.		13.59	SMC EMISSION LINE OBJECT
IC 1590	00 50 10.	+ 56 19			OPEN CLUSTER
LIN.CL 044	00 50 10.	- 73 12 35.	546	12.8	STAR CLUSTER IN SMC
RNGC 0288	00 50 11.	- 26 52		9.0	GLOBULAR CLUSTER
PHL 6800	00 50 12.	+ 02 24		17.6	BLUE STELLAR OBJECT
PHL 0877	00 50 12.	+ 13 00		17.8	BLUE STELLAR OBJECT
ZC 0050.2+2406	00 50 12.	+ 24 06	3700		CLUSTER OF GALAXIES
ZWG 501.031	00 50 12.	+ 28 45		14.1	GALAXY
UGC 00540	00 50 12.	+ 28 45	42	14.1	GALAXY PECULR
SHRV.73B 42	00 50 12.	+ 40 08	8	16.55	COMPACT GALAXY
ZC 0050.2-0102	00 50 12.	- 01 02	2820		CLUSTER OF GALAXIES
PHL 3054	00 50 12.	- 01 14		18.4	BLUE STELLAR OBJECT
PHL 6802	00 50 12.	- 02 31		18.1	BLUE STELLAR OBJECT
PHL 3053	00 50 12.	- 11 18		15.1	BLUE STELLAR OBJECT
PHL 6803	00 50 12.	- 12 50		18.2	BLUE STELLAR OBJECT
PHL 3056	00 50 12.	- 14 16		18.6	BLUE STELLAR OBJECT
PHL 3057	00 50 12.	- 18 40		18.1	BLUE STELLAR OBJECT
PHL 3055	00 50 12.	- 19 08		18.4	BLUE STELLAR OBJECT
PHL 6801	00 50 12.	- 22 53		17.9	BLUE STELLAR OBJECT
PHL 3058	00 50 12.	- 25 23		18.5	BLUE STELLAR OBJECT
GCL 002	00 50 12.	- 26 52	870	8.96	GLOBULAR STAR CLUSTER
ABC 0110	00 50 13.	+ 05 51		17.2	RICH CLUSTER OF GALAXIES
MCG+05-03-016	00 50 15.	+ 28 44 30.	42	14.5	GALAXY
LIN 195	00 50 15.	- 73 20 35.		12.60	SMC EMISSION LINE OBJECT
LIN 196	00 50 15.	- 73 37 17.		16.07	PLANETARY NEBULA IN SMC
LIN 194	00 50 16.	- 73 15 53.		14.45	SMC EMISSION LINE OBJECT
PHL 6805	00 50 18.	+ 11 56		16.7	BLUE STELLAR OBJECT
PHL 6806	00 50 18.	+ 13 12		18.0	BLUE STELLAR OBJECT
PHL 3059	00 50 18.	- 06 58		18.6	BLUE STELLAR OBJECT
PHL 6804	00 50 18.	- 18 10		15.7	BLUE STELLAR OBJECT
PHL 3060	00 50 18.	- 22 19		18.7	BLUE STELLAR OBJECT
MCG-04-03-022	00 50 18.	- 23 14	36	15.	GALAXY
MCG-04-03-023	00 50 18.	- 25 56 30.	12	15.	GALAXY
MCG-04-03-024	00 50 18.	- 26 00 30.	54	16.	GALAXY
PHL 3061	00 50 18.	- 27 00		17.6	BLUE STELLAR OBJECT
MCG-05-03-010	00 50 18.	- 31 30	180	12.	GALAXY
LB 00085	00 50 18.	- 34 15		16.0	FAINT BLUE STAR
LH115-N047	00 50 18.	- 73 37			EMISSION NEBULA IN SMC
CED 003	00 50 19.	+ 56 19	1620		DIFFUSE GALACTIC NEBULA
ABC 0111	00 50 19.	- 05 18		17.2	RICH CLUSTER OF GALAXIES
RNGC 0294	00 50 20.	- 73 37			DIFFUSE NEBULA IN SMC
PATH 1.018	00 50 21.	+ 00 07	14		NEBULA
LIN 199	00 50 21.	- 73 28 47.		14.49	SMC EMISSION LINE OBJECT
LIN 200	00 50 21.	- 73 30 29.		15.91	SMC EMISSION LINE OBJECT
PATH 1.019	00 50 23.	+ 00 05	11		NEBULA
LIN 198	00 50 23.	- 72 42 47.		14.54	SMC EMISSION LINE OBJECT
MCG+00-03-021	00 50 24.	+ 00 56	45	15.	GALAXY
ZWG 384.019	00 50 24.	+ 00 57		15.7	GALAXY
PHL 0878	00 50 24.	+ 02 48		18.0	BLUE STELLAR OBJECT
PHL 6809	00 50 24.	+ 03 35		18.4	BLUE STELLAR OBJECT
PHL 3063	00 50 24.	+ 14 46		18.5	BLUE STELLAR OBJECT
5ZW 037	00 50 24.	+ 50 46			COMPACT GALAXY
PHL 6807	00 50 24.	- 03 07		12.5	BLUE STELLAR OBJECT
PHL 0879	00 50 24.	- 08 50		18.3	BLUE STELLAR OBJECT
PHL 6811	00 50 24.	- 09 46		18.5	BLUE STELLAR OBJECT
PHL 6810	00 50 24.	- 12 22		18.1	BLUE STELLAR OBJECT
PHL 6812	00 50 24.	- 15 53		18.2	BLUE STELLAR OBJECT
PHL 3062	00 50 24.	- 18 40		18.6	BLUE STELLAR OBJECT
PHL 6808	00 50 24.	- 28 18		16.6	BLUE STELLAR OBJECT
RNGC 0289	00 50 24.	- 31 29		12.0	GALAXY
LIN 197	00 50 25.	- 72 01 35.		14.59	SMC EMISSION LINE OBJECT
LIN 203	00 50 26.	- 73 36 05.		13.57	SMC EMISSION LINE OBJECT
LIN 202	00 50 27.	- 73 17 47.		13.86	SMC EMISSION LINE OBJECT
LIN 201	00 50 29.	- 72 43 23.		14.44	SMC EMISSION LINE OBJECT
PHL 6814	00 50 30.	+ 11 09		18.1	BLUE STELLAR OBJECT
PHL 6813	00 50 30.	+ 12 21		18.3	BLUE STELLAR OBJECT

OBJECT NAME	RIGHT ASCEN.	DECLINATION	DIAM.	MAGN.	TYPE OF OBJECT
UGC 00541	00 50 30.	+ 21 39	72	16.0	GALAXY Sc
ZWG 550.017	00 50 30.	+ 47 10		15.7	GALAXY
MCG-01-03-028	00 50 30.	- 03 44	42	15.	GALAXY
LIN 200A	00 50 30.	- 72 24 41.		12.91	SMC EMISSION LINE OBJECT
LH115-N049	00 50 30.	- 73 52	16		EMISSION NEBULA IN SMC
LIN 205	00 50 32.	- 73 36 17.		14.68	SMC EMISSION LINE OBJECT
LIN 204	00 50 34.	- 73 01 53.		14.71	SMC EMISSION LINE OBJECT
PHL 6815	00 50 36.	+ 01 27		17.2	BLUE STELLAR OBJECT
ZCG 0050+23	00 50 36.	+ 23 21		17.3	COMPACT GALAXY
ZC 0050.6+4325	00 50 36.	+ 43 25	470		CLUSTER OF GALAXIES
ISS 0034	00 50 36.	+ 62 17	1039		STELLAR RING
PHL 3064	00 50 36.	- 03 26		18.4	BLUE STELLAR OBJECT
PHL 6816	00 50 36.	- 14 04		18.9	BLUE STELLAR OBJECT
PHL 6817	00 50 36.	- 17 39		18.1	BLUE STELLAR OBJECT
PHL 3065	00 50 36.	- 26 42		18.4	BLUE STELLAR OBJECT
PHL 3066	00 50 36.	- 26 50		16.5	BLUE STELLAR OBJECT
PHL 0880	00 50 36.	- 26 54		17.8	BLUE STELLAR OBJECT
TON-S 0174	00 50 36.	- 28 17		15.8	BLUE STAR
LH115-N048	00 50 36.	- 73 43	46		EMISSION NEBULA IN SMC
LIN 207	00 50 41.	- 72 33 59.		14.24	SMC EMISSION LINE OBJECT
PHL 0881	00 50 42.	+ 10 37		17.9	BLUE STELLAR OBJECT
MCG+05-03-017	00 50 42.	+ 28 59	114	14.5	GALAXY
ZWG 501.032	00 50 42.	+ 29 00		14.8	GALAXY
UGC 00542	00 50 42.	+ 29 00	138	14.8	GALAXY S
5ZW 038	00 50 42.	+ 44 22			COMPACT GALAXY
MCG+12-02-001A	00 50 42.	+ 72 48	6	18.	GALAXY
MCG+12-02-001	00 50 42.	+ 72 48	13	16.	GALAXY
PHL 3068	00 50 42.	- 07 28		18.4	BLUE STELLAR OBJECT
PHL 6818	00 50 42.	- 11 24		16.3	BLUE STELLAR OBJECT
MCG-02-03-031	00 50 42.	- 13 27	96	14.	GALAXY
PHL 6819	00 50 42.	- 16 16		18.3	BLUE STELLAR OBJECT
PHL 6821	00 50 42.	- 18 21		18.4	BLUE STELLAR OBJECT
PHL 6820	00 50 42.	- 18 47		17.5	BLUE STELLAR OBJECT
LB 00073	00 50 42.	- 23 07		16.5	FAINT BLUE STAR
PHL 3067	00 50 42.	- 26 48		17.5	BLUE STELLAR OBJECT
LIN 206	00 50 42.	- 72 27 11.		12.87	SMC EMISSION LINE OBJECT
RNGC 0283	00 50 45.	- 13 27		14.0	GALAXY
HSN 10	00 50 46.	- 70 45			BRIGHT GALAXY BEHIND SMC
KN 10.03	00 50 46.5	+ 05 29 51.			NEBULA
LIN 210	00 50 47.	- 72 44 11.		14.56	SMC EMISSION LINE OBJECT
ZWG 501.033	00 50 48.	+ 32 13		14.8	GALAXY
VB 096	00 50 48.	+ 62 20	1108		STELLAR RING
PHL 6823	00 50 48.	- 08 40		18.4	BLUE STELLAR OBJECT
PHL 6822	00 50 48.	- 13 20		15.7	BLUE STELLAR OBJECT
PHL 6824	00 50 48.	- 20 14		18.3	BLUE STELLAR OBJECT
PHL 3069	00 50 48.	- 27 44		18.5	BLUE STELLAR OBJECT
LB 00086	00 50 48.	- 35 36		15.4	FAINT BLUE STAR
LIN 208	00 50 48.	- 72 26 23.		14.70	SMC EMISSION LINE OBJECT
LIN 209	00 50 48.	- 72 28 47.		15.88	SMC EMISSION LINE OBJECT
RNGC 0287	00 50 50.	+ 32 13		15.0	GALAXY
LIN 212	00 50 50.	- 73 29 48.		14.37	SMC EMISSION LINE OBJECT
LIN 214	00 50 50.	- 73 37 00.		15.04	SMC EMISSION LINE OBJECT
HELW 258	00 50 51.	- 27 03 53.			NEBULA
KN 10.04	00 50 51.9	+ 05 30 02.			NEBULA
IC 1592	00 50 52.	+ 05 30 00.			NONSTELLAR OBJECT
LIN 211	00 50 52.	- 72 57 54.		14.17	SMC EMISSION LINE OBJECT
LIN 213	00 50 53.	- 72 36 12.		14.35	SMC EMISSION LINE OBJECT
PHL 6826	00 50 54.	+ 02 14		18.6	BLUE STELLAR OBJECT
MCG+00-03-022	00 50 54.	+ 02 29	33	15.	GALAXY
ZWG 384.020	00 50 54.	+ 02 30		15.4	GALAXY
ZWG 410.015	00 50 54.	+ 05 30		15.4	GALAXY
UGC 00543	00 50 54.	+ 05 30	66	15.4	GALAXY S
MRK 349	00 50 54.	+ 21 14	14	15.5	GALAXY WITH UV CONTINUUM
ZWG 501.034	00 50 54.	+ 31 09		15.5	GALAXY
ZWG 536.005	00 50 54.	+ 40 50		15.3	GALAXY
MRSL 123+02/1	00 50 54.	+ 65 26	2100		HII REGION
MCG-02-03-032	00 50 54.	- 13 25	12	15.	GALAXY
PHL 6825	00 50 54.	- 14 44		13.4	BLUE STELLAR OBJECT
PHL 6827	00 50 54.	- 19 02		18.2	BLUE STELLAR OBJECT
LIN.CL 045	00 50 54.	- 72 26 53.	678	12.7	STAR CLUSTER IN SMC
LH115-N051	00 50 54.	- 73 42	54		EMISSION NEBULA IN SMC
LIN 218	00 50 56.	- 73 35 18.		14.24	SMC EMISSION LINE OBJECT
LIN 219	00 50 56.	- 73 42 54.		14.33	SMC EMISSION LINE OBJECT
LIN 217	00 50 58.	- 72 58 42.		13.44	SMC EMISSION LINE OBJECT
LBN 0619	00 51	+ 34 20	3900		BRIGHT NEBULA
LBN 0618	00 51	+ 65 30	1920		BRIGHT NEBULA
ZWG 384.021	00 51 00.	+ 02 40		15.7	GALAXY
UGC 00544	00 51 00.	+ 02 40	102	15.7	GALAXY Sc
MCG+00-03-023	00 51 00.	+ 02 40	96	14.	GALAXY
ZWG 384.022	00 51 00.	+ 03 19		15.6	GALAXY
PHL 0882	00 51 00.	+ 10 03		17.2	BLUE STELLAR OBJECT
PHL 6830	00 51 00.	+ 12 05		17.2	BLUE STELLAR OBJECT
1ZW 001	00 51 00.	+ 12 25			COMPACT GALAXY
PHL 3072	00 51 00.	+ 12 25		13.9	BLUE STELLAR OBJECT
KW 62	00 51 00.	+ 12 25	22		SEYFERT GALAXY
VVI 05	00 51 00.	+ 12 25	12	14.26	SEYFERT GALAXY
UGC 00545	00 51 00.	+ 12 25	36	14.0	GALAXY VV CMPT
PHL 6829	00 51 00.	+ 13 02		17.9	BLUE STELLAR OBJECT
PHL 6828	00 51 00.	+ 13 08		17.9	BLUE STELLAR OBJECT
PHL 3070	00 51 00.	+ 14 23		18.3	BLUE STELLAR OBJECT
ZWG 458.007	00 51 00.	+ 17 48		15.7	GALAXY
UGC 00546	00 51 00.	+ 17 48	78	15.7	GALAXY
MCG+07-03-003	00 51 00.	+ 40 48 30.	48	15.	GALAXY
PHL 3071	00 51 00.	- 00 34		18.3	BLUE STELLAR OBJECT
MCG-02-03-035	00 51 00.	- 09 01	72	14.	GALAXY
PHL 3073	00 51 00.	- 09 46		18.4	BLUE STELLAR OBJECT
MCG-02-03-034	00 51 00.	- 13 22	48	14.5	GALAXY
MCG-02-03-033	00 51 00.	- 13 25	42	15.	GALAXY
PHL 6831	00 51 00.	- 18 24		17.5	BLUE STELLAR OBJECT
PHL 6832	00 51 00.	- 24 13		18.6	BLUE STELLAR OBJECT
LIN 216	00 51 00.	- 72 21 42.		14.97	SMC EMISSION LINE OBJECT
LIN 215	00 51 01.	- 71 59 54.		14.10	SMC EMISSION LINE OBJECT
HSN 11	00 51 01.	- 72 47			BRIGHT GALAXY BEHIND SMC
ARC 0113	00 51 02.	- 04 54		17.5	RICH CLUSTER OF GALAXIES
RNGC 0291	00 51 02.	- 09 01		14.0	GALAXY
RNGC 0292	00 51 02.	- 73 06			GALAXY
ARC 0112	00 51 03.	- 01 05		17.2	RICH CLUSTER OF GALAXIES
RNGC 0286	00 51 03.	- 13 22		14.0	GALAXY
RNGC 0284	00 51 03.	- 13 25		15.0	GALAXY
LIN 220	00 51 03.	- 73 10 18.		13.93	SMC EMISSION LINE OBJECT
PHL 6833	00 51 06.	+ 11 00		18.0	BLUE STELLAR OBJECT
PHL 3074	00 51 06.	+ 11 13		17.9	BLUE STELLAR OBJECT
MCG+03-03-005	00 51 06.	+ 17 50	60	15.	GALAXY
ZC 0051.1+2705	00 51 06.	+ 27 05	1610		CLUSTER OF GALAXIES
ZC 0051.1+3830	00 51 06.	+ 38 30	4570		CLUSTER OF GALAXIES
SHRV.73B 43	00 51 06.	+ 40 07	8	16.37	COMPACT GALAXY
SHRV.73B 44	00 51 06.	+ 42 40	8	16.34	COMPACT GALAXY
VB 019	00 51 06.	+ 69 33	410		STELLAR RING

OBJECT NAME	RIGHT ASCEN.	DECLINATION	DIAM.	MAGN.	TYPE OF OBJECT
PHL 0883	00 51 06.	- 11 42		17.8	BLUE STELLAR OBJECT
ARP 251	00 51 06.	- 14 09			PECULIAR GALAXY
LIN 221	00 51 06.	- 72 14 54.		14.88	SMC EMISSION LINE OBJECT
LIN 222	00 51 06.	- 72 22 06.		14.97	SMC EMISSION LINE OBJECT
LIN 223	00 51 06.	- 72 22 36.		14.20	SMC EMISSION LINE OBJECT
LH115-N050	00 51 06.	- 72 55	140		EMISSION NEBULA IN SMC
LIN 228	00 51 08.	- 73 33 54.		15.03	SMC EMISSION LINE OBJECT
RNGC 0285	00 51 09.	- 13 26			GALAXY
LIN 226	00 51 10.	- 72 55 18.		13.37	SMC EMISSION LINE OBJECT
LIN 227	00 51 10.	- 73 00 18.		13.22	SMC EMISSION LINE OBJECT
ARC 0114	00 51 11.	- 21 58		15.9	RICH CLUSTER OF GALAXIES
LIN 225	00 51 11.	- 72 29 30.		14.82	SMC EMISSION LINE OBJECT
PHL 0884	00 51 12.	+ 13 30		17.0	BLUE STELLAR OBJECT
PHL 3075	00 51 12.	+ 13 56		18.6	BLUE STELLAR OBJECT
ZWG 458.008	00 51 12.	+ 20 07		15.6	GALAXY
ZWG 480.019	00 51 12.	+ 24 30		15.3	GALAXY
UGC 00547	00 51 12.	+ 24 30	84	15.3	GALAXY S0
SHRV.73B 45	00 51 12.	+ 39 18	7	15.62	COMPACT GALAXY
ZWG 536.006	00 51 12.	+ 40 08		15.7	GALAXY
MCG+08-02-017	00 51 12.	+ 46 55	60	16.	GALAXY
PHL 3076	00 51 12.	- 04 06		18.4	BLUE STELLAR OBJECT
PHL 6836	00 51 12.	- 10 38		16.6	BLUE STELLAR OBJECT
PHL 3077	00 51 12.	- 11 20		18.5	BLUE STELLAR OBJECT
PHL 6835	00 51 12.	- 11 25		18.4	BLUE STELLAR OBJECT
PHL 6837	00 51 12.	- 12 01		18.5	BLUE STELLAR OBJECT
PHL 3080	00 51 12.	- 12 46		18.4	BLUE STELLAR OBJECT
PHL 6834	00 51 12.	- 16 58		17.3	BLUE STELLAR OBJECT
PHL 3078	00 51 12.	- 23 19		18.6	BLUE STELLAR OBJECT
PHL 3079	00 51 12.	- 27 16		18.6	BLUE STELLAR OBJECT
MCG-05-03-011	00 51 12.	- 27 19	66	15.	GALAXY
LIN 224	00 51 12.	- 72 16 06.		14.99	SMC EMISSION LINE OBJECT
LIN 229	00 51 14.	- 73 33 06.		14.61	SMC EMISSION LINE OBJECT
LIN 232	00 51 14.	- 73 34 42.		14.45	SMC EMISSION LINE OBJECT
LIN 233	00 51 14.	- 73 36 18.		16.32	SMC EMISSION LINE OBJECT
LIN.CL 046	00 51 15.	- 73 08 36.	678		STAR CLUSTER IN SMC
LIN 231	00 51 15.	- 73 19 30.		15.16	SMC EMISSION LINE OBJECT
HELW 259	00 51 16.	- 27 19 06.			NEBULA
LIN 230	00 51 16.	- 72 45 18.		15.71	SMC EMISSION LINE OBJECT
ZC 0051.3+1957	00 51 18.	+ 19 57	540		CLUSTER OF GALAXIES
MCG+03-03-006	00 51 18.	+ 20 07	27	15.	GALAXY
MCG+04-03-015	00 51 18.	+ 24 30	66	15.	GALAXY
PHL 3081	00 51 18.	- 01 48		18.2	BLUE STELLAR OBJECT
PHL 3082	00 51 18.	- 06 35		18.2	BLUE STELLAR OBJECT
MCG-02-03-036	00 51 18.	- 09 22	30	14.5	GALAXY
PHL 6838	00 51 18.	- 13 02		17.8	BLUE STELLAR OBJECT
MCG-02-03-037	00 51 18.	- 14 07	48	15.5	GALAXY
PHL 6839	00 51 18.	- 28 40		18.6	BLUE STELLAR OBJECT
MCG-06-03-003	00 51 18.	- 36 26	36	15.	GALAXY
FATH 1.020	00 51 19.	+ 00 43	5		NEBULA
LIN.CL 047	00 51 20.	- 73 37 00.	546	12.7	STAR CLUSTER IN SMC
IC 1597	00 51 21.	- 58 23			NONSTELLAR OBJECT
HN 0126	00 51 22.	- 58 23			NEBULA
LIN 234	00 51 22.	- 72 42 18.		15.02	SMC EMISSION LINE OBJECT
LIN 235	00 51 22.	- 72 43 24.		14.47	SMC EMISSION LINE OBJECT
LIN 236	00 51 22.	- 72 55 36.		13.97	SMC EMISSION LINE OBJECT
HN 0124	00 51 23.	- 47 55			NEBULA
ZWG 384.023	00 51 24.	+ 00 43		15.5	GALAXY
3ZW 015	00 51 24.	+ 16 38			COMPACT GALAXY
ZWG 501.035	00 51 24.	+ 30 05		15.3	GALAXY
KARA.73B 0037	00 51 24.	+ 30 05	54	15.3	ISOLATED GALAXY S
MCG+07-03-004	00 51 24.	+ 40 07	36	15.	GALAXY
SHRV.73B 46	00 51 24.	+ 41 18	7	17.06	COMPACT GALAXY
ZC 0051.4+4212	00 51 24.	+ 42 12	1010		CLUSTER OF GALAXIES
SHRV.73B 47	00 51 24.	+ 42 34	10	15.86	COMPACT GALAXY
PHL 3083	00 51 24.	- 07 20		18.5	BLUE STELLAR OBJECT
PHL 6841	00 51 24.	- 07 55		18.6	BLUE STELLAR OBJECT
PHL 6842	00 51 24.	- 08 32		18.4	BLUE STELLAR OBJECT
PHL 3084	00 51 24.	- 12 01		18.9	BLUE STELLAR OBJECT
PHL 6840	00 51 24.	- 19 19		17.1	BLUE STELLAR OBJECT
PHL 0885	00 51 24.	- 25 20		17.6	BLUE STELLAR OBJECT
MCG-05-03-012	00 51 24.	- 31 23	30	15.	GALAXY
IC 1594	00 51 24.	- 47 55			NONSTELLAR OBJECT
HSN 12	00 51 25.	- 72 51			BRIGHT GALAXY BEHIND SMC
LIN 238	00 51 27.	- 73 00 12.		15.13	SMC EMISSION LINE OBJECT
LIN 239	00 51 27.	- 73 00 36.		15.93	PLANETARY NEBULA IN SMC
PHL 6844	00 51 30.	+ 15 10		17.1	BLUE STELLAR OBJECT
MCG-02-03-038	00 51 30.	- 11 07	30	14.5	GALAXY
PHL 6843	00 51 30.	- 22 54		18.0	BLUE STELLAR OBJECT
HELW 322	00 51 31.	- 31 22 06.			NEBULA
HN 0125	00 51 31.	- 45 28			NEBULA
IC 1595	00 51 31.	- 45 28			NONSTELLAR OBJECT
LIN 237	00 51 31.	- 71 37 24.		14.28	SMC EMISSION LINE OBJECT
FATH 1.021	00 51 34.	+ 00 39	8		NEBULA
PHL 3085	00 51 36.	+ 02 17		18.4	BLUE STELLAR OBJECT
PHL 6846	00 51 36.	+ 13 53		16.7	BLUE STELLAR OBJECT
PHL 6847	00 51 36.	+ 14 10		18.3	BLUE STELLAR OBJECT
PHL 6845	00 51 36.	+ 14 52		15.8	BLUE STELLAR OBJECT
5ZW 039	00 51 36.	+ 50 21			COMPACT GALAXY
MCG-04-03-025	00 51 36.	- 21 56 30.	36	15.5	GALAXY
LIN.CL 048	00 51 37.	- 71 40 24.	474	11.9	STAR CLUSTER IN SMC
MCG-01-03-029	00 51 39.	- 07 30	12	18.5	GALAXY
FATH 1.022	00 51 40.	+ 00 48	8		NEBULA
SVEN 052	00 51 40.	- 37 37	30	15.1	GALAXY
PHL 6849	00 51 42.	+ 11 00		18.1	BLUE STELLAR OBJECT
PHL 0886	00 51 42.	+ 11 46		16.1	BLUE STELLAR OBJECT
PHL 6848	00 51 42.	+ 13 30		17.3	BLUE STELLAR OBJECT
MCG+05-03-018	00 51 42.	+ 31 23	66	14.5	GALAXY
ZWG 501.036	00 51 42.	+ 31 24		14.8	GALAXY
UGC 00548	00 51 42.	+ 31 24	84	14.8	GALAXY Sc
ZC 0051.7+3322	00 51 42.	+ 33 22	2890		CLUSTER OF GALAXIES
ZWG 384.024	00 51 42.	- 01 41		15.7	GALAXY
MCG-01-03-030	00 51 42.	- 07 30	60	14.	GALAXY
PHL 0887	00 51 42.	- 10 03		18.3	BLUE STELLAR OBJECT
PHL 0888	00 51 42.	- 11 37		18.3	BLUE STELLAR OBJECT
LB 00087	00 51 42.	- 36 19		15.8	FAINT BLUE STAR
LB 03167	00 51 42.	- 71 41		14.4	FAINT BLUE STAR
RNGC 0293	00 51 44.	- 07 30		14.0	GALAXY
MCG-04-03-026	00 51 45.	- 21 58 30.	12	15.5	GALAXY
LIN 243	00 51 45.	- 72 54 54.		14.63	SMC EMISSION LINE OBJECT
LIN 242	00 51 46.	- 72 50 06.		13.75	SMC EMISSION LINE OBJECT
PHL 3087	00 51 48.	+ 12 02		16.2	BLUE STELLAR OBJECT
PHL 3088	00 51 48.	+ 14 19		18.5	BLUE STELLAR OBJECT
PHL 6851	00 51 48.	+ 14 55		17.9	BLUE STELLAR OBJECT
MCG+07-03-005	00 51 48.	+ 41 59 30.	36	15.	GALAXY
5ZW 040	00 51 48.	+ 42 00			COMPACT GALAXY
ZWG 536.007	00 51 48.	+ 42 00		15.6	GALAXY
PHL 0889	00 51 48.	- 02 25		16.6	BLUE STELLAR OBJECT
PHL 3086	00 51 48.	- 02 56		10.0	BLUE STELLAR OBJECT

OBJECT NAME	RIGHT ASCEN.	DECLINATION	DIAM.	MAGN.	TYPE OF OBJECT
PHL 6852	00 51 48.	- 06 57		17.3	BLUE STELLAR OBJECT
PHL 6853	00 51 48.	- 16 10		18.7	BLUE STELLAR OBJECT
MCG-04-03-027	00 51 48.	- 23 50	24	16.	GALAXY
PHL 6850	00 51 48.	- 27 12		13.4	BLUE STELLAR OBJECT
LIN 241	00 51 48.	- 71 57 42.		14.22	SMC EMISSION LINE OBJECT
LIN 245	00 51 50.	- 73 26 13.		12.99	SMC EMISSION LINE OBJECT
LIN 244	00 51 51.	- 72 54 43.		14.61	SMC EMISSION LINE OBJECT
PHL 3090	00 51 54.	+ 10 52		18.2	BLUE STELLAR OBJECT
PHL 3089	00 51 54.	+ 14 33		18.6	BLUE STELLAR OBJECT
PHL 6855	00 51 54.	+ 14 57		13.7	BLUE STELLAR OBJECT
ZC 0051.9+2258	00 51 54.	+ 22 58	1410		CLUSTER OF GALAXIES
ZWG 520.002	00 51 54.	+ 36 30		15.7	GALAXY
UGC 00549	00 51 54.	+ 36 30	66	15.7	GALAXY Sc
KARA.73B 0038	00 51 54.	+ 36 30	72	15.7	ISOLATED GALAXY S
PHL 6857	00 51 54.	- 02 28		18.6	BLUE STELLAR OBJECT
MCG-01-03-031	00 51 54.	- 04 52	84	15.	GALAXY
PHL 6854	00 51 54.	- 12 04		18.7	BLUE STELLAR OBJECT
PHL 6856	00 51 54.	- 20 00		16.7	BLUE STELLAR OBJECT
PHL 0890	00 51 54.	- 22 22		18.6	BLUE STELLAR OBJECT
MCG-04-03-028	00 51 54.	- 23 49	36	15.5	GALAXY
PHL 3091	00 51 54.	- 24 10		18.3	BLUE STELLAR OBJECT
PHL 3092	00 51 54.	- 26 06		18.7	BLUE STELLAR OBJECT
TON-S 0175	00 51 54.	- 28 18		15.8	BLUE STAR
LB 01582	00 51 54.	- 58 48		14.8	FAINT BLUE STAR
LH115-N052A	00 51 54.	- 72 56	19		EMISSION NEBULA IN SMC
IC 1593	00 51 57.	+ 32 13			NONSTELLAR OBJECT
MCG-04-03-029	00 51 57.	- 25 43 30.	36	15.5	GALAXY
LIN 246	00 51 58.	- 72 28 25.		14.55	SMC EMISSION LINE OBJECT
KHAV 014	00 52	+ 61 34	5930		DARK NEBULA
MCG+01-03-007	00 52 00.	+ 05 30	42	14.5	GALAXY
PHL 6859	00 52 00.	+ 12 09		18.4	BLUE STELLAR OBJECT
PHL 0891	00 52 00.	+ 14 39		18.2	BLUE STELLAR OBJECT
ZWG 458.009	00 52 00.	+ 21 15		15.1	GALAXY
UGC 00550	00 52 00.	+ 21 15	150	15.1	GALAXY S
KARA.73B 0039	00 52 00.	+ 21 15	108	15.1	ISOLATED GALAXY S
MCG+03-03-007	00 52 00.	+ 21 15 30.	90	14.5	GALAXY
ZWG 480.020	00 52 00.	+ 24 36		15.5	GALAXY
MCG+05-03-019	00 52 00.	+ 28 28	18	16.	GALAXY
ZWG 501.037	00 52 00.	+ 28 30		15.3	GALAXY
UGC 00551	00 52 00.	+ 44 29	84	18.	GALAXY DWARF
ZWG 557.003	00 52 00.	+ 52 14		15.7	GALAXY
UGC 00552	00 52 00.	+ 52 14	90	15.7	GALAXY Sc
PHL 3093	00 52 00.	- 04 33		14.4	BLUE STELLAR OBJECT
PHL 3096	00 52 00.	- 06 43		18.3	BLUE STELLAR OBJECT
PHL 6862	00 52 00.	- 10 59		18.2	BLUE STELLAR OBJECT
PHL 3094	00 52 00.	- 12 37		18.9	BLUE STELLAR OBJECT
PHL 6860	00 52 00.	- 18 21		18.4	BLUE STELLAR OBJECT
PHL 6858	00 52 00.	- 22 06		17.5	BLUE STELLAR OBJECT
PHL 3095	00 52 00.	- 30 20		18.2	BLUE STELLAR OBJECT
PHL 6861	00 52 00.	- 32 58		17.4	BLUE STELLAR OBJECT
LB 00088	00 52 00.	- 34 01		16.2	FAINT BLUE STAR
LH115-N052B	00 52 00.	- 72 56	24		EMISSION NEBULA IN SMC
IC 1596	00 52 03.	+ 21 14 40.			NONSTELLAR OBJECT
MCG+05-03-020	00 52 03.	+ 28 25	48	15.	GALAXY
LIN 248	00 52 04.	- 72 27 43.		15.39	SMC EMISSION LINE OBJECT
PHL 3099	00 52 06.	+ 02 50		18.4	BLUE STELLAR OBJECT
PHL 3098	00 52 06.	+ 02 58		18.4	BLUE STELLAR OBJECT
ZWG 410.016	00 52 06.	+ 05 30		15.0	GALAXY
UGC 00553	00 52 06.	+ 05 30	78	15.0	GALAXY Sa
PHL 0893	00 52 06.	+ 09 48		18.4	BLUE STELLAR OBJECT
PHL 0892	00 52 06.	+ 14 30		18.3	BLUE STELLAR OBJECT
MCG+04-03-016	00 52 06.	+ 24 35 30.	36	15.5	GALAXY
ZWG 501.038	00 52 06.	+ 28 27		15.6	GALAXY
UGC 00554	00 52 06.	+ 28 27	72	15.6	GALAXY Sa-b
MCG+05-03-021	00 52 06.	+ 28 34 30.	54	15.	GALAXY
ZWG 501.039	00 52 06.	+ 28 35		14.9	GALAXY
UGC 00555	00 52 06.	+ 28 35	60	14.9	GALAXY S0-a
MCG+05-03-022	00 52 06.	+ 28 57	54	15.	GALAXY
ZWG 501.040	00 52 06.	+ 28 59		15.3	GALAXY
UGC 00556	00 52 06.	+ 28 59	72	15.3	GALAXY S
ZWG 501.041	00 52 06.	+ 31 05		14.9	GALAXY
UGC 00557	00 52 06.	+ 31 05	66	14.9	GALAXY S
MCG+05-03-023	00 52 06.	+ 31 05	66	15.	GALAXY
PHL 0894	00 52 06.	- 00 16		18.1	BLUE STELLAR OBJECT
PHL 3100	00 52 06.	- 01 06		18.3	BLUE STELLAR OBJECT
FEIG 009	00 52 06.	- 04 33		14.0	FAINT BLUE STAR
PHL 3097	00 52 06.	- 05 42		18.3	BLUE STELLAR OBJECT
PHL 3101	00 52 06.	- 10 09		17.3	BLUE STELLAR OBJECT
PHL 6863	00 52 06.	- 16 42		18.5	BLUE STELLAR OBJECT
MCG-04-03-030	00 52 06.	- 23 45	48	15.	GALAXY
LIN 247	00 52 06.	- 71 50 25.		14.02	SMC EMISSION LINE OBJECT
KN 10.05	00 52 06.6	+ 05 30 19.			NEBULA
IC 1598	00 52 07.	+ 05 30 10.			NONSTELLAR OBJECT
HN 0127	00 52 09.	- 23 46			NEBULA
IC 1599	00 52 09.	- 23 46			NONSTELLAR OBJECT
LIN.CL 049	00 52 10.	- 72 26 07.	342	11.7	STAR CLUSTER IN SMC
HOW 29	00 52 10.	- 74 24 56.	18		STAR CLUSTER IN SMC
3ZW 016	00 52 12.	+ 02 32			COMPACT GALAXY
ZWG 384.025	00 52 12.	+ 02 32		15.6	GALAXY
ZWG 435.018	00 52 12.	+ 10 16		15.7	GALAXY
UGC 00558	00 52 12.	+ 10 16	102	15.7	GALAXY S0
KARA.73B 0040	00 52 12.	+ 10 16	90	15.7	ISOLATED GALAXY S
MCG+02-03-010	00 52 12.	+ 11 33	36	15.	GALAXY
ZWG 435.019	00 52 12.	+ 11 34		15.7	GALAXY
UGC 00559	00 52 12.	+ 11 34	66	15.7	GALAXY S0?
UGC 00560	00 52 12.	+ 13 25	66	18.	GALAXY DWRF SP
PHL 3103	00 52 12.	+ 13 44		18.4	BLUE STELLAR OBJECT
PHL 3102	00 52 12.	+ 13 56		17.8	BLUE STELLAR OBJECT
ZWG 435.020	00 52 12.	+ 15 01		15.7	GALAXY
MCG+03-03-008	00 52 12.	+ 16 10	27	15.5	GALAXY
PHL 6864	00 52 12.	- 15 24		18.9	BLUE STELLAR OBJECT
PHL 3104	00 52 12.	- 25 00		16.9	BLUE STELLAR OBJECT
IC 0057	00 52 13.	+ 11 33 58.			NONSTELLAR OBJECT
RNGC 0299	00 52 13.	- 72 26		11.5	CLUSTER/NEBULOSITY IN SMC
MCG+02-03-011	00 52 15.	+ 10 15	90	15.	GALAXY
UGC 00561	00 52 18.	+ 30 32	90	16.0	GALAXY PECULR
4ZW 033	00 52 18.	+ 38 33			COMPACT GALAXY
MCG-02-03-039	00 52 18.	- 11 14	48	15.	GALAXY
PHL 6865	00 52 18.	- 21 18		18.9	BLUE STELLAR OBJECT
PHL 3105	00 52 18.	- 27 08		18.0	BLUE STELLAR OBJECT
MCG-06-03-004	00 52 18.	- 36 08	36	14.5	GALAXY
RNGC 0303	00 52 21.	- 16 56			GALAXY
LIN 250	00 52 21.	- 72 58 13.		13.31	SMC EMISSION LINE OBJECT
SVEN 053	00 52 22.	- 38 18 13.	12	16.1	GALAXY
LIN 249	00 52 22.	- 72 33 07.		15.41	SMC EMISSION LINE OBJECT
FATH 1.023	00 52 24.	+ 00 20	14		NEBULA
MCG+00-03-024	00 52 24.	+ 01 06	51	14.5	GALAXY
FATH 1.024	00 52 24.	+ 01 06	54		NEBULA
ZC 0052.4+0530	00 52 24.	+ 05 30	1280		CLUSTER OF GALAXIES
PHL 6870	00 52 24.	+ 09 50		18.0	BLUE STELLAR OBJECT
PHL 6869	00 52 24.	+ 13 00		17.9	BLUE STELLAR OBJECT
MCG+05-03-024	00 52 24.	+ 31 15 30.	132	13.5	GALAXY
ZWG 501.042	00 52 24.	+ 31 16		13.5	GALAXY
UGC 00562	00 52 24.	+ 31 16	138	13.5	GALAXY SB:b
ZWG 536.008	00 52 24.	+ 41 57		15.5	GALAXY
SHRV.73B 48	00 52 24.	+ 42 17	8	17.06	COMPACT GALAXY
ZWG 536.009	00 52 24.	+ 44 51		15.7	GALAXY
PHL 0896	00 52 24.	- 00 43		18.7	BLUE STELLAR OBJECT
PHL 0895	00 52 24.	- 01 00		18.1	BLUE STELLAR OBJECT
MCG-01-03-032	00 52 24.	- 03 20	18	14.5	GALAXY
PHL 3106	00 52 24.	- 06 01		18.6	BLUE STELLAR OBJECT
PHL 6871	00 52 24.	- 06 27		18.6	BLUE STELLAR OBJECT
MCG-01-03-033	00 52 24.	- 07 37	96	13.5	GALAXY
PHL 3108	00 52 24.	- 13 18		18.9	BLUE STELLAR OBJECT
PHL 0897	00 52 24.	- 14 43		15.8	BLUE STELLAR OBJECT
PHL 3109	00 52 24.	- 21 12		18.8	BLUE STELLAR OBJECT
PHL 6867	00 52 24.	- 25 08		18.6	BLUE STELLAR OBJECT
PHL 6868	00 52 24.	- 25 12		18.4	BLUE STELLAR OBJECT
PHL 6866	00 52 24.	- 26 11		18.5	BLUE STELLAR OBJECT
PHL 3107	00 52 24.	- 27 48		17.8	BLUE STELLAR OBJECT
LH115-N053	00 52 24.	- 71 51	23		EMISSION NEBULA IN SMC
RNGC 0298	00 52 26.	- 07 37		13.5	GALAXY
RNGC 0295	00 52 27.	+ 31 15		13.5	GALAXY
MCG+07-03-006	00 52 27.	+ 41 56	30	14.5	GALAXY
MCG-03-03-007	00 52 27.	- 19 18	150	13.5	GALAXY
LIN 252	00 52 27.	- 72 42 07.		14.53	SMC EMISSION LINE OBJECT
LIN 253	00 52 27.	- 72 42 49.		14.47	SMC EMISSION LINE OBJECT
LIN 251	00 52 28.	- 72 32 55.		13.77	SMC EMISSION LINE OBJECT
ZWG 384.026	00 52 30.	+ 01 07		15.1	GALAXY
UGC 00563	00 52 30.	+ 01 07	72	15.1	GALAXY S
ZWG 520.003	00 52 30.	+ 35 10		15.7	GALAXY
UGC 00564	00 52 30.	+ 35 10	60	15.7	GALAXY S
MCG-02-03-040	00 52 30.	- 11 19	48	15.5	GALAXY
PHL 0898	00 52 30.	- 18 32		18.4	BLUE STELLAR OBJECT
PHL 6872	00 52 30.	- 21 24		17.1	BLUE STELLAR OBJECT
PHL 6873	00 52 30.	- 30 00		18.2	BLUE STELLAR OBJECT
MCG-05-03-013	00 52 30.	- 32 20	12	15.5	GALAXY
MCG-06-03-005	00 52 30.	- 37 58	660	8.7	GALAXY
RNGC 0297	00 52 32.	- 07 37			NON-EXISTENT OBJECT
IC 0058	00 52 32.	- 13 57 09.			NONSTELLAR OBJECT
MCG-05-03-014	00 52 33.	- 32 19	30	15.	GALAXY
LIN 254	00 52 33.	- 72 45 55.		14.93	PLANETARY NEBULA IN SMC
PHL 3111	00 52 36.	+ 10 07		12.5	BLUE STELLAR OBJECT
ZWG 435.021	00 52 36.	+ 12 04		15.3	GALAXY
PHL 6874	00 52 36.	+ 12 10		18.4	BLUE STELLAR OBJECT
ZWG 501.043	00 52 36.	+ 30 13		15.7	GALAXY
ZWG 501.044	00 52 36.	+ 31 24		15.4	GALAXY
UGC 00565	00 52 36.	+ 31 24	72	15.4	GALAXY Sc
MCG+05-03-026	00 52 36.	+ 31 26 30.	60	15.5	GALAXY
ZWG 501.045	00 52 36.	+ 31 27		15.7	GALAXY
UGC 00566	00 52 36.	+ 31 27	90	15.7	GALAXY S IV
ZWG 501.046	00 52 36.	+ 31 28		14.9	GALAXY
UGC 00567	00 52 36.	+ 31 28	66	14.9	GALAXY S0
MCG+05-03-025	00 52 36.	+ 31 28	54	15.	GALAXY
ZC 0052.6+3453	00 52 36.	+ 34 53	1340		CLUSTER OF GALAXIES
MCG+00-03-025	00 52 36.	- 01 20	12	15.	GALAXY
PHL 3112	00 52 36.	- 02 04		18.5	BLUE STELLAR OBJECT
PHL 3113	00 52 36.	- 07 14		18.4	BLUE STELLAR OBJECT
PHL 3110	00 52 36.	- 12 02		18.5	BLUE STELLAR OBJECT
MCG-02-03-041	00 52 36.	- 13 57 30.	36	15.	GALAXY
PHL 6875	00 52 36.	- 16 38		18.6	BLUE STELLAR OBJECT
MCG-04-03-031	00 52 36.	- 23 47	42	15.	GALAXY
RNGC 0300	00 52 37.	- 37 58		10.0	GALAXY
KN 10.06	00 52 37.1	+ 09 10 52.			NEBULA
RNGC 0296	00 52 38.	+ 31 24		15.5	GALAXY
MCG+00-03-026	00 52 39.	- 01 22	36	15.	GALAXY
HN 0128	00 52 39.	- 23 48			NEBULA
IC 1600	00 52 39.	- 23 48			NONSTELLAR OBJECT
SVEN 054	00 52 40.	- 37 58	1140	10.5	GALAXY
UGC 00569	00 52 42.	+ 09 11	66	16.5	GALAXY Sc
MCG+02-03-012	00 52 42.	+ 12 02 30.	48	15.	GALAXY
MCG+05-03-027	00 52 42.	+ 31 23 30.	60	15.	GALAXY
MCG+08-02-016	00 52 42.	+ 48 47	60	16.	GALAXY
MCG+00-03-027	00 52 42.	- 01 12	36	15.5	GALAXY
ZWG 384.027	00 52 42.	- 01 18		15.7	GALAXY
UGC 00568	00 52 42.	- 01 18	90	15.7	GALAXY COMPACT
MCG+00-03-028	00 52 42.	- 01 34	12	15.	GALAXY
PHL 3114	00 52 42.	- 08 36		17.7	BLUE STELLAR OBJECT
PHL 3115	00 52 42.	- 20 02		18.6	BLUE STELLAR OBJECT
TON-S 0176	00 52 42.	- 27 05		15.9	BLUE STAR
TON-S 0177	00 52 42.	- 27 46		16.0	BLUE STAR
PHL 6876	00 52 42.	- 28 09		17.8	BLUE STELLAR OBJECT
PHL 6877	00 52 42.	- 32 00		18.4	BLUE STELLAR OBJECT
MCG-06-03-007	00 52 42.	- 35 36	36	14.5	GALAXY
MCG-06-03-006	00 52 42.	- 35 48	24	15.	GALAXY
LIN 258	00 52 44.	- 72 57 55.		14.65	SMC EMISSION LINE OBJECT
LIN 257	00 52 45.	- 72 43 13.		14.44	SMC EMISSION LINE OBJECT
LIN 256	00 52 46.	- 72 29 19.		14.92	SMC EMISSION LINE OBJECT
HOW 30	00 52 46.	- 73 53 14.	12		STAR CLUSTER IN SMC
PHL 3116	00 52 48.	+ 01 13		17.8	BLUE STELLAR OBJECT
PHL 3117	00 52 48.	+ 01 20		18.6	BLUE STELLAR OBJECT
PHL 6878	00 52 48.	+ 01 49		18.7	BLUE STELLAR OBJECT
PHL 0899	00 52 48.	+ 11 32		18.2	BLUE STELLAR OBJECT
PHL 6880	00 52 48.	+ 11 40		15.9	BLUE STELLAR OBJECT
ZWG 435.022	00 52 48.	+ 11 50		15.4	GALAXY
MCG+02-03-013	00 52 48.	+ 11 50	36	15.	GALAXY
ZWG 501.047	00 52 48.	+ 30 08		15.7	GALAXY
ZWG 550.018	00 52 48.	+ 48 45		15.7	GALAXY
ZWG 384.029	00 52 48.	- 01 20		15.7	GALAXY
ZWG 384.028	00 52 48.	- 01 31		15.2	GALAXY
PHL 6884	00 52 48.	- 11 02		18.3	BLUE STELLAR OBJECT
PHL 6883	00 52 48.	- 11 51		18.2	BLUE STELLAR OBJECT
PHL 6879	00 52 48.	- 16 08		18.6	BLUE STELLAR OBJECT
PHL 6881	00 52 48.	- 16 32		17.1	BLUE STELLAR OBJECT
PHL 6882	00 52 48.	- 20 29		17.9	BLUE STELLAR OBJECT
LB 00089	00 52 48.	- 36 36		15.2	FAINT BLUE STAR
LIN 255	00 52 48.	- 71 46 55.		15.37	SMC EMISSION LINE OBJECT
LIN 261	00 52 50.	- 72 54 43.		15.41	SMC EMISSION LINE OBJECT
LIN 259	00 52 51.	- 72 45 07.		13.49	SMC EMISSION LINE OBJECT
LIN 260	00 52 51.	- 72 46 01.		14.32	SMC EMISSION LINE OBJECT
PHL 6885	00 52 54.	+ 11 24		17.4	BLUE STELLAR OBJECT
PHL 3119	00 52 54.	+ 11 55		17.8	BLUE STELLAR OBJECT
PHL 0900	00 52 54.	+ 12 45		17.9	BLUE STELLAR OBJECT
PHL 3118	00 52 54.	- 00 57		18.4	BLUE STELLAR OBJECT
MCG-01-03-034	00 52 54.	- 03 42	36	15.5	GALAXY
PHL 0901	00 52 54.	- 08 08		9.8	BLUE STELLAR OBJECT

OBJECT NAME	RIGHT ASCEN.	DECLINATION	DIAM.	MAGN.	TYPE OF OBJECT
PHL 6886	00 52 54.	- 09 35		18.5	BLUE STELLAR OBJECT
LB 00090	00 52 54.	- 32 30		15.5	FAINT BLUE STAR
LIN 264	00 52 55.	- 73 21 08.		14.24	SMC EMISSION LINE OBJECT
LIN 263	00 52 56.	- 72 56 38.		14.31	SMC EMISSION LINE OBJECT
FATH 1.025	00 52 57.	+ 00 46	5		NEBULA
LIN 262	00 52 57.	- 72 42 56.		14.13	SMC EMISSION LINE OBJECT
FATH 1.026	00 52 59.	+ 00 10	8		NEBULA
PHL 0903	00 53 00.	+ 01 25		18.4	BLUE STELLAR OBJECT
PHL 3122	00 53 00.	+ 11 13		18.6	BLUE STELLAR OBJECT
PHL 0902	00 53 00.	+ 11 19		16.7	BLUE STELLAR OBJECT
ZWG 536.010	00 53 00.	+ 43 13		15.7	GALAXY
VB 097	00 53 00.	+ 61 34	517		STELLAR RING
URA 42	00 53 00.	+ 63 04	138		STELLAR RING
ASS 38	00 53 00.	+ 63 26			OB ASSOCIATION CAS OB7
MCG-01-03-035	00 53 00.	- 03 42	15	15.	GALAXY
PHL 3120	00 53 00.	- 08 48		17.0	BLUE STELLAR OBJECT
PHL 6887	00 53 00.	- 09 33		15.9	BLUE STELLAR OBJECT
MCG-02-03-043	00 53 00.	- 10 20	30	16.	GALAXY
MCG-02-03-044	00 53 00.	- 11 12	36	15.	GALAXY
PHL 6889	00 53 00.	- 13 18		18.1	BLUE STELLAR OBJECT
PHL 3121	00 53 00.	- 13 22		18.3	BLUE STELLAR OBJECT
MCG-02-03-042	00 53 00.	- 13 43	60	15.5	GALAXY
PHL 6888	00 53 00.	- 15 01		17.0	BLUE STELLAR OBJECT
MCG-04-03-032	00 53 00.	- 24 25 30.	48	14.	GALAXY
MCG-02-03-045	00 53 03.	- 10 20	36	15.	GALAXY
LIN.CL 050	00 53 03.	- 72 28 37.	270	12.5	STAR CLUSTER IN SMC
LIN 265	00 53 03.	- 72 38 20.		14.96	SMC EMISSION LINE OBJECT
SVEN 055	00 53 04.	- 38 27	6	15.9	GALAXY
4ZW 034	00 53 06.	+ 23 52			COMPACT GALAXY
ZWG 480.021	00 53 06.	+ 23 53		15.4	GALAXY
5ZW 041	00 53 06.	+ 47 09			COMPACT GALAXY
VB 098	00 53 06.	+ 62 41	262		STELLAR RING
PHL 3123	00 53 06.	- 22 04		18.7	BLUE STELLAR OBJECT
MCG-04-03-033	00 53 06.	- 24 25	24	16.	GALAXY
HELW 323	00 53 07.	- 10 22 31.			NEBULA
IC 1601	00 53 08.	- 24 26			NONSTELLAR OBJECT
RNGC 0306	00 53 08.	- 72 29		12.5	CLUSTER/NEBULOSITY IN SMC
LIN 268	00 53 08.	- 73 00 26.		13.97	SMC EMISSION LINE OBJECT
HN 0129	00 53 09.	- 24 26			NEBULA
LIN 266	00 53 09.	- 72 36 44.		13.92	SMC EMISSION LINE OBJECT
LIN 267	00 53 09.	- 72 39 56.		14.52	SMC EMISSION LINE OBJECT
FATH 1.027	00 53 10.	+ 00 26	5		NEBULA
SVEN 056	00 53 10.	- 37 41 14.	18	15.6	GALAXY
FATH 1.028	00 53 12.	+ 00 22	3		NEBULA
PHL 6896	00 53 12.	+ 01 22		18.4	BLUE STELLAR OBJECT
PHL 3124	00 53 12.	+ 01 36		16.6	BLUE STELLAR OBJECT
PHL 6895	00 53 12.	+ 09 52		17.8	BLUE STELLAR OBJECT
PHL 6894	00 53 12.	+ 14 05		16.2	BLUE STELLAR OBJECT
MCG+04-03-017	00 53 12.	+ 23 52 30.	36	15.5	GALAXY
PHL 3127	00 53 12.	- 00 12		18.5	BLUE STELLAR OBJECT
ZWG 384.030	00 53 12.	- 01 33		15.7	GALAXY
PHL 6893	00 53 12.	- 09 30		18.4	BLUE STELLAR OBJECT
PHL 6897	00 53 12.	- 10 17		18.2	BLUE STELLAR OBJECT
PHL 3125	00 53 12.	- 15 22		17.9	BLUE STELLAR OBJECT
PHL 6890	00 53 12.	- 17 00		16.9	BLUE STELLAR OBJECT
PHL 6892	00 53 12.	- 17 02		17.2	BLUE STELLAR OBJECT
PHL 6891	00 53 12.	- 20 16		16.8	BLUE STELLAR OBJECT
PHL 3126	00 53 12.	- 24 10		18.6	BLUE STELLAR OBJECT
PHL 3126	00 53 12.	- 24 10		18.6	BLUE STELLAR OBJECT
LIN 269	00 53 14.	- 72 50 08.		15.16	SMC EMISSION LINE OBJECT
LIN 270	00 53 14.	- 72 58 26.		15.95	SMC EMISSION LINE OBJECT
ARC 0116	00 53 16.	+ 00 23		15.7	RICH CLUSTER OF GALAXIES
SVEN 057	00 53 16.	- 37 39	18	16.1	GALAXY
LIN.CL 051	00 53 16.	- 72 22 56.	816	11.8	STAR CLUSTER IN SMC
ARC 0115	00 53 17.	+ 26 04		17.3	RICH CLUSTER OF GALAXIES
PHL 3130	00 53 18.	+ 03 24		18.6	BLUE STELLAR OBJECT
PHL 3129	00 53 18.	+ 11 28		17.2	BLUE STELLAR OBJECT
PHL 6909	00 53 18.	+ 12 03		17.9	BLUE STELLAR OBJECT
PHL 6908	00 53 18.	+ 12 24		17.5	BLUE STELLAR OBJECT
PHL 6907	00 53 18.	+ 12 58		17.9	BLUE STELLAR OBJECT
PHL 6899	00 53 18.	+ 13 50		18.2	BLUE STELLAR OBJECT
MCG+02-03-014	00 53 18.	+ 13 57 30.	48	15.	GALAXY
ZWG 480.022	00 53 18.	+ 26 02		15.7	GALAXY
ISS 0035	00 53 18.	+ 62 37	257		STELLAR RING
VB 020	00 53 18.	+ 64 25	490		STELLAR RING
PHL 0904	00 53 18.	- 09 02		16.4	BLUE STELLAR OBJECT
PHL 3131	00 53 18.	- 11 30		18.3	BLUE STELLAR OBJECT
PHL 6902	00 53 18.	- 11 44		16.0	BLUE STELLAR OBJECT
MCG-02-03-046	00 53 18.	- 14 25	30	15.	GALAXY
PHL 3132	00 53 18.	- 16 50		18.3	BLUE STELLAR OBJECT
PHL 6901	00 53 18.	- 16 53		13.8	BLUE STELLAR OBJECT
PHL 6900	00 53 18.	- 17 42		18.4	BLUE STELLAR OBJECT
PHL 6906	00 53 18.	- 18 56		18.3	BLUE STELLAR OBJECT
PHL 3133	00 53 18.	- 21 03		18.5	BLUE STELLAR OBJECT
PHL 6898	00 53 18.	- 26 38		17.4	BLUE STELLAR OBJECT
HELW 324	00 53 19.	- 10 35 50.			NEBULA
LIN 273	00 53 19.	- 73 12 38.		13.00	SMC EMISSION LINE OBJECT
ARC 0118	00 53 20.	- 26 41		16.5	RICH CLUSTER OF GALAXIES
MCG+00-03-029	00 53 21.	- 01 14	24	15.	GALAXY
IC 1602	00 53 21.	- 10 15 23.			NONSTELLAR OBJECT
SVEN 059	00 53 22.	- 37 40	24	15.9	GALAXY
SVEN 058	00 53 22.	- 37 41 14.	18	15.4	GALAXY
LIN 272	00 53 22.	- 72 20 02.		14.45	SMC EMISSION LINE OBJECT
LIN 271	00 53 23.	- 71 53 26.		15.28	SMC EMISSION LINE OBJECT
ZC 0053.4+0010	00 53 24.	+ 00 10	3900		CLUSTER OF GALAXIES
PHL 3128	00 53 24.	+ 03 20		17.9	BLUE STELLAR OBJECT
ZWG 435.023	00 53 24.	+ 11 50		15.4	GALAXY
RNGC 0305	00 53 24.	+ 11 50		15.5	GALAXY
UGC 00571	00 53 24.	+ 11 50	78	15.4	GALAXY SBb
MCG+02-03-015	00 53 24.	+ 11 50	48	15.	GALAXY
PHL 6904	00 53 24.	+ 12 12		17.2	BLUE STELLAR OBJECT
ZWG 435.024	00 53 24.	+ 13 57		15.6	GALAXY
UGC 00572	00 53 24.	+ 13 57	78	15.6	GALAXY Sc
PHL 6905	00 53 24.	+ 14 28		18.4	BLUE STELLAR OBJECT
ZWG 480.023	00 53 24.	+ 23 51		14.0	GALAXY
UGC 00573	00 53 24.	+ 23 51	84	14.0	GALAXY ERUPTVE
ZC 0053.4+2604	00 53 24.	+ 26 04	1610		CLUSTER OF GALAXIES
ZWG 501.048	00 53 24.	+ 30 48		15.6	GALAXY
SHRV.73B 49	00 53 24.	+ 39 55	8	16.1	COMPACT GALAXY
MCG+08-02-019	00 53 24.	+ 50 22	60	15.	GALAXY
ZWG 384.031	00 53 24.	- 01 10		15.3	GALAXY
UGC 00570	00 53 24.	- 01 10	66	15.3	GALAXY E
MCG-01-03-036	00 53 24.	- 06 31	60	16.	GALAXY
PHL 0905	00 53 24.	- 09 53		18.4	BLUE STELLAR OBJECT
MCG-02-03-047	00 53 24.	- 10 15	42	14.	GALAXY
PHL 6903	00 53 24.	- 16 04		18.2	BLUE STELLAR OBJECT
LIN 276	00 53 26.	- 72 44 56.		14.82	SMC EMISSION LINE OBJECT
RNGC 0302	00 53 27.	- 10 55			GALAXY
KN 10.07	00 53 27.3	+ 09 20 30.			NEBULA
RNGC 0304	00 53 28.	+ 23 51		14.0	GALAXY
SVEN 060	00 53 28.	- 37 39 14.	12	15.8	GALAXY
LIN 274	00 53 28.	- 72 14 08.		14.78	SMC EMISSION LINE OBJECT
LIN 275	00 53 28.	- 72 19 20.		15.71	PLANETARY NEBULA IN SMC
FATH 1.029	00 53 29.	+ 00 00	14		NEBULA
ARC 0117	00 53 29.	- 10 18		16.0	RICH CLUSTER OF GALAXIES
LIN 279	00 53 29.	- 73 43 26.		14.33	SMC EMISSION LINE OBJECT
MCG+00-03-030	00 53 30.	+ 00 21	24	15.5	GALAXY
ZWG 384.033	00 53 30.	+ 03 18		15.5	GALAXY
UGC 00574	00 53 30.	+ 04 08	78	17.	GALAXY DWARF
ZWG 410.017	00 53 30.	+ 09 21		15.5	GALAXY
MCG+04-03-018	00 53 30.	+ 23 51	60	14.5	GALAXY
MCG+04-03-019	00 53 30.	+ 26 01 30.	24	15.5	GALAXY
UGC 00575	00 53 30.	+ 30 49	60	16.5	GALAXY Sb-c
ZWG 550.019	00 53 30.	+ 50 21		15.3	GALAXY
UGC 00576	00 53 30.	+ 50 21	72	15.3	GALAXY Sc
UGC 00577	00 53 30.	+ 50 30	78	16.0	GALAXY E?
VB 021	00 53 30.	+ 66 42	389		STELLAR RING
ZWG 384.032	00 53 30.	- 01 35		15.4	GALAXY
MCG-06-03-008	00 53 30.	- 34 42	24	15.	GALAXY
MCG-06-03-009	00 53 30.	- 37 40	24	15.	GALAXY
MCG-06-03-010	00 53 30.	- 37 41	12	15.	GALAXY
LH115-N055	00 53 30.	- 72 19			EMISSION NEBULA IN SMC
LIN 278	00 53 30.	- 73 29 50.		15.25	SMC EMISSION LINE OBJECT
FATH 1.030	00 53 32.	+ 00 21	3		NEBULA
HSN 13	00 53 32.	- 72 57			BRIGHT GALAXY BEHIND SMC
MCG+03-03-009	00 53 33.	+ 19 42	45	15.	GALAXY
HELW 325	00 53 34.	- 10 12 02.			NEBULA
SVEN 061	00 53 34.	- 38 10 15.	12	16.4	GALAXY
LIN 277	00 53 34.	- 72 13 08.		14.74	SMC EMISSION LINE OBJECT
FATH 1.031	00 53 35.	+ 00 25	3		NEBULA
IC 0060	00 53 35.	- 13 38 23.			NONSTELLAR OBJECT
LIN 280	00 53 35.	- 73 38 08.		15.11	SMC EMISSION LINE OBJECT
ZWG 458.010	00 53 36.	+ 19 42		15.6	GALAXY
ZWG 536.011	00 53 36.	+ 39 33		14.8	GALAXY
UGC 00578	00 53 36.	+ 39 33	66	14.8	GALAXY
MCG+00-03-031	00 53 36.	- 01 25	15	15.5	GALAXY
MCG-02-03-049	00 53 36.	- 13 39	36	15.	GALAXY
MCG-02-03-048	00 53 36.	- 14 19 30.	36	16.	GALAXY
LH115-N056	00 53 36.	- 73 43			EMISSION NEBULA IN SMC
MCG-03-03-008	00 53 39.	- 14 33	48	15.	GALAXY
FATH 1.032	00 53 40.	+ 00 24	3		NEBULA
LIN.CL 052	00 53 41.	- 73 45 26.	204	13.7	STAR CLUSTER IN SMC
KN 10.08	00 53 41.8	+ 09 23 52.			NEBULA
PHL 6911	00 53 42.	+ 00 13		18.1	BLUE STELLAR OBJECT
MCG+02-03-016	00 53 42.	+ 13 35	18	15.5	GALAXY
UGC 00580	00 53 42.	+ 13 37	72	16.0	GALAXY
ZWG 480.024	00 53 42.	+ 25 20		15.6	GALAXY
MCG+07-03-007	00 53 42.	+ 39 33 30.	36	15.	GALAXY
ZWG 384.035	00 53 42.	- 01 22		15.6	GALAXY
ZWG 384.034	00 53 42.	- 01 30		15.0	GALAXY
UGC 00579	00 53 42.	- 01 30	90	15.0	GALAXY E
MCG+00-03-032	00 53 42.	- 01 33	60	15.	GALAXY
PHL 3134	00 53 42.	- 02 44		18.5	BLUE STELLAR OBJECT
PHL 6910	00 53 42.	- 03 14		17.2	BLUE STELLAR OBJECT
PHL 3136	00 53 42.	- 12 12		18.4	BLUE STELLAR OBJECT
PHL 3135	00 53 42.	- 25 26		18.4	BLUE STELLAR OBJECT
KN 10.09	00 53 44.8	+ 07 56 10.			NEBULA
MCG+02-03-017	00 53 45.	+ 13 35	9	15.5	GALAXY
MCG-01-03-037	00 53 45.	- 07 58	42	15.	GALAXY
HELW 326	00 53 46.	- 10 37 56.			NEBULA
UGC 00581	00 53 48.	+ 11 36	72	16.0	GALAXY Sb
PHL 6912	00 53 48.	+ 13 10		16.6	BLUE STELLAR OBJECT
UGC 00582	00 53 48.	+ 13 38	60	16.5	GALAXY S0?
PHL 3137	00 53 48.	+ 13 52		18.4	BLUE STELLAR OBJECT
PHL 6913	00 53 48.	+ 14 00		17.5	BLUE STELLAR OBJECT
ZC 0053.8+1459	00 53 48.	+ 14 59	3290		CLUSTER OF GALAXIES
MCG+04-03-020	00 53 48.	+ 25 19 30.	12	15.5	GALAXY
ZWG 536.012	00 53 48.	+ 42 23		15.6	GALAXY
DG 006	00 53 48.	+ 56 13	120		REFLECTION NEBULA
VB 022	00 53 48.	+ 66 52	309		STELLAR RING
ZWG 384.036	00 53 48.	- 01 28		15.3	GALAXY
MCG+00-03-033	00 53 48.	- 01 30	15	15.	GALAXY
MCG+00-03-034	00 53 48.	- 01 33	15	15.	GALAXY
PHL 3138	00 53 48.	- 01 34		18.5	BLUE STELLAR OBJECT
PHL 0906	00 53 48.	- 16 36		18.4	BLUE STELLAR OBJECT
PHL 6914	00 53 48.	- 26 55		17.0	BLUE STELLAR OBJECT
MCG-06-03-011	00 53 48.	- 36 27	24	15.	GALAXY
RNGC 0308	00 53 49.	- 02 03			GALAXY
LIN.CL 053	00 53 49.	- 73 06 56.	474	12.7	STAR CLUSTER IN SMC
LIN 282	00 53 50.	- 72 49 44.		14.49	SMC EMISSION LINE OBJECT
VDB.66N 005	00 53 51.	+ 60 30	7200		REFLECTION NEBULA
ARC 0119	00 53 51.	- 01 32		15.0	RICH CLUSTER OF GALAXIES
SVEN 062	00 53 51.	- 38 25 15.	24	15.7	GALAXY
LIN 280A	00 53 51.	- 72 22 38.		14.50	SMC EMISSION LINE OBJECT
LIN 281	00 53 51.	- 72 30 02.		14.82	SMC EMISSION LINE OBJECT
SHRV.73B 50	00 53 54.	+ 40 25	12	15.5	COMPACT GALAXY
MCG+07-03-008	00 53 54.	+ 42 22 30.	36	15.	GALAXY
URA 44	00 53 54.	+ 62 18	186		STELLAR RING
UGC 00583	00 53 54.	- 01 30	66	14.9	GALAXY E
ZWG 384.038	00 53 54.	- 01 31		14.9	GALAXY
ZWG 384.037	00 53 54.	- 01 32		15.6	GALAXY
MCG+00-03-035	00 53 54.	- 02 05	66	14.	GALAXY
PHL 3139	00 53 54.	- 13 12		16.7	BLUE STELLAR OBJECT
PHL 6915	00 53 54.	- 19 24		18.6	BLUE STELLAR OBJECT
PHL 3140	00 53 54.	- 23 07		18.1	BLUE STELLAR OBJECT
TON-S 0178	00 53 54.	- 26 52		15.7	BLUE STAR
LB 00091	00 53 54.	- 33 11		16.8	FAINT BLUE STAR
HOLM 027B	00 53 57.	- 10 12	18	15.0	PART OF MULTIPLE GALAXY
RNGC 0309	00 53 57.	- 10 13		12.5	GALAXY
LIN 288	00 53 58.	- 73 52 09.		14.76	SMC EMISSION LINE OBJECT
LIN 287	00 53 59.	- 73 27 39.		15.26	SMC EMISSION LINE OBJECT
HOW 31	00 53 59.	- 74 19 55.	18		STAR CLUSTER IN SMC
LBN 0621	00 54	+ 56 13	120		BRIGHT NEBULA
LBN 0620	00 54	+ 60 50	420		BRIGHT NEBULA
PHL 3141	00 54 00.	+ 02 00		18.3	BLUE STELLAR OBJECT
PHL 6919	00 54 00.	+ 11 08		17.2	BLUE STELLAR OBJECT
PHL 0907	00 54 00.	+ 14 30		18.4	BLUE STELLAR OBJECT
UGC 00585	00 54 00.	+ 26 44	66	16.0	GALAXY Sb-c
MCG+06-03-002	00 54 00.	+ 34 19	24	15.	GALAXY
MCG+06-03-001	00 54 00.	+ 34 21	18	16.	GALAXY
VB 023	00 54 00.	+ 65 03	295		STELLAR RING
MCG+00-03-036	00 54 00.	- 01 36	27	14.	GALAXY
ZWG 384.039	00 54 00.	- 02 02		14.1	GALAXY
UGC 00584	00 54 00.	- 02 02	102	14.1	GALAXY E-S0
MCG-01-03-038	00 54 00.	- 03 05	27	15.5	GALAXY
PHL 6917	00 54 00.	- 16 05		18.4	BLUE STELLAR OBJECT

OBJECT NAME	RIGHT ASCEN.	DECLINATION	DIAM.	MAGN.	TYPE OF OBJECT
PHL 0908	00 54 00.	- 20 49		17.7	BLUE STELLAR OBJECT
PHL 6916	00 54 00.	- 29 00		17.9	BLUE STELLAR OBJECT
PHL 6918	00 54 00.	- 31 28		12.0	BLUE STELLAR OBJECT
LB 00092	00 54 00.	- 34 05		16.7	FAINT BLUE STAR
LH115-N057	00 54 00.	- 72 33	35		EMISSION NEBULA IN SMC
RNGC 0307	00 54 01.	- 02 02		14.0	GALAXY
LIN 286	00 54 02.	- 72 40 57.		14.11	SMC EMISSION LINE OBJECT
LIN 284	00 54 03.	- 72 16 09.		14.46	SMC EMISSION LINE OBJECT
LIN 285	00 54 03.	- 72 20 21.		14.54	SMC EMISSION LINE OBJECT
LIN 283	00 54 04.	- 72 03 33.		15.34	SMC EMISSION LINE OBJECT
PHL 3142	00 54 06.	+ 00 26		18.5	BLUE STELLAR OBJECT
MCG+04-03-021	00 54 06.	+ 26 44	30	15.	GALAXY
ISS 0007	00 54 06.	+ 65 07	301		STELLAR RING
VB 024	00 54 06.	+ 66 36	342		STELLAR RING
ZWG 384.040	00 54 06.	- 01 33		15.5	GALAXY
PHL 6920	00 54 06.	- 21 32		18.2	BLUE STELLAR OBJECT
RNGC 0312	00 54 06.	- 53 03			UNVERIFIED SOUTHERN OBJECT
HSN 14	00 54 09.	- 72 59			BRIGHT GALAXY BEHIND SMC
LIN 295	00 54 10.	- 73 44 51.		15.87	SMC EMISSION LINE OBJECT
LIN 294	00 54 11.	- 73 39 57.		14.59	SMC EMISSION LINE OBJECT
PHL 6921	00 54 12.	+ 11 03		18.4	BLUE STELLAR OBJECT
MCG+04-03-022	00 54 12.	+ 26 43	48	15.	GALAXY
ZWG 384.041	00 54 12.	- 00 37		15.4	GALAXY
PHL 6922	00 54 12.	- 02 24		18.7	BLUE STELLAR OBJECT
PHL 3143	00 54 12.	- 06 28		18.5	BLUE STELLAR OBJECT
PHL 6923	00 54 12.	- 15 44		18.8	BLUE STELLAR OBJECT
LH115-N054	00 54 12.	- 70 36			EMISSION NEBULA IN SMC
LIN 293	00 54 12.	- 73 16 51.		14.96	SMC EMISSION LINE OBJECT
PATH 1.033	00 54 13.	+ 00 24			NEBULA
HOLM 027A	00 54 13.	- 10 11	120	12.6	PART OF MULTIPLE GALAXY
LIN 289	00 54 14.	- 70 35 51.		16.0	PLANETARY NEBULA IN SMC
MCG+00-03-037	00 54 15.	- 01 34	33	15.5	GALAXY
LIN 291	00 54 15.	- 72 14 57.		14.56	SMC EMISSION LINE OBJECT
LIN 292	00 54 15.	- 72 25 09.		13.06	SMC EMISSION LINE OBJECT
LIN 290	00 54 16.	- 71 57 33.		15.00	SMC EMISSION LINE OBJECT
LIN 299	00 54 16.	- 73 53 45.		15.27	SMC EMISSION LINE OBJECT
ZWG 384.044	00 54 18.	+ 00 05		15.6	GALAXY
UGC 00586	00 54 18.	+ 07 03	60	16.5	GALAXY S
PHL 6924	00 54 18.	+ 11 52		16.3	BLUE STELLAR OBJECT
ZWG 384.043	00 54 18.	- 00 15		15.4	GALAXY
MCG+00-03-038	00 54 18.	- 01 30	42	14.5	GALAXY
ZWG 384.042	00 54 18.	- 01 31		15.4	GALAXY
MCG-02-03-050	00 54 18.	- 10 10	180	12.	GALAXY
PHL 6925	00 54 18.	- 18 10		18.4	BLUE STELLAR OBJECT
RNGC 0310	00 54 19.	- 02 02			GALAXY
LIN 298	00 54 20.	- 72 44 21.		14.27	SMC EMISSION LINE OBJECT
LIN 296	00 54 21.	- 72 13 45.		14.24	SMC EMISSION LINE OBJECT
LIN 297	00 54 21.	- 72 19 15.		14.40	SMC EMISSION LINE OBJECT
KN 10.10	00 54 22.4	+ 08 45 25.			NEBULA
KN 10.11	00 54 23.7	+ 07 32 14.			NEBULA
PHL 3149	00 54 24.	+ 01 17		18.4	BLUE STELLAR OBJECT
MCG+01-03-008	00 54 24.	+ 07 12	24	15.	GALAXY
ZWG 410.018	00 54 24.	+ 08 46		15.2	GALAXY
PHL 6929	00 54 24.	+ 11 40		18.0	BLUE STELLAR OBJECT
PHL 3148	00 54 24.	+ 12 36		18.4	BLUE STELLAR OBJECT
PHL 3144	00 54 24.	+ 12 43		17.2	BLUE STELLAR OBJECT
PHL 3146	00 54 24.	+ 12 44		18.1	BLUE STELLAR OBJECT
PHL 6927	00 54 24.	+ 12 53		16.0	BLUE STELLAR OBJECT
PHL 3145	00 54 24.	+ 13 32		18.7	BLUE STELLAR OBJECT
PHL 3147	00 54 24.	+ 14 22		18.2	BLUE STELLAR OBJECT
ZC C054.4+2523	00 54 24.	+ 25 23	2550		CLUSTER OF GALAXIES
IC 0059	00 54 24.	+ 60 49	1980		EMISSION NEBULA
PHL 6928	00 54 24.	- 01 06		16.7	BLUE STELLAR OBJECT
MCG+00-03-039	00 54 24.	- 01 10	15	14.5	GALAXY
ZWG 384.045	00 54 24.	- 01 28		15.3	GALAXY
UGC 00587	00 54 24.	- 01 28	72	15.3	GALAXY
PHL 3150	00 54 24.	- 05 25		18.5	BLUE STELLAR OBJECT
MCG-04-03-034	00 54 24.	- 22 23	24	15.5	GALAXY
PHL 6931	00 54 24.	- 27 22		18.3	BLUE STELLAR OBJECT
PHL 6930	00 54 24.	- 31 00		18.1	BLUE STELLAR OBJECT
PHL 6926	00 54 24.	- 31 04		17.1	BLUE STELLAR OBJECT
RNGC 0314	00 54 24.	- 32 15		15.0	GALAXY
MCG-05-03-015	00 54 24.	- 32 15	18	15.	GALAXY
LB 03168	00 54 24.	- 73 12		12.9	FAINT BLUE STAR
CED 004A	00 54 25.	+ 60 49	1080		DIFFUSE GALACTIC NEBULA
SG 3.003	00 54 25.	+ 60 49	360		DIFFUSE EMISSION NEBULA
LIN 301	00 54 25.	- 72 53 21.		13.68	SMC EMISSION LINE OBJECT
KN 10.12	00 54 26.8	+ 07 11 31.			NEBULA
RNGC 0309A	00 54 27.	- 10 13			GALAXY
LIN 300	00 54 28.	- 71 56 15.		14.96	SMC EMISSION LINE OBJECT
ZWG 384.047	00 54 30.	+ 03 09		15.7	GALAXY
IC 0061	00 54 30.	+ 07 13 54.			NONSTELLAR OBJECT
ZWG 410.019	00 54 30.	+ 07 15		15.3	GALAXY
UGC 00589	00 54 30.	+ 07 15	90	15.3	GALAXY COMPACT
MCG+01-03-009	00 54 30.	+ 07 15	36	14.5	GALAXY
ZWG 435.025	00 54 30.	+ 14 38		15.7	GALAXY
UGC 00590	00 54 30.	+ 14 38	66	15.7	GALAXY S?
ZWG 384.046	00 54 30.	- 01 07		15.1	GALAXY
UGC 00588	00 54 30.	- 01 07	72	15.1	GALAXY
MCG-01-03-039	00 54 30.	- 03 02	36	15.5	GALAXY
LB 00093	00 54 30.	- 30 44		17.1	FAINT BLUE STAR
LH115-N058	00 54 30.	- 72 34	70		EMISSION NEBULA IN SMC
KN 10.13	00 54 30.5	+ 08 42 10.			NEBULA
KN 10.14	00 54 30.6	+ 07 17 43.			NEBULA
LIN 303	00 54 31.	- 72 44 21.		13.99	SMC EMISSION LINE OBJECT
LIN.CL 054	00 54 31.	- 72 44 45.	1224	7.4	STAR CLUSTER IN SMC
KN 10.15	00 54 31.6	+ 07 14 10.			NEBULA
RNGC 0331	00 54 32.	- 03 02		15.5	GALAXY
LIN 302	00 54 32.	- 72 22 57.		16.40	PLANETARY NEBULA IN SMC
RNGC 0330	00 54 33.	- 72 45		7.5	GLOBULAR CLUSTER IN SMC
PHL 0910	00 54 36.	+ 02 36		18.1	BLUE STELLAR OBJECT
PHL 6933	00 54 36.	+ 11 24		18.2	BLUE STELLAR OBJECT
PHL 0909	00 54 36.	+ 14 30		16.7	BLUE STELLAR OBJECT
PHL 6932	00 54 36.	+ 14 56		18.2	BLUE STELLAR OBJECT
ZWG 480.025	00 54 36.	+ 23 37		15.2	GALAXY
MRK 350	00 54 36.	+ 23 37	20	15.5	GALAXY WITH UV CONTINUUM
UGC 00591	00 54 36.	+ 23 37	60	15.2	GALAXY
ZC 0054.6-0127	00 54 36.	- 01 27	6050		CLUSTER OF GALAXIES
PHL 0911	00 54 36.	- 03 51		18.5	BLUE STELLAR OBJECT
PHL 3151	00 54 36.	- 04 42		18.5	BLUE STELLAR OBJECT
PHL 6934	00 54 36.	- 08 58		18.5	BLUE STELLAR OBJECT
PHL 3152	00 54 36.	- 11 31		18.7	BLUE STELLAR OBJECT
TON-S 0179	00 54 36.	- 27 38		14.8	BLUE STAR
MCG-05-03-016	00 54 36.	- 31 04	36	15.5	GALAXY
RNGC 0323	00 54 36.	- 53 15			UNVERIFIED SOUTHERN OBJECT
KN 10.16	00 54 36.5	+ 07 55 30.			NEBULA
ARC 0120	00 54 37.	- 16 41		17.2	RICH CLUSTER OF GALAXIES
LIN 306	00 54 41.	- 73 16 33.		14.99	SMC EMISSION LINE OBJECT

OBJECT NAME	RIGHT ASCEN.	DECLINATION	DIAM.	MAGN.	TYPE OF OBJECT
LIN 307	00 54 41.	- 73 18 39.		14.09	SMC EMISSION LINE OBJECT
LIN 308	00 54 41.	- 73 27 21.		14.53	SMC EMISSION LINE OBJECT
PHL 3153	00 54 42.	+ 12 16		18.4	BLUE STELLAR OBJECT
PHL 6936	00 54 42.	+ 13 29		17.7	BLUE STELLAR OBJECT
MCG+04-03-023	00 54 42.	+ 23 36 30.	48	15.	GALAXY
ZCG 0054+43	00 54 42.	+ 43 33		19.2	COMPACT GALAXY
VB C25	00 54 42.	+ 66 27	645		STELLAR RING
PHL 6937	00 54 42.	- 00 34		18.2	BLUE STELLAR OBJECT
MCG+00-03-040	00 54 42.	- 00 58	24	15.5	GALAXY
PHL 6938	00 54 42.	- 15 38		18.6	BLUE STELLAR OBJECT
SC 0052-1556.0	00 54 42.	- 15 39 44.	60		NEBULA
PHL 6935	00 54 42.	- 19 42		16.6	BLUE STELLAR OBJECT
PHL 0912	00 54 42.	- 22 38		13.8	BLUE STELLAR OBJECT
TON-S 0180	00 54 42.	- 22 38		14.2	BLUE STAR
HN 0130	00 54 42.	- 45 42			NEBULA
IC 1603	00 54 42.	- 45 42			NONSTELLAR OBJECT
LIN 305	00 54 43.	- 72 42 33.		15.76	SMC EMISSION LINE OBJECT
LIN 304	00 54 44.	- 72 34 39.		14.94	SMC EMISSION LINE OBJECT
MCG-04-03-035	00 54 45.	- 22 11 30.	24	15.5	GALAXY
PHL 3155	00 54 48.	+ 01 14		17.4	BLUE STELLAR OBJECT
ZC 0054.8+0635	00 54 48.	+ 06 35	1210		CLUSTER OF GALAXIES
3ZW 017	00 54 48.	+ 09 30			COMPACT GALAXY
ZWG 501.049	00 54 48.	+ 30 00		14.1	GALAXY
UGC 00592	00 54 48.	+ 30 00	90	14.1	GALAXY S0
MCG+05-03-028	00 54 48.	+ 30 00	30	14.5	GALAXY
ZWG 536.013	00 54 48.	+ 43 31		13.8	GALAXY
RNGC 0317A	00 54 48.	+ 43 31		15.0	GALAXY
UGC 00594	00 54 48.	+ 43 31	66	13.8	GALAXY SB
KARA.72 019B	00 54 48.	+ 43 31	60		PART OF DOUBLE GALAXY
5ZW 042	00 54 48.	+ 43 32			COMPACT GALAXY
KARA.72 019A	00 54 48.	+ 43 32	42	13.8	PART OF DOUBLE GALAXY
PHL 3156	00 54 48.	- 00 16		18.5	BLUE STELLAR OBJECT
PHL 6939	00 54 48.	- 10 48		18.4	BLUE STELLAR OBJECT
PHL 6940	00 54 48.	- 13 11		18.2	BLUE STELLAR OBJECT
LB 00094	00 54 48.	- 30 07		16.7	FAINT BLUE STAR
PHL 3154	00 54 48.	- 32 38		17.9	BLUE STELLAR OBJECT
LIN 309	00 54 48.	- 73 00 45.		13.52	SMC EMISSION LINE OBJECT
RNGC 0311	00 54 51.	+ 30 00		14.0	GALAXY
MCG+07-03-009	00 54 51.	+ 43 31 30.	15	15.	GALAXY
RNGC 0319	00 54 51.	- 44 08			GALAXY
LIN 311	00 54 51.	- 73 52 10.		14.27	SMC EMISSION LINE OBJECT
SVEN 063	00 54 52.	- 37 50	24	15.5	GALAXY
PHL 3158	00 54 54.	+ 01 12		18.5	BLUE STELLAR OBJECT
ZWG 536.014	00 54 54.	+ 43 25		14.9	GALAXY
MCG+07-03-011	00 54 54.	+ 43 25 30.	48	14.	GALAXY
RNGC 0317B	00 54 54.	+ 43 31		14.0	GALAXY
MCG+07-03-010	00 54 54.	+ 43 31	72	14.	GALAXY
PHL 3159	00 54 54.	- 19 07		18.1	BLUE STELLAR OBJECT
PHL 3157	00 54 54.	- 21 38		18.8	BLUE STELLAR OBJECT
LB 00095	00 54 54.	- 30 09		14.8	FAINT BLUE STAR
MCG-05-03-017	00 54 54.	- 31 13	12	15.5	GALAXY
RNGC 0328	00 54 54.	- 53 11			UNVERIFIED SOUTHERN OBJECT
LIN 310	00 54 56.	- 72 28 40.		14.22	SMC EMISSION LINE OBJECT
PATH 1.034	00 54 57.	+ 00 30	8		NEBULA
MCG-01-03-040	00 54 57.	- 04 25	30	16.	GALAXY
SN 1939D	00 54 59.	- 05 16		16.0	SUPERNOVA
LIN 314	00 54 59.	- 73 18 34.		13.25	SMC EMISSION LINE OBJECT
PHL 3162	00 55 00.	+ 01 42		18.4	BLUE STELLAR OBJECT
PHL 6942	00 55 00.	+ 02 07		18.4	BLUE STELLAR OBJECT
ZC 0055.0+0745	00 55 00.	+ 07 45	5040		CLUSTER OF GALAXIES
MCG+01-03-010	00 55 00.	+ 07 50	24	15.	GALAXY
ZC 0055.0+1212	00 55 00.	+ 12 12	6180		CLUSTER OF GALAXIES
ZWG 435.026	00 55 00.	+ 14 48		15.5	GALAXY
PHL 3160	00 55 00.	+ 15 28		18.6	BLUE STELLAR OBJECT
ZWG 501.050	00 55 00.	+ 32 17		15.7	GALAXY
MCG+05-03-029	00 55 00.	+ 33 04	24	15.	GALAXY
ZWG 501.051	00 55 00.	+ 33 05		15.0	GALAXY
MCG+00-03-041	00 55 00.	- 01 41	30	14.5	GALAXY
MCG-01-03-041	00 55 00.	- 05 16	42	14.5	GALAXY
ARC 0121	00 55 00.	- 07 17		16.0	RICH CLUSTER OF GALAXIES
PHL 3163	00 55 00.	- 09 50		18.5	BLUE STELLAR OBJECT
PHL 6941	00 55 00.	- 10 46		8.2	BLUE STELLAR OBJECT
MCG-03-03-009	00 55 00.	- 16 47	30	15.5	GALAXY
PHL 3161	00 55 00.	- 20 42		18.4	BLUE STELLAR OBJECT
MCG-05-03-018	00 55 00.	- 31 13	12	15.5	GALAXY
MCG-05-03-019	00 55 00.	- 31 15	60	15.5	GALAXY
LB 00096	00 55 00.	- 34 56		16.4	FAINT BLUE STAR
LIN 313	00 55 00.	- 73 00 04.		13.54	SMC EMISSION LINE OBJECT
ARC 0122	00 55 01.	- 26 33		17.1	RICH CLUSTER OF GALAXIES
LIN 312	00 55 01.	- 72 32 46.		15.09	SMC EMISSION LINE OBJECT
RNGC 0313	00 55 02.	+ 30 06			NON-EXISTENT OBJECT
RNGC 0321	00 55 02.	- 05 16		14.5	GALAXY
HOLM 029A	00 55 02.	- 05 17	36	14.3	PART OF MULTIPLE GALAXY
LIN.CL 055	00 55 02.	- 74 10 28.	474	14.9	STAR CLUSTER IN SMC
RNGC 0322	00 55 03.	- 44 00			GALAXY
KN 10.17	00 55 03.9	+ 07 49 27.			NEBULA
HOLM 028C	00 55 04.	+ 30 06	6	15.1	PART OF MULTIPLE GALAXY
HSN 15	00 55 05.	- 74 02			BRIGHT GALAXY BEHIND SMC
ZWG 410.020	00 55 06.	+ 07 49		15.6	GALAXY
UGC 00596	00 55 06.	+ 07 49	78	15.6	GALAXY COMPACT
PHL 6944	00 55 06.	+ 14 36		18.0	BLUE STELLAR OBJECT
PHL 6943	00 55 06.	+ 15 32		18.1	BLUE STELLAR OBJECT
MCG+05-03-031	00 55 06.	+ 30 04	48	13.	GALAXY
ZWG 501.052	00 55 06.	+ 30 05		12.5	GALAXY
UGC 00597	00 55 06.	+ 30 05	216	12.5	GALAXY E
MCG+05-03-030	00 55 06.	+ 31 12	54	14.5	GALAXY
ZWG 501.053	00 55 06.	+ 31 13		14.4	GALAXY
UGC 00598	00 55 06.	+ 31 13	60	14.4	GALAXY S0-a
MCG+07-03-012	00 55 06.	+ 42 03	48	16.	GALAXY
ZWG 550.020	00 55 06.	+ 45 59		15.6	GALAXY
ZWG 384.048	00 55 06.	- 01 38		14.8	GALAXY
UGC 00595	00 55 06.	- 01 38	90	14.8	GALAXY COMPACT
PHL 3164	00 55 06.	- 16 35		18.3	BLUE STELLAR OBJECT
LB 00097	00 55 06.	- 33 21		15.8	FAINT BLUE STAR
LB 01583	00 55 06.	- 48 40		13.7	FAINT BLUE STAR
KN 10.18	00 55 06.7	+ 07 44 13.			NEBULA
HOLM 028A	00 55 07.	+ 30 05	66	13.5	PART OF MULTIPLE GALAXY
LIN 315	00 55 07.	- 72 35 34.		15.00	SMC EMISSION LINE OBJECT
RNGC 0316	00 55 08.	+ 30 05			NON-EXISTENT OBJECT
RNGC 0315	00 55 08.	+ 30 05		12.5	GALAXY
HOLM 029B	00 55 09.	- 05 14	60	14.7	PART OF MULTIPLE GALAXY
HOLM 028B	00 55 11.	+ 30 05	12	14.9	PART OF MULTIPLE GALAXY
LIN 316	00 55 11.	- 73 17 40.		15.72	SMC EMISSION LINE OBJECT
PHL 0913	00 55 12.	+ 00 54		18.4	BLUE STELLAR OBJECT
PHL 3166	00 55 12.	+ 03 20		18.5	BLUE STELLAR OBJECT
MRK 351	00 55 12.	+ 28 06	10	16.	GALAXY WITH UV CONTINUUM
MCG+08-02-020	00 55 12.	+ 45 59	42	16.	GALAXY
ZWG 384.049	00 55 12.	- 00 40		14.7	GALAXY

OBJECT NAME	RIGHT ASCEN.	DECLINATION	DIAM.	MAGN.	TYPE OF OBJECT
UGC 00599	00 55 12.	- 00 40	78	14.7	GALAXY E?+COMP
MCG+00-03-042	00 55 12.	- 00 42 30.	21	15.	GALAXY
MCG-01-03-042	00 55 12.	- 05 13	36	14.5	GALAXY
PHL 3165	00 55 12.	- 06 04		18.5	BLUE STELLAR OBJECT
PHL 0914	00 55 12.	- 09 02		17.9	BLUE STELLAR OBJECT
PHL 6945	00 55 12.	- 11 50		17.6	BLUE STELLAR OBJECT
PHL 3167	00 55 12.	- 12 04		18.7	BLUE STELLAR OBJECT
PHL 6949	00 55 12.	- 13 46		18.3	BLUE STELLAR OBJECT
PHL 6946	00 55 12.	- 14 03		18.7	BLUE STELLAR OBJECT
SC 0052-1545.2	00 55 12.	- 15 28 57.	12		NEBULA
PHL 6948	00 55 12.	- 16 26		17.2	BLUE STELLAR OBJECT
PHL 6947	00 55 12.	- 22 02		18.0	BLUE STELLAR OBJECT
HELW 327	00 55 13.	- 10 34 57.			NEBULA
PHL 3169	00 55 18.	+ 15 05		18.3	BLUE STELLAR OBJECT
ZWG 501.054	00 55 18.	+ 30 09		15.2	GALAXY
ZWG 501.055	00 55 18.	+ 30 26		15.4	GALAXY
PHL 3168	00 55 18.	- 00 06		17.8	BLUE STELLAR OBJECT
MCG+00-03-043	00 55 18.	- 01 25	21	15.5	GALAXY
MCG-01-03-043	00 55 18.	- 05 24	15	16.	GALAXY
PHL 6950	00 55 18.	- 13 44		17.7	BLUE STELLAR OBJECT
PHL 6951	00 55 18.	- 14 02		18.2	BLUE STELLAR OBJECT
PHL 6952	00 55 18.	- 17 38		18.1	BLUE STELLAR OBJECT
MCG-05-03-020	00 55 18.	- 27 46	78	13.5	GALAXY
SG 3.004	00 55 19.	+ 60 43	1440		DIFFUSE EMISSION NEBULA
RNGC 0318	00 55 20.	+ 30 09		15.0	GALAXY
RNGC 0325	00 55 20.	- 05 24		16.0	GALAXY
RNGC 0301	00 55 21.	- 10 43			GALAXY
HN 0131	00 55 21.	- 49 11			NEBULA
IC 1605	00 55 21.	- 49 11			NONSTELLAR OBJECT
PHL 6953	00 55 24.	+ 10 45		15.1	BLUE STELLAR OBJECT
PHL 3170	00 55 24.	+ 13 50		17.9	BLUE STELLAR OBJECT
PHL 0915	00 55 24.	+ 15 38		18.4	BLUE STELLAR OBJECT
MCG+08-02-021	00 55 24.	+ 48 23	72	14.	GALAXY
ZWG 384.050	00 55 24.	- 01 03		15.7	GALAXY
MCG-01-03-044	00 55 24.	- 02 40	42	15.	GALAXY
PHL 3171	00 55 24.	- 03 36		18.3	BLUE STELLAR OBJECT
PHL 0916	00 55 24.	- 03 46		18.1	BLUE STELLAR OBJECT
MCG-01-03-045	00 55 24.	- 05 25	78	15.	GALAXY
PHL 0917	00 55 24.	- 13 42		17.6	BLUE STELLAR OBJECT
PHL 3172	00 55 24.	- 13 43		18.8	BLUE STELLAR OBJECT
PHL 6954	00 55 24.	- 16 30		17.5	BLUE STELLAR OBJECT
LB 03169	00 55 24.	- 69 11		14.2	FAINT BLUE STAR
LIN 318	00 55 24.	- 72 50 52.		14.94	SMC EMISSION LINE OBJECT
RNGC 0327	00 55 26.	- 05 24		13.0	GALAXY
HOLM 030A	00 55 26.	- 05 24	60	14.0	PART OF MULTIPLE GALAXY
LIN 317	00 55 27.	- 72 02 22.		15.17	SMC EMISSION LINE OBJECT
HELW 328	00 55 28.	- 10 22 10.			NEBULA
IC 1604	00 55 29.	- 16 31			NONSTELLAR OBJECT
PHL 6956	00 55 30.	+ 13 04		17.8	BLUE STELLAR OBJECT
MCG+04-03-024	00 55 30.	+ 26 36 30.	45	16.	GALAXY
ZWG 550.021	00 55 30.	+ 48 23		14.5	GALAXY
UGC 00600	00 55 30.	+ 48 23	96	14.5	GALAXY Sb/SBb
PHL 6957	00 55 30.	- 02 57		18.5	BLUE STELLAR OBJECT
MCG-01-03-046	00 55 30.	- 03 58	60	14.	GALAXY
MCG-01-03-048	00 55 30.	- 05 20	78	13.	GALAXY
MCG-01-03-047	00 55 30.	- 05 25	96	13.	GALAXY
PHL 0918	00 55 30.	- 08 40		17.7	BLUE STELLAR OBJECT
PHL 6958	00 55 30.	- 15 14		18.0	BLUE STELLAR OBJECT
PHL 6955	00 55 30.	- 29 46		17.5	BLUE STELLAR OBJECT
LH115-N059	00 55 30.	- 73 50	88		EMISSION NEBULA IN SMC
LIN 323	00 55 30.	- 74 29 46.		16.73	PLANETARY NEBULA IN SMC
RNGC 0329	00 55 32.	- 05 20		13.0	GALAXY
HOLM 030B	00 55 32.	- 05 21	78	14.3	PART OF MULTIPLE GALAXY
LIN 319	00 55 34.	- 71 41 58.		15.19	SMC EMISSION LINE OBJECT
LIN 322	00 55 34.	- 73 18 10.		15.14	SMC EMISSION LINE OBJECT
PHL 3174	00 55 36.	+ 00 50		18.2	BLUE STELLAR OBJECT
PHL 6959	00 55 36.	+ 02 32		18.1	BLUE STELLAR OBJECT
PHL 3173	00 55 36.	+ 02 44		18.1	BLUE STELLAR OBJECT
SHAH 031	00 55 36.	+ 13 38	60	17.6	GROUP OF COMPACT GALAXIES
MCG+08-02-022	00 55 36.	+ 49 47	60	15.	GALAXY
PHL 6960	00 55 36.	- 01 36		18.6	BLUE STELLAR OBJECT
MCG-01-03-050	00 55 36.	- 06 31	66	15.	GALAXY
MCG-01-03-049	00 55 36.	- 08 30	36	14.	GALAXY
HELW 329	00 55 36.	- 10 23 40.			NEBULA
PHL 0919	00 55 36.	- 13 05		18.3	BLUE STELLAR OBJECT
PHL 6961	00 55 36.	- 13 16		18.6	BLUE STELLAR OBJECT
LIN 321	00 55 36.	- 72 49 10.		14.08	PLANETARY NEBULA IN SMC
LH115-N060	00 55 36.	- 74 29			EMISSION NEBULA IN SMC
HOW 32	00 55 37.	- 71 26 36.	12		STAR CLUSTER IN SMC
LIN 320	00 55 38.	- 72 17 22.		12.95	SMC EMISSION LINE OBJECT
RNGC 0326	00 55 39.	+ 26 36		15.0	GALAXY
PHL 3176	00 55 42.	+ 00 57		18.6	BLUE STELLAR OBJECT
PHL 3175	00 55 42.	+ 09 41		18.2	BLUE STELLAR OBJECT
4ZW 035	00 55 42.	+ 26 35			COMPACT GALAXY
ZWG 480.026	00 55 42.	+ 26 36		14.9	GALAXY
UGC 00601	00 55 42.	+ 26 36	96	14.9	GALAXY PR CMPT
ZWG 520.004	00 55 42.	+ 36 28		14.5	GALAXY
UGC 00602	00 55 42.	+ 36 28	102	14.5	GALAXY Sc/SBc
KARA.73B 0041	00 55 42.	+ 36 28	96	14.5	ISOLATED GALAXY S
MCG+06-03-003	00 55 42.	+ 36 29	90	14.5	GALAXY
ZWG 551.001	00 55 42.	+ 49 46		15.4	GALAXY
ZWG 550.022	00 55 42.	+ 49 46		15.4	GALAXY
MCG-02-03-051	00 55 42.	- 10 22	36	14.5	GALAXY
HOW 33	00 55 43.	- 71 04 30.	12		STAR CLUSTER IN SMC
MCG+00-03-044	00 55 45.	- 02 13	9	15.	GALAXY
LIN 325	00 55 47.	- 72 56 40.		14.84	SMC EMISSION LINE OBJECT
PHL 0920	00 55 48.	+ 01 24		17.7	BLUE STELLAR OBJECT
MCG+00-03-045	00 55 48.	+ 01 47 30.	24	16.	GALAXY
ZWG 435.027	00 55 48.	+ 11 18		15.5	GALAXY
UGC 00603	00 55 48.	+ 11 18	60	15.5	GALAXY S
MCG+02-03-018	00 55 48.	+ 11 19	24	15.	GALAXY
ZWG 480.027	00 55 49.	+ 23 16		15.4	GALAXY
MCG+04-03-025	00 55 48.	+ 26 35	72	15.	GALAXY
UGC 00604	00 55 48.	+ 44 45	66	18.	GALAXY Sc
5ZW 043	00 55 48.	+ 51 21			COMPACT GALAXY
MCG-03-03-010	00 55 48.	- 15 40	60	14.	GALAXY
LH115-N061	00 55 48.	- 72 49			EMISSION NEBULA IN SMC
SG 1.02	00 55 49.	+ 60 38	720		DIFFUSE EMISSION NEBULA
LIN.CL 056	00 55 49.	- 72 32 46.	270	11.5	STAR CLUSTER IN SMC
LIN 324	00 55 49.	- 72 33 34.		15.61	SMC EMISSION LINE OBJECT
MCG+04-03-027	00 55 51.	+ 22 02 30.	30	15.5	GALAXY
MCG+04-03-026	00 55 51.	+ 23 16	42	15.	GALAXY
IC 1606	00 55 52.	- 12 26 57.			NONSTELLAR OBJECT
HSN 16	00 55 53.	- 74 03			BRIGHT GALAXY BEHIND SMC
ZWG 435.028	00 55 54.	+ 11 20		15.5	GALAXY
PHL 6962	00 55 54.	+ 12 54		14.5	BLUE STELLAR OBJECT
PHL 6963	00 55 54.	+ 13 11		15.4	BLUE STELLAR OBJECT
UGC 00605	00 55 54.	+ 22 02	66	16.0	GALAXY S
VB 099	00 55 54.	+ 57 25	168		STELLAR RING
ISS 0036	00 55 54.	+ 57 25	170		STELLAR RING
VB 100	00 55 54.	+ 59 10	577		STELLAR RING
URA 39	00 55 54.	+ 69 10	438		STELLAR RING
PHL 3177	00 55 54.	- 03 58		17.2	BLUE STELLAR OBJECT
MCG-02-03-052	00 55 54.	- 08 40	84	13.	GALAXY
PHL 6964	00 55 54.	- 15 08		18.8	BLUE STELLAR OBJECT
PHL 6965	00 55 54.	- 16 26		18.9	BLUE STELLAR OBJECT
PHL 3178	00 55 54.	- 20 44		18.3	BLUE STELLAR OBJECT
ARC 0123	00 55 56.	- 14 40		17.2	RICH CLUSTER OF GALAXIES
MCG+04-03-028	00 55 57.	+ 22 02	84	15.	GALAXY
HELW 330	00 55 59.	- 10 25 52.			NEBULA
LBN 0623	00 56	+ 60 40	4320		BRIGHT NEBULA
LBN 0622	00 56	+ 60 40	360		BRIGHT NEBULA
KHAV 015	00 56	+ 72 34	7500		DARK NEBULA
PHL 3179	00 56 00.	+ 01 24		18.0	BLUE STELLAR OBJECT
ZC 0056.0+0231	00 56 00.	+ 02 31	670		CLUSTER OF GALAXIES
PHL 6967	00 56 00.	+ 02 48		17.3	BLUE STELLAR OBJECT
MCG+02-03-021	00 56 00.	+ 11 32	48	14.5	GALAXY
MCG+02-03-019	00 56 00.	+ 12 25	24	16.	GALAXY
5ZW 044	00 56 00.	+ 47 06			COMPACT GALAXY
VB 026	00 56 00.	+ 63 52	477		STELLAR RING
ZC 0056.0+8450	00 56 00.	+ 84 50	1010		CLUSTER OF GALAXIES
PHL 6968	00 56 00.	- 00 52		17.9	BLUE STELLAR OBJECT
PHL 3181	00 56 00.	- 01 13		18.5	BLUE STELLAR OBJECT
MCG-02-03-053	00 56 00.	- 10 25	72	14.5	GALAXY
PHL 6966	00 56 00.	- 12 36		18.3	BLUE STELLAR OBJECT
PHL 6969	00 56 00.	- 14 12		18.8	BLUE STELLAR OBJECT
PHL 3180	00 56 00.	- 14 44		18.0	BLUE STELLAR OBJECT
MCG+00-03-046	00 56 03.	+ 02 45	27	15.5	GALAXY
LIN.CL 057	00 56 03.	- 73 40 59.	828		STAR CLUSTER IN SMC
PHL 6970	00 56 06.	+ 02 11		16.7	BLUE STELLAR OBJECT
IC 0062	00 56 06.	+ 11 32 15.			NONSTELLAR OBJECT
ZWG 435.029	00 56 06.	+ 11 33		15.0	GALAXY
UGC 00606	00 56 06.	+ 11 33	60	15.0	GALAXY S
ZWG 435.030	00 56 06.	+ 12 29		15.4	GALAXY
UGC 00607	00 56 06.	+ 12 29	78	15.4	GALAXY Sc
MCG+02-03-020	00 56 06.	+ 12 29	48	14.5	GALAXY
PHL 0921	00 56 06.	+ 12 40		18.0	BLUE STELLAR OBJECT
ZWG 551.002	00 56 06.	+ 47 45		15.6	GALAXY
ZWG 550.023	00 56 06.	+ 47 45		15.6	GALAXY
UGC 00608	00 56 06.	+ 47 45	120	15.6	GALAXY
MCG+08-03-001	00 56 06.	+ 47 46	108	15.	GALAXY
PHL 6971	00 56 06.	- 09 42		18.4	BLUE STELLAR OBJECT
MCG-02-03-054	00 56 06.	- 10 22	36	15.5	GALAXY
PHL 3183	00 56 06.	- 10 42		18.6	BLUE STELLAR OBJECT
PHL 3182	00 56 06.	- 10 44		18.2	BLUE STELLAR OBJECT
LB 03170	00 56 06.	- 63 01		14.4	FAINT BLUE STAR
LIN 326	00 56 08.	- 72 04 23.		15.00	SMC EMISSION LINE OBJECT
HSN 17	00 56 10.	- 73 13			BRIGHT GALAXY BEHIND SMC
LIN.CL 059	00 56 10.	- 74 44 47.	1020	12.0	STAR CLUSTER IN SMC
HELW 331	00 56 11.	- 07 57 58.			NEBULA
LIN 329	00 56 11.	- 72 53 23.		13.71	SMC EMISSION LINE OBJECT
HOW 34	00 56 11.	- 73 48 48.	54		STAR CLUSTER IN SMC
LIN.CL 058	00 56 11.	- 74 35 59.	342	14.1	STAR CLUSTER IN SMC
MCG+00-03-047	00 56 12.	+ 00 19	36	14.	GALAXY
ZWG 410.021	00 56 12.	+ 06 50		14.9	GALAXY
RNGC 0332	00 56 12.	+ 06 50		15.0	GALAXY
UGC 00609	00 56 12.	+ 06 50	96	14.9	GALAXY COMPACT
PHL 6974	00 56 12.	+ 12 09		18.1	BLUE STELLAR OBJECT
MCG+02-03-022	00 56 12.	+ 12 42	24	14.	GALAXY
ZWG 435.031	00 56 12.	+ 12 43		14.8	GALAXY
UGC 00610	00 56 12.	+ 12 43	108	14.8	GALAXY E
4ZW 036	00 56 12.	+ 27 20			COMPACT GALAXY
URA 46	00 56 12.	+ 60 46	1224		STELLAR RING
PHL 6972	00 56 12.	- 11 50		17.9	BLUE STELLAR OBJECT
PHL 6973	00 56 12.	- 16 48		18.3	BLUE STELLAR OBJECT
PHL 0922	00 56 12.	- 21 40		18.6	BLUE STELLAR OBJECT
PHL 3184	00 56 12.	- 29 38		4.3	BLUE STELLAR OBJECT
LH115-N065	00 56 12.	- 72 05			EMISSION NEBULA IN SMC
SVEN 064	00 56 13.	- 07 58	12	15.1	GALAXY
LIN 327	00 56 13.	- 72 12 17.		14.53	SMC EMISSION LINE OBJECT
LIN 328	00 56 13.	- 72 23 53.		15.00	SMC EMISSION LINE OBJECT
RNGC 0339	00 56 13.	- 74 45		12.0	GLOBULAR CLUSTER IN SMC
KN 10.19	00 56 13.4	+ 06 50 40.			NEBULA
RNGC 0324	00 56 14.	- 40 46			GALAXY
IC 1607	00 56 15.	+ 00 19 02.			NONSTELLAR OBJECT
MCG+08-03-002	00 56 15.	+ 50 01	36	16.	GALAXY
ZWG 384.051	00 56 18.	+ 00 20		14.5	GALAXY
UGC 00611	00 56 18.	+ 00 20	66	14.5	GALAXY PECULR
ZWG 480.028	00 56 18.	+ 23 35		14.5	GALAXY
UGC 00612	00 56 18.	+ 23 35	60	14.5	GALAXY
ZC 0056.3+4201	00 56 18.	+ 42 01	810		CLUSTER OF GALAXIES
MCG-04-03-036	00 56 18.	- 22 24	48	15.	GALAXY
LB 00098	00 56 18.	- 28 32		16.3	FAINT BLUE STAR
LB 00074	00 56 18.	- 36 24		15.4	FAINT BLUE STAR
LH115-N062	00 56 18.	- 72 56	65		EMISSION NEBULA IN SMC
RNGC 0333B	00 56 22.	- 16 45		14.0	GALAXY
RNGC 0333A	00 56 22.	- 16 45		16.0	GALAXY
RNGC 0336	00 56 22.	- 19 02		12.0	GALAXY
RNGC 0320	00 56 22.	- 21 06		15.0	GALAXY
PHL 6977	00 56 24.	+ 13 21		18.3	BLUE STELLAR OBJECT
3ZW 018	00 56 24.	+ 20 06			COMPACT GALAXY
MCG+04-03-029	00 56 24.	+ 23 34 30.	42	14.5	GALAXY
PHL 0923	00 56 24.	- 00 10		17.7	BLUE STELLAR OBJECT
PHL 6976	00 56 24.	- 03 36		17.0	BLUE STELLAR OBJECT
PHL 0924	00 56 24.	- 05 40		10.0	BLUE STELLAR OBJECT
PEIG 010	00 56 24.	- 05 41		9.8	FAINT BLUE STAR
PHL 6975	00 56 24.	- 09 06		17.4	BLUE STELLAR OBJECT
MCG-02-03-055	00 56 24.	- 11 12	48	15.	GALAXY
PHL 6978	00 56 24.	- 14 11		18.5	BLUE STELLAR OBJECT
MCG-03-03-013	00 56 24.	- 16 45 30.	15	14.5	GALAXY
MCG-03-03-012	00 56 24.	- 16 45 30.	4	16.	GALAXY
MCG-03-03-014	00 56 24.	- 17 08 30.	15	15.	GALAXY
MCG-03-03-011	00 56 24.	- 19 03	192	12.5	GALAXY
MCG-04-03-037	00 56 24.	- 21 06 30.	36	15.	GALAXY
PHL 3185	00 56 24.	- 24 30		18.4	BLUE STELLAR OBJECT
RNGC 0334	00 56 25.	- 35 24			GALAXY
MCG-04-03-038	00 56 27.	- 26 01 30.	30	14.5	GALAXY
LIN 331	00 56 29.	- 72 55 17.		14.64	SMC EMISSION LINE OBJECT
LIN 332	00 56 29.	- 73 02 11.		13.58	SMC EMISSION LINE OBJECT
PHL 3186	00 56 30.	+ 11 26		17.9	BLUE STELLAR OBJECT
ZWG 435.032	00 56 30.	+ 12 43		15.7	GALAXY
MCG+04-03-030	00 56 30.	+ 26 46	36	16.	GALAXY
ISS 0028	00 56 30.	+ 61 25	301		STELLAR RING
HELW 332	00 56 30.	- 07 21 59.			NEBULA
MCG-06-03-012	00 56 30.	- 35 24	48	14.	GALAXY
LH115-N066D	00 56 30.	- 72 27	32		EMISSION NEBULA IN SMC

OBJECT NAME	RIGHT ASCEN.	DECLINATION	DIAM.	MAGN.	TYPE OF OBJECT
HELW 333	00 56 31.	- 07 56 59.			NEBULA
SVEN 066	00 56 31.	- 07 57	18	16.7	GALAXY
SHB 014	00 56 31.7	- 00 09 16.		17.3	QUASI-STELLAR OBJECT
BC 4C-00.06	00 56 31.73	- 00 09 19.2		17.33	QUASI-STELLAR OBJECT
SVEN 065	00 56 32.	- 07 22	24	15.2	GALAXY
LIN 330	00 56 32.	- 72 01 17.		14.53	SMC EMISSION LINE OBJECT
PHL 6981	00 56 36.	+ 11 48		18.3	BLUE STELLAR OBJECT
PHL 3188	00 56 36.	+ 14 26		18.4	BLUE STELLAR OBJECT
PHL 3187	00 56 36.	+ 14 56		18.5	BLUE STELLAR OBJECT
ZWG 520.005	00 56 36.	+ 39 03		15.7	GALAXY
MCG+06-03-004	00 56 36.	+ 39 05	36	15.	GALAXY
ZC 0056.6+4000	00 56 36.	+ 40 00	340		CLUSTER OF GALAXIES
ARC 0124	00 56 36.	+ 42 07		17.5	RICH CLUSTER OF GALAXIES
ZC 0056.6+4302	00 56 36.	+ 43 02	1210		CLUSTER OF GALAXIES
PHL 6979	00 56 36.	- 00 32		17.5	BLUE STELLAR OBJECT
PHL 6982	00 56 36.	- 11 01		18.9	BLUE STELLAR OBJECT
PHL 6980	00 56 36.	- 17 19		18.2	BLUE STELLAR OBJECT
LH115-N063	00 56 36.	- 72 55	35		EMISSION NEBULA IN SMC
LIN 333	00 56 38.	- 71 51 29.		16.69	PLANETARY NEBULA IN SMC
RNGC 0335	00 56 40.	- 18 32		14.0	GALAXY
HSN 18	00 56 40.	- 72 55			BRIGHT GALAXY BEHIND SMC
HSN 19	00 56 41.	- 73 46			BRIGHT GALAXY BEHIND SMC
ZWG 410.022	00 56 42.	+ 06 40		15.7	GALAXY
MRK 559	00 56 42.	+ 06 40	12	15.5	GALAXY WITH UV CONTINUUM
PHL 3189	00 56 42.	+ 12 25		18.8	BLUE STELLAR OBJECT
PHL 6983	00 56 42.	+ 13 15		16.7	BLUE STELLAR OBJECT
ZWG 480.029	00 56 42.	+ 23 42		15.7	GALAXY
ZWG 480.030	00 56 42.	+ 26 47		15.6	GALAXY
UGC 00613	00 56 42.	+ 26 47	78	15.6	GALAXY
IC 0064	00 56 42.	+ 26 47 02.			NONSTELLAR OBJECT
MCG+00-03-048	00 56 42.	- 00 29	33	16.	GALAXY
HELW 334	00 56 42.	- 07 38 47.			NEBULA
PHL 3191	00 56 42.	- 11 45		18.9	BLUE STELLAR OBJECT
MCG-02-03-056	00 56 42.	- 14 02	72	15.5	GALAXY
MCG-03-03-015	00 56 42.	- 18 32	72	14.5	GALAXY
PHL 3190	00 56 42.	- 24 34		18.4	BLUE STELLAR OBJECT
PHL 0925	00 56 42.	- 24 34		17.3	BLUE STELLAR OBJECT
LB 00099	00 56 42.	- 28 40		17.1	FAINT BLUE STAR
LB 00100	00 56 42.	- 33 20		16.2	FAINT BLUE STAR
LIN 334	00 56 42.	- 72 29 47.		14.50	SMC EMISSION LINE OBJECT
LH115-N064	00 56 42.	- 72 56	80		EMISSION NEBULA IN SMC
KN 10.20	00 56 42.3	+ 06 39 09.			NEBULA
SVEN 067	00 56 43.	- 07 39	24	15.3	GALAXY
LIN 335	00 56 47.	- 72 56 30.		14.34	SMC EMISSION LINE OBJECT
MCG+04-03-031	00 56 48.	+ 26 47	60	15.5	GALAXY
ZWG 501.056	00 56 48.	+ 31 31		15.0	GALAXY
ZWG 520.006	00 56 48.	+ 35 18		14.0	GALAXY
UGC 00614	00 56 48.	+ 35 18	96	14.0	GALAXY SBc
VB 027	00 56 48.	+ 64 33	168		STELLAR RING
MCG-01-03-051	00 56 48.	- 05 05	126	14.5	GALAXY
LB 00101	00 56 48.	- 32 21		16.5	FAINT BLUE STAR
LB 00102	00 56 48.	- 32 35		16.3	FAINT BLUE STAR
LB 00075	00 56 48.	- 33 51		16.1	FAINT BLUE STAR
LH115-N064A	00 56 48.	- 72 56	24		EMISSION NEBULA IN SMC
HOLM 031A	00 56 50.	- 05 05	90	14.7	PART OF MULTIPLE GALAXY
MCG-01-03-052	00 56 51.	- 05 05	36	14.5	GALAXY
MCG-02-03-057	00 56 51.	- 13 58	30	15.	GALAXY
HOLM 031B	00 56 52.	- 05 04	30	15.0	PART OF MULTIPLE GALAXY
ARP 121	00 56 53.	- 05 03			PECULIAR GALAXY
IC 1608	00 56 53.	- 34 35 17.			NONSTELLAR OBJECT
LIN 337	00 56 53.	- 72 38 12.		15.10	SMC EMISSION LINE OBJECT
LIN 338	00 56 53.	- 72 39 06.		14.20	SMC EMISSION LINE OBJECT
LIN 339	00 56 53.	- 72 44 18.		15.71	PLANETARY NEBULA IN SMC
PHL 0926	00 56 54.	+ 00 56		18.6	BLUE STELLAR OBJECT
PHL 3193	00 56 54.	+ 01 18		18.5	BLUE STELLAR OBJECT
PHL 3192	00 56 54.	+ 13 43		18.6	BLUE STELLAR OBJECT
ZC 0056.9+1408	00 56 54.	+ 14 08	3020		CLUSTER OF GALAXIES
ZC 0056.9+2636	00 56 54.	+ 26 36	4700		CLUSTER OF GALAXIES
MCG+05-03-032	00 56 54.	+ 29 03 30.	30	15.	GALAXY
ZWG 501.057	00 56 54.	+ 29 04		15.1	GALAXY
MCG+06-03-005	00 56 54.	+ 35 18	84	14.	GALAXY
ZC 0056.9+4208	00 56 54.	+ 42 08	610		CLUSTER OF GALAXIES
MRSL 124-01/1	00 56 54.	+ 60 43	7200		HII REGION
PHL 0927	00 56 54.	- 07 50		17.9	BLUE STELLAR OBJECT
PHL 3194	00 56 54.	- 10 12		18.6	BLUE STELLAR OBJECT
PHL 0928	00 56 54.	- 10 52		18.2	BLUE STELLAR OBJECT
MCG-04-03-039	00 56 54.	- 20 50 30.	36	15.5	GALAXY
SC 0055-6745.1	00 56 54.	- 67 28 53.	6		NEBULA
LH115-N067	00 56 54.	- 71 52	19		EMISSION NEBULA IN SMC
LIN 336	00 56 54.	- 72 28 30.		13.79	SMC EMISSION LINE OBJECT
HELW 335	00 56 58.	- 08 04 11.			NEBULA
LIN 340	00 56 59.	- 72 37 06.		14.75	SMC EMISSION LINE OBJECT
LIN 341	00 56 59.	- 72 52 24.		14.42	SMC EMISSION LINE OBJECT
PHL 3195	00 57 00.	+ 09 46		17.7	BLUE STELLAR OBJECT
PHL 6988	00 57 00.	+ 12 04		16.0	BLUE STELLAR OBJECT
ZWG 435.033	00 57 00.	+ 12 43		15.7	GALAXY
PHL 3196	00 57 00.	+ 14 32		18.0	BLUE STELLAR OBJECT
MCG+02-03-023	00 57 00.	+ 15 03	54	14.	GALAXY
ZWG 435.034	00 57 00.	+ 15 04		14.2	GALAXY
UGC 00615	00 57 00.	+ 15 04	66	14.2	GALAXY Sa/SBb
ZWG 458.011	00 57 00.	+ 18 35		15.5	GALAXY
UGC 00616	00 57 00.	+ 18 35	90	15.5	GALAXY SBb
MCG+03-03-010	00 57 00.	+ 18 35	60	15.	GALAXY
MCG+08-03-003	00 57 00.	+ 47 32 30.	24	16.	GALAXY
PHL 6989	00 57 00.	- 00 21		18.3	BLUE STELLAR OBJECT
PHL 3198	00 57 00.	- 09 00		18.5	BLUE STELLAR OBJECT
PHL 3197	00 57 00.	- 10 17		18.9	BLUE STELLAR OBJECT
PHL 6987	00 57 00.	- 10 38		18.2	BLUE STELLAR OBJECT
PHL 6986	00 57 00.	- 14 37		17.8	BLUE STELLAR OBJECT
PHL 0929	00 57 00.	- 15 56		18.0	BLUE STELLAR OBJECT
PHL 6985	00 57 00.	- 16 19		16.5	BLUE STELLAR OBJECT
PHL 0930	00 57 00.	- 24 54		18.3	BLUE STELLAR OBJECT
PHL 6984	00 57 00.	- 32 14		10.0	BLUE STELLAR OBJECT
LB 00103	00 57 00.	- 36 11		15.3	FAINT BLUE STAR
LH115-N068	00 57 00.	- 72 44			EMISSION NEBULA IN SMC
SVEN 068	00 57 01.	- 08 04 18.	12	16.2	GALAXY
KARA.73 02A	00 57 03.	- 33 58			DWARF GALAXY
LIN 343	00 57 03.	- 73 13 48.		16.70	SMC EMISSION LINE OBJECT
MCG+00-03-049	00 57 06.	+ 00 38	48	15.	GALAXY
ZWG 458.012	00 57 06.	+ 17 45		15.2	GALAXY
UGC 00617	00 57 06.	+ 17 45	66	15.2	GALAXY SBa
ZWG 501.058	00 57 06.	+ 31 33		14.8	GALAXY
MRK 352	00 57 06.	+ 31 33	15	15.	GALAXY WITH UV CONTINUUM
KW 33	00 57 06.	+ 31 33	38		SEYFERT GALAXY
VVI 06	00 57 06.	+ 31 33	15	15.25	SEYFERT GALAXY
ZC 0057.1+4110	00 57 06.	+ 41 10	540		CLUSTER OF GALAXIES
PHL 0931	00 57 06.	- 09 00		18.1	BLUE STELLAR OBJECT
MCG-02-03-058	00 57 06.	- 10 35	42	15.	GALAXY
PHL 6990	00 57 06.	- 11 47		17.8	BLUE STELLAR OBJECT
MCG-02-03-059	00 57 06.	- 12 42	60	14.5	GALAXY
PHL 6991	00 57 06.	- 12 48		18.4	BLUE STELLAR OBJECT
PHL 6992	00 57 06.	- 19 27		18.8	BLUE STELLAR OBJECT
PHL 3199	00 57 06.	- 21 16		18.3	BLUE STELLAR OBJECT
MCG-06-03-013	00 57 06.	- 34 37	120	12.	GALAXY
MCG-06-03-014	00 57 06.	- 36 28	60	13.5	GALAXY
LIN 342	00 57 06.	- 72 27 18.		14.71	SMC EMISSION LINE OBJECT
HOW 35	00 57 06.	- 73 51 29.	36		STAR CLUSTER IN SMC
LIN 344	00 57 10.	- 72 54 36.		14.63	SMC EMISSION LINE OBJECT
ZWG 384.052	00 57 12.	+ 00 40		15.4	GALAXY
UGC 00618	00 57 12.	+ 00 40	72	15.4	GALAXY Sc
PHL 3202	00 57 12.	+ 01 28		18.1	BLUE STELLAR OBJECT
PHL 6994	00 57 12.	+ 10 50		16.8	BLUE STELLAR OBJECT
PHL 6998	00 57 12.	+ 12 49		18.1	BLUE STELLAR OBJECT
UGC 00619	00 57 12.	+ 14 27	78	16.5	GALAXY Sc
MCG+03-03-011	00 57 12.	+ 17 44	42	15.	GALAXY
ZWG 536.015	00 57 12.	+ 41 27		15.6	GALAXY
PHL 3201	00 57 12.	- 01 34		18.6	BLUE STELLAR OBJECT
PHL 6993	00 57 12.	- 02 37		15.8	BLUE STELLAR OBJECT
PHL 3203	00 57 12.	- 03 36		18.6	BLUE STELLAR OBJECT
PHL 6999	00 57 12.	- 09 22		18.4	BLUE STELLAR OBJECT
PHL 7000	00 57 12.	- 11 06		18.5	BLUE STELLAR OBJECT
PHL 6997	00 57 12.	- 11 41		17.5	BLUE STELLAR OBJECT
PHL 3200	00 57 12.	- 12 12		18.3	BLUE STELLAR OBJECT
PHL 6995	00 57 12.	- 16 52		17.5	BLUE STELLAR OBJECT
PHL 6696	00 57 12.	- 18 53		17.1	BLUE STELLAR OBJECT
PHL 3204	00 57 12.	- 24 52		18.6	BLUE STELLAR OBJECT
LB 00104	00 57 12.	- 33 06		16.0	FAINT BLUE STAR
CED 004B	00 57 14.	+ 60 34	660		DIFFUSE GALACTIC NEBULA
SG 3.005	00 57 14.	+ 60 34	600		DIFFUSE EMISSION NEBULA
RNGC 0337	00 57 14.	- 07 51		12.5	GALAXY
MCG-01-03-053	00 57 15.	- 07 52	156	11.	GALAXY
IC 0063	00 57 16.	+ 60 34			EMISSION NEBULA
PHL 7002	00 57 18.	+ 02 34		17.5	BLUE STELLAR OBJECT
PHL 7001	00 57 18.	+ 10 40		10.0	BLUE STELLAR OBJECT
PHL 0932	00 57 18.	+ 15 27		12.0	BLUE STELLAR OBJECT
ACK 125-47.1	00 57 18.	+ 15 28			PLANETARY NEBULA
ZC 0057.3+3107	00 57 18.	+ 31 07	270		CLUSTER OF GALAXIES
PHL 7003	00 57 18.	- 17 11		18.7	BLUE STELLAR OBJECT
PHL 3205	00 57 18.	- 25 02		18.6	BLUE STELLAR OBJECT
LIN 345	00 57 18.	- 72 26 36.		15.04	SMC EMISSION LINE OBJECT
SVEN 069	00 57 19.	- 07 51 18.	132	11.9	GALAXY
ARC 0126	00 57 20.	- 14 29		16.6	RICH CLUSTER OF GALAXIES
IC 1609	00 57 20.	- 40 37 42.			NONSTELLAR OBJECT
RNGC 0346	00 57 21.	- 72 27			CLUSTER/NEBULOSITY IN SMC
PHL 3209	00 57 24.	+ 00 00		17.9	BLUE STELLAR OBJECT
PHL 3206	00 57 24.	+ 10 06		17.9	BLUE STELLAR OBJECT
PHL 7005	00 57 24.	+ 15 34		17.9	BLUE STELLAR OBJECT
ZWG 520.007	00 57 24.	+ 35 01		15.4	GALAXY
PHL 3208	00 57 24.	- 03 14		18.2	BLUE STELLAR OBJECT
PHL 7006	00 57 24.	- 13 12		18.7	BLUE STELLAR OBJECT
PHL 3207	00 57 24.	- 26 11		17.2	BLUE STELLAR OBJECT
PHL 7004	00 57 24.	- 27 48		18.3	BLUE STELLAR OBJECT
LIN.CL 060	00 57 24.	- 72 27 18.	1698		STAR CLUSTER IN SMC
SVEN 070	00 57 25.	- 07 57 18.	12	16.2	GALAXY
SVEN 071	00 57 25.	- 08 04 18.	24	15.9	GALAXY
LIN 346	00 57 25.	- 72 00 30.		15.34	SMC EMISSION LINE OBJECT
HELW 336	00 57 26.	- 08 04 24.			NEBULA
ARC 0125	00 57 29.	+ 14 01		17.2	RICH CLUSTER OF GALAXIES
LIN 348	00 57 29.	- 72 43 12.		14.11	SMC EMISSION LINE OBJECT
PHL 3210	00 57 30.	+ 02 35		18.6	BLUE STELLAR OBJECT
UGC 00620	00 57 30.	+ 17 44	60	16.0	GALAXY Sb-c
ZWG 458.013	00 57 30.	+ 18 01		15.5	GALAXY
UGC 00621	00 57 30.	+ 18 01	72	15.5	GALAXY S
MCG+03-03-012	00 57 30.	+ 18 01	51	15.	GALAXY
MCG+06-03-006	00 57 30.	+ 35 02	24	16.	GALAXY
MCG+08-03-004	00 57 30.	+ 47 44	60	13.	GALAXY
MCG-02-03-061	00 57 30.	- 11 21	84	13.5	GALAXY
MCG-02-03-060	00 57 30.	- 14 16	36	15.	GALAXY
MCG-03-03-016	00 57 30.	- 18 55	66	14.	GALAXY
TON-S 0181	00 57 30.	- 27 51		15.8	BLUE STAR
LIN 347	00 57 30.	- 72 17 54.		16.07	PLANETARY NEBULA IN SMC
LH115-N066C	00 57 30.	- 72 26	19		EMISSION NEBULA IN SMC
LH115-N066B	00 57 30.	- 72 26	27		EMISSION NEBULA IN SMC
LH115-N066	00 57 30.	- 72 27	584		EMISSION NEBULA IN SMC
LH115-N069	00 57 30.	- 72 41	27		EMISSION NEBULA IN SMC
ARC 0127	00 57 32.	- 23 44		17.7	RICH CLUSTER OF GALAXIES
LIN 351	00 57 33.	- 73 08 42.		14.94	SMC EMISSION LINE OBJECT
LIN 350	00 57 35.	- 72 26 54.		13.90	SMC EMISSION LINE OBJECT
PHL 3211	00 57 36.	+ 13 19		18.1	BLUE STELLAR OBJECT
ZWG 551.003	00 57 36.	+ 47 43		14.6	GALAXY
UGC 00622	00 57 36.	+ 47 43	66	14.6	GALAXY Sc
VB 101	00 57 36.	+ 60 03	363		STELLAR RING
MCG+00-03-050	00 57 36.	- 02 18	48	15.5	GALAXY
MCG-01-03-054	00 57 36.	- 05 12	48	15.	GALAXY
HOLM 032A	00 57 36.	- 11 21	60	14.2	PART OF MULTIPLE GALAXY
PHL 3212	00 57 36.	- 12 52		18.9	BLUE STELLAR OBJECT
PHL 7008	00 57 36.	- 14 10		16.3	BLUE STELLAR OBJECT
PHL 7009	00 57 36.	- 17 52		18.4	BLUE STELLAR OBJECT
PHL 3213	00 57 36.	- 20 43		18.6	BLUE STELLAR OBJECT
PHL 3214	00 57 36.	- 25 18		18.0	BLUE STELLAR OBJECT
PHL 7007	00 57 36.	- 26 37		18.5	BLUE STELLAR OBJECT
LB 00105	00 57 36.	- 31 12		16.2	FAINT BLUE STAR
PHL 0933	00 57 36.	- 31 35		16.8	BLUE STELLAR OBJECT
MCG-06-03-015	00 57 36.	- 33 58	150	9.0	GALAXY
LH115-N070	00 57 36.	- 72 18			EMISSION NEBULA IN SMC
LIN 349	00 57 36.	- 72 25 18.		14.63	SMC EMISSION LINE OBJECT
LH115-N066A	00 57 36.	- 72 27			EMISSION NEBULA IN SMC
HOW 36	00 57 36.	- 74 07 17.	12		STAR CLUSTER IN SMC
SVEN 072	00 57 37.	- 07 59 18.	60	13.6	GALAXY
HELW 337	00 57 39.	- 07 59 24.			NEBULA
MCG-04-03-040	00 57 39.	- 21 45 30.	12	15.	GALAXY
HOW 37	00 57 39.	- 72 01 47.	18		STAR CLUSTER IN SMC
HOLM 032B	00 57 40.	- 11 21	18	14.9	PART OF MULTIPLE GALAXY
LIN 353	00 57 41.	- 72 25 54.		11.20	SMC EMISSION LINE OBJECT
PHL 7010	00 57 42.	+ 02 50		17.4	BLUE STELLAR OBJECT
ZWG 458.014	00 57 42.	+ 19 34		15.4	GALAXY
VB 102	00 57 42.	+ 60 00	161		STELLAR RING
ISS 0037	00 57 42.	+ 60 00	163		STELLAR RING
ASS 39	00 57 42.	+ 61 14	9600		OB ASSOCIATION CAS OB1
MCG-01-03-054A	00 57 42.	- 05 11	30		GALAXY
PHL 3215	00 57 42.	- 09 31		18.9	BLUE STELLAR OBJECT
MCG-02-03-062	00 57 42.	- 11 18	36	16.	GALAXY
PHL 3216	00 57 42.	- 13 42		18.6	BLUE STELLAR OBJECT
MCG-03-03-017	00 57 42.	- 15 34	36	15.	GALAXY
PHL 7054	00 57 42.	- 18 07		18.1	BLUE STELLAR OBJECT
PHL 0934	00 57 42.	- 28 24		17.8	BLUE STELLAR OBJECT

OBJECT NAME	RIGHT ASCEN.	DECLINATION	DIAM.	MAGN.	TYPE OF OBJECT
LIN 352	00 57 42.	- 72 17 00.		12.25	SMC EMISSION LINE OBJECT
HSN 20	00 57 43.	- 74 19			BRIGHT GALAXY BEHIND SMC
LIN 356	00 57 45.	- 73 08 49.		16.01	SMC EMISSION LINE OBJECT
IC 0067	00 57 47.	- 07 11			NONSTELLAR OBJECT
SHP 1	00 57 47.	- 33 58	4800	10.	SCULPTOR DWARF GALAXY
LIN 354	00 57 47.	- 72 26 25.		13.51	SMC EMISSION LINE OBJECT
LIN 355	00 57 47.	- 72 39 25.		14.25	SMC EMISSION LINE OBJECT
PHL 3217	00 57 48.	+ 13 13		18.4	BLUE STELLAR OBJECT
PHL 0935	00 57 48.	+ 13 38		18.4	BLUE STELLAR OBJECT
ZWG 480.031	00 57 48.	+ 26 45		15.7	GALAXY
MCG+05-03-033	00 57 48.	+ 30 30	48	15.	GALAXY
ZWG 501.059	00 57 48.	+ 30 32		15.0	GALAXY
UGC 00623	00 57 48.	+ 30 32	66	15.0	GALAXY Sa
UGC 00593	00 57 48.	+ 43 31	90		GALAXY
PHL 3220	00 57 48.	- 03 38		18.5	BLUE STELLAR OBJECT
PHL 3219	00 57 48.	- 09 22		18.8	BLUE STELLAR OBJECT
MCG-03-03-018	00 57 48.	- 15 26 30.	36	15.5	GALAXY
PHL 3218	00 57 48.	- 20 50		18.7	BLUE STELLAR OBJECT
PHL 7011	00 57 48.	- 22 40		18.8	BLUE STELLAR OBJECT
MCG-05-03-021	00 57 48.	- 31 06	54	15.	GALAXY
HOW 38	00 57 48.	- 74 05 41.	18		STAR CLUSTER IN SMC
IC 0066	00 57 49.	+ 30 31			NONSTELLAR OBJECT
BC PKS0057+07	00 57 49.8	+ 07 22 20.		18.5	QUASI-STELLAR OBJECT
RNGC 0338	00 57 50.	+ 30 23		14.0	GALAXY
MCG+04-03-032	00 57 51.	+ 26 45	30	15.5	GALAXY
MCG+05-03-034	00 57 51.	+ 30 23	108	14.	GALAXY
IC 0068	00 57 51.	- 07 13			NONSTELLAR OBJECT
RIC 947	00 57 52.34	+ 41 44 03.3		18.77	BLUE OBJECT
PHL 7012	00 57 54.	+ 12 54		17.0	BLUE STELLAR OBJECT
PHL 3221	00 57 54.	+ 15 16		17.0	BLUE STELLAR OBJECT
MCG+05-03-035	00 57 54.	+ 29 50	24	15.5	GALAXY
ZWG 501.060	00 57 54.	+ 29 51		15.4	GALAXY
ZWG 501.061	00 57 54.	+ 30 24		14.0	GALAXY
UGC 00624	00 57 54.	+ 30 24	132	14.0	GALAXY Sa-b
OCL 0314	00 57 54.	+ 63 41	600	13.	OPEN STAR CLUSTER
MCG+00-03-051	00 57 54.	- 01 59	27	15.0	GALAXY
LIN 358	00 57 54.	- 75 21 19.		15.8	PLANETARY NEBULA IN SMC
LIN 357	00 57 55.	- 71 53 55.		17.48	PLANETARY NEBULA IN SMC
MCG+04-03-033	00 57 57.	+ 26 50	30	15.5	GALAXY
HOW 39	00 57 57.	- 71 55 53.	24		STAR CLUSTER IN SMC
HOLM 033B	00 57 58.	- 01 57	54	15.0	PART OF MULTIPLE GALAXY
VDB.66G 007	00 58	+ 07 21	70		DWARF GALAXY
LBN 0624	00 58	+ 60 50	2400		BRIGHT NEBULA
PHL 0936	00 58 00.	+ 01 06		17.8	BLUE STELLAR OBJECT
PHL 7013	00 58 00.	+ 03 08		17.8	BLUE STELLAR OBJECT
ZWG 551.004	00 58 00.	+ 47 25		13.8	GALAXY
UGC 00625	00 58 00.	+ 47 25	264	13.8	GALAXY Sb/SBc
MCG+08-03-005	00 58 00.	+ 47 26	210	13.	GALAXY
PHL 7014	00 58 00.	- 00 34		17.5	BLUE STELLAR OBJECT
ZWG 384.053	00 58 00.	- 01 57		15.6	GALAXY
KARA.72 020A	00 58 00.	- 01 57	30	15.6	PART OF DOUBLE GALAXY
MCG+00-03-052	00 58 00.	- 02 00	42	14.9	GALAXY
MCG-01-03-055	00 58 00.	- 07 09	54	14.	GALAXY
MCG-01-03-056	00 58 00.	- 08 17	60	13.5	GALAXY
ARP 059	00 58 00.	- 09 25			PECULIAR GALAXY
PHL 7015	00 58 00.	- 16 10		18.2	BLUE STELLAR OBJECT
LB 01584	00 58 00.	- 53 25		13.9	FAINT BLUE STAR
IC 0065	00 58 01.	+ 47 25 00.			NONSTELLAR OBJECT
RNGC 0340	00 58 02.	- 07 09		14.0	GALAXY
HOLM 033A	00 58 03.	- 01 58	60	14.9	PART OF MULTIPLE GALAXY
ARC 0128	00 58 03.	- 13 14		17.0	RICH CLUSTER OF GALAXIES
LIN.CL 061	00 58 04.	- 72 36 25.	546	12.1	STAR CLUSTER IN SMC
HN 0132	00 58 04.	- 72 37			NEBULA
IC 1611	00 58 04.	- 72 37			NONSTELLAR OBJECT
PHL 3222	00 58 06.	+ 13 04		18.2	BLUE STELLAR OBJECT
ZC 0058.1+2020	00 58 06.	+ 20 20	2760		CLUSTER OF GALAXIES
ZWG 384.054	00 58 06.	- 01 58		15.0	GALAXY
UGC 00626	00 58 06.	- 01 58	60	15.0	GALAXY Sc
KARA.72 020B	00 58 06.	- 01 58	42	15.0	PART OF DOUBLE GALAXY
PHL 3223	00 58 06.	- 10 49		18.2	BLUE STELLAR OBJECT
PHL 3224	00 58 06.	- 26 04		18.5	BLUE STELLAR OBJECT
SVEN 074	00 58 07.	- 07 22	42	15.4	GALAXY
SVEN 073	00 58 07.	- 08 15	54	15.1	GALAXY
MCG+07-03-013	00 58 09.	+ 43 24	48	17.	GALAXY
SC 0056-6840.0	00 58 09.	- 68 23 48.	18		NEBULA
PHL 0938	00 58 12.	+ 01 56		16.9	BLUE STELLAR OBJECT
SHB 015	00 58 12.	+ 01 56		17.2	QUASI-STELLAR OBJECT
PHL 0937	00 58 12.	+ 02 48		18.1	BLUE STELLAR OBJECT
PHL 3226	00 58 12.	+ 09 41		18.3	BLUE STELLAR OBJECT
PHL 3227	00 58 12.	+ 12 50		18.0	BLUE STELLAR OBJECT
MCG+03-03-013	00 58 12.	+ 19 12	66	16.	GALAXY
MCG+05-03-036	00 58 12.	+ 29 18	63	15.	GALAXY
URA 43	00 58 12.	+ 63 18	216		STELLAR RING
MCG+00-03-053	00 58 12.	- 01 01	24	15.5	GALAXY
PHL 3225	00 58 12.	- 12 37		17.8	BLUE STELLAR OBJECT
PHL 7020	00 58 12.	- 13 00		17.0	BLUE STELLAR OBJECT
PHL 3229	00 58 12.	- 14 43		18.2	BLUE STELLAR OBJECT
PHL 7017	00 58 12.	- 15 04		18.3	BLUE STELLAR OBJECT
PHL 7016	00 58 12.	- 16 05		16.8	BLUE STELLAR OBJECT
PHL 3228	00 58 12.	- 16 48		18.7	BLUE STELLAR OBJECT
PHL 0939	00 58 12.	- 18 35		17.8	BLUE STELLAR OBJECT
PHL 7018	00 58 12.	- 25 55		18.1	BLUE STELLAR OBJECT
PHL 7019	00 58 12.	- 32 31		18.4	BLUE STELLAR OBJECT
LB 00106	00 58 12.	- 32 31		16.8	FAINT BLUE STAR
LIN 359	00 58 12.	- 72 14 25.		14.65	SMC EMISSION LINE OBJECT
RNGC 0342	00 58 14.	- 07 03		14.5	GALAXY
KN 10.21	00 58 14.8	+ 09 27 50.			NEBULA
MCG-01-03-058	00 58 15.	- 07 03	36	14.5	GALAXY
MCG-01-03-057	00 58 15.	- 07 12 30.	60	15.	GALAXY
RNGC 0341A	00 58 15.	- 09 27		13.0	GALAXY
MCG-02-03-063	00 58 15.	- 09 27 30.	60	13.	GALAXY
LIN 360	00 58 15.	- 72 56 49.		14.99	SMC EMISSION LINE OBJECT
LIN.CL 062	00 58 16.	- 72 38 43.	612	12.1	STAR CLUSTER IN SMC
PHL 3230	00 58 18.	+ 02 22		18.3	BLUE STELLAR OBJECT
PHL 7021	00 58 18.	+ 11 06		16.4	BLUE STELLAR OBJECT
ZWG 435.035	00 58 18.	+ 13 12		15.2	GALAXY
UGC 00627	00 58 18.	+ 13 12	78	15.2	GALAXY S
UGC 00628	00 58 18.	+ 19 13	138	17.	GALAXY DWRF SP
ZWG 501.062	00 58 18.	+ 29 20		14.9	GALAXY
UGC 00629	00 58 18.	+ 29 20	84	14.9	GALAXY Sc
UGC 00630	00 58 18.	+ 29 41	78	17.	GALAXY S IV
PHL 7022	00 58 18.	- 03 05		18.3	BLUE STELLAR OBJECT
PHL 7023	00 58 18.	- 05 36		18.6	BLUE STELLAR OBJECT
MCG-01-03-059	00 58 18.	- 07 20	42	15.	GALAXY
MCG-02-03-064	00 58 18.	- 09 27	18	15.	GALAXY
PHL 3231	00 58 18.	- 20 51		18.9	BLUE STELLAR OBJECT
SVEN 075	00 58 19.	- 07 18 19.	42	15.9	GALAXY
BC PHL938	00 58 19.69	+ 01 55 27.5		17.16	QUASI-STELLAR OBJECT
HELW 338	00 58 20.	- 07 18 19.			NEBULA
LIN 363	00 58 20.	- 73 11 25.		14.00	SMC EMISSION LINE OBJECT
MCG+05-03-037	00 58 21.	+ 29 39 30.	42	17.	GALAXY
RNGC 0341B	00 58 21.	- 09 27		15.0	GALAXY
MCG-02-03-065	00 58 21.	- 10 33	48	16.	GALAXY
HN 0133	00 58 22.	- 72 39			NEBULA
IC 1612	00 58 22.	- 72 39			NONSTELLAR OBJECT
LIN 362	00 58 22.	- 72 48 49.		14.78	SMC EMISSION LINE OBJECT
RNGC 0343	00 58 23.	- 23 30			NON-EXISTENT OBJECT
PHL 3236	00 58 24.	+ 03 09		18.3	BLUE STELLAR OBJECT
UGC 00631	00 58 24.	+ 09 27	102	16.0	GALAXY Sc
MCG+02-03-025	00 58 24.	+ 09 27	84	15.	GALAXY
MCG+02-03-024	00 58 24.	+ 13 12	48	15.	GALAXY
PHL 3232	00 58 24.	+ 13 56		16.7	BLUE STELLAR OBJECT
ZWG 458.015	00 58 24.	+ 16 25		15.5	GALAXY
MCG+05-03-038	00 58 24.	+ 29 50	60	15.	GALAXY
ZWG 501.063	00 58 24.	+ 29 52		14.8	GALAXY
UGC 00632	00 58 24.	+ 29 52	72	14.8	GALAXY SBa-b
PHL 3233	00 58 24.	- 03 23		16.5	BLUE STELLAR OBJECT
PHL 0940	00 58 24.	- 04 26		15.6	BLUE STELLAR OBJECT
PHL 3234	00 58 24.	- 06 50		18.4	BLUE STELLAR OBJECT
PHL 7024	00 58 24.	- 11 16		18.7	BLUE STELLAR OBJECT
PHL 7025	00 58 24.	- 13 46		16.6	BLUE STELLAR OBJECT
PHL 3235	00 58 24.	- 14 54		18.7	BLUE STELLAR OBJECT
LB 00107	00 58 24.	- 28 23		14.0	FAINT BLUE STAR
LB 00108	00 58 24.	- 28 51		16.7	FAINT BLUE STAR
LB 00109	00 58 24.	- 34 55		16.2	FAINT BLUE STAR
LB 01585	00 58 24.	- 47 30		13.8	FAINT BLUE STAR
LIN 361	00 58 25.	- 71 49 55.		14.72	SMC EMISSION LINE OBJECT
KN 10.22	00 58 26.0	+ 09 27 21.			NEBULA
MCG+05-03-039	00 58 27.	+ 29 52	24	15.	GALAXY
ARC 0129	00 58 28.	- 10 14		17.8	RICH CLUSTER OF GALAXIES
ZWG 501.064	00 58 30.	+ 29 53		14.9	GALAXY
IC 0070	00 58 30.	- 00 13 14.			NONSTELLAR OBJECT
MCG-01-03-060	00 58 30.	- 04 22	48	16.	GALAXY
HELW 339	00 58 30.	- 04 22 07.			NEBULA
MCG-01-03-061	00 58 30.	- 04 52	30	15.	GALAXY
PHL 3238	00 58 36.	+ 11 45		17.8	BLUE STELLAR OBJECT
PHL 3237	00 58 36.	+ 13 20		16.6	BLUE STELLAR OBJECT
MCG+05-03-040	00 58 36.	+ 31 13	90	14.5	GALAXY
ZWG 501.065	00 58 36.	+ 31 15		14.6	GALAXY
UGC 00633	00 58 36.	+ 31 15	102	14.6	GALAXY Sb
ZC 0058.6+4452	00 58 36.	+ 44 52	1410		CLUSTER OF GALAXIES
ZC 0058.6-0025	00 58 36.	- 00 25	2820		CLUSTER OF GALAXIES
PHL 3240	00 58 36.	- 02 36		18.2	BLUE STELLAR OBJECT
PHL 7027	00 58 36.	- 12 07		18.5	BLUE STELLAR OBJECT
PHL 7028	00 58 36.	- 14 35		18.7	BLUE STELLAR OBJECT
PHL 3239	00 58 36.	- 19 50		18.4	BLUE STELLAR OBJECT
PHL 7026	00 58 36.	- 26 34		16.7	BLUE STELLAR OBJECT
RNGC 0348	00 58 37.	- 53 30			UNVERIFIED SOUTHERN OBJECT
LIN 364	00 58 38.	- 73 04 14.		14.27	SMC EMISSION LINE OBJECT
MCG+05-03-041	00 58 39.	+ 30 45	48	14.5	GALAXY
LIN 367	00 58 39.	- 74 34 02.		16.05	SMC EMISSION LINE OBJECT
HELW 340	00 58 41.	- 04 42 13.			NEBULA
ZWG 384.055	00 58 42.	+ 00 35		15.5	GALAXY
PHL 0942	00 58 42.	+ 02 19		18.1	BLUE STELLAR OBJECT
PHL 0941	00 58 42.	+ 11 38		18.2	BLUE STELLAR OBJECT
ZWG 501.066	00 58 42.	+ 30 47		14.7	GALAXY
IC 0069	00 58 42.	+ 30 48 16.			NONSTELLAR OBJECT
MCG+00-03-054	00 58 42.	- 01 24	24	15.5	GALAXY
PHL 3241	00 58 42.	- 03 07		18.1	BLUE STELLAR OBJECT
MCG-01-03-062	00 58 42.	- 04 42	42	14.5	GALAXY
PHL 0943	00 58 42.	- 10 02		18.2	BLUE STELLAR OBJECT
MCG-04-03-041	00 58 42.	- 26 20	60	15.	GALAXY
MCG-04-03-042	00 58 42.	- 26 21	24	17.	GALAXY
TON-S 0182	00 58 42.	- 26 31		15.7	BLUE STAR
MCG-06-03-016	00 58 42.	- 35 32	30	15.	GALAXY
HOLM 034B	00 58 44.	+ 31 13	24	15.2	PART OF MULTIPLE GALAXY
HOLM 034A	00 58 44.	+ 31 15	150	13.9	PART OF MULTIPLE GALAXY
HOW 40	00 58 45.	- 71 33 22.	18		STAR CLUSTER IN SMC
LIN 365	00 58 46.	- 72 36 38.		14.22	SMC EMISSION LINE OBJECT
LIN 366	00 58 46.	- 72 39 02.		14.89	SMC EMISSION LINE OBJECT
IC 0071	00 58 47.	- 07 03			NONSTELLAR OBJECT
PHL 7030	00 58 48.	+ 00 31		18.2	BLUE STELLAR OBJECT
PHL 0944	00 58 48.	+ 01 22		17.9	BLUE STELLAR OBJECT
MCG+01-03-011	00 58 48.	+ 07 20	90	15.	GALAXY
ZWG 410.023	00 58 48.	+ 07 21		15.7	GALAXY
UGC 00634	00 58 48.	+ 07 21	120	15.7	GALAXY DWRF SP
ZC 0058.8+0831	00 58 48.	+ 08 31	610		CLUSTER OF GALAXIES
ZWG 435.036	00 58 48.	+ 11 13		15.7	GALAXY
PHL 3243	00 58 48.	+ 14 31		18.6	BLUE STELLAR OBJECT
MCG+07-03-014	00 58 48.	+ 43 47 30.	36	16.	GALAXY
UGC 00635	00 58 48.	+ 46 24	60	17.	GALAXY Sc
MCG+00-03-055	00 58 48.	- 01 35	42	15.5	GALAXY
PHL 3242	00 58 48.	- 02 29		18.4	BLUE STELLAR OBJECT
PHL 7029	00 58 48.	- 04 27		17.5	BLUE STELLAR OBJECT
MCG-01-03-063	00 58 48.	- 07 05	30	14.5	GALAXY
MCG-01-03-064	00 58 48.	- 07 10	60	13.5	GALAXY
PHL 0945	00 58 48.	- 09 44		17.5	BLUE STELLAR OBJECT
MCG-02-03-066	00 58 48.	- 10 07	36	14.5	GALAXY
PHL 3244	00 58 48.	- 13 31		18.8	BLUE STELLAR OBJECT
PHL 0946	00 58 48.	- 13 45		18.7	BLUE STELLAR OBJECT
TON-S 0183	00 58 48.	- 34 00		12.5	BLUE STAR
KN 10.23	00 58 49.5	+ 07 21 30.			NEBULA
RNGC 0347	00 58 50.	- 07 05		14.5	GALAXY
RNGC 0345	00 58 50.	- 07 10		13.5	GALAXY
LIN 368	00 58 51.	- 72 51 26.		14.79	SMC EMISSION LINE OBJECT
LIN.CL 063	00 58 52.	- 72 38 08.	270	12.3	STAR CLUSTER IN SMC
PHL 7031	00 58 54.	+ 11 12		17.7	BLUE STELLAR OBJECT
PHL 7033	00 58 54.	+ 12 56		18.2	BLUE STELLAR OBJECT
PHL 3245	00 58 54.	+ 13 15		18.3	BLUE STELLAR OBJECT
MCG+08-03-006	00 58 54.	+ 47 38	24	15.	GALAXY
ZC 0058.9-0304	00 58 54.	- 03 04	4370		CLUSTER OF GALAXIES
PHL 3246	00 58 54.	- 06 09		18.4	BLUE STELLAR OBJECT
PHL 0947	00 58 54.	- 07 50		18.4	BLUE STELLAR OBJECT
MCG-01-03-065	00 58 54.	- 07 52 30.	360	13.5	GALAXY
SVEN 076	00 58 55.	- 07 49	12	15.9	GALAXY
HOW 41	00 58 57.	- 71 43 58.	48		STAR CLUSTER IN SMC
LIN 369	00 58 57.	- 72 40 56.		14.86	SMC EMISSION LINE OBJECT
LIN 370	00 58 57.	- 72 41 56.		15.18	SMC EMISSION LINE OBJECT
ARC 0130	00 58 58.	- 00 25		17.2	RICH CLUSTER OF GALAXIES
LB 09774	00 59	- 84 14		14.6	FAINT BLUE STAR
PHL 7035	00 59 00.	+ 01 48		18.4	BLUE STELLAR OBJECT
PHL 3249	00 59 00.	+ 02 58		18.4	BLUE STELLAR OBJECT
PHL 7034	00 59 00.	+ 10 36		16.4	BLUE STELLAR OBJECT
PHL 3248	00 59 00.	+ 12 51		18.6	BLUE STELLAR OBJECT
PHL 3247	00 59 00.	+ 13 38		18.6	BLUE STELLAR OBJECT
ZWG 480.032	00 59 00.	+ 23 47		14.7	GALAXY

OBJECT NAME	RIGHT ASCEN.	DECLINATION	DIAM.	MAGN.	TYPE OF OBJECT
UGC 00636	00 59 00.	+ 23 47	90	14.7	GALAXY E
ZC 0059.0+3434	00 59 00.	+ 34 34	1480		CLUSTER OF GALAXIES
ZWG 384.056	00 59 00.	- 01 33		15.7	GALAXY
PHL 3250	00 59 00.	- 03 17		18.4	BLUE STELLAR OBJECT
MCG-01-03-066	00 59 00.	- 06 15	54	15.	GALAXY
PHL 7036	00 59 00.	- 17 06		18.0	BLUE STELLAR OBJECT
IC 0072	00 59 01.	- 07 02			NONSTELLAR OBJECT
SVEN 077	00 59 01.	- 07 51 20.	120	13.6	GALAXY
LIN 371	00 59 03.	- 72 53 26.		14.54	SMC EMISSION LINE OBJECT
HELW 341	00 59 04.	- 07 51 14.			NEBULA
PHL 7037	00 59 06.	+ 10 02		18.4	BLUE STELLAR OBJECT
PHL 3252	00 59 06.	+ 12 37		18.5	BLUE STELLAR OBJECT
MCG+04-03-034	00 59 06.	+ 23 47	57	15.	GALAXY
MCG+05-03-042	00 59 06.	+ 29 51	18	15.5	GALAXY
ZWG 501.067	00 59 06.	+ 29 53		15.0	GALAXY
ZWG 501.068	00 59 06.	+ 30 57		15.7	GALAXY
VB 103	00 59 06.	+ 58 03	497		STELLAR RING
PHL 3251	00 59 06.	- 00 45		18.3	BLUE STELLAR OBJECT
PHL 7038	00 59 06.	- 12 04		18.3	BLUE STELLAR OBJECT
PHL 7039	00 59 06.	- 15 10		18.0	BLUE STELLAR OBJECT
PHL 7041	00 59 06.	- 16 46		18.4	BLUE STELLAR OBJECT
PHL 7040	00 59 06.	- 16 46		18.4	BLUE STELLAR OBJECT
LB 00110	00 59 06.	- 32 17		15.8	FAINT BLUE STAR
LB 01586	00 59 06.	- 46 35		13.4	FAINT BLUE STAR
RNGC 0344	00 59 11.	- 23 30			GALAXY
LIN 375	00 59 11.	- 73 44 56.		14.70	SMC EMISSION LINE OBJECT
VB 028	00 59 12.	+ 66 48	208		STELLAR RING
SHAH 032	00 59 12.	- 01 51	198	16.6	GROUP OF COMPACT GALAXIES
PHL 7042	00 59 12.	- 03 30		17.9	BLUE STELLAR OBJECT
MCG-03-03-019	00 59 12.	- 15 33	30	15.	GALAXY
MCG-03-03-020	00 59 12.	- 15 50	72	14.	GALAXY
PHL 7043	00 59 12.	- 25 26		18.2	BLUE STELLAR OBJECT
IC 1610	00 59 13.	- 15 50 09.			NONSTELLAR OBJECT
KN 10.24	00 59 13.6	+ 07 17 38.			NEBULA
RNGC 0337A	00 59 14.	- 07 52		13.5	GALAXY
LIN 374	00 59 14.	- 72 53 26.		13.78	SMC EMISSION LINE OBJECT
MCG+00-03-056	00 59 15.	+ 02 47	54	15.	GALAXY
MCG-01-03-067	00 59 15.	- 05 08	42	15.5	GALAXY
LIN 373	00 59 16.	- 72 16 50.		14.34	SMC EMISSION LINE OBJECT
UGC 00637	00 59 18.	+ 02 47	78	16.0	GALAXY SBc
ZC 0059.3+1317	00 59 18.	+ 13 17	2080		CLUSTER OF GALAXIES
ZWG 480.033	00 59 18.	+ 26 13		15.6	GALAXY
MCG-01-03-068	00 59 18.	- 07 05	36	13.5	GALAXY
PHL 7044	00 59 18.	- 18 35		18.0	BLUE STELLAR OBJECT
LB 00112	00 59 18.	- 29 53		16.3	FAINT BLUE STAR
LB 00111	00 59 18.	- 30 56		16.6	FAINT BLUE STAR
LIN 372	00 59 18.	- 71 51 50.		15.08	PLANETARY NEBULA IN SMC
LH115-N071	00 59 18.	- 71 52	22		EMISSION NEBULA IN SMC
LIN.CL 064	00 59 18.	- 73 36 32.	270	13.5	STAR CLUSTER IN SMC
RNGC 0349	00 59 20.	- 07 05		13.5	GALAXY
PHL 3256	00 59 24.	+ 11 34		18.4	BLUE STELLAR OBJECT
PHL 3253	00 59 24.	+ 14 16		17.9	BLUE STELLAR OBJECT
ZWG 480.034	00 59 24.	+ 26 41		15.7	GALAXY
UGC 00638	00 59 24.	+ 26 41	102	15.7	GALAXY
ZWG 501.069	00 59 24.	+ 32 09		15.5	GALAXY
ZWG 501.070	00 59 24.	+ 32 11		15.1	GALAXY
5ZW 045	00 59 24.	+ 49 39			COMPACT GALAXY
PHL 3255	00 59 24.	- 00 04		17.7	BLUE STELLAR OBJECT
PHL 7045	00 59 24.	- 04 46		17.5	BLUE STELLAR OBJECT
MCG-01-03-070	00 59 24.	- 06 40	60	15.	GALAXY
MCG-01-03-069	00 59 24.	- 07 05	18	15.	GALAXY
PHL 0948	00 59 24.	- 07 18		18.3	BLUE STELLAR OBJECT
PHL 3254	00 59 24.	- 09 02		18.7	BLUE STELLAR OBJECT
PHL 7046	00 59 24.	- 17 36		18.2	BLUE STELLAR OBJECT
PHL 7047	00 59 24.	- 18 38		18.5	BLUE STELLAR OBJECT
PHL 7048	00 59 24.	- 26 25		18.9	BLUE STELLAR OBJECT
LIN 376	00 59 24.	- 71 43 50.		14.86	SMC EMISSION LINE OBJECT
RNGC 0350	00 59 26.	- 07 05		15.0	GALAXY
LIN.CL 065	00 59 26.	- 73 00 50.	546	13.4	STAR CLUSTER IN SMC
HOW 42	00 59 26.	- 74 21 28.	18		STAR CLUSTER IN SMC
LIN 377	00 59 27.	- 72 43 50.		15.46	SMC EMISSION LINE OBJECT
HOW 43	00 59 28.	- 72 01 04.	18		STAR CLUSTER IN SMC
PHL 3258	00 59 30.	+ 13 15		18.5	BLUE STELLAR OBJECT
MCG+04-03-035	00 59 30.	+ 26 40 30.	66	15.	GALAXY
ZC 0059.5+2908	00 59 30.	+ 29 08	3430		CLUSTER OF GALAXIES
PHL 0949	00 59 30.	- 02 08		18.1	BLUE STELLAR OBJECT
ZWG 384.057	00 59 30.	- 02 11		14.3	GALAXY
UGC 00639	00 59 30.	- 02 11	90	14.3	GALAXY SB0-a
MCG+00-03-057	00 59 30.	- 02 14	60	14.	GALAXY
PHL 3257	00 59 30.	- 07 19		13.8	BLUE STELLAR OBJECT
PHL 3259	00 59 30.	- 09 36		16.3	BLUE STELLAR OBJECT
PHL 0950	00 59 30.	- 12 41		18.0	BLUE STELLAR OBJECT
LB 00113	00 59 30.	- 30 32		15.7	FAINT BLUE STAR
RNGC 0351	00 59 31.	- 02 11		14.5	GALAXY
LIN 378	00 59 32.	- 73 03 57.		14.61	SMC EMISSION LINE OBJECT
PHL 7051	00 59 36.	+ 02 44		18.3	BLUE STELLAR OBJECT
PHL 7050	00 59 36.	+ 12 29		17.9	BLUE STELLAR OBJECT
PHL 7049	00 59 36.	- 03 14		17.9	BLUE STELLAR OBJECT
MCG-01-03-071	00 59 36.	- 04 31	138	12.5	GALAXY
PHL 0951	00 59 36.	- 10 55		18.6	BLUE STELLAR OBJECT
PHL 7052	00 59 36.	- 17 38		18.5	BLUE STELLAR OBJECT
MCG-04-03-043	00 59 36.	- 21 10 30.	42	15.	GALAXY
LB 00076	00 59 36.	- 31 37		14.0	FAINT BLUE STAR
LB 00114	00 59 36.	- 34 07		16.0	FAINT BLUE STAR
RNGC 0352	00 59 38.	- 04 31		12.5	GALAXY
KN 10.25	00 59 39.1	+ 08 49 57.			NEBULA
LIN 379	00 59 41.	- 72 02 03.		14.31	SMC EMISSION LINE OBJECT
ZWG 410.024	00 59 42.	+ 08 50		15.7	GALAXY
UGC 00640	00 59 42.	+ 08 50	66	15.7	GALAXY Sb
MCG+01-03-012	00 59 42.	+ 08 50	36	15.	GALAXY
PHL 3260	00 59 42.	+ 12 36		18.7	BLUE STELLAR OBJECT
ZWG 480.035	00 59 42.	+ 26 21		15.7	GALAXY
1ZW 002	00 59 42.	+ 30 27			COMPACT GALAXY
PHL 7053	00 59 42.	- 11 44		17.9	BLUE STELLAR OBJECT
PHL 3261	00 59 42.	- 14 16		18.3	BLUE STELLAR OBJECT
PHL 3262	00 59 42.	- 30 14		18.3	BLUE STELLAR OBJECT
PHL 3263	00 59 42.	- 30 34		18.2	BLUE STELLAR OBJECT
SVEN 078	00 59 43.	- 07 27 09.	12	15.9	GALAXY
ARC 0131	00 59 43.	- 15 03		17.2	RICH CLUSTER OF GALAXIES
LIN 382	00 59 44.	- 72 49 57.		14.38	SMC EMISSION LINE OBJECT
MCG-01-03-072	00 59 45.	- 04 47	48	14.5	GALAXY
MCG-01-03-073	00 59 45.	- 07 09	24	15.	GALAXY
LIN 381	00 59 45.	- 72 39 21.		13.25	SMC EMISSION LINE OBJECT
LIN 380	00 59 46.	- 72 08 33.		14.59	SMC EMISSION LINE OBJECT
HELW 342	00 59 47.	- 04 47 14.			NEBULA
PHL 7058	00 59 48.	+ 02 32		18.7	BLUE STELLAR OBJECT
ZWG 410.025	00 59 48.	+ 07 06		15.6	GALAXY
PHL 3264	00 59 48.	+ 13 10		17.0	BLUE STELLAR OBJECT

OBJECT NAME	RIGHT ASCEN.	DECLINATION	DIAM.	MAGN.	TYPE OF OBJECT
ZWG 520.008	00 59 48.	+ 34 00		14.9	GALAXY
PHL 0952	00 59 48.	- 00 54		16.6	BLUE STELLAR OBJECT
PHL 3266	00 59 48.	- 04 00		18.5	BLUE STELLAR OBJECT
PHL 7055	00 59 48.	- 09 18		17.8	BLUE STELLAR OBJECT
PHL 0953	00 59 48.	- 10 32		16.9	BLUE STELLAR OBJECT
PHL 7056	00 59 48.	- 12 36		18.4	BLUE STELLAR OBJECT
PHL 3265	00 59 48.	- 18 22		18.5	BLUE STELLAR OBJECT
PHL 7057	00 59 48.	- 26 40		18.5	BLUE STELLAR OBJECT
PHL 7059	00 59 48.	- 29 46		18.4	BLUE STELLAR OBJECT
LB 00115	00 59 48.	- 35 45		15.7	FAINT BLUE STAR
LH115-N072	00 59 48.	- 72 07	26		EMISSION NEBULA IN SMC
HOW 44	00 59 49.	- 74 04 10.	18		STAR CLUSTER IN SMC
HELW 343	00 59 50.	- 04 46 26.			NEBULA
HELW 344	00 59 52.	- 04 46 20.			NEBULA
HELW 345	00 59 52.	- 04 48 14.			NEBULA
LIN 383	00 59 52.	- 72 22 57.		14.67	SMC EMISSION LINE OBJECT
SHAH 139	00 59 54.	+ 03 17	84		GROUP OF COMPACT GALAXIES
PHL 7061	00 59 54.	+ 10 20		18.0	BLUE STELLAR OBJECT
PHL 0954	00 59 54.	+ 13 29		18.4	BLUE STELLAR OBJECT
5ZW 046	00 59 54.	+ 43 16			COMPACT GALAXY
MCG+00-03-058	00 59 54.	- 02 15	60	15.	GALAXY
PHL 7032	00 59 54.	- 10 22		17.8	BLUE STELLAR OBJECT
PHL 7060	00 59 54.	- 13 47		16.7	BLUE STELLAR OBJECT
PHL 7062	00 59 54.	- 14 01		18.4	BLUE STELLAR OBJECT
LB 03171	00 59 54.	- 60 36		11.0	FAINT BLUE STAR
LIN 384	00 59 58.	- 72 11 27.		13.81	SMC EMISSION LINE OBJECT
LBN 0617	01 00	+ 86 00	22500		BRIGHT NEBULA
PHL 3267	01 00 00.	+ 00 35		18.0	BLUE STELLAR OBJECT
PHL 7066	01 00 00.	+ 11 36		17.9	BLUE STELLAR OBJECT
ZC 0100.0+2147	01 00 00.	+ 21 47	1950		CLUSTER OF GALAXIES
LDN 1306	01 00 00.	+ 62 00	13440		DARK NEBULA
OCL 0315	01 00 00.	+ 62 32	180		OPEN STAR CLUSTER
LDN 1305	01 00 00.	+ 67 30	6600		DARK NEBULA
UGC 00642	01 00 00.	+ 80 24	78	16.0	GALAXY Sb
SHAH 140	01 00 00.	- 01 53	168		GROUP OF COMPACT GALAXIES
ZWG 384.058	01 00 00.	- 02 12		14.7	GALAXY
UGC 00641	01 00 00.	- 02 12	96	14.7	GALAXY SBa-b
MCG-01-03-074	01 00 00.	- 06 40	15	15.	GALAXY
PHL 7064	01 00 00.	- 11 14		18.5	BLUE STELLAR OBJECT
PHL 3268	01 00 00.	- 12 42		18.3	BLUE STELLAR OBJECT
PHL 7063	01 00 00.	- 13 33		18.3	BLUE STELLAR OBJECT
PHL 7065	01 00 00.	- 20 14		18.2	BLUE STELLAR OBJECT
RNGC 0353	01 00 01.	- 02 12		14.5	GALAXY
LIN 385	01 00 04.	- 72 09 21.		15.16	SMC EMISSION LINE OBJECT
KN 10.26	01 00 05.2	+ 09 31 37.			NEBULA
ZC 0100.1+0051	01 00 06.	+ 00 51	1410		CLUSTER OF GALAXIES
PHL 7067	01 00 06.	+ 01 35		17.7	BLUE STELLAR OBJECT
PHL 3270	01 00 06.	+ 10 00		18.7	BLUE STELLAR OBJECT
ZWG 520.009	01 00 06.	+ 37 24		15.6	GALAXY
MCG+06-03-007	01 00 06.	+ 37 26	42	15.	GALAXY
PHL 7089	01 00 06.	- 09 35		17.9	BLUE STELLAR OBJECT
PHL 0955	01 00 06.	- 12 36		18.1	BLUE STELLAR OBJECT
PHL 3269	01 00 06.	- 22 54		17.9	BLUE STELLAR OBJECT
LIN 388	01 00 07.	- 73 02 57.		14.68	SMC EMISSION LINE OBJECT
LIN 387	01 00 09.	- 72 26 51.		15.05	SMC EMISSION LINE OBJECT
LIN 386	01 00 10.	- 72 17 03.		14.21	SMC EMISSION LINE OBJECT
PHL 7070	01 00 12.	+ 00 24		18.4	BLUE STELLAR OBJECT
PHL 3274	01 00 12.	- 00 12		18.4	BLUE STELLAR OBJECT
MCG+00-03-059	01 00 12.	- 01 45 30.	42	15.	GALAXY
PHL 3271	01 00 12.	- 02 06		18.2	BLUE STELLAR OBJECT
MCG-01-03-075	01 00 12.	- 06 40	54	15.	GALAXY
PHL 7069	01 00 12.	- 09 00		17.8	BLUE STELLAR OBJECT
MCG-02-03-067	01 00 12.	- 13 08	36	15.	GALAXY
PHL 3272	01 00 12.	- 14 28		18.2	BLUE STELLAR OBJECT
PHL 3273	01 00 12.	- 23 29		18.3	BLUE STELLAR OBJECT
PHL 7068	01 00 12.	- 28 33		17.8	BLUE STELLAR OBJECT
LB 01587	01 00 12.	- 49 16		14.0	FAINT BLUE STAR
LIN 389	01 00 14.	- 72 39 21.		14.66	SMC EMISSION LINE OBJECT
LIN.CL 066	01 00 14.	- 72 49 33.	342	11.9	STAR CLUSTER IN SMC
HELW 346	01 00 15.	- 04 18 03.			NEBULA
MCG-04-03-045	01 00 15.	- 21 37	30	15.5	GALAXY
ARC 0133	01 00 15.	- 22 04		15.9	RICH CLUSTER OF GALAXIES
MCG-04-03-044	01 00 15.	- 22 10	36	15.	GALAXY
ZC 0100.3+0815	01 00 18.	+ 08 15	1410		CLUSTER OF GALAXIES
PHL 7071	01 00 18.	+ 11 59		15.8	BLUE STELLAR OBJECT
3ZW 019	01 00 18.	+ 16 09			COMPACT GALAXY
MCG+00-03-060	01 00 18.	- 01 51	27	15.5	GALAXY
PHL 3275	01 00 18.	- 05 00		12.0	BLUE STELLAR OBJECT
MCG-01-03-076	01 00 18.	- 05 02	42	14.5	GALAXY
PHL 7072	01 00 18.	- 10 06		18.4	BLUE STELLAR OBJECT
PHL 3276	01 00 18.	- 20 19		18.3	BLUE STELLAR OBJECT
ARC 0132	01 00 20.	+ 26 46		17.7	RICH CLUSTER OF GALAXIES
HELW 347	01 00 21.	- 04 47 57.			NEBULA
LIN 390	01 00 23.	- 71 45 45.		13.78	SMC EMISSION LINE OBJECT
PHL 3277	01 00 24.	+ 02 12		18.7	BLUE STELLAR OBJECT
PHL 3280	01 00 24.	+ 02 28		18.5	BLUE STELLAR OBJECT
PHL 7074	01 00 24.	+ 03 20		18.5	BLUE STELLAR OBJECT
PHL 7077	01 00 24.	+ 03 36		18.6	BLUE STELLAR OBJECT
PHL 7076	01 00 24.	+ 10 01		18.1	BLUE STELLAR OBJECT
PHL 7073	01 00 24.	+ 12 54		17.1	BLUE STELLAR OBJECT
ZWG 458.016	01 00 24.	+ 17 29		15.5	GALAXY
ZWG 480.036	01 00 24.	+ 24 42		15.7	GALAXY
UGC 00643	01 00 24.	+ 24 43	72	15.7	GALAXY Sc
ZWG 384.060	01 00 24.	- 01 43		15.0	GALAXY
ZWG 384.059	01 00 24.	- 01 49		15.7	GALAXY
PHL 7075	01 00 24.	- 16 49		17.6	BLUE STELLAR OBJECT
PHL 3279	01 00 24.	- 21 38		19.0	BLUE STELLAR OBJECT
PHL 3278	01 00 24.	- 22 38		18.3	BLUE STELLAR OBJECT
MCG-04-03-046	01 00 24.	- 24 30	60	15.	GALAXY
HELW 348	01 00 25.	- 04 33 39.			NEBULA
MCG+04-03-036	01 00 27.	+ 24 41 30.	54	15.5	GALAXY
HELW 349	01 00 28.	- 04 34 09.			NEBULA
MCG+00-03-061	01 00 30.	+ 01 22	45	16.	GALAXY
ZWG 410.026	01 00 30.	+ 06 34		15.5	GALAXY
4ZW 037	01 00 30.	+ 34 12			COMPACT GALAXY
PHL 0956	01 00 30.	- 09 24		18.3	BLUE STELLAR OBJECT
PHL 3281	01 00 30.	- 16 21		18.6	BLUE STELLAR OBJECT
TON-S 0184	01 00 30.	- 30 14		15.6	BLUE STAR
LB 03172	01 00 30.	- 62 55		12.9	FAINT BLUE STAR
KN 10.27	01 00 30.5	+ 06 50 32.			NEBULA
HELW 350	01 00 31.	- 04 50 39.			NEBULA
SVEN 080	01 00 31.	- 07 15	54	13.8	GALAXY
SVEN 079	01 00 31.	- 07 46 22.	42	15.2	GALAXY
KN 10.28	01 00 31.9	+ 06 33 49.			NEBULA
ARC 0134	01 00 32.	- 02 48		16.0	RICH CLUSTER OF GALAXIES
LIN 391	01 00 32.	- 72 38 10.		13.90	SMC EMISSION LINE OBJECT
LIN.CL 068	01 00 32.	- 74 10 52.	1020	11.9	STAR CLUSTER IN SMC
HELW 351	01 00 33.	- 04 43 15.			NEBULA

OBJECT NAME	RIGHT ASCEN.	DECLINATION	DIAM.	MAGN.	TYPE OF OBJECT
RNGC 0354	01 00 34.	+ 22 05		14.0	GALAXY
RNGC 0361	01 00 34.	- 71 53		12.0	GLOBULAR CLUSTER IN SMC
LIN.CL 067	01 00 35.	- 71 52 45.	1020	12.0	STAR CLUSTER IN SMC
PHL 0959	01 00 36.	+ 02 05		17.0	BLUE STELLAR OBJECT
MCG+01-03-013	01 00 36.	+ 08 33	48	15.	GALAXY
ZWG 410.027	01 00 36.	+ 08 34		15.6	GALAXY
PHL 0958	01 00 36.	+ 11 42		18.5	BLUE STELLAR OBJECT
PHL 3283	01 00 36.	+ 12 05		18.7	BLUE STELLAR OBJECT
PHL 0957	01 00 36.	+ 13 00		16.6	BLUE STELLAR OBJECT
BC PHL957	01 00 36.	+ 13 00		16.57	QUASI-STELLAR OBJECT
SHB 016	01 00 36.	+ 13 00 00.		16.6	QUASI-STELLAR OBJECT
ZWG 435.037	01 00 36.	+ 13 47		14.5	GALAXY
UGC 00644	01 00 36.	+ 13 47	60	14.5	GALAXY S-IRR
KARA.72 021A	01 00 36.	+ 13 47	48	14.5	PART OF DOUBLE GALAXY
PHL 3282	01 00 36.	+ 13 49		17.1	BLUE STELLAR OBJECT
ZWG 480.037	01 00 36.	+ 22 05		14.2	GALAXY
MRK 353	01 00 36.	+ 22 05	40	15.5	GALAXY WITH UV CONTINUUM
UGC 00645	01 00 36.	+ 22 05	60	14.2	GALAXY SB
ZWG 551.005	01 00 36.	+ 46 41		15.6	GALAXY
MCG+08-03-007	01 00 36.	+ 46 41	42	16.	GALAXY
ZWG 384.061	01 00 36.	- 01 22		15.6	GALAXY
MCG-01-03-077	01 00 36.	- 06 36	48	15.	GALAXY
PHL 3284	01 00 36.	- 10 54		18.6	BLUE STELLAR OBJECT
PHL 7078	01 00 36.	- 15 28		18.7	BLUE STELLAR OBJECT
PHL 7079	01 00 36.	- 24 08		18.7	BLUE STELLAR OBJECT
GCL 003	01 00 36.	- 71 07	1062	8.0	GLOBULAR STAR CLUSTER
LH115-N074	01 00 36.	- 72 07	53		EMISSION NEBULA IN SMC
RNGC 0355	01 00 38.	- 06 36		15.0	GALAXY
KN 10.29	01 00 38.3	+ 08 33 27.			NEBULA
MCG+04-03-037	01 00 39.	+ 22 03 30.	42	14.5	GALAXY
RNGC 0362	01 00 39.	- 71 07		8.0	GLOBULAR CLUSTER
LIN 392	01 00 39.	- 72 24 28.		14.69	SMC EMISSION LINE OBJECT
HELW 352	01 00 40.	- 04 36 21.			NEBULA
KARA.72 021B	01 00 42.	+ 13 46	42		PART OF DOUBLE GALAXY
MCG+02-03-026	01 00 42.	+ 13 46	48	15.	GALAXY
MCG+05-03-043	01 00 42.	+ 31 57	36	16.	GALAXY
ZWG 501.071	01 00 42.	+ 31 58		15.0	GALAXY
UGC 00646	01 00 42.	+ 31 58	120	15.0	GALAXY SB
PHL 0960	01 00 42.	- 02 23		17.9	BLUE STELLAR OBJECT
MCG-01-03-079	01 00 42.	- 03 52	60	14.	GALAXY
HELW 353	01 00 42.	- 04 43 33.			NEBULA
MCG-01-03-078	01 00 42.	- 07 16	78	13.5	GALAXY
LB 03173	01 00 42.	- 69 16		14.5	FAINT BLUE STAR
RNGC 0356	01 00 44.	- 07 16		13.5	GALAXY
HELW 354	01 00 45.	- 04 28 57.			NEBULA
HELW 355	01 00 45.	- 04 56 09.			NEBULA
HELW 356	01 00 46.	- 05 00 33.			NEBULA
PHL 0961	01 00 48.	+ 01 06		18.5	BLUE STELLAR OBJECT
PHL 3286	01 00 48.	+ 01 35		18.7	BLUE STELLAR OBJECT
MCG+00-03-062	01 00 48.	+ 01 35 30.	42	15.5	GALAXY
MCG+02-03-028	01 00 48.	+ 10 33	36	15.5	GALAXY
PHL 7083	01 00 48.	+ 12 54		18.1	BLUE STELLAR OBJECT
MCG+05-03-044	01 00 48.	+ 32 02	24	17.	GALAXY
ZWG 501.072	01 00 48.	+ 32 03		15.6	GALAXY
ZC 0100.8+3737	01 00 48.	+ 37 37	1280		CLUSTER OF GALAXIES
MCG+00-03-063	01 00 48.	- 01 48	30	15.5	GALAXY
PHL 3287	01 00 48.	- 03 40		17.0	BLUE STELLAR OBJECT
PHL 3285	01 00 48.	- 04 04		10.5	BLUE STELLAR OBJECT
MCG-01-03-080	01 00 48.	- 04 29	42	16.	GALAXY
MCG-01-03-081	01 00 48.	- 06 37	120	12.	GALAXY
PHL 0962	01 00 48.	- 06 48		13.0	BLUE STELLAR OBJECT
PHL 7082	01 00 48.	- 09 16		16.6	BLUE STELLAR OBJECT
PHL 0963	01 00 48.	- 10 16		18.5	BLUE STELLAR OBJECT
PHL 7080	01 00 48.	- 15 45		18.4	BLUE STELLAR OBJECT
MCG-04-03-047	01 00 48.	- 21 16 30.	36	15.	GALAXY
PHL 3288	01 00 48.	- 22 43		18.6	BLUE STELLAR OBJECT
PHL 7081	01 00 48.	- 30 00		12.0	BLUE STELLAR OBJECT
LH115-N075	01 00 48.	- 72 13	63		EMISSION NEBULA IN SMC
RNGC 0357	01 00 50.	- 06 37		13.0	GALAXY
ARC 0135	01 00 50.	- 22 51		17.7	RICH CLUSTER OF GALAXIES
SG 3.006	01 00 52.	+ 60 36	180		DIFFUSE EMISSION NEBULA
PHL 0964	01 00 54.	+ 09 54		18.2	BLUE STELLAR OBJECT
PHL 3289	01 00 54.	+ 13 08		18.1	BLUE STELLAR OBJECT
PHL 7084	01 00 54.	+ 14 52		18.3	BLUE STELLAR OBJECT
UGC 00647	01 00 54.	- 01 46	60	16.0	GALAXY Sb
PHL 7085	01 00 54.	- 02 40		18.1	BLUE STELLAR OBJECT
PHL 3290	01 00 54.	- 03 45		18.6	BLUE STELLAR OBJECT
MCG-01-03-082	01 00 54.	- 05 09	21	15.	GALAXY
MCG-02-03-068	01 00 54.	- 11 08	48	15.5	GALAXY
PHL 3291	01 00 54.	- 22 01		18.9	BLUE STELLAR OBJECT
KN 10.31	01 00 55.3	+ 08 51 47.			NEBULA
KN 10.30	01 00 55.5	+ 10 33 27.			NEBULA
RNGC 0360	01 00 57.	- 65 52			UNVERIFIED SOUTHERN OBJECT
LIN 393	01 00 57.	- 72 15 34.		12.73	SMC EMISSION LINE OBJECT
LIN 394	01 00 57.	- 72 24 40.		14.38	SMC EMISSION LINE OBJECT
KN 10.32	01 00 58.0	+ 08 18 26.			NEBULA
KN 10.34	01 00 59.8	+ 06 08 54.			NEBULA
KN 10.33	01 00 59.8	+ 08 35 36.			NEBULA
VDB.66G 008	01 01	+ 01 51	1010		DWARF GALAXY
ZWG 384.062	01 01 00.	+ 01 37		15.6	GALAXY
UGC 00648	01 01 00.	+ 01 37	66	15.6	GALAXY S
ZWG 410.028	01 01 00.	+ 06 09		15.7	GALAXY
UGC 00649	01 01 00.	+ 06 09	78	15.7	GALAXY Sa-b
UGC 00650	01 01 00.	+ 08 17	66	16.0	GALAXY S0-a
PHL 3293	01 01 00.	+ 11 00		18.7	BLUE STELLAR OBJECT
PHL 7086	01 01 00.	+ 12 30		16.1	BLUE STELLAR OBJECT
ZC 0101.0+2648	01 01 00.	+ 26 48	610		CLUSTER OF GALAXIES
VR 029	01 01 00.	+ 64 05	336		STELLAR RING
PHL 3292	01 01 00.	- 05 40		18.6	BLUE STELLAR OBJECT
PHL 7087	01 01 00.	- 11 12		18.2	BLUE STELLAR OBJECT
PHL 7088	01 01 00.	- 24 18		18.6	BLUE STELLAR OBJECT
PHL 3294	01 01 00.	- 25 16		18.5	BLUE STELLAR OBJECT
LIN 398	01 01 00.	- 73 06 40.		13.64	SMC EMISSION LINE OBJECT
LIN 396	01 01 01.	- 72 44 40.		14.29	SMC EMISSION LINE OBJECT
LIN 397	01 01 01.	- 72 45 52.		14.26	SMC EMISSION LINE OBJECT
LIN 395	01 01 03.	- 72 09 28.		15.26	SMC EMISSION LINE OBJECT
PHL 3298	01 01 06.	+ 01 09		18.4	BLUE STELLAR OBJECT
PHL 7117	01 01 06.	+ 11 32		18.5	BLUE STELLAR OBJECT
PHL 3295	01 01 06.	+ 13 07		18.4	BLUE STELLAR OBJECT
PHL 3297	01 01 06.	+ 13 16		13.6	BLUE STELLAR OBJECT
MCG+02-03-029	01 01 06.	+ 14 17 30.	24	15.5	GALAXY
UGC 00652	01 01 06.	+ 21 45	78	17.	GALAXY SB(C)
ZC 0101.1+2634	01 01 06.	+ 26 34	3830		CLUSTER OF GALAXIES
UGC 00653	01 01 06.	+ 30 56	60	16.5	GALAXY Sc
ZC 0101.1+3940	01 01 06.	+ 39 40	470		CLUSTER OF GALAXIES
UGC 00651	01 01 06.	- 00 45	72	16.0	GALAXY Sc
MCG+00-03-064	01 01 06.	- 00 47	54	15.5	GALAXY
PHL 3296	01 01 06.	- 05 39		18.4	BLUE STELLAR OBJECT

OBJECT NAME	RIGHT ASCEN.	DECLINATION	DIAM.	MAGN.	TYPE OF OBJECT
LB 03174	01 01 06.	- 60 49		12.6	FAINT BLUE STAR
LH115-N077B	01 01 06.	- 72 10	19		EMISSION NEBULA IN SMC
LIN 403	01 01 07.	- 72 56 10.		14.84	SMC EMISSION LINE OBJECT
KN 10.35	01 01 07.1	+ 07 12 13.			NEBULA
LIN 401	01 01 08.	- 72 23 58.		14.16	SMC EMISSION LINE OBJECT
LIN 402	01 01 08.	- 72 29 40.		13.99	SMC EMISSION LINE OBJECT
LIN 400	01 01 09.	- 72 09 16.		14.19	SMC EMISSION LINE OBJECT
LIN 399	01 01 10.	- 72 02 04.		14.24	SMC EMISSION LINE OBJECT
PHL 7091	01 01 12.	+ 00 06		18.6	BLUE STELLAR OBJECT
MCG+02-03-030	01 01 12.	+ 10 19	24	16.	GALAXY
PHL 7092	01 01 12.	+ 13 44		18.5	BLUE STELLAR OBJECT
ZWG 458.017	01 01 12.	+ 20 58		15.0	GALAXY
UGC 00654	01 01 12.	+ 20 58	120	15.0	GALAXY COMPACT
MCG+03-03-014	01 01 12.	+ 20 58	18	15.	GALAXY
MCG+04-03-038	01 01 12.	+ 21 44	72	16.	GALAXY
1ZW 003	01 01 12.	+ 29 52			COMPACT GALAXY
MCG+05-03-045	01 01 12.	+ 32 41	60	16.	GALAXY
MCG+07-03-015	01 01 12.	+ 41 33 30.	120	14.	GALAXY
ZWG 536.016	01 01 12.	+ 41 35		15.2	GALAXY
UGC 00655	01 01 12.	+ 41 35	180	15.2	GALAXY S IV-V
5ZW 047	01 01 12.	+ 45 52			COMPACT GALAXY
SHAH 033	01 01 12.	- 01 25	42	17.7	GROUP OF COMPACT GALAXIES
PHL 7090	01 01 12.	- 14 02		17.7	BLUE STELLAR OBJECT
PHL 3299	01 01 12.	- 14 40		17.9	BLUE STELLAR OBJECT
PHL 0965	01 01 12.	- 25 00		17.1	BLUE STELLAR OBJECT
LH115-N077A	01 01 12.	- 72 09	29		EMISSION NEBULA IN SMC
KN 10.36	01 01 13.6	+ 10 19 24.			NEBULA
LIN 404	01 01 14.	- 72 20 46.		12.62	SMC EMISSION LINE OBJECT
LIN.CL 069	01 01 14.	- 73 59 46.	204	14.9	STAR CLUSTER IN SMC
MCG+00-03-065	01 01 15.	+ 02 00	42	15.5	GALAXY
ZWG 384.063	01 01 18.	+ 02 01		15.6	GALAXY
UGC 00656	01 01 18.	+ 02 01	66	15.6	GALAXY SB
PHL 7093	01 01 18.	+ 10 37		18.2	BLUE STELLAR OBJECT
ZC 0101.3+2611	01 01 18.	+ 26 11	1680		CLUSTER OF GALAXIES
UGC 00657	01 01 18.	+ 32 42	72	17.	GALAXY DWRF SP
UGC 00658	01 01 18.	+ 35 41	72	16.0	GALAXY
5ZW 048	01 01 18.	+ 38 36			COMPACT GALAXY
PHL 7096	01 01 18.	- 13 57		18.2	BLUE STELLAR OBJECT
PHL 7094	01 01 18.	- 16 06		18.7	BLUE STELLAR OBJECT
PHL 7095	01 01 18.	- 18 04		18.3	BLUE STELLAR OBJECT
PHL 3300	01 01 18.	- 24 27		18.3	BLUE STELLAR OBJECT
TON-S 0186	01 01 18.	- 25 35		14.2	BLUE STAR
TON-S 0185	01 01 18.	- 27 06		14.6	BLUE STAR
PHL 0966	01 01 18.	- 27 08		15.7	BLUE STELLAR OBJECT
LIN 408	01 01 18.	- 72 59 41.		14.67	SMC EMISSION LINE OBJECT
LIN 406	01 01 19.	- 72 41 35.		13.16	SMC EMISSION LINE OBJECT
LIN 407	01 01 19.	- 72 46 23.		13.85	SMC EMISSION LINE OBJECT
LIN 405	01 01 20.	- 72 22 35.		14.13	SMC EMISSION LINE OBJECT
HOLM 035B	01 01 22.	+ 25 32	24	14.6	PART OF MULTIPLE GALAXY
LIN 409	01 01 23.	- 73 09 59.		14.82	SMC EMISSION LINE OBJECT
PHL 3302	01 01 24.	+ 00 58		18.4	BLUE STELLAR OBJECT
PHL 7101	01 01 24.	+ 01 05		18.5	BLUE STELLAR OBJECT
ZC 0101.4+0838	01 01 24.	+ 08 38	1480		CLUSTER OF GALAXIES
PHL 3301	01 01 24.	+ 09 26		18.4	BLUE STELLAR OBJECT
PHL 7098	01 01 24.	+ 11 13		18.2	BLUE STELLAR OBJECT
PHL 7097	01 01 24.	+ 13 22		15.9	BLUE STELLAR OBJECT
PHL 7100	01 01 24.	+ 14 42		18.8	BLUE STELLAR OBJECT
MRK 354	01 01 24.	+ 20 10	14	16.5	GALAXY WITH UV CONTINUUM
HOLM 035A	01 01 24.	+ 25 33	36	14.4	PART OF MULTIPLE GALAXY
MCG+06-03-008	01 01 24.	+ 35 41	42	16.	GALAXY
ISS 0008	01 01 24.	+ 63 28	223		STELLAR RING
PHL 7102	01 01 24.	- 04 20		18.4	BLUE STELLAR OBJECT
PHL 0967	01 01 24.	- 06 32		18.0	BLUE STELLAR OBJECT
PHL 7099	01 01 24.	- 15 45		17.4	BLUE STELLAR OBJECT
PHL 7103	01 01 24.	- 23 20		18.7	BLUE STELLAR OBJECT
PHL 3303	01 01 24.	- 26 11		18.3	BLUE STELLAR OBJECT
ARC 0136	01 01 25.	+ 24 49		17.5	RICH CLUSTER OF GALAXIES
MCG-01-03-083	01 01 30.	- 02 31	42	15.5	GALAXY
MCG-02-03-069	01 01 30.	- 11 55	9	15.	GALAXY
PHL 3304	01 01 30.	- 19 28		18.7	BLUE STELLAR OBJECT
MCG-04-03-048	01 01 30.	- 23 47	36	15.	GALAXY
TON-S 0187	01 01 30.	- 25 23		16.0	BLUE STAR
TON-S 0188	01 01 30.	- 29 41		15.4	BLUE STAR
LB 03175	01 01 30.	- 63 32		12.8	FAINT BLUE STAR
LH115-N076B	01 01 30.	- 72 23	31		EMISSION NEBULA IN SMC
LIN.CL 070	01 01 32.	- 72 32 29.	204	12.3	STAR CLUSTER IN SMC
PHL 7106	01 01 36.	+ 00 06		16.4	BLUE STELLAR OBJECT
PHL 7108	01 01 36.	+ 12 45		18.5	BLUE STELLAR OBJECT
PHL 0968	01 01 36.	+ 14 32		18.7	BLUE STELLAR OBJECT
ZWG 458.018	01 01 36.	+ 18 26		15.5	GALAXY
UGC 00659	01 01 36.	+ 18 26	66	15.5	GALAXY DBL SYS
MCG+03-03-015	01 01 36.	+ 18 26	48	15.	GALAXY
KARA.73B 0042	01 01 36.	+ 18 26	66	15.5	ISOLATED GALAXY S
MCG+00-03-066	01 01 36.	- 01 04	18	14.	GALAXY
MCG+00-03-067	01 01 36.	- 01 23	48	15.	GALAXY
PHL 7107	01 01 36.	- 09 52		17.9	BLUE STELLAR OBJECT
PHL 7105	01 01 36.	- 11 36		17.0	BLUE STELLAR OBJECT
PHL 7104	01 01 36.	- 18 51		9.3	BLUE STELLAR OBJECT
MCG-04-03-049	01 01 36.	- 21 39	48	16.	GALAXY
PHL 0969	01 01 36.	- 22 46		18.7	BLUE STELLAR OBJECT
RNGC 0371	01 01 41.	- 72 21			CLUSTER/NEBULOSITY IN SMC
FEIG 011	01 01 42.	+ 03 58		11.5	FAINT BLUE STAR
UGC 00660	01 01 42.	+ 06 21	60	18.	GALAXY DWARF
ZC 0101.7+4603	01 01 42.	+ 46 03	1210		CLUSTER OF GALAXIES
MCG+00-03-068	01 01 42.	- 01 24	24	16.	GALAXY
PHL 7109	01 01 42.	- 13 52		18.9	BLUE STELLAR OBJECT
PHL 0970	01 01 42.	- 18 19		16.3	BLUE STELLAR OBJECT
LIN 411	01 01 42.	- 72 59 35.		15.21	SMC EMISSION LINE OBJECT
LIN 410	01 01 44.	- 72 18 17.		13.77	SMC EMISSION LINE OBJECT
LIN.CL 071	01 01 44.	- 72 20 35.	4758		STAR CLUSTER IN SMC
LIN 413	01 01 47.	- 73 09 59.		14.36	SMC EMISSION LINE OBJECT
KN 10.37	01 01 47.7	+ 07 52 44.			NEBULA
ZC 0101.8+2329	01 01 48.	+ 23 29	1550		CLUSTER OF GALAXIES
ZWG 384.066	01 01 48.	- 01 01		14.8	GALAXY
UGC 00662	01 01 48.	- 01 01	126	14.8	GALAXY E-S0
ZWG 384.065	01 01 48.	- 01 19		14.9	GALAXY
UGC 00661	01 01 48.	- 01 19	72	14.9	GALAXY S0-a
ZWG 384.064	01 01 48.	- 01 22		15.7	GALAXY
PHL 7110	01 01 48.	- 04 32		17.7	BLUE STELLAR OBJECT
PHL 7111	01 01 48.	- 08 11		17.5	BLUE STELLAR OBJECT
ARC 0137	01 01 49.	+ 25 29		17.5	RICH CLUSTER OF GALAXIES
RNGC 0359	01 01 49.	- 01 01		15.0	GALAXY
LIN 412	01 01 50.	- 72 19 53.		14.99	SMC EMISSION LINE OBJECT
IC 1615	01 01 53.	- 51 25			NONSTELLAR OBJECT
PHL 0971	01 01 54.	+ 15 25		15.8	BLUE STELLAR OBJECT
ZC 0101.9+2532	01 01 54.	+ 25 32	1610		CLUSTER OF GALAXIES
MCG+07-03-016	01 01 54.	+ 41 23	36	17.	GALAXY
PHL 7112	01 01 54.	- 11 44		17.2	BLUE STELLAR OBJECT

OBJECT NAME	RIGHT ASCEN.	DECLINATION	DIAM.	MAGN.	TYPE OF OBJECT
PHL 7113	01 01 54.	- 15 12		18.5	BLUE STELLAR OBJECT
HN 0134	01 01 54.	- 51 25			NEBULA
LH115-N076	01 01 54.	- 72 20	322		EMISSION NEBULA IN SMC
LIN 416	01 01 55.	- 72 36 59.		15.18	SMC EMISSION LINE OBJECT
RNGC 0365	01 01 56.	- 35 25			GALAXY
LIN 415	01 01 56.	- 72 19 23.		15.65	SMC EMISSION LINE OBJECT
LIN 414	01 01 57.	- 72 05 53.		14.81	SMC EMISSION LINE OBJECT
PHL 3305	01 02 00.	+ 03 06		18.3	BLUE STELLAR OBJECT
SHAH 034	01 02 00.	+ 06 23	96	16.7	GROUP OF COMPACT GALAXIES
PHL 0972	01 02 00.	+ 09 33		13.8	BLUE STELLAR OBJECT
UGC 00664	01 02 00.	+ 09 38	78	16.0	GALAXY
ZWG 501.073	01 02 00.	+ 32 42		15.6	GALAXY
MCG+07-03-017	01 02 00.	+ 41 56	48	14.5	GALAXY
ZWG 536.017	01 02 00.	+ 41 57		15.4	GALAXY
UGC 00665	01 02 00.	+ 41 57	66	15.4	GALAXY SB
LDN 1307	01 02 00.	+ 65 00	6120		DARK NEBULA
UGC 00663	01 02 00.	- 00 26	72	16.0	GALAXY E
MCG+00-03-069	01 02 00.	- 01 07 30.	60	14.	GALAXY
PHL 7114	01 02 00.	- 16 12		18.4	BLUE STELLAR OBJECT
PHL 0973	01 02 00.	- 25 28		18.3	BLUE STELLAR OBJECT
SN 1970N	01 02 00.	- 35 23		18.8	SUPERNOVA
MCG-06-03-017	01 02 00.	- 35 24	48	12.5	GALAXY
RNGC 0364	01 02 01.	- 01 07		14.0	GALAXY
LIN 418	01 02 01.	- 72 29 23.		13.48	SMC EMISSION LINE OBJECT
RNGC 0358	01 02 02.	+ 61 46			NON-EXISTENT OBJECT
LIN 417	01 02 02.	- 72 21 41.		14.70	SMC EMISSION LINE OBJECT
IC 1617	01 02 05.	- 51 18			NONSTELLAR OBJECT
PHL 7118	01 02 06.	+ 02 35		18.0	BLUE STELLAR OBJECT
PHL 7116	01 02 06.	+ 14 00		17.2	BLUE STELLAR OBJECT
ZWG 480.038	01 02 06.	+ 22 03		15.6	GALAXY
MCG-01-03-084	01 02 06.	- 02 48	72	13.5	GALAXY
PHL 7115	01 02 06.	- 16 25		18.2	BLUE STELLAR OBJECT
PHL 7119	01 02 06.	- 17 16		18.3	BLUE STELLAR OBJECT
TON-S 0189	01 02 06.	- 26 47		14.6	BLUE STAR
HN 0135	01 02 06.	- 51 18			NEBULA
ARC 0140	01 02 07.	- 24 14		17.5	RICH CLUSTER OF GALAXIES
LIN 421	01 02 07.	- 72 28 53.		14.14	SMC EMISSION LINE OBJECT
KN 10.38	01 02 07.4	+ 09 36 54.			NEBULA
LIN 419	01 02 08.	- 72 06 23.		13.86	SMC EMISSION LINE OBJECT
LIN 420	01 02 08.	- 72 19 35.		15.65	SMC EMISSION LINE OBJECT
RNGC 0368	01 02 10.	- 43 32			GALAXY
HOW 45	01 02 11.	- 72 03 21.	48		STAR CLUSTER IN SMC
LIN 423	01 02 11.	- 72 59 36.		14.10	SMC EMISSION LINE OBJECT
LIN.CL 072	01 02 11.	- 73 05 23.	474	11.8	STAR CLUSTER IN SMC
PHL 7124	01 02 12.	+ 02 47		18.6	BLUE STELLAR OBJECT
UGC 00667	01 02 12.	+ 05 21	72	17.	GALAXY
PHL 7120	01 02 12.	+ 11 56		14.6	BLUE STELLAR OBJECT
PHL 7123	01 02 12.	+ 13 03		17.9	BLUE STELLAR OBJECT
PHL 7122	01 02 12.	+ 14 32		16.6	BLUE STELLAR OBJECT
MCG+04-03-039	01 02 12.	+ 22 01 30.	42	15.5	GALAXY
ZC 0102.2+2445	01 02 12.	+ 24 45	2080		CLUSTER OF GALAXIES
ZWG 384.067	01 02 12.	- 01 03		14.6	GALAXY
UGC 00666	01 02 12.	- 01 30	96	14.6	GALAXY S0
SHAH 141	01 02 12.	- 01 51	228		GROUP OF COMPACT GALAXIES
PHL 0974	01 02 12.	- 09 26		18.1	BLUE STELLAR OBJECT
PHL 3306	01 02 12.	- 10 20		18.7	BLUE STELLAR OBJECT
PHL 7125	01 02 12.	- 23 14		18.5	BLUE STELLAR OBJECT
TON-S 0190	01 02 12.	- 26 19		16.2	BLUE STAR
PHL 7121	01 02 12.	- 26 49		15.7	BLUE STELLAR OBJECT
PHL 7126	01 02 12.	- 27 14		18.0	BLUE STELLAR OBJECT
MCG-06-03-018	01 02 12.	- 33 56	36	13.	GALAXY
LB 01588	01 02 12.	- 46 57		13.9	FAINT BLUE STAR
LH115-N076A	01 02 12.	- 72 20	26		EMISSION NEBULA IN SMC
LIN 422	01 02 13.	- 72 24 36.		13.95	SMC EMISSION LINE OBJECT
RNGC 0376	01 02 13.	- 73 05			GLOBULAR CLUSTER IN SMC
IC 1613	01 02 13.4	+ 01 51 00.	1380	10.00	GALAXY IRR
KN 10.40	01 02 17.8	+ 04 30 06.			NEBULA
PHL 7128	01 02 18.	+ 00 44		17.8	BLUE STELLAR OBJECT
ZWG 410.029	01 02 18.	+ 04 30		15.4	GALAXY
ZWG 480.039	01 02 18.	+ 21 50		15.7	GALAXY
ZWG 501.074	01 02 18.	+ 32 33		15.7	GALAXY
ZC 0102.3+3418	01 02 18.	+ 34 18	2150		CLUSTER OF GALAXIES
PHL 7127	01 02 18.	- 05 19		17.6	BLUE STELLAR OBJECT
PHL 0975	01 02 18.	- 18 37		17.0	BLUE STELLAR OBJECT
MCG-06-03-019	01 02 18.	- 37 53	72	14.	GALAXY
LH115-N076C	01 02 18.	- 72 25			EMISSION NEBULA IN SMC
IC 0073	01 02 19.	+ 04 30 21.			NONSTELLAR OBJECT
KN 10.39	01 02 19.2	+ 07 44 43.			NEBULA
LIN 424	01 02 20.	- 72 12 24.		14.12	SMC EMISSION LINE OBJECT
HOW 46	01 02 20.	- 73 58 08.	12		STAR CLUSTER IN SMC
IC 1614	01 02 21.	+ 32 56 33.			NONSTELLAR OBJECT
SC 0059-1422.8	01 02 22.	- 14 06 41.	12		NEBULA
LIN 426	01 02 23.	- 73 01 06.		13.46	SMC EMISSION LINE OBJECT
ZWG 384.068	01 02 24.	+ 01 53		10.7	GALAXY
UGC 00668	01 02 24.	+ 01 53	1320	10.7	GALAXY DWRF IR
PHL 7129	01 02 24.	+ 03 05		18.2	BLUE STELLAR OBJECT
MCG+04-03-040	01 02 24.	+ 21 48	48	15.5	GALAXY
ZC 0102.4-0012	01 02 24.	- 00 12	1410		CLUSTER OF GALAXIES
PHL 0976	01 02 24.	- 01 48		17.9	BLUE STELLAR OBJECT
PHL 7130	01 02 24.	- 02 38		18.5	BLUE STELLAR OBJECT
MCG-02-03-070	01 02 24.	- 11 28	48	15.	GALAXY
PHL 3307	01 02 24.	- 13 01		18.3	BLUE STELLAR OBJECT
PHL 7131	01 02 24.	- 13 40		18.6	BLUE STELLAR OBJECT
PHL 3308	01 02 24.	- 15 00		18.8	BLUE STELLAR OBJECT
IC 1616	01 02 24.	- 27 40 16.			NONSTELLAR OBJECT
PHL 0977	01 02 24.	- 31 05		17.8	BLUE STELLAR OBJECT
LIN 425	01 02 24.	- 72 49 12.		12.68	SMC EMISSION LINE OBJECT
RNGC 0367	01 02 27.	- 12 36			GALAXY
ARC 0138	01 02 28.	+ 42 51		17.5	RICH CLUSTER OF GALAXIES
MCG+00-03-070	01 02 30.	+ 01 52	720	10.0	GALAXY
ZWG 501.075	01 02 30.	+ 31 25		15.7	GALAXY
UGC 00669	01 02 30.	+ 31 25	96	15.7	GALAXY Sc
ZC 0102.5+4251	01 02 30.	+ 42 51	1140		CLUSTER OF GALAXIES
MCG-01-03-087	01 02 30.	- 05 37	48	14.	GALAXY
MCG-01-03-085	01 02 30.	- 06 30	240	11.	GALAXY
MCG-01-03-088	01 02 30.	- 06 33	72	14.5	GALAXY
MCG-01-03-086	01 02 30.	- 07 01	42	15.	GALAXY
MCG-05-03-022	01 02 30.	- 27 41	84	14.	GALAXY
MCG+05-03-046	01 02 33.	+ 31 23 30.	84	15.	GALAXY
ARC 0139	01 02 34.	+ 36 08		17.5	RICH CLUSTER OF GALAXIES
RNGC 0369	01 02 34.	- 18 03		14.0	GALAXY
HSN 21	01 02 34.	- 74 29			BRIGHT GALAXY BEHIND SMC
HOW 47	01 02 35.	- 74 53 26.	12		STAR CLUSTER IN SMC
ZC 0102.6+0648	01 02 36.	+ 06 48	3630		CLUSTER OF GALAXIES
PHL 3309	01 02 36.	+ 10 00		9.5	BLUE STELLAR OBJECT
ZWG 480.040	01 02 36.	+ 23 46		15.6	GALAXY
SA 1	01 02 36.	- 06 29	210	12.8	GALAXY
MCG-03-03-021	01 02 36.	- 18 01	24	15.5	GALAXY

OBJECT NAME	RIGHT ASCEN.	DECLINATION	DIAM.	MAGN.	TYPE OF OBJECT
MCG-03-03-022	01 02 36.	- 18 03	60	14.	GALAXY
LIN 428	01 02 41.	- 73 04 00.		14.89	SMC EMISSION LINE OBJECT
PHL 7132	01 02 42.	+ 11 05		18.0	BLUE STELLAR OBJECT
ZC 0102.7+2733	01 02 42.	+ 27 33	1410		CLUSTER OF GALAXIES
LB 00077	01 02 42.	- 31 06		15.6	FAINT BLUE STAR
MOHR 2	01 02 42.	- 70 35 48.		16.5	STAR CLUSTER IN SMC
LIN 430	01 02 44.	- 73 37 42.		16.66	PLANETARY NEBULA IN SMC
MCG-01-03-089	01 02 45.	- 05 42	72	14.	GALAXY
MCG-01-03-090	01 02 45.	- 07 07	15	15.5	GALAXY
PHL 0978	01 02 48.	+ 01 10		18.1	BLUE STELLAR OBJECT
PHL 3310	01 02 48.	+ 01 50		17.2	BLUE STELLAR OBJECT
PHL 3312	01 02 48.	+ 13 10		18.3	BLUE STELLAR OBJECT
ZC 0102.8+1316	01 02 48.	+ 13 16	3090		CLUSTER OF GALAXIES
PHL 7133	01 02 48.	+ 15 34		14.2	BLUE STELLAR OBJECT
ZWG 501.076	01 02 48.	+ 31 42		15.5	GALAXY
ZWG 501.077	01 02 48.	+ 32 10		15.7	GALAXY
PHL 0979	01 02 48.	- 02 15		18.1	BLUE STELLAR OBJECT
MCG-02-03-071	01 02 48.	- 11 38	30	15.	GALAXY
PHL 7134	01 02 48.	- 12 05		18.7	BLUE STELLAR OBJECT
PHL 0980	01 02 48.	- 14 16		16.5	BLUE STELLAR OBJECT
PHL 7135	01 02 48.	- 17 44		18.2	BLUE STELLAR OBJECT
PHL 3311	01 02 48.	- 25 58		17.5	BLUE STELLAR OBJECT
MCG-05-03-023	01 02 48.	- 29 03	36	15.	GALAXY
LIN.CL 073	01 02 49.	- 70 36 48.	270	15.5	STAR CLUSTER IN SMC
LIN 429	01 02 49.	- 72 17 54.		13.89	SMC EMISSION LINE OBJECT
LIN.CL 074	01 02 49.	- 72 25 48.	342	12.3	STAR CLUSTER IN SMC
MCG+04-03-041	01 02 51.	+ 25 52 30.	36	16.	GALAXY
PHL 0981	01 02 54.	+ 01 24		17.8	BLUE STELLAR OBJECT
ZC 0102.9+0215	01 02 54.	+ 02 15	1480		CLUSTER OF GALAXIES
ZC 0102.9+2102	01 02 54.	+ 21 02	1080		CLUSTER OF GALAXIES
PHL 7136	01 02 54.	- 12 10		18.5	BLUE STELLAR OBJECT
LIN 432	01 02 54.	- 72 34 00.		14.46	SMC EMISSION LINE OBJECT
LIN 431	01 02 55.	- 72 16 06.		13.67	SMC EMISSION LINE OBJECT
KHAV 016	01 03	+ 60 58	4700		DARK NEBULA
PHL 3315	01 03 00.	+ 02 35		18.0	BLUE STELLAR OBJECT
ZWG 384.069	01 03 00.	+ 02 35		15.6	GALAXY
ZWG 435.038	01 03 00.	+ 14 29		15.5	GALAXY
MCG+02-03-031	01 03 00.	+ 14 29	48	15.	GALAXY
ZC 0103.0+1519	01 03 00.	+ 15 19	1610		CLUSTER OF GALAXIES
ZWG 520.010	01 03 00.	+ 34 44		15.4	GALAXY
ZWG 345.002	01 03 00.	+ 75 20		15.2	GALAXY
UGC 00670	01 03 00.	+ 75 20	108	15.2	GALAXY SB?b
MCG+12-02-002	01 03 00.	+ 75 20	57	14.	GALAXY
PHL 3317	01 03 00.	- 04 18		18.4	BLUE STELLAR OBJECT
PHL 3316	01 03 00.	- 11 18		18.0	BLUE STELLAR OBJECT
PHL 3313	01 03 00.	- 16 34		17.6	BLUE STELLAR OBJECT
PHL 7137	01 03 00.	- 17 24		18.2	BLUE STELLAR OBJECT
PHL 3314	01 03 00.	- 18 20		18.2	BLUE STELLAR OBJECT
MCG-04-03-050	01 03 00.	- 23 19	72	15.	GALAXY
LB 01589	01 03 00.	- 58 04		14.6	FAINT BLUE STAR
LIN 433	01 03 00.	- 72 35 43.		14.49	SMC EMISSION LINE OBJECT
MCG+06-03-009	01 03 03.	+ 34 44	30	15.	GALAXY
ZWG 384.070	01 03 06.	+ 01 20		15.7	GALAXY
PHL 7138	01 03 06.	+ 12 42		16.5	BLUE STELLAR OBJECT
PHL 3319	01 03 06.	+ 14 01		18.4	BLUE STELLAR OBJECT
ZC 0103.1-0038	01 03 06.	- 00 38	1080		CLUSTER OF GALAXIES
PHL 7139	01 03 06.	- 03 18		18.1	BLUE STELLAR OBJECT
PHL 3318	01 03 06.	- 14 33		16.7	BLUE STELLAR OBJECT
PHL 7140	01 03 06.	- 15 05		18.7	BLUE STELLAR OBJECT
LIN 434	01 03 06.	- 72 33 37.		14.35	SMC EMISSION LINE OBJECT
LIN 435	01 03 06.	- 72 41 43.		14.03	SMC EMISSION LINE OBJECT
IC 1618	01 03 09.	+ 32 09 55.			NONSTELLAR OBJECT
LIN 437	01 03 10.	- 73 02 31.		12.88	SMC EMISSION LINE OBJECT
PHL 3321	01 03 12.	+ 00 26		18.1	BLUE STELLAR OBJECT
PHL 7142	01 03 12.	+ 10 44		16.9	BLUE STELLAR OBJECT
PHL 7143	01 03 12.	+ 14 38		18.3	BLUE STELLAR OBJECT
ZWG 501.078	01 03 12.	+ 32 09		15.6	GALAXY
UGC 00671	01 03 12.	+ 32 09	72	15.6	GALAXY S0
VB 030	01 03 12.	+ 63 28	222		STELLAR RING
SHAH 035	01 03 12.	- 01 01	48	16.0	GROUP OF COMPACT GALAXIES
PHL 3320	01 03 12.	- 09 38		17.8	BLUE STELLAR OBJECT
PHL 7141	01 03 12.	- 10 32		18.3	BLUE STELLAR OBJECT
ARC 0141	01 03 12.	- 24 52		17.7	RICH CLUSTER OF GALAXIES
LB 00078	01 03 12.	- 32 53		15.8	FAINT BLUE STAR
LB 01590	01 03 12.	- 51 46		14.8	FAINT BLUE STAR
LIN 436	01 03 13.	- 72 21 43.		13.88	SMC EMISSION LINE OBJECT
PHL 7154	01 03 18.	+ 01 05		18.3	BLUE STELLAR OBJECT
PHL 7145	01 03 18.	+ 02 32		18.7	BLUE STELLAR OBJECT
ZWG 410.030	01 03 18.	+ 03 49		15.5	GALAXY
KARA.73B 0043	01 03 18.	+ 03 49	24	15.5	ISOLATED GALAXY E
PHL 3323	01 03 18.	+ 13 30		18.6	BLUE STELLAR OBJECT
5ZW 049	01 03 18.	+ 42 00			COMPACT GALAXY
UGC 00672	01 03 18.	+ 44 43	90	18.	GALAXY DWRF IR
OCL 0316	01 03 18.	+ 61 58	240	11.8	OPEN STAR CLUSTER
PHL 7155	01 03 18.	- 01 51		18.1	BLUE STELLAR OBJECT
ZC 0103.3-0210	01 03 18.	- 02 10	2150		CLUSTER OF GALAXIES
PHL 0982	01 03 18.	- 13 03		18.2	BLUE STELLAR OBJECT
PHL 7144	01 03 18.	- 15 44		15.9	BLUE STELLAR OBJECT
PHL 7146	01 03 18.	- 22 02		18.5	BLUE STELLAR OBJECT
RNGC 0366	01 03 20.	+ 61 58		12.0	OPEN CLUSTER
IC 0074	01 03 21.	+ 03 50 19.			NONSTELLAR OBJECT
MCG-04-03-051	01 03 21.	- 23 12 30.	48	15.5	GALAXY
HOW 48	01 03 21.	- 73 54 26.	12		STAR CLUSTER IN SMC
LIN 442	01 03 22.	- 73 02 25.		15.00	SMC EMISSION LINE OBJECT
LIN 440	01 03 23.	- 72 39 37.		13.60	SMC EMISSION LINE OBJECT
LIN 441	01 03 23.	- 72 40 43.		13.02	SMC EMISSION LINE OBJECT
PHL 7152	01 03 24.	+ 00 33		18.3	BLUE STELLAR OBJECT
PHL 7151	01 03 24.	+ 01 25		18.5	BLUE STELLAR OBJECT
ZWG 480.041	01 03 24.	+ 25 17		15.5	GALAXY
ZWG 501.079	01 03 24.	+ 31 08		15.5	GALAXY
UGC 00673	01 03 24.	+ 31 08	120	15.5	GALAXY Sc
SN 1960P	01 03 24.	+ 31 08		17.5	SUPERNOVA
PHL 7149	01 03 24.	- 01 34		18.2	BLUE STELLAR OBJECT
PHL 3322	01 03 24.	- 02 32		17.8	BLUE STELLAR OBJECT
PHL 7148	01 03 24.	- 04 49		15.5	BLUE STELLAR OBJECT
PHL 0983	01 03 24.	- 08 42		18.0	BLUE STELLAR OBJECT
PHL 7150	01 03 24.	- 18 00		18.4	BLUE STELLAR OBJECT
PHL 7147	01 03 24.	- 20 05		11.5	BLUE STELLAR OBJECT
PHL 0984	01 03 24.	- 27 54		15.5	BLUE STELLAR OBJECT
LH115-N073	01 03 24.	- 76 04			EMISSION NEBULA IN SMC
LIN 438	01 03 25.	- 72 14 49.		15.25	SMC EMISSION LINE OBJECT
LIN 439	01 03 25.	- 72 15 19.		14.28	SMC EMISSION LINE OBJECT
LIN 445A	01 03 26.	- 76 03 55.		17.0	PLANETARY NEBULA IN SMC
LIN 445	01 03 28.	- 73 03 49.		14.75	SMC EMISSION LINE OBJECT
LIN 443	01 03 28.	- 73 05 43.		14.21	SMC EMISSION LINE OBJECT
LIN 444	01 03 29.	- 72 40 07.		13.56	SMC EMISSION LINE OBJECT
MCG+00-03-071	01 03 30.	+ 00 30	48	15.5	GALAXY
ZC 0103.5+0032	01 03 30.	+ 00 32	670		CLUSTER OF GALAXIES

OBJECT NAME	RIGHT ASCEN.	DECLINATION	DIAM.	MAGN.	TYPE OF OBJECT
PHL 7153	01 03 30.	+ 02 19		18.2	BLUE STELLAR OBJECT
UGC 00674	01 03 30.	+ 21 07	78	18.	GALAXY DWARF
MCG+05-03-047	01 03 30.	+ 31 08	42	15.5	GALAXY
LH115-N078A	01 03 30.	- 72 15	19		EMISSION NEBULA IN SMC
LH115-N078B	01 03 30.	- 72 16	24		EMISSION NEBULA IN SMC
RNGC 0395	01 03 30.	- 72 16			CLUSTER/NEBULOSITY IN SMC
LIN.CL 075	01 03 31.	- 72 15 37.	1020		STAR CLUSTER IN SMC
MCG+04-03-042	01 03 33.	+ 25 17 30.	48	15.5	GALAXY
ABC 0142	01 03 34.	+ 00 42		18.0	RICH CLUSTER OF GALAXIES
UGC 00675	01 03 36.	+ 00 31	96	17.	GALAXY DWARF
ZC 0103.6+0403	01 03 36.	+ 04 03	1750		CLUSTER OF GALAXIES
ZWG 520.011	01 03 36.	+ 33 30		15.1	GALAXY
PHL 7156	01 03 36.	- 02 38		18.0	BLUE STELLAR OBJECT
PHL 7157	01 03 36.	- 03 52		18.6	BLUE STELLAR OBJECT
PHL 7158	01 03 36.	- 21 20		18.1	BLUE STELLAR OBJECT
LB 00079	01 03 36.	- 27 49		14.9	FAINT BLUE STAR
LH115-N078D	01 03 36.	- 72 14	33		EMISSION NEBULA IN SMC
BIGO 457	01 03 39.	+ 32 09			NEBULA
RNGC 0363	01 03 40.	- 16 50		15.0	GALAXY
KN 10.41	01 03 41.1	+ 07 01 25.			NEBULA
ZWG 384.071	01 03 42.	+ 00 58		15.6	GALAXY
UGC 00676	01 03 42.	+ 00 58	84	15.6	GALAXY S
MCG+00-03-072	01 03 42.	+ 03 17 30.	33	15.	GALAXY
SHAH 036	01 03 42.	+ 04 28	186	17.6	GROUP OF COMPACT GALAXIES
ZWG 435.039	01 03 42.	+ 14 19		15.4	GALAXY
UGC 00677	01 03 42.	+ 14 19	72	15.4	GALAXY S
3ZW 020	01 03 42.	+ 19 12			COMPACT GALAXY
ZCG 19.008	01 03 42.	+ 32 14		19.2	ELLIPTICAL GALAXY
MCG+06-03-010	01 03 42.	+ 33 30	36	15.	GALAXY
ZC 0103.7+3600	01 03 42.	+ 36 00	1340		CLUSTER OF GALAXIES
ZC 0103.7+3942	01 03 42.	+ 39 42	6120		CLUSTER OF GALAXIES
VB 031	01 03 42.	+ 69 34	208		STELLAR RING
PHL 7161	01 03 42.	- 00 12		18.1	BLUE STELLAR OBJECT
PHL 7163	01 03 42.	- 10 14		18.2	BLUE STELLAR OBJECT
PHL 7162	01 03 42.	- 12 07		16.7	BLUE STELLAR OBJECT
MCG-03-03-023	01 03 42.	- 16 50	15	15.	GALAXY
PHL 3324	01 03 42.	- 17 11		18.5	BLUE STELLAR OBJECT
MCG-03-03-024	01 03 42.	- 19 17	15	16.	GALAXY
PHL 7159	01 03 42.	- 20 28		18.4	BLUE STELLAR OBJECT
PHL 7160	01 03 42.	- 31 23		15.7	BLUE STELLAR OBJECT
HN 0137	01 03 42.	- 72 18			NEBULA
LIN.CL 076	01 03 42.	- 72 18 25.	408	12.5	STAR CLUSTER IN SMC
IC 1624	01 03 43.	- 72 18			NONSTELLAR OBJECT
MCG+02-03-032	01 03 45.	+ 14 18	48	15.	GALAXY
SN 19690	01 03 48.	+ 02 53		18.2	SUPERNOVA
ZWG 384.072	01 03 48.	+ 03 19		15.2	GALAXY
UGC 00678	01 03 48.	+ 03 19	120	15.2	GALAXY Sc
ZWG 410.031	01 03 48.	+ 06 01		15.4	GALAXY
ZWG 480.042	01 03 48.	+ 24 36		15.7	GALAXY
ZCG 19.009	01 03 48.	+ 32 16		18.4	ELLIPTICAL GALAXY
ZWG 520.012	01 03 48.	+ 36 38		15.7	GALAXY
5ZW 050	01 03 48.	+ 45 39			COMPACT GALAXY
PHL 0985	01 03 48.	- 05 06		18.3	BLUE STELLAR OBJECT
PHL 0986	01 03 48.	- 05 51		17.6	BLUE STELLAR OBJECT
MCG-01-03-091	01 03 48.	- 06 28	24	15.	GALAXY
PHL 3325	01 03 48.	- 17 38		18.7	BLUE STELLAR OBJECT
PHL 0987	01 03 48.	- 28 02		17.4	BLUE STELLAR OBJECT
PHL 7164	01 03 48.	- 29 29		18.5	BLUE STELLAR OBJECT
MCG-05-03-024	01 03 48.	- 30 27	84	14.5	GALAXY
LH115-N078	01 03 48.	- 72 17	365		EMISSION NEBULA IN SMC
RNGC 0370	01 03 49.	+ 32 09			NON-EXISTENT OBJECT
RNGC 0378	01 03 49.	- 30 27		14.0	GALAXY
MCG+04-03-043	01 03 51.	+ 24 36	48	15.	GALAXY
RNGC 0377	01 03 53.	- 20 19			NON-EXISTENT OBJECT
LIN 446	01 03 53.	- 72 44 32.		14.56	SMC EMISSION LINE OBJECT
PHL 0988	01 03 54.	+ 01 31		16.7	BLUE STELLAR OBJECT
ZWG 480.043	01 03 54.	+ 23 57		15.7	GALAXY
ZC 0103.9+2552	01 03 54.	+ 25 52	1480		CLUSTER OF GALAXIES
ZC 0103.9+4350	01 03 54.	+ 43 50	1340		CLUSTER OF GALAXIES
PHL 3329	01 03 54.	- 00 14		18.3	BLUE STELLAR OBJECT
ZWG 384.073	01 03 54.	- 00 23		15.7	GALAXY
MRK 560	01 03 54.	- 00 23	9	16.	GALAXY WITH UV CONTINUUM
PHL 3330	01 03 54.	- 10 32		18.6	BLUE STELLAR OBJECT
PHL 3327	01 03 54.	- 18 11		18.2	BLUE STELLAR OBJECT
PHL 3328	01 03 54.	- 19 00		18.4	BLUE STELLAR OBJECT
PHL 3326	01 03 54.	- 27 47		17.7	BLUE STELLAR OBJECT
MCG-05-03-025	01 03 54.	- 30 26	36	16.	GALAXY
LB 03176	01 03 54.	- 64 21		12.7	FAINT BLUE STAR
RNGC 0372	01 03 55.	+ 32 10			NON-EXISTENT OBJECT
SC 0101-1711.5	01 03 55.	- 16 55 25.	12		NEBULA
ABC 0144	01 03 57.	- 21 08		17.9	RICH CLUSTER OF GALAXIES
KN 10.42	01 03 58.8	+ 10 15 07.			NEBULA
KN 10.43	01 03 58.9	+ 08 55 30.			NEBULA
LBN 0625	01 04	+ 73 15	240		BRIGHT NEBULA
ZWG 410.032	01 04 00.	+ 07 16		15.7	GALAXY
ZWG 410.033	01 04 00.	+ 08 55		15.5	GALAXY
ZWG 435.040	01 04 00.	+ 10 16		14.8	GALAXY
KARA.73B 0044	01 04 00.	+ 10 16	48	14.8	ISOLATED GALAXY S
PHL 0989	01 04 00.	+ 10 38		18.5	BLUE STELLAR OBJECT
ZWG 536.018	01 04 00.	+ 44 01		15.5	GALAXY
MCG+07-03-018	01 04 00.	+ 44 01	48	14.5	GALAXY
LDN 1310	01 04 00.	+ 59 00	4260		DARK NEBULA
LDN 1309	01 04 00.	+ 62 00	2640		DARK NEBULA
MCG+00-03-073	01 04 00.	- 02 29 30.	42	15.5	GALAXY
PHL 3331	01 04 00.	- 18 04		18.4	BLUE STELLAR OBJECT
PHL 7165	01 04 00.	- 29 54		18.3	BLUE STELLAR OBJECT
ABC 0143	01 04 02.	+ 25 59		17.5	RICH CLUSTER OF GALAXIES
MCG-04-03-052	01 04 03.	- 23 58	48	14.5	GALAXY
LIN 448	01 04 05.	- 72 41 44.		13.28	SMC EMISSION LINE OBJECT
ZC 0104.1+4301	01 04 06.	+ 43 01	1880		CLUSTER OF GALAXIES
ZWG 384.074	01 04 06.	- 02 27		15.5	GALAXY
PHL 3332	01 04 06.	- 16 28		18.4	BLUE STELLAR OBJECT
PHL 7166	01 04 06.	- 30 28		18.2	BLUE STELLAR OBJECT
LH115-N078C	01 04 06.	- 72 20	32		EMISSION NEBULA IN SMC
LIN 447	01 04 07.	- 72 02 08.		12.80	SMC EMISSION LINE OBJECT
KN 10.44	01 04 07.2	+ 06 22 29.			NEBULA
HN 0136	01 04 08.	- 46 59			NEBULA
IC 1621	01 04 08.	- 46 59			NONSTELLAR OBJECT
KN 10.45	01 04 09.2	+ 06 11 36.			NEBULA
KN 10.46	01 04 09.8	+ 06 22 06.			NEBULA
PHL 3333	01 04 12.	+ 00 30		18.6	BLUE STELLAR OBJECT
PHL 0990	01 04 12.	+ 01 33		17.8	BLUE STELLAR OBJECT
PHL 7170	01 04 12.	+ 02 30		16.7	BLUE STELLAR OBJECT
ZWG 435.041	01 04 12.	+ 11 40		15.7	GALAXY
BIGO 458	01 04 12.	+ 32 03			NEBULA
ZCG 19.012	01 04 12.	+ 32 03		16.6	ELLIPTICAL GALAXY
ZCG 19.011	01 04 12.	+ 32 19		17.8	ELLIPTICAL GALAXY
ZCG 19.010	01 04 12.	+ 32 19		18.4	ELLIPTICAL GALAXY
ZCG 19.013	01 04 12.	+ 32 21		18.5	ELLIPTICAL GALAXY
PHL 3334	01 04 12.	- 02 34		18.4	BLUE STELLAR OBJECT
PHL 0991	01 04 12.	- 10 50		18.3	BLUE STELLAR OBJECT
PHL 7167	01 04 12.	- 12 21		18.6	BLUE STELLAR OBJECT
PHL 7168	01 04 12.	- 12 35		18.8	BLUE STELLAR OBJECT
PHL 7169	01 04 12.	- 12 50		18.5	BLUE STELLAR OBJECT
PHL 3335	01 04 12.	- 12 59		18.3	BLUE STELLAR OBJECT
MCG-04-03-053	01 04 12.	- 20 36	24	15.5	GALAXY
PHL 7171	01 04 12.	- 28 10		18.5	BLUE STELLAR OBJECT
RNGC 0373	01 04 13.	+ 32 02			GALAXY
ABC 0145	01 04 14.	- 02 43		17.6	RICH CLUSTER OF GALAXIES
LIN 449	01 04 16.	- 72 48 26.		14.44	SMC EMISSION LINE OBJECT
ZCG 19.014	01 04 18.	+ 32 04		15.9	ELLIPTICAL GALAXY
MCG+05-03-049	01 04 18.	+ 32 05	60	16.	GALAXY
UGC 00679	01 04 18.	+ 32 07	72	16.5	GALAXY
MCG+05-03-048	01 04 18.	+ 32 30 30.	60	14.5	GALAXY
ZWG 501.080	01 04 18.	+ 32 32		14.3	GALAXY
UGC 00680	01 04 18.	+ 32 32	78	14.3	GALAXY S0-a
PHL 7172	01 04 18.	- 15 28		18.6	BLUE STELLAR OBJECT
RNGC 0375	01 04 19.	+ 32 05		16.0	GALAXY
RNGC 0374	01 04 19.	+ 32 30		14.5	GALAXY
LIN 454	01 04 20.	- 74 45 14.		14.88	SMC EMISSION LINE OBJECT
ZC 0104.4+0048	01 04 24.	+ 00 48	1210		CLUSTER OF GALAXIES
PHL 3336	01 04 24.	+ 01 29		18.7	BLUE STELLAR OBJECT
ZWG 384.075	01 04 24.	+ 01 41		15.0	GALAXY
PHL 7173	01 04 24.	+ 09 37		18.8	BLUE STELLAR OBJECT
KARA.72 022A	01 04 24.	+ 13 42	36	14.2	PART OF DOUBLE GALAXY
ZCG 19.015	01 04 24.	+ 31 55		18.6	ELLIPTICAL GALAXY
ZC 0104.4+3628	01 04 24.	+ 36 28	1080		CLUSTER OF GALAXIES
PHL 3339	01 04 24.	- 05 48		17.7	BLUE STELLAR OBJECT
PHL 3338	01 04 24.	- 09 44		18.9	BLUE STELLAR OBJECT
PHL 3337	01 04 24.	- 09 44		18.9	BLUE STELLAR OBJECT
PHL 7174	01 04 24.	- 13 52		18.5	BLUE STELLAR OBJECT
PHL 0992	01 04 24.	- 19 15		17.6	BLUE STELLAR OBJECT
TON-S 0191	01 04 24.	- 33 39		12.9	BLUE STAR
LIN 450	01 04 24.	- 72 07 20.		14.34	SMC EMISSION LINE OBJECT
LIN 451	01 04 24.	- 72 16 32.		14.45	SMC EMISSION LINE OBJECT
MCG+02-03-033	01 04 27.	+ 13 41	36	15.	GALAXY
MCG+05-03-050	01 04 27.	+ 32 14	30	14.5	GALAXY
LIN 453	01 04 27.	- 73 05 20.		14.04	SMC EMISSION LINE OBJECT
HOW 49	01 04 27.	- 73 39 37.	18		STAR CLUSTER IN SMC
PHL 3343	01 04 28.	- 17 34		18.4	BLUE STELLAR OBJECT
LIN 452	01 04 29.	- 72 33 32.		13.00	SMC EMISSION LINE OBJECT
PHL 7175	01 04 30.	+ 00 00		17.0	BLUE STELLAR OBJECT
MCG+02-03-035	01 04 30.	+ 10 33	48	14.5	GALAXY
MCG+02-03-034	01 04 30.	+ 13 41	48	14.	GALAXY
ZWG 435.042	01 04 30.	+ 13 42		14.2	GALAXY
UGC 00681	01 04 30.	+ 13 42	72	14.2	GALAXY Sb/SBc
KARA.72 022B	01 04 30.	+ 13 42	60		PART OF DOUBLE GALAXY
ZWG 459.001	01 04 30.	+ 17 43		15.7	GALAXY
ZWG 458.019	01 04 30.	+ 17 43		15.7	GALAXY
MCG+03-03-016	01 04 30.	+ 17 44	36	15.5	GALAXY
ZWG 480.044	01 04 30.	+ 24 15		15.6	GALAXY
ZCG 19.016	01 04 30.	+ 31 56		17.6	ELLIPTICAL GALAXY
MCG+05-03-051	01 04 30.	+ 32 12	24	14.	GALAXY
ZWG 501.081	01 04 30.	+ 32 13		13.9	GALAXY
ZCG 19.002	01 04 30.	+ 32 13		13.9	ELLIPTICAL GALAXY
UGC 00682	01 04 30.	+ 32 13	90	13.9	GALAXY E
ZWG 501.082	01 04 30.	+ 32 15		14.0	GALAXY
ZCG 19.001	01 04 30.	+ 32 15		14.0	ELLIPTICAL GALAXY
UGC 00683	01 04 30.	+ 32 15	96	14.0	GALAXY S0
PHL 3340	01 04 30.	- 01 02		18.2	BLUE STELLAR OBJECT
HOW 50	01 04 30.	- 71 58 31.	54		STAR CLUSTER IN SMC
RNGC 0380	01 04 31.	+ 32 13		14.0	GALAXY
RNGC 0379	01 04 31.	+ 32 15		14.0	GALAXY
IC 0075	01 04 33.	+ 10 33 58.			NONSTELLAR OBJECT
MCG+05-03-052	01 04 33.	+ 32 07 30.	18	15.	GALAXY
HSN 22	01 04 34.	- 74 16			BRIGHT GALAXY BEHIND SMC
KN 10.47	01 04 34.7	+ 10 34 05.			NEBULA
IC 1620	01 04 35.	+ 13 41 22.			NONSTELLAR OBJECT
LIN 455	01 04 35.	- 72 28 20.		13.05	SMC EMISSION LINE OBJECT
ZWG 435.043	01 04 36.	+ 10 35		15.0	GALAXY
UGC 00684	01 04 36.	+ 10 35	96	15.0	GALAXY Sb
ZWG 436.001	01 04 36.	+ 14 00		15.7	GALAXY
ZWG 435.044	01 04 36.	+ 14 00		15.7	GALAXY
MCG+04-03-044	01 04 36.	+ 24 15	36	16.	GALAXY
4ZW 038	01 04 36.	+ 32 08			COMPACT GALAXY
VV 193B	01 04 36.	+ 32 08	21	14.	INTERACTING GALAXY
MCG+05-03-053	01 04 36.	+ 32 08	42	13.5	GALAXY
ARP 331	01 04 36.	+ 32 09			PECULIAR GALAXY
MCG+05-03-054	01 04 36.	+ 32 47	24	16.	GALAXY
ZWG 501.083	01 04 36.	+ 32 48		15.0	GALAXY
ZC 0104.6+4621	01 04 36.	+ 46 21	3230		CLUSTER OF GALAXIES
MCG-02-03-072	01 04 36.	- 09 06	18	16.	GALAXY
PHL 0993	01 04 36.	- 11 55		18.5	BLUE STELLAR OBJECT
PHL 3341	01 04 36.	- 20 07		18.5	BLUE STELLAR OBJECT
RNGC 0384	01 04 37.	+ 32 02		14.5	GALAXY
RNGC 0382	01 04 37.	+ 32 08		14.0	GALAXY
RNGC 0383	01 04 37.	+ 32 09		13.5	GALAXY
LIN.CL 077	01 04 37.	- 73 33 26.	342	13.0	STAR CLUSTER IN SMC
IC 1619	01 04 38.	+ 32 49 04.			NONSTELLAR OBJECT
LIN 458	01 04 38.	- 73 13 26.		15.29	SMC EMISSION LINE OBJECT
KN 10.48	01 04 38.7	+ 07 15 10.			NEBULA
MCG+05-03-055	01 04 39.	+ 32 00 30.	24	15.	GALAXY
VV 193A	01 04 39.	+ 32 09	42	13.6	INTERACTING GALAXY
LIN 457	01 04 40.	- 72 43 45.		12.66	SMC EMISSION LINE OBJECT
HSN 23	01 04 41.	- 74 24			BRIGHT GALAXY BEHIND SMC
ZC 0104.7+1149	01 04 42.	+ 11 49	1810		CLUSTER OF GALAXIES
PHL 3342	01 04 42.	+ 13 15		17.7	BLUE STELLAR OBJECT
ZWG 459.002	01 04 42.	+ 16 25		14.5	GALAXY
ZWG 458.020	01 04 42.	+ 16 25		14.5	GALAXY
UGC 00685	01 04 42.	+ 16 25	114	14.5	GALAXY IRR
KARA.73B 0045	01 04 42.	+ 16 25	66	14.5	ISOLATED GALAXY S
ZWG 501.084	01 04 42.	+ 32 01		14.3	GALAXY
ZCG 19.006	01 04 42.	+ 32 01		14.3	ELLIPTICAL GALAXY
UGC 00686	01 04 42.	+ 32 01	66	14.3	GALAXY S0
MCG+05-03-056	01 04 42.	+ 32 02	42	14.5	GALAXY
ZWG 501.085	01 04 42.	+ 32 03		14.3	GALAXY
ZCG 19.005	01 04 42.	+ 32 03		14.3	ELLIPTICAL GALAXY
UGC 00687	01 04 42.	+ 32 03	78	14.3	GALAXY E
MCG+05-03-057	01 04 42.	+ 32 05	18	16.5	GALAXY
ZWG 501.086	01 04 42.	+ 32 08		14.2	GALAXY
ZCG 19.017	01 04 42.	+ 32 08		14.2	ELLIPTICAL GALAXY
UGC 00688	01 04 42.	+ 32 08	15	14.2	GALAXY E?
KARA.72 023A	01 04 42.	+ 32 08	48	14.2	PART OF DOUBLE GALAXY
ZWG 501.087	01 04 42.	+ 32 09		13.6	GALAXY
ZCG 19.003	01 04 42.	+ 32 09		13.6	ELLIPTICAL GALAXY
UGC 00689	01 04 42.	+ 32 09	144	13.6	GALAXY S0

OBJECT NAME	RIGHT ASCEN.	DECLINATION	DIAM.	MAGN.	TYPE OF OBJECT
KARA.72 023B	01 04 42.	+ 32 09	114	13.6	PART OF DOUBLE GALAXY
ZWG 520.013	01 04 42.	+ 36 51		15.7	GALAXY
ZWG 520.014	01 04 42.	+ 39 08		13.8	GALAXY
UGC 00690	01 04 42.	+ 39 08	150	13.8	GALAXY Sc
ZWG 384.076	01 04 42.	- 00 53		15.1	GALAXY
MCG+00-03-074	01 04 42.	- 00 54	54	14.5	GALAXY
PHL 7176	01 04 42.	- 14 02		18.1	BLUE STELLAR OBJECT
PHL 0994	01 04 42.	- 24 22		17.4	BLUE STELLAR OBJECT
LIN 456	01 04 42.	- 72 16 27.		14.46	SMC EMISSION LINE OBJECT
IC 1626	01 04 42.	- 73 34			NONSTELLAR OBJECT
RNGC 0385	01 04 43.	+ 32 03		14.5	GALAXY
RNGC 0386	01 04 43.	+ 32 06		15.5	GALAXY
RNGC 0387	01 04 43.	+ 32 07			GALAXY
HN 0139	01 04 43.	- 73 34			NEBULA
MCG+03-03-017	01 04 45.	+ 16 25	60	14.	GALAXY
MCG-01-03-092	01 04 45.	- 03 23	42	14.5	GALAXY
LIN 459	01 04 45.	- 73 02 21.		14.97	SMC EMISSION LINE OBJECT
PHL 0995	01 04 48.	+ 09 54		14.4	BLUE STELLAR OBJECT
UGC 00691	01 04 48.	+ 31 49	84	16.0	GALAXY S
ZWG 501.088	01 04 48.	+ 32 05		15.4	GALAXY
ZCG 19.018	01 04 48.	+ 32 05		15.4	ELLIPTICAL GALAXY
ZCG 19.004	01 04 48.	+ 32 05		15.4	ELLIPTICAL GALAXY
ZCG 19.019	01 04 48.	+ 32 07		17.2	ELLIPTICAL GALAXY
MCG+05-03-058	01 04 48.	+ 32 39	48	15.	GALAXY
ZWG 501.089	01 04 48.	+ 32 40		15.1	GALAXY
UGC 00692	01 04 48.	+ 32 40	66	15.1	GALAXY SBc
ZWG 520.015	01 04 48.	+ 33 46		15.6	GALAXY
MCG+06-03-011	01 04 48.	+ 39 09	120	13.5	GALAXY
ZWG 551.006	01 04 48.	+ 48 30		15.7	GALAXY
MCG-01-03-093	01 04 48.	- 03 24 30.	24	15.	GALAXY
PHL 3344	01 04 48.	- 15 00		18.4	BLUE STELLAR OBJECT
PHL 7178	01 04 48.	- 17 06		18.3	BLUE STELLAR OBJECT
PHL 7177	01 04 48.	- 19 08		18.2	BLUE STELLAR OBJECT
PHL 3345	01 04 48.	- 30 56		18.1	BLUE STELLAR OBJECT
MCG+00-03-075	01 04 51.	+ 00 38	42	14.5	GALAXY
MCG+06-03-012	01 04 51.	+ 33 45	30	15.	GALAXY
LIN 461	01 04 51.	- 73 00 57.		15.19	SMC EMISSION LINE OBJECT
HN 0008	01 04 52.	+ 00 40			NEBULA
LIN 460	01 04 53.	- 72 24 15.		14.33	SMC EMISSION LINE OBJECT
ZWG 384.077	01 04 54.	+ 00 40		14.6	GALAXY
UGC 00693	01 04 54.	+ 00 40	72	14.6	GALAXY E
MCG+00-03-076	01 04 54.	+ 01 56	42	15.	GALAXY
ZC 0104.9+0715	01 04 54.	+ 07 15	1410		CLUSTER OF GALAXIES
ZWG 520.016	01 04 54.	+ 39 25		15.7	GALAXY
MCG-02-03-073	01 04 54.	- 13 37	30	16.	GALAXY
PHL 3346	01 04 54.	- 16 16		18.6	BLUE STELLAR OBJECT
LB 03177	01 04 54.	- 61 15		13.5	FAINT BLUE STAR
HOW 51	01 04 54.	- 74 54 37.	30		STAR CLUSTER IN SMC
RNGC 0391	01 04 55.	+ 00 40		14.5	GALAXY
MCG-02-03-074	01 04 57.	- 13 29	42	15.5	GALAXY
LIN 462	01 04 58.	- 72 33 09.		14.57	SMC EMISSION LINE OBJECT
KHAV 017	01 05	+ 58 34	4810		DARK NEBULA
KHAV 018	01 05	+ 58 58			DARK NEBULA
PHL 7179	01 05 00.	+ 01 54		15.4	BLUE STELLAR OBJECT
ZWG 385.001	01 05 00.	+ 01 55		15.5	GALAXY
ZWG 384.078	01 05 00.	+ 01 55		15.5	GALAXY
UGC 00694	01 05 00.	+ 01 55	66	15.5	GALAXY SBb
ZC 0105.0+0830	01 05 00.	+ 08 30	740		CLUSTER OF GALAXIES
PHL 7180	01 05 00.	+ 09 47		16.9	BLUE STELLAR OBJECT
PHL 7181	01 05 00.	+ 11 40		18.6	BLUE STELLAR OBJECT
MCG+05-03-059	01 05 00.	+ 32 01 30.	15	16.	GALAXY
ZWG 501.090	01 05 00.	+ 32 02		15.5	GALAXY
ZCG 19.020	01 05 00.	+ 32 02		15.5	ELLIPTICAL GALAXY
ZCG 19.007	01 05 00.	+ 32 02		15.5	ELLIPTICAL GALAXY
5ZW 051	01 05 00.	+ 41 43			COMPACT GALAXY
SN 1955B	01 05 00.	- 13 30		15.8	SUPERNOVA
RNGC 0388	01 05 01.	+ 32 03		15.5	GALAXY
LIN 463	01 05 01.	- 73 31 45.		13.75	SMC EMISSION LINE OBJECT
ARC 0146	01 05 03.	- 11 31		17.6	RICH CLUSTER OF GALAXIES
IC 1622	01 05 05.	- 17 46 27.			NONSTELLAR OBJECT
MCG+00-04-001	01 05 06.	+ 01 51	30	15.5	GALAXY
PHL 3347	01 05 06.	+ 11 58		18.5	BLUE STELLAR OBJECT
ZWG 436.002	01 05 06.	+ 12 02		15.7	GALAXY
ZWG 435.045	01 05 06.	+ 12 02		15.7	GALAXY
ZC 0105.1+2419	01 05 06.	+ 24 19	810		CLUSTER OF GALAXIES
MCG+06-03-013	01 05 06.	+ 33 33	36	15.5	GALAXY
LIN 466	01 05 07.	- 73 26 21.		13.05	SMC EMISSION LINE OBJECT
LIN 464	01 05 11.	- 72 14 33.		14.10	SMC EMISSION LINE OBJECT
LIN 465	01 05 11.	- 72 18 03.		14.18	SMC EMISSION LINE OBJECT
MCG+00-04-003	01 05 12.	+ 00 49	54	15.	GALAXY
MCG+00-04-002	01 05 12.	+ 01 55	33	16.	GALAXY
ZWG 459.003	01 05 12.	+ 16 11		15.5	GALAXY
ZWG 458.021	01 05 12.	+ 16 11		15.5	GALAXY
MCG+03-04-002	01 05 12.	+ 16 12 30.	30	16.5	GALAXY
MCG+03-04-001	01 05 12.	+ 18 30	24	16.	GALAXY
MCG+05-03-060	01 05 12.	+ 33 01	48	16.	GALAXY
ZWG 501.091	01 05 12.	+ 33 02		15.4	GALAXY
OCL 0317	01 05 12.	+ 61 19	840	9.4	OPEN STAR CLUSTER
ZC 0105.2-0007	01 05 12.	- 00 07	1210		CLUSTER OF GALAXIES
MCG-03-04-002	01 05 12.	- 17 30	30	15.5	GALAXY
SB 970	01 05 12.	- 17 44		15.	PEC EMISSION-LINE GALAXY
MCG-03-04-001	01 05 12.	- 17 49	30	14.5	GALAXY
PHL 7182	01 05 12.	- 19 35		18.4	BLUE STELLAR OBJECT
ARP 236	01 05 13.	- 17 44			PECULIAR GALAXY
IC 1623	01 05 13.	- 17 44 28.			NONSTELLAR OBJECT
RNGC 0381	01 05 14.	+ 61 19		9.5	OPEN CLUSTER
LIN.CL 079	01 05 16.	- 72 32 15.	342	12.0	STAR CLUSTER IN SMC
ZWG 385.002	01 05 18.	+ 00 48		15.4	GALAXY
UGC 00695	01 05 18.	+ 00 48	78	15.4	GALAXY PECULR
ZWG 459.004	01 05 18.	+ 20 50		15.0	GALAXY
ZWG 458.022	01 05 18.	+ 20 50		15.0	GALAXY
MRK 561	01 05 18.	+ 20 50	13	15.5	GALAXY WITH UV CONTINUUM
UGC 00696	01 05 18.	+ 20 50	66	15.0	GALAXY E-S0
ZWG 501.092	01 05 18.	+ 31 24		15.6	GALAXY
MCG+05-03-061	01 05 18.	+ 33 10	51	15.	GALAXY
ZWG 501.093	01 05 18.	+ 33 11		14.7	GALAXY
UGC 00697	01 05 18.	+ 33 11	66	14.7	GALAXY SB
ZC 0105.3+4422	01 05 18.	+ 44 22	1880		CLUSTER OF GALAXIES
PHL 7183	01 05 18.	- 01 46		15.3	BLUE STELLAR OBJECT
PHL 0996	01 05 18.	- 17 38		18.3	BLUE STELLAR OBJECT
VV 114B	01 05 18.	- 17 46	24	15.5	INTERACTING GALAXY
VV 114A	01 05 18.	- 17 46	48	14.5	INTERACTING GALAXY
VV 114	01 05 18.	- 17 46	48		INTERACTING GALAXY
MCG-06-03-020	01 05 18.	- 33 55	72	13.	GALAXY
LB 01591	01 05 18.	- 50 27		12.6	FAINT BLUE STAR
LIN.CL 078	01 05 18.	- 71 57 21.	678	13.0	STAR CLUSTER IN SMC
RNGC 0390	01 05 19.	+ 32 11			GALAXY
KN 10.49	01 05 19.8	+ 05 54 56.			NEBULA
ZWG 385.003	01 05 24.	+ 02 17		15.5	GALAXY
UGC 00698	01 05 24.	+ 02 17	90	15.5	GALAXY Sa-b
MCG+00-04-004	01 05 24.	+ 02 17 30.	54	15.5	GALAXY
PHL 7185	01 05 24.	- 14 32		18.3	BLUE STELLAR OBJECT
MCG-03-04-004	01 05 24.	- 17 48	24	15.5	GALAXY
MCG-03-04-003	01 05 24.	- 17 48	48	14.	GALAXY
PHL 7184	01 05 24.	- 20 39		15.4	BLUE STELLAR OBJECT
PHL 7186	01 05 24.	- 22 38		18.0	BLUE STELLAR OBJECT
PHL 3348	01 05 24.	- 25 20		17.1	BLUE STELLAR OBJECT
HN 0138	01 05 26.	- 47 11			NEBULA
IC 1625	01 05 26.	- 47 11			NONSTELLAR OBJECT
HOW 52	01 05 27.	- 73 29 43.	18		STAR CLUSTER IN SMC
HOW 53	01 05 28.	- 73 50 37.	18		STAR CLUSTER IN SMC
UGC 00699	01 05 30.	+ 20 49	60	19.	GALAXY DWRF SP
MCG+03-04-003	01 05 30.	+ 20 52	36	15.	GALAXY
ZC 0105.5+3020	01 05 30.	+ 30 20	940		CLUSTER OF GALAXIES
ZCG 19.021	01 05 30.	+ 32 16		18.8	ELLIPTICAL GALAXY
ZC 0105.5+3650	01 05 30.	+ 36 50	13240		CLUSTER OF GALAXIES
ZWG 551.007	01 05 30.	+ 49 08		15.6	GALAXY
ARC 0148	01 05 32.	- 13 27		17.2	RICH CLUSTER OF GALAXIES
LIN 427	01 05 33.	- 72 43 10.		14.17	SMC EMISSION LINE OBJECT
HOLM 036A	01 05 34.	+ 32 52	18	13.9	PART OF MULTIPLE GALAXY
LIN 467	01 05 34.	- 72 31 40.		14.26	SMC EMISSION LINE OBJECT
ARC 0147	01 05 35.	+ 01 55		15.0	RICH CLUSTER OF GALAXIES
MCG+00-04-005	01 05 36.	+ 01 57	12	16.	GALAXY
PHL 7187	01 05 36.	+ 11 58		18.3	BLUE STELLAR OBJECT
PHL 7190	01 05 36.	+ 13 22		18.5	BLUE STELLAR OBJECT
ZCG 19.022	01 05 36.	+ 32 10		17.8	ELLIPTICAL GALAXY
MCG+05-03-062	01 05 36.	+ 32 51	24	14.5	GALAXY
ZWG 501.094	01 05 36.	+ 32 52		13.9	GALAXY
UGC 00700	01 05 36.	+ 32 52	72	13.9	GALAXY E-S0
HOLM 036B	01 05 36.	+ 32 53	18	14.2	PART OF MULTIPLE GALAXY
MCG+06-03-014	01 05 36.	+ 39 27	42	14.5	GALAXY
MRSL 124+00/1	01 05 36.	+ 62 52	60		HII REGION
PHL 0997	01 05 36.	- 16 35		17.0	BLUE STELLAR OBJECT
PHL 7188	01 05 36.	- 18 18		18.5	BLUE STELLAR OBJECT
PHL 3349	01 05 36.	- 30 06		18.3	BLUE STELLAR OBJECT
PHL 7189	01 05 36.	- 31 25		18.5	BLUE STELLAR OBJECT
LB 03178	01 05 36.	- 65 44		13.2	FAINT BLUE STAR
RNGC 0392	01 05 37.	+ 32 51		14.0	GALAXY
RNGC 0394	01 05 37.	+ 32 52		15.0	GALAXY
MCG+00-04-006	01 05 39.	+ 01 54	9	16.	GALAXY
MCG+05-03-063	01 05 39.	+ 32 52	24	15.	GALAXY
BIGO 459	01 05 40.	+ 32 52			NEBULA
IC 0076	01 05 40.	- 04 49 22.			NONSTELLAR OBJECT
LIN 468	01 05 40.	- 72 36 04.		14.19	SMC EMISSION LINE OBJECT
RNGC 0389	01 05 41.	+ 39 26		15.0	GALAXY
HOW 54	01 05 41.	- 74 20 36.	18		STAR CLUSTER IN SMC
ZWG 385.004	01 05 42.	+ 01 56		15.5	GALAXY
UGC 00701	01 05 42.	+ 01 56	96	15.5	GALAXY DBL SYS
UGC 00702	01 05 42.	+ 08 08	66	18.	GALAXY DBL SYS
PHL 3350	01 05 42.	+ 13 59		18.4	BLUE STELLAR OBJECT
ZWG 459.005	01 05 42.	+ 16 49		15.7	GALAXY
MCG+03-04-004	01 05 42.	+ 16 50	15	16.	GALAXY
ZWG 459.006	01 05 42.	+ 21 28		15.7	GALAXY
ZWG 458.023	01 05 42.	+ 21 28		15.7	GALAXY
MCG+05-03-064	01 05 42.	+ 32 50	12	16.	GALAXY
ZWG 501.095	01 05 42.	+ 32 53		14.8	GALAXY
5ZW 052	01 05 42.	+ 39 23			COMPACT GALAXY
ZWG 520.017	01 05 42.	+ 39 26		15.0	GALAXY
UGC 00703	01 05 42.	+ 39 26	84	15.0	GALAXY S0
MCG+01-04-001	01 05 42.	- 04 48	42	15.5	GALAXY
MCG-03-04-005	01 05 42.	- 16 20	24	16.	GALAXY
MCG-05-03-026	01 05 42.	- 27 53	48	15.5	GALAXY
RNGC 0397	01 05 43.	+ 32 50		15.5	GALAXY
RNGC 0406	01 05 45.	- 70 09		12.5	GALAXY
KN 10.50	01 05 45.1	+ 05 29 40.			NEBULA
RNGC 0393	01 05 47.	+ 39 24		13.5	GALAXY
REIN 2.004	01 05 47.47	+ 39 22 38.2			NEBULA
ZC 0105.8+0352	01 05 48.	+ 03 52	940		CLUSTER OF GALAXIES
UGC 00704	01 05 48.	+ 05 29	66	16.0	GALAXY S
UGC 00705	01 05 48.	+ 06 12	78	16.0	GALAXY SBc
MCG+01-04-001	01 05 48.	+ 06 12 30.	36	15.	GALAXY
ZWG 411.001	01 05 48.	+ 08 04		15.4	GALAXY
UGC 00706	01 05 48.	+ 08 04	66	15.4	GALAXY SBb
PHL 3351	01 05 48.	+ 10 40		18.3	BLUE STELLAR OBJECT
ZWG 501.096	01 05 48.	+ 32 50		15.7	GALAXY
ZWG 520.018	01 05 48.	+ 39 23		13.3	GALAXY
UGC 00707	01 05 48.	+ 39 23	102	13.3	GALAXY E OR S0
MCG+06-03-015	01 05 48.	+ 39 24 30.	33	13.5	GALAXY
ZC 0105.8+3944	01 05 48.	+ 39 44	670		CLUSTER OF GALAXIES
MCG-03-04-006	01 05 48.	- 16 09	30	16.	GALAXY
KN 10.51	01 05 48.4	+ 08 04 55.			NEBULA
KN 10.52	01 05 49.6	+ 06 13 03.			NEBULA
LIN.CL 080	01 05 50.	- 73 01 58.	546	13.7	STAR CLUSTER IN SMC
HOW 55	01 05 51.	- 73 37 36.	36		STAR CLUSTER IN SMC
KN 10.53	01 05 52.0	+ 07 11 49.			NEBULA
SMB 017	01 05 53.4	- 00 53 23.		18.	QUASI-STELLAR OBJECT
ZWG 436.003	01 05 54.	+ 10 46		15.6	GALAXY
ZWG 459.007	01 05 54.	+ 16 37		14.9	GALAXY
UGC 00708	01 05 54.	+ 16 37	60	14.9	GALAXY SB:c
MCG+03-04-005	01 05 54.	+ 16 38	48	15.	GALAXY
ZWG 501.097	01 05 54.	+ 33 17		15.7	GALAXY
5ZW 053	01 05 54.	+ 38 33			COMPACT GALAXY
MCG+06-03-016	01 05 54.	+ 38 35	42	15.	GALAXY
MCG+01-04-002	01 05 54.	- 05 34	60	15.	GALAXY
PHL 0998	01 05 54.	- 13 41		14.9	BLUE STELLAR OBJECT
PHL 3352	01 05 54.	- 18 45		18.2	BLUE STELLAR OBJECT
LB 00080	01 05 54.	- 35 50		13.5	FAINT BLUE STAR
KN 10.54	01 05 56.7	+ 05 47 46.			NEBULA
HOW 56	01 05 59.	- 71 11 54.	12		STAR CLUSTER IN SMC
HMS 0106-1536	01 06	- 15 36			HAUFEN-A GALAXY CLSTR
MCG+00-04-009	01 06 00.	+ 01 04 30.	36	15.5	GALAXY
UGC 00709	01 06 00.	+ 01 05	84	16.0	GALAXY
MCG+00-04-008	01 06 00.	+ 01 23	222	13.	GALAXY
MCG+00-04-007	01 06 00.	+ 02 00	15	15.5	GALAXY
PHL 7191	01 06 00.	+ 12 26		16.6	BLUE STELLAR OBJECT
PHL 0999	01 06 00.	+ 15 00		16.7	BLUE STELLAR OBJECT
ZC 0106.0+2908	01 06 00.	+ 29 08	1880		CLUSTER OF GALAXIES
ZWG 501.098	01 06 00.	+ 33 12		15.6	GALAXY
UGC 00710	01 06 00.	+ 33 12	90	15.6	GALAXY Sb-c
ZWG 520.019	01 06 00.	+ 38 33		15.6	GALAXY
MCG-04-03-054	01 06 00.	- 23 48	48	14.5	GALAXY
LB 03179	01 06 00.	- 67 24		13.6	FAINT BLUE STAR
RNGC 0398	01 06 01.	+ 32 14		15.5	GALAXY
KARA.73 03	01 06 01.	- 38 32	34		DWARF GALAXY
HN 0140	01 06 02.	- 46 32			NEBULA
IC 1627	01 06 02.	- 46 32			NONSTELLAR OBJECT

OBJECT NAME	RIGHT ASCEN.	DECLINATION	DIAM.	MAGN.	TYPE OF OBJECT
LIN 471	01 06 02.	- 72 50 16.		13.67	SMC EMISSION LINE OBJECT
MCG+05-03-066	01 06 03.	+ 31 49	30	15.5	GALAXY
MCG+05-03-065	01 06 03.	+ 32 14	18	16.5	GALAXY
LIN 470	01 06 03.	- 72 35 40.		14.15	SMC EMISSION LINF OBJECT
LIN 469	01 06 04.	- 72 31 22.		13.46	SMC EMISSION LIFE OBJECT
SHB 018	01 06 04.4	+ 01 19 01.		18.4	QUASI-STELLAR OBJECT
BC PKS0106+01	01 06 04.48	+ 01 19 01.4		18.39	QUASI-STELLAR OBJECT
RNGC 0405	01 06 05.	- 46 56			NON-EXISTENT OBJECT
ZWG 385.005	01 06 06.	+ 01 23		14.8	GALAXY
UGC 00711	01 06 06.	+ 01 23	240	14.8	GALAXY Sc
ZWG 385.006	01 06 06.	+ 02 00		15.1	GALAXY
ZC 0106.1+0240	01 06 06.	+ 02 40	1810		CLUSTER OF GALAXIES
ZC 0106.1+1458	01 06 06.	+ 14 58	1140		CLUSTER OF GALAXIES
ZWG 501.099	01 06 06.	+ 31 50		15.0	GALAXY
ZWG 501.100	01 06 06.	+ 32 15		15.4	GALAXY
PHL 7192	01 06 06.	- 21 47		17.8	BLUE STELLAR OBJFCT
TON-S 0192	01 06 06.	- 33 01		12.5	BLUE STAR
LH115-N079	01 06 06.	- 72 51	18		EMISSION NEBULA IN SMC
HN 0141	01 06 07.	- 47 02			NEBULA
HOW 57	01 06 07.	- 72 08 48.	48		STAR CLUSTER IN SMC
IC 1630	01 06 08.	- 47 02			NONSTELLAR OBJFCT
MCG+06-03-017	01 06 09.	+ 37 15	48	15.5	GALAXY
LIN 472	01 06 09.	- 72 37 22.		14.15	SMC EMISSION LINE OBJECT
LIN.CL 081	01 06 10.	- 72 22 46.	342	14.1	STAR CLUSTER IN SMC
PHL 7193	01 06 12.	+ 12 38		17.4	BLUE STELLAR OBJECT
ZWG 459.008	01 06 12.	+ 21 28		15.7	GALAXY
SN 1972K	01 06 12.	+ 31 57		16.0	SUPERNOVA
MCG+05-03-067	01 06 12.	+ 32 21 30.	24	15.	GALAXY
ZWG 501.101	01 06 12.	+ 32 22		14.5	GALAXY
UGC 00712	01 06 12.	+ 32 22	72	14.5	GALAXY SBa
TON-S 0194	01 06 12.	- 27 07		11.5	BLUE STAR
PHL 1000	01 06 12.	- 27 10		12.5	BLUE STELLAR OBJECT
TON-S 0193	01 06 12.	- 32 56		14.5	BLUE STAR
LB 01592	01 06 12.	- 49 14		14.7	FAINT BLUE STAR
LB 01593	01 06 12.	- 54 09		13.8	FAINT BLUE STAR
RNGC 0399	01 06 13.	+ 32 21		14.5	GALAXY
RNGC 0400	01 06 13.	+ 32 28			NON-EXISTENT OBJECT
BIGO 460	01 06 14.	+ 32 22			NEBULA
IC 0077	01 06 15.2	- 15 41 08.			GALAXY SA(r)
LIN.CL 082	01 06 17.	- 72 01 46.	750	11.0	STAR CLUSTER IN SMC
SHAH 037	01 06 18.	+ 13 10	168	17.2	GROUP OF COMPACT GALAXIES
UGC 00713	01 06 18.	+ 37 35	66	16.5	GALAXY
MCG+00-04-010	01 06 18.	- 02 26	42	15.	GALAXY
MCG-02-04-001	01 06 18.	- 12 05	36	15.	GALAXY
MCG-02-04-003	01 06 18.	- 12 53	72	15.	GALAXY
MCG-02-04-002	01 06 18.	- 12 53	9	17.	GALAXY
MCG-03-04-007	01 06 18.	- 15 36	18	15.5	GALAXY
MCG-03-04-009	01 06 18.	- 15 40	18	15.	GALAXY
MCG-03-04-008	01 06 18.	- 15 40	30	15.	GALAXY
MCG-03-04-010	01 06 18.	- 16 07	96	14.	GALAXY
MCG-03-04-011	01 06 18.	- 16 12 30.	30	14.5	GALAXY
PHL 7194	01 06 18.	- 21 08		18.2	BLUE STELLAR OBJFCT
MCG-05-03-027	01 06 18.	- 28 51	36	14.5	GALAXY
MCG-05-03-028	01 06 18.	- 28 52	15	15.	GALAXY
MCG-06-03-021	01 06 18.	- 36 38	36	14.5	GALAXY
RNGC 0401	01 06 19.	+ 32 30			NON-EXISTENT OBJECT
RNGC 0411	01 06 19.	- 72 02		11.0	GLOBULAR CLUSTFR IN SMC
IC 0078	01 06 19.7	- 16 06 39.			GALAXY SA(r)
MCG-03-04-012	01 06 21.	- 15 41	48	15.5	GALAXY
IC 0079	01 06 21.3	- 16 12 58.			GALAXY SA0
IC 0080A	01 06 22.6	- 15 40 32.			GALAXY E2
IC 1628	01 06 23.	- 28 50 30.			NONSTELLAR OBJECT
IC 0080B	01 06 23.0	- 15 40 21.			GALAXY SA0
MCG+00-04-011	01 06 24.	+ 01 05 30.	42	15.	GALAXY
PHL 3353	01 06 24.	+ 09 29		7.2	BLUE STELLAR OBJECT
PHL 3354	01 06 24.	+ 11 46		18.2	BLUE STELLAR OBJECT
MCG+05-03-069	01 06 24.	+ 31 52	60	14.5	GALAXY
MCG+05-03-068	01 06 24.	+ 32 28 30.	90	14.5	GALAXY
ARC 0151	01 06 24.	- 15 41		15.0	RICH CLUSTER OF GALAXIFS
PHL 7195	01 06 24.	- 22 54		18.9	BLUE STELLAR OBJECT
RNGC 0402	01 06 25.	+ 32 33			NON-EXISTENT OBJECT
ZWG 385.007	01 06 30.	+ 01 07		15.1	GALAXY
PHL 3355	01 06 30.	+ 14 20		18.3	BLUF STELLAR OBJECT
ZWG 501.102	01 06 30.	+ 31 42		15.3	GALAXY
ZWG 501.103	01 06 30.	+ 31 53		14.5	GALAXY
UGC 00714	01 06 30.	+ 31 53	84	14.5	GALAXY Sc
ZCG 19.023	01 06 30.	+ 32 20		18.3	ELLIPTICAL GALAXY
ZWG 501.104	01 06 30.	+ 32 29		13.3	GALAXY
UGC 00715	01 06 30.	+ 32 29	120	13.3	GALAXY S0-a
MCG+05-03-070	01 06 30.	+ 33 12	30	17.	GALAXY
VB 032	01 06 30.	+ 69 30	551		STELLAR RING
MCG+01-04-003	01 06 30.	- 05 46	60	15.	GALAXY
RNGC 0403	01 06 31.	+ 32 29		13.5	GALAXY
ARC 0149	01 06 31.	+ 43 21		17.5	RICH CLUSTER OF GALAXIES
HN 0142	01 06 32.	- 46 44			NEBULA
IC 1631	01 06 32.	- 46 44			NONSTELLAR OBJFCT
MCG+05-03-071	01 06 33.	+ 32 26 30.	18	16.	GALAXY
ARC 0150	01 06 36.	+ 12 55		16.6	RICH CLUSTFR OF GALAXIES
ZWG 501.105	01 06 36.	+ 32 28		15.6	GALAXY
ZWG 501.106	01 06 36.	+ 33 13		15.5	GALAXY
MCG+06-03-018	01 06 36.	+ 35 29	138	10.	GALAXY
VB 104	01 06 36.	+ 60 38	322		STELLAR RING
VB 105	01 06 36.	+ 61 28	1074		STELLAR RING
MCG+00-04-012	01 06 36.	- 01 51 30.	30	16.	GALAXY
MCG-03-04-013	01 06 36.	- 16 16	36	14.5	GALAXY
PHL 3357	01 06 36.	- 20 52		18.4	BLUF STELLAR OBJFCT
PHL 3356	01 06 36.	- 23 22		18.6	BLUF STELLAR OBJECT
TON-S 0195	01 06 36.	- 33 26		12.5	BLUF STAR
IC 0082	01 06 37.3	- 16 15 56.			GALAXY SB(r)
RNGC 0416	01 06 38.	- 72 37		12.0	GLOBULAR CLUSTFR IN SMC
MCG+00-04-013	01 06 39.	+ 01 57	36	15.5	GALAXY
LIN.CL 083	01 06 39.	- 72 36 47.	678	12.0	STAR CLUSTER IN SMC
HOW 58	01 06 41.	- 73 57 42.	18		STAR CLUSTER IN SMC
ZWG 436.004	01 06 42.	+ 12 55		15.7	GALAXY
UGC 00716	01 06 42.	+ 12 55	84	15.7	GALAXY
ZWG 436.005	01 06 42.	+ 14 05		14.8	GALAXY
UGC 00717	01 06 42.	+ 14 05	96	14.8	GALAXY SBb
ZWG 520.020	01 06 42.	+ 35 27		11.3	GALAXY
RNGC 0404	01 06 42.	+ 35 27		12.0	GALAXY
UGC 00718	01 06 42.	+ 35 27	360	11.3	GALAXY E-S0
ZC 0106.7+4330	01 06 42.	+ 43 30	200		CLUSTER OF GALAXIES
ZC 0106.7-0112	01 06 42.	- 01 12	1480		CLUSTER OF GALAXIES
PHL 7196	01 06 42.	- 20 58		16.4	BLUE STELLAR OBJECT
PHL 7197	01 06 42.	- 21 00		18.5	BLUE STELLAR OBJECT
MCG-06-03-022	01 06 42.	- 37 35	66	13.	GALAXY
IC 1629	01 06 44.	+ 02 17 59.			NONSTELLAR OBJFCT
MCG+02-04-001	01 06 45.	+ 12 52	24	15.5	GALAXY
MCG+02-04-003	01 06 45.	+ 14 03	84	14.5	GALAXY
VV 348B	01 06 45.	+ 14 04 30.	42	14.5	INTERACTING GALAXY
MCG+02-04-002	01 06 45.	+ 14 07 30.	30	16.	GALAXY
LIN 473	01 06 45.	- 72 33 11.		14.06	SMC EMISSION LINE OBJECT
LIN.CL 086	01 06 47.	- 73 31 05.	342	14.5	STAR CLUSTER IN SMC
ZWG 385.008	01 06 48.	+ 01 57		15.4	GALAXY
UGC 00720	01 06 48.	+ 01 57	84	15.4	GALAXY DBL SYS
ZWG 385.009	01 06 48.	+ 02 18		15.7	GALAXY
MCG+02-04-004	01 06 48.	+ 14 02	24	16.	GALAXY
ZWG 436.006	01 06 48.	+ 14 06		15.2	GALAXY
UGC 00719	01 06 48.	+ 14 06	60	15.2	GALAXY SBb
ZWG 459.009	01 06 48.	+ 16 41		15.2	GALAXY
ZWG 459.010	01 06 48.	+ 19 35		15.1	GALAXY
ZWG 501.107	01 06 48.	+ 31 54		15.6	GALAXY
ZC 0106.8+4320	01 06 48.	+ 43 20	740		CLUSTER OF GALAXIES
PHL 7198	01 06 48.	- 32 50		18.1	BLUE STELLAR OBJFCT
LIN.CL 085	01 06 48.	- 73 08 41.	1362	10.2	STAR CLUSTER IN SMC
IC 0081	01 06 49.	- 01 57 13.			NONSTELLAR OBJECT
MCG+00-04-014	01 06 51.	+ 01 32	36	15.	GALAXY
MCG+02-04-005	01 06 51.	+ 14 04	36	15.	GALAXY
ARP 011	01 06 51.	+ 14 05			PECULIAR GALAXY
VV 348A	01 06 51.	+ 14 06	90	14.	INTFRACTING GALAXY
LIN.CL 084	01 06 52.	- 72 15 11.	678		STAR CLUSTER IN SMC
KN 10.55	01 06 53.2	+ 05 49 03.			NEBULA
ZC 0106.9+0028	01 06 54.	+ 00 28	11960		CLUSTER OF GALAXIES
ZWG 411.002	01 06 54.	+ 05 48		15.6	GALAXY
ZWG 480.045	01 06 54.	+ 23 02		15.7	GALAXY
MCG+05-03-072	01 06 54.	+ 32 10	30	15.5	GALAXY
DG 007	01 06 54.	+ 35 20			REFLECTION NFBULA
ZWG 385.010	01 06 54.	- 01 57		14.8	GALAXY
MCG+00-04-015	01 06 54.	- 01 59	54	14.	GALAXY
MCG+01-04-004	01 06 54.	- 03 06	54	16.	GALAXY
LH115-N080A	01 06 54.	- 72 16	32		EMISSION NEBULA IN SMC
LH115-N080	01 06 54.	- 72 16	198		EMISSION NEBULA IN SMC
RNGC 0419	01 06 55.	- 72 09		10.0	GLOBULAR CLUSTFR IN SMC
RNGC 0413	01 06 56.	- 03 06		16.0	GALAXY
LIN 477	01 06 56.	- 72 39 47.		13.52	SMC EMISSION LINE OBJECT
KN 10.56	01 06 56.3	+ 07 47 21.			NEBULA
MCG+05-03-073	01 06 57.	+ 30 51	24	16.	GALAXY
LIN 474	01 06 58.	- 72 14 05.		16.25	SMC EMISSION LINF OBJECT
LIN 475	01 06 58.	- 72 15 23.		15.82	SMC EMISSION LINE OBJECT
LIN 476	01 06 58.	- 72 15 41.		14.69	SMC EMISSION LINF OBJECT
VDB.66G 009	01 07	+ 49 20	100		DWARF GALAXY
MCG+00-04-016	01 07 00.	+ 00 52	24	16.	GALAXY
ZWG 385.011	01 07 00.	+ 01 34		15.6	GALAXY
UGC 00721	01 07 00.	+ 01 34	60	15.6	GALAXY S
ZC 0107.0+0459	01 07 00.	+ 04 59	1410		CLUSTER OF GALAXIES
4ZW 010	01 07 00.	+ 27 32			COMPACT GALAXY
ZWG 501.108	01 07 00.	+ 30 51		15.4	GALAXY
ZWG 501.109	01 07 00.	+ 32 10		14.9	GALAXY
NAB 0107-15	01 07 02.	- 15 37 40.		17.5	QUASI-STELLAR OBJECT
SHB 019	01 07 02.	- 15 37 40.		17.5	QUASI-STELLAR OBJECT
UGC 00722	01 07 06.	+ 13 02	120	16.0	GALAXY
MCG+05-03-074	01 07 06.	+ 30 10	15	15.5	GALAXY
PHL 3358	01 07 06.	- 24 06		18.5	BLUE STELLAR OBJFCT
ARC 0153	01 07 07.	+ 04 58		17.6	RICH CLUSTER OF GALAXIES
SN 1961R	01 07 11.	+ 32 38		17.0	SUPERNOVA
LIN 478	01 07 11.	- 73 31 06.		12.80	SMC EMISSION LINF OBJECT
MCG+02-04-006	01 07 12.	+ 13 00	108	16.	GALAXY
ARC 0152	01 07 12.	+ 13 43		17.2	RICH CLUSTER OF GALAXIES
ZWG 459.011	01 07 12.	+ 20 30		15.7	GALAXY
UGC 00723	01 07 12.	+ 20 30	108	15.7	GALAXY Sc
ZWG 501.110	01 07 12.	+ 30 12		15.7	GALAXY
ZWG 501.111	01 07 12.	+ 32 05		14.0	GALAXY
UGC 00724	01 07 12.	+ 32 05	138	14.0	GALAXY S
MCG+05-03-075	01 07 12.	+ 32 05	27	14.5	GALAXY
ZWG 501.112	01 07 12.	+ 32 38		15.5	GALAXY
MCG+05-03-076	01 07 12.	+ 32 39	36	16.	GALAXY
ZWG 551.008	01 07 12.	+ 49 55		15.7	GALAXY
MCG+01-04-005	01 07 12.	- 02 30	72	15.5	GALAXY
MCG-06-03-023	01 07 12.	- 36 06	60	12.	GALAXY
SN 1961M	01 07 13.	+ 32 06		17.0	SUPFRNOVA
MCG+07-03-019	01 07 15.	+ 42 50 30.	138	14.	GALAXY
ZC 0107.3+1558	01 07 18.	+ 15 58	940		CLUSTER OF GALAXIES
MCG+03-04-006	01 07 18.	+ 20 32	78	15.	GALAXY
ZWG 536.019	01 07 18.	+ 42 50		14.5	GALAXY
UGC 00725	01 07 18.	+ 42 50	132	14.5	GALAXY SB?c
KARA.72 024A	01 07 18.	+ 42 50	108	14.5	PART OF DOUBLF GALAXY
LB 03180	01 07 18.	- 76 19		13.3	FAINT BLUF STAF
RNGC 0409	01 07 20.	- 36 02			GALAXY
LIN 479	01 07 20.	- 72 39 36.		14.58	SMC EMISSION LINF OBJFCT
MCG+02-04-007	01 07 24.	+ 13 41	24	16.	GALAXY
UGC 00727	01 07 24.	+ 13 43	96	16.5	GALAXY MLT SYS
ZWG 480.046	01 07 24.	+ 22 41		15.6	GALAXY
KARA.73B 0046	01 07 24.	+ 22 41	36	15.4	ISOLATFD GALAXY S0
ZC 0107.4+3156	01 07 24.	+ 31 56	1280		CLUSTER OF GALAXIES
5ZW 065	01 07 24.	+ 40 18			COMPACT GALAXY
ZWG 385.012	01 07 24.	- 02 01		15.1	GALAXY
UGC 00726	01 07 24.	- 02 01	126	15.1	GALAXY DISTRBD
PHL 3359	01 07 24.	- 21 46		17.6	BLUF STELLAR OBJFCT
PHL 3360	01 07 24.	- 25 10		17.9	BLUE STELLAR OBJECT
HOW 59	01 07 28.	- 73 30 05.	12		STAR CLUSTER IN SMC
ZWG 385.013	01 07 30.	+ 00 09		15.0	GALAXY
MCG+00-04-019	01 07 30.	+ 00 09	30	15.	GALAXY
ZC 0107.5+0743	01 07 30.	+ 07 43	1280		CLUSTER OF GALAXIES
ZC 0107.5+3212	01 07 30.	+ 32 12	21500		CLUSTER OF GALAXIES
ZC 0107.5+4200	01 07 30.	+ 42 00	810		CLUSTER OF GALAXIES
MCG+00-04-017	01 07 30.	- 01 52	24	16.	GALAXY
MCG+00-04-018	01 07 30.	- 02 01	78	14.	GALAXY
MCG-02-04-004	01 07 30.	- 14 30 30.	36	15.	GALAXY
IC 1633	01 07 32.	- 46 13 51.			NONSTELLAR OBJFCT
HSN 24	01 07 32.	- 72 45			BRIGHT GALAXY BFHIND SMC
ZWG 501.113	01 07 36.	+ 29 55		15.5	GALAXY
ZWG 501.114	01 07 36.	+ 32 07		15.5	GALAXY
ZWG 536.020	01 07 36.	+ 43 01		14.2	GALAXY
UGC 00728	01 07 36.	+ 43 01	90	14.2	GALAXY Sc/SBc
KARA.72 024B	01 07 36.	+ 43 01	84	14.2	PART OF DOUBLF GALAXY
MCG+07-03-020	01 07 36.	+ 43 01	90	13.5	GALAXY
LIN 480	01 07 39.	- 72 23 18.		14.23	SMC EMISSION LINE OBJECT
HN 0083	01 07 40.	- 73 28			NEBULA
IC 1644	01 07 41.	- 73 28			NONSTELLAR OBJFCT
MCG-03-04-014	01 07 42.	- 17 07 30.	21	15.	GALAXY
MCG-03-04-015	01 07 42.	- 20 01	24	15.5	GALAXY
MCG-06-03-024	01 07 42.	- 35 47	48	13.	GALAXY
RNGC 0396	01 07 43.	+ 03 16		16.0	GALAXY
LIN 481	01 07 44.	- 72 27 54.		10.38	SMC FMISSION LINF OFJECT
MCG-03-04-016	01 07 45.	- 20 01	36	15.5	GALAXY
IC 1641	01 07 45.	- 72 02			SAME AS NGC 422
HN 0143	01 07 46.	- 72 02			NEBULA

OBJECT NAME	RIGHT ASCEN.	DECLINATION	DIAM.	MAGN.	TYPE OF OBJECT
ZWG 385.014	01 07 48.	+ 03 15		15.4	GALAXY
UGC 00729	01 07 48.	+ 03 15	66	15.4	GALAXY COMPACT
MCG+00-04-020	01 07 48.	+ 03 15	30	16.	GALAXY
ZC 0107.8+0439	01 07 48.	+ 04 39	670		CLUSTER OF GALAXIES
MCG+05-03-077	01 07 48.	+ 32 50 30.	90	14.5	GALAXY
ZWG 501.115	01 07 48.	+ 32 51		14.3	GALAXY
UGC 00730	01 07 48.	+ 32 51	120	14.3	GALAXY S0-a
ZC 0107.8+4342	01 07 48.	+ 43 42	1880		CLUSTER OF GALAXIES
UGC 00731	01 07 48.	+ 49 21	144	17.	GALAXY DWRF IR
MCG-02-04-005	01 07 48.	- 09 50 30.	36	14.5	GALAXY
LH115-N081	01 07 48.	- 73 28	30		EMISSION NEBULA IN SMC
RNGC 0407	01 07 49.	+ 32 51		14.5	GALAXY
RNGC 0415	01 07 50.	- 35 45			GALAXY
IC 0083	01 07 52.	+ 01 26 33.			NONSTELLAR OBJECT
LIN.CL 087	01 07 52.	- 72 01 24.	270	12.7	STAR CLUSTER IN SMC
RNGC 0412	01 07 53.	- 20 17			NON-EXISTENT OBJECT
ARC 0155	01 07 53.	- 25 06		17.5	RICH CLUSTER OF GALAXIES
MCG+00-04-021	01 07 54.	+ 01 25	30	15.	GALAXY
3ZW 021	01 07 54.	+ 10 23			COMPACT GALAXY
SN 1971O	01 07 54.	+ 31 18		18.5	SUPERNOVA
ZWG 501.116	01 07 54.	+ 33 18		14.7	GALAXY
UGC 00732	01 07 54.	+ 33 18	102	14.7	GALAXY Sc
ZC 0107.9+4017	01 07 54.	+ 40 17	2820		CLUSTER OF GALAXIES
MCG-03-04-017	01 07 54.	- 20 02 30.	36	15.5	GALAXY
RNGC 0422	01 07 55.	- 72 01		12.5	OPEN CLUSTER IN SMC
LIN 483	01 07 55.	- 72 40 13.		14.21	SMC EMISSION LINE OBJECT
LIN 482	01 07 56.	- 72 26 25.		15.09	SMC EMISSION LINE OBJECT
MCG+05-03-078	01 07 57.	+ 33 17	66	15.	GALAXY
KHAV 019	01 08	+ 76 28	9460		DARK NEBULA
LB 09775	01 08	- 85 20		12.5	FAINT BLUE STAR
ZWG 385.015	01 08 00.	+ 01 25		15.2	GALAXY
ZWG 459.012	01 08 00.	+ 16 20		15.7	GALAXY
UGC 00733	01 08 00.	+ 16 20	60	15.7	GALAXY SBb
MCG+03-04-007	01 08 00.	+ 16 20	30	15.	GALAXY
ZWG 536.021	01 08 00.	+ 42 31		15.7	GALAXY
MCG-02-04-006	01 08 00.	- 14 15 30.	36	15.	GALAXY
TON-S 0196	01 08 00.	- 26 39		15.5	BLUE STAR
PHL 7199	01 08 00.	- 26 42		16.2	BLUE STELLAR OBJECT
MCG-05-04-001	01 08 00.	- 31 08	72	15.5	GALAXY
MCG-06-03-025	01 08 00.	- 36 02	72	13.5	GALAXY
LIN 484	01 08 01.	- 72 38 55.		14.36	SMC EMISSION LINE OBJECT
HOW 60	01 08 02.	- 72 37 11.	12		STAR CLUSTER IN SMC
IC 1632	01 08 03.	+ 17 24 39.			NONSTELLAR OBJECT
MCG+05-03-079	01 08 03.	+ 32 19	30	16.	GALAXY
KW 10.57	01 08 04.1	+ 08 12 24.			NEBULA
ZWG 480.047	01 08 06.	+ 24 11		15.4	GALAXY
ZWG 501.117	01 08 06.	+ 32 19		15.4	GALAXY
ZWG 385.016	01 08 06.	- 00 32		15.6	GALAXY
UGC 00734	01 08 06.	- 00 32	72	15.6	GALAXY S
MCG+00-04-022	01 08 06.	- 00 32 30.	54	15.	GALAXY
MCG+01-04-006	01 08 06.	- 03 29	72	15.	GALAXY
MCG+01-04-007	01 08 06.	- 07 52	60	14.	GALAXY
PHL 7200	01 08 06.	- 25 58		14.0	BLUE STELLAR OBJECT
PHL 3361	01 08 06.	- 26 22		12.5	BLUE STELLAR OBJECT
RNGC 0408	01 08 07.	+ 32 50			GALAXY
HOW 61	01 08 08.	- 72 33 35.	18		STAR CLUSTER IN SMC
MCG+00-04-023	01 08 09.	+ 03 21	36	16.	GALAXY
MCG+05-03-080	01 08 09.	+ 32 52	42	13.5	GALAXY
3ZW 022	01 08 12.	+ 08 03			COMPACT GALAXY
ZC 0108.2+1008	01 08 12.	+ 10 08	2080		CLUSTER OF GALAXIES
ZC 0108.2+1726	01 08 12.	+ 17 26	3830		CLUSTER OF GALAXIES
ZWG 501.118	01 08 12.	+ 32 53		12.6	GALAXY
UGC 00735	01 08 12.	+ 32 53	144	12.6	GALAXY E
MCG-02-04-007	01 08 12.	- 12 19	36	15.	GALAXY
MCG-02-04-008	01 08 12.	- 13 55	48	15.	GALAXY
PHL 3362	01 08 12.	- 25 50		18.3	BLUE STELLAR OBJECT
MCG-05-04-002	01 08 12.	- 30 29	120	13.	GALAXY
RNGC 0410	01 08 13.	+ 32 53		12.5	GALAXY
RNGC 0418	01 08 13.	- 30 29		13.0	GALAXY
HOW 62	01 08 13.	- 72 01 41.	18		STAR CLUSTER IN SMC
SN 1967J	01 08 17.	+ 32 58		17.5	SUPERNOVA
SHAH 038	01 08 18.	+ 08 03	66	17.4	GROUP OF COMPACT GALAXIES
ZC 0108.3+1637	01 08 18.	+ 16 37	2290		CLUSTER OF GALAXIES
ZWG 501.119	01 08 18.	+ 32 58		15.7	GALAXY
ZWG 520.021	01 08 18.	+ 33 35		15.5	GALAXY
UGC 00738	01 08 18.	+ 33 35	78	15.5	GALAXY
SN 1966L	01 08 18.	+ 33 35		17.6	SUPERNOVA
5ZW 054	01 08 18.	+ 43 46			COMPACT GALAXY
ZWG 385.018	01 08 18.	- 00 04		15.7	GALAXY
UGC 00737	01 08 18.	- 00 04	78	15.7	GALAXY S
MCG+00-04-024	01 08 18.	- 00 05	27	15.	GALAXY
ZWG 385.017	01 08 18.	- 02 01		15.5	GALAXY
UGC 00736	01 08 18.	- 02 01	84	15.5	GALAXY Sc
MCG+00-04-025	01 08 18.	- 02 02 30.	66	15.	GALAXY
MCG+00-04-026	01 08 21.	+ 00 21 30.	24	15.5	GALAXY
ARC 0154	01 08 21.	+ 17 24		15.6	RICH CLUSTER OF GALAXIES
IC 1634	01 08 23.	+ 17 23 26.			NONSTELLAR OBJECT
ZWG 385.019	01 08 24.	+ 00 22		15.7	GALAXY
ZWG 411.003	01 08 24.	+ 08 18		15.7	GALAXY
ZWG 436.007	01 08 24.	+ 14 55		15.6	GALAXY
ZWG 459.013	01 08 24.	+ 17 22		15.7	GALAXY
UGC 00739	01 08 24.	+ 17 22	60	15.7	GALAXY MLT SYS
IC 1635	01 08 24.	+ 17 22 44.			NONSTELLAR OBJECT
ZWG 459.014	01 08 24.	+ 17 23		15.7	GALAXY
UGC 00740	01 08 24.	+ 17 23	60	15.7	GALAXY MLT SYS
MCG+03-04-009	01 08 24.	+ 17 23	18	16.	GALAXY
MCG+03-04-008	01 08 24.	+ 17 24	24	16.	GALAXY
ZWG 501.120	01 08 24.	+ 32 26		15.3	GALAXY
MCG+06-03-019	01 08 24.	+ 33 34	66	15.	GALAXY
ZWG 536.022	01 08 24.	+ 40 16		15.7	GALAXY
VB 033	01 08 24.	+ 65 36	1074		STELLAR RING
MCG+01-04-008	01 08 24.	- 06 04	132	14.	GALAXY
TON-S 0197	01 08 24.	- 27 02		15.9	BLUE STAR
PHL 7201	01 08 24.	- 28 00		15.9	BLUE STELLAR OBJECT
MCG+05-03-081	01 08 27.	+ 31 27	54	15.	GALAXY
MCG-02-04-009	01 08 27.	- 13 56	48	15.	GALAXY
MCG-03-04-018	01 08 27.	- 17 28	18	15.	GALAXY
ZWG 411.004	01 08 30.	+ 04 27		15.7	GALAXY
ZWG 411.005	01 08 30.	+ 08 30		15.5	GALAXY
UGC 00741	01 08 30.	+ 08 30	66	15.5	GALAXY E
ZWG 501.121	01 08 30.	+ 31 28		15.4	GALAXY
UGC 00742	01 08 30.	+ 31 28	78	15.4	GALAXY
ZWG 501.122	01 08 30.	+ 31 37		14.8	GALAXY
UGC 00743	01 08 30.	+ 31 37	90	14.8	GALAXY Sa
ZWG 501.123	01 08 30.	+ 32 50		14.5	GALAXY
UGC 00744	01 08 30.	+ 32 50	48	14.5	GALAXY DBL SYS
KARA.72 025B	01 08 30.	+ 32 50	24		PART OF DOUBLE GALAXY
KARA.72 025A	01 08 30.	+ 32 50	36	14.5	PART OF DOUBLE GALAXY
4ZW 039	01 08 30.	+ 32 51			COMPACT GALAXY
MCG+07-03-021	01 08 30.	+ 40 15 30.	18	16.	GALAXY
PHL 1001	01 08 30.	- 25 32		18.5	BLUE STELLAR OBJECT
RNGC 0414	01 08 31.	+ 32 50		14.5	GALAXY
KW 10.58	01 08 32.1	+ 08 30 43.			NEBULA
LIN 485	01 08 35.	- 73 00 55.		16.38	SMC EMISSION LINE OBJECT
LIN 486	01 08 35.	- 73 02 25.		16.50	SMC EMISSION LINE OBJECT
MCG+03-04-010	01 08 36.	+ 17 16	30	17.5	GALAXY
ZC 0108.6+1834	01 08 36.	+ 18 34	1610		CLUSTER OF GALAXIES
ZC 0108.6+2627	01 08 36.	+ 26 27	2820		CLUSTER OF GALAXIES
ZWG 501.124	01 08 36.	+ 27 39		15.5	GALAXY
VE 106	01 08 36.	+ 58 47	195		STELLAR RING
NAB 0108-14	01 08 36.	- 14 26 55.		17.5	QUASI-STELLAR OBJECT
SHB 020	01 08 36.	- 14 26 55.		17.5	QUASI-STELLAR OBJECT
ARC 0157	01 08 36.	- 14 41		16.9	RICH CLUSTER OF GALAXIES
MCG-05-04-003	01 08 36.	- 30 41	96	13.5	GALAXY
MCG-02-04-010	01 08 39.	- 14 14 30.	36	15.	GALAXY
RNGC 0417	01 08 40.	- 18 26		15.0	GALAXY
HN 0144	01 08 40.	- 30 42			NEBULA
IC 1637	01 08 40.	- 30 42			NONSTELLAR OBJECT
4ZW 040	01 08 42.	+ 29 51			COMPACT GALAXY
UGC 00745	01 08 42.	+ 36 25	60	16.5	GALAXY Sc
ZWG 551.009	01 08 42.	+ 48 51		14.8	GALAXY
UGC 00746	01 08 42.	+ 48 51	96	14.8	GALAXY E
MCG+08-03-008	01 08 42.	+ 48 51	30	14.	GALAXY
MCG+00-04-027	01 08 42.	- 00 44	30	16.	GALAXY
MCG-03-04-020	01 08 42.	- 18 25 30.	7	15.5	GALAXY
MCG-03-04-019	01 08 42.	- 18 26	24	15.	GALAXY
MCG-03-04-021	01 08 42.	- 19 48	30	15.	GALAXY
MCG+00-04-028	01 08 48.	+ 01 00	42	15.5	GALAXY
ZWG 385.020	01 08 48.	+ 01 01		15.7	GALAXY
UGC 00747	01 08 48.	+ 01 01	60	15.7	GALAXY
KARA.72 026A	01 08 48.	+ 01 01	48	15.7	PART OF DOUBLE GALAXY
IC 0084	01 08 48.	+ 01 23 31.			NONSTELLAR OBJECT
ZWG 480.048	01 08 48.	+ 22 57		15.5	GALAXY
ZWG 502.001	01 08 48.	+ 33 06		14.9	GALAXY
ZWG 501.125	01 08 48.	+ 33 06		14.9	GALAXY
MCG+01-04-009	01 08 49.	- 07 48	60	13.5	GALAXY
MCG-03-04-022	01 08 48.	- 17 20	36	15.5	GALAXY
PHL 7202	01 08 48.	- 21 55		12.5	BLUE STELLAR OBJECT
TON-S 0198	01 08 48.	- 28 31		15.4	BLUE STAR
LIN 487	01 08 48.	- 72 40 20.		14.67	SMC EMISSION LINE OBJECT
IC 1636	01 08 50.	+ 33 05 25.			NONSTELLAR OBJECT
MCG+00-04-029	01 08 51.	+ 01 21	36	15.	GALAXY
LIN 488	01 08 53.	- 73 02 38.		16.22	SMC EMISSION LINE OBJECT
HOW 63	01 08 53.	- 73 27 52.	18		STAR CLUSTER IN SMC
ZWG 385.021	01 08 54.	+ 01 23		15.0	GALAXY
ZC 0108.9+1128	01 08 54.	+ 11 28	1210		CLUSTER OF GALAXIES
ZC 0108.9+3300	01 08 54.	+ 33 00	2290		CLUSTER OF GALAXIES
ZWG 520.022	01 08 54.	+ 35 00		15.3	GALAXY
UGC 00748	01 08 54.	+ 35 00	120	15.3	GALAXY S
4ZW 041	01 08 54.	+ 35 26			COMPACT GALAXY
ZC 0108.9+3912	01 08 54.	+ 39 12	1410		CLUSTER OF GALAXIES
MCG+01-04-010	01 08 54.	- 07 04	45	14.5	GALAXY
LB 03181	01 08 54.	- 71 19		13.8	FAINT BLUE STAR
ARC 0156	01 08 57.	+ 33 11		16.9	RICH CLUSTER OF GALAXIES
MCG-03-04-023	01 08 57.	- 17 20	42	15.	GALAXY
SHB 021	01 09	+ 35			QUASI-STELLAR OBJECT
MCG+00-04-030	01 09 00.	+ 01 03	30	14.	GALAXY
ZWG 385.022	01 09 00.	+ 01 04		14.2	GALAXY
UGC 00749	01 09 00.	+ 01 04	78	14.2	GALAXY S-IRR
KARA.72 026B	01 09 00.	+ 01 04	66	14.2	PART OF DOUBLE GALAXY
MCG+06-03-020	01 09 00.	+ 35 00 30.	138	14.5	GALAXY
ZWG 361.001	01 09 00.	+ 84 54		15.6	GALAXY
ZWG 360.004	01 09 00.	+ 84 54		15.6	GALAXY
MCG-03-04-024	01 09 00.	- 18 30	30	16.	GALAXY
MCG-03-04-025	01 09 00.	- 18 34	24	15.5	GALAXY
MCG-05-04-004	01 09 00.	- 29 29	48	14.	GALAXY
MCG-06-03-026	01 09 00.	- 38 23	72	12.	GALAXY
LB 03189	01 09 00.	- 71 56		13.0	FAINT BLUE STAR
RNGC 0423	01 09 01.	- 29 29		14.0	GALAXY
RNGC 0424	01 09 03.	- 38 23			GALAXY
ZC 0109.1+0815	01 09 06.	+ 08 15	810		CLUSTER OF GALAXIES
ZC 0109.1+2415	01 09 06.	+ 24 15	2150		CLUSTER OF GALAXIES
ZWG 502.002	01 09 06.	+ 31 18		15.7	GALAXY
ZWG 501.126	01 09 06.	+ 31 18		15.7	GALAXY
ZC 0109.1+4432	01 09 06.	+ 44 32	2420		CLUSTER OF GALAXIES
OCL 0318	01 09 06.	+ 62 04	120	11.	OPEN STAR CLUSTER
VB 034	01 09 06.	+ 63 22	329		STELLAR RING
MCG-03-04-026	01 09 06.	- 17 20	24	15.5	GALAXY
ARC 0158	01 09 09.	+ 16 37		15.9	RICH CLUSTER OF GALAXIES
BC PKS0109+17	01 09 09.2	+ 17 37 55.		18.	QUASI-STELLAR OBJECT
ZC 0109.2+0445	01 09 12.	+ 04 45	1080		CLUSTER OF GALAXIES
ZC 0109.2+0947	01 09 12.	+ 09 47	670		CLUSTER OF GALAXIES
VB 035	01 09 12.	+ 66 58	316		STELLAR RING
ZWG 385.023	01 09 12.	- 00 55		14.2	GALAXY
MRK 563	01 09 12.	- 00 55	20	14.	GALAXY WITH UV CONTINUUM
UGC 00750	01 09 12.	- 00 55	54	14.2	GALAXY
IC 1639	01 09 12.	- 00 55		14.2	GALAXY
MCG+00-04-031	01 09 12.	- 00 56 30.	21	15.5	GALAXY
MCG+01-04-011	01 09 12.	- 03 35	24	15.	GALAXY
MCG-02-04-011	01 09 12.	- 14 20	36	15.5	GALAXY
LB 03183	01 09 12.	- 61 33		14.5	FAINT BLUE STAR
LIN 489	01 09 12.	- 72 37 26.		14.34	SMC EMISSION LINE OBJECT
IC 0085	01 09 13.	- 00 44			NONSTELLAR OBJECT
HOW 64	01 09 13.	- 71 36 16.	18		STAR CLUSTER IN SMC
MCG+00-04-032	01 09 15.	+ 01 47	36	15.	GALAXY
LIN.CL 088	01 09 16.	- 73 03 02.	204	13.7	STAR CLUSTER IN SMC
ZC 0109.3+0927	01 09 18.	+ 09 27	1340		CLUSTER OF GALAXIES
UGC 00751	01 09 18.	+ 17 02	90	16.0	GALAXY
MCG+03-04-011	01 09 18.	+ 17 02 30.	30	15.	GALAXY
MCG+05-03-083	01 09 18.	+ 31 50	30	14.	GALAXY
ZWG 502.003	01 09 18.	+ 31 52		13.4	GALAXY
ZWG 501.127	01 09 18.	+ 31 52		13.4	GALAXY
UGC 00752	01 09 18.	+ 31 52	120	13.4	GALAXY S0
5ZW 055	01 09 18.	+ 41 32			COMPACT GALAXY
ZWG 385.024	01 09 18.	- 00 53		15.1	GALAXY
IC 1640	01 09 18.	- 00 53		15.1	GALAXY
MRK 563	01 09 18.	- 01 55	18	15.5	GALAXY WITH UV CONTINUUM
PHL 1002	01 09 18.	- 22 42		18.1	BLUE STELLAR OBJECT
RNGC 0420	01 09 19.	+ 31 50		13.5	GALAXY
RNGC 0421	01 09 19.	+ 31 53			NON-EXISTENT OBJECT
ZC 0109.4+2724	01 09 24.	+ 27 24	2290		CLUSTER OF GALAXIES
ZWG 502.004	01 09 24.	+ 31 45		15.3	GALAXY
ZWG 501.128	01 09 24.	+ 31 45		15.3	GALAXY
MCG-05-04-005	01 09 24.	- 27 56	48	15.	GALAXY
LB 03184	01 09 24.	- 77 39		12.9	FAINT BLUE STAR
ZWG 436.008	01 09 30.	+ 12 00		15.5	GALAXY

OBJECT NAME	RIGHT ASCEN.	DECLINATION	DIAM.	MAGN.	TYPE OF OBJECT
5ZW 056	01 09 30.	+ 43 02			COMPACT GALAXY
ISS 0009	01 09 30.	+ 67 00	314		STELLAR RING
UGC 00753	01 09 30.	- 00 31	72	16.5	GALAXY
MCG+01-04-012	01 09 30.	- 06 00	48	15.5	GALAXY
ARC 0159	01 09 30.	- 15 23		17.2	RICH CLUSTER OF GALAXIES
TON-S 0199	01 09 30.	- 27 42		16.1	BLUE STAR
MCG+05-03-082	01 09 33.	+ 33 05	21	15.5	GALAXY
IC 1638	01 09 34.	+ 33 05 59.			NONSTELLAR OBJECT
ZWG 502.005	01 09 36.	+ 33 07		14.9	GALAXY
ZWG 501.129	01 09 36.	+ 33 07		14.9	GALAXY
UGC 00754	01 09 36.	+ 50 21	78	16.0	GALAXY SB0
ZWG 385.025	01 09 36.	- 00 40		15.4	GALAXY
IC 1643	01 09 36.	- 00 40		15.4	GALAXY
PHL 3363	01 09 36.	- 24 46		18.3	BLUE STELLAR OBJECT
TON-S 0200	01 09 36.	- 29 28		15.3	BLUE STAR
LB 01594	01 09 36.	- 58 16		14.3	FAINT BLUE STAR
LIN 490	01 09 38.	- 73 34 21.		13.41	SMC EMISSION LINE OBJECT
ZWG 459.015	01 09 42.	+ 16 15		15.2	GALAXY
MCG+00-04-033	01 09 42.	- 00 41	42	14.5	GALAXY
TON-S 0201	01 09 42.	- 26 29		11.3	BLUE STAR
IC 1642	01 09 43.	+ 15 29 35.			NONSTELLAR OBJECT
HOLM 037D	01 09 44.	- 04 24	48	14.4	PART OF MULTIPLE GALAXY
MCG+02-04-008	01 09 45.	+ 15 28	24	16.	GALAXY
MCG+08-03-009	01 09 45.	+ 50 20	60	16.	GALAXY
ZWG 436.009	01 09 48.	+ 15 29		15.5	GALAXY
ZWG 459.016	01 09 48.	+ 16 16		15.0	GALAXY
ZWG 520.023	01 09 48.	+ 38 14		14.6	GALAXY
UGC 00755	01 09 48.	+ 38 14	90	14.6	GALAXY Sc/SBc
MCG+06-03-021	01 09 48.	+ 38 16	72	14.5	GALAXY
PHL 1003	01 09 48.	- 26 31		12.0	BLUE STELLAR OBJECT
LIN 491	01 09 48.	- 72 38 39.		13.48	SMC EMISSION LINE OBJECT
IC 1645	01 09 49.	+ 15 28 35.			NONSTELLAR OBJECT
RNGC 0432	01 09 50.	- 61 52			UNVERIFIED SOUTHERN OBJECT
HN 0145	01 09 52.	- 56 08			NEBULA
IC 1649	01 09 52.	- 56 08			NONSTELLAR OBJECT
MCG+01-04-002	01 09 54.	+ 08 28	24	15.	GALAXY
MCG+05-03-084	01 09 54.	+ 32 41	48	15.	GALAXY
ZWG 502.006	01 09 54.	+ 32 43		14.7	GALAXY
ZWG 501.130	01 09 54.	+ 32 43		14.7	GALAXY
UGC 00756	01 09 54.	+ 32 43	60	14.7	GALAXY Sc
MCG+01-04-013	01 09 54.	- 03 01	66	14.	GALAXY
MCG-05-04-006	01 09 54.	- 32 30	36	15.	GALAXY
LB 01595	01 09 54.	- 51 13		14.5	FAINT BLUE STAR
LB 03185	01 09 54.	- 64 11		13.5	FAINT BLUE STAR
LIN 492	01 09 55.	- 72 16 27.		13.73	SMC EMISSION LINE OBJECT
HOLM 037A	01 09 56.	- 04 24	24	14.3	PART OF MULTIPLE GALAXY
HOLM 037C	01 09 56.	- 04 25	54	14.4	PART OF MULTIPLE GALAXY
MCG-02-04-012	01 09 57.	- 13 03	24	15.	GALAXY
HOLM 037B	01 09 59.	- 04 24	54	14.2	PART OF MULTIPLE GALAXY
UGC 00757	01 10 00.	+ 00 01	66	16.0	GALAXY SB
MCG+00-04-034	01 10 00.	+ 00 02	36	15.	GALAXY
MCG+02-04-009	01 10 00.	+ 15 25	30	15.5	GALAXY
MCG+05-04-001	01 10 00.	+ 32 43	48	14.5	GALAXY
ZWG 520.024	01 10 00.	+ 36 45		15.3	GALAXY
MCG+06-03-022	01 10 00.	+ 36 47	36	15.	GALAXY
VB 036	01 10 00.	+ 64 12	416		STELLAR RING
LDN 1308	01 10 00.	+ 72 40	5040		DARK NEBULA
MCG+01-04-014	01 10 00.	- 04 23	60	14.5	GALAXY
MCG+01-04-015	01 10 00.	- 04 24	54	14.5	GALAXY
LIN 493	01 10 02.	- 73 33 21.		14.20	SMC EMISSION LINE OBJECT
HSN 25	01 10 05.	- 73 18			BRIGHT GALAXY BEHIND SMC
IC 1646	01 10 06.	+ 15 25 58.			NONSTELLAR OBJECT
ZWG 436.010	01 10 06.	+ 15 27		15.3	GALAXY
ZWG 520.025	01 10 06.	+ 33 35		15.7	GALAXY
ZWG 536.023	01 10 06.	+ 39 45		15.2	GALAXY
MCG-03-04-027	01 10 06.	- 17 43	54	15.	GALAXY
MCG-05-04-007	01 10 06.	- 32 19	24	15.5	GALAXY
KN 10.59	01 10 07.6	+ 08 27 20.			NEBULA
RNGC 0427	01 10 08.	- 32 19		15.0	GALAXY
HN 0146	01 10 09.	- 50 39			NEBULA
IC 1650	01 10 09.	- 50 39			NONSTELLAR OBJECT
RNGC 0425	01 10 11.	+ 38 32		13.5	GALAXY
ZWG 411.006	01 10 12.	+ 08 27		15.7	GALAXY
ZWG 436.011	01 10 12.	+ 15 18		15.7	GALAXY
ZWG 520.026	01 10 12.	+ 38 30		13.5	GALAXY
UGC 00758	01 10 12.	+ 38 30	60	13.5	GALAXY S
MCG+06-03-023	01 10 12.	+ 38 32	66	14.	GALAXY
MCG+07-03-023	01 10 12.	+ 39 45	48	15.	GALAXY
MCG+07-03-022	01 10 12.	+ 41 59	48	15.	GALAXY
UGC 00759	01 10 12.	+ 49 24	60	16.5	GALAXY SB0-a
RNGC 0434	01 10 12.	- 58 31		13.5	GALAXY
ARC 0160	01 10 14.	+ 15 15		15.7	RICH CLUSTER OF GALAXIES
MCG+02-04-010	01 10 18.	+ 15 12 30.	36	15.5	GALAXY
ZWG 436.012	01 10 18.	+ 15 14		15.7	GALAXY
MCG+08-03-010	01 10 18.	+ 50 22 30.	48	15.	GALAXY
UGC 00761	01 10 18.	+ 50 25	72	16.0	GALAXY E?
ZWG 385.026	01 10 18.	- 00 33		14.4	GALAXY
UGC 00760	01 10 18.	- 00 33	108	14.4	GALAXY E
MCG+00-04-035	01 10 18.	- 00 33	66	14.	GALAXY
MCG-06-03-027	01 10 18.	- 33 57	15	15.	GALAXY
RNGC 0426	01 10 19.	- 00 33		14.5	GALAXY
HSN 26	01 10 20.	- 71 59			BRIGHT GALAXY BEHIND SMC
HN 0147	01 10 21.	- 71 36			NEBULA
IC 1655	01 10 21.	- 71 36			OPEN CLUSTER
MCG+00-04-036	01 10 24.	+ 00 42 30.	162	12.	GALAXY
ZWG 385.028	01 10 24.	+ 00 43		11.9	GALAXY
UGC 00763	01 10 24.	+ 00 43	306	11.9	GALAXY
UGC 00764	01 10 24.	+ 34 43	66	17.	GALAXY Sc
ZWG 520.027	01 10 24.	+ 33 37		15.2	GALAXY
MCG+06-03-024	01 10 24.	+ 38 39	42	15.	GALAXY
ZWG 385.027	01 10 24.	- 00 36		14.4	GALAXY
UGC 00762	01 10 24.	- 00 36	96	14.4	GALAXY S0
MCG+00-04-037	01 10 24.	- 00 37	54	14.5	GALAXY
MCG-05-04-008	01 10 24.	- 31 15	36	15.	GALAXY
RNGC 0428	01 10 25.	+ 00 43		12.0	GALAXY
RNGC 0429	01 10 25.	- 00 36		14.5	GALAXY
IC 1647	01 10 26.	+ 38 36 58.			NONSTELLAR OBJECT
MCG+00-04-038	01 10 27.	- 00 23 30.	15	16.	GALAXY
HOW 65	01 10 27.	- 72 30 51.	12		STAR CLUSTER IN SMC
ZC 0110.5+1515	01 10 30.	+ 15 15	6920		CLUSTER OF GALAXIES
UGC 00766	01 10 30.	+ 49 21	72	16.5	GALAXY Sc-IRR
VB 037	01 10 30.	+ 64 18	806		STELLAR RING
ZWG 385.029	01 10 30.	- 00 31		13.6	GALAXY
UGC 00765	01 10 30.	- 00 31	114	13.6	GALAXY E
MCG+00-04-039	01 10 30.	- 00 31	54	14.	GALAXY
MCG-05-04-009	01 10 30.	- 31 28	72	15.5	GALAXY
RNGC 0434A	01 10 30.	- 58 29			GALAXY
RNGC 0430	01 10 31.	- 00 31		13.5	GALAXY
MCG+00-04-040	01 10 33.	+ 02 01	30	15.5	GALAXY
ZWG 385.031	01 10 36.	+ 02 01		15.4	GALAXY
UGC 00768	01 10 36.	+ 02 01	102	15.4	GALAXY COMPACT
UGC 00769	01 10 36.	+ 12 30	66	16.0	GALAXY Sc/IRR
MCG+02-04-011	01 10 36.	+ 15 13	12	16.	GALAXY
ZWG 436.013	01 10 36.	+ 15 15		15.7	GALAXY
ZC 0110.6+2448	01 10 36.	+ 24 48	740		CLUSTER OF GALAXIES
ZWG 385.030	01 10 36.	- 01 59		15.7	GALAXY
UGC 00767	01 10 36.	- 01 59	66	15.7	GALAXY S
VV 258B	01 10 36.	- 19 16	9	16.5	INTERACTING GALAXY
VV 258A	01 10 36.	- 19 16	9	16.5	INTERACTING GALAXY
VV 258	01 10 36.	- 19 16	48		INTERACTING GALAXY
MCG-03-04-029	01 10 36.	- 19 17	12	16.	GALAXY
MCG-03-04-028	01 10 36.	- 19 17	8	15.5	GALAXY
KN 10.60	01 10 36.5	+ 08 35 59.			NEBULA
SHB 022	01 10 38.	+ 29 42 22.		17.	QUASI-STELLAR OBJECT
BC 4C29.02	01 10 38.2	+ 29 42 22.		17.	QUASI-STELLAR OBJECT
LIN 494	01 10 40.	- 72 51 52.		16.06	SMC EMISSION LINE OBJECT
ZWG 385.032	01 10 42.	+ 01 01		15.2	GALAXY
ZWG 411.007	01 10 42.	+ 08 36		15.7	GALAXY
MCG+02-04-012	01 10 42.	+ 12 28	54	16.	GALAXY
ZWG 459.017	01 10 42.	+ 15 31		15.7	GALAXY
5ZW 057	01 10 42.	+ 46 06			COMPACT GALAXY
MCG-03-04-030	01 10 42.	- 19 16 30.	10	15.5	GALAXY
MCG-05-04-010	01 10 42.	- 31 42	36	15.	GALAXY
MCG-03-04-031	01 10 45.	- 19 17	12	16.	GALAXY
MCG+01-04-003	01 10 48.	+ 03 56 30.	9	15.	GALAXY
ZWG 520.028	01 10 48.	+ 37 55		15.5	GALAXY
MCG+00-04-041	01 10 48.	- 02 01	48	15.5	GALAXY
MCG-06-03-028	01 10 48.	- 34 18	60	15.	GALAXY
LB 03186	01 10 48.	- 61 08		12.0	FAINT BLUE STAR
HOW 66	01 10 52.	- 75 28 27.	36		STAR CLUSTER IN SMC
UGC 00770	01 10 54.	+ 02 23	60	16.5	GALAXY S
ZC 0110.9+1751	01 10 54.	+ 17 51	2290		CLUSTER OF GALAXIES
IC 1648	01 10 54.	+ 32 57 14.			NONSTELLAR OBJECT
ZWG 502.007	01 10 54.	+ 32 58		15.0	GALAXY
ZWG 501.131	01 10 54.	+ 32 58		15.0	GALAXY
ZWG 385.033	01 10 54.	- 00 06		15.7	GALAXY
RNGC 0440	01 10 54.	- 58 33			GALAXY
IC 1651	01 10 55.	+ 01 50			NONSTELLAR OBJECT
MCG+02-04-013	01 11 00.	+ 11 13	24	16.	GALAXY
ZC 0111.0+1128	01 11 00.	+ 11 28	5980		CLUSTER OF GALAXIES
LDN 1312	01 11 00.	+ 61 50	1320		DARK NEBULA
MCG+00-04-042	01 11 00.	- 00 07 30.	36	15.	GALAXY
LH115-N082	01 11 00.	- 74 07			EMISSION NEBULA IN SMC
LIN.CL 089	01 11 01.	- 72 00 46.	546	12.7	STAR CLUSTER IN SMC
IC 0086	01 11 01.0	- 16 30 11.			GALAXY SBb
MCG+00-04-043	01 11 03.	+ 00 36	60	15.	GALAXY
UGC 00772	01 11 06.	+ 00 37	96	17.	GALAXY DWRF IR
ZWG 385.034	01 11 06.	- 00 21		15.0	GALAXY
UGC 00771	01 11 06.	- 00 21	84	15.0	GALAXY Sa-b
MCG+00-04-044	01 11 06.	- 00 22	48	15.	GALAXY
HN 0148	01 11 07.	- 72 01			NEBULA
IC 1660	01 11 07.	- 72 01			NONSTELLAR OBJECT
MCG-02-04-013	01 11 09.	- 10 51	36	15.	GALAXY
RNGC 0438	01 11 09.	- 38 12			GALAXY
LIN 495	01 11 10.	- 74 06 47.		14.24	SMC EMISSION LINE OBJECT
UGC 00773	01 11 12.	+ 02 07	90	16.0	GALAXY
MCG+01-04-004	01 11 12.	+ 07 09	48	14.5	GALAXY
MRK 564	01 11 12.	+ 07 31	11	16.	GALAXY WITH UV CONTINUUM
ZWG 436.014	01 11 12.	+ 13 00		15.0	GALAXY
UGC 00774	01 11 12.	+ 13 00	66	15.0	GALAXY S
UGC 00775	01 11 12.	+ 18 55	66	16.5	GALAXY Sb
ZWG 520.029	01 11 12.	+ 34 54		15.7	GALAXY
VB 107	01 11 12.	+ 58 40	275		STELLAR RING
MCG-05-04-011	01 11 12.	- 32 03	9	16.	GALAXY
LIN.CL 092	01 11 12.	- 73 42 23.	270	13.7	STAR CLUSTER IN SMC
HN 0149	01 11 12.	- 73 43			NEBULA
IC 1662	01 11 12.	- 73 43			OPEN CLUSTER
LIN.CL 091	01 11 13.	- 73 21 47.	342	14.8	STAR CLUSTER IN SMC
MCG+00-04-045	01 11 15.	+ 00 02 30.	36	16.	GALAXY
MCG+02-04-014	01 11 15.	+ 12 59	48	15.	GALAXY
MCG-06-03-029	01 11 15.	- 38 12	60	12.	GALAXY
ZWG 411.008	01 11 18.	+ 07 09		15.1	GALAXY
ZWG 502.008	01 11 18.	+ 33 27		14.0	GALAXY
ZWG 501.132	01 11 18.	+ 33 27		14.0	GALAXY
UGC 00776	01 11 18.	+ 33 27	84	14.0	GALAXY SB0
ZWG 536.024	01 11 18.	+ 41 59		15.4	GALAXY
UGC 00777	01 11 18.	+ 41 59	60	15.4	GALAXY Sc
UGC 00778	01 11 18.	+ 49 58	66	17.	GALAXY
MCG-05-04-012	01 11 18.	- 32 04	18	15.5	GALAXY
KN 10.61	01 11 18.6	+ 07 09 40.			NEBULA
LIN.CL 090	01 11 20.	- 71 35 17.	612	12.5	STAR CLUSTER IN SMC
MCG-05-04-013	01 11 21.	- 32 04	12	15.5	GALAXY
ZWG 385.035	01 11 24.	+ 01 48		15.0	GALAXY
UGC 00779	01 11 24.	+ 01 48	84	15.0	GALAXY S-IRR
MCG+00-04-046	01 11 24.	+ 01 49	54	14.	GALAXY
RNGC 0431	01 11 24.	+ 33 28		14.0	GALAXY
MCG+05-04-002	01 11 24.	+ 33 28	66	14.5	GALAXY
ZWG 520.030	01 11 24.	+ 37 41		15.5	GALAXY
UGC 00780	01 11 24.	+ 37 41	66	15.5	GALAXY Sb
UGC 00781	01 11 24.	+ 37 52	66	16.5	GALAXY Sb-c
ZWG 536.025	01 11 24.	+ 43 23		15.7	GALAXY
UGC 00782	01 11 24.	+ 50 54	84	16.5	GALAXY Sc?
LIN.CL 093	01 11 24.	- 73 43 35.	342		STAR CLUSTER IN SMC
RNGC 0435	01 11 25.	+ 01 48		15.0	GALAXY
HOW 67	01 11 25.	- 71 13 51.	12		STAR CLUSTER IN SMC
ZWG 536.026	01 11 30.	+ 42 17		14.6	GALAXY
UGC 00783	01 11 30.	+ 42 17	102	14.6	GALAXY Sc
MCG+07-03-024	01 11 30.	+ 42 17	78	14.	GALAXY
MCG-05-04-014	01 11 30.	- 31 19	60	15.	GALAXY
MCG-05-04-015	01 11 30.	- 32 00	36	13.	GALAXY
LB 03187	01 11 30.	- 73 03		13.0	FAINT BLUE STAR
KLEM 01	01 11 31.	- 32 03	900	14.	GROUP OF 8 GALAXIES
RNGC 0439	01 11 32.	- 32 00		13.0	GALAXY
ARC 0162	01 11 35.	+ 02 37		16.6	RICH CLUSTER OF GALAXIES
ZWG 436.015	01 11 36.	+ 12 24		15.6	GALAXY
ZWG 459.018	01 11 36.	+ 15 37		15.2	GALAXY
UGC 00785	01 11 36.	+ 15 37	66	15.2	GALAXY S0
MCG+03-04-012	01 11 36.	+ 15 37 30.	60	15.	GALAXY
ZWG 520.031	01 11 36.	+ 33 48		15.7	GALAXY
UGC 00786	01 11 36.	+ 50 12	60	17.	GALAXY Sc
ZWG 385.037	01 11 36.	- 02 00		15.0	GALAXY
UGC 00784	01 11 36.	- 02 00	90	15.0	GALAXY Sb
MCG+00-04-047	01 11 36.	- 02 01	54	14.5	GALAXY
ZWG 385.036	01 11 36.	- 02 07		15.1	GALAXY
MCG-02-04-014	01 11 36.	- 13 19	60	15.	GALAXY
MCG-05-04-016	01 11 36.	- 32 03	24	14.	GALAXY

OBJECT NAME	RIGHT ASCEN.	DECLINATION	DIAM.	MAGN.	TYPE OF OBJECT
RNGC 0441	01 11 38.	- 32 03		14.0	GALAXY
MCG+00-04-048	01 11 42.	+ 00 29 30.	30	15.	GALAXY
ZWG 385.038	01 11 42.	+ 00 30		15.4	GALAXY
IC 0087	01 11 42.	+ 00 30		15.4	GALAXY
UGC 00787	01 11 42.	+ 12 07	66	16.0	GALAXY Sb
MCG+06-03-025	01 11 42.	+ 33 47	48	15.	GALAXY
ZC 0111.7-0104	01 11 42.	- 01 04	1680		CLUSTER OF GALAXIES
MCG-05-04-017	01 11 42.	- 32 29	36	15.	GALAXY
MCG+00-04-049	01 11 45.	+ 00 38 30.	36	16.	GALAXY
MCG+01-04-005	01 11 45.	+ 05 39	66	13.	GALAXY
FATH 1.035	01 11 45.	- 14 48	11		NEBULA
FATH 1.036	01 11 45.	- 14 55	19		NEBULA
MCG-03-04-032	01 11 45.	- 16 56	54	15.	GALAXY
KN 10.62	01 11 46.0	+ 05 39 46.			NEBULA
ZWG 411.009	01 11 48.	+ 05 40		14.0	GALAXY
RNGC 0437	01 11 48.	+ 05 40		14.0	GALAXY
UGC 00788	01 11 48.	+ 05 40	114	14.0	GALAXY S0/Sa
MCG+02-04-015	01 11 48.	+ 12 05	36	15.	GALAXY
MCG-02-04-015	01 11 48.	- 10 22	36	15.	GALAXY
MCG-06-03-030	01 11 48.	- 32 56	162	12.	GALAXY
ABC 0164	01 11 49.	- 04 02		17.6	RICH CLUSTER OF GALAXIES
FATH 1.037	01 11 50.	- 14 46	11		NEBULA
HSN 27	01 11 50.	- 74 33			BRIGHT GALAXY BEHIND SMC
MCG+00-04-050	01 11 51.	+ 00 16	30	15.5	GALAXY
BIGO 461	01 11 51.	+ 04 06			NEBULA
MCG+00-04-051	01 11 54.	+ 00 03	27	16.	GALAXY
ZWG 385.039	01 11 54.	+ 00 16		15.6	GALAXY
MCG+01-04-016	01 11 54.	- 07 44	48	16.	GALAXY
MCG-05-04-018	01 11 54.	- 32 05	18	15.	GALAXY
IC 0088	01 11 56.	+ 00 31 54.			NONSTELLAR OBJECT
LIN 496	01 11 56.	- 73 10 24.		14.73	SMC EMISSION LINE OBJECT
ABC 0163	01 11 57.	+ 24 28		17.5	RICH CLUSTER OF GALAXIES
KHAV 021	01 12	+ 66 34	10870		DARK NEBULA
KHAV 020	01 12	+ 70 04	3610		DARK NEBULA
MCG+00-04-053	01 12 00.	+ 00 55	48	15.	GALAXY
MCG+00-04-052	01 12 00.	+ 01 34	60	15.	GALAXY
ZWG 385.040	01 12 00.	+ 03 28		15.5	GALAXY
ZC 0112.0+1941	01 12 00.	+ 19 41	1410		CLUSTER OF GALAXIES
ZC 0112.0+2732	01 12 00.	+ 27 32	1210		CLUSTER OF GALAXIES
MCG+01-04-017	01 12 00.	- 08 09	48	15.5	GALAXY
MCG-05-04-019	01 12 00.	- 32 31	30	15.	GALAXY
ABC 0161	01 12 02.	+ 37 09		16.4	RICH CLUSTER OF GALAXIES
MCG-04-04-001	01 12 03.	- 25 40	60	15.	GALAXY
ABC 0166	01 12 04.	- 16 33		16.3	RICH CLUSTER OF GALAXIES
ZWG 385.042	01 12 06.	+ 00 55		15.0	GALAXY
UGC 00790	01 12 06.	+ 00 55	60	15.0	GALAXY Sc
ZWG 385.043	01 12 06.	+ 01 34		15.0	GALAXY
UGC 00791	01 12 06.	+ 01 34	66	15.0	GALAXY SBa-b
ZWG 385.044	01 12 06.	+ 02 16		15.6	GALAXY
ZC 0112.1+3224	01 12 06.	+ 32 24	1340		CLUSTER OF GALAXIES
ZWG 551.010	01 12 06.	+ 50 05		15.7	GALAXY
OCL 0319	01 12 06.	+ 59 52	300	9.	OPEN STAR CLUSTER
ZWG 385.041	01 12 06.	- 01 17		14.5	GALAXY
UGC 00789	01 12 06.	- 01 17	102	14.5	GALAXY PECULR
MCG+00-04-054	01 12 06.	- 01 18	48	14.	GALAXY
IC 1657	01 12 06.	- 32 55 43.			NONSTELLAR OBJECT
RNGC 0433	01 12 07.	+ 59 52			OPEN CLUSTER
RNGC 0442	01 12 07.	- 01 17		14.5	GALAXY
LIN 497	01 12 08.	- 73 06 12.		16.22	SMC EMISSION LINE OBJECT
MCG+01-04-006	01 12 09.	+ 03 54	36	15.	GALAXY
IC 1652	01 12 09.	+ 31 40 36.			NONSTELLAR OBJECT
ZWG 385.045	01 12 12.	+ 00 51		15.1	GALAXY
BIGO 462	01 12 12.	+ 03 56			NEBULA
ZWG 502.009	01 12 12.	+ 31 42		14.3	GALAXY
UGC 00792	01 12 12.	+ 31 42	84	14.3	GALAXY S0-a
MCG+05-04-003	01 12 12.	+ 31 42	54	14.5	GALAXY
5ZW 058	01 12 12.	+ 50 05			COMPACT GALAXY
VB 108	01 12 12.	+ 59 56	457		STELLAR RING
HSN 28	01 12 15.	- 71 48			BRIGHT GALAXY BEHIND SMC
RNGC 0454	01 12 16.	- 55 40			UNVERIFIED SOUTHERN OBJECT
ZWG 385.047	01 12 18.	+ 01 40		15.5	GALAXY
ZWG 411.010	01 12 18.	+ 03 55		15.3	GALAXY
RNGC 0446	01 12 18.	+ 03 55		15.5	GALAXY
UGC 00794	01 12 18.	+ 03 55	78	15.3	GALAXY S
ZWG 411.011	01 12 18.	+ 04 46		15.7	GALAXY
UGC 00795	01 12 18.	+ 12 07	60	16.5	GALAXY Sb
4ZW 042	01 12 18.	+ 33 07			COMPACT GALAXY
ZWG 502.010	01 12 18.	+ 33 07		13.9	GALAXY
RNGC 0443	01 12 18.	+ 33 07		14.0	GALAXY
UGC 00796	01 12 18.	+ 33 07	54	13.9	GALAXY
ZWG 520.032	01 12 18.	+ 33 32		15.3	GALAXY
ZWG 385.046	01 12 18.	- 00 45		14.5	GALAXY
UGC 00793	01 12 18.	- 00 45	42	14.5	GALAXY DBL SYS
MCG+00-04-055	01 12 18.	- 00 46	24	14.	GALAXY
FATH 1.038	01 12 18.	- 14 14	14		NEBULA
MCG-05-04-021	01 12 18.	- 31 26	42	15.	GALAXY
MCG-05-04-020	01 12 18.	- 32 32	72	14.	GALAXY
RNGC 0445	01 12 19.	+ 01 40		15.5	GALAXY
IC 1653	01 12 20.	+ 33 06 47.			NONSTELLAR OBJECT
MCG+00-04-056	01 12 21.	- 00 09	24	16.	GALAXY
ZWG 385.048	01 12 24.	+ 00 10		15.5	GALAXY
UGC 00797	01 12 24.	+ 00 10	78	15.5	GALAXY E
MCG+00-04-057	01 12 24.	+ 00 10	27	15.	GALAXY
ZWG 385.049	01 12 24.	+ 01 25		15.5	GALAXY
ZWG 385.050	01 12 24.	+ 01 32		15.2	GALAXY
MCG+01-04-007	01 12 24.	+ 05 06	48	15.	GALAXY
MCG+02-04-016	01 12 24.	+ 12 05	24	16.	GALAXY
ZC 0112.4+2931	01 12 24.	+ 29 31	1610		CLUSTER OF GALAXIES
ZWG 502.011	01 12 24.	+ 29 56		14.3	GALAXY
UGC 00798	01 12 24.	+ 29 56	96	14.3	GALAXY SBa
ZC 0112.4+4049	01 12 24.	+ 40 49	5040		CLUSTER OF GALAXIES
LB 03188	01 12 24.	- 66 09		14.0	FAINT BLUE STAR
LIN 498	01 12 24.	- 73 27 48.		15.41	SMC EMISSION LINE OBJECT
LIN 499	01 12 24.	- 73 31 30.		13.35	SMC EMISSION LINE OBJECT
LH115-N083	01 12 24.	- 73 33	198		EMISSION NEBULA IN SMC
BIGO 463	01 12 25.	+ 04 02			NEBULA
FATH 1.039	01 12 25.	- 14 23	11		NEBULA
FATH 1.040	01 12 25.	- 14 29	16		NEBULA
IC 1654	01 12 26.	+ 29 55 47.			NONSTELLAR OBJECT
RNGC 0436	01 12 26.	+ 58 33		9.5	OPEN CLUSTER
FATH 1.041	01 12 26.	- 14 43	8		NEBULA
MCG+05-04-004	01 12 27.	+ 29 56	72	14.5	GALAXY
MCG+05-04-005	01 12 27.	+ 33 07 30.	48	14.5	GALAXY
MCG+06-03-026	01 12 27.	+ 33 31 30.	48	15.	GALAXY
LIN 500	01 12 29.	- 73 34 07.			SMC EMISSION LINE OBJECT
ZWG 411.012	01 12 30.	+ 05 06		15.1	GALAXY
UGC 00799	01 12 30.	+ 06 35	84	16.5	GALAXY Sc
ZC 0112.5+3703	01 12 30.	+ 37 03	1880		CLUSTER OF GALAXIES

OBJECT NAME	RIGHT ASCEN.	DECLINATION	DIAM.	MAGN.	TYPE OF OBJECT
OCL 0320	01 12 30.	+ 58 23	600	9.8	OPEN STAR CLUSTER
LH115-N083B	01 12 30.	- 73 32	24		EMISSION NEBULA IN SMC
LH115-N083A	01 12 30.	- 73 34	32		EMISSION NEBULA IN SMC
ABC 0165	01 12 34.	+ 32 20		17.5	RICH CLUSTER OF GALAXIES
ABC 0168	01 12 34.	- 00 02		15.4	RICH CLUSTER OF GALAXIES
FATH 1.042	01 12 35.	- 13 56	5		NEBULA
LIN 501	01 12 35.	- 73 32 01.		16.68	SMC EMISSION LINE OBJECT
LIN.CL 094	01 12 35.	- 73 32 13.	1362		STAR CLUSTER IN SMC
MCG+00-04-059	01 12 36.	+ 00 09	24	16.	GALAXY
MCG+00-04-058	01 12 36.	+ 00 40 30.	24	16.	GALAXY
ZWG 502.012	01 12 36.	+ 28 13		15.2	GALAXY
UGC 00800	01 12 36.	+ 28 13	78	15.2	GALAXY Sc
MCG+05-04-005A	01 12 36.	+ 28 13	48	15.	GALAXY
5ZW 059	01 12 36.	+ 56 13			COMPACT GALAXY
VB 109	01 12 36.	+ 59 53	1021		STELLAR RING
MCG-03-04-033	01 12 36.	- 17 28	30	16.	GALAXY
RNGC 0456	01 12 36.	- 73 32			CLUSTER/NEBULOSITY IN SMC
HOW 68	01 12 37.	- 73 40 20.	12		STAR CLUSTER IN SMC
FATH 1.043	01 12 40.	- 14 07	14		NEBULA
FATH 1.044	01 12 41.	- 14 20	14		NEBULA
LIN 502	01 12 41.	- 73 30 49.		15.54	SMC EMISSION LINE OBJECT
LIN 503	01 12 41.	- 73 32 01.		14.17	SMC EMISSION LINE OBJECT
SHB 023	01 12 41.9	- 01 42 54.		18.0	QUASI-STELLAR OBJECT
MCG+01-04-008	01 12 42.	+ 07 51	60	15.5	GALAXY
ZC 0112.7+1936	01 12 42.	+ 19 36	4570		CLUSTER OF GALAXIES
MCG-05-04-022	01 12 42.	- 32 30	30	15.	GALAXY
HN 0150	01 12 42.	- 70 04			NEBULA
LH115-N083C	01 12 42.	- 73 33	29		EMISSION NEBULA IN SMC
IC 1664	01 12 43.	- 70 04			MAY NOT EXIST
BC PKS0112-017	01 12 43.5	- 01 43 01.			QUASI-STELLAR OBJECT
RNGC 0448	01 12 44.	- 01 53		13.0	GALAXY
ABC 0167	01 12 45.	+ 24 13		17.5	RICH CLUSTER OF GALAXIES
MCG+00-04-060	01 12 45.	- 01 54	36	13.5	GALAXY
LIN 504	01 12 47.	- 73 31 37.		15.16	SMC EMISSION LINE OBJECT
UGC 00802	01 12 48.	+ 07 06	60	16.0	GALAXY Sc
UGC 00803	01 12 48.	+ 07 50	126	16.5	GALAXY
ZWG 459.019	01 12 48.	+ 19 46		15.5	GALAXY
ZC 0112.8+2505	01 12 48.	+ 25 05	540		CLUSTER OF GALAXIES
ZWG 502.013	01 12 48.	+ 32 48		14.0	GALAXY
RNGC 0449	01 12 48.	+ 32 48		14.0	GALAXY
UGC 00804	01 12 48.	+ 32 48	156	14.0	GALAXY SB0/SBa
VB 110	01 12 48.	+ 58 33	430		STELLAR RING
ISS 0038	01 12 48.	+ 59 51	517		STELLAR RING
ZWG 385.051	01 12 48.	- 01 53		13.2	GALAXY
UGC 00801	01 12 48.	- 01 53	114	13.2	GALAXY S0
MCG-03-04-034	01 12 48.	- 14 42	60	15.5	GALAXY
MCG-05-04-023	01 12 48.	- 32 02	24	15.	GALAXY
IC 1656	01 12 50.	+ 32 49 34.			SAME AS NGC 447
FATH 1.045	01 12 53.	- 14 44	16		NEBULA
ZC 0112.9+0231	01 12 54.	+ 02 31	740		CLUSTER OF GALAXIES
UGC 00805	01 12 54.	+ 06 38	84	16.5	GALAXY Sc
ZC 0112.9+2410	01 12 54.	+ 24 10	1680		CLUSTER OF GALAXIES
LH115-N084C	01 12 54.	- 73 32	33		EMISSION NEBULA IN SMC
RNGC 0450	01 12 55.	- 01 08		12.5	GALAXY
MCG+05-04-006	01 12 57.	+ 32 48 30.	120	14.	GALAXY
FATH 1.046	01 12 59.	- 14 42	24		NEBULA
LIN 505	01 12 59.	- 73 29 43.		15.06	SMC EMISSION LINE OBJECT
LBN 0639	01 13	+ 16 20	960		BRIGHT NEBULA
MCG+00-04-061	01 13 00.	+ 00 52	24	16.	GALAXY
ZWG 411.013	01 13 00.	+ 09 02		15.0	GALAXY
UGC 00808	01 13 00.	+ 09 02	168	15.0	GALAXY Sc
MCG+01-04-009	01 13 00.	+ 09 03	120	14.	GALAXY
ZWG 502.014	01 13 00.	+ 31 35		15.4	GALAXY
ZWG 520.033	01 13 00.	+ 33 33		14.8	GALAXY
UGC 00809	01 13 00.	+ 33 33	84	14.8	GALAXY Sc
URA 40	01 13 00.	+ 58 34	294		STELLAR RING
LDN 1313	01 13 00.	+ 61 25	720		DARK NEBULA
VB 038	01 13 00.	+ 63 55	416		STELLAR RING
VB 039	01 13 00.	+ 66 04	389		STELLAR RING
MCG+14-01-010	01 13 00.	+ 83 42	48	17.	GALAXY
MCG+14-01-011	01 13 00.	+ 83 57 30.	30	16.	GALAXY
MCG+14-01-012	01 13 00.	+ 84 02	12	17.	GALAXY
ZWG 385.052	01 13 00.	- 01 07		13.0	GALAXY
UGC 00807	01 13 00.	- 01 07	66	13.0	GALAXY S
UGC 00806	01 13 00.	- 01 07	210		GALAXY Sc
KARA.72 027A	01 13 00.	- 01 07	186	13.0	PART OF DOUBLE GALAXY
MCG+00-04-062	01 13 00.	- 01 10	162	12.6	GALAXY
HOW 69	01 13 01.	- 73 37 56.	18		STAR CLUSTER IN SMC
IC 1658	01 13 02.	+ 30 49 33.			SAME AS NGC 444
FATH 1.047	01 13 02.	- 14 14	8		NEBULA
MCG+04-04-001	01 13 03.	+ 24 07	42	16.	GALAXY
MCG+00-04-063	01 13 06.	+ 00 55 30.	24	16.	GALAXY
ZWG 385.053	01 13 06.	+ 00 56		15.7	GALAXY
ZWG 502.015	01 13 06.	+ 30 49		14.7	GALAXY
UGC 00810	01 13 06.	+ 30 49	126	14.7	GALAXY Sc
KARA.72 028A	01 13 06.	+ 30 49	84	14.7	PART OF DOUBLE GALAXY
MCG+05-04-007	01 13 06.	+ 30 49 30.	102	14.5	GALAXY
MCG+08-03-011	01 13 06.	+ 46 27 30.	48	15.	GALAXY
ZWG 551.011	01 13 06.	+ 46 28		15.1	GALAXY
KARA.72 027B	01 13 06.	- 01 06	60		PART OF DOUBLE GALAXY
MCG+00-04-064	01 13 06.	- 01 09	33	15.	GALAXY
MCG+01-04-019	01 13 06.	- 06 51	66	15.	GALAXY
MCG+01-04-018	01 13 06.	- 06 52	15	15.	GALAXY
LB 03190	01 13 06.	- 67 08		14.2	FAINT BLUE STAR
RNGC 0444	01 13 07.	+ 30 49		14.5	GALAXY
MCG+06-03-027	01 13 09.	+ 33 32	78	14.5	GALAXY
FATH 1.048	01 13 11.	- 14 17	8		NEBULA
MCG+01-04-010	01 13 12.	+ 04 47 30.	36	15.	GALAXY
ZWG 436.016	01 13 12.	+ 13 05		15.0	GALAXY
4ZW 043	01 13 12.	+ 30 43			COMPACT GALAXY
ZWG 502.016	01 13 12.	+ 33 24		15.5	GALAXY
ZWG 520.034	01 13 12.	+ 37 23		15.7	GALAXY
UGC 00811	01 13 12.	+ 37 23	66	15.7	GALAXY S0-a
ZC 0113.2-0004	01 13 12.	- 00 04	4970		CLUSTER OF GALAXIES
MCG+08-03-012	01 13 15.	+ 46 27 30.	60	14.	GALAXY
MCG-02-04-016	01 13 15.	- 14 17	24	16.	GALAXY
FATH 1.049	01 13 15.	- 14 47	19		NEBULA
LIN 506	01 13 16.	- 73 34 20.		11.60	SMC EMISSION LINE OBJECT
LIN 507	01 13 16.	- 73 35 20.		13.55	SMC EMISSION LINE OBJECT
ZWG 385.054	01 13 18.	+ 03 20		15.2	GALAXY
MCG+01-04-011	01 13 18.	+ 04 54 30.	96	13.	GALAXY
ZWG 411.014	01 13 18.	+ 05 04		15.0	GALAXY
ZWG 502.017	01 13 18.	+ 30 05		14.6	GALAXY
UGC 00812	01 13 18.	+ 30 05	96	14.6	GALAXY E
ZWG 502.018	01 13 18.	+ 32 50		15.2	GALAXY
RNGC 0447	01 13 18.	+ 32 50		15.0	GALAXY
MCG+06-03-028	01 13 18.	+ 37 25	42	15.	GALAXY
UGC 00814	01 13 18.	+ 40 54	60	17.	GALAXY

OBJECT NAME	RIGHT ASCEN.	DECLINATION	DIAM.	MAGN.	TYPE OF OBJECT
MCG+08-03-013	01 13 18.	+ 46 28	120	14.	GALAXY
ZWG 551.012	01 13 18.	+ 46 29		15.0	GALAXY
UGC 00813	01 13 18.	+ 46 29	84	15.0	GALAXY
MCG-05-04-024	01 13 18.	- 26 42	120	15.	GALAXY
LH115-N084A	01 13 18.	- 73 34	30		EMISSION NEBULA IN SMC
LH115-N084	01 13 18.	- 73 34	355		EMISSION NEBULA IN SMC
IC 1659	01 13 19.	+ 30 05 33.			NONSTELLAR OBJECT
LIN.CL 095	01 13 19.	- 71 36 07.	138	14.3	STAR CLUSTER IN SMC
MCG+05-04-008	01 13 21.	+ 30 05	24	15.	GALAXY
RNGC 0458	01 13 21.	- 71 48		10.5	GLOBULAR CLUSTER IN SMC
ARP 164	01 13 22.	+ 04 56			PECULIAR GALAXY
LIN 509	01 13 22.	- 73 36 20.		15.87	SMC EMISSION LINE OBJECT
RNGC 0460	01 13 23.	- 72 34			CLUSTER/NEBULOSITY IN SMC
ZWG 385.055	01 13 24.	+ 01 10		15.7	GALAXY
MCG+01-04-012	01 13 24.	+ 04 00	96	14.	GALAXY
ZWG 411.015	01 13 24.	+ 04 55		13.9	GALAXY
RNGC 0455	01 13 24.	+ 04 55		14.0	GALAXY
UGC 00815	01 13 24.	+ 04 55	162	13.9	GALAXY PECULR
ZC 0113.4+1247	01 13 24.	+ 12 47	940		CLUSTER OF GALAXIES
ZWG 459.020	01 13 24.	+ 19 45		15.5	GALAXY
ZWG 481.001	01 13 24.	+ 24 55		15.4	GALAXY
KARA.73B 0047	01 13 24.	+ 24 55	42	15.4	ISOLATED GALAXY S
ZWG 502.019	01 13 24.	+ 32 48		14.8	GALAXY
RNGC 0451	01 13 24.	+ 32 48		15.0	GALAXY
MRK 001	01 13 24.	+ 32 50	27	15.5	GALAXY WITH UV CONTINUUM
KW 01	01 13 24.	+ 32 50	34		SEYFERT GALAXY
VVI 07	01 13 24.	+ 32 50	36	16.64	SEYFERT GALAXY
MCG+05-04-009	01 13 24.	+ 32 50 30.	36	15.	GALAXY
ZWG 551.013	01 13 24.	+ 46 29		14.6	GALAXY
UGC 00816	01 13 24.	+ 46 29	108	14.6	GALAXY S
LIN.CL 096	01 13 24.	- 71 48 25.	750	10.5	STAR CLUSTER IN SMC
LH115-N084B	01 13 24.	- 73 36	21		EMISSION NEBULA IN SMC
LH115-N084D	01 13 24.	- 73 37	23		EMISSION NEBULA IN SMC
IC 1661	01 13 25.	+ 32 49 15.			SAME AS NGC 451
PATH 1.050	01 13 25.	- 14 05	10		NEBULA
IC 0089	01 13 27.	+ 04 01 38.			SAME AS NGC 446
LIN.CL 097	01 13 27.	- 72 33 56.	546		STAR CLUSTER IN SMC
LIN 508	01 13 27.	- 72 37 08.		13.93	SMC EMISSION LINE OBJECT
LIN 510	01 13 28.	- 73 35 44.		12.75	SMC EMISSION LINE OBJECT
HOLM 038B	01 13 29.	+ 19 30	12	15.2	PART OF MULTIPLE GALAXY
HOLM 038A	01 13 29.	+ 19 30	24	14.6	PART OF MULTIPLE GALAXY
ZWG 385.056	01 13 30.	+ 01 18		15.0	GALAXY
UGC 00817	01 13 30.	+ 01 18	72	15.0	GALAXY Sb/SBb
MCG+00-04-065	01 13 30.	+ 01 19	42	14.	GALAXY
ZWG 411.016	01 13 30.	+ 04 02		13.8	GALAXY
MRK 565	01 13 30.	+ 04 02	30	13.	GALAXY WITH UV CONTINUUM
UGC 00818	01 13 30.	+ 04 02	126	13.8	GALAXY SB0
MCG+01-04-013	01 13 30.	+ 04 53 30.	24	15.	GALAXY
UGC 00819	01 13 30.	+ 06 23	60	17.	GALAXY Sc-IRR
ZWG 502.020	01 13 30.	+ 30 46		14.0	GALAXY
UGC 00820	01 13 30.	+ 30 46	162	14.0	GALAXY SBa-b
KARA.72 028B	01 13 30.	+ 30 46	138	14.0	PART OF DOUBLE GALAXY
MCG+05-04-010	01 13 30.	+ 30 46	150	14.	GALAXY
RNGC 0453	01 13 30.	+ 32 47			NON-EXISTENT OBJECT
MCG+05-04-011	01 13 30.	+ 32 49	39	14.5	GALAXY
AND 2	01 13 30.	+ 33 09	210		DWARF SPHEROIDAL GALAXY
VB 040	01 13 30.	+ 65 45	389		STELLAR RING
MCG+00-04-066	01 13 30.	- 00 18 30.	48	16.	GALAXY
MCG+00-04-067	01 13 30.	- 00 23 30.	27	16.	GALAXY
MCG+01-04-020	01 13 30.	- 06 38	36	15.	GALAXY
MCG-05-04-025	01 13 30.	- 27 04	60	14.5	GALAXY
ARC 0170	01 13 31.	+ 12 50		16.6	RICH CLUSTER OF GALAXIES
RNGC 0452	01 13 31.	+ 30 46		14.0	GALAXY
PATH 1.051	01 13 31.	- 14 26	8		NEBULA
LIN 512	01 13 34.	- 73 35 08.		14.17	SMC EMISSION LINE OBJECT
HOW 70	01 13 35.	- 72 28 25.	12		STAR CLUSTER IN SMC
LIN 511	01 13 35.	- 73 27 20.		14.88	SMC EMISSION LINE OBJECT
ZWG 520.035	01 13 36.	+ 38 47		15.1	GALAXY
UGC 00822	01 13 36.	+ 38 47	78	15.1	GALAXY SBa
MCG+06-03-029	01 13 36.	+ 38 47 30.	66	14.5	GALAXY
UGC 00821	01 13 36.	- 00 17	66	16.0	GALAXY S
MCG-05-04-026	01 13 36.	- 31 51	42	15.	GALAXY
MCG-05-04-027	01 13 36.	- 32 44	48	14.5	GALAXY
ARC 0169	01 13 37.	+ 40 51		16.7	RICH CLUSTER OF GALAXIES
PATH 1.052	01 13 37.	- 14 39	14		NEBULA
HSN 29	01 13 39.	- 74 34			BRIGHT GALAXY BEHIND SMC
PATH 1.053	01 13 41.	- 14 38	5		NEBULA
ZC 0113.7+0645	01 13 42.	+ 06 45	1480		CLUSTER OF GALAXIES
VB 041	01 13 42.	+ 63 56	235		STELLAR RING
MCG+01-04-021	01 13 42.	- 06 52	48	14.5	GALAXY
MCG+01-04-022	01 13 42.	- 06 59	90	14.	GALAXY
MCG-03-04-036	01 13 42.	- 16 10	15	16.	GALAXY
MCG-03-04-035	01 13 42.	- 16 10 30.	30	16.	GALAXY
IC 1663	01 13 45.	- 30 55 29.			NONSTELLAR OBJECT
ZWG 502.021	01 13 48.	+ 28 43		15.5	GALAXY
LB 03191	01 13 48.	- 65 38		14.0	FAINT BLUE STAR
PATH 1.054	01 13 49.	- 14 01	5		NEBULA
MCG+05-04-012	01 13 51.	+ 32 38	39	15.5	GALAXY
BIGO 464	01 13 51.	+ 32 53			NEBULA
PATH 1.055	01 13 52.	- 14 11	27		NEBULA
LB 00441	01 13 53.	+ 31 46 30.		17.6	FAINT BLUE STAR
SN 1967A	01 13 54.	+ 02 45		17.0	SUPERNOVA
RNGC 0461	01 13 56.	- 34 13			GALAXY
LIN 513	01 13 57.	- 73 44 09.		13.87	SMC EMISSION LINE OBJECT
PATH 1.056	01 13 58.	- 14 11	8		NEBULA
PATH 1.057	01 13 58.	- 14 27	8		NEBULA
PATH 1.058	01 13 59.	- 14 18	5		NEBULA
HOW 71	01 13 59.	- 72 38 43.	18		STAR CLUSTER IN SMC
KHAV 022	01 14	+ 63 16	10000		DARK NEBULA
UGC 00823	01 14 00.	+ 00 57	60	18.	GALAXY DWRF SP
MCG+03-04-013	01 14 00.	+ 16 08	42	14.5	GALAXY
ZWG 536.027	01 14 00.	+ 43 21		15.7	GALAXY
VB 111	01 14 00.	+ 60 05	363		STELLAR RING
LDN 1314	01 14 00.	+ 61 10	360		DARK NEBULA
MCG+01-04-023	01 14 00.	- 08 14	30	14.	GALAXY
IC 0090	01 14 00.	- 08 14 24.			NONSTELLAR OBJECT
TON-S 0202	01 14 00.	- 27 10		14.8	BLUE STAR
PATH 1.059	01 14 02.	- 14 09	19		NEBULA
ZWG 459.021	01 14 06.	+ 16 08		14.7	GALAXY
ZWG 481.002	01 14 06.	+ 22 52		15.7	GALAXY
KARA.73B 0048	01 14 06.	+ 22 52	42	15.7	ISOLATED GALAXY S
MCG-05-04-028	01 14 06.	- 27 36	72	15.5	GALAXY
LIN.CL 098	01 14 07.	- 72 52 51.	270	14.9	STAR CLUSTER IN SMC
ARC 0171	01 14 09.	+ 16 00		15.9	RICH CLUSTER OF GALAXIES
PATH 1.060	01 14 09.	- 14 36	14		NEBULA
PATH 1.061	01 14 09.	- 14 52	11		NEBULA
LIN 516	01 14 09.	- 73 36 21.		14.21	SMC EMISSION LINE OBJECT
LIN 515	01 14 10.	- 73 35 03.		14.00	SMC EMISSION LINE OBJECT

OBJECT NAME	RIGHT ASCEN.	DECLINATION	DIAM.	MAGN.	TYPE OF OBJECT
ZC 0114.2+2308	01 14 12.	+ 23 08	1340		CLUSTER OF GALAXIES
ZWG 536.028	01 14 12.	+ 40 42		15.7	GALAXY
MCG+07-03-025	01 14 12.	+ 40 43	36	15.	GALAXY
PATH 1.062	01 14 13.	- 14 16	11		NEBULA
LIN 514	01 14 14.	- 72 36 09.		14.34	SMC EMISSION LINE OBJECT
VV 205B	01 14 15.	+ 14 27	18	17.	INTERACTING GALAXY
VV 205A	01 14 15.	+ 14 27	15	17.	INTERACTING GALAXY
LIN 517	01 14 16.	- 73 25 39.		15.33	PLANETARY NEBULA IN SMC
ZWG 436.017	01 14 18.	+ 12 45		14.9	GALAXY
UGC 00824	01 14 18.	+ 12 45	72	14.9	GALAXY Sa
ZC 0114.3+1717	01 14 18.	+ 17 17	4440		CLUSTER OF GALAXIES
MCG+00-04-068	01 14 18.	- 00 08	30	16.	GALAXY
PATH 1.063	01 14 18.	- 14 09	14		NEBULA
HOW 72	01 14 19.	- 73 25 19.	18		STAR CLUSTER IN SMC
ARC 0172	01 14 24.	+ 02 59		16.9	RICH CLUSTER OF GALAXIES
MCG+02-04-017	01 14 24.	+ 12 44	72	14.	GALAXY
ZC 0114.4+3944	01 14 24.	+ 39 44	2220		CLUSTER OF GALAXIES
LH115-N086	01 14 24.	- 73 26			EMISSION NEBULA IN SMC
LH115-N085	01 14 24.	- 73 36			EMISSION NEBULA IN SMC
RNGC 0465	01 14 25.	- 73 35			OPEN CLUSTER IN SMC
LIN.CL 099	01 14 27.	- 73 35 09.	5442		STAR CLUSTER IN SMC
LIN 518	01 14 27.	- 73 36 15.		15.47	SMC EMISSION LINE OBJECT
ZC 0114.5+0259	01 14 30.	+ 02 59	3290		CLUSTER OF GALAXIES
4ZW 044	01 14 30.	+ 27 32			COMPACT GALAXY
TON-S 0203	01 14 30.	- 28 03		14.7	BLUE STAR
MCG-05-04-029	01 14 30.	- 31 41	30	15.	GALAXY
MCG+00-04-069	01 14 33.	+ 02 01	18	16.	GALAXY
PATH 1.064	01 14 35.	- 14 28	8		NEBULA
ZWG 385.057	01 14 36.	+ 02 01		15.7	GALAXY
MCG+00-04-070	01 14 36.	+ 02 22	24	16.	GALAXY
ZWG 385.058	01 14 36.	+ 02 23		15.7	GALAXY
ZC 0114.6+1227	01 14 36.	+ 12 27	1480		CLUSTER OF GALAXIES
MCG+03-04-014	01 14 36.	+ 16 00	18	16.	GALAXY
5ZW 060	01 14 36.	+ 39 44			COMPACT GALAXY
UGC 00825	01 14 36.	+ 48 45	84	17.	GALAXY
LIN 519	01 14 38.	- 73 43 40.		14.25	SMC EMISSION LINE OBJECT
MCG+07-03-026	01 14 39.	+ 43 24	72	14.	GALAXY
MCG-02-04-017	01 14 39.	- 10 24	30	15.	GALAXY
MCG-02-04-018	01 14 39.	- 14 16	24	15.	GALAXY
PATH 1.065	01 14 40.	- 14 16	19		NEBULA
ZWG 536.029	01 14 42.	+ 43 22		14.8	GALAXY
UGC 00826	01 14 42.	+ 43 22	78	14.8	GALAXY S
5ZW 061	01 14 42.	+ 43 24			COMPACT GALAXY
MCG-04-04-002	01 14 42.	- 25 53 30.	30	15.5	GALAXY
MCG+02-04-018	01 14 45.	+ 14 26	36	15.5	GALAXY
MCG-03-04-037	01 14 45.	- 19 03	36	16.	GALAXY
MCG+01-04-015	01 14 48.	+ 03 53	6	15.5	GALAXY
MCG+01-04-014	01 14 48.	+ 03 53	24	15.5	GALAXY
ZWG 436.018	01 14 48.	+ 14 26		15.4	GALAXY
UGC 00827	01 14 48.	+ 14 26	66	15.4	GALAXY DBL SYS
ZC 0114.8+1604	01 14 48.	+ 16 04	4570		CLUSTER OF GALAXIES
MCG+00-04-071	01 14 48.	- 00 36 30.	24	16.	GALAXY
SHB 024	01 14 49.	+ 07 26 18.		18.	QUASI-STELLAR OBJECT
PATH 1.066	01 14 51.	- 14 32	8		NEBULA
MCG+01-04-016	01 14 54.	+ 03 52	24	15.	GALAXY
ZWG 411.017	01 14 54.	+ 03 55		15.7	GALAXY
MCG+03-04-015	01 14 54.	+ 19 42	30	15.5	GALAXY
MCG-02-04-019	01 14 54.	- 08 31	36	15.5	GALAXY
MCG-03-04-038	01 14 54.	- 16 19	36	15.	GALAXY
ARP 128	01 14 56.	+ 14 27			PECULIAR GALAXY
IC 1665	01 14 56.	+ 34 26 17.			NONSTELLAR OBJECT
ZWG 385.059	01 15 00.	+ 02 45		15.7	GALAXY
ZWG 459.022	01 15 00.	+ 19 41		15.7	GALAXY
ZWG 459.023	01 15 00.	+ 21 09		15.7	GALAXY
UGC 00828	01 15 00.	+ 21 09	72	15.7	GALAXY Sa-b
LB 03196	01 15 00.	- 71 09		12.2	FAINT BLUE STAR
PATH 1.067	01 15 02.	- 14 39	19		NEBULA
MCG+04-04-002	01 15 03.	+ 25 04	33	16.	GALAXY
HOW 73	01 15 03.	- 71 35 24.	18		STAR CLUSTER IN SMC
ZWG 436.019	01 15 06.	+ 09 56		15.7	GALAXY
UGC 00829	01 15 06.	+ 09 56	96	15.7	GALAXY PECULR
ZC 0115.1+1047	01 15 06.	+ 10 47	470		CLUSTER OF GALAXIES
ZWG 551.014	01 15 06.	+ 49 42		15.7	GALAXY
MCG+08-03-014	01 15 06.	+ 49 42	42	15.	GALAXY
5ZW 062	01 15 06.	+ 55 37			COMPACT GALAXY
MCG-06-04-002	01 15 06.	- 34 07	48	13.	GALAXY
MCG-06-04-001	01 15 06.	- 36 04	42	16.	GALAXY
PATH 1.068	01 15 07.	- 13 53	8		NEBULA
PATH 1.069	01 15 09.	- 14 17	11		NEBULA
SHAH 039	01 15 12.	+ 09 38	198	16.8	GROUP OF COMPACT GALAXIES
MCG+03-04-016	01 15 12.	+ 21 10	48	15.5	GALAXY
5ZW 063	01 15 12.	+ 50 33			COMPACT GALAXY
MCG-02-04-020	01 15 12.	- 08 52	96	16.	GALAXY
IC 1667	01 15 12.	- 17 22 27.			NONSTELLAR OBJECT
RNGC 0466	01 15 13.	- 59 11			UNVERIFIED SOUTHERN OBJECT
ZWG 385.061	01 15 18.	+ 03 02		15.6	GALAXY
ZWG 521.001	01 15 18.	+ 38 11		15.1	GALAXY
ZWG 520.036	01 15 18.	+ 38 11		15.1	GALAXY
UGC 00831	01 15 18.	+ 38 11	84	15.1	GALAXY S
ZWG 385.060	01 15 18.	- 02 13		15.6	GALAXY
MCG-02-04-021	01 15 18.	- 12 17	48	15.	GALAXY
RNGC 0459	01 15 22.	+ 17 17		15.5	GALAXY
LB 00442	01 15 22.	+ 29 51 12.		16.8	FAINT BLUE STAR
PATH 1.070	01 15 23.	- 13 56	11		NEBULA
ZWG 459.024	01 15 24.	+ 17 17		15.7	GALAXY
UGC 00832	01 15 24.	+ 17 17	66	15.7	GALAXY Sb-c
MCG+06-04-001	01 15 24.	+ 38 11 30.	48	15.	GALAXY
MCG+00-04-072	01 15 24.	- 02 13	114	15.	GALAXY
HOW 74	01 15 25.	- 73 24 30.	18		STAR CLUSTER IN SMC
PATH 1.072	01 15 26.	- 14 27	16		NEBULA
PATH 1.071	01 15 27.	- 13 58	16		NEBULA
RNGC 0462	01 15 30.	+ 03 58			GALAXY
ZWG 411.018	01 15 30.	+ 07 03		15.7	GALAXY
ZWG 436.020	01 15 30.	+ 11 07		14.3	GALAXY
UGC 00833	01 15 30.	+ 11 07	156	14.3	GALAXY SBc
KARA.73B 0049	01 15 30.	+ 11 07	132	14.3	ISOLATED GALAXY S
MCG+03-04-017	01 15 30.	+ 17 17	48	15.	GALAXY
ZC 0115.5+2054	01 15 30.	+ 20 54	2550		CLUSTER OF GALAXIES
VB 112	01 15 36.	+ 58 27	336		STELLAR RING
ISS 0039	01 15 36.	+ 58 27	342		STELLAR RING
MCG+01-04-024	01 15 36.	- 02 48	60	15.	GALAXY
LB 00176	01 15 36.	- 14 07 30.		14.4	FAINT BLUE STAR
MCG-06-04-003	01 15 36.	- 36 36	48	15.	GALAXY
LB 00177	01 15 38.	- 14 26 36.		15.8	FAINT BLUE STAR
MCG+02-04-019	01 15 42.	+ 11 06	144	13.	GALAXY
MCG+03-04-018	01 15 42.	+ 16 28	18	16.	GALAXY
ZC 0115.7+3849	01 15 42.	+ 38 49	1280		CLUSTER OF GALAXIES
VB 042	01 15 42.	+ 63 43	470		STELLAR RING

OBJECT NAME	RIGHT ASCEN.	DECLINATION	DIAM.	MAGN.	TYPE OF OBJECT
LB 01596	01 15 42.	- 46 30		14.2	FAINT BLUE STAR
BC PKS0115+02	01 15 42.8	+ 02 42 35.		17.5	QUASI-STELLAR OBJECT
SHB 025	01 15 43.6	+ 02 42 19.		17.5	QUASI-STELLAR OBJECT
IC 1666	01 15 44.	+ 32 12 27.			NONSTELLAR OBJECT
MCG+01-04-017	01 15 45.	+ 04 28	48	15.	GALAXY
FATH 1.073	01 15 47.	- 14 09	11		NEBULA
ZWG 411.019	01 15 48.	+ 04 29		15.7	GALAXY
UGC 00834	01 15 48.	+ 04 29	60	15.7	GALAXY Sb/SBc
ZC 0115.8+2719	01 15 48.	+ 27 19	870		CLUSTER OF GALAXIES
ZWG 502.022	01 15 48.	+ 32 00		15.6	GALAXY
LB 03192	01 15 48.	- 53 36		13.7	FAINT BLUE STAR
MCG+07-03-027	01 15 51.	+ 44 44	24	16.	GALAXY
MCG+05-04-013	01 15 54.	+ 30 46 30.	60	15.	GALAXY
ZWG 502.023	01 15 54.	+ 30 47		14.8	GALAXY
UGC 00835	01 15 54.	+ 30 47	72	14.8	GALAXY Sc
ZC 0115.9+3146	01 15 54.	+ 31 46	2220		CLUSTER OF GALAXIES
ARC 0173	01 15 54.	+ 38 44		17.4	RICH CLUSTER OF GALAXIES
MCG+07-03-028	01 15 54.	+ 44 02 30.	60	15.	GALAXY
OCL 0321	01 15 54.	+ 58 04	3600	8.4	OPEN STAR CLUSTER
TON-S 0204	01 15 54.	- 27 12		16.0	BLUE STAR
LIN 520	01 15 54.	- 74 00 11.		16.15	SMC EMISSION LINE OBJECT
RNGC 0457	01 15 56.	+ 58 04		8.0	OPEN CLUSTER
MCG+00-04-073	01 15 57.	- 01 28 30.	42	16.	GALAXY
LB 00443	01 15 58.	+ 31 44 12.		14.4	FAINT BLUE STAR
KHAV 023	01 16	+ 59 28	4700		DARK NEBULA
MCG+02-04-020	01 16 00.	+ 14 44	36	14.	GALAXY
MCG+05-04-014	01 16 00.	+ 32 00 30.	42	15.	GALAXY
ZWG 502.024	01 16 00.	+ 32 55		15.7	GALAXY
UGC 00836	01 16 00.	+ 44 02	84	16.0	GALAXY E
LDN 1316	01 16 00.	+ 61 10	840		DARK NEBULA
MCG+00-04-075	01 16 00.	- 00 30	21	16.5	GALAXY
MCG+00-04-074	01 16 00.	- 00 30 30.	18	15.5	GALAXY
AGU 05	01 16 00.	- 44 57 00.	66	12.5	PECULIAR GALAXY
IC 1668	01 16 04.	+ 32 55 02.			NONSTELLAR OBJECT
IC 0091	01 16 06.	+ 02 17 26.			NONSTELLAR OBJECT
ZWG 436.021	01 16 06.	+ 14 45		14.2	GALAXY
UGC 00838	01 16 06.	+ 14 45	39	14.2	GALAXY
ZWG 385.062	01 16 06.	- 00 30		15.1	GALAXY
UGC 00837	01 16 06.	- 01 28	66	16.5	GALAXY E-S
TON-S 0205	01 16 06.	- 23 10		15.5	BLUE STAR
MCG-06-04-004	01 16 06.	- 37 23	36	15.	GALAXY
MCG+00-04-076	01 16 09.	- 01 11 30.	36	15.5	GALAXY
RNGC 0464	01 16 12.	+ 34 42			GALAXY
ZWG 385.063	01 16 12.	- 01 10		15.4	GALAXY
MCG+01-04-025	01 16 12.	- 07 42	72	14.5	GALAXY
MCG-03-04-039	01 16 12.	- 17 19	54	14.5	GALAXY
MCG+01-04-018	01 16 15.	+ 05 17	36	15.	GALAXY
MCG+03-04-019	01 16 15.	+ 16 03	48	15.	GALAXY
HOW 75	01 16 15.	- 73 49 41.	30		STAR CLUSTER IN SMC
RNGC 0463	01 16 16.	+ 16 03		15.0	GALAXY
ZWG 411.020	01 16 18.	+ 04 05		15.7	GALAXY
SN 1968Q	01 16 18.	+ 04 40		18.5	SUPERNOVA
UGC 00839	01 16 18.	+ 12 41	60	16.5	GALAXY S
ZWG 459.025	01 16 18.	+ 16 03		15.2	GALAXY
UGC 00840	01 16 18.	+ 16 03	78	15.2	GALAXY DBL SYS
ZWG 502.025	01 16 18.	+ 32 47		15.0	GALAXY
UGC 00841	01 16 18.	+ 32 47	96	15.0	GALAXY Sb-c
ZC 0116.3+4322	01 16 18.	+ 43 22	5240		CLUSTER OF GALAXIES
MCG-03-04-040	01 16 21.	- 17 04	120	14.	GALAXY
MCG-04-04-003	01 16 21.	- 24 12 30.	54	14.	GALAXY
HOW 76	01 16 23.	- 74 36 41.	12		STAR CLUSTER IN SMC
UGC 00843	01 16 24.	+ 05 18	72	16.5	GALAXY
UGC 00844	01 16 24.	+ 09 46	60	16.5	GALAXY Sc
ZC 0116.4+2413	01 16 24.	+ 24 13	1810		CLUSTER OF GALAXIES
ZWG 385.064	01 16 24.	- 01 16		15.4	GALAXY
ZH 58	01 16 24.	- 01 16		15.0	GALAXY
UGC 00842	01 16 24.	- 01 16	114	15.4	GALAXY
MCG+00-04-077	01 16 24.	- 01 17 30.	54	15.	GALAXY
ZC 0116.4-0128	01 16 24.	- 01 28	2620		CLUSTER OF GALAXIES
MCG-03-04-041	01 16 24.	- 17 04	78	14.	GALAXY
MCG-03-04-042	01 16 24.	- 19 54	42	14.	GALAXY
LB 01597	01 16 24.	- 56 32		14.6	FAINT BLUE STAR
FATH 1.074	01 16 26.	+ 15 50	5		NEBULA
ARP 088	01 16 27.	+ 12 13			PECULIAR GALAXY
MCG+05-04-015	01 16 27.	+ 32 46	90	15.	GALAXY
ZWG 411.021	01 16 30.	+ 04 04		15.4	GALAXY
MRK 566	01 16 30.	+ 04 04	10	16.	GALAXY WITH UV CONTINUUM
ZWG 459.026	01 16 30.	+ 21 29		15.0	GALAXY
UGC 00845	01 16 30.	+ 21 29	72	15.0	GALAXY Sb
UGC 00846	01 16 30.	+ 48 56	72	16.5	GALAXY Sc/SBc
MCG+00-04-078	01 16 30.	- 00 25	72	15.	GALAXY
IC 0093	01 16 31.	- 17 19 48.			NONSTELLAR OBJECT
FATH 1.075	01 16 34.	- 14 47	14		NEBULA
ZWG 385.065	01 16 36.	+ 03 02		13.3	GALAXY
UGC 00848	01 16 36.	+ 03 02	150	13.3	GALAXY S0
MCG+00-04-079	01 16 36.	+ 03 02 30.	30	13.0	GALAXY
ZWG 459.027	01 16 36.	+ 20 31		15.6	GALAXY
ZC 0116.6+2947	01 16 36.	+ 29 47	2350		CLUSTER OF GALAXIES
ZWG 502.026	01 16 36.	+ 31 47		15.7	GALAXY
UGC 00847	01 16 36.	- 00 25	96	16.5	GALAXY
MCG-03-04-044	01 16 36.	- 17 18	42	15.	GALAXY
MCG-03-04-043	01 16 36.	- 17 20	72	14.	GALAXY
RNGC 0467	01 16 37.	+ 03 02		13.5	GALAXY
MCG+00-04-080	01 16 39.	+ 01 57	15	15.5	GALAXY
MCG+05-04-016	01 16 39.	+ 31 47	42	17.	GALAXY
ZWG 385.066	01 16 42.	+ 01 57		15.7	GALAXY
ZWG 411.022	01 16 42.	+ 04 19		14.9	GALAXY
MRK 567	01 16 42.	+ 04 19	13	15.	GALAXY WITH UV CONTINUUM
ZWG 436.022	01 16 42.	+ 12 11		15.1	GALAXY
UGC 00849	01 16 42.	+ 12 11	78	15.1	GALAXY Sc-IRR
KARA.72 029A	01 16 42.	+ 12 11	66	15.1	PART OF DOUBLE GALAXY
ZWG 436.023	01 16 42.	+ 12 12		15.2	GALAXY
KARA.72 029B	01 16 42.	+ 12 12	42	15.2	PART OF DOUBLE GALAXY
ZWG 459.028	01 16 42.	+ 20 53		15.6	GALAXY
UGC 00850	01 16 42.	+ 20 53	66	15.6	GALAXY Sb
MCG-03-04-045	01 16 42.	- 19 31	24	15.5	GALAXY
LB 00444	01 16 43.	+ 33 07 48.		16.2	FAINT BLUE STAR
MCG+00-04-081	01 16 45.	+ 01 59 30.	42	15.5	GALAXY
RNGC 0469	01 16 46.	+ 14 37		15.0	GALAXY
HELW 357	01 16 46.	- 34 22 06.			NEBULA
IC 1670	01 16 47.	- 17 06 43.			NONSTELLAR OBJECT
ZWG 385.067	01 16 48.	+ 01 59		15.7	GALAXY
VV 347B	01 16 48.	+ 12 10	30	15.5	INTERACTING GALAXY
VV 347A	01 16 48.	+ 12 10	66	15.	INTERACTING GALAXY
MCG+02-04-022	01 16 48.	+ 12 10	36	15.	GALAXY
MCG+02-04-021	01 16 48.	+ 12 10	84	15.	GALAXY
ZC 0116.8+1434	01 16 48.	+ 14 34	2490		CLUSTER OF GALAXIES
ZWG 436.024	01 16 48.	+ 14 36		15.0	GALAXY

OBJECT NAME	RIGHT ASCEN.	DECLINATION	DIAM.	MAGN.	TYPE OF OBJECT
MCG+02-04-023	01 16 48.	+ 14 37	36	15.	GALAXY
UGC 00851	01 16 48.	+ 19 28	66	16.5	GALAXY SBb
ZWG 502.027	01 16 48.	+ 31 55		15.7	GALAXY
ZH 81	01 16 48.	- 00 22		15.5	GALAXY
KN 14.001	01 16 48.1	+ 31 54 57.			NEBULA
IC 0095	01 16 50.	- 12 50 01.			NONSTELLAR OBJECT
ARP 119	01 16 51.	+ 12 12			PECULIAR GALAXY
MCG+05-04-017	01 16 51.	+ 31 55 30.	9	17.	GALAXY
MCG-02-04-022	01 16 51.	- 12 08 30.	60	15.	GALAXY
FATH 1.076	01 16 53.	- 14 10	16		NEBULA
SN 1972P	01 16 54.	+ 00 16		18.0	SUPERNOVA
MCG+00-04-082	01 16 54.	+ 02 57	54	15.5	GALAXY
ZWG 385.068	01 16 54.	+ 02 58		15.4	GALAXY
MCG+01-04-020	01 16 54.	+ 03 37	48	15.5	GALAXY
MCG+01-04-021	01 16 54.	+ 04 38	36	15.5	GALAXY
MCG+01-04-019	01 16 54.	+ 07 55	48	14.	GALAXY
UGC 00852	01 16 54.	+ 17 18	72	16.0	GALAXY
ZWG 459.029	01 16 54.	+ 18 10		14.7	GALAXY
MCG+03-04-020	01 16 54.	+ 19 27 30.	48	15.5	GALAXY
ZWG 502.028	01 16 54.	+ 31 24		15.6	GALAXY
UGC 00853	01 16 54.	+ 31 24	72	15.6	GALAXY SBb
MCG+05-04-018	01 16 54.	+ 31 56	12	16.	GALAXY
MCG-03-04-046	01 16 54.	- 15 57 30.	48	14.	GALAXY
HN 0151	01 16 55.	- 50 54			NEBULA
IC 1674	01 16 55.	- 50 55			NONSTELLAR OBJECT
LIN.CL 100	01 16 55.	- 72 15 24.	270	14.5	STAR CLUSTER IN SMC
ARC 0175	01 16 56.	+ 14 37		17.0	RICH CLUSTER OF GALAXIES
KHAV 024	01 17	+ 58 10	4260		DARK NEBULA
KHAV 025	01 17	+ 60 58	4380		DARK NEBULA
ZWG 385.069	01 17 00.	+ 00 12		15.7	GALAXY
ZH 85	01 17 00.	+ 00 12		15.6	GALAXY
UGC 00854	01 17 00.	+ 03 39	60	16.5	GALAXY Sc
ZWG 411.023	01 17 00.	+ 07 55		14.9	GALAXY
UGC 00855	01 17 00.	+ 07 55	72	14.9	GALAXY SBb
ZWG 502.029	01 17 00.	+ 32 31		15.1	GALAXY
RNGC 0468	01 17 00.	+ 32 31		15.0	GALAXY
5ZW 064	01 17 00.	+ 51 29			COMPACT GALAXY
LDN 1315	01 17 00.	+ 62 50	2100		DARK NEBULA
MCG-06-04-005	01 17 00.	- 34 22		12.5	GALAXY
HSN 30	01 17 00.	- 72 16			BRIGHT GALAXY BEHIND SMC
MCG+00-04-083	01 17 03.	+ 02 59 30.	54	16.	GALAXY
HOLM 039B	01 17 03.	+ 14 35	12	14.8	PART OF MULTIPLE GALAXY
HOLM 039A	01 17 03.	+ 14 36	18	14.4	PART OF MULTIPLE GALAXY
ARC 0174	01 17 03.	+ 35 33		15.8	RICH CLUSTER OF GALAXIES
MCG-04-04-004	01 17 03.	- 21 02	54	15.	GALAXY
ARC 0176	01 17 04.	- 08 25		17.6	RICH CLUSTER OF GALAXIES
FATH 1.077	01 17 04.	- 14 47	14		NEBULA
IC 0092	01 17 05.	+ 32 29			NONSTELLAR OBJECT
MCG+00-04-084	01 17 06.	+ 03 10	24	12.4	GALAXY
ZWG 436.025	01 17 06.	+ 09 54		15.5	GALAXY
ZWG 502.030	01 17 06.	+ 32 12		14.4	GALAXY
UGC 00857	01 17 06.	+ 32 12	72	14.4	GALAXY Sc
MCG+05-04-020	01 17 06.	+ 32 30 30.	39	15.	GALAXY
ZWG 502.031	01 17 06.	+ 32 42		15.7	GALAXY
UGC 00856	01 17 06.	- 01 58	78	16.5	GALAXY
IC 1671	01 17 06.	- 17 21 37.			NONSTELLAR OBJECT
MCG-03-04-047	01 17 06.	- 17 41	30	14.5	GALAXY
LIN 522	01 17 06.	- 73 47 19.		14.14	SMC EMISSION LINE OBJECT
KN 14.002	01 17 06.5	+ 33 12 25.			NEBULA
FATH 1.078	01 17 07.	- 13 59	19		NEBULA
MCG+05-04-019	01 17 09.	+ 32 12	60	14.5	GALAXY
MCG-03-04-048	01 17 09.	- 17 57	12	15.	GALAXY
RNGC 0473	01 17 10.	+ 16 17		13.5	GALAXY
LIN 521	01 17 10.	- 72 57 43.		16.92	SMC EMISSION LINE OBJECT
ZWG 385.070	01 17 12.	+ 03 08		12.4	GALAXY
UGC 00858	01 17 12.	+ 03 08	198	12.4	GALAXY Sb/Sc
ZWG 459.030	01 17 12.	+ 16 17		13.2	GALAXY
UGC 00859	01 17 12.	+ 16 17	156	13.2	GALAXY S0
MCG+03-04-022	01 17 12.	+ 16 17	54	13.5	GALAXY
ZWG 459.031	01 17 12.	+ 21 06		15.0	GALAXY
MCG+03-04-021	01 17 12.	+ 21 07 30.	39	15.5	GALAXY
MCG+04-04-003	01 17 12.	+ 23 43	27	17.	GALAXY
VB 043	01 17 12.	+ 64 28	1846		STELLAR RING
TON-S 0206	01 17 12.	- 28 45		16.0	BLUE STAR
LB 03193	01 17 12.	- 62 11		12.8	FAINT BLUE STAR
RNGC 0470	01 17 13.	+ 03 09		12.5	GALAXY
RNGC 0471	01 17 16.	+ 14 31		14.0	GALAXY
IC 1669	01 17 17.	+ 32 55 47.			NONSTELLAR OBJECT
ZWG 436.026	01 17 18.	+ 12 05		15.7	GALAXY
ZWG 436.027	01 17 18.	+ 12 22		15.7	GALAXY
ZWG 436.028	01 17 18.	+ 12 40		15.6	GALAXY
UGC 00860	01 17 18.	+ 12 40	66	15.6	GALAXY Sc
ZWG 436.029	01 17 18.	+ 14 31		14.0	GALAXY
UGC 00861	01 17 18.	+ 14 31	78	14.0	GALAXY S0
MCG+02-04-024	01 17 18.	+ 14 31	60	14.5	GALAXY
IC 0094	01 17 18.	+ 32 26			NONSTELLAR OBJECT
ZWG 502.032	01 17 18.	+ 32 56		15.7	GALAXY
MCG-02-04-023	01 17 18.	- 10 13	24	15.5	GALAXY
MCG-03-04-049	01 17 18.	- 16 31	36	15.5	GALAXY
TON-S 0207	01 17 18.	- 28 39		16.2	BLUE STAR
FATH 1.079	01 17 19.	+ 15 21	3		NEBULA
LIN 523	01 17 19.	- 73 33 31.		14.00	SMC EMISSION LINE OBJECT
IC 0097	01 17 21.	+ 14 35			NONSTELLAR OBJECT
RNGC 0475	01 17 22.	+ 14 35			GALAXY
MCG+02-04-026	01 17 24.	+ 12 39	48	14.5	GALAXY
MCG+02-04-025	01 17 24.	+ 14 05	30	14.5	GALAXY
ZWG 436.030	01 17 24.	+ 14 06		14.9	GALAXY
ZWG 502.033	01 17 24.	+ 33 15		14.5	GALAXY
UGC 00862	01 17 24.	+ 33 15	60	14.5	GALAXY
ZC 0117.4+3538	01 17 24.	+ 35 38	2620		CLUSTER OF GALAXIES
ZWG 551.015	01 17 24.	+ 49 53		15.0	GALAXY
UGC 00863	01 17 24.	+ 78 22	120	17.	GALAXY Sc?
SER 010.01	01 17 24.	- 68 19	40	15.	LOW SURFACE BRIGHT. GLXY
KN 14.003	01 17 24.9	+ 33 14 30.			NEBULA
ARP 048	01 17 27.	+ 12 05			PECULIAR GALAXY
ARP 227	01 17 28.	+ 03 10			PECULIAR GALAXY
ZWG 385.071	01 17 30.	+ 03 09		12.9	GALAXY
UGC 00864	01 17 30.	+ 03 09	600	12.9	GALAXY S0
MCG+02-04-027	01 17 30.	+ 14 17	24	16.	GALAXY
UGC 00865	01 17 30.	+ 14 18	90	16.0	GALAXY
ZWG 459.032	01 17 30.	+ 17 33		15.6	GALAXY
MCG+05-04-021	01 17 30.	+ 33 15 30.	36	14.5	GALAXY
MCG+08-03-015	01 17 30.	+ 49 52	24	14.	GALAXY
ZH 82	01 17 30.	- 00 44		15.7	GALAXY
RNGC 0474	01 17 31.	+ 03 09		12.0	GALAXY
IC 0096	01 17 32.	+ 29 24 28.			NONSTELLAR OBJECT
HELW 358	01 17 32.	- 34 30 37.			NEBULA
RNGC 0491A	01 17 33.	- 34 08			GALAXY

OBJECT NAME	RIGHT ASCEN.	DECLINATION	DIAM.	MAGN.	TYPE OF OBJECT
KN 14.004	01 17 33.0	+ 30 29 50.			NEBULA
RNGC 0476	01 17 34.	+ 15 45		15.0	GALAXY
HELW 359	01 17 34.	- 34 31 37.			NEBULA
MCG+00-04-085	01 17 36.	+ 03 10	30	13.0	GALAXY
ZWG 459.033	01 17 36.	+ 15 45		15.2	GALAXY
ZWG 459.034	01 17 36.	+ 21 13		15.2	GALAXY
ZWG 385.073	01 17 36.	- 00 11		15.3	GALAXY
ZH 79	01 17 36.	- 00 11		15.3	GALAXY
MCG+00-04-086	01 17 36.	- 00 11 30.	24	15.5	GALAXY
ZWG 385.072	01 17 36.	- 00 28		15.6	GALAXY
ZH 76	01 17 36.	- 00 28		15.6	GALAXY
UGC 00866	01 17 36.	- 00 28	102	15.6	GALAXY
MCG+00-04-087	01 17 36.	- 00 28	66	14.5	GALAXY
MCG+00-04-088	01 17 36.	- 00 35	27	15.5	GALAXY
ARC 0177	01 17 36.	- 21 18		17.5	RICH CLUSTER OF GALAXIES
TON-S 0208	01 17 36.	- 30 29		14.8	BLUE STAR
MCG-06-04-006	01 17 36.	- 33 22	66	14.	GALAXY
DV.56 N0491A	01 17 36.	- 34 08	216		SBc GALAXY
LIN 524	01 17 37.	- 73 31 44.		14.75	SMC EMISSION LINE OBJECT
RNGC 0484	01 17 38.	- 58 48			UNVERIFIED SOUTHERN OBJECT
MCG+03-04-023	01 17 39.	+ 15 45 30.	30	15.	GALAXY
HOLM 040B	01 17 40.	+ 15 45	60	15.2	PART OF MULTIPLE GALAXY
HOLM 040A	01 17 40.	+ 15 46	24	14.6	PART OF MULTIPLE GALAXY
KN 14.005	01 17 40.7	+ 32 26 49.			NEBULA
PATH 1.080	01 17 41.	+ 15 45	5		NEBULA
HELW 360	01 17 41.	- 34 09 37.			NEBULA
KN 14.077	01 17 41.6	+ 29 59 50.			
UGC 00868	01 17 42.	+ 09 03	72	16.0	GALAXY S
UGC 00869	01 17 42.	+ 30 30	66	16.0	GALAXY S
ZWG 502.034	01 17 42.	+ 32 27		14.2	GALAXY
RNGC 0472	01 17 42.	+ 32 27		14.0	GALAXY
UGC 00870	01 17 42.	+ 32 27	78	14.2	GALAXY VY CMPT
ZWG 521.002	01 17 42.	+ 37 54		14.9	GALAXY
ZWG 385.074	01 17 42.	- 00 36		15.5	GALAXY
ZH 77	01 17 42.	- 00 36		15.3	GALAXY
UGC 00867	01 17 42.	- 00 36	60	15.5	GALAXY Sb-c
MCG-06-04-007	01 17 42.	- 33 21	36	15.	GALAXY
AGU 06	01 17 42.	- 41 28 00.	42	13.	PECULIAR GALAXY
KN 14.078	01 17 42.0	+ 32 46 40.			
MCG+05-04-022	01 17 45.	+ 32 27	24	14.5	GALAXY
MCG-04-04-005	01 17 45.	- 22 38 30.	36	15.	GALAXY
KN 14.006	01 17 46.8	+ 29 21 26.			NEBULA
ZWG 385.075	01 17 48.	+ 01 37		15.7	GALAXY
MCG+01-04-022	01 17 48.	+ 05 34	60	16.	GALAXY
UGC 00871	01 17 48.	+ 05 35	120	17.	GALAXY DWRF IR
MCG+05-04-023	01 17 48.	+ 29 21	42	15.	GALAXY
ZWG 502.035	01 17 48.	+ 29 22		14.7	GALAXY
KARA.72 030A	01 17 48.	+ 29 22	54	14.7	PART OF DOUBLE GALAXY
RNGC 0478	01 17 48.	- 22 38		15.0	GALAXY
KN 14.007	01 17 48.8	+ 29 22 08.			NEBULA
KN 14.079	01 17 49.4	+ 31 25 26.			
KN 14.080	01 17 49.9	+ 31 34 48.			
IC 1672	01 17 50.	+ 29 26 04.			NONSTELLAR OBJECT
MCG-03-04-050	01 17 51.	- 18 28	36	15.5	GALAXY
KN 14.008	01 17 51.9	+ 29 26 20.			NEBULA
KN 14.081	01 17 53.5	+ 27 18 23.			
ZWG 411.024	01 17 54.	+ 07 05		15.3	GALAXY
ZWG 502.036	01 17 54.	+ 29 26		14.0	GALAXY
UGC 00872	01 17 54.	+ 29 26	108	14.0	GALAXY S
KARA.72 030B	01 17 54.	+ 29 26	72	14.0	PART OF DOUBLE GALAXY
MCG+05-04-024	01 17 54.	+ 29 26	66	14.5	GALAXY
ZWG 502.037	01 17 54.	+ 30 08		15.5	GALAXY
UGC 00873	01 17 54.	+ 30 08	66	15.5	GALAXY Sb-c
ZH 78	01 17 54.	- 00 20		15.4	GALAXY
MCG-02-04-024	01 17 54.	- 08 48 30.	48	15.5	GALAXY
MCG-06-04-008	01 17 54.	- 34 10	120	15.	GALAXY
KN 14.082	01 17 56.3	+ 30 22 42.			
KN 14.009	01 17 57.8	+ 32 46 53.			NEBULA
KN 14.083	01 17 57.9	+ 30 22 22.			
KN 14.084	01 17 59.0	+ 27 15 54.			
KN 14.085	01 17 59.2	+ 33 12 33.			
SN 1966H	01 18 00.	+ 03 10		19.0	SUPERNOVA
ZWG 459.035	01 18 00.	+ 18 15		15.6	GALAXY
IC 1673	01 18 00.	+ 32 45 39.			NONSTELLAR OBJECT
ZWG 502.038	01 18 00.	+ 32 48		14.6	GALAXY
ZWG 502.039	01 18 00.	+ 33 10		15.2	GALAXY
MCG+01-04-026	01 18 00.	- 07 05	48	16.5	GALAXY
KN 14.086	01 18 00.0	+ 29 23 17.			
MCG-03-04-051	01 18 03.	- 17 36	36	14.	GALAXY
MCG-03-04-052	01 18 03.	- 17 39	72	14.5	GALAXY
MCG-03-04-053	01 18 03.	- 17 39 30.	60	14.5	GALAXY
RNGC 0482	01 18 05.	- 41 13			GALAXY
KN 14.087	01 18 05.3	+ 31 23 17.			
ZWG 521.003	01 18 06.	+ 34 44		14.8	GALAXY
KN 14.088	01 18 08.0	+ 28 46 43.			
MCG+00-04-089	01 18 09.	+ 01 01	42	15.5	GALAXY
MCG+07-03-029	01 18 09.	+ 40 11	36	16.	GALAXY
MCG+07-03-030	01 18 09.	+ 42 36 30.	60	14.5	GALAXY
HELW 361	01 18 10.	- 34 23 02.			NEBULA
IC 1675	01 18 11.	+ 33 59 15.			NONSTELLAR OBJECT
KN 14.089	01 18 11.9	+ 29 02 13.			
ZWG 385.076	01 18 12.	+ 01 02		15.5	GALAXY
UGC 00874	01 18 12.	+ 01 02	78	15.5	GALAXY S0?
ZWG 385.077	01 18 12.	+ 01 10		15.2	GALAXY
UGC 00875	01 18 12.	+ 01 10	72	15.2	GALAXY Sc
MCG+00-04-090	01 18 12.	+ 01 11	42	15.	GALAXY
MCG+01-04-023	01 18 12.	+ 03 53 30.	60	16.	GALAXY
UGC 00876	01 18 12.	+ 03 56	66	17.	GALAXY
ZWG 436.031	01 18 12.	+ 11 38		15.4	GALAXY
ZWG 502.040	01 18 12.	+ 28 59		15.7	GALAXY
UGC 00877	01 18 12.	+ 28 59	60	15.7	GALAXY Sb-c
ZWG 502.041	01 18 12.	+ 30 00		15.7	GALAXY
IC 1676	01 18 12.	+ 30 00 09.			NONSTELLAR OBJECT
ZWG 521.004	01 18 12.	+ 33 38		14.7	GALAXY
UGC 00878	01 18 12.	+ 33 38	72	14.7	GALAXY S0?
ZWG 521.005	01 18 12.	+ 34 00		14.3	GALAXY
UGC 00879	01 18 12.	+ 34 00	48	14.3	GALAXY
MCG+06-04-002	01 18 12.	+ 34 42	33	14.5	GALAXY
ZWG 536.030	01 18 12.	+ 40 12		14.8	GALAXY
ZWG 536.031	01 18 12.	+ 42 35		15.0	GALAXY
UGC 00880	01 18 12.	+ 42 35	72	15.0	GALAXY S
VB 044	01 18 12.	+ 64 33	322		STELLAR RING
MCG-02-04-026	01 18 12.	- 08 45 30.	48	16.5	GALAXY
MCG-02-04-025	01 18 12.	- 08 45 30.	30	16.5	GALAXY
TON-S 0209	01 18 12.	- 27 16		16.1	BLUE STAR
LB 03194	01 18 12.	- 75 43		12.6	FAINT BLUE STAR
KN 14.010	01 18 12.1	+ 29 59 48.			NEBULA
KN 14.090	01 18 13.8	+ 30 20 29.			

OBJECT NAME	RIGHT ASCEN.	DECLINATION	DIAM.	MAGN.	TYPE OF OBJECT
MCG-03-04-054	01 18 14.	- 14 58	30	15.	GALAXY
MCG+01-04-024	01 18 15.	+ 04 32	18	15.	GALAXY
LIN 525	01 18 15.	- 72 53 09.		14.42	SMC EMISSION LINE OBJECT
KN 14.091	01 18 15.0	+ 32 22 42.			
KN 14.092	01 18 15.5	+ 31 11 32.			
KN 14.093	01 18 16.3	+ 33 22 22.			
KN 14.012	01 18 17.6	+ 31 02 14.			NEBULA
ZWG 411.025	01 18 18.	+ 04 32		15.4	GALAXY
UGC 00881	01 18 18.	+ 04 32	78	15.4	GALAXY E
ZWG 411.026	01 18 18.	+ 06 18		15.7	GALAXY
UGC 00882	01 18 18.	+ 06 18	120	15.7	GALAXY IRR
MCG+01-04-025	01 18 18.	+ 06 19	36	15.	GALAXY
MCG+02-04-028	01 18 18.	+ 11 37	12	15.	GALAXY
UGC 00883	01 18 18.	+ 16 49	90	16.5	GALAXY SB IV-V
MCG+03-04-024	01 18 18.	+ 16 49	66	15.	GALAXY
UGC 00884	01 18 18.	+ 25 18	90	16.5	GALAXY
IC 1677	01 18 18.	+ 32 57 38.			NONSTELLAR OBJECT
ZWG 502.042	01 18 18.	+ 32 58		15.2	GALAXY
ZWG 502.043	01 18 18.	+ 33 08		14.9	GALAXY
MCG+06-04-003	01 18 18.	+ 33 37	48	14.5	GALAXY
MCG+06-04-004	01 18 18.	+ 33 58	48	14.5	GALAXY
MCG+07-03-031	01 18 18.	+ 40 12 30.	48	14.5	GALAXY
MCG+00-04-091	01 18 18.	- 01 36	18	16.	GALAXY
KN 14.011	01 18 18.8	+ 32 57 05.			NEBULA
KN 14.094	01 18 19.0	+ 33 22 20.			
KN 14.013	01 18 19.7	+ 31 02 10.			NEBULA
KN 14.095	01 18 22.9	+ 29 55 07.			
UGC 00885	01 18 24.	+ 02 43	66	16.5	GALAXY Sc
MCG+01-04-026	01 18 24.	+ 04 18	48	15.5	GALAXY
ZWG 411.027	01 18 24.	+ 04 19		15.7	GALAXY
ZWG 411.028	01 18 24.	+ 05 18		15.5	GALAXY
ZWG 436.032	01 18 24.	+ 15 26		15.6	GALAXY
SN 1936B	01 18 24.	+ 15 26		14.0	SUPERNOVA
MCG+04-04-005	01 18 24.	+ 21 46	12	17.	GALAXY
MCG+04-04-004	01 18 24.	+ 21 46 30.	15	17.	GALAXY
3ZW 023	01 18 24.	+ 22 31			COMPACT GALAXY
ZWG 502.044	01 18 24.	+ 32 50		15.0	GALAXY
MCG+05-04-025	01 18 24.	+ 32 58	54	15.	GALAXY
ZWG 521.006	01 18 24.	+ 36 25		15.4	GALAXY
ZWG 536.032	01 18 24.	+ 40 13		14.0	GALAXY
UGC 00886	01 18 24.	+ 40 13	150	14.0	GALAXY Sc/SBc
KN 14.096	01 18 24.7	+ 30 50 44.			
KN 14.097	01 18 25.4	+ 30 57 45.			
IC 1678	01 18 26.	+ 05 17 32.			NONSTELLAR OBJECT
IC 0098	01 18 26.	- 12 52 11.			NONSTELLAR OBJECT
BC PKS0118+03	01 18 26.15	+ 03 28 29.3		18.	QUASI-STELLAR OBJECT
KN 14.098	01 18 26.5	+ 30 17 11.			
RNGC 0480	01 18 27.	- 10 08			NON-EXISTENT OBJECT
SHB 026	01 18 27.6	+ 03 28 19.		18.5	QUASI-STELLAR OBJECT
RNGC 0477	01 18 28.	+ 40 13		14.0	GALAXY
KN 14.014	01 18 29.2	+ 32 49 50.			NEBULA
ZWG 411.029	01 18 30.	+ 04 53		15.7	GALAXY
UGC 00887	01 18 30.	+ 04 53	72	15.7	GALAXY Sb-c
UGC 00888	01 18 30.	+ 05 10	90	17.	GALAXY DWRF IR
MCG+01-04-027	01 18 30.	+ 05 28	36	15.	GALAXY
ZWG 411.030	01 18 30.	+ 06 01		15.6	GALAXY
MCG+02-04-029	01 18 30.	+ 15 25	42	15.	GALAXY
UGC 00889	01 18 30.	+ 17 36	66	16.0	GALAXY SBc
3ZW 024	01 18 30.	+ 21 46			COMPACT GALAXY
VV 323A	01 18 30.	+ 39 11 30.	120	14.	INTERACTING GALAXY
MCG+07-03-032	01 18 30.	+ 40 13 30.	120	13.	GALAXY
MCG+00-04-092	01 18 30.	- 00 48 30.	30	16.	GALAXY
MCG-02-04-027	01 18 30.	- 12 53	36	15.	GALAXY
LIN 526	01 18 30.	- 73 25 39.		14.34	SMC EMISSION LINE OBJECT
KN 14.099	01 18 31.2	+ 31 29 12.			
KN 14.100	01 18 31.3	+ 33 01 25.			
VV 323B	01 18 33.	+ 39 10	90	14.5	INTERACTING GALAXY
RNGC 0481	01 18 33.	- 09 28		14.0	GALAXY
MCG-02-04-028	01 18 33.	- 12 50	12	15.	GALAXY
MCG-02-04-029	01 18 33.	- 14 07	36	15.	GALAXY
ZWG 385.078	01 18 36.	+ 01 06		14.3	GALAXY
UGC 00890	01 18 36.	+ 01 06	144	14.3	GALAXY Sb
MCG+00-04-093	01 18 36.	+ 01 07 30.	102	13.5	GALAXY
MCG+01-04-028	01 18 36.	+ 05 10	48	16.	GALAXY
MCG+01-04-029	01 18 36.	+ 06 11	24	16.	GALAXY
ZWG 436.033	01 18 36.	+ 12 09		15.7	GALAXY
UGC 00891	01 18 36.	+ 12 09	192	15.7	GALAXY DWARF
3ZW 025	01 18 36.	+ 16 46			COMPACT GALAXY
ZWG 459.036	01 18 36.	+ 16 46		15.6	GALAXY
ZC 0118.6+2035	01 18 36.	+ 20 35	1080		CLUSTER OF GALAXIES
MCG+06-04-005	01 18 36.	+ 36 24	48	15.	GALAXY
KN 14.015	01 18 36.0	+ 28 17 31.			NEBULA
MCG-02-04-030	01 18 39.	- 09 28 30.	24	14.5	GALAXY
LIN 527	01 18 40.	- 72 34 03.		14.62	SMC EMISSION LINE OBJECT
KN 14.101	01 18 40.2	+ 29 07 30.			
MCG+01-04-031	01 18 42.	+ 03 35	60	14.5	GALAXY
ZWG 411.031	01 18 42.	+ 03 36		15.1	GALAXY
RNGC 0479	01 18 42.	+ 03 36		15.0	GALAXY
UGC 00893	01 18 42.	+ 03 36	72	15.1	GALAXY Sb
MCG+01-04-030	01 18 42.	+ 04 53	72	15.	GALAXY
MCG+03-04-025	01 18 42.	+ 15 40	36	15.	GALAXY
ZC 0118.7+1946	01 18 42.	+ 19 46	1680		CLUSTER OF GALAXIES
ZWG 502.045	01 18 42.	+ 31 58		15.3	GALAXY
ZWG 385.079	01 18 42.	- 00 48		14.5	GALAXY
UGC 00892	01 18 42.	- 00 48	126	14.5	GALAXY SBa
MCG+00-04-094	01 18 42.	- 00 52	27	16.	GALAXY
MCG-05-04-030	01 18 42.	- 27 15	36	16.	GALAXY
LB 03195	01 18 42.	- 64 52		12.8	FAINT BLUE STAR
LB 03204	01 18 42.	- 73 21		12.8	FAINT BLUE STAR
KN 14.102	01 18 43.8	+ 31 11 31.			
HOW 77	01 18 44.	- 72 52 04.	36		STAR CLUSTER IN SMC
KN 14.103	01 18 44.1	+ 29 06 46.			
KN 14.016	01 18 44.4	+ 31 57 12.			NEBULA
KN 14.104	01 18 44.5	+ 28 54 46.			
MCG+02-04-030	01 18 45.	+ 12 08	108	14.5	GALAXY
MCG+00-04-095	01 18 45.	- 00 49	66	14.	GALAXY
KN 14.105	01 18 45.7	+ 32 41 33.			
ARP 067	01 18 47.	- 00 48			PECULIAR GALAXY
IC 1681	01 18 47.4	- 00 10 15.		14.8	GALAXY SB(s)
MCG+00-04-096	01 18 48.	+ 03 23	36	15.5	GALAXY
ZWG 411.032	01 18 48.	+ 06 45		14.2	GALAXY
RNGC 0485	01 18 48.	+ 06 45		14.0	GALAXY
UGC 00895	01 18 48.	+ 06 45	120	14.2	GALAXY S
MCG+01-04-032	01 18 48.	+ 06 45	90	13.	GALAXY
UGC 00896	01 18 48.	+ 08 55	72	16.5	GALAXY
ZWG 436.034	01 18 48.	+ 14 15		15.0	GALAXY
UGC 00897	01 18 48.	+ 15 41	60	16.0	GALAXY Sb-c
ZWG 459.037	01 18 48.	+ 19 36		15.7	GALAXY

OBJFCT NAME	RIGHT ASCEF.	DECLINATION	DIAM.	MAGN.	TYPE OF OBJECT
UGC 00898	01 18 48.	+ 19 36	60	15.7	GALAXY Sb
ZWG 459.038	01 18 48.	+ 20 39		15.2	GALAXY
UGC 00899	01 18 48.	+ 20 39	60	15.2	GALAXY Sb-c
ZWG 481.003	01 18 48.	+ 23 30		15.7	GALAXY
UGC 00900	01 18 48.	+ 23 30	96	15.7	GALAXY
MCG+04-04-006	01 18 48.	+ 23 30	42	16.	GALAXY
ZC 0118.8+2505	01 18 48.	+ 25 05	1280		CLUSTER OF GALAXIES
ZWG 502.046	01 18 48.	+ 32 21		14.5	GALAXY
UGC 00901	01 18 48.	+ 32 21	48	14.5	GALAXY
ZWG 502.047	01 18 48.	+ 33 21		15.5	GALAXY
ZWG 385.080	01 18 48.	- 00 10		14.8	GALAXY
ZH 75	01 18 48.	- 00 10		15.0	GALAXY
UGC 00894	01 18 48.	- 00 10	72	14.8	GALAXY S
MCG+00-04-097	01 18 48.	- 00 10	54	14.	GALAXY
ZH 57	01 18 48.	- 00 48		14.6	GALAXY
KN 14.017	01 18 48.4	+ 31 24 48.			NEBULA
MCG+05-04-026	01 18 51.	+ 33 21	36	15.	GALAXY
MCG-02-04-031	01 18 51.	- 10 53	72	15.	GALAXY
MCG-04-04-006	01 18 51.	- 23 03	108	14.	GALAXY
ZWG 385.081	01 18 54.	+ 03 22		15.5	GALAXY
ZWG 459.039	01 18 54.	+ 18 07		15.5	GALAXY
ZC 0118.9+2141	01 18 54.	+ 21 41	1080		CLUSTER OF GALAXIES
ZWG 502.048	01 18 54.	+ 33 15		15.5	GALAXY
ZWG 551.016	01 18 54.	+ 49 47		15.5	GALAXY
UGC 00902	01 18 54.	+ 49 47	96	15.5	GALAXY (E)
MCG+08-03-016	01 18 54.	+ 49 47	19	15.	GALAXY
MCG-05-04-031	01 18 54.	- 26 56	72	14.5	GALAXY
MCG-06-04-009	01 18 54.	- 36 23	90	14.	GALAXY
KN 14.106	01 18 54.9	+ 32 51 29.			
KN 14.018	01 18 56.1	+ 33 13 52.			NEBULA
KN 14.107	01 18 56.8	+ 32 51 21.			
IC 1679	01 18 57.	+ 33 13 43.			NONSTELLAR OBJECT
VDB.66G 010	01 19	+ 12 10	100		DWARF GALAXY
FEIG 012	01 19 00.	+ 12 26		12.7	FAINT BLUE STAR
MCG+03-04-026	01 19 00.	+ 17 20	102	14.	GALAXY
ZWG 459.040	01 19 00.	+ 18 18		15.0	GALAXY
MCG+04-04-007	01 19 00.	+ 23 32	60	17.	GALAXY
ZWG 502.049	01 19 00.	+ 23 02		14.9	GALAXY
MCG+05-04-027	01 19 00.	+ 33 12	30	15.	GALAXY
VB 113	01 19 00.	+ 60 45	282		STELLAR RING
ISS 0040	01 19 00.	+ 60 45	283		STELLAR RING
MCG-06-04-010	01 19 00.	- 36 45	36	14.5	GALAXY
KN 14.108	01 19 00.2	+ 27 10 26.			
IC 1680	01 19 02.	+ 33 01 37.			NONSTELLAR OBJECT
RNGC 0491	01 19 03.	- 34 19		13.0	GALAXY
KN 14.109	01 19 04.8	+ 32 32 54.			
MCG+01-04-033	01 19 06.	+ 05 00	300	11.	GALAXY
ZWG 436.035	01 19 06.	+ 11 28		15.1	GALAXY
ZWG 459.041	01 19 06.	+ 17 20		14.7	GALAXY
UGC 00903	01 19 06.	+ 17 20	120	14.7	GALAXY S
ZWG 459.042	01 19 06.	+ 18 00		15.3	GALAXY
UGC 00904	01 19 06.	+ 18 00	78	15.3	GALAXY SBa
MCG+03-04-027	01 19 06.	+ 18 00	60	14.5	GALAXY
ARC 0179	01 19 06.	+ 19 14		15.3	RICH CLUSTER OF GALAXIFS
ZC 0119.1+2204	01 19 06.	+ 22 04	6450		CLUSTER OF GALAXIES
UGC 00905	01 19 06.	+ 23 31	90	16.0	GALAXY S
MCG+05-04-028	01 19 06.	+ 33 02	36	15.	GALAXY
ZWG 502.050	01 19 06.	+ 33 16		14.0	GALAXY
UGC 00906	01 19 06.	+ 33 16	54	14.0	GALAXY
MCG-06-04-011	01 19 06.	- 34 20	78	13.0	GALAXY
KN 14.110	01 19 06.1	+ 28 10 37.			
ARC 0178	01 19 07.	+ 19 50		17.1	RICH CLUSTER OF GALAXIES
KN 14.111	01 19 08.0	+ 29 04 28.			
KN 14.019	01 19 08.2	+ 33 15 17.			NEBULA
KN 14.020	01 19 09.7	+ 30 55 52.			NEBULA
KN 14.112	01 19 10.0	+ 32 45 12.			
RNGC 0489	01 19 11.	+ 08 56		13.5	GALAXY
ZC 0119.2+0000	01 19 12.	+ 00 00	1750		CLUSTER OF GALAXIES
ZWG 411.033	01 19 12.	+ 05 00		11.6	GALAXY
RNGC 0488	01 19 12.	+ 05 00		11.5	GALAXY
UGC 00907	01 19 12.	+ 05 00	360	11.6	GALAXY Sb
ZWG 411.034	01 19 12.	+ 08 56		13.4	GALAXY
UGC 00908	01 19 12.	+ 08 56	108	13.4	GALAXY S
MCG+03-04-028	01 19 12.	+ 15 30	48	14.	GALAXY
ZC 0119.2+1932	01 19 12.	+ 19 32	4170		CLUSTER OF GALAXIES
RNGC 0483	01 19 12.	+ 33 16		14.0	GALAXY
MCG+05-04-029	01 19 12.	+ 33 16	42	14.5	GALAXY
ZWG 521.007	01 19 12.	+ 37 09		14.4	GALAXY
UGC 00909	01 19 12.	+ 37 09	102	14.4	GALAXY Sc
MCG+01-04-027	01 19 12.	- 07 36	18	15.5	GALAXY
LB 03197	01 19 12.	- 65 39		13.4	FAINT BLUE STAR
KN 14.113	01 19 14.8	+ 30 23 42.			
MCG+01-04-034	01 19 15.	+ 08 57	90	13.5	GALAXY
MCG+02-04-031	01 19 15.	+ 11 27	36	15.	GALAXY
MCG+06-04-006	01 19 15.	+ 37 08	90	14.	GALAXY
MCG+08-03-017	01 19 15.	+ 49 46	12	16.	GALAXY
MCG-03-04-055	01 19 15.	- 18 04	78	14.5	GALAXY
KN 14.114	01 19 15.2	+ 29 02 56.			
ARC 0180	01 19 18.	+ 02 46		17.0	RICH CLUSTER OF GALAXIFS
ZWG 459.043	01 19 18.	+ 15 31		15.0	GALAXY
UGC 00910	01 19 18.	+ 15 31	84	15.0	GALAXY Sc
ZWG 502.051	01 19 18.	+ 31 58		15.0	GALAXY
UGC 00911	01 19 18.	+ 31 58	66	15.0	GALAXY SBb
VB 114	01 19 18.	+ 58 06	188		STELLAR RING
ISS 0041	01 19 18.	+ 58 06	193		STELLAR RING
MCG+01-04-028	01 19 18.	- 03 46	42	15.	GALAXY
MCG-02-04-032	01 19 18.	- 12 02	60	14.	GALAXY
MCG-06-04-012	01 19 18.	- 36 49	78	15.	GALAXY
LB 00445	01 19 21.	+ 29 17 30.		16.2	FAINT BLUE STAR
MCG+06-04-007	01 19 21.	+ 34 22		16.	GALAXY
SG 3.007	01 19 21.	+ 61 45	180		DIFFUSF EMISSION NEBULA
LIN 528	01 19 21.	- 73 48 46.		14.41	SMC EMISSION LINE OBJECT
ARC 0181	01 19 22.	+ 00 03		17.2	RICH CLUSTER OF GALAXIFS
KN 14.022	01 19 22.1	+ 30 23 23.			NEBULA
KN 14.021	01 19 22.2	+ 31 57 11.			NEBULA
RNGC 0487	01 19 23.	- 16 37		14.0	GALAXY
MCG+01-04-035	01 19 24.	+ 05 06	36	15.	GALAXY
ZWG 502.052	01 19 24.	+ 31 27		15.7	GALAXY
MCG+05-04-030	01 19 24.	+ 31 58	42	15.	GALAXY
IC 1682	01 19 24.	+ 32 57 42.			NONSTELLAR OBJFCT
ZWG 502.053	01 19 24.	+ 33 00		14.3	GALAXY
UGC 00912	01 19 24.	+ 33 00	54	14.3	GALAXY
ZWG 521.008	01 19 24.	+ 34 25		14.4	GALAXY
UGC 00913	01 19 24.	+ 34 25	33	14.4	GALAXY PECULP
ZC 0119.4+4206	01 19 24.	+ 42 06	1080		CLUSTER OF GALAXIES
ZWG 385.083	01 19 24.	- 01 16		15.7	GALAXY
ZH 83	01 19 24.	- 01 16		15.5	GALAXY
ZWG 385.082	01 19 24.	- 01 18		15.1	GALAXY

OBJFCT NAME	RIGHT ASCEN.	DECLINATION	DIAM.	MAGN.	TYPE OF OBJECT
MCG+00-04-098	01 19 24.	- 01 18	36	16.5	GALAXY
MCG-03-04-056	01 19 24.	- 16 37	54	14.	GALAXY
MCG-06-04-013	01 19 24.	- 33 24		14.5	GALAXY
KN 14.023	01 19 24.8	+ 32 59 52.			NEBULA
MCG+08-03-018	01 19 27.	+ 49 45	12	16.	GALAXY
LIN 529	01 19 28.	- 73 47 05.		15.64	SMC EMISSION LINE OBJECT
ARC 0183	01 19 29.	- 22 10		16.5	RICH CLUSTER OF GALAXIES
ZWG 411.035	01 19 30.	+ 05 06		15.6	GALAXY
RNGC 0490	01 19 30.	+ 05 06		15.5	GALAXY
RNGC 0486	01 19 30.	+ 05 08		15.5	GALAXY
MCG+01-04-037	01 19 30.	+ 05 08 30.	18	15.5	GALAXY
MCG+01-04-036	01 19 30.	+ 05 31	24	17.	GALAXY
ZC 0119.5+1334	01 19 30.	+ 13 34	3700		CLUSTER OF GALAXIES
3ZW 026	01 19 30.	+ 16 02			COMPACT GALAXY
ZWG 502.054	01 19 30.	+ 28 32		15.6	GALAXY
MCG+05-04-031	01 19 30.	+ 28 32	54	15.	GALAXY
MCG+05-04-032	01 19 30.	+ 33 00 30.	36	15.	GALAXY
2ZW 001	01 19 30.	- 01 18			COMPACT GALAXY
KW 63	01 19 30.	- 01 18	15		SEYFERT GALAXY
ZH 84	01 19 30.	- 01 18		15.1	GALAXY
VVI 08	01 19 30.	- 01 18	30	15.56	SEYFERT GALAXY
MCG+01-04-029	01 19 30.	- 07 34	42	15.	GALAXY
MCG-03-04-057	01 19 30.	- 17 05	12	15.5	GALAXY
TON-S 0210	01 19 30.	- 28 37		15.0	BLUE STAR
BC TONS210	01 19 30.	- 28 37		14.83	QUASI-STFLLAR OFJECT
HELW 362	01 19 30.	- 34 24 10.			NEBULA
FATH 1.081	01 19 31.	+ 16 02	8		NEBULA
KN 14.024	01 19 31.1	+ 28 32 10.			NEBULA
MCG+01-04-038	01 19 33.	+ 05 09	36	15.	GALAXY
HELW 363	01 19 34.	- 34 27 28.			NEBULA
LIN 530	01 19 34.	- 73 36 47.		14.82	SMC EMISSION LINF OBJECT
ZWG 385.084	01 19 36.	+ 00 41		13.0	GALAXY
UGC 00914	01 19 36.	+ 00 41	258	13.0	GALAXY Sc
ZWG 411.036	01 19 36.	+ 05 09		15.5	GALAXY
RNGC 0492	01 19 36.	+ 05 09		15.5	GALAXY
ZC 0119.6+2428	01 19 36.	+ 24 28	2820		CLUSTER OF GALAXIES
ZC 0119.6+5035	01 19 36.	+ 50 35	14380		CLUSTER OF GALAXIFS
RNGC 0493	01 19 37.	+ 00 41		13.0	GALAXY
FATH 1.082	01 19 38.	+ 16 03	22		NEBULA
SN 1971S	01 19 39.	+ 00 42		15.5	SUPERNOVA
ARC 0182	01 19 40.	- 07 11		17.6	RICH CLUSTER OF GALAXIFS
LIN 531	01 19 41.	- 73 29 35.		13.78	SMC EMISSION LINE OBJECT
ZH 73	01 19 42.	+ 00 40		13.8	GALAXY
MCG+00-04-099	01 19 42.	+ 00 40	186	12.	GALAXY
ZWG 502.055	01 19 42.	+ 32 51		15.7	GALAXY
ZWG 521.009	01 19 42.	+ 38 57		15.2	GALAXY
ZH 60	01 19 42.	- 02 49		15.3	GALAXY
KN 14.025	01 19 42.4	+ 30 58 21.			NEBULA
MCG+01-04-039	01 19 45.	+ 06 12	42	15.5	GALAXY
LIN 532	01 19 47.	- 73 30 41.		15.94	PLANFTARY NEBULA IN SMC
KN 14.026	01 19 47.9	+ 32 16 11.			NEBULA
ZC 0119.8+0323	01 19 48.	+ 03 23	1410		CLUSTER OF GALAXIES
ZC 0119.8+0639	01 19 48.	+ 06 39	5240		CLUSTER OF GALAXIES
ZWG 411.037	01 19 48.	+ 08 47		15.3	GALAXY
MRK 568	01 19 48.	+ 08 47	17	15.5	GALAXY WITH UV CONTINUUM
ZWG 411.038	01 19 48.	+ 09 01		15.6	GALAXY
ZC 0119.8+0925	01 19 48.	+ 09 25	1210		CLUSTER OF GALAXIES
ZC 0119.8+1039	01 19 48.	+ 10 39	740		CLUSTER OF GALAXIES
ZWG 521.010	01 19 48.	+ 34 11		14.2	GALAXY
UGC 00916	01 19 48.	+ 34 11	102	14.2	GALAXY S
LDN 1318	01 19 48.	+ 61 35	180		DARK NEBULA
LDN 1317	01 19 48.	+ 61 35	180		DARK NEBULA
MRSL 126-00/1	01 19 48.	+ 61 36	600		HII REGION
ZWG 385.085	01 19 48.	- 01 08		14.1	GALAXY
ZH 22	01 19 48.	- 01 08		14.1	GALAXY
UGC 00915	01 19 48.	- 01 08	150	14.1	GALAXY SBb/SBc
MCG-06-04-014	01 19 48.	- 34 27	30	15.	GALAXY
LB 01598	01 19 48.	- 55 47		13.7	FAINT BLUE STAR
IC 1683	01 19 49.	+ 34 10 41.			NONSTELLAR OBJECT
RNGC 0497	01 19 49.	- 01 08		14.0	GALAXY
KN 14.027	01 19 51.1	+ 29 42 27.			NEBULA
ARP 008	01 19 53.	- 01 07			PECULIAR GALAXY
ZWG 481.004	01 19 54.	+ 26 36		15.2	GALAXY
MRK 356	01 19 54.	+ 26 36	12	16.	GALAXY WITH UV CONTINUUM
MRK 355	01 19 54.	+ 26 36	12	16.5	GALAXY WITH UV CONTINUUM
MCG+00-04-100	01 19 54.	- 01 09	114	13.	GALAXY
MCG-02-04-033	01 19 54.	- 09 54	36	16.	GALAXY
LW115-N087	01 19 54.	- 73 30			EMISSION NEBULA IN SMC
SHB 027	01 19 55.9	- 04 37 07.		16.9	QUASI-STELLAP OBJECT
BC PKS0119-04	01 19 55.91	- 04 37 07.0		16.88	QUASI-STELLAR OBJECT
MCG+06-04-008	01 19 57.	+ 34 10	36	14.	GALAXY
IC 0099	01 19 59.	- 13 12 45.			NONSTELLAR OBJFCT
ARC 0185	01 19 59.	- 21 44		17.1	RICH CLUSTFR OF GALAXIES
LBN 0644	01 20	+ 17 00	8700		BRIGHT NEBULA
KHAV 026	01 20	+ 51 34	12220		DARK NEBULA
LBN 0630	01 20	+ 61 35	120		BRIGHT NEBULA
LBN 0626	01 20	+ 73 00	5400		BRIGHT NEBULA
ZWG 385.087	01 20 00.	+ 01 37		14.9	GALAXY
MRK 569	01 20 00.	+ 01 37	14	15.5	GALAXY WITH UV CONTINUUM
UGC 00917	01 20 00.	+ 18 47	66	16.0	GALAXY S0
MCG+04-04-008	01 20 00.	+ 21 34	12	17.	GALAXY
MRK 357	01 20 00.	+ 22 54	10	16.	GALAXY WITH UV CONTINUUM
ZWG 521.011	01 20 00.	+ 33 42		15.5	GALAXY
ZWG 521.012	01 20 00.	+ 39 00		14.6	GALAXY
ZC 0120.0+3959	01 20 00.	+ 39 59	540		CLUSTER OF GALAXIES
MCG+08-03-019	01 20 00.	+ 49 40	16	16.	GALAXY
UGC 01000	01 20 00.	+ 50 22	66	16.5	GALAXY Sc
LDN 1319	01 20 00.	+ 61 00	1800		DARK NEBULA
ZWG 385.086	01 20 00.	- 00 50		15.6	GALAXY
MCG+00-04-101	01 20 00.	- 00 51	30	16.	GALAXY
ZH 59	01 20 00.	- 02 40		15.2	GALAXY
MCG-02-04-034	01 20 00.	- 13 13	30	15.	GALAXY
ARC 0184	01 20 01.	+ 12 48		17.8	RICH CLUSTER OF GALAXIES
MCG+00-04-102	01 20 03.	+ 03 20 30.	30	15.	GALAXY
KN 14.028	01 20 04.8	+ 33 09 12.			NEBULA
IC 1684	01 20 05.	+ 33 07 40.			NONSTELLAR OBJECT
MCG+00-04-103	01 20 06.	+ 00 27 30.	24	16.	GALAXY
ZWG 385.088	01 20 06.	+ 03 20		15.1	GALAXY
ZWG 411.039	01 20 06.	+ 05 08		15.2	GALAXY
RNGC 0500	01 20 06.	+ 05 08		15.0	GALAXY
MCG+01-04-040	01 20 06.	+ 05 08	36	15.	GALAXY
ZWG 459.044	01 20 06.	+ 19 02		15.6	GALAXY
UGC 00918	01 20 06.	+ 19 02	72	15.6	GALAXY Sc
MCG+03-04-029	01 20 06.	+ 19 02	48	15.	GALAXY
ZWG 502.056	01 20 06.	+ 28 35		15.2	GALAXY
ZWG 502.057	01 20 06.	+ 32 55		13.8	GALAXY
UGC 00919	01 20 06.	+ 32 55	126	13.8	GALAXY Sa-b
ZWG 502.058	01 20 06.	+ 33 13		14.0	GALAXY

OBJECT NAME	RIGHT ASCEN.	DECLINATION	DIAM.	MAGN.	TYPE OF OBJECT
RNGC 0495	01 20 06.	+ 33 13		14.0	GALAXY
UGC 00920	01 20 06.	+ 33 13	84	14.0	GALAXY SB0/SBa
MCG+06-04-010	01 20 06.	+ 33 40	36	16.	GALAXY
MCG+06-04-009	01 20 06.	+ 39 00	36	15.	GALAXY
5ZW 066	01 20 06.	+ 44 11			COMPACT GALAXY
ZH 80	01 20 06.	- 00 50		15.2	GALAXY
KN 14.029	01 20 06.2	+ 32 18 02.			NEBULA
KN 14.030	01 20 06.9	+ 32 54 45.			NEBULA
ARC 0187	01 20 07.	- 19 29		17.5	RICH CLUSTER OF GALAXIES
KN 14.031	01 20 07.8	+ 33 12 45.			NEBULA
HOW 78	01 20 09.	- 73 20 27.	18		STAR CLUSTER IN SMC
RNGC 0502	01 20 11.	+ 08 47		14.0	GALAXY
ZWG 385.090	01 20 12.	+ 01 27		15.7	GALAXY
MCG+01-04-042	01 20 12.	+ 08 35 30.	36	15.	GALAXY
ZWG 411.040	01 20 12.	+ 08 47		13.8	GALAXY
UGC 00922	01 20 12.	+ 08 47	114	13.8	GALAXY S0
MCG+01-04-041	01 20 12.	+ 09 12 30.	36	14.5	GALAXY
UGC 00923	01 20 12.	+ 12 52	60	16.5	GALAXY Sb-c
MCG+05-04-033	01 20 12.	+ 28 34	33	15.5	GALAXY
ZC 0120.2+3107	01 20 12.	+ 31 07	1480		CLUSTER OF GALAXIES
RNGC 0494	01 20 12.	+ 32 55		14.0	GALAXY
MCG+05-04-034	01 20 12.	+ 32 55	114	13.5	GALAXY
MCG+05-04-035	01 20 12.	+ 33 13 30.	66	14.	GALAXY
ZWG 521.013	01 20 12.	+ 33 52		15.7	GALAXY
ZWG 385.089	01 20 12.	- 01 39		14.6	GALAXY
ZH 23	01 20 12.	- 01 39		14.6	GALAXY
UGC 00921	01 20 12.	- 01 39	138	14.6	GALAXY S
KN 14.032	01 20 12.1	+ 29 55 54.			NEBULA
HELW 364	01 20 13.	- 34 07 53.			NEBULA
ARC 0186	01 20 14.	- 10 41		17.2	RICH CLUSTER OF GALAXIES
MCG+00-04-104	01 20 15.	+ 01 29 30.	36	15.	GALAXY
MCG+00-04-105	01 20 15.	- 01 39	90	14.	GALAXY
RNGC 0505	01 20 17.	+ 09 12		15.0	GALAXY
IC 1685	01 20 17.	+ 32 56 09.			NONSTELLAR OBJECT
MCG+01-04-043	01 20 18.	+ 08 47 30.	36	14.	GALAXY
ZWG 411.041	01 20 18.	+ 09 12		15.1	GALAXY
UGC 00924	01 20 18.	+ 09 12	72	15.1	GALAXY S0
ZWG 521.014	01 20 18.	+ 38 49		15.3	GALAXY
UGC 00925	01 20 18.	+ 42 43	60	16.5	GALAXY Sc
MCG+08-03-020	01 20 18.	+ 49 42	30	16.	GALAXY
ZH 74	01 20 18.	- 00 13		15.4	GALAXY
ARC 0188	01 20 18.	- 13 02		17.2	RICH CLUSTER OF GALAXIES
MCG-05-04-032	01 20 18.	- 30 13	54	16.	GALAXY
KN 14.033	01 20 18.3	+ 32 55 43.			NEBULA
KN 14.034	01 20 19.6	+ 32 55 01.			NEBULA
IC 0100	01 20 22.	- 04 54 10.			NONSTELLAR OBJECT
KN 14.035	01 20 22.4	+ 33 15 51.			NEBULA
KN 14.036	01 20 22.6	+ 33 11 59.			NEBULA
IC 1686	01 20 23.	+ 33 10 15.			SAME AS NGC 499
HOLM 041E	01 20 23.	- 06 57	36	14.8	PART OF MULTIPLE GALAXY
KN 14.037	01 20 23.0	+ 33 13 46.			NEBULA
ZWG 481.005	01 20 24.	+ 21 55		15.7	GALAXY
MCG+04-04-009	01 20 24.	+ 22 00	42	17.	GALAXY
RNGC 0499	01 20 24.	+ 33 12		13.0	GALAXY
ZWG 502.059	01 20 24.	+ 33 13		13.0	GALAXY
UGC 00926	01 20 24.	+ 33 13	120	13.0	GALAXY S0
ZWG 502.060	01 20 24.	+ 33 17		14.3	GALAXY
UGC 00927	01 20 24.	+ 33 17	102	14.3	GALAXY Sb-c
MCG+01-04-030	01 20 24.	- 04 52	60	14.	GALAXY
LB 03198	01 20 24.	- 72 46		14.2	FAINT BLUE STAR
BIGO 465	01 20 25.	+ 33 10			NEBULA
KN 14.038	01 20 25.5	+ 30 31 57.			NEBULA
ZWG 481.006	01 20 30.	+ 22 00		15.7	GALAXY
ZWG 502.061	01 20 30.	+ 33 02		14.8	GALAXY
MCG+05-04-038	01 20 30.	+ 33 12 30.	48	13.5	GALAXY
RNGC 0498	01 20 30.	+ 33 14		16.0	GALAXY
MCG+05-04-037	01 20 30.	+ 33 14	15	16.	GALAXY
RNGC 0496	01 20 30.	+ 33 17		14.5	GALAXY
MCG+05-04-036	01 20 30.	+ 33 17	90	14.	GALAXY
TON-S 0211	01 20 30.	- 30 10		14.5	BLUE STAR
KN 14.039	01 20 30.2	+ 33 00 56.			NEBULA
IC 1687	01 20 32.	+ 33 01 03.			NONSTELLAR OBJECT
HELW 365	01 20 32.	- 34 17 53.			NEBULA
HOLM 041B	01 20 33.	- 06 54	42	14.2	PART OF MULTIPLE GALAXY
KN 14.041	01 20 33.6	+ 30 39 44.			NEBULA
KN 14.040	01 20 34.1	+ 33 10 27.			NEBULA
MCG+00-04-106	01 20 36.	+ 02 51	36	16.	GALAXY
MCG+05-04-039	01 20 36.	+ 33 01	24	16.	GALAXY
ZWG 502.062	01 20 36.	+ 33 11		15.2	GALAXY
RNGC 0501	01 20 36.	+ 33 11		15.0	GALAXY
2ZW 002	01 20 36.	+ 34 19			COMPACT GALAXY
1ZW 004	01 20 36.	+ 34 19			COMPACT GALAXY
UGC 00930	01 20 36.	+ 49 54	84	16.0	GALAXY
ZWG 385.092	01 20 36.	- 00 39		14.9	GALAXY
UGC 00929	01 20 36.	- 00 39	96	14.9	GALAXY Sc
ZWG 385.091	01 20 36.	- 00 54		15.2	GALAXY
UGC 00928	01 20 36.	- 00 54	78	15.2	GALAXY S0
MCG+01-04-031	01 20 36.	- 06 53	36	15.5	GALAXY
LB 03199	01 20 36.	- 73 44		13.7	FAINT BLUE STAR
IC 1688	01 20 37.	+ 32 48 14.			NONSTELLAR OBJECT
KN 14.042	01 20 38.9	+ 32 56 32.			NEBULA
MCG-02-04-035	01 20 39.	- 09 18	24	15.	GALAXY
KN 14.043	01 20 39.5	+ 32 49 20.			NEBULA
KN 14.045	01 20 39.7	+ 30 30 43.			NEBULA
KN 14.044	01 20 40.4	+ 33 03 46.			NEBULA
ARP 070	01 20 41.	+ 30 31			PECULIAR GALAXY
KN 14.046	01 20 41.2	+ 30 31 45.			NEBULA
ZWG 411.042	01 20 42.	+ 07 32		15.2	GALAXY
MCG+01-04-044	01 20 42.	+ 07 32	36	14.5	GALAXY
ZWG 411.043	01 20 42.	+ 09 10		14.7	GALAXY
UGC 00932	01 20 42.	+ 09 10	96	14.7	GALAXY S0?
ZWG 436.036	01 20 42.	+ 14 57		15.5	GALAXY
UGC 00933	01 20 42.	+ 14 57	66	15.5	GALAXY SBa
MCG+02-04-032	01 20 42.	+ 14 57	48	15.	GALAXY
ZWG 502.063	01 20 42.	+ 30 32		14.5	GALAXY
UGC 00934	01 20 42.	+ 30 32	132	14.5	GALAXY S
ZWG 502.064	01 20 42.	+ 32 57		14.0	GALAXY
RNGC 0504	01 20 42.	+ 32 57		14.0	GALAXY
UGC 00935	01 20 42.	+ 32 57	96	14.0	GALAXY S0
MCG+05-04-041	01 20 42.	+ 32 57	90	14.	GALAXY
ZWG 502.065	01 20 42.	+ 33 05		15.1	GALAXY
RNGC 0503	01 20 42.	+ 33 05		15.0	GALAXY
MCG+05-04-040	01 20 42.	+ 33 05	18	15.5	GALAXY
ZWG 521.015	01 20 42.	+ 34 19		14.7	GALAXY
MCG+08-03-021	01 20 42.	+ 49 53	72	15.	GALAXY
ZH 71	01 20 42.	- 00 39		15.0	GALAXY
MCG+00-04-107	01 20 42.	- 00 39	54	14.	GALAXY
ZH 56	01 20 42.	- 00 54		15.3	GALAXY

OBJECT NAME	RIGHT ASCEN.	DECLINATION	DIAM.	MAGN.	TYPE OF OBJECT
MCG+00-04-108	01 20 42.	- 00 54	48	15.	GALAXY
ZWG 385.094	01 20 42.	- 00 57		15.7	GALAXY
UGC 00931	01 20 42.	- 00 57	78	15.7	GALAXY S
MCG+00-04-109	01 20 42.	- 00 59	66	15.5	GALAXY
ZH 36	01 20 42.	- 01 09		15.5	GALAXY
ZWG 385.093	01 20 42.	- 01 10		15.5	GALAXY
MCG-03-04-058	01 20 42.	- 17 47	24	15.5	GALAXY
MCG-05-04-033	01 20 42.	- 31 02	30	15.5	GALAXY
HOLM 041A	01 20 43.	- 06 54	42	14.2	PART OF MULTIPLE GALAXY
MCG+00-04-110	01 20 45.	+ 01 23	24	15.5	GALAXY
MCG+01-04-045	01 20 45.	+ 09 11	72	13.5	GALAXY
MCG+05-04-042	01 20 45.	+ 30 32	90	15.5	GALAXY
VV 341B	01 20 45.	+ 30 32 30.	21	16.	INTERACTING GALAXY
VV 341A	01 20 45.	+ 30 32 30.	78	14.5	INTERACTING GALAXY
KN 14.049	01 20 45.4	+ 30 06 22.			NEBULA
ARP 229	01 20 46.	+ 33 00			PECULIAR GALAXY
KN 14.047	01 20 46.3	+ 32 58 59.			NEBULA
KN 14.048	01 20 46.8	+ 33 00 37.			NEBULA
RNGC 0509	01 20 47.	+ 09 10		14.5	GALAXY
ZC 0120.8+0123	01 20 48.	+ 01 23	1680		CLUSTER OF GALAXIES
MCG+00-04-111	01 20 48.	+ 01 27	24	15.	GALAXY
ZWG 436.037	01 20 48.	+ 11 02		15.4	GALAXY
UGC 00936	01 20 48.	+ 11 02	72	15.4	GALAXY E
KARA.68 004	01 20 48.	+ 12 02	34		DWARF GALAXY
ZWG 502.066	01 20 48.	+ 32 23		15.0	GALAXY
UGC 00937	01 20 48.	+ 32 23	72	15.0	GALAXY PECULR
RNGC 0506	01 20 48.	+ 32 58			NON-EXISTENT OBJECT
ZWG 502.067	01 20 48.	+ 33 00		13.0	GALAXY
RNGC 0507	01 20 48.	+ 33 00		13.0	GALAXY
UGC 00938	01 20 48.	+ 33 00	240	13.0	GALAXY E
ZWG 502.068	01 20 48.	+ 33 02		14.5	GALAXY
RNGC 0508	01 20 48.	+ 33 02		14.5	GALAXY
UGC 00939	01 20 48.	+ 33 02	78	14.5	GALAXY E
ZWG 521.016	01 20 48.	+ 34 19		14.9	GALAXY
UGC 00940	01 20 48.	+ 34 19	66	14.9	GALAXY Sc
ZC 0120.8+3538	01 20 48.	+ 35 38	1140		CLUSTER OF GALAXIES
ZWG 385.095	01 20 48.	- 02 14		14.8	GALAXY
ZH 16	01 20 48.	- 02 14		14.8	GALAXY
MCG+00-04-112	01 20 48.	- 02 15	42	15.	GALAXY
MCG+01-04-032	01 20 48.	- 06 53	24	15.	GALAXY
MCG-05-04-034	01 20 48.	- 30 52	30	15.5	GALAXY
KN 14.050	01 20 48.8	+ 32 22 15.			NEBULA
KN 14.051	01 20 50.8	+ 32 59 44.			NEBULA
KN 14.052	01 20 51.8	+ 33 00 59.			NEBULA
LB 00446	01 20 52.	+ 29 47 30.		16.5	FAINT BLUE STAR
RNGC 0511	01 20 53.	+ 11 01		15.5	GALAXY
HOLM 041C	01 20 53.	- 06 52	18	14.6	PART OF MULTIPLE GALAXY
1ZW 005	01 20 54.	+ 01 25			COMPACT GALAXY
ZWG 385.096	01 20 54.	+ 01 26		15.5	GALAXY
MCG+01-04-046	01 20 54.	+ 06 41	48	14.5	GALAXY
MCG+05-04-043	01 20 54.	+ 32 22 30.	30	14.5	GALAXY
VV 207A	01 20 54.	+ 33 00 30.	120	12.8	INTERACTING GALAXY
RNGC 0510	01 20 54.	+ 33 10			GALAXY
MCG+06-04-011	01 20 54.	+ 34 18	54	15.	GALAXY
5ZW 067	01 20 54.	+ 49 43			COMPACT GALAXY
MCG+01-04-033	01 20 54.	- 07 03	24	15.5	GALAXY
MCG+01-04-034	01 20 54.	- 07 23	18	16.	GALAXY
MCG-06-04-015	01 20 54.	- 35 11	30	15.5	GALAXY
KN 14.053	01 20 54.3	+ 32 47 52.			NEBULA
MCG+02-04-033	01 20 57.	+ 11 01	18	15.	GALAXY
IC 1689	01 20 57.	+ 32 46 32.			NONSTELLAR OBJECT
MCG+05-04-044	01 20 57.	+ 33 00	48	13.	GALAXY
VV 207B	01 20 57.	+ 33 02 30.	60	13.	INTERACTING GALAXY
KN 14.054	01 20 58.0	+ 32 42 40.			NEBULA
BIGO 466	01 20 59.	+ 32 48			NEBULA
RNGC 0513	01 20 59.	+ 33 33		13.5	GALAXY
KN 14.055	01 20 59.0	+ 32 47 42.			NEBULA
ZWG 385.097	01 21 00.	+ 01 18		15.7	GALAXY
SN 1963S	01 21 00.	+ 01 18		15.0	SUPERNOVA
ZWG 411.044	01 21 00.	+ 06 42		15.7	GALAXY
UGC 00941	01 21 00.	+ 06 42	78	15.7	GALAXY IRR
ZWG 459.045	01 21 00.	+ 20 34		15.7	GALAXY
ZWG 502.069	01 21 00.	+ 32 25		15.7	GALAXY
ZWG 502.070	01 21 00.	+ 32 48		14.8	GALAXY
MCG+05-04-046	01 21 00.	+ 32 48	42	15.	GALAXY
ZWG 502.071	01 21 00.	+ 32 54		14.9	GALAXY
IC 1690	01 21 00.	+ 32 54 07.			NONSTELLAR OBJECT
MCG+05-04-045	01 21 00.	+ 33 02	30	14.	GALAXY
LDN 1322	01 21 00.	+ 61 00	240		DARK NEBULA
HOLM 041D	01 21 00.	- 06 51	12	14.7	PART OF MULTIPLE GALAXY
MCG-06-04-016	01 21 00.	- 35 00	84	14.5	GALAXY
LB 03200	01 21 00.	- 65 38		14.1	FAINT BLUE STAR
KN 14.056	01 21 00.4	+ 32 53 40.			NEBULA
KN 14.058	01 21 02.2	+ 30 33 08.			NEBULA
KN 14.057	01 21 03.2	+ 32 23 53.			NEBULA
KN 14.059	01 21 04.4	+ 32 41 00.			NEBULA
KN 14.060	01 21 04.7	+ 32 23 28.			NEBULA
ARC 0189	01 21 05.	+ 01 24		15.7	RICH CLUSTER OF GALAXIES
UGC 00942	01 21 06.	+ 06 27	72	16.0	GALAXY S IV-V
MCG+01-04-047	01 21 06.	+ 07 30 30.	30	15.	GALAXY
UGC 00943	01 21 06.	+ 11 15	78	16.0	GALAXY Sb
MCG+05-04-047	01 21 06.	+ 32 24 30.	15	16.5	GALAXY
ZWG 385.098	01 21 06.	- 02 05		15.5	GALAXY
ZH 52	01 21 06.	- 02 05		15.4	GALAXY
MCG-06-04-017	01 21 06.	- 33 06	72	14.	GALAXY
LB 03201	01 21 06.	- 63 35		13.7	FAINT BLUE STAR
LB 03202	01 21 06.	- 68 14		14.4	FAINT BLUE STAR
KN 14.061	01 21 06.8	+ 33 14 15.			NEBULA
KN 14.065	01 21 08.3	+ 28 34 31.			NEBULA
KEEL 027	01 21 08.4	+ 09 43 07.			NEBULA
KN 14.064	01 21 08.4	+ 30 51 42.			NEBULA
KN 14.062	01 21 09.6	+ 33 03 15.			NEBULA
KN 14.063	01 21 10.5	+ 33 38 41.			NEBULA
RNGC 0512	01 21 11.	+ 33 39		14.0	GALAXY
ZWG 411.045	01 21 12.	+ 07 31		15.6	GALAXY
MCG+02-04-034	01 21 12.	+ 11 14	72	15.5	GALAXY
MCG+03-04-030	01 21 12.	+ 20 35	24	16.	GALAXY
ZWG 502.072	01 21 12.	+ 33 04		15.0	GALAXY
ZWG 521.017	01 21 12.	+ 33 36		14.9	GALAXY
ZWG 521.018	01 21 12.	+ 33 39		14.0	GALAXY
UGC 00944	01 21 12.	+ 33 39	102	14.0	GALAXY Sa-b
ZWG 521.019	01 21 12.	+ 34 30		15.7	GALAXY
UGC 00945	01 21 12.	+ 42 58	78	16.0	GALAXY (S0)
ARC 0190	01 21 14.	- 10 07		17.2	RICH CLUSTER OF GALAXIES
MCG+05-04-048	01 21 15.	+ 33 04	24	16.	GALAXY
MCG+06-04-012	01 21 15.	+ 34 30	24	16.	GALAXY
ZWG 502.073	01 21 18.	+ 28 21		15.2	GALAXY
ZWG 502.074	01 21 18.	+ 32 31		15.6	GALAXY

OBJECT NAME	RIGHT ASCEN.	DECLINATION	DIAM.	MAGN.	TYPE OF OBJECT
MCG+06-04-013	01 21 18.	+ 33 38	90	14.5	GALAXY
KN 14.066	01 21 19.3	+ 32 24 17.			NEBULA
MCG+06-04-014	01 21 21.	+ 33 35	30	14.5	GALAXY
KN 14.067	01 21 22.2	+ 32 30 26.			NEBULA
RNGC 0514	01 21 23.	+ 12 40		12.5	GALAXY
KN 14.068	01 21 23.3	+ 32 19 58.			NEBULA
MCG+00-04-113	01 21 24.	+ 00 01	30	15.5	GALAXY
ZWG 411.046	01 21 24.	+ 09 18		14.3	GALAXY
UGC 00946	01 21 24.	+ 09 18	102	14.3	GALAXY S0
MCG+01-04-048	01 21 24.	+ 09 18	78	13.5	GALAXY
ZWG 436.038	01 21 24.	+ 12 39		12.8	GALAXY
UGC 00947	01 21 24.	+ 12 39	258	12.8	GALAXY Sc
3ZW 027	01 21 24.	+ 13 46			COMPACT GALAXY
MCG+03-04-031	01 21 24.	+ 17 00	24	16.	GALAXY
UGC 00948	01 21 24.	+ 26 48	66	17.	GALAXY DWARF
MCG+05-04-049	01 21 24.	+ 28 20	30	15.	GALAXY
ZH 31	01 21 24.	- 01 53		15.5	GALAXY
ZWG 385.099	01 21 24.	- 01 54		15.6	GALAXY
MCG-02-04-036	01 21 24.	- 09 50 30.	24	16.	GALAXY
MCG-02-04-037	01 21 24.	- 09 51 30.	36	16.	GALAXY
RNGC 0516	01 21 29.	+ 09 17		14.5	GALAXY
IC 1693	01 21 29.1	- 01 55 05.			GALAXY E4
ZC 0121.5+0113	01 21 30.	+ 01 13	8600		CLUSTER OF GALAXIES
ZWG 436.039	01 21 30.	+ 09 40		15.1	GALAXY
UGC 00949	01 21 30.	+ 09 40	96	15.1	GALAXY S
IC 0101	01 21 30.	+ 09 40 42.			NONSTELLAR OBJECT
MCG+02-04-035	01 21 30.	+ 12 40	180	12.	GALAXY
ZWG 502.075	01 21 30.	+ 31 58		15.2	GALAXY
UGC 00950	01 21 30.	+ 31 58	66	15.2	GALAXY Sa-b
MCG+05-04-050	01 21 30.	+ 32 20	24	17.	GALAXY
1ZW 006	01 21 30.	- 01 54			COMPACT GALAXY
ZH 32	01 21 30.	- 01 54		15.3	GALAXY
KEEL 028	01 21 31.0	+ 09 40 10.			SPIRAL NEBULA
MCG-02-04-038	01 21 33.	- 11 32	30	15.	GALAXY
SNO 13	01 21 33.	- 44 48 33.	360	17.	LINEAR GRP OF 4 GALAXIES
KN 14.069	01 21 33.5	+ 31 57 50.			NEBULA
ZWG 385.101	01 21 36.	+ 00 47		15.7	GALAXY
UGC 00951	01 21 36.	+ 00 47	78	15.7	GALAXY DISTRBD
ZWG 411.047	01 21 36.	+ 09 04		14.4	GALAXY
UGC 00952	01 21 36.	+ 09 04	108	14.4	GALAXY Sa
MCG+01-04-049	01 21 36.	+ 09 05	96	13.5	GALAXY
MCG+05-04-051	01 21 36.	+ 31 58	54	16.	GALAXY
MCG+06-04-016	01 21 36.	+ 33 31	36	14.	GALAXY
SN 1963T	01 21 36.	+ 33 47		17.5	SUPERNOVA
MCG+06-04-015	01 21 36.	+ 34 16 30.	15	16.	GALAXY
ACK 126+03.1	01 21 36.	+ 65 23			PLANETARY NEBULA
ZWG 385.100	01 21 36.	- 02 07		14.7	GALAXY
ZH 15	01 21 36.	- 02 07		14.7	GALAXY
MCG+00-04-114	01 21 36.	- 02 08	30	15.5	GALAXY
TON-S 0212	01 21 36.	- 26 12		15.5	BLUE STAR
MCG-05-04-035	01 21 36.	- 28 01	42	15.	GALAXY
MCG-06-04-018	01 21 36.	- 36 03	30	15.	GALAXY
LB 01599	01 21 36.	- 45 48		14.6	FAINT BLUE STAR
KN 14.070	01 21 37.2	+ 33 08 46.			NEBULA
KN 14.071	01 21 37.9	+ 33 32 29.			NEBULA
IC 1691	01 21 38.	+ 33 06 54.			NONSTELLAR OBJECT
RNGC 0526	01 21 39.	- 35 19			GALAXY
HOW 79	01 21 39.	- 75 16 07.	54		STAR CLUSTER IN SMC
KN 14.072	01 21 39.7	+ 32 20 00.			NEBULA
RNGC 0518	01 21 41.	+ 09 04		14.5	GALAXY
ZC 0121.7+0409	01 21 42.	+ 04 09	1410		CLUSTER OF GALAXIES
MCG+02-04-036	01 21 42.	+ 09 40	72	15.	GALAXY
ZWG 521.020	01 21 42.	+ 33 33		13.4	GALAXY
UGC 00953	01 21 42.	+ 33 33	45	13.4	GALAXY
ZC 0121.7+4055	01 21 42.	+ 40 55	1010		CLUSTER OF GALAXIES
MCG+00-04-115	01 21 42.	- 00 47	33	15.	GALAXY
MCG-03-04-059	01 21 42.	- 17 02 30.	30	15.	GALAXY
MCG-06-04-019	01 21 42.	- 35 20	90	14.5	GALAXY
ARC 0192	01 21 43.	+ 04 14		17.6	RICH CLUSTER OF GALAXIES
ARC 0191	01 21 44.	+ 20 40		17.5	RICH CLUSTER OF GALAXIES
MCG+03-04-032	01 21 45.	+ 15 30	66	15.	GALAXY
MCG-03-04-060	01 21 45.	- 16 45	42	15.	GALAXY
MCG-06-04-020	01 21 45.	- 35 20	90	14.5	GALAXY
RNGC 0527	01 21 45.	- 35 22			GALAXY
IC 0102	01 21 47.	+ 09 38 05.			NONSTELLAR OBJECT
RNGC 0515	01 21 47.	+ 33 13		14.5	GALAXY
KARA.68 005	01 21 48.	+ 03 36	54		DWARF GALAXY
ZWG 436.040	01 21 48.	+ 09 38		15.6	GALAXY
UGC 00954	01 21 48.	+ 09 38	60	15.6	GALAXY S0-a
ZWG 459.046	01 21 48.	+ 15 30		15.7	GALAXY
ZWG 436.041	01 21 48.	+ 15 30		15.7	GALAXY
UGC 00955	01 21 48.	+ 15 30	66	15.7	GALAXY SB?c
MCG+03-04-033	01 21 48.	+ 16 16	90	15.	GALAXY
ZWG 459.047	01 21 48.	+ 20 50		15.5	GALAXY
MCG+04-04-010	01 21 48.	+ 21 37	36	17.	GALAXY
ZWG 502.076	01 21 48.	+ 32 59		15.6	GALAXY
ZWG 502.077	01 21 48.	+ 33 13		14.3	GALAXY
UGC 00956	01 21 48.	+ 33 13	96	14.3	GALAXY S0
MRSL 126+01/1	01 21 48.	+ 63 41	18		HII REGION
ZWG 385.102	01 21 48.	- 02 00		14.9	GALAXY
ZH 14	01 21 48.	- 02 00		14.9	GALAXY
MCG+00-04-116	01 21 48.	- 02 00	36	15.	GALAXY
MCG+01-04-035	01 21 48.	- 06 13	60	15.	GALAXY
MCG-06-04-022	01 21 48.	- 35 23	36	15.5	GALAXY
MCG-06-04-021	01 21 48.	- 35 23	96	13.	GALAXY
KN 14.075	01 21 48.6	+ 30 52 39.			NEBULA
KEEL 029	01 21 48.8	+ 09 37 34.			SPIRAL NEBULA
KN 14.073	01 21 49.3	+ 33 12 57.			NEBULA
KN 14.074	01 21 50.3	+ 32 58 40.			NEBULA
KN 14.076	01 21 51.2	+ 30 24 28.			NEBULA
IC 1692	01 21 52.	+ 32 57 17.			NONSTELLAR OBJECT
RNGC 0517	01 21 53.	+ 33 11		13.5	GALAXY
ZWG 385.104	01 21 54.	+ 00 34		15.6	GALAXY
ZWG 385.105	01 21 54.	+ 00 52		15.2	GALAXY
UGC 00957	01 21 54.	+ 03 38	78	18.	GALAXY DWARF
ZWG 459.048	01 21 54.	+ 16 17		15.4	GALAXY
UGC 00958	01 21 54.	+ 16 17	108	15.4	GALAXY SB:c
ZWG 481.007	01 21 54.	+ 21 37		15.7	GALAXY
ZWG 502.078	01 21 54.	+ 31 55		14.2	GALAXY
UGC 00959	01 21 54.	+ 31 55	72	14.2	GALAXY Sa
VV 036C	01 21 54.	+ 33 10	3	19.	INTERACTING GALAXY
VV 036B	01 21 54.	+ 33 10	12	17.	INTERACTING GALAXY
VV 036A	01 21 54.	+ 33 10	78	13.	INTERACTING GALAXY
ZWG 502.079	01 21 54.	+ 33 11		13.6	GALAXY
UGC 00960	01 21 54.	+ 33 11	126	13.6	GALAXY S0
MCG+05-04-052	01 21 54.	+ 33 13 30.	66	14.	GALAXY
VB 115	01 21 54.	+ 57 32	269		STELLAR RING
ZH 28	01 21 54.	- 01 39		15.7	GALAXY
ZH 30	01 21 54.	- 01 53		15.3	GALAXY
ZWG 385.103	01 21 54.	- 01 54		15.3	GALAXY
MCG+01-04-036	01 21 54.	- 06 10	48	15.	GALAXY
MCG-02-04-039	01 21 54.	- 08 57	42	15.5	GALAXY
RNGC 0519	01 21 56.	- 01 54		15.5	GALAXY
ARP 157	01 21 58.	+ 03 32			PECULIAR GALAXY
IC 0103	01 21 59.	+ 01 47 34.			NONSTELLAR OBJECT
ZWG 385.106	01 22 00.	+ 01 28		12.9	GALAXY
UGC 00962	01 22 00.	+ 01 28	228	12.9	GALAXY SBb
MCG+00-04-118	01 22 00.	+ 01 29	180	13.0	GALAXY
ZWG 385.107	01 22 00.	+ 01 46		15.3	GALAXY
UGC 00963	01 22 00.	+ 01 46	84	15.3	GALAXY E-S0
MCG+00-04-117	01 22 00.	+ 01 47	30	15.	GALAXY
MCG+01-04-052	01 22 00.	+ 03 31 30.	240	12.5	GALAXY
RNGC 0520	01 22 00.	+ 03 32		12.5	GALAXY
VV 231C	01 22 00.	+ 03 32	30	15.	INTERACTING GALAXY
VV 231B	01 22 00.	+ 03 32	66	13.	INTERACTING GALAXY
VV 231A	01 22 00.	+ 03 32	102	12.5	INTERACTING GALAXY
VV 231	01 22 00.	+ 03 32	200	12.2	INTERACTING GALAXY
KARA.72 031A	01 22 00.	+ 03 33	114	12.4	PART OF DOUBLE GALAXY
MCG+01-04-050	01 22 00.	+ 07 27	60	15.	GALAXY
ZWG 411.048	01 22 00.	+ 07 28		15.5	GALAXY
UGC 00964	01 22 00.	+ 07 28	90	15.5	GALAXY S
ZWG 411.049	01 22 00.	+ 08 15		15.7	GALAXY
MCG+01-04-051	01 22 00.	+ 08 31	36	15.5	GALAXY
MCG+01-04-053	01 22 00.	+ 09 16 30.	150	11.	GALAXY
MCG+03-04-034	01 22 00.	+ 16 43	24	15.5	GALAXY
ZC 0122.0+2035	01 22 00.	+ 20 35	2020		CLUSTER OF GALAXIES
MCG+05-04-053	01 22 00.	+ 31 54 30.	36	15.	GALAXY
ZWG 502.080	01 22 00.	+ 31 59		15.1	GALAXY
MCG+05-04-054	01 22 00.	+ 33 10 30.	42	14.	GALAXY
LDN 1323	01 22 00.	+ 61 10	600		DARK NEBULA
UGC 00961	01 22 00.	- 00 18	72	16.0	GALAXY S
IC 0104	01 22 00.	- 01 43			DOUBLE STAR
MCG+01-04-037	01 22 00.	- 04 47	60	15.	GALAXY
MCG-03-04-061	01 22 00.	- 15 46	60	13.	GALAXY
MCG-04-04-007	01 22 00.	- 24 22	36	15.	GALAXY
MCG-06-04-023	01 22 00.	- 35 00	48	13.	GALAXY
RNGC 0521	01 22 01.	+ 01 28		12.5	GALAXY
SN 1966G	01 22 04.	+ 01 28		15.5	SUPERNOVA
BIGO 467	01 22 04.	+ 33 09			NEBULA
RNGC 0522	01 22 05.	+ 09 44		14.0	GALAXY
ZWG 385.109	01 22 06.	+ 01 49		15.6	GALAXY
KARA.72 031B	01 22 06.	+ 03 32	78		PART OF DOUBLE GALAXY
ZWG 411.050	01 22 06.	+ 03 33		12.4	GALAXY
UGC 00966	01 22 06.	+ 03 33	300	12.4	GALAXY PECULR
UGC 00967	01 22 06.	+ 08 32	66	16.0	GALAXY Sc
ZWG 411.051	01 22 06.	+ 09 17		11.5	GALAXY
UGC 00968	01 22 06.	+ 09 17	210	11.5	GALAXY S0
ZWG 436.042	01 22 06.	+ 09 30		15.7	GALAXY
ZWG 411.052	01 22 06.	+ 09 30		15.7	GALAXY
UGC 00969	01 22 06.	+ 09 30	78	15.7	GALAXY SBb
ZWG 436.043	01 22 06.	+ 09 44		14.2	GALAXY
UGC 00970	01 22 06.	+ 09 44	162	14.2	GALAXY Sb-c
ZWG 502.081	01 22 06.	+ 30 46		15.7	GALAXY
MCG+05-04-055	01 22 06.	+ 31 59	33	15.5	GALAXY
ZWG 385.108	01 22 06.	- 01 50		14.0	GALAXY
ZH 13	01 22 06.	- 01 50		14.5	GALAXY
UGC 00965	01 22 06.	- 01 50	108	14.0	GALAXY Sa
MCG+00-04-119	01 22 06.	- 01 50	66	14.	GALAXY
MCG-02-04-040	01 22 06.	- 09 06 30.	24	16.	GALAXY
MCG-06-04-024	01 22 06.	- 37 37	60	14.	GALAXY
IC 0105	01 22 08.	+ 01 49 28.			NONSTELLAR OBJECT
RNGC 0530	01 22 08.	- 01 50		14.0	GALAXY
IC 0106	01 22 08.	- 01 51			SAME AS NGC 530
KEEL 030	01 22 08.2	+ 09 29 59.			NEBULA
RNGC 0524	01 22 11.	+ 09 17		12.0	GALAXY
RNGC 0525	01 22 11.	+ 09 26		14.5	GALAXY
IC 0108	01 22 11.	- 12 53 39.			NONSTELLAR OBJECT
RNGC 0540	01 22 11.	- 20 10			GALAXY
HOW 80	01 22 11.	- 73 28 19.	6		STAR CLUSTER IN SMC
ZWG 385.110	01 22 12.	+ 00 39		15.2	GALAXY
UGC 00971	01 22 12.	+ 00 46	78	18.	GALAXY DWARF
ZWG 385.111	01 22 12.	+ 01 20		14.9	GALAXY
ZWG 411.053	01 22 12.	+ 09 26		14.5	GALAXY
UGC 00972	01 22 12.	+ 09 26	96	14.5	GALAXY S0
MCG+01-04-054	01 22 12.	+ 09 27	60	14.	GALAXY
VB 116	01 22 12.	+ 61 03	470		STELLAR RING
MCG-02-04-041	01 22 12.	- 12 54	48	15.	GALAXY
LIN 533	01 22 12.	- 72 47 09.		14.71	SMC EMISSION LINE OBJECT
IC 1694	01 22 14.	+ 01 21			NONSTELLAR OBJECT
MCG+02-04-037	01 22 15.	+ 09 30	48	15.	GALAXY
MCG+00-04-120	01 22 15.	- 01 58	24	15.5	GALAXY
MCG+02-04-038	01 22 18.	+ 09 44	156	13.	GALAXY
ZWG 459.049	01 22 18.	+ 15 35		15.4	GALAXY
MCG+05-04-056	01 22 18.	+ 29 10	30	15.	GALAXY
ZWG 385.115	01 22 18.	- 00 03		15.6	GALAXY
ZH 50	01 22 18.	- 01 30		16.4	GALAXY
ZWG 385.114	01 22 18.	- 01 45		15.0	GALAXY
ZH 27	01 22 18.	- 01 45		15.4	GALAXY
UGC 00974	01 22 18.	- 01 45	66	15.0	GALAXY S0-a
MCG+00-04-121	01 22 18.	- 01 46	36	15.	GALAXY
2ZW 003	01 22 18.	- 01 49			COMPACT GALAXY
ZWG 385.113	01 22 18.	- 01 52		14.7	GALAXY
ZH 24	01 22 18.	- 01 52		14.7	GALAXY
UGC 00973	01 22 18.	- 01 52	90	14.7	GALAXY E
MCG+00-04-122	01 22 18.	- 01 52 30.	30	15.	GALAXY
ZWG 385.112	01 22 18.	- 01 56		15.7	GALAXY
ZH 29	01 22 18.	- 01 56		15.5	GALAXY
MCG-02-04-042	01 22 18.	- 09 04	24	15.	GALAXY
MCG-04-04-008	01 22 18.	- 23 05 30.	36	15.	GALAXY
MCG-06-04-025	01 22 18.	- 33 26	72	14.5	GALAXY
LIN 534	01 22 19.	- 73 39 21.		13.83	SMC EMISSION LINE OBJECT
IC 1696	01 22 19.7	- 01 52 42.			GALAXY SA(s)
IC 1695	01 22 21.	+ 08 28 33.			NONSTELLAR OBJECT
ARP 158	01 22 22.	+ 33 46			PECULIAR GALAXY
RNGC 0534	01 22 23.	- 38 23			GALAXY
MCG+00-04-123	01 22 24.	+ 01 15	42	15.	GALAXY
ZC 0122.4+0813	01 22 24.	+ 08 13	3900		CLUSTER OF GALAXIES
SHAH 040	01 22 24.	+ 08 13			GROUP OF COMPACT GALAXIES
ZC 0122.4+3309	01 22 24.	+ 33 09	1080		CLUSTER OF GALAXIES
ZWG 521.021	01 22 24.	+ 34 06		15.0	GALAXY
UGC 00975	01 22 24.	+ 34 06	72	15.0	GALAXY S
VB 117	01 22 24.	+ 59 07	322		STELLAR RING
MCG-02-04-044	01 22 24.	- 11 51 30.	60	14.5	GALAXY
MCG-02-04-043	01 22 24.	- 12 55	30	15.	GALAXY
IC 0107	01 22 25.	+ 14 37 03.			NONSTELLAR OBJECT
ARC 0193	01 22 28.	+ 08 27		16.0	RICH CLUSTER OF GALAXIES

OBJECT NAME	RIGHT ASCEN.	DECLINATION	DIAM.	MAGN.	TYPE OF OBJECT
RNGC 0523	01 22 29.	+ 33 46		13.5	GALAXY
IC 1697	01 22 29.4	+ 00 11 04.		14.9	GALAXY SB0
ZWG 385.116	01 22 30.	+ 00 10		14.9	GALAXY
ZH 72	01 22 30.	+ 00 10		14.9	GALAXY
UGC 00976	01 22 30.	+ 00 10	72	14.9	GALAXY
MCG+00-04-125	01 22 30.	+ 00 11	30	15.	GALAXY
MCG+00-04-124	01 22 30.	+ 03 21 30.	30	16.	GALAXY
ZWG 411.054	01 22 30.	+ 08 26		15.4	GALAXY
UGC 00977	01 22 30.	+ 08 26	108	15.4	GALAXY COMPACT
MCG+01-04-055	01 22 30.	+ 08 26	36	15.	GALAXY
ZWG 436.044	01 22 30.	+ 14 37		15.4	GALAXY
UGC 00978	01 22 30.	+ 14 37	72	15.4	GALAXY Sc
MCG+02-04-039	01 22 30.	+ 14 37	60	15.	GALAXY
4ZW 045	01 22 30.	+ 33 46			COMPACT GALAXY
ZWG 521.022	01 22 30.	+ 33 46		13.5	GALAXY
UGC 00979	01 22 30.	+ 33 46	192	13.5	GALAXY PECULR
MCG+06-04-017	01 22 30.	+ 34 05	48	15.	GALAXY
VB 045	01 22 30.	+ 65 19	329		STELLAR RING
MCG+00-04-126	01 22 30.	- 00 03	36	15.	GALAXY
MCG-06-04-026	01 22 30.	- 38 25	60	14.	GALAXY
LB G3203	01 22 30.	- 53 15		12.8	FAINT BLUE STAR
MCG+00-04-127	01 22 33.	+ 02 01 30.	30	15.5	GALAXY
IC 0109	01 22 35.	+ 01 49 03.			NONSTELLAR OBJECT
ZWG 385.117	01 22 36.	+ 01 48		14.9	GALAXY
UGC 00980	01 22 36.	+ 01 48	78	14.9	GALAXY SB0
MCG+00-04-128	01 22 36.	+ 01 48	30	15.	GALAXY
ZWG 385.118	01 22 36.	+ 02 02		15.5	GALAXY
UGC 00981	01 22 36.	+ 02 02	66	15.5	GALAXY S
ZWG 411.055	01 22 36.	+ 09 00		13.5	GALAXY
UGC 00982	01 22 36.	+ 09 00	180	13.5	GALAXY Sa?
MCG+01-04-056	01 22 36.	+ 09 01	132	13.	GALAXY
ZWG 436.045	01 22 36.	+ 14 35		14.9	GALAXY
UGC 00983	01 22 36.	+ 14 35	120	14.9	GALAXY S0
VB 046	01 22 36.	+ 67 03	2786		STELLAR RING
MCG+01-04-038	01 22 36.	- 04 52	30	15.	GALAXY
MCG+01-04-039	01 22 36.	- 04 53	30	15.	GALAXY
MCG+01-04-040	01 22 36.	- 04 57	60	14.	GALAXY
MCG-02-04-045	01 22 36.	- 09 08	24	15.	GALAXY
MCG-05-04-036	01 22 36.	- 32 01	84	14.5	GALAXY
MCG+06-04-018	01 22 39.	+ 33 45	84	14.	GALAXY
LIN 536	01 22 40.	- 74 17 15.		17.12	PLANETARY NEBULA IN SMC
RNGC 0532	01 22 41.	+ 09 00		13.5	GALAXY
RNGC 0528	01 22 41.	+ 33 25		13.5	GALAXY
ZWG 436.046	01 22 42.	+ 14 16		15.2	GALAXY
UGC 00985	01 22 42.	+ 14 16	108	15.2	GALAXY Sb
MCG+02-04-040	01 22 42.	+ 14 35	90	15.	GALAXY
ZWG 436.047	01 22 42.	+ 14 36		14.3	GALAXY
UGC 00986	01 22 42.	+ 14 36	132	14.3	GALAXY E
ZWG 502.082	01 22 42.	+ 31 53		14.0	GALAXY
UGC 00987	01 22 42.	+ 31 53	156	14.0	GALAXY Sa
ZWG 502.083	01 22 42.	+ 33 25		13.7	GALAXY
UGC 00988	01 22 42.	+ 33 25	108	13.7	GALAXY S0
ZWG 385.119	01 22 42.	- 01 46		14.8	GALAXY
UGC 00984	01 22 42.	- 01 46	84	14.8	GALAXY S0
MCG+00-04-129	01 22 42.	- 01 47	42	15.	GALAXY
MCG-05-04-037	01 22 42.	- 30 52	36	15.5	GALAXY
IC 1698	01 22 44.	+ 14 34 56.			NONSTELLAR OBJFCT
LIN 535	01 22 44.	- 73 24 03.		14.05	SMC EMISSION LINE OBJECT
MCG+02-04-042	01 22 45.	+ 14 15	72	15.	GALAXY
MCG+02-04-041	01 22 45.	+ 14 37	36	15.	GALAXY
IC 1699	01 22 45.	+ 14 41 32.			NONSTELLAR OBJFCT
IC 1700	01 22 47.	+ 14 36 38.			NONSTELLAR OBJECT
RNGC 0549	01 22 47.	- 38 15			GALAXY
UGC 00989	01 22 48.	+ 07 17	96	16.0	GALAXY
MCG+01-04-057	01 22 48.	+ 07 17 30.	72	15.	GALAXY
MCG+01-04-058	01 22 48.	+ 07 43	48	15.	GALAXY
UGC 00990	01 22 48.	+ 10 33	90	17.	GALAXY DWARF
MCG+05-04-058	01 22 48.	+ 31 53	132	14.	GALAXY
MCG+05-04-057	01 22 48.	+ 33 25	72	14.	GALAXY
VB 118	01 22 48.	+ 57 49	289		STELLAR RING
ISS 0042	01 22 48.	+ 57 49	295		STELLAR RING
ZH 47	01 22 48.	- 01 26		15.6	GALAXY
ZH 08	01 22 48.	- 01 46		14.8	GALAXY
MCG-02-04-046	01 22 48.	- 12 11	24	15.	GALAXY
MCG-03-04-062	01 22 48.	- 16 42	48	15.	GALAXY
LIN 537	01 22 50.	- 73 23 52.		12.77	SMC EMISSION LINE OBJECT
MCG+02-04-043	01 22 51.	+ 09 57 30.	60	15.	GALAXY
RNGC 0529	01 22 53.	+ 34 28		13.0	GALAXY
RNGC 0539	01 22 53.	- 18 25		14.0	GALAXY
RNGC 0546	01 22 53.	- 38 19			GALAXY
RNGC 0544	01 22 53.	- 38 21			GALAXY
HOW 81	01 22 53.	- 73 24 24.	36		STAR CLUSTER IN SMC
ZWG 385.121	01 22 54.	+ 01 29		13.1	GALAXY
UGC 00992	01 22 54.	+ 01 29	300	13.1	GALAXY E
ZWG 411.056	01 22 54.	+ 07 44		15.6	GALAXY
UGC 00993	01 22 54.	+ 07 44	72	15.6	GALAXY DBL SYS
ZWG 436.048	01 22 54.	+ 09 52		15.5	GALAXY
MCG+02-04-046	01 22 54.	+ 09 54 30.	36	15.	GALAXY
MCG+02-04-045	01 22 54.	+ 09 55	18	15.	GALAXY
MCG+02-04-044	01 22 54.	+ 10 32	48	16.	GALAXY
UGC 00994	01 22 54.	+ 17 48	60	16.5	GALAXY Sc
MCG+06-04-019	01 22 54.	+ 34 27	36	13.5	GALAXY
ZWG 521.023	01 22 54.	+ 34 28		13.1	GALAXY
UGC 00995	01 22 54.	+ 34 28	144	13.1	GALAXY E-S0
ZWG 385.120	01 22 54.	- 01 48		14.7	GALAXY
ZH 06	01 22 54.	- 01 48		14.7	GALAXY
UGC 00991	01 22 54.	- 01 48	108	14.7	GALAXY Sa
MCG+00-04-130	01 22 54.	- 01 49	48	14.	GALAXY
MCG-03-04-063	01 22 54.	- 18 25	84	14.	GALAXY
LH115-N088	01 22 54.	- 73 24			EMISSION NEBULA IN SMC
BC PKS0122-00	01 22 55.13	- 00 21 31.5		16.70	QUASI-STELLAR OBJECT
SHB 028	01 22 55.5	- 00 21 34.		16.7	QUASI-STELLAR OBJECT
RNGC 0538	01 22 56.	- 01 48		14.5	GALAXY
ZC 0123.0+0013	01 23 00.	+ 00 13	1810		CLUSTER OF GALAXIES
MCG+00-04-132	01 23 00.	+ 00 54 30.	18	15.	GALAXY
ZC 0123.0+0058	01 23 00.	+ 00 58	810		CLUSTER OF GALAXIES
MCG+00-04-131	01 23 00.	+ 01 30	30	13.0	GALAXY
ZC 0123.0+0953	01 23 00.	+ 09 53	8400		CLUSTER OF GALAXIFS
UGC 00999	01 23 00.	+ 14 43	66	16.5	GALAXY Sc
MCG+03-04-035	01 23 00.	+ 17 54	48	15.	GALAXY
LDN 1324	01 23 00.	+ 61 20	720		DARK NEBULA
UGC 00998	01 23 00.	- 00 08	84	16.5	GALAXY IRR
ZWG 385.124	01 23 00.	- 01 39		14.9	GALAXY
ZH 04	01 23 00.	- 01 39		14.9	GALAXY
UGC 00997	01 23 00.	- 01 39	54	14.9	GALAXY S0
MCG+00-04-133	01 23 00.	- 01 40	60	14.5	GALAXY
ZWG 385.123	01 23 00.	- 01 45		14.8	GALAXY
ZH 07	01 23 00.	- 01 45		14.8	GALAXY
UGC 00996	01 23 00.	- 01 45	66	14.8	GALAXY S0-a
ZH 33	01 23 00.	- 02 03		15.8	GALAXY
ZWG 385.122	01 23 00.	- 02 04		15.7	GALAXY
TON-S 0213	01 23 00.	- 33 29		15.5	BLUE STAR
MCG-06-04-027	01 23 00.	- 33 42	36	15.	GALAXY
MCG-06-04-030	01 23 00.	- 36 05	72	15.	GALAXY
MCG-06-04-029	01 23 00.	- 38 22	66	14.	GALAXY
MCG-06-04-028	01 23 00.	- 38 23	60	13.5	GALAXY
RNGC 0533	01 23 01.	+ 01 30		13.0	GALAXY
RNGC 0535	01 23 02.	- 01 39		15.0	GALAXY
ARC 0194	01 23 02.	- 01 46		13.9	RICH CLUSTER OF GALAXIES
LIN 538	01 23 02.	- 73 29 10.		13.96	SMC EMISSION LINE OBJECT
MCG+00-04-134	01 23 03.	- 00 09	27	15.5	GALAXY
MCG-06-04-031	01 23 03.	- 36 05	36	15.5	GALAXY
MCG+00-04-136	01 23 06.	+ 00 10	18	15.5	GALAXY
UGC 01001	01 23 06.	+ 00 12	78	16.5	GALAXY S
ZWG 385.125	01 23 06.	+ 00 55		15.4	GALAXY
MCG+00-04-135	01 23 06.	+ 01 38	30	16.	GALAXY
MCG+03-04-036	01 23 06.	+ 16 20	48	14.	GALAXY
ZWG 459.050	01 23 06.	+ 17 55		15.3	GALAXY
UGC 01002	01 23 06.	+ 17 55	96	15.3	GALAXY VY CMPT
IC 0110	01 23 06.	+ 33 14			NONSTELLAR OBJECT
IC 0111	01 23 08.	+ 33 13			NONSTELLAR OBJECT
LIN.CL 101	01 23 08.	- 73 24 52.	4080		STAR CLUSTER IN SMC
IC 1701	01 23 09.	+ 17 56 13.			NONSTELLAR OBJFCT
IC 1702	01 23 11.	+ 16 20 19.			NONSTELLAR OBJFCT
LB 00447	01 23 11.	+ 33 08 36.		17.5	FAINT BLUE STAR
ARP 308	01 23 11.	- 01 37			PECULIAR GALAXY
ARP 133	01 23 11.	- 01 38			PECULIAR GALAXY
REIF 2.005	01 23 11.16	- 01 38 19.9			NEBULA
ZC 0123.2+1251	01 23 12.	+ 12 51	1210		CLUSTER OF GALAXIES
ZWG 459.051	01 23 12.	+ 16 20		14.5	GALAXY
UGC 01005	01 23 12.	+ 16 20	90	14.5	GALAXY Sc
ZWG 385.127	01 23 12.	- 01 35		15.2	GALAXY
ZH 09	01 23 12.	- 01 35		15.2	GALAXY
ZWG 385.128	01 23 12.	- 01 37		14.0	GALAXY
ZH 03	01 23 12.	- 01 37		14.0	GALAXY
UGC 01004	01 23 12.	- 01 37	156	14.0	GALAXY E
MCG+00-04-137	01 23 12.	- 01 39	24	13.	GALAXY
ZWG 385.126	01 23 12.	- 01 42		15.0	GALAXY
ZH 12	01 23 12.	- 01 42		14.9	GALAXY
UGC 01003	01 23 12.	- 01 42	60	15.0	GALAXY S0
ZH 48	01 23 12.	- 01 44		15.8	GALAXY
MCG+01-04-041	01 23 12.	- 08 00	48	15.	GALAXY
TON-S 0214	01 23 12.	- 28 11		15.7	BLUF STAR
RNGC 0541	01 23 13.	- 01 37		14.0	GALAXY
HOW 82	01 23 17.	- 73 25 30.	12		STAR CLUSTER IN SMC
ZWG 385.131	01 23 18.	+ 01 14		15.3	GALAXY
UGC 01006	01 23 18.	+ 01 14	72	15.3	GALAXY S0-a
ZWG 385.130	01 23 18.	- 01 32		15.0	GALAXY
ZH 10	01 23 18.	- 01 32		15.0	GALAXY
MCG+00-04-138	01 23 18.	- 01 33	30	15.	GALAXY
ZWG 385.129	01 23 18.	- 01 34		15.2	GALAXY
ZH 05	01 23 18.	- 01 34		15.2	GALAXY
MCG+00-04-140	01 23 18.	- 01 35	24	15.	GALAXY
MCG+00-04-139	01 23 18.	- 01 36	24	16.	GALAXY
MCG-03-04-064	01 23 18.	- 18 28 30.	30	15.	GALAXY
MCG-05-04-038	01 23 18.	- 30 37	36	15.	GALAXY
RNGC 0543	01 23 19.	- 01 32		15.0	GALAXY
MCG-02-04-047	01 23 21.	- 12 47	18	15.5	GALAXY
ZWG 436.049	01 23 24.	+ 11 11		14.2	GALAXY
UGC 01008	01 23 24.	+ 11 11	54	14.2	GALAXY S-IRR
IC 0112	01 23 24.	+ 11 11 31.			NONSTELLAR OBJECT
ZWG 502.084	01 23 24.	+ 33 09		15.0	GALAXY
ZWG 385.132	01 23 24.	- 01 35		13.7	GALAXY
ZH 02	01 23 24.	- 01 35		13.4	GALAXY
ZH 01	01 23 24.	- 01 35		13.7	GALAXY
UGC 01007	01 23 24.	- 01 25	180	13.7	GALAXY S0
KARA.72 032A	01 23 24.	- 01 35	144	13.7	PART OF DOUBLE GALAXY
MCG-02-04-048	01 23 24.	- 09 54 30.	36	15.	GALAXY
LB G1600	01 23 24.	- 48 29		15.0	FAINT BLUE STAR
LIN 539	01 23 24.	- 73 48 23.		14.58	SMC EMISSION LINF OBJECT
RNGC 0547A	01 23 25.	- 01 36			GALAXY
RNGC 0545	01 23 25.	- 01 36		13.5	GALAXY
REIN 2.006	01 23 25.97	- 01 35 58.4			NEBULA
HN 0033	01 23 27.	- 01 29			NEBULA
HOLM 042A	01 23 27.	- 01 36	30	13.7	PART OF MULTIPLE GALAXY
REIN 2.007	01 23 27.49	- 01 36 15.4			NEBULA
LB 00448	01 23 29.	+ 30 12 18.		16.9	FAINT BLUE STAR
RNGC 0537	01 23 29.	+ 33 49			NON-EXISTENT OBJFCT
RNGC 0531	01 23 29.	+ 34 29		15.0	GALAXY
HOLM 042B	01 23 29.	- 01 36	30	13.8	PART OF MULTIPLE GALAXY
HN 0152	01 23 29.	- 71 26			NEBULA
LIN.CL 102	01 23 29.	- 71 26 40.	270	14.3	STAR CLUSTER IN SMC
IC 1708	01 23 29.	- 71 27			NONSTELLAR OBJECT
UGC 01011	01 23 30.	+ 00 04	66	16.5	GALAXY S-IRR
ZC 0123.5+1021	01 23 30.	+ 10 21	940		CLUSTER OF GALAXIES
MCG+02-04-047	01 23 30.	+ 11 11	36	14.5	GALAXY
ZWG 521.024	01 23 30.	+ 34 30		14.9	GALAXY
UGC 01012	01 23 30.	+ 34 30	114	14.9	GALAXY SB:0-a
ZC 0123.5+3933	01 23 30.	+ 39 33	940		CLUSTER OF GALAXIES
MCG+08-03-022	01 23 30.	+ 48 08	30	15.	GALAXY
ZWG 385.134	01 23 30.	- 01 29		15.1	GALAXY
ZH 11	01 23 30.	- 01 29		15.1	GALAXY
UGC 01010	01 23 30.	- 01 29	78	15.1	GALAXY E-S0
MCG+00-04-141	01 23 30.	- 01 30	30	14.	GALAXY
ZWG 385.133	01 23 30.	- 01 36		13.4	GALAXY
UGC 01009	01 23 30.	- 01 36	90	13.4	GALAXY E
KARA.72 032B	01 23 30.	- 01 36	96	13.4	PART OF DOUBLE GALAXY
MCG+00-04-143	01 23 30.	- 01 37	30	13.8	GALAXY
MCG+00-04-142	01 23 30.	- 01 37	24	13.7	GALAXY
MCG-06-04-032	01 23 30.	- 35 07	30	15.	GALAXY
RNGC 0548	01 23 31.	- 01 29		15.0	GALAXY
RNGC 0547	01 23 31.	- 01 36		13.5	GALAXY
SN 1963N	01 23 34.	+ 34 27		17.7	SUPERNOVA
RNGC 0536	01 23 35.	+ 34 27		13.0	GALAXY
ZWG 521.025	01 23 36.	+ 34 27		13.2	GALAXY
UGC 01013	01 23 36.	+ 34 27	210	13.2	GALAXY SBb
MCG+06-04-020	01 23 36.	+ 34 29	108	15.	GALAXY
5ZW 068	01 23 36.	+ 48 08			COMPACT GALAXY
ZWG 551.017	01 23 36.	+ 48 08		15.3	GALAXY
ZC 0123.6-0133	01 23 36.	- 01 33	13570		CLUSTER OF GALAXIES
MCG+01-04-042	01 23 36.	- 04 12	108	14.5	GALAXY
MCG-02-04-049	01 23 36.	- 13 41 30.	24	15.	GALAXY
TON-S 0215	01 23 36.	- 22 15		15.6	BLUE STAR
SER 010.02	01 23 36.	- 68 53	240	16.5	COMPACT GROUP OF 9 GLXIES
KN 14.116	01 23 37.0	+ 29 26 20.			NEBULA
MCG+06-04-021	01 23 39.	+ 34 26	240	13.	GALAXY

OBJECT NAME	RIGHT ASCEN.	DECLINATION	DIAM.	MAGN.	TYPE OF OBJECT
MCG-06-04-033	01 23 39.	- 37 29	15	16.	GALAXY
RNGC 0542	01 23 41.	+ 34 25		15.5	GALAXY
HOLM 043B	01 23 41.	- 06 21	18	14.9	PART OF MULTIPLE GALAXY
KN 14.117	01 23 41.9	+ 31 24 56.			NEBULA
ZWG 411.057	01 23 42.	+ 06 01		15.2	GALAXY
UGC 01014	01 23 42.	+ 06 01	96	15.2	GALAXY IRR
KARA.73B 0050	01 23 42.	+ 06 01	48	15.2	ISOLATED GALAXY S
ZWG 436.050	01 23 42.	+ 09 40		15.7	GALAXY
UGC 01015	01 23 42.	+ 09 40	96	15.7	GALAXY S0
ZWG 502.085	01 23 42.	+ 31 22		15.0	GALAXY
MRK 358	01 23 42.	+ 31 22	12	16.	GALAXY WITH UV CONTINUUM
KW 34	01 23 42.	+ 31 22	42		SEYFERT GALAXY
SN 1969J	01 23 42.	+ 31 22		17.0	SUPERNOVA
ZWG 521.026	01 23 42.	+ 34 25		15.4	GALAXY
5ZW 069	01 23 42.	+ 39 37			COMPACT GALAXY
ZWG 537.001	01 23 42.	+ 39 37		15.7	GALAXY
ZH 34	01 23 42.	- 02 02		15.6	GALAXY
ZWG 385.135	01 23 42.	- 02 03		15.6	GALAXY
MCG+01-04-043	01 23 42.	- 04 04	60	16.	GALAXY
MCG-06-04-034	01 23 42.	- 37 31	48	14.5	GALAXY
IC 0113	01 23 43.	+ 18 56 00.			NONSTELLAR OBJECT
HOLM 043A	01 23 43.	- 06 21	162	13.1	PART OF MULTIPLE GALAXY
KN 14.118	01 23 43.3	+ 30 56 56.			NEBULA
KN 14.119	01 23 43.8	+ 29 36 49.			NEBULA
IC 0114	01 23 44.	+ 09 39 36.			NONSTELLAR OBJECT
KN 14.121	01 23 44.	+ 29 35 29.			NEBULA
MCG+01-04-059	01 23 45.	+ 06 01	60	14.	GALAXY
MCG+01-04-060	01 23 45.	+ 08 48	36	15.	GALAXY
MCG+06-04-022	01 23 45.	+ 34 24	54	15.	GALAXY
KN 14.120	01 23 45.3	+ 31 21 22.			NEBULA
LIN 540	01 23 47.	- 73 49 47.		14.11	SMC EMISSION LINE OBJECT
KN 14.124	01 23 47.0	+ 28 43 55.			NEBULA
KN 14.122	01 23 47.5	+ 30 15 38.			NEBULA
KN 14.125	01 23 47.7	+ 29 33 22.			NEBULA
3ZW 028	01 23 48.	+ 07 56			COMPACT GALAXY
MCG+02-04-048	01 23 48.	+ 09 39	84	15.	GALAXY
MCG+03-04-037	01 23 48.	+ 19 05	48	14.5	GALAXY
MCG+05-04-059	01 23 48.	+ 31 21	48	15.	GALAXY
ZC 0123.8+3904	01 23 48.	+ 39 04	1480		CLUSTER OF GALAXIES
UGC 01017	01 23 48.	+ 46 31	66	17.	GALAXY
ZWG 385.136	01 23 48.	- 01 54		14.9	GALAXY
UGC 01016	01 23 48.	- 01 54	120	14.9	GALAXY S0-a
SN 1963E	01 23 48.	- 01 54		16.5	SUPERNOVA
MCG+01-04-044	01 23 48.	- 06 20	180	12.5	GALAXY
MCG-02-04-050	01 23 48.	- 08 50	24	15.	GALAXY
MCG-02-04-051	01 23 48.	- 13 46 30.	18	15.5	GALAXY
MCG-03-04-065	01 23 48.	- 15 51 30.	36	16.	GALAXY
LB 03205	01 23 48.	- 60 52		13.4	FAINT BLUE STAR
KN 14.123	01 23 48.9	+ 31 22 34.			NEBULA
KN 14.126	01 23 49.4	+ 31 26 50.			NEBULA
KN 14.128	01 23 49.7	+ 28 11 41.			NEBULA
MCG+06-04-024	01 23 51.	+ 36 10	24	17.	GALAXY
MCG+06-04-023	01 23 51.	+ 36 10	24	16.	GALAXY
IC 1703	01 23 51.8	- 01 53 54.		14.9	GALAXY SB(rs)
KN 14.127	01 23 52.0	+ 31 47 01.			NEBULA
ZWG 385.137	01 23 54.	+ 01 11		15.6	GALAXY
MCG+10-03-002	01 23 54.	+ 58 14	24	18.	GALAXY
ZH 17	01 23 54.	- 01 53		14.9	GALAXY
MCG+00-04-144	01 23 54.	- 01 54	42		GALAXY
MCG+01-04-046	01 23 54.	- 07 42	9	15.5	GALAXY
MCG+01-04-045	01 23 54.	- 07 42	42	15.5	GALAXY
MCG+01-04-047	01 23 54.	- 08 20	60	15.	GALAXY
LIN.CL 103	01 23 55.	- 73 30 05.	2040		STAR CLUSTER IN SMC
KN 14.129	01 23 55.4	+ 32 33 24.			NEBULA
SHB 029	01 23 56.	+ 25 43 54.		17.5	QUASI-STELLAR OBJECT
RNGC 0557	01 23 56.	- 01 54		15.0	GALAXY
BC 4C25.05	01 23 57.	+ 25 43 33.			QUASI-STELLAR OBJECT
KHAV 027	01 24	+ 54 24	6230		DARK NEBULA
UGC 01018	01 24 00.	+ 00 18	66	18.	GALAXY DWARF
UGC 01019	01 24 00.	+ 10 01	78	16.5	GALAXY SB IV-V
ZWG 459.052	01 24 00.	+ 17 00		15.0	GALAXY
UGC 01020	01 24 00.	+ 17 00	84	15.0	GALAXY S
MCG+03-04-038	01 24 00.	+ 17 00	66	14.5	GALAXY
ZH 38	01 24 00.	- 01 21		15.6	GALAXY
MCG+00-04-145	01 24 00.	- 01 24	24	15.5	GALAXY
ZH 49	01 24 00.	- 01 49		15.9	GALAXY
ZH 35	01 24 00.	- 02 05		15.4	GALAXY
MCG-02-04-052	01 24 00.	- 13 45	30	15.	GALAXY
KN 14.130	01 24 00.1	+ 28 36 04.			NEBULA
KN 14.131	01 24 03.0	+ 31 31 39.			NEBULA
SN 1961Q	01 24 05.	+ 01 45		17.2	SUPERNOVA
KN 14.133	01 24 05.9	+ 31 25 41.			NEBULA
ZWG 385.139	01 24 06.	+ 01 45		13.6	GALAXY
UGC 01021	01 24 06.	+ 01 45	132	13.6	GALAXY S
MCG+00-04-146	01 24 06.	+ 01 47	66	13.	GALAXY
MCG+01-04-061	01 24 06.	+ 07 52 30.	15	15.5	GALAXY
MCG+02-04-050	01 24 06.	+ 10 02	48	16.	GALAXY
ZWG 459.053	01 24 06.	+ 21 21		15.7	GALAXY
ZC 0124.1+2703	01 24 06.	+ 27 03	1140		CLUSTER OF GALAXIES
ZWG 521.027	01 24 06.	+ 38 45		14.8	GALAXY
UGC 01022	01 24 06.	+ 38 45	72	14.8	GALAXY S
ZH 40	01 24 06.	- 01 33		15.6	GALAXY
ZWG 385.138	01 24 06.	- 01 34		15.2	GALAXY
LB 03206	01 24 06.	- 60 37		14.2	FAINT BLUE STAR
KN 14.132	01 24 06.9	+ 32 44 18.			NEBULA
RNGC 0550	01 24 07.	+ 01 46		13.5	GALAXY
LIN 541	01 24 07.	- 73 22 54.		14.48	SMC EMISSION LINE OBJECT
KN 14.134	01 24 07.9	+ 29 27 56.			NEBULA
ZWG 436.051	01 24 12.	+ 11 46		15.7	GALAXY
UGC 01023	01 24 12.	+ 11 46	96	15.7	GALAXY S0
ZWG 436.052	01 24 12.	+ 12 47		15.6	GALAXY
UGC 01024	01 24 12.	+ 12 47	72	15.6	GALAXY
ZWG 459.054	01 24 12.	+ 18 57		15.5	GALAXY
MCG+03-04-039	01 24 12.	+ 18 57	30	15.	GALAXY
IC 0115	01 24 12.	+ 18 57 23.			NONSTELLAR OBJECT
ZWG 521.028	01 24 12.	+ 33 42		15.6	GALAXY
MCG+06-04-025	01 24 12.	+ 38 45	36	16.	GALAXY
MRK 570	01 24 12.	- 01 04	16	14.5	GALAXY WITH UV CONTINUUM
ZH 37	01 24 12.	- 01 13		15.2	GALAXY
ZWG 385.140	01 24 12.	- 01 14		15.2	GALAXY
MCG+00-04-147	01 24 12.	- 01 35	30	15.5	GALAXY
MCG+01-04-048	01 24 12.	- 03 42	24	14.5	GALAXY
MCG-04-04-009	01 24 12.	- 23 28 30.	120	13.	GALAXY
MCG-04-04-010	01 24 12.	- 23 31	24	16.	GALAXY
LB 03207	01 24 12.	- 51 50		13.4	FAINT BLUE STAR
KN 14.135	01 24 12.1	+ 31 20 02.			NEBULA
KN 14.136	01 24 12.9	+ 31 17 55.			NEBULA
IC 1704	01 24 13.	+ 14 27 34.			NONSTELLAR OBJECT
ARC 0195	01 24 13.	+ 18 56		15.3	RICH CLUSTER OF GALAXIES
IC 1705	01 24 13.	- 03 45 38.			NONSTELLAR OBJECT
KN 14.137	01 24 14.3	+ 31 23 52.			NEBULA
MCG+02-04-049	01 24 15.	+ 11 46	48	15.	GALAXY
MCG+00-04-148	01 24 15.	- 01 15	18	15.	GALAXY
MCG+02-04-051	01 24 18.	+ 12 47	36	15.	GALAXY
ZWG 459.055	01 24 18.	+ 15 43		15.0	GALAXY
MCG+06-04-026	01 24 18.	+ 33 38 30.	24	16.	GALAXY
ZWG 385.141	01 24 18.	- 00 16		15.4	GALAXY
ZH 42	01 24 18.	- 01 45		16.3	GALAXY
MCG+01-04-049	01 24 18.	- 05 12	48	14.5	GALAXY
IC 0116	01 24 18.	- 05 14 32.			NONSTELLAR OBJECT
MCG-03-04-066	01 24 18.	- 17 17 30.	48	14.5	GALAXY
MCG-06-04-035	01 24 18.	- 37 45	48	15.5	GALAXY
LB 01601	01 24 18.	- 54 03		13.9	FAINT BLUE STAR
ARC 0197	01 24 19.	- 18 22		17.1	RICH CLUSTER OF GALAXIES
MCG+00-04-149	01 24 21.	- 00 17	24	15.	GALAXY
MCG-04-04-011	01 24 21.	- 25 33 30.	72	15.	GALAXY
KN 14.138	01 24 23.6	+ 31 19 43.			NEBULA
MCG+02-04-052	01 24 24.	+ 14 31	60	14.5	GALAXY
MCG+03-04-040	01 24 24.	+ 18 19	42	15.	GALAXY
ZWG 459.056	01 24 24.	+ 18 20		15.0	GALAXY
UGC 01025	01 24 24.	+ 18 20	72	15.0	GALAXY S0-a
ZC 0124.4+2256	01 24 24.	+ 22 56	1410		CLUSTER OF GALAXIES
ZWG 521.029	01 24 24.	+ 34 36		15.7	GALAXY
TON-S 0216	01 24 24.	- 25 45		15.2	BLUE STAR
LB 03208	01 24 24.	- 72 56		14.0	FAINT BLUE STAR
LIN.CL 104	01 24 24.	- 73 38 18.	3402		STAR CLUSTER IN SMC
RNGC 0567	01 24 27.	- 10 31		14.0	GALAXY
ARC 0196	01 24 29.	+ 22 58		17.5	RICH CLUSTER OF GALAXIES
ZWG 436.053	01 24 30.	+ 10 22		15.5	GALAXY
MCG+02-04-053	01 24 30.	+ 10 53	60	14.5	GALAXY
UGC 01026	01 24 30.	+ 13 20	120	17.	GALAXY DWRF SP
ZWG 436.054	01 24 30.	+ 14 31		14.2	GALAXY
UGC 01027	01 24 30.	+ 14 31	72	14.2	GALAXY S
ZC 0124.5+3147	01 24 30.	+ 31 47	810		CLUSTER OF GALAXIES
5ZW 070	01 24 30.	+ 41 48			COMPACT GALAXY
ZH 55	01 24 30.	- 00 16		15.6	GALAXY
MCG-02-04-053	01 24 30.	- 10 31 30.	48	14.5	GALAXY
MCG-03-04-067	01 24 30.	- 15 25	66	15.	GALAXY
LH115-N089	01 24 30.	- 73 38	283		EMISSION NEBULA IN SMC
MCG+02-04-054	01 24 33.	+ 10 53	72	15.	GALAXY
IC 1706	01 24 35.	+ 14 30 09.			NONSTELLAR OBJECT
ZWG 385.142	01 24 36.	+ 02 00		15.0	GALAXY
UGC 01028	01 24 36.	+ 02 00	72	15.0	GALAXY S
ZWG 436.055	01 24 36.	+ 11 44		15.7	GALAXY
UGC 01029	01 24 36.	+ 11 44	66	15.7	GALAXY Sa
ZC 0124.6+1610	01 24 36.	+ 16 10	2150		CLUSTER OF GALAXIES
ZC 0124.6+2853	01 24 36.	+ 28 53	6250		CLUSTER OF GALAXIES
MCG+01-04-050	01 24 36.	- 03 27	48	15.	GALAXY
MCG-03-04-068	01 24 36.	- 17 37	30	15.5	GALAXY
RNGC 0556	01 24 36.	- 23 03			GALAXY
KN 14.139	01 24 37.7	+ 32 11 21.			NEBULA
ARC 0198	01 24 38.	- 16 42		17.8	RICH CLUSTER OF GALAXIES
KN 14.140	01 24 39.1	+ 31 35 49.			NEBULA
RNGC 0563	01 24 41.	- 18 54		14.0	GALAXY
ZWG 436.056	01 24 42.	+ 13 48		15.6	GALAXY
ZWG 537.002	01 24 42.	+ 40 02		15.6	GALAXY
UGC 01031	01 24 42.	+ 40 02	78	15.6	GALAXY SB
MCG+08-03-023	01 24 42.	+ 48 33	66	15.	GALAXY
ZWG 385.144	01 24 42.	- 01 31		14.9	GALAXY
ZH 39	01 24 42.	- 01 31		14.9	GALAXY
UGC 01030	01 24 42.	- 01 31	66	14.9	GALAXY E
ZWG 385.143	01 24 42.	- 02 14		15.0	GALAXY
ZH 51	01 24 42.	- 02 16		15.1	GALAXY
MCG-03-04-069	01 24 42.	- 18 54 30.	18	14.5	GALAXY
ARC 0199	01 24 43.	- 18 03		17.5	RICH CLUSTER OF GALAXIES
RNGC 0558	01 24 44.	- 02 14		15.0	GALAXY
KN 14.141	01 24 44.2	+ 31 32 44.			NEBULA
MCG+03-04-041	01 24 45.	+ 18 54	30	14.5	GALAXY
MCG+00-04-150	01 24 45.	- 01 32	42	14.5	GALAXY
RNGC 0551	01 24 46.	+ 36 55		13.5	GALAXY
KN 14.142	01 24 46.6	+ 31 17 43.			NEBULA
ZWG 436.057	01 24 48.	+ 14 35		15.4	GALAXY
MRK 359	01 24 48.	+ 18 54	20	15.5	GALAXY WITH UV CONTINUUM
ZWG 459.057	01 24 48.	+ 18 55		13.8	GALAXY
UGC 01032	01 24 48.	+ 18 55	42	13.8	GALAXY PECULR
MCG+05-04-060	01 24 48.	+ 31 17 30.	180	14.	GALAXY
ZWG 502.086	01 24 48.	+ 31 18		14.4	GALAXY
UGC 01033	01 24 48.	+ 31 18	204	14.4	GALAXY Sc
ZWG 521.030	01 24 48.	+ 36 55		13.5	GALAXY
UGC 01034	01 24 48.	+ 36 55	120	13.5	GALAXY SBb-c
ZWG 551.018	01 24 48.	+ 48 33		15.4	GALAXY
UGC 01035	01 24 48.	+ 48 33	90	15.4	GALAXY Sb
MCG-04-04-015	01 24 48.	- 20 38 30.	48	15.	GALAXY
MCG-04-04-012	01 24 48.	- 22 01 30.	36	14.5	GALAXY
RNGC 0554	01 24 48.	- 22 59		15.0	GALAXY
MCG-04-04-013	01 24 48.	- 22 59 30.	24	15.	GALAXY
RNGC 0555	01 24 48.	- 23 01		15.0	GALAXY
MCG-04-04-014	01 24 48.	- 23 01 30.	24	15.	GALAXY
KN 14.143	01 24 49.	+ 27 59 27.			NEBULA
IC 0117	01 24 49.	- 02 07 22.			ONE STAR
SNO 14	01 24 50.	- 23 02 10.	600	16.	GROUP OF 7 GALAXIES
KN 14.144	01 24 51.1	+ 30 24 57.			NEBULA
MCG+06-04-027	01 24 54.	+ 36 54 30.	102	13.	GALAXY
ZWG 537.003	01 24 54.	+ 42 54		15.4	GALAXY
UGC 01037	01 24 54.	+ 42 54	72	15.4	GALAXY Sb
MCG+07-04-001	01 24 54.	+ 42 54	60	15.5	GALAXY
UGC 01038	01 24 54.	+ 43 00	84	17.	GALAXY DWARF
MCG+08-03-024	01 24 54.	+ 49 00	24	14.	GALAXY
ZH 43	01 24 54.	- 01 46		16.7	GALAXY
ZH 20	01 24 54.	- 02 09		14.2	GALAXY
ZWG 385.145	01 24 54.	- 02 10		14.0	GALAXY
UGC 01036	01 24 54.	- 02 10	132	14.0	GALAXY S0
MCG+00-04-151	01 24 54.	- 02 10	90	14.0	GALAXY
MCG+01-04-051	01 24 54.	- 06 22	30	15.	GALAXY
MCG-06-04-036	01 24 54.	- 35 53	18	15.	GALAXY
KN 14.145	01 24 55.8	+ 30 17 53.			NEBULA
RNGC 0560	01 24 56.	- 02 11		14.0	GALAXY
KN 14.146	01 24 57.8	+ 31 46 39.			NEBULA
RNGC 0553	01 24 59.	+ 33 12			NON-EXISTENT OBJECT
RNGC 0552	01 24 59.	+ 33 12			NON-EXISTENT OBJECT
ZWG 436.058	01 25 00.	+ 11 58		15.6	GALAXY
UGC 01041	01 25 00.	+ 11 58	102	15.6	GALAXY CHAIN
ZC 0125.0+1505	01 25 00.	+ 15 05	2620		CLUSTER OF GALAXIES
ZWG 551.019	01 25 00.	+ 48 59		14.9	GALAXY
UGC 01042	01 25 00.	+ 48 59	90	14.9	GALAXY S0
VB 047	01 25 00.	+ 64 10	450		STELLAR RING

OBJECT NAME	RIGHT ASCEN.	DECLINATION	DIAM.	MAGN.	TYPE OF OBJECT
VB 048	01 25 00.	+ 65 08	161		STELLAR RING
ZWG 361.002	01 25 00.	+ 84 46		15.4	GALAXY
ZWG 360.005	01 25 00.	+ 84 46		15.4	GALAXY
UGC 01039	01 25 00.	+ 84 46	102	15.4	GALAXY Sa-b
MCG+14-02-001	01 25 00.	+ 84 46	84	15.	GALAXY
ZH 19	01 25 00.	- 01 20		14.8	GALAXY
ZWG 385.146	01 25 00.	- 01 21		14.8	GALAXY
UGC 01040	01 25 00.	- 01 21	84	14.8	GALAXY S0-a
MCG+00-04-152	01 25 00.	- 01 22	48	14.5	GALAXY
MCG+01-04-052	01 25 00.	- 04 55	18	15.	GALAXY
LB 03209	01 25 00.	- 66 01		12.8	FAINT BLUE STAR
BC PKS0125-41	01 25 01.3	- 41 28 16.		17.	QUASI-STELLAR OBJECT
IC 0118	01 25 04.	- 05 15 16.			NONSTELLAR OBJECT
KN 14.150	01 25 05.9	+ 28 45 33.			NEBULA
MCG+02-04-055	01 25 06.	+ 11 57 30.	36	15.	GALAXY
4ZW 046	01 25 06.	+ 24 39			COMPACT GALAXY
ZH 41	01 25 06.	- 01 22		15.1	GALAXY
ZWG 385.147	01 25 06.	- 01 23		15.1	GALAXY
UGC 01043	01 25 06.	- 01 23	60	15.1	GALAXY E
MCG+01-04-053	01 25 06.	- 05 13	12	16.	GALAXY
MCG-02-04-054	01 25 06.	- 14 00 30.	36	15.	GALAXY
KN 14.147	01 25 06.1	+ 32 08 47.			NEBULA
KN 14.149	01 25 08.4	+ 32 11 20.			NEBULA
KN 14.148	01 25 08.8	+ 32 43 46.			NEBULA
KN 14.152	01 25 11.3	+ 28 02 56.			NEBULA
KN 14.151	01 25 11.8	+ 31 46 19.			NEBULA
ZC 0125.2+0702	01 25 12.	+ 07 02	1950		CLUSTER OF GALAXIES
ZC 0125.2+1954	01 25 12.	+ 19 54	1210		CLUSTER OF GALAXIES
ZWG 502.087	01 25 12.	+ 31 47		15.1	GALAXY
UGC 01045	01 25 12.	+ 31 47	78	15.1	GALAXY S
ZWG 502.088	01 25 12.	+ 32 45		15.7	GALAXY
KARA.73B 0051	01 25 12.	+ 32 45	36	15.7	ISOLATED GALAXY S
ZWG 521.031	01 25 12.	+ 37 58		15.7	GALAXY
UGC 01046	01 25 12.	+ 37 58	96	15.7	GALAXY S IV-V
MCG+00-04-153	01 25 12.	- 01 24	36	14.5	GALAXY
ZWG 385.148	01 25 12.	- 02 08		13.8	GALAXY
UGC 01044	01 25 12.	- 02 08	120	13.8	GALAXY E
HOLM 044B	01 25 13.	- 02 08	60	14.5	PART OF MULTIPLE GALAXY
RNGC 0564	01 25 14.	- 02 09		14.0	GALAXY
KN 14.153	01 25 14.6	+ 27 18 25.			NEBULA
ARC 0202	01 25 15.	+ 07 02		17.8	RICH CLUSTER OF GALAXIES
MCG+05-04-061	01 25 15.	+ 31 46	63	15.	GALAXY
HOLM 044A	01 25 15.	- 02 08	24	13.9	PART OF MULTIPLE GALAXY
MCG+00-04-154	01 25 15.	- 02 09	48	13.8	GALAXY
MCG-04-04-016	01 25 15.	- 21 53	30	16.	GALAXY
ARC 0200	01 25 16.	+ 14 58		17.1	RICH CLUSTER OF GALAXIES
MCG+00-04-155	01 25 18.	+ 01 52	15	15.	GALAXY
ZWG 385.150	01 25 18.	+ 01 53		15.0	GALAXY
ZC 0125.3+3159	01 25 18.	+ 31 59	870		CLUSTER OF GALAXIES
MCG+06-04-028	01 25 18.	+ 37 57	54	15.	GALAXY
ZH 21	01 25 18.	- 02 07		14.1	GALAXY
ZWG 385.149	01 25 18.	- 02 17		15.0	GALAXY
UGC 01047	01 25 18.	- 02 17	96	15.0	GALAXY S0-a
MCG-02-04-055	01 25 18.	- 08 38 30.	36	15.	GALAXY
MCG-03-04-070	01 25 18.	- 19 18	48	14.	GALAXY
TON-S 0217	01 25 18.	- 23 39		14.9	BLUE STAR
KN 14.154	01 25 20.7	+ 29 31 00.			NEBULA
MCG+00-04-156	01 25 21.	- 01 55	24	15.5	GALAXY
IC 0119	01 25 22.3	- 02 18 01.		15.0	GALAXY SB(r)
KN 14.155	01 25 22.8	+ 30 12 17.			NEBULA
ARC 0201	01 25 23.	+ 16 17		17.1	RICH CLUSTER OF GALAXIES
RNGC 0561	01 25 23.	+ 34 03		14.0	GALAXY
RNGC 0562	01 25 23.	+ 48 07		14.5	GALAXY
ZWG 521.032	01 25 24.	+ 34 03		14.1	GALAXY
UGC 01048	01 25 24.	+ 34 03	102	14.1	GALAXY SBa
ZWG 521.033	01 25 24.	+ 39 15		15.7	GALAXY
ZWG 551.020	01 25 24.	+ 48 07		14.5	GALAXY
UGC 01049	01 25 24.	+ 48 07	84	14.5	GALAXY Sc
MCG+08-03-025	01 25 24.	+ 48 08 30.	66	13.	GALAXY
ZH 44	01 25 24.	- 01 53		15.7	GALAXY
ZH 26	01 25 24.	- 02 17		15.0	GALAXY
MCG+00-04-157	01 25 24.	- 02 19	60	14.5	GALAXY
MCG-02-04-056	01 25 24.	- 09 44	42	14.5	GALAXY
ZC 0125.5+0133	01 25 30.	+ 01 33	2350		CLUSTER OF GALAXIES
UGC 01050	01 25 30.	+ 10 10	78	16.5	GALAXY Sc-IRR
UGC 01051	01 25 30.	+ 28 45	72	17.	GALAXY DWARF
ZWG 521.034	01 25 30.	+ 37 38		15.4	GALAXY
ZWG 537.004	01 25 30.	+ 39 58		15.5	GALAXY
ZWG 385.151	01 25 30.	- 01 00		15.6	GALAXY
ZH 53	01 25 30.	- 01 00		15.2	GALAXY
KN 14.156	01 25 30.6	+ 32 30 18.			NEBULA
MCG-03-04-071	01 25 33.	- 17 06	15	15.	GALAXY
KN 14.157	01 25 35.3	+ 31 15 45.			NEBULA
MCG+03-04-043	01 25 36.	+ 16 10	24	16.	GALAXY
MCG+03-04-042	01 25 36.	+ 16 11	24	16.	GALAXY
MCG+06-04-029	01 25 36.	+ 34 02	90	14.	GALAXY
MCG+08-03-026	01 25 36.	+ 48 04	24	16.	GALAXY
ZH 18	01 25 36.	- 01 32		14.6	GALAXY
ZWG 385.153	01 25 36.	- 01 33		14.5	GALAXY
UGC 01052	01 25 36.	- 01 33	90	14.5	GALAXY Sa
ZWG 385.152	01 25 36.	- 02 10		15.2	GALAXY
RNGC 0565	01 25 37.	- 01 33		14.5	GALAXY
KN 14.158	01 25 37.1	+ 32 04 34.			NEBULA
HN 0034	01 25 39.	- 01 34			NEBULA
IC 0120	01 25 39.9	- 02 10 33.		15.2	GALAXY E5
RNGC 0568	01 25 40.	- 35 59			GALAXY
ARC 0203	01 25 41.	+ 01 38		17.0	RICH CLUSTER OF GALAXIES
ZWG 459.058	01 25 42.	+ 16 11		15.2	GALAXY
KARA.72 033B	01 25 42.	+ 16 11	36	15.2	PART OF DOUBLE GALAXY
KARA.72 033A	01 25 42.	+ 16 12	36	15.5	PART OF DOUBLE GALAXY
ZWG 459.059	01 25 42.	+ 16 13		15.5	GALAXY
MCG+00-04-158	01 25 42.	- 01 34	66	14.	GALAXY
ZH 46	01 25 42.	- 02 09		15.3	GALAXY
MCG-03-04-072	01 25 42.	- 19 04	24	15.5	GALAXY
MCG-06-04-037	01 25 42.	- 36 00	90	12.5	GALAXY
LB 00449	01 25 43.	+ 29 19 48.		16.3	FAINT BLUE STAR
KN 14.159	01 25 43.8	+ 31 14 08.			NEBULA
MCG+00-04-159	01 25 45.	+ 02 15 30.	42	14.	GALAXY
MCG-03-04-073	01 25 45.	- 19 06	36	17.	GALAXY
IC 0121	01 25 46.	+ 02 16 30.			NONSTELLAR OBJECT
IC 0122	01 25 47.	- 15 05 43.			NONSTELLAR OBJECT
ZWG 385.154	01 25 48.	+ 02 15		14.3	GALAXY
UGC 01053	01 25 48.	+ 02 15	72	14.3	GALAXY Sb-c
ZWG 436.059	01 25 48.	+ 10 33		15.7	GALAXY
LB 01602	01 25 48.	- 58 14		14.0	FAINT BLUE STAR
LIN 542	01 25 48.	- 73 19 38.		13.94	SMC EMISSION LINE OBJECT
IC 1709	01 25 51.	- 36 01 25.			NONSTELLAR OBJECT
ZC 0125.9+2209	01 25 54.	+ 22 09	1410		CLUSTER OF GALAXIES
UGC 01054	01 25 54.	+ 34 05	84	17.	GALAXY Sc
ZWG 385.155	01 25 54.	- 00 15		15.4	GALAXY
MCG+01-04-054	01 25 54.	- 02 49	84	15.	GALAXY
ZH 61	01 25 54.	- 02 51		15.2	GALAXY
KN 14.160	01 25 58.0	+ 31 52 57.			NEBULA
ZWG 436.060	01 26 00.	+ 14 30		15.6	GALAXY
MCG+03-04-044	01 26 00.	+ 16 24	48	14.5	GALAXY
ZWG 502.089	01 26 00.	+ 31 27		15.5	GALAXY
SN 1971V	01 26 00.	+ 31 56		17.0	SUPERNOVA
IC 1707	01 26 00.	+ 33 20			NONSTELLAR OBJECT
ZC 0126.0+3435	01 26 00.	+ 34 35	1140		CLUSTER OF GALAXIES
5ZW 071	01 26 00.	+ 40 04			COMPACT GALAXY
MCG+14-02-002	01 26 00.	+ 84 05	39	17.	GALAXY
ZWG 385.156	01 26 00.	- 01 59		15.1	GALAXY
UGC 01055	01 26 00.	- 01 59	96	15.1	GALAXY SBa
MCG+00-04-160	01 26 00.	- 02 00	48	14.5	GALAXY
MCG-02-04-057	01 26 00.	- 13 50	42	15.	GALAXY
KN 14.161	01 26 00.8	+ 33 24 23.			NEBULA
KN 14.163	01 26 03.	+ 31 37 24.			NEBULA
KN 14.162	01 26 03.0	+ 31 25 58.			NEBULA
ARC 0204	01 26 04.	- 07 03		17.5	RICH CLUSTER OF GALAXIES
ZWG 411.058	01 26 06.	+ 06 36		15.6	GALAXY
ZWG 436.061	01 26 06.	+ 12 05		15.7	GALAXY
ZC 0126.1+1234	01 26 06.	+ 12 34	1750		CLUSTER OF GALAXIES
ZWG 459.060	01 26 06.	+ 16 25		15.0	GALAXY
UGC 01056	01 26 06.	+ 16 25	60	15.0	GALAXY IRR+CMP
MCG+03-04-045	01 26 06.	+ 18 42	24	15.5	GALAXY
ZWG 502.090	01 26 06.	+ 31 04		15.7	GALAXY
4ZW 047	01 26 06.	+ 39 09			COMPACT GALAXY
ISS 0043	01 26 06.	+ 60 35	333		STELLAR RING
OCL 0322	01 26 06.	+ 63 03	840	7.5	OPEN STAR CLUSTER
ZH 45	01 26 06.	- 01 58		15.0	GALAXY
ARC 0206	01 26 06.	- 25 51		17.5	RICH CLUSTER OF GALAXIES
RNGC 0559	01 26 07.	+ 63 03		7.5	OPEN CLUSTER
KN 14.166	01 26 07.1	+ 29 21 02.			NEBULA
KN 14.164	01 26 08.1	+ 32 29 24.			NEBULA
KN 14.167	01 26 08.7	+ 31 07 29.			NEBULA
KN 14.168	01 26 09.2	+ 31 03 28.			NEBULA
KN 14.165	01 26 09.7	+ 32 29 24.			NEBULA
FATH 1.083	01 26 11.	+ 30 38	11		NEBULA
RNGC 0566	01 26 11.	+ 32 05		14.5	GALAXY
KN 14.169	01 26 11.4	+ 30 38 55.			NEBULA
ZWG 385.157	01 26 12.	+ 02 10		15.0	GALAXY
ZWG 436.062	01 26 12.	+ 13 31		14.9	GALAXY
UGC 01057	01 26 12.	+ 13 31	96	14.9	GALAXY Sb-c
ZWG 459.061	01 26 12.	+ 18 43		14.9	GALAXY
MCG+03-04-046	01 26 12.	+ 19 18	27	14.5	GALAXY
ZWG 502.091	01 26 12.	+ 30 39		15.7	GALAXY
ZWG 502.092	01 26 12.	+ 32 05		14.6	GALAXY
UGC 01058	01 26 12.	+ 32 05	96	14.6	GALAXY S0
ZWG 521.035	01 26 12.	+ 39 09		15.4	GALAXY
5ZW 072	01 26 12.	+ 48 34			COMPACT GALAXY
VB 119	01 26 12.	+ 60 33	430		STELLAR RING
MCG-05-04-039	01 26 12.	- 28 57	36	15.	GALAXY
MCG-06-04-038	01 26 12.	- 36 16		14.	GALAXY
AGU 08	01 26 12.	- 43 49 00.		12.5	3 INTERACTING GALAXIES
LB 01603	01 26 12.	- 53 15		14.2	FAINT BLUE STAR
KN 14.170	01 26 13.3	+ 32 33 09.			NEBULA
KN 14.171	01 26 13.5	+ 32 04 29.			NEBULA
MCG+00-04-161	01 26 15.	+ 02 11 30.	24	15.	GALAXY
MCG+02-04-056	01 26 15.	+ 13 32 30.	84	15.	GALAXY
MCG+06-04-030	01 26 15.	+ 39 08	30	15.5	GALAXY
IC 0123	01 26 16.	+ 02 12 23.			NONSTELLAR OBJECT
ZWG 459.062	01 26 18.	+ 19 18		14.9	GALAXY
MCG+05-04-062	01 26 18.	+ 32 04 30.	78	15.	GALAXY
ZWG 502.093	01 26 18.	+ 33 25		15.3	GALAXY
ZWG 521.036	01 26 18.	+ 39 10		14.7	GALAXY
UGC 01059	01 26 18.	+ 39 10	96	14.7	GALAXY SB?
ZWG 521.037	01 26 18.	+ 39 18		15.1	GALAXY
ZWG 537.005	01 26 18.	+ 43 51		15.7	GALAXY
KN 14.172	01 26 22.7	+ 33 24 25.			NEBULA
RNGC 0572	01 26 23.	- 39 34			GALAXY
KN 14.173	01 26 23.4	+ 31 39 12.			NEBULA
ZWG 385.161	01 26 24.	+ 01 34		15.7	GALAXY
SHAH 041	01 26 24.	+ 07 25	54	17.2	GROUP OF COMPACT GALAXIES
ZWG 459.063	01 26 24.	+ 19 26		14.8	GALAXY
ZWG 502.094	01 26 24.	+ 31 30		15.7	GALAXY
ZWG 385.160	01 26 24.	- 00 49		14.0	GALAXY
ZH 54	01 26 24.	- 00 49		14.5	GALAXY
UGC 01062	01 26 24.	- 00 49	108	14.0	GALAXY S
ZH 25	01 26 24.	- 01 11		14.3	GALAXY
ZWG 385.159	01 26 24.	- 01 12		14.2	GALAXY
UGC 01061	01 26 24.	- 01 12	150	14.2	GALAXY SB0/SBa
MCG+00-04-162	01 26 24.	- 01 12 30.	54	12.	GALAXY
ZWG 385.158	01 26 24.	- 02 04		15.7	GALAXY
UGC 01060	01 26 24.	- 02 04	60	15.7	GALAXY S
MCG+00-04-161A	01 26 24.	- 02 26	48	16.	GALAXY
MCG-02-04-058	01 26 24.	- 09 51	42	15.	GALAXY
MCG-03-04-074	01 26 24.	- 16 01	36	15.5	GALAXY
MCG-04-04-017	01 26 24.	- 23 40	30	15.	GALAXY
KN 14.174	01 26 24.8	+ 31 30 17.			NEBULA
RNGC 0570	01 26 25.	- 01 12		14.0	GALAXY
ARC 0205	01 26 26.	+ 05 58		17.6	RICH CLUSTER OF GALAXIES
MCG+06-04-031	01 26 27.	+ 39 10	90	15.	GALAXY
HN 0032	01 26 27.	- 01 12			NEBULA
RNGC 0569	01 26 29.	+ 10 53		14.5	GALAXY
ZWG 436.063	01 26 30.	+ 10 53		14.7	GALAXY
UGC 01063	01 26 30.	+ 10 53	66	14.7	GALAXY
KARA.72 034A	01 26 30.	+ 10 53	60	14.7	PART OF DOUBLE GALAXY
MCG+03-04-047	01 26 30.	+ 19 25	42	16.	GALAXY
UGC 01064	01 26 30.	+ 50 08	90	18.	GALAXY DWRF SP
LDN 1325	01 26 30.	+ 62 50	180		DARK NEBULA
MCG+00-04-163	01 26 30.	- 00 50	54	14.	GALAXY
MCG+01-04-055	01 26 30.	- 02 40	66	15.5	GALAXY
MCG-03-04-075	01 26 30.	- 16 46	54	15.	GALAXY
KN 14.175	01 26 33.5	+ 32 31 37.			NEBULA
ZWG 436.064	01 26 36.	+ 10 53		15.7	GALAXY
UGC 01065	01 26 36.	+ 10 53	78	15.7	GALAXY Sc
KARA.72 034B	01 26 36.	+ 10 53	66	15.7	PART OF DOUBLE GALAXY
UGC 01066	01 26 36.	+ 31 49	84	16.5	GALAXY Sc
MCG-02-04-059	01 26 36.	- 09 29	48	15.	GALAXY
MCG-05-04-040	01 26 36.	- 32 17	30	15.	GALAXY
IC 0124	01 26 37.	- 02 11 33.			SINGLE STAR
ARC 0207	01 26 38.	- 02 45		17.2	RICH CLUSTER OF GALAXIES
FATH 1.084	01 26 39.	+ 30 22	16		NEBULA
KN 14.176	01 26 39.2	+ 30 53 21.			NEBULA
FATH 1.085	01 26 42.	+ 30 22	14		NEBULA
5ZW 073	01 26 42.	+ 43 50			COMPACT GALAXY

OBJECT NAME	RIGHT ASCEN.	DECLINATION	DIAM.	MAGN.	TYPE OF OBJECT
UGC 01067	01 26 42.	+ 51 41	72	16.5	GALAXY SB:a-b
VB 120	01 26 42.	+ 60 04	356		STELLAR RING
KN 14.178	01 26 43.6	+ 28 27 43.			NEBULA
MCG+08-03-027	01 26 45.	+ 45 20	96	14.	GALAXY
KN 14.177	01 26 45.3	+ 31 51 58.			NEBULA
RNGC 0574	01 26 46.	- 35 51			GALAXY
ZC 0126.8+0546	01 26 48.	+ 05 46	2020		CLUSTER OF GALAXIES
ZWG 502.095	01 26 48.	+ 32 34		15.7	GALAXY
ZWG 537.006	01 26 48.	+ 45 21		13.7	GALAXY
UGC 01068	01 26 48.	+ 45 21	114	13.7	GALAXY Sc
KARA.72 035A	01 26 48.	+ 45 21	102	13.7	PART OF DOUBLE GALAXY
MCG-02-04-060	01 26 48.	- 13 33	24	15.	GALAXY
KN 14.180	01 26 49.8	+ 32 34 12.			NEBULA
KN 14.179	01 26 50.0	+ 33 34 29.			NEBULA
IC 0125	01 26 51.	- 13 32 04.			NONSTELLAR OBJECT
MCG-03-04-076	01 26 51.	- 15 46	48	15.	GALAXY
SN 1967K	01 26 53.	+ 33 35		15.0	SUPERNOVA
SVEN 081	01 26 53.	- 22 59	12	16.2	GALAXY
RNGC 0576	01 26 53.	- 51 51			UNVERIFIED SOUTHRN OBJECT
KN 14.181	01 26 53.6	+ 32 32 26.			NEBULA
ZC 0126.9+2306	01 26 54.	+ 23 06	940		CLUSTER OF GALAXIES
ZWG 502.096	01 26 54.	+ 31 03		15.6	GALAXY
ZWG 502.097	01 26 54.	+ 31 46		15.6	GALAXY
ZWG 521.038	01 26 54.	+ 33 35		15.6	GALAXY
MCG-06-04-039	01 26 54.	- 35 53	90	14.	GALAXY
KN 14.182	01 26 55.4	+ 31 03 03.			NEBULA
KN 14.183	01 26 57.0	+ 31 45 34.			NEBULA
VDB.666 011	01 27	+ 25 35	100		DWARF GALAXY
LBN 0633	01 27	+ 58 07	360		BRIGHT NEBULA
VB 121	01 27 00.	+ 58 30	739		STELLAR RING
TON-S 0218	01 27 00.	- 28 17		16.0	BLUE STAR
FATH 1.086	01 27 03.	+ 30 37	5		NEBULA
MCG+06-04-032	01 27 03.	+ 33 33	36	16.	GALAXY
MCG+07-04-002	01 27 03.	+ 40 43	102	14.5	GALAXY
RNGC 0571	01 27 05.	+ 32 15		15.0	GALAXY
KN 14.184	01 27 05.1	+ 31 01 11.			NEBULA
SHAH 042	01 27 06.	+ 07 35	114	18.2	GROUP OF COMPACT GALAXIES
ZWG 459.064	01 27 06.	+ 16 40		15.0	GALAXY
ZWG 502.098	01 27 06.	+ 32 15		15.0	GALAXY
UGC 01069	01 27 06.	+ 32 15	108	15.0	GALAXY S
ZWG 537.007	01 27 06.	+ 40 43		14.3	GALAXY
UGC 01070	01 27 06.	+ 40 43	132	14.3	GALAXY Sc
MCG+08-03-028	01 27 06.	+ 45 22	42	16.	GALAXY
PHL 7203	01 27 06.	- 16 10		16.6	BLUE STELLAR OBJECT
KN 14.185	01 27 06.7	+ 32 14 49.			NEBULA
KN 14.186	01 27 09.7	+ 31 11 06.			NEBULA
SVEN 082	01 27 11.	- 22 34	18	15.5	GALAXY
SVEN 083	01 27 11.	- 22 35 57.	48	14.8	GALAXY
MCG+05-04-063	01 27 12.	+ 32 14 30.	66	14.5	GALAXY
ZWG 537.008	01 27 12.	+ 45 22		15.1	GALAXY
KARA.72 035B	01 27 12.	+ 45 22	48	15.1	PART OF DOUBLE GALAXY
ACK 127-01.1	01 27 12.	+ 60 16			PLANETARY NEBULA
ZWG 385.163	01 27 12.	- 01 30		14.7	GALAXY
ZH 64	01 27 12.	- 01 30		14.9	GALAXY
UGC 01072	01 27 12.	- 01 30	102	14.7	GALAXY S0
ZWG 385.162	01 27 12.	- 02 14		15.7	GALAXY
ZH 62	01 27 12.	- 02 14		15.4	GALAXY
UGC 01071	01 27 12.	- 02 14	90	15.7	GALAXY
KN 14.188	01 27 13.2	+ 28 31 06.			NEBULA
LB 00450	01 27 14.	+ 33 13 18.		16.0	FAINT BLUE STAR
KN 14.187	01 27 14.6	+ 33 03 32.			NEBULA
IC 0126	01 27 14.9	- 02 14 31.		15.7	GALAXY
MCG+00-04-164	01 27 15.	- 01 30	24	15.	GALAXY
MCG-04-04-018	01 27 15.	- 22 33	30	16.	GALAXY
BC 3CR43	01 27 15.18	+ 23 22 52.0		20.	QUASI-STELLAR OBJECT
SN 1968R	01 27 17.	- 02 57		15.5	SUPERNOVA
RNGC 0583	01 27 17.	- 18 35		15.0	GALAXY
IC 0127	01 27 17.0	- 07 14 18.			GALAXY
KN 14.189	01 27 17.3	+ 29 31 26.			NEBULA
ZC 0127.3+0258	01 27 18.	+ 02 58	1410		CLUSTER OF GALAXIES
MCG+03-04-048	01 27 18.	+ 18 53	36	16.	GALAXY
ZWG 481.008	01 27 18.	+ 25 36		15.3	GALAXY
UGC 01073	01 27 18.	+ 25 36	114	15.3	GALAXY DWRF SP
MCG+04-04-011	01 27 18.	+ 25 37 30.	96	14.5	GALAXY
ZWG 502.099	01 27 18.	+ 29 32		15.6	GALAXY
KARA.73B 0052	01 27 18.	+ 29 32	36	15.6	ISOLATED GALAXY S0
UGC 01074	01 27 18.	+ 33 04	72	17.	GALAXY DWRF IR
5ZW 074	01 27 18.	+ 44 31			COMPACT GALAXY
LDN 1326	01 27 18.	+ 62 50	180		DARK NEBULA
MCG+01-04-056	01 27 18.	- 02 55	60	15.	GALAXY
MCG+01-04-057	01 27 18.	- 07 13	90	14.	GALAXY
MCG-04-04-019	01 27 18.	- 22 34	48	15.5	GALAXY
MCG-03-04-077	01 27 21.	- 18 35 30.	24	15.	GALAXY
FATH 1.087	01 27 22.	+ 30 08	19		NEBULA
ZWG 459.065	01 27 24.	+ 18 55		15.4	GALAXY
ZC 0127.4+2011	01 27 24.	+ 20 11	1140		CLUSTER OF GALAXIES
ZWG 502.100	01 27 24.	+ 33 01		15.7	GALAXY
MRSL 128-04/1	01 27 24.	+ 58 07	540		HII REGION
LB C1604	01 27 24.	- 57 07		14.2	FAINT BLUE STAR
LB 03210	01 27 24.	- 73 04		13.3	FAINT BLUE STAR
KN 14.190	01 27 25.0	+ 33 00 35.			NEBULA
KN 14.191	01 27 26.2	+ 27 25 31.			NEBULA
UGC 01075	01 27 30.	+ 02 36	66	17.	GALAXY
MCG+03-04-049	01 27 30.	+ 16 43	60	16.	GALAXY
ZWG 459.066	01 27 30.	+ 20 20		14.9	GALAXY
ZC 0127.5+4457	01 27 30.	+ 44 57	1880		CLUSTER OF GALAXIES
MCG+01-04-058	01 27 30.	- 04 56	45	15.	GALAXY
PHL 3364	01 27 30.	- 16 03		18.1	BLUE STELLAR OBJECT
TON-S 0219	01 27 30.	- 31 09		15.5	BLUE STAR
KN 14.192	01 27 30.9	+ 28 21 11.			NEBULA
SG 2.003	01 27 31.	+ 58 03	720		DIFFUSE EMISSION NEBULA
ZWG 459.067	01 27 36.	+ 16 46		15.5	GALAXY
UGC 01076	01 27 36.	+ 16 46	66	15.5	GALAXY S
ZC 0127.6+1828	01 27 36.	+ 18 28	12630		CLUSTER OF GALAXIES
ZWG 459.068	01 27 36.	+ 19 02		15.4	GALAXY
PHL 7206	01 27 36.	- 08 55		16.6	BLUE STELLAR OBJECT
PHL 7205	01 27 36.	- 16 28		18.5	BLUE STELLAR OBJECT
PHL 1004	01 27 36.	- 17 16		17.9	BLUE STELLAR OBJECT
PHL 7204	01 27 36.	- 18 24		16.9	BLUE STELLAR OBJECT
LB 03211	01 27 36.	- 62 08		12.6	FAINT BLUE STAR
KN 14.193	01 27 37.0	+ 31 53 24.			NEBULA
KN 14.194	01 27 37.5	+ 29 44 59.			NEBULA
SVEN 084	01 27 41.	- 23 02	36	15.0	GALAXY
ZWG 385.164	01 27 42.	+ 02 09		15.4	GALAXY
ZWG 459.069	01 27 42.	+ 20 50		15.7	GALAXY
4ZW 048	01 27 42.	+ 33 58			COMPACT GALAXY
PHL 3365	01 27 42.	- 11 30		18.3	BLUE STELLAR OBJECT
PHL 7207	01 27 42.	- 13 08		16.8	BLUE STELLAR OBJECT
MCG+04-04-012	01 27 45.	+ 22 01	39	17.	GALAXY
KN 14.195	01 27 46.3	+ 31 58 28.			NEBULA
SN 1969N	01 27 47.	- 01 12		16.0	SUPERNOVA
ZC 0127.8+0505	01 27 48.	+ 05 05	1880		CLUSTER OF GALAXIES
ZWG 459.070	01 27 48.	+ 19 21		15.5	GALAXY
UGC 01077	01 27 48.	+ 19 21	78	15.5	GALAXY E-S0
FEIG 013	01 27 48.	- 09 55		13.3	FAINT BLUE STAR
PHL 3366	01 27 48.	- 10 04		18.7	BLUE STELLAR OBJECT
PHL 7208	01 27 48.	- 10 24		18.3	BLUE STELLAR OBJECT
LB 01605	01 27 48.	- 49 16		13.3	FAINT BLUE STAR
KN 14.196	01 27 50.1	+ 31 07 53.			NEBULA
MCG-02-04-062	01 27 51.	- 12 14	42	15.	GALAXY
MCG-02-04-061	01 27 51.	- 12 30	48	14.5	GALAXY
MCG+03-04-050	01 27 54.	+ 19 20	42	16.	GALAXY
ZWG 481.009	01 27 54.	+ 22 00		15.7	GALAXY
ZWG 537.009	01 27 54.	+ 40 25		15.5	GALAXY
ZWG 537.010	01 27 54.	+ 41 00		13.5	GALAXY
UGC 01078	01 27 54.	+ 41 00	27	13.5	GALAXY PECULR
UGC 01079	01 27 54.	+ 48 48	66	16.0	GALAXY S
VB 122	01 27 54.	+ 58 40	201		STELLAR RING
ISS 0045	01 27 54.	+ 58 40	202		STELLAR RING
VB 123	01 27 54.	+ 60 09	208		STELLAR RING
ISS 0044	01 27 54.	+ 60 09	193		STELLAR RING
PHL 7209	01 27 54.	- 10 00		12.5	BLUE STELLAR OBJECT
PHL 3367	01 27 54.	- 11 30		16.7	BLUE STELLAR OBJECT
PHL 7210	01 27 54.	- 12 46		18.0	BLUE STELLAR OBJECT
MCG-06-04-040	01 27 54.	- 36 13	12	15.5	GALAXY
LB 03212	01 27 54.	- 52 26		13.4	FAINT BLUE STAR
LB 03213	01 27 54.	- 76 01		14.2	FAINT BLUE STAR
KN 14.197	01 27 54.0	+ 30 43 34.			NEBULA
FATH 1.088	01 27 55.	+ 30 07	11		NEBULA
FATH 1.089	01 27 56.	+ 30 43	14		NEBULA
RNGC 0573	01 27 56.	+ 41 00		13.5	GALAXY
KN 14.201	01 27 56.4	+ 28 02 14.			NEBULA
KN 14.198	01 27 58.0	+ 32 44 08.			NEBULA
KN 14.199	01 27 58.2	+ 32 29 21.			NEBULA
SVEN 085	01 27 59.	- 22 56	282	12.0	GALAXY
KN 14.200	01 27 59.6	+ 32 27 21.			NEBULA
ZWG 436.065	01 28 00.	+ 12 58		15.6	GALAXY
ZWG 459.071	01 28 00.	+ 19 51		15.5	GALAXY
MCG+03-04-051	01 28 00.	+ 21 10	90	13.5	GALAXY
ZWG 481.010	01 28 00.	+ 25 39		15.6	GALAXY
MCG+04-04-013	01 28 00.	+ 25 40	30	16.5	GALAXY
ZWG 502.101	01 28 00.	+ 30 40		15.7	GALAXY
ZWG 521.039	01 28 00.	+ 33 49		15.7	GALAXY
PHL 7214	01 28 00.	- 08 59		18.2	BLUE STELLAR OBJECT
PHL 7215	01 28 00.	- 10 42		18.4	BLUE STELLAR OBJECT
PHL 7212	01 28 00.	- 11 02		17.2	BLUE STELLAR OBJECT
PHL 1005	01 28 00.	- 11 16		18.7	BLUE STELLAR OBJECT
PHL 7216	01 28 00.	- 12 16		18.1	BLUE STELLAR OBJECT
PHL 7213	01 28 00.	- 13 39		17.4	BLUE STELLAR OBJECT
PHL 3368	01 28 00.	- 14 00		12.5	BLUE STELLAR OBJECT
PHL 7211	01 28 00.	- 15 19		18.0	BLUE STELLAR OBJECT
PHL 3369	01 28 00.	- 17 26		18.2	BLUE STELLAR OBJECT
RNGC 0578	01 28 00.	- 22 56		11.5	GALAXY
LB 03214	01 28 00.	- 69 40		13.8	FAINT BLUE STAR
LB 03215	01 28 00.	- 75 24		12.9	FAINT BLUE STAR
RNGC 0575	01 28 02.	+ 21 10		14.0	GALAXY
IC 1710	01 28 03.	+ 21 10 54.			SAME AS NGC 575
KN 14.202	01 28 03.0	+ 30 39 41.			NEBULA
FATH 1.092	01 28 04.	+ 30 13	27		NEBULA
FATH 1.090	01 28 04.	+ 30 34	11		NEBULA
FATH 1.091	01 28 04.	+ 30 39	14		NEBULA
SCHO 0038	01 28 04.	+ 60 28 24.	550		ISOLATED DARK CLOUD
ZC 0128.1+1116	01 28 06.	+ 11 16	1340		CLUSTER OF GALAXIES
ZWG 459.072	01 28 06.	+ 21 11		13.8	GALAXY
UGC 01081	01 28 06.	+ 21 11	114	13.8	GALAXY SBc
KARA.73B 0053	01 28 06.	+ 21 11	102	13.8	ISOLATED GALAXY S
ZC 0128.1+3038	01 28 06.	+ 30 38	940		CLUSTER OF GALAXIES
ZC 0128.1+3519	01 28 06.	+ 35 19	1210		CLUSTER OF GALAXIES
ZWG 385.165	01 28 06.	- 02 15		14.2	GALAXY
ZH 63	01 28 06.	- 02 15		14.3	GALAXY
UGC 01080	01 28 06.	- 02 15	132	14.2	GALAXY SBa
MCG+01-04-059	01 28 06.	- 04 10	60	15.	GALAXY
PHL 3370	01 28 06.	- 11 57		18.0	BLUE STELLAR OBJECT
MCG-04-04-020	01 28 06.	- 22 56 30.	288	11.5	GALAXY
LIN 543	01 28 06.	- 72 58 18.		12.32	SMC EMISSION LINE OBJECT
RNGC 0577	01 28 08.	- 02 15		14.0	GALAXY
KN 14.204	01 28 09.3	+ 30 51 25.			NEBULA
KN 14.203	01 28 09.3	+ 33 13 44.			NEBULA
FATH 1.093	01 28 10.	+ 30 10	14		NEBULA
ZWG 385.166	01 28 12.	+ 02 20		15.5	GALAXY
PHL 3372	01 28 12.	+ 07 27		18.0	BLUE STELLAR OBJECT
MCG+03-04-052	01 28 12.	+ 16 53	150	14.	GALAXY
ZWG 459.073	01 28 12.	+ 16 56		14.8	GALAXY
UGC 01082	01 28 12.	+ 16 56	162	14.8	GALAXY Sb
4ZW 049	01 28 12.	+ 30 52			COMPACT GALAXY
VB 124	01 28 12.	+ 59 43	235		STELLAR RING
MCG+00-04-165	01 28 12.	- 02 16	54	13.	GALAXY
PHL 7217	01 28 12.	- 05 24		17.9	BLUE STELLAR OBJECT
PHL 3371	01 28 12.	- 11 14		16.6	BLUE STELLAR OBJECT
MCG-03-04-078	01 28 12.	- 16 42	18	15.	GALAXY
PHL 1006	01 28 12.	- 19 14		17.9	BLUE STELLAR OBJECT
MCG-05-04-041	01 28 12.	- 27 00	66	15.	GALAXY
MCG-06-04-041	01 28 12.	- 33 19	60	14.5	GALAXY
KN 14.205	01 28 12.8	+ 32 04 20.			NEBULA
IC 1711	01 28 14.	+ 16 56 17.			NONSTELLAR OBJECT
KN 14.206	01 28 14.4	+ 32 04 30.			NEBULA
KN 14.207	01 28 14.9	+ 30 51 27.			NEBULA
MCG+00-04-166	01 28 15.	+ 02 20 30.	48	15.5	GALAXY
KN 14.210	01 28 16.6	+ 30 51 25.			NEBULA
KN 14.211	01 28 16.7	+ 31 00 13.			NEBULA
SVEN 086	01 28 17.	- 22 38 59.	36	15.8	GALAXY
KN 14.208	01 28 17.2	+ 32 04 42.			NEBULA
ZWG 436.066	01 28 18.	+ 13 50		14.7	GALAXY
UGC 01083	01 28 18.	+ 13 50	96	14.7	GALAXY Sb
MCG+02-04-057	01 28 18.	+ 13 52	72	15.	GALAXY
PHL 3373	01 28 18.	- 11 24		18.3	BLUE STELLAR OBJECT
PHL 7218	01 28 18.	- 12 44		18.4	BLUE STELLAR OBJECT
LH115-N090	01 28 18.	- 73 49	199		EMISSION NEBULA IN SMC
KN 14.209	01 28 18.1	+ 32 57 37.			NEBULA
SVEN 087	01 28 23.	- 22 59	12	15.9	GALAXY
KN 14.212	01 28 23.2	+ 31 14 10.			NEBULA
PHL 3377	01 28 24.	+ 02 08		18.4	BLUE STELLAR OBJECT
PHL 1008	01 28 24.	+ 03 58		18.4	BLUE STELLAR OBJECT
PHL 7219	01 28 24.	+ 04 10		18.2	BLUE STELLAR OBJECT
PHL 3375	01 28 24.	+ 07 28		18.7	BLUE STELLAR OBJECT
BC PHL3375	01 28 24.	+ 07 28		18.02	QUASI-STELLAR OBJECT

OBJECT NAME	RIGHT ASCEN.	DECLINATION	DIAM.	MAGN.	TYPE OF OBJECT
SHB 030	01 28 24.	+ 07 28		18.0	QUASI-STELLAR OBJECT
PHL 1007	01 28 24.	+ 07 52		17.8	BLUE STELLAR OBJECT
PHL 7220	01 28 24.	+ 08 36		18.8	BLUE STELLAR OBJECT
PHL 3376	01 28 24.	+ 08 36		18.0	BLUE STELLAR OBJECT
ZWG 436.067	01 28 24.	+ 13 23		15.6	GALAXY
ZC 0128.4+3647	01 28 24.	+ 36 47	1080		CLUSTER OF GALAXIES
PHL 7221	01 28 24.	- 10 14		18.1	BLUE STELLAR OBJECT
PHL 3378	01 28 24.	- 11 01		18.3	BLUE STELLAR OBJECT
PHL 3374	01 28 24.	- 13 20		17.3	BLUE STELLAR OBJECT
MCG-04-04-021	01 28 24.	- 23 51	60	15.	GALAXY
KN 14.216	01 28 24.6	+ 28 39 44.			NEBULA
LIN.CL 105	01 28 25.	- 73 48 48.	2040		STAR CLUSTER IN SMC
KN 14.214	01 28 25.5	+ 30 11 59.			NEBULA
RNGC 0602	01 28 26.	- 73 49			CLUSTER/NEBULOSITY IN SMC
FATH 1.094	01 28 27.	+ 30 11	11		NEBULA
KN 14.217	01 28 27.6	+ 30 11 35.			NEBULA
KN 14.213	01 28 27.6	+ 32 53 10.			NEBULA
KN 14.215	01 28 28.2	+ 32 52 43.			NEBULA
KN 14.218	01 28 28.5	+ 30 12 53.			NEBULA
ZWG 459.074	01 28 30.	+ 17 28		15.6	GALAXY
5ZW 075	01 28 30.	+ 55 19			COMPACT GALAXY
VB 125	01 28 30.	+ 60 01	228		STELLAR RING
PHL 7222	01 28 30.	- 11 47		18.4	BLUE STELLAR OBJECT
PHL 3379	01 28 30.	- 12 38		18.2	BLUE STELLAR OBJECT
MCG-06-04-042	01 28 30.	- 38 01	60	14.5	GALAXY
RNGC 0580	01 28 32.	- 02 15			NON-EXISTENT OBJECT
LIN 544	01 28 32.	- 73 38 49.		14.13	SMC EMISSION LINE OBJECT
KN 14.220	01 28 32.7	+ 31 17 54.			NEBULA
KN 14.221	01 28 32.9	+ 30 07 04.			NEBULA
KN 14.219	01 28 33.4	+ 32 53 10.			NEBULA
KN 14.223	01 28 34.2	+ 28 10 51.			NEBULA
KN 14.222	01 28 35.1	+ 31 15 18.			NEBULA
PHL 3381	01 28 36.	+ 01 32		18.8	BLUE STELLAR OBJECT
PHL 3380	01 28 36.	+ 03 22		16.8	BLUE STELLAR OBJECT
ZWG 436.068	01 28 36.	+ 09 41		15.4	GALAXY
UGC 01084	01 28 36.	+ 23 42	84	17.	GALAXY DWRF SP
ZWG 502.102	01 28 36.	+ 31 15		15.6	GALAXY
ZWG 385.167	01 28 36.	- 01 45		14.7	GALAXY
ZH 65	01 28 36.	- 01 45		14.8	GALAXY
PHL 7224	01 28 36.	- 04 25		18.4	BLUE STELLAR OBJECT
PHL 7223	01 28 36.	- 07 44		17.5	BLUE STELLAR OBJECT
PHL 7225	01 28 36.	- 08 20		18.5	BLUE STELLAR OBJECT
PHL 7226	01 28 36.	- 11 48		18.5	BLUE STELLAR OBJECT
KN 14.225	01 28 37.2	+ 30 10 10.			NEBULA
KN 14.224	01 28 37.7	+ 31 12 22.			NEBULA
FATH 1.095	01 28 39.	+ 30 09	11		NEBULA
MCG+00-04-167	01 28 39.	- 01 45 30.	21	14.5	GALAXY
KN 14.226	01 28 40.6	+ 30 10 03.			NEBULA
ZC 0128.7+0022	01 28 42.	+ 00 22	3560		CLUSTER OF GALAXIES
UGC 01085	01 28 42.	+ 07 32	84	16.5	GALAXY IRR
PHL 7229	01 28 42.	+ 08 03		17.6	BLUE STELLAR OBJECT
ZWG 521.040	01 28 42.	+ 34 32		15.0	GALAXY
UGC 01086	01 28 42.	+ 34 32	60	15.0	GALAXY S
PHL 7227	01 28 42.	- 02 54		17.7	BLUE STELLAR OBJECT
PHL 7230	01 28 42.	- 09 50		18.4	BLUE STELLAR OBJECT
PHL 1009	01 28 42.	- 10 14		18.2	BLUE STELLAR OBJECT
PHL 7231	01 28 42.	- 11 24		18.0	BLUE STELLAR OBJECT
PHL 1010	01 28 42.	- 13 26		18.6	BLUE STELLAR OBJECT
PHL 7228	01 28 42.	- 14 08		15.4	BLUE STELLAR OBJECT
LIN 545	01 28 43.	- 73 39 19.		14.56	SMC EMISSION LINE OBJECT
LIN 546	01 28 44.	- 73 34 19.		13.25	SMC EMISSION LINE OBJECT
MCG+02-04-058	01 28 45.	+ 14 02	84	14.	GALAXY
RNGC 0584	01 28 45.	- 07 07		12.0	GALAXY
KN 14.227	01 28 45.4	+ 31 55 51.			NEBULA
LB 00451	01 28 47.	+ 32 22 06.		15.8	FAINT BLUE STAR
PHL 7233	01 28 48.	+ 05 58		17.4	BLUE STELLAR OBJECT
PHL 3383	01 28 48.	+ 07 17		16.5	BLUE STELLAR OBJECT
ZWG 437.001	01 28 48.	+ 14 01		15.1	GALAXY
ZWG 436.069	01 28 48.	+ 14 01		15.1	GALAXY
UGC 01087	01 28 48.	+ 14 01	96	15.1	GALAXY Sc
KN 14.228	01 28 48.	+ 33 05 17.			NEBULA
ZWG 521.041	01 28 48.	+ 36 35		15.7	GALAXY
UGC 01088	01 28 48.	+ 36 35	102	15.7	GALAXY Sc
ZWG 551.021	01 28 48.	+ 48 52		15.4	GALAXY
PHL 7232	01 28 48.	- 01 42		13.2	BLUE STELLAR OBJECT
PHL 3384	01 28 48.	- 04 37		18.8	BLUE STELLAR OBJECT
PHL 1011	01 28 48.	- 12 56		18.7	BLUE STELLAR OBJECT
PHL 3382	01 28 48.	- 14 24		15.7	BLUE STELLAR OBJECT
MCG-03-04-079	01 28 48.	- 17 56	36	15.5	GALAXY
PHL 1012	01 28 48.	- 19 29		18.4	BLUE STELLAR OBJECT
HOLM 045B	01 28 49.	- 07 08	84	11.2	PART OF MULTIPLE GALAXY
KN 14.229	01 28 50.6	+ 31 57 29.			NEBULA
IC 1712	01 28 51.	- 07 07 27.			SAME AS NGC 584
MCG-02-04-063	01 28 51.	- 12 53 30.	48	14.	GALAXY
MCG-04-04-022	01 28 51.	- 23 53	36	15.5	GALAXY
BIGO 468	01 28 53.	+ 33 14			NEBULA
RNGC 0579	01 28 53.	+ 33 22		13.5	GALAXY
SVEN 088	01 28 53.	- 22 42 00.	36	15.4	GALAXY
KN 14.233	01 28 53.0	+ 27 33 53.			NEBULA
PHL 7236	01 28 54.	+ 01 36		17.3	BLUE STELLAR OBJECT
PHL 7235	01 28 54.	+ 04 10		17.9	BLUE STELLAR OBJECT
PHL 7234	01 28 54.	+ 08 27		18.2	BLUE STELLAR OBJECT
3ZW 029	01 28 54.	+ 18 00			COMPACT GALAXY
ZWG 460.001	01 28 54.	+ 18 00		15.7	GALAXY
ZWG 459.075	01 28 54.	+ 18 00		15.7	GALAXY
4ZW 050	01 28 54.	+ 27 02			COMPACT GALAXY
ZWG 502.103	01 28 54.	+ 33 22		13.6	GALAXY
UGC 01089	01 28 54.	+ 33 22	102	13.6	GALAXY Sc
MCG+06-04-034	01 28 54.	+ 34 31	48	14.5	GALAXY
MCG+06-04-033	01 28 54.	+ 36 34	48	16.	GALAXY
ZWG 521.042	01 28 54.	+ 38 27		14.8	GALAXY
UGC 01090	01 28 54.	+ 38 27	72	14.8	GALAXY Sb
VB 126	01 28 54.	+ 61 10	470		STELLAR RING
VB 050	01 28 54.	+ 65 16	544		STELLAR RING
MCG+01-04-060	01 28 54.	- 07 07	132	12.	GALAXY
MCG-02-04-064	01 28 54.	- 11 06	48	14.5	GALAXY
KN 14.230	01 28 54.3	+ 31 30 47.			NEBULA
KN 14.232	01 28 55.	+ 31 45 17.			NEBULA
IC 0128	01 28 56.	- 12 52 58.			NONSTELLAR OBJECT
KN 14.231	01 28 56.0	+ 33 21 41.			NEBULA
ARC 0208	01 28 58.	+ 00 18		16.6	RICH CLUSTER OF GALAXIES
KN 14.234	01 28 59.5	+ 30 15 17.			NEBULA
PHL 3385	01 29 00.	+ 03 24		17.5	BLUE STELLAR OBJECT
PHL 3387	01 29 00.	+ 03 32		18.4	BLUE STELLAR OBJECT
PHL 3386	01 29 00.	+ 03 52		18.4	BLUE STELLAR OBJECT
UGC 01091	01 29 00.	+ 08 11	60	18.	GALAXY DWARF
3ZW 030	01 29 00.	+ 14 41			COMPACT GALAXY
FATH 1.096	01 29 00.	+ 30 14	16		NEBULA

OBJECT NAME	RIGHT ASCEN.	DECLINATION	DIAM.	MAGN.	TYPE OF OBJECT
MCG+05-04-064	01 29 00.	+ 33 22	72	14.	GALAXY
MCG+06-04-035	01 29 00.	+ 38 28	48	14.5	GALAXY
ZWG 551.022	01 29 00.	+ 45 33		15.7	GALAXY
ISS 0010	01 29 00.	+ 65 16	163		STELLAR RING
VB 051	01 29 00.	+ 65 16	161		STELLAR RING
PHL 7238	01 29 00.	- 03 16		18.5	BLUE STELLAR OBJECT
PHL 7241	01 29 00.	- 08 25		18.7	BLUE STELLAR OBJECT
PHL 3390	01 29 00.	- 08 28		18.5	BLUE STELLAR OBJECT
PHL 7240	01 29 00.	- 08 31		16.9	BLUE STELLAR OBJECT
PHL 3389	01 29 00.	- 08 43		17.9	BLUE STELLAR OBJECT
PHL 3388	01 29 00.	- 09 17		18.7	BLUE STELLAR OBJECT
PHL 1013	01 29 00.	- 11 36		18.3	BLUE STELLAR OBJECT
PHL 7237	01 29 00.	- 13 20		16.2	BLUE STELLAR OBJECT
PHL 1014	01 29 00.	- 14 10		18.6	BLUE STELLAR OBJECT
PHL 7239	01 29 00.	- 15 47		18.4	BLUE STELLAR OBJECT
MCG-03-05-001	01 29 00.	- 15 49	24	15.	GALAXY
PHL 7242	01 29 00.	- 16 02		18.2	BLUE STELLAR OBJECT
MCG-05-04-042	01 29 00.	- 27 04	15	15.	GALAXY
IC 0130	01 29 02.	- 15 50 52.			NONSTELLAR OBJECT
KN 14.235	01 29 02.7	+ 33 19 55.			NEBULA
RNGC 0586	01 29 03.	- 07 09		14.0	GALAXY
IC 0129	01 29 03.	- 12 54 46.			NONSTELLAR OBJECT
MCG-05-04-043	01 29 03.	- 26 41	18	15.5	GALAXY
KN 14.236	01 29 04.8	+ 33 18 47.			NEBULA
RNGC 0582	01 29 05.	+ 33 14		13.5	GALAXY
KN 14.237	01 29 05.9	+ 32 56 51.			NEBULA
MCG+03-05-002	01 29 06.	+ 17 17 30.	66	15.	GALAXY
ZWG 460.002	01 29 06.	+ 17 19		14.8	GALAXY
ZWG 459.076	01 29 06.	+ 17 19		14.8	GALAXY
UGC 01093	01 29 06.	+ 17 19	114	14.8	GALAXY
ZWG 460.003	01 29 06.	+ 18 20		15.0	GALAXY
ZWG 459.077	01 29 06.	+ 18 20		15.0	GALAXY
MCG+03-05-001	01 29 06.	+ 18 20	18	15.	GALAXY
ZWG 502.104	01 29 06.	+ 32 58		15.4	GALAXY
ZWG 502.105	01 29 06.	+ 33 14		13.7	GALAXY
UGC 01094	01 29 06.	+ 33 14	132	13.7	GALAXY SBb
ZWG 521.043	01 29 06.	+ 38 56		15.6	GALAXY
ZWG 386.005	01 29 06.	- 01 11		14.2	GALAXY
ZH 66	01 29 06.	- 01 11		14.4	GALAXY
UGC 01092	01 29 06.	- 01 11	162	14.2	GALAXY Sa
PHL 3391	01 29 06.	- 03 17		18.6	BLUE STELLAR OBJECT
HOLM 045A	01 29 06.	- 07 09	60	13.7	PART OF MULTIPLE GALAXY
MCG-02-05-001	01 29 06.	- 12 56	60	14.	GALAXY
PHL 3392	01 29 06.	- 13 10		18.4	BLUE STELLAR OBJECT
PHL 7243	01 29 06.	- 13 30		18.1	BLUE STELLAR OBJECT
RNGC 0585	01 29 07.	- 01 11		14.0	GALAXY
KN 14.239	01 29 07.1	+ 31 38 20.			NEBULA
KN 14.238	01 29 07.5	+ 33 13 03.			NEBULA
KN 14.242	01 29 08.9	+ 30 00 21.			NEBULA
KN 14.240	01 29 09.1	+ 33 18 36.			NEBULA
KN 14.241	01 29 11.1	+ 32 55 51.			NEBULA
KN 14.243	01 29 11.8	+ 32 57 02.			NEBULA
PHL 7249	01 29 12.	+ 02 24		17.9	BLUE STELLAR OBJECT
PHL 3394	01 29 12.	+ 04 10		16.9	BLUE STELLAR OBJECT
PHL 7244	01 29 12.	+ 05 08		18.0	BLUE STELLAR OBJECT
PHL 7248	01 29 12.	+ 08 34		17.7	BLUE STELLAR OBJECT
4ZW 051	01 29 12.	+ 31 51			COMPACT GALAXY
ZWG 502.106	01 29 12.	+ 32 55		15.1	GALAXY
MCG+05-04-065	01 29 12.	+ 33 14	138	14.	GALAXY
MCG+00-05-001	01 29 12.	- 01 12	114	13.5	GALAXY
PHL 7251	01 29 12.	- 02 44		18.5	BLUE STELLAR OBJECT
PHL 3393	01 29 12.	- 05 04		17.6	BLUE STELLAR OBJECT
PHL 7245	01 29 12.	- 05 15		18.4	BLUE STELLAR OBJECT
PHL 1015	01 29 12.	- 06 22		16.7	BLUE STELLAR OBJECT
MCG+01-05-001	01 29 12.	- 07 10 30.	60	14.	GALAXY
PHL 7250	01 29 12.	- 09 30		17.5	BLUE STELLAR OBJECT
PHL 7246	01 29 12.	- 12 45		18.5	BLUE STELLAR OBJECT
PHL 7247	01 29 12.	- 13 17		14.5	BLUE STELLAR OBJECT
KN 14.244	01 29 14.0	+ 32 55 13.			NEBULA
KN 14.245	01 29 17.	+ 31 32 16.			NEBULA
HOLM 045C	01 29 17.	- 07 12	36	14.4	PART OF MULTIPLE GALAXY
MCG+00-05-002	01 29 18.	+ 00 24 30.	27	15.	GALAXY
ZWG 386.006	01 29 18.	+ 00 25		15.7	GALAXY
PHL 3395	01 29 18.	+ 03 18		17.5	BLUE STELLAR OBJECT
PHL 7254	01 29 18.	+ 05 05		17.9	BLUE STELLAR OBJECT
PHL 7253	01 29 18.	+ 05 16		17.2	BLUE STELLAR OBJECT
ZWG 502.107	01 29 18.	+ 31 51		15.0	GALAXY
UGC 01095	01 29 18.	+ 31 51	60	15.0	GALAXY S
MCG+06-04-036	01 29 18.	+ 38 56 30.	42	14.5	GALAXY
5ZW 076	01 29 18.	+ 42 19			COMPACT GALAXY
PHL 1016	01 29 18.	- 06 15		17.7	BLUE STELLAR OBJECT
PHL 7252	01 29 18.	- 10 20		13.0	BLUE STELLAR OBJECT
PHL 1017	01 29 18.	- 20 36		15.7	BLUE STELLAR OBJECT
LB 03219	01 29 18.	- 73 19		12.5	FAINT BLUE STAR
KN 14.248	01 29 18.7	+ 32 09 54.			NEBULA
KN 14.247	01 29 18.7	+ 32 20 28.			NEBULA
KN 14.249	01 29 19.1	+ 31 50 50.			NEBULA
KN 14.246	01 29 19.2	+ 32 56 20.			NEBULA
SVEN 089	01 29 20.	- 29 13	18	15.8	GALAXY
KN 14.250	01 29 20.4	+ 31 50 06.			NEBULA
MCG+05-04-066	01 29 21.	+ 31 50	18	16.	GALAXY
ARP 098	01 29 22.	+ 31 51			PECULIAR GALAXY
KN 14.251	01 29 23.1	+ 32 49 47.			NEBULA
KN 14.252	01 29 23.8	+ 31 50 52.			NEBULA
PHL 3396	01 29 24.	+ 01 56		15.9	BLUE STELLAR OBJECT
ZC 0129.4+0625	01 29 24.	+ 06 25	2420		CLUSTER OF GALAXIES
VV 301A	01 29 24.	+ 31 51	12	16.	INTERACTING GALAXY
ZWG 386.007	01 29 24.	- 02 28		15.3	GALAXY
PHL 7255	01 29 24.	- 12 42		18.7	BLUE STELLAR OBJECT
PHL 3397	01 29 24.	- 14 32		17.2	BLUE STELLAR OBJECT
PHL 3398	01 29 24.	- 16 54		17.7	BLUE STELLAR OBJECT
KN 14.254	01 29 25.8	+ 30 42 41.			NEBULA
KN 14.256	01 29 26.8	+ 29 54 37.			NEBULA
KN 14.253	01 29 26.8	+ 32 56 45.			NEBULA
MCG+05-04-067	01 29 27.	+ 31 50	60	16.	GALAXY
KN 14.258	01 29 27.0	+ 28 03 11.			NEBULA
KN 14.255	01 29 27.0	+ 31 49 14.			NEBULA
ARC 0209	01 29 28.	- 13 51		17.8	RICH CLUSTER OF GALAXIES
UGC 01096	01 29 30.	+ 12 13	66	16.0	GALAXY S
UGC 01097	01 29 30.	+ 17 45	72	16.0	GALAXY S
MCG+03-05-003	01 29 30.	+ 17 45	48	16.	GALAXY
ZWG 460.004	01 29 30.	+ 21 10		14.8	GALAXY
ZWG 459.078	01 29 30.	+ 21 10		14.8	GALAXY
UGC 01098	01 29 30.	+ 21 10	66	14.8	GALAXY SBb-c
ZC 0129.5+2731	01 29 30.	+ 27 31	610		CLUSTER OF GALAXIES
VV 301B	01 29 30.	+ 31 51	60	16.	INTERACTING GALAXY
ZWG 521.044	01 29 30.	+ 38 36		15.6	GALAXY
ZWG 551.023	01 29 30.	+ 49 09		15.0	GALAXY

OBJECT NAME	RIGHT ASCEN.	DECLINATION	DIAM.	MAGN.	TYPE OF OBJECT
MCG+08-03-029	01 29 30.	+ 49 09	48	15.	GALAXY
PHL 7256	01 29 30.	- 01 46		18.6	BLUE STELLAR OBJECT
PHL 7257	01 29 30.	- 01 57		18.1	BLUE STELLAR OBJECT
PHL 3400	01 29 30.	- 02 12		18.7	BLUE STELLAR OBJECT
PHL 3399	01 29 30.	- 09 48		17.7	BLUE STELLAR OBJECT
KN 14.257	01 29 30.4	+ 32 29 59.			NEBULA
KN 14.259	01 29 32.4	+ 31 50 05.			NEBULA
MCG+05-04-068	01 29 33.	+ 32 30	42	15.	GALAXY
KN 14.260	01 29 33.8	+ 31 35 27.			NEBULA
SVEE 090	01 29 35.	- 22 43	18	15.8	GALAXY
KN 14.261	01 29 35.2	+ 32 29 30.			NEBULA
PHL 3401	01 29 36.	+ 03 52		16.9	BLUE STELLAR OBJECT
PHL 1019	01 29 36.	+ 05 09		16.4	BLUE STELLAR OBJECT
PHL 3402	01 29 36.	+ 05 38		17.1	BLUE STELLAR OBJECT
PHL 1018	01 29 36.	+ 07 20		17.8	BLUE STELLAR OBJECT
MCG+03-05-004	01 29 36.	+ 21 11	45	15.	GALAXY
5ZW 077	01 29 36.	+ 41 44			COMPACT GALAXY
PHL 7258	01 29 36.	- 00 02		16.6	BLUE STELLAR OBJECT
PHL 3404	01 29 36.	- 03 39		18.6	BLUE STELLAR OBJECT
PHL 3403	01 29 36.	- 04 54		18.9	BLUE STELLAR OBJECT
PHL 7259	01 29 36.	- 10 50		16.7	BLUE STELLAR OBJECT
LIN.CL 106	01 29 36.	- 76 19 15.	138	16.4	STAR CLUSTER IN SMC
BCL 0275	01 29 37.8	+ 30 20 48.	9		HII REGION IN M33
RNGC 0587	01 29 40.	+ 35 06		13.5	GALAXY
BCL 0274	01 29 40.2	+ 30 20 06.	57		HII REGION IN M33
KN 14.262	01 29 40.8	+ 32 02 40.			NEBULA
KN 14.263	01 29 41.3	+ 29 48 35.			NEBULA
MCG+01-05-001	01 29 42.	+ 03 36	36	14.5	GALAXY
ZWG 460.005	01 29 42.	+ 19 23		15.3	GALAXY
ZWG 459.079	01 29 42.	+ 19 23		15.3	GALAXY
UGC 01099	01 29 42.	+ 19 23	66	15.3	GALAXY COMPACT
ZC 0129.7+3127	01 29 42.	+ 31 27	340		CLUSTER OF GALAXIES
ZWG 502.108	01 29 42.	+ 32 03		15.7	GALAXY
SN 1968Y	01 29 42.	+ 32 03		18.5	SUPERNOVA
ZWG 521.045	01 29 42.	+ 35 06		13.7	GALAXY
UGC 01100	01 29 42.	+ 35 06	138	13.7	GALAXY Sb/SBb
ZWG 537.011	01 29 42.	+ 41 44		15.5	GALAXY
UGC 01101	01 29 42.	+ 41 44	72	15.5	GALAXY S0
PHL 7260	01 29 42.	- 03 23		18.6	BLUF STELLAR OBJECT
MCG-06-04-043	01 29 42.	- 33 23	36	14.5	GALAXY
MOHR 3	01 29 42.	- 72 13 36.		15.0	STAR CLUSTER IN SMC
BCL 0271	01 29 42.0	+ 30 11 36.	33		HII REGION IN M33
BCL 0272	01 29 43.2	+ 30 12 24.	15		HII REGION IN M33
BCL 0273	01 29 43.2	+ 30 15 06.	15		HII REGION IN M33
KN 14.264	01 29 43.3	+ 31 12 21.			NEBULA
KN 14.265	01 29 44.6	+ 32 03 26.			NEBULA
MCG+03-05-005	01 29 45.	+ 19 21	48	16.	GALAXY
KN 14.266	01 29 45.2	+ 31 10 38.			NEBULA
KN 14.268	01 29 45.6	+ 29 21 29.			NEBULA
BCL 0281	01 29 46.2	+ 30 24 42.	63		HII REGION IN M33
KN 14.267	01 29 46.3	+ 31 37 03.			NEBULA
BCL 0267	01 29 47.4	+ 30 09 18.	19		HII REGION IN M33
PHL 7266	01 29 48.	+ 00 46		19.0	BLUE STELLAR OBJECT
PHL 1020	01 29 48.	+ 01 20		17.5	BLUE STELLAR OBJECT
ZWG 412.001	01 29 48.	+ 03 37		15.3	GALAXY
PHL 7265	01 29 48.	+ 03 44		17.4	BLUE STELLAR OBJECT
PHL 3406	01 29 48.	+ 04 35		18.9	BLUF STELLAR OBJECT
PHL 3405	01 29 48.	+ 04 52		17.8	BLUF STELLAR OBJECT
MCG+06-04-037	01 29 48.	+ 35 05	132	13.5	GALAXY
PHL 7262	01 29 48.	- 01 58		18.0	BLUE STELLAR OBJECT
PHL 1021	01 29 48.	- 06 42		18.0	BLUE STELLAR OBJECT
PHL 7264	01 29 48.	- 08 46		16.6	BLUE STELLAR OBJECT
PHL 7261	01 29 48.	- 09 10		17.8	BLUE STELLAR OBJECT
MCG-02-05-002	01 29 48.	- 09 48	48	15.	GALAXY
PHL 7263	01 29 48.	- 11 14		18.3	BLUE STELLAR OBJECT
KN 14.269	01 29 48.9	+ 31 40 35.			NEBULA
BCL 0264	01 29 49.2	+ 30 06 54.	33		HII REGION IN M33
BCL 0266	01 29 49.2	+ 30 08 54.	15		HII REGION IN M33
BCL 0263	01 29 50.4	+ 30 05 24.	33		HII REGION IN M33
BCL 0269	01 29 50.4	+ 30 09 42.	18		HII REGION IN M33
MCG+01-05-002	01 29 51.	+ 04 20	72	14.	GALAXY
IC 1713	01 29 51.	+ 35 04			NONSTELLAR OBJECT
MCG-03-05-002	01 29 51.	- 16 22 30.	30	15.5	GALAXY
BCL 0268	01 29 51.0	+ 30 09 24.	21		HII REGION IN M33
BCL 0282	01 29 51.0	+ 30 25 12.	9		HII REGION IN M33
RNGC 0593	01 29 52.	- 12 37		14.0	GALAXY
ARC 0210	01 29 52.	- 26 15		17.5	RICH CLUSTER OF GALAXIES
BCL 0265	01 29 52.2	+ 30 08 36.	21		HII REGION IN M33
BCL 0262	01 29 53.4	+ 30 04 42.	32		HII REGION IN M33
BCL 0238	01 29 53.4	+ 30 19 36.	9		HII REGION IN M33
ZWG 386.008	01 29 54.	+ 01 44		15.7	GALAXY
VV 173B	01 29 54.	+ 04 20	24	15.	INTERACTING GALAXY
VV 173A	01 29 54.	+ 04 20	30	15.	INTERACTING GALAXY
VV 173	01 29 54.	+ 04 20	72	14.	INTEPACTING GALAXY
PHL 7268	01 29 54.	+ 05 36		18.0	BLUE STELLAR OBJECT
PHL 7267	01 29 54.	+ 05 54		17.6	BLUE STELLAR OBJECT
RNGC 0588	01 29 54.	+ 30 24			NEBULA IN EXTERNAL GALAXY
OCL 0326	01 29 54.	+ 60 27	4800	7.5	OPEN STAR CLUSTER
VB 127	01 29 54.	+ 61 08	322		STELLAR RING
MCG-02-05-003	01 29 54.	- 12 37 30.	72	14.5	GALAXY
PHL 3407	01 29 54.	- 13 51		18.0	BLUE STELLAR OBJECT
MCG-03-04-003	01 29 54.	- 15 21	30	15.	GALAXY
MCG-04-04-023	01 29 54.	- 21 04 30.	12	16.	GALAXY
LIN 547	01 29 54.	- 73 40 21.		12.12	SMC EMISSION LINE OBJECT
BCL 0280	01 29 54.6	+ 30 23 42.	39		HII REGION IN M33
KN 14.270	01 29 54.6	+ 31 29 31.			NEBULA
BCL 0270	01 29 55.2	+ 30 09 30.	60		HII REGION IN M33
RNGC 0581	01 29 56.	+ 60 27		7.0	OPEN CLUSTER
KN 14.271	01 29 56.8	+ 30 23 43.			NEBULA
MCG+01-05-002	01 29 57.	- 08 09	60	15.	GALAXY
ARP 306	01 29 58.	+ 04 21			PECULIAR GALAXY
RNGC 0597	01 29 58.	- 33 45			GALAXY
KARA.72 036A	01 30 00.	+ 04 19	54	14.6	PART OF DOUBLE GALAXY
ZWG 412.002	01 30 00.	+ 04 20		14.6	GALAXY
UGC 01102	01 30 00.	+ 04 20	96	14.6	GALAXY DBL SYS
KARA.72 036B	01 30 00.	+ 04 20	72		PART OF DOUBLE GALAXY
PHL 7270	01 30 00.	+ 04 22		17.0	BLUE STELLAR OBJECT
MCG+01-05-003	01 30 00.	+ 04 23	36	15.5	GALAXY
PHL 7269	01 30 00.	+ 06 28		17.7	BLUF STELLAR OBJECT
UGC 01103	01 30 00.	+ 11 35	66	17.	GALAXY Sc
MCG+03-05-006	01 30 00.	+ 18 02	48	14.5	GALAXY
ZWG 460.006	01 30 00.	+ 18 04		14.3	GALAXY
UGC 01104	01 30 00.	+ 18 04	72	14.3	GALAXY IRR
5ZW 078	01 30 00.	+ 39 57			COMPACT GALAXY
VB 128	01 30 00.	+ 60 02	416		STELLAR RING
PHL 1022	01 30 00.	- 00 42		17.2	BLUE STELLAR OBJECT
PHL 7272	01 30 00.	- 01 28		18.3	BLUE STELLAR OBJECT
PHL 3409	01 30 00.	- 02 16		18.7	BLUE STELLAR OBJECT
PHL 7273	01 30 00.	- 03 14		18.5	BLUE STELLAR OBJECT
PHL 7271	01 30 00.	- 06 46		17.9	BLUE STELLAR OBJECT
PHL 7274	01 30 00.	- 06 52		18.2	BLUE STELLAR OBJECT
MCG+01-05-003	01 30 00.	- 08 10	12	16.	GALAXY
PHL 3408	01 30 00.	- 09 12		18.6	BLUE STELLAR OBJECT
MCG-06-04-044	01 30 00.	- 33 45	60	14.5	GALAXY
LIN.CL 107	01 30 00.	- 73 40 09.	2718		STAR CLUSTER IN SMC
DV.56 N0643A	01 30 00.	- 76 19	96		GALAXY
RNGC 0643A	01 30 00.	- 76 19			GALAXY
BCL 0237	01 30 01.2	+ 30 19 36.	34		HII REGION IN M33
BCL 0236	01 30 01.2	+ 30 20 12.	21		HII REGION IN M33
BCL 0276	01 30 01.8	+ 30 22 42.	45		HII REGION IN M33
MCG+01-05-004	01 30 03.	- 08 10 30.	9	16.5	GALAXY
BCL 0261	01 30 03.0	+ 30 07 48.	20		HII REGION IN M33
BCL 0235	01 30 04.2	+ 30 20 36.	45		HII REGION IN M33
BCL 0279	01 30 04.2	+ 30 24 18.	14		HII REGION IN M33
BCL 0616	01 30 04.2	+ 30 35 12.	15		HII REGION IN M33
BCL 0240	01 30 04.8	+ 30 17 06.	32		HII REGION IN M33
BCL 0234	01 30 04.8	+ 30 20 12.	22		HII REGION IN M33
BCL 0283	01 30 04.8	+ 30 25 24.	45		HII REGION IN M33
KN 14.272	01 30 05.4	+ 27 24 13.			NEBULA
BCL 0243	01 30 05.4	+ 30 12 06.	27		HII REGION IN M33
BCL 0241	01 30 05.4	+ 30 16 30.	39		HII REGION IN M33
PHL 7277	01 30 06.	+ 00 15		18.8	BLUE STELLAR OBJECT
PHL 1023	01 30 06.	+ 03 36		18.5	BLUE STELLAR OBJECT
PHL 7275	01 30 06.	+ 03 40		18.6	BLUE STELLAR OBJECT
VV 174B	01 30 06.	+ 04 23	15	17.	INTERACTING GALAXY
VV 174A	01 30 06.	+ 04 23	15	17.	INTERACTING GALAXY
VV 174	01 30 06.	+ 04 23	60		INTERACTING GALAXY
PHL 7276	01 30 06.	+ 05 21		18.7	BLUE STELLAR OBJECT
VB 129	01 30 06.	+ 60 25	289		STELLAR RING
KARA.68 006	01 30 06.	- 07 14	34		DWARF GALAXY
PHL 7278	01 30 06.	- 07 25		18.4	BLUE STELLAR OBJECT
LB 03217	01 30 06.	- 78 05		13.5	FAINT BLUE STAR
BCL 0239	01 30 06.0	+ 30 17 36.	14		HII REGION IN M33
BCL 0232	01 30 06.6	+ 30 19 30.	21		HII REGION IN M33
BCL 0233	01 30 06.6	+ 30 20 42.	15		HII REGION IN M33
BCL 0289	01 30 07.8	+ 30 29 18.	26		HII REGION IN M33
BCL 0617	01 30 07.8	+ 30 37 06.	33		HII REGION IN M33
BCL 0231	01 30 08.4	+ 30 19 24.	4		HII REGION IN M33
LIN 548	01 30 09.	- 73 04 27.		12.23	SMC EMISSION LINE OBJECT
BCL 0218	01 30 09.6	+ 30 15 36.	27		HII REGION IN M33
KN 14.275	01 30 09.8	+ 28 00 31.			NEBULA
RNGC 0589	01 30 10.	- 12 18		14.0	GALAXY
BCL 0260	01 30 10.2	+ 30 09 06.	9		HII REGION IN M33
BCL 0219	01 30 10.2	+ 30 15 12.	21		HII REGION IN M33
BCL 0230	01 30 10.2	+ 30 18 36.	15		HII REGION IN M33
KN 14.276	01 30 10.3	+ 27 09 26.			NEBULA
KN 14.273	01 30 10.7	+ 31 55 02.			NEBULA
BCL 0229	01 30 10.8	+ 30 19 30.	8		HII REGION IN M33
PHL 7281	01 30 12.	+ 00 54		17.5	BLUE STELLAR OBJECT
PHL 1024	01 30 12.	+ 03 34		18.7	BLUE STELLAR OBJECT
UGC 01105	01 30 12.	+ 04 23	78	17.	GALAXY DWRF IR
ZC 0130.2+2404	01 30 12.	+ 24 04	1550		CLUSTER OF GALAXIES
PHL 7279	01 30 12.	- 10 12		18.7	BLUE STELLAR OBJECT
PHL 3410	01 30 12.	- 10 34		18.5	BLUE STELLAR OBJECT
PHL 7282	01 30 12.	- 10 40		17.8	BLUE STELLAR OBJECT
PHL 7283	01 30 12.	- 11 49		18.4	BLUE STELLAR OBJECT
PHL 7280	01 30 12.	- 13 44		18.1	BLUE STELLAR OBJECT
PHL 7284	01 30 12.	- 17 58		18.6	BLUE STELLAR OBJECT
LB 01606	01 30 12.	- 45 38		13.2	FAINT BLUE STAR
KN 14.274	01 30 12.4	+ 31 40 35.			NEBULA
BCL 0285	01 30 12.6	+ 30 26 00.	31		HII REGION IN M33
BCL 0284	01 30 13.2	+ 30 24 42.	12		HII REGION IN M33
KN 14.277	01 30 13.6	+ 30 44 46.			NEBULA
KN 14.281	01 30 13.8	+ 28 39 53.			NEBULA
KN 14.278	01 30 14.4	+ 29 56 17.			NEBULA
KN 14.282	01 30 14.8	+ 28 41 34.			NEBULA
MCG+01-05-004	01 30 15.	+ 05 28	36	15.	GALAXY
SCHO 0039	01 30 15.	+ 60 18 12.	450		ISOLATED DARK CLOUD
RNGC 0596	01 30 15.	- 07 17		12.5	GALAXY
MCG-02-05-004	01 30 15.	- 12 18	42	14.5	GALAXY
BCL 0258	01 30 15.0	+ 30 08 00.	21		HII REGION IN M33
BCL 0259	01 30 16.2	+ 30 07 36.	21		HII REGION IN M33
KN 14.280	01 30 16.5	+ 31 58 31.			NEBULA
KN 14.279	01 30 16.5	+ 32 03 42.			NEBULA
BCL 0288	01 30 16.8	+ 30 27 24.	21		HII REGION IN M33
BC PKS0130-17	01 30 17.	- 17 10 20.		18.44	QUASI-STELLAR OBJECT
ZWG 412.003	01 30 18.	+ 05 29		15.6	GALAXY
PHL 7287	01 30 18.	+ 09 28		17.5	BLUE STELLAR OBJECT
ZWG 502.109	01 30 18.	+ 32 00		15.7	GALAXY
5ZW 079	01 30 18.	+ 45 42			COMPACT GALAXY
VB 130	01 30 18.	+ 59 35	584		STELLAR RING
ISS 0046	01 30 18.	+ 59 35	590		STELLAR RING
PHL 3411	01 30 18.	- 02 36		18.5	BLUE STELLAR OBJECT
ARC 0211	01 30 18.	- 04 17		17.2	RICH CLUSTER OF GALAXIES
PHL 7286	01 30 18.	- 14 38		16.3	BLUE STELLAR OBJECT
PHL 7285	01 30 18.	- 17 20		12.9	BLUE STELLAR OBJECT
PHL 1025	01 30 18.	- 19 38		15.7	BLUE STELLAR OBJECT
TON-S 0220	01 30 18.	- 28 01		15.8	BLUE STAR
BCL 0257	01 30 18.0	+ 30 08 00.	15		HII REGION IN M33
BCL 0221	01 30 18.6	+ 30 12 00.	27		HII REGION IN M33
BCL 0217	01 30 18.6	+ 30 14 30.	15		HII REGION IN M33
BCL 0242	01 30 19.2	+ 30 17 42.	21		HII REGION IN M33
BCL 0255	01 30 19.8	+ 30 07 24.	21		HII REGION IN M33
BCL 0220	01 30 19.8	+ 30 11 42.	39		HII REGION IN M33
BCL 0605	01 30 19.8	+ 30 33 18.	56		HII REGION IN M33
KN 14.283	01 30 19.9	+ 32 27 29.			NEBULA
BCL 0256	01 30 20.4	+ 30 07 54.	27		HII REGION IN M33
BCL 0216	01 30 21.0	+ 30 14 30.	19		HII REGION IN M33
BCL 0227	01 30 21.0	+ 30 14 54.	36		HII REGION IN M33
BCL 0226	01 30 21.0	+ 30 15 06.	9		HII REGION IN M33
BCL 0228	01 30 21.0	+ 30 19 00.	15		HII REGION IN M33
BCL 0277	01 30 21.0	+ 30 23 30.	105		HII REGION IN M33
IC 0135	01 30 21.8	+ 30 12 09.			HII REGION IN M33
IC 0138	01 30 22.	- 00 56 01.			NONSTELLAR OBJECT
RNGC 0599	01 30 22.	- 12 27		13.0	GALAXY
IC 0131	01 30 22.3	+ 30 29 46.			HII REGION IN M33
IC 0136	01 30 23.0	+ 30 14 30.			HII REGION IN M33
KN 14.284	01 30 23.0	+ 33 20 47.			NEBULA
KN 14.285	01 30 23.3	+ 32 00 30.			NEBULA
BCL 0290B	01 30 23.4	+ 30 29 06.	9		HII REGION IN M33
BCL 0618	01 30 23.4	+ 30 35 18.	21		HII REGION IN M33
BCL 0621	01 30 23.4	+ 30 37 00.	27		HII REGION IN M33
KN 14.287	01 30 23.6	+ 28 36 22.			NEBULA
KN 14.288	01 30 23.8	+ 28 12 49.			NEBULA
PHL 1026	01 30 24.	+ 01 56		16.7	BLUE STELLAR OBJECT
PHL 3412	01 30 24.	+ 04 26		17.8	BLUF STELLAR OBJECT

OBJECT NAME	RIGHT ASCEN.	DECLINATION	DIAM.	MAGN.	TYPE OF OBJECT
RNGC 0592	01 30 24.	+ 30 23			NEBULA IN EXTERNAL GALAXY
ZWG 537.012	01 30 24.	+ 44 41		15.4	GALAXY
KARA.72 037A	01 30 24.	+ 44 41	36	15.4	PART OF DOUBLE GALAXY
ZWG 386.009	01 30 24.	- 00 57		14.9	GALAXY
ZH 70	01 30 24.	- 00 57		14.9	GALAXY
UGC 01106	01 30 24.	- 00 57	84	14.9	GALAXY Sc
PHL 7293	01 30 24.	- 02 06		18.9	BLUE STELLAR OBJECT
PHL 7288	01 30 24.	- 02 06		18.9	BLUE STELLAR OBJECT
PHL 7292	01 30 24.	- 02 36		16.6	BLUE STELLAR OBJECT
PHL 7294	01 30 24.	- 03 21		18.6	BLUE STELLAR OBJECT
PHL 7289	01 30 24.	- 04 44		18.5	BLUE STELLAR OBJECT
PHL 7290	01 30 24.	- 09 46		18.2	BLUE STELLAR OBJECT
MCG-02-05-005	01 30 24.	- 12 27	60	13.5	GALAXY
MCG-03-05-004	01 30 24.	- 15 04	60	14.	GALAXY
IC 0141	01 30 24.	- 15 04 14.			NONSTELLAR OBJECT
PHL 7291	01 30 24.	- 18 47		18.6	BLUE STELLAR OBJECT
BCL 0215	01 30 24.0	+ 30 17 12.	15		HII REGION IN M33
BCL 0286	01 30 24.0	+ 30 26 18.	15		HII REGION IN M33
BCL 0601	01 30 24.0	+ 30 30 42.	9		HII REGION IN M33
BCL 0619	01 30 24.0	+ 30 35 54.	9		HII REGION IN M33
BCL 0620	01 30 24.0	+ 30 36 24.	9		HII REGION IN M33
BCL 290C	01 30 24.6	+ 30 29 36.	69		HII REGION IN M33
BCL 0625	01 30 24.6	+ 30 38 24.	15		HII REGION IN M33
BCL 0290A	01 30 25.2	+ 30 29 36.	33		HII REGION IN M33
BCL 0622	01 30 25.2	+ 30 36 54.	9		HII REGION IN M33
BCL 0624	01 30 25.2	+ 30 38 06.	27		HII REGION IN M33
BCL 0638	01 30 25.8	+ 30 41 36.	40		HII REGION IN M33
KN 14.286	01 30 26.1	+ 33 17 55.			NEBULA
KN 14.289	01 30 26.2	+ 30 37 54.			NEBULA
BCL 0603	01 30 26.4	+ 30 31 30.	27		HII REGION IN M33
BCL 0623	01 30 26.4	+ 30 37 36.	27		HII REGION IN M33
BCL 0626	01 30 26.4	+ 30 38 48.	33		HII REGION IN M33
IC 0132	01 30 26.6	+ 30 41 23.			HII REGION IN M33
IC 0133	01 30 26.7	+ 30 37 57.			HII REGION IN M33
KN 14.290	01 30 26.7	+ 30 41 23.			NEBULA
MCG+01-05-005	01 30 27.	- 07 18 30.	90	11.5	GALAXY
IC 1714	01 30 27.	- 13 45 20.			NONSTELLAR OBJECT
BCL 0245	01 30 27.0	+ 30 11 30.	27		HII REGION IN M33
BCL 0602	01 30 27.0	+ 30 30 48.	26		HII REGION IN M33
BCL 0604	01 30 27.0	+ 30 31 48.	15		HII REGION IN M33
BCL 0287	01 30 27.6	+ 30 26 06.	15		HII REGION IN M33
BCL 0606	01 30 28.2	+ 30 32 42.	33		HII REGION IN M33
BCL 0244	01 30 28.8	+ 30 17 42.	20		HII REGION IN M33
BCL 0278	01 30 28.8	+ 30 24 18.	21		HII REGION IN M33
PHL 1027	01 30 30.	+ 03 22		17.8	BLUE STELLAR OBJECT
BC PHL1027	01 30 30.	+ 03 22		17.04	QUASI-STELLAR OBJECT
SHB 031	01 30 30.	+ 03 22		17.0	QUASI-STELLAR OBJECT
PHL 7295	01 30 30.	+ 05 06		17.1	BLUE STELLAR OBJECT
PHL 3413	01 30 30.	+ 08 03		17.4	BLUE STELLAR OBJECT
ZC 0130.5+3129	01 30 30.	+ 31 29	1210		CLUSTER OF GALAXIES
UGC 01107	01 30 30.	- 00 04	66	17.	GALAXY DWARF
MCG+00-05-003	01 30 30.	- 00 56 30.	48	15.	GALAXY
ZC 0130.5-0205	01 30 30.	- 02 05	810		CLUSTER OF GALAXIES
MCG+01-05-006	01 30 30.	- 05 17 30.	48	15.	GALAXY
MCG-02-05-006	01 30 30.	- 10 42 30.	54	14.5	GALAXY
BCL 0254	01 30 30.0	+ 30 02 12.	15		HII REGION IN M33
BCL 0224	01 30 30.0	+ 30 16 36.	12		HII REGION IN M33
BCL 0225	01 30 30.6	+ 30 16 06.	6		HII REGION IN M33
HELW 007	01 30 31.	- 29 34			NEBULA
BCL 0607	01 30 31.8	+ 30 32 24.	15		HII REGION IN M33
BCL 0222	01 30 33.0	+ 30 09 54.	12		HII REGION IN M33
KN 14.291	01 30 33.0	+ 30 53 47.			NEBULA
BCL 0223	01 30 33.6	+ 30 15 24.	18		HII REGION IN M33
BCL 0627	01 30 34.2	+ 30 35 12.	9		HII REGION IN M33
RNGC 0594	01 30 35.	- 16 48		14.0	GALAXY
BCL 0608	01 30 35.4	+ 30 32 24.	32		HII REGION IN M33
PHL 3415	01 30 36.	+ 03 22		18.2	BLUE STELLAR OBJECT
MCG+02-05-001	01 30 36.	+ 13 05	108	14.	GALAXY
UGC 01108	01 30 36.	+ 17 21	66	16.0	GALAXY S
IC 0134	01 30 36.	+ 30 38			SINGLE STAR
ZC 0130.6+3215	01 30 36.	+ 32 15	610		CLUSTER OF GALAXIES
ZWG 537.013	01 30 36.	+ 44 40		14.2	GALAXY
RNGC 0590	01 30 36.	+ 44 40		14.0	GALAXY
UGC 01109	01 30 36.	+ 44 40	216	14.2	GALAXY SB0/SBa
KARA.72 037B	01 30 36.	+ 44 40	144	14.2	PART OF DOUBLE GALAXY
ISS 0048	01 30 36.	+ 59 19	603		STELLAR RING
VB 131	01 30 36.	+ 59 23	571		STELLAR RING
VB 132	01 30 36.	+ 59 58	195		STELLAR RING
ISS 0047	01 30 36.	+ 59 58	200		STELLAR RING
PHL 7296	01 30 36.	- 00 05		16.1	BLUF STELLAR OBJECT
PHL 7298	01 30 36.	- 03 01		17.5	BLUE STELLAR OBJECT
PHL 7297	01 30 36.	- 04 48		17.3	BLUE STELLAR OBJECT
PHL 7299	01 30 36.	- 08 36		18.4	BLUE STELLAR OBJECT
PHL 3414	01 30 36.	- 13 31		16.8	BLUE STELLAR OBJECT
MCG-03-05-005	01 30 36.	- 16 48	78	14.	GALAXY
BCL 1007	01 30 36.0	+ 30 23 42.	21		HII REGION IN M33
BCL 0213	01 30 36.6	+ 30 16 24.	27		HII REGION IN M33
HELW 008	01 30 37.	- 29 34			NEBULA
BCL 0639	01 30 37.8	+ 30 45 18.	33		HII REGION IN M33
KN 14.292	01 30 38.3	+ 28 41 21.			NEBULA
BCL 0253	01 30 38.4	+ 30 02 24.	21		HII REGION IN M33
BCL 0045	01 30 38.4	+ 30 25 06.	27		HII REGION IN M33
BCL 0609	01 30 39.0	+ 30 32 30.	39		HII REGION IN M33
SHB 032	01 30 39.4	+ 24 12 07.		17.0	QUASI-STELLAR OBJECT
BCL 0252	01 30 39.6	+ 30 03 06.	9		HII REGION IN M33
BCL 0214	01 30 39.6	+ 30 16 18.	9		HII REGION IN M33
BCL 0212	01 30 39.6	+ 30 16 24.	33		HII REGION IN M33
BCL 0046	01 30 39.6	+ 30 25 30.	15		HII REGION IN M33
BC 4C24.02	01 30 39.7	+ 24 12 26.		16.8	QUASI-STELLAR OBJECT
RNGC 0591	01 30 40.	+ 35 25		14.0	GALAXY
BCL 0210	01 30 40.8	+ 30 18 06.	14		HII REGION IN M33
BCL 1006	01 30 40.8	+ 30 22 42.	21		HII REGION IN M33
BCL 0211	01 30 41.4	+ 30 16 30.	39		HII REGION IN M33
PHL 7304	01 30 42.	+ 01 22		18.6	BLUE STELLAR OBJECT
PHL 7300	01 30 42.	+ 01 51		15.7	BLUE STELLAR OBJECT
PHL 3416	01 30 42.	+ 08 22		18.7	BLUE STELLAR OBJECT
ZWG 437.002	01 30 42.	+ 13 05		15.3	GALAXY
UGC 01110	01 30 42.	+ 13 05	120	15.3	GALAXY Sc
MCG+03-05-007	01 30 42.	+ 17 09	42	15.5	GALAXY
RNGC 0595	01 30 42.	+ 30 26			NEBULA IN EXTERNAL GALAXY
ZWG 521.046	01 30 42.	+ 35 25		14.0	GALAXY
UGC 01111	01 30 42.	+ 35 25	84	14.5	GALAXY SB0/SBa
MCG+07-04-003	01 30 42.	+ 44 40	120	14.5	GALAXY
VB 133	01 30 42.	+ 59 46	329		STELLAR RING
PHL 7302	01 30 42.	- 02 35		18.4	BLUE STELLAR OBJECT
MCG+01-05-007	01 30 42.	- 07 35	240	12.5	GALAXY
PHL 7305	01 30 42.	- 09 58		18.3	BLUE STELLAR OBJECT
PHL 7301	01 30 42.	- 13 18		16.6	BLUE STELLAR OBJECT
PHL 7303	01 30 42.	- 18 20		18.5	BLUE STELLAR OBJECT
TON-S 0221	01 30 42.	- 32 00		15.8	BLUE STAR
LB 03218	01 30 42.	- 68 46		14.8	FAINT BLUE STAR
BCL 0034	01 30 42.0	+ 30 21 30.	7		HII REGION IN M33
BCL 0209	01 30 42.6	+ 30 18 18.	57		HII REGION IN M33
LIN.CL 108	01 30 43.	- 72 12 34.	342	14.8	STAR CLUSTER IN SMC
BCL 0208	01 30 43.2	+ 30 16 36.	33		HII REGION IN M33
BCL 1005	01 30 43.2	+ 30 22 12.	15		HII REGION IN M33
KN 14.294	01 30 43.5	+ 27 42 39.			NEBULA
BCL 0049	01 30 43.8	+ 30 26 18.	63		HII REGION IN M33
BCL 0610	01 30 43.8	+ 30 31 48.	21		HII REGION IN M33
BCL 0640	01 30 43.8	+ 30 45 42.	36		HII REGION IN M33
BCL 0611	01 30 44.4	+ 30 32 24.	15		HII REGION IN M33
BCL 0628	01 30 44.4	+ 30 34 54.	12		HII REGION IN M33
RNGC 0600	01 30 45.	- 07 35		12.5	GALAXY
BCL 0207	01 30 45.0	+ 30 14 24.	15		HII REGION IN M33
BCL 0032	01 30 45.0	+ 30 21 06.	24		HII REGION IN M33
BCL 0033	01 30 45.0	+ 30 21 42.	22		HII REGION IN M33
BCL 0041	01 30 45.0	+ 30 24 00.	21		HII REGION IN M33
BCL 0042	01 30 45.0	+ 30 24 12.	15		HII REGION IN M33
BCL 0630	01 30 45.0	+ 30 35 30.	38		HII REGION IN M33
BCL 0251	01 30 45.6	+ 30 04 42.	9		HII REGION IN M33
BCL 1008	01 30 45.6	+ 30 23 24.	15		HII REGION IN M33
BCL 0054	01 30 45.6	+ 30 27 18.	9		HII REGION IN M33
BCL 0057	01 30 45.6	+ 30 28 12.	3		HII REGION IN M33
BCL 0615	01 30 45.6	+ 30 33 18.	33		HII REGION IN M33
BCL 1004	01 30 46.2	+ 30 20 12.	9		HII REGION IN M33
BCL 0055	01 30 46.2	+ 30 27 48.	7		HII REGION IN M33
KN 14.293	01 30 46.2	+ 30 45 23.			NEBULA
BCL 0203	01 30 46.8	+ 30 16 36.	9		HII REGION IN M33
BCL 0031	01 30 46.8	+ 30 21 12.	15		HII REGION IN M33
BCL 0629	01 30 46.8	+ 30 35 18.	15		HII REGION IN M33
BCL 0206	01 30 47.4	+ 30 14 42.	21		HII REGION IN M33
PHL 7306	01 30 48.	+ 00 26		18.8	BLUE STELLAR OBJECT
MCG+00-05-004	01 30 48.	+ 02 49 30.	66	13.5	GALAXY
ZWG 386.010	01 30 48.	+ 02 50		15.0	GALAXY
UGC 01112	01 30 48.	+ 02 50	90	15.0	GALAXY Sc
PHL 3418	01 30 48.	+ 03 46		18.7	BLUE STELLAR OBJECT
PHL 7308	01 30 48.	+ 04 22		17.9	BLUE STELLAR OBJECT
PHL 1029	01 30 48.	+ 05 42		17.0	BLUE STELLAR OBJECT
PHL 1028	01 30 48.	+ 08 00		17.2	BLUE STELLAR OBJECT
UGC 01113	01 30 48.	+ 17 10	78	16.0	GALAXY
ZWG 460.007	01 30 48.	+ 18 46		15.6	GALAXY
UGC 01114	01 30 48.	+ 18 46	78	15.6	GALAXY Sc
MCG+03-05-008	01 30 48.	+ 18 47	66	15.	GALAXY
ZWG 460.008	01 30 48.	+ 20 52		15.5	GALAXY
MCG+06-04-038	01 30 48.	+ 35 24 30.	66	13.5	GALAXY
VB 134	01 30 48.	+ 61 08	376		STELLAR RING
MCG+00-05-005	01 30 48.	- 00 54	12	15.5	GALAXY
MCG+00-05-006	01 30 48.	- 00 55	24	15.5	GALAXY
PHL 7307	01 30 48.	- 06 16		18.4	BLUE STELLAR OBJECT
PHL 3417	01 30 48.	- 06 50		17.3	BLUE STELLAR OBJECT
PHL 7309	01 30 48.	- 10 56		18.5	BLUE STELLAR OBJECT
PHL 1030	01 30 48.	- 17 22		18.4	BLUE STELLAR OBJECT
TON-S 0222	01 30 48.	- 27 23		15.8	BLUE STAR
MOHR 4	01 30 48.	- 69 54 06.		15.5	STAR CLUSTER IN SMC
BCL 0250	01 30 48.0	+ 30 05 00.	48		HII REGION IN M33
BCL 0202	01 30 48.0	+ 30 16 36.	9		HII REGION IN M33
BCL 1003	01 30 48.0	+ 30 19 42.	9		HII REGION IN M33
BCL 0058	01 30 48.0	+ 30 27 42.	3		HII REGION IN M33
BCL 0249	01 30 48.6	+ 30 05 24.	27		HII REGION IN M33
BCL 0200	01 30 48.6	+ 30 17 12.	15		HII REGION IN M33
IC 0137	01 30 49.1	+ 30 15 54.			HII REGION IN M33
BCL 0205	01 30 49.2	+ 30 15 48.	39		HII REGION IN M33
BCL 0036	01 30 49.2	+ 30 22 42.	27		HII REGION IN M33
BCL 0035	01 30 49.2	+ 30 23 06.	39		HII REGION IN M33
BCL 0050	01 30 49.8	+ 30 25 48.	15		HII REGION IN M33
BCL 0246	01 30 50.4	+ 30 07 06.	27		HII REGION IN M33
BCL 0204	01 30 50.4	+ 30 16 00.	45		HII REGION IN M33
BCL 0030	01 30 50.4	+ 30 21 54.	25		HII REGION IN M33
BCL 0053	01 30 50.4	+ 30 26 18.	12		HII REGION IN M33
BCL 0059	01 30 50.4	+ 30 27 36.	15		HII REGION IN M33
BCL 0612	01 30 50.4	+ 30 30 42.	21		HII REGION IN M33
BCL 0201	01 30 51.0	+ 30 16 42.	27		HII REGION IN M33
BCL 0039	01 30 51.0	+ 30 23 36.	18		HII REGION IN M33
BCL 0048	01 30 51.0	+ 30 25 24.	15		HII REGION IN M33
BCL 0634	01 30 51.0	+ 30 37 06.	30		HII REGION IN M33
BCL 0024	01 30 51.6	+ 30 17 36.	7		HII REGION IN M33
BCL 0044	01 30 51.6	+ 30 24 24.	7		HII REGION IN M33
BCL 0056	01 30 51.6	+ 30 26 36.	22		HII REGION IN M33
BCL 0613	01 30 51.6	+ 30 32 18.	9		HII REGION IN M33
RNGC 0601	01 30 52.	- 12 28			NON-EXISTENT OBJECT
KN 14.295	01 30 52.0	+ 31 52 14.			NEBULA
BCL 0247	01 30 52.8	+ 30 07 12.	45		HII REGION IN M33
BCL 0023	01 30 52.8	+ 30 17 48.	11		HII REGION IN M33
BCL 1002	01 30 52.8	+ 30 20 06.	37		HII REGION IN M33
BCL 0038	01 30 52.8	+ 30 23 30.	19		HII REGION IN M33
BCL 0040	01 30 52.8	+ 30 23 48.	14		HII REGION IN M33
BCL 0061	01 30 52.8	+ 30 29 24.	15		HII REGION IN M33
IC 0139	01 30 53.	+ 30 13			SINGLE STAR
BCL 0028	01 30 53.4	+ 30 21 24.	13		HII REGION IN M33
BCL 0635	01 30 53.4	+ 30 38 00.	27		HII REGION IN M33
ZWG 437.003	01 30 54.	+ 12 19		14.5	GALAXY
UGC 01115	01 30 54.	+ 12 19	48	14.5	GALAXY IRR
KARA.73B 0054	01 30 54.	+ 12 19	42	14.5	ISOLATED GALAXY S
MCG+02-05-002	01 30 54.	+ 12 20	48	14.	GALAXY
MCG+03-05-009	01 30 54.	+ 18 19	48	16.	GALAXY
PHL 3419	01 30 54.	- 08 41		17.9	BLUE STELLAR OBJECT
PHL 7310	01 30 54.	- 08 52		18.0	BLUE STELLAR OBJFCT
PHL 3420	01 30 54.	- 12 42		18.6	BLUE STELLAR OBJECT
BCL 1502	01 30 54.0	+ 30 16 18.	51		HII REGION IN M33
BCL 0026	01 30 54.0	+ 30 20 36.	15		HII REGION IN M33
BCL 0051	01 30 54.0	+ 30 25 24.	21		HII REGION IN M33
BCL 0052	01 30 54.0	+ 30 25 36.	12		HII REGION IN M33
BCL 0614	01 30 54.0	+ 30 33 24.	6		HII REGION IN M33
BCL 0248	01 30 54.6	+ 30 06 00.	21		HII REGION IN M33
BCL 0062	01 30 54.6	+ 30 29 30.	33		HII REGION IN M33
BCL 0641	01 30 54.6	+ 30 46 36.	55		HII REGION IN M33
IC 0140	01 30 55.	+ 30 14			SINGLE STAR
BCL 0021	01 30 55.2	+ 30 17 12.	45		HII REGION IN M33
BCL 0022	01 30 55.2	+ 30 17 42.	7		HII REGION IN M33
BCL 0025	01 30 55.2	+ 30 21 00.	21		HII REGION IN M33
BCL 0027	01 30 55.2	+ 30 21 30.	21		HII REGION IN M33
KN 14.296	01 30 55.3	+ 30 46 54.			NEBULA
BCL 1001	01 30 55.8	+ 30 20 36.	9		HII REGION IN M33
BCL 0060	01 30 55.8	+ 30 27 24.	15		HII REGION IN M33
BCL 0633	01 30 56.4	+ 30 37 00.	12		HII REGION IN M33
KN 14.297	01 30 56.7	+ 27 09 11.			NEBULA

OBJECT NAME	RIGHT ASCEN.	DECLINATION	DIAM.	MAGN.	TYPE OF OBJECT
BCL 0019	01 30 57.0	+ 30 18 12.	13		HII REGION IN M33
BCL 0029	01 30 57.0	+ 30 23 18.	21		HII REGION IN M33
BCL 0017	01 30 57.6	+ 30 17 36.	27		HII REGION IN M33
BCL 0018	01 30 57.6	+ 30 18 12.	9		HII REGION IN M33
BCL 0063	01 30 57.6	+ 30 28 36.	27		HII REGION IN M33
BCL 0636	01 30 57.6	+ 30 38 54.	10		HII REGION IN M33
IC 1715	01 30 58.	+ 12 19 39.			NONSTELLAR OBJECT
BCL 0020	01 30 58.2	+ 30 22 42.	6		HII REGION IN M33
BCL 0037	01 30 58.2	+ 30 23 54.	7		HII REGION IN M33
BCL 0043	01 30 58.2	+ 30 24 12.	7		HII REGION IN M33
BCL 0631	01 30 58.2	+ 30 35 36.	21		HII REGION IN M33
BCL 0047	01 30 58.8	+ 30 24 30.	15		HII REGION IN M33
BCL 0049A	01 30 58.8	+ 30 26 18.	63		HII REGION IN M33
IC 1716	01 30 59.	- 12 34			NONSTELLAR OBJECT
VDB.66G 012	01 31	+ 04 09	100		DWARF GALAXY
KHAV 028	01 31	+ 56 51	6380		DARK NEBULA
PHL 1031	01 31 00.	+ 03 44		18.1	BLUE STELLAR OBJECT
PHL 7314	01 31 00.	+ 04 08		18.8	BLUE STELLAR OBJECT
PHL 3421	01 31 00.	+ 04 34		18.5	BLUE STELLAR OBJECT
PHL 7311	01 31 00.	+ 06 14		16.9	BLUE STELLAR OBJECT
ZWG 460.009	01 31 00.	+ 18 18		15.7	GALAXY
MCG+05-04-069	01 31 00.	+ 30 20	3840	6.1	GALAXY
ZWG 502.110	01 31 00.	+ 30 24		6.5	GALAXY
UGC 01117	01 31 00.	+ 30 24	4380	6.5	GALAXY Sc
5ZW 080	01 31 00.	+ 42 30			COMPACT GALAXY
LDN 1327	01 31 00.	+ 65 10	900		DARK NEBULA
ZWG 386.011	01 31 00.	- 01 21		15.2	GALAXY
ZH 67	01 31 00.	- 01 21		15.2	GALAXY
UGC 01116	01 31 00.	- 01 21	60	15.2	GALAXY Sc-IRR
MCG+00-05-007	01 31 00.	- 01 21	36	15.	GALAXY
PHL 7313	01 31 00.	- 03 33		15.8	BLUE STELLAR OBJECT
PHL 7312	01 31 00.	- 07 58		18.6	BLUE STELLAR OBJECT
PHL 7315	01 31 00.	- 08 30		18.0	BLUE STELLAR OBJECT
PHL 7316	01 31 00.	- 10 04		18.3	BLUE STELLAR OBJECT
PHL 3423	01 31 00.	- 12 05		18.8	BLUE STELLAR OBJECT
PHL 3422	01 31 00.	- 12 54		18.6	BLUE STELLAR OBJECT
PHL 1032	01 31 00.	- 19 22		18.2	BLUE STELLAR OBJECT
HN 0153	01 31 00.	- 67 48			NEBULA
BCL 1501	01 31 00.0	+ 30 17 06.	33		HII REGION IN M33
BCL 0015	01 31 00.0	+ 30 18 18.	15		HII REGION IN M33
BCL 0014	01 31 00.0	+ 30 21 48.	6		HII REGION IN M33
BCL 0016	01 31 00.0	+ 30 22 06.	12		HII REGION IN M33
BCL 0632	01 31 00.0	+ 30 36 24.	43		HII REGION IN M33
BCL 0069	01 31 00.6	+ 30 24 36.	15		HII REGION IN M33
BCL 0064	01 31 00.6	+ 30 26 18.	21		HII REGION IN M33
IC 1717	01 31 01.	- 67 47			NONSTELLAR OBJECT
BCL 0009	01 31 01.2	+ 30 23 30.	8		HII REGION IN M33
BCL 0070	01 31 01.2	+ 30 24 48.	21		HII REGION IN M33
BCL 0071	01 31 01.2	+ 30 24 54.	21		HII REGION IN M33
BCL 0637	01 31 01.2	+ 30 41 24.	27		HII REGION IN M33
BCL 0012	01 31 01.8	+ 30 21 12.	9		HII REGION IN M33
BCL 0066	01 31 01.8	+ 30 25 42.	15		HII REGION IN M33
PATH 2.003	01 31 02.	+ 30 24	3523		NEBULA
KN 14.298	01 31 02.0	+ 30 24 24.			NEBULA
BCL 0694	01 31 02.4	+ 30 32 00.	10		HII REGION IN M33
BCL 0093	01 31 03.0	+ 30 24 00.	27		HII REGION IN M33
BCL 0065	01 31 03.0	+ 30 28 36.	21		HII REGION IN M33
BCL 0697	01 31 03.0	+ 30 31 18.	3		HII REGION IN M33
BCL 0010	01 31 03.6	+ 30 20 06.	27		HII REGION IN M33
BCL 0099	01 31 03.6	+ 30 23 30.	21		HII REGION IN M33
KN 14.299	01 31 03.9	+ 29 58 30.			NEBULA
BCL 0013	01 31 04.2	+ 30 17 36.	21		HII REGION IN M33
BCL 0002	01 31 04.2	+ 30 22 42.	7		HII REGION IN M33
KN 14.300	01 31 04.3	+ 30 24 59.			NEBULA
BCL 0760	01 31 04.8	+ 30 04 18.	27		HII REGION IN M33
BCL 0003	01 31 04.8	+ 30 22 12.	21		HII REGION IN M33
BCL 0707	01 31 05.4	+ 30 18 36.	21		HII REGION IN M33
BCL 0011	01 31 05.4	+ 30 19 06.	27		HII REGION IN M33
BCL 0092	01 31 05.4	+ 30 23 36.	6		HII REGION IN M33
BCL 0086	01 31 05.4	+ 30 24 06.	6		HII REGION IN M33
MCG+00-05-008	01 31 06.	+ 03 17 30.	126	13.	GALAXY
PHL 7318	01 31 06.	+ 03 36		17.2	BLUE STELLAR OBJECT
RNGC 0598	01 31 06.	+ 30 24		7.0	GALAXY
PHL 7317	01 31 06.	- 03 24		17.9	BLUE STELLAR OBJECT
PHL 7319	01 31 06.	- 10 20		18.1	BLUE STELLAR OBJECT
BCL 0759	01 31 06.0	+ 30 06 48.	15		HII REGION IN M33
BCL 0758	01 31 06.0	+ 30 12 00.	39		HII REGION IN M33
BCL 0096	01 31 06.0	+ 30 23 06.	6		HII REGION IN M33
BCL 0301	01 31 06.0	+ 30 30 12.	33		HII REGION IN M33
IC 0142	01 31 06.2	+ 30 30 03.			HII REGION IN M33
BCL 0076	01 31 06.6	+ 30 25 42.	6		HII REGION IN M33
BCL 0075	01 31 06.6	+ 30 26 00.	6		HII REGION IN M33
BCL 0084	01 31 07.2	+ 30 24 06.	6		HII REGION IN M33
BCL 0704	01 31 07.8	+ 30 16 54.	21		HII REGION IN M33
BCL 0700	01 31 07.8	+ 30 17 48.	15		HII REGION IN M33
BCL 0073	01 31 07.8	+ 30 26 54.	9		HII REGION IN M33
BCL 0074	01 31 07.8	+ 30 27 12.	9		HII REGION IN M33
BC PHL3424	01 31 08.	+ 05 32 31.		18.25	QUASI-STELLAR OBJECT
BCL 0005	01 31 08.4	+ 30 20 00.	21		HII REGION IN M33
BCL 0696	01 31 08.4	+ 30 33 30.	45		HII REGION IN M33
BCL 0674	01 31 08.4	+ 30 36 00.	10		HII REGION IN M33
BCL 0008A	01 31 09.0	+ 30 18 12.	9		HII REGION IN M33
BCL 0007B	01 31 09.0	+ 30 19 00.	19		HII REGION IN M33
BCL 0006	01 31 09.0	+ 30 19 18.	21		HII REGION IN M33
BCL 0695	01 31 09.0	+ 30 34 06.	15		HII REGION IN M33
BCL 0669	01 31 09.0	+ 30 40 24.	9		HII REGION IN M33
BCL 0703	01 31 09.6	+ 30 17 24.	9		HII REGION IN M33
BCL 0008B	01 31 09.6	+ 30 18 30.	22		HII REGION IN M33
BCL 0004	01 31 09.6	+ 30 20 24.	24		HII REGION IN M33
BCL 0079	01 31 09.6	+ 30 25 36.	15		HII REGION IN M33
BCL 0007A	01 31 10.2	+ 30 19 00.	21		HII REGION IN M33
BCL 0094	01 31 10.2	+ 30 22 48.	14		HII REGION IN M33
BCL 0080	01 31 10.2	+ 30 25 30.	6		HII REGION IN M33
BCL 0699	01 31 10.8	+ 30 29 18.	3		HII REGION IN M33
KN 14.301	01 31 10.8	+ 32 03 24.			NEBULA
BCL 0001	01 31 11.4	+ 30 20 30.	9		HII REGION IN M33
BCL 0089A	01 31 11.4	+ 30 22 54.	57		HII REGION IN M33
BCL 0082	01 31 11.4	+ 30 24 48.	27		HII REGION IN M33
BCL 0081	01 31 11.4	+ 30 24 48.	15		HII REGION IN M33
BCL 0072	01 31 11.4	+ 30 28 36.	15		HII REGION IN M33
BCL 0673	01 31 11.4	+ 30 35 54.	10		HII REGION IN M33
PHL 3425	01 31 12.	+ 01 10		18.3	BLUE STELLAR OBJECT
ZWG 386.013	01 31 12.	+ 03 18		14.8	GALAXY
UGC 01118	01 31 12.	+ 03 18	150	14.8	GALAXY Sc
PHL 1033	01 31 12.	+ 03 40		18.7	BLUE STELLAR OBJECT
PHL 3424	01 31 12.	+ 05 32		18.7	BLUE STELLAR OBJECT
SHB 033	01 31 12.	+ 05 32		18.3	QUASI-STELLAR OBJECT
UGC 01119	01 31 12.	+ 16 58	108	16.5	GALAXY
ZC 0131.2+2752	01 31 12.	+ 27 52	3560		CLUSTER OF GALAXIES
PHL 7320	01 31 12.	- 01 45		16.6	BLUE STELLAR OBJECT
ZWG 386.012	01 31 12.	- 02 24		15.7	GALAXY
PHL 3428	01 31 12.	- 02 26		18.6	BLUE STELLAR OBJECT
PHL 1034	01 31 12.	- 04 20		18.0	BLUE STELLAR OBJECT
PHL 3426	01 31 12.	- 04 35		18.9	BLUE STELLAR OBJECT
PHL 7321	01 31 12.	- 09 50		18.1	BLUE STELLAR OBJECT
PHL 7322	01 31 12.	- 11 52		18.5	BLUE STELLAR OBJECT
PHL 3427	01 31 12.	- 13 08		18.4	BLUE STELLAR OBJECT
TON-S 0223	01 31 12.	- 27 27		14.2	BLUE STAR
BCL 0701	01 31 12.0	+ 30 15 42.	45		HII REGION IN M33
BCL 0100	01 31 12.0	+ 30 21 00.	27		HII REGION IN M33
BCL 0097	01 31 12.0	+ 30 22 00.	8		HII REGION IN M33
BCL 0698	01 31 12.0	+ 30 29 36.	9		HII REGION IN M33
BCL 0675	01 31 12.0	+ 30 35 18.	10		HII REGION IN M33
KN 14.302	01 31 12.1	+ 29 24 04.			NEBULA
BCL 0101	01 31 12.6	+ 30 20 42.	32		HII REGION IN M33
BCL 0089B	01 31 12.6	+ 30 22 36.	21		HII REGION IN M33
BCL 0087	01 31 12.6	+ 30 23 18.	39		HII REGION IN M33
BCL 0672	01 31 13.2	+ 30 36 12.	24		HII REGION IN M33
BCL 0693	01 31 13.8	+ 30 31 42.	33		HII REGION IN M33
BCL 0692	01 31 13.8	+ 30 32 42.	12		HII REGION IN M33
BCL 0670	01 31 13.8	+ 30 37 54.	9		HII REGION IN M33
BCL 0668	01 31 13.8	+ 30 39 48.	27		HII REGION IN M33
KN 14.303	01 31 14.2	+ 29 56 34.			NEBULA
BCL 0090	01 31 14.4	+ 30 22 48.	3		HII REGION IN M33
BCL 0098	01 31 15.6	+ 30 20 12.	3		HII REGION IN M33
BCL 0091	01 31 16.2	+ 30 22 30.	9		HII REGION IN M33
BCL 0077	01 31 16.8	+ 30 26 24.	27		HII REGION IN M33
BCL 0671	01 31 16.8	+ 30 36 30.	15		HII REGION IN M33
BCL 0646A	01 31 16.8	+ 30 48 36.	27		HII REGION IN M33
KN 14.306	01 31 16.9	+ 27 36 59.			NEBULA
KN 14.304	01 31 17.2	+ 28 54 43.			NEBULA
BCL 0751	01 31 17.4	+ 30 21 00.	15		HII REGION IN M33
BCL 0690	01 31 17.4	+ 30 31 42.	9		HII REGION IN M33
BCL 0648	01 31 17.4	+ 30 45 48.	45		HII REGION IN M33
KN 14.305	01 31 17.6	+ 28 50 41.			NEBULA
PHL 3429	01 31 18.	+ 03 40		17.2	BLUE STELLAR OBJECT
ZC 0131.3+2349	01 31 18.	+ 23 49	940		CLUSTER OF GALAXIES
PHL 7325	01 31 18.	- 02 22		18.2	BLUE STELLAR OBJECT
PHL 7324	01 31 18.	- 11 02		18.8	BLUE STELLAR OBJECT
PHL 7323	01 31 18.	- 14 44		16.6	BLUE STELLAR OBJECT
MCG-03-05-006	01 31 18.	- 17 38 30.	15	15.	GALAXY
BCL 0708	01 31 18.0	+ 30 18 36.	15		HII REGION IN M33
BCL 0085	01 31 18.0	+ 30 24 00.	36		HII REGION IN M33
IC 0143	01 31 18.1	+ 30 32 07.			HII REGION IN M33
BCL 0302	01 31 18.6	+ 30 32 24.	21		HII REGION IN M33
BCL 0702	01 31 19.8	+ 30 16 36.	21		HII REGION IN M33
BCL 0083	01 31 19.8	+ 30 23 48.	45		HII REGION IN M33
BCL 0647	01 31 19.8	+ 30 47 00.	74		HII REGION IN M33
BCL 0645	01 31 19.8	+ 30 48 24.	9		HII REGION IN M33
KN 14.307	01 31 20.2	+ 30 46 31.			NEBULA
BCL 0709	01 31 20.4	+ 30 19 12.	45		HII REGION IN M33
BCL 0646B	01 31 20.4	+ 30 49 54.	39		HII REGION IN M33
BCL 0095B	01 31 21.0	+ 30 20 24.	15		HII REGION IN M33
BCL 0078	01 31 21.0	+ 30 27 06.	10		HII REGION IN M33
BCL 0689	01 31 21.0	+ 30 31 12.	57		HII REGION IN M33
BCL 0095A	01 31 21.6	+ 30 20 54.	33		HII REGION IN M33
BCL 0642	01 31 22.2	+ 30 59 30.	27		HII REGION IN M33
BCL 0732	01 31 22.8	+ 30 24 42.	15		HII REGION IN M33
BCL 0733	01 31 22.8	+ 30 25 00.	10		HII REGION IN M33
BCL 0688	01 31 22.8	+ 30 31 00.	39		HII REGION IN M33
BCL 0744	01 31 23.4	+ 30 26 42.	6		HII REGION IN M33
BCL 0743	01 31 23.4	+ 30 27 12.	63		HII REGION IN M33
BCL 0747	01 31 23.4	+ 30 29 54.	9		HII REGION IN M33
BCL 0687	01 31 23.4	+ 30 33 12.	6		HII REGION IN M33
BCL 0667	01 31 23.4	+ 30 37 12.	33		HII REGION IN M33
BCL 0643	01 31 23.4	+ 30 55 00.	81		HII REGION IN M33
KN 14.310	01 31 23.5	+ 30 53 52.			NEBULA
PHL 7329	01 31 24.	+ 00 36		18.7	BLUE STELLAR OBJECT
PHL 7327	01 31 24.	+ 01 50		18.7	BLUE STELLAR OBJECT
PHL 3432	01 31 24.	+ 06 52		18.0	BLUE STELLAR OBJECT
PHL 3430	01 31 24.	+ 07 57		18.0	BLUE STELLAR OBJECT
MCG+02-05-004	01 31 24.	+ 13 14	24	16.	GALAXY
MCG+02-05-003	01 31 24.	+ 13 14	48	15.5	GALAXY
ZWG 502.111	01 31 24.	+ 31 48		14.9	GALAXY
ZWG 502.112	01 31 24.	+ 32 20		15.7	GALAXY
PHL 3431	01 31 24.	- 00 19		18.5	BLUE STELLAR OBJECT
PHL 7330	01 31 24.	- 05 28		18.0	BLUE STELLAR OBJECT
PHL 7328	01 31 24.	- 05 52		18.0	BLUE STELLAR OBJECT
PHL 7326	01 31 24.	- 14 00		15.7	BLUE STELLAR OBJECT
PHL 7331	01 31 24.	- 17 42		18.2	BLUE STELLAR OBJECT
MCG-06-04-045	01 31 24.	- 36 54		15.	GALAXY
BCL 0710	01 31 24.0	+ 30 18 18.	21		HII REGION IN M33
KN 14.308	01 31 24.2	+ 31 47 02.			NEBULA
KN 14.309	01 31 24.8	+ 32 19 41.			NEBULA
BCL 0714	01 31 25.2	+ 30 19 18.	45		HII REGION IN M33
BCL 0088	01 31 25.8	+ 30 21 48.	45		HII REGION IN M33
BCL 0720	01 31 26.4	+ 30 21 18.	21		HII REGION IN M33
BCL 0723	01 31 26.4	+ 30 24 00.	6		HII REGION IN M33
BCL 0666	01 31 26.4	+ 30 37 36.	39		HII REGION IN M33
BCL 0665	01 31 26.4	+ 30 38 30.	69		HII REGION IN M33
BCL 0660	01 31 26.4	+ 30 40 24.	81		HII REGION IN M33
MCG+05-04-070	01 31 27.	+ 31 47	36	15.	GALAXY
RNGC 0617	01 31 27.	- 10 03		15.0	GALAXY
BCL 0711	01 31 27.6	+ 30 18 12.	51		HII REGION IN M33
BCL 0686	01 31 27.6	+ 30 33 30.	15		HII REGION IN M33
BCL 0713	01 31 28.8	+ 30 19 00.	15		HII REGION IN M33
BCL 0722	01 31 28.8	+ 30 23 12.	9		HII REGION IN M33
KN 14.311	01 31 28.8	+ 32 27 04.			NEBULA
BCL 0712	01 31 29.4	+ 30 18 24.	9		HII REGION IN M33
BCL 0724	01 31 29.4	+ 30 23 54.	16		HII REGION IN M33
KN 14.312	01 31 29.9	+ 32 17 38.			NEBULA
UGC 01121	01 31 30.	+ 13 15	66	16.0	GALAXY Sb
UGC 01122	01 31 30.	+ 29 04	66	17.	GALAXY Sc
RNGC 0603	01 31 30.	+ 29 56			NON-EXISTENT OBJECT
ZWG 502.113	01 31 30.	+ 32 24		15.4	GALAXY
ZWG 558.001	01 31 30.	+ 55 10		15.3	GALAXY
ZWG 386.014	01 31 30.	- 01 20		14.9	GALAXY
ZH 68	01 31 30.	- 01 20		15.0	GALAXY
UGC 01120	01 31 30.	- 01 20	138	14.9	GALAXY SB:a
MCG+00-05-009	01 31 30.	- 01 20 30.	90	14.	GALAXY
MCG-02-05-007	01 31 30.	- 10 03	24	15.	GALAXY
PHL 3433	01 31 30.	- 15 08		18.5	BLUE STELLAR OBJECT
BCL 0746	01 31 30.0	+ 30 28 48.	81		HII REGION IN M33
BCL 0685	01 31 30.6	+ 30 33 42.	27		HII REGION IN M33
KN 14.313	01 31 30.8	+ 32 23 08.			NEBULA
KN 14.314	01 31 31.8	+ 29 00 23.			NEBULA

OBJECT NAME	RIGHT ASCEN.	DECLINATION	DIAM.	MAGN.	TYPE OF OBJECT
BCL 0721	01 31 31.8	+ 30 22 36.	21		HII REGION IN M33
BCL 0684	01 31 31.8	+ 30 33 24.	21		HII REGION IN M33
BCL 0661	01 31 31.8	+ 30 39 18.	9		HII REGION IN M33
BCL 0716	01 31 32.4	+ 30 17 12.	9		HII REGION IN M33
BCL 0715	01 31 33.0	+ 30 17 48.	33		HII REGION IN M33
BCL 0682	01 31 33.0	+ 30 31 42.	49		HII REGION IN M33
KN 14.316	01 31 33.5	+ 28 16 46.			NEBULA
BCL 0683	01 31 33.6	+ 30 33 36.	15		HII REGION IN M33
KN 14.317	01 31 34.2	+ 28 59 14.			NEBULA
BCL 0659	01 31 34.2	+ 30 41 30.	63		HII REGION IN M33
KN 14.315	01 31 34.4	+ 29 36 26.			NEBULA
BCL 0662	01 31 34.8	+ 30 39 42.	30		HII REGION IN M33
BCL 0731	01 31 35.4	+ 30 24 30.	15		HII REGION IN M33
PHL 7335	01 31 36.	+ 01 02		18.3	BLUE STELLAR OBJECT
PHL 7332	01 31 36.	+ 05 36		17.2	BLUE STELLAR OBJECT
MCG+05-04-071	01 31 36.	+ 32 23	30	15.	GALAXY
PHL 7333	01 31 36.	- 00 45		16.5	BLUE STELLAR OBJECT
ZWG 386.015	01 31 36.	- 01 17		14.4	GALAXY
ZH 69	01 31 36.	- 01 17		14.7	GALAXY
UGC 01123	01 31 36.	- 01 17	90	14.4	GALAXY Sa-b
MCG+00-05-010	01 31 36.	- 01 18	66	14.	GALAXY
PHL 1035	01 31 36.	- 03 26		18.3	BLUE STELLAR OBJECT
PHL 1036	01 31 36.	- 04 43		16.4	BLUE STELLAR OBJECT
PHL 7336	01 31 36.	- 06 51		18.6	BLUE STELLAR OBJECT
BCL 0664	01 31 36.6	+ 30 37 54.	93		HII REGION IN M33
KN 14.318	01 31 38.1	+ 32 17 18.			NEBULA
BCL 0644	01 31 39.0	+ 30 51 06.	33		HII REGION IN M33
BCL 0663	01 31 40.2	+ 30 38 00.	26		HII REGION IN M33
BCL 0734	01 31 40.8	+ 30 25 12.	28		HII REGION IN M33
BCL 0735	01 31 40.8	+ 30 26 00.	9		HII REGION IN M33
BCL 0651	01 31 40.8	+ 30 42 18.	57		HII REGION IN M33
RNGC 0604	01 31 41.	+ 30 32			NEBULA IN EXTERNAL GALAXY
RNGC 0612	01 31 41.	- 36 45			GALAXY
KN 14.322	01 31 41.6	+ 28 50 08.			NEBULA
PHL 1037	01 31 42.	+ 00 00		17.4	BLUE STELLAR OBJECT
PHL 7337	01 31 42.	+ 03 23		17.0	BLUE STELLAR OBJECT
ZWG 460.010	01 31 42.	+ 19 34		15.3	GALAXY
4ZW 052	01 31 42.	+ 37 43			COMPACT GALAXY
UGC 01124	01 31 42.	+ 55 10	78	17.	GALAXY Sc
PHL 3434	01 31 42.	- 01 11		18.3	BLUE STELLAR OBJECT
MCG-04-04-024	01 31 42.	- 21 48	48	16.	GALAXY
BCL 0719	01 31 42.6	+ 30 20 48.	12		HII REGION IN M33
BCL 0652	01 31 42.6	+ 30 41 54.	51		HII REGION IN M33
ARC 0212	01 31 43.	+ 04 14		17.2	RICH CLUSTER OF GALAXIES
KN 14.321	01 31 43.1	+ 31 20 56.			NEBULA
KN 14.319	01 31 43.1	+ 33 04 05.			NEBULA
BCL 0680	01 31 43.2	+ 30 31 54.	117		HII REGION IN M33
BCL 0676	01 31 43.2	+ 30 34 00.	21		HII REGION IN M33
KN 14.324	01 31 43.8	+ 30 32 00.			NEBULA
KN 14.323	01 31 43.8	+ 30 45 06.			NEBULA
LB 00452	01 31 44.	+ 31 58 06.		17.2	FAINT BLUE STAR
BCL 0745	01 31 44.4	+ 30 20 00.	21		HII REGION IN M33
BCL 0736	01 31 44.4	+ 30 25 42.	9		HII REGION IN M33
BCL 0650	01 31 44.4	+ 30 45 24.	45		HII REGION IN M33
KN 14.320	01 31 44.5	+ 33 03 27.			NEBULA
KN 14.325	01 31 44.8	+ 30 47 15.			NEBULA
BCL 0718	01 31 45.6	+ 30 20 30.	9		HII REGION IN M33
BCL 0748	01 31 46.8	+ 30 28 48.	16		HII REGION IN M33
BCL 0730	01 31 47.4	+ 30 22 48.	9		HII REGION IN M33
BCL 0679	01 31 47.4	+ 30 32 00.	9		HII REGION IN M33
PHL 7340	01 31 48.	+ 01 20		18.7	BLUE STELLAR OBJECT
PHL 7339	01 31 48.	+ 01 20		17.6	BLUE STELLAR OBJECT
PHL 7338	01 31 48.	+ 01 29		13.0	BLUE STELLAR OBJECT
PHL 1038	01 31 48.	+ 01 32		18.0	BLUE STELLAR OBJECT
ZC 0131.8+0515	01 31 48.	+ 05 15	810		CLUSTER OF GALAXIES
VB 135	01 31 48.	+ 60 20	242		STELLAR RING
PHL 7341	01 31 48.	- 01 42		18.9	BLUE STELLAR OBJECT
PHL 3435	01 31 48.	- 01 50		18.9	BLUE STELLAR OBJECT
PHL 1039	01 31 48.	- 12 32		18.1	BLUE STELLAR OBJECT
MCG-06-04-047	01 31 48.	- 34 40	78	14.	GALAXY
MCG-06-04-046	01 31 48.	- 36 47	72	14.	GALAXY
BCL 0717	01 31 48.0	+ 30 19 30.	21		HII REGION IN M33
BCL 0741	01 31 48.0	+ 30 26 24.	9		HII REGION IN M33
BCL 0681	01 31 48.0	+ 30 31 12.	8		HII REGION IN M33
BCL 0737	01 31 48.6	+ 30 25 30.	15		HII REGION IN M33
BCL 0729	01 31 49.2	+ 30 22 24.	21		HII REGION IN M33
BCL 0742	01 31 49.2	+ 30 26 00.	9		HII REGION IN M33
BCL 0653	01 31 49.2	+ 30 41 06.	26		HII REGION IN M33
BCL 0649	01 31 49.2	+ 30 47 36.	32		HII REGION IN M33
BCL 0738	01 31 49.8	+ 30 25 42.	9		HII REGION IN M33
BCL 0749	01 31 49.8	+ 30 28 36.	22		HII REGION IN M33
KN 14.326	01 31 49.8	+ 30 47 10.			NEBULA
BCL 0705	01 31 50.4	+ 30 15 54.	15		HII REGION IN M33
BCL 0740	01 31 50.4	+ 30 26 24.	51		HII REGION IN M33
RNGC 0607	01 31 51.	- 07 41			NON-EXISTENT OBJECT
MCG-04-04-025	01 31 51.	- 25 49	66	14.5	GALAXY
BCL 0678	01 31 51.0	+ 30 31 48.	9		HII REGION IN M33
BCL 0654	01 31 51.0	+ 30 41 00.	9		HII REGION IN M33
BCL 0728	01 31 51.6	+ 30 22 12.	27		HII REGION IN M33
BCL 0750	01 31 51.6	+ 30 28 12.	27		HII REGION IN M33
BCL 0677	01 31 51.6	+ 30 33 06.	27		HII REGION IN M33
BCL 0706	01 31 52.2	+ 30 16 12.	15		HII REGION IN M33
BCL 0739	01 31 52.8	+ 30 25 48.	9		HII REGION IN M33
BCL 0655	01 31 52.8	+ 30 40 42.	6		HII REGION IN M33
PHL 1040	01 31 54.	+ 01 49		13.9	BLUE STELLAR OBJECT
ZCG 0131+42	01 31 54.	+ 42 36		15.8	COMPACT GALAXY
PHL 3436	01 31 54.	- 01 41		18.6	BLUE STELLAR OBJECT
PHL 7342	01 31 54.	- 04 30		17.8	BLUE STELLAR OBJECT
PHL 1041	01 31 54.	- 05 05		18.8	BLUE STELLAR OBJECT
KARA.68 007	01 31 54.	- 07 37	34		DWARF GALAXY
RNGC 0611	01 31 54.	- 20 24			NON-EXISTENT OBJECT
RNGC 0610	01 31 54.	- 20 25			NON-EXISTENT OBJECT
SVEN 091	01 31 55.	- 29 40	300	10.7	GALAXY
BCL 0727	01 31 55.2	+ 30 22 30.	27		HII REGION IN M33
BCL 0757	01 31 55.2	+ 30 27 18.	15		HII REGION IN M33
BCL 0656	01 31 56.4	+ 30 40 18.	9		HII REGION IN M33
RNGC 0613	01 31 57.	- 29 40		11.0	GALAXY
BCL 0725	01 31 57.0	+ 30 22 48.	6		HII REGION IN M33
KN 14.327	01 31 57.5	+ 33 46 46.			NEBULA
BCL 0726	01 31 57.6	+ 30 22 30.	9		HII REGION IN M33
ARC 0214	01 31 58.	- 26 22		17.5	RICH CLUSTER OF GALAXIES
KN 14.328	01 31 59.7	+ 32 25 03.			NEBULA
LB 09961	01 32	- 87 28		14.5	FAINT BLUE STAR
PHL 3441	01 32 00.	+ 00 12		18.7	BLUE STELLAR OBJECT
PHL 7343	01 32 00.	+ 01 10		16.1	BLUE STELLAR OBJECT
PHL 1042	01 32 00.	+ 01 34		18.8	BLUE STELLAR OBJECT
PHL 7347	01 32 00.	+ 02 04		16.9	BLUE STELLAR OBJECT
PHL 7344	01 32 00.	+ 02 16		17.5	BLUE STELLAR OBJECT

OBJECT NAME	RIGHT ASCEN.	DECLINATION	DIAM.	MAGN.	TYPE OF OBJECT
PHL 7346	01 32 00.	+ 04 35		16.6	BLUE STELLAR OBJECT
PHL 3437	01 32 00.	+ 07 50		18.1	BLUE STELLAR OBJECT
PHL 3438	01 32 00.	+ 08 30		18.2	BLUE STELLAR OBJECT
ZWG 521.047	01 32 00.	+ 33 47		14.3	GALAXY
UGC 01125	01 32 00.	+ 33 47	39	14.3	GALAXY S
VB 136	01 32 00.	+ 59 35	349		STELLAR RING
ISS 0049	01 32 00.	+ 59 35	355		STELLAR RING
OCL 0327	01 32 00.	+ 61 11	180		OPEN STAR CLUSTER
LDN 1329	01 32 00.	+ 65 00	720		DARK NEBULA
7ZW 002	01 32 00.	+ 84 22			COMPACT GALAXY
PHL 3439	01 32 00.	- 00 06		18.4	BLUE STELLAR OBJECT
PHL 7348	01 32 00.	- 01 29		18.7	BLUE STELLAR OBJECT
PHL 3440	01 32 00.	- 02 22		18.7	BLUE STELLAR OBJECT
PHL 7349	01 32 00.	- 04 02		18.9	BLUE STELLAR OBJECT
PHL 7345	01 32 00.	- 13 25		14.5	BLUE STELLAR OBJECT
PHL 1043	01 32 00.	- 16 22		13.6	BLUE STELLAR OBJECT
PHL 3442	01 32 00.	- 17 30		18.6	BLUE STELLAR OBJECT
MCG-05-04-044	01 32 00.	- 29 39	300	11.	GALAXY
BCL 0657	01 32 00.0	+ 30 39 42.	33		HII REGION IN M33
BCL 0658	01 32 01.8	+ 30 39 06.	24		HII REGION IN M33
MCG+07-04-004	01 32 03.	+ 40 59	36	14.5	GALAXY
KN 14.329	01 32 03.0	+ 32 29 39.			NEBULA
KN 14.330	01 32 03.1	+ 31 58 46.			NEBULA
BCL 0752	01 32 05.4	+ 30 25 48.	27		HII REGION IN M33
PHL 1044	01 32 06.	+ 03 42		18.2	BLUE STELLAR OBJECT
PHL 3443	01 32 06.	+ 08 16		17.6	BLUE STELLAR OBJECT
ZWG 460.011	01 32 06.	+ 21 10		14.5	GALAXY
UGC 01126	01 32 06.	+ 21 10	108	14.5	GALAXY SBc
ZC 0132.1+2459	01 32 06.	+ 24 59	1080		CLUSTER OF GALAXIES
ZWG 521.048	01 32 06.	+ 33 40		15.5	GALAXY
ZWG 521.049	01 32 06.	+ 34 48		14.9	GALAXY
ZWG 521.050	01 32 06.	+ 35 46		15.6	GALAXY
4ZW 053	01 32 06.	+ 35 47			COMPACT GALAXY
UGC 01127	01 32 06.	+ 39 40	72	16.5	GALAXY
ZWG 537.014	01 32 06.	+ 41 00		14.3	GALAXY
UGC 01128	01 32 06.	+ 41 00	132	14.3	GALAXY S0
MCG-03-05-007	01 32 06.	- 16 05	48	14.	GALAXY
TON-S 0224	01 32 06.	- 28 19		16.2	BLUE STAR
MCG-06-04-048	01 32 06.	- 33 02	9	16.	GALAXY
RNGC 0606	01 32 08.	+ 21 10		14.5	GALAXY
RNGC 0605	01 32 08.	+ 41 00		14.5	GALAXY
MCG+06-04-039	01 32 09.	+ 33 47	30	14.5	GALAXY
BCL 0753	01 32 09.6	+ 30 26 12.	39		HII REGION IN M33
PHL 7350	01 32 12.	+ 02 16		16.4	BLUE STELLAR OBJECT
PHL 7355	01 32 12.	+ 04 54		17.8	BLUE STELLAR OBJECT
PHL 3444	01 32 12.	+ 07 28		18.4	BLUE STELLAR OBJECT
MCG+02-05-005	01 32 12.	+ 11 49	48	15.	GALAXY
ZWG 437.004	01 32 12.	+ 11 50		15.3	GALAXY
UGC 01129	01 32 12.	+ 11 50	72	15.3	GALAXY (S0)
MCG+03-05-010	01 32 12.	+ 21 11	24	14.	GALAXY
PHL 3445	01 32 12.	- 01 24		18.6	BLUE STELLAR OBJECT
PHL 1045	01 32 12.	- 01 50		18.6	BLUE STELLAR OBJECT
PHL 7356	01 32 12.	- 02 48		18.5	BLUE STELLAR OBJECT
PHL 7351	01 32 12.	- 03 10		18.2	BLUE STELLAR OBJECT
PHL 3446	01 32 12.	- 04 46		16.7	BLUE STELLAR OBJECT
PHL 7352	01 32 12.	- 07 24		18.6	BLUE STELLAR OBJECT
PHL 1046	01 32 12.	- 10 14		16.7	BLUE STELLAR OBJECT
PHL 7354	01 32 12.	- 12 36		16.9	BLUE STELLAR OBJECT
MCG-03-05-008	01 32 12.	- 15 46	90	15.	GALAXY
PHL 7353	01 32 12.	- 15 55		5.7	BLUE STELLAR OBJECT
MCG-06-04-049	01 32 12.	- 33 08	9	15.5	GALAXY
BCL 0754	01 32 12.0	+ 30 24 42.	57		HII REGION IN M33
KN 14.331	01 32 13.8	+ 32 15 07.			NEBULA
NAB 0132+20	01 32 15.	+ 20 30 25.		17.5	QUASI-STELLAR OBJECT
SHB 034	01 32 15.	+ 20 30 25.		17.5	QUASI-STELLAR OBJECT
BCL 0755	01 32 16.8	+ 30 25 42.	39		HII REGION IN M33
ZWG 481.011	01 32 18.	+ 21 39		15.3	GALAXY
KARA.73B 0055	01 32 18.	+ 21 39	66	15.3	ISOLATED GALAXY S
ZWG 502.114	01 32 18.	+ 30 47		15.5	GALAXY
ZWG 502.115	01 32 18.	+ 32 27		15.5	GALAXY
MCG+06-04-040	01 32 18.	+ 35 45 30.	42	16.	GALAXY
OCL 0328	01 32 18.	+ 61 02	660	9.0	OPEN STAR CLUSTER
PHL 7359	01 32 18.	- 03 18		18.1	BLUE STELLAR OBJECT
PHL 7357	01 32 18.	- 04 02		18.6	BLUE STELLAR OBJECT
PHL 7358	01 32 18.	- 05 10		17.9	BLUE STELLAR OBJECT
PHL 7360	01 32 18.	- 09 24		18.0	BLUE STELLAR OBJECT
PHL 1047	01 32 18.	- 09 26		18.5	BLUE STELLAR OBJECT
BCL 0756	01 32 18.0	+ 30 26 24.	39		HII REGION IN M33
KN 14.333	01 32 19.4	+ 30 46 59.			NEBULA
KN 14.332	01 32 20.0	+ 32 27 19.			NEBULA
PHL 7362	01 32 24.	+ 00 04		17.0	BLUE STELLAR OBJECT
ZWG 386.016	01 32 24.	+ 01 05		15.6	GALAXY
UGC 01130	01 32 24.	+ 01 05	84	15.6	GALAXY S
MCG+00-05-011	01 32 24.	+ 01 05	42	15.	GALAXY
MCG+05-04-072	01 32 24.	+ 32 26 30.	36	15.5	GALAXY
ZWG 502.116	01 32 24.	+ 32 48		15.2	GALAXY
ZWG 521.051	01 32 24.	+ 34 14		15.4	GALAXY
UGC 01131	01 32 24.	+ 34 14	84	15.4	GALAXY
ZWG 552.001	01 32 24.	+ 47 18		15.2	GALAXY
UGC 01132	01 32 24.	+ 47 18	114	15.2	GALAXY
MCG+08-04-001	01 32 24.	+ 47 18	90	15.	GALAXY
PHL 7363	01 32 24.	- 04 47		18.5	BLUE STELLAR OBJECT
PHL 1048	01 32 24.	- 12 10		18.0	BLUE STELLAR OBJECT
PHL 7361	01 32 24.	- 13 19		16.8	BLUE STELLAR OBJECT
PHL 3447	01 32 24.	- 13 20		18.6	BLUE STELLAR OBJECT
KN 14.334	01 32 24.8	+ 32 39 45.			NEBULA
KN 14.335	01 32 25.7	+ 32 37 45.			NEBULA
KN 14.336	01 32 26.7	+ 32 47 40.			NEBULA
MCG+01-05-005	01 32 27.	+ 04 06	120	16.	GALAXY
KN 14.337	01 32 27.8	+ 32 21 21.			NEBULA
LIN.CL 109	01 32 29.	- 74 25 01.	204	15.1	STAR CLUSTER IN SMC
UGC 01133	01 32 30.	+ 04 06	240	16.5	GALAXY DWRF IR
PHL 1049	01 32 30.	+ 07 42		18.2	BLUE STELLAR OBJECT
BC PHL1049	01 32 30.	+ 07 42		17.26	QUASI-STELLAR OBJECT
3ZW 031	01 32 30.	+ 17 27			COMPACT GALAXY
UGC 01134	01 32 30.	+ 36 04	66	17.	GALAXY DWRF SP
MCG+14-02-003	01 32 30.	+ 84 12	18	17.	GALAXY
MCG+14-02-004	01 32 30.	+ 84 13	30	17.	GALAXY
ZC 0132.5-0126	01 32 30.	- 01 26	740		CLUSTER OF GALAXIES
KEEL 031	01 32 31.4	+ 15 22 01.			NEBULA
MCG+06-04-041	01 32 33.	+ 34 13	54	15.	GALAXY
RNGC 0615	01 32 33.	- 07 35		12.5	GALAXY
RNGC 0608	01 32 34.	+ 33 25		14.0	GALAXY
RNGC 0619	01 32 35.	- 36 44			GALAXY
KEEL 032	01 32 35.4	+ 15 58 49.			NEBULA
PHL 7365	01 32 36.	+ 03 50		18.4	BLUE STELLAR OBJECT
PHL 1050	01 32 36.	+ 08 12		17.7	BLUE STELLAR OBJECT
PHL 3448	01 32 36.	+ 08 20		17.7	BLUE STELLAR OBJECT

OBJECT NAME	RIGHT ASCEN.	DECLINATION	DIAM.	MAGN.	TYPE OF OBJECT
ZWG 502.117	01 32 36.	+ 33 25		14.0	GALAXY
UGC 01135	01 32 36.	+ 33 25	54	14.0	GALAXY
KARA.72 038A	01 32 36.	+ 33 25	60	14.0	PART OF DOUBLE GALAXY
ZWG 521.052	01 32 36.	+ 35 25		15.5	GALAXY
UGC 01136	01 32 36.	+ 41 17	90	17.	GALAXY DWRF IR
PHL 1051	01 32 36.	- 01 46		18.7	BLUE STELLAR OBJECT
PHL 7366	01 32 36.	- 07 52		18.2	BLUE STELLAR OBJECT
PHL 7367	01 32 36.	- 11 12		18.2	BLUE STELLAR OBJECT
PHL 3449	01 32 36.	- 13 46		18.8	BLUE STELLAR OBJECT
PHL 7364	01 32 36.	- 15 32		17.2	BLUE STELLAR OBJECT
PHL 1052	01 32 36.	- 15 50		17.6	BLUE STELLAR OBJECT
MCG-06-04-050	01 32 36.	- 36 25	66	13.	GALAXY
KN 14.338	01 32 36.9	+ 33 24 14.			NEBULA
KN 14.339	01 32 37.0	+ 30 40 31.			NEBULA
KN 14.340	01 32 37.3	+ 30 57 48.			NEBULA
KN 14.341	01 32 38.7	+ 30 58 05.			NEBULA
ARC 0213	01 32 40.	+ 20 22		17.7	RICH CLUSTER OF GALAXIES
UGC 01137	01 32 42.	+ 05 14	78	18.	GALAXY DWARF
PHL 1053	01 32 42.	+ 06 38		18.7	BLUE STELLAR OBJECT
MCG+05-04-073	01 32 42.	+ 33 24	48	14.	GALAXY
PHL 7368	01 32 42.	- 05 08		17.1	BLUE STELLAR OBJECT
PHL 7370	01 32 42.	- 07 46		18.1	BLUE STELLAR OBJECT
PHL 7371	01 32 42.	- 11 50		18.3	BLUE STELLAR OBJECT
PHL 7369	01 32 42.	- 13 22		16.8	BLUE STELLAR OBJECT
MCG-06-04-051	01 32 42.	- 36 47	48	14.	GALAXY
MCG+01-05-006	01 32 45.	+ 04 37	72	15.	GALAXY
MCG+01-05-008	01 32 45.	- 07 37	228	12.	GALAXY
RNGC 0623	01 32 47.	- 36 45			GALAXY
PHL 7374	01 32 48.	+ 00 01		18.0	BLUE STELLAR OBJECT
PHL 7373	01 32 48.	+ 03 30		17.7	BLUE STELLAR OBJECT
UGC 01138	01 32 48.	+ 04 37	90	16.0	GALAXY IRR
MCG+02-05-006	01 32 48.	+ 11 41	36	14.5	GALAXY
ZC 0132.8+2022	01 32 48.	+ 20 22	810		CLUSTER OF GALAXIES
ZWG 521.053	01 32 48.	+ 35 13		15.2	GALAXY
VB 052	01 32 48.	+ 65 09	376		STELLAR RING
ISS 0011	01 32 48.	+ 65 15	264		STELLAR RING
PHL 7376	01 32 48.	- 00 15		17.2	BLUE STELLAR OBJECT
PHL 1054	01 32 48.	- 02 30		18.1	BLUE STELLAR OBJECT
PHL 7372	01 32 48.	- 05 52		16.6	BLUE STELLAR OBJECT
PHL 7375	01 32 48.	- 12 58		14.1	BLUE STELLAR OBJECT
PHL 3450	01 32 48.	- 13 07		18.8	BLUE STELLAR OBJECT
KN 14.342	01 32 53.1	+ 30 09 48.			NEBULA
PHL 3451	01 32 54.	+ 01 34		18.0	BLUE STELLAR OBJECT
ZWG 386.017	01 32 54.	+ 03 15		15.5	GALAXY
MCG+00-05-012	01 32 54.	+ 03 15	36	15.	GALAXY
ZWG 437.005	01 32 54.	+ 11 41		15.0	GALAXY
UGC 01139	01 32 54.	+ 11 41	66	15.0	GALAXY SB
MCG+06-04-042	01 32 54.	+ 35 12	24	15.5	GALAXY
ZC 0132.9+3638	01 32 54.	+ 36 38	1280		CLUSTER OF GALAXIES
PHL 7377	01 32 54.	- 10 49		14.6	BLUE STELLAR OBJECT
PHL 7378	01 32 54.	- 13 47		16.9	BLUE STELLAR OBJECT
MCG-06-04-052	01 32 54.	- 36 47	72	13.5	GALAXY
RNGC 0625	01 32 55.	- 41 41		12.5	GALAXY
KN 14.345	01 32 56.8	+ 30 11 40.			NEBULA
KN 14.343	01 32 56.9	+ 33 29 41.			NEBULA
RNGC 0614	01 32 58.	+ 33 26		14.0	GALAXY
KN 14.346	01 32 58.5	+ 31 34 34.			NEBULA
KN 14.344	01 32 58.6	+ 33 27 09.			NEBULA
KN 14.349	01 32 59.4	+ 31 36 56.			NEBULA
KN 14.347	01 32 59.4	+ 32 02 15.			NEBULA
KN 14.350	01 32 59.6	+ 29 49 14.			NEBULA
PHL 7379	01 33 00.	+ 03 38		17.8	BLUE STELLAR OBJECT
MCG+05-04-074	01 33 00.	+ 29 18	24	17.	GALAXY
ZWG 502.118	01 33 00.	+ 33 26		13.9	GALAXY
UGC 01140	01 33 00.	+ 33 26	96	13.9	GALAXY S0?
KARA.72 038B	01 33 00.	+ 33 26	84	13.9	PART OF DOUBLE GALAXY
LDN 1330	01 33 00.	+ 65 00	540		DARK NEBULA
UGC 01141	01 33 00.	+ 80 24	90	16.0	GALAXY Sb
PHL 1055	01 33 00.	- 04 42		16.9	BLUE STELLAR OBJECT
PHL 1056	01 33 00.	- 08 08		18.4	BLUE STELLAR OBJECT
PHL 1057	01 33 00.	- 08 28		18.2	BLUE STELLAR OBJECT
ARC 0215	01 33 00.	- 23 43		17.5	RICH CLUSTER OF GALAXIES
RNGC 0626	01 33 00.	- 39 24			GALAXY
KN 14.348	01 33 00.9	+ 33 25 31.			NEBULA
KEEL 033	01 33 01.7	+ 15 33 01.			NEBULA
KN 14.351	01 33 04.2	+ 29 18 58.			NEBULA
KEEL 034	01 33 05.6	+ 15 35 51.			NEBULA
PHL 3452	01 33 06.	+ 02 16		18.6	BLUE STELLAR OBJECT
PHL 7380	01 33 06.	+ 07 24		18.4	BLUE STELLAR OBJECT
MCG+05-04-075	01 33 06.	+ 33 26	60	14.	GALAXY
ZWG 386.018	01 33 06.	- 00 14		15.5	GALAXY
MCG+00-05-013	01 33 06.	- 00 15	30	14.	GALAXY
PHL 3453	01 33 06.	- 08 12		18.6	BLUE STELLAR OBJECT
MCG-02-05-008	01 33 06.	- 14 16	30	15.	GALAXY
PHL 7381	01 33 06.	- 31 02		18.4	BLUE STELLAR OBJECT
MCG-06-04-053	01 33 06.	- 33 03	15	15.	GALAXY
RNGC 0616	01 33 10.	+ 33 30			NON-EXISTENT OBJECT
KN 14.352	01 33 10.8	+ 31 30 59.			NEBULA
PHL 7383	01 33 12.	+ 00 11		18.9	BLUE STELLAR OBJECT
PHL 3454	01 33 12.	+ 01 50		18.9	BLUE STELLAR OBJECT
PHL 1058	01 33 12.	+ 07 32		16.6	BLUE STELLAR OBJECT
ZWG 521.054	01 33 12.	+ 35 40		15.4	GALAXY
PHL 7386	01 33 12.	- 00 52		16.1	BLUE STELLAR OBJECT
PHL 3455	01 33 12.	- 01 13		18.9	BLUE STELLAR OBJECT
PHL 7384	01 33 12.	- 01 44		18.3	BLUE STELLAR OBJECT
PHL 7382	01 33 12.	- 03 00		16.6	BLUE STELLAR OBJECT
PHL 7385	01 33 12.	- 09 24		18.6	BLUE STELLAR OBJECT
PHL 3457	01 33 12.	- 09 45		18.8	BLUE STELLAR OBJECT
MCG-02-05-009	01 33 12.	- 10 15	30	15.	GALAXY
PHL 3456	01 33 12.	- 11 40		18.5	BLUE STELLAR OBJECT
MCG+08-04-002	01 33 15.	+ 47 10	42	15.	GALAXY
RNGC 0624	01 33 16.	- 10 16		14.0	GALAXY
KEEL 035	01 33 16.9	+ 15 47 25.			NEBULA
PHL 3459	01 33 18.	+ 01 49		19.0	BLUE STELLAR OBJECT
ZWG 437.006	01 33 18.	+ 14 19		15.6	GALAXY
ZWG 552.002	01 33 18.	+ 47 08		15.7	GALAXY
UGC 01142	01 33 18.	+ 47 08	90	15.7	GALAXY COMPACT
PHL 7387	01 33 18.	- 01 35		16.4	BLUE STELLAR OBJECT
PHL 1059	01 33 18.	- 02 41		18.3	BLUE STELLAR OBJECT
PHL 3458	01 33 18.	- 04 02		18.7	BLUE STELLAR OBJECT
PHL 7389	01 33 18.	- 05 23		18.5	BLUE STELLAR OBJECT
MCG-02-05-010	01 33 18.	- 10 16	72	14.	GALAXY
PHL 7388	01 33 18.	- 12 45		17.3	BLUE STELLAR OBJECT
LIN.CL 110	01 33 18.	- 73 07 21.	678	14.3	STAR CLUSTER IN SMC
HOHE 5	01 33 18.	- 73 07 24.		16.0	STAR CLUSTER IN SMC
MCG+06-04-043	01 33 21.	+ 33 56	21	16.	GALAXY
KN 14.353	01 33 21.9	+ 33 10 38.			NEBULA
RNGC 0618	01 33 22.	+ 33 08			NON-EXISTENT OBJECT
PHL 7390	01 33 24.	+ 00 12		18.6	BLUE STELLAR OBJECT
PHL 7392	01 33 24.	+ 00 25		18.9	BLUE STELLAR OBJECT
ZWG 386.020	01 33 24.	+ 00 25		14.1	GALAXY
MRK 571	01 33 24.	+ 00 25	13	16.	GALAXY WITH UV CONTINUUM
UGC 01143	01 33 24.	+ 00 25	132	14.1	GALAXY SBb
MCG+00-05-014	01 33 24.	+ 00 25	90	13.	GALAXY
KARA.73B 0056	01 33 24.	+ 00 25	120	14.1	ISOLATED GALAXY S
PHL 3460	01 33 24.	+ 03 30		18.3	BLUE STELLAR OBJECT
ZWG 412.004	01 33 24.	+ 05 05		15.7	GALAXY
MCG+02-05-007	01 33 24.	+ 11 27	84	15.	GALAXY
MCG+06-04-044	01 33 24.	+ 35 40	36	15.	GALAXY
PHL 3461	01 33 24.	- 01 55		18.6	BLUE STELLAR OBJECT
ZWG 386.019	01 33 24.	- 02 29		15.7	GALAXY
PHL 3462	01 33 24.	- 05 18		18.3	BLUE STELLAR OBJECT
SN 1960Q	01 33 24.	- 05 45		17.5	SUPERNOVA
PHL 7391	01 33 24.	- 09 40		17.9	BLUE STELLAR OBJECT
MCG-02-05-011	01 33 24.	- 10 12	42	15.5	GALAXY
PHL 3463	01 33 24.	- 11 01		18.1	BLUE STELLAR OBJECT
PHL 1060	01 33 24.	- 31 44		18.0	BLUE STELLAR OBJECT
RNGC 0630	01 33 24.	- 39 37			GALAXY
RNGC 0622	01 33 25.	+ 00 25		14.0	GALAXY
KN 14.354	01 33 26.4	+ 32 58 24.			NEBULA
KN 14.357	01 33 26.6	+ 27 54 15.			NEBULA
KN 14.355	01 33 28.3	+ 33 09 01.			NEBULA
KN 14.356	01 33 28.4	+ 33 07 52.			NEBULA
REIN 2.008	01 33 29.41	+ 33 08 18.9			NEBULA
UGC 01144	01 33 30.	+ 11 26	108	16.0	GALAXY Sc
ZWG 460.012	01 33 30.	+ 19 51		15.6	GALAXY
MCG+04-04-014	01 33 30.	+ 23 00	24	17.	GALAXY
MCG+14-02-005	01 33 30.	+ 83 40	48	14.	GALAXY
PHL 7393	01 33 30.	- 07 08		16.6	BLUE STELLAR OBJECT
PHL 7394	01 33 30.	- 10 48		18.4	BLUE STELLAR OBJECT
PHL 3464	01 33 30.	- 14 42		18.8	BLUE STELLAR OBJECT
MCG-06-04-054	01 33 30.	- 36 26	36	14.5	GALAXY
KN 14.358	01 33 31.2	+ 31 00 46.			NEBULA
KN 14.359	01 33 34.6	+ 32 07 33.			NEBULA
KEEL 036	01 33 35.8	+ 15 58 23.			NEBULA
PHL 3469	01 33 36.	+ 00 14		18.2	BLUE STELLAR OBJECT
PHL 3468	01 33 36.	+ 03 58		18.4	BLUE STELLAR OBJECT
PHL 3467	01 33 36.	+ 08 42		17.7	BLUE STELLAR OBJECT
PHL 3466	01 33 36.	+ 08 56		17.1	BLUE STELLAR OBJECT
ZWG 460.013	01 33 36.	+ 17 34		15.5	GALAXY
MCG+07-04-005	01 33 36.	+ 39 39	60	15.	GALAXY
ZWG 537.015	01 33 36.	+ 39 40		15.2	GALAXY
UGC 01145	01 33 36.	+ 39 40	96	15.2	GALAXY SBb
PHL 7396	01 33 36.	- 02 51		17.8	BLUE STELLAR OBJECT
PHL 7395	01 33 36.	- 07 04		18.7	BLUE STELLAR OBJECT
PHL 3465	01 33 36.	- 19 44		15.9	BLUE STELLAR OBJECT
KN 14.360	01 33 36.0	+ 31 20 17.			NEBULA
REIN 2.009	01 33 38.35	+ 33 10 20.4			NEBULA
SHB 035	01 33 40.	+ 20 42 18.		18.1	QUASI-STELLAR OBJECT
BC 3CR47	01 33 40.30	+ 20 42 16.0		18.10	QUASI-STELLAR OBJECT
REIN 2.010	01 33 40.41	+ 33 12 52.9			NEBULA
PHL 3470	01 33 42.	+ 01 12		18.9	BLUE STELLAR OBJECT
UGC 01146	01 33 42.	+ 10 17	60	17.	GALAXY
OCL 0325	01 33 42.	+ 64 18	360	12.7	OPEN STAR CLUSTER
PHL 7397	01 33 42.	- 03 21		17.8	BLUE STELLAR OBJECT
MCG-02-05-012	01 33 42.	- 11 17 30.	60	14.5	GALAXY
PHL 7398	01 33 42.	- 12 50		18.0	BLUE STELLAR OBJECT
KN 14.361	01 33 43.4	+ 29 49 07.			NEBULA
RNGC 0609	01 33 45.	+ 64 17		12.5	OPEN CLUSTER
KEEL 037	01 33 46.1	+ 15 59 00.			NEBULA
ZWG 412.005	01 33 48.	+ 03 30		15.3	GALAXY
ZWG 386.021	01 33 48.	+ 03 30		15.3	GALAXY
VB 049	01 33 48.	+ 66 35	806		STELLAR RING
PHL 1061	01 33 48.	- 02 50		18.0	BLUE STELLAR OBJECT
PHL 7399	01 33 48.	- 06 18		18.2	BLUE STELLAR OBJECT
PHL 7400	01 33 48.	- 06 50		18.3	BLUE STELLAR OBJECT
PHL 7402	01 33 48.	- 07 36		18.7	BLUE STELLAR OBJECT
PHL 7401	01 33 48.	- 09 18		18.1	BLUE STELLAR OBJECT
PHL 1062	01 33 48.	- 11 34		14.0	BLUE STELLAR OBJECT
PHL 3471	01 33 48.	- 12 42		17.4	BLUE STELLAR OBJECT
PHL 7403	01 33 48.	- 13 42		18.6	BLUE STELLAR OBJECT
MCG-04-04-026	01 33 48.	- 22 47	30	16.	GALAXY
KN 14.362	01 33 50.5	+ 32 49 56.			NEBULA
RNGC 0621	01 33 52.	+ 35 16		14.0	GALAXY
PHL 7405	01 33 54.	+ 07 46		18.5	BLUE STELLAR OBJECT
MCG+03-05-011	01 33 54.	+ 15 31	660	9.	GALAXY
4ZW 054	01 33 54.	+ 35 16			COMPACT GALAXY
ZWG 521.055	01 33 54.	+ 35 16		14.2	GALAXY
UGC 01147	01 33 54.	+ 35 16	108	14.2	GALAXY SBO
PHL 7406	01 33 54.	- 00 24		18.6	BLUE STELLAR OBJECT
PHL 1063	01 33 54.	- 06 30		17.7	BLUE STELLAR OBJECT
PHL 7404	01 33 54.	- 08 44		16.8	BLUE STELLAR OBJECT
MCG-02-05-013	01 33 54.	- 13 57 30.	84	14.5	GALAXY
KN 14.364	01 33 54.4	+ 31 44 32.			NEBULA
KN 14.363	01 33 54.6	+ 32 45 44.			NEBULA
KN 14.365	01 33 56.7	+ 32 49 28.			NEBULA
KEEL 038	01 33 56.9	+ 15 46 06.			NEBULA
RNGC 0628	01 33 57.	+ 15 32		10.5	GALAXY
KN 14.366	01 33 58.0	+ 31 38 08.			NEBULA
KN 14.367	01 33 58.4	+ 31 44 09.			NEBULA
PHL 3472	01 34 00.	+ 05 10		16.9	BLUE STELLAR OBJECT
ZWG 460.014	01 34 00.	+ 15 32		10.5	GALAXY
UGC 01149	01 34 00.	+ 15 32	720	10.5	GALAXY Sc
5ZW 081	01 34 00.	+ 42 04			COMPACT GALAXY
ZWG 537.016	01 34 00.	+ 42 04		13.9	GALAXY
UGC 01150	01 34 00.	+ 42 04	60	13.9	GALAXY PECULR
MCG+07-04-006	01 34 00.	+ 42 04	60	13.	GALAXY
ZWG 361.003	01 34 00.	+ 83 43		15.5	GALAXY
ZWG 360.006	01 34 00.	+ 83 43		15.5	GALAXY
UGC 01148	01 34 00.	+ 83 43	72	15.5	GALAXY
PHL 7407	01 34 00.	- 03 18		17.1	BLUE STELLAR OBJECT
PHL 1064	01 34 00.	- 07 56		18.3	BLUE STELLAR OBJECT
MCG-06-04-055	01 34 00.	- 36 35	36	15.	GALAXY
RNGC 0620	01 34 01.	+ 42 04		14.0	GALAXY
KN 14.368	01 34 02.8	+ 31 38 17.			NEBULA
MCG+06-04-045	01 34 03.	+ 35 15	90	13.5	GALAXY
KN 14.369	01 34 05.7	+ 33 15 53.			NEBULA
PHL 7409	01 34 06.	+ 00 56		18.3	BLUE STELLAR OBJECT
PHL 3473	01 34 06.	+ 04 57		18.0	BLUE STELLAR OBJECT
PHL 7408	01 34 06.	+ 07 36		18.2	BLUE STELLAR OBJECT
MCG+05-04-076	01 34 06.	+ 31 43	54	15.	GALAXY
ZWG 502.119	01 34 06.	+ 31 44		14.6	GALAXY
UGC 01152	01 34 06.	+ 31 44	72	14.6	GALAXY Sb
ZWG 386.022	01 34 06.	- 01 38		15.1	GALAXY
UGC 01151	01 34 06.	- 01 38	66	15.1	GALAXY E-S0
MCG+00-05-015	01 34 06.	- 01 39	18	15.5	GALAXY

OBJECT NAME	RIGHT ASCEN.	DECLINATION	DIAM.	MAGN.	TYPE OF OBJECT
ARC 0218	01 34 06.	- 11 01		17.8	RICH CLUSTER OF GALAXIES
KN 14.370	01 34 06.0	+ 31 43 54.			NEBULA
KEEL 039	01 34 06.8	+ 15 59 55.			NEBULA
ARC 0217	01 34 08.	- 08 19		17.0	RICH CLUSTER OF GALAXIES
KN 14.371	01 34 08.1	+ 31 38 03.			NEBULA
MCG+01-05-007	01 34 09.	+ 05 35	18	14.5	GALAXY
RNGC 0627	01 34 10.	+ 33 19			NON-EXISTENT OBJECT
ARC 0216	01 34 10.	- 06 41		17.2	RICH CLUSTER OF GALAXIES
RNGC 0633	01 34 11.	- 37 34			GALAXY
KN 14.372	01 34 11.9	+ 31 39 37.			NEBULA
PHL 3474	01 34 12.	+ 00 50		18.6	BLUE STELLAR OBJECT
ZWG 412.006	01 34 12.	+ 05 35		15.0	GALAXY
RNGC 0631	01 34 12.	+ 05 35		15.0	GALAXY
UGC 01153	01 34 12.	+ 05 35	102	15.0	GALAXY E
PHL 1065	01 34 12.	+ 07 00		16.6	BLUE STELLAR OBJECT
ZC 0134.2+4050	01 34 12.	+ 40 50	1810		CLUSTER OF GALAXIES
PHL 3475	01 34 12.	- 05 36		18.8	BLUE STELLAR OBJECT
PHL 7410	01 34 12.	- 06 42		16.9	BLUE STELLAR OBJECT
PHL 7412	01 34 12.	- 10 06		18.3	BLUE STELLAR OBJECT
PHL 7411	01 34 12.	- 10 46		17.8	BLUE STELLAR OBJECT
MCG-06-04-056	01 34 12.	- 37 36	48	13.5	GALAXY
MCG-06-04-057	01 34 12.	- 37 37	24	15.	GALAXY
PK130-11.1	01 34 12.9	+ 50 12 57.	6		PLANETARY NEBULA
LIN.CL 111	01 34 17.	- 75 49 05.	1224	13.4	STAR CLUSTER IN SMC
KN 14.373	01 34 17.5	+ 33 23 18.			NEBULA
PHL 3478	01 34 18.	+ 06 40		18.4	BLUE STELLAR OBJECT
PHL 1066	01 34 18.	- 02 55		18.1	BLUE STELLAR OBJECT
PHL 3476	01 34 18.	- 03 50		18.4	BLUE STELLAR OBJECT
PHL 7413	01 34 18.	- 04 42		18.0	BLUE STELLAR OBJECT
PHL 7414	01 34 18.	- 05 07		18.1	BLUE STELLAR OBJECT
PHL 7415	01 34 18.	- 13 46		13.6	BLUE STELLAR OBJECT
MCG-02-05-014	01 34 18.	- 14 10	30	15.5	GALAXY
MCG-02-05-015	01 34 18.	- 14 19	36	15.	GALAXY
PHL 3477	01 34 18.	- 21 13		18.6	BLUE STELLAR OBJECT
MCG-06-04-058	01 34 18.	- 36 40	24	15.	GALAXY
RNGC 0643	01 34 18.	- 75 49		13.5	GLOBULAR CLUSTER IN LMC
MCG-02-05-016	01 34 21.	- 14 10 30.	24	15.5	GALAXY
PHL 7416	01 34 24.	+ 01 02		15.0	BLUE STELLAR OBJECT
PHL 7420	01 34 24.	+ 01 28		18.9	BLUE STELLAR OBJECT
PHL 7419	01 34 24.	+ 05 46		18.5	BLUE STELLAR OBJECT
PHL 7418	01 34 24.	+ 07 13		18.5	BLUE STELLAR OBJECT
ZC 0134.4+1140	01 34 24.	+ 11 40	3020		CLUSTER OF GALAXIES
MCG+04-04-015	01 34 24.	+ 23 00	24	17.	GALAXY
MCG+05-04-077	01 34 24.	+ 28 36	42	14.5	GALAXY
ZWG 502.120	01 34 24.	+ 28 38		14.4	GALAXY
UGC 01154	01 34 24.	+ 28 38	54	14.4	GALAXY S?
6ZW 002	01 34 24.	+ 30 28			COMPACT GALAXY
ZC 0134.4+3553	01 34 24.	+ 35 53	1950		CLUSTER OF GALAXIES
5ZW 082	01 34 24.	+ 42 36			COMPACT GALAXY
PHL 3479	01 34 24.	- 01 02		18.8	BLUE STELLAR OBJECT
PHL 3480	01 34 24.	- 01 30		18.3	BLUE STELLAR OBJECT
PHL 1067	01 34 24.	- 02 35		18.5	BLUE STELLAR OBJECT
MCG+01-05-009	01 34 24.	- 05 17 30.	30	15.5	GALAXY
PHL 3481	01 34 24.	- 05 34		18.7	BLUE STELLAR OBJECT
PHL 7421	01 34 24.	- 05 55		18.6	BLUE STELLAR OBJECT
PHL 7417	01 34 24.	- 08 11		18.6	BLUE STELLAR OBJECT
PHL 7422	01 34 24.	- 12 07		18.0	BLUE STELLAR OBJECT
PHL 7423	01 34 24.	- 13 09		18.8	BLUE STELLAR OBJECT
PHL 3482	01 34 24.	- 13 34		18.9	BLUE STELLAR OBJECT
PHL 3483	01 34 24.	- 14 28		18.2	BLUE STELLAR OBJECT
PHL 1068	01 34 24.	- 19 56		17.0	BLUE STELLAR OBJECT
PHL 7424	01 34 24.	- 24 50		18.5	BLUE STELLAR OBJECT
KEEL 041	01 34 25.6	+ 15 19 39.			NEBULA
KEEL 040	01 34 26.0	+ 16 02 10.			NEBULA
MCG+01-05-009	01 34 27.	+ 04 38	36	14.5	GALAXY
MCG+01-05-008	01 34 27.	+ 05 26	36	15.	GALAXY
KN 14.374	01 34 28.4	+ 28 38 11.			NEBULA
PHL 7425	01 34 30.	+ 07 38		17.9	BLUE STELLAR OBJECT
VB 137	01 34 30.	+ 60 09	248		STELLAR RING
MCG-02-05-018	01 34 30.	- 14 12 30.	24	15.	GALAXY
MCG-02-05-017	01 34 30.	- 14 16	48	14.5	GALAXY
PHL 3484	01 34 30.	- 14 26		18.0	BLUE STELLAR OBJECT
KN 14.375	01 34 34.4	+ 31 15 28.			NEBULA
PHL 7429	01 34 36.	+ 00 38		18.6	BLUE STELLAR OBJECT
ZWG 412.007	01 34 36.	+ 04 38		14.5	GALAXY
UGC 01155	01 34 36.	+ 04 38	60	14.5	GALAXY
PHL 7426	01 34 36.	+ 06 10		18.3	BLUE STELLAR OBJECT
PHL 3489	01 34 36.	+ 07 19		18.7	BLUE STELLAR OBJECT
PHL 7428	01 34 36.	+ 07 30		18.1	BLUE STELLAR OBJECT
ZC 0134.6+1246	01 34 36.	+ 12 46	2890		CLUSTER OF GALAXIES
UGC 01156	01 34 36.	+ 14 13	72	17.	GALAXY Sc
PHL 1069	01 34 36.	- 02 23		17.0	BLUE STELLAR OBJECT
PHL 3485	01 34 36.	- 02 41		18.8	BLUE STELLAR OBJECT
PHL 3486	01 34 36.	- 03 42		18.6	BLUE STELLAR OBJECT
PHL 7427	01 34 36.	- 05 26		16.1	BLUE STELLAR OBJECT
PHL 7430	01 34 36.	- 05 34		18.6	BLUE STELLAR OBJECT
PHL 3487	01 34 36.	- 08 35		18.4	BLUE STELLAR OBJECT
PHL 7431	01 34 36.	- 09 26		18.6	BLUE STELLAR OBJECT
MCG-02-05-019	01 34 36.	- 12 34	30	15.	GALAXY
PHL 3488	01 34 36.	- 12 39		18.6	BLUE STELLAR OBJECT
PHL 3490	01 34 36.	- 12 44		18.5	BLUE STELLAR OBJECT
ARC 0220	01 34 40.	+ 07 41		17.5	RICH CLUSTER OF GALAXIES
KN 14.376	01 34 40.1	+ 31 50 46.			NEBULA
PHL 3494	01 34 42.	+ 00 03		19.0	BLUE STELLAR OBJECT
ZWG 412.008	01 34 42.	+ 05 37		13.5	GALAXY
RNGC 0632	01 34 42.	+ 05 37		13.5	GALAXY
UGC 01157	01 34 42.	+ 05 37	102	13.5	GALAXY S0
MCG+01-05-010	01 34 42.	+ 05 37 30.	48	13.5	GALAXY
ARC 0219	01 34 42.	+ 08 54		17.2	RICH CLUSTER OF GALAXIES
6ZW 003	01 34 42.	+ 33 28			COMPACT GALAXY
ZWG 558.002	01 34 42.	+ 53 25		15.7	GALAXY
PHL 3491	01 34 42.	- 00 35		18.6	BLUE STELLAR OBJECT
PHL 7433	01 34 42.	- 02 50		18.6	BLUE STELLAR OBJECT
PHL 3492	01 34 42.	- 03 12		18.6	BLUE STELLAR OBJECT
PHL 7432	01 34 42.	- 09 56		18.1	BLUE STELLAR OBJECT
PHL 3493	01 34 42.	- 10 48		18.4	BLUE STELLAR OBJECT
MCG-03-04-009	01 34 42.	- 15 20	24	15.	GALAXY
KN 14.377	01 34 44.	+ 32 28 02.			NEBULA
KN 14.379	01 34 44.1	+ 28 01 32.			NEBULA
MCG-02-05-020	01 34 45.	- 09 27	18	15.	GALAXY
MCG-02-05-021	01 34 45.	- 13 54	30	15.5	GALAXY
KN 14.378	01 34 46.3	+ 32 11 42.			NEBULA
KEEL 042	01 34 46.9	+ 15 36 14.			NEBULA
ZC 0134.8+0242	01 34 48.	+ 02 42	3290		CLUSTER OF GALAXIES
PHL 1070	01 34 48.	+ 03 21		17.6	BLUE STELLAR OBJECT
BC PHL1070	01 34 48.	+ 03 21		17.6	QUASI-STELLAR OBJECT
SHB 036	01 34 48.	+ 03 21		17.6	QUASI-STELLAR OBJECT
ZC 0134.8+2257	01 34 48.	+ 22 57	1080		CLUSTER OF GALAXIES

OBJECT NAME	RIGHT ASCEN.	DECLINATION	DIAM.	MAGN.	TYPE OF OBJECT
6ZW 004	01 34 48.	+ 32 24			COMPACT GALAXY
UGC 01158	01 34 48.	+ 32 25	60	16.5	GALAXY S
ZWG 521.056	01 34 48.	+ 34 45		15.7	GALAXY
MCG+13-02-001	01 34 48.	+ 80 55	51	16.	GALAXY
ZC 0134.8-0045	01 34 48.	- 00 45	5170		CLUSTER OF GALAXIES
PHL 7436	01 34 48.	- 01 58		18.5	BLUE STELLAR OBJECT
PHL 7437	01 34 48.	- 04 42		18.8	BLUE STELLAR OBJECT
PHL 7435	01 34 48.	- 08 28		18.4	BLUE STELLAR OBJECT
MCG-02-05-022	01 34 48.	- 09 25	15	15.5	GALAXY
PHL 3495	01 34 48.	- 10 30		18.2	BLUE STELLAR OBJECT
PHL 7434	01 34 48.	- 23 04		17.4	BLUE STELLAR OBJECT
KN 14.380	01 34 48.7	+ 32 24 29.			NEBULA
KN 14.381	01 34 49.2	+ 32 24 05.			NEBULA
SHB 037	01 34 49.8	+ 32 54 18.		16.2	QUASI-STELLAR OBJECT
BC 3CR48	01 34 49.82	+ 32 54 20.2		16.20	QUASI-STELLAR OBJECT
KN 14.382	01 34 52.0	+ 31 11 20.			NEBULA
PHL 3496	01 34 54.	+ 06 16		17.4	BLUE STELLAR OBJECT
MCG-02-05-024	01 34 54.	- 09 31	24	15.5	GALAXY
MCG-02-05-023	01 34 54.	- 09 31	12	15.5	GALAXY
PHL 3497	01 34 54.	- 09 46		17.6	BLUE STELLAR OBJECT
PHL 3498	01 34 54.	- 12 56		18.0	BLUE STELLAR OBJECT
MCG-02-05-025	01 34 54.	- 13 48	36	15.	GALAXY
PHL 3499	01 34 54.	- 17 42		15.8	BLUE STELLAR OBJECT
ARC 0222	01 34 58.	- 13 15		17.6	RICH CLUSTER OF GALAXIES
PHL 7440	01 35 00.	+ 00 02		17.3	BLUE STELLAR OBJECT
PHL 3502	01 35 00.	+ 03 53		18.5	BLUE STELLAR OBJECT
MCG+01-05-012	01 35 00.	+ 06 40	36	14.5	GALAXY
MCG+01-05-011	01 35 00.	+ 06 40	24	17.	GALAXY
4ZW 055	01 35 00.	+ 23 53			COMPACT GALAXY
ZWG 481.012	01 35 00.	+ 27 17		15.7	GALAXY
VB 138	01 35 00.	+ 60 45	222		STELLAR RING
LDN 1331	01 35 00.	+ 64 15	960		DARK NEBULA
LDN 1328	01 35 00.	+ 67 00	4980		DARK NEBULA
ZWG 345.003	01 35 00.	+ 80 55		15.7	GALAXY
PHL 1071	01 35 00.	- 00 50		18.4	BLUE STELLAR OBJECT
PHL 3500	01 35 00.	- 02 52		18.6	BLUE STELLAR OBJECT
PHL 3503	01 35 00.	- 03 03		18.5	BLUE STELLAR OBJECT
PHL 7439	01 35 00.	- 03 27		17.3	BLUE STELLAR OBJECT
PHL 7438	01 35 00.	- 05 42		17.7	BLUE STELLAR OBJECT
MCG-02-05-026	01 35 00.	- 09 08	48	15.	GALAXY
PHL 7441	01 35 00.	- 11 16		18.0	BLUE STELLAR OBJECT
PHL 3501	01 35 00.	- 17 52		18.5	BLUE STELLAR OBJECT
ZWG 412.009	01 35 06.	+ 06 40		15.6	GALAXY
ZWG 521.057	01 35 06.	+ 34 35		15.5	GALAXY
VB 139	01 35 06.	+ 61 03	530		STELLAR RING
VB 054	01 35 06.	+ 63 54	504		STELLAR RING
VB 053	01 35 06.	+ 63 54	1007		STELLAR RING
ZWG 386.023	01 35 06.	- 00 12		15.5	GALAXY
UGC 01159	01 35 06.	- 00 12	66	15.5	GALAXY S
MCG+00-05-016	01 35 06.	- 00 13	36	15.	GALAXY
PHL 7442	01 35 06.	- 00 26		18.7	BLUE STELLAR OBJECT
PHL 7443	01 35 06.	- 10 08		18.3	BLUE STELLAR OBJECT
PHL 3505	01 35 06.	- 11 50		16.8	BLUE STELLAR OBJECT
PHL 3504	01 35 06.	- 11 50		16.8	BLUE STELLAR OBJECT
KN 14.383	01 35 07.0	+ 31 43 30.			NEBULA
KN 14.384	01 35 07.2	+ 30 21 55.			NEBULA
KN 14.387	01 35 07.9	+ 27 17 03.			NEBULA
MCG+01-05-010	01 35 09.	- 05 20 30.	18	15.	GALAXY
KN 14.385	01 35 09.0	+ 30 00 43.			NEBULA
KN 14.386	01 35 11.5	+ 32 14 23.			NEBULA
PHL 7446	01 35 12.	+ 01 58		18.4	BLUE STELLAR OBJECT
PHL 7448	01 35 12.	+ 04 40		17.7	BLUE STELLAR OBJECT
PHL 7447	01 35 12.	+ 05 19		17.0	BLUE STELLAR OBJECT
PHL 1072	01 35 12.	+ 05 39		18.3	BLUE STELLAR OBJECT
BC PHL1072	01 35 12.	+ 05 39		18.3	QUASI-STELLAR OBJECT
SHB 038	01 35 12.	+ 05 39		18.3	QUASI-STELLAR OBJECT
PHL 3508	01 35 12.	+ 09 08		18.5	BLUE STELLAR OBJECT
UGC 01160	01 35 12.	+ 32 14	114	16.0	GALAXY Sc
ZWG 503.001	01 35 12.	+ 32 15		15.7	GALAXY
ZWG 502.121	01 35 12.	+ 32 15		15.7	GALAXY
ZWG 521.058	01 35 12.	+ 36 42		15.4	GALAXY
PHL 1073	01 35 12.	- 05 02		18.2	BLUE STELLAR OBJECT
PHL 3510	01 35 12.	- 05 03		18.3	BLUE STELLAR OBJECT
PHL 3509	01 35 12.	- 05 13		18.2	BLUE STELLAR OBJECT
PHL 3511	01 35 12.	- 05 42		18.8	BLUE STELLAR OBJECT
PHL 3506	01 35 12.	- 07 08		17.8	BLUE STELLAR OBJECT
PHL 7444	01 35 12.	- 08 06		17.0	BLUE STELLAR OBJECT
PHL 7445	01 35 12.	- 08 42		17.7	BLUE STELLAR OBJECT
PHL 1074	01 35 12.	- 12 20		18.4	BLUE STELLAR OBJECT
MCG-02-05-027	01 35 12.	- 12 47	36	14.5	GALAXY
PHL 7449	01 35 12.	- 12 55		18.6	BLUE STELLAR OBJECT
PHL 3512	01 35 12.	- 21 12		18.4	BLUE STELLAR OBJECT
MCG-04-05-001	01 35 12.	- 21 29	54	15.	GALAXY
PHL 3507	01 35 12.	- 24 48		17.9	BLUE STELLAR OBJECT
IC 0144	01 35 14.	- 13 34 10.			NONSTELLAR OBJECT
MCG-02-05-028	01 35 15.	- 13 34	36	15.	GALAXY
KN 14.389	01 35 16.0	+ 28 03 13.			NEBULA
LIN.CL 112	01 35 17.	- 75 43 12.	204	16.2	STAR CLUSTER IN SMC
KN 14.390	01 35 17.2	+ 29 24 47.			NEBULA
KN 14.391	01 35 17.8	+ 29 25 13.			NEBULA
ZWG 386.024	01 35 18.	+ 02 03		15.5	GALAXY
KARA.73B 0057	01 35 18.	+ 02 03	24	15.5	ISOLATED GALAXY E
ZC 0135.3+1751	01 35 18.	+ 17 51	1210		CLUSTER OF GALAXIES
ZWG 521.059	01 35 18.	+ 36 21		15.0	GALAXY
UGC 01161	01 35 18.	+ 36 21	60	15.0	GALAXY COMPACT
MCG+07-04-007	01 35 18.	+ 41 23	90	15.	GALAXY
ZWG 537.017	01 35 18.	+ 41 24		15.1	GALAXY
UGC 01162	01 35 18.	+ 41 24	96	15.1	GALAXY SBb
ISS 0012	01 35 18.	+ 64 00	691		STELLAR RING
PHL 3513	01 35 18.	- 12 32		18.2	BLUE STELLAR OBJECT
KN 14.388	01 35 18.7	+ 32 15 37.			NEBULA
KN 14.392	01 35 19.3	+ 29 58 35.			NEBULA
RNGC 0634	01 35 22.	+ 35 07		14.0	GALAXY
KEEL 043	01 35 22.3	+ 15 38 41.			NEBULA
ZWG 386.025	01 35 24.	+ 00 45		15.5	GALAXY
UGC 01163	01 35 24.	+ 00 45	72	15.5	GALAXY COMPACT
MCG+00-05-017	01 35 24.	+ 00 45	15	15.	GALAXY
PHL 3516	01 35 24.	+ 01 49		18.9	BLUE STELLAR OBJECT
PHL 3514	01 35 24.	+ 06 14		18.1	BLUE STELLAR OBJECT
ZWG 521.060	01 35 24.	+ 35 07		14.0	GALAXY
UGC 01164	01 35 24.	+ 35 07	132	14.0	GALAXY Sa
MCG+06-04-046	01 35 24.	+ 36 41	57	15.	GALAXY
PHL 3515	01 35 24.	- 02 19		18.6	BLUE STELLAR OBJECT
PHL 3517	01 35 24.	- 05 04		18.5	BLUE STELLAR OBJECT
PHL 7450	01 35 24.	- 05 14		13.8	BLUE STELLAR OBJECT
MCG-03-05-010	01 35 24.	- 18 16	36	15.	GALAXY
PHL 7451	01 35 24.	- 30 18		18.3	BLUE STELLAR OBJECT
MCG-06-04-059	01 35 24.	- 34 12		14.	GALAXY

OBJECT NAME	RIGHT ASCEN.	DECLINATION	DIAM.	MAGN.	TYPE OF OBJECT
ARC 0221	01 35 26.	+ 17 53		17.7	RICH CLUSTER OF GALAXIES
IC 1719	01 35 26.	- 34 14 24.			NONSTELLAR OBJECT
ARC 0223	01 35 28.	- 13 03		17.6	RICH CLUSTER OF GALAXIES
BC 4C-05.06	01 35 29.	- 05 42 06.		18.25	QUASI-STELLAR OBJECT
SHB 039	01 35 29.	- 05 42 06.		18.3	QUASI-STELLAR OBJECT
SHB 039	01 35 29.	- 05 42 06.		18.3	QUASI-STELLAR OBJECT
KEEL 044	01 35 29.8	+ 15 27 46.			NEBULA
PHL 7453	01 35 30.	+ 01 21		18.6	BLUE STELLAR OBJECT
PHL 7458	01 35 30.	+ 01 54		18.9	BLUE STELLAR OBJECT
PHL 7454	01 35 30.	+ 04 08		18.2	BLUE STELLAR OBJECT
PHL 7455	01 35 30.	+ 04 44		18.2	BLUE STELLAR OBJECT
PHL 7456	01 35 30.	+ 04 58		18.1	BLUE STELLAR OBJECT
ZWG 460.015	01 35 30.	+ 20 10		15.5	GALAXY
MCG+06-04-048	01 35 30.	+ 35 06	120	14.	GALAXY
MCG+06-04-047	01 35 30.	+ 36 21	36	14.5	GALAXY
PHL 7457	01 35 30.	- 03 22		18.8	BLUE STELLAR OBJECT
PHL 7459	01 35 30.	- 06 12		18.0	BLUE STELLAR OBJECT
PHL 7452	01 35 30.	- 09 07		17.5	BLUE STELLAR OBJECT
IC 1718	01 35 35.	+ 33 05 32.			NONSTELLAR OBJECT
KN 14.393	01 35 35.4	+ 33 06 52.			NEBULA
PHL 7463	01 35 36.	+ 00 32		18.8	BLUE STELLAR OBJECT
PHL 1077	01 35 36.	+ 00 49		18.6	BLUE STELLAR OBJECT
PHL 3524	01 35 36.	+ 03 36		17.4	BLUE STELLAR OBJECT
PHL 3519	01 35 36.	+ 05 00		18.7	BLUE STELLAR OBJECT
PHL 3605	01 35 36.	+ 05 34		15.7	BLUE STELLAR OBJECT
PHL 7466	01 35 36.	+ 08 03		17.7	BLUE STELLAR OBJECT
PHL 1075	01 35 36.	+ 09 06		18.1	BLUE STELLAR OBJECT
ZWG 503.002	01 35 36.	+ 33 07		15.0	GALAXY
ZWG 502.122	01 35 36.	+ 33 07		15.0	GALAXY
PHL 1076	01 35 36.	- 01 24		17.4	BLUE STELLAR OBJECT
PHL 3520	01 35 36.	- 02 48		18.4	BLUE STELLAR OBJECT
PHL 7462	01 35 36.	- 03 56		17.4	BLUE STELLAR OBJECT
PHL 7461	01 35 36.	- 04 12		18.4	BLUE STELLAR OBJECT
PHL 3521	01 35 36.	- 04 33		18.9	BLUE STELLAR OBJECT
PHL 7464	01 35 36.	- 04 43		18.4	BLUE STELLAR OBJECT
PHL 1078	01 35 36.	- 05 44		18.6	BLUE STELLAR OBJECT
PHL 3518	01 35 36.	- 05 52		17.5	BLUE STELLAR OBJECT
PHL 3522	01 35 36.	- 05 57		18.8	BLUE STELLAR OBJECT
MCG+01-05-011	01 35 36.	- 06 00 30.	30	15.	GALAXY
MCG+01-05-012	01 35 36.	- 07 51	54	15.	GALAXY
PHL 7467	01 35 36.	- 09 34		18.4	BLUE STELLAR OBJECT
PHL 7465	01 35 36.	- 15 44		18.4	BLUE STELLAR OBJECT
PHL 7460	01 35 36.	- 24 31		18.5	BLUE STELLAR OBJECT
PHL 3523	01 35 36.	- 31 14		18.4	BLUE STELLAR OBJECT
MCG-06-04-060	01 35 36.	- 33 12	30	15.	GALAXY
KN 14.394	01 35 38.0	+ 28 53 20.			NEBULA
PHL 7470	01 35 42.	+ 04 52		17.5	BLUE STELLAR OBJECT
MCG+01-05-013	01 35 42.	+ 07 17	96	13.	GALAXY
PHL 3525	01 35 42.	+ 09 12		18.2	BLUE STELLAR OBJECT
ZWG 503.003	01 35 42.	+ 28 29		15.7	GALAXY
ZWG 502.123	01 35 42.	+ 28 29		15.7	GALAXY
UGC 01165	01 35 42.	+ 28 29	84	15.7	GALAXY Sc
KARA.73B 0058	01 35 42.	+ 28 29	78	15.7	ISOLATED GALAXY S
6ZW 005	01 35 42.	+ 30 31			COMPACT GALAXY
ZWG 521.061	01 35 42.	+ 34 45		14.0	GALAXY
UGC 01166	01 35 42.	+ 34 45	90	14.0	GALAXY S0-a
PHL 7469	01 35 42.	- 04 50		18.1	BLUE STELLAR OBJECT
PHL 7471	01 35 42.	- 07 02		18.6	BLUE STELLAR OBJECT
PHL 7468	01 35 42.	- 07 08		16.9	BLUE STELLAR OBJECT
KN 14.396	01 35 43.4	+ 28 28 23.			NEBULA
KN 14.397	01 35 43.7	+ 28 30 19.			NEBULA
KN 14.398	01 35 44.7	+ 28 38 58.			NEBULA
ARC 0224	01 35 45.	- 07 13		17.0	RICH CLUSTER OF GALAXIES
KN 14.395	01 35 45.3	+ 32 26 22.			NEBULA
KN 14.399	01 35 47.2	+ 28 29 30.			NEBULA
PHL 7474	01 35 48.	+ 01 46		17.9	BLUE STELLAR OBJECT
PHL 7472	01 35 48.	+ 02 26		18.2	BLUE STELLAR OBJECT
PHL 1079	01 35 48.	+ 03 23		12.4	BLUE STELLAR OBJECT
ZWG 412.010	01 35 48.	+ 07 16		14.8	GALAXY
UGC 01167	01 35 48.	+ 07 16	168	14.8	GALAXY Sc
KARA.73B 0059	01 35 48.	+ 07 16	192	14.8	ISOLATED GALAXY S
PHL 3527	01 35 48.	+ 07 41		17.9	BLUE STELLAR OBJECT
SHAH 043	01 35 48.	+ 08 16	210	16.5	GROUP OF COMPACT GALAXIES
ZC 0135.8+1830	01 35 48.	+ 18 30	3490		CLUSTER OF GALAXIES
ZWG 460.016	01 35 48.	+ 21 19		15.7	GALAXY
VB 140	01 35 48.	+ 58 02	168		STELLAR RING
PHL 3526	01 35 48.	- 04 32		18.9	BLUE STELLAR OBJECT
PHL 7473	01 35 48.	- 10 28		18.1	BLUE STELLAR OBJECT
PHL 7476	01 35 48.	- 12 26		18.3	BLUE STELLAR OBJECT
PHL 7475	01 35 48.	- 17 12		16.1	BLUE STELLAR OBJECT
RNGC 0646	01 35 48.	- 65 10			UNVERIFIED SOUTHERN OBJECT
MCG+06-04-049	01 35 51.	+ 34 44	78	14.5	GALAXY
KEEL 045	01 35 51.4	+ 15 32 07.			NEBULA
ZWG 460.017	01 35 54.	+ 20 33		15.7	GALAXY
MCG+00-05-018	01 35 54.	- 00 28	9	15.5	GALAXY
PHL 3528	01 35 54.	- 01 46		17.4	BLUE STELLAR OBJECT
PHL 7477	01 35 54.	- 02 32		18.5	BLUE STELLAR OBJECT
PHL 7478	01 35 54.	- 09 40		18.2	BLUE STELLAR OBJECT
MCG-02-05-029	01 35 54.	- 10 08	30	15.	GALAXY
RNGC 0635	01 35 54.	- 20 12			NON-EXISTENT OBJECT
MCG-06-04-061	01 35 54.	- 33 09	36	15.5	GALAXY
KEEL 046	01 35 54.2	+ 15 30 32.			NEBULA
KN 14.400	01 35 58.1	+ 31 32 14.			NEBULA
PHL 3529	01 36 00.	+ 08 41		17.8	BLUE STELLAR OBJECT
PHL 3531	01 36 00.	- 00 22		18.6	BLUE STELLAR OBJECT
PHL 7479	01 36 00.	- 03 03		17.0	BLUE STELLAR OBJECT
PHL 3532	01 36 00.	- 03 20		18.9	BLUE STELLAR OBJECT
PHL 7481	01 36 00.	- 05 10		17.3	BLUE STELLAR OBJECT
PHL 7480	01 36 00.	- 05 36		18.0	BLUE STELLAR OBJECT
PHL 3533	01 36 00.	- 07 42		18.7	BLUE STELLAR OBJECT
PHL 1080	01 36 00.	- 07 42		18.4	BLUE STELLAR OBJECT
PHL 7482	01 36 00.	- 08 13		18.6	BLUE STELLAR OBJECT
PHL 7484	01 36 00.	- 10 54		18.5	BLUE STELLAR OBJECT
PHL 7483	01 36 00.	- 12 54		16.8	BLUE STELLAR OBJECT
PHL 3530	01 36 00.	- 13 30		17.5	BLUE STELLAR OBJECT
PHL 1081	01 36 00.	- 21 18		18.4	BLUE STELLAR OBJECT
MCG-04-05-003	01 36 00.	- 21 22	36	15.5	GALAXY
MCG-04-05-002	01 36 00.	- 23 10 30.	30	15.	GALAXY
KN 14.401	01 36 00.4	+ 31 49 36.			NEBULA
KN 14.402	01 36 00.5	+ 31 11 21.			NEBULA
KN 14.403	01 36 03.7	+ 29 26 16.			NEBULA
IC 0145	01 36 04.	+ 00 29 23.			NONSTELLAR OBJECT
ZWG 386.026	01 36 06.	+ 00 30		15.6	GALAXY
PHL 7485	01 36 06.	+ 00 31		17.3	BLUE STELLAR OBJECT
MCG+02-05-007A	01 36 06.	+ 14 45	12	15.5	GALAXY
ZWG 437.007	01 36 06.	+ 14 48		15.7	GALAXY
ZWG 552.003	01 36 06.	+ 48 02		15.3	GALAXY
KN 14.406	01 36 08.1	+ 27 20 44.			NEBULA
MCG+00-05-020	01 36 09.	+ 00 29	36	15.	GALAXY
MCG+00-05-019	01 36 09.	- 00 25 30.	15	15.5	GALAXY
KN 14.404	01 36 09.0	+ 32 23 04.			NEBULA
KN 14.405	01 36 09.9	+ 32 22 59.			NEBULA
PHL 3534	01 36 12.	+ 03 39		16.9	BLUE STELLAR OBJECT
PHL 3535	01 36 12.	+ 08 04		18.2	BLUE STELLAR OBJECT
ZWG 437.008	01 36 12.	+ 14 46		15.5	GALAXY
MCG+02-05-008	01 36 12.	+ 14 47	66	15.	GALAXY
ZC 0136.2+2350	01 36 12.	+ 23 50	1340		CLUSTER OF GALAXIES
ZWG 552.004	01 36 12.	+ 48 31		15.7	GALAXY
UGC 01168	01 36 12.	+ 48 31	96	15.7	GALAXY Sb?
MCG+08-04-003	01 36 12.	+ 48 31	60	15.	GALAXY
ZWG 386.027	01 36 12.	- 00 24		15.7	GALAXY
PHL 1082	01 36 12.	- 03 32		17.9	BLUE STELLAR OBJECT
PHL 1083	01 36 12.	- 06 40		18.4	BLUE STELLAR OBJECT
PHL 3536	01 36 12.	- 12 04		16.5	BLUE STELLAR OBJECT
PHL 1084	01 36 12.	- 15 22		17.4	BLUE STELLAR OBJECT
PHL 1085	01 36 12.	- 15 24		18.5	BLUE STELLAR OBJECT
PHL 3537	01 36 12.	- 20 11		16.5	BLUE STELLAR OBJECT
LB 03220	01 36 12.	- 66 09		15.0	FAINT BLUE STAR
KN 14.407	01 36 13.3	+ 30 51 47.			NEBULA
KN 14.408	01 36 14.8	+ 31 47 09.			NEBULA
ARC 0225	01 36 15.	+ 18 38		15.9	RICH CLUSTER OF GALAXIES
IC 0146	01 36 15.	- 18 04 50.			NONSTELLAR OBJECT
RNGC 0629	01 36 17.	+ 72 37			NON-EXISTENT OBJECT
MCG+00-05-021	01 36 18.	+ 00 48 30.	48	14.	GALAXY
ZWG 386.028	01 36 18.	+ 00 50		14.2	GALAXY
UGC 01169	01 36 18.	+ 00 50	90	14.2	GALAXY S0
PHL 7488	01 36 18.	+ 04 00		18.1	BLUE STELLAR OBJECT
ZC 0136.3+0533	01 36 18.	+ 05 33	3290		CLUSTER OF GALAXIES
PHL 7486	01 36 18.	+ 06 38		18.0	BLUE STELLAR OBJECT
PHL 7487	01 36 18.	- 01 05		16.6	BLUE STELLAR OBJECT
PHL 3538	01 36 18.	- 01 15		17.7	BLUE STELLAR OBJECT
PHL 1086	01 36 18.	- 04 12		16.8	BLUE STELLAR OBJECT
RNGC 0648	01 36 18.	- 18 06		14.0	GALAXY
MCG-03-05-011	01 36 18.	- 18 06	48	14.5	GALAXY
PHL 7489	01 36 18.	- 25 06		18.5	BLUE STELLAR OBJECT
MCG-06-04-062	01 36 18.	- 33 52	60	15.	GALAXY
PHL 7497	01 36 24.	+ 01 10		18.0	BLUE STELLAR OBJECT
PHL 1087	01 36 24.	+ 03 38		18.4	BLUE STELLAR OBJECT
PHL 7496	01 36 24.	+ 04 14		18.0	BLUE STELLAR OBJECT
3ZW 032	01 36 24.	+ 21 28			COMPACT GALAXY
5ZW 083	01 36 24.	+ 32 25			COMPACT GALAXY
ZWG 503.004	01 36 24.	+ 32 25		15.7	GALAXY
ZWG 502.124	01 36 24.	+ 32 25		15.7	GALAXY
PHL 3539	01 36 24.	- 00 50		18.4	BLUE STELLAR OBJECT
PHL 7495	01 36 24.	- 03 21		17.3	BLUE STELLAR OBJECT
PHL 7498	01 36 24.	- 04 11		18.3	BLUE STELLAR OBJECT
PHL 3541	01 36 24.	- 04 58		17.5	BLUE STELLAR OBJECT
PHL 7492	01 36 24.	- 08 36		17.2	BLUE STELLAR OBJECT
PHL 7493	01 36 24.	- 10 11		18.4	BLUE STELLAR OBJECT
PHL 7494	01 36 24.	- 11 24		17.5	BLUE STELLAR OBJECT
PHL 3540	01 36 24.	- 12 40		18.7	BLUE STELLAR OBJECT
PHL 7490	01 36 24.	- 14 32		17.2	BLUE STELLAR OBJECT
PHL 7491	01 36 24.	- 27 38		16.7	BLUE STELLAR OBJECT
MCG-05-05-001	01 36 24.	- 32 05	36	15.5	GALAXY
KN 14.409	01 36 27.8	+ 32 24 25.			NEBULA
KN 14.410	01 36 29.0	+ 31 18 49.			NEBULA
PHL 7500	01 36 30.	+ 07 06		17.4	BLUE STELLAR OBJECT
5ZW 084	01 36 30.	+ 31 19			COMPACT GALAXY
PHL 3542	01 36 30.	- 00 49		17.4	BLUE STELLAR OBJECT
PHL 7499	01 36 30.	- 09 10		18.2	BLUE STELLAR OBJECT
PHL 1088	01 36 30.	- 09 59		17.2	BLUE STELLAR OBJECT
ARC 0226	01 36 30.	- 10 31		17.6	RICH CLUSTER OF GALAXIES
PHL 3543	01 36 30.	- 11 50		18.5	BLUE STELLAR OBJECT
MCG-06-04-063	01 36 30.	- 33 19	15	15.5	GALAXY
RNGC 0641	01 36 32.	- 42 47			GALAXY
RNGC 0636	01 36 33.	- 07 45		12.5	GALAXY
KN 14.411	01 36 35.4	+ 29 09 01.			NEBULA
PHL 7504	01 36 36.	+ 05 00		17.1	BLUE STELLAR OBJECT
PHL 7334	01 36 36.	+ 08 06		17.8	BLUE STELLAR OBJECT
6ZW 006	01 36 36.	+ 29 19			COMPACT GALAXY
ZWG 521.062	01 36 36.	+ 33 35		15.5	GALAXY
ZWG 521.063	01 36 36.	+ 34 55		15.4	GALAXY
VB 141	01 36 36.	+ 63 04	369		STELLAR RING
ISS 0050	01 36 36.	+ 63 04	371		STELLAR RING
PHL 7507	01 36 36.	- 00 01		17.7	BLUE STELLAR OBJECT
PHL 3544	01 36 36.	- 02 44		14.9	BLUE STELLAR OBJECT
PHL 7508	01 36 36.	- 04 22		18.3	BLUE STELLAR OBJECT
PHL 7501	01 36 36.	- 05 14		17.2	BLUE STELLAR OBJECT
PHL 7634	01 36 36.	- 08 18		18.5	BLUE STELLAR OBJECT
PHL 7502	01 36 36.	- 09 20		18.2	BLUE STELLAR OBJECT
MCG-02-05-030	01 36 36.	- 10 45	60	15.	GALAXY
PHL 7505	01 36 36.	- 12 14		17.2	BLUE STELLAR OBJECT
PHL 3546	01 36 36.	- 15 19		18.6	BLUE STELLAR OBJECT
PHL 3545	01 36 36.	- 15 32		18.6	BLUE STELLAR OBJECT
PHL 7506	01 36 36.	- 18 51		16.6	BLUE STELLAR OBJECT
PHL 7503	01 36 36.	- 23 10		18.4	BLUE STELLAR OBJECT
RNGC 0639	01 36 39.	- 30 11		15.0	GALAXY
ZC 0136.7+0053	01 36 42.	+ 00 53	3560		CLUSTER OF GALAXIES
PHL 3550	01 36 42.	+ 07 31		18.6	BLUE STELLAR OBJECT
PHL 3551	01 36 42.	- 01 18		18.7	BLUE STELLAR OBJECT
MCG+00-05-022	01 36 42.	- 01 55	24	15.	GALAXY
PHL 7511	01 36 42.	- 04 20		18.8	BLUE STELLAR OBJECT
PHL 3547	01 36 42.	- 05 38		18.0	BLUE STELLAR OBJECT
PHL 7510	01 36 42.	- 07 02		18.1	BLUE STELLAR OBJECT
PHL 3552	01 36 42.	- 08 14		18.7	BLUE STELLAR OBJECT
ARC 0228	01 36 42.	- 10 19		17.6	RICH CLUSTER OF GALAXIES
PHL 3548	01 36 42.	- 10 49		18.8	BLUE STELLAR OBJECT
PHL 7509	01 36 42.	- 13 49		17.9	BLUE STELLAR OBJECT
PHL 3549	01 36 42.	- 15 18		18.6	BLUE STELLAR OBJECT
MCG-05-05-002	01 36 42.	- 30 11	42	15.	GALAXY
RNGC 0644	01 36 44.	- 42 51			GALAXY
MCG+01-05-013	01 36 45.	- 07 47	66	12.5	GALAXY
RNGC 0642	01 36 45.	- 30 10		14.0	GALAXY
KN 14.412	01 36 47.3	+ 29 29 54.			NEBULA
PHL 7513	01 36 48.	+ 00 11		18.0	BLUE STELLAR OBJECT
PHL 7517	01 36 48.	+ 00 20		18.5	BLUE STELLAR OBJECT
PHL 7512	01 36 48.	+ 01 22		17.7	BLUE STELLAR OBJECT
PHL 7516	01 36 48.	+ 04 21		17.2	BLUE STELLAR OBJECT
4ZW 056	01 36 48.	+ 26 54			COMPACT GALAXY
MCG+06-04-050	01 36 48.	+ 34 54	48	15.	GALAXY
PHL 3554	01 36 48.	- 01 54		18.8	BLUE STELLAR OBJECT
PHL 7514	01 36 48.	- 05 57		16.6	BLUE STELLAR OBJECT
PHL 7515	01 36 48.	- 08 46		16.9	BLUE STELLAR OBJECT
PHL 3556	01 36 48.	- 10 51		18.7	BLUE STELLAR OBJECT
PHL 1089	01 36 48.	- 10 57		18.4	BLUE STELLAR OBJECT
PHL 3555	01 36 48.	- 11 17		17.7	BLUE STELLAR OBJECT

OBJECT NAME	RIGHT ASCEN.	DECLINATION	DIAM.	MAGN.	TYPE OF OBJECT
PHL 7518	01 36 48.	- 11 40		18.1	BLUE STELLAR OBJECT
MCG-03-05-012	01 36 48.	- 14 31	30	15.	GALAXY
PHL 3553	01 36 48.	- 19 16		16.8	BLUE STELLAR OBJECT
PHL 1090	01 36 48.	- 19 31		17.0	BLUE STELLAR OBJECT
MCG-05-05-004	01 36 48.	- 28 49	36	16.	GALAXY
MCG-05-05-003	01 36 48.	- 30 10	120	14.	GALAXY
RNGC 0640	01 36 51.	- 09 39		15.0	GALAXY
PHL 7519	01 36 54.	+ 03 32		18.7	BLUE STELLAR OBJECT
PHL 3557	01 36 54.	+ 06 16		17.0	BLUE STELLAR OBJECT
ZWG 537.018	01 36 54.	+ 42 38		15.7	GALAXY
ZWG 386.029	01 36 54.	- 00 12		15.7	GALAXY
MCG+00-05-023	01 36 54.	- 00 13	24	15.	GALAXY
ARC 0229	01 36 54.	- 03 54		16.6	RICH CLUSTER OF GALAXIES
MCG-02-05-031	01 36 54.	- 09 39	36	15.	GALAXY
PHL 1091	01 36 54.	- 28 00		16.6	BLUE STELLAR OBJECT
RNGC 0638	01 36 59.	+ 06 58		14.5	GALAXY
ARC 0230	01 36 59.	- 11 38		17.6	RICH CLUSTER OF GALAXIES
VDB.66G 013	01 37	+ 15 39	100		DWARF GALAXY
PHL 3563	01 37 00.	+ 01 00		18.4	BLUE STELLAR OBJECT
ZC 0137.0+0107	01 37 00.	+ 01 07	740		CLUSTER OF GALAXIES
PHL 3561	01 37 00.	+ 01 34		18.4	BLUE STELLAR OBJECT
PHL 3560	01 37 00.	+ 04 28		18.4	BLUE STELLAR OBJECT
PHL 3558	01 37 00.	+ 04 38		16.0	BLUE STELLAR OBJECT
MCG+01-05-015	01 37 00.	+ 05 30	48	15.	GALAXY
ZWG 412.011	01 37 00.	+ 06 58		14.4	GALAXY
UGC 01170	01 37 00.	+ 06 58	60	14.4	GALAXY
MCG+01-05-014	01 37 00.	+ 07 00	42	14.	GALAXY
UGC 01171	01 37 00.	+ 15 39	84	17.	GALAXY DWARF
ZC 0137.0+1721	01 37 00.	+ 17 21	3230		CLUSTER OF GALAXIES
MCG+06-04-051	01 37 00.	+ 34 08 30.	42	14.5	GALAXY
ZWG 521.064	01 37 00.	+ 34 10		14.9	GALAXY
ZWG 386.030	01 37 00.	- 00 16		15.6	GALAXY
MCG+00-05-024	01 37 00.	- 00 18	19	15.	GALAXY
PHL 3559	01 37 00.	- 06 27		17.0	BLUE STELLAR OBJECT
PHL 7522	01 37 00.	- 09 34		16.7	BLUE STELLAR OBJECT
PHL 7520	01 37 00.	- 10 46		18.4	BLUE STELLAR OBJECT
PHL 3562	01 37 00.	- 10 54		18.7	BLUE STELLAR OBJECT
PHL 7521	01 37 00.	- 10 59		17.8	BLUE STELLAR OBJECT
PHL 7523	01 37 00.	- 13 54		17.7	BLUE STELLAR OBJECT
PHL 7524	01 37 00.	- 14 16		18.6	BLUE STELLAR OBJECT
TON-S 0225	01 37 00.	- 28 01		15.7	BLUE STAR
KN 14.413	01 37 00.4	+ 32 04 17.			NEBULA
MCG-04-05-004	01 37 03.	- 20 41 30.	36	15.	GALAXY
KN 14.414	01 37 03.7	+ 28 45 04.			NEBULA
BB 7.10	01 37 03.95	+ 01 03 55.			GALAXY NEAR QSO PHL1093
BB 7.12	01 37 05.1	+ 01 08 58.5			GALAXY NEAR QSO PHL1093
ZWG 412.012	01 37 06.	+ 05 32		15.0	GALAXY
UGC 01172	01 37 06.	+ 05 32	60	15.0	GALAXY
UGC 01173	01 37 06.	+ 46 18	66	17.	GALAXY
MCG-02-05-032	01 37 06.	- 12 20	60	14.	GALAXY
PHL 3564	01 37 06.	- 16 17		18.6	BLUE STELLAR OBJECT
BB 7.11	01 37 06.45	+ 01 06 56.			GALAXY NEAR QSO PHL1093
ARC 0227	01 37 08.	+ 17 56		17.7	RICH CLUSTER OF GALAXIES
BB 7.04	01 37 08.55	+ 01 18 35.5			GALAXY NEAR QSO PHL1093
KN 14.415	01 37 09.8	+ 33 14 15.			NEBULA
KN 14.416	01 37 10.1	+ 30 46 11.			NEBULA
PHL 3565	01 37 12.	+ 05 50		18.2	BLUE STELLAR OBJECT
PHL 1092	01 37 12.	+ 06 03		17.0	BLUE STELLAR OBJECT
BC PHL1092	01 37 12.	+ 06 03		17.0	QUASI-STELLAR OBJECT
PHL 3566	01 37 12.	+ 07 42		18.3	BLUE STELLAR OBJECT
PHL 7527	01 37 12.	+ 08 12		18.6	BLUE STELLAR OBJECT
PHL 3567	01 37 12.	- 00 32		18.6	BLUE STELLAR OBJECT
PHL 7528	01 37 12.	- 02 44		18.8	BLUE STELLAR OBJECT
PHL 7526	01 37 12.	- 03 10		17.6	BLUE STELLAR OBJECT
PHL 7529	01 37 12.	- 03 12		18.7	BLUE STELLAR OBJECT
PHL 7525	01 37 12.	- 05 38		16.8	BLUE STELLAR OBJECT
PHL 7530	01 37 12.	- 15 31		18.7	BLUE STELLAR OBJECT
PHL 7531	01 37 12.	- 24 01		18.6	BLUE STELLAR OBJECT
BB 7.05	01 37 12.75	+ 01 17 45.			GALAXY NEAR QSO PHL1093
BB 7.06	01 37 14.05	+ 01 22 13.5			GALAXY NEAR QSO PHL1093
BB 7.03	01 37 17.35	+ 01 21 12.5			GALAXY NEAR QSO PHL1093
PHL 7532	01 37 18.	+ 00 50		17.8	BLUE STELLAR OBJECT
PHL 3570	01 37 18.	+ 05 03		18.7	BLUE STELLAR OBJECT
UGC 01175	01 37 18.	+ 10 51	72	17.	GALAXY S IV-V
PHL 7534	01 37 18.	- 01 06		18.6	BLUE STELLAR OBJECT
PHL 7536	01 37 18.	- 01 34		18.5	BLUE STELLAR OBJECT
ZWG 386.031	01 37 18.	- 02 17		15.1	GALAXY
UGC 01174	01 37 18.	- 02 17	78	15.1	GALAXY Sb-c
MCG+00-05-025	01 37 18.	- 02 18	48	14.	GALAXY
PHL 3569	01 37 18.	- 04 42		18.5	BLUE STELLAR OBJECT
PHL 7537	01 37 18.	- 07 26		18.7	BLUE STELLAR OBJECT
PHL 7533	01 37 18.	- 08 30		17.7	BLUE STELLAR OBJECT
PHL 7535	01 37 18.	- 09 30		17.9	BLUE STELLAR OBJECT
PHL 3568	01 37 18.	- 10 04		18.1	BLUE STELLAR OBJECT
PHL 3571	01 37 18.	- 13 29		18.7	BLUE STELLAR OBJECT
PHL 7538	01 37 18.	- 13 42		18.4	BLUE STELLAR OBJECT
KN 14.417	01 37 19.5	+ 30 17 40.			NEBULA
RNGC 0647	01 37 21.	- 09 29		14.0	GALAXY
BC 4C01.04	01 37 22.78	+ 01 16 35.2		17.07	QUASI-STELLAR OBJECT
SHB 040	01 37 23.	+ 01 16 18.		17.1	QUASI-STELLAR OBJECT
BB 7.01	01 37 23.2	+ 01 19 19.5			GALAXY NEAR QSO PHL1093
BC OC062	01 37 23.4	+ 01 16 57.		18.	QUASI-STELLAR OBJECT
PHL 1093	01 37 24.	+ 01 16		17.2	BLUE STELLAR OBJECT
PHL 7542	01 37 24.	+ 01 35		17.8	BLUE STELLAR OBJECT
PHL 3572	01 37 24.	+ 03 22		17.0	BLUE STELLAR OBJECT
UGC 01176	01 37 24.	+ 15 39	300	17.	GALAXY DWRF IR
ZWG 521.065	01 37 24.	+ 36 57		15.7	GALAXY
PHL 1094	01 37 24.	- 03 40		18.2	BLUE STELLAR OBJECT
PHL 7539	01 37 24.	- 03 47		18.2	BLUE STELLAR OBJECT
PHL 7540	01 37 24.	- 07 40		18.2	BLUE STELLAR OBJECT
MCG-02-05-033	01 37 24.	- 09 29	24	14.5	GALAXY
PHL 3573	01 37 24.	- 11 20		18.1	BLUE STELLAR OBJECT
PHL 3625	01 37 24.	- 12 03		17.7	BLUE STELLAR OBJECT
PHL 7541	01 37 24.	- 13 20		16.9	BLUE STELLAR OBJECT
PHL 3574	01 37 24.	- 19 27		18.6	BLUE STELLAR OBJECT
KN 14.418	01 37 24.7	+ 31 58 58.			NEBULA
BB 7.02	01 37 24.85	+ 01 20 02.5			GALAXY NEAR QSO PHL1093
KN 14.420	01 37 25.0	+ 29 18 55.			NEBULA
KN 14.419	01 37 25.5	+ 30 33 15.			NEBULA
BB 7.07	01 37 29.1	+ 01 16 57.5			GALAXY NEAR QSO PHL1093
PHL 7543	01 37 30.	+ 03 56		17.0	BLUE STELLAR OBJECT
PHL 3575	01 37 30.	+ 06 30		18.5	BLUE STELLAR OBJECT
KARA.68 008	01 37 30.	+ 15 38	168		DWARF GALAXY
PHL 7544	01 37 30.	- 04 10		17.0	BLUE STELLAR OBJECT
PHL 7545	01 37 30.	- 05 24		17.9	BLUE STELLAR OBJECT
SER 017.01	01 37 30.	- 51 15	40	16.	INTERACT. PAIR OF GALXIES
BB 7.09	01 37 32.2	+ 01 11 54.			GALAXY NEAR QSO PHL1093
MCG+01-05-016	01 37 33.	+ 05 28	138	13.	GALAXY
RNGC 0649	01 37 33.	- 09 30		15.0	GALAXY
IC 0147	01 37 33.	- 15 06 18.			NONSTELLAR OBJECT
KN 14.421	01 37 35.7	+ 31 59 36.			NEBULA
PHL 7551	01 37 36.	+ 01 20		18.0	BLUE STELLAR OBJECT
PHL 3576	01 37 36.	+ 01 39		18.9	BLUE STELLAR OBJECT
PHL 7550	01 37 36.	+ 05 20		18.6	BLUE STELLAR OBJECT
ZWG 412.013	01 37 36.	+ 05 28		13.8	GALAXY
RNGC 0645	01 37 36.	+ 05 28		14.0	GALAXY
UGC 01177	01 37 36.	+ 05 28	186	13.8	GALAXY SB:b
PHL 7547	01 37 36.	+ 06 24		17.6	BLUE STELLAR OBJECT
5ZW 085	01 37 36.	+ 32 00			COMPACT GALAXY
MCG+06-04-053	01 37 36.	+ 34 22	102	14.	GALAXY
ZWG 521.066	01 37 36.	+ 34 23		14.8	GALAXY
UGC 01178	01 37 36.	+ 34 23	120	14.8	GALAXY Sc
MCG+06-04-052	01 37 36.	+ 36 57	36	15.	GALAXY
ZWG 537.019	01 37 36.	+ 43 10		15.5	GALAXY
5ZW 086	01 37 36.	+ 43 37			COMPACT GALAXY
ZWG 552.005	01 37 36.	+ 45 35		15.1	GALAXY
PHL 7548	01 37 36.	- 09 19		17.8	BLUE STELLAR OBJECT
MCG-02-05-034	01 37 36.	- 09 30	48	15.	GALAXY
PHL 7552	01 37 36.	- 10 34		18.3	BLUE STELLAR OBJECT
ARC 0232	01 37 36.	- 10 38		17.6	RICH CLUSTER OF GALAXIES
PHL 1095	01 37 36.	- 12 07		18.5	BLUE STELLAR OBJECT
MCG-03-05-013	01 37 36.	- 15 07	48	14.5	GALAXY
PHL 7549	01 37 36.	- 15 46		17.5	BLUE STELLAR OBJECT
PHL 7546	01 37 36.	- 23 48		17.2	BLUE STELLAR OBJECT
KN 14.422	01 37 36.1	+ 31 59 49.			NEBULA
BB 7.08	01 37 39.1	+ 01 13 12.			GALAXY NEAR QSO PHL1093
PATH 1.097	01 37 40.	+ 45 35	55		NEBULA
PHL 3577	01 37 42.	+ 02 16		18.7	BLUE STELLAR OBJECT
5ZW 087	01 37 42.	+ 42 38			COMPACT GALAXY
UGC 01179	01 37 42.	+ 43 37	102	15.6	GALAXY COMPACT
5ZW 088	01 37 42.	+ 45 35			COMPACT GALAXY
ZWG 386.032	01 37 42.	- 00 45		15.2	GALAXY
KARA.73B 0060	01 37 42.	- 00 45	36	15.2	ISOLATED GALAXY S
MCG+00-05-026	01 37 42.	- 00 46 30.	30	15.	GALAXY
PHL 1096	01 37 42.	- 01 05		15.9	BLUE STELLAR OBJECT
PHL 7554	01 37 42.	- 05 01		18.2	BLUE STELLAR OBJECT
PHL 7553	01 37 42.	- 05 47		16.5	BLUE STELLAR OBJECT
PHL 1097	01 37 42.	- 11 56		18.8	BLUE STELLAR OBJECT
PHL 1103	01 37 42.	- 16 06		18.0	BLUE STELLAR OBJECT
MCG-05-05-005	01 37 42.	- 28 12	36	16.	GALAXY
MCG-05-05-006	01 37 42.	- 28 17	48	15.	GALAXY
MCG-05-05-007	01 37 42.	- 28 57	102	14.5	GALAXY
HOW 83	01 37 46.	- 74 47 54.	18		STAR CLUSTER IN SMC
ARC 0235	01 37 47.	- 17 41		17.5	RICH CLUSTER OF GALAXIES
IC 1720	01 37 47.	- 29 11 07.			NONSTELLAR OBJECT
PHL 3587	01 37 48.	+ 00 00		18.3	BLUE STELLAR OBJECT
PHL 3583	01 37 48.	+ 01 32		18.9	BLUE STELLAR OBJECT
PHL 3580	01 37 48.	+ 04 39		17.5	BLUE STELLAR OBJECT
PHL 7555	01 37 48.	+ 05 46		18.4	BLUE STELLAR OBJECT
PHL 3579	01 37 48.	+ 07 53		17.7	BLUE STELLAR OBJECT
UGC 01181	01 37 48.	+ 14 17	66	17.	GALAXY Sc
5ZW 089	01 37 48.	+ 42 00			COMPACT GALAXY
ZWG 552.006	01 37 48.	+ 49 00		15.6	GALAXY
UGC 01182	01 37 48.	+ 49 00	66	15.6	GALAXY Sc
UGC 01180	01 37 48.	- 00 33	66	16.0	GALAXY COMPACT
MCG+00-05-027	01 37 48.	- 00 33	12	15.	GALAXY
PHL 3581	01 37 48.	- 01 10		18.8	BLUE STELLAR OBJECT
PHL 3582	01 37 48.	- 01 54		18.7	BLUE STELLAR OBJECT
ZC 0137.8-0207	01 37 48.	- 02 07	1340		CLUSTER OF GALAXIES
PHL 3588	01 37 48.	- 02 36		18.4	BLUE STELLAR OBJECT
PHL 3584	01 37 48.	- 03 40		18.5	BLUE STELLAR OBJECT
PHL 3585	01 37 48.	- 18 06		18.2	BLUE STELLAR OBJECT
PHL 1098	01 37 48.	- 20 44		17.3	BLUE STELLAR OBJECT
PHL 3578	01 37 48.	- 22 30		15.7	BLUE STELLAR OBJECT
TON-S 0226	01 37 48.	- 22 56		15.2	BLUE STAR
PHL 3586	01 37 48.	- 22 57		16.6	BLUE STELLAR OBJECT
ARC 0233	01 37 50.	- 02 06		17.2	RICH CLUSTER OF GALAXIES
ARC 0231	01 37 51.	+ 24 16		17.9	RICH CLUSTER OF GALAXIES
PHL 3589	01 37 54.	+ 01 03		18.2	BLUE STELLAR OBJECT
UGC 01183	01 37 54.	+ 02 26	96	16.0	GALAXY S
MCG+00-05-028	01 37 54.	+ 02 26	24	15.	GALAXY
PHL 7556	01 37 54.	+ 06 09		18.4	BLUE STELLAR OBJECT
MCG+03-05-012	01 37 54.	+ 18 40	4	19.	GALAXY
ZC 0137.9+2418	01 37 54.	+ 24 18	1280		CLUSTER OF GALAXIES
VB 142	01 37 54.	+ 61 01	248		STELLAR RING
ISS 0051	01 37 54.	+ 61 01	253		STELLAR RING
PHL 7557	01 37 54.	- 01 24		18.6	BLUE STELLAR OBJECT
MCG+01-05-014	01 37 57.	- 05 47 30.	90	14.5	GALAXY
MCG+01-05-015	01 37 57.	- 08 12 30.	18	16.	GALAXY
ARC 0236	01 37 59.	- 12 07		17.2	RICH CLUSTER OF GALAXIES
HMS 0138+1840	01 38	+ 18 40			CLUSTER OF GALAXIES
LB 09962	01 38	- 87 01		14.6	FAINT BLUE STAR
PHL 7560	01 38 00.	+ 01 23		18.5	BLUE STELLAR OBJECT
PHL 7564	01 38 00.	+ 05 54		18.2	BLUE STELLAR OBJECT
MCG+01-05-017	01 38 00.	+ 07 44	42	14.	GALAXY
ZWG 460.018	01 38 00.	+ 20 35		14.9	GALAXY
6ZW 007	01 38 00.	+ 31 17			COMPACT GALAXY
PHL 7563	01 38 00.	- 01 00		17.0	BLUE STELLAR OBJECT
PHL 1099	01 38 00.	- 07 08		16.6	BLUE STELLAR OBJECT
PHL 7561	01 38 00.	- 09 32		18.2	BLUE STELLAR OBJECT
PHL 7559	01 38 00.	- 09 32		17.8	BLUE STELLAR OBJECT
PHL 7558	01 38 00.	- 09 46		16.9	BLUE STELLAR OBJECT
PHL 3590	01 38 00.	- 11 52		17.9	BLUE STELLAR OBJECT
PHL 7565	01 38 00.	- 13 12		18.1	BLUE STELLAR OBJECT
PHL 7562	01 38 00.	- 13 27		16.6	BLUE STELLAR OBJECT
PHL 1100	01 38 00.	- 23 36		17.5	BLUE STELLAR OBJECT
PHL 3591	01 38 00.	- 29 09		16.8	BLUE STELLAR OBJECT
MCG+01-05-016	01 38 03.	- 08 12	72	15.	GALAXY
RNGC 0652	01 38 05.	+ 07 43		14.5	GALAXY
KN 14.423	01 38 05.4	+ 31 35 20.			NEBULA
PHL 7568	01 38 06.	+ 03 51		18.0	BLUE STELLAR OBJECT
ZWG 412.014	01 38 06.	+ 07 43		14.7	GALAXY
UGC 01184	01 38 06.	+ 07 43	72	14.7	GALAXY S
ZC 0138.1+1839	01 38 06.	+ 18 39	670		CLUSTER OF GALAXIES
ZWG 521.067	01 38 06.	+ 38 23		15.6	GALAXY
PHL 3594	01 38 06.	- 03 06		19.0	BLUE STELLAR OBJECT
PHL 7566	01 38 06.	- 04 43		17.2	BLUE STELLAR OBJECT
PHL 7569	01 38 06.	- 10 54		18.4	BLUE STELLAR OBJECT
PHL 7567	01 38 06.	- 11 15		17.9	BLUE STELLAR OBJECT
PHL 7570	01 38 06.	- 12 18		18.5	BLUE STELLAR OBJECT
PHL 3592	01 38 06.	- 12 58		18.6	BLUE STELLAR OBJECT
PHL 3593	01 38 06.	- 20 40		18.6	BLUE STELLAR OBJECT
PHL 3595	01 38 06.	- 25 38		18.6	BLUE STELLAR OBJECT
MCG-05-05-008	01 38 06.	- 29 10	72	14.	GALAXY
MCG-06-04-064	01 38 06.	- 36 05		15.5	GALAXY
PHL 3597	01 38 12.	+ 07 38		17.0	BLUE STELLAR OBJECT

OBJECT NAME	RIGHT ASCEN.	DECLINATION	DIAM.	MAGN.	TYPE OF OBJECT
PHL 7575	01 38 12.	+ 07 42		18.3	BLUE STELLAR OBJECT
PHL 3596	01 38 12.	+ 07 52		17.7	BLUE STELLAR OBJECT
6ZW 008	01 38 12.	+ 28 00			COMPACT GALAXY
6ZW 009	01 38 12.	+ 32 38			COMPACT GALAXY
MCG+07-04-008	01 38 12.	+ 44 44	24	15.	GALAXY
ZWG 537.020	01 38 12.	+ 44 45		15.0	GALAXY
UGC 01185	01 38 12.	+ 44 45	90	15.0	GALAXY COMPACT
MCG+00-05-029	01 38 12.	- 01 33	36	15.5	GALAXY
PHL 7576	01 38 12.	- 01 44		18.1	BLUE STELLAR OBJECT
PHL 7574	01 38 12.	- 01 54		17.1	BLUE STELLAR OBJECT
PHL 7577	01 38 12.	- 04 00		18.4	BLUE STELLAR OBJECT
PHL 7571	01 38 12.	- 05 29		17.7	BLUE STELLAR OBJECT
PHL 3598	01 38 12.	- 05 33		18.5	BLUE STELLAR OBJECT
PHL 7572	01 38 12.	- 07 06		18.5	BLUE STELLAR OBJECT
PHL 1101	01 38 12.	- 07 57		18.2	BLUE STELLAR OBJECT
PHL 7578	01 38 12.	- 08 11		18.0	BLUE STELLAR OBJECT
PHL 7573	01 38 12.	- 11 50		18.1	BLUE STELLAR OBJECT
PHL 3599	01 38 12.	- 12 56		18.2	BLUE STELLAR OBJECT
PHL 3600	01 38 12.	- 30 59		18.2	BLUE STELLAR OBJECT
LIN 549	01 38 12.	- 73 25 30.		15.39	SMC EMISSION LINE OBJECT
LIN 550	01 38 12.	- 73 26 12.		15.97	SMC EMISSION LINE OBJECT
SCHO 0040	01 38 13.	+ 60 11 42.	670		ISOLATED DARK CLOUD
KN 14.424	01 38 13.9	+ 31 37 42.			NEBULA
ARC 0234	01 38 15.	+ 18 40		17.9	RICH CLUSTER OF GALAXIES
MCG+06-04-054	01 38 15.	+ 38 24	36	17.	GALAXY
SHAH 044	01 38 18.	+ 02 36	168	16.5	GROUP OF COMPACT GALAXIES
6ZW 010	01 38 18.	+ 28 03			COMPACT GALAXY
ZC 0138.3-0010	01 38 18.	- 00 10	2760		CLUSTER OF GALAXIES
PHL 7579	01 38 18.	- 07 18		18.2	BLUE STELLAR OBJECT
PHL 7581	01 38 18.	- 09 01		17.9	BLUE STELLAR OBJECT
PHL 7580	01 38 18.	- 09 02		17.9	BLUE STELLAR OBJECT
PHL 7583	01 38 18.	- 10 50		18.3	BLUE STELLAR OBJECT
PHL 7582	01 38 18.	- 18 50		17.7	BLUE STELLAR OBJECT
MCG-05-05-009	01 38 18.	- 32 05	60	15.	GALAXY
MCG-06-04-065	01 38 18.	- 32 58	24	15.5	GALAXY
ARC 0237	01 38 22.	+ 00 01		16.6	RICH CLUSTER OF GALAXIES
PHL 3602	01 38 24.	+ 05 35		18.2	BLUE STELLAR OBJECT
PHL 7586	01 38 24.	+ 08 08		17.5	BLUE STELLAR OBJECT
6ZW 011	01 38 24.	+ 27 29			COMPACT GALAXY
ZWG 521.068	01 38 24.	+ 34 34		14.7	GALAXY
PHL 7584	01 38 24.	- 00 36		16.9	BLUE STELLAR OBJECT
PHL 7587	01 38 24.	- 03 00		18.1	BLUE STELLAR OBJECT
PHL 1102	01 38 24.	- 03 00		17.3	BLUE STELLAR OBJECT
PHL 7588	01 38 24.	- 06 52		18.7	BLUE STELLAR OBJECT
MCG-02-05-035	01 38 24.	- 10 31	30	15.	GALAXY
PHL 3601	01 38 24.	- 12 36		16.5	BLUE STELLAR OBJECT
PHL 7589	01 38 24.	- 12 52		18.9	BLUE STELLAR OBJECT
PHL 7585	01 38 24.	- 13 17		18.7	BLUE STELLAR OBJECT
DV.56 N0643B	01 38 24.	- 75 16	84		S GALAXY
RNGC 0643B	01 38 24.	- 75 17			GALAXY
KN 14.425	01 38 24.3	+ 30 36 40.			NEBULA
MCG+01-05-018	01 38 27.	+ 08 12	36	15.	GALAXY
KN 14.426	01 38 28.2	+ 31 45 26.			NEBULA
BC 3CR49	01 38 28.56	+ 13 38 20.5		20.	QUASI-STELLAR OBJECT
ARC 0238	01 38 29.	- 23 18		17.5	RICH CLUSTER OF GALAXIES
PHL 7592	01 38 30.	+ 01 48		17.1	BLUE STELLAR OBJECT
PHL 7591	01 38 30.	+ 04 49		17.8	BLUE STELLAR OBJECT
PHL 7590	01 38 30.	+ 05 20		13.8	BLUE STELLAR OBJECT
MCG+06-04-055	01 38 30.	+ 34 33 30.	30	14.5	GALAXY
PHL 7593	01 38 30.	- 03 03		18.2	BLUE STELLAR OBJECT
PHL 3603	01 38 30.	- 11 11		18.6	BLUE STELLAR OBJECT
PHL 3604	01 38 30.	- 13 35		18.5	BLUE STELLAR OBJECT
LB 03221	01 38 30.	- 53 30		13.6	FAINT BLUE STAR
PHL 7594	01 38 36.	+ 01 26		15.7	BLUE STELLAR OBJECT
PHL 3606	01 38 36.	+ 06 04		17.1	BLUE STELLAR OBJECT
PHL 7596	01 38 36.	+ 07 21		17.0	BLUE STELLAR OBJECT
ZC 0138.6+2740	01 38 36.	+ 27 40	670		CLUSTER OF GALAXIES
VB 143	01 38 36.	+ 59 43	470		STELLAR RING
PHL 7597	01 38 36.	- 10 32		18.3	BLUE STELLAR OBJECT
PHL 7595	01 38 36.	- 14 25		17.9	BLUE STELLAR OBJECT
MCG+01-05-017	01 38 39.	- 05 50	156	13.5	GALAXY
KN 14.427	01 38 41.2	+ 30 16 46.			NEBULA
ZC 0138.7+1158	01 38 42.	+ 11 58	1480		CLUSTER OF GALAXIES
MCG+05-05-001	01 38 42.	+ 28 04	36	15.	GALAXY
ZWG 552.007	01 38 42.	+ 48 47		15.7	GALAXY
UGC 01186	01 38 42.	+ 48 47	96	15.7	GALAXY PECULR
MCG+08-04-004	01 38 42.	+ 48 48	60	16.	GALAXY
PHL 7600	01 38 42.	- 03 59		19.0	BLUE STELLAR OBJECT
PHL 7598	01 38 42.	- 10 00		18.6	BLUE STELLAR OBJECT
PHL 7599	01 38 42.	- 11 02		16.5	BLUE STELLAR OBJECT
KN 14.429	01 38 43.7	+ 28 05 05.			NEBULA
IC 1721	01 38 46.	+ 08 16 33.			NONSTELLAR OBJECT
KN 14.428	01 38 47.4	+ 32 07 59.			NEBULA
PHL 7602	01 38 48.	+ 04 56		17.8	BLUE STELLAR OBJECT
ZC 0138.8+0715	01 38 48.	+ 07 15	3830		CLUSTER OF GALAXIES
ZWG 412.015	01 38 48.	+ 08 15		14.3	GALAXY
UGC 01187	01 38 48.	+ 08 15	72	14.3	GALAXY S
MCG+01-05-019	01 38 48.	+ 08 18	36	14.	GALAXY
PHL 7601	01 38 48.	+ 09 35		17.9	BLUE STELLAR OBJECT
ZC 0138.8+0954	01 38 48.	+ 09 54	1680		CLUSTER OF GALAXIES
ZWG 503.005	01 38 48.	+ 28 05		15.7	GALAXY
ZC 0138.8+3605	01 38 48.	+ 36 05	940		CLUSTER OF GALAXIES
4ZW 057	01 38 48.	+ 37 18			COMPACT GALAXY
ZWG 521.069	01 38 48.	+ 37 53		15.7	GALAXY
PHL 7603	01 38 48.	- 06 25		18.5	BLUE STELLAR OBJECT
PHL 7604	01 38 48.	- 09 20		18.2	BLUE STELLAR OBJECT
MCG-02-05-036	01 38 48.	- 09 29 30.	36	16.	GALAXY
PHL 1104	01 38 48.	- 11 42		18.1	BLUE STELLAR OBJECT
PHL 3607	01 38 48.	- 13 19		16.6	BLUE STELLAR OBJECT
MCG-04-05-005	01 38 48.	- 26 17	9	15.5	GALAXY
LB 03222	01 38 48.	- 75 24		13.2	FAINT BLUE STAR
RNGC 0651	01 38 50.	+ 51 19		12.0	PLANETARY NEBULA
RNGC 0650	01 38 50.	+ 51 19		12.0	PLANETARY NEBULA
ARC 0239	01 38 53.	- 12 02		17.6	RICH CLUSTER OF GALAXIES
ZWG 482.001	01 38 54.	+ 27 15		15.7	GALAXY
MCG+06-04-057	01 38 54.	+ 37 53	24	17.	GALAXY
MCG+06-04-056	01 38 54.	+ 37 53	24	16.5	GALAXY
PHL 7605	01 38 54.	- 00 40		18.7	BLUE STELLAR OBJECT
ZWG 386.033	01 38 54.	- 02 21		15.7	GALAXY
MCG+00-05-030	01 38 54.	- 02 22	36	15.5	GALAXY
PHL 7606	01 38 54.	- 02 56		18.9	BLUE STELLAR OBJECT
PHL 3608	01 38 54.	- 17 33		18.7	BLUE STELLAR OBJECT
MCG-05-05-010	01 38 54.	- 26 32	84	16.	GALAXY
SER 017.05	01 38 54.	- 54 46	20	19.	CLOUD OF GALAXIES
MCG+01-05-018	01 38 57.	- 02 38	24	15.5	GALAXY
PHL 7608	01 39 00.	+ 04 05		17.5	BLUE STELLAR OBJECT
MCG+01-05-020	01 39 00.	+ 05 35	48	15.	GALAXY
PHL 1106	01 39 00.	+ 06 00		18.3	BLUE STELLAR OBJECT

OBJECT NAME	RIGHT ASCEN.	DECLINATION	DIAM.	MAGN.	TYPE OF OBJECT
PHL 1105	01 39 00.	+ 08 02		15.3	BLUE STELLAR OBJECT
MCG+07-04-009	01 39 00.	+ 42 41	15	16.5	GALAXY
LDN 1332	01 39 00.	+ 61 48	360		DARK NEBULA
PHL 1107	01 39 00.	- 00 26		18.6	BLUE STELLAR OBJECT
PHL 3609	01 39 00.	- 00 59		16.5	BLUE STELLAR OBJECT
PHL 3610	01 39 00.	- 01 10		18.9	BLUE STELLAR OBJECT
PHL 3612	01 39 00.	- 01 11		18.9	BLUE STELLAR OBJECT
PHL 7612	01 39 00.	- 01 28		18.0	BLUE STELLAR OBJECT
PHL 7613	01 39 00.	- 03 30		18.4	BLUE STELLAR OBJECT
PHL 7610	01 39 00.	- 05 46		18.4	BLUE STELLAR OBJECT
PHL 7611	01 39 00.	- 05 48		18.8	BLUE STELLAR OBJECT
PHL 1108	01 39 00.	- 06 34		18.6	BLUE STELLAR OBJECT
PHL 1109	01 39 00.	- 08 53		18.4	BLUE STELLAR OBJECT
PHL 7614	01 39 00.	- 14 10		18.1	BLUE STELLAR OBJECT
PHL 7609	01 39 00.	- 18 28		17.3	BLUE STELLAR OBJECT
PHL 1110	01 39 00.	- 20 14		17.9	BLUE STELLAR OBJECT
PHL 3611	01 39 00.	- 24 22		18.6	BLUE STELLAR OBJECT
PHL 7607	01 39 00.	- 29 08		15.4	BLUE STELLAR OBJECT
BC PKS0139-09	01 39 01.	- 09 42 18.			QUASI-STELLAR OBJECT
SHB 041	01 39 01.	- 09 42 18.		19.5	QUASI-STELLAR OBJECT
KN 14.430	01 39 04.6	+ 33 00 34.			NEBULA
ZWG 412.016	01 39 06.	+ 05 36		15.6	GALAXY
PHL 1111	01 39 06.	+ 05 58		15.7	BLUE STELLAR OBJECT
PHL 3613	01 39 06.	+ 07 26		18.2	BLUE STELLAR OBJECT
ZWG 503.006	01 39 06.	+ 33 00		15.7	GALAXY
PHL 3615	01 39 06.	- 03 12		18.7	BLUE STELLAR OBJECT
PHL 7615	01 39 06.	- 06 42		17.7	BLUE STELLAR OBJECT
PHL 3614	01 39 06.	- 10 08		18.5	BLUE STELLAR OBJECT
PHL 3616	01 39 06.	- 16 24		18.3	BLUE STELLAR OBJECT
KN 14.431	01 39 06.6	+ 30 43 09.			NEBULA
KN 14.432	01 39 06.9	+/ 30 04 40.			NEBULA
KN 14.433	01 39 07.3	+ 27 26 23.			NEBULA
REIN 2.011	01 39 08.47	+ 51 19 07.9			NEBULA
REIN 2.012	01 39 08.73	+ 51 18 46.3			NEBULA
PK130-10.1	01 39 10.16	+ 51 19 30.2	240	12.2	PLANETARY NEBULA
KN 14.434	01 39 11.0	+ 30 08 10.			NEBULA
REIN 2.013	01 39 11.05	+ 51 19 38.1			NEBULA
PHL 3617	01 39 12.	+ 03 46		17.8	BLUE STELLAR OBJECT
PHL 1112	01 39 12.	+ 06 56		16.7	BLUE STELLAR OBJECT
PHL 7620	01 39 12.	+ 08 27		16.9	BLUE STELLAR OBJECT
PHL 7616	01 39 12.	+ 08 57		17.7	BLUE STELLAR OBJECT
PHL 7617	01 39 12.	+ 09 20		18.2	BLUE STELLAR OBJECT
ZWG 482.002	01 39 12.	+ 22 26		15.7	GALAXY
UGC 01188	01 39 12.	+ 22 26	66	15.7	GALAXY S?
PHL 7621	01 39 12.	- 02 34		18.6	BLUE STELLAR OBJECT
MCG+01-05-019	01 39 12.	- 04 27	48	15.	GALAXY
PHL 7618	01 39 12.	- 10 18		18.1	BLUE STELLAR OBJECT
PHL 7622	01 39 12.	- 15 46		18.4	BLUE STELLAR OBJECT
MCG-03-05-014	01 39 12.	- 16 24	84	14.5	GALAXY
ARC 0241	01 39 12.	- 16 30		17.6	RICH CLUSTER OF GALAXIES
PHL 3619	01 39 12.	- 18 02		18.0	BLUE STELLAR OBJECT
PHL 3618	01 39 12.	- 18 14		18.4	BLUE STELLAR OBJECT
PHL 7619	01 39 12.	- 27 50		18.2	BLUE STELLAR OBJECT
LP 03223	01 39 12.	- 51 39		13.4	FAINT BLUE STAR
REIN 2.014	01 39 12.62	+ 51 19 46.9			NEBULA
ARC 0240	01 39 16.	+ 07 23		15.6	RICH CLUSTER OF GALAXIES
PHL 3623	01 39 18.	+ 04 10		17.1	BLUE STELLAR OBJECT
PHL 3620	01 39 18.	+ 04 16		18.9	BLUE STELLAR OBJECT
PHL 3622	01 39 18.	+ 04 27		18.6	BLUE STELLAR OBJECT
ZWG 412.017	01 39 18.	+ 06 52		15.5	GALAXY
UGC 01189	01 39 18.	+ 06 52	78	15.5	GALAXY S
MCG+01-05-021	01 39 18.	+ 06 53	48	15.	GALAXY
5ZW 090	01 39 18.	+ 29 20			COMPACT GALAXY
PHL 7626	01 39 18.	- 05 03		18.2	BLUE STELLAR OBJECT
PHL 7625	01 39 18.	- 05 32		18.0	BLUE STELLAR OBJECT
PHL 7623	01 39 18.	- 07 08		17.0	BLUE STELLAR OBJECT
PHL 7624	01 39 18.	- 08 31		17.1	BLUE STELLAR OBJECT
PHL 1113	01 39 18.	- 12 06		18.4	BLUE STELLAR OBJECT
PHL 3621	01 39 18.	- 21 20		18.6	BLUE STELLAR OBJECT
KN 14.436	01 39 20.5	+ 27 42 06.			NEBULA
MCG+01-05-021	01 39 21.	- 05 08	42	15.5	GALAXY
MCG+01-05-020	01 39 21.	- 05 09	30	15.5	GALAXY
KN 14.435	01 39 21.5	+ 29 35 42.			NEBULA
RNGC 0658	01 39 22.	+ 12 20		13.5	GALAXY
KN 14.437	01 39 23.3	+ 30 03 24.			NEBULA
PHL 7628	01 39 24.	+ 00 48		18.3	BLUE STELLAR OBJECT
PHL 1114	01 39 24.	+ 04 26		18.3	BLUE STELLAR OBJECT
MCG+02-05-009	01 39 24.	+ 12 20	156	13.	GALAXY
MCG+04-05-001	01 39 24.	+ 27 23	36	15.	GALAXY
ZWG 482.003	01 39 24.	+ 27 25		15.3	GALAXY
6ZW 012	01 39 24.	+ 32 03			COMPACT GALAXY
5ZW 091	01 39 24.	+ 33 27			COMPACT GALAXY
OCL 0329	01 39 24.	+ 63 45	540	7.5	OPEN STAR CLUSTER
PHL 3626	01 39 24.	- 12 59		18.8	BLUE STELLAR OBJECT
PHL 7627	01 39 24.	- 15 57		18.3	BLUE STELLAR OBJECT
PHL 1115	01 39 24.	- 21 24		18.2	BLUE STELLAR OBJECT
PHL 3624	01 39 24.	- 27 15		18.3	BLUE STELLAR OBJECT
KN 14.438	01 39 25.3	+ 27 25 24.			NEBULA
MCG+01-05-022	01 39 27.	+ 07 24	36	15.5	GALAXY
VV 177B	01 39 27.	+ 07 24 30.	12	17.	INTERACTING GALAXY
VV 177A	01 39 27.	+ 07 24 30.	15	17.	INTERACTING GALAXY
VV 177	01 39 27.	+ 07 24 30.	30		INTERACTING GALAXY
RNGC 0653	01 39 27.	+ 35 23		14.0	GALAXY
KN 14.439	01 39 28.8	+ 29 35 52.			NEBULA
KN 14.440	01 39 29.0	+ 29 14 51.			NEBULA
UGC 01190	01 39 30.	+ 02 33	60	16.5	GALAXY Sc
MCG+00-05-031	01 39 30.	+ 02 33	36	15.	GALAXY
PHL 7630	01 39 30.	+ 03 27		18.7	BLUE STELLAR OBJECT
MCG+01-05-023	01 39 30.	+ 07 25	36	15.	GALAXY
VV 176B	01 39 30.	+ 07 25 30.	12	17.	INTERACTING GALAXY
VV 176A	01 39 30.	+ 07 25 30.	12	17.	INTERACTING GALAXY
VV 176	01 39 30.	+ 07 25 30.	42		INTERACTING GALAXY
UGC 01191	01 39 30.	+ 07 26	60	16.0	GALAXY DBL SYS
ZWG 437.009	01 39 30.	+ 12 20		13.6	GALAXY
UGC 01192	01 39 30.	+ 12 20	210	13.6	GALAXY Sb
MCG+02-05-010	01 39 30.	+ 13 43	66	18.	GALAXY
ZWG 521.070	01 39 30.	+ 35 23		14.1	GALAXY
UGC 01193	01 39 30.	+ 35 23	96	14.1	GALAXY Sa-b
5ZW 092	01 39 30.	+ 49 48			COMPACT GALAXY
VB 144	01 39 30.	+ 59 52	222		STELLAR RING
ZC 0139.5-0122	01 39 30.	- 01 22	1880		CLUSTER OF GALAXIES
PHL 7629	01 39 30.	- 04 45		18.5	BLUE STELLAR OBJECT
PHL 7631	01 39 30.	- 09 54		18.1	BLUE STELLAR OBJECT
PHL 1116	01 39 30.	- 11 48		17.7	BLUE STELLAR OBJECT
RNGC 0637	01 39 32.	+ 63 45		7.5	OPEN CLUSTER
ARC 0242	01 39 32.	- 14 34		18.0	RICH CLUSTER OF GALAXIES
MCG-02-05-037	01 39 33.	- 13 20 30.	60	14.5	GALAXY
KN 14.441	01 39 33.1	+ 32 10 13.			NEBULA

OBJECT NAME	RIGHT ASCEN.	DECLINATION	DIAM.	MAGN.	TYPE OF OBJECT
KN 14.442	01 39 33.6	+ 30 04 57.			NEBULA
KN 14.443	01 39 34.9	+ 29 34 28.			NEBULA
RNGC 0655	01 39 35.	- 13 20		14.0	GALAXY
PHL 7633	01 39 36.	+ 01 28		16.7	BLUE STELLAR OBJECT
ZWG 482.004	01 39 36.	+ 25 53		13.5	GALAXY
RNGC 0656	01 39 36.	+ 25 53		13.5	GALAXY
UGC 01194	01 39 36.	+ 25 53	90	13.5	GALAXY SBO
KARA.73B 0061	01 39 36.	+ 25 53	90	13.5	ISOLATED GALAXY S
6ZW 013	01 39 36.	+ 30 04			COMPACT GALAXY
PHL 3627	01 39 36.	- 11 56		17.9	BLUE STELLAR OBJECT
PHL 3628	01 39 36.	- 30 20		18.0	BLUE STELLAR OBJECT
PHL 7638	01 39 42.	+ 00 12		17.0	BLUE STELLAR OBJECT
PHL 7637	01 39 42.	+ 05 49		17.8	BLUE STELLAR OBJECT
IC 0148	01 39 42.	+ 13 23 48.			NONSTELLAR OBJECT
ZWG 437.010	01 39 42.	+ 13 43		13.9	GALAXY
UGC 01195	01 39 42.	+ 13 43	210	13.9	GALAXY IRR
MCG+04-05-002	01 39 42.	+ 25 52	72	14.5	GALAXY
ZC 0139.7+2843	01 39 42.	+ 28 43	940		CLUSTER OF GALAXIES
MCG+06-04-058	01 39 42.	+ 35 23	84	14.	GALAXY
4ZW 058	01 39 42.	+ 38 32			COMPACT GALAXY
ZWG 386.034	01 39 42.	- 02 28		15.7	GALAXY
PHL 7635	01 39 42.	- 03 26		18.5	BLUE STELLAR OBJECT
PHL 7639	01 39 42.	- 05 14		17.2	BLUE STELLAR OBJECT
PHL 3629	01 39 42.	- 11 46		18.3	BLUE STELLAR OBJECT
PHL 7636	01 39 42.	- 25 42		18.6	BLUE STELLAR OBJECT
MCG-05-05-011	01 39 42.	- 31 17	60	15.	GALAXY
MCG+02-05-011	01 39 45.	+ 13 43	204	13.	GALAXY
PHL 7640	01 39 48.	+ 00 40		16.6	BLUE STELLAR OBJECT
UGC 01196	01 39 48.	+ 06 28	66	16.5	GALAXY Sb
PHL 7642	01 39 48.	+ 07 43		16.9	BLUE STELLAR OBJECT
ZWG 460.019	01 39 48.	+ 18 03		15.3	GALAXY
UGC 01197	01 39 48.	+ 18 03	120	15.3	GALAXY IRR
PHL 3630	01 39 48.	- 00 23		13.1	BLUE STELLAR OBJECT
ZWG 386.035	01 39 48.	- 02 23		14.9	GALAXY
KARA.72 039B	01 39 48.	- 02 23	24		PART OF DOUBLE GALAXY
KARA.72 039A	01 39 48.	- 02 23	24	14.9	PART OF DOUBLE GALAXY
PHL 7643	01 39 48.	- 03 56		18.6	BLUE STELLAR OBJECT
PHL 7644	01 39 48.	- 05 51		18.6	BLUE STELLAR OBJECT
PHL 7645	01 39 48.	- 06 09		18.5	BLUE STELLAR OBJECT
PHL 7641	01 39 48.	- 14 56		18.4	BLUE STELLAR OBJECT
KN 14.444	01 39 52.7	+ 28 41 33.			NEBULA
PHL 7648	01 39 54.	+ 01 22		17.9	BLUE STELLAR OBJECT
PHL 3631	01 39 54.	+ 03 39		17.8	BLUE STELLAR OBJECT
PHL 3632	01 39 54.	+ 06 10		18.6	BLUE STELLAR OBJECT
BC PHL3632	01 39 54.	+ 06 10		18.15	QUASI-STELLAR OBJECT
SHB 042	01 39 54.	+ 06 10		18.2	QUASI-STELLAR OBJECT
6ZW 014	01 39 54.	+ 31 52			COMPACT GALAXY
MCG+08-04-005	01 39 54.	+ 50 11	48	16.	GALAXY
PHL 7632	01 39 54.	- 05 15		18.4	BLUE STELLAR OBJECT
PHL 3633	01 39 54.	- 10 09		18.6	BLUE STELLAR OBJECT
PHL 7646	01 39 54.	- 16 14		16.3	BLUE STELLAR OBJECT
PHL 3634	01 39 54.	- 18 48		18.0	BLUE STELLAR OBJECT
PHL 7649	01 39 54.	- 28 22		18.4	BLUE STELLAR OBJECT
PHL 7647	01 39 54.	- 30 29		18.2	BLUE STELLAR OBJECT
KN 14.446	01 39 57.1	+ 28 28 52.			NEBULA
IC 0149	01 39 58.	- 16 32 50.			NONSTELLAR OBJECT
KN 14.445	01 39 59.7	+ 31 56 40.			NEBULA
PHL 7650	01 40 00.	+ 01 10		16.3	BLUE STELLAR OBJECT
KARA.68 009	01 40 00.	+ 02 42	34		DWARF GALAXY
PHL 3638	01 40 00.	+ 03 23		17.8	BLUE STELLAR OBJECT
PHL 7655	01 40 00.	+ 04 18		18.4	BLUE STELLAR OBJECT
PHL 3635	01 40 00.	+ 06 07		18.0	BLUE STELLAR OBJECT
MCG+01-05-024	01 40 00.	+ 07 57	48	15.	GALAXY
PHL 7652	01 40 00.	+ 08 12		18.5	BLUE STELLAR OBJECT
5ZW 093	01 40 00.	+ 25 22			COMPACT GALAXY
ZWG 503.007	01 40 00.	+ 31 14		15.6	GALAXY
MCG+05-05-002	01 40 00.	+ 31 14	30	15.5	GALAXY
5ZW 094	01 40 00.	+ 42 02			COMPACT GALAXY
LDN 1334	01 40 00.	+ 61 42	420		DARK NEBULA
7ZW 003	01 40 00.	+ 85 01			COMPACT GALAXY
ZWG 361.004	01 40 00.	+ 85 01		15.1	GALAXY
ZWG 360.007	01 40 00.	+ 85 01		15.1	GALAXY
UGC 01198	01 40 00.	+ 85 01	60	15.1	GALAXY ERUPTVE
ZWG 386.036	01 40 00.	- 01 04		15.3	GALAXY
PHL 3636	01 40 00.	- 02 01		18.9	BLUE STELLAR OBJECT
PHL 3639	01 40 00.	- 02 24		18.5	BLUE STELLAR OBJECT
PHL 7653	01 40 00.	- 05 44		16.9	BLUE STELLAR OBJECT
PHL 3640	01 40 00.	- 05 48		18.8	BLUE STELLAR OBJECT
PHL 7651	01 40 00.	- 09 14		16.5	BLUE STELLAR OBJECT
ABC 0243	01 40 00.	- 10 29		16.6	RICH CLUSTER OF GALAXIES
PHL 7656	01 40 00.	- 13 37		18.3	BLUE STELLAR OBJECT
PHL 7654	01 40 00.	- 14 34		17.1	BLUE STELLAR OBJECT
MCG-03-05-015	01 40 00.	- 16 33	60	14.	GALAXY
PHL 1117	01 40 00.	- 16 50		17.7	BLUE STELLAR OBJECT
PHL 3641	01 40 00.	- 18 18		18.2	BLUE STELLAR OBJECT
PHL 3637	01 40 00.	- 22 31		18.7	BLUE STELLAR OBJECT
MCG-06-04-066	01 40 00.	- 33 32	24	15.5	GALAXY
KN 14.447	01 40 00.4	+ 31 34 26.			NEBULA
KN 14.448	01 40 00.5	+ 31 13 34.			NEBULA
KN 14.449	01 40 02.0	+ 30 01 38.			NEBULA
KN 14.450	01 40 02.8	+ 29 53 25.			NEBULA
LIN 551	01 40 04.	- 73 31 45.		15.59	SMC EMISSION LINE OBJECT
PHL 7658	01 40 06.	+ 04 14		18.3	BLUE STELLAR OBJECT
PHL 3643	01 40 06.	+ 05 55		18.0	BLUE STELLAR OBJECT
PHL 7657	01 40 06.	+ 07 30		17.0	BLUE STELLAR OBJECT
ZWG 412.018	01 40 06.	+ 07 55		15.6	GALAXY
UGC 01199	01 40 06.	+ 07 55	66	15.6	GALAXY Sa-b
PHL 3642	01 40 06.	+ 07 58		18.1	BLUE STELLAR OBJECT
MCG+02-05-012	01 40 06.	+ 12 53	120	13.	GALAXY
ZWG 437.011	01 40 06.	+ 12 54		14.3	GALAXY
UGC 01200	01 40 06.	+ 12 54	126	14.3	GALAXY IRR
ZC 0140.1+3144	01 40 06.	+ 31 44	3700		CLUSTER OF GALAXIES
PHL 7659	01 40 06.	- 02 28		18.4	BLUE STELLAR OBJECT
PHL 1118	01 40 06.	- 16 23		18.6	BLUE STELLAR OBJECT
PHL 3644	01 40 06.	- 26 30		18.7	BLUE STELLAR OBJECT
KN 14.452	01 40 06.2	+ 27 58 06.			NEBULA
KN 14.451	01 40 07.	+ 31 37 48.			NEBULA
KN 14.454	01 40 10.2	+ 30 13 02.			NEBULA
KN 14.453	01 40 10.6	+ 32 05 26.			NEBULA
PHL 7664	01 40 12.	+ 01 20		17.8	BLUE STELLAR OBJECT
ZWG 386.037	01 40 12.	+ 01 30		15.7	GALAXY
PHL 3646	01 40 12.	+ 04 04		17.7	BLUE STELLAR OBJECT
PHL 7665	01 40 12.	+ 06 52		18.2	BLUE STELLAR OBJECT
PHL 7661	01 40 12.	+ 07 14		18.5	BLUE STELLAR OBJECT
ZWG 412.019	01 40 12.	+ 07 50		15.5	GALAXY
MCG+01-05-025	01 40 12.	+ 07 51	18	15.5	GALAXY
PHL 1119	01 40 12.	+ 08 06		18.2	BLUE STELLAR OBJECT
PHL 3645	01 40 12.	+ 08 16		16.6	BLUE STELLAR OBJECT

OBJECT NAME	RIGHT ASCEN.	DECLINATION	DIAM.	MAGN.	TYPE OF OBJECT
ZC 0140.2+1940	01 40 12.	+ 19 40	1410		CLUSTER OF GALAXIES
ZC 0140.2+3718	01 40 12.	+ 37 18	870		CLUSTER OF GALAXIES
PHL 3650	01 40 12.	- 00 10		18.3	BLUE STELLAR OBJECT
PHL 7666	01 40 12.	- 04 27		18.6	BLUE STELLAR OBJECT
PHL 7660	01 40 12.	- 04 38		16.6	BLUE STELLAR OBJECT
PHL 7662	01 40 12.	- 05 24		18.0	BLUE STELLAR OBJECT
PHL 7667	01 40 12.	- 05 40		18.6	BLUE STELLAR OBJECT
PHL 7668	01 40 12.	- 05 43		18.5	BLUE STELLAR OBJECT
PHL 3647	01 40 12.	- 05 59		18.8	BLUE STELLAR OBJECT
PHL 7663	01 40 12.	- 06 57		18.9	BLUE STELLAR OBJECT
PHL 7669	01 40 12.	- 10 08		18.2	BLUE STELLAR OBJECT
PHL 3651	01 40 12.	- 10 36		18.3	BLUE STELLAR OBJECT
PHL 3649	01 40 12.	- 12 06		17.9	BLUE STELLAR OBJECT
PHL 3648	01 40 12.	- 20 03		18.4	BLUE STELLAR OBJECT
KN 14.455	01 40 14.2	+ 31 35 51.			NEBULA
RNGC 0660	01 40 16.	+ 13 23		13.0	GALAXY
PHL 7673	01 40 18.	+ 09 12		18.3	BLUE STELLAR OBJECT
ZWG 437.012	01 40 18.	+ 13 23		12.8	GALAXY
UGC 01201	01 40 18.	+ 13 23	600	12.8	GALAXY SB(a)
MCG+02-05-013	01 40 18.	+ 13 23	456	12.	GALAXY
MCG+04-05-003	01 40 18.	+ 27 29	54	15.	GALAXY
ZWG 503.008	01 40 18.	+ 27 30		14.8	GALAXY
ZWG 482.005	01 40 18.	+ 27 30		14.8	GALAXY
PHL 7672	01 40 18.	- 03 21		18.7	BLUE STELLAR OBJECT
PHL 3652	01 40 18.	- 09 00		18.3	BLUE STELLAR OBJECT
PHL 7671	01 40 18.	- 15 40		16.1	BLUE STELLAR OBJECT
PHL 7670	01 40 18.	- 25 08		15.2	BLUE STELLAR OBJECT
IC 0150	01 40 19.	+ 03 56 04.			NONSTELLAR OBJECT
KN 14.457	01 40 19.3	+ 27 30 00.			NEBULA
MCG+01-05-026	01 40 21.	+ 03 56	60	14.5	GALAXY
MCG+01-05-022	01 40 21.	- 06 24	60	15.	GALAXY
MCG-03-05-016	01 40 21.	- 18 30	84	14.	GALAXY
KN 14.456	01 40 21.8	+ 30 26 18.			NEBULA
PHL 7679	01 40 24.	+ 02 18		18.3	BLUE STELLAR OBJECT
PHL 7676	01 40 24.	+ 03 57		16.6	BLUE STELLAR OBJECT
ZWG 412.020	01 40 24.	+ 03 57		14.9	GALAXY
UGC 01202	01 40 24.	+ 03 57	60	14.9	GALAXY Sb
PHL 3653	01 40 24.	+ 04 34		18.0	BLUE STELLAR OBJECT
UGC 01203	01 40 24.	+ 05 35	96	16.5	GALAXY
PHL 3654	01 40 24.	+ 06 14		18.2	BLUE STELLAR OBJECT
ZWG 521.071	01 40 24.	+ 36 20		15.3	GALAXY
PHL 7678	01 40 24.	- 00 04		18.7	BLUE STELLAR OBJECT
PHL 7674	01 40 24.	- 03 58		17.9	BLUE STELLAR OBJECT
PHL 7675	01 40 24.	- 07 14		18.1	BLUE STELLAR OBJECT
PHL 3655	01 40 24.	- 10 24		18.0	BLUE STELLAR OBJECT
PHL 7680	01 40 24.	- 11 13		18.0	BLUE STELLAR OBJECT
PHL 1120	01 40 24.	- 12 29		15.7	BLUE STELLAR OBJECT
PHL 3656	01 40 24.	- 16 22		18.6	BLUE STELLAR OBJECT
PHL 7677	01 40 24.	- 19 22		16.5	BLUE STELLAR OBJECT
KN 14.458	01 40 24.8	+ 31 03 26.			NEBULA
HOW 84	01 40 25.	- 71 25 03.	48		STAR CLUSTER IN SMC
KN 14.459	01 40 26.3	+ 31 03 39.			NEBULA
KN 14.460	01 40 27.4	+ 30 14 26.			NEBULA
RNGC 0657	01 40 28.	+ 55 37			OPEN CLUSTER
RNGC 0654	01 40 29.	+ 61 38		10.0	OPEN CLUSTER
PHL 3657	01 40 30.	+ 03 25		17.0	BLUE STELLAR OBJECT
OCL 0337	01 40 30.	+ 55 37			OPEN STAR CLUSTER
PHL 3658	01 40 30.	- 01 00		18.8	BLUE STELLAR OBJECT
PHL 7681	01 40 30.	- 23 22		17.9	BLUE STELLAR OBJECT
KN 14.461	01 40 30.9	+ 30 21 32.			NEBULA
MCG+01-05-027	01 40 33.	+ 04 01	72	15.	GALAXY
KN 14.462	01 40 35.7	+ 28 52 54.			NEBULA
PHL 3659	01 40 36.	+ 00 25		18.8	BLUE STELLAR OBJECT
PHL 7683	01 40 36.	+ 03 38		17.4	BLUE STELLAR OBJECT
ZWG 412.021	01 40 36.	+ 04 02		15.5	GALAXY
UGC 01204	01 40 36.	+ 04 02	96	15.5	GALAXY Sb-c
PHL 7682	01 40 36.	+ 08 30		16.7	BLUE STELLAR OBJECT
ZWG 412.022	01 40 36.	+ 08 38		14.2	GALAXY
UGC 01205	01 40 36.	+ 08 38	216	14.2	GALAXY Sb
ZWG 462.006	01 40 36.	+ 22 10		15.7	GALAXY
MCG+05-05-003	01 40 36.	+ 31 20	36	17.	GALAXY
4ZW 059	01 40 36.	+ 39 13			COMPACT GALAXY
5ZW 095	01 40 36.	+ 53 01			COMPACT GALAXY
VDB.66N 006	01 40 36.	+ 61 35	108		REFLECTION NEBULA
OCL 0330	01 40 36.	+ 61 38	660	10.0	OPEN STAR CLUSTER
MCG+01-05-023	01 40 36.	- 04 18	30	15.	GALAXY
MCG+01-05-024	01 40 36.	- 04 57 30.	54	15.	GALAXY
PHL 3660	01 40 36.	- 08 40		18.8	BLUE STELLAR OBJECT
PHL 1121	01 40 36.	- 10 38		18.4	BLUE STELLAR OBJECT
PHL 7684	01 40 36.	- 20 17		18.7	BLUE STELLAR OBJECT
IC 1722	01 40 38.	- 34 26 40.			NONSTELLAR OBJECT
HN 0154	01 40 38.	- 34 27			NEBULA
KN 14.463	01 40 38.1	+ 30 21 10.			NEBULA
MCG+01-05-028	01 40 39.	+ 06 40	180	13.	GALAXY
PHL 7685	01 40 42.	+ 04 57		18.5	BLUE STELLAR OBJECT
MCG+02-05-014	01 40 42.	+ 11 38	36	14.5	GALAXY
ZWG 437.013	01 40 42.	+ 11 39		15.3	GALAXY
UGC 01206	01 40 42.	+ 11 39	108	15.3	GALAXY
5ZW 096	01 40 42.	+ 28 42			COMPACT GALAXY
ZWG 503.009	01 40 42.	+ 30 21		15.5	GALAXY
PHL 7686	01 40 42.	- 08 50		18.3	BLUE STELLAR OBJECT
PHL 1122	01 40 42.	- 28 24		17.7	BLUE STELLAR OBJECT
PHL 3661	01 40 42.	- 30 54		18.1	BLUE STELLAR OBJECT
LB 03228	01 40 42.	- 73 21		13.5	FAINT BLUE STAR
MCG-03-05-017	01 40 45.	- 16 15	48	16.	GALAXY
KN 14.464	01 40 45.1	+ 28 42 18.			NEBULA
KN 14.465	01 40 46.9	+ 30 16 07.			NEBULA
PHL 7690	01 40 48.	+ 00 41		18.4	BLUE STELLAR OBJECT
PHL 3664	01 40 48.	+ 01 16		18.8	BLUE STELLAR OBJECT
PHL 7689	01 40 48.	+ 05 34		17.8	BLUE STELLAR OBJECT
PHL 3662	01 40 48.	+ 05 42		17.4	BLUE STELLAR OBJECT
ZC 0140.8+1200	01 40 48.	+ 12 00	1140		CLUSTER OF GALAXIES
ZWG 437.014	01 40 48.	+ 14 08		15.5	GALAXY
ZWG 503.010	01 40 48.	+ 28 42		15.4	GALAXY
OCL 0332	01 40 48.	+ 60 27	600	9.9	OPEN STAR CLUSTER
RNGC 0659	01 40 48.	+ 60 27		10.0	OPEN CLUSTER
PHL 7688	01 40 48.	- 03 02		18.2	BLUE STELLAR OBJECT
PHL 3663	01 40 48.	- 03 04		18.6	BLUE STELLAR OBJECT
PHL 7687	01 40 48.	- 06 48		16.9	BLUE STELLAR OBJECT
MCG-06-04-067	01 40 48.	- 34 26	78	15.	GALAXY
KN 14.467	01 40 49.7	+ 28 41 25.			NEBULA
KN 14.466	01 40 51.2	+ 31 21 07.			NEBULA
KN 14.468	01 40 52.6	+ 30 06 02.			NEBULA
LIN 552	01 40 53.	- 74 11 29.		14.99	SMC EMISSION LINE OBJECT
ZC 0140.9+2551	01 40 54.	+ 25 51	400		CLUSTER OF GALAXIES
ZWG 503.011	01 40 54.	+ 31 21		15.6	GALAXY
PHL 1123	01 40 54.	- 02 05		16.3	BLUE STELLAR OBJECT
PHL 7691	01 40 54.	- 02 52		18.5	BLUE STELLAR OBJECT

OBJECT NAME	RIGHT ASCEN.	DECLINATION	DIAM.	MAGN.	TYPE OF OBJECT
PHL 7692	01 40 54.	- 08 34		18.0	BLUE STELLAR OBJECT
MCG-02-05-038	01 40 54.	- 13 54	42	14.5	GALAXY
KN 14.469	01 40 55.9	+ 30 05 57.			NEBULA
HN 0155	01 40 56.	- 34 28			NEBULA
IC 1724	01 40 56.	- 34 28			NONSTELLAR OBJECT
KN 14.470	01 40 57.4	+ 28 55 01.			NEBULA
LBN 0637	01 41	+ 55 20	1320		BRIGHT NEBULA
MCG+00-05-032	01 41 00.	+ 01 48	42	15.	GALAXY
ZWG 386.038	01 41 00.	+ 01 50		15.7	GALAXY
UGC 01208	01 41 00.	+ 01 50	72	15.7	GALAXY Sa-b
PHL 3665	01 41 00.	+ 02 28		18.8	BLUE STELLAR OBJECT
PHL 7695	01 41 00.	+ 06 46		17.9	BLUE STELLAR OBJECT
MCG+02-05-015	01 41 00.	+ 11 53	60	14.5	GALAXY
KARA.68 010	01 41 00.	+ 15 25	60		DWARF GALAXY
ZWG 503.012	01 41 00.	+ 28 55		15.3	GALAXY
UGC 01207	01 41 00.	+ 81 58	108	16.5	GALAXY DWARF
PHL 7696	01 41 00.	- 00 54		17.3	BLUE STELLAR OBJECT
PHL 3666	01 41 00.	- 03 16		18.6	BLUE STELLAR OBJECT
PHL 7693	01 41 00.	- 04 27		16.7	BLUE STELLAR OBJECT
PHL 7694	01 41 00.	- 06 16		16.7	BLUE STELLAR OBJECT
PHL 7697	01 41 00.	- 07 42		18.6	BLUE STELLAR OBJECT
PHL 7698	01 41 00.	- 14 07		18.5	BLUE STELLAR OBJECT
MCG-06-04-068	01 41 00.	- 34 30	48	13.5	GALAXY
KN 14.471	01 41 01.9	+ 30 34 24.			NEBULA
PHL 1125	01 41 06.	+ 04 03		18.9	BLUE STELLAR OBJECT
PHL 1124	01 41 06.	+ 04 13		17.9	BLUE STELLAR OBJECT
ZWG 437.015	01 41 06.	+ 11 55		15.0	GALAXY
UGC 01209	01 41 06.	+ 11 55	90	15.0	GALAXY S-IRR
5ZW 097	01 41 06.	+ 28 40			COMPACT GALAXY
ISS 0052	01 41 06.	+ 58 59	425		STELLAR RING
PHL 7699	01 41 06.	- 06 06		17.8	BLUE STELLAR OBJECT
MCG-06-04-069	01 41 06.	- 34 28	36	15.	GALAXY
KN 14.472	01 41 07.4	+ 30 33 27.			NEBULA
RNGC 0643C	01 41 08.	- 75 31			GALAXY
KN 14.474	01 41 08.1	+ 28 26 51.			NEBULA
KN 14.473	01 41 08.9	+ 30 34 20.			NEBULA
MCG+01-05-029	01 41 09.	+ 03 57	72	13.	GALAXY
PHL 7702	01 41 12.	+ 00 16		18.1	BLUE STELLAR OBJECT
ZWG 412.023	01 41 12.	+ 03 59		13.9	GALAXY
RNGC 0664	01 41 12.	+ 03 59		14.0	GALAXY
UGC 01210	01 41 12.	+ 03 59	120	13.9	GALAXY Sb
PHL 3668	01 41 12.	+ 04 39		18.5	BLUE STELLAR OBJECT
UGC 01211	01 41 12.	+ 13 33	180	17.	GALAXY DWARF
3ZW 033	01 41 12.	+ 16 48			COMPACT GALAXY
ZWG 460.020	01 41 12.	+ 16 48		14.7	GALAXY
MRK 360	01 41 12.	+ 16 48	12	15.5	GALAXY WITH UV CONTINUUM
MCG+03-05-013	01 41 12.	+ 16 49	24	15.	GALAXY
MCG+06-04-059	01 41 12.	+ 34 08	72	14.5	GALAXY
ZWG 521.072	01 41 12.	+ 34 09		14.5	GALAXY
UGC 01212	01 41 12.	+ 34 09	102	14.5	GALAXY Sb
VB 145	01 41 12.	+ 59 02	423		STELLAR RING
MCG+14-02-006	01 41 12.	+ 85 01	30	15.	GALAXY
PHL 3667	01 41 12.	- 01 02		17.5	BLUE STELLAR OBJECT
MCG+01-05-025	01 41 12.	- 04 15	78	14.5	GALAXY
PHL 3670	01 41 12.	- 06 55		18.7	BLUE STELLAR OBJECT
PHL 7700	01 41 12.	- 07 11		16.8	BLUE STELLAR OBJECT
PHL 3669	01 41 12.	- 07 54		18.3	BLUE STELLAR OBJECT
MCG+01-05-026	01 41 12.	- 08 06	48	15.5	GALAXY
PHL 7701	01 41 12.	- 23 38		17.3	BLUE STELLAR OBJECT
DV.56 N0643C	01 41 12.	- 75 31	84		S GALAXY
KN 14.475	01 41 14.3	+ 34 08 05.			NEBULA
MCG+02-05-016	01 41 15.	+ 13 32 30.	36	16.	GALAXY
KN 14.476	01 41 15.5	+ 30 16 12.			NEBULA
KN 14.477	01 41 16.8	+ 30 33 32.			NEBULA
IC 0151	01 41 17.	+ 12 57 07.			NONSTELLAR OBJECT
PHL 3672	01 41 18.	+ 00 56		18.9	BLUE STELLAR OBJECT
PHL 3671	01 41 18.	+ 04 26		18.0	BLUE STELLAR OBJECT
ZWG 437.016	01 41 18.	+ 12 49		15.7	GALAXY
3ZW 034	01 41 18.	+ 15 00			COMPACT GALAXY
6ZW 015	01 41 18.	+ 27 40			COMPACT GALAXY
MCG+05-05-004	01 41 18.	+ 31 03	42	15.	GALAXY
ZWG 503.013	01 41 18.	+ 31 05		14.9	GALAXY
UGC 01213	01 41 18.	+ 31 05	78	14.9	GALAXY S0
6ZW 016	01 41 18.	+ 33 24			COMPACT GALAXY
BC 4C33.3	01 41 18.	+ 33 57		17.5	QUASI-STELLAR OBJECT
SHB 043	01 41 18.	+ 33 57 00.		17.5	QUASI-STELLAR OBJECT
MCG+07-04-010	01 41 18.	+ 42 46	48	17.	GALAXY
PHL 7703	01 41 18.	- 08 20		18.6	BLUE STELLAR OBJECT
PHL 7704	01 41 18.	- 12 04		17.0	BLUE STELLAR OBJECT
PHL 3673	01 41 18.	- 14 49		18.1	BLUE STELLAR OBJECT
KN 14.478	01 41 18.0	+ 31 04 05.			NEBULA
HOW 85	01 41 20.	- 71 32 09.	12		STAR CLUSTER IN SMC
RNGC 0661	01 41 23.	+ 28 27		13.0	GALAXY
ZWG 386.039	01 41 24.	+ 02 06		14.0	GALAXY
MRK 573	01 41 24.	+ 02 06	25	14.	GALAXY WITH UV CONTINUUM
UGC 01214	01 41 24.	+ 02 06	114	14.0	GALAXY S0
MCG+00-05-033	01 41 24.	+ 02 06	66	13.5	GALAXY
MRK 572	01 41 24.	+ 11 55	45	14.	GALAXY WITH UV CONTINUUM
MCG+05-05-005	01 41 24.	+ 28 26	36	14.	GALAXY
ZWG 503.014	01 41 24.	+ 28 28		13.0	GALAXY
UGC 01215	01 41 24.	+ 28 28	120	13.0	GALAXY E
UGC 01216	01 41 24.	+ 40 41	60	18.	GALAXY DWRF SP
ZWG 537.021	01 41 24.	+ 42 47		15.6	GALAXY
UGC 01217	01 41 24.	+ 42 47	66	15.6	GALAXY S
MCG+12-02-003	01 41 24.	+ 73 14	27	16.	GALAXY
PHL 3674	01 41 24.	- 00 57		18.8	BLUE STELLAR OBJECT
PHL 3676	01 41 24.	- 03 44		18.2	BLUE STELLAR OBJECT
PHL 7705	01 41 24.	- 09 33		16.5	BLUE STELLAR OBJECT
PHL 7706	01 41 24.	- 10 29		16.8	BLUE STELLAR OBJECT
MCG-03-05-018	01 41 24.	- 16 13	48	14.	GALAXY
PHL 3675	01 41 24.	- 17 42		18.3	BLUE STELLAR OBJECT
PHL 1126	01 41 24.	- 24 20		10.5	BLUE STELLAR OBJECT
PHL 3677	01 41 24.	- 29 05		18.4	BLUE STELLAR OBJECT
MCG-06-04-070	01 41 24.	- 33 58	24	15.	GALAXY
KN 14.479	01 41 25.2	+ 28 27 20.			NEBULA
ARC 0245	01 41 27.	+ 06 09		16.4	RICH CLUSTER OF GALAXIES
IC 0152	01 41 27.	+ 12 47 06.			NONSTELLAR OBJECT
ARC 0244	01 41 27.	+ 18 15		17.6	RICH CLUSTER OF GALAXIES
KN 14.480	01 41 29.6	+ 29 37 09.			NEBULA
PHL 1127	01 41 30.	+ 05 14		18.5	BLUE STELLAR OBJECT
SHB 044	01 41 30.	+ 05 14		18.3	QUASI-STELLAR OBJECT
6ZW 017	01 41 30.	+ 29 50			COMPACT GALAXY
TON-S 0227	01 41 30.	- 24 18		12.5	BLUE STAR
MCG-06-04-071	01 41 30.	- 36 21	90	12.5	GALAXY
BC PHL1127	01 41 33.	+ 05 15 17.		18.29	QUASI-STELLAR OBJECT
KN 14.481	01 41 33.9	+ 32 25 09.			NEBULA
IC 1723	01 41 35.	+ 08 32 54.			NONSTELLAR OBJECT
PHL 7708	01 41 36.	+ 07 10		18.2	BLUE STELLAR OBJECT

OBJECT NAME	RIGHT ASCEN.	DECLINATION	DIAM.	MAGN.	TYPE OF OBJECT
ZC 0141.6+0930	01 41 36.	+ 09 30	1210		CLUSTER OF GALAXIES
ZWG 437.017	01 41 36.	+ 12 00		15.5	GALAXY
UGC 01218	01 41 36.	+ 12 00	60	15.5	GALAXY SB:b-c
MCG+02-05-017	01 41 36.	+ 12 00	48	15.	GALAXY
ZWG 460.021	01 41 36.	+ 17 13		13.6	GALAXY
UGC 01219	01 41 36.	+ 17 13	84	13.6	GALAXY SB
MCG+03-05-014	01 41 36.	+ 17 14	66	13.5	GALAXY
ZWG 521.073	01 41 36.	+ 37 26		13.6	GALAXY
UGC 01220	01 41 36.	+ 37 26	54	13.6	GALAXY S
KARA.73B 0062	01 41 36.	+ 37 26	48	13.6	ISOLATED GALAXY S
6ZW 018	01 41 36.	+ 38 22			COMPACT GALAXY
PHL 7707	01 41 36.	- 02 48		17.5	BLUE STELLAR OBJECT
KN 14.483	01 41 37.9	+ 29 04 04.			NEBULA
RNGC 0662	01 41 38.	+ 37 26		13.5	GALAXY
KN 14.482	01 41 38.2	+ 31 45 46.			NEBULA
KN 14.484	01 41 39.5	+ 27 40 16.			NEBULA
HOW 86	01 41 41.	- 74 27 14.	36		STAR CLUSTER IN SMC
ZWG 521.074	01 41 42.	+ 34 25		15.0	GALAXY
ZWG 521.075	01 41 42.	+ 35 34		15.3	GALAXY
5ZW 098	01 41 42.	+ 37 27			COMPACT GALAXY
ZWG 521.076	01 41 42.	+ 37 57		15.0	GALAXY
UGC 01221	01 41 42.	+ 37 57	78	15.0	GALAXY Sb-c
ZWG 537.022	01 41 42.	+ 42 30		15.7	GALAXY
SCHO 0041	01 41 42.	+ 60 58 24.	350		ISOLATED DARK CLOUD
PHL 7711	01 41 42.	- 07 14		18.0	BLUE STELLAR OBJECT
PHL 7712	01 41 42.	- 08 27		18.7	BLUE STELLAR OBJECT
PHL 3678	01 41 42.	- 08 28		18.5	BLUE STELLAR OBJECT
PHL 7710	01 41 42.	- 11 24		17.1	BLUE STELLAR OBJECT
PHL 7709	01 41 42.	- 17 08		15.8	BLUE STELLAR OBJECT
MCG+02-05-018	01 41 45.	+ 12 10	48	15.	GALAXY
LIN 554	01 41 46.	- 74 07 36.		14.92	SMC EMISSION LINE OBJECT
PHL 7721	01 41 48.	+ 00 31		18.1	BLUE STELLAR OBJECT
PHL 1128	01 41 48.	+ 01 08		17.6	BLUE STELLAR OBJECT
PHL 7722	01 41 48.	+ 01 24		18.6	BLUE STELLAR OBJECT
PHL 7714	01 41 48.	+ 01 44		17.8	BLUE STELLAR OBJECT
PHL 3680	01 41 48.	+ 03 20		16.3	BLUE STELLAR OBJECT
PHL 7718	01 41 48.	+ 03 49		18.1	BLUE STELLAR OBJECT
ZWG 437.018	01 41 48.	+ 12 10		15.2	GALAXY
3ZW 035	01 41 48.	+ 16 50			COMPACT GALAXY
ZWG 482.007	01 41 48.	+ 21 40		15.3	GALAXY
MCG+06-04-060	01 41 48.	+ 37 27	48	14.	GALAXY
PHL 7719	01 41 48.	- 00 13		17.0	BLUE STELLAR OBJECT
PHL 7713	01 41 48.	- 00 48		15.8	BLUE STELLAR OBJECT
PHL 7716	01 41 48.	- 03 49		18.1	BLUE STELLAR OBJECT
PHL 7720	01 41 48.	- 06 10		17.1	BLUE STELLAR OBJECT
PHL 7717	01 41 48.	- 07 31		18.0	BLUE STELLAR OBJECT
PHL 3679	01 41 48.	- 08 04		18.8	BLUE STELLAR OBJECT
PHL 7715	01 41 48.	- 08 33		18.3	BLUE STELLAR OBJECT
PHL 7723	01 41 48.	- 11 02		18.6	BLUE STELLAR OBJECT
KN 14.485	01 41 48.0	+ 31 03 12.			NEBULA
LIN 553	01 41 51.	- 73 20 48.		15.55	SMC EMISSION LINE OBJECT
UGC 01222	01 41 54.	+ 04 25	72	16.5	GALAXY
5ZW 099	01 41 54.	+ 23 53			COMPACT GALAXY
6ZW 019	01 41 54.	+ 28 39			COMPACT GALAXY
MCG+06-04-061	01 41 54.	+ 37 58	48	15.	GALAXY
PHL 3681	01 41 54.	- 02 40		18.9	BLUE STELLAR OBJECT
LB 03224	01 41 54.	- 67 37		13.5	FAINT BLUE STAR
KN 14.486	01 41 55.5	+ 28 29 13.			NEBULA
IC 0153	01 41 56.	+ 12 22 41.			NONSTELLAR OBJECT
MCG-02-05-039	01 41 57.	- 09 01 30.	36	15.5	GALAXY
KN 14.487	01 41 57.3	+ 30 14 34.			NEBULA
LBN 0638	01 42	+ 54 50	1200		BRIGHT NEBULA
KHAV 029	01 42	+ 58 45	10280		DARK NEBULA
PHL 3685	01 42 00.	+ 01 57		18.9	BLUE STELLAR OBJECT
PHL 1129	01 42 00.	+ 04 17		18.8	BLUE STELLAR OBJECT
MCG+01-05-030	01 42 00.	+ 04 38	48	15.	GALAXY
PHL 3684	01 42 00.	+ 05 50		18.2	BLUE STELLAR OBJECT
PHL 3683	01 42 00.	+ 06 04		17.1	BLUE STELLAR OBJECT
ZC 0142.0+0607	01 42 00.	+ 06 07	3630		CLUSTER OF GALAXIES
MRK 574	01 42 00.	+ 11 28	7	16.5	GALAXY WITH UV CONTINUUM
6ZW 020	01 42 00.	+ 28 45			COMPACT GALAXY
LDN 1337	01 42 00.	+ 61 40	420		DARK NEBULA
PHL 7726	01 42 00.	- 01 20		18.8	BLUE STELLAR OBJECT
PHL 7724	01 42 00.	- 03 26		18.4	BLUE STELLAR OBJECT
PHL 7725	01 42 00.	- 06 25		18.4	BLUE STELLAR OBJECT
TON-S 0228	01 42 00.	- 22 12		14.8	BLUE STAR
PHL 3682	01 42 00.	- 22 14		15.1	BLUE STELLAR OBJECT
PHL 3686	01 42 00.	- 23 17		18.0	BLUE STELLAR OBJECT
KN 14.488	01 42 01.5	+ 31 40 44.			NEBULA
KN 14.489	01 42 02.6	+ 28 05 49.			NEBULA
PHL 7730	01 42 06.	+ 01 27		18.7	BLUE STELLAR OBJECT
ZC 0142.1+0308	01 42 06.	+ 03 08	1950		CLUSTER OF GALAXIES
PHL 3689	01 42 06.	+ 03 22		17.0	BLUE STELLAR OBJECT
ZWG 412.024	01 42 06.	+ 04 39		15.5	GALAXY
KARA.73B 0063	01 42 06.	+ 04 39	60	15.5	ISOLATED GALAXY S
PHL 7727	01 42 06.	+ 04 44		15.8	BLUE STELLAR OBJECT
PHL 7729	01 42 06.	+ 07 46		18.2	BLUE STELLAR OBJECT
MRK 361	01 42 06.	+ 16 52	8	16.	GALAXY WITH UV CONTINUUM
MCG+04-05-004	01 42 06.	+ 21 36	42	15.	GALAXY
ZWG 482.008	01 42 06.	+ 21 38		15.2	GALAXY
PHL 1130	01 42 06.	- 03 06		17.2	BLUE STELLAR OBJECT
PHL 3687	01 42 06.	- 03 28		18.8	BLUE STELLAR OBJECT
PHL 3690	01 42 06.	- 09 30		18.2	BLUE STELLAR OBJECT
PHL 7728	01 42 06.	- 09 54		17.7	BLUE STELLAR OBJECT
PHL 3688	01 42 06.	- 11 38		18.8	BLUE STELLAR OBJECT
LIN 555	01 42 06.	- 74 42 19.		16.18	SMC EMISSION LINE OBJECT
ARC 0246	01 42 09.	+ 05 34		16.4	RICH CLUSTER OF GALAXIES
MCG+01-05-027	01 42 09.	- 03 21	48	14.	GALAXY
RNGC 0665	01 42 11.	+ 10 10		13.5	GALAXY
PHL 7731	01 42 12.	+ 01 48		16.0	BLUE STELLAR OBJECT
PHL 1131	01 42 12.	+ 06 12		16.0	BLUE STELLAR OBJECT
5ZW 100	01 42 12.	+ 30 40			COMPACT GALAXY
PHL 7732	01 42 12.	- 04 05		16.6	BLUE STELLAR OBJECT
PHL 7734	01 42 12.	- 07 01		17.0	BLUE STELLAR OBJECT
PHL 7733	01 42 12.	- 09 06		18.5	BLUE STELLAR OBJECT
PHL 7735	01 42 12.	- 11 17		17.5	BLUE STELLAR OBJECT
PHL 3691	01 42 12.	- 18 47		17.2	BLUE STELLAR OBJECT
MCG+02-05-020	01 42 15.	+ 10 04	36	16.	GALAXY
MCG+02-05-019	01 42 15.	+ 10 10	48	13.	GALAXY
MCG+01-05-029	01 42 15.	- 04 23	42	15.	GALAXY
MCG+01-05-028	01 42 15.	- 07 54	72	15.	GALAXY
KN 14.490	01 42 16.7	+ 32 10 39.			NEBULA
PHL 1132	01 42 18.	+ 01 22		17.9	BLUE STELLAR OBJECT
PHL 3695	01 42 18.	+ 07 51		18.2	BLUE STELLAR OBJECT
PHL 3694	01 42 18.	+ 08 00		18.3	BLUE STELLAR OBJECT
ZWG 437.019	01 42 18.	+ 10 10		13.5	GALAXY
UGC 01223	01 42 18.	+ 10 10	156	13.5	GALAXY S0
UGC 01224	01 42 18.	+ 31 54	66	17.	GALAXY Sc

OBJECT NAME	RIGHT ASCEN.	DECLINATION	DIAM.	MAGN.	TYPE OF OBJECT
PHL 7736	01 42 18.	- 05 33		18.1	BLUE STELLAR OBJECT
PHL 3692	01 42 18.	- 06 10		17.6	BLUE STELLAR OBJECT
PHL 3693	01 42 18.	- 11 42		17.7	BLUE STELLAR OBJECT
KN 14.491	01 42 18.5	+ 31 52 21.			NEBULA
ARC 0248	01 42 19.	- 02 31		17.6	RICH CLUSTER OF GALAXIES
KN 14.492	01 42 23.6	+ 31 48 35.			NEBULA
PHL 3700	01 42 24.	+ 00 45		18.9	BLUE STELLAR OBJECT
PHL 3696	01 42 24.	+ 05 20		16.2	BLUE STELLAR OBJECT
PHL 3699	01 42 24.	+ 06 30		17.4	BLUE STELLAR OBJECT
PHL 3697	01 42 24.	+ 07 18		17.0	BLUE STELLAR OBJECT
PHL 7741	01 42 24.	+ 08 34		18.1	BLUE STELLAR OBJECT
MCG+02-05-021	01 42 24.	+ 10 06	60	15.5	GALAXY
UGC 01226	01 42 24.	+ 10 07	66	16.5	GALAXY S IV-V
MCG+02-05-022	01 42 24.	+ 10 10	24	17.	GALAXY
ZC 0142.4+1706	01 42 24.	+ 17 06	4700		CLUSTER OF GALAXIES
ZWG 482.009	01 42 24.	+ 21 31		15.3	GALAXY
ZWG 460.022	01 42 24.	+ 21 31		15.3	GALAXY
ZWG 482.010	01 42 24.	+ 25 32		15.4	GALAXY
MCG+05-05-006	01 42 24.	+ 31 48	30	15.5	GALAXY
ZWG 503.015	01 42 24.	+ 31 49		15.6	GALAXY
UGC 01227	01 42 24.	+ 31 49	60	15.6	GALAXY Sc
ZWG 521.077	01 42 24.	+ 34 01		15.2	GALAXY
UGC 01225	01 42 24.	- 00 32	60	16.5	GALAXY DBL SYS
PHL 7737	01 42 24.	- 01 30		18.7	BLUE STELLAR OBJECT
ZC 0142.4-0230	01 42 24.	- 02 30	810		CLUSTER OF GALAXIES
PHL 7740	01 42 24.	- 02 34		16.0	BLUE STELLAR OBJECT
PHL 7738	01 42 24.	- 03 11		18.6	BLUE STELLAR OBJECT
PHL 7739	01 42 24.	- 03 35		18.2	BLUE STELLAR OBJECT
MCG-02-05-040	01 42 24.	- 11 11 30.	36	15.	GALAXY
PHL 7742	01 42 24.	- 25 37		18.2	BLUE STELLAR OBJECT
PHL 3698	01 42 24.	- 27 05		18.2	BLUE STELLAR OBJECT
IC 1725	01 42 25.	+ 21 32 21.			NONSTELLAR OBJECT
MCG+05-05-007	01 42 30.	+ 28 27	66	14.5	GALAXY
ZWG 503.016	01 42 30.	+ 28 29		15.1	GALAXY
UGC 01228	01 42 30.	+ 28 29	108	15.1	GALAXY S-IRR
6ZW 021	01 42 30.	+ 34 45			COMPACT GALAXY
5ZW 101	01 42 30.	+ 38 28			COMPACT GALAXY
ZC 0142.5-0009	01 42 30.	- 00 09	200		CLUSTER OF GALAXIES
ARC 0247	01 42 32.	+ 17 24		16.9	RICH CLUSTER OF GALAXIES
KN 14.493	01 42 32.2	+ 28 28 31.			NEBULA
RNGC 0663	01 42 35.	+ 60 59		7.5	OPEN CLUSTER
ZC 0142.6+0120	01 42 36.	+ 01 20	1080		CLUSTER OF GALAXIES
PHL 7743	01 42 36.	+ 01 32		18.9	BLUE STELLAR OBJECT
ZC 0142.6+0805	01 42 36.	+ 08 05	4170		CLUSTER OF GALAXIES
MCG+02-05-024	01 42 36.	+ 10 05	36	15.	GALAXY
ZWG 437.020	01 42 36.	+ 10 06		15.5	GALAXY
MCG+02-05-023	01 42 36.	+ 10 22 30.	84	14.5	GALAXY
ZWG 437.021	01 42 36.	+ 10 24		14.8	GALAXY
UGC 01229	01 42 36.	+ 10 24	78	14.8	GALAXY Sb
OCL 0333	01 42 36.	+ 61 00	1980	7.9	OPEN STAR CLUSTER
MCG+00-05-034	01 42 36.	- 00 33	24	15.	GALAXY
MCG+01-05-030	01 42 36.	- 04 33 30.	60	15.5	GALAXY
PHL 3701	01 42 36.	- 06 19		18.7	BLUE STELLAR OBJECT
PHL 3702	01 42 36.	- 06 34		19.0	BLUE STELLAR OBJECT
IC 0154	01 42 38.	+ 10 24 08.			NONSTELLAR OBJECT
KN 14.494	01 42 39.7	+ 29 36 07.			NEBULA
KN 14.495	01 42 41.8	+ 28 12 12.			NEBULA
PHL 7744	01 42 42.	+ 01 32		17.4	BLUE STELLAR OBJECT
ZWG 412.025	01 42 42.	+ 04 22		14.8	GALAXY
UGC 01230	01 42 42.	+ 25 16	138	17.	GALAXY DWRF SP
ASS 40	01 42 42.	+ 61 04			OB ASSOCIATION CAS OB8
PHL 7745	01 42 42.	- 13 32		18.2	BLUE STELLAR OBJECT
TON-S 0229	01 42 42.	- 27 01		15.5	BLUE STAR
IC 1726	01 42 43.	+ 04 22 08.			NONSTELLAR OBJECT
RNGC 0667	01 42 43.	- 23 11			GALAXY
KN 14.496	01 42 44.2	+ 28 32 56.			NEBULA
KN 14.497	01 42 44.9	+ 28 42 28.			NEBULA
MCG+02-05-025	01 42 48.	+ 10 17	72	14.	GALAXY
ZWG 437.022	01 42 48.	+ 10 18		15.0	GALAXY
UGC 01231	01 42 48.	+ 10 18	108	15.0	GALAXY S
ZWG 482.011	01 42 48.	+ 23 11		15.7	GALAXY
MCG+04-05-005	01 42 48.	+ 25 15	120	16.	GALAXY
MCG+05-05-008	01 42 48.	+ 28 32	60	14.5	GALAXY
UGC 01232	01 42 48.	+ 40 58	66	17.	GALAXY Sc
PHL 3704	01 42 48.	- 04 22		18.7	BLUE STELLAR OBJECT
PHL 7747	01 42 48.	- 06 48		16.7	BLUE STELLAR OBJECT
PHL 7748	01 42 48.	- 06 55		18.5	BLUE STELLAR OBJECT
PHL 3703	01 42 48.	- 10 00		17.2	BLUE STELLAR OBJECT
PHL 7746	01 42 48.	- 13 36		13.7	BLUE STELLAR OBJECT
PHL 7749	01 42 48.	- 20 57		18.6	BLUE STELLAR OBJECT
IC 0156	01 42 50.	+ 10 18 14.			NONSTELLAR OBJECT
KN 14.498	01 42 51.9	+ 28 33 21.			NEBULA
PHL 3705	01 42 54.	+ 07 54		17.2	BLUE STELLAR OBJECT
ZWG 460.023	01 42 54.	+ 18 19		15.7	GALAXY
ZWG 503.017	01 42 54.	+ 28 34		14.7	GALAXY
UGC 01233	01 42 54.	+ 28 34	60	14.7	GALAXY Sc
ZWG 521.078	01 42 54.	+ 34 52		14.8	GALAXY
UGC 01234	01 42 54.	+ 34 52	72	14.8	GALAXY Sc/SBc
6ZW 022	01 42 54.	+ 37 54			COMPACT GALAXY
PHL 7751	01 42 54.	- 03 32		18.1	BLUE STELLAR OBJECT
MCG+01-05-031	01 42 54.	- 04 04	60	14.	GALAXY
PHL 7750	01 42 54.	- 04 51		16.7	BLUE STELLAR OBJECT
MCG-02-05-041	01 42 54.	- 10 20	24	15.	GALAXY
PHL 1133	01 42 54.	- 18 04		17.9	BLUE STELLAR OBJECT
LB 03225	01 42 54.	- 52 19		13.6	FAINT BLUE STAR
KN 14.499	01 42 57.0	+ 28 32 35.			NEBULA
PHL 7757	01 43 00.	+ 00 42		18.9	BLUE STELLAR OBJECT
PHL 3706	01 43 00.	+ 01 45		18.8	BLUE STELLAR OBJECT
ZWG 386.040	01 43 00.	+ 03 07		14.9	GALAXY
UGC 01235	01 43 00.	+ 03 07	66	14.9	GALAXY Sb-c
MCG+00-05-035	01 43 00.	+ 03 07	45	13.5	GALAXY
6ZW 023	01 43 00.	+ 37 53			COMPACT GALAXY
PHL 7756	01 43 00.	- 01 35		18.6	BLUE STELLAR OBJECT
PHL 7754	01 43 00.	- 05 40		17.1	BLUE STELLAR OBJECT
PHL 7755	01 43 00.	- 10 13		17.7	BLUE STELLAR OBJECT
PHL 7752	01 43 00.	- 12 48		17.8	BLUE STELLAR OBJECT
PHL 3707	01 43 00.	- 13 11		18.0	BLUE STELLAR OBJECT
PHL 7753	01 43 00.	- 19 55		18.6	BLUE STELLAR OBJECT
PHL 3708	01 43 00.	- 23 44		18.6	BLUE STELLAR OBJECT
SER 017.03	01 43 00.	- 53 16	100	19.	COMPACT CLSTR. OF GALXIES
IC 0157	01 43 02.	+ 12 37 25.			NONSTELLAR OBJECT
PHL 3710	01 43 06.	+ 08 19		6.6	BLUE STELLAR OBJECT
ZC 0143.1+1928	01 43 06.	+ 19 28	3290		CLUSTER OF GALAXIES
ZWG 503.018	01 43 06.	+ 32 08		15.6	GALAXY
MCG+06-05-001	01 43 06.	+ 34 51	60	14.	GALAXY
PHL 7758	01 43 06.	- 03 34		18.4	BLUE STELLAR OBJECT
PHL 3709	01 43 06.	- 14 45		16.1	BLUE STELLAR OBJECT
RNGC 0666	01 43 09.	+ 34 08		13.5	GALAXY

OBJECT NAME	RIGHT ASCEN.	DECLINATION	DIAM.	MAGN.	TYPE OF OBJECT
MCG+01-05-032	01 43 09.	- 04 56	78	16.	GALAXY
MCG-05-05-012	01 43 09.	- 26 33	24	17.	GALAXY
PHL 3712	01 43 12.	+ 08 23		17.8	BLUE STELLAR OBJECT
ZWG 521.079	01 43 12.	+ 34 08		13.6	GALAXY
UGC 01236	01 43 12.	+ 34 08	48	13.6	GALAXY
6ZW 024	01 43 12.	+ 37 28			COMPACT GALAXY
6ZW 025	01 43 12.	+ 39 17			COMPACT GALAXY
PHL 7760	01 43 12.	- 00 51		18.7	BLUE STELLAR OBJECT
ZC 0143.2-0214	01 43 12.	- 02 14	1480		CLUSTER OF GALAXIES
PHL 7759	01 43 12.	- 04 36		18.0	BLUE STELLAR OBJECT
PHL 7761	01 43 12.	- 07 10		18.8	BLUE STELLAR OBJECT
PHL 1134	01 43 12.	- 10 09		17.9	BLUE STELLAR OBJECT
PHL 3714	01 43 12.	- 10 23		18.3	BLUE STELLAR OBJECT
PHL 3713	01 43 12.	- 10 52		18.7	BLUE STELLAR OBJECT
PHL 3711	01 43 12.	- 13 24		16.8	BLUE STELLAR OBJECT
KN 14.500	01 43 14.1	+ 27 51 05.			NEBULA
MCG+02-05-026	01 43 15.	+ 09 33	72	14.5	GALAXY
LIN 556	01 43 15.	- 74 57 57.		14.29	SMC EMISSION LINE OBJECT
ZWG 437.023	01 43 18.	+ 09 34		15.5	GALAXY
UGC 01237	01 43 18.	+ 09 34	156	15.5	GALAXY Sa-b
VV 093	01 43 18.	+ 12 09	72	14.	INTERACTING GALAXY
5ZW 102	01 43 18.	+ 23 12			COMPACT GALAXY
MCG+04-05-007	01 43 18.	+ 23 13	18	15.5	GALAXY
MCG+04-05-006	01 43 18.	+ 23 24	42	15.5	GALAXY
5ZW 103	01 43 18.	+ 28 32			COMPACT GALAXY
MCG+06-05-002	01 43 18.	+ 34 07	42	14.	GALAXY
6ZW 026	01 43 18.	+ 34 08			COMPACT GALAXY
PHL 3716	01 43 18.	- 03 03		18.6	BLUE STELLAR OBJECT
PHL 7762	01 43 18.	- 05 28		18.7	BLUE STELLAR OBJECT
PHL 7763	01 43 18.	- 06 40		18.8	BLUE STELLAR OBJECT
PHL 7764	01 43 18.	- 08 15		18.7	BLUE STELLAR OBJECT
PHL 3715	01 43 18.	- 10 00		17.8	BLUE STELLAR OBJECT
PHL 7765	01 43 18.	- 12 28		18.1	BLUE STELLAR OBJECT
PHL 3717	01 43 18.	- 17 58		18.5	BLUE STELLAR OBJECT
LIN 557	01 43 21.	- 74 55 57.		11.92	SMC EMISSION LINE OBJECT
PHL 7772	01 43 24.	+ 03 41		18.5	BLUE STELLAR OBJECT
PHL 3718	01 43 24.	+ 04 11		18.2	BLUE STELLAR OBJECT
PHL 7767	01 43 24.	+ 04 50		18.1	BLUE STELLAR OBJECT
MCG+04-05-008	01 43 24.	+ 23 10	24	15.5	GALAXY
ZWG 522.001	01 43 24.	+ 36 12		13.5	GALAXY
ZWG 521.080	01 43 24.	+ 36 12		13.5	GALAXY
UGC 01238	01 43 24.	+ 36 12	132	13.5	GALAXY Sb
PHL 7770	01 43 24.	- 01 09		16.8	BLUE STELLAR OBJECT
PHL 7773	01 43 24.	- 02 27		18.9	BLUE STELLAR OBJECT
PHL 7766	01 43 24.	- 05 52		16.6	BLUE STELLAR OBJECT
IC 0158	01 43 24.	- 07 11 25.			NONSTELLAR OBJECT
PHL 7768	01 43 24.	- 07 53		18.7	BLUE STELLAR OBJECT
PHL 7771	01 43 24.	- 15 10		16.9	BLUE STELLAR OBJECT
PHL 3720	01 43 24.	- 16 49		18.2	BLUE STELLAR OBJECT
PHL 7769	01 43 24.	- 17 03		18.0	BLUE STELLAR OBJECT
PHL 3719	01 43 24.	- 19 01		18.3	BLUE STELLAR OBJECT
PHL 1135	01 43 24.	- 22 22		16.8	BLUE STELLAR OBJECT
RNGC 0668	01 43 26.	+ 36 12		13.5	GALAXY
PHL 3721	01 43 30.	+ 03 22		16.6	BLUE STELLAR OBJECT
ZWG 412.026	01 43 30.	+ 06 10		15.1	GALAXY
UGC 01239	01 43 30.	+ 06 10	78	15.1	GALAXY Sc-IRR
ZWG 522.002	01 43 30.	+ 34 41		14.7	GALAXY
ZWG 521.081	01 43 30.	+ 34 41		14.7	GALAXY
MCG+06-05-003	01 43 30.	+ 36 12	120	12.	GALAXY
VB 146	01 43 30.	+ 57 58	175		STELLAR RING
ISS 0053	01 43 30.	+ 57 58	177		STELLAR RING
PHL 1136	01 43 30.	- 07 58		18.2	BLUE STELLAR OBJECT
SER 017.04	01 43 30.	- 53 14	900		CLOUD OF GALAXIES
MCG+01-05-031	01 43 33.	+ 06 10	60	14.	GALAXY
PHL 3726	01 43 36.	+ 01 02		18.1	BLUE STELLAR OBJECT
PHL 7774	01 43 36.	+ 01 15		17.9	BLUE STELLAR OBJECT
PHL 3727	01 43 36.	+ 01 40		18.6	BLUE STELLAR OBJECT
PHL 3722	01 43 36.	+ 03 56		16.7	BLUE STELLAR OBJECT
PHL 7778	01 43 36.	- 02 36		18.0	BLUE STELLAR OBJECT
PHL 3725	01 43 36.	- 03 30		18.9	BLUE STELLAR OBJECT
PHL 7775	01 43 36.	- 06 12		17.8	BLUE STELLAR OBJECT
PHL 1137	01 43 36.	- 06 21		18.4	BLUE STELLAR OBJECT
PHL 3728	01 43 36.	- 09 26		17.6	BLUE STELLAR OBJECT
PHL 7776	01 43 36.	- 12 47		18.5	BLUE STELLAR OBJECT
PHL 7777	01 43 36.	- 13 04		17.9	BLUE STELLAR OBJECT
PHL 3724	01 43 36.	- 13 12		17.3	BLUE STELLAR OBJECT
PHL 3723	01 43 36.	- 16 16		16.8	BLUE STELLAR OBJECT
LIN 558	01 43 37.	- 73 29 28.		14.31	SMC EMISSION LINE OBJECT
KN 14.501	01 43 37.8	+ 29 06 01.			NEBULA
KN 14.504	01 43 39.7	+ 27 34 56.			NEBULA
KN 14.502	01 43 41.1	+ 30 50 45.			NEBULA
PHL 7779	01 43 42.	+ 02 10		15.2	BLUE STELLAR OBJECT
ZWG 412.027	01 43 42.	+ 04 01		14.9	GALAXY
UGC 01240	01 43 42.	+ 04 01	78	14.9	GALAXY S-IRR
5ZW 104	01 43 42.	+ 28 57			COMPACT GALAXY
VB 147	01 43 42.	+ 60 58	369		STELLAR RING
PHL 1138	01 43 42.	- 17 00		16.9	BLUE STELLAR OBJECT
PHL 3729	01 43 42.	- 20 25		18.5	BLUE STELLAR OBJECT
MCG-05-05-013	01 43 42.	- 29 18	12	15.	GALAXY
KN 14.503	01 43 42.0	+ 30 50 40.			NEBULA
KN 14.506	01 43 44.1	+ 28 28 38.			NEBULA
MCG+01-05-032	01 43 45.	+ 04 00	60	14.5	GALAXY
KN 14.505	01 43 46.6	+ 31 13 00.			NEBULA
PHL 1139	01 43 48.	+ 04 27		17.5	BLUE STELLAR OBJECT
PHL 3732	01 43 48.	+ 05 55		18.3	BLUE STELLAR OBJECT
PHL 3730	01 43 48.	+ 07 03		16.6	BLUE STELLAR OBJECT
PHL 3731	01 43 48.	+ 07 11		17.8	BLUE STELLAR OBJECT
ZC 0143.8+2323	01 43 48.	+ 23 23	7660		CLUSTER OF GALAXIES
ZWG 503.019	01 43 48.	+ 32 15		15.7	GALAXY
UGC 01241	01 43 48.	+ 48 10	78	17.	GALAXY
ZWG 559.001	01 43 48.	+ 52 42		15.6	GALAXY
PHL 1140	01 43 48.	- 01 07		18.8	BLUE STELLAR OBJECT
PHL 3733	01 43 48.	- 01 32		18.1	BLUE STELLAR OBJECT
PHL 7781	01 43 48.	- 03 16		18.8	BLUE STELLAR OBJECT
MCG+01-05-033	01 43 48.	- 06 07	14	16.	GALAXY
PHL 3735	01 43 48.	- 08 42		18.1	BLUE STELLAR OBJECT
MCG-02-05-042	01 43 48.	- 08 52	72	14.	GALAXY
PHL 7782	01 43 48.	- 11 35		17.4	BLUE STELLAR OBJECT
PHL 3734	01 43 48.	- 11 45		18.3	BLUE STELLAR OBJECT
PHL 7780	01 43 48.	- 18 52		17.4	BLUE STELLAR OBJECT
KN 14.507	01 43 48.1	+ 27 18 36.			NEBULA
KN 14.508	01 43 49.6	+ 28 43 35.			NEBULA
PHL 7785	01 43 54.	+ 01 12		18.8	BLUE STELLAR OBJECT
PHL 3737	01 43 54.	+ 05 53		15.4	BLUE STELLAR OBJECT
PHL 7784	01 43 54.	+ 06 37		18.2	BLUE STELLAR OBJECT
PHL 3736	01 43 54.	+ 07 33		17.2	BLUE STELLAR OBJECT
PHL 1141	01 43 54.	+ 07 44		18.9	BLUE STELLAR OBJECT
ZWG 437.024	01 43 54.	+ 14 26		15.7	GALAXY

OBJECT NAME	RIGHT ASCEN.	DECLINATION	DIAM.	MAGN.	TYPE OF OBJECT
UGC 01242	01 43 54.	+ 14 26	78	15.7	GALAXY
MCG+02-05-027	01 43 54.	+ 14 27	48	15.	GALAXY
MCG+03-05-015	01 43 54.	+ 18 20	48	15.	GALAXY
4ZW 060	01 43 54.	+ 34 37			COMPACT GALAXY
ZWG 537.023	01 43 54.	+ 43 07		15.6	GALAXY
PHL 7783	01 43 54.	- 09 42		16.8	BLUE STELLAR OBJECT
PHL 1142	01 43 54.	- 12 56		18.9	BLUE STELLAR OBJECT
ARC 0251	01 43 56.	- 07 33		17.2	RICH CLUSTER OF GALAXIES
IC 0159	01 43 57.	- 08 52 51.			NONSTELLAR OBJECT
ARC 0249	01 43 59.	+ 19 49		17.7	RICH CLUSTER OF GALAXIES
ZC 0144.0+1230	01 44 00.	+ 12 30	11020		CLUSTER OF GALAXIES
ZWG 460.024	01 44 00.	+ 18 19		15.0	GALAXY
UGC 01243	01 44 00.	+ 18 19	90	15.0	GALAXY S0?
ZWG 482.012	01 44 00.	+ 24 13		15.5	GALAXY
UGC 01244	01 44 00.	+ 24 13	84	15.5	GALAXY S
KARA.73B 0064	01 44 00.	+ 24 13	66	15.5	ISOLATED GALAXY S
MCG+05-05-009	01 44 00.	+ 31 36	12	15.5	GALAXY
ZWG 503.020	01 44 00.	+ 31 37		15.7	GALAXY
ZWG 503.021	01 44 00.	+ 33 09		15.2	GALAXY
KARA.73B 0065	01 44 00.	+ 33 09	36	15.2	ISOLATED GALAXY S
ZWG 522.003	01 44 00.	+ 34 32		15.2	GALAXY
5ZW 105	01 44 00.	+ 34 38			COMPACT GALAXY
ASS 41	01 44 00.	+ 55 25			OB ASSOCIATION CAS OB10
VB 148	01 44 00.	+ 57 58	154		STELLAR RING
ISS 0054	01 44 00.	+ 57 58	160		STELLAR RING
ZC 0144.0-0051	01 44 00.	- 00 51	3970		CLUSTER OF GALAXIES
PHL 1143	01 44 00.	- 04 31		18.1	BLUE STELLAR OBJECT
PHL 7786	01 44 00.	- 04 38		18.6	BLUE STELLAR OBJECT
MCG-02-05-043	01 44 00.	- 08 54	36	16.	GALAXY
PHL 1144	01 44 00.	- 11 48		18.4	BLUE STELLAR OBJECT
MCG-02-05-044	01 44 00.	- 13 30	48	14.5	GALAXY
MCG-03-05-019	01 44 00.	- 15 22	15	15.	GALAXY
LIN 559	01 44 02.	- 74 58 23.		16.83	SMC EMISSION LINE OBJECT
KN 14.509	01 44 02.5	+ 31 37 01.			NEBULA
IC 0155	01 44 03.	+ 59 32			NONSTELLAR OBJECT
CED 005	01 44 04.	+ 59 32	720		DIFFUSE GALACTIC NEBULA
IC 0160	01 44 04.	- 13 30 03.			NONSTELLAR OBJECT
KN 14.510	01 44 04.1	+ 31 51 25.			NEBULA
PHL 1145	01 44 06.	+ 00 54		18.2	BLUE STELLAR OBJECT
MCG+02-05-028	01 44 06.	+ 11 08	90	15.	GALAXY
ZWG 432.013	01 44 06.	+ 22 29		15.7	GALAXY
MCG+05-05-011	01 44 06.	+ 28 28	48	14.5	GALAXY
ZWG 503.022	01 44 06.	+ 28 30		15.2	GALAXY
ZWG 503.023	01 44 06.	+ 31 51		14.8	GALAXY
MCG+05-05-010	01 44 06.	+ 31 51	42	15.	GALAXY
5ZW 106	01 44 06.	+ 34 32			COMPACT GALAXY
MCG+01-05-034	01 44 06.	- 04 01	72	16.	GALAXY
MCG-02-05-045	01 44 06.	- 13 40	42	16.	GALAXY
PHL 7787	01 44 06.	- 13 54		18.2	BLUE STELLAR OBJECT
PHL 7788	01 44 06.	- 14 36		18.8	BLUE STELLAR OBJECT
ARC 0252	01 44 07.	- 02 57		17.8	RICH CLUSTER OF GALAXIES
KN 14.511	01 44 08.2	+ 28 30 32.			NEBULA
PHL 7739	01 44 12.	+ 05 30		16.5	BLUE STELLAR OBJECT
PHL 3739	01 44 12.	+ 06 30		18.7	BLUE STELLAR OBJECT
PHL 7791	01 44 12.	+ 06 57		18.2	BLUE STELLAR OBJECT
ZWG 437.025	01 44 12.	+ 11 10		15.0	GALAXY
UGC 01245	01 44 12.	+ 11 10	108	15.0	GALAXY S
6ZW 027	01 44 12.	+ 29 05			COMPACT GALAXY
ZWG 559.002	01 44 12.	+ 56 52		15.7	GALAXY
PHL 3738	01 44 12.	- 01 06		17.2	BLUE STELLAR OBJECT
PHL 3742	01 44 12.	- 02 38		19.0	BLUE STELLAR OBJECT
PHL 7790	01 44 12.	- 04 44		17.9	BLUE STELLAR OBJECT
PHL 7792	01 44 12.	- 05 17		16.8	BLUE STELLAR OBJECT
PHL 3740	01 44 12.	- 05 53		18.6	BLUE STELLAR OBJECT
PHL 7793	01 44 12.	- 10 04		17.7	BLUE STELLAR OBJECT
PHL 3741	01 44 12.	- 10 20		18.6	BLUE STELLAR OBJECT
PHL 7794	01 44 12.	- 11 42		18.3	BLUE STELLAR OBJECT
LB 03226	01 44 12.	- 74 45		12.2	FAINT BLUE STAR
LIN 560	01 44 13.	- 74 14 47.		14.97	SMC EMISSION LINE OBJECT
MCG-02-05-046	01 44 15.	- 13 39	48	15.	GALAXY
RNGC 0671	01 44 16.	+ 12 52		14.5	GALAXY
ARC 0250	01 44 17.	+ 19 27		17.7	RICH CLUSTER OF GALAXIES
KN 14.512	01 44 17.9	+ 29 56 15.			NEBULA
MCG+02-05-030	01 44 18.	+ 12 09	84	14.	GALAXY
ZWG 437.026	01 44 18.	+ 12 10		15.1	GALAXY
UGC 01246	01 44 18.	+ 12 10	102	15.1	GALAXY IRR
ZWG 437.027	01 44 18.	+ 12 52		14.3	GALAXY
UGC 01247	01 44 18.	+ 12 52	102	14.3	GALAXY S
MCG+02-05-029	01 44 18.	+ 12 52	84	14.	GALAXY
6ZW 028	01 44 18.	+ 28 30			COMPACT GALAXY
6ZW 029	01 44 18.	+ 28 56			COMPACT GALAXY
6ZW 030	01 44 18.	+ 33 46			COMPACT GALAXY
ZWG 522.004	01 44 18.	+ 35 18		12.9	GALAXY
UGC 01248	01 44 18.	+ 35 18	198	12.9	GALAXY Sa-b
OCL 0331	01 44 18.	+ 62 41	180	17.	OPEN STAR CLUSTER
OCL 0324	01 44 18.	+ 71 42	2190	5.8	OPEN STAR CLUSTER
PHL 3744	01 44 18.	- 08 36		18.0	BLUE STELLAR OBJECT
PHL 3743	01 44 18.	- 11 00		17.6	BLUE STELLAR OBJECT
PHL 7796	01 44 18.	- 12 20		18.5	BLUE STELLAR OBJECT
PHL 7795	01 44 18.	- 13 44		16.9	BLUE STELLAR OBJECT
PHL 3745	01 44 18.	- 14 28		18.3	BLUE STELLAR OBJECT
LB 03227	01 44 18.	- 66 05		15.3	FAINT BLUE STAR
LIN 561	01 44 18.	- 74 20 59.		15.23	SMC EMISSION LINE OBJECT
RNGC 0669	01 44 21.	+ 35 18		14.0	GALAXY
PHL 7801	01 44 24.	+ 01 03		18.2	BLUE STELLAR OBJECT
PHL 7797	01 44 24.	+ 04 23		16.8	BLUE STELLAR OBJECT
6ZW 031	01 44 24.	+ 29 07			COMPACT GALAXY
6ZW 032	01 44 24.	+ 32 55			COMPACT GALAXY
MCG+06-05-004	01 44 24.	+ 35 17	180	12.	GALAXY
VB 055	01 44 24.	+ 63 53	342		STELLAR RING
PHL 3748	01 44 24.	- 00 54		18.2	BLUE STELLAR OBJECT
PHL 7799	01 44 24.	- 02 24		16.8	BLUE STELLAR OBJECT
PHL 7802	01 44 24.	- 03 00		18.6	BLUE STELLAR OBJECT
PHL 3747	01 44 24.	- 03 28		18.9	BLUE STELLAR OBJECT
PHL 1146	01 44 24.	- 04 41		18.5	BLUE STELLAR OBJECT
PHL 7798	01 44 24.	- 06 00		16.7	BLUE STELLAR OBJECT
PHL 7803	01 44 24.	- 06 56		18.7	BLUE STELLAR OBJECT
PHL 7800	01 44 24.	- 08 38		17.7	BLUE STELLAR OBJECT
PHL 3746	01 44 24.	- 09 50		16.8	BLUE STELLAR OBJECT
MCG-02-05-047	01 44 24.	- 10 46	42	17.	GALAXY
PHL 7804	01 44 24.	- 12 36		18.4	BLUE STELLAR OBJECT
MCG-04-05-006	01 44 24.	- 25 36 30.	36	15.5	GALAXY
MCG-03-05-020	01 44 27.	- 14 42	36	14.5	GALAXY
PHL 3752	01 44 30.	+ 08 00		18.4	BLUE STELLAR OBJECT
PHL 1147	01 44 30.	+ 08 03		18.6	BLUE STELLAR OBJECT
6ZW 033	01 44 30.	+ 28 58			COMPACT GALAXY
6ZW 034	01 44 30.	+ 31 52			COMPACT GALAXY
5ZW 107	01 44 30.	+ 35 55			COMPACT GALAXY

OBJECT NAME	RIGHT ASCEN.	DECLINATION	DIAM.	MAGN.	TYPE OF OBJECT
PHL 3753	01 44 30.	- 08 56		18.1	BLUE STELLAR OBJECT
PHL 7805	01 44 30.	- 09 44		17.7	BLUE STELLAR OBJECT
PHL 1148	01 44 30.	- 12 14		18.3	BLUE STELLAR OBJECT
PHL 3749	01 44 30.	- 12 22		18.3	BLUE STELLAR OBJECT
PHL 7806	01 44 30.	- 13 04		18.1	BLUE STELLAR OBJECT
PHL 3751	01 44 30.	- 14 08		6.7	BLUE STELLAR OBJECT
PHL 3750	01 44 30.	- 16 06		18.0	BLUE STELLAR OBJECT
LIN 563	01 44 31.	- 74 59 18.		14.81	SMC EMISSION LINE OBJECT
LIN 562	01 44 33.	- 73 54 23.		14.51	SMC EMISSION LINE OBJECT
KN 14.513	01 44 34.5	+ 30 01 42.			NEBULA
RNGC 0670	01 44 35.	+ 27 38		13.0	GALAXY
KN 14.514	01 44 35.8	+ 27 38 06.			NEBULA
PHL 7807	01 44 36.	+ 00 27		16.7	BLUE STELLAR OBJECT
PHL 1149	01 44 36.	+ 02 23		15.8	BLUE STELLAR OBJECT
SHAH 142	01 44 36.	+ 03 10	162		GROUP OF COMPACT GALAXIES
PHL 3755	01 44 36.	+ 06 51		18.4	BLUE STELLAR OBJECT
MCG+03-05-016	01 44 36.	+ 16 06	72	16.	GALAXY
ZWG 482.014	01 44 36.	+ 27 05		12.2	GALAXY
UGC 01249	01 44 36.	+ 27 05	480	12.2	GALAXY
KARA.72 040A	01 44 36.	+ 27 05	420	12.2	PART OF DOUBLE GALAXY
MCG+05-05-012	01 44 36.	+ 27 36	90	14.	GALAXY
ZWG 503.024	01 44 36.	+ 27 38		13.1	GALAXY
UGC 01250	01 44 36.	+ 27 38	132	13.1	GALAXY S0
ZWG 522.005	01 44 36.	+ 35 47		15.0	GALAXY
UGC 01251	01 44 36.	+ 35 47	60	15.0	GALAXY PECULR
PHL 3756	01 44 36.	- 01 16		18.0	BLUE STELLAR OBJECT
PHL 1150	01 44 36.	- 02 39		17.6	BLUE STELLAR OBJECT
PHL 7808	01 44 36.	- 06 38		18.9	BLUE STELLAR OBJECT
PHL 7809	01 44 36.	- 07 07		18.0	BLUE STELLAR OBJECT
PHL 1151	01 44 36.	- 09 02		17.6	BLUE STELLAR OBJECT
PHL 3758	01 44 36.	- 09 10		18.7	BLUE STELLAR OBJECT
PHL 3757	01 44 36.	- 11 50		18.5	BLUE STELLAR OBJECT
PHL 7810	01 44 36.	- 14 02		18.6	BLUE STELLAR OBJECT
PHL 7811	01 44 36.	- 18 07		18.2	BLUE STELLAR OBJECT
PHL 3754	01 44 36.	- 19 13		17.8	BLUE STELLAR OBJECT
MCG-06-05-001	01 44 36.	- 34 09	48	14.	GALAXY
SER 019.06	01 44 36.	- 58 55	180	15.	IRR. MAGELLANIC GALAXY
HOLM 046B	01 44 38.	+ 27 17	210	13.0	PART OF MULTIPLE GALAXY
PHL 3759	01 44 42.	+ 03 40		18.6	BLUE STELLAR OBJECT
PHL 7812	01 44 42.	+ 04 02		17.0	BLUE STELLAR OBJECT
PHL 7813	01 44 42.	+ 08 26		18.6	BLUE STELLAR OBJECT
MCG+02-05-031	01 44 42.	+ 11 50	72	14.5	GALAXY
UGC 01252	01 44 42.	+ 16 03	72	16.5	GALAXY S(B)
MCG+04-05-009	01 44 42.	+ 27 04	360	12.1	GALAXY
IC 1727	01 44 42.	+ 27 04 50.	462	12.1	GALAXY SB(s)
VV 338A	01 44 42.	+ 27 05	300	12.	INTERACTING GALAXY
6ZW 035	01 44 42.	+ 37 23			COMPACT GALAXY
ARC 0254	01 44 42.	- 03 32		17.6	RICH CLUSTER OF GALAXIES
TON-S 0230	01 44 42.	- 25 43		14.2	BLUE STAR
KN 14.515	01 44 43.2	+ 27 04 39.			NEBULA
ZWG 386.041	01 44 48.	+ 02 35		14.9	GALAXY
PHL 7818	01 44 48.	+ 04 22		17.2	BLUE STELLAR OBJECT
PHL 3761	01 44 48.	+ 04 40		18.1	BLUE STELLAR OBJECT
ZWG 437.028	01 44 48.	+ 11 52		15.5	GALAXY
UGC 01253	01 44 48.	+ 11 52	96	15.5	GALAXY S
ZWG 522.006	01 44 48.	+ 34 46		15.0	GALAXY
MCG+06-05-005	01 44 48.	+ 34 46 30.	48	14.	GALAXY
PHL 7817	01 44 48.	- 00 16		16.4	BLUE STELLAR OBJECT
PHL 3760	01 44 48.	- 01 50		17.2	BLUE STELLAR OBJECT
PHL 3764	01 44 48.	- 02 16		18.5	BLUE STELLAR OBJECT
PHL 7814	01 44 48.	- 03 49		18.0	BLUE STELLAR OBJECT
PHL 3762	01 44 48.	- 04 12		18.7	BLUE STELLAR OBJECT
PHL 7815	01 44 48.	- 04 57		18.5	BLUE STELLAR OBJECT
PHL 7816	01 44 48.	- 05 14		15.9	BLUE STELLAR OBJECT
PHL 7819	01 44 48.	- 06 43		17.4	BLUE STELLAR OBJECT
PHL 1152	01 44 48.	- 08 39		16.8	BLUE STELLAR OBJECT
PHL 1153	01 44 48.	- 08 46		18.0	BLUE STELLAR OBJECT
PHL 3763	01 44 48.	- 10 58		18.7	BLUE STELLAR OBJECT
PHL 1154	01 44 48.	- 11 33		18.7	BLUE STELLAR OBJECT
PHL 1155	01 44 48.	- 20 01		16.6	BLUE STELLAR OBJECT
ARC 0255	01 44 50.	- 02 14		17.2	RICH CLUSTER OF GALAXIES
MCG+02-05-032	01 44 54.	+ 11 02	48	14.5	GALAXY
6ZW 036	01 44 54.	+ 35 40			COMPACT GALAXY
6ZW 037	01 44 54.	+ 37 06			COMPACT GALAXY
MCG+08-04-006	01 44 54.	+ 48 22	36	16.	GALAXY
UGC 01254	01 44 54.	+ 48 23	60	16.5	GALAXY Sc
PHL 7820	01 44 54.	- 00 15		18.0	BLUE STELLAR OBJECT
PHL 1156	01 44 54.	- 12 22		18.6	BLUE STELLAR OBJECT
MCG-02-05-048	01 44 54.	- 14 17 30.	48	14.5	GALAXY
VDB.66G 014	01 45	- 12 41	100		DWARF GALAXY
ZWG 437.029	01 45 00.	+ 11 06		15.7	GALAXY
UGC 01255	01 45 00.	+ 11 06	78	15.7	GALAXY
ZC 0145.0+1948	01 45 00.	+ 19 48	400		CLUSTER OF GALAXIES
MCG+04-05-010	01 45 00.	+ 25 18	36	15.5	GALAXY
ZWG 482.015	01 45 00.	+ 25 19		15.2	GALAXY
ZWG 482.016	01 45 00.	+ 27 11		11.4	GALAXY
UGC 01256	01 45 00.	+ 27 11	438	11.4	GALAXY SBc
VV 338B	01 45 00.	+ 27 11	300	11.4	INTERACTING GALAXY
KARA.72 040B	01 45 00.	+ 27 11	408	11.4	PART OF DOUBLE GALAXY
6ZW 038	01 45 00.	+ 33 07			COMPACT GALAXY
6ZW 039	01 45 00.	+ 33 22			COMPACT GALAXY
6ZW 040	01 45 00.	+ 37 04			COMPACT GALAXY
PHL 3765	01 45 00.	- 00 36		18.9	BLUE STELLAR OBJECT
PHL 7822	01 45 00.	- 04 36		17.7	BLUE STELLAR OBJECT
PHL 7821	01 45 00.	- 06 16		16.6	BLUE STELLAR OBJECT
PHL 1157	01 45 00.	- 10 30		18.0	BLUE STELLAR OBJECT
PHL 1158	01 45 00.	- 19 27		18.0	BLUE STELLAR OBJECT
PHL 1159	01 45 00.	- 22 11		15.3	BLUE STELLAR OBJECT
PHL 3766	01 45 00.	- 25 58		18.6	BLUE STELLAR OBJECT
HOLM 046A	01 45 01.	+ 27 11	360	12.1	PART OF MULTIPLE GALAXY
KN 14.516	01 45 03.7	+ 27 37 06.			NEBULA
KN 14.517	01 45 05.6	+ 27 11 01.			NEBULA
ARC 0253	01 45 06.	+ 20 25		17.9	RICH CLUSTER OF GALAXIES
5ZW 108	01 45 06.	+ 23 11			COMPACT GALAXY
MRK 362	01 45 06.	+ 23 24	7	14.5	GALAXY WITH UV CONTINUUM
MCG+04-05-011	01 45 06.	+ 27 10	420	11.	GALAXY
RNGC 0672	01 45 06.	+ 27 11		11.5	GALAXY
5ZW 109	01 45 06.	+ 53 03			COMPACT GALAXY
PHL 7824	01 45 06.	- 01 18		18.6	BLUE STELLAR OBJECT
ZC 0145.1-0218	01 45 06.	- 02 18	1810		CLUSTER OF GALAXIES
ARC 0256	01 45 06.	- 04 06		17.0	RICH CLUSTER OF GALAXIES
PHL 7823	01 45 06.	- 08 38		16.9	BLUE STELLAR OBJECT
ZC 0145.2+0002	01 45 12.	+ 00 02	1810		CLUSTER OF GALAXIES
PHL 3773	01 45 12.	+ 03 24		17.2	BLUE STELLAR OBJECT
PHL 3770	01 45 12.	+ 03 36		18.1	BLUE STELLAR OBJECT
PHL 3769	01 45 12.	+ 06 50		17.0	BLUE STELLAR OBJECT
PHL 3767	01 45 12.	+ 08 18		16.9	BLUE STELLAR OBJECT
ZC 0145.2+2146	01 45 12.	+ 21 46	1480		CLUSTER OF GALAXIES

OBJECT NAME	RIGHT ASCEN.	DECLINATION	DIAM.	MAGN.	TYPE OF OBJECT
ZC 0145.2+2902	01 45 12.	+ 29 02	940		CLUSTER OF GALAXIES
ZWG 522.007	01 45 12.	+ 36 12		15.0	GALAXY
UGC 01257	01 45 12.	+ 36 12	72	15.0	GALAXY Sa-b
MCG+06-05-006	01 45 12.	+ 36 12 30.	60	15.	GALAXY
PHL 7826	01 45 12.	- 01 30		18.6	BLUE STELLAR OBJECT
PHL 3768	01 45 12.	- 02 16		16.6	BLUF STELLAR OBJECT
PHL 7825	01 45 12.	- 02 42		18.7	BLUE STELLAR OBJECT
PHL 7827	01 45 12.	- 04 42		17.9	BLUE STELLAR OBJECT
PHL 3771	01 45 12.	- 11 38		18.2	BLUE STELLAR OBJECT
PHL 3772	01 45 12.	- 16 06		18.2	BLUE STELLAR OBJECT
SER 017.02	01 45 12.	- 52 10	780	17.	CHAIN OF 10 GALAXIES
PHL 1160	01 45 18.	+ 01 25		17.1	BLUE STELLAR OBJECT
UGC 01258	01 45 18.	+ 35 10	72	16.0	GALAXY DBL SYS
PHL 1161	01 45 18.	- 03 22		18.6	BLUE STELLAR OBJECT
PHL 7828	01 45 18.	- 04 50		18.4	BLUE STELLAR OBJECT
PHL 7829	01 45 18.	- 08 27		18.2	BLUE STELLAR OBJFCT
PHL 1162	01 45 18.	- 17 50		17.9	BLUE STELLAR OBJECT
PHL 7830	01 45 18.	- 22 08		18.7	BLUE STELLAR OBJECT
PHL 3774	01 45 18.	- 29 08		18.3	BLUE STELLAR OBJECT
LIN 564	01 45 23.	- 75 08 07.		15.97	SMC EMISSION LINF OBJECT
MCG+01-05-033	01 45 24.	+ 04 44	30	15.	GALAXY
PHL 3775	01 45 24.	+ 06 06		17.8	BLUE STELLAR OBJECT
PHL 3776	01 45 24.	+ 07 25		18.2	BLUE STELLAR OBJECT
ZWG 460.025	01 45 24.	+ 18 20		15.7	GALAXY
6ZW 041	01 45 24.	+ 32 22			COMPACT GALAXY
6ZW 042	01 45 24.	+ 32 37			COMPACT GALAXY
VB 149	01 45 24.	+ 57 50	195		STELLAR RING
ISS 0055	01 45 24.	+ 57 50	199		STELLAR RING
ZWG 326.001	01 45 24.	+ 75 01		15.7	GALAXY
PHL 7832	01 45 24.	- 06 11		17.6	BLUE STELLAR OBJECT
PHL 7831	01 45 24.	- 09 00		16.5	BLUE STELLAR OBJECT
PHL 7833	01 45 24.	- 11 00		18.5	BLUE STELLAR OBJECT
PHL 7834	01 45 24.	- 11 43		18.6	BLUE STELLAR OBJECT
MCG-02-05-049	01 45 24.	- 13 50	24	15.	GALAXY
RNGC 0690	01 45 24.	- 16 57		14.0	GALAXY
MCG-03-05-021	01 45 24.	- 16 57	72	14.	GALAXY
PHL 3778	01 45 24.	- 23 45		17.6	BLUE STELLAR OBJECT
PHL 7835	01 45 24.	- 28 12		18.0	BLUE STELLAR OBJECT
PHL 3777	01 45 24.	- 31 56		18.4	BLUE STELLAR OBJECT
MCG-06-05-002	01 45 24.	- 33 50	60	13.5	GALAXY
LB 03229	01 45 24.	- 51 48		13.2	FAINT BLUE STAR
HN 0156	01 45 26.	- 33 51			NEBULA
IC 1728	01 45 26.	- 33 51			NONSTELLAR OBJECT
PHL 7836	01 45 30.	+ 05 00		17.0	BLUE STELLAR OBJFCT
PHL 7837	01 45 30.	+ 06 50		18.7	BLUE STELLAR OBJECT
ZWG 460.026	01 45 30.	+ 18 30		15.2	GALAXY
6ZW 043	01 45 30.	+ 32 27			COMPACT GALAXY
PHL 3779	01 45 30.	- 01 58		16.5	BLUE STELLAR OBJECT
MCG-04-05-007	01 45 30.	- 22 04	36	15.	GALAXY
RNGC 0673	01 45 34.	+ 11 17		13.5	GALAXY
PHL 7839	01 45 36.	+ 01 00		18.8	BLUE STELLAR OBJECT
MCG+02-05-033	01 45 36.	+ 11 15	132	12.	GALAXY
6ZW 044	01 45 36.	+ 34 56			COMPACT GALAXY
VB 056	01 45 36.	+ 64 47	242		STELLAR RING
PHL 3782	01 45 36.	- 00 14		18.9	BLUE STELLAR OBJECT
PHL 1163	01 45 36.	- 00 26		18.9	BLUE STELLAR OBJECT
PHL 3780	01 45 36.	- 01 06		18.8	BLUE STELLAR OBJECT
PHL 7838	01 45 36.	- 05 14		16.5	BLUE STELLAR OBJECT
PHL 1164	01 45 36.	- 06 10		13.0	BLUE STELLAR OBJECT
FEIG 014	01 45 36.	- 06 11		12.5	FAINT BLUE STAR
PHL 7840	01 45 36.	- 23 57		18.6	BLUE STELLAR OBJECT
PHL 3781	01 45 36.	- 26 11		18.1	BLUE STELLAR OBJECT
IC 1729	01 45 37.	- 27 08 21.			NONSTELLAR OBJECT
PHL 7842	01 45 42.	+ 01 20		18.9	BLUE STELLAR OBJECT
PHL 3784	01 45 42.	+ 02 08		18.8	BLUE STELLAR OBJECT
PHL 3783	01 45 42.	+ 03 41		18.0	BLUE STELLAR OBJECT
ZWG 437.030	01 45 42.	+ 11 17		13.3	GALAXY
UGC 01259	01 45 42.	+ 11 17	138	13.3	GALAXY Sc
MCG+02-05-034	01 45 42.	+ 13 11	84	15.	GALAXY
ZWG 482.017	01 45 42.	+ 27 18		15.3	GALAXY
6ZW 045	01 45 42.	+ 32 27			COMPACT GALAXY
PHL 3786	01 45 42.	- 00 17		18.8	BLUE STELLAR OBJECT
PHL 7841	01 45 42.	- 04 36		18.5	BLUE STELLAR OBJECT
PHL 3785	01 45 42.	- 09 39		18.5	BLUE STELLAR OBJFCT
PHL 3787	01 45 42.	- 13 51		18.6	BLUE STELLAR OBJECT
MCG-05-05-014	01 45 42.	- 27 08	30	15.	GALAXY
MCG+02-05-035	01 45 45.	+ 10 18	60	16.	GALAXY
MCG+01-05-035	01 45 45.	- 04 38	42	15.5	GALAXY
PHL 3789	01 45 48.	+ 00 43		18.2	BLUE STELLAR OBJECT
PHL 7851	01 45 48.	+ 03 44		18.4	BLUE STELLAR OBJECT
PHL 7850	01 45 48.	+ 04 48		17.1	BLUE STELLAR OBJECT
PHL 7848	01 45 48.	+ 05 10		14.8	BLUE STELLAR OBJFCT
PHL 7849	01 45 48.	+ 06 44		16.9	BLUE STELLAR OBJECT
ZWG 437.031	01 45 48.	+ 12 21		14.0	GALAXY
MRK 575	01 45 48.	+ 12 21	16	15.5	GALAXY WITH UV CONTINUUM
UGC 01260	01 45 48.	+ 12 21	60	14.0	GALAXY SBa
ZWG 437.032	01 45 48.	+ 13 10		15.5	GALAXY
UGC 01261	01 45 48.	+ 13 10	84	15.5	GALAXY Sc
UGC 01262	01 45 48.	+ 13 27	84	16.0	GALAXY IRR
5ZW 110	01 45 48.	+ 32 43			COMPACT GALAXY
5ZW 111	01 45 48.	+ 43 38			COMPACT GALAXY
ZC 0145.8+4740	01 45 48.	+ 47 40	11220		CLUSTER OF GALAXIES
PHL 1165	01 45 48.	- 01 48		18.2	BLUE STELLAR OBJECT
PHL 7846	01 45 48.	- 03 16		18.0	BLUE STELLAR OBJECT
PHL 7847	01 45 48.	- 05 02		18.5	BLUE STELLAR OBJECT
PHL 7845	01 45 48.	- 05 44		17.0	BLUE STELLAR OBJECT
PHL 3788	01 45 48.	- 05 50		17.2	BLUE STELLAR OBJECT
PHL 1166	01 45 48.	- 06 10		17.1	BLUE STELLAR OBJECT
PHL 7844	01 45 48.	- 08 32		17.3	BLUE STELLAR OBJECT
PHL 7843	01 45 48.	- 12 38		16.7	BLUE STELLAR OBJECT
TON-S 0231	01 45 48.	- 25 47		14.4	BLUE STAR
PHL 7852	01 45 48.	- 29 53		18.4	BLUE STELLAR OBJECT
RNGC 0685	01 45 51.	- 53 01		12.0	GALAXY
YM 18	01 45 52.	+ 53 37	1500		SYMMETRIC GALACTIC NFBULA
PHL 7853	01 45 54.	+ 02 02		17.7	BLUE STELLAR OBJECT
UGC 01263	01 45 54.	+ 10 19	72	16.0	GALAXY S
UGC 01264	01 45 54.	+ 13 31	72	16.0	GALAXY Sc
6ZW 046	01 45 54.	+ 34 39			COMPACT GALAXY
6ZW 047	01 45 54.	+ 36 20			COMPACT GALAXY
PHL 7856	01 45 54.	- 03 58		18.0	BLUE STELLAR OBJECT
PHL 7854	01 45 54.	- 06 30		18.1	BLUE STELLAR OBJECT
PHL 7855	01 45 54.	- 06 58		17.7	BLUE STELLAR OBJECT
PHL 3790	01 45 54.	- 15 08		16.7	BLUE STELLAR OBJECT
VV 054C	01 45 57.	+ 10 14	6	20.	INTERACTING GALAXY
VV 054B	01 45 57.	+ 10 14	12	18.	INTERACTING GALAXY
VV 054A	01 45 57.	+ 10 14	36	15.	INTERACTING GALAXY
LBN 0640	01 46	+ 53 30	1800		BRIGHT NEBULA
VDB.66G 015	01 46	- 13 01	70		DWARF GALAXY

OBJECT NAME	RIGHT ASCEN.	DECLINATION	DIAM.	MAGN.	TYPE OF OBJECT
PHL 3792	01 46 00.	+ 05 25		18.0	BLUE STELLAR OBJECT
MCG+02-05-036	01 46 00.	+ 10 14	24	14.5	GALAXY
ZC 0146.0+1347	01 46 00.	+ 13 47	1610		CLUSTER OF GALAXIES
KARA.72 041B	01 46 00.	+ 19 59	36		PART OF DOUBLE GALAXY
ZWG 460.027	01 46 00.	+ 20 00		14.5	GALAXY
UGC 01265	01 46 00.	+ 20 00	66	14.5	GALAXY SB
KARA.72 041A	01 46 00.	+ 20 00	48	14.5	PART OF DOUBLE GALAXY
MCG+03-05-017	01 46 00.	+ 20 01	66	14.	GALAXY
ZC 0146.0+2246	01 46 00.	+ 22 46	1410		CLUSTER OF GALAXIES
PHL 3793	01 46 00.	- 02 02		17.1	BLUE STELLAR OBJECT
PHL 7858	01 46 00.	- 02 30		18.1	BLUE STELLAR OBJFCT
PHL 7857	01 46 00.	- 04 10		17.6	BLUE STELLAR OBJECT
PHL 1167	01 46 00.	- 09 04		18.3	BLUE STELLAR OBJECT
MCG-02-05-050	01 46 00.	- 12 38	168	14.	GALAXY
PHL 3791	01 46 00.	- 14 36		17.9	BLUE STELLAR OBJECT
PHL 7859	01 46 00.	- 22 16		18.6	BLUF STELLAR OPJFCT
ARP 004	01 46 01.	- 12 37			PECULIAR GALAXY
REIN 4.009	01 46 05.11	+ 05 44 57.2			NEBULA
IC 0161	01 46 06.	+ 10 06 45.			NONSTELLAR OBJECT
ZWG 437.033	01 46 06.	+ 10 15		14.2	GALAXY
UGC 01266	01 46 06.	+ 10 15	66	14.2	GALAXY
MCG+02-05-037	01 46 06.	+ 10 20	48	14.5	GALAXY
5ZW 112	01 46 06.	+ 32 52			COMPACT GALAXY
ZWG 522.008	01 46 06.	+ 34 59		15.6	GALAXY
6ZW 048	01 46 06.	+ 36 55			COMPACT GALAXY
PHL 7862	01 46 06.	- 02 36		18.1	BLUE STELLAR OBJFCT
PHL 7861	01 46 06.	- 09 18		17.1	BLUE STELLAR OBJECT
MCG-02-05-051	01 46 06.	- 10 33	60	14.5	GALAXY
PHL 7863	01 46 06.	- 12 28		18.4	BLUE STELLAR OBJECT
MCG-02-05-050A	01 46 06.	- 12 38	36	14.	GALAXY
PHL 7860	01 46 06.	- 27 08		18.0	BLUE STELLAR OBJECT
TON-S 0232	01 46 06.	- 28 15		16.1	BLUE STAR
MCG+02-05-038	01 46 09.	+ 10 15	72	13.	GALAXY
VV 053B	01 46 09.	+ 10 16 30.	9	18.	INTERACTING GALAXY
VV 053A	01 46 09.	+ 10 16 30.	60	13.5	INTERACTING GALAXY
REIN 4.010	01 46 10.90	+ 05 43 41.5			NEBULA
PHL 3794	01 46 12.	+ 05 50		17.2	BLUF STELLAR OFJECT
PHL 7870	01 46 12.	+ 06 21		18.4	BLUF STELLAR OBJFCT
PHL 7866	01 46 12.	+ 07 22		18.4	BLUE STELLAR OBJECT
MCG+02-05-039	01 46 12.	+ 10 15	36	15.	GALAXY
ZWG 437.034	01 46 12.	+ 10 16		14.8	GALAXY
UGC 01267	01 46 12.	+ 10 16	114	14.8	GALAXY S0
ZWG 437.035	01 46 12.	+ 10 20		15.3	GALAXY
UGC 01268	01 46 12.	+ 10 20	72	15.3	GALAXY S
MCG+04-05-012	01 46 12.	+ 25 41	27	16.5	GALAXY
MCG+06-05-007	01 46 12.	+ 34 43	36	15.	GALAXY
ZWG 522.009	01 46 12.	+ 34 44		15.4	GALAXY
UGC 01269	01 46 12.	+ 34 44	72	15.4	GALAXY E-S0
PHL 3797	01 46 12.	- 00 18		18.9	BLUE STELLAR OBJECT
PHL 3795	01 46 12.	- 01 50		18.8	BLUE STELLAR OBJECT
PHL 3796	01 46 12.	- 02 50		18.5	BLUE STELLAR OBJECT
PHL 7865	01 46 12.	- 03 14		17.6	BLUE STELLAR OBJECT
PHL 7869	01 46 12.	- 03 46		17.3	BLUE STELLAR OBJECT
PHL 7864	01 46 12.	- 04 36		16.6	BLUE STELLAR OBJECT
PHL 7867	01 46 12.	- 11 06		16.6	BLUE STELLAR OBJECT
PHL 7868	01 46 12.	- 13 32		16.8	BLUF STELLAR OBJFCT
PHL 1168	01 46 12.	- 14 26		18.4	BLUE STELLAR OBJECT
PHL 7871	01 46 12.	- 15 36		18.7	BLUE STELLAR OBJECT
SER 017.06	01 46 12.	- 52 17	20	15.5	COMPACT GROUP OF 6 GLXIES
MOHR 6	01 46 12.	- 74 01 24.		15.0	STAR CLUSTER IN SMC
IC 0162	01 46 13.	+ 10 16 38.			NONSTELLAR OBJECT
REIN 4.011	01 46 13.31	+ 05 39 44.0			NEBULA
REIN 4.012	01 46 13.34	+ 05 35 19.4			NEBULA
ARP 228	01 46 14.	+ 10 17			PFCULIAR GALAXY
MCG+02-05-040	01 46 15.	+ 12 57	72	14.	GALAXY
ARC 0257	01 46 17.	+ 13 45		16.9	RICH CLUSTER OF GALAXIES
PHL 1170	01 46 18.	+ 01 02		15.3	BLUE STELLAR OBJECT
PHL 1169	01 46 18.	+ C2 18		17.3	BLUE STELLAR OBJECT
ZWG 412.028	01 46 18.	+ 05 40		10.5	GALAXY
RNGC 0676	01 46 18.	+ 05 40		10.5	GALAXY
UGC 01270	01 46 18.	+ 05 40	300	10.5	GALAXY S0-a
PHL 3798	01 46 18.	+ 06 43		18.2	BLUF STELLAR OBJECT
PHL 7872	01 46 18.	+ 09 05		17.9	BLUE STELLAR OBJECT
PHL 7873	01 46 18.	+ 09 12		18.3	BLUE STELLAR OBJFCT
ZWG 437.036	01 46 18.	+ 12 57		14.6	GALAXY
UGC 01271	01 46 18.	+ 12 57	108	14.6	GALAXY SB0
ZWG 503.025	01 46 18.	+ 32 50		15.6	GALAXY
MCG+05-05-013	01 46 18.	+ 32 50	24	16.	GALAXY
ZWG 522.010	01 46 18.	+ 34 50		14.3	GALAXY
UGC 01272	01 46 18.	+ 34 50	90	14.3	GALAXY S0
ZWG 386.042	01 46 18.	- 01 22		15.7	GALAXY
PHL 7875	01 46 18.	- 07 27		17.0	BLUE STELLAR OBJECT
PHL 3799	01 46 18.	- 08 46		18.3	BLUE STELLAR OBJECT
PHL 7874	01 46 18.	- 10 24		18.4	BLUE STELLAR OBJECT
PHL 1171	01 46 18.	- 25 01		17.7	BLUE STELLAR OBJECT
REIN 2.015	01 46 20.36	+ 05 39 30.1			NEBULA
MCG+01-05-034	01 46 21.	+ 05 38 30.	210	12.5	GALAXY
RNGC 0675	01 46 22.	+ 12 48		15.5	GALAXY
REIN 4.013	01 46 22.24	+ 05 39 35.6			NEBULA
PHL 3800	01 46 24.	+ 04 26		16.6	BLUE STELLAR OBJECT
PHL 7880	01 46 24.	+ 05 02		16.8	BLUE STELLAR OBJECT
ZWG 437.037	01 46 24.	+ 12 48		15.5	GALAXY
UGC 01273	01 46 24.	+ 12 48	72	15.5	GALAXY S
MCG+02-05-041	01 46 24.	+ 12 48	48	14.5	GALAXY
MCG+04-05-013	01 46 24.	+ 22 49	30	16.5	GALAXY
6ZW 049	01 46 24.	+ 34 21			COMPACT GALAXY
MCG+06-05-008	01 46 24.	+ 34 42	15	16.	GALAXY
ZWG 522.011	01 46 24.	+ 34 43		15.4	GALAXY
MCG+06-05-009	01 46 24.	+ 34 48 30.	36	14.	GALAXY
ZWG 522.012	01 46 24.	+ 36 00		15.7	GALAXY
PHL 3801	01 46 24.	- 01 29		18.3	BLUE STELLAR OFJECT
PHL 7881	01 46 24.	- 01 50		18.2	BLUE STELLAR OPJFCT
PHL 7876	01 46 24.	- 05 34		16.5	BLUE STELLAR OBJECT
PHL 7878	01 46 24.	- 07 04		18.6	BLUE STELLAR OBJECT
PHL 7879	01 46 24.	- 12 42		15.5	BLUF STELLAR OBJECT
PHL 7877	01 46 24.	- 12 46		16.2	BLUE STELLAR OBJECT
PHL 1172	01 46 24.	- 14 14		17.7	BLUE STELLAR OBJECT
PHL 3802	01 46 24.	- 26 52		12.2	BLUE STELLAR OBJECT
RNGC 0674	01 46 25.	+ 22 06			NON-EXISTENT OBJECT
MCG+06-05-010	01 46 27.	+ 34 43	24	15.	GALAXY
RNGC 0677	01 46 28.	+ 12 48		14.5	GALAXY
LIN 565	01 46 28.	- 74 21 39.		16.70	SMC EMISSION LINE OBJECT
ZWG 412.029	01 46 30.	+ 05 23		15.2	GALAXY
MRK 576	01 46 30.	+ 05 23	16	15.5	GALAXY WITH UV CONTINUUM
MCG+02-05-043	01 46 30.	+ 12 35	72	14.5	GALAXY
ZWG 437.038	01 46 30.	+ 12 36		15.2	GALAXY
UGC 01274	01 46 30.	+ 12 36	102	15.2	GALAXY Sa
ZWG 437.039	01 46 30.	+ 12 48		14.3	GALAXY

OBJECT NAME	RIGHT ASCEN.	DECLINATION	DIAM.	MAGN.	TYPE OF OBJECT
UGC 01275	01 46 30.	+ 12 48	126	14.3	GALAXY E
MCG+02-05-042	01 46 30.	+ 12 48	24	13.5	GALAXY
FEIG 015	01 46 30.	+ 13 18		10.6	FAINT BLUE STAR
ZWG 460.028	01 46 30.	+ 20 27		13.8	GALAXY
UGC 01276	01 46 30.	+ 20 27	120	13.8	GALAXY
MCG+03-05-018	01 46 30.	+ 20 28	102	13.5	GALAXY
5ZW 113	01 46 30.	+ 34 44			COMPACT GALAXY
ZWG 522.013	01 46 30.	+ 34 44		15.5	GALAXY
MCG+06-05-011	01 46 30.	+ 35 11 30.	90	14.	GALAXY
ZWG 522.014	01 46 30.	+ 35 12		14.5	GALAXY
UGC 01277	01 46 30.	+ 35 12	120	14.5	GALAXY S0-a
6ZW 050	01 46 30.	+ 36 10			COMPACT GALAXY
MCG-05-05-015	01 46 30.	- 29 12	60	15.	GALAXY
REIN 4.014	01 46 31.01	+ 05 35 55.4			NEBULA
IC 0164	01 46 35.	- 04 10 12.			NONSTELLAR OBJECT
RNGC 0682	01 46 35.	- 15 13		13.0	GALAXY
PHL 7885	01 46 36.	+ 04 56		17.9	BLUE STELLAR OBJECT
MCG+02-05-044	01 46 36.	+ 12 27	60	14.5	GALAXY
ZWG 437.040	01 46 36.	+ 12 28		15.7	GALAXY
UGC 01278	01 46 36.	+ 12 28	66	15.7	GALAXY Sc
PHL 7883	01 46 36.	- 00 52		16.2	BLUE STELLAR OBJECT
PHL 7884	01 46 36.	- 03 56		16.8	BLUE STELLAR OBJECT
PHL 1173	01 46 36.	- 12 38		18.1	BLUE STELLAR OBJECT
MCG-03-05-022	01 46 36.	- 15 13	30	13.5	GALAXY
PHL 7882	01 46 36.	- 16 24		18.4	BLUE STELLAR OBJECT
MCG-05-05-016	01 46 36.	- 27 23	30	16.	GALAXY
RNGC 0678	01 46 37.	+ 21 45		13.5	GALAXY
IC 0163	01 46 38.	+ 20 28 01.			NONSTELLAR OBJECT
RNGC 0686	01 46 38.	- 24 02		13.0	GALAXY
MCG-04-05-008	01 46 39.	- 24 02	33	13.5	GALAXY
RNGC 0681	01 46 40.	- 10 40		13.0	GALAXY
PHL 3803	01 46 42.	+ 07 18		17.5	BLUE STELLAR OBJECT
ZWG 437.041	01 46 42.	+ 13 07		15.3	GALAXY
UGC 01279	01 46 42.	+ 13 07	60	15.3	GALAXY S
MCG+02-05-045	01 46 42.	+ 13 07	36	15.5	GALAXY
MCG+04-05-014	01 46 42.	+ 21 43	264	12.5	GALAXY
ZWG 482.018	01 46 42.	+ 21 45		13.3	GALAXY
UGC 01280	01 46 42.	+ 21 45	300	13.3	GALAXY Sb
6ZW 051	01 46 42.	+ 32 20			COMPACT GALAXY
ZWG 503.026	01 46 42.	+ 32 20		13.0	GALAXY
UGC 01281	01 46 42.	+ 32 20	282	13.0	GALAXY S IV
MCG+05-05-014	01 46 42.	+ 32 20	264	13.	GALAXY
PHL 3804	01 46 42.	- 03 40		18.3	BLUE STELLAR OBJECT
MCG-02-05-053	01 46 42.	- 10 18	150	13.	GALAXY
MCG-02-05-052	01 46 42.	- 10 40	138	12.	GALAXY
PHL 1175	01 46 42.	- 11 22		18.4	BLUE STELLAR OBJECT
PHL 1174	01 46 42.	- 11 22		17.3	BLUE STELLAR OBJECT
PHL 7886	01 46 42.	- 14 34		17.5	BLUE STELLAR OBJECT
MCG+02-05-046	01 46 45.	+ 12 14	60	14.	GALAXY
MCG+01-05-036	01 46 45.	- 03 55	60	15.	GALAXY
MCG+01-05-037	01 46 45.	- 04 08 30.	42	13.5	GALAXY
IC 1734	01 46 47.	- 33 00 19.			NONSTELLAR OBJECT
PHL 3807	01 46 48.	+ 01 50		18.6	BLUE STELLAR OBJECT
PHL 3811	01 46 48.	+ 04 02		18.7	BLUE STELLAR OBJECT
PHL 3806	01 46 48.	+ 04 12		18.0	BLUE STELLAR OBJECT
PHL 1176	01 46 48.	+ 07 24		18.1	BLUE STELLAR OBJECT
ZWG 437.042	01 46 48.	+ 12 15		14.2	GALAXY
MRK 577	01 46 48.	+ 12 15	9	15.	GALAXY WITH UV CONTINUUM
UGC 01282	01 46 48.	+ 12 15	96	14.2	GALAXY S0-a
ZC 0146.8+1325	01 46 48.	+ 13 25	1340		CLUSTER OF GALAXIES
ZWG 503.027	01 46 48.	+ 32 20		15.6	GALAXY
5ZW 114	01 46 48.	+ 35 32			COMPACT GALAXY
ZWG 522.015	01 46 48.	+ 35 32		13.1	GALAXY
UGC 01283	01 46 48.	+ 35 32	126	13.1	GALAXY E-S0
MCG+06-05-012	01 46 48.	+ 35 32	102	13.	GALAXY
VB 150	01 46 48.	+ 62 09	289		STELLAR RING
ZC 0146.8-0035	01 46 48.	- 00 35	270		CLUSTER OF GALAXIES
PHL 3805	01 46 48.	- 00 44		16.5	BLUE STELLAR OBJECT
PHL 7888	01 46 48.	- 02 10		16.9	BLUE STELLAR OBJECT
PHL 3813	01 46 48.	- 06 16		18.8	BLUE STELLAR OBJECT
PHL 3812	01 46 48.	- 06 25		19.0	BLUE STELLAR OBJECT
PHL 3808	01 46 48.	- 08 30		18.5	BLUE STELLAR OBJECT
MCG-02-05-054	01 46 48.	- 08 40 30.	36	15.	GALAXY
PHL 7889	01 46 48.	- 10 32		18.7	BLUE STELLAR OBJECT
PHL 3810	01 46 48.	- 11 44		18.1	BLUE STELLAR OBJECT
PHL 7890	01 46 48.	- 11 58		18.4	BLUE STELLAR OBJECT
PHL 3809	01 46 48.	- 13 07		18.2	BLUE STELLAR OBJECT
PHL 7887	01 46 48.	- 14 48		17.9	BLUE STELLAR OBJECT
RNGC 0692	01 46 48.	- 48 53			UNVERIFIED SOUTHERN OBJECT
RNGC 0679	01 46 50.	+ 35 32		13.0	GALAXY
UGC 01284	01 46 54.	+ 28 14	60	17.	GALAXY Sc
ZWG 537.024	01 46 54.	+ 43 16		15.7	GALAXY
PHL 3814	01 46 54.	- 00 06		17.7	BLUE STELLAR OBJECT
SHAH 143	01 46 54.	- 01 22	102		GROUP OF COMPACT GALAXIES
PHL 7891	01 46 54.	- 04 04		16.6	BLUE STELLAR OBJECT
PHL 3815	01 46 54.	- 11 25		18.4	BLUE STELLAR OBJECT
MCG-06-05-003	01 46 57.	- 32 58	84	13.	GALAXY
REIN 4.017	01 46 57.87	+ 05 55 36.1			NEBULA
RNGC 0683	01 46 58.	+ 11 27		15.0	GALAXY
KHAV 030	01 47	+ 66 09	10420		DARK NEBULA
PHL 1178	01 47 00.	+ 04 14		18.9	BLUE STELLAR OBJECT
PHL 1177	01 47 00.	+ 04 58		17.9	BLUE STELLAR OBJECT
PHL 3817	01 47 00.	+ 06 27		17.8	BLUE STELLAR OBJECT
PHL 7892	01 47 00.	+ 08 58		15.9	BLUE STELLAR OBJECT
PHL 3818	01 47 00.	+ 09 36		16.7	BLUE STELLAR OBJECT
MCG+02-05-047	01 47 00.	+ 11 27	60	14.	GALAXY
MCG+04-05-015	01 47 00.	+ 21 41	36	13.	GALAXY
ZWG 482.019	01 47 00.	+ 21 43		13.0	GALAXY
UGC 01286	01 47 00.	+ 21 43	162	13.0	GALAXY E
UGC 01287	01 47 00.	+ 22 08	72	18.	GALAXY DWARF
ZCG 0147+36.1	01 47 00.	+ 36 02		17.7	COMPACT GALAXY
ZWG 552.008	01 47 00.	+ 47 38		15.7	GALAXY
ZWG 370.001	01 47 00.	+ 86 27		14.5	GALAXY
ZWG 361.005	01 47 00.	+ 86 27		14.5	GALAXY
ZWG 360.008	01 47 00.	+ 86 27		14.5	GALAXY
UGC 01285	01 47 00.	+ 86 27	90	14.5	GALAXY Sb
KARA.73B 0066	01 47 00.	+ 86 27	66	14.5	ISOLATED GALAXY S
PHL 7893	01 47 00.	- 05 24		16.7	BLUE STELLAR OBJECT
PHL 7894	01 47 00.	- 06 27		16.8	BLUE STELLAR OBJECT
PHL 7896	01 47 00.	- 07 46		17.2	BLUE STELLAR OBJECT
PHL 3816	01 47 00.	- 08 54		18.4	BLUE STELLAR OBJECT
PHL 7898	01 47 00.	- 09 42		17.2	BLUE STELLAR OBJECT
MCG-02-05-055	01 47 00.	- 14 22	60	15.	GALAXY
PHL 1179	01 47 00.	- 14 42		18.4	BLUE STELLAR OBJECT
PHL 7895	01 47 00.	- 16 24		16.2	BLUE STELLAR OBJECT
PHL 7897	01 47 00.	- 16 46		13.6	BLUE STELLAR OBJECT
PHL 1180	01 47 00.	- 19 03		17.4	BLUE STELLAR OBJECT
RNGC 0680	01 47 01.	+ 21 43		13.0	GALAXY

OBJECT NAME	RIGHT ASCEN.	DECLINATION	DIAM.	MAGN.	TYPE OF OBJECT
MCG+02-05-048	01 47 03.	+ 11 27	48	16.	GALAXY
ZWG 412.030	01 47 06.	+ 03 32		15.7	GALAXY
ZWG 437.043	01 47 06.	+ 11 27		14.8	GALAXY
UGC 01288	01 47 06.	+ 11 27	60	14.8	GALAXY S
5ZW 115	01 47 06.	+ 34 07			COMPACT GALAXY
6ZW 052	01 47 06.	+ 35 51			COMPACT GALAXY
PHL 3819	01 47 06.	- 03 29		18.2	BLUE STELLAR OBJECT
PHL 7900	01 47 06.	- 05 34		17.8	BLUE STELLAR OBJECT
PHL 3821	01 47 06.	- 06 45		19.0	BLUE STELLAR OBJECT
PHL 3820	01 47 06.	- 10 21		18.5	BLUE STELLAR OBJECT
PHL 7899	01 47 06.	- 10 56		5.5	BLUE STELLAR OBJECT
IC 1730	01 47 09.	+ 21 43 47.			NONSTELLAR OBJECT
PHL 1181	01 47 12.	+ 00 00		16.6	BLUE STELLAR OBJECT
PHL 7903	01 47 12.	+ 01 29		18.0	BLUE STELLAR OBJECT
PHL 3828	01 47 12.	+ 01 57		18.3	BLUE STELLAR OBJECT
PHL 3822	01 47 12.	+ 04 56		17.3	BLUE STELLAR OBJECT
PHL 3827	01 47 12.	+ 04 58		18.0	BLUE STELLAR OBJECT
PHL 3823	01 47 12.	+ 05 10		17.4	BLUE STELLAR OBJECT
ZWG 412.031	01 47 12.	+ 07 07		15.7	GALAXY
UGC 01289	01 47 12.	+ 16 10	84	16.0	GALAXY Sc
MCG+04-05-015A	01 47 12.	+ 21 43	36	15.5	GALAXY
ZWG 482.020	01 47 12.	+ 21 46		15.5	GALAXY
MCG+04-05-016	01 47 12.	+ 26 50	39	16.	GALAXY
ZWG 522.016	01 47 12.	+ 37 06		15.7	GALAXY
PHL 3824	01 47 12.	- 02 08		18.5	BLUE STELLAR OBJECT
PHL 7901	01 47 12.	- 03 35		16.5	BLUE STELLAR OBJECT
PHL 7902	01 47 12.	- 04 18		17.8	BLUE STELLAR OBJECT
PHL 7904	01 47 12.	- 06 16		18.0	BLUE STELLAR OBJECT
PHL 1182	01 47 12.	- 10 48		17.9	BLUE STELLAR OBJECT
PHL 3825	01 47 12.	- 11 26		18.2	BLUE STELLAR OBJECT
PHL 1183	01 47 12.	- 12 14		18.0	BLUE STELLAR OBJECT
MCG-02-05-057	01 47 12.	- 13 04 30.	90	16.	GALAXY
MCG-02-05-056	01 47 12.	- 13 49	72	14.	GALAXY
PHL 3826	01 47 12.	- 14 06		18.1	BLUE STELLAR OBJECT
PHL 1184	01 47 12.	- 19 25		16.8	BLUE STELLAR OBJECT
MCG-06-05-004	01 47 12.	- 35 08	72	13.5	GALAXY
RNGC 0696	01 47 12.	- 35 09			GALAXY
ARC 0258	01 47 16.	+ 23 13		17.7	RICH CLUSTER OF GALAXIES
HARO 16	01 47 16.	- 27 59			BLUE EMISSION-LINE GALAXY
ZWG 386.043	01 47 18.	+ 03 25		15.5	GALAXY
ZC 0147.3+0803	01 47 18.	+ 08 03	4970		CLUSTER OF GALAXIES
MCG+02-05-049	01 47 18.	+ 11 33	30	14.5	GALAXY
ZWG 437.044	01 47 18.	+ 11 34		15.1	GALAXY
UGC 01290	01 47 18.	+ 11 34	60	15.1	GALAXY Sb-c
MCG+03-05-019	01 47 18.	+ 16 11	48	15.5	GALAXY
ZWG 460.029	01 47 18.	+ 16 38		15.6	GALAXY
ZWG 482.021	01 47 18.	+ 26 57		14.2	GALAXY
UGC 01291	01 47 18.	+ 26 57	102	14.2	GALAXY Sc
OCL 0371	01 47 18.	+ 27 00	360	8.4	OPEN STAR CLUSTER
ZWG 482.022	01 47 18.	+ 27 24		13.2	GALAXY
UGC 01292	01 47 18.	+ 27 24	210	13.2	GALAXY Sb
6ZW 053	01 47 18.	+ 29 30			COMPACT GALAXY
6ZW 054	01 47 18.	+ 35 48			COMPACT GALAXY
STOCK 01	01 47 18.	+ 35 59			BLUE KNOT NEAR ELLIP GLXY
5ZW 116	01 47 18.	+ 36 00			COMPACT GALAXY
6ZW 055	01 47 18.	+ 37 07			COMPACT GALAXY
PHL 3829	01 47 18.	- 00 03		17.6	BLUE STELLAR OBJECT
PHL 3831	01 47 18.	- 01 19		18.6	BLUE STELLAR OBJECT
MCG-05-05-017	01 47 18.	- 27 20	60	14.5	GALAXY
PHL 3830	01 47 18.	- 28 30		18.3	BLUE STELLAR OBJECT
MCG+04-05-017	01 47 21.	+ 27 23	210	13.5	GALAXY
IC 1731	01 47 22.	+ 26 57 17.			NONSTELLAR OBJECT
IC 0165	01 47 23.	+ 27 22 47.			MAY NOT EXIST
RNGC 0684	01 47 23.	+ 27 24		13.0	GALAXY
PHL 3832	01 47 24.	+ 01 30		18.5	BLUE STELLAR OBJECT
ZWG 386.044	01 47 24.	+ 01 50		15.0	GALAXY
UGC 01293	01 47 24.	+ 01 50	72	15.0	GALAXY Sa-b
UGC 01294	01 47 24.	+ 22 54	72	18.	GALAXY DWARF
MCG+04-05-018	01 47 24.	+ 26 56	84	14.	GALAXY
MCG+05-05-015	01 47 24.	+ 33 14	30	15.	GALAXY
ZWG 503.028	01 47 24.	+ 33 15		15.6	GALAXY
UGC 01295	01 47 24.	+ 33 15	84	15.6	GALAXY S0
5ZW 117	01 47 24.	+ 33 34			COMPACT GALAXY
6ZW 056	01 47 24.	+ 35 16			COMPACT GALAXY
5ZW 118	01 47 24.	+ 35 44			COMPACT GALAXY
ZCG 0147+36.0	01 47 24.	+ 36 00		18.2	COMPACT GALAXY
6ZW 057	01 47 24.	+ 36 57			COMPACT GALAXY
PHL 7905	01 47 24.	- 01 25		16.3	BLUE STELLAR OBJECT
PHL 7907	01 47 24.	- 03 08		18.5	BLUE STELLAR OBJECT
PHL 7906	01 47 24.	- 08 46		16.8	BLUE STELLAR OBJECT
PHL 7909	01 47 24.	- 08 49		18.5	BLUE STELLAR OBJECT
PHL 3833	01 47 24.	- 11 39		18.8	BLUE STELLAR OBJECT
PHL 1185	01 47 24.	- 12 42		17.7	BLUE STELLAR OBJECT
PHL 7908	01 47 24.	- 24 44		18.6	BLUE STELLAR OBJECT
PHL 7910	01 47 24.	- 31 26		18.0	BLUE STELLAR OBJECT
RNGC 0698	01 47 24.	- 35 04			GALAXY
MCG-06-05-005	01 47 24.	- 35 04	60	14.5	GALAXY
MCG-04-05-009	01 47 27.	- 22 04	18	17.	GALAXY
MCG+00-05-036	01 47 30.	+ 01 50	24	15.	GALAXY
ZWG 437.045	01 47 30.	+ 12 28		15.7	GALAXY
ZWG 386.045	01 47 30.	- 00 59		15.5	GALAXY
UGC 01296	01 47 30.	- 00 59	60	15.5	GALAXY S
PHL 3834	01 47 30.	- 07 25		15.3	BLUE STELLAR OBJECT
PHL 7911	01 47 30.	- 13 02		17.7	BLUE STELLAR OBJECT
MCG-04-05-010	01 47 30.	- 22 09	36	15.	GALAXY
MCG+01-05-038	01 47 33.	- 06 41	18	16.	GALAXY
MCG+01-05-039	01 47 33.	- 06 44	72	15.	GALAXY
HARO 17	01 47 34.	- 28 04			BLUE EMISSION-LINE GALAXY
ZWG 386.046	01 47 36.	+ 02 03		15.2	GALAXY
UGC 01297	01 47 36.	+ 02 03	78	15.2	GALAXY IRR
MCG+00-05-038	01 47 36.	+ 02 04 30.	54	14.	GALAXY
ZWG 412.032	01 47 36.	+ 03 46		15.6	GALAXY
PHL 7913	01 47 36.	+ 04 02		18.3	BLUE STELLAR OBJECT
PHL 1186	01 47 36.	+ 09 01		18.6	BLUE STELLAR OBJECT
BC PHL1186	01 47 36.	+ 09 01		18.6	QUASI-STELLAR OBJECT
SHB 046	01 47 36.	+ 09 01		18.6	QUASI-STELLAR OBJECT
3ZW 036	01 47 36.	+ 17 28			COMPACT GALAXY
ZWG 460.030	01 47 36.	+ 17 28		15.7	GALAXY
MCG+06-05-013	01 47 36.	+ 35 06	60	15.5	GALAXY
ZWG 522.017	01 47 36.	+ 36 07		13.3	GALAXY
UGC 01298	01 47 36.	+ 36 07	96	13.3	GALAXY S0
5ZW 119	01 47 36.	+ 46 38			COMPACT GALAXY
MCG+00-05-037	01 47 36.	- 00 59	48	15.	GALAXY
PHL 7914	01 47 36.	- 02 52		18.9	BLUE STELLAR OBJECT
PHL 1187	01 47 36.	- 02 54		18.9	BLUE STELLAR OBJECT
PHL 1188	01 47 36.	- 06 03		18.8	BLUE STELLAR OBJECT
PHL 7915	01 47 36.	- 07 52		18.7	BLUE STELLAR OBJECT
PHL 7912	01 47 36.	- 13 28		17.8	BLUE STELLAR OBJECT

OBJECT NAME	RIGHT ASCEN.	DECLINATION	DIAM.	MAGN.	TYPE OF OBJECT
PHL 3835	01 47 36.	- 14 57		18.7	BLUE STELLAR OBJECT
MCG-05-05-018	01 47 36.	- 26 31	48	15.	GALAXY
RNGC 0687	01 47 38.	+ 36 07		13.5	GALAXY
MCG+06-05-014	01 47 39.	+ 36 07	27	13.	GALAXY
RNGC 0689	01 47 39.	- 27 42		15.0	GALAXY
PHL 1189	01 47 42.	+ 07 42		18.8	BLUE STELLAR OBJECT
ZC 0147.7+1424	01 47 42.	+ 14 24	1410		CLUSTER OF GALAXIES
ZCG 0147+17	01 47 42.	+ 17 28		18.1	COMPACT GALAXY
6ZW 058	01 47 42.	+ 32 47			COMPACT GALAXY
ZWG 503.029	01 47 42.	+ 32 48		15.4	GALAXY
ZWG 503.030	01 47 42.	+ 33 23		15.3	GALAXY
ZWG 522.018	01 47 42.	+ 35 07		15.7	GALAXY
UGC 01299	01 47 42.	+ 35 07	60	15.7	GALAXY DWARF
MCG+08-04-007	01 47 42.	+ 48 06	60	15.	GALAXY
OCL 0335	01 47 42.	+ 60 50	360	14.	OPEN STAR CLUSTER
MCG-05-05-019	01 47 42.	- 27 42	48	15.	GALAXY
MCG-05-05-020	01 47 42.	- 28 02	18	15.	GALAXY
ARC 0259	01 47 45.	- 12 15		17.2	RICH CLUSTER OF GALAXIES
PHL 3840	01 47 48.	+ 07 37		16.8	BLUE STELLAR OBJECT
PHL 3837	01 47 48.	+ 08 04		18.9	BLUE STELLAR OBJECT
UGC 01300	01 47 48.	+ 08 09	90	16.5	GALAXY DBL SYS
PHL 3836	01 47 48.	+ 09 28		17.8	BLUE STELLAR OBJECT
ZWG 503.031	01 47 48.	+ 32 15		15.7	GALAXY
MCG+05-05-016	01 47 48.	+ 32 49	18	15.	GALAXY
ZWG 503.032	01 47 48.	+ 32 50		14.7	GALAXY
UGC 01301	01 47 48.	+ 32 50	102	14.7	GALAXY (E)
6ZW 059	01 47 48.	+ 32 51			COMPACT GALAXY
5ZW 120	01 47 48.	+ 33 30			COMPACT GALAXY
ZWG 522.019	01 47 48.	+ 33 30		15.0	GALAXY
ZWG 503.033	01 47 48.	+ 33 30		15.0	GALAXY
MCG+06-05-015	01 47 48.	+ 35 01 30.	120	13.	GALAXY
ZWG 522.020	01 47 48.	+ 35 02		13.3	GALAXY
UGC 01302	01 47 48.	+ 35 02	156	13.3	GALAXY SBb
6ZW 060	01 47 48.	+ 36 05			COMPACT GALAXY
6ZW 061	01 47 48.	+ 37 08			COMPACT GALAXY
ZWG 552.009	01 47 48.	+ 48 06		15.6	GALAXY
UGC 01303	01 47 48.	+ 48 06	96	15.6	GALAXY SBa
VB 057	01 47 48.	+ 64 17	658		STELLAR RING
PHL 7918	01 47 48.	- 03 27		17.6	BLUE STELLAR OBJECT
PHL 3841	01 47 48.	- 03 56		18.7	BLUE STELLAR OBJECT
PHL 1190	01 47 48.	- 04 30		18.5	BLUE STELLAR OBJECT
PHL 7916	01 47 48.	- 05 02		16.9	BLUE STELLAR OBJECT
PHL 7917	01 47 48.	- 07 39		16.7	BLUE STELLAR OBJECT
PHL 3838	01 47 48.	- 08 02		18.5	BLUE STELLAR OBJECT
MCG-02-05-058	01 47 48.	- 08 45	48	15.	GALAXY
PHL 3842	01 47 48.	- 09 36		18.4	BLUE STELLAR OBJECT
PHL 1191	01 47 48.	- 11 22		18.3	BLUE STELLAR OBJECT
PHL 3839	01 47 48.	- 17 04		18.7	BLUE STELLAR OBJECT
PHL 3843	01 47 48.	- 24 04		18.0	BLUE STELLAR OBJECT
RNGC 0688	01 47 50.	+ 35 02		13.5	GALAXY
IC 1733	01 47 51.	+ 32 50 45.			NONSTELLAR OBJECT
MCG+01-05-035	01 47 54.	+ 05 53 30.	120	13.	GALAXY
ZWG 412.033	01 47 54.	+ 05 54		13.5	GALAXY
RNGC 0693	01 47 54.	+ 05 54		13.5	GALAXY
UGC 01304	01 47 54.	+ 05 54	192	13.5	GALAXY S
PHL 7919	01 47 54.	+ 07 00		16.8	BLUE STELLAR OBJECT
PHL 3844	01 47 54.	+ 07 46		17.5	BLUE STELLAR OBJECT
ZC 0147.9+2009	01 47 54.	+ 20 09	540		CLUSTER OF GALAXIES
ZWG 482.023	01 47 54.	+ 21 30		13.5	GALAXY
ZWG 460.031	01 47 54.	+ 21 30		13.5	GALAXY
UGC 01305	01 47 54.	+ 21 30	228	13.5	GALAXY Sb/Sc
MCG+04-05-019	01 47 54.	+ 21 30	180	12.	GALAXY
MCG+05-05-017	01 47 54.	+ 32 17	60	15.	GALAXY
ZWG 503.034	01 47 54.	+ 32 18		15.0	GALAXY
UGC 01306	01 47 54.	+ 32 18	78	15.0	GALAXY S0
MCG+05-05-018	01 47 54.	+ 32 50	30	16.	GALAXY
6ZW 062	01 47 54.	+ 32 52			COMPACT GALAXY
ZWG 522.021	01 47 54.	+ 35 40		15.1	GALAXY
UGC 01307	01 47 54.	+ 35 40	102	15.1	GALAXY S
MCG+06-05-016	01 47 54.	+ 35 40	72	13.5	GALAXY
IC 1732	01 47 54.	+ 35 41			NONSTELLAR OBJECT
ZWG 522.022	01 47 54.	+ 36 01		14.5	GALAXY
UGC 01308	01 47 54.	+ 36 01	138	14.5	GALAXY E
KARA.72 042B	01 47 54.	+ 36 01	42		PART OF DOUBLE GALAXY
KARA.72 042A	01 47 54.	+ 36 01	24	14.5	PART OF DOUBLE GALAXY
MCG+06-05-017	01 47 54.	+ 36 01	24	13.5	GALAXY
PHL 3845	01 47 54.	- 06 59		18.8	BLUE STELLAR OBJECT
PHL 1192	01 47 54.	- 09 57		16.5	BLUE STELLAR OBJECT
PHL 7920	01 47 54.	- 12 56		18.6	BLUE STELLAR OBJECT
RNGC 0691	01 47 55.	+ 21 31		13.5	GALAXY
IC 1735	01 47 59.	+ 32 51 21.			NONSTELLAR OBJECT
IC 0168	01 47 59.	- 08 46 41.			NONSTELLAR OBJECT
PHL 3846	01 48 00.	+ 04 46		17.2	BLUE STELLAR OBJECT
6ZW 063	01 48 00.	+ 32 51			COMPACT GALAXY
LDN 1335	01 48 00.	+ 65 50	3180		DARK NEBULA
MCG+14-02-007	01 48 00.	+ 86 26	72	14.	GALAXY
PHL 7922	01 48 00.	- 02 36		18.0	BLUE STELLAR OBJECT
PHL 7925	01 48 00.	- 03 00		18.4	BLUE STELLAR OBJECT
PHL 3847	01 48 00.	- 03 05		18.7	BLUE STELLAR OBJECT
PHL 7924	01 48 00.	- 07 20		16.6	BLUE STELLAR OBJECT
PHL 3848	01 48 00.	- 12 06		18.4	BLUE STELLAR OBJECT
PHL 7921	01 48 00.	- 19 28		17.7	BLUE STELLAR OBJECT
PHL 7923	01 48 00.	- 21 27		18.5	BLUE STELLAR OBJECT
MCG-05-05-021	01 48 00.	- 29 08	36	15.5	GALAXY
MCG-06-05-006	01 48 00.	- 35 12	36	15.	GALAXY
PHL 3850	01 48 06.	+ 09 21		17.5	BLUE STELLAR OBJECT
ZC 0148.1+1340	01 48 06.	+ 13 40	1280		CLUSTER OF GALAXIES
MCG+03-05-020	01 48 06.	+ 18 03	48	15.	GALAXY
MCG+04-05-020	01 48 06.	+ 21 44	30	14.5	GALAXY
5ZW 121	01 48 06.	+ 28 11			COMPACT GALAXY
6ZW 064	01 48 06.	+ 32 43			COMPACT GALAXY
PHL 3851	01 48 06.	- 00 58		17.0	BLUE STELLAR OBJECT
PHL 7926	01 48 06.	- 03 17		18.2	BLUE STELLAR OBJECT
PHL 3849	01 48 06.	- 11 04		16.5	BLUE STELLAR OBJECT
LB 03230	01 48 06.	- 53 14		14.0	FAINT BLUE STAR
LB 03231	01 48 06.	- 70 46		12.5	FAINT BLUE STAR
MCG+01-05-040	01 48 09.	- 04 10	48	14.5	GALAXY
RNGC 0699	01 48 10.	- 12 17		14.0	GALAXY
IC 1736	01 48 11.	+ 18 02 44.			NONSTELLAR OBJECT
IC 0169	01 48 11.	- 12 55 29.			NONSTELLAR OBJECT
PHL 3852	01 48 12.	+ 03 58		17.2	BLUE STELLAR OBJECT
PHL 7928	01 48 12.	+ 05 31		17.2	BLUE STELLAR OBJECT
PHL 7930	01 48 12.	+ 06 41		18.1	BLUE STELLAR OBJECT
ZWG 437.046	01 48 12.	+ 12 18		15.2	GALAXY
ZWG 460.032	01 48 12.	+ 18 03		15.0	GALAXY
UGC 01309	01 48 12.	+ 18 03	84	15.0	GALAXY Sb-c
5ZW 122	01 48 12.	+ 21 45			COMPACT GALAXY
ZWG 482.024	01 48 12.	+ 21 45		13.9	GALAXY

OBJECT NAME	RIGHT ASCEN.	DECLINATION	DIAM.	MAGN.	TYPE OF OBJECT
MRK 363	01 48 12.	+ 21 45	25	15.5	GALAXY WITH UV CONTINUUM
UGC 01310	01 48 12.	+ 21 45	33	13.9	GALAXY PECULR
MCG+05-05-019	01 48 12.	+ 29 31	60	16.	GALAXY
UGC 01311	01 48 12.	+ 29 34	102	17.	GALAXY DWARF
6ZW 065	01 48 12.	+ 32 48			COMPACT GALAXY
PHL 3853	01 48 12.	- 01 04		17.3	BLUE STELLAR OBJECT
PHL 3854	01 48 12.	- 01 56		18.1	BLUE STELLAR OBJECT
PHL 3855	01 48 12.	- 03 28		18.6	BLUE STELLAR OBJECT
PHL 7927	01 48 12.	- 06 12		18.6	BLUE STELLAR OBJECT
PHL 7929	01 48 12.	- 11 19		17.8	BLUE STELLAR OBJECT
PHL 7931	01 48 12.	- 14 26		18.5	BLUE STELLAR OBJECT
PHL 3856	01 48 12.	- 16 55		18.4	BLUE STELLAR OBJECT
MCG-06-05-007	01 48 12.	- 34 17	48	15.	GALAXY
RNGC 0694	01 48 13.	+ 21 45		14.0	GALAXY
HN 0157	01 48 13.	- 34 18			NEBULA
IC 1739	01 48 14.	- 34 18			NONSTELLAR OBJECT
MCG+02-05-050	01 48 15.	+ 12 17 30.	24	15.5	GALAXY
MCG-02-05-059	01 48 15.	- 12 17	78	14.	GALAXY
MCG+02-05-052	01 48 18.	+ 12 20	72	15.5	GALAXY
ZWG 437.047	01 48 18.	+ 13 03		15.1	GALAXY
UGC 01312	01 48 18.	+ 13 03	90	15.1	GALAXY S
MCG+02-05-051	01 48 18.	+ 13 03	84	14.	GALAXY
ZWG 460.033	01 48 18.	+ 19 07		15.7	GALAXY
ARP 031	01 48 18.	+ 21 39			PECULIAR GALAXY
MCG+04-05-021	01 48 18.	+ 21 39	150	13.	GALAXY
ZWG 482.025	01 48 18.	+ 21 40		14.0	GALAXY
UGC 01313	01 48 18.	+ 21 40	180	14.0	GALAXY SB(c)
KARA.68 011	01 48 18.	+ 22 07	47		DWARF GALAXY
ZWG 503.035	01 48 18.	+ 32 23		15.5	GALAXY
ZCG 0148+32.6	01 48 18.	+ 32 34		17.9	COMPACT GALAXY
6ZW 066	01 48 18.	+ 32 49			COMPACT GALAXY
MCG+06-05-019	01 48 18.	+ 34 36	30	16.	GALAXY
MCG+06-05-018	01 48 18.	+ 34 38	48	15.	GALAXY
6ZW 067	01 48 18.	+ 36 01			COMPACT GALAXY
PHL 7932	01 48 18.	- 02 23		18.6	BLUE STELLAR OBJECT
PHL 3858	01 48 18.	- 05 12		18.8	BLUE STELLAR OBJECT
PHL 3857	01 48 18.	- 09 16		18.7	BLUE STELLAR OBJECT
PHL 7933	01 48 18.	- 20 15		18.4	BLUE STELLAR OBJECT
MCG-06-05-008	01 48 18.	- 34 17	42	16.	GALAXY
IC 0167	01 48 22.4	+ 21 39 59.			GALAXY
PHL 3859	01 48 24.	+ 00 14		18.0	BLUE STELLAR OBJECT
ZWG 412.034	01 48 24.	+ 05 28		15.6	GALAXY
UGC 01314	01 48 24.	+ 12 21	96	16.0	GALAXY SB IV
ZWG 482.026	01 48 24.	+ 22 20		13.7	GALAXY
UGC 01315	01 48 24.	+ 22 20	30	13.7	GALAXY PECULR
ZCG 0148+32.8	01 48 24.	+ 32 49		17.2	COMPACT GALAXY
6ZW 068	01 48 24.	+ 32 56			COMPACT GALAXY
UGC 01316	01 48 24.	+ 34 36	84	16.5	GALAXY S
ZWG 522.023	01 48 24.	+ 34 37		15.5	GALAXY
6ZW 069	01 48 24.	+ 36 02			COMPACT GALAXY
6ZW 070	01 48 24.	+ 36 35			COMPACT GALAXY
PHL 7937	01 48 24.	- 00 54		18.5	BLUE STELLAR OBJECT
PHL 7934	01 48 24.	- 01 34		18.9	BLUE STELLAR OBJECT
PHL 7938	01 48 24.	- 02 41		18.0	BLUE STELLAR OBJECT
PHL 7935	01 48 24.	- 02 54		18.3	BLUE STELLAR OBJECT
PHL 7939	01 48 24.	- 06 00		18.6	BLUE STELLAR OBJECT
PHL 7936	01 48 24.	- 08 20		18.6	BLUE STELLAR OBJECT
PHL 7940	01 48 24.	- 12 26		18.3	BLUE STELLAR OBJECT
PHL 1193	01 48 24.	- 14 58		17.1	BLUE STELLAR OBJECT
MCG+01-05-036	01 48 27.	+ 05 27	30	16.	GALAXY
HOLM 047B	01 48 27.	- 10 01	36	14.2	PART OF MULTIPLE GALAXY
LIN 566	01 48 27.	- 74 14 19.		14.83	SMC EMISSION LINE OBJECT
ZWG 386.047	01 48 30.	+ 01 46		15.7	GALAXY
ZWG 482.027	01 48 30.	+ 22 06		12.7	GALAXY
UGC 01317	01 48 30.	+ 22 06	306	12.7	GALAXY Sc
MCG+04-05-022	01 48 30.	+ 22 06	258	12.	GALAXY
5ZW 123	01 48 30.	+ 22 20			COMPACT GALAXY
MCG+05-05-020	01 48 30.	+ 32 46	30	15.5	GALAXY
ZWG 503.036	01 48 30.	+ 32 47		15.4	GALAXY
UGC 01318	01 48 30.	+ 32 47	90	15.4	GALAXY (E)
ZWG 522.024	01 48 30.	+ 35 49		14.5	GALAXY
UGC 01319	01 48 30.	+ 35 49	54	14.5	GALAXY S?
KARA.72 043B	01 48 30.	+ 35 49	36		PART OF DOUBLE GALAXY
KARA.72 043A	01 48 30.	+ 35 49	48	14.5	PART OF DOUBLE GALAXY
5ZW 124	01 48 30.	+ 37 52			COMPACT GALAXY
UGC 01320	01 48 30.	+ 41 35	84	16.5	GALAXY Sc
ZWG 537.025	01 48 30.	+ 41 36		15.7	GALAXY
PHL 7941	01 48 30.	- 06 29		17.5	BLUE STELLAR OBJECT
MCG-02-05-060	01 48 30.	- 09 56	150	12.5	GALAXY
MCG-06-05-009	01 48 30.	- 34 48	36	15.	GALAXY
LIN 568	01 48 30.	- 75 22 01.		16.48	SMC EMISSION LINE OBJECT
RNGC 0697	01 48 31.	+ 22 07		12.5	GALAXY
RNGC 0695	01 48 31.	+ 22 20		13.5	GALAXY
MCG+06-05-020	01 48 33.	+ 35 48 30.	48	13.5	GALAXY
MCG+01-05-041	01 48 33.	- 02 59	48	16.	GALAXY
LIN 567	01 48 33.	- 75 04 01.		17.26	SMC EMISSION LINE OBJECT
RNGC 0701	01 48 34.	- 09 56		13.0	GALAXY
HOLM 047A	01 48 34.	- 09 58	78	12.7	PART OF MULTIPLE GALAXY
REIN 2.016	01 48 35.44	- 09 56 59.4			NEBULA
PHL 3860	01 48 36.	+ 01 24		18.9	BLUE STELLAR OBJECT
PHL 7943	01 48 36.	+ 01 40		18.2	BLUE STELLAR OBJECT
PHL 7946	01 48 36.	+ 08 35		18.9	BLUE STELLAR OBJECT
PHL 7942	01 48 36.	+ 09 32		18.4	BLUE STELLAR OBJECT
5ZW 125	01 48 36.	+ 33 15			COMPACT GALAXY
5ZW 126	01 48 36.	+ 35 48			COMPACT GALAXY
ISS 0013	01 48 36.	+ 63 30	474		STELLAR RING
UGC 01321	01 48 36.	- 01 17	66	16.0	GALAXY S0
ZWG 386.048	01 48 36.	- 01 40		15.3	GALAXY
PHL 7947	01 48 36.	- 06 57		18.8	BLUE STELLAR OBJECT
PHL 3861	01 48 36.	- 08 14		18.5	BLUE STELLAR OBJECT
PHL 7944	01 48 36.	- 09 21		17.8	BLUE STELLAR OBJECT
MCG-02-05-061	01 48 36.	- 10 02	48	14.	GALAXY
PHL 7948	01 48 36.	- 10 08		18.1	BLUE STELLAR OBJECT
PHL 7945	01 48 36.	- 13 31		17.5	BLUE STELLAR OBJECT
MCG-06-05-010	01 48 36.	- 36 15	66	13.5	GALAXY
MCG+01-05-042	01 48 39.	- 03 43	72	14.5	GALAXY
IC 1738	01 48 39.4	- 10 02 19.			GALAXY SAB(r)
REIN 2.017	01 48 39.51	- 10 02 19.2			NEBULA
LIN.CL 113	01 48 41.	- 73 58 25.	1362	11.9	STAR CLUSTER IN SMC
PHL 3863	01 48 42.	+ 03 47		18.2	BLUE STELLAR OBJECT
PHL 1194	01 48 42.	+ 09 02		18.5	BLUE STELLAR OBJECT
SHB 047	01 48 42.	+ 09 02		17.5	QUASI-STELLAR OBJECT
ZWG 437.048	01 48 42.	+ 12 53		15.5	GALAXY
UGC 01322	01 48 42.	+ 12 53	102	15.5	GALAXY Sc
MCG+05-05-022	01 48 42.	+ 30 07	30	16.	GALAXY
6ZW 071	01 48 42.	+ 30 27			COMPACT GALAXY
5ZW 127	01 48 42.	+ 32 32			COMPACT GALAXY
MCG+05-05-021	01 48 42.	+ 32 57	48	16.	GALAXY

OBJECT NAME	RIGHT ASCEN.	DECLINATION	DIAM.	MAGN.	TYPE OF OBJECT
ZWG 503.037	01 48 42.	+ 32 58		15.3	GALAXY
ZWG 503.038	01 48 42.	+ 33 15		15.5	GALAXY
5ZW 128	01 48 42.	+ 33 16			COMPACT GALAXY
ZWG 386.049	01 48 42.	- 01 19		15.5	GALAXY
MCG+00-05-039	01 48 42.	- 01 39 30.	15	15.	GALAXY
PHL 7950	01 48 42.	- 03 35		18.0	BLUE STELLAR OBJECT
PHL 7951	01 48 42.	- 10 32		18.2	BLUE STELLAR OBJECT
PHL 3862	01 48 42.	- 11 56		16.5	BLUE STELLAR OBJECT
PHL 3864	01 48 42.	- 17 27		17.9	BLUE STELLAR OBJECT
PHL 7949	01 48 42.	- 24 36		18.6	BLUE STELLAR OBJECT
IC 1737	01 48 44.	+ 36 00			NONSTELLAR OBJECT
RNGC 0702	01 48 44.	- 04 17		14.0	GALAXY
MCG+02-05-053	01 48 45.	+ 12 52 30.	84	14.	GALAXY
MCG+01-05-043	01 48 45.	- 04 17	60	14.	GALAXY
MCG-05-05-022	01 48 45.	- 26 41	18	17.	GALAXY
RNGC 0707	01 48 46.	- 08 44		14.0	GALAXY
ARP 075	01 48 47.	- 04 18			PECULIAR GALAXY
PHL 7954	01 48 48.	+ 01 02		17.3	BLUE STELLAR OBJECT
PHL 7953	01 48 48.	+ 04 24		17.9	BLUE STELLAR OBJECT
PHL 7952	01 48 48.	+ 06 58		17.2	BLUE STELLAR OBJECT
PHL 3867	01 48 48.	+ 08 54		18.7	BLUE STELLAR OBJECT
UGC 01323	01 48 48.	+ 11 38	96	17.	GALAXY DWARF
ZWG 460.034	01 48 48.	+ 18 51		14.6	GALAXY
UGC 01324	01 48 48.	+ 18 51	96	14.6	GALAXY SBb
MCG+03-05-021	01 48 48.	+ 18 51	66	14.	GALAXY
MCG+05-05-023	01 48 48.	+ 30 16	24	16.	GALAXY
6ZW 072	01 48 48.	+ 33 02			COMPACT GALAXY
5ZW 129	01 48 48.	+ 33 17			COMPACT GALAXY
MCG+06-05-021	01 48 48.	+ 35 52 30.	36	15.5	GALAXY
VB 058	01 48 48.	+ 63 26	443		STELLAR RING
PHL 3866	01 48 48.	- 07 32		18.3	BLUE STELLAR OBJECT
PHL 7955	01 48 48.	- 08 39		17.2	BLUE STELLAR OBJECT
MCG-02-05-064	01 48 48.	- 08 40	30	14.5	GALAXY
MCG-02-05-063	01 48 48.	- 08 44	42	14.5	GALAXY
PHL 3868	01 48 48.	- 10 10		18.8	BLUE STELLAR OBJECT
PHL 7956	01 48 48.	- 12 16		18.0	BLUE STELLAR OBJECT
MCG-02-05-062	01 48 48.	- 12 46	42	14.5	GALAXY
PHL 1195	01 48 48.	- 20 14		16.3	BLUE STELLAR OBJECT
PHL 3865	01 48 48.	- 20 54		17.9	BLUE STELLAR OBJECT
BC PHL1194	01 48 52.	+ 09 02 35.		17.50	QUASI-STELLAR OBJECT
PHL 3872	01 48 54.	+ 03 56		17.3	BLUE STELLAR OBJECT
PHL 3869	01 48 54.	+ 07 42		18.7	BLUE STELLAR OBJECT
ZWG 412.035	01 48 54.	+ 08 00		14.2	GALAXY
UGC 01325	01 48 54.	+ 08 00	120	14.2	GALAXY E
KARA.72 044A	01 48 54.	+ 08 00	108	14.2	PART OF DOUBLE GALAXY
ZC 0148.9+1855	01 48 54.	+ 18 55	400		CLUSTER OF GALAXIES
ZWG 503.039	01 48 54.	+ 30 08		15.2	GALAXY
6ZW 073	01 48 54.	+ 32 58			COMPACT GALAXY
ZWG 503.040	01 48 54.	+ 33 17		15.7	GALAXY
6ZW 074	01 48 54.	+ 34 48			COMPACT GALAXY
MCG+06-05-022	01 48 54.	+ 35 24	24	16.	GALAXY
6ZW 075	01 48 54.	+ 36 39			COMPACT GALAXY
VB 151	01 48 54.	+ 61 42	416		STELLAR RING
MCG+00-05-040	01 48 54.	- 02 30 30.	12	15.	GALAXY
MCG-02-05-065	01 48 54.	- 08 37 30.	42	15.	GALAXY
PHL 3871	01 48 54.	- 19 12		17.9	BLUE STELLAR OBJECT
PHL 3870	01 48 54.	- 25 20		18.7	BLUE STELLAR OBJECT
ARC 0261	01 48 55.	- 02 30		17.2	RICH CLUSTER OF GALAXIES
ARC 0260	01 48 58.	+ 32 55		15.8	RICH CLUSTER OF GALAXIES
VDB.66G 016	01 49	+ 17 53	100		DWARF GALAXY
PHL 7962	01 49 00.	+ 03 32		17.5	BLUE STELLAR OBJECT
PHL 7961	01 49 00.	+ 04 05		18.9	BLUE STELLAR OBJECT
PHL 7959	01 49 00.	+ 04 46		17.2	BLUE STELLAR OBJECT
MRK 578	01 49 00.	+ 07 03	8	16.	GALAXY WITH UV CONTINUUM
ZWG 412.036	01 49 00.	+ 08 03		15.0	GALAXY
UGC 01326	01 49 00.	+ 08 03	72	15.0	GALAXY E
KARA.72 044B	01 49 00.	+ 08 03	66	15.0	PART OF DOUBLE GALAXY
MCG+01-05-037	01 49 00.	+ 08 04	30	15.	GALAXY
PHL 7958	01 49 00.	+ 08 22		17.9	BLUE STELLAR OBJECT
MCG+03-05-022	01 49 00.	+ 17 56	78	16.	GALAXY
ZWG 460.035	01 49 00.	+ 18 05		15.7	GALAXY
ZWG 503.041	01 49 00.	+ 30 18		14.9	GALAXY
UGC 01327	01 49 00.	+ 30 18	84	14.9	GALAXY E-S0
ZC 0149.0+3240	01 49 00.	+ 32 40	3020		CLUSTER OF GALAXIES
MCG+06-05-023	01 49 00.	+ 35 22 30.	48	16.	GALAXY
6ZW 076	01 49 00.	+ 36 30			COMPACT GALAXY
OCL 0334	01 49 00.	+ 61 35	270	17.	OPEN STAR CLUSTER
MCG+01-05-044	01 49 00.	- 06 45	84	15.	GALAXY
PHL 3873	01 49 00.	- 06 47		16.9	BLUE STELLAR OBJECT
PHL 3874	01 49 00.	- 07 18		18.8	BLUE STELLAR OBJECT
PHL 3875	01 49 00.	- 08 41		17.5	BLUE STELLAR OBJECT
PHL 7960	01 49 00.	- 14 17		17.5	BLUE STELLAR OBJECT
IC 1740	01 49 01.	- 30 11 27.			NONSTELLAR OBJECT
MCG+07-04-011	01 49 03.	+ 42 40	54	15.	GALAXY
BIGO 469	01 49 04.	+ 40 08			NEBULA
LIN 569	01 49 04.	- 74 51 51.		14.65	SMC EMISSION LINE OBJECT
IC 0166	01 49 05.	+ 61 35			OPEN CLUSTER
PHL 7963	01 49 06.	+ 03 42		17.5	BLUE STELLAR OBJECT
MCG+03-05-023	01 49 06.	+ 16 47	66	14.	GALAXY
ZWG 460.036	01 49 06.	+ 16 48		14.7	GALAXY
UGC 01328	01 49 06.	+ 16 48	96	14.7	GALAXY Sc
ZWG 460.037	01 49 06.	+ 17 55		15.7	GALAXY
UGC 01329	01 49 06.	+ 17 55	144	15.7	GALAXY SB:c
UGC 01330	01 49 06.	+ 34 48	78	16.0	GALAXY
MCG+06-05-024	01 49 06.	+ 35 52	42	15.	GALAXY
ZWG 522.025	01 49 06.	+ 35 53		15.6	GALAXY
ZWG 537.026	01 49 06.	+ 42 40		14.7	GALAXY
UGC 01331	01 49 06.	+ 42 40	60	14.7	GALAXY Sc
ZWG 552.010	01 49 06.	+ 47 50		15.7	GALAXY
UGC 01332	01 49 06.	+ 47 50	150	15.7	GALAXY E
MCG+08-04-008	01 49 06.	+ 47 50	22	15.	GALAXY
PHL 3878	01 49 06.	- 11 16		16.9	BLUE STELLAR OBJECT
PHL 3877	01 49 06.	- 19 13		18.0	BLUE STELLAR OBJECT
PHL 3876	01 49 06.	- 19 13		18.1	BLUE STELLAR OBJECT
PHL 3879	01 49 06.	- 23 02		17.7	BLUE STELLAR OBJECT
LB 03233	01 49 06.	- 73 57		14.4	FAINT BLUE STAR
MCG+01-05-038	01 49 09.	+ 08 02	24	14.	GALAXY
ZWG 386.050	01 49 12.	+ 00 01		15.7	GALAXY
UGC 01333	01 49 12.	+ 00 01	90	15.7	GALAXY S
ZC 0149.2+0136	01 49 12.	+ 01 36	1140		CLUSTER OF GALAXIES
PHL 1196	01 49 12.	+ 05 02		18.2	BLUE STELLAR OBJECT
ZWG 412.037	01 49 12.	+ 06 03		13.2	GALAXY
RNGC 0706	01 49 12.	+ 06 03		13.0	GALAXY
UGC 01334	01 49 12.	+ 06 03	132	13.2	GALAXY Sc
ZWG 412.038	01 49 12.	+ 08 11		15.7	GALAXY
UGC 01335	01 49 12.	+ 17 18	66	17.	GALAXY
5ZW 130	01 49 12.	+ 31 40			COMPACT GALAXY
6ZW 077	01 49 12.	+ 32 44			COMPACT GALAXY
6ZW 078	01 49 12.	+ 34 39			COMPACT GALAXY
5ZW 131	01 49 12.	+ 35 10			COMPACT GALAXY
ZWG 522.026	01 49 12.	+ 35 10		15.1	GALAXY
LB 03751	01 49 12.	+ 35 33 48.		17.3	FAINT BLUE STAR
ZWG 522.027	01 49 12.	+ 35 50		15.6	GALAXY
UGC 01336	01 49 12.	+ 35 50	72	15.6	GALAXY S0
5ZW 132	01 49 12.	+ 39 08			COMPACT GALAXY
ZWG 522.028	01 49 12.	+ 39 08		14.9	GALAXY
PHL 3880	01 49 12.	- 00 08		17.9	BLUE STELLAR OBJECT
PHL 3881	01 49 12.	- 03 23		18.3	BLUE STELLAR OBJECT
PHL 7966	01 49 12.	- 08 28		16.6	BLUE STELLAR OBJECT
PHL 7964	01 49 12.	- 08 34		18.1	BLUE STELLAR OBJECT
PHL 7967	01 49 12.	- 10 52		18.3	BLUE STELLAR OBJECT
PHL 1197	01 49 12.	- 12 38		18.4	BLUE STELLAR OBJECT
PHL 3882	01 49 12.	- 13 08		19.0	BLUE STELLAR OBJECT
PHL 7965	01 49 12.	- 14 34		18.0	BLUE STELLAR OBJECT
PHL 1198	01 49 12.	- 15 26		16.9	BLUE STELLAR OBJECT
RNGC 0700	01 49 14.	+ 35 50		15.5	GALAXY
MCG+00-05-041	01 49 18.	+ 00 01	30	15.	GALAXY
PHL 7968	01 49 18.	+ 00 28		18.0	BLUE STELLAR OBJECT
PHL 3883	01 49 18.	+ 03 46		17.3	BLUE STELLAR OBJECT
MCG+01-05-040	01 49 19.	+ 06 03	120	12.5	GALAXY
UGC 01337	01 49 18.	+ 08 35	66	16.5	GALAXY Sc
MCG+01-05-039	01 49 18.	+ 08 35 30.	60	15.	GALAXY
6ZW 079	01 49 18.	+ 32 45			COMPACT GALAXY
6ZW 080	01 49 18.	+ 32 46			COMPACT GALAXY
ZWG 522.029	01 49 18.	+ 34 55		15.6	GALAXY
ZWG 522.030	01 49 18.	+ 35 47		15.5	GALAXY
5ZW 133	01 49 18.	+ 35 51			COMPACT GALAXY
6ZW 081	01 49 18.	+ 37 59			COMPACT GALAXY
PHL 7969	01 49 18.	- 06 06		18.9	BLUE STELLAR OBJECT
MCG-02-05-066	01 49 18.	- 08 45	18	15.	GALAXY
PHL 7970	01 49 18.	- 11 29		18.0	BLUE STELLAR OBJECT
PHL 1199	01 49 18.	- 13 38		18.5	BLUE STELLAR OBJECT
LB 03752	01 49 20.	+ 35 42 42.		17.6	FAINT BLUE STAR
MCG+01-05-045	01 49 21.	- 05 44	102	15.	GALAXY
LB 03753	01 49 23.	+ 36 12 00.		19.0	FAINT BLUE STAR
PHL 3884	01 49 24.	+ 05 24		16.7	BLUE STELLAR OBJECT
MRK 579	01 49 24.	+ 07 01	9	16.5	GALAXY WITH UV CONTINUUM
MCG+05-05-024	01 49 24.	+ 31 16	42	16.	GALAXY
ZWG 503.042	01 49 24.	+ 31 17		15.1	GALAXY
ZWG 522.031	01 49 24.	+ 35 33		15.2	GALAXY
UGC 01338	01 49 24.	+ 35 33	60	15.2	GALAXY Sb
6ZW 082	01 49 24.	+ 36 50			COMPACT GALAXY
6ZW 083	01 49 24.	+ 37 57			COMPACT GALAXY
OCL 0342	01 49 24.	+ 56 49	1200	11.	OPEN STAR CLUSTER
PHL 3886	01 49 24.	- 00 44		18.8	BLUE STELLAR OBJECT
PHL 7972	01 49 24.	- 02 48		18.3	BLUE STELLAR OBJECT
PHL 7973	01 49 24.	- 03 58		18.5	BLUE STELLAR OBJECT
PHL 3885	01 49 24.	- 04 35		18.5	BLUE STELLAR OBJECT
MCG+01-05-046	01 49 24.	- 05 45	78	15.5	GALAXY
PHL 7971	01 49 24.	- 06 16		17.6	BLUE STELLAR OBJECT
PHL 7974	01 49 24.	- 08 54		18.6	BLUE STELLAR OBJECT
PHL 7975	01 49 24.	- 09 27		18.1	BLUE STELLAR OBJECT
PHL 7976	01 49 24.	- 13 38		18.3	BLUE STELLAR OBJECT
PHL 7977	01 49 24.	- 22 02		18.4	BLUE STELLAR OBJECT
MCG-04-05-012	01 49 24.	- 25 45 30.	18	16.5	GALAXY
MCG-04-05-011	01 49 24.	- 25 45 30.	36	16.	GALAXY
MCG-06-05-011	01 49 24.	- 36 26		14.5	GALAXY
HOLM 048A	01 49 25.	+ 35 36	12	14.6	PART OF MULTIPLE GALAXY
HOLM 048B	01 49 26.	+ 35 36	9	14.8	PART OF MULTIPLE GALAXY
MCG+06-05-025	01 49 27.	+ 35 32 30.	48	14.	GALAXY
LB 03754	01 49 29.	+ 35 41 00.		18.0	FAINT BLUE STAR
IC 0170	01 49 29.	- 08 46 28.			NONSTELLAR OBJECT
6ZW 084	01 49 30.	+ 32 57			COMPACT GALAXY
6ZW 085	01 49 30.	+ 32 58			COMPACT GALAXY
ZWG 522.032	01 49 30.	+ 35 36		15.0	GALAXY
UGC 01339	01 49 30.	+ 35 36	78	15.0	GALAXY SB0
MCG+06-05-026	01 49 30.	+ 35 36	48	14.	GALAXY
LB 03755	01 49 30.	+ 35 55 12.		19.0	FAINT BLUE STAR
UGC 01340	01 49 30.	+ 47 51	60	16.5	GALAXY E
MCG+08-04-009	01 49 30.	+ 47 51	22	16.	GALAXY
PHL 7979	01 49 30.	- 06 50		18.2	BLUE STELLAR OBJECT
PHL 7978	01 49 30.	- 08 18		16.1	BLUE STELLAR OBJECT
PHL 3887	01 49 30.	- 09 49		18.7	BLUE STELLAR OBJECT
MCG-03-05-023	01 49 30.	- 17 02	42	14.5	GALAXY
ARC 0264	01 49 30.	- 26 02		17.7	RICH CLUSTER OF GALAXIES
IC 1741	01 49 32.	- 17 01 58.			NONSTELLAR OBJECT
LB 03756	01 49 34.	+ 36 02 54.		18.1	FAINT BLUE STAR
LIN.CL 114	01 49 35.	- 74 36 03.	204	12.1	STAR CLUSTER IN SMC
ZC 0149.6+0117	01 49 36.	+ 01 17	1010		CLUSTER OF GALAXIES
PHL 3890	01 49 36.	+ 03 29		17.9	BLUE STELLAR OBJECT
PHL 3888	01 49 36.	+ 07 24		17.5	BLUE STELLAR OBJECT
PHL 7980	01 49 36.	+ 08 45		15.2	BLUE STELLAR OBJECT
MCG+05-05-025	01 49 36.	+ 31 43	72	15.	GALAXY
ZWG 503.043	01 49 36.	+ 31 45		15.6	GALAXY
UGC 01341	01 49 36.	+ 31 45	78	15.6	GALAXY Sc/SBc
6ZW 086	01 49 36.	+ 32 30			COMPACT GALAXY
6ZW 087	01 49 36.	+ 32 55			COMPACT GALAXY
6ZW 088	01 49 36.	+ 35 52			COMPACT GALAXY
ZWG 522.033	01 49 36.	+ 35 52		15.7	GALAXY
HOLM 049E	01 49 36.	+ 35 52	18	14.8	PART OF MULTIPLE GALAXY
MCG+06-05-027	01 49 36.	+ 36 15	96	13.5	GALAXY
PHL 1200	01 49 36.	- 01 44		16.8	BLUE STELLAR OBJECT
PHL 3889	01 49 36.	- 03 14		18.6	BLUE STELLAR OBJECT
PHL 7981	01 49 36.	- 03 46		19.0	BLUE STELLAR OBJECT
PHL 1201	01 49 36.	- 25 18		17.2	BLUE STELLAR OBJECT
LIN 570	01 49 37.	- 74 22 04.		13.29	SMC EMISSION LINE OBJECT
LB 03757	01 49 38.	+ 35 44 06.		18.8	FAINT BLUE STAR
MCG-04-05-013	01 49 39.	- 24 33 30.	54	14.5	GALAXY
LB 03758	01 49 40.	+ 35 45 48.		18.6	FAINT BLUE STAR
ZWG 460.038	01 49 42.	+ 17 16		14.5	GALAXY
UGC 01342	01 49 42.	+ 17 16	102	14.5	GALAXY S0
MCG+03-05-024	01 49 42.	+ 17 16 30.	66	14.5	GALAXY
6ZW 089	01 49 42.	+ 32 54			COMPACT GALAXY
ZWG 522.034	01 49 42.	+ 35 52		14.1	GALAXY
HOLM 049B	01 49 42.	+ 35 52	24	14.0	PART OF MULTIPLE GALAXY
UGC 01343	01 49 42.	+ 35 52	36	14.1	GALAXY DBL SYS
5ZW 134	01 49 42.	+ 35 53			COMPACT GALAXY
MCG+06-05-028	01 49 42.	+ 35 53	24	14.	GALAXY
LB 03759	01 49 42.	+ 35 53 42.		18.5	FAINT BLUE STAR
6ZW 090	01 49 42.	+ 35 54			COMPACT GALAXY
MCG+06-05-029	01 49 42.	+ 35 55	24	15.	GALAXY
ZWG 522.035	01 49 42.	+ 36 15		14.0	GALAXY
UGC 01344	01 49 42.	+ 36 15	114	14.0	GALAXY SBa
PHL 1202	01 49 42.	- 00 46		18.7	BLUE STELLAR OBJECT
PHL 7982	01 49 42.	- 07 45		16.7	BLUE STELLAR OBJECT
PHL 1203	01 49 42.	- 13 10		18.4	BLUE STELLAR OBJECT

OBJECT NAME	RIGHT ASCEN.	DECLINATION	DIAM.	MAGN.	TYPE OF OBJECT
PHL 3891	01 49 42.	- 18 18		18.5	BLUE STELLAR OBJECT
MCG-03-05-024	01 49 42.	- 20 26	84	14.	GALAXY
HOLM 049C	01 49 43.	+ 35 55	18	14.2	PART OF MULTIPLE GALAXY
LB 03760	01 49 43.	+ 35 59 36.		18.0	FAINT BLUE STAR
RNGC 0711	01 49 44.	+ 17 16		14.5	GALAXY
RNGC 0704	01 49 44.	+ 35 52		14.0	GALAXY
MCG+06-05-030	01 49 45.	+ 35 54	72	14.	GALAXY
LB 03761	01 49 46.	+ 35 48 12.		18.6	FAINT BLUE STAR
HOLM 049D	01 49 46.	+ 35 54	18	14.3	PART OF MULTIPLE GALAXY
PHL 3892	01 49 48.	+ 01 06		16.9	BLUE STELLAR OBJECT
PHL 3894	01 49 48.	+ 06 18		18.4	BLUE STELLAR OBJECT
ZC 0149.8+1927	01 49 48.	+ 19 27	2020		CLUSTER OF GALAXIES
ZC 0149.8+2925	01 49 48.	+ 29 25	470		CLUSTER OF GALAXIES
LB 03762	01 49 48.	+ 35 46 42.		16.0	FAINT BLUE STAR
ZWG 522.036	01 49 48.	+ 35 54		14.5	GALAXY
MCG+06-05-031	01 49 48.	+ 35 54 30.	24	13.	GALAXY
UGC 01345	01 49 48.	+ 35 55	72	14.5	GALAXY S0-a
ZWG 522.037	01 49 48.	+ 35 56		14.5	GALAXY
UGC 01346	01 49 48.	+ 35 56	72	14.5	GALAXY E-S0
ZWG 522.038	01 49 48.	+ 36 22		13.9	GALAXY
ZCG 0149+36	01 49 48.	+ 36 22		13.9	COMPACT GALAXY
UGC 01347	01 49 48.	+ 36 22	84	13.9	GALAXY Sc/SBc
SN 1952A	01 49 48.	+ 36 22		18.6	SUPERNOVA
MCG+06-05-032	01 49 48.	+ 36 22	78	13.	GALAXY
ZC 0149.8+4201	01 49 48.	+ 42 01	1810		CLUSTER OF GALAXIES
5ZW 135	01 49 48.	+ 53 58			COMPACT GALAXY
ZWG 386.051	01 49 48.	- 01 23		15.7	GALAXY
PHL 7987	01 49 48.	- 03 07		18.9	BLUE STELLAR OBJECT
PHL 7986	01 49 48.	- 06 35		17.5	BLUE STELLAR OBJECT
PHL 7985	01 49 48.	- 06 35		17.5	BLUE STELLAR OBJECT
PHL 7984	01 49 48.	- 06 37		16.6	BLUE STELLAR OBJECT
PHL 7988	01 49 48.	- 07 06		18.7	BLUE STELLAR OBJECT
PHL 1204	01 49 48.	- 12 27		16.6	BLUE STELLAR OBJECT
PHL 7983	01 49 48.	- 14 21		17.5	BLUE STELLAR OBJECT
PHL 3895	01 49 48.	- 14 21		18.6	BLUE STELLAR OBJECT
PHL 3893	01 49 48.	- 19 50		16.2	BLUE STELLAR OBJECT
PHL 7989	01 49 48.	- 26 36		18.5	BLUE STELLAR OBJECT
PHL 1205	01 49 48.	- 27 17		18.4	BLUE STELLAR OBJECT
LB 03763	01 49 49.	+ 35 39 54.		17.1	FAINT BLUE STAR
RNGC 0705	01 49 50.	+ 35 54		14.5	GALAXY
HOLM 049A	01 49 50.	+ 35 54	36	14.0	PART OF MULTIPLE GALAXY
RNGC 0703	01 49 50.	+ 35 56		14.5	GALAXY
LB 03764	01 49 51.	+ 36 19 42.		19.3	FAINT BLUE STAR
PHL 7957	01 49 54.	+ 07 54		18.3	BLUE STELLAR OBJECT
6ZW 091	01 49 54.	+ 34 42			COMPACT GALAXY
ZWG 522.039	01 49 54.	+ 35 55		14.8	GALAXY
UGC 01348	01 49 54.	+ 35 55	210	14.8	GALAXY E
ZWG 522.040	01 49 54.	+ 35 59		15.2	GALAXY
PHL 3897	01 49 54.	- 06 03		17.8	BLUE STELLAR OBJECT
PHL 3896	01 49 54.	- 18 56		16.5	BLUE STELLAR OBJECT
PHL 7990	01 49 54.	- 22 52		18.5	BLUE STELLAR OBJECT
LB 03765	01 49 55.	+ 35 41 54.		16.7	FAINT BLUE STAR
RNGC 0708	01 49 56.	+ 35 55		15.0	GALAXY
LB 03766	01 49 56.	+ 35 58 06.		18.4	FAINT BLUE STAR
RNGC 0709	01 49 56.	+ 35 59		15.0	GALAXY
ARC 0265	01 49 56.	- 07 17		17.6	RICH CLUSTER OF GALAXIES
ARC 0262	01 49 57.	+ 35 54		13.3	RICH CLUSTER OF GALAXIES
LB 03767	01 49 57.	+ 36 17 54		17.7	FAINT BLUE STAR
KHAV 031	01 50	+ 62 21	7640		DARK NEBULA
LB 09776	01 50	- 81 10		13.9	FAINT BLUE STAR
LB 09777	01 50	- 83 20		14.5	FAINT BLUE STAR
ZWG 386.052	01 50 00.	+ 01 05		15.3	GALAXY
FATH 1.098	01 50 00.	+ 01 06	16		NEBULA
MCG+00-05-042	01 50 00.	+ 01 06 30.	24	15.	GALAXY
PHL 7992	01 50 00.	+ 01 38		18.5	BLUE STELLAR OBJECT
PHL 7993	01 50 00.	+ 04 10		18.7	BLUE STELLAR OBJECT
PHL 3899	01 50 00.	+ 04 10		17.1	BLUE STELLAR OBJECT
PHL 3898	01 50 00.	+ 04 50		17.0	BLUE STELLAR OBJECT
ZWG 522.041	01 50 00.	+ 35 48		14.3	GALAXY
UGC 01349	01 50 00.	+ 35 48	102	14.3	GALAXY Sc
MCG+06-05-033	01 50 00.	+ 35 48	66	13.5	GALAXY
ZWG 522.042	01 50 00.	+ 36 15		14.5	GALAXY
UGC 01350	01 50 00.	+ 36 15	126	14.5	GALAXY SBb
MCG+06-05-034	01 50 00.	+ 36 15 30.	90	14.	GALAXY
ISS 0014	01 50 00.	+ 64 24	345		STELLAR RING
LDN 1336	01 50 00.	+ 66 30	4740		DARK NEBULA
ZC 0150.0-0106	01 50 00.	- 01 06	4230		CLUSTER OF GALAXIES
PHL 1206	01 50 00.	- 01 59		17.1	BLUE STELLAR OBJECT
PHL 3901	01 50 00.	- 04 03		18.3	BLUE STELLAR OBJECT
PHL 3902	01 50 00.	- 04 09		18.6	BLUE STELLAR OBJECT
PHL 1207	01 50 00.	- 06 44		18.7	BLUE STELLAR OBJECT
PHL 3900	01 50 00.	- 11 45		18.2	BLUE STELLAR OBJECT
PHL 1208	01 50 00.	- 18 48		17.7	BLUE STELLAR OBJECT
PHL 7994	01 50 00.	- 20 00		18.5	BLUE STELLAR OBJECT
PHL 7991	01 50 00.	- 25 04		17.9	BLUE STELLAR OBJECT
LB 03768	01 50 01.	+ 35 37 18.		18.5	FAINT BLUE STAR
RNGC 0710	01 50 02.	+ 35 48		14.5	GALAXY
ARC 0266	01 50 05.	- 04 24		17.2	RICH CLUSTER OF GALAXIES
PHL 3903	01 50 06.	+ 04 06		16.6	BLUE STELLAR OBJECT
ZWG 503.044	01 50 06.	+ 33 21		15.7	GALAXY
PHL 7995	01 50 06.	- 05 32		18.1	BLUE STELLAR OBJECT
PHL 3904	01 50 06.	- 07 01		18.5	BLUE STELLAR OBJECT
PHL 7997	01 50 06.	- 10 10		17.6	BLUE STELLAR OBJECT
PHL 7998	01 50 06.	- 14 02		18.7	BLUE STELLAR OBJECT
PHL 7996	01 50 06.	- 16 14		18.3	BLUE STELLAR OBJECT
MCG-02-05-067	01 50 09.	- 08 31	36	15.	GALAXY
PHL 3907	01 50 12.	+ 00 08		18.8	BLUE STELLAR OBJECT
PHL 7999	01 50 12.	+ 01 20		16.6	BLUE STELLAR OBJECT
PHL 3910	01 50 12.	+ 01 30		18.6	BLUE STELLAR OBJECT
PHL 3905	01 50 12.	+ 04 48		16.6	BLUE STELLAR OBJECT
PHL 8000	01 50 12.	+ 04 57		17.3	BLUE STELLAR OBJECT
PHL 3906	01 50 12.	+ 07 43		18.6	BLUE STELLAR OBJECT
PHL 3909	01 50 12.	+ 08 40		18.8	BLUE STELLAR OBJECT
ZWG 437.049	01 50 12.	+ 12 28		14.0	GALAXY
UGC 01351	01 50 12.	+ 12 28	126	14.0	GALAXY SB:a
MCG+06-05-035	01 50 12.	+ 36 33	33	14.	GALAXY
ZWG 522.043	01 50 12.	+ 36 34		13.9	GALAXY
UGC 01352	01 50 12.	+ 36 34	78	13.9	GALAXY S0
ZC 0150.2+3718	01 50 12.	+ 37 18	1080		CLUSTER OF GALAXIES
PHL 8001	01 50 12.	- 05 38		18.5	BLUE STELLAR OBJECT
PHL 1209	01 50 12.	- 07 12		18.0	BLUE STELLAR OBJECT
PHL 8002	01 50 12.	- 11 16		18.0	BLUE STELLAR OBJECT
RNGC 0725	01 50 12.	- 16 45		14.0	GALAXY
MCG-03-05-025	01 50 12.	- 16 45	42	14.5	GALAXY
MCG-03-05-026	01 50 12.	- 19 00	72	14.5	GALAXY
PHL 1210	01 50 12.	- 19 29		13.9	BLUE STELLAR OBJECT
PHL 3908	01 50 12.	- 20 18		17.1	BLUE STELLAR OBJECT
LB 03769	01 50 13.	+ 35 50 30.		18.1	FAINT BLUE STAR

OBJECT NAME	RIGHT ASCEN.	DECLINATION	DIAM.	MAGN.	TYPE OF OBJECT
RNGC 0712	01 50 14.	+ 36 34		14.0	GALAXY
MCG+02-05-054	01 50 15.	+ 12 27	96	13.	GALAXY
RNGC 0716	01 50 16.	+ 11 48			NON-EXISTENT OBJECT
ARC 0267	01 50 17.	+ 00 48		16.6	RICH CLUSTER OF GALAXIES
LB 03770	01 50 17.	+ 36 15 30.		18.4	FAINT BLUE STAR
PHL 3911	01 50 18.	+ 00 46		16.6	BLUE STELLAR OBJECT
PHL 8003	01 50 18.	+ 04 55		17.3	BLUE STELLAR OBJECT
MRK 580	01 50 18.	+ 06 43	8	16.5	GALAXY WITH UV CONTINUUM
ZWG 503.045	01 50 18.	+ 30 52		15.7	GALAXY
ZWG 522.044	01 50 18.	+ 36 05		15.7	GALAXY
PHL 3914	01 50 18.	- 03 00		18.8	BLUE STELLAR OBJECT
MCG+01-05-047	01 50 18.	- 03 40	180	13.	GALAXY
PHL 3912	01 50 18.	- 07 16		18.4	BLUE STELLAR OBJECT
PHL 3913	01 50 18.	- 07 32		18.7	BLUE STELLAR OBJECT
LB 03232	01 50 18.	- 67 26		15.4	FAINT BLUE STAR
IC 1743	01 50 20.	+ 12 28			NONSTELLAR OBJECT
MCG-03-05-027	01 50 21.	- 19 01	60	14.5	GALAXY
LB 03771	01 50 22.	+ 35 52 48.		18.2	FAINT BLUE STAR
ARC 0263	01 50 23.	+ 37 19		17.8	RICH CLUSTER OF GALAXIES
PHL 1211	01 50 24.	+ 01 28		17.4	BLUE STELLAR OBJECT
PHL 3915	01 50 24.	+ 06 00		17.0	BLUE STELLAR OBJECT
PHL 3916	01 50 24.	+ 07 28		18.5	BLUE STELLAR OBJECT
MCG+04-05-023	01 50 24.	+ 22 28	42	15.	GALAXY
ZWG 482.028	01 50 24.	+ 22 29		15.2	GALAXY
KARA.73B 0067	01 50 24.	+ 22 29	48	15.2	ISOLATED GALAXY S
6ZW 092	01 50 24.	+ 33 08			COMPACT GALAXY
ZWG 522.045	01 50 24.	+ 35 46		15.6	GALAXY
6ZW 093	01 50 24.	+ 36 43			COMPACT GALAXY
ZWG 522.046	01 50 24.	+ 36 43		14.4	GALAXY
UGC 01353	01 50 24.	+ 36 43	72	14.4	GALAXY E-S0
PK131-05.1	01 50 24.	+ 56 10			PLANETARY NEBULA
PHL 8005	01 50 24.	- 06 28		18.2	BLUE STELLAR OBJECT
PHL 8008	01 50 24.	- 08 38		17.5	BLUE STELLAR OBJECT
PHL 8006	01 50 24.	- 08 39		18.7	BLUE STELLAR OBJECT
PHL 8004	01 50 24.	- 14 20		16.3	BLUE STELLAR OBJECT
PHL 1212	01 50 24.	- 16 51		13.7	BLUE STELLAR OBJECT
PHL 8007	01 50 24.	- 23 25		18.7	BLUE STELLAR OBJECT
IC 1742	01 50 27.	+ 22 28 18.			NONSTELLAR OBJECT
MCG+06-05-036	01 50 27.	+ 36 42	13	15.	GALAXY
LB 03772	01 50 28.	+ 35 42 18.		20.2	FAINT BLUE STAR
PHL 3919	01 50 30.	+ 03 30		17.5	BLUE STELLAR OBJECT
PHL 8009	01 50 30.	+ 03 38		18.4	BLUE STELLAR OBJECT
ZWG 503.046	01 50 30.	+ 28 45		15.6	GALAXY
ZWG 538.001	01 50 30.	+ 43 43		14.2	GALAXY
ZWG 537.027	01 50 30.	+ 43 43		14.2	GALAXY
UGC 01355	01 50 30.	+ 43 43	102	14.2	GALAXY Sb/SBb
VB 059	01 50 30.	+ 64 20	537		STELLAR RING
ZWG 386.053	01 50 30.	- 01 20		15.7	GALAXY
UGC 01354	01 50 30.	- 01 20	60	15.7	GALAXY Sa-b
PHL 3918	01 50 30.	- 03 43		18.2	BLUE STELLAR OBJECT
PHL 3917	01 50 30.	- 29 53		18.2	BLUE STELLAR OBJECT
LB 03773	01 50 32.	+ 36 11 12.		18.5	FAINT BLUE STAR
MCG-04-05-014	01 50 33.	- 22 59	24	15.	GALAXY
LB 03774	01 50 34.	+ 36 07 00.		16.9	FAINT BLUE STAR
LB 03776	01 50 35.	+ 36 04 48.		16.0	FAINT BLUE STAR
LB 03775	01 50 35.	+ 36 21 36.		18.0	FAINT BLUE STAR
RNGC 0720	01 50 35.	- 13 58		11.5	GALAXY
IC 1745	01 50 35.	- 16 54 50.			NONSTELLAR OBJECT
ZWG 412.039	01 50 36.	+ 03 57		12.5	GALAXY
RNGC 0718	01 50 36.	+ 03 57		12.5	GALAXY
UGC 01356	01 50 36.	+ 03 57	180	12.5	GALAXY SBa
KARA.73B 0068	01 50 36.	+ 03 57	120	12.5	ISOLATED GALAXY S
PHL 3920	01 50 36.	+ 07 52		17.6	BLUE STELLAR OBJECT
PHL 1213	01 50 36.	+ 08 57		18.8	BLUE STELLAR OBJECT
ZWG 460.039	01 50 36.	+ 19 42		15.4	GALAXY
UGC 01357	01 50 36.	+ 19 42	78	15.4	GALAXY Sa
6ZW 094	01 50 36.	+ 33 09			COMPACT GALAXY
ZWG 522.047	01 50 36.	+ 35 58		13.9	GALAXY
UGC 01358	01 50 36.	+ 35 58	126	13.9	GALAXY S0-a
MCG+06-05-037	01 50 36.	+ 35 58	90	14.	GALAXY
MCG+07-05-001	01 50 36.	+ 43 42	60	14.5	GALAXY
OCL 0336	01 50 36.	+ 62 07	240	14.	OPEN STAR CLUSTER
MCG+00-05-043	01 50 36.	- 01 19	48	15.5	GALAXY
PHL 1214	01 50 36.	- 02 30		16.9	BLUE STELLAR OBJECT
PHL 3923	01 50 36.	- 02 45		18.7	BLUE STELLAR OBJECT
PHL 3921	01 50 36.	- 06 08		18.6	BLUE STELLAR OBJECT
PHL 8010	01 50 36.	- 06 42		17.5	BLUE STELLAR OBJECT
PHL 1215	01 50 36.	- 06 52		18.8	BLUE STELLAR OBJECT
PHL 1216	01 50 36.	- 07 44		16.8	BLUE STELLAR OBJECT
MCG-02-05-068	01 50 36.	- 13 59	96	11.5	GALAXY
PHL 3922	01 50 36.	- 19 23		18.0	BLUE STELLAR OBJECT
PHL 8011	01 50 36.	- 22 32		18.5	BLUE STELLAR OBJECT
RNGC 0714	01 50 38.	+ 35 58		14.0	GALAXY
MCG+01-05-041	01 50 39.	+ 03 56	150	12.5	GALAXY
FATH 1.099	01 50 40.	+ 00 48	11		NEBULA
LB 03779	01 50 40.	+ 35 33 54.		18.9	FAINT BLUE STAR
LB 03778	01 50 40.	+ 36 05 18.		17.9	FAINT BLUE STAR
LB 03777	01 50 40.	+ 36 13 54.		19.8	FAINT BLUE STAR
RNGC 0715	01 50 41.	- 13 07		15.0	GALAXY
MCG+03-05-025	01 50 42.	+ 19 42	48	15.5	GALAXY
LB 03780	01 50 42.	+ 35 51 24.		19.6	FAINT BLUE STAR
6ZW 095	01 50 42.	+ 36 04			COMPACT GALAXY
PHL 3925	01 50 42.	- 09 30		18.8	BLUE STELLAR OBJECT
PHL 3924	01 50 42.	- 09 37		18.6	BLUE STELLAR OBJECT
PHL 8012	01 50 42.	- 14 18		16.7	BLUE STELLAR OBJECT
MCG-02-05-069	01 50 45.	- 13 07	36	15.	GALAXY
MCG-04-05-015	01 50 45.	- 22 37	36	15.5	GALAXY
LB 03781	01 50 47.	+ 35 36 24.		18.7	FAINT BLUE STAR
PHL 1218	01 50 48.	+ 03 57		18.5	BLUE STELLAR OBJECT
PHL 3928	01 50 48.	+ 05 24		18.3	BLUE STELLAR OBJECT
PHL 1217	01 50 48.	+ 05 31		16.0	BLUE STELLAR OBJECT
PHL 8020	01 50 48.	+ 08 06		18.2	BLUE STELLAR OBJECT
PHL 3926	01 50 48.	+ 09 12		17.8	BLUE STELLAR OBJECT
MCG+03-05-026	01 50 48.	+ 19 35	24	14.5	GALAXY
MCG+05-05-026	01 50 48.	+ 29 39	45	14.5	GALAXY
ZWG 503.047	01 50 48.	+ 29 41		14.2	GALAXY
UGC 01359	01 50 48.	+ 29 41	54	14.2	GALAXY SB
KARA.73B 0069	01 50 48.	+ 29 41	42	14.2	ISOLATED GALAXY S
ZWG 503.048	01 50 48.	+ 31 14		15.3	GALAXY
ZC 0150.8+3615	01 50 48.	+ 36 15	12700		CLUSTER OF GALAXIES
VB 152	01 50 48.	+ 62 48	436		STELLAR RING
PHL 8022	01 50 48.	- 00 04		18.6	BLUE STELLAR OBJECT
PHL 3929	01 50 48.	- 00 04		18.3	BLUE STELLAR OBJECT
PHL 8021	01 50 48.	- 00 05		18.8	BLUE STELLAR OBJECT
ZC 0150.8-0124	01 50 48.	- 01 24	870		CLUSTER OF GALAXIES
PHL 8015	01 50 48.	- 04 40		17.8	BLUE STELLAR OBJECT
PHL 8017	01 50 48.	- 05 28		18.8	BLUE STELLAR OBJECT
PHL 8014	01 50 48.	- 06 04		16.9	BLUE STELLAR OBJECT

OBJECT NAME	RIGHT ASCEN.	DECLINATION	DIAM.	MAGN.	TYPE OF OBJECT
PHL 8018	01 50 48.	- 06 11		18.2	BLUE STELLAR OBJECT
PHL 8019	01 50 48.	- 07 49		18.4	BLUE STELLAR OBJECT
PHL 3927	01 50 48.	- 08 56		17.9	BLUE STELLAR OBJECT
PHL 3930	01 50 48.	- 11 00		18.6	BLUE STELLAR OBJECT
MCG-02-05-070	01 50 48.	- 12 37	48	15.	GALAXY
PHL 8013	01 50 48.	- 14 36		14.1	BLUE STELLAR OBJECT
PHL 8016	01 50 48.	- 19 06		17.8	BLUE STELLAR OBJECT
PHL 8023	01 50 48.	- 24 00		18.5	BLUE STELLAR OBJECT
ARC 0268	01 50 50.	- 01 23		16.6	RICH CLUSTER OF GALAXIES
MCG+06-05-038	01 50 51.	+ 36 32	24	14.5	GALAXY
MRK 581	01 50 54.	+ 06 25	15	16.5	GALAXY WITH UV CONTINUUM
ZWG 412.040	01 50 54.	+ 08 09		15.5	GALAXY
ZWG 460.040	01 50 54.	+ 19 35		14.7	GALAXY
UGC 01360	01 50 54.	+ 19 35	84	14.7	GALAXY S0?
IC 1744	01 50 54.	+ 19 36 22.			NONSTELLAR OBJFCT
ZC 0150.9+3050	01 50 54.	+ 30 50	18680		CLUSTER OF GALAXIFS
MCG+06-05-040	01 50 54.	+ 36 05 30.	12	15.	GALAXY
ZWG 522.048	01 50 54.	+ 36 06		15.3	GALAXY
ZWG 522.049	01 50 54.	+ 36 18		15.3	GALAXY
MCG+06-05-039	01 50 54.	+ 36 19	48	15.	GALAXY
ZWG 522.050	01 50 54.	+ 36 20		15.7	GALAXY
UGC 01361	01 50 54.	+ 36 20	72	15.7	GALAXY Sc
ZWG 522.051	01 50 54.	+ 36 32		15.1	GALAXY
MCG-03-05-028	01 50 54.	- 19 11	60	14.5	GALAXY
RNGC 0719	01 50 56.	+ 19 35		14.5	GALAXY
LB 03783	01 50 59.	+ 36 11 18.		18.8	FAINT BLUE STAR
LB 03782	01 50 59.	+ 36 16 42.		19.8	FAINT BLUE STAR
ARC 0269	01 50 59.	- 04 35		17.2	RICH CLUSTER OF GALAXIES
PHL 1219	01 51 00.	+ 04 10		17.5	BLUE STELLAR OBJECT
ZC 0151.0+0515	01 51 00.	+ 05 15	2350		CLUSTER OF GALAXIES
UGC 01362	01 51 00.	+ 14 32	60	18.	GALAXY DWRF SP
LB 03784	01 51 00.	+ 35 40 18.		19.1	FAINT BLUE STAR
MCG+06-05-041	01 51 00.	+ 35 58 30.	66	14.5	GALAXY
ZWG 522.052	01 51 00.	+ 35 59		14.7	GALAXY
UGC 01363	01 51 00.	+ 35 59	90	14.7	GALAXY S0-a
ZWG 522.053	01 51 00.	+ 36 23		15.4	GALAXY
6ZW 096	01 51 00.	+ 37 27			COMPACT GALAXY
LDN 1339	01 51 00.	+ 62 16	180		DARK NEBULA
LDN 1338	01 51 00.	+ 62 28	180		DARK NEBULA
VB 153	01 51 00.	+ 62 59	383		STELLAR RING
ZC 0151.0-0136	01 51 00.	- 01 36	1010		CLUSTER OF GALAXIES
PHL 1220	01 51 00.	- 10 15		18.3	BLUF STELLAR OBJECT
PHL 8024	01 51 00.	- 25 32		18.3	BLUE STELLAR OBJECT
RNGC 0717	01 51 02.	+ 35 59		14.5	GALAXY
LB 03786	01 51 05.	+ 36 19 18.		19.0	FAINT BLUE STAP
LB 03785	01 51 05.	+ 36 19 36.		17.0	FAINT BLUE STAR
PHL 8026	01 51 06.	+ 03 26		18.3	BLUE STELLAR OEJECT
PHL 8027	01 51 06.	+ 05 55		18.2	BLUE STELLAR OBJECT
6ZW 097	01 51 06.	+ 33 17			COMPACT GALAXY
LB 03787	01 51 06.	+ 35 40 06.		18.3	FAINT BLUE STAR
6ZW 098	01 51 06.	+ 37 34			COMPACT GALAXY
5ZW 136	01 51 06.	+ 40 02			COMPACT GALAXY
PHL 8025	01 51 06.	- 02 32		17.3	BLUE STELLAR OBJFCT
PHL 8028	01 51 06.	- 05 16		18.6	BLUE STELLAR OBJECT
PHL 1221	01 51 06.	- 11 27		18.7	BLUE STELLAR OBJECT
LB 03788	01 51 07.	+ 36 16 24.		18.9	FAINT BLUE STAR
LB 03789	01 51 08.	+ 35 34 24.		18.8	FAINT BLUE STAR
PHL 1222	01 51 12.	+ 04 48		18.5	BLUE STELLAR OBJECT
BC PHL1222	01 51 12.	+ 04 48		17.63	QUASI-STELLAR OBJECT
SHB 048	01 51 12.	+ 04 48		17.6	QUASI-STELLAR OBJECT
6ZW 099	01 51 12.	+ 31 38			COMPACT GALAXY
6ZW 100	01 51 12.	+ 33 55			COMPACT GALAXY
LB 03790	01 51 12.	+ 35 46 42.		18.8	FAINT BLUE STAR
PHL 8033	01 51 12.	- 00 19		18.3	BLUE STELLAR OBJECT
PHL 8029	01 51 12.	- 04 55		18.2	BLUF STELLAR OBJECT
PHL 8032	01 51 12.	- 05 00		17.9	BLUE STELLAR OBJECT
PHL 3933	01 51 12.	- 06 43		18.7	BLUE STELLAR OBJECT
PHL 8030	01 51 12.	- 06 52		18.4	BLUE STELLAR OBJECT
PHL 8031	01 51 12.	- 07 53		18.6	BLUE STELLAR OBJECT
PHL 3931	01 51 12.	- 09 52		18.0	BLUE STELLAR OBJECT
PHL 8034	01 51 12.	- 13 00		18.4	BLUE STELLAR OBJECT
PHL 3932	01 51 12.	- 13 54		18.0	BLUE STELLAR OBJECT
PHL 8035	01 51 12.	- 15 32		18.6	BLUE STELLAR OBJECT
PHL 8036	01 51 12.	- 29 10		18.4	BLUE STELLAR OBJECT
LB 03791	01 51 13.	+ 35 55 30.		17.0	FAINT BLUE STAR
FATH 1.100	01 51 17.	+ 00 29	5		NEBULA
PHL 3936	01 51 18.	+ 08 10		18.2	BLUE STELLAR OBJECT
PHL 8037	01 51 18.	+ 08 46		18.4	BLUE STELLAR OBJECT
ZC 0151.3+1334	01 51 18.	+ 13 34	1340		CLUSTER OF GALAXIES
UGC 01364	01 51 18.	+ 14 41	90	16.5	GALAXY SBc
ZWG 503.049	01 51 18.	+ 31 52		15.1	GALAXY
5ZW 137	01 51 18.	+ 34 04			COMPACT GALAXY
LB 03792	01 51 18.	+ 35 35 36.		17.9	FAINT BLUE STAR
ZWG 522.054	01 51 18.	+ 35 40		15.2	GALAXY
VB 060	01 51 18.	+ 66 37	336		STELLAR RING
PHL 8038	01 51 18.	- 04 08		18.4	BLUE STELLAR OBJECT
PHL 3934	01 51 18.	- 05 56		15.9	BLUF STELLAR OBJECT
PHL 3935	01 51 18.	- 12 12		18.9	BLUE STELLAR OBJECT
LB 00164	01 51 18.	- 15 10		17.2	FAINT BLUE STAR
ARC 0270	01 51 19.	- 03 05		17.6	RICH CLUSTER OF GALAXIES
LB 03793	01 51 20.	+ 36 08 00.		19.0	FAINT BLUE STAR
LB 03794	01 51 22.	+ 35 51 48.		18.4	FAINT BLUE STAR
PHL 8041	01 51 24.	+ 01 58		18.2	BLUE STELLAR OBJECT
PHL 8039	01 51 24.	+ 04 28		17.3	BLUE STELLAR OBJECT
MCG+03-05-027	01 51 24.	+ 17 50	60	14.5	GALAXY
LB 03795	01 51 24.	+ 35 36 00.		18.5	FAINT BLUE STAR
ZWG 522.055	01 51 24.	+ 36 22		14.7	GALAXY
UGC 01366	01 51 24.	+ 36 22	102	14.7	GALAXY SBc
MCG+06-05-042	01 51 24.	+ 36 23	108	14.5	GALAXY
ZWG 386.054	01 51 24.	- 01 00		14.6	GALAXY
UGC 01365	01 51 24.	- 01 00	72	14.6	GALAXY S
PHL 8040	01 51 24.	- 09 37		17.6	BLUE STELLAR OBJECT
PHL 1223	01 51 24.	- 12 50		18.8	BLUE STELLAR OBJECT
PHL 3937	01 51 24.	- 17 24		18.4	BLUE STELLAR OBJECT
RNGC 0724	01 51 26.	- 24 06			NON-EXISTENT OBJECT
LB 03796	01 51 28.	+ 36 11 06.		19.5	FAINT BLUE STAP
ARC 0271	01 51 29.	+ 01 31		16.6	RICH CLUSTER OF GALAXIFS
MCG+00-05-045	01 51 30.	+ 00 41 30.	24	15.	GALAXY
ZWG 386.055	01 51 30.	+ 00 42		15.7	GALAXY
ZC 0151.5+0052	01 51 30.	+ 00 52	4970		CLUSTER OF GALAXIES
ZWG 412.041	01 51 30.	+ 07 37		15.0	GALAXY
UGC 01368	01 51 30.	+ 07 37	96	15.0	GALAXY Sa-b
ZWG 460.041	01 51 30.	+ 17 51		14.9	GALAXY
UGC 01369	01 51 30.	+ 17 51	102	14.9	GALAXY Sc
LB 03797	01 51 30.	+ 36 04 18.		18.6	FAINT BLUE STAR
VB 154	01 51 30.	+ 58 16	638		STELLAR RING
UGC 01367	01 51 30.	- 01 00	66	17.	GALAXY DWARF
MCG+00-05-044	01 51 30.	- 01 00	48	14.	GALAXY
MCG+01-05-048	01 51 30.	- 02 51	42	17.	GALAXY
PHL 3938	01 51 30.	- 06 38		18.7	BLUE STELLAR OBJECT
PHL 8042	01 51 30.	- 10 15		18.2	BLUE STELLAR OBJECT
MCG-04-05-016	01 51 30.	- 24 00	60	13.5	GALAXY
LB 03798	01 51 31.	+ 36 16 24.		16.9	FAINT BLUE STAR
RNGC 0727	01 51 31.	- 36 06			GALAXY
FATH 1.101	01 51 32.	+ 00 43	14		NEBULA
LB 03799	01 51 32.	+ 35 56 24.		18.3	FAINT BLUE STAR
RNGC 0723	01 51 32.	- 24 00		13.0	GALAXY
PHL 3940	01 51 36.	+ 01 22		18.7	BLUE STELLAR OBJFCT
PHL 1224	01 51 36.	+ 01 47		14.8	BLUE STELLAR OBJECT
PHL 3942	01 51 36.	+ 04 36		16.6	BLUE STELLAR OBJECT
PHL 3943	01 51 36.	+ 07 28		18.7	BLUE STELLAR OBJECT
PHL 3939	01 51 36.	+ 08 48		17.8	BLUE STELLAR OBJECT
ZWG 437.050	01 51 36.	+ 13 17		15.6	GALAXY
UGC 01370	01 51 36.	+ 13 17	60	15.6	GALAXY Sa-b
6ZW 101	01 51 36.	+ 34 35			COMPACT GALAXY
LB 03802	01 51 36.	+ 35 48 30.		19.0	FAINT BLUE STAR
LB 03800	01 51 36.	+ 35 59 12.		19.2	FAINT BLUE STAR
LB 03801	01 51 36.	+ 36 19 48.		18.6	FAINT BLUE STAP
OCL 0338	01 51 36.	+ 61 05	480		OPEN STAR CLUSTER
PHL 8043	01 51 36.	- 01 42		18.0	BLUE STELLAR OBJECT
PHL 8044	01 51 36.	- 07 47		18.7	BLUE STELLAR OBJFCT
PHL 8045	01 51 36.	- 13 40		18.7	BLUE STELLAR OBJECT
PHL 3941	01 51 36.	- 23 54		18.4	BLUE STELLAR OBJECT
MCG-06-05-012	01 51 36.	- 36 05	60	14.5	GALAXY
LB 03803	01 51 38.	+ 36 19 48.		18.4	FAINT BLUE STAR
MCG+01-05-042	01 51 39.	+ 07 37 30.	72	14.	GALAXY
FATH 1.102	01 51 40.	+ 00 55	5		NEBULA
LB 03804	01 51 41.	+ 36 21 36.		18.5	FAINT BLUE STAR
ZWG 412.042	01 51 42.	+ 04 33		15.1	GALAXY
UGC 01371	01 51 42.	+ 04 33	90	15.1	GALAXY S0
ZWG 412.043	01 51 42.	+ 05 10		15.6	GALAXY
ZC 0151.7+0700	01 51 42.	+ 07 00	1410		CLUSTER OF GALAXIES
MCG+02-05-055	01 51 42.	+ 13 17	48	15.	GALAXY
MCG+03-05-030	01 51 42.	+ 16 37	48	15.	GALAXY
ZWG 460.042	01 51 42.	+ 17 25		15.1	GALAXY
UGC 01372	01 51 42.	+ 17 25	72	15.1	GALAXY S
MCG+03-05-029	01 51 42.	+ 17 25	42	15.	GALAXY
MCG+03-05-028	01 51 42.	+ 19 50	66	14.	GALAXY
PHL 8046	01 51 42.	- 06 21		18.2	BLUE STELLAR OBJECT
PHL 1225	01 51 42.	- 09 28		17.6	BLUF STELLAR OBJFCT
MCG-02-05-071	01 51 42.	- 09 57	48	14.5	GALAXY
MCG-02-05-072	01 51 42.	- 14 30	66	14.	GALAXY
LB 03805	01 51 45.	+ 35 41 00.		18.5	FAINT BLUE STAR
LB 03636	01 51 45.	+ 37 48 42.		19.7	FAINT BLUE STAR
FATH 1.103	01 51 48.	+ 00 23	5		NEBULA
PHL 1226	01 51 48.	+ 04 34		18.2	BLUE STELLAR OBJECT
MCG+01-05-043	01 51 48.	+ 04 34	84	15.	GALAXY
IC 1746	01 51 48.	+ 04 34		14.	GALAXY Sb
BC PHL1226	01 51 48.	+ 04 34		18.2	QUASI-STELLAR OBJECT
SHB 049	01 51 48.	+ 04 34		18.2	QUASI-STELLAR OBJECT
MCG+01-05-044	01 51 48.	+ 05 10	60	15.	GALAXY
UGC 01373	01 51 48.	+ 05 12	72	16.0	GALAXY S
ZWG 460.043	01 51 48.	+ 16 37		15.7	GALAXY
UGC 01374	01 51 48.	+ 16 37	72	15.7	GALAXY SBc
ZC 0151.8+1645	01 51 48.	+ 16 45	6180		CLUSTER OF GALAXIES
MCG+03-05-031	01 51 48.	+ 17 48	15	15.	GALAXY
ZWG 460.044	01 51 48.	+ 19 49		15.0	GALAXY
UGC 01375	01 51 48.	+ 19 49	84	15.0	GALAXY SBc
LB 03806	01 51 48.	+ 35 37 12.		17.8	FAINT BLUE STAR
ZWG 522.056	01 51 48.	+ 39 08		13.8	GALAXY
RNGC 0721	01 51 48.	+ 39 08		14.0	GALAXY
UGC 01376	01 51 48.	+ 39 09	108	13.8	GALAXY SBb
MCG+06-05-043	01 51 48.	+ 39 09	102	13.5	GALAXY
MCG+12-03-001	01 51 48.	+ 73 02 30.	150	14.	GALAXY
ZWG 386.057	01 51 48.	- 00 28		15.7	GALAXY
PHL 8048	01 51 48.	- 00 35		17.2	BLUE STELLAR OBJECT
PHL 3944	01 51 48.	- 00 46		18.7	BLUE STELLAR OBJECT
ZWG 386.056	01 51 48.	- 00 52		14.7	GALAXY
PHL 8051	01 51 48.	- 00 54		18.2	BLUE STELLAR OBJECT
MCG+01-05-049	01 51 48.	- 02 50	60	15.	GALAXY
PHL 8049	01 51 48.	- 07 18		19.0	BLUE STELLAR OBJFCT
PHL 3945	01 51 48.	- 09 47		18.1	BLUE STELLAR OBJECT
PHL 8050	01 51 48.	- 10 47		17.8	BLUE STELLAR OBJECT
PHL 8047	01 51 48.	- 19 41		14.0	BLUE STELLAR OBJECT
LB 03637	01 51 49.	+ 37 35 00.		18.8	FAINT BLUE STAR
RNGC 0729	01 51 49.	- 36 03			GALAXY
ARC 0273	01 51 50.	- 23 48		17.7	RICH CLUSTER OF GALAXIES
LB 03807	01 51 53.	+ 36 09 18.		20.0	FAINT BLUE STAR
PHL 3946	01 51 54.	+ 00 14		16.9	BLUE STELLAR OBJECT
PHL 3947	01 51 54.	+ 03 23		17.7	BLUE STELLAR OBJECT
KARA.68 012	01 51 54.	+ 05 15	27		DWARF GALAXY
UGC 01377	01 51 54.	+ 16 48	78	16.5	GALAXY SB:c
ZC 0151.9+2125	01 51 54.	+ 21 25	1280		CLUSTER OF GALAXIES
MRK 002	01 51 54.	+ 36 40	25	14.	GALAXY WITH UV CONTINUUM
ZWG 522.057	01 51 54.	+ 37 17		15.4	GALAXY
ZWG 326.002	01 51 54.	+ 73 02		14.8	GALAXY
UGC 01378	01 51 54.	+ 73 02	300	14.8	GALAXY SBa
MCG+00-05-047	01 51 54.	- 00 27	36	15.	GALAXY
MCG+00-05-046	01 51 54.	- 00 52 30.	24	15.	GALAXY
LB 03808	01 51 55.	+ 35 56 18.		18.3	FAINT BLUE STAR
LB 03810	01 51 57.	+ 35 45 12.		19.2	FAINT BLUE STAR
LB 03809	01 51 57.	+ 35 49 48.		17.1	FAINT BLUE STAR
MCG+06-05-044	01 51 57.	+ 36 40	36	14.5	GALAXY
LB 03812	01 51 58.	+ 36 09 18.		18.3	FAINT BLUE STAR
LB 03811	01 51 58.	+ 36 15 00.		20.5	FAINT BLUE STAR
LB 03638	01 51 58.	+ 37 35 12.		18.2	FAINT BLUE STAR
PHL 3949	01 52 00.	+ 04 30		18.7	BLUE STELLAR OBJECT
PHL 8053	01 52 00.	+ 05 01		18.0	BLUE STELLAR OBJECT
MCG+01-05-045	01 52 00.	+ 05 09	48	15.	GALAXY
ZWG 412.044	01 52 00.	+ 05 10		15.5	GALAXY
PHL 8052	01 52 00.	+ 05 48		17.2	BLUE STELLAR OBJECT
ZC 0152.0+0856	01 52 00.	+ 08 56	3090		CLUSTER OF GALAXIES
ZWG 460.045	01 52 00.	+ 17 48		15.0	GALAXY
ZWG 460.046	01 52 00.	+ 20 27		14.6	GALAXY
UGC 01379	01 52 00.	+ 20 27	126	14.6	GALAXY S
MCG+03-05-032	01 52 00.	+ 20 27	66	14.5	GALAXY
ZWG 503.050	01 52 00.	+ 28 18		15.7	GALAXY
KARA.73B 0070	01 52 00.	+ 28 18	24	15.7	ISOLATED GALAXY IR
ZC 0152.0+3337	01 52 00.	+ 33 37	4230		CLUSTER OF GALAXIFS
MCG+06-05-045	01 52 00.	+ 35 10	24	15.	GALAXY
MCG+06-05-046	01 52 00.	+ 36 08 30.	15	16.	GALAXY
MCG+06-05-047	01 52 00.	+ 36 37	30	14.5	GALAXY
ZWG 522.058	01 52 00.	+ 36 41		14.2	GALAXY
ZWG 522.059	01 52 00.	+ 37 05		15.6	GALAXY
UGC 01380	01 52 00.	+ 37 05	60	15.6	GALAXY S
UGC 01381	01 52 00.	+ 48 47	72	16.5	GALAXY Sc

OBJECT NAME	RIGHT ASCEN.	DECLINATION	DIAM.	MAGN.	TYPE OF OBJECT
VB 155	01 52 00.	+ 61 48	336		STELLAR RING
PHL 3948	01 52 00.	- 03 28		17.9	BLUE STELLAR OBJECT
PHL 3951	01 52 00.	- 06 04		17.8	BLUE STELLAR OBJECT
PHL 8054	01 52 00.	- 07 00		13.0	BLUE STELLAR OBJECT
PHL 8055	01 52 00.	- 08 54		18.2	BLUE STELLAR OBJECT
PHL 1227	01 52 00.	- 10 22		18.4	BLUE STELLAR OBJECT
PHL 3950	01 52 00.	- 12 02		18.8	BLUE STELLAR OBJECT
RNGC 0722	01 52 01.	+ 20 27		14.5	GALAXY
MCG+06-05-048	01 52 03.	+ 36 40	48	14.5	GALAXY
HOLM 050D	01 52 03.	- 10 03	30	14.8	PART OF MULTIPLE GALAXY
HOLM 050A	01 52 04.	- 10 03	30	14.2	PART OF MULTIPLE GALAXY
UGC 01383	01 52 06.	+ 05 47	102	16.0	GALAXY Sc
MCG+01-05-046	01 52 06.	+ 05 48	84	15.	GALAXY
UGC 01384	01 52 06.	+ 17 52	102	17.	GALAXY
6ZW 102	01 52 06.	+ 28 18			COMPACT GALAXY
5ZW 138	01 52 06.	+ 34 35			COMPACT GALAXY
ZWG 522.060	01 52 06.	+ 35 11		15.1	GALAXY
ZWG 522.061	01 52 06.	+ 36 38		15.2	GALAXY
ZWG 522.062	01 52 06.	+ 36 41		15.2	GALAXY
UGC 01385	01 52 06.	+ 36 41	48	14.2	GALAXY SBa
ZWG 538.002	01 52 06.	+ 39 40		15.6	GALAXY
ZWG 386.058	01 52 06.	- 00 24		14.7	GALAXY
UGC 01382	01 52 06.	- 00 24	96	14.7	GALAXY E?
PHL 8056	01 52 06.	- 06 00		18.3	BLUE STELLAR OBJECT
PHL 8057	01 52 06.	- 06 12		18.5	BLUE STELLAR OBJECT
FEIG 016	01 52 06.	- 07 01		11.0	FAINT BLUE STAR
RNGC 0756	01 52 06.	- 16 57		15.0	GALAXY
MCG-03-05-029	01 52 06.	- 16 57	24	15.	GALAXY
MCG-06-05-013	01 52 06.	- 38 02	48	14.5	GALAXY
HOLM 050C	01 52 08.	- 10 03	24	14.7	PART OF MULTIPLE GALAXY
MCG+00-05-048	01 52 09.	- 00 23 30.	36	14.	GALAXY
LB 03639	01 52 10.	+ 37 29 06.		18.0	FAINT BLUE STAR
PHL 8059	01 52 12.	+ 03 26		18.5	BLUE STELLAR OBJECT
PHL 3952	01 52 12.	+ 06 58		18.1	BLUE STELLAR OBJECT
ZWG 437.051	01 52 12.	+ 13 14		15.7	GALAXY
UGC 01386	01 52 12.	+ 13 14	60	15.7	GALAXY Sb-c
6ZW 103	01 52 12.	+ 27 58			COMPACT GALAXY
5ZW 139	01 52 12.	+ 33 39			COMPACT GALAXY
MCG+06-05-049	01 52 12.	+ 36 00	48	15.	GALAXY
ZWG 522.063	01 52 12.	+ 36 01		15.4	GALAXY
UGC 01387	01 52 12.	+ 36 01	66	15.4	GALAXY S-IRR
ZWG 552.011	01 52 12.	+ 46 19		15.2	GALAXY
LDN 1342	01 52 12.	+ 62 27	180		DARK NEBULA
LDN 1341	01 52 12.	+ 62 27	180		DARK NEBULA
PHL 8060	01 52 12.	- 02 18		18.5	BLUE STELLAR OBJECT
PHL 8058	01 52 12.	- 13 45		16.8	BLUE STELLAR OBJECT
LB 03640	01 52 14.	+ 37 53 30.		20.2	FAINT BLUE STAR
MCG+02-05-058	01 52 15.	+ 10 35	48	16.	GALAXY
MCG+02-05-057	01 52 15.	+ 13 13	48	15.5	GALAXY
MCG+02-05-056	01 52 15.	+ 13 14	36	15.5	GALAXY
MCG+06-05-050	01 52 15.	+ 35 02	42	13.5	GALAXY
LB 03641	01 52 15.	+ 37 53 48.		19.5	FAINT BLUE STAR
ABC 0274	01 52 15.	- 06 32		16.3	RICH CLUSTER OF GALAXIES
LB 03642	01 52 16.	+ 37 57 48.		18.2	FAINT BLUE STAR
HOLM 050B	01 52 16.	- 10 02	36	14.6	PART OF MULTIPLE GALAXY
LB 03813	01 52 17.	+ 35 34 36.		16.8	FAINT BLUE STAR
ZWG 386.060	01 52 18.	+ 00 33		14.6	GALAXY
ZC 0152.3+1947	01 52 18.	+ 19 47	1810		CLUSTER OF GALAXIES
ZWG 482.029	01 52 18.	+ 23 48		15.6	GALAXY
5ZW 140	01 52 18.	+ 29 55			COMPACT GALAXY
ZWG 522.064	01 52 18.	+ 35 02		13.8	GALAXY
UGC 01388	01 52 18.	+ 35 02	150	13.8	GALAXY
ZWG 522.065	01 52 18.	+ 35 13		15.6	GALAXY
ZWG 552.012	01 52 18.	+ 47 43		15.6	GALAXY
UGC 01389	01 52 18.	+ 47 43	120	15.6	GALAXY E
MCG+08-04-010	01 52 18.	+ 47 43	22	15.	GALAXY
ZWG 386.059	01 52 18.	- 01 46		15.7	GALAXY
PHL 3956	01 52 18.	- 02 41		18.7	BLUE STELLAR OBJECT
PHL 3954	01 52 18.	- 02 41		18.7	BLUE STELLAR OBJECT
PHL 8062	01 52 18.	- 03 16		18.4	BLUE STELLAR OBJECT
PHL 3955	01 52 18.	- 05 56		10.8	BLUE STELLAR OBJECT
MCG-02-05-073	01 52 18.	- 09 14	36	13.5	GALAXY
PHL 8063	01 52 18.	- 13 06		18.3	BLUE STELLAR OBJECT
PHL 8061	01 52 18.	- 13 20		16.6	BLUE STELLAR OBJECT
PHL 3953	01 52 18.	- 27 22		15.6	BLUE STELLAR OBJECT
SER 019.04	01 52 18.	- 56 56	480	15.	COMPACT GROUP OF GALAXIES
RNGC 0745	01 52 19.	- 56 56			UNVERIFIED SOUTHERN OBJECT
IC 0172	01 52 20.	+ 00 34	8		BRIGHT NEBULA
FATH 2.004	01 52 20.	+ 00 34	8		NEBULA
LB 03814	01 52 22.	+ 35 35 36.		18.0	FAINT BLUE STAR
LB 03815	01 52 23.	+ 36 07 18.		16.8	FAINT BLUE STAR
MCG+00-05-049	01 52 24.	+ 00 34	24	15.	GALAXY
PHL 3958	01 52 24.	+ 09 30		17.2	BLUE STELLAR OBJECT
ZWG 522.066	01 52 24.	+ 36 03		15.5	GALAXY
UGC 01390	01 52 24.	+ 36 03	60	15.5	GALAXY S
MCG+06-05-051	01 52 24.	+ 36 03 30.	42	15.5	GALAXY
5ZW 141	01 52 24.	+ 37 03			COMPACT GALAXY
ZWG 552.013	01 52 24.	+ 47 19		15.4	GALAXY
MCG+00-05-050	01 52 24.	- 01 45 30.	30	15.	GALAXY
PHL 3959	01 52 24.	- 01 48		18.8	BLUE STELLAR OBJECT
PHL 3957	01 52 24.	- 02 34		12.9	BLUE STELLAR OBJECT
PHL 8065	01 52 24.	- 05 23		17.9	BLUE STELLAR OBJECT
PHL 8064	01 52 24.	- 05 44		18.6	BLUE STELLAR OBJECT
PHL 1228	01 52 24.	- 07 26		16.7	BLUE STELLAR OBJECT
RNGC 0734	01 52 24.	- 17 17			GALAXY
PHL 1229	01 52 24.	- 22 22		17.0	BLUE STELLAR OBJECT
PHL 1230	01 52 24.	- 23 44		18.3	BLUE STELLAR OBJECT
TON-S 0233	01 52 24.	- 27 25		15.6	BLUE STAR
ABC 0272	01 52 25.	+ 33 42		16.8	RICH CLUSTER OF GALAXIES
IC 0171	01 52 25.	+ 35 02 59.			NONSTELLAR OBJECT
LB 03816	01 52 25.	+ 35 55 30.		19.5	FAINT BLUE STAR
LB 03644	01 52 25.	+ 37 23 18.		18.8	FAINT BLUE STAR
LB 03643	01 52 25.	+ 37 52 48.		18.5	FAINT BLUE STAR
SHB 050	01 52 25.	+ 43 32		19.0	QUASI-STELLAR OBJECT
LB 03645	01 52 26.	+ 37 29 54.		17.9	FAINT BLUE STAR
LB 03646	01 52 27.	+ 37 41 36.		17.9	FAINT BLUE STAR
LB 03647	01 52 29.	+ 37 08 18.		18.4	FAINT BLUE STAR
RNGC 0728	01 52 30.	+ 03 58			NON-EXISTENT OBJECT
ZWG 412.045	01 52 30.	+ 05 08		15.5	GALAXY
MCG+02-05-059	01 52 30.	+ 09 45	84	14.5	GALAXY
ZWG 437.052	01 52 30.	+ 09 46		15.5	GALAXY
UGC 01391	01 52 30.	+ 09 46	90	15.5	GALAXY Sc
KARA.73B 0071	01 52 30.	+ 09 46	96	15.5	ISOLATED GALAXY S
UGC 01392	01 52 30.	+ 36 19	72	16.0	GALAXY S0?
MCG+08-04-011	01 52 30.	+ 46 33	144	14.	GALAXY
ZWG 559.003	01 52 30.	+ 52 03		15.2	GALAXY
LDN 1344	01 52 30.	+ 62 10	540		DARK NEBULA
LDN 1343	01 52 30.	+ 62 10	540		DARK NEBULA
PHL 3960	01 52 30.	- 02 50		18.9	BLUE STELLAR OBJECT
PHL 8066	01 52 30.	- 02 52		19.0	BLUE STELLAR OBJECT
PHL 1231	01 52 30.	- 10 10		18.0	BLUE STELLAR OBJECT
PHL 3961	01 52 30.	- 11 08		18.4	BLUE STELLAR OBJECT
PHL 3962	01 52 30.	- 12 59		18.7	BLUE STELLAR OBJECT
LB 03817	01 52 31.	+ 35 59 42.		20.3	FAINT BLUE STAR
LB 03649	01 52 31.	+ 37 24 48.		18.2	FAINT BLUE STAR
LB 03648	01 52 31.	+ 37 38 00.		17.7	FAINT BLUE STAR
LB 03818	01 52 32.	+ 35 49 42.		18.7	FAINT BLUE STAR
MCG+01-05-047	01 52 33.	+ 05 07	48	15.	GALAXY
LB 03819	01 52 33.	+ 36 04 18.		18.9	FAINT BLUE STAR
MCG+06-05-052	01 52 33.	+ 36 18 30.	36	16.	GALAXY
LB 03650	01 52 33.	+ 37 36 24.		16.5	FAINT BLUE STAR
RNGC 0731	01 52 34.	- 09 15		13.0	GALAXY
PHL 1232	01 52 36.	+ 01 36		17.2	BLUE STELLAR OBJECT
PHL 8068	01 52 36.	+ 04 32		16.1	BLUE STELLAR OBJECT
PHL 3963	01 52 36.	+ 06 34		18.3	BLUE STELLAR OBJECT
ZWG 461.001	01 52 36.	+ 21 02		15.6	GALAXY
ZWG 460.047	01 52 36.	+ 21 02		15.6	GALAXY
UGC 01393	01 52 36.	+ 21 02	60	15.6	GALAXY E-S0
KARA.72 045A	01 52 36.	+ 21 02	54	15.6	PART OF DOUBLE GALAXY
MCG+03-05-033	01 52 36.	+ 21 03	12	16.	GALAXY
ZWG 482.030	01 52 36.	+ 23 54		15.7	GALAXY
6ZW 104	01 52 36.	+ 32 24			COMPACT GALAXY
ZWG 552.014	01 52 36.	+ 46 34		15.0	GALAXY
UGC 01394	01 52 36.	+ 46 34	138	15.0	GALAXY Sa
PHL 8067	01 52 36.	- 03 59		17.7	BLUE STELLAR OBJECT
PHL 1233	01 52 36.	- 10 38		17.8	BLUE STELLAR OBJECT
MCG-02-05-074	01 52 36.	- 13 55	36	14.	GALAXY
PHL 1235	01 52 36.	- 20 32		17.0	BLUE STELLAR OBJECT
PHL 1234	01 52 36.	- 20 36		17.0	BLUE STELLAR OBJECT
MCG-06-05-014	01 52 36.	- 35 24	15	15.	GALAXY
RNGC 0754	01 52 37.	- 57 00			UNVERIFIED SOUTHERN OBJECT
LB 03652	01 52 38.	+ 37 23 30.		18.6	FAINT BLUE STAR
LB 03651	01 52 38.	+ 38 00 00.		18.4	FAINT BLUE STAR
LB 03821	01 52 39.	+ 35 36 42.		19.4	FAINT BLUE STAR
LB 03820	01 52 39.	+ 36 18 06.		18.6	FAINT BLUE STAR
RNGC 0713	01 52 40.	- 09 19		15.0	GALAXY
LB 03823	01 52 41.	+ 35 38 24.		20.0	FAINT BLUE STAR
LB 03822	01 52 41.	+ 36 23 30.		18.8	FAINT BLUE STAR
3ZW 037	01 52 42.	+ 01 14			COMPACT GALAXY
PHL 8071	01 52 42.	+ 01 18		14.0	BLUE STELLAR OBJECT
RNGC 0730	01 52 42.	+ 05 23			NON-EXISTENT OBJECT
PHL 8070	01 52 42.	+ 06 10		8.1	BLUE STELLAR OBJECT
ZWG 412.046	01 52 42.	+ 06 21		14.5	GALAXY
UGC 01395	01 52 42.	+ 06 21	102	14.5	GALAXY Sb
KARA.73B 0072	01 52 42.	+ 06 21	66	14.5	ISOLATED GALAXY S
ZWG 412.047	01 52 42.	+ 08 26		15.3	GALAXY
MCG+06-05-053	01 52 42.	+ 36 48	30	15.5	GALAXY
ZWG 522.067	01 52 42.	+ 37 09		15.5	GALAXY
LB 03653	01 52 42.	+ 37 47 48.		19.0	FAINT BLUE STAR
VB 061	01 52 42.	+ 65 07	584		STELLAR RING
PHL 8072	01 52 42.	- 05 37		18.6	BLUE STELLAR OBJECT
PHL 8069	01 52 42.	- 12 18		18.4	BLUE STELLAR OBJECT
PHL 3964	01 52 42.	- 12 18		16.9	BLUE STELLAR OBJECT
LB 03824	01 52 44.	+ 35 54 30.		20.3	FAINT BLUE STAR
MCG-02-05-075	01 52 45.	- 09 19 30.	48	15.	GALAXY
MCG-04-05-017	01 52 45.	- 26 16	30	15.	GALAXY
MCG+01-05-048	01 52 48.	+ 06 22	72	13.5	GALAXY
PHL 1237	01 52 48.	+ 06 28		18.5	BLUE STELLAR OBJECT
PHL 1236	01 52 48.	+ 07 29		17.9	BLUE STELLAR OBJECT
ZWG 461.002	01 52 48.	+ 21 05		15.1	GALAXY
ZWG 460.048	01 52 48.	+ 21 05		15.1	GALAXY
UGC 01396	01 52 48.	+ 21 05	66	15.1	GALAXY S0
KARA.72 045B	01 52 48.	+ 21 05	78	15.1	PART OF DOUBLE GALAXY
MCG+03-05-034	01 52 48.	+ 21 05	54	15.	GALAXY
6ZW 105	01 52 48.	+ 27 29			COMPACT GALAXY
ZWG 503.051	01 52 48.	+ 31 10		15.7	GALAXY
ZCG 0152+33	01 52 48.	+ 33 13		18.0	COMPACT GALAXY
LB 03825	01 52 48.	+ 35 54 42.		17.8	FAINT BLUE STAR
ZWG 522.068	01 52 48.	+ 36 48		15.6	GALAXY
PHL 8074	01 52 48.	- 02 44		18.3	BLUE STELLAR OBJECT
PHL 3965	01 52 48.	- 03 26		18.2	BLUE STELLAR OBJECT
PHL 8073	01 52 48.	- 04 48		17.9	BLUE STELLAR OBJECT
PHL 8075	01 52 48.	- 10 52		18.4	BLUE STELLAR OBJECT
LIN 571	01 52 51.	- 74 31 10.		14.15	SMC EMISSION LINE OBJECT
LB 03826	01 52 53.	+ 35 47 12.		18.7	FAINT BLUE STAR
LB 03654	01 52 53.	+ 37 16 30.		19.2	FAINT BLUE STAR
PHL 1238	01 52 54.	+ 03 44		18.9	BLUE STELLAR OBJECT
ZWG 482.031	01 52 54.	+ 22 50		15.7	GALAXY
KARA.73B 0073	01 52 54.	+ 22 50	36	15.7	ISOLATED GALAXY S
MCG+05-05-027	01 52 54.	+ 32 43	48	15.	GALAXY
5ZW 142	01 52 54.	+ 34 23			COMPACT GALAXY
LB 03655	01 52 54.	+ 37 20 48.		15.8	FAINT BLUE STAR
PHL 8076	01 52 54.	- 07 10		17.1	BLUE STELLAR OBJECT
LB 00165	01 52 54.	- 14 44		17.7	FAINT BLUE STAR
LB 03656	01 52 57.	+ 37 12 12.		19.8	FAINT BLUE STAR
MCG-02-05-076	01 52 57.	- 13 09	60	15.	GALAXY
LB 03828	01 52 59.	+ 35 47 00.		18.7	FAINT BLUE STAR
LB 03827	01 52 59.	+ 36 23 18.		17.6	FAINT BLUE STAR
LIN 572	01 52 59.	- 74 16 28.		13.93	SMC EMISSION LINE OBJECT
LB 09778	01 53	- 82 36		14.8	FAINT BLUE STAR
ZC 0153.0+0430	01 53 00.	+ 04 30	1610		CLUSTER OF GALAXIES
ZC 0153.0+1426	01 53 00.	+ 14 26	1210		CLUSTER OF GALAXIES
ZWG 461.003	01 53 00.	+ 17 45		15.2	GALAXY
ZWG 460.049	01 53 00.	+ 17 45		15.2	GALAXY
ZWG 503.052	01 53 00.	+ 32 45		15.0	GALAXY
UGC 01397	01 53 00.	+ 32 45	66	15.0	GALAXY S0
ZWG 522.069	01 53 00.	+ 36 53		14.9	GALAXY
UGC 01398	01 53 00.	+ 36 53	84	14.9	GALAXY Sc
MCG+06-05-054	01 53 00.	+ 36 53	60	14.5	GALAXY
MCG-02-06-002	01 53 00.	- 10 42 30.	48	15.	GALAXY
PHL 1239	01 53 00.	- 10 54		16.8	BLUE STELLAR OBJECT
MCG-02-06-001	01 53 00.	- 14 23 30.	36	15.5	GALAXY
LIN 576	01 53 02.	- 75 17 41.		16.87	SMC EMISSION LINE OBJECT
MCG+03-06-001	01 53 03.	+ 17 48	18	15.	GALAXY
LIN 574	01 53 03.	- 74 28 11.		15.63	SMC EMISSION LINE OBJECT
LIN 575	01 53 03.	- 75 15 23.		17.8	SMC EMISSION LINE OBJECT
LIN 573	01 53 05.	- 74 11 59.		14.03	SMC EMISSION LINE OBJECT
ZWG 438.001	01 53 06.	+ 14 42		15.2	GALAXY
ZWG 437.053	01 53 06.	+ 14 42		15.2	GALAXY
ZWG 461.004	01 53 06.	+ 17 48		15.1	GALAXY
ZWG 460.050	01 53 06.	+ 17 48		15.1	GALAXY
UGC 01399	01 53 06.	+ 17 48	60	15.1	GALAXY S0
ZC 0153.1+2600	01 53 06.	+ 26 00	1080		CLUSTER OF GALAXIES
ZWG 503.053	01 53 06.	+ 32 53		15.7	GALAXY
ZWG 522.070	01 53 06.	+ 39 25		15.7	GALAXY
PHL 8078	01 53 06.	- 08 34		17.9	BLUE STELLAR OBJECT

OBJECT NAME	RIGHT ASCEN.	DECLINATION	DIAM.	MAGN.	TYPE OF OBJECT
PHL 8077	01 53 06.	- 14 16		14.7	BLUE STELLAR OBJECT
PHL 1240	01 53 06.	- 22 12		18.0	BLUE STELLAR OBJECT
LB 00210	01 53 07.	+ 00 13 18.		16.7	FAINT BLUE STAR
ARC 0275	01 53 07.	+ 14 25		17.1	RICH CLUSTER OF GALAXIES
LB 03829	01 53 08.	+ 36 10 48.		19.4	FAINT BLUE STAR
MCG-04-05-018	01 53 09.	- 26 11	54	16.	GALAXY
LB 03657	01 53 10.	+ 37 11 36.		17.3	FAINT BLUE STAR
RNGC 0726	01 53 10.	- 11 03		14.0	GALAXY
PHL 1242	01 53 12.	+ 00 14		16.6	BLUE STELLAR OBJECT
PHL 1241	01 53 12.	+ 03 54		18.5	BLUE STELLAR OBJECT
SHAH 045	01 53 12.	+ 14 35	132	17.3	GROUP OF COMPACT GALAXIES
ZWG 522.071	01 53 12.	+ 35 53		13.8	GALAXY
UGC 01400	01 53 12.	+ 35 53	156	13.8	GALAXY Sb
MCG+06-05-055	01 53 12.	+ 35 53	150	13.5	GALAXY
6ZW 106	01 53 12.	+ 35 56			COMPACT GALAXY
6ZW 107	01 53 12.	+ 36 04			COMPACT GALAXY
VB 062	01 53 12.	+ 63 33	436		STELLAR RING
PHL 8079	01 53 12.	- 02 36		18.5	BLUE STELLAR OBJECT
PHL 8080	01 53 12.	- 04 08		18.5	BLUE STELLAR OBJECT
PHL 1243	01 53 12.	- 09 12		17.2	BLUE STELLAR OBJECT
PHL 8081	01 53 12.	- 21 00		18.0	BLUE STELLAR OBJECT
LB 03830	01 53 14.	+ 35 45 12.		19.1	FAINT BLUE STAR
MCG-02-06-003	01 53 15.	- 11 02	66	14.	GALAXY
FATH 1.104	01 53 16.	+ 00 28	5		NEBULA
LB 03658	01 53 16.	+ 37 16 30.		17.7	FAINT BLUE STAR
LB 03659	01 53 17.	+ 37 35 06.		17.9	FAINT BLUE STAR
PHL 8084	01 53 18.	+ 01 49		13.6	BLUE STELLAR OBJECT
ZWG 461.005	01 53 18.	+ 17 17		15.7	GALAXY
ZWG 460.051	01 53 18.	+ 17 17		15.7	GALAXY
ZWG 503.054	01 53 18.	+ 31 15		15.7	GALAXY
5ZW 143	01 53 18.	+ 34 46			COMPACT GALAXY
LB 03831	01 53 18.	+ 36 11 00.		20.2	FAINT BLUE STAR
VB 156	01 53 18.	+ 59 44	833		STELLAR RING
PHL 8082	01 53 18.	- 04 46		18.1	BLUE STELLAR OBJECT
PHL 8083	01 53 18.	- 06 42		18.5	BLUE STELLAR OBJECT
PHL 1244	01 53 18.	- 22 40		18.6	BLUE STELLAR OBJECT
ARC 0277	01 53 20.	- 07 38		15.6	RICH CLUSTER OF GALAXIES
IC 0173	01 53 21.	+ 01 02	27		FAINT NEBULA
FATH 2.005	01 53 21.	+ 01 02	27		NEBULA
MCG+00-06-001	01 53 21.	+ 01 04	48	15.	GALAXY
RNGC 0749	01 53 22.	- 30 11		14.0	GALAXY
LB 03660	01 53 23.	+ 37 25 12.		17.7	FAINT BLUE STAR
ZWG 386.001	01 53 24.	+ 01 02		14.9	GALAXY
UGC 01402	01 53 24.	+ 01 02	66	14.9	GALAXY SBb
ZWG 461.006	01 53 24.	+ 17 24		14.7	GALAXY
ZWG 460.052	01 53 24.	+ 17 24		14.7	GALAXY
UGC 01403	01 53 24.	+ 17 24	72	14.7	GALAXY Sb/SBc
5ZW 144	01 53 24.	+ 35 20			COMPACT GALAXY
ZWG 522.072	01 53 24.	+ 35 20		15.2	GALAXY
6ZW 108	01 53 24.	+ 36 02			COMPACT GALAXY
6ZW 109	01 53 24.	+ 36 14			COMPACT GALAXY
MCG+06-05-056	01 53 24.	+ 36 58	48	15.	GALAXY
ZWG 522.073	01 53 24.	+ 36 59		15.6	GALAXY
UGC 01404	01 53 24.	+ 36 59	84	15.6	GALAXY SBb
ZWG 522.074	01 53 24.	+ 37 12		15.7	GALAXY
UGC 01405	01 53 24.	+ 37 12	72	15.7	GALAXY Sc
ZWG 522.075	01 53 24.	+ 37 15		15.7	GALAXY
UGC 01401	01 53 24.	- 00 34	60	16.5	GALAXY Sc
PHL 8085	01 53 24.	- 06 02		16.6	BLUE STELLAR OBJECT
PHL 8086	01 53 24.	- 10 24		16.6	BLUE STELLAR OBJECT
PHL 3966	01 53 24.	- 12 10		16.8	BLUE STELLAR OBJECT
PHL 8087	01 53 24.	- 19 48		15.4	BLUE STELLAR OBJECT
PHL 1245	01 53 24.	- 24 29		18.1	BLUE STELLAR OBJECT
PHL 8088	01 53 24.	- 26 44		18.3	BLUE STELLAR OBJECT
MCG-05-05-023	01 53 24.	- 30 11	120	14.	GALAXY
LB 03832	01 53 25.	+ 35 52 30.		18.4	FAINT BLUE STAR
IC 1748	01 53 25.	+ 17 23 49.			NONSTELLAR OBJECT
LB 03661	01 53 26.	+ 37 12 48.		16.9	FAINT BLUE STAR
RNGC 0758	01 53 26.	- 03 20			GALAXY
RNGC 0733	01 53 27.	+ 32 50			GALAXY
LB 03663	01 53 28.	+ 37 19 30.		17.1	FAINT BLUE STAR
LB 03662	01 53 28.	+ 37 53 00.		16.8	FAINT BLUE STAR
ZWG 413.001	01 53 30.	+ 05 32		15.5	GALAXY
ZWG 438.002	01 53 30.	+ 10 33		15.0	GALAXY
MCG+02-06-001	01 53 30.	+ 10 34	30	15.5	GALAXY
MCG+03-06-002	01 53 30.	+ 17 23	54	14.5	GALAXY
5ZW 145	01 53 30.	+ 33 17			COMPACT GALAXY
LB 03833	01 53 30.	+ 36 13 00.		18.0	FAINT BLUE STAR
STOCK 02	01 53 30.	+ 36 32			BLUE KNOT NEAR ELLIP GLXY
MCG+06-05-057	01 53 30.	+ 36 33 30.	24	15.	GALAXY
ZWG 522.076	01 53 30.	+ 36 34		14.9	GALAXY
UGC 01406	01 53 30.	+ 36 34	102	14.9	GALAXY S0
LDN 1346	01 53 30.	+ 62 23	480		DARK NEBULA
LDN 1345	01 53 30.	+ 62 24	480		DARK NEBULA
VB 157	01 53 30.	+ 63 06	349		STELLAR RING
PHL 8089	01 53 30.	- 06 32		17.8	BLUE STELLAR OBJECT
LB 00166	01 53 30.	- 13 51		17.4	FAINT BLUE STAR
LB 00081	01 53 30.	- 14 15		16.7	FAINT BLUE STAR
LB 03834	01 53 31.	+ 35 50 12.		17.9	FAINT BLUE STAR
RNGC 0732	01 53 31.	+ 36 34		15.0	GALAXY
LB 03664	01 53 32.	+ 37 24 18.		17.2	FAINT BLUE STAR
IC 1749	01 53 33.	+ 06 29 54.			NONSTELLAR OBJECT
HOLM 051A	01 53 34.	+ 10 42	54	13.8	PART OF MULTIPLE GALAXY
LB 03665	01 53 34.	+ 37 59 48.		17.7	FAINT BLUE STAR
LB 03835	01 53 35.	+ 35 55 30.		18.6	FAINT BLUE STAR
LB 03666	01 53 35.	+ 37 42 42.		17.6	FAINT BLUE STAR
FATH 1.105	01 53 36.	+ 00 01	5		NEBULA
PHL 8092	01 53 36.	+ 02 16		16.8	BLUE STELLAR OBJECT
MCG+01-06-002	01 53 36.	+ 05 18	12	15.	GALAXY
ZWG 413.002	01 53 36.	+ 05 21		15.0	GALAXY
ZCG 0153+05.3	01 53 36.	+ 05 21		15.0	COMPACT GALAXY
MCG+01-06-001	01 53 36.	+ 06 29	36	14.5	GALAXY
ZWG 413.003	01 53 36.	+ 06 30		14.8	GALAXY
UGC 01407	01 53 36.	+ 06 30	72	14.8	GALAXY S0
HOLM 051B	01 53 36.	+ 10 43	18	14.8	PART OF MULTIPLE GALAXY
ZWG 438.003	01 53 36.	+ 12 55		15.7	GALAXY
UGC 01408	01 53 36.	+ 12 55	72	15.7	GALAXY Sc
MCG+02-06-002	01 53 36.	+ 12 57	48	15.	GALAXY
LB 03836	01 53 36.	+ 35 44 54.		18.9	FAINT BLUE STAR
ZWG 522.077	01 53 36.	+ 37 05		15.5	GALAXY
LB 03667	01 53 36.	+ 37 25 30.		19.8	FAINT BLUE STAR
MCG+01-06-001	01 53 36.	- 02 55 30.	21	15.	GALAXY
PHL 3967	01 53 36.	- 04 14		17.8	BLUE STELLAR OBJECT
PHL 8091	01 53 36.	- 04 28		17.9	BLUE STELLAR OBJECT
PHL 8093	01 53 36.	- 06 45		17.9	BLUE STELLAR OBJECT
PHL 8090	01 53 36.	- 07 33		16.5	BLUE STELLAR OBJECT
MCG-02-06-004	01 53 36.	- 10 12	72	14.	GALAXY
PHL 3968	01 53 36.	- 21 35		18.4	BLUE STELLAR OBJECT
PHL 1246	01 53 36.	- 22 52		17.5	BLUE STELLAR OBJECT
PHL 3969	01 53 36.	- 24 22		18.3	BLUE STELLAR OBJECT
PHL 1247	01 53 36.	- 28 58		17.1	BLUE STELLAR OBJECT
LB 03234	01 53 36.	- 79 45		14.1	FAINT BLUE STAR
LB 03668	01 53 37.	+ 37 25 24.		20.0	FAINT BLUE STAR
LB 03837	01 53 38.	+ 35 38 00.		20.0	FAINT BLUE STAR
IC 0174	01 53 39.	+ 03 30 30.			NONSTELLAR OBJECT
MCG+01-06-002	01 53 39.	- 02 57	24	15.	GALAXY
LB C3669	01 53 40.	+ 37 24 48.		18.0	FAINT BLUE STAR
MCG+00-06-002	01 53 42.	+ 00 49	9	16.	GALAXY
ZWG 386.002	01 53 42.	+ 03 15		15.7	GALAXY
MCG+01-06-008	01 53 42.	+ 03 30	48	14.	GALAXY
ZWG 413.004	01 53 42.	+ 03 31		14.6	GALAXY
UGC 01409	01 53 42.	+ 03 31	102	14.6	GALAXY S0?
IC 1750	01 53 42.	+ 03 50 05.			NONSTELLAR OBJECT
MCG+01-06-007	01 53 42.	+ 04 22	96	15.	GALAXY
ZWG 413.005	01 53 42.	+ 04 24		15.5	GALAXY
UGC 01410	01 53 42.	+ 04 24	108	15.5	GALAXY Sc
MCG+01-06-005	01 53 42.	+ 05 19	12	17.	GALAXY
MCG+01-06-004	01 53 42.	+ 05 21	18	14.	GALAXY
MCG+01-06-003	01 53 42.	+ 05 21	36	12.	GALAXY
MCG+01-06-006	01 53 42.	+ 05 22	12	15.5	GALAXY
3ZW 038	01 53 42.	+ 05 23			COMPACT GALAXY
RNGC 0741	01 53 42.	+ 05 23		13.0	GALAXY
ZWG 413.006	01 53 42.	+ 05 25		15.1	GALAXY
ZCG 0153+05.4	01 53 42.	+ 05 25		15.1	COMPACT GALAXY
6ZW 110	01 53 42.	+ 31 10			COMPACT GALAXY
MCG+05-05-028	01 53 42.	+ 32 47	72	14.	GALAXY
6ZW 111	01 53 42.	+ 32 48			COMPACT GALAXY
MCG+06-05-058	01 53 42.	+ 33 55	108	13.	GALAXY
5ZW 146	01 53 42.	+ 33 56			COMPACT GALAXY
ZWG 522.078	01 53 42.	+ 33 56		13.9	GALAXY
UGC 01411	01 53 42.	+ 33 56	108	13.9	GALAXY Sb
ZWG 522.079	01 53 42.	+ 35 21		15.3	GALAXY
LB 03670	01 53 42.	+ 37 46 24.		18.8	FAINT BLUE STAR
VB 164	01 53 42.	+ 60 48	269		STELLAR RING
PHL 8094	01 53 42.	- 07 01		18.3	BLUE STELLAR OBJECT
LB 00167	01 53 42.	- 15 03		16.8	FAINT BLUE STAR
IC 0175	01 53 44.	+ 01 05	11		FAINT NEBULA
FATH 2.006	01 53 44.	+ 01 05	14		NEBULA
RNGC 0735	01 53 44.	+ 33 56		14.0	GALAXY
SN 1572L	01 53 44.	+ 33 56		15.0	SUPERNOVA
REIN 2.018	01 53 44.28	+ 05 23 05.0			NEBULA
IC 1751	01 53 45.	+ 05 22 59.			SAME AS NGC 741
MCG+06-05-059	01 53 45.	+ 36 38	66	15.	GALAXY
MCG+01-06-003	01 53 45.	- 04 26 30.	48	15.	GALAXY
RNGC 0755	01 53 46.	- 09 18		13.0	GALAXY
RNGC 0747	01 53 46.	- 09 42			NON-EXISTENT OBJECT
ARC 0279	01 53 47.	+ 00 49		17.2	RICH CLUSTER OF GALAXIES
HN 0103	01 53 47.	+ 63 04			NEBULA
REIN 2.019	01 53 47.43	+ 05 22 57.6			NEBULA
ZWG 386.003	01 53 48.	+ 01 05		15.7	GALAXY
ZWG 413.007	01 53 48.	+ 03 50		15.4	GALAXY
UGC 01412	01 53 48.	+ 03 50	66	15.4	GALAXY S0-a
ZWG 413.008	01 53 48.	+ 05 23		13.2	GALAXY
UGC 01413	01 53 48.	+ 05 23	180	13.2	GALAXY E
ZC 0153.8+0706	01 53 48.	+ 07 06	1550		CLUSTER OF GALAXIES
ZWG 461.007	01 53 48.	+ 16 40		15.7	GALAXY
6ZW 112	01 53 48.	+ 31 12			COMPACT GALAXY
ZWG 503.055	01 53 48.	+ 32 48		13.6	GALAXY
UGC 01414	01 53 48.	+ 32 48	120	13.6	GALAXY E
6ZW 113	01 53 48.	+ 32 49			COMPACT GALAXY
6ZW 114	01 53 48.	+ 33 56			COMPACT GALAXY
MCG+06-05-061	01 53 48.	+ 35 20	42	14.5	GALAXY
ZWG 522.080	01 53 48.	+ 36 08		14.5	GALAXY
UGC 01415	01 53 48.	+ 36 08	84	14.5	GALAXY S0-a
MCG+06-05-060	01 53 48.	+ 36 08	66	14.5	GALAXY
ZWG 522.081	01 53 48.	+ 36 39		14.9	GALAXY
UGC 01416	01 53 48.	+ 36 39	78	14.9	GALAXY S
IC 1747	01 53 48.	+ 63 04	13	13.9	PLANETARY NEBULA
PHL 3971	01 53 48.	- 03 18		17.6	BLUE STELLAR OBJECT
PHL 8095	01 53 48.	- 06 46		17.6	BLUE STELLAR OBJECT
PHL 3970	01 53 48.	- 06 46		18.4	BLUE STELLAR OBJECT
LB 03671	01 53 49.	+ 37 28 36.		19.2	FAINT BLUE STAR
LB 03672	01 53 50.	+ 37 48 48.		19.4	FAINT BLUE STAR
RNGC 0748	01 53 50.	- 04 43		12.0	GALAXY
REIN 2.020	01 53 50.56	- 04 42 41.8			NEBULA
RNGC 0736	01 53 51.	+ 32 48		13.5	GALAXY
RNGC 0737	01 53 51.	+ 32 49			NON-EXISTENT OBJECT
MCG+06-05-062	01 53 51.	+ 35 45	48	16.	GALAXY
LB 03673	01 53 51.	+ 37 39 24.		16.0	FAINT BLUE STAR
MCG+01-06-004	01 53 51.	- 04 43 30.	132	12.	GALAXY
ZWG 413.009	01 53 54.	+ 05 23		14.8	GALAXY
RNGC 0742	01 53 54.	+ 05 23		15.0	GALAXY
UGC 01417	01 53 54.	+ 17 28	78	18.	GALAXY
MCG+05-05-029	01 53 54.	+ 31 26	48	15.	GALAXY
ZWG 503.056	01 53 54.	+ 31 27		14.9	GALAXY
ZWG 503.057	01 53 54.	+ 32 49		15.5	GALAXY
ZWG 522.082	01 53 54.	+ 35 45		15.3	GALAXY
ZWG 538.003	01 53 54.	+ 40 06		14.5	GALAXY
UGC 01418	01 53 54.	+ 40 06	114	14.5	GALAXY S0
VB 165	01 53 54.	+ 60 59	356		STELLAR RING
MCG-02-06-005	01 53 54.	- 09 18	192	13.	GALAXY
PHL 8096	01 53 54.	- 22 46		18.5	BLUE STELLAR OBJECT
LB 03235	01 53 54.	- 68 09		14.4	FAINT BLUE STAR
LB 03676	01 53 55.	+ 37 24 00.		18.8	FAINT BLUE STAR
LB 03675	01 53 55.	+ 37 46 30.		18.0	FAINT BLUE STAR
LB 03674	01 53 55.	+ 37 49 12.		18.7	FAINT BLUE STAR
LB 03678	01 53 56.	+ 37 17 12.		19.0	FAINT BLUE STAR
LB 03677	01 53 56.	+ 37 40 54.		18.4	FAINT BLUE STAR
RNGC 0738	01 53 57.	+ 32 49		15.5	GALAXY
LB 03679	01 53 57.	+ 37 57 30.		19.3	FAINT BLUE STAR
PK130+01.1	01 53 57.7	+ 63 04 42.	13	13.6	PLANETARY NEBULA
RNGC 0757	01 53 58.	- 09 09			NON-EXISTENT OBJECT
ZC 0154.0+0048	01 54 00.	+ 00 48	1010		CLUSTER OF GALAXIES
MCG+01-06-009	01 54 00.	+ 05 32	12	15.	GALAXY
UGC 01419	01 54 00.	+ 10 03	72	17.	GALAXY
ZWG 438.004	01 54 00.	+ 14 46		14.9	GALAXY
UGC 01420	01 54 00.	+ 14 46	60	14.9	GALAXY S
MCG+02-06-003	01 54 00.	+ 14 48	48	14.5	GALAXY
6ZW 115	01 54 00.	+ 32 45			COMPACT GALAXY
ZWG 503.058	01 54 00.	+ 32 46		14.9	GALAXY
UGC 01421	01 54 00.	+ 32 46	102	14.9	GALAXY S
MCG+05-05-030	01 54 00.	+ 33 00	24	15.5	GALAXY
ZWG 503.059	01 54 00.	+ 33 01		14.8	GALAXY
MCG+06-05-063	01 54 00.	+ 35 41 30.	36	15.5	GALAXY
ZWG 522.083	01 54 00.	+ 35 43		15.7	GALAXY
VB 063	01 54 00.	+ 65 27	410		STELLAR RING

OBJECT NAME	RIGHT ASCEN.	DECLINATION	DIAM.	MAGN.	TYPE OF OBJECT
ZWG 386.004	01 54 00.	- 00 25		15.5	GALAXY
ZC 0154.0-0205	01 54 00.	- 02 05	1140		CLUSTER OF GALAXIES
MCG-04-05-019	01 54 00.	- 23 10	120	14.5	GALAXY
PHL 3972	01 54 00.	- 28 34		18.4	BLUE STELLAR OBJECT
ARC 0280	01 54 02.	- 02 02		17.4	RICH CLUSTER OF GALAXIES
RNGC 0740	01 54 03.	+ 32 46		15.0	GALAXY
RNGC 0739	01 54 03.	+ 33 00		15.0	GALAXY
LB 03680	01 54 03.	+ 37 06 06.		18.7	FAINT BLUE STAR
VV 175A	01 54 06.	+ 05 23	42	13.	INTERACTING GALAXY
ZWG 413.010	01 54 06.	+ 05 34		15.4	GALAXY
ZC 0154.1+1120	01 54 06.	+ 11 20	1410		CLUSTER OF GALAXIES
ZWG 461.008	01 54 06.	+ 19 21		15.7	GALAXY
MCG+05-05-031	01 54 06.	+ 32 45	60	14.	GALAXY
LB 03682	01 54 06.	+ 37 25 54.		19.7	FAINT BLUE STAR
LB 03681	01 54 06.	+ 37 49 42.		19.5	FAINT BLUE STAR
MCG+07-05-002	01 54 06.	+ 40 04 30.	60	14.5	GALAXY
ARC 0276	01 54 07.	+ 41 07		16.3	RICH CLUSTER OF GALAXIES
LB 03683	01 54 08.	+ 37 54 24.		20.4	FAINT BLUE STAR
BIGO 470	01 54 11.	+ 32 49			NEBULA
LIN 577	01 54 11.	- 74 49 37.		17.11	SMC EMISSION LINE OBJECT
ZC 0154.2+0033	01 54 12.	+ 00 33	870		CLUSTER OF GALAXIES
ZWG 438.005	01 54 12.	+ 14 38		15.6	GALAXY
MCG+03-06-003	01 54 12.	+ 19 20	30	16.	GALAXY
ZWG 503.060	01 54 12.	+ 32 33		14.3	GALAXY
UGC 01422	01 54 12.	+ 32 33	78	14.3	GALAXY
6ZW 116	01 54 12.	+ 33 00			COMPACT GALAXY
UGC 01423	01 54 12.	+ 74 52	78	17.	GALAXY S
MCG+00-06-003	01 54 12.	- 00 28	30	16.	GALAXY
MCG-03-06-001	01 54 12.	- 20 20	78	15.	GALAXY
LB 03684	01 54 13.	+ 37 48 30.		20.2	FAINT BLUE STAR
IC 1754	01 54 14.	+ 03 46 57.			NONSTELLAR OBJECT
MCG+01-06-005	01 54 15.	- 03 59	78	14.5	GALAXY
LB 03685	01 54 16.	+ 37 52 00.		18.6	FAINT BLUE STAR
ZWG 413.011	01 54 18.	+ 03 47		15.4	GALAXY
UGC 01424	01 54 18.	+ 03 47	66	15.4	GALAXY COMPACT
ZWG 413.012	01 54 18.	+ 05 32		14.6	GALAXY
UGC 01425	01 54 18.	+ 05 32	102	14.6	GALAXY
ZC 0154.3+1043	01 54 18.	+ 10 43	1210		CLUSTER OF GALAXIES
MCG+05-05-032	01 54 18.	+ 28 20	15	17.	GALAXY
5ZW 147	01 54 18.	+ 31 46			COMPACT GALAXY
ZC 0154.3+3201	01 54 18.	+ 32 01	2960		CLUSTER OF GALAXIES
6ZW 117	01 54 18.	+ 32 18			COMPACT GALAXY
5ZW 148	01 54 18.	+ 54 43			COMPACT GALAXY
ISS 0015	01 54 18.	+ 63 41	524		STELLAR RING
VB 064	01 54 18.	+ 65 13	477		STELLAR RING
LB 03686	01 54 19.	+ 37 51 24.		18.8	FAINT BLUE STAR
MCG+01-06-011	01 54 21.	+ 05 30	24	16.	GALAXY
MCG+01-06-010	01 54 21.	+ 05 30	18	15.	GALAXY
MCG+00-06-004	01 54 21.	- 02 16 30.	66	14.	GALAXY
LB 03687	01 54 22.	+ 37 27 30.		20.2	FAINT BLUE STAR
IC 0176	01 54 22.	- 02 15 33.			NONSTELLAR OBJECT
ARC 0278	01 54 23.	+ 31 59		15.6	RICH CLUSTER OF GALAXIES
LB 03688	01 54 23.	+ 37 11 54.		17.9	FAINT BLUE STAR
MCG+01-06-012	01 54 24.	+ 05 46	36	15.	GALAXY
ZWG 413.013	01 54 24.	+ 05 48		14.7	GALAXY
UGC 01427	01 54 24.	+ 05 48	66	14.7	GALAXY S0
MCG+02-06-004	01 54 24.	+ 11 18	36	15.5	GALAXY
ZWG 461.009	01 54 24.	+ 17 32		15.2	GALAXY
MCG+05-05-033	01 54 24.	+ 28 18	18	16.	GALAXY
5ZW 149	01 54 24.	+ 28 20			COMPACT GALAXY
ZWG 503.061	01 54 24.	+ 28 20		15.0	GALAXY
KARA.73B 0074	01 54 24.	+ 28 20	36	15.0	ISOLATED GALAXY E
6ZW 118	01 54 24.	+ 28 23			COMPACT GALAXY
6ZW 119	01 54 24.	+ 29 35			COMPACT GALAXY
5ZW 150	01 54 24.	+ 31 45			COMPACT GALAXY
6ZW 120	01 54 24.	+ 32 12			COMPACT GALAXY
6ZW 121	01 54 24.	+ 32 26			COMPACT GALAXY
ZWG 387.002	01 54 24.	- 00 53		15.5	GALAXY
ZWG 387.001	01 54 24.	- 02 16		15.0	GALAXY
UGC 01426	01 54 24.	- 02 16	144	15.0	GALAXY Sc
PHL 3973	01 54 24.	- 26 21		18.6	BLUE STELLAR OBJECT
IC 1752	01 54 26.	+ 28 22 39.			NONSTELLAR OBJECT
IC 1755	01 54 28.	+ 14 17 33.			NONSTELLAR OBJECT
IC 1758	01 54 28.	- 16 46 58.			NONSTELLAR OBJECT
IC 1756	01 54 29.	- 00 43 04.			NONSTELLAR OBJECT
ARC 0282	01 54 29.	- 10 22		17.8	RICH CLUSTER OF GALAXIES
MCG+01-06-013	01 54 30.	+ 04 17	36	14.5	GALAXY
VV 175B	01 54 30.	+ 05 23	18	15.	INTERACTING GALAXY
ZWG 438.006	01 54 30.	+ 11 17		15.7	GALAXY
KARA.73B 0075	01 54 30.	+ 11 17	42	15.7	ISOLATED GALAXY S
ZWG 438.007	01 54 30.	+ 14 19		14.8	GALAXY
UGC 01428	01 54 30.	+ 14 19	96	14.8	GALAXY Sa
MCG+02-06-005	01 54 30.	+ 14 20	72	14.	GALAXY
ZWG 438.008	01 54 30.	+ 14 30		15.7	GALAXY
6ZW 122	01 54 30.	+ 28 21			COMPACT GALAXY
IC 1753	01 54 30.	+ 28 21 15.			NONSTELLAR OBJECT
5ZW 151	01 54 30.	+ 29 13			COMPACT GALAXY
6ZW 123	01 54 30.	+ 32 58			COMPACT GALAXY
LDN 1347	01 54 30.	+ 62 05	840		DARK NEBULA
ZWG 387.003	01 54 30.	- 00 20		15.4	GALAXY
MCG+00-06-005	01 54 30.	- 00 42 30.	72	14.5	GALAXY
MCG+01-06-006	01 54 30.	- 05 40 30.	66	13.5	GALAXY
ARP 056	01 54 31.	+ 16 58			PECULIAR GALAXY
LB 03689	01 54 32.	+ 37 15 06.		20.2	FAINT BLUE STAR
IC 0177	01 54 33.	- 00 23 34.			NONSTELLAR OBJECT
IC 1757	01 54 33.	- 00 43 28.			NONSTELLAR OBJECT
RNGC 0762	01 54 33.	- 05 40		13.5	GALAXY
ARC 0281	01 54 33.	- 06 06		17.0	RICH CLUSTER OF GALAXIES
ARC 0283	01 54 33.	- 22 18		17.1	RICH CLUSTER OF GALAXIES
ZWG 413.014	01 54 36.	+ 04 19		14.6	GALAXY
5ZW 152	01 54 36.	+ 28 44			COMPACT GALAXY
MCG+05-05-035	01 54 36.	+ 32 56	120	13.5	GALAXY
KARA.72 046B	01 54 36.	+ 32 57	78	13.9	PART OF DOUBLE GALAXY
ZWG 503.062	01 54 36.	+ 32 58		12.9	GALAXY
UGC 01431	01 54 36.	+ 32 58	150		GALAXY E+E
UGC 01430	01 54 36.	+ 32 58	150		GALAXY E+E
VV 189B	01 54 36.	+ 32 58	12	14.1	INTERACTING GALAXY
VV 189A	01 54 36.	+ 32 58	15	13.7	INTERACTING GALAXY
VV 189	01 54 36.	+ 32 58	210	12.9	INTERACTING GALAXY
KARA.72 046A	01 54 36.	+ 32 58	84	13.5	PART OF DOUBLE GALAXY
MCG+05-05-034	01 54 36.	+ 32 58	120	13.5	GALAXY
MCG+06-05-064	01 54 36.	+ 35 58	66	15.	GALAXY
LB 03690	01 54 36.	+ 37 34 48.		19.1	FAINT BLUE STAR
5ZW 153	01 54 36.	+ 57 11			COMPACT GALAXY
VB 065	01 54 36.	+ 63 37	537		STELLAR RING
ZWG 387.004	01 54 36.	- 00 43		15.5	GALAXY
UGC 01429	01 54 36.	- 00 43	90	15.5	GALAXY Sc
PHL 3974	01 54 36.	- 21 16		18.5	BLUE STELLAR OBJECT
MCG-04-05-020	01 54 36.	- 25 46	24	17.	GALAXY
ARP 166	01 54 38.	+ 32 58			PECULIAR GALAXY
RNGC 0751	01 54 39.	+ 32 58		14.0	GALAXY
RNGC 0750	01 54 39.	+ 32 58		13.0	GALAXY
LB 03691	01 54 39.	+ 37 41 12.		18.9	FAINT BLUE STAR
BIGO 471	01 54 39.	+ 44 33			NEBULA
MCG-04-05-021	01 54 39.	- 25 30	36	16.	GALAXY
LB 03692	01 54 40.	+ 37 53 06.		18.2	FAINT BLUE STAR
ZWG 461.010	01 54 42.	+ 16 58		14.8	GALAXY
UGC 01432	01 54 42.	+ 16 58	60	14.8	GALAXY Sb-c
KARA.73B 0076	01 54 42.	+ 16 58	60	14.8	ISOLATED GALAXY S
ZWG 461.011	01 54 42.	+ 19 41		15.7	GALAXY
UGC 01433	01 54 42.	+ 19 41	66	15.7	GALAXY Sc
ZWG 522.084	01 54 42.	+ 36 00		15.4	GALAXY
UGC 01434	01 54 42.	+ 36 00	90	15.4	GALAXY S0-a
ZWG 522.085	01 54 42.	+ 37 12		15.6	GALAXY
LB 03693	01 54 42.	+ 37 34 30.		18.0	FAINT BLUE STAR
MCG-02-06-006	01 54 42.	- 12 01	150	14.	GALAXY
YC 0154-44	01 54 42.	- 44 13 00.			UNUSUAL SOUTHERN GALAXY
LB 03694	01 54 43.	+ 37 09 36.		20.6	FAINT BLUE STAR
LB 03695	01 54 44.	+ 37 30 42.		18.7	FAINT BLUE STAR
RNGC 0746	01 54 45.	+ 44 41		14.0	GALAXY
SN 1954E	01 54 46.	+ 35 40		18.5	SUPERNOVA
LB 03696	01 54 46.	+ 37 16 06.		15.8	FAINT BLUE STAR
MCG+01-06-015	01 54 48.	+ 04 04	36	15.	GALAXY
ZWG 413.015	01 54 48.	+ 04 06		15.7	GALAXY
MCG+01-06-014	01 54 48.	+ 05 20	48	14.5	GALAXY
ZWG 413.016	01 54 48.	+ 05 22		15.0	GALAXY
UGC 01435	01 54 48.	+ 05 22	84	15.0	GALAXY S
ZWG 461.012	01 54 48.	+ 16 32		14.7	GALAXY
KARA.73B 0077	01 54 48.	+ 16 32	48	14.7	ISOLATED GALAXY S
VV 012C	01 54 48.	+ 16 58	18	17.	INTERACTING GALAXY
VV 012B	01 54 48.	+ 16 58	12	17.	INTERACTING GALAXY
VV 012A	01 54 48.	+ 16 58	54	15.	INTERACTING GALAXY
MCG+03-06-004	01 54 48.	+ 16 58	48	14.	GALAXY
UGC 01436	01 54 48.	+ 19 39	66	16.0	GALAXY Sc
MCG+03-06-005	01 54 48.	+ 19 40	24	15.	GALAXY
5ZW 154	01 54 48.	+ 31 13			COMPACT GALAXY
ZWG 522.086	01 54 48.	+ 35 40		12.6	GALAXY
UGC 01437	01 54 48.	+ 35 40	198	12.6	GALAXY Sc
MCG+06-05-066	01 54 48.	+ 35 40	150	13.	GALAXY
MCG+06-05-065	01 54 48.	+ 37 07	66	15.	GALAXY
OCL 0363	01 54 48.	+ 37 26	3480	7.0	OPEN STAR CLUSTER
LB 03697	01 54 48.	+ 37 36 00.		18.3	FAINT BLUE STAR
MCG+07-05-003	01 54 48.	+ 44 40	96	14.	GALAXY
ZWG 538.004	01 54 48.	+ 44 41		13.8	GALAXY
UGC 01438	01 54 48.	+ 44 41	132	13.8	GALAXY IRR
ACK 129+04.1	01 54 48.	+ 66 19			PLANETARY NEBULA
RNGC 0764	01 54 48.	- 16 16			NON-EXISTENT OBJECT
PHL 1248	01 54 48.	- 24 06		16.6	BLUE STELLAR OBJECT
MCG-04-05-022	01 54 48.	- 25 31	15	16.	GALAXY
LB 03698	01 54 49.	+ 37 17 00.		14.6	FAINT BLUE STAR
RNGC 0752	01 54 49.	+ 37 26		6.5	OPEN CLUSTER
RNGC 0753	01 54 50.	+ 35 40		13.0	GALAXY
LB 03700	01 54 50.	+ 37 08 00.		18.8	FAINT BLUE STAR
LB 03699	01 54 50.	+ 37 14 54.		16.6	FAINT BLUE STAR
RNGC 0760	01 54 51.	+ 33 07			NON-EXISTENT OBJECT
MCG+06-05-067	01 54 51.	+ 36 06	30	14.	GALAXY
LB 03702	01 54 51.	+ 37 27 00.		17.0	FAINT BLUE STAR
LB 03701	01 54 51.	+ 37 28 12.		16.5	FAINT BLUE STAR
LB 03705	01 54 52.	+ 37 12 06.		17.8	FAINT BLUE STAR
LB 03704	01 54 52.	+ 37 30 36.		16.3	FAINT BLUE STAR
LB 03703	01 54 52.	+ 37 43 42.		16.9	FAINT BLUE STAR
5ZW 155	01 54 54.	+ 27 37			COMPACT GALAXY
ZWG 503.063	01 54 54.	+ 27 37		15.2	GALAXY
MRK 364	01 54 54.	+ 27 37	18	15.5	GALAXY WITH UV CONTINUUM
6ZW 124	01 54 54.	+ 30 33			COMPACT GALAXY
MCG+05-05-036	01 54 54.	+ 33 07	66	14.5	GALAXY
ZWG 503.064	01 54 54.	+ 33 08		14.5	GALAXY
UGC 01439	01 54 54.	+ 33 08	114	14.5	GALAXY SB0-a
ZWG 522.087	01 54 54.	+ 36 05		13.7	GALAXY
UGC 01440	01 54 54.	+ 36 05	108	13.7	GALAXY E
ZWG 522.088	01 54 54.	+ 37 07		15.5	GALAXY
UGC 01441	01 54 54.	+ 37 07	66	15.5	GALAXY Sb
LB 03707	01 54 54.	+ 37 40 12.		19.3	FAINT BLUE STAR
LB 03706	01 54 54.	+ 37 46 48.		16.9	FAINT BLUE STAR
RNGC 0759	01 54 55.	+ 36 06		13.5	GALAXY
RNGC 0761	01 54 57.	+ 33 08		14.5	GALAXY
LIN 578	01 54 57.	- 74 13 09.		15.01	SMC EMISSION LINE OBJECT
RNGC 0763	01 54 58.	- 09 13			NON-EXISTENT OBJECT
LB 03709	01 54 59.	+ 37 17 54.		17.6	FAINT BLUE STAR
LB 03708	01 54 59.	+ 37 43 48.		16.3	FAINT BLUE STAR
LB 09779	01 55	- 82 51		14.2	FAINT BLUE STAR
ZWG 387.006	01 55 00.	+ 03 28		15.7	GALAXY
UGC 01443	01 55 00.	+ 03 29	72	16.5	GALAXY S
MCG+01-06-016	01 55 00.	+ 03 44	60	16.5	GALAXY
ZC 0155.0+1257	01 55 00.	+ 12 57	5240		CLUSTER OF GALAXIES
MCG+03-06-008	01 55 00.	+ 16 31	48	14.5	GALAXY
MCG+03-06-007	01 55 00.	+ 19 37	30	16.	GALAXY
MCG+03-06-006	01 55 00.	+ 19 38	18	18.	GALAXY
6ZW 125	01 55 00.	+ 27 38			COMPACT GALAXY
5ZW 156	01 55 00.	+ 35 39			COMPACT GALAXY
MCG+06-05-068	01 55 00.	+ 36 40	12	16.	GALAXY
ZWG 522.089	01 55 00.	+ 36 41		15.7	GALAXY
LB 03710	01 55 00.	+ 37 12 48.		16.3	FAINT BLUE STAR
LDN 1348	01 55 00.	+ 62 10	840		DARK NEBULA
ZWG 387.005	01 55 00.	- 02 20		15.7	GALAXY
UGC 01442	01 55 00.	- 02 20	96	15.7	GALAXY
MCG+00-06-006	01 55 00.	- 02 21	54	15.	GALAXY
MCG-03-06-002	01 55 00.	- 15 31	24	15.	GALAXY
LB 03711	01 55 01.	+ 37 49 00.		19.2	FAINT BLUE STAR
LB 03712	01 55 02.	+ 37 11 06.		17.3	FAINT BLUE STAR
LIN 580	01 55 03.	- 75 01 33.		15.47	SMC EMISSION LINE OBJECT
LIN 579	01 55 04.	- 74 08 09.		14.97	SMC EMISSION LINE OBJECT
LB 03714	01 55 05.	+ 37 12 42.		17.4	FAINT BLUE STAR
LB 03713	01 55 05.	+ 37 23 00.		19.2	FAINT BLUE STAR
FEIG 017	01 55 06.	+ 06 57		15.0	FAINT BLUE STAR
ZWG 413.017	01 55 06.	+ 08 45		15.6	GALAXY
ZWG 522.090	01 55 06.	+ 34 03		15.7	GALAXY
6ZW 126	01 55 06.	+ 36 03			COMPACT GALAXY
5ZW 157	01 55 06.	+ 37 20			COMPACT GALAXY
ZWG 522.091	01 55 06.	+ 37 20		15.3	GALAXY
ZWG 538.005	01 55 06.	+ 41 43		15.6	GALAXY
OCL 0345	01 55 06.	+ 55 14	1080	9.4	OPEN STAR CLUSTER
LB 00168	01 55 06.	- 14 30		16.0	FAINT BLUE STAR
LB 03715	01 55 07.	+ 37 57 18.		18.0	FAINT BLUE STAR
LB 03716	01 55 08.	+ 37 18 24.		17.6	FAINT BLUE STAR
RNGC 0743	01 55 09.	+ 59 56			OPEN CLUSTER

OBJECT NAME	RIGHT ASCEN.	DECLINATION	DIAM.	MAGN.	TYPE OF OBJECT
MCG-03-06-003	01 55 09.	- 15 16	24	15.	GALAXY
LB 03717	01 55 10.	+ 37 24 12.		18.7	FAINT BLUE STAR
LB 03719	01 55 11.	+ 37 22 06.		16.0	FAINT BLUE STAR
LB 03718	01 55 11.	+ 37 43 24.		18.1	FAINT BLUE STAR
ZC 0155.2+4051	01 55 12.	+ 40 51	4910		CLUSTER OF GALAXIES
ZWG 538.006	01 55 12.	+ 42 02		15.6	GALAXY
OCL 0343	01 55 12.	+ 59 56	318	9.5	OPEN STAR CLUSTER
ZWG 387.007	01 55 12.	- 01 26		15.6	GALAXY
SHAH 144	01 55 12.	- 02 05	132		GROUP OF COMPACT GALAXIES
MCG+01-06-007	01 55 12.	- 05 56	90	13.5	GALAXY
PHL 8097	01 55 12.	- 23 00		17.3	BLUE STELLAR OBJECT
RNGC 0744	01 55 14.	+ 55 14		9.0	OPEN CLUSTER
BC PKS0155-10	01 55 14.06	- 10 58 16.6		17.09	QUASI-STELLAR OBJECT
SHB 051	01 55 14.1	- 10 58 17.		17.1	QUASI-STELLAR OBJECT
MCG+00-06-007	01 55 15.	- 01 31	12	15.5	GALAXY
MCG-02-06-007	01 55 15.	- 09 42	60	14.	GALAXY
LB 03720	01 55 16.	+ 37 34 12.		19.2	FAINT BLUE STAR
IC 1759	01 55 17.	- 33 16 50.			NONSTELLAR OBJECT
LB 03722	01 55 18.	+ 37 16 30.		17.3	FAINT BLUE STAR
LB 03721	01 55 18.	+ 37 40 24.		19.3	FAINT BLUE STAR
VB 158	01 55 18.	+ 62 22	316		STELLAR RING
ZWG 387.008	01 55 18.	- 01 30		15.7	GALAXY
PHL 3975	01 55 18.	- 22 28		18.3	BLUE STELLAR OBJECT
PHL 3976	01 55 18.	- 26 54		18.3	BLUE STELLAR OBJECT
PHL 8098	01 55 18.	- 28 04		17.9	BLUE STELLAR OBJECT
IC 1760	01 55 18.	- 32 15 02.			NONSTELLAR OBJECT
LB 03723	01 55 19.	+ 37 19 00.		16.4	FAINT BLUE STAR
HOLM 052B	01 55 19.	- 00 54	30	15.3	PART OF MULTIPLE GALAXY
MCG+00-06-008	01 55 21.	+ 03 08 30.	66	13.5	GALAXY
LB 03724	01 55 21.	+ 37 05 12.		19.2	FAINT BLUE STAR
ARC 0284	01 55 21.	- 00 53		17.5	RICH CLUSTER OF GALAXIES
HOLM 052A	01 55 21.	- 00 54	18	14.8	PART OF MULTIPLE GALAXY
LB 03726	01 55 22.	+ 37 07 30.		18.5	FAINT BLUE STAR
LB 03725	01 55 22.	+ 37 10 30.		17.1	FAINT BLUE STAR
LB 03727	01 55 23.	+ 37 58 54.		20.3	FAINT BLUE STAR
VV 122C	01 55 24.	+ 02 51 30.	6	18.	INTERACTING GALAXY
VV 122B	01 55 24.	+ 02 51 30.	30	16.	INTERACTING GALAXY
VV 122A	01 55 24.	+ 02 51 30.	30	14.5	INTERACTING GALAXY
VV 122	01 55 24.	+ 02 51 30.	90	15.	INTERACTING GALAXY
MCG+01-06-017	01 55 24.	+ 04 07	108	15.	GALAXY
ZWG 413.018	01 55 24.	+ 04 09		15.6	GALAXY
UGC 01444	01 55 24.	+ 04 09	126	15.6	GALAXY S?
ZWG 461.013	01 55 24.	+ 18 52		15.1	GALAXY
UGC 01445	01 55 24.	+ 18 52	72	15.1	GALAXY S0
ZWG 461.014	01 55 24.	+ 21 06		14.8	GALAXY
LB 03728	01 55 24.	+ 37 27 00.		17.8	FAINT BLUE STAR
ARC 0285	01 55 24.	- 03 59		17.2	RICH CLUSTER OF GALAXIES
MCG-06-05-015	01 55 24.	- 33 2	90	13.	GALAXY
LB 00453	01 55 25.	- 04 03 00.		17.5	FAINT BLUE STAR
MCG+00-06-009	01 55 27.	+ 02 51 30.	42	15.	GALAXY
MCG+00-06-010	01 55 30.	+ 02 51 30.	24	14.5	GALAXY
ZWG 387.009	01 55 30.	+ 03 08		14.9	GALAXY
UGC 01446	01 55 30.	+ 03 08	90	14.9	GALAXY IRR+CMP
ZWG 538.007	01 55 30.	+ 44 20		15.1	GALAXY
UGC 01447	01 55 30.	+ 44 20	72	15.1	GALAXY Sc
MCG-02-06-008	01 55 30.	- 08 47	30	16.	GALAXY
LIN 581	01 55 31.	- 74 23 16.		15.29	SMC EMISSION LINE OBJECT
LB 03729	01 55 32.	+ 37 42 06.		20.3	FAINT BLUE STAR
LIN.CL 116	01 55 32.	- 77 53 58.	138		STAR CLUSTER IN SMC
PK133-08.1	01 55 32.9	+ 52 39 15.	1		PLANETARY NEBULA
MCG+00-06-011	01 55 33.	+ 01 50 30.	72	15.	GALAXY
MCG+07-05-004	01 55 33.	+ 44 19	60	15.	GALAXY
ARP 126	01 55 34.	+ 02 51			PECULIAR GALAXY
LB 03732	01 55 34.	+ 37 11 00.		18.5	FAINT BLUE STAR
LB 03731	01 55 34.	+ 37 14 24.		17.8	FAINT BLUE STAR
LB 03730	01 55 34.	+ 37 57 48.		17.8	FAINT BLUE STAR
LB 03733	01 55 35.	+ 37 33 36.		19.6	FAINT BLUE STAR
ZWG 387.010	01 55 36.	+ 01 50		15.1	GALAXY
UGC 01448	01 55 36.	+ 01 50	78	15.1	GALAXY Sc
KARA.73B 0078	01 55 36.	+ 01 50	90	15.1	ISOLATED GALAXY S
ZWG 387.011	01 55 36.	+ 02 50		14.0	GALAXY
MRK 582	01 55 36.	+ 02 50	14	15.5	GALAXY WITH UV CONTINUUM
UGC 01449	01 55 36.	+ 02 50	132	14.0	GALAXY IRR
KARA.72 047B	01 55 36.	+ 02 50	30		PART OF DOUBLE GALAXY
KARA.72 047A	01 55 36.	+ 02 50	36	14.0	PART OF DOUBLE GALAXY
ZWG 438.009	01 55 36.	+ 14 03		15.6	GALAXY
UGC 01450	01 55 36.	+ 14 03	78	15.6	GALAXY
MCG+02-06-006	01 55 36.	+ 14 04	18	15.5	GALAXY
MCG+03-06-009	01 55 36.	+ 18 51	42	15.	GALAXY
ZWG 461.015	01 55 36.	+ 20 44		15.0	GALAXY
KARA.73B 0079	01 55 36.	+ 20 44	24	15.0	ISOLATED GALAXY S0
MCG+04-05-024	01 55 36.	+ 25 06	72	15.	GALAXY
ZWG 482.032	01 55 36.	+ 25 07		14.3	GALAXY
UGC 01451	01 55 36.	+ 25 07	84	14.3	GALAXY SBb?
ZWG 503.065	01 55 36.	+ 31 50		15.6	GALAXY
ZWG 522.092	01 55 36.	+ 36 15		15.7	GALAXY
5ZW 158	01 55 36.	+ 36 32			COMPACT GALAXY
LB 03734	01 55 36.	+ 37 36 42.		18.4	FAINT BLUE STAR
ZWG 522.093	01 55 36.	+ 38 29		15.3	GALAXY
MCG+06-05-069	01 55 36.	+ 38 29 30.	36	15.	GALAXY
ZC 0155.6-0055	01 55 36.	- 00 55	1080		CLUSTER OF GALAXIES
MCG-06-05-016	01 55 36.	- 33 13	60	13.	GALAXY
LB 03735	01 55 37.	+ 37 39 42.		20.0	FAINT BLUE STAR
UGC 01452	01 55 42.	+ 21 54	84	17.	GALAXY
5ZW 159	01 55 42.	+ 37 28			COMPACT GALAXY
LB 01607	01 55 42.	- 45 46		14.4	FAINT BLUE STAR
LB 03736	01 55 44.	+ 37 59 36.		17.6	FAINT BLUE STAR
IC 1762	01 55 47.	- 33 31 04.			NONSTELLAR OBJECT
MCG-06-05-017	01 55 48.	- 35 48	42	14.5	GALAXY
MCG-06-05-018	01 55 48.	- 38 17	60	15.	GALAXY
LB 03737	01 55 49.	+ 37 19 42.		19.5	FAINT BLUE STAR
LB 03738	01 55 52.	+ 37 10 12.		18.7	FAINT BLUE STAR
MCG+01-06-018	01 55 54.	+ 09 02	48	15.	GALAXY
UGC 01453	01 55 54.	+ 24 25	102	17.	GALAXY DWARF
MCG+06-05-070	01 55 54.	+ 36 27	66	14.	GALAXY
VB 166	01 55 54.	+ 60 25	295		STELLAR RING
MCG+00-06-012	01 55 54.	- 01 42	30	15.5	GALAXY
PHL 8099	01 55 54.	- 30 30		17.2	BLUE STELLAR OBJECT
LB 03739	01 55 55.	+ 37 45 24.		19.3	FAINT BLUE STAR
ARC 0286	01 55 56.	- 02 01		17.2	RICH CLUSTER OF GALAXIES
MCG+00-06-014	01 55 57.	+ 03 01 30.	78	15.	GALAXY
MCG+00-06-013	01 55 57.	+ 03 12	48	15.5	GALAXY
MCG+01-06-008	01 55 57.	- 07 25 30.	60	15.5	GALAXY
MCG-02-06-009	01 55 57.	- 10 51 30.	36	15.	GALAXY
IC 0178	01 55 58.	+ 36 22 04.			NONSTELLAR OBJECT
LB 03740	01 55 58.	+ 37 06 48.		16.9	FAINT BLUE STAR
LB 03741	01 55 59.	+ 37 44 00.		17.6	FAINT BLUE STAR
LB 09780	01 56	- 81 40		13.5	FAINT BLUE STAR

OBJECT NAME	RIGHT ASCEN.	DECLINATION	DIAM.	MAGN.	TYPE OF OBJECT
UGC 01454	01 56 00.	+ 03 01	114	16.5	GALAXY
MCG+01-06-020	01 56 00.	+ 08 02	30	15.	GALAXY
MCG+01-06-019	01 56 00.	+ 08 06	24	14.	GALAXY
ZWG 482.033	01 56 00.	+ 24 39		14.2	GALAXY
RNGC 0765	01 56 00.	+ 24 39		14.0	GALAXY
UGC 01455	01 56 00.	+ 24 39	180	14.2	GALAXY SBb/Sc
MCG+04-05-025	01 56 00.	+ 24 39	150	14.	GALAXY
ZWG 522.094	01 56 00.	+ 36 26		14.0	GALAXY
UGC 01456	01 56 00.	+ 36 26	84	14.0	GALAXY Sa-b
ZWG 522.095	01 56 00.	+ 37 30		15.6	GALAXY
ZWG 538.008	01 56 00.	+ 41 25		15.7	GALAXY
VB 159	01 56 00.	+ 59 04	154		STELLAR RING
ISS 0056	01 56 00.	+ 59 04	160		STELLAR RING
VB 160	01 56 00.	+ 60 26	295		STELLAR RING
ZWG 387.012	01 56 00.	- 01 42		15.6	GALAXY
KARA.72 048A	01 56 00.	- 01 42	42	15.6	PART OF DOUBLE GALAXY
ZC 0156.0-0202	01 56 00.	- 02 02	940		CLUSTER OF GALAXIES
MCG+01-06-009	01 56 00.	- 03 01	36	16.	GALAXY
PHL 1249	01 56 00.	- 25 28		17.7	BLUE STELLAR OBJECT
LB 03743	01 56 03.	+ 37 09 06.		17.8	FAINT BLUE STAR
LB 03742	01 56 03.	+ 37 28 12.		18.1	FAINT BLUE STAR
MCG+00-06-015	01 56 03.	- 01 41 30.	27	15.	GALAXY
MCG+01-06-010	01 56 03.	- 03 00	30	16.	GALAXY
RNGC 0782	01 56 03.	- 58 01		13.0	GALAXY
RNGC 0766	01 56 05.	+ 08 06		14.5	GALAXY
RNGC 0796	01 56 05.	- 74 27			CLUSTER/NEBULOSITY IN SMC
ZWG 387.014	01 56 06.	+ 00 17		14.3	GALAXY
UGC 01457	01 56 06.	+ 00 17	126	14.3	GALAXY Sb
KARA.72 049A	01 56 06.	+ 00 17	84	14.3	PART OF DOUBLE GALAXY
MCG+00-06-016	01 56 06.	+ 00 18 30.	78	13.5	GALAXY
ZWG 413.019	01 56 06.	+ 08 06		14.4	GALAXY
UGC 01458	01 56 06.	+ 08 06	120	14.4	GALAXY E
ZWG 522.096	01 56 06.	+ 35 49		15.4	GALAXY
UGC 01459	01 56 06.	+ 35 49	96	15.4	GALAXY Sc
ZWG 522.097	01 56 06.	+ 36 01		15.0	GALAXY
UGC 01460	01 56 06.	+ 36 01	90	15.0	GALAXY Sa
MCG+06-05-071	01 56 06.	+ 36 01	90	14.	GALAXY
ZCG 0156+36.2	01 56 06.	+ 36 34		17.8	COMPACT GALAXY
ZWG 387.013	01 56 06.	- 01 42		14.9	GALAXY
KARA.72 048B	01 56 06.	- 01 42	42	14.9	PART OF DOUBLE GALAXY
PHL 3977	01 56 06.	- 21 32		18.1	BLUE STELLAR OBJECT
LIN.CL 115	01 56 06.	- 74 27 29.	204		STAR CLUSTER IN SMC
RNGC 0768	01 56 07.	+ 00 17		14.5	GALAXY
MCG+06-05-072	01 56 09.	+ 35 49	84	15.	GALAXY
LB 03744	01 56 11.	+ 37 34 18.		18.3	FAINT BLUE STAR
MCG+01-06-024	01 56 12.	+ 05 06	36	15.	GALAXY
MCG+01-06-023	01 56 12.	+ 05 09	36	15.	GALAXY
MCG+01-06-022	01 56 12.	+ 05 19	60	15.	GALAXY
UGC 01461	01 56 12.	+ 05 21	78	16.0	GALAXY Sc
MCG+01-06-021	01 56 12.	+ 08 06	24	16.5	GALAXY
6ZW 127	01 56 12.	+ 32 20			COMPACT GALAXY
ZCG 0156+36.3	01 56 12.	+ 36 34		18.0	COMPACT GALAXY
MCG+01-06-011	01 56 12.	- 07 58	42	15.	GALAXY
LB 03746	01 56 14.	+ 37 25 06.		18.6	FAINT BLUE STAR
LB 03745	01 56 14.	+ 37 40 00.		20.6	FAINT BLUE STAR
RNGC 0775	01 56 15.	- 26 32		14.0	GALAXY
LB 03747	01 56 17.	+ 37 08 18.		19.6	FAINT BLUE STAR
ZWG 387.015	01 56 18.	+ 00 19		15.5	GALAXY
KARA.72 049B	01 56 18.	+ 00 19	42	15.5	PART OF DOUBLE GALAXY
IC 1761	01 56 18.	+ 00 20 38.			NONSTELLAR OBJECT
ZWG 413.020	01 56 18.	+ 05 10		15.7	GALAXY
MCG+04-05-026	01 56 18.	+ 25 08	90	15.	GALAXY
ZWG 482.034	01 56 18.	+ 25 09		15.6	GALAXY
UGC 01462	01 56 18.	+ 25 09	102	15.6	GALAXY SBc
ZCG 0156+36.1	01 56 18.	+ 36 33		18.2	COMPACT GALAXY
5ZW 160	01 56 18.	+ 36 36			COMPACT GALAXY
FEIG 018	01 56 18.	- 05 03		9.3	FAINT BLUE STAR
MCG-05-05-024	01 56 18.	- 26 32	90	14.	GALAXY
MCG-06-05-019	01 56 18.	- 35 06	66	14.5	GALAXY
LIN 582	01 56 18.	- 74 30 17.		15.32	SMC EMISSION LINE OBJECT
MCG+01-06-025	01 56 24.	+ 06 16	30	15.	GALAXY
ZWG 413.021	01 56 24.	+ 06 17		15.4	GALAXY
ZWG 461.016	01 56 24.	+ 18 43		14.2	GALAXY
UGC 01463	01 56 24.	+ 18 43	66	14.2	GALAXY
ZWG 482.035	01 56 24.	+ 24 10		14.8	GALAXY
6ZW 128	01 56 24.	+ 36 02			COMPACT GALAXY
ZWG 522.098	01 56 24.	+ 36 35		15.0	GALAXY
MCG+06-05-073	01 56 24.	+ 36 35 30.	12	15.	GALAXY
OCL 0323	01 56 24.	+ 75 20	480	14.	OPEN STAR CLUSTER
PHL 3979	01 56 24.	- 25 04		18.6	BLUE STELLAR OBJECT
PHL 3978	01 56 24.	- 25 18		18.4	BLUE STELLAR OBJECT
RNGC 0767	01 56 28.	- 09 49		14.0	GALAXY
UGC 01464	01 56 30.	+ 01 40	108	17.	GALAXY DWRF IR
UGC 01465	01 56 30.	+ 17 47	60	17.	GALAXY SB?
ZWG 461.017	01 56 30.	+ 18 02		15.7	GALAXY
MCG+03-06-010	01 56 30.	+ 18 43	27	14.5	GALAXY
ZWG 461.018	01 56 30.	+ 18 46		11.3	GALAXY
UGC 01466	01 56 30.	+ 18 46	480	11.3	GALAXY Sb
KARA.73B 0080	01 56 30.	+ 18 46	468	11.3	ISOLATED GALAXY S
MCG-02-06-010	01 56 30.	- 09 49	72	14.5	GALAXY
LB 01608	01 56 30.	- 46 31		13.8	FAINT BLUE STAR
RNGC 0770	01 56 32.	+ 18 43		14.0	GALAXY
IC 1763	01 56 32.	- 28 01 54.			NONSTELLAR OBJECT
MCG+01-06-012	01 56 33.	- 08 25 30.	66	14.	GALAXY
RNGC 0773	01 56 35.	- 11 45		14.0	GALAXY
MRK 583	01 56 36.	+ 09 41	12	16.	GALAXY WITH UV CONTINUUM
ARP 078	01 56 36.	+ 18 46			PECULIAR GALAXY
ZWG 461.019	01 56 36.	+ 21 20		15.6	GALAXY
MCG+05-05-037	01 56 36.	+ 30 38	48	14.	GALAXY
ZWG 387.016	01 56 36.	- 00 08		15.0	GALAXY
MCG+01-06-013	01 56 36.	- 03 23 30.	42	14.5	GALAXY
MCG-02-06-011	01 56 36.	- 11 45	72	14.	GALAXY
SER 019.02	01 56 36.	- 56 32	120	15.	COMPACT GROUP OF GALAXIES
RNGC 0772	01 56 38.	+ 18 46		11.5	GALAXY
RNGC 0769	01 56 39.	+ 30 40		13.5	GALAXY
MCG+01-06-014	01 56 39.	- 03 24	24	16.	GALAXY
MCG+03-06-011	01 56 42.	+ 18 45	450	11.	GALAXY
ZWG 503.066	01 56 42.	+ 30 40		13.4	GALAXY
UGC 01467	01 56 42.	+ 30 40	60	13.4	GALAXY S
PHL 8100	01 56 42.	- 23 08		17.9	BLUE STELLAR OBJECT
LB 03749	01 56 44.	+ 37 06 12.		17.2	FAINT BLUE STAR
LB 03748	01 56 44.	+ 37 15 48.		19.7	FAINT BLUE STAR
RNGC 0774	01 56 45.	+ 13 47		14.5	GALAXY
LB 03750	01 56 47.	+ 37 47 36.		16.0	FAINT BLUE STAR
UGC 01468	01 56 48.	+ 13 42	102	16.0	GALAXY Sc
MCG+02-06-007	01 56 48.	+ 13 42 30.	84	15.5	GALAXY
MCG+04-05-027	01 56 48.	+ 27 11	48	15.	GALAXY
ZWG 482.036	01 56 48.	+ 27 12		15.2	GALAXY

OBJECT NAME	RIGHT ASCEN.	DECLINATION	DIAM.	MAGN.	TYPE OF OBJECT
LB 00169	01 56 48.	- 17 08		16.5	FAINT BLUE STAR
MCG+02-06-008	01 56 51.	+ 13 47	48	14.	GALAXY
ZWG 438.010	01 56 54.	+ 13 46		14.4	GALAXY
UGC 01469	01 56 54.	+ 13 46	126	14.4	GALAXY S0
UGC 01470	01 56 54.	+ 31 50	102	16.5	GALAXY
5ZW 161	01 56 54.	+ 39 17			COMPACT GALAXY
MCG-05-05-025	01 56 54.	- 28 03	36	15.	GALAXY
MCG+04-05-028	01 57 00.	+ 23 23	96	14.	GALAXY
ZC 0157.0+3315	01 57 00.	+ 33 15	810		CLUSTER OF GALAXIES
MCG+06-05-074	01 57 00.	+ 36 20 30.	30	15.5	GALAXY
ZWG 522.099	01 57 00.	+ 36 21		15.4	GALAXY
LB 00454	01 57 04.	- 03 15 06.		17.1	FAINT BLUE STAR
IC 0183	01 57 04.	- 05 35 20.			NONSTELLAR OBJECT
MCG+04-05-029	01 57 06.	+ 23 21	45	15.	GALAXY
ZWG 482.037	01 57 06.	+ 23 24		13.4	GALAXY
RNGC 0776	01 57 06.	+ 23 24		13.5	GALAXY
UGC 01471	01 57 06.	+ 23 24	114	13.4	GALAXY SBb/Sb
MCG+04-05-030	01 57 06.	+ 23 24	24	15.5	GALAXY
ZWG 482.038	01 57 06.	+ 24 04		15.6	GALAXY
6ZW 129	01 57 06.	+ 30 34			COMPACT GALAXY
UGC 01472	01 57 06.	+ 34 06	78	18.	GALAXY DWARF
MCG+01-06-015	01 57 06.	- 05 36 30.	60	15.	GALAXY
LB 00170	01 57 06.	- 13 25		16.8	FAINT BLUE STAR
LIN 583	01 57 07.	- 74 56 01.		15.69	SMC EMISSION LINE OBJECT
SCHO 0042	01 57 09.	+ 52 48 48.	530		ISOLATED DARK CLOUD
RNGC 0779	01 57 09.	- 06 12		12.0	GALAXY
IC 0179	01 57 11.	+ 37 47 30.			NONSTELLAR OBJECT
MCG+01-06-026	01 57 12.	+ 07 09	36	14.	GALAXY
ZWG 413.022	01 57 12.	+ 07 10		14.6	GALAXY
UGC 01473	01 57 12.	+ 07 10	120	14.6	GALAXY SBb
MCG+02-06-009	01 57 12.	+ 09 40	36	15.	GALAXY
ZWG 482.039	01 57 12.	+ 23 22		15.3	GALAXY
ZWG 482.040	01 57 12.	+ 23 25		15.7	GALAXY
MCG+04-05-031	01 57 12.	+ 23 30	42	15.	GALAXY
5ZW 162	01 57 12.	+ 23 31			COMPACT GALAXY
6ZW 130	01 57 12.	+ 30 38			COMPACT GALAXY
ZWG 522.100	01 57 12.	+ 37 21		15.0	GALAXY
UGC 01474	01 57 12.	+ 37 21	96	15.0	GALAXY SB III
MCG+06-05-076	01 57 12.	+ 37 22	60	14.	GALAXY
ZWG 522.101	01 57 12.	+ 37 47		13.4	GALAXY
UGC 01475	01 57 12.	+ 37 47	108	13.4	GALAXY E
MCG+06-05-075	01 57 12.	+ 37 47 30.	24	14.	GALAXY
ZWG 387.017	01 57 12.	- 00 31		15.6	GALAXY
PHL 8101	01 57 12.	- 22 52		18.4	BLUE STELLAR OBJECT
LB 01609	01 57 12.	- 47 56		14.4	FAINT BLUE STAR
REIN 2.021	01 57 12.39	- 06 12 22.4			NEBULA
IC 0182	01 57 13.	+ 07 09 04.			NONSTELLAR OBJECT
IC 0180	01 57 14.	+ 23 21 41.			NONSTELLAR OBJECT
IC 0181	01 57 16.	+ 23 24 59.			NONSTELLAR OBJECT
ZC 0157.3+1003	01 57 18.	+ 10 03	810		CLUSTER OF GALAXIES
ZWG 482.041	01 57 18.	+ 23 31		15.6	GALAXY
6ZW 131	01 57 18.	+ 30 38			COMPACT GALAXY
ZWG 503.067	01 57 18.	+ 31 10		12.7	GALAXY
UGC 01476	01 57 18.	+ 31 10	174	12.7	GALAXY E
MCG+05-05-038	01 57 18.	+ 31 12	42	13.	GALAXY
ZWG 503.068	01 57 18.	+ 31 48		15.4	GALAXY
6ZW 132	01 57 18.	+ 34 23			COMPACT GALAXY
MCG+01-06-016	01 57 18.	- 06 13	240	12.	GALAXY
HOLM 053B	01 57 20.	- 07 17	30	14.3	PART OF MULTIPLE GALAXY
RNGC 0778	01 57 21.	+ 31 03		14.0	GALAXY
RNGC 0777	01 57 21.	+ 31 11		13.0	GALAXY
MCG+01-06-017	01 57 21.	- 07 14	18	17.	GALAXY
IC 0184	01 57 22.	- 07 04 57.			NONSTELLAR OBJECT
HOLM 053A	01 57 23.	- 07 19	30	14.3	PART OF MULTIPLE GALAXY
ZWG 413.023	01 57 24.	+ 06 43		15.7	GALAXY
UGC 01477	01 57 24.	+ 06 43	72	15.7	GALAXY
ZWG 482.042	01 57 24.	+ 24 00		14.6	GALAXY
UGC 01478	01 57 24.	+ 24 00	60	14.6	GALAXY SBc
MCG+04-05-032	01 57 24.	+ 24 00	48	15.	GALAXY
ZWG 482.043	01 57 24.	+ 24 14		14.8	GALAXY
UGC 01479	01 57 24.	+ 24 14	84	14.8	GALAXY S
6ZW 133	01 57 24.	+ 30 35			COMPACT GALAXY
MCG+05-05-039	01 57 24.	+ 31 02	48	14.5	GALAXY
ZWG 503.069	01 57 24.	+ 31 03		14.2	GALAXY
UGC 01480	01 57 24.	+ 31 03	66	14.2	GALAXY S0
MCG+01-06-019	01 57 24.	- 07 16 30.	30	16.	GALAXY
MCG+01-06-018	01 57 24.	- 08 06	60	14.	GALAXY
PHL 8102	01 57 24.	- 24 41		18.2	BLUE STELLAR OBJECT
PHL 8103	01 57 24.	- 25 05		18.3	BLUE STELLAR OBJECT
MCG+04-05-034	01 57 27.	+ 24 14	51	15.	GALAXY
MCG+04-05-033	01 57 27.	+ 24 20	72	14.5	GALAXY
MCG+01-06-021	01 57 27.	- 07 05	60	14.5	GALAXY
MCG+01-06-020	01 57 27.	- 07 18	90	15.	GALAXY
RNGC 0781	01 57 28.	+ 12 24		14.0	GALAXY
ZWG 438.011	01 57 30.	+ 12 24		14.0	GALAXY
UGC 01482	01 57 30.	+ 12 24	102	14.0	GALAXY S
KARA.73B 0081	01 57 30.	+ 12 24	90	14.0	ISOLATED GALAXY S
MCG+02-06-010	01 57 30.	+ 12 26	84	14.	GALAXY
ZWG 461.020	01 57 30.	+ 21 04		15.7	GALAXY
UGC 01483	01 57 30.	+ 22 39	78	18.	GALAXY DWARF
MCG+05-05-040	01 57 30.	+ 32 12	60	16.	GALAXY
ZWG 387.018	01 57 30.	- 00 31		15.6	GALAXY
UGC 01481	01 57 30.	- 00 31	72	15.6	GALAXY Sa-b
MCG+00-06-017	01 57 30.	- 01 14 30.	36	16.	GALAXY
MCG+01-06-022	01 57 30.	- 07 20	48	15.	GALAXY
IC 1767	01 57 32.	- 11 21 46.			NONSTELLAR OBJECT
MCG+00-06-018	01 57 33.	- 01 15	48	15.5	GALAXY
IC 1764	01 57 34.	+ 24 20 16.			NONSTELLAR OBJECT
IC 0185	01 57 34.	- 01 45 57.			NONSTELLAR OBJECT
UGC 01484	01 57 36.	+ 15 44	60	17.	GALAXY
ZWG 461.021	01 57 36.	+ 20 52		14.6	GALAXY
UGC 01485	01 57 36.	+ 20 52	84	14.6	GALAXY PECULR
ZWG 482.044	01 57 36.	+ 24 20		14.5	GALAXY
UGC 01486	01 57 36.	+ 24 20	102	14.5	GALAXY SBb
ZWG 482.045	01 57 36.	+ 24 50		15.7	GALAXY
UGC 01487	01 57 36.	+ 24 50	72	15.7	GALAXY (S0)
MCG+04-05-035	01 57 36.	+ 24 50	48	15.	GALAXY
MCG+05-05-041	01 57 36.	+ 27 56	66	14.5	GALAXY
ZWG 503.070	01 57 36.	+ 29 39		15.5	GALAXY
6ZW 134	01 57 36.	+ 29 46			COMPACT GALAXY
ZWG 503.071	01 57 36.	+ 32 13		15.7	GALAXY
KARA.73B 0082	01 57 36.	+ 32 13	42	15.7	ISOLATED GALAXY S
5ZW 163	01 57 36.	+ 46 33			COMPACT GALAXY
MCG+00-06-019	01 57 36.	- 01 45 30.	30	16.	GALAXY
PHL 3980	01 57 36.	- 22 42		16.7	BLUE STELLAR OBJECT
RNGC 0780	01 57 40.	+ 27 59		14.5	GALAXY
MCG+03-06-012	01 57 42.	+ 20 53	66	14.	GALAXY
5ZW 164	01 57 42.	+ 27 59			COMPACT GALAXY
ZWG 503.072	01 57 42.	+ 27 59		14.6	GALAXY
UGC 01488	01 57 42.	+ 27 59	114	14.6	GALAXY
6ZW 135	01 57 42.	+ 33 54			COMPACT GALAXY
UGC 01489	01 57 42.	+ 42 14	72	17.	GALAXY Sc
MCG+08-04-012	01 57 42.	+ 45 54	78	15.	GALAXY
IC 1765	01 57 44.	+ 31 36 09.			SAME AS NGC 783
MCG+03-06-013	01 57 45.	+ 16 19	39	15.	GALAXY
HELW 366	01 57 45.	- 06 19 55.			NEBULA
MCG-02-06-012	01 57 45.	- 11 18	96	14.	GALAXY
REIN 2.022	01 57 45.36	- 06 19 59.3			NEBULA
ZWG 461.022	01 57 48.	+ 16 19		14.9	GALAXY
ZWG 461.023	01 57 48.	+ 21 03		14.1	GALAXY
UGC 01490	01 57 48.	+ 21 03	30	14.1	GALAXY PECULR
UGC 01491	01 57 48.	+ 29 29	60	18.	GALAXY DWRF SP
6ZW 136	01 57 48.	+ 36 30			COMPACT GALAXY
ZWG 552.015	01 57 48.	+ 45 55		14.9	GALAXY
UGC 01492	01 57 48.	+ 45 55	90	14.9	GALAXY S0-a
MCG+00-06-021	01 57 48.	- 01 47 30.	24	15.5	GALAXY
MCG+00-06-020	01 57 48.	- 01 47 30.	18	15.	GALAXY
MCG+01-06-023	01 57 48.	- 06 21	36	14.5	GALAXY
PHL 3981	01 57 48.	- 22 10		18.0	BLUE STELLAR OBJECT
PHL 8104	01 57 48.	- 31 00		18.4	BLUE STELLAR OBJECT
LB 01610	01 57 48.	- 45 05		14.0	FAINT BLUE STAR
LB 03236	01 57 48.	- 51 26		14.0	FAINT BLUE STAR
MCG+06-05-077	01 57 51.	+ 37 58 30.	72	14.	GALAXY
IC 0186	01 57 52.	- 01 47 23.			NONSTELLAR OBJECT
ZWG 387.020	01 57 54.	+ 02 25		15.3	GALAXY
MRK 584	01 57 54.	+ 02 25	11	16.	GALAXY WITH UV CONTINUUM
ZWG 522.102	01 57 54.	+ 37 58		14.0	GALAXY
UGC 01493	01 57 54.	+ 37 58	138	14.0	GALAXY SB?a-b
ZWG 552.016	01 57 54.	+ 50 16		14.4	GALAXY
UGC 01493A	01 57 54.	+ 50 16	48	14.4	GALAXY
ZWG 387.019	01 57 54.	- 01 47		15.3	GALAXY
RNGC 0802	01 57 54.	- 68 06			UNVERIFIED SOUTHERN OBJECT
ARC 0287	01 57 55.	- 08 06		17.2	RICH CLUSTER OF GALAXIES
KHAV 032	01 58	+ 60 27	8420		DARK NEBULA
ZC 0158.0+0018	01 58 00.	+ 00 18	1210		CLUSTER OF GALAXIES
3ZW 039	01 58 00.	+ 11 56			COMPACT GALAXY
ZWG 522.103	01 58 00.	+ 37 40		15.6	GALAXY
ZWG 522.104	01 58 00.	+ 38 33		15.2	GALAXY
ZWG 559.004	01 58 00.	+ 55 00		15.5	GALAXY
MCG-04-05-023	01 58 00.	- 24 48	36	16.	GALAXY
PHL 1250	01 58 00.	- 30 14		18.0	BLUE STELLAR OBJECT
MCG-06-05-020	01 58 00.	- 34 33		13.5	GALAXY
RNGC 0795	01 58 02.	- 56 03			UNVERIFIED SOUTHERN OBJECT
ZWG 461.024	01 58 06.	+ 16 06		15.7	GALAXY
MCG+03-06-014	01 58 06.	+ 17 03	60	16.	GALAXY
UGC 01494	01 58 06.	+ 17 04	78	16.0	GALAXY S
MCG+03-06-015	01 58 06.	+ 17 28	48	15.	GALAXY
ZWG 461.025	01 58 06.	+ 17 29		15.6	GALAXY
UGC 01495	01 58 06.	+ 17 29	60	15.6	GALAXY Sc
ZC 0158.1+2630	01 58 06.	+ 26 30	1750		CLUSTER OF GALAXIES
VB 161	01 58 06.	+ 61 41	322		STELLAR RING
ISS 0057	01 58 06.	+ 61 41	326		STELLAR RING
LB 00171	01 58 06.	- 17 43		16.4	FAINT BLUE STAR
LIN 584	01 58 07.	- 74 08 21.		15.41	SMC EMISSION LINE OBJECT
ZWG 438.012	01 58 12.	+ 14 56		14.6	GALAXY
UGC 01496	01 58 12.	+ 14 56	132	14.6	GALAXY S0
ZWG 482.046	01 58 12.	+ 23 11		15.2	GALAXY
MCG+05-05-043	01 58 12.	+ 31 31	24	18.	GALAXY
MCG+05-05-042	01 58 12.	+ 31 37	90	13.5	GALAXY
ZWG 503.073	01 58 12.	+ 31 38		12.8	GALAXY
UGC 01497	01 58 12.	+ 31 38	102	12.8	GALAXY Sc
PHL 8105	01 58 12.	- 23 46		18.5	BLUE STELLAR OBJECT
PHL 8106	01 58 12.	- 28 30		18.4	BLUE STELLAR OBJECT
MCG-06-05-021	01 58 12.	- 34 29		13.5	GALAXY
RNGC 0783	01 58 15.	+ 31 38		13.0	GALAXY
IC 1766	01 58 17.	+ 31 32 07.			SAME AS NGC 785
ZWG 413.024	01 58 18.	+ 08 04		14.4	GALAXY
UGC 01498	01 58 18.	+ 08 04	60	14.4	GALAXY S
MCG+01-06-027	01 58 18.	+ 08 04	60	14.5	GALAXY
MCG+02-06-011	01 58 18.	+ 14 58	36	14.5	GALAXY
ZWG 482.047	01 58 18.	+ 26 36		14.8	GALAXY
MCG+04-05-036	01 58 18.	+ 26 39	30	15.	GALAXY
UGC 01499	01 58 18.	+ 31 43	66	16.5	GALAXY Sa-b
6ZW 137	01 58 18.	+ 34 27			COMPACT GALAXY
MCG-02-06-013	01 58 18.	- 10 56	60	15.	GALAXY
PHL 8107	01 58 18.	- 29 44		18.1	BLUE STELLAR OBJECT
MCG-04-05-024	01 58 21.	- 25 17 30.	54	15.	GALAXY
RNGC 0784	01 58 22.	+ 28 36		12.0	GALAXY
ARC 0289	01 58 23.	- 24 52		17.5	RICH CLUSTER OF GALAXIES
MCG+01-06-028	01 58 24.	+ 06 16	60	15.	GALAXY
UGC 01500	01 58 24.	+ 19 26	72	16.5	GALAXY Sc
MCG+05-05-045	01 58 24.	+ 28 34	360	12.	GALAXY
ZWG 503.074	01 58 24.	+ 28 35		12.1	GALAXY
UGC 01501	01 58 24.	+ 28 35	408	12.1	GALAXY
5ZW 165	01 58 24.	+ 29 57			COMPACT GALAXY
UGC 01502	01 58 24.	+ 30 07	120	18.	GALAXY DWARF
MCG+05-05-044	01 58 24.	+ 33 04	30	15.	GALAXY
ZWG 503.075	01 58 24.	+ 33 05		14.4	GALAXY
UGC 01503	01 58 24.	+ 33 05	60	14.4	GALAXY E
KARA.73B 0083	01 58 24.	+ 33 05	36	14.4	ISOLATED GALAXY S0
ZWG 538.009	01 58 24.	+ 44 46		14.8	GALAXY
UGC 01504	01 58 24.	+ 44 46	84	14.8	GALAXY S0-a
OCL 0340	01 58 24.	+ 62 38	180		OPEN STAR CLUSTER
MCG-02-06-014	01 58 24.	- 13 44 30.	30	15.	GALAXY
MCG-04-05-025	01 58 24.	- 20 38	48	15.	GALAXY
LIN 585	01 58 26.	- 74 45 22.		15.10	SMC EMISSION LINE OBJECT
IC 1768	01 58 27.	- 25 20 08.			NONSTELLAR OBJECT
ZWG 413.025	01 58 30.	+ 06 17		14.7	GALAXY
UGC 01505	01 58 30.	+ 06 17	72	14.7	GALAXY S0
MCG+03-06-016	01 58 30.	+ 19 26	60	16.	GALAXY
5ZW 166	01 58 30.	+ 28 30			COMPACT GALAXY
MCG+07-05-005	01 58 30.	+ 44 44 30.	60	15.	GALAXY
ZWG 552.017	01 58 30.	+ 46 10		15.5	GALAXY
VB 162	01 58 30.	+ 59 05	369		STELLAR RING
MCG+01-06-024	01 58 30.	- 05 46	60	15.	GALAXY
SER 019.01	01 58 30.	- 56 27	70	16.	INTERACTING PEC. GALAXIES
RNGC 0788	01 58 33.	- 07 03		13.5	GALAXY
MCG-04-05-026	01 58 33.	- 25 16	36	14.5	GALAXY
RNGC 0787	01 58 34.	- 09 14		13.0	GALAXY
ZWG 438.013	01 58 36.	+ 15 24		14.3	GALAXY
UGC 01506	01 58 36.	+ 15 24	54	14.3	GALAXY S
KARA.72 050B	01 58 36.	+ 15 25	30		PART OF DOUBLE GALAXY
KARA.72 050A	01 58 36.	+ 15 25	36	14.3	PART OF DOUBLE GALAXY
MCG-02-06-016	01 58 36.	- 09 03	90	14.5	GALAXY
MCG-02-06-015	01 58 36.	- 09 14	60	13.5	GALAXY
PHL 1251	01 58 36.	- 22 42		16.6	BLUE STELLAR OBJECT

OBJECT NAME	RIGHT ASCEN.	DECLINATION	DIAM.	MAGN.	TYPE OF OBJECT
MCG-06-05-022	01 58 36.	- 34 58	36	15.	GALAXY
LIN 586	01 58 36.	- 74 16 34.		15.35	SMC EMISSION LINE OBJECT
MCG+02-06-012	01 58 42.	+ 15 26	36	14.	GALAXY
ZWG 482.048	01 58 42.	+ 26 15		13.9	GALAXY
UGC 01507	01 58 42.	+ 26 15	126	13.9	GALAXY SBa
MCG+04-05-037	01 58 42.	+ 26 15	120	14.	GALAXY
6ZW 138	01 58 42.	+ 29 20			COMPACT GALAXY
6ZW 139	01 58 42.	+ 30 35			COMPACT GALAXY
MCG+05-05-046	01 58 42.	+ 31 33	66	14.5	GALAXY
6ZW 140	01 58 42.	+ 33 32			COMPACT GALAXY
ZWG 538.010	01 58 42.	+ 44 40		15.6	GALAXY
UGC 01508	01 58 42.	+ 44 40	132	15.6	GALAXY
ZWG 552.018	01 58 42.	+ 46 05		15.5	GALAXY
SER 019.08	01 58 42.	- 56 17	20	16.	INTERACTING GALAXIES
HN 0158	01 58 43.	- 32 09			NEBULA
IC 1769	01 58 43.	- 32 10			NONSTELLAR OBJECT
RNGC 0786	01 58 45.	+ 15 24		14.5	GALAXY
MCG+07-05-006	01 58 45.	+ 44 36 30.	12	16.	GALAXY
MCG+07-05-007	01 58 45.	+ 44 38	96	15.5	GALAXY
MCG+01-06-025	01 58 45.	- 07 04	78	13.	GALAXY
MCG+01-06-029	01 58 48.	+ 08 15	36	16.	GALAXY
ZWG 461.026	01 58 48.	+ 18 25		15.7	GALAXY
MCG+03-06-017	01 58 48.	+ 18 25	30	15.	GALAXY
ZWG 503.076	01 58 48.	+ 31 35		13.9	GALAXY
UGC 01509	01 58 48.	+ 31 35	108	13.9	GALAXY E-S0
6ZW 141	01 58 48.	+ 32 43			COMPACT GALAXY
OCL 0341	01 58 48.	+ 62 01	300		OPEN STAR CLUSTER
ZC 0158.8-0040	01 58 48.	- 00 40	940		CLUSTER OF GALAXIES
MCG+00-06-022	01 58 48.	- 01 32	12	15.5	GALAXY
ZWG 387.021	01 58 48.	- 01 33		15.5	GALAXY
PHL 3982	01 58 48.	- 25 08		18.2	BLUE STELLAR OBJECT
RNGC 0785	01 58 51.	+ 31 35		14.0	GALAXY
MCG+07-05-008	01 58 51.	+ 44 40	48	17.	GALAXY
MCG+01-06-030	01 58 54.	+ 05 40	30	15.	GALAXY
ZWG 482.049	01 58 54.	+ 26 18		14.4	GALAXY
UGC 01510	01 58 54.	+ 26 18	36	14.4	GALAXY
MCG+04-05-038	01 58 54.	+ 26 19	36	15.5	GALAXY
6ZW 142	01 58 54.	+ 37 16			COMPACT GALAXY
MCG+01-06-026	01 58 54.	- 05 37 30.	24	13.5	GALAXY
MCG-06-05-023	01 58 54.	- 33 26	54	13.	GALAXY
SHB 052	01 58 54.4	+ 18 21 40.		17.5	QUASI-STELLAR OBJECT
LIN 587	01 58 55.	- 74 46 05.		15.70	SMC EMISSION LINE OBJECT
BC PKS0158+18	01 58 55.3	+ 18 22 01.			QUASI-STELLAR OBJECT
RNGC 0790	01 58 57.	- 05 37		13.5	GALAXY
MCG-06-05-024	01 58 57.	- 34 36	60	15.	GALAXY
KHAV 033	01 59	+ 77 03	14470		DARK NEBULA
ZWG 413.026	01 59 00.	+ 05 40		15.7	GALAXY
ZWG 413.027	01 59 00.	+ 08 14		15.7	GALAXY
MCG+01-06-032	01 59 00.	+ 08 15	30	16.	GALAXY
MCG+01-06-031	01 59 00.	+ 08 16	22	14.	GALAXY
MCG+03-06-018	01 59 00.	+ 17 40	42	15.	GALAXY
ZWG 461.027	01 59 00.	+ 17 41		15.6	GALAXY
ARC 0288	01 59 00.	+ 18 12		17.5	RICH CLUSTER OF GALAXIES
ZWG 482.050	01 59 00.	+ 23 27		15.5	GALAXY
6ZW 143	01 59 00.	+ 36 00			COMPACT GALAXY
ZC 0159.0+3932	01 59 00.	+ 39 32	4440		CLUSTER OF GALAXIES
ZWG 361.006	01 59 00.	+ 85 08		15.7	GALAXY
ZWG 360.009	01 59 00.	+ 85 08		15.7	GALAXY
ZC 0159.0-0121	01 59 00.	- 01 21	5650		CLUSTER OF GALAXIES
PHL 3983	01 59 00.	- 21 58		18.3	BLUE STELLAR OBJECT
MCG-05-05-026	01 59 00.	- 31 58	30	15.5	GALAXY
IC 0187	01 59 01.	+ 26 13 52.			NONSTELLAR OBJECT
LB 00455	01 59 04.	- 05 12 12.		15.7	FAINT BLUE STAR
RNGC 0791	01 59 05.	+ 08 15		15.0	GALAXY
ZWG 413.028	01 59 06.	+ 08 15		14.8	GALAXY
UGC 01511	01 59 06.	+ 08 15	96	14.8	GALAXY E
ZWG 461.028	01 59 06.	+ 16 01		14.6	GALAXY
UGC 01512	01 59 06.	+ 16 01	90	14.6	GALAXY S0-a
ZC 0159.1+1812	01 59 06.	+ 18 12	740		CLUSTER OF GALAXIES
5ZW 167	01 59 06.	+ 23 17			COMPACT GALAXY
MCG+04-05-039	01 59 06.	+ 23 18 30.	42	15.	GALAXY
ZWG 482.051	01 59 06.	+ 23 19		14.9	GALAXY
ZC 0159.1+3232	01 59 06.	+ 32 32	1080		CLUSTER OF GALAXIES
IC 0189	01 59 07.	+ 23 18 34.			NONSTELLAR OBJECT
IC 0188	01 59 07.	+ 26 47 16.			NONSTELLAR OBJECT
ARC 0291	01 59 07.	- 02 25		17.8	RICH CLUSTER OF GALAXIES
MCG+03-06-019	01 59 09.	+ 16 00	42	15.	GALAXY
RNGC 0771	01 59 09.	+ 72 11			NON-EXISTENT OBJECT
MCG+01-06-027	01 59 09.	- 07 31	60	15.	GALAXY
ZC 0159.2+0330	01 59 12.	+ 03 30	870		CLUSTER OF GALAXIES
ZWG 413.029	01 59 12.	+ 08 14		15.7	GALAXY
UGC 01513	01 59 12.	+ 08 14	60	15.7	GALAXY SB0-a
ZWG 461.029	01 59 12.	+ 20 52		14.7	GALAXY
UGC 01514	01 59 12.	+ 20 52	90	14.7	GALAXY
6ZW 144	01 59 12.	+ 34 25			COMPACT GALAXY
ZWG 387.022	01 59 12.	- 01 35		15.7	GALAXY
ZC 0159.2-0226	01 59 12.	- 02 26	470		CLUSTER OF GALAXIES
MCG-04-05-027	01 59 15.	- 25 09	60	14.5	GALAXY
UGC 01515	01 59 18.	+ 11 29	78	18.	GALAXY DWARF
MCG+03-06-020	01 59 18.	+ 20 52 30.	24	14.5	GALAXY
MCG+04-05-040	01 59 18.	+ 23 18	15	15.	GALAXY
ZWG 482.052	01 59 18.	+ 23 19		15.1	GALAXY
PHL 8108	01 59 18.	- 22 48		18.4	BLUE STELLAR OBJECT
ARC 0290	01 59 21.	+ 20 48		16.5	RICH CLUSTER OF GALAXIES
IC 0190	01 59 21.	+ 23 18 21.			NONSTELLAR OBJECT
UGC 01516	01 59 24.	+ 44 44	66	16.5	GALAXY Sc
MCG+01-06-028	01 59 24.	- 03 04	24	16.	GALAXY
ARC 0293	01 59 26.	+ 03 33		17.3	RICH CLUSTER OF GALAXIES
ZWG 413.030	01 59 30.	+ 08 12		15.2	GALAXY
ZWG 438.014	01 59 30.	+ 15 28		14.6	GALAXY
UGC 01517	01 59 30.	+ 15 28	108	14.6	GALAXY S0
ZC 0159.5+1850	01 59 30.	+ 18 50	3020		CLUSTER OF GALAXIES
UGC 01518	01 59 30.	+ 18 50	120	16.0	GALAXY
UGC 01519	01 59 30.	+ 18 57	96	17.	GALAXY
MCG+05-05-047	01 59 30.	+ 31 48	33	14.5	GALAXY
ZWG 503.077	01 59 30.	+ 31 50		14.0	GALAXY
UGC 01520	01 59 30.	+ 31 50	48	14.0	GALAXY PECULR
6ZW 145	01 59 30.	+ 39 28			COMPACT GALAXY
ZWG 538.011	01 59 30.	+ 39 53		15.7	GALAXY
UGC 01521	01 59 30.	+ 39 53	72	16.5	GALAXY S
BC 3C57	01 59 30.	- 11 47 00.		16.40	QUASI-STELLAR OBJECT
SHB 053	01 59 30.4	- 11 47 00.		16.4	QUASI-STELLAR OBJECT
ARC 0294	01 59 33.	+ 05 11		17.6	RICH CLUSTER OF GALAXIES
MCG+02-06-014	01 59 33.	+ 09 44	23	14.5	GALAXY
MCG+02-06-013	01 59 33.	+ 09 45	36	14.5	GALAXY
RNGC 0792	01 59 33.	+ 15 28		14.5	GALAXY
RNGC 0789	01 59 33.	+ 31 50		14.0	GALAXY
IC 1770	01 59 34.	+ 09 44 37.			NONSTELLAR OBJECT

OBJECT NAME	RIGHT ASCEN.	DECLINATION	DIAM.	MAGN.	TYPE OF OBJECT
ZWG 438.015	01 59 36.	+ 09 43		15.4	GALAXY
KARA.72 051B	01 59 36.	+ 09 43	48	15.4	PART OF DOUBLE GALAXY
IC 1771	01 59 36.	+ 09 43 49.			NONSTELLAR OBJECT
ZWG 438.016	01 59 36.	+ 09 44		15.4	GALAXY
UGC 01522	01 59 36.	+ 09 44	90	15.4	GALAXY S0
KARA.72 051A	01 59 36.	+ 09 44	54	15.4	PART OF DOUBLE GALAXY
MCG+02-06-015	01 59 36.	+ 15 29	72	14.5	GALAXY
MCG+03-06-023	01 59 36.	+ 18 50	36	15.	GALAXY
MCG+03-06-022	01 59 36.	+ 18 56	48	16.	GALAXY
ZWG 461.030	01 59 36.	+ 19 25		15.7	GALAXY
UGC 01523	01 59 36.	+ 19 25	72	15.7	GALAXY Sb
MCG+03-06-021	01 59 36.	+ 19 25	27	16.	GALAXY
ZC 0159.6+2043	01 59 36.	+ 20 43	2220		CLUSTER OF GALAXIES
UGC 01524	01 59 36.	+ 23 24	78	16.5	GALAXY
6ZW 146	01 59 36.	+ 34 05			COMPACT GALAXY
ZC 0159.6+3833	01 59 36.	+ 38 33	1140		CLUSTER OF GALAXIES
VB 066	01 59 36.	+ 65 12	275		STELLAR RING
MCG+01-06-029	01 59 36.	- 03 04	30	14.5	GALAXY
MCG-02-06-017	01 59 36.	- 10 42	72	14.	GALAXY
PHL 8109	01 59 36.	- 28 09		18.3	BLUE STELLAR OBJECT
HOLM 054A	01 59 39.	- 00 19	30	14.4	PART OF MULTIPLE GALAXY
MCG+00-06-023	01 59 39.	- 00 20 30.	102	14.4	GALAXY
HOLM 054B	01 59 39.	- 00 21	54	14.0	PART OF MULTIPLE GALAXY
MCG+00-06-024	01 59 39.	- 00 22	42	14.0	GALAXY
MCG+01-06-030	01 59 39.	- 08 30	60	15.	GALAXY
HOLM 055B	01 59 40.	- 06 19	30	14.6	PART OF MULTIPLE GALAXY
ZWG 461.031	01 59 42.	+ 18 08		14.0	GALAXY
UGC 01528	01 59 42.	+ 18 08	84	14.0	GALAXY E-S0
ZWG 482.053	01 59 42.	+ 22 05		15.7	GALAXY
5ZW 168	01 59 42.	+ 23 24			COMPACT GALAXY
MCG+07-05-009	01 59 42.	+ 39 51	54	16.	GALAXY
HOLM 054C	01 59 42.	- 00 19	24	14.8	PART OF MULTIPLE GALAXY
ZWG 387.025	01 59 42.	- 00 20		14.2	GALAXY
UGC 01527	01 59 42.	- 00 20	168	14.2	GALAXY SBa
KARA.72 052B	01 59 42.	- 00 20	126	14.2	PART OF DOUBLE GALAXY
ZWG 387.024	01 59 42.	- 00 22		14.7	GALAXY
UGC 01526	01 59 42.	- 00 22	84	14.7	GALAXY Sc
KARA.72 052A	01 59 42.	- 00 22	72	14.7	PART OF DOUBLE GALAXY
ZWG 387.023	01 59 42.	- 01 21		15.7	GALAXY
UGC 01525	01 59 42.	- 01 21	120	15.7	GALAXY E
HOLM 055A	01 59 42.	- 06 18	48	14.3	PART OF MULTIPLE GALAXY
MCG-05-05-027	01 59 42.	- 31 02	24	16.	GALAXY
LB 01611	01 59 42.	- 58 11		14.1	FAINT BLUE STAR
IC 0191	01 59 43.	+ 18 07 43.			SAME AS NGC 794
ARC 0292	01 59 43.	+ 18 52		16.5	RICH CLUSTER OF GALAXIES
RNGC 0799	01 59 43.	- 00 20		14.0	GALAXY
RNGC 0800	01 59 43.	- 00 22		14.5	GALAXY
RNGC 0794	01 59 44.	+ 18 08		14.0	GALAXY
MCG+00-06-025	01 59 45.	- 01 21 30.	36	15.	GALAXY
MCG+01-06-031	01 59 45.	- 06 19 30.	60	15.	GALAXY
ARC 0297	01 59 45.	- 25 49		17.5	RICH CLUSTER OF GALAXIES
SCHO 0043	01 59 47.	+ 54 31 00.	590		ISOLATED DARK CLOUD
ZWG 438.017	01 59 48.	+ 10 51		14.7	GALAXY
UGC 01529	01 59 48.	+ 10 51	120	14.7	GALAXY Sc
KARA.73B 0084	01 59 48.	+ 10 51	90	14.7	ISOLATED GALAXY S
MCG+03-06-025	01 59 48.	+ 15 45	36	15.	GALAXY
ZWG 461.032	01 59 48.	+ 15 47		14.8	GALAXY
UGC 01530	01 59 48.	+ 15 47	66	14.8	GALAXY S0
ZWG 461.033	01 59 48.	+ 16 55		15.1	GALAXY
UGC 01531	01 59 48.	+ 16 55	84	15.1	GALAXY Sb
MCG+03-06-024	01 59 48.	+ 18 08	24	13.5	GALAXY
ZWG 482.054	01 59 48.	+ 22 08		15.7	GALAXY
MCG+04-05-041	01 59 48.	+ 23 24 30.	9	15.5	GALAXY
ZWG 482.055	01 59 48.	+ 27 20		15.0	GALAXY
ZWG 482.056	01 59 48.	+ 27 26		15.6	GALAXY
6ZW 147	01 59 48.	+ 29 46			COMPACT GALAXY
ZWG 538.012	01 59 48.	+ 44 47		15.3	GALAXY
UGC 01532	01 59 48.	+ 44 47	60	15.3	GALAXY E-S
TON-S 0234	01 59 48.	- 22 50		15.2	BLUE STAR
IC 0192	01 59 50.	+ 15 46 55.			NONSTELLAR OBJECT
MCG+02-06-016	01 59 51.	+ 10 52	108	14.	GALAXY
MCG+01-06-032	01 59 51.	- 07 15	12	16.	GALAXY
MCG+03-06-026	01 59 54.	+ 16 55	66	15.	GALAXY
ZWG 482.057	01 59 54.	+ 26 20		15.4	GALAXY
UGC 01533	01 59 54.	+ 26 20	66	15.4	GALAXY SBc/Sc
MCG+04-05-042	01 59 54.	+ 26 21	48	15.	GALAXY
ZWG 482.058	01 59 54.	+ 27 26		15.3	GALAXY
6ZW 148	01 59 54.	+ 35 07			COMPACT GALAXY
UGC 01534	01 59 54.	+ 46 14	60	16.5	GALAXY SB0
ACK 130-03.1	01 59 54.	+ 64 43			PLANETARY NEBULA
ARC 0296	01 59 54.	- 03 22		17.6	RICH CLUSTER OF GALAXIES
MCG+01-06-034	01 59 54.	- 04 25	60	15.	GALAXY
MCG+01-06-033	01 59 54.	- 07 14 30.	9	16.	GALAXY
MCG-05-05-028	01 59 54.	- 31 16	30	15.	GALAXY
ARC 0295	01 59 56.	- 01 19		16.6	RICH CLUSTER OF GALAXIES
RNGC 0793	01 59 57.	+ 31 46			NON-EXISTENT OBJECT
VDB.66G 017	02 00	+ 21 43	100		DWARF GALAXY
LB 09781	02 00	- 81 23		15.3	FAINT BLUE STAR
ZWG 413.031	02 00 00.	+ 03 55		15.6	GALAXY
MCG+01-06-033	02 00 00.	+ 07 30	36	15.	GALAXY
DG 008	02 00 00.	+ 19 55	3600		REFLECTION NEBULA
ZC 0200.0+2032	02 00 00.	+ 20 32	6520		CLUSTER OF GALAXIES
6ZW 149	02 00 00.	+ 38 34			COMPACT GALAXY
MCG+08-04-013	02 00 00.	+ 48 14	72	17.	GALAXY
PHL 3984	02 00 00.	- 21 31		18.0	BLUE STELLAR OBJECT
MCG-04-06-001	02 00 00.	- 22 00	66	14.5	GALAXY
MCG-05-06-001	02 00 00.	- 28 54	96	14.5	GALAXY
LB 03237	02 00 00.	- 54 00		13.4	FAINT BLUE STAR
IC 0193	02 00 02.	+ 10 50 12.			NONSTELLAR OBJECT
IC 1772	02 00 05.	+ 07 30 23.			NONSTELLAR OBJECT
5ZW 169	02 00 06.	+ 30 01			COMPACT GALAXY
6ZW 150	02 00 06.	+ 31 44			COMPACT GALAXY
6ZW 151	02 00 06.	+ 34 32			COMPACT GALAXY
PHL 8110	02 00 06.	- 26 34		17.4	BLUE STELLAR OBJECT
SNO 15	02 00 07.	- 45 01 26.	1200	17.	CLUSTER OF 25 GALAXIES
MCG+01-06-034	02 00 12.	+ 07 17	48	15.	GALAXY
6ZW 153	02 00 12.	+ 30 18			COMPACT GALAXY
6ZW 152	02 00 12.	+ 31 44			COMPACT GALAXY
UGC 01535	02 00 12.	+ 36 04	66	17.	GALAXY DWRF IR
ZC 0200.2+4013	02 00 12.	+ 40 13	2690		CLUSTER OF GALAXIES
MCG+01-06-035	02 00 12.	- 06 29	60	15.5	GALAXY
SER 019.03	02 00 12.	- 58 10	20	16.	PECULIAR GALAXY
ARC 0298	02 00 17.	- 09 35		17.2	RICH CLUSTER OF GALAXIES
ZWG 413.032	02 00 18.	+ 07 17		15.4	GALAXY
UGC 01536	02 00 18.	+ 07 17	72	15.4	GALAXY Sb-c
3ZW 040	02 00 18.	+ 20 46			COMPACT GALAXY
BIGO 472	02 00 18.	+ 37 59			NEBULA
UGC 01537	02 00 18.	+ 48 13	84	16.5	GALAXY Sc

OBJECT NAME	RIGHT ASCEN.	DECLINATION	DIAM.	MAGN.	TYPE OF OBJECT
MCG-04-06-002	02 00 18.	- 24 42 30.	30	15.5	GALAXY
PHL 3985	02 00 18.	- 28 30		17.9	BLUE STELLAR OBJECT
UGC 01538	02 00 24.	+ 23 31	84	17.	GALAXY DWRF IR
MCG+05-05-048	02 00 24.	+ 31 48	27	15.	GALAXY
ZWG 503.078	02 00 24.	+ 31 50		14.7	GALAXY
UGC 01539	02 00 24.	+ 31 50	78	14.7	GALAXY E
UGC 01540	02 00 24.	+ 33 24	78	17.	GALAXY DWRF SP
5ZW 170	02 00 24.	+ 37 52			COMPACT GALAXY
ZWG 522.105	02 00 24.	+ 37 53		13.1	GALAXY
RNGC 0797	02 00 24.	+ 37 53		13.0	GALAXY
UGC 01541	02 00 24.	+ 37 53	114	13.1	GALAXY Sa/SBa
MCG+06-05-078	02 00 24.	+ 37 53 30.	84	13.5	GALAXY
SER 019.05	02 00 24.	- 58 40	480	15.	LOOSE GROUP OF GALAXIES
RNGC 0813	02 00 26.	- 68 41			UNVERIFIED SOUTHRN OBJECT
MCG+00-06-026	02 00 27.	+ 02 22 30.	72	14.	GALAXY
RNGC 0798	02 00 27.	+ 31 50		14.5	GALAXY
MCG-03-06-004	02 00 27.	- 14 56	54	16.	GALAXY
IC 0194	02 00 29.	+ 02 22 10.			NONSTELLAR OBJECT
ZWG 387.026	02 00 30.	+ 02 22		15.4	GALAXY
UGC 01542	02 00 30.	+ 02 22	102	15.4	GALAXY Sb
ZWG 413.033	02 00 30.	+ 04 00		15.6	GALAXY
MCG+01-06-035	02 00 30.	+ 05 24	48	15.	GALAXY
ZWG 461.034	02 00 30.	+ 19 29		15.7	GALAXY
UGC 01543	02 00 30.	+ 19 29	66	15.7	GALAXY E
MCG+03-06-027	02 00 30.	+ 19 29	15	15.	GALAXY
UGC 01544	02 00 30.	+ 47 52	90	16.5	GALAXY SBc
MCG+08-04-014	02 00 30.	+ 48 07	48	16.	GALAXY
RNGC 0815	02 00 30.	- 14 52		16.0	GALAXY
MCG-03-06-005	02 00 30.	- 14 52	48	16.	GALAXY
RNGC 0814	02 00 30.	- 14 56		16.0	GALAXY
ZC 0200.6+0324	02 00 36.	+ 03 24	1550		CLUSTER OF GALAXIES
ZWG 413.034	02 00 36.	+ 05 25		15.0	GALAXY
UGC 01545	02 00 36.	+ 05 25	66	15.0	GALAXY S0-a
ZWG 461.035	02 00 36.	+ 18 24		14.8	GALAXY
UGC 01546	02 00 36.	+ 18 24	72	14.8	GALAXY Sc
ZWG 461.036	02 00 36.	+ 21 03		15.7	GALAXY
ZWG 482.059	02 00 36.	+ 21 48		15.0	GALAXY
UGC 01547	02 00 36.	+ 21 48	132	15.0	GALAXY IRR
MCG+04-05-043	02 00 36.	+ 21 48	138	15.	GALAXY
KARA.73B 0085	02 00 36.	+ 21 48	168	15.0	ISOLATED GALAXY SPEC
6ZW 154	02 00 36.	+ 37 34			COMPACT GALAXY
UGC 01548	02 00 36.	+ 42 46	66	16.0	GALAXY Sc
MCG+08-04-015	02 00 36.	+ 47 52	66	16.	GALAXY
LB 00456	02 00 37.	- 04 19 42.		17.7	FAINT BLUE STAR
ZWG 482.060	02 00 42.	+ 26 04		14.6	GALAXY
UGC 01549	02 00 42.	+ 26 04	60	14.6	GALAXY S0-a
MCG+04-05-044	02 00 42.	+ 26 04 30.	48	15.	GALAXY
ZC 0200.7+3725	02 00 42.	+ 37 25	1880		CLUSTER OF GALAXIES
ZWG 522.106	02 00 42.	+ 38 01		13.5	GALAXY
RNGC 0801	02 00 42.	+ 38 01		13.5	GALAXY
UGC 01550	02 00 42.	+ 38 01	204	13.5	GALAXY Sc
MCG+06-05-079	02 00 42.	+ 38 02	180	13.5	GALAXY
ZWG 538.013	02 00 42.	+ 42 46		15.2	GALAXY
PHL 3986	02 00 42.	- 29 24		17.8	BLUE STELLAR OBJECT
TON-S 0235	02 00 42.	- 29 55		16.2	BLUE STAR
BIGO 473	02 00 43.	+ 38 00			NEBULA
MCG-02-06-019	02 00 45.	- 09 52 30.	174	14.	GALAXY
MCG-02-06-018	02 00 45.	- 12 47	54	15.	GALAXY
ZWG 482.061	02 00 48.	+ 23 50		14.1	GALAXY
UGC 01551	02 00 48.	+ 23 50	180	14.1	GALAXY SB IV-V
MCG+04-05-045	02 00 48.	+ 23 50	150	14.	GALAXY
ZWG 482.062	02 00 48.	+ 26 02		14.9	GALAXY
6ZW 155	02 00 48.	+ 35 34			COMPACT GALAXY
6ZW 156	02 00 48.	+ 39 02			COMPACT GALAXY
MCG+07-05-010	02 00 48.	+ 42 46	48	15.	GALAXY
MCG+10-04-001	02 00 48.	+ 57 55	9	17.	GALAXY
OCL 0339	02 00 48.	+ 64 12	900	7.	OPEN STAR CLUSTER
MCG+04-05-046	02 00 51.	+ 26 03	33	15.	GALAXY
MRK 585	02 00 54.	+ 02 19	10	15.5	GALAXY WITH UV CONTINUUM
MCG+01-06-036	02 00 54.	+ 04 31	42	14.5	GALAXY
ZWG 461.037	02 00 54.	+ 15 30		15.7	GALAXY
ZWG 438.018	02 00 54.	+ 15 30		15.7	GALAXY
6ZW 157	02 00 54.	+ 31 03			COMPACT GALAXY
ZWG 552.019	02 00 54.	+ 47 44		14.7	GALAXY
UGC 01552	02 00 54.	+ 47 44	84	14.7	GALAXY SB0
MCG+08-04-016	02 00 54.	+ 47 44	66	15.	GALAXY
VB 163	02 00 54.	+ 62 24	248		STELLAR RING
LIN 588	02 00 58.	- 74 55 22.		15.93	SMC EMISSION LINE OBJECT
ZWG 413.035	02 01 00.	+ 04 33		15.5	GALAXY
UGC 01553	02 01 00.	+ 04 33	84	15.5	GALAXY Sc
IC 0195	02 01 00.	+ 14 27 38.			NONSTELLAR OBJECT
ZWG 438.019	02 01 00.	+ 14 28		14.3	GALAXY
UGC 01555	02 01 00.	+ 14 28	90	14.3	GALAXY S0
KARA.72 053A	02 01 00.	+ 14 28	84	14.3	PART OF DOUBLE GALAXY
MCG+03-06-028	02 01 00.	+ 15 47	180	13.	GALAXY
ZWG 461.038	02 01 00.	+ 15 48		13.5	GALAXY
UGC 01554	02 01 00.	+ 15 48	210	13.5	GALAXY Sc
ZWG 461.039	02 01 00.	+ 18 04		15.5	GALAXY
5ZW 171	02 01 00.	+ 25 41			COMPACT GALAXY
ZWG 482.063	02 01 00.	+ 25 41		14.7	GALAXY
MCG+04-05-047	02 01 00.	+ 25 41 30.	24	15.	GALAXY
5ZW 172	02 01 00.	+ 47 44			COMPACT GALAXY
MCG-02-06-020	02 01 00.	- 12 51 30.	48	15.	GALAXY
VV 309B	02 01 00.	- 13 35	54	14.	INTERACTING GALAXY
RNGC 0803	02 01 02.	+ 15 48		13.5	GALAXY
RNGC 0804	02 01 03.	+ 30 35		14.5	GALAXY
MCG+08-04-017	02 01 03.	+ 45 32	36	15.	GALAXY
MCG+08-04-018	02 01 04.	+ 45 32	21	16.	GALAXY
MCG+02-06-017	02 01 06.	+ 14 29	72	14.5	GALAXY
ZWG 438.020	02 01 06.	+ 14 30		14.2	GALAXY
UGC 01556	02 01 06.	+ 14 30	180	14.2	GALAXY SB(b)
KARA.72 053B	02 01 06.	+ 14 30	138	14.2	PART OF DOUBLE GALAXY
ZWG 461.040	02 01 06.	+ 19 25		15.5	GALAXY
MCG+05-05-049	02 01 06.	+ 30 33	60	15.	GALAXY
ZWG 504.001	02 01 06.	+ 30 35		14.7	GALAXY
ZWG 503.079	02 01 06.	+ 30 35		14.7	GALAXY
UGC 01557	02 01 06.	+ 30 35	84	14.7	GALAXY S0
6ZW 158	02 01 06.	+ 34 48			COMPACT GALAXY
VV 037A	02 01 06.	+ 45 31	42	16.	INTERACTING GALAXY
MCG+08-04-019	02 01 06.	+ 45 32	24	15.	GALAXY
IC 1773	02 01 09.	+ 30 34			NONSTELLAR OBJECT
VV 037B	02 01 09.	+ 45 31	42	15.	INTERACTING GALAXY
RNGC 0806	02 01 10.	- 10 10		14.0	GALAXY
UGC 01558	02 01 12.	+ 11 43	72	16.0	GALAXY DWARF
MCG+02-06-018	02 01 12.	+ 14 31	156	13.5	GALAXY
ZWG 438.021	02 01 12.	+ 15 04		15.2	GALAXY
UGC 01559	02 01 12.	+ 15 04	126	15.2	GALAXY Sc/SBc
ZWG 461.041	02 01 12.	+ 19 25		14.9	GALAXY
UGC 01560	02 01 12.	+ 19 25	84	14.9	GALAXY SBb
MCG+03-06-029	02 01 12.	+ 19 25	42	16.	GALAXY
5ZW 173	02 01 12.	+ 23 58			COMPACT GALAXY
ZWG 482.064	02 01 12.	+ 23 58		14.7	GALAXY
UGC 01561	02 01 12.	+ 23 58	90	14.7	GALAXY IRR
5ZW 174	02 01 12.	+ 34 55			COMPACT GALAXY
6ZW 159	02 01 12.	+ 36 22			COMPACT GALAXY
5ZW 175	02 01 12.	+ 45 31			COMPACT GALAXY
ZWG 552.020	02 01 12.	+ 45 32		15.0	GALAXY
UGC 01562	02 01 12.	+ 45 32	84	15.0	GALAXY E
MCG+08-04-020	02 01 12.	+ 45 32	42	16.	GALAXY
LB 03238	02 01 12.	- 51 30		14.0	FAINT BLUE STAR
ARP 290	02 01 13.	+ 14 29			PECULIAR GALAXY
MCG+02-06-019	02 01 15.	+ 11 43	72	15.	GALAXY
IC 0196	02 01 15.	+ 14 29 01.			NONSTELLAR OBJECT
IC 1774	02 01 15.	+ 15 04 25.			NONSTELLAR OBJECT
MCG-02-06-021	02 01 15.	- 10 10	72	14.	GALAXY
VV 309A	02 01 15.	- 13 37	150	14.	INTERACTING GALAXY
LB 00457	02 01 16.	- 05 36 24.		17.3	FAINT BLUE STAR
MCG+02-06-021	02 01 18.	+ 14 05	24	15.	GALAXY
MCG+02-06-020	02 01 18.	+ 15 06	108	14.	GALAXY
MCG+03-06-030	02 01 18.	+ 19 26	66	15.	GALAXY
MCG+08-04-021	02 01 18.	+ 47 42	21	15.	GALAXY
MCG+04-05-048	02 01 21.	+ 23 58 30.	48	15.	GALAXY
ZWG 504.002	02 01 24.	+ 28 25		14.7	GALAXY
ZWG 503.080	02 01 24.	+ 28 25		14.7	GALAXY
MRK 365	02 01 24.	+ 28 25	25	15.	GALAXY WITH UV CONTINUUM
ZWG 552.021	02 01 24.	+ 45 32		15.4	GALAXY
ZWG 552.022	02 01 24.	+ 47 42		15.5	GALAXY
UGC 01563	02 01 24.	+ 47 42	78	15.5	GALAXY E
MCG+01-06-036	02 01 24.	- 07 40	42	15.	GALAXY
MCG+00-06-027	02 01 27.	+ 01 33	54	13.	GALAXY
LB 00458	02 01 27.	- 02 45 54.		18.2	FAINT BLUE STAR
ZWG 387.027	02 01 30.	+ 02 33		14.3	GALAXY
UGC 01564	02 01 30.	+ 02 33	60	14.3	GALAXY SB:b-c
IC 0197	02 01 30.	+ 02 33 48.			NONSTELLAR OBJECT
MCG+05-05-051	02 01 30.	+ 27 38	66	15.5	GALAXY
MCG+05-05-050	02 01 30.	+ 28 32	54	15.	GALAXY
MCG+01-06-037	02 01 30.	- 08 23	30	15.5	GALAXY
RNGC 0808	02 01 33.	- 23 32		14.0	GALAXY
RNGC 0805	02 01 34.	+ 28 34		14.5	GALAXY
ZWG 504.003	02 01 36.	+ 27 41		15.4	GALAXY
ZWG 503.081	02 01 36.	+ 27 41		15.4	GALAXY
UGC 01565	02 01 36.	+ 27 41	96	15.4	GALAXY
ZWG 504.004	02 01 36.	+ 28 34		14.7	GALAXY
ZWG 503.082	02 01 36.	+ 28 34		14.7	GALAXY
UGC 01566	02 01 36.	+ 28 34	66	14.7	GALAXY SB0
6ZW 160	02 01 36.	+ 36 59			COMPACT GALAXY
ZC 0201.7+2557	02 01 42.	+ 25 57	1280		CLUSTER OF GALAXIES
6ZW 161	02 01 42.	+ 34 53			COMPACT GALAXY
UGC 01567	02 01 42.	+ 42 55	78	16.0	GALAXY Sc
MCG-04-06-003	02 01 42.	- 23 33	72	14.5	GALAXY
MCG-05-06-002	02 01 42.	- 32 02	42	15.	GALAXY
MCG+03-06-031	02 01 48.	+ 21 22	27	17.	GALAXY
6ZW 162	02 01 48.	+ 33 34			COMPACT GALAXY
UGC 01568	02 01 48.	+ 35 26	66	17.	GALAXY
MCG+06-05-080	02 01 48.	+ 35 45	12	15.5	GALAXY
MCG+06-05-081	02 01 51.	+ 34 53	66	16.	GALAXY
MCG+06-05-082	02 01 51.	+ 34 54	36	16.	GALAXY
MCG+01-06-038	02 01 51.	- 05 09	60	14.5	GALAXY
MCG+05-05-052	02 01 54.	+ 32 00	48	15.	GALAXY
ZWG 504.005	02 01 54.	+ 32 02		15.5	GALAXY
ZWG 503.083	02 01 54.	+ 32 02		15.5	GALAXY
UGC 01569	02 01 54.	+ 34 55	72	16.0	GALAXY Sb
5ZW 176	02 01 54.	+ 39 27			COMPACT GALAXY
ZWG 552.023	02 01 54.	+ 46 08		15.6	GALAXY
MCG-02-06-022	02 01 54.	- 10 18	48	15.	GALAXY
RNGC 0807	02 01 58.	+ 28 45		14.0	GALAXY
RNGC 0809	02 01 58.	- 08 57		14.0	GALAXY
LBN 0688	02 02	+ 20 00	2700		BRIGHT NEBULA
MCG+01-06-037	02 02 00.	+ 08 19	72	14.5	GALAXY
ZC 0202.0+0940	02 02 00.	+ 09 40	1550		CLUSTER OF GALAXIES
ZWG 461.042	02 02 00.	+ 21 12		15.5	GALAXY
UGC 01570	02 02 00.	+ 21 12	66	15.5	GALAXY Sa-b
ZWG 504.006	02 02 00.	+ 28 45		13.8	GALAXY
ZWG 503.084	02 02 00.	+ 28 45		13.8	GALAXY
UGC 01571	02 02 00.	+ 28 45	150	13.8	GALAXY E
MCG-02-06-023	02 02 00.	- 08 57 30.	60	14.	GALAXY
SCHO 0044	02 02 02.	+ 54 36 00.	470		ISOLATED DARK CLOUD
ZC 0202.1+0307	02 02 06.	+ 03 07	1750		CLUSTER OF GALAXIES
ZWG 413.036	02 02 06.	+ 08 18		14.3	GALAXY
UGC 01572	02 02 06.	+ 08 18	90	14.3	GALAXY S
MCG+02-06-022	02 02 06.	+ 11 11	48	14.5	GALAXY
MCG+05-06-001	02 02 06.	+ 28 44	66	14.	GALAXY
5ZW 177	02 02 06.	+ 33 10			COMPACT GALAXY
ZCG 0202+51	02 02 06.	+ 51 45		16.6	COMPACT GALAXY
MCG+01-06-039	02 02 06.	- 06 26	162	13.5	GALAXY
ARC 0299	02 02 10.	+ 00 06		17.6	RICH CLUSTER OF GALAXIES
SHB 054	02 02 10.	+ 31 57 18.		18.	QUASI-STELLAR OBJECT
BC DW0202+31	02 02 10.	+ 31 57 38.		18.	QUASI-STELLAR OBJECT
UGC 01573	02 02 12.	+ 01 40	60	16.5	GALAXY Sc
ZWG 438.022	02 02 12.	+ 11 11		15.5	GALAXY
ZWG 522.107	02 02 12.	+ 37 24		15.7	GALAXY
UGC 01574	02 02 12.	+ 37 24	72	15.7	GALAXY Sc
5ZW 178	02 02 12.	+ 51 46			COMPACT GALAXY
MCG+09-04-001	02 02 12.	+ 55 49	36	17.	GALAXY
MCG+01-06-040	02 02 12.	- 02 38	60	16.	GALAXY
MCG-02-06-024	02 02 15.	- 10 20	54	14.5	GALAXY
RNGC 0811	02 02 16.	- 10 20		14.0	GALAXY
MCG+00-06-028	02 02 18.	+ 01 40	48	15.	GALAXY
ZWG 461.043	02 02 18.	+ 15 37		15.7	GALAXY
UGC 01575	02 02 18.	+ 24 26	72	18.	GALAXY DWARF
MCG+06-05-083	02 02 18.	+ 34 32	48	15.	GALAXY
ZWG 522.108	02 02 18.	+ 34 34		15.4	GALAXY
MCG+09-04-002	02 02 18.	+ 55 55	30	17.	GALAXY
LB 00459	02 02 19.	- 03 37 30.		17.4	FAINT BLUE STAR
UGC 01576	02 02 24.	+ 29 45	72	17.	GALAXY Sc
MCG+09-04-003	02 02 24.	+ 55 47	18	19.	GALAXY
ZC 0202.5+1741	02 02 30.	+ 17 41	740		CLUSTER OF GALAXIES
ZWG 483.001	02 02 30.	+ 24 52		15.5	GALAXY
ZWG 504.007	02 02 30.	+ 30 56		14.0	GALAXY
ZWG 503.085	02 02 30.	+ 30 56		14.0	GALAXY
UGC 01577	02 02 30.	+ 30 56	144	14.0	GALAXY SBb
KARA.73B 0086	02 02 30.	+ 30 56	132	14.0	ISOLATED GALAXY S
ZWG 522.109	02 02 30.	+ 34 42		15.5	GALAXY
6ZW 163	02 02 30.	+ 37 22			COMPACT GALAXY
5ZW 179	02 02 30.	+ 37 22			COMPACT GALAXY
ZWG 522.110	02 02 30.	+ 37 23		15.7	GALAXY

OBJECT NAME	RIGHT ASCEN.	DECLINATION	DIAM.	MAGN.	TYPE OF OBJECT
UGC 01578	02 02 30.	+ 50 30	72	17.	GALAXY Sc
MCG+12-03-002	02 02 30.	+ 74 56	138	13.	GALAXY
MCG-05-06-003	02 02 30.	- 29 31	42	15.	GALAXY
SHB 055	02 02 34.4	- 17 15 37.		18.	QUASI-STELLAR OBJECT
BC PKS0202-17	02 02 34.55	- 17 15 39.1		18.	QUASI-STELLAR OBJECT
IC 1775	02 02 35.	+ 13 16 38.			NONSTELLAR OBJECT
SHAH 046	02 02 36.	+ 03 05	402	17.3	GROUP OF COMPACT GALAXIES
MCG+01-06-038	02 02 36.	+ 05 51	96	13.5	GALAXY
ZWG 413.037	02 02 36.	+ 05 52		14.4	GALAXY
UGC 01579	02 02 36.	+ 05 52	138	14.4	GALAXY SBc
ZWG 438.023	02 02 36.	+ 09 41		15.2	GALAXY
UGC 01580	02 02 36.	+ 09 41	90	15.2	GALAXY S
MCG+02-06-024	02 02 36.	+ 09 41	72	14.	GALAXY
MCG+02-06-023	02 02 36.	+ 13 17	36	15.	GALAXY
ARC 0300	02 02 36.	+ 17 37		17.7	RICH CLUSTER OF GALAXIFS
ZC 0202.6+1852	02 02 36.	+ 18 52	8200		CLUSTER OF GALAXIES
MCG+04-06-001	02 02 36.	+ 24 53	36	15.5	GALAXY
MCG+05-06-002	02 02 36.	+ 30 56	120	14.	GALAXY
MCG+06-05-084	02 02 36.	+ 34 38	90	14.	GALAXY
ZWG 522.111	02 02 36.	+ 34 40		15.2	GALAXY
UGC 01581	02 02 36.	+ 34 40	108	15.2	GALAXY S-IRR
ZWG 538.014	02 02 36.	+ 39 36		14.3	GALAXY
UGC 01582	02 02 36.	+ 39 36	84	14.3	GALAXY Sc
ZWG 538.015	02 02 36.	+ 44 58		15.6	GALAXY
5ZW 180	02 02 36.	+ 51 47			COMPACT GALAXY
TON-S 0236	02 02 36.	- 23 04		15.2	BLUE STAR
IC 1776	02 02 39.	+ 05 52 43.			NONSTELLAR OBJFCT
RNGC 0810	02 02 39.	+ 13 00		15.5	GALAXY
ZWG 413.038	02 02 42.	+ 08 17		15.6	GALAXY
ZWG 438.024	02 02 42.	+ 13 00		15.4	GALAXY
UGC 01583	02 02 42.	+ 13 00	108	15.4	GALAXY E?
UGC 01584	02 02 42.	+ 13 08	90	16.0	GALAXY Sc
MCG+01-06-041	02 02 42.	- 06 44 30.	60	14.5	GALAXY
MCG+02-06-026	02 02 45.	+ 13 01	24	14.	GALAXY
MCG+02-06-025	02 02 45.	+ 13 07	84	15.5	GALAXY
MCG-02-06-025	02 02 45.	- 13 07 30.	48	15.	GALAXY
3ZW 041	02 02 48.	+ 18 42			COMPACT GALAXY
MCG+05-06-003	02 02 48.	+ 28 22	36	16.5	GALAXY
MCG+07-05-011	02 02 48.	+ 39 34	84	14.	GALAXY
MCG-04-06-004	02 02 48.	- 22 01	48	15.	GALAXY
MCG-05-06-004	02 02 48.	- 32 36	30	15.	GALAXY
ARC 0301	02 02 49.	- 02 21		18.0	RICH CLUSTER OF GALAXIFS
MCG+01-06-039	02 02 54.	+ 06 31	60	14.5	GALAXY
ZWG 483.002	02 02 54.	+ 24 00		15.4	GALAXY
5ZW 181	02 02 54.	+ 27 10			COMPACT GALAXY
6ZW 164	02 02 54.	+ 33 18			COMPACT GALAXY
MCG+07-05-012	02 02 54.	+ 44 56	48	14.5	GALAXY
ZWG 538.016	02 02 54.	+ 44 58		14.6	GALAXY
UGC 01585	02 02 54.	+ 44 58	72	14.6	GALAXY Sa/SBb
ZWG 552.024	02 02 54.	+ 49 40		15.6	GALAXY
UGC 01586	02 02 54.	+ 49 40	72	15.6	GALAXY Sb/SBb
MCG+01-06-042	02 02 54.	- 06 41	72	14.5	GALAXY
MCG+04-06-002	02 02 57.	+ 24 01	48	15.	GALAXY
ZWG 413.039	02 03 00.	+ 06 32		14.6	GALAXY
UGC 01587	02 03 00.	+ 06 32	66	14.6	GALAXY S
KARA.73B 0087	02 03 00.	+ 06 32	60	14.6	ISOLATED GALAXY S
6ZW 165	02 03 00.	+ 31 57			COMPACT GALAXY
ZC 0203.0-0220	02 03 00.	- 02 20	1010		CLUSTER OF GALAXIES
VB 067	02 03 06.	+ 64 07	309		STELLAR RING
MCG+00-06-029	02 03 06.	- 00 55 30.	66	15.	GALAXY
ZWG 387.028	02 03 06.	- 00 56		15.7	GALAXY
UGC 01588	02 03 06.	- 00 56	84	15.7	GALAXY Sc
TON-S 0237	02 03 06.	- 29 17		16.1	BLUE STAR
ARC 0302	02 03 10.	- 25 01		17.9	RICH CLUSTFR OF GALAXIFS
ZWG 438.025	02 03 12.	+ 14 41		15.6	GALAXY
UGC 01589	02 03 12.	+ 14 41	60	15.6	GALAXY Sa-b
ZWG 461.044	02 03 12.	+ 18 13		15.6	GALAXY
MCG+03-06-032	02 03 12.	+ 18 13	39	15.	GALAXY
ZWG 504.008	02 03 12.	+ 29 33		14.0	GALAXY
UGC 01590	02 03 12.	+ 29 33	120	14.0	GALAXY S0-E
6ZW 166	02 03 12.	+ 39 06			COMPACT GALAXY
LB 03239	02 03 12.	- 67 28		14.6	FAINT BLUE STAR
MCG+05-06-004	02 03 15.	+ 29 33	60	14.	GALAXY
MCG+01-06-040	02 03 18.	+ 09 05	60	14.5	GALAXY
MCG+02-06-027	02 03 18.	+ 14 42	48	15.	GALAXY
5ZW 182	02 03 18.	+ 29 29			COMPACT GALAXY
MCG+05-06-005	02 03 18.	+ 29 43	78	14.5	GALAXY
ZWG 504.009	02 03 18.	+ 29 44		14.5	GALAXY
UGC 01591	02 03 18.	+ 29 44	102	14.5	GALAXY S
5ZW 183	02 03 18.	+ 32 43			COMPACT GALAXY
MCG+01-06-043	02 03 18.	- 05 31	60	15.	GALAXY
LB 00460	02 03 21.	- 03 30 00.		17.2	FAINT BLUE STAR
ZWG 387.029	02 03 24.	+ 00 31		15.7	GALAXY
SHAH 047	02 03 24.	+ 02 40	78	18.0	GROUP OF COMPACT GALAXIES
ZWG 413.040	02 03 24.	+ 09 03		14.8	GALAXY
UGC 01592	02 03 24.	+ 09 03	72	14.8	GALAXY S
IC 0198	02 03 24.	+ 09 04 22.			NONSTELLAR OBJECT
ZWG 438.026	02 03 24.	+ 13 02		15.6	GALAXY
UGC 01593	02 03 24.	+ 13 02	78	15.6	GALAXY Sc
MCG+02-06-028	02 03 24.	+ 13 03	36	15.	GALAXY
ZWG 438.027	02 03 24.	+ 14 58		15.7	GALAXY
ZWG 538.017	02 03 24.	+ 44 08		15.7	GALAXY
ZC 0203.4-0016	02 03 24.	- 00 16	3970		CLUSTER OF GALAXIES
MCG-03-06-006	02 03 24.	- 18 05	36	15.	GALAXY
MCG-06-05-025	02 03 24.	- 32 54	60	13.	GALAXY
IC 1777	02 03 26.	+ 14 57 59.			NONSTELLAR OBJECT
MCG+01-06-041	02 03 30.	+ 09 00	72	14.5	GALAXY
MCG+02-06-029	02 03 30.	+ 15 00	36	15.5	GALAXY
MCG+07-05-013	02 03 30.	+ 44 07	60	16.	GALAXY
MCG-02-06-026	02 03 30.	- 10 02 30.	102	14.5	GALAXY
ZWG 413.041	02 03 36.	+ 08 59		15.4	GALAXY
UGC 01594	02 03 36.	+ 08 59	96	15.4	GALAXY Sa-b
ZWG 413.042	02 03 36.	+ 09 24		15.4	GALAXY
ZC 0203.6+1008	02 03 36.	+ 10 08	1550		CLUSTER OF GALAXIES
ZWG 483.003	02 03 36.	+ 26 48		15.5	GALAXY
UGC 01595	02 03 36.	+ 26 48	84	15.5	GALAXY Sc
KARA.73B 0088	02 03 36.	+ 26 48	66	15.5	ISOLATED GALAXY S
ZWG 504.010	02 03 36.	+ 27 42		15.7	GALAXY
ZWG 504.011	02 03 36.	+ 29 45		14.8	GALAXY
UGC 01596	02 03 36.	+ 29 45	78	14.8	GALAXY S0
MCG+05-06-006	02 03 36.	+ 29 45	42	14.5	GALAXY
5ZW 184	02 03 36.	+ 38 16			COMPACT GALAXY
ZWG 538.018	02 03 36.	+ 45 22		15.6	GALAXY
ZC 0203.6-0154	02 03 36.	- 01 54	1410		CLUSTER OF GALAXIES
IC 1778	02 03 39.	+ 08 59 03.			NONSTELLAR OBJFCT
MCG+01-06-044	02 03 39.	- 02 34	30	15.5	GALAXY
IC 0199	02 03 40.	+ 09 00 21.			NONSTELLAR OBJFCT
5ZW 185	02 03 42.	+ 32 45			COMPACT GALAXY
5ZW 186	02 03 42.	+ 33 01			COMPACT GALAXY
ZWG 538.019	02 03 42.	+ 44 20		12.8	GALAXY
UGC 01598	02 03 42.	+ 44 20	192	12.8	GALAXY S
ZWG 552.025	02 03 42.	+ 45 27		15.7	GALAXY
ZWG 538.020	02 03 42.	+ 45 27		15.7	GALAXY
VB 068	02 03 42.	+ 64 40	692		STELLAR RING
MCG+00-06-030	02 03 42.	- 00 31	48	15.	GALAXY
ZWG 387.030	02 03 42.	- 00 32		14.6	GALAXY
UGC 01597	02 03 42.	- 00 32	78	14.6	GALAXY S0
ARC 0303	02 03 42.	- 03 34		16.6	RICH CLUSTER OF GALAXIFS
MCG-06-05-026	02 03 42.	- 38 21	18	15.	GALAXY
RNGC 0812	02 03 44.	+ 44 20		13.0	GALAXY
MCG+07-05-014	02 03 45.	+ 44 19	780	13.	GALAXY
IC 1779	02 03 48.	+ 03 27 21.			NONSTELLAR OBJECT
ZWG 461.045	02 03 48.	+ 16 46		15.4	GALAXY
UGC 01599	02 03 48.	+ 37 11	66	16.5	GALAXY Sc-IRR
ZWG 387.031	02 03 54.	+ 03 27		15.2	GALAXY
ZWG 438.028	02 03 54.	+ 14 59		15.3	GALAXY
ZWG 504.012	02 03 54.	+ 30 55		15.2	GALAXY
ARC 0304	02 03 56.	+ 04 15		17.6	RICH CLUSTER OF GALAXIFS
MCG+00-06-031	02 03 57.	+ 01 17 30.	66	15.	GALAXY
LB 09782	02 04	- 81 05		14.7	FAINT BLUE STAR
UGC 01600	02 04 00.	+ 01 17	78	17.	GALAXY Sc
ZC 0204.0+0502	02 04 00.	+ 05 02	1550		CLUSTER OF GALAXIES
ZC 0204.0+1452	02 04 00.	+ 14 52	1610		CLUSTER OF GALAXIES
MCG+02-06-031	02 04 00.	+ 15 00	36	15.5	GALAXY
MCG+02-06-030	02 04 00.	+ 15 00	36	15.5	GALAXY
ZWG 522.112	02 04 00.	+ 34 40		15.1	GALAXY
ZWG 538.021	02 04 00.	+ 42 44		15.5	GALAXY
UGC 01601	02 04 00.	+ 42 44	78	15.5	GALAXY SBb-c
ZC 0204.0-0132	02 04 00.	- 01 32	540		CLUSTER OF GALAXIES
MCG+01-06-045	02 04 00.	- 02 33 30.	42	15.5	GALAXY
MCG-02-06-027	02 04 00.	- 09 05	60	15.	GALAXY
MCG-05-06-005	02 04 00.	- 28 48	30	15.	GALAXY
MCG-06-05-027	02 04 00.	- 37 35	42	15.	GALAXY
MCG+01-06-046	02 04 03.	- 02 33	36	16.	GALAXY
ARC 0305	02 04 04.	- 15 10		17.8	RICH CLUSTER OF GALAXIES
ZWG 438.029	02 04 06.	+ 14 29		15.5	GALAXY
UGC 01602	02 04 06.	+ 43 32	90	16.5	GALAXY
IC 1780	02 04 09.	+ 14 29 26.			NONSTELLAR OBJFCT
MCG+07-05-015	02 04 09.	+ 42 43	60	16.	GALAXY
MCG+00-06-032	02 04 09.	- 01 06	36	14.5	GALAXY
LB 00461	02 04 11.	- 02 26 48.		16.2	FAINT BLUE STAR
ZWG 504.013	02 04 12.	+ 32 44		14.9	GALAXY
KARA.72 054A	02 04 12.	+ 32 44	42	14.9	PART OF DOUBLE GALAXY
6ZW 167	02 04 12.	+ 38 54			COMPACT GALAXY
6ZW 168	02 04 12.	+ 39 13			COMPACT GALAXY
ZWG 387.032	02 04 12.	- 01 06		15.6	GALAXY
UGC 01603	02 04 12.	- 01 06	96	15.6	GALAXY
MCG+02-06-032	02 04 15.	+ 14 30	36	15.	GALAXY
LB 00462	02 04 15.	- 05 21 42.		17.6	FAINT BLUE STAR
MCG+01-06-042	02 04 18.	+ 07 54	72	15.	GALAXY
6ZW 169	02 04 18.	+ 36 52			COMPACT GALAXY
ZWG 522.113	02 04 18.	+ 36 52		15.0	GALAXY
UGC 01604	02 04 18.	+ 36 52	90	15.0	GALAXY S0?
MCG+06-05-085	02 04 18.	+ 36 52	27	14.	GALAXY
ZCG 0204+38	02 04 18.	+ 38 46		18.3	COMPACT GALAXY
ZWG 387.033	02 04 18.	- 00 45		15.3	GALAXY
MCG-05-06-006	02 04 18.	- 32 01	72	15.	GALAXY
IC 1781	02 04 22.	- 00 46 00.			NONSTELLAR OBJECT
ZWG 413.043	02 04 24.	+ 07 54		15.4	GALAXY
UGC 01605	02 04 24.	+ 07 54	96	15.4	GALAXY
ZWG 504.014	02 04 24.	+ 32 44		14.8	GALAXY
KARA.72 054B	02 04 24.	+ 32 44	42	14.8	PART OF DOUBLE GALAXY
MCG+08-05-001	02 04 24.	+ 45 23	72	14.	GALAXY
LB 00463	02 04 25.	- 03 46 48.		17.8	FAINT BLUE STAP
UGC 01606	02 04 30.	+ 08 45	72	16.5	GALAXY Sc
5ZW 187	02 04 30.	+ 24 15			COMPACT GALAXY
MCG+05-06-007	02 04 30.	+ 32 43	36	15.	GALAXY
ZWG 538.022	02 04 30.	+ 45 23		14.2	GALAXY
UGC 01607	02 04 30.	+ 45 23	126	14.2	GALAXY SBb
ACK 132-00.1	02 04 30.	+ 60 31			PLANETARY NEBULA
TON-S 0238	02 04 30.	- 23 28		15.0	BLUE STAR
IC 0200	02 04 31.	+ 30 56 01.			NONSTELLAR OBJFCT
RNGC 0822	02 04 34.	- 41 23			GALAXY
IC 0201	02 04 36.	+ 08 52 06.			NONSTELLAR OBJFCT
UGC 01608	02 04 36.	+ 30 13	66	16.5	GALAXY
MCG+05-06-008	02 04 36.	+ 32 43	42	15.	GALAXY
6ZW 170	02 04 36.	+ 35 12			COMPACT GALAXY
RNGC 0824	02 04 38.	- 36 42			GALAXY
LB 00464	02 04 39.	- 03 07 12.		15.7	FAINT BLUE STAP
MCG+01-06-043	02 04 42.	+ 08 57	72	15.	GALAXY
ZWG 461.046	02 04 42.	+ 18 57		15.6	GALAXY
MCG+07-05-016	02 04 42.	+ 44 35	72	15.	GALAXY
ZWG 538.023	02 04 42.	+ 44 37		15.3	GALAXY
UGC 01609	02 04 42.	+ 44 37	78	15.3	GALAXY Sb
MCG-06-05-028	02 04 42.	- 36 41	72	13.	GALAXY
ARC 0306	02 04 44.	- 12 03		17.8	RICH CLUSTER OF GALAXIES
ZWG 413.044	02 04 48.	+ 08 56		15.3	GALAXY
UGC 01610	02 04 48.	+ 08 56	96	15.3	GALAXY Sb
MCG+03-06-033	02 04 48.	+ 16 57	42	14.5	GALAXY
ZWG 461.047	02 04 48.	+ 16 58		13.9	GALAXY
UGC 01611	02 04 48.	+ 16 58	48	13.9	GALAXY PECULR
6ZW 171	02 04 48.	+ 38 17			COMPACT GALAXY
ZWG 538.024	02 04 48.	+ 43 21		15.5	GALAXY
UGC 01612	02 04 48.	+ 43 21	84	15.5	GALAXY Sc
IC 0202	02 04 49.	+ 08 55 23.			NONSTELLAR OBJECT
RNGC 0817	02 04 50.	+ 16 58		14.0	GALAXY
MCG+00-06-033	02 04 51.	+ 01 52 30.	12	15.	GALAXY
IC 0203	02 04 51.	+ 08 52 29.			NONSTELLAR OBJECT
IC 0204	02 04 52.	- 01 37 44.			NONSTELLAR OBJFCT
ZWG 387.036	02 04 54.	+ 01 52		15.5	GALAXY
ZWG 438.030	02 04 54.	+ 15 07		15.7	GALAXY
UGC 01614	02 04 54.	+ 15 07	90	15.7	GALAXY S0-a
ZWG 522.114	02 04 54.	+ 34 48		15.7	GALAXY
5ZW 188	02 04 54.	+ 34 53			COMPACT GALAXY
UGC 01615	02 04 54.	+ 41 05	66	17.	GALAXY
MCG+07-05-017	02 04 54.	+ 43 20	48	15.	GALAXY
ZWG 387.035	02 04 54.	- 01 40		15.7	GALAXY
ZWG 387.034	02 04 54.	- 02 20		14.8	GALAXY
UGC 01613	02 04 54.	- 02 20	66	14.8	GALAXY SBa
MCG+00-06-034	02 04 54.	- 02 20	54	14.	GALAXY
LB 03240	02 04 54.	- 73 35		14.2	FAINT BLUE STAR
IC 0205	02 04 56.	- 02 20 02.			NONSTELLAR OBJECT
MCG+00-06-035	02 04 57.	+ 01 52 30.	15	14.5	GALAXY
LB 09783	02 05	- 82 25		14.5	FAINT BLUE STAR
ZWG 387.037	02 05 00.	+ 01 52		15.1	GALAXY
UGC 01617	02 05 00.	+ 01 52	72	15.1	GALAXY E-S0

OBJECT NAME	RIGHT ASCEN.	DECLINATION	DIAM.	MAGN.	TYPE OF OBJECT
ZWG 387.038	02 05 00.	+ 01 56		15.3	GALAXY
UGC 01618	02 05 00.	+ 01 56	72	15.3	GALAXY E-SO
MCG+00-06-037	02 05 00.	+ 01 56 30.	18	15.	GALAXY
MCG+00-06-036	02 05 00.	+ 01 58	27	15.	GALAXY
MCG+02-06-033	02 05 00.	+ 15 09	84	15.5	GALAXY
5ZW 189	02 05 00.	+ 34 57			COMPACT GALAXY
UGC 01619	02 05 00.	+ 34 58	66	16.0	GALAXY DBL SYS
ZC 0205.0+8450	02 05 00.	+ 84 50	2020		CLUSTER OF GALAXIES
UGC 01616	02 05 00.	+ 85 05	90	16.5	GALAXY SB
LB 00465	02 05 01.	- 06 32 06.		17.1	FAINT BLUE STAR
IC 0206	02 05 02.	- 07 15 51.			NONSTELLAR OBJECT
RNGC 0823	02 05 04.	- 25 41		14.0	GALAXY
ZWG 387.039	02 05 06.	+ 01 54		15.1	GALAXY
UGC 01620	02 05 06.	+ 01 54	66	15.1	GALAXY E-SO
MCG+00-06-038	02 05 06.	+ 01 54 30.	18	15.	GALAXY
MCG+03-06-034	02 05 06.	+ 18 57	42	14.5	GALAXY
ZWG 504.015	02 05 06.	+ 28 38		15.6	GALAXY
MCG-04-06-005	02 05 06.	- 25 41	21	14.	GALAXY
RNGC 0816	02 05 09.	+ 29 01		15.5	GALAXY
ARP 074	02 05 10.	+ 41 14			PECULIAR GALAXY
UGC 01621	02 05 12.	+ 02 27	72	16.5	GALAXY Sc
MAB 0205+02	02 05 12.	+ 02 28 55.		15.4	QUASI-STELLAR OBJECT
SHB 056	02 05 12.	+ 02 28 55.		15.4	QUASI-STELLAR OBJECT
ZWG 461.048	02 05 12.	+ 15 59		15.0	GALAXY
UGC 01622	02 05 12.	+ 15 59	72	15.0	GALAXY Sc
MCG+03-06-035	02 05 12.	+ 15 59	60	14.5	GALAXY
ZWG 461.049	02 05 12.	+ 18 56		15.7	GALAXY
UGC 01623	02 05 12.	+ 18 56	66	15.7	GALAXY SB
ZWG 504.016	02 05 12.	+ 29 01		15.3	GALAXY
IC 0207	02 05 12.	- 07 13 09.			NONSTELLAR OBJECT
IC 1782	02 05 12.	- 25 43 10.			NONSTELLAR OBJECT
MCG-06-05-029	02 05 12.	- 35 22	30	15.5	GALAXY
MCG+00-06-039	02 05 15.	+ 02 27 30.	36	16.	GALAXY
MCG+05-06-009	02 05 15.	+ 28 03	60	15.5	GALAXY
ZWG 387.040	02 05 18.	+ 01 55		15.2	GALAXY
UGC 01624	02 05 18.	+ 01 55	90	15.2	GALAXY SO
MCG+00-06-040	02 05 18.	+ 01 56 30.	36	14.5	GALAXY
MRK 586	02 05 18.	+ 02 28	7	16.	GALAXY WITH UV CONTINUUM
ZC 0205.3+1012	02 05 18.	+ 10 12	1010		CLUSTER OF GALAXIES
MCG+03-06-036	02 05 18.	+ 20 08	24	16.	GALAXY
UGC 01625	02 05 18.	+ 28 06	66	16.5	GALAXY Sb-c
ZWG 538.025	02 05 18.	+ 41 15		14.5	GALAXY
UGC 01626	02 05 18.	+ 41 15	108	14.5	GALAXY Sc/SBc
MCG-04-06-006	02 05 18.	- 21 39 30.	60	15.	GALAXY
ARC 0307	02 05 21.	+ 10 14		17.5	RICH CLUSTER OF GALAXIES
MCG+07-05-018	02 05 21.	+ 41 13	84	14.	GALAXY
HELW 367	02 05 26.	- 07 42 23.			NEBULA
MCG+01-06-047	02 05 27.	- 06 46	48	14.5	GALAXY
ZC 0205.5+0110	02 05 30.	+ 01 10	7800		CLUSTER OF GALAXIES
ZWG 387.041	02 05 30.	+ 01 39		14.9	GALAXY
UGC 01627	02 05 30.	+ 01 39	66	14.9	GALAXY S
MCG+00-06-041	02 05 30.	+ 01 40	30	15.	GALAXY
ZC 0205.5+3342	02 05 30.	+ 33 42	1680		CLUSTER OF GALAXIES
5ZW 190	02 05 30.	+ 35 55			COMPACT GALAXY
6ZW 172	02 05 30.	+ 37 25			COMPACT GALAXY
ZWG 538.026	02 05 30.	+ 42 46		15.4	GALAXY
UGC 01628	02 05 30.	+ 42 46	60	15.4	GALAXY S
ARC 0308	02 05 30.	- 03 36		17.2	RICH CLUSTER OF GALAXIES
MCG-03-06-007	02 05 30.	- 16 56	42	15.5	GALAXY
ZWG 438.031	02 05 36.	+ 14 06		13.7	GALAXY
UGC 01629	02 05 36.	+ 14 06	90	13.7	GALAXY Sb
ZWG 438.032	02 05 36.	+ 14 44		14.3	GALAXY
UGC 01630	02 05 36.	+ 14 44	72	14.3	GALAXY S
TON-S 0239	02 05 36.	- 24 06		14.8	BLUE STAR
MCG-06-05-030	02 05 36.	- 35 26	72	14.	GALAXY
MCG+00-06-042	02 05 39.	+ 00 58	15	15.	GALAXY
RNGC 0820	02 05 39.	+ 14 06		13.5	GALAXY
RNGC 0819	02 05 39.	+ 29 00		14.0	GALAXY
MCG+07-05-019	02 05 39.	+ 42 46	60	15.	GALAXY
RNGC 0821	02 05 40.	+ 10 45		12.5	GALAXY
RNGC 0818	02 05 41.	+ 38 32		12.5	GALAXY
ZWG 387.042	02 05 42.	+ 00 57		15.7	GALAXY
KARA.72 055A	02 05 42.	+ 00 57	42	15.7	PART OF DOUBLE GALAXY
ZWG 438.033	02 05 42.	+ 10 45		12.6	GALAXY
UGC 01631	02 05 42.	+ 10 45	210	12.6	GALAXY E
KARA.73B 0089	02 05 42.	+ 10 45	156	12.6	ISOLATED GALAXY E
MCG+02-06-034	02 05 42.	+ 10 46	66	12.	GALAXY
ZWG 461.050	02 05 42.	+ 18 45		15.7	GALAXY
ZWG 504.017	02 05 42.	+ 29 00		14.1	GALAXY
UGC 01632	02 05 42.	+ 29 00	39	14.1	GALAXY PECULR
6ZW 173	02 05 42.	+ 33 17			COMPACT GALAXY
ZWG 522.115	02 05 42.	+ 35 41		15.7	GALAXY
ZWG 522.116	02 05 42.	+ 38 32		12.7	GALAXY
UGC 01633	02 05 42.	+ 38 32	210	12.7	GALAXY SB:b-c
MCG+06-05-086	02 05 42.	+ 38 34	156	13.5	GALAXY
ZWG 553.001	02 05 42.	+ 46 59		15.0	GALAXY
UGC 01634	02 05 42.	+ 46 59	192	15.0	GALAXY Sc
MCG+08-05-002	02 05 42.	+ 46 59	120	13.	GALAXY
MCG-02-06-028	02 05 42.	- 10 35	54	14.5	GALAXY
REIN 2.023	02 05 42.41	+ 14 06 46.4			NEBULA
ARC 0309	02 05 43.	+ 02 46		17.8	RICH CLUSTER OF GALAXIES
MCG+00-06-043	02 05 45.	+ 00 57	36	15.	GALAXY
MCG+02-06-036	02 05 45.	+ 14 07 30.	72	14.	GALAXY
MCG+02-06-035	02 05 45.	+ 14 45	42	14.5	GALAXY
ZWG 387.043	02 05 48.	+ 00 56		15.5	GALAXY
KARA.72 055B	02 05 48.	+ 00 56	42	15.5	PART OF DOUBLE GALAXY
KARA.72 056A	02 05 48.	+ 06 09	96	14.8	PART OF DOUBLE GALAXY
MCG+01-06-044	02 05 48.	+ 06 09	90	13.5	GALAXY
ZWG 413.045	02 05 48.	+ 06 10		14.8	GALAXY
UGC 01635	02 05 48.	+ 06 10	126	14.8	GALAXY Sb
MCG-06-05-031	02 05 48.	- 34 34	30	15.	GALAXY
MCG-06-05-032	02 05 48.	- 34 35	18	15.	GALAXY
IC 0208	02 05 51.	+ 06 08			NONSTELLAR OBJECT
ARC 0310	02 05 52.	+ 05 14		17.6	RICH CLUSTER OF GALAXIES
RNGC 0825	02 05 53.	+ 06 05		14.5	GALAXY
ZC 0205.9+0248	02 05 54.	+ 02 48	1080		CLUSTER OF GALAXIES
ZWG 413.046	02 05 54.	+ 06 05		14.5	GALAXY
UGC 01636	02 05 54.	+ 06 05	150	14.5	GALAXY Sa
KARA.72 056B	02 05 54.	+ 06 05	132	14.5	PART OF DOUBLE GALAXY
MCG+01-06-045	02 05 54.	+ 06 05	120	13.5	GALAXY
ZWG 461.051	02 05 54.	+ 16 00		15.6	GALAXY
LDN 1350	02 05 54.	+ 60 45	240		DARK NEBULA
MCG+01-06-048	02 05 54.	- 03 42	48	15.5	GALAXY
MCG-05-06-007	02 05 54.	- 28 51	84	14.5	GALAXY
MCG-06-05-034	02 05 54.	- 34 30		15.	GALAXY
MCG-06-05-033	02 05 54.	- 34 35	18	15.5	GALAXY
KHAV 034	02 06	+ 56 26	3170		DARK NEBULA
MCG+07-05-020	02 06 00.	+ 44 02	60	16.	GALAXY

OBJECT NAME	RIGHT ASCEN.	DECLINATION	DIAM.	MAGN.	TYPE OF OBJECT
ZWG 538.027	02 06 00.	+ 44 03		15.6	GALAXY
UGC 01637	02 06 00.	+ 44 03	60	15.6	GALAXY Sc
MCG-03-06-008	02 06 00.	- 17 40	48	15.5	GALAXY
LIN 589	02 06 00.	- 74 51 51.		15.40	SMC EMISSION LINE OBJECT
HELW 368	02 06 02.	- 07 46 55.			NEBULA
MCG+06-05-087	02 06 03.	+ 34 51	42	15.	GALAXY
ZWG 483.004	02 06 06.	+ 25 48		15.4	GALAXY
UGC 01638	02 06 06.	+ 25 48	60	15.4	GALAXY Sc
KARA.73B 0090	02 06 06.	+ 25 48	60	15.4	ISOLATED GALAXY S
ZWG 483.005	02 06 06.	+ 27 18		15.4	GALAXY
ZWG 522.117	02 06 06.	+ 34 52		15.7	GALAXY
UGC 01639	02 06 06.	+ 34 52	60	15.7	GALAXY S
MCG-04-06-007	02 06 06.	- 24 47 30.	60	15.	GALAXY
RNGC 0829	02 06 10.	- 08 02		14.0	GALAXY
BIGO 474	02 06 11.	+ 07 39			NEBULA
RNGC 0827	02 06 11.	+ 07 44		14.0	GALAXY
ZWG 413.047	02 06 12.	+ 07 44		14.0	GALAXY
UGC 01640	02 06 12.	+ 07 44	162	14.0	GALAXY S
MCG+01-06-046	02 06 12.	+ 07 44	120	13.5	GALAXY
UGC 01644	02 06 12.	+ 30 06	60	18.	GALAXY DWARF
ZWG 504.018	02 06 12.	+ 31 46		14.7	GALAXY
UGC 01641	02 06 12.	+ 31 46	102	14.7	GALAXY
UGC 01642	02 06 12.	+ 36 58	84	17.	GALAXY
MCG+07-05-021	02 06 12.	+ 44 12	36	17.	GALAXY
KARA.68 013	02 06 12.	- 08 06	27		DWARF GALAXY
MCG+00-06-044	02 06 15.	+ 02 29	24	15.5	GALAXY
HELW 369	02 06 15.	- 08 04 01.			NEBULA
MCG-02-06-029	02 06 15.	- 08 51	36	15.	GALAXY
ZWG 387.044	02 06 18.	+ 00 32		15.5	GALAXY
MCG+00-06-045	02 06 18.	+ 00 34	24	15.5	GALAXY
ZWG 387.045	02 06 18.	+ 02 28		15.4	GALAXY
UGC 01643	02 06 18.	+ 02 28	78	15.4	GALAXY SBb
MCG+05-06-010	02 06 18.	+ 31 46	72	14.5	GALAXY
UGC 01645	02 06 18.	+ 34 02	60	17.	GALAXY S
6ZW 174	02 06 18.	+ 38 32			COMPACT GALAXY
MCG+01-06-049	02 06 21.	- 08 02	54	14.	GALAXY
ZWG 413.048	02 06 24.	+ 04 53		15.0	GALAXY
UGC 01646	02 06 24.	+ 04 53	132	15.0	GALAXY SBb-c
UGC 01647	02 06 24.	+ 05 54	78	16.0	GALAXY SB:c
ZWG 483.006	02 06 24.	+ 25 20		14.5	GALAXY
UGC 01648	02 06 24.	+ 25 20	72	14.5	GALAXY
KARA.73B 0091	02 06 24.	+ 25 20	66	14.5	ISOLATED GALAXY SO
ZWG 504.019	02 06 24.	+ 30 30		15.4	GALAXY
6ZW 175	02 06 24.	+ 35 25			COMPACT GALAXY
ZWG 538.028	02 06 24.	+ 44 15		15.5	GALAXY
MCG-06-05-035	02 06 24.	- 34 36	48	15.	GALAXY
HOLM 056B	02 06 26.	- 07 16	30	14.4	PART OF MULTIPLE GALAXY
ARC 0311	02 06 27.	+ 19 29		16.5	RICH CLUSTER OF GALAXIES
RNGC 0826	02 06 27.	+ 30 30		15.5	GALAXY
IC 0209	02 06 28.	- 07 17 51.			NONSTELLAR OBJECT
HOLM 056A	02 06 29.	- 07 18	48	13.5	PART OF MULTIPLE GALAXY
MCG+01-06-048	02 06 30.	+ 04 51	96	14.	GALAXY
MCG+01-06-047	02 06 30.	+ 06 21	48	15.5	GALAXY
UGC 01649	02 06 30.	+ 06 23	108	16.5	GALAXY
ZWG 504.020	02 06 30.	+ 33 15		15.7	GALAXY
UGC 01650	02 06 30.	+ 37 01	132	17.	GALAXY S
RNGC 0830	02 06 34.	- 08 00		14.0	GALAXY
ZC 0206.6+2306	02 06 36.	+ 23 06	3020		CLUSTER OF GALAXIES
5ZW 191	02 06 36.	+ 35 33			COMPACT GALAXY
MAI 001	02 06 36.	+ 82 30	40		DWARF SPHEROIDAL GALAXY
MCG+01-06-051	02 06 36.	- 07 17	108	14.	GALAXY
MCG+01-06-050	02 06 36.	- 08 00	42	14.	GALAXY
SNO 16	02 06 36.	- 42 44 23.	18	19.	GROUP OF 8 GALAXIES
MCG+06-05-088	02 06 39.	+ 35 33 30.	21	14.5	GALAXY
ZC 0206.7+0905	02 06 42.	+ 09 05	1410		CLUSTER OF GALAXIES
MCG+03-06-037	02 06 42.	+ 19 33	12	15.5	GALAXY
ZWG 461.052	02 06 42.	+ 21 22		15.7	GALAXY
ZWG 522.118	02 06 42.	+ 35 34		14.9	GALAXY
UGC 01651	02 06 42.	+ 35 34	138	14.9	GALAXY E?
6ZW 176	02 06 42.	+ 35 40			COMPACT GALAXY
MCG+03-06-038	02 06 45.	+ 21 02	78	14.5	GALAXY
ZC 0206.8+1025	02 06 48.	+ 10 25	1810		CLUSTER OF GALAXIES
ZWG 461.053	02 06 48.	+ 21 00		15.0	GALAXY
UGC 01652	02 06 48.	+ 21 00	96	15.0	GALAXY Sc
MCG+06-05-089	02 06 48.	+ 35 30 30.	36	16.	GALAXY
ZWG 553.002	02 06 48.	+ 45 47		15.7	GALAXY
5ZW 192	02 06 48.	+ 54 27			COMPACT GALAXY
VB 069	02 06 48.	+ 64 25	336		STELLAR RING
TON-S 0240	02 06 48.	- 27 49		15.1	BLUE STAR
RNGC 0831	02 06 53.	+ 05 52		15.0	GALAXY
ZC 0206.9+0437	02 06 54.	+ 04 37	1480		CLUSTER OF GALAXIES
ZWG 413.049	02 06 54.	+ 05 52		15.2	GALAXY
ZWG 522.119	02 06 54.	+ 37 57		15.6	GALAXY
ARP 318	02 06 55.	- 10 22			PECULIAR GALAXY
MCG+06-05-091	02 06 57.	+ 36 28	48	14.	GALAXY
IC 0210	02 06 57.	- 09 54 35.			NONSTELLAR OBJECT
MCG-04-06-008	02 06 57.	- 22 55	54	15.5	GALAXY
RNGC 0835	02 06 58.	- 10 22		13.0	GALAXY
RNGC 0833	02 06 58.	- 10 22		14.0	GALAXY
HOLM 057A	02 06 59.	- 07 13	48	13.5	PART OF MULTIPLE GALAXY
VDB.66G 018	02 07	+ 06 32	70		DWARF GALAXY
ZC 0207.0+0529	02 07 00.	+ 05 29	2620		CLUSTER OF GALAXIES
MCG+02-06-037	02 07 00.	+ 13 45	36	16.	GALAXY
ZC 0207.0+2830	02 07 00.	+ 28 30	870		CLUSTER OF GALAXIES
ZWG 504.021	02 07 00.	+ 29 17		15.5	GALAXY
KARA.73B 0092	02 07 00.	+ 29 17	36	15.5	ISOLATED GALAXY S
ZWG 522.120	02 07 00.	+ 33 55		15.7	GALAXY
ZWG 522.121	02 07 00.	+ 34 43		15.4	GALAXY
ZWG 522.122	02 07 00.	+ 35 29		14.9	GALAXY
ZWG 522.123	02 07 00.	+ 36 47		15.6	GALAXY
ZC 0207.0-0113	02 07 00.	- 01 13	2150		CLUSTER OF GALAXIES
MCG+01-06-052	02 07 00.	- 06 26	72	14.5	GALAXY
MCG-02-06-030	02 07 00.	- 10 21	84	14.	GALAXY
ARC 0312	02 07 03.	+ 04 38		17.5	RICH CLUSTER OF GALAXIES
MCG-02-06-031	02 07 03.	- 10 21	60	13.5	GALAXY
RNGC 0828	02 07 05.	+ 38 57		13.0	GALAXY
ZWG 413.050	02 07 06.	+ 07 05		15.6	GALAXY
UGC 01653	02 07 06.	+ 07 05	60	15.6	GALAXY PECULR
ZC 0207.1+0940	02 07 06.	+ 09 40	2420		CLUSTER OF GALAXIES
ZWG 522.124	02 07 06.	+ 36 28		15.1	GALAXY
UGC 01654	02 07 06.	+ 36 28	66	15.1	GALAXY Sc
6ZW 177	02 07 06.	+ 38 57			COMPACT GALAXY
ZWG 522.125	02 07 06.	+ 38 57		13.0	GALAXY
UGC 01655	02 07 06.	+ 38 57	210	13.0	GALAXY PECULR
MCG+06-05-092	02 07 06.	+ 38 58	138	13.	GALAXY
UGC 01656	02 07 06.	+ 41 20	84	18.	GALAXY
UGC 01657	02 07 06.	+ 45 22	90	16.0	GALAXY
LDN 1351	02 07 06.	+ 60 35	180		DARK NEBULA

OBJECT NAME	RIGHT ASCEN.	DECLINATION	DIAM.	MAGN.	TYPE OF OBJECT
MCG+01-06-053	02 07 06.	- 07 11 30.	42	14.5	GALAXY
MCG-02-06-032	02 07 06.	- 09 54	120	13.5	GALAXY
MCG-04-06-009	02 07 06.	- 23 39	102	13.	GALAXY
RNGC 0832	02 07 07.	+ 35 18			GALAXY
HOLM 057B	02 07 08.	- 07 10	60	13.7	PART OF MULTIPLE GALAXY
RNGC 0838	02 07 10.	- 10 23		13.0	GALAXY
ZWG 438.034	02 07 12.	+ 10 44		15.4	GALAXY
UGC 01658	02 07 12.	+ 10 44	60	15.4	GALAXY S
ZWG 438.035	02 07 12.	+ 11 07		15.7	GALAXY
ZWG 461.054	02 07 12.	+ 15 48		14.9	GALAXY
UGC 01659	02 07 12.	+ 15 48	102	14.9	GALAXY SBc
6ZW 178	02 07 12.	+ 34 46			COMPACT GALAXY
ZWG 522.126	02 07 12.	+ 38 21		15.4	GALAXY
UGC 01660	02 07 12.	+ 38 21	60	15.4	GALAXY Sc
MCG+06-05-093	02 07 12.	+ 38 22	48	15.	GALAXY
ZWG 538.029	02 07 12.	+ 41 17		15.5	GALAXY
UGC 01661	02 07 12.	+ 41 17	84	15.5	GALAXY (E)
ZWG 559.005	02 07 12.	+ 52 05		15.3	GALAXY
VB 070	02 07 12.	+ 65 06	591		STELLAR RING
ABC 0314	02 07 12.	- 13 09		17.6	RICH CLUSTER OF GALAXIES
MCG-03-06-009	02 07 12.	- 17 02	15	15.	GALAXY
MCG+02-06-039	02 07 15.	+ 10 32	48	15.	GALAXY
MCG+02-06-038	02 07 15.	+ 10 43	42	15.	GALAXY
MCG+03-06-039	02 07 15.	+ 15 47 30.	90	14.	GALAXY
MCG+01-06-054	02 07 15.	- 07 09	66	14.5	GALAXY
LIN 590	02 07 15.	- 75 02 24.		15.96	SMC EMISSION LINE OBJECT
RNGC 0839	02 07 16.	- 10 26		13.0	GALAXY
RNGC 0852	02 07 16.	- 56 58			GALAXY
ZWG 438.036	02 07 18.	+ 10 32		15.3	GALAXY
UGC 01662	02 07 18.	+ 10 32	60	15.3	GALAXY Sa-b
6ZW 179	02 07 18.	+ 34 26			COMPACT GALAXY
ZC 0207.3+3755	02 07 18.	+ 37 55	1480		CLUSTER OF GALAXIES
MCG+07-05-022	02 07 18.	+ 41 16	18	15.	GALAXY
MCG-02-06-033	02 07 18.	- 10 22	60	13.5	GALAXY
MCG-04-06-010	02 07 18.	- 23 33	54	17.	GALAXY
RNGC 0842	02 07 22.	- 08 00		14.0	GALAXY
ZC 0207.4+0242	02 07 24.	+ 02 42	2080		CLUSTER OF GALAXIES
ZWG 413.051	02 07 24.	+ 07 25		15.1	GALAXY
UGC 01663	02 07 24.	+ 07 25	126	15.1	GALAXY Sc
ZC 0207.4+0754	02 07 24.	+ 07 54	1280		CLUSTER OF GALAXIES
ZWG 483.007	02 07 24.	+ 25 26		15.3	GALAXY
ZC 0207.4+2700	02 07 24.	+ 27 00	1680		CLUSTER OF GALAXIES
MCG-02-06-034	02 07 24.	- 10 24	72	13.5	GALAXY
MCG+01-06-050	02 07 30.	+ 07 24	72	14.5	GALAXY
MCG+01-06-049	02 07 30.	+ 07 36	120	14.	GALAXY
MCG+01-06-055	02 07 30.	- 08 00	54	14.	GALAXY
ABC 0316	02 07 30.	- 13 44		17.2	RICH CLUSTER OF GALAXIES
ABC 0313	02 07 31.	+ 02 40		17.6	RICH CLUSTER OF GALAXIES
ABC 0315	02 07 32.	- 01 15		17.5	RICH CLUSTER OF GALAXIES
IC 0211	02 07 33.	+ 03 36 30.			NONSTELLAR OBJECT
HELW 370	02 07 33.	- 07 51 46.			NEBULA
RNGC 0844	02 07 35.	+ 05 49		15.0	GALAXY
RNGC 0840	02 07 35.	+ 07 36		14.5	GALAXY
ZWG 413.052	02 07 36.	+ 05 49		15.0	GALAXY
ZWG 413.053	02 07 36.	+ 07 36		14.7	GALAXY
UGC 01664	02 07 36.	+ 07 36	156	14.7	GALAXY SBb
ZWG 461.055	02 07 36.	+ 21 09		15.1	GALAXY
MCG+03-06-040	02 07 36.	+ 21 10	24	15.	GALAXY
ZWG 538.030	02 07 36.	+ 41 17		15.4	GALAXY
MCG+00-06-046	02 07 36.	- 01 37 30.	36	15.	GALAXY
ZWG 387.046	02 07 36.	- 01 38		15.5	GALAXY
MCG+07-05-023	02 07 39.	+ 41 15	48	16.	GALAXY
ABC 0317	02 07 41.	- 08 47		17.6	RICH CLUSTER OF GALAXIES
ZWG 504.022	02 07 42.	+ 31 06		15.0	GALAXY
MCG+06-05-095	02 07 42.	+ 34 53	48	16.	GALAXY
MCG+06-05-094	02 07 42.	+ 34 57	15	14.5	GALAXY
5ZW 193	02 07 42.	+ 34 58			COMPACT GALAXY
MCG+01-06-056	02 07 42.	- 07 51 30.	30	15.	GALAXY
MCG-06-05-036	02 07 42.	- 35 37	36	15.5	GALAXY
LIN 591	02 07 43.	- 74 32 01.		14.81	SMC EMISSION LINE OBJECT
LB 00466	02 07 45.	- 03 55 06.		15.8	FAINT BLUE STAR
HELW 009	02 07 45.	- 09 57			NEBULA
UGC 01665	02 07 48.	+ 00 58	60	18.	GALAXY DWARF
ZWG 413.054	02 07 48.	+ 08 10		15.7	GALAXY
MCG+06-05-097	02 07 48.	+ 34 43	36	16.	GALAXY
MCG+06-05-096	02 07 48.	+ 34 44	12	15.	GALAXY
UGC 01666	02 07 48.	+ 34 45	66	16.0	GALAXY DISTRBD
UGC 01667	02 07 48.	+ 34 54	60	16.5	GALAXY S IV
ZWG 522.127	02 07 48.	+ 34 59		15.7	GALAXY
UGC 01668	02 07 48.	+ 34 59	66	15.7	GALAXY PECULR
ZWG 538.031	02 07 48.	+ 41 07		15.7	GALAXY
MCG-06-05-037	02 07 48.	- 33 09	90	12.5	GALAXY
IC 1783	02 07 48.	- 33 13	72	13.1	GALAXY SA
RNGC 0849	02 07 51.	- 22 33			GALAXY
RNGC 0848	02 07 52.	- 10 34		13.0	GALAXY
MCG+01-06-051	02 07 54.	+ 05 38	36	15.	GALAXY
VB 071	02 07 54.	+ 63 38	2276		STELLAR RING
ISS 0016	02 07 54.	+ 63 38	2279		STELLAR RING
MCG-02-06-035	02 07 54.	- 09 56	108	13.5	GALAXY
MCG+06-05-098	02 07 57.	+ 37 35	60	15.	GALAXY
LBN 0641	02 08	+ 62 20	1320		BRIGHT NEBULA
ZWG 387.047	02 08 00.	+ 00 30		15.7	GALAXY
ZWG 413.055	02 08 00.	+ 05 38		14.5	GALAXY
MRK 587	02 08 00.	+ 05 38	20	15.	GALAXY WITH UV CONTINUUM
UGC 01669	02 08 00.	+ 05 38	66	14.5	GALAXY S
UGC 01670	02 08 00.	+ 06 31	150	16.0	GALAXY DWRF SP
MCG+01-06-052	02 08 00.	+ 06 31	120	15.5	GALAXY
ZC 0208.0+1515	02 08 00.	+ 15 15	18140		CLUSTER OF GALAXIES
ZWG 504.023	02 08 00.	+ 31 03		15.1	GALAXY
ZWG 504.024	02 08 00.	+ 32 28		15.5	GALAXY
UGC 01671	02 08 00.	+ 32 28	66	15.5	GALAXY SB
KARA.72 057A	02 08 00.	+ 32 28	66	15.5	PART OF DOUBLE GALAXY
ZWG 504.025	02 08 00.	+ 32 30		15.6	GALAXY
KARA.72 057B	02 08 00.	+ 32 30	54	15.6	PART OF DOUBLE GALAXY
6ZW 180	02 08 00.	+ 33 58			COMPACT GALAXY
ZWG 522.128	02 08 00.	+ 37 26		13.2	GALAXY
RNGC 0834	02 08 00.	+ 37 26		13.0	GALAXY
UGC 01672	02 08 00.	+ 37 26	72	13.2	GALAXY S
MCG+06-05-099	02 08 00.	+ 37 26	60	14.	GALAXY
ZWG 522.129	02 08 00.	+ 37 35		15.2	GALAXY
UGC 01673	02 08 00.	+ 37 35	66	15.2	GALAXY S
MCG+14-02-008	02 08 00.	+ 82 22	48	16.	GALAXY
MCG-02-06-036	02 08 00.	- 10 32	84	13.5	GALAXY
MCG-04-06-011	02 08 00.	- 22 40	48	15.	GALAXY
RNGC 0837	02 08 03.	- 22 40		15.0	GALAXY
6ZW 181	02 08 06.	+ 34 51			COMPACT GALAXY
ZWG 522.130	02 08 06.	+ 38 31		15.1	GALAXY
UGC 01674	02 08 06.	+ 38 31	66	15.1	GALAXY Sc
MCG+01-06-058	02 08 06.	- 06 50 30.	12	16.	GALAXY
MCG+01-06-057	02 08 06.	- 06 51	6	18.	GALAXY
RNGC 0843	02 08 08.	+ 31 52			NON-EXISTENT OBJECT
RNGC 0836	02 08 08.	- 22 18		14.0	GALAXY
MCG+06-05-100	02 08 09.	+ 38 32	54	15.	GALAXY
MCG+01-06-059	02 08 09.	- 06 50	12	17.	GALAXY
MCG-04-06-012	02 08 09.	- 22 18	42	14.5	GALAXY
ZWG 461.056	02 08 12.	+ 18 55		15.6	GALAXY
MCG+03-06-041	02 08 12.	+ 18 55	30	15.	GALAXY
MCG+05-06-011	02 08 12.	+ 32 28	54	15.	GALAXY
MCG+05-06-012	02 08 12.	+ 32 30	48	15.	GALAXY
5ZW 194	02 08 12.	+ 37 15			COMPACT GALAXY
5ZW 195	02 08 12.	+ 50 06			COMPACT GALAXY
ZC 0208.2-0050	02 08 12.	- 00 50	610		CLUSTER OF GALAXIES
MCG+01-06-060	02 08 12.	- 06 49 30.	18	17.	GALAXY
LB 01612	02 08 12.	- 56 28		14.2	FAINT BLUE STAR
MCG+01-06-061	02 08 14.	- 06 48 30.	24	16.	GALAXY
MCG+01-06-063	02 08 15.	- 03 00	78	15.	GALAXY
MCG+01-06-062	02 08 15.	- 06 48	15	16.	GALAXY
MCG-03-06-010	02 08 15.	- 16 02	78	14.5	GALAXY
FEIG 019	02 08 18.	+ 01 34		14.4	FAINT BLUE STAR
MCG+03-06-042	02 08 18.	+ 16 57 30.	30	15.	GALAXY
ZWG 461.057	02 08 18.	+ 16 58		15.7	GALAXY
ZWG 483.008	02 08 18.	+ 25 35		15.5	GALAXY
UGC 01675	02 08 18.	+ 35 35	72	15.5	GALAXY SB
ZWG 522.131	02 08 18.	+ 37 16		12.8	GALAXY
RNGC 0841	02 08 18.	+ 37 16		13.0	GALAXY
UGC 01676	02 08 18.	+ 37 16	126	12.8	GALAXY SBa/Sb
MCG+06-05-101	02 08 18.	+ 37 16	96	13.	GALAXY
HELW 371	02 08 19.	- 07 47 42.			NEBULA
UGC 01677	02 08 24.	+ 06 27	72	18.	GALAXY
ZWG 413.056	02 08 24.	+ 07 26		15.4	GALAXY
ZC 0208.4+4140	02 08 24.	+ 41 40	1340		CLUSTER OF GALAXIES
PK131+02.1	02 08 24.	+ 63 55	66	18.2	PLANETARY NEBULA
MCG+01-06-064	02 08 24.	- 06 46	18	16.	GALAXY
LB 00173	02 08 24.	- 15 34		17.0	FAINT BLUE STAR
MCG-03-06-011	02 08 24.	- 20 02	24	15.	GALAXY
MCG-04-06-013	02 08 24.	- 22 53	72	16.	GALAXY
MCG+01-06-053	02 08 30.	+ 03 36	120	13.	GALAXY
ZWG 413.057	02 08 30.	+ 03 38		14.5	GALAXY
UGC 01678	02 08 30.	+ 03 38	156	14.5	GALAXY Sc
KARA.72 059A	02 08 30.	+ 03 38	162	14.5	PART OF DOUBLE GALAXY
ZWG 438.037	02 08 30.	+ 13 53		14.8	GALAXY
ZWG 522.132	02 08 30.	+ 34 36		15.6	GALAXY
MCG+00-06-047	02 08 30.	- 00 53	30	15.	GALAXY
ZWG 387.048	02 08 30.	- 00 54		14.7	GALAXY
KARA.72 058B	02 08 30.	- 00 54	42		PART OF DOUBLE GALAXY
KARA.72 058A	02 08 30.	- 00 54	48	14.7	PART OF DOUBLE GALAXY
MCG+00-06-048	02 08 33.	- 00 53	42	15.	GALAXY
ZWG 387.051	02 08 36.	+ 00 50		15.6	GALAXY
MCG+01-06-054	02 08 36.	+ 03 31	48	14.5	GALAXY
ZWG 413.058	02 08 36.	+ 03 33		14.7	GALAXY
MRK 588	02 08 36.	+ 03 33	22	14.5	GALAXY WITH UV CONTINUUM
RNGC 0851	02 08 36.	+ 03 33		14.5	GALAXY
UGC 01680	02 08 36.	+ 03 33	78	14.7	GALAXY S0
KARA.72 059B	02 08 36.	+ 03 33	78	14.7	PART OF DOUBLE GALAXY
MCG+02-06-040	02 08 36.	+ 13 53	24	15.	GALAXY
UGC 01681	02 08 36.	+ 14 16	60	16.5	GALAXY Sc
ZWG 461.058	02 08 36.	+ 15 40		15.4	GALAXY
ZWG 504.026	02 08 36.	+ 31 17		14.7	GALAXY
UGC 01682	02 08 36.	+ 31 17	72	14.7	GALAXY Sc
5ZW 196	02 08 36.	+ 38 05			COMPACT GALAXY
5ZW 197	02 08 36.	+ 53 08			COMPACT GALAXY
ZWG 387.050	02 08 36.	- 01 04		15.3	GALAXY
ZWG 387.049	02 08 36.	- 01 44		14.1	GALAXY
UGC 01679	02 08 36.	- 01 44	102	14.1	GALAXY S0-a
MCG+01-06-065	02 08 36.	- 06 46	18	17.	GALAXY
RNGC 0850	02 08 38.	- 01 44		14.0	GALAXY
MCG+06-05-102	02 08 39.	+ 33 47	60	14.5	GALAXY
MCG-04-06-014	02 08 39.	- 21 26	36	15.	GALAXY
UGC 01683	02 08 42.	+ 15 41	60	16.5	GALAXY Sb-c
MCG+03-06-043	02 08 42.	+ 15 43	78	14.	GALAXY
ZWG 461.059	02 08 42.	+ 15 44		14.8	GALAXY
UGC 01684	02 08 42.	+ 15 44	102	14.8	GALAXY SB0
MCG+05-06-013	02 08 42.	+ 31 16	54	14.5	GALAXY
ZWG 522.133	02 08 42.	+ 33 49		14.9	GALAXY
UGC 01685	02 08 42.	+ 33 49	66	14.9	GALAXY Sa-b
MCG+00-06-049	02 08 42.	- 01 42 30.	60	14.	GALAXY
MCG+01-06-066	02 08 45.	- 06 39	54	15.	GALAXY
ZWG 438.038	02 08 48.	+ 11 36		15.7	GALAXY
KARA.73B 0093	02 08 48.	+ 11 36	30	15.7	ISOLATED GALAXY E
ZC 0208.8+1145	02 08 48.	+ 11 45	1810		CLUSTER OF GALAXIES
VV 044B	02 08 48.	+ 14 03 30.	4	19.	INTERACTING GALAXY
VV 044A	02 08 48.	+ 14 03 30.	24	16.	INTERACTING GALAXY
ZWG 438.039	02 08 48.	+ 14 04		15.6	GALAXY
MCG+02-06-041	02 08 48.	+ 14 04	24	16.	GALAXY
VB 072	02 08 48.	+ 63 46	342		STELLAR RING
3ZW 042	02 08 54.	+ 13 40			COMPACT GALAXY
ZWG 438.040	02 08 54.	+ 13 40		15.0	GALAXY
MRK 366	02 08 54.	+ 13 40	12	15.5	GALAXY WITH UV CONTINUUM
ZWG 504.027	02 08 54.	+ 30 17		15.7	GALAXY
5ZW 198	02 08 54.	+ 30 23			COMPACT GALAXY
MCG+08-05-003	02 08 54.	+ 46 07	42	16.	GALAXY
UGC 01686	02 08 54.	+ 46 08	72	17.	GALAXY
ZWG 438.041	02 09 00.	+ 14 04		15.3	GALAXY
UGC 01687	02 09 00.	+ 14 04	66	15.3	GALAXY S0
MCG+02-06-042	02 09 00.	+ 14 04	18	14.5	GALAXY
5ZW 199	02 09 00.	+ 30 17			COMPACT GALAXY
ZWG 538.032	02 09 00.	+ 44 20		13.2	GALAXY
UGC 01688	02 09 00.	+ 44 20	138	13.2	GALAXY SBb
MCG+07-05-024	02 09 00.	+ 44 20	96	13.	GALAXY
MCG-02-06-037	02 09 00.	- 10 20	36	15.	GALAXY
RNGC 0846	02 09 02.	+ 44 20		13.0	GALAXY
RNGC 0847	02 09 02.	+ 44 26			GALAXY
ZC 0209.1+0008	02 09 06.	+ 00 08	670		CLUSTER OF GALAXIES
UGC 01689	02 09 06.	+ 13 46	84	16.5	GALAXY DWARF
ZC 0209.1+2317	02 09 06.	+ 23 17	870		CLUSTER OF GALAXIES
ZWG 504.028	02 09 06.	+ 29 05		15.7	GALAXY
UGC 01690	02 09 06.	+ 29 05	96	15.7	GALAXY S
6ZW 182	02 09 06.	+ 35 28			COMPACT GALAXY
6ZW 183	02 09 06.	+ 39 00			COMPACT GALAXY
ZWG 522.134	02 09 06.	+ 39 00		14.6	GALAXY
UGC 01691	02 09 06.	+ 39 00	108	14.6	GALAXY S0
UGC 01692	02 09 06.	+ 42 09	108	16.5	GALAXY Sb-c
LB 00467	02 09 08.	- 05 20 36.		14.9	FAINT BLUE STAR
MCG+06-05-103	02 09 09.	+ 39 01	42	14.	GALAXY
UGC 01693	02 09 12.	+ 13 52	96	17.	GALAXY DWRF SP
5ZW 200	02 09 12.	+ 30 51			COMPACT GALAXY

OBJECT NAME	RIGHT ASCEN.	DECLINATION	DIAM.	MAGN.	TYPE OF OBJECT
ZWG 504.029	02 09 12.	+ 31 21		15.0	GALAXY
MCG+07-05-025	02 09 12.	+ 42 09	36	17.	GALAXY
MCG-03-06-012	02 09 12.	- 20 18 30.	24	15.5	GALAXY
MCG+01-06-067	02 09 15.	- 06 42 30.	84	15.	GALAXY
RNGC 0845	02 09 17.	+ 37 15		14.5	GALAXY
UGC 01694	02 09 18.	+ 09 19	96	17.	GALAXY
ZWG 438.042	02 09 18.	+ 15 21		15.2	GALAXY
ZC 0209.3+2125	02 09 18.	+ 21 25	1480		CLUSTER OF GALAXIES
MCG+05-06-014	02 09 18.	+ 31 20	30	15.	GALAXY
MCG+06-05-104	02 09 18.	+ 37 14	90	14.	GALAXY
ZWG 522.135	02 09 18.	+ 37 15		14.5	GALAXY
UGC 01695	02 09 18.	+ 37 15	108	14.5	GALAXY Sb
MCG+01-06-068	02 09 18.	- 06 07 30.	54	15.	GALAXY
LB 00082	02 09 18.	- 16 05		16.7	FAINT BLUE STAR
MCG-06-05-038	02 09 18.	- 36 04 30.	120	13.	GALAXY
LB 01613	02 09 18.	- 48 06		14.4	FAINT BLUE STAR
MCG+02-06-043	02 09 21.	+ 13 47 30.	60	16.	GALAXY
RNGC 0853	02 09 22.	- 09 32		13.0	GALAXY
MCG+00-06-050	02 09 24.	+ 00 59	30	16.	GALAXY
FBIG 020	02 09 24.	+ 08 33		14.0	FAINT BLUE STAR
MCG+02-06-044	02 09 24.	+ 13 52 30.	84	17.	GALAXY
MCG-02-06-038	02 09 24.	- 09 32	96	13.5	GALAXY
MCG-03-06-013	02 09 24.	- 18 32 30.	78	14.5	GALAXY
MCG-03-06-014	02 09 24.	- 20 12	60	15.	GALAXY
ARC 0319	02 09 25.	- 12 20		17.6	RICH CLUSTER OF GALAXIES
RNGC 0854	02 09 26.	- 36 05			GALAXY
MCG+02-06-045	02 09 27.	+ 15 22	48	15.	GALAXY
B 201	02 09 31.	+ 56 51	600		DARK OBJECT
SCHO 0045	02 09 33.	+ 58 06 00.	590		ISOLATED DARK CLOUD
ZWG 504.030	02 09 36.	+ 29 37		15.6	GALAXY
UGC 01696	02 09 36.	+ 29 37	78	15.6	GALAXY
MCG+05-06-015	02 09 36.	+ 29 37	60	15.	GALAXY
6ZW 184	02 09 36.	+ 34 59			COMPACT GALAXY
MCG+01-06-069	02 09 36.	- 02 41	36	16.	GALAXY
LB 00174	02 09 36.	- 15 06		16.2	FAINT BLUE STAR
MCG-04-06-015	02 09 39.	- 20 36 30.	60	14.	GALAXY
ZWG 413.059	02 09 42.	+ 08 23		15.3	GALAXY
6ZW 185	02 09 42.	+ 39 01			COMPACT GALAXY
MCG+09-04-004	02 09 42.	+ 53 10	96	14.	GALAXY
UGC 01699	02 09 42.	+ 53 12	90	15.1	GALAXY S0-a
ZWG 387.053	02 09 42.	- 01 03		15.3	GALAXY
UGC 01698	02 09 42.	- 01 03	78	15.3	GALAXY SBa
ZWG 387.052	02 09 42.	- 02 23		15.0	GALAXY
UGC 01697	02 09 42.	- 02 23	84	15.0	GALAXY Sb
MCG+06-05-105	02 09 45.	+ 36 03 30.	78	15.	GALAXY
MCG+00-06-051	02 09 45.	- 01 02 30.	36	15.	GALAXY
ARC 0321	02 09 46.	+ 00 10		17.6	RICH CLUSTER OF GALAXIES
ZWG 461.060	02 09 48.	+ 16 37		15.7	GALAXY
UGC 01700	02 09 48.	+ 16 37	60	15.7	GALAXY Sc
ZWG 522.136	02 09 48.	+ 36 05		15.0	GALAXY
ZCG 0209+36	02 09 48.	+ 36 05		15.0	COMPACT GALAXY
UGC 01701	02 09 48.	+ 36 05	78	15.0	GALAXY Sb
ZWG 522.137	02 09 48.	+ 37 35		15.7	GALAXY
MCG+00-06-052	02 09 48.	- 02 22 30.	60	15.	GALAXY
MCG-02-06-039	02 09 51.	- 13 56	12	15.5	GALAXY
ZWG 438.043	02 09 54.	+ 14 08		15.2	GALAXY
VV 027B	02 09 54.	+ 14 08	6	18.	INTERACTING GALAXY
VV 027A	02 09 54.	+ 14 08	42	15.	INTERACTING GALAXY
MCG+02-06-046	02 09 54.	+ 14 09	36	15.	GALAXY
SCHO 0046	02 09 54.	+ 56 45 24.	660		ISOLATED DARK CLOUD
UGC 01702	02 10 00.	+ 13 44	60	16.0	GALAXY
MCG+02-06-047	02 10 00.	+ 13 44	60	15.	GALAXY
ZWG 461.061	02 10 00.	+ 18 57		15.7	GALAXY
MCG+03-06-044	02 10 00.	+ 18 57 30.	42	15.	GALAXY
UGC 01703	02 10 00.	+ 32 37	108	18.	GALAXY DWARF
6ZW 186	02 10 00.	+ 36 12			COMPACT GALAXY
ZC 0210.0+3852	02 10 00.	+ 38 52	4700		CLUSTER OF GALAXIES
ZWG 559.006	02 10 00.	+ 53 12		15.1	GALAXY
LDN 1349	02 10 00.	+ 66 00	4380		DARK NEBULA
ARC 0318	02 10 01.	+ 26 13		16.9	RICH CLUSTER OF GALAXIES
ZC 0210.1+1857	02 10 06.	+ 18 57	4910		CLUSTER OF GALAXIES
ZC 0210.2+2255	02 10 12.	+ 22 55	1080		CLUSTER OF GALAXIES
ZWG 538.033	02 10 12.	+ 41 01		15.5	GALAXY
UGC 01704	02 10 12.	+ 41 01	96	15.5	GALAXY Sc
ZC 0210.2+5130	02 10 12.	+ 51 30	1010		CLUSTER OF GALAXIES
RNGC 0858	02 10 15.	- 22 41		13.0	GALAXY
MCG-04-06-016	02 10 15.	- 22 41 30.	72	13.5	GALAXY
ARC 0320	02 10 17.	+ 25 10		17.7	RICH CLUSTER OF GALAXIES
ZWG 504.031	02 10 18.	+ 27 44		15.7	GALAXY
ZC 0210.3+3049	02 10 18.	+ 30 49	1010		CLUSTER OF GALAXIES
MCG+07-05-026	02 10 21.	+ 41 00	60	16.	GALAXY
ZWG 413.060	02 10 24.	+ 09 15		15.2	GALAXY
ZWG 413.061	02 10 24.	+ 09 16		15.6	GALAXY
UGC 01705	02 10 24.	+ 09 16	60	15.6	GALAXY SBb
6ZW 187	02 10 24.	+ 35 56			COMPACT GALAXY
6ZW 188	02 10 24.	+ 38 12			COMPACT GALAXY
MCG-05-06-008	02 10 24.	- 32 11	72	13.5	GALAXY
RNGC 0857	02 10 25.	- 32 11		13.0	GALAXY
MCG+00-06-053	02 10 30.	+ 01 10	36	15.	GALAXY
6ZW 189	02 10 30.	+ 38 13			COMPACT GALAXY
MCG-03-06-015	02 10 33.	- 15 21	72	14.5	GALAXY
MCG-03-06-016	02 10 33.	- 19 35	66	15.	GALAXY
6ZW 190	02 10 36.	+ 36 43			COMPACT GALAXY
LB 00175	02 10 36.	- 16 00		16.5	FAINT BLUE STAR
ZWG 483.009	02 10 42.	+ 25 37		14.7	GALAXY
UGC 01706	02 10 42.	+ 25 37	72	14.7	GALAXY Sc
KARA.73B 0094	02 10 42.	+ 25 37	60	14.7	ISOLATED GALAXY S
ZWG 504.032	02 10 42.	+ 30 58		15.7	GALAXY
ZWG 522.138	02 10 42.	+ 36 05		15.7	GALAXY
6ZW 191	02 10 42.	+ 36 44			COMPACT GALAXY
ARC 0325	02 10 43.	- 25 31		17.9	RICH CLUSTER OF GALAXIES
MCG+01-06-073	02 10 45.	- 02 31	36	15.	GALAXY
MCG+01-06-072	02 10 45.	- 03 16 30.	48	15.	GALAXY
MCG+01-06-071	02 10 45.	- 06 33	15	15.	GALAXY
MCG+01-06-070	02 10 45.	- 07 53	84	13.5	GALAXY
ZWG 413.062	02 10 48.	+ 04 31		15.5	GALAXY
UGC 01707	02 10 48.	+ 04 31	78	15.5	GALAXY SBb
UGC 01708	02 10 48.	+ 09 23	78	16.0	GALAXY S
5ZW 201	02 10 48.	+ 39 24			COMPACT GALAXY
VVI 09	02 10 48.	+ 86 05		19.	SEYFERT GALAXY
ARC 0327	02 10 48.	- 26 21		17.9	RICH CLUSTER OF GALAXIES
LB 01614	02 10 48.	- 58 48		13.2	FAINT BLUE STAR
SHB 057	02 10 49.	+ 86 05 06.		19.	QUASI-STELLAR OBJECT
BC 3CR61.1.BS	02 10 49.	+ 86 05 10.		19.	QUASI-STELLAR OBJECT
IC 0212	02 10 52.	+ 16 21 40.			NONSTELLAR OBJECT
UGC 01709	02 10 54.	+ 03 45	66	16.0	GALAXY
MCG+01-06-055	02 10 54.	+ 04 31	48	15.	GALAXY
ARC 0322	02 10 54.	+ 07 17		17.6	RICH CLUSTER OF GALAXIES
UGC 01710	02 10 54.	+ 13 29	60	16.0	GALAXY SBa
MCG+02-06-048	02 10 54.	+ 13 30	36	16.	GALAXY
ZWG 461.062	02 10 54.	+ 16 22		15.6	GALAXY
ZWG 461.063	02 10 54.	+ 16 51		15.5	GALAXY
MRK 367	02 10 54.	+ 16 51	20	16.	GALAXY WITH UV CONTINUUM
UGC 01711	02 10 54.	+ 24 40	60	16.5	GALAXY DWARF
UGC 01712	02 10 54.	+ 33 23	66	17.	GALAXY
ZWG 538.034	02 10 54.	+ 41 39		15.0	GALAXY
ZC 0210.9+5038	02 10 54.	+ 50 38	8000		CLUSTER OF GALAXIES
LB 01615	02 10 54.	- 47 50		14.8	FAINT BLUE STAR
UGC 01714	02 11 00.	+ 10 06	78	17.	GALAXY
MCG+03-06-045	02 11 00.	+ 16 33	48	15.	GALAXY
ZWG 461.064	02 11 00.	+ 16 52		15.7	GALAXY
ZWG 461.065	02 11 00.	+ 19 01		15.6	GALAXY
ZC 0211.0+2550	02 11 00.	+ 25 50	1680		CLUSTER OF GALAXIES
ZWG 504.033	02 11 00.	+ 28 39		15.6	GALAXY
5ZW 202	02 11 00.	+ 31 38			COMPACT GALAXY
ZWG 504.034	02 11 00.	+ 31 38		15.4	GALAXY
MCG+07-05-027	02 11 00.	+ 41 39	48	15.	GALAXY
UGC 01715	02 11 00.	+ 49 47	72	16.0	GALAXY Sc/SBc
ASS 42	02 11 00.	+ 57 05	21600		OB ASSOCIATION PER OB1
ZWG 387.054	02 11 00.	- 00 58		14.4	GALAXY
UGC 01713	02 11 00.	- 00 58	84	14.4	GALAXY Sa
MCG+01-06-074	02 11 00.	- 07 57	42	15.5	GALAXY
MCG-02-06-040	02 11 00.	- 14 13	12	15.5	GALAXY
MCG-03-06-017	02 11 00.	- 19 05 30.	48	15.	GALAXY
RNGC 0862	02 11 00.	- 42 18			GALAXY
RNGC 0856	02 11 01.	- 00 58		14.5	GALAXY
MCG+06-05-106	02 11 03.	+ 35 29 30.	15	15.5	GALAXY
MCG+00-06-054	02 11 03.	- 00 57 30.	66	13.5	GALAXY
ZWG 413.063	02 11 06.	+ 03 53		14.3	GALAXY
MRK 589	02 11 06.	+ 03 53	11	14.5	GALAXY WITH UV CONTINUUM
UGC 01716	02 11 06.	+ 03 53	54	14.3	GALAXY EX CMPT
ZWG 461.066	02 11 06.	+ 16 34		15.7	GALAXY
UGC 01717	02 11 06.	+ 16 34	66	15.7	GALAXY SBc
ZWG 522.139	02 11 06.	+ 35 30		15.5	GALAXY
ZWG 538.035	02 11 06.	+ 39 50		15.7	GALAXY
ARC 0323	02 11 06.	- 03 06		17.5	RICH CLUSTER OF GALAXIES
MCG+01-06-075	02 11 06.	- 07 55	36	15.	GALAXY
TON-S 0241	02 11 06.	- 24 07		14.6	BLUE STAR
RNGC 0855	02 11 10.	+ 27 39		13.0	GALAXY
ZC 0211.2+0009	02 11 12.	+ 00 09	1340		CLUSTER OF GALAXIES
MCG+01-06-056	02 11 12.	+ 03 51	30	15.	GALAXY
3ZW 043	02 11 12.	+ 03 53			COMPACT GALAXY
MCG+05-06-016	02 11 12.	+ 27 37 30.	36	13.5	GALAXY
ZWG 504.035	02 11 12.	+ 27 39		13.0	GALAXY
UGC 01718	02 11 12.	+ 27 39	198	13.0	GALAXY E
ARC 0326	02 11 13.	- 07 21		17.0	RICH CLUSTER OF GALAXIES
ARC 0324	02 11 14.	- 01 46		17.5	RICH CLUSTER OF GALAXIES
MCG+03-06-046	02 11 18.	+ 16 12 30.	90	14.	GALAXY
ZWG 461.067	02 11 18.	+ 16 14		15.4	GALAXY
UGC 01719	02 11 18.	+ 16 14	138	15.4	GALAXY Sb
6ZW 192	02 11 18.	+ 36 18			COMPACT GALAXY
ZC 0211.3-0145	02 11 18.	- 01 45	1680		CLUSTER OF GALAXIES
IC 0213	02 11 19.	+ 16 13 02.			NONSTELLAR OBJECT
RNGC 0859	02 11 19.	- 00 58			NON-EXISTENT OBJECT
MCG+07-05-028	02 11 21.	+ 39 49	24	16.	GALAXY
ZWG 413.064	02 11 24.	+ 04 57		14.4	GALAXY
UGC 01720	02 11 24.	+ 04 57	54	14.4	GALAXY PECULR
ZWG 483.010	02 11 24.	+ 23 24		15.7	GALAXY
6ZW 193	02 11 24.	+ 36 20			COMPACT GALAXY
OCL 0348	02 11 24.	+ 59 02	3600	13.	OPEN STAR CLUSTER
ARC 0328	02 11 25.	- 07 06		17.3	RICH CLUSTER OF GALAXIES
IC 0214	02 11 28.	+ 04 56 01.			NONSTELLAR OBJECT
MCG+01-06-057	02 11 30.	+ 04 56	48	14.	GALAXY
ZCG 0211+36	02 11 30.	+ 36 21		17.9	COMPACT GALAXY
ZWG 523.001	02 11 30.	+ 37 10		14.5	GALAXY
ZWG 522.140	02 11 30.	+ 37 10		14.5	GALAXY
UGC 01721	02 11 30.	+ 37 10	138	14.5	GALAXY SBb/SBc
UGC 01722	02 11 30.	+ 43 09	66	16.5	GALAXY
UGC 01723	02 11 36.	+ 07 20	84	16.0	GALAXY Sc
MCG+01-06-058	02 11 36.	+ 07 20	72	15.	GALAXY
MCG+03-06-047	02 11 36.	+ 16 57	18	15.	GALAXY
MCG+06-05-107	02 11 36.	+ 37 10	108	14.	GALAXY
LB 03241	02 11 36.	- 49 59		12.7	FAINT BLUE STAR
IC 0215	02 11 40.	- 07 02 12.			NONSTELLAR OBJECT
ZWG 413.065	02 11 42.	+ 07 37		15.7	GALAXY
UGC 01724	02 11 42.	+ 07 37	72	15.7	GALAXY Sc
MCG+01-06-059	02 11 42.	+ 07 38	60	14.5	GALAXY
ZC 0211.7+1138	02 11 42.	+ 11 38	1010		CLUSTER OF GALAXIES
6ZW 194	02 11 42.	+ 36 58			COMPACT GALAXY
6ZW 195	02 11 42.	+ 38 42			COMPACT GALAXY
TON-S 0242	02 11 42.	- 28 23		15.0	BLUE STAR
MCG+01-06-076	02 11 45.	- 07 01	60	15.	GALAXY
MCG-02-06-041	02 11 45.	- 13 02 30.	12	15.5	GALAXY
VB 073	02 11 48.	+ 64 36	148		STELLAR RING
MCG+00-06-055	02 11 51.	+ 01 14 30.	48	15.	GALAXY
UGC 01725	02 11 54.	+ 01 15	66	16.5	GALAXY Sc-IRR
ZWG 504.036	02 11 54.	+ 31 15		14.6	GALAXY
UGC 01726	02 11 54.	+ 31 15	114	14.6	GALAXY Sb-c
ZWG 523.002	02 11 54.	+ 36 20		14.9	GALAXY
ZWG 522.141	02 11 54.	+ 36 20		14.9	GALAXY
LB 03242	02 11 54.	- 69 25		14.4	FAINT BLUE STAR
SCHO 0047	02 11 55.	+ 58 36 12.	370		ISOLATED DARK CLOUD
MCG+01-06-060	02 12 00.	+ 07 12	48	15.	GALAXY
ZC 0212.0+1020	02 12 00.	+ 10 20	3560		CLUSTER OF GALAXIES
MCG+05-06-017	02 12 00.	+ 31 14	72	14.5	GALAXY
ZC 0212.0+3701	02 12 00.	+ 37 01	810		CLUSTER OF GALAXIES
UGC 01728	02 12 00.	+ 49 38	108	17.	GALAXY Sc
MCG+00-06-056	02 12 00.	- 00 59 30.	66	13.	GALAXY
ZWG 387.055	02 12 00.	- 01 00		14.0	GALAXY
MRK 590	02 12 00.	- 01 00	18	14.5	GALAXY WITH UV CONTINUUM
UGC 01727	02 12 00.	- 01 00	84	14.0	GALAXY Sa
MCG+01-06-077	02 12 00.	- 07 34 30.	132	14.	GALAXY
MCG-05-06-009	02 12 00.	- 31 23	72	15.	GALAXY
RNGC 0863	02 12 01.	- 01 00		14.0	GALAXY
ZWG 504.037	02 12 06.	+ 30 33		15.1	GALAXY
ZWG 504.038	02 12 06.	+ 32 30		15.6	GALAXY
UGC 01729	02 12 06.	+ 32 30	102	15.6	GALAXY SBc
5ZW 203	02 12 06.	+ 35 43			COMPACT GALAXY
TON-S 0243	02 12 06.	- 23 08		15.6	BLUE STAR
RNGC 0860	02 12 08.	+ 30 33		15.0	GALAXY
LIN 592	02 12 08.	- 74 36 29.		15.79	SMC EMISSION LINE OBJECT
ARC 0329	02 12 10.	- 04 46		17.2	RICH CLUSTER OF GALAXIES
5ZW 204	02 12 12.	+ 30 31			COMPACT GALAXY
6ZW 196	02 12 12.	+ 33 33			COMPACT GALAXY
MCG+01-06-078	02 12 12.	- 07 31	60	15.5	GALAXY
MCG-02-06-042	02 12 12.	- 13 29	18	15.5	GALAXY

OBJECT NAME	RIGHT ASCEN.	DECLINATION	DIAM.	MAGN.	TYPE OF OBJECT
ZWG 504.039	02 12 18.	+ 30 57		15.6	GALAXY
UGC 01730	02 12 18.	+ 32 28	66	16.0	GALAXY SB
MCG+05-06-018	02 12 18.	+ 32 29	72	16.	GALAXY
MCG+06-06-001	02 12 18.	+ 33 31	36	15.	GALAXY
ZWG 523.003	02 12 18.	+ 33 32		15.3	GALAXY
ZWG 522.142	02 12 18.	+ 33 32		15.3	GALAXY
5ZW 205	02 12 18.	+ 35 02			COMPACT GALAXY
SN 1961C	02 12 18.	+ 40 53		18.2	SUPERNOVA
MCG+07-05-029	02 12 18.	+ 40 53	30	16.5	GALAXY
ZWG 538.036	02 12 24.	+ 43 52		15.6	GALAXY
SER 019.07	02 12 24.	- 59 55	80	16.	PECULIAR GALAXY
ZWG 461.068	02 12 30.	+ 17 46		14.9	GALAXY
MCG+03-06-049	02 12 30.	+ 17 46	15	14.5	GALAXY
ZWG 461.069	02 12 30.	+ 18 05		15.0	GALAXY
UGC 01731	02 12 30.	+ 18 05	84	15.0	GALAXY E?
MCG+03-06-048	02 12 30.	+ 18 05	36	14.5	GALAXY
UGC 01732	02 12 30.	+ 18 27	66	16.5	GALAXY Sc
ZWG 483.011	02 12 30.	+ 21 47		15.6	GALAXY
UGC 01733	02 12 30.	+ 21 47	108	15.6	GALAXY Sc
KARA.73B 0095	02 12 30.	+ 21 47	108	15.6	ISOLATED GALAXY S
ZWG 504.040	02 12 30.	+ 31 40		15.7	GALAXY
UGC 01734	02 12 30.	+ 31 40	60	15.7	GALAXY Sa-b
6ZW 197	02 12 30.	+ 38 30			COMPACT GALAXY
MCG-04-06-017	02 12 33.	- 25 04	42	14.5	GALAXY
MCG+06-06-002	02 12 36.	+ 35 16	54	15.	GALAXY
ZWG 523.004	02 12 36.	+ 35 17		14.0	GALAXY
ZWG 522.143	02 12 36.	+ 35 17		14.0	GALAXY
UGC 01735	02 12 36.	+ 35 17	84	14.0	GALAXY E-S0
MCG-03-06-018	02 12 36.	- 20 29	120	13.	GALAXY
TON-S 0244	02 12 36.	- 25 53		14.8	BLUE STAR
ARC 0330	02 12 40.	+ 10 07		17.2	RICH CLUSTER OF GALAXIES
ZC 0212.7+0655	02 12 42.	+ 06 55	5380		CLUSTER OF GALAXIES
6ZW 198	02 12 42.	+ 35 18			COMPACT GALAXY
RNGC 0864	02 12 47.	+ 05 46		12.0	GALAXY
ZWG 413.066	02 12 48.	+ 05 46		12.0	GALAXY
UGC 01736	02 12 48.	+ 05 46	312	12.0	GALAXY Sc
KARA.73B 0096	02 12 48.	+ 05 46	270	12.0	ISOLATED GALAXY S
MCG+01-06-061	02 12 48.	+ 05 47	300	11.	GALAXY
ZC 0212.8+1109	02 12 48.	+ 11 09	810		CLUSTER OF GALAXIES
ZWG 438.044	02 12 48.	+ 13 38		15.6	GALAXY
ZWG 523.005	02 12 48.	+ 35 41		14.8	GALAXY
RNGC 0861	02 12 48.	+ 35 41		15.0	GALAXY
UGC 01737	02 12 48.	+ 35 41	90	14.8	GALAXY Sb
VV 094B	02 12 48.	+ 37 08 30.	9	18.	INTERACTING GALAXY
VV 094A	02 12 48.	+ 37 08 30.	12	16.5	INTERACTING GALAXY
VV 094	02 12 48.	+ 37 08 30.	36		INTERACTING GALAXY
ZWG 538.037	02 12 48.	+ 42 35		15.6	GALAXY
UGC 01738	02 12 48.	+ 42 35	90	16.5	GALAXY Sc
KARA.72 060A	02 12 48.	+ 42 35	48	15.6	PART OF DOUBLE GALAXY
MCG+07-05-030	02 12 48.	+ 42 35	54	15.5	GALAXY
MCG+06-06-003	02 12 51.	+ 35 40	90	14.5	GALAXY
ARC 0331	02 12 53.	+ 11 09		17.1	RICH CLUSTER OF GALAXIES
ZWG 483.012	02 12 54.	+ 24 59		14.9	GALAXY
UGC 01739	02 12 54.	+ 24 59	78	14.9	GALAXY S
ZWG 523.006	02 12 54.	+ 33 35		14.7	GALAXY
MCG+06-06-004	02 12 54.	+ 33 35	48	14.5	GALAXY
UGC 01740	02 12 54.	+ 43 16	60	17.	GALAXY DWARF
LBN 0653	02 13	+ 55 10	300		BRIGHT NEBULA
ZWG 387.056	02 13 00.	+ 01 25		15.1	GALAXY
MCG+00-06-057	02 13 00.	+ 01 26	42	15.	GALAXY
UGC 01741	02 13 00.	+ 01 28	60	15.1	GALAXY SBb-c
ZC 0213.0+0825	02 13 00.	+ 08 25	1140		CLUSTER OF GALAXIES
ZWG 438.045	02 13 00.	+ 15 10		15.7	GALAXY
UGC 01742	02 13 00.	+ 15 10	60	15.7	GALAXY Sc
MCG+02-06-049	02 13 00.	+ 15 10	48	15.	GALAXY
MCG+03-06-050	02 13 00.	+ 20 29	24	15.	GALAXY
6ZW 199	02 13 00.	+ 35 50			COMPACT GALAXY
MCG-03-06-019	02 13 00.	- 18 02	72	14.	GALAXY
RNGC 0872	02 13 01.	- 18 02		14.0	GALAXY
MCG-05-06-010	02 13 03.	- 26 36	36	15.	GALAXY
ARC 0332	02 13 05.	- 13 50		17.6	RICH CLUSTER OF GALAXIES
ZWG 538.038	02 13 06.	+ 42 35		15.7	GALAXY
UGC 01743	02 13 06.	+ 42 35	78	15.7	GALAXY SBb
KARA.72 060B	02 13 06.	+ 42 35	42	15.7	PART OF DOUBLE GALAXY
MCG+07-05-031	02 13 06.	+ 42 35	60	16.	GALAXY
MCG-02-06-043	02 13 09.	- 13 17	36	15.	GALAXY
ZWG 504.041	02 13 12.	+ 32 25		14.5	GALAXY
UGC 01744	02 13 12.	+ 32 25	138	14.5	GALAXY Sb-c
KARA.72 061A	02 13 12.	+ 32 25	108	14.5	PART OF DOUBLE GALAXY
UGC 01745	02 13 12.	+ 50 27	84	16.5	GALAXY
RNGC 0866	02 13 13.	- 01 00			NON-EXISTENT OBJECT
RNGC 0865	02 13 15.	+ 28 22		14.0	GALAXY
IC 1784	02 13 16.	+ 32 24 01.			NONSTELLAR OBJECT
ZWG 387.058	02 13 18.	+ 01 32		15.2	GALAXY
UGC 01746	02 13 18.	+ 01 32	90	15.2	GALAXY Sc
MCG+00-06-058	02 13 18.	+ 01 33	54	15.	GALAXY
5ZW 206	02 13 18.	+ 25 00			COMPACT GALAXY
ZC 0213.3+2500	02 13 18.	+ 25 00	2150		CLUSTER OF GALAXIES
ZWG 504.042	02 13 18.	+ 28 22		14.0	GALAXY
UGC 01747	02 13 18.	+ 28 22	120	14.0	GALAXY S
MCG+05-06-019	02 13 18.	+ 32 25	60	14.	GALAXY
5ZW 207	02 13 18.	+ 36 04			COMPACT GALAXY
ZWG 387.057	02 13 18.	- 02 15		15.7	GALAXY
MCG-02-06-044	02 13 18.	- 13 57	48	15.	GALAXY
LB 01616	02 13 18.	- 55 53		14.3	FAINT BLUE STAR
LB 03243	02 13 18.	- 63 28		14.2	FAINT BLUE STAR
LB 03244	02 13 18.	- 72 29		13.7	FAINT BLUE STAR
RNGC 0867	02 13 19.	+ 00 49			NON-EXISTENT OBJECT
ZWG 413.067	02 13 24.	+ 04 55		15.5	GALAXY
ZWG 461.070	02 13 24.	+ 17 44		14.8	GALAXY
ZWG 461.071	02 13 24.	+ 18 05		15.7	GALAXY
UGC 01749	02 13 24.	+ 18 05	72	15.7	GALAXY Sa-b
ZWG 483.013	02 13 24.	+ 23 25		15.6	GALAXY
MCG+05-06-020	02 13 24.	+ 28 20	84	14.	GALAXY
ZWG 504.043	02 13 24.	+ 31 47		14.6	GALAXY
UGC 01750	02 13 24.	+ 31 47	84	14.6	GALAXY Sa-b
IC 1785	02 13 24.	+ 32 25 00.			NONSTELLAR OBJECT
ZWG 504.044	02 13 24.	+ 32 26		15.6	GALAXY
KARA.72 061B	02 13 24.	+ 32 26	42	15.6	PART OF DOUBLE GALAXY
ZWG 387.059	02 13 24.	- 00 58		15.6	GALAXY
UGC 01748	02 13 24.	- 00 58	84	15.6	GALAXY
IC 0216	02 13 24.	- 02 14 37.			NONSTELLAR OBJECT
RNGC 0868	02 13 25.	- 00 58		15.5	GALAXY
IC 1786	02 13 26.	+ 04 55 05.			NONSTELLAR OBJECT
MCG+07-05-032	02 13 27.	+ 39 46	12	16.5	GALAXY
MCG-03-06-020	02 13 27.	- 16 56 30.	54	14.5	GALAXY
MCG-04-06-018	02 13 27.	- 23 45	48	15.	GALAXY
UGC 01751	02 13 30.	+ 16 21	60	16.0	GALAXY SBb

OBJECT NAME	RIGHT ASCEN.	DECLINATION	DIAM.	MAGN.	TYPE OF OBJECT
MCG+03-06-051	02 13 30.	+ 18 05	66	14.5	GALAXY
UGC 01752	02 13 30.	+ 24 40	102	16.5	GALAXY Sc
MCG+05-06-022	02 13 30.	+ 31 46	72	15.	GALAXY
MCG+05-06-021	02 13 30.	+ 32 26	33	15.	GALAXY
MCG-02-06-045	02 13 30.	- 14 02 30.	15	15.5	GALAXY
MCG-03-06-021	02 13 30.	- 19 10	36	15.	GALAXY
SCHO 0048	02 13 33.	+ 57 43 06.	770		ISOLATED DARK CLOUD
ZWG 504.045	02 13 36.	+ 27 59		15.1	GALAXY
UGC 01753	02 13 36.	+ 27 59	72	15.1	GALAXY IRR
ZC 0213.6-0303	02 13 36.	- 03 03	3360		CLUSTER OF GALAXIES
MCG-05-06-011	02 13 36.	- 31 26	120	13.5	GALAXY
IC 1788	02 13 38.8	+ 31 25 56.	96	13.2	GALAXY SA
ZC 0213.7+1637	02 13 42.	+ 16 37	1210		CLUSTER OF GALAXIES
ZWG 483.014	02 13 42.	+ 26 06		15.5	GALAXY
MCG+05-06-023	02 13 42.	+ 27 57 30.	60	15.	GALAXY
MCG-04-06-019	02 13 42.	- 23 32	24	15.5	GALAXY
ARC 0335	02 13 43.	- 12 23		17.6	RICH CLUSTER OF GALAXIES
ARC 0334	02 13 46.	- 04 26		17.5	RICH CLUSTER OF GALAXIES
IC 0217	02 13 46.	- 12 09 39.			NONSTELLAR OBJECT
IC 1787	02 13 46.	- 12 10 57.			NONSTELLAR OBJECT
ZC 0213.8+1859	02 13 48.	+ 18 59	1210		CLUSTER OF GALAXIES
MCG-02-06-046	02 13 48.	- 12 08	132	13.5	GALAXY
KARA.68 014	02 13 48.	- 20 42	47		DWARF GALAXY
MCG+06-06-005	02 13 51.	+ 37 08	24	15.	GALAXY
RNGC 0874	02 13 51.	- 23 25			NON-EXISTENT OBJECT
MCG+02-06-050	02 13 54.	+ 12 45	36	15.	GALAXY
ARC 0333	02 13 54.	+ 16 38		17.6	RICH CLUSTER OF GALAXIES
ZC 0213.9+1734	02 13 54.	+ 17 34	740		CLUSTER OF GALAXIES
VMT 03	02 14	+ 62 18			SUPERNOVA REMNANT
MIL 03	02 14	+ 63 15	4800		SUPERNOVA REMNANT
ZWG 504.046	02 14 00.	+ 30 43		15.6	GALAXY
UGC 01754	02 14 00.	+ 30 43	90	15.6	GALAXY SB?b-c
ZWG 538.039	02 14 00.	+ 39 54		15.7	GALAXY
MCG-02-06-047	02 14 00.	- 11 12 30.	36	15.	GALAXY
MCG+02-06-051	02 14 06.	+ 11 34	48	15.5	GALAXY
5ZW 208	02 14 06.	+ 27 22			COMPACT GALAXY
MCG+07-05-033	02 14 06.	+ 39 53 30.	36	16.	GALAXY
ZC 0214.1-0214	02 14 06.	- 02 14	1280		CLUSTER OF GALAXIES
SHAH 145	02 14 06.	- 02 23	396		GROUP OF COMPACT GALAXIES
ARC 0336	02 14 07.	- 02 21		17.2	RICH CLUSTER OF GALAXIES
RNGC 0873	02 14 11.	- 11 34		13.0	GALAXY
ZC 0214.2+1106	02 14 12.	+ 11 06	2350		CLUSTER OF GALAXIES
UGC 01755	02 14 12.	+ 11 34	72	16.0	GALAXY Sb-c
ZC 0214.2+2045	02 14 12.	+ 20 45	1010		CLUSTER OF GALAXIES
ZWG 504.047	02 14 12.	+ 30 21		15.7	GALAXY
MCG-02-06-048	02 14 12.	- 11 34	84	13.	GALAXY
MCG+00-06-059	02 14 15.	+ 01 59	24	15.	GALAXY
LB 00468	02 14 15.	- 04 49 12.		15.4	FAINT BLUE STAR
ZWG 387.060	02 14 18.	+ 01 57		15.2	GALAXY
UGC 01756	02 14 18.	+ 01 57	66	15.2	GALAXY S
5ZW 209	02 14 18.	+ 27 30			COMPACT GALAXY
5ZW 210	02 14 18.	+ 27 37			COMPACT GALAXY
5ZW 211	02 14 18.	+ 36 29			COMPACT GALAXY
ZWG 523.007	02 14 18.	+ 36 52		15.1	GALAXY
ZWG 523.008	02 14 18.	+ 38 11		13.6	GALAXY
UGC 01757	02 14 18.	+ 38 11	54	13.6	GALAXY S
UGC 01758	02 14 18.	+ 46 49	60	17.	GALAXY Sc/SBc
MCG+01-06-079	02 14 18.	- 03 23	39	15.5	GALAXY
RNGC 0879	02 14 22.	- 09 12			GALAXY
ZWG 413.068	02 14 24.	+ 05 04		15.4	GALAXY
ZWG 438.046	02 14 24.	+ 14 19		13.6	GALAXY
UGC 01759	02 14 24.	+ 14 19	66	13.6	GALAXY
MCG+02-06-052	02 14 24.	+ 14 19	12	16.	GALAXY
MCG+02-06-053	02 14 24.	+ 14 20	72	14.	GALAXY
MCG+06-06-006	02 14 24.	+ 38 12	42	15.	GALAXY
MCG-05-06-012	02 14 24.	- 30 10	60	15.5	GALAXY
LIN 593	02 14 25.	- 74 35 41.		13.82	SMC EMISSION LINE OBJECT
RNGC 0870	02 14 26.	+ 14 19		16.0	GALAXY
SHB 058	02 14 27.	+ 10 50 24.		17.	QUASI-STELLAR OBJECT
BC PKS0214+10	02 14 27.	+ 10 50 4		17.	QUASI-STELLAR OBJECT
ZWG 387.061	02 14 30.	+ 01 00		14.2	GALAXY
UGC 01760	02 14 30.	+ 01 00	90	14.2	GALAXY S0-a
KARA.72 062A	02 14 30.	+ 01 00	72	14.2	PART OF DOUBLE GALAXY
MCG+00-06-060	02 14 30.	+ 01 01 30.	36	14.	GALAXY
ZWG 387.062	02 14 30.	+ 01 02		15.7	GALAXY
KARA.72 062B	02 14 30.	+ 01 02	48	15.7	PART OF DOUBLE GALAXY
MCG+00-06-061	02 14 30.	+ 01 04	42	15.	GALAXY
RNGC 0875	02 14 31.	+ 01 00		14.0	GALAXY
RNGC 0871	02 14 32.	+ 14 19		13.5	GALAXY
IC 0218	02 14 33.	+ 01 03 12.			NONSTELLAR OBJECT
ZWG 387.063	02 14 36.	+ 01 27		15.2	GALAXY
MRK 591	02 14 36.	+ 01 27	12	15.5	GALAXY WITH UV CONTINUUM
ZWG 438.047	02 14 36.	+ 09 35		15.4	GALAXY
ZCG 0214+27	02 14 36.	+ 27 28		18.2	COMPACT GALAXY
5ZW 212	02 14 36.	+ 29 18			COMPACT GALAXY
LB 03245	02 14 36.	- 70 49		14.3	FAINT BLUE STAR
ZWG 438.048	02 14 42.	+ 09 38		15.4	GALAXY
ZWG 438.049	02 14 42.	+ 10 02		15.7	GALAXY
UGC 01761	02 14 42.	+ 14 21	72	16.5	GALAXY DWRF IR
MCG+02-06-054	02 14 42.	+ 14 21	60	16.	GALAXY
ZWG 504.048	02 14 42.	+ 29 18		15.2	GALAXY
MCG+01-06-080	02 14 45.	- 05 40 30.	72	15.	GALAXY
MCG+01-06-081	02 14 45.	- 06 16	36	15.5	GALAXY
FEIG 021	02 14 48.	+ 04 01		14.1	FAINT BLUE STAR
ZWG 523.009	02 14 48.	+ 35 35		15.2	GALAXY
MCG+06-06-007	02 14 48.	+ 35 35 30.	30	15.	GALAXY
VB 074	02 14 48.	+ 64 51	510		STELLAR RING
TON-S 0245	02 14 48.	- 23 29		15.7	BLUE STAR
ZWG 438.050	02 14 54.	+ 12 17		15.7	GALAXY
UGC 01762	02 14 54.	+ 12 17	66	15.7	GALAXY S
KARA.72 063A	02 14 54.	+ 12 17	60	15.7	PART OF DOUBLE GALAXY
MCG+02-06-055	02 14 54.	+ 12 17	60	15.	GALAXY
ZWG 504.049	02 14 54.	+ 32 10		14.8	GALAXY
UGC 01763	02 14 54.	+ 32 10	162	14.8	GALAXY Sa
5ZW 213	02 14 54.	+ 34 17			COMPACT GALAXY
ZC 0214.9+3935	02 14 54.	+ 39 35	1610		CLUSTER OF GALAXIES
MCG-06-08-002	02 14 54.	- 35 43	90	15.	GALAXY
IC 1790	02 14 56.	+ 12 16 35.			NONSTELLAR OBJECT
MCG+02-06-056	02 14 57.	+ 12 14	24	14.5	GALAXY
LBN 0703	02 15	+ 19 00	16200		BRIGHT NEBULA
IC 1791	02 15 00.	+ 12 14 17.			NONSTELLAR OBJECT
ZWG 438.051	02 15 00.	+ 12 15		14.6	GALAXY
UGC 01764	02 15 00.	+ 12 15	72	14.6	GALAXY S0
KARA.72 063B	02 15 00.	+ 12 15	78	14.6	PART OF DOUBLE GALAXY
ZC 0215.0+1716	02 15 00.	+ 17 16	1280		CLUSTER OF GALAXIES
5ZW 214	02 15 00.	+ 22 01			COMPACT GALAXY
ZWG 504.050	02 15 00.	+ 31 40		15.4	GALAXY
MCG+05-06-024	02 15 00.	+ 32 10	120	15.	GALAXY

OBJECT NAME	RIGHT ASCEN.	DECLINATION	DIAM.	MAGN.	TYPE OF OBJECT
ZWG 523.010	02 15 00.	+ 35 32		15.4	GALAXY
UGC 01765	02 15 00.	+ 35 32	72	15.4	GALAXY SB?c-IR
MCG+06-06-008	02 15 00.	+ 35 32 30.	60	15.	GALAXY
ZC 0215.0+8630	02 15 00.	+ 86 30	2690		CLUSTER OF GALAXIES
ZWG 387.064	02 15 00.	- 00 44		15.6	GALAXY
KARA.73 04	02 15 01.	- 41 35	34		DWARF GALAXY
UGC 01766	02 15 06.	+ 14 18	102	16.5	GALAXY S
ZWG 523.011	02 15 06.	+ 37 45		15.7	GALAXY
ZWG 523.012	02 15 06.	+ 37 51		14.0	GALAXY
UGC 01767	02 15 06.	+ 37 51	72	14.0	GALAXY IRR
MCG+06-06-009	02 15 06.	+ 37 53	54	14.	GALAXY
MCG-02-06-049	02 15 06.	- 11 55	36	15.	GALAXY
ARC 0337	02 15 07.	+ 17 18		17.6	RICH CLUSTER OF GALAXIES
ARC 0338	02 15 07.	- 11 30		17.6	RICH CLUSTER OF GALAXIES
MCG+02-06-057	02 15 12.	+ 14 17	120	14.5	GALAXY
ZWG 438.052	02 15 12.	+ 14 19		12.5	GALAXY
UGC 01768	02 15 12.	+ 14 19	138	12.5	GALAXY Sc
ZWG 523.013	02 15 12.	+ 36 51		13.9	GALAXY
UGC 01769	02 15 12.	+ 36 51	60	13.9	GALAXY Sb/Sc
MCG+06-06-010	02 15 12.	+ 36 53	60	14.	GALAXY
ZC 0215.2+3905	02 15 12.	+ 39 05	670		CLUSTER OF GALAXIES
BA 10	02 15 12.	+ 58 05	288		STELLAR GROUP
MCG+01-06-082	02 15 12.	- 07 54 30.	54	15.5	GALAXY
MCG-02-06-051	02 15 12.	- 09 24	15	15.5	GALAXY
MCG-02-06-050	02 15 12.	- 11 44	72	15.	GALAXY
LB 01617	02 15 12.	- 58 22		14.0	FAINT BLUE STAR
RNGC 0876	02 15 14.	+ 14 17		14.5	GALAXY
MCG-04-06-020	02 15 15.	- 23 08	84	14.5	GALAXY
ARC 0339	02 15 16.	- 09 20		17.8	RICH CLUSTER OF GALAXIES
IC 1789	02 15 17.	+ 32 10 22.			NONSTELLAR OBJECT
ZWG 387.065	02 15 18.	+ 00 00		15.7	GALAXY
MCG+02-06-058	02 15 18.	+ 14 19	120	12.	GALAXY
ZWG 504.051	02 15 18.	+ 32 45		15.4	GALAXY
UGC 01770	02 15 18.	+ 36 17	66	16.5	GALAXY S0-a
MCG+06-06-012	02 15 18.	+ 37 16	24	15.5	GALAXY
UGC 01771	02 15 18.	+ 37 41	132	17.	GALAXY DWRF SP
MCG+06-06-011	02 15 18.	+ 37 41	78	16.	GALAXY
MCG+07-05-034	02 15 18.	+ 44 03	36	17.	GALAXY
PHL 3987	02 15 18.	- 11 48		17.5	BLUE STELLAR OBJECT
RNGC 0877	02 15 20.	+ 14 19		12.5	GALAXY
MCG+01-06-083	02 15 21.	- 07 02	60	15.5	GALAXY
MCG-03-06-022	02 15 21.	- 16 51	48	15.	GALAXY
ZWG 523.014	02 15 24.	+ 37 48		13.7	GALAXY
UGC 01772	02 15 24.	+ 37 48	66	13.7	GALAXY
ZWG 438.053	02 15 30.	+ 12 58		15.3	GALAXY
UGC 01773	02 15 30.	+ 12 58	96	15.3	GALAXY SB
OCL 0350	02 15 30.	+ 56 55	4680	5.3	OPEN STAR CLUSTER
RNGC 0869	02 15 32.	+ 56 55		4.5	OPEN CLUSTER
ARP 010	02 15 33.	+ 05 24			PECULIAR GALAXY
RNGC 0880	02 15 33.	- 04 27			GALAXY
RNGC 0878	02 15 33.	- 23 36		15.0	GALAXY
PHL 8111	02 15 36.	- 11 16		17.5	BLUE STELLAR OBJECT
PHL 8112	02 15 36.	- 18 24		17.7	BLUE STELLAR OBJECT
PHL 1252	02 15 36.	- 20 14		17.7	BLUE STELLAR OBJECT
MCG-03-06-023	02 15 39.	- 15 42 30.	66	15.	GALAXY
MCG-04-06-021	02 15 39.	- 23 36 30.	30	15.	GALAXY
MCG-04-06-022	02 15 39.	- 23 58	36	15.	GALAXY
UGC 01774	02 15 42.	+ 05 19	60	16.0	GALAXY Sb-c
ZWG 413.069	02 15 42.	+ 05 25		14.2	GALAXY
UGC 01775	02 15 42.	+ 05 25	120	14.2	GALAXY S
ZWG 523.015	02 15 42.	+ 35 14		15.3	GALAXY
UGC 01776	02 15 42.	+ 35 14	96	15.3	GALAXY SBb
MCG+06-06-013	02 15 42.	+ 35 15	60	15.	GALAXY
ZWG 523.016	02 15 42.	+ 36 54		15.5	GALAXY
6ZW 200	02 15 42.	+ 38 25			COMPACT GALAXY
PHL 8113	02 15 42.	- 10 00		17.6	BLUE STELLAR OBJECT
MCG-06-06-001	02 15 42.	- 35 04	78	15.	GALAXY
MCG+01-06-084	02 15 45.	- 07 01	48	15.5	GALAXY
VV 302A	02 15 45.	- 12 27 30.	54	15.	INTERACTING GALAXY
ZC 0215.8+0140	02 15 48.	+ 01 40	810		CLUSTER OF GALAXIES
MCG+01-06-062	02 15 48.	+ 05 27	60	14.	GALAXY
MCG+02-06-059	02 15 48.	+ 12 59	84	15.	GALAXY
ZWG 504.052	02 15 48.	+ 30 17		15.7	GALAXY
UGC 01777	02 15 48.	+ 30 17	60	15.7	GALAXY SBc
ZWG 523.017	02 15 48.	+ 33 30		14.6	GALAXY
ZWG 504.053	02 15 48.	+ 33 30		14.6	GALAXY
UGC 01778	02 15 48.	+ 33 30	84	15.6	GALAXY
UGC 01779	02 15 48.	+ 38 28	66	17.	GALAXY
SCHO 0049	02 15 48.	+ 57 04 54.	450		ISOLATED DARK CLOUD
OCL 0346	02 15 48.	+ 63 31	240	15.	OPEN STAR CLUSTER
ZC 0215.8-0038	02 15 48.	- 00 38	1010		CLUSTER OF GALAXIES
MCG+01-06-085	02 15 48.	- 06 54	30	15.	GALAXY
PHL 3989	02 15 48.	- 08 48		18.4	BLUE STELLAR OBJECT
PHL 8114	02 15 48.	- 09 19		18.5	BLUE STELLAR OBJECT
PHL 3990	02 15 48.	- 09 36		19.0	BLUE STELLAR OBJECT
PHL 3991	02 15 48.	- 10 29		18.5	BLUE STELLAR OBJECT
PHL 1253	02 15 48.	- 11 08		18.5	BLUE STELLAR OBJECT
PHL 3988	02 15 48.	- 11 20		18.8	BLUE STELLAR OBJECT
MCG-05-06-013	02 15 48.	- 28 02	36	16.	GALAXY
LB 03246	02 15 48.	- 66 54		15.5	FAINT BLUE STAR
MCG+06-06-014	02 15 51.	+ 33 30	60	14.	GALAXY
MCG+01-06-086	02 15 51.	- 06 59	24	15.5	GALAXY
ZC 0215.9+2819	02 15 54.	+ 28 19	1280		CLUSTER OF GALAXIES
ZWG 504.054	02 15 54.	+ 29 27		15.7	GALAXY
ZWG 538.040	02 15 54.	+ 40 20		15.6	GALAXY
UGC 01780	02 15 54.	+ 40 20	120	15.6	GALAXY IRR
MCG+07-05-035	02 15 54.	+ 40 20	96	16.	GALAXY
PHL 8115	02 15 54.	- 09 52		17.2	BLUE STELLAR OBJECT
PHL 8154	02 15 54.	- 10 33		18.4	BLUE STELLAR OBJECT
MCG-02-06-052	02 15 54.	- 12 27	54	14.5	GALAXY
VV 302B	02 15 57.	- 12 27 12.	72	15.5	INTERACTING GALAXY
RNGC 0888	02 15 58.	- 60 05			UNVERIFIED SOUTHERN OBJECT
ZC 0216.0+2746	02 16 00.	+ 27 46	1080		CLUSTER OF GALAXIES
ZWG 523.018	02 16 00.	+ 34 15		15.0	GALAXY
UGC 01781	02 16 00.	+ 34 15	102	15.0	GALAXY COMPACT
MCG+06-06-015	02 16 00.	+ 34 15	60	14.	GALAXY
ZC 0216.0+3625	02 16 00.	+ 36 25	16200		CLUSTER OF GALAXIES
UGC 01782	02 16 00.	+ 42 31	108	17.	GALAXY S IV
MCG+07-05-036	02 16 00.	+ 43 03	24	16.	GALAXY
UGC 01783	02 16 00.	+ 43 54	90	17.	GALAXY DWARF
LDN 1354	02 16 00.	+ 61 30	480		DARK NEBULA
MCG-02-06-053	02 16 00.	- 09 31	48	15.	GALAXY
ARC 0340	02 16 00.	- 12 55		17.6	RICH CLUSTER OF GALAXIES
PHL 1254	02 16 00.	- 14 19		16.6	BLUE STELLAR OBJECT
MCG-02-06-054	02 16 03.	- 12 24	54	15.	GALAXY
IC 1792	02 16 04.	+ 34 13 49.			NONSTELLAR OBJECT
ZC 0216.1+0629	02 16 06.	+ 06 29	810		CLUSTER OF GALAXIES
ZWG 438.054	02 16 06.	+ 14 56		15.7	GALAXY
MCG+05-06-025	02 16 06.	+ 33 28	42	14.5	GALAXY
5ZW 215	02 16 06.	+ 34 17			COMPACT GALAXY
ZWG 538.041	02 16 06.	+ 41 30		15.1	GALAXY
MCG+07-05-037	02 16 06.	+ 42 30	84	16.	GALAXY
PHL 1255	02 16 06.	- 09 34		18.1	BLUE STELLAR OBJECT
RNGC 0881	02 16 09.	- 06 49		12.5	GALAXY
IC 0219	02 16 09.	- 07 08 19.			NONSTELLAR OBJECT
UGC 01784	02 16 12.	+ 36 27	72	17.	GALAXY IRR
MCG+01-06-088	02 16 12.	- 07 06	22	15.	GALAXY
MCG+01-06-087	02 16 12.	- 07 06	24	17.	GALAXY
PHL 8116	02 16 12.	- 08 44		18.0	BLUE STELLAR OBJECT
PHL 3992	02 16 12.	- 11 25		18.6	BLUE STELLAR OBJECT
PHL 8117	02 16 12.	- 12 02		18.7	BLUE STELLAR OBJECT
MCG-02-06-055	02 16 12.	- 13 02	60	15.	GALAXY
MCG+01-06-089	02 16 15.	- 06 49	132	12.5	GALAXY
ZWG 438.055	02 16 18.	+ 15 02		15.7	GALAXY
ZC 0216.3+2505	02 16 18.	+ 25 05	1810		CLUSTER OF GALAXIES
ZWG 504.055	02 16 18.	+ 31 10		15.7	GALAXY
ZWG 523.019	02 16 18.	+ 33 35		15.6	GALAXY
5ZW 216	02 16 18.	+ 34 46			COMPACT GALAXY
PHL 3993	02 16 18.	- 19 50		17.3	BLUE STELLAR OBJECT
FATH 1.106	02 16 21.	+ 15 02	14		NEBULA
MCG-04-06-023	02 16 21.	- 25 59	60	14.5	GALAXY
ZWG 387.066	02 16 24.	+ 00 05		15.6	GALAXY
UGC 01785	02 16 24.	+ 04 35	84	16.0	GALAXY Sc
6ZW 201	02 16 24.	+ 33 58			COMPACT GALAXY
5ZW 217	02 16 24.	+ 36 21			COMPACT GALAXY
PHL 8119	02 16 24.	- 08 43		18.4	BLUE STELLAR OBJECT
PHL 8120	02 16 24.	- 09 27		18.5	BLUE STELLAR OBJECT
MCG-02-06-056	02 16 24.	- 11 42	48	15.	GALAXY
PHL 8118	02 16 24.	- 13 42		16.9	BLUE STELLAR OBJECT
MCG-04-06-024	02 16 24.	- 22 43 30.	36	15.	GALAXY
MCG-04-06-025	02 16 27.	- 22 44 30.	24	16.	GALAXY
PHL 3994	02 16 30.	+ 01 38		13.9	BLUE STELLAR OBJECT
ZWG 461.072	02 16 30.	+ 15 48		15.6	GALAXY
ZWG 504.056	02 16 30.	+ 31 49		15.6	GALAXY
6ZW 202	02 16 30.	+ 34 53			COMPACT GALAXY
ZWG 523.020	02 16 30.	+ 36 53		15.6	GALAXY
UGC 01786	02 16 30.	+ 36 53	60	15.6	GALAXY E
MCG+06-06-016	02 16 30.	+ 36 55	15	15.5	GALAXY
ZC 0216.5+4353	02 16 30.	+ 43 53	1950		CLUSTER OF GALAXIES
RNGC 0883	02 16 33.	- 06 59		13.0	GALAXY
5ZW 218	02 16 36.	+ 29 33			COMPACT GALAXY
ZWG 523.021	02 16 36.	+ 37 24		15.7	GALAXY
ZWG 523.022	02 16 36.	+ 37 43		14.7	GALAXY
UGC 01787	02 16 36.	+ 37 43	84	14.7	GALAXY S-IRR
MCG+06-06-017	02 16 36.	+ 37 44	78	15.	GALAXY
ZWG 538.042	02 16 36.	+ 43 46		15.7	GALAXY
MCG+01-06-090	02 16 36.	- 06 59	42	13.	GALAXY
PHL 8121	02 16 36.	- 08 00		13.7	BLUE STELLAR OBJECT
PHL 3996	02 16 36.	- 11 36		18.5	BLUE STELLAR OBJECT
PHL 3995	02 16 36.	- 14 12		18.4	BLUE STELLAR OBJECT
ARC 0341	02 16 36.	- 17 03		17.8	RICH CLUSTER OF GALAXIES
TON-S 0246	02 16 36.	- 22 52		15.2	BLUE STAR
MCG+06-06-019	02 16 39.	+ 36 25	36	15.	GALAXY
MCG+06-06-018	02 16 39.	+ 37 03 30.	36	16.	GALAXY
PHL 1256	02 16 42.	+ 03 13		14.2	BLUE STELLAR OBJECT
UGC 01788	02 16 42.	+ 36 24	72	16.0	GALAXY E
MCG+06-06-020	02 16 42.	+ 36 52 30.	24	15.5	GALAXY
SCHO 0050	02 16 42.	+ 52 01 18.	350		ISOLATED DARK CLOUD
IC 0220	02 16 47.	- 13 00 46.			NONSTELLAR OBJECT
ARC 0343	02 16 47.	- 22 06		17.7	RICH CLUSTER OF GALAXIES
PHL 8122	02 16 48.	+ 01 00		16.5	BLUE STELLAR OBJECT
MCG+03-06-052	02 16 48.	+ 15 34	60	14.5	GALAXY
ZC 0216.8+3857	02 16 48.	+ 38 57	2490		CLUSTER OF GALAXIES
VB 076	02 16 48.	+ 67 58	430		STELLAR RING
PHL 3998	02 16 48.	- 09 12		19.0	BLUE STELLAR OBJECT
PHL 3997	02 16 48.	- 13 30		18.7	BLUE STELLAR OBJECT
PHL 8123	02 16 48.	- 14 17		17.9	BLUE STELLAR OBJECT
MCG-03-06-024	02 16 48.	- 19 12	132	14.	GALAXY
ZWG 462.001	02 16 54.	+ 15 35		14.9	GALAXY
ZWG 461.073	02 16 54.	+ 15 35		14.9	GALAXY
UGC 01789	02 16 54.	+ 15 35	72	14.9	GALAXY S0
ZWG 538.043	02 16 54.	+ 41 03		15.0	GALAXY
ZWG 538.044	02 16 54.	+ 41 29		15.4	GALAXY
UGC 01790	02 16 54.	+ 41 29	78	15.4	GALAXY
MCG-02-06-057	02 16 54.	- 13 00	48	14.5	GALAXY
FATH 2.007	02 16 55.	+ 15 35	5		NEBULA
RNGC 0882	02 16 56.	+ 15 35		15.0	GALAXY
LBN 0642	02 17	+ 61 40	900		BRIGHT NEBULA
UGC 01791	02 17 00.	+ 28 02	66	17.	GALAXY DWARF
MCG+05-06-026	02 17 00.	+ 28 47	96	14.	GALAXY
ZWG 504.057	02 17 00.	+ 28 49		14.3	GALAXY
UGC 01792	02 17 00.	+ 28 49	156	14.3	GALAXY SBc
ZWG 504.058	02 17 00.	+ 33 00		15.7	GALAXY
ZWG 523.023	02 17 00.	+ 37 41		15.7	GALAXY
UGC 01793	02 17 00.	+ 37 41	66	15.7	GALAXY Sc
MCG+00-07-001	02 17 00.	- 00 30	36	15.	GALAXY
PHL 8124	02 17 00.	- 07 39		18.7	BLUE STELLAR OBJECT
PHL 3999	02 17 00.	- 10 42		17.9	BLUE STELLAR OBJECT
MCG-02-07-001	02 17 00.	- 12 00	54	15.	GALAXY
PHL 4001	02 17 00.	- 14 07		18.9	BLUE STELLAR OBJECT
PHL 4000	02 17 00.	- 16 20		18.5	BLUE STELLAR OBJECT
6ZW 203	02 17 06.	+ 33 45			COMPACT GALAXY
VB 077	02 17 06.	+ 64 37	356		STELLAR RING
OCL 0344	02 17 06.	+ 65 40	240	14.	OPEN STAR CLUSTER
RNGC 0889	02 17 06.	- 41 59			GALAXY
ZC 0217.2+0231	02 17 12.	+ 02 31	1340		CLUSTER OF GALAXIES
5ZW 219	02 17 12.	+ 29 48			COMPACT GALAXY
ZWG 388.001	02 17 12.	- 00 29		14.6	GALAXY
MRK 592	02 17 12.	- 00 29	10	15.	GALAXY WITH UV CONTINUUM
UGC 01794	02 17 12.	- 00 29	102	14.6	GALAXY SBb
PHL 4002	02 17 12.	- 01 02		18.8	BLUE STELLAR OBJECT
PHL 8125	02 17 12.	- 07 55		15.2	BLUE STELLAR OBJECT
PHL 8127	02 17 12.	- 11 59		17.5	BLUE STELLAR OBJECT
PHL 8126	02 17 12.	- 14 22		16.8	BLUE STELLAR OBJECT
MCG-03-07-001	02 17 12.	- 16 18	120	13.	GALAXY
MCG-04-06-026	02 17 12.	- 21 39	84	15.	GALAXY
ARC 0342	02 17 13.	+ 02 30		17.7	RICH CLUSTER OF GALAXIES
RNGC 0885	02 17 13.	- 01 00			NON-EXISTENT OBJECT
RNGC 0887	02 17 13.	- 16 18		13.0	GALAXY
KEEL 047	02 17 17.4	+ 42 04 00.			NEBULA
ZC 0217.3+1707	02 17 18.	+ 17 07	1340		CLUSTER OF GALAXIES
UGC 01795	02 17 18.	+ 35 41	60	16.5	GALAXY SBc
ZWG 523.024	02 17 18.	+ 37 38		15.1	GALAXY
ZWG 538.045	02 17 18.	+ 40 34		15.5	GALAXY
UGC 01796	02 17 18.	+ 40 34	96	15.5	GALAXY S IV
PHL 1257	02 17 18.	- 00 36		16.8	BLUE STELLAR OBJECT

OBJECT NAME	RIGHT ASCEN.	DECLINATION	DIAM.	MAGN.	TYPE OF OBJECT
MCG-04-06-027	02 17 18.	- 23 05 30.	15	15.	GALAXY
MCG+00-07-002	02 17 21.	+ 01 42 30.	18	16.	GALAXY
MCG+06-06-021	02 17 21.	+ 36 47 30.	15	15.	GALAXY
ZC 0217.4+0117	02 17 24.	+ 01 17	470		CLUSTER OF GALAXIES
MCG+00-07-003	02 17 24.	+ 01 42	30	15.	GALAXY
ZWG 388.002	02 17 24.	+ 01 43		15.2	GALAXY
UGC 01797	02 17 24.	+ 01 43	84	15.2	GALAXY S0-a
ZWG 483.015	02 17 24.	+ 23 15		15.6	GALAXY
ZWG 523.025	02 17 24.	+ 36 46		15.7	GALAXY
ZWG 523.026	02 17 24.	+ 38 18		15.6	GALAXY
UGC 01798	02 17 24.	+ 38 18	60	15.6	GALAXY Sa-b
MCG+07-05-038	02 17 24.	+ 40 33	48	17.	GALAXY
ZWG 538.046	02 17 24.	+ 41 20		15.3	GALAXY
UGC 01799	02 17 24.	+ 48 13	78	16.0	GALAXY SBb/Sc
PHL 1258	02 17 24.	- 04 54		18.4	BLUE STELLAR OBJECT
PHL 8128	02 17 24.	- 10 12		17.9	BLUE STELLAR OBJECT
LB 03247	02 17 24.	- 76 47		13.8	FAINT BLUE STAR
ZWG 504.059	02 17 30.	+ 31 27		15.5	GALAXY
UGC 01800	02 17 30.	+ 34 59	114	16.0	GALAXY S
ISS 0017	02 17 30.	+ 64 12	264		STELLAR RING
PHL 8129	02 17 30.	- 02 46		17.4	BLUE STELLAR OBJECT
PHL 8130	02 17 30.	- 06 20		17.5	BLUE STELLAR OBJECT
LB 01618	02 17 30.	- 45 30		15.2	FAINT BLUE STAR
MCG+07-05-039	02 17 33.	+ 41 20	36	15.	GALAXY
KEEL 048	02 17 33.6	+ 42 02 52.			NEBULA
PHL 8131	02 17 36.	+ 00 00		18.6	BLUE STELLAR OBJECT
MCG+01-07-001	02 17 36.	+ 07 45	48	15.	GALAXY
ZWG 414.001	02 17 36.	+ 07 58		15.6	GALAXY
5ZW 220	02 17 36.	+ 28 56			COMPACT GALAXY
VB 075	02 17 36.	+ 64 09	369		STELLAR RING
PHL 4003	02 17 36.	- 03 55		18.0	BLUE STELLAR OBJECT
MCG-06-06-002	02 17 36.	- 38 04	108	12.5	GALAXY
KEEL 049	02 17 40.9	+ 41 51 22.			NEBULA
ZWG 414.002	02 17 42.	+ 07 48		15.7	GALAXY
UGC 01801	02 17 42.	+ 07 48	72	15.7	GALAXY SBb
MCG+08-05-004	02 17 42.	+ 48 42	24	14.	GALAXY
PHL 8133	02 17 42.	- 01 12		16.4	BLUE STELLAR OBJECT
PHL 8132	02 17 42.	- 08 08		16.4	BLUE STELLAR OBJECT
PHL 8134	02 17 42.	- 09 14		17.6	BLUE STELLAR OBJECT
PHL 4004	02 17 42.	- 11 35		18.6	BLUE STELLAR OBJECT
KEEL 050	02 17 43.7	+ 41 52 15.			NEBULA
PHL 4005	02 17 48.	+ 01 00		18.5	BLUE STELLAR OBJECT
MCG+01-07-002	02 17 48.	+ 06 34	138	14.5	GALAXY
5ZW 221	02 17 48.	+ 28 01			COMPACT GALAXY
ZWG 553.003	02 17 48.	+ 48 44		15.6	GALAXY
UGC 01802	02 17 48.	+ 48 44	102	15.6	GALAXY E
MCG+00-07-004	02 17 48.	- 00 33	33	15.5	GALAXY
PHL 4007	02 17 48.	- 05 48		18.5	BLUE STELLAR OBJECT
PHL 8135	02 17 48.	- 05 57		17.9	BLUE STELLAR OBJECT
PHL 8138	02 17 48.	- 08 22		17.9	BLUE STELLAR OBJECT
PHL 8136	02 17 48.	- 10 00		18.2	BLUE STELLAR OBJECT
PHL 4006	02 17 48.	- 13 15		19.0	BLUE STELLAR OBJECT
PHL 8137	02 17 48.	- 14 09		18.9	BLUE STELLAR OBJECT
MCG-03-07-002	02 17 48.	- 20 00	150	13.5	GALAXY
ZWG 414.003	02 17 54.	+ 06 35		15.1	GALAXY
UGC 01803	02 17 54.	+ 06 35	156	15.1	GALAXY IRR
5ZW 222	02 17 54.	+ 34 40			COMPACT GALAXY
ZWG 523.027	02 17 54.	+ 38 26		15.5	GALAXY
UGC 01804	02 17 54.	+ 38 26	102	15.5	GALAXY
MCG+06-06-022	02 17 54.	+ 38 28	48	16.	GALAXY
ISS 0058	02 17 54.	+ 58 59	2614		STELLAR RING
ZWG 388.003	02 17 54.	- 00 32		15.7	GALAXY
PHL 4008	02 17 54.	- 06 31		18.1	BLUE STELLAR OBJECT
PHL 8139	02 17 54.	- 10 02		19.0	BLUE STELLAR OBJECT
AGU 11	02 17 54.	- 41 39 00.	66	12.5	PECULIAR GALAXY
LBN 0715	02 18	+ 12 40	2040		BRIGHT NEBULA
LBN 0668	02 18	+ 50 10	6300		BRIGHT NEBULA
LBN 0666	02 18	+ 50 20	960		BRIGHT NEBULA
LBN 0648	02 18	+ 59 42	120		BRIGHT NEBULA
PHL 4011	02 18 00.	+ 00 51		18.5	BLUE STELLAR OBJECT
PHL 8141	02 18 00.	+ 00 53		18.7	BLUE STELLAR OBJECT
PHL 4010	02 18 00.	+ 01 22		18.3	BLUE STELLAR OBJECT
PHL 4009	02 18 00.	+ 03 00		18.4	BLUE STELLAR OBJECT
ZWG 504.060	02 18 00.	+ 32 37		14.5	GALAXY
UGC 01805	02 18 00.	+ 32 37	60	14.5	GALAXY
UGC 01806	02 18 00.	+ 33 10	72	18.	GALAXY DWARF
ZWG 538.047	02 18 00.	+ 41 35		15.6	GALAXY
MCG+07-05-040	02 18 00.	+ 42 32	60	16.	GALAXY
UGC 01807	02 18 00.	+ 42 33	150	16.5	GALAXY DWARF
MCG+07-05-041	02 18 00.	+ 42 38	42	16.	GALAXY
PHL 8142	02 18 00.	- 06 07		18.1	BLUE STELLAR OBJECT
PHL 1259	02 18 00.	- 07 52		15.8	BLUE STELLAR OBJECT
PHL 8140	02 18 00.	- 14 53		16.6	BLUE STELLAR OBJECT
PHL 8143	02 18 00.	- 16 26		18.1	BLUE STELLAR OBJECT
RNGC 0893	02 18 00.	- 41 37			GALAXY
KEEL 051	02 18 03.0	+ 42 38 10.			NEBULA
ZWG 462.002	02 18 06.	+ 16 07		15.7	GALAXY
VB 167	02 18 06.	+ 58 57	2585		STELLAR RING
MCG+03-07-001	02 18 12.	+ 16 08	30	15.5	GALAXY
ZWG 483.016	02 18 12.	+ 23 22		14.7	GALAXY
UGC 01808	02 18 12.	+ 23 22	72	14.7	GALAXY Sb
KARA.73B 0097	02 18 12.	+ 23 22	60	14.7	ISOLATED GALAXY S
ZWG 538.048	02 18 12.	+ 42 39		15.3	GALAXY
MCG+07-05-042	02 18 12.	+ 42 39	54	15.	GALAXY
ZWG 553.004	02 18 12.	+ 46 20		15.7	GALAXY
URA 25	02 18 12.	+ 64 47	198		STELLAR RING
URA 28	02 18 12.	+ 68 24	156		STELLAR RING
PHL 4013	02 18 12.	- 08 57		18.7	BLUE STELLAR OBJECT
PHL 8144	02 18 12.	- 13 44		16.7	BLUE STELLAR OBJECT
PHL 4012	02 18 12.	- 16 55		18.3	BLUE STELLAR OBJECT
TON-S 0247	02 18 12.	- 25 51		16.0	BLUE STAR
FATH 1.107	02 18 14.	- 14 29	24		NEBULA
KEEL 052	02 18 14.3	+ 42 38 55.			SPIRAL NEBULA
ZWG 388.004	02 18 18.	+ 00 20		15.1	GALAXY
UGC 01809	02 18 18.	+ 00 20	72	15.1	GALAXY S0-a
MCG+00-07-005	02 18 18.	+ 00 20	48	14.5	GALAXY
PHL 8145	02 18 18.	- 09 40		18.6	BLUE STELLAR OBJECT
ARP 273	02 18 22.	+ 39 09			PECULIAR GALAXY
KEEL 053	02 18 22.8	+ 42 30 33.			NEBULA
PHL 8146	02 18 24.	+ 01 22		17.0	BLUE STELLAR OBJECT
MCG+02-07-001	02 18 24.	+ 13 54	36	16.	GALAXY
5ZW 223	02 18 24.	+ 39 09			COMPACT GALAXY
ZWG 523.028	02 18 24.	+ 39 09		13.7	GALAXY
UGC 01810	02 18 24.	+ 39 09	126	13.7	GALAXY S(b)
KARA.72 064A	02 18 24.	+ 39 09	108	13.7	PART OF DOUBLE GALAXY
PHL 4016	02 18 24.	- 05 03		18.5	BLUE STELLAR OBJECT
PHL 4014	02 18 24.	- 09 08		18.9	BLUE STELLAR OBJECT
PHL 8147	02 18 24.	- 12 54		18.9	BLUE STELLAR OBJECT
PHL 4015	02 18 24.	- 14 24		18.7	BLUE STELLAR OBJECT
MCG-03-07-003	02 18 24.	- 18 52 30.	48	14.5	GALAXY
KEEL 054	02 18 24.0	+ 42 27 53.			NEBULA
MCG+06-06-023	02 18 27.	+ 39 11	120	13.	GALAXY
RNGC 0892	02 18 27.	- 23 21			GALAXY
MCG+02-07-002	02 18 30.	+ 13 28	36	15.5	GALAXY
UGC 01811	02 18 30.	+ 13 30	78	16.0	GALAXY Sc
ZWG 483.017	02 18 30.	+ 25 11		14.6	GALAXY
UGC 01812	02 18 30.	+ 25 11	78	14.6	GALAXY Sc
ZWG 523.029	02 18 30.	+ 39 08		15.3	GALAXY
UGC 01813	02 18 30.	+ 39 08	84	15.3	GALAXY SB
KARA.72 064B	02 18 30.	+ 39 08	66	15.3	PART OF DOUBLE GALAXY
MCG+06-06-024	02 18 30.	+ 39 10	78	14.	GALAXY
MCG-03-07-004	02 18 30.	- 20 02	42	15.	GALAXY
SG 3.008	02 18 31.	+ 61 38	840		DIFFUSE EMISSION NEBULA
IC 1793	02 18 34.	+ 32 18 08.			NONSTELLAR OBJECT
PHL 1260	02 18 36.	+ 00 53		18.1	BLUE STELLAR OBJECT
ZWG 462.003	02 18 36.	+ 16 20		14.0	GALAXY
UGC 01814	02 18 36.	+ 16 20	162	14.0	GALAXY Sb/SBc
KARA.72 065B	02 18 36.	+ 16 20	78		PART OF DOUBLE GALAXY
KARA.72 065A	02 18 36.	+ 16 20	78	14.0	PART OF DOUBLE GALAXY
MCG+03-07-002	02 18 36.	+ 16 20	144	14.	GALAXY
ZWG 504.061	02 18 36.	+ 30 49		15.7	GALAXY
UGC 01815	02 18 36.	+ 30 49	84	15.7	GALAXY SB IV
KARA.73B 0098	02 18 36.	+ 30 49	66	15.7	ISOLATED GALAXY S
MCG+05-06-027	02 18 36.	+ 32 19	66	14.5	GALAXY
ZWG 504.062	02 18 36.	+ 32 20		14.8	GALAXY
UGC 01816	02 18 36.	+ 32 20	108	14.8	GALAXY Sa-b
MCG+08-05-005	02 18 36.	+ 50 23	72	14.	GALAXY
PHL 8148	02 18 36.	- 09 58		17.6	BLUE STELLAR OBJECT
PHL 4017	02 18 36.	- 12 58		17.6	BLUE STELLAR OBJECT
PHL 8149	02 18 36.	- 20 08		18.4	BLUE STELLAR OBJECT
ZWG 462.004	02 18 42.	+ 15 31		15.0	GALAXY
MCG+03-07-003	02 18 42.	+ 15 32	42	15.	GALAXY
MCG+05-06-028	02 18 42.	+ 30 49	66	16.	GALAXY
5ZW 224	02 18 42.	+ 44 08			COMPACT GALAXY
MCG+07-05-043	02 18 42.	+ 44 10	36	15.5	GALAXY
URA 86	02 18 42.	+ 58 38	540		STELLAR RING
PHL 1261	02 18 42.	- 14 00		18.7	BLUE STELLAR OBJECT
MCG-04-06-028	02 18 42.	- 21 02	36	15.5	GALAXY
ARC 0345	02 18 45.	+ 13 22		17.5	RICH CLUSTER OF GALAXIES
MCG+02-07-003	02 18 45.	+ 13 57	144	14.5	GALAXY
IC 1794	02 18 47.	+ 15 32			NEBULA
FATH 2.008	02 18 47.	+ 15 32	5		NEBULA
KEEL 055	02 18 47.2	+ 43 08 48.			NEBULA
ZWG 439.001	02 18 48.	+ 13 59		14.8	GALAXY
UGC 01817	02 18 48.	+ 13 59	162	14.8	GALAXY Sc
UGC 01818	02 18 48.	+ 15 07	60	16.5	GALAXY SB
UGC 01819	02 18 48.	+ 16 09	60	18.	GALAXY DWARF
ZC 0218.8+2106	02 18 48.	+ 21 06	1610		CLUSTER OF GALAXIES
ZWG 504.063	02 18 48.	+ 32 48		15.1	GALAXY
UGC 01820	02 18 48.	+ 32 48	108	15.1	GALAXY Sc
ZWG 538.049	02 18 48.	+ 44 08		15.6	GALAXY
PHL 8150	02 18 48.	- 04 25		17.0	BLUE STELLAR OBJECT
PHL 8151	02 18 48.	- 08 30		18.2	BLUE STELLAR OBJECT
PHL 8152	02 18 48.	- 09 28		18.2	BLUE STELLAR OBJECT
PHL 8153	02 18 48.	- 12 54		17.5	BLUE STELLAR OBJECT
PHL 1262	02 18 48.	- 18 48		18.2	BLUE STELLAR OBJECT
TON-S 0248	02 18 48.	- 25 31		15.9	BLUE STAR
MCG-05-06-014	02 18 48.	- 32 11	60	15.	GALAXY
LB 03248	02 18 48.	- 65 16		14.7	FAINT BLUE STAR
KEEL 056	02 18 51.8	+ 42 27 46.			NEBULA
MCG+05-06-029	02 18 54.	+ 32 46	96	14.5	GALAXY
OCL 0353	02 18 54.	+ 56 53	4680	5.94	OPEN STAR CLUSTER
PHL 4018	02 18 54.	- 04 42		18.3	BLUE STELLAR OBJECT
PHL 8156	02 18 54.	- 10 22		19.0	BLUE STELLAR OBJECT
PHL 8155	02 18 54.	- 10 32		18.0	BLUE STELLAR OBJECT
PHL 1263	02 18 54.	- 10 44		18.7	BLUE STELLAR OBJECT
PHL 1264	02 18 54.	- 20 36		17.1	BLUE STELLAR OBJECT
MCG-06-06-003	02 18 54.	- 33 58		11.	GALAXY
ARC 0344	02 18 55.	+ 21 08		17.4	RICH CLUSTER OF GALAXIES
RNGC 0884	02 18 56.	+ 56 53		4.5	OPEN CLUSTER
RNGC 0897	02 18 56.	- 33 56			GALAXY
LBN 0671	02 19	+ 50 00	840		BRIGHT NEBULA
KHAV 035	02 19	+ 60 50	9980		DARK NEBULA
PHL 1265	02 19 00.	+ 02 30		18.5	BLUE STELLAR OBJECT
UGC 01821	02 19 00.	+ 06 33	66	17.	GALAXY DWARF
ZWG 462.005	02 19 00.	+ 16 39		15.7	GALAXY
UGC 01822	02 19 00.	+ 16 39	78	15.7	GALAXY Sc
ZWG 504.064	02 19 00.	+ 33 03		12.5	GALAXY
UGC 01823	02 19 00.	+ 33 03	174	12.5	GALAXY S0
MCG+07-05-044	02 19 00.	+ 43 20	36	15.	GALAXY
LDN 1356	02 19 00.	+ 61 30	1140		DARK NEBULA
PHL 8158	02 19 00.	- 00 57		18.9	BLUE STELLAR OBJECT
MCG+01-07-001	02 19 00.	- 04 38	72	15.	GALAXY
PHL 8157	02 19 00.	- 07 33		17.0	BLUE STELLAR OBJECT
PHL 8159	02 19 00.	- 08 58		18.7	BLUE STELLAR OBJECT
RNGC 0890	02 19 01.	+ 33 02		13.0	GALAXY
MCG+06-06-025	02 19 03.	+ 33 42 30.	60	14.	GALAXY
RNGC 0894	02 19 03.	- 05 44			NON-EXISTENT OBJECT
RNGC 0895	02 19 03.	- 05 45		12.0	GALAXY
KARA.73 05	02 19 05.	- 40 07	27		DWARF GALAXY
KEEL 057	02 19 05.6	+ 43 19 11.			NEBULA
MCG+00-07-006	02 19 06.	+ 00 01	42	15.	GALAXY
ZWG 388.005	02 19 06.	+ 00 02		15.6	GALAXY
UGC 01824	02 19 06.	+ 00 02	66	15.6	GALAXY S0-a
MCG+03-07-004	02 19 06.	+ 16 39	66	15.	GALAXY
ZWG 504.065	02 19 06.	+ 32 00		15.0	GALAXY
UGC 01825	02 19 06.	+ 32 00	78	15.0	GALAXY Sc
MCG+05-06-031	02 19 06.	+ 32 00	48	15.	GALAXY
MCG+05-06-030	02 19 06.	+ 33 02	48	13.	GALAXY
ZWG 523.030	02 19 06.	+ 33 43		14.6	GALAXY
UGC 01826	02 19 06.	+ 33 43	72	14.6	GALAXY SB
5ZW 225	02 19 06.	+ 37 03			COMPACT GALAXY
ZWG 538.050	02 19 06.	+ 42 35		15.7	GALAXY
ZWG 538.051	02 19 06.	+ 43 19		15.7	GALAXY
UGC 01827	02 19 06.	+ 43 19	60	15.7	GALAXY S-IRR
MCG+01-07-002	02 19 06.	- 05 45	210	11.5	GALAXY
PHL 8160	02 19 06.	- 06 02		18.2	BLUE STELLAR OBJECT
MCG-06-06-004	02 19 06.	- 34 35	36	15.	GALAXY
KEEL 058	02 19 09.2	+ 42 34 42.			NEBULA
MCG+08-05-006	02 19 12.	+ 47 36	78	15.	GALAXY
PHL 8161	02 19 12.	- 05 00		18.4	BLUE STELLAR OBJECT
PHL 4019	02 19 12.	- 05 48		18.1	BLUE STELLAR OBJECT
PHL 8162	02 19 12.	- 08 42		19.0	BLUE STELLAR OBJECT
PHL 8163	02 19 12.	- 13 14		18.5	BLUE STELLAR OBJECT
FATH 1.109	02 19 12.	- 14 31	5		NEBULA
KEEL 059	02 19 13.9	+ 41 58 38.			NEBULA

OBJECT NAME	RIGHT ASCEN.	DECLINATION	DIAM.	MAGN.	TYPE OF OBJECT
KEEL 060	02 19 17.2	+ 41 59 13.			NEBULA
PATH 1.108	02 19 18.	+ 15 20	2		NEBULA
ZC 0219.3+2606	02 19 18.	+ 26 06	1750		CLUSTER OF GALAXIES
MCG+05-06-032	02 19 18.	+ 28 29	48	15.	GALAXY
ZWG 504.066	02 19 18.	+ 28 31		15.5	GALAXY
UGC 01828	02 19 18.	+ 28 31	84	15.5	GALAXY S
KARA.72 066A	02 19 18.	+ 28 31	78	15.5	PART OF DOUBLE GALAXY
ZWG 523.031	02 19 18.	+ 33 31		15.7	GALAXY
UGC 01829	02 19 18.	+ 33 31	66	15.7	GALAXY IRR+CMP
5ZW 226	02 19 18.	+ 36 15			COMPACT GALAXY
ZWG 523.032	02 19 18.	+ 38 54		15.2	GALAXY
MCG+07-05-045	02 19 18.	+ 42 50	48	15.	GALAXY
5ZW 227	02 19 18.	+ 47 37			COMPACT GALAXY
ZWG 553.005	02 19 18.	+ 47 37		14.8	GALAXY
UGC 01830	02 19 18.	+ 47 37	192	14.8	GALAXY SB0/SBa
PHL 8164	02 19 18.	- 03 16		18.6	BLUE STELLAR OBJECT
PHL 1266	02 19 18.	- 08 57		18.8	BLUE STELLAR OBJECT
PHL 8165	02 19 18.	- 10 46		18.6	BLUE STELLAR OBJECT
MCG-04-06-029	02 19 18.	- 20 48 30.	24	15.	GALAXY
TON-S 0249	02 19 18.	- 26 42		15.0	BLUE STAR
MCG-02-07-002	02 19 21.	- 10 15	54	14.	GALAXY
KEEL 061	02 19 21.6	+ 42 50 19.			SPIRAL NEBULA
MCG+05-06-033	02 19 24.	+ 28 28	48	15.	GALAXY
MCG+06-06-026	02 19 24.	+ 33 31 30.	30	15.	GALAXY
ZWG 538.052	02 19 24.	+ 42 07		10.8	GALAXY
UGC 01831	02 19 24.	+ 42 07	840	10.8	GALAXY Sb
ZWG 538.053	02 19 24.	+ 42 50		15.4	GALAXY
UGC 01832	02 19 24.	+ 42 50	102	15.4	GALAXY Sa
5ZW 228	02 19 24.	+ 43 42			COMPACT GALAXY
PHL 8166	02 19 24.	- 00 38		18.4	BLUE STELLAR OBJECT
PHL 4021	02 19 24.	- 06 26		13.8	BLUE STELLAR OBJECT
PHL 4020	02 19 24.	- 09 14		19.0	BLUE STELLAR OBJECT
LB 03249	02 19 24.	- 68 12		14.6	FAINT BLUE STAR
RNGC 0891	02 19 26.	+ 42 07		11.5	GALAXY
ZWG 504.067	02 19 30.	+ 28 30		15.6	GALAXY
UGC 01833	02 19 30.	+ 28 30	96	15.6	GALAXY SBc
KARA.72 066B	02 19 30.	+ 28 30	66	15.6	PART OF DOUBLE GALAXY
MCG+07-05-046	02 19 30.	+ 42 07 30.	840	10.	GALAXY
PHL 8167	02 19 30.	- 03 42		18.1	BLUE STELLAR OBJECT
PHL 1267	02 19 30.	- 05 23		18.6	BLUE STELLAR OBJECT
PHL 4022	02 19 30.	- 07 04		17.9	BLUE STELLAR OBJECT
PHL 4023	02 19 30.	- 09 39		18.5	BLUE STELLAR OBJECT
PATH 1.110	02 19 30.	- 14 19			NEBULA
MCG-05-06-015	02 19 30.	- 27 29	36	15.	GALAXY
SHB 059	02 19 30.0	+ 42 48 29.		15.2	QUASI-STELLAR OBJECT
ZWG 439.002	02 19 36.	+ 11 49		15.3	GALAXY
UGC 01834	02 19 36.	+ 11 49	90	15.3	GALAXY SB(c)
ZWG 462.006	02 19 36.	+ 17 26		15.4	GALAXY
PHL 1268	02 19 36.	- 01 14		18.4	BLUE STELLAR OBJECT
PHL 8168	02 19 36.	- 01 55		17.7	BLUE STELLAR OBJECT
PHL 4024	02 19 36.	- 04 42		18.7	BLUE STELLAR OBJECT
PHL 8171	02 19 36.	- 09 32		18.4	BLUE STELLAR OBJECT
PHL 8170	02 19 36.	- 09 34		17.0	BLUE STELLAR OBJECT
PHL 8169	02 19 36.	- 13 40		18.1	BLUE STELLAR OBJECT
MCG-04-06-030	02 19 36.	- 21 02 30.	96	13.	GALAXY
ARC 0346	02 19 39.	+ 26 10		16.9	RICH CLUSTER OF GALAXIES
RNGC 0899	02 19 39.	- 21 02		13.0	GALAXY
KEEL 062	02 19 39.8	+ 42 53 13.			NEBULA
RNGC 0886	02 19 40.	+ 63 33			NON-EXISTENT OBJECT
ZC 0219.7+0440	02 19 42.	+ 04 40	940		CLUSTER OF GALAXIES
MCG+03-07-005	02 19 42.	+ 17 27	39	15.	GALAXY
ZWG 538.054	02 19 42.	+ 41 56		15.7	GALAXY
ZWG 553.006	02 19 42.	+ 46 20		15.7	GALAXY
PHL 4025	02 19 42.	- 01 58		18.1	BLUE STELLAR OBJECT
MCG-06-06-005	02 19 42.	- 34 34	9	16.	GALAXY
LB 01619	02 19 42.	- 46 01		14.0	FAINT BLUE STAR
KEEL 063	02 19 42.0	+ 41 55 52.			NEBULA
SVEN 092	02 19 43.	- 21 02	48	13.2	GALAXY
IC 0223	02 19 44.	- 20 58			NONSTELLAR OBJECT
IC 0221	02 19 45.	+ 28 02 21.			NONSTELLAR OBJECT
ARP 145	02 19 45.	+ 40 59			PECULIAR GALAXY
MCG-04-06-031	02 19 45.	- 20 58	60	14.5	GALAXY
ZWG 414.004	02 19 48.	+ 05 37		15.5	GALAXY
KARA.73B 0099	02 19 48.	+ 05 37	36	15.5	ISOLATED GALAXY S0
MCG+05-06-034	02 19 48.	+ 28 00	90	14.	GALAXY
ZWG 504.068	02 19 48.	+ 28 02		13.9	GALAXY
UGC 01835	02 19 48.	+ 28 02	132	13.9	GALAXY Sc
ZWG 504.069	02 19 48.	+ 28 15		15.0	GALAXY
UGC 01836	02 19 48.	+ 37 52	66	18.	GALAXY DWARF
ZWG 539.001	02 19 48.	+ 42 47		15.2	GALAXY
ZWG 538.055	02 19 48.	+ 42 47		15.2	GALAXY
UGC 01837	02 19 48.	+ 42 47	78	15.2	GALAXY S0
OCL 0347	02 19 48.	+ 63 38	1200	11.	OPEN STAR CLUSTER
VB 078	02 19 48.	+ 64 09	551		STELLAR RING
PHL 1269	02 19 48.	- 06 40		17.7	BLUE STELLAR OBJECT
PHL 4026	02 19 48.	- 09 42		17.8	BLUE STELLAR OBJECT
PHL 4027	02 19 48.	- 12 54		18.8	BLUE STELLAR OBJECT
MCG-04-06-032	02 19 48.	- 21 21	72	16.	GALAXY
TON-S 0250	02 19 48.	- 23 03		15.2	BLUE STAR
KEEL 064	02 19 48.9	+ 42 47 07.			NEBULA
SVEN 093	02 19 49.	- 20 57	18	14.1	GALAXY
KEEL 065	02 19 50.8	+ 41 51 59.			NEBULA
MCG-04-06-033	02 19 51.	- 20 37	30	15.	GALAXY
MCG+00-07-008	02 19 54.	+ 00 08 30.	12	18.	GALAXY
MCG+00-07-007	02 19 54.	+ 00 09	24	16.	GALAXY
UGC 01838	02 19 54.	+ 24 08	78	16.5	GALAXY Sc
6ZW 204	02 19 54.	+ 36 59			COMPACT GALAXY
ZWG 523.033	02 19 54.	+ 37 10		15.7	GALAXY
ZWG 559.007	02 19 54.	+ 53 10		15.1	GALAXY
MCG+00-07-008A	02 19 54.	- 00 51	90	14.	GALAXY
PHL 4029	02 19 54.	- 03 08		18.6	BLUE STELLAR OBJECT
PHL 4028	02 19 54.	- 06 41		18.5	BLUE STELLAR OBJECT
MCG-06-06-006	02 19 54.	- 34 41	24	15.5	GALAXY
MCG-06-06-007	02 19 54.	- 37 43	72	15.	GALAXY
SVEN 094	02 19 55.	- 21 30	18	14.7	GALAXY
LBN 0634	02 20	+ 75 10	900		BRIGHT NEBULA
ZWG 388.007	02 20 00.	+ 00 09		15.6	GALAXY
PHL 1270	02 20 00.	+ 03 10		16.9	BLUE STELLAR OBJECT
ZC 0220.0+1255	02 20 00.	+ 12 55	540		CLUSTER OF GALAXIES
ZWG 483.018	02 20 00.	+ 25 05		15.7	GALAXY
ZWG 483.019	02 20 00.	+ 25 43		15.7	GALAXY
5ZW 229	02 20 00.	+ 41 08			COMPACT GALAXY
ZWG 539.002	02 20 00.	+ 41 09		14.1	GALAXY
ZWG 538.056	02 20 00.	+ 41 09		14.1	GALAXY
UGC 01840	02 20 00.	+ 41 09	108	14.1	GALAXY PECULR
5ZW 230	02 20 00.	+ 42 45			COMPACT GALAXY
ZWG 539.003	02 20 00.	+ 42 46		15.0	GALAXY
ZWG 538.057	02 20 00.	+ 42 46		15.0	GALAXY
UGC 01841	02 20 00.	+ 42 46	180	15.0	GALAXY E
MCG+07-06-001	02 20 00.	+ 42 48	54	15.	GALAXY
ZWG 361.007	02 20 00.	+ 83 48		15.7	GALAXY
ZWG 388.006	02 20 00.	- 00 51		15.6	GALAXY
UGC 01839	02 20 00.	- 00 51	180	15.6	GALAXY
PHL 8173	02 20 00.	- 09 22		18.5	BLUE STELLAR OBJECT
PHL 8172	02 20 00.	- 15 40		18.2	BLUE STELLAR OBJECT
PHL 4030	02 20 00.	- 16 22		18.0	BLUE STELLAR OBJECT
MCG-03-07-005	02 20 00.	- 16 54 30.	30	14.5	GALAXY
RNGC 0902	02 20 01.	- 16 54		14.0	GALAXY
KEEL 066	02 20 01.8	+ 42 45 57.			NEBULA
MCG+01-07-003	02 20 03.	+ 05 17 30.	36	15.	GALAXY
MCG+02-07-004	02 20 03.	+ 11 24	36	14.5	GALAXY
SG 3.009	02 20 03.	+ 62 06	960		DIFFUSE EMISSION NEBULA
ZWG 414.005	02 20 06.	+ 05 19		15.4	GALAXY
ZWG 439.003	02 20 06.	+ 11 25		15.2	GALAXY
ZWG 462.007	02 20 06.	+ 17 04		15.7	GALAXY
5ZW 231	02 20 06.	+ 36 08			COMPACT GALAXY
PHL 1271	02 20 06.	- 05 20		16.9	BLUE STELLAR OBJECT
PHL 8176	02 20 06.	- 10 26		19.0	BLUE STELLAR OBJECT
PHL 8175	02 20 06.	- 12 43		17.6	BLUE STELLAR OBJECT
LB 01620	02 20 06.	- 46 18		14.2	FAINT BLUE STAR
KEEL 067	02 20 06.1	+ 42 30 33.			NEBULA
IC 0222	02 20 08.	+ 11 24 42.			NONSTELLAR OBJECT
MCG+07-06-002	02 20 09.	+ 41 09	84	14.	GALAXY
RNGC 0905	02 20 10.	- 08 57			GALAXY
PHL 8180	02 20 12.	+ 00 33		18.2	BLUE STELLAR OBJECT
PHL 8179	02 20 12.	+ 02 44		18.8	BLUE STELLAR OBJECT
PHL 4034	02 20 12.	+ 03 22		17.9	BLUE STELLAR OBJECT
ZWG 439.004	02 20 12.	+ 14 15		15.7	GALAXY
MCG+03-07-006	02 20 12.	+ 17 05	30	15.	GALAXY
ZWG 483.020	02 20 12.	+ 25 12		15.7	GALAXY
ZWG 539.004	02 20 12.	+ 41 44		13.8	GALAXY
ZWG 538.058	02 20 12.	+ 41 44		13.8	GALAXY
UGC 01842	02 20 12.	+ 41 44	120	13.8	GALAXY Sa
MCG+07-06-003	02 20 12.	+ 42 47	24	14.5	GALAXY
PHL 4033	02 20 12.	- 01 31		14.4	BLUE STELLAR OBJECT
PHL 4035	02 20 12.	- 01 58		18.4	BLUE STELLAR OBJECT
MCG+01-07-003	02 20 12.	- 04 11	54	16.	GALAXY
PHL 8177	02 20 12.	- 05 34		15.9	BLUE STELLAR OBJECT
PHL 8178	02 20 12.	- 06 30		17.9	BLUE STELLAR OBJECT
PHL 1272	02 20 12.	- 08 50		18.5	BLUE STELLAR OBJECT
PHL 8181	02 20 12.	- 11 54		18.0	BLUE STELLAR OBJECT
PHL 4031	02 20 12.	- 14 13		18.8	BLUE STELLAR OBJECT
PHL 4032	02 20 12.	- 15 38		18.1	BLUE STELLAR OBJECT
ZWG 483.021	02 20 18.	+ 24 31		15.7	GALAXY
ZWG 504.070	02 20 18.	+ 31 57		15.2	GALAXY
KARA.72 067A	02 20 18.	+ 31 57	36	15.2	PART OF DOUBLE GALAXY
MCG+05-06-035	02 20 18.	+ 31 58	21	15.5	GALAXY
MCG+07-06-004	02 20 18.	+ 41 44	102	14.	GALAXY
5ZW 232	02 20 18.	+ 45 35			COMPACT GALAXY
PHL 8182	02 20 18.	- 12 04		18.5	BLUE STELLAR OBJECT
LB 03250	02 20 18.	- 65 23		14.0	FAINT BLUE STAR
LB 03251	02 20 18.	- 72 24		13.7	FAINT BLUE STAR
BV 09	02 20 20.	- 19 46 18.		18.0	FAINT BLUE VARIABLE
KEEL 068	02 20 22.3	+ 42 36 21.			NEBULA
5ZW 233	02 20 24.	+ 31 58			COMPACT GALAXY
ZWG 504.071	02 20 24.	+ 31 58		14.9	GALAXY
KARA.72 067B	02 20 24.	+ 31 58	36	14.9	PART OF DOUBLE GALAXY
MCG+05-06-036	02 20 24.	+ 31 59	24	15.	GALAXY
PHL 4036	02 20 24.	- 10 13		18.6	BLUE STELLAR OBJECT
PHL 8183	02 20 24.	- 12 20		18.5	BLUE STELLAR OBJECT
PHL 8184	02 20 24.	- 13 06		18.5	BLUE STELLAR OBJECT
MCG-03-07-006	02 20 24.	- 15 37	24	15.	GALAXY
PHL 4037	02 20 24.	- 19 48		18.4	BLUE STELLAR OBJECT
LB 01621	02 20 24.	- 44 55		14.4	FAINT BLUE STAR
KEEL 069	02 20 27.7	+ 42 49 56.			NEBULA
PHL 1273	02 20 30.	- 09 45		18.2	BLUE STELLAR OBJECT
ZC 0220.6+1716	02 20 36.	+ 17 16	2150		CLUSTER OF GALAXIES
PHL 4038	02 20 36.	- 01 02		18.5	BLUE STELLAR OBJECT
ZWG 388.008	02 20 36.	- 01 59		15.2	GALAXY
PHL 8185	02 20 36.	- 03 36		17.5	BLUE STELLAR OBJECT
LB 01622	02 20 36.	- 56 11		13.3	FAINT BLUE STAR
KEEL 070	02 20 36.0	+ 41 46 46.			NEBULA
KEEL 071	02 20 37.1	+ 42 23 18.			NEBULA
RNGC 0907	02 20 39.	- 20 56		13.0	GALAXY
RNGC 0900	02 20 40.	+ 26 17		15.0	GALAXY
KEEL 072	02 20 41.7	+ 42 34 00.			NEBULA
ZWG 483.022	02 20 42.	+ 25 15		15.7	GALAXY
ZWG 483.023	02 20 42.	+ 26 17		15.0	GALAXY
UGC 01843	02 20 42.	+ 26 17	66	15.0	GALAXY S0
ZWG 483.024	02 20 42.	+ 27 05		15.6	GALAXY
6ZW 205	02 20 42.	+ 34 14			COMPACT GALAXY
MCG+01-07-004	02 20 42.	- 04 43	36	15.	GALAXY
PHL 4039	02 20 42.	- 06 14		18.3	BLUE STELLAR OBJECT
SVEN 095	02 20 43.	- 20 56	78	12.8	GALAXY
MCG+08-05-007	02 20 45.	+ 47 43	48	16.	GALAXY
MCG-04-06-034	02 20 45.	- 20 56 30.	90	13.	GALAXY
RNGC 0908	02 20 45.	- 21 27		11.0	GALAXY
KEEL 074	02 20 45.0	+ 41 58 45.			NEBULA
KEEL 073	02 20 45.6	+ 43 17 08.			NEBULA
RNGC 0901	02 20 46.	+ 26 20			NON-EXISTENT OBJECT
IC 1796	02 20 47.	- 41 36			NONSTELLAR OBJECT
ZC 0220.8+0624	02 20 48.	+ 06 24	1610		CLUSTER OF GALAXIES
MCG+03-07-007	02 20 48.	+ 16 45	42	16.	GALAXY
ZWG 483.025	02 20 48.	+ 26 56		15.7	GALAXY
UGC 01844	02 20 48.	+ 26 56	96	15.7	GALAXY
ZWG 504.072	02 20 48.	+ 28 29		15.7	GALAXY
ZWG 539.005	02 20 48.	+ 41 59		15.7	GALAXY
ZWG 538.059	02 20 48.	+ 41 59		15.7	GALAXY
ZWG 553.007	02 20 48.	+ 47 45		15.7	GALAXY
UGC 01845	02 20 48.	+ 47 45	96	15.7	GALAXY Sa-b
VB 217	02 20 48.	+ 53 36	403		STELLAR RING
PHL 1274	02 20 48.	- 02 58		18.2	BLUE STELLAR OBJECT
PHL 4040	02 20 48.	- 05 14		17.0	BLUE STELLAR OBJECT
PHL 8186	02 20 48.	- 11 00		16.5	BLUE STELLAR OBJECT
PHL 4041	02 20 48.	- 11 32		18.8	BLUE STELLAR OBJECT
PHL 4042	02 20 48.	- 12 03		18.6	BLUE STELLAR OBJECT
MCG-03-07-007	02 20 48.	- 15 23	36	15.	GALAXY
MCG-04-06-035	02 20 48.	- 21 27 30.	318	11.	GALAXY
HN 0159	02 20 48.	- 41 36			NEBULA
SVEN 096	02 20 49.	- 21 27	276	10.6	GALAXY
KEEL 075	02 20 50.2	+ 42 28 51.			NEBULA
KEEL 076	02 20 51.6	+ 42 19 04.			NEBULA
KEEL 077	02 20 53.1	+ 42 32 11.			NEBULA
ZC 0220.9+2510	02 20 54.	+ 25 10	1410		CLUSTER OF GALAXIES
ZWG 504.073	02 20 54.	+ 28 38		15.5	GALAXY
UGC 01846	02 20 54.	+ 28 38	78	15.5	GALAXY COMPACT

OBJECT NAME	RIGHT ASCEN.	DECLINATION	DIAM.	MAGN.	TYPE OF OBJECT
ZWG 523.034	02 20 54.	+ 36 50		15.6	GALAXY
ZWG 539.006	02 20 54.	+ 41 46		14.8	GALAXY
ZWG 538.060	02 20 54.	+ 41 46		14.8	GALAXY
MCG+07-06-005	02 20 54.	+ 43 05	42	15.	GALAXY
KEEL 078	02 20 54.1	+ 42 36 28.			NEBULA
RNGC 0898	02 20 56.	+ 41 46		14.0	GALAXY
RNGC 0903	02 20 57.	+ 27 08			GALAXY
MCG+07-06-006	02 20 57.	+ 41 46 30.	36	15.	GALAXY
MCG-03-07-008	02 20 57.	- 19 04	36	15.	GALAXY
KEEL 080	02 20 59.0	+ 42 00 04.			NEBULA
LBN 0707	02 21	+ 19 30	2100		BRIGHT NEBULA
VDB.66G 019	02 21	+ 35 46	100		DWARF GALAXY
ZWG 388.009	02 21 00.	+ 02 18		15.7	GALAXY
UGC 01847	02 21 00.	+ 02 18	66	15.7	GALAXY DBL SYS
PHL 1275	02 21 00.	+ 02 34		18.8	BLUE STELLAR OBJECT
ZWG 483.026	02 21 00.	+ 25 19		15.0	GALAXY
ZC 0221.0+2537	02 21 00.	+ 25 37	1610		CLUSTER OF GALAXIES
ZWG 483.027	02 21 00.	+ 27 16		15.5	GALAXY
UGC 01848	02 21 00.	+ 27 16	72	15.5	GALAXY Sb
UGC 01849	02 21 00.	+ 43 04	96	16.0	GALAXY S
MRSL 133+01/1	02 21 00.	+ 61 40	1500		HII REGION
LDN 1333	02 21 00.	+ 75 15	300		DARK NEBULA
PHL 4043	02 21 00.	- 12 15		18.9	BLUE STELLAR OBJECT
KEEL 079	02 21 00.0	+ 43 04 02.			NEBULA
UGC 01850	02 21 06.	+ 26 53	66	16.5	GALAXY DWRF SP
MCG+07-06-007	02 21 06.	+ 43 03	60	17.	GALAXY
UGC 01851	02 21 06.	+ 49 17	72	17.	GALAXY
MCG+01-07-005	02 21 06.	- 07 30	75	15.5	GALAXY
PHL 8187	02 21 06.	- 10 32		17.6	BLUE STELLAR OBJECT
TON-S 0251	02 21 06.	- 26 40		16.0	BLUE STAR
KEEL 082	02 21 08.8	+ 42 02 37.			NEBULA
RNGC 0904	02 21 09.	+ 27 07		15.0	GALAXY
KEEL 081	02 21 09.4	+ 43 02 12.			NEBULA
ZWG 483.028	02 21 12.	+ 27 07		15.0	GALAXY
UGC 01852	02 21 12.	+ 27 07	102	15.0	GALAXY E
PHL 4044	02 21 12.	- 08 51		18.7	BLUE STELLAR OBJECT
PHL 8188	02 21 12.	- 10 03		18.8	BLUE STELLAR OBJECT
PHL 4045	02 21 12.	- 11 11		18.8	BLUE STELLAR OBJECT
KEEL 083	02 21 12.7	+ 42 27 42.			NEBULA
UGC 01853	02 21 18.	+ 33 25	72	17.	GALAXY Sc
UGC 01854	02 21 18.	+ 43 02	114	16.5	GALAXY S
PHL 8189	02 21 18.	- 03 21		17.3	BLUE STELLAR OBJECT
MCG-02-07-003	02 21 21.	- 10 36	36	15.	GALAXY
KEEL 084	02 21 23.2	+ 42 21 08.			SPIRAL NEBULA
MCG+07-06-008	02 21 24.	+ 40 38 30.	48	15.	GALAXY
ZWG 539.007	02 21 24.	+ 40 39		15.1	GALAXY
ZWG 538.061	02 21 24.	+ 40 39		15.1	GALAXY
UGC 01855	02 21 24.	+ 40 39	90	15.1	GALAXY SBa
MCG+01-07-006	02 21 24.	- 07 35	21	15.	GALAXY
PHL 8190	02 21 24.	- 07 36		17.9	BLUE STELLAR OBJECT
ARC 0348	02 21 28.	- 08 50		18.0	RICH CLUSTER OF GALAXIES
MCG+01-07-007	02 21 30.	- 04 55	60	15.	GALAXY
PHL 1276	02 21 30.	- 05 36		15.7	BLUE STELLAR OBJECT
PHL 8191	02 21 30.	- 09 38		18.0	BLUE STELLAR OBJECT
PHL 8192	02 21 30.	- 11 28		18.0	BLUE STELLAR OBJECT
MCG-04-06-036	02 21 30.	- 21 14 30.	48	15.5	GALAXY
LB 02725	02 21 31.	+ 30 37 48.		15.5	FAINT BLUE STAR
MCG+01-07-004	02 21 33.	+ 05 07 30.	48	15.	GALAXY
KEEL 085	02 21 33.0	+ 43 09 51.			NEBULA
KEEL 086	02 21 35.2	+ 42 23 49.			NEBULA
ZWG 388.010	02 21 36.	+ 03 02		15.5	GALAXY
ZWG 414.006	02 21 36.	+ 05 08		15.2	GALAXY
MCG+01-07-005	02 21 36.	+ 06 14	36	14.5	GALAXY
MCG+05-06-038	02 21 36.	+ 31 22	132	14.5	GALAXY
ZWG 504.074	02 21 36.	+ 31 23		14.9	GALAXY
UGC 01856	02 21 36.	+ 31 23	132	14.9	GALAXY Sc
MCG+05-06-037	02 21 36.	+ 32 55	66	15.	GALAXY
ZWG 504.075	02 21 36.	+ 32 57		15.3	GALAXY
UGC 01857	02 21 36.	+ 32 57	144	15.3	GALAXY
ZWG 539.008	02 21 36.	+ 41 28		15.7	GALAXY
ZWG 538.062	02 21 36.	+ 41 28		15.7	GALAXY
UGC 01858	02 21 36.	+ 41 28	108	15.7	GALAXY SB
ZWG 539.009	02 21 36.	+ 41 48		15.7	GALAXY
ZWG 538.063	02 21 36.	+ 41 48		15.7	GALAXY
ZWG 539.010	02 21 36.	+ 42 24		14.3	GALAXY
ZWG 538.064	02 21 36.	+ 42 24		14.3	GALAXY
UGC 01859	02 21 36.	+ 42 24	102	14.3	GALAXY COMPACT
PHL 4046	02 21 36.	- 02 54		18.6	BLUE STELLAR OBJECT
MCG+01-07-008	02 21 36.	- 03 04	54	15.	GALAXY
PHL 8193	02 21 36.	- 06 14		18.6	BLUE STELLAR OBJECT
PHL 8194	02 21 36.	- 08 14		18.2	BLUE STELLAR OBJECT
PHL 8197	02 21 36.	- 09 21		18.6	BLUE STELLAR OBJECT
PHL 8195	02 21 36.	- 13 23		18.0	BLUE STELLAR OBJECT
PHL 8196	02 21 36.	- 17 18		18.0	BLUE STELLAR OBJECT
TON-S 0252	02 21 36.	- 27 42		14.8	BLUE STAR
TON-S 0253	02 21 36.	- 32 52		14.7	BLUE STAR
LB 01623	02 21 36.	- 46 27		13.4	FAINT BLUE STAR
KEEL 087	02 21 38.5	+ 41 47 54.			NEBULA
ZWG 414.007	02 21 42.	+ 06 15		15.1	GALAXY
ZWG 462.008	02 21 42.	+ 20 46		15.7	GALAXY
ZWG 504.076	02 21 42.	+ 28 23		15.7	GALAXY
MCG+07-06-009	02 21 42.	+ 42 25	24	15.	GALAXY
ZWG 539.011	02 21 42.	+ 43 06		15.6	GALAXY
ZWG 538.065	02 21 42.	+ 43 06		15.6	GALAXY
PHL 8200	02 21 42.	- 03 43		18.8	BLUE STELLAR OBJECT
PHL 4047	02 21 42.	- 09 40		19.1	BLUE STELLAR OBJECT
PHL 8198	02 21 42.	- 16 06		13.5	BLUE STELLAR OBJECT
PHL 8199	02 21 42.	- 19 46		18.3	BLUE STELLAR OBJECT
KEEL 089	02 21 42.6	+ 42 28 25.			NEBULA
KEEL 088	02 21 42.9	+ 43 06 00.			NEBULA
MCG+07-06-010	02 21 45.	+ 41 28	42	16.	GALAXY
SG 1.03	02 21 45.	+ 61 44	180		DIFFUSE EMISSION NEBULA
KEEL 090	02 21 47.1	+ 43 18 07.			NEBULA
ZWG 483.029	02 21 48.	+ 25 20		14.7	GALAXY
UGC 01860	02 21 48.	+ 25 20	90	14.7	GALAXY SBb
UGC 01861	02 21 48.	+ 30 37	66	17.	GALAXY DWRF IR
6ZW 206	02 21 48.	+ 34 51			COMPACT GALAXY
5ZW 234	02 21 48.	+ 41 18			COMPACT GALAXY
URA 87	02 21 48.	+ 60 16	300		STELLAR RING
MCG+00-07-009	02 21 48.	- 02 24	60	13.	GALAXY
PHL 8201	02 21 48.	- 03 10		15.0	BLUE STELLAR OBJECT
PHL 4048	02 21 48.	- 11 08		18.5	BLUE STELLAR OBJECT
PHL 8202	02 21 48.	- 11 36		18.9	BLUE STELLAR OBJECT
ZWG 388.012	02 21 54.	+ 01 37		15.3	GALAXY
UGC 01863	02 21 54.	+ 01 37	66	15.3	GALAXY Sc
MCG+00-07-010	02 21 54.	+ 01 37	27	14.5	GALAXY
KARA.73B 0100	02 21 54.	+ 01 37	48	15.3	ISOLATED GALAXY S
UGC 01862	02 21 54.	+ 02 23	138	14.0	GALAXY Sc

OBJECT NAME	RIGHT ASCEN.	DECLINATION	DIAM.	MAGN.	TYPE OF OBJECT
ZWG 483.030	02 21 54.	+ 25 18		15.7	GALAXY
MCG+06-06-027	02 21 54.	+ 35 05	36	15.5	GALAXY
ZWG 523.035	02 21 54.	+ 37 15		15.7	GALAXY
UGC 01864	02 21 54.	+ 37 15	84	15.7	GALAXY Sb
BIGO 475	02 21 54.	+ 41 37			NEBULA
ZWG 388.011	02 21 54.	- 02 23		14.0	GALAXY
PHL 8203	02 21 54.	- 06 30		18.0	BLUE STELLAR OBJECT
MCG-03-07-009	02 21 54.	- 20 03 30.	36	15.	GALAXY
LBN 0645	02 22	+ 61 45	720		BRIGHT NEBULA
LB 09784	02 22	- 81 13		15.2	FAINT BLUE STAR
PHL 8204	02 22 00.	+ 01 40		17.8	BLUE STELLAR OBJECT
PHL 8205	02 22 00.	+ 02 38		16.0	BLUE STELLAR OBJECT
3ZW 044	02 22 00.	+ 29 10			COMPACT GALAXY
UGC 01865	02 22 00.	+ 35 49	210	16.5	GALAXY DWRF SP
MCG+06-06-028	02 22 00.	+ 35 50	150	15.	GALAXY
ZWG 539.012	02 22 00.	+ 41 38		14.9	GALAXY
ZWG 538.066	02 22 00.	+ 41 38		14.9	GALAXY
UGC 01866	02 22 00.	+ 41 38	66	14.9	GALAXY SBa
ZWG 539.013	02 22 00.	+ 45 15		15.5	GALAXY
ZWG 538.067	02 22 00.	+ 45 15		15.5	GALAXY
UGC 01867	02 22 00.	+ 45 15	132	15.5	GALAXY Sc
LDN 1359	02 22 00.	+ 61 50	960		DARK NEBULA
MCG+14-02-009	02 22 00.	+ 83 48	36	15.	GALAXY
ARP 054	02 22 00.	- 04 52			PECULIAR GALAXY
PHL 4050	02 22 00.	- 05 43		17.9	BLUE STELLAR OBJECT
PHL 4051	02 22 00.	- 09 34		18.9	BLUE STELLAR OBJECT
PHL 8206	02 22 00.	- 11 28		18.3	BLUE STELLAR OBJECT
PHL 4049	02 22 00.	- 13 23		18.8	BLUE STELLAR OBJECT
PHL 8207	02 22 00.	- 20 03		18.2	BLUE STELLAR OBJECT
LB 03252	02 22 00.	- 50 20		13.5	FAINT BLUE STAR
RNGC 0896	02 22 01.	+ 61 45			DIFFUSE NEBULA
ZWG 483.031	02 22 06.	+ 25 49		15.7	GALAXY
5ZW 235	02 22 06.	+ 26 08			COMPACT GALAXY
ZWG 483.032	02 22 06.	+ 27 18		15.6	GALAXY
MCG+07-06-011	02 22 06.	+ 41 38	48	15.	GALAXY
ZWG 539.014	02 22 06.	+ 41 52		14.4	GALAXY
UGC 01868	02 22 06.	+ 41 52	138	14.4	GALAXY SBa
5ZW 236	02 22 06.	+ 45 26			COMPACT GALAXY
MCG+08-05-008	02 22 06.	+ 50 34	60	16.	GALAXY
ZWG 553.008	02 22 06.	+ 50 36		15.7	GALAXY
UGC 01869	02 22 06.	+ 50 36	60	15.7	GALAXY Sa-b
PHL 8174	02 22 06.	- 01 39		17.0	BLUE STELLAR OBJECT
PHL 4052	02 22 06.	- 13 37		18.7	BLUE STELLAR OBJECT
RNGC 0906	02 22 08.	+ 41 52		14.5	GALAXY
SG 1.04	02 22 09.	+ 61 49	780		DIFFUSE EMISSION NEBULA
MCG-03-07-010	02 22 09.	- 14 44	54	15.	GALAXY
ZC 0222.2+0104	02 22 12.	+ 01 04	3020		CLUSTER OF GALAXIES
PHL 8208	02 22 12.	+ 01 07		16.1	BLUE STELLAR OBJECT
MCG+03-07-008	02 22 12.	+ 15 57	42	15.	GALAXY
UGC 01870	02 22 12.	+ 19 27	90	18.	GALAXY S
ZWG 483.033	02 22 12.	+ 22 00		14.9	GALAXY
UGC 01871	02 22 12.	+ 22 00	72	14.9	GALAXY S
ZWG 523.036	02 22 12.	+ 36 35		15.7	GALAXY
ZWG 539.015	02 22 12.	+ 41 30		15.7	GALAXY
ZWG 539.016	02 22 12.	+ 41 49		14.5	GALAXY
UGC 01872	02 22 12.	+ 41 49	78	14.5	GALAXY E
MCG+07-06-012	02 22 12.	+ 41 52 30.	84	14.5	GALAXY
PHL 4056	02 22 12.	- 02 18		18.7	BLUE STELLAR OBJECT
PHL 4053	02 22 12.	- 03 16		18.7	BLUE STELLAR OBJECT
PHL 4055	02 22 12.	- 09 13		18.8	BLUE STELLAR OBJECT
PHL 4054	02 22 12.	- 09 48		18.5	BLUE STELLAR OBJECT
PHL 1277	02 22 12.	- 12 04		18.8	BLUE STELLAR OBJECT
PHL 8209	02 22 12.	- 12 16		18.8	BLUE STELLAR OBJECT
TON-S 0254	02 22 12.	- 23 35		15.5	BLUE STAR
RNGC 0909	02 22 14.	+ 41 49		14.5	GALAXY
MCG-02-07-004	02 22 15.	- 12 00	54	14.5	GALAXY
LB 02726	02 22 15.	+ 31 24 24.		15.8	FAINT BLUE STAR
MCG+01-07-009	02 22 15.	- 08 12	54	15.	GALAXY
SG 1.05	02 22 16.	+ 62 04	600		DIFFUSE EMISSION NEBULA
MCG+02-07-005	02 22 18.	+ 14 50	18	15.	GALAXY
UGC 01873	02 22 18.	+ 16 52	60	16.0	GALAXY
ZWG 483.034	02 22 18.	+ 23 37		15.6	GALAXY
UGC 01874	02 22 18.	+ 40 29	84	17.	GALAXY
ZWG 539.017	02 22 18.	+ 41 36		14.5	GALAXY
UGC 01875	02 22 18.	+ 41 36	240	14.5	GALAXY E
ZWG 539.018	02 22 18.	+ 41 41		15.7	GALAXY
MCG+07-06-013	02 22 18.	+ 41 49	18	15.	GALAXY
SCHO 0051	02 22 18.	+ 53 20 48.	300		ISOLATED DARK CLOUD
PHL 1278	02 22 18.	- 08 04		18.3	BLUE STELLAR OBJECT
PHL 8210	02 22 18.	- 11 08		16.7	BLUE STELLAR OBJECT
PHL 4057	02 22 18.	- 11 16		17.9	BLUE STELLAR OBJECT
MCG-02-07-005	02 22 18.	- 12 48	30	14.5	GALAXY
PHL 8211	02 22 18.	- 16 18		18.3	BLUE STELLAR OBJECT
RNGC 0910	02 22 20.	+ 41 36		14.5	GALAXY
IC 0224	02 22 21.	- 12 47 41.			NONSTELLAR OBJECT
PHL 8213	02 22 24.	+ 02 09		15.9	BLUE STELLAR OBJECT
PHL 4058	02 22 24.	+ 02 56		18.5	BLUE STELLAR OBJECT
PHL 8214	02 22 24.	+ 03 32		17.7	BLUE STELLAR OBJECT
MCG+03-07-009	02 22 24.	+ 16 51	18	16.	GALAXY
ZWG 462.009	02 22 24.	+ 20 44		15.7	GALAXY
ZWG 483.035	02 22 24.	+ 24 02		15.0	GALAXY
ZWG 523.037	02 22 24.	+ 36 57		15.0	GALAXY
MCG+06-06-029	02 22 24.	+ 36 58	18	15.	GALAXY
ZWG 523.038	02 22 24.	+ 37 57		15.6	GALAXY
ZWG 553.009	02 22 24.	+ 45 55		15.7	GALAXY
PHL 1279	02 22 24.	- 01 52		17.0	BLUE STELLAR OBJECT
PHL 4059	02 22 24.	- 07 18		18.3	BLUE STELLAR OBJECT
PHL 8212	02 22 24.	- 10 40		18.2	BLUE STELLAR OBJECT
PHL 8215	02 22 24.	- 13 30		17.8	BLUE STELLAR OBJECT
MCG-03-07-011	02 22 24.	- 19 23	60	14.	GALAXY
MCG+07-06-014	02 22 27.	+ 41 36	24	14.5	GALAXY
ZWG 439.005	02 22 30.	+ 14 50		15.6	GALAXY
ZWG 483.036	02 22 30.	+ 23 36		15.5	GALAXY
ACK 132+04.1	02 22 30.	+ 65 34			PLANETARY NEBULA
UGC 01876	02 22 30.	- 00 49	66	16.5	GALAXY S
PHL 8216	02 22 30.	- 07 40		18.6	BLUE STELLAR OBJECT
PHL 1280	02 22 30.	- 16 24		18.1	BLUE STELLAR OBJECT
LB 03253	02 22 30.	- 70 53		13.5	FAINT BLUE STAR
MIN.47 01	02 22 33.	+ 61 51			DIFFUSE NEBULA
MCG-03-07-012	02 22 33.	- 19 22 30.	36	15.	GALAXY
ZC 0222.6+1033	02 22 36.	+ 10 33	1140		CLUSTER OF GALAXIES
ZWG 523.039	02 22 36.	+ 36 45		15.0	GALAXY
UGC 01877	02 22 36.	+ 36 45	102	15.0	GALAXY E
MCG+06-06-030	02 22 36.	+ 36 46	48	14.5	GALAXY
ZWG 539.019	02 22 36.	+ 40 55		15.5	GALAXY
ZWG 539.020	02 22 36.	+ 41 33		15.0	GALAXY
5ZW 237	02 22 36.	+ 41 39			COMPACT GALAXY
ZWG 539.021	02 22 36.	+ 41 45		14.0	GALAXY

OBJECT NAME	RIGHT ASCEN.	DECLINATION	DIAM.	MAGN.	TYPE OF OBJECT
UGC 01878	02 22 36.	+ 41 45	108	14.0	GALAXY E
ZWG 539.022	02 22 36.	+ 43 23		15.6	GALAXY
PHL 8217	02 22 36.	- 00 38		16.6	BLUE STELLAR OBJECT
PHL 8218	02 22 36.	- 06 11		18.6	BLUE STELLAR OBJECT
PHL 4060	02 22 36.	- 06 36		18.8	BLUE STELLAR OBJECT
RNGC 0912	02 22 38.	+ 41 33		15.0	GALAXY
RNGC 0913	02 22 38.	+ 41 34			GALAXY
RNGC 0911	02 22 38.	+ 41 45		14.0	GALAXY
SG 3.010	02 22 38.	+ 61 26	2400		DIFFUSE EMISSION NEBULA
ARC 0350	02 22 39.	- 10 03		17.5	RICH CLUSTER OF GALAXIES
IC 1797	02 22 41.	+ 20 11 07.			NONSTELLAR OBJECT
PHL 1282	02 22 42.	+ 00 16		18.7	BLUE STELLAR OBJECT
PHL 4062	02 22 42.	+ 01 28		18.7	BLUE STELLAR OBJECT
PHL 1281	02 22 42.	+ 02 58		16.7	BLUE STELLAR OBJECT
PHL 1283	02 22 42.	+ 03 31		18.4	BLUE STELLAR OBJECT
MCG+02-07-006	02 22 42.	+ 10 51	72	14.	GALAXY
ZWG 439.006	02 22 42.	+ 10 52		15.6	GALAXY
UGC 01879	02 22 42.	+ 10 52	84	15.6	GALAXY Sc
ZWG 462.010	02 22 42.	+ 20 10		15.3	GALAXY
UGC 01880	02 22 42.	+ 20 10	72	15.3	GALAXY SB
MCG+03-07-010	02 22 42.	+ 20 11	27	15.	GALAXY
ZWG 483.037	02 22 42.	+ 26 31		14.8	GALAXY
UGC 01881	02 22 42.	+ 26 31	60	14.8	GALAXY Sb+COMP
ZWG 483.038	02 22 42.	+ 27 05		15.6	GALAXY
ZWG 523.040	02 22 42.	+ 37 00		14.9	GALAXY
UGC 01882	02 22 42.	+ 37 00	96	14.9	GALAXY Sc
MCG+06-06-031	02 22 42.	+ 37 02	66	15.	GALAXY
MCG+07-06-015	02 22 42.	+ 41 34	15	15.	GALAXY
MCG+07-06-016	02 22 42.	+ 41 45	24	15.	GALAXY
IC 1795	02 22 42.	+ 61 51	1200		HII REGION
PHL 4061	02 22 42.	- 05 10		17.9	BLUE STELLAR OBJECT
PHL 4063	02 22 42.	- 12 58		18.	BLUE STELLAR OBJECT
ARC 0347	02 22 44.	+ 41 39		13.3	RICH CLUSTER OF GALAXIES
RNGC 0915	02 22 45.	+ 27 00		15.0	GALAXY
PHL 8219	02 22 48.	+ 03 26		18.4	BLUE STELLAR OBJECT
ZWG 414.008	02 22 48.	+ 04 13		15.2	GALAXY
KARA.73B 0101	02 22 48.	+ 04 13	24	15.2	ISOLATED GALAXY E
MCG+02-07-007	02 22 48.	+ 11 14	108	14.5	GALAXY
ZWG 439.007	02 22 48.	+ 11 15		15.7	GALAXY
UGC 01883	02 22 48.	+ 11 15	150	15.7	GALAXY
ZWG 483.039	02 22 48.	+ 22 47		15.6	GALAXY
5ZW 238	02 22 48.	+ 24 35			COMPACT GALAXY
ZWG 483.040	02 22 48.	+ 24 45		15.5	GALAXY
ZWG 483.041	02 22 48.	+ 27 00		15.0	GALAXY
ZWG 523.041	02 22 48.	+ 37 40		15.7	GALAXY
UGC 01884	02 22 48.	+ 37 40	90	15.7	GALAXY SBa
MCG+06-06-032	02 22 48.	+ 37 42	36	16.	GALAXY
6ZW 207	02 22 48.	+ 38 17			COMPACT GALAXY
VB 218	02 22 48.	+ 55 22	477		STELLAR RING
PHL 4064	02 22 48.	- 05 00		18.6	BLUE STELLAR OBJECT
PHL 1284	02 22 48.	- 14 49		16.2	BLUE STELLAR OBJECT
PHL 8220	02 22 48.	- 15 58		18.1	BLUE STELLAR OBJECT
MCG-04-06-037	02 22 48.	- 25 00 30.	96	12.	GALAXY
RNGC 0916	02 22 51.	+ 27 01		15.0	GALAXY
ARC 0351	02 22 52.	- 08 57		16.9	RICH CLUSTER OF GALAXIES
RNGC 0922	02 22 52.	- 25 01		12.5	GALAXY
ZC 0222.9+1457	02 22 54.	+ 14 57	1280		CLUSTER OF GALAXIES
ZWG 483.042	02 22 54.	+ 24 33		15.7	GALAXY
ZWG 483.043	02 22 54.	+ 27 01		14.9	GALAXY
ZWG 483.044	02 22 54.	+ 27 11		15.6	GALAXY
UGC 01885	02 22 54.	+ 27 11	72	15.6	GALAXY SBb
ZC 0222.9+3633	02 22 54.	+ 36 33	1480		CLUSTER OF GALAXIES
ZWG 523.042	02 22 54.	+ 39 15		13.1	GALAXY
UGC 01886	02 22 54.	+ 39 15	270	13.1	GALAXY Sb
KARA.73B 0102	02 22 54.	+ 39 15	204	13.1	ISOLATED GALAXY S
MCG+06-06-033	02 22 54.	+ 39 17	216	13.5	GALAXY
ZWG 539.023	02 22 54.	+ 41 55		13.9	GALAXY
UGC 01887	02 22 54.	+ 41 55	126	13.9	GALAXY Sc
PHL 8221	02 22 54.	- 04 00		18.4	BLUE STELLAR OBJECT
PHL 1285	02 22 54.	- 11 05		19.0	BLUE STELLAR OBJECT
RNGC 0914	02 22 56.	+ 41 55		14.0	GALAXY
LBN 0650	02 23	+ 61 20	1920		BRIGHT NEBULA
LBN 0646	02 23	+ 62 00	600		BRIGHT NEBULA
VDB.66G 020	02 23	- 10 04	130		DWARF GALAXY
VDB.66G 021	02 23	- 21 39	100		DWARF GALAXY
ZWG 483.045	02 23 00.	+ 24 38		15.6	GALAXY
MCG+07-06-017	02 23 00.	+ 41 56	90	14.5	GALAXY
LDN 1360	02 23 00.	+ 61 00	420		DARK NEBULA
PHL 8223	02 23 00.	- 01 40		18.5	BLUE STELLAR OBJECT
PHL 8222	02 23 00.	- 02 41		16.3	BLUE STELLAR OBJECT
PHL 4065	02 23 00.	- 03 55		18.5	BLUE STELLAR OBJECT
PHL 8224	02 23 00.	- 14 35		18.5	BLUE STELLAR OBJECT
SCHO 0052	02 23 02.	+ 56 27 00.	400		ISOLATED DARK CLOUD
CED 006C	02 23 03.	+ 61 51	1620		DIFFUSE GALACTIC NEBULA
CED 006B	02 23 03.	+ 61 51	60		DIFFUSE GALACTIC NEBULA
CED 006A	02 23 03.	+ 61 51	360		DIFFUSE GALACTIC NEBULA
ZWG 462.011	02 23 06.	+ 18 16		14.3	GALAXY
UGC 01888	02 23 06.	+ 18 16	228	14.3	GALAXY Sc
KARA.73B 0103	02 23 06.	+ 18 16	234	14.3	ISOLATED GALAXY S
ZWG 504.077	02 23 06.	+ 29 57		15.0	GALAXY
UGC 01889	02 23 06.	+ 29 57	78	15.0	GALAXY SB0
5ZW 239	02 23 06.	+ 36 34			COMPACT GALAXY
ZWG 553.010	02 23 06.	+ 46 02		15.6	GALAXY
MCG+08-05-009	02 23 06.	+ 49 48	36	15.	GALAXY
PHL 1286	02 23 06.	- 05 23		18.6	BLUE STELLAR OBJECT
AGU 12	02 23 06.	- 40 41 00.		12.5	2 INTERACTING GALAXIES
BIGO 476	02 23 07.	+ 18 14			NEBULA
RNGC 0918	02 23 07.	+ 18 17		14.5	GALAXY
RNGC 0917	02 23 07.	+ 32 00			NON-EXISTENT OBJECT
SN 1967E	02 23 11.	+ 42 38		18.0	SUPERNOVA
PHL 1287	02 23 12.	+ 01 13		18.2	BLUE STELLAR OBJECT
PHL 8227	02 23 12.	+ 02 03		17.9	BLUE STELLAR OBJECT
MCG+03-07-011	02 23 12.	+ 18 17	180	13.	GALAXY
ZWG 504.078	02 23 12.	+ 28 46		15.7	GALAXY
MCG+05-06-040	02 23 12.	+ 29 57	48	14.5	GALAXY
MCG+05-06-039	02 23 12.	+ 31 40	90	14.	GALAXY
ZWG 504.079	02 23 12.	+ 31 41		14.5	GALAXY
UGC 01890	02 23 12.	+ 31 41	150	14.5	GALAXY Sa-b
6ZW 208	02 23 12.	+ 36 46			COMPACT GALAXY
MCG+06-06-034	02 23 12.	+ 36 47	36	16.	GALAXY
5ZW 240	02 23 12.	+ 37 00			COMPACT GALAXY
MCG+07-06-018	02 23 12.	+ 42 38	18	15.5	GALAXY
PHL 8226	02 23 12.	- 02 14		16.9	BLUE STELLAR OBJECT
PHL 4066	02 23 12.	- 03 16		18.7	BLUE STELLAR OBJECT
PHL 8229	02 23 12.	- 09 14		18.5	BLUE STELLAR OBJECT
PHL 4067	02 23 12.	- 11 12		18.8	BLUE STELLAR OBJECT
PHL 8228	02 23 12.	- 13 16		17.6	BLUE STELLAR OBJECT
PHL 8225	02 23 12.	- 17 22		18.0	BLUE STELLAR OBJECT

OBJECT NAME	RIGHT ASCEN.	DECLINATION	DIAM.	MAGN.	TYPE OF OBJECT
MCG-04-06-038	02 23 12.	- 25 51 30.	60	14.	GALAXY
ARC 0349	02 23 16.	+ 36 37		17.3	RICH CLUSTER OF GALAXIES
ZWG 483.046	02 23 18.	+ 25 50		15.5	GALAXY
UGC 01891	02 23 18.	+ 25 50	102	15.5	GALAXY E
ZWG 483.047	02 23 18.	+ 25 53		15.1	GALAXY
ZWG 483.048	02 23 18.	+ 27 23		15.3	GALAXY
UGC 01892	02 23 18.	+ 27 23	78	15.3	GALAXY SBb
ZWG 553.011	02 23 18.	+ 49 50		15.5	GALAXY
UGC 01893	02 23 18.	+ 49 50	102	15.5	GALAXY SB0/SBa
URA 24	02 23 18.	+ 64 31	228		STELLAR RING
PHL 8230	02 23 18.	- 05 28		16.8	BLUE STELLAR OBJECT
MCG-03-07-013	02 23 18.	- 16 20	36	15.	GALAXY
RNGC 0919	02 23 21.	+ 26 59		15.5	GALAXY
PHL 4069	02 23 24.	+ 00 38		18.8	BLUE STELLAR OBJECT
ZC 0223.4+1326	02 23 24.	+ 13 26	940		CLUSTER OF GALAXIES
ZWG 483.049	02 23 24.	+ 26 59		15.5	GALAXY
UGC 01894	02 23 24.	+ 26 59	84	15.5	GALAXY Sa-b
ZWG 504.080	02 23 24.	+ 28 17		15.7	GALAXY
UGC 01895	02 23 24.	+ 28 17	90	15.7	GALAXY SBc
ZWG 504.081	02 23 24.	+ 30 13		14.6	GALAXY
UGC 01896	02 23 24.	+ 30 13	90	14.6	GALAXY S0
6ZW 209	02 23 24.	+ 36 49			COMPACT GALAXY
MCG+06-06-035	02 23 24.	+ 37 12	66	16.	GALAXY
PHL 4068	02 23 24.	- 05 08		16.7	BLUE STELLAR OBJECT
PHL 1288	02 23 24.	- 06 16		16.7	BLUE STELLAR OBJECT
PHL 8231	02 23 24.	- 08 18		17.6	BLUE STELLAR OBJECT
MCG-02-07-006	02 23 24.	- 11 25	48	14.5	GALAXY
PHL 4070	02 23 24.	- 11 55		19.0	BLUE STELLAR OBJECT
ZWG 388.013	02 23 30.	+ 00 48		15.7	GALAXY
MCG+02-07-008	02 23 30.	+ 12 12 30.	36	15.	GALAXY
UGC 01897	02 23 30.	+ 12 14	66	16.0	GALAXY Sc
UGC 01898	02 23 30.	+ 22 47	90	16.5	GALAXY Sc
ZWG 483.050	02 23 30.	+ 27 26		15.0	GALAXY
UGC 01899	02 23 30.	+ 27 26	78	15.0	GALAXY DISTRBD
MCG+05-06-041	02 23 30.	+ 30 11	39	14.5	GALAXY
ZWG 504.082	02 23 30.	+ 30 22		15.6	GALAXY
UGC 01900	02 23 30.	+ 37 11	90	16.0	GALAXY Sa-b
SN 19610	02 23 30.	+ 43 24		17.0	SUPERNOVA
MCG+07-06-019	02 23 30.	+ 43 24	36	16.5	GALAXY
MCG+00-07-011	02 23 30.	- 00 33	66	13.	GALAXY
MCG+01-07-010	02 23 30.	- 04 32 30.	42	14.5	GALAXY
PHL 4071	02 23 30.	- 07 48		17.2	BLUE STELLAR OBJECT
IC 1798	02 23 33.	+ 13 13 15.			NONSTELLAR OBJECT
SN 1964N	02 23 35.	+ 29 36		16.0	SUPERNOVA
PHL 8235	02 23 36.	+ 00 38		17.9	BLUE STELLAR OBJECT
ZWG 439.008	02 23 36.	+ 10 11		15.5	GALAXY
ZWG 483.051	02 23 36.	+ 21 55		15.7	GALAXY
UGC 01902	02 23 36.	+ 27 35	72	16.5	GALAXY Sc
MCG+05-06-042	02 23 36.	+ 29 35	60	16.	GALAXY
ZWG 504.083	02 23 36.	+ 29 36		15.7	GALAXY
UGC 01903	02 23 36.	+ 29 36	60	15.7	GALAXY Sb
KARA.73B 0104	02 23 36.	+ 29 36	72	15.7	ISOLATED GALAXY S
UGC 01904	02 23 36.	+ 36 56	84	16.0	GALAXY Sb-c
ZWG 539.024	02 23 36.	+ 41 37		15.0	GALAXY
ZWG 388.014	02 23 36.	- 00 33		13.9	GALAXY
UGC 01901	02 23 36.	- 00 33	150	13.9	GALAXY SBb
PHL 8234	02 23 36.	- 01 39		16.8	BLUE STELLAR OBJECT
PHL 1289	02 23 36.	- 11 48		19.1	BLUE STELLAR OBJECT
PHL 8233	02 23 36.	- 12 39		18.0	BLUE STELLAR OBJECT
PHL 8232	02 23 36.	- 18 10		17.7	BLUE STELLAR OBJECT
LB 03254	02 23 36.	- 65 12		13.4	FAINT BLUE STAR
RNGC 0926	02 23 37.	- 00 33		14.0	GALAXY
PHL 1290	02 23 42.	+ 00 02		17.5	BLUE STELLAR OBJECT
ZWG 483.052	02 23 42.	+ 25 35		15.6	GALAXY
MCG+05-06-043	02 23 42.	+ 27 34	36	17.	GALAXY
ZWG 539.025	02 23 42.	+ 41 28		15.3	GALAXY
ZWG 539.026	02 23 42.	+ 41 48		15.7	GALAXY
URA 26	02 23 42.	+ 64 44	78		STELLAR RING
MCG+00-07-012	02 23 42.	- 01 47 30.	30	15.	GALAXY
PHL 8236	02 23 42.	- 04 14		17.6	BLUE STELLAR OBJECT
MCG-04-06-039	02 23 42.	- 24 55 30.	48	15.	GALAXY
MCG+07-06-020	02 23 45.	+ 41 37	54	15.5	GALAXY
REIN 7.001	02 23 46.74	- 01 46 34.5			NEBULA
PHL 8237	02 23 48.	+ 01 16		16.8	BLUE STELLAR OBJECT
UGC 01906	02 23 48.	+ 22 32	84	16.5	GALAXY Sb
6ZW 210	02 23 48.	+ 36 30			COMPACT GALAXY
ZWG 539.027	02 23 48.	+ 42 35		15.7	GALAXY
PHL 8239	02 23 48.	- 00 17		17.8	BLUE STELLAR OBJECT
ZWG 388.015	02 23 48.	- 01 46		15.6	GALAXY
UGC 01905	02 23 48.	- 01 46	60	15.6	GALAXY S0
PHL 8238	02 23 48.	- 07 06		16.6	BLUE STELLAR OBJECT
MCG-04-06-040	02 23 48.	- 21 39	120	16.	GALAXY
LB 02727	02 23 49.	+ 31 26 24.		16.2	FAINT BLUE STAR
IC 0225	02 23 52.	+ 00 56 43.			NONSTELLAR OBJECT
ZWG 388.016	02 23 54.	+ 00 56		14.7	GALAXY
UGC 01907	02 23 54.	+ 00 56	108	14.7	GALAXY E
MCG+00-07-013	02 23 54.	+ 00 56	18	14.	GALAXY
MCG+02-07-009	02 23 54.	+ 11 55	72	13.5	GALAXY
ZWG 439.009	02 23 54.	+ 11 56		14.5	GALAXY
MRK 593	02 23 54.	+ 11 56	10	16.5	GALAXY WITH UV CONTINUUM
UGC 01908	02 23 54.	+ 11 56	84	14.5	GALAXY SBc
ZWG 504.084	02 23 54.	+ 31 52		15.4	GALAXY
UGC 01909	02 23 54.	+ 31 52	72	15.4	GALAXY Sc
6ZW 211	02 23 54.	+ 34 18			COMPACT GALAXY
ZWG 523.043	02 23 54.	+ 34 58		14.9	GALAXY
UGC 01910	02 23 54.	+ 34 58	138	14.8	GALAXY S
MCG+06-06-036	02 23 54.	+ 34 58	90	14.	GALAXY
MCG-02-07-007	02 23 54.	- 10 03 30.	180	14.	GALAXY
PHL 8240	02 23 54.	- 13 08		18.3	BLUE STELLAR OBJECT
LB 01624	02 23 54.	- 49 07		14.4	FAINT BLUE STAR
RNGC 0927	02 23 57.	+ 11 56		14.5	GALAXY
LBN 0730	02 24	+ 11 30	5760		BRIGHT NEBULA
LBN 0646	02 24	+ 61 50	900		BRIGHT NEBULA
LBN 0635	02 24	+ 72 50	600		BRIGHT NEBULA
LB 09785	02 24	- 83 56		14.6	FAINT BLUE STAR
UGC 01911	02 24 00.	+ 00 25	60	16.5	GALAXY Sc
MCG+00-07-014	02 24 00.	+ 00 25	48	15.5	GALAXY
PHL 8241	02 24 00.	+ 01 12		18.7	BLUE STELLAR OBJECT
ZWG 462.012	02 24 00.	+ 20 16		13.8	GALAXY
RNGC 0924	02 24 00.	+ 20 16		14.0	GALAXY
UGC 01912	02 24 00.	+ 20 16	138	13.8	GALAXY S0
ZWG 483.053	02 24 00.	+ 24 48		15.1	GALAXY
5ZW 241	02 24 00.	+ 27 58			COMPACT GALAXY
MCG+05-06-044	02 24 00.	+ 31 52	48	15.	GALAXY
LDN 1363	02 24 00.	+ 60 50	420		DARK NEBULA
LDN 1362	02 24 00.	+ 61 00	240		DARK NEBULA
PHL 8242	02 24 00.	- 01 04		18.5	BLUE STELLAR OBJECT
MCG-03-07-014	02 24 00.	- 15 27	18	15.	GALAXY

OBJECT NAME	RIGHT ASCEN.	DECLINATION	DIAM.	MAGN.	TYPE OF OBJECT
PHL 4072	02 24 06.	+ 02 20		18.5	BLUE STELLAR OBJECT
PHL 1291	02 24 06.	+ 03 28		18.5	BLUE STELLAR OBJECT
MCG+03-07-012	02 24 06.	+ 20 17	30	14.	GALAXY
ZWG 462.013	02 24 06.	+ 20 18		15.7	GALAXY
PHL 1292	02 24 06.	- 02 52		17.3	BLUE STELLAR OBJECT
PHL 8243	02 24 06.	- 05 06		15.7	BLUE STELLAR OBJECT
PHL 8244	02 24 06.	- 12 21		16.6	BLUE STELLAR OBJECT
LB 02728	02 24 08.	+ 30 02 06.		16.8	FAINT BLUE STAR
PHL 4073	02 24 12.	+ 00 44		18.0	BLUE STELLAR OBJECT
ZC 0224.2+1352	02 24 12.	+ 13 52	1010		CLUSTER OF GALAXIES
ZWG 539.028	02 24 12.	+ 40 50		15.2	GALAXY
VB 219	02 24 12.	+ 56 16	356		STELLAR RING
PHL 4074	02 24 12.	- 07 57		18.4	BLUE STELLAR OBJECT
PHL 8245	02 24 12.	- 15 30		18.2	BLUE STELLAR OBJECT
MCG-03-07-015	02 24 12.	- 16 04 30.	72	14.5	GALAXY
MCG-04-06-041	02 24 12.	- 24 31	150	12.	GALAXY
LB 03255	02 24 12.	- 73 47		13.4	FAINT BLUE STAR
RNGC 0921	02 24 13.	- 16 04		14.0	GALAXY
MCG+03-07-013	02 24 18.	+ 20 19	45	15.	GALAXY
ZWG 504.085	02 24 18.	+ 33 21		10.5	GALAXY
RNGC 0925	02 24 18.	+ 33 21		11.5	GALAXY
UGC 01913	02 24 18.	+ 33 21	780	10.5	GALAXY Sc/SBc
KARA.73B 0105	02 24 18.	+ 33 21	636	10.5	ISOLATED GALAXY S
MCG+05-06-045	02 24 18.	+ 33 22	600	11.	GALAXY
6ZW 212	02 24 18.	+ 36 28			COMPACT GALAXY
ZWG 523.044	02 24 18.	+ 37 25		15.7	GALAXY
UGC 01914	02 24 18.	+ 40 50	84	15.2	GALAXY (E)
ZWG 539.029	02 24 18.	+ 41 42		15.7	GALAXY
MCG+08-05-010	02 24 18.	+ 50 15	18	15.	GALAXY
RNGC 0944	02 24 18.	- 14 44		14.0	GALAXY
MCG-04-06-042	02 24 18.	- 23 13	48	15.5	GALAXY
IC 0228	02 24 19.	- 14 43 56.			NONSTELLAR OBJECT
MCG-03-07-016	02 24 21.	- 14 44	48	14.5	GALAXY
LB 02729	02 24 24.	+ 31 08 54.		16.2	FAINT BLUE STAR
ZWG 523.045	02 24 24.	+ 36 21		15.7	GALAXY
MCG+06-06-037	02 24 24.	+ 36 22	12	16.	GALAXY
ZWG 539.030	02 24 24.	+ 41 45		14.4	GALAXY
UGC 01915	02 24 24.	+ 41 45	54	14.4	GALAXY S(b)
ZWG 539.031	02 24 24.	+ 41 47		15.0	GALAXY
UGC 01916	02 24 24.	+ 50 16	78	16.0	GALAXY E
PHL 8248	02 24 24.	- 10 40		17.7	BLUE STELLAR OBJECT
PHL 8247	02 24 24.	- 10 48		18.5	BLUE STELLAR OBJECT
PHL 8246	02 24 24.	- 10 48		18.6	BLUE STELLAR OBJECT
MCG-04-06-043	02 24 24.	- 23 39	10	15.5	GALAXY
RNGC 0923	02 24 25.	+ 41 45		14.5	GALAXY
SG 3.011	02 24 25.	+ 62 53	900		DIFFUSE EMISSION NEBULA
MCG+07-06-021	02 24 27.	+ 41 43	36	16.	GALAXY
RNGC 0939	02 24 27.	- 44 40			GALAXY
3ZW 045	02 24 30.	+ 12 58			COMPACT GALAXY
5ZW 242	02 24 30.	+ 22 52			COMPACT GALAXY
ZWG 483.054	02 24 30.	+ 22 52		15.7	GALAXY
MCG+07-06-022	02 24 30.	+ 41 46 30.	45	14.5	GALAXY
LDN 1361	02 24 30.	+ 61 25	420		DARK NEBULA
DG 009	02 24 30.	+ 72 46	840		REFLECTION NEBULA
PHL 8249	02 24 30.	- 01 32		16.2	BLUE STELLAR OBJECT
MCG+01-07-011	02 24 30.	- 04 49	24	16.	GALAXY
MCG-02-07-008	02 24 30.	- 09 30	48	14.5	GALAXY
MCG+06-06-038	02 24 33.	+ 35 56 30.	66	14.5	GALAXY
MCG+07-06-023	02 24 33.	+ 41 48	24	15.5	GALAXY
RNGC 0920	02 24 35.	+ 45 44		15.5	GALAXY
PHL 8251	02 24 36.	+ 01 58		18.2	BLUE STELLAR OBJECT
PHL 8252	02 24 36.	+ 02 36		16.5	BLUE STELLAR OBJECT
ZC 0224.6+0546	02 24 36.	+ 05 46	1410		CLUSTER OF GALAXIES
5ZW 243	02 24 36.	+ 21 46			COMPACT GALAXY
ZWG 483.055	02 24 36.	+ 24 02		15.5	GALAXY
UGC 01917	02 24 36.	+ 24 02	66	15.5	GALAXY SBb
ZWG 483.056	02 24 36.	+ 25 26		14.7	GALAXY
UGC 01918	02 24 36.	+ 25 26	102	14.7	GALAXY SBa-b
ZWG 523.046	02 24 36.	+ 35 56		15.0	GALAXY
UGC 01919	02 24 36.	+ 35 56	96	15.0	GALAXY SBb
MCG+08-05-011	02 24 36.	+ 45 42	60	15.	GALAXY
ZWG 553.012	02 24 36.	+ 45 44		15.6	GALAXY
UGC 01920	02 24 36.	+ 45 44	114	15.6	GALAXY SBb
PHL 8250	02 24 36.	- 02 58		16.0	BLUE STELLAR OBJECT
MCG+01-07-012	02 24 36.	- 03 05	60		GALAXY
PHL 4075	02 24 36.	- 03 38		15.9	BLUE STELLAR OBJECT
PHL 8253	02 24 36.	- 03 57		16.9	BLUE STELLAR OBJECT
PHL 8254	02 24 36.	- 05 36		17.5	BLUE STELLAR OBJECT
HELW 095	02 24 39.	- 10 28 10.			NEBULA
6ZW 213	02 24 42.	+ 33 51			COMPACT GALAXY
VB 168	02 24 42.	+ 63 31	141		STELLAR RING
ISS 0059	02 24 42.	+ 63 31	142		STELLAR RING
PHL 8255	02 24 42.	- 12 22		16.0	BLUE STELLAR OBJECT
MCG-03-07-017	02 24 42.	- 19 30	30	14.5	GALAXY
RNGC 0928	02 24 45.	+ 27 00		14.5	GALAXY
MCG-04-06-044	02 24 45.	- 23 09	36	15.5	GALAXY
PHL 8256	02 24 48.	+ 00 36		18.1	BLUE STELLAR OBJECT
PHL 1293	02 24 48.	+ 01 10		18.0	BLUE STELLAR OBJECT
MCG+00-07-015	02 24 48.	+ 01 53	30	16.	GALAXY
ZWG 483.057	02 24 48.	+ 21 56		15.7	GALAXY
3ZW 046	02 24 48.	+ 21 57			COMPACT GALAXY
ZWG 483.058	02 24 48.	+ 26 00		14.9	GALAXY
5ZW 244	02 24 48.	+ 26 22			COMPACT GALAXY
ZWG 483.059	02 24 48.	+ 26 22		14.6	GALAXY
UGC 01921	02 24 48.	+ 26 22	66	14.6	GALAXY SBb
ZWG 483.060	02 24 48.	+ 27 00		14.7	GALAXY
UGC 01922	02 24 48.	+ 27 58	138	16.0	GALAXY S
MCG+05-06-046	02 24 48.	+ 27 58	30	15.	GALAXY
PHL 8259	02 24 48.	- 05 00		17.7	BLUE STELLAR OBJECT
PHL 4076	02 24 48.	- 05 01		18.2	BLUE STELLAR OBJECT
PHL 8258	02 24 48.	- 08 32		16.0	BLUE STELLAR OBJECT
PHL 8257	02 24 48.	- 08 50		18.4	BLUE STELLAR OBJECT
MCG-03-07-018	02 24 48.	- 14 35	48	15.	GALAXY
PHL 4077	02 24 48.	- 16 56		17.5	BLUE STELLAR OBJECT
PHL 4078	02 24 48.	- 17 12		17.5	BLUE STELLAR OBJECT
IC 0226	02 24 51.	+ 27 59 16.			NONSTELLAR OBJECT
RNGC 0929	02 24 53.	- 12 19		15.0	GALAXY
UGC 01923	02 24 54.	+ 01 54	66	17.	GALAXY S-IRR
PHL 4080	02 24 54.	+ 02 17		18.8	BLUE STELLAR OBJECT
5ZW 245	02 24 54.	+ 22 36			COMPACT GALAXY
ZWG 504.086	02 24 54.	+ 31 30		15.4	GALAXY
UGC 01924	02 24 54.	+ 31 30	114	15.4	GALAXY Sc
MCG+05-06-047	02 24 54.	+ 31 30	90	15.	GALAXY
UGC 01925	02 24 54.	+ 37 14	66	16.5	GALAXY DBL SYS
VB 220	02 24 54.	+ 53 07	1175		STELLAR RING
PHL 4079	02 24 54.	- 03 53		16.6	BLUE STELLAR OBJECT
MCG+01-07-013	02 24 54.	- 04 10	42	15.5	GALAXY
MCG-02-07-009	02 24 54.	- 12 19	54	15.	GALAXY

OBJECT NAME	RIGHT ASCEN.	DECLINATION	DIAM.	MAGN.	TYPE OF OBJECT
MCG-03-07-019	02 24 54.	- 15 39	78	14.5	GALAXY
MCG-03-07-020	02 24 57.	- 19 26	30	14.5	GALAXY
ARP 276	02 24 59.	+ 19 22			PECULIAR GALAXY
LBN 0649	02 25	+ 62 00	5400		BRIGHT NEBULA
LBN 0636	02 25	+ 72 50	300		BRIGHT NEBULA
PHL 8262	02 25 00.	+ 02 36		18.4	BLUE STELLAR OBJECT
UGC 01927	02 25 00.	+ 43 23	72	17.	GALAXY
UGC 01928	02 25 00.	+ 43 38	108	16.5	GALAXY S
MCG+09-05-001	02 25 00.	+ 51 12	72	15.	GALAXY
ZWG 553.013	02 25 00.	+ 51 14		15.2	GALAXY
UGC 01930	02 25 00.	+ 51 14	138	15.2	GALAXY SB0
LDN 1365	02 25 00.	+ 60 35	1200		DARK NEBULA
ZWG 388.017	02 25 00.	- 00 28		14.4	GALAXY
UGC 01926	02 25 00.	- 00 28	108	14.4	GALAXY S0
MCG+00-07-016	02 25 00.	- 00 29	48	14.	GALAXY
PHL 8261	02 25 00.	- 02 26		17.4	BLUE STELLAR OBJECT
PHL 1294	02 25 00.	- 03 42		18.2	BLUE STELLAR OBJECT
PHL 4081	02 25 00.	- 09 02		19.0	BLUE STELLAR OBJECT
PHL 8264	02 25 00.	- 13 10		16.3	BLUE STELLAR OBJECT
PHL 8263	02 25 00.	- 13 51		18.9	BLUE STELLAR OBJECT
PHL 8260	02 25 00.	- 31 32		16.4	BLUE STELLAR OBJECT
LB 01625	02 25 00.	- 45 14		13.4	FAINT BLUE STAR
RNGC 0934	02 25 01.	- 00 28		14.5	GALAXY
MCG+00-07-017	02 25 03.	- 01 23	180	11.2	GALAXY
REIN 2.024	02 25 04.71	- 01 22 46.6			NEBULA
ZC 0225.1+1419	02 25 06.	+ 14 19	1340		CLUSTER OF GALAXIES
ZWG 462.014	02 25 06.	+ 20 06		13.7	GALAXY
RNGC 0930	02 25 06.	+ 20 06		13.5	GALAXY
UGC 01931	02 25 06.	+ 20 06	114	13.7	GALAXY Sa
RNGC 0932	02 25 06.	+ 20 07			NON-EXISTENT OBJECT
ZWG 483.061	02 25 06.	+ 21 35		15.5	GALAXY
MCG+05-06-048	02 25 06.	+ 27 55	24	15.	GALAXY
ZWG 504.087	02 25 06.	+ 27 57		15.5	GALAXY
UGC 01932	02 25 06.	+ 27 57	138	15.5	GALAXY E
OCL 0349	02 25 06.	+ 61 34	150	16.	OPEN STAR CLUSTER
ZWG 388.018	02 25 06.	- 01 23		11.3	GALAXY
UGC 01929	02 25 06.	- 01 23	348	11.3	GALAXY SB0
PHL 8265	02 25 06.	- 04 16		17.9	BLUE STELLAR OBJECT
MCG+01-07-014	02 25 06.	- 04 16	60	15.5	GALAXY
MCG-02-07-010	02 25 06.	- 10 23	72	14.	GALAXY
PHL 4082	02 25 06.	- 10 40		18.9	BLUE STELLAR OBJECT
IC 0229	02 25 07.	- 24 02 36.			NONSTELLAR OBJECT
RNGC 0936	02 25 08.	- 01 23		11.0	GALAXY
HELW 096	02 25 08.	- 10 23 23.			NEBULA
IC 0227	02 25 09.	+ 27 57 14.			NONSTELLAR OBJECT
ARC 0353	02 25 09.	- 22 17		17.7	RICH CLUSTER OF GALAXIES
MCG+03-07-014	02 25 12.	+ 20 07	120	13.5	GALAXY
ZC 0225.2+2313	02 25 12.	+ 23 13	1410		CLUSTER OF GALAXIES
ZWG 504.088	02 25 12.	+ 32 08		15.1	GALAXY
UGC 01933	02 25 12.	+ 38 13	84	16.0	GALAXY S
MCG+06-06-039	02 25 12.	+ 38 14	9	16.	GALAXY
MCG-02-07-011	02 25 12.	- 10 09	36	15.5	GALAXY
PHL 4083	02 25 12.	- 12 44		16.7	BLUE STELLAR OBJECT
PHL 4084	02 25 12.	- 14 19		18.6	BLUE STELLAR OBJECT
PHL 8266	02 25 12.	- 19 48		16.6	BLUE STELLAR OBJECT
MCG+05-06-049	02 25 15.	+ 31 05	240	14.	GALAXY
ZWG 388.019	02 25 18.	+ 00 16		15.5	GALAXY
UGC 01934	02 25 18.	+ 00 16	114	15.5	GALAXY Sb-c
MCG+00-07-018	02 25 18.	+ 00 16 30.	54	14.5	GALAXY
ZWG 483.062	02 25 18.	+ 23 42		15.2	GALAXY
ZWG 504.089	02 25 18.	+ 31 05		13.9	GALAXY
UGC 01935	02 25 18.	+ 31 05	270	13.9	GALAXY Sb
PHL 4085	02 25 18.	- 25 00		17.7	BLUE STELLAR OBJECT
MCG-05-06-016	02 25 18.	- 28 42	24	17.	GALAXY
RNGC 0931	02 25 19.	+ 31 05		14.0	GALAXY
PHL 4088	02 25 24.	+ 01 44		18.3	BLUE STELLAR OBJECT
PHL 4086	02 25 24.	+ 03 29		17.7	BLUE STELLAR OBJECT
ZC 0225.4+1157	02 25 24.	+ 11 57	3290		CLUSTER OF GALAXIES
ZWG 462.015	02 25 24.	+ 19 21		14.8	GALAXY
UGC 01936	02 25 24.	+ 19 21	84	14.8	GALAXY SBb
KARA.72 068B	02 25 24.	+ 19 21	72	14.8	PART OF DOUBLE GALAXY
ZWG 462.016	02 25 24.	+ 19 22		13.9	GALAXY
RNGC 0935	02 25 24.	+ 19 22		14.0	GALAXY
UGC 01937	02 25 24.	+ 19 22	114	13.9	GALAXY Sc
KARA.72 068A	02 25 24.	+ 19 22	108	13.9	PART OF DOUBLE GALAXY
ZWG 483.063	02 25 24.	+ 23 00		15.0	GALAXY
UGC 01938	02 25 24.	+ 23 00	90	15.0	GALAXY Sb-c
ZWG 483.064	02 25 24.	+ 26 05		14.9	GALAXY
UGC 01939	02 25 24.	+ 26 05	72	14.9	GALAXY SBa
ZWG 504.090	02 25 24.	+ 29 15		15.5	GALAXY
KARA.73B 0106	02 25 24.	+ 29 15	24	15.5	ISOLATED GALAXY S
ZWG 504.091	02 25 24.	+ 30 29		15.6	GALAXY
MCG+08-05-012	02 25 24.	+ 45 44	60	16.	GALAXY
URA 16	02 25 24.	+ 63 26	108		STELLAR RING
ZC 0225.4-0227	02 25 24.	- 02 27	810		CLUSTER OF GALAXIES
PHL 8268	02 25 24.	- 09 46		17.5	BLUE STELLAR OBJECT
PHL 8267	02 25 24.	- 14 12		17.6	BLUE STELLAR OBJECT
PHL 4087	02 25 24.	- 16 20		18.2	BLUE STELLAR OBJECT
PHL 1295	02 25 24.	- 19 14		16.0	BLUE STELLAR OBJECT
IC 1801	02 25 25.	+ 19 21 07.			NONSTELLAR OBJECT
IC 1799	02 25 29.	+ 45 44			NONSTELLAR OBJECT
VV 238B	02 25 30.	+ 19 22	78	13.5	INTERACTING GALAXY
VV 238A	02 25 30.	+ 19 22	102	13.	INTERACTING GALAXY
MCG+03-07-016	02 25 30.	+ 19 23	66	14.	GALAXY
MCG+03-07-015	02 25 30.	+ 19 24	96	13.	GALAXY
UGC 01940	02 25 30.	+ 27 55	72	16.5	GALAXY S0
UGC 01941	02 25 30.	+ 37 45	96	16.0	GALAXY
MCG+06-06-040	02 25 30.	+ 37 45 30.	72	16.	GALAXY
UGC 01942	02 25 30.	+ 38 24	66	16.5	GALAXY Sb
ZWG 553.014	02 25 30.	+ 45 45		15.0	GALAXY
UGC 01943	02 25 30.	+ 45 45	78	15.0	GALAXY S
PHL 8269	02 25 30.	- 06 00		16.5	BLUE STELLAR OBJECT
SHB 060	02 25 35.0	- 01 29 06.		18.	QUASI-STELLAR OBJECT
BC PKS0225-014	02 25 35.0	- 01 29 07.		18.	QUASI-STELLAR OBJECT
MCG+00-07-019	02 25 36.	+ 00 34	18	15.	GALAXY
ZWG 388.020	02 25 36.	+ 00 35		15.0	GALAXY
UGC 01944	02 25 36.	+ 29 47	90	17.	GALAXY DWRF SP
IC 1800	02 25 36.	+ 31 11			OPEN CLUSTER
MCG+00-07-020	02 25 36.	- 01 35	66	13.	GALAXY
PHL 4089	02 25 36.	- 08 48		17.5	BLUE STELLAR OBJECT
PHL 4090	02 25 36.	- 09 28		18.4	BLUE STELLAR OBJECT
PHL 8270	02 25 36.	- 27 45		13.7	BLUE STELLAR OBJECT
ARC 0352	02 25 37.	- 02 24		17.6	RICH CLUSTER OF GALAXIES
HELW 010	02 25 38.	- 01 35			NEBULA
REIN 7.002B	02 25 39.90	- 01 34 10.4			NEBULA
REIN 7.002A	02 25 39.93	- 01 34 10.4			NEBULA
MCG+02-07-010	02 25 42.	+ 10 09	36	14.5	GALAXY
ZWG 439.010	02 25 42.	+ 10 10		15.1	GALAXY

OBJECT NAME	RIGHT ASCEN.	DECLINATION	DIAM.	MAGN.	TYPE OF OBJECT
UGC 01946	02 25 42.	+ 15 25	114	16.0	GALAXY Sc
ZC 0225.7+1706	02 25 42.	+ 17 06	340		CLUSTER OF GALAXIES
ZWG 462.017	02 25 42.	+ 20 03		13.8	GALAXY
RNGC 0938	02 25 42.	+ 20 03		14.0	GALAXY
UGC 01947	02 25 42.	+ 20 03	102	13.8	GALAXY E
ZWG 523.047	02 25 42.	+ 37 57		15.2	GALAXY
UGC 01948	02 25 42.	+ 37 57	60	15.2	GALAXY Sa-b
ZWG 388.021	02 25 42.	- 01 34		14.4	GALAXY
UGC 01945	02 25 42.	- 01 34	144	14.4	GALAXY Sc-IRR
KARA.68 015	02 25 42.	- 01 54	40		DWARF GALAXY
PHL 4091	02 25 42.	- 11 48		18.7	BLUE STELLAR OBJECT
LB 02730	02 25 43.	+ 31 39 00.		17.2	FAINT BLUE STAR
MCG+01-07-015	02 25 45.	- 06 03	72	15.	GALAXY
ZWG 388.022	02 25 48.	+ 00 28		15.4	GALAXY
MCG+00-07-021	02 25 48.	+ 00 28	24	15.5	GALAXY
MCG+03-07-017	02 25 48.	+ 20 05	30	14.	GALAXY
5ZW 247	02 25 48.	+ 22 48			COMPACT GALAXY
ZWG 483.065	02 25 48.	+ 23 35		15.7	GALAXY
UGC 01950	02 25 48.	+ 23 35	66	15.7	GALAXY
ZC 0225.8+3347	02 25 48.	+ 33 47	1680		CLUSTER OF GALAXIES
UGC 01951	02 25 48.	+ 37 30	72	16.0	GALAXY E
MCG+06-06-041	02 25 48.	+ 37 57 30.	54	16.	GALAXY
ZWG 553.015	02 25 48.	+ 48 13		15.5	GALAXY
UGC 01949	02 25 48.	- 00 48	60	18.	GALAXY DWARF
PHL 4093	02 25 48.	- 04 26		18.3	BLUE STELLAR OBJECT
PHL 4092	02 25 48.	- 09 57		18.9	BLUE STELLAR OBJECT
PHL 8271	02 25 48.	- 10 14		16.9	BLUE STELLAR OBJECT
PHL 1296	02 25 48.	- 10 18		18.1	BLUE STELLAR OBJECT
PHL 1297	02 25 48.	- 11 48		18.8	BLUE STELLAR OBJECT
ARC 0355	02 25 48.	- 12 09		18.0	RICH CLUSTER OF GALAXIES
PHL 1298	02 25 48.	- 13 40		18.5	BLUE STELLAR OBJECT
PHL 8272	02 25 48.	- 24 54		18.2	BLUE STELLAR OBJECT
IC 1802	02 25 50.	+ 22 53 47.			NONSTELLAR OBJECT
HELW 372	02 25 53.	- 03 22 31.			NEBULA
MCG+00-07-023	02 25 54.	+ 00 33	12	15.5	GALAXY
PHL 1299	02 25 54.	+ 01 49		18.6	BLUE STELLAR OBJECT
ZWG 504.092	02 25 54.	+ 30 40		15.7	GALAXY
MCG+06-06-042	02 25 54.	+ 35 10 30.	54	16.	GALAXY
ZWG 523.048	02 25 54.	+ 37 53		15.5	GALAXY
VV 107B	02 25 54.	+ 37 54	6	17.	INTERACTING GALAXY
VV 107A	02 25 54.	+ 37 54	12	16.	INTERACTING GALAXY
MCG+06-06-043	02 25 54.	+ 37 54	15	15.	GALAXY
UGC 01952	02 25 54.	+ 40 15	72	16.5	GALAXY Sc
UGC 01953	02 25 54.	+ 44 51	72	16.5	GALAXY S
OCL 0351	02 25 54.	+ 60 26	270	9.	OPEN STAR CLUSTER
MCG+00-07-022	02 25 54.	- 01 22 30.	120	12.8	GALAXY
PHL 1300	02 25 54.	- 01 46		15.5	BLUE STELLAR OBJECT
PHL 1301	02 25 54.	- 04 15		17.9	BLUE STELLAR OBJECT
VV 217B	02 25 54.	- 11 02	48	14.	INTERACTING GALAXY
VV 217A	02 25 54.	- 11 03	72	14.	INTERACTING GALAXY
SVEN 097	02 25 55.	- 03 23	30	14.5	GALAXY
REIN 7.003	02 25 55.04	- 01 22 28.2			NEBULA
RNGC 0941	02 25 56.	- 01 22		13.0	GALAXY
ARP 309	02 25 56.	- 11 03			PECULIAR GALAXY
ARC 0354	02 25 59.	+ 01 13		17.6	RICH CLUSTER OF GALAXIES
RNGC 0933	02 25 59.	+ 45 42		15.5	GALAXY
LDN 0711	02 26	+ 19 20	1200		BRIGHT NEBULA
LB 09786	02 26	- 80 55		14.7	FAINT BLUE STAR
LB 09787	02 26	- 85 45		14.7	FAINT BLUE STAR
PHL 8274	02 26 00.	+ 02 45		17.7	BLUE STELLAR OBJECT
ZWG 483.066	02 26 00.	+ 25 07		15.4	GALAXY
UGC 01955	02 26 00.	+ 25 07	96	15.4	GALAXY S
ZC 0226.0+2600	02 26 00.	+ 26 00	33600		CLUSTER OF GALAXIES
MCG+08-05-013	02 26 00.	+ 45 40	36	14.	GALAXY
ZWG 553.016	02 26 00.	+ 45 42		15.5	GALAXY
UGC 01956	02 26 00.	+ 45 42	132	15.5	GALAXY
ZWG 553.017	02 26 00.	+ 47 17		15.7	GALAXY
UGC 01957	02 26 00.	+ 47 17	102	15.7	GALAXY Sc
MCG+08-05-014	02 26 00.	+ 48 15	30	15.	GALAXY
LDN 1364	02 26 00.	+ 61 20	480		DARK NEBULA
LDN 1340	02 26 00.	+ 72 44	120		DARK NEBULA
ZWG 388.023	02 26 00.	- 01 22		13.4	GALAXY
UGC 01954	02 26 00.	- 01 22	210	13.4	GALAXY Sc
PHL 4094	02 26 00.	- 14 02		17.5	BLUE STELLAR OBJECT
PHL 1302	02 26 00.	- 17 15		17.0	BLUE STELLAR OBJECT
PHL 8273	02 26 00.	- 19 20		13.5	BLUE STELLAR OBJECT
PHL 8275	02 26 00.	- 26 58		18.4	BLUE STELLAR OBJECT
LB 03256	02 26 00.	- 78 40		14.5	FAINT BLUE STAR
HELW 097	02 26 01.	- 10 36 13.			NEBULA
ZC 0226.1+0111	02 26 06.	+ 01 11	870		CLUSTER OF GALAXIES
ZWG 504.093	02 26 06.	+ 27 55		15.7	GALAXY
UGC 01958	02 26 06.	+ 27 55	90	15.7	GALAXY Sc
UGC 01959	02 26 06.	+ 34 19	72	18.	GALAXY DWARF
ZWG 523.049	02 26 06.	+ 38 10		15.5	GALAXY
UGC 01960	02 26 06.	+ 38 10	66	15.5	GALAXY
PHL 8276	02 26 06.	- 05 26		14.0	BLUE STELLAR OBJECT
MCG-02-07-012	02 26 06.	- 11 30 30.	36	14.5	GALAXY
PHL 8277	02 26 06.	- 23 45		18.4	BLUE STELLAR OBJECT
MCG-05-07-001	02 26 06.	- 32 07	36	15.	GALAXY
HOLM 058A	02 26 08.	- 10 46	108	12.9	PART OF MULTIPLE GALAXY
MCG+06-06-044	02 26 09.	+ 38 12	30	15.	GALAXY
MCG-02-07-013	02 26 09.	- 10 45 30.	132	12.5	GALAXY
RNGC 0945	02 26 11.	- 10 45		12.0	GALAXY
PHL 8281	02 26 12.	+ 01 42		18.2	BLUE STELLAR OBJECT
PHL 8280	02 26 12.	- 01 14		17.3	BLUE STELLAR OBJECT
PHL 1303	02 26 12.	- 04 48		17.8	BLUE STELLAR OBJECT
PHL 1304	02 26 12.	- 05 58		17.9	BLUE STELLAR OBJECT
PHL 8279	02 26 12.	- 08 18		16.7	BLUE STELLAR OBJECT
PHL 8282	02 26 12.	- 08 46		18.7	BLUE STELLAR OBJECT
PHL 4095	02 26 12.	- 09 26		18.7	BLUE STELLAR OBJECT
MCG-02-07-014	02 26 12.	- 10 54	36	15.	GALAXY
MCG-03-07-021	02 26 12.	- 19 13 30.	54	15.	GALAXY
MCG-03-07-022	02 26 12.	- 19 17	120	12.5	GALAXY
PHL 8278	02 26 12.	- 21 46		14.1	BLUE STELLAR OBJECT
RNGC 0947	02 26 14.	- 19 17		12.0	GALAXY
HOLM 058B	02 26 16.	- 10 44	72	13.7	PART OF MULTIPLE GALAXY
HELW 098	02 26 16.	- 10 54 32.			NEBULA
RNGC 0948	02 26 17.	- 10 44		14.0	GALAXY
MCG+00-07-024	02 26 18.	+ 00 09	54	13.5	GALAXY
ZWG 483.067	02 26 18.	+ 22 51		15.0	GALAXY
ZWG 539.032	02 26 18.	+ 42 02		15.0	GALAXY
UGC 01961	02 26 18.	+ 42 02	66	15.0	GALAXY SB?c
PHL 4096	02 26 18.	- 05 25		18.5	BLUE STELLAR OBJECT
MCG-05-07-002	02 26 18.	- 28 11	36	15.5	GALAXY
LB 03257	02 26 18.	- 61 23		14.3	FAINT BLUE STAR
RNGC 0937	02 26 19.	+ 42 02		15.0	GALAXY
MCG+07-06-024	02 26 21.	+ 42 02	63	15.	GALAXY
SHB 061	02 26 21.	- 03 52		17.0	QUASI-STELLAR OBJECT
BC PHL1305	02 26 21.	- 03 54 18.		16.96	QUASI-STELLAR OBJECT
MCG-02-07-015	02 26 21.	- 10 44	60	14.	GALAXY
BC PKS0226-038	02 26 22.5	- 03 50 58.		17.5	QUASI-STELLAR OBJECT
SN 1965K	02 26 23.	+ 31 15		16.0	SUPERNOVA
IC 0230	02 26 23.	- 11 03 35.			NONSTELLAR OBJECT
ZWG 388.024	02 26 24.	+ 00 09		14.6	GALAXY
UGC 01962	02 26 24.	+ 00 09	96	14.6	GALAXY Sc
PHL 4098	02 26 24.	+ 00 51		16.8	BLUE STELLAR OBJECT
PHL 8285	02 26 24.	+ 01 49		16.7	BLUE STELLAR OBJECT
ZC 0226.4+0420	02 26 24.	+ 04 20	1210		CLUSTER OF GALAXIES
5ZW 248	02 26 24.	+ 27 47			COMPACT GALAXY
ZWG 504.094	02 26 24.	+ 31 15		14.5	GALAXY
UGC 01963	02 26 24.	+ 31 15	84	14.5	GALAXY Sa/SBb
ZWG 504.095	02 26 24.	+ 31 25		13.4	GALAXY
UGC 01964	02 26 24.	+ 31 25	114	13.4	GALAXY S0
MCG+06-06-045	02 26 24.	+ 37 30	36	16.	GALAXY
PHL 8286	02 26 24.	- 00 08		18.0	BLUE STELLAR OBJECT
PHL 1305	02 26 24.	- 03 52		17.5	BLUE STELLAR OBJECT
PHL 4097	02 26 24.	- 04 51		18.0	BLUE STELLAR OBJECT
PHL 8284	02 26 24.	- 09 08		17.6	BLUE STELLAR OBJECT
PHL 8283	02 26 24.	- 09 28		13.0	BLUE STELLAR OBJECT
MCG-02-07-016	02 26 24.	- 11 02	12	15.5	GALAXY
RNGC 0940	02 26 25.	+ 31 25		13.5	GALAXY
ARC 0356	02 26 27.	+ 04 16		17.8	RICH CLUSTER OF GALAXIES
IC 1803	02 26 27.	+ 22 55 44.			NONSTELLAR OBJECT
HELW 099	02 26 29.	- 10 48 39.			NEBULA
ZWG 414.009	02 26 30.	+ 06 38		15.6	GALAXY
ZC 0226.5+3111	02 26 30.	+ 31 11	1210		CLUSTER OF GALAXIES
MCG+05-06-051	02 26 30.	+ 31 14	66	14.5	GALAXY
MCG+05-06-050	02 26 30.	+ 31 25	54	14.	GALAXY
MCG-02-07-017	02 26 30.	- 08 44	48	15.	GALAXY
IC 1804	02 26 32.	+ 22 56 14.			NONSTELLAR OBJECT
HELW 100	02 26 33.	- 10 47 33.			NEBULA
RNGC 0951	02 26 33.	- 22 34		15.0	GALAXY
RNGC 0943	02 26 35.	- 11 03		14.0	GALAXY
RNGC 0942	02 26 35.	- 11 03		14.0	GALAXY
PHL 8287	02 26 36.	+ 01 38		18.1	BLUE STELLAR OBJECT
ZWG 462.018	02 26 36.	+ 20 00		15.7	GALAXY
UGC 01965	02 26 36.	+ 20 00	102	15.7	GALAXY S
MCG+03-07-018	02 26 36.	+ 20 00	84	15.	GALAXY
PHL 4100	02 26 36.	- 06 36		18.5	BLUE STELLAR OBJECT
PHL 4099	02 26 36.	- 16 06		18.4	BLUE STELLAR OBJECT
MCG-04-07-001	02 26 36.	- 22 34	48	15.	GALAXY
MCG-04-07-002	02 26 36.	- 22 36	30	16.	GALAXY
SVEN 098	02 26 38.	- 02 59 03.	12	15.4	GALAXY
HOLM 059A	02 26 38.	- 11 04	24	14.0	PART OF MULTIPLE GALAXY
HELW 101	02 26 38.	- 11 14 45.			NEBULA
HOLM 059B	02 26 39.	- 11 05	48	14.0	PART OF MULTIPLE GALAXY
ZWG 439.011	02 26 42.	+ 09 57		14.9	GALAXY
UGC 01966	02 26 42.	+ 09 57	96	14.9	GALAXY Sb
MCG-02-07-018	02 26 42.	- 11 03	24	14.	GALAXY
MCG-02-07-019	02 26 42.	- 11 03 30.	24	14.	GALAXY
MCG-02-07-020	02 26 42.	- 11 15	36	15.	GALAXY
PHL 4101	02 26 42.	- 12 25		16.7	BLUE STELLAR OBJECT
PHL 1306	02 26 42.	- 18 42		15.8	BLUE STELLAR OBJECT
IC 1806	02 26 43.	+ 22 43 43.			NONSTELLAR OBJECT
SVEN 099	02 26 43.	- 03 31 03.	24	14.9	GALAXY
MCG+02-07-011	02 26 45.	+ 09 57	72	14.5	GALAXY
ARC 0357	02 26 45.	+ 13 02		16.8	RICH CLUSTER OF GALAXIES
LB 02731	02 26 45.	+ 30 35 18.		16.2	FAINT BLUE STAR
HELW 373	02 26 45.	- 03 31 03.			NEBULA
MCG-04-07-003	02 26 45.	- 23 59	36	15.	GALAXY
RNGC 0950	02 26 47.	- 11 15		14.0	GALAXY
PHL 8288	02 26 48.	+ 00 25		17.3	BLUE STELLAR OBJECT
3ZW 047	02 26 48.	+ 22 02			COMPACT GALAXY
ZWG 504.096	02 26 48.	+ 28 21		15.4	GALAXY
ZWG 504.097	02 26 48.	+ 31 13		15.3	GALAXY
UGC 01967	02 26 48.	+ 44 57	78	16.5	GALAXY
PHL 4106	02 26 48.	- 00 14		18.3	BLUE STELLAR OBJECT
PHL 4102	02 26 48.	- 00 40		18.6	BLUE STELLAR OBJECT
PHL 1307	02 26 48.	- 04 48		18.2	BLUE STELLAR OBJECT
PHL 8289	02 26 48.	- 10 38		14.0	BLUE STELLAR OBJECT
MCG-02-07-021	02 26 48.	- 11 15	72	14.5	GALAXY
PHL 1308	02 26 48.	- 12 12		18.5	BLUE STELLAR OBJECT
PHL 4103	02 26 48.	- 15 08		18.1	BLUE STELLAR OBJECT
PHL 4104	02 26 48.	- 15 16		18.1	BLUE STELLAR OBJECT
PHL 4105	02 26 48.	- 19 05		18.4	BLUE STELLAR OBJECT
MCG-05-07-003	02 26 48.	- 31 13	36	16.	GALAXY
RNGC 0954	02 26 49.	- 41 37			GALAXY
UGC 01968	02 26 54.	+ 29 32	90	17.	GALAXY DWARF
PHL 1309	02 26 54.	- 10 42		18.5	BLUE STELLAR OBJECT
PHL 1310	02 26 54.	- 12 06		18.1	BLUE STELLAR OBJECT
ZWG 388.025	02 27 00.	+ 00 33		15.7	GALAXY
UGC 01969	02 27 00.	+ 15 50	60	18.	GALAXY DWARF
ZWG 483.068	02 27 00.	+ 22 52		15.7	GALAXY
ZWG 483.069	02 27 00.	+ 25 02		15.3	GALAXY
UGC 01970	02 27 00.	+ 25 02	132	15.3	GALAXY Sc
MCG+05-06-052	02 27 00.	+ 28 23	45	14.5	GALAXY
ZWG 504.098	02 27 00.	+ 28 25		14.9	GALAXY
UGC 01971	02 27 00.	+ 28 25	84	14.9	GALAXY
ZWG 523.050	02 27 00.	+ 37 53		15.7	GALAXY
UGC 01972	02 27 00.	+ 37 53	96	15.7	GALAXY (E)
ISS 0060	02 27 00.	+ 57 57	235		STELLAR RING
7ZW 004	02 27 00.	+ 84 38			COMPACT GALAXY
ZWG 361.008	02 27 00.	+ 84 40		15.6	GALAXY
PHL 8292	02 27 00.	- 02 28		18.1	BLUE STELLAR OBJECT
HELW 374	02 27 00.	- 03 25 52.			NEBULA
MCG+01-07-016	02 27 00.	- 03 26	66	14.	GALAXY
PHL 4107	02 27 00.	- 10 24		18.8	BLUE STELLAR OBJECT
PHL 8290	02 27 00.	- 14 08		17.1	BLUE STELLAR OBJECT
PHL 8291	02 27 00.	- 24 40		16.1	BLUE STELLAR OBJECT
SVEN 100	02 27 01.	- 03 26 04.	48	13.2	GALAXY
MCG+02-07-012	02 27 03.	+ 09 57 30.	36	15.	GALAXY
LB 02732	02 27 05.	+ 32 00 42.		16.3	FAINT BLUE STAR
UGC 01973	02 27 06.	+ 06 40	66	16.0	GALAXY
ZWG 439.012	02 27 06.	+ 09 59		15.5	GALAXY
ZC 0227.1+1301	02 27 06.	+ 13 01	1680		CLUSTER OF GALAXIES
PHL 1311	02 27 06.	- 06 06		17.9	BLUE STELLAR OBJECT
PHL 4108	02 27 06.	- 11 29		18.4	BLUE STELLAR OBJECT
SVEN 101	02 27 07.	- 03 15 04.	12	15.9	GALAXY
MCG-04-07-004	02 27 09.	- 22 44	48	16.	GALAXY
MCG+02-07-013	02 27 12.	+ 09 37	60	15.5	GALAXY
ZWG 439.013	02 27 12.	+ 09 38		15.7	GALAXY
UGC 01974	02 27 12.	+ 09 38	72	15.7	GALAXY Sa
ZWG 504.099	02 27 12.	+ 32 55		15.3	GALAXY
UGC 01975	02 27 12.	+ 32 55	78	15.3	GALAXY S
KARA.73B 0107	02 27 12.	+ 32 55	60	15.3	ISOLATED GALAXY S
PHL 8294	02 27 12.	- 02 03		18.1	BLUE STELLAR OBJECT

OBJECT NAME	RIGHT ASCEN.	DECLINATION	DIAM.	MAGN.	TYPE OF OBJECT
PHL 8293	02 27 12.	- 03 26		17.7	BLUE STELLAR OBJECT
PHL 8295	02 27 12.	- 09 42		18.1	BLUE STELLAR OBJECT
PHL 4110	02 27 12.	- 20 34		18.2	BLUE STELLAR OBJECT
PHL 4109	02 27 14.	- 12 53		18.1	BLUE STELLAR OBJECT
MCG+00-07-025	02 27 18.	+ 00 57 30.	42	15.	GALAXY
ZWG 523.051	02 27 18.	+ 35 07		15.0	GALAXY
UGC 01976	02 27 18.	+ 35 07	102	15.0	GALAXY Sb
ZWG 523.052	02 27 18.	+ 38 09		15.5	GALAXY
ZWG 539.033	02 27 18.	+ 41 14		15.7	GALAXY
MCG+07-06-025	02 27 18.	+ 41 14	45	16.	GALAXY
UGC 01977	02 27 18.	+ 49 55	72	16.0	GALAXY Sb/SBc
MCG-03-07-023	02 27 18.	- 17 17 30.	42	14.5	GALAXY
PHL 4112	02 27 18.	- 17 46		18.5	BLUE STELLAR OBJECT
PHL 4111	02 27 18.	- 19 50		18.4	BLUE STELLAR OBJECT
MCG-05-07-004	02 27 18.	- 26 45	36	15.	GALAXY
IC 0231	02 27 19.	+ 00 58 15.			NONSTELLAR OBJECT
LB 02733	02 27 21.	+ 31 53 06.		16.4	FAINT BLUE STAR
MCG+06-06-046	02 27 21.	+ 35 07	96	14.5	GALAXY
MCG+06-06-047	02 27 21.	+ 38 10	36	15.	GALAXY
REIN 7.004	02 27 22.02	+ 00 57 24.2			NEBULA
ZWG 388.026	02 27 24.	+ 00 57		15.0	GALAXY
UGC 01978	02 27 24.	+ 00 57	66	15.0	GALAXY (S0)
PHL 1312	02 27 24.	+ 03 07		17.7	BLUE STELLAR OBJECT
ZWG 483.070	02 27 24.	+ 22 55		15.5	GALAXY
PHL 4113	02 27 24.	- 05 16		18.6	BLUE STELLAR OBJECT
PHL 8296	02 27 24.	- 10 15		17.1	BLUE STELLAR OBJECT
PHL 8297	02 27 24.	- 18 31		18.2	BLUE STELLAR OBJECT
MCG-02-07-022	02 27 27.	- 11 20	54	14.	GALAXY
ZWG 388.027	02 27 30.	+ 01 12		15.6	GALAXY
ZC 0227.5+1853	02 27 30.	+ 18 53	610		CLUSTER OF GALAXIES
KARA.68 016	02 27 30.	+ 27 43	34		DWARF GALAXY
ZWG 539.034	02 27 30.	+ 42 01		14.5	GALAXY
UGC 01979	02 27 30.	+ 42 01	102	14.5	GALAXY (S0)
MCG+07-06-026	02 27 30.	+ 42 01	60	15.	GALAXY
PHL 4114	02 27 30.	- 11 55		16.8	BLUE STELLAR OBJECT
MCG-03-07-024	02 27 30.	- 15 54	36	15.	GALAXY
RNGC 0946	02 27 31.	+ 42 01		14.5	GALAXY
HN 0160	02 27 32.	- 43 19			NEBULA
IC 1810	02 27 32.	- 43 19			NONSTELLAR OBJECT
MCG-02-07-023	02 27 33.	- 13 29	36	15.	GALAXY
PHL 8298	02 27 36.	+ 03 20		16.9	BLUE STELLAR OBJECT
ZWG 483.071	02 27 36.	+ 22 44		15.7	GALAXY
MCG+05-06-053	02 27 36.	+ 28 30	24	16.	GALAXY
ZWG 504.100	02 27 36.	+ 28 31		15.4	GALAXY
ZWG 504.101	02 27 36.	+ 30 40		15.0	GALAXY
ZWG 504.102	02 27 36.	+ 31 57		14.5	GALAXY
UGC 01980	02 27 36.	+ 31 57	66	14.5	GALAXY S
PHL 4116	02 27 36.	- 04 28		18.6	BLUE STELLAR OBJECT
PHL 4117	02 27 36.	- 07 18		18.1	BLUE STELLAR OBJECT
MCG-02-07-024	02 27 36.	- 09 13	54	15.	GALAXY
HN 0161	02 27 38.	- 43 03			NEBULA
IC 1812	02 27 38.	- 43 03			NONSTELLAR OBJECT
MCG+01-07-006	02 27 39.	+ 08 29	48	15.	GALAXY
IC 1807	02 27 39.	+ 22 44 03.			NONSTELLAR OBJECT
RNGC 0949	02 27 40.	+ 36 55		13.0	GALAXY
HELW 102	02 27 41.	- 10 41 00.			NEBULA
MCG+00-07-026	02 27 42.	+ 00 42	48	15.	GALAXY
UGC 01981	02 27 42.	+ 00 44	102	17.	GALAXY DWARF
PHL 4118	02 27 42.	+ 02 26		18.7	BLUE STELLAR OBJECT
FEIG 022	02 27 42.	+ 05 03		12.4	FAINT BLUE STAR
ZWG 483.072	02 27 42.	+ 26 56		15.7	GALAXY
KARA.73B 0108	02 27 42.	+ 26 56	42	15.7	ISOLATED GALAXY S
MCG+05-06-054	02 27 42.	+ 31 57	66	14.5	GALAXY
MCG+06-06-048	02 27 42.	+ 36 56	156	13.	GALAXY
MCG+00-07-027	02 27 42.	- 00 53	48	16.	GALAXY
MCG-02-07-025	02 27 42.	- 10 40	60	14.	GALAXY
MCG-04-07-005	02 27 42.	- 20 34	48	16.	GALAXY
PHL 8299	02 27 42.	- 32 00		17.4	BLUE STELLAR OBJECT
LB 03258	02 27 42.	- 68 07		13.5	FAINT BLUE STAR
REIN 7.005	02 27 45.07	+ 01 04 45.1			NEBULA
ARC 0358	02 27 46.	- 13 25		15.6	RICH CLUSTER OF GALAXIES
ZWG 388.028	02 27 48.	+ 01 05		15.7	GALAXY
PHL 4119	02 27 48.	+ 02 28		18.7	BLUE STELLAR OBJECT
PHL 8300	02 27 48.	+ 03 03		17.8	BLUE STELLAR OBJECT
UGC 01982	02 27 48.	+ 08 29	72	16.5	GALAXY IRR
ZWG 523.053	02 27 48.	+ 36 55		12.0	GALAXY
UGC 01983	02 27 48.	+ 36 55	216	12.0	GALAXY S
KARA.73B 0109	02 27 48.	+ 36 55	204	12.0	ISOLATED GALAXY S
UGC 01984	02 27 48.	+ 43 08	72	16.5	GALAXY Sc
SN 1963W	02 27 48.	+ 43 44		17.0	SUPERNOVA
PHL 8301	02 27 48.	- 09 04		18.7	BLUE STELLAR OBJECT
PHL 8302	02 27 48.	- 10 43		18.9	BLUE STELLAR OBJECT
PHL 4120	02 27 48.	- 14 04		18.5	BLUE STELLAR OBJECT
PHL 1313	02 27 48.	- 21 06		18.5	BLUE STELLAR OBJECT
LB 01626	02 27 48.	- 48 40		13.5	FAINT BLUE STAR
SVEN 102	02 27 49.	- 03 13	6	15.9	GALAXY
IC 1811	02 27 51.	- 34 28 38.			NONSTELLAR OBJECT
ZWG 483.073	02 27 54.	+ 25 58		15.7	GALAXY
ZCG 0227+32	02 27 54.	+ 32 31		18.2	COMPACT GALAXY
5ZW 246	02 27 54.	+ 37 13			COMPACT GALAXY
PHL 8303	02 27 54.	- 04 37		17.6	BLUE STELLAR OBJECT
LB 03259	02 27 54.	- 74 01		13.5	FAINT BLUE STAR
HELW 375	02 27 55.	- 03 19 36.			NEBULA
SVEN 103	02 27 55.	- 03 20	18	15.0	GALAXY
VDB.66G 023	02 28	- 11 00	70		DWARF GALAXY
LB 09788	02 28	- 83 33		14.6	FAINT BLUE STAR
PHL 8306	02 28 00.	+ 01 00		18.5	BLUE STELLAR OBJECT
PHL 8304	02 28 00.	+ 01 40		17.9	BLUE STELLAR OBJECT
5ZW 249	02 28 00.	+ 22 15			COMPACT GALAXY
ZWG 483.074	02 28 00.	+ 25 11		15.5	GALAXY
MCG+04-07-001	02 28 00.	+ 25 12	30	15.	GALAXY
MCG+04-07-002	02 28 00.	+ 26 00	36	15.5	GALAXY
5ZW 250	02 28 00.	+ 41 45			COMPACT GALAXY
ZWG 539.035	02 28 00.	+ 43 07		15.7	GALAXY
UGC 01987	02 28 00.	+ 43 07	78	15.7	GALAXY E?
LDN 1370	02 28 00.	+ 60 15	720		DARK NEBULA
LDN 1367	02 28 00.	+ 61 05	480		DARK NEBULA
UGC 01985	02 28 00.	+ 81 59	60	16.5	GALAXY SB
ZWG 388.029	02 28 00.	- 01 20		13.0	GALAXY
UGC 01986	02 28 00.	- 01 20	204	13.0	GALAXY Sa-b
MCG+00-07-027A	02 28 00.	- 01 20	156	13.	GALAXY
PHL 1315	02 28 00.	- 02 44		18.7	BLUE STELLAR OBJECT
PHL 1314	02 28 00.	- 03 00		17.9	BLUE STELLAR OBJECT
HELW 376	02 28 00.	- 03 06 06.			NEBULA
IC 1808	02 28 00.	- 04 26 01.			NONSTELLAR OBJECT
MCG+01-07-017	02 28 00.	- 04 27	36	14.5	GALAXY
PHL 8305	02 28 00.	- 06 12		17.5	BLUE STELLAR OBJECT
MCG-02-07-026	02 28 00.	- 10 58	120	14.5	GALAXY

OBJECT NAME	RIGHT ASCEN.	DECLINATION	DIAM.	MAGN.	TYPE OF OBJECT
HELW 103	02 28 00.	- 10 58 36.			NEBULA
PHL 4121	02 28 00.	- 15 34		18.5	BLUE STELLAR OBJECT
PHL 4122	02 28 00.	- 28 44		18.6	BLUE STELLAR OBJECT
REIN 2.025	02 28 00.39	- 01 19 48.6			NEBULA
SVEN 104	02 28 01.	- 03 06 07.	42	14.5	GALAXY
RNGC 0955	02 28 02.	- 01 20		13.5	GALAXY
RNGC 0958	02 28 02.	- 03 09		13.0	GALAXY
MCG-03-07-025	02 28 03.	- 17 32	54	15.	GALAXY
PHL 8308	02 28 06.	+ 01 27		18.2	BLUE STELLAR OBJECT
ZWG 539.036	02 28 06.	+ 40 10		14.7	GALAXY
UGC 01988	02 28 06.	+ 40 10	72	14.7	GALAXY Sa-b
MCG+07-06-027	02 28 06.	+ 40 10	54	15.	GALAXY
UGC 01989	02 28 06.	+ 42 40	96	16.0	GALAXY S
MCG+07-06-028	02 28 06.	+ 43 07	24	15.5	GALAXY
ZWG 539.037	02 28 06.	+ 43 15		15.0	GALAXY
VB 169	02 28 06.	+ 57 53	235		STELLAR RING
PHL 8307	02 28 06.	- 00 19		17.8	BLUE STELLAR OBJECT
PHL 8309	02 28 06.	- 00 28		18.2	BLUE STELLAR OBJECT
MCG+01-07-018	02 28 06.	- 04 50	18	15.5	GALAXY
PHL 1316	02 28 06.	- 08 29		16.0	BLUE STELLAR OBJECT
IC 1813	02 28 06.	- 34 28 21.			NONSTELLAR OBJECT
PHL 4123	02 28 12.	+ 02 54		18.3	BLUE STELLAR OBJECT
ZWG 504.103	02 28 12.	+ 27 29		15.7	GALAXY
UGC 01990	02 28 12.	+ 27 29	78	15.7	GALAXY PECULR
KARA.68 017	02 28 12.	+ 27 43	27		DWARF GALAXY
ZWG 505.001	02 28 12.	+ 29 22		14.5	GALAXY
ZWG 504.104	02 28 12.	+ 29 22		14.5	GALAXY
UGC 01991	02 28 12.	+ 29 22	108	14.5	GALAXY E
MCG+05-07-001	02 28 12.	+ 29 23	21	14.5	GALAXY
UGC 01992	02 28 12.	+ 42 42	66	18.	GALAXY DWARF
MCG+01-07-019	02 28 12.	- 03 09	150	13.0	GALAXY
MCG+01-07-020	02 28 12.	- 04 01	180	15.	GALAXY
PHL 8311	02 28 12.	- 10 42		16.7	BLUE STELLAR OBJECT
PHL 8310	02 28 12.	- 27 02		15.6	BLUE STELLAR OBJECT
HELW 377	02 28 13.	- 03 07 01.			NEBULA
SVEN 106	02 28 13.	- 03 07 07.	12	15.0	GALAXY
SVEN 105	02 28 13.	- 03 10	138	11.5	GALAXY
SVEN 107	02 28 13.	- 03 10 07.	12	14.9	GALAXY
RNGC 0953	02 28 14.	+ 29 22		14.5	GALAXY
HELW 378	02 28 14.	- 03 10 13.			NEBULA
CED 007	02 28 16.	+ 61 15	3000		DIFFUSE GALACTIC NEBULA
RNGC 0952	02 28 17.	+ 34 31			NON-EXISTENT OBJECT
ZC 0228.3+0235	02 28 18.	+ 02 35	740		CLUSTER OF GALAXIES
MCG+01-07-007	02 28 18.	+ 03 36	36	15.	GALAXY
ZWG 414.010	02 28 18.	+ 03 38		15.7	GALAXY
ZC 0228.3+3153	02 28 18.	+ 31 53	1480		CLUSTER OF GALAXIES
ZWG 539.038	02 28 18.	+ 40 02		15.7	GALAXY
PHL 8312	02 28 18.	- 04 07		17.4	BLUE STELLAR OBJECT
PHL 8313	02 28 18.	- 12 50		17.9	BLUE STELLAR OBJECT
PHL 1317	02 28 18.	- 14 36		17.3	BLUE STELLAR OBJECT
MCG-05-07-005	02 28 18.	- 31 50	84	15.	GALAXY
LB 03260	02 28 18.	- 72 58		14.5	FAINT BLUE STAR
ARC 0359	02 28 19.	+ 02 37		18.0	RICH CLUSTER OF GALAXIES
HOLM 060B	02 28 19.	+ 38 50	24	14.8	PART OF MULTIPLE GALAXY
SVEN 108	02 28 19.	- 03 08	6	15.9	GALAXY
SVEN 109	02 28 19.	- 03 10 07.	12	15.9	GALAXY
IC 1814	02 28 19.	- 36 15 28.			NONSTELLAR OBJECT
PHL 8314	02 28 24.	+ 01 15		18.0	BLUE STELLAR OBJECT
HOLM 060A	02 28 24.	+ 38 50	18	14.6	PART OF MULTIPLE GALAXY
ZWG 539.039	02 28 24.	+ 41 20		15.6	GALAXY
PHL 1318	02 28 24.	- 03 34		18.4	BLUE STELLAR OBJECT
PHL 4124	02 28 24.	- 05 38		17.4	BLUE STELLAR OBJECT
PHL 1319	02 28 24.	- 22 48		17.1	BLUE STELLAR OBJECT
MCG-04-07-006	02 28 24.	- 24 18	36	15.	GALAXY
PHL 1320	02 28 24.	- 28 04		17.3	BLUE STELLAR OBJECT
MCG-06-06-008	02 28 24.	- 34 30	42	14.	GALAXY
SCHO 0053	02 28 27.	+ 53 56 42.	270		ISOLATED DARK CLOUD
SN 1973P	02 28 29.	+ 39 10		18.0	SUPERNOVA
PHL 8315	02 28 30.	+ 00 16		17.7	BLUE STELLAR OBJECT
ZWG 523.054	02 28 30.	+ 39 10		14.3	GALAXY
UGC 01993	02 28 30.	+ 39 10	144	14.3	GALAXY Sb
MCG+14-02-010	02 28 30.	+ 82 00	51	16.	GALAXY
PHL 1321	02 28 30.	- 03 41		18.5	BLUE STELLAR OBJECT
LB 01627	02 28 30.	- 48 49		14.0	FAINT BLUE STAR
IC 0232	02 28 31.	+ 01 02 51.			NONSTELLAR OBJECT
ZWG 388.030	02 28 36.	+ 01 02		14.2	GALAXY
UGC 01994	02 28 36.	+ 01 02	102	14.2	GALAXY S0
MCG+00-07-028	02 28 36.	+ 01 02 30.	24	14.5	GALAXY
PHL 1322	02 28 36.	+ 01 28		18.6	BLUE STELLAR OBJECT
PHL 1323	02 28 36.	- 06 50		17.9	BLUE STELLAR OBJECT
PHL 4125	02 28 36.	- 10 02		18.0	BLUE STELLAR OBJECT
PHL 1324	02 28 36.	- 10 10		18.9	BLUE STELLAR OBJECT
PHL 1325	02 28 36.	- 13 35		18.7	BLUE STELLAR OBJECT
PHL 8317	02 28 36.	- 16 45		16.2	BLUE STELLAR OBJECT
PHL 8318	02 28 36.	- 20 48		17.8	BLUE STELLAR OBJECT
PHL 8316	02 28 36.	- 20 51		15.8	BLUE STELLAR OBJECT
PHL 1326	02 28 36.	- 21 16		17.7	BLUE STELLAR OBJECT
TON-S 0255	02 28 36.	- 27 01		15.0	BLUE STAR
PHL 1327	02 28 36.	- 28 49		17.3	BLUE STELLAR OBJECT
MCG-06-06-009	02 28 36.	- 34 27	66	14.	GALAXY
SVEN 110	02 28 37.	- 03 10	18	14.5	GALAXY
REIN 7.006	02 28 37.20	+ 01 02 39.2			NEBULA
REIN 7.007	02 28 38.54	+ 01 23 50.2			NEBULA
MCG+06-06-049	02 28 39.	+ 39 11	138	15.	GALAXY
HELW 379	02 28 40.	- 03 09 50.			NEBULA
MCG+00-07-029	02 28 42.	+ 01 36 30.	42	15.	GALAXY
ZWG 388.031	02 28 42.	+ 01 37		15.7	GALAXY
MCG+03-07-020	02 28 42.	+ 18 34	57	15.	GALAXY
MCG+03-07-019	02 28 42.	+ 20 28	27	15.	GALAXY
MCG+04-07-004	02 28 42.	+ 22 44	48	15.	GALAXY
ZWG 484.001	02 28 42.	+ 25 58		15.6	GALAXY
MCG+04-07-003	02 28 42.	+ 25 58	30	15.	GALAXY
PHL 8319	02 28 42.	- 09 28		18.3	BLUE STELLAR OBJECT
PHL 4126	02 28 42.	- 12 28		18.8	BLUE STELLAR OBJECT
MCG-03-07-026	02 28 42.	- 19 34	78	14.	GALAXY
PHL 4127	02 28 42.	- 31 30		18.6	BLUE STELLAR OBJECT
ARC 0360	02 28 43.	+ 06 47		17.8	RICH CLUSTER OF GALAXIES
LB 03611	02 28 45.	+ 43 39 00.		20.0	FAINT BLUE STAR
RNGC 0961	02 28 46.	- 07 08			NON-EXISTENT OBJECT
REIN 7.008	02 28 47.62	+ 01 25 50.9			NEBULA
ZWG 388.032	02 28 48.	+ 01 07		14.6	GALAXY
UGC 01995	02 28 48.	+ 01 07	102	14.6	GALAXY Sc
MCG+00-07-030	02 28 48.	+ 01 07 30.	90	13.5	GALAXY
ZC 0228.8+0244	02 28 48.	+ 02 44	1410		CLUSTER OF GALAXIES
ZC 0228.8+0644	02 28 48.	+ 06 44	1010		CLUSTER OF GALAXIES
ZWG 484.002	02 28 48.	+ 22 42		15.0	GALAXY
UGC 01996	02 28 48.	+ 22 42	72	15.0	GALAXY SBb
3ZW 048	02 28 48.	+ 22 43			COMPACT GALAXY

OBJECT NAME	RIGHT ASCEN.	DECLINATION	DIAM.	MAGN.	TYPE OF OBJECT
PHL 8322	02 28 48.	- 03 27		18.7	BLUE STELLAR OBJECT
PHL 8321	02 28 48.	- 05 28		16.6	BLUE STELLAR OBJECT
PHL 8325	02 28 48.	- 06 16		18.5	BLUE STELLAR OBJECT
PHL 8326	02 28 48.	- 12 28		18.4	BLUE STELLAR OBJECT
PHL 8320	02 28 48.	- 12 29		15.3	BLUE STELLAR OBJECT
PHL 8324	02 28 48.	- 12 41		15.6	BLUE STELLAR OBJECT
PHL 8323	02 28 48.	- 14 17		18.2	BLUE STELLAR OBJECT
PHL 4128	02 28 48.	- 16 20		18.1	BLUE STELLAR OBJECT
IC 1809	02 28 49.	+ 22 41 27.			NONSTELLAR OBJECT
REIN 7.009	02 28 51.49	+ 01 07 28.8			NEBULA
ZC 0228.9+0319	02 28 54.	+ 03 19	1680		CLUSTER OF GALAXIES
OCL 0352	02 28 54.	+ 61 14	2820	7.0	OPEN STAR CLUSTER
PHL 1328	02 28 54.	- 03 12		17.9	BLUE STELLAR OBJECT
PHL 8327	02 28 54.	- 11 59		17.7	BLUE STELLAR OBJECT
MCG-04-07-007	02 28 54.	- 20 44	30	15.	GALAXY
PHL 4129	02 28 54.	- 21 38		18.0	BLUE STELLAR OBJECT
TON-S 0256	02 28 54.	- 28 46		16.0	BLUE STAR
ARC 0361	02 28 55.	+ 02 40		17.6	RICH CLUSTER OF GALAXIES
LB 03612	02 28 55.	+ 43 31 12.		19.8	FAINT BLUE STAR
SVEN 111	02 28 55.	- 03 06 09.	18	15.4	GALAXY
SVEN 112	02 28 55.	- 03 20 09.	24	14.9	GALAXY
HELW 380	02 28 56.	- 03 20 09.			NEBULA
VDB.66G 022	02 29	+ 38 25	70		DWARF GALAXY
LBN 0657	02 29	+ 60 15	720		BRIGHT NEBULA
LBN 0656	02 29	+ 60 20	2220		BRIGHT NEBULA
ZWG 388.033	02 29 00.	+ 02 36		14.9	GALAXY
KARA.72 069A	02 29 00.	+ 02 36	24	14.9	PART OF DOUBLE GALAXY
5ZW 251	02 29 00.	+ 32 29			COMPACT GALAXY
ZWG 539.040	02 29 00.	+ 43 14		15.4	GALAXY
UGC 01997	02 29 00.	+ 43 14	84	15.4	GALAXY Sb
LDN 1368	02 29 00.	+ 61 10	300		DARK NEBULA
LDN 1366	02 29 00.	+ 61 30	1860		DARK NEBULA
PHL 8329	02 29 00.	- 01 45		18.6	BLUE STELLAR OBJECT
PHL 8328	02 29 00.	- 01 53		16.7	BLUE STELLAR OBJECT
PHL 4134	02 29 00.	- 03 11		18.4	BLUE STELLAR OBJECT
PHL 8331	02 29 00.	- 06 08		16.9	BLUE STELLAR OBJECT
PHL 4135	02 29 00.	- 08 02		17.5	BLUE STELLAR OBJECT
PHL 4130	02 29 00.	- 14 40		18.2	BLUE STELLAR OBJECT
PHL 4131	02 29 00.	- 15 31		18.4	BLUE STELLAR OBJECT
MCG-03-07-027	02 29 00.	- 17 53 30.	36	15.	GALAXY
PHL 4132	02 29 00.	- 20 16		18.1	BLUE STELLAR OBJECT
PHL 8330	02 29 00.	- 26 48		18.5	BLUE STELLAR OBJECT
PHL 4133	02 29 00.	- 27 06		18.7	BLUE STELLAR OBJECT
MCG-06-06-010	02 29 00.	- 36 15	120	12.5	GALAXY
IC 0234	02 29 02.	- 00 21 29.			NONSTELLAR OBJECT
BC PKS0229+13	02 29 02.35	+ 13 09 41.0		17.71	QUASI-STELLAR OBJECT
SHB 062	02 29 02.5	+ 13 09 39.		17.7	QUASI-STELLAR OBJECT
IC 0233	02 29 04.	+ 02 35 19.			NONSTELLAR OBJECT
RNGC 0964	02 29 04.	- 36 15			GALAXY
UGC 01998	02 29 06.	+ 00 05	78	16.0	GALAXY Sc
MCG+00-07-031	02 29 06.	+ 00 40	24	15.	GALAXY
ZWG 388.035	02 29 06.	+ 00 41		15.7	GALAXY
ZWG 388.036	02 29 06.	+ 02 35		15.7	GALAXY
KARA.72 069B	02 29 06.	+ 02 35	18	15.7	PART OF DOUBLE GALAXY
ZWG 462.019	02 29 06.	+ 18 55		15.1	GALAXY
UGC 01999	02 29 06.	+ 18 55	186	15.1	GALAXY Sc
MCG+04-07-005	02 29 06.	+ 23 34	42	15.	GALAXY
ZWG 388.034	02 29 06.	- 00 21		15.2	GALAXY
PHL 1329	02 29 06.	- 03 24		18.1	BLUE STELLAR OBJECT
LB 01628	02 29 06.	- 48 09		13.5	FAINT BLUE STAR
LB 02735	02 29 07.	+ 29 09 12.		16.9	FAINT BLUE STAR
LB 02734	02 29 07.	+ 31 12 48.		16.3	FAINT BLUE STAR
REIN 7.010	02 29 08.99	+ 00 41 00.5			NEBULA
MCG+00-07-032	02 29 09.	+ 00 04	48	15.	GALAXY
MCG+07-06-029	02 29 09.	+ 43 14 30.	72	16.	GALAXY
RNGC 0963	02 29 09.	- 04 14			GALAXY
ARC 0362	02 29 09.	- 05 06		17.7	RICH CLUSTER OF GALAXIES
REIN 7.011	02 29 09.48	+ 00 04 25.4			NEBULA
FATH 1.111	02 29 11.	+ 30 24	14		NEBULA
RNGC 0956	02 29 11.	+ 44 25		9.0	OPEN CLUSTER
RNGC 0960	02 29 11.	- 09 31		14.0	GALAXY
PHL 8332	02 29 12.	+ 00 02		18.2	BLUE STELLAR OBJECT
ZWG 388.038	02 29 12.	+ 00 05		15.6	GALAXY
ZWG 388.039	02 29 12.	+ 01 02		15.4	GALAXY
MCG+00-07-033	02 29 12.	+ 01 02	12	16.	GALAXY
ZWG 388.040	02 29 12.	+ 01 50		15.7	GALAXY
UGC 02000	02 29 12.	+ 05 24	84	16.5	GALAXY
ZWG 414.011	02 29 12.	+ 09 15		15.7	GALAXY
ZWG 439.014	02 29 12.	+ 09 31		15.4	GALAXY
ZWG 414.012	02 29 12.	+ 09 31		15.4	GALAXY
MCG+03-07-021	02 29 12.	+ 18 56	186	14.	GALAXY
ZWG 484.003	02 29 12.	+ 23 32		15.4	GALAXY
5ZW 252	02 29 12.	+ 37 18			COMPACT GALAXY
ZWG 539.041	02 29 12.	+ 41 59		14.6	GALAXY
UGC 02001	02 29 12.	+ 41 59	102	14.6	GALAXY Sa-b
OCL 0377	02 29 12.	+ 44 26	480	9.2	OPEN STAR CLUSTER
VB 221	02 29 12.	+ 54 27	463		STELLAR RING
ZWG 388.037	02 29 12.	- 01 10		15.7	GALAXY
PHL 1330	02 29 12.	- 02 18		18.6	BLUE STELLAR OBJECT
PHL 1331	02 29 12.	- 05 26		18.2	BLUE STELLAR OBJECT
PHL 8333	02 29 12.	- 05 54		18.5	BLUE STELLAR OBJECT
PHL 4136	02 29 12.	- 06 46		18.4	BLUE STELLAR OBJECT
MCG-02-07-028	02 29 12.	- 09 31	60	14.	GALAXY
MCG-02-07-027	02 29 12.	- 13 39	48	16.	GALAXY
PHL 4137	02 29 12.	- 28 46		18.8	BLUE STELLAR OBJECT
MCG-05-07-006	02 29 12.	- 29 47	15	15.	GALAXY
LB 03261	02 29 12.	- 52 52		14.4	FAINT BLUE STAR
SVEN 113	02 29 13.	- 03 23	18	14.9	GALAXY
REIN 7.012	02 29 14.79	+ 01 02 24.5			NEBULA
MCG+00-07-034	02 29 15.	+ 00 59	12	16.	GALAXY
MCG+07-06-030	02 29 15.	+ 42 00	78	15.	GALAXY
MCG-04-07-008	02 29 15.	- 23 10	21	15.	GALAXY
REIN 7.013	02 29 15.98	+ 00 01 46.6			NEBULA
HELW 381	02 29 16.	- 03 22 52.			NEBULA
RNGC 0959	02 29 17.	+ 35 17		12.5	GALAXY
ZWG 388.041	02 29 18.	+ 02 31		15.6	GALAXY
5ZW 253	02 29 18.	+ 25 08			COMPACT GALAXY
ZWG 523.055	02 29 18.	+ 35 17		12.5	GALAXY
UGC 02002	02 29 18.	+ 35 17	168	12.5	GALAXY Sc/IRR
MCG+06-06-050	02 29 18.	+ 39 11	42	15.5	GALAXY
ZWG 523.056	02 29 18.	+ 39 12		15.0	GALAXY
UGC 02003	02 29 18.	+ 39 12	66	15.0	GALAXY S
ZWG 539.042	02 29 18.	+ 41 44		15.6	GALAXY
LB 03613	02 29 18.	+ 43 48 00.		19.0	FAINT BLUE STAR
OCL 0359	02 29 18.	+ 58 31	420		OPEN STAR CLUSTER
MRSL 134+01/1	02 29 18.	+ 61 15	9000		HII REGION
PHL 8334	02 29 18.	- 00 48		17.3	BLUE STELLAR OBJECT
PHL 8335	02 29 18.	- 02 58		18.7	BLUE STELLAR OBJECT
PHL 4138	02 29 18.	- 05 04		18.6	BLUE STELLAR OBJECT
MCG-03-07-028	02 29 18.	- 15 41	12	15.	GALAXY
MCG-04-07-009	02 29 18.	- 20 46	36	15.	GALAXY
PHL 1332	02 29 18.	- 23 21		16.4	BLUE STELLAR OBJECT
FATH 1.112	02 29 19.	+ 30 56	27		NEBULA
REIN 7.014	02 29 19.10	+ 00 59 19.9			NEBULA
LB 02736	02 29 21.	+ 31 31 54.		16.6	FAINT BLUE STAR
MCG+06-06-051	02 29 21.	+ 35 17	96	13.	GALAXY
SG 1.06	02 29 22.	+ 61 13	5400		DIFFUSE EMISSION NEBULA
REIN 7.015	02 29 23.29	+ 01 01 32.1			NEBULA
MCG+00-07-036	02 29 24.	+ 00 40	60	14.	GALAXY
ZWG 388.042	02 29 24.	+ 00 42		14.8	GALAXY
UGC 02004	02 29 24.	+ 00 42	84	14.8	GALAXY SBc
PHL 8337	02 29 24.	+ 00 54		18.6	BLUE STELLAR OBJECT
PHL 1333	02 29 24.	+ 00 54		18.1	BLUE STELLAR OBJECT
ZWG 388.043	02 29 24.	+ 01 01		14.8	GALAXY
UGC 02005	02 29 24.	+ 01 01	180	14.8	GALAXY (E)
MCG+00-07-035	02 29 24.	+ 01 01	30	14.	GALAXY
ZC 0229.4+2556	02 29 24.	+ 25 56	1480		CLUSTER OF GALAXIES
ZWG 539.043	02 29 24.	+ 42 12		15.7	GALAXY
UGC 02006	02 29 24.	+ 42 12	90	15.7	GALAXY E
ZWG 553.018	02 29 24.	+ 49 25		15.7	GALAXY
PHL 8336	02 29 24.	- 03 00		18.5	BLUE STELLAR OBJECT
PHL 4141	02 29 24.	- 03 58		18.2	BLUE STELLAR OBJECT
PHL 4142	02 29 24.	- 05 04		18.0	BLUE STELLAR OBJECT
PHL 4139	02 29 24.	- 13 52		18.5	BLUE STELLAR OBJECT
PHL 4140	02 29 24.	- 15 24		17.9	BLUE STELLAR OBJECT
PHL 8338	02 29 24.	- 22 30		17.7	BLUE STELLAR OBJECT
REIN 7.016	02 29 25.04	+ 00 41 22.9			NEBULA
REIN 7.017	02 29 25.84	+ 00 41 11.0			NEBULA
RNGC 0966	02 29 27.	- 20 07		14.0	GALAXY
MCG-03-07-029	02 29 27.	- 20 07 30.	48	14.5	GALAXY
PHL 4143	02 29 30.	+ 00 20		18.7	BLUE STELLAR OBJECT
ZWG 439.015	02 29 30.	+ 09 30		15.3	GALAXY
ZWG 414.013	02 29 30.	+ 09 30		15.3	GALAXY
UGC 02007	02 29 30.	+ 09 30	78	15.3	GALAXY SB0
UGC 02008	02 29 30.	+ 31 23	60	16.5	GALAXY
IC 1805	02 29 30.	+ 61 13	9000		HII REGION
MCG+00-07-037	02 29 30.	- 01 36	66	14.	GALAXY
TON-S 0257	02 29 30.	- 23 18		15.1	BLUE STAR
MCG+07-06-031	02 29 33.	+ 42 13	30	16.5	GALAXY
UGC 02010	02 29 36.	+ 01 35	102	14.3	GALAXY SBb
UGC 02009	02 29 36.	+ 01 36	66	16.0	GALAXY S
PHL 4144	02 29 36.	+ 02 47		12.3	BLUE STELLAR OBJECT
ZWG 505.002	02 29 36.	+ 31 20		15.6	GALAXY
UGC 02011	02 29 36.	+ 31 20	102	15.6	GALAXY Sb
MCG+05-07-002	02 29 36.	+ 31 24	36	17.	GALAXY
PHL 1334	02 29 36.	- 00 20		18.1	BLUE STELLAR OBJECT
ZWG 388.044	02 29 36.	- 01 35		14.3	GALAXY
MCG+00-07-038	02 29 36.	- 01 37	36	15.	GALAXY
PHL 4145	02 29 36.	- 04 14		17.7	BLUE STELLAR OBJECT
PHL 8702	02 29 36.	- 21 28		18.5	BLUE STELLAR OBJECT
MCG-02-07-029	02 29 39.	- 13 33	72	15.	GALAXY
RNGC 0979	02 29 39.	- 44 45			GALAXY
SCHO 0054	02 29 40.	+ 55 41 24.	430		ISOLATED DARK CLOUD
PHL 8339	02 29 42.	+ 00 59		17.3	BLUE STELLAR OBJECT
FEIG 023	02 29 42.	+ 02 47		11.9	FAINT BLUE STAR
3ZW 049	02 29 42.	+ 29 26			COMPACT GALAXY
MCG+05-07-003	02 29 42.	+ 31 22	66	15.5	GALAXY
PHL 8340	02 29 42.	- 05 07		18.8	BLUE STELLAR OBJECT
PHL 4146	02 29 42.	- 20 32		16.6	BLUE STELLAR OBJECT
MCG+05-07-004	02 29 45.	+ 27 51	27	14.5	GALAXY
ZC 0229.8+0903	02 29 48.	+ 09 03	2820		CLUSTER OF GALAXIES
ZWG 462.020	02 29 48.	+ 20 48		15.1	GALAXY
UGC 02012	02 29 48.	+ 20 48	72	15.1	GALAXY Sb-c
ZWG 505.003	02 29 48.	+ 27 51		14.2	GALAXY
UGC 02013	02 29 48.	+ 27 51	102	14.2	GALAXY E
FATH 1.113	02 29 48.	+ 30 36	16		NEBULA
MCG+06-06-052	02 29 48.	+ 34 38 30.	15	15.	GALAXY
UGC 02014	02 29 48.	+ 38 28	144	17.	GALAXY DWARF
PHL 8342	02 29 48.	- 09 11		18.7	BLUE STELLAR OBJECT
PHL 8341	02 29 48.	- 13 17		16.9	BLUE STELLAR OBJECT
PHL 4147	02 29 48.	- 19 16		18.3	BLUE STELLAR OBJECT
PHL 1335	02 29 48.	- 23 28		18.1	BLUE STELLAR OBJECT
MCG-06-06-011	02 29 48.	- 36 54	72	13.	GALAXY
SVEN 114	02 29 49.	- 03 09 11.	48	14.9	GALAXY
RNGC 0962	02 29 50.	+ 27 51		14.0	GALAXY
RNGC 0967	02 29 50.	- 17 27		14.0	GALAXY
MCG-03-07-030	02 29 51.	- 17 27	24	14.	GALAXY
IC 1816	02 29 53.	- 36 58 29.			NONSTELLAR OBJECT
ZWG 462.021	02 29 54.	+ 18 15		15.6	GALAXY
ZWG 505.004	02 29 54.	+ 30 31		15.7	GALAXY
MCG+05-07-005	02 29 54.	+ 30 32	42	15.	GALAXY
UGC 02015	02 29 54.	+ 34 39	78	16.0	GALAXY S0?
OCL 0355	02 29 54.	+ 59 40	360		OPEN STAR CLUSTER
PHL 8344	02 29 54.	- 16 26		17.8	BLUE STELLAR OBJECT
PHL 8343	02 29 54.	- 32 40		16.5	BLUE STELLAR OBJECT
REIN 7.018	02 29 56.78	+ 00 00 50.9			NEBULA
FATH 1.114	02 29 57.	+ 30 32	27		NEBULA
LB 03614	02 29 57.	+ 43 54 24.		20.6	FAINT BLUE STAR
RNGC 0957	02 29 59.	+ 57 18		7.0	OPEN CLUSTER
VDB.66G 025	02 30	+ 33 15	130		DWARF GALAXY
VDB.66G 024	02 30	+ 40 15	100		DWARF GALAXY
KHAV 036	02 30	+ 59 01	6190		DARK NEBULA
LBN 0655	02 30	+ 60 50	3300		BRIGHT NEBULA
LBN 0654	02 30	+ 61 00	3600		BRIGHT NEBULA
LB 09789	02 30	- 80 19		15.0	FAINT BLUE STAR
ZWG 388.045	02 30 00.	+ 00 22		15.7	GALAXY
ZC 0230.0+0624	02 30 00.	+ 06 24	1950		CLUSTER OF GALAXIES
MCG+03-07-023	02 30 00.	+ 18 15	36	14.5	GALAXY
ZWG 462.022	02 30 00.	+ 20 25		14.5	GALAXY
MRK 368	02 30 00.	+ 20 25	22	15.	GALAXY WITH UV CONTINUUM
UGC 02016	02 30 00.	+ 20 25	27	14.5	GALAXY PECULR
MCG+03-07-022	02 30 00.	+ 20 50	57	14.5	GALAXY
MCG+04-07-008	02 30 00.	+ 21 44	36	15.	GALAXY
MCG+04-07-007	02 30 00.	+ 23 09	72	15.	GALAXY
ZWG 484.004	02 30 00.	+ 24 53		14.9	GALAXY
MCG+04-07-006	02 30 00.	+ 24 53 30.	30	15.	GALAXY
KARA.68 018	02 30 00.	+ 24 59	27		DWARF GALAXY
UGC 02017	02 30 00.	+ 28 36	150	17.	GALAXY DWRF IR
ZWG 523.057	02 30 00.	+ 35 18		15.6	GALAXY
MCG+06-06-053	02 30 00.	+ 35 18	24	15.5	GALAXY
MCG+06-06-054	02 30 00.	+ 38 27	108	15.5	GALAXY
OCL 0362	02 30 00.	+ 57 19	1140	7.2	OPEN STAR CLUSTER
LDN 1369	02 30 00.	+ 61 10	420		DARK NEBULA
PHL 8345	02 30 00.	- 07 44		18.5	BLUE STELLAR OBJECT
PHL 8346	02 30 00.	- 16 01		18.3	BLUE STELLAR OBJECT
PHL 1336	02 30 00.	- 17 12		16.6	BLUE STELLAR OBJECT

OBJECT NAME	RIGHT ASCEN.	DECLINATION	DIAM.	MAGN.	TYPE OF OBJECT
MCG+00-07-040	02 30 03.	+ 00 01	48	13.	GALAXY
MCG+00-07-039	02 30 03.	+ 00 23 30.	27	15.	GALAXY
LB 03615	02 30 03.	+ 43 52 00.		19.8	FAINT BLUE STAR
ARC 0363	02 30 04.	+ 09 14		17.1	RICH CLUSTER OF GALAXIES
IC 0235	02 30 04.	+ 20 25 27.			NONSTELLAR OBJECT
REIN 7.019	02 30 04.10	+ 00 22 26.9			NEBULA
REIN 7.020	02 30 05.31	+ 00 23 50.6			NEBULA
ZWG 388.046	02 30 06.	+ 00 03		14.3	GALAXY
UGC 02018	02 30 06.	+ 00 03	96	14.3	GALAXY S
ZWG 388.047	02 30 06.	+ 00 24		14.4	GALAXY
UGC 02019	02 30 06.	+ 00 24	33	14.4	GALAXY PECULR
ZWG 462.023	02 30 06.	+ 19 02		14.9	GALAXY
ZWG 484.005	02 30 06.	+ 23 07		15.6	GALAXY
UGC 02020	02 30 06.	+ 23 07	96	15.6	GALAXY Sc
ZWG 523.058	02 30 06.	+ 34 49		15.7	GALAXY
OCL 0354	02 30 06.	+ 59 58	180		OPEN STAR CLUSTER
PHL 4148	02 30 06.	- 06 39		18.8	BLUE STELLAR OBJECT
PHL 8347	02 30 06.	- 11 35		18.3	BLUE STELLAR OBJECT
MCG-03-07-031	02 30 06.	- 18 53	72	14.5	GALAXY
PHL 1337	02 30 06.	- 31 02		18.4	BLUE STELLAR OBJECT
MCG-06-06-012	02 30 06.	- 33 26	30	14.5	GALAXY
REIN 7.021	02 30 06.41	+ 00 02 24.6			NEBULA
REIN 7.022	02 30 07.06	+ 00 55 56.1			NEBULA
RNGC 0965	02 30 08.	- 18 53		14.0	GALAXY
MCG+04-07-009	02 30 09.	+ 25 02 30.	33	16.	GALAXY
FATH 1.115	02 30 09.	+ 30 18	19		NEBULA
LB 02737	02 30 09.	+ 30 43 12.		15.8	FAINT BLUE STAR
REIN 7.023	02 30 09.81	+ 01 03 50.4			NEBULA
REIN 7.024	02 30 11.36	+ 00 05 42.7			NEBULA
PHL 8349	02 30 12.	+ 02 20		16.6	BLUE STELLAR OBJECT
ZC 0230.2+0404	02 30 12.	+ 04 04	4300		CLUSTER OF GALAXIES
ZWG 414.014	02 30 12.	+ 04 35		15.7	GALAXY
ZWG 414.015	02 30 12.	+ 05 29		14.9	GALAXY
UGC 02021	02 30 12.	+ 05 29	78	14.9	GALAXY S0
ZWG 484.006	02 30 12.	+ 21 40		14.8	GALAXY
LB 02739	02 30 12.	+ 30 17 48.		15.7	FAINT BLUE STAR
ZC 0230.2+3050	02 30 12.	+ 30 50	1140		CLUSTER OF GALAXIES
LB 02738	02 30 12.	+ 31 34 12.		16.7	FAINT BLUE STAR
6ZW 214	02 30 12.	+ 39 28			COMPACT GALAXY
ZWG 523.059	02 30 12.	+ 39 28		15.4	GALAXY
VB 170	02 30 12.	+ 58 13	255		STELLAR RING
PHL 8348	02 30 12.	- 05 34		16.6	BLUE STELLAR OBJECT
PHL 1338	02 30 12.	- 09 09		18.8	BLUE STELLAR OBJECT
PHL 1339	02 30 12.	- 13 23		18.8	BLUE STELLAR OBJECT
MCG-06-06-013	02 30 12.	- 35 16	90	12.5	GALAXY
REIN 7.025	02 30 14.54	+ 00 56 37.9			NEBULA
MCG+05-07-006	02 30 15.	+ 32 32	30	15.	GALAXY
MCG+01-07-008	02 30 18.	+ 05 28 30.	36	15.	GALAXY
ZWG 505.005	02 30 18.	+ 32 32		15.2	GALAXY
UGC 02022	02 30 18.	+ 32 32	66	15.2	GALAXY E
ZWG 505.006	02 30 18.	+ 33 17		14.9	GALAXY
UGC 02023	02 30 18.	+ 33 17	180	14.9	GALAXY DWARF
ZC 0230.3+3711	02 30 18.	+ 37 11	1480		CLUSTER OF GALAXIES
LB 03616	02 30 19.	+ 43 35 30.		18.4	FAINT BLUE STAR
IC 0236	02 30 20.	- 00 20 59.			NONSTELLAR OBJECT
KARA.68 019	02 30 21.	+ 27 57	27		DWARF GALAXY
ZWG 388.048	02 30 24.	+ 00 12		14.5	GALAXY
UGC 02024	02 30 24.	+ 00 12	72	14.5	GALAXY Sa-b
MCG+00-07-041	02 30 24.	+ 00 12	48	15.	GALAXY
PHL 8350	02 30 24.	+ 02 28		16.9	BLUE STELLAR OBJECT
ZC 0230.4+2002	02 30 24.	+ 20 02	1280		CLUSTER OF GALAXIES
ZWG 462.024	02 30 24.	+ 21 17		15.7	GALAXY
MCG+04-07-011	02 30 24.	+ 22 12	66	14.5	GALAXY
ZWG 484.007	02 30 24.	+ 25 17		15.7	GALAXY
UGC 02025	02 30 24.	+ 25 17	84	15.7	GALAXY Sc
MCG+04-07-010	02 30 24.	+ 25 18 30.	84	15.5	GALAXY
UGC 02026	02 30 24.	+ 27 05	90	17.	GALAXY DWARF
MCG+05-07-007	02 30 24.	+ 33 18	90	15.5	GALAXY
PHL 1340	02 30 24.	- 02 32		17.7	BLUE STELLAR OBJECT
PHL 8352	02 30 24.	- 08 32		18.3	BLUE STELLAR OBJECT
PHL 1341	02 30 24.	- 10 11		18.0	BLUE STELLAR OBJECT
PHL 8351	02 30 24.	- 26 48		16.6	BLUE STELLAR OBJECT
FATH 1.116	02 30 27.	+ 30 38	27		NEBULA
REIN 7.026	02 30 27.33	+ 00 12 03.4			NEBULA
LB 00178	02 30 28.	+ 30 28 12.		15.8	FAINT BLUE STAR
FATH 1.117	02 30 28.	+ 30 39	24		NEBULA
RNGC 0981	02 30 29.	- 11 11		14.0	GALAXY
UGC 02027	02 30 30.	+ 08 42	90	16.0	GALAXY
ZWG 414.016	02 30 30.	+ 09 28		15.5	GALAXY
MCG+03-07-024	02 30 30.	+ 21 17	30	15.5	GALAXY
ZWG 484.008	02 30 30.	+ 22 10		14.6	GALAXY
UGC 02028	02 30 30.	+ 22 10	90	14.6	GALAXY Sc
UGC 02029	02 30 30.	+ 22 32	66	17.	GALAXY IV?
ZWG 539.044	02 30 30.	+ 41 08		15.3	GALAXY
PHL 8353	02 30 30.	- 02 18		17.2	BLUE STELLAR OBJECT
PHL 4150	02 30 30.	- 06 53		18.8	BLUE STELLAR OBJECT
MCG-02-07-030	02 30 30.	- 11 11 30.	48	14.	GALAXY
PHL 4149	02 30 30.	- 27 02		17.9	BLUE STELLAR OBJECT
FATH 1.118	02 30 32.	+ 29 59	16		NEBULA
RNGC 0975	02 30 34.	+ 09 23		14.0	GALAXY
RNGC 0977	02 30 35.	- 10 58		13.0	GALAXY
PHL 4151	02 30 36.	+ 03 15		18.7	BLUE STELLAR OBJECT
ZWG 414.017	02 30 36.	+ 09 23		14.2	GALAXY
UGC 02030	02 30 36.	+ 09 23	78	14.2	GALAXY S0-a
ZWG 462.025	02 30 36.	+ 20 03		15.5	GALAXY
UGC 02031	02 30 36.	+ 20 03	90	15.5	GALAXY S0
MCG+03-07-025	02 30 36.	+ 20 04	18	15.	GALAXY
UGC 02032	02 30 36.	+ 20 22	78	16.5	GALAXY Sc
ZWG 505.007	02 30 36.	+ 32 19		15.7	GALAXY
ZWG 505.008	02 30 36.	+ 32 48		15.2	GALAXY
ZWG 523.060	02 30 36.	+ 37 27		14.7	GALAXY
UGC 02033	02 30 36.	+ 37 27	96	14.7	GALAXY SBb
MCG+06-06-055	02 30 36.	+ 37 28	66	15.	GALAXY
ZWG 539.045	02 30 36.	+ 39 53		15.3	GALAXY
MCG+07-06-032	02 30 36.	+ 40 18	120	14.	GALAXY
ZWG 539.046	02 30 36.	+ 40 19		15.0	GALAXY
UGC 02034	02 30 36.	+ 40 19	210	15.0	GALAXY DWRF IR
PHL 8354	02 30 36.	- 03 22		17.4	BLUE STELLAR OBJECT
MCG-02-07-031	02 30 36.	- 10 58 30.	120	13.5	GALAXY
PHL 1342	02 30 36.	- 20 50		13.4	BLUE STELLAR OBJECT
PHL 1343	02 30 36.	- 28 46		17.8	BLUE STELLAR OBJECT
SG 3.012	02 30 38.	+ 62 09	2400		DIFFUSE EMISSION NEBULA
MCG+01-07-009	02 30 39.	+ 09 23	48	14.5	GALAXY
ZWG 388.049	02 30 42.	+ 01 06		15.2	GALAXY
ZWG 462.026	02 30 42.	+ 21 16		15.7	GALAXY
ZWG 484.009	02 30 42.	+ 21 56		15.7	GALAXY
ZWG 539.047	02 30 42.	+ 44 08		14.3	GALAXY
UGC 02035	02 30 42.	+ 44 08	126	14.3	GALAXY Sb

OBJECT NAME	RIGHT ASCEN.	DECLINATION	DIAM.	MAGN.	TYPE OF OBJECT
PHL 8355	02 30 42.	- 08 38		18.4	BLUE STELLAR OBJECT
PHL 1344	02 30 42.	- 20 35		10.0	BLUE STELLAR OBJECT
DV.56 N0986A	02 30 42.	- 39 32	102		S GALAXY
RNGC 0986A	02 30 42.	- 39 32			GALAXY
REIN 7.027	02 30 43.80	+ 01 06 01.2			NEBULA
PHL 1345	02 30 48.	+ 00 16		18.7	BLUE STELLAR OBJECT
MCG+03-07-026	02 30 48.	+ 21 16	39	15.5	GALAXY
ZWG 505.009	02 30 48.	+ 29 58		15.7	GALAXY
LB 02740	02 30 48.	+ 30 38 42.		16.5	FAINT BLUE STAR
5ZW 254	02 30 48.	+ 33 23			COMPACT GALAXY
UGC 02036	02 30 48.	+ 41 15	60	17.	GALAXY
ZWG 539.048	02 30 48.	+ 42 28		15.7	GALAXY
PHL 1346	02 30 48.	- 05 08		17.5	BLUE STELLAR OBJECT
PHL 4152	02 30 48.	- 09 23		18.7	BLUE STELLAR OBJECT
PHL 4153	02 30 48.	- 12 50		18.2	BLUE STELLAR OBJECT
FATH 1.119	02 30 51.	+ 29 58	22		NEBULA
LB 03617	02 30 51.	+ 43 42 30.		19.8	FAINT BLUE STAR
MCG-02-07-032	02 30 51.	- 11 58	90	14.	GALAXY
ZWG 388.050	02 30 54.	+ 00 55		14.6	GALAXY
MCG+00-07-042	02 30 54.	+ 00 55	30	15.	GALAXY
ZWG 414.018	02 30 54.	+ 05 55		15.5	GALAXY
ZWG 414.019	02 30 54.	+ 06 19		15.2	GALAXY
FATH 1.120	02 30 54.	+ 29 57	19		NEBULA
MCG+07-06-034	02 30 54.	+ 42 27 30.	96	14.	GALAXY
MCG+07-06-033	02 30 54.	+ 44 07	78	15.	GALAXY
PHL 4154	02 30 54.	- 01 08		16.4	BLUE STELLAR OBJECT
TON-S 0258	02 30 54.	- 28 57		13.6	BLUE STAR
TON-S 0259	02 30 54.	- 31 39		15.2	BLUE STAR
LB 01629	02 30 54.	- 56 03		14.2	FAINT BLUE STAR
IC 0237	02 30 57.	+ 00 54 46.			NONSTELLAR OBJECT
REIN 7.028	02 30 57.08	+ 00 55 14.3			NEBULA
VDB.66G 026	02 31	+ 29 30	70		DWARF GALAXY
PHL 4155	02 31 00.	+ 00 51		18.1	BLUE STELLAR OBJECT
PHL 8357	02 31 00.	+ 02 32		17.4	BLUE STELLAR OBJECT
ZWG 414.020	02 31 00.	+ 06 15		15.6	GALAXY
UGC 02037	02 31 00.	+ 06 15	78	15.6	GALAXY S
MCG+01-07-010	02 31 00.	+ 06 15	60	15.	GALAXY
5ZW 255	02 31 00.	+ 31 45			COMPACT GALAXY
MCG+06-06-056	02 31 00.	+ 34 16	33	14.	GALAXY
MCG+07-06-035	02 31 00.	+ 40 17	36	17.	GALAXY
UGC 02038	02 31 00.	+ 40 18	66	17.	GALAXY
LDN 1374	02 31 00.	+ 58 10	420		DARK NEBULA
LDN 1371	02 31 00.	+ 60 30	1800		DARK NEBULA
VB 079	02 31 00.	+ 63 16	309		STELLAR RING
LDN 1352	02 31 00.	+ 67 10	480		DARK NEBULA
PHL 1347	02 31 00.	- 01 22		18.4	BLUE STELLAR OBJECT
PHL 1348	02 31 00.	- 09 36		18.3	BLUE STELLAR OBJECT
PHL 4156	02 31 00.	- 09 45		19.0	BLUE STELLAR OBJECT
PHL 8356	02 31 00.	- 10 36		17.9	BLUE STELLAR OBJECT
PHL 1349	02 31 00.	- 13 58		17.6	BLUE STELLAR OBJECT
MCG-03-07-032	02 31 00.	- 15 32	18	15.	GALAXY
PHL 1350	02 31 00.	- 19 47		17.7	BLUE STELLAR OBJECT
MCG+02-07-015	02 31 03.	+ 10 57 30.	36	15.5	GALAXY
MCG+02-07-014	02 31 03.	+ 10 57 30.	18	15.5	GALAXY
RNGC 0968	02 31 05.	+ 34 16		14.0	GALAXY
ZWG 505.010	02 31 06.	+ 32 44		13.5	GALAXY
RNGC 0969	02 31 06.	+ 32 44		13.5	GALAXY
UGC 02039	02 31 06.	+ 32 44	120	13.5	GALAXY S0
MCG+05-07-008	02 31 06.	+ 32 45	78	14.5	GALAXY
MCG+05-07-009	02 31 06.	+ 32 47	36	16.	GALAXY
MCG+06-06-057	02 31 06.	+ 34 15	18	15.	GALAXY
ZWG 523.061	02 31 06.	+ 34 16		13.8	GALAXY
UGC 02040	02 31 06.	+ 34 16	258	13.8	GALAXY E
URA 17	02 31 06.	+ 63 18	198		STELLAR RING
PHL 1351	02 31 06.	- 12 30		17.4	BLUE STELLAR OBJECT
PHL 8358	02 31 06.	- 30 50		15.9	BLUE STELLAR OBJECT
MCG-06-06-014	02 31 06.	- 37 04	48	14.5	GALAXY
IC 1817	02 31 08.	+ 10 59 15.			NONSTELLAR OBJECT
RNGC 0976	02 31 11.	+ 20 46		13.5	GALAXY
PHL 1352	02 31 12.	+ 02 17		17.2	BLUE STELLAR OBJECT
ZWG 414.021	02 31 12.	+ 09 25		15.3	GALAXY
UGC 02041	02 31 12.	+ 09 25	84	15.3	GALAXY Sa-b
MCG+01-07-011	02 31 12.	+ 09 25	60	15.	GALAXY
ZWG 439.016	02 31 12.	+ 11 00		14.9	GALAXY
KARA.72 070B	02 31 12.	+ 11 00	30		PART OF DOUBLE GALAXY
KARA.72 070A	02 31 12.	+ 11 00	18	14.9	PART OF DOUBLE GALAXY
ZWG 462.027	02 31 12.	+ 20 45		12.9	GALAXY
UGC 02042	02 31 12.	+ 20 45	108	12.9	GALAXY Sb
LB 02741	02 31 12.	+ 28 52 24.		16.4	FAINT BLUE STAR
RNGC 0971	02 31 12.	+ 32 45			GALAXY
ZWG 505.011	02 31 12.	+ 32 46		15.7	GALAXY
RNGC 0970	02 31 12.	+ 32 46		15.5	GALAXY
ZWG 523.062	02 31 12.	+ 34 15		15.6	GALAXY
MCG+07-06-036	02 31 12.	+ 40 19	18	17.5	GALAXY
ZWG 539.049	02 31 12.	+ 44 48		15.0	GALAXY
UGC 02043	02 31 12.	+ 44 48	66	15.0	GALAXY SBc
PHL 4157	02 31 12.	- 05 58		18.6	BLUE STELLAR OBJECT
PHL 1353	02 31 12.	- 06 40		17.9	BLUE STELLAR OBJECT
PHL 4158	02 31 12.	- 10 52		18.5	BLUE STELLAR OBJECT
PHL 8359	02 31 12.	- 12 14		17.2	BLUE STELLAR OBJECT
PHL 4159	02 31 12.	- 19 00		18.3	BLUE STELLAR OBJECT
MCG-03-07-033	02 31 12.	- 20 26	42	14.5	GALAXY
LB 03262	02 31 12.	- 62 54		14.5	FAINT BLUE STAR
LB 03263	02 31 12.	- 73 57		14.0	FAINT BLUE STAR
BC PKS0231+022	02 31 15.	+ 02 16 18.		18.	QUASI-STELLAR OBJECT
SHB 063	02 31 15.	+ 02 16 18.		18.	QUASI-STELLAR OBJECT
MCG+05-07-010	02 31 15.	+ 29 07	162	13.	GALAXY
MCG+05-07-011	02 31 15.	+ 29 49	48	15.5	GALAXY
REIN 7.029	02 31 17.15	+ 00 55 04.0			NEBULA
ZWG 388.051	02 31 18.	+ 00 42		15.7	GALAXY
UGC 02044	02 31 18.	+ 00 42	60	15.7	GALAXY SBc
ZWG 388.052	02 31 18.	+ 00 55		15.5	GALAXY
PHL 8360	02 31 18.	+ 02 10		17.6	BLUE STELLAR OBJECT
MCG+03-07-027	02 31 18.	+ 20 46	90	12.7	GALAXY
ZWG 505.012	02 31 18.	+ 29 05		12.1	GALAXY
UGC 02045	02 31 18.	+ 29 05	240	12.1	GALAXY Sa-b
UGC 02046	02 31 18.	+ 29 47	84	16.5	GALAXY
ZWG 505.013	02 31 18.	+ 32 13		14.3	GALAXY
UGC 02047	02 31 18.	+ 32 13	120	14.3	GALAXY SB0
MCG+05-07-014	02 31 18.	+ 32 13	72	15.	GALAXY
ZWG 505.014	02 31 18.	+ 32 17		13.7	GALAXY
RNGC 0973	02 31 18.	+ 32 17		13.5	GALAXY
UGC 02048	02 31 18.	+ 32 17	240	13.7	GALAXY Sb
MCG+05-07-013	02 31 18.	+ 32 18	210	14.	GALAXY
MCG+05-07-012	02 31 18.	+ 32 45	90	14.	GALAXY
ZWG 539.050	02 31 18.	+ 40 56		15.2	GALAXY
PHL 4160	02 31 18.	- 07 02		18.6	BLUE STELLAR OBJECT
PHL 1354	02 31 18.	- 07 29		13.9	BLUE STELLAR OBJECT

OBJECT NAME	RIGHT ASCEN.	DECLINATION	DIAM.	MAGN.	TYPE OF OBJECT
RNGC 0972	02 31 20.	+ 29 06		12.5	GALAXY
MCG+00-07-043	02 31 21.	+ 00 42	36	15.5	GALAXY
LB 02742	02 31 21.	+ 31 39 18.		16.8	FAINT BLUE STAR
IC 1815	02 31 21.	+ 32 12 40.			NONSTELLAR OBJECT
REIN 7.030	02 31 21.96	+ 00 42 16.3			NEBULA
PHL 4162	02 31 24.	+ 00 36		18.0	BLUE STELLAR OBJECT
ZWG 505.015	02 31 24.	+ 32 45		13.9	GALAXY
RNGC 0974	02 31 24.	+ 32 45		14.0	GALAXY
UGC 02049	02 31 24.	+ 32 45	240	13.9	GALAXY S
ZWG 539.051	02 31 24.	+ 44 42		15.2	GALAXY
UGC 02050	02 31 24.	+ 44 42	72	15.2	GALAXY Sc/SBc
PHL 1355	02 31 24.	- 04 36		17.4	BLUE STELLAR OBJECT
PHL 4163	02 31 24.	- 05 17		18.4	BLUE STELLAR OBJECT
PHL 1356	02 31 24.	- 06 02		18.1	BLUE STELLAR OBJECT
PHL 8361	02 31 24.	- 06 22		18.4	BLUE STELLAR OBJECT
PHL 4161	02 31 24.	- 08 32		18.7	BLUE STELLAR OBJECT
MCG-03-07-034	02 31 24.	- 16 44 30.	15	15.	GALAXY
PHL 8362	02 31 24.	- 21 04		18.4	BLUE STELLAR OBJECT
LB 03264	02 31 24.	- 64 14		12.2	FAINT BLUE STAR
RNGC 0989	02 31 25.	- 16 44		15.0	GALAXY
ZWG 388.053	02 31 30.	+ 01 08		14.4	GALAXY
UGC 02051	02 31 30.	+ 01 08	96	14.4	GALAXY S0
MCG+00-07-044	02 31 30.	+ 01 08	16	15.	GALAXY
UGC 02052	02 31 30.	+ 25 01	90	18.	GALAXY DWARF?
ZWG 505.016	02 31 30.	+ 29 32		15.7	GALAXY
UGC 02053	02 31 30.	+ 29 32	138	15.7	GALAXY DWRF IR
MCG+05-07-015	02 31 30.	+ 29 32 30.	108	15.	GALAXY
MCG+06-06-058	02 31 30.	+ 33 43	72	15.	GALAXY
ZWG 523.063	02 31 30.	+ 33 44		15.3	GALAXY
UGC 02054	02 31 30.	+ 33 44	102	15.3	GALAXY S-IRR
UGC 02055	02 31 30.	+ 36 56	84	18.	GALAXY DWRF SP
PHL 1357	02 31 30.	- 01 40		18.4	BLUE STELLAR OBJECT
PHL 8363	02 31 30.	- 05 48		17.0	BLUE STELLAR OBJECT
PHL 4164	02 31 30.	- 08 56		18.9	BLUE STELLAR OBJECT
TON-S 0260	02 31 30.	- 24 24		14.8	BLUE STAR
REIN 7.031	02 31 30.35	+ 01 07 55.9			NEBULA
MCG+06-06-059	02 31 33.	+ 34 39	42	15.	GALAXY
PHL 8367	02 31 36.	+ 01 26		18.1	BLUE STELLAR OBJECT
PHL 8366	02 31 36.	+ 01 26		18.0	BLUE STELLAR OBJECT
PHL 8365	02 31 36.	+ 02 00		17.5	BLUE STELLAR OBJECT
ZWG 388.054	02 31 36.	+ 02 24		15.0	GALAXY
UGC 02056	02 31 36.	+ 02 24	66	15.0	GALAXY SBb/Sc
ZWG 439.017	02 31 36.	+ 10 20		15.4	GALAXY
5ZW 256	02 31 36.	+ 29 45			COMPACT GALAXY
ZWG 505.017	02 31 36.	+ 29 45		15.0	GALAXY
ZWG 523.064	02 31 36.	+ 34 40		15.0	GALAXY
PHL 1358	02 31 36.	- 05 25		13.7	BLUE STELLAR OBJECT
PHL 1359	02 31 36.	- 12 57		13.7	BLUE STELLAR OBJECT
MCG-03-07-035	02 31 36.	- 15 34 30.	24	15.	GALAXY
MCG-03-07-036	02 31 36.	- 20 27	36	15.	GALAXY
PHL 8364	02 31 36.	- 24 22		14.4	BLUE STELLAR OBJECT
RNGC 0986	02 31 36.	- 39 16		12.0	GALAXY
MCG-04-07-010	02 31 39.	- 21 15	48	15.	GALAXY
MCG+00-07-045	02 31 42.	+ 02 23 30.	36	15.	GALAXY
PHL 8368	02 31 42.	- 02 19		14.2	BLUE STELLAR OBJECT
PHL 1360	02 31 42.	- 04 20		16.6	BLUE STELLAR OBJECT
IC 1818	02 31 42.	- 11 15 30.			NONSTELLAR OBJECT
PHL 8369	02 31 42.	- 28 27		5.0	BLUE STELLAR OBJECT
FATH 1.121	02 31 47.	+ 30 32	14		NEBULA
LB 02743	02 31 47.	+ 31 34 18.		16.8	FAINT BLUE STAR
MRK 594	02 31 48.	+ 07 31	16	16.	GALAXY WITH UV CONTINUUM
MCG+04-07-012	02 31 48.	+ 23 14	24	14.	GALAXY
ZWG 505.018	02 31 48.	+ 32 38		13.3	GALAXY
RNGC 0978B	02 31 48.	+ 32 38		15.0	GALAXY
RNGC 0978A	02 31 48.	+ 32 38		14.0	GALAXY
UGC 02057	02 31 48.	+ 32 38	120	13.3	GALAXY E-S0
KARA.72 071B	02 31 48.	+ 32 38	42		PART OF DOUBLE GALAXY
KARA.72 071A	02 31 48.	+ 32 38	96	13.3	PART OF DOUBLE GALAXY
MCG+05-07-017	02 31 48.	+ 32 39	24	15.5	GALAXY
MCG+05-07-016	02 31 48.	+ 32 39	30	14.5	GALAXY
ZWG 523.065	02 31 48.	+ 34 17		15.7	GALAXY
ZWG 539.052	02 31 48.	+ 40 55		15.6	GALAXY
UGC 02058	02 31 48.	+ 40 55	102	15.6	GALAXY Sb/SBc
VB 171	02 31 48.	+ 61 10	376		STELLAR RING
ZC 0231.8-0030	02 31 48.	- 00 30	740		CLUSTER OF GALAXIES
PHL 1361	02 31 48.	- 13 10		18.6	BLUE STELLAR OBJECT
PHL 4166	02 31 48.	- 19 49		18.4	BLUE STELLAR OBJECT
PHL 8370	02 31 48.	- 25 38		18.3	BLUE STELLAR OBJECT
PHL 4165	02 31 48.	- 26 07		18.0	BLUE STELLAR OBJECT
PHL 1362	02 31 48.	- 28 52		18.1	BLUE STELLAR OBJECT
RNGC 0984	02 31 52.	+ 23 12		14.5	GALAXY
FATH 1.122	02 31 52.	+ 29 57	11		NEBULA
FATH 1.123	02 31 52.	+ 30 01	19		NEBULA
PHL 1363	02 31 54.	+ 01 10		18.1	BLUE STELLAR OBJECT
5ZW 257	02 31 54.	+ 23 11			COMPACT GALAXY
ZWG 484.010	02 31 54.	+ 23 12		14.5	GALAXY
UGC 02059	02 31 54.	+ 23 12	180	14.5	GALAXY S0
MCG+07-06-037	02 31 54.	+ 41 08	54	15.	GALAXY
ZWG 539.053	02 31 54.	+ 41 09		14.7	GALAXY
UGC 02060	02 31 54.	+ 41 09	96	14.7	GALAXY SBa-b
VB 080	02 31 54.	+ 63 55	363		STELLAR RING
PHL 1364	02 31 54.	- 05 08		16.3	BLUE STELLAR OBJECT
MCG-02-07-033	02 31 54.	- 11 03	180	13.5	GALAXY
PHL 4167	02 31 54.	- 12 56		18.6	BLUE STELLAR OBJECT
PHL 1365	02 31 54.	- 13 27		18.3	BLUE STELLAR OBJECT
ZWG 388.055	02 32 00.	+ 01 08		15.0	GALAXY
UGC 02062	02 32 00.	+ 01 08	60	15.0	GALAXY Sb-c
MCG+00-07-046	02 32 00.	+ 01 08	48	14.5	GALAXY
PHL 1367	02 32 00.	+ 02 00		18.3	BLUE STELLAR OBJECT
PHL 1366	02 32 00.	+ 02 30		18.8	BLUE STELLAR OBJECT
OCL 0361	02 32 00.	+ 58 47	180	16.	OPEN STAR CLUSTER
UGC 02061	02 32 00.	- 01 11	102	16.0	GALAXY S
MCG+00-07-047	02 32 00.	- 01 12 30.	30	15.5	GALAXY
MCG+01-07-021	02 32 00.	- 04 42	60	15.	GALAXY
SN 1963V	02 32 00.	- 06 18		16.0	SUPERNOVA
PHL 4168	02 32 00.	- 08 54		18.5	BLUE STELLAR OBJECT
PHL 1368	02 32 00.	- 11 47		18.6	BLUE STELLAR OBJECT
PHL 4169	02 32 00.	- 13 30		18.7	BLUE STELLAR OBJECT
PHL 1369	02 32 00.	- 30 40		17.9	BLUE STELLAR OBJECT
RNGC 0983	02 32 01.	+ 31 17			NON-EXISTENT OBJECT
REIN 7.032	02 32 01.21	+ 01 07 43.8			NEBULA
MCG-02-07-034	02 32 03.	- 14 30	36	14.5	GALAXY
RNGC 0985	02 32 04.	- 09 00		14.0	GALAXY
LB 02744	02 32 06.	+ 32 44 12.		16.2	FAINT BLUE STAR
VB 172	02 32 06.	+ 59 04	322		STELLAR RING
ISS 0061	02 32 06.	+ 59 04	323		STELLAR RING
OCL 0358	02 32 06.	+ 59 25	480		OPEN STAR CLUSTER
ISS 0018	02 32 06.	+ 63 55	324		STELLAR RING
VV 285	02 32 06.	- 09 00	42	14.	INTERACTING GALAXY
MCG-02-07-035	02 32 06.	- 09 00	48	14.5	GALAXY
PHL 4170	02 32 06.	- 11 15		16.7	BLUE STELLAR OBJECT
PHL 8371	02 32 06.	- 14 44		18.7	BLUE STELLAR OBJECT
MCG-03-07-037	02 32 06.	- 19 50	72	15.5	GALAXY
PHL 4171	02 32 06.	- 32 38		18.3	BLUE STELLAR OBJECT
LB 01630	02 32 06.	- 59 26		14.4	FAINT BLUE STAR
MCG+07-06-038	02 32 12.	+ 40 42 30.	30	14.5	GALAXY
ZWG 539.054	02 32 12.	+ 40 43		14.3	GALAXY
UGC 02063	02 32 12.	+ 40 43	102	14.3	GALAXY S0
URA 34	02 32 12.	+ 63 57	84		STELLAR RING
PHL 1371	02 32 12.	- 00 11		18.9	BLUE STELLAR OBJECT
PHL 1372	02 32 12.	- 01 31		16.8	BLUE STELLAR OBJECT
PHL 1370	02 32 12.	- 01 47		17.5	BLUE STELLAR OBJECT
PHL 4173	02 32 12.	- 07 47		18.6	BLUE STELLAR OBJECT
PHL 1373	02 32 12.	- 13 48		18.9	BLUE STELLAR OBJECT
PHL 4172	02 32 12.	- 14 44		17.9	BLUE STELLAR OBJECT
PHL 8372	02 32 12.	- 27 48		16.8	BLUE STELLAR OBJECT
PHL 8373	02 32 12.	- 28 44		18.8	BLUE STELLAR OBJECT
RNGC 0982	02 32 13.	+ 40 43		14.5	GALAXY
FATH 1.124	02 32 14.	+ 30 10	5		NEBULA
HOLM 061B	02 32 15.	+ 37 17	30	14.3	PART OF MULTIPLE GALAXY
HOLM 061A	02 32 17.	+ 37 16	48	13.8	PART OF MULTIPLE GALAXY
ZWG 462.028	02 32 18.	+ 20 38		15.0	GALAXY
UGC 02064	02 32 18.	+ 20 38	138	15.0	GALAXY SBb/Sc
ZWG 523.066	02 32 18.	+ 37 16		14.6	GALAXY
UGC 02065	02 32 18.	+ 37 16	102	14.6	GALAXY
MCG+06-06-060	02 32 18.	+ 37 17	36	14.	GALAXY
ZWG 539.055	02 32 18.	+ 40 05		15.5	GALAXY
MCG+07-06-039	02 32 18.	+ 40 39	72	14.	GALAXY
ZWG 539.056	02 32 18.	+ 40 40		13.2	GALAXY
UGC 02066	02 32 18.	+ 40 40	126	13.2	GALAXY Sa
URA 33	02 32 18.	+ 63 57	156		STELLAR RING
ZWG 388.056	02 32 18.	- 01 00		15.7	GALAXY
PHL 4174	02 32 18.	- 17 44		18.5	BLUE STELLAR OBJECT
RNGC 0980	02 32 19.	+ 40 40		13.0	GALAXY
LB 00179	02 32 20.	+ 30 08 06.		17.1	FAINT BLUE STAR
LB 02745	02 32 22.	+ 31 55 00.		16.0	FAINT BLUE STAR
LB 02746	02 32 23.	+ 32 53 30.		16.0	FAINT BLUE STAR
PHL 4175	02 32 24.	+ 03 14		17.2	BLUE STELLAR OBJECT
ZC 0232.4+0828	02 32 24.	+ 08 28	1550		CLUSTER OF GALAXIES
MCG+03-07-028	02 32 24.	+ 20 39	108	14.	GALAXY
ZWG 523.067	02 32 24.	+ 37 18		14.8	GALAXY
HOLM 061C	02 32 24.	+ 37 18	30	14.3	PART OF MULTIPLE GALAXY
UGC 02067	02 32 24.	+ 37 18	150	14.8	GALAXY Sa-b
MCG+06-06-061	02 32 24.	+ 37 19	96	15.	GALAXY
MCG+07-06-040	02 32 24.	+ 40 40 30.	15	18.	GALAXY
UGC 02068	02 32 24.	+ 40 42	72	17.	GALAXY
PHL 1374	02 32 24.	- 06 22		15.8	BLUE STELLAR OBJECT
MCG+01-07-022	02 32 24.	- 07 55	60	14.	GALAXY
PHL 8374	02 32 24.	- 21 20		12.5	BLUE STELLAR OBJECT
PHL 1375	02 32 24.	- 24 40		16.7	BLUE STELLAR OBJECT
REIN 7.033	02 32 24.07	+ 01 02 26.4			NEBULA
ARC 0364	02 32 27.	+ 08 28		17.1	RICH CLUSTER OF GALAXIES
MCG-02-07-036	02 32 27.	- 13 52 30.	120	14.	GALAXY
PHL 1376	02 32 30.	+ 03 31		12.7	BLUE STELLAR OBJECT
FEIG 024	02 32 30.	+ 03 31		12.3	FAINT BLUE STAR
5ZW 258	02 32 30.	+ 31 25			COMPACT GALAXY
5ZW 259	02 32 30.	+ 32 07			COMPACT GALAXY
SN 1961P	02 32 30.	+ 37 24		14.3	SUPERNOVA
ZWG 523.068	02 32 30.	+ 37 25		13.2	GALAXY
UGC 02069	02 32 30.	+ 37 25	162	13.2	GALAXY SBc
KARA.72 072B	02 32 30.	+ 37 25	18		PART OF DOUBLE GALAXY
KARA.72 072A	02 32 30.	+ 37 25	18	13.2	PART OF DOUBLE GALAXY
VV 096	02 32 30.	+ 37 26	138	13.	INTERACTING GALAXY
MCG+06-06-062	02 32 30.	+ 37 26	120	13.5	GALAXY
PHL 4176	02 32 30.	- 00 42		18.5	BLUE STELLAR OBJECT
PHL 4177	02 32 30.	- 05 01		18.7	BLUE STELLAR OBJECT
IC 0238	02 32 31.	+ 12 36 51.			NONSTELLAR OBJECT
LB 02747	02 32 32.	+ 32 06 30.		14.9	FAINT BLUE STAR
LB 02748	02 32 35.	+ 31 20 42.		16.7	FAINT BLUE STAR
PHL 8375	02 32 36.	+ 01 32		18.1	BLUE STELLAR OBJECT
ZWG 439.018	02 32 36.	+ 12 36		14.1	GALAXY
UGC 02070	02 32 36.	+ 12 36	108	14.1	GALAXY S0
MCG+02-07-016	02 32 36.	+ 12 37	72	14.5	GALAXY
ZWG 462.029	02 32 36.	+ 19 26		15.4	GALAXY
ZWG 462.030	02 32 36.	+ 19 43		15.6	GALAXY
UGC 02071	02 32 36.	+ 19 43	78	15.6	GALAXY Sb-c
MCG+03-07-029	02 32 36.	+ 19 43	66	14.5	GALAXY
VB 222	02 32 36.	+ 52 58	363		STELLAR RING
MAFFEI 1	02 32 36.	+ 59 25 48.	50		INFRARED OBJECT
PHL 1377	02 32 36.	- 04 14		16.6	BLUE STELLAR OBJECT
SHB 064	02 32 36.	- 04 15 12.		16.5	QUASI-STELLAR OBJECT
PHL 4178	02 32 36.	- 07 56		15.5	BLUE STELLAR OBJECT
PHL 8377	02 32 36.	- 14 57		18.2	BLUE STELLAR OBJECT
PHL 8376	02 32 36.	- 22 49		18.5	BLUE STELLAR OBJECT
BC 4C-04.06	02 32 36.6	- 04 15 11.		16.46	QUASI-STELLAR OBJECT
MCG+03-07-030	02 32 42.	+ 19 27	24	16.	GALAXY
URA 18	02 32 42.	+ 63 17	738		STELLAR RING
PHL 4179	02 32 42.	- 00 20		16.8	BLUE STELLAR OBJECT
PHL 4180	02 32 42.	- 04 10		18.4	BLUE STELLAR OBJECT
PHL 8378	02 32 42.	- 28 06		18.5	BLUE STELLAR OBJECT
LB 02749	02 32 43.	+ 32 27 36.		16.9	FAINT BLUE STAR
LB 02750	02 32 45.	+ 33 08 18.		16.4	FAINT BLUE STAR
LB 02751	02 32 47.	+ 32 27 42.		16.2	FAINT BLUE STAR
RNGC 0988	02 32 47.	- 09 34		11.0	GALAXY
SN 1960A	02 32 48.	+ 01 53		16.0	SUPERNOVA
UGC 02072	02 32 48.	+ 45 28	72	16.5	GALAXY S
PHL 1378	02 32 48.	- 04 00		17.8	BLUE STELLAR OBJECT
PHL 1379	02 32 48.	- 07 38		18.6	BLUE STELLAR OBJECT
PHL 1380	02 32 48.	- 09 48		18.2	BLUE STELLAR OBJECT
PHL 1381	02 32 48.	- 10 17		18.7	BLUE STELLAR OBJECT
PHL 8379	02 32 48.	- 10 32		18.6	BLUE STELLAR OBJECT
PHL 8380	02 32 48.	- 12 16		18.8	BLUE STELLAR OBJECT
PHL 4181	02 32 48.	- 20 46		17.5	BLUE STELLAR OBJECT
MCG-02-07-037	02 32 51.	- 09 34	300	11.	GALAXY
ZWG 462.031	02 32 54.	+ 20 11		15.7	GALAXY
MCG+04-07-013	02 32 54.	+ 24 15 30.	36	15.	GALAXY
LB 02752	02 32 54.	+ 32 42 30.		16.8	FAINT BLUE STAR
ZWG 539.057	02 32 54.	+ 42 12		14.0	GALAXY
UGC 02073	02 32 54.	+ 42 12	126	14.0	GALAXY S0?
ZWG 539.058	02 32 54.	+ 42 24		15.0	GALAXY
UGC 02074	02 32 54.	+ 42 24	72	15.0	GALAXY VY CMPT
PHL 1382	02 32 54.	- 08 45		17.3	BLUE STELLAR OBJECT
PHL 1383	02 32 54.	- 12 01		18.7	BLUE STELLAR OBJECT
MCG-05-07-007	02 32 54.	- 29 49	30	15.5	GALAXY
LB 03265	02 32 54.	- 52 11		13.7	FAINT BLUE STAR
LB 02753	02 32 56.	+ 31 27 36.		16.5	FAINT BLUE STAR

OBJECT NAME	RIGHT ASCEN.	DECLINATION	DIAM.	MAGN.	TYPE OF OBJECT
SHB 065	02 32 59.9	- 02 32 23.		19.5	QUASI-STELLAR OBJECT
LBN 0659	02 33	+ 59 25	120		BRIGHT NEBULA
ZC 0233.0+0124	02 33 00.	+ 01 24	9880		CLUSTER OF GALAXIES
ZWG 414.022	02 33 00.	+ 06 03		15.0	GALAXY
UGC 02075	02 33 00.	+ 06 03	90	15.0	GALAXY S0-a
MCG+01-07-012	02 33 00.	+ 06 03	60	14.5	GALAXY
UGC 02076	02 33 00.	+ 11 15	72	16.0	GALAXY Sb-c
MCG+03-07-031	02 33 00.	+ 20 12	24	15.	GALAXY
ZWG 484.011	02 33 00.	+ 24 15		15.5	GALAXY
MCG+07-06-041	02 33 00.	+ 42 12	54	15.	GALAXY
PHL 8382	02 33 00.	- 01 39		15.7	BLUE STELLAR OBJECT
PHL 8381	02 33 00.	- 03 03		14.8	BLUE STELLAR OBJECT
PHL 1384	02 33 00.	- 03 22		18.4	BLUE STELLAR OBJECT
PHL 1385	02 33 00.	- 06 49		18.3	BLUE STELLAR OBJECT
MCG+01-07-023	02 33 00.	- 07 22	168	12.7	GALAXY
PHL 8383	02 33 00.	- 20 36		17.2	BLUE STELLAR OBJECT
LB 01631	02 33 00.	- 45 35		13.7	FAINT BLUE STAR
PATH 1.125	02 33 04.	+ 30 19	14		NEBULA
IC 1819	02 33 05.	+ 03 49 24.			NONSTELLAR OBJECT
LB 02754	02 33 05.	+ 32 57 48.		16.3	FAINT BLUE STAR
PHL 8384	02 33 06.	+ 00 43		18.4	BLUE STELLAR OBJECT
ZWG 388.057	02 33 06.	+ 02 22		15.7	GALAXY
ZWG 414.023	02 33 06.	+ 03 50		15.0	GALAXY
KARA.73B 0110	02 33 06.	+ 03 50	96	15.0	ISOLATED GALAXY S0
ZWG 523.069	02 33 06.	+ 34 23		15.5	GALAXY
UGC 02077	02 33 06.	+ 34 23	66	15.5	GALAXY Sc
ZWG 539.059	02 33 06.	+ 40 00		15.7	GALAXY
ZC 0233.1-0132	02 33 06.	- 01 32	1680		CLUSTER OF GALAXIES
TON-S 0261	02 33 06.	- 31 24		15.2	BLUE STAR
ARC 0365	02 33 07.	- 01 39		17.4	RICH CLUSTER OF GALAXIES
MCG+06-06-063	02 33 09.	+ 34 22	36	15.	GALAXY
RNGC 0991	02 33 10.	- 07 22		12.5	GALAXY
PHL 8385	02 33 12.	+ 00 08		18.3	BLUE STELLAR OBJECT
PHL 4185	02 33 12.	+ 01 06		16.7	BLUE STELLAR OBJECT
PHL 1386	02 33 12.	+ 02 36		17.7	BLUE STELLAR OBJECT
ZWG 414.024	02 33 12.	+ 05 50		15.4	GALAXY
ZC 0233.2+1035	02 33 12.	+ 10 35	7390		CLUSTER OF GALAXIES
SN 1963R	02 33 12.	+ 35 44		16.8	SUPERNOVA
MCG+06-06-064	02 33 12.	+ 35 44	36	16.	GALAXY
ZWG 523.070	02 33 12.	+ 35 45		15.7	GALAXY
URA 19	02 33 12.	+ 63 22	120		STELLAR RING
PHL 1387	02 33 12.	- 06 08		17.4	BLUE STELLAR OBJECT
PHL 4186	02 33 12.	- 07 25		18.6	BLUE STELLAR OBJECT
PHL 8387	02 33 12.	- 09 18		17.6	BLUE STELLAR OBJECT
PHL 8386	02 33 12.	- 09 54		18.6	BLUE STELLAR OBJECT
MCG-02-07-039	02 33 12.	- 10 45	30	15.	GALAXY
MCG-02-07-038	02 33 12.	- 12 30	78	14.5	GALAXY
PHL 4184	02 33 12.	- 13 34		18.6	BLUE STELLAR OBJECT
PHL 4182	02 33 12.	- 13 35		16.2	BLUE STELLAR OBJECT
PHL 4183	02 33 12.	- 13 46		18.9	BLUE STELLAR OBJECT
PHL 1388	02 33 12.	- 21 41		17.7	BLUE STELLAR OBJECT
MCG-04-07-011	02 33 12.	- 25 44	54	14.5	GALAXY
ARC 0366	02 33 14.	- 05 40		17.6	RICH CLUSTER OF GALAXIES
IC 1822	02 33 14.	- 08 47			NONSTELLAR OBJECT
IC 1820	02 33 15.	+ 05 49 41.			NONSTELLAR OBJECT
BV 10	02 33 17.	- 13 34 06.		18.1	FAINT BLUE VARIABLE
UGC 02078	02 33 18.	+ 10 13	90	16.0	GALAXY
MCG+02-07-017	02 33 18.	+ 10 13	36	15.5	GALAXY
ZWG 484.012	02 33 18.	+ 23 41		14.8	GALAXY
UGC 02079	02 33 18.	+ 23 41	132	14.8	GALAXY Sc/SBc
MCG+04-07-014	02 33 18.	+ 23 42	96	14.	GALAXY
ZWG 484.013	02 33 18.	+ 26 59		15.7	GALAXY
KARA.73B 0111	02 33 18.	+ 26 59	36	15.7	ISOLATED GALAXY E
5ZW 260	02 33 18.	+ 33 22			COMPACT GALAXY
ZWG 523.071	02 33 18.	+ 38 45		12.1	GALAXY
UGC 02080	02 33 18.	+ 38 45	360	12.1	GALAXY Sc
VB 081	02 33 18.	+ 66 52	497		STELLAR RING
MCG-02-07-040	02 33 18.	- 13 53 30.	36	15.	GALAXY
LB 03618	02 33 19.	+ 40 40 18.		20.5	FAINT BLUE STAR
MCG+04-07-015	02 33 21.	+ 26 59 30.	24	15.	GALAXY
MCG-02-07-041	02 33 21.	- 13 52	72	14.	GALAXY
IC 0239	02 33 23.	+ 38 45 55.	3120	11.8	GALAXY SAB(rs)
ZWG 388.058	02 33 24.	+ 00 12		15.1	GALAXY
UGC 02081	02 33 24.	+ 00 12	180	15.1	GALAXY Sc
MCG+00-07-048	02 33 24.	+ 00 12	72	14.5	GALAXY
ZWG 484.014	02 33 24.	+ 25 12		14.0	GALAXY
UGC 02082	02 33 24.	+ 25 12	378	14.0	GALAXY Sc
KARA.73B 0112	02 33 24.	+ 25 12	318	14.0	ISOLATED GALAXY S
MCG+04-07-016	02 33 24.	+ 25 14	300	13.	GALAXY
5ZW 261	02 33 24.	+ 31 30			COMPACT GALAXY
ZWG 505.019	02 33 24.	+ 31 30		14.6	GALAXY
MCG+05-07-018	02 33 24.	+ 31 30	30	16.	GALAXY
ZWG 505.020	02 33 24.	+ 32 30		15.0	GALAXY
UGC 02083	02 33 24.	+ 32 30	102	15.0	GALAXY Sb-c
ZWG 505.021	02 33 24.	+ 33 22		15.7	GALAXY
6ZW 215	02 33 24.	+ 33 48			COMPACT GALAXY
ZWG 523.072	02 33 24.	+ 35 55		15.7	GALAXY
UGC 02084	02 33 24.	+ 35 55	66	15.7	GALAXY S
MCG+06-06-065	02 33 24.	+ 38 47	240	13.	GALAXY
PHL 4187	02 33 24.	- 05 04		18.6	BLUE STELLAR OBJECT
PHL 8389	02 33 24.	- 15 32		18.3	BLUE STELLAR OBJECT
PHL 4188	02 33 24.	- 19 11		18.2	BLUE STELLAR OBJECT
PHL 1389	02 33 24.	- 29 52		16.1	BLUE STELLAR OBJECT
PHL 8388	02 33 24.	- 31 20		15.2	BLUE STELLAR OBJECT
LB 02755	02 33 26.	+ 33 10 18.		16.0	FAINT BLUE STAR
ZWG 388.059	02 33 30.	+ 01 20		15.7	GALAXY
UGC 02085	02 33 30.	+ 01 20	78	15.7	GALAXY Sc
ZWG 388.060	02 33 30.	+ 01 34		15.7	GALAXY
PHL 4189	02 33 30.	+ 03 25		18.4	BLUE STELLAR OBJECT
MCG+02-07-018	02 33 30.	+ 11 24	21	13.5	GALAXY
UGC 02086	02 33 30.	+ 13 52	66	16.0	GALAXY SB:a-b
ZC 0233.5+1833	02 33 30.	+ 18 33	470		CLUSTER OF GALAXIES
ZWG 505.022	02 33 30.	+ 31 23		15.1	GALAXY
UGC 02087	02 33 30.	+ 31 23	66	15.1	GALAXY Sc
MCG+05-07-019	02 33 30.	+ 32 31	90	15.	GALAXY
5ZW 262	02 33 30.	+ 45 10			COMPACT GALAXY
PHL 4190	02 33 30.	- 12 30		18.3	BLUE STELLAR OBJECT
REIN 6.001	02 33 31.98	+ 01 20 09.2			NEBULA
MCG+06-06-066	02 33 33.	+ 34 23	48	15.	GALAXY
LB 02756	02 33 34.	+ 33 13 54.		16.8	FAINT BLUE STAR
LB 03619	02 33 34.	+ 40 59 36.		19.6	FAINT BLUE STAR
MCG+00-07-049	02 33 36.	+ 01 20	36	15.	GALAXY
ZWG 439.019	02 33 36.	+ 11 26		13.9	GALAXY
UGC 02089	02 33 36.	+ 11 26	120	13.9	GALAXY E
MCG+05-07-020	02 33 36.	+ 31 24	54	15.	GALAXY
ZWG 523.073	02 33 36.	+ 34 24		15.4	GALAXY
UGC 02090	02 33 36.	+ 34 24	66	15.4	GALAXY Sb
6ZW 216	02 33 36.	+ 34 42			COMPACT GALAXY
UGC 02088	02 33 36.	- 00 55	66	16.5	GALAXY Sc
PHL 8390	02 33 36.	- 02 01		16.4	BLUE STELLAR OBJECT
PHL 8391	02 33 36.	- 08 21		18.0	BLUE STELLAR OBJECT
PHL 1390	02 33 36.	- 08 56		17.7	BLUE STELLAR OBJECT
PHL 4191	02 33 36.	- 28 14		18.6	BLUE STELLAR OBJECT
MCG+01-07-013	02 33 39.	+ 07 18	42	14.5	GALAXY
RNGC 0990	02 33 39.	+ 11 26		14.0	GALAXY
MCG-03-07-038	02 33 39.	- 17 28 30.	24	15.	GALAXY
IC 1821	02 33 41.	+ 13 34 39.			NONSTELLAR OBJECT
ZWG 414.025	02 33 42.	+ 07 19		15.4	GALAXY
ZWG 439.020	02 33 42.	+ 13 33		15.4	GALAXY
5ZW 263	02 33 42.	+ 31 14			COMPACT GALAXY
MCG+06-06-067	02 33 42.	+ 35 54	102	13.5	GALAXY
OCL 0365	02 33 42.	+ 55 46	1440	9.0	OPEN STAR CLUSTER
MCG-02-07-042	02 33 42.	- 09 31	24	15.	GALAXY
MCG-03-07-040	02 33 42.	- 17 30	60	15.	GALAXY
MCG-03-07-039	02 33 42.	- 17 30	24	15.	GALAXY
PHL 8392	02 33 42.	- 26 40		16.7	BLUE STELLAR OBJECT
ZWG 388.061	02 33 48.	+ 00 30		15.1	GALAXY
UGC 02091	02 33 48.	+ 00 30	108	15.1	GALAXY Sc
MCG+00-07-050	02 33 48.	+ 00 30	48	15.	GALAXY
ZWG 414.026	02 33 48.	+ 07 05		15.6	GALAXY
UGC 02092	02 33 48.	+ 07 05	198	15.6	GALAXY Sc
ZWG 505.023	02 33 48.	+ 33 06		13.4	GALAXY
RNGC 0987	02 33 48.	+ 33 06		13.5	GALAXY
UGC 02093	02 33 48.	+ 33 06	108	13.4	GALAXY SB0/SBa
MCG+05-07-021	02 33 48.	+ 33 08	78	14.5	GALAXY
ZWG 523.074	02 33 48.	+ 35 54		13.8	GALAXY
UGC 02094	02 33 48.	+ 35 54	138	13.8	GALAXY SBc
ZWG 523.075	02 33 48.	+ 37 40		15.5	GALAXY
PHL 8396	02 33 48.	- 00 09		18.2	BLUE STELLAR OBJECT
PHL 4193	02 33 48.	- 00 47		18.8	BLUE STELLAR OBJECT
PHL 4192	02 33 48.	- 01 46		18.0	BLUE STELLAR OBJECT
PHL 4194	02 33 48.	- 03 16		18.7	BLUE STELLAR OBJECT
PHL 8394	02 33 48.	- 04 46		17.3	BLUE STELLAR OBJECT
PHL 8395	02 33 48.	- 12 16		17.0	BLUE STELLAR OBJECT
PHL 8393	02 33 48.	- 16 30		18.5	BLUE STELLAR OBJECT
MCG+01-07-014	02 33 51.	+ 07 05	180	14.	GALAXY
MCG-02-07-043	02 33 54.	- 09 34	12	15.	GALAXY
PHL 4195	02 33 54.	- 17 08		17.7	BLUE STELLAR OBJECT
LB 02758	02 33 55.	+ 30 39 18.		13.2	FAINT BLUE STAR
LB 02757	02 33 55.	+ 32 34 24.		15.8	FAINT BLUE STAR
MCG+00-07-051	02 33 57.	+ 00 53	30	15.5	GALAXY
LB 02759	02 33 57.	+ 32 21 42.		16.1	FAINT BLUE STAR
LBN 0693	02 34	+ 31 50	3120		BRIGHT NEBULA
KHAV 037	02 34	+ 62 25	2470		DARK NEBULA
ZWG 388.062	02 34 00.	+ 00 54		15.5	GALAXY
PHL 4196	02 34 00.	+ 01 32		16.6	BLUE STELLAR OBJECT
PHL 1391	02 34 00.	+ 01 47		18.4	BLUE STELLAR OBJECT
PHL 4197	02 34 00.	+ 02 36		16.6	BLUE STELLAR OBJECT
SN 1966M	02 34 00.	+ 37 56		18.5	SUPERNOVA
LB 03620	02 34 00.	+ 40 41 30.		20.3	FAINT BLUE STAR
LDN 1373	02 34 00.	+ 61 00	14040		DARK NEBULA
LDN 1372	02 34 00.	+ 61 10	660		DARK NEBULA
PHL 1392	02 34 00.	- 15 20		18.0	BLUE STELLAR OBJECT
LB 03621	02 34 04.	+ 40 52 36.		19.2	FAINT BLUE STAR
LB 02760	02 34 05.	+ 28 28 18.		15.9	FAINT BLUE STAR
PHL 4198	02 34 06.	+ 00 47		18.4	BLUE STELLAR OBJECT
ZWG 388.063	02 34 06.	+ 01 50		14.9	GALAXY
RNGC 0994	02 34 06.	+ 01 50			NON-EXISTENT OBJECT
RNGC 0993	02 34 06.	+ 01 50		15.0	GALAXY
UGC 02095	02 34 06.	+ 01 50	66	14.9	GALAXY E?
MCG+00-07-052	02 34 06.	+ 01 50	42	14.5	GALAXY
MCG+02-07-019	02 34 06.	+ 10 51	60	15.5	GALAXY
ZWG 484.015	02 34 06.	+ 25 14		14.8	GALAXY
MCG+04-07-017	02 34 06.	+ 25 15	39	15.	GALAXY
PHL 8397	02 34 06.	- 08 02		18.5	BLUE STELLAR OBJECT
LB 03266	02 34 06.	- 51 13		14.1	FAINT BLUE STAR
REIN 6.002	02 34 09.98	+ 01 50 28.8			NEBULA
REIN 6.003	02 34 10.98	+ 01 49 58.5			NEBULA
PHL 4203	02 34 12.	+ 01 29		18.7	BLUE STELLAR OBJECT
UGC 02096	02 34 12.	+ 10 51	72	16.0	GALAXY Sb-c
ZWG 462.032	02 34 12.	+ 20 13		15.7	GALAXY
VB 082	02 34 12.	+ 66 48	389		STELLAR RING
PHL 8398	02 34 12.	- 08 39		18.6	BLUE STELLAR OBJECT
PHL 4204	02 34 12.	- 11 08		18.8	BLUE STELLAR OBJECT
PHL 4200	02 34 12.	- 14 00		17.8	BLUE STELLAR OBJECT
PHL 4201	02 34 12.	- 14 24		17.3	BLUE STELLAR OBJECT
PHL 4199	02 34 12.	- 14 42		16.6	BLUE STELLAR OBJECT
PHL 1393	02 34 12.	- 15 33		18.0	BLUE STELLAR OBJECT
PHL 4202	02 34 12.	- 17 48		18.5	BLUE STELLAR OBJECT
PHL 8399	02 34 12.	- 20 54		17.9	BLUE STELLAR OBJECT
REIN 6.004	02 34 13.00	+ 01 52 34.0			NEBULA
LB 03622	02 34 17.	+ 40 56 12.		19.8	FAINT BLUE STAR
SCHO 0055	02 34 17.	+ 53 33 24.	360		ISOLATED DARK CLOUD
UGC 02097	02 34 18.	+ 05 14	66	17.	GALAXY Sc
MCG+03-07-032	02 34 18.	+ 20 14	30	15.	GALAXY
PHL 1394	02 34 18.	- 01 16		18.7	BLUE STELLAR OBJECT
ARC 0367	02 34 18.	- 19 37		16.5	RICH CLUSTER OF GALAXIES
PHL 8400	02 34 18.	- 26 36		17.5	BLUE STELLAR OBJECT
MCG+00-07-053	02 34 21.	- 02 26 30.	48	15.5	GALAXY
LB 02761	02 34 23.	+ 29 35 42.		16.3	FAINT BLUE STAR
UGC 02098	02 34 24.	+ 19 44	78	16.5	GALAXY S
MCG+03-07-034	02 34 24.	+ 19 45	66	15.5	GALAXY
ZWG 462.033	02 34 24.	+ 21 21		15.5	GALAXY
UGC 02099	02 34 24.	+ 21 21	84	15.5	GALAXY S0
MCG+03-07-033	02 34 24.	+ 21 22	60	15.	GALAXY
ZWG 505.024	02 34 24.	+ 33 25		15.6	GALAXY
UGC 02100	02 34 24.	+ 33 25	66	15.6	GALAXY
ZWG 539.060	02 34 24.	+ 42 25		15.1	GALAXY
UGC 02101	02 34 24.	+ 42 25	126	15.1	GALAXY Sb
ZWG 388.064	02 34 24.	- 02 25		15.7	GALAXY
PHL 8402	02 34 24.	- 05 40		17.1	BLUE STELLAR OBJECT
PHL 8403	02 34 24.	- 10 18		17.8	BLUE STELLAR OBJECT
PHL 1395	02 34 24.	- 11 50		18.4	BLUE STELLAR OBJECT
PHL 8401	02 34 24.	- 28 22		13.8	BLUE STELLAR OBJECT
PHL 1396	02 34 24.	- 29 04		18.2	BLUE STELLAR OBJECT
PHL 8404	02 34 30.	+ 00 16		18.0	BLUE STELLAR OBJECT
SN 1958B	02 34 30.	+ 01 07		18.0	SUPERNOVA
ZWG 462.034	02 34 30.	+ 20 12		15.0	GALAXY
MCG+04-07-018	02 34 30.	+ 23 07	48	15.	GALAXY
ZWG 539.061	02 34 30.	+ 40 05		15.7	GALAXY
MCG+07-06-042	02 34 30.	+ 42 25	96	15.	GALAXY
PHL 4205	02 34 30.	- 05 06		15.7	BLUE STELLAR OBJECT
MCG+01-07-024	02 34 30.	- 05 34	60	13.	GALAXY
PHL 8405	02 34 30.	- 11 02		18.4	BLUE STELLAR OBJECT
PHL 8407	02 34 30.	- 28 42		18.8	BLUE STELLAR OBJECT
PHL 8406	02 34 30.	- 29 05		17.0	BLUE STELLAR OBJECT

OBJECT NAME	RIGHT ASCEN.	DECLINATION	DIAM.	MAGN.	TYPE OF OBJECT
LB 03267	02 34 30.	- 65 28		14.7	FAINT BLUE STAR
RNGC 1025	02 34 31.	- 55 06			UNVERIFIED SOUTHRN OBJECT
MCG+03-07-036	02 34 33.	+ 20 13	30	15.	GALAXY
MCG+03-07-035	02 34 33.	+ 20 54	51	14.5	GALAXY
RNGC 0998	02 34 34.	+ 07 13		15.5	GALAXY
RNGC 0997	02 34 35.	+ 07 05		14.5	GALAXY
RNGC 0992	02 34 35.	+ 20 53		13.5	GALAXY
PHL 1397	02 34 36.	+ 01 22		17.8	BLUE STELLAR OBJECT
ZWG 414.027	02 34 36.	+ 07 05		14.6	GALAXY
UGC 02102	02 34 36.	+ 07 05	78	14.6	GALAXY E
MCG+01-07-015	02 34 36.	+ 07 06	30	14.	GALAXY
ZWG 414.028	02 34 36.	+ 07 13		15.6	GALAXY
ZWG 462.035	02 34 36.	+ 20 53		13.5	GALAXY
UGC 02103	02 34 36.	+ 20 53	54	13.5	GALAXY S?
2ZW 004	02 34 36.	+ 20 56			COMPACT GALAXY
GCL 004	02 34 36.	+ 20 56	6	9.7	GLOBULAR STAR CLUSTER
ZWG 484.016	02 34 36.	+ 23 05		15.5	GALAXY
UGC 02104	02 34 36.	+ 23 05	60	15.5	GALAXY Sc
HMS 1.06	02 34 36.	+ 34 12			SBa GALAXY
MCG+06-06-068	02 34 36.	+ 34 12	48	14.5	GALAXY
ZWG 523.076	02 34 36.	+ 34 14		13.9	GALAXY
UGC 02105	02 34 36.	+ 34 14	102	13.9	GALAXY SBa
SN 1938A	02 34 36.	+ 34 14		15.2	SUPERNOVA
PHL 8409	02 34 36.	- 03 08		17.5	BLUE STELLAR OBJECT
PHL 4206	02 34 36.	- 10 30		18.8	BLUE STELLAR OBJECT
PHL 1398	02 34 36.	- 11 01		17.7	BLUE STELLAR OBJECT
PHL 4208	02 34 36.	- 12 22		18.3	BLUE STELLAR OBJECT
PHL 8408	02 34 36.	- 12 44		17.4	BLUE STELLAR OBJECT
PHL 4207	02 34 36.	- 27 44		18.8	BLUE STELLAR OBJECT
LB 02762	02 34 38.	+ 31 23 48.		15.2	FAINT BLUE STAR
MCG+01-07-016	02 34 39.	+ 07 04 30.	36	15.	GALAXY
PHL 8410	02 34 42.	+ 02 35		15.5	BLUE STELLAR OBJECT
3ZW 050	02 34 42.	+ 20 56			COMPACT GALAXY
ZWG 462.036	02 34 42.	+ 20 56		15.0	GALAXY
MRK 369	02 34 42.	+ 20 56	12	16.	GALAXY WITH UV CONTINUUM
MCG+04-07-019	02 34 42.	+ 25 34	36	15.5	GALAXY
MCG-05-07-008	02 34 42.	- 29 24	48	15.5	GALAXY
ZC 0234.8+0302	02 34 48.	+ 03 02	3020		CLUSTER OF GALAXIES
UGC 02107	02 34 48.	+ 12 58	60	16.0	GALAXY S0-a
ZCG 0234+20	02 34 48.	+ 20 58		16.4	COMPACT GALAXY
UGC 02108	02 34 48.	+ 42 21	84	17.	GALAXY
ZWG 388.065	02 34 48.	- 02 02		15.4	GALAXY
UGC 02106	02 34 48.	- 02 02	60	15.4	GALAXY S
PHL 1399	02 34 48.	- 13 52		16.7	BLUE STELLAR OBJECT
PHL 4209	02 34 48.	- 26 58		17.0	BLUE STELLAR OBJECT
MCG-03-07-041	02 34 51.	- 16 02	48	15.	GALAXY
REIN 7.034	02 34 51.86	- 02 02 14.8			NEBULA
LB 03623	02 34 52.	+ 41 00 18.		19.2	FAINT BLUE STAR
ZWG 388.066	02 34 54.	+ 01 25		15.5	GALAXY
MCG+06-06-069	02 34 54.	+ 34 01	90	14.5	GALAXY
ZWG 523.077	02 34 54.	+ 34 02		14.5	GALAXY
UGC 02109	02 34 54.	+ 34 02	132	14.5	GALAXY Sc
UGC 02110	02 34 54.	+ 37 56	66	16.5	GALAXY SBb-c
ZWG 539.062	02 34 54.	+ 41 35		14.8	GALAXY
UGC 02111	02 34 54.	+ 41 35	168	14.8	GALAXY Sa-b
URA 20	02 34 54.	+ 63 25	168		STELLAR RING
PHL 8411	02 34 54.	- 05 22		16.6	BLUE STELLAR OBJECT
KARA.68 020	02 34 54.	- 07 40	27		DWARF GALAXY
LB 03268	02 34 54.	- 79 24		13.7	FAINT BLUE STAR
RNGC 1031	02 34 55.	- 55 05			UNVERIFIED SOUTHERN OBJECT
KEEL 091	02 34 57.8	+ 38 29 36.			NEBULA
REIN 6.005	02 34 58.14	+ 01 26 06.8			NEBULA
RNGC 1006	02 34 59.	- 11 14			NON-EXISTENT OBJECT
ZWG 388.067	02 35 00.	+ 01 23		15.6	GALAXY
MCG+00-07-054	02 35 00.	+ 01 25	24	15.5	GALAXY
ZWG 388.068	02 35 00.	+ 01 45		14.3	GALAXY
UGC 02112	02 35 00.	+ 01 45	114	14.3	GALAXY E
MCG+07-06-043	02 35 00.	+ 41 34	66	15.	GALAXY
PHL 1400	02 35 00.	- 12 35		15.0	BLUE STELLAR OBJECT
PHL 8412	02 35 00.	- 13 10		16.0	BLUE STELLAR OBJECT
TON-S 0262	02 35 00.	- 26 57		16.0	BLUE STAR
HOD.61 1	02 35 00.	- 34 23	36	19.7	GLOB CLSTR IN FORNAX GALX
LB 03269	02 35 00.	- 66 31		14.4	FAINT BLUE STAR
RNGC 1004	02 35 01.	+ 01 45		14.5	GALAXY
REIN 6.006	02 35 01.94	+ 01 23 19.1			NEBULA
LB 02763	02 35 02.	+ 30 00 42.		16.5	FAINT BLUE STAR
MCG+00-07-055	02 35 03.	+ 01 22 30.	18	15.5	GALAXY
REIN 7.035	02 35 04.80	- 01 07 17.5			NEBULA
LB 02764	02 35 05.	+ 32 32 06.		16.0	FAINT BLUE STAR
RNGC 1010	02 35 05.	- 11 14		14.0	GALAXY
REIN 6.008	02 35 05.18	+ 01 45 23.2			NEBULA
LB 03624	02 35 06.	+ 40 50 48.		19.7	FAINT BLUE STAR
ISS 0019	02 35 06.	+ 66 47	302		STELLAR RING
MCG-02-07-044	02 35 06.	- 11 14	48	14.	GALAXY
PHL 4210	02 35 06.	- 28 20		18.2	BLUE STELLAR OBJECT
REIN 6.009	02 35 06.74	+ 01 45 31.9			NEBULA
RNGC 1011	02 35 11.	- 11 13		15.0	GALAXY
MCG+00-07-057	02 35 12.	+ 01 45	54	14.	GALAXY
ZWG 388.069	02 35 12.	+ 01 56		15.7	GALAXY
RNGC 1007	02 35 12.	+ 01 56		15.5	GALAXY
MCG+00-07-056	02 35 12.	+ 02 01	42	15.	GALAXY
PHL 8414	02 35 12.	+ 02 40		18.6	BLUE STELLAR OBJECT
PHL 8413	02 35 12.	+ 03 08		16.8	BLUE STELLAR OBJECT
MCG+05-07-022	02 35 12.	+ 30 39	66	15.	GALAXY
ZWG 505.025	02 35 12.	+ 32 56		15.7	GALAXY
OCL 0360	02 35 12.	+ 60 12	300	13.	OPEN STAR CLUSTER
PHL 1401	02 35 12.	- 03 15		16.9	BLUE STELLAR OBJECT
MCG-02-07-045	02 35 12.	- 11 13	36	15.	GALAXY
PHL 1402	02 35 12.	- 12 11		18.2	BLUE STELLAR OBJECT
PHL 4211	02 35 12.	- 13 28		18.6	BLUE STELLAR OBJECT
MCG-04-07-012	02 35 12.	- 20 35	24	15.5	GALAXY
ARC 0368	02 35 12.	- 26 43		17.5	RICH CLUSTER OF GALAXIES
HOLM 062A	02 35 13.	- 11 15	36	13.6	PART OF MULTIPLE GALAXY
MCG+00-07-059	02 35 15.	+ 02 00	36	15.	GALAXY
MCG+00-07-058	02 35 15.	+ 02 07	24	15.	GALAXY
PATH 1.126	02 35 15.	+ 30 37	54		NEBULA
HOLM 062B	02 35 17.	- 11 14	18	14.1	PART OF MULTIPLE GALAXY
REIN 6.010	02 35 17.05	+ 01 56 21.5			NEBULA
ZWG 388.070	02 35 18.	+ 01 52		14.9	GALAXY
RNGC 1008	02 35 18.	+ 01 52		15.0	GALAXY
UGC 02114	02 35 18.	+ 01 52	60	14.9	GALAXY E
ZWG 388.071	02 35 18.	+ 02 07		14.5	GALAXY
UGC 02115	02 35 18.	+ 02 07	72	14.5	GALAXY
ZC 0235.3+1832	02 35 18.	+ 18 32	2290		CLUSTER OF GALAXIES
UGC 02116	02 35 18.	+ 30 37	120	16.0	GALAXY S IV
UGC 02117	02 35 18.	+ 32 59	66	16.5	GALAXY Sc-IRR
ZWG 539.063	02 35 18.	+ 41 20		14.9	GALAXY
UGC 02118	02 35 18.	+ 41 20	114	14.9	GALAXY S0
UGC 02113	02 35 18.	- 01 54	78	16.5	GALAXY
PHL 8416	02 35 18.	- 04 39		17.7	BLUE STELLAR OBJECT
PHL 8417	02 35 18.	- 15 07		18.1	BLUE STELLAR OBJECT
PHL 8415	02 35 18.	- 22 40		18.3	BLUE STELLAR OBJECT
MCG-05-07-009	02 35 18.	- 27 39	36	16.	GALAXY
RNGC 0995	02 35 19.	+ 41 20		15.0	GALAXY
REIN 6.011	02 35 19.16	+ 02 06 41.6			NEBULA
REIN 7.036	02 35 19.20	- 01 21 28.6			NEBULA
IC 0241	02 35 20.	+ 02 07			NONSTELLAR OBJECT
REIN 6.012	02 35 20.11	+ 01 51 47.4			NEBULA
KARA.68 021	02 35 21.	+ 29 43	67		DWARF GALAXY
REIN 7.037	02 35 22.82	- 01 57 48.4			NEBULA
RNGC 1014	02 35 23.	- 09 44			NON-EXISTENT OBJECT
RNGC 1017	02 35 23.	- 11 13		14.0	GALAXY
ZWG 388.073	02 35 24.	+ 01 28		15.0	GALAXY
UGC 02121	02 35 24.	+ 01 28	102	15.0	GALAXY Sc
MCG+00-07-060	02 35 24.	+ 01 52	36	15.	GALAXY
MCG+05-07-023	02 35 24.	+ 29 33	48	14.5	GALAXY
LB 02765	02 35 24.	+ 30 51 54.		16.9	FAINT BLUE STAR
MCG+07-06-044	02 35 24.	+ 41 18	72	15.	GALAXY
PHL 4213	02 35 24.	- 01 50		18.0	BLUE STELLAR OBJECT
ZWG 388.072	02 35 24.	- 02 04		14.5	GALAXY
UGC 02119	02 35 24.	- 02 04	132	14.5	GALAXY SBb
MCG+00-07-061	02 35 24.	- 02 05	90	14.	GALAXY
PHL 4216	02 35 24.	- 04 02		18.4	BLUE STELLAR OBJECT
PHL 4214	02 35 24.	- 05 00		18.8	BLUE STELLAR OBJECT
PHL 8419	02 35 24.	- 09 26		18.5	BLUE STELLAR OBJECT
RNGC 1013	02 35 24.	- 11 43		14.0	GALAXY
MCG-02-07-046	02 35 24.	- 11 43	24	14.5	GALAXY
MCG-03-07-042	02 35 24.	- 16 00	72	14.5	GALAXY
PHL 1403	02 35 24.	- 20 22		18.0	BLUE STELLAR OBJECT
PHL 4215	02 35 24.	- 22 16		18.0	BLUE STELLAR OBJECT
PHL 4212	02 35 24.	- 23 04		17.8	BLUE STELLAR OBJECT
PHL 8418	02 35 24.	- 28 58		16.1	BLUE STELLAR OBJECT
MCG-06-06-015	02 35 24.	- 33 08	90	13.	GALAXY
RNGC 1037	02 35 26.	- 02 03		14.5	GALAXY
REIN 7.038	02 35 26.29	- 02 03 34.3			NEBULA
MCG+00-07-062	02 35 27.	+ 01 27 30.	66	14.5	GALAXY
MCG-02-07-047	02 35 27.	- 11 13 30.	36	14.5	GALAXY
HOLM 062C	02 35 28.	- 11 14	18	14.5	PART OF MULTIPLE GALAXY
REIN 6.014	02 35 28.88	+ 01 28 16.6			NEBULA
ZWG 388.074	02 35 30.	+ 01 10		15.1	GALAXY
UGC 02120	02 35 30.	+ 01 10	96	15.1	GALAXY SB:b-c
MCG+00-07-063	02 35 30.	+ 01 54	36	15.	GALAXY
ZWG 505.026	02 35 30.	+ 29 32		14.7	GALAXY
UGC 02122	02 35 30.	+ 29 32	72	14.7	GALAXY Sc/SBc
ZWG 523.078	02 35 30.	+ 34 53		15.7	GALAXY
6ZW 217	02 35 30.	+ 37 22			COMPACT GALAXY
MCG+07-06-045	02 35 30.	+ 41 25 30.	21	14.5	GALAXY
ZWG 539.064	02 35 30.	+ 41 27		14.5	GALAXY
UGC 02123	02 35 30.	+ 41 28	102	14.5	GALAXY E
1ZW 007	02 35 30.	- 09 38			COMPACT GALAXY
RNGC 0996	02 35 31.	+ 41 27		14.5	GALAXY
LB 03625	02 35 32.	+ 40 56 24.		18.8	FAINT BLUE STAR
SVEN 115	02 35 33.	- 08 17	24	14.8	GALAXY
RNGC 1018	02 35 35.	- 09 46		15.0	GALAXY
PHL 4217	02 35 36.	+ 01 06		18.6	BLUE STELLAR OBJECT
MCG+00-07-064	02 35 36.	+ 01 10	48	15.	GALAXY
UGC 02124	02 35 36.	+ 01 32	198	13.5	GALAXY SBa
MCG+00-07-065	02 35 36.	+ 02 06	90	14.5	GALAXY
ZWG 505.027	02 35 36.	+ 31 51		14.8	GALAXY
UGC 02125	02 35 36.	+ 31 51	138	14.8	GALAXY SBc
MCG+05-07-024	02 35 36.	+ 31 52 30.	120	15.	GALAXY
ZWG 539.065	02 35 36.	+ 40 30		15.4	GALAXY
UGC 02126	02 35 36.	+ 40 30	102	15.4	GALAXY
ZWG 539.066	02 35 36.	+ 41 28		14.5	GALAXY
UGC 02127	02 35 36.	+ 41 28	66	14.5	GALAXY Sa/SBa
VB 223	02 35 36.	+ 53 26	289		STELLAR RING
OCL 0367	02 35 36.	+ 54 43	180		OPEN STAR CLUSTER
ZWG 388.075	02 35 36.	- 01 32		13.5	GALAXY
MCG+00-07-066	02 35 36.	- 01 33	126	13.	GALAXY
PHL 8420	02 35 36.	- 02 17		16.5	BLUE STELLAR OBJECT
HELW 382	02 35 36.	- 08 17 03.			NEBULA
PHL 4218	02 35 36.	- 08 44		18.8	BLUE STELLAR OBJECT
PHL 8421	02 35 36.	- 20 29		18.4	BLUE STELLAR OBJECT
PHL 1404	02 35 36.	- 20 56		18.4	BLUE STELLAR OBJECT
IC 1823	02 35 37.	+ 31 50 49.			NONSTELLAR OBJECT
RNGC 0999	02 35 37.	+ 41 28		14.5	GALAXY
RNGC 1015	02 35 38.	- 01 32		13.5	GALAXY
REIN 7.039	02 35 38.81	- 01 32 03.5			NEBULA
MCG+07-06-046	02 35 39.	+ 40 28	48	15.	GALAXY
MCG+07-06-047	02 35 39.	+ 41 27	45	15.	GALAXY
SVEN 116	02 35 39.	- 08 35	24	14.5	GALAXY
HELW 383	02 35 39.	- 08 35 03.			NEBULA
MCG-02-07-048	02 35 39.	- 09 46 30.	66	15.	GALAXY
LB 02766	02 35 41.	+ 33 05 00.		15.9	FAINT BLUE STAR
ZWG 388.076	02 35 42.	+ 01 55		13.3	GALAXY
RNGC 1016	02 35 42.	+ 01 55		13.5	GALAXY
UGC 02128	02 35 42.	+ 01 55	168	13.3	GALAXY E
ZWG 388.077	02 35 42.	+ 02 06		15.4	GALAXY
RNGC 1009	02 35 42.	+ 02 06		15.5	GALAXY
UGC 02129	02 35 42.	+ 02 06	102	15.4	GALAXY Sb
ZWG 414.029	02 35 42.	+ 07 47		15.4	GALAXY
UGC 02130	02 35 42.	+ 07 47	72	15.4	GALAXY S
KARA.73B 0113	02 35 42.	+ 07 47	66	15.4	ISOLATED GALAXY S
ZWG 505.028	02 35 42.	+ 33 14		15.6	GALAXY
UGC 02131	02 35 42.	+ 33 14	102	15.6	GALAXY SBc
MCG+05-07-025	02 35 42.	+ 33 15	48	15.5	GALAXY
MCG+07-06-048	02 35 42.	+ 41 14	36	16.5	GALAXY
ZWG 539.067	02 35 42.	+ 41 15		15.6	GALAXY
URA 21A	02 35 42.	+ 63 24	72		STELLAR RING
PHL 8422	02 35 42.	- 29 52		18.4	BLUE STELLAR OBJECT
REIN 6.015	02 35 42.98	+ 01 43 13.6			NEBULA
RNGC 1000	02 35 43.	+ 41 15		15.5	GALAXY
REIN 6.016	02 35 43.75	+ 02 05 37.8			NEBULA
REIN 6.017	02 35 44.42	+ 01 54 12.2			NEBULA
MCG+00-07-067	02 35 45.	+ 01 54	36	12.	GALAXY
MCG+00-07-068	02 35 48.	+ 01 41	54	14.	GALAXY
ZWG 388.079	02 35 48.	+ 01 42		14.6	GALAXY
UGC 02132	02 35 48.	+ 01 42	78	14.6	GALAXY SBb
ZWG 388.080	02 35 48.	+ 01 45		15.6	GALAXY
MCG+06-06-070	02 35 48.	+ 34 24	60	14.	GALAXY
UGC 02169	02 35 48.	+ 37 01	120	15.5	GALAXY S0
URA 21B	02 35 48.	+ 63 23	174		STELLAR RING
VB 083	02 35 48.	+ 63 30	604		STELLAR RING
ZWG 388.078	02 35 48.	- 00 24		15.6	GALAXY
PHL 8423	02 35 48.	- 10 44		18.0	BLUE STELLAR OBJECT
PHL 4220	02 35 48.	- 14 40		18.7	BLUE STELLAR OBJECT

OBJECT NAME	RIGHT ASCEN.	DECLINATION	DIAM.	MAGN.	TYPE OF OBJECT
MCG-03-07-043	02 35 48.	- 16 01 30.	36	14.	GALAXY
MCG-03-07-044	02 35 48.	- 20 25	24	13.5	GALAXY
PHL 4221	02 35 48.	- 27 56		18.3	BLUE STELLAR OBJECT
PHL 4219	02 35 48.	- 30 54		17.7	BLUE STELLAR OBJECT
LB 03270	02 35 48.	- 52 27		15.4	FAINT BLUE STAR
RNGC 1019	02 35 49.	+ 01 42		14.5	GALAXY
RNGC 1034	02 35 49.	- 16 01		14.0	GALAXY
REIN 6.018	02 35 50.29	+ 01 45 06.1			NEBULA
REIN 6.019	02 35 52.41	+ 01 41 31.7			NEBULA
RNGC 1002	02 35 53.	+ 34 25		14.0	GALAXY
IC 0240	02 35 53.	+ 41 31			NONSTELLAR OBJECT
MCG+05-07-026	02 35 54.	+ 27 38	90	14.5	GALAXY
ZWG 523.079	02 35 54.	+ 34 25		14.0	GALAXY
UGC 02133	02 35 54.	+ 34 25	96	14.0	GALAXY SBb
ZWG 523.080	02 35 54.	+ 35 23		15.5	GALAXY
PHL 8425	02 35 54.	- 00 39		18.7	BLUE STELLAR OBJECT
MCG+00-07-069	02 35 54.	- 01 16	12	16.	GALAXY
PHL 4222	02 35 54.	- 10 41		18.5	BLUE STELLAR OBJECT
PHL 8424	02 35 54.	- 10 53		17.7	BLUE STELLAR OBJECT
MCG-03-07-045	02 35 54.	- 20 24	60	15.	GALAXY
PHL 1405	02 35 54.	- 29 46		18.0	BLUE STELLAR OBJECT
TON-S 0263	02 35 54.	- 32 24		15.9	BLUE STAR
REIN 7.040	02 35 55.15	- 01 15 43.2			NEBULA
REIN 7.041	02 35 55.81	- 01 15 58.3			NEBULA
IC 0242	02 35 56.	- 07 08 45.			NONSTELLAR OBJECT
LBN 0663	02 36	+ 59 22	60		BRIGHT NEBULA
ZWG 414.030	02 36 00.	+ 03 50		15.3	GALAXY
KARA.73B 0114	02 36 00.	+ 03 50	60	15.3	ISOLATED GALAXY S
FEIG 025	02 36 00.	+ 05 15		11.6	FAINT BLUE STAR
ZWG 505.029	02 36 00.	+ 27 37		14.4	GALAXY
UGC 02134	02 36 00.	+ 27 37	108	14.4	GALAXY Sb
MCG+07-06-049	02 36 00.	+ 41 01	15	15.	GALAXY
ZWG 539.068	02 36 00.	+ 41 02		15.7	GALAXY
UGC 02135	02 36 00.	+ 41 02	90	15.7	GALAXY Sc
ZWG 539.069	02 36 00.	+ 41 28		14.7	GALAXY
RNGC 1001	02 36 00.	+ 41 28		14.5	GALAXY
MCG+01-07-025	02 36 00.	- 06 54	132	12.0	GALAXY
MCG+01-07-026	02 36 00.	- 07 08	54	15.	GALAXY
PHL 8426	02 36 00.	- 09 52		18.6	BLUE STELLAR OBJECT
PHL 4224	02 36 00.	- 11 26		16.7	BLUE STELLAR OBJECT
MCG-05-07-010	02 36 00.	- 27 27	24	16.	GALAXY
PHL 4223	02 36 00.	- 29 22		18.7	BLUE STELLAR OBJECT
HELW 384	02 36 02.	- 08 19 22.			NEBULA
MCG+07-06-050	02 36 03.	+ 41 26 30.	39	15.	GALAXY
SVEN 117	02 36 03.	- 08 19 28.	18	15.1	GALAXY
RNGC 1022	02 36 04.	- 06 52		12.5	GALAXY
IC 0243	02 36 04.	- 07 06 45.			NONSTELLAR OBJECT
REIN 2.026	02 36 04.11	- 06 53 35.3			NEBULA
ZWG 388.081	02 36 06.	+ 02 01		15.0	GALAXY
RNGC 1020	02 36 06.	+ 02 01		15.0	GALAXY
UGC 02136	02 36 06.	+ 06 42	66	16.0	GALAXY Sb
MCG+07-06-051	02 36 06.	+ 40 39	270	13.	GALAXY
ZWG 539.070	02 36 06.	+ 40 40		12.1	GALAXY
UGC 02137	02 36 06.	+ 40 40	420	12.1	GALAXY Sc
PHL 1406	02 36 06.	- 13 44		17.9	BLUE STELLAR OBJECT
PHL 1407	02 36 06.	- 29 26		18.7	BLUE STELLAR OBJECT
RNGC 1003	02 36 07.	+ 40 39		12.0	GALAXY
SN 1937D	02 36 08.	+ 40 40		12.8	SUPERNOVA
SS 05	02 36 08.	+ 59 23			DIFFUSE GALACTIC NEBULA
REIN 6.020	02 36 09.09	+ 02 00 56.3			NEBULA
MCG+00-07-070	02 36 12.	+ 00 18 30.	9	17.	GALAXY
ZWG 388.082	02 36 12.	+ 01 43		15.7	GALAXY
ZWG 388.083	02 36 12.	+ 01 50		15.7	GALAXY
ZWG 388.084	02 36 12.	+ 02 00		15.1	GALAXY
RNGC 1021	02 36 12.	+ 02 00		15.0	GALAXY
PHL 1408	02 36 12.	+ 02 42		18.6	BLUE STELLAR OBJECT
MCG+01-07-017	02 36 12.	+ 07 52	72	14.	GALAXY
ZWG 414.031	02 36 12.	+ 08 53		14.9	GALAXY
UGC 02138	02 36 12.	+ 08 53	84	14.9	GALAXY Sc
KARA.73B 0115	02 36 12.	+ 08 53	72	14.9	ISOLATED GALAXY S
ZWG 439.021	02 36 12.	+ 10 05		15.5	GALAXY
ZC 0236.2+3249	02 36 12.	+ 32 49	12100		CLUSTER OF GALAXIES
ZC 0236.2+3417	02 36 12.	+ 34 17	1080		CLUSTER OF GALAXIES
UGC 02139	02 36 12.	+ 36 12	60	18.	GALAXY DWARF
5ZW 264	02 36 12.	+ 45 48			COMPACT GALAXY
PHL 8428	02 36 12.	- 00 37		18.6	BLUE STELLAR OBJECT
PHL 8427	02 36 12.	- 08 59		17.0	BLUE STELLAR OBJECT
PHL 8429	02 36 12.	- 09 38		18.3	BLUE STELLAR OBJECT
PHL 4225	02 36 12.	- 10 06		18.9	BLUE STELLAR OBJECT
PHL 8433	02 36 12.	- 13 55		18.3	BLUE STELLAR OBJECT
PHL 8434	02 36 12.	- 16 48		18.5	BLUE STELLAR OBJECT
PHL 4226	02 36 12.	- 18 12		18.4	BLUE STELLAR OBJECT
MCG-05-07-011	02 36 12.	- 27 25	24	15.5	GALAXY
PHL 8430	02 36 12.	- 27 56		18.9	BLUE STELLAR OBJECT
PHL 8431	02 36 12.	- 28 24		18.6	BLUE STELLAR OBJECT
PHL 8435	02 36 12.	- 28 42		18.9	BLUE STELLAR OBJECT
PHL 8432	02 36 12.	- 31 46		18.0	BLUE STELLAR OBJECT
REIN 6.021	02 36 12.81	+ 02 00 05.0			NEBULA
MCG+00-07-071	02 36 15.	+ 00 17 30.	24	16.	GALAXY
IC 1825	02 36 15.	+ 08 52 57.			NONSTELLAR OBJECT
VV 143	02 36 15.	+ 18 10	108	14.	INTERACTING GALAXY
MCG+05-07-027	02 36 15.	+ 29 57 30.	132	14.	GALAXY
IC 1826	02 36 15.	- 27 39 23.			NONSTELLAR OBJECT
ARP 258	02 36 17.	+ 18 10			PECULIAR GALAXY
ZWG 414.032	02 36 18.	+ 08 29		15.7	GALAXY
ZWG 462.037	02 36 18.	+ 18 10		14.6	GALAXY
UGC 02140	02 36 18.	+ 18 10	120	14.6	GALAXY IRR
MCG+03-07-037	02 36 18.	+ 18 11	90	14.5	GALAXY
ZWG 505.030	02 36 18.	+ 29 56		13.1	GALAXY
UGC 02141	02 36 18.	+ 29 56	174	13.1	GALAXY S-IRR
MCG+07-06-052	02 36 18.	+ 41 16	18	15.	GALAXY
ZWG 539.071	02 36 18.	+ 41 17		14.7	GALAXY
5ZW 265	02 36 18.	+ 42 19			COMPACT GALAXY
MRSL 136-00/1	02 36 18.	+ 59 26	180		HII REGION
LB 01632	02 36 18.	- 56 14		13.4	FAINT BLUE STAR
RNGC 1012	02 36 19.	+ 29 56		13.0	GALAXY
RNGC 1005	02 36 19.	+ 41 17		14.5	GALAXY
SVEN 118	02 36 21.	- 08 46	12	15.1	GALAXY
HELW 385	02 36 23.	- 08 46 05.			NEBULA
PHL 4228	02 36 24.	+ 01 44		18.4	BLUE STELLAR OBJECT
BIGO 477	02 36 24.	+ 06 23			NEBULA
MCG+03-07-038	02 36 24.	+ 18 10	48	15.5	GALAXY
ZWG 539.072	02 36 24.	+ 40 35		15.7	GALAXY
MCG+07-06-053	02 36 24.	+ 41 47	30	15.	GALAXY
ZWG 539.073	02 36 24.	+ 41 48		15.1	GALAXY
PHL 4229	02 36 24.	- 06 25		18.4	BLUE STELLAR OBJECT
PHL 8436	02 36 24.	- 12 40		17.0	BLUE STELLAR OBJECT
PHL 8437	02 36 24.	- 14 16		18.3	BLUE STELLAR OBJECT

OBJECT NAME	RIGHT ASCEN.	DECLINATION	DIAM.	MAGN.	TYPE OF OBJECT
PHL 4230	02 36 24.	- 26 53		17.1	BLUE STELLAR OBJECT
PHL 4227	02 36 24.	- 26 56		17.1	BLUE STELLAR OBJECT
RNGC 1024	02 36 27.	+ 10 38		14.0	GALAXY
ZWG 439.022	02 36 30.	+ 10 38		13.8	GALAXY
UGC 02142	02 36 30.	+ 10 38	300	13.8	GALAXY Sb
MCG+02-07-020	02 36 30.	+ 10 38	258	14.	GALAXY
5ZW 266	02 36 30.	+ 35 52			COMPACT GALAXY
ZWG 523.081	02 36 30.	+ 35 52		14.0	GALAXY
UGC 02143	02 36 30.	+ 35 52	33	14.0	GALAXY PECULR
ZWG 388.085	02 36 30.	- 01 46		15.7	GALAXY
MCG-03-07-046	02 36 30.	- 14 31	42	14.5	GALAXY
ARP 333	02 36 31.	+ 10 38			PECULIAR GALAXY
IC 0245	02 36 32.	- 14 30 54.			NONSTELLAR OBJECT
REIN 7.042	02 36 33.69	- 01 46 42.7			NEBULA
MCG+02-07-021	02 36 36.	+ 12 27	18	17.	GALAXY
UGC 02144	02 36 36.	+ 29 01	72	17.	GALAXY DWRF SP
PHL 1409	02 36 36.	- 09 40		18.9	BLUE STELLAR OBJECT
MCG-03-07-047	02 36 36.	- 14 32	84	15.	GALAXY
MCG-03-07-048	02 36 36.	- 19 34	24	15.	GALAXY
TON-S 0264	02 36 36.	- 26 55		15.9	BLUE STAR
HOD.61 2	02 36 36.	- 35 01	36	19.6	GLOB CLSTR IN FORNAX GAL
MCG-06-06-016	02 36 36.	- 35 01	72	14.5	GALAXY
RNGC 1026	02 36 41.	+ 06 20		14.0	GALAXY
PHL 8441	02 36 42.	+ 01 09		18.6	BLUE STELLAR OBJECT
ZWG 414.033	02 36 42.	+ 06 20		14.1	GALAXY
UGC 02145	02 36 42.	+ 06 20	120	14.1	GALAXY S0
MCG+01-07-018	02 36 42.	+ 06 20	24	13.5	GALAXY
MCG+02-07-022	02 36 42.	+ 12 27	72	16.	GALAXY
3ZW 051	02 36 42.	+ 12 28			COMPACT GALAXY
ZWG 439.023	02 36 42.	+ 12 28		15.7	GALAXY
ZWG 462.038	02 36 42.	+ 17 25		15.5	GALAXY
UGC 02146	02 36 42.	+ 42 53	102	17.	GALAXY DWARF
PHL 8439	02 36 42.	- 03 02		16.9	BLUE STELLAR OBJECT
PHL 8438	02 36 42.	- 08 02		15.8	BLUE STELLAR OBJECT
PHL 4231	02 36 42.	- 15 22		18.0	BLUE STELLAR OBJECT
PHL 8440	02 36 42.	- 28 10		17.0	BLUE STELLAR OBJECT
MCG+00-07-073	02 36 45.	+ 00 52 30.	180	12.	GALAXY
MCG+00-07-072	02 36 45.	+ 01 45	48	15.	GALAXY
PHL 1410	02 36 48.	+ 00 48		18.6	BLUE STELLAR OBJECT
ZWG 388.086	02 36 48.	+ 00 53		13.2	GALAXY
UGC 02147	02 36 48.	+ 00 53	246	13.2	GALAXY Sa
ZWG 388.087	02 36 48.	+ 01 45		15.4	GALAXY
IC 0244	02 36 48.	+ 02 29 42.			NONSTELLAR OBJECT
MCG+00-07-074	02 36 48.	+ 02 30 30.	36	15.	GALAXY
ZWG 388.088	02 36 48.	+ 02 31		15.3	GALAXY
MCG+02-07-024	02 36 48.	+ 10 34	72	14.	GALAXY
MCG+02-07-023	02 36 48.	+ 10 37 30.	48	14.5	GALAXY
UGC 02148	02 36 48.	+ 12 28	66	16.0	GALAXY S
MCG+05-07-028	02 36 48.	+ 28 07	72	15.	GALAXY
SG 3.013	02 36 48.	+ 60 46	2700		DIFFUSE EMISSION NEBULA
URA 32	02 36 48.	+ 89 32	60		STELLAR RING
PHL 1412	02 36 48.	- 00 14		16.1	BLUE STELLAR OBJECT
PHL 1411	02 36 48.	- 01 28		18.3	BLUE STELLAR OBJECT
PHL 8443	02 36 48.	- 06 48		16.5	BLUE STELLAR OBJECT
PHL 8442	02 36 48.	- 07 49		15.5	BLUE STELLAR OBJECT
PHL 4233	02 36 48.	- 10 34		18.5	BLUE STELLAR OBJECT
MCG-04-07-013	02 36 48.	- 22 52 30.	102	14.	GALAXY
PHL 8444	02 36 48.	- 24 14		17.9	BLUE STELLAR OBJECT
MCG-05-07-012	02 36 48.	- 27 39	78	13.5	GALAXY
PHL 4232	02 36 48.	- 28 28		17.9	BLUE STELLAR OBJECT
PHL 8445	02 36 48.	- 30 08		18.8	BLUE STELLAR OBJECT
RNGC 1032	02 36 49.	+ 00 53		13.0	GALAXY
REIN 2.027A	02 36 49.07	+ 00 52 42.8			NEBULA
REIN 2.027B	02 36 49.14	+ 00 52 42.2			NEBULA
LB 02767	02 36 50.	+ 32 45 18.		16.5	FAINT BLUE STAR
RNGC 1029	02 36 51.	+ 10 35		14.0	GALAXY
RNGC 1028	02 36 51.	+ 10 38		15.5	GALAXY
HARO 18	02 36 52.	- 27 38			BLUE EMISSION-LINE GALAXY
IC 1830	02 36 52.3	- 27 39 31.			GALAXY
ZL 018	02 36 53.	- 01 56 12.		16.0	ULTRAFAINT BLUE STAR
PHL 1413	02 36 54.	+ 00 07		4.0	BLUE STELLAR OBJECT
ZWG 439.024	02 36 54.	+ 10 35		14.1	GALAXY
UGC 02149	02 36 54.	+ 10 35	96	14.9	GALAXY S0-a
ZWG 439.025	02 36 54.	+ 10 38		15.3	GALAXY
MCG+06-06-071	02 36 54.	+ 38 18	96	14.5	GALAXY
URA 31	02 36 54.	+ 69 28	72		STELLAR RING
PHL 8446	02 36 54.	- 02 17		8.1	BLUE STELLAR OBJECT
PHL 1414	02 36 54.	- 11 40		17.6	BLUE STELLAR OBJECT
PHL 4234	02 36 54.	- 12 11		17.4	BLUE STELLAR OBJECT
HN 0162	02 36 54.	- 27 40			NEBULA
REIN 7.043	02 36 55.31	- 01 29 28.7			NEBULA
RNGC 1035	02 36 58.	- 08 19		13.0	GALAXY
VDB.66G 027	02 37	+ 01 05	70		DWARF GALAXY
LBN 0658	02 37	+ 61 00	3900		BRIGHT NEBULA
KHAV 038	02 37	+ 61 13	1990		DARK NEBULA
PHL 4236	02 37 00.	+ 01 09		8.6	BLUE STELLAR OBJECT
UGC 02150	02 37 00.	+ 09 41	72	18.	GALAXY DWRF SP
MCG+03-07-039	02 37 00.	+ 17 49	90	14.	GALAXY
ZWG 505.031	02 37 00.	+ 28 06		14.8	GALAXY
UGC 02151	02 37 00.	+ 28 06	120	14.8	GALAXY Sc
5ZW 267	02 37 00.	+ 35 57			COMPACT GALAXY
MCG+11-04-001	02 37 00.	+ 66 16	48	16.	GALAXY
PHL 4235	02 37 00.	- 03 28		18.5	BLUE STELLAR OBJECT
PHL 8447	02 37 00.	- 03 35		16.6	BLUE STELLAR OBJECT
MCG+01-07-027	02 37 00.	- 08 21	120	12.8	GALAXY
PHL 8448	02 37 00.	- 13 50		16.6	BLUE STELLAR OBJECT
PHL 4237	02 37 00.	- 19 20		18.1	BLUE STELLAR OBJECT
PHL 8449	02 37 00.	- 23 50		16.6	BLUE STELLAR OBJECT
PHL 8450	02 37 00.	- 27 33		16.3	BLUE STELLAR OBJECT
REIN 7.044	02 37 00.53	- 01 51 00.5			NEBULA
KEEL 092	02 37 00.8	+ 38 32 27.			NEBULA
LB 00469	02 37 01.	+ 23 41 30.		15.6	FAINT BLUE STAR
ZL 019	02 37 01.	- 01 42 42.		21.8	ULTRAFAINT BLUE STAR
ZL 020	02 37 02.	- 01 46 12.		22.0	ULTRAFAINT BLUE STAR
SVEN 119	02 37 03.	- 08 21	114	12.4	GALAXY
KARA.73 05A	02 37 03.	- 34 44	336		DWARF GALAXY
KEEL 093	02 37 04.2	+ 39 02 15.			NEBULA
ZL 021	02 37 05.	- 01 38 30.		21.5	ULTRAFAINT BLUE STAR
ZWG 388.089	02 37 06.	+ 01 20		14.5	GALAXY
UGC 02152	02 37 06.	+ 01 20	72	14.5	GALAXY Sa
ZC 0237.1+0232	02 37 06.	+ 02 32	1340		CLUSTER OF GALAXIES
ZWG 462.039	02 37 06.	+ 17 49		14.5	GALAXY
RNGC 1030	02 37 06.	+ 17 49		14.5	GALAXY
UGC 02153	02 37 06.	+ 17 49	102	14.5	GALAXY S?
ZC 0237.1+1927	02 37 06.	+ 19 27	1410		CLUSTER OF GALAXIES
ZWG 523.082	02 37 06.	+ 38 08		15.6	GALAXY
MCG+06-06-072	02 37 06.	+ 38 09	18	15.5	GALAXY
VB 084	02 37 06.	+ 69 14	530		STELLAR RING

OBJECT NAME	RIGHT ASCEN.	DECLINATION	DIAM.	MAGN.	TYPE OF OBJECT
ISS 0020	02 37 06.	+ 69 14	53		STELLAR RING
PHL 8451	02 37 06.	- 01 08		19.0	BLUE STELLAR OBJECT
PHL 4238	02 37 06.	- 12 12		18.5	BLUE STELLAR OBJECT
PHL 4239	02 37 06.	- 17 00		18.3	BLUE STELLAR OBJECT
MCG-03-07-049	02 37 06.	- 20 05	48	15.	GALAXY
TON-S 0265	02 37 06.	- 32 11		15.8	BLUE STAR
LB 03271	02 37 06.	- 67 23		16.0	FAINT BLUE STAR
ZL 022	02 37 07.	- 01 36 00.		21.6	ULTRAFAINT BLUE STAR
IC 1827	02 37 09.	+ 01 20 46.			NONSTELLAR OBJECT
ARC 0369	02 37 10.	- 03 45		17.8	RICH CLUSTER OF GALAXIES
MCG+00-07-075	02 37 12.	+ 01 20	60	15.	GALAXY
ZC 0237.2+2514	02 37 12.	+ 25 14	340		CLUSTER OF GALAXIES
ZWG 523.083	02 37 12.	+ 38 51		10.5	GALAXY
UGC 02154	02 37 12.	+ 38 51	510	10.5	GALAXY SB0
LB 03565	02 37 12.	+ 42 18 06.		19.5	FAINT BLUE STAR
ZC 0237.2-0146	02 37 12.	- 01 46	670		CLUSTER OF GALAXIES
PHL 4240	02 37 12.	- 14 51		18.1	BLUE STELLAR OBJECT
PHL 1415	02 37 12.	- 18 10		16.8	BLUE STELLAR OBJECT
PHL 8453	02 37 12.	- 18 28		18.5	BLUE STELLAR OBJECT
PHL 8452	02 37 12.	- 22 33		16.6	BLUE STELLAR OBJECT
ZL 023	02 37 13.	- 01 46 00.		21.5	ULTRAFAINT BLUE STAR
KEEL 094	02 37 14.9	+ 38 56 04.			NEBULA
ARP 135	02 37 15.	+ 38 52			PECULIAR GALAXY
KEEL 095	02 37 16.6	+ 38 56 10.			NEBULA
ZC 0237.3+0338	02 37 18.	+ 03 38	1080		CLUSTER OF GALAXIES
UGC 02155	02 37 18.	+ 14 10	66	16.0	GALAXY SBb
ZWG 505.032	02 37 18.	+ 32 02		14.5	GALAXY
UGC 02156	02 37 18.	+ 32 02	132	14.5	GALAXY Sc
MCG+05-07-029	02 37 18.	+ 32 05	90	14.	GALAXY
ZWG 523.084	02 37 18.	+ 38 21		14.9	GALAXY
UGC 02157	02 37 18.	+ 38 21	120	14.9	GALAXY
URA 88	02 37 18.	+ 61 28	540		STELLAR RING
PHL 4241	02 37 18.	- 00 44		18.5	BLUE STELLAR OBJECT
ZL 024	02 37 18.	- 01 52 00.		21.3	ULTRAFAINT BLUE STAR
MCG-02-07-049	02 37 18.	- 10 49 30.	54	15.	GALAXY
PHL 4242	02 37 18.	- 15 40		18.3	BLUE STELLAR OBJECT
TON-S 0266	02 37 18.	- 27 31		15.5	BLUE STAR
PHL 8455	02 37 18.	- 28 14		18.8	BLUE STELLAR OBJECT
PHL 8454	02 37 18.	- 32 09		16.2	BLUE STELLAR OBJECT
KEEL 096	02 37 18.5	+ 38 20 39.			SPIRAL NEBULA
ARC 0370	02 37 19.	- 01 48		17.8	RICH CLUSTER OF GALAXIES
RNGC 1023	02 37 20.	+ 38 51		11.0	GALAXY
MCG-03-07-050	02 37 21.	- 16 13 30.	60	15.	GALAXY
ZL 025	02 37 22.	- 01 47 24.		22.8	ULTRAFAINT BLUE STAR
REIN 7.045	02 37 22.03	- 01 45 25.9			NEBULA
LB 03566	02 37 23.	+ 42 45 30.		19.8	FAINT BLUE STAR
IC 1824	02 37 23.	+ 61 24			OPEN CLUSTER
REIN 7.046	02 37 23.77	- 01 44 28.0			NEBULA
ZWG 462.040	02 37 24.	+ 17 40		14.9	GALAXY
MCG+03-07-040	02 37 24.	+ 17 40	45	15.	GALAXY
MCG+06-06-073	02 37 24.	+ 38 52	210	10.	GALAXY
PHL 8456	02 37 24.	- 07 12		16.5	BLUE STELLAR OBJECT
PHL 8457	02 37 24.	- 07 26		16.1	BLUE STELLAR OBJECT
PHL 1416	02 37 24.	- 13 28		17.8	BLUE STELLAR OBJECT
PHL 4244	02 37 24.	- 13 50		19.0	BLUE STELLAR OBJECT
PHL 1417	02 37 24.	- 17 52		14.5	BLUE STELLAR OBJECT
PHL 1418	02 37 24.	- 20 42		18.0	BLUE STELLAR OBJECT
MCG-04-07-014	02 37 24.	- 21 26	42	14.5	GALAXY
PHL 1419	02 37 24.	- 23 12		17.3	BLUE STELLAR OBJECT
PHL 4243	02 37 24.	- 27 44		17.9	BLUE STELLAR OBJECT
PHL 8458	02 37 24.	- 28 15		18.3	BLUE STELLAR OBJECT
LB 03567	02 37 27.	+ 42 39 42.		19.6	FAINT BLUE STAR
MCG+00-07-076	02 37 30.	+ 01 17 30.	72	14.	GALAXY
ZWG 388.090	02 37 30.	+ 01 18		14.4	GALAXY
UGC 02158	02 37 30.	+ 01 18	84	14.4	GALAXY Sa
FEIG 026	02 37 30.	+ 03 43		12.9	FAINT BLUE STAR
MCG+01-07-019	02 37 30.	+ 06 45	24	16.	GALAXY
UGC 02159	02 37 30.	+ 29 52	66	16.5	GALAXY Sc
LB 03568	02 37 30.	+ 42 33 36.		18.7	FAINT BLUE STAR
PHL 4246	02 37 30.	- 03 07		18.6	BLUE STELLAR OBJECT
PHL 4245	02 37 30.	- 05 47		18.6	BLUE STELLAR OBJECT
PHL 4247	02 37 30.	- 13 48		18.8	BLUE STELLAR OBJECT
RNGC 1038	02 37 31.	+ 01 18		14.5	GALAXY
REIN 6.022A	02 37 31.47	+ 01 17 39.9			NEBULA
REIN 6.022B	02 37 31.53	+ 01 17 40.3			NEBULA
MCG-02-07-050	02 37 33.	- 14 25	36	15.	GALAXY
ZL 026	02 37 34.	- 01 40 12.		20.5	ULTRAFAINT BLUE STAR
PHL 4248	02 37 36.	+ 00 23		18.2	BLUE STELLAR OBJECT
PHL 1420	02 37 36.	+ 01 40		18.5	BLUE STELLAR OBJECT
FEIG 027	02 37 36.	+ 08 48		11.4	FAINT BLUE STAR
ZWG 462.041	02 37 36.	+ 19 05		13.5	GALAXY
MRK 370	02 37 36.	+ 19 05	30	14.5	GALAXY WITH UV CONTINUUM
RNGC 1036	02 37 36.	+ 19 05		13.5	GALAXY
UGC 02160	02 37 36.	+ 19 05	96	13.5	GALAXY PECULR
MCG+03-07-041	02 37 36.	+ 19 05	90	13.5	GALAXY
5ZW 268	02 37 36.	+ 35 06			COMPACT GALAXY
ZWG 523.085	02 37 36.	+ 35 06		15.7	GALAXY
ZC 0237.6+3618	02 37 36.	+ 36 18	870		CLUSTER OF GALAXIES
ZWG 539.074	02 37 36.	+ 43 36		15.4	GALAXY
UGC 02161	02 37 36.	+ 43 36	132	15.4	GALAXY S0
VB 085	02 37 36.	+ 63 43	457		STELLAR RING
PHL 4249	02 37 36.	- 00 33		18.5	BLUE STELLAR OBJECT
PHL 1421	02 37 36.	- 05 48		18.0	BLUE STELLAR OBJECT
PHL 4250	02 37 36.	- 11 18		18.8	BLUE STELLAR OBJECT
PHL 8459	02 37 36.	- 25 27		17.6	BLUE STELLAR OBJECT
PHL 8460	02 37 36.	- 31 56		17.9	BLUE STELLAR OBJECT
PHL 1422	02 37 36.	- 32 02		17.2	BLUE STELLAR OBJECT
MCG-06-06-017	02 37 36.	- 34 29	72		GALAXY
IC 1828	02 37 37.	+ 19 06 51.			NONSTELLAR OBJECT
REIN 7.047	02 37 38.61	- 01 59 18.3			NEBULA
LB 00470	02 37 39.	+ 24 49 36.		17.6	FAINT BLUE STAR
MCG+07-06-054	02 37 39.	+ 43 36	54	15.5	GALAXY
RNGC 1049	02 37 40.	- 34 29			GLOB.CL. IN EXTRNL GALAXY
RNGC 1033	02 37 41.	- 09 00		14.0	GALAXY
MCG+00-07-077	02 37 42.	+ 01 01	78	17.	GALAXY
ZWG 388.091	02 37 42.	+ 02 13		15.6	GALAXY
5ZW 269	02 37 42.	+ 32 30			COMPACT GALAXY
MCG+01-07-028	02 37 42.	- 02 56	120	15.5	GALAXY
PHL 1423	02 37 42.	- 03 45		17.9	BLUE STELLAR OBJECT
MCG-02-07-052	02 37 42.	- 11 57	72	14.	GALAXY
MCG-02-07-051	02 37 42.	- 12 43	24	15.	GALAXY
IC 0247	02 37 43.	- 11 56 24.			NONSTELLAR OBJECT
KARA.68 022	02 37 45.	+ 01 00	67		DWARF GALAXY
ZL 027	02 37 45.	- 01 33 36.		17.8	ULTRAFAINT BLUE STAR
MCG-02-07-053	02 37 45.	- 09 00	72	14.	GALAXY
ZL 028	02 37 47.	- 01 49 48.		18.7	ULTRAFAINT BLUE STAR
REIN 7.048	02 37 47.90	- 01 39 33.0			NEBULA
UGC 02162	02 37 48.	+ 01 01	120	18.	GALAXY DWRF IR

OBJECT NAME	RIGHT ASCEN.	DECLINATION	DIAM.	MAGN.	TYPE OF OBJECT
ZWG 388.093	02 37 48.	+ 02 16		15.5	GALAXY
MCG+00-07-078	02 37 48.	+ 02 16	36	15.5	GALAXY
ZWG 439.026	02 37 48.	+ 14 05		15.7	GALAXY
MCG+02-07-025	02 37 48.	+ 15 29	60	15.5	GALAXY
5ZW 270	02 37 48.	+ 38 29			COMPACT GALAXY
ZWG 523.086	02 37 48.	+ 38 30		15.7	GALAXY
ZWG 539.075	02 37 48.	+ 41 01		15.6	GALAXY
ZC 0237.8-0120	02 37 48.	- 01 20	810		CLUSTER OF GALAXIES
ZWG 388.092	02 37 48.	- 01 40		15.7	GALAXY
MCG-02-07-054	02 37 48.	- 08 38	330	12.	GALAXY
PHL 8461	02 37 48.	- 10 08		16.5	BLUE STELLAR OBJECT
PHL 4251	02 37 48.	- 15 08		18.2	BLUE STELLAR OBJECT
PHL 8462	02 37 48.	- 23 22		17.1	BLUE STELLAR OBJECT
HOD.61 3	02 37 48.	- 34 29	36	19.6	GLOB CLSTR IN FORNAX GALX
LB 03272	02 37 48.	- 50 50		14.3	FAINT BLUE STAR
REIN 7.049	02 37 48.13	- 01 39 58.7			NEBULA
HELW 386	02 37 49.	- 08 59 39.			NEBULA
SG 3.014	02 37 51.	+ 61 43	6600		DIFFUSE EMISSION NEBULA
SG 1.07	02 37 51.	+ 61 43			DIFFUSE EMISSION NEBULA
RNGC 1041	02 37 51.	- 05 40		14.0	GALAXY
SVEN 120	02 37 51.	- 08 59 33.	36	14.4	GALAXY
KEEL 098	02 37 51.4	+ 38 29 14.			NEBULA
KEEL 097	02 37 51.7	+ 38 51 25.			NEBULA
IC 1829	02 37 52.	+ 19 05 55.			NONSTELLAR OBJECT
SHB 066	02 37 52.7	- 23 22 06.		16.6	QUASI-STELLAR OBJECT
BC PKS0237-23	02 37 52.72	- 23 22 08.5		16.63	QUASI-STELLAR OBJECT
IC 0246	02 37 53.	+ 02 16 12.			NONSTELLAR OBJECT
ZWG 414.034	02 37 54.	+ 08 38		15.5	GALAXY
MCG+01-07-020	02 37 54.	+ 08 38 30.	60	15.	GALAXY
UGC 02163	02 37 54.	+ 15 30	60	16.0	GALAXY Sb?
MCG+03-07-042	02 37 54.	+ 21 24	42	15.5	GALAXY
MCG+05-07-030	02 37 54.	+ 31 59	36	16.	GALAXY
ZWG 539.076	02 37 54.	+ 43 28		15.6	GALAXY
UGC 02164	02 37 54.	+ 43 28	102	15.6	GALAXY Sb-c
ZL 029	02 37 54.	- 01 51 18.		20.6	ULTRAFAINT BLUE STAR
MCG+01-07-029	02 37 54.	- 02 56	15	15.5	GALAXY
PHL 4252	02 37 54.	- 05 00		18.4	BLUE STELLAR OBJECT
MCG+01-07-030	02 37 54.	- 05 40	30	14.	GALAXY
PHL 1424	02 37 54.	- 05 55		18.5	BLUE STELLAR OBJECT
MCG-02-07-055	02 37 54.	- 12 41 30.	30	16.	GALAXY
MCG-02-07-056	02 37 54.	- 12 42	24	15.5	GALAXY
MCG-04-07-015	02 37 54.	- 25 26	42	15.	GALAXY
LB 03569	02 37 55.	+ 42 35 24.		19.3	FAINT BLUE STAR
SS 06	02 37 56.	+ 59 24			DIFFUSE GALACTIC NEBULA
SVEN 121	02 37 57.	- 08 39	186	13.0	GALAXY
MCG-02-07-057	02 37 57.	- 13 06	15	15.5	GALAXY
LB 03570	02 37 58.	+ 42 31 12.		20.5	FAINT BLUE STAR
RNGC 1042	02 37 58.	- 08 39		12.0	GALAXY
RNGC 1045	02 37 59.	- 11 30		13.0	GALAXY
LBN 0665	02 38	+ 59 21	120		BRIGHT NEBULA
PHL 8463	02 38 00.	+ 02 52		18.5	BLUE STELLAR OBJECT
ZWG 462.042	02 38 00.	+ 21 23		15.5	GALAXY
ZWG 505.033	02 38 00.	+ 31 58		15.7	GALAXY
UGC 02165	02 38 00.	+ 38 32	114	16.0	GALAXY
5ZW 271	02 38 00.	+ 45 38			COMPACT GALAXY
PHL 1425	02 38 00.	- 00 01		18.7	BLUE STELLAR OBJECT
MCG+01-07-031	02 38 00.	- 06 20	60	16.	GALAXY
MCG+01-07-032	02 38 00.	- 08 22	60	15.	GALAXY
MCG-02-07-058	02 38 00.	- 08 45 30.	36	14.5	GALAXY
PHL 1426	02 38 00.	- 09 50		17.9	BLUE STELLAR OBJECT
MCG-02-07-059	02 38 00.	- 11 30	42	13.	GALAXY
VV 200B	02 38 00.	- 13 06	12	16.	INTERACTING GALAXY
VV 200A	02 38 00.	- 13 06	15	15.5	INTERACTING GALAXY
MCG-02-07-060	02 38 00.	- 13 06	18	15.5	GALAXY
MCG-02-07-061	02 38 00.	- 13 06 30.	18	15.5	GALAXY
PHL 1427	02 38 00.	- 16 44		16.2	BLUE STELLAR OBJECT
SVEN 122	02 38 03.	- 08 21 34.	42	14.2	GALAXY
MCG-02-07-062	02 38 03.	- 08 45 30.	54	14.5	GALAXY
LB 02768	02 38 05.	+ 32 20 42.		16.5	FAINT BLUE STAR
ZWG 388.094	02 38 06.	+ 01 08		15.7	GALAXY
ZWG 523.087	02 38 06.	+ 35 39		15.7	GALAXY
UGC 02166	02 38 06.	+ 35 39	84	15.7	GALAXY Sc
PHL 4253	02 38 06.	- 07 32		18.6	BLUE STELLAR OBJECT
PHL 8464	02 38 06.	- 07 34		16.5	BLUE STELLAR OBJECT
PHL 1428	02 38 06.	- 08 04		18.1	BLUE STELLAR OBJECT
PHL 8465	02 38 06.	- 24 56		18.3	BLUE STELLAR OBJECT
HMS 1.07	02 38 06.	- 34 41			Ep GALAXY
MCG-06-07-001	02 38 06.	- 34 41			GALAXY
HOD.61 4	02 38 06.	- 34 45	24	19.7	GLOB CLSTR IN FORNAX GALX
RNGC 1043	02 38 07.	+ 01 08		15.5	GALAXY
LB 03571	02 38 07.	+ 42 35 36.		20.0	FAINT BLUE STAR
KEEL 099	02 38 08.1	+ 38 31 42.			NEBULA
MCG+00-07-079	02 38 09.	+ 00 19	36	16.	GALAXY
MCG+06-06-074	02 38 09.	+ 38 33	60	15.	GALAXY
MCG+06-06-074	02 38 09.	+ 38 33	60	15.	GALAXY
SVEN 124	02 38 09.	- 08 45	54	14.1	GALAXY
SVEN 123	02 38 09.	- 08 46	48	14.8	GALAXY
HELW 387	02 38 10.	- 05 36 52.			NEBULA
RNGC 1047	02 38 10.	- 08 22		15.0	GALAXY
RNGC 1048B	02 38 10.	- 08 45		14.0	GALAXY
RNGC 1048A	02 38 10.	- 08 45		14.0	GALAXY
REIN 7.050	02 38 10.76	- 01 54 16.6			NEBULA
MCG+00-07-080	02 38 12.	+ 00 06 30.	36	15.5	GALAXY
PHL 8466	02 38 12.	+ 00 56		18.5	BLUE STELLAR OBJECT
BIGO 478	02 38 12.	+ 01 08			NEBULA
PHL 1429	02 38 12.	+ 02 16		17.7	BLUE STELLAR OBJECT
ZWG 414.035	02 38 12.	+ 08 23		15.1	GALAXY
UGC 02167	02 38 12.	+ 08 23	78	15.1	GALAXY SB:c
MCG+01-07-021	02 38 12.	+ 08 23 30.	60	15.	GALAXY
MCG+03-07-043	02 38 12.	+ 17 13	72	15.	GALAXY
LB 03572	02 38 12.	+ 42 17 36.		19.6	FAINT BLUE STAR
MAFFEI 2	02 38 12.	+ 59 23 42.			INFRARED OBJECT
PHL 4254	02 38 12.	- 05 04		18.1	BLUE STELLAR OBJECT
PHL 1430	02 38 12.	- 06 52		18.1	BLUE STELLAR OBJECT
PHL 8467	02 38 12.	- 16 36		17.9	BLUE STELLAR OBJECT
TON-S 0267	02 38 12.	- 24 49		15.2	BLUE STAR
REIN 6.023	02 38 14.27	+ 01 05 57.8			NEBULA
HELW 388	02 38 16.	- 05 54 40.			NEBULA
ZC 0238.3+0321	02 38 18.	+ 03 21	1480		CLUSTER OF GALAXIES
ZWG 414.036	02 38 18.	+ 08 32		15.7	GALAXY
UGC 02168	02 38 18.	+ 17 13	84	16.5	GALAXY DWARF
ZWG 523.088	02 38 18.	+ 36 45		15.4	GALAXY
PHL 4255	02 38 18.	- 00 53		18.3	BLUE STELLAR OBJECT
PHL 8468	02 38 18.	- 04 37		17.8	BLUE STELLAR OBJECT
MCG-03-07-051	02 38 18.	- 16 00	30	15.5	GALAXY
PHL 1431	02 38 18.	- 19 06		15.9	BLUE STELLAR OBJECT
MCG-04-07-016	02 38 18.	- 20 35	30	15.	GALAXY
PHL 8469	02 38 18.	- 23 11		18.0	BLUE STELLAR OBJECT

OBJECT NAME	RIGHT ASCEN.	DECLINATION	DIAM.	MAGN.	TYPE OF OBJECT
MCG-04-07-017	02 38 20.	- 20 35 30.	12	15.5	GALAXY
SVEN 125	02 38 21.	- 08 11 35.	24	14.5	GALAXY
MCG-04-07-018	02 38 21.	- 20 36 30.	36	14.	GALAXY
RNGC 1044	02 38 22.	+ 08 31		15.0	GALAXY
LB 03573	02 38 23.	+ 42 36 12.		20.5	FAINT BLUE STAR
HELW 389	02 38 23.	- 08 11 28.			NEBULA
PHL 4256	02 38 24.	+ 00 54		18.6	BLUE STELLAR OBJECT
PHL 8471	02 38 24.	+ 03 12		17.2	BLUE STELLAR OBJECT
ZWG 414.037	02 38 24.	+ 08 00		15.5	GALAXY
ZWG 414.038	02 38 24.	+ 08 31		14.8	GALAXY
MCG+01-07-023	02 38 24.	+ 08 31	30	15.	GALAXY
MCG+01-07-022	02 38 24.	+ 08 31 30.	12	16.	GALAXY
6ZW 218	02 38 24.	+ 37 01			COMPACT GALAXY
PHL 8470	02 38 24.	- 08 03		16.2	BLUE STELLAR OBJECT
MCG-02-07-063	02 38 24.	- 13 37	54	15.	GALAXY
MCG-03-07-052	02 38 24.	- 15 21 30.	72	14.	GALAXY
MCG-04-07-019	02 38 24.	- 21 12	60	14.	GALAXY
PHL 8472	02 38 24.	- 27 56		18.1	BLUE STELLAR OBJECT
LB 03574	02 38 25.	+ 42 42 24.		17.8	FAINT BLUE STAR
MCG-03-07-053	02 38 27.	- 17 20	15	15.	GALAXY
RNGC 1046	02 38 28.	+ 08 30		15.0	GALAXY
RNGC 1051	02 38 28.	- 07 10		13.0	GALAXY
LB 03575	02 38 29.	+ 42 30 18.		20.6	FAINT BLUE STAR
ZWG 414.039	02 38 30.	+ 08 30		14.9	GALAXY
MCG+01-07-024	02 38 30.	+ 08 30	18	15.	GALAXY
ZWG 523.089	02 38 30.	+ 37 01		15.5	GALAXY
MCG+06-06-075	02 38 30.	+ 37 01	36	15.5	GALAXY
ZWG 539.077	02 38 30.	+ 43 52		15.1	GALAXY
MCG+01-07-033	02 38 30.	- 07 10	102	13.	GALAXY
MCG+01-07-034	02 38 30.	- 08 29	78	12.	GALAXY
MCG-02-07-064	02 38 30.	- 13 31 30.	48	15.	GALAXY
IC 0250	02 38 30.	- 13 31 46.			NONSTELLAR OBJECT
IC 0249	02 38 33.	- 07 09 04.			NONSTELLAR OBJECT
SVEN 126	02 38 33.	- 08 44	42	15.0	GALAXY
HELW 390	02 38 34.	- 05 45 59.			NEBULA
RNGC 1052	02 38 34.	- 08 27		12.0	GALAXY
ZWG 462.043	02 38 36.	+ 17 35		14.4	GALAXY
UGC 02170	02 38 36.	+ 17 35	78	14.4	GALAXY Sa
MCG+03-07-044	02 38 36.	+ 17 35	48	14.	GALAXY
PHL 1432	02 38 36.	- 11 53		18.7	BLUE STELLAR OBJECT
PHL 4257	02 38 36.	- 24 40		18.2	BLUE STELLAR OBJECT
PHL 8473	02 38 36.	- 27 10		17.2	BLUE STELLAR OBJECT
PHL 8474	02 38 36.	- 28 18		18.5	BLUE STELLAR OBJECT
IC 0248	02 38 37.	+ 17 36 04.			NONSTELLAR OBJECT
REIN 2.028	02 38 37.42	- 08 28 09.4			NEBULA
SVEN 127	02 38 39.	- 08 28 36.	66	11.7	GALAXY
MCG-02-07-065	02 38 39.	- 13 40	48	15.	GALAXY
ARC 0371	02 38 41.	- 11 26		17.0	RICH CLUSTER OF GALAXIES
2ZW 005	02 38 42.	+ 04 00			COMPACT GALAXY
PHL 1433	02 38 42.	- 01 41		18.1	BLUE STELLAR OBJECT
PHL 4258	02 38 42.	- 09 56		18.9	BLUE STELLAR OBJECT
PHL 8475	02 38 42.	- 28 42		16.2	BLUE STELLAR OBJECT
LB 01633	02 38 42.	- 48 37		13.9	FAINT BLUE STAR
LB 03576	02 38 44.	+ 42 40 42.		20.6	FAINT BLUE STAR
RNGC 1027	02 38 46.	+ 61 20		7.5	OPEN CLUSTER
RNGC 1039	02 38 47.	+ 42 34		6.0	OPEN CLUSTER
LB 03577	02 38 47.	+ 42 42 24.		19.5	FAINT BLUE STAR
PHL 8476	02 38 48.	+ 01 12		16.4	BLUE STELLAR OBJECT
OCL 0382	02 38 48.	+ 42 34	4200	6.8	OPEN STAR CLUSTER
VB 173	02 38 48.	+ 59 41	376		STELLAR RING
OCL 0357	02 38 48.	+ 61 20	2460	7.5	OPEN STAR CLUSTER
PHL 4260	02 38 48.	- 08 12		15.1	BLUE STELLAR OBJECT
MCG-02-07-066	02 38 48.	- 10 09	66	14.	GALAXY
PHL 8477	02 38 48.	- 11 22		18.6	BLUE STELLAR OBJECT
PHL 4259	02 38 48.	- 14 32		18.9	BLUE STELLAR OBJECT
MCG-03-07-054	02 38 48.	- 15 11	24	15.	GALAXY
PHL 1434	02 38 48.	- 19 14		12.5	BLUE STELLAR OBJECT
LB 03578	02 38 49.	+ 42 30 00.		17.8	FAINT BLUE STAR
LB 03579	02 38 51.	+ 42 27 36.		17.6	FAINT BLUE STAR
IC 0251	02 38 52.	- 15 10 18.			NONSTELLAR OBJECT
ZWG 414.040	02 38 54.	+ 06 58		15.0	GALAXY
MRK 595	02 38 54.	+ 06 58	15	16.	GALAXY WITH UV CONTINUUM
UGC 02171	02 38 54.	+ 31 52	90	17.	GALAXY Sc
6ZW 219	02 38 54.	+ 34 49			COMPACT GALAXY
ZWG 539.078	02 38 54.	+ 43 09		14.6	GALAXY
UGC 02172	02 38 54.	+ 43 09	108	14.6	GALAXY IRR
PHL 8478	02 38 54.	- 00 20		16.7	BLUE STELLAR OBJECT
MCG-02-07-067	02 38 54.	- 13 40	60	15.	GALAXY
PHL 8480	02 38 54.	- 26 48		17.3	BLUE STELLAR OBJECT
PHL 8479	02 38 54.	- 26 48		18.6	BLUE STELLAR OBJECT
TON-S 0268	02 38 54.	- 30 04		14.3	BLUE STAR
CED 008	02 39	+ 63	7200		DIFFUSE GALACTIC NEBULA
PHL 8482	02 39 00.	+ 01 48		17.5	BLUE STELLAR OBJECT
MCG+05-07-031	02 39 00.	+ 32 10	120	15.	GALAXY
ZWG 523.090	02 39 00.	+ 35 30		15.5	GALAXY
RNGC 1040	02 39 00.	+ 41 18			NON-EXISTENT OBJECT
MCG+07-06-055	02 39 00.	+ 43 08	54	15.	GALAXY
URA 27	02 39 00.	+ 66 39	306		STELLAR RING
PHL 4261	02 39 00.	- 03 59		18.7	BLUE STELLAR OBJECT
PHL 4262	02 39 00.	- 05 44		18.1	BLUE STELLAR OBJECT
PHL 4263	02 39 00.	- 06 40		18.3	BLUE STELLAR OBJECT
MCG-02-07-068	02 39 00.	- 13 20 30.	60	15.	GALAXY
PHL 4264	02 39 00.	- 14 55		18.0	BLUE STELLAR OBJECT
MCG-03-07-055	02 39 00.	- 18 10	54	15.	GALAXY
MCG-03-07-056	02 39 00.	- 18 11	12	15.	GALAXY
PHL 8483	02 39 00.	- 27 24		18.7	BLUE STELLAR OBJECT
PHL 8481	02 39 00.	- 28 04		17.8	BLUE STELLAR OBJECT
MCG+07-06-056	02 39 03.	+ 42 10	39	15.5	GALAXY
IC 1832	02 39 05.	+ 18 48 37.			NONSTELLAR OBJECT
ZWG 388.095	02 39 06.	+ 00 13		12.5	GALAXY
UGC 02173	02 39 06.	+ 00 13	480	12.5	GALAXY Sb
PHL 8485	02 39 06.	+ 02 28		16.7	BLUE STELLAR OBJECT
ZWG 462.044	02 39 06.	+ 18 49		15.1	GALAXY
MCG+03-07-045	02 39 06.	+ 18 49	24	15.	GALAXY
5ZW 272	02 39 06.	+ 29 56			COMPACT GALAXY
ZWG 505.034	02 39 06.	+ 32 09		15.2	GALAXY
UGC 02174	02 39 06.	+ 32 09	138	15.2	GALAXY Sc/SBc
6ZW 220	02 39 06.	+ 35 31			COMPACT GALAXY
ZWG 523.091	02 39 06.	+ 35 32		15.7	GALAXY
ZWG 539.079	02 39 06.	+ 42 08		15.7	GALAXY
ZWG 539.080	02 39 06.	+ 42 11		15.2	GALAXY
UGC 02175	02 39 06.	+ 42 11	84	15.2	GALAXY Sb-c
PHL 4266	02 39 06.	- 05 14		18.9	BLUE STELLAR OBJECT
PHL 4265	02 39 06.	- 20 40		15.3	BLUE STELLAR OBJECT
PHL 8484	02 39 06.	- 22 08		13.8	BLUE STELLAR OBJECT
LB 03273	02 39 06.	- 67 32		14.4	FAINT BLUE STAR
LB 03580	02 39 07.	+ 42 45 00.		20.6	FAINT BLUE STAR
HELW 391	02 39 08.	- 08 23 30.			NEBULA
SVEN 128	02 39 09.	- 08 23 37.	18	15.3	GALAXY
REIN 2.029	02 39 11.22	+ 00 13 45.0			NEBULA
ZWG 414.041	02 39 12.	+ 05 45		14.6	GALAXY
UGC 02176	02 39 12.	+ 05 45	84	14.6	GALAXY Sb
PHL 1435	02 39 12.	- 00 08		18.8	BLUE STELLAR OBJECT
PHL 4267	02 39 12.	- 03 13		18.7	BLUE STELLAR OBJECT
PHL 8487	02 39 12.	- 05 46		18.6	BLUE STELLAR OBJECT
PHL 4268	02 39 12.	- 06 44		18.5	BLUE STELLAR OBJECT
PHL 4269	02 39 12.	- 09 26		18.2	BLUE STELLAR OBJECT
PHL 1436	02 39 12.	- 11 11		19.0	BLUE STELLAR OBJECT
MCG-02-07-069	02 39 12.	- 12 34	48	14.5	GALAXY
MCG-03-07-057	02 39 12.	- 16 54	42	14.5	GALAXY
PHL 8486	02 39 12.	- 21 07		17.6	BLUE STELLAR OBJECT
RNGC 1055	02 39 13.	+ 00 14		11.5	GALAXY
LB 02769	02 39 13.	+ 31 05 42.		16.8	FAINT BLUE STAR
MCG+00-07-081	02 39 15.	+ 00 13	420	11.3	GALAXY
ZWG 388.096	02 39 18.	+ 01 01		15.0	GALAXY
ZC 0239.3+2550	02 39 18.	+ 25 50	670		CLUSTER OF GALAXIES
VB 224	02 39 18.	+ 54 04	416		STELLAR RING
ASS 43	02 39 18.	+ 61 10	28800		OB ASSOCIATION CAS OB6
PHL 4270	02 39 18.	- 10 11		18.1	BLUE STELLAR OBJECT
PHL 8488	02 39 18.	- 28 50		18.6	BLUE STELLAR OBJECT
IC 0252	02 39 23.	- 15 03 38.			NONSTELLAR OBJECT
PHL 8489	02 39 24.	+ 00 02		18.3	BLUE STELLAR OBJECT
ZC 0239.4+1402	02 39 24.	+ 14 02	1010		CLUSTER OF GALAXIES
ZWG 462.045	02 39 24.	+ 18 00		14.6	GALAXY
RNGC 1054	02 39 24.	+ 18 00		14.5	GALAXY
MCG+03-07-046	02 39 24.	+ 18 00	48	14.5	GALAXY
SN 1970K	02 39 24.	+ 36 20		18.5	SUPERNOVA
PHL 1437	02 39 24.	- 02 39		15.7	BLUE STELLAR OBJECT
PHL 4272	02 39 24.	- 04 00		17.7	BLUE STELLAR OBJECT
PHL 1439	02 39 24.	- 06 42		15.8	BLUE STELLAR OBJECT
PHL 1438	02 39 24.	- 07 46		18.4	BLUE STELLAR OBJECT
PHL 4273	02 39 24.	- 13 31		18.5	BLUE STELLAR OBJECT
PHL 4271	02 39 24.	- 21 47		18.6	BLUE STELLAR OBJECT
PHL 4274	02 39 24.	- 28 16		18.8	BLUE STELLAR OBJECT
MCG-05-07-013	02 39 24.	- 28 22	60	14.5	GALAXY
KEEL 101	02 39 26.7	- 00 11 55.			NEBULA
RNGC 1050	02 39 28.	+ 34 34		13.5	GALAXY
IC 1833	02 39 29.	- 28 22 58.			NONSTELLAR OBJECT
2ZW 006	02 39 30.	+ 04 03			COMPACT GALAXY
ZWG 524.001	02 39 30.	+ 34 34		13.5	GALAXY
ZWG 523.092	02 39 30.	+ 34 34		13.5	GALAXY
UGC 02178	02 39 30.	+ 34 34	120	13.5	GALAXY SBa
KARA.73B 0116	02 39 30.	+ 34 34	102	13.5	ISOLATED GALAXY S
MCG+06-06-076	02 39 30.	+ 34 59	78	14.5	GALAXY
ZWG 524.002	02 39 30.	+ 35 00		15.1	GALAXY
ZWG 523.093	02 39 30.	+ 35 00		15.1	GALAXY
UGC 02179	02 39 30.	+ 35 00	102	15.1	GALAXY SBc
ZWG 524.003	02 39 30.	+ 36 00		15.7	GALAXY
ZWG 523.094	02 39 30.	+ 36 00		15.7	GALAXY
MCG+06-06-077	02 39 30.	+ 36 20	24	15.5	GALAXY
ZWG 524.004	02 39 30.	+ 39 20		14.6	GALAXY
ZWG 523.095	02 39 30.	+ 39 20		14.6	GALAXY
UGC 02180	02 39 30.	+ 39 20	60	14.6	GALAXY S
UGC 02177	02 39 30.	- 01 06	66	16.0	GALAXY Sc
MCG+01-07-035	02 39 30.	- 07 00	60	16.	GALAXY
MCG-04-07-020	02 39 30.	- 21 00	60	15.	GALAXY
PHL 4275	02 39 30.	- 23 28		18.5	BLUE STELLAR OBJECT
MCG-05-07-014	02 39 30.	- 27 30	48	16.	GALAXY
MCG+06-06-078	02 39 33.	+ 34 32 30.	60	14.	GALAXY
MCG-04-07-021	02 39 33.	- 21 01	60	14.5	GALAXY
LB 03581	02 39 34.	+ 42 22 48.		20.2	FAINT BLUE STAR
MCG+00-07-082	02 39 36.	+ 02 12	24	15.5	GALAXY
ZWG 388.097	02 39 36.	+ 02 13		15.2	GALAXY
UGC 02181	02 39 36.	+ 02 13	66	15.2	GALAXY S
ZWG 462.046	02 39 36.	+ 17 57		15.5	GALAXY
MCG+03-07-047	02 39 36.	+ 17 57	18	15.5	GALAXY
ZWG 505.035	02 39 36.	+ 32 31		15.7	GALAXY
MCG+06-06-079	02 39 36.	+ 39 20	36	14.5	GALAXY
PHL 1440	02 39 36.	- 01 37		16.5	BLUE STELLAR OBJECT
MCG+01-07-036	02 39 36.	- 05 48	72	14.	GALAXY
PHL 8490	02 39 36.	- 09 22		17.1	BLUE STELLAR OBJECT
PHL 4276	02 39 36.	- 10 22		18.9	BLUE STELLAR OBJECT
PHL 4277	02 39 36.	- 10 24		19.1	BLUE STELLAR OBJECT
PHL 1441	02 39 36.	- 17 32		15.9	BLUE STELLAR OBJECT
MCG-04-07-022	02 39 36.	- 21 10	60	14.	GALAXY
PHL 1442	02 39 36.	- 28 18		18.6	BLUE STELLAR OBJECT
LB 02770	02 39 38.	+ 01 09 42.		16.6	FAINT BLUE STAR
RNGC 1063	02 39 39.	- 05 46		14.0	GALAXY
KEEL 102	02 39 39.7	+ 00 10 09.			NEBULA
BIGO 479	02 39 40.	- 15 16			NEBULA
KEEL 100	02 39 40.5	+ 38 43 19.			NEBULA
ZWG 414.042	02 39 42.	+ 08 00		15.4	GALAXY
RNGC 1059	02 39 42.	+ 17 49			NON-EXISTENT OBJECT
MCG-02-07-070	02 39 42.	- 09 44	42	15.	GALAXY
PHL 8491	02 39 42.	- 11 40		18.1	BLUE STELLAR OBJECT
IC 0253	02 39 42.	- 15 15 40.			NONSTELLAR OBJECT
MCG-03-07-058	02 39 42.	- 15 16	18	14.5	GALAXY
MCG-03-07-059	02 39 42.	- 15 18	30	14.5	GALAXY
IC 0254	02 39 42.	- 15 19 16.			NONSTELLAR OBJECT
MCG-04-07-023	02 39 42.	- 21 29	54	15.	GALAXY
RNGC 1065	02 39 43.	- 15 18		14.0	GALAXY
LB 03582	02 39 44.	+ 42 16 24.		20.6	FAINT BLUE STAR
BIGO 480	02 39 45.	- 15 21			NEBULA
RNGC 1064	02 39 47.	- 09 34		14.0	GALAXY
MCG+05-07-032	02 39 48.	+ 28 22	132	14.	GALAXY
UGC 02182	02 39 48.	+ 32 28	84	17.	GALAXY DWRF IR
MCG+07-06-057	02 39 48.	+ 40 15	30	17.	GALAXY
MCG+01-07-037	02 39 48.	- 03 19	18	15.	GALAXY
PHL 8492	02 39 48.	- 14 50		17.9	BLUE STELLAR OBJECT
PHL 8493	02 39 48.	- 19 33		18.1	BLUE STELLAR OBJECT
MCG-03-07-060	02 39 48.	- 20 04	48	14.5	GALAXY
PHL 4278	02 39 48.	- 23 45		17.5	BLUE STELLAR OBJECT
MCG-02-07-071	02 39 51.	- 09 34	60	14.5	GALAXY
MCG-04-07-024	02 39 51.	- 24 20	72	14.	GALAXY
LB 02771	02 39 52.	+ 01 00 18.		15.0	FAINT BLUE STAR
ZC 0239.9+2147	02 39 54.	+ 21 47	470		CLUSTER OF GALAXIES
ZWG 505.036	02 39 54.	+ 28 21		13.5	GALAXY
UGC 02183	02 39 54.	+ 28 21	150	13.5	GALAXY Sa
5ZW 273	02 39 54.	+ 34 53			COMPACT GALAXY
6ZW 221	02 39 54.	+ 37 36			COMPACT GALAXY
MCG+07-06-058	02 39 54.	+ 40 51	60	15.	GALAXY
MCG+00-07-083	02 39 54.	- 00 13 30.	420	9.6	GALAXY
PHL 8494	02 39 54.	- 02 27		15.8	BLUE STELLAR OBJECT
PHL 8495	02 39 54.	- 14 56		18.2	BLUE STELLAR OBJECT
PHL 4279	02 39 54.	- 20 24		18.1	BLUE STELLAR OBJECT
RNGC 1056	02 39 56.	+ 28 21		13.5	GALAXY

OBJECT NAME	RIGHT ASCEN.	DECLINATION	DIAM.	MAGN.	TYPE OF OBJECT
MCG+07-06-059	02 39 57.	+ 40 12	150	14.	GALAXY
MCG+00-07-084	02 39 57.	- 01 14	36	15.	GALAXY
SHB 067	02 40	- 06			QUASI-STELLAR OBJECT
PHL 1443	02 40 00.	+ 00 45		16.8	BLUE STELLAR OBJECT
PHL 8496	02 40 00.	+ 02 35		18.2	BLUE STELLAR OBJECT
5ZW 274	02 40 00.	+ 32 06			COMPACT GALAXY
ZWG 505.037	02 40 00.	+ 32 17		15.7	GALAXY
RNGC 1057	02 40 00.	+ 32 17		15.5	GALAXY
UGC 02184	02 40 00.	+ 32 17	72	15.7	GALAXY S0
MCG+05-07-033	02 40 00.	+ 32 18	48	15.	GALAXY
6ZW 222	02 40 00.	+ 37 36			COMPACT GALAXY
ZWG 539.081	02 40 00.	+ 40 14		13.8	GALAXY
UGC 02185	02 40 00.	+ 40 14	210	13.8	GALAXY Sc
ZWG 539.082	02 40 00.	+ 40 51		14.5	GALAXY
UGC 02186	02 40 00.	+ 40 51	102	14.5	GALAXY S0-a
MCG+07-06-060	02 40 00.	+ 41 16 30.	72	15.	GALAXY
ZWG 539.083	02 40 00.	+ 41 18		14.0	GALAXY
RNGC 1053	02 40 00.	+ 41 18		14.0	GALAXY
UGC 02187	02 40 00.	+ 41 18	102	14.0	GALAXY S0
LDN 1376	02 40 00.	+ 60 20	540		DARK NEBULA
LDN 1353	02 40 00.	+ 68 00	6420		DARK NEBULA
MCG-02-07-072	02 40 00.	- 09 40	48	15.	GALAXY
PHL 8497	02 40 00.	- 13 18		18.8	BLUE STELLAR OBJECT
PHL 8498	02 40 00.	- 22 12		18.6	BLUE STELLAR OBJECT
PHL 4280	02 40 00.	- 28 06		18.6	BLUE STELLAR OBJECT
ARP 037	02 40 05.	- 00 13			PECULIAR GALAXY
REIN 2.030	02 40 05.41	- 00 14 04.6			NEBULA
PHL 4281	02 40 06.	+ 02 02		18.8	BLUE STELLAR OBJECT
ZWG 388.099	02 40 06.	+ 02 52		15.2	GALAXY
UGC 02189	02 40 06.	+ 02 52	66	15.2	GALAXY Sb
UGC 02190	02 40 06.	+ 37 36	60	16.0	GALAXY S0
ZWG 388.098	02 40 06.	- 00 13		9.7	GALAXY
UGC 02188	02 40 06.	- 00 13	540	9.7	GALAXY Sb
VVI 10	02 40 06.	- 00 14	420	9.63	SEYFERT GALAXY
MCG-05-07-015	02 40 06.	- 30 32	30	14.	GALAXY
IC 1831	02 40 07.	+ 63 23			NONSTELLAR OBJECT
RNGC 1068	02 40 07.	- 00 14		10.5	GALAXY
REIN 2.031	02 40 07.12	- 00 13 31.1			NEBULA
MCG-02-07-073	02 40 09.	- 12 38	42	14.5	GALAXY
MCG+00-07-085	02 40 12.	+ 02 52	42		GALAXY
MCG+01-07-025	02 40 12.	+ 07 22	42	14.5	GALAXY
ZWG 414.043	02 40 12.	+ 07 23		15.0	GALAXY
MRK 596	02 40 12.	+ 07 23	15	16.	GALAXY WITH UV CONTINUUM
MCG+03-07-048	02 40 12.	+ 16 59	42	16.	GALAXY
ZWG 462.047	02 40 12.	+ 17 00		15.7	GALAXY
ZC 0240.2+2750	02 40 12.	+ 27 50	1010		CLUSTER OF GALAXIES
MCG+05-07-034	02 40 12.	+ 32 04	36	17.	GALAXY
ZWG 505.038	02 40 12.	+ 32 12		13.4	GALAXY
RNGC 1060	02 40 12.	+ 32 12		13.5	GALAXY
UGC 02191	02 40 12.	+ 32 12	150	13.4	GALAXY E-S0
MCG+05-07-035	02 40 12.	+ 32 13	30	13.	GALAXY
ZWG 505.039	02 40 12.	+ 32 15		15.2	GALAXY
RNGC 1061	02 40 12.	+ 32 15		15.0	GALAXY
MCG+05-07-036	02 40 12.	+ 32 16	42	15.	GALAXY
MCG+07-06-061	02 40 12.	+ 41 11	72	15.	GALAXY
ZC 0240.2+4140	02 40 12.	+ 41 40	2350		CLUSTER OF GALAXIES
ZWG 554.001	02 40 12.	+ 47 37		15.2	GALAXY
UGC 02192	02 40 12.	+ 47 37	66	15.2	GALAXY S0-a
PHL 4283	02 40 12.	- 02 16		18.3	BLUE STELLAR OBJECT
PHL 8499	02 40 12.	- 02 37		15.7	BLUE STELLAR OBJECT
PHL 4282	02 40 12.	- 03 09		17.9	BLUE STELLAR OBJECT
PHL 1444	02 40 12.	- 08 12		18.0	BLUE STELLAR OBJECT
MCG-02-07-074	02 40 12.	- 12 38	48	17.5	GALAXY
PHL 4284	02 40 12.	- 19 34		18.2	BLUE STELLAR OBJECT
PHL 8500	02 40 12.	- 28 52		18.4	BLUE STELLAR OBJECT
IC 1834	02 40 14.	+ 02 53 19.			NONSTELLAR OBJECT
MCG-04-07-025	02 40 15.	- 21 48	48	14.5	GALAXY
LB 03583	02 40 16.	+ 42 34 18.		19.3	FAINT BLUE STAR
ZC 0240.3+0105	02 40 18.	+ 01 05	1410		CLUSTER OF GALAXIES
ZWG 414.044	02 40 18.	+ 07 55		15.7	GALAXY
ZWG 524.005	02 40 18.	+ 37 08		11.8	GALAXY
ZWG 523.096	02 40 18.	+ 37 08		11.8	GALAXY
UGC 02193	02 40 18.	+ 37 08	228	11.8	GALAXY Sc
ZWG 539.084	02 40 18.	+ 41 13		14.6	GALAXY
UGC 02194	02 40 18.	+ 41 13	102	14.6	GALAXY SB:b
ZWG 539.085	02 40 18.	+ 41 44		15.7	GALAXY
UGC 02195	02 40 18.	+ 41 44	66	15.7	GALAXY Sc
MCG+07-06-062	02 40 18.	+ 44 08	42	15.5	GALAXY
ZWG 539.086	02 40 18.	+ 44 10		15.6	GALAXY
UGC 02196	02 40 18.	+ 44 10	66	15.6	GALAXY S
PHL 1445	02 40 18.	- 12 00		18.4	BLUE STELLAR OBJECT
PHL 4285	02 40 18.	- 13 26		17.7	BLUE STELLAR OBJECT
PHL 8501	02 40 18.	- 19 33		18.0	BLUE STELLAR OBJECT
MCG-06-07-002	02 40 18.	- 34 18	72	14.5	GALAXY
LB 01634	02 40 18.	- 49 03		14.0	FAINT BLUE STAR
MCG-02-07-075	02 40 21.	- 12 37	60	15.	GALAXY
SN 1961V	02 40 22.	+ 37 08		12.2	SUPERNOVA
RNGC 1069	02 40 22.	- 08 30		14.5	GALAXY
MCG+03-07-049	02 40 24.	+ 16 23	24	15.5	GALAXY
ZWG 462.048	02 40 24.	+ 16 24		15.3	GALAXY
ZWG 505.040	02 40 24.	+ 31 15		15.6	GALAXY
UGC 02197	02 40 24.	+ 31 15	120	15.6	GALAXY Sc
MCG+05-07-037	02 40 24.	+ 31 17	72	15.5	GALAXY
MCG+06-07-001	02 40 24.	+ 37 08	138	11.	GALAXY
MCG+01-07-038	02 40 24.	- 08 30	60	14.5	GALAXY
PHL 4286	02 40 24.	- 11 00		18.7	BLUE STELLAR OBJECT
PHL 4287	02 40 24.	- 19 54		18.3	BLUE STELLAR OBJECT
PHL 8502	02 40 24.	- 24 22		18.4	BLUE STELLAR OBJECT
HOD.61 5	02 40 24.	- 34 21	30	19.6	GLOB CLSTR IN FORNAX GALX
LB 01635	02 40 24.	- 57 38		13.7	FAINT BLUE STAR
LB 03274	02 40 24.	- 58 20		14.0	FAINT BLUE STAR
RNGC 1058	02 40 27.	+ 37 08		12.5	GALAXY
ARC 0372	02 40 27.	+ 41 39		15.7	RICH CLUSTER OF GALAXIES
ARC 0373	02 40 28.	+ 27 48		17.7	RICH CLUSTER OF GALAXIES
SN 1969L	02 40 28.	+ 37 06		12.8	SUPERNOVA
ZC 0240.5+1037	02 40 30.	+ 10 37	2420		CLUSTER OF GALAXIES
MCG+03-07-050	02 40 30.	+ 16 27	42	15.	GALAXY
KARA.68 023	02 40 30.	+ 16 31	67		DWARF GALAXY
ZWG 505.041	02 40 30.	+ 31 35		15.7	GALAXY
UGC 02198	02 40 30.	+ 31 35	66	15.7	GALAXY Sc
MCG+05-07-038	02 40 30.	+ 31 35	66	15.	GALAXY
ZWG 505.042	02 40 30.	+ 32 02		15.6	GALAXY
PHL 8503	02 40 30.	- 07 42		18.2	BLUE STELLAR OBJECT
PHL 4288	02 40 30.	- 07 42		18.3	BLUE STELLAR OBJECT
MCG-02-07-076	02 40 30.	- 11 53	30	14.5	GALAXY
SVEN 129	02 40 33.	- 08 30 41.	60	13.7	GALAXY
MCG-04-07-026	02 40 33.	- 21 40	60	13.5	GALAXY
RNGC 1071	02 40 35.	- 08 58		15.0	GALAXY

OBJECT NAME	RIGHT ASCEN.	DECLINATION	DIAM.	MAGN.	TYPE OF OBJECT
PHL 8504	02 40 36.	+ 00 24		16.5	BLUE STELLAR OBJECT
ZWG 388.100	02 40 36.	+ 03 25		15.5	GALAXY
ZC 0240.6+0740	02 40 36.	+ 07 40	13980		CLUSTER OF GALAXIES
KARA.72 073B	02 40 36.	+ 16 26	30		PART OF DOUBLE GALAXY
ZWG 463.001	02 40 36.	+ 16 27		14.8	GALAXY
ZWG 462.049	02 40 36.	+ 16 27		14.8	GALAXY
VV 278C	02 40 36.	+ 16 27	12	16.	INTERACTING GALAXY
VV 278B	02 40 36.	+ 16 27	12	16.	INTERACTING GALAXY
VV 278A	02 40 36.	+ 16 27	12	16.	INTERACTING GALAXY
VV 278	02 40 36.	+ 16 27	54		INTERACTING GALAXY
KARA.72 073A	02 40 36.	+ 16 27	36	14.8	PART OF DOUBLE GALAXY
ZC 0240.6+1933	02 40 36.	+ 19 33	1950		CLUSTER OF GALAXIES
MCG+05-07-039	02 40 36.	+ 32 03	36	15.5	GALAXY
PHL 4289	02 40 36.	- 00 21		17.0	BLUE STELLAR OBJECT
MCG-02-07-077	02 40 36.	- 08 58 30.	60	15.	GALAXY
PHL 1446	02 40 36.	- 10 13		17.5	BLUE STELLAR OBJECT
PHL 8505	02 40 36.	- 25 27		18.3	BLUE STELLAR OBJECT
MCG-04-07-027	02 40 36.	- 25 47	24	15.5	GALAXY
PHL 4290	02 40 36.	- 26 24		18.5	BLUE STELLAR OBJECT
LB 01636	02 40 36.	- 51 15		14.3	FAINT BLUE STAR
LB 02772	02 40 39.	+ 01 07 00.		16.8	FAINT BLUE STAR
SVEN 130	02 40 39.	- 08 59 41.	18	14.2	GALAXY
MCG-04-07-028	02 40 39.	- 25 40	60	14.	GALAXY
RNGC 1070	02 40 41.	+ 04 46		13.0	GALAXY
RNGC 1062	02 40 41.	+ 32 18		15.0	GALAXY
PHL 1447	02 40 42.	+ 01 08		16.9	BLUE STELLAR OBJECT
ZWG 388.101	02 40 42.	+ 01 33		14.8	GALAXY
UGC 02199	02 40 42.	+ 01 33	78	14.8	GALAXY S0-a
PHL 4291	02 40 42.	+ 02 44		17.9	BLUE STELLAR OBJECT
ZWG 388.102	02 40 42.	+ 02 53		15.6	GALAXY
ZWG 414.045	02 40 42.	+ 04 46		13.0	GALAXY
UGC 02200	02 40 42.	+ 04 46	168	13.0	GALAXY Sb
ZC 0240.7+1259	02 40 42.	+ 12 59	2550		CLUSTER OF GALAXIES
ZWG 505.043	02 40 42.	+ 32 17		15.7	GALAXY
UGC 02201	02 40 42.	+ 32 17	102	15.7	GALAXY Sc
ZWG 539.087	02 40 42.	+ 41 15		15.6	GALAXY
RLWT 064	02 40 42.	+ 57 40		20.	FAINT VERY BLUE STAR
PHL 8506	02 40 42.	- 17 26		16.0	BLUE STELLAR OBJECT
MCG-04-07-029	02 40 42.	- 25 47	36	15.5	GALAXY
TON-S 0269	02 40 42.	- 27 00		16.0	BLUE STAR
TON-S 0270	02 40 42.	- 28 36		15.8	BLUE STAR
LB 01637	02 40 42.	- 52 38		14.5	FAINT BLUE STAR
REIN 2.032	02 40 44.80	+ 04 45 25.5			NEBULA
MCG+00-07-086	02 40 45.	+ 01 32	27	15.	GALAXY
MCG+00-07-087	02 40 45.	+ 02 53		15.	GALAXY
MCG+01-07-026	02 40 45.	+ 04 45 30.	120	13.	GALAXY
MCG+05-07-040	02 40 45.	+ 32 11	36	16.	GALAXY
MCG+05-07-041	02 40 45.	+ 32 18	72	15.	GALAXY
RNGC 1066	02 40 47.	+ 32 15		15.0	GALAXY
RNGC 1067	02 40 47.	+ 32 18		14.5	GALAXY
ZC 0240.8+2825	02 40 48.	+ 28 25	1340		CLUSTER OF GALAXIES
UGC 02202	02 40 48.	+ 32 10	60	16.5	GALAXY DWARF
ZWG 505.044	02 40 48.	+ 32 15		14.9	GALAXY
UGC 02203	02 40 48.	+ 32 15	120	14.9	GALAXY (E)
MCG+05-07-042	02 40 48.	+ 32 17 30.	21	14.5	GALAXY
ZWG 505.045	02 40 48.	+ 32 18		14.6	GALAXY
UGC 02204	02 40 48.	+ 32 18	72	14.6	GALAXY Sc
MCG+05-07-043	02 40 48.	+ 32 19	60	14.5	GALAXY
ZWG 505.046	02 40 48.	+ 32 57		15.3	GALAXY
UGC 02205	02 40 48.	+ 32 57	72	15.3	GALAXY Sc
ZWG 505.047	02 40 48.	+ 33 08		15.0	GALAXY
UGC 02206	02 40 48.	+ 33 08	66	15.0	GALAXY
MCG+05-07-044	02 40 48.	+ 33 10	48	15.	GALAXY
UGC 02207	02 40 48.	+ 38 36	66	16.5	GALAXY SB
OCL 0356	02 40 48.	+ 62 08	360		OPEN STAR CLUSTER
ZC 0240.8-0110	02 40 48.	- 01 10	1680		CLUSTER OF GALAXIES
MCG+01-07-039	02 40 48.	- 04 06	48	15.5	GALAXY
MCG-03-07-061	02 40 48.	- 15 29	18	15.	GALAXY
PHL 4292	02 40 48.	- 22 11		18.6	BLUE STELLAR OBJECT
IC 1836	02 40 49.	+ 02 54 22.			NONSTELLAR OBJECT
ARC 0374	02 40 51.	+ 04 02		17.8	RICH CLUSTER OF GALAXIES
MCG+00-07-088	02 40 54.	+ 00 05	96	14.	GALAXY
ZC 0240.9+1905	02 40 54.	+ 19 05	1280		CLUSTER OF GALAXIES
MCG+05-07-045	02 40 54.	+ 32 59	48	15.	GALAXY
KARA.68 024	02 40 54.	- 01 22	27		DWARF GALAXY
MCG+01-07-040	02 40 54.	- 08 22	18	15.5	GALAXY
REIN 2.033	02 40 57.28	+ 00 05 43.8			NEBULA
IC 1837	02 40 58.	- 00 05 57.			SAME AS NGC 1072
ZWG 389.001	02 41 00.	+ 00 05		14.3	GALAXY
ZWG 388.103	02 41 00.	+ 00 05		14.3	GALAXY
UGC 02208	02 41 00.	+ 00 05	90	14.3	GALAXY Sa-b
ZC 0241.0+0402	02 41 00.	+ 04 02	1280		CLUSTER OF GALAXIES
ZWG 415.001	02 41 00.	+ 07 29		15.4	GALAXY
ZWG 414.046	02 41 00.	+ 07 29		15.4	GALAXY
ZC 0241.0+3617	02 41 00.	+ 36 17	1680		CLUSTER OF GALAXIES
5ZW 275	02 41 00.	+ 37 18			COMPACT GALAXY
UGC 02209	02 41 00.	+ 49 22	60	16.5	GALAXY S
LDN 1375	02 41 00.	+ 60 40	660		DARK NEBULA
MCG-06-07-003	02 41 00.	- 33 00	60	14.5	GALAXY
MCG+00-08-001	02 41 03.	+ 01 10	300	11.4	GALAXY
IC 1835	02 41 03.	+ 14 40 33.			NONSTELLAR OBJECT
SVEN 131	02 41 03.	- 00 20	30	15.8	GALAXY
SVEN 132	02 41 03.	- 00 44 42.	30	14.8	GALAXY
REIN 6.024	02 41 05.89	+ 01 09 53.8			NEBULA
LB 02773	02 41 06.	+ 00 36 48.		16.8	FAINT BLUE STAR
ZWG 389.002	02 41 06.	+ 01 10		12.5	GALAXY
UGC 02210	02 41 06.	+ 01 10	330	12.5	GALAXY SBc
ZWG 440.001	02 41 06.	+ 14 41		15.7	GALAXY
ZWG 439.027	02 41 06.	+ 14 41		15.7	GALAXY
ZCG 0241+14	02 41 06.	+ 14 41		15.7	COMPACT GALAXY
PHL 4293	02 41 06.	- 06 50		15.7	BLUE STELLAR OBJECT
PHL 8507	02 41 06.	- 23 59		18.5	BLUE STELLAR OBJECT
RNGC 1073	02 41 07.	+ 01 10		11.5	GALAXY
SN 1962L	02 41 07.	+ 01 11		13.9	SUPERNOVA
UGC 02211	02 41 12.	+ 06 26	78	16.5	GALAXY Sc
ZC 0241.2+3558	02 41 12.	+ 35 58	9070		CLUSTER OF GALAXIES
ZWG 539.088	02 41 12.	+ 44 20		15.6	GALAXY
PHL 8508	02 41 12.	- 29 36		16.1	BLUE STELLAR OBJECT
PHL 4294	02 41 12.	- 30 14		18.5	BLUE STELLAR OBJECT
PHL 8509	02 41 12.	- 30 30		18.8	BLUE STELLAR OBJECT
LB 02774	02 41 13.	+ 00 22 18.		16.4	FAINT BLUE STAR
LB 02775	02 41 15.	+ 00 22 24.		16.3	FAINT BLUE STAR
KEEL 103	02 41 17.7	- 00 03 32.			NEBULA
ZWG 415.002	02 41 18.	+ 07 36		15.6	GALAXY
ZWG 463.002	02 41 18.	+ 16 20		15.4	GALAXY
ZWG 524.006	02 41 18.	+ 37 17		15.7	GALAXY
MCG+01-08-001	02 41 18.	- 06 53	60	16.	GALAXY
MCG-03-08-003	02 41 18.	- 14 57	120	13.	GALAXY

OBJECT NAME	RIGHT ASCEN.	DECLINATION	DIAM.	MAGN.	TYPE OF OBJECT
MCG-03-08-002	02 41 18.	- 16 23	36	15.	GALAXY
MCG-03-08-001	02 41 18.	- 16 30	42	14.5	GALAXY
TON-S 0271	02 41 18.	- 29 32		15.8	BLUE STAR
LB 03275	02 41 18.	- 63 30		13.1	FAINT BLUE STAR
RNGC 1076	02 41 19.	- 14 57		13.0	GALAXY
IC 1840	02 41 19.	- 15 54 54.			NONSTELLAR OBJECT
RNGC 1075	02 41 20.	- 16 23		15.0	GALAXY
RNGC 1074	02 41 20.	- 16 30		14.0	GALAXY
MCG+05-07-046	02 41 21.	+ 32 10	42	17.	GALAXY
SVEN 133	02 41 21.	- 00 49 43.	24	15.2	GALAXY
LB 02776	02 41 22.	+ 01 33 24.		16.3	FAINT BLUE STAR
PHL 4295	02 41 24.	+ 03 00		16.6	BLUE STELLAR OBJECT
ZWG 415.003	02 41 24.	+ 04 51		15.4	GALAXY
ZWG 415.004	02 41 24.	+ 05 13		14.9	GALAXY
KARA.72 074A	02 41 24.	+ 05 13	30	14.9	PART OF DOUBLE GALAXY
ZWG 415.005	02 41 24.	+ 05 14		15.7	GALAXY
MCG+06-07-002	02 41 24.	+ 33 30	48	15.5	GALAXY
ZWG 524.007	02 41 24.	+ 33 31		15.4	GALAXY
UGC 02212	02 41 24.	+ 33 31	66	15.4	GALAXY
ZWG 524.008	02 41 24.	+ 37 47		15.7	GALAXY
UGC 02213	02 41 24.	+ 37 47	66	15.7	GALAXY Sb
KW 48	02 41 24.	- 00 08	380		SEYFERT GALAXY
PHL 4296	02 41 24.	- 09 47		17.8	BLUE STELLAR OBJECT
PHL 8510	02 41 24.	- 10 12		17.8	BLUE STELLAR OBJECT
PHL 1448	02 41 24.	- 10 57		16.1	BLUE STELLAR OBJECT
MCG-03-08-004	02 41 24.	- 15 54 30.	36	15.	GALAXY
PHL 8511	02 41 24.	- 29 18		18.7	BLUE STELLAR OBJECT
PHL 1449	02 41 24.	- 29 54		18.0	BLUE STELLAR OBJECT
PHL 1450	02 41 24.	- 30 43		18.0	BLUE STELLAR OBJECT
PHL 1451	02 41 24.	- 31 16		18.3	BLUE STELLAR OBJECT
MCG-05-07-016	02 41 24.	- 32 10	84	15.	GALAXY
LB 02777	02 41 25.	+ 01 34 00.		16.0	FAINT BLUE STAR
LB 02778	02 41 26.	+ 01 21 12.		16.9	FAINT BLUE STAR
MCG-04-07-030	02 41 27.	- 24 24	72	15.	GALAXY
MCG+00-08-002	02 41 30.	+ 00 34	36	16.	GALAXY
KARA.72 074B	02 41 30.	+ 05 14	36	15.7	PART OF DOUBLE GALAXY
MCG+06-07-003	02 41 30.	+ 37 46	48	15.	GALAXY
TON-S 0272	02 41 30.	- 27 23		16.0	BLUE STAR
MCG-05-07-017	02 41 30.	- 29 12	108	12.5	GALAXY
SN 1961A	02 41 31.	+ 00 34		19.0	SUPERNOVA
MCG-04-07-031	02 41 33.	- 26 19	48	14.	GALAXY
ZWG 463.003	02 41 36.	+ 17 22		15.5	GALAXY
ZWG 524.009	02 41 36.	+ 35 15		15.7	GALAXY
PHL 4297	02 41 36.	- 28 22		18.7	BLUE STELLAR OBJECT
RNGC 1079	02 41 37.	- 29 13		12.5	GALAXY
SVEN 134	02 41 39.	- 00 50 43.	12	15.5	GALAXY
KARA.68 025	02 41 39.	- 06 57	34		DWARF GALAXY
UGC 02214	02 41 42.	+ 09 31	78	16.5	GALAXY
MCG+03-08-001	02 41 42.	+ 17 20	30	16.	GALAXY
ZWG 524.010	02 41 42.	+ 37 27		15.7	GALAXY
MCG+07-06-063	02 41 42.	+ 41 46	36	16.	GALAXY
ZWG 539.089	02 41 42.	+ 41 47		14.8	GALAXY
UGC 02215	02 41 42.	+ 41 47	60	14.8	GALAXY
MCG-03-08-005	02 41 42.	- 15 00	24	15.	GALAXY
PHL 4299	02 41 42.	- 16 36		17.9	BLUE STELLAR OBJECT
PHL 4298	02 41 42.	- 21 16		18.6	BLUE STELLAR OBJECT
PHL 8512	02 41 42.	- 22 25		18.0	BLUE STELLAR OBJECT
PHL 8513	02 41 42.	- 29 18		18.6	BLUE STELLAR OBJECT
RNGC 1078	02 41 47.	- 09 38		15.0	GALAXY
PHL 8515	02 41 48.	- 14 04		4.3	BLUE STELLAR OBJECT
PHL 8514	02 41 48.	- 25 30		18.1	BLUE STELLAR OBJECT
PHL 4300	02 41 48.	- 28 52		18.7	BLUE STELLAR OBJECT
LB 02779	02 41 49.	+ 00 44 18.		15.6	FAINT BLUE STAR
MCG+00-08-003	02 41 51.	+ 00 28	60	15.5	GALAXY
MCG-02-08-001	02 41 51.	- 09 38	18	15.5	GALAXY
UGC 02216	02 41 54.	+ 00 28	108	16.0	GALAXY IRR
MCG+02-08-001	02 41 54.	+ 15 01	60	15.	GALAXY
ZWG 463.004	02 41 54.	+ 16 31		15.6	GALAXY
UGC 02217	02 41 54.	+ 16 31	84	15.6	GALAXY Sb
IC 1838	02 41 54.	+ 19 13 59.			NONSTELLAR OBJECT
MCG+03-08-002	02 41 54.	+ 19 14	36	16.	GALAXY
ZWG 463.005	02 41 54.	+ 19 15		15.2	GALAXY
ARC 0375	02 41 54.	+ 28 51		17.8	RICH CLUSTER OF GALAXIES
ZC 0241.9+2852	02 41 54.	+ 28 52	1340		CLUSTER OF GALAXIES
ZWG 505.048	02 41 54.	+ 32 30		15.4	GALAXY
UGC 02218	02 41 54.	+ 32 30	66	15.4	GALAXY VY CMPT
MCG+05-07-047	02 41 54.	+ 32 30	36	15.	GALAXY
UGC 02219	02 41 54.	+ 38 22	78	16.5	GALAXY Sc
MCG+07-06-064	02 41 54.	+ 42 32 30.	36	16.	GALAXY
ZWG 539.090	02 41 54.	+ 42 34		15.7	GALAXY
PHL 8516	02 41 54.	- 28 44		18.8	BLUE STELLAR OBJECT
TON-S 0273	02 41 54.	- 31 39		16.0	BLUE STAR
IC 1839	02 41 57.	+ 15 02 53.			NONSTELLAR OBJECT
ZWG 440.002	02 42 00.	+ 15 02		15.3	GALAXY
UGC 02220	02 42 00.	+ 15 02	66	15.3	GALAXY Sb-c
MCG+03-08-003	02 42 00.	+ 16 29	66	15.	GALAXY
UGC 02221	02 42 00.	+ 30 10	102	17.	GALAXY
ZWG 524.011	02 42 00.	+ 37 50		15.6	GALAXY
VB 174	02 42 00.	+ 58 54	618		STELLAR RING
MCG-02-08-002	02 42 00.	- 09 55	24	15.5	GALAXY
MCG-02-08-003	02 42 03.	- 09 55	42	15.5	GALAXY
LB 02780	02 42 05.	+ 00 55 18.		16.9	FAINT BLUE STAR
LB 00471	02 42 05.	+ 25 22 48.		16.8	FAINT BLUE STAR
ZWG 415.006	02 42 06.	+ 05 08		15.0	GALAXY
MCG+03-08-004	02 42 06.	+ 16 57	36	16.	GALAXY
ZWG 505.049	02 42 06.	+ 32 47		14.9	GALAXY
UGC 02222	02 42 06.	+ 32 47	90	14.9	GALAXY
KARA.72 075A	02 42 06.	+ 32 47	84	14.9	PART OF DOUBLE GALAXY
MCG+05-07-048	02 42 06.	+ 32 48	24	16.	GALAXY
ZWG 524.012	02 42 06.	+ 34 59		14.8	GALAXY
UGC 02223	02 42 06.	+ 34 59	72	14.8	GALAXY Sc
UGC 02224	02 42 06.	+ 45 47	72	16.5	GALAXY S
ZWG 389.003	02 42 06.	- 01 40		15.2	GALAXY
LB 02781	02 42 07.	+ 00 56 12.		16.2	FAINT BLUE STAR
MCG+05-07-049	02 42 09.	+ 32 47 30.	48	16.	GALAXY
ARC 0380	02 42 11.	- 26 29		16.9	RICH CLUSTER OF GALAXIES
ZC 0242.2+2652	02 42 12.	+ 26 52	1480		CLUSTER OF GALAXIES
ZWG 505.050	02 42 12.	+ 32 04		15.5	GALAXY
ZWG 505.051	02 42 12.	+ 32 46		15.2	GALAXY
UGC 02225	02 42 12.	+ 32 46	66	15.2	GALAXY S
KARA.72 075B	02 42 12.	+ 32 46	60	15.2	PART OF DOUBLE GALAXY
ZC 0242.2+3323	02 42 12.	+ 33 23	540		CLUSTER OF GALAXIES
PK144-15.1	02 42 12.	+ 42 20	22	16.7	PLANETARY NEBULA
ZWG 539.091	02 42 12.	+ 42 21		15.6	GALAXY
PHL 8517	02 42 12.	- 22 12		18.1	BLUE STELLAR OBJECT
TON-S 0274	02 42 12.	- 25 10		15.1	BLUE STAR
PHL 8518	02 42 12.	- 27 27		18.7	BLUE STELLAR OBJECT
LB 01638	02 42 12.	- 52 00		13.7	FAINT BLUE STAR
SVEN 135	02 42 15.	- 00 42 45.	18	15.8	GALAXY
5ZW 276	02 42 18.	+ 41 21			COMPACT GALAXY
MCG+07-06-065	02 42 18.	+ 41 56	30	16.	GALAXY
ZWG 539.092	02 42 18.	+ 41 57		15.6	GALAXY
KARA.72 076A	02 42 18.	+ 41 57	54	15.6	PART OF DOUBLE GALAXY
LB 02782	02 42 24.	+ 01 01 48.		16.6	FAINT BLUE STAR
MCG+02-08-002	02 42 24.	+ 13 57 30.	36	15.5	GALAXY
MCG+07-06-066	02 42 24.	+ 41 56	36	16.	GALAXY
ZWG 539.093	02 42 24.	+ 41 57		15.7	GALAXY
UGC 02226	02 42 24.	+ 41 57	90	15.7	GALAXY E
KARA.72 076B	02 42 24.	+ 41 57	60	15.7	PART OF DOUBLE GALAXY
MCG-03-08-007	02 42 24.	- 17 54 30.	30	15.5	GALAXY
MCG-03-08-006	02 42 24.	- 17 55	30	15.5	GALAXY
PHL 8519	02 42 24.	- 22 13		16.9	BLUE STELLAR OBJECT
MCG+01-08-002	02 42 27.	- 08 24 30.	60	16.	GALAXY
ARC 0378	02 42 29.	- 03 19		17.6	RICH CLUSTER OF GALAXIES
LB 02783	02 42 36.	+ 00 34 54.		16.7	FAINT BLUE STAR
ZWG 463.006	02 42 36.	+ 17 55		15.4	GALAXY
MCG-04-07-032	02 42 36.	- 24 42 30.	30	14.5	GALAXY
LB 02784	02 42 39.	+ 00 25 06.		16.8	FAINT BLUE STAR
SVEN 136	02 42 39.	- 00 57	12	15.4	GALAXY
RNGC 1080	02 42 39.	- 04 56		14.5	GALAXY
RNGC 1096	02 42 39.	- 60 07			UNVERIFIED SOUTHERN OBJECT
ARC 0376	02 42 40.	+ 36 39		15.4	RICH CLUSTER OF GALAXIES
ZWG 440.003	02 42 42.	+ 11 15		15.7	GALAXY
KARA.73B 0117	02 42 42.	+ 11 15	66	15.7	ISOLATED GALAXY
IC 1842	02 42 42.	+ 11 15 01.			NONSTELLAR OBJECT
MCG+03-08-005	02 42 42.	+ 17 54	42	16.	GALAXY
ZWG 539.094	02 42 42.	+ 42 36		14.7	GALAXY
UGC 02227	02 42 42.	+ 42 36	138	14.7	GALAXY SBb
MCG+07-06-067	02 42 42.	+ 42 36	90	15.	GALAXY
MCG+01-08-003	02 42 42.	- 04 56	60	14.5	GALAXY
MCG-02-08-004	02 42 42.	- 13 26	84	15.5	GALAXY
MCG-03-08-009	02 42 42.	- 17 30	60	15.	GALAXY
MCG-03-08-008	02 42 42.	- 17 51	24	14.	GALAXY
LB 03276	02 42 42.	- 75 15		13.0	FAINT BLUE STAR
RNGC 1098	02 42 44.	- 17 51		14.0	GALAXY
MCG+00-08-004	02 42 45.	+ 02 41	30	14.5	GALAXY
IC 1841	02 42 46.	+ 18 42 19.			NONSTELLAR OBJECT
IC 1845	02 42 46.	- 28 09 56.			NONSTELLAR OBJECT
MCG+00-08-005	02 42 48.	+ 00 44	36	16.	GALAXY
ZWG 389.004	02 42 48.	+ 02 40		14.2	GALAXY
UGC 02228	02 42 48.	+ 02 40	102	14.2	GALAXY SBa
ZWG 463.007	02 42 48.	+ 18 43		15.1	GALAXY
ZC 0242.8+2916	02 42 48.	+ 29 16	1080		CLUSTER OF GALAXIES
MCG+07-06-069	02 42 48.	+ 39 52	48	14.5	GALAXY
MCG+07-06-068	02 42 48.	+ 39 52	30	16.5	GALAXY
VB 175	02 42 48.	+ 58 02	470		STELLAR RING
VB 176	02 42 48.	+ 60 17	295		STELLAR RING
MCG-03-08-010	02 42 48.	- 15 47	96	13.5	GALAXY
RNGC 1081	02 42 49.	- 15 47		13.0	GALAXY
IC 1843	02 42 50.	+ 02 40 36.			NONSTELLAR OBJECT
MCG+00-08-006	02 42 54.	+ 00 42	36	15.	GALAXY
ZWG 389.005	02 42 54.	+ 00 43		14.8	GALAXY
UGC 02229	02 42 54.	+ 00 43	102	14.8	GALAXY E-S
ZC 0242.9+0311	02 42 54.	+ 03 11	1210		CLUSTER OF GALAXIES
ZC 0242.9+2734	02 42 54.	+ 27 34	1410		CLUSTER OF GALAXIES
ZWG 539.095	02 42 54.	+ 39 53		14.6	GALAXY
UGC 02230	02 42 54.	+ 39 53	78	14.6	GALAXY Sb
UGC 02231	02 42 54.	+ 41 54	72	17.	GALAXY
RNGC 1077B	02 42 55.	+ 39 53		14.0	GALAXY
RNGC 1077A	02 42 55.	+ 39 53		16.0	GALAXY
ARC 0379	02 42 56.	+ 03 12		17.6	RICH CLUSTER OF GALAXIES
SVEN 147	02 42 56.	- 30 00	30	16.0	GALAXY
SVEN 137	02 42 57.	- 01 04 47.	12	16.0	GALAXY
ZWG 463.008	02 43 00.	+ 16 12		15.6	GALAXY
MCG+05-07-050	02 43 00.	+ 27 49	48	15.	GALAXY
MCG+06-07-004	02 43 00.	+ 36 41	36	16.	GALAXY
UGC 02232	02 43 00.	+ 36 42	84	16.0	GALAXY
ZWG 539.096	02 43 00.	+ 44 45		15.5	GALAXY
UGC 02233	02 43 00.	+ 44 45	102	15.5	GALAXY Sb
MCG-03-08-012	02 43 00.	- 17 30	18	15.	GALAXY
MCG-03-08-011	02 43 00.	- 17 53	108	13.5	GALAXY
PHL 8520	02 43 00.	- 21 22		18.1	BLUE STELLAR OBJECT
PHL 4301	02 43 00.	- 28 05		18.5	BLUE STELLAR OBJECT
TON-S 0275	02 43 00.	- 31 07		14.4	BLUE STAR
RNGC 1099	02 43 02.	- 17 53		13.0	GALAXY
SVEN 148	02 43 02.	- 30 42	12	15.9	GALAXY
LB 02785	02 43 03.	+ 00 26 42.		16.4	FAINT BLUE STAR
ARC 0377	02 43 04.	+ 27 36		17.7	RICH CLUSTER OF GALAXIES
MCG+03-08-006	02 43 06.	+ 16 11	24	15.5	GALAXY
ZWG 505.052	02 43 06.	+ 27 49		15.6	GALAXY
ZWG 524.013	02 43 06.	+ 39 10		15.0	GALAXY
ZWG 539.097	02 43 06.	+ 44 56		15.4	GALAXY
UGC 02234	02 43 06.	+ 44 56	84	15.4	GALAXY Sc
PHL 8521	02 43 06.	- 31 06		14.7	BLUE STELLAR OBJECT
MCG+00-08-007	02 43 09.	+ 03 02	42	15.	GALAXY
SVEN 139	02 43 09.	- 00 44 48.	30	14.9	GALAXY
SVEN 138	02 43 09.	- 00 45 48.	18	15.9	GALAXY
SVEN 140	02 43 09.	- 00 51	6	16.1	GALAXY
MCG-04-07-033	02 43 09.	- 21 02	36	15.	GALAXY
ZWG 389.006	02 43 12.	+ 03 01		15.0	GALAXY
UGC 02235	02 43 12.	+ 45 00	72	17.	GALAXY S:IV-V
ZC 0243.2-0054	02 43 12.	- 00 54	1340		CLUSTER OF GALAXIES
MCG-03-08-013	02 43 12.	- 17 43	36	14.	GALAXY
PHL 4302	02 43 12.	- 22 20		18.3	BLUE STELLAR OBJECT
PHL 4303	02 43 12.	- 23 00		18.1	BLUE STELLAR OBJECT
TON-S 0276	02 43 12.	- 26 49		15.7	BLUE STAR
PHL 8522	02 43 12.	- 26 56		16.0	BLUE STELLAR OBJECT
PHL 8523	02 43 12.	- 27 50		18.2	BLUE STELLAR OBJECT
PHL 4314	02 43 12.	- 29 42		18.6	BLUE STELLAR OBJECT
RNGC 1091	02 43 14.	- 17 43		14.0	GALAXY
MCG+04-07-020	02 43 15.	+ 26 50	66	15.	GALAXY
SVEN 141	02 43 15.	- 00 26 48.	18	16.5	GALAXY
IC 1844	02 43 16.	+ 03 02 03.			NONSTELLAR OBJECT
RNGC 1082	02 43 16.	- 08 25		15.5	GALAXY
ZWG 463.009	02 43 18.	+ 15 38		14.8	GALAXY
MRK 597	02 43 18.	+ 15 38	9	16.5	GALAXY WITH UV CONTINUUM
ZWG 484.017	02 43 18.	+ 26 50		15.5	GALAXY
UGC 02236	02 43 18.	+ 26 50	78	15.5	GALAXY S
ZWG 524.014	02 43 18.	+ 35 01		15.5	GALAXY
UGC 02237	02 43 18.	+ 35 01	78	15.5	GALAXY S
5ZW 277	02 43 18.	+ 35 14			COMPACT GALAXY
MRSL 136+02/1	02 43 18.	+ 61 47	60		HII REGION
MCG-03-08-014	02 43 18.	- 17 44	54	14.	GALAXY
MCG-05-07-018	02 43 18.	- 28 10	48	15.	GALAXY
LB 02786	02 43 19.	+ 00 57 54.		16.8	FAINT BLUE STAR
RNGC 1092	02 43 20.	- 17 44		14.0	GALAXY

OBJECT NAME	RIGHT ASCEN.	DECLINATION	DIAM.	MAGN.	TYPE OF OBJECT
LB 02787	02 43 21.	- 00 12 36.		17.0	FAINT BLUE STAR
SVEN 143	02 43 21.	- 00 28	12	16.3	GALAXY
SVEN 145	02 43 21.	- 00 39	6	14.7	GALAXY
SVEN 142	02 43 21.	- 00 43 48.	12	15.8	GALAXY
SVEN 146	02 43 21.	- 00 48	24	15.2	GALAXY
SVEN 144	02 43 21.	- 00 57 48.	18	15.4	GALAXY
MCG+01-08-004	02 43 21.	- 08 25	6	15.5	GALAXY
ZWG 389.007	02 43 24.	+ 03 08		15.5	GALAXY
MCG+03-08-007	02 43 24.	+ 15 36	39	15.	GALAXY
ZWG 463.010	02 43 24.	+ 15 38		15.7	GALAXY
LB 00472	02 43 24.	+ 26 58 00.		16.5	FAINT BLUE STAR
MCG+06-07-005	02 43 24.	+ 35 01	48	15.	GALAXY
MRSL 136+02/3	02 43 24.	+ 61 44	120		HII REGION
MCG-03-08-017	02 43 24.	- 15 23	30	15.5	GALAXY
MCG-03-08-015	02 43 24.	- 15 32 30.	84	14.	GALAXY
MCG-03-08-016	02 43 24.	- 17 52	90	13.5	GALAXY
MCG-04-07-034	02 43 24.	- 25 01	48	14.	GALAXY
PHL 8524	02 43 24.	- 26 46		18.1	BLUE STELLAR OBJECT
MCG-05-07-019	02 43 24.	- 28 07	24	16.	GALAXY
LB 02788	02 43 25.	+ 00 41 12.		16.4	FAINT BLUE STAR
RNGC 1083	02 43 25.	- 15 32		14.0	GALAXY
ARC 0381	02 43 26.	- 00 52		17.6	RICH CLUSTER OF GALAXIES
RNGC 1100	02 43 26.	- 17 52		13.0	GALAXY
MCG+05-07-051	02 43 27.	+ 32 16	60	15.	GALAXY
RNGC 1084	02 43 28.	- 07 47		12.0	GALAXY
MCG+00-08-008	02 43 30.	+ 03 07	24	16.	GALAXY
KARA.68 026	02 43 30.	+ 03 20	27		DWARF GALAXY
ZWG 524.015	02 43 30.	+ 34 50		15.2	GALAXY
MCG-03-08-018	02 43 30.	- 15 42	48	15.	GALAXY
MCG-04-07-035	02 43 30.	- 25 05	36	15.	GALAXY
PHL 4304	02 43 30.	- 27 55		16.0	BLUE STELLAR OBJECT
SN 1963P	02 43 32.	- 07 47		14.0	SUPERNOVA
REIN 2.034	02 43 32.213	- 07 47 15.8			NEBULA
SVEN 149	02 43 33.	- 00 32	12	14.9	GALAXY
SVEN 150	02 43 33.	- 00 48	6	16.0	GALAXY
MCG+02-08-003	02 43 36.	+ 12 52 30.	72	14.5	GALAXY
ZWG 440.004	02 43 36.	+ 12 53		15.2	GALAXY
UGC 02238	02 43 36.	+ 12 53	108	15.2	GALAXY IRR?
ZWG 505.053	02 43 36.	+ 32 14		15.6	GALAXY
UGC 02239	02 43 36.	+ 32 14	114	15.6	GALAXY SB?b-c
ZWG 539.098	02 43 36.	+ 41 50		15.7	GALAXY
MCG+08-06-001	02 43 36.	+ 47 59	42	16.	GALAXY
ZWG 554.002	02 43 36.	+ 48 00		15.4	GALAXY
UGC 02240	02 43 36.	+ 48 00	102	15.4	GALAXY SB0
MRSL 136+02/2	02 43 36.	+ 61 48	120		HII REGION
MCG+01-08-005	02 43 36.	- 03 32 30.	15	17.	GALAXY
MCG+01-08-006	02 43 36.	- 03 33	12	17.	GALAXY
TON-S 0277	02 43 36.	- 27 51		15.8	BLUE STAR
TON-S 0278	02 43 36.	- 30 23		15.9	BLUE STAR
MCG+01-08-007	02 43 39.	- 07 48 30.	156	12.	GALAXY
KEEL 104	02 43 39.2	- 07 50 36.			NEBULA
ZWG 440.005	02 43 42.	+ 10 53		15.4	GALAXY
MCG-05-07-020	02 43 42.	- 28 28	60	15.	GALAXY
LB 01639	02 43 42.	- 47 43		12.7	FAINT BLUE STAR
YC 0243-55	02 43 42.	- 55 55 00.			UNUSUAL SOUTHERN OBJECT
MCG+00-08-010	02 43 45.	+ 03 25	54	13.5	GALAXY
MCG+00-08-009	02 43 45.	- 00 43	180	11.4	GALAXY
SVEN 151	02 43 45.	- 00 48 49.	12	15.3	GALAXY
ZWG 389.008	02 43 48.	+ 03 24		13.6	GALAXY
RNGC 1085	02 43 48.	+ 03 24		13.5	GALAXY
UGC 02241	02 43 48.	+ 03 24	150	13.6	GALAXY Sb
UGC 02242	02 43 48.	+ 20 10	72	17.	GALAXY DWARF
ZWG 524.016	02 43 48.	+ 37 25		15.7	GALAXY
ZWG 524.017	02 43 48.	+ 38 49		14.9	GALAXY
UGC 02243	02 43 48.	+ 38 49	84	14.9	GALAXY SBb-c
MCG+06-07-006	02 43 48.	+ 39 24 30.	30	16.	GALAXY
ZWG 524.018	02 43 48.	+ 39 25		15.4	GALAXY
UGC 02244	02 43 48.	+ 39 25	60	15.4	GALAXY
MCG+06-07-007	02 43 48.	+ 39 25	15	16.	GALAXY
ZWG 539.099	02 43 48.	+ 41 35		15.7	GALAXY
MCG-05-07-021	02 43 48.	- 32 23	60	14.5	GALAXY
LB 01640	02 43 48.	- 46 13		13.2	FAINT BLUE STAR
REIN 2.035	02 43 50.76	- 00 42 28.7			NEBULA
SVEN 152	02 43 51.	- 00 42	180	11.3	GALAXY
REIN 2.036	02 43 52.07	- 00 42 29.3			NEBULA
MCG+06-07-008	02 43 54.	+ 38 48 30.	54	15.	GALAXY
UGC 02246	02 43 54.	+ 43 00	72	17.	GALAXY Sc
SG 3.015	02 43 54.	+ 61 39	60		DIFFUSE EMISSION NEBULA
ZWG 389.010	02 43 54.	- 00 42		11.4	GALAXY
UGC 02245	02 43 54.	- 00 42	240	11.4	GALAXY Sc
ZWG 389.009	02 43 54.	- 01 57		15.4	GALAXY
MCG-03-08-019	02 43 54.	- 17 18	48	15.	GALAXY
PHL 8525	02 43 54.	- 27 00		18.3	BLUE STELLAR OBJECT
MCG-05-07-022	02 43 54.	- 30 27	36	14.5	GALAXY
RNGC 1087	02 43 55.	- 00 42		11.5	GALAXY
MCG+04-07-021	02 43 57.	+ 23 23 30.	54	15.	GALAXY
SN 1962K	02 43 57.	- 00 27		18.2	SUPERNOVA
SVEN 154	02 43 57.	- 00 33	18	15.5	GALAXY
SVEN 153	02 43 57.	- 00 51 50.	30	15.0	GALAXY
SN 1971T	02 43 59.	- 00 27		16.0	SUPERNOVA
VDB.66G 028	02 44	+ 03 39	200		DWARF GALAXY
LBN 0662	02 44	+ 61 40	120		BRIGHT NEBULA
LBN 0661	02 44	+ 61 43	60		BRIGHT NEBULA
LBN 0660	02 44	+ 61 44	180		BRIGHT NEBULA
ZWG 389.012	02 44 00.	+ 01 14		15.3	GALAXY
ZC 0244.0+0519	02 44 00.	+ 05 19	4700		CLUSTER OF GALAXIES
ZWG 484.018	02 44 00.	+ 23 24		15.7	GALAXY
UGC 02248	02 44 00.	+ 23 24	72	15.7	GALAXY S0
UGC 02249	02 44 00.	+ 45 20	120	16.0	GALAXY S0
MCG+08-06-002	02 44 00.	+ 45 20	54	16.	GALAXY
UGC 02250	02 44 00.	+ 46 08	78	18.	GALAXY DWARF
ZWG 389.011	02 44 00.	- 00 27		12.8	GALAXY
UGC 02247	02 44 00.	- 00 27	270	12.8	GALAXY SBb
MCG+00-08-011	02 44 00.	- 00 27 30.	222	13.	GALAXY
MCG-03-08-020	02 44 00.	- 15 16	24	14.5	GALAXY
MCG-05-07-023	02 44 00.	- 26 31	36	14.5	GALAXY
REIN 6.025	02 44 00.35	- 00 28 09.1			NEBULA
REIN 6.026A	02 44 00.41	- 00 27 23.6			NEBULA
REIN 6.026B	02 44 00.41	- 00 27 24.1			NEBULA
RNGC 1090	02 44 01.	- 00 27		12.5	GALAXY
RNGC 1089	02 44 01.	- 15 16		14.0	GALAXY
SVEN 174	02 44 02.	- 30 25	30	14.2	GALAXY
SVEN 173	02 44 02.	- 30 31 51.	12	16.2	GALAXY
SVEN 155	02 44 03.	- 00 27	180	12.2	GALAXY
SVEN 157	02 44 03.	- 00 39 50.	24	15.1	GALAXY
SVEN 156	02 44 03.	- 00 51	6	14.8	GALAXY
SVEN 158	02 44 03.	- 01 02 50.	12	15.6	GALAXY
UGC 02251	02 44 06.	+ 24 39	66	17.	GALAXY DWARF?
MCG+04-07-022	02 44 06.	+ 24 39	36	16.	GALAXY
ZWG 554.003	02 44 06.	+ 47 35		15.5	GALAXY
MCG+08-06-003	02 44 06.	+ 47 35	48	16.	GALAXY
MCG+00-08-012	02 44 06.	- 00 40	24	16.	GALAXY
MCG-03-08-021	02 44 06.	- 14 58	36	15.	GALAXY
MCG-03-08-022	02 44 06.	- 18 58	30	15.	GALAXY
MCG-04-07-036	02 44 06.	- 21 23	30	15.5	GALAXY
MCG-05-07-024	02 44 06.	- 30 30	480	10.5	GALAXY
AGU 13	02 44 06.	- 38 42 00.		16.	2 INTERACTING GALAXIES
SVEN 159	02 44 09.	- 01 01 50.	12	16.0	GALAXY
ZWG 389.013	02 44 12.	+ 00 13		15.7	GALAXY
ZC 0244.2+0343	02 44 12.	+ 03 43	1210		CLUSTER OF GALAXIES
ZWG 440.006	02 44 12.	+ 15 13		14.8	GALAXY
UGC 02252	02 44 12.	+ 15 13	66	14.8	GALAXY COMPACT
ZWG 463.011	02 44 12.	+ 15 59		14.8	GALAXY
UGC 02253	02 44 12.	+ 15 59	78	14.8	GALAXY S0-a
ZWG 463.012	02 44 12.	+ 16 04		15.7	GALAXY
ZC 0244.2+3344	02 44 12.	+ 33 44	1140		CLUSTER OF GALAXIES
ZC 0244.2+3503	02 44 12.	+ 35 03	1080		CLUSTER OF GALAXIES
UGC 02254	02 44 12.	+ 37 19	66	17.	GALAXY DWRF IR
PHL 8526	02 44 12.	- 21 08		15.3	BLUE STELLAR OBJECT
PHL 8528	02 44 12.	- 27 10		18.5	BLUE STELLAR OBJECT
PHL 8527	02 44 12.	- 28 30		17.1	BLUE STELLAR OBJECT
LB 02789	02 44 13.	+ 00 44 18.		16.8	FAINT BLUE STAR
RNGC 1088	02 44 13.	+ 15 59		15.0	GALAXY
IC 0255	02 44 14.	+ 16 03 23.			NONSTELLAR OBJECT
SVEN 177	02 44 14.	- 30 28	570	10.8	GALAXY
MCG+02-08-004	02 44 15.	+ 15 12 30.	60	15.	GALAXY
SVEN 160	02 44 15.	- 00 38	12	16.0	GALAXY
SVEN 162	02 44 15.	- 00 46	12	14.8	GALAXY
SVEN 161	02 44 15.	- 00 55 51.	42	15.1	GALAXY
MCG+00-08-013	02 44 18.	+ 00 14	27	16.	GALAXY
ZWG 463.013	02 44 18.	+ 15 43		15.7	GALAXY
MCG+03-08-008	02 44 18.	+ 16 03	36	16.	GALAXY
ZWG 389.014	02 44 18.	- 02 04		15.7	GALAXY
MCG-03-08-023	02 44 18.	- 18 57 30.	30	15.5	GALAXY
PHL 1452	02 44 18.	- 27 16		18.4	BLUE STELLAR OBJECT
RNGC 1097A	02 44 20.	- 30 29		14.0	GALAXY
RNGC 1097	02 44 20.	- 30 29		10.5	GALAXY
MCG+03-08-009	02 44 21.	+ 15 58	48	15.	GALAXY
SVEN 168	02 44 21.	- 00 21	24	15.8	GALAXY
SVEN 167	02 44 21.	- 00 26 51.	6	15.9	GALAXY
SVEN 164	02 44 21.	- 00 39	12	15.6	GALAXY
SVEN 166	02 44 21.	- 00 45 51.	18	15.1	GALAXY
SVEN 165	02 44 21.	- 01 01 51.	12	16.1	GALAXY
SVEN 163	02 44 21.	- 01 05 51.	18	15.5	GALAXY
MCG-04-07-037	02 44 21.	- 26 27	72	13.5	GALAXY
ARP 077	02 44 21.	- 30 29			PECULIAR GALAXY
KEEL 105	02 44 21.6	- 07 25 30.			NEBULA
ZWG 415.007	02 44 24.	+ 08 26		14.3	GALAXY
UGC 02255	02 44 24.	+ 08 26	78	14.3	GALAXY S0-a
ZC 0244.4+2755	02 44 24.	+ 27 55	1010		CLUSTER OF GALAXIES
VB 177	02 44 24.	+ 61 28	530		STELLAR RING
PHL 1453	02 44 24.	- 22 28		18.2	BLUE STELLAR OBJECT
MCG-05-07-025	02 44 24.	- 27 10	42	14.5	GALAXY
SVEN 169	02 44 27.	- 00 19 51.	30	15.3	GALAXY
SVEN 172	02 44 27.	- 00 38 51.	18	15.5	GALAXY
SVEN 171	02 44 27.	- 00 39	6	15.3	GALAXY
SVEN 170	02 44 27.	- 00 47 51.	12	14.5	GALAXY
ARC 0382	02 44 28.	+ 04 06		17.2	RICH CLUSTER OF GALAXIES
ZC 0244.5+0412	02 44 30.	+ 04 12	1210		CLUSTER OF GALAXIES
ZWG 440.007	02 44 30.	+ 15 11		15.6	GALAXY
MCG+02-08-005	02 44 30.	+ 15 11	24	15.	GALAXY
5ZW 278	02 44 30.	+ 21 42			COMPACT GALAXY
ZWG 524.019	02 44 30.	+ 34 42		15.6	GALAXY
MCG+07-06-070	02 44 30.	+ 40 16	18	15.5	GALAXY
ZWG 539.100	02 44 30.	+ 40 18		15.1	GALAXY
UGC 02256	02 44 30.	+ 40 18	84	15.1	GALAXY S0
ZWG 389.015	02 44 30.	- 00 48		15.7	GALAXY
MCG-04-07-038	02 44 30.	- 25 33	60	14.	GALAXY
LB 02790	02 44 31.	+ 01 32 30.		16.6	FAINT BLUE STAR
MRK 598	02 44 36.	+ 07 11	10	17.	GALAXY WITH UV CONTINUUM
OCL 0287	02 44 36.	+ 16 59	720		OPEN STAR CLUSTER
UGC 02257	02 44 36.	+ 48 34	78	16.5	GALAXY S0?
MCG-04-07-039	02 44 36.	- 22 52	72	13.5	GALAXY
PHL 4305	02 44 36.	- 25 19		17.7	BLUE STELLAR OBJECT
PHL 4306	02 44 36.	- 30 12		18.4	BLUE STELLAR OBJECT
LB 02792	02 44 39.	+ 00 09 12.		16.6	FAINT BLUE STAR
LB 02791	02 44 39.	+ 01 22 24.		16.5	FAINT BLUE STAR
SVEN 176	02 44 39.	- 00 35	6	15.0	GALAXY
SVEN 175	02 44 39.	- 01 14 52.	18	15.2	GALAXY
MCG+01-08-008	02 44 39.	- 03 11	72	15.5	GALAXY
VDB.66N 007	02 44 40.	+ 69 27	372		REFLECTION NEBULA
ZWG 463.014	02 44 42.	+ 15 54		15.4	GALAXY
MCG+07-06-071	02 44 42.	+ 41 02	84	14.	GALAXY
ZWG 539.101	02 44 42.	+ 41 03		13.6	GALAXY
RNGC 1086	02 44 42.	+ 41 03		13.5	GALAXY
UGC 02258	02 44 42.	+ 41 03	102	13.6	GALAXY Sc
MCG+08-06-004	02 44 42.	+ 47 20	9	17.	GALAXY
SVEN 178	02 44 45.	- 00 39 52.	30	15.2	GALAXY
SVEN 179	02 44 45.	- 00 56 52.	12	15.2	GALAXY
KEEL 106	02 44 46.5	- 07 49 49.			NEBULA
SN 1963L	02 44 47.	+ 37 20		15.0	SUPERNOVA
ZWG 463.015	02 44 48.	+ 15 40		15.7	GALAXY
ZWG 524.020	02 44 48.	+ 37 20		14.3	GALAXY
UGC 02259	02 44 48.	+ 37 20	180	14.3	GALAXY
KARA.72 077A	02 44 48.	+ 37 20	132	14.3	PART OF DOUBLE GALAXY
ZWG 524.021	02 44 48.	+ 37 22		15.6	GALAXY
KARA.72 077B	02 44 48.	+ 37 22	54	15.6	PART OF DOUBLE GALAXY
UGC 02260	02 44 48.	+ 42 40	60	16.0	GALAXY COMPACT
MCG+08-06-005	02 44 48.	+ 48 31	36	16.	GALAXY
ZWG 554.004	02 44 48.	+ 50 36		15.2	GALAXY
UGC 02261	02 44 48.	+ 50 36	120	15.2	GALAXY E
KARA.72 078A	02 44 48.	+ 50 36	66	15.2	PART OF DOUBLE GALAXY
PHL 8529	02 44 48.	- 21 37		18.6	BLUE STELLAR OBJECT
PHL 4307	02 44 48.	- 22 30		18.1	BLUE STELLAR OBJECT
PHL 4308	02 44 48.	- 24 59		18.1	BLUE STELLAR OBJECT
MCG-05-07-026	02 44 48.	- 31 42	60	15.	GALAXY
LB 01641	02 44 48.	- 48 03		13.4	FAINT BLUE STAR
MCG+06-07-009	02 44 51.	+ 37 20	144	14.	GALAXY
MCG+06-07-010	02 44 51.	+ 37 22	15	16.	GALAXY
SVEN 181	02 44 51.	- 00 27 52.	30	14.3	GALAXY
SVEN 180	02 44 51.	- 00 28	72	13.0	GALAXY
RNGC 1102	02 44 52.	- 22 26		14.0	GALAXY
MCG+03-08-010	02 44 54.	+ 16 36	42	17.	GALAXY
UGC 02263	02 44 54.	+ 16 38	60	16.0	GALAXY S0?
MCG+04-07-023	02 44 54.	+ 23 11 30.	48	15.	GALAXY
MCG+08-06-006	02 44 54.	+ 50 36	24	15.	GALAXY

OBJECT NAME	RIGHT ASCEN.	DECLINATION	DIAM.	MAGN.	TYPE OF OBJECT
ZWG 389.017	02 44 54.	- 00 28		15.6	GALAXY
MCG+00-08-014	02 44 54.	- 00 28	30	15.	GALAXY
MCG+00-08-015	02 44 54.	- 00 29	66	13.5	GALAXY
ZWG 389.016	02 44 54.	- 00 30		13.5	GALAXY
PHL 4309	02 44 54.	- 22 09		18.5	BLUE STELLAR OBJECT
MCG-04-07-040	02 44 54.	- 22 26	42	14.5	GALAXY
LB 03277	02 44 54.	- 70 48		15.0	FAINT BLUE STAR
REIN 6.027	02 44 54.52	- 00 29 37.8			NEBULA
REIN 6.028	02 44 54.90	- 00 28 36.3			NEBULA
RNGC 1094	02 44 55.	- 00 30		13.5	GALAXY
MCG+03-08-011	02 44 57.	+ 17 01	18	15.5	GALAXY
ARP 131	02 44 57.	- 14 59			PECULIAR GALAXY
RNGC 1095	02 44 59.	+ 04 26		14.0	GALAXY
ZWG 415.008	02 45 00.	+ 04 26		14.2	GALAXY
UGC 02264	02 45 00.	+ 04 26	84	14.2	GALAXY SBc
IC 1846	02 45 00.	+ 13 01 43.			NONSTELLAR OBJECT
MCG+02-08-006	02 45 00.	+ 13 02	15	15.	GALAXY
ZWG 440.008	02 45 00.	+ 13 03		15.4	GALAXY
UGC 02265	02 45 00.	+ 13 03	66	15.4	GALAXY COMPACT
UGC 02266	02 45 00.	+ 14 11	72	17.	GALAXY
ZWG 440.009	02 45 00.	+ 15 20		15.6	GALAXY
ZWG 484.019	02 45 00.	+ 23 12		15.6	GALAXY
UGC 02267	02 45 00.	+ 23 12	126	15.6	GALAXY SBb
UGC 02268	02 45 00.	+ 25 53	66	17.	GALAXY DWARF
UGC 02269	02 45 00.	+ 44 27	102	18.	GALAXY DWARF
ZWG 554.005	02 45 00.	+ 50 33		14.8	GALAXY
UGC 02270	02 45 00.	+ 50 33	132	14.8	GALAXY Sb
KARA.72 078B	02 45 00.	+ 50 33	96	14.8	PART OF DOUBLE GALAXY
RLWT 065	02 45 00.	+ 61 42		13.5	FAINT VERY BLUE STAR
MCG-03-08-024	02 45 00.	- 15 36	24	16.	GALAXY
SVEN 183	02 45 02.	- 30 16	12	15.8	GALAXY
MCG+01-08-001	02 45 03.	+ 04 25 30.	72	14.	GALAXY
VV 336A	02 45 03.	- 14 58 30.	36	15.	INTERACTING GALAXY
VV 336B	02 45 03.	- 14 59	27	15.	INTERACTING GALAXY
IC 1847	02 45 05.	+ 14 18 01.			NONSTELLAR OBJECT
UGC 02271	02 45 06.	+ 00 13	96	15.6	GALAXY SBb
ZWG 389.018	02 45 06.	+ 02 57		15.0	GALAXY
MRK 599	02 45 06.	+ 02 57	12	16.	GALAXY WITH UV CONTINUUM
ZWG 415.009	02 45 06.	+ 09 09		15.5	GALAXY
KARA.73B 0118	02 45 06.	+ 09 09	60	15.5	ISOLATED GALAXY E
ZWG 440.010	02 45 06.	+ 14 18		15.7	GALAXY
ZWG 484.020	02 45 06.	+ 26 54		14.7	GALAXY
UGC 02272	02 45 06.	+ 26 54	120	14.7	GALAXY Sb-c
ZWG 539.102	02 45 06.	+ 41 25		15.4	GALAXY
ZWG 554.006	02 45 06.	+ 47 47		15.7	GALAXY
UGC 02273	02 45 06.	+ 47 47	90	15.7	GALAXY SBc
MCG+08-06-007	02 45 06.	+ 47 47	66	16.	GALAXY
MCG+08-06-008	02 45 06.	+ 50 33	102	15.	GALAXY
MCG-03-08-025	02 45 06.	- 15 00	48	14.5	GALAXY
LB 03278	02 45 06.	- 68 13		13.5	FAINT BLUE STAR
LB 02793	02 45 07.	+ 01 28 30.		16.4	FAINT BLUE STAR
MCG+04-07-024	02 45 09.	+ 26 55	96	14.	GALAXY
RNGC 1093	02 45 10.	+ 34 13		14.5	GALAXY
MCG+00-08-016	02 45 12.	+ 00 12	66	15.	GALAXY
ZWG 389.019	02 45 12.	+ 00 13		15.6	GALAXY
MCG+01-08-002	02 45 12.	+ 09 09 30.	24	15.	GALAXY
MCG+02-08-007	02 45 12.	+ 10 11	24	16.	GALAXY
ZWG 524.022	02 45 12.	+ 34 13		14.3	GALAXY
UGC 02274	02 45 12.	+ 34 13	132	14.3	GALAXY S
MRK 371	02 45 12.	- 01 07	6	16.	GALAXY WITH UV CONTINUUM
MCG-03-08-026	02 45 12.	- 14 59	36	15.	GALAXY
MCG-04-07-041	02 45 12.	- 22 29	18	15.5	GALAXY
PHL 4310	02 45 12.	- 26 14		17.9	BLUE STELLAR OBJECT
TON-S 0279	02 45 12.	- 33 16		15.8	BLUE STAR
LB 02794	02 45 17.	- 00 27 24.		16.6	FAINT BLUE STAR
MCG+00-08-017	02 45 18.	+ 02 58	27	15.	GALAXY
MCG+06-07-011	02 45 18.	+ 34 12	78	14.	GALAXY
LB 03279	02 45 18.	- 68 56		13.4	FAINT BLUE STAR
LB 02795	02 45 19.	+ 00 37 18.		16.9	FAINT BLUE STAR
ZWG 415.010	02 45 24.	+ 03 41		15.7	GALAXY
UGC 02275	02 45 24.	+ 03 41	360	15.7	GALAXY DWRF SP
MCG+02-08-008	02 45 24.	+ 13 30	36	16.	GALAXY
ZC 0245.4-0040	02 45 24.	- 00 40	340		CLUSTER OF GALAXIES
MCG-04-07-042	02 45 24.	- 20 49	18	16.	GALAXY
PHL 8530	02 45 24.	- 21 10		17.1	BLUE STELLAR OBJECT
ABC 0385	02 45 24.	- 22 02		17.9	RICH CLUSTER OF GALAXIES
MCG-04-07-043	02 45 24.	- 25 22	36	14.5	GALAXY
SVEN 182	02 45 27.	- 00 56	12	15.2	GALAXY
IC 1849	02 45 30.	+ 09 09 16.			NONSTELLAR OBJECT
UGC 02276	02 45 30.	+ 17 32	84	16.0	GALAXY S
URA 29	02 45 30.	+ 68 47	360		STELLAR RING
PHL 8531	02 45 30.	- 28 14		18.4	BLUE STELLAR OBJECT
SVEN 184	02 45 33.	- 00 56 54.	18	14.6	GALAXY
ABC 0383	02 45 34.	- 03 43		17.6	RICH CLUSTER OF GALAXIES
MCG+03-08-013	02 45 36.	+ 15 49	30	16.	GALAXY
MCG+03-08-012	02 45 36.	+ 17 31	42	15.	GALAXY
MCG+07-06-072	02 45 36.	+ 40 27 30.	18	15.5	GALAXY
ZWG 539.103	02 45 36.	+ 40 29		15.5	GALAXY
UGC 02277	02 45 36.	+ 40 29	66	15.5	GALAXY COMPACT
MCG+00-08-018	02 45 36.	- 00 58	15	16.	GALAXY
TON-S 0280	02 45 36.	- 29 31		14.2	BLUE STAR
PHL 8532	02 45 36.	- 29 34		14.0	BLUE STELLAR OBJECT
RNGC 1103	02 45 37.	- 14 09		13.0	GALAXY
MCG-02-08-005	02 45 39.	- 14 09	120	13.5	GALAXY
MCG-02-08-006	02 45 39.	- 14 11	48	14.	GALAXY
RNGC 1101	02 45 41.	+ 04 22		15.0	GALAXY
ZWG 415.011	02 45 42.	+ 04 22		14.8	GALAXY
UGC 02278	02 45 42.	+ 04 22	96	14.8	GALAXY S0
MCG+01-08-003	02 45 42.	+ 04 22	48	14.	GALAXY
UGC 02279	02 45 42.	+ 18 45	60	16.5	GALAXY S
MCG+07-06-073	02 45 42.	+ 41 33	102	15.5	GALAXY
ZWG 539.104	02 45 42.	+ 41 35		15.0	GALAXY
UGC 02280	02 45 42.	+ 41 35	144	15.0	GALAXY Sb
UGC 02281	02 45 42.	+ 46 57	78	16.0	GALAXY S
ABC 0384	02 45 42.	- 02 30		18.0	RICH CLUSTER OF GALAXIES
IC 1853	02 45 42.	- 14 11 58.			NONSTELLAR OBJECT
MCG-04-07-045	02 45 42.	- 22 13	42	15.5	GALAXY
MCG-04-07-044	02 45 42.	- 26 15	48	14.	GALAXY
ZWG 440.011	02 45 48.	+ 11 48		15.5	GALAXY
ZWG 440.012	02 45 48.	+ 14 03		15.5	GALAXY
MCG+02-08-009	02 45 48.	+ 14 03	36	15.	GALAXY
MCG+08-06-009	02 45 48.	+ 46 55	48	16.	GALAXY
IZW 008	02 45 48.	- 09 08			COMPACT GALAXY
MCG-04-07-046	02 45 48.	- 22 26	72	15.	GALAXY
ZWG 440.013	02 45 54.	+ 14 06		14.8	GALAXY
UGC 02282	02 45 54.	+ 14 06	72	14.8	GALAXY SBc
MCG+02-08-010	02 45 54.	+ 14 06	60	14.	GALAXY
ZWG 440.014	02 45 54.	+ 15 19		15.7	GALAXY
ZWG 463.016	02 45 54.	+ 17 00		15.3	GALAXY
UGC 02283	02 45 54.	+ 17 00	60	15.3	GALAXY SB:c
MCG+03-08-014	02 45 54.	+ 17 00	45	15.	GALAXY
ZC 0245.9+2845	02 45 54.	+ 28 45	1340		CLUSTER OF GALAXIES
MCG-02-08-007	02 45 54.	- 10 26	66	15.	GALAXY
TON-S 0281	02 45 54.	- 30 16		13.8	BLUE STAR
IC 1850	02 45 56.	+ 13 02 02.			NONSTELLAR OBJECT
MCG+01-08-004	02 45 57.	+ 06 19	78	15.	GALAXY
MCG+03-08-015	02 45 57.	+ 20 40	24	15.	GALAXY
MCG-02-08-008	02 45 57.	- 09 07	30	15.5	GALAXY
VDB.66G 029	02 46	+ 01 54	200		DWARF GALAXY
LBN 0669	02 46	+ 59 30	600		BRIGHT NEBULA
LBN 0652	02 46	+ 65 20	1620		BRIGHT NEBULA
ZWG 415.012	02 46 00.	+ 06 19		15.3	GALAXY
UGC 02285	02 46 00.	+ 06 19	84	15.3	GALAXY Sc-IRR
KARA.73B 0119	02 46 00.	+ 06 19	72	15.3	ISOLATED GALAXY S
ZWG 463.017	02 46 00.	+ 20 39		15.4	GALAXY
UGC 02286	02 46 00.	+ 20 39	66	15.4	GALAXY COMPACT
ZWG 505.054	02 46 00.	+ 28 04		15.7	GALAXY
ZWG 524.023	02 46 00.	+ 36 15		15.4	GALAXY
SG 1.08	02 46 00.	+ 59 32	420		DIFFUSE EMISSION NEBULA
UGC 02284	02 46 00.	- 02 18	84	17.	GALAXY
PHL 4311	02 46 00.	- 23 46		18.3	BLUE STELLAR OBJECT
PHL 8533	02 46 00.	- 30 14		13.5	BLUE STELLAR OBJECT
SVEN 187	02 46 03.	- 00 29	48	14.0	GALAXY
SVEN 185	02 46 03.	- 00 44 56.	18	15.8	GALAXY
SVEN 186	02 46 03.	- 01 11	54	15.4	GALAXY
REIN 6.029	02 46 05.18	- 00 28 45.0			NEBULA
ZWG 524.024	02 46 06.	+ 35 08		15.5	GALAXY
UGC 02288	02 46 06.	+ 35 08	78	15.5	GALAXY S0-a
ZWG 389.020	02 46 06.	- 00 28		14.8	GALAXY
UGC 02287	02 46 06.	- 00 28	90	14.8	GALAXY SBa
MCG+00-08-019	02 46 06.	- 00 29	42	14.5	GALAXY
ZC 0246.1-0045	02 46 06.	- 00 45	8940		CLUSTER OF GALAXIES
PHL 4312	02 46 06.	- 21 16		18.7	BLUE STELLAR OBJECT
LB 03280	02 46 06.	- 73 14		15.3	FAINT BLUE STAR
RNGC 1104	02 46 07.	- 00 29		15.0	GALAXY
RNGC 1108	02 46 10.	- 08 10			GALAXY
ZWG 463.018	02 46 12.	+ 15 31		15.4	GALAXY
UGC 02289	02 46 12.	+ 15 31	60	15.4	GALAXY S
UGC 02290	02 46 12.	+ 22 48	102	18.	GALAXY DWRF SP
MCG+06-07-012	02 46 12.	+ 35 08	42	15.	GALAXY
MRSL 137+00/1	02 46 12.	+ 59 30	540		HII REGION
MCG+01-08-009	02 46 12.	- 04 50	30	15.5	GALAXY
PHL 4313	02 46 12.	- 22 48		18.4	BLUE STELLAR OBJECT
MCG-05-07-027	02 46 12.	- 29 48	36	15.	GALAXY
TON-S 0282	02 46 12.	- 31 09		15.9	BLUE STAR
PHL 1454	02 46 12.	- 31 46		16.8	BLUE STELLAR OBJECT
LB 03281	02 46 12.	- 57 18		14.2	FAINT BLUE STAR
MCG+00-08-020	02 46 15.	+ 00 47	42	15.	GALAXY
IC 1852	02 46 15.	+ 12 59 49.			NONSTELLAR OBJECT
SVEN 188	02 46 15.	- 00 56	54	14.0	GALAXY
BAK 1.001	02 46 17.	- 29 48 03.	23	15.2	EXTRAGALACTIC NEBULA
IC 1856	02 46 17.9	+ 00 58 30.			GALAXY
ZWG 389.022	02 46 18.	+ 00 47		15.4	GALAXY
UGC 02292	02 46 18.	+ 00 47	72	15.4	GALAXY Sa
MCG+02-08-011	02 46 18.	+ 13 00	72	14.	GALAXY
ZWG 440.015	02 46 18.	+ 13 01		14.9	GALAXY
UGC 02293	02 46 18.	+ 13 01	78	14.9	GALAXY SBc
MCG+03-08-016	02 46 18.	+ 15 30	45	15.	GALAXY
5ZW 279	02 46 18.	+ 35 12			COMPACT GALAXY
ZWG 539.105	02 46 18.	+ 40 21		15.7	GALAXY
UGC 02294	02 46 18.	+ 46 57	72	16.0	GALAXY S
1ZW 009	02 46 18.	- 00 44			COMPACT GALAXY
ZWG 389.021	02 46 18.	- 00 58		14.5	GALAXY
UGC 02291	02 46 18.	- 00 58	72	14.5	GALAXY SB:b
MCG+00-08-021	02 46 18.	- 00 58 30.	66	14.5	GALAXY
TON-S 0283	02 46 18.	- 31 48		15.7	BLUE STAR
RNGC 1109	02 46 20.	+ 13 03		15.0	GALAXY
MCG+01-08-005	02 46 21.	+ 08 13	36	15.	GALAXY
MCG+03-08-017	02 46 21.	+ 18 07	27	15.	GALAXY
SVEN 189	02 46 21.	- 01 09 57.	24	15.0	GALAXY
ZWG 389.023	02 46 24.	+ 02 58		14.0	GALAXY
UGC 02295	02 46 24.	+ 02 58	96	14.0	GALAXY S
MCG+00-08-022	02 46 24.	+ 02 59	72	14.	GALAXY
ZWG 463.019	02 46 24.	+ 18 07		13.1	GALAXY
UGC 02296	02 46 24.	+ 18 07	60	13.1	GALAXY EX CMPT
UGC 02297	02 46 24.	+ 30 27	66	16.0	GALAXY S0-a
MCG+08-06-010	02 46 24.	+ 46 22	42	16.	GALAXY
ZWG 554.007	02 46 24.	+ 46 23		15.7	GALAXY
5ZW 280	02 46 24.	+ 46 46			COMPACT GALAXY
MCG+08-06-011	02 46 24.	+ 46 46	30	15.	GALAXY
ZWG 554.008	02 46 24.	+ 46 47		14.5	GALAXY
UGC 02298	02 46 24.	+ 46 47	150	14.5	GALAXY E-S0
MCG-05-07-028	02 46 24.	- 27 39	18	15.	GALAXY
MCG-05-07-029	02 46 24.	- 31 46	120	14.	GALAXY
IC 0256	02 46 25.	+ 46 46 21.			NONSTELLAR OBJECT
SVEN 190	02 46 27.	- 01 04	12	15.8	GALAXY
LB 02796	02 46 29.	+ 01 08 12.		16.4	FAINT BLUE STAR
HOLM 063B	02 46 30.	+ 07 54	24	13.3	PART OF MULTIPLE GALAXY
UGC 02299	02 46 30.	+ 10 54	60	17.	GALAXY
MCG+02-08-012	02 46 30.	+ 14 19	48	15.5	GALAXY
ZWG 463.020	02 46 30.	+ 19 05		15.0	GALAXY
MRK 372	02 46 30.	+ 19 05	20	15.5	GALAXY WITH UV CONTINUUM
KW 35	02 46 30.	+ 19 05	23		SEYFERT GALAXY
VVI 11	02 46 30.	+ 19 05	20	15.86	SEYFERT GALAXY
UGC 02300	02 46 30.	+ 30 29	66	16.0	GALAXY Sa-b
UGC 02301	02 46 30.	+ 38 04	60	18.	GALAXY DWARF
ZWG 539.106	02 46 30.	+ 40 51		15.5	GALAXY
KARA.72 079A	02 46 30.	+ 40 51	36	15.5	PART OF DOUBLE GALAXY
IC 0257	02 46 30.	+ 46 46 38.			NONSTELLAR OBJECT
TON-S 0284	02 46 30.	- 27 00		15.7	BLUE STAR
MCG-05-07-030	02 46 30.	- 32 34	36	14.5	GALAXY
SVEN 194	02 46 31.	- 30 58	12	15.9	GALAXY
IC 1855	02 46 32.	+ 13 09 29.			NONSTELLAR OBJECT
IC 1854	02 46 32.	+ 19 06 24.			SEYFERT GALAXY
LB 02797	02 46 33.	+ 01 19 00.		16.6	FAINT BLUE STAR
MCG+04-07-025	02 46 33.	+ 21 56	72	15.5	GALAXY
IC 0258	02 46 33.	+ 40 51 13.			NONSTELLAR OBJECT
MCG+07-06-074	02 46 33.	+ 41 14 30.	30	15.5	GALAXY
SVEN 191	02 46 33.	- 01 05 57.	30	15.0	GALAXY
BAK 1.002	02 46 34.	- 31 44 26.	35	14.8	EXTRAGALACTIC NEBULA
ZWG 389.024	02 46 36.	+ 01 56		15.3	GALAXY
UGC 02302	02 46 36.	+ 01 56	360	15.3	GALAXY DWRF SP
MCG+00-08-023	02 46 36.	+ 01 56	180	15.	GALAXY
ZWG 463.021	02 46 36.	+ 17 27		14.5	GALAXY
UGC 02303	02 46 36.	+ 17 27	84	14.5	GALAXY Sb
MCG+03-08-018	02 46 36.	+ 17 27	66	14.	GALAXY

OBJECT NAME	RIGHT ASCEN.	DECLINATION	DIAM.	MAGN.	TYPE OF OBJECT
ZC 0246.6+1942	02 46 36.	+ 19 42	1880		CLUSTER OF GALAXIES
UGC 02304	02 46 36.	+ 21 55	84	16.0	GALAXY S
ZWG 524.025	02 46 36.	+ 37 51		15.5	GALAXY
UGC 02305	02 46 36.	+ 37 51	90	15.5	GALAXY Sc
ZWG 539.107	02 46 36.	+ 40 51		15.6	GALAXY
UGC 02306	02 46 36.	+ 40 51	102	15.6	GALAXY SB0
KARA.72 079B	02 46 36.	+ 40 51	54	15.6	PART OF DOUBLE GALAXY
ZWG 539.108	02 46 36.	+ 41 16		14.7	GALAXY
ZWG 539.109	02 46 36.	+ 42 20		15.7	GALAXY
MCG-03-08-027	02 46 36.	- 20 10	72	14.5	GALAXY
PHL 4315	02 46 36.	- 22 49		18.4	BLUE STELLAR OBJECT
PHL 8534	02 46 36.	- 31 32		16.9	BLUE STELLAR OBJECT
IC 1858	02 46 37.	- 31 29 58.			NONSTELLAR OBJECT
LB 02798	02 46 38.	- 00 52 00.		16.2	FAINT BLUE STAR
SVEN 192	02 46 39.	- 00 56 58.	18	15.4	GALAXY
IC 1859	02 46 39.	- 31 23 58.			NONSTELLAR OBJECT
RNGC 1107	02 46 40.	+ 07 54		14.0	GALAXY
HOLM 063A	02 46 40.	+ 07 54	42	13.0	PART OF MULTIPLE GALAXY
ZWG 415.013	02 46 42.	+ 07 54		14.1	GALAXY
UGC 02307	02 46 42.	+ 07 54	138	14.1	GALAXY S0
MCG+01-08-006	02 46 42.	+ 07 54	24	13.5	GALAXY
ZWG 463.022	02 46 42.	+ 15 35		15.6	GALAXY
MCG+06-07-013	02 46 42.	+ 37 51 30.	48	15.	GALAXY
UGC 02308	02 46 42.	+ 46 52	90	16.0	GALAXY MLT SYS
IC 0261	02 46 42.	- 14 40 39.			NONSTELLAR OBJECT
MCG-05-07-031	02 46 42.	- 28 44	36	15.	GALAXY
MCG-05-07-032	02 46 42.	- 31 23	72	14.	GALAXY
PHL 4316	02 46 42.	- 31 26		18.0	BLUE STELLAR OBJECT
MCG-06-07-004	02 46 42.	- 36 54	54	15.	GALAXY
LB 02799	02 46 43.	+ 02 18 42.		16.0	FAINT BLUE STAR
RNGC 1110	02 46 46.	- 08 04		14.0	GALAXY
MCG+02-08-013	02 46 48.	+ 14 25	60	15.	GALAXY
ZWG 463.023	02 46 48.	+ 21 00		15.6	GALAXY
UGC 02309	02 46 48.	+ 21 00	102	15.6	GALAXY S0
MCG+03-08-019	02 46 48.	+ 21 00	30	15.	GALAXY
ZWG 524.026	02 46 48.	+ 34 47		15.4	GALAXY
UGC 02310	02 46 48.	+ 35 40	66	16.5	GALAXY
MCG-02-08-009	02 46 48.	- 10 51 30.	36	14.5	GALAXY
MCG-03-08-028	02 46 48.	- 14 39	84	14.	GALAXY
MCG-03-08-029	02 46 48.	- 17 12	120	13.	GALAXY
PHL 8535	02 46 48.	- 23 24		18.2	BLUE STELLAR OBJECT
MCG-05-07-033	02 46 48.	- 31 30	24	14.5	GALAXY
RNGC 1120	02 46 49.	- 14 39		14.0	GALAXY
BAK 1.003	02 46 49.	- 28 44 49.	19	15.1	EXTRAGALACTIC NEBULA
SVEN 197	02 46 49.	- 30 51 59.	12	15.0	GALAXY
RNGC 1114	02 46 50.	- 17 12		13.0	GALAXY
MCG+01-08-010	02 46 51.	- 08 04 30.	168	14.	GALAXY
ZC 0246.9+0834	02 46 54.	+ 08 34	1280		CLUSTER OF GALAXIES
IC 1857	02 46 54.	+ 14 24 57.			NONSTELLAR OBJECT
ZWG 440.016	02 46 54.	+ 14 25		15.1	GALAXY
UGC 02312	02 46 54.	+ 14 25	60	15.1	GALAXY S
ZWG 539.110	02 46 54.	+ 41 11		15.7	GALAXY
UGC 02313	02 46 54.	+ 41 11	90	15.7	GALAXY Sc
UGC 02314	02 46 54.	+ 47 20	96	18.	GALAXY DWARF
MCG+11-04-002	02 46 54.	+ 67 36	39	12.	GALAXY
MCG+11-04-002	02 46 54.	+ 67 36	39	12.	GALAXY
ZWG 389.025	02 46 54.	- 01 05		13.9	GALAXY
UGC 02311	02 46 54.	- 01 05	102	13.9	GALAXY SBb
MCG+00-08-024	02 46 54.	- 01 05	66	13.5	GALAXY
TON-S 0285	02 46 54.	- 31 04		15.9	BLUE STAR
LB 02800	02 46 55.	- 00 11 42.		16.0	FAINT BLUE STAR
RNGC 1111	02 46 56.	+ 12 56			GALAXY
IC 0259	02 46 56.	+ 40 51 17.			NONSTELLAR OBJECT
SVEN 193	02 46 57.	- 01 03 59.	36	13.6	GALAXY
LB 02801	02 46 58.	- 00 04 06.		16.5	FAINT BLUE STAR
LBN 0670	02 47	+ 59 45	900		BRIGHT NEBULA
LBN 0667	02 47	+ 60 05	1500		BRIGHT NEBULA
LBN 0664	02 47	+ 62 00	360		BRIGHT NEBULA
LBN 0643	02 47	+ 68 40	120		BRIGHT NEBULA
LBN 0820	02 47	- 04 00	1440		BRIGHT NEBULA
UGC 02315	02 47 00.	+ 22 31	66	17.	GALAXY Sb-c
UGC 02316	02 47 00.	+ 39 00	66	16.0	GALAXY S0
ZWG 554.009	02 47 00.	+ 46 30		15.6	GALAXY
UGC 02317	02 47 00.	+ 46 30	108	15.6	GALAXY Sa
MCG+08-06-012	02 47 00.	+ 46 30	84	16.	GALAXY
IC 1848	02 47 00.	+ 60 10	7200		HII REGION
MCG+01-08-011	02 47 00.	- 08 25	42	15.	GALAXY
PHL 8536	02 47 00.	- 27 49		17.9	BLUE STELLAR OBJECT
LB 02802	02 47 02.	+ 01 06 12.		16.4	FAINT BLUE STAR
SVEN 196	02 47 03.	- 00 15 59.	30	14.4	GALAXY
SVEN 195	02 47 03.	- 00 54 59.	30	14.9	GALAXY
ZWG 440.017	02 47 06.	+ 14 12		15.7	GALAXY
5ZW 281	02 47 06.	+ 38 59			COMPACT GALAXY
ZWG 389.027	02 47 06.	- 00 16		14.7	GALAXY
ZWG 389.026	02 47 06.	- 00 56		15.7	GALAXY
SVEN 198	02 47 07.	- 30 46 00.	30	14.3	GALAXY
BAK 1.004	02 47 07.	- 31 15 06.	23	15.7	EXTRAGALACTIC NEBULA
IC 1860	02 47 11.	- 31 24 01.			NONSTELLAR OBJECT
ZC 0247.2+0229	02 47 12.	+ 02 29	1680		CLUSTER OF GALAXIES
ZWG 539.111	02 47 12.	+ 43 27		15.7	GALAXY
UGC 02318	02 47 12.	+ 43 27	66	15.7	GALAXY Sb-c
VB 178	02 47 12.	+ 57 28	1007		STELLAR RING
VDB.66N 008	02 47 12.	+ 67 40	276		REFLECTION NEBULA
MCG+00-08-025	02 47 12.	- 00 43	48	15.	GALAXY
ZWG 389.028	02 47 12.	- 00 44		15.1	GALAXY
SVEN 199	02 47 13.	- 30 30 00.	18	15.7	GALAXY
RNGC 1112	02 47 14.	+ 13 01			NON-EXISTENT OBJECT
SN 1959F	02 47 14.	- 00 44		18.5	SUPERNOVA
MCG+07-06-075	02 47 15.	+ 43 26	60	16.	GALAXY
MCG+01-08-012	02 47 15.	- 06 26	42	15.5	GALAXY
LB 02803	02 47 18.	+ 02 04 18.		16.5	FAINT BLUE STAR
OCL 0364	02 47 18.	+ 60 14	2880	7.6	OPEN STAR CLUSTER
MRSL 136+02/4	02 47 18.	+ 62 02	240		HII REGION
MCG+00-08-026	02 47 18.	- 01 13	66	15.	GALAXY
MCG-02-08-010	02 47 18.	- 12 26 30.	30	15.	GALAXY
MCG-03-08-030	02 47 18.	- 14 40	24	15.5	GALAXY
PHL 8538	02 47 18.	- 26 44		16.4	BLUE STELLAR OBJECT
PHL 1455	02 47 18.	- 27 32		18.2	BLUE STELLAR OBJECT
PHL 8537	02 47 18.	- 30 47		17.8	BLUE STELLAR OBJECT
MCG-05-07-034	02 47 18.	- 30 47	30	14.	GALAXY
MCG-05-07-035	02 47 18.	- 31 23 30.	36	13.	GALAXY
MCG-05-07-036	02 47 18.	- 31 24	36	15.5	GALAXY
KLEM 02	02 47 22.	- 31 29	2280	14.	CLUSTER OF 35 GALAXIES
RNGC 1106	02 47 23.	+ 41 29		13.5	GALAXY
DG 010	02 47 24.	+ 68 43	360		REFLECTION NEBULA
ZWG 440.018	02 47 24.	+ 12 38		15.4	GALAXY
ZWG 440.019	02 47 24.	+ 12 40		15.2	GALAXY
UGC 02320	02 47 24.	+ 12 40	72	15.2	GALAXY S
ZWG 463.024	02 47 24.	+ 18 09		15.7	GALAXY
UGC 02321	02 47 24.	+ 18 09	78	15.7	GALAXY Sa-b
MCG+03-08-020	02 47 24.	+ 18 09	48	15.	GALAXY
MCG+07-06-076	02 47 24.	+ 41 26 30.	90	14.	GALAXY
ZWG 539.112	02 47 24.	+ 41 29		13.7	GALAXY
UGC 02322	02 47 24.	+ 41 29	114	13.7	GALAXY S0
ZC 0247.4+4157	02 47 24.	+ 41 57	1750		CLUSTER OF GALAXIES
ZWG 554.010	02 47 24.	+ 46 34		15.7	GALAXY
ZWG 389.029	02 47 24.	- 01 12		15.6	GALAXY
UGC 02319	02 47 24.	- 01 12	84	15.6	GALAXY Sc
MCG-05-07-037	02 47 24.	- 31 12	24	15.	GALAXY
PHL 8539	02 47 24.	- 31 18		16.8	BLUE STELLAR OBJECT
RNGC 1113	02 47 26.	+ 13 06			NON-EXISTENT OBJECT
MCG+02-08-014	02 47 27.	+ 12 37 30.	18	15.	GALAXY
CED 009A	02 47 27.	+ 60 14	3600		DIFFUSE GALACTIC NEBULA
BAK 1.005	02 47 27.	- 31 23 53.	12	16.0	EXTRAGALACTIC NEBULA
MCG+02-08-015	02 47 30.	+ 12 40	30	15.	GALAXY
MCG+03-08-021	02 47 30.	+ 18 54	45	15.	GALAXY
ZWG 463.025	02 47 30.	+ 18 55		14.8	GALAXY
MCG+08-06-013	02 47 30.	+ 46 33	60	16.	GALAXY
PHL 8540	02 47 30.	- 25 02		18.2	BLUE STELLAR OBJECT
LB 02804	02 47 31.	+ 01 31 18.		16.9	FAINT BLUE STAR
ARP 190	02 47 31.	+ 12 41			PECULIAR GALAXY
HELW 049	02 47 31.	- 30 43			NEBULA
BAK 1.006	02 47 31.	- 31 11 47.	23	15.6	EXTRAGALACTIC NEBULA
LB 02805	02 47 34.	+ 02 58 36.		16.4	FAINT BLUE STAR
ZWG 415.014	02 47 36.	+ 07 58		15.7	GALAXY
UGC 02323	02 47 36.	+ 07 58	66	15.7	GALAXY SB?c
VV 221B	02 47 36.	+ 12 41	12	16.	INTERACTING GALAXY
VV 221A	02 47 36.	+ 12 41	12	16.	INTERACTING GALAXY
VV 221	02 47 36.	+ 12 41	48		INTERACTING GALAXY
ZWG 539.113	02 47 36.	+ 39 45		15.7	GALAXY
RNGC 1118	02 47 36.	- 12 21		14.0	GALAXY
MCG-02-08-011	02 47 36.	- 12 21	84	14.	GALAXY
BAK 1.007	02 47 39.	- 31 21 23.	19	15.4	EXTRAGALACTIC NEBULA
LB 00473	02 47 40.	+ 24 02 48.		17.8	FAINT BLUE STAR
IC 0260	02 47 40.	+ 46 45 20.			NONSTELLAR OBJECT
HN 0163	02 47 41.	+ 23 23			NEBULA
VDB.66N 009	02 47 41.	+ 68 39	444		REFLECTION NEBULA
MCG+01-08-007	02 47 42.	+ 07 58	36	15.	GALAXY
ZWG 440.020	02 47 42.	+ 13 03		15.6	GALAXY
MCG+02-08-016	02 47 42.	+ 13 03	36	16.	GALAXY
ZWG 505.055	02 47 42.	+ 33 18		15.7	GALAXY
ZWG 554.011	02 47 42.	+ 46 45		15.2	GALAXY
UGC 02325	02 47 42.	+ 46 45	108	15.2	GALAXY E
MCG+08-06-014	02 47 42.	+ 46 45	24	15.	GALAXY
ZWG 389.030	02 47 42.	- 00 06		14.9	GALAXY
UGC 02324	02 47 42.	- 00 06	66	14.9	GALAXY SBb
MCG+00-08-027	02 47 42.	- 01 05	18	16.	GALAXY
ARC 0386	02 47 43.	- 17 22		17.2	RICH CLUSTER OF GALAXIES
RNGC 1115	02 47 44.	+ 13 03		15.5	GALAXY
MCG+03-08-022	02 47 45.	+ 20 29	42	14.5	GALAXY
MCG+00-08-028	02 47 45.	- 00 07 30.	36	14.5	GALAXY
RNGC 1119	02 47 45.	- 18 14			NON-EXISTENT OBJECT
LB 02806	02 47 47.	+ 00 46 30.		16.0	FAINT BLUE STAR
ZC 0247.8+0506	02 47 48.	+ 05 06	940		CLUSTER OF GALAXIES
ZWG 440.021	02 47 48.	+ 13 07		15.4	GALAXY
UGC 02326	02 47 48.	+ 13 07	90	15.4	GALAXY Sa-b
MCG+02-08-017	02 47 48.	+ 13 07	72	15.	GALAXY
ZWG 463.026	02 47 48.	+ 15 51		15.7	GALAXY
ZC 0247.8+1754	02 47 48.	+ 17 54	870		CLUSTER OF GALAXIES
ZWG 463.027	02 47 48.	+ 20 29		15.0	GALAXY
UGC 02327	02 47 48.	+ 20 29	72	15.0	GALAXY Sb
ZWG 554.012	02 47 48.	+ 50 40		15.3	GALAXY
VB 179	02 47 48.	+ 57 25	671		STELLAR RING
MCG+00-08-029	02 47 48.	- 00 07 30.	48	14.5	GALAXY
PHL 1456	02 47 48.	- 28 38		18.5	BLUE STELLAR OBJECT
RNGC 1116	02 47 50.	+ 13 07		15.5	GALAXY
MCG+02-08-018	02 47 51.	+ 14 47	24	15.5	GALAXY
MCG+04-07-026	02 47 51.	+ 22 34	48	15.5	GALAXY
MCG+06-07-014	02 47 51.	+ 37 16	27	14.5	GALAXY
ZWG 389.031	02 47 54.	+ 00 53		15.7	GALAXY
ZWG 484.021	02 47 54.	+ 22 35		15.5	GALAXY
ZWG 524.027	02 47 54.	+ 37 15		13.8	GALAXY
UGC 02328	02 47 54.	+ 37 15	102	13.8	GALAXY E
ISS 0062	02 47 54.	+ 62 24	396		STELLAR RING
PHL 4317	02 47 54.	- 23 10		18.0	BLUE STELLAR OBJECT
LB 03282	02 47 54.	- 68 03		13.0	FAINT BLUE STAR
CED 010	02 47 57.	+ 58 07	360		DIFFUSE GALACTIC NEBULA
KARA.68 027	02 47 57.	- 08 23	27		DWARF GALAXY
LB 02807	02 47 58.	+ 00 59 48.		16.5	FAINT BLUE STAR
VDB.66G 030	02 48	- 01 21	100		DWARF GALAXY
ZC 0248.0+1307	02 48 00.	+ 13 07	9270		CLUSTER OF GALAXIES
ZWG 463.028	02 48 00.	+ 15 48		15.0	GALAXY
UGC 02329	02 48 00.	+ 15 48	78	15.0	GALAXY Sa-b
MCG+03-08-023	02 48 00.	+ 15 52	42	15.	GALAXY
ZWG 463.029	02 48 00.	+ 15 57		15.7	GALAXY
PHL 8541	02 48 00.	+ 24 59		14.5	BLUE STELLAR OBJECT
UGC 02330	02 48 00.	+ 41 22	84	16.0	GALAXY E
UGC 02331	02 48 00.	+ 47 00	66	16.5	GALAXY Sa
IC 1851	02 48 00.	+ 58 06 32.			MAY NOT EXIST
VB 086	02 48 00.	+ 67 09	1296		STELLAR RING
LDN 1357	02 48 00.	+ 68 40	240		DARK NEBULA
MCG-02-08-012	02 48 00.	- 08 46	132	14.5	GALAXY
PHL 4318	02 48 00.	- 25 12		18.5	BLUE STELLAR OBJECT
PHL 4319	02 48 00.	- 28 04		17.9	BLUE STELLAR OBJECT
LB 02808	02 48 01.	+ 01 40 48.		16.4	FAINT BLUE STAR
IC 0263	02 48 01.	- 00 19 27.			NONSTELLAR OBJECT
MCG+07-06-077	02 48 03.	+ 41 20	15	16.	GALAXY
SS 07	02 48 09.	+ 62 00			DIFFUSE GALACTIC NEBULA
MCG+01-08-013	02 48 09.	- 03 11	42	16.	GALAXY
LB 02809	02 48 10.	+ 00 38 18.		16.8	FAINT BLUE STAR
VB 180	02 48 12.	+ 62 22	376		STELLAR RING
LDN 1355	02 48 12.	+ 68 50	300		DARK NEBULA
ZWG 389.032	02 48 12.	- 01 56		13.7	GALAXY
UGC 02332	02 48 12.	- 01 56	66	13.7	GALAXY S
MCG+00-08-030	02 48 12.	- 01 57	30	15.	GALAXY
PHL 4321	02 48 12.	- 26 08		17.2	BLUE STELLAR OBJECT
PHL 4320	02 48 12.	- 29 46		18.2	BLUE STELLAR OBJECT
MCG-05-07-038	02 48 12.	- 31 46	48	15.	GALAXY
LB 03283	02 48 12.	- 57 54		13.6	FAINT BLUE STAR
RNGC 1121	02 48 14.	- 01 56		13.5	GALAXY
MCG+04-07-027	02 48 15.	+ 25 13	42	15.5	GALAXY
ZWG 484.022	02 48 18.	+ 25 12		15.7	GALAXY
UGC 02333	02 48 18.	+ 25 12	60	15.7	GALAXY Sb-c
UGC 02334	02 48 18.	+ 40 32	84	16.0	GALAXY
MCG+01-08-014	02 48 18.	- 06 58	72	15.	GALAXY
MCG-03-08-031	02 48 18.	- 19 10	36	15.	GALAXY

OBJECT NAME	RIGHT ASCEN.	DECLINATION	DIAM.	MAGN.	TYPE OF OBJECT
PHL 4322	02 48 18.	- 22 08		18.5	BLUE STELLAR OBJECT
MCG-05-07-040	02 48 18.	- 31 37	36	14.5	GALAXY
MCG-05-07-039	02 48 18.	- 31 44	36	14.5	GALAXY
MCG-06-07-005	02 48 18.	- 35 01	12	16.	GALAXY
IC 0264	02 48 20.	- 00 21 53.			NONSTELLAR OBJECT
BAK 1.008	02 48 20.	- 31 45 45.	24	15.0	EXTRAGALACTIC NEBULA
MCG+00-08-031	02 48 21.	+ 03 20	27	16.	GALAXY
IC 0262	02 48 23.	+ 42 37 16.			NONSTELLAR OBJECT
ZWG 389.033	02 48 24.	+ 03 19		15.2	GALAXY
MCG+00-08-032	02 48 24.	+ 03 20 30.	24	16.	GALAXY
ZWG 524.028	02 48 24.	+ 37 38		15.5	GALAXY
MCG+07-06-078	02 48 24.	+ 40 31	30	16.	GALAXY
ZWG 539.114	02 48 24.	+ 42 38		14.8	GALAXY
UGC 02335	02 48 24.	+ 42 38	138	14.8	GALAXY SB0
BAK 1.009	02 48 24.	- 31 36 02.	34	14.5	EXTRAGALACTIC NEBULA
MCG-06-07-006	02 48 24.	- 35 02	12	16.	GALAXY
LB 01642	02 48 24.	- 45 20		12.0	FAINT BLUE STAR
MCG+07-06-079	02 48 27.	+ 40 31 30.	45	17.	GALAXY
MCG+07-06-080	02 48 27.	+ 42 36	54	15.	GALAXY
SG 2.004	02 48 27.	+ 62 03	300		DIFFUSE EMISSION NEBULA
BAK 1.010	02 48 29.	- 31 43 44.	28	14.7	EXTRAGALACTIC NEBULA
MCG+00-08-033	02 48 30.	+ 02 24	72	14.	GALAXY
MCG+01-08-008	02 48 30.	+ 04 14 30.	24	15.	GALAXY
ZWG 415.015	02 48 30.	+ 04 15		15.2	GALAXY
MRK 600	02 48 30.	+ 04 15	15	16.	GALAXY WITH UV CONTINUUM
ZWG 415.016	02 48 30.	+ 07 52		15.7	GALAXY
UGC 02336	02 48 30.	+ 07 52	114	15.7	GALAXY SBc
KARA.72 080B	02 48 30.	+ 12 58	36		PART OF DOUBLE GALAXY
MCG+02-08-020	02 48 30.	+ 12 58	18	15.5	GALAXY
MCG+02-08-019	02 48 30.	+ 12 58	24	15.5	GALAXY
ZWG 440.022	02 48 30.	+ 12 59		14.9	GALAXY
UGC 02337	02 48 30.	+ 12 59	60	14.9	GALAXY TRP SYS
KARA.72 080A	02 48 30.	+ 12 59	36	14.9	PART OF DOUBLE GALAXY
MCG+06-07-015	02 48 30.	+ 37 37	36	15.	GALAXY
MCG-06-07-007	02 48 30.	- 35 03	36	14.	GALAXY
LB 01643	02 48 30.	- 54 28		13.0	FAINT BLUE STAR
LB 02810	02 48 31.	- 00 23 48.		16.8	FAINT BLUE STAR
RNGC 1117	02 48 32.	+ 12 59		15.0	GALAXY
KLEM 03	02 48 33.	- 35 06	480	14.	LINEAR GRP OF 5 GALAXIES
KN 11.01	02 48 34.2	+ 02 23 06.			NEBULA
ZWG 389.034	02 48 36.	+ 02 23		14.6	GALAXY
UGC 02338	02 48 36.	+ 02 23	102	14.6	GALAXY Sc
MCG+01-08-009	02 48 36.	+ 07 53	84	14.5	GALAXY
MCG+03-08-024	02 48 36.	+ 15 49 30.	42	15.5	GALAXY
UGC 02339	02 48 36.	+ 15 50	96	16.5	GALAXY
PHL 4323	02 48 36.	- 22 10		18.3	BLUE STELLAR OBJECT
PHL 1457	02 48 36.	- 23 50		16.2	BLUE STELLAR OBJECT
PHL 1458	02 48 36.	- 27 24		18.0	BLUE STELLAR OBJECT
LB 03284	02 48 36.	- 75 53		13.2	FAINT BLUE STAR
LB 02811	02 48 39.	+ 01 27 24.		16.6	FAINT BLUE STAR
MCG+02-08-021	02 48 42.	+ 13 22	60	15.	GALAXY
5ZW 282	02 48 42.	+ 27 00			COMPACT GALAXY
MCG-06-07-008	02 48 42.	- 35 04	24	16.	GALAXY
LB 02812	02 48 44.	+ 03 17 12.		16.8	FAINT BLUE STAR
MCG+03-08-025	02 48 45.	+ 17 26	45	17.	GALAXY
SHAP250-5330.7	02 48 45.	- 53 43 02.	42	16.75	GALAXY
UGC 02340	02 48 48.	+ 13 22	72	16.0	GALAXY SBc
ZWG 524.029	02 48 48.	+ 38 06		15.6	GALAXY
MCG+08-06-015	02 48 48.	+ 47 23	36	16.	GALAXY
5ZW 283	02 48 48.	+ 47 24			COMPACT GALAXY
PK141-07.1	02 48 48.	+ 50 24	138	16.0	PLANETARY NEBULA
ISS 0063	02 48 48.	+ 58 49	353		STELLAR RING
MCG-03-08-032	02 48 48.	- 14 55	60	14.5	GALAXY
MCG-05-07-041	02 48 48.	- 27 07	30	15.	GALAXY
MCG-06-07-009	02 48 48.	- 35 05	24	16.5	GALAXY
LB 02813	02 48 51.	+ 00 48 18.		15.8	FAINT BLUE STAR
SG 1.09	02 48 51.	+ 60 04	3600		DIFFUSE EMISSION NEBULA
KN 11.02	02 48 53.6	+ 02 25 53.			NEBULA
UGC 02341	02 48 54.	+ 17 28	66	16.5	GALAXY Sb/SBb
MCG+08-06-016	02 48 54.	+ 46 57	42	17.	GALAXY
VB 181	02 48 54.	+ 58 48	356		STELLAR RING
MCG-03-08-033	02 48 54.	- 14 52	36	14.5	GALAXY
LB 03285	02 48 54.	- 55 43		12.8	FAINT BLUE STAR
MCG+06-07-016	02 48 57.	+ 38 06	48	16.	GALAXY
ZC 0249.0+0314	02 49 00.	+ 03 14	1410		CLUSTER OF GALAXIES
ZWG 539.115	02 49 00.	+ 43 52		15.2	GALAXY
MCG+08-06-017	02 49 00.	+ 46 53	42	17.	GALAXY
MCG+00-08-034	02 49 00.	- 00 56	15	15.5	GALAXY
MCG-02-08-013	02 49 00.	- 13 30	12	15.5	GALAXY
CED 009B	02 49 03.	+ 60 15			DIFFUSE GALACTIC NEBULA
ARC 0388	02 49 04.	- 03 58		17.6	RICH CLUSTER OF GALAXIES
LB 02814	02 49 06.	+ 00 05 42.		16.8	FAINT BLUE STAR
ZWG 389.035	02 49 06.	- 00 57		15.2	GALAXY
UGC 02343	02 49 06.	- 00 57	66	15.2	GALAXY Sa
PHL 8542	02 49 06.	- 22 30		18.2	BLUE STELLAR OBJECT
ARC 0389	02 49 06.	- 25 07		15.9	RICH CLUSTER OF GALAXIES
MCG-05-07-043	02 49 06.	- 27 09	24	15.	GALAXY
MCG-05-07-042	02 49 06.	- 28 23	72	14.5	GALAXY
SHAP250-5321.3	02 49 06.	- 53 33 37.	54	15.75	GALAXY
LB 02815	02 49 07.	- 01 26 54.		16.4	FAINT BLUE STAR
BAK 1.012	02 49 08.	- 28 24 12.	46	14.8	EXTRAGALACTIC NEBULA
BAK 1.011	02 49 10.	- 31 31 48.	20	15.6	EXTRAGALACTIC NEBULA
ZWG 415.017	02 49 12.	+ 07 34		15.7	GALAXY
ZC 0249.2+1058	02 49 12.	+ 10 58	4100		CLUSTER OF GALAXIES
UGC 02344	02 49 12.	+ 13 43	60	16.5	GALAXY Sb-c
5ZW 284	02 49 12.	+ 36 32			COMPACT GALAXY
ZWG 524.030	02 49 12.	+ 36 32		15.4	GALAXY
5ZW 285	02 49 12.	+ 46 46			COMPACT GALAXY
URA 35	02 49 12.	+ 64 09	246		STELLAR RING
PHL 4324	02 49 12.	- 25 51		18.0	BLUE STELLAR OBJECT
BAK 1.013	02 49 13.	- 30 59 36.	13	15.6	EXTRAGALACTIC NEBULA
IC 1862	02 49 14.	- 33 34 24.			NONSTELLAR OBJECT
RNGC 1135	02 49 15.	- 55 09			UNVERIFIED SOUTHRN OBJECT
SHAP250-5246.0	02 49 17.	- 52 58 19.	42	16.75	GALAXY
UGC 02346	02 49 18.	+ 00 23	78	15.7	GALAXY Sb-c
ZWG 389.036	02 49 18.	+ 00 27		15.7	GALAXY
UGC 02342	02 49 18.	+ 43 52	78	15.2	GALAXY VY CMPT
VB 087	02 49 18.	+ 67 07	1034		STELLAR RING
ISS 0021	02 49 18.	+ 67 07	1039		STELLAR RING
UGC 02345	02 49 18.	- 01 22	270	16.0	GALAXY DWRF SP
MCG+00-08-035	02 49 18.	- 01 23	90	15.	GALAXY
SHAP250-5236.8	02 49 18.	- 52 49 07.	42	17.0	GALAXY
LB 02816	02 49 19.	+ 02 41 42.		15.9	FAINT BLUE STAR
RNGC 1136	02 49 22.	- 55 16			GALAXY
MCG+00-08-036	02 49 24.	+ 00 28	48	15.	GALAXY
MCG+02-08-022	02 49 24.	+ 11 55	48	15.	GALAXY
UGC 02347	02 49 24.	+ 11 56	78	16.0	GALAXY S
ZWG 440.023	02 49 24.	+ 13 47		15.2	GALAXY
UGC 02348	02 49 24.	+ 13 47	96	15.2	GALAXY S0?
MCG+02-08-023	02 49 24.	+ 13 47	18	15.	GALAXY
UGC 02349	02 49 24.	+ 41 07	108	16.0	GALAXY S0
MCG+07-06-081	02 49 24.	+ 41 10	42	15.	GALAXY
ZWG 539.116	02 49 24.	+ 41 12		14.9	GALAXY
UGC 02350	02 49 24.	+ 41 12	72	14.9	GALAXY Sb
MCG+08-06-018	02 49 24.	+ 46 44	72	15.	GALAXY
ZWG 554.013	02 49 24.	+ 46 45		14.9	GALAXY
UGC 02351	02 49 24.	+ 46 45	138	14.9	GALAXY SBb
ZWG 389.037	02 49 24.	- 01 02		14.9	GALAXY
MCG+00-08-037	02 49 24.	- 01 02 30.	30	15.	GALAXY
MCG-03-08-035	02 49 24.	- 16 50	96	13.5	GALAXY
MCG-03-08-034	02 49 24.	- 16 50	18	15.	GALAXY
RNGC 1124	02 49 24.	- 25 54		15.0	GALAXY
MCG-04-07-047	02 49 24.	- 25 54	60	15.	GALAXY
PHL 8543	02 49 24.	- 26 00		18.2	BLUE STELLAR OBJECT
PHL 8544	02 49 24.	- 28 29		18.2	BLUE STELLAR OBJECT
MCG+07-06-082	02 49 26.	+ 41 06	36	15.	GALAXY
RNGC 1125	02 49 26.	- 16 50		13.0	GALAXY
BAK 1.014	02 49 26.	- 30 25 17.	53	15.5	EXTRAGALACTIC NEBULA
KN 11.03	02 49 27.1	+ 03 01 03.			NEBULA
UGC 02352	02 49 30.	+ 04 11	66	17.	GALAXY DWARF
ARC 0387	02 49 30.	+ 27 55		17.6	RICH CLUSTER OF GALAXIES
RNGC 1122	02 49 35.	+ 42 00		13.0	GALAXY
RNGC 1123	02 49 35.	+ 42 01			NON-EXISTENT OBJECT
ZC 0249.6+0530	02 49 36.	+ 05 30	870		CLUSTER OF GALAXIES
MCG+03-08-026	02 49 36.	+ 17 34	24	16.	GALAXY
ZC 0249.6+2753	02 49 36.	+ 27 53	1410		CLUSTER OF GALAXIES
ZWG 539.117	02 49 36.	+ 42 00		13.0	GALAXY
UGC 02353	02 49 36.	+ 42 00	132	13.0	GALAXY Sb/SBb
MCG+07-06-083	02 49 39.	+ 41 59 30.	72	13.5	GALAXY
CED 009C	02 49 40.	+ 60 16	240		DIFFUSE GALACTIC NEBULA
CED 009D	02 49 40.	+ 60 21			DIFFUSE GALACTIC NEBULA
UGC 02354	02 49 42.	+ 42 02	60	16.5	GALAXY Sc
MCG+00-08-038	02 49 45.	- 01 30	36	15.	GALAXY
LB 02819	02 49 46.	+ 01 59 18.		16.3	FAINT BLUE STAR
LB 02818	02 49 46.	+ 02 16 24.		15.6	FAINT BLUE STAR
LB 02817	02 49 46.	+ 03 17 42.		17.0	FAINT BLUE STAR
LB 00474	02 49 46.	+ 25 13 18.		17.6	FAINT BLUE STAR
SHAP251-5354.2	02 49 47.	- 54 06 29.	78	16.25	GALAXY
UGC 02355	02 49 48.	+ 01 47	90	18.	GALAXY DWARF
ZWG 539.118	02 49 48.	+ 41 20		15.6	GALAXY
ZWG 389.038	02 49 48.	- 01 30		15.4	GALAXY
MCG-06-07-010	02 49 48.	- 33 32	96	15.	GALAXY
RNGC 1126	02 49 50.	- 01 30		15.5	GALAXY
SHAP251-5327.8	02 49 50.	- 53 40 05.	36	17.75	GALAXY
MCG+00-08-039	02 49 51.	- 01 30	36	15.	GALAXY
SHAP251-5325.2	02 49 53.	- 53 37 29.	42	17.5	GALAXY
MCG+08-06-019	02 49 54.	+ 47 22	42	16.	GALAXY
MCG+12-03-003	02 49 54.	+ 74 59	66	16.	GALAXY
ZWG 389.039	02 49 54.	- 01 29		15.7	GALAXY
SHAP251-5227.6	02 49 56.	- 52 39 53.	60	17.0	GALAXY
SHAP251-5335.4	02 49 57.	- 53 47 41.	42	16.75	GALAXY
LB 00475	02 49 58.	+ 24 54 30.		17.4	FAINT BLUE STAR
RNGC 1128	02 49 59.	+ 05 51			NON-EXISTENT OBJECT
LBN 0672	02 50	+ 60 00	5400		BRIGHT NEBULA
VB 225	02 50 00.	+ 56 20	369		STELLAR RING
MCG-03-08-036	02 50 00.	- 15 54	24	15.	GALAXY
PHL 4325	02 50 00.	- 20 53		18.4	BLUE STELLAR OBJECT
SHAP251-5207.1	02 50 01.	- 52 19 22.	36	16.5	GALAXY
RNGC 1105	02 50 02.	- 15 54		15.0	GALAXY
MCG-02-08-014	02 50 03.	- 08 41	84	14.5	GALAXY
BAK 1.015	02 50 04.	- 30 20 03.	10	16.6	EXTRAGALACTIC NEBULA
ZC 0250.1+0051	02 50 06.	+ 00 51	1280		CLUSTER OF GALAXIES
ZWG 440.024	02 50 06.	+ 13 03		15.7	GALAXY
UGC 02356	02 50 06.	+ 13 03	60	15.7	GALAXY SBa-b
ZWG 484.023	02 50 06.	+ 25 17		14.7	GALAXY
UGC 02357	02 50 06.	+ 25 17	108	14.7	GALAXY S0
KARA.73B 0120	02 50 06.	+ 25 17	78	14.7	ISOLATED GALAXY S
MCG-03-08-037	02 50 06.	- 14 44	60	15.	GALAXY
PHL 8545	02 50 06.	- 27 42		18.5	BLUE STELLAR OBJECT
RNGC 1127	02 50 08.	+ 13 03		15.5	GALAXY
KLEM 04	02 50 10.	- 31 05	1800	16.	GROUP OF 6 GALAXIES
BAK 1.017	02 50 11.	- 30 21 03.	17	15.8	EXTRAGALACTIC NEBULA
BAK 1.016	02 50 11.	- 31 09 03.	17	16.3	EXTRAGALACTIC NEBULA
MCG+02-08-024	02 50 12.	+ 13 02 30.	30	15.	GALAXY
MCG+04-07-028	02 50 12.	+ 25 18	30	14.	GALAXY
ZWG 327.001	02 50 12.	+ 74 57		15.7	GALAXY
ZWG 326.003	02 50 12.	+ 74 57		15.7	GALAXY
UGC 02358	02 50 12.	+ 74 57	132	15.7	GALAXY SBb
KARA.72 081A	02 50 12.	+ 74 57	78	15.7	PART OF DOUBLE GALAXY
PHL 1459	02 50 12.	- 20 45		17.2	BLUE STELLAR OBJECT
PHL 8546	02 50 12.	- 22 13		16.6	BLUE STELLAR OBJECT
PHL 8547	02 50 12.	- 23 29		18.3	BLUE STELLAR OBJECT
MCG-05-07-044	02 50 12.	- 30 59	36	15.	GALAXY
BAK 1.018	02 50 16.	- 31 08 27.	12	16.4	EXTRAGALACTIC NEBULA
RNGC 1133	02 50 17.	- 08 59		15.0	GALAXY
MCG+00-08-041	02 50 18.	+ 02 52	30	16.	GALAXY
UGC 02360	02 50 18.	+ 31 28	78	16.0	GALAXY
ZC 0250.3+3903	02 50 18.	+ 39 03	2350		CLUSTER OF GALAXIES
MCG+07-06-084	02 50 18.	+ 41 40	60	14.	GALAXY
ZWG 540.001	02 50 18.	+ 41 42		14.4	GALAXY
ZWG 539.119	02 50 18.	+ 41 42		14.4	GALAXY
UGC 02361	02 50 18.	+ 41 42	96	14.4	GALAXY SBb
ZWG 389.040	02 50 18.	- 01 28		13.9	GALAXY
UGC 02359	02 50 18.	- 01 28	174	13.9	GALAXY E
MCG+00-08-040	02 50 18.	- 01 29	24	13.	GALAXY
PHL 8548	02 50 18.	- 21 06		18.3	BLUE STELLAR OBJECT
BAK 1.020	02 50 18.	- 32 01 15.	22	15.0	EXTRAGALACTIC NEBULA
BAK 1.019	02 50 19.	- 30 59 03.	31	14.6	EXTRAGALACTIC NEBULA
RNGC 1132	02 50 20.	- 01 28		14.0	GALAXY
SHAP251-5050.8	02 50 20.	- 51 03 03.	36	17.5	GALAXY
SHAP251-5330.4	02 50 20.	- 53 42 40.	36	16.25	GALAXY
MCG-02-08-015	02 50 21.	- 08 59	36	15.	GALAXY
IC 1861	02 50 22.	+ 25 16 34.			NONSTELLAR OBJECT
ARC 0390	02 50 23.	- 15 10		18.0	RICH CLUSTER OF GALAXIES
ZWG 440.025	02 50 24.	+ 12 50		15.5	GALAXY
UGC 02362	02 50 24.	+ 12 50	102	15.5	GALAXY IRR
MCG+02-08-025	02 50 24.	+ 12 50	60	14.5	GALAXY
ZWG 540.002	02 50 24.	+ 41 32		15.7	GALAXY
ZWG 539.120	02 50 24.	+ 41 32		15.7	GALAXY
PHL 1460	02 50 24.	- 21 54		16.6	BLUE STELLAR OBJECT
TON-S 0286	02 50 24.	- 28 18		16.0	BLUE STAR
SHAP251-5338.4	02 50 24.	- 53 50 39.	48	17.5	GALAXY
LB 03286	02 50 24.	- 66 19		11.8	FAINT BLUE STAR
KN 11.04	02 50 27.02	+ 03 23 48.			NEBULA
MCG-03-08-038	02 50 30.	- 14 43	72	15.	GALAXY
PHL 8549	02 50 30.	- 27 02		17.1	BLUE STELLAR OBJECT

OBJECT NAME	RIGHT ASCEN.	DECLINATION	DIAM.	MAGN.	TYPE OF OBJECT
BAK 1.021	02 50 30.	- 29 51 08.	17	16.0	EXTRAGALACTIC NEBULA
LB 03287	02 50 30.	- 70 41		15.2	FAINT BLUE STAR
RNGC 1139	02 50 31.	- 14 43		15.0	GALAXY
KN 11.05	02 50 31.8	+ 03 59 20.			NEBULA
ZWG 440.026	02 50 36.	+ 11 45		15.7	GALAXY
MCG+04-07-029	02 50 36.	+ 25 45 30.	24	16.	GALAXY
ZWG 524.031	02 50 36.	+ 39 23		14.6	GALAXY
MRSL 137+01/1	02 50 36.	+ 60 12	7200		HII REGION
PHL 1461	02 50 36.	- 31 10		17.4	BLUE STELLAR OBJECT
SHAP252-5258.6	02 50 37.	- 53 10 51.	36	16.5	GALAXY
SHAP252-5202.5	02 50 39.	- 52 14 45.	36	16.5	GALAXY
SHAP252-5402.1	02 50 39.	- 54 14 21.	48	17.25	GALAXY
UGC 02364	02 50 42.	+ 06 20	96	16.5	GALAXY
ZWG 415.018	02 50 42.	+ 09 14		15.4	GALAXY
MCG+06-07-017	02 50 42.	+ 39 25	21	15.	GALAXY
MCG+11-04-003	02 50 42.	+ 66 10	12	18.	GALAXY
LB 02820	02 50 42.	- 00 26 24.		16.3	FAINT BLUE STAR
MCG-03-08-040	02 50 42.	- 18 00	48	16.	GALAXY
MCG-03-08-039	02 50 42.	- 18 00	12	16.	GALAXY
MCG-04-07-049	02 50 42.	- 24 59	24	14.5	GALAXY
MCG-04-07-048	02 50 42.	- 25 04	72	15.5	GALAXY
PHL 4326	02 50 42.	- 25 46		17.8	BLUE STELLAR OBJECT
BAK 1.022	02 50 42.	- 29 21 26.	13	16.6	EXTRAGALACTIC NEBULA
SHAP252-5221.5	02 50 42.	- 52 33 44.	30	16.75	GALAXY
LB 03288	02 50 42.	- 58 29		14.4	FAINT BLUE STAR
LB 02821	02 50 44.	- 00 01 36.		16.7	FAINT BLUE STAR
KN 11.06	02 50 44.5	+ 03 37 18.			NEBULA
MCG+01-08-010	02 50 45.	+ 03 36	36	15.5	GALAXY
MCG+01-08-011	02 50 45.	+ 06 19 30.	60	15.	GALAXY
MCG+02-08-026	02 50 45.	+ 12 26	48	15.	GALAXY
BAK 1.023	02 50 46.	- 29 27 25.	17	16.5	EXTRAGALACTIC NEBULA
BAK 1.025	02 50 47.	- 29 50 01.	13	16.4	EXTRAGALACTIC NEBULA
BAK 1.024	02 50 47.	- 31 03 37.	31	15.7	EXTRAGALACTIC NEBULA
ZWG 415.019	02 50 48.	+ 03 38		15.5	GALAXY
LB 02822	02 50 48.	- 01 30 48.		16.7	FAINT BLUE STAR
PHL 1462	02 50 48.	- 22 56		18.4	BLUE STELLAR OBJECT
SHAP252-5049.2	02 50 48.	- 51 01 26.	48	16.75	GALAXY
SHAP252-5222.8	02 50 48.	- 52 35 02.	30	17.25	GALAXY
LB 02823	02 50 50.	+ 00 35 00.		14.5	FAINT BLUE STAR
ZWG 440.027	02 50 54.	+ 12 48		13.2	GALAXY
UGC 02365	02 50 54.	+ 12 48	156	13.2	GALAXY S
ARP 200	02 50 54.	+ 12 48			PECULIAR GALAXY
MCG+02-08-027	02 50 54.	+ 12 49	156	13.	GALAXY
ZWG 524.032	02 50 54.	+ 36 13		15.7	GALAXY
UGC 02366	02 50 54.	+ 36 13	84	15.7	GALAXY S0
ZWG 327.002	02 50 54.	+ 74 53		15.7	GALAXY
ZWG 326.004	02 50 54.	+ 74 53		15.7	GALAXY
KARA.72 081B	02 50 54.	+ 74 53	42	15.7	PART OF DOUBLE GALAXY
PHL 8550	02 50 54.	- 27 55		17.5	BLUE STELLAR OBJECT
SHAP252-5400.0	02 50 55.	- 54 12 14.	48	17.75	GALAXY
LB 02824	02 50 56.	+ 01 27 18.		17.0	FAINT BLUE STAR
RNGC 1134	02 50 56.	+ 12 48		13.0	GALAXY
MCG+01-08-011A	02 50 57.	+ 06 15	36	15.	GALAXY
KN 11.07	02 50 59.4	+ 05 51 25.			NEBULA
ZWG 415.020	02 51 00.	+ 06 04		15.3	GALAXY
UGC 02367	02 51 00.	+ 06 04	138	15.3	GALAXY S0
MCG+06-07-018	02 51 00.	+ 36 13	48	16.	GALAXY
ZWG 540.003	02 51 00.	+ 42 32		14.7	GALAXY
ZWG 539.121	02 51 00.	+ 42 32		14.7	GALAXY
LDN 1358	02 51 00.	+ 69 10	300		DARK NEBULA
ABC 0391	02 51 00.	- 02 42		17.6	RICH CLUSTER OF GALAXIES
PHL 4327	02 51 00.	- 20 48		18.4	BLUE STELLAR OBJECT
PHL 4328	02 51 00.	- 21 26		18.6	BLUE STELLAR OBJECT
PHL 8551	02 51 00.	- 31 36		8.1	BLUE STELLAR OBJECT
KN 11.08	02 51 02.6	+ 06 03 49.			NEBULA
MCG-03-08-041	02 51 03.	- 17 49	30	15.	GALAXY
LB 02825	02 51 05.	+ 02 29 00.		16.5	FAINT BLUE STAR
MCG+01-08-012	02 51 06.	+ 06 03 30.	72	14.	GALAXY
ZWG 440.028	02 51 06.	+ 12 38		14.1	GALAXY
UGC 02368	02 51 06.	+ 12 38	126	14.1	GALAXY SBb
WEED 1	02 51 06.	+ 12 39		17.5	VERY BLUE STELLAR OBJECT
MCG+02-08-028	02 51 06.	+ 12 39	120	13.5	GALAXY
MCG+07-07-002	02 51 06.	+ 41 22	24	15.	GALAXY
VV 085D	02 51 06.	+ 41 23	6	18.	INTERACTING GALAXY
VV 085C	02 51 06.	+ 41 23	6	18.	INTERACTING GALAXY
VV 085B	02 51 06.	+ 41 23	18	15.	INTERACTING GALAXY
VV 085A	02 51 06.	+ 41 23	120	13.	INTERACTING GALAXY
VV 085	02 51 06.	+ 41 23	150		INTERACTING GALAXY
MCG+07-07-001	02 51 06.	+ 42 32	42	14.5	GALAXY
LB 03289	02 51 06.	- 72 47		15.0	FAINT BLUE STAR
IC 1864	02 51 08.	- 34 24 34.			NONSTELLAR OBJECT
SHAP252-5336.0	02 51 08.	- 53 48 13.	36	16.75	GALAXY
MCG+07-07-003	02 51 09.	+ 41 21	8	15.	GALAXY
SHAP252-5354.0	02 51 09.	- 54 06 13.	30	17.5	GALAXY
RNGC 1130	02 51 11.	+ 41 25		13.0	GALAXY
LB 02826	02 51 12.	+ 00 32 54.		16.9	FAINT BLUE STAR
ZWG 440.029	02 51 12.	+ 13 24		15.7	GALAXY
ZWG 440.030	02 51 12.	+ 14 46		14.6	GALAXY
UGC 02369	02 51 12.	+ 14 46	60	14.6	GALAXY DBL SYS
KARA.72 082A	02 51 12.	+ 14 46	36	14.6	PART OF DOUBLE GALAXY
KARA.72 082B	02 51 12.	+ 14 47	42		PART OF DOUBLE GALAXY
MCG+02-08-030	02 51 12.	+ 14 47	36	15.5	GALAXY
MCG+02-08-029	02 51 12.	+ 14 47	24	15.5	GALAXY
5ZW 286	02 51 12.	+ 41 22			COMPACT GALAXY
MCG+07-07-004	02 51 12.	+ 41 22 30.	18	13.	GALAXY
ZWG 540.004	02 51 12.	+ 41 25		15.6	GALAXY
ZWG 539.122	02 51 12.	+ 41 25		15.6	GALAXY
ZWG 540.005	02 51 12.	+ 42 27		15.7	GALAXY
ZWG 539.123	02 51 12.	+ 42 27		15.7	GALAXY
UGC 02370	02 51 12.	+ 42 27	180	15.7	GALAXY PLAT
ZC 0251.2-0247	02 51 12.	- 02 47	2420		CLUSTER OF GALAXIES
MCG+05-07-052	02 51 15.	+ 31 07	48	15.	GALAXY
MCG+07-07-005	02 51 15.	+ 41 19	12	15.	GALAXY
RNGC 1129	02 51 17.	+ 41 23		14.5	GALAXY
ZWG 389.041	02 51 18.	+ 02 05		15.7	GALAXY
UGC 02371	02 51 18.	+ 02 05	90	15.7	GALAXY Sa-b
MCG+00-08-042	02 51 18.	+ 02 06	60	15.	GALAXY
MCG+00-08-043	02 51 18.	+ 02 47	120	14.	GALAXY
ZWG 415.021	02 51 18.	+ 05 47		15.5	GALAXY
UGC 02372	02 51 18.	+ 05 47	66	15.5	GALAXY Sc
ZWG 440.031	02 51 18.	+ 11 37		15.7	GALAXY
1ZW 010	02 51 18.	+ 13 25			COMPACT GALAXY
ZWG 540.006	02 51 18.	+ 41 23		14.6	GALAXY
ZWG 539.124	02 51 18.	+ 41 23		14.6	GALAXY
UGC 02373	02 51 18.	+ 41 23	300	14.6	GALAXY E
PHL 8552	02 51 18.	- 21 54		14.1	BLUE STELLAR OBJECT
MCG-05-07-045	02 51 18.	- 31 04	42	14.5	GALAXY
KN 11.09	02 51 20.2	+ 05 46 55.			NEBULA
BAK 1.026	02 51 22.	- 31 03 36.	49	14.8	EXTRAGALACTIC NEBULA
SHAP252-5249.5	02 51 22.	- 53 01 42.	36	17.0	GALAXY
IC 0267	02 51 23.	+ 12 38 16.			NONSTELLAR OBJECT
RNGC 1131	02 51 23.	+ 41 22		15.5	GALAXY
ZWG 389.042	02 51 24.	+ 02 45		13.5	GALAXY
RNGC 1137	02 51 24.	+ 02 45		13.5	GALAXY
UGC 02374	02 51 24.	+ 02 45	162	13.5	GALAXY Sb
MCG+01-08-013	02 51 24.	+ 05 47	48	15.	GALAXY
ZWG 415.022	02 51 24.	+ 06 03		15.5	GALAXY
UGC 02375	02 51 24.	+ 06 03	66	15.5	GALAXY S
MCG+02-08-031	02 51 24.	+ 11 37	36	15.	GALAXY
UGC 02376	02 51 24.	+ 31 05	120	16.0	GALAXY Sc
ZWG 540.007	02 51 24.	+ 41 22		15.6	GALAXY
ZWG 539.125	02 51 24.	+ 41 22		15.6	GALAXY
MCG+07-07-006	02 51 24.	+ 41 25	12	15.	GALAXY
MCG+07-07-007	02 51 24.	+ 41 56	12	15.5	GALAXY
PHL 1463	02 51 24.	- 26 22		8.6	BLUE STELLAR OBJECT
PHL 4329	02 51 24.	- 32 46		18.4	BLUE STELLAR OBJECT
LB 02827	02 51 25.	- 00 35 00.		16.8	FAINT BLUE STAR
SHAP253-5055.0	02 51 25.	- 51 07 12.	42	17.0	GALAXY
MCG-02-08-016	02 51 27.	- 14 04	24	15.	GALAXY
KN 11.10	02 51 27.0	+ 02 45 30.			NEBULA
LB 02828	02 51 28.	+ 00 46 48.		17.0	FAINT BLUE STAR
LB 02829	02 51 28.	- 01 14 12.		16.2	FAINT BLUE STAR
LB 02830	02 51 29.	+ 02 43 48.		15.8	FAINT BLUE STAR
ABC 0394	02 51 29.	- 14 52		17.8	RICH CLUSTER OF GALAXIES
MCG+01-08-014	02 51 30.	+ 06 03	60	15.	GALAXY
ZWG 524.033	02 51 30.	+ 39 11		15.6	GALAXY
UGC 02377	02 51 30.	+ 39 11	72	15.6	GALAXY Sa-b
ZWG 540.008	02 51 30.	+ 41 20		14.8	GALAXY
ZWG 539.126	02 51 30.	+ 41 20		14.8	GALAXY
MCG+07-07-008	02 51 30.	+ 41 20	24	15.	GALAXY
ZWG 540.009	02 51 30.	+ 41 28		15.7	GALAXY
ZWG 539.127	02 51 30.	+ 41 28		15.7	GALAXY
ZWG 540.010	02 51 30.	+ 41 40		15.7	GALAXY
ZWG 539.128	02 51 30.	+ 41 40		15.7	GALAXY
SHAP253-5241.0	02 51 30.	- 52 53 12.	36	16.0	GALAXY
IC 0265	02 51 33.	+ 41 27 29.			NONSTELLAR OBJECT
ZWG 540.011	02 51 36.	+ 41 07		15.1	GALAXY
ZWG 539.129	02 51 36.	+ 41 07		15.1	GALAXY
MCG-02-08-017	02 51 36.	- 13 04	72	15.	GALAXY
PHL 8553	02 51 36.	- 23 12		18.3	BLUE STELLAR OBJECT
TON-S 0287	02 51 36.	- 27 03		16.2	BLUE STAR
MCG-06-07-011	02 51 36.	- 34 23	60	13.5	GALAXY
SHAP253-5114.0	02 51 39.	- 51 26 12.	36	17.25	GALAXY
LB 02831	02 51 41.	+ 01 14 24.		16.2	FAINT BLUE STAR
ABC 0392	02 51 41.	+ 04 44		17.4	RICH CLUSTER OF GALAXIES
ZC 0251.7+0114	02 51 42.	+ 01 14	1810		CLUSTER OF GALAXIES
UGC 02378	02 51 42.	+ 11 32	96	16.0	GALAXY Sc
ZWG 440.032	02 51 42.	+ 13 57		15.5	GALAXY
UGC 02379	02 51 42.	+ 13 57	84	15.5	GALAXY S+COMP
MCG+08-06-020	02 51 42.	+ 47 03	36	17.	GALAXY
UGC 02380	02 51 42.	+ 51 43	96	16.0	GALAXY Sb
LB 02832	02 51 42.	- 00 01 12.		16.6	FAINT BLUE STAR
PHL 8554	02 51 42.	- 23 05		18.6	BLUE STELLAR OBJECT
LB 02833	02 51 43.	+ 02 29 54.		17.1	FAINT BLUE STAR
ABC 0393	02 51 45.	+ 03 43		17.1	RICH CLUSTER OF GALAXIES
MCG+02-08-032	02 51 45.	+ 11 32	84	15.	GALAXY
MCG+07-07-009	02 51 45.	+ 41 22 30.	9	16.	GALAXY
MCG+07-07-010	02 51 45.	+ 42 02	18	15.5	GALAXY
SHAP253-5100.4	02 51 45.	- 51 12 35.	42	16.5	GALAXY
ZC 0251.8+0440	02 51 48.	+ 04 40	1010		CLUSTER OF GALAXIES
ZWG 415.023	02 51 48.	+ 09 15		15.5	GALAXY
ZWG 463.030	02 51 48.	+ 15 38		15.7	GALAXY
ZWG 524.034	02 51 48.	+ 39 03		15.6	GALAXY
IC 0266	02 51 48.	+ 42 03 28.			NONSTELLAR OBJECT
ZWG 540.012	02 51 48.	+ 42 05		15.7	GALAXY
ZWG 539.130	02 51 48.	+ 42 05		15.7	GALAXY
ZWG 540.013	02 51 48.	+ 43 20		15.7	GALAXY
ZWG 539.131	02 51 48.	+ 43 20		15.7	GALAXY
SHAP253-5200.2	02 51 49.	- 52 12 23.	42	16.75	GALAXY
MCG+03-08-027	02 51 51.	+ 15 38	33	15.5	GALAXY
MCG+06-07-019	02 51 51.	+ 39 04	18	15.	GALAXY
SHAP253-5402.3	02 51 53.	- 54 14 29.	42	16.75	GALAXY
MCG+02-08-033	02 51 54.	+ 11 48	60	15.5	GALAXY
UGC 02381	02 51 54.	+ 11 49	60	16.0	GALAXY S
5ZW 287	02 51 54.	+ 36 01			COMPACT GALAXY
MCG+07-07-011	02 51 54.	+ 43 20	36	17.	GALAXY
MCG-02-08-018	02 51 54.	- 08 40	18	15.	GALAXY
SHAP253-5249.1	02 51 55.	- 53 01 17.	30	16.5	GALAXY
MCG+02-08-034	02 51 57.	+ 15 23	36	15.5	GALAXY
SHAP253-5317.0	02 51 58.	- 53 29 11	30	17.5	GALAXY
LBN 0673	02 52	+ 60 20	960		BRIGHT NEBULA
ZC 0252.0+0346	02 52 00.	+ 03 46	1280		CLUSTER OF GALAXIES
UGC 02382	02 52 00.	+ 09 10	66	17.	GALAXY Sc
ZWG 440.033	02 52 00.	+ 15 23		15.7	GALAXY
ZWG 440.034	02 52 00.	+ 15 27		15.7	GALAXY
ZC 0252.0+2452	02 52 00.	+ 24 52	1010		CLUSTER OF GALAXIES
UGC 02383	02 52 00.	+ 46 09	72	18.	GALAXY
PHL 1464	02 52 00.	- 26 30		17.7	BLUE STELLAR OBJECT
TON-S 0288	02 52 00.	- 27 25		16.0	BLUE STAR
LB 03290	02 52 00.	- 62 30		12.7	FAINT BLUE STAR
LB 02834	02 52 02.	+ 03 10 12.		16.2	FAINT BLUE STAR
SHAP253-5329.0	02 52 04.	- 53 41 10.	36	16.75	GALAXY
ABC 0395	02 52 06.	- 10 36		17.6	RICH CLUSTER OF GALAXIES
SHAP253-5141.0	02 52 06.	- 51 53 10.	36	17.5	GALAXY
MCG-02-08-019	02 52 09.	- 10 12	84	13.5	GALAXY
IC 1863	02 52 10.	+ 08 33 47.			NONSTELLAR OBJECT
SHAP253-5155.6	02 52 10.	- 52 07 46.	30	17.0	GALAXY
RNGC 1140	02 52 11.	- 10 14		13.5	GALAXY
ZWG 415.024	02 52 12.	+ 08 35		15.7	GALAXY
ZWG 463.031	02 52 12.	+ 15 49		15.1	GALAXY
UGC 02384	02 52 12.	+ 15 49	60	15.1	GALAXY S
ZWG 484.024	02 52 12.	+ 24 26		15.6	GALAXY
MCG+08-06-021	02 52 12.	+ 45 52 30.	36	16.	GALAXY
SHAP253-5247.2	02 52 12.	- 52 59 22.	42	17.0	GALAXY
RNGC 1145	02 52 15.	- 18 50		13.0	GALAXY
MCG-03-08-042	02 52 15.	- 18 50	210	13.	GALAXY
MCG+01-08-015	02 52 18.	+ 08 36 30.	24	15.5	GALAXY
MCG+03-08-028	02 52 18.	+ 15 48	45	15.	GALAXY
ZWG 524.035	02 52 18.	+ 36 00		15.6	GALAXY
5ZW 288	02 52 18.	+ 45 30			COMPACT GALAXY
ZWG 389.043	02 52 18.	- 01 34		15.6	GALAXY
KARA.68 028	02 52 18.	- 12 05	34		DWARF GALAXY
PHL 8555	02 52 18.	- 26 31		18.4	BLUE STELLAR OBJECT
SHAP253-5249.2	02 52 18.	- 53 01 22.	30	16.5	GALAXY
BAK 1.027	02 52 19.	- 30 30 33.	24	16.4	EXTRAGALACTIC NEBULA
MCG+00-08-044	02 52 21.	- 00 10	36	16.	GALAXY

OBJECT NAME	RIGHT ASCEN.	DECLINATION	DIAM.	MAGN.	TYPE OF OBJECT
MCG-02-08-020	02 52 21.	- 13 39 30.	30	15.	GALAXY
LB 02835	02 52 22.	+ 00 31 48.		16.3	FAINT BLUE STAR
MCG+06-07-020	02 52 24.	+ 35 59 30.	36	16.	GALAXY
SHAP253-5231.5	02 52 25.	- 52 43 39.	30	17.5	GALAXY
SHAP254-5247.4	02 52 28.	- 52 59 33.	30	17.0	GALAXY
UGC 02386	02 52 30.	+ 09 09	72	16.0	GALAXY S
ZWG 440.035	02 52 30.	+ 12 01		15.7	GALAXY
UGC 02387	02 52 30.	+ 12 01	102	15.7	GALAXY Sc
ZWG 463.032	02 52 30.	+ 15 58		15.5	GALAXY
5ZW 289	02 52 30.	+ 35 53			COMPACT GALAXY
ZWG 389.045	02 52 30.	- 01 23		15.5	GALAXY
UGC 02385	02 52 30.	- 01 23	66	15.5	GALAXY Sc
ZWG 389.044	02 52 30.	- 02 04		15.6	GALAXY
FEIG 028	02 52 30.	- 04 01		11.5	FAINT BLUE STAR
MCG+01-08-015	02 52 30.	- 06 34 30.	90	14.	GALAXY
IC 1866	02 52 32.	- 15 51 16.			NONSTELLAR OBJECT
MCG+01-08-016	02 52 33.	+ 09 10	48	15.5	GALAXY
MCG+03-08-029	02 52 33.	+ 15 57	18	15.	GALAXY
MCG+00-08-045	02 52 33.	- 02 05	36	15.	GALAXY
MCG-03-08-043	02 52 33.	- 19 05	60	15.5	GALAXY
MCG-04-08-001	02 52 33.	- 25 22	24	15.	GALAXY
SHAP254-5230.8	02 52 34.	- 52 42 57.	30	17.5	GALAXY
SHAP254-5230.8	02 52 34.	- 52 42 57.	30	17.5	GALAXY
SHAP254-5240.1	02 52 34.	- 52 52 15.	24	17.0	GALAXY
ARP 118	02 52 35.	- 00 22			PECULIAR GALAXY
LB 02836	02 52 35.	- 00 42 54.		14.9	FAINT BLUE STAR
BAK 1.028	02 52 35.	- 30 37 32.	24	15.6	EXTRAGALACTIC NEBULA
ZC 0252.6+1038	02 52 36.	+ 10 38	3290		CLUSTER OF GALAXIES
MCG+02-08-035	02 52 36.	+ 12 00	72	15.	GALAXY
UGC 02390	02 52 36.	+ 36 29	66	16.0	GALAXY SBb
ZWG 554.014	02 52 36.	+ 47 36		15.7	GALAXY
MCG+08-06-022	02 52 36.	+ 47 36	36	16.	GALAXY
ZWG 389.046	02 52 36.	- 00 23		13.2	GALAXY
UGC 02389	02 52 36.	- 00 23	84	13.2	GALAXY E
UGC 02388	02 52 36.	- 00 23	84	13.2	GALAXY E
VV 331B	02 52 36.	- 00 23	36	15.	INTERACTING GALAXY
VV 331A	02 52 36.	- 00 23	42	14.5	INTERACTING GALAXY
KARA.72 083A	02 52 36.	- 00 23	48	13.2	PART OF DOUBLE GALAXY
MCG+00-08-046	02 52 36.	- 01 24	42	15.	GALAXY
MCG-03-08-044	02 52 36.	- 15 50	24	14.5	GALAXY
MCG-05-08-001	02 52 36.	- 30 38	90	14.	GALAXY
SHAP254-5155.5	02 52 36.	- 52 07 39.	30	16.75	GALAXY
LB 03291	02 52 36.	- 79 02		14.5	FAINT BLUE STAR
RNGC 1142	02 52 37.	+ 00 16			NON-EXISTENT OBJECT
RNGC 1141	02 52 37.	+ 00 16			NON-EXISTENT OBJECT
RNGC 1144	02 52 37.	- 00 23		13.0	GALAXY
RNGC 1143	02 52 37.	- 00 23		14.0	GALAXY
SHAP254-5246.0	02 52 38.	- 52 58 09.	42	16.5	GALAXY
MCG-04-08-002	02 52 39.	- 22 06	48	14.	GALAXY
SHAP254-5254.4	02 52 39.	- 53 06 33.	30	16.0	GALAXY
IC 1865	02 52 40.	+ 08 36 27.			NONSTELLAR OBJECT
KN 11.11	02 52 41.6	+ 05 55 21.			NEBULA
ZWG 415.025	02 52 42.	+ 05 55		15.7	GALAXY
ZWG 415.026	02 52 42.	+ 08 37		15.0	GALAXY
UGC 02391	02 52 42.	+ 08 37	96	15.0	GALAXY Sb-c
5ZW 290	02 52 42.	+ 30 55			COMPACT GALAXY
MCG+06-07-021	02 52 42.	+ 33 33	108	15.	GALAXY
ZWG 524.036	02 52 42.	+ 33 35		15.0	GALAXY
UGC 02392	02 52 42.	+ 33 35	120	15.0	GALAXY Sc?
MCG+06-07-022	02 52 42.	+ 36 28	72	16.	GALAXY
5ZW 291	02 52 42.	+ 41 21			COMPACT GALAXY
KARA.72 083B	02 52 42.	- 00 23	42		PART OF DOUBLE GALAXY
MCG-02-08-021	02 52 42.	- 10 56 30.	30	15.	GALAXY
MCG-03-08-045	02 52 42.	- 17 15	90	15.	GALAXY
MCG-05-08-002	02 52 42.	- 30 53	60	15.	GALAXY
SHAP254-5022.0	02 52 42.	- 50 34 08.	54	17.0	GALAXY
SHAP254-5156.0	02 52 42.	- 52 08 08.	30	17.25	GALAXY
BAK 1.029	02 52 43.	- 30 52 32.	17	15.5	EXTRAGALACTIC NEBULA
SHAP254-5148.1	02 52 44.	- 52 00 14.	24	17.25	GALAXY
MCG+01-08-017	02 52 45.	+ 08 38 30.	48	14.5	GALAXY
MCG+00-08-047	02 52 45.	- 00 23	30	14.	GALAXY
SHAP254-5346.8	02 52 46.	- 53 58 56.	36	17.25	GALAXY
ZWG 505.056	02 52 48.	+ 32 08		15.5	GALAXY
UGC 02393	02 52 48.	+ 32 08	66	15.5	GALAXY Sc
UGC 02394	02 52 48.	+ 37 56	84	18.	GALAXY Sc
MCG+00-08-048	02 52 48.	- 00 23	42	13.	GALAXY
MCG+01-08-016	02 52 48.	- 04 47	45	17.	GALAXY
BAK 1.030	02 52 48.	- 30 36 31.	23	16.4	EXTRAGALACTIC NEBULA
LB 01644	02 52 48.	- 47 31		12.5	FAINT BLUE STAR
SHAP254-5415.0	02 52 48.	- 54 27 08.	36	16.75	GALAXY
SHAP254-5341.6	02 52 50.	- 53 53 44.	36	16.5	GALAXY
MCG+01-08-017	02 52 51.	- 04 46 30.	24	17.	GALAXY
MCG-02-08-022	02 52 51.	- 13 43 30.	36	15.	GALAXY
RNGC 1147	02 52 53.	- 09 19			NON-EXISTENT OBJECT
ZWG 540.014	02 52 54.	+ 41 26		15.2	GALAXY
UGC 02395	02 52 54.	+ 41 26	66	15.2	GALAXY S0-a
MCG+00-08-049	02 52 54.	- 01 45 30.	27	16.	GALAXY
LB 02837	02 52 56.	+ 01 19 12.		17.1	FAINT BLUE STAR
SHAP254-5356.2	02 52 58.	- 54 08 20.	24	17.0	GALAXY
SHAP254-5413.4	02 52 59.	- 54 25 32.	30	17.75	GALAXY
LBN 0753	02 53	+ 18 00	4800		BRIGHT NEBULA
LBN 0743	02 53	+ 20 00	6300		BRIGHT NEBULA
LDN 1457	02 53 00.	+ 19 20	2040		DARK NEBULA
UGC 02396	02 53 00.	+ 36 52	84	16.5	GALAXY
UGC 02397	02 53 00.	+ 36 57	66	17.	GALAXY PECULR
MCG-02-08-023	02 53 00.	- 14 16	60	15.	GALAXY
MCG-02-08-024	02 53 00.	- 14 18	36	15.5	GALAXY
PHL 8556	02 53 00.	- 22 38		18.2	BLUE STELLAR OBJECT
BAK 1.031	02 53 00.	- 32 22 55.	17	16.5	EXTRAGALACTIC NEBULA
SHAP254-5406.1	02 53 00.	- 54 18 14.	54	17.5	GALAXY
IC 0268	02 53 03.	- 14 18 19.			NONSTELLAR OBJECT
SHAP254-5355.6	02 53 03.	- 54 07 44.	36	17.5	GALAXY
IC 0269	02 53 04.	- 14 16 01.			NONSTELLAR OBJECT
KN 11.12	02 53 04.0	+ 06 08 44.			NEBULA
BAK 1.032	02 53 05.	- 28 50 00.	17	16.3	EXTRAGALACTIC NEBULA
SHAP254-5133.9	02 53 05.	- 51 46 01.	30	17.25	GALAXY
SHAP254-5153.0	02 53 05.	- 52 05 07.	24	17.0	GALAXY
SHAP254-5153.9	02 53 05.	- 52 06 01.	18	18.0	GALAXY
ZWG 415.027	02 53 06.	+ 06 08		15.6	GALAXY
ZWG 463.033	02 53 06.	+ 15 46		15.4	GALAXY
MCG+08-06-023	02 53 06.	+ 48 29	18	16.	GALAXY
ZWG 554.015	02 53 06.	+ 48 30		15.7	GALAXY
UGC 02398	02 53 06.	+ 48 30	72	15.7	GALAXY E
PHL 4330	02 53 06.	- 24 28		16.9	BLUE STELLAR OBJECT
SHAP254-5129.8	02 53 06.	- 51 41 55.	30	16.25	GALAXY
SHAP254-5410.0	02 53 06.	- 54 22 07.	42	18.0	GALAXY
SHAP254-5205.4	02 53 07.	- 52 17 31.	66	15.75	GALAXY
SHAP254-5320.6	02 53 08.	- 53 32 43.	30	17.25	GALAXY

OBJECT NAME	RIGHT ASCEN.	DECLINATION	DIAM.	MAGN.	TYPE OF OBJECT
KN 11.13	02 53 08.3	+ 06 01 04.			NEBULA
MCG-04-08-003	02 53 09.	- 24 57 30.	42	15.5	GALAXY
IC 1867	02 53 10.	+ 09 07 36.			NONSTELLAR OBJECT
BAK 1.033	02 53 10.	- 30 02 18.	46	15.6	EXTRAGALACTIC NEBULA
SHAP254-5158.9	02 53 10.	- 52 11 01.	36	16.25	GALAXY
KN 11.14	02 53 10.9	+ 05 15 59.			NEBULA
SHAP254-5422.2	02 53 11.	- 54 34 19.	30	17.5	GALAXY
ZWG 415.028	02 53 12.	+ 06 00		15.3	GALAXY
UGC 02399	02 53 12.	+ 06 00	96	15.3	GALAXY Sc
MCG+01-08-018	02 53 12.	+ 06 00 30.	60	14.5	GALAXY
ZWG 415.029	02 53 12.	+ 09 06		15.0	GALAXY
UGC 02400	02 53 12.	+ 09 06	132	15.0	GALAXY
LDN 1453	02 53 12.	+ 20 05	1020		DARK NEBULA
MCG+05-07-053	02 53 12.	+ 31 41	48	16.	GALAXY
5ZW 292	02 53 12.	+ 35 20			COMPACT GALAXY
UGC 02401	02 53 12.	+ 36 49	72	16.0	GALAXY S0?
UGC 02402	02 53 12.	+ 45 55	72	16.5	GALAXY SB
ZWG 554.016	02 53 12.	+ 47 20		14.9	GALAXY
MCG-02-08-025	02 53 12.	- 10 11	30	15.5	GALAXY
MCG-02-08-026	02 53 12.	- 10 20	24	15.	GALAXY
MCG-04-08-004	02 53 12.	- 24 46 30.	54	15.	GALAXY
MCG-05-08-003	02 53 12.	- 30 03	48	15.	GALAXY
SHAP254-5335.3	02 53 12.	- 53 47 25.	42	16.75	GALAXY
SHAP254-5408.0	02 53 12.	- 54 20 07.	54	18.0	GALAXY
MCG-02-08-027	02 53 15.	- 10 40	24	15.5	GALAXY
MCG-04-08-005	02 53 15.	- 23 15	36	15.	GALAXY
RNGC 1138	02 53 16.	+ 42 51		14.0	GALAXY
ZWG 389.047	02 53 18.	+ 00 29		14.9	GALAXY
UGC 02403	02 53 18.	+ 00 29	108	14.9	GALAXY SB
MCG+00-08-050	02 53 18.	+ 00 30	60	14.	GALAXY
UGC 02404	02 53 18.	+ 00 54	78	16.5	GALAXY Sc
ZWG 415.030	02 53 18.	+ 06 17		15.1	GALAXY
UGC 02405	02 53 18.	+ 06 17	108	15.1	GALAXY Sc
MCG+01-08-019	02 53 18.	+ 09 08	12	15.	GALAXY
ZWG 463.034	02 53 18.	+ 15 35		15.7	GALAXY
UGC 02406	02 53 18.	+ 15 35	72	15.7	GALAXY Sa
ZC 0253.3+2317	02 53 18.	+ 23 17	1210		CLUSTER OF GALAXIES
UGC 02407	02 53 18.	+ 31 42	60		GALAXY SB?a-b
ZWG 540.015	02 53 18.	+ 42 51		14.1	GALAXY
UGC 02408	02 53 18.	+ 42 51	126	14.1	GALAXY SB0
UGC 02409	02 53 18.	+ 50 26	66	17.	GALAXY S
ZC 0253.3+7900	02 53 18.	+ 79 00	940		CLUSTER OF GALAXIES
MCG-02-08-028	02 53 18.	- 14 24	72	14.5	GALAXY
KN 11.15	02 53 18.7	+ 06 17 35.			NEBULA
MCG+07-07-012	02 53 21.	+ 42 51	36	14.	GALAXY
SHAP254-5151.9	02 53 21.	- 52 04 00.	36	17.0	GALAXY
FATH 1.127	02 53 22.	+ 00 30	60		NEBULA
IC 0270	02 53 22.	- 14 24 21.			NONSTELLAR OBJECT
SHAP254-5353.5	02 53 22.	- 54 05 37.	36	16.5	GALAXY
MCG+01-08-020	02 53 24.	+ 06 17	84	15.	GALAXY
ZCG 0253+15	02 53 24.	+ 15 02		15.9	COMPACT GALAXY
MCG+03-08-030	02 53 24.	+ 15 35	48	15.5	GALAXY
5ZW 293	02 53 24.	+ 26 14			COMPACT GALAXY
ZWG 540.016	02 53 24.	+ 41 08		15.6	GALAXY
UGC 02410	02 53 24.	+ 41 08	72	15.6	GALAXY S
UGC 02411	02 53 24.	+ 75 33	270	16.5	GALAXY
LB 02838	02 53 26.	- 00 28 24.		16.7	FAINT BLUE STAR
SHAP255-5023.8	02 53 26.	- 50 35 54.	42	16.5	GALAXY
SHAP255-5226.6	02 53 26.	- 52 38 42.	24	17.25	GALAXY
MCG+01-08-021	02 53 27.	+ 07 50	48	16.	GALAXY
IC 1868	02 53 27.	+ 09 08 41.			NONSTELLAR OBJECT
SHAP255-5248.1	02 53 27.	- 53 00 12.	36	17.0	GALAXY
SHAP254-5313.8	02 53 27.	- 53 25 54.	30	16.75	GALAXY
LB 02839	02 53 30.	- 00 37 48.		16.6	FAINT BLUE STAR
UGC 02412	02 53 30.	- 01 55	66	16.5	GALAXY Sc
MCG-05-08-004	02 53 30.	- 27 37	90	13.5	GALAXY
SHAP255-5406.2	02 53 30.	- 54 18 18.	42	16.75	GALAXY
BAK 1.034	02 53 32.	- 27 37 41.	49	15.1	EXTRAGALACTIC NEBULA
SHAP255-5331.9	02 53 33.	- 53 44 00.	36	16.75	GALAXY
ZWG 463.035	02 53 36.	+ 15 43		15.2	GALAXY
UGC 02413	02 53 36.	+ 15 43	72	15.2	GALAXY COMPACT
ZWG 463.036	02 53 36.	+ 15 49		15.7	GALAXY
UGC 02363	02 53 36.	+ 39 23	96	14.6	GALAXY COMPACT
MCG-02-08-029	02 53 36.	- 12 12	48	15.	GALAXY
IC 0271	02 53 37.	- 12 12 52.			NONSTELLAR OBJECT
SHAP255-5244.8	02 53 39.	- 52 56 54.	24	17.0	GALAXY
KN 11.17	02 53 40.4	+ 04 19 44.			NEBULA
ZWG 415.031	02 53 42.	+ 04 20		15.5	GALAXY
UGC 02414	02 53 42.	+ 04 20	72	15.5	GALAXY Sc
ZWG 415.032	02 53 42.	+ 05 57		15.5	GALAXY
UGC 02415	02 53 42.	+ 05 57	66	15.5	GALAXY SBb-c
MCG+03-08-031	02 53 42.	+ 15 42	42	15.	GALAXY
ZWG 463.037	02 53 42.	+ 15 43		15.7	GALAXY
ZWG 463.038	02 53 42.	+ 16 41		15.7	GALAXY
VB 182	02 53 42.	+ 63 16	289		STELLAR RING
MCG-02-08-030	02 53 42.	- 14 23	54	15.	GALAXY
LB 01645	02 53 42.	- 44 58		12.2	FAINT BLUE STAR
KN 11.16	02 53 42.8	+ 06 15 24.			NEBULA
IC 0272	02 53 44.	- 14 22 58.			NONSTELLAR OBJECT
KN 11.18	02 53 44.1	+ 05 57 08.			NEBULA
MCG+01-08-022	02 53 45.	+ 04 19	72	15.	GALAXY
MCG+00-08-051	02 53 48.	+ 00 41	54	15.	GALAXY
ZWG 389.048	02 53 48.	+ 01 47		15.3	GALAXY
ZWG 415.033	02 53 48.	+ 04 25		15.2	GALAXY
ZWG 415.034	02 53 48.	+ 06 07		15.7	GALAXY
MCG+03-08-032	02 53 48.	+ 15 41 30.	12	15.5	GALAXY
UGC 02416	02 53 48.	+ 18 30	60	17.	GALAXY Sb
ZC 0253.8+2706	02 53 48.	+ 27 06	1480		CLUSTER OF GALAXIES
UGC 02417	02 53 48.	+ 36 08	72	18.	GALAXY DWRF IR
MCG-02-08-031	02 53 48.	- 13 53	24	15.	GALAXY
MCG-03-08-046	02 53 48.	- 15 35	78	14.5	GALAXY
KN 11.20	02 53 49.0	+ 01 46 42.			NEBULA
LB 00180	02 53 50.	+ 00 24 00.		17.3	FAINT BLUE STAR
KN 11.19	02 53 50.8	+ 04 24 40.			NEBULA
ARC 0398	02 53 51.	- 15 54		17.8	RICH CLUSTER OF GALAXIES
MCG-04-08-006	02 53 51.	- 22 27 30.	54	15.	GALAXY
ZWG 389.049	02 53 54.	+ 00 40		15.2	GALAXY
UGC 02418	02 53 54.	+ 00 40	78	15.2	GALAXY SBb/Sc
RLWT 066	02 53 54.	+ 60 18		16.	FAINT VERY BLUE STAR
MCG-03-08-047	02 53 54.	- 19 51	48	14.5	GALAXY
BAK 1.035	02 53 58.	- 30 17 40.	17	16.8	EXTRAGALACTIC NEBULA
BAK 1.036	02 53 58.	- 30 26 40.	14	16.2	EXTRAGALACTIC NEBULA
ZWG 389.050	02 54 00.	+ 02 02		15.7	GALAXY
ZWG 415.035	02 54 00.	+ 07 08		14.8	GALAXY
UGC 02419	02 54 00.	+ 07 08	78	14.8	GALAXY SBa
ZWG 415.036	02 54 00.	+ 09 13		15.2	GALAXY
UGC 02420	02 54 00.	+ 15 46	84	16.0	GALAXY Sb
LDN 1454	02 54 00.	+ 20 00	3720		DARK NEBULA

OBJECT NAME	RIGHT ASCEN.	DECLINATION	DIAM.	MAGN.	TYPE OF OBJECT
UGC 02421	02 54 00.	+ 44 33	78	17.	GALAXY Sc
UGC 02422	02 54 00.	+ 44 57	84	17.	GALAXY
KN 11.21	02 54 00.8	+ 02 02 05.			NEBULA
BAK 1.037	02 54 04.	- 28 14 39.	17	15.2	EXTRAGALACTIC NEBULA
ZWG 415.037	02 54 06.	+ 04 47		15.7	GALAXY
UGC 02423	02 54 06.	+ 04 47	90	15.7	GALAXY Sc
VB 183	02 54 06.	+ 63 14	248		STELLAR RING
MRK 601	02 54 06.	- 02 59	11	15.	GALAXY WITH UV CONTINUUM
MCG-05-08-005	02 54 06.	- 28 14	36	15.	GALAXY
MCG-05-08-006	02 54 06.	- 32 24	60	14.5	GALAXY
RNGC 1146	02 54 07.	+ 46 15			NON-EXISTENT OBJECT
KN 11.22	02 54 07.0	+ 04 05 34.			NEBULA
SHAP255-5410.7	02 54 08.	- 54 22 46.	42	17.0	GALAXY
KN 11.23	02 54 09.3	+ 04 46 39.			NEBULA
ABC 0397	02 54 10.	+ 15 45		15.1	RICH CLUSTER OF GALAXIES
MCG+01-08-023	02 54 12.	+ 07 08	78	15.	GALAXY
VB 226	02 54 12.	+ 56 07	269		STELLAR RING
TON-S 0289	02 54 12.	- 26 20		15.8	BLUE STAR
TON-S 0290	02 54 12.	- 30 18		13.8	BLUE STAR
MCG+01-08-024	02 54 15.	+ 04 46	48	15.	GALAXY
ZWG 415.038	02 54 18.	+ 06 00		15.3	GALAXY
ZWG 463.039	02 54 18.	+ 17 19		14.6	GALAXY
UGC 02424	02 54 18.	+ 17 19	150	14.6	GALAXY Sb-c
MCG+03-08-033	02 54 18.	+ 17 19	120	14.	GALAXY
ZWG 540.017	02 54 18.	+ 41 20		15.3	GALAXY
TON-S 0291	02 54 18.	- 27 50		17.0	BLUE STAR
KN 11.24	02 54 18.1	+ 06 00 21.			NEBULA
ABC 0396	02 54 19.	+ 41 24		16.6	RICH CLUSTER OF GALAXIES
BAK 1.038	02 54 19.	- 32 22 39.	18	14.5	EXTRAGALACTIC NEBULA
SHAP255-5151.8	02 54 21.	- 52 03 51.	18	17.0	GALAXY
ZC 0254.4+1240	02 54 24.	+ 12 40	3360		CLUSTER OF GALAXIES
ZWG 389.051	02 54 24.	- 01 30		15.6	GALAXY
MCG-02-08-032	02 54 24.	- 11 26	48	15.5	GALAXY
LB 03292	02 54 24.	- 59 23		14.3	FAINT BLUE STAR
LB 02840	02 54 25.	+ 00 17 06.		16.9	FAINT BLUE STAR
ZWG 389.052	02 54 30.	+ 02 35		14.4	GALAXY
UGC 02425	02 54 30.	+ 02 35	108	14.4	GALAXY SB:0-a
MCG+00-08-052	02 54 30.	+ 02 36	66	13.5	GALAXY
ZWG 415.039	02 54 30.	+ 05 07		15.1	GALAXY
UGC 02426	02 54 30.	+ 05 07	66	15.1	GALAXY Sb
UGC 02427	02 54 30.	+ 46 42	60	16.5	GALAXY Sb-c
MCG+00-08-053	02 54 30.	- 01 30	30	15.5	GALAXY
KN 11.25	02 54 31.4	+ 05 07 21.			NEBULA
BAK 1.040	02 54 32.	- 27 48 38.	40	15.8	EXTRAGALACTIC NEBULA
BAK 1.039	02 54 32.	- 28 38 08.	13	16.6	EXTRAGALACTIC NEBULA
LB 00181	02 54 33.	+ 00 35 00.		17.1	FAINT BLUE STAR
IC 0273	02 54 35.	+ 02 35 28.			NONSTELLAR OBJECT
SHAP256-5137.0	02 54 35.	- 51 49 03.	36	15.5	GALAXY
KN 11.26	02 54 35.0	+ 02 34 30.			NEBULA
UGC 02429	02 54 36.	+ 01 09	84	16.0	GALAXY IRR
MCG+01-08-025	02 54 36.	+ 05 08	48	15.	GALAXY
ZWG 440.036	02 54 36.	+ 15 16		15.6	GALAXY
MCG+03-08-034	02 54 36.	+ 15 53	18	17.	GALAXY
UGC 02428	02 54 36.	- 00 25	90	16.5	GALAXY
MCG+00-08-054	02 54 36.	- 01 33	18	16.	GALAXY
MCG-05-08-007	02 54 36.	- 27 48	42	15.5	GALAXY
SHAP256-5154.1	02 54 36.	- 52 06 09.	24	17.0	GALAXY
SHAP256-5302.4	02 54 38.	- 53 14 27.	18	17.25	GALAXY
MCG+00-08-055	02 54 39.	+ 01 08	54	15.5	GALAXY
MCG+00-08-056	02 54 39.	- 00 25	36	15.	GALAXY
KN 11.27	02 54 39.4	+ 05 46 37.			NEBULA
RNGC 1148	02 54 40.	- 07 54		15.0	GALAXY
BAK 1.041	02 54 40.	- 32 11 20.	17	16.1	EXTRAGALACTIC NEBULA
ZC 0254.7+0555	02 54 42.	+ 05 55	9410		CLUSTER OF GALAXIES
ZWG 440.037	02 54 42.	+ 10 17		15.3	GALAXY
UGC 02430	02 54 42.	+ 10 17	66	15.3	GALAXY Sa
ZC 0254.7+1606	02 54 42.	+ 16 06	7060		CLUSTER OF GALAXIES
ZWG 463.040	02 54 42.	+ 16 26		15.6	GALAXY
MCG+03-08-035	02 54 42.	+ 16 26	30	15.5	GALAXY
ZWG 554.017	02 54 42.	+ 46 18		15.0	GALAXY
UGC 02431	02 54 42.	+ 47 58	72	17.	GALAXY
PK136+04.1	02 54 42.	+ 64 18	198	15.5	PLANETARY NEBULA
ZWG 389.053	02 54 42.	- 01 38		15.7	GALAXY
MCG+01-08-018	02 54 42.	- 07 54 30.	60	15.	GALAXY
MCG-03-08-048	02 54 42.	- 15 14	42	15.	GALAXY
RNGC 1151	02 54 43.	- 15 12			GALAXY
RNGC 1150	02 54 43.	- 15 15		15.0	GALAXY
MCG+02-08-037	02 54 45.	+ 10 16	12	15.	GALAXY
MCG+02-08-036	02 54 45.	+ 10 16	48	14.5	GALAXY
SHAP256-5235.6	02 54 45.	- 52 47 38.	24	17.0	GALAXY
BAK 1.042	02 54 46.	- 28 47 01.	17	16.6	EXTRAGALACTIC NEBULA
UGC 02432	02 54 48.	+ 09 55	72	17.	GALAXY DWRF SP
MCG+02-08-038	02 54 48.	+ 10 15	36	14.5	GALAXY
UGC 02433	02 54 48.	+ 10 16	84	16.0	GALAXY SBa-b
ZWG 463.041	02 54 48.	+ 15 52		15.7	GALAXY
ZWG 485.001	02 54 48.	+ 27 23		15.7	GALAXY
ZWG 484.025	02 54 48.	+ 27 23		15.7	GALAXY
ZWG 389.054	02 54 48.	- 00 30		15.1	GALAXY
MCG+00-08-058	02 54 48.	- 00 30	30	15.5	GALAXY
MCG+00-08-057	02 54 48.	- 01 37 30.	18	16.	GALAXY
LB 02841	02 54 49.	- 00 24 48.		17.2	FAINT BLUE STAR
RNGC 1149	02 54 49.	- 00 30		15.0	GALAXY
KN 11.28	02 54 50.7	+ 05 01 17.			NEBULA
MCG+03-08-036	02 54 51.	+ 15 51	24	15.5	GALAXY
SHAP256-5448.0	02 54 52.	- 55 00 02.	42	15.75	GALAXY
ZWG 415.040	02 54 54.	+ 05 46		15.1	GALAXY
UGC 02434	02 54 54.	+ 07 55	60	16.0	GALAXY Sc
ZWG 524.037	02 54 54.	+ 35 04		14.6	GALAXY
UGC 02435	02 54 54.	+ 35 04	150	14.6	GALAXY Sc
SG 1.10	02 54 54.	+ 60 20	2400		DIFFUSE EMISSION NEBULA
URA 30	02 54 54.	+ 68 32	192		STELLAR RING
MCG-03-08-049	02 54 54.	- 14 35	36	15.	GALAXY
MCG-03-08-050	02 54 54.	- 16 52	36	15.5	GALAXY
LB 02842	02 54 55.	- 00 17 12.		16.1	FAINT BLUE STAR
RNGC 1158	02 54 55.	- 14 35		15.0	GALAXY
KN 11.29	02 54 55.2	+ 05 46 38.			NEBULA
MCG+06-07-023	02 54 57.	+ 35 02	108	14.	GALAXY
SHAP256-5245.2	02 54 58.	- 52 57 14.	24	17.0	GALAXY
3ZW 052	02 55 00.	+ 05 49			COMPACT GALAXY
ZWG 415.041	02 55 00.	+ 05 50		14.7	GALAXY
KARA.72 084B	02 55 00.	+ 05 50	30		PART OF DOUBLE GALAXY
KARA.72 084A	02 55 00.	+ 05 50	30	14.7	PART OF DOUBLE GALAXY
MCG+01-08-026	02 55 00.	+ 07 54	48	15.	GALAXY
LDN 1458	02 55 00.	+ 20 00	960		DARK NEBULA
MCG+04-08-001	02 55 00.	+ 25 13	36	15.5	GALAXY
MCG+08-06-024	02 55 00.	+ 46 26	24	17.	GALAXY
TON-S 0292	02 55 00.	- 23 36		15.3	BLUE STAR
MCG-06-07-012	02 55 00.	- 34 43	15	15.	GALAXY
ABC 0400	02 55 01.	+ 05 50		13.9	RICH CLUSTER OF GALAXIES
KN 11.30	02 55 02.2	+ 05 42 14.			NEBULA
KN 11.31	02 55 03.0	+ 05 49 42.			NEBULA
KN 11.32	02 55 03.1	+ 05 49 19.			NEBULA
SHAP256-5235.2	02 55 04.	- 52 47 13.	30	17.5	GALAXY
KN 11.33	02 55 04.1	+ 05 13 52.			NEBULA
KN 11.34	02 55 04.9	+ 05 40 01.			NEBULA
ZWG 415.042	02 55 06.	+ 05 45		15.7	GALAXY
MCG+03-08-037	02 55 06.	+ 15 32	36	16.	GALAXY
ZWG 485.002	02 55 06.	+ 25 14		15.6	GALAXY
ZWG 484.026	02 55 06.	+ 25 14		15.6	GALAXY
5ZW 294	02 55 06.	+ 31 25			COMPACT GALAXY
UGC 02436	02 55 06.	+ 36 18	84	17.	GALAXY
KN 11.35	02 55 06.2	+ 05 28 20.			NEBULA
KN 11.36	02 55 06.8	+ 05 37 36.			NEBULA
KN 11.37	02 55 07.3	+ 05 45 13.			NEBULA
SHAP256-5127.0	02 55 08.	- 51 39 01.	24	16.5	GALAXY
MCG+01-08-027	02 55 09.	+ 05 49	42	14.	GALAXY
KN 11.38	02 55 09.1	+ 05 49 41.			NEBULA
RNGC 1152	02 55 10.	- 07 59		15.0	GALAXY
ABC 0399	02 55 11.	+ 12 49		15.6	RICH CLUSTER OF GALAXIES
ZWG 440.038	02 55 12.	+ 10 59		15.4	GALAXY
UGC 02437	02 55 12.	+ 10 59	96	15.4	GALAXY Sc
UGC 02438	02 55 12.	+ 12 50	72	16.5	GALAXY
MCG+02-08-039	02 55 12.	+ 12 50	48	15.	GALAXY
ZWG 485.003	02 55 12.	+ 23 50		15.5	GALAXY
MCG+04-08-002	02 55 12.	+ 25 15	36	16.	GALAXY
MCG+01-08-019	02 55 12.	- 07 59	42	15.	GALAXY
SHAP256-5254.0	02 55 12.	- 53 06 01.	18	17.0	GALAXY
SHAP256-5244.6	02 55 14.	- 52 56 37.	24	17.0	GALAXY
MCG+01-08-028	02 55 15.	+ 05 45	48	15.	GALAXY
ZWG 415.043	02 55 18.	+ 05 35		15.5	GALAXY
MCG+02-08-040	02 55 18.	+ 10 58	96	14.	GALAXY
MCG+04-08-003	02 55 18.	+ 23 51	36	15.	GALAXY
ZWG 506.001	02 55 18.	+ 32 08		15.5	GALAXY
ZWG 505.057	02 55 18.	+ 32 08		15.5	GALAXY
FEIG 029	02 55 18.	- 02 12		10.3	FAINT BLUE STAR
TON-S 0293	02 55 18.	- 22 33		15.2	BLUE STAR
MCG-06-07-013	02 55 18.	- 35 45	72	14.	GALAXY
KN 11.39	02 55 19.4	+ 05 35 15.			NEBULA
KN 11.40	02 55 20.4	+ 05 35 02.			NEBULA
IC 1870	02 55 21.	- 02 31 42.			NONSTELLAR OBJECT
LB 02843	02 55 22.	+ 02 17 48.		16.3	FAINT BLUE STAR
ABC 0402	02 55 22.	- 22 20		17.9	RICH CLUSTER OF GALAXIES
YC 0255-54	02 55 23.	- 54 46 00.			UNUSUAL SOUTHERN GALAXY
ZWG 485.004	02 55 24.	+ 25 15		15.7	GALAXY
MCG+04-08-004	02 55 24.	+ 25 15	36	15.	GALAXY
OCL 0368	02 55 24.	+ 60 12	4800	6.2	OPEN STAR CLUSTER
MCG-02-08-033	02 55 24.	- 10 21	150	14.	GALAXY
MCG-03-08-051	02 55 24.	- 16 46	54	14.	GALAXY
SHAP256-5301.0	02 55 24.	- 53 13 00.	18	17.0	GALAXY
SHAP256-5239.3	02 55 25.	- 52 51 18.	24	17.0	GALAXY
SHAP256-5242.3	02 55 26.	- 52 54 18.	18	17.0	GALAXY
MCG+01-08-020	02 55 27.	- 02 33	144	13.	GALAXY
SHAP256-5434.9	02 55 28.	- 54 46 54.	510	13.5	GALAXY
ZWG 389.055	02 55 30.	+ 03 10		13.5	GALAXY
RNGC 1153	02 55 30.	+ 03 10		13.5	GALAXY
UGC 02439	02 55 30.	+ 03 10	108	13.5	GALAXY E-S
MCG+00-08-059	02 55 30.	+ 03 10	24	13.	GALAXY
ZWG 415.044	02 55 30.	+ 05 38		15.6	GALAXY
IC 1869	02 55 30.	+ 05 38 17.			NONSTELLAR OBJECT
ZWG 440.039	02 55 30.	+ 10 55		15.4	GALAXY
5ZW 295	02 55 30.	+ 34 45			COMPACT GALAXY
MCG-03-08-052	02 55 30.	- 15 05	48	15.	GALAXY
SHAP257-5258.3	02 55 32.	- 53 10 18.	24	16.5	GALAXY
SPA 2	02 55 32.	- 54 47	360	12.	PECULIAR SPIRAL GALAXY
KN 11.41	02 55 33.4	+ 05 38 07.			NEBULA
SHAP257-5059.4	02 55 34.	- 51 11 24.	30	16.5	GALAXY
KN 11.42	02 55 34.2	+ 03 09 46.			NEBULA
MCG+02-08-041	02 55 36.	+ 10 55	15	15.	GALAXY
ZC 0255.6+2222	02 55 36.	+ 22 22	1280		CLUSTER OF GALAXIES
5ZW 296	02 55 36.	+ 35 50			COMPACT GALAXY
MCG+01-08-021	02 55 36.	- 04 56	48	16.	GALAXY
SHAP257-5242.1	02 55 37.	- 52 54 06.	24	17.5	GALAXY
MCG+02-08-042	02 55 39.	+ 10 54	36	15.5	GALAXY
MCG+01-08-022	02 55 39.	- 05 16	36	16.	GALAXY
RNGC 1154	02 55 41.	- 10 33		14.0	GALAXY
ZWG 415.045	02 55 42.	+ 03 40		15.0	GALAXY
UGC 02441	02 55 42.	+ 03 40	138	15.0	GALAXY Sc
ZWG 415.046	02 55 42.	+ 05 54		15.5	GALAXY
ZWG 415.047	02 55 42.	+ 06 23		15.4	GALAXY
MCG+04-08-005	02 55 42.	+ 25 05	48	15.5	GALAXY
UGC 02442	02 55 42.	+ 25 06	78	16.0	GALAXY
5ZW 297	02 55 42.	+ 41 06			COMPACT GALAXY
ZWG 389.056	02 55 42.	- 02 28		15.7	GALAXY
UGC 02440	02 55 42.	- 02 28	60	15.7	GALAXY Sb-c
MCG-02-08-034	02 55 42.	- 10 33	42	14.	GALAXY
TON-S 0294	02 55 42.	- 29 14		13.8	BLUE STAR
KN 11.43	02 55 42.5	+ 05 53 48.			NEBULA
HOLM 064A	02 55 43.	- 10 34	36	13.8	PART OF MULTIPLE GALAXY
LB 02844	02 55 45.	+ 02 01 24.		16.4	FAINT BLUE STAR
KN 11.45	02 55 45.3	+ 03 39 28.			NEBULA
KN 11.44	02 55 45.6	+ 06 23 32.			NEBULA
SHAP257-5335.0	02 55 46.	- 53 46 59.	36	16.5	GALAXY
RNGC 1155	02 55 47.	- 10 32		14.0	GALAXY
MCG+01-08-029	02 55 48.	+ 03 37	132	14.	GALAXY
ZWG 415.048	02 55 48.	+ 06 06		15.2	GALAXY
UGC 02444	02 55 48.	+ 06 06	66	15.2	GALAXY S
5ZW 298	02 55 48.	+ 25 12			COMPACT GALAXY
ZWG 485.005	02 55 48.	+ 25 12		15.7	GALAXY
UGC 02445	02 55 48.	+ 25 34	72	17.	GALAXY DWARF?
ZWG 540.018	02 55 48.	+ 41 05		15.5	GALAXY
ZWG 540.019	02 55 48.	+ 42 09		14.9	GALAXY
5ZW 299	02 55 48.	+ 46 18			COMPACT GALAXY
VB 184	02 55 48.	+ 60 04	363		STELLAR RING
ZWG 389.057	02 55 48.	- 02 14		14.6	GALAXY
UGC 02443	02 55 48.	- 02 14	90	14.6	GALAXY Sc
MCG+00-08-060	02 55 48.	- 02 15	60	14.	GALAXY
MCG-02-08-035	02 55 48.	- 10 32	48	14.5	GALAXY
MCG-06-07-014	02 55 48.	- 36 54	96	13.	GALAXY
SHAP257-5343.0	02 55 48.	- 53 54 59.	30	16.5	GALAXY
HOLM 064B	02 55 49.	- 10 33	24	14.0	PART OF MULTIPLE GALAXY
SHAP257-5238.2	02 55 49.	- 52 50 11.	36	16.5	GALAXY
PATH 1.128	02 55 50.	+ 00 26	19		NEBULA
RNGC 1157	02 55 50.	- 15 19			GALAXY
KN 11.46	02 55 50.8	+ 06 06 27.			NEBULA
MCG+01-08-031	02 55 51.	+ 06 06	36	15.	GALAXY
MCG+01-08-030	02 55 51.	+ 06 23 30.	18	14.5	GALAXY

OBJECT NAME	RIGHT ASCEN.	DECLINATION	DIAM.	MAGN.	TYPE OF OBJECT
MCG+01-08-023	02 55 51.	- 05 01	30	15.5	GALAXY
KARA.68 029	02 55 54.	+ 09 32	47		DWARF GALAXY
MCG+01-08-024	02 55 54.	- 04 29	120	14.	GALAXY
MCG+06-07-024	02 55 57.	+ 33 41 30.	15	15.	GALAXY
SHAP257-5410.1	02 55 59.	- 54 22 05.	36	16.75	GALAXY
LB 09790	02 56	- 84 02		10.7	FAINT BLUE STAR
ZWG 389.058	02 56 00.	+ 03 14		14.4	GALAXY
UGC 02446	02 56 00.	+ 03 14	33	14.4	GALAXY PECULR
UGC 02447	02 56 00.	+ 32 26	72	17.	GALAXY Sb-c
MCG+06-07-025	02 56 00.	+ 33 41 30.	48	15.	GALAXY
ZWG 524.038	02 56 00.	+ 33 43		15.6	GALAXY
UGC 02448	02 56 00.	+ 33 43	78	15.6	GALAXY
ZWG 554.018	02 56 00.	+ 46 53		15.5	GALAXY
UGC 02449	02 56 00.	+ 46 53	108	15.5	GALAXY PECULR
VB 227	02 56 00.	+ 56 22	369		STELLAR RING
SHAP257-5016.5	02 56 01.	- 50 28 28.	36	17.0	GALAXY
BAK 1.043	02 56 02.	- 29 38 34.	17	16.5	EXTRAGALACTIC NEBULA
SHAP257-5102.0	02 56 02.	- 51 13 58.	30	17.0	GALAXY
KN 11.48	02 56 05.2	+ 03 14 11.			NEBULA
KN 11.47	02 56 05.8	+ 06 13 27.			NEBULA
5ZW 300	02 56 06.	+ 39 28			COMPACT GALAXY
MCG-04-08-007	02 56 06.	- 23 15	18	15.5	GALAXY
SHAP257-5034.8	02 56 08.	- 50 46 46.	30	16.5	GALAXY
MCG+00-08-061	02 56 09.	+ 03 15	18	15.	GALAXY
SHAP257-5243.4	02 56 10.	- 52 55 22.	24	17.25	GALAXY
ARC 0401	02 56 12.	+ 13 23		15.6	RICH CLUSTER OF GALAXIES
UGC 02450	02 56 12.	+ 13 24	108	16.5	GALAXY
ZC 0256.2+1327	02 56 12.	+ 13 27	3630		CLUSTER OF GALAXIES
MCG+06-07-026	02 56 12.	+ 33 53 30.	30	16.	GALAXY
ZWG 524.039	02 56 12.	+ 33 55		15.7	GALAXY
5ZW 301	02 56 12.	+ 44 08			COMPACT GALAXY
MCG-04-08-008	02 56 12.	- 23 16	12	16.	GALAXY
TON-S 0295	02 56 12.	- 30 29		15.2	BLUE STAR
KN 11.49	02 56 12.3	+ 06 10 26.			NEBULA
SHAP257-5240.6	02 56 14.	- 52 52 34.	24	16.5	GALAXY
SHAP257-5250.0	02 56 14.	- 53 01 58.	30	16.25	GALAXY
SHAP257-5348.2	02 56 14.	- 54 00 10.	24	17.5	GALAXY
SHAP257-5238.4	02 56 16.	- 52 50 22.	24	16.5	GALAXY
SHAP257-5240.0	02 56 16.	- 52 51 58.	18	17.0	GALAXY
ZWG 415.049	02 56 18.	+ 06 04		15.5	GALAXY
MCG+02-08-044	02 56 18.	+ 13 23	60	15.5	GALAXY
MCG+02-08-043	02 56 18.	+ 13 24	18	17.	GALAXY
UGC 02451	02 56 18.	+ 36 49	78	16.5	GALAXY Sc
ZWG 540.020	02 56 18.	+ 41 11		15.4	GALAXY
UGC 02452	02 56 18.	+ 44 20	78	19.	GALAXY DWARF
MCG-03-08-053	02 56 18.	- 18 54	72	15.	GALAXY
SHAP257-5058.0	02 56 18.	- 51 09 57.	30	17.0	GALAXY
IC 0276	02 56 21.	- 15 54 25.			NONSTELLAR OBJECT
KN 11.50	02 56 22.2	+ 06 04 16.			NEBULA
SHAP257-5316.1	02 56 23.	- 53 28 03.	24	17.25	GALAXY
ZC 0256.4+0316	02 56 24.	+ 03 16	1080		CLUSTER OF GALAXIES
ZWG 463.042	02 56 24.	+ 15 50		15.6	GALAXY
UGC 02453	02 56 24.	+ 15 50	96	15.6	GALAXY Sb
5ZW 302	02 56 24.	+ 31 11			COMPACT GALAXY
MCG-03-08-054	02 56 24.	- 15 53	120	14.	GALAXY
SHAP258-5002.1	02 56 24.	- 50 14 03.	36	17.0	GALAXY
SHAP257-5217.8	02 56 25.	- 52 29 45.	18	16.75	GALAXY
SHAP257-5316.0	02 56 25.	- 53 27 57.	18	17.0	GALAXY
SHAP257-5246.0	02 56 26.	- 52 57 57.	18	17.0	GALAXY
SHAP257-5237.6	02 56 27.	- 52 49 33.	18	17.25	GALAXY
MCG+03-08-038	02 56 30.	+ 15 50	48	15.	GALAXY
RNGC 1162	02 56 30.	- 12 36		13.0	GALAXY
MCG-02-08-036	02 56 30.	- 12 36	30	13.5	GALAXY
MCG-05-08-008	02 56 30.	- 27 00	42	15.	GALAXY
TON-S 0296	02 56 30.	- 27 28		16.0	BLUE STAR
LB 01646	02 56 30.	- 45 48		11.6	FAINT BLUE STAR
ARC 0406	02 56 32.	- 19 49		17.7	RICH CLUSTER OF GALAXIES
SHAP258-5245.2	02 56 32.	- 52 57 09.	24	16.75	GALAXY
RNGC 1165	02 56 34.	- 32 19		13.0	GALAXY
ZWG 415.050	02 56 36.	+ 05 56		15.0	GALAXY
DG 011	02 56 36.	+ 25 01	120		REFLECTION NEBULA
5ZW 303	02 56 36.	+ 32 36			COMPACT GALAXY
ZWG 506.002	02 56 36.	+ 32 36		15.7	GALAXY
5ZW 304	02 56 36.	+ 35 28			COMPACT GALAXY
TON-S 0297	02 56 36.	- 26 34		15.8	BLUE STAR
MCG-05-08-009	02 56 36.	- 32 19	144	13.5	GALAXY
SHAP258-5222.2	02 56 36.	- 52 34 09.	30	17.0	GALAXY
SHAP258-5240.8	02 56 36.	- 52 52 45.	24	16.5	GALAXY
SHAP259-5244.4	02 56 36.	- 52 56 21.	18	16.75	GALAXY
SS 08	02 56 37.	+ 25 01			DIFFUSE GALACTIC NEBULA
KN 11.51	02 56 37.6	+ 05 56 11.			NEBULA
ARC 0403	02 56 39.	+ 03 18		17.5	RICH CLUSTER OF GALAXIES
UGC 02454	02 56 42.	+ 07 06	78	16.0	GALAXY Sc
MCG+01-08-032	02 56 42.	+ 07 06 30.	72	15.	GALAXY
5ZW 305	02 56 42.	+ 41 21			COMPACT GALAXY
MCG+01-08-025	02 56 42.	- 04 40	60	15.	GALAXY
SHAP258-5256.2	02 56 43.	- 53 08 08.	24	16.75	GALAXY
SHAP258-5243.4	02 56 45.	- 52 55 20.	24	16.75	GALAXY
SHAP258-5408.0	02 56 45.	- 54 19 56.	36	17.5	GALAXY
IC 0274	02 56 46.	+ 44 01 02.			NONSTELLAR OBJECT
SHAP258-5127.9	02 56 46.	- 51 39 50.	24	16.5	GALAXY
MCG+04-08-006	02 56 48.	+ 25 02 30.	198	12.5	GALAXY
ZWG 485.006	02 56 48.	+ 25 03		12.0	GALAXY
UGC 02455	02 56 48.	+ 25 03	222	12.0	GALAXY IRR
KARA.73B 0121	02 56 48.	+ 25 03	174	12.0	ISOLATED GALAXY IR
5ZW 306	02 56 48.	+ 26 14			COMPACT GALAXY
ZWG 524.040	02 56 48.	+ 36 37		13.8	GALAXY
UGC 02456	02 56 48.	+ 36 37	120	13.8	GALAXY SB0
ZWG 524.041	02 56 48.	+ 39 18		14.8	GALAXY
PHL 4331	02 56 48.	- 21 39		18.1	BLUE STELLAR OBJECT
SHAP258-4937.2	02 56 49.	- 49 49 08.	42	16.5	GALAXY
RNGC 1156	02 56 50.	+ 25 02		12.5	GALAXY
SHAP258-5202.9	02 56 50.	- 52 14 50.	30	17.25	GALAXY
SHAP258-5403.5	02 56 50.	- 54 15 26.	30	16.25	GALAXY
MCG+00-08-062	02 56 51.	+ 01 21	36	16.	GALAXY
MCG+06-07-027	02 56 51.	+ 36 37	72	14.5	GALAXY
ZC 0256.9+0002	02 56 54.	+ 00 02	610		CLUSTER OF GALAXIES
MCG+03-08-039	02 56 54.	+ 16 51	48	17.	GALAXY
MCG+03-08-040	02 56 54.	+ 16 52	12	16.	GALAXY
ZC 0256.9+2440	02 56 54.	+ 24 40	1010		CLUSTER OF GALAXIES
LB 03293	02 56 54.	- 73 04		14.9	FAINT BLUE STAR
SHB 068	02 56 54.4	- 00 31 54.		17.5	QUASI-STELLAR OBJECT
BC PKS0256-005	02 56 54.8	- 00 31 52.			QUASI-STELLAR OBJECT
BAK 1.044	02 56 56.	- 28 53 31.	22	17.0	EXTRAGALACTIC NEBULA
LBN 0762	02 57	+ 17 00	1800		BRIGHT NEBULA
LB 09791	02 57	- 83 31		14.0	FAINT BLUE STAR
ZWG 440.040	02 57 00.	+ 15 05		15.6	GALAXY
MCG+02-08-045	02 57 00.	+ 15 07	48	15.	GALAXY
ZWG 485.007	02 57 00.	+ 24 02		15.5	GALAXY
UGC 02457	02 57 00.	+ 24 02	78	15.5	GALAXY Sc
MCG+04-08-007	02 57 00.	+ 24 02	48	15.	GALAXY
5ZW 307	02 57 00.	+ 39 18			COMPACT GALAXY
OCL 0369	02 57 00.	+ 60 13	1500	7.1	OPEN STAR CLUSTER
MCG-04-08-009	02 57 00.	- 23 52	72	15.	GALAXY
SHAP258-5233.4	02 57 00.	- 52 45 19.	18	17.0	GALAXY
SHAP258-5405.6	02 57 01.	- 54 17 31.	30	16.75	GALAXY
SHAP258-4957.0	02 57 02.	- 50 08 55.	36	17.5	GALAXY
LB 02845	02 57 04.	+ 00 26 00.		16.3	FAINT BLUE STAR
BAK 1.045	02 57 04.	- 29 19 30.	12	16.7	EXTRAGALACTIC NEBULA
LB 02846	02 57 06.	+ 00 44 12.		16.9	FAINT BLUE STAR
ZWG 389.059	02 57 06.	+ 02 53		15.6	GALAXY
UGC 02458	02 57 06.	+ 44 27	72	17.	GALAXY Sc
UGC 02459	02 57 06.	+ 48 52	180	18.	GALAXY
MCG-02-08-037	02 57 06.	- 13 50	36	15.5	GALAXY
BAK 1.046	02 57 06.	- 29 16 54.	18	17.4	EXTRAGALACTIC NEBULA
BAK 1.047	02 57 07.	- 27 26 30.	22	16.6	EXTRAGALACTIC NEBULA
SHAP258-5226.8	02 57 08.	- 52 38 43.	24	17.0	GALAXY
KN 11.52	02 57 10.4	+ 05 32 48.			NEBULA
BAK 1.048	02 57 11.	- 31 44 48.	19	16.8	EXTRAGALACTIC NEBULA
SHAP258-5327.5	02 57 11.	- 53 39 25.	18	17.0	GALAXY
ZWG 389.060	02 57 12.	+ 02 34		13.8	GALAXY
MRK 602	02 57 12.	+ 02 34	12	15.	GALAXY WITH UV CONTINUUM
UGC 02460	02 57 12.	+ 02 34	102	13.8	GALAXY SBb-c
ZWG 415.051	02 57 12.	+ 06 20		15.5	GALAXY
ZWG 524.042	02 57 12.	+ 33 53		15.7	GALAXY
UGC 02461	02 57 12.	+ 33 53	72	15.7	GALAXY Sc
KN 11.54	02 57 12.5	+ 05 30 11.			NEBULA
KN 11.53	02 57 13.3	+ 06 20 06.			NEBULA
KN 11.56	02 57 13.7	+ 02 53 05.			NEBULA
SHAP258-4951.4	02 57 14.	- 50 03 19.	42	17.0	GALAXY
KN 11.57	02 57 14.8	+ 02 34 23.			NEBULA
IC 0277	02 57 15.	+ 02 34 20.			NONSTELLAR OBJECT
MCG+00-08-064	02 57 15.	+ 02 35	60	13.5	GALAXY
MCG+00-08-063	02 57 15.	+ 02 54	24	15.5	GALAXY
BAK 1.050	02 57 15.	- 27 25 42.	28	16.6	EXTRAGALACTIC NEBULA
KN 11.55	02 57 17.2	+ 06 05 12.			NEBULA
ZWG 463.043	02 57 18.	+ 18 42		15.5	GALAXY
MCG+03-08-041	02 57 18.	+ 18 42	30	15.	GALAXY
UGC 02462	02 57 18.	+ 33 51	60	16.0	GALAXY SBc
MCG+06-07-028	02 57 18.	+ 33 51	54	16.	GALAXY
MCG-03-08-055	02 57 18.	- 15 28	36	15.5	GALAXY
BAK 1.049	02 57 18.	- 31 59 00.	16	16.4	EXTRAGALACTIC NEBULA
SHAP258-5326.9	02 57 18.	- 53 38 49.	24	17.0	GALAXY
BAK 1.051	02 57 19.	- 31 25 30.	36	16.6	EXTRAGALACTIC NEBULA
SHAP258-5240.5	02 57 19.	- 52 52 24.	24	16.75	GALAXY
MCG+01-08-033	02 57 21.	+ 06 19	36	15.	GALAXY
SHAP258-5404.8	02 57 21.	- 54 16 42.	30	16.75	GALAXY
KN 11.58	02 57 21.4	+ 05 46 15.			NEBULA
LB 02847	02 57 22.	+ 00 32 54.		16.0	FAINT BLUE STAR
KN 11.59	02 57 23.0	+ 03 10 49.			NEBULA
ZC 0257.4+2235	02 57 24.	+ 22 35	870		CLUSTER OF GALAXIES
MCG+06-07-029	02 57 24.	+ 33 49	42	16.	GALAXY
5ZW 308	02 57 24.	+ 34 59			COMPACT GALAXY
ZWG 540.021	02 57 24.	+ 40 03		15.0	GALAXY
UGC 02463	02 57 24.	+ 40 03	180	15.0	GALAXY S IV-V
MCG+07-07-013	02 57 24.	+ 40 03	60	15.	GALAXY
ZWG 540.022	02 57 24.	+ 40 41		15.6	GALAXY
KARA.68 030	02 57 24.	+ 44 53	34		DWARF GALAXY
TON-S 0298	02 57 24.	- 22 51		14.8	BLUE STAR
LB 01647	02 57 24.	- 46 12		12.6	FAINT BLUE STAR
SHAP258-5121.4	02 57 24.	- 51 33 18.	24	17.5	GALAXY
SHAP258-5412.2	02 57 24.	- 54 24 06.	42	16.5	GALAXY
BAK 1.052	02 57 26.	- 30 55 05.	24	16.2	EXTRAGALACTIC NEBULA
SHAP259-5159.2	02 57 26.	- 52 11 06.	24	16.5	GALAXY
LB 02848	02 57 27.	+ 00 52 36.		15.9	FAINT BLUE STAR
RNGC 1159	02 57 27.	+ 42 58		14.0	GALAXY
SHAP258-5404.2	02 57 27.	- 54 16 06.	36	16.75	GALAXY
SHAP259-5143.0	02 57 29.	- 51 54 54.	24	16.75	GALAXY
ZWG 415.052	02 57 30.	+ 05 36		15.3	GALAXY
MCG+01-08-034	02 57 30.	+ 05 46	30	15.	GALAXY
ZWG 463.044	02 57 30.	+ 18 33		15.5	GALAXY
MCG+05-08-001	02 57 30.	+ 32 26	36	16.	GALAXY
UGC 02464	02 57 30.	+ 32 28	66	16.0	GALAXY S0-a
MCG+06-07-031	02 57 30.	+ 34 58	54	15.	GALAXY
ZWG 524.043	02 57 30.	+ 34 59		14.7	GALAXY
UGC 02465	02 57 30.	+ 34 59	78	14.7	GALAXY Sa
MCG+06-07-030	02 57 30.	+ 35 25	60	16.	GALAXY
UGC 02466	02 57 30.	+ 35 27	102	16.5	GALAXY IRR
ZWG 540.023	02 57 30.	+ 42 58		14.2	GALAXY
UGC 02467	02 57 30.	+ 42 58	30	14.2	GALAXY
UGC 02468	02 57 30.	+ 44 15	120	16.5	GALAXY E-S
MCG-04-08-010	02 57 30.	- 24 30	24	15.	GALAXY
KN 11.60	02 57 30.3	+ 05 36 21.			NEBULA
KN 11.61	02 57 31.1	+ 05 30 50.			NEBULA
SHAP259-5414.2	02 57 34.	- 54 26 06.	30	16.5	GALAXY
KN 11.62	02 57 35.1	+ 04 14 25.			NEBULA
MCG+01-08-036	02 57 36.	+ 05 30	30	15.	GALAXY
MCG+01-08-035	02 57 36.	+ 05 35 30.	30	15.	GALAXY
5ZW 309	02 57 36.	+ 44 09			COMPACT GALAXY
PHL 4115	02 57 36.	- 04 57		18.0	BLUE STELLAR OBJECT
LB 02849	02 57 37.	+ 00 28 24.		16.8	FAINT BLUE STAR
IC 0275	02 57 37.	+ 44 08 57.			NONSTELLAR OBJECT
LB 02850	02 57 38.	+ 01 44 06.		16.2	FAINT BLUE STAR
SHAP259-5219.0	02 57 38.	- 52 30 53.	30	16.5	GALAXY
SHAP259-5411.8	02 57 40.	- 54 23 42.	30	17.0	GALAXY
ZWG 415.053	02 57 42.	+ 05 31		15.2	GALAXY
UGC 02469	02 57 42.	+ 05 31	90	15.2	GALAXY PECULR?
ZC 0257.7+3729	02 57 42.	+ 37 29	1280		CLUSTER OF GALAXIES
ZWG 540.024	02 57 42.	+ 42 50		14.7	GALAXY
UGC 02470	02 57 42.	+ 42 50	84	14.7	GALAXY Sa
ZWG 554.019	02 57 42.	+ 46 06		15.7	GALAXY
MCG-02-08-038	02 57 42.	- 11 01	48	15.	GALAXY
TON-S 0299	02 57 42.	- 23 27		14.7	BLUE STAR
MCG-05-08-010	02 57 42.	- 31 57	36	15.5	GALAXY
KN 11.63	02 57 42.3	+ 05 30 51.			NEBULA
SHAP259-5349.0	02 57 43.	- 54 00 53.	30	17.0	GALAXY
SHAP259-5110.4	02 57 45.	- 51 22 17.	30	17.0	GALAXY
SHAP259-5123.2	02 57 46.	- 51 35 05.	24	17.75	GALAXY
SHAP259-5215.0	02 57 46.	- 52 26 53.	18	17.0	GALAXY
ARC 0405	02 57 47.	+ 37 34		17.5	RICH CLUSTER OF GALAXIES
BAK 1.053	02 57 47.	- 31 55 40.	36	15.6	EXTRAGALACTIC NEBULA
MCG+01-08-037	02 57 48.	+ 05 30 30.	36	14.5	GALAXY
ZWG 440.041	02 57 48.	+ 11 39		15.4	GALAXY
UGC 02471	02 57 48.	+ 11 39	78	15.4	GALAXY Sa-b
KARA.72 085A	02 57 48.	+ 11 39	66	15.4	PART OF DOUBLE GALAXY
MCG+02-08-046	02 57 48.	+ 11 39	72	14.	GALAXY

OBJECT NAME	RIGHT ASCEN.	DECLINATION	DIAM.	MAGN.	TYPE OF OBJECT
UGC 02472	02 57 48.	+ 22 59	66	16.0	GALAXY Sa
ZC 0257.8+3542	02 57 48.	+ 35 42	3900		CLUSTER OF GALAXIES
SHAP259-5246.6	02 57 49.	- 52 58 29.	18	17.0	GALAXY
SHAP259-5349.2	02 57 49.	- 54 01 05.	18	17.0	GALAXY
RNGC 1166	02 57 50.	+ 11 39		15.5	GALAXY
SHAP259-5202.2	02 57 50.	- 52 14 05.	18	16.25	GALAXY
SHAP259-5111.8	02 57 51.	- 51 23 41.	24	17.0	GALAXY
ZWG 540.025	02 57 54.	+ 41 12		15.4	GALAXY
UGC 02473	02 57 54.	+ 41 12	102	15.4	GALAXY S(B:c)
ZWG 540.026	02 57 54.	+ 44 43		12.6	GALAXY
UGC 02474	02 57 54.	+ 44 43	180	12.6	GALAXY S0
KARA.72 086B	02 57 54.	+ 44 43	138	12.6	PART OF DOUBLE GALAXY
ZWG 540.027	02 57 54.	+ 44 46		13.0	GALAXY
UGC 02475	02 57 54.	+ 44 46	102	13.0	GALAXY Sc
KARA.72 086A	02 57 54.	+ 44 46	96	13.0	PART OF DOUBLE GALAXY
SHAP259-5240.5	02 57 55.	- 52 52 23.	18	16.75	GALAXY
ARC 0404	02 57 56.	+ 41 13		16.8	RICH CLUSTER OF GALAXIES
RNGC 1161	02 57 56.	+ 44 43		12.5	GALAXY
RNGC 1160	02 57 56.	+ 44 46		13.0	GALAXY
KN 11.64	02 57 56.0	+ 03 51 29.			NEBULA
SHAP259-5154.8	02 57 57.	- 52 06 40.	24	17.25	GALAXY
ZWG 440.042	02 58 00.	+ 11 35		15.4	GALAXY
UGC 02476	02 58 00.	+ 11 35	90	15.4	GALAXY Sb
KARA.72 085B	02 58 00.	+ 11 35	66	15.4	PART OF DOUBLE GALAXY
ZCG 102A02	02 58 00.	+ 35 38		17.8	FUZZY ELLIPTICAL GALAXY
ZCG 102A01	02 58 00.	+ 35 42		17.9	FUZZY ELLIPTICAL GALAXY
UGC 02477	02 58 00.	+ 41 31	102	16.0	GALAXY S
MCG+07-07-015	02 58 00.	+ 44 43	42	13.	GALAXY
MCG+07-07-014	02 58 00.	+ 44 46	78	13.	GALAXY
ZWG 327.003	02 58 00.	+ 74 28		15.6	GALAXY
ZWG 326.005	02 58 00.	+ 74 28		15.6	GALAXY
UGC 02478	02 58 00.	+ 74 28	84	15.6	GALAXY S0-a
MCG+14-02-011	02 58 00.	+ 86 41	39	16.	GALAXY
MCG-03-08-056	02 58 00.	- 17 21	162	14.	GALAXY
TON-S 0300	02 58 00.	- 21 59		14.5	BLUE STAR
BAK 1.054	02 58 00.	- 30 29 22.	11	16.5	EXTRAGALACTIC NEBULA
SHAP259-5027.2	02 58 00.	- 50 39 04.	30	16.5	GALAXY
RNGC 1168	02 58 02.	+ 11 35		15.5	GALAXY
RNGC 1163	02 58 02.	- 17 21		14.0	GALAXY
SHAP259-5111.6	02 58 03.	- 51 23 28.	24	17.0	GALAXY
MCG+02-08-047	02 58 06.	+ 11 34	60	15.	GALAXY
MCG-02-08-039	02 58 06.	- 11 37	60	14.5	GALAXY
MCG-04-08-012	02 58 06.	- 22 29 30.	36	15.5	GALAXY
MCG-04-08-011	02 58 06.	- 23 30	24	15.	GALAXY
SHAP259-5150.2	02 58 06.	- 52 02 04.	18	16.5	GALAXY
SHAP259-5239.6	02 58 06.	- 52 51 28.	18	16.75	GALAXY
MCG+00-08-065	02 58 09.	- 00 10 30.	60	15.	GALAXY
MCG-03-08-057	02 58 09.	- 15 55	78	13.	GALAXY
UGC 02480	02 58 12.	+ 23 08	66	16.5	GALAXY
UGC 02479	02 58 12.	- 00 10	108	16.0	GALAXY
FEIG 030	02 58 12.	- 11 38		13.0	FAINT BLUE STAR
SHAP259-5241.9	02 58 12.	- 52 53 46.	18	16.75	GALAXY
SVEN 200	02 58 14.	- 22 29	48	15.0	GALAXY
BAK 1.055	02 58 15.	- 28 41 51.	31	16.7	EXTRAGALACTIC NEBULA
MCG+00-08-066	02 58 18.	+ 00 37	30	15.	GALAXY
ZWG 524.044	02 58 18.	+ 37 35		14.8	GALAXY
UGC 02481	02 58 18.	+ 37 35	138	14.8	GALAXY (E)
VB 228	02 58 18.	+ 52 43	363		STELLAR RING
ISS 0109	02 58 18.	+ 52 43	363		STELLAR RING
TON-S 0301	02 58 18.	- 24 05		15.0	BLUE STAR
TON-S 0302	02 58 18.	- 26 40		16.0	BLUE STAR
TON-S 0303	02 58 18.	- 31 56		14.8	BLUE STAR
IC 0278	02 58 19.	+ 37 34 05.			NONSTELLAR OBJECT
SHAP259-5020.5	02 58 19.	- 50 32 21.	36	17.0	GALAXY
SHAP300-4947.2	02 58 22.	- 49 59 03.	36	17.5	GALAXY
SHAP300-4948.1	02 58 22.	- 49 59 57.	30	17.0	GALAXY
ZWG 463.045	02 58 24.	+ 16 01		15.2	GALAXY
IC 0279	02 58 24.	+ 16 01 20.			NONSTELLAR OBJECT
ZCG 102A03	02 58 24.	+ 35 32		18.0	FUZZY ELLIPTICAL GALAXY
ZCG 102A05	02 58 24.	+ 35 44		18.1	SPHERICAL GALAXY
ZCG 102A04	02 58 24.	+ 35 44		17.6	SPHERICAL GALAXY
MCG+06-07-032	02 58 24.	+ 37 33	48	15.	GALAXY
VB 230	02 58 24.	+ 52 38	369		STELLAR RING
TON-S 0304	02 58 24.	- 23 36		14.8	BLUE STAR
SHAP259-5158.1	02 58 25.	- 52 09 57.	18	16.5	GALAXY
SHAP300-5134.2	02 58 26.	- 51 46 03.	24	17.5	GALAXY
SHAP300-5127.2	02 58 29.	- 51 39 03.	18	17.0	GALAXY
SHAP300-5301.1	02 58 29.	- 53 12 57.	24	17.0	GALAXY
UGC 02483	02 58 30.	+ 31 39	96	16.0	GALAXY
ZCG 102A06	02 58 30.	+ 35 34		18.2	SPHERICAL GALAXY
UGC 02484	02 58 30.	+ 41 55	60	17.	GALAXY Sc
UGC 02485	02 58 30.	+ 74 14	96	17.	GALAXY Sc
UGC 02482	02 58 30.	- 00 56	72	16.5	GALAXY
MCG+00-08-067	02 58 30.	- 00 56	48	15.5	GALAXY
MCG-04-08-013	02 58 30.	- 22 20	54	14.	GALAXY
LB 03294	02 58 30.	- 74 17		14.8	FAINT BLUE STAR
BC PKS0258-344	02 58 34.7	- 34 25 55.		18.	QUASI-STELLAR OBJECT
SHAP300-5102.0	02 58 35.	- 51 13 51.	24	16.5	GALAXY
ZWG 463.046	02 58 36.	+ 17 39		14.8	GALAXY
UGC 02486	02 58 36.	+ 17 39	66	14.8	GALAXY Sb/Sc
MCG+06-07-033	02 58 36.	+ 34 59	108	14.	GALAXY
ZWG 524.045	02 58 36.	+ 35 01		14.0	GALAXY
UGC 02487	02 58 36.	+ 35 01	210	14.0	GALAXY S0
ZCG 102A08	02 58 36.	+ 35 41		18.2	FUZZY SPHERICAL GALAXY
ZCG 102A07	02 58 36.	+ 35 43		18.0	ELLIPTICAL GALAXY
5ZW 310	02 58 36.	+ 44 04			COMPACT GALAXY
VB 229	02 58 36.	+ 52 45	289		STELLAR RING
ZWG 389.061	02 58 36.	- 02 07		15.7	GALAXY
ARC 0407	02 58 37.	+ 35 39		14.7	RICH CLUSTER OF GALAXIES
RNGC 1167	02 58 39.	+ 35 01		14.0	GALAXY
SHAP300-5154.0	02 58 39.	- 52 05 50.	24	17.25	GALAXY
SHAP300-5439.3	02 58 39.	- 54 51 09.	36	16.75	GALAXY
RNGC 1164	02 58 40.	+ 42 24		14.5	GALAXY
ZWG 463.047	02 58 42.	+ 16 29		15.4	GALAXY
MCG+03-08-042	02 58 42.	+ 17 40	60	14.	GALAXY
UGC 02488	02 58 42.	+ 28 32	72	18.	GALAXY DWARF
MCG+05-08-002	02 58 42.	+ 28 59	12	15.	GALAXY
MCG+06-07-034	02 58 42.	+ 35 38	48	15.	GALAXY
5ZW 311	02 58 42.	+ 35 39			COMPACT GALAXY
UGC 02489	02 58 42.	+ 35 39	66	16.0	GALAXY CMPT GP
ZCG 102A09	02 58 42.	+ 35 40		18.2	SPHERICAL GALAXY
ZWG 540.028	02 58 42.	+ 42 24		14.4	GALAXY
UGC 02490	02 58 42.	+ 42 24	96	14.4	GALAXY SBa
ISS 0064	02 58 42.	+ 62 15	724		STELLAR RING
MCG+07-07-016	02 58 45.	+ 42 23 30.	66	14.5	GALAXY
MCG+00-08-068	02 58 45.	- 02 07 30.	48	15.5	GALAXY
MCG-03-08-058	02 58 45.	- 15 22	48	15.	GALAXY
SHAP300-5200.0	02 58 47.	- 52 11 50.	24	16.5	GALAXY
ZWG 506.003	02 58 48.	+ 29 00		15.7	GALAXY
MCG+06-07-035	02 58 48.	+ 35 31	78	15.	GALAXY
ZWG 524.046	02 58 48.	+ 35 33		14.8	GALAXY
UGC 02491	02 58 48.	+ 35 33	132	14.8	GALAXY SBa
ZCG 102A10	02 58 48.	+ 35 40		18.6	FUZZY SPHERICAL GALAXY
UGC 02492	02 58 48.	+ 47 47	60	17.	GALAXY
PHL 8557	02 58 48.	- 21 53		18.3	BLUE STELLAR OBJECT
SHAP300-5142.2	02 58 48.	- 51 54 02.	36	17.5	GALAXY
SHAP300-5146.4	02 58 48.	- 51 58 14.	24	16.5	GALAXY
SHAP300-5413.2	02 58 48.	- 54 25 02.	24	17.0	GALAXY
SHAP300-5453.6	02 58 48.	- 55 05 26.	36	16.5	GALAXY
SVEN 201	02 58 50.	- 22 59	6	15.4	GALAXY
MCG+07-07-017	02 58 51.	+ 41 23 30.	15	15.	GALAXY
SHAP300-5250.6	02 58 51.	- 53 02 26.	24	17.75	GALAXY
SHAP300-5150.2	02 58 53.	- 52 02 02.	18	16.5	GALAXY
ZC 0258.9+0142	02 58 54.	+ 01 42	1140		CLUSTER OF GALAXIES
MCG+06-07-036	02 58 54.	+ 35 33	72	15.	GALAXY
ZWG 524.047	02 58 54.	+ 35 35		15.5	GALAXY
ZCG 102A13	02 58 54.	+ 35 35		16.2	COMPACT CORE GALAXY
ZCG 102A12	02 58 54.	+ 35 35		16.2	COMPACT CORE GALAXY
UGC 02493	02 58 54.	+ 35 35	84	15.5	GALAXY DBL SYS
ZCG 102A11	02 58 54.	+ 35 39		18.7	SPHERICAL GALAXY
MCG+06-07-037	02 58 54.	+ 35 53	54	15.	GALAXY
ZWG 524.048	02 58 54.	+ 35 55		14.7	GALAXY
UGC 02494	02 58 54.	+ 35 55	66	14.7	GALAXY S
ZWG 540.029	02 58 54.	+ 41 25		14.9	GALAXY
UGC 02495	02 58 54.	+ 41 25	78	14.9	GALAXY S0
MCG+09-06-001	02 58 54.	+ 51 50	36	19.	GALAXY
VB 185	02 58 54.	+ 62 12	739		STELLAR RING
SHAP300-5141.7	02 58 54.	- 51 53 32.	24	16.75	GALAXY
SHAP300-5148.6	02 58 57.	- 52 00 25.	18	17.0	GALAXY
LBN 0675	02 59	+ 60 14	240		BRIGHT NEBULA
LB 09792	02 59	- 85 37		13.8	FAINT BLUE STAR
ZC 0259.0+0414	02 59 00.	+ 04 14	1280		CLUSTER OF GALAXIES
ZCG 102A14	02 59 00.	+ 35 35		18.0	SPHERICAL GALAXY
ZC 0259.0+3838	02 59 00.	+ 38 38	1810		CLUSTER OF GALAXIES
TON-S 0305	02 59 00.	- 24 50		15.8	BLUE STAR
SHAP300-5038.0	02 59 04.	- 50 49 49.	42	17.5	GALAXY
ZCG 102A15	02 59 06.	+ 35 34		17.8	FUZZY ELLIPTICAL GALAXY
UGC 02496	02 59 06.	+ 46 15	66	18.	GALAXY
ZWG 389.062	02 59 06.	- 00 47		15.2	GALAXY
KN 11.65	02 59 06.9	+ 03 17 39.			NEBULA
SG 2.005	02 59 07.	+ 60 18	180		DIFFUSE EMISSION NEBULA
BAK 1.056	02 59 08.	- 28 39 54.	41	15.4	EXTRAGALACTIC NEBULA
LB 02851	02 59 10.	+ 01 03 42.		17.0	FAINT BLUE STAR
LB 02852	02 59 11.	+ 00 30 12.		16.9	FAINT BLUE STAR
MCG+05-08-003	02 59 12.	+ 28 53	132	15.	GALAXY
ZWG 506.004	02 59 12.	+ 28 55		14.8	GALAXY
UGC 02497	02 59 12.	+ 28 55	204	14.8	GALAXY
5ZW 312	02 59 12.	+ 28 57			COMPACT GALAXY
ZCG 102A16	02 59 12.	+ 35 32		17.8	FUZZY ELLIPTICAL GALAXY
MRSL 138+01/1	02 59 12.	+ 60 17	300		HII REGION
MCG+00-08-069	02 59 12.	- 00 47	36	14.5	GALAXY
MCG-05-08-011	02 59 12.	- 28 40	60	14.	GALAXY
SHAP300-5143.4	02 59 14.	- 51 55 13.	24	17.25	GALAXY
MCG-03-08-059	02 59 15.	- 15 01	36	13.	GALAXY
SHAP300-5055.7	02 59 15.	- 51 07 30.	24	16.75	GALAXY
BAK 1.057	02 59 17.	- 28 51 35.	19	16.4	EXTRAGALACTIC NEBULA
ZWG 463.048	02 59 18.	+ 17 09		15.7	GALAXY
UGC 02498	02 59 18.	+ 17 09	102	15.7	GALAXY SBc
KARA.73B 0122	02 59 18.	+ 17 09	96	15.7	ISOLATED GALAXY S
UGC 02499	02 59 18.	+ 39 35	90	18.	GALAXY DWRF IR
ARP 179	02 59 18.	- 04 54			PECULIAR GALAXY
BAK 1.058	02 59 18.	- 28 51 23.	22	16.6	EXTRAGALACTIC NEBULA
RNGC 1172	02 59 19.	- 15 02		13.0	GALAXY
SHAP300-5347.6	02 59 22.	- 53 59 24.	18	17.75	GALAXY
SHAP300-5102.6	02 59 23.	- 51 14 24.	30	17.0	GALAXY
ZC 0259.4+0248	02 59 24.	+ 02 48	1340		CLUSTER OF GALAXIES
MCG+03-08-043	02 59 24.	+ 17 09	90	15.	GALAXY
MCG+00-08-070	02 59 24.	- 01 57	48	16.	GALAXY
TON-S 0306	02 59 24.	- 24 29		15.0	BLUE STAR
TON-S 0307	02 59 24.	- 30 41		14.9	BLUE STAR
HN 0038	02 59 27.	+ 26 53			NEBULA
SHAP301-5150.4	02 59 27.	- 52 02 12.	24	17.25	GALAXY
SHAP300-5353.8	02 59 28.	- 54 05 36.	24	16.25	GALAXY
BAK 1.059	02 59 29.	- 29 28 11.	14	16.6	EXTRAGALACTIC NEBULA
ZWG 540.030	02 59 30.	+ 41 26		15.4	GALAXY
UGC 02500	02 59 30.	+ 41 26	102	15.4	GALAXY SBa
SHAP301-5113.4	02 59 30.	- 51 25 12.	30	16.5	GALAXY
RNGC 1170	02 59 31.	+ 26 52			NON-EXISTENT OBJECT
SHAP301-5331.4	02 59 31.	- 53 43 12.	24	17.25	GALAXY
BAK 1.060	02 59 33.	- 27 58 59.	22	16.3	EXTRAGALACTIC NEBULA
ARC 0408	02 59 35.	+ 32 14		17.4	RICH CLUSTER OF GALAXIES
SHAP301-5333.0	02 59 35.	- 53 44 48.	42	17.25	GALAXY
ZC 0259.6+1303	02 59 36.	+ 13 03	3360		CLUSTER OF GALAXIES
MCG-02-08-040	02 59 36.	- 10 49	12	15.5	GALAXY
MCG-04-08-014	02 59 36.	- 25 30	15	15.	GALAXY
LB 03295	02 59 36.	- 77 12		14.8	FAINT BLUE STAR
SVEN 202	02 59 37.	- 23 13 37.	24	15.9	GALAXY
RNGC 1179	02 59 39.	- 19 06		12.0	GALAXY
BAK 1.061	02 59 39.	- 27 35 10.	17	16.4	EXTRAGALACTIC NEBULA
BAK 1.062	02 59 41.	- 28 40 22.	14	17.0	EXTRAGALACTIC NEBULA
SHAP301-5058.6	02 59 41.	- 51 10 23.	18	16.75	GALAXY
SHAP301-4947.6	02 59 43.	- 49 59 23.	36	17.5	GALAXY
SHAP301-5115.4	02 59 44.	- 51 27 11.	30	17.0	GALAXY
SHAP301-5339.4	02 59 44.	- 53 51 11.	18	17.25	GALAXY
SHAP301-5112.2	02 59 47.	- 51 23 59.	18	16.5	GALAXY
ZWG 389.063	02 59 48.	+ 00 47		15.7	GALAXY
SHAP301-5353.6	02 59 48.	- 54 05 23.	18	17.0	GALAXY
RNGC 1180	02 59 50.	- 15 13			NON-EXISTENT OBJECT
RNGC 1181	02 59 50.	- 15 15			NON-EXISTENT OBJECT
SHAP301-5404.8	02 59 52.	- 54 16 35.	24	16.25	GALAXY
SHAP301-5021.9	02 59 53.	- 50 33 40.	66	17.0	GALAXY
SHAP301-5329.1	02 59 53.	- 53 40 53.	18	16.5	GALAXY
MCG+00-08-071	02 59 54.	+ 00 48	48	15.	GALAXY
ZWG 389.064	02 59 54.	+ 02 00		14.0	GALAXY
UGC 02501	02 59 54.	+ 02 00	84	14.0	GALAXY IRR
BAK 1.064	02 59 54.	- 28 45 34.	17	16.6	EXTRAGALACTIC NEBULA
BAK 1.063	02 59 56.	- 31 49 46.	14	16.6	EXTRAGALACTIC NEBULA
SHAP301-5112.2	02 59 56.	- 51 23 58.	24	17.0	GALAXY
KN 11.66	02 59 58.2	+ 01 59 53.			NEBULA
SHAP301-5339.1	02 59 59.	- 53 50 52.	12	16.75	GALAXY
LB 09793	03 00	- 85 33		14.0	FAINT BLUE STAR
UGC 02502	03 00 00.	+ 00 47	60	16.0	GALAXY S
ZC 0300.0+1422	03 00 00.	+ 14 22	1680		CLUSTER OF GALAXIES
VB 088	03 00 00.	+ 66 17	369		STELLAR RING
BAK 1.065	03 00 01.	- 28 47 21.	16	16.5	EXTRAGALACTIC NEBULA
SHAP301-5151.7	03 00 01.	- 52 03 28.	24	16.5	GALAXY

OBJECT NAME	RIGHT ASCEN.	DECLINATION	DIAM.	MAGN.	TYPE OF OBJECT
SHAP301-5251.8	03 00 01.	- 53 03 34.	18	17.0	GALAXY
SHAP301-5319.2	03 00 01.	- 53 30 58.	12	17.5	GALAXY
KN 11.67	03 00 01.7	+ 02 44 48.			NEBULA
SHAP301-5319.6	03 00 02.	- 53 31 22.	12	16.5	GALAXY
MCG+00-08-072	03 00 03.	+ 02 01	66	14.	GALAXY
IC 0280	03 00 03.	+ 42 09 38.			GROUP OF STARS
SHAP301-5338.4	03 00 03.	- 53 50 10.	24	16.25	GALAXY
SG 3.016	03 00 04.	+ 46 08	120		DIFFUSE EMISSION NEBULA
BAK 1.066	03 00 05.	- 27 16 51.	13	16.3	EXTRAGALACTIC NEBULA
SHAP301-5043.8	03 00 05.	- 50 55 34.	48	16.75	GALAXY
SHAP301-5218.6	03 00 05.	- 52 30 22.	36	16.5	GALAXY
ZWG 554.020	03 00 06.	+ 46 12		13.2	GALAXY
UGC 02503	03 00 06.	+ 46 12	312	13.2	GALAXY Sb/SBb
ZWG 554.021	03 00 06.	+ 46 45		15.5	GALAXY
UGC 02504	03 00 06.	+ 46 45	102	15.5	GALAXY PECULR
TON-S 0308	03 00 06.	- 26 13		12.8	BLUE STAR
SVEN 204	03 00 08.	- 22 30 38.	30	15.2	GALAXY
SVEN 203	03 00 08.	- 22 52	6	15.8	GALAXY
SHAP301-5406.2	03 00 08.	- 54 17 58.	18	17.0	GALAXY
SHAP301-5110.0	03 00 09.	- 51 21 46.	18	17.5	GALAXY
SHAP301-5056.9	03 00 10.	- 51 08 40.	24	17.0	GALAXY
SHAP301-5339.8	03 00 10.	- 53 51 34.	24	17.0	GALAXY
5ZW 313	03 00 12.	+ 27 22			COMPACT GALAXY
MCG+08-06-025	03 00 12.	+ 46 11	240	12.9	GALAXY
MCG+08-06-026	03 00 12.	+ 46 44	78	15.	GALAXY
TON-S 0309	03 00 12.	- 25 20		16.0	BLUE STAR
SVEN 205	03 00 14.	- 22 59	30	15.2	GALAXY
SHAP301-5132.8	03 00 14.	- 51 44 33.	24	16.5	GALAXY
MCG-04-08-015	03 00 15.	- 23 00	12	16.	GALAXY
SHAP301-5045.3	03 00 15.	- 50 57 03.	24	17.25	GALAXY
BAK 1.067	03 00 16.	- 32 22 33.	16	16.6	EXTRAGALACTIC NEBULA
SHAP301-5042.4	03 00 16.	- 50 54 09.	24	17.0	GALAXY
SHAP301-5055.0	03 00 16.	- 51 06 45.	24	17.5	GALAXY
SVEN 206	03 00 19.	- 23 19	12	14.8	GALAXY
SHAP301-5311.0	03 00 19.	- 53 22 45.	18	17.25	GALAXY
MCG-03-08-060	03 00 21.	- 19 06	240	13.	GALAXY
SHAP301-5115.5	03 00 22.	- 51 27 15.	24	17.0	GALAXY
RNGC 1187	03 00 23.	- 23 04		11.0	GALAXY
ZWG 415.054	03 00 24.	+ 04 12		15.3	GALAXY
OCL 0373	03 00 24.	+ 58 34	420	16.	OPEN STAR CLUSTER
UGC 02505	03 00 24.	- 01 17	66	18.	GALAXY DWRF IR
ZWG 389.065	03 00 24.	- 01 34		15.6	GALAXY
MCG-04-08-016	03 00 24.	- 23 04	360	11.	GALAXY
TON-S 0310	03 00 24.	- 26 15		16.0	BLUE STAR
SVEN 207	03 00 25.	- 23 03	240	11.3	GALAXY
SHAP302-5100.8	03 00 25.	- 51 12 33.	18	17.0	GALAXY
SHAP301-5315.6	03 00 25.	- 53 27 21.	24	16.0	GALAXY
SHAP301-5317.5	03 00 26.	- 53 29 15.	18	17.0	GALAXY
KN 11.68	03 00 26.7	+ 04 11 40.			NEBULA
SHAP302-5148.2	03 00 27.	- 51 59 57.	18	16.5	GALAXY
SHAP302-5213.4	03 00 28.	- 52 25 09.	18	17.5	GALAXY
SHAP301-5301.9	03 00 28.	- 53 13 39.	18	17.0	GALAXY
SHAP301-5314.1	03 00 28.	- 53 25 51.	42	16.5	GALAXY
SHAP301-5336.6	03 00 29.	- 53 48 21.	18	16.75	GALAXY
ZWG 506.005	03 00 30.	+ 27 30		15.0	GALAXY
ZWG 485.008	03 00 30.	+ 27 30		15.0	GALAXY
UGC 02506	03 00 30.	+ 27 30	84	15.0	GALAXY COMPACT
ISS 0022	03 00 30.	+ 66 14	344		STELLAR RING
SHAP302-5243.4	03 00 30.	- 52 55 09.	12	17.0	GALAXY
SHAP302-5341.4	03 00 31.	- 53 53 09.	18	16.75	GALAXY
MCG+00-08-073	03 00 33.	- 01 16 30.	48	16.	GALAXY
BAK 1.068	03 00 33.	- 29 18 50.	34	16.9	EXTRAGALACTIC NEBULA
RNGC 1185	03 00 35.	- 09 20		15.0	GALAXY
5ZW 314	03 00 36.	+ 27 15			COMPACT GALAXY
UGC 02507	03 00 36.	+ 30 26	66	16.5	GALAXY SB?c
MCG-02-08-041	03 00 36.	- 09 20	66	15.	GALAXY
SHAP302-5149.0	03 00 37.	- 52 00 44.	18	17.0	GALAXY
SHAP302-5248.8	03 00 40.	- 53 00 32.	60	16.5	GALAXY
SHAP302-5257.5	03 00 40.	- 53 09 14.	12	17.0	GALAXY
BAK 1.069	03 00 41.	- 30 58 19.	23	16.5	EXTRAGALACTIC NEBULA
SHAP302-5104.9	03 00 41.	- 51 16 38.	18	17.75	GALAXY
KN 11.69	03 00 41.0	+ 04 04 58.			NEBULA
ARC 0409	03 00 42.	+ 01 42		17.8	RICH CLUSTER OF GALAXIES
ZC 0300.7+0148	03 00 42.	+ 01 48	1280		CLUSTER OF GALAXIES
ZWG 415.055	03 00 42.	+ 04 05		15.5	GALAXY
ZWG 415.056	03 00 42.	+ 04 18		15.3	GALAXY
UGC 02509	03 00 42.	+ 04 18	60	15.3	GALAXY S
MCG+04-08-008	03 00 42.	+ 27 29	12	15.	GALAXY
5ZW 315	03 00 42.	+ 30 25			COMPACT GALAXY
ZWG 540.031	03 00 42.	+ 43 12		13.6	GALAXY
UGC 02510	03 00 42.	+ 43 12	210	13.6	GALAXY Sc
MCG+07-07-018	03 00 42.	+ 43 12	120	13.5	GALAXY
UGC 02508	03 00 42.	- 02 10	72	16.5	GALAXY Sc
SHAP302-5312.6	03 00 42.	- 53 24 20.	30	16.5	GALAXY
SHAP302-5424.6	03 00 44.	- 54 36 20.	54	16.5	GALAXY
RNGC 1171	03 00 45.	+ 43 12		13.5	GALAXY
MCG+00-08-074	03 00 45.	- 02 09	54	15.5	GALAXY
SHAP302-5301.7	03 00 45.	- 53 13 26.	18	17.0	GALAXY
SHAP302-5320.8	03 00 46.	- 53 32 32.	18	17.0	GALAXY
SHAP302-5153.2	03 00 47.	- 52 04 56.	18	16.5	GALAXY
KN 11.70	03 00 47.0	+ 04 17 40.			NEBULA
HN 0164	03 00 48.	+ 17 24			NEBULA
ZC 0300.8+2528	03 00 48.	+ 25 28	1280		CLUSTER OF GALAXIES
UGC 02511	03 00 48.	+ 48 05	90	16.0	GALAXY SB:a-b
TON-S 0311	03 00 48.	- 22 30		14.8	BLUE STAR
SHAP302-5210.4	03 00 48.	- 52 22 08.	24	16.75	GALAXY
RNGC 1168	03 00 50.	- 15 42			NON-EXISTENT OBJECT
MCG+01-08-038	03 00 51.	+ 04 17	54	15.	GALAXY
VV 346B	03 00 51.	- 22 24	18	16.5	INTERACTING GALAXY
VV 346A	03 00 51.	- 22 24	48	16.	INTERACTING GALAXY
MCG-04-08-017	03 00 51.	- 22 25	24	16.	GALAXY
ZC 0300.9+1533	03 00 54.	+ 15 33	2350		CLUSTER OF GALAXIES
UGC 02512	03 00 54.	+ 30 04	66	16.5	GALAXY Sc
ZC 0300.9+3150	03 00 54.	+ 31 50	1410		CLUSTER OF GALAXIES
5ZW 316	03 00 54.	+ 47 19			COMPACT GALAXY
MCG+08-06-027	03 00 54.	+ 48 04	60	16.	GALAXY
ZWG 389.066	03 00 54.	- 02 26		14.7	GALAXY
MCG-04-08-018	03 00 54.	- 22 23	60	15.	GALAXY
MCG-04-08-019	03 00 54.	- 23 23	54	14.	GALAXY
SVEN 208	03 00 55.	- 23 22 41.	48	14.0	GALAXY
SHAP302-5118.4	03 00 55.	- 51 30 07.	18	16.5	GALAXY
RNGC 1173	03 00 58.	+ 41 10			NON-EXISTENT OBJECT
SHAP302-5346.8	03 00 58.	- 53 58 31.	30	16.25	GALAXY
SHAP302-5245.9	03 00 59.	- 52 57 37.	12	17.25	GALAXY
VDB.66G 227	03 01	- 25 27	70		DWARF GALAXY
ZWG 389.067	03 01 00.	+ 01 45		15.0	GALAXY
UGC 02513	03 01 00.	+ 01 45	96	15.0	GALAXY Sa:
5ZW 317	03 01 00.	+ 31 11			COMPACT GALAXY

OBJECT NAME	RIGHT ASCEN.	DECLINATION	DIAM.	MAGN.	TYPE OF OBJECT
VVI 12	03 01 00.	+ 31 11	11	16.5	SEYFERT GALAXY
RNGC 1195	03 01 00.	- 12 14		14.0	GALAXY
MCG-03-08-061	03 01 00.	- 15 49	108	14.5	GALAXY
ARP 108	03 01 00.	- 22 24			PECULIAR GALAXY
SHAP302-5019.4	03 01 00.	- 50 31 07.	24	17.0	GALAXY
SHAP302-5107.6	03 01 00.	- 51 19 19.	24	17.0	GALAXY
SHAP302-5246.4	03 01 00.	- 52 58 07.	18	16.75	GALAXY
SHAP302-5247.7	03 01 00.	- 52 59 25.	24	16.5	GALAXY
BAK 1.070	03 01 01.	- 32 32 18.	25	16.6	EXTRAGALACTIC NEBULA
SHAP302-5155.5	03 01 01.	- 52 07 13.	18	16.75	GALAXY
SHAP302-5340.1	03 01 01.	- 53 51 49.	18	16.75	GALAXY
RNGC 1189	03 01 02.	- 15 49		14.0	GALAXY
MCG+00-08-075	03 01 03.	- 02 26 30.	42	15.5	GALAXY
MCG-02-08-042A	03 01 03.	- 12 14 30.	24	14.5	GALAXY
SHAP302-5029.3	03 01 03.	- 50 41 01.	72	16.5	GALAXY
KN 11.71	03 01 04.2	+ 01 45 13.			NEBULA
RNGC 1182	03 01 05.	- 09 52			GALAXY
SHAP302-5117.5	03 01 05.	- 51 29 13.	18	16.5	GALAXY
SHAP302-5423.3	03 01 05.	- 54 35 01.	24	16.5	GALAXY
5ZW 318	03 01 06.	+ 45 19			COMPACT GALAXY
MCG+09-06-002	03 01 06.	+ 53 58	30	18.	GALAXY
RNGC 1196	03 01 06.	- 12 17		14.0	GALAXY
MCG-02-08-042B	03 01 06.	- 12 17	72	14.5	GALAXY
MCG-03-08-062	03 01 06.	- 15 51	54	15.	GALAXY
SHAP302-5104.8	03 01 06.	- 51 16 31.	18	17.0	GALAXY
IC 1873	03 01 08.	+ 09 19 41.			NONSTELLAR OBJECT
HOLM 065B	03 01 08.	- 12 14	36	14.2	PART OF MULTIPLE GALAXY
RNGC 1190	03 01 08.	- 15 51		15.0	GALAXY
RNGC 1174	03 01 09.	+ 42 38			NON-EXISTENT OBJECT
SHAP302-5228.4	03 01 10.	- 52 40 07.	24	17.25	GALAXY
SHAP302-5315.1	03 01 10.	- 53 26 49.	18	16.75	GALAXY
KN 11.72	03 01 10.5	+ 01 31 59.			NEBULA
BIGO 481	03 01 11.	+ 42 38			NEBULA
HOLM 065A	03 01 11.	- 12 16	48	13.9	PART OF MULTIPLE GALAXY
ZWG 415.057	03 01 12.	+ 09 25		15.3	GALAXY
MCG+02-08-048	03 01 12.	+ 11 53	36	15.	GALAXY
RNGC 1169	03 01 12.	+ 46 11		12.5	GALAXY
ZWG 389.069	03 01 12.	- 00 24		15.0	GALAXY
ZWG 389.068	03 01 12.	- 01 17		14.7	GALAXY
UGC 02514	03 01 12.	- 01 17	138	14.7	GALAXY S0-a
MCG-03-08-064	03 01 12.	- 15 52 30.	24	15.	GALAXY
MCG-03-08-063	03 01 12.	- 20 23	84	15.	GALAXY
SHAP302-5106.8	03 01 12.	- 51 18 30.	24	17.0	GALAXY
SHAP302-5312.7	03 01 12.	- 53 24 25.	18	17.0	GALAXY
SVEN 209	03 01 13.	- 23 19 42.	6	15.7	GALAXY
SHAP302-5342.9	03 01 13.	- 53 54 36.	24	16.75	GALAXY
RNGC 1194	03 01 14.	- 01 17		14.5	GALAXY
RNGC 1191	03 01 14.	- 15 52		15.0	GALAXY
ARC 0410	03 01 15.	+ 03 37		16.9	RICH CLUSTER OF GALAXIES
MCG+07-07-019	03 01 15.	+ 42 09	84	13.5	GALAXY
IC 0281	03 01 15.	+ 42 09 31.			SAME AS NGC 1177
MCG-03-08-065	03 01 15.	- 15 52	12	15.	GALAXY
MCG-03-08-066	03 01 15.	- 18 34	60	15.	GALAXY
SHAP302-5351.4	03 01 15.	- 54 03 06.	24	17.0	GALAXY
RNGC 1176	03 01 16.	+ 41 11			NON-EXISTENT OBJECT
RNGC 1175	03 01 16.	+ 42 09		13.5	GALAXY
BAK 1.071	03 01 16.	- 28 30 17.	26	16.4	EXTRAGALACTIC NEBULA
SHAP302-5246.8	03 01 16.	- 52 58 30.	12	16.75	GALAXY
IC 1872	03 01 17.	+ 42 37 01.			OPEN CLUSTER
ZWG 389.070	03 01 18.	+ 00 13		15.7	GALAXY
ZC 0301.3+0838	03 01 18.	+ 08 38	1020		CLUSTER OF GALAXIES
MCG+01-08-039	03 01 18.	+ 09 25	24	15.	GALAXY
ZWG 540.032	03 01 18.	+ 42 09		13.8	GALAXY
UGC 02515	03 01 18.	+ 42 09	138	13.8	GALAXY S
UGC 02516	03 01 18.	+ 47 07	90	17.	GALAXY
IC 0283	03 01 18.	- 00 24 21.			NONSTELLAR OBJECT
MCG-03-08-067	03 01 18.	- 15 48	48	12.5	GALAXY
SHAP302-5003.8	03 01 18.	- 50 15 30.	30	17.5	GALAXY
SHAP302-5220.9	03 01 18.	- 52 32 36.	24	16.75	GALAXY
SHAP302-5133.0	03 01 19.	- 51 44 42.	24	17.5	GALAXY
SHAP302-5137.1	03 01 19.	- 51 48 48.	24	17.0	GALAXY
RNGC 1199	03 01 20.	- 15 48		13.0	GALAXY
RNGC 1192	03 01 20.	- 15 52		15.0	GALAXY
SHAP302-5435.7	03 01 20.	- 54 47 24.	24	17.25	GALAXY
MCG+00-08-077	03 01 21.	+ 00 14	12	15.	GALAXY
MCG+07-07-020	03 01 21.	+ 42 10 30.	24	15.5	GALAXY
MCG+00-08-076	03 01 21.	- 00 24	24	15.	GALAXY
MCG+00-08-078	03 01 21.	- 01 18	66	14.5	GALAXY
SHAP302-5003.5	03 01 21.	- 50 15 12.	24	17.5	GALAXY
SHAP302-5132.0	03 01 21.	- 51 43 42.	24	16.5	GALAXY
RNGC 1178	03 01 22.	+ 41 06			NON-EXISTENT OBJECT
RNGC 1177	03 01 22.	+ 42 10		15.5	GALAXY
IC 1875	03 01 22.	- 39 40 48.			NONSTELLAR OBJECT
SHAP302-5404.3	03 01 22.	- 54 16 00.	18	17.5	GALAXY
ZC 0301.4+0338	03 01 24.	+ 03 38	1810		CLUSTER OF GALAXIES
ZC 0301.4+0608	03 01 24.	+ 06 08	610		CLUSTER OF GALAXIES
ZC 0301.4+2705	03 01 24.	+ 27 05	1410		CLUSTER OF GALAXIES
ZWG 540.033	03 01 24.	+ 42 10		15.7	GALAXY
MCG+00-08-079	03 01 24.	- 01 23	18	16.	GALAXY
RNGC 1200	03 01 24.	- 12 11		13.0	GALAXY
MCG-03-08-068	03 01 24.	- 15 40	60	14.5	GALAXY
TON-S 0312	03 01 24.	- 23 59		14.9	BLUE STAR
SVEN 210	03 01 26.	- 22 26	54	14.2	GALAXY
SHAP302-5433.9	03 01 26.	- 54 45 36.	24	17.25	GALAXY
MCG-02-08-043	03 01 27.	- 12 11 30.	24	13.5	GALAXY
BAK 1.072	03 01 27.	- 29 31 59.	23	16.4	EXTRAGALACTIC NEBULA
RNGC 1183	03 01 28.	+ 41 10			NON-EXISTENT OBJECT
SHAP303-5104.4	03 01 28.	- 51 16 06.	24	16.5	GALAXY
SHAP303-5138.4	03 01 29.	- 51 50 06.	12	17.0	GALAXY
MCG-04-08-020	03 01 30.	- 22 27	48	15.	GALAXY
BAK 1.073	03 01 30.	- 31 05 17.	14	16.4	EXTRAGALACTIC NEBULA
SHAP303-5304.1	03 01 30.	- 53 15 48.	12	17.5	GALAXY
SHAP303-5107.4	03 01 32.	- 51 19 05.	24	16.75	GALAXY
SHAP303-5143.6	03 01 32.	- 51 55 17.	18	16.5	GALAXY
SHAP302-5408.2	03 01 32.	- 54 19 54.	18	16.75	GALAXY
SHAP303-5030.2	03 01 34.	- 50 41 53.	24	17.0	GALAXY
SHAP303-5404.0	03 01 34.	- 54 15 41.	18	17.0	GALAXY
SHAP303-5030.4	03 01 35.	- 50 42 05.	54	16.0	GALAXY
SHAP303-5244.4	03 01 35.	- 52 56 05.	18	17.0	GALAXY
ZC 0301.6+0053	03 01 36.	+ 00 53	2690		CLUSTER OF GALAXIES
ZC 0301.6+2244	03 01 36.	+ 22 44	3760		CLUSTER OF GALAXIES
ZC 0301.6+3809	03 01 36.	+ 38 09	2150		CLUSTER OF GALAXIES
ZWG 389.071	03 01 36.	- 01 23		14.9	GALAXY
UGC 02517	03 01 36.	- 01 23	78	14.9	GALAXY SB
MCG-03-08-069	03 01 36.	- 14 31	84	14.5	GALAXY
BAK 1.074	03 01 36.	- 27 32 16.	19	16.4	EXTRAGALACTIC NEBULA
TON-S 0313	03 01 36.	- 30 38		15.1	BLUE STAR
HN 0165	03 01 36.	- 50 42			NEBULA

OBJECT NAME	RIGHT ASCEN.	DECLINATION	DIAM.	MAGN.	TYPE OF OBJECT
IC 1877	03 01 36.	- 50 42			NONSTELLAR OBJECT
SHAP303-5144.6	03 01 36.	- 51 56 17.	24	17.25	GALAXY
SHAP303-5358.8	03 01 36.	- 54 10 29.	18	17.5	GALAXY
SHAP303-5359.7	03 01 36.	- 54 11 23.	24	17.0	GALAXY
SHAP303-5158.1	03 01 39.	- 52 09 47.	24	17.5	GALAXY
ZWG 389.073	03 01 42.	+ 01 34		15.6	GALAXY
UGC 02518	03 01 42.	+ 01 34	72	15.6	GALAXY Sc
ZWG 463.049	03 01 42.	+ 17 29		15.5	GALAXY
MCG+00-08-080	03 01 42.	- 01 23	48	15.	GALAXY
ZWG 389.072	03 01 42.	- 02 30		15.5	GALAXY
MCG-02-08-044	03 01 42.	- 12 13	60	15.	GALAXY
MCG-04-08-021	03 01 42.	- 25 27 30.	120	15.	GALAXY
TON-S 0314	03 01 42.	- 29 01		14.8	BLUE STAR
SHAP303-5029.8	03 01 42.	- 50 41 29.	96	15.25	GALAXY
KN 11.73	03 01 42.7	+ 01 33 50.			NEBULA
IC 0285	03 01 43.	- 12 13 00.			NONSTELLAR OBJECT
SHAP303-5230.4	03 01 43.	- 52 42 05.	24	17.0	GALAXY
SHAP303-5421.4	03 01 43.	- 54 33 05.	18	17.75	GALAXY
MCG+00-08-081	03 01 45.	+ 01 34 30.	54	15.	GALAXY
MCG-04-08-022	03 01 45.	- 24 46	48	15.	GALAXY
SHAP303-5145.9	03 01 47.	- 51 57 35.	18	17.25	GALAXY
SHAP303-5244.2	03 01 47.	- 52 55 53.	30	17.25	GALAXY
ZC 0301.9+1617	03 01 54.	+ 16 17	1280		CLUSTER OF GALAXIES
IC 0282	03 01 54.	+ 41 39 39.			NONSTELLAR OBJECT
LB 01648	03 01 54.	- 53 42		13.8	FAINT BLUE STAR
SHAP303-5408.8	03 01 54.	- 54 20 28.	24	17.25	GALAXY
LB 03296	03 01 54.	- 57 33		13.9	FAINT BLUE STAR
SHAP303-5032.8	03 01 55.	- 50 44 28.	18	17.25	GALAXY
SHAP303-5050.0	03 01 55.	- 51 01 40.	24	18.0	GALAXY
SHAP303-5220.5	03 01 57.	- 52 32 10.	12	17.5	GALAXY
ARC 0411	03 01 59.	+ 00 49		17.6	RICH CLUSTER OF GALAXIES
SHAP303-5131.8	03 01 59.	- 51 43 28.	12	17.0	GALAXY
FEIG 031	03 02 00.	+ 02 46		15.0	FAINT BLUE STAR
ZWG 415.058	03 02 00.	+ 05 15		15.7	GALAXY
MCG+13-03-001	03 02 00.	+ 79 56 30.	66	13.	GALAXY
ZWG 346.001	03 02 00.	+ 79 57		14.3	GALAXY
UGC 02519	03 02 00.	+ 79 57	102	14.3	GALAXY Sc
MCG+00-08-082	03 02 00.	- 00 13	30	15.5	GALAXY
TON-S 0315	03 02 00.	- 22 28		13.5	BLUE STAR
MCG-04-08-023	03 02 00.	- 26 16	84	12.	GALAXY
MCG-05-08-012	03 02 00.	- 26 47	48	15.	GALAXY
RNGC 1201	03 02 01.	- 26 15		12.0	GALAXY
KN 11.74	03 02 01.0	+ 05 14 54.			NEBULA
BAK 1.075	03 02 04.	- 30 19 45.	17	16.4	EXTRAGALACTIC NEBULA
SHAP303-5349.8	03 02 04.	- 54 01 28.	18	17.25	GALAXY
SHAP303-5353.1	03 02 04.	- 54 04 46.	24	17.5	GALAXY
SHAP303-5018.0	03 02 05.	- 50 29 40.	24	16.5	GALAXY
UGC 02520	03 02 06.	+ 21 12	90	16.5	GALAXY SB:c
BAK 1.076	03 02 06.	- 27 32 57.	22	16.5	EXTRAGALACTIC NEBULA
SHAP303-5105.8	03 02 06.	- 51 17 28.	18	17.25	GALAXY
SHAP303-5206.0	03 02 06.	- 52 17 40.	30	15.7	GALAXY
IC 1878	03 02 07.	- 52 18			NONSTELLAR OBJECT
HN 0166	03 02 08.	- 52 18			NEBULA
SHAP303-5206.5	03 02 08.	- 52 18 10.	42	15.7	GALAXY
RNGC 1186	03 02 09.	+ 42 38		12.5	GALAXY
SHAP303-5216.8	03 02 10.	- 52 28 27.	18	17.0	GALAXY
SHAP303-5118.8	03 02 11.	- 51 30 27.	30	16.75	GALAXY
MCG+07-07-021	03 02 12.	+ 42 38	138	12.5	GALAXY
ZWG 540.034	03 02 12.	+ 42 39		12.5	GALAXY
UGC 02521	03 02 12.	+ 42 39	270	12.5	GALAXY SBc
MCG+09-06-003	03 02 12.	+ 54 06	4	18.	GALAXY
RNGC 1204	03 02 12.	- 12 32		15.0	GALAXY
MCG-02-08-045	03 02 12.	- 12 32	54	15.	GALAXY
SHAP303-5430.1	03 02 12.	- 54 41 46.	24	17.0	GALAXY
BAK 1.077	03 02 13.	- 30 34 02.	14	16.6	EXTRAGALACTIC NEBULA
SHAP303-5033.5	03 02 15.	- 50 45 09.	24	17.25	GALAXY
RNGC 1206	03 02 17.	- 08 58			GALAXY
SHAP303-5038.3	03 02 17.	- 50 49 57.	18	17.0	GALAXY
UGC 02522	03 02 18.	+ 00 01	102	16.0	GALAXY Sb
MCG+00-08-083	03 02 18.	+ 00 02	30	15.5	GALAXY
IC 0286	03 02 19.	- 06 41			NONSTELLAR OBJECT
IC 1879	03 02 19.	- 52 18			NONSTELLAR OBJECT
HN 0167	03 02 20.	- 52 18			NEBULA
SHAP303-5206.5	03 02 20.	- 52 18 09.	66	16.0	GALAXY
CED 011	03 02 21.	+ 60 29	60		DIFFUSE GALACTIC NEBULA
SHAP303-5225.4	03 02 21.	- 52 37 03.	12	17.25	GALAXY
IC 1871	03 02 22.	+ 60 29 10.			MAY NOT EXIST
ZC 0302.4+0240	03 02 24.	+ 02 40	1480		CLUSTER OF GALAXIES
MCG+05-08-004	03 02 24.	+ 33 12	42	15.	GALAXY
URA 23	03 02 24.	+ 65 13	240		STELLAR RING
MCG-05-08-013	03 02 24.	- 27 38	36	14.5	GALAXY
SHAP303-5150.0	03 02 25.	- 52 01 39.	12	17.5	GALAXY
SHAP304-5006.4	03 02 26.	- 50 18 03.	24	17.5	GALAXY
SHAP304-5011.6	03 02 27.	- 50 23 14.	30	17.0	GALAXY
SHAP303-5239.0	03 02 27.	- 52 50 39.	12	17.0	GALAXY
SHAP303-5411.4	03 02 27.	- 54 23 03.	24	16.75	GALAXY
RNGC 1202	03 02 28.	- 06 41			GALAXY
IC 1876	03 02 28.	- 27 40 53.			NONSTELLAR OBJECT
BAK 1.078	03 02 29.	- 30 38 08.	17	16.4	EXTRAGALACTIC NEBULA
SHAP303-5259.0	03 02 29.	- 53 10 39.	18	17.0	GALAXY
ZWG 389.074	03 02 30.	+ 00 09		15.2	GALAXY
MCG+00-08-084	03 02 30.	+ 00 10	36	15.	GALAXY
ZWG 389.075	03 02 30.	+ 00 54		15.3	GALAXY
UGC 02523	03 02 30.	+ 00 54	114	15.3	GALAXY Sc
ZWG 415.059	03 02 30.	+ 05 40		15.2	GALAXY
ZWG 463.050	03 02 30.	+ 19 10		15.5	GALAXY
OCL 0390	03 02 30.	+ 44 11	300	12.6	OPEN STAR CLUSTER
SHAP304-5211.4	03 02 30.	- 52 23 02.	18	16.5	GALAXY
SHAP304-5254.2	03 02 30.	- 53 05 50.	12	16.5	GALAXY
LB 01649	03 02 30.	- 55 22		13.3	FAINT BLUE STAR
SHAP304-5123.4	03 02 31.	- 51 35 02.	12	17.0	GALAXY
SHAP304-5317.6	03 02 31.	- 53 29 14.	12	16.5	GALAXY
RNGC 1193	03 02 32.	+ 44 11		12.5	OPEN CLUSTER
SHAP304-5212.2	03 02 32.	- 52 23 50.	12	16.5	GALAXY
SHAP304-5403.1	03 02 33.	- 54 14 44.	18	17.25	GALAXY
SHAP304-5412.8	03 02 33.	- 54 24 26.	24	17.0	GALAXY
KN 11.76	03 02 34.7	+ 00 53 58.			NEBULA
IC 0287	03 02 35.	- 12 16 23.			NONSTELLAR OBJECT
SHAP304-5314.0	03 02 35.	- 53 25 38.	18	18.0	GALAXY
KN 11.75	03 02 35.0	+ 05 39 39.			NEBULA
MCG+00-08-085	03 02 36.	+ 00 55	78	15.	GALAXY
UGC 02524	03 02 36.	+ 05 03	60	18.	GALAXY DWARF
ZWG 506.006	03 02 36.	+ 33 12		15.6	GALAXY
UGC 02525	03 02 36.	+ 33 12	60	15.6	GALAXY S
MCG+06-07-038	03 02 36.	+ 36 34	180	13.	GALAXY
ZWG 524.049	03 02 36.	+ 36 35		13.5	GALAXY
UGC 02526	03 02 36.	+ 36 35	270	13.5	GALAXY Sb
TON-S 0317	03 02 36.	- 22 22		15.1	BLUE STAR
TON-S 0316	03 02 36.	- 23 05		14.7	BLUE STAR
TON-S 0318	03 02 36.	- 33 27		14.8	BLUE STAR
SHAP304-4957.2	03 02 38.	- 50 08 50.	30	16.5	GALAXY
SHAP304-5101.4	03 02 38.	- 51 13 02.	24	16.25	GALAXY
IC 0284	03 02 40.	+ 42 10 23.			NONSTELLAR OBJECT
SHAP304-5147.4	03 02 40.	- 51 59 02.	30	17.0	GALAXY
BAK 1.079	03 02 41.	- 28 14 13.	14	16.2	EXTRAGALACTIC NEBULA
SHAP304-5029.4	03 02 41.	- 50 41 02.	24	16.5	GALAXY
SHAP304-5124.0	03 02 41.	- 51 35 38.	18	17.0	GALAXY
SHAP304-5313.2	03 02 41.	- 53 24 50.	12	17.0	GALAXY
ZWG 389.077	03 02 42.	+ 02 11		14.9	GALAXY
UGC 02527	03 02 42.	+ 02 11	72	14.9	GALAXY SB
MCG+01-08-040	03 02 42.	+ 05 39	30	15.	GALAXY
MCG+07-07-022	03 02 42.	+ 41 23	54	14.5	GALAXY
ZWG 540.035	03 02 42.	+ 41 25		14.8	GALAXY
UGC 02528	03 02 42.	+ 41 25	120	14.8	GALAXY S0-a
ZWG 389.076	03 02 42.	- 00 21		15.2	GALAXY
MCG+00-08-086	03 02 42.	- 00 30	42	16.	GALAXY
MCG-02-08-046	03 02 42.	- 13 13	60	16.	GALAXY
HARO 19	03 02 43.	- 27 43			BLUE EMISSION-LINE GALAXY
SHAP304-5122.1	03 02 43.	- 51 33 44.	18	17.25	GALAXY
KN 11.77	03 02 44.4	+ 02 11 14.			NEBULA
MCG+00-08-087	03 02 45.	+ 02 12	42	15.	GALAXY
MCG+00-08-088	03 02 45.	- 01 16	36	16.	GALAXY
BAK 1.080	03 02 46.	- 27 41 55.	18	15.9	EXTRAGALACTIC NEBULA
BAK 1.081	03 02 47.	- 27 41 43.	22	15.8	EXTRAGALACTIC NEBULA
SHAP304-5138.0	03 02 47.	- 51 49 38.	18	16.75	GALAXY
MCG+00-08-089	03 02 48.	- 00 21	27	16.	GALAXY
MCG-03-08-071	03 02 48.	- 14 34	6	16.	GALAXY
MCG-03-08-070	03 02 48.	- 14 34	21	15.5	GALAXY
MCG-05-08-015	03 02 48.	- 27 41	24	15.	GALAXY
MCG-05-08-014	03 02 48.	- 27 41	21	15.	GALAXY
SHAP304-5214.6	03 02 48.	- 52 26 13.	18	17.5	GALAXY
BC PKS0302-623	03 02 48.3	- 62 23 05.		18.5	QUASI-STELLAR OBJECT
RNGC 1203B	03 02 49.	- 14 34		16.0	GALAXY
RNGC 1203A	03 02 49.	- 14 34		15.0	GALAXY
RNGC 1197	03 02 50.	+ 43 52			NON-EXISTENT OBJECT
MCG+01-08-026	03 02 51.	- 02 32	24	15.5	GALAXY
SHAP304-5114.4	03 02 51.	- 51 26 01.	24	16.5	GALAXY
SHAP304-5106.6	03 02 52.	- 51 18 13.	18	17.75	GALAXY
SHAP304-5310.2	03 02 52.	- 53 21 49.	12	17.25	GALAXY
ZWG 415.060	03 02 54.	+ 05 56		15.2	GALAXY
UGC 02530	03 02 54.	+ 22 00	108	16.5	GALAXY
ZWG 540.036	03 02 54.	+ 41 33		15.7	GALAXY
MCG+07-07-024	03 02 54.	+ 41 39	60	14.5	GALAXY
MCG+07-07-023	03 02 54.	+ 42 10 30.	180	14.5	GALAXY
5ZW 319	03 02 54.	+ 42 11			COMPACT GALAXY
ZWG 540.037	03 02 54.	+ 42 11		13.8	GALAXY
UGC 02531	03 02 54.	+ 42 11	312	13.8	GALAXY
UGC 02529	03 02 54.	- 00 34	60	16.0	GALAXY S
MCG+00-08-090	03 02 54.	- 00 34	60	15.5	GALAXY
MCG+00-08-091	03 02 54.	- 00 35	15	15.	GALAXY
ZWG 389.078	03 02 54.	- 00 36		14.9	GALAXY
SHAP304-5216.2	03 02 57.	- 52 27 49.	24	16.75	GALAXY
RNGC 1198	03 02 58.	+ 41 40		14.0	GALAXY
BAK 1.082	03 02 58.	- 30 11 36.	17	16.5	EXTRAGALACTIC NEBULA
KN 11.78	03 02 58.8	+ 04 48 07.			NEBULA
RNGC 1205	03 02 59.	- 09 53			NON-EXISTENT OBJECT
BAK 1.083	03 02 59.	- 27 31 48.	26	15.2	EXTRAGALACTIC NEBULA
LBN 0679	03 03	+ 56 30	3420		BRIGHT NEBULA
MCG+01-08-041	03 03 00.	+ 05 57	24	15.	GALAXY
UGC 02532	03 03 00.	+ 15 55	90	16.5	GALAXY S0?
ZC 0303.0+4125	03 03 00.	+ 41 25	51810		CLUSTER OF GALAXIES
ZWG 540.038	03 03 00.	+ 41 40		14.0	GALAXY
UGC 02533	03 03 00.	+ 41 40	168	14.0	GALAXY E-S0
MCG-05-08-016	03 03 00.	- 27 31	60	15.	GALAXY
SHAP304-5256.8	03 03 00.	- 53 08 25.	12	17.0	GALAXY
LB 01650	03 03 00.	- 53 33		12.4	FAINT BLUE STAR
LB 03297	03 03 00.	- 58 49		13.4	FAINT BLUE STAR
SHAP304-5004.0	03 03 02.	- 50 15 37.	30	17.25	GALAXY
BAK 1.085	03 03 05.	- 27 53 12.	25	16.4	EXTRAGALACTIC NEBULA
BAK 1.084	03 03 05.	- 31 42 42.	19	16.6	EXTRAGALACTIC NEBULA
ZWG 389.079	03 03 06.	+ 01 07		15.6	GALAXY
BAK 1.086	03 03 06.	- 29 29 54.	17	16.4	EXTRAGALACTIC NEBULA
BAK 1.087	03 03 10.	- 29 05 59.	14	16.4	EXTRAGALACTIC NEBULA
SHAP304-5332.2	03 03 11.	- 53 43 48.	12	17.25	GALAXY
ZWG 415.061	03 03 12.	+ 04 40		15.3	GALAXY
ZWG 524.050	03 03 12.	+ 35 50		14.8	GALAXY
ZWG 389.080	03 03 12.	- 00 22		14.7	GALAXY
KN 11.79	03 03 12.1	+ 01 07 02.			NEBULA
IC 1874	03 03 13.	+ 35 49 26.			NONSTELLAR OBJECT
SHAP304-5447.4	03 03 14.	- 54 59 00.	24	17.0	GALAXY
MCG+06-07-039	03 03 15.	+ 35 48	54	14.5	GALAXY
SHAP304-5010.8	03 03 15.	- 50 22 24.	24	17.5	GALAXY
SHAP304-5108.0	03 03 16.	- 51 19 36.	18	17.5	GALAXY
KN 11.80	03 03 17.7	+ 04 39 53.			NEBULA
MCG+00-08-092	03 03 18.	+ 01 08	36	15.5	GALAXY
ZC 0303.3+1821	03 03 18.	+ 18 21	3020		CLUSTER OF GALAXIES
ZWG 540.039	03 03 18.	+ 41 17		15.7	GALAXY
UGC 02534	03 03 18.	+ 41 17	84	15.7	GALAXY PECULR?
MCG+07-07-025	03 03 18.	+ 41 17	9	15.5	GALAXY
TON-S 0319	03 03 18.	- 23 25		14.6	BLUE STAR
SHAP304-5415.2	03 03 18.	- 54 26 46.	36	16.5	GALAXY
MCG-03-08-072	03 03 21.	- 16 55	36	15.	GALAXY
SHAP304-5158.7	03 03 21.	- 52 10 18.	12	17.5	GALAXY
MCG+01-08-042	03 03 24.	+ 04 38 30.	18	15.	GALAXY
5ZW 320	03 03 24.	+ 47 17			COMPACT GALAXY
PHL 1465	03 03 24.	- 11 06		17.8	BLUE STELLAR OBJECT
BAK 1.088	03 03 25.	- 27 53 35.	10	16.4	EXTRAGALACTIC NEBULA
SHAP305-5110.0	03 03 25.	- 51 21 35.	24	16.5	GALAXY
SHAP304-5307.7	03 03 26.	- 53 19 18.	24	16.5	GALAXY
SHAP304-5330.2	03 03 26.	- 53 41 48.	12	17.5	GALAXY
BAK 1.089	03 03 28.	- 27 37 34.	13	16.4	EXTRAGALACTIC NEBULA
UGC 02535	03 03 30.	+ 01 25	60	18.	GALAXY DWRF IR
MCG+01-08-027	03 03 30.	- 05 56	36	15.5	GALAXY
SHAP305-5200.6	03 03 31.	- 52 12 11.	18	17.0	GALAXY
SHAP305-5221.0	03 03 34.	- 52 32 35.	24	17.0	GALAXY
SHAP305-5235.2	03 03 34.	- 52 46 47.	12	17.0	GALAXY
SHAP305-5350.8	03 03 34.	- 54 02 23.	18	16.5	GALAXY
SHAP305-5201.7	03 03 35.	- 52 13 17.	24	17.0	GALAXY
ZWG 540.040	03 03 36.	+ 41 20		15.7	GALAXY
UGC 02536	03 03 36.	+ 41 20	96	15.7	GALAXY S0
ZWG 540.041	03 03 36.	+ 42 20		15.6	GALAXY
MCG+08-06-028	03 03 36.	+ 46 25	60	15.	GALAXY
ZWG 554.022	03 03 36.	+ 46 26		15.1	GALAXY
UGC 02537	03 03 36.	+ 46 26	84	15.1	GALAXY Sc
PHL 4332	03 03 36.	- 12 21		13.9	BLUE STELLAR OBJECT
SHAP305-5047.0	03 03 36.	- 50 58 35.	18	17.5	GALAXY

OBJECT NAME	RIGHT ASCEN.	DECLINATION	DIAM.	MAGN.	TYPE OF OBJECT
SHAP305-5212.9	03 03 36.	- 52 24 29.	18	17.0	GALAXY
SHAP305-5331.4	03 03 36.	- 53 42 59.	12	17.25	GALAXY
SHAP305-5121.8	03 03 37.	- 51 33 23.	18	16.5	GALAXY
BAK 1.090	03 03 38.	- 28 54 46.	42	16.4	EXTRAGALACTIC NEBULA
ARC 0414	03 03 40.	- 14 40		17.5	RICH CLUSTER OF GALAXIES
SHAP305-5103.6	03 03 40.	- 51 15 11.	24	17.0	GALAXY
ZC 0303.7-0020	03 03 42.	- 00 20	1280		CLUSTER OF GALAXIES
MCG-03-08-073	03 03 42.	- 15 48	60	12.5	GALAXY
TON-S 0320	03 03 42.	- 21 03		13.5	BLUE STAR
MCG-05-08-017	03 03 42.	- 27 25	60	15.	GALAXY
BAK 1.091	03 03 42.	- 27 52 22.	12	16.6	EXTRAGALACTIC NEBULA
LB 01651	03 03 42.	- 47 30		12.6	FAINT BLUE STAR
ARC 0412	03 03 45.	- 00 23		17.5	RICH CLUSTER OF GALAXIES
BAK 1.092	03 03 45.	- 27 25 22.	20	15.4	EXTRAGALACTIC NEBULA
BAK 1.093	03 03 45.	- 27 30 10.	11	17.0	EXTRAGALACTIC NEBULA
SHAP305-5213.0	03 03 45.	- 52 24 35.	12	17.0	GALAXY
SHAP305-5409.9	03 03 45.	- 54 21 29.	18	16.75	GALAXY
REIN 2.037	03 03 46.36	- 09 43 59.7			NEBULA
RNGC 1208	03 03 47.	- 09 44		14.0	GALAXY
ZWG 540.042	03 03 48.	+ 41 34		15.6	GALAXY
UGC 02538	03 03 48.	+ 41 34	102	15.6	GALAXY SBa
MCG-02-08-047	03 03 48.	- 09 44 30.	84	14.	GALAXY
PHL 1466	03 03 48.	- 14 43		15.7	BLUE STELLAR OBJECT
PHL 1467	03 03 48.	- 20 40		16.6	BLUE STELLAR OBJECT
TON-S 0321	03 03 48.	- 24 57		15.1	BLUE STAR
LB 01652	03 03 48.	- 47 13		11.1	FAINT BLUE STAR
SHAP305-5223.0	03 03 49.	- 52 34 34.	18	17.0	GALAXY
SHAP305-5454.4	03 03 49.	- 55 05 58.	24	16.5	GALAXY
RNGC 1209	03 03 50.	- 15 48		12.5	GALAXY
SHAP305-5100.5	03 03 50.	- 51 12 04.	24	17.0	GALAXY
SHAP305-5105.6	03 03 50.	- 51 17 10.	18	17.25	GALAXY
SHAP305-5059.4	03 03 51.	- 51 10 58.	18	17.5	GALAXY
SHAP305-5354.4	03 03 51.	- 54 05 58.	18	17.0	GALAXY
UGC 02539	03 03 54.	+ 07 39	60	17.	GALAXY Sc
MCG+06-07-040	03 03 54.	+ 35 57	90	15.	GALAXY
ZWG 524.051	03 03 54.	+ 35 59		15.7	GALAXY
UGC 02540	03 03 54.	+ 35 59	114	15.7	GALAXY Sa-b
MCG+07-07-026	03 03 54.	+ 41 32 30.	24	15.5	GALAXY
PHL 8558	03 03 54.	- 15 27		17.8	BLUE STELLAR OBJECT
TON-S 0322	03 03 54.	- 27 38		13.3	BLUE STAR
SHAP305-5141.7	03 03 54.	- 51 53 16.	12	17.0	GALAXY
SHAP305-5236.1	03 03 54.	- 52 47 40.	18	17.5	GALAXY
SHAP305-5237.2	03 03 54.	- 52 48 46.	18	17.5	GALAXY
BAK 1.094	03 03 55.	- 28 07 27.	16	17.2	EXTRAGALACTIC NEBULA
SHAP305-5141.2	03 03 56.	- 51 52 46.	12	17.25	GALAXY
BAK 1.095	03 03 57.	- 27 34 33.	22	16.6	EXTRAGALACTIC NEBULA
SHAP305-4947.7	03 03 58.	- 49 59 16.	30	16.5	GALAXY
UGC 02541	03 04 00.	+ 34 10	72	17.	GALAXY
ZWG 554.023	03 04 00.	+ 47 20		15.7	GALAXY
PHL 1468	03 04 00.	- 03 23		18.0	BLUE STELLAR OBJECT
MCG-02-08-048	03 04 00.	- 10 01	24	18.	GALAXY
PHL 4333	03 04 00.	- 10 06		18.8	BLUE STELLAR OBJECT
SHAP305-5103.0	03 04 00.	- 51 14 34.	18	17.25	GALAXY
ARC 0413	03 04 01.	+ 02 04		17.5	RICH CLUSTER OF GALAXIES
SHAP305-4946.0	03 04 02.	- 49 57 34.	30	16.5	GALAXY
MCG-02-08-049	03 04 03.	- 09 55	60	15.	GALAXY
IC 1880	03 04 03.	- 09 55 13.			NONSTELLAR OBJECT
SHAP305-5119.1	03 04 04.	- 51 30 39.	18	17.5	GALAXY
SHAP305-5308.4	03 04 04.	- 53 19 58.	18	17.5	GALAXY
BAK 1.097	03 04 05.	- 28 26 08.	19	17.2	EXTRAGALACTIC NEBULA
SHAP305-5121.1	03 04 05.	- 51 32 39.	24	17.5	GALAXY
ZC 0304.1+2609	03 04 06.	+ 26 09	940		CLUSTER OF GALAXIES
UGC 02542	03 04 06.	+ 70 22	96	17.	GALAXY Sb
PHL 8559	03 04 06.	- 06 51		18.2	BLUE STELLAR OBJECT
TON-S 0323	03 04 06.	- 28 02		16.0	BLUE STAR
SHAP305-5047.4	03 04 06.	- 50 58 57.	24	17.0	GALAXY
SHAP305-5342.4	03 04 06.	- 53 53 57.	12	17.0	GALAXY
SHAP305-5336.5	03 04 07.	- 53 48 03.	12	17.25	GALAXY
SHAP305-5354.9	03 04 07.	- 54 06 27.	12	16.5	GALAXY
SHB 069	03 04 07.6	- 04 01 30.			QUASI-STELLAR OBJECT
BAK 1.096	03 04 08.	- 31 22 44.	23	16.7	EXTRAGALACTIC NEBULA
SHAP305-5111.8	03 04 09.	- 51 23 21.	12	17.25	GALAXY
SHAP305-5133.4	03 04 09.	- 51 44 57.	12	17.0	GALAXY
SHAP305-5335.5	03 04 09.	- 53 47 03.	24	16.0	GALAXY
BAK 1.098	03 04 10.	- 30 21 50.	13	16.9	EXTRAGALACTIC NEBULA
SHAP305-5427.2	03 04 10.	- 54 38 45.	24	17.5	GALAXY
IC 0288	03 04 11.	+ 42 10 27.			OPEN CLUSTER
ZC 0304.2+0142	03 04 12.	+ 01 42	3360		CLUSTER OF GALAXIES
ZWG 524.052	03 04 12.	+ 37 39		14.9	GALAXY
UGC 02543	03 04 12.	+ 37 39	126	14.9	GALAXY
ZWG 540.043	03 04 12.	+ 42 12		15.0	GALAXY
UGC 02544	03 04 12.	+ 42 12	78	15.0	GALAXY S
SHAP305-5255.8	03 04 12.	- 53 07 21.	24	17.75	GALAXY
SHAP305-5332.8	03 04 13.	- 53 44 21.	12	16.5	GALAXY
MCG+07-07-027	03 04 15.	+ 42 11	48	15.	GALAXY
MCG-03-08-074	03 04 15.	- 15 45 30.	48	14.5	GALAXY
RNGC 1217	03 04 15.	- 39 13			GALAXY
BAK 1.099	03 04 16.	- 27 54 02.	12	16.6	EXTRAGALACTIC NEBULA
BAK 1.100	03 04 17.	- 28 15 38.	26	16.9	EXTRAGALACTIC NEBULA
SHAP305-5219.4	03 04 17.	- 52 30 57.	12	17.25	GALAXY
ZWG 389.082	03 04 18.	+ 03 05		15.4	GALAXY
KARA.68 031	03 04 18.	+ 04 09	27		DWARF GALAXY
ZWG 524.053	03 04 18.	+ 39 05		15.6	GALAXY
UGC 02546	03 04 18.	+ 39 05	60	15.6	GALAXY SBb
ZWG 389.081	03 04 18.	- 00 59		13.5	GALAXY
UGC 02545	03 04 18.	- 00 59	174	13.5	GALAXY SB0/SBa
MCG+00-08-093	03 04 18.	- 00 59 30.	66	14.	GALAXY
BAK 1.103	03 04 18.	- 28 31 32.	14	17.2	EXTRAGALACTIC NEBULA
SHAP305-5109.3	03 04 18.	- 51 20 51.	18	16.5	GALAXY
SHAP305-5202.8	03 04 18.	- 52 14 21.	30	17.25	GALAXY
BAK 1.104	03 04 19.	- 27 55 02.	23	16.2	EXTRAGALACTIC NEBULA
BAK 1.102	03 04 19.	- 28 46 38.	20	16.5	EXTRAGALACTIC NEBULA
BAK 1.101	03 04 19.	- 28 48 44.	11	16.6	EXTRAGALACTIC NEBULA
RNGC 1211	03 04 20.	- 00 59		13.5	GALAXY
SHAP305-5329.1	03 04 20.	- 53 40 39.	24	18.0	GALAXY
MCG+06-07-041	03 04 21.	+ 37 37	102	15.5	GALAXY
SHAP305-4934.2	03 04 21.	- 49 45 44.	24	16.5	GALAXY
SHAP305-5038.4	03 04 21.	- 50 49 57.	24	17.0	GALAXY
SHAP305-5118.1	03 04 21.	- 51 29 39.	12	16.5	GALAXY
SHAP305-5318.9	03 04 21.	- 53 30 27.	12	16.75	GALAXY
SHAP305-5335.7	03 04 21.	- 53 47 15.	18	16.5	GALAXY
SHAP305-4958.7	03 04 23.	- 50 10 14.	36	17.0	GALAXY
KN 11.81	03 04 23.5	+ 03 04 44.			NEBULA
PHL 1469	03 04 24.	- 12 24		16.6	BLUE STELLAR OBJECT
PHL 4335	03 04 24.	- 13 14		18.4	BLUE STELLAR OBJECT
PHL 4334	03 04 24.	- 17 34		17.2	BLUE STELLAR OBJECT
PHL 1470	03 04 24.	- 19 13		17.9	BLUE STELLAR OBJECT
SHAP305-5042.4	03 04 24.	- 50 53 56.	24	18.0	GALAXY

OBJECT NAME	RIGHT ASCEN.	DECLINATION	DIAM.	MAGN.	TYPE OF OBJECT
BAK 1.105	03 04 25.	- 30 03 43.	13	17.2	EXTRAGALACTIC NEBULA
ARC 0415	03 04 26.	- 12 15		16.3	RICH CLUSTER OF GALAXIES
SHAP305-5318.0	03 04 26.	- 53 29 32.	18	17.5	GALAXY
SHAP306-5126.6	03 04 27.	- 51 38 08.	18	16.5	GALAXY
SHAP305-5206.8	03 04 27.	- 52 18 20.	12	16.5	GALAXY
BAK 1.106	03 04 28.	- 27 44 55.	13	16.4	EXTRAGALACTIC NEBULA
SHAP306-5021.5	03 04 28.	- 50 33 02.	24	17.5	GALAXY
RNGC 1214	03 04 29.	- 09 44		14.0	GALAXY
ZWG 524.054	03 04 30.	+ 38 10		14.9	GALAXY
MCG-02-08-051	03 04 30.	- 09 44	72	14.5	GALAXY
MCG-02-08-050	03 04 30.	- 09 50	36	15.	GALAXY
MCG-02-08-052	03 04 30.	- 12 11	48	15.	GALAXY
LB 03298	03 04 30.	- 65 59		12.8	FAINT BLUE STAR
HOLM 066A	03 04 31.	- 09 43	36	13.7	PART OF MULTIPLE GALAXY
SHAP306-5124.2	03 04 31.	- 51 35 44.	18	17.25	GALAXY
SHAP306-5315.6	03 04 31.	- 53 27 08.	18	17.5	GALAXY
MCG+06-07-042	03 04 33.	+ 39 04	24	15.5	GALAXY
BAK 1.107	03 04 33.	- 30 02 43.	18	16.0	EXTRAGALACTIC NEBULA
SHAP306-5320.2	03 04 33.	- 53 31 44.	18	17.0	GALAXY
SHAP306-5236.0	03 04 34.	- 52 47 32.	24	16.75	GALAXY
SHAP306-5219.2	03 04 35.	- 52 30 44.	12	17.75	GALAXY
SHAP306-5408.4	03 04 35.	- 54 19 56.	24	16.75	GALAXY
PHL 8560	03 04 36.	- 03 48		17.0	BLUE STELLAR OBJECT
MCG-02-08-053	03 04 36.	- 14 13 30.	60	16.	GALAXY
MCG-04-08-024	03 04 36.	- 25 54	24	13.	GALAXY
BAK 1.108	03 04 36.	- 27 45 43.	8	17.2	EXTRAGALACTIC NEBULA
SHAP306-5141.4	03 04 36.	- 51 52 56.	18	16.75	GALAXY
RNGC 1210	03 04 37.	- 25 54		13.0	GALAXY
SHAP306-5030.8	03 04 37.	- 50 42 20.	18	17.5	GALAXY
SHAP306-5039.0	03 04 37.	- 50 50 32.	24	17.0	GALAXY
HN 0168	03 04 38.	- 33 02			NEBULA
IC 1885	03 04 38.	- 33 02			NONSTELLAR OBJECT
RNGC 1215	03 04 41.	- 09 47		14.0	GALAXY
SHAP306-5010.6	03 04 41.	- 50 22 07.	54	16.75	GALAXY
SZW 321	03 04 42.	+ 29 25			COMPACT GALAXY
PHL 8561	03 04 42.	- 04 37		18.8	BLUE STELLAR OBJECT
PHL 8562	03 04 42.	- 11 26		18.7	BLUE STELLAR OBJECT
SHAP306-5400.4	03 04 42.	- 54 11 56.	24	17.75	GALAXY
KN 11.82	03 04 42.2	+ 04 16 37.			NEBULA
SHAP306-5445.8	03 04 43.	- 54 57 20.	24	16.25	GALAXY
SHAP306-5043.6	03 04 44.	- 50 55 07.	18	18.0	GALAXY
HOLM 066B	03 04 45.	- 09 46	60	14.0	PART OF MULTIPLE GALAXY
MCG-02-08-054	03 04 45.	- 09 47	24	17.	GALAXY
MCG-02-08-055	03 04 45.	- 09 47 30.	72	14.5	GALAXY
SHAP306-5251.6	03 04 45.	- 53 03 07.	18	16.75	GALAXY
SHAP306-5446.5	03 04 45.	- 54 58 02.	24	16.75	GALAXY
KN 11.83	03 04 45.3	+ 04 13 48.			NEBULA
SHAP306-5158.2	03 04 46.	- 52 09 43.	24	17.5	GALAXY
SHAP306-5437.7	03 04 47.	- 54 49 13.	18	17.25	GALAXY
ZWG 389.083	03 04 48.	+ 02 20		15.4	GALAXY
UGC 02547	03 04 48.	+ 02 20	60	15.4	GALAXY SBa-b
ZC 0304.8+0250	03 04 48.	+ 02 50	1280		CLUSTER OF GALAXIES
SHAP306-5321.2	03 04 48.	- 53 32 43.	12	16.5	GALAXY
SHAP306-5034.0	03 04 49.	- 50 45 31.	24	17.0	GALAXY
BAK 1.109	03 04 51.	- 30 51 12.	24	16.6	EXTRAGALACTIC NEBULA
KN 11.84	03 04 52.4	+ 02 20 15.			NEBULA
RNGC 1216	03 04 53.	- 09 48		15.0	GALAXY
SHAP306-5224.4	03 04 53.	- 52 35 55.	24	16.75	GALAXY
MCG+00-08-094	03 04 54.	+ 02 20	60	16.	GALAXY
MCG+00-08-095	03 04 54.	- 01 11	18	14.	GALAXY
MCG-02-08-056	03 04 54.	- 09 48 30.	36	15.	GALAXY
ARC 0416	03 04 54.	- 16 56		17.7	RICH CLUSTER OF GALAXIES
SHAP306-5229.0	03 04 54.	- 52 40 31.	12	16.75	GALAXY
SHAP306-4934.4	03 04 55.	- 49 45 55.	36	16.5	GALAXY
SHAP306-5425.6	03 04 56.	- 54 37 07.	18	17.0	GALAXY
SHAP306-5114.8	03 04 57.	- 51 26 19.	18	15.5	GALAXY
SHAP306-5405.5	03 04 57.	- 54 17 01.	12	17.0	GALAXY
SHAP306-5139.2	03 04 58.	- 51 50 43.	12	17.5	GALAXY
BAK 1.110	03 04 59.	- 30 52 36.	13	16.8	EXTRAGALACTIC NEBULA
SHAP306-5214.2	03 04 59.	- 52 25 43.	30	16.75	GALAXY
MCG+05-08-005	03 05 00.	+ 30 16	30	16.	GALAXY
ZWG 524.055	03 05 00.	+ 38 11		13.7	GALAXY
RNGC 1207	03 05 00.	+ 38 11		13.5	GALAXY
UGC 02548	03 05 00.	+ 38 11	168	13.7	GALAXY Sb
KARA.72 087B	03 05 00.	+ 38 11	36		PART OF DOUBLE GALAXY
KARA.72 087A	03 05 00.	+ 38 11	30	13.7	PART OF DOUBLE GALAXY
ZWG 390.001	03 05 00.	- 01 09		14.7	GALAXY
ZWG 389.084	03 05 00.	- 01 09		14.7	GALAXY
MCG+00-09-001	03 05 00.	- 01 09	48	15.	GALAXY
PHL 4336	03 05 00.	- 15 32		18.5	BLUE STELLAR OBJECT
MCG-05-08-018	03 05 00.	- 31 38	156	14.	GALAXY
SHAP306-5141.2	03 05 00.	- 51 52 43.	48	16.5	GALAXY
SHAP306-5216.1	03 05 00.	- 52 27 37.	12	17.25	GALAXY
IC 0291	03 05 03.	- 12 47 07.			NONSTELLAR OBJECT
MCG-02-09-001	03 05 03.	- 12 48	48	14.5	GALAXY
SHAP306-4846.9	03 05 03.	- 48 58 24.	42	17.0	GALAXY
SHAP306-5159.4	03 05 03.	- 52 10 54.	18	17.0	GALAXY
SHAP306-4949.5	03 05 04.	- 50 01 00.	36	16.5	GALAXY
SHAP306-5046.4	03 05 04.	- 50 57 54.	18	17.7	GALAXY
SHAP306-5154.0	03 05 05.	- 52 05 30.	12	17.25	GALAXY
SHAP306-5258.9	03 05 05.	- 53 10 24.	24	16.25	GALAXY
UGC 02549	03 05 06.	+ 23 27	72	16.0	GALAXY Sb
MCG+04-08-009	03 05 06.	+ 23 27	42	16.	GALAXY
ZWG 524.056	03 05 06.	+ 36 15		15.1	GALAXY
UGC 02550	03 05 06.	+ 36 15	126	15.1	GALAXY Sc
FEIG 033	03 05 06.	- 10 57		13.5	FAINT BLUE STAR
PHL 8563	03 05 06.	- 11 56		12.5	BLUE STELLAR OBJECT
FEIG 032	03 05 06.	- 11 56		14.0	FAINT BLUE STAR
BAK 1.111	03 05 06.	- 31 27 29.	18	16.4	EXTRAGALACTIC NEBULA
SHAP306-5054.5	03 05 07.	- 51 06 00.	12	17.25	GALAXY
SHAP306-5205.2	03 05 07.	- 52 16 42.	12	17.0	GALAXY
MCG+03-09-001	03 05 09.	+ 17 36	30	15.5	GALAXY
MCG+06-07-043	03 05 09.	+ 38 08	96	14.	GALAXY
BAK 1.112	03 05 09.	- 31 35 29.	98	14.0	EXTRAGALACTIC NEBULA
ARC 0417	03 05 10.	- 14 46		17.8	RICH CLUSTER OF GALAXIES
ZWG 390.003	03 05 12.	+ 02 58		15.1	GALAXY
UGC 02551	03 05 12.	+ 02 58	60	15.1	GALAXY S
MCG+00-09-002	03 05 12.	+ 02 58	54	14.5	GALAXY
ZWG 524.057	03 05 12.	+ 34 11		15.7	GALAXY
MCG+06-07-044	03 05 12.	+ 36 14	84	15.	GALAXY
ZWG 390.002	03 05 12.	- 01 00		15.6	GALAXY
MCG+00-09-003	03 05 12.	- 01 01	18	16.	GALAXY
PHL 8564	03 05 12.	- 14 33		17.3	BLUE STELLAR OBJECT
SHAP306-5143.8	03 05 12.	- 51 55 18.	12	17.5	GALAXY
SHAP306-5425.8	03 05 12.	- 54 37 18.	24	17.5	GALAXY
KN 11.85	03 05 13."3	+ 02 57 19.			NEBULA
IC 1882	03 05 14.	+ 02 57 36.			NONSTELLAR OBJECT
SHAP306-5445.0	03 05 15.	- 54 56 30.	24	17.5	GALAXY

OBJECT NAME	RIGHT ASCEN.	DECLINATION	DIAM.	MAGN.	TYPE OF OBJECT
SHAP306-5403.5	03 05 16.	- 54 15 00.	18	17.0	GALAXY
ZWG 390.004	03 05 18.	- 00 58		15.2	GALAXY
PHL 8565	03 05 18.	- 03 54		18.5	BLUE STELLAR OBJECT
SHAP306-5117.8	03 05 19.	- 51 29 18.	18	17.5	GALAXY
SHAP306-5306.5	03 05 19.	- 53 18 00.	12	16.5	GALAXY
SHAP306-5348.2	03 05 19.	- 53 59 42.	24	17.5	GALAXY
SHAP306-5143.1	03 05 20.	- 51 54 35.	12	16.5	GALAXY
SHAP306-5439.6	03 05 21.	- 54 51 06.	24	16.75	GALAXY
RNGC 1212	03 05 22.	+ 40 39			GALAXY
SHAP306-5407.9	03 05 22.	- 54 19 24.	18	16.75	GALAXY
KN 11.86	03 05 22.3	+ 05 46 25.			NEBULA
SHAP306-5457.5	03 05 23.	- 55 09 00.	30	17.0	GALAXY
ZWG 416.001	03 05 24.	+ 05 46		15.4	GALAXY
UGC 02552	03 05 24.	+ 42 41	66	16.5	GALAXY S
PHL 1471	03 05 24.	- 03 56		18.0	BLUE STELLAR OBJECT
SHAP306-5430.5	03 05 24.	- 54 41 59.	24	17.75	GALAXY
SHAP306-5457.4	03 05 24.	- 55 08 53.	24	16.5	GALAXY
KN 11.87	03 05 25.4	+ 05 44 18.			NEBULA
SHAP306-5143.2	03 05 26.	- 51 54 41.	18	16.5	GALAXY
SHAP306-5409.9	03 05 27.	- 54 21 23.	18	17.0	GALAXY
SHAP306-5445.4	03 05 27.	- 54 56 53.	18	17.5	GALAXY
SHAP306-5342.0	03 05 28.	- 53 53 29.	18	17.5	GALAXY
SHAP307-4930.7	03 05 29.	- 49 42 11.	42	16.0	GALAXY
MCG+01-09-001	03 05 30.	- 04 35	15	15.5	GALAXY
SHAP306-5256.7	03 05 30.	- 53 08 11.	12	17.0	GALAXY
SHAP307-5205.6	03 05 31.	- 52 17 05.	30	17.25	GALAXY
IC 1886	03 05 32.	- 04 35			NONSTELLAR OBJECT
SHAP306-5424.4	03 05 32.	- 54 35 53.	18	16.5	GALAXY
SHAP306-5427.8	03 05 32.	- 54 39 17.	18	16.5	GALAXY
MCG+00-09-004	03 05 33.	- 01 26 30.	30	16.	GALAXY
SHAP306-5458.2	03 05 34.	- 55 09 41.	30	17.0	GALAXY
BAK 1.114	03 05 35.	- 28 11 58.	11	16.0	EXTRAGALACTIC NEBULA
SHAP307-5319.5	03 05 35.	- 53 30 59.	18	16.5	GALAXY
SHAP307-5450.6	03 05 35.	- 55 02 05.	24	17.5	GALAXY
ZWG 441.001	03 05 36.	+ 11 43		15.5	GALAXY
ZWG 464.001	03 05 36.	+ 20 35		15.1	GALAXY
ZWG 463.051	03 05 36.	+ 20 35		15.1	GALAXY
UGC 02553	03 05 36.	+ 20 35	78	15.1	GALAXY SBb
MCG+07-07-028	03 05 36.	+ 41 16	15	15.	GALAXY
ZWG 540.044	03 05 36.	+ 41 17		15.7	GALAXY
UGC 02554	03 05 36.	+ 41 17	96	15.7	GALAXY S0?
ZWG 540.045	03 05 36.	+ 41 37		15.7	GALAXY
MCG+00-09-005	03 05 36.	- 01 00	36	15.	GALAXY
SHAP307-4950.4	03 05 36.	- 50 01 53.	24	17.5	GALAXY
SHAP307-5258.2	03 05 36.	- 53 09 41.	12	17.0	GALAXY
BAK 1.116	03 05 37.	- 27 40 16.	17	17.0	EXTRAGALACTIC NFBULA
BAK 1.113	03 05 37.	- 30 40 04.	14	16.4	EXTRAGALACTIC NEBULA
BAK 1.115	03 05 38.	- 30 52 04.	17	17.0	EXTRAGALACTIC NEBULA
RNGC 1221	03 05 39.	- 04 27		15.0	GALAXY
BAK 1.117	03 05 39.	- 27 15 15.	16	15.7	EXTRAGALACTIC NFBULA
SHAP307-5409.9	03 05 40.	- 54 21 23.	18	16.75	GALAXY
BAK 1.118	03 05 41.	- 27 26 33.	20	16.0	EXTRAGALACTIC NEBULA
SHAP307-5440.0	03 05 41.	- 54 51 29.	24	17.25	GALAXY
MCG+04-08-010	03 05 42.	+ 22 55	36	15.	GALAXY
ZWG 554.024	03 05 42.	+ 47 09		15.7	GALAXY
ZWG 390.005	03 05 42.	- 00 59		15.6	GALAXY
MCG+01-09-002	03 05 42.	- 04 27	48	15.	GALAXY
TON-S 0324	03 05 42.	- 27 14		14.6	BLUE STAR
MCG-05-08-019	03 05 42.	- 27 15	30	15.5	GALAXY
SHAP307-5059.0	03 05 43.	- 51 10 28.	18	17.25	GALAXY
MCG+03-09-002	03 05 45.	+ 20 37	48	15.	GALAXY
BIGO 482	03 05 45.	- 04 23			NEBULA
RNGC 1218	03 05 47.	+ 03 55		14.0	GALAXY
SHAP307-5100.4	03 05 47.	- 51 11 52.	18	17.0	GALAXY
MCG+00-09-006	03 05 48.	+ 01 55	48	13.	GALAXY
ZWG 416.002	03 05 48.	+ 03 55		14.0	GALAXY
UGC 02555	03 05 48.	+ 03 55	84	14.0	GALAXY S0-a
5ZW 322	03 05 48.	+ 34 21			COMPACT GALAXY
MCG+01-09-003	03 05 48.	- 04 20	18	15.	GALAXY
PHL 8567	03 05 48.	- 11 45		17.0	BLUE STELLAR OBJECT
PHL 8566	03 05 48.	- 13 40		15.9	BLUE STELLAR OBJECT
PHL 4337	03 05 48.	- 16 27		18.5	BLUE STELLAR OBJECT
VV 337B	03 05 48.	- 23 12	90	15.	INTERACTING GALAXY
LB 01653	03 05 48.	- 51 59		12.8	FAINT BLUE STAR
SHAP307-5130.0	03 05 49.	- 51 41 28.	18	17.0	GALAXY
KN 11.88	03 05 49.2	+ 03 55 14.			NEBULA
SHAP307-5313.2	03 05 50.	- 53 24 40.	18	17.5	GALAXY
RNGC 1225	03 05 51.	- 04 20		15.0	GALAXY
VV 337A	03 05 51.	- 23 10	72	14.	INTERACTING GALAXY
SHAP307-5440.2	03 05 51.	- 54 51 40.	36	17.0	GALAXY
SHAP307-5329.8	03 05 52.	- 53 41 16.	18	17.75	GALAXY
RNGC 1244	03 05 52.	- 66 58			UNVERIFIED SOUTHPN OBJECT
KN 11.89	03 05 52.8	+ 01 55 06.			NEBULA
ARC 0418	03 05 53.	- 13 56		17.8	RICH CLUSTER OF GALAXIFS
SHAP307-5316.7	03 05 53.	- 53 28 10.	18	17.0	GALAXY
SHAP307-5332.0	03 05 53.	- 53 43 28.	12	17.0	GALAXY
MCG+00-09-007	03 05 54.	+ 00 12	48	15.	GALAXY
ZWG 390.006	03 05 54.	+ 01 57		13.5	GALAXY
RNGC 1219	03 05 54.	+ 01 57		13.5	GALAXY
UGC 02556	03 05 54.	+ 01 57	78	13.5	GALAXY Sc
MCG+01-09-001	03 05 54.	+ 03 52	48	14.	GALAXY
SHAP307-5311.4	03 05 55.	- 53 22 52.	18	16.5	GALAXY
SHAP307-5057.2	03 05 56.	- 51 08 40.	18	17.25	GALAXY
SHAP307-5331.2	03 05 56.	- 53 42 40.	18	17.0	GALAXY
SHAP307-5026.0	03 05 57.	- 50 37 27.	24	17.5	GALAXY
SER 028.01	03 06	- 66 59	100	15.	LOOSE GROUP OF GALAXIES
ZWG 390.007	03 06 00.	+ 00 13		15.6	GALAXY
ZWG 524.058	03 06 00.	+ 38 27		15.7	GALAXY
RNGC 1213	03 06 00.	+ 38 27		15.5	GALAXY
UGC 02557	03 06 00.	+ 38 27	180	15.7	GALAXY
UGC 02558	03 06 00.	+ 39 30	66	18.	GALAXY DWARF
MCG+07-07-029	03 06 00.	+ 42 46	36	15.	GALAXY
ZWG 540.046	03 06 00.	+ 42 47		14.5	GALAXY
UGC 02559	03 06 00.	+ 42 47	132	14.5	GALAXY S0
URA 22	03 06 00.	+ 64 05	330		STELLAR RING
MCG-02-09-002	03 06 00.	- 14 07	54	14.5	GALAXY
RNGC 1228	03 06 00.	- 23 07		14.0	GALAXY
RNGC 1229	03 06 00.	- 23 09		14.0	GALAXY
MCG-04-08-025	03 06 00.	- 23 09	84	14.	GALAXY
TON-S 0325	03 06 00.	- 30 49		14.7	BLUE STAR
SHAP307-5204.3	03 06 02.	- 52 15 45.	18	17.0	GALAXY
MCG-04-08-026	03 06 03.	- 23 07 30.	72	14.	GALAXY
SHAP307-5149.0	03 06 03.	- 52 00 27.	12	16.0	GALAXY
SHAP307-5352.0	03 06 03.	- 54 03 27.	12	16.5	GALAXY
KARA.73 06	03 06 04.	- 34 13	67		DWARF GALAXY
IC 1881	03 06 05.	+ 38 29			NONSTELLAR OBJECT
BAK 1.119	03 06 05.	- 31 25 38.	26	16.4	EXTRAGALACTIC NFBULA
SHAP307-5423.5	03 06 05.	- 54 34 57.	18	16.75	GALAXY

OBJECT NAME	RIGHT ASCEN.	DECLINATION	DIAM.	MAGN.	TYPE OF OBJECT
SHAP307-5428.0	03 06 05.	- 54 39 27.	24	17.0	GALAXY
PHL 4338	03 06 06.	- 03 51		18.5	BLUE STELLAR OBJECT
RNGC 1230	03 06 06.	- 23 12		15.0	GALAXY
MCG-04-08-027	03 06 06.	- 23 12	18	15.	GALAXY
LB 01654	03 06 06.	- 53 36		13.2	FAINT BLUE STAR
SHAP307-5440.3	03 06 06.	- 54 51 45.	30	17.25	GALAXY
SHAP307-5426.8	03 06 07.	- 54 38 15.	30	15.5	GALAXY
SHAP307-5440.7	03 06 07.	- 54 52 09.	18	17.0	GALAXY
SHAP307-4941.4	03 06 08.	- 49 52 51.	24	17.5	GALAXY
SHAP307-5054.8	03 06 08.	- 51 06 15.	18	16.75	GALAXY
SHAP307-5405.6	03 06 08.	- 54 17 03.	12	17.5	GALAXY
MCG-04-08-028	03 06 09.	- 24 51 30.	36	15.	GALAXY
BAK 1.120	03 06 09.	- 31 35 26.	17	17.0	EXTRAGALACTIC NEBULA
SHAP307-5205.0	03 06 09.	- 52 16 27.	18	17.5	GALAXY
SHAP307-5434.4	03 06 09.	- 54 45 51.	18	16.75	GALAXY
BAK 1.122	03 06 10.	- 30 40 50.	17	17.0	EXTRAGALACTIC NEBULA
BAK 1.123	03 06 10.	- 30 41 44.	14	16.7	EXTRAGALACTIC NEBULA
BAK 1.121	03 06 10.	- 31 07 26.	17	16.3	EXTRAGALACTIC NEBULA
SHAP307-5107.0	03 06 11.	- 51 18 27.	18	16.75	GALAXY
ZWG 441.002	03 06 12.	+ 11 43		15.7	GALAXY
MCG+06-07-045	03 06 12.	+ 38 27	54	16.	GALAXY
MCG+00-09-008	03 06 12.	- 00 56	42	15.	GALAXY
PHL 1472	03 06 12.	- 03 08		16.6	BLUE STELLAR OBJECT
PHL 8568	03 06 12.	- 09 34		17.3	BLUE STELLAR OBJECT
MCG-04-08-029	03 06 12.	- 23 14	24	16.	GALAXY
SHAP307-5325.8	03 06 12.	- 53 37 15.	78	16.5	GALAXY
SHAP307-5422.6	03 06 12.	- 54 34 03.	42	16.25	GALAXY
IC 1892	03 06 14.	- 23 14 44.			NONSTELLAR OBJECT
BAK 1.124	03 06 14.	- 31 25 50.	31	16.8	EXTRAGALACTIC NEBULA
VV 260B	03 06 15.	- 23 14 30.	36	15.	INTERACTING GALAXY
VV 260A	03 06 15.	- 23 14 30.	36	15.	INTERACTING GALAXY
VV 260	03 06 15.	- 23 14 30.	102		INTERACTING GALAXY
MCG-04-08-030	03 06 15.	- 23 15	120	13.	GALAXY
SHAP307-5104.7	03 06 15.	- 51 16 09.	24	16.5	GALAXY
SHAP307-5107.8	03 06 15.	- 51 19 15.	18	16.5	GALAXY
SHAP307-5407.8	03 06 15.	- 54 19 15.	18	17.0	GALAXY
SHAP307-5429.4	03 06 15.	- 54 40 51.	18	17.0	GALAXY
BAK 1.125	03 06 16.	- 28 59 20.	26	16.6	EXTRAGALACTIC NEBULA
SHAP307-5403.0	03 06 16.	- 54 14 27.	36	16.5	GALAXY
PK138+02.1	03 06 16.4	+ 61 07 39.	42	12.3	PLANETARY NEBULA
IC 0289	03 06 16.4	+ 61 08 39.	18	12.3	PLANETARY NEBULA
ARC 0419	03 06 17.	- 23 51		15.7	RICH CLUSTER OF GALAXIES
RLWT 067	03 06 18.	+ 61 03		16.	FAINT VERY BLUE STAR
ZWG 390.008	03 06 18.	- 00 55		15.5	GALAXY
MCG+01-09-004	03 06 18.	- 04 18	36	15.	GALAXY
MCG-03-09-001	03 06 18.	- 17 54	36	15.5	GALAXY
ARP 332	03 06 19.	- 23 15			PECULIAR GALAXY
BAK 1.127	03 06 19.	- 23 48 13.	26	15.8	EXTRAGALACTIC NEBULA
BAK 1.126	03 06 20.	- 31 53 13.	20	16.8	EXTRAGALACTIC NEBULA
RNGC 1223	03 06 21.	- 04 18		15.0	GALAXY
SHAP307-5317.5	03 06 21.	- 53 28 56.	12	16.25	GALAXY
SHAP307-5411.0	03 06 22.	- 54 22 26.	18	16.75	GALAXY
SHAP307-5219.4	03 06 23.	- 52 30 50.	24	16.5	GALAXY
UGC 02560	03 06 24.	+ 40 43	66	16.0	GALAXY S0-a
ZWG 540.047	03 06 24.	+ 40 48		15.5	GALAXY
UGC 02561	03 06 24.	+ 40 48	72	15.5	GALAXY Sb
MCG+00-09-009	03 06 24.	- 01 11	48	15.	GALAXY
PHL 1474	03 06 24.	- 03 08		18.0	BLUE STELLAR OBJECT
MRK 603	03 06 24.	- 03 09	8	14.5	GALAXY WITH UV CONTINUUM
MCG+01-09-005	03 06 24.	- 03 09	48	14.	GALAXY
PHL 1473	03 06 24.	- 07 11		18.0	BLUE STELLAR OBJECT
MCG+01-09-006	03 06 24.	- 07 14	66	15.	GALAXY
PHL 4339	03 06 24.	- 09 15		18.6	BLUE STELLAR OBJECT
IC 1883	03 06 25.	+ 40 42 08.			NONSTELLAR OBJECT
SHAP307-5259.1	03 06 25.	- 53 10 32.	24	16.5	GALAXY
SHAP307-5421.9	03 06 25.	- 54 33 20.	18	17.5	GALAXY
IC 0290	03 06 26.	+ 40 47 38.			NONSTELLAR OBJECT
RNGC 1222	03 06 26.	- 03 09		14.0	GALAXY
SHAP307-5332.6	03 06 26.	- 53 44 02.	18	17.5	GALAXY
IC 1884	03 06 27.	+ 40 47 08.			NONSTELLAR OBJECT
BAK 1.128	03 06 28.	- 31 55 01.	25	16.2	EXTRAGALACTIC NEBULA
BAK 1.129	03 06 29.	- 29 44 01.	34	16.8	EXTRAGALACTIC NEBULA
SHAP308-5102.0	03 06 29.	- 51 13 26.	12	17.5	GALAXY
RNGC 1246	03 06 29.	- 67 09			UNVERIFIED SOUTHERN OBJECT
ZWG 540.048	03 06 30.	+ 41 04		15.7	GALAXY
MCG+00-09-010	03 06 30.	- 01 07	12	16.	GALAXY
MCG+01-09-007	03 06 30.	- 05 19	60	15.5	GALAXY
SHAP307-5332.3	03 06 30.	- 53 43 44.	18	17.5	GALAXY
BAK 1.130	03 06 31.	- 28 14 55.	19	16.5	EXTRAGALACTIC NEBULA
SHAP307-5333.8	03 06 31.	- 53 45 14.	12	17.5	GALAXY
SHAP307-5413.8	03 06 31.	- 54 25 14.	42	15.25	GALAXY
BAK 1.134	03 06 32.	- 24 12 01.	14	16.1	EXTRAGALACTIC NFBULA
SHAP308-5310.2	03 06 32.	- 53 21 38.	18	17.0	GALAXY
HN 0169	03 06 32.	- 54 25			NEBULA
IC 1896	03 06 32.	- 54 25			NONSTELLAR OBJECT
SHAP308-5318.1	03 06 33.	- 53 29 32.	12	17.0	GALAXY
SHAP308-5400.3	03 06 34.	- 54 11 44.	18	16.5	GALAXY
SHAP308-5413.6	03 06 34.	- 54 25 02.	18	16.5	GALAXY
ZWG 464.002	03 06 36.	+ 15 55		15.7	GALAXY
MCG+03-09-003	03 06 36.	+ 15 56	12	16.	GALAXY
5ZW 323	03 06 36.	+ 41 22			COMPACT GALAXY
MRSL 138+04/1	03 06 36.	+ 62 38	360		HII REGION
PHL 4340	03 06 36.	- 14 54		17.8	BLUE STELLAR OBJECT
BAK 1.133	03 06 36.	- 28 51 55.	12	16.6	EXTRAGALACTIC NEBULA
BAK 1.131	03 06 36.	- 29 34 07.	14	17.0	EXTRAGALACTIC NEBULA
SHAP308-5434.4	03 06 36.	- 54 45 50.	54	16.5	GALAXY
BAK 1.132	03 06 37.	- 30 21 13.	14	16.4	EXTRAGALACTIC NEBULA
KLEE 05	03 06 38.	- 36 55	1080	16.	CLUSTER OF 35 GALAXIFS
SHAP308-5401.2	03 06 38.	- 54 12 38.	12	17.5	GALAXY
SHAP308-5231.8	03 06 39.	- 52 43 13.	18	17.0	GALAXY
SHAP308-5108.4	03 06 40.	- 51 19 49.	18	16.5	GALAXY
SHAP308-5407.8	03 06 40.	- 54 19 13.	12	17.25	GALAXY
SHAP308-5440.2	03 06 41.	- 54 51 37.	24	17.5	GALAXY
UGC 02562	03 06 42.	+ 41 55	72	16.0	GALAXY S0
MCG+01-09-009	03 06 42.	- 04 07	48	16.	GALAXY
MCG+01-09-008	03 06 42.	- 04 07	36	16.	GALAXY
MCG-02-09-003	03 06 42.	- 10 28	78	14.	GALAXY
SHAP308-5316.0	03 06 42.	- 53 27 25.	18	16.0	GALAXY
BAK 1.138	03 06 46.	- 23 25 48.	24	16.4	EXTRAGALACTIC NEBULA
BAK 1.136	03 06 46.	- 28 07 00.	17	16.2	EXTRAGALACTIC NFBULA
BAK 1.135	03 06 47.	- 31 06 48.	17	16.6	EXTRAGALACTIC NEBULA
SHAP308-5410.8	03 06 47.	- 54 22 13.	18	17.0	GALAXY
ZWG 464.003	03 06 48.	+ 18 19		15.5	GALAXY
UGC 02563	03 06 48.	+ 18 19	96	15.5	GALAXY Sc
ZWG 390.009	03 06 48.	- 01 05		15.7	GALAXY
PHL 4341	03 06 48.	- 13 07		18.5	BLUE STELLAR OBJECT
SHAP308-5425.8	03 06 48.	- 54 37 13.	36	17.0	GALAXY
BAK 1.137	03 06 49.	- 30 46 54.	16	16.8	EXTRAGALACTIC NFBULA

OBJECT NAME	RIGHT ASCEN.	DECLINATION	DIAM.	MAGN.	TYPE OF OBJECT
MCG-02-09-004	03 06 51.	- 10 59 30.	48	15.	GALAXY
MCG+07-07-030	03 06 54.	+ 40 34	48	14.	GALAXY
IC 1887	03 06 54.	+ 40 34 05.			NONSTELLAR OBJECT
UGC 02564	03 06 54.	+ 41 59	60	16.0	GALAXY S0
SHAP308-4957.4	03 06 54.	- 50 08 48.	24	17.75	GALAXY
SHAP308-5408.7	03 06 54.	- 54 20 07.	12	17.0	GALAXY
SHAP308-5449.8	03 06 54.	- 55 01 13.	24	17.0	GALAXY
SHAP308-5334.6	03 06 55.	- 53 46 01.	18	16.5	GALAXY
SHAP308-5442.4	03 06 55.	- 54 53 49.	24	17.25	GALAXY
ARC 0420	03 06 57.	- 11 43		16.8	RICH CLUSTER OF GALAXIES
BAK 1.140	03 06 59.	- 26 08 29.	20	16.2	EXTRAGALACTIC NEBULA
BAK 1.139	03 06 59.	- 27 19 59.	6	16.3	EXTRAGALACTIC NEBULA
KN 11.90	03 06 59.9	+ 02 06 03.			NEBULA
LBN 0674	03 07	+ 62 35	480		BRIGHT NEBULA
ZWG 390.010	03 07 00.	+ 02 07		15.6	GALAXY
UGC 02565	03 07 00.	+ 31 44	90	16.5	GALAXY
MCG+05-08-006	03 07 00.	+ 31 44	66	16.	GALAXY
UGC 02566	03 07 00.	+ 33 40	66	17.	GALAXY
IC 0292	03 07 00.	+ 40 34 23.			NONSTELLAR OBJECT
ZWG 540.049	03 07 00.	+ 40 35		14.3	GALAXY
CR 01	03 07 00.	+ 40 35		14.3	GALAXY IN PERSEUS CLUSTER
UGC 02567	03 07 00.	+ 40 35	78	14.3	GALAXY S-IRR
ZWG 540.050	03 07 00.	+ 42 02		15.5	GALAXY
UGC 02568	03 07 00.	+ 42 02	90	15.5	GALAXY S0
PHL 8569	03 07 00.	- 03 35		16.5	BLUE STELLAR OBJECT
MCG+01-09-010	03 07 00.	- 05 07	66	15.	GALAXY
MCG-04-08-031	03 07 00.	- 25 43	72	15.	GALAXY
SHAP308-5133.1	03 07 00.	- 51 44 30.	18	16.5	GALAXY
SHAP308-5248.8	03 07 01.	- 53 00 12.	12	17.25	GALAXY
SHAP308-5317.4	03 07 01.	- 53 28 48.	18	16.7	GALAXY
SHAP308-5427.4	03 07 01.	- 54 38 48.	24	17.5	GALAXY
SHAP308-4944.0	03 07 02.	- 49 55 24.	36	17.0	GALAXY
SHAP308-4945.0	03 07 02.	- 49 56 24.	30	16.5	GALAXY
SHAP308-5201.3	03 07 02.	- 52 12 42.	30	16.5	GALAXY
MCG+00-09-011	03 07 03.	- 01 28 30.	36	16.	GALAXY
BAK 1.141	03 07 03.	- 24 00 05.	22	16.0	EXTRAGALACTIC NEBULA
SHAP308-5031.7	03 07 03.	- 50 43 06.	48	17.0	GALAXY
SHAP308-5300.0	03 07 03.	- 53 11 24.	24	17.25	GALAXY
SHAP308-5425.0	03 07 03.	- 54 36 24.	18	16.75	GALAXY
ZWG 540.051	03 07 06.	+ 42 14		15.7	GALAXY
LB 00476	03 07 06.	- 01 00 48.		16.3	FAINT BLUE STAR
ZWG 390.011	03 07 06.	- 02 12		15.5	GALAXY
SHAP308-5102.6	03 07 06.	- 51 14 00.	24	17.0	GALAXY
BAK 1.142	03 07 07.	- 23 37 35.	23	16.1	EXTRAGALACTIC NEBULA
IC 1890	03 07 09.	+ 19 00 26.			NONSTELLAR OBJECT
SHAP308-5153.1	03 07 10.	- 52 04 30.	18	15.75	GALAXY
RNGC 1234	03 07 11.	- 08 03		15.0	GALAXY
SHAP308-5245.6	03 07 11.	- 52 57 00.	18	16.5	GALAXY
ZWG 464.004	03 07 12.	+ 15 50		15.6	GALAXY
ZWG 464.005	03 07 12.	+ 19 00		15.6	GALAXY
MCG+03-09-004	03 07 12.	+ 19 02	12	15.	GALAXY
MCG+01-09-011	03 07 12.	- 08 03	60	15.	GALAXY
MCG-02-09-005	03 07 12.	- 10 00 30.	30	15.5	GALAXY
PHL 8570	03 07 12.	- 10 26		18.5	BLUE STELLAR OBJECT
PHL 4342	03 07 12.	- 12 20		18.4	BLUE STELLAR OBJECT
MCG-03-09-002	03 07 12.	- 17 37	78	16.	GALAXY
SHAP308-4952.4	03 07 12.	- 50 03 47.	24	17.5	GALAXY
SHAP308-5405.0	03 07 12.	- 54 16 24.	18	17.5	GALAXY
SHAP308-5328.9	03 07 13.	- 53 40 18.	18	16.5	GALAXY
SHAP308-5424.7	03 07 13.	- 54 36 06.	18	17.5	GALAXY
RNGC 1231	03 07 14.	- 15 47			NON-EXISTENT OBJECT
MCG+00-09-012	03 07 15.	+ 00 02	24	15.5	GALAXY
SHAP308-5019.1	03 07 15.	- 50 30 29.	60	17.25	GALAXY
SHAP308-5403.2	03 07 16.	- 54 14 35.	24	16.7	GALAXY
BAK 1.143	03 07 17.	- 29 14 10.	30	16.6	EXTRAGALACTIC NEBULA
UGC 02569	03 07 18.	+ 41 20	66	16.5	GALAXY S0?
SHAP308-5303.1	03 07 18.	- 53 14 29.	12	17.5	GALAXY
SHAP308-5426.0	03 07 18.	- 54 37 23.	18	18.0	GALAXY
LB 03299	03 07 18.	- 58 29		14.4	FAINT BLUE STAR
IC 1895	03 07 19.	- 25 31 02.			NONSTELLAR OBJECT
SHAP308-5321.0	03 07 19.	- 53 32 23.	18	16.75	GALAXY
SHAP308-5426.9	03 07 19.	- 54 38 17.	12	17.0	GALAXY
SHAP308-5413.6	03 07 20.	- 54 24 59.	18	17.25	GALAXY
IC 1891	03 07 21.	+ 19 25 07.			NONSTELLAR OBJECT
SHAP308-5156.2	03 07 21.	- 52 07 35.	42	16.5	GALAXY
BAK 1.144	03 07 23.	- 31 19 58.	23	16.0	EXTRAGALACTIC NEBULA
SHAP308-5427.8	03 07 23.	- 54 39 11.	18	17.0	GALAXY
MCG+04-08-011	03 07 24.	+ 25 28	72	15.	GALAXY
MCG+06-07-046	03 07 24.	+ 34 52	42	15.5	GALAXY
ZWG 524.059	03 07 24.	+ 34 55		15.4	GALAXY
PHL 8571	03 07 24.	- 13 22		18.9	BLUE STELLAR OBJECT
MCG-03-09-003	03 07 24.	- 16 03	60	15.5	GALAXY
SHAP308-5315.2	03 07 24.	- 53 26 35.	12	17.0	GALAXY
SHAP308-5328.6	03 07 24.	- 53 39 59.	12	16.75	GALAXY
BAK 1.145	03 07 25.	- 28 55 10.	19	17.0	EXTRAGALACTIC NEBULA
IC 1893	03 07 26.	+ 19 25 54.			NONSTELLAR OBJECT
SHAP309-4950.8	03 07 26.	- 50 02 11.	24	17.0	GALAXY
SHAP309-5029.2	03 07 26.	- 50 40 35.	18	17.25	GALAXY
SHAP308-5138.4	03 07 26.	- 51 49 47.	18	16.5	GALAXY
SHAP309-4936.6	03 07 27.	- 49 47 59.	36	16.0	GALAXY
SHAP308-5147.8	03 07 27.	- 51 59 11.	18	17.5	GALAXY
RNGC 1232	03 07 28.	- 20 46		10.5	GALAXY
BAK 1.147	03 07 28.	- 23 37 10.	20	15.9	EXTRAGALACTIC NEBULA
ZWG 416.003	03 07 30.	+ 04 40		15.5	GALAXY
UGC 02570	03 07 30.	+ 20 13	96	16.5	GALAXY S
UGC 02571	03 07 30.	+ 22 42	66	16.0	GALAXY Sb
UGC 02572	03 07 30.	+ 44 00	66	16.5	GALAXY SB:b-c
VB 186	03 07 30.	+ 61 09	215		STELLAR RING
MCG-02-09-006	03 07 30.	- 10 14 30.	36	14.5	GALAXY
MCG-04-08-032	03 07 30.	- 20 45	510	10.	GALAXY
ARP 041	03 07 30.	- 20 46			PECULIAR GALAXY
MCG-04-08-033	03 07 30.	- 25 26	24	13.5	GALAXY
MCG-05-08-020	03 07 30.	- 31 55	36	15.	GALAXY
KN 11.91	03 07 31.6	+ 04 39 29.			NEBULA
BAK 1.146	03 07 32.	- 31 54 16.	16	16.4	EXTRAGALACTIC NEBULA
SHAP308-5355.7	03 07 32.	- 54 07 05.	12	16.75	GALAXY
MCG-02-09-007	03 07 33.	- 12 10	54	15.	GALAXY
SVEN 211	03 07 33.	- 20 47	402	10.6	GALAXY
SHAP309-5318.6	03 07 34.	- 53 29 58.	12	16.75	GALAXY
IC 1894	03 07 35.	+ 19 25 12.			NONSTELLAR OBJECT
IC 0293	03 07 35.	+ 40 57 08.			NONSTELLAR OBJECT
SHAP309-5009.2	03 07 35.	- 50 20 34.	18	17.25	GALAXY
SHAP309-5147.1	03 07 35.	- 51 58 28.	12	17.0	GALAXY
SHAP309-5214.8	03 07 35.	- 52 26 10.	18	17.0	GALAXY
ZWG 464.006	03 07 36.	+ 19 25		15.6	GALAXY
MCG+07-07-031	03 07 36.	+ 40 56	12	15.	GALAXY
MCG+07-07-032	03 07 36.	+ 44 49	36	15.5	GALAXY
ZWG 540.052	03 07 36.	+ 44 50		15.7	GALAXY
UGC 02573	03 07 36.	+ 44 50	102	15.7	GALAXY SBb
OCL 0366	03 07 36.	+ 63 04	1680	7.2	OPEN STAR CLUSTER
PHL 8572	03 07 36.	- 05 50		18.3	BLUE STELLAR OBJECT
BAK 1.148	03 07 38.	- 31 53 03.	35	15.7	EXTRAGALACTIC NEBULA
SHAP309-5401.4	03 07 38.	- 54 12 46.	18	17.25	GALAXY
SHAP309-4951.0	03 07 39.	- 50 02 22.	24	16.5	GALAXY
SHAP309-4955.6	03 07 39.	- 50 06 58.	30	17.25	GALAXY
SHAP309-5306.4	03 07 39.	- 53 17 46.	42	17.0	GALAXY
IC 0294	03 07 40.	+ 40 26 37.			NONSTELLAR OBJECT
BAK 1.149	03 07 40.	- 24 39 45.	26	15.6	EXTRAGALACTIC NEBULA
ZC 0307.7+0934	03 07 42.	+ 09 34	1480		CLUSTER OF GALAXIES
ZC 0307.7+1907	03 07 42.	+ 19 07	7060		CLUSTER OF GALAXIES
IC 1888	03 07 42.	+ 40 56 31.			NONSTELLAR OBJECT
ZWG 540.053	03 07 42.	+ 40 58		15.7	GALAXY
VB 187	03 07 42.	+ 58 00	322		STELLAR RING
ISS 0065	03 07 42.	+ 58 00	326		STELLAR RING
PHL 8573	03 07 42.	- 13 03		18.5	BLUE STELLAR OBJECT
MCG-04-08-034	03 07 42.	- 24 40	42	14.5	GALAXY
MCG-05-08-021	03 07 42.	- 31 22	54	14.5	GALAXY
AGU 14	03 07 42.	- 40 11 30.		13.	2 INTERACTING GALAXIES
SHAP309-5345.3	03 07 42.	- 53 56 40.	18	17.75	GALAXY
SHAP309-5118.5	03 07 44.	- 51 29 52.	18	17.5	GALAXY
SHAP309-5449.0	03 07 44.	- 55 00 22.	24	16.75	GALAXY
MCG+07-07-033	03 07 45.	+ 40 25	48	15.	GALAXY
IC 1889	03 07 45.	+ 40 25 25.			NONSTELLAR OBJECT
IC 0295	03 07 45.	+ 40 25 37.			NONSTELLAR OBJECT
SVEN 212	03 07 45.	- 20 48 02.	42	14.8	GALAXY
SHAP309-5115.8	03 07 45.	- 51 27 10.	18	17.0	GALAXY
SHAP309-5253.9	03 07 47.	- 53 05 16.	12	16.75	GALAXY
ZWG 525.001	03 07 48.	+ 38 28		15.6	GALAXY
ZWG 524.060	03 07 48.	+ 38 28		15.6	GALAXY
ZWG 540.054	03 07 48.	+ 40 26		15.6	GALAXY
UGC 02574	03 07 48.	+ 40 26	168	15.6	GALAXY SB0/SBa
5ZW 324	03 07 48.	+ 47 09			COMPACT GALAXY
PHL 8575	03 07 48.	- 13 35		18.5	BLUE STELLAR OBJECT
PHL 8574	03 07 48.	- 15 32		17.4	BLUE STELLAR OBJECT
SHAP309-5204.4	03 07 48.	- 52 15 46.	18	17.0	GALAXY
SHAP309-5408.0	03 07 48.	- 54 19 22.	12	17.25	GALAXY
SHAP309-5409.3	03 07 48.	- 54 20 40.	18	17.25	GALAXY
IC 0296	03 07 49.	+ 40 26 18.			NONSTELLAR OBJECT
SHAP309-5306.0	03 07 49.	- 53 17 22.	18	16.25	GALAXY
SHAP309-5125.5	03 07 50.	- 51 36 52.	18	17.25	GALAXY
MCG+06-07-047	03 07 51.	+ 38 27	42	15.	GALAXY
ARC 0422	03 07 51.	- 11 14		17.6	RICH CLUSTER OF GALAXIES
MCG-04-08-035	03 07 51.	- 21 46 30.	42	15.	GALAXY
SHAP309-5113.6	03 07 51.	- 51 24 57.	18	17.0	GALAXY
BAK 1.152	03 07 52.	- 23 23 44.	26	15.8	EXTRAGALACTIC NEBULA
BAK 1.150	03 07 52.	- 29 42 20.	13	16.6	EXTRAGALACTIC NEBULA
ZWG 390.012	03 07 54.	+ 00 55		15.6	GALAXY
ZWG 525.002	03 07 54.	+ 35 12		14.5	GALAXY
ZWG 524.061	03 07 54.	+ 35 12		14.5	GALAXY
UGC 02575	03 07 54.	+ 35 12	156	14.5	GALAXY E
MCG+07-07-034	03 07 54.	+ 41 10	18	14.5	GALAXY
PHL 8576	03 07 54.	- 05 00		18.8	BLUE STELLAR OBJECT
PHL 4343	03 07 54.	- 12 52		18.3	BLUE STELLAR OBJECT
SHAP309-5035.4	03 07 54.	- 50 46 45.	12	17.5	GALAXY
SHAP309-5133.1	03 07 54.	- 51 44 27.	12	17.5	GALAXY
SHAP309-5134.0	03 07 54.	- 51 45 21.	18	16.5	GALAXY
SHAP309-5210.0	03 07 54.	- 52 21 21.	24	17.25	GALAXY
SHAP309-5358.2	03 07 54.	- 54 09 33.	12	17.0	GALAXY
ARC 0421	03 07 55.	+ 09 38		17.1	RICH CLUSTER OF GALAXIES
BAK 1.151	03 07 55.	- 31 19 44.	42	14.7	EXTRAGALACTIC NEBULA
SHAP309-5056.0	03 07 55.	- 51 07 21.	18	17.0	GALAXY
SHAP309-5339.0	03 07 55.	- 53 50 21.	30	17.5	GALAXY
RNGC 1226	03 07 56.	+ 35 12		14.5	GALAXY
SHAP309-5130.4	03 07 56.	- 51 41 45.	12	17.5	GALAXY
SHAP309-5226.4	03 07 56.	- 52 37 45.	18	17.0	GALAXY
MCG-02-09-008	03 07 57.	- 10 57	36	15.	GALAXY
SHAP309-5143.1	03 07 57.	- 51 54 27.	24	16.5	GALAXY
SHAP309-5413.1	03 07 57.	- 54 24 27.	24	17.25	GALAXY
RNGC 1224	03 07 58.	+ 41 11		15.5	GALAXY
RNGC 1220	03 07 58.	+ 53 09		12.0	OPEN CLUSTER
RNGC 1232A	03 07 58.	- 20 47			GALAXY
SHAP309-5356.1	03 07 58.	- 54 07 27.	18	17.5	GALAXY
SHAP309-5035.4	03 07 59.	- 50 46 45.	12	16.5	GALAXY
SHAP309-5149.2	03 07 59.	- 52 00 33.	18	16.75	GALAXY
LBN 0725	03 08	+ 29 30	7200		BRIGHT NEBULA
LB 09795	03 08	- 81 26		14.1	FAINT BLUE STAR
LB 09794	03 08	- 84 50		14.8	FAINT BLUE STAR
5ZW 325	03 08 00.	+ 22 16			COMPACT GALAXY
ZWG 525.003	03 08 00.	+ 35 08		15.7	GALAXY
ZWG 524.062	03 08 00.	+ 35 08		15.7	GALAXY
UGC 02577	03 08 00.	+ 35 08	78	15.7	GALAXY SB0/SBa
ZWG 540.055	03 08 00.	+ 41 11		15.5	GALAXY
CR 02	03 08 00.	+ 41 11		15.5	GALAXY IN PERSEUS CLUSTER
UGC 02578	03 08 00.	+ 41 11	108	15.5	GALAXY E-S0
OCL 0380	03 08 00.	+ 53 09	420	11.8	OPEN STAR CLUSTER
UGC 02576	03 08 00.	- 01 28	60	16.5	GALAXY Sb-c
PHL 8577	03 08 00.	- 11 02		18.5	BLUE STELLAR OBJECT
PHL 8578	03 08 00.	- 19 19		18.5	BLUE STELLAR OBJECT
SHAP309-5354.3	03 08 01.	- 54 05 39.	18	17.25	GALAXY
RNGC 1227	03 08 02.	+ 35 08		15.5	GALAXY
SHAP309-5409.0	03 08 02.	- 54 20 21.	24	16.75	GALAXY
SHAP309-5451.2	03 08 02.	- 55 02 33.	24	16.25	GALAXY
SHAP309-5141.0	03 08 03.	- 51 52 21.	12	16.5	GALAXY
SHAP309-5319.5	03 08 03.	- 53 30 51.	12	16.5	GALAXY
SHAP309-5214.3	03 08 04.	- 52 25 39.	12	16.75	GALAXY
SHAP309-5258.5	03 08 04.	- 53 09 51.	30	16.0	GALAXY
SHAP309-5312.1	03 08 04.	- 53 23 27.	24	16.5	GALAXY
SHAP309-5350.6	03 08 04.	- 54 01 57.	24	17.0	GALAXY
LB 00477	03 08 05.	- 01 23 36.		17.6	FAINT BLUE STAR
SHAP309-5025.8	03 08 05.	- 50 37 09.	24	17.75	GALAXY
SHAP309-5357.8	03 08 05.	- 54 09 09.	18	17.0	GALAXY
ZC 0308.1+2501	03 08 06.	+ 25 01	870		CLUSTER OF GALAXIES
MCG+06-08-001	03 08 06.	+ 35 09	90	15.	GALAXY
MCG-04-08-036	03 08 06.	- 22 35	216	13.	GALAXY
SHAP309-5411.2	03 08 06.	- 54 22 33.	18	17.5	GALAXY
BV 01	03 08 08.	- 09 56 00.		17.0	FAINT BLUE VARIABLE
SHAP309-5452.2	03 08 08.	- 55 03 33.	24	16.25	GALAXY
SHAP309-5136.0	03 08 09.	- 51 47 20.	36	16.5	GALAXY
SHAP309-5149.4	03 08 09.	- 52 00 45.	18	17.25	GALAXY
SHAP309-5202.5	03 08 09.	- 52 13 51.	24	17.25	GALAXY
SHAP309-5250.4	03 08 09.	- 53 01 45.	12	17.25	GALAXY
SHAP309-5123.4	03 08 10.	- 51 34 44.	12	17.25	GALAXY
ZC 0308.2+1839	03 08 12.	+ 18 39	1410		CLUSTER OF GALAXIES
PHL 1475	03 08 12.	- 09 57		16.6	BLUE STELLAR OBJECT
HN 0170	03 08 13.	- 22 36			NEBULA
SHAP309-5331.8	03 08 13.	- 53 43 08.	12	17.5	GALAXY

OBJECT NAME	RIGHT ASCEN.	DECLINATION	DIAM.	MAGN.	TYPE OF OBJECT
SHAP309-5402.4	03 08 13.	- 54 13 44.	30	16.75	GALAXY
IC 1898	03 08 14.	- 22 35			NONSTELLAR OBJECT
SHAP309-5125.6	03 08 14.	- 51 36 56.	18	17.25	GALAXY
BAK 1.155	03 08 15.	- 25 54 19.	18	16.2	EXTRAGALACTIC NEBULA
BAK 1.154	03 08 15.	- 26 42 31.	40	16.0	EXTRAGALACTIC NEBULA
SHAP309-5020.8	03 08 16.	- 50 32 08.	18	17.0	GALAXY
SHAP309-5408.8	03 08 16.	- 54 20 08.	18	16.0	GALAXY
RNGC 1237	03 08 17.	- 08 52			NON-EXISTENT OBJECT
BAK 1.153	03 08 17.	- 30 25 01.	17	16.8	EXTRAGALACTIC NEBULA
ZWG 525.004	03 08 18.	+ 34 48		15.7	GALAXY
ZWG 524.063	03 08 18.	+ 34 48		15.7	GALAXY
ZWG 525.005	03 08 18.	+ 35 17		15.7	GALAXY
ZWG 524.064	03 08 18.	+ 35 17		15.7	GALAXY
UGC 02579	03 08 18.	+ 35 17	72	15.7	GALAXY E?
PHL 1476	03 08 18.	- 04 27		18.0	BLUE STELLAR OBJECT
PHL 8579	03 08 18.	- 11 24		18.2	BLUE STELLAR OBJECT
SHAP309-5212.6	03 08 18.	- 52 23 56.	18	17.0	GALAXY
SHAP309-5322.4	03 08 18.	- 53 33 44.	24	17.0	GALAXY
SHAP309-5213.8	03 08 20.	- 52 25 08.	12	16.75	GALAXY
IC 1897	03 08 21.	- 10 59 13.			NONSTELLAR OBJECT
BAK 1.156	03 08 22.	- 27 32 31.	14	16.6	EXTRAGALACTIC NEBULA
SHAP309-5211.5	03 08 23.	- 52 22 50.	12	17.25	GALAXY
SHAP309-5307.4	03 08 23.	- 53 18 44.	12	16.5	GALAXY
SHAP309-5335.8	03 08 23.	- 53 47 08.	18	17.25	GALAXY
MCG+00-09-013	03 08 24.	+ 01 35	24	16.	GALAXY
MCG+01-09-012	03 08 24.	- 02 43	15	15.	GALAXY
HOLM 067A	03 08 24.	- 10 57	66	13.5	PART OF MULTIPLE GALAXY
MCG-02-09-009	03 08 24.	- 11 00	30	14.5	GALAXY
TON-S 0326	03 08 24.	- 22 00		15.2	BLUE STAR
TON-S 0327	03 08 24.	- 27 54		15.8	BLUE STAR
SHAP309-5223.6	03 08 24.	- 52 34 56.	18	17.25	GALAXY
SHAP309-5256.8	03 08 24.	- 53 08 08.	12	16.5	GALAXY
SHAP309-5341.6	03 08 24.	- 53 52 56.	24	17.5	GALAXY
SHAP309-5443.1	03 08 24.	- 54 54 26.	24	16.5	GALAXY
RNGC 1239	03 08 26.	- 02 43		15.0	GALAXY
HOLM 067B	03 08 26.	- 10 55	48	13.8	PART OF MULTIPLE GALAXY
SHAP309-5152.8	03 08 27.	- 52 04 08.	12	17.5	GALAXY
SHAP309-5352.2	03 08 28.	- 54 03 32.	18	17.25	GALAXY
BAK 1.157	03 08 29.	- 29 27 07.	19	17.2	EXTRAGALACTIC NEBULA
ZC 0308.5+2330	03 08 30.	+ 23 30	3560		CLUSTER OF GALAXIES
MCG+06-08-002	03 08 30.	+ 35 14	36	15.5	GALAXY
ZWG 390.013	03 08 30.	- 02 24		15.7	GALAXY
RNGC 1238	03 08 30.	- 10 57		14.0	GALAXY
MCG-02-09-010	03 08 30.	- 10 57	24	14.	GALAXY
PHL 4344	03 08 30.	- 13 30		17.9	BLUE STELLAR OBJECT
SHAP309-5300.0	03 08 30.	- 53 11 19.	12	16.5	GALAXY
LB 03300	03 08 30.	- 63 33		13.4	FAINT BLUE STAR
SHAP310-5251.6	03 08 31.	- 53 02 55.	12	17.5	GALAXY
ARP 147	03 08 34.	+ 01 07			PECULIAR GALAXY
RNGC 1249	03 08 34.	- 53 32		12.0	GALAXY
SHAP310-5334.2	03 08 34.	- 53 45 31.	12	17.5	GALAXY
ZWG 390.015	03 08 36.	+ 00 04		15.6	GALAXY
MCG+00-09-014	03 08 36.	+ 02 22 30.	30	15.5	GALAXY
ZWG 390.014	03 08 36.	- 00 43		15.7	GALAXY
PHL 8580	03 08 36.	- 03 36		18.6	BLUE STELLAR OBJECT
PHL 4345	03 08 36.	- 04 47		16.8	BLUE STELLAR OBJECT
SHAP310-5100.8	03 08 37.	- 51 12 07.	12	17.25	GALAXY
SHAP310-5111.0	03 08 37.	- 51 22 19.	36	17.25	GALAXY
SHAP310-5218.4	03 08 38.	- 52 29 43.	18	17.25	GALAXY
SHAP310-5135.2	03 08 39.	- 51 46 31.	18	16.5	GALAXY
SHAP310-5137.4	03 08 39.	- 51 48 43.	12	16.75	GALAXY
SHAP310-5134.6	03 08 40.	- 51 45 55.	12	17.0	GALAXY
SHAP310-5320.3	03 08 40.	- 53 31 37.	486	12.3	GALAXY
IC 0299	03 08 41.	- 13 17 57.			NONSTELLAR OBJECT
SHAP310-5133.8	03 08 41.	- 51 45 07.	12	17.5	GALAXY
SHAP310-5230.8	03 08 41.	- 52 42 07.	24	16.5	GALAXY
SHAP310-5342.2	03 08 41.	- 53 53 31.	12	17.25	GALAXY
1ZW 011	03 08 42.	+ 01 08			COMPACT GALAXY
ZWG 390.016	03 08 42.	+ 01 08		15.1	GALAXY
MCG+01-09-013	03 08 42.	- 04 26	90	16.	GALAXY
PHL 1477	03 08 42.	- 13 12		17.5	BLUE STELLAR OBJECT
LB 01655	03 08 42.	- 46 34		14.4	FAINT BLUE STAR
IC 0298	03 08 43.	+ 01 08 16.			NONSTELLAR OBJECT
SHAP310-5107.0	03 08 44.	- 51 18 19.	12	17.75	GALAXY
KN 11.92	03 08 44.5	+ 01 07 21.			NEBULA
MCG+00-09-015	03 08 45.	+ 01 08	24	15.	GALAXY
VV 334B	03 08 45.	- 09 06 30.	48	14.	INTERACTING GALAXY
VV 334A	03 08 45.	- 09 06 30.	138	12.	INTERACTING GALAXY
RNGC 1241	03 08 47.	- 09 07		13.0	GALAXY
ZWG 441.003	03 08 48.	+ 10 37		15.7	GALAXY
PHL 8582	03 08 48.	- 10 50		16.6	BLUE STELLAR OBJECT
PHL 8581	03 08 48.	- 12 18		15.0	BLUE STELLAR OBJECT
PHL 8583	03 08 48.	- 17 58		16.9	BLUE STELLAR OBJECT
SHAP310-5300.1	03 08 48.	- 53 11 25.	18	17.25	GALAXY
SHAP310-5330.4	03 08 48.	- 53 41 43.	24	17.25	GALAXY
SHAP310-5435.0	03 08 48.	- 54 46 19.	30	17.25	GALAXY
REIN 2.038	03 08 48.77	- 09 06 35.2			NEBULA
SHAP310-5139.2	03 08 49.	- 51 50 30.	12	16.5	GALAXY
SHAP310-5231.4	03 08 49.	- 52 42 42.	18	17.5	GALAXY
HOLM 068A	03 08 50.	- 09 07	72	13.6	PART OF MULTIPLE GALAXY
ARP 304	03 08 50.	- 09 07			PECULIAR GALAXY
RNGC 1236	03 08 51.	+ 10 37		15.5	GALAXY
MCG-02-09-012	03 08 51.	- 09 03	60	14.	GALAXY
MCG-02-09-011	03 08 51.	- 09 07	198	13.	GALAXY
BAK 1.158	03 08 51.	- 29 36 41.	13	17.0	EXTRAGALACTIC NEBULA
RNGC 1242	03 08 53.	- 09 06		14.0	GALAXY
SHAP310-5100.0	03 08 53.	- 51 11 18.	12	17.5	GALAXY
SHAP310-5127.4	03 08 53.	- 51 38 42.	12	17.25	GALAXY
SHAP310-5204.9	03 08 53.	- 52 16 12.	12	17.25	GALAXY
SHAP310-5319.1	03 08 53.	- 53 30 24.	18	16.5	GALAXY
MCG+00-09-016	03 08 54.	+ 01 06 30.	30	15.5	GALAXY
UGC 02580	03 08 54.	+ 06 31	72	18.	GALAXY PECULR
UGC 02581	03 08 54.	+ 40 04	78	17.	GALAXY Sc
MCG+00-09-017	03 08 54.	- 00 56	30	15.5	GALAXY
HOLM 068C	03 08 54.	- 09 06	30	14.1	PART OF MULTIPLE GALAXY
PHL 1478	03 08 54.	- 10 04		18.5	BLUE STELLAR OBJECT
LB 01656	03 08 54.	- 46 12		12.2	FAINT BLUE STAR
SHAP310-5355.1	03 08 54.	- 54 06 24.	24	17.5	GALAXY
ARC 0423	03 08 55.	- 12 18		16.6	RICH CLUSTER OF GALAXIES
KN 11.93	03 08 55.5	+ 04 19 16.			NEBULA
SHAP310-5308.8	03 08 56.	- 53 20 06.	24	17.0	GALAXY
SHAP310-5003.8	03 08 57.	- 50 15 06.	18	17.0	GALAXY
RNGC 1243	03 08 59.	- 09 09			NON-EXISTENT OBJECT
BAK 1.159	03 08 59.	- 25 30 41.	25	15.8	EXTRAGALACTIC NEBULA
BAK 1.160	03 08 59.	- 25 49 41.	20	16.6	EXTRAGALACTIC NEBULA
LB 09796	03 09	- 83 40		13.4	FAINT BLUE STAR
LB 09963	03 09	- 87 42		13.5	FAINT BLUE STAR
UGC 02582	03 09 00.	+ 27 56	60	16.5	GALAXY Sc

OBJECT NAME	RIGHT ASCEN.	DECLINATION	DIAM.	MAGN.	TYPE OF OBJECT
MCG+13-03-002	03 09 00.	+ 80 36 30.	156	13.	GALAXY
ZWG 346.002	03 09 00.	+ 80 37		13.4	GALAXY
UGC 02583	03 09 00.	+ 80 37	216	13.4	GALAXY S0-a
ZWG 390.017	03 09 00.	- 00 55		15.7	GALAXY
MCG-04-08-037	03 09 00.	- 25 30	42	15.	GALAXY
SHAP310-5318.9	03 09 00.	- 53 30 12.	12	16.75	GALAXY
RNGC 1184	03 09 01.	+ 80 37		13.5	GALAXY
HOLM 068B	03 09 01.	- 09 08	24	13.7	PART OF MULTIPLE GALAXY
BAK 1.161	03 09 01.	- 26 18 05.	19	16.6	EXTRAGALACTIC NEBULA
SHAP310-4958.1	03 09 01.	- 50 09 24.	18	17.0	GALAXY
SHAP310-5312.0	03 09 01.	- 53 23 18.	18	17.25	GALAXY
SHAP310-5315.4	03 09 04.	- 53 26 42.	18	17.0	GALAXY
SHAP310-5432.2	03 09 05.	- 54 43 30.	24	17.0	GALAXY
UGC 02584	03 09 06.	+ 00 53	66	18.	GALAXY DWARF?
SHAP310-5204.1	03 09 08.	- 52 15 23.	24	17.25	GALAXY
BAK 1.164	03 09 10.	- 25 18 10.	34	16.1	EXTRAGALACTIC NEBULA
BAK 1.163	03 09 10.	- 26 07 34.	14	16.1	EXTRAGALACTIC NEBULA
BAK 1.162	03 09 10.	- 28 27 28.	11	17.3	EXTRAGALACTIC NEBULA
SHAP310-5404.0	03 09 10.	- 54 15 17.	18	17.25	GALAXY
ZWG 540.056	03 09 12.	+ 42 30		15.6	GALAXY
ZWG 390.018	03 09 12.	- 00 35		14.4	GALAXY
UGC 02585	03 09 12.	- 00 35	108	14.4	GALAXY SBb
MCG+00-09-018	03 09 12.	- 00 36	90	13.	GALAXY
MCG-02-09-013	03 09 12.	- 10 49	60	15.	GALAXY
PHL 1479	03 09 12.	- 11 07		18.8	BLUE STELLAR OBJECT
MCG-04-08-038	03 09 12.	- 25 17	36	15.	GALAXY
SHAP310-5230.0	03 09 13.	- 52 41 17.	12	17.5	GALAXY
MCG+00-09-019	03 09 15.	- 00 02 30.	42	15.	GALAXY
SHAP310-5023.6	03 09 15.	- 50 34 53.	30	16.0	GALAXY
SHAP310-5424.0	03 09 15.	- 54 35 17.	24	17.0	GALAXY
RNGC 1233	03 09 17.	+ 39 08		14.0	GALAXY
SHAP310-5406.9	03 09 17.	- 54 18 11.	18	17.75	GALAXY
SHAP310-5421.0	03 09 17.	- 54 32 17.	12	17.5	GALAXY
RNGC 1252	03 09 17.	- 58 20			UNVERIFIED SOUTHERN OBJECT
ZWG 525.006	03 09 18.	+ 39 08		13.9	GALAXY
ZWG 524.065	03 09 18.	+ 39 08		13.9	GALAXY
UGC 02586	03 09 18.	+ 39 08	138	13.9	GALAXY Sb
ZWG 390.019	03 09 18.	- 00 01		15.7	GALAXY
PHL 1480	03 09 18.	- 08 50		17.9	BLUE STELLAR OBJECT
SHAP310-5125.2	03 09 18.	- 51 36 29.	18	17.0	GALAXY
SHAP310-5235.8	03 09 19.	- 52 47 05.	12	16.25	GALAXY
SHAP310-5425.7	03 09 19.	- 54 36 59.	24	16.5	GALAXY
RNGC 1313A	03 09 23.	- 66 52			GALAXY
ZC 0309.4+2235	03 09 24.	+ 22 35	870		CLUSTER OF GALAXIES
MCG+06-08-003	03 09 24.	+ 39 07	90	14.	GALAXY
MCG+00-09-020	03 09 24.	- 00 17	12	14.5	GALAXY
UGC 02587	03 09 24.	- 01 21	90	16.0	GALAXY Sb
MCG+00-09-021	03 09 24.	- 01 22	90	15.5	GALAXY
MCG+01-09-014	03 09 24.	- 03 09	30	16.	GALAXY
SHAP310-5208.8	03 09 24.	- 52 20 05.	18	16.5	GALAXY
SHAP310-5300.1	03 09 24.	- 53 11 23.	18	16.5	GALAXY
DV.56 N1313A	03 09 24.	- 66 52	72		S GALAXY
BAK 1.165	03 09 25.	- 28 10 03.	13	16.4	EXTRAGALACTIC NEBULA
MCG-03-09-004	03 09 27.	- 17 07	10	15.5	GALAXY
BAK 1.166	03 09 27.	- 27 05 27.	22	16.0	EXTRAGALACTIC NEBULA
SHAP310-5155.8	03 09 27.	- 52 07 04.	24	17.0	GALAXY
SHAP310-5406.4	03 09 27.	- 54 17 41.	24	17.5	GALAXY
SHAP311-5150.8	03 09 29.	- 52 02 04.	18	17.25	GALAXY
ZWG 390.020	03 09 30.	- 00 15		14.8	GALAXY
SHAP310-5259.3	03 09 30.	- 53 10 34.	18	16.7	GALAXY
BAK 1.167	03 09 31.	- 30 23 27.	16	16.7	EXTRAGALACTIC NEBULA
SHAP311-5026.7	03 09 32.	- 50 37 58.	36	16.5	GALAXY
BAK 1.168	03 09 34.	- 25 06 09.	17	16.0	EXTRAGALACTIC NEBULA
SHAP311-5218.7	03 09 34.	- 52 29 58.	18	17.5	GALAXY
RNGC 1235	03 09 35.	+ 38 44			NON-EXISTENT OBJECT
SHAP311-4835.8	03 09 35.	- 48 47 04.	36	17.0	GALAXY
ZWG 464.007	03 09 36.	+ 19 03		15.1	GALAXY
MCG+03-09-005	03 09 36.	+ 19 05	30	15.	GALAXY
MCG+00-09-022	03 09 36.	- 01 17	36	15.	GALAXY
PHL 1481	03 09 36.	- 12 40		18.5	BLUE STELLAR OBJECT
SHAP311-5139.2	03 09 37.	- 51 50 28.	24	16.5	GALAXY
BAK 1.170	03 09 38.	- 24 28 09.	18	16.5	EXTRAGALACTIC NEBULA
SHAP311-5433.8	03 09 38.	- 54 45 04.	18	17.0	GALAXY
SHAP311-4801.4	03 09 39.	- 48 12 39.	60	16.75	GALAXY
IC 1899	03 09 40.	- 25 30 40.			NONSTELLAR OBJECT
SHAP311-5422.4	03 09 40.	- 54 33 40.	12	17.25	GALAXY
ARC 0424	03 09 41.	- 02 50		17.5	RICH CLUSTER OF GALAXIES
BAK 1.169	03 09 41.	- 31 10 39.	13	16.4	EXTRAGALACTIC NEBULA
SHAP311-5046.3	03 09 41.	- 50 57 34.	36	16.0	GALAXY
SHAP311-5251.6	03 09 41.	- 53 02 52.	12	17.0	GALAXY
UGC 02588	03 09 42.	+ 14 14	60	17.	GALAXY IRR
5ZW 326	03 09 42.	+ 36 05			COMPACT GALAXY
ZC 0309.7+7759	03 09 42.	+ 77 59	1210		CLUSTER OF GALAXIES
ZWG 390.021	03 09 42.	- 01 14		14.9	GALAXY
PHL 1482	03 09 42.	- 11 56		17.5	BLUE STELLAR OBJECT
BAK 1.171	03 09 46.	- 28 26 26.	12	16.3	EXTRAGALACTIC NEBULA
BAK 1.175	03 09 47.	- 23 15 14.	18	16.7	EXTRAGALACTIC NEBULA
BAK 1.172	03 09 47.	- 29 02 26.	24	17.0	EXTRAGALACTIC NEBULA
ZWG 390.023	03 09 48.	+ 02 34		15.6	GALAXY
UGC 02589	03 09 48.	+ 12 13	84	16.0	GALAXY Sb-c
ZC 0309.8+3305	03 09 48.	+ 33 05	4030		CLUSTER OF GALAXIES
ZWG 540.057	03 09 48.	+ 42 16		15.6	GALAXY
UGC 02590	03 09 48.	+ 42 16	72	15.6	GALAXY
ZWG 390.022	03 09 48.	- 00 21		15.1	GALAXY
PHL 8584	03 09 48.	- 03 25		15.4	BLUE STELLAR OBJECT
PHL 8586	03 09 48.	- 08 22		18.5	BLUE STELLAR OBJECT
PHL 8585	03 09 48.	- 08 58		16.3	BLUE STELLAR OBJECT
RNGC 1247	03 09 48.	- 10 40		13.0	GALAXY
MCG-02-09-014	03 09 48.	- 10 40	210	13.	GALAXY
PHL 4347	03 09 48.	- 18 08		18.5	BLUE STELLAR OBJECT
BAK 1.173	03 09 48.	- 30 26 02.	14	16.8	EXTRAGALACTIC NEBULA
LB 01657	03 09 48.	- 49 05		14.2	FAINT BLUE STAR
SHAP311-5200.8	03 09 48.	- 52 12 03.	12	17.25	GALAXY
SHAP311-5207.2	03 09 48.	- 52 18 27.	18	17.25	GALAXY
LB 01658	03 09 48.	- 52 51		14.0	FAINT BLUE STAR
SHAP311-5425.8	03 09 48.	- 54 37 03.	12	17.0	GALAXY
BAK 1.174	03 09 49.	- 26 59 02.	19	16.4	EXTRAGALACTIC NEBULA
SVEN 213	03 09 50.	- 21 14 09.	24	14.8	GALAXY
SHAP311-5216.2	03 09 50.	- 52 27 27.	12	17.5	GALAXY
SVEN 214	03 09 51.	- 21 13	48	14.8	GALAXY
SHAP311-5420.0	03 09 51.	- 54 31 15.	24	17.5	GALAXY
RNGC 1240	03 09 53.	+ 30 22			NON-EXISTENT OBJECT
MCG+02-09-001	03 09 54.	+ 12 11	60	15.	GALAXY
5ZW 327	03 09 54.	+ 34 02			COMPACT GALAXY
5ZW 328	03 09 54.	+ 34 48			COMPACT GALAXY
UGC 02591	03 09 54.	+ 42 35	78	16.5	GALAXY Sb-c
ZWG 540.058	03 09 54.	+ 42 49		15.7	GALAXY
ZC 0309.9-0157	03 09 54.	- 01 57	1680		CLUSTER OF GALAXIES

OBJECT NAME	RIGHT ASCEN.	DECLINATION	DIAM.	MAGN.	TYPE OF OBJECT
PHL 4346	03 09 54.	- 06 15		17.8	BLUE STELLAR OBJECT
PHL 8587	03 09 54.	- 15 14		18.5	BLUE STELLAR OBJECT
SHAP311-5311.0	03 09 54.	- 53 22 15.	12	16.5	GALAXY
LB 03301	03 09 54.	- 79 34		14.2	FAINT BLUE STAR
BAK 1.176	03 09 55.	- 27 17 50.	34	16.0	EXTRAGALACTIC NEBULA
IC 0297	03 09 57.	+ 41 55 06.			TWO STARS
BAK 1.177	03 09 57.	- 24 26 32.	18	16.6	EXTRAGALACTIC NEBULA
SHAP311-5300.6	03 09 57.	- 53 11 51.	24	16.5	GALAXY
SHAP311-4933.4	03 09 58.	- 49 44 39.	36	16.0	GALAXY
SHAP311-5307.7	03 09 58.	- 53 18 57.	18	16.5	GALAXY
LBN 0763	03 10	+ 19 40	1080		BRIGHT NEBULA
LBN 0726	03 10	+ 30 00	20700		BRIGHT NEBULA
LBN 0677	03 10	+ 59 30	10800		BRIGHT NEBULA
LBN 0856	03 10	- 09 25	900		BRIGHT NEBULA
MCG+00-09-023	03 10 00.	+ 00 49 30.	36	16.	GALAXY
ZWG 464.008	03 10 00.	+ 18 36		15.7	GALAXY
UGC 02592	03 10 00.	+ 18 36	66	15.7	GALAXY Sb
MCG+03-09-006	03 10 00.	+ 18 38	54	15.	GALAXY
ZWG 485.009	03 10 00.	+ 22 20		15.6	GALAXY
UGC 02593	03 10 00.	+ 28 27	72	16.0	GALAXY S
ZC 0310.0-0130	03 10 00.	- 01 30	20160		CLUSTER OF GALAXIES
PHL 1483	03 10 00.	- 16 06		15.1	BLUE STELLAR OBJECT
MCG-04-08-040	03 10 00.	- 21 13 30.	24	14.5	GALAXY
BAK 1.180	03 10 00.	- 23 49 07.	30	15.9	EXTRAGALACTIC NEBULA
MCG-04-08-039	03 10 00.	- 23 50	24	15.5	GALAXY
MCG-04-08-042	03 10 00.	- 25 19	90	15.	GALAXY
BAK 1.179	03 10 00.	- 25 19 02.	19	15.7	EXTRAGALACTIC NEBULA
MCG-04-08-041	03 10 00.	- 25 30	48	14.	GALAXY
SHAP311-5137.2	03 10 01.	- 51 48 27.	24	17.75	GALAXY
BAK 1.178	03 10 02.	- 29 39 01.	11	16.3	EXTRAGALACTIC NEBULA
SHAP311-5017.5	03 10 02.	- 50 28 44.	30	16.75	GALAXY
MCG-04-08-043	03 10 03.	- 25 22 30.	24	15.	GALAXY
BAK 1.185	03 10 05.	- 23 20 31.	13	16.6	EXTRAGALACTIC NEBULA
BAK 1.183	03 10 05.	- 24 59 43.	19	16.6	EXTRAGALACTIC NEBULA
SHAP311-4942.4	03 10 05.	- 49 53 38.	36	16.25	GALAXY
SHAP311-5324.0	03 10 05.	- 53 35 14.	24	16.5	GALAXY
SHAP311-5329.0	03 10 05.	- 53 40 14.	18	17.5	GALAXY
SHB 070	03 10 06.	+ 01 26			QUASI-STELLAR OBJECT
MCG+04-08-012	03 10 06.	+ 22 20	30	15.5	GALAXY
MCG+05-08-007	03 10 06.	+ 28 27	48	15.	GALAXY
MCG+07-07-035	03 10 06.	+ 43 56	120	15.	GALAXY
MCG-04-08-044	03 10 06.	- 24 48	48	15.	GALAXY
MCG-04-08-045	03 10 06.	- 25 20	18	15.	GALAXY
SHAP311-5127.4	03 10 06.	- 51 38 38.	18	17.25	GALAXY
BAK 1.181	03 10 07.	- 28 25 07.	7	17.2	EXTRAGALACTIC NEBULA
SHAP311-5444.5	03 10 07.	- 54 55 44.	18	17.0	GALAXY
BAK 1.187	03 10 08.	- 24 48 25.	19	15.9	EXTRAGALACTIC NEBULA
BAK 1.186	03 10 08.	- 25 28 01.	23	16.0	EXTRAGALACTIC NEBULA
BAK 1.188	03 10 09.	- 25 20 01.	19	16.0	EXTRAGALACTIC NEBULA
BAK 1.184	03 10 09.	- 28 04 49.	17	16.4	EXTRAGALACTIC NEBULA
BAK 1.182	03 10 09.	- 29 04 13.	14	16.7	EXTRAGALACTIC NEBULA
SHAP311-5037.8	03 10 09.	- 50 49 02.	18	17.5	GALAXY
BAK 1.189	03 10 11.	- 26 24 49.	18	16.2	EXTRAGALACTIC NEBULA
MCG+01-09-002	03 10 12.	+ 04 30	96	14.	GALAXY
ZWG 416.004	03 10 12.	+ 04 31		14.0	GALAXY
UGC 02595	03 10 12.	+ 04 31	150	14.0	GALAXY SBb/SBc
KARA.73B 0123	03 10 12.	+ 04 31	114	14.0	ISOLATED GALAXY S
ZWG 540.059	03 10 12.	+ 40 54		15.0	GALAXY
ZWG 540.060	03 10 12.	+ 43 57		14.5	GALAXY
UGC 02596	03 10 12.	+ 43 57	132	14.5	GALAXY S IV-V
ZWG 554.025	03 10 12.	+ 46 52		15.7	GALAXY
ZWG 390.025	03 10 12.	- 00 30		14.5	GALAXY
UGC 02594	03 10 12.	- 00 30	66	14.5	GALAXY SB:0-a
MCG+00-09-024	03 10 12.	- 00 32	48	14.	GALAXY
ZWG 390.024	03 10 12.	- 02 28		15.7	GALAXY
PHL 4348	03 10 12.	- 04 14		18.6	BLUE STELLAR OBJECT
PHL 8588	03 10 12.	- 14 16		15.2	BLUE STELLAR OBJECT
MCG-04-08-047	03 10 12.	- 21 11	90	14.5	GALAXY
MCG-04-08-046	03 10 12.	- 25 28	30	15.	GALAXY
SHAP311-5003.8	03 10 12.	- 50 15 02.	24	16.5	GALAXY
IC 0302	03 10 13.	+ 04 31 07.			NONSTELLAR OBJECT
KN 11.94	03 10 13.5	+ 04 31 09.			NEBULA
BAK 1.191	03 10 14.	- 24 43 55.	77	15.6	EXTRAGALACTIC NEBULA
RNGC 1248	03 10 15.	- 05 25		13.5	GALAXY
MCG-04-08-048	03 10 15.	- 24 43	72	15.	GALAXY
SHAP311-5436.1	03 10 15.	- 54 47 20.	18	17.0	GALAXY
BAK 1.190	03 10 16.	- 27 59 37.	14	17.0	EXTRAGALACTIC NEBULA
SHAP311-5108.6	03 10 16.	- 51 19 50.	12	17.0	GALAXY
IC 0303	03 10 17.	- 11 52 48.			NONSTELLAR OBJECT
SHAP311-5023.4	03 10 17.	- 50 34 38.	30	16.5	GALAXY
ZWG 390.026	03 10 18.	+ 02 35		15.5	GALAXY
5ZW 329	03 10 18.	+ 30 03			COMPACT GALAXY
BAK 1.192	03 10 18.	- 27 19 25.	16	16.5	EXTRAGALACTIC NEBULA
BAK 1.193	03 10 18.	- 27 19 49.	16	16.2	EXTRAGALACTIC NEBULA
SHAP311-5316.8	03 10 18.	- 53 28 02.	12	17.0	GALAXY
LB 03302	03 10 18.	- 55 26		13.0	FAINT BLUE STAR
SHAP311-5023.3	03 10 19.	- 50 34 31.	24	16.25	GALAXY
BAK 1.195	03 10 20.	- 26 08 36.	17	16.6	EXTRAGALACTIC NEBULA
SHAP311-5436.4	03 10 20.	- 54 47 38.	24	17.5	GALAXY
BAK 1.194	03 10 21.	- 28 28 36.	17	16.4	EXTRAGALACTIC NEBULA
SHAP311-5042.3	03 10 21.	- 50 53 31.	24	16.25	GALAXY
BAK 1.196	03 10 23.	- 28 25 06.	16	17.2	EXTRAGALACTIC NEBULA
UGC 02597	03 10 24.	+ 17 48	90	16.0	GALAXY Sb
VB 188	03 10 24.	+ 59 41	423		STELLAR RING
VB 189	03 10 24.	+ 59 44	336		STELLAR RING
MCG+01-09-015	03 10 24.	- 03 05	22	15.	GALAXY
PHL 8589	03 10 24.	- 04 20		17.9	BLUE STELLAR OBJECT
MCG+01-09-016	03 10 24.	- 05 24	30	13.5	GALAXY
MRK 604	03 10 24.	- 05 26	8	15.5	GALAXY WITH UV CONTINUUM
MCG-03-09-005	03 10 24.	- 16 53	30	15.	GALAXY
MCG-03-09-006	03 10 24.	- 17 18	60	15.	GALAXY
BAK 1.197	03 10 24.	- 29 00 00.	12	16.6	EXTRAGALACTIC NEBULA
LB 03303	03 10 24.	- 68 48		12.2	FAINT BLUE STAR
BAK 1.198	03 10 25.	- 28 25 30.	17	16.9	EXTRAGALACTIC NEBULA
SHAP311-5309.8	03 10 25.	- 53 21 01.	12	16.5	GALAXY
SHAP312-4941.4	03 10 28.	- 49 52 37.	30	16.75	GALAXY
SHAP311-5213.1	03 10 28.	- 52 24 19.	24	16.75	GALAXY
SHAP311-5310.2	03 10 28.	- 53 21 25.	18	16.5	GALAXY
BAK 1.201	03 10 29.	- 28 28 12.	12	16.6	EXTRAGALACTIC NEBULA
BAK 1.200	03 10 29.	- 29 05 00.	11	17.0	EXTRAGALACTIC NEBULA
BAK 1.199	03 10 29.	- 29 50 48.	17	16.2	EXTRAGALACTIC NEBULA
MCG+03-09-007	03 10 30.	+ 17 49	48	16.	GALAXY
PHL 8590	03 10 30.	- 18 41		18.3	BLUE STELLAR OBJECT
SHAP311-5256.2	03 10 31.	- 53 07 25.	24	16.5	GALAXY
SHAP311-5326.2	03 10 31.	- 53 37 25.	54	16.5	GALAXY
SHAP312-5310.0	03 10 33.	- 53 21 13.	12	16.0	GALAXY
SHAP312-5325.0	03 10 33.	- 53 36 13.	24	17.0	GALAXY
SHAP312-5105.7	03 10 34.	- 51 16 55.	18	16.5	GALAXY

OBJECT NAME	RIGHT ASCEN.	DECLINATION	DIAM.	MAGN.	TYPE OF OBJECT
BAK 1.203	03 10 35.	- 27 18 24.	19	16.4	EXTRAGALACTIC NEBULA
BAK 1.204	03 10 35.	- 27 50 36.	12	16.9	EXTRAGALACTIC NEBULA
SHAP312-4949.2	03 10 35.	- 50 00 25.	24	17.0	GALAXY
5ZW 330	03 10 36.	+ 41 04			COMPACT GALAXY
IC 0306	03 10 36.	- 11 54 20.			NONSTELLAR OBJECT
MCG-05-08-022	03 10 36.	- 31 42	96	15.	GALAXY
TON-S 0328	03 10 36.	- 31 59		14.6	BLUE STAR
SHAP312-5426.9	03 10 36.	- 54 38 07.	18	16.5	GALAXY
BAK 1.207	03 10 37.	- 25 32 48.	19	16.5	EXTRAGALACTIC NEBULA
SHAP312-4743.6	03 10 37.	- 47 54 48.	42	16.5	GALAXY
SHAP312-5448.2	03 10 37.	- 54 59 25.	24	16.75	GALAXY
BAK 1.202	03 10 38.	- 30 51 00.	11	16.8	EXTRAGALACTIC NEBULA
BAK 1.206	03 10 39.	- 28 05 23.	7	16.7	EXTRAGALACTIC NEBULA
BAK 1.208	03 10 40.	- 27 15 35.	14	16.8	EXTRAGALACTIC NEBULA
SHAP312-5249.2	03 10 40.	- 53 00 24.	12	17.25	GALAXY
BAK 1.205	03 10 41.	- 31 08 59.	17	16.4	EXTRAGALACTIC NEBULA
SHAP312-5245.9	03 10 41.	- 52 57 06.	12	17.0	GALAXY
5ZW 331	03 10 42.	+ 41 05			COMPACT GALAXY
MCG-02-09-015	03 10 42.	- 11 54	30	15.	GALAXY
MCG-03-09-007	03 10 42.	- 18 06	60	15.5	GALAXY
SHAP312-5013.2	03 10 44.	- 50 24 24.	24	17.5	GALAXY
SHAP312-5248.2	03 10 44.	- 52 59 24.	12	17.0	GALAXY
BAK 1.209	03 10 45.	- 31 40 11.	24	16.0	EXTRAGALACTIC NEBULA
SHAP312-5325.4	03 10 45.	- 53 36 36.	30	17.25	GALAXY
BAK 1.210	03 10 46.	- 29 09 35.	13	16.8	EXTRAGALACTIC NEBULA
SHAP312-5113.0	03 10 46.	- 51 24 12.	24	17.25	GALAXY
LB 00478	03 10 47.	- 02 15 24.		18.0	FAINT BLUE STAR
BAK 1.211	03 10 47.	- 28 02 47.	16	16.8	EXTRAGALACTIC NEBULA
ZWG 525.007	03 10 48.	+ 33 47		15.7	GALAXY
ZWG 540.061	03 10 48.	+ 41 06		15.1	GALAXY
ZCG 0310+41	03 10 48.	+ 41 06		15.1	COMPACT GALAXY
UGC 02598	03 10 48.	+ 41 06	90	15.1	GALAXY (S0)
ZWG 540.062	03 10 48.	+ 42 34		14.9	GALAXY
MCG+08-06-029	03 10 48.	+ 45 50	24	17.	GALAXY
PHL 1484	03 10 48.	- 07 19		16.6	BLUE STELLAR OBJECT
PHL 8593	03 10 48.	- 08 52		18.5	BLUE STELLAR OBJECT
PHL 8592	03 10 48.	- 09 22		15.9	BLUE STELLAR OBJECT
PHL 8591	03 10 48.	- 12 08		15.2	BLUE STELLAR OBJECT
SHAP312-5022.2	03 10 48.	- 50 33 24.	24	17.0	GALAXY
SHAP312-5246.0	03 10 48.	- 52 57 12.	12	18.0	GALAXY
LB 03304	03 10 48.	- 58 17		14.2	FAINT BLUE STAR
BAK 1.212	03 10 50.	- 29 41 47.	13	16.5	EXTRAGALACTIC NEBULA
SHAP312-5313.7	03 10 50.	- 53 24 54.	18	16.5	GALAXY
SHAP312-5104.2	03 10 52.	- 51 15 24.	12	16.5	GALAXY
SHAP312-4947.3	03 10 53.	- 49 58 30.	24	17.5	GALAXY
SHAP312-5312.8	03 10 53.	- 53 24 00.	12	17.0	GALAXY
MCG+00-09-025	03 10 54.	+ 02 33 30.	48	15.	GALAXY
LDN 1378	03 10 54.	+ 60 09	120		DARK NEBULA
PHL 4349	03 10 54.	- 11 36		18.0	BLUE STELLAR OBJECT
PHL 8594	03 10 54.	- 12 49		18.4	BLUE STELLAR OBJECT
PHL 1485	03 10 54.	- 13 54		18.2	BLUE STELLAR OBJECT
MCG-03-09-008	03 10 54.	- 18 06	30	17.	GALAXY
MCG-03-09-009	03 10 54.	- 20 24	36	16.	GALAXY
SHAP312-5024.1	03 10 54.	- 50 35 18.	24	16.75	GALAXY
SHAP312-5303.8	03 10 54.	- 53 15 00.	12	17.0	GALAXY
GCL 005	03 10 54.	- 55 25	240	9.5	GLOBULAR STAR CLUSTER
SHAP312-5020.9	03 10 55.	- 50 32 06.	12	17.0	GALAXY
RNGC 1261	03 10 55.	- 55 25		9.5	GLOBULAR CLUSTER
IC 0300	03 10 56.9	+ 42 13 50.			GALAXY S0
SHAP312-5323.0	03 10 57.	- 53 34 12.	18	17.5	GALAXY
BAK 1.213	03 10 58.	- 28 27 46.	14	17.0	EXTRAGALACTIC NEBULA
SHAP312-4940.9	03 10 58.	- 49 52 05.	36	16.0	GALAXY
LBN 0676	03 11	+ 60 40	720		BRIGHT NEBULA
VDB.66G 031	03 11	- 02 58	100		DWARF GALAXY
VDB.66G 032	03 11	- 04 58	70		DWARF GALAXY
OCL 0384	03 11 00.	+ 52 32	450	13.	OPEN STAR CLUSTER
VB 231	03 11 00.	+ 56 57	423		STELLAR RING
LDN 1377	03 11 00.	+ 60 50	600		DARK NEBULA
MCG+13-03-003	03 11 00.	+ 80 36 30.	27	17.	GALAXY
MCG+01-09-017	03 11 00.	- 08 25	42	15.	GALAXY
PHL 1486	03 11 00.	- 09 03		17.9	BLUE STELLAR OBJECT
PHL 4350	03 11 00.	- 15 52		18.6	BLUE STELLAR OBJECT
PHL 4351	03 11 00.	- 16 22		18.5	BLUE STELLAR OBJECT
PHL 4352	03 11 00.	- 18 31		18.2	BLUE STELLAR OBJECT
BAK 1.215	03 11 00.	- 23 34 04.	16	16.4	EXTRAGALACTIC NEBULA
BAK 1.214	03 11 00.	- 24 17 34.	17	16.6	EXTRAGALACTIC NEBULA
SHAP312-5026.2	03 11 00.	- 50 37 23.	36	17.0	GALAXY
SHAP312-5331.4	03 11 00.	- 53 42 35.	18	17.0	GALAXY
SHAP312-5020.2	03 11 01.	- 50 31 23.	24	16.5	GALAXY
SHAP312-5304.1	03 11 01.	- 53 15 17.	30	15.0	GALAXY
SHAP312-4950.9	03 11 02.	- 50 02 05.	24	16.5	GALAXY
SHAP312-5251.9	03 11 02.	- 53 03 05.	12	17.5	GALAXY
BAK 1.216	03 11 03.	- 23 37 28.	13	16.8	EXTRAGALACTIC NEBULA
SHAP312-5354.4	03 11 03.	- 54 05 35.	12	17.5	GALAXY
SHAP312-5146.2	03 11 04.	- 51 57 23.	18	17.5	GALAXY
ZWG 390.027	03 11 06.	+ 02 35		15.6	GALAXY
UGC 02599	03 11 06.	+ 02 35	72	15.6	GALAXY Sb
ZWG 525.008	03 11 06.	+ 35 21		15.6	GALAXY
PHL 4353	03 11 06.	- 08 02		18.6	BLUE STELLAR OBJECT
PHL 8595	03 11 06.	- 10 21		17.5	BLUE STELLAR OBJECT
MCG-05-08-023	03 11 06.	- 31 53	36	15.	GALAXY
SHAP312-4847.6	03 11 06.	- 48 58 47.	36	17.0	GALAXY
SHAP312-5116.2	03 11 06.	- 51 27 23.	12	16.5	GALAXY
SHAP312-5119.6	03 11 06.	- 51 30 47.	42	16.5	GALAXY
SHAP312-4933.0	03 11 07.	- 49 44 11.	24	17.0	GALAXY
MCG+00-09-026	03 11 09.	+ 00 30 30.	36	14.5	GALAXY
SHAP312-4828.9	03 11 09.	- 48 40 05.	42	16.5	GALAXY
SHAP312-5037.8	03 11 09.	- 50 48 59.	18	17.5	GALAXY
SHAP312-5312.3	03 11 09.	- 53 23 29.	18	16.0	GALAXY
RNGC 1245	03 11 10.	+ 47 03		9.0	OPEN CLUSTER
KLEM 06	03 11 11.	- 53 13	900		GROUP OF 12 GALAXIES
ZC 0311.2+0223	03 11 12.	+ 02 23	4910		CLUSTER OF GALAXIES
5ZW 332	03 11 12.	+ 42 58			COMPACT GALAXY
OCL 0389	03 11 12.	+ 47 04	1800	9.0	OPEN STAR CLUSTER
ZWG 390.028	03 11 12.	- 00 24		14.6	GALAXY
UGC 02600	03 11 12.	- 00 24	114	14.6	GALAXY Sa
MCG+00-09-027	03 11 12.	- 00 26	90	13.	GALAXY
PHL 4354	03 11 12.	- 12 00		18.4	BLUE STELLAR OBJECT
PHL 1487	03 11 12.	- 12 16		17.8	BLUE STELLAR OBJECT
BAK 1.219	03 11 12.	- 25 03 22.	28	16.0	EXTRAGALACTIC NEBULA
BAK 1.217	03 11 12.	- 31 49 58.	17	15.0	EXTRAGALACTIC NEBULA
SHAP312-5313.8	03 11 12.	- 53 24 59.	18	17.0	GALAXY
LB 01659	03 11 12.	- 56 00		13.8	FAINT BLUE STAR
IC 0307	03 11 13.	- 00 23 41.			NONSTELLAR OBJECT
BAK 1.218	03 11 14.	- 28 25 10.	17	16.4	EXTRAGALACTIC NEBULA
SHAP312-5014.6	03 11 14.	- 50 25 46.	12	17.5	GALAXY
SHAP312-5356.6	03 11 14.	- 54 07 47.	12	17.0	GALAXY
MCG-04-08-049	03 11 15.	- 25 03	48	15.	GALAXY

OBJECT NAME	RIGHT ASCEN.	DECLINATION	DIAM.	MAGN.	TYPE OF OBJECT
SHAP312-5013.6	03 11 16.	- 50 24 46.	30	17.25	GALAXY
SHAP312-5205.6	03 11 16.	- 52 16 46.	42	17.0	GALAXY
BAK 1.220	03 11 17.	- 28 39 45.	13	16.9	EXTRAGALACTIC NEBULA
ZWG 390.029	03 11 18.	+ 00 31		14.8	GALAXY
MCG+03-09-008	03 11 18.	+ 16 18	66	14.5	GALAXY
UGC 02601	03 11 18.	+ 39 36	78	17.	GALAXY DWARF
PHL 8596	03 11 18.	- 09 16		18.4	BLUE STELLAR OBJECT
MCG-03-09-010	03 11 18.	- 15 26 30.	36	14.5	GALAXY
SHAP312-5136.6	03 11 18.	- 51 47 46.	18	17.0	GALAXY
LB 01660	03 11 18.	- 52 11		14.0	FAINT BLUE STAR
BAK 1.221	03 11 19.	- 25 34 09.	22	15.8	EXTRAGALACTIC NEBULA
ZWG 464.009	03 11 24.	+ 16 18		15.0	GALAXY
UGC 02602	03 11 24.	+ 16 18	96	15.0	GALAXY SBc
MCG+03-09-009	03 11 24.	+ 16 21	15	16.	GALAXY
MCG+06-08-004	03 11 24.	+ 39 25	48	14.5	GALAXY
MCG+07-07-036	03 11 24.	+ 42 02	15	14.5	GALAXY
UGC 02603	03 11 24.	+ 81 10	114	17.	GALAXY DWRF IR
MCG+00-09-028	03 11 24.	- 01 31	30	16.	GALAXY
PHL 4355	03 11 24.	- 17 40		17.9	BLUE STELLAR OBJECT
MCG-04-08-050	03 11 24.	- 25 55	270	11.	GALAXY
RNGC 1255	03 11 25.	- 25 59		12.0	GALAXY
SHAP312-5056.5	03 11 26.	- 51 07 40.	12	17.0	GALAXY
BAK 1.222	03 11 27.	- 24 42 57.	18	16.8	EXTRAGALACTIC NEBULA
IC 0301	03 11 28.9	+ 42 02 12.			GALAXY SAO
BAK 1.224	03 11 29.	- 25 21 57.	25	14.6	EXTRAGALACTIC NEBULA
BAK 1.223	03 11 29.	- 26 47 21.	18	16.8	EXTRAGALACTIC NEBULA
ZWG 525.009	03 11 30.	+ 39 27		14.8	GALAXY
UGC 02604	03 11 30.	+ 39 27	96	14.8	GALAXY SBc
UGC 02605	03 11 30.	+ 40 06	90	18.	GALAXY DWARF
ZWG 540.063	03 11 30.	+ 42 02		15.2	GALAXY
UGC 02606	03 11 30.	+ 42 02	102	15.2	GALAXY E
LDN 1379	03 11 30.	+ 60 10	180		DARK NEBULA
TON-S 0329	03 11 30.	- 23 04		14.1	BLUE STAR
SHAP312-5346.2	03 11 30.	- 53 57 22.	30	16.5	GALAXY
RNGC 1251	03 11 31.	+ 01 16			NON-EXISTENT OBJECT
SG 3.017	03 11 31.	+ 60 38	1200		DIFFUSE EMISSION NEBULA
MAI 002	03 11 31.	+ 81 11	47		DWARF SPHEROIDAL GALAXY
MCG+04-08-013	03 11 33.	+ 23 58	30	16.	GALAXY
MCG+00-09-029	03 11 33.	- 01 49 30.	66	14.5	GALAXY
SHAP313-5035.3	03 11 33.	- 50 46 27.	60	16.25	GALAXY
SHAP313-5035.6	03 11 33.	- 50 46 45.	18	17.5	GALAXY
HN 0171	03 11 33.	- 50 47			NEBULA
IC 1903	03 11 33.	- 50 47			NONSTELLAR OBJECT
SHAP313-4928.4	03 11 34.	- 49 39 33.	24	16.5	GALAXY
SHAP312-5430.6	03 11 34.	- 54 41 46.	18	17.5	GALAXY
SHAP312-5438.7	03 11 34.	- 54 49 52.	18	17.0	GALAXY
REIN 2.039A	03 11 34.90	- 03 00 46.2			NEBULA
REIN 2.039B	03 11 34.92	- 03 00 57.6			NEBULA
HN 0024	03 11 35.	+ 01 16			NEBULA
SHAP313-5406.2	03 11 35.	- 54 17 22.	24	17.0	GALAXY
5ZW 333	03 11 36.	+ 25 24			COMPACT GALAXY
ZWG 390.030	03 11 36.	- 01 49		14.8	GALAXY
UGC 02607	03 11 36.	- 01 49	102	14.8	GALAXY SBb
MCG+01-09-018	03 11 36.	- 03 00	264	13.	GALAXY
PHL 4356	03 11 36.	- 08 28		18.6	BLUE STELLAR OBJECT
PHL 1488	03 11 36.	- 10 18		18.6	BLUE STELLAR OBJECT
MCG-03-09-011	03 11 36.	- 18 24	48	15.	GALAXY
MCG-04-08-051	03 11 36.	- 25 22 30.	258	13.	GALAXY
SHAP313-5033.1	03 11 36.	- 50 44 15.	24	16.5	GALAXY
RNGC 1253	03 11 38.	- 03 00		13.0	GALAXY
SHAP313-5021.2	03 11 38.	- 50 32 21.	18	16.5	GALAXY
REIN 2.040A	03 11 38.15	- 03 00 31.8			NEBULA
REIN 2.040B	03 11 38.44	- 03 00 32.9			NEBULA
RNGC 1258	03 11 41.	- 21 57		13.0	GALAXY
RNGC 1256	03 11 41.	- 22 10		14.0	GALAXY
MCG+00-09-030	03 11 42.	+ 00 37	30	16.	GALAXY
MCG+07-07-037	03 11 42.	+ 41 50	48	14.5	GALAXY
ZWG 540.064	03 11 42.	+ 41 51		14.0	GALAXY
UGC 02608	03 11 42.	+ 41 51	66	14.0	GALAXY SBb
LDN 1380	03 11 42.	+ 59 58	120		DARK NEBULA
MCG+00-09-031	03 11 42.	- 01 30	30	15.	GALAXY
ARP 279	03 11 42.	- 02 59			PECULIAR GALAXY
MCG-04-08-052	03 11 42.	- 22 10	60	14.5	GALAXY
SHAP313-5004.7	03 11 42.	- 50 15 51.	30	16.25	GALAXY
SHAP313-5217.2	03 11 42.	- 52 28 21.	24	16.5	GALAXY
FATH 1.129	03 11 44.	+ 14 44	27		NEBULA
BAK 1.225	03 11 44.	- 30 20 20.	26	16.0	EXTRAGALACTIC NEBULA
SHAP313-5321.5	03 11 44.	- 53 32 39.	18	17.5	GALAXY
MCG-04-08-053	03 11 45.	- 21 57 30.	78	13.	GALAXY
SHAP313-5113.0	03 11 45.	- 51 24 09.	12	16.5	GALAXY
BAK 1.226	03 11 46.	- 24 43 20.	18	16.9	EXTRAGALACTIC NEBULA
SHAP313-5018.4	03 11 46.	- 50 29 33.	30	17.25	GALAXY
SHAP313-5414.6	03 11 46.	- 54 25 45.	36	16.25	GALAXY
BAK 1.227	03 11 47.	- 23 30 26.	14	17.0	EXTRAGALACTIC NEBULA
ZWG 390.032	03 11 48.	+ 02 30		15.5	GALAXY
RNGC 1254	03 11 48.	+ 02 30		15.5	GALAXY
ZC 0311.8+0940	03 11 48.	+ 09 40	4370		CLUSTER OF GALAXIES
ZWG 464.010	03 11 48.	+ 16 05		15.5	GALAXY
ZWG 525.010	03 11 48.	+ 37 43		14.8	GALAXY
UGC 02609	03 11 48.	+ 37 43	78	14.8	GALAXY Sb
ZWG 525.011	03 11 48.	+ 39 11		15.7	GALAXY
UGC 02610	03 11 48.	+ 39 11	72	15.7	GALAXY Sb
5ZW 334	03 11 48.	+ 49 05			COMPACT GALAXY
7ZW 005	03 11 48.	+ 76 48			COMPACT GALAXY
ZWG 390.031	03 11 48.	- 01 29		15.4	GALAXY
PHL 8597	03 11 48.	- 13 37		15.7	BLUE STELLAR OBJECT
SHAP313-5022.9	03 11 48.	- 50 34 03.	30	16.25	GALAXY
SHAP313-5320.2	03 11 49.	- 53 31 21.	18	17.5	GALAXY
IC 0304	03 11 50.	+ 37 41 49.			NONSTELLAR OBJECT
MCG+00-09-032	03 11 51.	- 02 06 30.	15	14.5	GALAXY
LB 00479	03 11 51.	- 02 21 06.		16.5	FAINT BLUE STAR
MCG+01-09-019	03 11 51.	- 02 59	54	15.	GALAXY
MCG+01-09-020	03 11 51.	- 06 17	42	15.	GALAXY
SHAP313-5225.0	03 11 51.	- 52 36 09.	18	17.5	GALAXY
IC 0305	03 11 52.	+ 37 40 31.			NONSTELLAR OBJECT
SHAP313-5018.0	03 11 52.	- 50 29 08.	18	17.25	GALAXY
MCG+00-09-033	03 11 54.	+ 02 33	21	14.	GALAXY
MCG+06-08-005	03 11 54.	+ 37 40	60	15.	GALAXY
ZWG 525.012	03 11 54.	+ 37 41		15.7	GALAXY
MCG+07-07-038	03 11 54.	+ 41 47	36	15.	GALAXY
ZWG 540.065	03 11 54.	+ 41 48		15.4	GALAXY
UGC 02612	03 11 54.	+ 41 48	66	15.4	GALAXY Sc
MCG+07-07-039	03 11 54.	+ 42 30	48	15.	GALAXY
ZWG 390.033	03 11 54.	- 02 05		14.8	GALAXY
UGC 02611	03 11 54.	- 02 05	72	14.8	GALAXY S0
PHL 4357	03 11 54.	- 10 01		18.7	BLUE STELLAR OBJECT
TON-S 0330	03 11 54.	- 22 52		14.0	BLUE STAR
SHAP313-4753.2	03 11 54.	- 48 04 20.	48	17.0	GALAXY
BAK 1.229	03 11 56.	- 23 50 19.	22	16.8	EXTRAGALACTIC NEBULA
BAK 1.228	03 11 56.	- 25 44 31.	23	16.7	EXTRAGALACTIC NEBULA
SHAP313-5323.8	03 11 56.	- 53 34 56.	18	17.5	GALAXY
SHAP313-5431.1	03 11 56.	- 54 42 14.	18	17.5	GALAXY
RNGC 1250	03 11 57.	+ 41 10		14.0	GALAXY
SHAP313-5007.3	03 11 57.	- 50 18 26.	36	16.5	GALAXY
SHAP313-4847.6	03 11 58.	- 48 58 44.	66	17.5	GALAXY
SHAP313-5225.6	03 11 59.	- 52 36 44.	18	17.0	GALAXY
SHAP313-5325.8	03 11 59.	- 53 36 56.	18	17.0	GALAXY
LBN 0680	03 12	+ 56 55	420		BRIGHT NEBULA
MCG+06-08-006	03 12 00.	+ 37 39	24	15.	GALAXY
MCG+07-07-040	03 12 00.	+ 41 09	60	14.5	GALAXY
ZWG 540.066	03 12 00.	+ 41 10		14.2	GALAXY
CR 03	03 12 00.	+ 41 10		14.2	GALAXY IN PERSEUS CLUSTER
UGC 02613	03 12 00.	+ 41 10	150	14.2	GALAXY S0
ZWG 540.067	03 12 00.	+ 41 25		15.3	GALAXY
ZWG 540.068	03 12 00.	+ 42 30		14.8	GALAXY
UGC 02614	03 12 00.	+ 42 30	120	14.8	GALAXY S0-a
MCG-02-09-017	03 12 00.	- 09 38	48	15.5	GALAXY
MCG-02-09-016	03 12 00.	- 10 57	48	15.	GALAXY
BAK 1.230	03 12 00.	- 25 42 43.	23	16.8	EXTRAGALACTIC NEBULA
SHAP313-5133.5	03 12 01.	- 51 44 38.	18	17.0	GALAXY
SHAP313-5212.6	03 12 01.	- 52 23 44.	12	16.5	GALAXY
SHAP313-5306.7	03 12 01.	- 53 17 50.	18	17.75	GALAXY
SHAP313-5437.1	03 12 01.	- 54 48 14.	18	17.25	GALAXY
FATH 1.130	03 12 02.	+ 15 01	11		NEBULA
SHAP313-5011.3	03 12 02.	- 50 22 26.	18	17.25	GALAXY
SHAP313-4748.4	03 12 03.	- 47 59 32.	48	16.5	GALAXY
SHAP313-5112.0	03 12 03.	- 51 23 08.	18	16.75	GALAXY
SHAP313-5225.0	03 12 03.	- 52 36 08.	12	18.0	GALAXY
SHAP313-5044.6	03 12 04.	- 50 55 44.	18	16.5	GALAXY
SHAP313-5059.2	03 12 04.	- 51 10 20.	24	17.0	GALAXY
MCG+00-09-035	03 12 06.	+ 01 34	36	16.	GALAXY
MCG+00-09-034	03 12 06.	+ 03 00 30.	42	15.	GALAXY
UGC 02615	03 12 06.	+ 30 45	72	16.0	GALAXY SB0-a
MCG+00-09-036	03 12 06.	- 00 56	30	15.	GALAXY
MCG-02-09-018	03 12 06.	- 09 08	42	15.	GALAXY
PHL 8598	03 12 06.	- 09 24		18.5	BLUE STELLAR OBJECT
SHAP313-5109.1	03 12 06.	- 51 20 14.	12	16.5	GALAXY
BAK 1.231	03 12 07.	- 30 37 49.	12	16.4	EXTRAGALACTIC NEBULA
SHAP313-5133.5	03 12 07.	- 51 44 38.	18	16.0	GALAXY
SHAP313-5207.2	03 12 07.	- 52 18 20.	36	17.0	GALAXY
VDB.66N 010	03 12 09.	+ 56 57	960		REFLECTION NEBULA
SHAP313-5021.0	03 12 09.	- 50 32 08.	18	16.5	GALAXY
ZWG 390.035	03 12 12.	+ 03 00		15.6	GALAXY
ZC 0312.2+1551	03 12 12.	+ 15 51	9410		CLUSTER OF GALAXIES
5ZW 336	03 12 12.	+ 30 45			COMPACT GALAXY
ZWG 390.034	03 12 12.	- 00 55		15.6	GALAXY
PHL 8600	03 12 12.	- 03 18		16.9	BLUE STELLAR OBJECT
PHL 8599	03 12 12.	- 04 17		16.3	BLUE STELLAR OBJECT
MCG+01-09-021	03 12 12.	- 04 58	108	15.	GALAXY
PHL 4358	03 12 12.	- 12 31		17.5	BLUE STELLAR OBJECT
SHAP313-4717.1	03 12 12.	- 47 28 13.	36	17.5	GALAXY
SHAP313-5022.4	03 12 13.	- 50 33 31.	24	17.0	GALAXY
SHAP313-5317.0	03 12 13.	- 53 28 07.	18	17.5	GALAXY
SHAP313-5002.5	03 12 16.	- 50 13 37.	84	16.0	GALAXY
SHAP313-5221.0	03 12 17.	- 52 32 07.	18	17.0	GALAXY
OCL 0375	03 12 18.	+ 59 51	300		OPEN STAR CLUSTER
PHL 4359	03 12 18.	- 16 07		18.4	BLUE STELLAR OBJECT
BAK 1.232	03 12 19.	- 29 01 18.	14	16.6	EXTRAGALACTIC NEBULA
SHAP314-4719.1	03 12 21.	- 47 30 13.	42	17.0	GALAXY
SHAP313-5132.9	03 12 23.	- 51 44 01.	24	16.5	GALAXY
SHAP313-5333.5	03 12 24.	- 53 44 37.	12	17.0	GALAXY
SHAP313-5333.4	03 12 26.	- 53 44 31.	18	17.25	GALAXY
BAK 1.234	03 12 27.	- 23 39 17.	22	16.4	EXTRAGALACTIC NEBULA
SHAP314-5004.8	03 12 27.	- 50 15 55.	30	16.5	GALAXY
ZWG 464.011	03 12 30.	+ 17 46		15.6	GALAXY
UGC 02616	03 12 30.	+ 41 40	78	16.5	GALAXY E-S0
LDN 1381	03 12 30.	+ 60 02	300		DARK NEBULA
PHL 8601	03 12 30.	- 06 27		18.4	BLUE STELLAR OBJECT
PHL 8602	03 12 30.	- 08 02		18.5	BLUE STELLAR OBJECT
BAK 1.233	03 12 31.	- 30 12 17.	14	16.8	EXTRAGALACTIC NEBULA
SHAP314-5045.1	03 12 32.	- 50 56 12.	18	16.0	GALAXY
SHAP314-5109.4	03 12 33.	- 51 20 30.	18	16.5	GALAXY
BAK 1.235	03 12 34.	- 29 06 53.	18	16.2	EXTRAGALACTIC NEBULA
SHAP314-4721.2	03 12 35.	- 47 32 18.	36	17.5	GALAXY
ZWG 390.036	03 12 36.	+ 02 21		15.7	GALAXY
MCG+03-09-010	03 12 36.	+ 16 17	42	16.	GALAXY
DG 012	03 12 36.	+ 57 04	600		REFLECTION NEBULA
PHL 4360	03 12 36.	- 07 08		17.8	BLUE STELLAR OBJECT
PHL 1489	03 12 36.	- 08 26		17.0	BLUE STELLAR OBJECT
PHL 8603	03 12 36.	- 12 49		16.1	BLUE STELLAR OBJECT
BAK 1.236	03 12 36.	- 25 02 47.	18	16.8	EXTRAGALACTIC NEBULA
BAK 1.237	03 12 36.	- 25 03 05.	18	16.5	EXTRAGALACTIC NEBULA
SHAP314-5046.6	03 12 37.	- 50 57 42.	18	16.5	GALAXY
SHAP314-5055.1	03 12 38.	- 51 06 12.	12	16.75	GALAXY
SHAP314-5141.5	03 12 38.	- 51 52 36.	12	17.25	GALAXY
MCG+03-09-011	03 12 39.	+ 15 59	24	17.	GALAXY
SHAP314-5209.5	03 12 39.	- 52 20 36.	12	17.0	GALAXY
SHAP314-5322.3	03 12 39.	- 53 33 24.	18	16.75	GALAXY
BAK 1.238	03 12 40.	- 26 29 11.	25	16.8	EXTRAGALACTIC NEBULA
MCG+07-07-041	03 12 42.	+ 40 41	120	15.	GALAXY
ZWG 540.069	03 12 42.	+ 40 43		14.3	GALAXY
UGC 02617	03 12 42.	+ 40 43	180	14.3	GALAXY Sc
MCG+07-07-042	03 12 42.	+ 41 52 30.	60	15.	GALAXY
ZWG 540.070	03 12 42.	+ 41 53		14.9	GALAXY
UGC 02618	03 12 42.	+ 41 53	90	14.9	GALAXY Sa-b
ZWG 540.071	03 12 42.	+ 42 44		15.6	GALAXY
BAK 1.239	03 12 43.	- 24 20 05.	19	17.1	EXTRAGALACTIC NEBULA
SHAP314-5041.4	03 12 43.	- 50 52 30.	36	16.5	GALAXY
SHAP314-5044.4	03 12 43.	- 50 55 30.	24	16.5	GALAXY
IC 1900	03 12 44.	+ 36 58 08.			NONSTELLAR OBJECT
BAK 1.240	03 12 44.	- 25 37 35.	17	16.9	EXTRAGALACTIC NEBULA
SHAP314-5414.6	03 12 44.	- 54 25 42.	18	17.0	GALAXY
FATH 1.131	03 12 45.	+ 15 21	27		NEBULA
BAK 1.241	03 12 45.	- 26 44 29.	19	16.7	EXTRAGALACTIC NEBULA
LB 02853	03 12 46.	+ 00 50 24.		16.6	FAINT BLUE STAR
SHAP314-5321.5	03 12 46.	- 53 32 36.	12	17.5	GALAXY
SHAP314-5054.3	03 12 47.	- 51 05 23.	18	17.0	GALAXY
MCG+06-08-007	03 12 48.	+ 36 57	36	15.	GALAXY
ZWG 525.013	03 12 48.	+ 36 59		15.0	GALAXY
MCG+07-07-043	03 12 48.	+ 40 36	36	15.	GALAXY
ZWG 540.072	03 12 48.	+ 40 38		14.9	GALAXY
CR 04	03 12 48.	+ 40 38		14.9	GALAXY IN PERSEUS CLUSTER
VB 190	03 12 48.	+ 61 01	295		STELLAR RING
MCG+13-03-004	03 12 48.	+ 80 04	42	15.	GALAXY
PHL 8604	03 12 48.	- 03 20		18.0	BLUE STELLAR OBJECT
PHL 8605	03 12 48.	- 05 48		18.7	BLUE STELLAR OBJECT

OBJECT NAME	RIGHT ASCEN.	DECLINATION	DIAM.	MAGN.	TYPE OF OBJECT
SHAP314-5357.0	03 12 48.	- 54 08 06.	12	16.5	GALAXY
SHAP314-4801.5	03 12 49.	- 48 12 35.	42	17.0	GALAXY
SHAP314-5354.7	03 12 49.	- 54 05 48.	12	17.0	GALAXY
IC 0309	03 12 49.2	+ 40 37 11.			GALAXY SA(s)
SHAP314-5048.3	03 12 51.	- 50 59 23.	18	17.0	GALAXY
IC 1901	03 12 52.	+ 36 55 37.			NONSTELLAR OBJECT
HN 0172	03 12 52.	- 30 53			NEBULA
IC 1904	03 12 52.	- 30 53			NONSTELLAR OBJECT
SHAP314-5337.8	03 12 52.	- 53 48 53.	18	17.5	GALAXY
ZWG 525.014	03 12 54.	+ 36 56		15.6	GALAXY
MCG-05-08-024	03 12 54.	- 30 55	72	14.	GALAXY
TON-S 0331	03 12 54.	- 31 35		15.8	BLUE STAR
SHAP314-5406.9	03 12 54.	- 54 17 59.	12	17.0	GALAXY
BAK 1.242	03 12 55.	- 29 43 04.	18	16.8	EXTRAGALACTIC NEBULA
SHAP314-5113.8	03 12 55.	- 51 24 53.	12	16.5	GALAXY
FATH 1.132	03 12 56.	+ 14 52	14		NEBULA
BAK 1.243	03 12 56.	- 28 22 10.	17	16.7	EXTRAGALACTIC NEBULA
IC 0308	03 12 58.6	+ 40 59 46.			GALAXY SB(s)
BAK 1.244	03 12 59.	- 28 22 04.	22	16.5	EXTRAGALACTIC NEBULA
RNGC 1269	03 12 59.	- 41 16			NON-EXISTENT OBJECT
ZWG 390.037	03 13 00.	+ 02 20		15.6	GALAXY
MCG+01-09-003	03 13 00.	+ 09 00	60	15.	GALAXY
MCG+06-08-008	03 13 00.	+ 36 54 30.	36	16.	GALAXY
ZWG 525.015	03 13 00.	+ 37 00		15.7	GALAXY
CR 05	03 13 00.	+ 41 00			GALAXY IN PERSEUS CLUSTER
UGC 02619	03 13 00.	+ 41 00	96	16.5	GALAXY S0
OCL 0376	03 13 00.	+ 58 25	420		OPEN STAR CLUSTER
LDN 1382	03 13 00.	+ 59 53	360		DARK NEBULA
ZWG 346.003	03 13 00.	+ 80 04		15.5	GALAXY
UGC 02620	03 13 00.	+ 80 04	138	15.5	GALAXY S(B)IV?
MCG+00-09-037	03 13 00.	- 01 13	42	15.	GALAXY
PHL 1490	03 13 00.	- 04 35		18.0	BLUE STELLAR OBJECT
MCG-03-09-012	03 13 00.	- 18 06	30	15.5	GALAXY
PHL 4361	03 13 00.	- 18 32		18.6	BLUE STELLAR OBJECT
MCG-03-09-013	03 13 00.	- 19 10	36	15.	GALAXY
BAK 1.245	03 13 00.	- 29 20 52.	13	16.6	EXTRAGALACTIC NEBULA
SHAP314-5213.6	03 13 00.	- 52 24 41.	24	17.25	GALAXY
IC 1902	03 13 01.	+ 36 59 30.			NONSTELLAR OBJECT
SHAP314-4723.2	03 13 01.	- 47 34 17.	48	17.25	GALAXY
RNGC 1257	03 13 03.	+ 41 20		14.5	GALAXY
BAK 1.246	03 13 03.	- 29 26 04.	17	16.7	EXTRAGALACTIC NEBULA
SHAP314-5258.2	03 13 05.	- 53 09 17.	12	17.5	GALAXY
MCG+07-07-044	03 13 06.	+ 41 20	60	15.	GALAXY
ZWG 390.038	03 13 06.	- 01 12		15.5	GALAXY
MRK 605	03 13 06.	- 03 38	10	16.5	GALAXY WITH UV CONTINUUM
SHAP314-5057.8	03 13 07.	- 51 08 52.	18	15.5	GALAXY
SHAP314-5214.7	03 13 07.	- 52 25 46.	12	17.0	GALAXY
FATH 1.133	03 13 08.	+ 15 43	11		NEBULA
SHAP314-5322.7	03 13 08.	- 53 33 47.	12	17.0	GALAXY
WK 001	03 13 08.23	+ 41 20 49.3			NEBULA
SHAP314-5323.4	03 13 09.	- 53 34 28.	12	16.25	GALAXY
WK 002	03 13 09.63	+ 41 26 35.9			NEBULA
SHAP314-5225.8	03 13 10.	- 52 36 52.	30	16.75	GALAXY
BAK 1.249	03 13 11.	- 23 24 39.	13	17.4	EXTRAGALACTIC NEBULA
ZWG 540.073	03 13 12.	+ 41 21		14.7	GALAXY
CR 06A	03 13 12.	+ 41 21		14.7	GALAXY IN PERSEUS CLUSTER
UGC 02621	03 13 12.	+ 41 21	102	14.7	GALAXY Sa
ZWG 540.074	03 13 12.	+ 41 27		15.7	GALAXY
PHL 1491	03 13 12.	- 04 05		18.2	BLUE STELLAR OBJECT
PHL 4362	03 13 12.	- 08 39		17.6	BLUE STELLAR OBJECT
PHL 1492	03 13 12.	- 09 23		16.9	BLUE STELLAR OBJECT
PHL 8708	03 13 12.	- 23 47		18.1	BLUE STELLAR OBJECT
SHAP314-5002.2	03 13 12.	- 50 13 16.	24	16.75	GALAXY
SHAP314-5057.6	03 13 12.	- 51 08 40.	18	17.5	GALAXY
BAK 1.247	03 13 13.	- 29 14 03.	18	16.8	EXTRAGALACTIC NEBULA
SHAP314-5321.4	03 13 13.	- 53 32 28.	24	16.25	GALAXY
SHAP314-5337.4	03 13 13.	- 53 48 28.	18	17.0	GALAXY
BAK 1.248	03 13 14.	- 27 21 39.	13	16.4	EXTRAGALACTIC NEBULA
FATH 1.134	03 13 15.	+ 15 25	11		NEBULA
SHAP314-5204.7	03 13 15.	- 52 15 46.	18	16.5	GALAXY
SHAP314-5311.8	03 13 15.	- 53 22 52.	12	17.0	GALAXY
SHAP314-5139.7	03 13 16.	- 51 50 46.	12	16.5	GALAXY
MCG+00-09-038	03 13 18.	+ 01 27 30.	36	15.	GALAXY
UGC 02622	03 13 18.	+ 08 59	96	16.0	GALAXY SBb
ZWG 441.004	03 13 18.	+ 15 25		15.7	GALAXY
ZWG 525.016	03 13 18.	+ 34 53		15.7	GALAXY
UGC 02623	03 13 18.	+ 34 53	162	15.7	GALAXY SBc
MCG+01-09-022	03 13 18.	- 03 48	42	14.	GALAXY
PHL 4363	03 13 18.	- 09 06		18.6	BLUE STELLAR OBJECT
MCG-03-09-014	03 13 18.	- 16 04	48	14.5	GALAXY
BAK 1.250	03 13 19.	- 22 59 51.	16	16.1	EXTRAGALACTIC NEBULA
RNGC 1262	03 13 20.	- 16 04		14.0	GALAXY
BAK 1.251	03 13 20.	- 23 00 09.	8	16.2	EXTRAGALACTIC NEBULA
SHAP314-5302.2	03 13 23.	- 53 13 16.	24	17.0	GALAXY
SHAP314-5442.6	03 13 23.	- 54 53 40.	18	17.25	GALAXY
SHAP314-5443.8	03 13 23.	- 54 54 52.	24	17.0	GALAXY
ZWG 390.039	03 13 24.	+ 01 28		15.7	GALAXY
ZWG 441.005	03 13 24.	+ 15 25		15.6	GALAXY
IC 0311	03 13 24.	+ 39 49 28.			NONSTELLAR OBJECT
MCG+07-07-045	03 13 24.	+ 41 07 30.	15	14.5	GALAXY
ZWG 540.075	03 13 24.	+ 41 08		14.3	GALAXY
CR 06B	03 13 24.	+ 41 08		14.3	GALAXY IN PERSEUS CLUSTER
UGC 02624	03 13 24.	+ 41 08	108	14.3	GALAXY S0
BIGO 483	03 13 24.	+ 41 09			NEBULA
ZWG 346.004	03 13 24.	+ 75 54		15.6	GALAXY
MCG+13-03-005	03 13 24.	+ 75 54	36	16.	GALAXY
MCG+01-09-023	03 13 24.	- 02 35	36	14.	GALAXY
MCG-02-09-019	03 13 24.	- 12 12	120	14.	GALAXY
PHL 1493	03 13 24.	- 15 12		18.4	BLUE STELLAR OBJECT
MCG-03-09-015	03 13 24.	- 15 17	36	15.	GALAXY
PHL 1494	03 13 24.	- 16 55		17.3	BLUE STELLAR OBJECT
IC 0310	03 13 24.8	+ 41 08 33.			GALAXY SA(r)
FATH 1.135	03 13 25.	+ 15 25	14		NEBULA
WK 003	03 13 25.22	+ 41 08 30.7			NEBULA
RNGC 1266	03 13 26.	- 02 35		14.0	GALAXY
RNGC 1263	03 13 26.	- 15 17		15.0	GALAXY
SHAP314-5140.3	03 13 26.	- 51 51 21.	12	17.5	GALAXY
BAK 1.252	03 13 27.	- 23 51 26.	23	17.2	EXTRAGALACTIC NEBULA
SHAP314-5442.8	03 13 27.	- 54 53 52.	24	17.25	GALAXY
BAK 1.253	03 13 28.	- 23 35 50.	23	16.3	EXTRAGALACTIC NEBULA
SHAP314-5332.8	03 13 28.	- 53 43 51.	12	17.0	GALAXY
ZWG 540.076	03 13 30.	+ 39 50		15.7	GALAXY
UGC 02625	03 13 30.	+ 39 50	114	15.7	GALAXY S?
ZWG 540.077	03 13 30.	+ 40 31		15.5	GALAXY
MCG+00-09-039	03 13 30.	- 02 17	36	15.5	GALAXY
MCG+01-09-024	03 13 30.	- 05 41	102	13.5	GALAXY
PHL 8606	03 13 30.	- 06 06		6.0	BLUE STELLAR OBJECT
PHL 4364	03 13 30.	- 17 01		7.6	BLUE STELLAR OBJECT
BAK 1.254	03 13 31.	- 24 21 38.	13	16.4	EXTRAGALACTIC NEBULA
SHAP314-5257.2	03 13 31.	- 53 08 15.	18	17.0	GALAXY
ARC 0425	03 13 32.	- 11 56		17.8	RICH CLUSTER OF GALAXIES
MCG+06-08-009	03 13 33.	+ 34 51	120	15.	GALAXY
ZWG 390.040	03 13 36.	- 02 15		15.6	GALAXY
TON-S 0332	03 13 36.	- 23 15		14.2	BLUE STAR
BAK 1.255	03 13 36.	- 25 26 44.	17	16.4	EXTRAGALACTIC NEBULA
SHAP315-5127.2	03 13 37.	- 51 38 15.	12	16.5	GALAXY
SHAP315-5337.9	03 13 37.	- 53 48 57.	24	17.0	GALAXY
SHAP315-5037.4	03 13 38.	- 50 48 27.	12	17.25	GALAXY
RNGC 1259	03 13 39.	+ 41 10		15.5	GALAXY
BAK 1.256	03 13 40.	- 26 55 08.	25	16.2	EXTRAGALACTIC NEBULA
BAK 1.258	03 13 41.	- 27 21 38.	19	16.2	EXTRAGALACTIC NEBULA
SHAP315-5034.3	03 13 41.	- 50 45 20.	12	17.5	GALAXY
SHAP315-5241.3	03 13 41.	- 52 52 21.	12	17.5	GALAXY
WK 004	03 13 41.86	+ 41 10 26.6			NEBULA
ZWG 540.078	03 13 42.	+ 41 10		15.7	GALAXY
CR 07	03 13 42.	+ 41 10		15.7	GALAXY IN PERSEUS CLUSTER
UGC 02626	03 13 42.	+ 41 10	84	15.7	GALAXY Sa?
ZWG 540.079	03 13 42.	+ 41 27		15.0	GALAXY
CR 08	03 13 42.	+ 41 27		15.0	GALAXY IN PERSEUS CLUSTER
ZWG 540.080	03 13 42.	+ 43 20		15.7	GALAXY
BAK 1.257	03 13 42.	- 29 10 38.	17	16.6	EXTRAGALACTIC NEBULA
HPW 01	03 13 42.	- 41 19			DWARF GALAXY IN FORNAX
SHAP315-5241.0	03 13 42.	- 52 52 03.	12	17.75	GALAXY
ARC 0428	03 13 43.	- 19 18		16.5	RICH CLUSTER OF GALAXIES
SHAP315-5011.8	03 13 43.	- 50 22 50.	24	17.0	GALAXY
SHAP315-5034.0	03 13 43.	- 50 45 02.	12	17.5	GALAXY
SHAP315-5009.2	03 13 44.	- 50 20 14.	18	17.5	GALAXY
LB 00480	03 13 45.	+ 00 34 42.		15.8	FAINT BLUE STAR
SHAP315-4944.8	03 13 45.	- 49 55 50.	18	17.25	GALAXY
WK 005	03 13 45.25	+ 41 27 01.5			NEBULA
SHAP315-5017.9	03 13 46.	- 50 28 56.	18	17.25	GALAXY
HN 0174	03 13 46.	- 55 01			NEBULA
IC 1908	03 13 46.	- 55 01			NONSTELLAR OBJECT
BAK 1.259	03 13 47.	- 29 43 13.	19	16.2	EXTRAGALACTIC NEBULA
ZC 0313.8+0448	03 13 48.	+ 04 48	2220		CLUSTER OF GALAXIES
MCG+03-09-012	03 13 48.	+ 16 12	24	17.	GALAXY
ZC 0313.8+2118	03 13 48.	+ 21 18	1280		CLUSTER OF GALAXIES
MCG+05-08-008	03 13 48.	+ 31 23	84	14.5	GALAXY
PHL 8607	03 13 48.	- 08 27		17.7	BLUE STELLAR OBJECT
MCG-02-09-020	03 13 48.	- 10 51 30.	60	15.	GALAXY
MCG-02-09-021	03 13 48.	- 13 13 30.	66	14.5	GALAXY
PHL 1495	03 13 48.	- 18 56		18.5	BLUE STELLAR OBJECT
SHAP315-5009.2	03 13 48.	- 50 20 14.	24	17.0	GALAXY
SHAP315-5045.2	03 13 48.	- 50 56 14.	18	17.5	GALAXY
SHAP315-5102.6	03 13 48.	- 51 13 38.	12	16.5	GALAXY
SHAP315-5449.7	03 13 48.	- 55 00 44.	78	14.8	GALAXY
SHAP315-4953.1	03 13 50.	- 50 04 08.	18	17.0	GALAXY
BAK 1.260	03 13 51.	- 29 20 49.	19	16.6	EXTRAGALACTIC NEBULA
SHAP315-5258.8	03 13 51.	- 53 09 50.	24	16.5	GALAXY
SHAP315-5321.0	03 13 52.	- 53 32 02.	12	17.0	GALAXY
SHAP315-4936.0	03 13 53.	- 49 47 02.	30	17.0	GALAXY
SHAP315-5025.6	03 13 53.	- 50 36 38.	24	16.75	GALAXY
ZWG 506.007	03 13 54.	+ 31 23		14.9	GALAXY
UGC 02627	03 13 54.	+ 31 23	120	14.9	GALAXY Sc
KARA.72 088A	03 13 54.	+ 31 23	102	14.9	PART OF DOUBLE GALAXY
CR 09	03 13 54.	+ 41 28			GALAXY IN PERSEUS CLUSTER
PHL 4365	03 13 54.	- 09 11		18.8	BLUE STELLAR OBJECT
WK 006	03 13 54.76	+ 41 28 04.3			NEBULA
SHAP315-5443.0	03 13 56.	- 54 54 02.	42	16.0	GALAXY
MCG+07-07-046	03 13 57.	+ 41 11	9	16.	GALAXY
SHAP315-5044.2	03 13 57.	- 50 55 14.	12	16.5	GALAXY
SHAP315-5130.2	03 13 57.	- 51 41 14.	18	16.5	GALAXY
SHAP315-4940.4	03 13 58.	- 49 51 26.	18	16.75	GALAXY
BAK 1.262	03 13 59.	- 23 51 48.	14	17.0	EXTRAGALACTIC NEBULA
WK 007	03 13 59.28	+ 41 12 10.7			NEBULA
WK 008	03 13 59.67	+ 41 14 20.0			NEBULA
WK 009	03 13 59.97	+ 41 18 46.6			NEBULA
LBN 0686	03 14	+ 53 52	240		BRIGHT NEBULA
ZWG 506.008	03 14 00.	+ 31 24		15.6	GALAXY
UGC 02629	03 14 00.	+ 31 24	60	15.6	GALAXY SBc
KARA.72 088B	03 14 00.	+ 31 24	48	15.6	PART OF DOUBLE GALAXY
MCG+05-08-009	03 14 00.	+ 31 25	42	15.	GALAXY
ZWG 525.017	03 14 00.	+ 36 57		15.5	GALAXY
UGC 02630	03 14 00.	+ 36 57	60	15.5	GALAXY Sb-c
CR 10	03 14 00.	+ 41 12			GALAXY IN PERSEUS CLUSTER
LDN 1383	03 14 00.	+ 59 55	180		DARK NEBULA
ZWG 390.041	03 14 00.	- 00 39		14.9	GALAXY
UGC 02628	03 14 00.	- 00 39	102	14.9	GALAXY Sb-c
MCG+00-09-040	03 14 00.	- 00 40	90	14.5	GALAXY
PHL 4366	03 14 00.	- 07 18		15.4	BLUE STELLAR OBJECT
PHL 4367	03 14 00.	- 09 08		18.7	BLUE STELLAR OBJECT
LB 02854	03 14 01.	- 00 23 24.		15.4	FAINT BLUE STAR
SHAP315-4628.2	03 14 01.	- 46 39 13.	48	17.25	GALAXY
SHAP315-5350.6	03 14 01.	- 54 01 38.	18	17.25	GALAXY
LB 02855	03 14 02.	+ 00 01 00.		16.9	FAINT BLUE STAR
SHAP315-4524.8	03 14 03.	- 45 35 49.	42	17.25	GALAXY
SHAP315-5057.0	03 14 03.	- 51 08 01.	18	17.0	GALAXY
SHAP315-5258.1	03 14 03.	- 53 09 07.	36	17.0	GALAXY
SHAP315-5207.4	03 14 04.	- 52 18 25.	24	16.5	GALAXY
FATH 1.136	03 14 05.	+ 15 04	16		NEBULA
MCG+07-07-047	03 14 05.	+ 41 12 30.	42	15.	GALAXY
BAK 1.261	03 14 05.	- 29 29 24.	18	16.2	EXTRAGALACTIC NEBULA
HN 0173	03 14 05.	- 34 33			NEBULA
SHAP315-4629.4	03 14 05.	- 46 40 25.	42	17.5	GALAXY
SHAP315-5247.6	03 14 05.	- 52 58 37.	12	17.5	GALAXY
SHAP315-5439.9	03 14 05.	- 54 50 55.	24	16.5	GALAXY
UGC 02631	03 14 06.	+ 05 50	78	17.	GALAXY S
ZCG 0314+23.5	03 14 06.	+ 23 29		18.5	COMPACT GALAXY
MCG+06-08-010	03 14 06.	+ 36 56	27	17.	GALAXY
MCG+07-07-048	03 14 06.	+ 41 10 30.	8	17.	GALAXY
ZC 0314.1+4716	03 14 06.	+ 47 16	5580		CLUSTER OF GALAXIES
PHL 1496	03 14 06.	- 03 04		17.6	BLUE STELLAR OBJECT
MCG-06-08-001	03 14 06.	- 34 32	60	15.	GALAXY
IC 1906	03 14 06.	- 34 33			NONSTELLAR OBJECT
RNGC 1260	03 14 09.	+ 41 13		14.0	GALAXY
SHAP315-5130.5	03 14 09.	- 51 41 31.	24	15.75	GALAXY
WK 010	03 14 09.13	+ 41 13 20.4			NEBULA
WK 011	03 14 09.49	+ 41 25 47.8			NEBULA
WK 012	03 14 09.71	+ 41 18 19.6			NEBULA
LB 02856	03 14 10.	+ 01 13 06.		16.7	FAINT BLUE STAR
BAK 1.263	03 14 10.	- 30 17 48.	14	16.6	EXTRAGALACTIC NEBULA
SHAP315-5327.5	03 14 10.	- 53 38 31.	12	17.5	GALAXY
BAK 1.266	03 14 11.	- 22 40 30.	22	17.2	EXTRAGALACTIC NEBULA
BAK 1.264	03 14 11.	- 29 25 00.	18	16.6	EXTRAGALACTIC NEBULA
SHAP315-5257.9	03 14 11.	- 53 08 55.	12	16.25	GALAXY
WK 013	03 14 11.15	+ 41 11 34.6			NEBULA

OBJECT NAME	RIGHT ASCEN.	DECLINATION	DIAM.	MAGN.	TYPE OF OBJECT
ZCG 0314+23.3	03 14 12.	+ 23 27		18.8	COMPACT GALAXY
ZWG 525.018	03 14 12.	+ 36 02		15.7	GALAXY
UGC 02633	03 14 12.	+ 36 23	96	17.	GALAXY Sc
CR 13	03 14 12.	+ 41 11			GALAXY IN PERSEUS CLUSTER
ZWG 540.081	03 14 12.	+ 41 13		14.2	GALAXY
CR 12	03 14 12.	+ 41 13		14.2	GALAXY IN PERSEUS CLUSTER
UGC 02634	03 14 12.	+ 41 13	84	14.2	GALAXY S0-a
CR 11	03 14 12.	+ 41 18			GALAXY IN PERSEUS CLUSTER
MCG+00-09-041	03 14 12.	- 01 42 30.	42	16.	GALAXY
UGC 02632	03 14 12.	- 02 13	60	16.5	GALAXY
PHL 4369	03 14 12.	- 04 19		18.5	BLUE STELLAR OBJECT
PHL 8608	03 14 12.	- 05 50		16.0	BLUE STELLAR OBJECT
PHL 4368	03 14 12.	- 06 36		15.7	BLUE STELLAR OBJECT
MCG-04-08-054	03 14 12.	- 23 59	36	15.	GALAXY
MCG-04-08-055	03 14 12.	- 26 02	24	14.5	GALAXY
BAK 1.265	03 14 12.	- 26 02 30.	36	15.0	EXTRAGALACTIC NEBULA
SHAP315-5153.3	03 14 12.	- 52 04 19.	24	17.0	GALAXY
BAK 1.267	03 14 13.	- 23 59 24.	28	16.2	EXTRAGALACTIC NEBULA
WK 014	03 14 13.91	+ 41 13 45.5			NEBULA
SHAP315-5402.1	03 14 15.	- 54 13 07.	18	17.75	GALAXY
WK 015	03 14 15.02	+ 41 06 53.1			NEBULA
BAK 1.268	03 14 16.	- 25 06 36.	26	16.3	EXTRAGALACTIC NEBULA
LB 02857	03 14 17.	+ 00 30 36.		15.8	FAINT BLUE STAR
FATH 1.137	03 14 17.	+ 14 54	14		NEBULA
ZWG 441.006	03 14 18.	+ 15 18		14.7	GALAXY
MCG+02-09-002	03 14 18.	+ 15 18	42	14.5	GALAXY
FATH 1.138	03 14 18.	+ 15 19	35		NEBULA
ZC 0314.3+2355	03 14 18.	+ 23 55	1340		CLUSTER OF GALAXIES
MCG+06-08-011	03 14 18.	+ 36 00	36	16.	GALAXY
UGC 02636	03 14 18.	+ 36 52	108	17.	GALAXY Sc
UGC 02637	03 14 18.	+ 37 52	90	17.	GALAXY Sc
CR 14	03 14 18.	+ 41 13			GALAXY IN PERSEUS CLUSTER
UGC 02635	03 14 18.	- 01 43	60	16.5	GALAXY Sb-c
PHL 1497	03 14 18.	- 04 06		16.6	BLUE STELLAR OBJECT
PHL 8609	03 14 18.	- 09 30		16.1	BLUE STELLAR OBJECT
PHL 8610	03 14 18.	- 18 24		18.5	BLUE STELLAR OBJECT
BAK 1.269	03 14 18.	- 26 35 59.	25	16.7	EXTRAGALACTIC NEBULA
SHAP315-5148.6	03 14 20.	- 51 59 36.	12	16.75	GALAXY
MCG+00-09-042	03 14 21.	+ 01 04	54	15.5	GALAXY
FATH 1.139	03 14 22.	+ 14 43	11		NEBULA
BAK 1.272	03 14 22.	- 23 28 11.	25	16.7	EXTRAGALACTIC NEBULA
SHAP315-5325.8	03 14 22.	- 53 36 48.	30	16.25	GALAXY
SHAP315-5050.1	03 14 23.	- 51 01 06.	18	16.75	GALAXY
ZWG 390.043	03 14 24.	+ 03 29		15.7	GALAXY
ZWG 540.082	03 14 24.	+ 42 11		15.7	GALAXY
ZWG 390.042	03 14 24.	- 00 13		15.4	GALAXY
BAK 1.273	03 14 24.	- 24 30 23.	18	17.2	EXTRAGALACTIC NEBULA
SHAP315-5124.2	03 14 24.	- 51 35 12.	18	17.0	GALAXY
SHAP315-5318.8	03 14 24.	- 53 29 48.	18	17.5	GALAXY
SHAP315-5327.2	03 14 24.	- 53 38 12.	18	17.75	GALAXY
SHAP315-5116.8	03 14 25.	- 51 27 48.	18	17.0	GALAXY
SHAP315-5246.7	03 14 25.	- 52 57 42.	18	15.5	GALAXY
BAK 1.270	03 14 26.	- 31 18 23.	22	16.6	EXTRAGALACTIC NEBULA
WK 016	03 14 27.87	+ 41 20 19.4			NEBULA
SHAP316-4627.4	03 14 28.	- 46 38 24.	36	17.5	GALAXY
SHAP316-5056.4	03 14 28.	- 51 07 24.	12	17.25	GALAXY
WK 017	03 14 28.37	+ 41 20 27.7			NEBULA
UGC 02638	03 14 30.	+ 01 05	60	16.0	GALAXY Sa-b
MCG+00-09-043	03 14 30.	+ 03 25	18	15.	GALAXY
5ZW 336	03 14 30.	+ 23 29			COMPACT GALAXY
ZWG 540.083	03 14 30.	+ 41 47		15.6	GALAXY
UGC 02639	03 14 30.	+ 41 47	78	15.6	GALAXY Sa-b
ZWG 540.084	03 14 30.	+ 43 07		14.8	GALAXY
UGC 02640	03 14 30.	+ 43 07	78	14.8	GALAXY SBb
MCG+07-07-049	03 14 30.	+ 43 07	48	15.	GALAXY
PHL 1498	03 14 30.	- 12 39		18.4	BLUE STELLAR OBJECT
BAK 1.271	03 14 30.	- 31 09 59.	14	16.6	EXTRAGALACTIC NEBULA
SHAP316-5101.4	03 14 31.	- 51 12 24.	18	17.25	GALAXY
WK 018	03 14 31.19	+ 41 47 05.2			NEBULA
FATH 1.140	03 14 32.	+ 14 39	14		NEBULA
SHAP316-4629.0	03 14 32.	- 46 39 59.	42	17.0	GALAXY
SHAP315-5447.8	03 14 32.	- 54 58 48.	30	16.5	GALAXY
WK 019	03 14 32.87	+ 41 16 06.1			NEBULA
RNGC 1264	03 14 33.	+ 41 16		15.5	GALAXY
BAK 1.274	03 14 34.	- 31 09 59.	16	16.1	EXTRAGALACTIC NEBULA
SHAP316-5140.9	03 14 34.	- 51 51 54.	18	17.0	GALAXY
SHAP316-5257.9	03 14 35.	- 53 08 54.	12	16.5	GALAXY
UGC 02641	03 14 36.	+ 03 25	72	17.	GALAXY
ZCG 0314+23.6	03 14 36.	+ 23 29		18.2	COMPACT GALAXY
ZWG 540.085	03 14 36.	+ 41 16		15.4	GALAXY
CR 15	03 14 36.	+ 41 16		15.4	GALAXY IN PERSEUS CLUSTER
PHL 1499	03 14 36.	- 04 33		18.5	BLUE STELLAR OBJECT
SHAP316-5139.4	03 14 36.	- 51 50 24.	18	17.0	GALAXY
SHAP315-5449.2	03 14 36.	- 55 00 12.	24	17.5	GALAXY
SHAP316-5024.2	03 14 37.	- 50 35 11.	24	17.0	GALAXY
SHAP315-5447.7	03 14 37.	- 54 58 42.	30	18.0	GALAXY
MCG+07-07-050	03 14 39.	+ 41 19	36	15.	GALAXY
SHAP316-5057.0	03 14 39.	- 51 07 59.	24	16.0	GALAXY
SHAP316-5057.8	03 14 39.	- 51 08 47.	24	15.5	GALAXY
SHAP316-5437.6	03 14 39.	- 54 48 36.	24	17.0	GALAXY
WK 020	03 14 39.25	+ 41 18 52.7			NEBULA
FATH 1.141	03 14 40.	+ 15 48	14		NEBULA
WK 021	03 14 41.31	+ 41 20 17.0			NEBULA
ZWG 390.044	03 14 42.	+ 03 26		15.2	GALAXY
ZCG 0314+23.1	03 14 42.	+ 23 25		17.4	COMPACT GALAXY
UGC 02642	03 14 42.	+ 40 45	60	16.5	GALAXY SBb
CR 16	03 14 42.	+ 41 19			GALAXY IN PERSEUS CLUSTER
CR 17	03 14 42.	+ 41 20			GALAXY IN PERSEUS CLUSTER
UGC 02643	03 14 42.	+ 41 20	72	16.0	GALAXY SBa-b
VB 232	03 14 42.	+ 52 34	1141		STELLAR RING
MCG+00-09-044	03 14 42.	- 00 16	21	16.	GALAXY
WK 022	03 14 42.30	+ 41 40 14.7			NEBULA
WK 024	03 14 42.82	+ 41 07 02.2			NEBULA
BAK 1.276	03 14 43.	- 29 07 16.	11	16.7	EXTRAGALACTIC NEBULA
WK 023	03 14 43.37	+ 41 39 26.1			NEBULA
SHAP316-5057.4	03 14 44.	- 51 08 23.	12	17.0	GALAXY
SHAP316-5325.8	03 14 44.	- 53 36 47.	18	17.0	GALAXY
WK 025	03 14 44.47	+ 41 11 01.7			NEBULA
MCG+07-07-051	03 14 45.	+ 41 34	36	15.	GALAXY
SN 1964B	03 14 47.	+ 40 11		16.7	SUPERNOVA
SHAP316-5203.0	03 14 47.	- 52 13 59.	12	17.5	GALAXY
WK 026	03 14 47.71	+ 41 36 59.1			NEBULA
ZCG 0314+23.2	03 14 48.	+ 23 26		17.9	COMPACT GALAXY
ZWG 540.086	03 14 48.	+ 41 34		14.9	GALAXY
CR 18	03 14 48.	+ 41 34		14.9	GALAXY IN PERSEUS CLUSTER
UGC 02644	03 14 48.	+ 41 34	96	14.9	GALAXY E
MCG+09-06-004	03 14 48.	+ 51 13	42	19.	GALAXY
MRSL 140+01/1	03 14 48.	+ 59 27	10200		HII REGION
ZWG 390.045	03 14 48.	- 00 14		15.7	GALAXY
MCG+00-09-045	03 14 48.	- 00 18	60	14.5	GALAXY
MCG-04-08-056	03 14 48.	- 23 03	120	13.5	GALAXY
TON-S 0333	03 14 48.	- 23 59		15.1	BLUE STAR
SHAP316-5408.6	03 14 48.	- 54 19 35.	18	17.75	GALAXY
SHAP316-5100.0	03 14 49.	- 51 10 59.	12	17.25	GALAXY
SHAP316-5139.9	03 14 49.	- 51 50 53.	24	16.75	GALAXY
WK 027	03 14 49.86	+ 41 34 19.6			NEBULA
SHAP316-5053.1	03 14 50.	- 51 04 05.	12	16.5	GALAXY
IC 0312	03 14 50.1	+ 41 34 18.			GALAXY E6
SHAP316-5055.6	03 14 51.	- 51 06 35.	30	16.5	GALAXY
SHAP316-5207.8	03 14 51.	- 52 18 47.	60	16.5	GALAXY
BAK 1.280	03 14 53.	- 23 02 46.	37	14.5	EXTRAGALACTIC NEBULA
SHAP316-5100.5	03 14 53.	- 51 11 29.	6	17.5	GALAXY
SHAP316-5343.6	03 14 53.	- 53 54 35.	12	17.0	GALAXY
WK 028	03 14 53.33	+ 41 10 49.7			NEBULA
ZCG 0314+23.4	03 14 54.	+ 23 27		18.6	COMPACT GALAXY
UGC 02646	03 14 54.	+ 40 11	72	16.5	GALAXY S0
MCG+07-07-052	03 14 54.	+ 41 40	18	14.	GALAXY
SG 3.018	03 14 54.	+ 60 01	1800		DIFFUSE EMISSION NEBULA
ZWG 390.046	03 14 54.	- 00 15		15.7	GALAXY
UGC 02645	03 14 54.	- 00 15	78	15.7	GALAXY Sb
PHL 4370	03 14 54.	- 06 34		13.9	BLUE STELLAR OBJECT
SHAP316-5054.4	03 14 54.	- 51 05 23.	12	17.25	GALAXY
BAK 1.277	03 14 55.	- 31 07 10.	23	16.6	EXTRAGALACTIC NEBULA
SHAP316-5201.0	03 14 56.	- 52 11 59.	12	17.5	GALAXY
SHAP316-5404.2	03 14 56.	- 54 15 11.	24	17.25	GALAXY
BAK 1.278	03 14 57.	- 31 33 58.	19	16.1	EXTRAGALACTIC NEBULA
SHAP316-4832.6	03 14 57.	- 48 43 34.	30	16.5	GALAXY
WK 028	03 14 57.07	+ 41 40 33.4			NEBULA
WK 029	03 14 57.76	+ 41 33 09.3			NEBULA
SHAP316-5054.6	03 14 58.	- 51 05 34.	12	16.5	GALAXY
BAK 1.279	03 14 59.	- 30 52 09.	16	16.4	EXTRAGALACTIC NEBULA
SHAP316-4808.0	03 14 59.	- 48 18 58.	54	17.75	GALAXY
SHAP316-5054.2	03 14 59.	- 51 05 10.	12	17.25	GALAXY
SHAP316-5054.4	03 14 59.	- 51 05 22.	18	16.5	GALAXY
WK 030	03 14 59.31	+ 41 29 35.8			NEBULA
ZC 0315.0+3419	03 15 00.	+ 34 19	940		CLUSTER OF GALAXIES
UGC 02647	03 15 00.	+ 38 55	66	16.5	GALAXY SBc
CR 19A	03 15 00.	+ 41 17			GALAXY IN PERSEUS CLUSTER
UGC 02648	03 15 00.	+ 47 15	90	16.5	GALAXY SBc
MCG+00-09-046	03 15 00.	- 01 53	54	15.	GALAXY
MCG+00-09-047	03 15 00.	- 01 55	36	15.	GALAXY
MCG+00-09-048	03 15 00.	- 02 01	18	14.	GALAXY
PHL 4371	03 15 00.	- 09 54		18.7	BLUE STELLAR OBJECT
PHL 4372	03 15 00.	- 12 52		16.6	BLUE STELLAR OBJECT
PHL 1500	03 15 00.	- 13 00		17.2	BLUE STELLAR OBJECT
BAK 1.282	03 15 00.	- 24 15 09.	17	16.8	EXTRAGALACTIC NEBULA
MCG-05-08-025	03 15 00.	- 32 48	120	13.	GALAXY
HPW 48	03 15 00.	- 35 42			IRREG GALAXY IN FORNAX
SHAP316-5021.9	03 15 00.	- 50 32 52.	30	16.75	GALAXY
SHAP316-5326.6	03 15 00.	- 53 37 34.	12	17.5	GALAXY
SHAP316-5330.4	03 15 00.	- 53 41 22.	18	17.0	GALAXY
WK 031	03 15 00.59	+ 41 17 15.1			NEBULA
SHAP316-5117.7	03 15 01.	- 51 28 40.	24	17.0	GALAXY
WK 032	03 15 02.64	+ 41 16 53.2			NEBULA
SHAP316-4814.4	03 15 03.	- 48 25 22.	36	17.5	GALAXY
SHAP316-5053.0	03 15 04.	- 51 03 58.	18	17.0	GALAXY
SHAP316-5311.6	03 15 04.	- 53 22 34.	12	17.5	GALAXY
WK 033	03 15 04.26	+ 41 13 40.9			NEBULA
SHAP316-4624.6	03 15 05.	- 46 35 34.	30	17.0	GALAXY
WK 034	03 15 05.03	+ 41 27 37.0			NEBULA
WK 035	03 15 05.09	+ 41 05 13.6			NEBULA
ZWG 525.019	03 15 06.	+ 38 50		15.7	GALAXY
ZWG 540.087	03 15 06.	+ 41 14		15.3	GALAXY
CR 20	03 15 06.	+ 41 14		15.3	GALAXY IN PERSEUS CLUSTER
CR 21	03 15 06.	+ 41 17			GALAXY IN PERSEUS CLUSTER
ZWG 540.088	03 15 06.	+ 41 40		14.7	GALAXY
CR 19B	03 15 06.	+ 41 40		14.7	GALAXY IN PERSEUS CLUSTER
UGC 02651	03 15 06.	+ 41 40	138	14.7	GALAXY E
ZWG 390.049	03 15 06.	- 01 52		15.6	GALAXY
UGC 02650	03 15 06.	- 01 52	72	15.6	GALAXY Sa-b
ZWG 390.048	03 15 06.	- 01 54		15.6	GALAXY
ZWG 390.047	03 15 06.	- 02 00		14.5	GALAXY
UGC 02649	03 15 06.	- 02 00	84	14.5	GALAXY S0
TON-S 0334	03 15 06.	- 24 14		15.2	BLUE STAR
SHAP316-5314.6	03 15 06.	- 53 25 34.	24	17.75	GALAXY
BAK 1.281	03 15 07.	- 31 09 51.	19	16.2	EXTRAGALACTIC NEBULA
SHAP316-4758.8	03 15 07.	- 48 09 46.	48	17.0	GALAXY
SHAP316-5133.6	03 15 07.	- 51 44 34.	12	16.5	GALAXY
ARC 0427	03 15 08.	+ 34 16		17.7	RICH CLUSTER OF GALAXIES
SHAP316-5105.8	03 15 08.	- 51 16 46.	24	17.25	GALAXY
SHAP316-5135.6	03 15 08.	- 51 46 34.	12	16.5	GALAXY
RNGC 1265	03 15 09.	+ 41 40		14.5	GALAXY
SHAP316-5102.4	03 15 09.	- 51 13 22.	12	16.75	GALAXY
SHAP316-4631.8	03 15 10.	- 46 42 45.	42	17.25	GALAXY
SHAP316-5102.8	03 15 10.	- 51 13 46.	24	17.25	GALAXY
BAK 1.283	03 15 11.	- 24 10 09.	38	16.4	EXTRAGALACTIC NEBULA
SHAP316-4603.4	03 15 11.	- 46 14 21.	30	17.25	GALAXY
MCG+08-07-001	03 15 12.	+ 47 14	36	17.	GALAXY
MCG+00-09-049	03 15 12.	- 01 53	24	15.	GALAXY
RNGC 1284	03 15 12.	- 10 28		14.0	GALAXY
MCG-02-09-022	03 15 12.	- 10 28	72	14.	GALAXY
PHL 1501	03 15 12.	- 10 42		15.5	BLUE STELLAR OBJECT
IC 1912	03 15 13.	- 50 50			NONSTELLAR OBJECT
SHAP316-5327.4	03 15 13.	- 53 38 22.	12	17.0	GALAXY
WK 036	03 15 13.70	+ 41 05 34.1			NEBULA
SHAP316-4746.9	03 15 14.	- 47 57 51.	36	16.75	GALAXY
SHAP316-5038.8	03 15 14.	- 50 49 45.	66	16.25	GALAXY
HN 0177	03 15 14.	- 50 50			NEBULA
SHAP316-5153.6	03 15 15.	- 52 04 33.	24	17.5	GALAXY
SHAP316-5302.4	03 15 15.	- 53 13 22.	18	16.0	GALAXY
SHAP316-5302.8	03 15 16.	- 53 13 46.	30	17.0	GALAXY
RNGC 1288	03 15 17.	- 32 45		13.0	GALAXY
WK 037	03 15 17.56	+ 41 17 44.4			NEBULA
ZWG 540.089	03 15 18.	+ 41 17		15.7	GALAXY
CR 22	03 15 18.	+ 41 17		15.7	GALAXY IN PERSEUS CLUSTER
MCG+07-07-053	03 15 18.	+ 42 07	60	14.5	GALAXY
MCG+00-09-050	03 15 18.	- 00 22	42	13.	GALAXY
ZWG 390.050	03 15 18.	- 00 44		15.7	GALAXY
BAK 1.284	03 15 18.	- 26 01 08.	12	16.8	EXTRAGALACTIC NEBULA
HN 0175	03 15 18.	- 33 52			NEBULA
IC 1909	03 15 18.	- 33 52			NONSTELLAR OBJECT
SHAP316-4631.1	03 15 18.	- 46 42 03.	42	17.25	GALAXY
LB 03305	03 15 18.	- 66 45		14.2	FAINT BLUE STAR
ARC 0426	03 15 20.	+ 41 20		12.5	RICH CLUSTER OF GALAXIES
WK 038	03 15 20.22	+ 41 29 13.8			NEBULA
MCG+07-07-054	03 15 21.	+ 43 03	90	15.	GALAXY

OBJECT NAME	RIGHT ASCEN.	DECLINATION	DIAM.	MAGN.	TYPE OF OBJECT
RNGC 1285	03 15 22.	- 07 30		14.0	GALAXY
RNGC 1286	03 15 23.	- 07 49		15.0	GALAXY
SHAP316-5300.4	03 15 23.	- 53 11 21.	12	16.5	GALAXY
UGC 02653	03 15 24.	+ 37 26	138	16.5	GALAXY Sc
MCG+07-07-055	03 15 24.	+ 41 16	12	15.	GALAXY
MCG+07-07-056	03 15 24.	+ 41 17	48	15.	GALAXY
ZWG 540.090	03 15 24.	+ 42 07		14.6	GALAXY
UGC 02654	03 15 24.	+ 42 07	96	14.6	GALAXY PECULR
ZWG 540.091	03 15 24.	+ 43 03		14.1	GALAXY
UGC 02655	03 15 24.	+ 43 03	126	14.1	GALAXY SBc
ZWG 390.051	03 15 24.	- 00 20		13.9	GALAXY
UGC 02652	03 15 24.	- 00 20	66	13.9	GALAXY SB
PHL 1502	03 15 24.	- 07 06		17.8	BLUE STELLAR OBJECT
MCG+01-09-026	03 15 24.	- 07 30	90	14.	GALAXY
MCG+01-09-025	03 15 24.	- 07 49	42	15.	GALAXY
MCG-02-09-023	03 15 24.	- 13 38	54	14.5	GALAXY
MCG-06-08-003	03 15 24.	- 33 52	72	14.	GALAXY
LB 03306	03 15 24.	- 65 56		13.7	FAINT BLUE STAR
RNGC 1280	03 15 25.	- 00 20		14.0	GALAXY
SHAP316-4942.4	03 15 25.	- 49 53 21.	24	17.0	GALAXY
SHAP316-5324.2	03 15 25.	- 53 35 09.	18	17.0	GALAXY
SHAP316-5412.5	03 15 25.	- 54 23 27.	18	17.75	GALAXY
WK 040	03 15 25.84	+ 40 53 30.0			NEBULA
SHAP316-5049.2	03 15 26.	- 51 00 09.	12	17.25	GALAXY
WK 039	03 15 26.36	+ 41 19 47.3			NEBULA
WK 041	03 15 26.45	+ 41 17 10.7			NEBULA
WK 042	03 15 26.76	+ 41 18 26.2			NEBULA
IC 1905	03 15 28.	+ 41 11			THREE STARS
WK 043	03 15 28.99	+ 41 14 49.6			NEBULA
RNGC 1291	03 15 29.	- 41 16		10.0	GALAXY
SHAP316-5418.0	03 15 29.	- 54 28 57.	36	17.0	GALAXY
MCG+06-08-012	03 15 30.	+ 37 25 30.	48	16.	GALAXY
UGC 02656	03 15 30.	+ 40 54	66	16.0	GALAXY E
ZWG 540.092	03 15 30.	+ 41 17		15.4	GALAXY
CR 23	03 15 30.	+ 41 17		15.4	GALAXY IN PERSEUS CLUSTER
UGC 02657	03 15 30.	+ 41 17	66	15.4	GALAXY E?
ZWG 540.093	03 15 30.	+ 41 18		14.5	GALAXY
CR 24	03 15 30.	+ 41 18		14.5	GALAXY IN PERSEUS CLUSTER
UGC 02658	03 15 30.	+ 41 18	66	14.5	GALAXY Sb
ZWG 390.052	03 15 30.	- 00 43		15.7	GALAXY
MCG+00-09-051	03 15 30.	- 01 32 30.	30	15.5	GALAXY
TON-S 0335	03 15 30.	- 22 20		14.9	BLUE STAR
SHAP317-5028.0	03 15 30.	- 50 38 57.	24	17.5	GALAXY
LB 03307	03 15 30.	- 63 29		14.3	FAINT BLUE STAR
WK 045	03 15 32.64	+ 41 12 39.0			NEBULA
RNGC 1267	03 15 33.	+ 41 17		15.5	GALAXY
RNGC 1268	03 15 33.	+ 41 18		14.5	GALAXY
SHAP316-5344.8	03 15 33.	- 53 55 45.	18	17.0	GALAXY
WK 044	03 15 33.27	+ 41 41 59.0			NEBULA
SHAP317-5028.3	03 15 35.	- 50 39 14.	18	17.5	GALAXY
WK 046	03 15 35.40	+ 41 37 18.5			NEBULA
ZWG 540.094	03 15 36.	+ 40 25		14.9	GALAXY
UGC 02659	03 15 36.	+ 40 25	84	14.9	GALAXY Sb-c
MCG+07-07-057	03 15 36.	+ 41 16	18	15.	GALAXY
ZWG 390.053	03 15 36.	- 01 30		15.7	GALAXY
PHL 4373	03 15 36.	- 17 53		18.5	BLUE STELLAR OBJECT
BAK 1.285	03 15 36.	- 28 36 19.	17	16.2	EXTRAGALACTIC NEBULA
SHAP317-4510.1	03 15 38.	- 45 21 02.	54	17.25	GALAXY
SHAP317-5419.1	03 15 38.	- 54 30 02.	18	17.75	GALAXY
WK 047	03 15 39.27	+ 41 27 14.6			NEBULA
WK 048	03 15 39.45	+ 41 31 15.5			NEBULA
WK 049	03 15 39.65	+ 41 17 19.5			NEBULA
SHAP317-5332.1	03 15 41.	- 53 43 02.	24	17.0	GALAXY
ZWG 540.095	03 15 42.	+ 41 17		14.4	GALAXY
CR 25	03 15 42.	+ 41 17		14.4	GALAXY IN PERSEUS CLUSTER
UGC 02660	03 15 42.	+ 41 17	78	14.4	GALAXY E
PHL 8612	03 15 42.	- 16 15		18.2	BLUE STELLAR OBJECT
PHL 4374	03 15 42.	- 17 05		18.0	BLUE STELLAR OBJECT
SHAP317-5433.0	03 15 42.	- 54 43 56.	18	17.25	GALAXY
BAK 1.286	03 15 43.	- 27 29 55.	16	16.6	EXTRAGALACTIC NEBULA
SHAP317-5103.0	03 15 43.	- 51 13 56.	12	17.25	GALAXY
SHAP317-5121.4	03 15 43.	- 51 32 20.	24	16.5	GALAXY
WK 050	03 15 43.83	+ 41 22 08.6			NEBULA
HN 0176	03 15 44.	- 21 37			NEBULA
IC 1910	03 15 44.	- 21 37			NONSTELLAR OBJECT
SHAP317-5021.6	03 15 44.	- 50 32 32.	24	16.25	GALAXY
RNGC 1270	03 15 45.	+ 41 17			GALAXY
MCG+00-09-052	03 15 45.	- 01 01	21	15.5	GALAXY
SHAP317-5146.3	03 15 45.	- 51 57 14.	18	16.0	GALAXY
SHAP317-5148.9	03 15 45.	- 51 59 50.	24	17.75	GALAXY
SHAP317-5330.2	03 15 45.	- 53 41 08.	18	17.0	GALAXY
SHAP317-5416.6	03 15 45.	- 54 27 32.	18	18.0	GALAXY
BAK 1.287	03 15 46.	- 25 57 49.	14	16.8	EXTRAGALACTIC NEBULA
WK 051	03 15 46.25	+ 41 17 15.2			NEBULA
ZWG 390.054	03 15 48.	- 01 00		15.6	GALAXY
PHL 4375	03 15 48.	- 07 25		17.7	BLUE STELLAR OBJECT
WK 053	03 15 48.52	+ 41 11 49.9			NEBULA
SHAP317-5104.9	03 15 49.	- 51 15 50.	18	17.25	GALAXY
WK 052	03 15 49.09	+ 41 40 24.5			NEBULA
WK 054	03 15 49.26	+ 40 57 44.4			NEBULA
BAK 1.289	03 15 50.	- 26 56 18.	18	16.6	EXTRAGALACTIC NEBULA
RNGC 1271	03 15 51.	+ 41 10		15.5	GALAXY
MCG-03-09-016	03 15 51.	- 16 28	24	15.5	GALAXY
SHAP317-5244.8	03 15 51.	- 52 55 44.	18	17.25	GALAXY
BAK 1.288	03 15 52.	- 29 49 36.	14	17.0	EXTRAGALACTIC NEBULA
SHAP317-5101.8	03 15 52.	- 51 12 43.	18	16.5	GALAXY
LB 02858	03 15 53.	- 00 06 18.		16.5	FAINT BLUE STAR
SHAP317-4636.9	03 15 53.	- 46 47 49.	36	17.0	GALAXY
WK 055	03 15 53.05	+ 41 10 19.6			NEBULA
MCG+00-09-053	03 15 54.	+ 00 43	30	15.	GALAXY
ZWG 540.096	03 15 54.	+ 41 10		15.4	GALAXY
CR 26	03 15 54.	+ 41 10		15.4	GALAXY IN PERSEUS CLUSTER
PHL 8613	03 15 54.	- 09 55		8.2	BLUE STELLAR OBJECT
PHL 1503	03 15 54.	- 15 13		18.5	BLUE STELLAR OBJECT
SHAP317-4829.9	03 15 54.	- 48 40 49.	42	17.75	GALAXY
SHAP317-5109.0	03 15 54.	- 51 19 55.	18	17.0	GALAXY
SHAP317-5126.1	03 15 54.	- 51 37 01.	12	17.5	GALAXY
SHAP317-5310.3	03 15 54.	- 53 21 13.	18	17.0	GALAXY
SHAP317-5451.1	03 15 54.	- 55 02 02.	24	16.5	GALAXY
BAK 1.290	03 15 55.	- 30 18 18.	20	16.4	EXTRAGALACTIC NEBULA
SHAP317-5422.0	03 15 55.	- 54 32 55.	24	17.0	GALAXY
SHAP317-5431.9	03 15 55.	- 54 42 49.	24	17.25	GALAXY
RNGC 1292	03 15 56.	- 27 48		13.0	GALAXY
SHAP317-4951.1	03 15 56.	- 50 02 01.	36	16.75	GALAXY
SHAP317-5146.4	03 15 56.	- 51 57 19.	24	17.0	GALAXY
WK 056	03 15 56.64	+ 40 57 40.0			NEBULA
MCG+07-07-058	03 15 57.	+ 41 18	15	14.5	GALAXY
SHAP317-5055.0	03 15 57.	- 51 05 55.	12	17.0	GALAXY

OBJECT NAME	RIGHT ASCEN.	DECLINATION	DIAM.	MAGN.	TYPE OF OBJECT
BAK 1.293	03 15 58.	- 24 56 18.	14	16.9	EXTRAGALACTIC NEBULA
BAK 1.291	03 15 58.	- 29 48 54.	14	16.2	EXTRAGALACTIC NEBULA
SHAP317-4545.0	03 15 58.	- 45 55 55.	36	16.75	GALAXY
SHAP317-4637.2	03 15 58.	- 46 48 07.	30	17.5	GALAXY
SHAP317-4630.0	03 15 59.	- 46 40 55.	36	17.5	GALAXY
WK 057	03 15 59.04	+ 41 27 48.3			NEBULA
LBN 0685	03 16	+ 54 30	4800		BRIGHT NEBULA
ZWG 525.020	03 16 00.	+ 39 17		14.7	GALAXY
UGC 02661	03 16 00.	+ 39 17	84	14.7	GALAXY S0
CR 27	03 16 00.	+ 40 58			GALAXY IN PERSEUS CLUSTER
CR 28	03 16 00.	+ 41 28			GALAXY IN PERSEUS CLUSTER
ZWG 555.001	03 16 00.	+ 45 41		15.7	GALAXY
LDN 1384	03 16 00.	+ 59 50	1020		DARK NEBULA
PHL 4376	03 16 00.	- 09 54		18.7	BLUE STELLAR OBJECT
PHL 8614	03 16 00.	- 09 56		17.8	BLUE STELLAR OBJECT
LB 03308	03 16 00.	- 66 21		13.2	FAINT BLUE STAR
RNGC 1287	03 16 02.	- 02 55			GALAXY
SHAP317-5132.0	03 16 02.	- 51 42 55.	6	17.25	GALAXY
SHAP317-5449.0	03 16 02.	- 54 59 55.	30	16.5	GALAXY
WK 058	03 16 02.73	+ 41 18 35.4			NEBULA
MCG+07-07-059	03 16 03.	+ 41 20 30.	15	14.5	GALAXY
MCG+07-07-060	03 16 03.	+ 41 26	60	15.	GALAXY
BAK 1.294	03 16 03.	- 28 38 18.	17	17.0	EXTRAGALACTIC NEBULA
SHAP317-5411.6	03 16 03.	- 54 22 31.	24	17.5	GALAXY
BAK 1.292	03 16 04.	- 31 59 54.	18	16.4	EXTRAGALACTIC NEBULA
SHAP317-4940.1	03 16 04.	- 49 51 01.	42	16.75	GALAXY
SHAP317-5337.5	03 16 04.	- 53 48 25.	18	18.0	GALAXY
BAK 1.295	03 16 05.	- 28 37 06.	14	16.3	EXTRAGALACTIC NEBULA
SHAP317-4543.0	03 16 05.	- 45 53 54.	36	16.75	GALAXY
SHAP317-5123.6	03 16 05.	- 51 34 31.	12	17.25	GALAXY
SHAP317-5327.6	03 16 05.	- 53 38 31.	18	17.0	GALAXY
SHAP317-5421.6	03 16 05.	- 54 32 31.	48	14.8	GALAXY
5ZW 337	03 16 06.	+ 27 07			COMPACT GALAXY
ZWG 540.097	03 16 06.	+ 40 19		15.4	GALAXY
ZWG 540.098	03 16 06.	+ 41 18		14.5	GALAXY
CR 29	03 16 06.	+ 41 18		14.5	GALAXY IN PERSEUS CLUSTER
UGC 02662	03 16 06.	+ 41 18	180	14.5	GALAXY E
MCG-05-08-026	03 16 06.	- 27 48	198	12.	GALAXY
SHAP317-5210.2	03 16 06.	- 52 21 07.	18	15.75	GALAXY
SHAP317-5213.1	03 16 06.	- 52 24 01.	18	17.0	GALAXY
SHAP317-5340.4	03 16 06.	- 53 51 19.	18	17.0	GALAXY
SHAP317-4834.1	03 16 08.	- 48 45 00.	36	17.5	GALAXY
WK 059	03 16 08.14	+ 41 21 34.2			NEBULA
WK 060	03 16 08.61	+ 41 27 15.8			NEBULA
RNGC 1272	03 16 09.	+ 41 19		14.5	GALAXY
RNGC 1273	03 16 09.	+ 41 22		14.5	GALAXY
MCG+07-07-061	03 16 09.	+ 41 23	12	15.	GALAXY
SHAP317-5420.2	03 16 09.	- 54 31 07.	24	17.25	GALAXY
UGC 02663	03 16 12.	+ 08 37	66	17.	GALAXY Sc:
UGC 02664	03 16 12.	+ 40 34	84	18.	GALAXY DWARF
ZWG 540.099	03 16 12.	+ 41 21		14.7	GALAXY
CR 30	03 16 12.	+ 41 21		14.7	GALAXY IN PERSEUS CLUSTER
ZWG 540.100	03 16 12.	+ 41 27		15.5	GALAXY
CR 31	03 16 12.	+ 41 27		15.5	GALAXY IN PERSEUS CLUSTER
UGC 02665	03 16 12.	+ 41 27	66	15.5	GALAXY Sc?
MCG+00-09-054	03 16 12.	- 02 10	36	13.	GALAXY
PHL 1504	03 16 12.	- 12 30		18.6	BLUE STELLAR OBJECT
PHL 4377	03 16 12.	- 14 37		18.4	BLUE STELLAR OBJECT
SHAP317-5126.0	03 16 13.	- 51 36 54.	18	17.0	GALAXY
SHAP317-5429.8	03 16 13.	- 54 40 42.	18	17.75	GALAXY
MCG+07-07-062	03 16 15.	+ 41 21	24	15.5	GALAXY
SHAP317-5339.6	03 16 15.	- 53 50 30.	6	17.3	GALAXY
WK 061	03 16 15.57	+ 41 23 59.8			NEBULA
BAK 1.296	03 16 17.	- 30 16 53.	16	16.4	EXTRAGALACTIC NEBULA
SHAP317-5442.2	03 16 17.	- 54 53 06.	24	17.0	GALAXY
WK 063	03 16 17.97	+ 40 58 15.5			NEBULA
ZWG 416.005	03 16 18.	+ 06 20		15.6	GALAXY
KARA.73B 0124	03 16 18.	+ 06 20	42	15.6	ISOLATED GALAXY S0
ZWG 540.101	03 16 18.	+ 41 24		15.2	GALAXY
CR 32	03 16 18.	+ 41 24		15.2	GALAXY IN PERSEUS CLUSTER
ZWG 390.055	03 16 18.	- 02 09		13.8	GALAXY
UGC 02666	03 16 18.	- 02 09	126	13.8	GALAXY E-S0
IC 0314	03 16 18.	- 02 09			NONSTELLAR OBJECT
PHL 4378	03 16 18.	- 03 56		18.6	BLUE STELLAR OBJECT
TON-S 0336	03 16 18.	- 23 48		15.0	BLUE STAR
SHAP318-4541.4	03 16 18.	- 45 52 18.	42	17.0	GALAXY
SHAP317-4657.0	03 16 18.	- 47 07 54.	24	17.0	GALAXY
WK 062	03 16 18.68	+ 41 27 08.5			NEBULA
WK 064	03 16 18.72	+ 41 18 19.7			NEBULA
SHAP317-5140.0	03 16 19.	- 51 50 54.	24	17.25	GALAXY
SHAP317-5422.8	03 16 19.	- 54 33 42.	24	16.75	GALAXY
LB 02859	03 16 20.	+ 00 32 30.		16.1	FAINT BLUE STAR
RNGC 1289	03 16 20.	- 02 09		14.0	GALAXY
SHAP317-5122.0	03 16 20.	- 51 32 54.	12	17.0	GALAXY
SHAP318-4719.0	03 16 21.	- 47 29 54.	30	17.5	GALAXY
SHAP317-5104.5	03 16 21.	- 51 15 24.	18	17.0	GALAXY
SHAP317-5130.3	03 16 21.	- 51 41 12.	24	17.0	GALAXY
SHAP317-5337.0	03 16 21.	- 53 47 54.	24	17.9	GALAXY
SHAP317-5349.2	03 16 21.	- 54 00 06.	18	17.5	GALAXY
WK 066	03 16 21.92	+ 41 08 55.3			NEBULA
WK 065	03 16 21.94	+ 41 22 05.4			NEBULA
SHAP318-4537.1	03 16 22.	- 45 47 59.	48	17.0	GALAXY
BAK 1.297	03 16 23.	- 26 00 53.	84	14.7	EXTRAGALACTIC NEBULA
SHAP317-5254.0	03 16 23.	- 53 04 54.	12	17.0	GALAXY
ZC 0316.4+3636	03 16 24.	+ 36 36	1080		CLUSTER OF GALAXIES
UGC 02667	03 16 24.	+ 38 15	66	17.	GALAXY Sc
MCG+07-07-063	03 16 24.	+ 41 19	72	12.5	GALAXY
ZWG 540.102	03 16 24.	+ 41 22		15.1	GALAXY
CR 34	03 16 24.	+ 41 22		15.1	GALAXY IN PERSEUS CLUSTER
CR 33	03 16 24.	+ 41 27			GALAXY IN PERSEUS CLUSTER
PHL 4379	03 16 24.	- 13 48		18.4	BLUE STELLAR OBJECT
PHL 8615	03 16 24.	- 14 14		18.7	BLUE STELLAR OBJECT
PHL 1505	03 16 24.	- 14 54		18.2	BLUE STELLAR OBJECT
PHL 1506	03 16 24.	- 15 40		18.1	BLUE STELLAR OBJECT
MCG-04-08-057	03 16 24.	- 26 01	180	13.	GALAXY
SHAP317-5057.8	03 16 24.	- 51 08 42.	30	16.5	GALAXY
WK 067	03 16 25.24	+ 41 16 35.1			NEBULA
SHAP317-5333.2	03 16 26.	- 53 44 06.	18	17.8	GALAXY
WK 068	03 16 26.16	+ 41 15 57.6			NEBULA
WK 069	03 16 26.35	+ 41 05 52.4			NEBULA
RNGC 1274	03 16 27.	+ 41 22		15.0	GALAXY
MCG+07-07-064	03 16 27.	+ 41 22 30.	36	15.	GALAXY
MCG-02-09-024	03 16 27.	- 13 49	36	15.5	GALAXY
SHAP318-4958.0	03 16 27.	- 50 08 53.	30	16.5	GALAXY
SHAP317-5352.3	03 16 27.	- 54 03 12.	16	17.4	GALAXY
SHAP317-5406.4	03 16 27.	- 54 17 18.	12	17.9	GALAXY
SHAP318-4649.0	03 16 28.	- 46 59 53.	36	17.25	GALAXY
SHAP317-5318.0	03 16 28.	- 53 28 54.	12	17.5	GALAXY

OBJECT NAME	RIGHT ASCEN.	DECLINATION	DIAM.	MAGN.	TYPE OF OBJECT
SHAP317-5453.0	03 16 28.	- 55 03 54.	24	16.5	GALAXY
WK 071	03 16 28.63	+ 41 05 25.5			NEBULA
SHAP318-5003.0	03 16 29.	- 50 13 53.	18	16.75	GALAXY
SHAP317-5339.0	03 16 29.	- 53 49 53.	12	17.3	GALAXY
WK 070	03 16 29.05	+ 41 24 57.3			NEBULA
WK 072	03 16 29.60	+ 41 19 52.0			NEBULA
LB 00481	03 16 30.	+ 00 32 18.		16.8	FAINT BLUE STAR
ZWG 416.006	03 16 30.	+ 03 51		15.1	GALAXY
ZWG 464.012	03 16 30.	+ 18 01		15.7	GALAXY
UGC 02668	03 16 30.	+ 40 17	84	16.5	GALAXY IV?
CR 35A	03 16 30.	+ 41 06			GALAXY IN PERSEUS CLUSTER
ZWG 540.103	03 16 30.	+ 41 20		13.0	GALAXY
CR 35B	03 16 30.	+ 41 20		13.0	GALAXY IN PERSEUS CLUSTER
VVI 13	03 16 30.	+ 41 20	72	13.14	SEYFERT GALAXY
UGC 02669	03 16 30.	+ 41 20	210	13.0	GALAXY PECULR
SN 1968A	03 16 30.	+ 41 20		15.5	SUPERNOVA
MCG+07-07-065	03 16 30.	+ 41 22	18	14.5	GALAXY
CR 36	03 16 30.	+ 41 25			GALAXY IN PERSEUS CLUSTER
MCG-02-09-025	03 16 30.	- 13 14 30.	54	14.5	GALAXY
RNGC 1296	03 16 31.	- 13 14		14.0	GALAXY
BAK 1.299	03 16 31.	- 25 56 16.	18	16.5	EXTRAGALACTIC NEBULA
BAK 1.298	03 16 31.	- 28 39 52.	16	17.1	EXTRAGALACTIC NEBULA
SHAP318-5105.6	03 16 31.	- 51 16 29.	12	16.5	GALAXY
SHAP318-5151.7	03 16 31.	- 52 02 35.	18	17.0	GALAXY
SHAP318-5151.8	03 16 31.	- 52 02 41.	24	16.75	GALAXY
SHAP318-5151.8	03 16 31.	- 52 02 41.	18	16.75	GALAXY
SHAP317-5423.4	03 16 31.	- 54 34 17.	30	16.5	GALAXY
IC 0315	03 16 32.	+ 03 50 30.			NONSTELLAR OBJECT
ARC 0429	03 16 32.	+ 36 38		17.7	RICH CLUSTER OF GALAXIES
SHAP318-5128.4	03 16 32.	- 51 39 17.	42	16.25	GALAXY
SHAP318-5155.8	03 16 32.	- 52 06 41.	24	17.0	GALAXY
WK 073	03 16 32.76	+ 41 23 35.4			NEBULA
RNGC 1275	03 16 33.	+ 41 20		12.5	GALAXY
RNGC 1276	03 16 33.	+ 41 23			GALAXY
SHAP318-5057.9	03 16 33.	- 51 08 47.	12	17.5	GALAXY
SHAP318-5223.2	03 16 33.	- 52 34 05.	12	16.5	GALAXY
WK 074	03 16 33.79	+ 41 22 09.8			NEBULA
IC 1907	03 16 33.8	+ 41 22 07.			GALAXY E4
IC 0317	03 16 34.	- 12 55 37.			NONSTELLAR OBJECT
SHAP318-5028.0	03 16 34.	- 50 38 53.	18	17.0	GALAXY
SHAP318-5309.9	03 16 34.	- 53 20 47.	24	17.0	GALAXY
SHAP317-5343.6	03 16 34.	- 53 54 29.	18	18.0	GALAXY
WK 075	03 16 34.67	+ 41 07 19.3			NEBULA
SHAP317-5355.4	03 16 35.	- 54 06 17.	18	17.3	GALAXY
WK 076	03 16 35.48	+ 41 22 59.2			NEBULA
ZWG 390.056	03 16 36.	+ 00 47		15.6	GALAXY
CR 39	03 16 36.	+ 41 07			GALAXY IN PERSEUS CLUSTER
ZWG 540.104	03 16 36.	+ 41 23		14.9	GALAXY
CR 37	03 16 36.	+ 41 23		14.9	GALAXY IN PERSEUS CLUSTER
KARA.68 032	03 16 36.	- 10 43	81		DWARF GALAXY
PHL 8616	03 16 36.	- 14 49		15.9	BLUE STELLAR OBJECT
PHL 8617	03 16 36.	- 15 08		17.9	BLUE STELLAR OBJECT
TON-S 0337	03 16 36.	- 22 28		14.9	BLUE STAR
SHAP318-5008.1	03 16 36.	- 50 18 59.	18	16.75	GALAXY
WK 078	03 16 36.95	+ 41 20 33.9			NEBULA
SHAP318-5010.2	03 16 37.	- 50 21 05.	18	16.5	GALAXY
SHAP318-5350.0	03 16 37.	- 54 00 53.	12	17.8	GALAXY
WK 077	03 16 37.21	+ 41 36 50.5			NEBULA
SHAP318-5126.0	03 16 38.	- 51 36 53.	12	17.25	GALAXY
MCG+07-07-066	03 16 39.	+ 41 03	12	15.	GALAXY
RNGC 1278	03 16 39.	+ 41 23		14.5	GALAXY
RNGC 1277	03 16 39.	+ 41 24		15.0	GALAXY
MCG-02-09-026	03 16 39.	- 12 56	48	14.5	GALAXY
BAK 1.300	03 16 39.	- 25 42 28.	26	17.0	EXTRAGALACTIC NEBULA
SHAP318-5326.9	03 16 39.	- 53 37 47.	30	16.0	GALAXY
SHAP318-5212.4	03 16 40.	- 52 23 17.	12	17.0	GALAXY
SHAP318-5432.9	03 16 40.	- 54 43 47.	24	17.5	GALAXY
WK 079	03 16 40.54	+ 41 17 56.9			NEBULA
BAK 1.301	03 16 41.	- 26 45 28.	26	16.3	EXTRAGALACTIC NEBULA
SHAP318-5318.8	03 16 41.	- 53 29 41.	18	17.0	GALAXY
ZC 0316.7+0153	03 16 42.	+ 01 53	270		CLUSTER OF GALAXIES
CR 41	03 16 42.	+ 41 18			GALAXY IN PERSEUS CLUSTER
5ZW 338	03 16 42.	+ 41 20			COMPACT GALAXY
ZWG 540.105	03 16 42.	+ 41 22		14.4	GALAXY
CR 38	03 16 42.	+ 41 22		14.4	GALAXY IN PERSEUS CLUSTER
UGC 02670	03 16 42.	+ 41 22	108	14.4	GALAXY E
5ZW 339	03 16 42.	+ 41 23			COMPACT GALAXY
MCG-02-09-027	03 16 42.	- 12 08	48	15.5	GALAXY
BAK 1.304	03 16 43.	- 23 39 15.	20	16.8	EXTRAGALACTIC NEBULA
BAK 1.302	03 16 43.	- 24 09 15.	26	16.8	EXTRAGALACTIC NEBULA
SHAP318-4608.0	03 16 43.	- 46 18 52.	36	17.0	GALAXY
SHAP318-4613.2	03 16 43.	- 46 24 04.	42	17.5	GALAXY
SHAP318-4715.7	03 16 43.	- 47 26 34.	24	17.75	GALAXY
SHAP318-5052.6	03 16 43.	- 51 03 29.	18	17.0	GALAXY
SHAP318-5409.8	03 16 43.	- 54 20 41.	24	17.3	GALAXY
WK 080	03 16 43.32	+ 41 04 17.7			NEBULA
SHAP318-5028.4	03 16 44.	- 50 39 16.	24	17.5	GALAXY
WK 081	03 16 44.10	+ 41 19 44.9			NEBULA
RNGC 1279	03 16 45.	+ 41 18			GALAXY
MCG+07-07-067	03 16 45.	+ 41 26 30.	30	15.	GALAXY
BAK 1.306	03 16 45.	- 22 54 15.	19	16.6	EXTRAGALACTIC NEBULA
BAK 1.303	03 16 45.	- 26 25 15.	17	16.3	EXTRAGALACTIC NEBULA
SHAP318-5356.0	03 16 45.	- 54 06 53.	18	16.4	GALAXY
WK 082	03 16 45.48	+ 41 01 01.7			NEBULA
SHAP318-5215.0	03 16 46.	- 52 25 52.	24	16.5	GALAXY
SHAP318-5306.6	03 16 46.	- 53 17 29.	12	16.5	GALAXY
SHAP318-5354.0	03 16 46.	- 54 04 53.	18	17.3	GALAXY
SHAP318-5359.2	03 16 46.	- 54 10 05.	18	16.9	GALAXY
SHAP318-5010.8	03 16 47.	- 50 21 40.	24	16.75	GALAXY
SHAP318-5214.4	03 16 47.	- 52 25 16.	12	17.0	GALAXY
SHAP318-5257.2	03 16 47.	- 53 08 04.	12	17.0	GALAXY
SHAP318-5338.8	03 16 47.	- 53 49 40.	24	16.9	GALAXY
WK 083	03 16 47.26	+ 41 26 58.9			NEBULA
WK 084	03 16 47.99	+ 41 39 07.5			NEBULA
UGC 02671	03 16 48.	+ 07 58	72	17.	GALAXY
ZWG 540.106	03 16 48.	+ 40 44		15.7	GALAXY
UGC 02672	03 16 48.	+ 40 44	66	15.7	GALAXY Sa?
ZWG 540.107	03 16 48.	+ 41 04		15.6	GALAXY
CR 40	03 16 48.	+ 41 04		15.6	GALAXY IN PERSEUS CLUSTER
UGC 02673	03 16 48.	+ 41 04	102	15.6	GALAXY S0
MCG+07-07-068	03 16 48.	+ 41 10	15	15.	GALAXY
ZWG 540.108	03 16 48.	+ 41 27		15.0	GALAXY
CR 42	03 16 48.	+ 41 27		15.0	GALAXY IN PERSEUS CLUSTER
MCG-02-09-028	03 16 48.	- 11 31	78	14.5	GALAXY
PHL 8619	03 16 48.	- 13 28		18.4	BLUE STELLAR OBJECT
PHL 4380	03 16 48.	- 17 18		18.0	BLUE STELLAR OBJECT
PHL 8618	03 16 48.	- 17 34		16.8	BLUE STELLAR OBJECT
TON-S 0338	03 16 48.	- 23 36		15.3	BLUE STAR

OBJECT NAME	RIGHT ASCEN.	DECLINATION	DIAM.	MAGN.	TYPE OF OBJECT
SHAP318-5408.6	03 16 49.	- 54 19 28.	30	17.3	GALAXY
BAK 1.305	03 16 50.	- 29 18 03.	14	17.1	EXTRAGALACTIC NEBULA
SHAP318-5149.0	03 16 50.	- 51 59 52.	36	16.75	GALAXY
SHAP318-5304.4	03 16 50.	- 53 15 16.	12	17.5	GALAXY
MCG+07-07-069	03 16 51.	+ 41 12	12	15.5	GALAXY
RNGC 1281	03 16 51.	+ 41 27		15.0	GALAXY
MCG-03-09-017	03 16 51.	- 19 18	30	13.	GALAXY
SHAP318-5117.1	03 16 51.	- 51 27 58.	18	16.75	GALAXY
WK 085	03 16 51.70	+ 41 10 16.3			NEBULA
SHAP318-5358.8	03 16 52.	- 54 09 40.	18	18.2	GALAXY
SHAP318-4707.2	03 16 53.	- 47 18 04.	36	17.0	GALAXY
SHAP318-5327.4	03 16 53.	- 53 38 16.	42	17.5	GALAXY
WK 087	03 16 53.76	+ 41 03 00.3			NEBULA
WK 086	03 16 53.78	+ 41 11 12.9			NEBULA
ZC 0316.9+2048	03 16 54.	+ 20 48	1610		CLUSTER OF GALAXIES
MCG+00-09-056	03 16 54.	- 02 14	24	15.	GALAXY
MCG+00-09-055	03 16 54.	- 02 18	24	15.	GALAXY
MCG+01-09-027	03 16 54.	- 06 18	48	15.	GALAXY
WK 088	03 16 54.28	+ 41 19 19.5			NEBULA
WK 089	03 16 54.96	+ 41 06 24.9			NEBULA
HELW 392	03 16 55.	- 02 17 37.			NEBULA
SHAP318-5438.2	03 16 55.	- 54 49 04.	24	16.5	GALAXY
BAK 1.307	03 16 56.	- 28 23 51.	30	17.1	EXTRAGALACTIC NEBULA
SHAP318-5343.0	03 16 56.	- 53 53 52.	12	17.9	GALAXY
SHAP318-5343.5	03 16 56.	- 53 54 22.	18	18.0	GALAXY
WK 090	03 16 56.95	+ 41 13 07.0			NEBULA
RNGC 1282	03 16 57.	+ 41 11		14.5	GALAXY
MCG-02-09-029	03 16 57.	- 12 18	78	14.	GALAXY
WK 091	03 16 57.47	+ 41 18 08.4			NEBULA
RNGC 1297	03 16 58.	- 19 15		13.0	GALAXY
BAK 1.308	03 16 58.	- 23 35 39.	24	16.7	EXTRAGALACTIC NEBULA
SHAP318-5258.8	03 16 58.	- 53 09 40.	18	17.5	GALAXY
SHAP318-5332.3	03 16 58.	- 53 43 10.	24	17.2	GALAXY
WK 092	03 16 58.11	+ 41 11 55.1			NEBULA
WK 093	03 16 58.70	+ 41 11 30.3			NEBULA
WK 094	03 16 58.86	+ 41 10 08.9			NEBULA
HELW 393	03 16 59.	- 02 13 20.			NEBULA
SHAP318-5004.6	03 16 59.	- 50 15 28.	30	16.25	GALAXY
WK 095	03 16 59.49	+ 41 11 15.8			NEBULA
ZWG 441.007	03 17 00.	+ 09 34		15.3	GALAXY
UGC 02674	03 17 00.	+ 09 34	66	15.3	GALAXY S0?
ZWG 525.021	03 17 00.	+ 39 23		15.5	GALAXY
ZWG 540.109	03 17 00.	+ 41 11		14.3	GALAXY
CR 43	03 17 00.	+ 41 11		14.3	GALAXY IN PERSEUS CLUSTER
UGC 02675	03 17 00.	+ 41 11	96	14.3	GALAXY E
ZWG 540.110	03 17 00.	+ 41 13		15.6	GALAXY
CR 44	03 17 00.	+ 41 13		15.6	GALAXY IN PERSEUS CLUSTER
UGC 02676	03 17 00.	+ 41 13	72	15.6	GALAXY COMPACT
MCG+07-07-070	03 17 00.	+ 41 27	30	15.5	GALAXY
LDN 1385	03 17 00.	+ 59 38	240		DARK NEBULA
ZWG 390.057	03 17 00.	- 02 12		15.6	GALAXY
PHL 8620	03 17 00.	- 05 50		18.6	BLUE STELLAR OBJECT
PHL 4382	03 17 00.	- 07 38		18.0	BLUE STELLAR OBJECT
PHL 4381	03 17 00.	- 16 30		16.8	BLUE STELLAR OBJECT
SHAP318-4708.0	03 17 00.	- 47 18 51.	24	17.25	GALAXY
SHAP318-5205.0	03 17 00.	- 52 15 52.	18	17.5	GALAXY
SHAP318-5108.5	03 17 01.	- 51 19 22.	30	16.75	GALAXY
SHAP318-4602.8	03 17 02.	- 46 13 39.	30	17.5	GALAXY
WK 096	03 17 02.42	+ 41 11 38.6			NEBULA
RNGC 1283	03 17 03.	+ 41 13		15.5	GALAXY
MCG+07-07-071	03 17 03.	+ 42 21	12	15.	GALAXY
WK 097	03 17 03.00	+ 41 27 36.9			NEBULA
SHAP318-5440.6	03 17 04.	- 54 51 28.	24	17.5	GALAXY
SHAP318-5141.0	03 17 05.	- 51 51 51.	24	16.5	GALAXY
KLEM 07	03 17 05.	- 54 08	3000	15.	CLUSTER OF 50 GALAXIES
SHAP318-5359.6	03 17 05.	- 54 10 28.	18	17.9	GALAXY
MCG+00-09-057	03 17 06.	+ 00 22	42	15.	GALAXY
ZWG 390.058	03 17 06.	+ 00 23		15.6	GALAXY
KARA.72 089A	03 17 06.	+ 00 23	48	15.6	PART OF DOUBLE GALAXY
TON-S 0339	03 17 06.	- 28 51		15.8	BLUE STAR
RNGC 1295	03 17 08.	- 14 11			GALAXY
BAK 1.310	03 17 08.	- 26 41 26.	26	16.8	EXTRAGALACTIC NEBULA
BAK 1.309	03 17 08.	- 28 57 50.	29	16.1	EXTRAGALACTIC NEBULA
SHAP318-5051.1	03 17 08.	- 51 01 57.	12	17.25	GALAXY
SHAP318-5404.4	03 17 08.	- 54 15 15.	24	17.6	GALAXY
BAK 1.311	03 17 09.	- 27 18 32.	22	16.2	EXTRAGALACTIC NEBULA
SHAP318-5408.7	03 17 09.	- 54 19 33.	12	17.4	GALAXY
WK 098	03 17 09.94	+ 41 18 32.4			NEBULA
WK 099	03 17 10.62	+ 41 20 09.4			NEBULA
MCG+00-09-058	03 17 12.	+ 00 23	48	14.	GALAXY
ZWG 390.059	03 17 12.	+ 00 24		14.9	GALAXY
KARA.72 089B	03 17 12.	+ 00 24	48	14.9	PART OF DOUBLE GALAXY
MRK 606	03 17 12.	+ 03 58	8	16.5	GALAXY WITH UV CONTINUUM
PHL 8621	03 17 12.	- 07 36		18.5	BLUE STELLAR OBJECT
PHL 4383	03 17 12.	- 11 00		18.6	BLUE STELLAR OBJECT
PHL 8623	03 17 12.	- 11 48		18.5	BLUE STELLAR OBJECT
PHL 8622	03 17 12.	- 16 58		16.9	BLUE STELLAR OBJECT
BAK 1.313	03 17 12.	- 25 23 02.	18	16.6	EXTRAGALACTIC NEBULA
PHL 4384	03 17 12.	- 28 38		18.0	BLUE STELLAR OBJECT
SHAP318-5104.2	03 17 12.	- 51 15 03.	36	17.75	GALAXY
WK 100	03 17 12.25	+ 41 19 45.2			NEBULA
WK 101	03 17 12.89	+ 41 34 41.6			NEBULA
SHAP318-5413.5	03 17 13.	- 54 24 21.	18	17.0	GALAXY
WK 103	03 17 13.48	+ 41 06 05.8			NEBULA
WK 102	03 17 13.95	+ 41 33 06.8			NEBULA
SHAP318-5341.3	03 17 14.	- 53 52 09.	12	16.8	GALAXY
BAK 1.312	03 17 16.	- 30 45 26.	12	16.8	EXTRAGALACTIC NEBULA
SHAP318-5022.2	03 17 16.	- 50 33 03.	36	16.75	GALAXY
SHAP318-5217.0	03 17 16.	- 52 27 51.	24	16.0	GALAXY
SHAP318-5331.9	03 17 17.	- 53 42 45.	12	17.7	GALAXY
SHAP318-5340.4	03 17 17.	- 53 51 15.	24	17.3	GALAXY
ZWG 390.060	03 17 18.	+ 03 24		15.2	GALAXY
UGC 02677	03 17 18.	+ 03 24	102	15.2	GALAXY Sb
MCG+00-09-059	03 17 18.	+ 03 24	24	14.	GALAXY
UGC 02678	03 17 18.	+ 37 19	132	16.0	GALAXY SBb
MCG-05-08-027	03 17 18.	- 32 42	84	14.5	GALAXY
SHAP318-5226.4	03 17 18.	- 52 37 15.	18	17.0	GALAXY
SHAP318-5408.2	03 17 19.	- 54 19 03.	18	16.9	GALAXY
BAK 1.317	03 17 20.	- 27 18 37.	13	16.4	EXTRAGALACTIC NEBULA
SHAP318-5414.2	03 17 20.	- 54 25 03.	24	15.7	GALAXY
SHAP318-5237.8	03 17 21.	- 52 48 39.	12	17.5	GALAXY
SHAP318-5239.6	03 17 21.	- 52 50 27.	12	16.5	GALAXY
SHAP318-5305.0	03 17 21.	- 53 15 51.	24	17.0	GALAXY
SHAP318-5335.6	03 17 21.	- 53 46 27.	18	18.0	GALAXY
SHAP318-5348.2	03 17 21.	- 53 59 03.	12	17.6	GALAXY
SHAP318-5403.6	03 17 21.	- 54 14 27.	24	17.4	GALAXY
BAK 1.316	03 17 22.	- 30 27 01.	14	16.4	EXTRAGALACTIC NEBULA
BAK 1.314	03 17 22.	- 32 38 50.	17	16.8	EXTRAGALACTIC NEBULA

OBJECT NAME	RIGHT ASCEN.	DECLINATION	DIAM.	MAGN.	TYPE OF OBJECT
SHAP318-5405.6	03 17 22.	- 54 16 27.	12	18.0	GALAXY
SHAP318-5409.8	03 17 22.	- 54 20 39.	24	17.9	GALAXY
BAK 1.318	03 17 23.	- 28 19 31.	17	17.2	EXTRAGALACTIC NEBULA
WK 104	03 17 23.49	+ 41 13 27.7			NEBULA
WK 105	03 17 23.98	+ 41 30 20.4			NEBULA
MCG+00-09-060	03 17 24.	+ 01 10	114	13.5	GALAXY
MCG+07-07-072	03 17 24.	+ 42 36	36	17.	GALAXY
PHL 4385	03 17 24.	- 11 54		18.4	BLUE STELLAR OBJECT
MCG-03-09-018	03 17 24.	- 19 36 30.	360	11.	GALAXY
PHL 8624	03 17 24.	- 22 48		16.5	BLUE STELLAR OBJECT
TON-S 0340	03 17 24.	- 26 38		16.1	BLUE STAR
BAK 1.315	03 17 24.	- 32 50 01.	28	16.0	EXTRAGALACTIC NEBULA
SHAP318-5217.8	03 17 24.	- 52 28 38.	18	17.5	GALAXY
SHAP318-5413.0	03 17 24.	- 54 23 51.	24	17.9	GALAXY
SHAP318-5440.9	03 17 24.	- 54 51 45.	24	16.75	GALAXY
SHAP318-5317.6	03 17 25.	- 53 28 26.	24	17.0	GALAXY
SHAP318-5333.8	03 17 25.	- 53 44 38.	18	17.7	GALAXY
SHAP318-5334.9	03 17 25.	- 53 45 44.	18	17.8	GALAXY
SHAP319-4914.8	03 17 26.	- 49 25 38.	24	17.25	GALAXY
SHAP318-5158.2	03 17 26.	- 52 09 02.	18	17.0	GALAXY
WK 106	03 17 26.65	+ 41 33 18.0			NEBULA
SHAP318-5426.9	03 17 27.	- 54 37 44.	24	17.9	GALAXY
RNGC 1300	03 17 28.	- 19 34		11.0	GALAXY
BAK 1.321	03 17 28.	- 25 14 01.	12	16.7	EXTRAGALACTIC NEBULA
BAK 1.320	03 17 28.	- 25 14 37.	12	17.1	EXTRAGALACTIC NEBULA
BAK 1.319	03 17 28.	- 26 20 01.	18	16.6	EXTRAGALACTIC NEBULA
SHAP318-5442.6	03 17 28.	- 54 53 26.	30	16.5	GALAXY
BAK 1.322	03 17 29.	- 26 12 01.	18	16.8	EXTRAGALACTIC NEBULA
SHAP318-5400.4	03 17 29.	- 54 11 14.	18	17.3	GALAXY
ZWG 390.061	03 17 30.	+ 01 11		15.4	GALAXY
UGC 02679	03 17 30.	+ 01 11	138	15.4	GALAXY Sb-c
MCG+06-08-013	03 17 30.	+ 37 18	42	16.	GALAXY
MCG+00-09-061	03 17 30.	- 02 05	17	14.5	GALAXY
PHL 8625	03 17 30.	- 05 45		15.6	BLUE STELLAR OBJECT
WK 107	03 17 30.82	+ 41 11 28.5			NEBULA
BAK 1.323	03 17 31.	- 27 40 25.	22	16.4	EXTRAGALACTIC NEBULA
HN 0178	03 17 31.	- 32 39			NEBULA
SHAP319-5038.8	03 17 31.	- 50 49 38.	24	17.5	GALAXY
SHAP319-5104.8	03 17 31.	- 51 15 38.	18	17.25	GALAXY
SHAP318-5406.8	03 17 31.	- 54 17 38.	12	17.75	GALAXY
SHAP318-5420.9	03 17 31.	- 54 31 44.	18	17.7	GALAXY
WK 108	03 17 31.79	+ 41 25 14.9			NEBULA
LB 02860	03 17 32.	+ 00 05 12.		16.5	FAINT BLUE STAR
SHAP319-5114.2	03 17 32.	- 51 25 02.	18	16.5	GALAXY
SHAP318-5344.0	03 17 32.	- 53 54 50.	12	17.6	GALAXY
SHAP319-4935.0	03 17 33.	- 49 45 50.	252	15.0	GALAXY
SHAP319-5154.6	03 17 33.	- 52 05 26.	18	17.25	GALAXY
IC 1913	03 17 33.1	- 32 38 49.			GALAXY
LB 00482	03 17 34.	+ 00 05 12.		17.7	FAINT BLUE STAR
HN 0179	03 17 34.	- 49 46			NEBULA
IC 1914	03 17 34.	- 49 46			NONSTELLAR OBJECT
WK 109	03 17 34.65	+ 41 30 08.9			NEBULA
HELW 394	03 17 35.	- 02 03 58.			NEBULA
SHAP319-5316.1	03 17 35.	- 53 26 56.	12	17.0	GALAXY
ZC 0317.6+3411	03 17 36.	+ 34 11	1140		CLUSTER OF GALAXIES
UGC 02681	03 17 36.	+ 38 55	66	17.	GALAXY
ZWG 540.111	03 17 36.	+ 41 42		15.1	GALAXY
CR 45	03 17 36.	+ 41 42		15.1	GALAXY IN PERSEUS CLUSTER
UGC 02682	03 17 36.	+ 41 42	84	15.1	GALAXY E
MCG+07-07-073	03 17 36.	+ 41 42	24	15.	GALAXY
ZWG 390.062	03 17 36.	- 02 03		14.9	GALAXY
UGC 02680	03 17 36.	- 02 03	84	14.9	GALAXY PECULR?
PHL 4386	03 17 36.	- 10 02		18.8	BLUE STELLAR OBJECT
MCG-02-09-030	03 17 36.	- 14 11	48	15.	GALAXY
PHL 1507	03 17 36.	- 14 38		17.7	BLUE STELLAR OBJECT
BAK 1.325	03 17 36.	- 27 11 13.	20	16.0	EXTRAGALACTIC NEBULA
SHAP319-5343.2	03 17 36.	- 53 54 02.	24	17.9	GALAXY
SHAP319-5358.4	03 17 36.	- 54 09 14.	12	17.2	GALAXY
RNGC 1313	03 17 36.	- 66 39		10.5	GALAXY
SN 1962M	03 17 36.	- 66 40		11.7	SUPERNOVA
BAK 1.324	03 17 37.	- 29 11 25.	17	16.7	EXTRAGALACTIC NEBULA
SHAP319-4550.0	03 17 37.	- 46 00 49.	36	17.25	GALAXY
SHAP319-5336.8	03 17 37.	- 53 47 38.	18	17.3	GALAXY
IC 1911	03 17 38.	+ 35 07			NONSTELLAR OBJECT
RNGC 1290	03 17 38.	- 14 11		15.0	GALAXY
SHAP319-5351.6	03 17 38.	- 54 02 26.	24	17.2	GALAXY
WK 110	03 17 38.39	+ 41 42 47.9			NEBULA
WK 111	03 17 38.95	+ 41 19 37.3			NEBULA
SHAP319-4543.5	03 17 39.	- 45 54 19.	48	17.5	GALAXY
SHAP319-4942.8	03 17 39.	- 49 53 37.	30	17.0	GALAXY
SHAP319-5103.0	03 17 39.	- 51 13 49.	24	16.75	GALAXY
IC 0313	03 17 39.2	+ 41 42 49.			GALAXY SA0
RNGC 1299	03 17 40.	- 06 27		14.0	GALAXY
BAK 1.326	03 17 40.	- 28 14 00.	24	16.4	EXTRAGALACTIC NEBULA
SHAP319-5422.8	03 17 40.	- 54 33 38.	18	17.8	GALAXY
SHAP319-5425.1	03 17 40.	- 54 35 56.	18	17.0	GALAXY
LB 02861	03 17 41.	+ 01 36 30.		17.0	FAINT BLUE STAR
BAK 1.327	03 17 41.	- 29 10 36.	23	15.8	EXTRAGALACTIC NEBULA
SHAP319-5412.0	03 17 41.	- 54 22 50.	42	17.2	GALAXY
UGC 02684	03 17 42.	+ 17 08	120	18.	GALAXY DWARF?
ZWG 525.022	03 17 42.	+ 37 41		15.4	GALAXY
ZWG 525.023	03 17 42.	+ 38 05		14.8	GALAXY
UGC 02685	03 17 42.	+ 38 05	132	14.8	GALAXY SBb
UGC 02686	03 17 42.	+ 40 37	96	16.0	GALAXY SBa
ZWG 390.063	03 17 42.	- 02 17		14.2	GALAXY
UGC 02683	03 17 42.	- 02 17	78	14.2	GALAXY E
MCG+00-09-062	03 17 42.	- 02 18	24	13.	GALAXY
PHL 8627	03 17 42.	- 06 11		18.4	BLUE STELLAR OBJECT
MCG+01-09-028	03 17 42.	- 06 27	60	14.	GALAXY
PHL 8628	03 17 42.	- 12 24		18.5	BLUE STELLAR OBJECT
PHL 1508	03 17 42.	- 12 53		18.3	BLUE STELLAR OBJECT
PHL 8626	03 17 42.	- 22 11		18.4	BLUE STELLAR OBJECT
PHL 1509	03 17 42.	- 24 22		18.2	BLUE STELLAR OBJECT
BAK 1.328	03 17 42.	- 29 22 12.	22	17.4	EXTRAGALACTIC NEBULA
BAK 1.329	03 17 42.	- 29 25 06.	12	17.0	EXTRAGALACTIC NEBULA
SHAP319-5352.4	03 17 42.	- 54 03 13.	12	16.2	GALAXY
SHAP319-5353.0	03 17 42.	- 54 03 49.	18	16.9	GALAXY
SHAP319-4923.3	03 17 43.	- 49 34 07.	36	17.0	GALAXY
SHAP319-5402.2	03 17 43.	- 54 13 01.	18	17.2	GALAXY
WK 112	03 17 43.51	+ 41 13 49.2			NEBULA
RNGC 1298	03 17 44.	- 02 17		14.0	GALAXY
RNGC 1302	03 17 44.	- 26 14		12.0	GALAXY
SHAP319-4915.1	03 17 44.	- 49 25 55.	24	17.0	GALAXY
SHAP319-5326.0	03 17 44.	- 53 36 49.	18	16.5	GALAXY
MCG+06-08-014	03 17 45.	+ 38 04	72	15.	GALAXY
MCG+00-09-063	03 17 45.	- 02 17	30	14.5	GALAXY
MCG-04-08-058	03 17 45.	- 26 14	240	12.	GALAXY
BAK 1.330	03 17 45.	- 28 45 30.	12	17.0	EXTRAGALACTIC NEBULA
SHAP319-4658.2	03 17 45.	- 47 09 01.	30	17.25	GALAXY
WK 113	03 17 45.28	+ 41 22 34.9			NEBULA
BAK 1.331	03 17 46.	- 29 19 24.	17	17.0	EXTRAGALACTIC NEBULA
SHAP319-5110.2	03 17 46.	- 51 21 01.	18	17.0	GALAXY
BAK 1.332	03 17 47.	- 24 32 36.	12	17.2	EXTRAGALACTIC NEBULA
SHAP319-4932.2	03 17 47.	- 49 43 01.	24	17.0	GALAXY
SHAP319-5347.6	03 17 47.	- 53 58 25.	18	16.4	GALAXY
SHAP319-5358.2	03 17 47.	- 54 09 01.	24	17.3	GALAXY
MCG+06-08-015	03 17 48.	+ 37 40 30.	36	15.	GALAXY
ZWG 390.064	03 17 48.	- 02 15		15.2	GALAXY
UGC 02687	03 17 48.	- 02 15	78	15.2	GALAXY SBa
PHL 4387	03 17 48.	- 10 15		18.6	BLUE STELLAR OBJECT
PHL 8629	03 17 48.	- 11 11		18.6	BLUE STELLAR OBJECT
PHL 1510	03 17 48.	- 21 22		18.0	BLUE STELLAR OBJECT
TON-S 0341	03 17 48.	- 21 44		14.9	BLUE STAR
PHL 1511	03 17 48.	- 31 45		18.1	BLUE STELLAR OBJECT
SHAP319-4955.9	03 17 48.	- 50 06 43.	18	17.25	GALAXY
SHAP319-5141.1	03 17 48.	- 51 51 55.	36	16.0	GALAXY
SHAP319-5055.8	03 17 49.	- 51 06 37.	18	17.25	GALAXY
SHAP319-5140.7	03 17 49.	- 51 51 31.	12	17.0	GALAXY
SHAP319-5356.8	03 17 49.	- 54 07 37.	24	15.4	GALAXY
HELW 395	03 17 50.	- 02 15 52.			NEBULA
SHAP319-4558.0	03 17 50.	- 46 08 49.	36	17.5	GALAXY
SHAP319-5345.3	03 17 50.	- 53 56 07.	18	16.7	GALAXY
SHAP319-5346.8	03 17 50.	- 53 57 37.	12	17.8	GALAXY
SHAP319-5445.2	03 17 50.	- 54 56 01.	24	16.5	GALAXY
WK 114	03 17 50.15	+ 41 04 26.2			NEBULA
MCG-02-09-031	03 17 51.	- 11 02	30	15.5	GALAXY
SHAP319-5040.8	03 17 51.	- 50 51 37.	12	17.0	GALAXY
SHAP319-5350.8	03 17 52.	- 54 01 37.	18	17.7	GALAXY
SHAP319-5130.0	03 17 53.	- 51 40 49.	36	17.0	GALAXY
SHAP319-5134.8	03 17 53.	- 51 45 37.	12	17.0	GALAXY
MCG+07-07-074	03 17 54.	+ 41 44	24	16.	GALAXY
MCG+00-09-064	03 17 54.	- 00 31	36	15.	GALAXY
MCG+00-09-065	03 17 54.	- 02 12	48	14.	GALAXY
BAK 1.333	03 17 55.	- 27 11 36.	18	15.9	EXTRAGALACTIC NEBULA
SHAP319-5157.5	03 17 55.	- 52 08 19.	12	16.5	GALAXY
SHAP319-5400.8	03 17 55.	- 54 11 37.	18	16.1	GALAXY
SHAP319-5133.1	03 17 56.	- 51 43 55.	18	17.0	GALAXY
SHB 071	03 17 56.5	- 02 19 24.		19.5	QUASI-STELLAR OBJECT
BAK 1.335	03 17 57.	- 27 06 23.	20	16.6	EXTRAGALACTIC NEBULA
SHAP319-5345.4	03 17 57.	- 53 56 13.	18	16.0	GALAXY
SHAP319-5347.6	03 17 57.	- 53 58 25.	24	17.8	GALAXY
SHAP319-5233.0	03 17 58.	- 52 43 49.	12	17.5	GALAXY
SHAP319-5405.2	03 17 58.	- 54 16 01.	18	17.4	GALAXY
BAK 1.336	03 17 59.	- 28 49 47.	22	17.2	EXTRAGALACTIC NEBULA
LB 09797	03 18	- 83 54		14.2	FAINT BLUE STAR
3ZW 053	03 18 00.	+ 15 46			COMPACT GALAXY
ZWG 464.013	03 18 00.	+ 15 46		15.4	GALAXY
ZWG 540.112	03 18 00.	+ 41 45		15.0	GALAXY
UGC 02688	03 18 00.	+ 41 45	90	15.0	GALAXY
ZWG 390.066	03 18 00.	- 00 31		15.4	GALAXY
ZC 0318.0-0131	03 18 00.	- 01 31	1480		CLUSTER OF GALAXIES
ZWG 390.065	03 18 00.	- 02 10		15.5	GALAXY
MCG-02-09-032	03 18 00.	- 11 02	24	15.5	GALAXY
MCG-03-09-020	03 18 00.	- 15 36 30.	36	14.5	GALAXY
MCG-03-09-019	03 18 00.	- 17 16	30	16.	GALAXY
PHL 8630	03 18 00.	- 21 30		15.3	BLUE STELLAR OBJECT
BAK 1.334	03 18 00.	- 31 19 59.	24	17.0	EXTRAGALACTIC NEBULA
SHAP319-5148.1	03 18 00.	- 51 58 54.	12	17.5	GALAXY
SHAP319-5400.5	03 18 00.	- 54 11 19.	24	17.3	GALAXY
WK 115	03 18 00.22	+ 41 45 07.8			NEBULA
IC 0316A	03 18 00.7	+ 41 45 13.			GALAXY E1p
IC 0316B	03 18 00.9	+ 41 44 56.			GALAXY S0p
HELW 396	03 18 01.	- 02 11 17.			NEBULA
BAK 1.337	03 18 01.	- 29 26 23.	26	16.6	EXTRAGALACTIC NEBULA
LB 02862	03 18 02.	- 00 26 18.		15.8	FAINT BLUE STAR
SHAP319-5350.7	03 18 02.	- 54 01 30.	12	17.1	GALAXY
WK 116	03 18 02.10	+ 41 16 52.8			NEBULA
SHAP319-5341.8	03 18 03.	- 53 52 36.	18	17.7	GALAXY
SHAP319-5401.6	03 18 03.	- 54 12 24.	18	17.8	GALAXY
SHAP319-5144.5	03 18 05.	- 51 55 18.	36	16.7	GALAXY
SHAP319-5154.6	03 18 05.	- 52 05 24.	36	17.5	GALAXY
SHAP319-5413.0	03 18 05.	- 54 23 48.	24	17.3	GALAXY
UGC 02689	03 18 06.	+ 40 38	102	16.0	GALAXY S0?
ZWG 540.113	03 18 06.	+ 41 17		15.1	GALAXY
ZWG 390.067	03 18 06.	- 01 12		15.7	GALAXY
MCG+00-09-066	03 18 06.	- 01 18	78	14.	GALAXY
TON-S 0342	03 18 06.	- 21 28		14.8	BLUE STAR
KARA.68 033	03 18 06.	- 27 00	27		DWARF GALAXY
HPW 02	03 18 06.	- 39 09			DWARF GALAXY IN FORNAX
SHAP319-5339.4	03 18 06.	- 53 50 12.	12	17.8	GALAXY
SHAP319-4543.5	03 18 07.	- 45 54 18.	36	16.75	GALAXY
SHAP319-5414.0	03 18 07.	- 54 24 48.	18	17.8	GALAXY
BAK 1.338	03 18 08.	- 26 36 47.	17	16.7	EXTRAGALACTIC NEBULA
MCG+07-07-075	03 18 09.	+ 41 11 30.	12	15.	GALAXY
BAK 1.339	03 18 09.	- 26 24 17.	14	16.2	EXTRAGALACTIC NEBULA
SHAP319-5349.8	03 18 09.	- 54 00 36.	18	16.2	GALAXY
SHAP319-5350.3	03 18 09.	- 54 01 06.	24	16.7	GALAXY
SHAP319-5448.8	03 18 09.	- 54 59 36.	24	16.5	GALAXY
WK 117	03 18 09.82	+ 41 21 28.7			NEBULA
BAK 1.342	03 18 10.	- 26 38 41.	90	15.8	EXTRAGALACTIC NEBULA
SHAP319-5102.2	03 18 10.	- 51 13 00.	36	16.0	GALAXY
SHAP319-5102.7	03 18 10.	- 51 13 30.	36	18.0	GALAXY
MCG+07-07-076	03 18 12.	+ 41 09 30.	12	15.	GALAXY
KW 49	03 18 12.	+ 41 25	68		SEYFERT GALAXY
MRSL 143-01/1	03 18 12.	+ 54 42	2700		HII REGION
LDN 1386	03 18 12.	+ 59 30	360		DARK NEBULA
ZWG 390.070	03 18 12.	- 01 10		15.1	GALAXY
ZWG 390.069	03 18 12.	- 01 13		14.9	GALAXY
UGC 02691	03 18 12.	- 01 13	102	14.9	GALAXY DBL SYS
MCG+00-09-067	03 18 12.	- 01 15	54	14.	GALAXY
ZWG 390.068	03 18 12.	- 01 17		14.8	GALAXY
UGC 02690	03 18 12.	- 01 17	90	14.8	GALAXY Sb-c
PHL 8631	03 18 12.	- 10 31		16.5	BLUE STELLAR OBJECT
PHL 4388	03 18 12.	- 13 44		18.9	BLUE STELLAR OBJECT
MCG-03-09-021	03 18 12.	- 15 24 30.	48	15.	GALAXY
MCG-05-09-001	03 18 12.	- 26 38	72	16.	GALAXY
BAK 1.340	03 18 12.	- 29 34 17.	17	16.6	EXTRAGALACTIC NEBULA
SHAP319-5044.4	03 18 12.	- 50 55 12.	24	17.75	GALAXY
SHAP319-5411.7	03 18 12.	- 54 22 30.	36	17.8	GALAXY
HN 0180	03 18 13.	- 50 52			NEBULA
IC 1915	03 18 13.	- 50 52			NONSTELLAR OBJECT
SHAP319-5420.8	03 18 13.	- 54 31 36.	24	17.3	GALAXY
BAK 1.343	03 18 14.	- 28 03 22.	16	17.0	EXTRAGALACTIC NEBULA
BAK 1.341	03 18 14.	- 31 20 23.	12	16.8	EXTRAGALACTIC NEBULA
SHAP319-5041.2	03 18 14.	- 50 52 00.	36	16.25	GALAXY
SHAP319-5342.8	03 18 14.	- 53 53 36.	18	17.2	GALAXY

OBJECT NAME	RIGHT ASCEN.	DECLINATION	DIAM.	MAGN.	TYPE OF OBJECT
SHAP319-5325.0	03 18 15.	- 53 35 48.	12	17.0	GALAXY
WK 118	03 18 15.23	+ 41 19 37.1			NEBULA
RNGC 1301	03 18 16.	- 18 55		13.0	GALAXY
BAK 1.344	03 18 16.	- 28 38 58.	19	17.1	EXTRAGALACTIC NEBULA
SHAP319-5109.8	03 18 16.	- 51 20 35.	24	17.25	GALAXY
SHAP319-5423.0	03 18 16.	- 54 33 48.	30	17.2	GALAXY
RNGC 1303	03 18 17.	- 07 35		15.0	GALAXY
BAK 1.348	03 18 17.	- 27 17 22.	14	16.4	EXTRAGALACTIC NEBULA
SHAP319-5039.2	03 18 17.	- 50 49 59.	42	17.25	GALAXY
WK 119	03 18 17.75	+ 41 12 50.5			NEBULA
5ZW 340	03 18 18.	+ 32 30			COMPACT GALAXY
ZWG 540.114	03 18 18.	+ 40 15		15.6	GALAXY
ZWG 540.115	03 18 18.	+ 41 19		15.6	GALAXY
VB 233	03 18 18.	+ 51 25	551		STELLAR RING
VB 234	03 18 18.	+ 51 52	470		STELLAR RING
MCG+00-09-068	03 18 18.	- 00 33	66	13.5	GALAXY
MCG+01-09-029	03 18 18.	- 07 35	30	15.	GALAXY
PHL 1512	03 18 18.	- 10 16		18.1	BLUE STELLAR OBJECT
PHL 8632	03 18 18.	- 12 41		16.0	BLUE STELLAR OBJECT
MCG-03-09-022	03 18 18.	- 18 55	108	13.5	GALAXY
BAK 1.346	03 18 18.	- 29 24 16.	13	17.0	EXTRAGALACTIC NEBULA
BAK 1.347	03 18 18.	- 29 24 28.	14	16.6	EXTRAGALACTIC NEBULA
BAK 1.345	03 18 19.	- 31 15 46.	16	17.0	EXTRAGALACTIC NEBULA
SHAP319-5403.6	03 18 19.	- 54 14 23.	24	17.0	GALAXY
SHAP319-5338.4	03 18 20.	- 53 49 11.	42	16.2	GALAXY
RNGC 1293	03 18 21.	+ 41 13		15.0	GALAXY
SHAP319-4805.0	03 18 21.	- 48 15 47.	24	16.5	GALAXY
SHAP319-5111.1	03 18 21.	- 51 21 53.	24	18.0	GALAXY
WK 120	03 18 21.24	+ 41 10 54.7			NEBULA
LB 02863	03 18 22.	+ 01 43 12.		16.8	FAINT BLUE STAR
BAK 1.349	03 18 22.	- 31 14 10.	14	16.9	EXTRAGALACTIC NEBULA
IC 0318	03 18 23.	- 14 44 42.			NONSTELLAR OBJECT
SHAP319-5120.9	03 18 23.	- 51 31 41.	24	16.5	GALAXY
ZWG 540.116	03 18 24.	+ 41 13		15.0	GALAXY
CR 46	03 18 24.	+ 41 13		15.0	GALAXY IN PERSEUS CLUSTER
ZWG 390.071	03 18 24.	- 00 32		14.1	GALAXY
UGC 02692	03 18 24.	- 00 32	84	14.1	GALAXY Sc
PHL 4391	03 18 24.	- 02 47		18.0	BLUE STELLAR OBJECT
PHL 4389	03 18 24.	- 06 16		17.1	BLUE STELLAR OBJECT
MCG-04-09-001	03 18 24.	- 23 07 30.	36	15.	GALAXY
PHL 4390	03 18 24.	- 28 46		17.7	BLUE STELLAR OBJECT
BAK 1.350	03 18 24.	- 31 24 58.	14	16.3	EXTRAGALACTIC NEBULA
SHAP319-5347.2	03 18 24.	- 53 57 59.	18	17.7	GALAXY
SHAP320-4623.5	03 18 25.	- 46 34 17.	30	17.5	GALAXY
SHAP319-5329.6	03 18 25.	- 53 40 23.	18	17.5	GALAXY
LB 00483	03 18 26.	- 00 58 06.		17.4	FAINT BLUE STAR
BAK 1.351	03 18 26.	- 30 17 58.	17	16.6	EXTRAGALACTIC NEBULA
SHAP319-5046.6	03 18 26.	- 50 57 23.	18	16.5	GALAXY
SHAP319-5403.6	03 18 26.	- 54 14 23.	24	17.8	GALAXY
RNGC 1294	03 18 27.	+ 41 11		15.0	GALAXY
MCG-03-09-023	03 18 27.	- 14 45	48	14.5	GALAXY
BAK 1.356	03 18 27.	- 24 48 46.	12	17.2	EXTRAGALACTIC NEBULA
SHAP319-5142.0	03 18 27.	- 51 52 47.	12	16.5	GALAXY
BAK 1.354	03 18 28.	- 27 19 16.	14	17.0	EXTRAGALACTIC NEBULA
BAK 1.352	03 18 28.	- 31 26 34.	18	16.6	EXTRAGALACTIC NEBULA
SHAP320-4713.9	03 18 28.	- 47 24 41.	24	17.0	GALAXY
BAK 1.358	03 18 29.	- 23 06 28.	13	15.6	EXTRAGALACTIC NEBULA
BAK 1.353	03 18 29.	- 30 22 58.	19	17.4	EXTRAGALACTIC NEBULA
SHAP319-5101.4	03 18 29.	- 51 12 11.	30	17.5	GALAXY
SHAP319-5201.2	03 18 29.	- 52 11 59.	24	17.5	GALAXY
SHAP319-5410.2	03 18 29.	- 54 20 59.	18	17.8	GALAXY
UGC 02693	03 18 30.	+ 17 41	66	16.5	GALAXY
ZWG 540.117	03 18 30.	+ 41 11		15.1	GALAXY
CR 47	03 18 30.	+ 41 11		15.1	GALAXY IN PERSEUS CLUSTER
UGC 02694	03 18 30.	+ 41 11	102	15.1	GALAXY S0?
OCL 0392	03 18 30.	+ 48 26	15600	2.3	OPEN STAR CLUSTER
PHL 1513	03 18 30.	- 19 06		17.4	BLUE STELLAR OBJECT
BAK 1.355	03 18 30.	- 28 46 40.	22	17.4	EXTRAGALACTIC NEBULA
TON-S 0343	03 18 30.	- 30 25		15.2	BLUE STAR
SHAP319-5223.2	03 18 30.	- 52 33 59.	12	16.75	GALAXY
LB 03309	03 18 30.	- 70 21		13.3	FAINT BLUE STAR
SHAP320-4805.1	03 18 31.	- 48 15 52.	24	16.5	GALAXY
LB 02864	03 18 32.	+ 01 01 12.		16.1	FAINT BLUE STAR
SHAP319-5345.6	03 18 32.	- 53 56 23.	12	17.9	GALAXY
RNGC 1304	03 18 33.	- 04 45		14.5	GALAXY
SHAP320-4912.2	03 18 33.	- 49 22 58.	24	16.5	GALAXY
SHAP319-5340.3	03 18 33.	- 53 51 05.	18	16.8	GALAXY
SHAP320-4808.0	03 18 34.	- 48 18 46.	30	16.75	GALAXY
BAK 1.359	03 18 35.	- 28 38 09.	18	17.0	EXTRAGALACTIC NEBULA
MCG+01-09-030	03 18 36.	- 04 45	21	14.5	GALAXY
PHL 8611	03 18 36.	- 17 17		18.2	BLUE STELLAR OBJECT
MCG-03-09-024	03 18 36.	- 18 05	48	15.	GALAXY
PHL 8633	03 18 36.	- 25 03		18.5	BLUE STELLAR OBJECT
MCG-05-09-002	03 18 36.	- 31 01	48	15.5	GALAXY
SHAP320-5122.0	03 18 36.	- 51 32 46.	24	18.0	GALAXY
SHAP320-5345.8	03 18 36.	- 53 56 34.	18	16.5	GALAXY
SHAP320-5210.5	03 18 37.	- 52 21 16.	300	13.6	GALAXY
SHAP319-5447.1	03 18 37.	- 54 57 53.	24	15.75	GALAXY
BAK 1.357	03 18 39.	- 32 22 39.	34	16.4	EXTRAGALACTIC NEBULA
BAK 1.360	03 18 40.	- 29 18 15.	14	16.8	EXTRAGALACTIC NEBULA
SHAP320-5103.2	03 18 40.	- 51 13 58.	36	17.5	GALAXY
RNGC 1311	03 18 40.	- 52 20			GALAXY
SHAP320-5410.8	03 18 40.	- 54 21 34.	18	17.7	GALAXY
BAK 1.361	03 18 41.	- 24 17 45.	14	17.1	EXTRAGALACTIC NEBULA
SHAP320-4804.4	03 18 41.	- 48 15 10.	24	17.0	GALAXY
UGC 02695	03 18 42.	+ 07 15	96	16.5	GALAXY Sc?
MCG+07-07-077	03 18 42.	+ 40 40	12	15.	GALAXY
ZWG 540.118	03 18 42.	+ 42 00		15.7	GALAXY
UGC 02696	03 18 42.	+ 42 00	60	15.7	GALAXY S?
SHAP320-5224.6	03 18 42.	- 52 35 22.	24	17.75	GALAXY
SHAP320-5421.0	03 18 42.	- 54 31 46.	24	18.1	GALAXY
RNGC 1305	03 18 44.	- 02 29		15.0	GALAXY
SHAP320-5325.9	03 18 44.	- 53 36 40.	18	16.5	GALAXY
MCG+00-09-069	03 18 45.	- 02 30	24	13.	GALAXY
SHAP320-4902.2	03 18 46.	- 49 12 58.	54	15.75	GALAXY
IC 1916	03 18 46.	- 49 13			NONSTELLAR OBJECT
SHAP320-5342.5	03 18 46.	- 53 53 16.	12	16.9	GALAXY
SHAP320-5354.2	03 18 46.	- 54 04 58.	24	17.7	GALAXY
SHAP320-5414.5	03 18 46.	- 54 25 16.	24	16.6	GALAXY
HN 0181	03 18 47.	- 49 13			NEBULA
ZWG 441.008	03 18 48.	+ 13 22		15.7	GALAXY
MCG+07-07-078	03 18 48.	+ 40 32	36	17.	GALAXY
ZWG 540.119	03 18 48.	+ 40 42		14.4	GALAXY
UGC 02698	03 18 48.	+ 40 42	90	14.4	GALAXY E
ZWG 390.072	03 18 48.	- 02 30		14.8	GALAXY
UGC 02697	03 18 48.	- 02 30	102	14.8	GALAXY S0?
PHL 8634	03 18 48.	- 10 11		15.3	BLUE STELLAR OBJECT
PHL 8635	03 18 48.	- 13 10		18.6	BLUE STELLAR OBJECT
PHL 4392	03 18 48.	- 17 36		18.2	BLUE STELLAR OBJECT
SHAP320-5247.6	03 18 48.	- 52 58 22.	18	17.5	GALAXY
BAK 1.362	03 18 49.	- 30 57 57.	37	16.2	EXTRAGALACTIC NEBULA
SHAP320-4910.9	03 18 49.	- 49 21 39.	24	16.0	GALAXY
SHAP320-5443.4	03 18 49.	- 54 54 10.	36	16.0	GALAXY
SHAP320-4836.2	03 18 50.	- 48 46 57.	36	17.5	GALAXY
SHAP320-5113.2	03 18 50.	- 51 23 58.	24	17.5	GALAXY
SHAP320-5338.2	03 18 50.	- 53 48 58.	18	17.8	GALAXY
BAK 1.363	03 18 51.	- 24 22 20.	14	17.1	EXTRAGALACTIC NEBULA
SHAP320-5357.2	03 18 51.	- 54 07 58.	18	18.0	GALAXY
SHAP320-5101.8	03 18 52.	- 51 12 33.	24	16.5	GALAXY
SHAP320-5150.6	03 18 52.	- 52 01 21.	24	17.5	GALAXY
SHAP320-4804.8	03 18 53.	- 48 15 33.	24	17.25	GALAXY
SHAP320-5136.9	03 18 53.	- 51 47 39.	24	17.5	GALAXY
SHAP320-5319.6	03 18 53.	- 53 30 22.	18	17.0	GALAXY
MCG+02-09-004	03 18 54.	+ 13 22	36	16.	GALAXY
MCG+02-09-003	03 18 54.	+ 13 22	24	16.	GALAXY
ZWG 540.120	03 18 54.	+ 40 34		15.6	GALAXY
PHL 4393	03 18 54.	- 09 50		18.1	BLUE STELLAR OBJECT
PHL 8636	03 18 54.	- 26 42		17.6	BLUE STELLAR OBJECT
MCG-05-09-003	03 18 54.	- 31 13	30	15.	GALAXY
SHAP320-5108.3	03 18 54.	- 51 19 03.	24	17.0	GALAXY
SHAP320-5233.2	03 18 54.	- 52 43 57.	24	16.8	GALAXY
SHAP320-5306.1	03 18 54.	- 53 16 51.	24	16.5	GALAXY
SHAP320-5416.0	03 18 56.	- 54 26 45.	24	16.7	GALAXY
SHAP320-5442.0	03 18 57.	- 54 52 45.	30	16.75	GALAXY
SHAP320-5408.4	03 18 58.	- 54 19 09.	12	17.9	GALAXY
SHAP320-4538.4	03 18 59.	- 45 49 09.	36	17.25	GALAXY
SHAP320-5324.6	03 18 59.	- 53 35 21.	18	17.0	GALAXY
SHAP320-5440.5	03 18 59.	- 54 51 15.	24	16.75	GALAXY
ZWG 525.024	03 19 00.	+ 36 17		15.6	GALAXY
MCG-02-09-033	03 19 00.	- 11 20	30	15.	GALAXY
PHL 1514	03 19 00.	- 21 00		16.9	BLUE STELLAR OBJECT
MCG-06-08-004	03 19 00.	- 37 18	120	12.5	GALAXY
SHAP320-5416.4	03 19 00.	- 54 27 09.	30	14.9	GALAXY
WK 121	03 19 00.87	+ 41 04 40.8			NEBULA
RNGC 1306	03 19 01.	- 25 42			GALAXY
SHAP320-4551.2	03 19 01.	- 46 01 57.	36	17.0	GALAXY
SHAP320-5210.6	03 19 01.	- 52 21 21.	18	17.5	GALAXY
BAK 1.364	03 19 02.	- 30 43 08.	14	16.7	EXTRAGALACTIC NEBULA
SHAP320-5322.6	03 19 02.	- 53 33 21.	18	16.25	GALAXY
BAK 1.365	03 19 03.	- 31 20 26.	17	16.8	EXTRAGALACTIC NEBULA
RNGC 1310	03 19 03.	- 37 18			GALAXY
SHAP320-5151.8	03 19 03.	- 52 02 33.	24	17.5	GALAXY
SHAP320-5259.7	03 19 03.	- 53 10 27.	24	17.5	GALAXY
SHAP320-5203.9	03 19 04.	- 52 14 39.	18	17.0	GALAXY
HELW 104	03 19 05.	- 36 54 22.			NEBULA
SHAP320-4547.8	03 19 05.	- 45 58 32.	36	16.5	GALAXY
WK 122	03 19 05.41	+ 41 18 06.5			NEBULA
SHAP320-5211.2	03 19 06.	- 52 21 57.	24	17.75	GALAXY
SHAP320-5417.2	03 19 06.	- 54 27 57.	24	16.5	GALAXY
BAK 1.366	03 19 07.	- 29 59 20.	18	17.0	EXTRAGALACTIC NEBULA
SN 1965J	03 19 07.	- 37 19		18.0	SUPERNOVA
SHAP320-5437.6	03 19 07.	- 54 48 21.	24	16.5	GALAXY
SHAP320-4529.4	03 19 08.	- 45 40 08.	36	17.0	GALAXY
BAK 1.371	03 19 09.	- 24 56 43.	24	15.3	EXTRAGALACTIC NEBULA
BAK 1.367	03 19 09.	- 31 09 20.	29	15.8	EXTRAGALACTIC NEBULA
SHAP320-4826.1	03 19 09.	- 48 36 50.	24	17.25	GALAXY
SHAP320-5310.6	03 19 09.	- 53 21 21.	24	17.0	GALAXY
SHAP320-5442.6	03 19 09.	- 54 53 21.	24	16.75	GALAXY
BAK 1.368	03 19 10.	- 31 32 32.	19	16.8	EXTRAGALACTIC NEBULA
SHAP320-5049.9	03 19 10.	- 51 00 38.	30	17.25	GALAXY
SHAP320-5258.7	03 19 11.	- 53 09 26.	36	18.0	GALAXY
SHAP320-5350.0	03 19 11.	- 54 00 45.	18	18.0	GALAXY
SHAP320-5437.8	03 19 11.	- 54 48 33.	36	16.5	GALAXY
MCG-03-09-025	03 19 12.	- 16 55 30.	12	15.	GALAXY
PHL 1515	03 19 12.	- 19 34		17.2	BLUE STELLAR OBJECT
PHL 4394	03 19 12.	- 20 50		16.5	BLUE STELLAR OBJECT
BAK 1.369	03 19 12.	- 30 28 49.	13	16.4	EXTRAGALACTIC NEBULA
ARC 0430	03 19 13.	- 15 31		17.7	RICH CLUSTER OF GALAXIES
SHAP320-5141.5	03 19 13.	- 51 52 14.	24	16.5	GALAXY
BAK 1.370	03 19 14.	- 31 34 31.	14	16.8	EXTRAGALACTIC NEBULA
HELW 105	03 19 14.	- 36 20 52.			NEBULA
SHAP320-4514.5	03 19 14.	- 45 25 14.	42	16.75	GALAXY
SHAP320-5141.6	03 19 15.	- 51 52 20.	24	16.75	GALAXY
SHAP320-5247.2	03 19 15.	- 52 57 56.	18	17.5	GALAXY
SHAP320-5413.2	03 19 16.	- 54 23 56.	24	17.5	GALAXY
SHAP320-5417.0	03 19 16.	- 54 27 44.	90	14.8	GALAXY
WK 123	03 19 16.61	+ 41 13 22.5			NEBULA
ARC 0431	03 19 17.	- 16 44		17.4	RICH CLUSTER OF GALAXIES
SHAP320-5440.5	03 19 17.	- 54 51 14.	24	17.0	GALAXY
BAK 1.372	03 19 18.	- 31 32 31.	16	16.7	EXTRAGALACTIC NEBULA
SHAP320-5049.1	03 19 18.	- 50 59 50.	24	17.25	GALAXY
SHAP320-5103.6	03 19 18.	- 51 14 20.	24	18.0	GALAXY
BAK 1.373	03 19 19.	- 27 19 07.	17	16.8	EXTRAGALACTIC NEBULA
SHAP320-5411.6	03 19 20.	- 54 22 20.	30	17.0	GALAXY
SHAP320-5313.8	03 19 21.	- 53 24 32.	24	17.75	GALAXY
SHAP320-4819.9	03 19 22.	- 48 30 38.	24	17.25	GALAXY
SHAP320-5146.4	03 19 22.	- 51 57 08.	24	17.5	GALAXY
SHAP320-5217.2	03 19 23.	- 52 27 56.	24	16.75	GALAXY
SHAP320-5221.7	03 19 23.	- 52 32 26.	30	17.0	GALAXY
SHAP320-5222.0	03 19 23.	- 52 32 44.	18	17.5	GALAXY
MCG-03-09-026	03 19 24.	- 16 56	15	15.5	GALAXY
PHL 8637	03 19 24.	- 17 23		18.1	BLUE STELLAR OBJECT
BAK 1.374	03 19 24.	- 26 42 13.	17	17.0	EXTRAGALACTIC NEBULA
PHL 1516	03 19 24.	- 31 14		18.5	BLUE STELLAR OBJECT
HPW 03	03 19 24.	- 35 03			DWARF GALAXY IN FORNAX
SHAP320-5115.2	03 19 24.	- 51 25 56.	18	17.5	GALAXY
SHAP320-5218.1	03 19 24.	- 52 28 50.	18	17.0	GALAXY
BAK 1.375	03 19 25.	- 24 46 12.	10	17.3	EXTRAGALACTIC NEBULA
SHAP320-5310.8	03 19 25.	- 53 21 32.	18	17.0	GALAXY
SHAP320-5430.0	03 19 25.	- 54 40 44.	24	17.0	GALAXY
SHAP321-4748.3	03 19 26.	- 47 59 01.	18	17.5	GALAXY
MCG+07-07-079	03 19 27.	+ 41 21	60	15.	GALAXY
SHAP320-5242.6	03 19 27.	- 52 53 20.	18	17.5	GALAXY
BAK 1.380	03 19 28.	- 24 13 42.	14	16.7	EXTRAGALACTIC NEBULA
BAK 1.377	03 19 28.	- 26 57 36.	26	16.6	EXTRAGALACTIC NEBULA
SHAP320-5418.1	03 19 28.	- 54 28 50.	18	17.4	GALAXY
SHAP321-4523.8	03 19 29.	- 45 34 31.	36	17.25	GALAXY
SHAP320-5313.9	03 19 29.	- 53 24 37.	24	17.0	GALAXY
SHAP320-5440.2	03 19 29.	- 54 50 56.	24	16.75	GALAXY
LB 02865	03 19 30.	+ 00 00 48.		16.2	FAINT BLUE STAR
OCL 0386	03 19 30.	+ 52 04	180		OPEN STAR CLUSTER
MCG+00-09-070	03 19 30.	- 01 02	42	14.	GALAXY
UGC 02699	03 19 30.	- 01 14	66	16.5	GALAXY S?
PHL 8638	03 19 30.	- 09 12		16.9	BLUE STELLAR OBJECT
PHL 4395	03 19 30.	- 20 41		18.6	BLUE STELLAR OBJECT
SHAP321-5107.7	03 19 30.	- 51 18 25.	24	17.5	GALAXY

OBJECT NAME	RIGHT ASCEN.	DECLINATION	DIAM.	MAGN.	TYPE OF OBJECT
SHAP320-5123.7	03 19 30.	- 51 34 25.	24	17.25	GALAXY
SER 028.02	03 19 30.	- 66 54	180		VERY COMPACT GALAXY
SHAP320-5226.4	03 19 31.	- 52 37 07.	24	17.5	GALAXY
SHAP320-5306.9	03 19 31.	- 53 17 37.	24	17.75	GALAXY
SHAP320-5351.3	03 19 31.	- 54 02 01.	24	16.5	GALAXY
YM 37	03 19 32.	+ 36 53	210		SYMMETRIC GALACTIC NEBULA
BAK 1.381	03 19 32.	- 24 37 00.	24	16.4	EXTRAGALACTIC NEBULA
BAK 1.376	03 19 32.	- 30 50 12.	18	17.0	EXTRAGALACTIC NEBULA
SHAP321-4955.8	03 19 33.	- 50 06 31.	24	17.0	GALAXY
SHAP321-5229.8	03 19 33.	- 52 40 31.	30	17.5	GALAXY
SHAP321-4712.8	03 19 34.	- 47 23 31.	30	16.5	GALAXY
SHAP321-4722.1	03 19 34.	- 47 32 49.	18	17.5	GALAXY
SHAP320-5322.6	03 19 34.	- 53 33 19.	24	17.75	GALAXY
BAK 1.378	03 19 35.	- 31 22 24.	26	17.2	EXTRAGALACTIC NEBULA
SHAP320-5433.7	03 19 35.	- 54 44 25.	36	16.5	GALAXY
ZC 0319.6+2320	03 19 36.	+ 23 20	2690		CLUSTER OF GALAXIES
ZWG 540.121	03 19 36.	+ 42 22		15.5	GALAXY
UGC 02700	03 19 36.	+ 42 22	120	15.5	GALAXY SB?b
VB 235	03 19 36.	+ 54 33	470		STELLAR RING
ZWG 390.073	03 19 36.	- 01 00		15.4	GALAXY
MCG-02-09-034	03 19 36.	- 12 30	60	15.	GALAXY
MCG-03-09-027	03 19 36.	- 15 54	84	15.5	GALAXY
BAK 1.384	03 19 36.	- 24 20 42.	17	16.4	EXTRAGALACTIC NEBULA
BAK 1.379	03 19 36.	- 31 39 24.	25	16.4	EXTRAGALACTIC NEBULA
SHAP321-5059.5	03 19 36.	- 51 10 13.	24	17.75	GALAXY
SHAP321-5103.6	03 19 37.	- 51 14 19.	30	18.0	GALAXY
BAK 1.382	03 19 38.	- 29 35 18.	17	16.4	EXTRAGALACTIC NEBULA
SHAP321-5101.2	03 19 38.	- 51 11 55.	24	17.5	GALAXY
SHAP321-5141.8	03 19 39.	- 51 52 31.	18	16.75	GALAXY
SHAP321-5231.2	03 19 39.	- 52 41 55.	18	17.75	GALAXY
LB 02866	03 19 40.	+ 00 44 48.		16.4	FAINT BLUE STAR
BAK 1.386	03 19 40.	- 23 21 54.	30	16.9	EXTRAGALACTIC NEBULA
BAK 1.385	03 19 40.	- 25 49 54.	22	16.8	EXTRAGALACTIC NEBULA
SHAP321-5017.0	03 19 40.	- 50 27 43.	18	17.25	GALAXY
SHAP321-5107.4	03 19 40.	- 51 18 07.	18	18.0	GALAXY
SHAP321-5153.8	03 19 40.	- 52 04 31.	24	17.5	GALAXY
SHAP321-4530.0	03 19 41.	- 45 40 42.	48	17.0	GALAXY
SHAP321-4749.1	03 19 41.	- 47 59 48.	18	17.5	GALAXY
SHAP321-5227.3	03 19 41.	- 52 38 01.	18	17.0	GALAXY
SHAP321-5437.8	03 19 41.	- 54 48 31.	24	17.0	GALAXY
UGC 02701	03 19 42.	+ 09 16	84	16.0	GALAXY Sb-c
ZWG 525.025	03 19 42.	+ 36 51		15.5	GALAXY
UGC 02702	03 19 42.	+ 36 51	180	15.5	GALAXY
MCG-02-09-035	03 19 42.	- 13 50	60	15.5	GALAXY
BAK 1.388	03 19 42.	- 23 00 23.	14	16.8	EXTRAGALACTIC NEBULA
BAK 1.383	03 19 42.	- 30 56 42.	22	16.4	EXTRAGALACTIC NEBULA
LB 01661	03 19 42.	- 46 42		14.8	FAINT BLUE STAR
SHAP321-5111.4	03 19 42.	- 51 22 07.	30	18.0	GALAXY
SHAP321-5157.0	03 19 42.	- 52 07 43.	24	17.5	GALAXY
SHAP321-5321.3	03 19 42.	- 53 32 01.	24	17.5	GALAXY
SHAP321-5327.2	03 19 42.	- 53 37 55.	18	17.0	GALAXY
SHAP321-5157.8	03 19 43.	- 52 08 31.	18	17.5	GALAXY
SHAP321-5232.9	03 19 43.	- 52 43 37.	18	17.5	GALAXY
SHAP321-5328.4	03 19 43.	- 53 39 07.	24	17.5	GALAXY
RNGC 1309	03 19 44.	- 15 35		12.5	GALAXY
SHAP321-5223.4	03 19 44.	- 52 34 07.	18	17.5	GALAXY
RNGC 1307	03 19 45.	- 04 43			NON-EXISTENT OBJECT
SHAP321-5043.8	03 19 45.	- 50 54 30.	18	16.0	GALAXY
BAK 1.390	03 19 46.	- 23 13 53.	11	16.9	EXTRAGALACTIC NEBULA
BAK 1.389	03 19 46.	- 24 11 29.	12	16.6	EXTRAGALACTIC NEBULA
BAK 1.387	03 19 47.	- 29 15 23.	19	17.2	EXTRAGALACTIC NEBULA
SHAP321-5204.1	03 19 47.	- 52 14 48.	24	17.5	GALAXY
ZWG 441.009	03 19 48.	+ 14 44		15.7	GALAXY
UGC 02703	03 19 48.	+ 14 44	90	15.7	GALAXY Sb
MCG+02-09-005	03 19 48.	+ 14 44	84	15.	GALAXY
ZWG 441.010	03 19 48.	+ 14 47		15.7	GALAXY
PHL 4398	03 19 48.	- 05 57		18.0	BLUE STELLAR OBJECT
PHL 4397	03 19 48.	- 06 10		17.0	BLUE STELLAR OBJECT
MCG+01-09-031	03 19 48.	- 07 16	60	13.5	GALAXY
MCG-03-09-028	03 19 48.	- 15 35 30.	150	12.	GALAXY
PHL 8639	03 19 48.	- 21 16		18.2	BLUE STELLAR OBJECT
BAK 1.391	03 19 48.	- 23 58 53.	12	16.2	EXTRAGALACTIC NEBULA
SHAP321-5103.9	03 19 48.	- 51 14 36.	24	17.25	GALAXY
SHAP321-5326.7	03 19 48.	- 53 37 24.	24	17.25	GALAXY
SHAP321-5035.5	03 19 49.	- 50 46 12.	24	17.5	GALAXY
SHAP321-5246.6	03 19 50.	- 52 57 18.	18	17.5	GALAXY
MCG+06-08-016	03 19 51.	+ 36 51 30.	150	15.	GALAXY
BAK 1.392	03 19 52.	- 27 16 17.	23	16.3	EXTRAGALACTIC NEBULA
BAK 1.396	03 19 53.	- 24 02 41.	12	16.8	EXTRAGALACTIC NEBULA
SHAP321-5438.4	03 19 53.	- 54 49 06.	24	17.5	GALAXY
SHAP321-5444.2	03 19 53.	- 54 54 54.	24	17.0	GALAXY
PHL 1517	03 19 54.	- 02 52		17.6	BLUE STELLAR OBJECT
MCG+01-09-032	03 19 54.	- 02 55	54	15.	GALAXY
PHL 4399	03 19 54.	- 10 46		18.5	BLUE STELLAR OBJECT
PHL 4396	03 19 54.	- 18 08		17.1	BLUE STELLAR OBJECT
LB 01662	03 19 54.	- 51 05		12.6	FAINT BLUE STAR
SHAP321-5328.6	03 19 54.	- 53 39 18.	24	16.2	GALAXY
BAK 1.397	03 19 55.	- 24 21 53.	22	16.6	EXTRAGALACTIC NEBULA
SHAP321-5309.2	03 19 55.	- 53 19 54.	18	17.3	GALAXY
SHAP321-5312.5	03 19 55.	- 53 23 12.	18	17.7	GALAXY
RNGC 1308	03 19 56.	- 02 55		15.0	GALAXY
SHAP321-5219.8	03 19 56.	- 52 30 30.	18	17.5	GALAXY
SHAP321-5220.0	03 19 56.	- 52 30 42.	24	16.25	GALAXY
SHAP321-5235.0	03 19 56.	- 52 45 42.	18	18.0	GALAXY
BAK 1.393	03 19 57.	- 30 48 05.	16	17.3	EXTRAGALACTIC NEBULA
SHAP321-5043.2	03 19 58.	- 50 53 54.	18	16.5	GALAXY
BAK 1.394	03 19 59.	- 31 09 17.	16	16.6	EXTRAGALACTIC NEBULA
SHAP321-4453.1	03 19 59.	- 45 03 47.	48	16.75	GALAXY
LBN 0678	03 20	+ 61 20	120		BRIGHT NEBULA
5ZW 341	03 20 00.	+ 24 41			COMPACT GALAXY
HUB C01	03 20 00.	+ 30 07			DIFFUSE NEBULA
DG 013	03 20 00.	+ 61 21	180		REFLECTION NEBULA
PHL 4400	03 20 00.	- 13 26		14.0	BLUE STELLAR OBJECT
PHL 8640	03 20 00.	- 13 52		18.4	BLUE STELLAR OBJECT
MCG-03-09-029	03 20 00.	- 17 23	30	15.5	GALAXY
PHL 8641	03 20 00.	- 24 52		18.5	BLUE STELLAR OBJECT
PHL 1518	03 20 00.	- 26 37		18.4	BLUE STELLAR OBJECT
PHL 8642	03 20 00.	- 26 50		18.5	BLUE STELLAR OBJECT
BAK 1.400	03 20 00.	- 27 45 29.	13	16.7	EXTRAGALACTIC NEBULA
BAK 1.399	03 20 00.	- 27 45 47.	13	15.8	EXTRAGALACTIC NEBULA
SHAP321-5326.4	03 20 00.	- 53 37 06.	18	16.5	GALAXY
BAK 1.395	03 20 01.	- 31 21 17.	23	17.0	EXTRAGALACTIC NEBULA
SHAP321-4644.0	03 20 01.	- 46 54 41.	24	18.0	GALAXY
BAK 1.403	03 20 02.	- 20 56 40.	19	16.2	EXTRAGALACTIC NEBULA
BAK 1.401	03 20 02.	- 27 58 29.	20	16.5	EXTRAGALACTIC NEBULA
SHAP321-4721.0	03 20 02.	- 47 31 41.	24	17.5	GALAXY
SHAP321-4933.8	03 20 02.	- 49 44 29.	24	16.25	GALAXY
BAK 1.402	03 20 03.	- 27 59 40.	16	16.9	EXTRAGALACTIC NEBULA

OBJECT NAME	RIGHT ASCEN.	DECLINATION	DIAM.	MAGN.	TYPE OF OBJECT
BAK 1.398	03 20 03.	- 30 51 53.	20	16.8	EXTRAGALACTIC NEBULA
SHAP321-5206.5	03 20 04.	- 52 17 11.	24	17.0	GALAXY
SHAP321-5128.6	03 20 05.	- 51 39 17.	24	17.25	GALAXY
SHAP321-5406.4	03 20 05.	- 54 17 06.	18	17.75	GALAXY
5ZW 342	03 20 06.	+ 32 40			COMPACT GALAXY
SS 09	03 20 06.	+ 61 22			DIFFUSE GALACTIC NEBULA
VDB.66N 011	03 20 06.	+ 61 24	372		REFLECTION NEBULA
ZWG 390.074	03 20 06.	- 02 14		15.6	GALAXY
MCG+00-09-071	03 20 06.	- 02 14 30.	24	15.	GALAXY
PHL 4401	03 20 06.	- 04 34		18.6	BLUE STELLAR OBJECT
PHL 8643	03 20 06.	- 08 52		16.9	BLUE STELLAR OBJECT
SHAP321-5405.8	03 20 06.	- 54 16 29.	12	17.0	GALAXY
SHAP321-4531.2	03 20 07.	- 45 41 53.	36	17.0	GALAXY
BAK 1.405	03 20 08.	- 25 42 16.	14	16.4	EXTRAGALACTIC NEBULA
SHAP321-5332.5	03 20 08.	- 53 43 11.	24	17.25	GALAXY
IC 0319	03 20 08.0	+ 41 13 38.			GALAXY SB0
RNGC 1314	03 20 09.	- 04 21		15.5	GALAXY
BAK 1.404	03 20 09.	- 27 48 04.	13	16.4	EXTRAGALACTIC NEBULA
SHAP321-5036.0	03 20 09.	- 50 46 41.	30	17.25	GALAXY
SHAP321-5048.2	03 20 11.	- 50 58 53.	24	17.5	GALAXY
SHAP321-5103.9	03 20 11.	- 51 14 35.	18	18.0	GALAXY
SHAP321-5258.6	03 20 11.	- 53 09 17.	18	16.7	GALAXY
MCG+07-07-080	03 20 12.	+ 40 30	12	15.	GALAXY
1ZW 012	03 20 12.	+ 43 22			COMPACT GALAXY
ZWG 390.076	03 20 12.	- 00 01		15.0	GALAXY
UGC 02705	03 20 12.	- 00 01	66	15.0	GALAXY Sc
MCG+00-09-072	03 20 12.	- 00 02 30.	54	14.	GALAXY
ZWG 390.075	03 20 12.	- 02 05		14.3	GALAXY
UGC 02704	03 20 12.	- 02 05	60	14.3	GALAXY
MCG+00-09-073	03 20 12.	- 02 12	21	15.	GALAXY
MCG+01-09-033	03 20 12.	- 04 22	60	15.5	GALAXY
MCG-03-09-030	03 20 12.	- 17 23	30	15.5	GALAXY
PHL 1519	03 20 12.	- 18 34		18.6	BLUE STELLAR OBJECT
PHL 4402	03 20 12.	- 19 16		16.3	BLUE STELLAR OBJECT
PHL 8644	03 20 12.	- 26 18		18.6	BLUE STELLAR OBJECT
SHAP321-5054.7	03 20 12.	- 51 05 23.	18	16.0	GALAXY
SHAP321-5139.4	03 20 12.	- 51 50 05.	54	16.5	GALAXY
BAK 1.406	03 20 15.	- 30 57 58.	26	16.3	EXTRAGALACTIC NEBULA
SHAP321-5037.8	03 20 15.	- 50 48 29.	30	17.75	GALAXY
SHAP321-5042.9	03 20 15.	- 50 53 35.	24	17.5	GALAXY
SHAP321-4526.4	03 20 16.	- 45 37 04.	36	17.0	GALAXY
SHAP321-4948.0	03 20 17.	- 49 58 41.	12	16.5	GALAXY
SHAP321-5447.4	03 20 17.	- 54 58 05.	24	16.5	GALAXY
MCG+00-09-074	03 20 18.	+ 01 11	48	15.	GALAXY
ZWG 390.077	03 20 18.	- 02 12		15.6	GALAXY
PHL 1520	03 20 18.	- 05 54		17.6	BLUE STELLAR OBJECT
PHL 4403	03 20 18.	- 08 28		17.9	BLUE STELLAR OBJECT
TON-S 0344	03 20 18.	- 22 11		15.5	BLUE STAR
BAK 1.407	03 20 18.	- 30 53 34.	18	17.4	EXTRAGALACTIC NEBULA
BAK 1.408	03 20 18.	- 30 56 28.	14	17.7	EXTRAGALACTIC NEBULA
SER 031.04	03 20 18.	- 51 52	30	17.	PEC. COMPACT GALAXY
SHAP321-5120.8	03 20 19.	- 51 31 29.	24	17.75	GALAXY
SHAP321-5258.4	03 20 19.	- 53 09 05.	24	16.4	GALAXY
SHAP321-5354.7	03 20 19.	- 54 05 23.	24	17.75	GALAXY
BAK 1.409	03 20 22.	- 30 21 21.	17	17.4	EXTRAGALACTIC NEBULA
SHAP321-5229.6	03 20 22.	- 52 40 16.	18	17.0	GALAXY
SHAP321-5433.6	03 20 22.	- 54 44 17.	24	16.5	GALAXY
SHAP322-4657.6	03 20 23.	- 47 08 16.	36	16.5	GALAXY
SHAP321-5245.2	03 20 23.	- 52 55 52.	18	17.0	GALAXY
ZWG 390.078	03 20 24.	+ 01 12		15.0	GALAXY
ZWG 525.026	03 20 24.	+ 37 57		15.6	GALAXY
MCG+07-07-081	03 20 24.	+ 40 22	48	15.5	GALAXY
PHL 8645	03 20 24.	- 09 24		16.9	BLUE STELLAR OBJECT
PHL 4404	03 20 24.	- 11 30		17.5	BLUE STELLAR OBJECT
PHL 4405	03 20 24.	- 22 08		18.2	BLUE STELLAR OBJECT
SHAP321-5041.4	03 20 24.	- 50 52 04.	24	17.5	GALAXY
LB 02867	03 20 25.	+ 02 31 12.		16.8	FAINT BLUE STAR
SHAP322-4759.7	03 20 25.	- 48 10 22.	24	17.0	GALAXY
SHAP321-5136.8	03 20 25.	- 51 47 28.	24	17.5	GALAXY
SHAP321-5217.0	03 20 25.	- 52 27 40.	18	17.5	GALAXY
SHAP321-5310.0	03 20 25.	- 53 20 40.	18	17.5	GALAXY
BAK 1.410	03 20 26.	- 24 03 09.	14	17.0	EXTRAGALACTIC NEBULA
HELW 106	03 20 26.	- 37 34 08.			NEBULA
SHAP321-5119.0	03 20 26.	- 51 29 40.	18	16.5	GALAXY
SHAP321-5149.6	03 20 26.	- 52 00 16.	18	17.5	GALAXY
SHAP321-5202.2	03 20 26.	- 52 12 52.	24	16.75	GALAXY
LB 00484	03 20 27.	- 00 04 18.		17.6	FAINT BLUE STAR
SHAP321-5314.6	03 20 27.	- 53 25 16.	24	17.6	GALAXY
SHAP321-5437.6	03 20 27.	- 54 48 16.	24	16.75	GALAXY
SHAP321-5131.5	03 20 28.	- 51 42 10.	24	17.0	GALAXY
SHAP321-5157.7	03 20 28.	- 52 08 22.	24	16.75	GALAXY
SHAP321-5204.1	03 20 28.	- 52 14 46.	18	17.5	GALAXY
SHAP321-5322.6	03 20 28.	- 53 33 16.	24	17.1	GALAXY
SHAP322-4831.4	03 20 29.	- 48 42 04.	18	17.5	GALAXY
SHAP321-5134.4	03 20 29.	- 51 45 04.	24	17.75	GALAXY
SHAP321-5304.4	03 20 29.	- 53 15 04.	18	17.5	GALAXY
ZWG 525.027	03 20 30.	+ 36 45		15.7	GALAXY
UGC 02706	03 20 30.	+ 36 45	84	15.7	GALAXY S-IRR
UGC 02707	03 20 30.	+ 36 52	84	18.	GALAXY
PHL 8646	03 20 30.	- 10 07		16.2	BLUE STELLAR OBJECT
BAK 1.413	03 20 30.	- 24 58 15.	17	16.7	EXTRAGALACTIC NEBULA
SER 031.03	03 20 30.	- 51 30	180	18.	CLUSTER WITH cD GALAXY
SHAP321-5140.1	03 20 30.	- 51 50 46.	18	17.25	GALAXY
SHAP321-5144.5	03 20 30.	- 51 55 10.	18	17.5	GALAXY
SHAP321-5201.8	03 20 30.	- 52 12 28.	24	17.5	GALAXY
SHAP321-5214.5	03 20 31.	- 52 25 10.	18	17.5	GALAXY
BAK 1.411	03 20 33.	- 27 55 27.	20	17.2	EXTRAGALACTIC NEBULA
BAK 1.412	03 20 33.	- 27 59 51.	12	16.8	EXTRAGALACTIC NEBULA
SHAP322-5056.6	03 20 33.	- 51 07 16.	24	16.5	GALAXY
SHAP322-5117.5	03 20 33.	- 51 28 10.	24	18.0	GALAXY
SHAP321-5345.8	03 20 33.	- 53 56 28.	24	17.75	GALAXY
BAK 1.414	03 20 34.	- 26 00 51.	19	16.6	EXTRAGALACTIC NEBULA
SHAP322-5101.8	03 20 34.	- 51 12 28.	24	17.5	GALAXY
SHAP321-5318.3	03 20 34.	- 53 28 58.	24	16.7	GALAXY
SHAP322-4832.2	03 20 35.	- 48 42 52.	18	17.5	GALAXY
SHAP322-4840.6	03 20 35.	- 48 51 16.	24	17.5	GALAXY
SHAP321-5434.8	03 20 35.	- 54 45 28.	36	17.25	GALAXY
ZWG 541.001	03 20 36.	+ 40 23		15.7	GALAXY
ZWG 540.122	03 20 36.	+ 40 23		15.7	GALAXY
UGC 02708	03 20 36.	+ 40 23	84	15.7	GALAXY S0
5ZW 343	03 20 36.	+ 42 07			COMPACT GALAXY
MCG-02-09-036	03 20 36.	- 11 22 30.	138	14.	GALAXY
PHL 4406	03 20 36.	- 31 59		18.1	BLUE STELLAR OBJECT
SHAP322-5336.8	03 20 36.	- 53 47 28.	24	17.25	GALAXY
SHAP321-5344.9	03 20 36.	- 53 55 34.	18	17.75	GALAXY
SHAP322-5258.6	03 20 37.	- 53 09 16.	18	16.5	GALAXY
SHAP322-5321.2	03 20 37.	- 53 31 52.	24	16.8	GALAXY
SHAP322-5021.5	03 20 38.	- 50 32 09.	24	17.0	GALAXY

OBJECT NAME	RIGHT ASCEN.	DECLINATION	DIAM.	MAGN.	TYPE OF OBJECT
SHAP322-5138.2	03 20 38.	- 51 48 52.	24	16.0	GALAXY
SHAP322-5212.6	03 20 38.	- 52 23 16.	24	17.0	GALAXY
SHAP322-5422.8	03 20 38.	- 54 33 28.	24	16.75	GALAXY
MCG+06-08-017	03 20 39.	+ 36 44	42	15.5	GALAXY
RNGC 1316	03 20 39.	- 37 25		10.5	GALAXY
SHAP322-5033.2	03 20 39.	- 50 43 51.	24	17.5	GALAXY
SHAP322-5257.1	03 20 39.	- 53 07 46.	18	16.6	GALAXY
SHAP322-5247.9	03 20 40.	- 52 58 33.	18	16.6	GALAXY
BAK 1.419	03 20 41.	- 22 37 08.	17	16.6	EXTRAGALACTIC NEBULA
SHAP322-5121.8	03 20 41.	- 51 32 27.	24	17.75	GALAXY
SHAP322-5128.2	03 20 41.	- 51 38 51.	30	17.5	GALAXY
SHAP322-5153.0	03 20 41.	- 52 03 39.	18	18.0	GALAXY
SHAP322-5421.9	03 20 41.	- 54 32 34.	24	16.75	GALAXY
ZWG 416.007	03 20 42.	+ 08 43		15.4	GALAXY
KARA.73B 0125	03 20 42.	+ 08 43	36	15.4	ISOLATED GALAXY S
ZWG 525.028	03 20 42.	+ 37 47		15.7	GALAXY
ZWG 525.029	03 20 42.	+ 38 30		14.8	GALAXY
UGC 02709	03 20 42.	+ 38 30	210	14.8	GALAXY Sb/SBb
PHL 4407	03 20 42.	- 14 12		18.4	BLUE STELLAR OBJECT
BAK 1.415	03 20 42.	- 29 26 32.	35	15.8	EXTRAGALACTIC NEBULA
MCG-06-08-005	03 20 42.	- 37 24	420		GALAXY
SHAP322-4911.6	03 20 42.	- 49 22 15.	30	16.5	GALAXY
SHAP322-5259.5	03 20 42.	- 53 10 09.	24	17.6	GALAXY
LB 03310	03 20 42.	- 70 45		14.9	FAINT BLUE STAR
BAK 1.416	03 20 43.	- 27 54 56.	16	16.9	EXTRAGALACTIC NEBULA
SHAP322-5200.6	03 20 43.	- 52 11 15.	18	17.5	GALAXY
SHAP322-5223.6	03 20 43.	- 52 34 15.	18	17.25	GALAXY
SHAP322-5435.0	03 20 43.	- 54 45 39.	36	17.75	GALAXY
ARP 154	03 20 44.	- 37 24			PECULIAR GALAXY
MCG+06-08-018	03 20 45.	+ 38 30	102	14.	GALAXY
BAK 1.421	03 20 45.	- 22 13 38.	12	16.2	EXTRAGALACTIC NEBULA
RNGC 1317	03 20 45.	- 37 17		12.5	GALAXY
MCG-06-08-006	03 20 45.	- 37 18	180		GALAXY
SHAP322-5106.0	03 20 45.	- 51 16 39.	24	17.5	GALAXY
SHAP322-5329.4	03 20 45.	- 53 40 03.	24	17.25	GALAXY
SHAP322-5156.4	03 20 46.	- 52 07 03.	18	17.0	GALAXY
SHAP322-5244.4	03 20 46.	- 52 55 03.	18	17.75	GALAXY
SHAP322-5312.4	03 20 46.	- 53 23 03.	24	17.3	GALAXY
SHAP322-4903.6	03 20 47.	- 49 14 15.	18	16.5	GALAXY
SHAP322-5113.0	03 20 47.	- 51 23 39.	24	18.0	GALAXY
SHAP322-5126.2	03 20 47.	- 51 36 51.	18	16.5	GALAXY
SHAP322-5145.6	03 20 47.	- 51 56 15.	18	16.0	GALAXY
SHAP322-5309.4	03 20 47.	- 53 20 03.	18	17.4	GALAXY
SHAP322-5340.7	03 20 47.	- 53 51 21.	24	17.75	GALAXY
SHAP322-5352.5	03 20 47.	- 54 03 09.	30	17.25	GALAXY
ZWG 390.079	03 20 48.	+ 00 16		15.5	GALAXY
MCG+00-09-076	03 20 48.	+ 00 17 30.	36	15.	GALAXY
MCG+00-09-075	03 20 48.	+ 01 09	30	15.5	GALAXY
ZC 0320.8+0602	03 20 48.	+ 06 02	1340		CLUSTER OF GALAXIES
ZWG 525.030	03 20 48.	+ 37 35		15.0	GALAXY
UGC 02710	03 20 48.	+ 37 35	102	15.0	GALAXY S0
PHL 1521	03 20 48.	- 12 28		18.5	BLUE STELLAR OBJECT
PHL 1522	03 20 48.	- 19 52		18.4	BLUE STELLAR OBJECT
PHL 4408	03 20 48.	- 23 22		18.4	BLUE STELLAR OBJECT
PHL 8647	03 20 48.	- 24 17		11.0	BLUE STELLAR OBJECT
BAK 1.418	03 20 48.	- 32 17 26.	14	17.0	EXTRAGALACTIC NEBULA
BAK 1.417	03 20 48.	- 32 18 14.	14	16.8	EXTRAGALACTIC NEBULA
HELW 107	03 20 48.	- 36 36 51.			NEBULA
SHAP322-4502.8	03 20 48.	- 45 13 27.	48	17.0	GALAXY
SHAP322-5110.8	03 20 48.	- 51 21 27.	18	17.0	GALAXY
SHAP322-5129.6	03 20 48.	- 51 40 15.	24	17.0	GALAXY
SHAP322-5311.3	03 20 48.	- 53 21 57.	66	16.9	GALAXY
IC 1917	03 20 48.	- 53 23			NONSTELLAR OBJECT
SHAP322-4917.3	03 20 49.	- 49 27 57.	24	16.25	GALAXY
SHAP322-5217.1	03 20 49.	- 52 27 45.	24	18.0	GALAXY
SHAP322-5300.9	03 20 49.	- 53 11 33.	12	17.2	GALAXY
HN 0182	03 20 49.	- 53 22			NEBULA
SHAP322-4646.0	03 20 50.	- 46 56 39.	24	16.75	GALAXY
SHAP322-5338.6	03 20 50.	- 53 49 15.	24	17.25	GALAXY
MCG+06-08-019	03 20 51.	+ 37 34 30.	60	15.	GALAXY
RNGC 1318	03 20 51.	- 37 17			NON-EXISTENT OBJECT
SHAP322-5159.9	03 20 51.	- 52 10 33.	18	17.5	GALAXY
BAK 1.424	03 20 52.	- 23 21 50.	36	16.4	EXTRAGALACTIC NEBULA
BAK 1.422	03 20 52.	- 28 01 38.	14	16.2	EXTRAGALACTIC NEBULA
SHAP322-5200.8	03 20 52.	- 52 11 27.	18	17.5	GALAXY
RNGC 1315	03 20 53.	- 21 33		13.0	GALAXY
BAK 1.420	03 20 53.	- 32 12 14.	24	16.6	EXTRAGALACTIC NEBULA
SHAP322-5124.9	03 20 53.	- 51 35 33.	30	18.0	GALAXY
ZWG 390.080	03 20 54.	+ 01 09		15.7	GALAXY
UGC 02711	03 20 54.	+ 01 09	90	15.7	GALAXY SB:b
PHL 8649	03 20 54.	- 04 34		16.8	BLUE STELLAR OBJECT
PHL 8648	03 20 54.	- 13 44		15.5	BLUE STELLAR OBJECT
TON-S 0345	03 20 54.	- 29 57		14.8	BLUE STAR
LB 01663	03 20 54.	- 53 57		13.6	FAINT BLUE STAR
RNGC 1312	03 20 55.	+ 01 09		15.5	GALAXY
SHAP322-5245.0	03 20 55.	- 52 55 39.	18	17.5	GALAXY
BAK 1.427	03 20 57.	- 24 00 43.	17	16.0	EXTRAGALACTIC NEBULA
HELW 108	03 20 57.	- 36 56 40.			NEBULA
SHAP322-4633.8	03 20 57.	- 46 44 26.	30	16.5	GALAXY
SHAP322-5123.6	03 20 57.	- 51 34 14.	24	18.0	GALAXY
LB 02868	03 20 58.	- 00 20 48.		16.3	FAINT BLUE STAR
BAK 1.423	03 20 58.	- 31 24 25.	22	16.6	EXTRAGALACTIC NEBULA
SHAP322-5103.3	03 20 58.	- 51 13 56.	54	16.5	GALAXY
LB 02869	03 20 59.	- 00 20 54.		16.1	FAINT BLUE STAR
BAK 1.425	03 20 59.	- 28 56 13.	13	16.8	EXTRAGALACTIC NEBULA
SHAP322-5212.2	03 20 59.	- 52 22 50.	18	17.25	GALAXY
LB 09798	03 21	- 80 03		14.0	FAINT BLUE STAR
LB 09799	03 21	- 80 28		14.5	FAINT BLUE STAR
LB 09800	03 21	- 82 50		14.5	FAINT BLUE STAR
UGC 02712	03 21 00.	+ 06 24	102	17.	GALAXY
MCG+01-09-004	03 21 00.	+ 06 24	72	15.	GALAXY
ZC 0321.0+0648	03 21 00.	+ 06 48	10820		CLUSTER OF GALAXIES
MCG+02-09-006	03 21 00.	+ 14 00	48	16.	GALAXY
ZC 0321.0+8124	03 21 00.	+ 81 24	1140		CLUSTER OF GALAXIES
MCG-03-09-031	03 21 00.	- 15 50	48	15.5	GALAXY
MCG-04-09-002	03 21 00.	- 21 31 30.	72	13.5	GALAXY
PHL 4409	03 21 00.	- 21 51		18.4	BLUE STELLAR OBJECT
PHL 1523	03 21 00.	- 29 56		14.6	BLUE STELLAR OBJECT
BAK 1.426	03 21 00.	- 30 18 31.	25	16.7	EXTRAGALACTIC NEBULA
SHAP322-5020.2	03 21 01.	- 50 30 50.	24	17.0	GALAXY
SHAP322-5211.0	03 21 01.	- 52 21 38.	18	17.75	GALAXY
SHAP322-5219.8	03 21 01.	- 52 30 26.	18	17.0	GALAXY
KARA.73 07	03 21 03.	- 40 50	27		DWARF GALAXY
SHAP322-5042.6	03 21 03.	- 50 53 14.	18	18.0	GALAXY
SHAP322-5126.5	03 21 03.	- 51 37 08.	24	17.25	GALAXY
SHAP322-5350.0	03 21 03.	- 54 00 38.	24	17.5	GALAXY
HN 0023	03 21 04.	+ 01 01			NEBULA
SHAP322-5140.6	03 21 04.	- 51 51 14.	18	17.75	GALAXY

OBJECT NAME	RIGHT ASCEN.	DECLINATION	DIAM.	MAGN.	TYPE OF OBJECT
SHAP322-5150.8	03 21 04.	- 52 01 26.	18	17.75	GALAXY
SHAP322-5224.0	03 21 04.	- 52 34 38.	18	17.25	GALAXY
SHAP322-5226.6	03 21 04.	- 52 37 14.	24	17.0	GALAXY
SHAP322-5326.0	03 21 04.	- 53 36 38.	18	17.2	GALAXY
SHAP322-5203.1	03 21 05.	- 52 13 44.	18	17.75	GALAXY
SHAP322-5210.8	03 21 05.	- 52 21 26.	24	17.75	GALAXY
SHAP322-5243.1	03 21 05.	- 52 53 44.	18	17.0	GALAXY
SHAP322-5314.0	03 21 05.	- 53 24 38.	24	17.0	GALAXY
MCG+03-09-013	03 21 06.	+ 17 35	6	18.	GALAXY
UGC 02714	03 21 06.	+ 37 51	60	17.	GALAXY
UGC 02715	03 21 06.	+ 40 38	60	16.0	GALAXY Sb-c
ZWG 390.081	03 21 06.	- 00 17		15.3	GALAXY
UGC 02713	03 21 06.	- 00 17	72	15.3	GALAXY TRP SYS
MCG+00-09-077	03 21 06.	- 00 19	12	16.	GALAXY
SHAP322-5119.0	03 21 06.	- 51 29 38.	30	17.25	GALAXY
SHAP322-5151.9	03 21 06.	- 52 02 32.	48	17.0	GALAXY
SHAP322-4818.1	03 21 07.	- 48 28 44.	24	16.5	GALAXY
SHAP322-5157.0	03 21 07.	- 52 07 38.	18	17.0	GALAXY
SHAP322-5232.7	03 21 07.	- 52 43 20.	24	16.5	GALAXY
SHAP322-5351.1	03 21 07.	- 54 01 44.	30	17.0	GALAXY
MCG+00-09-078	03 21 09.	- 00 19	9	16.	GALAXY
MCG-02-09-037	03 21 09.	- 10 40	36	15.5	GALAXY
SHAP322-4624.0	03 21 09.	- 46 34 38.	24	17.5	GALAXY
SHAP322-5210.0	03 21 09.	- 52 20 38.	18	17.5	GALAXY
SHAP322-5223.4	03 21 09.	- 52 34 02.	24	17.0	GALAXY
SHAP322-4818.0	03 21 10.	- 48 28 38.	48	16.0	GALAXY
SHAP322-5042.9	03 21 10.	- 50 53 32.	30	18.0	GALAXY
SHAP322-5315.8	03 21 10.	- 53 26 26.	24	17.3	GALAXY
BAK 1.429	03 21 11.	- 27 50 01.	65	16.1	EXTRAGALACTIC NEBULA
SHAP322-5119.0	03 21 11.	- 51 29 38.	24	17.0	GALAXY
SHAP322-5141.9	03 21 11.	- 51 52 32.	24	17.5	GALAXY
SHAP322-5151.7	03 21 11.	- 52 02 20.	30	17.5	GALAXY
SHAP322-5204.8	03 21 11.	- 52 15 26.	30	18.0	GALAXY
ZWG 464.014	03 21 12.	+ 16 00		14.8	GALAXY
KARA.73B 0126	03 21 12.	+ 16 00	54	14.8	ISOLATED GALAXY S
ZWG 464.015	03 21 12.	+ 17 34		15.4	GALAXY
UGC 02716	03 21 12.	+ 17 34	120	15.4	GALAXY S-IRR
MCG+03-09-014	03 21 12.	+ 17 35	66	14.5	GALAXY
MCG+07-08-001	03 21 12.	+ 39 51	12	16.5	GALAXY
PHL 8650	03 21 12.	- 03 24		18.6	BLUE STELLAR OBJECT
MCG-03-09-032	03 21 12.	- 16 58	60	14.5	GALAXY
BAK 1.428	03 21 12.	- 31 21 37.	23	16.6	EXTRAGALACTIC NEBULA
SHAP322-5148.1	03 21 12.	- 51 58 44.	18	17.5	GALAXY
SHAP322-5415.7	03 21 12.	- 54 26 20.	24	16.5	GALAXY
SN 1962O	03 21 13.	+ 39 51		19.8	SUPERNOVA
SHAP322-5140.3	03 21 14.	- 51 50 56.	18	17.75	GALAXY
SHAP322-5220.1	03 21 14.	- 52 30 44.	18	17.25	GALAXY
SHAP322-5303.6	03 21 14.	- 53 14 14.	18	17.5	GALAXY
BAK 1.431	03 21 15.	- 28 22 54.	18	16.9	EXTRAGALACTIC NEBULA
SHAP322-4739.8	03 21 15.	- 47 50 25.	36	17.0	GALAXY
SHAP322-4843.0	03 21 15.	- 48 53 37.	24	17.0	GALAXY
BAK 1.430	03 21 16.	- 31 35 48.	20	17.0	EXTRAGALACTIC NEBULA
SHAP322-5020.0	03 21 16.	- 50 30 37.	18	17.25	GALAXY
SHAP322-5020.2	03 21 16.	- 50 30 49.	18	17.25	GALAXY
SHAP322-5153.6	03 21 16.	- 52 04 13.	24	17.5	GALAXY
SHAP322-5154.6	03 21 16.	- 52 05 13.	18	17.5	GALAXY
SHAP322-5227.8	03 21 16.	- 52 38 25.	24	17.5	GALAXY
SHAP322-5325.2	03 21 16.	- 53 35 50.	24	17.5	GALAXY
MCG+07-08-001A	03 21 18.	+ 40 30	18	14.5	GALAXY
MCG-02-09-038	03 21 18.	- 11 22	96	15.	GALAXY
TON-S 0346	03 21 18.	- 32 02		16.0	BLUE STAR
TON-S 0347	03 21 18.	- 32 41		16.0	BLUE STAR
DV.56 N1316A	03 21 18.	- 37 06			S GALAXY
HPW 04	03 21 18.	- 40 50			DWARF GALAXY IN FORNAX
SHAP322-5217.1	03 21 18.	- 52 27 43.	18	17.5	GALAXY
SHAP322-5224.9	03 21 18.	- 52 35 31.	30	17.0	GALAXY
SHAP322-5040.0	03 21 19.	- 50 50 37.	24	18.0	GALAXY
SHAP322-5219.9	03 21 19.	- 52 30 31.	18	17.0	GALAXY
HELW 109	03 21 20.	- 37 30 05.			NEBULA
SHAP322-5235.4	03 21 21.	- 52 46 01.	18	17.5	GALAXY
SHAP322-5144.0	03 21 22.	- 51 54 37.	12	17.5	GALAXY
SHAP322-5151.0	03 21 22.	- 52 01 37.	18	18.0	GALAXY
BAK 1.437	03 21 23.	- 22 11 30.	16	16.2	EXTRAGALACTIC NEBULA
BAK 1.433	03 21 23.	- 26 23 00.	12	16.7	EXTRAGALACTIC NEBULA
BAK 1.432	03 21 23.	- 28 31 36.	12	16.8	EXTRAGALACTIC NEBULA
ZC 0321.4+0317	03 21 24.	+ 03 17	1410		CLUSTER OF GALAXIES
ZWG 541.002	03 21 24.	+ 40 31		15.0	GALAXY
ZWG 540.123	03 21 24.	+ 40 31		15.0	GALAXY
UGC 02717	03 21 24.	+ 40 31	60	15.0	GALAXY E
PHL 1524	03 21 24.	- 09 19		15.1	BLUE STELLAR OBJECT
PHL 1525	03 21 24.	- 15 48		17.3	BLUE STELLAR OBJECT
LB 01664	03 21 24.	- 51 19		13.0	FAINT BLUE STAR
SHAP322-5141.0	03 21 24.	- 51 51 37.	24	17.5	GALAXY
SHAP322-5152.8	03 21 24.	- 52 03 25.	24	17.0	GALAXY
SHAP322-5239.2	03 21 24.	- 52 49 49.	24	17.0	GALAXY
SHAP322-5311.9	03 21 26.	- 53 22 31.	18	17.4	GALAXY
MCG-03-09-033	03 21 27.	- 19 58	78	14.	GALAXY
SHAP322-5413.0	03 21 27.	- 54 23 37.	36	16.75	GALAXY
HELW 110	03 21 28.	- 37 28 48.			NEBULA
SHAP322-5414.8	03 21 28.	- 54 25 25.	30	17.25	GALAXY
SHAP322-5202.9	03 21 29.	- 52 13 31.	24	16.25	GALAXY
SHAP322-5230.3	03 21 29.	- 52 40 55.	30	16.5	GALAXY
MCG+04-09-001	03 21 30.	+ 25 30	7	16.	GALAXY
UGC 02718	03 21 30.	+ 41 48	72	16.0	GALAXY Sc
VB 236	03 21 30.	+ 52 17	490		STELLAR RING
ZWG 390.082	03 21 30.	- 01 13		15.7	GALAXY
MCG+00-09-079	03 21 30.	- 01 14	48	15.	GALAXY
PHL 8651	03 21 30.	- 03 26		15.9	BLUE STELLAR OBJECT
BAK 1.436	03 21 30.	- 29 02 36.	12	16.7	EXTRAGALACTIC NEBULA
BAK 1.434	03 21 30.	- 30 25 36.	14	16.9	EXTRAGALACTIC NEBULA
SHAP322-5415.0	03 21 30.	- 54 25 37.	36	17.25	GALAXY
LB 02870	03 21 31.	- 01 38 42.		15.9	FAINT BLUE STAR
BAK 1.435	03 21 31.	- 30 36 06.	22	17.0	EXTRAGALACTIC NEBULA
SHAP323-5112.4	03 21 31.	- 51 23 01.	24	17.0	GALAXY
SHAP322-5201.8	03 21 31.	- 52 12 25.	24	16.25	GALAXY
SHAP322-5206.5	03 21 31.	- 52 17 07.	18	18.0	GALAXY
SHAP322-5236.8	03 21 32.	- 52 47 25.	18	18.0	GALAXY
SHAP322-5306.9	03 21 33.	- 53 17 31.	18	17.0	GALAXY
BAK 1.440	03 21 34.	- 25 04 29.	17	16.9	EXTRAGALACTIC NEBULA
ARC 0432	03 21 35.	- 06 00		17.8	RICH CLUSTER OF GALAXIES
SHAP323-4933.5	03 21 35.	- 49 44 06.	24	16.5	GALAXY
SHAP323-5113.0	03 21 35.	- 51 23 36.	24	17.0	GALAXY
UGC 02719	03 21 36.	+ 38 46	66	16.5	GALAXY Sc
KARA.68 034	03 21 36.	- 19 27	67		DWARF GALAXY
BAK 1.444	03 21 36.	- 23 38 53.	12	16.4	EXTRAGALACTIC NEBULA
MCG-06-08-007	03 21 36.	- 37 05	18	15.5	GALAXY
MCG-06-08-008	03 21 36.	- 37 06	36	15.	GALAXY
MCG-06-08-009	03 21 36.	- 37 06 30.	24	15.5	GALAXY

OBJECT NAME	RIGHT ASCEN.	DECLINATION	DIAM.	MAGN.	TYPE OF OBJECT
SHAP323-4930.4	03 21 36.	- 49 41 00.	24	16.0	GALAXY
SHAP323-5304.0	03 21 36.	- 53 14 36.	24	16.3	GALAXY
SHAP322-5427.0	03 21 36.	- 54 37 36.	30	17.75	GALAXY
BAK 1.445	03 21 37.	- 24 25 35.	42	16.5	EXTRAGALACTIC NEBULA
SHAP323-4845.4	03 21 37.	- 48 56 00.	30	16.0	GALAXY
SHAP323-5045.9	03 21 37.	- 50 56 30.	24	17.5	GALAXY
SHAP323-5131.1	03 21 37.	- 51 41 42.	18	17.5	GALAXY
SHAP323-5214.6	03 21 37.	- 52 25 12.	18	17.5	GALAXY
BAK 1.438	03 21 38.	- 31 51 23.	18	17.0	EXTRAGALACTIC NEBULA
SHAP323-4851.8	03 21 38.	- 49 02 24.	24	17.0	GALAXY
BAK 1.441	03 21 39.	- 29 04 47.	22	17.0	EXTRAGALACTIC NEBULA
HELW 111	03 21 39.	- 37 13 48.			NEBULA
SHAP323-4746.6	03 21 39.	- 47 57 12.	18	17.0	GALAXY
SHAP323-5344.8	03 21 39.	- 53 55 24.	18	17.5	GALAXY
BAK 1.446	03 21 40.	- 26 21 53.	14	16.2	EXTRAGALACTIC NEBULA
SHAP323-5224.6	03 21 40.	- 52 35 12.	24	17.5	GALAXY
RNGC 1319	03 21 41.	- 21 41		14.0	GALAXY
BAK 1.439	03 21 41.	- 31 46 23.	19	16.6	EXTRAGALACTIC NEBULA
SHAP323-4642.2	03 21 41.	- 46 52 48.	24	16.5	GALAXY
SHAP323-5128.3	03 21 41.	- 51 38 54.	18	17.0	GALAXY
SHAP323-5343.3	03 21 41.	- 53 53 54.	24	17.0	GALAXY
ZWG 525.031	03 21 42.	+ 38 57		15.5	GALAXY
UGC 02720	03 21 42.	+ 38 57	66	15.5	GALAXY Sa-b
PHL 8652	03 21 42.	- 07 54		16.4	BLUE STELLAR OBJECT
PHL 8654	03 21 42.	- 10 14		16.1	BLUE STELLAR OBJECT
PHL 8653	03 21 42.	- 24 23		18.2	BLUE STELLAR OBJECT
BAK 1.442	03 21 42.	- 31 48 53.	18	16.8	EXTRAGALACTIC NEBULA
HELW 112	03 21 42.	- 37 03 30.			NEBULA
HELW 113	03 21 42.	- 37 04 36.			NEBULA
SHAP323-4959.4	03 21 42.	- 50 10 00.	30	16.5	GALAXY
SHAP323-4724.2	03 21 43.	- 47 34 48.	24	17.5	GALAXY
SHAP323-5159.0	03 21 43.	- 52 09 36.	18	17.0	GALAXY
SHAP323-5200.1	03 21 43.	- 52 10 42.	18	17.0	GALAXY
BAK 1.443	03 21 44.	- 32 42 35.	24	15.8	EXTRAGALACTIC NEBULA
HELW 114	03 21 44.	- 37 04 54.			NEBULA
SHAP323-5414.2	03 21 44.	- 54 24 48.	18	16.5	GALAXY
BAK 1.448	03 21 45.	- 26 41 23.	25	16.3	EXTRAGALACTIC NEBULA
RNGC 1316A	03 21 45.	- 37 07			GALAXY
RNGC 1316B	03 21 45.	- 37 08			GALAXY
SHAP323-5049.8	03 21 45.	- 51 00 24.	18	17.0	GALAXY
SHAP323-5159.6	03 21 45.	- 52 10 12.	18	17.5	GALAXY
SHAP323-5304.4	03 21 45.	- 53 15 00.	18	17.1	GALAXY
SHAP323-4930.4	03 21 46.	- 49 41 00.	18	16.5	GALAXY
SHAP323-5104.0	03 21 46.	- 51 14 36.	18	16.5	GALAXY
BAK 1.447	03 21 47.	- 29 45 23.	17	16.6	EXTRAGALACTIC NEBULA
SHAP323-5108.8	03 21 47.	- 51 19 24.	24	16.0	GALAXY
SHAP323-5114.5	03 21 47.	- 51 25 06.	24	17.0	GALAXY
SHAP323-5221.6	03 21 47.	- 52 32 12.	24	17.75	GALAXY
MCG+06-08-020	03 21 48.	+ 38 57	54	16.5	GALAXY
MCG+07-08-002	03 21 48.	+ 42 12	30	16.	GALAXY
UGC 02721	03 21 48.	- 00 04	66	17.	GALAXY
PHL 1526	03 21 48.	- 02 38		17.8	BLUE STELLAR OBJECT
PHL 8655	03 21 48.	- 08 50		15.2	BLUE STELLAR OBJECT
PHL 4410	03 21 48.	- 12 46		18.6	BLUE STELLAR OBJECT
MCG-03-09-034	03 21 48.	- 16 30	48	15.	GALAXY
MCG-04-09-003	03 21 48.	- 21 41	36	14.	GALAXY
PHL 8656	03 21 48.	- 24 14		18.6	BLUE STELLAR OBJECT
MCG-06-08-010	03 21 48.	- 37 42	48	14.	GALAXY
SHAP323-5127.0	03 21 49.	- 51 37 36.	18	17.0	GALAXY
SHAP323-5157.2	03 21 49.	- 52 07 48.	12	17.25	GALAXY
MCG+00-09-080	03 21 51.	- 00 03 30.	36	16.	GALAXY
MCG+00-09-081	03 21 51.	- 00 15 30.	54	15.	GALAXY
SHAP323-5253.1	03 21 51.	- 53 03 42.	24	16.9	GALAXY
SHAP323-5434.8	03 21 51.	- 54 45 24.	30	17.5	GALAXY
LB 02871	03 21 52.	+ 01 05 18.		16.5	FAINT BLUE STAR
SHAP323-5207.2	03 21 53.	- 52 17 47.	18	17.5	GALAXY
UGC 02722	03 21 54.	+ 42 14	78	16.5	GALAXY Sc
UGC 02723	03 21 54.	+ 42 30	96	16.5	GALAXY SBc
ZWG 390.083	03 21 54.	- 00 15		15.7	GALAXY
PHL 4411	03 21 54.	- 32 01		18.2	BLUE STELLAR OBJECT
SHAP323-5108.3	03 21 55.	- 51 18 53.	18	17.25	GALAXY
SHAP323-5109.1	03 21 55.	- 51 19 41.	18	18.0	GALAXY
SHAP323-5157.7	03 21 55.	- 52 08 17.	24	17.5	GALAXY
SHAP323-5215.9	03 21 55.	- 52 26 29.	18	17.0	GALAXY
SHAP323-5252.2	03 21 55.	- 53 02 47.	18	17.6	GALAXY
RNGC 1326	03 21 56.	- 36 39		11.5	GALAXY
SHAP323-4952.0	03 21 56.	- 50 02 35.	30	17.0	GALAXY
LB 02872	03 21 57.	+ 02 21 30.		16.0	FAINT BLUE STAR
SHAP323-5108.6	03 21 58.	- 51 19 11.	24	18.0	GALAXY
SHAP323-5429.4	03 21 58.	- 54 39 59.	24	17.25	GALAXY
SHAP323-5447.5	03 21 58.	- 54 58 05.	42	17.25	GALAXY
5ZW 344	03 22 00.	+ 41 44			COMPACT GALAXY
MCG+07-08-003	03 22 00.	+ 42 29	48	17.	GALAXY
PHL 4412	03 22 00.	- 03 36		14.1	BLUE STELLAR OBJECT
MCG+01-09-034	03 22 00.	- 05 00	42	15.	GALAXY
PHL 4413	03 22 00.	- 11 54		16.2	BLUE STELLAR OBJECT
MCG-06-08-011	03 22 00.	- 36 40	210		GALAXY
HELW 115	03 22 00.	- 37 40 55.			NEBULA
SHAP323-4721.4	03 22 01.	- 47 31 59.	24	17.5	GALAXY
SHAP323-5127.7	03 22 01.	- 51 38 17.	12	18.0	GALAXY
BAK 1.449	03 22 02.	- 26 41 46.	14	16.0	EXTRAGALACTIC NEBULA
SHAP323-4944.9	03 22 02.	- 49 55 29.	18	16.5	GALAXY
MCG+00-09-082	03 22 03.	- 00 51	36	15.	GALAXY
MCG-02-09-039	03 22 03.	- 11 51	36	14.5	GALAXY
ARC 0433	03 22 04.	- 06 59		17.8	RICH CLUSTER OF GALAXIES
SHAP323-5125.6	03 22 04.	- 51 36 11.	12	17.5	GALAXY
SHAP323-5228.9	03 22 04.	- 52 39 29.	24	17.0	GALAXY
SHAP323-5314.0	03 22 04.	- 53 24 35.	30	16.5	GALAXY
SHAP323-5345.6	03 22 05.	- 53 56 11.	36	16.5	GALAXY
LB 02873	03 22 06.	+ 00 24 12.		16.3	FAINT BLUE STAR
UGC 02724	03 22 06.	+ 40 28	66	16.0	GALAXY Sc-IRR
ZWG 390.084	03 22 06.	- 00 51		15.7	GALAXY
SHAP323-5054.3	03 22 06.	- 51 04 53.	18	17.25	GALAXY
SHAP323-5132.0	03 22 06.	- 51 42 35.	18	17.0	GALAXY
LB 01665	03 22 06.	- 52 19		13.0	FAINT BLUE STAR
SHAP323-5232.8	03 22 06.	- 52 43 23.	18	17.0	GALAXY
SHAP323-5417.8	03 22 06.	- 54 28 23.	24	17.25	GALAXY
BAK 1.450	03 22 07.	- 27 55 45.	16	17.2	EXTRAGALACTIC NEBULA
SHAP323-5308.2	03 22 07.	- 53 18 47.	24	17.6	GALAXY
SHAP323-5402.6	03 22 08.	- 54 13 11.	18	17.5	GALAXY
SHAP323-5444.0	03 22 08.	- 54 54 35.	36	18.0	GALAXY
IC 0321	03 22 09.	- 15 09 41.			NONSTELLAR OBJECT
SHAP323-4950.0	03 22 09.	- 50 00 34.	12	17.0	GALAXY
SHAP323-5031.3	03 22 09.	- 50 41 52.	30	18.0	GALAXY
SHAP323-5455.2	03 22 09.	- 55 05 47.	42	17.5	GALAXY
SHAP323-5054.1	03 22 10.	- 51 04 40.	24	17.5	GALAXY
SHAP323-5413.0	03 22 10.	- 54 23 35.	18	17.75	GALAXY
SHAP323-5443.3	03 22 11.	- 54 53 53.	36	17.5	GALAXY

OBJECT NAME	RIGHT ASCEN.	DECLINATION	DIAM.	MAGN.	TYPE OF OBJECT
MCG+07-08-004	03 22 12.	+ 40 20	48	15.	GALAXY
ZWG 541.003	03 22 12.	+ 40 21		14.9	GALAXY
ZWG 541.004	03 22 12.	+ 41 04		15.6	GALAXY
UGC 02725	03 22 12.	+ 41 04	84	15.6	GALAXY S0
PHL 8657	03 22 12.	- 02 54		18.0	BLUE STELLAR OBJECT
PHL 8659	03 22 12.	- 11 06		19.0	BLUE STELLAR OBJECT
MCG-03-09-035	03 22 12.	- 15 11	21	15.	GALAXY
PHL 8658	03 22 12.	- 24 14		18.6	BLUE STELLAR OBJECT
SHAP323-5205.0	03 22 12.	- 52 15 34.	24	18.0	GALAXY
SHAP323-5324.2	03 22 12.	- 53 34 46.	24	17.3	GALAXY
SHAP323-4741.1	03 22 13.	- 47 51 40.	18	17.5	GALAXY
SHAP323-5242.2	03 22 13.	- 52 52 46.	24	17.1	GALAXY
SHAP323-4952.9	03 22 14.	- 50 03 28.	24	16.75	GALAXY
MCG+07-08-005	03 22 15.	+ 41 03 30.	36	15.	GALAXY
SHAP323-5317.6	03 22 15.	- 53 28 10.	18	17.2	GALAXY
SHAP323-5421.6	03 22 15.	- 54 32 10.	30	17.75	GALAXY
CED 013	03 22 16.	+ 30 46	240		DIFFUSE GACACTIC NEBULA
SHAP323-4748.8	03 22 16.	- 47 59 22.	30	17.5	GALAXY
SHAP323-4839.0	03 22 16.	- 48 49 34.	18	16.5	GALAXY
SHAP323-4936.0	03 22 16.	- 49 46 34.	30	16.5	GALAXY
SHAP323-5058.4	03 22 16.	- 51 08 58.	24	17.25	GALAXY
SHAP323-5242.5	03 22 16.	- 52 53 04.	18	17.2	GALAXY
CED 012	03 22 17.	+ 31 34	300		DIFFUSE GALACTIC NEBULA
RNGC 1325	03 22 17.	- 21 43		12.5	GALAXY
ZWG 390.085	03 22 18.	+ 02 44		15.7	GALAXY
KARA.72 090A	03 22 18.	+ 02 44	24	15.7	PART OF DOUBLE GALAXY
UGC 02726	03 22 18.	+ 07 15	90	16.5	GALAXY S+COMP
DG G14	03 22 18.	+ 31 34	180		REFLECTION NEBULA
MRK 608	03 22 18.	- 03 10	7	14.5	GALAXY WITH UV CONTINUUM
MCG+01-09-035	03 22 18.	- 03 10	30	14.	GALAXY
MRK 607	03 22 18.	- 03 12	45	14.	GALAXY WITH UV CONTINUUM
MCG+01-09-036	03 22 18.	- 03 12	78	14.	GALAXY
PHL 4414	03 22 18.	- 11 07		18.7	BLUE STELLAR OBJECT
PHL 4415	03 22 18.	- 25 54		18.5	BLUE STELLAR OBJECT
PHL 8660	03 22 18.	- 28 33		18.5	BLUE STELLAR OBJECT
SHAP323-5318.9	03 22 18.	- 53 29 28.	18	17.4	GALAXY
SHAP323-5249.9	03 22 19.	- 53 00 28.	24	17.0	GALAXY
SHAP323-5435.8	03 22 19.	- 54 46 22.	18	17.75	GALAXY
SHAP323-5311.4	03 22 20.	- 53 21 58.	18	16.3	GALAXY
SHAP323-5316.8	03 22 20.	- 53 27 22.	24	16.4	GALAXY
SHAP323-5454.2	03 22 20.	- 55 04 46.	36	17.5	GALAXY
RNGC 1321	03 22 21.	- 03 10		14.0	GALAXY
RNGC 1320	03 22 21.	- 03 12		14.0	GALAXY
MCG-04-09-004	03 22 21.	- 21 41	300	12.5	GALAXY
SHAP323-5216.8	03 22 21.	- 52 27 22.	18	17.0	GALAXY
SHAP323-5315.7	03 22 21.	- 53 26 16.	24	17.7	GALAXY
SHAP324-4620.2	03 22 22.	- 46 30 45.	24	17.25	GALAXY
SHAP323-4749.8	03 22 22.	- 48 00 21.	24	17.0	GALAXY
SHAP323-5145.2	03 22 22.	- 51 55 46.	24	16.5	GALAXY
SHAP323-5219.0	03 22 22.	- 52 29 34.	18	17.75	GALAXY
SHAP323-5314.9	03 22 22.	- 53 25 28.	24	17.4	GALAXY
SHAP323-5320.1	03 22 22.	- 53 30 40.	24	16.5	GALAXY
SHAP324-4730.7	03 22 23.	- 47 41 15.	18	17.5	GALAXY
ZWG 390.086	03 22 24.	+ 02 43		15.7	GALAXY
KARA.72 090B	03 22 24.	+ 02 43	24	15.7	PART OF DOUBLE GALAXY
UGC 02727	03 22 24.	+ 05 02	78	17.	GALAXY Sc
ZC 0322.4+2325	03 22 24.	+ 23 25	2150		CLUSTER OF GALAXIES
UGC 02728	03 22 24.	+ 41 54	72	17.	GALAXY Sc
PHL 8661	03 22 24.	- 02 46		18.7	BLUE STELLAR OBJECT
MCG+01-09-037	03 22 24.	- 03 04	18	15.	GALAXY
KARA.68 035	03 22 24.	- 03 14	27		DWARF GALAXY
PHL 4417	03 22 24.	- 05 06		18.5	BLUE STELLAR OBJECT
PHL 4416	03 22 24.	- 12 40		18.5	BLUE STELLAR OBJECT
MCG-04-09-005	03 22 24.	- 22 06	48	17.	GALAXY
SHAP323-5142.0	03 22 24.	- 51 52 34.	24	18.0	GALAXY
SHAP323-5318.9	03 22 24.	- 53 29 28.	18	17.1	GALAXY
SHAP323-5327.4	03 22 24.	- 53 37 58.	24	17.5	GALAXY
SHAP323-5031.8	03 22 25.	- 50 42 21.	24	17.5	GALAXY
SHAP323-5305.1	03 22 25.	- 53 15 40.	18	17.6	GALAXY
SHAP323-5244.0	03 22 26.	- 52 54 34.	24	17.6	GALAXY
SHAP323-5439.6	03 22 26.	- 54 50 10.	30	17.75	GALAXY
RNGC 1323	03 22 27.	- 02 59			GALAXY
RNGC 1322	03 22 27.	- 03 04		15.0	GALAXY
SHAP323-5441.6	03 22 27.	- 54 52 10.	30	18.0	GALAXY
BAK 1.452	03 22 28.	- 26 09 44.	24	17.1	EXTRAGALACTIC NEBULA
BAK 1.453	03 22 28.	- 26 11 44.	14	16.6	EXTRAGALACTIC NEBULA
SHAP323-5135.2	03 22 28.	- 51 45 45.	12	17.5	GALAXY
SHAP324-5012.0	03 22 29.	- 50 22 33.	12	17.5	GALAXY
VDB.66N 012	03 22 30.	+ 31 35	276		REFLECTION NEBULA
UGC 02729	03 22 30.	+ 68 23	210	18.	GALAXY
PHL 4418	03 22 30.	- 14 37		18.4	BLUE STELLAR OBJECT
SER 031.06	03 22 30.	- 53 29	1200	17.5	CHAIN OF 12 GALAXIES
SHAP324-5032.8	03 22 31.	- 50 43 21.	24	17.5	GALAXY
SHAP324-5105.3	03 22 31.	- 51 15 51.	18	17.25	GALAXY
RNGC 1327	03 22 32.	- 25 50			NON-EXISTENT OBJECT
SHAP324-4754.6	03 22 32.	- 48 05 09.	24	17.5	GALAXY
SHAP324-5044.1	03 22 32.	- 50 54 39.	18	17.5	GALAXY
SHAP324-5046.7	03 22 32.	- 50 57 15.	24	17.5	GALAXY
BAK 1.451	03 22 33.	- 31 44 56.	19	16.3	EXTRAGALACTIC NEBULA
RNGC 1324	03 22 34.	- 05 55		14.0	GALAXY
SHAP324-4604.6	03 22 34.	- 46 15 09.	30	17.5	GALAXY
SHAP324-5010.0	03 22 34.	- 50 20 33.	48	17.0	GALAXY
SHAP324-5049.4	03 22 34.	- 50 59 57.	24	17.5	GALAXY
B 202	03 22 35.	+ 30 06	1980		DARK OBJECT
ARC 0434	03 22 35.	- 09 39		17.6	RICH CLUSTER OF GALAXIES
SHAP324-5033.6	03 22 35.	- 50 44 09.	30	17.25	GALAXY
SHAP324-5205.8	03 22 35.	- 52 16 21.	12	17.5	GALAXY
SHAP323-5325.4	03 22 35.	- 53 35 57.	30	17.6	GALAXY
ZWG 541.005	03 22 36.	+ 40 35		15.3	GALAXY
UGC 02730	03 22 36.	+ 40 35	114	15.3	GALAXY Sb
PHL 1527	03 22 36.	- 05 26		17.5	BLUE STELLAR OBJECT
MCG+01-09-038	03 22 36.	- 05 55	120	14.	GALAXY
BAK 1.455	03 22 36.	- 31 01 20.	22	16.5	EXTRAGALACTIC NEBULA
SHAP324-4927.4	03 22 36.	- 49 37 57.	18	17.25	GALAXY
SHAP324-5144.0	03 22 36.	- 51 54 33.	18	17.0	GALAXY
SHAP323-5307.9	03 22 36.	- 53 18 27.	18	16.5	GALAXY
HELW 011	03 22 37.	- 21 30			NEBULA
BAK 1.454	03 22 37.	- 32 44 32.	30	16.6	EXTRAGALACTIC NEBULA
SHAP324-5013.0	03 22 37.	- 50 23 33.	24	17.25	GALAXY
SHAP324-5108.8	03 22 37.	- 51 19 21.	24	17.5	GALAXY
SHAP324-5148.1	03 22 37.	- 51 58 39.	18	16.5	GALAXY
SHAP323-5243.9	03 22 37.	- 52 54 27.	18	17.2	GALAXY
SHAP324-5347.2	03 22 37.	- 53 57 45.	36	17.25	GALAXY
BAK 1.458	03 22 38.	- 26 42 20.	34	16.6	EXTRAGALACTIC NEBULA
BAK 1.456	03 22 38.	- 31 51 32.	17	17.0	EXTRAGALACTIC NEBULA
SHAP324-4953.0	03 22 38.	- 50 03 33.	24	16.5	GALAXY
SHAP324-5312.1	03 22 38.	- 53 22 39.	18	17.1	GALAXY
MCG+07-08-006	03 22 39.	+ 40 33	72	15.	GALAXY

OBJECT NAME	RIGHT ASCEN.	DECLINATION	DIAM.	MAGN.	TYPE OF OBJECT
BAK 1.457	03 22 40.	- 31 01 08.	23	16.5	EXTRAGALACTIC NEBULA
SHAP324-5146.6	03 22 40.	- 51 57 09.	18	17.0	GALAXY
SHAP324-5151.4	03 22 40.	- 52 01 57.	30	17.0	GALAXY
SHAP324-5249.6	03 22 40.	- 53 00 09.	24	17.1	GALAXY
IC 0320	03 22 40.5	+ 40 36 48.			GALAXY SB(rs)
VDB.66N 013	03 22 41.	+ 30 42	276		REFLECTION NEBULA
HOLM 069B	03 22 41.	- 06 21	60	13.5	PART OF MULTIPLE GALAXY
RNGC 1325A	03 22 41.	- 21 31		13.5	GALAXY
MCG+00-09-083	03 22 42.	+ 02 50	30	16.	GALAXY
MCG+07-08-007	03 22 42.	+ 40 35 30.	36	14.	GALAXY
UGC 02731	03 22 42.	+ 41 52	72	17.	GALAXY Sc
MCG+01-09-039	03 22 42.	- 06 21	48	14.5	GALAXY
PHL 8662	03 22 42.	- 18 44		16.4	BLUE STELLAR OBJECT
BAK 1.460	03 22 42.	- 25 41 55.	17	16.6	EXTRAGALACTIC NEBULA
SHAP324-4957.6	03 22 42.	- 50 08 08.	24	15.5	GALAXY
SHAP324-5222.7	03 22 42.	- 52 33 15.	18	17.5	GALAXY
SHAP324-5320.4	03 22 42.	- 53 30 57.	48	16.5	GALAXY
SHAP324-5409.3	03 22 42.	- 54 19 51.	42	17.0	GALAXY
LB 03311	03 22 42.	- 69 17		13.5	FAINT BLUE STAR
BAK 1.461	03 22 43.	- 26 10 43.	20	16.8	EXTRAGALACTIC NEBULA
BAK 1.459	03 22 43.	- 31 04 08.	22	17.2	EXTRAGALACTIC NEBULA
SHAP324-5101.4	03 22 44.	- 51 11 56.	24	17.0	GALAXY
SHAP324-5216.4	03 22 44.	- 52 26 57.	12	17.5	GALAXY
SHAP324-5311.6	03 22 44.	- 53 22 09.	18	17.4	GALAXY
B 203	03 22 46.	+ 30 36			DARK OBJECT
HOLM 069A	03 22 46.	- 06 24	72	13.5	PART OF MULTIPLE GALAXY
SHAP324-4952.7	03 22 47.	- 50 03 14.	30	16.75	GALAXY
DG 015	03 22 48.	+ 30 46	120		REFLECTION NEBULA
ZWG 541.006	03 22 48.	+ 40 37		15.4	GALAXY
UGC 02732	03 22 48.	+ 40 37	96	15.4	GALAXY SBb
ZWG 541.007	03 22 48.	+ 41 05		15.3	GALAXY
UGC 02733	03 22 48.	+ 41 05	72	15.3	GALAXY E?
UGC 02734	03 22 48.	+ 46 27	78	17.	GALAXY
PHL 8663	03 22 48.	- 06 38		17.7	BLUE STELLAR OBJECT
MCG-02-09-040	03 22 48.	- 12 29 30.	54	15.	GALAXY
MCG-03-09-036	03 22 48.	- 15 15	48	15.5	GALAXY
MCG-04-09-006	03 22 48.	- 21 29 30.	120	13.5	GALAXY
SHAP324-5056.3	03 22 48.	- 51 06 50.	24	17.25	GALAXY
SHAP324-5306.3	03 22 48.	- 53 16 50.	18	17.5	GALAXY
SHAP324-5137.9	03 22 49.	- 51 48 26.	18	18.0	GALAXY
HN 0183	03 22 49.	- 52 53			NEBULA
SHAP324-5242.7	03 22 49.	- 52 53 14.	42	16.5	GALAXY
IC 1920	03 22 49.	- 52 54			NONSTELLAR OBJECT
SHAP324-5252.0	03 22 50.	- 53 02 32.	24	17.7	GALAXY
MCG+07-08-008	03 22 51.	+ 41 04	45	15.	GALAXY
MCG-03-09-038	03 22 51.	- 16 16	60	17.	GALAXY
MCG-03-09-037	03 22 51.	- 16 21	60	17.	GALAXY
BAK 1.462	03 22 51.	- 30 38 49.	26	16.4	EXTRAGALACTIC NEBULA
SHAP324-5113.2	03 22 51.	- 51 23 44.	24	17.0	GALAXY
SHAP324-5251.5	03 22 51.	- 53 02 02.	18	17.0	GALAXY
SHAP324-5410.6	03 22 51.	- 54 21 08.	36	17.0	GALAXY
SHAP324-5104.8	03 22 52.	- 51 15 20.	18	17.5	GALAXY
SHAP324-5209.0	03 22 52.	- 52 19 32.	18	18.0	GALAXY
SHAP324-4640.4	03 22 53.	- 46 50 56.	24	17.25	GALAXY
SHAP324-4726.8	03 22 53.	- 47 37 20.	24	17.5	GALAXY
MRK 609	03 22 54.	- 06 19	13	14.5	GALAXY WITH UV CONTINUUM
MCG-04-09-007	03 22 54.	- 21 55	72	16.	GALAXY
TON-S 0348	03 22 54.	- 27 00		15.7	BLUE STAR
SHAP324-5150.9	03 22 54.	- 52 01 26.	18	18.0	GALAXY
SHAP324-5346.0	03 22 55.	- 53 56 32.	24	17.5	GALAXY
LB 02874	03 22 56.	+ 02 48 24.		17.2	FAINT BLUE STAR
BAK 1.463	03 22 56.	- 28 39 49.	13	16.2	EXTRAGALACTIC NEBULA
SHAP324-5206.3	03 22 56.	- 52 16 50.	24	16.5	GALAXY
SHAP324-5237.5	03 22 56.	- 52 48 02.	24	17.5	GALAXY
RNGC 1316C	03 22 57.	- 37 12			GALAXY
SHAP324-5200.0	03 22 57.	- 52 10 32.	18	17.75	GALAXY
SHAP324-5209.6	03 22 57.	- 52 20 08.	18	17.5	GALAXY
SHAP324-5409.3	03 22 57.	- 54 19 50.	30	17.25	GALAXY
SHAP324-4957.8	03 22 58.	- 50 08 20.	24	16.5	GALAXY
SHAP324-5237.7	03 22 58.	- 52 48 14.	36	17.4	GALAXY
SHAP324-5242.5	03 22 59.	- 52 53 02.	42	16.7	GALAXY
LBN 0734	03 23	+ 31 30	360		BRIGHT NEBULA
LBN 0716	03 23	+ 36 40	540		BRIGHT NEBULA
LDN 1451	03 23 00.	+ 30 00	1500		DARK NEBULA
LDN 1448	03 23 00.	+ 30 30	900		DARK NEBULA
ZWG 525.032	03 23 00.	+ 37 54		15.5	GALAXY
VB 237	03 23 00.	+ 54 07	651		STELLAR RING
ISS 0110	03 23 00.	+ 54 07	657		STELLAR RING
MCG+00-09-084	03 23 00.	- 01 12	54	15.	GALAXY
MRK 610	03 23 00.	- 06 18	12	16.5	GALAXY WITH UV CONTINUUM
PHL 4419	03 23 00.	- 09 22		18.5	BLUE STELLAR OBJECT
PHL 8664	03 23 00.	- 10 20		16.9	BLUE STELLAR OBJECT
MCG-03-09-040	03 23 00.	- 15 34 30.	30	16.	GALAXY
MCG-03-09-039	03 23 00.	- 17 51 30.	48	15.5	GALAXY
DV.56 N1316C	03 23 00.	- 37 12	72		S GALAXY
MCG-06-08-012	03 23 00.	- 37 12	60	14.	GALAXY
SHAP324-4955.9	03 23 00.	- 50 06 25.	18	16.25	GALAXY
LB 03312	03 23 00.	- 66 01		14.0	FAINT BLUE STAR
SHAP324-5016.2	03 23 01.	- 50 26 43.	18	17.0	GALAXY
SHAP324-5132.9	03 23 01.	- 51 43 26.	36	16.75	GALAXY
SHAP324-5243.7	03 23 01.	- 52 54 14.	24	17.3	GALAXY
BAK 1.464	03 23 02.	- 24 31 42.	17	16.6	EXTRAGALACTIC NEBULA
RNGC 1326A	03 23 02.	- 36 31			GALAXY
SHAP324-4949.6	03 23 02.	- 50 00 07.	12	17.0	GALAXY
HELW 116	03 23 03.	- 37 10 53.			NEBULA
SHAP324-5144.5	03 23 03.	- 51 55 01.	24	17.0	GALAXY
SHAP324-5201.4	03 23 03.	- 52 11 55.	18	17.0	GALAXY
BAK 1.465	03 23 04.	- 24 10 00.	12	16.5	EXTRAGALACTIC NEBULA
SHAP324-5107.1	03 23 04.	- 51 17 37.	18	17.25	GALAXY
SHAP324-5136.0	03 23 04.	- 51 46 31.	18	17.75	GALAXY
SHAP324-5237.1	03 23 04.	- 52 47 37.	24	17.5	GALAXY
SHAP324-5455.3	03 23 04.	- 55 05 50.	36	17.25	GALAXY
SHAP324-4642.2	03 23 05.	- 46 52 43.	18	17.0	GALAXY
SHAP324-5021.0	03 23 05.	- 50 31 31.	18	17.0	GALAXY
ZWG 390.087	03 23 06.	- 01 12		15.6	GALAXY
UGC 02735	03 23 06.	- 01 12	66	15.6	GALAXY SB
PHL 1528	03 23 06.	- 10 12		18.4	BLUE STELLAR OBJECT
MCG-03-09-041	03 23 06.	- 16 25	168	14.	GALAXY
DV.56 N1326A	03 23 06.	- 36 31	84		SBmp GALAXY
MCG-06-08-013	03 23 06.	- 36 33	60	14.5	GALAXY
SHAP324-4455.2	03 23 06.	- 45 05 43.	36	17.5	GALAXY
BAK 1.467	03 23 07.	- 25 31 30.	22	16.4	EXTRAGALACTIC NEBULA
SHAP324-4942.0	03 23 07.	- 49 52 31.	24	16.5	GALAXY
SHAP324-5048.5	03 23 07.	- 50 59 01.	18	17.0	GALAXY
HELW 117	03 23 08.	- 37 05 47.			NEBULA
RNGC 1328	03 23 09.	- 04 18			GALAXY
BAK 1.466	03 23 09.	- 28 26 42.	12	16.6	EXTRAGALACTIC NEBULA
HELW 118	03 23 09.	- 37 05 53.			NEBULA

OBJECT NAME	RIGHT ASCEN.	DECLINATION	DIAM.	MAGN.	TYPE OF OBJECT
SHAP324-5152.0	03 23 09.	- 52 02 31.	18	18.0	GALAXY
SHAP324-5159.9	03 23 09.	- 52 10 25.	18	16.5	GALAXY
SHAP324-5202.6	03 23 09.	- 52 13 07.	18	17.5	GALAXY
SHAP324-5138.7	03 23 10.	- 51 49 13.	18	16.75	GALAXY
BAK 1.468	03 23 11.	- 26 28 30.	22	16.8	EXTRAGALACTIC NEBULA
SHAP324-4637.9	03 23 11.	- 46 48 25.	18	17.0	GALAXY
SHAP324-5101.1	03 23 11.	- 51 11 37.	18	17.25	GALAXY
SHAP324-5200.7	03 23 11.	- 52 11 13.	18	16.75	GALAXY
MCG+07-08-009	03 23 12.	+ 40 18	78	14.5	GALAXY
ZWG 541.008	03 23 12.	+ 40 20		14.7	GALAXY
UGC 02736	03 23 12.	+ 40 20	108	14.7	GALAXY Sa-b
PHL 1529	03 23 12.	- 12 57		17.7	BLUE STELLAR OBJECT
PHL 4421	03 23 12.	- 14 15		17.7	BLUE STELLAR OBJECT
KARA.68 036	03 23 12.	- 21 06	54		DWARF GALAXY
PHL 4420	03 23 12.	- 23 45		18.7	BLUE STELLAR OBJECT
HN 0184	03 23 12.	- 50 52			NEBULA
SHAP324-5041.7	03 23 12.	- 50 52 13.	24	16.5	GALAXY
IC 1921	03 23 12.	- 50 53			NONSTELLAR OBJECT
HN 0185	03 23 12.	- 50 54			NEBULA
SHAP324-5043.7	03 23 12.	- 50 54 13.	24	16.5	GALAXY
IC 1922	03 23 12.	- 50 55			NONSTELLAR OBJECT
SHAP324-5440.6	03 23 12.	- 54 51 07.	30	18.0	GALAXY
SHAP324-4639.0	03 23 13.	- 46 49 30.	36	17.0	GALAXY
SHAP324-4952.4	03 23 13.	- 50 02 55.	18	17.0	GALAXY
SHAP324-5108.0	03 23 13.	- 51 18 31.	18	17.75	GALAXY
BAK 1.470	03 23 14.	- 25 51 00.	38	15.9	EXTRAGALACTIC NEBULA
RNGC 1326B	03 23 14.	- 36 31			GALAXY
LB 02875	03 23 15.	- 00 49 54.		17.0	FAINT BLUE STAR
MCG-04-09-008	03 23 15.	- 25 50	36	15.5	GALAXY
BAK 1.473	03 23 16.	- 24 34 05.	16	16.8	EXTRAGALACTIC NEBULA
SHAP324-4829.2	03 23 16.	- 48 39 42.	18	17.25	GALAXY
SHAP324-5259.0	03 23 16.	- 53 09 31.	24	17.6	GALAXY
BAK 1.474	03 23 17.	- 25 49 35.	12	16.8	EXTRAGALACTIC NEBULA
SHAP324-5153.3	03 23 17.	- 52 03 49.	24	18.0	GALAXY
SHAP324-5325.7	03 23 17.	- 53 36 13.	48	16.6	GALAXY
LB 02876	03 23 18.	+ 01 31 54.		16.5	FAINT BLUE STAR
ZWG 416.008	03 23 18.	+ 03 30		15.4	GALAXY
ZWG 390.089	03 23 18.	+ 03 30		15.4	GALAXY
ZWG 525.033	03 23 18.	+ 38 30		15.6	GALAXY
ZWG 390.088	03 23 18.	- 00 50		15.6	GALAXY
PHL 1530	03 23 18.	- 18 46		17.8	BLUE STELLAR OBJECT
DV.56 N1326B	03 23 18.	- 36 31	222		SBc GALAXY
MCG-06-08-014	03 23 18.	- 36 34	168	13.	GALAXY
HN 0186	03 23 18.	- 50 44			NEBULA
SHAP324-5033.7	03 23 18.	- 50 44 12.	30	16.75	GALAXY
IC 1923	03 23 18.	- 50 45			NONSTELLAR OBJECT
SHAP324-5139.4	03 23 18.	- 51 49 55.	18	17.5	GALAXY
LB 03313	03 23 18.	- 69 17		13.3	FAINT BLUE STAR
BAK 1.471	03 23 19.	- 28 51 35.	18	17.1	EXTRAGALACTIC NEBULA
BAK 1.469	03 23 19.	- 30 37 05.	13	17.2	EXTRAGALACTIC NEBULA
SHAP324-4636.5	03 23 19.	- 46 47 00.	36	17.0	GALAXY
SHAP324-5003.1	03 23 19.	- 50 13 36.	30	16.5	GALAXY
SHAP324-5158.2	03 23 19.	- 52 08 43.	12	17.25	GALAXY
SHAP324-5235.8	03 23 20.	- 52 46 18.	18	17.4	GALAXY
SHAP324-5407.2	03 23 20.	- 54 17 43.	18	17.25	GALAXY
BAK 1.472	03 23 21.	- 30 38 17.	24	17.2	EXTRAGALACTIC NEBULA
HELW 119	03 23 21.	- 36 32 48.			NEBULA
SHAP324-5035.6	03 23 21.	- 50 46 06.	24	17.5	GALAXY
SHAP324-5233.2	03 23 21.	- 52 43 42.	30	17.2	GALAXY
SHAP324-5441.8	03 23 22.	- 54 52 19.	36	17.5	GALAXY
SHAP324-5154.0	03 23 23.	- 52 04 30.	12	18.0	GALAXY
SHAP324-5205.2	03 23 23.	- 52 15 42.	18	17.5	GALAXY
IC 0322	03 23 24.	+ 03 30 07.			NONSTELLAR OBJECT
PHL 4422	03 23 24.	- 03 20		18.6	BLUE STELLAR OBJECT
PHL 4423	03 23 24.	- 03 31		18.5	BLUE STELLAR OBJECT
PHL 4424	03 23 24.	- 08 26		18.0	BLUE STELLAR OBJECT
PHL 1531	03 23 24.	- 22 59		18.5	BLUE STELLAR OBJECT
BAK 1.476	03 23 24.	- 23 00 41.	22	16.4	EXTRAGALACTIC NEBULA
SHAP324-5024.6	03 23 24.	- 50 35 06.	12	17.5	GALAXY
SHAP324-5318.6	03 23 24.	- 53 29 06.	30	17.7	GALAXY
LB 03314	03 23 24.	- 72 54		14.8	FAINT BLUE STAR
IC 1919	03 23 25.	- 33 04 14.			NONSTELLAR OBJECT
SHAP324-5017.1	03 23 25.	- 50 27 36.	18	17.0	GALAXY
SHAP324-5106.6	03 23 25.	- 51 17 06.	24	16.5	GALAXY
SHAP324-5321.9	03 23 25.	- 53 32 24.	18	17.5	GALAXY
BAK 1.475	03 23 26.	- 28 50 05.	17	16.4	EXTRAGALACTIC NEBULA
ZWG 390.090	03 23 30.	- 00 47		15.6	GALAXY
ARC 0435	03 23 30.	- 05 49		17.8	RICH CLUSTER OF GALAXIES
PHL 4425	03 23 30.	- 29 24		17.4	BLUE STELLAR OBJECT
SHAP325-4914.2	03 23 30.	- 49 24 42.	42	15.5	GALAXY
SHAP324-5327.4	03 23 30.	- 53 37 54.	18	17.0	GALAXY
SHAP325-5036.3	03 23 31.	- 50 46 48.	30	17.25	GALAXY
SHAP325-5102.6	03 23 31.	- 51 13 06.	24	16.75	GALAXY
SHAP324-5418.3	03 23 31.	- 54 28 48.	24	16.5	GALAXY
BAK 1.477	03 23 32.	- 27 57 17.	34	16.8	EXTRAGALACTIC NEBULA
SHAP325-5141.7	03 23 32.	- 51 52 12.	36	17.0	GALAXY
SHAP324-5338.5	03 23 32.	- 53 49 00.	60	16.7	GALAXY
SHAP324-5438.6	03 23 32.	- 54 49 06.	30	17.5	GALAXY
SHAP325-5105.9	03 23 33.	- 51 16 24.	24	16.75	GALAXY
HN 0187	03 23 34.	- 51 52			NEBULA
IC 1924	03 23 34.	- 51 52			NONSTELLAR OBJECT
SHAP324-5420.0	03 23 35.	- 54 30 30.	24	17.5	GALAXY
LB 02877	03 23 36.	+ 01 06 36.		16.5	FAINT BLUE STAR
ZWG 416.009	03 23 36.	+ 04 22		15.2	GALAXY
ZWG 555.002	03 23 36.	+ 46 49		15.7	GALAXY
UGC 02737	03 23 36.	+ 46 49	66	15.7	GALAXY S0-a
MRK 611	03 23 36.	- 00 22	6	16.5	GALAXY WITH UV CONTINUUM
PHL 1532	03 23 36.	- 20 30		9.0	BLUE STELLAR OBJECT
PHL 8665	03 23 36.	- 21 44		17.6	BLUE STELLAR OBJECT
PHL 8666	03 23 36.	- 26 04		18.6	BLUE STELLAR OBJECT
MCG-04-09-009	03 23 36.	- 26 33 30.	30	15.5	GALAXY
BAK 1.479	03 23 36.	- 26 33 40.	22	16.0	EXTRAGALACTIC NEBULA
BAK 1.478	03 23 36.	- 28 20 40.	17	16.8	EXTRAGALACTIC NEBULA
SHAP325-4943.1	03 23 36.	- 49 53 35.	30	17.25	GALAXY
SHAP325-5248.2	03 23 37.	- 52 58 42.	12	17.5	GALAXY
SHAP325-5324.9	03 23 37.	- 53 35 24.	24	16.3	GALAXY
SHAP325-5054.2	03 23 38.	- 51 04 41.	18	16.5	GALAXY
SHAP324-5433.2	03 23 38.	- 54 43 42.	30	17.0	GALAXY
SHAP324-5433.4	03 23 38.	- 54 43 54.	30	17.0	GALAXY
RNGC 1329	03 23 39.	- 17 47		14.0	GALAXY
SHAP325-4909.3	03 23 39.	- 49 19 47.	24	16.75	GALAXY
SHAP325-5252.0	03 23 39.	- 53 02 29.	30	16.9	GALAXY
SHAP325-5337.5	03 23 39.	- 53 47 59.	24	17.4	GALAXY
LB 00485	03 23 40.	- 00 21 36.		17.3	FAINT BLUE STAR
SHAP325-5208.9	03 23 40.	- 52 19 23.	18	17.0	GALAXY
SHAP325-5314.1	03 23 40.	- 53 24 35.	18	17.4	GALAXY
SHAP325-5314.2	03 23 40.	- 53 24 41.	24	17.3	GALAXY
IC 1918	03 23 41.	+ 04 21 48.			NONSTELLAR OBJECT

OBJECT NAME	RIGHT ASCEN.	DECLINATION	DIAM.	MAGN.	TYPE OF OBJECT
ZWG 441.011	03 23 42.	+ 09 30		15.7	GALAXY
ZWG 416.010	03 23 42.	+ 09 30		15.7	GALAXY
MCG+06-08-021	03 23 42.	+ 39 21	36	15.	GALAXY
UGC 02738	03 23 42.	+ 39 23	84	16.5	GALAXY Sc
PHL 1533	03 23 42.	- 11 08		18.0	BLUE STELLAR OBJECT
MCG-03-09-043	03 23 42.	- 17 51 30.	48	15.	GALAXY
SHAP325-4749.0	03 23 42.	- 47 59 29.	12	17.0	GALAXY
SHAP325-5109.0	03 23 42.	- 51 19 29.	18	16.25	GALAXY
SHAP325-5135.0	03 23 42.	- 51 45 29.	24	17.0	GALAXY
SHAP325-5319.0	03 23 42.	- 53 29 29.	18	17.5	GALAXY
LB 03315	03 23 42.	- 71 39		13.8	FAINT BLUE STAR
LB 02878	03 23 43.	- 00 21 00.		16.6	FAINT BLUE STAR
SHAP325-5338.2	03 23 43.	- 53 48 41.	18	17.4	GALAXY
SHAP325-5141.7	03 23 44.	- 51 52 11.	42	16.25	GALAXY
SHAP325-5143.7	03 23 44.	- 51 54 11.	24	17.0	GALAXY
SHAP325-5338.0	03 23 44.	- 53 48 29.	24	17.2	GALAXY
SHAP325-5355.8	03 23 44.	- 54 06 17.	18	17.0	GALAXY
MCG-03-09-042	03 23 45.	- 17 47	60	14.	GALAXY
BAK 1.481	03 23 45.	- 25 58 58.	17	16.8	EXTRAGALACTIC NEBULA
BAK 1.480	03 23 45.	- 30 35 04.	17	16.8	EXTRAGALACTIC NEBULA
SHAP325-5115.7	03 23 45.	- 51 26 11.	84	15.2	GALAXY
IC 1926	03 23 45.	- 51 52			NONSTELLAR OBJECT
HN 0190	03 23 45.	- 51 54			NEBULA
IC 1927	03 23 45.	- 51 54			NONSTELLAR OBJECT
SHAP325-5414.4	03 23 45.	- 54 24 53.	24	17.25	GALAXY
HN 0189	03 23 46.	- 51 52			NEBULA
BAK 1.482	03 23 47.	- 27 47 22.	22	16.4	EXTRAGALACTIC NEBULA
HN 0188	03 23 47.	- 51 26			NEBULA
IC 1925	03 23 47.	- 51 26			NONSTELLAR OBJECT
SHAP325-5219.8	03 23 47.	- 52 30 17.	18	17.25	GALAXY
UGC 02739	03 23 48.	+ 49 49	66	17.	GALAXY Sc
PHL 8668	03 23 48.	- 02 55		18.5	BLUE STELLAR OBJECT
PHL 8667	03 23 48.	- 07 10		12.2	BLUE STELLAR OBJECT
PHL 4426	03 23 48.	- 09 55		18.6	BLUE STELLAR OBJECT
PHL 4427	03 23 48.	- 13 41		18.6	BLUE STELLAR OBJECT
SHAP325-5403.5	03 23 48.	- 54 13 59.	24	17.0	GALAXY
SHAP325-4713.6	03 23 49.	- 47 24 05.	18	17.5	GALAXY
SHAP325-5055.4	03 23 49.	- 51 05 53.	24	17.75	GALAXY
SHAP325-5216.5	03 23 49.	- 52 26 59.	24	17.0	GALAXY
SHAP325-5246.6	03 23 49.	- 52 57 05.	18	17.3	GALAXY
SHAP325-5056.7	03 23 51.	- 51 07 11.	18	18.0	GALAXY
SHAP325-5248.4	03 23 51.	- 52 58 53.	24	17.1	GALAXY
SHAP325-4623.9	03 23 52.	- 46 34 22.	24	17.0	GALAXY
SHAP325-4804.0	03 23 52.	- 48 14 28.	12	17.5	GALAXY
SHAP325-5231.6	03 23 52.	- 52 42 05.	18	16.7	GALAXY
SHAP325-5335.3	03 23 52.	- 53 45 47.	18	17.0	GALAXY
RNGC 1331	03 23 53.	- 21 32		14.0	GALAXY
BAK 1.483	03 23 53.	- 25 46 39.	12	16.0	EXTRAGALACTIC NEBULA
SHAP325-5243.2	03 23 53.	- 52 53 41.	24	17.1	GALAXY
5ZW 345	03 23 54.	+ 22 53			COMPACT GALAXY
MCG-04-09-010	03 23 54.	- 25 54	24	15.5	GALAXY
BAK 1.484	03 23 54.	- 25 54 51.	34	15.6	EXTRAGALACTIC NEBULA
SHAP325-5102.4	03 23 55.	- 51 12 52.	48	17.0	GALAXY
SHAP325-5212.0	03 23 55.	- 52 22 29.	12	16.75	GALAXY
LB 02879	03 23 56.	+ 02 13 06.		15.2	FAINT BLUE STAR
BAK 1.485	03 23 56.	- 28 28 27.	14	16.9	EXTRAGALACTIC NEBULA
SHAP325-5321.0	03 23 56.	- 53 31 28.	24	16.8	GALAXY
SHAP325-5056.5	03 23 57.	- 51 06 58.	24	17.25	GALAXY
SHAP325-5115.7	03 23 57.	- 51 26 10.	36	17.25	GALAXY
SHAP325-5206.6	03 23 57.	- 52 17 04.	18	17.5	GALAXY
SHAP325-5045.9	03 23 58.	- 50 56 22.	24	17.0	GALAXY
HN 0191	03 23 58.	- 51 26			NEBULA
SHAP325-5018.9	03 23 59.	- 50 29 22.	18	17.0	GALAXY
IC 1929	03 23 59.	- 51 26			NONSTELLAR OBJECT
SHAP325-5341.2	03 23 59.	- 53 51 40.	18	17.6	GALAXY
LBN 0740	03 24	+ 30 44	900		BRIGHT NEBULA
KHAV 039	03 24	+ 32 59	19540		DARK NEBULA
LB 09801	03 24	- 80 40		13.7	FAINT BLUE STAR
ZWG 416.011	03 24 00.	+ 07 32		15.2	GALAXY
UGC 02740	03 24 00.	+ 07 32	90	15.2	GALAXY Sb
MCG+01-09-005	03 24 00.	+ 07 33	72	14.5	GALAXY
ZWG 390.091	03 24 00.	- 00 46		15.6	GALAXY
PHL 4428	03 24 00.	- 20 28		18.2	BLUE STELLAR OBJECT
MCG-06-08-015	03 24 00.	- 33 04	78	13.	GALAXY
SHAP325-5044.9	03 24 00.	- 50 55 22.	24	17.75	GALAXY
SHAP325-5046.4	03 24 00.	- 50 56 52.	18	16.5	GALAXY
SHAP325-5324.2	03 24 01.	- 53 34 40.	24	17.2	GALAXY
SHAP325-5032.6	03 24 02.	- 50 43 04.	30	17.5	GALAXY
BAK 1.488	03 24 03.	- 22 35 57.	14	16.6	EXTRAGALACTIC NEBULA
BAK 1.486	03 24 03.	- 27 42 03.	18	16.4	EXTRAGALACTIC NEBULA
SHAP325-4757.0	03 24 03.	- 48 07 28.	30	17.75	GALAXY
SHAP325-5045.7	03 24 03.	- 50 56 10.	24	16.75	GALAXY
SHAP325-4909.4	03 24 04.	- 49 19 52.	24	16.75	GALAXY
RNGC 1332	03 24 05.	- 21 31		11.5	GALAXY
BAK 1.487	03 24 05.	- 27 50 15.	22	16.8	EXTRAGALACTIC NEBULA
KARA.73 08	03 24 05.	- 34 27	27		DWARF GALAXY
SHAP325-5017.2	03 24 05.	- 50 27 40.	18	17.5	GALAXY
SHAP325-5335.1	03 24 05.	- 53 45 34.	24	17.4	GALAXY
ZC 0324.1+2441	03 24 06.	+ 24 41	1550		CLUSTER OF GALAXIES
PHL 1534	03 24 06.	- 25 29		13.7	BLUE STELLAR OBJECT
SHAP325-5013.6	03 24 06.	- 50 24 04.	30	17.5	GALAXY
SHAP325-5340.4	03 24 07.	- 53 50 52.	18	16.9	GALAXY
BAK 1.489	03 24 08.	- 26 18 51.	17	16.8	EXTRAGALACTIC NEBULA
SHAP325-4659.0	03 24 08.	- 47 09 27.	18	17.0	GALAXY
SHAP325-5329.6	03 24 08.	- 53 40 04.	18	17.0	GALAXY
SHAP325-5232.6	03 24 10.	- 52 43 04.	18	17.7	GALAXY
SHAP325-5041.0	03 24 11.	- 50 51 27.	24	17.5	GALAXY
SHAP325-5052.6	03 24 11.	- 51 03 04.	24	18.0	GALAXY
UGC 02741	03 24 12.	+ 06 56	78	16.5	GALAXY S
ASS 46	03 24 12.	+ 49 44			OB ASSOCIATION PER OB3
ZC 0324.2-0224	03 24 12.	- 02 24	1210		CLUSTER OF GALAXIES
PHL 1535	03 24 12.	- 06 45		15.7	BLUE STELLAR OBJECT
PHL 1536	03 24 12.	- 16 14		15.2	BLUE STELLAR OBJECT
MCG-04-09-011	03 24 12.	- 21 29	132	12.	GALAXY
TON-S 0349	03 24 12.	- 22 30		15.0	BLUE STAR
PHL 1537	03 24 12.	- 22 34		16.4	BLUE STELLAR OBJECT
PHL 8669	03 24 12.	- 24 58		18.5	BLUE STELLAR OBJECT
BAK 1.491	03 24 12.	- 26 38 14.	12	16.7	EXTRAGALACTIC NEBULA
SHAP325-5120.9	03 24 13.	- 51 31 21.	18	17.0	GALAXY
SHAP325-5156.3	03 24 13.	- 52 06 45.	18	17.75	GALAXY
SHAP325-5345.4	03 24 13.	- 53 55 52.	24	17.2	GALAXY
SHAP325-5348.4	03 24 13.	- 53 58 52.	42	17.6	GALAXY
BAK 1.493	03 24 14.	- 23 01 38.	16	16.6	EXTRAGALACTIC NEBULA
BAK 1.490	03 24 14.	- 28 30 02.	16	16.4	EXTRAGALACTIC NEBULA
SHAP325-5153.2	03 24 15.	- 52 03 39.	36	17.0	GALAXY
IC 0324	03 24 16.	- 21 32			SAME AS NGC 1331
BAK 1.492	03 24 16.	- 26 40 50.	35	16.6	EXTRAGALACTIC NEBULA
SHAP325-4845.4	03 24 16.	- 48 55 51.	24	16.5	GALAXY
SHAP325-5232.7	03 24 16.	- 52 43 09.	36	16.2	GALAXY
SHAP325-5349.4	03 24 16.	- 53 59 51.	24	17.3	GALAXY
ARC 0437	03 24 17.	- 02 55		16.5	RICH CLUSTER OF GALAXIES
BAK 1.494	03 24 17.	- 21 10 56.	22	16.4	EXTRAGALACTIC NEBULA
ZWG 441.012	03 24 18.	+ 12 22		15.7	GALAXY
MCG+08-07-002	03 24 18.	+ 46 45	36	16.	GALAXY
ZWG 555.003	03 24 18.	+ 46 47		15.5	GALAXY
PHL 8670	03 24 18.	- 08 53		17.1	BLUE STELLAR OBJECT
PHL 4429	03 24 18.	- 21 14		18.5	BLUE STELLAR OBJECT
PHL 1538	03 24 18.	- 22 06		17.8	BLUE STELLAR OBJECT
SHAP325-5121.8	03 24 18.	- 51 32 15.	24	17.5	GALAXY
SHAP325-5232.2	03 24 18.	- 52 42 39.	18	16.4	GALAXY
IC 1933	03 24 18.	- 52 57	138	13.2	GALAXY S
SHAP325-5246.8	03 24 18.	- 52 57 15.	174	13.2	GALAXY
SHAP325-5346.8	03 24 18.	- 53 57 15.	18	17.5	GALAXY
ARC 0436	03 24 19.	+ 18 59		17.1	RICH CLUSTER OF GALAXIES
HN 0193	03 24 19.	- 52 57			NEBULA
SHAP325-5251.3	03 24 19.	- 53 01 45.	24	17.6	GALAXY
SHAP325-4700.2	03 24 20.	- 47 10 39.	12	17.0	GALAXY
SHAP325-5122.2	03 24 20.	- 51 32 39.	18	17.5	GALAXY
SHAP325-5401.2	03 24 20.	- 54 11 39.	24	17.6	GALAXY
MCG-04-09-012	03 24 21.	- 21 30	24	14.	GALAXY
SHAP325-5017.0	03 24 21.	- 50 27 27.	18	16.25	GALAXY
SHAP325-5115.2	03 24 21.	- 51 25 39.	18	18.0	GALAXY
SHAP325-5119.8	03 24 21.	- 51 30 15.	48	15.5	GALAXY
SHAP325-5253.1	03 24 21.	- 53 03 33.	24	17.3	GALAXY
SHAP325-5335.3	03 24 21.	- 53 45 45.	18	17.4	GALAXY
SHAP325-5417.6	03 24 21.	- 54 28 03.	18	17.5	GALAXY
HN 0192	03 24 22.	- 51 30			NEBULA
SHAP325-5230.6	03 24 22.	- 52 41 03.	66	17.1	GALAXY
IC 1932	03 24 23.	- 51 30			NONSTELLAR OBJECT
SHAP325-5319.7	03 24 23.	- 53 30 09.	42	16.5	GALAXY
MCG+02-09-008	03 24 24.	+ 12 22	6	15.	GALAXY
MCG+02-09-007	03 24 24.	+ 12 22	48	14.5	GALAXY
MCG+07-08-010	03 24 24.	+ 40 42 30.	42	15.	GALAXY
ZWG 541.009	03 24 24.	+ 40 44		15.5	GALAXY
UGC 02742	03 24 24.	+ 40 44	66	15.5	GALAXY SBc
BAK 1.495	03 24 24.	- 23 06 50.	25	16.3	EXTRAGALACTIC NEBULA
SHAP325-4807.2	03 24 24.	- 48 17 39.	12	17.0	GALAXY
SHAP325-5229.3	03 24 24.	- 52 39 45.	18	17.0	GALAXY
SHAP326-4706.2	03 24 25.	- 47 16 38.	18	17.5	GALAXY
SHAP325-5247.4	03 24 25.	- 52 57 51.	24	16.5	GALAXY
SHAP325-5328.1	03 24 25.	- 53 38 33.	18	17.1	GALAXY
SHAP325-5333.9	03 24 25.	- 53 44 21.	18	17.5	GALAXY
SHAP326-4848.0	03 24 27.	- 48 58 26.	24	17.5	GALAXY
SHAP325-5059.0	03 24 27.	- 51 09 27.	18	16.5	GALAXY
SHAP325-5216.8	03 24 27.	- 52 27 15.	18	17.25	GALAXY
SHAP325-5319.1	03 24 27.	- 53 29 33.	18	16.6	GALAXY
SHAP325-5212.7	03 24 28.	- 52 23 09.	12	17.5	GALAXY
SHAP325-5322.6	03 24 28.	- 53 33 03.	30	17.0	GALAXY
SHAP325-5323.8	03 24 28.	- 53 34 15.	18	17.4	GALAXY
BAK 1.497	03 24 29.	- 25 26 25.	14	16.4	EXTRAGALACTIC NEBULA
SHAP326-5010.3	03 24 29.	- 50 20 44.	30	17.0	GALAXY
SHAP325-5042.0	03 24 29.	- 50 52 26.	24	17.5	GALAXY
SHAP325-5114.5	03 24 29.	- 51 24 56.	18	17.5	GALAXY
SHAP325-5338.2	03 24 29.	- 53 48 39.	24	17.4	GALAXY
UGC 02743	03 24 30.	+ 39 52	72	16.0	GALAXY Sc
MCG-06-08-016	03 24 30.	- 35 54	120	12.5	GALAXY
LB 01666	03 24 30.	- 46 22		15.0	FAINT BLUE STAR
SHAP325-5124.8	03 24 30.	- 51 35 14.	48	17.0	GALAXY
SHAP325-5216.2	03 24 30.	- 52 26 39.	18	17.5	GALAXY
BAK 1.498	03 24 31.	- 23 40 19.	13	16.4	EXTRAGALACTIC NEBULA
BAK 1.496	03 24 32.	- 28 50 49.	17	16.8	EXTRAGALACTIC NEBULA
SHAP326-5128.6	03 24 32.	- 51 39 02.	24	17.5	GALAXY
SHAP325-5229.8	03 24 32.	- 52 40 14.	18	17.5	GALAXY
SHAP325-5347.5	03 24 32.	- 53 57 57.	24	16.7	GALAXY
LB 02880	03 24 34.	- 01 12 12.		16.4	FAINT BLUE STAR
SHAP326-5116.0	03 24 34.	- 51 26 26.	24	18.0	GALAXY
SHAP326-5138.4	03 24 34.	- 51 48 50.	18	17.75	GALAXY
SHAP326-5013.9	03 24 35.	- 50 24 20.	24	17.0	GALAXY
SHAP326-5140.0	03 24 35.	- 51 50 26.	24	16.5	GALAXY
SHAP325-5412.8	03 24 35.	- 54 23 14.	18	17.25	GALAXY
MCG+00-09-085	03 24 36.	+ 02 25	48	14.	GALAXY
PHL 8671	03 24 36.	- 03 18		15.7	BLUE STELLAR OBJECT
PHL 4430	03 24 36.	- 04 58		18.4	BLUE STELLAR OBJECT
PHL 1539	03 24 36.	- 31 14		13.8	BLUE STELLAR OBJECT
SHAP325-5347.6	03 24 36.	- 53 58 02.	24	16.6	GALAXY
LB 03316	03 24 36.	- 65 13		12.7	FAINT BLUE STAR
HN 0195	03 24 37.	- 50 11			NEBULA
IC 1935	03 24 37.	- 50 11			NONSTELLAR OBJECT
SHAP326-5000.8	03 24 37.	- 50 11 14.	66	15.25	GALAXY
SHAP326-5016.3	03 24 37.	- 50 26 44.	30	17.75	GALAXY
SHAP326-5050.6	03 24 37.	- 51 01 02.	30	18.0	GALAXY
SHAP326-5140.0	03 24 37.	- 51 50 26.	24	16.75	GALAXY
SHAP325-5347.4	03 24 37.	- 53 57 50.	18	17.1	GALAXY
RNGC 1336	03 24 38.	- 35 54			GALAXY
SHAP326-5252.8	03 24 38.	- 53 03 14.	18	17.4	GALAXY
BAK 1.499	03 24 39.	- 26 15 25.	24	16.6	EXTRAGALACTIC NEBULA
SHAP326-5235.4	03 24 39.	- 52 45 50.	18	16.9	GALAXY
SHAP326-5249.1	03 24 39.	- 52 59 32.	24	16.8	GALAXY
SHAP326-5014.7	03 24 40.	- 50 25 08.	36	16.5	GALAXY
SHAP326-5208.7	03 24 40.	- 52 19 08.	12	17.5	GALAXY
SHAP326-5257.2	03 24 40.	- 53 07 38.	24	17.7	GALAXY
SHAP326-5325.4	03 24 41.	- 53 35 50.	24	16.5	GALAXY
ZWG 390.092	03 24 42.	+ 02 26		15.5	GALAXY
UGC 02744	03 24 42.	+ 02 26	66	15.5	GALAXY Sc
ZWG 525.034	03 24 42.	+ 36 13		15.5	GALAXY
UGC 02745	03 24 42.	+ 36 13	90	15.5	GALAXY S
MCG+07-08-011	03 24 42.	+ 39 43	42	15.	GALAXY
ZC 0324.7+4203	03 24 42.	+ 42 03	1340		CLUSTER OF GALAXIES
ZWG 327.004	03 24 42.	+ 74 45		15.6	GALAXY
PHL 1540	03 24 42.	- 12 18		18.4	BLUE STELLAR OBJECT
SHAP326-4742.2	03 24 42.	- 47 52 38.	12	17.5	GALAXY
SHAP326-5015.3	03 24 42.	- 50 25 44.	24	16.5	GALAXY
SHAP326-5118.8	03 24 42.	- 51 29 14.	18	17.25	GALAXY
SHAP326-5211.4	03 24 42.	- 52 21 50.	18	17.5	GALAXY
BAK 1.500	03 24 43.	- 26 40 13.	16	17.5	EXTRAGALACTIC NEBULA
SHAP326-5318.0	03 24 43.	- 53 28 26.	24	16.0	GALAXY
SHAP326-5209.1	03 24 44.	- 52 19 32.	12	17.5	GALAXY
SHAP326-5258.4	03 24 44.	- 53 08 50.	18	17.0	GALAXY
SHAP326-5305.6	03 24 44.	- 53 16 02.	18	17.7	GALAXY
SHAP326-5428.4	03 24 44.	- 54 38 50.	24	17.0	GALAXY
SHAP326-5257.0	03 24 45.	- 53 07 26.	18	17.5	GALAXY
BAK 1.501	03 24 46.	- 26 57 54.	22	16.4	EXTRAGALACTIC NEBULA
SHAP326-5227.6	03 24 46.	- 52 38 02.	12	17.6	GALAXY
SHAP326-5136.4	03 24 47.	- 51 46 49.	12	17.0	GALAXY
MCG+06-08-022	03 24 48.	+ 36 12 30.	36	15.	GALAXY
ZWG 541.010	03 24 48.	+ 39 44		14.9	GALAXY

OBJECT NAME	RIGHT ASCEN.	DECLINATION	DIAM.	MAGN.	TYPE OF OBJECT
PHL 4431	03 24 48.	- 10 21		18.6	BLUE STELLAR OBJECT
PHL 1541	03 24 48.	- 19 42		16.2	BLUE STELLAR OBJECT
MCG-06-08-017	03 24 48.	- 35 13	42	15.5	GALAXY
SHAP326-5310.5	03 24 48.	- 53 20 56.	18	17.3	GALAXY
SHAP326-5400.2	03 24 48.	- 54 10 38.	24	17.3	GALAXY
SHAP326-5405.6	03 24 48.	- 54 16 02.	24	17.7	GALAXY
SHAP326-5123.8	03 24 49.	- 51 34 13.	18	17.0	GALAXY
SHAP326-5131.2	03 24 49.	- 51 41 37.	18	17.0	GALAXY
SHAP326-5122.8	03 24 50.	- 51 33 13.	18	17.5	GALAXY
SHAP326-5131.6	03 24 50.	- 51 42 01.	18	16.25	GALAXY
SHAP326-5234.2	03 24 50.	- 52 44 37.	24	17.1	GALAXY
SHAP326-5258.1	03 24 50.	- 53 08 31.	24	17.0	GALAXY
SHAP326-5313.6	03 24 50.	- 53 24 01.	18	17.3	GALAXY
SHAP326-5346.3	03 24 50.	- 53 56 43.	36	17.4	GALAXY
BAK 1.503	03 24 51.	- 27 37 24.	23	17.4	EXTRAGALACTIC NEBULA
SHAP326-5124.3	03 24 51.	- 51 34 43.	24	16.75	GALAXY
SHAP326-5136.0	03 24 52.	- 51 46 25.	18	17.5	GALAXY
SHAP326-5332.4	03 24 52.	- 53 42 49.	24	17.1	GALAXY
BAK 1.502	03 24 53.	- 30 53 24.	19	16.6	EXTRAGALACTIC NEBULA
SHAP326-5106.6	03 24 53.	- 51 17 01.	18	17.5	GALAXY
SHAP326-5145.2	03 24 53.	- 51 55 37.	12	17.5	GALAXY
SHAP326-5228.0	03 24 53.	- 52 38 25.	18	16.8	GALAXY
ZWG 416.012	03 24 54.	+ 05 34		15.3	GALAXY
5ZW 346	03 24 54.	+ 21 10			COMPACT GALAXY
PHL 1542	03 24 54.	- 03 50		17.5	BLUE STELLAR OBJECT
PHL 1543	03 24 54.	- 07 22		17.7	BLUE STELLAR OBJECT
PHL 8672	03 24 54.	- 15 28		16.9	BLUE STELLAR OBJECT
SHAP326-5320.1	03 24 54.	- 53 30 31.	18	17.0	GALAXY
SHAP326-5239.4	03 24 56.	- 52 49 49.	30	16.7	GALAXY
SHAP326-5254.5	03 24 56.	- 53 04 55.	18	17.7	GALAXY
SHAP326-5432.0	03 24 56.	- 54 42 25.	30	17.5	GALAXY
SHAP326-5118.8	03 24 57.	- 51 29 13.	42	15.5	GALAXY
SHAP326-5155.4	03 24 57.	- 52 05 49.	18	17.25	GALAXY
SHAP326-5045.6	03 24 58.	- 50 56 01.	24	17.75	GALAXY
HN 0196	03 24 58.	- 51 30			NEBULA
SHAP326-5229.7	03 24 58.	- 52 40 07.	18	17.2	GALAXY
IC 1936	03 24 59.	- 51 29			NONSTELLAR OBJECT
LBN 0682	03 25	+ 58 40	1620		BRIGHT NEBULA
ZC 0325.0+2613	03 25 00.	+ 26 13	1340		CLUSTER OF GALAXIES
LDN 1455	03 25 00.	+ 30 00	600		DARK NEBULA
LDN 1452	03 25 00.	+ 30 30	5220		DARK NEBULA
5ZW 347	03 25 00.	+ 43 00			COMPACT GALAXY
7ZW 006	03 25 00.	+ 84 12			COMPACT GALAXY
PHL 1544	03 25 00.	- 11 46		18.2	BLUE STELLAR OBJECT
PHL 8673	03 25 00.	- 13 22		18.6	BLUE STELLAR OBJECT
MCG-02-09-041	03 25 00.	- 13 55 30.	72	15.	GALAXY
KARA.73 09	03 25 01.	- 33 39	60		DWARF GALAXY
SHAP326-5332.3	03 25 01.	- 53 42 43.	36	17.5	GALAXY
SHAP326-5149.8	03 25 02.	- 52 00 13.	18	18.0	GALAXY
SHAP326-5213.0	03 25 02.	- 52 23 25.	24	17.25	GALAXY
KARA.73 10	03 25 03.	- 36 19	27		DWARF GALAXY
SHAP326-5347.4	03 25 04.	- 53 57 49.	24	15.2	GALAXY
SHAP326-5436.0	03 25 04.	- 54 46 25.	42	16.75	GALAXY
BAK 1.505	03 25 05.	- 28 26 35.	22	17.0	EXTRAGALACTIC NEBULA
SHAP326-5328.7	03 25 05.	- 53 39 07.	18	16.6	GALAXY
VDB.66N 014	03 25 06.	+ 59 47	760		REFLECTION NEBULA
PHL 4432	03 25 06.	- 23 13		15.7	BLUE STELLAR OBJECT
BAK 1.506	03 25 06.	- 24 15 29.	13	16.8	EXTRAGALACTIC NEBULA
SHAP326-5123.4	03 25 06.	- 51 33 48.	18	17.5	GALAXY
SHAP326-5259.1	03 25 06.	- 53 09 31.	18	17.5	GALAXY
SHAP326-5320.0	03 25 06.	- 53 30 25.	24	16.9	GALAXY
SHAP326-5347.8	03 25 06.	- 53 58 13.	18	16.7	GALAXY
LB 01667	03 25 06.	- 59 32		14.3	FAINT BLUE STAR
BAK 1.504	03 25 07.	- 31 38 11.	26	16.8	EXTRAGALACTIC NEBULA
SHAP326-4536.6	03 25 08.	- 45 47 00.	24	17.5	GALAXY
SHAP326-5338.6	03 25 08.	- 53 49 00.	24	16.8	GALAXY
SHAP326-5345.2	03 25 08.	- 53 55 36.	54	17.4	GALAXY
VDB.66N 016	03 25 09.	+ 29 33	540		REFLECTION NEBULA
SHAP326-5337.6	03 25 09.	- 53 48 00.	24	16.7	GALAXY
BAK 1.508	03 25 10.	- 26 10 59.	14	16.2	EXTRAGALACTIC NEBULA
BAK 1.507	03 25 10.	- 28 11 11.	17	17.0	EXTRAGALACTIC NEBULA
SHAP326-5306.5	03 25 10.	- 53 16 54.	18	16.8	GALAXY
SHAP326-4914.0	03 25 11.	- 49 24 24.	24	17.5	GALAXY
SHAP326-5032.1	03 25 11.	- 50 42 30.	24	18.0	GALAXY
SHAP326-5305.8	03 25 11.	- 53 16 12.	30	16.4	GALAXY
SHAP326-5426.0	03 25 11.	- 54 36 24.	30	18.0	GALAXY
UGC 02746	03 25 12.	+ 36 37	108	16.0	GALAXY E-S0
UGC 02747	03 25 12.	+ 37 51	72	16.0	GALAXY E?
ZWG 541.011	03 25 12.	+ 39 59		15.0	GALAXY
PHL 8674	03 25 12.	- 17 46		18.0	BLUE STELLAR OBJECT
HPW 49	03 25 12.	- 33 30			IRREG GALAXY IN FORNAX
MCG-06-08-018	03 25 12.	- 33 40	96	17.	GALAXY
SHAP326-4831.2	03 25 12.	- 48 41 36.	18	17.25	GALAXY
LB 01668	03 25 12.	- 53 08		14.2	FAINT BLUE STAR
SHAP326-5335.7	03 25 12.	- 53 46 06.	30	16.8	GALAXY
SHAP326-5358.4	03 25 12.	- 54 08 48.	24	17.7	GALAXY
SHAP326-4535.4	03 25 13.	- 45 45 48.	24	17.0	GALAXY
SHAP326-5251.9	03 25 13.	- 53 02 18.	24	16.8	GALAXY
SHAP326-5429.1	03 25 13.	- 54 39 30.	30	17.0	GALAXY
MCG+00-09-086	03 25 15.	+ 02 22 30.	48	15.	GALAXY
CED 015	03 25 15.	+ 29 40	660		DIFFUSE GALACTIC NEBULA
SHAP326-4841.8	03 25 15.	- 48 52 12.	54	15.25	GALAXY
SHAP326-5140.1	03 25 15.	- 51 50 30.	18	18.0	GALAXY
SHAP326-5200.5	03 25 15.	- 52 10 54.	18	17.5	GALAXY
SHAP326-5239.2	03 25 15.	- 52 49 36.	18	17.7	GALAXY
SHAP326-4833.4	03 25 16.	- 48 43 48.	24	17.0	GALAXY
IC 1937	03 25 16.	- 48 52			NONSTELLAR OBJECT
HN 0197	03 25 16.	- 48 53			NEBULA
SHAP326-5305.6	03 25 16.	- 53 16 00.	24	17.6	GALAXY
SHAP326-5436.0	03 25 16.	- 54 46 24.	30	16.5	GALAXY
IC 1928	03 25 16.0	- 21 43 54.			GALAXY S0
SHAP326-5335.6	03 25 17.	- 53 46 00.	24	16.5	GALAXY
ZWG 390.093	03 25 18.	+ 02 23		15.5	GALAXY
UGC 02748	03 25 18.	+ 02 23	102	15.5	GALAXY E
DG 016	03 25 18.	+ 29 40	300		REFLECTION NEBULA
ZWG 525.035	03 25 18.	+ 36 31		15.5	GALAXY
UGC 02749	03 25 18.	+ 39 54	78	16.0	GALAXY S-IRR
MCG-06-08-019	03 25 18.	- 34 43	48	13.5	GALAXY
HN 0194	03 25 19.	- 21 44			NEBULA
SHAP326-5001.9	03 25 20.	- 50 12 18.	24	17.25	GALAXY
SHAP326-5345.8	03 25 20.	- 53 56 12.	30	17.7	GALAXY
BAK 1.511	03 25 21.	- 21 23 46.	96	15.4	EXTRAGALACTIC NEBULA
MCG-04-09-013	03 25 21.	- 21 42 30.	84	14.	GALAXY
SHAP326-5202.9	03 25 21.	- 52 13 18.	12	17.25	GALAXY
SHAP326-5305.2	03 25 22.	- 53 15 36.	18	17.6	GALAXY
BAK 1.509	03 25 23.	- 26 43 10.	24	17.0	EXTRAGALACTIC NEBULA
BAK 1.510	03 25 23.	- 26 44 58.	26	16.8	EXTRAGALACTIC NEBULA
SHAP326-5016.4	03 25 23.	- 50 26 47.	18	16.75	GALAXY

OBJECT NAME	RIGHT ASCEN.	DECLINATION	DIAM.	MAGN.	TYPE OF OBJECT
MCG+06-08-024	03 25 24.	+ 36 30	12	17.	GALAXY
MCG+06-08-023	03 25 24.	+ 36 30	21	17.	GALAXY
PHL 8676	03 25 24.	- 06 46		18.1	BLUE STELLAR OBJECT
PHL 4433	03 25 24.	- 07 15		18.4	BLUE STELLAR OBJECT
PHL 4434	03 25 24.	- 08 09		18.4	BLUE STELLAR OBJECT
MCG-03-09-044	03 25 24.	- 17 12	48	15.	GALAXY
PHL 8675	03 25 24.	- 18 20		16.4	BLUE STELLAR OBJECT
SHAP326-5019.8	03 25 24.	- 50 30 11.	12	17.0	GALAXY
SHAP326-5330.6	03 25 24.	- 53 41 00.	30	16.4	GALAXY
B 204	03 25 25.	+ 30 01	840		DARK OBJECT
SHAP327-4609.2	03 25 25.	- 46 19 35.	18	17.5	GALAXY
SHAP327-4805.4	03 25 25.	- 48 15 47.	18	17.75	GALAXY
SHAP326-5402.4	03 25 25.	- 54 12 48.	18	17.3	GALAXY
SHAP326-5108.6	03 25 26.	- 51 18 59.	24	16.0	GALAXY
B 205	03 25 27.	+ 30 56	900		DARK OBJECT
MCG+06-08-025	03 25 27.	+ 36 36	15	15.5	GALAXY
SHAP326-5323.8	03 25 27.	- 53 34 11.	18	17.6	GALAXY
SHAP326-5335.8	03 25 27.	- 53 46 11.	24	17.0	GALAXY
SHAP326-5321.4	03 25 29.	- 53 31 47.	18	17.2	GALAXY
UGC 02750	03 25 30.	+ 36 22	96	16.5	GALAXY Sb-c
DG 017	03 25 30.	+ 58 43	1800		REFLECTION NEBULA
MCG-04-09-014	03 25 30.	- 21 22	108	14.	GALAXY
PHL 8677	03 25 30.	- 21 32		9.4	BLUE STELLAR OBJECT
SHAP326-5251.4	03 25 30.	- 53 01 47.	18	17.2	GALAXY
SHAP326-5439.8	03 25 30.	- 54 50 11.	30	17.25	GALAXY
SHAP326-5252.8	03 25 32.	- 53 03 11.	18	16.9	GALAXY
SHAP326-5439.0	03 25 34.	- 54 49 23.	24	16.75	GALAXY
CED 014	03 25 35.	+ 58 43	1800		DIFFUSE GALACTIC NEBULA
RNGC 1337	03 25 35.	- 08 34		12.0	GALAXY
SHAP326-5255.4	03 25 35.	- 53 05 47.	24	17.5	GALAXY
SHAP326-5331.3	03 25 35.	- 53 41 41.	24	17.0	GALAXY
SHAP326-5404.7	03 25 35.	- 54 15 05.	24	16.6	GALAXY
ZC 0325.6+0036	03 25 36.	+ 00 36	1140		CLUSTER OF GALAXIES
LB 02881	03 25 36.	+ 01 17 54.		16.2	FAINT BLUE STAR
ZC 0325.6+0838	03 25 36.	+ 08 38	4500		CLUSTER OF GALAXIES
UGC 02751	03 25 36.	+ 39 52	66	16.0	GALAXY S0?
ZWG 541.012	03 25 36.	+ 41 46		15.6	GALAXY
MCG+01-09-040	03 25 36.	- 04 59	60	15.	GALAXY
MCG-02-09-042	03 25 36.	- 09 34	420	12.	GALAXY
TON-S 0350	03 25 36.	- 23 38		15.3	BLUE STAR
BAK 1.513	03 25 36.	- 26 41 22.	17	16.5	EXTRAGALACTIC NEBULA
BAK 1.512	03 25 36.	- 28 33 28.	35	16.3	EXTRAGALACTIC NEBULA
LB 01669	03 25 36.	- 51 29		14.2	FAINT BLUE STAR
SHAP327-5220.3	03 25 36.	- 52 30 41.	18	17.4	GALAXY
SHAP327-5309.2	03 25 36.	- 53 19 35.	24	17.5	GALAXY
SHAP326-5352.4	03 25 36.	- 54 02 47.	24	16.9	GALAXY
SHAP327-5153.4	03 25 37.	- 52 03 47.	30	16.5	GALAXY
SHAP327-5317.2	03 25 37.	- 53 27 35.	18	17.1	GALAXY
SHAP327-5309.0	03 25 38.	- 53 19 23.	24	17.4	GALAXY
LB 02882	03 25 39.	+ 00 55 18.		15.4	FAINT BLUE STAR
SHAP327-5320.9	03 25 39.	- 53 31 17.	42	15.9	GALAXY
BAK 1.515	03 25 40.	- 27 07 51.	13	16.4	EXTRAGALACTIC NEBULA
SHAP327-4545.6	03 25 40.	- 45 55 58.	18	18.0	GALAXY
SHAP327-5106.8	03 25 40.	- 51 17 10.	24	16.5	GALAXY
SHAP327-5424.2	03 25 40.	- 54 34 35.	48	17.0	GALAXY
BAK 1.518	03 25 41.	- 22 56 09.	14	16.9	EXTRAGALACTIC NEBULA
BAK 1.514	03 25 41.	- 28 33 15.	17	16.3	EXTRAGALACTIC NEBULA
SHAP327-4558.5	03 25 41.	- 46 08 52.	24	17.0	GALAXY
SHAP327-4558.8	03 25 41.	- 46 09 10.	18	17.75	GALAXY
ZC 0325.7+1758	03 25 42.	+ 17 58	1880		CLUSTER OF GALAXIES
GCL 006	03 25 42.	+ 79 28	78		GLOBULAR STAR CLUSTER
BAK 1.516	03 25 42.	- 25 29 09.	18	16.6	EXTRAGALACTIC NEBULA
SHAP327-5125.6	03 25 42.	- 51 35 58.	18	17.5	GALAXY
SHAP327-5212.7	03 25 42.	- 52 23 04.	18	16.25	GALAXY
BAK 1.517	03 25 43.	- 26 21 45.	14	16.2	EXTRAGALACTIC NEBULA
SHAP327-4619.0	03 25 43.	- 46 29 22.	24	17.0	GALAXY
SHAP327-4622.9	03 25 43.	- 46 33 16.	18	17.0	GALAXY
SHAP327-5012.8	03 25 43.	- 50 23 10.	18	17.0	GALAXY
SHAP327-5127.6	03 25 43.	- 51 37 58.	18	18.0	GALAXY
SHAP327-5155.6	03 25 43.	- 52 05 58.	18	17.5	GALAXY
SHAP327-5328.2	03 25 43.	- 53 38 34.	24	17.5	GALAXY
SHAP327-5353.4	03 25 43.	- 54 03 46.	24	17.4	GALAXY
SHAP327-5220.0	03 25 44.	- 52 30 22.	12	17.5	GALAXY
SHAP327-5421.8	03 25 44.	- 54 32 10.	24	16.75	GALAXY
HH 14C	03 25 44.1	+ 30 50 29.			HERBIG-HARO OBJECT
HH 14E	03 25 44.4	+ 30 50 56.			HERBIG-HARO OBJECT
HH 14D	03 25 44.8	+ 30 51 05.			HERBIG-HARO OBJECT
BAK 1.520	03 25 45.	- 24 03 51.	14	16.4	EXTRAGALACTIC NEBULA
SHAP327-5141.0	03 25 45.	- 51 51 22.	24	17.25	GALAXY
HH 14B	03 25 45.0	+ 30 50 50.			HERBIG-HARO OBJECT
SHAP327-4743.0	03 25 46.	- 47 53 22.	12	17.0	GALAXY
SHAP327-4809.9	03 25 46.	- 48 20 16.	12	17.25	GALAXY
SHAP327-5210.1	03 25 46.	- 52 20 28.	24	17.25	GALAXY
SHAP327-5211.0	03 25 46.	- 52 21 22.	12	17.5	GALAXY
SHAP327-5301.0	03 25 46.	- 53 11 22.	42	16.1	GALAXY
SHAP327-5127.6	03 25 47.	- 51 37 58.	24	16.5	GALAXY
SHAP327-5219.6	03 25 47.	- 52 29 58.	18	17.6	GALAXY
SHAP327-5236.1	03 25 47.	- 52 46 28.	18	16.5	GALAXY
SHAP327-5321.2	03 25 47.	- 53 31 34.	24	17.1	GALAXY
SHAP327-5336.9	03 25 47.	- 53 47 16.	24	17.0	GALAXY
SHAP327-5338.9	03 25 47.	- 53 49 16.	18	16.7	GALAXY
SHAP327-5410.2	03 25 47.	- 54 20 34.	72	17.0	GALAXY
MCG+07-08-012	03 25 48.	+ 40 38 30.	90	15.	GALAXY
ZWG 541.013	03 25 48.	+ 40 39		15.6	GALAXY
UGC 02752	03 25 48.	+ 40 39	120	15.6	GALAXY S0
PHL 8678	03 25 48.	- 03 28		17.3	BLUE STELLAR OBJECT
BAK 1.519	03 25 48.	- 28 36 09.	16	16.7	EXTRAGALACTIC NEBULA
SHAP327-5044.2	03 25 48.	- 50 54 34.	42	17.75	GALAXY
SHAP327-5210.0	03 25 48.	- 52 20 22.	12	17.5	GALAXY
HN 0198	03 25 48.	- 53 12			NEBULA
SHAP327-5302.2	03 25 48.	- 53 12 34.	18	17.4	GALAXY
SHAP327-5339.8	03 25 48.	- 53 50 10.	18	17.1	GALAXY
SHAP327-5044.0	03 25 49.	- 50 54 22.	24	17.5	GALAXY
IC 1938	03 25 49.	- 53 11			NONSTELLAR OBJECT
HARO 20	03 25 50.	- 17 33			BLUE EMISSION-LINE GALAXY
SHAP327-5308.5	03 25 50.	- 53 18 52.	24	17.3	GALAXY
SHAP327-5334.2	03 25 50.	- 53 44 34.	24	17.1	GALAXY
SHAP327-5234.4	03 25 51.	- 52 44 46.	18	17.6	GALAXY
BAK 1.521	03 25 52.	- 28 49 21.	18	17.0	EXTRAGALACTIC NEBULA
SHAP327-5021.0	03 25 52.	- 50 31 22.	36	17.75	GALAXY
SHAP327-5039.4	03 25 52.	- 50 49 46.	30	17.75	GALAXY
SHAP327-5127.2	03 25 52.	- 51 37 34.	12	17.5	GALAXY
HH 12D	03 25 52.1	+ 31 09 51.			HERBIG-HARO OBJECT
HH 12C	03 25 52.4	+ 31 10 04.			HERBIG-HARO OBJECT
BAK 1.522	03 25 53.	- 27 53 03.	18	17.0	EXTRAGALACTIC NEBULA
HH 12G	03 25 53.1	+ 31 10 26.			HERBIG-HARO OBJECT
HH 12B	03 25 53.4	+ 31 10 11.			HERBIG-HARO OBJECT
HH 15	03 25 53.5	+ 30 57 43.	10		HERBIG-HARO OBJECT

OBJECT NAME	RIGHT ASCEN.	DECLINATION	DIAM.	MAGN.	TYPE OF OBJECT
HH 12E	03 25 53.6	+ 31 09 49.			HERBIG-HARO OBJECT
HH 12F	03 25 53.7	+ 31 09 29.			HERBIG-HARO OBJECT
MCG+07-08-013	03 25 54.	+ 41 12	21	15.5	GALAXY
ZWG 541.014	03 25 54.	+ 41 13		15.5	GALAXY
7ZW 007	03 25 54.	+ 79 25			COMPACT GALAXY
MCG-03-09-045	03 25 54.	- 17 36	24	15.	GALAXY
BAK 1.523	03 25 54.	- 27 34 09.	17	16.8	EXTRAGALACTIC NEBULA
SHAP327-5247.0	03 25 54.	- 52 57 22.	18	17.6	GALAXY
SHAP327-5241.0	03 25 55.	- 52 51 22.	12	17.3	GALAXY
RNGC 1330	03 25 56.	+ 41 12		15.5	GALAXY
SHAP327-4924.4	03 25 56.	- 49 34 45.	30	17.0	GALAXY
SHAP327-5125.4	03 25 56.	- 51 35 46.	18	17.25	GALAXY
BAK 1.524	03 25 57.	- 28 26 56.	24	16.3	EXTRAGALACTIC NEBULA
SHAP327-4857.4	03 25 57.	- 49 07 45.	24	17.25	GALAXY
SHAP327-5050.6	03 25 57.	- 51 00 57.	18	17.5	GALAXY
SHAP327-5238.5	03 25 57.	- 52 48 52.	24	17.7	GALAXY
SHAP327-5406.4	03 25 57.	- 54 16 46.	24	17.0	GALAXY
BAK 1.525	03 25 58.	- 28 25 20.	26	16.6	EXTRAGALACTIC NEBULA
SHAP327-5024.6	03 25 58.	- 50 34 57.	24	16.5	GALAXY
SHAP327-5255.5	03 25 58.	- 53 05 52.	12	17.1	GALAXY
SHAP327-5309.7	03 25 58.	- 53 20 04.	18	16.9	GALAXY
SHAP327-5324.8	03 25 58.	- 53 35 10.	18	16.7	GALAXY
SHAP327-5328.2	03 25 58.	- 53 38 34.	30	16.9	GALAXY
SHAP327-4855.2	03 25 59.	- 49 05 33.	18	17.0	GALAXY
SHAP327-5308.0	03 25 59.	- 53 18 21.	24	16.9	GALAXY
HH 11	03 25 59.0	+ 31 05 35.	2		HERBIG-HARO OBJECT
HH 10	03 25 59.8	+ 31 05 28.	8		HERBIG-HARO OBJECT
LBN 0746	03 26	+ 29 40	600		BRIGHT NEBULA
LBN 0741	03 26	+ 31 13	480		BRIGHT NEBULA
LBN 0681	03 26	+ 59 40	840		BRIGHT NEBULA
LBN 0870	03 26	- 09 30	1980		BRIGHT NEBULA
LB 09802	03 26	- 85 44		13.9	FAINT BLUE STAR
UGC 02753	03 26 00.	+ 82 15	60	16.0	GALAXY Sb-c
MCG+14-02-012	03 26 00.	+ 82 15	57	16.	GALAXY
PHL 8679	03 26 00.	- 04 30		18.5	BLUE STELLAR OBJECT
BAK 1.526	03 26 00.	- 24 08 20.	16	16.4	EXTRAGALACTIC NEBULA
PHL 4435	03 26 00.	- 25 48		13.2	BLUE STELLAR OBJECT
MCG-06-08-020	03 26 00.	- 37 20	72		GALAXY
SHAP327-4625.6	03 26 00.	- 46 35 57.	30	17.5	GALAXY
SHAP327-4924.0	03 26 00.	- 49 34 21.	24	16.75	GALAXY
SHAP327-4926.4	03 26 00.	- 49 36 45.	12	17.5	GALAXY
SHAP327-5046.5	03 26 00.	- 50 56 51.	24	17.5	GALAXY
SHAP327-5229.0	03 26 00.	- 52 39 21.	30	17.4	GALAXY
HH 08	03 26 00.7	+ 31 05 19.	9		HERBIG-HARO OBJECT
HH 09	03 26 00.9	+ 31 05 35.	2		HERBIG-HARO OBJECT
SHAP327-4930.2	03 26 01.	- 49 40 33.	18	16.5	GALAXY
SHAP327-5234.4	03 26 01.	- 52 44 45.	12	16.7	GALAXY
SHAP327-5114.6	03 26 02.	- 51 24 57.	66	15.0	GALAXY
SHAP327-5237.6	03 26 02.	- 52 47 57.	18	17.4	GALAXY
HH 07C	03 26 02.3	+ 31 05 08.			HERBIG-HARO OBJECT
HH 07B	03 26 02.5	+ 31 05 10.			HERBIG-HARO OBJECT
HH 07	03 26 02.5	+ 31 05 13.	10		HERBIG-HARO OBJECT
HH 07D	03 26 02.7	+ 31 05 11.			HERBIG-HARO OBJECT
HH 16	03 26 02.8	+ 30 58 52.	7		HERBIG-HARO OBJECT
RNGC 1333	03 26 03.	+ 31 12			DIFFUSE NEBULA
KARA.73 11	03 26 03.	- 33 54	27		DWARF GALAXY
RNGC 1341	03 26 03.	- 37 19		13.5	GALAXY
SHAP327-5240.5	03 26 03.	- 52 50 51.	12	17.4	GALAXY
SHAP327-5325.3	03 26 03.	- 53 35 39.	18	16.7	GALAXY
SHAP327-5401.8	03 26 03.	- 54 12 09.	24	17.3	GALAXY
SHAP327-5304.2	03 26 04.	- 53 14 33.	48	16.2	GALAXY
B 206	03 26 05.	+ 30 01	300		DARK OBJECT
SHAP327-5240.1	03 26 05.	- 52 50 27.	18	17.1	GALAXY
SHAP327-5252.0	03 26 05.	- 53 02 21.	24	17.5	GALAXY
SHAP327-5403.8	03 26 05.	- 54 14 09.	24	17.7	GALAXY
HH 06D	03 26 05.8	+ 31 08 10.			HERBIG-HARO OBJECT
RNGC 1339	03 26 06.	- 32 27		13.0	GALAXY
SHAP327-5255.5	03 26 06.	- 53 05 51.	18	17.2	GALAXY
SHAP327-5342.4	03 26 06.	- 53 52 45.	24	17.2	GALAXY
SHAP327-5424.0	03 26 06.	- 54 34 21.	24	17.5	GALAXY
HH 06C	03 26 06.5	+ 31 08 15.			HERBIG-HARO OBJECT
HH 06E	03 26 06.6	+ 31 08 24.			HERBIG-HARO OBJECT
VDB.66N 015	03 26 07.	+ 58 44	3240		REFLECTION NEBULA
SHAP327-5400.2	03 26 07.	- 54 10 33.	18	17.6	GALAXY
HH 06F	03 26 07.0	+ 31 08 23.			HERBIG-HARO OBJECT
HH 06B	03 26 07.2	+ 31 08 28.			HERBIG-HARO OBJECT
BAK 1.527	03 26 08.	- 26 26 14.	26	16.2	EXTRAGALACTIC NEBULA
SHAP327-4641.5	03 26 08.	- 46 51 51.	18	18.25	GALAXY
SHAP327-5310.6	03 26 08.	- 53 20 57.	18	16.2	GALAXY
SHAP327-5323.5	03 26 08.	- 53 33 51.	24	17.4	GALAXY
IC 1930	03 26 09.	+ 04 13 33.			NONSTELLAR OBJECT
IC 0323	03 26 10.	+ 41 41 24.			NONSTELLAR OBJECT
ARC 0438	03 26 10.	- 10 01		17.2	RICH CLUSTER OF GALAXIES
BAK 1.528	03 26 10.	- 28 12 08.	24	16.3	EXTRAGALACTIC NEBULA
SHAP327-5421.8	03 26 10.	- 54 32 09.	24	16.75	GALAXY
CED 016	03 26 11.	+ 31 12	540		DIFFUSE GALACTIC NEBULA
SHAP327-5320.2	03 26 11.	- 53 30 33.	36	17.0	GALAXY
DG 018	03 26 12.	+ 31 13	420		REFLECTION NEBULA
UGC 02754	03 26 12.	+ 37 56	66	16.5	GALAXY Sb-c
MCG+07-08-014	03 26 12.	+ 39 36 30.	18	15.	GALAXY
ZWG 541.015	03 26 12.	+ 39 37		14.9	GALAXY
UGC 02755	03 26 12.	+ 39 37	90	14.9	GALAXY
PHL 8680	03 26 12.	- 11 18		18.6	BLUE STELLAR OBJECT
PHL 1545	03 26 12.	- 12 29		17.0	BLUE STELLAR OBJECT
TON-S 0351	03 26 12.	- 23 12		14.8	BLUE STAR
BAK 1.529	03 26 12.	- 27 52 08.	19	16.5	EXTRAGALACTIC NEBULA
LB 01670	03 26 12.	- 45 40		15.5	FAINT BLUE STAR
SHAP327-5139.7	03 26 12.	- 51 50 03.	18	16.5	GALAXY
SHAP327-5208.0	03 26 13.	- 52 18 21.	54	15.5	GALAXY
SHAP327-5435.5	03 26 13.	- 54 45 51.	36	17.25	GALAXY
VDB.66N 017	03 26 14.	+ 31 15	480		REFLECTION NEBULA
IC 1940	03 26 14.	- 52 18			NONSTELLAR OBJECT
HN 0200	03 26 14.	- 52 19			NEBULA
SHAP327-5340.2	03 26 14.	- 53 50 33.	18	17.3	GALAXY
HH 17	03 26 14.7	+ 31 08 17.	10		HERBIG-HARO OBJECT
HH 05	03 26 14.8	+ 31 02 34.	5		HERBIG-HARO OBJECT
MCG+00-09-087	03 26 15.	+ 01 33	48	14.	GALAXY
SHAP327-4722.9	03 26 15.	- 47 33 14.	12	18.0	GALAXY
SHAP327-4855.6	03 26 15.	- 49 05 56.	18	17.5	GALAXY
SHAP327-5139.9	03 26 15.	- 51 50 14.	24	18.0	GALAXY
SHAP327-5232.0	03 26 15.	- 52 42 21.	18	17.1	GALAXY
SHAP327-5240.9	03 26 15.	- 52 51 15.	18	17.5	GALAXY
BAK 1.531	03 26 16.	- 26 02 19.	17	16.6	EXTRAGALACTIC NEBULA
SHAP327-4739.0	03 26 16.	- 47 49 20.	18	17.5	GALAXY
SHAP327-5058.0	03 26 16.	- 51 08 20.	24	17.5	GALAXY
HN 0199	03 26 16.	- 51 15			NEBULA
SHAP327-5327.1	03 26 16.	- 53 37 27.	18	17.2	GALAXY
BAK 1.530	03 26 17.	- 29 36 55.	14	16.5	EXTRAGALACTIC NEBULA
RNGC 1340	03 26 17.	- 31 04			NON-EXISTENT OBJECT
SHAP327-4855.4	03 26 17.	- 49 05 44.	24	17.25	GALAXY
IC 1939	03 26 17.	- 51 14			NONSTELLAR OBJECT
SHAP327-5143.0	03 26 17.	- 51 53 20.	24	16.5	GALAXY
SHAP327-5354.0	03 26 17.	- 54 04 21.	24	17.5	GALAXY
ZWG 390.094	03 26 18.	+ 01 35		15.5	GALAXY
MCG+07-08-016	03 26 18.	+ 40 41	36	15.	GALAXY
MCG+07-08-015	03 26 18.	+ 40 41	39	15.	GALAXY
5ZW 348	03 26 18.	+ 40 42			COMPACT GALAXY
ZWG 541.016	03 26 18.	+ 40 42		15.0	GALAXY
UGC 02756	03 26 18.	+ 40 42	90	15.0	GALAXY DBL SYS
UGC 02757	03 26 18.	+ 43 08	72	17.	GALAXY Sc
PHL 1546	03 26 18.	- 21 08		16.3	BLUE STELLAR OBJECT
BAK 1.533	03 26 18.	- 23 30 31.	17	16.6	EXTRAGALACTIC NEBULA
PHL 8681	03 26 18.	- 23 36		18.1	BLUE STELLAR OBJECT
SHAP327-5355.0	03 26 18.	- 54 05 21.	24	17.1	GALAXY
SHAP327-5410.8	03 26 18.	- 54 21 09.	30	17.3	GALAXY
HH 04	03 26 18.6	+ 31 09 41.	8		HERBIG-HARO OBJECT
BAK 1.532	03 26 20.	- 29 29 01.	17	16.6	EXTRAGALACTIC NEBULA
SHAP327-5230.6	03 26 20.	- 52 40 56.	18	17.4	GALAXY
SHAP327-5333.2	03 26 20.	- 53 43 32.	30	17.4	GALAXY
HH 18B	03 26 20.8	+ 30 57 00.			HERBIG-HARO OBJECT
BAK 1.534	03 26 21.	- 23 41 31.	16	16.6	EXTRAGALACTIC NEBULA
SHAP327-4706.7	03 26 21.	- 47 17 02.	24	16.25	GALAXY
SHAP327-5238.4	03 26 21.	- 52 48 44.	30	16.4	GALAXY
SHAP327-5400.9	03 26 21.	- 54 11 14.	24	17.4	GALAXY
HH 18A	03 26 21.0	+ 30 57 21.			HERBIG-HARO OBJECT
IC 1931	03 26 22.	+ 01 34 37.			NONSTELLAR OBJECT
SHAP327-4806.0	03 26 22.	- 48 16 20.	18	17.5	GALAXY
SHAP327-5255.6	03 26 22.	- 53 05 56.	18	17.3	GALAXY
SHAP327-5307.2	03 26 22.	- 53 17 32.	18	17.1	GALAXY
SHAP327-5420.8	03 26 22.	- 54 31 08.	24	17.25	GALAXY
SHAP327-4738.2	03 26 23.	- 47 48 32.	12	17.5	GALAXY
PHL 8682	03 26 24.	- 08 58		17.8	BLUE STELLAR OBJECT
PHL 4436	03 26 24.	- 10 10		18.8	BLUE STELLAR OBJECT
PHL 1547	03 26 24.	- 19 49		18.4	BLUE STELLAR OBJECT
MCG-05-09-004	03 26 24.	- 32 29	90	13.5	GALAXY
SHAP328-4550.2	03 26 24.	- 46 00 32.	36	17.5	GALAXY
SHAP327-5222.8	03 26 24.	- 52 33 08.	24	17.0	GALAXY
SHAP327-5241.3	03 26 24.	- 52 51 38.	24	16.9	GALAXY
SHAP327-5334.1	03 26 24.	- 53 44 26.	24	16.5	GALAXY
SHAP327-5336.4	03 26 24.	- 53 46 44.	24	17.6	GALAXY
SHAP327-5338.2	03 26 24.	- 53 48 32.	18	17.0	GALAXY
SHAP327-5323.2	03 26 25.	- 53 33 32.	90	16.9	GALAXY
SHAP327-5325.3	03 26 25.	- 53 35 38.	24	16.5	GALAXY
SHAP327-5326.5	03 26 25.	- 53 36 50.	12	17.3	GALAXY
SHAP328-4614.2	03 26 26.	- 46 24 32.	18	17.0	GALAXY
SHAP327-5222.2	03 26 26.	- 52 32 32.	18	17.3	GALAXY
SHAP328-4754.4	03 26 27.	- 48 04 44.	24	17.5	GALAXY
SHAP327-5236.2	03 26 27.	- 52 46 32.	12	17.0	GALAXY
SHAP327-5240.2	03 26 27.	- 52 50 32.	18	16.6	GALAXY
SHAP327-5305.3	03 26 27.	- 53 15 38.	24	16.9	GALAXY
SHAP327-5308.4	03 26 27.	- 53 18 44.	24	17.6	GALAXY
SHAP327-5321.6	03 26 27.	- 53 31 56.	24	16.4	GALAXY
BAK 1.536	03 26 29.	- 25 02 18.	14	17.0	EXTRAGALACTIC NEBULA
SHAP328-4807.2	03 26 29.	- 48 17 31.	18	17.75	GALAXY
SHAP327-5234.2	03 26 29.	- 52 44 32.	12	17.1	GALAXY
SHAP327-5321.4	03 26 29.	- 53 31 44.	18	17.5	GALAXY
SHAP327-5325.8	03 26 29.	- 53 36 08.	18	16.7	GALAXY
LDN 1450	03 26 30.	+ 31 10	360		DARK NEBULA
MCG-05-09-005	03 26 30.	- 31 15	90	12.	GALAXY
LB 03317	03 26 30.	- 64 04		13.8	FAINT BLUE STAR
SN 1969M	03 26 31.	+ 39 51		17.5	SUPERNOVA
BAK 1.535	03 26 32.	- 28 52 18.	17	17.0	EXTRAGALACTIC NEBULA
SHAP327-5415.8	03 26 32.	- 54 26 08.	36	17.25	GALAXY
MCG-02-09-043	03 26 33.	- 13 59	36	15.	GALAXY
SHAP328-4741.2	03 26 33.	- 47 51 31.	12	17.75	GALAXY
SHAP328-5144.2	03 26 33.	- 51 54 31.	18	17.25	GALAXY
SHAP327-5310.0	03 26 33.	- 53 20 20.	18	17.1	GALAXY
SHAP327-5319.1	03 26 33.	- 53 29 26.	18	17.5	GALAXY
BAK 1.537	03 26 34.	- 29 49 30.	16	16.9	EXTRAGALACTIC NEBULA
SHAP328-5042.3	03 26 35.	- 50 52 37.	18	16.5	GALAXY
ZWG 441.013	03 26 36.	+ 13 35		15.5	GALAXY
MCG+07-08-017	03 26 36.	+ 39 39	36	15.5	GALAXY
MCG-02-09-044	03 26 36.	- 12 20	72	13.5	GALAXY
IC 1942	03 26 36.	- 52 50			NONSTELLAR OBJECT
SHAP328-5240.0	03 26 36.	- 52 50 19.	30	16.4	GALAXY
HN 0201	03 26 36.	- 52 51			NEBULA
SHAP328-5302.2	03 26 36.	- 53 12 31.	18	16.7	GALAXY
SHAP327-5335.2	03 26 36.	- 53 45 31.	24	17.4	GALAXY
LB 03318	03 26 36.	- 69 16		13.6	FAINT BLUE STAR
RNGC 1338	03 26 37.	- 12 20		13.0	GALAXY
BAK 1.540	03 26 37.	- 19 42 36.	12	16.0	EXTRAGALACTIC NEBULA
BAK 1.538	03 26 37.	- 23 33 42.	14	16.4	EXTRAGALACTIC NEBULA
SHAP327-5347.1	03 26 37.	- 53 57 25.	18	16.6	GALAXY
SHAP327-5439.0	03 26 37.	- 54 49 19.	24	17.5	GALAXY
SHAP328-5214.4	03 26 38.	- 52 24 43.	24	16.9	GALAXY
SHAP328-5236.1	03 26 39.	- 52 46 25.	18	17.6	GALAXY
SHAP328-5320.0	03 26 39.	- 53 30 19.	24	17.6	GALAXY
SHAP328-5406.3	03 26 39.	- 54 16 37.	24	17.3	GALAXY
BAK 1.539	03 26 40.	- 23 38 54.	14	15.4	EXTRAGALACTIC NEBULA
SHAP328-5429.9	03 26 40.	- 54 40 13.	24	17.75	GALAXY
RNGC 1344	03 26 41.	- 31 14		12.0	GALAXY
SHAP328-5404.0	03 26 41.	- 54 14 19.	24	17.0	GALAXY
ZWG 525.036	03 26 42.	+ 38 54		15.7	GALAXY
UGC 02758	03 26 42.	+ 38 54	90	15.7	GALAXY SB0
MCG+07-08-018	03 26 42.	+ 41 39	72	14.	GALAXY
ZWG 541.017	03 26 42.	+ 41 40		14.8	GALAXY
UGC 02759	03 26 42.	+ 41 40	108	14.8	GALAXY PECULR
ZC 0326.7-0051	03 26 42.	- 00 51	3090		CLUSTER OF GALAXIES
PHL 1548	03 26 42.	- 11 52		12.5	BLUE STELLAR OBJECT
BAK 1.545	03 26 42.	- 20 34 06.	35	16.1	EXTRAGALACTIC NEBULA
SHAP328-4518.7	03 26 42.	- 45 29 01.	24	17.0	GALAXY
SHAP328-5244.3	03 26 42.	- 52 54 37.	30	17.8	GALAXY
SHAP328-5303.0	03 26 42.	- 53 13 19.	24	17.0	GALAXY
BAK 1.542	03 26 43.	- 23 54 06.	18	16.9	EXTRAGALACTIC NEBULA
SHAP328-5407.0	03 26 43.	- 54 17 19.	24	17.2	GALAXY
RNGC 1334	03 26 44.	+ 41 40		15.0	GALAXY
BAK 1.543	03 26 44.	- 23 10 54.	50	16.3	EXTRAGALACTIC NEBULA
SHAP328-5043.7	03 26 44.	- 50 54 01.	30	17.0	GALAXY
SHAP328-5313.2	03 26 44.	- 53 23 31.	18	17.0	GALAXY
SHAP328-5333.0	03 26 44.	- 53 43 19.	24	17.5	GALAXY
BAK 1.546	03 26 45.	- 22 04 17.	26	16.1	EXTRAGALACTIC NEBULA
BAK 1.541	03 26 45.	- 26 46 54.	25	16.0	EXTRAGALACTIC NEBULA
SHAP328-5230.2	03 26 45.	- 52 40 31.	24	17.7	GALAXY
SHAP328-5248.4	03 26 46.	- 52 58 43.	18	17.5	GALAXY
SHAP328-5337.0	03 26 46.	- 53 47 19.	24	17.5	GALAXY
SHAP328-5434.3	03 26 46.	- 54 44 37.	24	17.5	GALAXY

OBJECT NAME	RIGHT ASCEN.	DECLINATION	DIAM.	MAGN.	TYPE OF OBJECT
SHAP328-5140.1	03 26 47.	- 51 50 25.	18	17.25	GALAXY
SHAP328-5402.2	03 26 47.	- 54 12 31.	18	16.5	GALAXY
SHAP328-5404.2	03 26 47.	- 54 14 31.	18	17.4	GALAXY
UGC 02760	03 26 48.	+ 42 47	66	16.5	GALAXY Sc
MCG-06-08-021	03 26 48.	- 35 21	150	13.	GALAXY
SHAP328-5213.3	03 26 48.	- 52 23 37.	18	17.7	GALAXY
SHAP328-5402.6	03 26 48.	- 54 12 55.	36	17.6	GALAXY
SHAP328-5428.2	03 26 48.	- 54 38 31.	18	17.25	GALAXY
BAK 1.547	03 26 49.	- 22 18 35.	17	15.9	EXTRAGALACTIC NEBULA
BAK 1.544	03 26 49.	- 27 33 05.	24	16.6	EXTRAGALACTIC NEBULA
SHAP328-4918.0	03 26 49.	- 49 28 18.	12	17.25	GALAXY
SHAP328-5139.6	03 26 49.	- 51 49 55.	12	16.5	GALAXY
SHAP328-5410.0	03 26 49.	- 54 20 19.	18	17.5	GALAXY
SHAP328-5337.6	03 26 50.	- 53 47 55.	18	17.4	GALAXY
SHAP328-5409.4	03 26 50.	- 54 19 43.	24	17.5	GALAXY
ARC 0440	03 26 51.	- 10 47		17.2	RICH CLUSTER OF GALAXIES
MCG-04-09-015	03 26 51.	- 22 18 30.	72	15.	GALAXY
SHAP328-4633.7	03 26 51.	- 46 44 00.	18	18.0	GALAXY
SHAP328-4857.4	03 26 51.	- 49 07 42.	24	17.5	GALAXY
SHAP328-5408.8	03 26 51.	- 54 19 07.	24	17.4	GALAXY
BAK 2.0001	03 26 52.	- 26 37 17.	25	14.1	GALAXY
SHAP328-5027.7	03 26 52.	- 50 38 00.	18	17.5	GALAXY
SHAP328-4556.4	03 26 53.	- 46 06 42.	24	16.5	GALAXY
SHAP328-4701.0	03 26 53.	- 47 11 18.	18	18.0	GALAXY
SHAP328-4816.9	03 26 53.	- 48 27 12.	12	17.5	GALAXY
SHAP328-4820.9	03 26 53.	- 48 31 12.	24	16.5	GALAXY
SHAP328-5129.6	03 26 53.	- 51 39 54.	18	17.25	GALAXY
SHAP328-5241.5	03 26 53.	- 52 51 48.	18	17.5	GALAXY
SHAP328-5400.0	03 26 53.	- 54 10 18.	18	16.4	GALAXY
ZWG 525.037	03 26 54.	+ 38 17		15.4	GALAXY
UGC 02761	03 26 54.	+ 38 17	84	15.4	GALAXY S0-a
HPW 05	03 26 54.	- 34 02			DWARF GALAXY IN FORNAX
DV.56 N1351A	03 26 54.	- 35 21			S GALAXY
SHAP328-5403.5	03 26 54.	- 54 13 48.	24	16.9	GALAXY
SHAP328-4600.9	03 26 55.	- 46 11 12.	36	18.0	GALAXY
SHAP328-4946.7	03 26 55.	- 49 57 00.	24	17.25	GALAXY
SHAP328-5306.9	03 26 55.	- 53 17 12.	18	17.4	GALAXY
SHAP328-5410.2	03 26 55.	- 54 20 30.	24	17.3	GALAXY
BAK 1.548	03 26 56.	- 26 36 53.	19	14.7	EXTRAGALACTIC NEBULA
RNGC 1351A	03 26 56.	- 35 21			GALAXY
SHAP328-5305.1	03 26 56.	- 53 15 24.	24	17.3	GALAXY
SHAP328-5338.4	03 26 56.	- 53 48 42.	18	17.6	GALAXY
SHAP328-5158.6	03 26 57.	- 52 08 54.	18	18.0	GALAXY
SHAP328-5318.3	03 26 57.	- 53 28 36.	24	17.5	GALAXY
SHAP328-5348.4	03 26 57.	- 53 58 42.	36	17.1	GALAXY
SHAP328-5158.8	03 26 58.	- 52 09 06.	18	17.5	GALAXY
SHAP328-5338.7	03 26 58.	- 53 49 00.	18	17.4	GALAXY
SC 0325-5455.0	03 26 58.	- 54 44 37.	72		NEBULA
SHAP328-4617.2	03 26 59.	- 46 27 30.	18	16.5	GALAXY
SHAP328-5340.4	03 26 59.	- 53 50 42.	18	17.5	GALAXY
MCG+06-08-026	03 27 00.	+ 38 17	66	15.5	GALAXY
MCG+07-08-019	03 27 00.	+ 41 23	42	15.5	GALAXY
ZWG 541.018	03 27 00.	+ 41 25		15.7	GALAXY
UGC 02762	03 27 00.	+ 41 25	78	15.7	GALAXY E-S0
ZWG 390.095	03 27 00.	- 01 42		15.7	GALAXY
KARA.73B 0127	03 27 00.	- 01 42	42	15.7	ISOLATED GALAXY E
MCG+00-09-088	03 27 00.	- 01 43	12	15.	GALAXY
PHL 1549	03 27 00.	- 02 43		17.7	BLUE STELLAR OBJECT
PHL 4437	03 27 00.	- 14 36		18.3	BLUE STELLAR OBJECT
PHL 8683	03 27 00.	- 22 50		16.6	BLUE STELLAR OBJECT
BAK 1.551	03 27 00.	- 26 23 17.	26	16.4	EXTRAGALACTIC NEBULA
BAK 1.549	03 27 00.	- 28 12 41.	14	16.8	EXTRAGALACTIC NEBULA
PHL 8684	03 27 00.	- 28 40		18.3	BLUE STELLAR OBJECT
BAK 1.550	03 27 01.	- 28 51 41.	17	16.6	EXTRAGALACTIC NEBULA
SHAP328-5155.1	03 27 01.	- 52 05 24.	18	16.5	GALAXY
SHAP328-5327.6	03 27 01.	- 53 37 54.	18	17.5	GALAXY
RNGC 1335	03 27 02.	+ 41 25		15.5	GALAXY
SHAP328-5155.2	03 27 02.	- 52 05 30.	18	17.0	GALAXY
SHAP328-5204.6	03 27 02.	- 52 14 54.	18	17.5	GALAXY
SHAP328-5243.8	03 27 02.	- 52 54 06.	18	17.4	GALAXY
SHAP328-5334.0	03 27 03.	- 53 44 18.	132	16.5	GALAXY
SHAP328-5359.3	03 27 03.	- 54 09 36.	30	17.6	GALAXY
LB 02883	03 27 04.	- 00 01 42.		16.3	FAINT BLUE STAR
RNGC 1345	03 27 04.	- 17 58		14.0	GALAXY
SHAP328-4555.9	03 27 04.	- 46 06 11.	24	17.0	GALAXY
SHAP328-5342.5	03 27 04.	- 53 52 48.	24	17.6	GALAXY
ZC 0327.1+0002	03 27 06.	+ 00 02	810		CLUSTER OF GALAXIES
OCL 0387	03 27 06.	+ 52 29	540		OPEN STAR CLUSTER
MCG-03-09-046	03 27 06.	- 17 58	54	14.	GALAXY
HPW 06	03 27 06.	- 35 30			DWARF GALAXY IN FORNAX
SHAP328-4713.9	03 27 06.	- 47 24 11.	30	17.0	GALAXY
SHAP328-4858.6	03 27 06.	- 49 08 53.	18	16.75	GALAXY
SHAP328-4601.8	03 27 07.	- 46 12 05.	24	17.0	GALAXY
SHAP328-5001.6	03 27 07.	- 50 11 53.	24	16.75	GALAXY
SHAP328-5056.5	03 27 07.	- 51 06 47.	18	17.5	GALAXY
SHAP328-5247.3	03 27 07.	- 52 57 36.	12	17.5	GALAXY
SHAP328-5256.6	03 27 07.	- 53 06 54.	24	16.8	GALAXY
SHAP328-5413.6	03 27 08.	- 54 23 54.	24	16.7	GALAXY
SHAP328-5353.7	03 27 09.	- 54 04 00.	42	17.5	GALAXY
SHAP328-5347.0	03 27 10.	- 53 57 17.	24	17.3	GALAXY
BAK 1.552	03 27 11.	- 27 49 16.	29	15.5	EXTRAGALACTIC NEBULA
SHAP328-5323.4	03 27 11.	- 53 33 41.	30	17.5	GALAXY
ZWG 525.038	03 27 12.	+ 36 47		15.6	GALAXY
UGC 02763	03 27 12.	+ 36 47	96	15.6	GALAXY S0-a
PHL 8685	03 27 12.	- 05 30		18.1	BLUE STELLAR OBJECT
MCG-03-09-047	03 27 12.	- 15 25	120	15.	GALAXY
PHL 8687	03 27 12.	- 21 20		18.2	BLUE STELLAR OBJECT
PHL 8686	03 27 12.	- 27 30		13.8	BLUE STELLAR OBJECT
SHAP328-5149.6	03 27 12.	- 51 59 53.	18	17.0	GALAXY
SHAP328-5334.3	03 27 12.	- 53 44 35.	24	17.5	GALAXY
SHAP328-5430.2	03 27 12.	- 54 40 29.	24	17.25	GALAXY
HARO 21	03 27 13.	- 17 57			BLUE EMISSION-LINE GALAXY
BAK 1.553	03 27 13.	- 28 55 04.	13	17.2	EXTRAGALACTIC NEBULA
SHAP328-4614.8	03 27 13.	- 46 25 05.	18	17.75	GALAXY
SHAP328-4944.2	03 27 13.	- 49 54 29.	18	17.25	GALAXY
SHAP328-5434.8	03 27 13.	- 54 45 05.	24	17.25	GALAXY
SHAP328-5301.8	03 27 14.	- 53 12 05.	12	16.3	GALAXY
BAK 2.0002	03 27 15.	- 26 35 40.	11	16.3	GALAXY
BAK 1.555	03 27 15.	- 28 18 04.	48	15.6	EXTRAGALACTIC NEBULA
BAK 1.556	03 27 16.	- 25 10 52.	17	16.2	EXTRAGALACTIC NEBULA
BAK 1.557	03 27 17.	- 22 34 04.	24	17.2	EXTRAGALACTIC NEBULA
BAK 1.554	03 27 17.	- 31 51 34.	24	16.7	EXTRAGALACTIC NEBULA
SHAP328-5307.0	03 27 17.	- 53 17 17.	12	17.4	GALAXY
SHAP328-5325.1	03 27 17.	- 53 35 23.	24	16.9	GALAXY
SHAP328-5400.2	03 27 17.	- 54 10 29.	24	17.5	GALAXY
MCG-05-09-006	03 27 18.	- 27 49	18	15.	GALAXY
SHAP328-5320.0	03 27 19.	- 53 30 17.	24	17.5	GALAXY
BAK 1.561	03 27 20.	- 19 51 39.	31	16.0	EXTRAGALACTIC NEBULA

OBJECT NAME	RIGHT ASCEN.	DECLINATION	DIAM.	MAGN.	TYPE OF OBJECT
SHAP328-5237.6	03 27 20.	- 52 47 53.	18	17.0	GALAXY
SHAP328-5326.6	03 27 20.	- 53 36 53.	24	17.0	GALAXY
SHAP328-4840.2	03 27 21.	- 48 50 28.	18	17.5	GALAXY
SHAP328-5226.4	03 27 21.	- 52 36 41.	18	17.2	GALAXY
SHAP328-5233.2	03 27 21.	- 52 43 29.	18	17.7	GALAXY
SHAP328-5234.2	03 27 21.	- 52 44 29.	24	17.6	GALAXY
SHAP328-5238.0	03 27 21.	- 52 48 17.	24	17.6	GALAXY
SHAP328-5303.1	03 27 21.	- 53 13 23.	18	17.1	GALAXY
SHAP329-4551.2	03 27 23.	- 46 01 28.	24	17.5	GALAXY
SHAP328-5310.4	03 27 23.	- 53 20 41.	18	17.6	GALAXY
SHAP328-5357.6	03 27 23.	- 54 07 53.	24	17.4	GALAXY
5ZW 349	03 27 24.	+ 24 29			COMPACT GALAXY
5ZW 350	03 27 24.	+ 32 38			COMPACT GALAXY
UGC 02764	03 27 24.	+ 41 51	84	16.5	GALAXY S0-a
PHL 8688	03 27 24.	- 18 57		16.9	BLUE STELLAR OBJECT
VV 023B	03 27 24.	- 22 26	24	16.	INTERACTING GALAXY
VV 023A	03 27 24.	- 22 26	84	13.	INTERACTING GALAXY
VV 023	03 27 24.	- 22 26	102		INTERACTING GALAXY
PHL 4438	03 27 24.	- 24 47		18.6	BLUE STELLAR OBJECT
BAK 1.558	03 27 24.	- 28 52 27.	14	16.5	EXTRAGALACTIC NEBULA
SHAP328-5104.5	03 27 24.	- 51 14 47.	18	17.5	GALAXY
SHAP328-5354.0	03 27 24.	- 54 04 17.	18	17.0	GALAXY
SHAP328-5146.4	03 27 25.	- 51 56 40.	30	16.25	GALAXY
SHAP328-5237.6	03 27 25.	- 52 47 53.	12	17.7	GALAXY
SHAP328-5257.3	03 27 25.	- 53 07 35.	18	16.7	GALAXY
BAK 1.559	03 27 26.	- 28 19 39.	22	16.8	EXTRAGALACTIC NEBULA
SHAP328-5113.0	03 27 26.	- 51 23 16.	30	17.5	GALAXY
BAK 1.562	03 27 27.	- 23 31 39.	60	15.9	EXTRAGALACTIC NEBULA
SHAP328-5309.2	03 27 27.	- 53 19 28.	24	17.5	GALAXY
SHAP328-5357.2	03 27 27.	- 54 07 29.	24	17.5	GALAXY
SHAP328-5302.8	03 27 28.	- 53 13 04.	18	17.4	GALAXY
SHAP328-5329.5	03 27 28.	- 53 39 46.	18	17.2	GALAXY
BAK 1.564	03 27 29.	- 22 27 51.	26	16.4	EXTRAGALACTIC NEBULA
BAK 1.560	03 27 29.	- 28 07 03.	12	16.5	EXTRAGALACTIC NEBULA
SHAP329-4558.2	03 27 29.	- 46 08 28.	18	16.5	GALAXY
SHAP328-5212.7	03 27 29.	- 52 22 58.	24	16.5	GALAXY
ZC 0327.5+2440	03 27 30.	+ 24 40	2020		CLUSTER OF GALAXIES
UGC 02765	03 27 30.	+ 68 12	240	18.	GALAXY
PHL 8689	03 27 30.	- 04 08		18.5	BLUE STELLAR OBJECT
PHL 8690	03 27 30.	- 07 28		17.3	BLUE STELLAR OBJECT
MCG-02-09-045	03 27 30.	- 13 32	42	15.	GALAXY
RNGC 1347	03 27 30.	- 22 26		13.0	GALAXY
MCG-04-09-016	03 27 30.	- 23 30	60	15.	GALAXY
PHL 4439	03 27 30.	- 23 49		18.1	BLUE STELLAR OBJECT
MCG-05-09-007	03 27 30.	- 28 17	48	15.	GALAXY
SHAP329-4915.6	03 27 30.	- 49 25 52.	18	17.0	GALAXY
SHAP328-5301.4	03 27 30.	- 53 11 40.	18	17.2	GALAXY
SHAP328-5431.3	03 27 30.	- 54 41 34.	18	17.5	GALAXY
ARC 0439	03 27 31.	+ 24 38		17.0	RICH CLUSTER OF GALAXIES
BAK 1.563	03 27 31.	- 27 15 27.	14	16.5	EXTRAGALACTIC NEBULA
SHAP329-4512.0	03 27 31.	- 45 22 16.	30	17.5	GALAXY
SHAP328-5145.2	03 27 31.	- 51 55 28.	18	17.0	GALAXY
BAK 1.566	03 27 32.	- 22 34 03.	17	16.2	EXTRAGALACTIC NEBULA
SHAP328-5211.8	03 27 32.	- 52 22 04.	24	17.4	GALAXY
SHAP328-5337.3	03 27 32.	- 53 47 34.	30	16.2	GALAXY
MCG-04-09-017	03 27 33.	- 22 26	96	13.5	GALAXY
SHAP328-5327.7	03 27 33.	- 53 37 58.	24	17.2	GALAXY
SHAP329-5212.8	03 27 35.	- 52 23 04.	24	16.9	GALAXY
SHAP328-5305.8	03 27 35.	- 53 16 04.	24	17.5	GALAXY
ASS 44	03 27 36.	+ 58 28			OB ASSOCIATION CAM OB1
MCG+01-09-041	03 27 36.	- 05 42	60	16.	GALAXY
PHL 4440	03 27 36.	- 20 43		16.6	BLUE STELLAR OBJECT
SHAP329-5250.0	03 27 36.	- 53 00 16.	12	16.9	GALAXY
SHAP328-5423.8	03 27 36.	- 54 34 04.	18	16.75	GALAXY
ARP 039	03 27 37.	- 22 26			PECULIAR GALAXY
SHAP328-5335.6	03 27 37.	- 53 45 52.	24	17.7	GALAXY
SHAP328-5403.2	03 27 37.	- 54 13 28.	30	17.3	GALAXY
SHAP329-4947.4	03 27 38.	- 49 57 40.	18	17.0	GALAXY
MCG+07-08-021	03 27 39.	+ 40 37	30	15.	GALAXY
MCG+07-08-020	03 27 39.	+ 43 04	42	15.	GALAXY
BAK 1.565	03 27 39.	- 29 13 51.	14	17.2	EXTRAGALACTIC NEBULA
SHAP329-4506.6	03 27 39.	- 45 16 51.	30	17.0	GALAXY
SHAP329-5240.7	03 27 39.	- 52 50 58.	18	16.9	GALAXY
SHAP328-5431.8	03 27 39.	- 54 42 04.	18	17.25	GALAXY
RNGC 1346	03 27 40.	- 05 43		15.0	GALAXY
SHAP329-4823.5	03 27 40.	- 48 33 45.	18	18.0	GALAXY
SHAP329-5232.0	03 27 40.	- 52 42 16.	18	17.0	GALAXY
SHAP329-5322.2	03 27 40.	- 53 32 28.	18	17.2	GALAXY
UGC 02766	03 27 42.	+ 40 39	72	16.0	GALAXY S0
ZWG 541.019	03 27 42.	+ 43 05		15.7	GALAXY
ZWG 390.096	03 27 42.	- 01 05		15.1	GALAXY
MCG+00-09-089	03 27 42.	- 01 07	30	15.	GALAXY
MCG+01-09-043	03 27 42.	- 04 24	60	14.	GALAXY
MCG+01-09-042	03 27 42.	- 05 43	42	15.	GALAXY
LB 01671	03 27 42.	- 49 31		14.7	FAINT BLUE STAR
SHAP329-5245.6	03 27 42.	- 52 55 52.	12	17.4	GALAXY
BAK 1.567	03 27 43.	- 29 11 50.	19	16.8	EXTRAGALACTIC NEBULA
SHAP329-5424.8	03 27 43.	- 54 35 04.	18	16.75	GALAXY
BAK 1.570	03 27 44.	- 25 35 32.	22	16.1	EXTRAGALACTIC NEBULA
BAK 1.569	03 27 44.	- 29 11 44.	13	17.1	EXTRAGALACTIC NEBULA
SHAP329-5327.7	03 27 44.	- 53 37 58.	18	16.7	GALAXY
SHAP329-5314.0	03 27 45.	- 53 24 15.	18	17.3	GALAXY
BAK 1.571	03 27 46.	- 25 49 50.	17	17.2	EXTRAGALACTIC NEBULA
BAK 1.568	03 27 46.	- 31 57 26.	18	16.2	EXTRAGALACTIC NEBULA
SHAP329-5324.2	03 27 46.	- 53 34 27.	18	17.5	GALAXY
SHAP329-5327.4	03 27 46.	- 53 37 39.	18	16.1	GALAXY
SHAP329-5329.0	03 27 46.	- 53 39 15.	18	17.3	GALAXY
SHAP329-4555.6	03 27 47.	- 46 05 51.	18	17.5	GALAXY
SHAP329-5311.8	03 27 47.	- 53 22 03.	18	17.4	GALAXY
SHAP329-5325.4	03 27 47.	- 53 35 39.	18	17.3	GALAXY
MCG+00-09-090	03 27 48.	+ 00 34	30	15.	GALAXY
MCG+00-09-090	03 27 48.	+ 00 38	36	15.	GALAXY
MCG+07-08-022	03 27 48.	+ 43 43 30.	54	15.	GALAXY
5ZW 351	03 27 48.	+ 46 50			COMPACT GALAXY
UGC 02767	03 27 48.	+ 79 56	102	18.	GALAXY DWRF SP
BAK 2.0003	03 27 48.	- 17 51 02.	15	15.9	GALAXY
SHAP329-5307.0	03 27 48.	- 53 17 15.	24	17.3	GALAXY
SHAP329-5228.6	03 27 49.	- 52 38 51.	18	16.8	GALAXY
SHAP329-5311.2	03 27 49.	- 53 21 27.	12	17.7	GALAXY
SHAP329-5314.0	03 27 49.	- 53 24 15.	18	17.6	GALAXY
SHAP329-4446.9	03 27 50.	- 44 57 09.	36	17.25	GALAXY
SHAP329-5329.6	03 27 50.	- 53 39 51.	24	17.5	GALAXY
BAK 1.572	03 27 51.	- 28 28 38.	22	16.4	EXTRAGALACTIC NEBULA
BAK 1.573	03 27 51.	- 28 56 20.	14	16.2	EXTRAGALACTIC NEBULA
IC 1934	03 27 52.	+ 42 37 32.			NONSTELLAR OBJECT
BAK 1.574	03 27 52.	- 28 56 32.	48	15.5	EXTRAGALACTIC NEBULA
SHAP329-5326.3	03 27 52.	- 53 36 33.	24	17.7	GALAXY
SHAP329-4558.3	03 27 53.	- 46 08 33.	24	17.0	GALAXY

OBJECT NAME	RIGHT ASCEN.	DECLINATION	DIAM.	MAGN.	TYPE OF OBJECT
ZWG 390.097	03 27 54.	+ 00 35		15.4	GALAXY
ZWG 390.098	03 27 54.	+ 00 39		15.6	GALAXY
ZWG 525.039	03 27 54.	+ 39 28		15.4	GALAXY
PHL 1550	03 27 54.	- 11 50		18.5	BLUE STELLAR OBJECT
BAK 1.575	03 27 54.	- 28 53 44.	13	16.2	EXTRAGALACTIC NEBULA
SHAP329-5238.2	03 27 54.	- 52 48 27.	36	16.3	GALAXY
HN 0204	03 27 54.	- 52 49			NEBULA
SHAP329-5325.6	03 27 54.	- 53 35 51.	30	17.4	GALAXY
SHAP329-5328.6	03 27 54.	- 53 38 51.	24	17.6	GALAXY
IC 1945	03 27 55.	- 52 48			NONSTELLAR OBJECT
SHAP329-5246.9	03 27 56.	- 52 57 09.	18	17.3	GALAXY
MCG+06-08-027	03 27 57.	+ 39 28	48	15.	GALAXY
SHAP329-4912.6	03 27 57.	- 49 22 50.	42	17.25	GALAXY
SHAP329-5225.1	03 27 57.	- 52 35 21.	18	17.3	GALAXY
SHAP329-5359.2	03 27 57.	- 54 09 27.	24	17.0	GALAXY
SHAP329-5256.0	03 27 58.	- 53 06 15.	24	17.3	GALAXY
ARC 0442	03 27 59.	- 13 05		17.6	RICH CLUSTER OF GALAXIES
SHAP329-4600.9	03 27 59.	- 46 11 08.	12	17.5	GALAXY
LBN 0719	03 28	+ 37 25	1920		BRIGHT NEBULA
LBN 0717	03 28	+ 37 30	7200		BRIGHT NEBULA
LBN 0684	03 28	+ 58 20	1800		BRIGHT NEBULA
UGC 02768	03 28 00.	+ 39 15	66	17.	GALAXY Sb-c
UGC 02769	03 28 00.	+ 42 38	78	16.0	GALAXY S0
CED 017	03 28 00.	+ 43 44			DIFFUSE GALACTIC NEBULA
PHL 8691	03 28 00.	- 05 15		4.8	BLUE STELLAR OBJECT
MCG-02-09-046	03 28 00.	- 14 12	36	15.	GALAXY
PHL 8692	03 28 00.	- 30 02		18.6	BLUE STELLAR OBJECT
SHAP329-4510.4	03 28 00.	- 45 20 38.	24	17.0	GALAXY
SHAP329-5230.7	03 28 00.	- 52 40 57.	24	17.7	GALAXY
SHAP329-5232.0	03 28 00.	- 52 42 15.	18	16.9	GALAXY
SHAP329-5237.2	03 28 00.	- 52 47 27.	72	16.2	GALAXY
HN 0205	03 28 00.	- 52 48			NEBULA
SHAP329-4449.9	03 28 01.	- 45 00 08.	24	17.25	GALAXY
IC 1946	03 28 01.	- 52 47			NONSTELLAR OBJECT
IC 1943	03 28 02.	- 44 16 38.			NONSTELLAR OBJECT
SHAP329-5211.0	03 28 02.	- 52 21 14.	12	17.6	GALAXY
SHAP329-5415.4	03 28 02.	- 54 25 39.	18	17.75	GALAXY
ARC 0441	03 28 03.	- 07 08		17.6	RICH CLUSTER OF GALAXIES
SHAP329-4800.2	03 28 04.	- 48 10 26.	36	15.75	GALAXY
SHAP329-5239.8	03 28 04.	- 52 50 02.	24	17.6	GALAXY
SHAP329-5244.4	03 28 04.	- 52 54 38.	24	16.8	GALAXY
SHAP329-5256.2	03 28 04.	- 53 06 26.	30	17.6	GALAXY
BAK 1.580	03 28 05.	- 24 31 25.	22	15.4	EXTRAGALACTIC NEBULA
BAK 1.579	03 28 05.	- 25 49 25.	17	17.4	EXTRAGALACTIC NEBULA
BAK 1.578	03 28 05.	- 28 52 37.	12	17.2	EXTRAGALACTIC NEBULA
BAK 1.577	03 28 05.	- 29 12 25.	13	16.9	EXTRAGALACTIC NEBULA
SHAP329-4658.0	03 28 05.	- 47 08 14.	24	16.5	GALAXY
IC 1944	03 28 05.	- 48 10			NONSTELLAR OBJECT
HN 0203	03 28 05.	- 48 11			NEBULA
SHAP329-5304.8	03 28 05.	- 53 15 02.	18	17.5	GALAXY
ZWG 525.040	03 28 06.	+ 35 18		14.9	GALAXY
UGC 02770	03 28 06.	+ 35 18	102	14.9	GALAXY SBb
MCG+07-08-023	03 28 06.	+ 39 33	78	15.	GALAXY
ZWG 541.020	03 28 06.	+ 39 35		15.4	GALAXY
UGC 02771	03 28 06.	+ 39 35	132	15.4	GALAXY S0
BAK 2.0004	03 28 06.	- 17 24 31.	34	16.0	GALAXY
MCG-05-09-009	03 28 06.	- 28 56	42	15.	GALAXY
MCG-05-09-008	03 28 06.	- 28 56	12	15.	GALAXY
TON-S 0352	03 28 06.	- 29 29		15.7	BLUE STAR
SHAP329-4510.2	03 28 06.	- 45 20 26.	18	17.75	GALAXY
SHAP329-4558.1	03 28 06.	- 46 08 20.	18	17.0	GALAXY
SHAP329-5215.8	03 28 06.	- 52 26 02.	18	16.6	GALAXY
SHAP329-5226.4	03 28 06.	- 52 36 38.	18	17.4	GALAXY
SHAP329-5240.2	03 28 06.	- 52 50 26.	18	17.5	GALAXY
SHAP329-5245.4	03 28 06.	- 52 55 38.	18	17.3	GALAXY
SHAP329-5249.2	03 28 06.	- 52 59 26.	18	16.8	GALAXY
SHAP329-5226.0	03 28 07.	- 52 36 14.	18	17.2	GALAXY
SHAP329-5239.4	03 28 07.	- 52 49 38.	18	16.9	GALAXY
BAK 1.581	03 28 08.	- 24 30 49.	19	16.0	EXTRAGALACTIC NEBULA
BAK 1.576	03 28 08.	- 32 38 25.	25	16.0	EXTRAGALACTIC NEBULA
SHAP329-5242.6	03 28 08.	- 52 52 50.	24	17.3	GALAXY
SHAP329-5245.4	03 28 08.	- 52 55 38.	18	17.2	GALAXY
SHAP329-5245.4	03 28 08.	- 52 55 38.	18	17.5	GALAXY
SHAP329-5250.0	03 28 08.	- 53 00 14.	18	17.0	GALAXY
SHAP329-4602.4	03 28 09.	- 46 12 38.	24	16.25	GALAXY
SHAP329-5228.8	03 28 09.	- 52 39 02.	18	17.5	GALAXY
SHAP329-5245.7	03 28 09.	- 52 55 56.	24	17.6	GALAXY
SHAP329-5216.8	03 28 10.	- 52 27 02.	12	16.9	GALAXY
SHAP329-5419.0	03 28 10.	- 54 29 14.	30	16.5	GALAXY
SHAP329-5241.1	03 28 11.	- 52 51 20.	18	16.6	GALAXY
SHAP329-5313.4	03 28 11.	- 53 23 38.	18	16.4	GALAXY
SHAP329-5344.5	03 28 11.	- 53 54 44.	24	17.6	GALAXY
ZC 0328.2+1812	03 28 12.	+ 18 12	670		CLUSTER OF GALAXIES
MCG+06-08-028	03 28 12.	+ 35 17	90	14.5	GALAXY
MRK 612	03 28 12.	- 03 16	10	15.5	GALAXY WITH UV CONTINUUM
MCG+01-09-044	03 28 12.	- 05 01	18	15.	GALAXY
PHL 8693	03 28 12.	- 20 40		16.9	BLUE STELLAR OBJECT
TON-S 0353	03 28 12.	- 26 09		15.7	BLUE STAR
SHAP329-5247.8	03 28 13.	- 52 58 02.	18	17.6	GALAXY
SHAP329-5444.4	03 28 13.	- 54 54 38.	30	17.25	GALAXY
SHAP329-5241.2	03 28 14.	- 52 51 26.	18	17.6	GALAXY
SHAP329-5259.9	03 28 14.	- 53 10 08.	12	17.6	GALAXY
BAK 1.583	03 28 15.	- 24 34 00.	20	16.0	EXTRAGALACTIC NEBULA
SHAP329-4518.8	03 28 15.	- 45 29 01.	36	17.5	GALAXY
SHAP329-5211.8	03 28 15.	- 52 22 02.	18	17.2	GALAXY
SHAP329-5244.0	03 28 16.	- 52 54 14.	24	17.2	GALAXY
IC 0326	03 28 17.	- 14 35 54.			NONSTELLAR OBJECT
SHAP329-5229.6	03 28 17.	- 52 39 50.	12	17.6	GALAXY
UGC 02772	03 28 18.	+ 72 01	72	17.	GALAXY Sc
MCG-03-09-048	03 28 18.	- 18 07	72	14.	GALAXY
PHL 8694	03 28 18.	- 29 42		18.4	BLUE STELLAR OBJECT
SHAP329-5245.6	03 28 18.	- 52 55 50.	18	17.5	GALAXY
SHAP329-4605.8	03 28 19.	- 46 16 01.	24	16.5	GALAXY
SHAP329-5335.3	03 28 19.	- 53 45 32.	18	17.6	GALAXY
SHAP329-5345.1	03 28 19.	- 53 55 20.	30	16.9	GALAXY
BAK 2.0005	03 28 20.	- 18 06 48.	48	14.5	GALAXY
BAK 1.582	03 28 20.	- 29 38 24.	19	16.9	EXTRAGALACTIC NEBULA
MCG-03-09-049	03 28 21.	- 14 36	24	15.5	GALAXY
BAK 1.585	03 28 21.	- 26 10 42.	23	16.7	EXTRAGALACTIC NEBULA
BAK 1.584	03 28 21.	- 29 15 24.	24	16.8	EXTRAGALACTIC NEBULA
SHAP329-5405.1	03 28 21.	- 54 15 19.	18	17.2	GALAXY
IC 0325	03 28 22.	- 07 13 06.			NONSTELLAR OBJECT
SHAP329-4930.1	03 28 22.	- 49 40 19.	24	17.5	GALAXY
RNGC 1342	03 28 23.	+ 37 10		7.0	OPEN CLUSTER
SHAP330-4453.8	03 28 23.	- 45 04 01.	42	17.5	GALAXY
SHAP330-4603.2	03 28 23.	- 46 13 25.	18	17.5	GALAXY
SHAP330-4653.1	03 28 23.	- 47 03 19.	30	17.5	GALAXY
SHAP329-5257.6	03 28 23.	- 53 07 49.	24	17.5	GALAXY

OBJECT NAME	RIGHT ASCEN.	DECLINATION	DIAM.	MAGN.	TYPE OF OBJECT
SHAP329-5439.0	03 28 23.	- 54 49 13.	30	17.5	GALAXY
OCL 0401	03 28 24.	+ 37 10	1800	7.2	OPEN STAR CLUSTER
PHL 8695	03 28 24.	- 08 46		16.5	BLUE STELLAR OBJECT
PHL 4441	03 28 24.	- 10 42		18.5	BLUE STELLAR OBJECT
PHL 1551	03 28 24.	- 13 54		18.6	BLUE STELLAR OBJECT
SHAP329-5217.6	03 28 24.	- 52 27 49.	18	17.7	GALAXY
SHAP329-5234.2	03 28 24.	- 52 44 25.	24	17.4	GALAXY
SHAP329-5431.6	03 28 25.	- 54 41 49.	24	17.25	GALAXY
SHAP329-5436.9	03 28 25.	- 54 47 07.	24	17.75	GALAXY
SHAP329-5233.6	03 28 26.	- 52 43 49.	24	16.5	GALAXY
SHAP330-4746.9	03 28 27.	- 47 57 07.	24	17.5	GALAXY
SHAP329-5245.1	03 28 27.	- 52 55 19.	24	17.6	GALAXY
SHAP330-4514.3	03 28 28.	- 45 24 30.	24	17.0	GALAXY
SHAP330-4750.3	03 28 28.	- 48 00 31.	48	14.2	GALAXY
SHAP329-5234.2	03 28 28.	- 52 44 25.	18	16.5	GALAXY
SHAP329-5021.8	03 28 29.	- 50 32 01.	24	17.5	GALAXY
SHAP329-5308.0	03 28 29.	- 53 18 13.	24	16.6	GALAXY
SHAP329-5401.8	03 28 29.	- 54 12 01.	24	17.6	GALAXY
ZC 0328.5-0205	03 28 30.	- 02 05	1340		CLUSTER OF GALAXIES
TON-S 0354	03 28 30.	- 27 58		16.0	BLUE STAR
HPW 07	03 28 30.	- 34 26			DWARF GALAXY IN FORNAX
SHAP330-4558.9	03 28 31.	- 46 09 06.	24	17.5	GALAXY
SHAP329-5231.3	03 28 31.	- 52 41 31.	30	16.8	GALAXY
SHAP329-5301.0	03 28 31.	- 53 11 13.	12	17.0	GALAXY
SHAP329-5405.5	03 28 31.	- 54 15 43.	24	17.5	GALAXY
BAK 1.590	03 28 32.	- 20 41 23.	42	16.2	EXTRAGALACTIC NEBULA
HELW 397	03 28 32.	- 20 41 47.			NEBULA
SHAP330-4608.2	03 28 32.	- 46 18 24.	18	17.5	GALAXY
SHAP329-5436.8	03 28 32.	- 54 47 01.	18	17.5	GALAXY
SHAP330-4656.6	03 28 33.	- 47 06 48.	24	17.5	GALAXY
SHAP329-5435.8	03 28 33.	- 54 46 01.	30	16.5	GALAXY
BAK 1.593	03 28 34.	- 21 13 35.	66	15.5	EXTRAGALACTIC NEBULA
BAK 1.586	03 28 34.	- 28 31 11.	12	17.0	EXTRAGALACTIC NEBULA
SHAP330-4620.8	03 28 34.	- 46 31 00.	18	17.25	GALAXY
SHAP329-5327.7	03 28 34.	- 53 37 55.	24	16.7	GALAXY
SHAP330-5219.2	03 28 35.	- 52 29 25.	24	17.5	GALAXY
SHAP329-5348.4	03 28 35.	- 53 58 37.	24	17.6	GALAXY
SHAP329-5824.2	03 28 35.	- 58 34 25.	42	17.0	GALAXY
MCG+08-07-003	03 28 36.	+ 47 37	66	14.	GALAXY
ZWG 555.004	03 28 36.	+ 47 38		15.0	GALAXY
UGC 02773	03 28 36.	+ 47 38	102	15.0	GALAXY DBL SYS
PHL 8696	03 28 36.	- 04 38		18.0	BLUE STELLAR OBJECT
PHL 1552	03 28 36.	- 19 12		18.1	BLUE STELLAR OBJECT
PHL 4442	03 28 36.	- 25 41		18.4	BLUE STELLAR OBJECT
MCG-06-08-022	03 28 36.	- 35 02	180		GALAXY
SHAP330-4945.7	03 28 36.	- 49 55 54.	24	16.5	GALAXY
SHAP330-5226.0	03 28 36.	- 52 36 12.	24	17.6	GALAXY
SHAP329-5343.2	03 28 36.	- 53 53 25.	24	15.8	GALAXY
SHAP329-5414.2	03 28 36.	- 54 24 25.	18	18.0	GALAXY
HELW 398	03 28 37.	- 21 14 17.			NEBULA
SHAP330-4851.6	03 28 37.	- 49 01 48.	30	17.5	GALAXY
SHAP330-5217.6	03 28 37.	- 52 27 48.	18	17.0	GALAXY
SHAP330-5224.5	03 28 37.	- 52 34 42.	18	17.8	GALAXY
SHAP329-5401.0	03 28 37.	- 54 11 12.	24	17.5	GALAXY
SHAP329-5438.8	03 28 37.	- 54 49 01.	24	17.0	GALAXY
RNGC 1351	03 28 38.	- 35 02		13.0	GALAXY
SHAP330-4610.5	03 28 38.	- 46 20 42.	18	17.5	GALAXY
BAK 1.588	03 28 39.	- 28 31 11.	22	16.9	EXTRAGALACTIC NEBULA
BAK 1.587	03 28 39.	- 32 06 11.	25	16.2	EXTRAGALACTIC NEBULA
SHAP330-5309.8	03 28 39.	- 53 20 00.	18	17.4	GALAXY
SHAP330-5341.4	03 28 39.	- 53 51 36.	24	17.7	GALAXY
BAK 1.589	03 28 40.	- 29 16 35.	17	17.2	EXTRAGALACTIC NEBULA
SHAP330-5328.0	03 28 40.	- 53 38 12.	24	17.5	GALAXY
SHAP330-5400.8	03 28 40.	- 54 11 00.	30	17.4	GALAXY
HELW 399	03 28 41.	- 21 15 18.			NEBULA
BAK 1.598	03 28 41.	- 23 17 11.	12	16.8	EXTRAGALACTIC NEBULA
BAK 1.595	03 28 41.	- 24 43 05.	29	16.4	EXTRAGALACTIC NEBULA
SHAP330-5214.6	03 28 41.	- 52 24 48.	18	17.6	GALAXY
SHAP330-5224.4	03 28 41.	- 52 34 36.	24	17.1	GALAXY
SHAP330-5225.0	03 28 41.	- 52 35 12.	24	17.8	GALAXY
SHAP330-5311.0	03 28 41.	- 53 21 12.	18	17.8	GALAXY
SHAP330-5314.9	03 28 41.	- 53 25 06.	12	17.6	GALAXY
SHAP330-5326.6	03 28 41.	- 53 36 48.	36	17.3	GALAXY
SHAP330-5401.9	03 28 41.	- 54 12 06.	24	17.3	GALAXY
VB 191	03 28 42.	+ 60 07	638		STELLAR RING
MCG-04-09-018	03 28 42.	- 21 12	60	14.	GALAXY
SHAP330-5248.0	03 28 42.	- 52 58 12.	30	16.4	GALAXY
BAK 1.597	03 28 43.	- 26 06 35.	19	16.4	EXTRAGALACTIC NEBULA
SHAP330-5214.2	03 28 43.	- 52 24 24.	18	16.5	GALAXY
BAK 1.594	03 28 44.	- 29 58 47.	31	16.9	EXTRAGALACTIC NEBULA
BAK 1.592	03 28 44.	- 30 41 35.	18	17.2	EXTRAGALACTIC NEBULA
BAK 1.591	03 28 44.	- 30 42 11.	25	16.4	EXTRAGALACTIC NEBULA
SHAP330-5121.0	03 28 44.	- 51 31 12.	24	16.5	GALAXY
SHAP330-5233.5	03 28 44.	- 52 43 42.	24	16.3	GALAXY
SHAP330-5229.8	03 28 45.	- 52 40 00.	18	17.7	GALAXY
BAK 1.596	03 28 46.	- 28 59 11.	19	17.2	EXTRAGALACTIC NEBULA
SHAP330-5152.2	03 28 46.	- 52 02 24.	24	17.5	GALAXY
SHAP330-5322.4	03 28 46.	- 53 32 36.	18	17.5	GALAXY
RNGC 1349	03 28 47.	+ 04 12		15.0	GALAXY
ZWG 416.013	03 28 48.	+ 04 12		15.0	GALAXY
UGC 02774	03 28 48.	+ 04 12	60	15.0	GALAXY S0
KARA.73B 0128	03 28 48.	+ 04 12	48	15.0	ISOLATED GALAXY E
MCG+07-08-024	03 28 48.	+ 41 24	84	16.	GALAXY
UGC 02775	03 28 48.	+ 41 26	108	16.0	GALAXY Sc
MCG+01-09-045	03 28 48.	- 05 17	48	15.	GALAXY
MCG-03-09-050	03 28 48.	- 14 53	36	15.	GALAXY
BAK 2.0006	03 28 48.	- 18 45 10.	17	15.0	GALAXY
BAK 1.602	03 28 48.	- 22 52 46.	11	17.3	EXTRAGALACTIC NEBULA
PHL 4443	03 28 48.	- 27 58		17.9	BLUE STELLAR OBJECT
HPW 08	03 28 48.	- 35 44			DWARF GALAXY IN FORNAX
SHAP330-5244.2	03 28 48.	- 52 54 24.	18	17.4	GALAXY
SHAP330-5311.2	03 28 48.	- 53 21 24.	30	17.7	GALAXY
SHAP330-5323.6	03 28 48.	- 53 33 48.	18	16.5	GALAXY
IC 0327	03 28 49.	- 14 52 03.			NONSTELLAR OBJECT
BAK 1.599	03 28 49.	- 29 03 35.	17	17.6	EXTRAGALACTIC NEBULA
SHAP330-5312.4	03 28 49.	- 53 22 36.	36	17.0	GALAXY
SHAP330-5345.5	03 28 49.	- 53 55 42.	36	16.4	GALAXY
SHAP330-4843.4	03 28 50.	- 48 53 35.	36	17.0	GALAXY
SHAP330-5113.5	03 28 50.	- 51 23 42.	24	16.5	GALAXY
IC 0328	03 28 51.	- 14 48 40.			NONSTELLAR OBJECT
MCG-03-09-051	03 28 51.	- 14 50	30	14.5	GALAXY
SHAP330-5219.2	03 28 51.	- 52 29 24.	18	17.6	GALAXY
SHAP330-5233.2	03 28 51.	- 52 43 24.	18	16.5	GALAXY
SHAP330-5308.5	03 28 51.	- 53 18 42.	18	17.5	GALAXY
SHAP330-5232.8	03 28 52.	- 52 43 00.	18	17.3	GALAXY
BAK 1.600	03 28 53.	- 28 20 10.	14	16.5	EXTRAGALACTIC NEBULA
BAK 1.601	03 28 53.	- 28 41 46.	23	17.0	EXTRAGALACTIC NEBULA
SHAP330-5254.7	03 28 53.	- 53 04 54.	18	17.6	GALAXY

OBJECT NAME	RIGHT ASCEN.	DECLINATION	DIAM.	MAGN.	TYPE OF OBJECT
SHAP330-5314.8	03 28 53.	- 53 25 00.	18	17.1	GALAXY
MCG+01-09-006	03 28 54.	+ 04 13	36	15.	GALAXY
SCHO 0056	03 28 54.	+ 37 58 12.	360		ISOLATED DARK CLOUD
PHL 8697	03 28 54.	- 12 10		18.2	BLUE STELLAR OBJECT
MCG-04-09-019	03 28 54.	- 22 46	36	15.5	GALAXY
SHAP330-5013.7	03 28 54.	- 50 23 53.	36	17.5	GALAXY
SHAP330-5230.6	03 28 54.	- 52 40 47.	18	17.1	GALAXY
BAK 1.606	03 28 55.	- 22 46 28.	24	15.8	EXTRAGALACTIC NEBULA
SHAP330-5229.6	03 28 55.	- 52 39 47.	18	16.9	GALAXY
SHAP330-5333.2	03 28 55.	- 53 43 23.	42	17.6	GALAXY
SHAP330-5336.6	03 28 55.	- 53 46 47.	18	16.6	GALAXY
SHAP330-5431.6	03 28 55.	- 54 41 48.	24	17.25	GALAXY
BAK 1.603	03 28 56.	- 28 21 58.	10	17.2	EXTRAGALACTIC NEBULA
SHAP330-5227.8	03 28 56.	- 52 37 59.	24	17.1	GALAXY
SHAP330-5230.2	03 28 56.	- 52 40 23.	18	17.3	GALAXY
SHAP330-5301.8	03 28 56.	- 53 11 59.	18	17.7	GALAXY
SHAP330-5344.5	03 28 56.	- 53 54 41.	30	17.1	GALAXY
BAK 1.604	03 28 57.	- 28 16 58.	13	17.1	EXTRAGALACTIC NEBULA
SHAP330-4831.0	03 28 57.	- 48 41 11.	12	17.5	GALAXY
SHAP330-4851.1	03 28 57.	- 49 01 17.	18	17.5	GALAXY
SHAP330-5403.0	03 28 58.	- 54 13 11.	18	16.7	GALAXY
SHAP330-5253.5	03 28 59.	- 53 03 41.	30	17.5	GALAXY
SHAP330-5301.9	03 28 59.	- 53 12 05.	24	17.5	GALAXY
SER 031.05	03 29	- 52 42	4500	17.	CLOUD, ABOUT 500 GALAXIES
UGC 02776	03 29 00.	+ 10 21	60	17.	GALAXY Sc
MCG+08-07-004	03 29 00.	+ 45 17	24	18.	GALAXY
MCG+08-07-005	03 29 00.	+ 47 18	30	17.	GALAXY
OCL 0388	03 29 00.	+ 52 32	300	15.	OPEN STAR CLUSTER
KARA.68 037	03 29 00.	+ 67 55	107		DWARF GALAXY
KARA.68 038	03 29 00.	+ 68 12	107		DWARF GALAXY
PHL 4444	03 29 00.	- 25 35		18.2	BLUE STELLAR OBJECT
TON-S 0355	03 29 00.	- 29 32		15.6	BLUE STAR
PHL 8698	03 29 00.	- 31 29		18.5	BLUE STELLAR OBJECT
SHAP330-4718.8	03 29 00.	- 47 28 59.	12	17.0	GALAXY
SHAP330-5246.3	03 29 00.	- 52 56 29.	18	16.4	GALAXY
SHAP330-5248.9	03 29 00.	- 52 59 05.	24	16.9	GALAXY
SHAP330-5332.6	03 29 00.	- 53 42 47.	24	17.6	GALAXY
SHAP330-5350.7	03 29 00.	- 54 00 53.	36	17.1	GALAXY
SHAP330-5357.0	03 29 00.	- 54 07 11.	24	17.6	GALAXY
BAK 1.608	03 29 01.	- 24 06 10.	13	17.5	EXTRAGALACTIC NEBULA
SHAP330-4540.2	03 29 01.	- 45 50 23.	30	17.0	GALAXY
SHAP330-5228.0	03 29 01.	- 52 38 11.	18	17.0	GALAXY
SHAP330-5301.3	03 29 01.	- 53 11 29.	18	17.4	GALAXY
SHAP330-5336.2	03 29 01.	- 53 46 23.	24	17.2	GALAXY
BAK 1.607	03 29 02.	- 27 07 46.	16	16.7	EXTRAGALACTIC NEBULA
SHAP330-5243.3	03 29 02.	- 52 53 29.	24	17.6	GALAXY
SHAP330-5315.9	03 29 02.	- 53 26 05.	66	15.5	GALAXY
SHAP330-5317.6	03 29 02.	- 53 27 47.	24	17.5	GALAXY
BAK 1.605	03 29 03.	- 29 57 58.	14	17.0	EXTRAGALACTIC NEBULA
SHAP330-5401.2	03 29 03.	- 54 11 23.	18	17.7	GALAXY
SHAP330-5436.8	03 29 03.	- 54 46 59.	18	18.0	GALAXY
ARC 0443	03 29 04.	- 06 25		17.9	RICH CLUSTER OF GALAXIES
SHAP330-5019.3	03 29 04.	- 50 29 29.	36	15.25	GALAXY
SHAP330-5317.8	03 29 04.	- 53 27 59.	18	17.7	GALAXY
SHAP330-5406.0	03 29 04.	- 54 16 11.	24	17.6	GALAXY
KARA.73 12	03 29 05.	- 34 30	27		DWARF GALAXY
SHAP330-5352.9	03 29 05.	- 54 03 05.	24	16.8	GALAXY
5ZW 352	03 29 06.	+ 24 41			COMPACT GALAXY
ISS 0066	03 29 06.	+ 60 07	640		STELLAR RING
PHL 1553	03 29 06.	- 10 32		17.2	BLUE STELLAR OBJECT
MCG-03-10-001	03 29 06.	- 15 52	30	16.	GALAXY
MCG-06-08-023	03 29 06.	- 33 48	270	10.	GALAXY
SER 031.03	03 29 06.	- 50 27	600	15.	COMPACT AND SPINDLE GLXYS
IC 1947	03 29 06.	- 50 29			NONSTELLAR OBJECT
HN 0206	03 29 06.	- 50 30			NEBULA
KLEM 08	03 29 06.	- 52 45	3000	15.	RICH CLUSTER OF GALAXIES
SHAP330-5237.5	03 29 06.	- 52 47 41.	18	17.1	GALAXY
SHAP330-5400.2	03 29 06.	- 54 10 23.	24	16.9	GALAXY
BAK 1.614	03 29 07.	- 23 38 57.	16	16.4	EXTRAGALACTIC NEBULA
BAK 1.611	03 29 07.	- 27 00 21.	17	16.6	EXTRAGALACTIC NEBULA
SN 1959A	03 29 07.	- 33 46		16.0	SUPERNOVA
RNGC 1350	03 29 07.	- 33 47		12.0	GALAXY
SHAP330-5032.1	03 29 07.	- 50 42 17.	24	16.0	GALAXY
SHAP330-5344.8	03 29 07.	- 53 54 59.	18	17.0	GALAXY
BAK 1.609	03 29 08.	- 28 58 33.	19	16.8	EXTRAGALACTIC NEBULA
SHAP330-5246.0	03 29 08.	- 52 56 11.	18	17.0	GALAXY
SHAP330-5334.8	03 29 08.	- 53 44 59.	24	17.7	GALAXY
SHAP330-5336.0	03 29 08.	- 53 46 11.	24	17.8	GALAXY
BAK 1.610	03 29 09.	- 28 58 51.	12	17.6	EXTRAGALACTIC NEBULA
SHAP330-4745.0	03 29 09.	- 47 55 10.	18	17.75	GALAXY
SHAP330-5234.5	03 29 09.	- 52 44 41.	24	16.2	GALAXY
SHAP330-5301.8	03 29 09.	- 53 11 59.	18	17.0	GALAXY
SHAP330-5301.8	03 29 09.	- 53 11 59.	18	17.0	GALAXY
SHAP330-5018.6	03 29 10.	- 50 28 46.	30	16.25	GALAXY
BAK 1.615	03 29 11.	- 26 43 33.	58	14.8	EXTRAGALACTIC NEBULA
BAK 1.613	03 29 11.	- 28 27 15.	12	17.0	EXTRAGALACTIC NEBULA
BAK 1.612	03 29 11.	- 28 57 03.	23	17.5	EXTRAGALACTIC NEBULA
SHAP330-4932.9	03 29 11.	- 49 43 04.	24	17.5	GALAXY
SHAP330-5017.0	03 29 11.	- 50 27 10.	132	15.0	GALAXY
ZC 0329.2-0056	03 29 12.	- 00 56	810		CLUSTER OF GALAXIES
PHL 8699	03 29 12.	- 10 17		15.3	BLUE STELLAR OBJECT
PHL 4445	03 29 12.	- 20 39		18.5	BLUE STELLAR OBJECT
PHL 8700	03 29 12.	- 23 23		17.9	BLUE STELLAR OBJECT
MCG-04-09-020	03 29 12.	- 26 15	24	16.	GALAXY
PHL 4446	03 29 12.	- 29 32		16.5	BLUE STELLAR OBJECT
MCG-06-08-024	03 29 12.	- 36 29	24	15.	GALAXY
SHAP330-4615.8	03 29 12.	- 46 25 58.	24	17.5	GALAXY
BAK 1.616	03 29 13.	- 28 37 33.	19	17.2	EXTRAGALACTIC NEBULA
SHAP330-5221.4	03 29 13.	- 52 31 34.	18	17.2	GALAXY
SHAP330-5226.8	03 29 13.	- 52 36 58.	18	16.9	GALAXY
SHAP330-5250.4	03 29 13.	- 53 00 34.	18	16.3	GALAXY
SHAP330-5302.2	03 29 13.	- 53 12 22.	24	17.5	GALAXY
SHAP330-5225.5	03 29 14.	- 52 35 40.	18	17.5	GALAXY
HN 0202	03 29 15.	+ 24 15			NEBULA
IC 1941	03 29 15.	+ 24 16			NONSTELLAR OBJECT
SCHO 0057	03 29 15.	+ 36 49 24.	720		ISOLATED DARK CLOUD
BAK 2.0007	03 29 15.	- 17 42 45.	20	15.6	GALAXY
HELW 120	03 29 15.	- 26 16 32.			NEBULA
SHAP330-4758.3	03 29 15.	- 48 08 28.	60	16.5	GALAXY
RNGC 1356	03 29 15.	- 50 27			GALAXY
SHAP330-5235.0	03 29 15.	- 52 45 10.	24	17.1	GALAXY
SHAP330-5319.1	03 29 15.	- 53 29 16.	18	17.0	GALAXY
RNGC 1352	03 29 16.	- 19 26		15.0	GALAXY
SHAP330-4523.0	03 29 16.	- 45 33 10.	30	16.5	GALAXY
BAK 1.617	03 29 17.	- 28 32 27.	17	16.8	EXTRAGALACTIC NEBULA
IC 1948	03 29 17.	- 48 08			NONSTELLAR OBJECT
HN 0207	03 29 17.	- 48 09			NEBULA
ZWG 391.001	03 29 18.	+ 02 07		15.3	GALAXY
PHL 8701	03 29 18.	- 09 34		18.1	BLUE STELLAR OBJECT
MCG-03-10-002	03 29 18.	- 19 26	24	15.	GALAXY
PHL 4448	03 29 18.	- 24 32		18.8	BLUE STELLAR OBJECT
PHL 4447	03 29 18.	- 25 44		18.1	BLUE STELLAR OBJECT
MCG-05-09-010	03 29 18.	- 26 42	60	14.5	GALAXY
SHAP330-5018.6	03 29 18.	- 50 28 46.	24	17.0	GALAXY
SHAP330-5046.7	03 29 18.	- 50 56 52.	24	17.75	GALAXY
SHAP330-5240.8	03 29 18.	- 52 50 58.	24	17.8	GALAXY
SHAP330-5444.6	03 29 18.	- 54 54 46.	30	16.5	GALAXY
SHAP331-4440.1	03 29 19.	- 44 50 16.	36	16.75	GALAXY
SHAP330-5226.4	03 29 19.	- 52 36 34.	18	16.8	GALAXY
SHAP330-5811.0	03 29 19.	- 58 21 10.	36	17.0	GALAXY
HELW 121	03 29 20.	- 25 47 56.			NEBULA
BAK 1.618	03 29 20.	- 29 59 51.	17	17.4	EXTRAGALACTIC NEBULA
SHAP330-5222.6	03 29 20.	- 52 32 46.	24	16.8	GALAXY
SHAP330-5231.0	03 29 20.	- 52 41 10.	18	17.3	GALAXY
SHAP330-5231.1	03 29 20.	- 52 41 16.	18	17.7	GALAXY
SHAP330-5444.4	03 29 20.	- 54 54 34.	18	17.75	GALAXY
SHAP330-4759.3	03 29 21.	- 48 09 28.	72	14.4	GALAXY
BAK 1.622	03 29 22.	- 22 46 44.	13	16.6	EXTRAGALACTIC NEBULA
SHAP330-5318.2	03 29 22.	- 53 28 22.	24	17.7	GALAXY
SHAP330-5434.4	03 29 22.	- 54 44 34.	24	16.25	GALAXY
IC 1949	03 29 23.	- 48 09			NONSTELLAR OBJECT
HN 0208	03 29 23.	- 48 10			NEBULA
SHAP330-5223.0	03 29 23.	- 52 33 10.	18	17.4	GALAXY
SHAP330-5433.8	03 29 23.	- 54 43 58.	36	16.75	GALAXY
SHAP330-5437.3	03 29 23.	- 54 47 28.	18	17.5	GALAXY
UGC 02777	03 29 24.	+ 40 10	84	16.5	GALAXY
PHL 1554	03 29 24.	- 15 54		18.2	BLUE STELLAR OBJECT
PHL 4450	03 29 24.	- 24 04		18.2	BLUE STELLAR OBJECT
TON-S 0356	03 29 24.	- 26 52		16.0	BLUE STAR
PHL 4449	03 29 24.	- 31 00		17.0	BLUE STELLAR OBJECT
MCG-06-08-025	03 29 24.	- 35 30	48	15.	GALAXY
SHAP330-5315.2	03 29 24.	- 53 25 22.	18	17.4	GALAXY
BAK 1.619	03 29 25.	- 27 38 32.	26	16.0	EXTRAGALACTIC NEBULA
SHAP331-4731.2	03 29 25.	- 47 41 21.	36	17.0	GALAXY
SHAP330-5229.2	03 29 26.	- 52 39 22.	24	16.4	GALAXY
SHAP330-5235.6	03 29 26.	- 52 45 46.	24	16.4	GALAXY
SHAP330-5430.0	03 29 26.	- 54 40 10.	24	17.5	GALAXY
IC 0329	03 29 27.	+ 00 07 12.			NONSTELLAR OBJECT
SVEN 215	03 29 27.	- 19 27 14.	24	14.0	GALAXY
BAK 1.624	03 29 27.	- 22 52 32.	13	16.6	EXTRAGALACTIC NEBULA
BAK 1.620	03 29 27.	- 29 05 14.	14	17.3	EXTRAGALACTIC NEBULA
SHAP331-4553.4	03 29 27.	- 46 03 33.	24	17.75	GALAXY
SHAP330-5233.4	03 29 27.	- 52 43 34.	18	17.4	GALAXY
SHAP330-5241.4	03 29 27.	- 52 51 34.	18	16.9	GALAXY
BAK 1.621	03 29 28.	- 30 22 56.	62	14.6	EXTRAGALACTIC NEBULA
SHAP331-4814.8	03 29 28.	- 48 24 57.	36	17.0	GALAXY
LB 02884	03 29 29.	+ 00 38 54.		16.3	FAINT BLUE STAR
BAK 1.625	03 29 29.	- 25 10 38.	41	14.9	EXTRAGALACTIC NEBULA
SHAP330-5453.6	03 29 29.	- 55 03 46.	54	16.5	GALAXY
MCG+00-10-001	03 29 30.	+ 00 05	30	14.5	GALAXY
ZWG 391.002	03 29 30.	+ 00 06		15.1	GALAXY
ZWG 391.003	03 29 30.	+ 01 51		15.2	GALAXY
UGC 02778	03 29 30.	+ 36 05	66	16.5	GALAXY SBb
MCG-04-09-021	03 29 30.	- 25 09	60	14.	GALAXY
BAK 1.623	03 29 30.	- 27 23 08.	24	17.5	EXTRAGALACTIC NEBULA
SHAP330-5241.4	03 29 30.	- 52 51 33.	18	17.1	GALAXY
SHAP330-5410.0	03 29 30.	- 54 20 09.	24	17.25	GALAXY
LB 03319	03 29 30.	- 64 00		13.8	FAINT BLUE STAR
SHAP330-5223.5	03 29 31.	- 52 33 39.	30	16.8	GALAXY
SHAP330-5224.5	03 29 31.	- 52 34 39.	24	16.5	GALAXY
SHAP330-5339.2	03 29 31.	- 53 49 21.	24	17.7	GALAXY
SHAP330-5344.7	03 29 31.	- 53 54 51.	24	16.9	GALAXY
SHAP330-5215.5	03 29 33.	- 52 25 39.	18	17.5	GALAXY
SHAP331-4815.0	03 29 34.	- 48 25 09.	12	17.5	GALAXY
SHAP331-5137.8	03 29 34.	- 51 47 57.	18	16.75	GALAXY
SHAP330-5234.4	03 29 34.	- 52 44 33.	24	16.4	GALAXY
SHAP330-5248.2	03 29 34.	- 52 58 21.	24	16.5	GALAXY
SHAP330-5351.9	03 29 34.	- 54 02 03.	24	16.5	GALAXY
IC 0330	03 29 35.	+ 00 11 47.			NONSTELLAR OBJECT
SHAP331-5025.5	03 29 35.	- 50 35 39.	108	15.5	GALAXY
HN 0209	03 29 35.	- 50 36			NEBULA
IC 1950	03 29 35.	- 50 36			NONSTELLAR OBJECT
SHAP330-5232.8	03 29 35.	- 52 42 57.	24	17.0	GALAXY
MCG+00-10-002	03 29 36.	+ 00 09	54	14.	GALAXY
ZWG 391.004	03 29 36.	+ 00 11		15.0	GALAXY
UGC 02779	03 29 36.	+ 00 11	60	15.0	GALAXY Sa-b
ZWG 442.001	03 29 36.	+ 15 14		15.7	GALAXY
PHL 1555	03 29 36.	- 16 26		16.7	BLUE STELLAR OBJECT
PHL 8703	03 29 36.	- 22 07		18.5	BLUE STELLAR OBJECT
SHAP330-5250.1	03 29 36.	- 53 00 15.	18	17.0	GALAXY
BAK 1.626	03 29 37.	- 29 58 56.	19	16.6	EXTRAGALACTIC NEBULA
SHAP331-5212.5	03 29 37.	- 52 22 39.	24	16.5	GALAXY
BAK 1.628	03 29 38.	- 28 45 56.	17	17.0	EXTRAGALACTIC NEBULA
BAK 1.627	03 29 38.	- 29 11 20.	12	17.6	EXTRAGALACTIC NEBULA
SHAP331-5215.6	03 29 38.	- 52 25 45.	16	17.4	GALAXY
SHAP331-5230.8	03 29 38.	- 52 40 57.	36	16.9	GALAXY
SHAP331-5248.4	03 29 38.	- 52 58 33.	18	16.9	GALAXY
SHAP330-5336.5	03 29 38.	- 53 46 39.	18	18.0	GALAXY
SHAP331-4930.7	03 29 39.	- 49 40 51.	12	17.5	GALAXY
SHAP331-5203.9	03 29 39.	- 52 14 03.	24	16.4	GALAXY
SHAP331-5228.8	03 29 39.	- 52 38 57.	18	17.4	GALAXY
SHAP331-5308.5	03 29 39.	- 53 18 39.	42	16.8	GALAXY
SHAP331-5339.6	03 29 39.	- 53 49 45.	30	17.5	GALAXY
BAK 1.629	03 29 40.	- 23 59 31.	12	17.2	EXTRAGALACTIC NEBULA
SHAP331-5104.3	03 29 40.	- 51 14 27.	18	17.0	GALAXY
SHAP331-5221.2	03 29 40.	- 52 31 21.	12	17.6	GALAXY
IC 1951	03 29 40.	- 53 19			NONSTELLAR OBJECT
SHAP331-4757.4	03 29 41.	- 48 07 32.	12	17.5	GALAXY
SHAP331-5226.0	03 29 41.	- 52 36 09.	24	17.7	GALAXY
SHAP331-5246.0	03 29 41.	- 52 56 09.	24	16.7	GALAXY
SHAP331-5305.0	03 29 41.	- 53 15 09.	18	17.6	GALAXY
HN 0210	03 29 41.	- 53 19			NEBULA
ZC 0329.7+1822	03 29 42.	+ 18 22	1410		CLUSTER OF GALAXIES
ZC 0329.7+2452	03 29 42.	+ 24 52	1010		CLUSTER OF GALAXIES
MCG-05-09-011	03 29 42.	- 30 23	54	14.	GALAXY
HPW 09	03 29 42.	- 37 52			DWARF GALAXY IN FORNAX
SHAP331-5059.6	03 29 42.	- 51 09 45.	18	17.75	GALAXY
SHAP331-5234.4	03 29 42.	- 52 44 33.	24	17.2	GALAXY
SHAP331-5210.7	03 29 43.	- 52 20 51.	24	17.4	GALAXY
SHAP331-5230.0	03 29 44.	- 52 40 09.	18	17.7	GALAXY
MCG+00-10-003	03 29 45.	+ 00 05	36	14.5	GALAXY
IC 0331	03 29 45.	+ 00 07 22.			NONSTELLAR OBJECT
PHL 4451	03 29 45.	- 21 24		8.5	BLUE STELLAR OBJECT
SHAP331-4522.1	03 29 45.	- 45 32 14.	30	18.0	GALAXY
SHAP331-5224.1	03 29 45.	- 52 34 14.	18	16.9	GALAXY
SHAP331-5229.6	03 29 45.	- 52 39 44.	18	17.1	GALAXY

OBJECT NAME	RIGHT ASCEN.	DECLINATION	DIAM.	MAGN.	TYPE OF OBJECT
SHAP331-5320.2	03 29 45.	- 53 30 21.	24	17.6	GALAXY
SHAP331-5341.5	03 29 45.	- 53 51 39.	24	16.9	GALAXY
SHAP331-5210.0	03 29 46.	- 52 20 08.	24	17.4	GALAXY
SHAP331-5235.9	03 29 46.	- 52 46 02.	18	17.3	GALAXY
SHAP331-5245.8	03 29 46.	- 52 55 56.	24	17.6	GALAXY
RNGC 1353	03 29 47.	- 21 00		12.5	GALAXY
SHAP331-4750.7	03 29 47.	- 48 00 50.	18	17.5	GALAXY
SHAP331-5231.3	03 29 47.	- 52 41 26.	18	17.3	GALAXY
SHAP331-5238.6	03 29 47.	- 52 48 44.	30	15.9	GALAXY
SHAP331-5241.5	03 29 47.	- 52 51 38.	30	16.5	GALAXY
SHAP331-5348.2	03 29 47.	- 53 58 20.	18	17.2	GALAXY
ZWG 391.005	03 29 48.	+ 00 07		14.9	GALAXY
MCG+06-08-029	03 29 48.	+ 36 02	30	17.	GALAXY
HPW 10	03 29 48.	- 35 14			DWARF GALAXY IN FORNAX
SHAP331-5225.3	03 29 48.	- 52 35 26.	24	17.4	GALAXY
SHAP331-5436.1	03 29 48.	- 54 46 14.	24	17.75	GALAXY
BAK 1.632	03 29 49.	- 23 28 31.	12	16.2	EXTRAGALACTIC NEBULA
BAK 1.630	03 29 49.	- 27 31 37.	23	17.0	EXTRAGALACTIC NEBULA
SHAP331-4538.2	03 29 49.	- 45 48 20.	18	17.5	GALAXY
SHAP331-5221.6	03 29 49.	- 52 31 44.	18	16.4	GALAXY
SHAP331-5229.0	03 29 49.	- 52 39 08.	24	17.2	GALAXY
BAK 1.633	03 29 50.	- 23 29 55.	16	16.6	EXTRAGALACTIC NEBULA
SHAP331-5318.4	03 29 50.	- 53 28 32.	18	17.6	GALAXY
SHAP331-5403.6	03 29 50.	- 54 13 44.	18	17.0	GALAXY
SHAP331-5255.4	03 29 51.	- 53 05 32.	24	17.1	GALAXY
B 001	03 29 52.	+ 30 59	1800		DARK OBJECT
SHAP331-5235.2	03 29 52.	- 52 45 20.	30	16.3	GALAXY
SHAP331-5347.2	03 29 52.	- 53 57 20.	24	16.2	GALAXY
SHAP331-5228.8	03 29 53.	- 52 38 56.	24	17.4	GALAXY
SHAP331-5440.8	03 29 53.	- 54 50 56.	24	17.5	GALAXY
ZC 0329.9+0242	03 29 54.	+ 02 42	4910		CLUSTER OF GALAXIES
UGC 02780	03 29 54.	+ 36 00	108	16.5	GALAXY
MCG+06-08-030	03 29 54.	+ 36 04	24	17.	GALAXY
MCG-03-10-003	03 29 54.	- 17 53	72	16.	GALAXY
MCG-04-09-022	03 29 54.	- 20 58	210	12.	GALAXY
TON-S 0357	03 29 54.	- 28 46		15.6	BLUE STAR
SHAP331-4551.4	03 29 54.	- 46 01 32.	18	17.0	GALAXY
SHAP331-4639.0	03 29 54.	- 46 49 08.	12	17.5	GALAXY
SHAP331-5231.4	03 29 54.	- 52 41 32.	24	17.7	GALAXY
SHAP331-5231.8	03 29 54.	- 52 41 56.	18	17.2	GALAXY
BAK 2.0008	03 29 55.	- 18 02 42.	15	15.6	GALAXY
BAK 1.631	03 29 55.	- 31 30 07.	68	14.7	EXTRAGALACTIC NEBULA
SHAP331-5406.9	03 29 55.	- 54 17 02.	24	17.75	GALAXY
BAK 1.634	03 29 56.	- 28 31 07.	12	16.8	EXTRAGALACTIC NEBULA
BAK 1.636	03 29 57.	- 27 10 25.	36	16.7	EXTRAGALACTIC NEBULA
BAK 1.635	03 29 57.	- 28 31 37.	17	16.6	EXTRAGALACTIC NEBULA
SHAP331-5245.6	03 29 58.	- 52 55 44.	18	17.4	GALAXY
BAK 1.638	03 29 59.	- 24 03 42.	24	16.0	EXTRAGALACTIC NEBULA
HELW 122	03 29 59.	- 26 08 58.			NEBULA
SHAP331-5217.2	03 29 59.	- 52 27 20.	18	17.1	GALAXY
SHAP331-5300.8	03 29 59.	- 53 10 56.	18	17.3	GALAXY
SHAP331-5346.0	03 29 59.	- 53 56 08.	24	17.7	GALAXY
LBN 0651	03 30	+ 73 00	12600		BRIGHT NEBULA
MCG+00-10-004	03 30 00.	+ 01 12	60	15.	GALAXY
MCG-03-10-004	03 30 00.	- 15 23	96	14.5	GALAXY
PHL 4452	03 30 00.	- 26 06		18.3	BLUE STELLAR OBJECT
HPW 11	03 30 00.	- 37 14			DWARF GALAXY IN FORNAX
SHAP331-4521.5	03 30 00.	- 45 31 37.	30	16.5	GALAXY
SHAP331-4546.7	03 30 00.	- 45 56 49.	30	16.25	GALAXY
LB 01672	03 30 00.	- 47 58		16.4	FAINT BLUE STAR
SHAP331-4543.0	03 30 01.	- 45 53 07.	24	17.0	GALAXY
RNGC 1354	03 30 02.	- 15 23		14.0	GALAXY
SHAP331-4650.2	03 30 02.	- 47 00 19.	30	16.0	GALAXY
SHAP331-5441.1	03 30 02.	- 54 51 14.	30	18.0	GALAXY
SHAP331-4943.3	03 30 03.	- 49 53 25.	30	16.5	GALAXY
SHAP331-5219.2	03 30 03.	- 52 29 19.	12	17.7	GALAXY
SHAP331-5243.2	03 30 03.	- 52 53 19.	18	17.1	GALAXY
IC 0332	03 30 04.	+ 01 13 44.			NONSTELLAR OBJECT
BAK 1.637	03 30 04.	- 28 55 54.	20	17.0	EXTRAGALACTIC NEBULA
SHAP331-5231.0	03 30 04.	- 52 41 07.	18	17.6	GALAXY
SHAP331-5235.2	03 30 04.	- 52 45 19.	24	17.4	GALAXY
SHAP331-5345.8	03 30 04.	- 53 55 55.	18	17.6	GALAXY
BAK 1.642	03 30 05.	- 26 17 06.	26	16.6	EXTRAGALACTIC NEBULA
BAK 1.639	03 30 05.	- 28 43 12.	13	17.4	EXTRAGALACTIC NEBULA
SHAP331-4941.2	03 30 05.	- 49 51 19.	30	16.75	GALAXY
SHAP331-5219.8	03 30 05.	- 52 29 55.	30	16.5	GALAXY
ZWG 391.006	03 30 06.	+ 01 13		15.0	GALAXY
KARA.73B 0129	03 30 06.	+ 01 13	60	15.0	ISOLATED GALAXY S
SCHO 0058	03 30 06.	+ 39 39 06.	440		ISOLATED DARK CLOUD
MCG+07-08-025	03 30 06.	+ 39 58	42	15.5	GALAXY
MCG+01-10-001	03 30 06.	- 04 56	36	15.	GALAXY
BAK 1.643	03 30 06.	- 24 29 06.	14	17.0	EXTRAGALACTIC NEBULA
BAK 1.640	03 30 06.	- 28 56 24.	16	17.3	EXTRAGALACTIC NEBULA
SHAP331-5244.2	03 30 06.	- 52 54 19.	24	17.4	GALAXY
SHAP331-5245.8	03 30 06.	- 52 55 55.	18	17.4	GALAXY
SHAP331-5313.9	03 30 06.	- 53 24 01.	24	17.1	GALAXY
LB 03320	03 30 06.	- 72 51		15.0	FAINT BLUE STAR
SHAP331-4937.2	03 30 07.	- 49 47 19.	24	17.0	GALAXY
SHAP331-5246.6	03 30 07.	- 52 56 43.	18	17.7	GALAXY
SHAP331-4624.5	03 30 08.	- 46 34 37.	24	18.0	GALAXY
SHAP331-5430.5	03 30 08.	- 54 40 37.	30	17.0	GALAXY
SHAP331-5140.4	03 30 09.	- 51 50 31.	30	16.75	GALAXY
SHAP331-5300.8	03 30 09.	- 53 10 55.	18	17.1	GALAXY
SHAP331-5713.5	03 30 09.	- 57 23 37.	60	16.5	GALAXY
RNGC 1348	03 30 10.	+ 51 16			OPEN CLUSTER
BAK 1.641	03 30 10.	- 31 38 54.	18	16.8	EXTRAGALACTIC NEBULA
SHAP331-5239.5	03 30 10.	- 52 49 37.	42	16.6	GALAXY
SHAP331-4927.2	03 30 11.	- 49 37 19.	36	17.5	GALAXY
HN 0211	03 30 11.	- 57 24			NEBULA
IC 1955	03 30 11.	- 57 24			NONSTELLAR OBJECT
MCG+06-08-031	03 30 12.	+ 36 00	102	15.	GALAXY
OCL 0391	03 30 12.	+ 51 16			OPEN STAR CLUSTER
ZWG 391.007	03 30 12.	- 01 15		15.0	GALAXY
BAK 1.649	03 30 12.	- 23 29 24.	18	16.8	EXTRAGALACTIC NEBULA
TON-S 0358	03 30 12.	- 27 15		15.8	BLUE STAR
MCG-05-09-012	03 30 12.	- 31 31	66	14.	GALAXY
HPW 12	03 30 12.	- 37 34			DWARF GALAXY IN FORNAX
IC 1954	03 30 12.	- 52 05	156	12.2	GALAXY SB(s)
SHAP331-5232.9	03 30 12.	- 52 43 01.	18	16.6	GALAXY
LB 03321	03 30 12.	- 66 48		13.6	FAINT BLUE STAR
BAK 1.644	03 30 13.	- 30 02 48.	17	16.5	EXTRAGALACTIC NEBULA
SHAP331-5308.6	03 30 13.	- 53 18 43.	24	17.0	GALAXY
SHAP331-5350.6	03 30 13.	- 54 00 43.	36	16.0	GALAXY
SHAP331-5418.8	03 30 13.	- 54 28 55.	18	18.0	GALAXY
BAK 2.0009	03 30 14.	- 19 07 29.	23	14.6	GALAXY
BAK 1.645	03 30 14.	- 29 17 18.	13	16.8	EXTRAGALACTIC NEBULA
SHAP331-5205.9	03 30 14.	- 52 16 01.	24	17.5	GALAXY
SHAP331-5247.1	03 30 14.	- 52 57 13.	24	17.5	GALAXY
MCG-03-10-005	03 30 15.	- 19 06	24	15.	GALAXY
BAK 1.652	03 30 15.	- 22 17 59.	12	16.6	EXTRAGALACTIC NEBULA
BAK 1.646	03 30 15.	- 29 47 24.	22	17.0	EXTRAGALACTIC NEBULA
LB 02885	03 30 16.	+ 00 19 00.		16.8	FAINT BLUE STAR
BAK 1.647	03 30 16.	- 30 54 48.	19	17.1	EXTRAGALACTIC NEBULA
SHAP331-5057.0	03 30 16.	- 51 07 07.	18	17.5	GALAXY
SHAP331-5154.5	03 30 16.	- 52 04 37.	222	12.2	GALAXY
SHAP331-5221.6	03 30 16.	- 52 31 43.	18	17.4	GALAXY
SHAP331-5228.4	03 30 16.	- 52 38 31.	24	16.8	GALAXY
SHAP331-4657.1	03 30 17.	- 47 07 12.	18	17.0	GALAXY
SHAP331-5259.2	03 30 17.	- 53 09 19.	18	17.2	GALAXY
UGC 02781	03 30 18.	+ 15 42	90	17.	GALAXY Sc
ZC 0330.3+3859	03 30 18.	+ 38 59	1140		CLUSTER OF GALAXIES
UGC 02782	03 30 18.	+ 40 12	84	17.	GALAXY
BAK 2.0010	03 30 18.	- 22 34 29.	24	16.0	GALAXY
BAK 1.655	03 30 18.	- 23 33 11.	20	16.4	EXTRAGALACTIC NEBULA
BAK 1.648	03 30 18.	- 29 49 23.	18	16.2	EXTRAGALACTIC NEBULA
SHAP331-4524.5	03 30 18.	- 45 34 36.	18	17.0	GALAXY
SHAP331-5243.2	03 30 18.	- 52 53 19.	18	17.5	GALAXY
LB 03322	03 30 18.	- 71 53		15.0	FAINT BLUE STAR
BAK 1.657	03 30 19.	- 22 33 35.	54	16.0	EXTRAGALACTIC NEBULA
BAK 1.654	03 30 19.	- 24 34 05.	14	16.4	EXTRAGALACTIC NEBULA
SHAP331-5259.2	03 30 19.	- 53 09 19.	24	17.0	GALAXY
BAK 2.0011	03 30 21.	- 17 11 41.	17	15.9	GALAXY
SHAP331-5125.7	03 30 21.	- 51 35 48.	24	17.75	GALAXY
BAK 1.651	03 30 22.	- 29 27 05.	26	16.6	EXTRAGALACTIC NEBULA
BAK 1.653	03 30 23.	- 28 59 17.	23	16.8	EXTRAGALACTIC NEBULA
SHAP331-5307.0	03 30 23.	- 53 17 06.	18	17.5	GALAXY
B 002	03 30 24.	+ 32 09	1200		DARK OBJECT
BAK 1.650	03 30 24.	- 31 39 53.	20	16.3	EXTRAGALACTIC NEBULA
HPW 13	03 30 24.	- 36 23			DWARF GALAXY IN FORNAX
SHAP331-5117.4	03 30 24.	- 51 27 30.	24	17.0	GALAXY
SHAP331-5200.4	03 30 25.	- 52 10 30.	30	17.4	GALAXY
BAK 1.656	03 30 27.	- 29 47 59.	22	16.4	EXTRAGALACTIC NEBULA
SHAP331-5206.7	03 30 28.	- 52 16 48.	36	16.2	GALAXY
SHAP331-5217.4	03 30 28.	- 52 27 30.	18	17.0	GALAXY
SHAP332-4555.4	03 30 29.	- 46 05 29.	24	17.0	GALAXY
SHAP331-5104.6	03 30 29.	- 51 14 42.	24	16.5	GALAXY
SHAP331-5224.5	03 30 29.	- 52 34 36.	24	16.5	GALAXY
PHL 8704	03 30 30.	- 23 48		9.7	BLUE STELLAR OBJECT
SHAP332-4620.7	03 30 30.	- 46 30 47.	18	16.5	GALAXY
SHAP332-4537.4	03 30 31.	- 45 47 29.	24	17.0	GALAXY
SHAP332-4744.6	03 30 31.	- 47 54 41.	18	18.0	GALAXY
SHAP332-5024.2	03 30 31.	- 50 34 18.	24	17.0	GALAXY
SHAP331-5242.2	03 30 31.	- 52 52 18.	30	16.4	GALAXY
BAK 1.659	03 30 33.	- 28 05 29.	17	16.9	EXTRAGALACTIC NEBULA
SHAP331-5432.5	03 30 33.	- 54 42 36.	30	17.5	GALAXY
BAK 1.666	03 30 34.	- 21 06 04.	19	16.3	EXTRAGALACTIC NEBULA
BAK 1.661	03 30 34.	- 25 33 10.	18	16.6	EXTRAGALACTIC NEBULA
BAK 1.658	03 30 34.	- 29 36 53.	19	16.6	EXTRAGALACTIC NEBULA
SHAP332-5054.3	03 30 34.	- 51 04 24.	18	18.0	GALAXY
BAK 1.662	03 30 35.	- 26 15 22.	16	16.8	EXTRAGALACTIC NEBULA
SHAP332-4729.0	03 30 35.	- 47 39 05.	12	17.5	GALAXY
SHAP332-5217.8	03 30 35.	- 52 27 54.	24	17.0	GALAXY
ZWG 391.008	03 30 36.	- 02 08		15.1	GALAXY
PHL 8705	03 30 36.	- 24 41		16.0	BLUE STELLAR OBJECT
BAK 1.660	03 30 36.	- 29 37 28.	16	17.0	EXTRAGALACTIC NEBULA
HPW 14	03 30 36.	- 35 44			DWARF GALAXY IN FORNAX
SHAP332-4657.9	03 30 36.	- 47 07 59.	30	17.75	GALAXY
SHAP332-4729.6	03 30 36.	- 47 39 41.	18	17.25	GALAXY
SHAP331-5255.8	03 30 36.	- 53 05 54.	18	16.6	GALAXY
BAK 1.667	03 30 37.	- 24 24 28.	14	16.6	EXTRAGALACTIC NEBULA
BAK 1.664	03 30 37.	- 25 27 52.	22	16.2	EXTRAGALACTIC NEBULA
SHAP332-4544.6	03 30 38.	- 45 54 41.	24	16.75	GALAXY
SHAP332-4905.6	03 30 38.	- 49 15 41.	18	17.0	GALAXY
SHAP332-5257.0	03 30 38.	- 53 07 05.	24	17.6	GALAXY
MCG+06-08-032	03 30 39.	+ 36 08 30.	42	17.	GALAXY
SHAP332-4842.9	03 30 39.	- 48 52 59.	18	17.5	GALAXY
SHAP332-5233.4	03 30 39.	- 52 43 29.	18	17.4	GALAXY
LB 02886	03 30 41.	+ 00 14 36.		16.5	FAINT BLUE STAR
BAK 1.665	03 30 41.	- 29 16 40.	19	17.2	EXTRAGALACTIC NEBULA
SHAP332-4534.5	03 30 41.	- 45 44 35.	24	15.5	GALAXY
SHAP332-4546.1	03 30 41.	- 45 56 11.	24	17.0	GALAXY
SHAP332-5417.8	03 30 41.	- 54 27 53.	24	17.75	GALAXY
PHL 8706	03 30 42.	- 21 36		18.5	BLUE STELLAR OBJECT
PHL 4453	03 30 42.	- 26 13		18.5	BLUE STELLAR OBJECT
BAK 1.663	03 30 42.	- 31 45 16.	18	16.5	EXTRAGALACTIC NEBULA
HPW 15	03 30 42.	- 38 15			DWARF GALAXY IN FORNAX
LB 01673	03 30 42.	- 46 46		15.8	FAINT BLUE STAR
SHAP332-4737.5	03 30 43.	- 47 47 35.	18	17.75	GALAXY
SHAP332-4842.0	03 30 43.	- 48 52 05.	12	17.5	GALAXY
SHAP332-5214.4	03 30 43.	- 52 24 29.	24	16.7	GALAXY
SHAP332-5243.8	03 30 43.	- 52 53 53.	18	16.9	GALAXY
SHAP332-5246.4	03 30 43.	- 52 56 29.	12	17.4	GALAXY
BAK 1.672	03 30 44.	- 21 15 28.	20	15.7	EXTRAGALACTIC NEBULA
SHAP332-5229.4	03 30 44.	- 52 39 29.	24	17.1	GALAXY
BAK 1.668	03 30 45.	- 28 07 40.	18	17.3	EXTRAGALACTIC NEBULA
SHAP332-4600.0	03 30 45.	- 46 10 05.	24	17.5	GALAXY
SHAP332-5232.0	03 30 45.	- 52 42 05.	24	16.8	GALAXY
SHAP332-5256.2	03 30 45.	- 53 06 17.	24	16.9	GALAXY
SHAP332-5242.8	03 30 46.	- 52 52 53.	18	17.6	GALAXY
SHAP332-5312.3	03 30 46.	- 53 22 23.	24	17.4	GALAXY
LB 02887	03 30 47.	+ 00 34 30.		16.6	FAINT BLUE STAR
BAK 1.669	03 30 47.	- 29 39 52.	17	17.2	EXTRAGALACTIC NEBULA
SHAP332-4547.0	03 30 47.	- 45 57 04.	24	16.5	GALAXY
SHAP332-5226.5	03 30 47.	- 52 36 35.	66	16.6	GALAXY
SHAP332-5231.2	03 30 47.	- 52 41 17.	24	16.8	GALAXY
ZC 0330.8-0151	03 30 48.	- 01 51	810		CLUSTER OF GALAXIES
MCG-04-09-023	03 30 48.	- 21 14	72	14.	GALAXY
BAK 1.675	03 30 48.	- 22 23 39.	12	16.3	EXTRAGALACTIC NEBULA
SHAP332-5221.6	03 30 48.	- 52 31 41.	24	17.0	GALAXY
HN 0214	03 30 48.	- 52 37			NEBULA
IC 1957	03 30 48.	- 52 37			NONSTELLAR OBJECT
BAK 1.670	03 30 49.	- 28 50 28.	14	16.8	EXTRAGALACTIC NEBULA
BAK 1.671	03 30 49.	- 29 06 52.	17	17.0	EXTRAGALACTIC NEBULA
SHAP332-5304.7	03 30 49.	- 53 14 47.	24	17.7	GALAXY
SHAP332-5336.4	03 30 49.	- 53 46 29.	24	16.5	GALAXY
SHAP332-4803.2	03 30 50.	- 48 13 16.	12	17.0	GALAXY
MCG-04-09-024	03 30 51.	- 24 18	156	15.	GALAXY
SHAP332-4558.9	03 30 51.	- 46 08 58.	30	17.5	GALAXY
RNGC 1355	03 30 52.	- 05 10		15.0	GALAXY
BAK 1.674	03 30 52.	- 26 38 03.	17	16.2	EXTRAGALACTIC NEBULA
SHAP332-5032.3	03 30 52.	- 50 42 22.	18	16.75	GALAXY
BAK 1.676	03 30 53.	- 24 18 03.	10	15.7	EXTRAGALACTIC NEBULA
SHAP332-4531.6	03 30 53.	- 45 41 40.	24	16.75	GALAXY
SHAP332-4751.1	03 30 53.	- 48 01 10.	18	17.5	GALAXY
SHAP332-4834.2	03 30 53.	- 48 44 16.	36	17.75	GALAXY
ZC 0330.9+0204	03 30 54.	+ 02 04	1550		CLUSTER OF GALAXIES

OBJECT NAME	RIGHT ASCEN.	DECLINATION	DIAM.	MAGN.	TYPE OF OBJECT
5ZW 353	03 30 54.	+ 40 45			COMPACT GALAXY
MCG+01-10-002	03 30 54.	- 05 12	72	15.	GALAXY
PHL 8707	03 30 54.	- 27 30		18.3	BLUE STELLAR OBJECT
BAK 2.0012	03 30 55.	- 27 08 15.	16	16.3	GALAXY
BAK 1.673	03 30 55.	- 30 02 39.	20	16.6	EXTRAGALACTIC NEBULA
SHAP332-4750.7	03 30 55.	- 48 00 46.	18	17.5	GALAXY
RNGC 1357	03 30 56.	- 13 50		12.5	GALAXY
SHAP332-5107.4	03 30 56.	- 51 17 28.	24	17.0	GALAXY
SHAP332-5218.0	03 30 56.	- 52 28 04.	18	17.4	GALAXY
SHAP332-5224.0	03 30 56.	- 52 34 04.	36	17.0	GALAXY
SHAP332-5106.2	03 30 57.	- 51 16 16.	18	17.5	GALAXY
SHAP332-5341.4	03 30 57.	- 53 51 28.	24	17.5	GALAXY
BAK 1.677	03 30 58.	- 27 07 39.	24	16.3	EXTRAGALACTIC NEBULA
BAK 1.680	03 30 59.	- 24 02 15.	17	16.4	EXTRAGALACTIC NEBULA
SHAP332-4546.8	03 30 59.	- 45 56 52.	18	17.25	GALAXY
SHAP332-4552.8	03 30 59.	- 46 02 52.	24	17.75	GALAXY
SHAP332-5825.7	03 30 59.	- 58 35 47.	30	17.25	GALAXY
LBN 0720	03 31	+ 37 40	2820		BRIGHT NEBULA
LB 09803	03 31	- 80 30		14.2	FAINT BLUE STAR
ZWG 525.041	03 31 00.	+ 39 12		14.2	GALAXY
UGC 02783	03 31 00.	+ 39 12	102	14.2	GALAXY E?
ZWG 525.042	03 31 00.	+ 39 23		14.3	GALAXY
UGC 02784	03 31 00.	+ 39 23	102	14.3	GALAXY S0
ZWG 525.043	03 31 00.	+ 39 26		15.0	GALAXY
ZC 0331.0+8451	03 31 00.	+ 84 51	1480		CLUSTER OF GALAXIES
MCG-02-10-001	03 31 00.	- 13 50	120	12.5	GALAXY
BAK 1.681	03 31 00.	- 24 50 39.	19	16.2	EXTRAGALACTIC NEBULA
SHAP332-4635.7	03 31 00.	- 46 45 46.	30	17.25	GALAXY
SHAP332-4914.3	03 31 00.	- 49 24 22.	24	17.0	GALAXY
SHAP332-5218.7	03 31 00.	- 52 28 46.	18	17.0	GALAXY
SHAP332-5224.2	03 31 00.	- 52 34 16.	24	17.5	GALAXY
LB 03323	03 31 00.	- 79 03		14.4	FAINT BLUE STAR
BAK 1.678	03 31 01.	- 29 14 39.	19	16.7	EXTRAGALACTIC NEBULA
KARA.73 13	03 31 01.	- 35 55	27		DWARF GALAXY
SHAP332-4820.6	03 31 01.	- 48 30 40.	18	17.75	GALAXY
SHAP332-5327.9	03 31 01.	- 53 37 58.	30	17.25	GALAXY
BAK 1.679	03 31 02.	- 29 38 21.	17	17.2	EXTRAGALACTIC NEBULA
SHAP332-5412.2	03 31 02.	- 54 22 16.	24	17.0	GALAXY
SHAP332-5222.4	03 31 04.	- 52 32 28.	18	16.9	GALAXY
KARA.73 14	03 31 05.	- 36 46	27		DWARF GALAXY
SHAP332-4732.4	03 31 05.	- 47 42 28.	18	17.5	GALAXY
SHAP332-5212.8	03 31 05.	- 52 22 52.	30	17.4	GALAXY
SHAP332-5244.0	03 31 05.	- 52 54 04.	24	17.5	GALAXY
SHAP332-5323.8	03 31 05.	- 53 33 52.	12	17.5	GALAXY
UGC 02785	03 31 06.	+ 40 45	84	16.5	GALAXY Sc
BAK 1.682	03 31 06.	- 25 38 51.	16	16.4	EXTRAGALACTIC NEBULA
HELW 123	03 31 06.	- 26 22 32.			NEBULA
SHAP332-5201.0	03 31 06.	- 52 11 04.	24	17.2	GALAXY
SHAP332-5216.0	03 31 06.	- 52 26 04.	24	16.6	GALAXY
SHAP332-5403.4	03 31 06.	- 54 13 28.	54	16.5	GALAXY
SHAP332-5414.4	03 31 06.	- 54 24 28.	18	17.0	GALAXY
PK220-53.1	03 31 07.	- 26 02 20.	540		PLANETARY NEBULA
SHAP332-4537.2	03 31 07.	- 45 47 15.	24	17.25	GALAXY
SHAP332-4730.7	03 31 07.	- 47 40 45.	12	17.0	GALAXY
SHAP332-5243.4	03 31 07.	- 52 53 28.	24	17.4	GALAXY
SHAP332-4915.2	03 31 08.	- 49 25 15.	24	16.75	GALAXY
SHAP332-5229.0	03 31 08.	- 52 39 04.	30	17.6	GALAXY
SHAP332-5308.3	03 31 08.	- 53 18 22.	18	17.4	GALAXY
MCG+06-08-034	03 31 09.	+ 39 11	48	15.	GALAXY
MCG+06-08-033	03 31 09.	+ 39 23	48	15.	GALAXY
SHAP332-5329.7	03 31 09.	- 53 39 46.	18	17.25	GALAXY
RNGC 1358	03 31 10.	- 05 16		12.5	GALAXY
SHAP332-4727.6	03 31 11.	- 47 37 39.	24	18.0	GALAXY
SHAP332-5337.2	03 31 11.	- 53 47 16.	24	17.5	GALAXY
MCG+01-10-003	03 31 12.	- 05 17	120	14.	GALAXY
MCG-03-10-006	03 31 12.	- 18 18	84	14.5	GALAXY
PHL 1556	03 31 12.	- 26 02		9.0	BLUE STELLAR OBJECT
SHAP332-5019.2	03 31 12.	- 50 29 15.	18	16.5	GALAXY
SHAP332-5038.8	03 31 12.	- 50 48 51.	18	16.5	GALAXY
BAK 2.0014	03 31 13.	- 18 19 02.	39	14.8	GALAXY
BAK 1.683	03 31 13.	- 29 50 08.	17	16.6	EXTRAGALACTIC NEBULA
SHAP332-4638.6	03 31 13.	- 46 48 39.	18	17.0	GALAXY
RNGC 1360	03 31 14.	- 26 01			PLANETARY NEBULA
SHAP332-4606.9	03 31 14.	- 46 16 57.	18	17.5	GALAXY
BAK 1.686	03 31 15.	- 21 43 44.	36	14.7	EXTRAGALACTIC NEBULA
BAK 1.684	03 31 15.	- 27 01 38.	12	17.2	EXTRAGALACTIC NEBULA
HN 0212	03 31 16.	- 23 53			NEBULA
IC 1952	03 31 16.	- 23 53			NONSTELLAR OBJECT
SHAP332-5439.1	03 31 16.	- 54 49 09.	24	17.25	GALAXY
SHAP332-5057.8	03 31 17.	- 51 07 51.	24	17.0	GALAXY
MCG-04-09-025	03 31 18.	- 23 53	150	13.	GALAXY
SHAP332-4757.6	03 31 18.	- 48 07 39.	18	17.25	GALAXY
SHAP332-5326.0	03 31 18.	- 53 36 03.	18	16.75	GALAXY
BAK 2.0013	03 31 19.	- 26 44 26.	13	16.7	GALAXY
BAK 1.685	03 31 19.	- 29 57 02.	19	16.8	EXTRAGALACTIC NEBULA
SHAP332-5126.5	03 31 19.	- 51 36 33.	30	16.5	GALAXY
SHAP332-5344.2	03 31 19.	- 53 54 15.	18	17.5	GALAXY
HN 0215	03 31 20.	- 51 37			NEBULA
IC 1958	03 31 20.	- 51 37			NONSTELLAR OBJECT
SHAP332-5206.9	03 31 20.	- 52 16 57.	24	17.1	GALAXY
SHAP332-5216.8	03 31 20.	- 52 26 51.	24	17.6	GALAXY
SHAP332-5227.8	03 31 20.	- 52 37 51.	24	17.5	GALAXY
SHAP332-5233.3	03 31 20.	- 52 43 21.	36	17.1	GALAXY
SHAP332-5350.0	03 31 20.	- 54 00 03.	24	17.25	GALAXY
SHAP332-5711.7	03 31 21.	- 57 21 45.	54	16.0	GALAXY
BAK 2.0015	03 31 22.	- 18 36 43.	16	15.8	GALAXY
SHAP332-5221.8	03 31 22.	- 52 31 51.	36	17.2	GALAXY
HN 0217	03 31 22.	- 57 22			NEBULA
SHAP333-4550.6	03 31 23.	- 46 00 38.	24	17.5	GALAXY
SHAP332-5146.1	03 31 23.	- 51 56 09.	30	17.0	GALAXY
IC 1960	03 31 23.	- 57 22			NONSTELLAR OBJECT
MCG+11-05-001	03 31 24.	+ 67 24 30.	42	16.	GALAXY
PHL 1557	03 31 24.	- 30 14		16.7	BLUE STELLAR OBJECT
HPW 16	03 31 24.	- 35 57			DWARF GALAXY IN FORNAX
SHAP333-4554.6	03 31 24.	- 46 04 38.	18	17.0	GALAXY
SHAP332-5222.0	03 31 24.	- 52 32 03.	18	17.6	GALAXY
SHAP332-5325.2	03 31 24.	- 53 35 15.	18	16.5	GALAXY
SHAP332-5344.3	03 31 24.	- 53 54 21.	18	17.75	GALAXY
SHAP333-4637.1	03 31 25.	- 46 47 08.	30	17.0	GALAXY
SHAP333-4739.2	03 31 25.	- 47 49 14.	24	18.0	GALAXY
SHAP333-4739.7	03 31 26.	- 47 49 44.	18	17.5	GALAXY
SHAP332-5135.8	03 31 26.	- 51 45 51.	18	17.5	GALAXY
SHAP332-5221.6	03 31 26.	- 52 31 39.	24	17.7	GALAXY
BAK 1.687	03 31 27.	- 22 57 07.	24	16.1	EXTRAGALACTIC NEBULA
SHAP332-5248.2	03 31 27.	- 52 58 15.	18	17.3	GALAXY
SHAP332-5328.0	03 31 27.	- 53 38 03.	18	17.5	GALAXY
BAK 1.688	03 31 28.	- 24 20 43.	22	16.5	EXTRAGALACTIC NEBULA
SHAP333-4721.2	03 31 28.	- 47 31 14.	30	18.25	GALAXY
SHAP333-4738.8	03 31 28.	- 47 48 50.	24	16.75	GALAXY
SHAP333-4755.8	03 31 28.	- 48 05 50.	24	17.5	GALAXY
RNGC 1359	03 31 29.	- 19 41		13.0	GALAXY
SHAP333-4636.0	03 31 29.	- 46 46 02.	18	17.75	GALAXY
SHAP332-5242.0	03 31 29.	- 52 52 02.	24	17.5	GALAXY
SHAP332-5257.8	03 31 29.	- 53 07 50.	18	17.75	GALAXY
SHAP332-5326.2	03 31 29.	- 53 36 15.	24	17.5	GALAXY
IC 1953	03 31 29.3	- 21 38 43.	156	12.5	GALAXY SB(rs)
UGC 02786	03 31 30.	+ 02 56	84	17.	GALAXY
UGC 02787	03 31 30.	+ 41 10	66	16.0	GALAXY
MCG-02-10-002	03 31 30.	- 10 24	66	14.5	GALAXY
MCG-03-10-007	03 31 30.	- 19 40	90	13.5	GALAXY
HN 0213	03 31 30.	- 21 39			NEBULA
TON-S 0359	03 31 30.	- 22 57		13.8	BLUE STAR
BAK 1.690	03 31 30.	- 24 20 07.	17	16.6	EXTRAGALACTIC NEBULA
SHAP333-4532.2	03 31 31.	- 45 42 14.	24	17.0	GALAXY
SHAP333-4904.8	03 31 31.	- 49 14 50.	24	16.5	GALAXY
SHAP332-5327.9	03 31 31.	- 53 37 56.	24	17.5	GALAXY
SHAP332-5350.5	03 31 32.	- 54 00 32.	24	17.0	GALAXY
SHAP332-5418.6	03 31 32.	- 54 28 38.	18	18.0	GALAXY
SHAP332-5437.5	03 31 32.	- 54 47 32.	24	17.5	GALAXY
IC 0333	03 31 33.	- 05 17			NONSTELLAR OBJECT
SVEN 216	03 31 33.	- 19 39 21.	108	11.7	GALAXY
SVEN 217	03 31 33.	- 19 44	36	14.4	GALAXY
BAK 1.691	03 31 33.	- 24 22 01.	19	16.4	EXTRAGALACTIC NEBULA
SHAP332-5418.0	03 31 33.	- 54 28 02.	24	16.75	GALAXY
BAK 1.689	03 31 34.	- 29 07 19.	22	16.2	EXTRAGALACTIC NEBULA
SHAP333-4720.6	03 31 34.	- 47 30 38.	42	17.25	GALAXY
HELW 400	03 31 35.	- 19 43 52.			NEBULA
RNGC 1362	03 31 35.	- 20 27		14.0	GALAXY
SHAP333-4542.1	03 31 35.	- 45 52 08.	24	16.25	GALAXY
SHAP333-4824.1	03 31 35.	- 48 34 08.	18	17.0	GALAXY
5ZW 354	03 31 36.	+ 28 34			COMPACT GALAXY
ZWG 525.044	03 31 36.	+ 36 03		15.7	GALAXY
UGC 02788	03 31 36.	+ 36 03	126	15.7	GALAXY E?
MCG-03-10-009	03 31 36.	- 16 31	36	15.5	GALAXY
BAK 1.693	03 31 36.	- 20 17 19.	16	16.6	EXTRAGALACTIC NEBULA
MCG-03-10-008	03 31 36.	- 20 27	30	14.5	GALAXY
HELW 401	03 31 36.	- 21 13 10.			NEBULA
MCG-04-09-026	03 31 36.	- 21 37	180	12.5	GALAXY
PHL 8709	03 31 36.	- 21 48		4.3	BLUE STELLAR OBJECT
PHL 4454	03 31 36.	- 22 54		13.3	BLUE STELLAR OBJECT
SHAP332-5302.8	03 31 36.	- 53 12 50.	24	16.75	GALAXY
HELW 402	03 31 37.	- 21 37 22.			NEBULA
HELW 124	03 31 37.	- 26 20 34.			NEBULA
SHAP333-4825.8	03 31 37.	- 48 35 50.	24	18.0	GALAXY
SHAP333-5146.4	03 31 37.	- 51 56 26.	18	16.5	GALAXY
SHAP333-5151.7	03 31 37.	- 52 01 44.	24	17.5	GALAXY
BAK 1.695	03 31 38.	- 21 13 12.	25	15.5	EXTRAGALACTIC NEBULA
SHAP333-4638.2	03 31 38.	- 46 48 14.	24	17.5	GALAXY
MCG+01-10-004	03 31 39.	- 07 19 30.	102	15.	GALAXY
BAK 1.696	03 31 39.	- 21 37 00.	19	16.2	EXTRAGALACTIC NEBULA
SHAP333-4901.4	03 31 39.	- 49 11 26.	24	17.5	GALAXY
SHAP333-4909.5	03 31 39.	- 49 19 32.	18	17.5	GALAXY
SHAP333-5046.4	03 31 39.	- 50 56 26.	24	17.0	GALAXY
BAK 1.697	03 31 40.	- 20 10 06.	16	16.2	EXTRAGALACTIC NEBULA
SHAP333-5024.7	03 31 40.	- 50 34 44.	276	14.0	GALAXY
HN 0216	03 31 41.	- 50 35			NEBULA
IC 1959	03 31 41.	- 50 35			NONSTELLAR OBJECT
SHAP333-5152.4	03 31 41.	- 52 02 26.	18	17.5	GALAXY
SHAP333-5425.5	03 31 41.	- 54 35 32.	24	17.5	GALAXY
SHAP332-5745.0	03 31 41.	- 57 55 02.	36	16.75	GALAXY
UGC 02789	03 31 42.	+ 67 24	138	17.	GALAXY SB
BAK 2.0018	03 31 42.	- 15 45 00.	18	15.8	GALAXY
BAK 2.0017	03 31 42.	- 16 30 54.	27	16.0	GALAXY
MCG-04-09-027	03 31 42.	- 21 12 30.	36	15.	GALAXY
PHL 8710	03 31 42.	- 23 41		18.6	BLUE STELLAR OBJECT
MCG-06-08-026	03 31 42.	- 36 20	420		GALAXY
SHAP333-4631.4	03 31 42.	- 46 41 25.	24	17.75	GALAXY
LB 01674	03 31 42.	- 48 40		16.2	FAINT BLUE STAR
SHAP333-5440.0	03 31 42.	- 54 50 02.	30	17.0	GALAXY
SVEN 221	03 31 43.	- 36 18	318	10.9	GALAXY
SHAP333-5233.4	03 31 43.	- 52 43 26.	18	16.7	GALAXY
BAK 1.692	03 31 44.	- 31 42 42.	24	16.6	EXTRAGALACTIC NEBULA
SHAP333-4620.0	03 31 44.	- 46 30 01.	18	17.5	GALAXY
MCG+06-08-035	03 31 45.	+ 36 02	24	15.	GALAXY
SVEN 218	03 31 45.	- 19 42 22.	36	15.2	GALAXY
SN 1957C	03 31 45.	- 36 17		16.5	SUPERNOVA
RNGC 1365	03 31 45.	- 36 18		10.5	GALAXY
SHAP333-4714.9	03 31 45.	- 47 24 55.	18	17.0	GALAXY
SHAP333-4907.4	03 31 45.	- 49 17 25.	24	16.0	GALAXY
SHAP333-5341.8	03 31 45.	- 53 51 50.	12	16.75	GALAXY
SHAP333-5433.0	03 31 45.	- 54 43 02.	30	17.75	GALAXY
RNGC 1361	03 31 46.	- 06 25			GALAXY
BAK 2.0019	03 31 46.	- 17 39 24.	24	16.0	GALAXY
BAK 1.694	03 31 46.	- 28 59 36.	18	17.0	EXTRAGALACTIC NEBULA
SHAP333-4907.8	03 31 46.	- 49 17 49.	24	16.5	GALAXY
SHAP333-4945.4	03 31 46.	- 49 55 25.	24	16.25	GALAXY
SHAP333-5131.1	03 31 46.	- 51 41 07.	18	18.0	GALAXY
SHAP333-5213.8	03 31 46.	- 52 23 49.	30	16.7	GALAXY
SHAP333-5216.6	03 31 46.	- 52 26 37.	18	17.7	GALAXY
SHAP333-5217.1	03 31 46.	- 52 27 07.	18	16.8	GALAXY
VDB.66N 018	03 31 47.	+ 37 53	852		REFLECTION NEBULA
BAK 1.698	03 31 47.	- 24 45 12.	17	16.1	EXTRAGALACTIC NEBULA
SHAP333-4737.4	03 31 47.	- 47 47 25.	18	17.5	GALAXY
MCG-04-09-028	03 31 48.	- 21 36	78	15.	GALAXY
PHL 8711	03 31 48.	- 25 46		18.3	BLUE STELLAR OBJECT
SHAP333-4910.2	03 31 48.	- 49 20 13.	24	16.5	GALAXY
SHAP333-5215.6	03 31 48.	- 52 25 37.	24	17.6	GALAXY
BAK 2.0016	03 31 49.	- 27 21 24.	15	16.7	GALAXY
SHAP333-4721.2	03 31 49.	- 47 31 13.	18	18.0	GALAXY
SHAP333-4822.4	03 31 49.	- 48 32 25.	30	16.5	GALAXY
SHAP333-5211.2	03 31 49.	- 52 21 13.	24	17.6	GALAXY
SHAP333-5328.8	03 31 49.	- 53 38 49.	12	17.25	GALAXY
BAK 2.0020	03 31 50.	- 16 17 48.	17	16.6	GALAXY
HELW 125	03 31 50.	- 25 47 41.			NEBULA
SHAP333-5326.2	03 31 51.	- 53 36 13.	18	17.0	GALAXY
SHAP333-5341.2	03 31 51.	- 53 51 13.	24	17.25	GALAXY
SHAP333-5420.8	03 31 52.	- 54 30 49.	36	16.75	GALAXY
SHAP333-4842.4	03 31 53.	- 48 52 25.	42	16.5	GALAXY
SHAP333-5144.0	03 31 53.	- 51 54 01.	24	16.25	GALAXY
SHAP333-5147.1	03 31 53.	- 51 57 07.	18	16.5	GALAXY
MCG+01-10-005	03 31 54.	- 06 27 30.	24	15.	GALAXY
SHAP333-4823.8	03 31 54.	- 48 33 49.	18	16.75	GALAXY
SHAP333-5201.6	03 31 54.	- 52 11 37.	24	17.3	GALAXY
SHAP333-5802.4	03 31 54.	- 58 12 25.	30	17.25	GALAXY
SHAP333-4744.8	03 31 55.	- 47 54 49.	18	17.25	GALAXY
SHAP333-5228.6	03 31 55.	- 52 38 37.	18	17.0	GALAXY

OBJECT NAME	RIGHT ASCEN.	DECLINATION	DIAM.	MAGN.	TYPE OF OBJECT
HELW 403	03 31 56.	- 19 43 17.			NEBULA
SHAP333-5005.0	03 31 56.	- 50 15 01.	24	17.0	GALAXY
SHAP333-5226.8	03 31 56.	- 52 36 49.	24	16.9	GALAXY
SHAP333-5328.6	03 31 56.	- 53 38 37.	18	17.25	GALAXY
SVEN 219	03 31 57.	- 19 43 22.	24	14.7	GALAXY
BAK 1.699	03 31 57.	- 30 23 12.	14	17.1	EXTRAGALACTIC NEBULA
SHAP333-5331.4	03 31 57.	- 53 41 25.	12	17.5	GALAXY
SHAP333-5232.3	03 31 58.	- 52 42 19.	18	17.1	GALAXY
SHAP333-5632.7	03 31 58.	- 56 42 43.	60	16.25	GALAXY
RNGC 1366	03 31 59.	- 31 23		14.0	GALAXY
SHAP333-4721.4	03 31 59.	- 47 31 24.	18	18.0	GALAXY
SHAP333-5132.8	03 31 59.	- 51 42 49.	18	17.25	GALAXY
LBN 0721	03 32	+ 37 50	660		BRIGHT NEBULA
ZWG 391.009	03 32 00.	+ 03 25		15.3	GALAXY
LB 00486	03 32 00.	+ 16 36 24.		16.4	FAINT BLUE STAR
UGC 02790	03 32 00.	+ 41 10	72	16.5	GALAXY Sc
ACK 142+03.1	03 32 00.	+ 59 54			PLANETARY NEBULA
MCG-03-10-010	03 32 00.	- 19 35	54	15.	GALAXY
PHL 4455	03 32 00.	- 24 28		18.0	BLUE STELLAR OBJECT
SHAP333-4728.1	03 32 00.	- 47 38 06.	36	17.0	GALAXY
SHAP333-4857.2	03 32 00.	- 49 07 12.	24	16.5	GALAXY
SHAP333-5229.4	03 32 00.	- 52 39 25.	24	17.4	GALAXY
HN 0219	03 32 00.	- 56 43			NEBULA
SHAP333-4856.7	03 32 01.	- 49 06 42.	36	15.5	GALAXY
IC 1965	03 32 01.	- 56 43			NONSTELLAR OBJECT
HN 0218	03 32 02.	- 49 07			NEBULA
IC 1961	03 32 02.	- 49 07			NONSTELLAR OBJECT
SHAP333-5443.7	03 32 02.	- 54 53 43.	24	16.5	GALAXY
MCG+01-10-006	03 32 03.	- 04 59 30.	60	15.	GALAXY
SVEN 220	03 32 03.	- 19 35 23.	60	13.7	GALAXY
SHAP333-4841.6	03 32 03.	- 48 51 36.	24	17.0	GALAXY
BAK 1.703	03 32 04.	- 19 35 35.	54	15.2	EXTRAGALACTIC NEBULA
BAK 1.701	03 32 04.	- 23 25 47.	14	16.2	EXTRAGALACTIC NEBULA
SHAP333-5312.8	03 32 04.	- 53 22 48.	18	17.5	GALAXY
SHAP333-5436.5	03 32 04.	- 54 46 31.	30	17.0	GALAXY
HELW 404	03 32 05.	- 19 35 17.			NEBULA
SHAP333-4910.6	03 32 05.	- 49 20 36.	24	17.0	GALAXY
ZWG 391.010	03 32 06.	- 02 06		15.1	GALAXY
PHL 8712	03 32 06.	- 22 01		18.4	BLUE STELLAR OBJECT
PHL 8713	03 32 06.	- 29 24		18.0	BLUE STELLAR OBJECT
MCG-05-09-013	03 32 06.	- 31 22	120	14.	GALAXY
LB 01675	03 32 06.	- 46 34		16.0	FAINT BLUE STAR
SHAP333-4709.0	03 32 06.	- 47 19 00.	18	17.0	GALAXY
SHAP333-4951.5	03 32 06.	- 50 01 30.	24	17.25	GALAXY
SHAP333-4530.3	03 32 07.	- 45 40 18.	18	17.5	GALAXY
SHAP333-4726.0	03 32 07.	- 47 36 00.	18	17.25	GALAXY
SHAP333-5344.4	03 32 07.	- 53 54 24.	12	16.75	GALAXY
BAK 2.0021	03 32 08.	- 26 13 11.	15	16.6	GALAXY
SHAP333-4506.0	03 32 08.	- 45 16 00.	24	16.75	GALAXY
SHAP333-4805.8	03 32 08.	- 48 15 48.	18	17.5	GALAXY
SHAP333-4850.8	03 32 08.	- 49 00 48.	18	17.5	GALAXY
SHAP333-5312.0	03 32 08.	- 53 22 00.	30	17.75	GALAXY
BAK 1.700	03 32 09.	- 29 43 05.	14	16.9	EXTRAGALACTIC NEBULA
RNGC 1369	03 32 09.	- 36 27			NON-EXISTENT OBJECT
SHAP333-4718.9	03 32 09.	- 47 28 54.	24	17.5	GALAXY
SHAP333-4913.0	03 32 09.	- 49 23 00.	18	17.0	GALAXY
SHAP333-5138.9	03 32 09.	- 51 48 54.	18	17.5	GALAXY
SHAP333-4602.0	03 32 10.	- 46 12 00.	18	17.5	GALAXY
SHAP333-5323.0	03 32 10.	- 53 33 00.	18	17.5	GALAXY
SHAP333-4620.3	03 32 11.	- 46 30 18.	24	18.0	GALAXY
SHAP333-4633.8	03 32 11.	- 46 43 48.	24	18.0	GALAXY
UGC 02791	03 32 12.	+ 41 15	90	17.	GALAXY
BAK 2.0022	03 32 12.	- 15 06 34.	17	15.7	GALAXY
BAK 1.705	03 32 12.	- 20 55 17.	22	16.8	EXTRAGALACTIC NEBULA
PHL 4456	03 32 12.	- 23 52		18.7	BLUE STELLAR OBJECT
BAK 2.0024	03 32 13.	- 15 49 46.	14	16.6	GALAXY
BAK 1.704	03 32 13.	- 27 02 59.	84	16.6	EXTRAGALACTIC NEBULA
BAK 1.702	03 32 14.	- 30 53 17.	24	15.6	EXTRAGALACTIC NEBULA
SHAP333-4958.0	03 32 14.	- 50 08 00.	18	16.2	GALAXY
SHAP333-5210.6	03 32 14.	- 52 20 36.	18	16.7	GALAXY
BAK 1.706	03 32 15.	- 23 17 58.	14	16.0	EXTRAGALACTIC NEBULA
BAK 1.707	03 32 15.	- 23 21 28.	24	16.4	EXTRAGALACTIC NEBULA
SHAP333-5309.7	03 32 15.	- 53 19 42.	36	16.5	GALAXY
SHAP333-5422.0	03 32 15.	- 54 32 00.	24	17.0	GALAXY
BAK 2.0023	03 32 16.	- 18 49 40.	11	16.2	GALAXY
SHAP333-4858.0	03 32 16.	- 49 07 59.	24	17.5	GALAXY
SHAP333-5134.7	03 32 16.	- 51 44 42.	18	17.0	GALAXY
HN 0220	03 32 16.	- 53 20			NEBULA
IC 1964	03 32 16.	- 53 20			NONSTELLAR OBJECT
ZC 0332.3+7750	03 32 18.	+ 77 50	1140		CLUSTER OF GALAXIES
PHL 8714	03 32 18.	- 22 30		16.7	BLUE STELLAR OBJECT
HPW 17	03 32 18.	- 36 37			DWARF GALAXY IN FORNAX
SHAP333-4447.8	03 32 18.	- 44 57 47.	30	17.5	GALAXY
SHAP333-4959.4	03 32 18.	- 50 09 23.	18	17.0	GALAXY
BAK 2.0052	03 32 20.	- 15 49 52.	15	16.2	GALAXY
SHAP333-4906.0	03 32 20.	- 49 15 59.	24	16.5	GALAXY
SVEN 222	03 32 21.	- 19 12	48	14.2	GALAXY
SHAP333-5217.4	03 32 21.	- 52 27 23.	18	17.5	GALAXY
SHAP333-5229.4	03 32 21.	- 52 39 23.	18	17.7	GALAXY
SHAP333-5710.8	03 32 21.	- 57 20 48.	36	16.25	GALAXY
BAK 2.0053	03 32 22.	- 17 38 46.	23	16.0	GALAXY
SHAP333-5426.0	03 32 23.	- 54 35 59.	18	17.5	GALAXY
7ZW 008	03 32 24.	+ 72 24			COMPACT GALAXY
ZWG 327.005	03 32 24.	+ 72 24		14.1	GALAXY
RNGC 1343	03 32 24.	+ 72 24		14.0	GALAXY
UGC 02792	03 32 24.	+ 72 24	162	14.1	GALAXY PECULR
MCG+12-04-001	03 32 24.	+ 72 24	150	14.	GALAXY
MCG+00-10-005	03 32 24.	- 01 57	12	15.	GALAXY
RNGC 1363	03 32 24.	- 10 00			GALAXY
MCG-03-10-011	03 32 24.	- 19 11 30.	72	14.5	GALAXY
PHL 8715	03 32 24.	- 22 14		18.5	BLUE STELLAR OBJECT
PHL 4457	03 32 24.	- 23 18		18.7	BLUE STELLAR OBJECT
BAK 1.708	03 32 24.	- 29 06 10.	19	16.8	EXTRAGALACTIC NEBULA
MCG-05-09-014	03 32 24.	- 30 54	84	15.	GALAXY
SHAP333-5225.0	03 32 24.	- 52 34 59.	18	17.2	GALAXY
LB 01676	03 32 24.	- 53 01		13.6	FAINT BLUE STAR
SHAP333-5301.2	03 32 24.	- 53 11 11.	12	17.5	GALAXY
SHAP333-5400.7	03 32 24.	- 54 10 41.	24	17.0	GALAXY
SHAP333-5426.8	03 32 24.	- 54 36 47.	24	17.75	GALAXY
SHAP334-4622.5	03 32 25.	- 46 32 29.	18	17.75	GALAXY
BC PKS0332-403	03 32 25.3	- 40 18 24.		18.5	QUASI-STELLAR OBJECT
SHAP333-5135.6	03 32 26.	- 51 45 35.	18	17.25	GALAXY
SHAP333-5201.7	03 32 26.	- 52 11 41.	24	17.6	GALAXY
SHAP333-5440.3	03 32 26.	- 54 50 17.	30	17.25	GALAXY
SHAP333-5728.6	03 32 26.	- 57 38 36.	30	17.25	GALAXY
SHAP333-5208.0	03 32 27.	- 52 17 59.	18	17.4	GALAXY
SHAP333-5209.6	03 32 27.	- 52 19 35.	24	17.1	GALAXY
SHAP333-5210.4	03 32 28.	- 52 20 23.	18	17.8	GALAXY

OBJECT NAME	RIGHT ASCEN.	DECLINATION	DIAM.	MAGN.	TYPE OF OBJECT
BAK 2.0054	03 32 29.	- 19 11 39.	37	14.1	GALAXY
BAK 1.709	03 32 29.	- 21 00 46.	24	16.3	EXTRAGALACTIC NEBULA
MCG-06-08-027	03 32 30.	- 35 44	48	16.	GALAXY
SHAP334-4455.2	03 32 30.	- 45 05 10.	30	17.0	GALAXY
SHAP333-5213.0	03 32 30.	- 52 22 59.	18	16.9	GALAXY
SHAP333-5434.1	03 32 31.	- 54 44 05.	24	17.25	GALAXY
BAK 2.0055	03 32 32.	- 16 24 57.	14	16.6	GALAXY
RNGC 1367	03 32 32.	- 25 06			NON-EXISTENT OBJECT
SHAP334-5006.6	03 32 32.	- 50 16 35.	24	16.5	GALAXY
SHAP334-5006.8	03 32 32.	- 50 16 47.	18	17.0	GALAXY
SHAP333-5750.1	03 32 32.	- 58 00 05.	30	16.25	GALAXY
SHAP334-4604.2	03 32 33.	- 46 14 10.	24	17.75	GALAXY
SHAP333-5213.8	03 32 34.	- 52 23 47.	24	17.6	GALAXY
BAK 2.0056	03 32 35.	- 18 37 09.	20	16.2	GALAXY
SHAP334-5208.5	03 32 35.	- 52 18 29.	24	17.6	GALAXY
SHAP333-5332.3	03 32 35.	- 53 42 17.	24	17.5	GALAXY
RNGC 1364	03 32 36.	- 10 00			GALAXY
MCG-03-10-012	03 32 36.	- 15 48 30.	36	15.	GALAXY
PHL 4458	03 32 36.	- 23 07		18.3	BLUE STELLAR OBJECT
HOD.60 01	03 32 36.	- 34 40	42	17.5	E0 GALAXY IN FORNAX CLSTR
HPW 18	03 32 36.	- 40 38			DWARF GALAXY IN FORNAX
SER 031.02	03 32 36.	- 51 28	60	16.5	INTERACTING GALAXIES
BAK 2.0057	03 32 37.	- 16 23 15.	12	16.6	GALAXY
SHAP334-5118.7	03 32 37.	- 51 28 40.	42	16.0	GALAXY
SHAP333-5325.2	03 32 37.	- 53 35 11.	18	17.5	GALAXY
SHAP333-5444.2	03 32 37.	- 54 54 11.	36	16.75	GALAXY
SHAP333-5738.8	03 32 37.	- 57 48 47.	30	15.75	GALAXY
HN 0222	03 32 38.	- 51 29			NEBULA
IC 1966	03 32 38.	- 51 29			NONSTELLAR OBJECT
SHAP334-5209.2	03 32 38.	- 52 19 10.	18	17.0	GALAXY
SHAP333-5356.0	03 32 38.	- 54 05 59.	30	16.5	GALAXY
RNGC 1368	03 32 39.	- 15 48		15.0	GALAXY
BAK 1.710	03 32 39.	- 29 20 21.	24	17.0	EXTRAGALACTIC NEBULA
SHAP334-4534.6	03 32 39.	- 45 44 34.	18	17.5	GALAXY
BAK 2.0058	03 32 40.	- 17 12 45.	19	16.2	GALAXY
SHAP334-5117.6	03 32 40.	- 51 27 34.	24	16.5	GALAXY
SHAP334-5307.6	03 32 40.	- 53 17 34.	18	18.0	GALAXY
SHAP333-5748.8	03 32 40.	- 57 58 47.	42	16.5	GALAXY
BAK 1.711	03 32 41.	- 28 56 45.	26	16.4	EXTRAGALACTIC NEBULA
SHAP334-5024.4	03 32 41.	- 50 34 22.	12	17.25	GALAXY
ZWG 391.011	03 32 42.	- 01 23		15.6	GALAXY
KARA.72 091A	03 32 42.	- 01 23	30	15.6	PART OF DOUBLE GALAXY
KARA.68 039	03 32 42.	- 21 25	34		DWARF GALAXY
BAK 1.712	03 32 42.	- 29 22 27.	17	17.0	EXTRAGALACTIC NEBULA
BAK 1.714	03 32 43.	- 24 57 51.	19	16.1	EXTRAGALACTIC NEBULA
SHAP334-4453.6	03 32 43.	- 45 03 34.	30	17.0	GALAXY
ARC 0444	03 32 45.	+ 03 08		17.5	RICH CLUSTER OF GALAXIES
BAK 1.713	03 32 45.	- 28 40 03.	19	16.7	EXTRAGALACTIC NEBULA
SHAP334-5221.1	03 32 46.	- 52 31 04.	18	17.6	GALAXY
UGC 02793	03 32 48.	+ 36 30	72	16.5	GALAXY
MCG+08-07-006	03 32 48.	+ 45 28	42	16.	GALAXY
ZWG 391.012	03 32 48.	- 01 24		15.7	GALAXY
KARA.72 091B	03 32 48.	- 01 24	30	15.7	PART OF DOUBLE GALAXY
BAK 2.0059	03 32 48.	- 18 05 44.	20	16.4	GALAXY
PHL 8717	03 32 48.	- 22 30		18.5	BLUE STELLAR OBJECT
PHL 8716	03 32 48.	- 24 02		8.8	BLUE STELLAR OBJECT
SHAP334-5220.0	03 32 48.	- 52 29 58.	18	17.8	GALAXY
BAK 2.0060	03 32 49.	- 15 59 44.	19	15.0	GALAXY
BAK 1.717	03 32 49.	- 20 12 26.	16	16.5	EXTRAGALACTIC NEBULA
SHAP334-5129.8	03 32 49.	- 51 39 46.	18	18.0	GALAXY
SHAP334-5420.8	03 32 49.	- 54 30 46.	18	18.0	GALAXY
SHAP334-5424.4	03 32 49.	- 54 34 22.	24	17.5	GALAXY
RNGC 1371	03 32 50.	- 25 06		12.0	GALAXY
SHAP334-4604.5	03 32 50.	- 46 14 27.	30	17.5	GALAXY
SHAP334-5115.4	03 32 50.	- 51 25 22.	18	17.75	GALAXY
SHAP334-5307.2	03 32 50.	- 53 17 10.	18	17.5	GALAXY
SHAP334-5432.1	03 32 50.	- 54 42 04.	30	17.25	GALAXY
SHAP334-5434.6	03 32 50.	- 54 44 34.	24	17.75	GALAXY
MCG-04-09-029	03 32 51.	- 25 05 30.	330	12.	GALAXY
SVEN 223	03 32 51.	- 25 06 26.	48	13.1	GALAXY
SHAP334-5323.0	03 32 51.	- 53 32 58.	12	17.5	GALAXY
IC 1956	03 32 53.	+ 04 54 09.			NONSTELLAR OBJECT
SHAP333-5955.6	03 32 53.	- 60 05 34.	48	15.0	GALAXY
UGC 02794	03 32 54.	+ 48 20	96	18.	GALAXY S
PHL 4459	03 32 54.	- 21 18		18.5	BLUE STELLAR OBJECT
HOD.60 02	03 32 54.	- 36 03	48	17.5	E0 GALAXY IN FORNAX CLSTR
SHAP334-4509.8	03 32 54.	- 45 19 45.	24	16.5	GALAXY
SHAP334-5307.0	03 32 54.	- 53 16 58.	18	17.25	GALAXY
SHAP334-5433.8	03 32 54.	- 54 43 46.	30	17.0	GALAXY
BAK 1.716	03 32 55.	- 31 28 32.	23	17.0	EXTRAGALACTIC NEBULA
BAK 1.715	03 32 55.	- 32 49 26.	32	15.8	EXTRAGALACTIC NEBULA
SHAP334-5703.2	03 32 55.	- 57 13 10.	30	16.75	GALAXY
SHAP334-5311.1	03 32 56.	- 53 21 03.	18	17.75	GALAXY
SHAP334-5352.0	03 32 56.	- 54 01 57.	24	17.0	GALAXY
MCG+01-10-007	03 32 57.	- 04 29 30.	42	15.5	GALAXY
IC 1963	03 32 57.	- 34 36 49.			NONSTELLAR OBJECT
SHAP334-5022.6	03 32 57.	- 50 32 33.	18	17.25	GALAXY
SHAP334-5103.1	03 32 58.	- 51 13 03.	30	16.0	GALAXY
SHAP334-5323.2	03 32 58.	- 53 33 09.	18	17.5	GALAXY
RNGC 1370	03 32 59.	- 20 32		14.0	GALAXY
BAK 2.0061	03 32 59.	- 25 40 02.	14	16.6	GALAXY
SHAP334-5722.6	03 32 59.	- 57 32 34.	30	16.0	GALAXY
LB 09804	03 33	- 83 08		14.6	FAINT BLUE STAR
ZC 0333.0+0300	03 33 00.	+ 03 00	1340		CLUSTER OF GALAXIES
ZWG 417.001	03 33 00.	+ 04 55		15.6	GALAXY
UGC 02795	03 33 00.	+ 04 55	84	15.6	GALAXY SB:b-c
MCG+01-10-001	03 33 00.	+ 04 55	72	15.	GALAXY
KARA.73B 0130	03 33 00.	+ 04 55	126	15.6	ISOLATED GALAXY S
LDN 1434	03 33 00.	+ 37 30	3360		DARK NEBULA
MCG-03-10-013	03 33 00.	- 20 32	30	14.	GALAXY
PHL 4460	03 33 00.	- 24 30		15.5	BLUE STELLAR OBJECT
BAK 2.0062	03 33 00.	- 25 30 56.	14	16.8	GALAXY
PHL 8718	03 33 00.	- 29 11		16.6	BLUE STELLAR OBJECT
MCG-06-08-028	03 33 00.	- 35 21	30	15.	GALAXY
SHAP334-5720.1	03 33 00.	- 57 30 04.	36	16.5	GALAXY
SHAP334-5233.0	03 33 01.	- 52 42 57.	12	17.5	GALAXY
BAK 2.0063	03 33 02.	- 18 03 32.	16	16.3	GALAXY
HELW 050	03 33 02.	- 35 20			NEBULA
BAK 1.720	03 33 03.	- 24 39 44.	18	16.2	EXTRAGALACTIC NEBULA
BAK 1.718	03 33 03.	- 25 46 32.	17	16.7	EXTRAGALACTIC NEBULA
SHAP334-4540.8	03 33 03.	- 45 50 45.	36	16.5	GALAXY
SHAP334-5440.6	03 33 03.	- 54 50 33.	24	18.0	GALAXY
SHAP334-5451.2	03 33 03.	- 55 01 09.	84	17.0	GALAXY
SHAP334-5029.0	03 33 04.	- 50 38 57.	18	17.5	GALAXY
BAK 2.0067	03 33 05.	- 18 42 07.	20	16.1	GALAXY
BAK 1.723	03 33 05.	- 23 29 08.	14	15.9	EXTRAGALACTIC NEBULA
IC 0335	03 33 05.	- 34 36 50.			NONSTELLAR OBJECT
KARA.73 15	03 33 05.	- 35 28	27		DWARF GALAXY

OBJECT NAME	RIGHT ASCEN.	DECLINATION	DIAM.	MAGN.	TYPE OF OBJECT
SHAP334-5300.0	03 33 05.	- 53 09 57.	18	18.0	GALAXY
PHL 8719	03 33 06.	- 28 44		18.4	BLUE STELLAR OBJECT
MCG-05-09-015	03 33 06.	- 32 50	48	15.	GALAXY
1ZW 013	03 33 06.	- 35 20			COMPACT GALAXY
SHAP334-4504.6	03 33 06.	- 45 14 32.	30	17.25	GALAXY
SHAP334-5241.1	03 33 06.	- 52 51 03.	18	17.75	GALAXY
SHAP334-5040.8	03 33 07.	- 50 50 45.	24	18.0	GALAXY
SHAP334-5144.0	03 33 07.	- 51 53 57.	18	17.5	GALAXY
BAK 1.721	03 33 08.	- 28 55 19.	14	16.8	EXTRAGALACTIC NEBULA
RNGC 1373	03 33 08.	- 35 21			GALAXY
SHAP334-4549.1	03 33 08.	- 45 59 02.	18	17.5	GALAXY
SHAP334-5229.1	03 33 08.	- 52 39 03.	18	17.0	GALAXY
SHAP334-5311.0	03 33 08.	- 53 20 57.	18	18.0	GALAXY
BAK 1.724	03 33 09.	- 27 03 43.	16	16.7	EXTRAGALACTIC NEBULA
SHAP334-4501.4	03 33 09.	- 45 11 20.	30	17.5	GALAXY
SHAP334-4935.2	03 33 09.	- 49 45 08.	48	16.75	GALAXY
BAK 1.722	03 33 10.	- 29 08 01.	11	16.6	EXTRAGALACTIC NEBULA
BAK 1.719	03 33 10.	- 31 27 43.	16	16.8	EXTRAGALACTIC NEBULA
SHAP334-4503.1	03 33 10.	- 45 13 02.	24	17.0	GALAXY
SHAP334-5756.4	03 33 10.	- 58 06 21.	24	16.75	GALAXY
BAK 2.0069	03 33 11.	- 18 53 19.	18	16.2	GALAXY
BAK 2.0064	03 33 11.	- 24 48 55.	20	16.9	GALAXY
PHL 8720	03 33 12.	- 20 42		18.5	BLUE STELLAR OBJECT
BAK 2.0068	03 33 12.	- 24 39 07.	22	16.9	GALAXY
BAK 2.0065	03 33 12.	- 25 13 55.	20	16.8	GALAXY
PHL 4461	03 33 12.	- 25 31		18.3	BLUE STELLAR OBJECT
TON-S 0360	03 33 12.	- 27 03		16.0	BLUE STAR
LB 01677	03 33 12.	- 55 55		13.5	FAINT BLUE STAR
BAK 2.0066	03 33 13.	- 26 26 07.	15	16.7	GALAXY
BAK 1.725	03 33 13.	- 28 35 31.	17	16.6	EXTRAGALACTIC NEBULA
SHAP334-4601.9	03 33 13.	- 46 11 50.	24	17.75	GALAXY
SHAP334-5420.4	03 33 13.	- 54 30 21.	36	17.0	GALAXY
SHAP334-5037.8	03 33 15.	- 50 47 44.	30	16.5	GALAXY
BAK 1.730	03 33 16.	- 21 22 55.	29	15.9	EXTRAGALACTIC NEBULA
BAK 1.727	03 33 16.	- 27 55 07.	16	16.8	EXTRAGALACTIC NEBULA
HN 0223	03 33 16.	- 50 48			NEBULA
IC 1968	03 33 16.	- 50 48			NONSTELLAR OBJECT
SHAP334-5343.6	03 33 16.	- 53 53 32.	18	17.5	GALAXY
SHAP334-5424.6	03 33 16.	- 54 34 32.	18	17.75	GALAXY
BAK 1.728	03 33 17.	- 24 40 55.	16	16.2	EXTRAGALACTIC NEBULA
SHAP334-5336.0	03 33 17.	- 53 45 56.	18	17.5	GALAXY
5ZW 355	03 33 18.	+ 34 03			COMPACT GALAXY
VB 192	03 33 18.	+ 60 26	275		STELLAR RING
ISS 0067	03 33 18.	+ 60 26	277		STELLAR RING
ZWG 391.013	03 33 18.	- 02 09		15.0	GALAXY
MCG+01-10-008	03 33 18.	- 03 01 30.	15	15.	GALAXY
PHL 4462	03 33 18.	- 21 00		17.5	BLUE STELLAR OBJECT
PHL 1558	03 33 18.	- 23 47		18.3	BLUE STELLAR OBJECT
BAK 1.729	03 33 18.	- 24 38 25.	13	16.3	EXTRAGALACTIC NEBULA
PHL 4463	03 33 18.	- 27 06		18.2	BLUE STELLAR OBJECT
PHL 4464	03 33 18.	- 28 18		18.5	BLUE STELLAR OBJECT
BAK 1.726	03 33 18.	- 32 47 01.	30	15.2	EXTRAGALACTIC NEBULA
MCG-06-08-029	03 33 18.	- 35 24	150		GALAXY
MCG-06-08-030	03 33 18.	- 35 26	90	11.5	GALAXY
SHAP334-5326.4	03 33 18.	- 53 36 20.	12	17.25	GALAXY
SHAP334-4727.4	03 33 19.	- 47 37 20.	18	17.0	GALAXY
BAK 2.0070	03 33 20.	- 26 53 07.	16	17.0	GALAXY
RNGC 1375	03 33 20.	- 35 26			GALAXY
MCG-04-09-030	03 33 21.	- 21 22 30.	30	15.	GALAXY
SHAP334-4550.1	03 33 21.	- 46 00 02.	18	17.5	GALAXY
SHAP334-5719.1	03 33 22.	- 57 29 02.	30	16.5	GALAXY
BC NRAO140	03 33 22.32	+ 32 08 36.1		18.5	QUASI-STELLAR OBJECT
SHB 072	03 33 22.4	+ 32 08 36.		18.5	QUASI-STELLAR OBJECT
SHAP335-4513.0	03 33 23.	- 45 22 55.	24	17.5	GALAXY
SHAP334-5707.4	03 33 23.	- 57 17 20.	36	16.5	GALAXY
ZWG 391.014	03 33 24.	+ 00 56		15.4	GALAXY
BAK 2.0071	03 33 24.	- 26 46 54.	18	16.8	GALAXY
SHAP334-5325.2	03 33 24.	- 53 35 08.	18	17.5	GALAXY
BAK 2.0072	03 33 25.	- 19 16 42.	10	16.4	GALAXY
HN 0221	03 33 25.	- 21 28			NEBULA
IC 1962	03 33 25.	- 21 28			NONSTELLAR OBJECT
SHAP335-4729.5	03 33 25.	- 47 39 25.	24	18.0	GALAXY
SHAP335-5140.0	03 33 25.	- 51 49 56.	18	17.0	GALAXY
SHAP334-5347.2	03 33 25.	- 53 57 08.	18	17.0	GALAXY
SHAP334-5749.9	03 33 25.	- 57 59 50.	30	16.0	GALAXY
RNGC 1374	03 33 26.	- 35 24		12.5	GALAXY
SHAP334-5208.2	03 33 26.	- 52 18 08.	24	17.25	GALAXY
SHAP334-5258.0	03 33 27.	- 53 07 56.	18	17.25	GALAXY
SHAP334-5423.5	03 33 27.	- 54 33 26.	18	18.0	GALAXY
SHAP334-5313.4	03 33 28.	- 53 23 20.	18	17.0	GALAXY
BAK 2.0074	03 33 29.	- 14 57 48.	22	16.2	GALAXY
MCG-04-09-031	03 33 30.	- 21 27	180	14.	GALAXY
MCG-05-09-016	03 33 30.	- 32 47	90	14.5	GALAXY
SHAP334-5431.8	03 33 30.	- 54 41 44.	24	17.25	GALAXY
LB 01678	03 33 30.	- 58 00		13.7	FAINT BLUE STAR
BAK 2.0075	03 33 31.	- 17 32 18.	14	16.9	GALAXY
BAK 1.733	03 33 31.	- 26 49 06.	18	18.9	EXTRAGALACTIC NEBULA
BAK 1.731	03 33 31.	- 32 38 30.	40	15.2	EXTRAGALACTIC NEBULA
SHAP335-4605.6	03 33 31.	- 46 15 31.	18	17.5	GALAXY
SHAP334-5441.5	03 33 31.	- 54 51 25.	18	17.5	GALAXY
BAK 1.732	03 33 32.	- 28 29 18.	22	17.5	EXTRAGALACTIC NEBULA
SHAP334-5339.0	03 33 32.	- 53 48 55.	18	17.5	GALAXY
SHAP334-5443.5	03 33 32.	- 54 53 25.	30	16.75	GALAXY
SHAP334-5816.1	03 33 32.	- 58 26 02.	30	16.25	GALAXY
BAK 1.734	03 33 33.	- 28 28 54.	19	17.3	EXTRAGALACTIC NEBULA
SHAP335-4629.8	03 33 33.	- 46 39 43.	18	16.75	GALAXY
SHAP335-5106.2	03 33 33.	- 51 16 07.	24	18.0	GALAXY
SHAP334-5440.8	03 33 34.	- 54 50 43.	24	18.0	GALAXY
SHAP334-5712.0	03 33 34.	- 57 21 56.	30	16.75	GALAXY
BAK 1.741	03 33 35.	- 22 21 54.	18	15.9	EXTRAGALACTIC NEBULA
BAK 1.738	03 33 35.	- 23 27 54.	14	16.8	EXTRAGALACTIC NEBULA
SHAP334-5240.5	03 33 35.	- 52 50 25.	18	17.0	GALAXY
SHAP334-5324.7	03 33 35.	- 53 34 37.	18	17.0	GALAXY
ZWG 525.045	03 33 36.	+ 38 32		15.7	GALAXY
ZWG 391.015	03 33 36.	- 02 07		15.0	GALAXY
BAK 2.0073	03 33 36.	- 24 04 54.	18	16.8	GALAXY
SHAP334-5332.3	03 33 36.	- 53 42 13.	18	17.5	GALAXY
SHAP334-5342.5	03 33 36.	- 53 52 25.	24	17.0	GALAXY
SHAP334-5345.6	03 33 36.	- 53 55 31.	24	17.5	GALAXY
SHAP334-5753.5	03 33 36.	- 58 03 25.	24	17.0	GALAXY
BAK 1.737	03 33 37.	- 26 49 54.	22	16.4	EXTRAGALACTIC NEBULA
SHAP335-4513.0	03 33 37.	- 45 22 55.	18	16.75	GALAXY
SHAP335-5151.5	03 33 37.	- 52 01 25.	12	17.0	GALAXY
BAK 2.0077	03 33 38.	- 17 29 23.	17	16.7	GALAXY
BAK 1.739	03 33 38.	- 25 25 42.	18	16.6	EXTRAGALACTIC NEBULA
BAK 1.736	03 33 38.	- 30 47 06.	11	16.6	EXTRAGALACTIC NEBULA
BAK 1.735	03 33 38.	- 32 14 30.	24	16.8	EXTRAGALACTIC NEBULA
BAK 2.0078	03 33 39.	- 16 11 53.	17	16.0	GALAXY
SHAP335-4513.7	03 33 39.	- 45 23 36.	24	16.5	GALAXY
SHAP335-5326.2	03 33 39.	- 53 36 07.	18	17.25	GALAXY
SHAP335-4632.8	03 33 40.	- 46 42 42.	24	18.0	GALAXY
SHAP335-5042.2	03 33 40.	- 50 52 07.	24	17.5	GALAXY
SHAP335-5144.8	03 33 40.	- 51 54 43.	12	17.5	GALAXY
SHAP335-5300.0	03 33 40.	- 53 09 55.	18	17.5	GALAXY
SHAP335-5351.0	03 33 40.	- 54 00 55.	18	16.75	GALAXY
BAK 1.742	03 33 41.	- 26 10 17.	14	17.1	EXTRAGALACTIC NEBULA
SHAP335-5322.0	03 33 41.	- 53 31 55.	18	17.0	GALAXY
SHAP335-5334.4	03 33 41.	- 53 44 19.	30	16.5	GALAXY
SHAP335-5336.6	03 33 41.	- 53 46 31.	18	17.5	GALAXY
SHAP334-5613.8	03 33 41.	- 56 23 43.	42	16.5	GALAXY
BAK 1.747	03 33 42.	- 22 18 17.	16	16.2	EXTRAGALACTIC NEBULA
MCG-05-09-017	03 33 42.	- 32 39	24	14.5	GALAXY
SHAP335-4827.6	03 33 42.	- 48 37 30.	12	17.25	GALAXY
SHAP335-5340.8	03 33 42.	- 53 50 43.	18	17.5	GALAXY
SHAP335-5422.0	03 33 42.	- 54 31 55.	30	17.75	GALAXY
SHAP334-5903.6	03 33 42.	- 59 13 31.	36	17.0	GALAXY
SHAP335-4508.6	03 33 43.	- 45 18 30.	24	17.0	GALAXY
SHAP335-5352.0	03 33 43.	- 54 01 55.	18	16.5	GALAXY
BAK 2.0025	03 33 44.	- 16 06 05.	11	16.4	GALAXY
BAK 2.0076	03 33 44.	- 25 23 53.	16	17.0	GALAXY
BAK 1.743	03 33 44.	- 26 08 41.	22	16.8	EXTRAGALACTIC NEBULA
BAK 1.740	03 33 44.	- 30 13 05.	41	16.2	EXTRAGALACTIC NEBULA
SHAP335-4437.8	03 33 44.	- 44 47 42.	30	17.5	GALAXY
SHAP335-5259.2	03 33 44.	- 53 09 07.	30	16.75	GALAXY
BAK 1.744	03 33 45.	- 26 10 17.	18	16.8	EXTRAGALACTIC NEBULA
BAK 1.745	03 33 45.	- 26 22 29.	17	16.8	EXTRAGALACTIC NEBULA
SHAP335-5250.4	03 33 45.	- 53 00 19.	18	17.5	GALAXY
BAK 1.746	03 33 46.	- 26 09 41.	12	16.9	EXTRAGALACTIC NEBULA
SHAP335-5343.2	03 33 46.	- 53 53 07.	18	17.0	GALAXY
BAK 2.0028	03 33 47.	- 16 20 11.	15	16.8	GALAXY
PHL 4465	03 33 48.	- 22 00		15.6	BLUE STELLAR OBJECT
SHAP335-4506.4	03 33 48.	- 45 16 18.	30	17.0	GALAXY
SHAP335-4603.1	03 33 48.	- 46 13 00.	18	17.0	GALAXY
SHAP335-5237.9	03 33 48.	- 52 47 48.	18	17.5	GALAXY
SHAP335-5335.5	03 33 48.	- 53 45 24.	18	17.25	GALAXY
SHAP335-5346.8	03 33 48.	- 53 56 42.	18	17.5	GALAXY
SHAP335-5511.6	03 33 48.	- 55 21 31.	60	16.0	GALAXY
BAK 2.0032	03 33 49.	- 16 42 41.	16	16.8	GALAXY
SHAP335-5232.8	03 33 49.	- 52 42 42.	24	17.5	GALAXY
SHAP335-5325.2	03 33 49.	- 53 35 06.	12	17.25	GALAXY
SHAP335-5444.5	03 33 49.	- 54 54 24.	24	17.0	GALAXY
SHAP335-5505.4	03 33 49.	- 55 15 19.	54	16.0	GALAXY
SHAP334-5752.7	03 33 49.	- 58 02 37.	24	17.0	GALAXY
BAK 2.0033	03 33 50.	- 16 42 59.	13	16.6	GALAXY
BAK 2.0079	03 33 50.	- 26 02 53.	18	17.0	GALAXY
SHAP335-4509.8	03 33 50.	- 45 19 42.	30	16.0	GALAXY
SHAP335-5148.3	03 33 50.	- 51 58 12.	30	15.5	GALAXY
SHAP334-5804.3	03 33 50.	- 58 14 13.	36	17.25	GALAXY
BAK 2.0030	03 33 51.	- 19 23 05.	12	16.0	GALAXY
BAK 1.748	03 33 51.	- 26 26 05.	24	17.1	EXTRAGALACTIC NEBULA
SHAP335-4535.8	03 33 51.	- 45 45 42.	36	17.5	GALAXY
SHAP335-4847.8	03 33 51.	- 48 57 42.	24	17.5	GALAXY
SHAP335-5130.2	03 33 52.	- 51 40 06.	24	17.0	GALAXY
SHAP335-5150.3	03 33 52.	- 52 00 12.	18	15.5	GALAXY
SHAP335-5422.7	03 33 52.	- 54 32 36.	24	17.5	GALAXY
SHAP335-5715.3	03 33 52.	- 57 25 12.	36	17.0	GALAXY
SHAP335-4508.1	03 33 53.	- 45 18 00.	30	17.25	GALAXY
SHAP335-5031.7	03 33 53.	- 50 41 36.	18	17.5	GALAXY
SHAP335-5325.8	03 33 53.	- 53 35 42.	24	17.75	GALAXY
SHAP335-5333.1	03 33 53.	- 53 43 00.	18	16.75	GALAXY
ZWG 465.001	03 33 54.	+ 20 58		15.5	GALAXY
MCG+03-10-001	03 33 54.	+ 20 58	42	15.	GALAXY
KARA.73B 0131	03 33 54.	+ 20 58	36	15.5	ISOLATED GALAXY S
MRK 613	03 33 54.	- 04 52	11	16.	GALAXY WITH UV CONTINUUM
BAK 2.0034	03 33 54.	- 19 52 17.	14	15.8	GALAXY
MCG-06-08-031	03 33 54.	- 34 37	120	13.	GALAXY
LB 01679	03 33 54.	- 46 31		15.3	FAINT BLUE STAR
SHAP335-5510.6	03 33 54.	- 55 20 30.	48	16.5	GALAXY
LB 03324	03 33 54.	- 64 12		13.5	FAINT BLUE STAR
SHAP335-4548.9	03 33 55.	- 45 58 48.	18	17.0	GALAXY
SHAP335-5051.2	03 33 55.	- 51 01 06.	24	16.75	GALAXY
SHAP335-5445.8	03 33 55.	- 54 55 42.	24	18.0	GALAXY
BAK 2.0027	03 33 56.	- 25 38 47.	36	16.6	GALAXY
BAK 2.0026	03 33 56.	- 26 08 11.	18	16.7	GALAXY
BAK 1.752	03 33 56.	- 26 09 11.	17	16.3	EXTRAGALACTIC NEBULA
BAK 1.751	03 33 56.	- 27 50 11.	13	16.6	EXTRAGALACTIC NEBULA
BAK 1.750	03 33 56.	- 29 09 29.	19	16.6	EXTRAGALACTIC NEBULA
RNGC 1378	03 33 56.	- 35 22			NON-EXISTENT OBJECT
SHAP335-5013.2	03 33 56.	- 50 23 06.	18	17.0	GALAXY
SHAP335-5258.9	03 33 56.	- 53 08 48.	18	17.25	GALAXY
SHAP335-5331.8	03 33 56.	- 53 41 42.	18	17.0	GALAXY
BAK 2.0037	03 33 57.	- 19 26 16.	11	16.4	GALAXY
BAK 2.0029	03 33 57.	- 25 35 34.	22	16.6	GALAXY
BAK 1.749	03 33 57.	- 31 15 05.	17	16.2	EXTRAGALACTIC NEBULA
ABC 0445	03 33 58.	- 03 03		17.5	RICH CLUSTER OF GALAXIES
BAK 1.754	03 33 58.	- 25 42 16.	17	16.8	EXTRAGALACTIC NEBULA
SHAP335-5521.2	03 33 58.	- 55 31 06.	42	16.75	GALAXY
BAK 2.0041	03 33 59.	- 18 13 52.	17	16.3	GALAXY
BAK 2.0041	03 33 59.	- 18 13 52.	17	16.3	GALAXY
BAK 1.755	03 33 59.	- 26 46 22.	14	16.6	EXTRAGALACTIC NEBULA
SHAP335-5248.5	03 33 59.	- 52 58 24.	12	17.5	GALAXY
SHAP335-5325.3	03 33 59.	- 53 35 12.	36	17.5	GALAXY
SHAP335-5353.1	03 33 59.	- 54 03 00.	18	17.0	GALAXY
UGC 02796	03 34 00.	+ 13 15	96	16.5	GALAXY Sb-c
MCG+01-10-009	03 34 00.	- 06 54	72	15.5	GALAXY
PHL 8721	03 34 00.	- 20 54		18.5	BLUE STELLAR OBJECT
PHL 8722	03 34 00.	- 26 23		18.7	BLUE STELLAR OBJECT
BAK 2.0031	03 34 00.	- 26 57 40.	18	16.8	GALAXY
KARA.73 16	03 34 00.	- 36 10	27		DWARF GALAXY
SHAP335-5352.4	03 34 00.	- 54 02 18.	18	17.5	GALAXY
SHAP335-5403.7	03 34 00.	- 54 13 36.	24	17.5	GALAXY
SHAP335-5413.0	03 34 00.	- 54 22 54.	24	17.5	GALAXY
SHAP335-5655.0	03 34 00.	- 57 04 54.	42	16.5	GALAXY
BAK 2.0036	03 34 01.	- 24 03 46.	22	16.7	GALAXY
SHAP335-4825.2	03 34 01.	- 48 35 05.	18	17.25	GALAXY
SHAP335-5344.0	03 34 01.	- 53 53 54.	12	17.0	GALAXY
SHAP335-5413.3	03 34 01.	- 54 23 12.	18	17.5	GALAXY
SHAP335-5724.4	03 34 01.	- 57 34 18.	30	16.75	GALAXY
BAK 2.0035	03 34 02.	- 26 58 28.	17	17.0	GALAXY
BAK 2.0039	03 34 03.	- 24 03 58.	17	16.8	GALAXY
BAK 1.753	03 34 03.	- 30 32 04.	18	16.2	EXTRAGALACTIC NEBULA
SHAP335-4922.2	03 34 03.	- 49 32 05.	36	17.0	GALAXY
SHAP335-5307.0	03 34 03.	- 53 16 53.	18	16.75	GALAXY
SHAP335-5413.4	03 34 03.	- 54 23 18.	18	17.5	GALAXY
BAK 2.0040	03 34 04.	- 24 09 04.	19	16.7	GALAXY
BAK 1.756	03 34 04.	- 29 47 40.	14	16.6	EXTRAGALACTIC NEBULA

OBJECT NAME	RIGHT ASCEN.	DECLINATION	DIAM.	MAGN.	TYPE OF OBJECT
BAK 2.0038	03 34 05.	- 26 38 52.	21	17.0	GALAXY
BAK 1.757	03 34 05.	- 26 51 40.	16	16.8	EXTRAGALACTIC NEBULA
BAK 1.758	03 34 05.	- 27 03 10.	18	15.9	EXTRAGALACTIC NEBULA
SHAP335-5348.4	03 34 05.	- 53 58 17.	24	17.75	GALAXY
SHAP335-5445.6	03 34 05.	- 54 55 30.	36	17.0	GALAXY
SHAP335-5844.6	03 34 05.	- 58 54 30.	30	17.0	GALAXY
5ZW 356	03 34 06.	+ 47 15			COMPACT GALAXY
MCG-06-09-001	03 34 06.	- 35 37	240	12.3	GALAXY
SHAP335-5050.2	03 34 06.	- 51 00 05.	24	18.0	GALAXY
BAK 2.0043	03 34 07.	- 23 35 58.	20	16.8	GALAXY
RNGC 1379	03 34 08.	- 35 37		12.5	GALAXY
SHAP335-5222.5	03 34 08.	- 52 32 23.	18	18.0	GALAXY
SHAP335-5252.3	03 34 08.	- 53 02 11.	18	16.75	GALAXY
HELW 126	03 34 09.	- 25 46 07.			NEBULA
BAK 2.0042	03 34 09.	- 26 37 04.	21	17.1	GALAXY
SHAP335-4535.1	03 34 09.	- 45 44 59.	24	17.5	GALAXY
SHAP335-5814.0	03 34 10.	- 58 23 54.	42	16.0	GALAXY
SHAP335-5015.1	03 34 11.	- 50 24 59.	24	16.5	GALAXY
SHAP335-5335.4	03 34 11.	- 53 45 17.	12	17.5	GALAXY
SHAP335-5405.4	03 34 11.	- 54 15 17.	18	17.0	GALAXY
ZWG 541.021	03 34 12.	+ 39 33		15.7	GALAXY
VB 193	03 34 12.	+ 60 37	154		STELLAR RING
ISS 0068	03 34 12.	+ 60 37	158		STELLAR RING
MCG+13-03-006	03 34 12.	+ 79 21	36	16.	GALAXY
MCG-04-09-032	03 34 12.	- 25 46	84	14.	GALAXY
BAK 2.0044	03 34 12.	- 26 48 04.	17	17.2	GALAXY
SHAP335-5342.0	03 34 12.	- 53 51 53.	18	17.0	GALAXY
SHAP335-5342.6	03 34 12.	- 53 52 29.	18	17.0	GALAXY
SHAP335-5343.8	03 34 12.	- 53 53 41.	18	16.75	GALAXY
SHAP335-4906.6	03 34 13.	- 49 16 29.	30	16.75	GALAXY
SHAP335-5008.1	03 34 13.	- 50 17 59.	12	16.75	GALAXY
SHAP335-5047.5	03 34 13.	- 50 57 23.	30	17.0	GALAXY
SHAP335-5841.4	03 34 13.	- 58 51 17.	30	16.5	GALAXY
BAK 1.761	03 34 14.	- 20 37 03.	22	16.4	EXTRAGALACTIC NEBULA
SHAP335-4828.2	03 34 14.	- 48 38 05.	12	16.25	GALAXY
MCG+00-10-006	03 34 15.	+ 00 10	30	15.	GALAXY
SHAP335-4510.4	03 34 15.	- 45 20 16.	24	16.75	GALAXY
BAK 1.760	03 34 16.	- 25 49 21.	16	16.2	EXTRAGALACTIC NEBULA
BAK 2.0045	03 34 16.	- 26 05 51.	19	16.9	GALAXY
SHAP335-5020.4	03 34 16.	- 50 30 17.	12	17.25	GALAXY
BAK 2.0049	03 34 17.	- 16 04 09.	17	15.6	GALAXY
ZWG 391.016	03 34 18.	+ 00 52		15.4	GALAXY
BAK 2.0048	03 34 18.	- 19 26 03.	13	16.4	GALAXY
SVEN 224	03 34 18.	- 36 09	48	12.5	GALAXY
SHAP335-5017.0	03 34 18.	- 50 26 52.	42	17.0	GALAXY
SHAP335-5631.5	03 34 18.	- 56 41 23.	36	16.0	GALAXY
BAK 2.0046	03 34 19.	- 25 12 09.	14	17.2	GALAXY
BAK 1.762	03 34 19.	- 25 49 09.	16	16.6	EXTRAGALACTIC NEBULA
BAK 2.0047	03 34 20.	- 22 15 33.	18	16.5	GALAXY
SHAP335-4545.2	03 34 20.	- 45 55 04.	24	17.0	GALAXY
SHAP335-4742.0	03 34 20.	- 47 51 52.	30	17.0	GALAXY
SHAP335-5339.8	03 34 20.	- 53 49 41.	24	16.75	GALAXY
SHAP335-5344.9	03 34 20.	- 53 54 47.	18	17.0	GALAXY
MCG+01-10-010	03 34 21.	- 07 58	36	15.	GALAXY
BAK 2.0051	03 34 22.	- 18 31 09.	14	16.4	GALAXY
BAK 1.759	03 34 22.	- 31 54 39.	17	16.2	EXTRAGALACTIC NEBULA
BAK 1.764	03 34 23.	- 22 05 03.	23	16.2	EXTRAGALACTIC NEBULA
SHAP335-4810.6	03 34 23.	- 48 20 28.	12	17.25	GALAXY
SHAP335-5404.1	03 34 23.	- 54 13 58.	18	17.0	GALAXY
SHAP335-5712.1	03 34 23.	- 57 21 59.	24	17.0	GALAXY
SHAP335-5740.2	03 34 23.	- 57 50 05.	24	17.0	GALAXY
DG 019	03 34 24.	+ 37 54	780		REFLECTION NEBULA
PHL 4466	03 34 24.	- 23 00		18.5	BLUE STELLAR OBJECT
SHAP335-4951.0	03 34 24.	- 50 00 52.	18	17.5	GALAXY
SHAP335-5339.9	03 34 24.	- 53 49 46.	18	17.25	GALAXY
SHAP335-4806.9	03 34 25.	- 48 16 46.	12	16.75	GALAXY
SHAP335-5053.9	03 34 25.	- 51 03 46.	24	16.0	GALAXY
SHAP336-4701.4	03 34 26.	- 47 11 16.	12	17.5	GALAXY
SHAP335-5433.4	03 34 26.	- 54 43 16.	36	17.5	GALAXY
SHAP335-5017.8	03 34 27.	- 50 27 40.	36	17.25	GALAXY
SHAP335-5138.6	03 34 27.	- 51 48 28.	30	18.0	GALAXY
SHAP335-5143.8	03 34 27.	- 51 53 40.	18	17.75	GALAXY
SHAP335-5457.0	03 34 27.	- 55 06 52.	72	17.25	GALAXY
SHAP335-5806.4	03 34 27.	- 58 16 17.	24	16.75	GALAXY
SHAP335-5322.8	03 34 28.	- 53 32 40.	18	17.75	GALAXY
SHAP335-5856.1	03 34 28.	- 59 05 59.	30	17.25	GALAXY
BAK 2.0050	03 34 29.	- 25 21 57.	24	17.2	GALAXY
RNGC 1377	03 34 30.	- 21 04		14.0	GALAXY
MCG-04-09-033	03 34 30.	- 21 04	48	14.	GALAXY
BAK 1.763	03 34 30.	- 28 50 27.	19	16.6	EXTRAGALACTIC NEBULA
TON-S 0361	03 34 30.	- 32 09		15.2	BLUE STAR
MCG-06-09-002	03 34 30.	- 35 09	300		GALAXY
SHAP336-4743.0	03 34 31.	- 47 52 52.	30	16.5	GALAXY
SHAP335-5325.2	03 34 31.	- 53 35 04.	12	17.75	GALAXY
SHAP335-5403.4	03 34 31.	- 54 13 16.	24	17.75	GALAXY
BAK 1.765	03 34 32.	- 28 51 09.	14	16.8	EXTRAGALACTIC NEBULA
SHAP335-5212.4	03 34 32.	- 52 22 16.	12	17.5	GALAXY
SHAP336-4510.8	03 34 33.	- 45 20 39.	102	16.5	GALAXY
HN 0224	03 34 33.	- 45 21			NEBULA
IC 1969	03 34 33.	- 45 21			NONSTELLAR OBJECT
SHAP335-5413.6	03 34 33.	- 54 23 28.	30	17.5	GALAXY
SHAP335-5844.7	03 34 33.	- 58 54 34.	30	17.0	GALAXY
SHAP336-4548.1	03 34 34.	- 45 57 57.	24	17.25	GALAXY
SHAP335-5211.6	03 34 34.	- 52 21 28.	18	17.5	GALAXY
SHAP335-5238.0	03 34 34.	- 52 47 52.	36	16.0	GALAXY
SHAP335-5427.8	03 34 34.	- 54 37 40.	30	17.25	GALAXY
BAK 2.0081	03 34 35.	- 16 28 14.	14	16.5	GALAXY
BAK 1.766	03 34 35.	- 28 27 26.	18	16.9	EXTRAGALACTIC NEBULA
SHAP335-5208.6	03 34 35.	- 52 18 28.	18	17.5	GALAXY
HN 0225	03 34 35.	- 52 48			NEBULA
IC 1971	03 34 35.	- 52 48			NONSTELLAR OBJECT
SHAP335-5440.7	03 34 35.	- 54 50 34.	42	17.25	GALAXY
SHAP335-5442.4	03 34 35.	- 54 52 16.	24	18.0	GALAXY
SHAP335-5513.1	03 34 35.	- 55 22 58.	36	16.75	GALAXY
ZWG 391.017	03 34 36.	+ 01 42		15.1	GALAXY
UGC 02797	03 34 36.	+ 23 07	72	16.0	GALAXY
MCG+01-10-011	03 34 36.	- 05 13 30.	102	13.	GALAXY
BAK 2.0080	03 34 36.	- 18 29 44.	14	16.2	GALAXY
PHL 8724	03 34 36.	- 23 56		17.5	BLUE STELLAR OBJECT
PHL 8725	03 34 36.	- 24 16		18.6	BLUE STELLAR OBJECT
PHL 8723	03 34 36.	- 30 44		17.4	BLUE STELLAR OBJECT
MCG-06-09-003	03 34 36.	- 35 28	132	12.6	GALAXY
IC 0337	03 34 37.	- 06 53 09.			NONSTELLAR OBJECT
SHAP336-4917.0	03 34 37.	- 49 26 51.	24	17.0	GALAXY
SHAP336-4921.4	03 34 37.	- 49 31 15.	24	17.5	GALAXY
RNGC 1380	03 34 38.	- 35 09		11.5	GALAXY
RNGC 1381	03 34 38.	- 35 28		13.0	GALAXY
SHAP336-4502.4	03 34 38.	- 45 12 15.	30	16.5	GALAXY

OBJECT NAME	RIGHT ASCEN.	DECLINATION	DIAM.	MAGN.	TYPE OF OBJECT
MCG+04-09-002	03 34 39.	+ 23 07	30	15.5	GALAXY
RNGC 1372	03 34 39.	- 16 03			GALAXY
SHAP336-5201.5	03 34 39.	- 52 11 21.	24	16.5	GALAXY
SHAP336-5229.4	03 34 39.	- 52 39 15.	18	17.0	GALAXY
RNGC 1376	03 34 40.	- 05 12		13.0	GALAXY
BAK 1.768	03 34 40.	- 28 08 02.	14	16.8	EXTRAGALACTIC NEBULA
BAK 1.769	03 34 40.	- 28 26 02.	18	16.3	EXTRAGALACTIC NEBULA
SHAP336-5345.1	03 34 40.	- 53 54 57.	24	16.25	GALAXY
SHAP336-5347.6	03 34 40.	- 53 57 27.	18	17.25	GALAXY
ABC 0446	03 34 41.	- 02 36		17.5	RICH CLUSTER OF GALAXIES
BAK 1.770	03 34 41.	- 26 35 14.	20	16.2	EXTRAGALACTIC NEBULA
SHAP336-5234.0	03 34 41.	- 52 43 51.	18	16.75	GALAXY
BAK 2.0082	03 34 42.	- 19 13 14.	15	16.4	GALAXY
SHAP336-5338.1	03 34 42.	- 53 47 57.	24	17.5	GALAXY
SHAP336-5340.9	03 34 42.	- 53 50 45.	42	16.75	GALAXY
SHAP336-5435.5	03 34 42.	- 54 45 21.	24	18.0	GALAXY
LB 03325	03 34 42.	- 64 38		13.6	FAINT BLUE STAR
BAK 2.0083	03 34 43.	- 19 41 02.	15	16.4	GALAXY
BAK 1.767	03 34 43.	- 30 54 14.	34	15.6	EXTRAGALACTIC NEBULA
SVEN 225	03 34 44.	- 25 04 33.	30	14.4	GALAXY
BAK 1.771	03 34 45.	- 25 55 02.	17	16.7	EXTRAGALACTIC NEBULA
BAK 1.772	03 34 46.	- 25 04 26.	34	15.3	EXTRAGALACTIC NEBULA
HELW 405	03 34 46.	- 25 04 39.			NEBULA
SHAP336-5215.0	03 34 46.	- 52 24 51.	24	18.0	GALAXY
SHAP336-5238.6	03 34 46.	- 52 48 27.	24	17.5	GALAXY
BAK 2.0087	03 34 47.	- 17 39 19.	13	16.6	GALAXY
BAK 2.0084	03 34 47.	- 23 54 13.	19	16.7	GALAXY
BAK 1.773	03 34 47.	- 26 02 02.	12	16.3	EXTRAGALACTIC NEBULA
MCG+01-10-002	03 34 48.	+ 05 12	36	16.	GALAXY
MCG+07-08-026	03 34 48.	+ 40 48 30.	96	14.5	GALAXY
ZWG 541.022	03 34 48.	+ 40 50		14.8	GALAXY
UGC 02798	03 34 48.	+ 40 50	138	14.8	GALAXY Sb/SBc
OCL 0370	03 34 48.	+ 66 22	720	14.	OPEN STAR CLUSTER
ZC 0334.8-0237	03 34 48.	- 02 37	1010		CLUSTER OF GALAXIES
MCG-04-09-034	03 34 48.	- 25 04	48	15.	GALAXY
MCG-05-09-018	03 34 48.	- 30 54	72	15.	GALAXY
HOD.60 04	03 34 48.	- 34 05	36	14.8	E0 GALAXY IN FORNAX CLSTR
HOD.60 03	03 34 48.	- 34 27	54	17.5	E0 GALAXY IN FORNAX CLSTR
DV.56 N1380A	03 34 48.	- 34 53			S GALAXY
IC 1970	03 34 48.	- 44 07	180	13.4	GALAXY S
BAK 2.0085	03 34 49.	- 25 06 13.	17	17.0	GALAXY
BAK 1.774	03 34 49.	- 28 06 02.	24	16.2	EXTRAGALACTIC NEBULA
SHAP336-4813.9	03 34 49.	- 48 23 44.	18	17.5	GALAXY
BAK 2.0088	03 34 50.	- 17 40 37.	14	17.0	GALAXY
RNGC 1380A	03 34 50.	- 34 53			GALAXY
HELW 051	03 34 50.	- 34 54			NEBULA
SHAP335-5908.5	03 34 50.	- 59 18 21.	36	17.0	GALAXY
BAK 1.775	03 34 51.	- 24 08 49.	14	16.2	EXTRAGALACTIC NEBULA
MCG-06-09-004	03 34 51.	- 36 25	78	13.	GALAXY
SHAP336-5004.2	03 34 51.	- 50 14 02.	18	16.75	GALAXY
SHAP336-5236.0	03 34 51.	- 52 45 51.	24	17.25	GALAXY
SHAP336-5359.0	03 34 51.	- 54 08 51.	18	17.5	GALAXY
SHAP336-5407.1	03 34 51.	- 54 16 57.	18	16.75	GALAXY
SHAP336-5411.6	03 34 51.	- 54 21 27.	24	18.0	GALAXY
SHAP336-5434.2	03 34 52.	- 54 44 03.	24	17.25	GALAXY
BAK 2.0086	03 34 53.	- 26 16 37.	19	17.2	GALAXY
SHAP336-5234.6	03 34 53.	- 52 44 27.	18	17.25	GALAXY
SHAP336-5748.2	03 34 53.	- 57 58 03.	30	16.25	GALAXY
MCG+00-10-007	03 34 54.	+ 02 57	48	14.5	GALAXY
UGC 02799	03 34 54.	+ 39 45	60	16.5	GALAXY S
UGC 02800	03 34 54.	+ 71 14	180	16.5	GALAXY DWARF
MCG+12-04-002	03 34 54.	+ 71 15	114	17.	GALAXY
KARA.68 040	03 34 54.	- 20 43	34		DWARF GALAXY
PHL 8726	03 34 54.	- 23 09		15.9	BLUE STELLAR OBJECT
MCG-04-09-035	03 34 54.	- 25 24	24	15.	GALAXY
BAK 1.776	03 34 54.	- 26 36 55.	14	16.8	EXTRAGALACTIC NEBULA
TON-S 0362	03 34 54.	- 31 17		14.7	BLUE STAR
MCG-06-09-006	03 34 54.	- 34 54	138	12.5	GALAXY
MCG-06-09-005	03 34 54.	- 36 10	210	11.8	GALAXY
HELW 012	03 34 54.	- 36 25			NEBULA
SN 1969A	03 34 54.	- 36 25		17.	SUPERNOVA
LB 01680	03 34 54.	- 45 59		14.2	FAINT BLUE STAR
SHAP336-5205.9	03 34 54.	- 52 15 44.	12	17.5	GALAXY
SHAP336-5337.8	03 34 54.	- 53 47 39.	18	17.5	GALAXY
SHAP336-5753.6	03 34 54.	- 58 03 27.	24	16.5	GALAXY
SHAP336-4724.0	03 34 55.	- 47 33 50.	24	17.0	GALAXY
SHAP336-5443.5	03 34 55.	- 54 53 21.	24	16.5	GALAXY
BAK 2.0092	03 34 56.	- 17 03 07.	12	16.4	GALAXY
BAK 1.778	03 34 56.	- 25 24 43.	36	15.6	EXTRAGALACTIC NEBULA
BAK 2.0095	03 34 57.	- 16 34 13.	18	16.3	GALAXY
RNGC 1386	03 34 57.	- 36 10		12.5	GALAXY
RNGC 1392	03 34 57.	- 36 26			GALAXY
SHAP336-4558.0	03 34 57.	- 46 07 50.	30	17.0	GALAXY
SHAP336-5226.5	03 34 57.	- 52 36 20.	24	17.0	GALAXY
SHAP336-5254.0	03 34 57.	- 53 03 50.	24	17.25	GALAXY
SHAP336-5444.0	03 34 57.	- 54 53 50.	30	16.75	GALAXY
SHAP336-5725.9	03 34 57.	- 57 35 45.	24	16.75	GALAXY
SHAP336-5726.6	03 34 57.	- 57 36 27.	24	17.0	GALAXY
BAK 1.777	03 34 58.	- 28 17 07.	12	17.1	EXTRAGALACTIC NEBULA
SHAP336-5735.4	03 34 58.	- 57 45 15.	24	16.5	GALAXY
SHAP336-4535.6	03 34 59.	- 45 45 26.	24	17.5	GALAXY
CED 019A	03 35	+ 22	7200		DIFFUSE GALACTIC NEBULA
LB 09805	03 35	- 80 57		13.4	FAINT BLUE STAR
ZWG 391.018	03 35 00.	+ 02 58		15.0	GALAXY
ZWG 391.019	03 35 00.	+ 03 06		15.7	GALAXY
UGC 02801	03 35 00.	+ 41 07	90	16.0	GALAXY Sc
MCG+14-02-013	03 35 00.	+ 82 09	27	17.	GALAXY
BAK 2.0089	03 35 00.	- 25 26 49.	17	16.8	GALAXY
HPW 19	03 35 00.	- 35 07			DWARF GALAXY IN FORNAX
MCG-06-09-008	03 35 00.	- 35 32	36	15.	GALAXY
MCG-06-09-007	03 35 00.	- 35 40	240	12.1	GALAXY
HOD.60 13	03 35 00.	- 35 46	36	17.0	IRRG GLXY IN FORNAX CLSTR
SHAP336-5158.0	03 35 00.	- 52 07 50.	42	16.5	GALAXY
HN 0226	03 35 00.	- 52 08			NEBULA
SHAP336-5159.0	03 35 00.	- 52 08 50.	24	16.75	GALAXY
HN 0027	03 35 00.	- 52 09			NEBULA
BC PKS0335-364	03 35 00.0	- 36 25 55.		18.	QUASI-STELLAR OBJECT
BAK 1.779	03 35 01.	- 25 26 25.	14	16.2	EXTRAGALACTIC NEBULA
HELW 052	03 35 01.	- 35 32			NEBULA
IC 1972	03 35 01.	- 52 08			NONSTELLAR OBJECT
IC 1973	03 35 01.	- 52 09			NONSTELLAR OBJECT
SHAP336-5402.6	03 35 01.	- 54 12 26.	24	17.5	GALAXY
IC 0338	03 35 02.	+ 02 58 25.			NONSTELLAR OBJECT
BAK 2.0096	03 35 02.	- 17 39 24.	13	16.6	GALAXY
BAK 2.0090	03 35 02.	- 25 26 43.	20	16.8	GALAXY
RNGC 1382	03 35 02.	- 35 21			GALAXY
RNGC 1380B	03 35 02.	- 35 21			NON-EXISTENT OBJECT
SHAP336-4531.8	03 35 02.	- 45 41 38.	18	17.5	GALAXY

OBJECT NAME	RIGHT ASCEN.	DECLINATION	DIAM.	MAGN.	TYPE OF OBJECT
SHAP336-5238.6	03 35 02.	- 52 48 26.	18	17.0	GALAXY
SHAP336-5324.8	03 35 02.	- 53 34 38.	18	17.25	GALAXY
SHAP336-5432.0	03 35 02.	- 54 41 50.	24	17.0	GALAXY
RNGC 1387	03 35 03.	- 35 41		12.5	GALAXY
SHAP336-5325.7	03 35 03.	- 53 35 32.	12	17.5	GALAXY
BAK 2.0094	03 35 04.	- 23 45 36.	29	16.8	GALAXY
SHAP336-4814.5	03 35 04.	- 48 24 20.	18	16.75	GALAXY
SHAP336-5028.6	03 35 04.	- 50 38 26.	18	17.0	GALAXY
SHAP336-5344.5	03 35 04.	- 53 54 20.	24	17.75	GALAXY
SHAP336-5459.9	03 35 04.	- 55 09 44.	42	17.25	GALAXY
BAK 2.0091	03 35 05.	- 26 37 55.	18	16.5	GALAXY
SHAP336-4814.4	03 35 05.	- 48 24 14.	12	16.75	GALAXY
SHAP336-5228.6	03 35 05.	- 52 38 26.	24	17.0	GALAXY
SHAP336-5346.2	03 35 05.	- 53 56 02.	24	17.0	GALAXY
SHAP336-5402.0	03 35 05.	- 54 11 50.	24	17.0	GALAXY
MCG+00-10-008	03 35 06.	+ 03 07	48	14.5	GALAXY
ZWG 417.002	03 35 06.	+ 04 48		15.7	GALAXY
KARA.72 092A	03 35 06.	+ 04 48	36	15.7	PART OF DOUBLE GALAXY
ZC 0335.1+0956	03 35 06.	+ 09 56	5240		CLUSTER OF GALAXIES
SCHO 0059	03 35 06.	+ 39 21 12.	400		ISOLATED DARK CLOUD
BAK 2.0093	03 35 06.	- 26 17 36.	17	16.8	GALAXY
BAK 1.780	03 35 06.	- 29 16 55.	19	16.6	EXTRAGALACTIC NEBULA
BAK 1.781	03 35 06.	- 29 19 49.	14	16.7	EXTRAGALACTIC NEBULA
DV.56 N1380B	03 35 06.	- 35 21			GALAXY
MCG-06-09-009	03 35 06.	- 35 21	90	13.	GALAXY
SHAP336-5030.2	03 35 06.	- 50 40 02.	18	17.5	GALAXY
BAK 2.0100	03 35 07.	- 15 12 42.	10	15.9	GALAXY
SHAP336-5411.7	03 35 08.	- 54 21 32.	24	17.5	GALAXY
SHAP336-5818.9	03 35 08.	- 58 28 44.	30	17.75	GALAXY
SHAP336-5344.8	03 35 09.	- 53 54 38.	18	18.0	GALAXY
BAK 2.0101	03 35 10.	- 17 54 12.	12	16.7	GALAXY
BAK 2.0102	03 35 10.	- 18 17 48.	22	16.5	GALAXY
IC 1967	03 35 11.	+ 03 06 25.			NONSTELLAR OBJECT
SHAP336-4934.0	03 35 11.	- 49 43 49.	54	15.75	GALAXY
SHAP336-5452.4	03 35 11.	- 55 02 14.	30	17.25	GALAXY
ZWG 391.020	03 35 12.	+ 03 06		15.0	GALAXY
MCG+01-10-003	03 35 12.	+ 04 48	30	16.	GALAXY
PHL 4467	03 35 12.	- 23 10		17.5	BLUE STELLAR OBJECT
BAK 1.783	03 35 12.	- 30 10 24.	16	16.5	EXTRAGALACTIC NEBULA
BAK 1.782	03 35 12.	- 32 09 24.	32	15.8	EXTRAGALACTIC NEBULA
HPW 20	03 35 12.	- 35 46			DWARF GALAXY IN FORNAX
HN 0228	03 35 12.	- 49 44			NEBULA
IC 1974	03 35 12.	- 49 44			NONSTELLAR OBJECT
SHAP336-5345.2	03 35 12.	- 53 55 01.	18	18.0	GALAXY
SHAP336-5413.6	03 35 12.	- 54 23 26.	30	17.0	GALAXY
SHAP336-5442.8	03 35 12.	- 54 52 38.	24	17.25	GALAXY
SHAP336-5846.8	03 35 12.	- 58 56 38.	24	17.75	GALAXY
HELW 053	03 35 13.	- 35 21			NEBULA
SHAP336-5356.0	03 35 13.	- 54 05 49.	24	17.75	GALAXY
RNGC 1385	03 35 14.	- 24 40		12.0	GALAXY
BAK 2.0097	03 35 14.	- 26 20 48.	14	17.1	GALAXY
SHAP336-5518.5	03 35 14.	- 55 28 20.	60	17.0	GALAXY
BAK 2.0107	03 35 15.	- 18 16 24.	22	16.4	GALAXY
BAK 2.0099	03 35 15.	- 24 24 12.	16	16.8	GALAXY
RNGC 1389	03 35 15.	- 35 55		13.0	GALAXY
SHAP336-4912.9	03 35 15.	- 49 22 43.	24	16.5	GALAXY
SHAP336-5430.5	03 35 15.	- 54 40 19.	18	17.25	GALAXY
BAK 2.0098	03 35 16.	- 25 55 18.	17	16.7	GALAXY
SHAP336-5246.2	03 35 16.	- 52 56 01.	30	17.75	GALAXY
SHAP336-5557.6	03 35 16.	- 56 07 25.	36	16.5	GALAXY
LB 01465	03 35 17.	+ 23 09 18.		17.6	FAINT BLUE STAR
BAK 2.0104	03 35 17.	- 22 51 48.	16	16.9	GALAXY
BAK 2.0105	03 35 17.	- 22 55 36.	24	16.7	GALAXY
BAK 2.0103	03 35 17.	- 24 05 30.	20	16.9	GALAXY
ZWG 417.003	03 35 18.	+ 04 47		15.6	GALAXY
UGC 02802	03 35 18.	+ 04 47	84	15.6	GALAXY S
KARA.72 092B	03 35 18.	+ 04 47	48	15.6	PART OF DOUBLE GALAXY
MCG+01-10-004	03 35 18.	+ 04 48	60	16.	GALAXY
UGC 02803	03 35 18.	+ 12 04	60	17.	GALAXY SBb
IC 0336	03 35 18.	+ 23 18			NONSTELLAR OBJECT
ISS 0142	03 35 18.	+ 45 15	247		STELLAR RING
ZWG 346.005	03 35 18.	+ 76 51		15.7	GALAXY
MCG-03-10-014	03 35 18.	- 15 52	60	16.	GALAXY
MCG-06-09-010	03 35 18.	- 35 54	150	12.8	GALAXY
BAK 1.784	03 35 19.	- 25 57 24.	19	16.9	EXTRAGALACTIC NEBULA
SHAP336-4615.6	03 35 19.	- 46 25 25.	24	17.5	GALAXY
SHAP336-5411.6	03 35 19.	- 54 21 25.	24	17.0	GALAXY
BAK 2.0108	03 35 20.	- 15 52 47.	21	15.7	GALAXY
BAK 2.0109	03 35 20.	- 16 28 47.	24	16.6	GALAXY
BAK 2.0106	03 35 20.	- 24 43 24.	23	16.6	GALAXY
SHAP336-5342.8	03 35 20.	- 53 52 37.	24	17.25	GALAXY
SHAP336-5413.5	03 35 20.	- 54 23 19.	18	17.5	GALAXY
SHAP336-5521.2	03 35 20.	- 55 31 01.	48	16.0	GALAXY
SCHO 0060	03 35 21.	+ 38 30 36.	370		ISOLATED DARK CLOUD
MCG-03-10-015	03 35 21.	- 18 30	36	14.5	GALAXY
SVEN 226	03 35 21.	- 24 40	150	11.5	GALAXY
RNGC 1383	03 35 22.	- 18 30		14.0	GALAXY
SHAP336-4648.4	03 35 22.	- 46 58 12.	42	17.5	GALAXY
SHAP336-5036.2	03 35 22.	- 50 46 01.	24	17.0	GALAXY
BAK 1.787	03 35 23.	- 25 00 11.	17	15.8	EXTRAGALACTIC NEBULA
SHAP336-5100.9	03 35 23.	- 51 10 43.	18	17.5	GALAXY
SHAP336-5416.2	03 35 23.	- 54 26 01.	24	17.5	GALAXY
ZWG 465.002	03 35 24.	+ 19 13		15.7	GALAXY
MCG+03-10-002	03 35 24.	+ 19 14	18	15.	GALAXY
ZWG 526.001	03 35 24.	+ 37 52		15.7	GALAXY
ZWG 525.046	03 35 24.	+ 37 52		15.7	GALAXY
MCG-03-10-016	03 35 24.	- 15 02	60	15.	GALAXY
PHL 4468	03 35 24.	- 23 00		18.6	BLUE STELLAR OBJECT
MCG-04-09-036	03 35 24.	- 24 39 30.	180	12.	GALAXY
HELW 406	03 35 24.	- 25 00 23.			NEBULA
SHAP336-4848.3	03 35 24.	- 48 58 06.	18	17.5	GALAXY
SHAP337-4642.6	03 35 25.	- 46 52 24.	24	17.5	GALAXY
SHAP336-5327.6	03 35 25.	- 53 37 25.	24	17.25	GALAXY
SVEN 227	03 35 26.	- 25 00 35.	24	15.0	GALAXY
BAK 1.785	03 35 26.	- 28 30 59.	23	16.4	EXTRAGALACTIC NEBULA
SHAP337-4540.2	03 35 26.	- 45 50 00.	24	17.5	GALAXY
SHAP336-5252.4	03 35 26.	- 53 02 13.	12	17.25	GALAXY
SHAP336-5436.4	03 35 27.	- 54 46 13.	18	17.0	GALAXY
SHAP336-5437.2	03 35 27.	- 54 47 01.	24	17.5	GALAXY
SHAP336-5429.5	03 35 28.	- 54 39 19.	42	17.0	GALAXY
SHAP336-5528.2	03 35 28.	- 55 38 01.	48	17.0	GALAXY
BAK 1.786	03 35 29.	- 30 39 11.	18	17.0	EXTRAGALACTIC NEBULA
UGC 02804	03 35 30.	+ 01 46	66	16.5	GALAXY Sc
ZWG 541.023	03 35 30.	+ 40 50		15.7	GALAXY
UGC 02805	03 35 30.	+ 40 50	72	15.7	GALAXY Sc/SBc
ARC 0447	03 35 30.	- 05 17		17.7	RICH CLUSTER OF GALAXIES
BAK 2.0111	03 35 30.	- 16 31 59.	13	16.3	GALAXY
TON-S 0363	03 35 30.	- 23 20		15.3	BLUE STAR

OBJECT NAME	RIGHT ASCEN.	DECLINATION	DIAM.	MAGN.	TYPE OF OBJECT
SHAP337-4905.6	03 35 30.	- 49 15 24.	24	17.25	GALAXY
BAK 2.0115	03 35 32.	- 15 02 59.	55	14.8	GALAXY
BAK 2.0110A	03 35 32.	- 25 15 47.	14	16.9	GALAXY
SHAP336-5250.6	03 35 32.	- 53 00 24.	18	17.25	GALAXY
SHAP336-5341.0	03 35 32.	- 53 50 48.	24	17.75	GALAXY
SHAP336-5458.7	03 35 32.	- 55 08 30.	30	17.25	GALAXY
MCG+07-08-027	03 35 33.	+ 40 49	60	15.	GALAXY
BAK 2.0114	03 35 33.	- 17 09 41.	22	16.6	GALAXY
BAK 2.0112	03 35 33.	- 19 30 47.	15	16.6	GALAXY
BAK 1.791	03 35 33.	- 23 03 59.	50	15.4	EXTRAGALACTIC NEBULA
BAK 1.790	03 35 33.	- 25 53 23.	18	16.6	EXTRAGALACTIC NEBULA
SHAP337-4726.0	03 35 33.	- 47 35 48.	66	16.25	GALAXY
SHAP337-4843.5	03 35 33.	- 48 53 18.	24	17.5	GALAXY
SHAP337-5009.0	03 35 33.	- 50 18 48.	72	16.0	GALAXY
SHAP336-5410.5	03 35 33.	- 54 20 18.	24	17.5	GALAXY
BAK 2.0110B	03 35 34.	- 24 01 41.	14	16.8	GALAXY
BAK 1.789	03 35 34.	- 29 12 23.	19	16.5	EXTRAGALACTIC NEBULA
SHAP336-5415.0	03 35 34.	- 54 24 48.	24	17.75	GALAXY
BAK 1.788	03 35 35.	- 29 12 59.	17	16.4	EXTRAGALACTIC NEBULA
HN 0229	03 35 35.	- 47 36			NEBULA
IC 1976	03 35 35.	- 47 36			NONSTELLAR OBJECT
HN 0230	03 35 35.	- 50 19			NEBULA
IC 1978	03 35 35.	- 50 19			NONSTELLAR OBJECT
SHAP336-5721.0	03 35 35.	- 57 30 48.	36	16.5	GALAXY
SHAP336-5756.2	03 35 35.	- 58 06 01.	72	16.5	GALAXY
MCG+01-10-012	03 35 36.	- 06 27	78	14.5	GALAXY
MCG-04-09-037	03 35 36.	- 23 04	72	14.5	GALAXY
BAK 1.792	03 35 36.	- 25 08 59.	13	16.6	EXTRAGALACTIC NEBULA
PHL 4470	03 35 36.	- 25 32		18.1	BLUE STELLAR OBJECT
PHL 4469	03 35 36.	- 27 33		18.3	BLUE STELLAR OBJECT
HOD.60 14	03 35 36.	- 35 05	42	17.0	IRRG GLXY IN FORNAX CLSTR
SHAP337-5147.4	03 35 36.	- 51 57 12.	30	16.5	GALAXY
SHAP336-5617.0	03 35 36.	- 56 26 48.	36	16.5	GALAXY
SHAP337-5104.1	03 35 37.	- 51 13 54.	18	17.5	GALAXY
HN 0231	03 35 37.	- 58 06			NEBULA
IC 1979	03 35 37.	- 58 06			NONSTELLAR OBJECT
SHAP336-5512.7	03 35 38.	- 55 22 30.	42	16.5	GALAXY
LB 01466	03 35 39.	+ 23 16 06.		19.8	FAINT BLUE STAR
BAK 2.0119	03 35 39.	- 16 23 22.	31	16.2	GALAXY
SHAP337-4959.0	03 35 39.	- 50 08 48.	18	17.0	GALAXY
BAK 1.794	03 35 40.	- 26 24 34.	23	16.8	EXTRAGALACTIC NEBULA
SHAP337-5102.6	03 35 40.	- 51 12 24.	18	17.75	GALAXY
SHAP336-5632.4	03 35 40.	- 56 42 12.	30	17.0	GALAXY
RNGC 1390	03 35 41.	- 19 10		15.0	GALAXY
BAK 2.0117	03 35 41.	- 22 56 46.	12	16.6	GALAXY
BAK 2.0113	03 35 41.	- 25 42 40.	18	17.0	GALAXY
SHAP336-5416.4	03 35 41.	- 54 26 12.	24	17.5	GALAXY
ZWG 391.021	03 35 42.	+ 01 00		15.0	GALAXY
UGC 02806	03 35 42.	+ 07 28	66	16.0	GALAXY S
MCG-03-10-017	03 35 42.	- 19 10	36	15.	GALAXY
BAK 2.0116	03 35 42.	- 24 49 58.	24	17.0	GALAXY
BAK 1.793	03 35 42.	- 30 22 46.	13	16.5	EXTRAGALACTIC NEBULA
SHAP337-4949.1	03 35 42.	- 49 58 53.	24	16.5	GALAXY
SHAP337-5327.2	03 35 42.	- 53 37 00.	30	17.25	GALAXY
SHAP337-5417.2	03 35 42.	- 54 27 00.	24	17.5	GALAXY
BAK 2.0118	03 35 43.	- 22 46 10.	15	16.6	GALAXY
SHAP337-4847.4	03 35 43.	- 48 57 11.	18	17.0	GALAXY
SHAP337-5022.2	03 35 43.	- 50 31 59.	18	17.5	GALAXY
SHAP336-5537.5	03 35 43.	- 55 47 18.	60	16.5	GALAXY
SHAP336-5636.9	03 35 43.	- 56 46 42.	30	17.25	GALAXY
BAK 2.0123	03 35 44.	- 16 23 10.	14	16.7	GALAXY
SHAP337-5021.8	03 35 44.	- 50 31 35.	24	17.0	GALAXY
SHAP337-5344.4	03 35 44.	- 53 54 12.	24	16.5	GALAXY
SHAP337-5142.6	03 35 45.	- 51 52 23.	18	17.5	GALAXY
SHAP337-5310.8	03 35 45.	- 53 20 36.	42	16.0	GALAXY
SHAP337-5336.6	03 35 45.	- 53 46 24.	24	17.5	GALAXY
BAK 1.797	03 35 46.	- 21 49 22.	24	16.1	EXTRAGALACTIC NEBULA
SHAP337-5345.4	03 35 46.	- 53 55 11.	24	18.0	GALAXY
BAK 1.795	03 35 47.	- 27 52 10.	24	16.7	EXTRAGALACTIC NEBULA
SHAP337-5239.2	03 35 47.	- 52 48 59.	18	18.0	GALAXY
SHAP337-5250.8	03 35 47.	- 53 00 35.	48	15.25	GALAXY
SHAP336-5758.2	03 35 47.	- 58 08 00.	96	16.0	GALAXY
MCG-03-10-018	03 35 48.	- 19 55	90	15.5	GALAXY
HOD.60 15	03 35 48.	- 33 38	36	16.5	IRRG GLXY IN FORNAX CLSTR
HOD.60 05	03 35 48.	- 35 57	36	17.0	E0 GALAXY IN FORNAX CLSTR
SHAP337-5412.8	03 35 48.	- 54 22 35.	24	17.5	GALAXY
LB 03326	03 35 48.	- 70 15		13.2	FAINT BLUE STAR
CED 018A	03 35 49.	+ 32 05			DIFFUSE GALACTIC NEBULA
SG 3.019	03 35 49.	+ 32 05	3000		DIFFUSE EMISSION NEBULA
BAK 2.0120	03 35 49.	- 25 38 46.	14	16.6	GALAXY
BAK 1.796	03 35 49.	- 26 30 10.	24	16.8	EXTRAGALACTIC NEBULA
SHAP337-5310.2	03 35 49.	- 53 19 59.	18	16.5	GALAXY
SHAP337-5351.1	03 35 49.	- 54 00 53.	36	16.75	GALAXY
HN 0232	03 35 49.	- 58 08			NEBULA
IC 1980	03 35 49.	- 58 08			NONSTELLAR OBJECT
BAK 2.0121	03 35 50.	- 23 19 22.	18	16.7	GALAXY
SHAP337-4642.4	03 35 50.	- 46 52 11.	18	17.0	GALAXY
BAK 2.0126	03 35 51.	- 16 16 09.	13	16.9	GALAXY
IC 0339	03 35 51.	- 18 32 36.			SINGLE STAR
BAK 2.0122	03 35 51.	- 24 34 04.	23	16.8	GALAXY
SHAP337-5139.0	03 35 51.	- 51 48 47.	18	16.75	GALAXY
SHAP337-5303.4	03 35 51.	- 53 13 11.	36	17.5	GALAXY
BAK 2.0127	03 35 52.	- 17 03 27.	12	16.7	GALAXY
SHAP337-4950.5	03 35 53.	- 50 00 17.	24	17.0	GALAXY
SHAP337-5140.0	03 35 53.	- 51 49 47.	18	16.5	GALAXY
SHAP337-5336.7	03 35 53.	- 53 46 29.	18	17.5	GALAXY
UGC 02807	03 35 54.	+ 30 06	96	17.	GALAXY DWARF
MCG+01-10-013	03 35 54.	- 04 28 30.	72	15.	GALAXY
BAK 2.0129	03 35 54.	- 16 44 09.	29	16.2	GALAXY
PHL 4472	03 35 54.	- 24 22		18.3	BLUE STELLAR OBJECT
PHL 4471	03 35 54.	- 25 24		18.6	BLUE STELLAR OBJECT
HPW 21	03 35 54.	- 35 05			DWARF GALAXY IN FORNAX
SHAP337-5457.6	03 35 54.	- 55 07 23.	36	16.5	GALAXY
BAK 2.0132	03 35 55.	- 17 36 57.	14	16.0	GALAXY
SHAP337-5137.0	03 35 55.	- 51 46 47.	18	17.0	GALAXY
SHAP337-5140.2	03 35 55.	- 51 49 59.	18	16.5	GALAXY
BAK 2.0128	03 35 56.	- 19 53 33.	51	15.8	GALAXY
BAK 2.0124	03 35 56.	- 23 09 45.	16	16.8	GALAXY
RNGC 1388	03 35 57.	- 16 03			GALAXY
SHAP337-5019.2	03 35 57.	- 50 28 59.	18	17.0	GALAXY
SHAP337-5236.8	03 35 57.	- 52 46 35.	24	17.25	GALAXY
BAK 1.798	03 35 58.	- 29 13 22.	14	16.6	EXTRAGALACTIC NEBULA
BAK 2.0136	03 35 59.	- 17 13 27.	14	17.0	GALAXY
DG U20	03 36	+ 22 10	5400		REFLECTION NEBULA
LBN 0761	03 36	+ 29 00	5100		BRIGHT NEBULA
LBN 0749	03 36	+ 32 00	6300		BRIGHT NEBULA
ZWG 465.003	03 36 00.	+ 20 15		15.1	GALAXY
PHL 8727	03 36 00.	- 25 13		18.5	BLUE STELLAR OBJECT

OBJECT NAME	RIGHT ASCEN.	DECLINATION	DIAM.	MAGN.	TYPE OF OBJECT
BAK 2.0125	03 36 00.	- 25 40 39.	18	17.0	GALAXY
PHL 8728	03 36 00.	- 29 50		18.7	BLUE STELLAR OBJECT
HPW 22	03 36 00.	- 35 57			DWARF GALAXY IN FORNAX
SHAP337-5349.2	03 36 00.	- 53 58 59.	24	17.5	GALAXY
SHAP337-5824.8	03 36 00.	- 58 34 35.	30	18.0	GALAXY
SHAP337-5924.5	03 36 00.	- 59 34 17.	36	17.0	GALAXY
BAK 2.0138	03 36 01.	- 15 17 21.	10	16.0	GALAXY
BAK 2.0130	03 36 01.	- 23 21 27.	16	16.7	GALAXY
KARA.73 17	03 36 01.	- 37 26	27		DWARF GALAXY
SHAP337-5424.0	03 36 01.	- 54 33 47.	30	16.5	GALAXY
LB 01467	03 36 02.	+ 24 15 36.		18.0	FAINT BLUE STAR
BAK 2.0131	03 36 02.	- 24 54 39.	13	16.4	GALAXY
KARA.73 18	03 36 02.	- 37 27	40		DWARF GALAXY
BAK 1.799	03 36 03.	- 28 58 57.	19	16.4	EXTRAGALACTIC NEBULA
BAK 2.0134	03 36 04.	- 22 38 39.	24	16.5	GALAXY
SHAP337-5140.0	03 36 04.	- 51 49 46.	24	16.5	GALAXY
BAK 2.0137	03 36 05.	- 23 01 03.	24	16.6	GALAXY
SHAP337-4915.1	03 36 05.	- 49 24 52.	30	17.5	GALAXY
SHAP337-5138.2	03 36 05.	- 51 47 58.	24	17.0	GALAXY
SHAP337-5450.4	03 36 05.	- 55 00 10.	24	16.75	GALAXY
ZC 0336.1+0338	03 36 06.	+ 03 38	670		CLUSTER OF GALAXIES
MCG+01-10-014	03 36 06.	- 05 39	72	15.	GALAXY
BAK 2.0139	03 36 06.	- 17 43 33.	14	16.4	GALAXY
MCG-06-09-011	03 36 06.	- 33 15	48	15.	GALAXY
SHAP337-5843.7	03 36 06.	- 58 53 29.	24	17.75	GALAXY
BAK 2.0140	03 36 07.	- 16 44 33.	14	16.8	GALAXY
BAK 1.800	03 36 07.	- 23 35 09.	26	15.4	EXTRAGALACTIC NEBULA
BAK 2.0135	03 36 07.	- 25 47 45.	24	16.7	GALAXY
BAK 2.0133	03 36 07.	- 27 01 21.	16	16.7	GALAXY
SHAP337-5039.0	03 36 07.	- 50 48 46.	24	18.0	GALAXY
LB 01468	03 36 08.	+ 22 16 00.		19.9	FAINT BLUE STAR
SHAP337-5111.1	03 36 08.	- 51 20 52.	66	15.0	GALAXY
SHAP337-5459.6	03 36 08.	- 55 09 22.	48	16.0	GALAXY
RNGC 1396	03 36 09.	- 35 50			NON-EXISTENT OBJECT
SHAP337-5127.5	03 36 09.	- 51 37 16.	24	17.0	GALAXY
SHAP337-5324.8	03 36 09.	- 53 34 34.	12	17.5	GALAXY
MCG+01-10-015	03 36 12.	- 05 31	72	15.	GALAXY
MCG-04-09-038	03 36 12.	- 23 35	36	15.	GALAXY
HPW 23	03 36 12.	- 37 26			DWARF GALAXY IN FORNAX
SHAP337-5151.6	03 36 12.	- 52 01 22.	18	17.75	GALAXY
SHAP337-5245.0	03 36 12.	- 52 54 46.	18	16.75	GALAXY
SHAP337-4945.6	03 36 13.	- 49 55 22.	30	16.5	GALAXY
SHAP337-5222.5	03 36 13.	- 52 32 16.	18	18.0	GALAXY
BAK 2.0144	03 36 14.	- 16 01 50.	19	15.8	GALAXY
MCG+07-08-028	03 36 15.	+ 40 55	48	14.5	GALAXY
SHAP337-5230.8	03 36 16.	- 52 40 34.	18	17.0	GALAXY
BAK 2.0145	03 36 17.	- 17 05 44.	22	16.4	GALAXY
MCG+03-10-003	03 36 18.	+ 15 39	42	15.	GALAXY
LB 01469	03 36 18.	+ 25 05 06.		20.0	FAINT BLUE STAR
5ZW 357	03 36 18.	+ 28 00			COMPACT GALAXY
ZWG 526.002	03 36 18.	+ 38 32		15.2	GALAXY
ZWG 525.047	03 36 18.	+ 38 32		15.2	GALAXY
UGC 02808	03 36 18.	+ 38 32	66	15.2	GALAXY Sc
ZWG 541.024	03 36 18.	+ 40 55		15.1	GALAXY
MCG-02-10-003	03 36 18.	- 09 53	42	15.	GALAXY
TON-S 0364	03 36 18.	- 25 15		15.1	BLUE STAR
PHL 4473	03 36 18.	- 26 00		13.8	BLUE STELLAR OBJECT
SHAP337-5224.3	03 36 18.	- 52 34 04.	48	16.75	GALAXY
SHAP337-5810.3	03 36 18.	- 58 20 04.	30	17.0	GALAXY
ARC 0448	03 36 19.	- 11 17		17.6	RICH CLUSTER OF GALAXIES
RNGC 1395	03 36 19.	- 23 11		11.5	GALAXY
BAK 2.0142	03 36 19.	- 24 22 02.	24	16.6	GALAXY
BAK 2.0143	03 36 20.	- 23 31 56.	24	16.7	GALAXY
BAK 2.0141	03 36 20.	- 26 54 20.	11	16.8	GALAXY
SHAP337-5240.4	03 36 20.	- 52 50 09.	24	18.0	GALAXY
SHAP338-4514.2	03 36 21.	- 45 23 57.	30	17.25	GALAXY
SHAP337-5053.7	03 36 21.	- 51 03 27.	18	17.0	GALAXY
SHAP337-5539.2	03 36 21.	- 55 48 58.	48	17.75	GALAXY
SHAP337-5551.2	03 36 21.	- 56 00 58.	30	17.75	GALAXY
RNGC 1393	03 36 22.	- 18 36		14.0	GALAXY
SHAP338-4558.2	03 36 23.	- 46 07 57.	18	17.5	GALAXY
SHAP337-4656.0	03 36 23.	- 47 05 45.	24	17.25	GALAXY
ZWG 465.004	03 36 24.	+ 15 40		15.6	GALAXY
RNGC 1384	03 36 24.	+ 15 40		15.5	GALAXY
MCG+06-09-001	03 36 24.	+ 38 33	48	15.	GALAXY
MCG-03-10-019	03 36 24.	- 18 35 30.	72	14.	GALAXY
MCG-04-09-039	03 36 24.	- 23 12	96	12.	GALAXY
PHL 8729	03 36 24.	- 25 40		18.6	BLUE STELLAR OBJECT
SHAP337-5459.6	03 36 24.	- 55 09 21.	48	16.5	GALAXY
BAK 2.0146	03 36 25.	- 24 20 14.	19	17.0	GALAXY
SHAP338-4530.6	03 36 25.	- 45 40 21.	30	17.5	GALAXY
BAK 1.801	03 36 26.	- 26 30 32.	14	16.8	EXTRAGALACTIC NEBULA
SHAP338-4513.0	03 36 26.	- 45 22 45.	24	17.0	GALAXY
SHAP337-5116.8	03 36 26.	- 51 26 33.	42	16.5	GALAXY
SHAP337-5212.2	03 36 26.	- 52 21 57.	18	17.0	GALAXY
SHAP337-5819.5	03 36 26.	- 58 29 16.	24	17.25	GALAXY
SHAP337-4854.6	03 36 27.	- 49 04 21.	36	17.0	GALAXY
BAK 2.0147	03 36 28.	- 24 59 20.	14	16.5	GALAXY
BAK 1.803	03 36 29.	- 26 40 56.	34	15.9	EXTRAGALACTIC NEBULA
HELW 407	03 36 29.	- 26 40 57.			NEBULA
BAK 1.802	03 36 29.	- 28 49 20.	25	16.2	EXTRAGALACTIC NEBULA
VB 194	03 36 30.	+ 61 48	477		STELLAR RING
MCG+01-10-016	03 36 30.	- 03 47	22	14.5	GALAXY
BAK 2.0148	03 36 30.	- 17 50 55.	23	16.4	GALAXY
PHL 4474	03 36 30.	- 22 25		18.3	BLUE STELLAR OBJECT
MCG-05-09-019	03 36 30.	- 26 41	36	16.	GALAXY
TON-S 0365	03 36 30.	- 31 25		13.8	BLUE STAR
HPW 25	03 36 30.	- 35 41			DWARF GALAXY IN FORNAX
HPW 24	03 36 30.	- 37 26			DWARF GALAXY IN FORNAX
SHAP338-4902.4	03 36 30.	- 49 12 09.	24	16.5	GALAXY
SHAP337-5746.3	03 36 30.	- 57 56 03.	36	16.0	GALAXY
SHAP337-5358.1	03 36 31.	- 54 07 51.	18	17.0	GALAXY
SHAP337-5725.0	03 36 31.	- 57 34 45.	24	17.25	GALAXY
SHAP337-5728.0	03 36 31.	- 57 37 45.	24	17.5	GALAXY
HN 0233	03 36 31.	- 57 56			NEBULA
SHAP337-5422.1	03 36 32.	- 54 31 51.	24	17.5	GALAXY
IC 1982	03 36 32.	- 57 56			NONSTELLAR OBJECT
MCG-03-10-020	03 36 33.	- 18 31 30.	36	15.	GALAXY
RNGC 1399	03 36 33.	- 35 37		11.5	GALAXY
SHAP337-5353.2	03 36 33.	- 54 02 57.	18	17.0	GALAXY
BAK 2.0150	03 36 34.	- 16 59 13.	19	16.6	GALAXY
RNGC 1391	03 36 34.	- 18 31		15.0	GALAXY
BAK 1.805	03 36 34.	- 22 32 25.	23	16.1	EXTRAGALACTIC NEBULA
SHAP338-5013.6	03 36 34.	- 50 23 20.	30	17.0	GALAXY
SHAP338-5116.2	03 36 34.	- 51 25 56.	18	17.5	GALAXY
SHAP338-5127.2	03 36 35.	- 51 36 56.	12	17.5	GALAXY
MCG+03-10-004	03 36 36.	+ 19 37	66	15.	GALAXY
ZWG 465.005	03 36 36.	+ 20 22		15.7	GALAXY

OBJECT NAME	RIGHT ASCEN.	DECLINATION	DIAM.	MAGN.	TYPE OF OBJECT
PHL 1559	03 36 36.	- 21 13		17.2	BLUE STELLAR OBJECT
PHL 4475	03 36 36.	- 23 33		18.8	BLUE STELLAR OBJECT
PHL 8730	03 36 36.	- 24 18		18.6	BLUE STELLAR OBJECT
BAK 1.804	03 36 36.	- 29 05 01.	24	16.9	EXTRAGALACTIC NEBULA
PHL 4476	03 36 36.	- 30 30		18.6	BLUE STELLAR OBJECT
MCG-06-09-012	03 36 36.	- 35 37	360	10.9	GALAXY
BAK 2.0149	03 36 37.	- 23 36 55.	19	16.7	GALAXY
SHAP338-4722.1	03 36 37.	- 47 31 50.	18	17.25	GALAXY
SHAP338-5220.5	03 36 37.	- 52 30 14.	18	17.75	GALAXY
BAK 2.0153	03 36 38.	- 17 14 31.	12	16.8	GALAXY
BAK 2.0152	03 36 38.	- 18 03 07.	12	16.7	GALAXY
SHAP338-4723.7	03 36 38.	- 47 33 26.	18	17.5	GALAXY
SHAP337-5447.6	03 36 38.	- 54 57 20.	30	17.5	GALAXY
SHAP337-5500.5	03 36 38.	- 55 10 15.	30	17.0	GALAXY
BAK 2.0154	03 36 39.	- 16 53 19.	12	16.2	GALAXY
BAK 2.0155	03 36 39.	- 17 18 43.	19	16.4	GALAXY
BAK 1.806	03 36 39.	- 25 09 19.	14	16.0	EXTRAGALACTIC NEBULA
SHAP338-4813.1	03 36 39.	- 48 22 50.	18	16.75	GALAXY
SHAP337-5529.6	03 36 39.	- 55 39 20.	42	17.5	GALAXY
BAK 2.0158	03 36 40.	- 16 15 25.	17	16.4	GALAXY
BAK 2.0156	03 36 40.	- 17 41 55.	17	16.6	GALAXY
SHAP337-5459.6	03 36 40.	- 55 09 20.	48	16.0	GALAXY
SHAP338-5134.8	03 36 41.	- 51 44 32.	18	17.5	GALAXY
LB 01470	03 36 42.	+ 25 40 24.		17.7	FAINT BLUE STAR
ZC 0336.7+2754	03 36 42.	+ 27 54	1480		CLUSTER OF GALAXIES
ZWG 391.022	03 36 42.	- 02 10		15.2	GALAXY
BAK 2.0157	03 36 42.	- 19 28 43.	19	16.4	GALAXY
MCG-04-09-040	03 36 42.	- 26 30	360	11.	GALAXY
TON-S 0366	03 36 42.	- 31 19		16.0	BLUE STAR
LB 01681	03 36 42.	- 51 28		11.4	FAINT BLUE STAR
SHAP338-5126.5	03 36 42.	- 51 36 14.	24	17.5	GALAXY
SHAP338-5130.2	03 36 42.	- 51 39 56.	30	17.0	GALAXY
SHAP338-5421.7	03 36 42.	- 54 31 26.	24	17.5	GALAXY
BAK 2.0159	03 36 43.	- 18 27 24.	12	16.7	GALAXY
BAK 1.807	03 36 43.	- 28 52 07.	22	17.0	EXTRAGALACTIC NEBULA
BAK 2.0151	03 36 44.	- 25 34 49.	14	17.0	GALAXY
SHAP338-5235.8	03 36 44.	- 52 45 32.	18	17.5	GALAXY
SHAP338-5424.3	03 36 44.	- 54 34 02.	18	17.25	GALAXY
RNGC 1398	03 36 45.	- 26 30		11.0	GALAXY
SHAP338-4721.4	03 36 45.	- 47 31 08.	24	17.0	GALAXY
IC 1975	03 36 46.	- 15 39 47.			NONSTELLAR OBJECT
RNGC 1394	03 36 46.	- 18 27		14.0	GALAXY
BAK 1.810	03 36 46.	- 25 36 01.	14	16.6	EXTRAGALACTIC NEBULA
SHAP337-5845.7	03 36 46.	- 58 55 26.	30	17.0	GALAXY
BAK 2.0164	03 36 47.	- 14 46 30.	25	15.8	GALAXY
BAK 2.0162	03 36 47.	- 17 04 06.	26	16.6	GALAXY
BAK 2.0161	03 36 47.	- 18 01 30.	16	15.6	GALAXY
SHAP338-5400.3	03 36 47.	- 54 10 02.	30	17.75	GALAXY
5ZW 358	03 36 48.	+ 23 58			COMPACT GALAXY
5ZW 359	03 36 48.	+ 38 30			COMPACT GALAXY
ZWG 526.003	03 36 48.	+ 39 27		15.0	GALAXY
UGC 02810	03 36 48.	+ 39 27	72	15.0	GALAXY Sc
5ZW 360	03 36 48.	+ 42 17			COMPACT GALAXY
RLWT 068	03 36 48.	+ 56 13		13.	FAINT VERY BLUE STAR
MCG-03-10-021	03 36 48.	- 18 28	54	14.5	GALAXY
BAK 1.808	03 36 48.	- 28 48 43.	23	16.5	EXTRAGALACTIC NEBULA
SHAP338-5423.7	03 36 48.	- 54 33 26.	18	17.25	GALAXY
LB 01682	03 36 48.	- 58 21		13.5	FAINT BLUE STAR
SHAP337-5844.0	03 36 48.	- 58 53 44.	24	16.0	GALAXY
SHAP338-5422.8	03 36 49.	- 54 32 32.	24	18.0	GALAXY
SHAP337-5732.2	03 36 49.	- 57 41 56.	24	17.0	GALAXY
BAK 2.0163	03 36 50.	- 18 46 42.	20	16.5	GALAXY
SHAP338-4842.6	03 36 50.	- 48 52 19.	30	17.5	GALAXY
SHAP337-5726.1	03 36 50.	- 57 35 50.	36	16.25	GALAXY
BAK 1.811	03 36 51.	- 25 09 30.	11	16.5	EXTRAGALACTIC NEBULA
SHAP338-4900.4	03 36 51.	- 49 10 07.	36	17.0	GALAXY
SHAP337-5728.2	03 36 51.	- 57 37 56.	30	17.0	GALAXY
BAK 2.0165	03 36 52.	- 17 52 54.	12	16.8	GALAXY
BAK 2.0160	03 36 52.	- 25 10 54.	14	16.6	GALAXY
BAK 1.809	03 36 52.	- 30 50 18.	23	16.9	EXTRAGALACTIC NEBULA
SHAP338-4611.6	03 36 53.	- 46 21 19.	24	17.25	GALAXY
ZWG 391.023	03 36 54.	+ 02 25		15.2	GALAXY
B 003	03 36 54.	+ 31 49	1200		DARK OBJECT
UGC 02811	03 36 54.	+ 43 27	72	17.	GALAXY S IV?
BAK 2.0168	03 36 54.	- 16 52 12.	13	16.7	GALAXY
BAK 2.0169	03 36 54.	- 16 54 54.	12	16.8	GALAXY
BAK 2.0166	03 36 54.	- 17 41 30.	13	17.0	GALAXY
MCG-06-09-013	03 36 54.	- 35 45	210	11.5	GALAXY
SHAP338-5816.2	03 36 54.	- 58 25 56.	30	18.0	GALAXY
BAK 2.0170	03 36 56.	- 17 08 30.	19	16.5	GALAXY
SHAP338-4828.0	03 36 56.	- 48 37 43.	18	17.5	GALAXY
SHAP338-5535.8	03 36 56.	- 55 45 32.	42	16.5	GALAXY
BAK 2.0171	03 36 57.	- 16 52 36.	24	16.1	GALAXY
RNGC 1404	03 36 57.	- 35 45		12.0	GALAXY
SHAP338-5834.4	03 36 57.	- 58 44 08.	24	17.5	GALAXY
BAK 2.0172	03 36 58.	- 17 45 18.	12	16.0	GALAXY
SHAP338-5334.9	03 36 58.	- 53 44 37.	24	16.7	GALAXY
SHAP338-5354.2	03 36 58.	- 54 03 55.	18	17.75	GALAXY
SHAP338-5730.1	03 36 58.	- 57 39 50.	36	17.5	GALAXY
BC PKS0336-01	03 36 58.97	- 01 56 16.6		18.41	QUASI-STELLAR OBJECT
SHAP338-4720.0	03 36 59.	- 47 29 43.	30	17.75	GALAXY
SHB 073	03 36 59.2	- 01 56 19.		18.4	QUASI-STELLAR OBJECT
LDN 1468	03 37 00.	+ 31 15	1380		DARK NEBULA
BAK 2.0174	03 37 00.	- 18 53 41.	14	16.8	GALAXY
PHL 1560	03 37 00.	- 23 32		18.2	BLUE STELLAR OBJECT
PHL 4477	03 37 00.	- 23 57		18.2	BLUE STELLAR OBJECT
LB 01683	03 37 00.	- 52 18		12.9	FAINT BLUE STAR
LB 01471	03 37 01.	+ 23 24 00.		17.0	FAINT BLUE STAR
BAK 1.812	03 37 01.	- 26 21 12.	23	17.0	EXTRAGALACTIC NEBULA
SHAP338-5326.2	03 37 01.	- 53 35 55.	24	17.1	GALAXY
BAK 2.0167	03 37 02.	- 25 19 18.	17	17.0	GALAXY
HUB C02	03 37 03.	+ 31 42			DIFFUSE NEBULA
KARA.73 19	03 37 03.	- 35 53	27		DWARF GALAXY
SHAP338-5440.6	03 37 03.	- 54 50 19.	30	17.0	GALAXY
BAK 2.0176	03 37 04.	- 18 10 41.	14	16.0	GALAXY
RNGC 1411	03 37 04.	- 44 15		12.0	GALAXY
SHAP338-5154.5	03 37 04.	- 52 04 13.	18	17.0	GALAXY
SHAP338-5334.3	03 37 04.	- 53 44 01.	24	16.8	GALAXY
SHAP338-5439.1	03 37 05.	- 54 48 49.	30	16.5	GALAXY
VB 195	03 37 06.	+ 58 14	269		STELLAR RING
MCG-02-10-004	03 37 06.	- 09 46	48	15.	GALAXY
BAK 2.0178	03 37 06.	- 15 58 41.	18	16.3	GALAXY
RNGC 1403	03 37 06.	- 22 33		14.0	GALAXY
MCG-04-09-041	03 37 06.	- 22 33	33	14.	GALAXY
SHAP338-4719.1	03 37 06.	- 47 28 48.	12	16.5	GALAXY
SHAP338-5325.2	03 37 06.	- 53 34 55.	24	17.2	GALAXY
SHAP338-5733.0	03 37 06.	- 57 42 43.	30	17.0	GALAXY
SHAP338-4718.4	03 37 07.	- 47 28 06.	18	17.5	GALAXY

OBJECT NAME	RIGHT ASCEN.	DECLINATION	DIAM.	MAGN.	TYPE OF OBJECT
SHAP338-4720.8	03 37 07.	- 47 30 30.	24	16.5	GALAXY
BAK 1.813	03 37 08.	- 21 34 35.	19	15.8	EXTRAGALACTIC NEBULA
BAK 2.0173	03 37 08.	- 26 47 41.	16	17.0	GALAXY
RNGC 1397	03 37 09.	- 04 50		14.5	GALAXY
IC 0340	03 37 09.	- 13 16 38.			NONSTELLAR OBJECT
MCG-02-10-005	03 37 09.	- 13 17	84	15.	GALAXY
BAK 2.0179	03 37 09.	- 18 55 05.	19	16.6	GALAXY
RNGC 1402	03 37 10.	- 18 41		15.0	GALAXY
RNGC 1400	03 37 10.	- 18 51		13.0	GALAXY
BAK 2.0175	03 37 10.	- 25 35 59.	20	16.8	GALAXY
SHAP338-5030.2	03 37 11.	- 50 39 54.	18	17.0	GALAXY
SHAP338-5421.8	03 37 11.	- 54 31 31.	42	16.5	GALAXY
VB 196	03 37 12.	+ 61 48	295		STELLAR RING
ISS 0069	03 37 12.	+ 61 48	299		STELLAR RING
MCG-03-10-023	03 37 12.	- 18 41 30.	24	15.	GALAXY
MCG-03-10-022	03 37 12.	- 18 51	42	12.5	GALAXY
BAK 2.0183	03 37 12.	- 19 22 05.	14	16.4	GALAXY
HPW 27	03 37 12.	- 35 25			DWARF GALAXY IN FORNAX
HOD.60 06	03 37 12.	- 35 35	36	16.0	E0 GALAXY IN FORNAX CLSTR
HPW 26	03 37 12.	- 38 21			DWARF GALAXY IN FORNAX
BAK 2.0189	03 37 13.	- 15 36 53.	16	16.5	GALAXY
BAK 2.0184	03 37 13.	- 18 09 17.	17	16.5	GALAXY
RNGC 1401	03 37 13.	- 22 53		13.0	GALAXY
SHAP338-4749.6	03 37 13.	- 47 59 18.	18	17.5	GALAXY
SHAP338-5356.2	03 37 13.	- 54 05 54.	24	16.75	GALAXY
SHAP338-5733.2	03 37 13.	- 57 42 55.	30	16.5	GALAXY
SHAP338-4635.5	03 37 14.	- 46 45 12.	24	17.25	GALAXY
MCG+01-10-017	03 37 15.	- 04 50	108	14.5	GALAXY
MCG-04-09-043	03 37 15.	- 21 34	48	15.	GALAXY
MCG-04-09-042	03 37 15.	- 22 53 30.	120	13.5	GALAXY
BAK 2.0180	03 37 15.	- 24 21 17.	18	17.2	GALAXY
BAK 2.0177	03 37 15.	- 27 08 53.	14	16.8	GALAXY
BAK 2.0190	03 37 16.	- 17 49 17.	19	16.0	GALAXY
BAK 1.814	03 37 16.	- 25 12 17.	14	16.2	EXTRAGALACTIC NEBULA
SHAP338-5411.2	03 37 16.	- 54 20 54.	24	17.0	GALAXY
SHAP338-5310.1	03 37 17.	- 53 19 48.	18	17.5	GALAXY
ZWG 465.006	03 37 18.	+ 17 15		15.7	GALAXY
MCG-03-10-024	03 37 18.	- 14 42 30.	84	15.5	GALAXY
MCG-03-10-025	03 37 18.	- 20 11	60	15.	GALAXY
HPW 28	03 37 18.	- 35 54			DWARF GALAXY IN FORNAX
BAK 2.0181	03 37 19.	- 26 57 41.	18	16.4	GALAXY
SHAP338-5036.2	03 37 19.	- 50 45 54.	24	17.0	GALAXY
SHAP338-5054.4	03 37 19.	- 51 04 06.	24	17.5	GALAXY
SHAP338-5353.8	03 37 19.	- 54 03 30.	30	17.75	GALAXY
BAK 2.0182	03 37 20.	- 26 40 59.	17	17.0	GALAXY
MCG-02-10-006	03 37 21.	- 12 51 30.	48	15.	GALAXY
BAK 2.0186	03 37 21.	- 24 35 34.	12	17.0	GALAXY
BAK 2.0185	03 37 21.	- 24 59 04.	18	16.8	GALAXY
SHAP338-4607.1	03 37 21.	- 46 16 47.	42	18.0	GALAXY
SHAP338-5315.2	03 37 21.	- 53 24 54.	18	17.6	GALAXY
SHAP338-5327.0	03 37 21.	- 53 36 42.	24	16.8	GALAXY
BAK 2.0193	03 37 22.	- 14 43 28.	73	14.9	GALAXY
BAK 2.0192	03 37 22.	- 17 26 28.	12	16.6	GALAXY
BAK 1.815	03 37 22.	- 20 09 52.	71	15.6	EXTRAGALACTIC NEBULA
SHAP338-5336.1	03 37 22.	- 53 45 48.	18	17.4	GALAXY
SHAP338-5412.4	03 37 22.	- 54 22 06.	18	17.25	GALAXY
ARP 219	03 37 23.	- 02 17			PECULIAR GALAXY
BAK 2.0187	03 37 23.	- 26 26 16.	17	16.9	GALAXY
SHAP338-5434.3	03 37 23.	- 54 44 00.	24	17.5	GALAXY
ZWG 391.024	03 37 24.	- 02 16		14.9	GALAXY
UGC 02812	03 37 24.	- 02 16	60	14.9	GALAXY SB+COMP
PHL 8731	03 37 24.	- 24 59		18.3	BLUE STELLAR OBJECT
BAK 2.0188	03 37 24.	- 26 09 34.	19	17.0	GALAXY
SHAP338-5304.2	03 37 24.	- 53 13 54.	30	17.1	GALAXY
SHAP338-5047.5	03 37 25.	- 50 57 11.	24	17.0	GALAXY
BAK 2.0191	03 37 27.	- 24 40 22.	12	16.8	GALAXY
RNGC 1408	03 37 27.	- 35 41			NON-EXISTENT OBJECT
SHAP338-5440.0	03 37 27.	- 54 49 42.	30	18.0	GALAXY
SHAP338-5757.2	03 37 27.	- 58 06 54.	24	16.0	GALAXY
SHAP338-5052.6	03 37 29.	- 51 02 17.	18	17.5	GALAXY
SHAP338-5853.6	03 37 29.	- 59 03 18.	30	17.0	GALAXY
UGC 02813	03 37 30.	+ 71 09	150	16.5	GALAXY DWARF
MCG+00-10-009	03 37 30.	- 02 17 30.	48	14.5	GALAXY
RNGC 1406	03 37 30.	- 31 28		13.0	GALAXY
MCG-05-09-020	03 37 30.	- 31 30	228	11.5	GALAXY
BAK 2.0197	03 37 31.	- 19 09 28.	14	15.8	GALAXY
LB 01472	03 37 32.	+ 24 20 30.		16.4	FAINT BLUE STAR
SHAP338-5153.5	03 37 32.	- 52 03 11.	30	17.0	GALAXY
SHAP338-5812.9	03 37 32.	- 58 22 36.	24	17.0	GALAXY
BAK 1.817	03 37 33.	- 25 35 16.	32	16.5	EXTRAGALACTIC NEBULA
BAK 1.816	03 37 33.	- 28 52 22.	14	16.4	EXTRAGALACTIC NEBULA
BAK 2.0194	03 37 34.	- 25 10 46.	18	16.5	GALAXY
SHAP339-4630.2	03 37 34.	- 46 39 53.	30	17.5	GALAXY
SHAP339-4730.0	03 37 34.	- 47 39 41.	24	17.25	GALAXY
SHAP338-5329.0	03 37 34.	- 53 38 41.	24	17.4	GALAXY
SHAP338-5345.0	03 37 34.	- 53 54 41.	24	17.1	GALAXY
SHAP338-5357.8	03 37 34.	- 54 07 29.	18	17.0	GALAXY
SHAP338-5618.2	03 37 34.	- 56 27 53.	24	16.5	GALAXY
SHAP339-4518.1	03 37 35.	- 45 27 47.	30	17.5	GALAXY
SHAP338-5618.6	03 37 35.	- 56 28 17.	24	16.5	GALAXY
ZWG 391.025	03 37 36.	- 02 14		15.0	GALAXY
UGC 02814	03 37 36.	- 02 14	60	15.0	GALAXY
BAK 2.0196	03 37 36.	- 24 00 34.	38	16.7	GALAXY
BAK 2.0195	03 37 36.	- 25 31 04.	19	16.8	GALAXY
BAK 1.818	03 37 36.	- 25 35 04.	11	16.4	EXTRAGALACTIC NEBULA
SHAP338-5429.1	03 37 37.	- 54 38 47.	24	17.0	GALAXY
SHAP338-5330.2	03 37 38.	- 53 39 53.	24	17.3	GALAXY
MCG-03-10-026	03 37 39.	- 19 34	30	15.	GALAXY
RNGC 1427A	03 37 39.	- 35 47			GALAXY
SHAP339-5117.0	03 37 39.	- 51 26 41.	18	17.5	GALAXY
BAK 2.0199	03 37 40.	- 22 51 27.	17	16.8	GALAXY
BAK 2.0198	03 37 40.	- 24 41 51.	12	16.8	GALAXY
BAK 2.0201	03 37 41.	- 16 58 51.	22	16.2	GALAXY
BAK 2.0202	03 37 41.	- 17 02 39.	17	16.6	GALAXY
SHAP339-5341.4	03 37 41.	- 53 51 05.	18	17.5	GALAXY
ZWG 391.026	03 37 42.	+ 00 56		15.1	GALAXY
LB 01473	03 37 42.	+ 22 54 18.		17.9	FAINT BLUE STAR
ZC 0337.7+2631	03 37 42.	+ 26 31	2220		CLUSTER OF GALAXIES
MCG+00-10-010	03 37 42.	- 02 15	42	14.5	GALAXY
PHL 8732	03 37 42.	- 25 08		13.4	BLUE STELLAR OBJECT
TON-S 0367	03 37 42.	- 25 09		13.9	BLUE STAR
PHL 4478	03 37 42.	- 28 04		17.9	BLUE STELLAR OBJECT
DV.56 N1427A	03 37 42.	- 35 47	108		I GALAXY
SHAP339-4541.1	03 37 42.	- 45 50 46.	24	17.5	GALAXY
SHAP339-5022.2	03 37 42.	- 50 31 52.	18	17.0	GALAXY
SHAP339-5139.8	03 37 42.	- 51 49 28.	18	17.25	GALAXY
SHAP338-5857.0	03 37 42.	- 59 06 41.	24	16.5	GALAXY
BAK 1.821	03 37 43.	- 28 54 51.	16	16.9	EXTRAGALACTIC NEBULA
BAK 1.820	03 37 43.	- 29 07 15.	17	16.7	EXTRAGALACTIC NEBULA
SHAP339-5049.1	03 37 43.	- 50 58 46.	24	16.5	GALAXY
BAK 1.825	03 37 44.	- 25 29 15.	12	16.6	EXTRAGALACTIC NEBULA
BAK 1.822	03 37 44.	- 27 45 27.	20	16.8	EXTRAGALACTIC NEBULA
SHAP339-4525.4	03 37 44.	- 45 35 04.	24	17.25	GALAXY
SHAP339-5056.0	03 37 44.	- 51 05 40.	24	17.5	GALAXY
SHAP339-5233.4	03 37 44.	- 52 43 04.	24	17.0	GALAXY
SHAP339-5422.1	03 37 44.	- 54 31 47.	24	16.75	GALAXY
MCG-03-10-027	03 37 45.	- 19 31	60	15.5	GALAXY
BAK 1.823	03 37 45.	- 27 44 45.	19	16.4	EXTRAGALACTIC NEBULA
SHAP339-5050.2	03 37 45.	- 50 59 52.	18	17.5	GALAXY
SHAP339-5216.0	03 37 45.	- 52 25 40.	12	17.5	GALAXY
IC 1981	03 37 46.	- 27 01 25.			NONSTELLAR OBJECT
BAK 2.0205	03 37 47.	- 15 50 51.	22	16.7	GALAXY
BAK 1.827	03 37 47.	- 19 35 09.	26	15.5	EXTRAGALACTIC NEBULA
BAK 1.824	03 37 47.	- 29 42 57.	22	16.5	EXTRAGALACTIC NEBULA
SHAP338-5700.0	03 37 47.	- 57 09 41.	24	17.0	GALAXY
SHAP338-5738.8	03 37 47.	- 57 48 29.	24	17.0	GALAXY
ZC 0337.8+1536	03 37 48.	+ 15 36	6720		CLUSTER OF GALAXIES
MCG+03-10-005	03 37 48.	+ 17 35	66	14.5	GALAXY
MCG-03-10-028	03 37 48.	- 15 40	72	16.	GALAXY
BAK 1.826	03 37 48.	- 28 53 15.	14	16.8	EXTRAGALACTIC NEBULA
SHAP339-5151.9	03 37 48.	- 52 01 34.	18	17.75	GALAXY
SHAP339-5336.2	03 37 48.	- 53 45 52.	24	17.5	GALAXY
BAK 2.0200	03 37 49.	- 25 01 51.	38	16.6	GALAXY
SHAP339-4752.4	03 37 49.	- 48 02 04.	12	17.5	GALAXY
BAK 2.0203	03 37 50.	- 23 32 15.	12	17.0	GALAXY
SHAP339-5203.8	03 37 50.	- 52 13 28.	18	16.5	GALAXY
SHAP339-5323.0	03 37 50.	- 53 32 40.	36	17.6	GALAXY
RNGC 1405	03 37 51.	- 15 40		16.0	GALAXY
BAK 2.0207	03 37 51.	- 17 12 50.	14	16.6	GALAXY
MCG-03-10-029	03 37 51.	- 18 36 30.	48	14.5	GALAXY
BAK 2.0206	03 37 51.	- 19 31 08.	53	15.7	GALAXY
SHAP339-5150.2	03 37 51.	- 51 59 52.	18	17.0	GALAXY
SHAP339-5321.4	03 37 51.	- 53 31 04.	24	16.5	GALAXY
SHAP339-5332.0	03 37 51.	- 53 41 40.	24	17.6	GALAXY
IC 0343	03 37 51.7	- 18 36 12.			GALAXY SB(s)
RNGC 1407	03 37 52.	- 18 44		12.0	GALAXY
SHAP339-5235.0	03 37 52.	- 52 44 40.	18	17.5	GALAXY
SHAP339-5333.3	03 37 52.	- 53 42 58.	18	17.7	GALAXY
IC 1977	03 37 53.	+ 17 35 33.			NONSTELLAR OBJECT
SHAP339-5408.1	03 37 53.	- 54 17 46.	24	17.5	GALAXY
ZWG 465.007	03 37 54.	+ 17 35		14.6	GALAXY
UGC 02815	03 37 54.	+ 17 35	102	14.6	GALAXY SBb
ZC 0337.9+7820	03 37 54.	+ 78 20	1280		CLUSTER OF GALAXIES
BAK 2.0204	03 37 54.	- 24 12 44.	20	17.0	GALAXY
PHL 8733	03 37 54.	- 28 52		18.4	BLUE STELLAR OBJECT
TON-S 0368	03 37 54.	- 31 10		15.8	BLUE STAR
MCG-06-09-014	03 37 54.	- 37 57	9	15.	GALAXY
SHAP339-5327.2	03 37 54.	- 53 36 52.	24	17.6	GALAXY
SHAP339-5343.0	03 37 54.	- 53 52 40.	24	17.3	GALAXY
SHAP339-4656.0	03 37 55.	- 47 05 39.	18	17.5	GALAXY
SHAP339-5339.8	03 37 55.	- 53 49 28.	24	16.7	GALAXY
BAK 1.828	03 37 56.	- 27 28 02.	26	16.8	EXTRAGALACTIC NEBULA
SHAP339-5131.9	03 37 56.	- 51 41 34.	18	17.5	GALAXY
MCG+01-10-018	03 37 57.	- 02 46	48	15.5	GALAXY
RNGC 1413	03 37 57.	- 15 45			GALAXY
SHAP339-5152.8	03 37 57.	- 52 02 28.	18	17.0	GALAXY
SHAP339-5320.9	03 37 57.	- 53 30 34.	18	17.5	GALAXY
BAK 2.0212	03 37 58.	- 15 48 50.	19	16.4	GALAXY
BAK 2.0209	03 37 58.	- 17 50 14.	46	16.5	GALAXY
BAK 1.829	03 37 58.	- 25 24 44.	12	16.2	EXTRAGALACTIC NEBULA
SHAP339-5619.9	03 37 58.	- 56 29 34.	24	18.0	GALAXY
SHAP339-5700.0	03 37 58.	- 57 09 40.	24	16.5	GALAXY
BAK 2.0211	03 37 59.	- 18 29 08.	14	16.6	GALAXY
SHAP339-5641.6	03 37 59.	- 56 51 16.	24	17.25	GALAXY
PK147-02.1	03 37 59.1	+ 52 07 26.	4		PLANETARY NEBULA
CED 019B	03 38	+ 21 40	8100		DIFFUSE GALACTIC NEBULA
IC 0341	03 38	+ 21 48			NONSTELLAR OBJECT
LDN 1472	03 38 00.	+ 31 00	7680		DARK NEBULA
RLWT 069	03 38 00.	+ 52 10		16.	FAINT VERY BLUE STAR
ZC 0338.0+8356	03 38 00.	+ 83 56	3630		CLUSTER OF GALAXIES
MCG-03-10-031	03 38 00.	- 19 05	54	14.5	GALAXY
MCG-06-09-015	03 38 00.	- 37 58	60	15.	GALAXY
SHAP339-5551.1	03 38 00.	- 56 00 46.	54	17.0	GALAXY
BAK 2.0208	03 38 01.	- 26 25 14.	17	16.8	GALAXY
SHAP339-5505.9	03 38 02.	- 55 15 34.	48	16.0	GALAXY
BAK 2.0214	03 38 03.	- 19 14 14.	22	16.0	GALAXY
BAK 1.831	03 38 03.	- 22 31 02.	37	16.8	EXTRAGALACTIC NEBULA
SHAP339-4512.6	03 38 04.	- 45 22 15.	30	16.5	GALAXY
BAK 2.0217	03 38 05.	- 16 46 26.	13	16.6	GALAXY
BAK 2.0215	03 38 05.	- 19 05 14.	48	15.2	GALAXY
BAK 1.830	03 38 05.	- 26 03 38.	14	16.5	EXTRAGALACTIC NEBULA
SHAP339-4713.6	03 38 05.	- 47 23 15.	30	16.5	GALAXY
SHAP339-4906.6	03 38 05.	- 49 16 15.	30	16.25	GALAXY
SHAP339-4912.0	03 38 05.	- 49 21 39.	30	17.5	GALAXY
SHAP339-5122.5	03 38 05.	- 51 32 09.	18	17.5	GALAXY
SHAP339-5312.2	03 38 05.	- 53 21 51.	24	17.4	GALAXY
ZCG 112.01	03 38 06.	+ 15 09		18.0	RED COMPACT GALAXY
LB 01474	03 38 06.	+ 25 02 12.		19.9	FAINT BLUE STAR
ZWG 391.027	03 38 06.	- 01 32		15.0	GALAXY
PHL 4479	03 38 06.	- 21 35		17.7	BLUE STELLAR OBJECT
BAK 2.0213	03 38 06.	- 23 29 32.	17	16.8	GALAXY
PHL 8734	03 38 06.	- 29 22		17.6	BLUE STELLAR OBJECT
HPW 29	03 38 06.	- 34 56			DWARF GALAXY IN FORNAX
BAK 2.0210	03 38 07.	- 26 25 38.	17	17.0	GALAXY
SHAP339-5213.2	03 38 07.	- 52 22 51.	24	17.5	GALAXY
SHAP339-5315.6	03 38 07.	- 53 25 15.	18	17.5	GALAXY
SHAP339-5421.3	03 38 07.	- 54 30 57.	24	17.25	GALAXY
SHAP339-5542.8	03 38 07.	- 55 52 27.	36	16.5	GALAXY
SHAP339-5309.8	03 38 08.	- 53 19 27.	24	16.8	GALAXY
SHAP339-5609.4	03 38 08.	- 56 19 03.	24	17.0	GALAXY
BAK 2.0218	03 38 09.	- 16 44 37.	12	16.0	GALAXY
SHAP339-4807.0	03 38 09.	- 48 16 39.	18	17.0	GALAXY
SHAP339-5302.8	03 38 09.	- 53 12 27.	24	16.7	GALAXY
SHAP339-5351.2	03 38 09.	- 54 00 51.	30	16.4	GALAXY
BAK 2.0219	03 38 10.	- 16 43 31.	17	16.2	GALAXY
SHAP339-4808.7	03 38 11.	- 48 18 21.	18	17.5	GALAXY
SHAP339-5033.4	03 38 11.	- 50 43 03.	24	17.75	GALAXY
SHAP339-5301.6	03 38 11.	- 53 11 15.	24	16.4	GALAXY
SHAP339-5423.5	03 38 11.	- 54 33 09.	24	17.0	GALAXY
MCG+01-10-005	03 38 12.	+ 08 05	48	16.	GALAXY
ZWG 486.001	03 38 12.	+ 23 51		15.7	GALAXY
UGC 02816	03 38 12.	+ 23 51	84	15.7	GALAXY
BAK 2.0220	03 38 12.	- 16 18 49.	29	16.3	GALAXY
PHL 1561	03 38 12.	- 22 44		18.2	BLUE STELLAR OBJECT
BAK 2.0216	03 38 12.	- 24 50 01.	14	16.8	GALAXY
HOD.60 16	03 38 12.	- 35 48	108	15.0	IRRG GLXY IN FORNAX CLSTR

OBJECT NAME	RIGHT ASCEN.	DECLINATION	DIAM.	MAGN.	TYPE OF OBJECT
SHAP339-4716.2	03 38 12.	- 47 25 50.	12	17.5	GALAXY
BAK 2.0221	03 38 13.	- 17 32 07.	19	16.8	GALAXY
SHAP339-4650.0	03 38 13.	- 46 59 38.	24	16.25	GALAXY
SHAP339-5132.6	03 38 13.	- 51 42 15.	18	17.5	GALAXY
BAK 1.832	03 38 14.	- 28 51 25.	14	16.8	EXTRAGALACTIC NEBULA
SHAP339-4536.8	03 38 14.	- 45 46 26.	24	17.5	GALAXY
BAK 1.833	03 38 15.	- 29 06 31.	31	15.8	EXTRAGALACTIC NEBULA
YC 0338-35	03 38 15.	- 35 47		18.	UNUSUAL SOUTHERN GALAXY
SHAP339-4740.1	03 38 15.	- 47 49 44.	18	17.75	GALAXY
SHAP339-5224.0	03 38 15.	- 52 33 39.	24	18.0	GALAXY
SHAP339-4704.3	03 38 16.	- 47 13 56.	48	16.5	GALAXY
IC 1984	03 38 16.	- 47 14			NONSTELLAR OBJECT
SHAP339-5405.2	03 38 16.	- 54 14 51.	24	17.5	GALAXY
BAK 1.837	03 38 17.	- 25 06 31.	34	16.4	EXTRAGALACTIC NEBULA
HN 0234	03 38 17.	- 47 14			NEBULA
SHAP339-5435.0	03 38 17.	- 54 44 39.	24	16.5	GALAXY
3ZW 054	03 38 18.	+ 15 10			COMPACT GALAXY
ZCG 112.02	03 38 18.	+ 15 18		17.5	RED COMPACT GALAXY
MCG+04-09-003	03 38 18.	+ 23 50	60	15.5	GALAXY
MCG+01-10-019	03 38 18.	- 06 35	96	15.	GALAXY
PHL 8735	03 38 18.	- 30 24		16.6	BLUE STELLAR OBJECT
TON-S 0369	03 38 18.	- 30 26		15.7	BLUE STAR
MCG-06-09-016	03 38 18.	- 35 47	120	14.	GALAXY
SHAP339-5234.6	03 38 18.	- 52 44 14.	18	17.0	GALAXY
BAK 1.834	03 38 19.	- 28 40 55.	22	17.0	EXTRAGALACTIC NEBULA
BAK 1.835	03 38 19.	- 28 49 37.	17	16.8	EXTRAGALACTIC NEBULA
BAK 1.836	03 38 19.	- 28 51 01.	12	16.8	EXTRAGALACTIC NEBULA
SHAP339-4443.0	03 38 19.	- 44 52 38.	36	17.5	GALAXY
SHAP339-5503.2	03 38 19.	- 55 12 51.	48	15.75	GALAXY
SHAP339-5417.3	03 38 20.	- 54 26 56.	24	17.25	GALAXY
SHAP339-5643.3	03 38 20.	- 56 52 57.	24	17.5	GALAXY
BAK 1.838	03 38 21.	- 28 49 37.	18	16.4	EXTRAGALACTIC NEBULA
BAK 1.839	03 38 21.	- 28 51 13.	14	17.4	EXTRAGALACTIC NEBULA
SHAP339-5705.0	03 38 21.	- 57 14 39.	18	17.5	GALAXY
SHAP339-5707.9	03 38 21.	- 57 17 33.	18	17.0	GALAXY
SHAP339-5731.0	03 38 21.	- 57 40 39.	18	17.25	GALAXY
SHAP339-4800.4	03 38 22.	- 48 10 02.	24	16.5	GALAXY
BAK 2.0223	03 38 23.	- 23 46 19.	22	16.9	GALAXY
BAK 1.840	03 38 23.	- 26 12 25.	18	16.6	EXTRAGALACTIC NEBULA
SHAP339-5703.5	03 38 23.	- 57 13 09.	24	17.5	GALAXY
SHAP339-5710.0	03 38 23.	- 57 19 39.	24	17.5	GALAXY
SHAP339-5822.4	03 38 23.	- 58 32 03.	36	16.5	GALAXY
SHAP339-5823.6	03 38 23.	- 58 33 15.	24	16.5	GALAXY
ZCG 112.03	03 38 24.	+ 15 12		17.5	RED COMPACT GALAXY
ZCG 112.04	03 38 24.	+ 15 13		17.5	RED COMPACT GALAXY
ZCG 112.05	03 38 24.	+ 15 14		17.0	RED COMPACT GALAXY
ZCG 112.06	03 38 24.	+ 15 14		17.9	RED COMPACT GALAXY
MCG+04-09-004	03 38 24.	+ 25 31	36	16.	GALAXY
UGC 02817	03 38 24.	+ 37 03	150	17.	GALAXY
ZWG 526.004	03 38 24.	+ 39 11		14.5	GALAXY
UGC 02818	03 38 24.	+ 39 11	78	14.5	GALAXY S
MCG+07-08-029	03 38 24.	+ 39 47	27	16.	GALAXY
MCG-04-09-044	03 38 24.	- 22 26	84	15.	GALAXY
PHL 4480	03 38 24.	- 23 42		18.6	BLUE STELLAR OBJECT
BAK 2.0222	03 38 24.	- 26 49 37.	23	16.9	GALAXY
HELW 408	03 38 24.	- 27 01 10.			NEBULA
SHAP340-4554.8	03 38 25.	- 46 04 26.	24	17.5	GALAXY
SHAP339-5352.0	03 38 25.	- 54 01 38.	24	17.0	GALAXY
SHAP339-5646.8	03 38 25.	- 56 56 26.	30	17.5	GALAXY
SHAP339-4759.1	03 38 26.	- 48 08 44.	12	17.0	GALAXY
SHAP339-5235.2	03 38 26.	- 52 44 50.	24	17.5	GALAXY
SHAP339-5407.8	03 38 26.	- 54 17 26.	24	17.5	GALAXY
SHAP339-5438.2	03 38 26.	- 54 47 50.	30	16.5	GALAXY
RNGC 1412	03 38 27.	- 26 23			NON-EXISTENT OBJECT
SHAP339-4851.2	03 38 27.	- 49 00 50.	30	17.5	GALAXY
BAK 2.0224	03 38 28.	- 23 26 54.	14	16.8	GALAXY
BAK 1.841	03 38 28.	- 28 50 25.	14	17.0	EXTRAGALACTIC NEBULA
BAK 2.0227	03 38 29.	- 18 48 18.	24	15.7	GALAXY
BAK 1.844	03 38 29.	- 22 26 24.	72	15.8	EXTRAGALACTIC NEBULA
SHAP339-5341.8	03 38 29.	- 53 51 26.	24	17.1	GALAXY
SHAP339-5621.4	03 38 29.	- 56 31 02.	24	17.0	GALAXY
ZC 0338.5+0946	03 38 30.	+ 09 46	2150		CLUSTER OF GALAXIES
ZCG 112.07	03 38 30.	+ 15 16		17.2	RED COMPACT GALAXY
MCG+06-09-002	03 38 30.	+ 39 11	48	14.5	GALAXY
UGC 02819	03 38 30.	+ 39 48	84	16.0	GALAXY E
MCG+13-03-007	03 38 30.	+ 76 29	114	14.	GALAXY
PHL 8736	03 38 30.	- 21 25		16.6	BLUE STELLAR OBJECT
RNGC 1414	03 38 30.	- 21 51		14.0	GALAXY
MCG-04-09-045	03 38 30.	- 21 51	72	14.	GALAXY
BAK 1.847	03 38 31.	- 22 48 48.	34	16.2	EXTRAGALACTIC NEBULA
SHAP340-4653.4	03 38 31.	- 47 03 01.	18	17.75	GALAXY
BAK 1.842	03 38 32.	- 28 47 18.	14	16.9	EXTRAGALACTIC NEBULA
SHAP340-4905.8	03 38 32.	- 49 15 25.	36	17.5	GALAXY
BAK 2.0228	03 38 33.	- 18 53 30.	14	16.6	GALAXY
BAK 1.848	03 38 33.	- 22 49 24.	19	17.0	EXTRAGALACTIC NEBULA
SHAP340-5024.0	03 38 33.	- 50 33 37.	36	16.75	GALAXY
BAK 2.0225	03 38 34.	- 25 02 12.	14	17.2	GALAXY
BAK 1.845	03 38 34.	- 26 56 36.	72	15.8	EXTRAGALACTIC NEBULA
SHAP340-4723.6	03 38 34.	- 47 33 13.	18	17.5	GALAXY
SHAP339-5420.6	03 38 35.	- 54 30 14.	24	17.75	GALAXY
UGC 02820	03 38 36.	+ 05 28	84	16.5	GALAXY S
MCG+01-10-006	03 38 36.	+ 05 29	36	16.	GALAXY
ZCG 112.08	03 38 36.	+ 15 14		17.8	RED COMPACT GALAXY
ZCG 112.09	03 38 36.	+ 15 15		17.8	RED COMPACT GALAXY
MCG-04-09-046	03 38 36.	- 22 50	12	15.	GALAXY
BAK 2.0226	03 38 36.	- 26 58 30.	21	17.2	GALAXY
MCG-05-09-021	03 38 36.	- 27 00	96	14.	GALAXY
HPW 30	03 38 36.	- 35 28			DWARF GALAXY IN FORNAX
HOD.60 07	03 38 36.	- 35 28	42	17.5	E0 GALAXY IN FORNAX CLSTR
SHAP339-5448.0	03 38 36.	- 54 57 38.	30	16.75	GALAXY
SHAP339-5603.0	03 38 36.	- 56 12 38.	30	16.5	GALAXY
SCHO 0061	03 38 37.	+ 37 09 18.	670		ISOLATED DARK CLOUD
SHAP340-4558.7	03 38 37.	- 46 08 19.	24	17.5	GALAXY
SHAP340-4744.0	03 38 37.	- 47 53 37.	24	17.75	GALAXY
SHAP340-4746.0	03 38 37.	- 47 55 37.	24	17.75	GALAXY
SHAP339-5320.8	03 38 38.	- 53 30 25.	18	17.2	GALAXY
SHAP339-5348.0	03 38 38.	- 53 57 37.	30	17.0	GALAXY
SHAP339-5449.3	03 38 38.	- 54 58 55.	30	16.5	GALAXY
BAK 2.0231	03 38 39.	- 18 38 30.	18	16.2	GALAXY
SHAP339-5329.5	03 38 39.	- 53 39 07.	18	17.5	GALAXY
SEY 001	03 38 40.	+ 68 09 48.		14.9	FAINT GALAXY
BAK 2.0232	03 38 40.	- 18 50 24.	12	16.8	GALAXY
SHAP340-5317.4	03 38 40.	- 53 27 01.	24	17.2	GALAXY
SHAP340-4936.6	03 38 41.	- 49 46 13.	30	17.0	GALAXY
SHAP340-5013.2	03 38 41.	- 50 22 49.	30	17.0	GALAXY
SHAP340-5238.2	03 38 41.	- 52 47 49.	24	17.5	GALAXY
SHAP340-5338.1	03 38 41.	- 53 47 43.	18	17.5	GALAXY
ABC 0450	03 38 42.	+ 23 21		16.4	RICH CLUSTER OF GALAXIES

OBJECT NAME	RIGHT ASCEN.	DECLINATION	DIAM.	MAGN.	TYPE OF OBJECT
UGC 02822	03 38 42.	+ 40 59	66	16.5	GALAXY Sb-c
3ZW 055	03 38 42.	- 01 27			COMPACT GALAXY
ZWG 391.028	03 38 42.	- 01 27		14.7	GALAXY
KW 66	03 38 42.	- 01 27	13		SEYFERT GALAXY
VVI 14	03 38 42.	- 01 27	12	14.	SEYFERT GALAXY
UGC 02821	03 38 42.	- 01 27	72	14.7	GALAXY DBL SYS
KARA.72 093B	03 38 42.	- 01 27	30		PART OF DOUBLE GALAXY
KARA.72 093A	03 38 42.	- 01 27	42	14.7	PART OF DOUBLE GALAXY
BAK 2.0233	03 38 42.	- 18 46 47.	20	16.5	GALAXY
BAK 2.0229	03 38 42.	- 23 10 48.	22	16.8	GALAXY
BAK 1.851	03 38 42.	- 25 22 48.	17	16.8	EXTRAGALACTIC NEBULA
BAK 1.849	03 38 42.	- 28 52 24.	12	16.6	EXTRAGALACTIC NEBULA
HPW 31	03 38 42.	- 36 39			DWARF GALAXY IN FORNAX
SHAP340-4510.6	03 38 42.	- 45 20 12.	24	17.25	GALAXY
SHAP340-5159.6	03 38 42.	- 52 09 13.	12	16.75	GALAXY
SHAP340-5216.5	03 38 42.	- 52 26 07.	18	17.0	GALAXY
RNGC 1415	03 38 43.	- 22 43		12.5	GALAXY
SHAP340-5338.8	03 38 43.	- 53 48 25.	24	17.6	GALAXY
SHAP339-5618.2	03 38 43.	- 56 27 49.	36	17.5	GALAXY
SHAP339-5701.2	03 38 43.	- 57 10 49.	24	17.0	GALAXY
RNGC 1410	03 38 44.	- 01 27		14.0	GALAXY
RNGC 1409	03 38 44.	- 01 27		14.0	GALAXY
BAK 2.0230	03 38 44.	- 24 57 48.	12	16.8	GALAXY
BAK 1.853	03 38 44.	- 25 53 00.	17	17.3	EXTRAGALACTIC NEBULA
SHAP340-5013.6	03 38 44.	- 50 23 13.	24	17.25	GALAXY
IC 0334	03 38 45.	+ 76 30			NONSTELLAR OBJECT
MCG+00-10-012	03 38 45.	- 01 28	12	14.	GALAXY
MCG+00-10-011	03 38 45.	- 01 28	48	14.	GALAXY
BAK 1.850	03 38 45.	- 28 53 54.	14	16.2	EXTRAGALACTIC NEBULA
BAK 1.854	03 38 46.	- 25 30 29.	14	16.3	EXTRAGALACTIC NEBULA
RNGC 1419	03 38 46.	- 37 41			GALAXY
SHAP340-4536.1	03 38 46.	- 45 45 42.	24	17.25	GALAXY
SHAP340-5126.8	03 38 46.	- 51 36 25.	18	16.5	GALAXY
SHAP340-5402.2	03 38 46.	- 54 11 49.	24	17.1	GALAXY
BAK 2.0239	03 38 47.	- 15 35 29.	22	15.9	GALAXY
SHAP340-5130.6	03 38 47.	- 51 40 13.	12	17.75	GALAXY
UGC 02823	03 38 48.	+ 15 51	96	16.0	GALAXY Sb
ZC 0338.8+2327	03 38 48.	+ 23 27	3020		CLUSTER OF GALAXIES
LF 01475	03 38 48.	+ 24 10 12.		20.0	FAINT BLUE STAR
MCG+11-05-002	03 38 48.	+ 68 09	54	15.	GALAXY
ZWG 346.006	03 38 48.	+ 76 29		13.2	GALAXY
UGC 02824	03 38 48.	+ 76 29	360	13.2	GALAXY PECULR
BAK 2.0236	03 38 48.	- 19 37 17.	29	16.3	GALAXY
MCG-04-09-047	03 38 48.	- 22 44	180	13.	GALAXY
MCG-06-09-017	03 38 48.	- 37 40	60	12.	GALAXY
SHAP340-5438.6	03 38 48.	- 54 48 13.	30	17.5	GALAXY
BAK 2.0240	03 38 49.	- 18 27 11.	22	16.6	GALAXY
BAK 2.0238	03 38 49.	- 19 14 47.	35	16.1	GALAXY
RNGC 1416	03 38 49.	- 22 56		14.0	GALAXY
BAK 1.855	03 38 49.	- 28 05 35.	22	16.8	EXTRAGALACTIC NEBULA
BAK 2.0234	03 38 50.	- 25 10 05.	14	16.8	GALAXY
SHAP340-5048.0	03 38 50.	- 50 57 36.	24	16.5	GALAXY
SHAP340-5647.8	03 38 50.	- 56 57 25.	24	17.0	GALAXY
SHAP340-5653.6	03 38 50.	- 57 03 13.	24	17.25	GALAXY
BAK 1.856	03 38 51.	- 28 07 05.	36	16.9	EXTRAGALACTIC NEBULA
BAK 2.0235	03 38 52.	- 25 33 35.	12	16.9	GALAXY
SHAP340-5713.4	03 38 52.	- 57 23 01.	18	17.25	GALAXY
BAK 2.0243	03 38 53.	- 17 58 35.	12	16.4	GALAXY
BAK 2.0242	03 38 53.	- 19 35 47.	21	16.8	GALAXY
BAK 2.0241	03 38 53.	- 20 37 47.	17	16.4	GALAXY
BAK 1.857	03 38 53.	- 24 45 59.	16	16.0	EXTRAGALACTIC NEBULA
SHAP340-5354.8	03 38 53.	- 54 04 24.	24	17.7	GALAXY
ZC 0338.9+2631	03 38 54.	+ 26 31	200		CLUSTER OF GALAXIES
ZWG 391.029	03 38 54.	- 01 46		15.5	GALAXY
UGC 02825	03 38 54.	- 01 46	78	15.5	GALAXY S
MCG+01-10-020	03 38 54.	- 04 49 30.	54	15.	GALAXY
MCG-03-10-032	03 38 54.	- 18 28 30.	24	15.	GALAXY
MCG-04-09-048	03 38 54.	- 22 53 30.	54	14.	GALAXY
MCG-04-09-049	03 38 54.	- 24 46 30.	24	15.5	GALAXY
SHAP340-5047.6	03 38 54.	- 50 57 12.	18	17.0	GALAXY
SHAP340-5331.2	03 38 54.	- 53 40 48.	24	16.5	GALAXY
BAK 2.0245	03 38 55.	- 16 06 11.	13	16.8	GALAXY
BAK 2.0244	03 38 55.	- 18 52 23.	19	16.4	GALAXY
BAK 2.0237	03 38 55.	- 24 58 11.	24	16.8	GALAXY
SHAP340-4521.3	03 38 55.	- 45 30 54.	72	15.5	GALAXY
SHAP340-4757.0	03 38 55.	- 48 06 36.	18	17.5	GALAXY
SHAP340-5451.7	03 38 55.	- 55 01 18.	30	16.5	GALAXY
SHAP340-5653.6	03 38 55.	- 57 03 13.	18	17.5	GALAXY
IC 1983	03 38 56.	- 22 45 56.			MAY NOT EXIST
HN 0235	03 38 56.	- 45 31			NEBULA
IC 1986	03 38 56.	- 45 31			NONSTELLAR OBJECT
SHAP340-5049.2	03 38 56.	- 50 58 48.	18	17.5	GALAXY
SHAP340-5403.8	03 38 56.	- 54 13 24.	24	17.1	GALAXY
BAK 2.0246	03 38 57.	- 18 26 28.	16	17.0	GALAXY
SHAP340-5134.9	03 38 57.	- 51 44 30.	18	17.0	GALAXY
SHAP340-5309.3	03 38 57.	- 53 18 54.	24	16.6	GALAXY
SHAP340-5309.5	03 38 57.	- 53 19 06.	30	17.7	GALAXY
SHAP340-5342.5	03 38 57.	- 53 52 06.	24	17.0	GALAXY
SHAP340-5405.0	03 38 57.	- 54 14 36.	18	17.6	GALAXY
SHAP340-5515.5	03 38 57.	- 55 25 06.	48	17.5	GALAXY
SHAP340-5544.8	03 38 57.	- 55 54 24.	36	16.75	GALAXY
BAK 1.858	03 38 58.	- 27 30 11.	14	16.6	EXTRAGALACTIC NEBULA
BAK 1.859	03 38 58.	- 27 55 47.	17	16.8	EXTRAGALACTIC NEBULA
SHAP340-5326.8	03 38 58.	- 53 36 24.	18	17.0	GALAXY
SHAP340-4541.8	03 38 59.	- 45 51 24.	24	17.5	GALAXY
SHAP340-4751.1	03 38 59.	- 48 00 42.	24	18.0	GALAXY
ZCG 112.10	03 39 00.	+ 15 12		17.9	RED COMPACT GALAXY
ASS 47	03 39 00.	+ 33 16	28800		OB ASSOCIATION PER OB2
UGC 02826	03 39 00.	+ 58 08	96	15.0	GALAXY
ZWG 305.001	03 39 00.	+ 68 08		15.0	GALAXY
UGC 02827	03 39 00.	+ 69 10	90	17.	GALAXY SB
MCG+12-04-003	03 39 00.	+ 69 10	45	18.	GALAXY
MCG+00-10-013	03 39 00.	- 01 47	66	14.5	GALAXY
IC 0344	03 39 00.	- 04 49			GALAXY Sa
BAK 2.0247	03 39 00.	- 18 53 04.	17	16.3	GALAXY
PHL 8737	03 39 00.	- 24 52		18.5	BLUE STELLAR OBJECT
BAK 1.860	03 39 00.	- 28 50 35.	17	16.6	EXTRAGALACTIC NEBULA
TON-S 0370	03 39 00.	- 31 04		15.8	BLUE STAR
MCG-06-09-018	03 39 00.	- 33 55	90	15.	GALAXY
SHAP340-4538.2	03 39 00.	- 45 47 47.	24	17.5	GALAXY
SHAP340-5346.4	03 39 01.	- 53 56 00.	30	16.8	GALAXY
SHAP340-5145.1	03 39 02.	- 51 54 42.	18	17.5	GALAXY
SHAP340-5412.4	03 39 02.	- 54 22 00.	24	17.5	GALAXY
SHAP340-5503.5	03 39 02.	- 55 13 06.	60	16.0	GALAXY
BAK 2.0248	03 39 03.	- 18 53 34.	16	16.6	GALAXY
SHAP340-5337.2	03 39 03.	- 53 46 48.	24	16.5	GALAXY
HN 0236	03 39 03.	- 55 13			NEBULA
IC 1987	03 39 03.	- 55 14			NONSTELLAR OBJECT

OBJECT NAME	RIGHT ASCEN.	DECLINATION	DIAM.	MAGN.	TYPE OF OBJECT
BAK 2.0250	03 39 04.	- 18 12 10.	14	16.6	GALAXY
SHAP340-4540.2	03 39 04.	- 45 49 47.	24	16.5	GALAXY
SHAP340-4810.4	03 39 04.	- 48 19 59.	18	18.0	GALAXY
SHAP340-4837.2	03 39 04.	- 48 46 47.	24	17.0	GALAXY
SHAP340-5011.4	03 39 04.	- 50 21 00.	24	17.5	GALAXY
SHAP340-5144.8	03 39 04.	- 51 54 24.	18	17.25	GALAXY
BAK 2.0252	03 39 05.	- 17 55 34.	36	15.8	GALAXY
BAK 2.0251	03 39 05.	- 18 49 10.	12	16.6	GALAXY
BAK 1.862	03 39 05.	- 24 00 04.	61	15.3	EXTRAGALACTIC NEBULA
BAK 1.861	03 39 05.	- 28 06 34.	34	16.0	EXTRAGALACTIC NEBULA
SHAP340-5340.5	03 39 05.	- 53 50 06.	30	17.2	GALAXY
SHAP340-5352.1	03 39 05.	- 54 01 42.	24	17.7	GALAXY
SHAP340-5700.0	03 39 05.	- 57 09 36.	24	17.5	GALAXY
SHAP340-5702.8	03 39 05.	- 57 12 24.	24	16.75	GALAXY
ZWG 526.005	03 39 06.	+ 39 05		15.3	GALAXY
UGC 02828	03 39 06.	+ 39 05	90	15.3	GALAXY SBb
MCG-03-10-033	03 39 06.	- 17 54 30.	60	14.5	GALAXY
BAK 2.0253	03 39 06.	- 18 37 22.	18	16.6	GALAXY
PHL 8738	03 39 06.	- 26 07		13.4	BLUE STELLAR OBJECT
HPW 32	03 39 06.	- 35 08			DWARF GALAXY IN FORNAX
HOD.60 08	03 39 06.	- 35 08	36	16.5	E0 GALAXY IN FORNAX CLSTR
HPW 33	03 39 06.	- 37 33			DWARF GALAXY IN FORNAX
SHAP340-5320.7	03 39 06.	- 53 30 18.	24	17.4	GALAXY
SHAP340-5451.5	03 39 06.	- 55 01 06.	30	17.0	GALAXY
SHAP340-5514.4	03 39 06.	- 55 24 00.	42	16.25	GALAXY
BAK 2.0249	03 39 07.	- 21 59 22.	21	16.2	GALAXY
SHAP340-4552.0	03 39 07.	- 46 01 35.	24	17.75	GALAXY
SHAP340-5011.5	03 39 07.	- 50 21 05.	30	16.75	GALAXY
SHAP340-5150.5	03 39 08.	- 52 00 05.	18	17.25	GALAXY
SHAP340-5700.4	03 39 08.	- 57 10 00.	30	16.5	GALAXY
MCG-04-09-050	03 39 09.	- 24 00	120	13.	GALAXY
SHAP340-5256.0	03 39 09.	- 53 05 35.	18	17.5	GALAXY
SHAP340-5240.2	03 39 10.	- 52 49 47.	24	17.25	GALAXY
SHAP340-5727.9	03 39 10.	- 57 37 30.	30	16.5	GALAXY
SHAP340-4551.2	03 39 11.	- 46 00 47.	30	18.0	GALAXY
SHAP340-5120.1	03 39 11.	- 51 29 41.	18	17.5	GALAXY
SHAP340-5700.0	03 39 11.	- 57 09 36.	24	17.0	GALAXY
ZWG 417.004	03 39 12.	+ 08 00		15.0	GALAXY
UGC 02829	03 39 12.	+ 08 00	84	15.0	GALAXY S0
ZWG 465.008	03 39 12.	+ 15 50		15.7	GALAXY
UGC 02830	03 39 12.	+ 15 50	72	15.7	GALAXY (S0)
5ZW 361	03 39 12.	+ 26 25			COMPACT GALAXY
ZWG 526.006	03 39 12.	+ 38 48		15.2	GALAXY
MCG+06-09-003	03 39 12.	+ 39 06	60	15.	GALAXY
VB 238	03 39 12.	+ 55 21	1041		STELLAR RING
PHL 4481	03 39 12.	- 26 17		18.8	BLUE STELLAR OBJECT
SHAP340-4539.6	03 39 12.	- 45 49 11.	24	16.5	GALAXY
SHAP340-5330.8	03 39 12.	- 53 40 23.	24	17.1	GALAXY
BV 11	03 39 13.	- 26 45 42.		18.2	FAINT BLUE VARIABLE
SHAP340-5506.3	03 39 13.	- 55 15 53.	42	16.0	GALAXY
SHAP340-5511.1	03 39 13.	- 55 20 41.	42	16.0	GALAXY
SHAP340-5132.4	03 39 14.	- 51 41 59.	18	17.0	GALAXY
BAK 2.0255	03 39 15.	- 18 25 39.	24	16.4	GALAXY
BAK 1.863	03 39 15.	- 22 37 52.	14	16.0	EXTRAGALACTIC NEBULA
SHAP340-5201.2	03 39 15.	- 52 10 47.	24	18.0	GALAXY
SHAP340-5247.6	03 39 15.	- 52 57 11.	18	17.0	GALAXY
SHAP340-5707.8	03 39 15.	- 57 17 23.	24	18.0	GALAXY
SHAP340-5847.6	03 39 15.	- 58 57 12.	30	17.0	GALAXY
SHAP340-5643.0	03 39 16.	- 56 52 35.	30	17.25	GALAXY
SHAP340-5805.3	03 39 16.	- 58 14 53.	18	17.5	GALAXY
SHAP340-5826.2	03 39 16.	- 58 35 48.	24	17.0	GALAXY
SHAP340-5417.2	03 39 17.	- 54 26 47.	24	18.0	GALAXY
SHAP340-5709.5	03 39 17.	- 57 19 05.	30	17.75	GALAXY
SHAP340-5725.4	03 39 17.	- 57 34 59.	36	16.0	GALAXY
SHAP340-5820.8	03 39 17.	- 58 30 23.	24	17.25	GALAXY
MCG+01-10-007	03 39 18.	+ 08 02	48	15.	GALAXY
MCG+06-09-004	03 39 18.	+ 38 49 30.	30	15.	GALAXY
RLWT 070	03 39 18.	+ 52 21		16.	FAINT VERY BLUE STAR
BAK 2.0258	03 39 18.	- 17 29 39.	14	16.1	GALAXY
MCG-03-10-034	03 39 18.	- 20 03 30.	24	15.5	GALAXY
RNGC 1422	03 39 18.	- 21 51		13.0	GALAXY
PHL 4482	03 39 18.	- 26 46		18.7	BLUE STELLAR OBJECT
HPW 34	03 39 18.	- 33 55			DWARF GALAXY IN FORNAX
LB 01684	03 39 18.	- 51 27		13.7	FAINT BLUE STAR
SHAP340-5713.9	03 39 18.	- 57 23 29.	18	17.0	GALAXY
BAK 1.864	03 39 19.	- 20 03 57.	24	15.6	EXTRAGALACTIC NEBULA
SHAP340-4751.2	03 39 19.	- 48 00 46.	18	17.0	GALAXY
SHAP340-5638.5	03 39 19.	- 56 48 05.	36	18.0	GALAXY
SHAP340-5700.0	03 39 19.	- 57 09 35.	24	16.25	GALAXY
SHAP340-5231.6	03 39 20.	- 52 41 11.	18	16.5	GALAXY
SHAP340-5659.8	03 39 20.	- 57 09 23.	24	16.5	GALAXY
SHAP340-5200.0	03 39 21.	- 52 09 35.	18	17.25	GALAXY
SHAP340-5034.0	03 39 22.	- 50 43 34.	18	16.75	GALAXY
SHAP340-5637.4	03 39 22.	- 56 46 59.	24	17.75	GALAXY
BAK 2.0263	03 39 23.	- 15 26 45.	22	16.4	GALAXY
BAK 2.0260	03 39 23.	- 18 47 15.	13	16.4	GALAXY
BAK 2.0254	03 39 23.	- 26 29 21.	17	16.8	GALAXY
SHAP341-4540.5	03 39 23.	- 45 50 04.	18	16.75	GALAXY
SHAP340-5336.4	03 39 23.	- 53 45 59.	24	17.4	GALAXY
5ZW 362	03 39 24.	+ 25 55			COMPACT GALAXY
MCG+01-10-021	03 39 24.	- 04 52	144	12.	GALAXY
MCG-04-09-051	03 39 24.	- 21 50	96	13.5	GALAXY
PHL 4483	03 39 24.	- 22 03		18.5	BLUE STELLAR OBJECT
SHAP341-4539.4	03 39 24.	- 45 48 58.	30	17.5	GALAXY
SHAP340-5327.6	03 39 24.	- 53 37 11.	24	17.1	GALAXY
SHAP340-5651.1	03 39 24.	- 57 00 41.	24	18.0	GALAXY
BAK 2.0259	03 39 25.	- 23 27 57.	12	17.2	GALAXY
BAK 2.0256	03 39 25.	- 25 15 09.	22	16.8	GALAXY
SHAP340-5201.2	03 39 25.	- 52 10 46.	18	17.25	GALAXY
SHAP340-5359.2	03 39 25.	- 54 08 47.	18	17.5	GALAXY
SHAP340-5545.9	03 39 25.	- 55 55 29.	24	17.0	GALAXY
SHAP340-5448.0	03 39 26.	- 54 57 35.	30	17.0	GALAXY
SHAP340-5803.2	03 39 26.	- 58 12 47.	24	16.5	GALAXY
RNGC 1417	03 39 27.	- 04 52		13.0	GALAXY
MCG-03-10-035	03 39 27.	- 18 26	84	14.5	GALAXY
BAK 2.0257	03 39 27.	- 26 27 45.	13	17.0	GALAXY
SHAP340-5847.0	03 39 27.	- 58 56 35.	18	16.75	GALAXY
SHAP340-5201.7	03 39 28.	- 52 11 16.	18	17.25	GALAXY
HOLM 070A	03 39 29.	- 04 51	78	13.3	PART OF MULTIPLE GALAXY
BAK 1.865	03 39 29.	- 28 18 57.	19	16.3	EXTRAGALACTIC NEBULA
SHAP340-5027.8	03 39 29.	- 50 37 22.	24	16.75	GALAXY
SHAP340-5404.2	03 39 29.	- 54 13 46.	24	17.4	GALAXY
IC 0346	03 39 29.,4	- 18 25 34.			GALAXY SB(s)
UGC 02831	03 39 30.	+ 03 02	96	17.	GALAXY
MCG+00-10-014	03 39 30.	+ 03 02	66	15.	GALAXY
ZWG 417.005	03 39 30.	+ 04 05		15.1	GALAXY
KARA.73B 0132	03 39 30.	+ 04 05	36	15.1	ISOLATED GALAXY S
ZC 0339.5+0521	03 39 30.	+ 05 21	2490		CLUSTER OF GALAXIES

OBJECT NAME	RIGHT ASCEN.	DECLINATION	DIAM.	MAGN.	TYPE OF OBJECT
SHAH 146	03 39 30.	- 02 06	96		GROUP OF COMPACT GALAXIES
IC 0346	03 39 30.	- 18 25 29.			NONSTELLAR OBJECT
BAK 1.866	03 39 30.	- 27 58 57.	17	16.0	EXTRAGALACTIC NEBULA
SHAP341-4540.7	03 39 31.	- 45 50 16.	30	17.5	GALAXY
SHAP340-5203.4	03 39 31.	- 52 12 58.	18	17.5	GALAXY
BAK 2.0264	03 39 32.	- 19 29 20.	23	16.3	GALAXY
BAK 2.0261	03 39 32.	- 25 32 27.	14	17.0	GALAXY
SHAP340-5409.5	03 39 32.	- 54 19 04.	30	16.7	GALAXY
SHAP340-5708.0	03 39 32.	- 57 17 34.	30	16.75	GALAXY
BAK 2.0262	03 39 33.	- 26 13 15.	23	16.8	GALAXY
SHAP340-5054.8	03 39 33.	- 51 04 22.	18	17.0	GALAXY
SHAP341-5104.0	03 39 34.	- 51 13 34.	18	17.0	GALAXY
SHAP340-5314.9	03 39 34.	- 53 24 28.	24	17.5	GALAXY
SHAP340-5339.8	03 39 34.	- 53 49 22.	24	16.8	GALAXY
ARC 0451	03 39 35.	- 02 35		17.5	RICH CLUSTER OF GALAXIES
SHAP341-4821.0	03 39 35.	- 48 30 34.	18	17.5	GALAXY
SHAP340-5645.4	03 39 35.	- 56 54 58.	24	17.0	GALAXY
MCG-03-10-036	03 39 36.	- 19 03	36	15.	GALAXY
BAK 2.0266	03 39 36.	- 19 44 26.	19	16.4	GALAXY
BAK 1.867	03 39 36.	- 25 50 20.	13	16.9	EXTRAGALACTIC NEBULA
MCG-06-09-019	03 39 36.	- 35 02	72	15.	GALAXY
HPW 35	03 39 36.	- 37 49			DWARF GALAXY IN FORNAX
SHAP341-4820.2	03 39 36.	- 48 29 45.	36	16.75	GALAXY
SHAP340-5202.9	03 39 36.	- 52 12 28.	18	17.5	GALAXY
SHAP340-5555.6	03 39 36.	- 56 05 10.	36	17.5	GALAXY
SHAP340-5727.5	03 39 36.	- 57 37 04.	24	17.0	GALAXY
SHAP341-4547.5	03 39 37.	- 45 57 03.	24	17.25	GALAXY
SHAP341-4744.8	03 39 37.	- 47 54 21.	18	17.5	GALAXY
BAK 2.0267	03 39 38.	- 16 48 20.	13	16.5	GALAXY
BAK 2.0265	03 39 38.	- 24 34 44.	19	17.2	GALAXY
SHAP341-4731.6	03 39 38.	- 47 41 09.	24	16.75	GALAXY
SHAP341-5146.8	03 39 39.	- 51 56 22.	30	17.0	GALAXY
SHAP340-5741.5	03 39 39.	- 57 51 04.	24	17.0	GALAXY
SHAP341-5105.9	03 39 40.	- 51 15 27.	18	17.5	GALAXY
SHAP340-5411.7	03 39 40.	- 54 21 16.	24	16.5	GALAXY
SHAP340-5445.2	03 39 40.	- 54 54 46.	24	17.5	GALAXY
SHAP340-5702.2	03 39 40.	- 57 11 46.	24	16.5	GALAXY
SHAP340-5757.5	03 39 40.	- 58 07 04.	36	16.5	GALAXY
BAK 2.0269	03 39 41.	- 17 37 50.	12	16.6	GALAXY
BAK 2.0268	03 39 41.	- 19 03 14.	27	15.8	GALAXY
BAK 1.868	03 39 41.	- 28 16 38.	18	16.2	EXTRAGALACTIC NEBULA
SHAP340-5512.0	03 39 41.	- 55 21 34.	48	15.5	GALAXY
SHAP340-5821.9	03 39 41.	- 58 31 28.	24	16.75	GALAXY
MCG+06-09-005	03 39 42.	+ 38 06 30.	42	17.	GALAXY
MCG-02-10-007	03 39 42.	- 09 46	42	14.5	GALAXY
BAK 2.0271	03 39 42.	- 16 31 32.	16	16.5	GALAXY
BAK 2.0270	03 39 42.	- 17 41 08.	12	16.9	GALAXY
MCG-03-10-037	03 39 42.	- 18 39	72	15.	GALAXY
PHL 1562	03 39 42.	- 27 05		16.6	BLUE STELLAR OBJECT
HOD.60 09	03 39 42.	- 36 09	48	17.5	E0 GALAXY IN FORNAX CLSTR
SHAP341-5221.0	03 39 42.	- 52 30 33.	30	18.0	GALAXY
SHAP341-5313.8	03 39 42.	- 53 23 21.	24	17.5	GALAXY
SHAP340-5519.1	03 39 42.	- 55 28 40.	36	16.25	GALAXY
SHAP340-5734.5	03 39 42.	- 57 44 04.	24	16.5	GALAXY
SHAP341-5340.5	03 39 43.	- 53 50 03.	24	16.9	GALAXY
SHAP341-4533.0	03 39 44.	- 45 42 33.	30	17.5	GALAXY
SHAP341-4552.8	03 39 44.	- 46 02 21.	24	17.5	GALAXY
SHAP341-4818.0	03 39 44.	- 48 27 33.	30	18.0	GALAXY
MCG+00-10-015	03 39 45.	- 00 30	54	15.	GALAXY
RNGC 1418	03 39 45.	- 04 53		14.5	GALAXY
MCG+01-10-022	03 39 45.	- 04 53	60	14.5	GALAXY
BAK 2.0272	03 39 45.	- 18 52 26.	16	16.2	GALAXY
SHAP341-4758.1	03 39 45.	- 48 07 39.	24	17.75	GALAXY
SHAP341-5012.0	03 39 45.	- 50 21 33.	36	16.5	GALAXY
SHAP340-5641.4	03 39 45.	- 56 50 58.	18	17.5	GALAXY
SHAP340-5805.3	03 39 45.	- 58 14 52.	24	17.0	GALAXY
BAK 2.0274	03 39 46.	- 17 27 08.	14	15.8	GALAXY
BAK 2.0273	03 39 46.	- 17 44 56.	14	16.5	GALAXY
SHAP341-5137.0	03 39 46.	- 51 46 33.	18	17.0	GALAXY
SHAP341-5348.1	03 39 46.	- 53 57 39.	18	17.5	GALAXY
BAK 2.0275	03 39 47.	- 17 40 56.	23	15.6	GALAXY
SHAP341-5511.7	03 39 47.	- 55 21 15.	36	16.25	GALAXY
UGC 02832	03 39 48.	- 00 30	60	16.5	GALAXY Sc-IRR
HOLM 070B	03 39 48.	- 04 53	78	13.9	PART OF MULTIPLE GALAXY
MCG+01-10-023	03 39 48.	- 06 56	18	15.	GALAXY
PHL 1563	03 39 48.	- 27 03		18.5	BLUE STELLAR OBJECT
TON-S 0371	03 39 48.	- 27 03		16.0	BLUE STAR
HPW 36	03 39 48.	- 35 23			DWARF GALAXY IN FORNAX
LB 01685	03 39 48.	- 50 28		13.2	FAINT BLUE STAR
SHAP341-5309.2	03 39 48.	- 53 18 45.	24	16.9	GALAXY
SHAP341-5355.6	03 39 48.	- 54 05 09.	30	17.1	GALAXY
SHAP341-5408.2	03 39 48.	- 54 17 45.	24	17.4	GALAXY
SHAP341-5312.6	03 39 49.	- 53 22 09.	36	17.2	GALAXY
SHAP341-5314.8	03 39 49.	- 53 24 21.	24	16.7	GALAXY
SHAP340-5705.2	03 39 49.	- 57 14 45.	24	16.75	GALAXY
LB 00487	03 39 50.	+ 16 23 54.		16.2	FAINT BLUE STAR
LB 01476	03 39 50.	+ 25 23 54.		19.5	FAINT BLUE STAR
BAK 2.0276	03 39 50.	- 18 38 37.	45	16.0	GALAXY
BAK 1.871	03 39 50.	- 21 24 25.	38	12.9	EXTRAGALACTIC NEBULA
SHAP341-4528.1	03 39 50.	- 45 37 38.	24	17.0	GALAXY
SHAP341-4530.2	03 39 50.	- 45 39 44.	24	17.5	GALAXY
SHAP341-5332.4	03 39 50.	- 53 41 57.	24	16.6	GALAXY
SHAP341-5448.0	03 39 50.	- 54 57 33.	30	17.75	GALAXY
SHAP340-5719.2	03 39 50.	- 57 28 45.	42	16.75	GALAXY
SHAP340-5801.8	03 39 50.	- 58 11 21.	18	17.0	GALAXY
MCG-04-09-052	03 39 51.	- 21 24	60	14.	GALAXY
BAK 1.869	03 39 51.	- 29 04 32.	17	16.9	EXTRAGALACTIC NEBULA
SHAP341-5330.8	03 39 51.	- 53 40 21.	24	17.2	GALAXY
SHAP341-5336.2	03 39 51.	- 53 45 45.	24	17.1	GALAXY
SHAP341-5351.8	03 39 51.	- 54 01 21.	18	17.3	GALAXY
SHAP341-5542.2	03 39 51.	- 55 51 45.	36	16.5	GALAXY
SHAP340-5810.6	03 39 51.	- 58 20 09.	18	17.0	GALAXY
SHAP340-5858.2	03 39 52.	- 59 07 45.	36	16.0	GALAXY
BAK 1.870	03 39 53.	- 28 05 38.	19	16.4	EXTRAGALACTIC NEBULA
SHAP341-5651.9	03 39 53.	- 57 01 27.	24	16.0	GALAXY
SHAP341-5708.3	03 39 53.	- 57 17 51.	18	16.25	GALAXY
SHAP341-5708.6	03 39 53.	- 57 18 09.	18	16.75	GALAXY
ZC 0339.9+1816	03 39 54.	+ 18 16	2420		CLUSTER OF GALAXIES
DG 021	03 39 54.	+ 31 49	120		REFLECTION NEBULA
MCG+08-07-007	03 39 54.	+ 46 00	24	17.	GALAXY
VB 197	03 39 54.	+ 63 02	477		STELLAR RING
ISS 0070	03 39 54.	+ 63 02	482		STELLAR RING
BAK 2.0280	03 39 54.	- 15 53 55.	23	16.6	GALAXY
BAK 2.0279	03 39 54.	- 17 29 43.	20	16.2	GALAXY
HPW 37	03 39 54.	- 33 56			DWARF GALAXY IN FORNAX
HPW 38	03 39 54.	- 34 16			DWARF GALAXY IN FORNAX
MCG-06-09-020	03 39 54.	- 38 12	36	15.	GALAXY
SHAP341-5238.5	03 39 54.	- 52 48 03.	18	18.0	GALAXY

OBJECT NAME	RIGHT ASCEN.	DECLINATION	DIAM.	MAGN.	TYPE OF OBJECT
SHAP341-5356.0	03 39 54.	- 54 05 33.	30	17.4	GALAXY
SHAP340-5903.5	03 39 54.	- 59 13 03.	24	16.0	GALAXY
SS 10	03 39 55.	+ 31 49			DIFFUSE GALACTIC NEBULA
BAK 2.0281	03 39 55.	- 16 58 19.	24	16.7	GALAXY
BAK 1.874	03 39 55.	- 21 41 19.	13	16.4	EXTRAGALACTIC NEBULA
SHAP341-5245.9	03 39 55.	- 52 55 27.	18	17.5	GALAXY
SHAP341-5328.8	03 39 55.	- 53 38 21.	24	17.2	GALAXY
LB 01477	03 39 56.	+ 23 07 12.		18.4	FAINT BLUE STAR
SHAP341-4620.9	03 39 56.	- 46 30 26.	18	17.5	GALAXY
SHAP341-5247.3	03 39 56.	- 52 56 51.	24	16.75	GALAXY
SHAP341-5400.4	03 39 56.	- 54 09 57.	24	16.0	GALAXY
SHAP341-5637.5	03 39 56.	- 56 47 03.	36	16.75	GALAXY
BAK 2.0283	03 39 57.	- 18 04 19.	16	16.5	GALAXY
BAK 2.0282	03 39 57.	- 18 39 55.	29	16.5	GALAXY
BAK 2.0277	03 39 57.	- 25 13 13.	14	17.1	GALAXY
BAK 1.872	03 39 57.	- 27 21 55.	14	16.6	EXTRAGALACTIC NEBULA
SHAP341-4748.0	03 39 57.	- 47 57 32.	18	16.25	GALAXY
SHAP341-5243.5	03 39 57.	- 52 53 03.	18	17.5	GALAXY
SHAP341-5344.1	03 39 57.	- 53 53 39.	30	17.1	GALAXY
SHAP341-5500.1	03 39 57.	- 55 09 39.	48	17.0	GALAXY
SHAP341-5109.1	03 39 58.	- 51 18 38.	24	17.25	GALAXY
SHAP341-5322.8	03 39 58.	- 53 32 21.	18	17.4	GALAXY
SHAP341-5323.1	03 39 58.	- 53 32 39.	30	17.7	GALAXY
SHAP341-5326.0	03 39 58.	- 53 35 33.	18	17.1	GALAXY
SHAP341-5400.8	03 39 58.	- 54 10 21.	24	17.1	GALAXY
SHAP341-5336.8	03 39 59.	- 53 46 20.	24	17.0	GALAXY
SHAP341-5917.6	03 39 59.	- 59 27 09.	24	16.5	GALAXY
LBN 0773	03 40	+ 22 30	8100		BRIGHT NEBULA
KHAV 040	03 40	+ 54 34	11500		DARK NEBULA
UGC 02833	03 40 00.	+ 15 50	60	16.5	GALAXY Sc
LDN 1470	03 40 00.	+ 31 40	2520		DARK NEBULA
MCG+01-10-024	03 40 00.	- 04 27	42	13.5	GALAXY
PHL 4484	03 40 00.	- 21 32		17.0	BLUE STELLAR OBJECT
PHL 8739	03 40 00.	- 23 21		17.5	BLUE STELLAR OBJECT
BAK 2.0278	03 40 00.	- 25 35 13.	16	16.8	GALAXY
BAK 1.873	03 40 00.	- 28 58 55.	17	16.8	EXTRAGALACTIC NEBULA
SHAP341-4834.0	03 40 00.	- 48 43 32.	24	18.0	GALAXY
LB 01686	03 40 00.	- 51 13		13.5	FAINT BLUE STAR
SHAP341-5325.0	03 40 00.	- 53 34 32.	24	17.4	GALAXY
SHAP341-5641.6	03 40 00.	- 56 51 09.	18	17.25	GALAXY
SHAP341-5748.6	03 40 00.	- 57 58 09.	24	17.5	GALAXY
SHAP341-5651.6	03 40 01.	- 57 01 09.	24	17.5	GALAXY
SHAP341-5206.4	03 40 02.	- 52 15 56.	24	17.5	GALAXY
SHAP341-5356.7	03 40 02.	- 54 06 14.	30	17.4	GALAXY
BAK 2.0288	03 40 03.	- 18 01 55.	24	16.6	GALAXY
BAK 2.0285	03 40 03.	- 20 25 55.	13	16.6	GALAXY
SHAP341-5057.2	03 40 04.	- 51 06 44.	24	17.25	GALAXY
IC 0347	03 40 05.	- 04 28 25.			NONSTELLAR OBJECT
RNGC 1425	03 40 05.	- 30 04		12.0	GALAXY
SHAP341-4537.6	03 40 05.	- 45 47 08.	30	17.5	GALAXY
UGC 02834	03 40 06.	+ 42 37	60	18.	GALAXY DWRF IR
MCG-03-10-038	03 40 06.	- 16 09	24	15.	GALAXY
PHL 4485	03 40 06.	- 22 44		18.7	BLUE STELLAR OBJECT
BAK 1.876	03 40 06.	- 24 49 07.	18	15.8	EXTRAGALACTIC NEBULA
EOD.60 10	03 40 06.	- 34 19	42	17.0	EO GALAXY IN FORNAX CLSTR
SHAP341-5704.0	03 40 06.	- 57 13 32.	24	16.5	GALAXY
LB 01478	03 40 07.	+ 22 52 36.		20.4	FAINT BLUE STAR
BAK 2.0284	03 40 07.	- 26 22 37.	12	17.2	GALAXY
BAK 1.875	03 40 07.	- 28 01 37.	48	14.7	EXTRAGALACTIC NEBULA
BAK 2.0287	03 40 08.	- 23 54 06.	15	16.5	GALAXY
SHAP341-5543.2	03 40 08.	- 55 52 44.	30	16.5	GALAXY
MCG-02-10-008	03 40 09.	- 13 38	192	12.	GALAXY
BAK 1.877	03 40 09.	- 28 02 01.	23	16.6	EXTRAGALACTIC NEBULA
SHAP341-5325.6	03 40 09.	- 53 35 08.	18	17.4	GALAXY
SHAP341-5627.8	03 40 09.	- 56 37 20.	30	17.5	GALAXY
RNGC 1423	03 40 10.	- 06 32		15.0	GALAXY
BAK 2.0292	03 40 10.	- 19 28 18.	38	16.4	GALAXY
BAK 2.0286	03 40 10.	- 26 39 36.	17	17.0	GALAXY
BAK 1.878	03 40 10.	- 28 04 00.	30	16.9	EXTRAGALACTIC NEBULA
SHAP341-4530.5	03 40 11.	- 45 40 01.	30	17.25	GALAXY
SHAP341-5507.0	03 40 11.	- 55 16 32.	42	16.5	GALAXY
MCG+07-08-030	03 40 12.	+ 40 54	66	16.	GALAXY
UGC 02835	03 40 12.	+ 40 55	84	16.0	GALAXY Sb
MCG+08-07-008	03 40 12.	+ 46 00	18	17.	GALAXY
BAK 2.0294	03 40 12.	- 17 17 54.	25	16.4	GALAXY
MCG-04-09-053	03 40 12.	- 22 56	120	16.	GALAXY
BAK 1.880	03 40 12.	- 28 50 42.	18	16.6	EXTRAGALACTIC NEBULA
BAK 1.879	03 40 12.	- 29 02 00.	19	16.6	EXTRAGALACTIC NEBULA
SHAP341-5056.8	03 40 12.	- 51 06 19.	24	17.5	GALAXY
SHAP341-5402.9	03 40 12.	- 54 12 26.	24	17.3	GALAXY
SHAP341-5645.1	03 40 12.	- 56 54 38.	24	17.5	GALAXY
SHAP341-5707.2	03 40 12.	- 57 16 44.	24	17.75	GALAXY
SHAP341-5712.4	03 40 12.	- 57 21 56.	24	16.5	GALAXY
BAK 2.0295	03 40 13.	- 16 09 42.	34	15.1	GALAXY
BAK 2.0290	03 40 13.	- 23 41 42.	19	16.8	GALAXY
BAK 2.0289	03 40 13.	- 24 51 42.	20	16.8	GALAXY
SHAP341-5328.8	03 40 13.	- 53 38 20.	24	17.1	GALAXY
SHAP341-5542.0	03 40 13.	- 55 51 32.	42	16.5	GALAXY
RNGC 1421	03 40 14.	- 13 40		12.5	GALAXY
BAK 2.0296	03 40 14.	- 15 38 30.	18	16.4	GALAXY
SHAP341-5254.8	03 40 14.	- 53 04 20.	24	17.5	GALAXY
SHAP341-5535.2	03 40 14.	- 55 44 44.	42	17.0	GALAXY
SHAP341-5333.4	03 40 15.	- 53 42 56.	9	17.2	GALAXY
SHAP341-5341.1	03 40 15.	- 53 50 38.	18	16.2	GALAXY
RNGC 1420	03 40 16.	- 06 01			NON-EXISTENT OBJECT
BAK 1.882	03 40 16.	- 21 03 36.	19	16.5	EXTRAGALACTIC NEBULA
BAK 2.0291	03 40 16.	- 25 51 00.	19	17.1	GALAXY
SHAP341-5056.0	03 40 16.	- 51 05 31.	18	17.5	GALAXY
SHAP341-5642.2	03 40 16.	- 56 51 44.	36	16.5	GALAXY
SHAP341-5806.2	03 40 16.	- 58 15 44.	24	16.75	GALAXY
SHAP341-5309.6	03 40 17.	- 53 19 07.	24	16.7	GALAXY
SHAP341-5705.2	03 40 17.	- 57 14 44.	18	16.5	GALAXY
PHL 1564	03 40 18.	- 24 20		14.2	BLUE STELLAR OBJECT
TON-S 0372	03 40 18.	- 24 20		14.7	BLUE STAR
BAK 2.0293	03 40 18.	- 25 29 00.	16	17.0	GALAXY
PHL 4486	03 40 18.	- 27 03		17.8	BLUE STELLAR OBJECT
MCG-05-09-022	03 40 18.	- 28 01	90	14.	GALAXY
MCG-05-09-023	03 40 18.	- 30 03	360	12.	GALAXY
SHAP341-5013.2	03 40 18.	- 50 22 43.	30	17.5	GALAXY
BAK 1.881	03 40 19.	- 26 44 18.	14	16.4	EXTRAGALACTIC NEBULA
SHAP341-5012.9	03 40 19.	- 50 22 25.	24	17.0	GALAXY
SHAP341-5313.0	03 40 20.	- 53 22 31.	24	17.2	GALAXY
SHAP341-5357.9	03 40 20.	- 54 07 25.	24	17.3	GALAXY
SHAP341-5634.6	03 40 20.	- 56 44 08.	18	17.0	GALAXY
BAK 2.0297	03 40 21.	- 19 06 05.	13	16.8	GALAXY
RNGC 1427	03 40 21.	- 35 34		12.5	GALAXY
SHAP341-5202.2	03 40 21.	- 52 11 43.	18	17.5	GALAXY
SHAP341-5708.8	03 40 21.	- 57 18 20.	24	16.25	GALAXY

OBJECT NAME	RIGHT ASCEN.	DECLINATION	DIAM.	MAGN.	TYPE OF OBJECT
SHAP341-5509.2	03 40 22.	- 55 18 43.	36	16.0	GALAXY
SHAP341-5633.8	03 40 22.	- 56 43 19.	18	17.25	GALAXY
SHAP341-5954.6	03 40 22.	- 60 04 08.	30	16.0	GALAXY
ZC 0340.4+2732	03 40 24.	+ 27 32	870		CLUSTER OF GALAXIES
BAK 2.0301	03 40 24.	- 16 07 41.	23	16.4	GALAXY
PHL 1565	03 40 24.	- 22 22		17.9	BLUE STELLAR OBJECT
MCG-06-09-022	03 40 24.	- 35 18	72	13.	GALAXY
MCG-06-09-021	03 40 24.	- 35 32	180	12.4	GALAXY
HPW 39	03 40 24.	- 35 43			DWARF GALAXY IN FORNAX
LB 01687	03 40 24.	- 49 28		16.1	FAINT BLUE STAR
RNGC 1433	03 40 25.	- 47 24		11.0	GALAXY
SHAP341-4935.6	03 40 25.	- 49 45 07.	30	17.25	GALAXY
SHAP341-5154.9	03 40 25.	- 52 04 25.	24	17.25	GALAXY
SHAP341-5213.5	03 40 25.	- 52 23 01.	18	17.75	GALAXY
SHAP341-5442.2	03 40 25.	- 54 51 43.	24	16.5	GALAXY
SHAP341-5736.5	03 40 25.	- 57 46 01.	24	16.5	GALAXY
SHAP341-5911.5	03 40 25.	- 59 21 01.	24	16.75	GALAXY
LB 01479	03 40 26.	+ 24 17 54.		19.6	FAINT BLUE STAR
BAK 2.0298	03 40 26.	- 23 41 53.	13	16.7	GALAXY
BAK 1.884	03 40 26.	- 25 49 05.	26	16.1	EXTRAGALACTIC NEBULA
SHAP342-4635.0	03 40 26.	- 46 44 30.	18	17.5	GALAXY
SHAP341-5658.5	03 40 26.	- 57 08 01.	24	17.75	GALAXY
SHAP341-5939.4	03 40 26.	- 59 48 55.	30	17.25	GALAXY
BAK 2.0303	03 40 27.	- 18 05 41.	25	16.4	GALAXY
BAK 1.885	03 40 27.	- 25 08 17.	14	15.9	EXTRAGALACTIC NEBULA
BAK 1.883	03 40 27.	- 28 07 41.	19	16.4	EXTRAGALACTIC NEBULA
RNGC 1428	03 40 27.	- 35 19			GALAXY
SHAP342-4713.5	03 40 27.	- 47 23 00.	462	11.4	GALAXY
SHAP341-5049.6	03 40 27.	- 50 59 07.	48	17.0	GALAXY
SHAP341-5339.5	03 40 27.	- 53 49 01.	18	17.1	GALAXY
SHAP341-5357.8	03 40 27.	- 54 07 19.	30	17.2	GALAXY
SHAP341-5947.0	03 40 27.	- 59 56 31.	36	16.75	GALAXY
BAK 2.0304	03 40 28.	- 17 40 05.	36	16.4	GALAXY
SHAP341-5109.7	03 40 28.	- 51 19 13.	24	17.25	GALAXY
BAK 2.0299	03 40 29.	- 22 59 29.	16	16.5	GALAXY
SHAP341-5323.6	03 40 29.	- 53 33 07.	30	17.1	GALAXY
SHAP341-5331.4	03 40 29.	- 53 40 55.	24	17.4	GALAXY
BAK 2.0306	03 40 30.	- 15 24 05.	16	16.4	GALAXY
BAK 1.886	03 40 30.	- 26 43 53.	22	16.1	EXTRAGALACTIC NEBULA
SHAP341-5704.8	03 40 30.	- 57 14 19.	24	16.5	GALAXY
SHAP341-5708.2	03 40 30.	- 57 17 43.	24	16.75	GALAXY
BAK 2.0300	03 40 31.	- 23 43 47.	14	16.9	GALAXY
SHAP341-5057.7	03 40 31.	- 51 07 12.	48	15.25	GALAXY
SHAP341-5214.2	03 40 31.	- 52 23 42.	18	17.5	GALAXY
SHAP341-5342.6	03 40 31.	- 53 52 07.	24	17.4	GALAXY
SHAP341-5417.1	03 40 31.	- 54 26 37.	24	17.25	GALAXY
SHAP341-5545.2	03 40 31.	- 55 54 43.	30	17.5	GALAXY
SHAP341-5912.5	03 40 31.	- 59 22 01.	42	16.25	GALAXY
LB 01480	03 40 32.	+ 25 21 12.		17.8	FAINT BLUE STAR
BAK 2.0305	03 40 32.	- 18 04 41.	11	16.2	GALAXY
BAK 2.0302	03 40 32.	- 23 46 23.	12	16.6	GALAXY
HN 0237	03 40 32.	- 51 07			NEBULA
IC 1999	03 40 32.	- 51 08			NONSTELLAR OBJECT
SHAP341-5354.8	03 40 32.	- 54 04 19.	30	17.7	GALAXY
SHAP341-5556.2	03 40 32.	- 56 05 43.	24	17.5	GALAXY
BAK 1.887	03 40 33.	- 28 47 47.	20	16.6	EXTRAGALACTIC NEBULA
SHAP341-5335.0	03 40 33.	- 53 44 30.	24	17.3	GALAXY
SHAP341-5433.2	03 40 33.	- 54 42 43.	36	17.75	GALAXY
SHAP341-5434.0	03 40 33.	- 54 43 31.	24	17.5	GALAXY
SHAP341-5555.3	03 40 33.	- 56 04 49.	36	17.25	GALAXY
SHAP341-5736.2	03 40 33.	- 57 45 43.	24	15.5	GALAXY
BAK 2.0307	03 40 34.	- 18 58 53.	12	16.8	GALAXY
SHAP341-5400.5	03 40 34.	- 54 10 00.	30	17.2	GALAXY
BAK 2.0308	03 40 35.	- 19 16 17.	14	16.7	GALAXY
SHAP342-4649.2	03 40 35.	- 46 58 42.	24	17.5	GALAXY
SHAP342-4747.9	03 40 35.	- 47 57 24.	24	17.0	GALAXY
SHAP341-5334.6	03 40 35.	- 53 44 06.	24	17.1	GALAXY
SHAP341-5457.2	03 40 35.	- 55 06 42.	36	16.75	GALAXY
ZWG 526.007	03 40 36.	+ 39 09		13.8	GALAXY
UGC 02836	03 40 36.	+ 39 09	84	13.8	GALAXY E-S0
MCG+07-08-031	03 40 36.	+ 39 50 30.	72	15.	GALAXY
ZWG 541.025	03 40 36.	+ 39 52		15.1	GALAXY
UGC 02837	03 40 36.	+ 39 52	108	15.1	GALAXY SBb
MCG-02-10-009	03 40 36.	- 13 04	180	13.5	GALAXY
MCG-03-10-039	03 40 36.	- 19 11	36	15.5	GALAXY
RNGC 1426	03 40 36.	- 22 16		13.0	GALAXY
PHL 8740	03 40 36.	- 24 01		18.2	BLUE STELLAR OBJECT
PHL 1566	03 40 36.	- 24 54		17.5	BLUE STELLAR OBJECT
BAK 1.889	03 40 36.	- 25 37 23.	12	16.1	EXTRAGALACTIC NEBULA
PHL 4487	03 40 36.	- 29 36		17.9	BLUE STELLAR OBJECT
SHAP341-5554.8	03 40 36.	- 56 04 18.	36	17.0	GALAXY
SHAP341-5659.1	03 40 36.	- 57 08 37.	24	17.5	GALAXY
SHAP341-5251.2	03 40 37.	- 53 00 42.	18	17.0	GALAXY
BAK 1.888	03 40 38.	- 28 53 29.	18	16.8	EXTRAGALACTIC NEBULA
SHAP342-4748.4	03 40 38.	- 47 57 54.	24	17.25	GALAXY
SHAP341-5307.2	03 40 38.	- 53 16 42.	24	17.6	GALAXY
SHAP341-5316.5	03 40 38.	- 53 26 00.	36	16.5	GALAXY
SHAP341-5558.0	03 40 38.	- 56 07 30.	24	17.75	GALAXY
SHAP341-5810.8	03 40 38.	- 58 20 19.	18	17.0	GALAXY
SHAP341-5347.6	03 40 39.	- 53 57 06.	24	16.7	GALAXY
SHAP341-5357.5	03 40 39.	- 54 07 00.	30	17.0	GALAXY
BAK 2.0310	03 40 40.	- 16 30 04.	16	16.8	GALAXY
SHAP342-5311.0	03 40 40.	- 53 20 30.	24	17.0	GALAXY
SHAP341-5330.6	03 40 40.	- 53 40 06.	24	17.7	GALAXY
SHAP341-5435.6	03 40 40.	- 54 45 06.	24	17.0	GALAXY
SHAP341-5647.6	03 40 40.	- 56 57 06.	30	17.5	GALAXY
LB 01481	03 40 41.	+ 25 07 18.		19.5	FAINT BLUE STAR
BAK 2.0312	03 40 41.	- 16 29 04.	13	16.6	GALAXY
SHAP342-5304.6	03 40 41.	- 53 14 06.	18	17.4	GALAXY
SHAP342-5326.8	03 40 41.	- 53 36 18.	24	17.6	GALAXY
SHAP342-5329.5	03 40 41.	- 53 39 00.	24	16.9	GALAXY
SHAP341-5518.2	03 40 41.	- 55 27 42.	42	16.5	GALAXY
SHAP341-5614.4	03 40 41.	- 56 23 54.	24	17.5	GALAXY
SHAP341-5640.1	03 40 41.	- 56 49 36.	18	17.25	GALAXY
SHAP341-5752.4	03 40 41.	- 58 01 54.	24	17.0	GALAXY
ZC 0340.7+2429	03 40 42.	+ 24 29	1410		CLUSTER OF GALAXIES
MCG+06-09-006	03 40 42.	+ 39 08	36	14.5	GALAXY
ZWG 391.030	03 40 42.	- 01 10		15.3	GALAXY
MCG-03-10-040	03 40 42.	- 17 21 30.	60	16.	GALAXY
BAK 2.0311	03 40 42.	- 19 11 04.	40	16.0	GALAXY
MCG-04-09-054	03 40 42.	- 22 17	54	13.	GALAXY
MCG-06-09-023	03 40 42.	- 34 05	30	15.	GALAXY
SHAP341-5722.9	03 40 42.	- 57 32 24.	24	16.5	GALAXY
BAK 2.0314	03 40 43.	- 16 30 22.	16	16.8	GALAXY
SHAP342-5040.0	03 40 43.	- 50 49 30.	30	16.75	GALAXY
SHAP342-5424.2	03 40 43.	- 54 33 42.	30	17.5	GALAXY
SHAP342-5225.9	03 40 44.	- 52 35 24.	24	18.0	GALAXY
SHAP342-5327.8	03 40 44.	- 53 37 18.	24	17.6	GALAXY

OBJECT NAME	RIGHT ASCEN.	DECLINATION	DIAM.	MAGN.	TYPE OF OBJECT
SHAP342-5402.2	03 40 44.	- 54 11 42.	24	16.8	GALAXY
RNGC 1424	03 40 45.	- 04 53		14.5	GALAXY
MCG+01-10-026	03 40 45.	- 04 53	96	14.5	GALAXY
MCG+01-10-025	03 40 45.	- 06 32 30.	54	15.	GALAXY
BAK 2.0309	03 40 45.	- 26 11 46.	26	17.0	GALAXY
SHAP342-5103.2	03 40 45.	- 51 12 42.	24	17.0	GALAXY
SHAP342-5111.3	03 40 45.	- 51 20 48.	24	18.0	GALAXY
SHAP342-5342.4	03 40 45.	- 53 51 54.	24	16.9	GALAXY
SHAP342-4657.6	03 40 46.	- 47 07 05.	24	17.25	GALAXY
SHAP342-5307.6	03 40 46.	- 53 17 06.	18	17.6	GALAXY
SHAP342-5332.2	03 40 46.	- 53 41 42.	24	17.2	GALAXY
BAK 2.0315	03 40 47.	- 18 50 10.	20	17.0	GALAXY
BAK 1.891	03 40 47.	- 25 40 58.	14	16.8	EXTRAGALACTIC NEBULA
SHAP342-5337.6	03 40 47.	- 53 47 06.	24	17.5	GALAXY
UGC 02838	03 40 48.	+ 23 53	84	16.0	GALAXY Sc
MCG+04-09-005	03 40 48.	+ 23 53	78	17.	GALAXY
PHL 4488	03 40 48.	- 21 30		18.5	BLUE STELLAR OBJECT
SHAP342-5218.0	03 40 48.	- 52 27 29.	18	17.25	GALAXY
SHAP342-5346.6	03 40 48.	- 53 56 06.	30	16.8	GALAXY
SHAP341-5745.2	03 40 48.	- 57 54 42.	18	17.0	GALAXY
SHAP342-5057.2	03 40 49.	- 51 06 41.	18	17.75	GALAXY
BAK 2.0317	03 40 50.	- 17 20 16.	54	16.1	GALAXY
BAK 2.0313	03 40 50.	- 24 18 10.	41	16.6	GALAXY
SHAP342-5223.8	03 40 50.	- 52 33 17.	18	17.5	GALAXY
SHAP342-5305.0	03 40 50.	- 53 14 29.	24	16.7	GALAXY
SHAP342-5315.6	03 40 50.	- 53 25 05.	36	17.5	GALAXY
SHAP341-5902.2	03 40 50.	- 59 11 42.	42	16.5	GALAXY
SHAP342-5417.6	03 40 51.	- 54 27 05.	24	17.0	GALAXY
SHAP342-5506.2	03 40 51.	- 55 15 41.	42	17.25	GALAXY
SHAP342-5643.1	03 40 51.	- 56 52 36.	24	17.0	GALAXY
BC 3C093	03 40 51.47	+ 04 48 21.6		18.09	QUASI-STELLAR OBJECT
SHB 074	03 40 51.5	+ 04 48 21.		18.1	QUASI-STELLAR OBJECT
SHAP342-5136.0	03 40 52.	- 51 45 29.	36	17.0	GALAXY
SHAP342-5309.0	03 40 52.	- 53 18 29.	24	17.5	GALAXY
SHAP342-5321.2	03 40 52.	- 53 30 41.	24	17.5	GALAXY
SHAP342-5355.5	03 40 52.	- 54 04 59.	30	17.6	GALAXY
SHAP342-5322.7	03 40 53.	- 53 32 11.	24	16.2	GALAXY
SHAP342-5332.5	03 40 53.	- 53 41 59.	24	17.2	GALAXY
SHAP342-5350.0	03 40 53.	- 53 59 29.	18	17.6	GALAXY
SHAP342-5433.4	03 40 53.	- 54 42 53.	30	16.5	GALAXY
MCG+08-07-009	03 40 54.	+ 46 06	60	16.	GALAXY
KARA.68 041	03 40 54.	- 21 30	47		DWARF GALAXY
HOD.60 11	03 40 54.	- 34 04	36	15.0	E0 GALAXY IN FORNAX CLSTR
SHAP342-4606.5	03 40 54.	- 46 15 59.	24	17.5	GALAXY
SHAP342-5055.6	03 40 54.	- 51 05 05.	24	17.25	GALAXY
SHAP342-5428.2	03 40 54.	- 54 37 41.	24	17.75	GALAXY
SHAP341-5759.9	03 40 54.	- 58 09 24.	12	17.0	GALAXY
B 004	03 40 55.	+ 31 39			DARK OBJECT
SHAP342-4747.8	03 40 55.	- 47 57 17.	24	17.5	GALAXY
SHAP342-5040.8	03 40 55.	- 50 50 17.	24	17.5	GALAXY
SHAP342-5338.4	03 40 55.	- 53 47 53.	24	17.0	GALAXY
BAK 2.0318	03 40 56.	- 16 49 39.	28	15.6	GALAXY
SHAP342-5107.6	03 40 56.	- 51 17 05.	24	17.75	GALAXY
BAK 2.0320	03 40 57.	- 15 31 57.	18	16.7	GALAXY
BAK 2.0316	03 40 57.	- 25 43 46.	19	17.0	GALAXY
IC 1988	03 40 57.	- 40 02 41.			NONSTELLAR OBJECT
SHAP342-5328.0	03 40 57.	- 53 37 29.	24	16.8	GALAXY
SHAP342-4605.6	03 40 58.	- 46 15 04.	24	17.0	GALAXY
SHAP342-5302.8	03 40 58.	- 53 12 17.	24	16.6	GALAXY
SHAP342-5026.1	03 40 59.	- 50 35 35.	24	17.5	GALAXY
SHAP342-5252.0	03 40 59.	- 53 01 29.	18	18.0	GALAXY
SHAP342-5647.2	03 40 59.	- 56 56 41.	18	16.5	GALAXY
LBN 0758	03 41	+ 31 59	600		BRIGHT NEBULA
UGC 02839	03 41 00.	+ 14 10	60	18.	GALAXY Sc-IRR
UGC 02840	03 41 00.	+ 22 29	108	16.0	GALAXY
MCG+04-09-006	03 41 00.	+ 22 29	36	16.	GALAXY
UGC 02841	03 41 00.	+ 39 11	66	16.0	GALAXY S0
ZC 0341.0+8235	03 41 00.	+ 82 35	1210		CLUSTER OF GALAXIES
PHL 8741	03 41 00.	- 21 34		18.4	BLUE STELLAR OBJECT
PHL 4489	03 41 00.	- 26 07		18.4	BLUE STELLAR OBJECT
SHAP342-5041.4	03 41 00.	- 50 50 53.	24	17.0	GALAXY
BAK 2.0321	03 41 01.	- 18 06 15.	15	16.3	GALAXY
SHAP342-4614.2	03 41 02.	- 46 23 40.	24	16.75	GALAXY
SHAP342-4709.4	03 41 02.	- 47 18 52.	24	16.5	GALAXY
SHAP342-5317.2	03 41 02.	- 53 26 41.	24	16.2	GALAXY
SHAP342-5348.0	03 41 02.	- 53 57 29.	24	17.4	GALAXY
BAK 2.0322	03 41 03.	- 17 51 51.	22	16.3	GALAXY
BAK 2.0323	03 41 03.	- 18 03 51.	13	16.2	GALAXY
RNGC 1437A	03 41 03.	- 36 26			GALAXY
SHAP342-5340.8	03 41 03.	- 53 50 17.	36	16.3	GALAXY
SHAP342-5728.2	03 41 03.	- 57 37 41.	18	17.0	GALAXY
BAK 2.0326	03 41 04.	- 15 41 39.	17	17.0	GALAXY
SHAP342-5501.1	03 41 04.	- 55 10 35.	54	16.5	GALAXY
BAK 2.0324	03 41 05.	- 18 31 51.	34	16.4	GALAXY
BAK 1.892	03 41 05.	- 27 45 03.	17	16.6	EXTRAGALACTIC NEBULA
SHAP342-5207.2	03 41 05.	- 52 16 40.	12	17.5	GALAXY
DG 022	03 41 06.	+ 32 08	660		REFLECTION NEBULA
MCG+08-07-010	03 41 06.	+ 46 13	23	17.	GALAXY
BAK 2.0325	03 41 06.	- 18 38 15.	16	16.3	GALAXY
PHL 8742	03 41 06.	- 23 10		17.4	BLUE STELLAR OBJECT
BAK 2.0319	03 41 06.	- 25 44 21.	18	17.0	GALAXY
BAK 1.893	03 41 06.	- 25 52 03.	12	16.7	EXTRAGALACTIC NEBULA
HPW 40	03 41 06.	- 34 33			DWARF GALAXY IN FORNAX
DV.56 N1437A	03 41 06.	- 36 26			GALAXY
BAK 1.894	03 41 07.	- 23 06 51.	14	16.2	EXTRAGALACTIC NEBULA
SHAP342-4655.1	03 41 07.	- 47 04 34.	18	16.75	GALAXY
SHAP342-5109.4	03 41 07.	- 51 18 52.	24	17.0	GALAXY
CED 018B	03 41 08.	+ 32 08	9000		DIFFUSE GALACTIC NEBULA
BAK 1.896	03 41 08.	- 22 58 15.	13	16.4	EXTRAGALACTIC NEBULA
SHAP342-5155.9	03 41 08.	- 52 05 22.	24	17.5	GALAXY
SHAP342-5753.0	03 41 08.	- 58 02 29.	18	17.5	GALAXY
TC 0341-36	03 41 09.	- 36 25 30.			UNUSUAL SOUTHERN GALAXY
SHAP342-5348.4	03 41 09.	- 53 57 52.	24	17.4	GALAXY
SHAP342-5615.8	03 41 09.	- 56 25 17.	18	18.0	GALAXY
RNGC 1430	03 41 10.	- 18 24			NON-EXISTENT OBJECT
BAK 1.895	03 41 10.	- 26 08 33.	17	17.0	EXTRAGALACTIC NEBULA
SHAP342-4609.6	03 41 10.	- 46 19 04.	24	17.0	GALAXY
SHAP342-5213.4	03 41 10.	- 52 22 52.	24	16.5	GALAXY
BAK 2.0329	03 41 11.	- 18 58 26.	22	16.6	GALAXY
SHAP342-5331.6	03 41 11.	- 53 41 04.	24	17.6	GALAXY
SHAP342-5749.0	03 41 11.	- 57 58 29.	18	18.0	GALAXY
ZWG 417.006	03 41 12.	+ 06 22		15.2	GALAXY
BLWT 071	03 41 12.	+ 52 14		12.5	FAINT VERY BLUE STAR
MCG-03-10-041	03 41 12.	- 16 10	120	15.	GALAXY
PHL 4492	03 41 12.	- 22 22		18.0	BLUE STELLAR OBJECT
PHL 4490	03 41 12.	- 23 18		18.6	BLUE STELLAR OBJECT
PHL 8743	03 41 12.	- 23 58		18.6	BLUE STELLAR OBJECT
MCG-06-09-024	03 41 12.	- 36 25	84	15.5	GALAXY

OBJECT NAME	RIGHT ASCEN.	DECLINATION	DIAM.	MAGN.	TYPE OF OBJECT
HPW 50	03 41 12.	- 36 26			IRREG GALAXY IN FORNAX
SHAP342-5338.4	03 41 12.	- 53 47 52.	24	16.6	GALAXY
SHAP342-5653.2	03 41 12.	- 57 02 40.	18	17.5	GALAXY
SHAP342-5718.1	03 41 12.	- 57 27 34.	18	16.75	GALAXY
SHAP342-5333.0	03 41 13.	- 53 42 28.	24	17.0	GALAXY
BAK 2.0330	03 41 14.	- 21 29 14.	22	16.5	GALAXY
BAK 2.0327	03 41 14.	- 25 23 15.	37	16.8	GALAXY
SHAP342-5444.6	03 41 14.	- 54 54 04.	36	16.5	GALAXY
SHAP342-5635.1	03 41 14.	- 56 44 34.	30	17.75	GALAXY
SHAP342-5741.4	03 41 14.	- 57 50 52.	12	17.5	GALAXY
SHAP342-5113.4	03 41 15.	- 51 22 52.	18	17.0	GALAXY
SHAP342-5705.6	03 41 15.	- 57 15 04.	12	17.5	GALAXY
SHAP342-5740.9	03 41 15.	- 57 50 22.	18	16.5	GALAXY
BAK 2.0328	03 41 16.	- 24 20 08.	24	16.8	GALAXY
SHAP342-5055.2	03 41 16.	- 51 04 40.	18	17.5	GALAXY
SHAP342-5124.2	03 41 17.	- 51 33 40.	18	17.0	GALAXY
SHAP342-5340.1	03 41 17.	- 53 49 34.	24	17.4	GALAXY
SHAP342-5757.8	03 41 17.	- 58 07 16.	24	17.0	GALAXY
ZWG 391.031	03 41 18.	+ 00 27		15.3	GALAXY
UGC 02842	03 41 18.	+ 00 27	66	15.3	GALAXY Sc
MCG+00-10-016	03 41 18.	+ 00 28	36	14.5	GALAXY
MCG+01-10-008	03 41 18.	+ 06 23	36	15.	GALAXY
MCG+08-07-011	03 41 18.	+ 45 41	60	17.	GALAXY
UGC 02843	03 41 18.	+ 46 04	66	16.5	GALAXY Sb-c
BAK 2.0336	03 41 18.	- 16 10 14.	94	15.2	GALAXY
PHL 4493	03 41 18.	- 29 16		18.5	BLUE STELLAR OBJECT
SHAP342-5218.0	03 41 18.	- 52 27 28.	12	17.25	GALAXY
SHAP342-5337.6	03 41 18.	- 53 47 04.	24	16.6	GALAXY
BAK 2.0335	03 41 19.	- 18 26 02.	13	16.8	GALAXY
SHAP342-5730.6	03 41 19.	- 57 40 04.	18	16.75	GALAXY
SHAP342-5217.7	03 41 20.	- 52 27 10.	18	18.0	GALAXY
SHAP342-5611.8	03 41 20.	- 56 21 16.	30	16.75	GALAXY
SHAP342-5635.8	03 41 20.	- 56 45 16.	24	17.25	GALAXY
BAK 2.0339	03 41 21.	- 15 15 02.	14	16.2	GALAXY
BAK 2.0333	03 41 21.	- 22 12 02.	19	16.3	GALAXY
SHAP342-5341.0	03 41 21.	- 53 50 28.	24	16.9	GALAXY
SHAP342-5630.7	03 41 21.	- 56 40 10.	24	17.25	GALAXY
PHL 4491	03 41 22.	- 23 09		17.2	BLUE STELLAR OBJECT
BAK 2.0332	03 41 22.	- 24 39 50.	17	16.8	GALAXY
BAK 2.0331	03 41 22.	- 25 33 14.	12	16.6	GALAXY
SHAP342-5626.9	03 41 22.	- 56 36 22.	24	16.5	GALAXY
BAK 1.897	03 41 23.	- 25 40 02.	13	15.8	EXTRAGALACTIC NEBULA
SHAP342-5107.6	03 41 23.	- 51 17 03.	18	17.5	GALAXY
SHAP342-5231.4	03 41 23.	- 52 40 51.	24	17.0	GALAXY
SHAP342-5336.6	03 41 23.	- 53 46 03.	24	17.5	GALAXY
SHAP342-5651.3	03 41 23.	- 57 00 46.	18	17.0	GALAXY
LB 01482	03 41 24.	+ 25 06 24.		15.0	FAINT BLUE STAR
DG 023	03 41 24.	+ 32 01	180		REFLECTION NEBULA
OCL 0409	03 41 24.	+ 32 08		10.	OPEN STAR CLUSTER
BAK 2.0340	03 41 24.	- 16 26 14.	26	15.6	GALAXY
PHL 1567	03 41 24.	- 24 49		14.0	BLUE STELLAR OBJECT
TON-S 0373	03 41 24.	- 26 26		16.0	BLUE STAR
HOD.60 17	03 41 24.	- 36 15	84	15.0	IRRG GLXY IN FORNAX CLSTR
SHAP342-5929.9	03 41 24.	- 59 39 22.	30	15.75	GALAXY
IC 0348	03 41 25.	+ 32 00 36.	360		OPEN CLUSTER
SHAP342-5248.2	03 41 25.	- 52 57 39.	24	17.25	GALAXY
SHAP342-5323.0	03 41 25.	- 53 32 27.	24	16.9	GALAXY
IC 1985	03 41 26.	+ 32 00 24.			NONSTELLAR OBJECT
CED 020	03 41 26.	+ 32 01	500		DIFFUSE GALACTIC NEBULA
BAK 2.0338	03 41 26.	- 21 23 38.	55	15.8	GALAXY
BAK 2.0337	03 41 26.	- 22 46 38.	19	16.7	GALAXY
BAK 2.0334	03 41 26.	- 26 26 14.	17	16.9	GALAXY
SHAP343-4724.0	03 41 26.	- 47 33 27.	24	18.0	GALAXY
SHAP342-5053.4	03 41 26.	- 51 02 51.	18	17.5	GALAXY
SHAP342-5226.4	03 41 26.	- 52 35 51.	18	17.25	GALAXY
SHAP342-5442.9	03 41 26.	- 54 52 21.	30	18.0	GALAXY
VDB.66N 019	03 41 27.	+ 32 00	336		REFLECTION NEBULA
RNGC 1436	03 41 27.	- 36 17			NON-EXISTENT OBJECT
SHAP343-4720.8	03 41 27.	- 47 30 15.	18	17.5	GALAXY
SHAP343-4746.6	03 41 27.	- 47 56 03.	24	17.5	GALAXY
SHAP342-5151.3	03 41 27.	- 52 00 45.	24	17.5	GALAXY
SHAP342-5335.8	03 41 27.	- 53 45 15.	30	17.1	GALAXY
BAK 2.0341	03 41 28.	- 19 58 49.	16	16.2	GALAXY
SHAP342-5556.3	03 41 28.	- 56 05 45.	30	17.5	GALAXY
SHAP342-5322.5	03 41 29.	- 53 31 57.	24	17.4	GALAXY
SHAP342-5337.6	03 41 29.	- 53 47 03.	30	16.3	GALAXY
DG 024	03 41 30.	+ 32 11	120		REFLECTION NEBULA
MCG+08-07-012	03 41 30.	+ 45 49	36	16.	GALAXY
ZWG 555.005	03 41 30.	+ 45 50		15.3	GALAXY
UGC 02844	03 41 30.	+ 45 50	84	15.3	GALAXY E
MCG+13-03-008	03 41 30.	+ 79 31	24	17.	GALAXY
MCG+01-10-027	03 41 30.	- 04 10 30.	60	15.	GALAXY
MCG-03-10-042	03 41 30.	- 14 31	114	13.5	GALAXY
MCG-04-09-055	03 41 30.	- 21 24	78	15.5	GALAXY
TON-S 0374	03 41 30.	- 24 49		14.6	BLUE STAR
SHAP343-4724.2	03 41 30.	- 47 33 39.	24	18.0	GALAXY
SHAP342-5234.2	03 41 30.	- 52 43 39.	18	17.0	GALAXY
SHAP342-5358.9	03 41 30.	- 54 08 21.	24	16.9	GALAXY
SHAP342-5514.0	03 41 30.	- 55 23 27.	30	17.5	GALAXY
LB 03327	03 41 30.	- 63 28		13.5	FAINT BLUE STAR
CED 021	03 41 32.	+ 32 11	360		DIFFUSE GALACTIC NEBULA
SHAP342-5305.0	03 41 32.	- 53 14 27.	24	17.5	GALAXY
SHAP342-5307.7	03 41 32.	- 53 17 09.	24	17.5	GALAXY
SHAP342-5350.8	03 41 32.	- 54 00 15.	24	17.2	GALAXY
SHAP342-5508.0	03 41 32.	- 55 17 27.	36	17.25	GALAXY
SHAP342-5707.6	03 41 32.	- 57 17 03.	18	17.0	GALAXY
BAK 2.0342	03 41 33.	- 22 17 07.	19	15.8	GALAXY
SHAP342-5147.6	03 41 33.	- 51 57 03.	12	17.75	GALAXY
SHAP342-5305.2	03 41 33.	- 53 14 39.	24	17.8	GALAXY
SHAP342-5307.5	03 41 33.	- 53 16 57.	24	17.7	GALAXY
LB 01483	03 41 34.	+ 25 49 42.		19.6	FAINT BLUE STAR
RNGC 1429	03 41 34.	- 04 53			NON-EXISTENT OBJECT
BAK 2.0343	03 41 34.	- 23 27 37.	19	16.8	GALAXY
SHAP342-5105.5	03 41 34.	- 51 14 57.	18	16.5	GALAXY
SHAP342-5353.2	03 41 34.	- 54 02 39.	30	17.3	GALAXY
SHAP342-5711.6	03 41 34.	- 57 21 03.	18	16.75	GALAXY
SHAP342-5825.9	03 41 34.	- 58 35 21.	30	16.25	GALAXY
SHAP342-5337.2	03 41 35.	- 53 46 39.	24	17.2	GALAXY
3ZW 056	03 41 36.	+ 01 33			COMPACT GALAXY
ZWG 391.032	03 41 36.	+ 01 33		15.4	GALAXY
MCG+13-03-009	03 41 36.	+ 79 31	8	17.	GALAXY
HPW 41	03 41 36.	- 34 50			DWARF GALAXY IN FORNAX
SHAP343-4724.7	03 41 36.	- 47 34 08.	24	17.5	GALAXY
SHAP342-5354.4	03 41 36.	- 54 03 51.	24	17.3	GALAXY
SHAP342-5706.4	03 41 36.	- 57 15 51.	24	17.0	GALAXY
LB 03328	03 41 36.	- 74 56		13.2	FAINT BLUE STAR
BAK 1.898	03 41 37.	- 27 40 13.	18	16.7	EXTRAGALACTIC NEBULA
SHAP343-4742.4	03 41 37.	- 47 51 50.	24	17.5	GALAXY

OBJECT NAME	RIGHT ASCEN.	DECLINATION	DIAM.	MAGN.	TYPE OF OBJECT
SHAP343-4948.6	03 41 37.	- 49 58 02.	24	17.25	GALAXY
SHAP342-5245.2	03 41 37.	- 52 54 39.	18	17.0	GALAXY
SHAP342-5538.8	03 41 37.	- 55 48 15.	30	17.0	GALAXY
BAK 2.0346	03 41 38.	- 17 49 37.	18	16.0	GALAXY
SHAP342-5543.0	03 41 38.	- 55 52 27.	30	18.0	GALAXY
BAK 2.0347	03 41 39.	- 18 41 31.	14	17.1	GALAXY
BAK 2.0345	03 41 39.	- 20 09 43.	14	16.2	GALAXY
RNGC 1437	03 41 39.	- 36 01		13.0	GALAXY
SHAP342-5315.2	03 41 39.	- 53 24 38.	24	17.2	GALAXY
SHAP342-5321.6	03 41 39.	- 53 31 03.	24	16.8	GALAXY
SHAP342-5337.6	03 41 39.	- 53 47 02.	24	17.6	GALAXY
SHAP342-5341.2	03 41 39.	- 53 50 39.	24	17.6	GALAXY
SHAP342-5513.0	03 41 39.	- 55 22 27.	36	16.5	GALAXY
SHAP342-5737.6	03 41 39.	- 57 47 03.	24	18.0	GALAXY
SHAP342-5757.6	03 41 39.	- 58 07 03.	24	17.25	GALAXY
BAK 1.899	03 41 40.	- 25 28 43.	22	16.2	EXTRAGALACTIC NEBULA
SHAP343-5322.4	03 41 41.	- 53 31 50.	30	17.5	GALAXY
SHAP343-5334.6	03 41 41.	- 53 44 02.	30	16.2	GALAXY
SHAP343-5337.8	03 41 41.	- 53 47 14.	24	17.5	GALAXY
SHAP342-5652.6	03 41 41.	- 57 02 03.	24	17.0	GALAXY
SHAP343-5049.6	03 41 42.	- 50 59 02.	18	16.5	GALAXY
SHAP343-5051.4	03 41 42.	- 51 00 50.	24	17.5	GALAXY
SHAP343-5341.0	03 41 42.	- 53 50 26.	18	17.4	GALAXY
SHAP342-5433.1	03 41 42.	- 54 42 32.	60	17.25	GALAXY
SHAP342-5608.8	03 41 42.	- 56 18 15.	24	17.5	GALAXY
BAK 2.0344	03 41 43.	- 25 43 25.	18	17.2	GALAXY
YC 0342-36	03 41 43.	- 36 00 36.			UNUSUAL SOUTHERN GALAXY
SHAP343-5328.0	03 41 43.	- 53 37 26.	24	17.2	GALAXY
SHAP343-5328.2	03 41 43.	- 53 37 38.	24	17.5	GALAXY
SHAP343-5337.6	03 41 43.	- 53 47 02.	24	17.2	GALAXY
SHAP342-5653.2	03 41 43.	- 57 02 39.	24	17.75	GALAXY
SHAP343-4546.9	03 41 44.	- 45 56 20.	42	17.0	GALAXY
SHAP343-5424.5	03 41 44.	- 54 33 56.	24	17.5	GALAXY
SHAP343-5323.6	03 41 45.	- 53 33 02.	24	16.5	GALAXY
SHAP343-5337.9	03 41 45.	- 53 47 20.	24	17.1	GALAXY
SHAP343-5434.0	03 41 45.	- 54 43 26.	36	17.5	GALAXY
SHAP342-5650.0	03 41 45.	- 56 59 26.	18	17.5	GALAXY
BAK 2.0354	03 41 46.	- 16 01 36.	13	16.4	GALAXY
BAK 2.0355	03 41 46.	- 16 04 48.	19	16.8	GALAXY
BAK 2.0349	03 41 46.	- 19 48 24.	30	16.4	GALAXY
BAK 2.0348	03 41 46.	- 24 52 55.	13	17.4	GALAXY
BAK 2.0352	03 41 47.	- 18 38 00.	20	16.4	GALAXY
SHAP343-5207.6	03 41 47.	- 52 17 02.	18	17.75	GALAXY
SHAP343-5240.2	03 41 47.	- 52 49 38.	18	16.75	GALAXY
SHAP343-5435.9	03 41 47.	- 54 45 20.	30	17.0	GALAXY
SHAP343-5523.0	03 41 47.	- 55 32 26.	30	17.0	GALAXY
VDB.66N 020	03 41 48.	+ 23 54	1320		REFLECTION NEBULA
BAK 2.0353	03 41 48.	- 18 46 30.	14	16.4	GALAXY
PHL 8744	03 41 48.	- 24 59		18.4	BLUE STELLAR OBJECT
MCG-06-09-025	03 41 48.	- 36 00	132	12.9	GALAXY
SHAP343-4618.4	03 41 48.	- 46 27 49.	24	17.5	GALAXY
SHAP343-5330.9	03 41 48.	- 53 40 20.	30	17.7	GALAXY
SHAP343-5341.6	03 41 48.	- 53 51 02.	24	17.6	GALAXY
SHAP343-5514.8	03 41 48.	- 55 24 14.	30	16.5	GALAXY
BAK 1.900	03 41 49.	- 23 41 00.	16	15.6	EXTRAGALACTIC NEBULA
SHAP343-4540.0	03 41 49.	- 45 49 25.	42	16.5	GALAXY
SHAP343-5213.6	03 41 50.	- 52 23 02.	18	17.0	GALAXY
SHAP343-5613.4	03 41 50.	- 56 22 50.	30	16.5	GALAXY
SHAP342-5825.0	03 41 50.	- 58 34 26.	24	17.5	GALAXY
BAK 2.0356	03 41 51.	- 18 55 18.	14	16.7	GALAXY
BAK 2.0350	03 41 51.	- 24 18 18.	13	16.7	GALAXY
SHAP343-5250.4	03 41 51.	- 52 59 50.	24	17.5	GALAXY
SHAP343-5339.6	03 41 51.	- 53 49 02.	24	16.9	GALAXY
SHAP343-5437.4	03 41 51.	- 54 46 50.	36	17.5	GALAXY
SHAP343-5515.6	03 41 51.	- 55 25 02.	36	16.25	GALAXY
SHAP343-5622.8	03 41 51.	- 56 32 14.	24	16.25	GALAXY
SHAP343-5700.2	03 41 51.	- 57 09 38.	24	17.25	GALAXY
CED 019C	03 41 52.	+ 24 09	960		DIFFUSE GALACTIC NEBULA
BAK 2.0351	03 41 52.	- 25 23 36.	17	17.2	GALAXY
BAK 1.901	03 41 52.	- 25 24 54.	14	16.6	EXTRAGALACTIC NEBULA
SHAP343-5049.8	03 41 52.	- 50 59 13.	24	16.75	GALAXY
BAK 1.902	03 41 53.	- 25 26 36.	16	16.5	EXTRAGALACTIC NEBULA
SHAP343-5444.0	03 41 53.	- 54 53 26.	36	17.5	GALAXY
SHAP343-5525.0	03 41 53.	- 55 34 26.	48	18.0	GALAXY
MCG+11-05-003	03 41 54.	+ 67 57	1200	10.7	GALAXY
SHAP343-5239.5	03 41 54.	- 52 48 56.	18	17.5	GALAXY
BAK 2.0357	03 41 55.	- 22 05 48.	22	16.2	GALAXY
BAK 1.903	03 41 55.	- 25 24 36.	6	17.0	EXTRAGALACTIC NEBULA
IC 0342	03 41 56.	+ 67 57	1500	12.7	GALAXY Sc
SHAP343-5306.8	03 41 56.	- 53 16 13.	24	17.6	GALAXY
SHAP343-5337.2	03 41 56.	- 53 46 37.	24	17.5	GALAXY
SHAP343-5537.2	03 41 56.	- 55 46 38.	24	17.5	GALAXY
SHAP343-4740.1	03 41 57.	- 47 49 31.	24	17.0	GALAXY
SHAP343-5337.2	03 41 57.	- 53 46 37.	24	17.3	GALAXY
SHAP343-5829.5	03 41 57.	- 58 38 56.	24	17.0	GALAXY
CED 019D	03 41 58.	+ 23 58	1200		DIFFUSE GALACTIC NEBULA
BAK 2.0360	03 41 58.	- 19 28 36.	29	16.4	GALAXY
BAK 2.0361	03 41 58.	- 19 34 18.	11	16.8	GALAXY
BAK 2.0362	03 41 58.	- 19 35 24.	18	16.8	GALAXY
SHAP343-4611.8	03 41 58.	- 46 21 13.	30	17.5	GALAXY
SHAP343-5124.1	03 41 58.	- 51 33 31.	18	17.5	GALAXY
SHAP343-5339.6	03 41 58.	- 53 49 01.	24	16.7	GALAXY
SHAP343-5342.0	03 41 58.	- 53 51 25.	18	17.0	GALAXY
SHAP343-5350.1	03 41 58.	- 53 59 31.	24	17.0	GALAXY
BAK 2.0364	03 41 59.	- 18 50 18.	13	16.3	GALAXY
SHAP343-4542.8	03 41 59.	- 45 52 13.	36	16.5	GALAXY
CED 019G	03 42	+ 23 54	24000		DIFFUSE GALACTIC NEBULA
KHAV 041	03 42	+ 32 04	11910		DARK NEBULA
LBN 0601	03 42	+ 32 10	60		BRIGHT NEBULA
LBN 0714	03 42	+ 43 20	2700		BRIGHT NEBULA
LBN 0713	03 42	+ 44 00	7200		BRIGHT NEBULA
ZWG 391.033	03 42 00.	+ 02 41		15.5	GALAXY
RNGC 1431	03 42 00.	+ 02 41		15.5	GALAXY
UGC 02845	03 42 00.	+ 02 41	60	15.5	GALAXY S0?
MCG+00-10-017	03 42 00.	+ 02 42	48	15.	GALAXY
UGC 02846	03 42 00.	+ 38 36	60	17.	GALAXY Sb-c
ZWG 305.002	03 42 00.	+ 67 57		10.5	GALAXY
UGC 02847	03 42 00.	+ 67 57	1680	10.5	GALAXY Sc
UGC 02848	03 42 00.	+ 72 30	72	16.5	GALAXY Sc-IRR
RNGC 1434	03 42 00.	- 09 51			NON-EXISTENT OBJECT
BAK 2.0366	03 42 00.	- 18 22 00.	23	16.4	GALAXY
SHAP343-5342.1	03 42 00.	- 53 51 31.	30	17.0	GALAXY
SHAP343-5353.8	03 42 00.	- 54 03 13.	30	16.5	GALAXY
SHAP343-5828.3	03 42 00.	- 58 37 44.	18	17.0	GALAXY
BAK 2.0367	03 42 01.	- 17 35 11.	17	16.8	GALAXY
SHAP343-5828.8	03 42 01.	- 58 38 14.	18	17.25	GALAXY
BAK 2.0359	03 42 02.	- 24 43 12.	12	17.2	GALAXY
BAK 2.0358	03 42 02.	- 26 24 48.	17	17.0	GALAXY
SHAP343-5300.6	03 42 02.	- 53 10 01.	24	17.25	GALAXY
SHAP343-4438.5	03 42 03.	- 44 47 54.	492	11.8	GALAXY
SHAP343-4728.5	03 42 03.	- 47 37 55.	24	16.75	GALAXY
SHAP343-5333.1	03 42 03.	- 53 42 31.	24	17.2	GALAXY
SHAP343-5619.5	03 42 04.	- 56 28 55.	24	16.5	GALAXY
BAK 2.0363	03 42 05.	- 25 39 05.	16	17.0	GALAXY
MCG+03-10-006	03 42 06.	+ 20 37	48	15.5	GALAXY
UGC 02849	03 42 06.	+ 44 43	120	17.	GALAXY Sc
MCG+08-07-013	03 42 06.	+ 46 31	78	15.	GALAXY
VB 198	03 42 06.	+ 57 52	497		STELLAR RING
ISS 0071	03 42 06.	+ 57 52	503		STELLAR RING
VB 199	03 42 06.	+ 62 55	436		STELLAR RING
SHAP343-5140.4	03 42 06.	- 51 49 49.	24	17.0	GALAXY
BAK 2.0371	03 42 07.	- 20 09 59.	21	16.6	GALAXY
BAK 2.0369	03 42 07.	- 21 24 35.	13	16.2	GALAXY
BAK 2.0368	03 42 07.	- 22 07 41.	15	16.0	GALAXY
BAK 2.0365	03 42 07.	- 26 08 11.	17	17.0	GALAXY
BAK 1.905	03 42 07.	- 26 45 11.	16	16.2	EXTRAGALACTIC NEBULA
SHAP343-5329.6	03 42 07.	- 53 39 01.	24	17.2	GALAXY
SHAP343-5335.6	03 42 07.	- 53 45 01.	24	17.0	GALAXY
BAK 2.0374	03 42 08.	- 18 49 35.	12	16.6	GALAXY
SHAP343-4612.0	03 42 08.	- 46 21 24.	24	17.75	GALAXY
SHAP343-5141.8	03 42 08.	- 51 51 13.	24	17.5	GALAXY
SHAP343-5341.1	03 42 08.	- 53 50 31.	24	17.6	GALAXY
SHAP343-5216.8	03 42 09.	- 52 26 13.	18	16.5	GALAXY
LB 01484	03 42 10.	+ 24 58 30.		19.2	FAINT BLUE STAR
BAK 2.0375	03 42 10.	- 20 53 47.	24	16.5	GALAXY
BAK 2.0370	03 42 10.	- 23 45 23.	13	17.0	GALAXY
SHAP343-5126.2	03 42 10.	- 51 35 36.	18	17.5	GALAXY
SHAP343-5203.0	03 42 10.	- 52 12 25.	24	17.5	GALAXY
SHAP343-5259.0	03 42 10.	- 53 08 25.	18	17.5	GALAXY
SHAP343-5359.0	03 42 10.	- 54 08 25.	24	17.5	GALAXY
SHAP343-5626.3	03 42 10.	- 56 35 43.	36	17.0	GALAXY
BAK 2.0372	03 42 11.	- 23 46 17.	17	16.8	GALAXY
SHAP343-4544.8	03 42 11.	- 45 54 12.	36	17.25	GALAXY
SHAP343-5315.0	03 42 11.	- 53 24 25.	24	17.6	GALAXY
ZWG 391.034	03 42 12.	+ 03 11		15.3	GALAXY
UGC 02850	03 42 12.	+ 20 34	84	16.5	GALAXY IRR
ZWG 555.006	03 42 12.	+ 46 32		15.4	GALAXY
UGC 02851	03 42 12.	+ 46 32	126	15.4	GALAXY Sb
MCG+08-07-014	03 42 12.	+ 46 32	90	17.	GALAXY
OCL 0381	03 42 12.	+ 56 17	690	10.	OPEN STAR CLUSTER
KARA.68 042	03 42 12.	- 25 04	34		DWARF GALAXY
BAK 2.0373	03 42 12.	- 25 24 11.	14	17.1	GALAXY
HOD.60 12	03 42 12.	- 34 47	36	17.0	E0 GALAXY IN FORNAX CLSTR
SHAP343-5341.6	03 42 12.	- 53 51 01.	30	16.2	GALAXY
SHAP343-5350.0	03 42 12.	- 53 59 25.	24	17.2	GALAXY
SHAP343-5600.6	03 42 12.	- 56 10 01.	24	17.0	GALAXY
SHAP343-5645.2	03 42 12.	- 56 54 37.	30	16.25	GALAXY
SHAP343-5250.4	03 42 13.	- 52 59 48.	18	17.5	GALAXY
SHAP343-5333.2	03 42 13.	- 53 42 36.	24	17.5	GALAXY
SHAP343-5608.2	03 42 13.	- 56 17 37.	24	16.0	GALAXY
IC 0350	03 42 14.	- 11 56 58.			NONSTELLAR OBJECT
SHAP343-5037.0	03 42 14.	- 50 46 24.	24	17.25	GALAXY
SHAP343-5340.5	03 42 14.	- 53 49 54.	30	16.9	GALAXY
SHAP343-5713.1	03 42 14.	- 57 22 31.	18	17.25	GALAXY
SHAP343-4732.4	03 42 15.	- 47 41 48.	24	18.25	GALAXY
CED 019E	03 42 16.	+ 24 20	1260		DIFFUSE GALACTIC NEBULA
BAK 2.0376	03 42 16.	- 23 46 35.	12	16.8	GALAXY
SHAP343-4623.5	03 42 16.	- 46 32 54.	24	17.25	GALAXY
SHAP343-5054.0	03 42 16.	- 51 03 24.	30	17.0	GALAXY
SHAP343-5334.8	03 42 16.	- 53 44 12.	24	17.0	GALAXY
SHAP343-5749.0	03 42 16.	- 57 58 25.	18	17.25	GALAXY
SHAP343-5249.2	03 42 17.	- 52 58 36.	24	17.0	GALAXY
SHAP343-5302.8	03 42 17.	- 53 12 12.	18	17.5	GALAXY
SHAP343-5330.8	03 42 17.	- 53 40 12.	24	17.5	GALAXY
UGC 02852	03 42 18.	+ 05 45	102	16.0	GALAXY
MCG-02-10-010	03 42 18.	- 11 58	48	14.5	GALAXY
PHL 4494	03 42 18.	- 23 32		18.1	BLUE STELLAR OBJECT
SHAP343-5209.2	03 42 18.	- 52 18 36.	24	17.5	GALAXY
SHAP343-5315.1	03 42 18.	- 53 24 30.	24	16.9	GALAXY
SHAP343-5628.9	03 42 19.	- 56 38 18.	24	16.5	GALAXY
SHAP343-5639.3	03 42 19.	- 56 48 42.	18	16.25	GALAXY
SHAP343-5719.5	03 42 19.	- 57 28 54.	18	16.5	GALAXY
SHAP343-5127.7	03 42 20.	- 51 37 06.	18	17.5	GALAXY
SHAP343-5217.0	03 42 20.	- 52 26 24.	24	17.5	GALAXY
SHAP343-5217.3	03 42 20.	- 52 26 42.	24	17.5	GALAXY
SHAP343-5614.0	03 42 20.	- 56 23 24.	18	16.75	GALAXY
BAK 2.0379	03 42 21.	- 18 48 04.	14	16.8	GALAXY
SHAP343-5600.4	03 42 21.	- 56 09 48.	24	17.5	GALAXY
SHAP343-5609.0	03 42 21.	- 56 18 24.	24	16.5	GALAXY
SHAP343-5806.6	03 42 21.	- 58 16 00.	18	16.75	GALAXY
BAK 2.0380	03 42 22.	- 19 17 46.	24	16.6	GALAXY
BAK 2.0377	03 42 22.	- 21 20 34.	27	16.2	GALAXY
SHAP343-5625.4	03 42 22.	- 56 34 48.	18	16.5	GALAXY
BAK 2.0382	03 42 23.	- 18 30 34.	37	16.6	GALAXY
SHAP343-4750.5	03 42 24.	- 47 59 53.	24	17.0	GALAXY
SHAP343-5257.4	03 42 24.	- 53 06 48.	24	17.5	GALAXY
SER 031.07	03 42 24.	- 53 50	1020	17.	CLUSTER OF GALAXIES
SHAP343-5333.8	03 42 25.	- 53 43 12.	24	17.5	GALAXY
SHAP343-5520.2	03 42 25.	- 55 29 36.	36	16.25	GALAXY
SHAP343-5356.8	03 42 26.	- 54 06 12.	24	16.5	GALAXY
SHAP343-5622.1	03 42 26.	- 56 31 30.	24	17.5	GALAXY
SHAP343-5629.8	03 42 26.	- 56 39 12.	18	18.0	GALAXY
BAK 2.0387	03 42 27.	- 18 19 10.	12	17.0	GALAXY
BAK 2.0388	03 42 27.	- 18 50 10.	22	16.9	GALAXY
BAK 2.0384	03 42 27.	- 20 10 10.	20	16.6	GALAXY
BAK 2.0385	03 42 27.	- 20 57 58.	12	16.6	GALAXY
BAK 2.0381	03 42 27.	- 23 45 22.	17	16.6	GALAXY
BAK 2.0378	03 42 27.	- 25 26 46.	19	16.9	GALAXY
SHAP343-5200.2	03 42 27.	- 52 09 35.	30	17.5	GALAXY
SHAP343-5818.5	03 42 27.	- 58 27 54.	24	16.75	GALAXY
BAK 2.0390	03 42 28.	- 15 38 22.	12	16.6	GALAXY
BAK 1.906	03 42 28.	- 24 16 22.	22	16.0	EXTRAGALACTIC NEBULA
SHAP343-5400.2	03 42 28.	- 54 09 36.	24	17.5	GALAXY
SHAP344-4530.4	03 42 29.	- 45 39 47.	24	16.5	GALAXY
SHAP344-4634.8	03 42 29.	- 46 44 11.	24	18.0	GALAXY
SHAP344-4711.8	03 42 29.	- 47 21 11.	24	17.0	GALAXY
SHAP343-5532.2	03 42 29.	- 55 41 36.	30	16.25	GALAXY
SHAP343-5617.8	03 42 29.	- 56 27 12.	24	16.5	GALAXY
SHAP343-5945.6	03 42 29.	- 59 55 00.	36	17.25	GALAXY
RNGC 1445	03 42 30.	- 10 00			GALAXY
BAK 2.0386	03 42 30.	- 23 41 10.	22	16.8	GALAXY
SHAP343-5342.3	03 42 30.	- 53 51 41.	24	17.5	GALAXY
SHAP343-5357.9	03 42 30.	- 54 07 17.	30	17.5	GALAXY
SHAP343-5617.2	03 42 30.	- 56 26 36.	24	17.5	GALAXY
ARC 0453	03 42 31.	- 20 12		17.9	RICH CLUSTER OF GALAXIES
SHAP344-4729.9	03 42 31.	- 47 39 17.	18	17.5	GALAXY

OBJECT NAME	RIGHT ASCEN.	DECLINATION	DIAM.	MAGN.	TYPE OF OBJECT
SHAP343-5338.4	03 42 31.	- 53 47 47.	24	17.4	GALAXY
BAK 2.0383	03 42 32.	- 27 01 40.	22	16.6	GALAXY
SHAP343-5158.8	03 42 32.	- 52 08 11.	24	17.0	GALAXY
SHAP343-5214.5	03 42 32.	- 52 23 53.	18	17.25	GALAXY
SHAP343-5819.0	03 42 33.	- 58 28 24.	18	17.25	GALAXY
SHAP343-5128.6	03 42 34.	- 51 37 59.	24	17.5	GALAXY
SHAP343-5538.4	03 42 34.	- 55 47 47.	36	16.5	GALAXY
SHAP343-5620.2	03 42 34.	- 56 29 35.	24	17.25	GALAXY
SHAP343-5747.2	03 42 34.	- 57 56 36.	18	17.5	GALAXY
BAK 2.0397	03 42 36.	- 17 17 27.	26	16.6	GALAXY
RNGC 1439	03 42 36.	- 22 05		13.0	GALAXY
HPW 42	03 42 36.	- 35 22			DWARF GALAXY IN FORNAX
SHAP343-5339.0	03 42 36.	- 53 48 23.	24	17.5	GALAXY
BAK 2.0389	03 42 37.	- 27 04 58.	17	16.6	GALAXY
BAK 1.907	03 42 37.	- 27 05 10.	17	17.0	EXTRAGALACTIC NEBULA
SHAP343-5337.8	03 42 37.	- 53 47 11.	30	16.3	GALAXY
SHAP344-5214.1	03 42 38.	- 52 23 29.	18	17.0	GALAXY
SHAP343-5809.9	03 42 38.	- 58 19 17.	25	16.5	GALAXY
BAK 2.0393	03 42 39.	- 21 47 33.	12	16.5	GALAXY
MCG-04-09-056	03 42 39.	- 22 06	48	13.	GALAXY
SHAP344-5235.3	03 42 39.	- 52 44 41.	18	17.25	GALAXY
SHAP343-5300.0	03 42 39.	- 53 09 23.	24	18.0	GALAXY
SHAP343-5314.9	03 42 39.	- 53 24 17.	24	17.0	GALAXY
SHAP343-5845.4	03 42 39.	- 58 54 47.	24	18.0	GALAXY
BAK 2.0395	03 42 40.	- 21 47 21.	10	16.6	GALAXY
BAK 2.0396	03 42 40.	- 21 48 03.	11	16.5	GALAXY
BAK 2.0392	03 42 40.	- 24 12 39.	17	17.0	GALAXY
BAK 2.0391	03 42 40.	- 24 46 57.	16	16.4	GALAXY
SHAP344-5313.3	03 42 40.	- 53 22 41.	24	17.6	GALAXY
SHAP343-5713.8	03 42 40.	- 57 23 11.	18	16.5	GALAXY
SHAP343-5808.2	03 42 40.	- 58 17 35.	24	16.5	GALAXY
BAK 2.0398	03 42 41.	- 19 47 51.	17	16.8	GALAXY
SHAP344-5334.8	03 42 41.	- 53 44 11.	24	17.6	GALAXY
SHAP344-5335.6	03 42 41.	- 53 44 59.	24	17.5	GALAXY
MCG+08-07-015	03 42 42.	+ 45 41	42	18.	GALAXY
ISS 0072	03 42 42.	+ 62 56	421		STELLAR RING
MCG-03-10-043	03 42 42.	- 18 26	96	13.5	GALAXY
MCG-05-09-024	03 42 42.	- 26 51	30	16.	GALAXY
SHAP344-5345.0	03 42 42.	- 53 54 23.	30	17.4	GALAXY
SHAP343-5703.0	03 42 42.	- 57 12 23.	18	17.5	GALAXY
BAK 2.0394	03 42 43.	- 24 46 33.	26	16.7	GALAXY
BAK 1.908	03 42 43.	- 26 18 21.	22	16.4	EXTRAGALACTIC NEBULA
SHAP344-5331.8	03 42 43.	- 53 41 11.	24	17.4	GALAXY
SHAP343-5528.6	03 42 43.	- 55 37 59.	36	17.0	GALAXY
SHAP344-5340.8	03 42 44.	- 53 50 11.	30	17.3	GALAXY
SHAP343-5702.0	03 42 44.	- 57 11 23.	18	18.0	GALAXY
BAK 2.0401	03 42 45.	- 20 27 57.	15	16.6	GALAXY
SHAP344-4638.6	03 42 45.	- 46 47 58.	24	17.0	GALAXY
SHAP344-5358.4	03 42 45.	- 54 07 47.	18	18.0	GALAXY
SHAP343-5615.5	03 42 45.	- 56 24 53.	42	16.5	GALAXY
BAK 2.0405	03 42 46.	- 16 28 15.	12	17.0	GALAXY
RNGC 1440	03 42 46.	- 18 27		13.0	GALAXY
BAK 1.909	03 42 46.	- 23 51 21.	28	16.6	EXTRAGALACTIC NEBULA
SHAP344-5433.9	03 42 46.	- 54 43 17.	30	17.5	GALAXY
SHAP343-5745.9	03 42 46.	- 57 55 17.	24	17.0	GALAXY
SHAP343-5920.4	03 42 46.	- 59 29 47.	30	16.25	GALAXY
RNGC 1442	03 42 47.	- 19 25			NON-EXISTENT OBJECT
BAK 2.0402	03 42 47.	- 19 58 21.	15	16.0	GALAXY
SHAP344-5029.9	03 42 47.	- 50 39 16.	24	17.75	GALAXY
SHAP344-5228.8	03 42 47.	- 52 38 10.	18	17.5	GALAXY
SHAP343-5651.4	03 42 47.	- 57 00 47.	24	16.75	GALAXY
SHAP343-5758.4	03 42 47.	- 58 07 47.	30	16.5	GALAXY
SHAP343-5912.9	03 42 47.	- 59 22 17.	48	16.5	GALAXY
ZC 0342.8+0052	03 42 48.	+ 00 52	1010		CLUSTER OF GALAXIES
BAK 1.910	03 42 48.	- 22 35 09.	22	16.5	EXTRAGALACTIC NEBULA
MCG-04-09-057	03 42 48.	- 23 52	30	15.5	GALAXY
BAK 2.0399	03 42 48.	- 23 52 15.	18	16.8	GALAXY
BAK 2.0400	03 42 48.	- 24 00 33.	13	17.0	GALAXY
LB 01688	03 42 48.	- 46 51		15.5	FAINT BLUE STAR
SHAP344-5259.2	03 42 48.	- 53 08 34.	24	17.5	GALAXY
SHAP344-5345.8	03 42 48.	- 53 55 10.	24	17.5	GALAXY
SHAP344-5237.1	03 42 49.	- 52 46 28.	18	17.75	GALAXY
BAK 2.0407	03 42 50.	- 17 58 08.	14	16.4	GALAXY
SHAP344-5306.7	03 42 50.	- 53 16 04.	30	17.0	GALAXY
ARC 0454	03 42 51.	- 13 09		17.5	RICH CLUSTER OF GALAXIES
BAK 2.0403	03 42 51.	- 22 51 09.	18	17.1	GALAXY
SHAP344-5316.0	03 42 51.	- 53 25 22.	24	17.4	GALAXY
CED 019F	03 42 52.	+ 24 13	1800		DIFFUSE GALACTIC NEBULA
BAK 2.0404	03 42 52.	- 23 54 21.	12	16.8	GALAXY
SHAP344-4645.4	03 42 52.	- 46 54 46.	24	17.25	GALAXY
SHAP344-5206.6	03 42 52.	- 52 15 58.	18	17.0	GALAXY
SHAP344-5210.0	03 42 52.	- 52 19 22.	18	17.5	GALAXY
SHAP344-5214.4	03 42 52.	- 52 23 46.	24	17.5	GALAXY
SHAP344-5340.8	03 42 52.	- 53 50 10.	30	17.3	GALAXY
SHAP344-5400.0	03 42 52.	- 54 09 22.	42	17.0	GALAXY
SHAP344-5406.4	03 42 52.	- 54 15 46.	30	16.5	GALAXY
HUB C03	03 42 53.	+ 24 35			DIFFUSE NEBULA
RNGC 1457	03 42 53.	- 44 48			NON-EXISTENT OBJECT
RNGC 1448	03 42 53.	- 44 48		11.5	GALAXY
SHAP344-5545.3	03 42 53.	- 55 54 40.	30	16.0	GALAXY
SHAP344-5551.6	03 42 53.	- 56 00 58.	30	16.0	GALAXY
BAK 2.0409	03 42 54.	- 20 03 26.	19	16.4	GALAXY
PHL 4495	03 42 54.	- 23 30		18.6	BLUE STELLAR OBJECT
PHL 8745	03 42 54.	- 25 26		13.6	BLUE STELLAR OBJECT
TON-S 0375	03 42 54.	- 25 26		13.9	BLUE STAR
SHAP344-5125.0	03 42 54.	- 51 34 22.	24	17.5	GALAXY
SHAP344-5358.1	03 42 54.	- 54 07 28.	24	18.0	GALAXY
SHAP344-5604.9	03 42 54.	- 56 14 16.	24	17.25	GALAXY
SHAP344-5706.0	03 42 54.	- 57 15 22.	24	16.75	GALAXY
BAK 1.911	03 42 55.	- 26 50 56.	31	15.9	EXTRAGALACTIC NEBULA
SHAP344-5113.2	03 42 55.	- 51 22 34.	24	17.5	GALAXY
SHAP344-5143.7	03 42 55.	- 51 53 04.	24	17.5	GALAXY
BAK 2.0408	03 42 56.	- 22 45 56.	17	16.2	GALAXY
SHAP344-5134.8	03 42 56.	- 51 44 10.	18	17.25	GALAXY
BAK 2.0406	03 42 57.	- 26 57 44.	17	16.8	GALAXY
SHAP344-5148.1	03 42 57.	- 51 57 28.	18	17.0	GALAXY
SHAP344-5230.5	03 42 57.	- 52 39 52.	18	17.5	GALAXY
SHAP344-5313.7	03 42 57.	- 53 23 04.	24	16.9	GALAXY
VDB.66N 021	03 42 58.	+ 24 12	3120		REFLECTION NEBULA
BAK 2.0411	03 42 58.	- 20 42 02.	19	16.4	GALAXY
SHAP344-4619.9	03 42 58.	- 46 29 15.	24	17.25	GALAXY
SHAP344-5419.0	03 42 58.	- 54 28 22.	30	17.0	GALAXY
SHAP344-5422.3	03 42 58.	- 54 31 40.	30	17.75	GALAXY
CED 019H	03 42 59.	+ 24 24	900		DIFFUSE GALACTIC NEBULA
SHAP344-5154.1	03 42 59.	- 52 03 28.	18	17.5	GALAXY
SHAP344-5229.0	03 42 59.	- 52 38 22.	36	17.25	GALAXY
SHAP344-5544.8	03 42 59.	- 55 54 10.	24	17.75	GALAXY
SHAP344-5901.4	03 42 59.	- 59 10 46.	36	17.25	GALAXY

OBJECT NAME	RIGHT ASCEN.	DECLINATION	DIAM.	MAGN.	TYPE OF OBJECT
LBN 0771	03 43	+ 24 00	3000		BRIGHT NEBULA
LBN 0770	03 43	+ 24 00	5400		BRIGHT NEBULA
LBN 0772	03 43	+ 24 00	10800		BRIGHT NEBULA
ZC 0343.0+2520	03 43 00.	+ 25 20	1080		CLUSTER OF GALAXIES
LDN 1446	03 43 00.	+ 37 00	2160		DARK NEBULA
UGC 02853	03 43 00.	+ 41 44	66	18.	GALAXY DWARF
PHL 4496	03 43 00.	- 26 58		18.4	BLUE STELLAR OBJECT
SHAP344-5245.4	03 43 00.	- 52 54 46.	18	17.25	GALAXY
RNGC 1432	03 43 01.	+ 24 00			DIFFUSE NEBULA
SHAP344-5203.2	03 43 01.	- 52 12 33.	18	17.5	GALAXY
SHAP344-5730.5	03 43 01.	- 57 39 52.	18	17.0	GALAXY
SHAP344-5839.5	03 43 01.	- 58 48 52.	24	17.25	GALAXY
SHAP344-5941.4	03 43 01.	- 59 50 46.	24	16.75	GALAXY
BAK 2.0413	03 43 02.	- 19 31 50.	17	16.6	GALAXY
SHAP344-5258.0	03 43 02.	- 53 07 21.	24	17.5	GALAXY
SHAP344-5344.4	03 43 02.	- 53 53 46.	24	17.6	GALAXY
SHAP344-5403.8	03 43 02.	- 54 13 10.	30	16.75	GALAXY
SHAP344-5707.1	03 43 02.	- 57 16 28.	18	17.25	GALAXY
SHAP344-5952.2	03 43 02.	- 60 01 34.	36	16.0	GALAXY
BAK 2.0415	03 43 03.	- 18 50 02.	14	16.6	GALAXY
SHAP344-5301.4	03 43 03.	- 53 10 45.	24	17.0	GALAXY
SEY 002	03 43 04.	+ 69 59 33.		15.0	FAINT GALAXY
SHAP344-5342.5	03 43 04.	- 53 51 51.	24	17.3	GALAXY
RNGC 1452	03 43 05.	- 18 47		13.0	GALAXY
BAK 2.0410	03 43 05.	- 27 24 56.	24	16.4	GALAXY
ZC 0343.1+2445	03 43 06.	+ 24 45	2420		CLUSTER OF GALAXIES
BAK 2.0412	03 43 06.	- 25 37 44.	28	16.6	GALAXY
MCG-06-09-026	03 43 06.	- 36 07	60	15.	GALAXY
SHAP344-4936.2	03 43 06.	- 49 45 33.	48	17.0	GALAXY
SHAP344-5102.0	03 43 06.	- 51 11 21.	24	17.5	GALAXY
SHAP344-5122.5	03 43 06.	- 51 31 51.	24	17.0	GALAXY
SHAP344-5122.5	03 43 06.	- 51 31 51.	24	17.0	GALAXY
SHAP344-5631.4	03 43 06.	- 56 40 46.	24	17.0	GALAXY
BAK 2.0417	03 43 07.	- 16 55 55.	12	17.1	GALAXY
BAK 2.0416	03 43 07.	- 20 55 08.	17	16.8	GALAXY
SHAP344-5355.9	03 43 07.	- 54 05 15.	30	17.5	GALAXY
SHAP344-5614.6	03 43 07.	- 56 23 57.	24	16.75	GALAXY
SHAP344-5128.2	03 43 08.	- 51 37 33.	30	17.25	GALAXY
SHAP344-5215.6	03 43 08.	- 52 24 57.	18	18.0	GALAXY
RNGC 1441	03 43 09.	- 04 15		14.0	GALAXY
MCG+01-10-028	03 43 09.	- 07 38 30.	12	15.5	GALAXY
BAK 2.0414	03 43 09.	- 26 02 02.	13	16.8	GALAXY
SHAP344-5246.0	03 43 09.	- 52 55 21.	48	17.5	GALAXY
SHAP344-5713.8	03 43 09.	- 57 23 09.	18	17.8	GALAXY
SHAP344-5909.8	03 43 09.	- 59 19 10.	24	15.5	GALAXY
SHAP344-5057.4	03 43 10.	- 51 06 45.	18	17.25	GALAXY
SHAP344-5915.8	03 43 10.	- 59 25 10.	24	17.25	GALAXY
SHAP344-5005.4	03 43 11.	- 50 14 45.	30	17.0	GALAXY
SHAP344-5400.4	03 43 11.	- 54 09 45.	24	17.5	GALAXY
SHAP344-5625.4	03 43 11.	- 56 34 45.	24	16.5	GALAXY
SHAP344-5633.7	03 43 11.	- 56 43 03.	24	16.5	GALAXY
ARC 0452	03 43 12.	+ 01 32		17.9	RICH CLUSTER OF GALAXIES
MCG+01-10-009	03 43 12.	+ 07 10	12	14.	GALAXY
MCG+01-10-029	03 43 12.	- 04 14	66	14.	GALAXY
RNGC 1450	03 43 12.	- 09 23			GALAXY
MCG-03-10-044	03 43 12.	- 18 46 30.	60	14.	GALAXY
ARC 0456	03 43 12.	- 20 54		17.5	RICH CLUSTER OF GALAXIES
MCG-04-09-058	03 43 12.	- 23 10	120	13.	GALAXY
KARA.68 043	03 43 12.	- 24 45	27		DWARF GALAXY
SHAP344-5246.6	03 43 12.	- 52 55 57.	30	16.75	GALAXY
SHAP344-5333.8	03 43 12.	- 53 43 09.	24	17.1	GALAXY
SHAP344-5357.3	03 43 12.	- 54 06 39.	24	17.0	GALAXY
RNGC 1435	03 43 13.	+ 23 37			DIFFUSE NEBULA
BAK 2.0418	03 43 13.	- 19 19 01.	32	17.0	GALAXY
RNGC 1438	03 43 13.	- 23 10		13.0	GALAXY
SHAP344-5158.2	03 43 13.	- 52 07 33.	24	16.5	GALAXY
IC 0349	03 43 14.	+ 23 36			NONSTELLAR OBJECT
SHAP344-5630.9	03 43 14.	- 56 40 15.	18	16.5	GALAXY
RNGC 1443	03 43 15.	- 04 11			NON-EXISTENT OBJECT
BAK 2.0419	03 43 15.	- 19 53 19.	15	16.6	GALAXY
CED 019I	03 43 16.	+ 23 36	30		DIFFUSE GALACTIC NEBULA
BAK 2.0420	03 43 16.	- 19 24 43.	22	17.0	GALAXY
SHAP344-4617.8	03 43 16.	- 46 27 08.	24	16.5	GALAXY
SHAP344-5158.6	03 43 16.	- 52 07 57.	24	17.0	GALAXY
SHAP344-5312.6	03 43 16.	- 53 21 57.	30	16.8	GALAXY
SHAP344-5400.0	03 43 16.	- 54 09 21.	24	16.25	GALAXY
SHAP344-5859.0	03 43 16.	- 59 08 21.	18	16.25	GALAXY
VDB.66N 022	03 43 17.	+ 23 44	3120		REFLECTION NEBULA
BAK 2.0421	03 43 17.	- 19 35 37.	13	16.6	GALAXY
SHAP344-4643.8	03 43 17.	- 46 53 08.	42	17.5	GALAXY
SHAP344-4745.0	03 43 17.	- 47 54 20.	24	17.0	GALAXY
SHAP344-5132.0	03 43 17.	- 51 41 20.	66	16.5	GALAXY
IC 1991	03 43 17.	- 51 42			NONSTELLAR OBJECT
UGC 02854	03 43 18.	+ 36 35	72	17.	GALAXY Sc
VB 200	03 43 18.	+ 57 24	631		STELLAR RING
ZWG 327.006	03 43 18.	+ 70 00		14.6	GALAXY
UGC 02855	03 43 18.	+ 70 00	276	14.6	GALAXY Sc/SBc
MCG-06-09-027	03 43 18.	- 38 33	24	15.5	GALAXY
HN 0238	03 43 18.	- 51 42			NEBULA
SHAP344-5143.6	03 43 19.	- 51 52 56.	18	17.5	GALAXY
SHAP344-5726.9	03 43 19.	- 57 36 15.	24	16.75	GALAXY
SHAP344-5342.6	03 43 20.	- 53 51 56.	24	17.0	GALAXY
SHAP344-5710.2	03 43 20.	- 57 19 33.	24	17.8	GALAXY
SHAP344-5912.8	03 43 20.	- 59 22 09.	18	17.25	GALAXY
BAK 1.912	03 43 21.	- 27 41 37.	17	17.2	EXTRAGALACTIC NEBULA
SHAP344-4841.6	03 43 21.	- 48 50 56.	24	18.25	GALAXY
CED 019J	03 43 22.	+ 23 37	1800		DIFFUSE GALACTIC NEBULA
BAK 1.913	03 43 22.	- 27 39 43.	22	16.2	EXTRAGALACTIC NEBULA
SHAP344-5245.9	03 43 22.	- 52 55 14.	24	17.0	GALAXY
SHAP344-5200.0	03 43 23.	- 52 09 20.	18	16.75	GALAXY
SHAP344-5755.0	03 43 23.	- 58 04 21.	18	17.0	GALAXY
ZWG 442.002	03 43 24.	+ 11 06		15.7	GALAXY
ISS 0143	03 43 24.	+ 45 13	347		STELLAR RING
MCG+12-04-004	03 43 24.	+ 69 49	252	14.	GALAXY
MCG+01-10-030	03 43 24.	- 07 40 30.	60	15.	GALAXY
RNGC 1447	03 43 24.	- 09 09			GALAXY
MCG-06-09-028	03 43 24.	- 37 05	30	16.	GALAXY
SHAP344-5133.1	03 43 24.	- 51 42 26.	30	16.0	GALAXY
SHAP344-5302.1	03 43 24.	- 53 11 26.	24	17.5	GALAXY
ARC 0457	03 43 25.	- 20 18		17.9	RICH CLUSTER OF GALAXIES
SHAP344-4834.3	03 43 25.	- 48 43 38.	24	18.0	GALAXY
SHAP344-5524.1	03 43 25.	- 55 33 26.	30	17.0	GALAXY
SHAP344-5711.0	03 43 26.	- 57 20 20.	24	16.9	GALAXY
SHAP344-5817.9	03 43 26.	- 58 27 14.	30	16.25	GALAXY
MCG+01-10-031	03 43 27.	- 03 20 30.	60	15.5	GALAXY
RNGC 1446	03 43 27.	- 04 13			NON-EXISTENT OBJECT
RNGC 1449	03 43 27.	- 04 18		14.5	GALAXY
BAK 2.0423	03 43 27.	- 18 10 18.	14	16.8	GALAXY

OBJECT NAME	RIGHT ASCEN.	DECLINATION	DIAM.	MAGN.	TYPE OF OBJECT
SHAP344-5240.8	03 43 27.	- 52 50 08.	18	17.5	GALAXY
SHAP344-5246.6	03 43 28.	- 52 55 56.	24	17.5	GALAXY
SHAP344-5257.2	03 43 28.	- 53 06 32.	18	17.5	GALAXY
SHAP344-5715.8	03 43 28.	- 57 25 08.	12	18.1	GALAXY
SHAP344-5800.2	03 43 29.	- 58 09 32.	18	16.5	GALAXY
MCG+01-10-032	03 43 30.	- 04 17	36	14.6	GALAXY
SHAP344-5710.2	03 43 30.	- 57 19 32.	24	16.8	GALAXY
BAK 2.0426	03 43 31.	- 18 51 30.	21	17.0	GALAXY
BAK 2.0427	03 43 31.	- 19 04 00.	14	16.7	GALAXY
SHAP344-5258.8	03 43 31.	- 53 08 08.	18	17.5	GALAXY
SHAP344-5640.5	03 43 31.	- 56 49 50.	24	17.9	GALAXY
SHAP344-5341.9	03 43 32.	- 53 51 14.	24	17.1	GALAXY
ARC 0449	03 43 33.	+ 75 02		16.2	RICH CLUSTER OF GALAXIES
MCG+01-10-033	03 43 33.	- 04 13	24	14.5	GALAXY
RNGC 1451	03 43 33.	- 04 14		14.5	GALAXY
BAK 2.0430	03 43 33.	- 15 00 18.	20	16.2	GALAXY
BAK 2.0428	03 43 33.	- 20 27 06.	13	16.8	GALAXY
SHAP344-5111.6	03 43 33.	- 51 20 55.	24	17.0	GALAXY
SHAP344-5447.3	03 43 33.	- 54 56 38.	48	16.5	GALAXY
SHAP344-5652.6	03 43 33.	- 57 01 56.	30	17.2	GALAXY
BAK 2.0422	03 43 34.	- 27 22 06.	20	16.8	GALAXY
BAK 2.0432	03 43 35.	- 15 27 06.	13	16.4	GALAXY
BAK 2.0425	03 43 35.	- 23 51 12.	25	16.7	GALAXY
SHAP344-5146.0	03 43 35.	- 51 55 19.	18	17.0	GALAXY
UGC 02856	03 43 36.	+ 15 17	60	17.	GALAXY Sb-c
BAK 2.0429	03 43 36.	- 19 27 42.	22	16.8	GALAXY
PHL 4497	03 43 36.	- 22 30		18.6	BLUE STELLAR OBJECT
BAK 2.0424	03 43 36.	- 25 40 42.	14	16.7	GALAXY
LB 01689	03 43 36.	- 47 49		14.2	FAINT BLUE STAR
SHAP344-5659.1	03 43 36.	- 57 08 26.	24	17.0	GALAXY
BAK 2.0431	03 43 37.	- 18 50 06.	18	16.2	GALAXY
SHAP344-5257.5	03 43 38.	- 53 06 49.	42	16.0	GALAXY
SHAP344-5631.6	03 43 38.	- 56 40 56.	18	17.0	GALAXY
CED 019K	03 43 39.	+ 23 28	300		DIFFUSE GALACTIC NEBULA
ARC 0458	03 43 40.	- 24 27		17.2	RICH CLUSTER OF GALAXIES
SHAP344-5641.9	03 43 40.	- 56 51 14.	24	17.1	GALAXY
SHAP344-5844.5	03 43 40.	- 58 53 50.	36	16.75	GALAXY
SHAP344-5909.7	03 43 40.	- 59 19 02.	24	15.0	GALAXY
SHAP345-4627.8	03 43 41.	- 46 37 07.	24	17.5	GALAXY
ZC 0343.7+0736	03 43 42.	+ 07 36	1680		CLUSTER OF GALAXIES
ZWG 526.008	03 43 42.	+ 38 29		15.0	GALAXY
UGC 02857	03 43 42.	+ 38 29	90	15.0	GALAXY SBb
OCL 0385	03 43 42.	+ 58 54	1020	14.	OPEN STAR CLUSTER
RNGC 1454	03 43 42.	- 20 51			NON-EXISTENT OBJECT
BAK 2.0433	03 43 42.	- 21 03 05.	21	16.4	GALAXY
SHAP345-5100.0	03 43 42.	- 51 09 19.	24	16.5	GALAXY
SHAP344-5631.0	03 43 42.	- 56 40 19.	30	17.5	GALAXY
LB 01485	03 43 43.	+ 22 49 24.		19.3	FAINT BLUE STAR
HN 0239	03 43 43.	- 51 10			NEBULA
IC 1992	03 43 43.	- 51 10			NONSTELLAR OBJECT
SHAP345-5145.0	03 43 43.	- 51 54 19.	24	17.0	GALAXY
SHAP344-5423.8	03 43 43.	- 54 33 07.	30	17.5	GALAXY
SHAP344-5431.9	03 43 43.	- 54 41 13.	30	17.0	GALAXY
BAK 2.0435	03 43 44.	- 19 24 53.	14	17.0	GALAXY
SHAP345-4944.5	03 43 44.	- 49 53 49.	30	17.25	GALAXY
SHAP345-5101.5	03 43 44.	- 51 10 49.	24	17.5	GALAXY
SHAP345-5211.0	03 43 44.	- 52 20 19.	24	17.0	GALAXY
SHAP344-5742.8	03 43 44.	- 57 52 07.	24	17.0	GALAXY
BAK 2.0434	03 43 45.	- 24 15 05.	14	17.2	GALAXY
SHAP345-5221.8	03 43 46.	- 52 31 07.	18	17.0	GALAXY
SHAP344-5536.1	03 43 46.	- 55 45 25.	24	17.0	GALAXY
SHAP344-5907.2	03 43 46.	- 59 16 31.	66	15.5	GALAXY
SHAP345-4547.0	03 43 47.	- 45 56 18.	30	17.25	GALAXY
SHAP345-5522.6	03 43 47.	- 55 31 55.	36	17.25	GALAXY
MCG+06-09-007	03 43 48.	+ 38 30	60	14.5	GALAXY
MCG+08-07-016	03 43 48.	+ 45 49	60	17.	GALAXY
PHL 4498	03 43 48.	- 21 53		18.3	BLUE STELLAR OBJECT
SHAP345-4555.1	03 43 48.	- 46 04 24.	30	17.5	GALAXY
SHAP345-4732.4	03 43 48.	- 47 41 42.	36	17.0	GALAXY
SHAP345-5135.2	03 43 48.	- 51 44 31.	24	16.5	GALAXY
SHAP345-5208.0	03 43 48.	- 52 17 19.	24	17.5	GALAXY
SHAP344-5743.2	03 43 48.	- 57 52 31.	18	17.0	GALAXY
HN 0241	03 43 48.	- 59 17			NEBULA
IC 1997	03 43 48.	- 59 17			NONSTELLAR OBJECT
BAK 2.0439	03 43 49.	- 18 42 35.	16	17.0	GALAXY
SHAP344-5855.2	03 43 49.	- 59 04 31.	24	16.75	GALAXY
SHAP344-5859.8	03 43 49.	- 59 09 07.	18	17.5	GALAXY
BAK 2.0438	03 43 50.	- 20 59 11.	14	16.8	GALAXY
BAK 2.0436	03 43 50.	- 23 38 05.	26	16.2	GALAXY
BAK 2.0437	03 43 50.	- 24 26 59.	24	16.6	GALAXY
SHAP345-5036.6	03 43 50.	- 50 45 54.	24	17.25	GALAXY
SHAP344-5702.5	03 43 50.	- 57 11 49.	18	16.4	GALAXY
MCG+01-10-034	03 43 51.	- 04 06 30.	60	13.	GALAXY
BAK 2.0440	03 43 51.	- 18 15 53.	13	16.9	GALAXY
BAK 1.914	03 43 51.	- 27 59 47.	18	16.6	EXTRAGALACTIC NEBULA
SHAP345-4719.6	03 43 51.	- 47 28 54.	24	17.75	GALAXY
SHAP345-5356.6	03 43 51.	- 54 05 55.	24	17.5	GALAXY
SHAP345-5517.2	03 43 51.	- 55 26 31.	36	16.25	GALAXY
SHAP345-5629.7	03 43 51.	- 56 39 01.	24	17.5	GALAXY
SHAP345-5635.8	03 43 51.	- 56 45 07.	24	16.9	GALAXY
SHAP344-5715.9	03 43 51.	- 57 25 13.	18	17.9	GALAXY
SHAP345-5314.1	03 43 52.	- 53 23 24.	30	17.5	GALAXY
SHAP345-5432.5	03 43 52.	- 54 41 49.	36	17.25	GALAXY
SHAP345-5702.4	03 43 52.	- 57 11 43.	18	17.5	GALAXY
BAK 2.0443	03 43 53.	- 17 05 29.	13	16.8	GALAXY
RNGC 1455	03 43 53.	- 18 48			GALAXY
SHAP345-4718.5	03 43 53.	- 47 27 48.	24	17.5	GALAXY
SHAP345-5719.2	03 43 53.	- 57 28 31.	48	16.3	GALAXY
ARC 0455	03 43 54.	+ 07 43		17.9	RICH CLUSTER OF GALAXIES
MCG+07-08-032	03 43 54.	+ 40 42	48	15.	GALAXY
UGC 02858	03 43 54.	+ 48 48	120	16.5	GALAXY
HPW 43	03 43 54.	- 37 13			DWARF GALAXY IN FORNAX
LB 01690	03 43 54.	- 52 10		13.0	FAINT BLUE STAR
SHAP345-5708.6	03 43 54.	- 57 17 55.	24	17.9	GALAXY
SHAP344-5818.8	03 43 54.	- 58 28 07.	18	16.5	GALAXY
ARC 0459	03 43 55.	- 20 28		17.7	RICH CLUSTER OF GALAXIES
SHAP345-5102.8	03 43 55.	- 51 12 06.	24	17.0	GALAXY
SHAP345-5642.0	03 43 55.	- 56 51 19.	18	16.8	GALAXY
HN 0240	03 43 55.	- 57 29			NEBULA
IC 1996	03 43 55.	- 57 29			NONSTELLAR OBJECT
BAK 2.0444	03 43 56.	- 19 54 17.	16	17.0	GALAXY
SHAP345-5158.6	03 43 56.	- 52 07 54.	30	16.0	GALAXY
SHAP345-5652.4	03 43 56.	- 57 01 43.	18	17.7	GALAXY
RNGC 1453	03 43 57.	- 04 08		13.0	GALAXY
SHAP345-5133.2	03 43 57.	- 51 42 30.	18	17.5	GALAXY
SHAP345-5237.9	03 43 57.	- 52 47 12.	18	17.0	GALAXY
SHAP345-5321.1	03 43 57.	- 53 30 24.	24	17.0	GALAXY
SHAP345-5353.4	03 43 57.	- 54 02 42.	24	17.5	GALAXY

OBJECT NAME	RIGHT ASCEN.	DECLINATION	DIAM.	MAGN.	TYPE OF OBJECT
SHAP345-5717.9	03 43 57.	- 57 27 13.	18	17.0	GALAXY
SHAP345-5826.1	03 43 57.	- 58 35 25.	18	18.0	GALAXY
BAK 2.0445	03 43 58.	- 20 31 04.	18	16.8	GALAXY
BAK 2.0441	03 43 58.	- 24 11 35.	19	17.2	GALAXY
RNGC 1437B	03 43 58.	- 36 31			GALAXY
SHAP345-5217.6	03 43 58.	- 52 26 54.	30	18.0	GALAXY
SHAP345-5547.9	03 43 58.	- 55 57 12.	24	16.0	GALAXY
SHAP345-5702.9	03 43 58.	- 57 12 12.	18	17.6	GALAXY
BAK 2.0449	03 43 59.	- 16 09 04.	22	15.8	GALAXY
SHAP345-4621.0	03 43 59.	- 46 30 18.	30	16.5	GALAXY
SHAP345-5832.0	03 43 59.	- 58 41 19.	18	17.5	GALAXY
OCL 0421	03 44 00.	+ 23 58	21600	1.6	OPEN STAR CLUSTER
UGC 02859	03 44 00.	+ 40 43	84	16.0	GALAXY Sc
UGC 02860	03 44 00.	+ 78 10	78	16.0	GALAXY Sc
BAK 2.0442	03 44 00.	- 24 28 05.	19	17.0	GALAXY
TON-S 0376	03 44 00.	- 28 19		15.8	BLUE STAR
MCG-06-09-029	03 44 00.	- 36 30	90	13.	GALAXY
DV.56 N1437B	03 44 00.	- 36 31			GALAXY
SHAP345-5122.8	03 44 00.	- 51 32 06.	24	17.0	GALAXY
SHAP345-5640.4	03 44 00.	- 56 49 42.	24	17.0	GALAXY
SHAP345-5704.9	03 44 00.	- 57 14 12.	18	17.6	GALAXY
SHAP345-5848.0	03 44 00.	- 58 57 19.	18	17.5	GALAXY
BAK 2.0448	03 44 01.	- 19 45 52.	22	16.6	GALAXY
SHAP345-5137.3	03 44 01.	- 51 46 36.	18	17.0	GALAXY
BAK 2.0446	03 44 02.	- 21 50 40.	13	16.2	GALAXY
SHAP345-5708.1	03 44 02.	- 57 17 24.	18	17.7	GALAXY
MCG+01-10-035	03 44 03.	- 04 36	138	14.	GALAXY
BAK 2.0447	03 44 03.	- 23 05 22.	17	16.8	GALAXY
SHAP345-5102.9	03 44 03.	- 51 12 12.	24	17.0	GALAXY
SHAP345-5245.1	03 44 03.	- 52 54 24.	24	17.5	GALAXY
SHAP345-5418.0	03 44 03.	- 54 27 18.	30	17.25	GALAXY
SHAP345-5517.2	03 44 03.	- 55 26 30.	42	18.0	GALAXY
SHAP345-5711.6	03 44 03.	- 57 20 54.	18	17.6	GALAXY
BAK 2.0454	03 44 04.	- 18 01 40.	14	16.4	GALAXY
BAK 2.0453	03 44 04.	- 18 42 40.	12	16.9	GALAXY
BAK 2.0450	03 44 04.	- 19 13 10.	14	16.9	GALAXY
BAK 2.0451	03 44 04.	- 19 13 28.	16	16.8	GALAXY
SHAP345-5458.3	03 44 04.	- 55 07 36.	36	16.25	GALAXY
SHAP345-5706.0	03 44 04.	- 57 15 18.	18	17.3	GALAXY
SHAP345-5831.6	03 44 04.	- 58 40 54.	18	18.0	GALAXY
KARA.73 20	03 44 05.	- 37 19	27		DWARF GALAXY
SHAP345-5236.2	03 44 05.	- 52 45 30.	24	17.5	GALAXY
SHAP345-5647.6	03 44 05.	- 56 56 54.	18	17.4	GALAXY
SHAP345-5704.0	03 44 05.	- 57 13 18.	18	17.1	GALAXY
VB 201	03 44 06.	+ 58 56	275		STELLAR RING
BAK 2.0455	03 44 06.	- 19 12 04.	24	16.4	GALAXY
SHAP345-5039.6	03 44 06.	- 50 48 53.	30	17.0	GALAXY
SHAP345-5658.9	03 44 06.	- 57 08 12.	30	18.0	GALAXY
SHAP345-5135.0	03 44 07.	- 51 44 17.	18	17.5	GALAXY
SHAP345-5313.5	03 44 07.	- 53 22 48.	18	17.0	GALAXY
SHAP345-5322.9	03 44 07.	- 53 32 12.	24	17.5	GALAXY
SHAP345-5710.0	03 44 07.	- 57 19 18.	24	17.1	GALAXY
SHAP345-5754.0	03 44 07.	- 58 03 18.	24	16.5	GALAXY
BAK 2.0457	03 44 08.	- 19 11 28.	22	16.4	GALAXY
BAK 2.0456	03 44 08.	- 20 41 16.	19	16.8	GALAXY
SHAP345-4804.0	03 44 09.	- 48 13 17.	18	17.0	GALAXY
SHAP345-5518.0	03 44 09.	- 55 27 18.	30	17.25	GALAXY
SHAP345-5519.1	03 44 09.	- 55 28 24.	48	17.75	GALAXY
SHAP345-5645.5	03 44 09.	- 56 54 48.	18	17.1	GALAXY
SHAP345-5804.2	03 44 09.	- 58 13 30.	18	17.25	GALAXY
VDB.66N 023	03 44 10.	+ 23 58	2040		REFLECTION NEBULA
SHAP345-4550.4	03 44 10.	- 45 59 41.	24	17.5	GALAXY
SHAP345-4931.5	03 44 10.	- 49 40 47.	36	17.25	GALAXY
SHAP345-5624.8	03 44 10.	- 56 34 06.	24	17.0	GALAXY
BAK 2.0452	03 44 11.	- 25 29 40.	17	16.6	GALAXY
SHAP345-5324.1	03 44 11.	- 53 33 23.	30	16.25	GALAXY
SHAP345-5635.6	03 44 11.	- 56 44 54.	24	18.2	GALAXY
SHAP345-5934.8	03 44 11.	- 59 44 06.	36	16.75	GALAXY
BAK 2.0459	03 44 12.	- 18 52 28.	19	16.6	GALAXY
BAK 2.0458	03 44 12.	- 22 57 04.	22	17.1	GALAXY
SHAP345-5636.5	03 44 12.	- 56 45 48.	18	18.2	GALAXY
SHAP345-5700.8	03 44 12.	- 57 10 06.	24	17.8	GALAXY
SHAP345-5708.1	03 44 12.	- 57 17 24.	18	16.9	GALAXY
SHAP345-5754.1	03 44 12.	- 58 03 24.	24	17.75	GALAXY
SHAP345-5644.2	03 44 13.	- 56 53 30.	18	17.1	GALAXY
LB 01486	03 44 14.	+ 24 39 12.		14.6	FAINT BLUE STAR
BAK 2.0460	03 44 14.	- 20 11 52.	15	16.6	GALAXY
SHAP345-5318.9	03 44 14.	- 53 28 11.	30	16.5	GALAXY
SHAP345-5339.9	03 44 14.	- 53 49 11.	30	16.5	GALAXY
SHAP345-5700.0	03 44 14.	- 57 09 17.	24	16.9	GALAXY
BAK 2.0466	03 44 15.	- 15 38 15.	18	16.7	GALAXY
BAK 2.0461	03 44 15.	- 18 59 57.	16	16.6	GALAXY
BAK 2.0462	03 44 15.	- 19 00 51.	20	16.8	GALAXY
SHAP345-5457.6	03 44 15.	- 55 06 53.	36	16.25	GALAXY
LB 01487	03 44 16.	+ 23 10 06.		19.7	FAINT BLUE STAR
RNGC 1460	03 44 16.	- 36 51			GALAXY
SHAP345-4615.9	03 44 16.	- 46 25 11.	42	16.0	GALAXY
SHAP345-5214.5	03 44 16.	- 52 23 47.	24	16.5	GALAXY
SHAP345-5329.0	03 44 16.	- 53 38 17.	24	17.0	GALAXY
SHAP345-5709.0	03 44 16.	- 57 18 17.	24	16.8	GALAXY
SHAP345-5842.8	03 44 16.	- 58 52 06.	36	18.0	GALAXY
BAK 2.0463	03 44 17.	- 20 12 39.	10	16.8	GALAXY
SHAP345-5517.2	03 44 17.	- 55 26 29.	36	17.0	GALAXY
SHAP345-5633.4	03 44 17.	- 56 42 41.	18	17.7	GALAXY
ZC 0344.3+2836	03 44 18.	+ 28 36	940		CLUSTER OF GALAXIES
BAK 2.0469	03 44 18.	- 16 11 57.	14	17.0	GALAXY
MCG-03-10-045	03 44 18.	- 16 41	60	14.	GALAXY
TON-S 0377	03 44 18.	- 30 26		15.6	BLUE STAR
SHAP345-5234.0	03 44 18.	- 52 43 17.	24	17.5	GALAXY
SHAP345-5431.2	03 44 18.	- 54 40 29.	132	16.25	GALAXY
SHAP345-5550.1	03 44 18.	- 55 59 23.	24	17.25	GALAXY
SHAP345-5827.6	03 44 18.	- 58 36 53.	12	17.5	GALAXY
SHAP345-5233.2	03 44 19.	- 52 42 29.	24	17.0	GALAXY
SHAP345-5235.4	03 44 19.	- 52 44 41.	30	17.5	GALAXY
SHAP345-5711.0	03 44 19.	- 57 20 17.	18	17.7	GALAXY
ARC 0460	03 44 20.	- 13 51		17.5	RICH CLUSTER OF GALAXIES
BAK 1.915	03 44 20.	- 28 16 51.	19	16.1	EXTRAGALACTIC NEBULA
PK159-15.1	03 44 20.2	+ 34 53 35.	8	12.4	PLANETARY NEBULA
IC 0351	03 44 20.2	+ 34 53 35.	8	12.4	PLANETARY NEBULA
MCG+01-10-036	03 44 21.	- 03 35	78	15.	GALAXY
BAK 2.0470	03 44 21.	- 16 42 15.	55	14.5	GALAXY
BAK 2.0464	03 44 21.	- 23 15 33.	19	16.7	GALAXY
SHAP345-4537.1	03 44 21.	- 45 46 22.	36	17.5	GALAXY
SHAP345-5048.1	03 44 21.	- 50 57 23.	24	17.25	GALAXY
SHAP345-5626.8	03 44 21.	- 56 36 05.	24	17.75	GALAXY
BAK 2.0471	03 44 22.	- 15 10 27.	67	15.8	GALAXY
BAK 2.0468	03 44 22.	- 20 12 39.	34	16.5	GALAXY
SHAP345-5645.4	03 44 22.	- 56 54 41.	24	17.3	GALAXY

OBJECT NAME	RIGHT ASCEN.	DECLINATION	DIAM.	MAGN.	TYPE OF OBJECT
SHAP345-5654.4	03 44 22.	- 57 03 41.	18	17.4	GALAXY
SHAP345-5658.8	03 44 22.	- 57 08 05.	24	16.4	GALAXY
SHAP345-5743.0	03 44 22.	- 57 52 17.	24	17.25	GALAXY
LB 01488	03 44 23.	+ 25 46 54.		18.4	FAINT BLUE STAR
SHAP345-5139.2	03 44 23.	- 51 48 28.	48	16.5	GALAXY
HN 0242	03 44 23.	- 51 49			NEBULA
IC 1994	03 44 23.	- 51 49			NONSTELLAR OBJECT
ZWG 391.035	03 44 24.	+ 01 15		15.2	GALAXY
UGC 02861	03 44 24.	+ 39 12	138	16.0	GALAXY Sb
MCG+06-09-008	03 44 24.	+ 39 13	60	15.	GALAXY
VB 202	03 44 24.	+ 57 32	242		STELLAR RING
BAK 2.0472	03 44 24.	- 16 04 03.	16	16.8	GALAXY
BAK 1.917	03 44 24.	- 26 32 51.	13	16.6	EXTRAGALACTIC NEBULA
BAK 1.916	03 44 24.	- 26 33 39.	10	16.8	EXTRAGALACTIC NEBULA
PHL 8746	03 44 24.	- 30 26		16.0	BLUE STELLAR OBJECT
MCG-06-09-030	03 44 24.	- 35 04	180	11.	GALAXY
MCG-06-09-031	03 44 24.	- 36 52	84	12.	GALAXY
SHAP345-5700.6	03 44 24.	- 57 09 53.	24	16.7	GALAXY
SHAP345-5716.8	03 44 24.	- 57 26 05.	18	18.3	GALAXY
SHAP346-4604.2	03 44 25.	- 46 13 28.	30	17.5	GALAXY
SHAP345-5313.4	03 44 25.	- 53 22 40.	24	17.5	GALAXY
BAK 2.0467	03 44 26.	- 24 45 51.	14	17.0	GALAXY
SHAP345-5621.5	03 44 26.	- 56 30 47.	24	17.5	GALAXY
BAK 2.0465	03 44 27.	- 27 29 51.	17	16.9	GALAXY
BAK 1.918	03 44 27.	- 27 50 21.	13	16.5	EXTRAGALACTIC NEBULA
SHAP345-5220.2	03 44 27.	- 52 29 28.	24	17.0	GALAXY
SHAP345-5238.2	03 44 27.	- 52 47 28.	18	17.5	GALAXY
SHAP345-5315.5	03 44 27.	- 53 24 46.	18	17.75	GALAXY
SHAP345-5342.8	03 44 27.	- 53 52 04.	24	16.8	GALAXY
SHAP345-5514.2	03 44 27.	- 55 23 29.	30	16.0	GALAXY
SHAP345-5638.6	03 44 27.	- 56 47 53.	24	18.2	GALAXY
SHAP345-5653.9	03 44 27.	- 57 03 11.	18	17.5	GALAXY
CED 019L	03 44 28.	+ 23 57	1620		DIFFUSE GALACTIC NEBULA
BAK 1.919	03 44 28.	- 27 50 15.	13	16.8	EXTRAGALACTIC NEBULA
SHAP346-4614.5	03 44 28.	- 46 23 46.	36	17.0	GALAXY
SHAP345-5615.1	03 44 28.	- 56 24 23.	24	17.5	GALAXY
SHAP345-5624.2	03 44 28.	- 56 33 29.	24	17.0	GALAXY
SHAP345-5231.2	03 44 29.	- 52 40 28.	18	16.75	GALAXY
SHAP345-5314.1	03 44 29.	- 53 23 22.	42	17.0	GALAXY
SHAP345-5743.2	03 44 29.	- 57 52 29.	24	17.75	GALAXY
SHAP345-5744.1	03 44 29.	- 57 53 23.	24	17.5	GALAXY
RLWT 072	03 44 30.	+ 52 01		13.	FAINT VERY BLUE STAR
SHAP345-5118.4	03 44 30.	- 51 27 40.	18	17.5	GALAXY
SHAP345-5533.4	03 44 30.	- 55 42 40.	30	15.25	GALAXY
SHAP345-5839.0	03 44 30.	- 58 48 17.	18	18.0	GALAXY
BAK 2.0478	03 44 31.	- 16 25 02.	42	16.2	GALAXY
BAK 2.0476	03 44 31.	- 18 00 08.	10	17.2	GALAXY
BAK 2.0477	03 44 31.	- 18 00 20.	10	17.0	GALAXY
SHAP345-5321.6	03 44 31.	- 53 30 52.	24	17.5	GALAXY
SHAP345-5742.9	03 44 31.	- 57 52 11.	24	17.75	GALAXY
BAK 2.0474	03 44 32.	- 19 13 02.	19	16.4	GALAXY
BAK 2.0475	03 44 32.	- 19 43 26.	15	16.9	GALAXY
SHAP345-5236.3	03 44 32.	- 52 45 34.	24	17.5	GALAXY
SHAP345-5645.0	03 44 32.	- 56 54 16.	18	16.2	GALAXY
IC 1990	03 44 33.	+ 24 28			NONSTELLAR OBJECT
BAK 2.0473	03 44 33.	- 20 58 08.	18	17.0	GALAXY
SHAP345-5639.0	03 44 33.	- 56 48 16.	24	16.9	GALAXY
SHAP345-5649.2	03 44 33.	- 56 58 28.	18	18.0	GALAXY
SHAP345-5701.2	03 44 33.	- 57 10 28.	18	17.7	GALAXY
BAK 2.0480	03 44 34.	- 19 04 08.	26	16.2	GALAXY
SHAP345-5124.5	03 44 34.	- 51 33 46.	24	17.5	GALAXY
SHAP345-5241.6	03 44 34.	- 52 50 52.	30	16.0	GALAXY
SHAP345-5419.0	03 44 34.	- 54 28 16.	36	18.0	GALAXY
SHAP345-5659.2	03 44 34.	- 57 08 28.	24	17.2	GALAXY
SHAP345-5704.4	03 44 34.	- 57 13 40.	24	16.9	GALAXY
SHAP345-5814.6	03 44 34.	- 58 23 52.	18	16.5	GALAXY
CED 019M	03 44 35.	+ 24 28			DIFFUSE GALACTIC NEBULA
BAK 2.0481	03 44 35.	- 19 27 56.	24	16.6	GALAXY
BAK 2.0479	03 44 35.	- 20 47 02.	16	16.7	GALAXY
SHAP345-5644.0	03 44 35.	- 56 53 16.	12	17.1	GALAXY
SHAP345-5707.1	03 44 35.	- 57 16 22.	24	17.9	GALAXY
SHAP345-5710.1	03 44 35.	- 57 19 22.	18	16.6	GALAXY
SHAP345-5824.0	03 44 35.	- 58 33 16.	18	17.0	GALAXY
BAK 2.0482	03 44 36.	- 20 07 02.	17	17.1	GALAXY
PHL 1568	03 44 36.	- 24 45		17.0	BLUE STELLAR OBJECT
SHAP345-5815.7	03 44 36.	- 58 24 58.	54	16.0	GALAXY
SHAP345-5312.6	03 44 37.	- 53 21 52.	24	17.5	GALAXY
SHAP345-5711.2	03 44 37.	- 57 20 28.	18	17.5	GALAXY
RNGC 1466	03 44 37.	- 71 46		11.5	GLOBULAR CLUSTER IN LMC
SL 001	03 44 37.	- 71 46	135		STAR CLUSTER IN LMC
BAK 2.0484	03 44 38.	- 20 22 26.	16	16.8	GALAXY
BAK 2.0483	03 44 38.	- 20 55 02.	17	16.8	GALAXY
SHAP346-5125.2	03 44 38.	- 51 34 28.	24	18.0	GALAXY
SHAP345-5650.5	03 44 38.	- 56 59 46.	24	18.1	GALAXY
SHAP345-5729.6	03 44 38.	- 57 38 52.	24	16.75	GALAXY
MCG+07-08-033	03 44 39.	+ 42 03 30.	36	16.	GALAXY
SHAP345-5443.5	03 44 39.	- 54 52 46.	36	17.0	GALAXY
SHAP345-5550.9	03 44 39.	- 56 00 10.	36	16.75	GALAXY
SHAP345-5637.8	03 44 39.	- 56 47 04.	24	17.4	GALAXY
SHAP345-5640.2	03 44 39.	- 56 49 28.	24	17.8	GALAXY
SHAP345-5853.8	03 44 39.	- 59 03 04.	24	17.5	GALAXY
SHAP346-4752.0	03 44 40.	- 48 01 15.	24	17.25	GALAXY
SHAP346-5127.2	03 44 40.	- 51 36 27.	24	17.0	GALAXY
SHAP345-5655.2	03 44 40.	- 57 04 28.	18	16.6	GALAXY
SHAP345-5659.8	03 44 41.	- 57 09 04.	24	17.4	GALAXY
SHAP345-5701.4	03 44 41.	- 57 10 40.	36	15.9	GALAXY
SHAP345-5703.2	03 44 41.	- 57 12 28.	24	16.6	GALAXY
ZC 0344.7+2409	03 44 42.	+ 24 09	1410		CLUSTER OF GALAXIES
BAK 2.0485	03 44 42.	- 21 32 26.	17	16.8	GALAXY
SHAP345-5422.3	03 44 42.	- 54 31 34.	30	18.0	GALAXY
SHAP345-5704.0	03 44 42.	- 57 13 16.	18	17.4	GALAXY
B 005	03 44 43.	+ 32 44	3600		DARK OBJECT
SHAP345-5444.4	03 44 43.	- 54 53 40.	42	17.5	GALAXY
SHAP345-5702.8	03 44 43.	- 57 12 04.	18	17.1	GALAXY
SHAP345-5639.5	03 44 44.	- 56 48 46.	30	17.3	GALAXY
SHAP345-5700.1	03 44 44.	- 57 09 22.	18	17.6	GALAXY
SHAP345-5701.3	03 44 44.	- 57 10 34.	24	18.2	GALAXY
SHAP345-5618.7	03 44 45.	- 56 27 58.	24	17.2	GALAXY
BAK 2.0487	03 44 46.	- 16 26 37.	24	16.6	GALAXY
RNGC 1458	03 44 46.	- 18 24			NON-EXISTENT OBJECT
BAK 1.920	03 44 46.	- 30 05 26.	37	14.7	EXTRAGALACTIC NEBULA
SHAP345-5627.4	03 44 46.	- 56 36 39.	24	17.75	GALAXY
SHAP345-5701.8	03 44 46.	- 57 11 04.	30	17.5	GALAXY
SHAP345-5707.9	03 44 46.	- 57 17 10.	24	18.3	GALAXY
SHAP345-5659.5	03 44 47.	- 57 08 45.	18	17.9	GALAXY
SHAP345-5814.1	03 44 47.	- 58 23 22.	18	17.25	GALAXY
ZWG 442.003	03 44 48.	+ 13 05		15.6	GALAXY
UGC 02862	03 44 48.	+ 13 05	96	15.6	GALAXY Sa

OBJECT NAME	RIGHT ASCEN.	DECLINATION	DIAM.	MAGN.	TYPE OF OBJECT
KARA.72 094A	03 44 48.	+ 13 05	84	15.6	PART OF DOUBLE GALAXY
LDN 1471	03 44 48.	+ 32 45	1260		DARK NEBULA
ISS 0073	03 44 48.	+ 57 33	231		STELLAR RING
MCG-05-10-001	03 44 49.	- 30 06	54	15.	GALAXY
SHAP346-5252.6	03 44 48.	- 53 01 51.	18	18.0	GALAXY
SHAP346-5333.2	03 44 48.	- 53 42 27.	42	17.25	GALAXY
SHAP345-5651.2	03 44 48.	- 57 00 27.	18	16.9	GALAXY
SHAP345-5704.8	03 44 48.	- 57 14 03.	18	16.9	GALAXY
BAK 2.0486	03 44 50.	- 21 49 01.	11	16.4	GALAXY
SHAP345-5645.9	03 44 50.	- 56 55 09.	18	17.7	GALAXY
SHAP345-5711.4	03 44 50.	- 57 20 39.	24	15.6	GALAXY
LW 001	03 44 50.	- 71 50			STAR CLUSTER IN LMC
RNGC 1459	03 44 51.	- 25 41		14.0	GALAXY
MCG-04-10-001	03 44 51.	- 25 41	84	14.	GALAXY
SHAP346-5639.4	03 44 51.	- 56 48 39.	30	16.7	GALAXY
SHAP345-5648.5	03 44 51.	- 56 57 45.	18	16.8	GALAXY
SHAP345-5650.8	03 44 51.	- 57 00 03.	18	17.0	GALAXY
SHAP346-5241.2	03 44 52.	- 52 50 27.	24	17.75	GALAXY
SHAP346-5652.5	03 44 52.	- 57 01 45.	18	17.1	GALAXY
BAK 2.0490	03 44 53.	- 19 19 25.	16	17.1	GALAXY
SHAP346-5200.0	03 44 53.	- 52 09 15.	24	17.5	GALAXY
SHAP346-5312.2	03 44 53.	- 53 21 27.	24	17.5	GALAXY
SHAP346-5316.5	03 44 53.	- 53 25 45.	24	17.0	GALAXY
SHAP346-5707.4	03 44 53.	- 57 16 39.	18	17.6	GALAXY
MCG+02-10-001	03 44 54.	+ 13 06	96	15.	GALAXY
UGC 02863	03 44 54.	+ 41 46	102	16.0	GALAXY SBb
MCG+07-08-034	03 44 54.	+ 41 46	48	16.	GALAXY
ZWG 327.007	03 44 54.	+ 73 58		15.2	GALAXY
UGC 02864	03 44 54.	+ 73 58	72	15.2	GALAXY Sa?
UGC 02262	03 44 54.	- 00 30	96	13.5	GALAXY Sa/Sb
MCG-05-10-002	03 44 54.	- 27 06	30	15.5	GALAXY
TON-S 0378	03 44 54.	- 31 30		15.6	BLUE STAR
SHAP346-5625.0	03 44 54.	- 56 34 15.	24	17.0	GALAXY
SHAP346-5657.0	03 44 54.	- 57 06 15.	18	18.2	GALAXY
SHAP346-5130.4	03 44 55.	- 51 39 38.	30	17.0	GALAXY
SHAP346-5621.8	03 44 55.	- 56 31 03.	30	18.0	GALAXY
SHAP346-5735.9	03 44 55.	- 57 45 09.	24	17.25	GALAXY
SHAP346-5739.9	03 44 55.	- 57 49 09.	18	17.0	GALAXY
SHAP345-5915.4	03 44 55.	- 59 24 39.	30	17.0	GALAXY
IC 1993	03 44 56.	- 33 52 36.			NONSTELLAR OBJECT
SHAP346-5118.4	03 44 56.	- 51 27 38.	24	18.0	GALAXY
SHAP346-5725.8	03 44 56.	- 57 35 03.	18	17.0	GALAXY
SHAP345-5927.6	03 44 56.	- 59 36 51.	24	17.0	GALAXY
SHAP346-5357.9	03 44 57.	- 54 07 09.	36	17.0	GALAXY
SHAP346-5633.3	03 44 57.	- 56 42 33.	30	16.6	GALAXY
SHAP346-4652.9	03 44 58.	- 47 02 08.	36	17.25	GALAXY
SHAP346-5315.8	03 44 58.	- 53 25 02.	36	16.75	GALAXY
SHAP346-5651.4	03 44 58.	- 57 00 39.	30	16.9	GALAXY
BAK 2.0492	03 44 59.	- 23 06 01.	18	17.0	GALAXY
SHAP346-5230.0	03 44 59.	- 52 39 14.	24	17.0	GALAXY
SHAP346-5309.4	03 44 59.	- 53 18 38.	30	16.5	GALAXY
SHAP346-5604.5	03 44 59.	- 56 13 45.	18	17.0	GALAXY
LBN 0738	03 45	+ 36 30	12600		BRIGHT NEBULA
LBN 0631	03 45	+ 85 00	3300		BRIGHT NEBULA
ZWG 327.008	03 45 00.	+ 72 52		15.1	GALAXY
UGC 02865	03 45 00.	+ 72 52	102	15.1	GALAXY SB
MCG+12-04-005	03 45 00.	+ 72 52	78	15.	GALAXY
MCG-02-10-011	03 45 00.	- 11 52	54	15.	GALAXY
BAK 2.0499	03 45 00.	- 16 41 49.	16	16.8	GALAXY
BAK 2.0497	03 45 00.	- 18 22 01.	19	17.1	GALAXY
BAK 2.0495	03 45 00.	- 21 31 25.	12	16.4	GALAXY
BAK 2.0491	03 45 00.	- 24 40 55.	33	16.0	GALAXY
BAK 2.0488	03 45 00.	- 26 11 01.	18	17.0	GALAXY
BAK 2.0489	03 45 00.	- 26 20 37.	19	16.7	GALAXY
MCG-06-09-032	03 45 00.	- 33 50	120	11.	GALAXY
HPW 44	03 45 00.	- 37 06			DWARF GALAXY IN FORNAX
SHAP346-5159.9	03 45 00.	- 52 09 06.	24	17.75	GALAXY
SHAP346-5627.2	03 45 00.	- 56 36 27.	24	17.5	GALAXY
SHAP346-5801.4	03 45 00.	- 58 10 39.	18	17.25	GALAXY
SHAP346-5915.4	03 45 00.	- 59 24 39.	18	17.25	GALAXY
BAK 2.0501	03 45 01.	- 16 06 48.	14	17.0	GALAXY
SHAP346-5629.6	03 45 01.	- 56 38 51.	24	17.0	GALAXY
SHAP346-5638.8	03 45 01.	- 56 48 03.	42	17.2	GALAXY
SHAP346-5645.6	03 45 01.	- 56 54 51.	18	17.1	GALAXY
SHAP346-5759.5	03 45 01.	- 58 08 45.	30	16.0	GALAXY
BAK 2.0498	03 45 02.	- 20 52 31.	17	16.8	GALAXY
BAK 2.0493	03 45 02.	- 23 49 49.	23	17.0	GALAXY
SHAP346-5649.9	03 45 02.	- 56 59 09.	18	17.1	GALAXY
SHAP346-5708.0	03 45 02.	- 57 17 15.	24	17.1	GALAXY
SHAP346-5819.5	03 45 02.	- 58 28 45.	24	16.75	GALAXY
BAK 2.0494	03 45 03.	- 24 39 49.	14	16.7	GALAXY
SHAP346-5123.2	03 45 03.	- 51 32 26.	24	17.5	GALAXY
SHAP346-5429.9	03 45 03.	- 54 39 08.	30	17.5	GALAXY
SHAP346-5654.5	03 45 03.	- 57 03 45.	18	17.2	GALAXY
SHAP346-5702.6	03 45 03.	- 57 11 50.	24	17.8	GALAXY
BAK 2.0496	03 45 04.	- 24 41 43.	22	16.3	GALAXY
SHAP346-5615.0	03 45 04.	- 56 24 14.	24	17.5	GALAXY
SHAP346-5732.6	03 45 04.	- 57 41 51.	18	17.5	GALAXY
BAK 1.921	03 45 05.	- 27 05 49.	14	15.8	EXTRAGALACTIC NEBULA
SHAP346-5631.1	03 45 05.	- 56 40 20.	36	17.0	GALAXY
SHAP346-5709.0	03 45 05.	- 57 18 14.	18	17.3	GALAXY
SHAP346-5714.2	03 45 05.	- 57 23 26.	24	17.3	GALAXY
SHAP346-5730.0	03 45 05.	- 57 39 14.	24	16.5	GALAXY
SHAP346-5903.4	03 45 05.	- 59 12 39.	24	17.5	GALAXY
MCG+12-04-006	03 45 06.	+ 69 57	66	16.	GALAXY
SHAP346-5336.6	03 45 06.	- 53 45 50.	30	17.5	GALAXY
SHAP346-5451.5	03 45 06.	- 55 00 44.	36	17.0	GALAXY
SHAP346-5810.4	03 45 06.	- 58 19 38.	18	17.0	GALAXY
SHAP346-4816.2	03 45 07.	- 48 25 26.	24	17.25	GALAXY
SHAP346-4927.3	03 45 07.	- 49 36 32.	30	16.75	GALAXY
SHAP346-5158.8	03 45 07.	- 52 08 02.	30	17.5	GALAXY
SHAP346-5711.4	03 45 07.	- 57 20 38.	24	16.9	GALAXY
CED 019N	03 45 08.	+ 22 24			DIFFUSE GALACTIC NEBULA
SHAP346-4458.6	03 45 08.	- 45 07 49.	48	16.75	GALAXY
SHAP346-5759.9	03 45 08.	- 58 09 08.	24	17.25	GALAXY
MCG+01-10-037	03 45 09.	- 04 29 30.	84	15.	GALAXY
BAK 2.0502	03 45 09.	- 17 51 24.	18	16.8	GALAXY
SHAP346-5224.6	03 45 09.	- 52 33 50.	42	16.75	GALAXY
SHAP346-5228.4	03 45 09.	- 52 37 38.	36	18.0	GALAXY
SHAP346-5346.2	03 45 09.	- 53 55 26.	24	17.5	GALAXY
BAK 2.0504	03 45 10.	- 15 57 12.	14	16.6	GALAXY
SHAP346-5457.5	03 45 10.	- 55 06 44.	36	16.0	GALAXY
SHAP346-5651.0	03 45 10.	- 57 00 14.	18	18.2	GALAXY
SHAP346-5738.1	03 45 10.	- 57 47 20.	18	16.75	GALAXY
BAK 2.0503	03 45 11.	- 19 30 00.	19	16.6	GALAXY
BAK 2.0500	03 45 11.	- 26 23 00.	17	17.1	GALAXY
SHAP346-4718.0	03 45 11.	- 47 27 13.	48	17.25	GALAXY
SHAP346-5641.1	03 45 11.	- 56 50 20.	12	17.2	GALAXY

OBJECT NAME	RIGHT ASCEN.	DECLINATION	DIAM.	MAGN.	TYPE OF OBJECT
ZWG 327.009	03 45 12.	+ 69 57		15.5	GALAXY
UGC 02866	03 45 12.	+ 69 57	90	15.5	GALAXY
HPW 45	03 45 12.	- 36 35			DWARF GALAXY IN FORNAX
SHAP346-5304.4	03 45 12.	- 53 13 38.	18	17.5	GALAXY
SHAP346-5630.6	03 45 12.	- 56 39 50.	24	18.0	GALAXY
SHAP346-5814.1	03 45 12.	- 58 23 20.	24	17.0	GALAXY
IC 0352	03 45 13.	- 08 53 29.			NONSTELLAR OBJECT
SHAP346-4714.8	03 45 13.	- 47 24 01.	30	16.75	GALAXY
SHAP346-5309.8	03 45 13.	- 53 19 02.	18	18.0	GALAXY
SHAP346-5659.0	03 45 13.	- 57 08 14.	18	15.9	GALAXY
RNGC 1463	03 45 13.	- 59 58			GALAXY
RNGC 1456	03 45 14.	+ 22 24			NON-EXISTENT OBJECT
SHAP346-5255.6	03 45 14.	- 53 04 49.	48	16.5	GALAXY
SHAP346-5409.5	03 45 14.	- 54 18 44.	48	17.0	GALAXY
BAK 2.0505	03 45 15.	- 15 43 06.	14	16.4	GALAXY
SHAP346-5814.2	03 45 15.	- 58 23 26.	24	18.0	GALAXY
SHAP346-5948.4	03 45 15.	- 59 57 38.	42	15.5	GALAXY
SHAP346-5616.5	03 45 16.	- 56 25 44.	24	16.5	GALAXY
SHAP346-5651.8	03 45 16.	- 57 01 02.	18	16.9	GALAXY
SHAP346-5733.2	03 45 16.	- 57 42 26.	18	17.0	GALAXY
UGC 02867	03 45 18.	+ 41 43	84	16.0	GALAXY S0-a
ZWG 391.036	03 45 18.	- 02 18		14.8	GALAXY
HPW 46	03 45 18.	- 36 28			DWARF GALAXY IN FORNAX
SHAP346-5013.5	03 45 18.	- 50 22 43.	24	16.5	GALAXY
SHAP346-5305.6	03 45 18.	- 53 14 49.	30	17.5	GALAXY
SHAP346-5424.2	03 45 18.	- 54 33 25.	42	17.0	GALAXY
BAK 2.0507	03 45 19.	- 17 50 59.	14	17.0	GALAXY
SHAP346-5608.2	03 45 20.	- 56 17 25.	30	17.5	GALAXY
SHAP346-5642.8	03 45 20.	- 56 52 01.	18	18.3	GALAXY
SHAP346-5745.2	03 45 20.	- 57 54 26.	24	17.5	GALAXY
SHAP346-5915.6	03 45 20.	- 59 24 50.	24	17.25	GALAXY
BAK 2.0508	03 45 21.	- 18 50 59.	23	16.8	GALAXY
SHAP346-4716.2	03 45 21.	- 47 25 25.	30	17.25	GALAXY
SHAP346-4958.6	03 45 21.	- 50 07 49.	42	16.25	GALAXY
SHAP346-5302.1	03 45 21.	- 53 11 19.	24	17.5	GALAXY
SHAP346-5359.5	03 45 21.	- 54 08 43.	24	17.5	GALAXY
SHAP346-5716.2	03 45 21.	- 57 25 25.	18	17.5	GALAXY
SHAP346-5929.0	03 45 21.	- 59 38 14.	24	17.25	GALAXY
SHAP346-5640.2	03 45 22.	- 56 49 25.	24	17.1	GALAXY
SHAP346-5153.2	03 45 23.	- 52 02 25.	18	17.0	GALAXY
SHAP346-5524.6	03 45 23.	- 55 33 49.	36	16.5	GALAXY
SHAP346-5641.4	03 45 23.	- 56 50 37.	24	17.1	GALAXY
ZC 0345.4+0519	03 45 24.	+ 05 19	1680		CLUSTER OF GALAXIES
UGC 02868	03 45 24.	+ 34 59	108	17.	GALAXY Sb-c
BLWT 073	03 45 24.	+ 49 55		18.	FAINT VERY BLUE STAR
MCG+00-10-018	03 45 24.	- 02 17	36	14.5	GALAXY
PHL 1569	03 45 24.	- 32 44		18.5	BLUE STELLAR OBJECT
LB 01691	03 45 24.	- 46 35		14.8	FAINT BLUE STAR
SHAP346-4759.8	03 45 24.	- 48 09 00.	24	17.0	GALAXY
BAK 2.0509	03 45 25.	- 21 48 23.	18	16.8	GALAXY
SHAP346-5720.8	03 45 25.	- 57 30 01.	18	17.8	GALAXY
SHAP346-5724.9	03 45 25.	- 57 34 07.	24	17.0	GALAXY
SHAP346-5729.2	03 45 25.	- 57 38 25.	18	17.25	GALAXY
SHAP346-5753.8	03 45 25.	- 58 03 01.	24	17.5	GALAXY
SHAP346-5825.6	03 45 25.	- 58 34 49.	24	17.0	GALAXY
BAK 2.0506	03 45 26.	- 25 56 59.	16	16.8	GALAXY
SHAP346-4715.5	03 45 26.	- 47 24 42.	30	17.25	GALAXY
SHAP346-5608.6	03 45 26.	- 56 17 49.	24	16.0	GALAXY
SHAP347-4711.0	03 45 27.	- 47 20 12.	24	17.5	GALAXY
SHAP346-5638.6	03 45 27.	- 56 47 49.	24	18.1	GALAXY
SHAP346-5704.9	03 45 27.	- 57 14 07.	18	17.7	GALAXY
SHAP346-5711.4	03 45 27.	- 57 20 37.	18	17.7	GALAXY
SHAP346-5837.3	03 45 27.	- 58 46 31.	24	16.25	GALAXY
SHAP346-5901.0	03 45 28.	- 59 10 13.	18	17.25	GALAXY
BAK 2.0510	03 45 29.	- 19 29 05.	14	16.4	GALAXY
BAK 2.0511	03 45 29.	- 19 32 17.	15	16.9	GALAXY
SHAP346-4957.2	03 45 29.	- 50 06 24.	48	16.75	GALAXY
ZC 0345.5+3448	03 45 30.	+ 34 48	3760		CLUSTER OF GALAXIES
UGC 02869	03 45 30.	+ 42 12	84	17.	GALAXY Sa-b
SHAP346-5623.8	03 45 30.	- 56 33 01.	18	17.0	GALAXY
SHAP346-5644.6	03 45 30.	- 56 53 49.	30	18.2	GALAXY
SHAP346-5659.8	03 45 30.	- 57 09 01.	18	17.8	GALAXY
SHAP346-5837.8	03 45 30.	- 58 47 01.	24	16.25	GALAXY
SHAP346-5903.4	03 45 30.	- 59 12 37.	18	17.5	GALAXY
BAK 2.0515	03 45 31.	- 15 20 29.	13	16.8	GALAXY
SHAP346-5629.8	03 45 31.	- 56 39 01.	18	18.1	GALAXY
SHAP346-5708.4	03 45 31.	- 57 17 37.	18	17.2	GALAXY
SHAP346-5609.3	03 45 32.	- 56 18 31.	18	17.25	GALAXY
BAK 2.0519	03 45 33.	- 15 51 59.	20	16.9	GALAXY
BAK 2.0516	03 45 33.	- 17 29 35.	12	16.8	GALAXY
BAK 2.0517	03 45 33.	- 17 41 11.	25	16.8	GALAXY
BAK 2.0512	03 45 34.	- 21 35 23.	14	17.0	GALAXY
SHAP347-4733.8	03 45 34.	- 47 43 00.	24	17.0	GALAXY
SHAP346-5420.2	03 45 34.	- 54 29 24.	30	18.0	GALAXY
SHAP346-5727.2	03 45 34.	- 57 36 25.	18	18.0	GALAXY
BAK 2.0518	03 45 35.	- 18 39 47.	28	16.4	GALAXY
BAK 2.0514	03 45 35.	- 20 11 11.	23	15.7	GALAXY
SHAP347-4533.1	03 45 35.	- 45 42 18.	36	16.5	GALAXY
SHAP347-5103.0	03 45 35.	- 51 12 12.	30	17.0	GALAXY
SHAP346-5215.0	03 45 35.	- 52 24 12.	18	17.0	GALAXY
SHAP346-5400.2	03 45 35.	- 54 09 24.	60	16.5	GALAXY
SHAP346-5707.2	03 45 35.	- 57 16 25.	18	18.0	GALAXY
SHAP346-5711.3	03 45 35.	- 57 20 31.	24	17.0	GALAXY
SHAP346-5719.8	03 45 35.	- 57 29 01.	18	17.6	GALAXY
SHAP346-5752.9	03 45 35.	- 58 02 07.	18	17.25	GALAXY
ZC 0345.6+0400	03 45 36.	+ 04 00	1480		CLUSTER OF GALAXIES
UGC 02870	03 45 36.	+ 42 11	84	16.5	GALAXY S0
OCL 0394	03 45 36.	+ 52 31	660	6.4	OPEN STAR CLUSTER
MCG-03-10-046	03 45 36.	- 20 12	30	15.5	GALAXY
LB 01692	03 45 36.	- 47 12		15.4	FAINT BLUE STAR
SHAP347-5127.1	03 45 36.	- 51 36 18.	24	17.5	GALAXY
SHAP346-5328.5	03 45 36.	- 53 37 42.	24	17.25	GALAXY
SHAP346-5656.8	03 45 36.	- 57 06 00.	18	16.6	GALAXY
SHAP346-5854.2	03 45 36.	- 59 03 25.	18	16.75	GALAXY
RNGC 1444	03 45 37.	+ 52 30		6.5	OPEN CLUSTER
BAK 2.0513	03 45 37.	- 24 16 23.	31	16.0	GALAXY
SHAP346-5451.4	03 45 37.	- 55 00 36.	60	17.5	GALAXY
SHAP347-4834.0	03 45 38.	- 48 43 12.	30	17.5	GALAXY
SHAP347-4547.0	03 45 39.	- 45 56 11.	36	17.0	GALAXY
SHAP346-5614.2	03 45 39.	- 56 23 24.	24	17.25	GALAXY
SHAP347-5150.0	03 45 40.	- 51 59 12.	12	17.0	GALAXY
SHAP347-5233.8	03 45 41.	- 52 43 00.	24	17.5	GALAXY
SHAP346-5650.2	03 45 41.	- 56 59 24.	18	17.3	GALAXY
SHAP346-5743.5	03 45 41.	- 57 52 42.	36	17.75	GALAXY
BAK 2.0522	03 45 42.	- 19 36 16.	15	17.0	GALAXY
SHAP347-5341.6	03 45 42.	- 53 50 48.	30	17.5	GALAXY
SHAP346-5405.8	03 45 42.	- 54 15 00.	24	17.5	GALAXY
SHAP346-5658.5	03 45 42.	- 57 07 42.	18	16.7	GALAXY

OBJECT NAME	RIGHT ASCEN.	DECLINATION	DIAM.	MAGN.	TYPE OF OBJECT
SHAP346-5710.5	03 45 42.	- 57 19 42.	18	17.8	GALAXY
SHAP347-4452.7	03 45 43.	- 45 01 53.	60	16.5	GALAXY
SHAP347-5100.7	03 45 43.	- 51 09 54.	30	16.5	GALAXY
SHAP347-5305.4	03 45 43.	- 53 14 36.	24	17.5	GALAXY
SHAP347-5312.0	03 45 43.	- 53 21 12.	24	17.0	GALAXY
SHAP346-5754.6	03 45 43.	- 58 03 48.	24	17.75	GALAXY
SHAP347-5403.3	03 45 44.	- 54 12 30.	24	17.75	GALAXY
SHAP346-5611.2	03 45 44.	- 56 20 24.	24	17.0	GALAXY
BAK 2.0526	03 45 45.	- 18 28 22.	20	16.6	GALAXY
SHAP347-5309.9	03 45 45.	- 53 19 06.	36	17.0	GALAXY
SHAP346-5602.8	03 45 45.	- 56 12 00.	24	17.5	GALAXY
SHAP346-5849.5	03 45 45.	- 58 58 42.	18	17.25	GALAXY
BAK 2.0528	03 45 46.	- 18 02 34.	13	17.1	GALAXY
SHAP347-4535.6	03 45 46.	- 45 44 47.	36	17.0	GALAXY
SHAP347-5203.0	03 45 46.	- 52 12 11.	18	17.5	GALAXY
SHAP347-5216.6	03 45 46.	- 52 25 48.	24	17.5	GALAXY
SHAP347-5233.6	03 45 46.	- 52 42 47.	30	17.5	GALAXY
SHAP347-5221.4	03 45 47.	- 52 30 35.	24	17.5	GALAXY
SHAP346-5821.2	03 45 47.	- 58 30 24.	36	17.0	GALAXY
BAK 2.0532	03 45 48.	- 16 28 10.	14	16.8	GALAXY
BAK 2.0520	03 45 48.	- 26 11 58.	32	16.6	GALAXY
SHAP347-4652.2	03 45 48.	- 47 01 23.	24	16.0	GALAXY
SHAP347-5130.4	03 45 48.	- 51 39 35.	24	17.0	GALAXY
SHAP346-5654.2	03 45 48.	- 57 03 24.	18	17.5	GALAXY
BAK 2.0524	03 45 49.	- 25 06 58.	16	16.8	GALAXY
BAK 2.0523	03 45 49.	- 25 20 16.	12	16.8	GALAXY
BAK 2.0521	03 45 49.	- 26 11 22.	20	16.8	GALAXY
SHAP347-4454.9	03 45 49.	- 45 04 05.	60	16.5	GALAXY
SHAP347-4749.0	03 45 49.	- 47 58 11.	30	18.0	GALAXY
SHAP347-5101.2	03 45 49.	- 51 10 23.	30	17.0	GALAXY
SHAP347-5401.0	03 45 49.	- 54 10 11.	36	17.5	GALAXY
SHAP347-5435.5	03 45 49.	- 54 44 41.	60	15.7	GALAXY
SHAP346-5814.1	03 45 49.	- 58 23 18.	18	16.5	GALAXY
SHAP346-5901.1	03 45 49.	- 59 10 18.	24	17.0	GALAXY
BAK 2.0534	03 45 50.	- 17 54 34.	46	16.6	GALAXY
SHAP347-5617.1	03 45 50.	- 56 26 18.	36	17.5	GALAXY
SHAP346-5644.2	03 45 50.	- 56 53 24.	18	17.3	GALAXY
SHAP346-5651.5	03 45 50.	- 57 00 42.	24	16.6	GALAXY
ARC 0461	03 45 51.	+ 27 00		17.6	RICH CLUSTER OF GALAXIES
BAK 2.0530	03 45 51.	- 20 30 22.	14	17.2	GALAXY
BAK 2.0527	03 45 51.	- 23 59 34.	14	16.8	GALAXY
BAK 2.0525	03 45 51.	- 26 04 04.	12	17.0	GALAXY
SHAP347-5129.8	03 45 51.	- 51 38 59.	18	17.5	GALAXY
SHAP347-5226.6	03 45 51.	- 52 35 47.	24	17.5	GALAXY
SHAP346-5854.8	03 45 51.	- 59 04 00.	18	17.0	GALAXY
BAK 2.0531	03 45 52.	- 21 35 34.	10	17.0	GALAXY
SHAP346-5821.4	03 45 52.	- 58 30 36.	30	17.75	GALAXY
MCG+08-07-017	03 45 54.	+ 46 50	30	17.	GALAXY
MCG-02-10-012	03 45 54.	- 12 37	48	15.	GALAXY
SHAP347-5246.2	03 45 54.	- 52 55 23.	18	17.0	GALAXY
LB 01489	03 45 55.	+ 23 26 48.		18.6	FAINT BLUE STAR
SHAP347-5237.8	03 45 55.	- 52 46 59.	18	17.5	GALAXY
SHAP347-5248.2	03 45 56.	- 52 57 23.	24	17.5	GALAXY
SHAP347-5359.6	03 45 56.	- 54 08 47.	24	17.5	GALAXY
SHAP347-5619.2	03 45 56.	- 56 28 23.	24	17.75	GALAXY
SHAP347-5642.2	03 45 56.	- 56 51 23.	18	16.8	GALAXY
SHAP347-5658.3	03 45 56.	- 57 07 29.	30	16.0	GALAXY
BAK 2.0533	03 45 57.	- 25 08 57.	14	16.7	GALAXY
BAK 2.0529	03 45 57.	- 27 07 03.	19	16.6	GALAXY
SHAP347-5309.2	03 45 57.	- 53 18 23.	18	17.0	GALAXY
SHAP347-5752.2	03 45 57.	- 58 01 23.	24	18.0	GALAXY
SHAP347-5224.2	03 45 58.	- 52 33 23.	18	17.5	GALAXY
SHAP347-5254.2	03 45 58.	- 53 03 23.	30	18.0	GALAXY
BAK 2.0537	03 45 59.	- 19 36 57.	20	16.2	GALAXY
SHAP347-4533.5	03 45 59.	- 45 42 40.	36	17.5	GALAXY
LB 09806	03 46	- 81 40		13.6	FAINT BLUE STAR
ZWG 442.004	03 46 00.	+ 12 58		15.5	GALAXY
UGC 02871	03 46 00.	+ 12 58	78	15.5	GALAXY Sb
KARA.72 094B	03 46 00.	+ 12 58	66	15.5	PART OF DOUBLE GALAXY
ZWG 526.009	03 46 00.	+ 37 08		15.6	GALAXY
ISS 0144	03 46 00.	+ 46 52	231		STELLAR RING
ZC 0346.0+8509	03 46 00.	+ 85 09	2080		CLUSTER OF GALAXIES
MCG+01-10-038	03 46 00.	- 06 46 30.	48	15.	GALAXY
ARC 0462	03 46 00.	- 17 49		17.5	RICH CLUSTER OF GALAXIES
BAK 2.0535	03 46 00.	- 21 31 03.	11	17.0	GALAXY
MCG-06-09-033	03 46 00.	- 36 37	30	17.	GALAXY
SHAP347-5305.8	03 46 00.	- 53 14 59.	30	17.5	GALAXY
SHAP347-5816.6	03 46 00.	- 58 25 47.	24	17.5	GALAXY
SHAP347-4622.9	03 46 01.	- 46 32 04.	24	17.5	GALAXY
SHAP347-5222.6	03 46 01.	- 52 31 47.	18	17.0	GALAXY
SHAP347-5308.6	03 46 01.	- 53 17 47.	30	17.0	GALAXY
SHAP347-5636.9	03 46 01.	- 56 46 05.	24	17.8	GALAXY
LB 01491	03 46 02.	+ 23 26 48.		19.1	FAINT BLUE STAR
LB 01490	03 46 02.	+ 23 26 48.		18.8	FAINT BLUE STAR
BAK 2.0539	03 46 02.	- 21 37 09.	147	14.6	GALAXY
BAK 2.0538	03 46 02.	- 21 59 21.	12	16.7	GALAXY
SHAP347-5237.9	03 46 02.	- 52 47 05.	24	17.0	GALAXY
SHAP347-5659.6	03 46 02.	- 57 08 47.	18	16.9	GALAXY
SHAP347-5708.5	03 46 02.	- 57 17 41.	24	16.4	GALAXY
RNGC 1461	03 46 03.	- 16 33		13.0	GALAXY
BAK 2.0541	03 46 03.	- 18 09 09.	14	16.4	GALAXY
SHAP347-4530.1	03 46 03.	- 45 39 16.	36	16.5	GALAXY
SHAP347-4900.1	03 46 03.	- 49 09 16.	36	17.0	GALAXY
BAK 2.0536	03 46 04.	- 24 50 09.	22	16.6	GALAXY
SHAP347-5134.4	03 46 04.	- 51 43 34.	36	17.5	GALAXY
SHAP347-5231.5	03 46 05.	- 52 40 40.	24	17.25	GALAXY
SHAP347-5609.9	03 46 05.	- 56 19 05.	24	17.25	GALAXY
MCG+02-10-002	03 46 06.	+ 13 00	48	15.	GALAXY
DG 025	03 46 06.	+ 32 00	420		REFLECTION NEBULA
MCG+06-09-009	03 46 06.	+ 37 09 30.	24	15.	GALAXY
BAK 2.0542	03 46 06.	- 21 22 45.	14	16.6	GALAXY
MCG-04-10-002	03 46 06.	- 21 36 30.	144	14.	GALAXY
BAK 2.0540	03 46 06.	- 21 52 57.	10	16.6	GALAXY
BAK 2.0549	03 46 07.	- 16 23 09.	14	17.0	GALAXY
SHAP347-5225.9	03 46 07.	- 52 35 04.	18	17.75	GALAXY
BAK 2.0546	03 46 08.	- 19 23 15.	14	16.7	GALAXY
BAK 2.0544	03 46 08.	- 20 54 39.	19	17.2	GALAXY
SHAP347-5201.2	03 46 08.	- 52 10 22.	18	17.5	GALAXY
SHAP347-5520.8	03 46 08.	- 55 29 58.	30	16.25	GALAXY
BAK 2.0550	03 46 09.	- 18 59 02.	14	17.0	GALAXY
BAK 2.0547	03 46 09.	- 19 36 32.	14	16.4	GALAXY
SHAP347-5543.8	03 46 09.	- 55 52 58.	24	16.75	GALAXY
CED 0190	03 46 10.	+ 23 54	660		DIFFUSE GALACTIC NEBULA
BAK 2.0543	03 46 10.	- 25 02 15.	19	15.6	GALAXY
SHAP347-5244.5	03 46 10.	- 52 53 40.	24	17.5	GALAXY
SHAP347-5254.2	03 46 10.	- 53 03 22.	24	17.25	GALAXY
BAK 2.0554	03 46 11.	- 19 30 50.	18	16.4	GALAXY
SHAP347-5552.2	03 46 11.	- 56 01 22.	30	17.5	GALAXY

OBJECT NAME	RIGHT ASCEN.	DECLINATION	DIAM.	MAGN.	TYPE OF OBJECT
SHAP347-5646.2	03 46 11.	- 56 55 22.	30	17.1	GALAXY
SHAP347-5711.6	03 46 11.	- 57 20 46.	12	17.5	GALAXY
SS 11	03 46 12.	+ 38 48			DIFFUSE GALACTIC NEBULA
ZC 0346.2-0158	03 46 12.	- 01 58	2150		CLUSTER OF GALAXIES
MCG-03-10-047	03 46 12.	- 16 32	150	13.	GALAXY
BAK 2.0551	03 46 12.	- 21 34 44.	47	16.1	GALAXY
SHAP347-5648.8	03 46 12.	- 56 57 58.	24	16.6	GALAXY
SHAP347-5649.1	03 46 12.	- 56 58 16.	30	18.0	GALAXY
BAK 2.0559	03 46 13.	- 18 54 20.	72	15.9	GALAXY
BAK 2.0552	03 46 13.	- 21 48 08.	11	17.1	GALAXY
BAK 2.0553	03 46 13.	- 21 56 56.	34	16.0	GALAXY
BAK 2.0545	03 46 13.	- 25 20 14.	19	16.8	GALAXY
SHAP347-5230.0	03 46 13.	- 52 39 10.	24	17.75	GALAXY
SHAP347-5641.8	03 46 13.	- 56 50 58.	30	16.7	GALAXY
BAK 2.0555	03 46 14.	- 21 58 44.	10	16.8	GALAXY
BAK 2.0548	03 46 14.	- 24 12 20.	19	16.8	GALAXY
SHAP347-5718.2	03 46 14.	- 57 27 22.	24	17.1	GALAXY
MCG-03-10-048	03 46 15.	- 18 54	66	15.	GALAXY
MCG-04-10-003	03 46 15.	- 21 34	60	16.	GALAXY
SHAP347-5629.6	03 46 15.	- 56 38 46.	24	17.1	GALAXY
SHAP347-5703.3	03 46 15.	- 57 12 28.	24	16.6	GALAXY
SHAP347-5804.8	03 46 15.	- 58 13 58.	24	17.0	GALAXY
CED 019P	03 46 16.	+ 24 00			DIFFUSE GALACTIC NEBULA
BAK 2.0561	03 46 16.	- 17 21 20.	14	16.8	GALAXY
BAK 2.0562	03 46 16.	- 17 38 44.	12	17.0	GALAXY
BAK 2.0558	03 46 16.	- 22 06 14.	15	16.6	GALAXY
BAK 2.0557	03 46 16.	- 22 42 08.	17	16.7	GALAXY
BAK 1.922	03 46 16.	- 28 48 44.	22	15.6	EXTRAGALACTIC NEBULA
SHAP347-5155.4	03 46 16.	- 52 04 34.	42	16.75	GALAXY
SHAP347-5218.4	03 46 16.	- 52 27 34.	18	17.75	GALAXY
SHAP347-5610.2	03 46 16.	- 56 19 22.	18	17.0	GALAXY
SHAP347-5644.4	03 46 16.	- 56 53 34.	18	16.7	GALAXY
SHAP347-5646.8	03 46 16.	- 56 55 58.	24	16.8	GALAXY
SHAP347-5724.1	03 46 16.	- 57 33 16.	24	18.0	GALAXY
MCG+00-10-019	03 46 18.	+ 01 04	54	15.5	GALAXY
ZC 0346.3+2702	03 46 18.	+ 27 02	1080		CLUSTER OF GALAXIES
VDB.66N 024	03 46 18.	+ 38 50	540		REFLECTION NEBULA
VB 203	03 46 18.	+ 58 30	658		STELLAR RING
BAK 2.0566	03 46 18.	- 18 07 20.	22	16.4	GALAXY
SHAP347-4847.0	03 46 18.	- 48 56 09.	30	16.75	GALAXY
SHAP347-5324.0	03 46 18.	- 53 33 10.	24	17.0	GALAXY
SHAP347-5805.4	03 46 18.	- 58 14 34.	24	17.0	GALAXY
SHAP347-5816.8	03 46 18.	- 58 25 58.	18	16.5	GALAXY
BAK 2.0563	03 46 19.	- 20 47 20.	12	16.6	GALAXY
BAK 2.0556	03 46 19.	- 26 40 20.	18	16.0	GALAXY
SHAP347-4533.6	03 46 19.	- 45 42 45.	36	17.5	GALAXY
SHAP347-4746.1	03 46 19.	- 47 55 15.	24	17.5	GALAXY
SHAP347-5633.5	03 46 19.	- 56 42 40.	24	18.1	GALAXY
SHAP347-5717.2	03 46 19.	- 57 26 22.	18	17.1	GALAXY
SHAP347-5744.4	03 46 19.	- 57 53 34.	18	17.75	GALAXY
BAK 2.0567	03 46 20.	- 18 28 08.	22	16.9	GALAXY
BAK 2.0564	03 46 20.	- 21 32 32.	12	17.0	GALAXY
BAK 2.0560	03 46 20.	- 24 36 02.	24	16.8	GALAXY
SHAP347-5022.2	03 46 20.	- 50 31 21.	48	17.5	GALAXY
SHAP347-5649.6	03 46 20.	- 56 58 46.	30	16.6	GALAXY
SHAP347-5903.2	03 46 20.	- 59 12 22.	30	17.75	GALAXY
BAK 2.0573	03 46 21.	- 16 36 20.	17	16.2	GALAXY
SHAP347-5607.0	03 46 21.	- 56 16 10.	24	17.25	GALAXY
BAK 2.0565	03 46 22.	- 23 22 44.	17	16.8	GALAXY
SHAP347-5216.6	03 46 22.	- 52 25 45.	24	17.5	GALAXY
SHAP347-5243.6	03 46 22.	- 52 52 45.	24	17.25	GALAXY
SHAP347-5259.0	03 46 22.	- 53 08 09.	18	17.5	GALAXY
SHAP347-5807.0	03 46 22.	- 58 16 10.	24	17.25	GALAXY
BAK 2.0568	03 46 23.	- 19 55 56.	18	16.2	GALAXY
SHAP347-5230.0	03 46 23.	- 52 39 09.	24	17.75	GALAXY
SHAP347-5702.0	03 46 23.	- 57 11 10.	24	18.2	GALAXY
ZC 0346.4+2558	03 46 24.	+ 25 58	670		CLUSTER OF GALAXIES
BAK 2.0570	03 46 24.	- 20 45 44.	13	16.4	GALAXY
BAK 2.0571	03 46 24.	- 20 47 32.	18	17.0	GALAXY
BAK 2.0569	03 46 24.	- 21 23 50.	14	16.8	GALAXY
SHAP348-4523.8	03 46 24.	- 45 32 57.	42	16.5	GALAXY
SHAP347-5749.0	03 46 24.	- 57 58 10.	18	17.5	GALAXY
SHAP347-5624.6	03 46 26.	- 56 33 45.	24	17.25	GALAXY
SHAP347-5716.4	03 46 26.	- 57 25 34.	12	17.2	GALAXY
SHAP348-4626.1	03 46 27.	- 46 35 15.	30	18.0	GALAXY
BAK 2.0577	03 46 28.	- 18 44 31.	19	17.0	GALAXY
SHAP348-4541.2	03 46 28.	- 45 50 20.	36	17.5	GALAXY
BAK 2.0574	03 46 29.	- 23 15 49.	13	16.7	GALAXY
SHAP347-5657.5	03 46 29.	- 57 06 39.	36	16.1	GALAXY
ZWG 391.037	03 46 30.	+ 01 00		14.8	GALAXY
UGC 02872	03 46 30.	+ 01 02	66	17.	GALAXY
MCG+00-10-020	03 46 30.	- 01 16	36	15.	GALAXY
MCG+01-10-039	03 46 30.	- 03 11	72	15.	GALAXY
BAK 2.0576	03 46 30.	- 21 07 31.	13	16.9	GALAXY
SHAP348-4625.7	03 46 30.	- 46 34 50.	54	17.75	GALAXY
BAK 2.0579	03 46 31.	- 17 39 43.	20	16.1	GALAXY
BAK 2.0580	03 46 31.	- 17 40 31.	18	15.8	GALAXY
BAK 2.0581	03 46 31.	- 17 45 07.	22	16.9	GALAXY
BAK 2.0572	03 46 31.	- 26 46 43.	25	16.7	GALAXY
SHAP347-5134.6	03 46 31.	- 51 43 45.	42	17.0	GALAXY
HN 0243	03 46 31.	- 57 07			NEBULA
BAK 2.0575	03 46 32.	- 25 39 07.	22	16.4	GALAXY
IC 1999	03 46 32.	- 57 06			NONSTELLAR OBJECT
MCG+00-10-021	03 46 33.	+ 01 03	33	14.5	GALAXY
BAK 2.0578	03 46 33.	- 20 40 01.	17	17.2	GALAXY
SHAP347-5659.5	03 46 33.	- 57 08 39.	18	18.3	GALAXY
BAK 2.0582	03 46 34.	- 18 34 55.	14	17.0	GALAXY
SHAP347-5144.0	03 46 34.	- 51 53 09.	30	17.5	GALAXY
SHAP347-5752.0	03 46 34.	- 58 01 09.	30	17.5	GALAXY
SHAP347-5629.8	03 46 35.	- 56 38 57.	24	18.1	GALAXY
SHAP347-5632.4	03 46 35.	- 56 41 33.	24	17.2	GALAXY
SHAP347-5745.0	03 46 35.	- 57 54 09.	30	16.5	GALAXY
MCG-03-10-049	03 46 36.	- 19 08	48	15.	GALAXY
LB 01693	03 46 36.	- 46 07		14.1	FAINT BLUE STAR
SHAP348-5014.6	03 46 36.	- 50 23 44.	36	17.5	GALAXY
SHAP348-5014.7	03 46 36.	- 50 23 50.	30	17.5	GALAXY
SHAP347-5335.1	03 46 36.	- 53 44 15.	36	17.5	GALAXY
SHAP347-5444.6	03 46 36.	- 54 53 45.	42	18.0	GALAXY
BAK 2.0585	03 46 37.	- 18 24 43.	58	15.6	GALAXY
SHAP347-5236.2	03 46 37.	- 52 45 20.	24	16.75	GALAXY
SHAP347-5745.6	03 46 37.	- 57 54 45.	30	17.75	GALAXY
BAK 2.0583	03 46 38.	- 21 13 13.	14	17.0	GALAXY
SHAP348-5033.6	03 46 38.	- 50 42 44.	60	15.75	GALAXY
BAK 2.0584	03 46 39.	- 21 29 43.	16	16.8	GALAXY
SHAP348-4553.4	03 46 39.	- 46 02 32.	30	17.5	GALAXY
SHAP347-5755.2	03 46 40.	- 58 04 21.	24	17.5	GALAXY
SHAP347-5938.8	03 46 40.	- 59 47 57.	24	17.25	GALAXY
BAK 2.0587	03 46 41.	- 19 38 07.	12	17.0	GALAXY

OBJECT NAME	RIGHT ASCEN.	DECLINATION	DIAM.	MAGN.	TYPE OF OBJECT
ZWG 508.001	03 46 42.	+ 28 00		15.6	GALAXY
ZWG 507.001	03 46 42.	+ 28 00		15.6	GALAXY
VB 239	03 46 42.	+ 56 03	342		STELLAR RING
RLWT 074	03 46 42.	+ 57 04		12.	FAINT VERY BLUE STAR
SHAP347-5344.6	03 46 42.	- 53 53 44.	42	16.5	GALAXY
SHAP348-5351.8	03 46 43.	- 54 00 56.	24	17.5	GALAXY
SHAP347-5737.8	03 46 43.	- 57 46 57.	18	17.5	GALAXY
SHAP347-5738.8	03 46 43.	- 57 47 57.	18	17.75	GALAXY
SHAP347-5755.7	03 46 43.	- 58 04 51.	24	17.75	GALAXY
LB 01492	03 46 45.	+ 23 23 18.		18.1	FAINT BLUE STAR
BAK 2.0589	03 46 45.	- 18 34 12.	12	16.8	GALAXY
SHAP348-5156.4	03 46 45.	- 52 05 32.	30	17.5	GALAXY
BAK 2.0592	03 46 46.	- 18 36 18.	14	16.6	GALAXY
BAK 2.0588	03 46 46.	- 22 17 06.	25	14.6	GALAXY
SHAP348-5446.3	03 46 46.	- 54 55 26.	42	17.5	GALAXY
SHAP347-5639.6	03 46 46.	- 56 48 44.	24	18.2	GALAXY
SHAP347-5652.4	03 46 46.	- 57 01 32.	18	17.2	GALAXY
BAK 2.0586	03 46 47.	- 26 02 48.	22	16.7	GALAXY
VB 204	03 46 48.	+ 61 02	302		STELLAR RING
BAK 2.0590	03 46 48.	- 21 30 06.	18	17.0	GALAXY
SHAP347-5731.6	03 46 48.	- 57 40 44.	24	17.5	GALAXY
SHAP347-5833.0	03 46 48.	- 58 42 08.	24	17.5	GALAXY
BAK 2.0596	03 46 49.	- 16 13 30.	16	17.1	GALAXY
SHAP347-5705.1	03 46 49.	- 57 14 14.	18	17.8	GALAXY
BAK 2.0593	03 46 50.	- 22 24 06.	39	15.8	GALAXY
BAK 2.0591	03 46 50.	- 23 03 30.	19	16.8	GALAXY
SHAP348-5558.6	03 46 50.	- 56 07 44.	18	17.25	GALAXY
SHAP347-5638.6	03 46 50.	- 56 47 44.	24	18.1	GALAXY
SHAP347-5654.8	03 46 50.	- 57 03 56.	18	17.7	GALAXY
BAK 2.0597	03 46 51.	- 18 42 42.	13	16.8	GALAXY
BAK 2.0594	03 46 51.	- 20 59 18.	24	16.0	GALAXY
BAK 2.0595	03 46 51.	- 21 00 36.	24	16.2	GALAXY
MCG-04-10-004	03 46 51.	- 22 17	30	14.5	GALAXY
SHAP348-5345.5	03 46 51.	- 53 54 38.	24	17.5	GALAXY
SHAP348-5352.0	03 46 51.	- 54 01 08.	24	18.0	GALAXY
BAK 2.0598	03 46 52.	- 19 07 06.	20	17.0	GALAXY
SHAP348-4530.9	03 46 52.	- 45 40 01.	30	16.5	GALAXY
BAK 2.0600	03 46 53.	- 18 20 48.	22	17.0	GALAXY
SHAP348-4624.4	03 46 53.	- 46 33 31.	42	17.25	GALAXY
SHAP347-5726.4	03 46 53.	- 57 35 32.	24	17.25	GALAXY
MCG-04-10-005	03 46 54.	- 22 23 30.	36	16.	GALAXY
SHAP348-5330.5	03 46 54.	- 53 39 37.	36	18.0	GALAXY
SHAP348-5401.9	03 46 54.	- 54 11 02.	24	17.75	GALAXY
SHAP348-5141.5	03 46 55.	- 51 50 37.	24	17.5	GALAXY
SHAP348-5329.0	03 46 55.	- 53 38 07.	42	17.5	GALAXY
BAK 2.0602	03 46 56.	- 20 16 06.	14	17.0	GALAXY
SHAP348-5642.6	03 46 56.	- 56 51 44.	18	18.0	GALAXY
SHAP348-5701.6	03 46 57.	- 57 10 44.	18	17.7	GALAXY
BAK 2.0605	03 46 58.	- 18 39 41.	16	17.1	GALAXY
ARC 0463	03 46 58.	- 21 44		17.9	RICH CLUSTER OF GALAXIES
BAK 2.0599	03 46 58.	- 23 37 06.	24	17.2	GALAXY
SHAP348-5701.4	03 46 58.	- 57 10 32.	18	18.1	GALAXY
BAK 2.0606	03 46 59.	- 19 03 41.	14	17.2	GALAXY
BAK 2.0603	03 46 59.	- 22 00 06.	10	16.8	GALAXY
LB 09807	03 47	- 82 28		14.4	FAINT BLUE STAR
LDN 1442	03 47 00.	+ 38 50	2280		DARK NEBULA
MAI 003	03 47 00.	+ 88 43	40		DWARF SPHEROIDAL GALAXY
BAK 2.0601	03 47 00.	- 24 52 24.	16	17.0	GALAXY
SHAP348-5610.7	03 47 00.	- 56 19 49.	30	16.25	GALAXY
SHAP348-5612.0	03 47 00.	- 56 21 07.	24	16.25	GALAXY
BAK 2.0608	03 47 01.	- 17 58 53.	18	17.0	GALAXY
BAK 2.0611	03 47 02.	- 16 45 17.	18	16.8	GALAXY
BAK 2.0609	03 47 02.	- 18 54 23.	14	17.1	GALAXY
SHAP348-5315.8	03 47 02.	- 53 24 55.	24	17.5	GALAXY
SHAP348-5703.5	03 47 02.	- 57 12 37.	24	18.1	GALAXY
SHAP348-5707.1	03 47 02.	- 57 16 13.	24	17.4	GALAXY
SHAP348-5821.0	03 47 02.	- 58 30 07.	24	18.0	GALAXY
MCG+03-10-007	03 47 03.	+ 17 41	51	16.	GALAXY
BAK 2.0610	03 47 03.	- 18 47 41.	30	16.6	GALAXY
SHAP348-5553.0	03 47 03.	- 56 02 07.	30	17.75	GALAXY
SHAP348-5646.1	03 47 03.	- 56 55 13.	24	16.7	GALAXY
BAK 2.0613	03 47 04.	- 16 08 05.	41	16.2	GALAXY
BAK 2.0607	03 47 04.	- 23 40 41.	24	16.8	GALAXY
BAK 2.0604	03 47 04.	- 24 38 41.	28	16.7	GALAXY
SHAP348-5440.7	03 47 04.	- 54 49 49.	42	17.0	GALAXY
SHAP348-5529.4	03 47 04.	- 55 38 31.	36	17.25	GALAXY
SHAP348-5600.8	03 47 04.	- 56 09 55.	18	16.75	GALAXY
SHAP348-5604.0	03 47 04.	- 56 13 07.	24	18.0	GALAXY
SHAP348-5601.4	03 47 05.	- 56 10 31.	18	17.5	GALAXY
SHAP348-5654.5	03 47 05.	- 57 03 37.	24	17.2	GALAXY
SHAP348-5745.2	03 47 05.	- 57 54 19.	18	17.5	GALAXY
UGC 02873	03 47 06.	+ 17 41	84	16.0	GALAXY Sb
LB 01694	03 47 06.	- 56 50		13.3	FAINT BLUE STAR
SHAP348-5642.2	03 47 06.	- 56 51 19.	24	18.1	GALAXY
SHAP348-5653.3	03 47 06.	- 57 02 25.	24	17.1	GALAXY
SHAP348-5702.8	03 47 07.	- 57 11 55.	18	17.5	GALAXY
SHAP348-5602.0	03 47 08.	- 56 11 07.	24	17.0	GALAXY
SHAP348-5800.8	03 47 08.	- 58 09 55.	24	18.0	GALAXY
BAK 2.0615	03 47 09.	- 18 19 05.	14	17.2	GALAXY
SHAP348-5405.2	03 47 09.	- 54 14 19.	30	17.25	GALAXY
BAK 2.0616	03 47 10.	- 18 55 29.	16	16.9	GALAXY
ARC 0464	03 47 11.	- 17 58		17.7	RICH CLUSTER OF GALAXIES
BAK 2.0617	03 47 11.	- 18 42 41.	16	16.4	GALAXY
BAK 2.0612	03 47 11.	- 23 51 59.	23	16.7	GALAXY
SHAP348-4824.7	03 47 11.	- 48 33 48.	48	17.75	GALAXY
SHAP348-5053.9	03 47 11.	- 51 03 00.	24	17.5	GALAXY
SHAP348-5344.6	03 47 11.	- 53 53 42.	42	17.0	GALAXY
SHAP348-5558.4	03 47 11.	- 56 07 31.	24	16.25	GALAXY
SHAP348-5627.4	03 47 11.	- 56 36 31.	24	17.0	GALAXY
SHAP348-5648.4	03 47 11.	- 56 57 31.	18	17.7	GALAXY
SHAP348-5740.6	03 47 11.	- 57 49 43.	12	17.25	GALAXY
RNGC 1473	03 47 11.	- 68 22			GALAXY
ZWG 391.038	03 47 12.	- 01 55		14.6	GALAXY
MCG-02-10-013	03 47 12.	- 14 03 30.	48	15.5	GALAXY
BAK 2.0618	03 47 12.	- 17 41 35.	14	16.8	GALAXY
MCG-05-10-003	03 47 12.	- 26 52	42	15.5	GALAXY
LB 01695	03 47 12.	- 50 39		12.6	FAINT BLUE STAR
SHAP348-5605.7	03 47 12.	- 56 14 49.	24	17.5	GALAXY
SHAP348-5611.2	03 47 12.	- 56 20 19.	24	18.0	GALAXY
SHAP348-5721.5	03 47 12.	- 57 30 37.	24	17.75	GALAXY
BAK 2.0614	03 47 13.	- 22 51 53.	14	16.8	GALAXY
SHAP348-5344.0	03 47 13.	- 53 53 06.	30	18.0	GALAXY
SHAP348-5402.8	03 47 13.	- 54 11 54.	24	17.5	GALAXY
SHAP348-5419.8	03 47 13.	- 54 28 54.	54	16.5	GALAXY
SHAP348-5701.3	03 47 13.	- 57 10 25.	48	17.6	GALAXY
LB 01494	03 47 14.	+ 25 10 30.		19.7	FAINT BLUE STAR
SHAP348-5651.2	03 47 14.	- 57 00 19.	18	17.8	GALAXY
SHAP348-5556.8	03 47 15.	- 56 05 54.	18	16.75	GALAXY

OBJECT NAME	RIGHT ASCEN.	DECLINATION	DIAM.	MAGN.	TYPE OF OBJECT
BAK 2.0619	03 47 16.	- 18 20 28.	14	17.0	GALAXY
SHAP348-5602.5	03 47 16.	- 56 11 36.	24	17.5	GALAXY
SHAP348-5140.4	03 47 17.	- 51 49 30.	36	17.5	GALAXY
SHAP348-5732.8	03 47 17.	- 57 41 54.	18	17.5	GALAXY
ZWG 465.009	03 47 18.	+ 17 32		15.7	GALAXY
UGC 02874	03 47 18.	+ 17 32	66	15.7	GALAXY S0-a
CED 022	03 47 18.	+ 25 26			DIFFUSE GALACTIC NEBULA
IC 1995	03 47 18.	+ 25 26 03.			NONSTELLAR OBJECT
UGC 02875	03 47 18.	+ 36 23	66	16.5	GALAXY S
MCG+07-08-035	03 47 18.	+ 42 07	42	15.5	GALAXY
BAK 2.0621	03 47 18.	- 18 53 22.	50	16.3	GALAXY
BAK 1.923	03 47 20.	- 26 52 52.	42	15.5	EXTRAGALACTIC NEBULA
SHAP348-5609.1	03 47 20.	- 56 18 12.	24	17.25	GALAXY
MCG+03-10-008	03 47 21.	+ 18 04	33	16.	GALAXY
BAK 2.0622	03 47 21.	- 19 30 16.	16	17.0	GALAXY
SHAP348-5753.6	03 47 22.	- 58 02 42.	36	17.75	GALAXY
SHAP348-5755.7	03 47 22.	- 58 04 48.	30	16.5	GALAXY
BAK 2.0623	03 47 23.	- 18 26 28.	17	16.7	GALAXY
SHAP348-5621.8	03 47 23.	- 56 30 54.	18	17.5	GALAXY
SHAP348-5920.4	03 47 23.	- 59 29 30.	18	17.25	GALAXY
ZWG 541.026	03 47 24.	+ 42 08		15.7	GALAXY
MCG+00-10-022	03 47 24.	- 01 53	42	13.5	GALAXY
BAK 2.0620	03 47 24.	- 25 49 52.	24	16.7	GALAXY
MCG-05-10-004	03 47 24.	- 27 08	156	13.5	GALAXY
SHAP348-5614.0	03 47 24.	- 56 23 06.	24	18.0	GALAXY
SHAP348-5327.8	03 47 25.	- 53 36 54.	60	16.5	GALAXY
SHAP348-5958.8	03 47 26.	- 60 07 54.	42	17.25	GALAXY
BAK 2.0624	03 47 27.	- 18 33 16.	23	16.9	GALAXY
SHAP348-4900.0	03 47 27.	- 49 09 05.	36	17.5	GALAXY
SHAP348-5623.1	03 47 27.	- 56 32 12.	24	17.25	GALAXY
SHAP348-5916.2	03 47 27.	- 59 25 18.	24	18.0	GALAXY
SHAP348-5910.2	03 47 28.	- 59 19 18.	24	17.25	GALAXY
SHAP348-5334.6	03 47 29.	- 53 43 41.	36	17.5	GALAXY
BAK 2.0627	03 47 30.	- 16 54 39.	17	16.6	GALAXY
SHAP348-5603.1	03 47 30.	- 56 12 11.	24	16.75	GALAXY
SHAP348-5611.6	03 47 30.	- 56 20 41.	18	18.0	GALAXY
SHAP349-4713.5	03 47 31.	- 47 22 35.	30	17.0	GALAXY
SHAP348-5223.3	03 47 31.	- 52 32 23.	24	16.5	GALAXY
BAK 2.0626	03 47 32.	- 19 15 15.	14	16.5	GALAXY
SHAP348-6000.4	03 47 32.	- 60 09 30.	36	17.25	GALAXY
MCG+01-10-040	03 47 33.	- 04 25	18	16.	GALAXY
BAK 2.0625	03 47 34.	- 22 42 03.	23	16.6	GALAXY
SHAP348-5610.8	03 47 34.	- 56 19 53.	24	16.75	GALAXY
SHAP348-5348.8	03 47 35.	- 53 57 53.	30	16.75	GALAXY
TON-S 0379	03 47 36.	- 22 56		14.8	BLUE STAR
SHAP348-6004.2	03 47 36.	- 60 13 18.	36	17.0	GALAXY
BAK 2.0629	03 47 37.	- 22 10 09.	10	16.6	GALAXY
BAK 2.0628	03 47 37.	- 23 13 57.	21	16.8	GALAXY
BAK 1.925	03 47 38.	- 26 36 39.	16	16.5	EXTRAGALACTIC NEBULA
BAK 1.924	03 47 38.	- 27 08 39.	182	14.4	EXTRAGALACTIC NEBULA
SHAP348-5524.9	03 47 38.	- 55 33 59.	36	16.0	GALAXY
SHAP348-5734.9	03 47 38.	- 57 43 59.	18	17.5	GALAXY
SHAP348-5734.2	03 47 39.	- 57 43 17.	30	16.25	GALAXY
RNGC 1462	03 47 40.	+ 06 49		15.5	GALAXY
BAK 2.0631	03 47 40.	- 21 53 15.	12	17.2	GALAXY
SHAP349-5217.4	03 47 40.	- 52 26 29.	24	17.0	GALAXY
SHAP348-5336.5	03 47 40.	- 53 45 35.	60	17.25	GALAXY
SHAP348-5611.4	03 47 40.	- 56 20 29.	18	17.0	GALAXY
BAK 2.0633	03 47 41.	- 22 09 27.	11	16.6	GALAXY
BAK 2.0630	03 47 41.	- 23 32 51.	19	16.8	GALAXY
IC 2000	03 47 41.	- 49 00			NONSTELLAR OBJECT
SHAP349-4851.5	03 47 41.	- 49 00 34.	276	13.8	GALAXY
HN 0244	03 47 41.	- 49 01			NEBULA
SHAP348-5554.9	03 47 41.	- 56 03 59.	30	16.25	GALAXY
ZWG 417.007	03 47 42.	+ 06 49		15.3	GALAXY
MCG+01-10-010	03 47 42.	+ 06 50	48	14.5	GALAXY
UGC 02876	03 47 42.	+ 37 28	84	17.	GALAXY S
SHAP348-5650.2	03 47 42.	- 56 59 17.	36	18.2	GALAXY
BAK 2.0644	03 47 43.	- 15 47 39.	24	16.6	GALAXY
BAK 2.0639	03 47 43.	- 18 45 39.	22	16.6	GALAXY
BAK 2.0634	03 47 43.	- 23 21 03.	18	16.9	GALAXY
SHAP349-5355.0	03 47 43.	- 54 04 04.	36	17.5	GALAXY
BAK 2.0636	03 47 44.	- 23 08 09.	16	16.2	GALAXY
BAK 2.0635	03 47 44.	- 23 22 15.	16	16.9	GALAXY
BAK 2.0632	03 47 44.	- 25 09 21.	20	16.4	GALAXY
BAK 2.0640	03 47 45.	- 20 20 39.	14	17.2	GALAXY
SHAP348-5642.8	03 47 45.	- 56 51 53.	24	16.5	GALAXY
BAK 2.0645	03 47 46.	- 15 26 44.	12	16.5	GALAXY
BAK 2.0641	03 47 46.	- 20 21 03.	12	16.4	GALAXY
BAK 2.0638	03 47 46.	- 22 09 39.	15	16.9	GALAXY
BAK 2.0637	03 47 46.	- 24 16 03.	19	16.8	GALAXY
BAK 2.0646	03 47 47.	- 16 51 14.	19	16.6	GALAXY
SHAP349-5300.9	03 47 47.	- 53 09 58.	24	17.5	GALAXY
ZC 0347.8+3258	03 47 48.	+ 32 58	1480		CLUSTER OF GALAXIES
VB 205	03 47 48.	+ 57 11	403		STELLAR RING
BAK 2.0647	03 47 48.	- 17 54 26.	42	16.2	GALAXY
SHAP349-5114.2	03 47 48.	- 51 23 16.	42	17.25	GALAXY
SHAP348-5559.4	03 47 48.	- 56 08 28.	24	17.5	GALAXY
SHAP349-5352.2	03 47 49.	- 54 01 16.	36	17.0	GALAXY
SHAP349-5525.7	03 47 49.	- 55 34 46.	30	16.0	GALAXY
SHAP348-5617.0	03 47 49.	- 56 26 04.	30	16.5	GALAXY
SHAP348-5844.7	03 47 49.	- 58 53 47.	36	16.75	GALAXY
SHAP349-4914.4	03 47 50.	- 49 23 28.	36	17.5	GALAXY
SHAP348-5858.4	03 47 50.	- 59 07 29.	18	17.75	GALAXY
BAK 2.0643	03 47 51.	- 24 01 50.	19	16.8	GALAXY
BAK 2.0642	03 47 51.	- 24 32 26.	17	16.8	GALAXY
SHAP349-5626.6	03 47 51.	- 56 35 40.	12	17.5	GALAXY
LB 01495	03 47 52.	+ 25 09 24.		18.6	FAINT BLUE STAR
LB 01493	03 47 52.	+ 25 56 00.		18.0	FAINT BLUE STAR
SHAP349-5151.1	03 47 53.	- 52 00 10.	24	17.75	GALAXY
SHAP348-5845.3	03 47 53.	- 58 54 22.	36	16.5	GALAXY
UGC 02877	03 47 54.	+ 36 45	96	17.	GALAXY
PK149-01.1	03 47 54.	+ 51 20	84		PLANETARY NEBULA
BAK 2.0649	03 47 54.	- 18 51 26.	17	17.2	GALAXY
BAK 2.0652	03 47 55.	- 17 55 26.	16	16.6	GALAXY
SHAP349-5637.9	03 47 55.	- 56 46 58.	18	17.0	GALAXY
SHAP349-5606.6	03 47 56.	- 56 15 40.	24	16.25	GALAXY
SHAP349-5712.2	03 47 56.	- 57 21 16.	18	17.25	GALAXY
BAK 2.0658	03 47 57.	- 15 34 26.	19	16.6	GALAXY
BAK 2.0654	03 47 57.	- 17 44 44.	19	16.8	GALAXY
SHAP349-5518.8	03 47 57.	- 55 27 52.	24	17.0	GALAXY
BAK 2.0656	03 47 59.	- 18 11 32.	14	17.0	GALAXY
SHAP349-5225.5	03 47 59.	- 52 34 33.	24	17.75	GALAXY
SHAP349-5617.2	03 47 59.	- 56 26 16.	18	16.75	GALAXY
SHAP348-5922.6	03 47 59.	- 59 31 40.	24	17.0	GALAXY
HMS 0348+0613	03 48	+ 06 13			CLUSTER OF GALAXIES
LBN 0748	03 48	+ 35 20	2100		BRIGHT NEBULA
BAK 2.0655	03 48 00.	- 20 18 02.	17	16.0	GALAXY

OBJECT NAME	RIGHT ASCEN.	DECLINATION	DIAM.	MAGN.	TYPE OF OBJECT
SHAP349-5323.5	03 48 00.	- 53 32 33.	24	17.5	GALAXY
SHAP349-5324.6	03 48 00.	- 53 33 39.	30	17.0	GALAXY
SHAP349-5748.2	03 48 00.	- 57 57 16.	18	18.0	GALAXY
SHAP349-5841.4	03 48 00.	- 58 50 28.	24	16.5	GALAXY
SHAP348-5931.2	03 48 00.	- 59 40 16.	24	16.5	GALAXY
BAK 2.0650	03 48 01.	- 24 16 38.	22	16.6	GALAXY
SHAP349-5017.4	03 48 01.	- 50 26 27.	36	16.5	GALAXY
BAK 2.0648	03 48 02.	- 26 08 14.	19	17.0	GALAXY
SHAP349-5342.2	03 48 02.	- 53 51 15.	36	17.5	GALAXY
SHAP349-5714.0	03 48 02.	- 57 23 04.	24	17.75	GALAXY
BAK 2.0660	03 48 03.	- 19 24 02.	14	17.2	GALAXY
BAK 2.0651	03 48 03.	- 25 50 08.	18	17.0	GALAXY
SHAP349-5913.6	03 48 03.	- 59 22 40.	24	17.75	GALAXY
SHAP349-5917.6	03 48 03.	- 59 26 40.	24	17.5	GALAXY
BAK 2.0653	03 48 04.	- 24 14 50.	17	16.8	GALAXY
SHAP349-6007.0	03 48 04.	- 60 16 04.	36	17.0	GALAXY
BAK 2.0661	03 48 05.	- 18 40 49.	20	17.1	GALAXY
SHAP349-5225.2	03 48 05.	- 52 34 15.	30	17.5	GALAXY
5ZW 363	03 48 06.	+ 37 30			COMPACT GALAXY
OCL 0378	03 48 06.	+ 61 51	30C		OPEN STAR CLUSTER
MCG+01-10-041	03 48 06.	- 02 59	60	15.	GALAXY
BAK 2.0666	03 48 06.	- 16 40 49.	25	16.3	GALAXY
BAK 2.0664	03 48 06.	- 17 53 01.	18	17.1	GALAXY
SHAP349-5714.2	03 48 06.	- 57 23 15.	18	17.5	GALAXY
BAK 2.0662	03 48 07.	- 20 16 01.	13	16.8	GALAXY
BAK 2.0657	03 48 07.	- 25 50 13.	25	17.0	GALAXY
BAK 2.0667	03 48 08.	- 18 40 49.	26	17.0	GALAXY
ARC 0467	03 48 08.	- 22 25		17.5	RICH CLUSTER OF GALAXIES
SHAP349-5055.6	03 48 08.	- 51 04 39.	66	16.5	GALAXY
SHAP349-5524.5	03 48 08.	- 55 33 33.	30	17.0	GALAXY
SHAP349-5900.0	03 48 08.	- 59 09 03.	18	17.25	GALAXY
BAK 2.0659	03 48 09.	- 25 09 25.	12	16.9	GALAXY
SHAP349-5715.1	03 48 09.	- 57 24 09.	24	17.25	GALAXY
SHAP349-5717.4	03 48 10.	- 57 26 27.	24	18.0	GALAXY
BAK 2.0670	03 48 11.	- 17 57 49.	20	16.5	GALAXY
BAK 2.0668	03 48 11.	- 20 29 31.	13	16.8	GALAXY
SHAP349-5623.3	03 48 11.	- 56 32 21.	18	16.5	GALAXY
MCG+01-10-011	03 48 12.	+ 06 10	4	19.5	GALAXY
ZC 0348.2+1030	03 48 12.	+ 10 30	340		CLUSTER OF GALAXIES
SCHO 0062	03 48 12.	+ 36 59 24.	390		ISOLATED DARK CLOUD
SHAP349-5226.7	03 48 12.	- 52 35 45.	30	17.5	GALAXY
SHAP349-5548.1	03 48 12.	- 55 57 09.	30	17.5	GALAXY
BAK 2.0665	03 48 13.	- 23 42 13.	13	16.7	GALAXY
SHAP349-5300.9	03 48 13.	- 53 09 57.	30	17.5	GALAXY
ARC 0465	03 48 15.	+ 06 10		17.7	RICH CLUSTER OF GALAXIES
BAK 2.0663	03 48 15.	- 26 53 37.	21	16.6	GALAXY
SHAP349-5409.5	03 48 15.	- 54 18 33.	48	17.25	GALAXY
BAK 2.0673	03 48 16.	- 19 20 43.	13	16.8	GALAXY
SHAP349-5556.6	03 48 16.	- 56 05 39.	24	17.5	GALAXY
SHAP349-5716.5	03 48 16.	- 57 25 33.	24	17.25	GALAXY
SHAP349-5906.6	03 48 16.	- 59 15 39.	24	16.0	GALAXY
BAK 2.0674	03 48 17.	- 19 54 19.	14	16.8	GALAXY
SHAP349-5224.2	03 48 17.	- 52 33 14.	24	17.5	GALAXY
SHAP349-5326.4	03 48 17.	- 53 35 26.	30	18.0	GALAXY
SHAP349-5734.8	03 48 17.	- 57 43 51.	18	17.5	GALAXY
SHAP349-5738.2	03 48 17.	- 57 47 15.	18	17.5	GALAXY
SHAP349-5742.0	03 48 17.	- 57 51 03.	18	17.75	GALAXY
ZWG 526.010	03 48 18.	+ 36 56		15.7	GALAXY
MCG+06-09-010	03 48 18.	+ 36 56	30	15.5	GALAXY
BAK 2.0680	03 48 18.	- 17 53 37.	22	16.2	GALAXY
BAK 2.0681	03 48 19.	- 18 19 19.	16	16.9	GALAXY
BAK 2.0669	03 48 20.	- 26 09 01.	22	16.8	GALAXY
SHAP349-5553.0	03 48 20.	- 56 02 02.	24	17.5	GALAXY
SHAP349-5902.0	03 48 20.	- 59 11 03.	18	17.0	GALAXY
BAK 2.0685	03 48 21.	- 16 12 36.	17	16.8	GALAXY
BAK 2.0677	03 48 21.	- 21 17 24.	14	16.4	GALAXY
BAK 2.0678	03 48 21.	- 21 18 06.	13	17.0	GALAXY
BAK 2.0675	03 48 21.	- 23 00 37.	17	16.6	GALAXY
BAK 2.0671	03 48 21.	- 26 32 37.	13	16.7	GALAXY
SHAP349-5223.8	03 48 21.	- 52 32 50.	42	17.0	GALAXY
SHAP349-5945.8	03 48 21.	- 59 54 51.	24	17.5	GALAXY
BAK 2.0679	03 48 22.	- 21 54 00.	19	16.8	GALAXY
BAK 2.0672	03 48 22.	- 25 10 25.	18	16.8	GALAXY
SHAP349-5325.7	03 48 22.	- 53 34 44.	30	18.0	GALAXY
SHAP349-5400.1	03 48 22.	- 54 09 08.	48	17.5	GALAXY
SHAP349-5943.5	03 48 22.	- 59 52 33.	24	17.5	GALAXY
BAK 2.0684	03 48 23.	- 19 46 36.	17	16.6	GALAXY
BAK 2.0676	03 48 23.	- 24 19 30.	24	16.8	GALAXY
SHAP349-4856.8	03 48 23.	- 49 05 50.	36	17.5	GALAXY
SHAP349-5403.2	03 48 24.	- 54 12 14.	30	17.5	GALAXY
SHAP349-5724.8	03 48 24.	- 57 33 50.	24	17.5	GALAXY
BAK 2.0689	03 48 25.	- 17 12 42.	16	17.2	GALAXY
SHAP349-5110.2	03 48 25.	- 51 19 14.	24	18.0	GALAXY
SHAP349-5701.8	03 48 25.	- 57 10 50.	18	17.5	GALAXY
BAK 2.0682	03 48 27.	- 24 20 30.	19	17.1	GALAXY
BAK 2.0691	03 48 28.	- 18 10 12.	17	16.9	GALAXY
BAK 2.0683	03 48 28.	- 25 44 36.	19	16.8	GALAXY
SHAP349-5538.8	03 48 28.	- 55 47 50.	30	16.75	GALAXY
SHAP349-5557.4	03 48 28.	- 56 06 26.	30	18.0	GALAXY
BAK 2.0690	03 48 29.	- 19 57 12.	14	16.6	GALAXY
BAK 2.0688	03 48 29.	- 23 05 00.	15	16.8	GALAXY
SHAP349-5548.8	03 48 29.	- 55 57 50.	54	16.0	GALAXY
ZWG 417.008	03 48 30.	+ 03 39		15.4	GALAXY
KARA.73B 0133	03 48 30.	+ 03 39	42	15.4	ISOLATED GALAXY S
ZC 0348.5+0607	03 48 30.	+ 06 07	1810		CLUSTER OF GALAXIES
UGC 02878	03 48 30.	+ 42 40	72	17.	GALAXY Sc
MCG+01-10-042	03 48 30.	- 07 12 30.	42	15.5	GALAXY
BAK 2.0692	03 48 30.	- 17 05 00.	12	16.4	GALAXY
MCG-04-10-006	03 48 30.	- 25 26	90	14.	GALAXY
SHAP349-5642.4	03 48 30.	- 56 51 26.	24	17.25	GALAXY
SHAP349-5943.8	03 48 30.	- 59 52 50.	24	17.5	GALAXY
BAK 2.0686	03 48 31.	- 25 15 36.	18	16.8	GALAXY
BAK 2.0687	03 48 31.	- 25 25 54.	66	14.1	GALAXY
MCG+01-10-043	03 48 33.	- 03 14	60	16.	GALAXY
BAK 2.0696	03 48 33.	- 16 37 00.	12	17.0	GALAXY
BAK 2.0694	03 48 33.	- 18 08 36.	19	17.3	GALAXY
BAK 2.0697	03 48 34.	- 17 49 36.	14	16.9	GALAXY
BAK 2.0699	03 48 35.	- 16 59 59.	13	16.9	GALAXY
BAK 2.0698	03 48 35.	- 17 04 35.	12	16.8	GALAXY
BAK 2.0693	03 48 35.	- 20 58 00.	16	17.0	GALAXY
SHAP349-5350.2	03 48 35.	- 53 59 13.	36	17.5	GALAXY
SHAP349-5652.1	03 48 35.	- 57 01 08.	24	17.5	GALAXY
UGC 02879	03 48 36.	+ 32 52	90	18.	GALAXY DWARF
MCG-02-10-014	03 48 36.	- 08 40	36	15.	GALAXY
BAK 2.0700	03 48 36.	- 17 24 35.	15	16.5	GALAXY
BAK 2.0695	03 48 36.	- 19 55 06.	13	16.7	GALAXY
SHAP349-5553.6	03 48 36.	- 56 02 37.	30	17.75	GALAXY
SHAP349-5740.3	03 48 36.	- 57 49 20.	24	17.75	GALAXY

OBJECT NAME	RIGHT ASCEN.	DECLINATION	DIAM.	MAGN.	TYPE OF OBJECT
SHB 075	03 48 37.	+ 33 04 54.		18.5	QUASI-STELLAR OBJECT
BC 3C93.1/113	03 48 37.	+ 33 05			QUASI-STELLAR OBJECT
SHAP349-5345.0	03 48 37.	- 53 54 01.	24	17.5	GALAXY
NAB 0348+06	03 48 38.	+ 06 10 30.		17.6	QUASI-STELLAR OBJECT
SHB 076	03 48 38.	+ 06 10 30.		17.6	QUASI-STELLAR OBJECT
SHAP350-4516.0	03 48 38.	- 45 25 00.	60	17.25	GALAXY
BAK 2.0704	03 48 39.	- 17 54 35.	20	16.6	GALAXY
BAK 2.0701	03 48 39.	- 19 02 11.	16	16.9	GALAXY
SHAP349-5442.0	03 48 39.	- 54 51 01.	36	16.25	GALAXY
SHAP349-5408.2	03 48 40.	- 54 17 13.	36	17.5	GALAXY
SHAP349-5726.5	03 48 41.	- 57 35 31.	18	17.0	GALAXY
SHAP349-5739.3	03 48 41.	- 57 48 19.	24	17.75	GALAXY
SHAP349-5739.6	03 48 41.	- 57 48 37.	24	17.75	GALAXY
SHAP349-5712.6	03 48 42.	- 57 21 37.	18	17.5	GALAXY
BAK 2.0705	03 48 43.	- 17 55 17.	30	16.6	GALAXY
BAK 2.0702	03 48 43.	- 22 58 59.	17	16.9	GALAXY
SHAP349-5643.9	03 48 43.	- 56 52 55.	12	17.5	GALAXY
SHAP349-5913.6	03 48 43.	- 59 22 37.	24	16.25	GALAXY
SHAP349-5729.6	03 48 44.	- 57 38 37.	24	17.0	GALAXY
BAK 2.0703	03 48 45.	- 23 41 41.	13	16.8	GALAXY
SS 12	03 48 46.	+ 32 59			DIFFUSE GALACTIC NEBULA
BAK 2.0706	03 48 46.	- 18 15 11.	14	16.8	GALAXY
SHAP349-5634.4	03 48 46.	- 56 43 25.	24	17.5	GALAXY
SHAP349-5721.2	03 48 46.	- 57 30 13.	30	16.5	GALAXY
SHAP349-5929.1	03 48 46.	- 59 38 07.	54	16.0	GALAXY
BAK 2.0707	03 48 47.	- 17 37 59.	30	16.4	GALAXY
MCG-06-09-034	03 48 48.	- 36 03	36	13.	GALAXY
SHAP350-5016.6	03 48 48.	- 50 25 36.	54	17.0	GALAXY
SHAP349-5756.5	03 48 48.	- 58 05 31.	18	17.75	GALAXY
SHAP349-5929.8	03 48 48.	- 59 38 49.	18	17.0	GALAXY
SHAP349-5610.2	03 48 49.	- 56 19 13.	48	16.5	GALAXY
SHAP349-5933.0	03 48 49.	- 59 42 01.	24	17.75	GALAXY
BAK 2.0708	03 48 50.	- 20 16 59.	18	16.0	GALAXY
SHAP349-5730.9	03 48 50.	- 57 39 55.	24	17.0	GALAXY
BAK 2.0710	03 48 51.	- 20 17 23.	16	16.8	GALAXY
SHAP350-5233.2	03 48 52.	- 52 42 12.	24	17.5	GALAXY
SHAP349-5929.5	03 48 52.	- 59 38 31.	18	17.0	GALAXY
SHAP350-5326.7	03 48 53.	- 53 35 42.	36	16.5	GALAXY
SHAP349-5727.5	03 48 53.	- 57 36 31.	24	16.75	GALAXY
SHAP349-5736.0	03 48 53.	- 57 45 01.	24	16.25	GALAXY
ZWG 391.039	03 48 54.	+ 01 02		15.7	GALAXY
UGC 02880	03 48 54.	+ 24 46	84	16.0	GALAXY Sc
MCG+04-10-001	03 48 54.	+ 24 48	66	16.	GALAXY
BAK 2.0709	03 48 54.	- 23 36 35.	13	16.8	GALAXY
SHAP350-5226.6	03 48 54.	- 52 35 36.	30	18.0	GALAXY
IC 1998	03 48 55.	+ 01 03 14.			NONSTELLAR OBJECT
LB 01496	03 48 55.	+ 24 57 24.		16.2	FAINT BLUE STAR
BAK 2.0711	03 48 55.	- 21 54 52.	10	17.0	GALAXY
BAK 2.0712	03 48 55.	- 21 56 22.	14	16.8	GALAXY
SHAP350-5708.2	03 48 55.	- 57 17 12.	24	17.5	GALAXY
YM 19	03 48 57.	+ 53 22	1800		SYMMETRIC GALACTIC NEBULA
SHAP350-5454.2	03 48 57.	- 55 03 12.	36	17.75	GALAXY
SHAP350-5148.1	03 48 58.	- 51 57 06.	24	17.5	GALAXY
SHAP350-5319.6	03 48 58.	- 53 28 36.	24	18.0	GALAXY
SHAP350-5342.8	03 48 58.	- 53 51 48.	42	18.0	GALAXY
SHAP350-5735.3	03 48 58.	- 57 44 18.	18	16.5	GALAXY
BAK 2.0713	03 48 59.	- 25 06 16.	18	16.4	GALAXY
SHAP350-5757.7	03 48 59.	- 58 06 42.	36	16.75	GALAXY
CED 0190	03 49	+ 23			DIFFUSE GALACTIC NEBULA
LBN 0757	03 49	+ 34 00	1800		BRIGHT NEBULA
LBN 0696	03 49	+ 53 10	2040		BRIGHT NEBULA
LB 09808	03 49	- 84 06		14.8	FAINT BLUE STAR
LB 09809	03 49	- 85 29		14.5	FAINT BLUE STAR
ZWG 526.011	03 49 00.	+ 36 05		15.0	GALAXY
UGC 02881	03 49 00.	+ 36 05	90	15.0	GALAXY S0
MCG+06-09-011	03 49 00.	+ 36 05 30.	72	15.	GALAXY
MCG+00-10-023	03 49 00.	- 00 35	24	15.	GALAXY
ZWG 391.040	03 49 00.	- 00 37		15.3	GALAXY
BAK 2.0716	03 49 00.	- 20 24 34.	14	17.0	GALAXY
BAK 2.0714	03 49 00.	- 22 20 04.	15	16.6	GALAXY
SHAP350-5704.6	03 49 00.	- 57 13 36.	24	17.0	GALAXY
BAK 2.0715	03 49 01.	- 22 30 46.	13	16.4	GALAXY
SHAP350-4553.0	03 49 01.	- 46 01 59.	48	17.5	GALAXY
SHAP350-5018.0	03 49 02.	- 50 26 59.	42	16.5	GALAXY
SHAP350-5859.6	03 49 02.	- 59 08 36.	24	16.5	GALAXY
MCG+01-10-044	03 49 03.	- 07 49 30.	108	15.	GALAXY
SHAP350-5419.1	03 49 03.	- 54 28 06.	42	17.0	GALAXY
BAK 2.0719	03 49 04.	- 19 28 04.	16	16.8	GALAXY
HZ 05	03 49 05.	+ 07 00		14.7	DECIDEDLY BLUE STAR
BAK 2.0724	03 49 05.	- 15 52 46.	38	16.6	GALAXY
BAK 2.0723	03 49 05.	- 18 27 34.	30	15.4	GALAXY
LB 01497	03 49 06.	+ 24 47 12.		15.4	FAINT BLUE STAR
ZWG 391.041	03 49 06.	- 00 40		15.0	GALAXY
BAK 2.0720	03 49 06.	- 20 21 22.	17	17.1	GALAXY
BAK 2.0721	03 49 06.	- 20 22 34.	17	17.1	GALAXY
BAK 2.0726	03 49 07.	- 16 58 03.	19	16.4	GALAXY
SHAP350-5019.0	03 49 07.	- 50 27 59.	42	16.5	GALAXY
BAK 2.0727	03 49 08.	- 17 28 57.	18	16.4	GALAXY
BAK 2.0718	03 49 08.	- 24 35 34.	18	16.7	GALAXY
BAK 2.0717	03 49 08.	- 25 08 04.	17	17.0	GALAXY
SHAP350-5240.1	03 49 08.	- 52 49 05.	30	17.5	GALAXY
SHAP350-5606.3	03 49 08.	- 56 15 17.	24	17.25	GALAXY
SHAP350-5932.6	03 49 08.	- 59 41 36.	24	16.75	GALAXY
RNGC 1464	03 49 09.	- 15 33			GALAXY
SHAP350-4613.2	03 49 09.	- 46 22 11.	48	17.5	GALAXY
SHAP350-5504.4	03 49 09.	- 55 13 23.	30	16.75	GALAXY
SHAP350-5732.7	03 49 09.	- 57 41 42.	18	16.0	GALAXY
SHAP350-5757.2	03 49 09.	- 58 06 12.	24	17.0	GALAXY
BC 3C95	03 49 09.5	- 14 38 03.		16.24	QUASI-STELLAR OBJECT
SHB 077	03 49 09.5	- 14 38 07.		16.2	QUASI-STELLAR OBJECT
LB 01498	03 49 10.	+ 22 57 48.		19.0	FAINT BLUE STAR
BAK 2.0722	03 49 11.	- 24 25 04.	17	17.0	GALAXY
SHAP350-5315.4	03 49 11.	- 53 24 23.	46	17.0	GALAXY
SHAP350-5723.2	03 49 11.	- 57 32 11.	30	16.75	GALAXY
MCG+00-10-024	03 49 12.	- 00 37	24	15.	GALAXY
TON-S 0380	03 49 12.	- 25 49		15.2	BLUE STAR
LB 01696	03 49 12.	- 47 51		15.4	FAINT BLUE STAR
SHAP350-5149.0	03 49 12.	- 51 57 59.	30	17.75	GALAXY
SHAP350-5418.6	03 49 13.	- 54 27 35.	48	17.0	GALAXY
SHAP350-5620.8	03 49 13.	- 56 29 47.	24	17.0	GALAXY
SHAP350-5722.9	03 49 14.	- 57 31 53.	18	17.75	GALAXY
BAK 2.0731	03 49 15.	- 16 13 45.	22	16.6	GALAXY
SHAP350-5244.8	03 49 15.	- 52 53 47.	24	17.5	GALAXY
SHAP350-5800.0	03 49 15.	- 58 08 59.	18	18.0	GALAXY
BAK 2.0730	03 49 16.	- 18 56 33.	12	16.8	GALAXY
BAK 2.0728	03 49 16.	- 23 23 45.	26	16.9	GALAXY
SHAP350-5234.8	03 49 16.	- 52 43 47.	24	17.5	GALAXY
ARC 0466	03 49 17.	+ 25 05		17.5	RICH CLUSTER OF GALAXIES
SHAP350-5549.6	03 49 17.	- 55 58 35.	30	17.0	GALAXY
SHAP350-5801.0	03 49 17.	- 58 09 59.	18	17.75	GALAXY
BAK 2.0725	03 49 18.	- 26 54 21.	19	16.7	GALAXY
TON-S 0381	03 49 18.	- 32 01		15.6	BLUE STAR
BAK 2.0729	03 49 20.	- 23 22 57.	15	16.0	GALAXY
SHAP350-5211.2	03 49 21.	- 52 20 10.	30	17.5	GALAXY
SHAP350-4802.5	03 49 22.	- 48 11 26.	30	17.5	GALAXY
HN 0245	03 49 23.	- 48 47			NEBULA
ZC 0349.4+2458	03 49 24.	+ 24 58	3290		CLUSTER OF GALAXIES
IC 2001	03 49 24.	- 48 47			NONSTELLAR OBJECT
SHAP350-5317.8	03 49 24.	- 53 26 46.	36	17.5	GALAXY
BAK 2.0732	03 49 25.	- 25 25 45.	14	16.8	GALAXY
SHAP350-5023.1	03 49 25.	- 50 32 04.	48	17.0	GALAXY
SHAP350-5023.9	03 49 25.	- 50 32 52.	48	17.0	GALAXY
SHAP350-5258.6	03 49 25.	- 53 07 34.	30	17.75	GALAXY
SHAP350-5727.9	03 49 25.	- 57 36 53.	18	17.25	GALAXY
YM 20	03 49 26.	+ 52 31	1320		SYMMETRIC GALACTIC NEBULA
BAK 2.0734	03 49 26.	- 22 26 33.	15	16.4	GALAXY
SHAP350-5252.8	03 49 26.	- 53 01 46.	36	17.0	GALAXY
SHAP350-5634.2	03 49 26.	- 56 43 10.	18	17.5	GALAXY
BAK 2.0733	03 49 27.	- 25 29 45.	17	16.3	GALAXY
BAK 2.0735	03 49 29.	- 18 54 56.	12	17.1	GALAXY
UGC 02882	03 49 30.	+ 34 32	108	17.	GALAXY S IV
RNGC 1467	03 49 30.	- 08 59		15.0	GALAXY
MCG-02-10-015	03 49 30.	- 08 59 30.	30	15.	GALAXY
SHAP350-5854.1	03 49 30.	- 59 03 04.	24	17.5	GALAXY
BAK 2.0736	03 49 31.	- 20 07 20.	14	17.2	GALAXY
BAK 2.0737	03 49 31.	- 20 15 32.	22	16.8	GALAXY
SHAP350-5749.0	03 49 31.	- 57 57 58.	24	16.5	GALAXY
SHAP350-5726.5	03 49 32.	- 57 35 28.	36	17.0	GALAXY
SHAP350-5740.6	03 49 32.	- 57 49 34.	24	17.75	GALAXY
SHAP350-5928.4	03 49 33.	- 59 37 22.	24	17.0	GALAXY
BAK 2.0740	03 49 34.	- 20 35 26.	12	17.0	GALAXY
BAK 2.0738	03 49 34.	- 22 08 44.	12	16.8	GALAXY
SHAP350-5117.8	03 49 35.	- 51 26 45.	42	16.5	GALAXY
SHAP350-5718.6	03 49 36.	- 57 27 34.	18	17.5	GALAXY
SHAP350-5158.4	03 49 38.	- 52 07 21.	48	16.25	GALAXY
SHAP350-5511.5	03 49 38.	- 55 20 28.	24	16.75	GALAXY
SHAP350-5553.1	03 49 38.	- 56 02 04.	24	17.0	GALAXY
BAK 2.0739	03 49 39.	- 25 55 32.	24	16.6	GALAXY
SHAP350-5337.9	03 49 39.	- 53 46 51.	36	17.25	GALAXY
SHAP350-5713.2	03 49 39.	- 57 22 10.	18	17.0	GALAXY
LB 01499	03 49 41.	+ 23 42 00.		20.5	FAINT BLUE STAR
BAK 2.0743	03 49 41.	- 19 26 01.	14	17.0	GALAXY
BAK 2.0742	03 49 41.	- 21 24 56.	14	16.8	GALAXY
BAK 2.0741	03 49 41.	- 22 03 08.	15	17.0	GALAXY
SHAP351-5058.1	03 49 41.	- 51 07 03.	42	16.75	GALAXY
SHAP350-5646.8	03 49 41.	- 56 55 46.	18	17.0	GALAXY
UGC 02884	03 49 42.	+ 02 14	72	16.5	GALAXY Sc
ZWG 391.042	03 49 42.	- 01 40		15.0	GALAXY
UGC 02883	03 49 42.	- 01 40	72	15.0	GALAXY
KARA.73B 0134	03 49 42.	- 01 40	78	15.0	ISOLATED GALAXY S
LB 01697	03 49 42.	- 51 58		14.0	FAINT BLUE STAR
SHAP350-5949.8	03 49 42.	- 59 58 46.	30	17.5	GALAXY
SHAP350-5639.6	03 49 43.	- 56 48 33.	18	17.75	GALAXY
SHAP350-5710.9	03 49 43.	- 57 19 51.	12	17.7	GALAXY
BAK 2.0749	03 49 45.	- 17 35 07.	14	17.0	GALAXY
BAK 2.0744	03 49 45.	- 21 59 19.	21	16.9	GALAXY
SHAP350-5803.2	03 49 45.	- 58 12 09.	24	17.0	GALAXY
RNGC 1468	03 49 46.	- 06 29		15.0	GALAXY
BAK 2.0750	03 49 46.	- 17 56 25.	20	16.8	GALAXY
SHAP351-5303.1	03 49 46.	- 53 12 03.	36	18.0	GALAXY
SHAP351-5311.8	03 49 46.	- 53 20 45.	36	17.5	GALAXY
SHAP350-5506.3	03 49 46.	- 55 15 15.	30	16.25	GALAXY
SHAP350-5634.4	03 49 46.	- 56 43 21.	18	17.25	GALAXY
SHAP350-5615.0	03 49 47.	- 56 23 57.	24	18.0	GALAXY
ZC 0349.8+2117	03 49 48.	+ 21 17	1340		CLUSTER OF GALAXIES
MCG+06-09-012	03 49 48.	+ 35 26 30.	180	14.	GALAXY
ZWG 526.012	03 49 48.	+ 35 27		14.4	GALAXY
UGC 02885	03 49 48.	+ 35 27	330	14.4	GALAXY Sc
VB 206	03 49 48.	+ 57 07	537		STELLAR RING
SHAH 147	03 49 48.	- 00 03	78		GROUP OF COMPACT GALAXIES
ZWG 391.043	03 49 48.	- 01 07		15.0	GALAXY
MCG+00-10-025	03 49 48.	- 01 37	54	13.5	GALAXY
MCG+01-10-045	03 49 48.	- 06 29	21	15.	GALAXY
RNGC 1470	03 49 48.	- 09 09		15.0	GALAXY
TON-S 0382	03 49 48.	- 27 03		15.7	BLUE STAR
HPW 47	03 49 48.	- 35 45			DWARF GALAXY IN FORNAX
SHAP351-5400.2	03 49 48.	- 54 09 09.	42	17.25	GALAXY
SHAP350-5610.5	03 49 48.	- 56 19 27.	24	17.25	GALAXY
BAK 2.0745	03 49 49.	- 24 53 19.	14	17.0	GALAXY
SHAP351-5133.3	03 49 49.	- 51 42 15.	30	16.75	GALAXY
ARC 0469	03 49 50.	- 22 20		17.9	RICH CLUSTER OF GALAXIES
ARC 0468	03 49 51.	+ 21 16		17.4	RICH CLUSTER OF GALAXIES
MCG-02-10-016	03 49 51.	- 09 09	72	15.	GALAXY
MCG-02-10-017	03 49 51.	- 09 16 30.	48	15.	GALAXY
BAK 2.0746	03 49 51.	- 24 39 19.	38	16.7	GALAXY
SHAP351-5126.3	03 49 51.	- 51 35 14.	36	17.0	GALAXY
SHAP351-5552.6	03 49 51.	- 56 01 33.	30	17.5	GALAXY
BAK 2.0752	03 49 52.	- 19 22 19.	12	16.6	GALAXY
SHAP351-5625.1	03 49 52.	- 56 34 03.	24	17.0	GALAXY
SHAP351-5636.2	03 49 52.	- 56 45 09.	24	17.25	GALAXY
BAK 2.0754	03 49 53.	- 15 36 49.	12	16.7	GALAXY
BAK 2.0747	03 49 53.	- 25 56 55.	20	16.7	GALAXY
SHAP351-5638.8	03 49 53.	- 56 47 45.	24	17.5	GALAXY
UGC 02886	03 49 54.	+ 34 49	102	18.	GALAXY S IV-V
BAK 2.0748	03 49 54.	- 26 38 31.	21	16.3	GALAXY
SHAP350-5741.9	03 49 54.	- 57 50 51.	18	17.0	GALAXY
SHAP350-5956.9	03 49 54.	- 60 05 51.	24	16.5	GALAXY
BAK 2.0751	03 49 55.	- 23 22 43.	12	16.0	GALAXY
SCHO 0063	03 49 56.	+ 34 58 06.	400		ISOLATED DARK CLOUD
BAK 2.0755	03 49 56.	- 18 16 43.	25	17.0	GALAXY
SHAP351-5620.4	03 49 56.	- 56 29 21.	18	17.0	GALAXY
SHAP351-5634.2	03 49 56.	- 56 43 09.	12	17.25	GALAXY
SHAP351-5651.0	03 49 57.	- 56 59 57.	18	17.5	GALAXY
SHAP351-5246.5	03 49 59.	- 52 55 26.	36	17.5	GALAXY
IC 0354	03 50	+ 23			NONSTELLAR OBJECT
LBN 0776	03 50	+ 24 00	7800		BRIGHT NEBULA
IC 0353	03 50	+ 25 46			NONSTELLAR OBJECT
SIV 04	03 50	+ 32 45	32400		FAINT H EMISSION REGION
MCG+05-10-001	03 50 00.	+ 32 09	30	15.	GALAXY
UGC 02887	03 50 00.	+ 34 48	66	18.	GALAXY DWARF
LDN 1449	03 50 00.	+ 37 30	6120		DARK NEBULA
MCG+14-03-001	03 50 00.	+ 83 36	42	17.	GALAXY
UGC 02886A	03 50 00.	+ 88 37	72	17.	GALAXY Sc
BAK 2.0758	03 50 00.	- 18 16 30.	16	17.2	GALAXY
KARA.73 21	03 50 00.	- 38 36	67		DWARF GALAXY
BAK 2.0753	03 50 01.	- 26 37 55.	19	16.2	GALAXY

OBJECT NAME	RIGHT ASCEN.	DECLINATION	DIAM.	MAGN.	TYPE OF OBJECT
SHAP351-5741.8	03 50 01.	- 57 50 44.	24	17.0	GALAXY
SHAP351-4913.8	03 50 02.	- 49 22 44.	54	17.25	GALAXY
SHAP351-5916.6	03 50 02.	- 59 25 33.	18	17.25	GALAXY
BAK 2.0763	03 50 03.	- 18 00 18.	14	16.8	GALAXY
BAK 2.0759	03 50 03.	- 20 53 18.	17	16.4	GALAXY
BAK 2.0756	03 50 03.	- 21 45 54.	13	16.8	GALAXY
BC 3C94	03 50 04.	- 07 19 50.		16.49	QUASI-STELLAR OBJECT
BAK 2.0764	03 50 04.	- 19 12 18.	18	17.1	GALAXY
SHAP351-5627.4	03 50 04.	- 56 36 20.	24	17.5	GALAXY
SHAP351-5636.6	03 50 04.	- 56 45 32.	18	18.0	GALAXY
SHB 078	03 50 04.1	- 07 19 55.		16.5	QUASI-STELLAR OBJECT
BAK 2.0765	03 50 05.	- 19 24 18.	16	16.6	GALAXY
BAK 2.0760	03 50 05.	- 20 56 30.	19	16.6	GALAXY
BAK 2.0761	03 50 05.	- 20 59 06.	17	16.4	GALAXY
SHAP351-4819.4	03 50 05.	- 48 28 19.	48	18.0	GALAXY
SHAP351-5253.0	03 50 05.	- 53 01 56.	48	17.0	GALAXY
SHAP351-5343.1	03 50 05.	- 53 52 02.	42	17.75	GALAXY
ZWG 508.002	03 50 06.	+ 32 09		15.7	GALAXY
MCG+05-10-002	03 50 06.	+ 32 11 30.	60	14.5	GALAXY
BAK 2.0768	03 50 06.	- 16 17 42.	19	16.2	GALAXY
BAK 2.0766	03 50 06.	- 19 12 06.	17	17.1	GALAXY
BAK 2.0757	03 50 06.	- 24 43 30.	23	16.8	GALAXY
SHAP351-5133.0	03 50 06.	- 51 41 56.	36	17.5	GALAXY
SHAP351-5524.1	03 50 07.	- 55 33 02.	30	17.0	GALAXY
SHAP351-5642.5	03 50 07.	- 56 51 26.	18	16.75	GALAXY
SHAP351-5735.4	03 50 07.	- 57 44 20.	18	18.0	GALAXY
SHAP351-5907.6	03 50 07.	- 59 16 32.	36	16.0	GALAXY
BAK 2.0762	03 50 08.	- 23 48 06.	14	16.8	GALAXY
SHAP351-4635.2	03 50 09.	- 46 44 07.	42	16.5	GALAXY
SHAP351-5803.9	03 50 09.	- 58 12 50.	18	17.5	GALAXY
SHAP351-5907.5	03 50 09.	- 59 16 26.	24	17.0	GALAXY
BAK 2.0767	03 50 10.	- 21 05 06.	22	16.7	GALAXY
SHAP351-5420.8	03 50 11.	- 54 29 43.	48	16.5	GALAXY
SHAP351-5647.6	03 50 11.	- 56 56 32.	18	17.0	GALAXY
SHAP351-5714.5	03 50 11.	- 57 23 26.	24	18.0	GALAXY
SHAP351-5744.8	03 50 11.	- 57 53 44.	30	17.5	GALAXY
ZWG 417.009	03 50 12.	+ 06 54		15.6	GALAXY
ZWG 508.003	03 50 12.	+ 32 10		15.2	GALAXY
UGC 02888	03 50 12.	+ 32 10	66	15.2	GALAXY S
SZW 264	03 50 12.	+ 35 45			COMPACT GALAXY
ISS 0074	03 50 12.	+ 57 08	524		STELLAR RING
KARA.68 044	03 50 12.	- 02 49	27		DWARF GALAXY
SHAP351-5359.2	03 50 12.	- 54 08 07.	36	17.5	GALAXY
SHAP351-5259.3	03 50 13.	- 53 08 13.	30	17.5	GALAXY
BAK 2.0772	03 50 14.	- 18 04 35.	17	16.7	GALAXY
BAK 2.0770	03 50 14.	- 19 08 05.	12	17.2	GALAXY
SHAP351-5408.4	03 50 14.	- 54 17 19.	36	16.5	GALAXY
SHAP351-5633.0	03 50 14.	- 56 41 55.	18	17.25	GALAXY
SHAP351-5909.4	03 50 15.	- 59 18 20.	24	17.0	GALAXY
BAK 2.0775	03 50 16.	- 16 19 29.	16	16.6	GALAXY
BAK 2.0769	03 50 16.	- 23 10 59.	53	16.0	GALAXY
SHAP351-5348.4	03 50 16.	- 53 57 19.	42	17.25	GALAXY
BAK 2.0771	03 50 17.	- 21 55 53.	12	16.8	GALAXY
RNGC 1476	03 50 17.	- 44 40			GALAXY
SHAP351-5656.6	03 50 17.	- 57 05 31.	18	17.25	GALAXY
SHAP351-5726.6	03 50 17.	- 57 35 31.	24	17.0	GALAXY
MCG+01-10-012	03 50 18.	+ 06 55	36	16.	GALAXY
ZWG 526.013	03 50 18.	+ 37 06		15.7	GALAXY
UGC 02889	03 50 18.	+ 37 06	96	15.7	GALAXY Sb/SBb
VB 207	03 50 18.	+ 57 29	235		STELLAR RING
ZWG 327.010	03 50 18.	+ 72 46		14.7	GALAXY
UGC 02890	03 50 18.	+ 72 46	156	14.7	GALAXY S-IRR
BAK 2.0774	03 50 18.	- 19 05 53.	14	16.8	GALAXY
MCG-04-10-007	03 50 18.	- 23 11 30.	36	16.	GALAXY
TON-S 0383	03 50 18.	- 24 23		15.2	BLUE STAR
SHAP351-5543.6	03 50 18.	- 55 52 31.	24	16.25	GALAXY
SHAP351-5342.5	03 50 19.	- 53 51 25.	24	17.75	GALAXY
SHAP351-5507.0	03 50 19.	- 55 15 55.	24	16.75	GALAXY
BAK 2.0773	03 50 20.	- 21 59 53.	11	16.9	GALAXY
SHAP351-4924.8	03 50 20.	- 49 33 42.	60	15.5	GALAXY
SHAP351-5620.1	03 50 20.	- 56 29 01.	12	17.0	GALAXY
SHAP351-5732.7	03 50 20.	- 57 41 37.	24	16.5	GALAXY
MCG+05-10-003	03 50 21.	+ 32 21	96	14.5	GALAXY
BAK 2.0779	03 50 21.	- 16 23 23.	13	16.7	GALAXY
BAK 2.0776	03 50 21.	- 17 25 29.	16	17.2	GALAXY
HN 0246	03 50 21.	- 49 34			NEBULA
IC 2004	03 50 21.	- 49 34			NONSTELLAR OBJECT
SHAP351-5643.8	03 50 21.	- 56 52 43.	36	17.0	GALAXY
BAK 2.0777	03 50 22.	- 18 19 47.	16	16.7	GALAXY
SHAP351-5844.0	03 50 22.	- 58 52 55.	18	17.5	GALAXY
BAK 2.0778	03 50 23.	- 19 10 35.	24	17.0	GALAXY
SHAP351-5724.8	03 50 23.	- 57 33 43.	24	17.0	GALAXY
ZWG 508.004	03 50 24.	+ 32 20		14.9	GALAXY
UGC 02891	03 50 24.	+ 32 20	132	14.9	GALAXY S0-a
MCG+06-09-013	03 50 24.	+ 37 07 30.	90	15.	GALAXY
VB 208	03 50 24.	+ 57 05	282		STELLAR RING
KARA.68 045	03 50 24.	- 02 55	34		DWARF GALAXY
RNGC 1465	03 50 25.	+ 32 20		15.0	GALAXY
SHAP351-5639.4	03 50 25.	- 56 48 19.	18	17.5	GALAXY
SHAP351-5321.4	03 50 26.	- 53 30 18.	36	17.0	GALAXY
SHAP351-5723.4	03 50 26.	- 57 32 19.	24	17.0	GALAXY
SHAP351-5804.8	03 50 26.	- 58 13 43.	18	17.0	GALAXY
BAK 2.0788	03 50 27.	- 15 38 05.	23	16.4	GALAXY
SHAP351-5804.4	03 50 27.	- 58 13 19.	18	17.75	GALAXY
BAK 2.0787	03 50 28.	- 17 11 17.	17	16.8	GALAXY
BAK 2.0782	03 50 28.	- 18 33 41.	13	16.6	GALAXY
BAK 2.0780	03 50 28.	- 21 29 47.	24	16.8	GALAXY
SHAP351-5949.6	03 50 28.	- 59 58 31.	36	16.25	GALAXY
BAK 2.0783	03 50 29.	- 18 53 59.	13	16.0	GALAXY
SHAP351-5932.0	03 50 29.	- 59 40 55.	24	17.0	GALAXY
CED 024	03 50 30.	+ 25 32			DIFFUSE GALACTIC NEBULA
BAK 2.0784	03 50 30.	- 20 08 52.	17	16.8	GALAXY
SHAP351-5839.0	03 50 30.	- 58 47 55.	18	17.0	GALAXY
BAK 2.0791	03 50 31.	- 17 34 04.	22	17.0	GALAXY
BAK 2.0790	03 50 31.	- 18 41 52.	16	16.8	GALAXY
BAK 2.0785	03 50 31.	- 20 52 04.	17	16.9	GALAXY
SHAP351-5032.5	03 50 31.	- 50 41 24.	42	16.5	GALAXY
SHAP351-5932.6	03 50 31.	- 59 41 31.	36	16.25	GALAXY
BAK 2.0789	03 50 32.	- 20 17 40.	12	16.9	GALAXY
SHAP352-4916.8	03 50 32.	- 49 25 42.	48	17.0	GALAXY
SHAP351-5318.6	03 50 32.	- 53 27 30.	42	17.0	GALAXY
SHAP351-5626.8	03 50 32.	- 56 35 42.	24	17.5	GALAXY
PK171-25.1	03 50 33.	+ 19 19	40	13.9	PLANETARY NEBULA
CED 023	03 50 33.	+ 32 21			DIFFUSE GALACTIC NEBULA
SHAP351-5803.2	03 50 33.	- 58 12 06.	24	17.0	GALAXY
BAK 2.0781	03 50 34.	- 24 24 10.	16	16.8	GALAXY
SHAP351-5425.8	03 50 34.	- 54 34 42.	54	17.5	GALAXY
SHAP351-5732.0	03 50 34.	- 57 40 54.	18	17.0	GALAXY
SHAP351-5817.0	03 50 34.	- 58 25 54.	18	17.75	GALAXY
BAK 2.0792	03 50 35.	- 17 59 16.	14	16.8	GALAXY
MCG+12-04-007	03 50 36.	+ 72 47	186	14.	GALAXY
LB 01698	03 50 36.	- 53 15		14.0	FAINT BLUE STAR
SHAP351-5307.8	03 50 36.	- 53 16 42.	36	17.5	GALAXY
SHAP351-5350.2	03 50 37.	- 53 59 06.	48	17.5	GALAXY
BAK 2.0795	03 50 38.	- 16 35 04.	13	16.2	GALAXY
BAK 2.0786	03 50 38.	- 26 36 16.	30	16.4	GALAXY
MCG+03-10-009	03 50 39.	+ 18 57	66	15.5	GALAXY
SHAP351-5316.2	03 50 39.	- 53 25 06.	42	17.25	GALAXY
SHAP351-5937.4	03 50 39.	- 59 46 18.	30	16.5	GALAXY
SHAP352-4429.7	03 50 40.	- 44 38 35.	72	15.5	GALAXY
SHAP351-5704.2	03 50 40.	- 57 13 06.	12	17.0	GALAXY
SHAP351-5842.2	03 50 40.	- 58 51 06.	18	17.5	GALAXY
SHAP351-5914.4	03 50 40.	- 59 23 18.	30	17.5	GALAXY
UGC 02892	03 50 42.	+ 18 57	96	16.0	GALAXY SBb
MCG+12-04-008	03 50 42.	+ 71 47	51	15.	GALAXY
BAK 2.0797	03 50 42.	- 18 10 04.	16	16.6	GALAXY
BAK 2.0794	03 50 42.	- 20 40 52.	12	17.0	GALAXY
SHAP351-5621.8	03 50 42.	- 56 30 42.	24	17.5	GALAXY
SHAP351-5635.2	03 50 42.	- 56 44 06.	30	16.0	GALAXY
LB 03329	03 50 42.	- 69 29		14.4	FAINT BLUE STAR
BAK 2.0796	03 50 43.	- 20 57 28.	17	16.9	GALAXY
SHAP351-5755.8	03 50 43.	- 58 04 42.	24	17.5	GALAXY
SHAP351-5929.9	03 50 43.	- 59 38 48.	30	17.5	GALAXY
SHAP351-5934.8	03 50 43.	- 59 43 42.	24	17.0	GALAXY
SHAP352-5358.8	03 50 44.	- 54 07 41.	42	17.5	GALAXY
SHAP351-5755.0	03 50 44.	- 58 03 54.	12	17.75	GALAXY
RNGC 1471	03 50 45.	- 15 32			NON-EXISTENT OBJECT
BAK 2.0799	03 50 45.	- 18 09 40.	14	16.6	GALAXY
SHAP351-5515.0	03 50 45.	- 55 23 53.	36	16.0	GALAXY
SHAP351-5540.2	03 50 45.	- 55 49 05.	24	17.5	GALAXY
SHAP352-5433.0	03 50 46.	- 54 41 53.	36	18.0	GALAXY
SHAP351-5543.2	03 50 46.	- 55 52 05.	30	17.0	GALAXY
SHAP351-5923.0	03 50 46.	- 59 31 54.	24	16.75	GALAXY
SHAP351-5945.8	03 50 46.	- 59 54 42.	24	17.0	GALAXY
SHAP351-5752.5	03 50 47.	- 58 01 24.	24	17.0	GALAXY
SHAP351-5846.2	03 50 47.	- 58 55 06.	18	17.0	GALAXY
HZ 03	03 50 48.	+ 10 36		13.0	VERY BLUE STAR
ZWG 465.010	03 50 48.	+ 19 49		15.4	GALAXY
BAK 2.0793	03 50 48.	- 26 49 22.	17	16.4	GALAXY
SHAP351-5541.5	03 50 48.	- 55 50 23.	24	17.0	GALAXY
SHAP351-5753.0	03 50 48.	- 58 01 54.	18	16.75	GALAXY
SHAP351-5947.2	03 50 48.	- 59 56 06.	30	17.5	GALAXY
SHAP351-5614.6	03 50 50.	- 56 23 29.	18	17.5	GALAXY
SHAP351-5620.2	03 50 50.	- 56 29 05.	24	17.75	GALAXY
SHAP351-5628.8	03 50 50.	- 56 37 41.	18	16.5	GALAXY
MCG+03-10-010	03 50 51.	+ 19 50	30	15.5	GALAXY
SHAP351-5935.8	03 50 51.	- 59 44 42.	24	16.5	GALAXY
IC 0355	03 50 52.	+ 19 51 33.			NONSTELLAR OBJECT
SHAP352-5331.0	03 50 52.	- 53 39 53.	36	17.25	GALAXY
SHAP352-5431.8	03 50 52.	- 54 40 41.	60	18.0	GALAXY
BAK 2.0801	03 50 53.	- 18 05 45.	24	16.6	GALAXY
BAK 2.0798	03 50 53.	- 26 17 57.	20	16.8	GALAXY
SHAP352-5330.2	03 50 53.	- 53 39 05.	36	17.25	GALAXY
SHAP352-5650.0	03 50 53.	- 56 58 53.	18	17.5	GALAXY
BAK 2.0800	03 50 54.	- 23 51 57.	16	16.8	GALAXY
SHAP351-5736.2	03 50 54.	- 57 45 05.	18	17.5	GALAXY
SHAP351-5757.0	03 50 54.	- 58 05 53.	18	17.25	GALAXY
SHAP351-5916.8	03 50 54.	- 59 25 41.	24	17.75	GALAXY
SHAP352-4708.2	03 50 55.	- 47 17 04.	48	17.25	GALAXY
SHAP352-5609.9	03 50 55.	- 56 18 47.	24	17.75	GALAXY
SHAP351-5752.7	03 50 55.	- 58 01 35.	24	17.25	GALAXY
BAK 2.0807	03 50 56.	- 16 45 45.	17	16.2	GALAXY
BAK 2.0806	03 50 56.	- 17 12 15.	33	16.8	GALAXY
SHAP352-5707.9	03 50 56.	- 57 16 47.	18	17.0	GALAXY
SHAP352-5740.5	03 50 56.	- 57 49 23.	18	16.75	GALAXY
SHAP351-5913.6	03 50 56.	- 59 22 29.	24	16.75	GALAXY
BAK 2.0802	03 50 57.	- 19 49 39.	18	17.3	GALAXY
SHAP352-5452.8	03 50 57.	- 55 01 41.	42	16.25	GALAXY
BAK 2.0803	03 50 58.	- 19 48 27.	13	17.2	GALAXY
BAK 2.0804	03 50 58.	- 20 25 15.	14	16.8	GALAXY
SHAP352-5509.0	03 50 58.	- 55 17 53.	24	17.0	GALAXY
SHAP352-5812.6	03 50 58.	- 58 21 29.	18	17.0	GALAXY
SHAP352-5446.2	03 50 59.	- 54 55 05.	54	17.0	GALAXY
SHAP352-5607.0	03 50 59.	- 56 15 53.	24	17.5	GALAXY
LBN 0689	03 51	+ 57 10	1200		BRIGHT NEBULA
ZC 0351.0+2820	03 51 00.	+ 28 20	1750		CLUSTER OF GALAXIES
UGC 02893	03 51 00.	+ 36 22	72	17.	GALAXY SBc
MCG-02-10-018	03 51 00.	- 10 37	54	15.	GALAXY
SHAP352-5850.1	03 51 00.	- 58 58 59.	24	17.75	GALAXY
BAK 2.0809	03 51 01.	- 16 20 44.	17	16.8	GALAXY
SHAP352-5857.2	03 51 01.	- 59 06 05.	18	16.75	GALAXY
SHAP352-5708.6	03 51 02.	- 57 17 29.	12	17.25	GALAXY
SHAP352-5915.1	03 51 02.	- 59 23 59.	24	17.0	GALAXY
SHAP352-5721.6	03 51 03.	- 57 30 29.	30	17.0	GALAXY
SHAP352-5913.8	03 51 04.	- 59 22 41.	24	16.75	GALAXY
BAK 2.0805	03 51 05.	- 26 04 33.	23	16.8	GALAXY
SHAP352-5410.4	03 51 05.	- 54 19 16.	48	17.75	GALAXY
VB 209	03 51 06.	+ 57 17	302		STELLAR RING
LB 01699	03 51 06.	- 49 40		14.8	FAINT BLUE STAR
SHAP352-5955.0	03 51 06.	- 60 03 53.	60	15.5	GALAXY
SHAP352-5453.0	03 51 07.	- 55 01 52.	72	16.0	GALAXY
SHAP352-5750.0	03 51 07.	- 57 58 52.	24	17.75	GALAXY
SHAP352-5930.5	03 51 07.	- 59 39 23.	24	17.5	GALAXY
HN 0247	03 51 07.	- 60 04			NEBULA
IC 2010	03 51 07.	- 60 04			NONSTELLAR OBJECT
BAK 2.0812	03 51 08.	- 19 49 32.	13	16.9	GALAXY
BAK 2.0813	03 51 08.	- 20 22 44.	12	17.0	GALAXY
MCG-03-10-050	03 51 09.	- 17 15	60	15.5	GALAXY
BAK 2.0814	03 51 09.	- 18 05 26.	23	16.5	GALAXY
SHAP352-5713.9	03 51 09.	- 57 22 46.	24	17.5	GALAXY
BAK 2.0808	03 51 10.	- 26 18 14.	32	16.6	GALAXY
SHAP352-5826.3	03 51 10.	- 58 35 10.	24	16.25	GALAXY
SHAP352-5634.2	03 51 11.	- 56 43 04.	18	17.0	GALAXY
MCG+03-10-011	03 51 12.	+ 15 50	48	15.	GALAXY
SHAP352-5841.3	03 51 12.	- 58 50 10.	24	17.0	GALAXY
BAK 2.0817	03 51 13.	- 17 16 44.	43	15.8	GALAXY
SHAP352-4946.5	03 51 13.	- 49 55 21.	66	17.0	GALAXY
SHAP352-5456.0	03 51 13.	- 55 04 52.	42	16.0	GALAXY
FATH 1.142	03 51 14.	- 00 17	11		NEBULA
BAK 2.0811	03 51 14.	- 26 08 56.	18	16.8	GALAXY
BAK 2.0810	03 51 14.	- 26 43 50.	22	16.2	GALAXY
RNGC 1483	03 51 14.	- 47 37			GALAXY
MCG-02-10-019	03 51 15.	- 09 37	36	15.	GALAXY
BAK 2.0819	03 51 15.	- 16 53 14.	20	16.4	GALAXY
BAK 2.0818	03 51 15.	- 17 56 08.	26	16.7	GALAXY
BAK 2.0815	03 51 15.	- 22 04 20.	47	16.6	GALAXY

OBJECT NAME	RIGHT ASCEN.	DECLINATION	DIAM.	MAGN.	TYPE OF OBJECT
SHAP352-4728.7	03 51 15.	- 47 37 33.	90	14.2	GALAXY
RNGC 1472	03 51 17.	- 08 43			GALAXY
SHAP352-5826.0	03 51 17.	- 58 34 52.	24	17.0	GALAXY
ZWG 465.011	03 51 18.	+ 15 50		15.6	GALAXY
UGC 02894	03 51 18.	+ 15 50	72	15.6	GALAXY Sc
KARA.72 095A	03 51 18.	+ 15 50		15.6	PART OF DOUBLE GALAXY
MCG-02-10-020	03 51 18.	- 12 30 30.	36	15.5	GALAXY
SHAP352-5712.4	03 51 18.	- 57 21 16.	42	16.75	GALAXY
BAK 2.0820	03 51 20.	- 17 40 01.	14	16.6	GALAXY
BAK 2.0816	03 51 20.	- 26 18 50.	26	16.2	GALAXY
BAK 2.0821	03 51 21.	- 18 43 01.	13	17.0	GALAXY
SHAP352-5749.6	03 51 21.	- 57 58 28.	18	17.75	GALAXY
BAK 2.0823	03 51 22.	- 16 55 07.	19	16.8	GALAXY
SHAP352-5707.0	03 51 22.	- 57 15 51.	24	17.0	GALAXY
BAK 2.0822	03 51 23.	- 18 36 31.	19	17.0	GALAXY
ZWG 465.012	03 51 24.	+ 15 47		15.3	GALAXY
KARA.72 095B	03 51 24.	+ 15 47		15.3	PART OF DOUBLE GALAXY
UGC 02895	03 51 24.	+ 17 26	78	16.0	GALAXY S
ZWG 347.001	03 51 24.	+ 79 25		15.4	GALAXY
ZWG 346.007	03 51 24.	+ 79 25		15.4	GALAXY
UGC 02896	03 51 24.	+ 79 25	102	15.4	GALAXY Sc
MCG+13-04-001	03 51 24.	+ 79 25	57	16.	GALAXY
SHAP352-5345.2	03 51 24.	- 53 54 03.	36	17.5	GALAXY
BAK 2.0825	03 51 25.	- 16 17 13.	15	16.4	GALAXY
SHAP352-5410.1	03 51 26.	- 54 18 57.	54	17.0	GALAXY
SHAP352-5729.0	03 51 26.	- 57 37 51.	72	16.25	GALAXY
SHAP352-5540.1	03 51 28.	- 55 48 57.	42	18.0	GALAXY
HN 0248	03 51 28.	- 57 38			NEBULA
IC 2011	03 51 28.	- 57 38			NONSTELLAR OBJECT
LB 01500	03 51 29.	+ 23 16 00.		19.2	FAINT BLUE STAR
LB 01201	03 51 30.	+ 22 11 24.		15.7	FAINT BLUE STAR
UGC 02897	03 51 30.	+ 42 05	60	17.	GALAXY
HUB E03	03 51 30.	+ 50 17			DIFFUSE NEBULA
SHAP352-5907.6	03 51 30.	- 59 16 27.	24	17.75	GALAXY
SHAP352-5205.6	03 51 31.	- 52 14 26.	42	17.5	GALAXY
SHAP352-5615.2	03 51 31.	- 56 24 03.	24	17.75	GALAXY
SHAP352-5926.0	03 51 31.	- 59 34 51.	24	17.0	GALAXY
BAK 2.0826	03 51 32.	- 22 10 37.	13	16.8	GALAXY
SHAP352-5447.0	03 51 32.	- 54 55 51.	48	16.0	GALAXY
BAK 2.0827	03 51 33.	- 22 00 49.	12	16.8	GALAXY
SHAP352-5506.5	03 51 33.	- 55 15 20.	30	16.25	GALAXY
SHAP352-5928.1	03 51 33.	- 59 36 57.	30	17.75	GALAXY
BAK 2.0824	03 51 34.	- 26 43 49.	34	15.5	GALAXY
SHAP352-5349.2	03 51 34.	- 53 58 02.	54	17.0	GALAXY
SHAP352-5440.6	03 51 34.	- 54 49 26.	84	16.75	GALAXY
SHAP352-5813.0	03 51 34.	- 58 21 51.	12	17.5	GALAXY
SHAP352-5933.2	03 51 34.	- 59 42 03.	24	17.75	GALAXY
ZC 0351.6+7940	03 51 36.	+ 79 40	1810		CLUSTER OF GALAXIES
RNGC 1477	03 51 36.	- 08 43			GALAXY
BAK 2.0831	03 51 36.	- 18 01 48.	20	16.5	GALAXY
BAK 2.0830	03 51 36.	- 18 01 48.	14	17.2	GALAXY
SHAP352-5519.2	03 51 36.	- 55 28 02.	30	17.5	GALAXY
BAK 2.0833	03 51 37.	- 17 44 54.	47	15.7	GALAXY
BAK 2.0828	03 51 37.	- 25 33 25.	20	16.6	GALAXY
BAK 2.0829	03 51 38.	- 22 24 48.	11	16.5	GALAXY
MCG-03-10-051	03 51 39.	- 17 43 30.	48	15.	GALAXY
BAK 2.0835	03 51 39.	- 18 34 24.	20	17.0	GALAXY
SHAP352-5447.8	03 51 39.	- 54 56 38.	42	16.0	GALAXY
BAK 2.0834	03 51 40.	- 19 50 06.	13	17.2	GALAXY
RNGC 1475	03 51 41.	- 08 15			NON-EXISTENT OBJECT
RNGC 1478	03 51 41.	- 08 42			GALAXY
BAK 2.0832	03 51 41.	- 22 28 18.	16	16.6	GALAXY
ZWG 442.005	03 51 42.	+ 10 34		15.1	GALAXY
UGC 02898	03 51 42.	+ 10 34	84	15.1	GALAXY SBa
KARA.73B 0135	03 51 42.	+ 10 34	60	15.1	ISOLATED GALAXY S
MRSL 145+03/1	03 51 42.	+ 57 17	2400		HII REGION
BAK 2.0836	03 51 42.	- 20 05 48.	19	17.0	GALAXY
SHAP353-5322.6	03 51 43.	- 53 31 26.	42	17.0	GALAXY
SHAP352-5635.4	03 51 43.	- 56 44 14.	30	16.0	GALAXY
IC 2002	03 51 44.	+ 10 34 08.			NONSTELLAR OBJECT
SHAP352-5701.6	03 51 44.	- 57 10 26.	18	17.5	GALAXY
BAK 2.0837	03 51 45.	- 18 12 54.	15	17.1	GALAXY
BAK 2.0839	03 51 46.	- 17 26 12.	33	17.0	GALAXY
SHAP353-5343.6	03 51 46.	- 53 52 26.	42	16.5	GALAXY
SHAP352-5736.4	03 51 46.	- 57 45 14.	24	17.0	GALAXY
SHAP352-5838.2	03 51 47.	- 58 47 02.	84	16.5	GALAXY
UGC 02899	03 51 48.	+ 06 27	90	16.0	GALAXY
MCG+01-10-013	03 51 48.	+ 06 28	72	15.	GALAXY
MCG+02-10-003	03 51 48.	+ 10 33	60	14.	GALAXY
RNGC 1474	03 51 48.	- 10 33		14.0	GALAXY
HN 0250	03 51 48.	- 58 47			NEBULA
IC 2012	03 51 48.	- 58 47			NONSTELLAR OBJECT
SHAP352-5654.4	03 51 49.	- 57 03 14.	18	18.0	GALAXY
SHAP352-5848.0	03 51 49.	- 58 56 50.	36	16.75	GALAXY
SHAP353-5440.4	03 51 50.	- 54 49 13.	54	17.25	GALAXY
MCG-02-10-021	03 51 51.	- 08 53	36	15.	GALAXY
BAK 2.0838	03 51 51.	- 23 58 00.	14	16.7	GALAXY
SHAP352-5736.6	03 51 51.	- 57 45 26.	12	17.5	GALAXY
SHAP353-5331.1	03 51 53.	- 53 39 55.	48	16.75	GALAXY
MCG+01-10-046	03 51 54.	- 03 01 30.	24	15.	GALAXY
SHAP353-5345.0	03 51 54.	- 53 53 49.	36	17.5	GALAXY
SHAP353-5623.9	03 51 54.	- 56 32 43.	30	16.0	GALAXY
SHAP352-5849.8	03 51 54.	- 58 58 38.	42	16.75	GALAXY
SHAP353-5636.2	03 51 55.	- 56 45 01.	24	16.5	GALAXY
SHAP353-5641.9	03 51 55.	- 56 50 43.	18	18.0	GALAXY
SHAP353-5706.2	03 51 56.	- 57 15 01.	24	17.25	GALAXY
BAK 2.0841	03 51 57.	- 18 23 11.	12	16.7	GALAXY
BAK 2.0842	03 51 58.	- 18 35 53.	17	16.6	GALAXY
SHAP353-4659.3	03 51 59.	- 47 08 06.	54	17.5	GALAXY
LBN 0774	03 52	+ 25 20	6300		BRIGHT NEBULA
KHAV 042	03 52	+ 49 15	10690		DARK NEBULA
LB 09810	03 52	- 81 18		15.1	FAINT BLUE STAR
LB 09964	03 52	- 88 57		14.7	FAINT BLUE STAR
UGC 02900	03 52 00.	+ 18 14	66	16.0	GALAXY S
LDN 1456	03 52 00.	+ 37 00	840		DARK NEBULA
RNGC 1479	03 52 00.	- 10 20			NON-EXISTENT OBJECT
BAK 2.0843	03 52 00.	- 17 28 53.	20	16.1	GALAXY
MCG-06-09-035	03 52 00.	- 35 40	30	15.5	GALAXY
KARA.73 22	03 52 02.	- 36 12	34		DWARF GALAXY
MCG-03-10-052	03 52 03.	- 19 20	60	15.	GALAXY
BAK 2.0840	03 52 03.	- 26 58 23.	19	16.4	GALAXY
BAK 2.0846	03 52 04.	- 17 07 47.	17	16.6	GALAXY
BAK 2.0845	03 52 04.	- 18 12 17.	12	17.1	GALAXY
BAK 2.0844	03 52 04.	- 20 46 59.	14	17.2	GALAXY
SHAP353-4859.0	03 52 04.	- 49 07 48.	96	16.25	GALAXY
HN 0249	03 52 04.	- 49 08			NEBULA
IC 2009	03 52 04.	- 49 08			NONSTELLAR OBJECT
SHAP353-5629.4	03 52 04.	- 56 38 13.	30	16.5	GALAXY
ZWG 417.010	03 52 06.	+ 06 07		15.5	GALAXY
ZWG 487.001	03 52 06.	+ 27 27		15.4	GALAXY
VB 240	03 52 06.	+ 51 58	430		STELLAR RING
SHAP353-5600.6	03 52 07.	- 56 09 24.	18	17.0	GALAXY
BAK 2.0847	03 52 08.	- 19 40 46.	12	17.1	GALAXY
SHAP353-5543.5	03 52 08.	- 55 52 18.	42	17.0	GALAXY
SHAP353-5331.1	03 52 09.	- 53 39 54.	42	16.75	GALAXY
BAK 2.0848	03 52 10.	- 21 43 58.	14	17.0	GALAXY
SHAP353-5542.5	03 52 10.	- 55 51 18.	24	16.5	GALAXY
SHAP353-5953.5	03 52 10.	- 60 02 19.	18	17.0	GALAXY
BAK 2.0851	03 52 11.	- 19 20 16.	30	15.4	GALAXY
BAK 2.0852	03 52 11.	- 19 20 34.	19	16.8	GALAXY
SHAP353-5910.2	03 52 11.	- 59 19 01.	24	18.0	GALAXY
RLWT 075	03 52 12.	+ 46 34		14.	FAINT VERY BLUE STAR
RLWT 076	03 52 12.	+ 52 07		13.5	FAINT VERY BLUE STAR
RNGC 1481	03 52 12.	- 20 33		15.0	GALAXY
MCG-03-10-053	03 52 12.	- 20 33	12	15.	GALAXY
BAK 2.0849	03 52 12.	- 22 23 28.	12	16.0	GALAXY
IC 2006	03 52 12.	- 36 08	78	12.8	GALAXY E1
BAK 2.0850	03 52 13.	- 22 02 58.	22	16.8	GALAXY
SHAP353-5831.8	03 52 13.	- 58 40 36.	30	17.25	GALAXY
SHAP353-5934.6	03 52 13.	- 59 43 24.	24	17.25	GALAXY
BIGO 484	03 52 15.	- 20 38			NEBULA
SHAP353-5648.9	03 52 15.	- 56 57 42.	24	17.0	GALAXY
BAK 2.0854	03 52 16.	- 17 26 58.	16	16.8	GALAXY
SHAP353-5516.6	03 52 16.	- 55 25 24.	24	17.25	GALAXY
SHAP353-5553.5	03 52 16.	- 56 02 18.	30	16.5	GALAXY
BAK 2.0853	03 52 17.	- 19 19 34.	23	16.3	GALAXY
SHAP353-5315.9	03 52 17.	- 53 24 42.	48	17.5	GALAXY
MRSL 148-00/1	03 52 18.	+ 53 03	7200		HII REGION
MCG-03-10-054	03 52 18.	- 20 38	60	14.	GALAXY
SHAP353-5933.0	03 52 19.	- 59 41 48.	42	17.0	GALAXY
SHAP353-5750.6	03 52 21.	- 57 59 24.	24	17.25	GALAXY
SHAP353-5708.0	03 52 22.	- 57 16 48.	54	17.5	GALAXY
SN 1937E	03 52 23.	- 20 38		15.0	SUPERNOVA
RNGC 1484	03 52 23.	- 37 08			GALAXY
BAK 2.0860	03 52 24.	- 17 47 09.	18	16.6	GALAXY
BAK 2.0858	03 52 24.	- 17 56 33.	16	17.0	GALAXY
BAK 2.0855	03 52 24.	- 20 03 45.	13	17.0	GALAXY
MCG-06-09-037	03 52 24.	- 36 08	150	12.8	GALAXY
MCG-06-09-036	03 52 24.	- 37 08	120	12.5	GALAXY
SHAP353-5515.2	03 52 24.	- 55 23 59.	24	16.25	GALAXY
SHAP353-5706.0	03 52 24.	- 57 14 48.	12	18.0	GALAXY
LB 03330	03 52 24.	- 65 57		13.6	FAINT BLUE STAR
B 006	03 52 25.	+ 55 59			DARK OBJECT
SHAP353-5341.5	03 52 25.	- 53 50 17.	48	16.5	GALAXY
SHAP353-5648.2	03 52 25.	- 56 56 59.	18	17.0	GALAXY
BAK 2.0856	03 52 27.	- 22 04 15.	35	15.6	GALAXY
SHAP353-5648.7	03 52 27.	- 56 57 29.	30	17.0	GALAXY
SHAP353-5931.2	03 52 27.	- 59 40 00.	48	16.5	GALAXY
BAK 2.0862	03 52 29.	- 17 35 45.	23	16.8	GALAXY
SHAP353-5544.8	03 52 29.	- 55 53 35.	24	17.5	GALAXY
SHAH 148	03 52 30.	+ 00 50	234		GROUP OF COMPACT GALAXIES
LDN 1387	03 52 30.	+ 55 55	660		DARK NEBULA
RNGC 1482	03 52 30.	- 20 38		14.0	GALAXY
BAK 2.0861	03 52 30.	- 20 54 57.	24	16.7	GALAXY
SHAP353-5710.2	03 52 30.	- 57 18 59.	24	18.0	GALAXY
BAK 2.0857	03 52 31.	- 25 16 39.	19	16.8	GALAXY
BAK 2.0859	03 52 32.	- 24 53 21.	24	17.1	GALAXY
HZ 04	03 52 34.	+ 09 37		14.8	BLUE STAR
BAK 2.0864	03 52 35.	- 17 46 45.	22	16.8	GALAXY
SHAP353-5452.4	03 52 36.	- 55 01 11.	48	15.75	GALAXY
SHAP353-5537.9	03 52 36.	- 55 46 41.	36	17.0	GALAXY
BAK 2.0863	03 52 37.	- 19 59 45.	16	17.2	GALAXY
SHAP354-4936.2	03 52 37.	- 49 44 58.	54	17.5	GALAXY
SHAP353-5532.1	03 52 37.	- 55 40 53.	24	16.25	GALAXY
MCG+01-10-047	03 52 39.	- 06 21	42	15.	GALAXY
BAK 2.0867	03 52 39.	- 17 25 26.	19	16.3	GALAXY
SHAP353-5755.2	03 52 39.	- 58 03 59.	24	17.0	GALAXY
BAK 2.0865	03 52 40.	- 21 31 21.	17	16.8	GALAXY
SHAP353-5854.6	03 52 40.	- 59 03 23.	24	17.75	GALAXY
VB 241	03 52 42.	+ 55 46	2182		STELLAR RING
MCG-03-10-055	03 52 42.	- 20 30	30	17.	GALAXY
ARC 0470	03 52 43.	- 04 50		16.9	RICH CLUSTER OF GALAXIES
BAK 2.0870	03 52 43.	- 17 16 08.	14	16.2	GALAXY
BAK 2.0866	03 52 43.	- 22 04 14.	26	16.8	GALAXY
IC 2007	03 52 43.	- 28 17 27.			NONSTELLAR OBJECT
SHAP353-5341.4	03 52 43.	- 53 50 10.	42	17.0	GALAXY
SHAP353-5711.4	03 52 43.	- 57 20 10.	30	18.0	GALAXY
SHAP353-5815.8	03 52 43.	- 58 24 34.	24	17.75	GALAXY
BAK 2.0869	03 52 45.	- 20 12 08.	14	16.6	GALAXY
BAK 2.0872	03 52 46.	- 17 17 44.	14	16.2	GALAXY
BAK 2.0871	03 52 46.	- 20 15 44.	23	16.2	GALAXY
BAK 2.0868	03 52 46.	- 22 07 32.	43	16.9	GALAXY
SHAP353-5731.5	03 52 46.	- 57 40 16.	24	18.0	GALAXY
SHAP353-5507.4	03 52 47.	- 55 16 10.	36	16.5	GALAXY
MCG-03-10-056	03 52 48.	- 17 35 30.	36	15.	GALAXY
SHAP354-5218.4	03 52 48.	- 52 27 10.	48	17.5	GALAXY
SHAP353-5548.2	03 52 48.	- 55 56 58.	24	17.0	GALAXY
SHAP353-5930.2	03 52 48.	- 59 38 58.	36	16.25	GALAXY
BAK 2.0874	03 52 49.	- 17 49 50.	20	16.8	GALAXY
BAK 2.0873	03 52 51.	- 20 31 50.	31	16.2	GALAXY
BAK 2.0876	03 52 52.	- 17 37 08.	23	15.4	GALAXY
BAK 2.0875	03 52 52.	- 17 49 56.	19	17.1	GALAXY
SHAP354-5533.0	03 52 52.	- 55 41 46.	30	16.5	GALAXY
SHAP353-5908.8	03 52 52.	- 59 17 34.	18	18.0	GALAXY
ZC 0352.9+7214	03 52 54.	+ 72 14	1610		CLUSTER OF GALAXIES
MCG+13-04-002	03 52 54.	+ 79 17	30	16.	GALAXY
PATH 1.143	03 52 54.	- 00 19	14		NEBULA
LB 01700	03 52 54.	- 48 25		15.2	FAINT BLUE STAR
LB 01202	03 52 55.	+ 09 43 36.		15.3	FAINT BLUE STAR
BAK 2.0880	03 52 55.	- 17 27 01.	23	15.8	GALAXY
SHAP354-5344.6	03 52 56.	- 53 53 21.	48	17.0	GALAXY
SHAP354-5729.6	03 52 56.	- 57 38 22.	24	18.0	GALAXY
SHAP353-5944.8	03 52 56.	- 59 53 34.	30	17.75	GALAXY
BAK 2.0877	03 52 57.	- 21 59 31.	16	16.8	GALAXY
BAK 2.0878	03 52 58.	- 22 23 13.	23	16.1	GALAXY
SHAP354-5825.5	03 52 59.	- 58 34 16.	18	17.25	GALAXY
SHAP353-5836.8	03 52 59.	- 58 45 34.	18	16.5	GALAXY
SHB 079	03 52 59.3	+ 12 23 03.		19.5	QUASI-STELLAR OBJECT
BAK 2.0879	03 53 00.	- 24 28 37.	19	16.8	GALAXY
BAK 2.0881	03 53 01.	- 22 00 37.	13	16.6	GALAXY
SHAP354-5546.1	03 53 01.	- 55 54 51.	30	17.5	GALAXY
SHAP354-5556.0	03 53 01.	- 56 04 45.	30	17.5	GALAXY
SCHO 0064	03 53 02.	+ 34 05 42.	420		ISOLATED DARK CLOUD
SHAP354-5603.6	03 53 02.	- 56 12 21.	24	17.5	GALAXY
SHAP354-5613.6	03 53 02.	- 56 22 21.	18	17.25	GALAXY
MCG+03-11-001	03 53 03.	+ 16 20	114	15.	GALAXY

OBJECT NAME	RIGHT ASCEN.	DECLINATION	DIAM.	MAGN.	TYPE OF OBJECT
SHAP354-5817.8	03 53 03.	- 58 26 33.	18	17.0	GALAXY
BAK 2.0882	03 53 04.	- 21 59 43.	14	15.8	GALAXY
IC 2008	03 53 04.	- 28 21 35.			NONSTELLAR OBJECT
SHAP354-5324.9	03 53 04.	- 53 33 39.	42	16.0	GALAXY
SHAP354-5349.2	03 53 04.	- 53 57 57.	54	17.25	GALAXY
SHAP354-5942.5	03 53 04.	- 59 51 15.	36	16.0	GALAXY
SHAP354-5325.5	03 53 05.	- 53 34 15.	42	16.5	GALAXY
RNGC 1490	03 53 05.	- 66 10			GALAXY
MCG+03-11-002	03 53 06.	+ 16 22	42	15.	GALAXY
RLWT 077	03 53 06.	+ 56 59		14.	FAINT VERY BLUF STAR
VB 210	03 53 06.	+ 57 14	383		STELLAR RING
SHAP354-5632.5	03 53 07.	- 56 41 15.	12	17.5	GALAXY
SHAP354-5647.9	03 53 08.	- 56 56 39.	18	17.5	GALAXY
BAK 2.0884	03 53 09.	- 19 19 25.	14	17.0	GALAXY
BAK 2.0883	03 53 09.	- 21 59 31.	12	16.1	GALAXY
SHAP354-5218.8	03 53 10.	- 52 27 32.	42	17.0	GALAXY
SHAP354-5535.0	03 53 10.	- 55 43 45.	30	17.25	GALAXY
PK161-14.1	03 53 10.1	+ 33 43 52.	7	12.6	PLANETARY NEBULA
IC 2003	03 53 10.1	+ 33 43 52.	6	12.6	PLANETARY NEBULA
SHAP354-5527.3	03 53 11.	- 55 36 02.	30	17.75	GALAXY
MCG-02-11-001	03 53 12.	- 09 41	42	15.	GALAXY
BAK 2.0885	03 53 12.	- 18 08 54.	18	16.1	GALAXY
MCG-05-10-005	03 53 12.	- 28 18	72	14.	GALAXY
SHAP354-5331.0	03 53 14.	- 53 39 44.	42	16.5	GALAXY
SHAP354-5629.0	03 53 14.	- 56 37 44.	18	17.25	GALAXY
BAK 2.0887	03 53 15.	- 19 23 06.	22	16.8	GALAXY
SHAP354-5544.2	03 53 15.	- 55 52 56.	36	17.0	GALAXY
SHAP354-5718.8	03 53 15.	- 57 27 32.	24	17.5	GALAXY
BAK 2.0886	03 53 16.	- 21 51 54.	18	16.8	GALAXY
SHAP354-5723.0	03 53 16.	- 57 31 44.	18	17.5	GALAXY
SHAP354-5929.6	03 53 16.	- 59 38 21.	18	18.0	GALAXY
VB 242	03 53 18.	+ 57 16	342		STELLAR RING
SHAP354-5545.0	03 53 18.	- 55 53 44.	24	16.75	GALAXY
BAK 2.0888	03 53 19.	- 21 58 12.	21	16.8	GALAXY
SHAP354-5939.4	03 53 20.	- 59 48 08.	30	16.5	GALAXY
SHAP354-5527.6	03 53 21.	- 55 36 20.	42	16.25	GALAXY
SHAP354-5608.9	03 53 21.	- 56 17 38.	24	18.0	GALAXY
SHAP354-5814.2	03 53 21.	- 58 22 56.	24	16.75	GALAXY
BAK 2.0891	03 53 23.	- 18 33 24.	14	17.3	GALAXY
BAK 2.0889	03 53 23.	- 21 23 42.	16	17.0	GALAXY
BAK 2.0890	03 53 23.	- 21 54 12.	17	17.0	GALAXY
SHAP354-5348.6	03 53 23.	- 53 57 20.	78	16.75	GALAXY
SHAP354-5259.2	03 53 24.	- 53 07 55.	36	16.5	GALAXY
SHAP354-5624.7	03 53 24.	- 56 33 26.	18	16.75	GALAXY
SHAP354-5635.8	03 53 24.	- 56 44 32.	18	17.5	GALAXY
SHAP354-5647.2	03 53 24.	- 56 55 56.	24	17.5	GALAXY
SHAP354-5723.4	03 53 24.	- 57 32 08.	18	17.5	GALAXY
SHAP354-4936.4	03 53 25.	- 49 45 07.	60	17.5	GALAXY
SHAP354-5630.1	03 53 25.	- 56 38 50.	18	17.75	GALAXY
SHAP354-5856.0	03 53 25.	- 59 04 44.	18	17.75	GALAXY
SHAP354-5944.4	03 53 25.	- 59 53 08.	30	17.0	GALAXY
SHAP354-5348.5	03 53 26.	- 53 57 13.	48	17.5	GALAXY
MCG+06-09-014	03 53 27.	+ 34 42	60	15.	GALAXY
BAK 2.0892	03 53 29.	- 17 15 47.	21	16.2	GALAXY
SHAP354-5601.8	03 53 29.	- 56 10 31.	24	17.5	GALAXY
UGC 02901	03 53 30.	+ 34 43	66	16.0	GALAXY S
TON-S 0384	03 53 30.	- 24 56		15.3	BLUE STAR
LB 03331	03 53 30.	- 63 33		14.5	FAINT BLUE STAR
BAK 2.0893	03 53 33.	- 19 38 53.	14	17.0	GALAXY
SHAP354-5909.2	03 53 33.	- 59 17 55.	12	17.75	GALAXY
SHAP354-5952.3	03 53 33.	- 60 01 02.	36	16.5	GALAXY
SHAP354-5330.8	03 53 34.	- 53 39 31.	36	17.5	GALAXY
LB 01203	03 53 35.	+ 09 32 30.		15.5	FAINT BLUE STAR
BAK 2.0894	03 53 35.	- 16 25 53.	17	16.6	GALAXY
5ZW 365	03 53 36.	+ 27 13			COMPACT GALAXY
ZWG 487.002	03 53 36.	+ 27 13		15.6	GALAXY
KARA.73B 0136	03 53 36.	+ 27 13	24	15.6	ISOLATED GALAXY E
SHAP354-5746.2	03 53 37.	- 57 54 55.	18	17.25	GALAXY
SHAP354-5752.6	03 53 37.	- 58 01 19.	12	17.5	GALAXY
SHAP354-5453.6	03 53 38.	- 55 02 19.	36	16.25	GALAXY
SHAP354-5603.5	03 53 39.	- 56 12 13.	24	17.5	GALAXY
SHAP354-5746.2	03 53 39.	- 57 54 55.	30	17.0	GALAXY
RLWT 078	03 53 42.	+ 56 48		15.	FAINT VERY BLUF STAR
BAK 2.0897	03 53 42.	- 17 15 10.	14	16.6	GALAXY
BAK 2.0895	03 53 44.	- 22 06 23.	14	16.1	GALAXY
SHAP354-5746.8	03 53 45.	- 57 55 31.	24	18.0	GALAXY
BAK 2.0896	03 53 46.	- 21 41 16.	14	15.5	GALAXY
SHAP355-5401.0	03 53 47.	- 54 09 42.	48	17.0	GALAXY
TON-S 0385	03 53 48.	- 22 15		15.2	BLUE STAR
BAK 2.0899	03 53 49.	- 18 09 52.	24	16.6	GALAXY
SHAP354-5747.1	03 53 49.	- 57 55 48.	18	17.5	GALAXY
SHAP354-5832.6	03 53 49.	- 58 41 18.	12	16.75	GALAXY
SHAP355-5218.3	03 53 50.	- 52 27 00.	42	15.75	GALAXY
BAK 2.0900	03 53 51.	- 19 28 40.	19	16.3	GALAXY
SHAP355-5607.4	03 53 51.	- 56 16 06.	24	17.5	GALAXY
SHAP354-5738.4	03 53 52.	- 57 47 06.	24	18.0	GALAXY
BAK 2.0901	03 53 53.	- 21 09 40.	12	16.8	GALAXY
SHAP355-5534.0	03 53 53.	- 55 42 42.	24	17.25	GALAXY
ZWG 526.014	03 53 54.	+ 34 01		15.7	GALAXY
UGC 02902	03 53 54.	+ 34 01	102	15.7	GALAXY S0
FATH 1.144	03 53 54.	- 00 12	14		NEBULA
LB 03332	03 53 54.	- 67 18		14.5	FAINT BLUE STAR
BAK 2.0906	03 53 55.	- 15 14 28.	16	15.9	GALAXY
BAK 2.0898	03 53 55.	- 26 22 22.	25	16.0	GALAXY
SHAP355-5132.4	03 53 55.	- 51 41 05.	54	16.25	GALAXY
SHAP355-5550.6	03 53 55.	- 55 59 18.	24	16.75	GALAXY
SHAP354-5923.0	03 53 55.	- 59 31 42.	18	18.0	GALAXY
BAK 2.0903	03 53 56.	- 20 01 16.	30	16.8	GALAXY
SHAP354-5912.6	03 53 56.	- 59 21 18.	18	17.75	GALAXY
MCG+06-09-015	03 53 57.	+ 34 00	60	14.5	GALAXY
BAK 2.0905	03 53 57.	- 19 28 16.	16	17.2	GALAXY
BAK 2.0907	03 53 59.	- 18 22 51.	19	17.0	GALAXY
BAK 2.0904	03 53 59.	- 21 23 04.	11	16.8	GALAXY
SHAP355-5732.6	03 53 59.	- 57 41 18.	18	18.0	GALAXY
LB 09611	03 54	- 83 07		13.6	FAINT BLUE STAR
LDN 1391	03 54 00.	+ 53 00	4740		DARK NEBULA
ZC 0354.0+7900	03 54 00.	+ 79 00	9340		CLUSTER OF GALAXIES
BAK 2.0908	03 54 01.	- 20 23 57.	22	15.6	GALAXY
BAK 2.0902	03 54 01.	- 25 59 28.	19	16.6	GALAXY
SHAP355-5553.0	03 54 01.	- 56 01 41.	24	17.5	GALAXY
SHAP355-5607.6	03 54 01.	- 56 16 17.	24	18.0	GALAXY
SHAP355-5741.6	03 54 01.	- 57 50 18.	12	16.0	GALAXY
SHAP355-5839.4	03 54 02.	- 58 48 06.	24	17.75	GALAXY
RNGC 1487	03 54 03.	- 42 31			GALAXY
ZWG 418.001	03 54 06.	+ 07 57		15.5	GALAXY
UGC 02903	03 54 06.	+ 07 57	66	15.5	GALAXY S0-a
MCG-04-10-008	03 54 06.	- 21 57 30.	36	15.5	GALAXY
VV 078B	03 54 06.	- 42 31	30		INTFRACTING GALAXY

OBJECT NAME	RIGHT ASCEN.	DECLINATION	DIAM.	MAGN.	TYPE OF OBJECT
VV 078A	03 54 06.	- 42 31	132		INTERACTING GALAXY
VV 078	03 54 06.	- 42 31	150	12.6	INTERACTING GALAXY
AGU 15	03 54 06.	- 42 31 00.	126	12.5	PECULIAR GALAXY
SHAP355-5651.4	03 54 06.	- 57 00 05.	18	17.5	GALAXY
RNGC 1486	03 54 07.	- 21 57		15.0	GALAXY
SHAP355-5458.2	03 54 07.	- 55 06 53.	24	16.0	GALAXY
SHAP355-5810.4	03 54 08.	- 58 19 05.	36	16.75	GALAXY
SHAP355-5351.8	03 54 09.	- 54 00 29.	48	17.0	GALAXY
BAK 2.0909	03 54 10.	- 18 09 39.	12	17.0	GALAXY
BAK 2.0910	03 54 10.	- 18 10 03.	14	17.1	GALAXY
BAK 2.0911	03 54 11.	- 19 23 57.	14	16.9	GALAXY
SHAP355-5741.1	03 54 11.	- 57 49 47.	18	17.0	GALAXY
ZWG 466.001	03 54 12.	+ 16 20		15.7	GALAXY
UGC 02904	03 54 12.	+ 16 20	96	15.7	GALAXY Sb
KARA.72 096A	03 54 12.	+ 16 20	60	15.7	PART OF DOUBLE GALAXY
ZWG 466.002	03 54 12.	+ 16 22		15.7	GALAXY
UGC 02905	03 54 12.	+ 16 22	66	15.7	GALAXY IRR
KARA.72 096B	03 54 12.	+ 16 22	54	15.7	PART OF DOUBLE GALAXY
ZC 0354.2+2045	03 54 12.	+ 20 45	1410		CLUSTER OF GALAXIES
RLWT 079	03 54 12.	+ 55 38		17.	FAINT VERY BLUF STAR
SHAP355-5747.0	03 54 13.	- 57 55 41.	24	17.5	GALAXY
SHAP355-5823.2	03 54 13.	- 58 31 53.	18	17.5	GALAXY
SHAP355-5941.5	03 54 14.	- 59 50 11.	24	17.25	GALAXY
SHAP355-5454.5	03 54 15.	- 55 03 10.	36	16.5	GALAXY
SHAP355-5608.9	03 54 16.	- 56 17 35.	18	17.25	GALAXY
SHAP355-5657.2	03 54 16.	- 57 05 53.	18	16.25	GALAXY
SHAP355-5828.9	03 54 16.	- 58 37 35.	24	17.25	GALAXY
BAK 2.0912	03 54 17.	- 22 07 20.	15	16.9	GALAXY
SHAP355-5518.3	03 54 17.	- 55 26 58.	24	16.0	GALAXY
MCG+00-11-001	03 54 18.	+ 00 40	30	15.5	GALAXY
ZWG 418.002	03 54 18.	+ 08 22		14.9	GALAXY
ZWG 526.015	03 54 18.	+ 36 39		15.7	GALAXY
MCG-03-11-001	03 54 18.	- 18 11	42	14.5	GALAXY
BAK 2.0914	03 54 18.	- 18 47 50.	14	16.2	GALAXY
SHAP355-5459.5	03 54 18.	- 55 08 10.	36	16.5	GALAXY
SHAP355-5644.3	03 54 18.	- 56 52 58.	66	16.25	GALAXY
HN 0252	03 54 18.	- 56 53			NEBULA
IC 2014	03 54 18.	- 56 53			NONSTELLAR OBJECT
SCHO 0065	03 54 19.	+ 36 54 48.	400		ISOLATED DARK CLOUD
BAK 2.0913	03 54 19.	- 21 22 02.	12	17.1	GALAXY
SHAP355-5857.4	03 54 19.	- 59 06 05.	18	17.0	GALAXY
SHAP355-5200.4	03 54 20.	- 52 09 04.	36	17.0	GALAXY
SHAP355-5414.4	03 54 20.	- 54 23 04.	54	16.25	GALAXY
BAK 2.0916	03 54 21.	- 18 12 26.	41	15.6	GALAXY
IC 2005	03 54 23.	+ 36 39 42.			NONSTELLAR OBJECT
BAK 2.0915	03 54 23.	- 20 28 14.	16	16.7	GALAXY
SHAP355-5632.6	03 54 23.	- 56 41 16.	18	17.75	GALAXY
MCG+01-11-001	03 54 24.	+ 08 23 30.	48	14.5	GALAXY
TON-S 0387	03 54 24.	- 23 55		15.4	BLUE STAR
TON-S 0386	03 54 24.	- 31 57		14.6	BLUE STAR
SHAP355-5613.0	03 54 24.	- 56 21 40.	30	17.0	GALAXY
SHAP355-5901.6	03 54 24.	- 59 10 16.	18	17.5	GALAXY
BAK 2.0918	03 54 26.	- 19 54 44.	13	17.2	GALAXY
BAK 2.0917	03 54 26.	- 20 38 14.	17	16.6	GALAXY
SHAP355-5129.6	03 54 26.	- 51 38 15.	66	16.5	GALAXY
BAK 2.0919	03 54 27.	- 18 46 08.	34	15.8	GALAXY
HN 0251	03 54 28.	- 17 15			NEBULA
IC 2013	03 54 29.	- 17 15			NONSTELLAR OBJECT
SHAP355-5743.5	03 54 29.	- 57 52 10.	52	17.75	GALAXY
LDN 1440	03 54 30.	+ 41 20	560		DARK NEBULA
BAK 2.0921	03 54 30.	- 19 54 38.	13	16.9	GALAXY
SHAP355-5831.5	03 54 30.	- 58 40 10.	24	17.25	GALAXY
SHAP355-5844.9	03 54 31.	- 58 53 34.	18	17.25	GALAXY
SHAP355-5358.2	03 54 33.	- 54 06 51.	42	16.75	GALAXY
BAK 2.0922	03 54 34.	- 19 15 43.	35	16.6	GALAXY
SHAP356-4929.0	03 54 34.	- 49 37 39.	72	16.0	GALAXY
SHAP355-5914.9	03 54 34.	- 59 23 34.	24	18.0	GALAXY
SHAP355-5705.9	03 54 35.	- 57 14 33.	24	17.75	GALAXY
BAK 2.0920	03 54 36.	- 26 10 13.	21	16.4	GALAXY
SHAP355-5627.8	03 54 37.	- 56 36 27.	18	17.25	GALAXY
SHAP355-5837.8	03 54 38.	- 58 46 27.	24	18.0	GALAXY
BAK 2.0923	03 54 39.	- 19 44 01.	17	17.0	GALAXY
SHAP355-5325.4	03 54 40.	- 53 34 03.	36	16.5	GALAXY
BAK 2.0924	03 54 41.	- 19 27 49.	12	17.2	GALAXY
SHAP355-5609.0	03 54 41.	- 56 17 39.	24	17.0	GALAXY
SHAP355-5815.6	03 54 41.	- 58 24 15.	42	16.25	GALAXY
MCG+07-09-001	03 54 42.	+ 43 24	36	15.	GALAXY
SHAP355-5647.2	03 54 42.	- 56 55 51.	24	17.0	GALAXY
SHAP355-5934.9	03 54 42.	- 59 43 33.	30	17.0	GALAXY
SHAP356-5232.5	03 54 45.	- 52 41 08.	48	16.0	GALAXY
SHAP356-5343.2	03 54 45.	- 53 51 50.	30	16.0	GALAXY
SHAP355-5849.2	03 54 45.	- 58 57 51.	36	17.25	GALAXY
BAK 2.0925	03 54 46.	- 20 06 01.	14	17.2	GALAXY
RLWT 080	03 54 48.	+ 46 20		15.5	FAINT VERY BLUF STAR
VB 243	03 54 48.	+ 57 14	235		STELLAR RING
MCG-03-11-002	03 54 48.	- 18 54	72	14.5	GALAXY
SHAP355-5629.4	03 54 49.	- 56 38 02.	36	18.0	GALAXY
SHAP355-5915.8	03 54 49.	- 59 24 27.	24	18.0	GALAXY
SHAP356-5231.4	03 54 50.	- 52 40 02.	36	16.0	GALAXY
SHAP356-5530.1	03 54 50.	- 55 38 44.	30	17.0	GALAXY
SHAP355-5942.8	03 54 50.	- 59 51 27.	18	17.5	GALAXY
SHAP356-5630.4	03 54 53.	- 56 39 02.	24	18.0	GALAXY
SHAP355-5824.0	03 54 55.	- 58 32 38.	18	17.5	GALAXY
BAK 2.0926	03 54 56.	- 19 22 24.	14	17.0	GALAXY
BAK 2.0927	03 54 57.	- 18 54 54.	50	14.6	GALAXY
SHAP355-5819.5	03 54 57.	- 58 28 08.	18	18.0	GALAXY
SHAP356-5326.2	03 54 58.	- 53 34 50.	30	16.5	GALAXY
SHAP356-5548.8	03 54 59.	- 55 57 26.	24	16.0	GALAXY
SHAP356-5817.6	03 54 59.	- 58 26 14.	24	17.75	GALAXY
LBN 0701	03 55	+ 52 50	5400		BRIGHT NEBULA
RLWT 081	03 55 00.	+ 54 39		15.5	FAINT VERY BLUF STAR
MCG-05-10-006	03 55 00.	- 29 01	108	14.5	GALAXY
SHAP356-5717.0	03 55 00.	- 57 25 38.	12	17.5	GALAXY
BAK 2.0928	03 55 01.	- 22 23 24.	23	16.4	GALAXY
BAK 2.0929	03 55 01.	- 22 24 36.	12	16.6	GALAXY
SHAP356-5231.9	03 55 01.	- 52 40 31.	48	16.5	GALAXY
MCG-04-10-009	03 55 03.	- 25 32 30.	84	15.	GALAXY
SHAP356-4959.4	03 55 03.	- 50 08 01.	66	16.5	GALAXY
SHAP356-5736.0	03 55 04.	- 57 44 38.	18	17.75	GALAXY
SHAP356-5938.6	03 55 04.	- 59 47 14.	24	17.5	GALAXY
BAK 2.0931	03 55 05.	- 21 26 23.	13	16.7	GALAXY
SHAP356-5713.0	03 55 05.	- 57 21 38.	24	17.25	GALAXY
LB 01204	03 55 06.	+ 21 54 42.		16.9	FAINT BLUE STAR
ZWG 327.011	03 55 06.	+ 73 56		14.3	GALAXY
UGC 02906	03 55 06.	+ 73 56	198	14.3	GALAXY Sb
BAK 2.0930	03 55 06.	- 25 32 06.	36	14.8	GALAXY
SHAP356-5731.3	03 55 07.	- 57 39 55.	18	17.0	GALAXY
SHAP356-5134.6	03 55 08.	- 51 43 13.	36	15.75	GALAXY

OBJECT NAME	RIGHT ASCEN.	DECLINATION	DIAM.	MAGN.	TYPE OF OBJECT
BAK 2.0932	03 55 09.	- 19 01 11.	13	16.8	GALAXY
SHAP356-5442.3	03 55 09.	- 54 50 55.	36	16.25	GALAXY
SHAP356-5630.1	03 55 09.	- 56 38 43.	18	17.0	GALAXY
SHAP356-5640.2	03 55 11.	- 56 48 49.	24	17.25	GALAXY
SHAP356-5848.9	03 55 11.	- 58 57 31.	48	16.5	GALAXY
ZWG 487.003	03 55 12.	+ 24 07		15.6	GALAXY
VB 244	03 55 12.	+ 51 38	269		STELLAR RING
OCL 0395	03 55 12.	+ 51 39	360	16.	OPEN STAR CLUSTER
7ZW 009	03 55 12.	+ 66 58			COMPACT GALAXY
MCG+12-04-009	03 55 12.	+ 73 57	132	14.	GALAXY
SHAP356-5606.0	03 55 12.	- 56 14 37.	24	17.5	GALAXY
BAK 2.0934	03 55 13.	- 19 36 47.	19	16.8	GALAXY
BAK 2.0933	03 55 13.	- 20 58 59.	14	17.3	GALAXY
SHAP356-5235.6	03 55 13.	- 52 44 13.	36	16.0	GALAXY
SHAP356-5512.9	03 55 14.	- 55 21 31.	36	16.0	GALAXY
SHAP356-5353.3	03 55 15.	- 54 01 55.	60	16.0	GALAXY
SHAP356-5401.5	03 55 15.	- 54 10 07.	42	16.0	GALAXY
RNGC 1488	03 55 16.	+ 18 26		15.5	GALAXY
SHAP356-5052.6	03 55 16.	- 51 01 12.	36	16.75	GALAXY
SHAP356-5359.0	03 55 16.	- 54 07 37.	72	16.5	GALAXY
SHAP356-5934.6	03 55 16.	- 59 43 13.	30	17.0	GALAXY
SHAP356-5546.8	03 55 17.	- 55 55 25.	18	15.75	GALAXY
SHAP356-5942.9	03 55 17.	- 59 51 31.	18	17.0	GALAXY
ZWG 466.003	03 55 18.	+ 18 26		15.5	GALAXY
MCG-03-11-003	03 55 18.	- 19 20	54	14.5	GALAXY
SHAP356-5636.4	03 55 18.	- 56 45 01.	24	17.0	GALAXY
SHAP356-5548.5	03 55 19.	- 55 57 07.	30	17.0	GALAXY
SHAP356-5551.0	03 55 19.	- 55 59 37.	36	16.5	GALAXY
BAK 2.0937	03 55 20.	- 19 24 10.	11	16.8	GALAXY
BAK 2.0936	03 55 20.	- 21 16 11.	17	16.6	GALAXY
SHAP356-5830.6	03 55 20.	- 58 39 13.	24	16.75	GALAXY
BAK 2.0935	03 55 21.	- 23 19 23.	10	16.4	GALAXY
BAK 2.0939	03 55 23.	- 19 17 22.	11	17.0	GALAXY
RNGC 1489	03 55 23.	- 19 22		14.0	GALAXY
SHAP356-5236.9	03 55 23.	- 52 45 30.	48	16.25	GALAXY
SHAP356-5307.5	03 55 23.	- 53 16 06.	54	17.25	GALAXY
SHAP356-5636.9	03 55 23.	- 56 45 30.	24	17.5	GALAXY
RLWT 082	03 55 24.	+ 51 31		12.5	FAINT VERY BLUE STAR
MCG+11-05-004	03 55 24.	+ 68 25 30.	66	15.	GALAXY
7ZW 010	03 55 24.	+ 78 08			COMPACT GALAXY
ZWG 347.002	03 55 24.	+ 78 08		15.2	GALAXY
ZWG 346.008	03 55 24.	+ 78 08		15.2	GALAXY
UGC 02907	03 55 24.	+ 78 08	108	15.2	GALAXY PECULR
MCG+00-11-002	03 55 24.	- 00 20	24	15.	GALAXY
SHAP356-5050.1	03 55 24.	- 50 58 42.	42	16.25	GALAXY
SHAP356-5919.9	03 55 25.	- 59 28 31.	36	16.5	GALAXY
BAK 2.0940	03 55 26.	- 20 55 46.	12	16.7	GALAXY
SHAP356-5708.0	03 55 26.	- 57 16 36.	24	16.5	GALAXY
SHAP356-5807.8	03 55 27.	- 58 16 24.	18	18.0	GALAXY
BAK 2.0942	03 55 28.	- 19 22 52.	23	16.8	GALAXY
BAK 2.0941	03 55 28.	- 20 22 10.	12	16.5	GALAXY
BAK 2.0938	03 55 28.	- 25 32 58.	23	16.3	GALAXY
LB 01206	03 55 30.	+ 19 12 06.		15.8	FAINT BLUE STAR
LB 01205	03 55 30.	+ 19 36 06.		17.2	FAINT BLUE STAR
MCG+07-09-002	03 55 30.	+ 43 11	72	15.5	GALAXY
UGC 02908	03 55 30.	+ 43 13	120	16.0	GALAXY S
RLWT 083	03 55 30.	+ 52 28		11.5	FAINT VERY BLUE STAR
ZWG 305.003	03 55 30.	+ 68 26		14.5	GALAXY
RNGC 1469	03 55 30.	+ 68 26		14.5	GALAXY
UGC 02909	03 55 30.	+ 68 26	150	14.5	GALAXY S0
SHAP356-5324.4	03 55 30.	- 53 33 00.	42	16.5	GALAXY
SHAP356-5221.5	03 55 31.	- 52 30 05.	36	16.25	GALAXY
BAK 2.0943	03 55 32.	- 20 39 22.	13	16.7	GALAXY
LB 01207	03 55 33.	+ 19 47 24.		15.6	FAINT BLUE STAR
SHAP356-5743.6	03 55 35.	- 57 52 12.	24	17.5	GALAXY
TON-S 0388	03 55 36.	- 27 01		15.7	BLUE STAR
BAK 2.0945	03 55 37.	- 21 06 21.	14	16.7	GALAXY
LB 01208	03 55 38.	+ 23 52 42.		16.2	FAINT BLUE STAR
SHAP356-5830.4	03 55 38.	- 58 39 00.	12	17.5	GALAXY
BAK 2.0946	03 55 39.	- 18 59 21.	12	16.8	GALAXY
BAK 2.0947	03 55 39.	- 19 05 21.	18	16.4	GALAXY
BAK 2.0944	03 55 39.	- 26 49 34.	23	15.7	GALAXY
BAK 2.0948	03 55 40.	- 19 17 51.	14	17.2	GALAXY
BAK 2.0949	03 55 41.	- 18 24 57.	23	16.8	GALAXY
SHAP356-5537.3	03 55 41.	- 55 45 53.	30	18.0	GALAXY
UGC 02910	03 55 42.	+ 06 35	60	16.5	GALAXY Sb-c
ZWG 508.005	03 55 42.	+ 27 37		15.6	GALAXY
MCG+05-10-004	03 55 42.	+ 27 37	24	16.5	GALAXY
MCG-03-11-004	03 55 42.	- 18 54	48	16.	GALAXY
SHAP356-5354.2	03 55 42.	- 54 02 47.	30	16.5	GALAXY
SHAP356-5822.9	03 55 42.	- 58 31 29.	18	17.75	GALAXY
SHAP356-5923.7	03 55 42.	- 59 32 17.	54	16.0	GALAXY
SHAP356-5840.4	03 55 43.	- 58 48 59.	18	18.0	GALAXY
HN 0254	03 55 44.	- 59 32			NEBULA
IC 2017	03 55 45.	- 59 32			NONSTELLAR OBJECT
SHAP356-5800.0	03 55 46.	- 58 08 35.	30	17.25	GALAXY
MCG-02-11-002	03 55 48.	- 10 04 30.	54	15.	GALAXY
BAK 2.0950	03 55 48.	- 18 55 33.	41	16.2	GALAXY
SHAP357-5211.8	03 55 48.	- 52 20 22.	30	15.5	GALAXY
SHAP356-5541.8	03 55 49.	- 55 50 23.	24	16.75	GALAXY
SHAP357-5548.1	03 55 51.	- 55 56 41.	30	17.0	GALAXY
SHAP357-5601.6	03 55 52.	- 56 10 11.	24	17.5	GALAXY
SHAP356-5747.2	03 55 52.	- 57 55 47.	18	18.0	GALAXY
SHAP356-5759.7	03 55 52.	- 58 08 17.	18	18.0	GALAXY
BAK 2.0952	03 55 53.	- 18 25 56.	14	17.0	GALAXY
BAK 2.0954	03 55 54.	- 16 09 44.	17	16.1	GALAXY
BAK 2.0951	03 55 54.	- 20 08 26.	12	16.4	GALAXY
SHAP356-5826.6	03 55 54.	- 58 35 11.	24	18.0	GALAXY
BAK 2.0953	03 55 55.	- 19 56 08.	12	17.0	GALAXY
RNGC 1493	03 55 55.	- 46 21		12.0	GALAXY
SHAP356-5838.2	03 55 58.	- 58 46 46.	30	16.5	GALAXY
SHAP357-5220.2	03 55 59.	- 52 28 46.	42	16.5	GALAXY
UGC 02911	03 56 00.	+ 43 11	66	16.5	GALAXY SB
MCG-04-10-010	03 56 00.	- 22 43	36	16.	GALAXY
SHAP357-5241.5	03 56 00.	- 52 50 04.	42	16.5	GALAXY
SHAP357-5646.2	03 56 00.	- 56 54 46.	36	16.0	GALAXY
BAK 2.0955	03 56 01.	- 22 08 20.	12	16.4	GALAXY
BAK 2.0956	03 56 02.	- 22 42 32.	32	15.6	GALAXY
SHAP357-5646.8	03 56 02.	- 56 55 22.	24	16.5	GALAXY
BAK 2.0957	03 56 03.	- 18 26 20.	13	17.0	GALAXY
SHAP357-5625.1	03 56 04.	- 56 33 40.	18	17.25	GALAXY
HUB E04	03 56 05.	+ 36 13			DIFFUSE NEBULA
RLWT 084	03 56 06.	+ 51 25		12.	FAINT VERY BLUE STAR
TON-S 0389	03 56 06.	- 30 27		15.2	BLUE STAR
SHAP357-5142.8	03 56 06.	- 51 51 21.	36	16.5	GALAXY
RNGC 1503	03 56 06.	- 66 10			GALAXY
BAK 2.0960	03 56 09.	- 18 11 49.	19	17.2	GALAXY
BAK 2.0958	03 56 09.	- 19 12 19.	19	16.9	GALAXY
BAK 2.0959	03 56 09.	- 19 23 01.	17	16.6	GALAXY
SHAP357-5142.4	03 56 09.	- 51 50 57.	54	16.75	GALAXY
BAK 2.0963	03 56 10.	- 18 30 31.	13	17.2	GALAXY
BAK 2.0961	03 56 10.	- 19 23 31.	19	17.0	GALAXY
RNGC 1494	03 56 10.	- 49 03		12.5	GALAXY
SHAP357-5218.6	03 56 10.	- 52 27 09.	30	15.75	GALAXY
SHAP357-5656.1	03 56 11.	- 57 04 39.	24	18.0	GALAXY
LB 01209	03 56 12.	+ 16 23 00.		13.7	FAINT BLUE STAR
UGC 02912	03 56 12.	+ 42 29	84	16.0	GALAXY S0-a
TON-S 0390	03 56 12.	- 25 13		14.7	BLUE STAR
BAK 2.0962	03 56 13.	- 21 36 55.	22	16.8	GALAXY
SHAP357-5536.4	03 56 13.	- 55 44 57.	30	18.0	GALAXY
SHAP357-5626.4	03 56 13.	- 56 34 57.	18	17.5	GALAXY
MAI 004	03 56 14.	+ 79 57	40		DWARF SPHEROIDAL GALAXY
IC 2015	03 56 14.	- 40 35			NONSTELLAR OBJECT
HN 0253	03 56 15.	- 40 35			NEBULA
SHAP357-5837.0	03 56 15.	- 58 45 33.	18	17.0	GALAXY
BAK 2.0964	03 56 16.	- 19 50 37.	19	16.8	GALAXY
RNGC 1492	03 56 16.	- 35 35			GALAXY
BAK 2.0966	03 56 17.	- 19 11 01.	18	17.2	GALAXY
BAK 2.0965	03 56 17.	- 20 09 07.	12	16.8	GALAXY
MCG+13-04-003	03 56 18.	+ 78 09	114	15.	GALAXY
SHAP357-5720.7	03 56 18.	- 57 29 15.	24	17.5	GALAXY
LB 01210	03 56 20.	+ 20 31 36.		16.4	FAINT BLUE STAR
MCG-02-11-003	03 56 21.	- 10 26	36	15.5	GALAXY
BAK 2.0967	03 56 21.	- 18 36 31.	14	16.6	GALAXY
SHAP357-5609.4	03 56 22.	- 56 17 57.	24	17.25	GALAXY
RNGC 1495	03 56 23.	- 44 37			GALAXY
SHAP357-5303.6	03 56 23.	- 53 12 08.	30	17.0	GALAXY
MCG+00-11-003	03 56 24.	+ 01 13	48	15.	GALAXY
UGC 02913	03 56 24.	+ 01 14	60	16.0	GALAXY SBa
VB 211	03 56 24.	+ 57 42	235		STELLAR RING
SHAP357-5211.0	03 56 25.	- 52 19 32.	24	16.0	GALAXY
SHAP357-5349.3	03 56 25.	- 53 57 50.	36	16.5	GALAXY
SHAP357-5900.2	03 56 25.	- 59 08 45.	24	16.75	GALAXY
BAK 2.0968	03 56 26.	- 18 18 18.	16	16.7	GALAXY
SHAP357-5903.8	03 56 26.	- 59 12 21.	30	17.5	GALAXY
SHAP357-5626.6	03 56 27.	- 56 35 08.	24	16.25	GALAXY
LB 01211	03 56 28.	+ 18 48 24.		16.2	FAINT BLUE STAR
BAK 2.0970	03 56 28.	- 19 02 06.	13	16.4	GALAXY
BAK 2.0971	03 56 28.	- 19 06 42.	14	16.6	GALAXY
BAK 2.0969	03 56 28.	- 19 26 00.	14	17.2	GALAXY
SHAP357-5903.6	03 56 29.	- 59 12 08.	36	16.0	GALAXY
MCG+00-11-004	03 56 30.	+ 01 05	30	15.5	GALAXY
SHAP357-5206.2	03 56 30.	- 52 14 44.	66	16.75	GALAXY
BAK 2.0972	03 56 31.	- 21 44 54.	15	16.8	GALAXY
SHAP357-5010.1	03 56 31.	- 50 18 38.	60	16.5	GALAXY
SHAP357-5342.1	03 56 31.	- 53 50 38.	60	17.0	GALAXY
SHAP357-5913.5	03 56 31.	- 59 22 02.	24	16.75	GALAXY
SHAP357-5228.6	03 56 33.	- 52 37 08.	24	16.5	GALAXY
BAK 2.0975	03 56 34.	- 19 02 30.	17	16.4	GALAXY
BAK 2.0974	03 56 34.	- 20 11 12.	12	17.2	GALAXY
BAK 2.0973	03 56 34.	- 21 45 00.	16	16.9	GALAXY
SHAP357-5246.7	03 56 35.	- 52 55 14.	48	15.75	GALAXY
3ZW 057	03 56 36.	+ 07 48			COMPACT GALAXY
ZWG 487.004	03 56 36.	+ 21 40		15.3	GALAXY
ZWG 346.009	03 56 36.	+ 75 44		15.6	GALAXY
BAK 2.0976	03 56 36.	- 19 00 30.	54	16.2	GALAXY
TON-S 0391	03 56 36.	- 30 08		14.9	BLUE STAR
HN 0255	03 56 36.	- 52 55			NEBULA
IC 2018	03 56 36.	- 52 55			NONSTELLAR OBJECT
SHAP357-5425.6	03 56 36.	- 54 34 08.	48	17.0	GALAXY
SHAP357-5900.9	03 56 36.	- 59 09 26.	30	17.25	GALAXY
SHAP357-5427.4	03 56 37.	- 54 35 56.	48	17.0	GALAXY
SHAP357-5607.0	03 56 37.	- 56 15 32.	24	17.0	GALAXY
SHAP357-5931.6	03 56 37.	- 59 40 08.	18	18.0	GALAXY
BAK 2.0977	03 56 38.	- 20 19 06.	12	16.5	GALAXY
MCG+01-11-002	03 56 39.	+ 06 33	36	14.5	GALAXY
SHAP358-5013.0	03 56 40.	- 50 21 31.	48	17.0	GALAXY
SHAP357-5312.9	03 56 40.	- 53 21 25.	30	16.0	GALAXY
SHAP357-5648.2	03 56 40.	- 56 56 44.	24	18.0	GALAXY
SHAP358-5221.9	03 56 41.	- 52 30 25.	42	16.25	GALAXY
SHAP358-5230.0	03 56 41.	- 52 38 31.	24	16.75	GALAXY
MCG+01-11-004A	03 56 42.	+ 06 32 30.	9	15.5	GALAXY
MCG+01-11-003	03 56 42.	+ 06 32 30.	24	15.	GALAXY
ZWG 418.003	03 56 42.	+ 06 33		14.8	GALAXY
UGC 02914	03 56 42.	+ 06 33	78	14.8	GALAXY SB0
LDN 1441	03 56 42.	+ 41 25	660		DARK NEBULA
RLWT 085	03 56 42.	+ 52 29		11.	FAINT VERY BLUE STAR
SHAP357-5731.9	03 56 43.	- 57 40 25.	24	17.0	GALAXY
SHAP357-5912.0	03 56 43.	- 59 20 32.	18	17.0	GALAXY
SHAP357-5402.7	03 56 44.	- 54 11 13.	36	16.5	GALAXY
SHAP357-5512.4	03 56 44.	- 55 20 55.	30	17.75	GALAXY
SHAP357-5733.0	03 56 44.	- 57 41 31.	18	17.75	GALAXY
SHAP358-5224.1	03 56 45.	- 52 32 37.	24	16.0	GALAXY
SHAP357-5720.6	03 56 46.	- 57 29 07.	18	17.0	GALAXY
SHAP357-5931.0	03 56 46.	- 59 39 32.	18	17.0	GALAXY
SHAP357-5534.4	03 56 47.	- 55 42 55.	30	17.0	GALAXY
SHAP357-5912.2	03 56 47.	- 59 20 43.	30	17.5	GALAXY
MCG+01-11-004	03 56 48.	+ 06 30	18	15.	GALAXY
ZWG 418.004	03 56 48.	+ 06 32		15.4	GALAXY
ZWG 418.005	03 56 48.	+ 06 33		15.6	GALAXY
UGC 02915	03 56 48.	+ 32 29	84	17.	GALAXY Sc
ISS 0075	03 56 48.	+ 57 42	240		STELLAR RING
SHAP358-5152.8	03 56 48.	- 52 01 19.	24	17.0	GALAXY
SHAP357-5649.2	03 56 48.	- 56 57 43.	24	18.0	GALAXY
BAK 2.0978	03 56 49.	- 20 40 05.	17	16.6	GALAXY
SHAP358-5222.6	03 56 49.	- 52 31 07.	36	17.0	GALAXY
SHAP358-5436.5	03 56 49.	- 54 45 01.	42	17.0	GALAXY
SHAP357-5703.7	03 56 49.	- 57 12 13.	24	16.25	GALAXY
RNGC 1500	03 56 50.	- 52 27			GALAXY
BAK 2.0979	03 56 52.	- 19 12 17.	19	16.4	GALAXY
SHAP358-5438.1	03 56 52.	- 54 46 37.	42	16.5	GALAXY
SHAP358-5545.0	03 56 53.	- 55 53 31.	24	16.5	GALAXY
SHAP358-5615.0	03 56 53.	- 56 23 31.	24	18.0	GALAXY
SHAP357-5834.2	03 56 53.	- 58 42 43.	30	17.5	GALAXY
SHAP357-5714.2	03 56 54.	- 57 22 43.	36	16.5	GALAXY
LB 01212	03 56 55.	+ 20 51 48.		15.2	FAINT BLUE STAR
SHAP358-5219.3	03 56 55.	- 52 27 48.	36	15.0	GALAXY
SHAP358-5118.8	03 56 56.	- 51 27 18.	30	16.5	GALAXY
SHAP358-5612.4	03 56 56.	- 56 20 54.	24	17.25	GALAXY
LB 01213	03 56 58.	+ 09 41 36.		15.0	FAINT BLUE STAR
SHAP358-5212.8	03 56 59.	- 52 21 18.	30	16.5	GALAXY
SHAP357-5901.8	03 56 59.	- 59 10 19.	30	17.25	GALAXY
LBN 0835	03 57	+ 04 50	780		BRIGHT NEBULA
LBN 0829	03 57	+ 06 30	960		BRIGHT NEBULA
LBN 0861	03 57	- 04 00	3120		BRIGHT NEBULA
LDN 1462	03 57 00.	+ 36 50	660		DARK NEBULA

OBJECT NAME	RIGHT ASCEN.	DECLINATION	DIAM.	MAGN.	TYPE OF OBJECT
LDN 1443	03 57 00.	+ 40 40	5160		DARK NEBULA
TON-S 0392	03 57 00.	- 23 18		15.0	BLUE STAR
SHAP358-5437.6	03 57 00.	- 54 46 06.	36	16.0	GALAXY
BAK 2.0980	03 57 01.	- 19 35 58.	25	17.2	GALAXY
SHAP358-5212.1	03 57 01.	- 52 20 36.	30	16.5	GALAXY
SHAP358-5619.0	03 57 01.	- 56 27 30.	24	17.5	GALAXY
BAK 2.0981	03 57 02.	- 18 40 16.	17	16.8	GALAXY
SHAP358-5225.2	03 57 02.	- 52 33 42.	36	16.0	GALAXY
SHAP357-5922.6	03 57 02.	- 59 31 07.	36	16.75	GALAXY
SHAP358-5221.2	03 57 03.	- 52 29 42.	30	16.0	GALAXY
SHAP358-5615.6	03 57 03.	- 56 24 06.	24	16.75	GALAXY
SHAP358-5246.1	03 57 04.	- 52 54 36.	30	16.5	GALAXY
SHAP358-5533.6	03 57 04.	- 55 42 06.	24	16.0	GALAXY
SHAP358-5930.8	03 57 04.	- 59 39 18.	42	16.5	GALAXY
SHAP358-5506.2	03 57 05.	- 55 14 42.	30	17.0	GALAXY
LB 01214	03 57 06.	+ 13 31 48.		17.4	FAINT BLUE STAR
VB 212	03 57 06.	+ 60 21	483		STELLAR RING
ZWG 327.012	03 57 06.	+ 71 34		14.9	GALAXY
UGC 02916	03 57 06.	+ 71 34	114	14.9	GALAXY Sa-b
LB 01215	03 57 07.	+ 19 43 54.		15.8	FAINT BLUE STAR
BAK 2.0983	03 57 07.	- 20 18 52.	14	16.8	GALAXY
SHAP358-5615.7	03 57 07.	- 56 24 12.	18	16.5	GALAXY
BAK 2.0982	03 57 08.	- 22 01 58.	10	16.6	GALAXY
SHAP358-5256.6	03 57 08.	- 53 05 05.	42	17.0	GALAXY
SHAP358-5617.2	03 57 09.	- 56 25 42.	18	17.0	GALAXY
LB 01216	03 57 10.	+ 23 37 18.		16.0	FAINT BLUE STAR
BAK 2.0984	03 57 10.	- 21 22 16.	22	16.9	GALAXY
SHAP358-5152.4	03 57 10.	- 52 00 53.	30	16.75	GALAXY
SHAP358-5619.6	03 57 11.	- 56 28 06.	18	17.25	GALAXY
ZC 0357.2+1850	03 57 12.	+ 18 50	5710		CLUSTER OF GALAXIES
UGC 02917	03 57 12.	+ 79 43	72	18.	GALAXY DWRF SP
ARC 0471	03 57 13.	- 13 48		17.6	RICH CLUSTER OF GALAXIES
SHAP358-5325.8	03 57 13.	- 53 34 17.	24	16.25	GALAXY
SHAP358-5621.2	03 57 13.	- 56 29 41.	18	18.0	GALAXY
SHAP358-5805.2	03 57 13.	- 58 13 42.	24	17.25	GALAXY
SHAP358-5834.6	03 57 13.	- 58 43 06.	24	17.0	GALAXY
BAK 2.0985	03 57 14.	- 21 28 39.	19	16.8	GALAXY
SHAP358-5913.0	03 57 14.	- 59 21 30.	24	17.0	GALAXY
MCG+04-10-002	03 57 15.	+ 22 47	60	16.	GALAXY
BAK 2.0986	03 57 15.	- 21 28 51.	13	16.9	GALAXY
SHAP358-5205.0	03 57 15.	- 52 13 29.	30	16.0	GALAXY
SHAP358-5433.6	03 57 15.	- 54 42 05.	42	16.25	GALAXY
BAK 2.0987	03 57 16.	- 22 23 09.	15	16.2	GALAXY
SHAP358-5349.0	03 57 16.	- 53 57 29.	42	16.5	GALAXY
SHAP358-5928.2	03 57 16.	- 59 36 42.	30	17.0	GALAXY
SHAP358-5621.2	03 57 17.	- 56 29 41.	24	16.5	GALAXY
ZWG 487.005	03 57 18.	+ 22 45		15.7	GALAXY
UGC 02918	03 57 18.	+ 22 45	90	15.7	GALAXY S IV
BAK 2.0988	03 57 18.	- 18 05 15.	36	16.6	GALAXY
SHAP358-5705.4	03 57 19.	- 57 13 53.	24	17.25	GALAXY
SHAP358-5427.8	03 57 20.	- 54 36 17.	60	17.0	GALAXY
SHAP358-5438.8	03 57 20.	- 54 47 17.	36	16.75	GALAXY
SHAP358-5843.0	03 57 20.	- 58 51 29.	18	16.5	GALAXY
BAK 2.0989	03 57 21.	- 19 58 27.	17	17.2	GALAXY
SHAP358-5533.0	03 57 22.	- 55 41 29.	24	17.0	GALAXY
SHAP358-5544.2	03 57 22.	- 55 52 41.	24	16.5	GALAXY
SHAP358-5924.6	03 57 22.	- 59 33 05.	30	16.25	GALAXY
SHAP358-5213.2	03 57 23.	- 52 21 40.	36	17.0	GALAXY
SHAP358-5258.5	03 57 23.	- 53 06 59.	30	17.0	GALAXY
UGC 02919	03 57 24.	+ 05 34	72	16.0	GALAXY Sc
MCG+01-11-005	03 57 24.	+ 05 34	48	16.	GALAXY
MRSL 160-12/1	03 57 24.	+ 36 29	19200		HII REGION
LDN 1464	03 57 24.	+ 36 32	240		DARK NEBULA
BAK 2.0990	03 57 24.	- 22 11 27.	16	16.6	GALAXY
BAK 2.0991	03 57 25.	- 21 15 51.	16	16.2	GALAXY
SHAP358-5134.8	03 57 25.	- 51 43 16.	30	17.0	GALAXY
SHAP358-5657.8	03 57 25.	- 57 06 17.	36	17.5	GALAXY
BAK 2.0992	03 57 26.	- 21 04 51.	17	16.7	GALAXY
SHAP358-5854.5	03 57 26.	- 59 02 59.	30	17.5	GALAXY
SHAP358-5131.0	03 57 27.	- 51 39 28.	30	16.0	GALAXY
LB 01217	03 57 28.	+ 19 15 12.		17.0	FAINT BLUE STAR
SHAP358-5159.5	03 57 28.	- 52 07 58.	24	17.0	GALAXY
SHAP358-5749.0	03 57 28.	- 57 57 29.	30	17.25	GALAXY
BAK 2.0993	03 57 29.	- 19 02 02.	14	17.0	GALAXY
SHAP358-5439.4	03 57 29.	- 54 47 52.	48	16.5	GALAXY
ZWG 526.016	03 57 30.	+ 34 53		15.3	GALAXY
UGC 02920	03 57 30.	+ 34 53	138	15.3	GALAXY Sc
SHAP358-5433.2	03 57 32.	- 54 41 40.	72	16.25	GALAXY
SHAP358-5438.0	03 57 32.	- 54 46 28.	42	16.0	GALAXY
LB 01218	03 57 33.	+ 24 18 18.		15.9	FAINT BLUE STAR
BAK 2.0994	03 57 33.	- 21 54 02.	11	16.7	GALAXY
SHAP358-5158.0	03 57 33.	- 52 06 28.	30	16.5	GALAXY
SHAP358-5656.9	03 57 33.	- 57 05 22.	24	16.0	GALAXY
SHAP358-5210.2	03 57 34.	- 52 18 40.	54	16.5	GALAXY
SHAP358-5316.2	03 57 34.	- 53 24 40.	30	16.5	GALAXY
SHAP358-5935.7	03 57 35.	- 59 44 10.	30	16.25	GALAXY
MCG+00-11-005	03 57 36.	+ 00 36 30.	66	15.5	GALAXY
UGC 02921	03 57 36.	+ 00 37	90	17.	GALAXY
ZWG 466.004	03 57 36.	+ 20 10		15.6	GALAXY
MCG+03-11-003	03 57 36.	+ 20 12	45	16.	GALAXY
MCG+06-09-016	03 57 36.	+ 34 51 30.	138	14.5	GALAXY
SCHO 0066	03 57 37.	+ 36 33 12.	350		ISOLATED DARK CLOUD
SHAP358-5935.4	03 57 37.	- 59 43 52.	24	15.75	GALAXY
SHAP359-5119.8	03 57 38.	- 51 28 15.	36	16.75	GALAXY
SHAP358-5328.6	03 57 38.	- 53 37 04.	36	17.0	GALAXY
BAK 2.0996	03 57 40.	- 18 38 38.	25	16.7	GALAXY
BAK 2.0995	03 57 40.	- 20 03 38.	14	16.7	GALAXY
SHAP359-5132.8	03 57 40.	- 51 41 15.	30	16.0	GALAXY
SHAP358-5602.5	03 57 40.	- 56 10 58.	18	16.5	GALAXY
SHAP358-5934.6	03 57 40.	- 59 43 04.	36	16.5	GALAXY
SHAP359-5217.5	03 57 41.	- 52 25 57.	42	17.25	GALAXY
SHAP358-5631.2	03 57 42.	- 56 39 40.	18	17.0	GALAXY
SHAP358-5402.8	03 57 43.	- 54 11 15.	42	16.0	GALAXY
BAK 2.0997	03 57 44.	- 21 16 26.	29	15.6	GALAXY
HN 0256	03 57 44.	- 54 11			NEBULA
IC 2020	03 57 44.	- 54 12			NONSTELLAR OBJECT
SHAP358-5901.8	03 57 44.	- 59 10 16.	168	16.5	GALAXY
SHAP358-5935.0	03 57 44.	- 59 43 28.	42	16.0	GALAXY
HN 0258	03 57 45.	- 59 10			NEBULA
IC 2022	03 57 45.	- 59 11			NONSTELLAR OBJECT
SHAP358-5924.0	03 57 45.	- 59 32 28.	18	17.0	GALAXY
SHAP358-5827.8	03 57 46.	- 58 36 16.	24	17.0	GALAXY
SHAP358-5828.6	03 57 46.	- 58 37 04.	18	17.25	GALAXY
SHAP359-5215.4	03 57 47.	- 52 23 51.	36	17.0	GALAXY
SHAP358-5808.2	03 57 47.	- 58 16 40.	18	17.0	GALAXY
SHAP358-5809.4	03 57 47.	- 58 17 52.	24	17.75	GALAXY
SHAP358-5931.2	03 57 47.	- 59 39 40.	30	16.5	GALAXY
MCG+03-11-005	03 57 48.	+ 17 26	60	15.	GALAXY

OBJECT NAME	RIGHT ASCEN.	DECLINATION	DIAM.	MAGN.	TYPE OF OBJECT
MCG+03-11-004	03 57 48.	+ 20 54	30	16.	GALAXY
ZWG 466.005	03 57 48.	+ 20 55		15.5	GALAXY
KARA.73B 0137	03 57 48.	+ 20 55	24	15.5	ISOLATED GALAXY E
ZC 0357.8+3154	03 57 48.	+ 31 54	1750		CLUSTER OF GALAXIES
VB 213	03 57 48.	+ 59 04	517		STELLAR RING
SHAP358-5600.0	03 57 48.	- 56 08 27.	18	17.5	GALAXY
SHAP358-5758.8	03 57 49.	- 58 07 15.	42	17.0	GALAXY
BAK 2.0998	03 57 51.	- 22 11 25.	10	16.3	GALAXY
SHAP359-5217.0	03 57 51.	- 52 25 27.	18	17.0	GALAXY
BAK 2.0999	03 57 52.	- 20 29 01.	12	16.6	GALAXY
BAK 2.1000	03 57 53.	- 19 57 25.	13	16.8	GALAXY
SHAP359-5308.5	03 57 53.	- 53 16 57.	24	16.0	GALAXY
SHAP358-5806.2	03 57 53.	- 58 14 39.	18	17.5	GALAXY
SHAP358-5831.6	03 57 53.	- 58 40 03.	24	17.0	GALAXY
UGC 02922	03 57 54.	+ 17 27	66	16.5	GALAXY Sb
ZC 0357.9+3432	03 57 54.	+ 34 32	4100		CLUSTER OF GALAXIES
ZWG 327.013	03 57 54.	+ 71 35		15.7	GALAXY
SHAP359-5159.6	03 57 54.	- 52 08 02.	24	17.0	GALAXY
LB 01220	03 57 56.	+ 17 09 30.		15.9	FAINT BLUE STAR
LB 01219	03 57 56.	+ 22 01 24.		17.1	FAINT BLUE STAR
SHAP359-5131.7	03 57 56.	- 51 40 08.	30	17.25	GALAXY
SHAP359-5443.1	03 57 56.	- 54 51 33.	42	16.25	GALAXY
SHAP358-5934.5	03 57 56.	- 59 42 57.	48	16.5	GALAXY
SHAP359-5132.6	03 57 58.	- 51 41 02.	54	16.75	GALAXY
LBN 0853	03 58	+ 00 00	31500		BRIGHT NEBULA
LBN 0779	03 58	+ 25 00	6300		BRIGHT NEBULA
LBN 0756	03 58	+ 36 30	6000		BRIGHT NEBULA
LBN 0752	03 58	+ 36 50	10800		BRIGHT NEBULA
KHAV 043	03 58	+ 39 39	22430		DARK NEBULA
LDN 1463	03 58 00.	+ 36 45	660		DARK NEBULA
ZWG 542.001	03 58 00.	+ 42 42		15.7	GALAXY
RLWT 086	03 58 00.	+ 56 50		14.	FAINT VERY BLUE STAR
UGC 02923	03 58 00.	+ 78 32	66	16.0	GALAXY Sb/SBb
SHAP358-5849.8	03 58 00.	- 58 58 15.	24	17.5	GALAXY
BAK 2.1001	03 58 02.	- 20 15 24.	14	16.8	GALAXY
SHAP359-5921.6	03 58 04.	- 59 30 03.	24	17.25	GALAXY
SHAP359-5239.8	03 58 05.	- 52 48 14.	36	15.5	GALAXY
VB 245	03 58 06.	+ 57 21	195		STELLAR RING
ISS 0111	03 58 06.	+ 57 21	195		STELLAR RING
MCG-02-11-004	03 58 06.	- 09 19	36	15.5	GALAXY
BAK 2.1003	03 58 06.	- 19 15 18.	13	16.8	GALAXY
HN 0257	03 58 06.	- 52 48			NEBULA
IC 2021	03 58 06.	- 52 49			NONSTELLAR OBJECT
SHAP359-5957.6	03 58 06.	- 60 06 03.	24	17.75	GALAXY
LB 03333	03 58 06.	- 61 05		11.7	FAINT BLUE STAR
RNGC 1498	03 58 07.	- 12 09			NON-EXISTENT OBJECT
BAK 2.1002	03 58 07.	- 21 32 06.	16	17.0	GALAXY
SHAP359-5213.2	03 58 07.	- 52 21 38.	24	16.0	GALAXY
LB 01221	03 58 08.	+ 19 06 54.		16.6	FAINT BLUE STAR
BAK 2.1004	03 58 08.	- 20 40 36.	22	16.7	GALAXY
SHAP359-5238.4	03 58 08.	- 52 46 50.	18	16.5	GALAXY
SHAP359-5530.9	03 58 08.	- 55 39 20.	24	17.25	GALAXY
SHAP359-5646.8	03 58 08.	- 56 55 14.	18	17.75	GALAXY
SHAP359-5546.8	03 58 09.	- 55 55 14.	30	17.5	GALAXY
SHAP359-5724.9	03 58 09.	- 57 33 20.	18	17.0	GALAXY
SHAP359-5528.2	03 58 10.	- 55 36 38.	36	16.5	GALAXY
BAK 2.1005	03 58 11.	- 20 46 12.	14	16.3	GALAXY
BAK 2.1006	03 58 11.	- 20 50 54.	25	17.0	GALAXY
SHAP359-5804.8	03 58 11.	- 58 13 14.	24	17.5	GALAXY
RLWT 088	03 58 12.	+ 51 51		12.5	FAINT VERY BLUE STAR
RLWT 087	03 58 12.	+ 52 24		11.5	FAINT VERY BLUE STAR
ISS 0076	03 58 12.	+ 59 05	513		STELLAR RING
BAK 2.1007	03 58 12.	- 21 16 00.	19	16.8	GALAXY
TON-S 0393	03 58 12.	- 22 55		14.9	BLUE STAR
SHAP359-5844.5	03 58 12.	- 58 52 56.	36	17.75	GALAXY
SHAP359-5853.1	03 58 12.	- 59 01 32.	24	18.0	GALAXY
SHAP359-5338.9	03 58 13.	- 53 47 19.	42	16.5	GALAXY
SHAP359-5443.2	03 58 13.	- 54 51 38.	42	16.0	GALAXY
SHAP359-5930.8	03 58 14.	- 59 39 14.	18	17.25	GALAXY
BAK 2.1008	03 58 15.	- 20 55 00.	25	15.6	GALAXY
SHAP359-5427.4	03 58 15.	- 54 35 49.	48	16.5	GALAXY
HN 0268	03 58 15.	- 84 00			NEBULA
IC 2051	03 58 16.	- 83 59			NONSTELLAR OBJECT
UGC 02924	03 58 18.	+ 26 42	60	16.5	GALAXY S
SHAP359-5739.3	03 58 20.	- 57 47 43.	18	17.0	GALAXY
BAK 2.1009	03 58 21.	- 19 48 17.	14	17.4	GALAXY
BAK 2.1010	03 58 21.	- 20 06 41.	23	16.4	GALAXY
MCG-03-11-005	03 58 21.	- 20 55	30	15.	GALAXY
MCG-04-10-011	03 58 21.	- 25 19	150	14.5	GALAXY
SHAP359-5530.1	03 58 21.	- 55 38 31.	24	17.0	GALAXY
SHAP359-5847.8	03 58 21.	- 58 56 13.	24	16.5	GALAXY
SHAP359-5240.8	03 58 23.	- 52 49 13.	24	15.5	GALAXY
IC 2023	03 58 23.	- 52 50			NONSTELLAR OBJECT
ZWG 487.006	03 58 24.	+ 23 07		15.4	GALAXY
MCG+04-10-004	03 58 24.	+ 23 07	42	15.	GALAXY
MCG+04-10-003	03 58 24.	+ 26 41 30.	36	15.5	GALAXY
ASS 45	03 58 24.	+ 56 29			OB ASSOCIATION CAM OB3
VB 246	03 58 24.	+ 57 20	383		STELLAR RING
BAK 2.1011	03 58 24.	- 21 23 35.	16	16.7	GALAXY
HN 0259	03 58 24.	- 52 50			NEBULA
SHAP359-5837.8	03 58 25.	- 58 46 13.	18	17.5	GALAXY
SHAP359-5112.2	03 58 26.	- 51 20 36.	36	15.25	GALAXY
SHAP359-5844.0	03 58 26.	- 58 52 25.	18	18.0	GALAXY
MCG+00-11-006	03 58 27.	- 00 51	60	15.	GALAXY
SHAP359-5142.2	03 58 27.	- 51 50 36.	42	17.25	GALAXY
SHAP359-5154.5	03 58 27.	- 52 02 54.	24	17.0	GALAXY
SHAP359-5606.0	03 58 28.	- 56 14 25.	18	16.5	GALAXY
LDN 1569	03 58 30.	+ 00 30	3180		DARK NEBULA
UGC 02926	03 58 30.	+ 05 25	72	15.5	GALAXY SBb
UGC 02925	03 58 30.	- 00 51	102	17.	GALAXY
SHAP359-5336.4	03 58 30.	- 53 44 48.	24	16.5	GALAXY
BAK 2.1013	03 58 31.	- 20 55 23.	19	16.0	GALAXY
BAK 2.1012	03 58 31.	- 21 49 23.	13	16.7	GALAXY
BAK 2.1014	03 58 32.	- 21 50 58.	10	16.9	GALAXY
SHAP359-5059.9	03 58 32.	- 51 08 18.	30	16.25	GALAXY
SHAP359-5237.0	03 58 32.	- 52 45 24.	24	16.0	GALAXY
SHAP359-5340.5	03 58 32.	- 53 48 54.	24	16.5	GALAXY
SHAP359-5850.1	03 58 32.	- 58 58 31.	24	17.75	GALAXY
MCG+01-11-006	03 58 33.	+ 05 25	54	15.5	GALAXY
LB 00221	03 58 34.	+ 19 18 00.		16.2	FAINT BLUE STAR
SHAP359-5102.0	03 58 34.	- 51 10 24.	24	16.25	GALAXY
SHAP359-5158.1	03 58 34.	- 52 06 30.	24	16.0	GALAXY
SHAP359-5537.2	03 58 35.	- 55 45 36.	24	16.75	GALAXY
ZWG 418.006	03 58 36.	+ 05 25		15.5	GALAXY
MCG+04-10-006	03 58 36.	+ 22 59	36	15.	GALAXY
MCG-02-11-005	03 58 36.	- 10 23	54	14.5	GALAXY
TON-S 0394	03 58 36.	- 23 02		15.5	BLUE STAR
MCG-05-10-007	03 58 36.	- 30 58	60	14.5	GALAXY

OBJECT NAME	RIGHT ASCEN.	DECLINATION	DIAM.	MAGN.	TYPE OF OBJECT
SHAP359-5956.9	03 58 37.	- 60 05 19.	24	16.25	GALAXY
BAK 2.1015	03 58 38.	- 18 04 40.	22	16.6	GALAXY
MCG+04-10-005	03 58 39.	+ 23 05	54	15.5	GALAXY
SHAP359-5311.2	03 58 40.	- 53 19 36.	30	16.75	GALAXY
SHAP359-5610.2	03 58 41.	- 56 18 36.	24	17.0	GALAXY
ZWG 487.007	03 58 42.	+ 22 58		15.2	GALAXY
UGC 02927	03 58 42.	+ 22 58	144	15.2	GALAXY SBa
ZWG 487.008	03 58 42.	+ 23 04		15.5	GALAXY
UGC 02928	03 58 42.	+ 23 04	60	15.5	GALAXY Sa/SBa
MCG+04-10-007	03 58 42.	+ 24 40 30.	60	16.	GALAXY
LB 03334	03 58 42.	- 68 50		13.5	FAINT BLUE STAR
BAK 2.1016	03 58 43.	- 22 01 34.	19	16.4	GALAXY
SHAP400-5107.1	03 58 44.	- 51 15 29.	24	16.0	GALAXY
SHAP400-5157.1	03 58 44.	- 52 05 29.	36	16.75	GALAXY
SHAP359-5714.6	03 58 45.	- 57 23 00.	30	16.5	GALAXY
BAK 2.1017	03 58 46.	- 21 10 28.	14	16.8	GALAXY
SHAP359-5804.0	03 58 47.	- 58 12 24.	24	17.25	GALAXY
LB 01222	03 58 48.	+ 12 57 54.		17.3	FAINT BLUE STAR
ZC 0358.8+1511	03 58 48.	+ 15 11	4570		CLUSTER OF GALAXIES
BAK 2.1018	03 58 50.	- 21 12 45.	13	16.8	GALAXY
SHAP400-5210.8	03 58 50.	- 52 19 11.	18	17.0	GALAXY
LB 01223	03 58 51.	+ 13 19 54.		16.3	FAINT BLUE STAR
SHAP400-5101.9	03 58 51.	- 51 10 17.	36	17.0	GALAXY
SHAP400-5322.0	03 58 51.	- 53 30 23.	84	16.0	GALAXY
SHAP359-5608.2	03 58 51.	- 56 16 35.	24	17.5	GALAXY
SHAP359-5614.1	03 58 51.	- 56 22 29.	18	17.0	GALAXY
SHAP359-5756.4	03 58 51.	- 58 04 48.	24	17.5	GALAXY
SHAP400-5100.5	03 58 52.	- 51 08 53.	24	16.75	GALAXY
HN 0260	03 58 52.	- 53 31			NEBULA
IC 2024	03 58 52.	- 53 31			NONSTELLAR OBJECT
SHAP400-5548.2	03 58 52.	- 55 56 35.	24	17.5	GALAXY
SHAP359-5608.9	03 58 52.	- 56 17 17.	24	17.25	GALAXY
SHAP400-5235.8	03 58 53.	- 52 44 11.	18	16.25	GALAXY
SHAP400-5156.2	03 58 54.	- 52 04 35.	30	16.5	GALAXY
LB 01224	03 58 55.	+ 13 34 36.		15.2	FAINT BLUE STAR
SHAP400-5048.2	03 58 55.	- 50 56 35.	36	17.25	GALAXY
SHAP400-5156.6	03 58 55.	- 52 04 59.	30	16.5	GALAXY
LB 01225	03 58 56.	+ 23 54 18.		16.9	FAINT BLUE STAR
SHAP400-5612.4	03 58 56.	- 56 20 47.	18	17.0	GALAXY
SHAP400-5600.8	03 58 57.	- 56 09 11.	24	16.75	GALAXY
SHAP359-5832.1	03 58 57.	- 58 40 29.	24	17.0	GALAXY
SHAP400-5609.6	03 58 58.	- 56 17 59.	24	16.75	GALAXY
SHAP400-5609.0	03 58 59.	- 56 17 23.	24	17.25	GALAXY
SHAP359-5835.1	03 58 59.	- 58 43 29.	48	16.0	GALAXY
LBN 0836	03 59	+ 03 40	2760		BRIGHT NEBULA
LBN 0704	03 59	+ 51 11	420		BRIGHT NEBULA
ZWG 466.006	03 59 00.	+ 20 05		15.7	GALAXY
VB 247	03 59 00.	+ 54 01	530		STELLAR RING
LDN 1388	03 59 00.	+ 56 40	300		DARK NEBULA
MCG+14-03-002	03 59 00.	+ 83 19	39	17.	GALAXY
LB 01226	03 59 01.	+ 10 27 36.		16.1	FAINT BLUE STAR
SHAP400-5220.2	03 59 01.	- 52 28 34.	24	16.0	GALAXY
SHAP400-5553.4	03 59 02.	- 56 01 47.	30	17.5	GALAXY
RNGC 1506	03 59 03.	- 52 42			GALAXY
SHAP400-5304.0	03 59 03.	- 53 12 22.	30	16.25	GALAXY
SHAP400-5900.8	03 59 03.	- 59 09 11.	36	18.0	GALAXY
IC 2016	03 59 04.	+ 20 06 44.			NONSTELLAR OBJECT
SHAP400-5234.3	03 59 04.	- 52 42 40.	36	15.0	GALAXY
HN 0261	03 59 04.	- 53 13			NEBULA
IC 2025	03 59 04.	- 53 13			NONSTELLAR OBJECT
SHAP400-5341.4	03 59 05.	- 53 49 46.	30	16.75	GALAXY
RNGC 1520	03 59 05.	- 76 58			NON-EXISTENT OBJECT
ZWG 487.009	03 59 06.	+ 23 00		14.5	GALAXY
UGC 02929	03 59 06.	+ 23 00	120	14.5	GALAXY S0
MCG+04-10-008	03 59 06.	+ 23 00	30	14.5	GALAXY
5ZW 366	03 59 06.	+ 46 29			COMPACT GALAXY
BAK 2.1019	03 59 06.	- 20 14 20.	22	16.6	GALAXY
SHAP400-5237.3	03 59 06.	- 52 45 40.	24	17.0	GALAXY
SHAP400-5703.6	03 59 06.	- 57 11 58.	30	17.5	GALAXY
RNGC 1497	03 59 07.	+ 23 00		14.5	GALAXY
SHAP400-5114.0	03 59 07.	- 51 22 22.	24	16.75	GALAXY
SHAP400-5833.1	03 59 07.	- 58 41 29.	18	17.0	GALAXY
SHAP400-5941.8	03 59 07.	- 59 50 11.	36	17.0	GALAXY
LB 01227	03 59 08.	+ 13 01 18.		17.2	FAINT BLUE STAR
BAK 2.1021	03 59 08.	- 18 24 56.	18	16.4	GALAXY
SHAP400-5831.1	03 59 08.	- 58 39 28.	24	17.0	GALAXY
SHAP400-5939.5	03 59 08.	- 59 47 53.	30	16.75	GALAXY
MCG+04-10-009	03 59 09.	+ 21 48 30.	30	15.5	GALAXY
BAK 2.1022	03 59 10.	- 20 25 50.	17	16.2	GALAXY
BAK 2.1023	03 59 11.	- 20 23 08.	22	16.8	GALAXY
BAK 2.1020	03 59 11.	- 22 03 44.	17	15.6	GALAXY
ZWG 487.010	03 59 12.	+ 21 49		15.7	GALAXY
MCG-02-11-006	03 59 12.	- 10 32	36	15.	GALAXY
MCG-03-11-006	03 59 12.	- 19 22	60	15.	GALAXY
SHAP400-5205.9	03 59 12.	- 52 14 16.	24	16.0	GALAXY
SHAP400-5606.5	03 59 12.	- 56 14 52.	36	17.0	GALAXY
BAK 2.1024	03 59 13.	- 21 32 20.	15	15.8	GALAXY
SHAP400-5602.0	03 59 14.	- 56 10 22.	24	17.5	GALAXY
SHAP400-5938.6	03 59 14.	- 59 46 58.	30	16.0	GALAXY
MCG+01-11-007	03 59 15.	+ 05 30	24	15.5	GALAXY
BAK 2.1026	03 59 15.	- 18 27 08.	17	16.7	GALAXY
SHAP400-5831.5	03 59 15.	- 58 39 52.	24	17.75	GALAXY
SHAP400-5906.4	03 59 15.	- 59 14 46.	36	17.5	GALAXY
SHAP400-5939.9	03 59 15.	- 59 48 16.	30	16.5	GALAXY
BAK 2.1027	03 59 16.	- 19 52 32.	16	16.8	GALAXY
BAK 2.1025	03 59 16.	- 21 48 14.	17	16.8	GALAXY
SHAP400-5224.8	03 59 17.	- 52 33 09.	18	16.5	GALAXY
SHAP400-5318.2	03 59 17.	- 53 26 33.	30	17.0	GALAXY
IC 2019	03 59 18.	+ 05 29 05.			NONSTELLAR OBJECT
ZWG 418.007	03 59 18.	+ 05 30		15.3	GALAXY
UGC 02930	03 59 18.	+ 05 30	66	15.3	GALAXY DBL SYS
ZWG 487.011	03 59 18.	+ 25 40		14.8	GALAXY
UGC 02931	03 59 18.	+ 25 40	84	14.8	GALAXY Sc
VB 214	03 59 18.	+ 59 01	168		STELLAR RING
ISS 0077	03 59 18.	+ 59 01	174		STELLAR RING
BAK 2.1028	03 59 18.	- 19 22 20.	59	15.5	GALAXY
SHAP400-5801.2	03 59 18.	- 58 09 34.	18	16.5	GALAXY
SHAP400-5838.2	03 59 18.	- 58 46 34.	42	16.5	GALAXY
RNGC 1511	03 59 18.	- 67 46		12.5	GALAXY
SHAP400-5154.6	03 59 19.	- 52 02 57.	24	17.0	GALAXY
SHAP400-5711.4	03 59 19.	- 57 19 46.	18	16.5	GALAXY
SHAP400-5935.1	03 59 19.	- 59 43 28.	36	16.5	GALAXY
LB 01228	03 59 20.	+ 23 02 36.		17.1	FAINT BLUE STAR
SHAP400-5857.1	03 59 20.	- 59 05 28.	30	17.25	GALAXY
MCG+04-10-010	03 59 21.	+ 25 40 30.	48	15.	GALAXY
SHAP400-5243.5	03 59 21.	- 52 51 51.	18	16.0	GALAXY
SHAP400-5719.0	03 59 22.	- 57 27 21.	24	16.5	GALAXY
SHAP400-5301.9	03 59 23.	- 53 10 15.	24	16.5	GALAXY
SHAP400-5341.3	03 59 23.	- 53 49 39.	36	16.5	GALAXY
ZWG 487.012	03 59 24.	+ 26 41		15.6	GALAXY
MRSL 150-00/1	03 59 24.	+ 51 12	3000		HII REGION
BAK 2.1029	03 59 25.	- 19 19 55.	19	17.2	GALAXY
SHAP400-5835.8	03 59 25.	- 58 44 09.	18	17.25	GALAXY
SHAP400-5203.6	03 59 26.	- 52 11 57.	24	16.0	GALAXY
SHAP400-5351.3	03 59 26.	- 53 59 39.	30	17.25	GALAXY
SHAP400-5456.4	03 59 26.	- 55 04 45.	42	16.0	GALAXY
MCG+04-10-011	03 59 27.	+ 21 44 30.	42	15.5	GALAXY
SHAP400-5853.8	03 59 27.	- 59 02 09.	48	16.25	GALAXY
SHAP400-5317.9	03 59 28.	- 53 26 15.	24	16.75	GALAXY
SHAP400-5714.0	03 59 28.	- 57 22 21.	78	17.75	GALAXY
SHAP400-5324.4	03 59 29.	- 53 32 45.	30	17.0	GALAXY
REIN 2.041	03 59 29.87	+ 51 11 02.7			NEBULA
SHAP400-5857.2	03 59 30.	- 59 05 33.	24	18.0	GALAXY
LB 01229	03 59 31.	+ 19 43 06.		16.4	FAINT BLUE STAR
RNGC 1491	03 59 31.	+ 51 10			DIFFUSE NEBULA
CED 025	03 59 33.	+ 51 10	180		DIFFUSE GALACTIC NEBULA
SG 2.006	03 59 33.	+ 51 10	180		DIFFUSE EMISSION NEBULA
SHAP400-5211.4	03 59 33.	- 52 19 44.	30	16.75	GALAXY
SHAP400-5606.8	03 59 33.	- 56 15 09.	24	17.25	GALAXY
SHAP400-5359.4	03 59 34.	- 54 07 44.	24	16.5	GALAXY
REIN 2.042	03 59 34.84	+ 51 10 36.5			NEBULA
SHAP400-5341.4	03 59 35.	- 53 49 44.	30	16.5	GALAXY
MCG+04-10-012	03 59 36.	+ 21 25	36	15.	GALAXY
MCG+04-10-012	03 59 36.	+ 21 25	36	15.	GALAXY
ZWG 487.013	03 59 36.	+ 21 45		15.5	GALAXY
SHAP400-5651.1	03 59 38.	- 56 59 26.	24	17.0	GALAXY
MCG-02-11-007	03 59 39.	- 09 36	72	15.	GALAXY
SHAP400-5653.9	03 59 39.	- 57 02 14.	24	17.0	GALAXY
SHAP400-5828.0	03 59 39.	- 58 36 21.	18	17.25	GALAXY
BAK 2.1030	03 59 40.	- 20 24 42.	14	16.8	GALAXY
SHAP400-5210.9	03 59 40.	- 52 19 14.	24	16.25	GALAXY
SHAP400-5651.1	03 59 40.	- 56 59 26.	24	17.25	GALAXY
SHAP401-5125.4	03 59 41.	- 51 33 44.	30	16.75	GALAXY
SHAP400-5328.0	03 59 41.	- 53 36 20.	30	16.75	GALAXY
ZWG 466.007	03 59 42.	+ 21 25		15.7	GALAXY
MCG+04-10-013	03 59 42.	+ 25 06	30	16.	GALAXY
UGC 02932	03 59 42.	+ 33 44	66	18.	GALAXY
ZWG 327.014	03 59 42.	+ 70 52		13.6	GALAXY
UGC 02933	03 59 42.	+ 70 52	162	13.6	GALAXY S
MCG+12-04-010	03 59 42.	+ 70 52	114	13.	GALAXY
BAK 2.1031	03 59 42.	- 19 20 42.	16	17.1	GALAXY
SHAP400-5258.5	03 59 42.	- 53 06 50.	24	15.75	GALAXY
SHAP400-5414.1	03 59 42.	- 54 22 26.	36	16.75	GALAXY
SHAP400-5420.0	03 59 42.	- 54 28 20.	36	16.0	GALAXY
RNGC 1485	03 59 43.	+ 70 52		13.5	GALAXY
SHAP401-5257.8	03 59 43.	- 53 06 08.	24	16.75	GALAXY
SHAP400-5413.0	03 59 43.	- 54 21 20.	36	16.75	GALAXY
SHAP400-5809.4	03 59 43.	- 58 17 44.	18	17.5	GALAXY
SHAP401-5140.5	03 59 44.	- 51 48 50.	24	17.0	GALAXY
SHAP401-5222.6	03 59 44.	- 52 30 56.	24	16.75	GALAXY
SHAP400-5652.2	03 59 45.	- 57 00 32.	24	17.5	GALAXY
SHAP400-5550.1	03 59 46.	- 55 58 26.	42	16.5	GALAXY
LDN 1469	03 59 48.	+ 06 30	300		DARK NEBULA
MCG+04-10-014	03 59 48.	+ 21 20	48	15.5	GALAXY
ZC 0359.8+2859	03 59 48.	+ 28 59	1480		CLUSTER OF GALAXIES
MCG-03-11-007	03 59 48.	- 16 34	66	15.	GALAXY
SHAP400-5724.1	03 59 48.	- 57 32 26.	48	17.75	GALAXY
LB 03335	03 59 48.	- 63 06		13.5	FAINT BLUE STAR
SHAP401-5205.5	03 59 49.	- 52 13 49.	18	17.25	GALAXY
SHAP401-5258.0	03 59 49.	- 53 06 19.	24	15.5	GALAXY
SHAP400-5907.4	03 59 49.	- 59 15 44.	18	17.25	GALAXY
BAK 2.1032	03 59 50.	- 18 19 53.	14	16.6	GALAXY
SHAP401-5329.2	03 59 50.	- 53 37 31.	24	16.5	GALAXY
SHAP400-5649.2	03 59 51.	- 56 57 32.	24	17.25	GALAXY
SHAP400-5814.9	03 59 52.	- 58 23 14.	24	16.75	GALAXY
BAK 2.1033	03 59 53.	- 20 58 29.	23	16.9	GALAXY
SHAP401-5111.6	03 59 53.	- 51 19 55.	54	16.75	GALAXY
SHAP401-5238.4	03 59 53.	- 52 46 43.	30	17.25	GALAXY
SHAP400-5830.1	03 59 53.	- 58 38 26.	18	17.5	GALAXY
MRSL 186-34/1	03 59 54.	+ 04 00	43200		HII REGION
UGC 02934	03 59 54.	+ 30 55	102	17.	GALAXY S
SHAP401-5227.6	03 59 54.	- 52 35 55.	24	16.5	GALAXY
SHAP401-5244.0	03 59 55.	- 52 52 19.	18	17.25	GALAXY
SHAP401-5312.4	03 59 55.	- 53 20 43.	18	16.5	GALAXY
SHAP401-5228.5	03 59 58.	- 52 36 49.	24	17.0	GALAXY
SHAP401-5230.1	03 59 58.	- 52 38 25.	24	16.75	GALAXY
SHAP401-5248.0	03 59 58.	- 52 56 19.	42	16.0	GALAXY
SHAP401-5301.9	03 59 58.	- 53 10 13.	24	16.75	GALAXY
SHAP401-5634.9	03 59 58.	- 56 43 13.	18	17.0	GALAXY
SHAP401-5242.0	03 59 59.	- 52 50 19.	36	15.0	GALAXY
HN 0262	03 59 59.	- 52 51			NEBULA
IC 2028	03 59 59.	- 52 51			NONSTELLAR OBJECT
HN 0263	03 59 59.	- 52 57			NEBULA
IC 2029	03 59 59.	- 52 57			NONSTELLAR OBJECT
SHAP401-5707.9	03 59 59.	- 57 16 13.	24	17.0	GALAXY
LBN 0839	04 00	+ 03 00	18900		BRIGHT NEBULA
LBN 0775	04 00	+ 26 30	2100		BRIGHT NEBULA
LBN 0736	04 00	+ 40 30	9900		BRIGHT NEBULA
LBN 0724	04 00	+ 43 50	1440		BRIGHT NEBULA
LBN 0709	04 00	+ 50 00	1920		BRIGHT NEBULA
LBN 0706	04 00	+ 50 50	1620		BRIGHT NEBULA
LBN 0705	04 00	+ 51 10	1500		BRIGHT NEBULA
ZWG 466.008	04 00 00.	+ 21 20		15.5	GALAXY
LDN 1484	04 00 00.	+ 29 00	2700		DARK NEBULA
LDN 1447	04 00 00.	+ 40 40	3240		DARK NEBULA
UGC 02935	04 00 00.	+ 83 20	66	17.	GALAXY Sc
LDN 1320	04 00 00.	+ 85 30	2760		DARK NEBULA
LB 01230	04 00 01.	+ 08 22 24.		13.3	FAINT BLUE STAR
SHAP401-5124.6	04 00 01.	- 51 32 54.	48	14.4	GALAXY
SHAP401-5241.8	04 00 01.	- 52 50 07.	18	17.0	GALAXY
SHAP401-5408.7	04 00 01.	- 54 17 01.	30	16.5	GALAXY
SHAP400-5910.6	04 00 02.	- 59 18 55.	24	17.5	GALAXY
SHAP401-5415.0	04 00 03.	- 54 23 19.	24	15.5	GALAXY
SHAP401-5814.4	04 00 03.	- 58 22 43.	12	17.5	GALAXY
LB 01231	04 00 04.	+ 23 31 42.		15.8	FAINT BLUE STAR
RNGC 1499	04 00 04.	+ 36 17			DIFFUSE NEBULA
SHAP401-5300.0	04 00 04.	- 53 08 18.	24	16.75	GALAXY
SHAP401-5518.2	04 00 04.	- 55 26 31.	30	17.0	GALAXY
SHAP401-5725.0	04 00 04.	- 57 33 19.	18	17.5	GALAXY
LB 01232	04 00 05.	+ 14 07 18.		16.6	FAINT BLUE STAR
SHAP401-5754.4	04 00 05.	- 58 02 43.	18	17.0	GALAXY
RNGC 1504	04 00 06.	- 09 27		15.0	GALAXY
MCG-02-11-008	04 00 06.	- 09 27	30	15.5	GALAXY
SHAP401-5230.4	04 00 06.	- 52 38 42.	24	17.0	GALAXY
DV.56 N1511A	04 00 06.	- 67 56	72		S GALAXY
RNGC 1511A	04 00 06.	- 67 56			GALAXY

OBJECT NAME	RIGHT ASCEN.	DECLINATION	DIAM.	MAGN.	TYPE OF OBJECT
SHAP401-5220.0	04 00 07.	- 52 28 18.	18	16.0	GALAXY
SHAP401-5416.2	04 00 07.	- 54 24 30.	36	16.5	GALAXY
SHAP401-5721.3	04 00 07.	- 57 29 37.	24	17.25	GALAXY
BAK 2.1034	04 00 09.	- 20 20 16.	17	16.9	GALAXY
SHAP401-5227.9	04 00 09.	- 52 36 12.	24	17.0	GALAXY
SHAP401-5255.8	04 00 09.	- 53 04 06.	24	16.75	GALAXY
SHAP401-5709.6	04 00 09.	- 57 17 54.	24	17.5	GALAXY
CED 026	04 00 10.	+ 36 16	9420		DIFFUSE GALACTIC NEBULA
SHAP401-5744.1	04 00 10.	- 57 52 24.	18	17.5	GALAXY
SHAP401-5902.8	04 00 11.	- 59 11 07.	30	17.5	GALAXY
SHAP401-5906.4	04 00 11.	- 59 14 43.	24	17.0	GALAXY
ZWG 392.001	04 00 12.	+ 01 49		15.7	GALAXY
UGC 02936	04 00 12.	+ 01 49	168	15.7	GALAXY Sc
KARA.73B 0138	04 00 12.	+ 01 49	204	15.7	ISOLATED GALAXY S
UGC 02937	04 00 12.	+ 46 35	84	16.0	GALAXY S0-a
MCG+08-08-001	04 00 12.	+ 46 35	72	16.	GALAXY
MCG+01-11-001	04 00 12.	- 04 42	48	15.	GALAXY
MCG-03-11-008	04 00 12.	- 18 11	36	14.5	GALAXY
SHAP401-5506.9	04 00 12.	- 55 15 12.	42	16.5	GALAXY
SHAP401-5910.2	04 00 12.	- 59 18 31.	30	17.5	GALAXY
BAK 2.1035	04 00 13.	- 19 41 40.	22	16.0	GALAXY
SHAP401-5901.5	04 00 13.	- 59 09 48.	36	17.0	GALAXY
SHAP401-5110.1	04 00 14.	- 51 18 24.	24	16.25	GALAXY
MCG+00-11-007	04 00 15.	+ 01 50	144	14.	GALAXY
SHAP401-5758.8	04 00 15.	- 58 07 06.	24	16.75	GALAXY
SHAP401-5908.1	04 00 16.	- 59 16 24.	24	17.0	GALAXY
ZWG 487.014	04 00 18.	+ 26 52		15.7	GALAXY
RNGC 1505	04 00 18.	- 09 26		15.0	GALAXY
MCG-02-11-009	04 00 18.	- 09 26	36	15.	GALAXY
BAK 2.1036	04 00 19.	- 22 26 28.	10	15.7	GALAXY
LB 01233	04 00 20.	+ 24 00 24.		16.4	FAINT BLUE STAR
SG 2.007	04 00 21.	+ 51 09	1440		DIFFUSE EMISSION NEBULA
SHAP401-5218.4	04 00 21.	- 52 26 41.	30	17.25	GALAXY
SHAP401-5204.2	04 00 22.	- 52 12 29.	24	16.75	GALAXY
SHAP401-5517.2	04 00 22.	- 55 25 29.	18	17.0	GALAXY
SHAP401-5905.1	04 00 22.	- 59 13 24.	30	17.25	GALAXY
SHAP401-5444.9	04 00 23.	- 54 53 11.	42	15.75	GALAXY
SHAP401-5648.2	04 00 23.	- 56 56 30.	24	17.25	GALAXY
SHAP401-5830.0	04 00 23.	- 58 38 18.	24	17.5	GALAXY
ZWG 418.008	04 00 24.	+ 04 25		15.7	GALAXY
UGC 02938	04 00 24.	+ 04 25	78	15.7	GALAXY SBa-b
MCG+01-11-008	04 00 24.	+ 04 25	60	15.	GALAXY
VB 248	04 00 24.	+ 56 52	436		STELLAR RING
SHAP401-5720.2	04 00 24.	- 57 28 30.	18	17.5	GALAXY
BAK 2.1037	04 00 25.	- 21 41 03.	15	16.0	GALAXY
SHAP401-5038.6	04 00 25.	- 50 46 53.	30	17.25	GALAXY
SHAP401-5220.6	04 00 25.	- 52 28 53.	30	16.5	GALAXY
SHAP401-5548.6	04 00 25.	- 55 56 53.	24	17.0	GALAXY
SHAP401-5647.0	04 00 25.	- 56 55 17.	24	16.75	GALAXY
SHAP401-5657.1	04 00 26.	- 57 05 23.	42	16.0	GALAXY
SHAP401-5039.1	04 00 27.	- 50 47 23.	24	16.75	GALAXY
SHAP401-5643.1	04 00 27.	- 56 51 23.	66	16.0	GALAXY
SHAP401-5659.0	04 00 27.	- 57 07 17.	24	16.75	GALAXY
SHAP401-5848.4	04 00 27.	- 58 56 42.	24	17.25	GALAXY
BAK 2.1038	04 00 28.	- 19 04 27.	34	16.4	GALAXY
SHAP401-5850.5	04 00 28.	- 58 58 47.	36	16.75	GALAXY
SHAP401-5917.6	04 00 28.	- 59 25 54.	24	17.0	GALAXY
SHAP401-5746.4	04 00 29.	- 57 54 41.	24	18.0	GALAXY
ZC 0400.5+1335	04 00 30.	+ 13 35	7860		CLUSTER OF GALAXIES
ZWG 487.015	04 00 30.	+ 26 13		14.9	GALAXY
5ZW 367	04 00 30.	+ 32 00			COMPACT GALAXY
UGC 02939	04 00 30.	+ 46 05	84	17.	GALAXY Sc-IRR
LDN 1389	04 00 30.	+ 56 46	180		DARK NEBULA
SHAP401-5144.8	04 00 30.	- 51 53 05.	18	16.25	GALAXY
BAK 2.1040	04 00 31.	- 18 13 21.	13	16.1	GALAXY
BAK 2.1039	04 00 31.	- 20 29 39.	17	17.0	GALAXY
SHAP401-5439.0	04 00 32.	- 54 47 17.	30	16.5	GALAXY
SHAP401-5908.8	04 00 32.	- 59 17 05.	24	17.25	GALAXY
SHAP401-5042.6	04 00 33.	- 50 50 52.	30	16.75	GALAXY
SHAP401-5522.4	04 00 33.	- 55 30 41.	30	17.0	GALAXY
SHAP401-5605.0	04 00 33.	- 56 13 17.	24	17.25	GALAXY
SHAP401-5334.5	04 00 34.	- 53 42 47.	36	16.0	GALAXY
SHAP401-5647.4	04 00 34.	- 56 55 41.	24	17.0	GALAXY
RNGC 1496	04 00 35.	+ 52 29		9.5	OPEN CLUSTER
5ZW 368	04 00 36.	+ 23 51			COMPACT GALAXY
MCG+04-10-015	04 00 36.	+ 26 14	45	15.	GALAXY
OCL 0396	04 00 36.	+ 52 29	390	9.6	OPEN STAR CLUSTER
MCG+00-11-008	04 00 36.	- 02 45 30.	30	16.	GALAXY
MCG-02-11-010	04 00 36.	- 09 10	60	15.	GALAXY
BAK 2.1041	04 00 36.	- 21 15 21.	45	14.6	GALAXY
SHAP401-5552.6	04 00 37.	- 56 00 53.	24	17.25	GALAXY
SHAP401-5808.9	04 00 38.	- 58 17 11.	18	17.5	GALAXY
MCG+04-10-016	04 00 39.	+ 22 00 30.	72	14.5	GALAXY
SHAP401-5605.8	04 00 39.	- 56 14 04.	24	17.5	GALAXY
SG 3.020	04 00 40.	+ 36 11	12000		DIFFUSE EMISSION NEBULA
MCG-04-10-012	04 00 42.	- 21 16	48	15.	GALAXY
TON-S 0395	04 00 42.	- 29 39		15.5	BLUE STAR
DV.56 W1511B	04 00 42.	- 67 45	84		S GALAXY
RNGC 1511B	04 00 42.	- 67 45			GALAXY
SHAP402-5220.0	04 00 43.	- 52 28 16.	30	17.0	GALAXY
SHAP401-5609.8	04 00 43.	- 56 18 04.	24	17.5	GALAXY
SHAP401-5647.4	04 00 43.	- 56 55 40.	18	16.75	GALAXY
SHAP401-5734.0	04 00 44.	- 57 42 16.	24	18.0	GALAXY
MCG+03-11-006	04 00 45.	+ 19 45	48	15.5	GALAXY
MCG+05-10-005	04 00 45.	+ 30 42	36	15.	GALAXY
SHAP402-5258.2	04 00 45.	- 53 06 28.	36	16.75	GALAXY
SHAP401-5656.4	04 00 46.	- 57 04 40.	18	16.75	GALAXY
SHAP401-5852.0	04 00 46.	- 59 00 16.	24	17.25	GALAXY
IC 0358	04 00 47.	+ 19 46 15.			NONSTELLAR OBJECT
SHAP402-5206.2	04 00 47.	- 52 14 28.	30	16.75	GALAXY
SHAP401-5757.9	04 00 47.	- 58 06 10.	18	18.0	GALAXY
ZWG 466.009	04 00 48.	+ 19 45		15.3	GALAXY
UGC 02940	04 00 48.	+ 19 45	84	15.3	GALAXY S0
ZWG 487.016	04 00 48.	+ 22 01		14.3	GALAXY
UGC 02941	04 00 48.	+ 22 01	78	14.3	GALAXY SBb
ZWG 508.006	04 00 48.	+ 30 42		15.1	GALAXY
SHAP401-5707.3	04 00 48.	- 57 15 34.	18	17.25	GALAXY
IC 0357	04 00 49.	+ 22 01 09.			NONSTELLAR OBJECT
SHAP401-5605.0	04 00 49.	- 56 13 16.	18	17.75	GALAXY
SHAP401-5624.2	04 00 50.	- 56 32 28.	24	16.5	GALAXY
MCG+04-10-017	04 00 51.	+ 24 15	30	15.5	GALAXY
SHAP402-5534.3	04 00 52.	- 55 42 34.	24	17.0	GALAXY
BAK 2.1042	04 00 53.	- 21 37 08.	15	16.4	GALAXY
ZWG 487.017	04 00 54.	+ 24 15		15.6	GALAXY
MCG-02-11-011	04 00 54.	- 14 26 30.	36	15.	GALAXY
SHAP402-5303.1	04 00 58.	- 53 11 21.	42	17.0	GALAXY
SHAP401-5810.1	04 00 58.	- 58 18 22.	30	18.0	GALAXY
LB 01234	04 01 00.	+ 23 48 24.		16.9	FAINT BLUE STAR
SHAP402-5054.5	04 01 00.	- 51 02 45.	24	15.75	GALAXY
SHAP402-5244.4	04 01 01.	- 52 52 39.	24	16.5	GALAXY
SHAP402-5744.6	04 01 01.	- 57 52 51.	54	16.75	GALAXY
BC PKS0402-362	04 01 02.2	- 36 13 16.		16.	QUASI-STELLAR OBJECT
LB 01235	04 01 03.	+ 12 41 24.		17.0	FAINT BLUE STAR
BAK 2.1043	04 01 03.	- 20 25 25.	15	16.8	GALAXY
BAK 2.1044	04 01 03.	- 20 31 37.	17	16.0	GALAXY
SHAP402-5243.8	04 01 03.	- 52 52 03.	30	16.75	GALAXY
SHAP402-5739.8	04 01 03.	- 57 48 03.	24	17.5	GALAXY
BAK 2.1045	04 01 04.	- 20 01 31.	18	16.9	GALAXY
SHAP402-5431.3	04 01 04.	- 54 39 33.	48	17.0	GALAXY
SHAP402-5907.9	04 01 04.	- 59 16 09.	24	16.5	GALAXY
LB 01236	04 01 06.	+ 09 48 24.		15.7	FAINT BLUE STAR
MCG+04-10-018	04 01 06.	+ 21 59	72	15.5	GALAXY
DG 026	04 01 06.	+ 26 12	1680		REFLECTION NEBULA
ZC 0401.1+3247	04 01 06.	+ 32 47	3230		CLUSTER OF GALAXIES
ZC 0401.1-0135	04 01 06.	- 01 35	1010		CLUSTER OF GALAXIES
SHAP402-5056.1	04 01 06.	- 51 04 20.	24	16.25	GALAXY
BAK 2.1046	04 01 07.	- 20 45 13.	16	16.3	GALAXY
SS 13	04 01 08.	+ 26 12			DIFFUSE GALACTIC NEBULA
ZWG 487.018	04 01 12.	+ 22 00		15.7	GALAXY
UGC 02943	04 01 12.	+ 22 00	102	15.7	GALAXY
UGC 02942	04 01 12.	+ 22 00	84	15.7	GALAXY S
MCG-05-10-008	04 01 12.	- 30 56	30	15.	GALAXY
SHAP402-5652.6	04 01 12.	- 57 00 51.	24	16.75	GALAXY
SHAP402-5737.0	04 01 12.	- 57 45 15.	24	16.5	GALAXY
SHAP402-5403.5	04 01 13.	- 54 11 44.	36	16.75	GALAXY
BAK 2.1048	04 01 14.	- 18 06 48.	17	16.1	GALAXY
BAK 2.1047	04 01 15.	- 22 01 24.	19	15.4	GALAXY
SHAP402-5913.4	04 01 15.	- 59 21 39.	24	16.5	GALAXY
SHAP402-5410.0	04 01 16.	- 54 18 14.	24	16.5	GALAXY
SHAP402-5708.9	04 01 16.	- 57 17 08.	24	16.75	GALAXY
LB 01237	04 01 17.	+ 23 28 54.		16.5	FAINT BLUE STAR
SHAP402-5033.8	04 01 17.	- 50 42 02.	30	16.75	GALAXY
VB 249	04 01 18.	+ 54 56	282		STELLAR RING
TON-S 0396	04 01 18.	- 23 57		14.8	BLUE STAR
TON-S 0397	04 01 18.	- 29 30		13.6	BLUE STAR
SHAP402-5454.1	04 01 18.	- 55 02 20.	30	17.0	GALAXY
SHAP402-5807.0	04 01 18.	- 58 15 14.	24	17.5	GALAXY
SHAP402-5908.0	04 01 18.	- 59 16 14.	36	16.25	GALAXY
BAK 2.1049	04 01 19.	- 20 39 48.	11	16.8	GALAXY
SHAP402-5425.1	04 01 19.	- 54 33 20.	30	15.5	GALAXY
SHAP402-5645.5	04 01 19.	- 56 53 44.	24	17.25	GALAXY
SHAP402-5309.0	04 01 21.	- 53 17 14.	30	16.25	GALAXY
SHAP402-5822.0	04 01 21.	- 58 30 14.	18	17.25	GALAXY
SHAP402-5213.2	04 01 22.	- 52 21 25.	30	17.25	GALAXY
SHAP402-5617.2	04 01 22.	- 56 25 26.	24	17.0	GALAXY
SHAP402-5811.6	04 01 22.	- 58 19 50.	24	17.75	GALAXY
SHAP402-5103.0	04 01 23.	- 51 11 13.	24	17.0	GALAXY
SHAP402-5830.0	04 01 23.	- 58 38 14.	24	17.5	GALAXY
SHAP402-6001.2	04 01 23.	- 60 09 26.	36	17.0	GALAXY
UGC 02944	04 01 24.	+ 33 10	102	17.	GALAXY S
ZWG 526.017	04 01 24.	+ 33 40		15.2	GALAXY
UGC 02945	04 01 24.	+ 33 40	96	15.2	GALAXY S0-a
ARC 0472	04 01 24.	- 17 14		17.5	RICH CLUSTER OF GALAXIES
SHAP402-5349.4	04 01 24.	- 53 57 37.	24	16.5	GALAXY
LB 01238	04 01 25.	+ 18 55 24.		16.7	FAINT BLUE STAR
SHAP402-5856.2	04 01 25.	- 59 04 26.	18	17.25	GALAXY
LB 01239	04 01 28.	+ 18 54 54.		16.7	FAINT BLUE STAR
BAK 2.1050	04 01 28.	- 20 50 59.	19	16.4	GALAXY
SHAP402-5342.4	04 01 28.	- 53 50 37.	18	17.0	GALAXY
SHAP402-5110.2	04 01 29.	- 51 18 25.	30	16.5	GALAXY
UGC 02946	04 01 30.	+ 25 50	66	16.5	GALAXY
MCG+06-09-017	04 01 30.	+ 33 39 30.	36	14.5	GALAXY
ISS 0112	04 01 30.	+ 54 59	284		STELLAR RING
MCG-02-11-012	04 01 30.	- 11 20	30	15.5	GALAXY
SHAP402-5632.5	04 01 30.	- 56 40 43.	18	17.25	GALAXY
SHAP402-5858.0	04 01 31.	- 59 06 14.	24	17.5	GALAXY
BAK 2.1051	04 01 32.	- 20 43 59.	15	16.9	GALAXY
SHAP402-5629.6	04 01 32.	- 56 37 49.	18	17.0	GALAXY
SHAP402-5800.2	04 01 32.	- 58 08 25.	30	17.75	GALAXY
LB 01240	04 01 33.	+ 25 01 24.		14.8	FAINT BLUE STAR
B 207	04 01 33.	+ 26 13			DARK OBJECT
IC 2026	04 01 33.	- 11 19			NONSTELLAR OBJECT
SHAP402-5747.9	04 01 33.	- 57 56 07.	18	18.0	GALAXY
SHAP402-5757.6	04 01 33.	- 58 05 49.	60	17.0	GALAXY
SHAP402-5321.0	04 01 35.	- 53 29 13.	36	17.0	GALAXY
SHAP402-5523.6	04 01 35.	- 55 31 49.	30	17.25	GALAXY
SHAP402-5632.6	04 01 35.	- 56 40 49.	24	18.0	GALAXY
SHAP402-5644.1	04 01 35.	- 56 52 19.	18	17.5	GALAXY
SHAP402-5851.1	04 01 35.	- 58 59 19.	66	17.25	GALAXY
SHAP402-5942.3	04 01 35.	- 59 50 31.	24	17.5	GALAXY
LDN 1491	04 01 36.	+ 26 09	240		DARK NEBULA
MCG-02-11-013	04 01 36.	- 11 20	48	15.	GALAXY
SHAP402-5634.4	04 01 36.	- 56 42 37.	24	17.25	GALAXY
LB 01241	04 01 37.	+ 07 52 48.		15.4	FAINT BLUE STAR
RNGC 1509	04 01 37.	- 11 20		15.0	GALAXY
SHAP402-5545.2	04 01 38.	- 55 53 25.	24	17.25	GALAXY
SHAP402-5659.5	04 01 38.	- 57 07 43.	24	17.75	GALAXY
SHAP402-5617.9	04 01 40.	- 56 26 07.	24	17.5	GALAXY
SHAP402-5810.9	04 01 40.	- 58 19 07.	18	17.25	GALAXY
SHAP403-5125.6	04 01 41.	- 51 33 48.	18	17.25	GALAXY
SHAP402-5631.7	04 01 41.	- 56 39 55.	36	16.5	GALAXY
LDN 1489	04 01 42.	+ 26 20	660		DARK NEBULA
SHAP403-5249.6	04 01 43.	- 52 57 48.	18	16.25	GALAXY
RNGC 1507	04 01 44.	- 02 20		13.0	GALAXY
SHAP403-5317.4	04 01 45.	- 53 25 36.	24	16.0	GALAXY
SHAP402-5744.8	04 01 45.	- 57 53 01.	18	17.5	GALAXY
SHAP402-5427.4	04 01 46.	- 54 35 36.	30	15.5	GALAXY
LB 01242	04 01 47.	+ 14 53 24.		15.6	FAINT BLUE STAR
ZC 0401.8+0219	04 01 48.	+ 02 19	810		CLUSTER OF GALAXIES
SHAP403-5230.2	04 01 48.	- 52 38 24.	24	17.0	GALAXY
SHAP403-5313.0	04 01 48.	- 53 21 12.	30	16.5	GALAXY
SHAP402-5932.7	04 01 48.	- 59 40 55.	30	16.0	GALAXY
LB 00222	04 01 50.	+ 13 14 30.		16.6	FAINT BLUE STAR
SHAP403-5039.4	04 01 50.	- 50 47 36.	30	17.25	GALAXY
SHAP403-5315.2	04 01 50.	- 53 23 24.	24	16.5	GALAXY
BAK 2.1053	04 01 51.	- 17 41 46.	14	16.7	GALAXY
SHAP403-5524.6	04 01 51.	- 55 32 48.	24	17.25	GALAXY
SHAP402-5754.5	04 01 51.	- 58 02 42.	48	18.0	GALAXY
BAK 2.1052	04 01 52.	- 20 57 46.	12	16.4	GALAXY
LB 01243	04 01 53.	+ 13 24 30.		16.6	FAINT BLUE STAR
RNGC 1510	04 01 53.	- 43 33			GALAXY
ZWG 392.002	04 01 54.	- 02 19		13.1	GALAXY
UGC 02947	04 01 54.	- 02 19	216	13.1	GALAXY
KARA.72 097B	04 01 54.	- 02 19	84		PART OF DOUBLE GALAXY
KARA.73B 0139	04 01 54.	- 02 19	246	13.1	ISOLATED GALAXY S
KARA.72 097A	04 01 54.	- 02 20	150	13.1	PART OF DOUBLE GALAXY

OBJECT NAME	RIGHT ASCEN.	DECLINATION	DIAM.	MAGN.	TYPE OF OBJECT
MCG+00-11-009	04 01 54.	- 02 20	204	12.9	GALAXY
LB 01701	04 01 54.	- 46 00		14.8	FAINT BLUE STAR
SHAP402-5634.6	04 01 54.	- 56 42 48.	18	16.75	GALAXY
SHAP403-5224.5	04 01 57.	- 52 32 41.	18	16.75	GALAXY
SHAP403-5311.6	04 01 59.	- 53 19 47.	24	17.25	GALAXY
SHAP403-5354.5	04 01 59.	- 54 02 41.	24	17.0	GALAXY
LBN 0777	04 02	+ 26 15	1140		BRIGHT NEBULA
ZWG 466.010	04 02 00.	+ 20 41		15.7	GALAXY
UGC 02948	04 02 00.	+ 20 41	84	15.7	GALAXY Sa-b
ZWG 487.019	04 02 00.	+ 25 07		14.6	GALAXY
UGC 02949	04 02 00.	+ 25 07	72	14.6	GALAXY SB:a-b
UGC 02950	04 02 00.	+ 30 54	114	17.	GALAXY S
MAI 005	04 02 00.	+ 89 05	67		DWARF SPHEROIDAL GALAXY
TON-S 0398	04 02 00.	- 29 20		13.5	BLUE STAR
SHAP403-5214.8	04 02 00.	- 52 22 59.	30	17.0	GALAXY
SHAP403-5215.6	04 02 00.	- 52 23 47.	48	16.75	GALAXY
LB 01702	04 02 00.	- 54 50		13.1	FAINT BLUE STAR
SHAP403-5628.2	04 02 00.	- 56 36 23.	18	17.5	GALAXY
BAK 2.1054	04 02 01.	- 20 45 21.	19	16.4	GALAXY
SHAP403-5051.5	04 02 01.	- 50 59 41.	30	17.0	GALAXY
SHAP403-5243.0	04 02 01.	- 52 51 11.	24	16.0	GALAXY
SHAP403-5600.0	04 02 01.	- 56 08 11.	24	17.25	GALAXY
MCG+04-10-019	04 02 03.	+ 25 07 30.	45	14.5	GALAXY
BAK 2.1055	04 02 03.	- 19 33 09.	16	15.7	GALAXY
SHAP403-5258.0	04 02 03.	- 53 06 11.	18	16.25	GALAXY
SHAP403-5557.9	04 02 03.	- 56 06 05.	24	17.25	GALAXY
SHAP403-5220.5	04 02 04.	- 52 28 41.	18	17.0	GALAXY
MCG+00-11-010	04 02 06.	+ 02 18	48	15.	GALAXY
MCG-02-11-014	04 02 06.	- 10 20	48	15.	GALAXY
SHAP403-5658.1	04 02 06.	- 57 06 17.	24	17.25	GALAXY
SHAP403-5824.8	04 02 08.	- 58 32 59.	24	17.25	GALAXY
SHAP403-5237.1	04 02 09.	- 52 45 16.	24	17.0	GALAXY
SHAP403-5748.5	04 02 09.	- 57 56 41.	24	18.0	GALAXY
ARC 0473	04 02 11.	- 17 36		17.7	RICH CLUSTER OF GALAXIES
BAK 2.1056	04 02 11.	- 20 20 33.	20	16.8	GALAXY
SHAP403-5236.8	04 02 11.	- 52 44 58.	24	17.0	GALAXY
SHAP403-5248.3	04 02 11.	- 52 56 28.	24	16.5	GALAXY
SHAP403-5552.9	04 02 11.	- 56 01 05.	24	17.5	GALAXY
UGC 02951	04 02 12.	+ 02 18	60	17.	GALAXY Sc
5ZW 369	04 02 12.	+ 51 18			COMPACT GALAXY
RLWT 089	04 02 12.	+ 54 19		15.5	FAINT VERY BLUE STAR
MCG-02-11-015	04 02 12.	- 10 27	48	15.5	GALAXY
MCG-03-11-009	04 02 12.	- 18 01	72	15.5	GALAXY
BAK 2.1057	04 02 12.	- 20 56 09.	12	16.2	GALAXY
SHAP403-5146.1	04 02 12.	- 51 54 16.	42	15.5	GALAXY
SHAP403-5625.3	04 02 12.	- 56 33 29.	24	17.0	GALAXY
SHAP403-5755.8	04 02 12.	- 58 03 59.	18	17.5	GALAXY
SHAP403-5100.0	04 02 13.	- 51 08 10.	18	16.25	GALAXY
SHAP403-5645.4	04 02 14.	- 56 53 35.	24	18.0	GALAXY
LB 01244	04 02 15.	+ 12 56 24.		16.5	FAINT BLUE STAR
BAK 2.1058	04 02 15.	- 18 02 08.	25	15.4	GALAXY
SHAP403-5007.2	04 02 16.	- 50 15 22.	36	16.75	GALAXY
SHAP403-5630.0	04 02 16.	- 56 38 10.	24	17.25	GALAXY
RNGC 1512	04 02 17.	- 43 29		11.5	GALAXY
SHAP403-5743.8	04 02 17.	- 57 51 58.	42	16.25	GALAXY
TON-S 0399	04 02 18.	- 30 21		14.8	BLUE STAR
SHAP403-5422.4	04 02 18.	- 54 30 34.	36	17.0	GALAXY
SHAP403-5623.6	04 02 18.	- 56 31 46.	48	17.0	GALAXY
SHAP403-5151.8	04 02 20.	- 51 59 58.	18	16.5	GALAXY
MCG+04-10-020	04 02 21.	+ 22 06	36	16.	GALAXY
BAK 2.1058A	04 02 21.	- 21 57 08.	30	16.1	GALAXY
SHAP403-5005.8	04 02 21.	- 50 13 57.	36	17.0	GALAXY
SHAP403-5844.0	04 02 21.	- 58 52 10.	24	17.25	GALAXY
LB 00223	04 02 22.	+ 19 08 36.		17.0	FAINT BLUE STAR
BAK 2.1059	04 02 22.	- 21 12 20.	22	16.4	GALAXY
SHAP403-5049.8	04 02 22.	- 50 57 57.	18	17.25	GALAXY
SHAP403-5624.6	04 02 22.	- 56 32 46.	24	16.5	GALAXY
SHAP403-5813.1	04 02 22.	- 58 21 16.	24	17.5	GALAXY
SHAP403-5548.1	04 02 23.	- 55 56 16.	24	17.25	GALAXY
SHAP403-5643.4	04 02 23.	- 56 51 34.	36	17.0	GALAXY
MCG+12-04-011	04 02 24.	+ 69 40	210	13.	GALAXY
SHAP403-5138.3	04 02 24.	- 51 46 27.	18	16.75	GALAXY
SHAP403-5730.3	04 02 24.	- 57 38 28.	24	17.75	GALAXY
SHAP403-5005.0	04 02 25.	- 50 13 09.	42	16.75	GALAXY
SHAP403-5719.6	04 02 25.	- 57 27 46.	24	17.5	GALAXY
SHAP403-5922.1	04 02 27.	- 59 30 16.	24	16.5	GALAXY
SHAP403-5339.3	04 02 28.	- 53 47 27.	48	16.5	GALAXY
SHAP403-5403.8	04 02 28.	- 54 11 57.	30	17.0	GALAXY
LB 01245	04 02 29.	+ 14 51 30.		17.0	FAINT BLUE STAR
SHAP403-5145.9	04 02 29.	- 51 54 03.	18	16.5	GALAXY
SHAP403-5406.8	04 02 29.	- 54 14 57.	54	16.0	GALAXY
SHAP403-5758.9	04 02 29.	- 58 07 04.	30	16.5	GALAXY
LB 01246	04 02 30.	+ 14 46 00.		17.1	FAINT BLUE STAR
BAK 2.1060	04 02 30.	- 19 26 43.	15	16.8	GALAXY
RNGC 1515A	04 02 30.	- 54 14			GALAXY
SHAP403-5437.0	04 02 30.	- 54 45 09.	30	16.25	GALAXY
ARP 213	04 02 31.	+ 69 42			PECULIAR GALAXY
SHAP403-5843.0	04 02 32.	- 58 51 10.	24	17.25	GALAXY
IC 0356	04 02 33.	+ 69 40 52.	60	13.28	GALAXY Sa
SHAP403-5235.8	04 02 33.	- 52 43 57.	24	17.0	GALAXY
SHAP403-5248.4	04 02 33.	- 52 56 33.	24	16.5	GALAXY
SHAP403-5440.0	04 02 33.	- 54 48 09.	42	17.0	GALAXY
SHAP403-5621.5	04 02 34.	- 56 29 39.	24	16.75	GALAXY
3ZW 058	04 02 36.	+ 19 19			COMPACT GALAXY
UGC 02952	04 02 36.	+ 36 56	78	16.0	GALAXY S
ZWG 327.015	04 02 36.	+ 69 41		13.3	GALAXY
UGC 02953	04 02 36.	+ 69 41	300	13.3	GALAXY Sb-c
SHAP403-5403.2	04 02 37.	- 54 11 21.	18	17.0	GALAXY
SHAP403-5636.0	04 02 37.	- 56 44 09.	30	16.5	GALAXY
SHAP403-5944.4	04 02 37.	- 59 52 33.	42	17.0	GALAXY
SHAP403-5236.0	04 02 38.	- 52 44 09.	24	16.0	GALAXY
MCG+06-09-018	04 02 39.	+ 36 56	60	15.5	GALAXY
RNGC 1501	04 02 39.	+ 60 47		13.5	PLANETARY NEBULA
SHAP403-5320.9	04 02 39.	- 53 29 03.	24	17.0	GALAXY
SHAP403-5617.2	04 02 39.	- 56 25 21.	18	17.5	GALAXY
SHAP403-5712.6	04 02 39.	- 57 20 45.	18	16.75	GALAXY
SHAP403-5802.1	04 02 40.	- 58 10 15.	24	18.0	GALAXY
PK144+06.1	04 02 40.76	+ 60 47 10.5	60	13.3	PLANETARY NEBULA
SHAP404-5032.1	04 02 41.	- 50 40 14.	30	16.25	GALAXY
SHAP404-5204.7	04 02 41.	- 52 12 50.	18	16.5	GALAXY
SHAP403-5235.0	04 02 41.	- 52 43 08.	18	16.0	GALAXY
ZWG 487.020	04 02 42.	+ 25 21		15.3	GALAXY
BAK 2.1061	04 02 42.	- 19 34 19.	17	16.9	GALAXY
RNGC 1515	04 02 42.	- 54 14		12.0	GALAXY
SHAP403-5733.1	04 02 42.	- 57 41 15.	24	17.25	GALAXY
SHAP403-5756.2	04 02 42.	- 58 04 21.	24	17.25	GALAXY
HN 0264	04 02 43.	- 19 22			NEBULA
SHAP403-5307.2	04 02 43.	- 53 15 20.	24	17.0	GALAXY
SHAP403-5713.0	04 02 43.	- 57 21 09.	18	17.0	GALAXY
IC 2030	04 02 44.	- 19 22			NONSTELLAR OBJECT
SHAP403-5648.9	04 02 44.	- 56 57 03.	30	17.75	GALAXY
SHAP403-5628.6	04 02 46.	- 56 36 45.	24	18.0	GALAXY
RNGC 1508	04 02 47.	+ 25 16		15.0	GALAXY
SHAP403-5532.2	04 02 47.	- 55 40 20.	48	17.0	GALAXY
ZWG 487.021	04 02 48.	+ 25 16		15.2	GALAXY
MCG+04-10-021	04 02 48.	+ 25 16	24	14.5	GALAXY
SHAP404-5219.4	04 02 48.	- 52 27 32.	24	16.5	GALAXY
SHAP403-5800.1	04 02 48.	- 58 08 15.	24	17.75	GALAXY
SHAP404-5406.3	04 02 49.	- 54 14 26.	72	12.1	GALAXY
SHAP403-5526.5	04 02 49.	- 55 34 38.	30	17.25	GALAXY
SHAP403-5738.8	04 02 49.	- 57 46 56.	24	17.0	GALAXY
SHAP403-5616.0	04 02 50.	- 56 24 08.	24	17.0	GALAXY
SHAP403-5658.2	04 02 50.	- 57 06 20.	36	17.25	GALAXY
MCG+01-11-010	04 02 51.	+ 04 16	36	15.	GALAXY
MCG+01-11-009	04 02 51.	+ 04 18	48	15.	GALAXY
SHAP403-5742.2	04 02 51.	- 57 50 20.	18	17.75	GALAXY
SHAP404-5556.2	04 02 53.	- 56 04 20.	24	16.5	GALAXY
ZWG 418.009	04 02 54.	+ 04 17		15.4	GALAXY
KARA.72 098B	04 02 54.	+ 04 17	36	15.4	PART OF DOUBLE GALAXY
ZWG 418.010	04 02 54.	+ 04 19		15.4	GALAXY
UGC 02954	04 02 54.	+ 04 19	60	15.4	GALAXY Sc
KARA.72 098A	04 02 54.	+ 04 19	54	15.4	PART OF DOUBLE GALAXY
LB 01247	04 02 54.	+ 21 11 00.		16.7	FAINT BLUE STAR
ZC 0402.9-0031	04 02 54.	- 00 31	1010		CLUSTER OF GALAXIES
SHAP404-5125.1	04 02 54.	- 51 33 14.	30	16.0	GALAXY
SHAP404-5256.8	04 02 54.	- 53 04 56.	24	16.75	GALAXY
SHAP404-5552.5	04 02 55.	- 56 00 38.	24	16.5	GALAXY
SHAP404-5632.0	04 02 55.	- 56 40 08.	24	18.0	GALAXY
SHAP404-5050.8	04 02 57.	- 50 58 55.	24	16.75	GALAXY
SHAP404-5716.9	04 02 57.	- 57 25 02.	24	17.5	GALAXY
SHAP404-5341.5	04 02 58.	- 53 49 37.	24	16.5	GALAXY
SHAP404-5556.8	04 02 58.	- 56 04 56.	24	17.0	GALAXY
SHAP404-5406.8	04 02 59.	- 54 14 55.	30	16.75	GALAXY
SHAP404-5434.4	04 02 59.	- 54 42 31.	36	16.5	GALAXY
KHAV 044	04 03	+ 54 50	8340		DARK NEBULA
RNGC 1502	04 03 00.	+ 62 11		5.5	OPEN CLUSTER
LB 01703	04 03 00.	- 51 54		14.7	FAINT BLUE STAR
SHAP403-5828.6	04 03 01.	- 58 36 44.	24	17.0	GALAXY
SHAP404-5559.5	04 03 02.	- 56 07 37.	24	16.75	GALAXY
SHAP404-5334.4	04 03 03.	- 53 42 31.	24	16.5	GALAXY
LB 01248	04 03 04.	+ 14 29 00.		15.6	FAINT BLUE STAR
SHAP404-5352.0	04 03 04.	- 54 00 07.	36	16.25	GALAXY
SHAP404-5544.0	04 03 05.	- 55 52 07.	24	17.0	GALAXY
SHAP404-5633.4	04 03 05.	- 56 41 31.	24	17.25	GALAXY
ZC 0403.1+3040	04 03 06.	+ 30 40	12370		CLUSTER OF GALAXIES
VB 250	04 03 06.	+ 56 12	383		STELLAR RING
SHAP404-5910.8	04 03 07.	- 59 18 55.	24	16.5	GALAXY
SHAP404-5609.8	04 03 08.	- 56 17 55.	24	16.75	GALAXY
LB 01249	04 03 09.	+ 20 40 36.		15.6	FAINT BLUE STAR
SHAP404-5131.7	04 03 09.	- 51 39 49.	24	16.25	GALAXY
SHAP404-5524.4	04 03 09.	- 55 32 31.	36	17.0	GALAXY
SHAP404-5813.8	04 03 09.	- 58 21 55.	24	17.5	GALAXY
LB 01251	04 03 11.	+ 14 30 24.		16.2	FAINT BLUE STAR
LB 01250	04 03 11.	+ 22 37 30.		16.3	FAINT BLUE STAR
SHAP404-5315.2	04 03 11.	- 53 23 19.	24	16.0	GALAXY
SHAP404-5355.2	04 03 11.	- 54 03 19.	30	17.0	GALAXY
SHAP404-5704.8	04 03 11.	- 57 12 55.	18	17.0	GALAXY
UGC 02955	04 03 12.	+ 69 32	78	17.	GALAXY S
SHAP404-5432.0	04 03 12.	- 54 40 07.	30	15.75	GALAXY
SHAP404-5758.2	04 03 12.	- 58 06 19.	24	17.5	GALAXY
SHAP404-5852.6	04 03 13.	- 59 00 43.	24	17.0	GALAXY
BC PKS0403-13	04 03 13.98	- 13 16 17.9		17.09	QUASI-STELLAR OBJECT
SHAP404-5757.2	04 03 14.	- 58 05 19.	18	17.5	GALAXY
SHB 080	04 03 14.0	- 13 16 16.		17.2	QUASI-STELLAR OBJECT
MCG+05-10-006	04 03 15.	+ 31 10	54	15.5	GALAXY
BAK 2.1062	04 03 16.	- 21 38 22.	26	15.8	GALAXY
SHAP404-5215.4	04 03 16.	- 52 23 30.	18	17.25	GALAXY
SHAP404-5138.4	04 03 17.	- 51 46 30.	24	16.5	GALAXY
SHAP404-5314.2	04 03 17.	- 53 22 18.	18	16.75	GALAXY
ZWG 508.007	04 03 18.	+ 31 09		15.7	GALAXY
UGC 02956	04 03 18.	+ 31 10	132	17.	GALAXY Sc
OCL 0383	04 03 18.	+ 62 12	900	6.1	OPEN STAR CLUSTER
SHAP404-5402.4	04 03 19.	- 54 10 30.	24	16.0	GALAXY
SHAP404-5530.8	04 03 19.	- 55 38 54.	36	16.5	GALAXY
SHAP404-5657.9	04 03 19.	- 57 06 00.	24	17.0	GALAXY
SHAP404-5607.6	04 03 20.	- 56 15 42.	30	16.75	GALAXY
SHAP404-5807.8	04 03 20.	- 58 15 55.	18	17.75	GALAXY
BAK 2.1063	04 03 22.	- 19 37 58.	10	16.7	GALAXY
IC 2027	04 03 23.	+ 37 00 29.			NONSTELLAR OBJECT
UGC 02957	04 03 24.	+ 36 59	72	16.0	GALAXY E
MCG+06-09-019	04 03 24.	+ 36 59	24	15.5	GALAXY
MCG-05-10-009	04 03 24.	- 30 22	18	16.	GALAXY
SHAP404-5306.0	04 03 24.	- 53 14 06.	18	16.75	GALAXY
SHAP404-5215.0	04 03 25.	- 52 23 06.	18	17.0	GALAXY
IC 2031	04 03 26.	- 05 45			NONSTELLAR OBJECT
MCG+05-10-007	04 03 27.	+ 30 15 30.	24	17.	GALAXY
LB 01252	04 03 29.	+ 12 04 48.		16.3	FAINT BLUE STAR
SHAP404-5144.2	04 03 29.	- 51 52 17.	48	16.0	GALAXY
OCL 0425	04 03 30.	+ 27 18			OPEN STAR CLUSTER
RLWT 090	04 03 30.	+ 51 58		11.5	FAINT VERY BLUE STAR
TON-S 0400	04 03 30.	- 24 15		15.4	BLUE STAR
SHAP404-5121.6	04 03 30.	- 51 29 41.	12	17.0	GALAXY
SHAP404-5338.6	04 03 30.	- 53 46 41.	18	17.5	GALAXY
SHAP407_5312.8	04 03 31.	+ 53 04 48.	24	17.0	GALAXY
SHAP404-5057.2	04 03 31.	- 51 05 17.	24	17.0	GALAXY
SHAP404-5819.1	04 03 31.	- 58 27 12.	24	17.75	GALAXY
SEY 003	04 03 32.	+ 70 31 47.		15.4	FAINT GALAXY
SHAP404-5322.3	04 03 33.	- 53 30 23.	24	17.5	GALAXY
SHAP404-5542.6	04 03 33.	- 55 50 41.	42	16.25	GALAXY
SHAP404-5543.2	04 03 33.	- 55 51 17.	36	16.25	GALAXY
CED 027	04 03 34.	+ 27 29			DIFFUSE GALACTIC NEBULA
MCG-03-11-010	04 03 36.	- 15 16	36	15.	GALAXY
SHAP405-5025.5	04 03 37.	- 50 33 35.	30	17.25	GALAXY
LB 01253	04 03 39.	+ 20 52 36.		16.4	FAINT BLUE STAR
MCG-03-11-011	04 03 39.	- 17 54	150	14.5	GALAXY
SHAP405-5012.0	04 03 41.	- 50 20 04.	30	17.5	GALAXY
SHAP405-5144.9	04 03 41.	- 51 52 59.	18	17.0	GALAXY
SHAP404-5253.2	04 03 41.	- 53 01 17.	12	17.25	GALAXY
VB 251	04 03 42.	+ 51 14	1209		STELLAR RING
SHAP404-5614.9	04 03 42.	- 56 22 59.	30	16.75	GALAXY
LB 01254	04 03 44.	+ 11 31 42.		16.7	FAINT BLUE STAR
BAK 2.1064	04 03 44.	- 17 54 45.	106	16.6	GALAXY
SHAP404-5645.7	04 03 44.	- 56 53 47.	36	17.5	GALAXY
SHAP405-5232.9	04 03 45.	- 52 40 58.	18	17.0	GALAXY
SHAP404-5415.4	04 03 46.	- 54 23 28.	42	16.0	GALAXY
SHAP404-5602.0	04 03 46.	- 56 10 05.	30	17.25	GALAXY

OBJECT NAME	RIGHT ASCEN.	DECLINATION	DIAM.	MAGN.	TYPE OF OBJECT
SHAP405-5100.8	04 03 47.	- 51 08 52.	18	16.75	GALAXY
SHAP405-5152.6	04 03 47.	- 52 00 40.	24	17.0	GALAXY
SHAP405-5313.0	04 03 47.	- 53 21 04.	18	16.0	GALAXY
SHAP405-5347.8	04 03 47.	- 53 55 52.	24	17.5	GALAXY
SHAP405-5406.0	04 03 47.	- 54 14 04.	18	16.75	GALAXY
MCG+04-10-022	04 03 48.	+ 22 43	72	15.	GALAXY
RLWT 091	04 03 48.	+ 52 24		12.5	FAINT VERY BLUE STAR
BAK 2.1065	04 03 49.	- 21 08 50.	20	16.5	GALAXY
SHAP405-5043.8	04 03 49.	- 50 51 52.	24	16.75	GALAXY
SHAP405-5430.1	04 03 49.	- 54 38 10.	42	17.5	GALAXY
SHAP405-5107.6	04 03 50.	- 51 15 40.	18	17.5	GALAXY
SHAP405-5204.1	04 03 50.	- 52 12 10.	12	16.75	GALAXY
SHAP405-5402.5	04 03 51.	- 54 10 34.	36	17.0	GALAXY
SHAP405-5523.4	04 03 51.	- 55 31 28.	24	16.5	GALAXY
SHAP404-5552.0	04 03 51.	- 56 00 04.	24	17.0	GALAXY
SHAP405-5406.1	04 03 52.	- 54 14 10.	18	16.0	GALAXY
LB 00224	04 03 54.	+ 20 18 42.		15.3	FAINT BLUE STAR
ZWG 487.022	04 03 54.	+ 22 44		15.0	GALAXY
UGC 02958	04 03 54.	+ 22 44	90	15.0	GALAXY Sb
MCG+06-09-020	04 03 54.	+ 34 18	30	15.5	GALAXY
SHAP405-5253.6	04 03 54.	- 53 01 40.	18	16.5	GALAXY
SHAP404-5832.2	04 03 55.	- 58 40 16.	18	17.25	GALAXY
SHAP404-5917.2	04 03 55.	- 59 25 16.	18	17.5	GALAXY
SHAP405-5249.7	04 03 57.	- 52 57 46.	12	16.5	GALAXY
BAK 2.1066	04 03 58.	- 20 08 14.	65	15.8	GALAXY
BAK 2.1067	04 03 58.	- 20 21 38.	21	16.1	GALAXY
SHAP405-5722.5	04 03 58.	- 57 30 34.	24	18.0	GALAXY
SHAP405-5358.0	04 03 59.	- 54 06 04.	24	16.25	GALAXY
SHAP405-5400.0	04 03 59.	- 54 08 04.	24	16.5	GALAXY
SHAP405-5602.3	04 03 59.	- 56 10 22.	30	17.25	GALAXY
UGC 02959	04 04 00.	+ 34 19	84	16.0	GALAXY S
ZC 0404.0+8304	04 04 00.	+ 83 04	610		CLUSTER OF GALAXIES
MCG+01-11-002	04 04 00.	- 08 52	60	15.	GALAXY
SHAP405-5300.0	04 04 00.	- 53 08 03.	18	17.0	GALAXY
SHAP404-5905.2	04 04 02.	- 59 13 16.	42	17.0	GALAXY
SHAP404-5939.5	04 04 03.	- 59 47 34.	24	17.25	GALAXY
MCG+00-11-011	04 04 06.	+ 00 09	42	15.5	GALAXY
BAK 2.1068	04 04 07.	- 21 39 25.	22	16.3	GALAXY
SHAP405-5157.7	04 04 07.	- 52 05 45.	12	16.75	GALAXY
SHAP405-5341.8	04 04 07.	- 53 49 51.	24	17.25	GALAXY
SHAP405-5121.2	04 04 08.	- 51 29 15.	18	17.0	GALAXY
SHAP405-5122.0	04 04 08.	- 51 30 03.	24	16.75	GALAXY
SHAP405-5136.5	04 04 08.	- 51 44 33.	18	17.0	GALAXY
MCG+01-11-011	04 04 09.	+ 06 49 30.	36	15.	GALAXY
LB 01255	04 04 10.	+ 11 23 54.		16.2	FAINT BLUE STAR
SHAP405-5248.4	04 04 10.	- 52 56 27.	12	16.75	GALAXY
ZWG 418.011	04 04 12.	+ 06 49		15.4	GALAXY
SHAP405-5204.1	04 04 12.	- 52 12 09.	30	16.75	GALAXY
SHAP405-5632.2	04 04 12.	- 56 40 15.	18	16.75	GALAXY
SHAP405-5640.6	04 04 12.	- 56 48 39.	24	17.5	GALAXY
SHAP405-5211.8	04 04 15.	- 52 19 50.	18	17.0	GALAXY
SHAP405-5556.1	04 04 15.	- 56 04 09.	48	17.0	GALAXY
LB 00225	04 04 16.	+ 19 34 18.		15.5	FAINT BLUE STAR
UGC 02960	04 04 18.	+ 24 28	84	16.5	GALAXY
UGC 02961	04 04 18.	+ 26 35	90	17.	GALAXY DWARF
BAK 2.1069	04 04 18.	- 20 12 36.	22	16.9	GALAXY
SHAP405-5434.0	04 04 18.	- 54 42 02.	24	17.0	GALAXY
SHAP405-5934.2	04 04 19.	- 59 42 15.	24	17.25	GALAXY
SHAP405-5640.6	04 04 20.	- 56 48 39.	24	17.5	GALAXY
MCG+04-10-023	04 04 21.	+ 24 28	48	15.5	GALAXY
SHAP405-5110.9	04 04 21.	- 51 18 56.	18	17.5	GALAXY
SHAP405-5712.4	04 04 21.	- 57 20 26.	24	16.25	GALAXY
SHAP405-5849.4	04 04 21.	- 58 57 27.	24	17.5	GALAXY
SHAP405-5643.2	04 04 22.	- 56 51 14.	24	17.5	GALAXY
SHAP405-5025.6	04 04 23.	- 50 33 38.	24	16.75	GALAXY
SHAP405-5419.7	04 04 24.	- 54 27 44.	30	17.0	GALAXY
SHAP405-5530.7	04 04 24.	- 55 38 44.	30	16.75	GALAXY
SHAP405-5916.4	04 04 24.	- 59 24 27.	30	17.75	GALAXY
LB 01256	04 04 25.	+ 14 25 48.		16.2	FAINT BLUE STAR
SHAP405-5028.0	04 04 25.	- 50 36 02.	24	17.0	GALAXY
SHAP405-5221.6	04 04 26.	- 52 29 38.	24	16.25	GALAXY
SHAP405-5906.1	04 04 26.	- 59 14 08.	24	17.5	GALAXY
SHAP405-5939.6	04 04 27.	- 59 47 38.	42	16.0	GALAXY
BAK 2.1070	04 04 29.	- 21 43 30.	22	16.4	GALAXY
SHAP405-5433.2	04 04 29.	- 54 41 14.	24	16.5	GALAXY
SHAP405-5241.5	04 04 30.	- 52 49 31.	18	17.0	GALAXY
SHAP405-5821.2	04 04 30.	- 58 29 14.	18	17.25	GALAXY
SHAP405-4959.9	04 04 32.	- 50 07 55.	36	17.5	GALAXY
SHAP405-5726.5	04 04 32.	- 57 34 32.	24	16.25	GALAXY
SHAP405-5822.6	04 04 34.	- 58 30 38.	30	17.0	GALAXY
BAK 2.1071	04 04 36.	- 20 46 05.	23	16.0	GALAXY
MCG-05-10-010	04 04 36.	- 27 55	48	15.	GALAXY
MCG-05-10-011	04 04 36.	- 27 56	36	16.	GALAXY
SHAP405-5857.8	04 04 36.	- 59 05 50.	30	16.25	GALAXY
SHAP405-5514.4	04 04 37.	- 55 22 25.	24	17.0	GALAXY
LB 01257	04 04 39.	+ 18 27 24.		17.0	FAINT BLUE STAR
SHAP405-5757.4	04 04 41.	- 58 05 25.	24	17.5	GALAXY
UGC 02962	04 04 42.	+ 25 39	102	18.	GALAXY DWARF?
ZC 0404.7+7947	04 04 42.	+ 79 47	2550		CLUSTER OF GALAXIES
MCG-04-10-013	04 04 42.	- 21 19	156	12.5	GALAXY
SHAP405-5800.0	04 04 42.	- 58 08 01.	36	17.0	GALAXY
RNGC 1518	04 04 43.	- 21 18		12.5	GALAXY
SHAP405-5709.9	04 04 43.	- 57 17 55.	30	17.0	GALAXY
SHAP405-5458.0	04 04 44.	- 55 06 01.	24	16.5	GALAXY
SHAP406-5312.6	04 04 46.	- 53 20 37.	18	15.75	GALAXY
SHAP405-5640.6	04 04 46.	- 56 48 37.	18	17.0	GALAXY
RLWT 092	04 04 48.	+ 51 08		16.	FAINT VERY BLUE STAR
ZC 0404.8-0005	04 04 48.	- 00 05	940		CLUSTER OF GALAXIES
MCG-02-11-016	04 04 48.	- 10 22 30.	36	15.	GALAXY
MCG-03-11-012	04 04 48.	- 17 20 30.	120	15.	GALAXY
SHAP406-5247.9	04 04 48.	- 52 55 54.	24	16.75	GALAXY
LB 03336	04 04 48.	- 61 29		12.9	FAINT BLUE STAR
RNGC 1526	04 04 49.	- 65 58			GALAXY
SHAP406-5214.5	04 04 50.	- 52 22 30.	18	16.25	GALAXY
SHAP406-5340.2	04 04 50.	- 53 48 12.	18	17.0	GALAXY
SHAP406-5239.8	04 04 52.	- 52 47 48.	36	14.0	GALAXY
RNGC 1522	04 04 52.	- 52 48			GALAXY
SHAP406-5417.5	04 04 52.	- 54 25 30.	24	16.5	GALAXY
SHAP406-5303.7	04 04 53.	- 53 11 42.	18	16.75	GALAXY
MCG+00-11-012	04 04 54.	+ 01 38	42	14.5	GALAXY
SHAP406-5242.4	04 04 54.	- 52 50 24.	18	17.0	GALAXY
SHAP406-5258.6	04 04 55.	- 53 06 36.	36	16.75	GALAXY
SHAP406-5555.8	04 04 55.	- 56 03 48.	30	16.75	GALAXY
SHAP406-5620.2	04 04 55.	- 56 28 12.	24	17.75	GALAXY
SHAP406-5639.8	04 04 56.	- 56 47 48.	24	17.5	GALAXY
MCG+04-10-024	04 04 57.	+ 23 08 30.	36	14.5	GALAXY
SHAP406-5352.6	04 04 57.	- 54 00 36.	24	17.25	GALAXY
SHAP405-5817.2	04 04 57.	- 58 25 12.	24	17.5	GALAXY
LB 01258	04 04 58.	+ 11 36 06.		17.1	FAINT BLUE STAR
BAK 2.1072	04 04 59.	- 17 20 34.	28	15.2	GALAXY
SHAP406-5307.8	04 04 59.	- 53 15 48.	30	16.0	GALAXY
SHAP406-5655.0	04 04 59.	- 57 03 00.	18	17.25	GALAXY
KHAV 045	04 05	+ 30 38	6990		DARK NEBULA
LB 09812	04 05	- 80 42		14.0	FAINT BLUE STAR
MCG+01-11-012	04 05 00.	+ 03 50	108	15.	GALAXY
LB 01259	04 05 00.	+ 10 42 48.		15.0	FAINT BLUE STAR
ZWG 487.023	04 05 00.	+ 23 10		14.9	GALAXY
LDN 1486	04 05 00.	+ 29 00	2760		DARK NEBULA
LDN 1459	04 05 00.	+ 39 30	12660		DARK NEBULA
LDN 1390	04 05 00.	+ 55 00	3960		DARK NEBULA
MCG+00-11-013	04 05 00.	- 02 34 30.	15	15.5	GALAXY
RNGC 1523	04 05 00.	- 54 14			NON-EXISTENT OBJECT
SHAP406-5419.6	04 05 00.	- 54 27 36.	24	16.5	GALAXY
SHAP406-5636.8	04 05 02.	- 56 44 48.	24	17.25	GALAXY
B 218	04 05 03.	+ 26 12	900		DARK OBJECT
SHAP406-5057.5	04 05 03.	- 51 05 29.	18	16.25	GALAXY
SHAP406-5350.2	04 05 03.	- 53 58 11.	18	17.5	GALAXY
SHAP406-5638.8	04 05 04.	- 56 46 48.	24	17.5	GALAXY
SHAP406-5715.2	04 05 04.	- 57 23 12.	24	17.5	GALAXY
SHAP406-5813.9	04 05 05.	- 58 21 54.	30	16.5	GALAXY
ZWG 418.012	04 05 06.	+ 03 51		15.3	GALAXY
UGC 02963	04 05 06.	+ 03 51	144	15.3	GALAXY Sc
SHAP406-5335.7	04 05 07.	- 53 43 41.	24	16.5	GALAXY
SHAP406-5558.6	04 05 08.	- 56 06 35.	36	17.0	GALAXY
SHAP406-5911.1	04 05 09.	- 59 19 06.	36	18.0	GALAXY
SHAP406-5109.5	04 05 11.	- 51 17 29.	18	17.0	GALAXY
BC PKS0405-385	04 05 11.4	- 38 34 14.		17.5	QUASI-STELLAR OBJECT
SHAP406-5458.6	04 05 12.	- 55 06 35.	24	16.5	GALAXY
SHAP406-5411.4	04 05 14.	- 54 19 23.	30	16.25	GALAXY
SHAP406-5444.9	04 05 14.	- 54 52 53.	42	16.5	GALAXY
SHAP406-5729.4	04 05 14.	- 57 37 23.	18	16.0	GALAXY
SHAP406-5109.2	04 05 16.	- 51 17 10.	30	16.75	GALAXY
SHAP406-5637.8	04 05 16.	- 56 45 47.	24	17.75	GALAXY
SHAP406-5659.2	04 05 16.	- 57 07 11.	24	16.75	GALAXY
SHAP406-5057.5	04 05 17.	- 51 05 28.	24	17.0	GALAXY
SHAP406-5623.7	04 05 17.	- 56 31 41.	30	16.75	GALAXY
SHAP406-5702.0	04 05 17.	- 57 09 59.	36	17.0	GALAXY
SHAP406-5614.0	04 05 18.	- 56 21 59.	24	17.75	GALAXY
LB 03337	04 05 18.	- 61 46		13.4	FAINT BLUE STAR
SHAP406-5106.9	04 05 21.	- 51 14 52.	42	17.25	GALAXY
SHAP406-5406.6	04 05 22.	- 54 14 34.	36	16.75	GALAXY
MCG+00-11-014	04 05 24.	- 02 41 30.	36	15.5	GALAXY
BAK 2.1073	04 05 24.	- 21 33 56.	20	14.4	GALAXY
AGU 16	04 05 24.	- 40 18 12.		12.5	2 INTERACTING GALAXIES
SHAP406-5117.6	04 05 25.	- 51 25 34.	12	16.75	GALAXY
SHAP406-5701.0	04 05 25.	- 57 08 58.	24	17.75	GALAXY
SHAP406-5726.0	04 05 26.	- 57 33 58.	30	17.75	GALAXY
SHAP406-5932.5	04 05 26.	- 59 40 29.	24	17.25	GALAXY
MCG-04-10-014	04 05 27.	- 21 34	15	15.5	GALAXY
SHB 081	04 05 27.4	- 12 19 34.		17.1	QUASI-STELLAR OBJECT
BC PKS0405-12	04 05 27.46	- 12 19 32.3		14.79	QUASI-STELLAR OBJECT
SHAP406-5232.8	04 05 28.	- 52 40 46.	18	17.0	GALAXY
SHAP406-5320.0	04 05 28.	- 53 27 58.	36	17.0	GALAXY
SHAP406-5454.0	04 05 31.	- 55 01 58.	36	16.25	GALAXY
SHAP406-5827.4	04 05 32.	- 58 35 22.	30	17.75	GALAXY
SHAP406-5022.0	04 05 33.	- 50 29 57.	30	17.5	GALAXY
SHAP406-5159.3	04 05 33.	- 52 07 15.	18	17.0	GALAXY
SHAP406-5824.8	04 05 33.	- 58 32 46.	24	16.5	GALAXY
BAK 2.1074	04 05 34.	- 21 04 56.	21	15.9	GALAXY
SHAP406-5710.6	04 05 34.	- 57 18 34.	24	17.0	GALAXY
SHAP406-5552.8	04 05 35.	- 56 00 46.	42	16.25	GALAXY
SHAP406-5836.8	04 05 35.	- 58 44 46.	30	17.75	GALAXY
ARC 0474	04 05 36.	- 16 50		17.1	RICH CLUSTER OF GALAXIES
SHAP406-5126.1	04 05 36.	- 51 34 03.	18	17.5	GALAXY
SHAP406-5658.3	04 05 36.	- 57 06 16.	30	17.0	GALAXY
SHAP406-5343.4	04 05 37.	- 53 51 21.	18	17.0	GALAXY
SHAP407-5033.0	04 05 38.	- 50 40 57.	30	17.0	GALAXY
SHAP406-5143.6	04 05 39.	- 51 51 33.	24	17.25	GALAXY
SHAP406-5338.8	04 05 40.	- 53 46 45.	36	16.5	GALAXY
SHAP406-5756.8	04 05 40.	- 58 04 46.	102	16.25	GALAXY
LB 01260	04 05 41.	+ 23 13 18.		15.3	FAINT BLUE STAR
SHAP407-5139.5	04 05 41.	- 51 47 27.	18	16.75	GALAXY
HN 0266	04 05 41.	- 58 05			NEBULA
ZC 0405.7-0153	04 05 42.	- 01 53	1010		CLUSTER OF GALAXIES
RNGC 1516B	04 05 42.	- 09 01		15.0	GALAXY
RNGC 1516A	04 05 42.	- 09 01		15.0	GALAXY
MCG-02-11-017	04 05 42.	- 09 01	24	15.	GALAXY
MCG-02-11-018	04 05 42.	- 09 01 30.	30	15.	GALAXY
MCG-05-10-012	04 05 42.	- 29 58	120	14.	GALAXY
IC 2034	04 05 42.	- 58 05			NONSTELLAR OBJECT
SHAP407-5224.0	04 05 43.	- 52 31 57.	18	16.0	GALAXY
SHAP407-5100.6	04 05 44.	- 51 08 33.	18	17.25	GALAXY
SHAP406-5359.4	04 05 44.	- 54 07 21.	24	17.0	GALAXY
SHAP406-5423.5	04 05 45.	- 54 31 27.	24	16.5	GALAXY
SHAP406-5626.4	04 05 45.	- 56 34 21.	24	18.0	GALAXY
LB 01261	04 05 46.	+ 21 29 12.		17.0	FAINT BLUE STAR
RNGC 1519	04 05 46.	- 17 20		13.0	GALAXY
LB 01262	04 05 47.	+ 12 56 00.		17.0	FAINT BLUE STAR
SHAP407-5111.9	04 05 47.	- 51 19 50.	24	16.25	GALAXY
SHAP407-5317.2	04 05 47.	- 53 25 09.	24	15.25	GALAXY
RLWT 093	04 05 48.	+ 51 44 30.		13.5	FAINT VERY BLUE STAR
RLWT 094	04 05 48.	+ 51 44 42.		13.5	FAINT VERY BLUE STAR
MCG-03-11-013	04 05 48.	- 17 20	102	13.5	GALAXY
SHAP406-5518.8	04 05 48.	- 55 26 45.	90	16.25	GALAXY
SHAP406-5738.2	04 05 48.	- 57 46 09.	30	17.0	GALAXY
SHAP407-5110.5	04 05 49.	- 51 18 26.	18	17.0	GALAXY
SHAP407-5342.0	04 05 49.	- 53 49 57.	24	17.0	GALAXY
SHAP406-5704.4	04 05 49.	- 57 12 21.	30	16.25	GALAXY
SHAP407-5246.1	04 05 50.	- 52 54 02.	24	16.5	GALAXY
HN 0265	04 05 50.	- 55 27			NEBULA
IC 2032	04 05 50.	- 55 27			NONSTELLAR OBJECT
SHAP406-5911.3	04 05 51.	- 59 19 15.	30	18.0	GALAXY
SHAP407-5306.4	04 05 52.	- 53 14 20.	18	17.0	GALAXY
BAK 2.1075	04 05 53.	- 21 30 42.	24	16.1	GALAXY
SHAP407-5400.2	04 05 53.	- 54 08 08.	18	16.5	GALAXY
SHAP406-5815.1	04 05 53.	- 58 23 03.	30	17.75	GALAXY
ZWG 487.024	04 05 54.	+ 27 04		15.5	GALAXY
UGC 02964	04 05 54.	+ 27 04	78	15.5	GALAXY Sc
KARA.73B 0140	04 05 54.	+ 27 04	48	15.5	ISOLATED GALAXY S
MCG+00-11-015	04 05 54.	- 01 13 30.	48	15.	GALAXY
SHAP407-5309.6	04 05 55.	- 53 17 32.	24	17.5	GALAXY
SHAP407-5343.2	04 05 55.	- 53 51 08.	18	17.0	GALAXY
SHAP406-5652.8	04 05 55.	- 57 00 44.	30	17.25	GALAXY
SHAP407-5200.1	04 05 56.	- 52 08 02.	42	16.0	GALAXY
SHAP407-5332.0	04 05 57.	- 53 39 56.	18	17.5	GALAXY
SHAP407-5644.4	04 05 57.	- 56 52 20.	24	17.0	GALAXY

OBJECT NAME	RIGHT ASCEN.	DECLINATION	DIAM.	MAGN.	TYPE OF OBJECT
SHAP407-5128.4	04 05 58.	- 51 36 20.	18	17.0	GALAXY
SHAP407-5233.2	04 05 59.	- 52 41 08.	18	17.5	GALAXY
SHAP407-5234.4	04 05 59.	- 52 42 20.	24	17.0	GALAXY
SHAP407-5325.6	04 05 59.	- 53 33 32.	12	16.5	GALAXY
KHAV 046	04 06	+ 24 44	3400		DARK NEBULA
KHAV 047	04 06	+ 25 56	3540		DARK NEBULA
LB 09813	04 06	- 81 05		13.7	FAINT BLUE STAR
UGC 02965	04 06 00.	+ 03 00	60	18.	GALAXY DWRF IR
UGC 02966	04 06 00.	+ 08 25	66	16.5	GALAXY Sc
MCG+04-10-025	04 06 00.	+ 27 04	66	15.	GALAXY
VB 215	04 06 00.	+ 58 40	289		STELLAR RING
SHAP407-5340.8	04 06 00.	- 53 48 44.	30	15.5	GALAXY
SHAP406-5826.8	04 06 00.	- 58 34 44.	24	17.75	GALAXY
HN 0267	04 06 01.	- 53 49			NEBULA
SHAP407-5450.5	04 06 01.	- 54 58 26.	24	16.25	GALAXY
SHAP407-5754.3	04 06 02.	- 58 02 14.	42	16.5	GALAXY
SHAP407-5810.1	04 06 02.	- 58 18 02.	24	17.25	GALAXY
SHAP407-5134.9	04 06 03.	- 51 42 49.	18	16.25	GALAXY
LB 00226	04 06 04.	+ 11 46 36.		15.3	FAINT BLUE STAR
MCG+00-11-016	04 06 06.	- 01 28	24	15.5	GALAXY
SHAP407-5330.4	04 06 06.	- 53 38 19.	24	16.75	GALAXY
IC 2033	04 06 06.	- 53 48			GALAXY S0
SHAP407-5647.0	04 06 07.	- 56 54 56.	24	17.75	GALAXY
SHAP407-5946.5	04 06 07.	- 59 54 26.	36	17.25	GALAXY
CED 028	04 06 08.	+ 30 31	240		DIFFUSE GALACTIC NEBULA
RNGC 1514	04 06 08.	+ 30 38			PLANETARY NEBULA
SHAP407-5645.2	04 06 08.	- 56 53 08.	18	17.5	GALAXY
PK165-15.1	04 06 08.26	+ 30 38 42.2	180	10.	PLANETARY NEBULA
BAK 2.1076	04 06 09.	- 21 09 23.	13	16.2	GALAXY
SHAP407-5424.7	04 06 09.	- 54 32 37.	24	17.5	GALAXY
SHAP407-5704.1	04 06 09.	- 57 12 02.	24	17.5	GALAXY
SHAP407-5321.6	04 06 11.	- 53 29 31.	42	16.0	GALAXY
MCG+00-11-017	04 06 12.	+ 00 05	30	15.5	GALAXY
MCG+03-11-007	04 06 12.	+ 16 58	72	17.	GALAXY
UGC 02967	04 06 12.	+ 36 52	66	17.	GALAXY SB
MCG-02-11-019	04 06 12.	- 10 20 30.	42	15.	GALAXY
MCG-02-11-020	04 06 12.	- 14 08 30.	30	14.5	GALAXY
MCG-03-11-014	04 06 12.	- 16 53	60	15.	GALAXY
MCG-04-10-015	04 06 12.	- 21 11 30.	120	13.	GALAXY
SHAP407-5609.4	04 06 12.	- 56 17 19.	30	16.5	GALAXY
SHAP407-5925.6	04 06 12.	- 59 33 32.	36	17.0	GALAXY
RNGC 1521	04 06 13.	- 21 11		13.0	GALAXY
SHAP407-5600.4	04 06 14.	- 56 08 19.	24	16.5	GALAXY
SHAP407-5026.8	04 06 15.	- 50 34 43.	24	16.0	GALAXY
SHAP407-5334.2	04 06 15.	- 53 42 07.	18	17.5	GALAXY
SHAP407-5603.0	04 06 15.	- 56 10 55.	30	16.75	GALAXY
SHAP407-5339.3	04 06 17.	- 53 47 13.	18	16.75	GALAXY
UGC 02968	04 06 18.	+ 16 58	126	16.0	GALAXY Sb
5ZW 370	04 06 18.	+ 44 13			COMPACT GALAXY
OCL 0398	04 06 18.	+ 49 23	1320	9.0	OPEN STAR CLUSTER
ZC 0406.3-0009	04 06 18.	- 00 09	1210		CLUSTER OF GALAXIES
LB 01704	04 06 18.	- 53 16		14.7	FAINT BLUE STAR
SHAP407-5723.1	04 06 18.	- 57 31 01.	24	17.5	GALAXY
SHAP407-5425.5	04 06 19.	- 54 33 25.	24	16.75	GALAXY
RNGC 1513	04 06 21.	+ 49 23		9.0	OPEN CLUSTER
SHAP407-5326.5	04 06 21.	- 53 34 24.	30	16.5	GALAXY
SHAP407-5332.2	04 06 22.	- 53 40 06.	24	16.0	GALAXY
SHAP407-5812.2	04 06 23.	- 58 20 07.	24	17.25	GALAXY
BAK 2.1077	04 06 25.	- 21 12 04.	19	15.7	GALAXY
SHAP407-5358.4	04 06 25.	- 54 06 18.	18	17.5	GALAXY
SHAP407-5705.2	04 06 25.	- 57 13 07.	24	17.5	GALAXY
SHAP407-5157.3	04 06 26.	- 52 05 12.	42	16.5	GALAXY
SHAP407-5344.0	04 06 26.	- 53 51 54.	18	17.5	GALAXY
SHAP407-5713.2	04 06 26.	- 57 21 07.	24	17.0	GALAXY
LB 01263	04 06 27.	+ 15 41 48.		16.8	FAINT BLUE STAR
SHAP407-5047.6	04 06 27.	- 50 55 30.	18	17.0	GALAXY
ZWG 418.013	04 06 30.	+ 08 31		14.3	GALAXY
UGC 02970	04 06 30.	+ 08 31	72	14.3	GALAXY Sc
UGC 02971	04 06 30.	+ 36 53	108	16.5	GALAXY
MCG+06-10-001	04 06 30.	+ 36 54	84	17.	GALAXY
ZWG 392.003	04 06 30.	- 01 17		14.8	GALAXY
UGC 02969	04 06 30.	- 01 17	90	14.8	GALAXY Sb-c
MCG+00-11-018	04 06 30.	- 01 18	66	14.	GALAXY
SHAP407-5112.9	04 06 30.	- 51 20 48.	12	16.75	GALAXY
LB 01264	04 06 31.	+ 08 21 36.		13.8	FAINT BLUE STAR
SHAP407-5026.0	04 06 32.	- 50 33 54.	24	17.25	GALAXY
RNGC 1517	04 06 33.	+ 08 31		14.5	GALAXY
ARC 0475	04 06 33.	- 09 32		17.9	RICH CLUSTER OF GALAXIES
LB 01265	04 06 34.	+ 11 28 48.		14.5	FAINT BLUE STAR
SHAP407-5820.4	04 06 34.	- 58 28 18.	24	18.0	GALAXY
SHAP407-5303.6	04 06 35.	- 53 11 30.	12	17.0	GALAXY
SHAP407-5304.1	04 06 35.	- 53 12 00.	30	17.25	GALAXY
SHAP407-5310.4	04 06 35.	- 53 18 18.	36	17.25	GALAXY
SHAP407-5648.0	04 06 35.	- 56 55 54.	24	16.5	GALAXY
RNGC 1529	04 06 35.	- 63 02			GALAXY
LB 00227	04 06 36.	+ 17 00 12.		15.6	FAINT BLUE STAR
UGC 02972	04 06 36.	+ 36 23	96	17.	GALAXY
LB 01705	04 06 36.	- 49 30		14.2	FAINT BLUE STAR
SHAP407-5025.4	04 06 36.	- 50 33 17.	18	17.5	GALAXY
SHAP407-5452.9	04 06 36.	- 55 00 48.	36	16.25	GALAXY
SHAP408-5017.8	04 06 37.	- 50 25 41.	18	16.75	GALAXY
SHAP407-5500.3	04 06 37.	- 55 08 12.	30	16.25	GALAXY
SHAP408-5020.8	04 06 38.	- 50 28 41.	18	17.0	GALAXY
MCG+00-11-019	04 06 39.	- 01 36	12	15.5	GALAXY
SHAP407-5629.0	04 06 39.	- 56 36 54.	30	17.75	GALAXY
SHAP407-5357.5	04 06 41.	- 54 05 23.	18	17.0	GALAXY
SHAP407-5737.2	04 06 41.	- 57 45 06.	24	17.0	GALAXY
MCG+00-11-020	04 06 42.	- 01 36	24	15.	GALAXY
ZWG 392.004	04 06 42.	- 01 37		15.7	GALAXY
MCG+01-11-003	04 06 42.	- 04 21	54	16.	GALAXY
MCG-04-10-016	04 06 42.	- 21 52	24	15.5	GALAXY
SHAP408-5026.6	04 06 42.	- 50 34 29.	24	17.5	GALAXY
SHAP407-5302.0	04 06 42.	- 53 09 53.	18	17.5	GALAXY
BAK 2.1078	04 06 43.	- 17 09 33.	24	15.5	GALAXY
FATH 1.145	04 06 44.	+ 74 52	8		NEBULA
SHAP407-5722.4	04 06 45.	- 57 30 17.	24	17.5	GALAXY
LB 01266	04 06 46.	+ 18 52 48.		17.2	FAINT BLUE STAR
TON-S 0401	04 06 48.	- 29 55		12.1	BLUE STAR
SHAP408-5031.8	04 06 48.	- 50 39 40.	24	17.0	GALAXY
SHAP408-5059.1	04 06 48.	- 51 06 58.	12	17.25	GALAXY
SHAP408-5341.8	04 06 48.	- 53 49 41.	24	16.75	GALAXY
SHAP407-5724.2	04 06 49.	- 57 32 05.	30	17.25	GALAXY
MCG+00-11-021	04 06 51.	- 00 19 30.	24	16.	GALAXY
RNGC 1527	04 06 52.	- 48 01		12.5	GALAXY
SHAP408-5107.1	04 06 53.	- 51 14 58.	24	16.75	GALAXY
ZWG 347.003	04 06 54.	+ 74 53		15.0	GALAXY
ZWG 327.016	04 06 54.	+ 74 53		15.0	GALAXY
KARA.73B 0141	04 06 54.	+ 74 53	48	15.0	ISOLATED GALAXY S0

OBJECT NAME	RIGHT ASCEN.	DECLINATION	DIAM.	MAGN.	TYPE OF OBJECT
MCG-03-11-015	04 06 54.	- 20 22	6	16.	GALAXY
SHAP408-5347.0	04 06 54.	- 53 54 52.	24	17.25	GALAXY
SHAP408-5517.6	04 06 57.	- 55 25 28.	24	16.75	GALAXY
SHAP408-5030.2	04 06 58.	- 50 38 04.	24	17.0	GALAXY
SHAP408-5621.9	04 06 58.	- 56 29 46.	30	17.0	GALAXY
SHAP408-5646.5	04 06 58.	- 56 54 22.	24	16.5	GALAXY
LBN 0710	04 07	+ 51 05	420		BRIGHT NEBULA
LB 01267	04 07 00.	+ 11 00 18.		16.5	FAINT BLUE STAR
SHAP408-5722.1	04 07 01.	- 57 29 58.	30	16.75	GALAXY
SHAP408-5022.0	04 07 03.	- 50 29 52.	18	17.25	GALAXY
SHAP408-5051.2	04 07 03.	- 50 59 04.	42	17.0	GALAXY
SHAP408-5301.8	04 07 03.	- 53 09 40.	12	17.25	GALAXY
SHAP408-5402.4	04 07 03.	- 54 10 16.	42	15.5	GALAXY
SHAP408-5826.9	04 07 03.	- 58 34 46.	30	17.75	GALAXY
BAK 2.1079	04 07 04.	- 20 21 02.	30	15.6	GALAXY
SHAP408-5111.0	04 07 04.	- 51 18 52.	18	16.75	GALAXY
SHAP408-5244.7	04 07 04.	- 52 52 34.	24	17.0	GALAXY
SHAP408-5239.4	04 07 05.	- 52 47 16.	24	16.5	GALAXY
SHAP408-5401.0	04 07 05.	- 54 08 52.	18	17.25	GALAXY
SHAP407-5934.6	04 07 05.	- 59 42 28.	36	16.5	GALAXY
TON-S 0402	04 07 06.	- 25 30		16.0	BLUE STAR
SHAP408-5531.1	04 07 06.	- 55 38 58.	42	17.0	GALAXY
SHAP408-5906.8	04 07 07.	- 59 14 40.	24	17.5	GALAXY
SHAP408-5558.0	04 07 08.	- 56 05 52.	24	17.0	GALAXY
SHAP408-5217.8	04 07 11.	- 52 25 39.	18	17.5	GALAXY
SHAP408-5646.2	04 07 11.	- 56 54 04.	42	17.0	GALAXY
MCG-05-10-013	04 07 12.	- 30 32	72	14.	GALAXY
LB 01706	04 07 12.	- 48 52		14.2	FAINT BLUE STAR
SHAP408-5131.8	04 07 12.	- 51 39 39.	24	16.5	GALAXY
SHAP408-5633.8	04 07 12.	- 56 41 40.	36	17.0	GALAXY
SHAP408-5352.5	04 07 13.	- 54 00 21.	24	17.5	GALAXY
SHAP408-5522.2	04 07 13.	- 55 30 03.	24	17.0	GALAXY
SHAP408-4935.2	04 07 14.	- 49 43 03.	72	14.5	GALAXY
SHAP408-5157.2	04 07 14.	- 52 05 03.	18	16.0	GALAXY
SHAP408-5320.3	04 07 14.	- 53 28 09.	18	17.5	GALAXY
SHAP408-5458.9	04 07 14.	- 55 06 45.	24	17.0	GALAXY
LB 01268	04 07 15.	+ 10 39 06.		15.5	FAINT BLUE STAR
SHAP408-5021.2	04 07 15.	- 50 29 03.	42	17.0	GALAXY
SHAP408-5652.6	04 07 15.	- 57 00 27.	42	17.0	GALAXY
HZ 10	04 07 16.	+ 17 54		14.4	DECIDEDLY BLUE STAR
SHAP408-5402.0	04 07 16.	- 54 09 51.	36	16.5	GALAXY
SHAP408-5041.9	04 07 17.	- 50 49 45.	18	16.75	GALAXY
SHAP408-5932.8	04 07 17.	- 59 40 40.	36	17.5	GALAXY
MRSL 151-00/1	04 07 18.	+ 51 02	840		HII REGION
VB 252	04 07 18.	+ 53 27	336		STELLAR RING
VB 216	04 07 18.	+ 61 00	779		STELLAR RING
SHAP408-5808.2	04 07 19.	- 58 16 03.	36	17.75	GALAXY
SHAP408-5839.8	04 07 20.	- 58 47 39.	66	16.0	GALAXY
SHAP408-5530.2	04 07 22.	- 55 38 03.	24	17.25	GALAXY
SHAP408-5622.4	04 07 23.	- 56 30 15.	36	17.0	GALAXY
SHAP408-5802.6	04 07 23.	- 58 10 27.	30	17.0	GALAXY
5ZW 371	04 07 24.	+ 29 41			COMPACT GALAXY
ZWG 392.005	04 07 24.	- 01 30		15.3	GALAXY
UGC 02973	04 07 24.	- 01 30	66	15.3	GALAXY Sa-b
MCG+00-11-022	04 07 24.	- 01 31	48	14.5	GALAXY
MCG-04-10-017	04 07 24.	- 21 55	90	16.	GALAXY
LB 01707	04 07 24.	- 51 15		12.5	FAINT BLUE STAR
SHAP408-5824.8	04 07 24.	- 58 32 39.	24	17.5	GALAXY
SHAP408-5756.1	04 07 25.	- 58 03 57.	30	17.75	GALAXY
SHAP408-5844.0	04 07 25.	- 58 51 51.	102	16.0	GALAXY
SHAP408-5734.9	04 07 26.	- 57 42 45.	24	17.0	GALAXY
HN 0269	04 07 26.	- 58 52			NEBULA
IC 2037	04 07 26.	- 58 52			NONSTELLAR OBJECT
SHAP408-5226.4	04 07 27.	- 52 34 14.	24	17.0	GALAXY
LDN 1499	04 07 30.	+ 24 40	1080		DARK NEBULA
MCG-02-11-021	04 07 30.	- 10 24 30.	36	15.	GALAXY
SHAP408-5453.2	04 07 30.	- 55 01 02.	24	16.75	GALAXY
SHAP408-5628.2	04 07 30.	- 56 36 02.	54	17.75	GALAXY
SHAP408-5644.9	04 07 31.	- 56 52 44.	30	17.0	GALAXY
SHAP408-5804.4	04 07 31.	- 58 12 14.	30	17.5	GALAXY
SHAP408-5459.8	04 07 33.	- 55 07 38.	24	16.25	GALAXY
SHAP408-5256.5	04 07 35.	- 53 04 20.	24	17.0	GALAXY
SHAP408-5354.0	04 07 35.	- 54 01 50.	18	17.0	GALAXY
IC 2035	04 07 36.	- 45 38	36	12.6	GALAXY S0
SHAP409-4930.2	04 07 36.	- 49 38 01.	30	17.25	GALAXY
SHAP408-5254.8	04 07 37.	- 53 02 38.	18	17.0	GALAXY
SHAP408-5310.0	04 07 37.	- 53 17 50.	24	16.75	GALAXY
SHAP408-5805.8	04 07 37.	- 58 13 38.	36	18.0	GALAXY
SHAP408-5835.0	04 07 37.	- 58 42 50.	42	17.0	GALAXY
SHAP409-5036.4	04 07 38.	- 50 44 13.	18	17.25	GALAXY
SHAP408-5135.3	04 07 38.	- 51 43 07.	12	17.0	GALAXY
SHAP408-5902.8	04 07 39.	- 59 10 38.	30	17.0	GALAXY
SG 2.008	04 07 40.	+ 51 00	360		DIFFUSE EMISSION NEBULA
LB 01269	04 07 43.	+ 15 56 24.		16.8	FAINT BLUE STAR
SHAP408-5309.2	04 07 43.	- 53 17 01.	42	16.5	GALAXY
SHAP408-5805.1	04 07 44.	- 58 12 56.	24	17.25	GALAXY
SHAP408-5939.5	04 07 45.	- 59 47 20.	42	16.5	GALAXY
SHAP408-5559.0	04 07 47.	- 56 06 49.	240	15.5	GALAXY
TON-S 0403	04 07 48.	- 28 30		13.0	BLUE STAR
SHAP409-5140.9	04 07 48.	- 51 48 43.	24	16.5	GALAXY
HN 0270	04 07 48.	- 56 07			NEBULA
IC 2038	04 07 48.	- 56 07			NONSTELLAR OBJECT
SHAP409-5425.9	04 07 49.	- 54 33 43.	24	16.5	GALAXY
SHAP408-5937.2	04 07 49.	- 59 45 01.	36	17.5	GALAXY
SHAP408-5630.0	04 07 50.	- 56 37 49.	24	17.5	GALAXY
HOLM 071A	04 07 53.	- 15 25	48	14.4	PART OF MULTIPLE GALAXY
SHAP409-5007.2	04 07 53.	- 50 15 00.	24	17.0	GALAXY
SHAP409-5358.8	04 07 53.	- 54 06 37.	24	17.0	GALAXY
SHAP408-5600.2	04 07 53.	- 56 08 01.	60	15.5	GALAXY
SHAP408-5748.2	04 07 53.	- 57 56 01.	24	18.0	GALAXY
SHAP408-5917.2	04 07 53.	- 59 25 01.	42	16.5	GALAXY
MCG+00-11-023	04 07 54.	- 00 47	36	16.	GALAXY
RNGC 1525	04 07 54.	- 08 56			NON-EXISTENT OBJECT
RNGC 1524	04 07 54.	- 08 56			NON-EXISTENT OBJECT
MCG-03-11-016	04 07 54.	- 15 25	8	16.5	GALAXY
SHAP409-5527.2	04 07 54.	- 55 35 01.	24	16.0	GALAXY
HN 0271	04 07 54.	- 56 08			NEBULA
IC 2039	04 07 54.	- 56 08			NONSTELLAR OBJECT
HOLM 071B	04 07 55.	- 15 26	24	14.9	PART OF MULTIPLE GALAXY
SHAP409-5602.2	04 07 56.	- 56 10 01.	24	16.25	GALAXY
SHAP409-5259.3	04 07 57.	- 53 07 06.	12	17.5	GALAXY
SHAP409-5429.8	04 07 57.	- 54 37 36.	48	17.25	GALAXY
SHAP409-5435.2	04 07 58.	- 54 43 00.	30	17.0	GALAXY
SHAP409-5040.6	04 07 59.	- 50 48 24.	18	17.0	GALAXY
SHAP409-5350.9	04 07 59.	- 53 58 42.	30	17.0	GALAXY
LDN 1498	04 08 00.	+ 24 50	1380		DARK NEBULA
UGC 02974	04 08 00.	+ 26 02	66	17.	GALAXY
UGC 02975	04 08 00.	+ 26 30	84	17.	GALAXY

OBJECT NAME	RIGHT ASCEN.	DECLINATION	DIAM.	MAGN.	TYPE OF OBJECT
LDN 1473	04 08 00.	+ 38 00	1560		DARK NEBULA
MCG+14-03-003	04 08 00.	+ 86 04	24	15.	GALAXY
SHAP409-5005.6	04 08 00.	- 50 13 24.	24	16.75	GALAXY
SHAP409-5146.8	04 08 00.	- 51 54 36.	18	17.5	GALAXY
SHAP409-5153.0	04 08 01.	- 52 00 48.	18	17.5	GALAXY
SHAP409-5504.4	04 08 01.	- 55 12 12.	24	17.0	GALAXY
SHAP409-5204.4	04 08 02.	- 52 12 12.	18	17.5	GALAXY
SHAP409-5359.1	04 08 02.	- 54 06 54.	24	16.5	GALAXY
SHAP409-5223.6	04 08 03.	- 52 31 24.	18	17.0	GALAXY
SHAP409-5207.2	04 08 04.	- 52 15 00.	18	17.0	GALAXY
SHAP409-5126.0	04 08 05.	- 51 33 48.	18	17.0	GALAXY
SHAP409-5926.6	04 08 07.	- 59 34 24.	36	16.5	GALAXY
LB 01270	04 08 08.	+ 15 40 12.		16.1	FAINT BLUE STAR
SHAP409-5139.2	04 08 08.	- 51 46 59.	12	17.5	GALAXY
RNGC 1534	04 08 11.	- 62 55			GALAXY
MCG+01-11-004	04 08 12.	- 07 22	114	14.	GALAXY
AGU 17	04 08 12.	- 39 48 24.	60	12.5	N GALAXY
SHAP409-5232.0	04 08 12.	- 52 39 47.	12	17.5	GALAXY
LB 01271	04 08 13.	+ 16 43 48.		16.2	FAINT BLUE STAR
SHAP409-5905.0	04 08 13.	- 59 12 48.	36	16.0	GALAXY
IC 2036	04 08 14.	- 39 50 08.			NONSTELLAR OBJECT
SHAP409-5328.2	04 08 14.	- 53 35 59.	18	17.0	GALAXY
SHAP409-5331.8	04 08 14.	- 53 39 35.	24	16.5	GALAXY
SHAP409-5400.0	04 08 14.	- 54 07 47.	30	16.5	GALAXY
LB 01273	04 08 15.	+ 09 41 18.		16.0	FAINT BLUE STAR
LB 01272	04 08 15.	+ 19 33 42.		14.8	FAINT BLUE STAR
MCG-04-10-018	04 08 15.	- 23 46	36	14.5	GALAXY
SHAP409-5633.0	04 08 16.	- 56 40 47.	30	17.25	GALAXY
SHAP409-5439.5	04 08 17.	- 54 47 17.	24	17.0	GALAXY
ZWG 487.025	04 08 18.	+ 26 45		14.6	GALAXY
UGC 02976	04 08 18.	+ 26 45	108	14.6	GALAXY Sc
RLWT 095	04 08 18.	+ 53 10		11.5	FAINT VERY BLUE STAR
SHAP409-5108.6	04 08 18.	- 51 16 23.	18	17.5	GALAXY
SHAP409-5311.9	04 08 18.	- 53 19 41.	12	17.25	GALAXY
SHAP409-5040.5	04 08 19.	- 50 48 17.	24	17.0	GALAXY
SHAP409-5505.0	04 08 19.	- 55 12 47.	24	16.0	GALAXY
SHAP409-5936.8	04 08 19.	- 59 44 36.	54	15.75	GALAXY
SHAP409-5630.3	04 08 20.	- 56 38 05.	36	17.0	GALAXY
MCG-02-11-022	04 08 24.	- 09 08	48	15.	GALAXY
LB 01708	04 08 24.	- 47 14		15.6	FAINT BLUE STAR
SHAP409-5032.8	04 08 24.	- 50 40 34.	24	17.0	GALAXY
SHAP409-5807.6	04 08 24.	- 58 15 23.	18	17.5	GALAXY
MCG+04-10-026	04 08 27.	+ 26 46	66	14.5	GALAXY
MCG-02-11-023	04 08 27.	- 09 46	84	14.5	GALAXY
SHAP409-5519.2	04 08 28.	- 55 26 58.	36	16.5	GALAXY
SHAP409-5927.0	04 08 28.	- 59 34 47.	36	16.75	GALAXY
SHAP409-5420.6	04 08 29.	- 54 28 22.	24	17.0	GALAXY
SHAP409-5935.2	04 08 29.	- 59 42 59.	42	16.75	GALAXY
MRK 614	04 08 30.	- 07 30	12	15.	GALAXY WITH UV CONTINUUM
MCG+01-11-005	04 08 30.	- 07 30	54	14.5	GALAXY
SHAP409-5129.6	04 08 30.	- 51 37 22.	18	17.0	GALAXY
SHAP409-5401.8	04 08 30.	- 54 09 34.	24	17.0	GALAXY
SHAP409-5431.5	04 08 30.	- 54 39 16.	18	16.5	GALAXY
LB 01274	04 08 31.	+ 12 33 42.		15.1	FAINT BLUE STAR
B 208	04 08 31.	+ 25 02			DARK OBJECT
SHAP409-5342.5	04 08 31.	- 53 50 16.	18	17.0	GALAXY
SHAP409-4944.6	04 08 32.	- 49 52 22.	36	17.5	GALAXY
SHAP409-5520.0	04 08 32.	- 55 27 46.	24	16.75	GALAXY
SHAP409-5923.0	04 08 32.	- 59 30 47.	36	16.75	GALAXY
SHAP409-5456.5	04 08 33.	- 55 04 16.	30	16.25	GALAXY
SHAP409-5749.0	04 08 33.	- 57 56 46.	42	17.25	GALAXY
SHAP409-5241.0	04 08 36.	- 52 48 46.	30	17.5	GALAXY
SHAP409-5323.0	04 08 36.	- 53 30 46.	18	16.5	GALAXY
SHAP409-5444.3	04 08 37.	- 54 52 04.	30	17.0	GALAXY
SHAP409-5758.4	04 08 37.	- 58 06 10.	24	17.0	GALAXY
SHAP409-5930.6	04 08 37.	- 59 38 22.	48	15.5	GALAXY
SHAP409-5801.0	04 08 38.	- 58 08 46.	24	17.0	GALAXY
SHAP409-5440.0	04 08 39.	- 54 47 46.	24	17.0	GALAXY
SHAP409-5736.4	04 08 40.	- 57 44 10.	30	17.5	GALAXY
LB 01275	04 08 41.	+ 09 32 42.		15.2	FAINT BLUE STAR
SHAP409-5255.1	04 08 42.	- 53 02 51.	18	16.0	GALAXY
RNGC 1533	04 08 45.	- 56 15		12.0	GALAXY
SHAP409-5833.2	04 08 45.	- 58 40 58.	30	17.0	GALAXY
SHAP410-4935.2	04 08 46.	- 49 42 57.	30	17.5	GALAXY
SHAP409-5719.0	04 08 46.	- 57 26 46.	24	17.5	GALAXY
SHAP410-5350.2	04 08 47.	- 53 57 57.	18	17.0	GALAXY
SHAP409-5757.8	04 08 47.	- 58 05 34.	24	17.25	GALAXY
VB 253	04 08 48.	+ 51 01	604		STELLAR RING
MCG-03-11-017	04 08 48.	- 15 01	48	14.5	GALAXY
MCG-05-10-014	04 08 48.	- 30 02	72	14.5	GALAXY
LB 01709	04 08 48.	- 50 10		14.4	FAINT BLUE STAR
SHAP409-5607.1	04 08 48.	- 56 14 51.	138	12.3	GALAXY
SHAP410-5045.4	04 08 49.	- 50 53 09.	18	17.0	GALAXY
SHAP410-5216.7	04 08 49.	- 52 24 27.	24	16.0	GALAXY
SHAP409-5743.1	04 08 49.	- 57 50 51.	24	17.5	GALAXY
SHAP409-5756.5	04 08 49.	- 58 04 15.	24	17.25	GALAXY
HZ 06	04 08 52.	+ 15 15		14.6	BLUE STAR
SHAP410-5332.7	04 08 52.	- 53 40 27.	36	17.0	GALAXY
SHAP409-5935.0	04 08 53.	- 59 42 45.	54	16.25	GALAXY
SHAP410-5443.7	04 08 56.	- 54 51 27.	18	17.0	GALAXY
SHAP410-5303.5	04 08 57.	- 53 11 14.	24	17.0	GALAXY
SHAP410-5441.5	04 08 58.	- 54 49 14.	30	17.0	GALAXY
SHAP410-5624.4	04 08 58.	- 56 32 09.	24	16.75	GALAXY
SHAP410-5639.4	04 08 58.	- 56 47 09.	36	16.75	GALAXY
ARC 0476	04 08 59.	- 11 21		17.6	RICH CLUSTER OF GALAXIES
SHAP410-5048.4	04 08 59.	- 50 56 08.	30	17.5	GALAXY
SHAP410-5303.0	04 08 59.	- 53 10 44.	30	17.0	GALAXY
KHAV 048	04 09	+ 28 20	8950		DARK NEBULA
MCG+05-10-008	04 09 00.	+ 27 48	48	15.5	GALAXY
UGC 02977	04 09 00.	+ 27 49	72	16.0	GALAXY Sa-b
UGC 02978	04 09 00.	+ 34 47	96	17.	GALAXY SB?
LDN 1393	04 09 00.	+ 54 40	540		DARK NEBULA
ZWG 362.001	04 09 00.	+ 86 05		15.5	GALAXY
ZWG 361.009	04 09 00.	+ 86 05		15.5	GALAXY
KARA.73B 0142	04 09 00.	+ 86 05	36	15.5	ISOLATED GALAXY S
MCG+01-11-006	04 09 00.	- 04 00	60	17.	GALAXY
MCG-03-11-018	04 09 00.	- 16 20	84	15.	GALAXY
MCG-05-10-015	04 09 00.	- 31 31	90	14.	GALAXY
LB 01276	04 09 02.	+ 15 21 12.		16.5	FAINT BLUE STAR
SHAP410-5132.5	04 09 03.	- 51 40 14.	18	16.75	GALAXY
SHAP410-5316.4	04 09 03.	- 53 24 08.	12	17.25	GALAXY
SHAP410-5348.4	04 09 03.	- 53 56 08.	12	17.0	GALAXY
SHAP410-5444.6	04 09 06.	- 54 52 20.	24	16.0	GALAXY
SHAP410-5630.6	04 09 06.	- 56 38 20.	48	16.5	GALAXY
SHAP410-5442.8	04 09 08.	- 54 50 32.	24	16.5	GALAXY
SHAP410-5608.0	04 09 09.	- 56 15 44.	24	17.0	GALAXY
SHAP410-5810.6	04 09 09.	- 58 18 20.	24	17.5	GALAXY
SHAP410-5312.6	04 09 10.	- 53 20 20.	12	16.75	GALAXY
SHAP410-5936.5	04 09 10.	- 59 44 14.	48	16.5	GALAXY
PK147+04.1	04 09 10.2	+ 56 49 20.	12		PLANETARY NEBULA
HUB C04	04 09 11.	+ 27 55			DIFFUSE NEBULA
UGC 02979	04 09 12.	+ 06 20	84	16.0	GALAXY S
RLWT 096	04 09 12.	+ 53 10		16.	FAINT VERY BLUE STAR
SHAP410-5930.8	04 09 13.	- 59 38 32.	36	16.5	GALAXY
SG 3.021	04 09 16.	+ 27 29	60		DIFFUSE EMISSION NEBULA
SHAP410-5428.4	04 09 17.	- 54 36 07.	24	17.0	GALAXY
B 209	04 09 18.	+ 28 12			DARK OBJECT
ISS 0113	04 09 18.	+ 51 04	562		STELLAR RING
TON-S 0404	04 09 18.	- 30 32		15.2	BLUE STAR
SHAP410-5117.5	04 09 19.	- 51 25 13.	24	17.5	GALAXY
SHAP410-5420.0	04 09 19.	- 54 27 43.	18	17.0	GALAXY
SHAP410-5856.0	04 09 19.	- 59 03 44.	30	17.75	GALAXY
SHAP410-5304.6	04 09 20.	- 53 12 19.	24	16.5	GALAXY
MCG+05-10-009	04 09 21.	+ 27 35	42	15.	GALAXY
SHAP410-5440.3	04 09 22.	- 54 48 01.	24	15.75	GALAXY
SHAP410-5423.0	04 09 23.	- 54 30 43.	12	17.0	GALAXY
SHAP410-5826.0	04 09 23.	- 58 33 43.	30	16.75	GALAXY
VDB.66N 025	04 09 24.	+ 23 30	516		REFLECTION NEBULA
ZWG 508.008	04 09 24.	+ 27 35		15.4	GALAXY
UGC 02980	04 09 24.	+ 27 35	138	15.4	GALAXY E-SO
LB 03338	04 09 24.	- 73 32		12.8	FAINT BLUE STAR
LB 01277	04 09 25.	+ 13 24 12.		16.0	FAINT BLUE STAR
SHAP410-5320.6	04 09 25.	- 53 28 19.	18	16.5	GALAXY
HN 0277	04 09 25.	- 78 23			NEBULA
IC 2054	04 09 25.	- 78 23			NONSTELLAR OBJECT
SHAP410-5507.2	04 09 26.	- 55 14 55.	24	17.25	GALAXY
SHAP410-5716.2	04 09 26.	- 57 23 55.	30	17.5	GALAXY
MCG-02-11-024	04 09 27.	- 12 36	54	15.	GALAXY
SHAP410-5336.2	04 09 27.	- 53 43 55.	24	17.25	GALAXY
SHAP410-5714.6	04 09 27.	- 57 22 19.	42	17.5	GALAXY
SHAP410-5738.9	04 09 27.	- 57 46 37.	30	17.0	GALAXY
SHAP410-5939.4	04 09 27.	- 59 47 07.	42	16.0	GALAXY
SS 14	04 09 29.	+ 23 28			DIFFUSE GALACTIC NEBULA
SHAP410-5338.0	04 09 29.	- 53 45 42.	30	17.0	GALAXY
MCG+00-11-024	04 09 30.	+ 02 42 30.	36	15.5	GALAXY
DG 027	04 09 30.	+ 23 28	300		REFLECTION NEBULA
UGC 02981	04 09 30.	+ 36 46	78	18.	GALAXY S
LB 03339	04 09 30.	- 63 49		14.6	FAINT BLUE STAR
SHAP410-5044.0	04 09 31.	- 50 51 42.	18	17.5	GALAXY
SHAP410-5359.6	04 09 33.	- 54 07 18.	18	17.0	GALAXY
SHAP410-5425.6	04 09 33.	- 54 33 18.	24	17.5	GALAXY
SHAP410-5938.6	04 09 33.	- 59 46 19.	42	16.0	GALAXY
SHAP410-5043.2	04 09 34.	- 50 50 54.	30	16.5	GALAXY
SHAP410-5437.2	04 09 34.	- 54 44 54.	24	16.5	GALAXY
SHAP410-5329.2	04 09 35.	- 53 36 54.	24	17.0	GALAXY
SHAP410-5711.9	04 09 35.	- 57 19 36.	36	17.0	GALAXY
ACK 151+00.1	04 09 36.	+ 51 43			PLANETARY NEBULA
VB 254	04 09 36.	+ 52 51	161		STELLAR RING
ARC 0477	04 09 36.	- 02 00		17.5	RICH CLUSTER OF GALAXIES
ZC 0409.6-0201	04 09 36.	- 02 01	1140		CLUSTER OF GALAXIES
TON-S 0405	04 09 36.	- 32 31		15.3	BLUE STAR
SHAP410-5009.2	04 09 36.	- 50 16 54.	18	17.25	GALAXY
SHAP410-5044.5	04 09 36.	- 50 52 12.	12	17.5	GALAXY
SHAP410-5519.4	04 09 36.	- 55 27 06.	24	17.0	GALAXY
SHAP410-5903.2	04 09 36.	- 59 10 54.	36	17.75	GALAXY
SHAP411-5004.6	04 09 37.	- 50 12 18.	30	17.25	GALAXY
SHAP410-5045.4	04 09 37.	- 50 53 06.	18	17.5	GALAXY
SHAP410-5354.6	04 09 37.	- 54 02 18.	18	17.0	GALAXY
SHAP410-5402.8	04 09 37.	- 54 10 30.	18	17.5	GALAXY
MCG+00-11-025	04 09 39.	- 00 05	42	15.	GALAXY
SHAP410-5336.8	04 09 39.	- 53 44 30.	24	16.5	GALAXY
LB 01279	04 09 40.	+ 09 40 30.		16.4	FAINT BLUE STAR
LB 01278	04 09 40.	+ 22 55 06.		16.3	FAINT BLUE STAR
SHAP410-5706.6	04 09 40.	- 57 14 18.	30	16.5	GALAXY
ZWG 418.014	04 09 42.	+ 05 26		15.5	GALAXY
UGC 02982	04 09 42.	+ 05 26	60	15.5	GALAXY S
KARA.73B 0143	04 09 42.	+ 05 26	54	15.5	ISOLATED GALAXY S
STOCK 03	04 09 42.	- 03 10			BLUE KNOT NEAR ELLIP GLXY
LB 01710	04 09 42.	- 50 58		14.6	FAINT BLUE STAR
SHAP411-5109.6	04 09 42.	- 51 17 17.	18	16.75	GALAXY
LB 01711	04 09 42.	- 52 00		15.8	FAINT BLUE STAR
SHAP410-5358.7	04 09 42.	- 54 06 24.	72	16.5	GALAXY
SHAP410-5341.8	04 09 43.	- 53 49 29.	24	16.5	GALAXY
SHAP410-5342.2	04 09 44.	- 53 49 53.	24	17.25	GALAXY
SHAP410-5839.0	04 09 44.	- 58 46 42.	36	16.0	GALAXY
MCG+01-11-013	04 09 45.	+ 05 25	48	15.	GALAXY
SHAP410-5340.8	04 09 45.	- 53 48 29.	30	17.0	GALAXY
LB 01280	04 09 46.	+ 15 16 48.		16.7	FAINT BLUE STAR
SHAP411-5047.9	04 09 47.	- 50 55 35.	24	17.5	GALAXY
SHAP411-5048.1	04 09 47.	- 50 55 47.	24	17.25	GALAXY
SHAP410-5418.5	04 09 47.	- 54 26 11.	18	17.25	GALAXY
MCG+01-11-007	04 09 48.	- 06 55	42	15.5	GALAXY
SHAP411-5250.9	04 09 49.	- 52 58 35.	18	17.5	GALAXY
SHAP411-5320.6	04 09 50.	- 53 28 17.	36	16.5	GALAXY
SHAP411-5050.8	04 09 51.	- 50 58 29.	24	16.25	GALAXY
SHAP411-5129.6	04 09 52.	- 51 37 17.	18	17.5	GALAXY
SHAP411-5211.0	04 09 52.	- 52 18 41.	12	17.0	GALAXY
SHAP411-5419.4	04 09 52.	- 54 27 05.	18	17.25	GALAXY
SHAP411-5350.9	04 09 53.	- 53 58 35.	18	17.25	GALAXY
SHAP411-5218.9	04 09 55.	- 52 26 35.	24	17.25	GALAXY
SHAP411-5354.1	04 09 55.	- 54 01 47.	30	16.75	GALAXY
SHAP411-5410.2	04 09 55.	- 54 17 53.	18	17.0	GALAXY
SHAP411-5422.6	04 09 55.	- 54 30 17.	18	17.5	GALAXY
SHAP411-5359.8	04 09 56.	- 54 07 29.	30	17.0	GALAXY
SHAP411-5328.0	04 09 57.	- 53 35 41.	18	17.5	GALAXY
SHAP411-5346.9	04 09 58.	- 53 54 35.	54	17.5	GALAXY
SHAP411-5359.8	04 09 58.	- 54 07 29.	36	17.5	GALAXY
SHAP411-5620.7	04 09 58.	- 56 28 23.	24	16.0	GALAXY
RNGC 1536	04 09 58.	- 56 36		13.5	GALAXY
SHAP411-5628.5	04 09 58.	- 56 36 11.	96	13.2	GALAXY
LB 01281	04 09 59.	+ 12 05 06.		16.0	FAINT BLUE STAR
SHAP411-5341.3	04 09 59.	- 53 48 58.	84	15.5	GALAXY
LBN 0786	04 10	+ 25 30	8400		BRIGHT NEBULA
LBN 0628	04 10	+ 85 50	660		BRIGHT NEBULA
ISS 0145	04 10 00.	+ 49 25	527		STELLAR RING
LDN 1395	04 10 00.	+ 54 00	4860		DARK NEBULA
LDN 1392	04 10 00.	+ 55 00	720		DARK NEBULA
ZC 0410.0+8359	04 10 00.	+ 83 59	1550		CLUSTER OF GALAXIES
SHAP411-5145.0	04 10 00.	- 51 52 40.	12	17.5	GALAXY
SHAP411-5305.8	04 10 00.	- 53 13 28.	30	17.0	GALAXY
IC 2043	04 10 00.	- 53 49			NONSTELLAR OBJECT
MAI 006	04 10 01.	+ 85 58	40		DWARF SPHEROIDAL GALAXY
HN 0272	04 10 01.	- 53 49			NEBULA
SHAP411-5759.2	04 10 01.	- 58 06 53.	30	17.25	GALAXY
RNGC 1531	04 10 02.	- 32 59		13.0	GALAXY
SHAP411-5520.4	04 10 02.	- 55 28 04.	18	17.0	GALAXY

OBJECT NAME	RIGHT ASCEN.	DECLINATION	DIAM.	MAGN.	TYPE OF OBJECT
HZ 02	04 10 03.	+ 11 45		13.7	VERY BLUE STAR
SHAP411-5431.3	04 10 03.	- 54 38 58.	24	16.5	GALAXY
SHAP411-5636.8	04 10 03.	- 56 44 29.	30	17.0	GALAXY
SHAP411-5750.0	04 10 03.	- 57 57 41.	30	16.75	GALAXY
SHAP411-5049.4	04 10 04.	- 50 57 04.	18	16.5	GALAXY
SHAP411-5055.0	04 10 04.	- 51 02 40.	24	17.0	GALAXY
SHAP411-5352.2	04 10 04.	- 53 59 52.	30	16.75	GALAXY
HN 0273	04 10 04.	- 54 39			NEBULA
IC 2044	04 10 04.	- 54 39			NONSTELLAR OBJECT
SHAP411-5624.0	04 10 04.	- 56 31 40.	24	17.0	GALAXY
MCG-05-11-001	04 10 06.	- 32 59	36	13.	GALAXY
SHAP411-5832.8	04 10 06.	- 58 40 29.	36	17.0	GALAXY
SHAP411-5253.2	04 10 07.	- 53 00 52.	18	17.0	GALAXY
RNGC 1532	04 10 08.	- 33 00		11.5	GALAXY
SHAP411-5754.1	04 10 11.	- 58 01 46.	24	17.25	GALAXY
ZWG 392.006	04 10 12.	+ 02 14		15.3	GALAXY
UGC 02983	04 10 12.	+ 02 14	168	15.3	GALAXY Sb
MCG-02-11-025	04 10 12.	- 13 18	60	15.	GALAXY
MCG-05-11-002	04 10 12.	- 33 00	240	11.	GALAXY
LB 03340	04 10 12.	- 76 56		13.3	FAINT BLUE STAR
SHAP411-5503.2	04 10 13.	- 55 10 52.	18	17.5	GALAXY
LB 00228	04 10 14.	+ 18 51 54.		15.4	FAINT BLUE STAR
IC 2042	04 10 14.	- 47 23 17.			NONSTELLAR OBJECT
SHAP411-5440.5	04 10 14.	- 54 48 10.	36	15.5	GALAXY
HN 0274	04 10 15.	- 54 48			NEBULA
IC 2046	04 10 16.	- 54 48			NONSTELLAR OBJECT
SHAP411-5747.9	04 10 16.	- 57 55 34.	36	17.0	GALAXY
SHAP411-5256.6	04 10 18.	- 53 04 15.	18	16.5	GALAXY
SHAP411-5348.9	04 10 18.	- 53 56 33.	54	15.5	GALAXY
SHAP411-5739.8	04 10 18.	- 57 47 28.	30	17.0	GALAXY
SHAP411-5545.8	04 10 20.	- 55 53 27.	18	17.0	GALAXY
MCG+05-10-010	04 10 21.	+ 27 28	48	17.	GALAXY
SHAP411-5248.9	04 10 21.	- 52 56 33.	18	17.0	GALAXY
IC 0359	04 10 23.	+ 27 34 22.	24	13.5	REFLECTION NEBULA
MCG+00-11-026	04 10 24.	+ 02 15	114	14.	GALAXY
UGC 02984	04 10 24.	+ 13 19	108	17.	GALAXY
UGC 02985	04 10 24.	+ 27 26	84	16.0	GALAXY Sc
SHAP411-5800.1	04 10 24.	- 58 07 45.	36	16.5	GALAXY
SHAP411-5222.2	04 10 25.	- 52 29 51.	18	16.5	GALAXY
SHAP411-5319.2	04 10 25.	- 53 26 51.	24	16.5	GALAXY
IC 2040	04 10 26.	- 32 42 04.			MAY NOT EXIST
MCG+05-10-011	04 10 27.	+ 27 25	48	15.	GALAXY
SHAP411-4945.3	04 10 27.	- 49 52 56.	24	17.0	GALAXY
ZWG 418.015	04 10 30.	+ 06 26		15.7	GALAXY
ZWG 487.026	04 10 30.	+ 27 25		15.2	GALAXY
UGC 02986	04 10 30.	+ 27 28	60	16.5	GALAXY IRR
UGC 02987	04 10 30.	+ 79 41	96	16.5	GALAXY
SHAP411-5112.4	04 10 32.	- 51 20 02.	18	17.25	GALAXY
MCG+01-11-014	04 10 33.	+ 06 25 30.	36	15.	GALAXY
SHAP411-4950.1	04 10 33.	- 49 57 44.	24	17.0	GALAXY
LB 01282	04 10 34.	+ 07 48 30.		16.1	FAINT BLUE STAR
SHAP411-5307.6	04 10 34.	- 53 15 14.	36	17.0	GALAXY
ARC 0478	04 10 36.	+ 10 22		17.4	RICH CLUSTER OF GALAXIES
ZWG 487.027	04 10 36.	+ 25 21		15.5	GALAXY
UGC 02988	04 10 36.	+ 25 21	180	15.5	GALAXY Sb
KARA.73E 0144	04 10 36.	+ 25 21	204	15.5	ISOLATED GALAXY S
MCG-04-11-001	04 10 36.	- 23 18	24	14.	GALAXY
SHAP411-5011.8	04 10 36.	- 50 19 26.	24	16.75	GALAXY
SHAP411-5309.6	04 10 36.	- 53 17 14.	30	17.0	GALAXY
SHAP411-5348.4	04 10 37.	- 53 56 02.	18	17.5	GALAXY
SHAP411-5103.5	04 10 38.	- 51 11 08.	18	16.0	GALAXY
MCG+04-10-027	04 10 39.	+ 25 20 30.	84	15.	GALAXY
IC 2041	04 10 40.	- 33 00 06.			MAY NOT EXIST
SHAP411-5400.2	04 10 40.	- 54 07 50.	30	17.25	GALAXY
SHAP411-5400.6	04 10 40.	- 54 08 14.	24	16.5	GALAXY
SHAP412-5028.1	04 10 41.	- 50 35 43.	18	17.0	GALAXY
DG 028	04 10 42.	+ 10 06	660		REFLECTION NEBULA
UGC 02989	04 10 42.	+ 29 03	102	16.0	GALAXY SB?
UGC 02990	04 10 42.	+ 31 20	84	16.5	GALAXY Sa-b
UGC 02991	04 10 42.	+ 36 44	132	16.5	GALAXY SBb
SHAP412-5056.2	04 10 42.	- 51 03 49.	24	16.75	GALAXY
SHAP412-5150.8	04 10 42.	- 51 58 25.	12	17.5	GALAXY
SHAP411-5403.8	04 10 43.	- 54 11 26.	18	16.5	GALAXY
SHAP411-5542.7	04 10 43.	- 55 50 20.	24	17.0	GALAXY
SHAP412-5015.2	04 10 44.	- 50 22 49.	42	15.5	GALAXY
MCG+05-10-012	04 10 45.	+ 29 02	48	15.	GALAXY
SHAP412-5317.3	04 10 46.	- 53 24 55.	24	17.0	GALAXY
SHAP411-5348.2	04 10 46.	- 53 55 49.	30	17.0	GALAXY
SHAP412-5015.3	04 10 47.	- 50 22 55.	30	16.0	GALAXY
ZC 0410.8+1023	04 10 48.	+ 10 23	2620		CLUSTER OF GALAXIES
5ZW 372	04 10 48.	+ 29 02			COMPACT GALAXY
MCG+06-10-002	04 10 48.	+ 36 45	90	17.	GALAXY
VB 255	04 10 48.	+ 51 52	403		STELLAR RING
MCG-05-11-003	04 10 48.	- 30 59	30	15.	GALAXY
SHAP412-4951.3	04 10 48.	- 49 58 55.	30	17.0	GALAXY
SHAP411-5411.5	04 10 48.	- 54 19 07.	24	17.0	GALAXY
SHAP411-5935.6	04 10 48.	- 59 43 14.	48	16.0	GALAXY
SHAP412-5046.2	04 10 49.	- 50 53 49.	18	17.0	GALAXY
SHAP412-5041.2	04 10 51.	- 50 48 49.	12	17.5	GALAXY
SHAP412-5320.6	04 10 52.	- 53 28 13.	12	17.5	GALAXY
SHAP412-5355.8	04 10 52.	- 54 03 25.	18	17.0	GALAXY
SHAP412-5357.8	04 10 52.	- 54 05 25.	24	16.75	GALAXY
SHAP411-5638.2	04 10 52.	- 56 45 49.	30	17.5	GALAXY
VDB.66N 026	04 10 55.	+ 10 05	684		REFLECTION NEBULA
SHAP412-5337.6	04 10 55.	- 53 45 13.	36	17.0	GALAXY
SHAP412-5338.4	04 10 55.	- 53 46 01.	24	17.0	GALAXY
SHAP412-5310.8	04 10 58.	- 53 18 25.	30	16.25	GALAXY
SHAP412-5311.8	04 10 58.	- 53 19 25.	24	16.5	GALAXY
SHAP412-5719.8	04 10 59.	- 57 27 25.	42	17.25	GALAXY
LBN 0828	04 11	+ 10 05	300		BRIGHT NEBULA
CED 029	04 11	+ 25 48			DIFFUSE GALACTIC NEBULA
IC 0360	04 11	+ 25 54			NONSTELLAR OBJECT
LBN 0697	04 11	+ 56 45	300		BRIGHT NEBULA
SHAP412-5503.1	04 11 00.	- 55 10 43.	30	17.5	GALAXY
SHAP412-5542.4	04 11 00.	- 55 50 01.	18	17.0	GALAXY
SHAP412-5335.8	04 11 01.	- 53 43 24.	18	17.5	GALAXY
HELW 013	04 11 02.	- 32 40			NEBULA
SHAP412-5356.6	04 11 02.	- 54 04 12.	24	16.5	GALAXY
SHAP412-5401.5	04 11 02.	- 54 09 06.	12	17.5	GALAXY
SHAP412-5439.1	04 11 02.	- 54 46 42.	18	17.5	GALAXY
SHAP412-5648.0	04 11 02.	- 56 55 37.	24	16.5	GALAXY
SHAP412-5416.3	04 11 05.	- 54 23 54.	24	17.0	GALAXY
ZWG 392.007	04 11 06.	+ 01 38		15.4	GALAXY
UGC 02992	04 11 06.	+ 01 38	90	15.4	GALAXY DBL SYS
ISS 0114	04 11 06.	+ 51 56	389		STELLAR RING
MCG-05-11-004	04 11 06.	- 32 40	60	14.5	GALAXY
SHAP412-5017.5	04 11 06.	- 50 25 06.	18	17.5	GALAXY
SHAP412-5356.2	04 11 06.	- 54 03 48.	30	17.0	GALAXY

OBJECT NAME	RIGHT ASCEN.	DECLINATION	DIAM.	MAGN.	TYPE OF OBJECT
SHAP412-5502.0	04 11 06.	- 55 09 36.	24	16.25	GALAXY
SHAP412-5328.6	04 11 07.	- 53 36 12.	18	17.25	GALAXY
SHAP412-5337.8	04 11 07.	- 53 45 24.	12	17.0	GALAXY
SHAP412-5832.5	04 11 07.	- 58 40 07.	66	16.25	GALAXY
SHAP412-5423.8	04 11 08.	- 54 31 24.	18	17.0	GALAXY
SHAP412-5502.3	04 11 08.	- 55 09 54.	18	17.25	GALAXY
HN 0275	04 11 08.	- 58 40			NEBULA
IC 2049	04 11 08.	- 58 40			NONSTELLAR OBJECT
SHAP412-5909.8	04 11 08.	- 59 17 25.	36	16.25	GALAXY
SHAP412-5412.0	04 11 09.	- 54 19 36.	24	16.5	GALAXY
SHAP412-5447.5	04 11 09.	- 54 55 06.	18	17.0	GALAXY
ZWG 487.028	04 11 12.	+ 24 32		15.5	GALAXY
MCG-06-10-001	04 11 12.	- 34 33	60	15.	GALAXY
SHAP412-5230.9	04 11 12.	- 52 38 30.	24	17.25	GALAXY
SHAP412-5415.6	04 11 13.	- 54 23 12.	18	17.0	GALAXY
SHAP412-5038.2	04 11 14.	- 50 45 47.	12	17.5	GALAXY
SHAP412-5348.8	04 11 14.	- 53 56 24.	18	15.5	GALAXY
MCG+04-10-028	04 11 15.	+ 24 31 30.	30	15.5	GALAXY
SHAP412-5551.4	04 11 16.	- 55 59 00.	18	17.5	GALAXY
SHAP412-5623.2	04 11 16.	- 56 30 48.	24	16.5	GALAXY
SHAP412-5850.6	04 11 17.	- 58 58 12.	30	17.25	GALAXY
SHAP412-5255.0	04 11 19.	- 53 02 35.	18	17.0	GALAXY
SHAP412-5339.8	04 11 19.	- 53 47 23.	36	17.0	GALAXY
SHAP412-5028.1	04 11 20.	- 50 35 41.	30	17.25	GALAXY
SHAP412-5351.0	04 11 20.	- 53 58 35.	12	16.5	GALAXY
SHAP412-5128.0	04 11 21.	- 51 35 35.	12	17.0	GALAXY
SHAP412-5242.2	04 11 21.	- 52 49 47.	18	17.5	GALAXY
SHAP412-5059.8	04 11 22.	- 51 07 23.	18	17.0	GALAXY
SHAP412-5306.4	04 11 22.	- 53 13 59.	12	16.5	GALAXY
SHAP412-5410.2	04 11 23.	- 54 17 47.	36	15.5	GALAXY
SHAP412-5411.2	04 11 23.	- 54 18 47.	24	16.0	GALAXY
SHAP412-5416.2	04 11 23.	- 54 23 47.	18	17.5	GALAXY
LB 01712	04 11 24.	- 45 18		14.9	FAINT BLUE STAR
LB 01713	04 11 24.	- 50 59		15.2	FAINT BLUE STAR
SHAP412-5254.2	04 11 25.	- 53 01 47.	18	17.0	GALAXY
SHAP412-5409.1	04 11 25.	- 54 16 41.	18	17.0	GALAXY
SHAP412-5411.0	04 11 25.	- 54 18 35.	24	17.25	GALAXY
SHAP412-4957.8	04 11 27.	- 50 05 22.	18	17.5	GALAXY
SHAP412-5053.8	04 11 27.	- 51 01 23.	12	17.25	GALAXY
SHAP412-5407.6	04 11 27.	- 54 15 11.	18	17.25	GALAXY
SHAP412-5350.7	04 11 28.	- 53 58 17.	24	17.0	GALAXY
SHAP412-5521.8	04 11 29.	- 55 29 23.	18	16.5	GALAXY
UGC 02993	04 11 30.	+ 35 03	84	17.	GALAXY Sc
MCG-04-11-002	04 11 30.	- 24 10	18	15.5	GALAXY
SHAP412-5101.4	04 11 30.	- 51 08 58.	18	17.0	GALAXY
SHAP412-5614.6	04 11 30.	- 56 22 11.	24	16.75	GALAXY
SHAP412-5728.8	04 11 31.	- 57 36 23.	36	17.0	GALAXY
SHAP412-5251.8	04 11 32.	- 52 59 22.	12	17.0	GALAXY
SHAP412-5522.1	04 11 33.	- 55 29 41.	18	17.5	GALAXY
SHAP412-5357.1	04 11 34.	- 54 04 40.	12	17.0	GALAXY
SHAP412-5329.6	04 11 35.	- 53 37 10.	24	16.5	GALAXY
SHAP412-5344.2	04 11 35.	- 53 51 46.	18	17.0	GALAXY
OCL 0397	04 11 36.	+ 51 07	2340	6.8	OPEN STAR CLUSTER
RNGC 1528	04 11 36.	+ 51 07		6.5	OPEN CLUSTER
VB 256	04 11 36.	+ 53 02	302		STELLAR RING
MCG-04-11-003	04 11 36.	- 24 09	30	15.	GALAXY
SHAP412-5343.6	04 11 36.	- 53 51 10.	54	16.0	GALAXY
SHAP412-5122.5	04 11 37.	- 51 30 04.	18	16.25	GALAXY
SHAP412-5215.9	04 11 37.	- 52 23 28.	48	17.0	GALAXY
SHAP412-5343.1	04 11 38.	- 53 50 40.	60	16.5	GALAXY
MCG+00-11-027	04 11 39.	+ 02 44	18	16.	GALAXY
SHAP412-5310.0	04 11 39.	- 53 17 34.	12	16.5	GALAXY
SHAP412-5408.2	04 11 39.	- 54 15 46.	24	17.25	GALAXY
SHAP412-5415.4	04 11 39.	- 54 22 58.	36	17.0	GALAXY
SHAP413-5002.4	04 11 40.	- 50 09 58.	18	17.25	GALAXY
SHAP412-5255.6	04 11 40.	- 53 03 10.	18	17.0	GALAXY
SHAP412-5556.0	04 11 40.	- 56 03 34.	18	17.0	GALAXY
REIN 4.019A	04 11 40.66	- 12 51 28.3			NEBULA
REIN 4.019B	04 11 40.67	- 12 51 28.3			NEBULA
SHAP412-5334.4	04 11 41.	- 53 41 58.	30	17.0	GALAXY
SHAP412-5507.5	04 11 41.	- 55 15 04.	18	17.0	GALAXY
ZC 0411.7-0157	04 11 42.	- 01 57	1340		CLUSTER OF GALAXIES
MCG-02-11-026	04 11 42.	- 13 32 30.	60	15.	GALAXY
SHAP412-5313.5	04 11 42.	- 53 21 04.	12	17.25	GALAXY
SHAP412-5313.6	04 11 42.	- 53 21 10.	18	16.75	GALAXY
SHAP412-5410.2	04 11 42.	- 54 17 46.	18	17.0	GALAXY
RNGC 1543	04 11 42.	- 57 52		12.0	GALAXY
SHAP412-5417.6	04 11 43.	- 54 25 10.	30	17.25	GALAXY
HOLM 072A	04 11 44.	- 12 49	18	14.5	PART OF MULTIPLE GALAXY
SHAP413-5126.0	04 11 44.	- 51 33 33.	18	16.25	GALAXY
SHAP412-5511.6	04 11 44.	- 55 19 10.	18	17.0	GALAXY
SHAP412-5744.1	04 11 44.	- 57 51 40.	210	12.0	GALAXY
SHAP412-5410.2	04 11 45.	- 54 17 46.	30	16.5	GALAXY
SHAP413-5131.6	04 11 46.	- 51 39 09.	24	17.0	GALAXY
HOLM 072B	04 11 47.	- 12 50	12	15.3	PART OF MULTIPLE GALAXY
SHAP413-5327.2	04 11 47.	- 53 34 45.	18	16.25	GALAXY
ZWG 392.008	04 11 48.	+ 02 35		15.7	GALAXY
UGC 02994	04 11 48.	+ 02 35	102	15.7	GALAXY Sc
MCG+00-11-028	04 11 48.	+ 02 36	90	14.	GALAXY
TON-S 0406	04 11 48.	- 31 32		15.3	BLUE STAR
MCG-05-11-005	04 11 48.	- 31 45	60	13.	GALAXY
SHAP413-5102.2	04 11 48.	- 51 09 45.	18	17.25	GALAXY
LB 01283	04 11 49.	+ 09 16 12.		15.3	FAINT BLUE STAR
SHAP412-5528.8	04 11 49.	- 55 36 21.	18	17.5	GALAXY
RNGC 1537	04 11 50.	- 31 41		13.0	GALAXY
SHAP413-5041.0	04 11 50.	- 50 48 33.	18	17.25	GALAXY
ARC 0479	04 11 51.	- 03 33		17.5	RICH CLUSTER OF GALAXIES
SHAP413-5337.2	04 11 51.	- 53 44 45.	12	17.0	GALAXY
SHAP413-5350.6	04 11 51.	- 53 58 09.	18	17.5	GALAXY
SHAP413-5358.9	04 11 51.	- 54 06 27.	18	15.0	GALAXY
SHAP413-5403.1	04 11 51.	- 54 10 39.	18	16.25	GALAXY
REIN 4.021	04 11 51.14	- 12 50 30.1			NEBULA
SHAP413-5048.6	04 11 52.	- 50 56 09.	18	17.0	GALAXY
SHAP413-5406.8	04 11 52.	- 54 14 21.	12	16.75	GALAXY
SHAP413-5305.2	04 11 53.	- 53 12 45.	12	17.0	GALAXY
RLWT 097	04 11 54.	+ 56 02		12.5	FAINT VERY BLUE STAR
MCG-05-11-006	04 11 54.	- 32 07	36	14.5	GALAXY
SHAP413-5016.8	04 11 54.	- 50 24 21.	12	17.0	GALAXY
SHAP413-5122.4	04 11 54.	- 51 29 57.	18	17.25	GALAXY
SHAP413-5224.2	04 11 54.	- 52 31 45.	18	17.25	GALAXY
SHAP413-5340.4	04 11 54.	- 53 47 57.	24	16.5	GALAXY
SHAP413-5326.7	04 11 55.	- 53 34 15.	18	17.0	GALAXY
SHAP413-5329.3	04 11 55.	- 53 36 51.	18	16.0	GALAXY
SHAP413-5417.8	04 11 55.	- 54 25 21.	30	17.0	GALAXY
REIN 2.043	04 11 55.60	- 12 51 52.6			NEBULA
PK206-40.1	04 11 55.67	- 12 51 52.0	48	10.4	PLANETARY NEBULA
RNGC 1535	04 11 56.	- 12 52			PLANETARY NEBULA
SHAP413-5239.6	04 11 56.	- 52 47 09.	18	16.0	GALAXY

OBJECT NAME	RIGHT ASCEN.	DECLINATION	DIAM.	MAGN.	TYPE OF OBJECT
SHAP413-5416.2	04 11 56.	- 54 23 45.	18	17.5	GALAXY
SHAP413-5102.8	04 11 57.	- 51 10 21.	18	17.5	GALAXY
SHAP413-5246.6	04 11 57.	- 52 54 09.	12	17.5	GALAXY
SHAP413-5404.4	04 11 57.	- 54 11 57.	30	16.5	GALAXY
LB 01284	04 11 58.	+ 16 02 42.		16.2	FAINT BLUE STAR
SHAP413-5105.8	04 11 59.	- 51 13 20.	12	17.0	GALAXY
SHAP413-5303.8	04 11 59.	- 53 11 21.	18	17.5	GALAXY
KHAV 049	04 12	+ 53 50	5110		DARK NEBULA
LDN 1501	04 12 00.	+ 25 00	4440		DARK NEBULA
ZWG 362.002	04 12 00.	+ 82 18		15.7	GALAXY
ZWG 361.010	04 12 00.	+ 82 18		15.7	GALAXY
MCG-06-10-002	04 12 00.	- 38 15	30	15.5	GALAXY
SHAP413-5104.8	04 12 00.	- 51 12 20.	18	17.0	GALAXY
SHAP413-5149.0	04 12 00.	- 51 56 32.	12	17.5	GALAXY
SHAP413-5301.0	04 12 00.	- 53 08 33.	12	17.5	GALAXY
LB 01285	04 12 01.	+ 16 49 18.		16.8	FAINT BLUE STAR
SHAP413-5033.9	04 12 01.	- 50 41 26.	18	17.25	GALAXY
SHAP413-5320.8	04 12 01.	- 53 28 21.	18	17.5	GALAXY
SHAP413-5050.0	04 12 02.	- 50 57 32.	18	16.0	GALAXY
SHAP413-5340.6	04 12 02.	- 53 48 09.	18	17.25	GALAXY
LB 01286	04 12 03.	+ 08 36 48.		15.4	FAINT BLUE STAR
SHAP413-5138.9	04 12 03.	- 51 46 26.	12	17.5	GALAXY
VB 257	04 12 06.	+ 52 38	336		STELLAR RING
MCG+01-11-008	04 12 06.	- 06 50	60	15.5	GALAXY
SHAP413-5245.4	04 12 06.	- 52 52 56.	18	17.0	GALAXY
SHAP413-5303.0	04 12 07.	- 53 10 32.	36	16.5	GALAXY
LB 01287	04 12 08.	+ 15 39 18.		15.9	FAINT BLUE STAR
SHAP413-5349.9	04 12 08.	- 53 57 26.	12	17.0	GALAXY
SHAP413-5054.0	04 12 09.	- 51 01 32.	18	17.0	GALAXY
SHAP413-5100.8	04 12 09.	- 51 08 20.	24	16.25	GALAXY
SHAP413-5253.2	04 12 09.	- 53 00 44.	48	15.75	GALAXY
SHAP413-5355.6	04 12 09.	- 54 03 08.	18	17.5	GALAXY
SHAP413-5453.0	04 12 09.	- 55 00 32.	18	17.0	GALAXY
SHAP413-5100.7	04 12 11.	- 51 08 14.	18	17.5	GALAXY
SHAP413-5428.8	04 12 11.	- 54 36 20.	18	16.75	GALAXY
ACK 153-01.1	04 12 12.	+ 48 42			PLANETARY NEBULA
SHAP413-5107.0	04 12 12.	- 51 14 32.	12	16.5	GALAXY
SHAP413-5143.0	04 12 12.	- 51 50 32.	12	17.5	GALAXY
SHAP413-5150.7	04 12 14.	- 51 58 14.	24	16.5	GALAXY
SHAP413-5246.5	04 12 15.	- 52 54 02.	18	16.75	GALAXY
SHAP413-5402.6	04 12 15.	- 54 10 08.	24	16.5	GALAXY
IC 2045	04 12 16.	- 13 17 50.			NONSTELLAR OBJECT
SHAP413-5311.2	04 12 16.	- 53 18 44.	24	17.0	GALAXY
SHAP413-5354.5	04 12 16.	- 54 02 02.	12	17.5	GALAXY
REIN 4.024B	04 12 16.42	- 13 17 57.7			NEBULA
REIN 4.024A	04 12 16.47	- 13 17 57.7			NEBULA
SHAP413-5322.2	04 12 17.	- 53 29 43.	24	17.0	GALAXY
SHAP413-5052.9	04 12 18.	- 51 00 25.	18	16.5	GALAXY
SHAP413-5353.9	04 12 18.	- 54 01 25.	18	17.0	GALAXY
SHAP413-5109.8	04 12 19.	- 51 17 19.	18	16.5	GALAXY
SHAP413-5413.2	04 12 19.	- 54 20 43.	12	17.5	GALAXY
RNGC 1538	04 12 20.	- 13 20		15.0	GALAXY
SHAP413-5012.0	04 12 20.	- 50 19 31.	18	16.75	GALAXY
SHAP413-5455.8	04 12 20.	- 55 03 19.	18	17.25	GALAXY
SHAP413-5500.0	04 12 20.	- 55 07 31.	18	17.25	GALAXY
MCG-02-11-027	04 12 21.	- 13 20	24	15.	GALAXY
SHAP413-5351.6	04 12 21.	- 53 59 07.	24	16.25	GALAXY
RLWT 098	04 12 24.	+ 52 08		12.5	FAINT VERY BLUE STAR
SHAP413-5008.8	04 12 24.	- 50 16 19.	18	17.5	GALAXY
SHAP413-5246.0	04 12 25.	- 52 53 31.	24	17.25	GALAXY
SHAP413-5337.1	04 12 25.	- 53 44 37.	30	16.5	GALAXY
IC 2046	04 12 26.	- 33 14 48.			MAY NOT EXIST
SHAP413-5619.0	04 12 26.	- 56 26 31.	18	17.0	GALAXY
SHAP413-5100.6	04 12 28.	- 51 08 07.	12	16.5	GALAXY
SHAP413-5425.9	04 12 28.	- 54 33 25.	18	17.25	GALAXY
ARC 0480	04 12 29.	+ 00 53		17.6	RICH CLUSTER OF GALAXIES
SHAP413-5248.3	04 12 29.	- 52 55 49.	18	16.5	GALAXY
MCG+00-11-029	04 12 30.	- 01 03 30.	39	16.	GALAXY
SHAP413-5101.4	04 12 30.	- 51 08 54.	18	16.0	GALAXY
SHAP413-5303.2	04 12 31.	- 53 10 43.	18	15.5	GALAXY
SHAP413-5439.6	04 12 31.	- 54 47 07.	18	17.0	GALAXY
B 210	04 12 32.	+ 24 57			DARK OBJECT
SHAP413-5045.6	04 12 32.	- 50 53 06.	18	17.0	GALAXY
SHAP413-5737.6	04 12 32.	- 57 45 07.	36	17.0	GALAXY
SHAP413-5128.5	04 12 34.	- 51 36 00.	12	16.5	GALAXY
SHAP413-5355.1	04 12 35.	- 54 02 36.	12	16.5	GALAXY
SHAP413-5402.0	04 12 35.	- 54 09 30.	72	16.5	GALAXY
SHAP413-5412.2	04 12 35.	- 54 19 42.	18	16.5	GALAXY
MCG+00-11-030	04 12 36.	- 00 59	12	16.	GALAXY
IC 2047	04 12 36.	- 13 18 53.			NONSTELLAR OBJECT
SHAP413-5247.5	04 12 36.	- 52 55 00.	12	16.0	GALAXY
SHAP413-5503.8	04 12 36.	- 55 11 18.	18	17.0	GALAXY
SHAP413-5621.2	04 12 36.	- 56 28 43.	18	17.0	GALAXY
REIN 4.025	04 12 36.51	- 13 18 56.7			NEBULA
SHAP413-5214.5	04 12 37.	- 52 22 00.	18	16.25	GALAXY
SHAP413-5243.3	04 12 37.	- 52 50 48.	18	16.5	GALAXY
SHAP413-5101.6	04 12 38.	- 51 09 06.	18	16.25	GALAXY
REIN 4.026	04 12 39.46	- 13 17 57.2			NEBULA
LB 01288	04 12 41.	+ 24 02 12.		13.9	FAINT BLUE STAR
SHAP413-5325.9	04 12 41.	- 53 33 24.	18	17.0	GALAXY
REIN 4.027B	04 12 41.26	- 13 11 04.2			NEBULA
REIN 4.027A	04 12 41.26	- 13 11 04.2			NEBULA
ZC 0412.7+0054	04 12 42.	+ 00 54	1080		CLUSTER OF GALAXIES
MCG+00-11-031	04 12 42.	- 00 50 30.	54	15.5	GALAXY
SHAP414-5102.7	04 12 42.	- 51 10 12.	18	17.5	GALAXY
SHAP413-5327.7	04 12 42.	- 53 35 12.	48	15.5	GALAXY
SHAP413-5348.7	04 12 42.	- 53 56 12.	12	17.0	GALAXY
SHAP413-5936.8	04 12 42.	- 59 44 19.	48	16.5	GALAXY
HOLM 073A	04 12 43.	- 13 28	12	15.0	PART OF MULTIPLE GALAXY
HN 0276	04 12 43.	- 53 35			NEBULA
IC 2050	04 12 43.	- 53 35			NONSTELLAR OBJECT
SHAP413-5350.8	04 12 44.	- 53 58 18.	18	17.0	GALAXY
SHAP413-5427.6	04 12 44.	- 54 35 06.	18	17.25	GALAXY
SHAP413-5637.5	04 12 44.	- 56 45 00.	24	17.0	GALAXY
REIN 4.028B	04 12 45.58	- 13 21 22.1			NEBULA
REIN 4.028A	04 12 45.61	- 13 21 22.1			NEBULA
HOLM 073B	04 12 46.	- 13 28	12	15.4	PART OF MULTIPLE GALAXY
SHAP414-5225.8	04 12 46.	- 52 33 17.	18	16.75	GALAXY
SHAP413-5444.2	04 12 46.	- 54 51 42.	24	17.5	GALAXY
SHAP413-5340.1	04 12 47.	- 53 47 36.	18	16.0	GALAXY
ZC 0412.8+0950	04 12 48.	+ 09 50	1340		CLUSTER OF GALAXIES
ZWG 392.009	04 12 48.	- 01 06		15.5	GALAXY
SHAP414-5248.9	04 12 48.	- 52 56 23.	24	16.75	GALAXY
SHAP413-5356.6	04 12 48.	- 54 04 06.	24	17.0	GALAXY
SHAP413-5418.9	04 12 48.	- 54 26 24.	24	17.5	GALAXY
REIN 4.029	04 12 48.15	- 13 21 03.2			NEBULA
SHAP414-5241.4	04 12 49.	- 52 48 53.	18	17.25	GALAXY
SHAP414-5256.0	04 12 49.	- 53 03 29.	24	16.0	GALAXY
SHAP414-5340.3	04 12 50.	- 53 47 47.	12	17.0	GALAXY
SHAP413-5458.8	04 12 50.	- 55 06 18.	18	17.25	GALAXY
MCG+00-11-032	04 12 51.	- 01 06	24	15.5	GALAXY
SHAP414-5342.2	04 12 51.	- 53 49 41.	24	16.0	GALAXY
SHAP414-5417.4	04 12 51.	- 54 24 53.	18	17.5	GALAXY
SHAP414-5049.4	04 12 52.	- 50 56 53.	42	16.5	GALAXY
SHAP414-5430.2	04 12 52.	- 54 37 41.	12	17.0	GALAXY
SHAP414-5250.1	04 12 53.	- 52 57 35.	12	17.0	GALAXY
7ZW 011	04 12 54.	+ 75 27			COMPACT GALAXY
TON-S 0407	04 12 54.	- 32 17		16.0	BLUE STAR
SHAP414-5500.9	04 12 55.	- 55 08 23.	36	17.0	GALAXY
FATH 1.147	04 12 56.	- 15 26	19		NEBULA
MCG+00-11-033	04 12 57.	- 01 06	9	17.	GALAXY
SHAP414-4944.0	04 12 57.	- 49 51 29.	24	16.5	GALAXY
SHAP414-5553.4	04 12 57.	- 56 00 53.	24	16.5	GALAXY
SHAP414-5012.8	04 12 58.	- 50 20 17.	18	17.5	GALAXY
SHAP414-5022.4	04 12 58.	- 50 29 53.	12	17.0	GALAXY
SHAP414-5504.9	04 12 58.	- 55 12 23.	24	17.5	GALAXY
FATH 1.146	04 12 59.	+ 75 26	8		NEBULA
FATH 1.148	04 12 59.	- 14 28	14		NEBULA
SHAP414-5118.8	04 12 59.	- 51 26 17.	12	17.25	GALAXY
SHAP414-5239.8	04 12 59.	- 52 47 17.	18	17.25	GALAXY
SHAP414-5636.8	04 12 59.	- 56 44 17.	36	17.25	GALAXY
SHAP414-5642.0	04 12 59.	- 56 49 29.	18	17.25	GALAXY
LBN 0834	04 13	+ 08 00	4920		BRIGHT NEBULA
LBN 0832	04 13	+ 08 40	240		BRIGHT NEBULA
LBN 0831	04 13	+ 08 40	1320		BRIGHT NEBULA
KHAV 052	04 13	+ 17 50			DARK NEBULA
KHAV 050	04 13	+ 42 14	8420		DARK NEBULA
KHAV 051	04 13	+ 55 02	3810		DARK NEBULA
UGC 02995	04 13 00.	+ 00 22	66	17.	GALAXY Sc
ZWG 392.010	04 13 00.	+ 01 02		15.3	GALAXY
UGC 02996	04 13 00.	+ 01 02	72	15.3	GALAXY S0-a
MCG+00-11-034	04 13 00.	+ 01 04 30.	54	14.5	GALAXY
LDN 1394	04 13 00.	+ 55 10	1740		DARK NEBULA
SHAP414-5044.8	04 13 00.	- 50 52 16.	18	17.25	GALAXY
SHAP414-5307.2	04 13 00.	- 53 14 41.	24	16.75	GALAXY
SHAP414-5308.4	04 13 00.	- 53 15 53.	24	16.5	GALAXY
SHAP414-5500.8	04 13 00.	- 55 08 17.	18	17.0	GALAXY
SHAP414-5502.5	04 13 00.	- 55 09 59.	18	17.0	GALAXY
SHAP414-5052.6	04 13 01.	- 51 00 04.	36	17.0	GALAXY
SHAP414-5208.0	04 13 01.	- 52 15 28.	24	16.0	GALAXY
SHAP414-5307.0	04 13 01.	- 53 14 29.	24	17.25	GALAXY
SHAP414-5452.6	04 13 01.	- 55 00 05.	18	17.25	GALAXY
SHAP414-5047.3	04 13 02.	- 50 54 46.	24	15.5	GALAXY
SHAP414-5553.4	04 13 03.	- 56 00 53.	18	17.0	GALAXY
SHAP414-5809.2	04 13 05.	- 58 16 41.	36	17.0	GALAXY
SHAP414-5347.8	04 13 06.	- 53 55 16.	18	16.75	GALAXY
SHAP414-5633.2	04 13 06.	- 56 40 41.	24	17.25	GALAXY
SHAP414-5552.4	04 13 07.	- 55 59 53.	18	17.25	GALAXY
SHAP413-5938.0	04 13 07.	- 59 45 29.	42	16.25	GALAXY
ARC 0481	04 13 08.	- 10 04		17.9	RICH CLUSTER OF GALAXIES
SHAP414-5037.1	04 13 08.	- 50 44 34.	12	17.25	GALAXY
SHAP414-5421.6	04 13 08.	- 54 29 04.	24	16.5	GALAXY
SHAP414-5541.6	04 13 08.	- 55 49 04.	30	17.0	GALAXY
SHAP414-5037.8	04 13 09.	- 50 45 16.	12	17.0	GALAXY
SHAP414-5354.3	04 13 09.	- 54 01 46.	24	16.5	GALAXY
FATH 1.149	04 13 12.	- 15 24	11		NEBULA
MCG-04-11-004	04 13 12.	- 24 47 30.	48	16.	GALAXY
SHAP414-5109.4	04 13 12.	- 51 16 52.	18	17.0	GALAXY
SHAP414-5459.3	04 13 14.	- 55 06 46.	24	17.0	GALAXY
SHAP414-5953.6	04 13 14.	- 60 01 05.	48	17.0	GALAXY
SHAP414-5240.6	04 13 16.	- 52 48 04.	18	16.0	GALAXY
ZWG 418.016	04 13 18.	+ 08 04		15.0	GALAXY
UGC 02997	04 13 18.	+ 08 04	72	15.0	GALAXY S0-a
SHAP414-5420.8	04 13 18.	- 54 28 16.	18	17.0	GALAXY
SHAP414-5048.3	04 13 19.	- 50 55 45.	18	17.0	GALAXY
SHAP414-5123.8	04 13 19.	- 51 31 15.	24	17.0	GALAXY
LB 01289	04 13 20.	+ 14 06 48.		16.2	FAINT BLUE STAR
SHAP414-4944.3	04 13 21.	- 49 51 45.	18	17.0	GALAXY
SHAP414-5038.6	04 13 21.	- 50 46 03.	12	16.75	GALAXY
SHAP414-5111.5	04 13 21.	- 51 18 57.	18	16.5	GALAXY
SHAP414-5305.6	04 13 21.	- 53 13 03.	24	16.0	GALAXY
SHAP414-5406.6	04 13 21.	- 54 14 03.	18	17.0	GALAXY
SHAP414-5057.1	04 13 22.	- 51 04 33.	12	17.0	GALAXY
FATH 1.150	04 13 23.	- 14 46			NEBULA
SNO 17	04 13 23.	- 24 46 06.	420	17.	GROUP OF 6 GALAXIES
RNGC 1540	04 13 23.	- 28 32			GALAXY
SHAP414-5326.6	04 13 23.	- 53 34 03.	18	17.0	GALAXY
SHAP414-5409.4	04 13 24.	- 54 16 51.	18	16.5	GALAXY
SHAP414-5811.9	04 13 24.	- 58 19 22.	42	16.5	GALAXY
SHAP414-5158.6	04 13 25.	- 52 06 03.	18	17.0	GALAXY
LB 01290	04 13 26.	+ 10 17 12.		16.8	FAINT BLUE STAR
SHAP414-5027.4	04 13 27.	- 50 34 51.	24	16.25	GALAXY
SHAP414-5528.4	04 13 27.	- 55 35 51.	18	17.0	GALAXY
SHAP414-5009.4	04 13 28.	- 50 16 51.	24	17.0	GALAXY
SHAP414-5050.6	04 13 28.	- 50 58 03.	18	17.0	GALAXY
SHAP414-5055.8	04 13 28.	- 51 03 15.	18	16.0	GALAXY
SHAP414-5122.8	04 13 29.	- 51 30 15.	18	16.75	GALAXY
ZWG 392.011	04 13 30.	+ 01 57		15.4	GALAXY
MCG+00-11-035	04 13 30.	+ 01 57	30	15.5	GALAXY
MCG+01-11-015	04 13 30.	+ 08 03 30.	48	14.5	GALAXY
ZC 0413.5+0907	04 13 30.	+ 09 07	2220		CLUSTER OF GALAXIES
ZC 0413.5-0213	04 13 30.	- 02 13	1750		CLUSTER OF GALAXIES
MCG-02-11-028	04 13 30.	- 13 38	36	14.5	GALAXY
MCG-06-10-003	04 13 30.	- 38 11	24	16.	GALAXY
SHAP414-5255.6	04 13 30.	- 53 03 03.	18	17.5	GALAXY
SHAP414-5347.8	04 13 30.	- 53 55 15.	18	17.0	GALAXY
SHAP414-4933.5	04 13 31.	- 49 40 56.	24	16.5	GALAXY
SHAP414-5649.5	04 13 31.	- 56 56 57.	36	17.25	GALAXY
LB 01291	04 13 32.	+ 19 34 00.		16.5	FAINT BLUE STAR
SHAP414-5603.8	04 13 32.	- 56 11 15.	180	12.5	GALAXY
RNGC 1557	04 13 32.	- 70 33			NON-EXISTENT OBJECT
SHAP414-5002.5	04 13 34.	- 50 09 56.	24	17.5	GALAXY
SHAP414-5348.0	04 13 34.	- 53 55 26.	24	16.5	GALAXY
SHAP414-5350.8	04 13 34.	- 53 58 14.	24	17.25	GALAXY
RNGC 1546	04 13 34.	- 56 11		12.5	GALAXY
SHAP414-5048.6	04 13 36.	- 50 56 02.	18	17.0	GALAXY
SHAP414-5857.5	04 13 36.	- 59 04 57.	54	16.0	GALAXY
LB 03341	04 13 36.	- 63 57		12.6	FAINT BLUE STAR
SHAP414-5207.8	04 13 39.	- 52 15 14.	12	17.0	GALAXY
SHAP414-5349.6	04 13 39.	- 53 57 02.	30	16.75	GALAXY
LB 01292	04 13 40.	+ 15 00 06.		16.5	FAINT BLUE STAR
ARC 0483	04 13 40.	- 11 40		17.9	RICH CLUSTER OF GALAXIES
SHAP414-5137.3	04 13 40.	- 51 44 44.	12	17.0	GALAXY
SHAP414-5311.8	04 13 40.	- 53 19 14.	12	17.0	GALAXY
SHAP414-5335.4	04 13 40.	- 53 42 50.	18	16.5	GALAXY
MCG-04-11-005	04 13 42.	- 24 55 30.	54	15.	GALAXY

OBJECT NAME	RIGHT ASCEN.	DECLINATION	DIAM.	MAGN.	TYPE OF OBJECT
LB 01293	04 13 43.	+ 09 42 48.		17.0	FAINT BLUE STAR
SHAP414-5349.1	04 13 43.	- 53 56 32.	42	16.5	GALAXY
SHAP414-5639.6	04 13 43.	- 56 47 02.	30	17.5	GALAXY
SHAP415-5046.5	04 13 44.	- 50 53 56.	18	17.25	GALAXY
SHAP414-5356.7	04 13 45.	- 54 04 08.	18	16.25	GALAXY
LB 01294	04 13 47.	+ 14 37 24.		16.3	FAINT BLUE STAR
SHAP415-4936.3	04 13 47.	- 49 43 43.	18	16.75	GALAXY
SHAP415-5047.8	04 13 48.	- 50 55 13.	36	17.0	GALAXY
LB 03342	04 13 48.	- 74 16		13.1	FAINT BLUE STAR
MCG+00-11-036	04 13 51.	+ 00 31	24	15.5	GALAXY
B 008	04 13 51.	+ 55 07	9360		DARK OBJECT
MCG-03-11-019	04 13 51.	- 16 52	84	14.5	GALAXY
SHAP415-5021.9	04 13 51.	- 50 29 19.	30	17.5	GALAXY
SHAP415-5122.9	04 13 51.	- 51 30 19.	18	15.5	GALAXY
SHAP415-5419.8	04 13 51.	- 54 27 13.	66	15.75	GALAXY
SHAP415-5444.2	04 13 51.	- 54 51 37.	24	17.0	GALAXY
IC 2052	04 13 52.	- 54 27			NONSTELLAR OBJECT
HN 0278	04 13 52.	- 54 28			NEBULA
SHAP415-5339.2	04 13 53.	- 53 46 37.	24	16.5	GALAXY
SHAP415-5345.4	04 13 53.	- 53 52 49.	30	16.5	GALAXY
SHAP415-5354.6	04 13 53.	- 54 02 01.	24	16.5	GALAXY
ZWG 392.012	04 13 54.	+ 02 37		14.9	GALAXY
UGC 02998	04 13 54.	+ 02 37	108	14.9	GALAXY SBb
UGC 02999	04 13 54.	+ 23 23	66	16.0	GALAXY S?
ARC 0484	04 13 54.	- 07 48		16.9	RICH CLUSTER OF GALAXIES
SHAP415-5350.8	04 13 54.	- 53 58 13.	24	16.75	GALAXY
SHAP415-5411.8	04 13 56.	- 54 19 13.	18	17.25	GALAXY
SHAP415-4950.5	04 13 57.	- 49 57 55.	18	16.5	GALAXY
SHAP415-5354.7	04 13 57.	- 54 02 07.	24	17.5	GALAXY
SHAP415-5057.3	04 13 58.	- 51 04 43.	30	15.5	GALAXY
ARC 0482	04 13 59.	- 02 16		17.5	RICH CLUSTER OF GALAXIES
SHAP415-5445.1	04 13 59.	- 54 52 31.	18	17.0	GALAXY
B 009	04 14	+ 54 56			DARK OBJECT
LB 09814	04 14	- 83 13		13.3	FAINT BLUE STAR
MCG+00-11-037	04 14 00.	+ 02 38	90	13.	GALAXY
ISS 0146	04 14 00.	+ 48 09	248		STELLAR RING
SHAP415-5054.9	04 14 00.	- 51 02 19.	24	17.5	GALAXY
SHAP415-5353.3	04 14 01.	- 54 00 43.	18	16.0	GALAXY
SHAP415-5811.2	04 14 03.	- 58 18 37.	36	17.0	GALAXY
MRSL 158-05/1	04 14 06.	+ 42 30	120		HII REGION
SHAP415-5128.8	04 14 06.	- 51 36 12.	18	16.5	GALAXY
B 211	04 14 07.	+ 27 41			DARK OBJECT
SHAP415-5255.2	04 14 08.	- 53 02 36.	48	17.0	GALAXY
SHAP415-5105.2	04 14 09.	- 51 12 36.	12	17.0	GALAXY
SHAP415-5244.2	04 14 09.	- 52 51 36.	24	16.5	GALAXY
SHAP415-5324.3	04 14 09.	- 53 31 42.	30	16.25	GALAXY
SHAP415-5341.4	04 14 10.	- 53 48 48.	18	17.0	GALAXY
MCG+00-11-038	04 14 12.	+ 02 26 30.	30	15.5	GALAXY
MCG-02-11-029	04 14 12.	- 11 24 30.	36	15.	GALAXY
SHAP415-5241.6	04 14 13.	- 52 49 00.	18	17.0	GALAXY
SHAP415-5343.8	04 14 13.	- 53 51 12.	24	16.75	GALAXY
SHAP415-4958.5	04 14 14.	- 50 05 54.	12	17.0	GALAXY
SHAP415-5242.4	04 14 16.	- 52 49 48.	18	15.5	GALAXY
SHAP415-5225.4	04 14 17.	- 52 32 48.	18	16.0	GALAXY
SHAP415-5238.6	04 14 17.	- 52 46 00.	24	16.5	GALAXY
SHAP415-5554.4	04 14 17.	- 56 01 48.	18	16.5	GALAXY
SHAP415-5654.8	04 14 17.	- 57 02 12.	30	16.75	GALAXY
ZC 0414.3+0103	04 14 18.	+ 01 03	740		CLUSTER OF GALAXIES
UGC 03000	04 14 18.	+ 36 10	90	16.0	GALAXY S0-a
B 007	04 14 19.	+ 28 26			DARK OBJECT
SHAP415-5242.4	04 14 19.	- 52 49 47.	30	15.5	GALAXY
SHAP415-5107.9	04 14 21.	- 51 15 17.	18	17.0	GALAXY
SHAP415-5238.5	04 14 21.	- 52 45 53.	24	16.5	GALAXY
IC 0362	04 14 22.	- 12 19 23.			NONSTELLAR OBJECT
SHAP415-4958.4	04 14 22.	- 50 05 47.	18	17.0	GALAXY
SHAP415-5236.0	04 14 23.	- 52 43 23.	18	16.0	GALAXY
ZWG 392.013	04 14 24.	+ 00 42		14.9	GALAXY
UGC 03001	04 14 24.	+ 00 42	90	14.9	GALAXY S0-a
UGC 03002	04 14 24.	+ 02 58	72	16.0	GALAXY SBc
MCG+00-11-039	04 14 24.	+ 02 58	60	15.	GALAXY
MCG+06-10-003	04 14 24.	+ 36 11	36	16.5	GALAXY
MCG-02-11-031	04 14 24.	- 12 20 30.	18	14.	GALAXY
MCG-02-11-030	04 14 24.	- 12 32	102	14.5	GALAXY
IC 2053	04 14 24.	- 49 29			NONSTELLAR OBJECT
HN 0279	04 14 24.	- 49 30			NEBULA
SHAP415-5226.1	04 14 24.	- 52 33 29.	12	17.0	GALAXY
RNGC 1541	04 14 25.	+ 00 42		15.0	GALAXY
SHAP415-5327.9	04 14 25.	- 53 35 17.	24	17.25	GALAXY
SHAP415-5755.2	04 14 25.	- 58 02 36.	42	17.5	GALAXY
SHAP415-5634.9	04 14 29.	- 56 42 17.	30	17.25	GALAXY
MCG+00-11-040	04 14 30.	+ 00 42	66	14.5	GALAXY
ZC 0414.5+0216	04 14 30.	+ 02 16	9340		CLUSTER OF GALAXIES
MCG-06-10-004	04 14 30.	- 35 53	60	15.5	GALAXY
SHAP415-5208.2	04 14 30.	- 52 15 35.	12	17.0	GALAXY
SHAP415-5300.9	04 14 30.	- 53 08 17.	24	17.0	GALAXY
SHAP415-5400.2	04 14 31.	- 54 07 35.	90	16.5	GALAXY
SHAP415-5558.6	04 14 33.	- 56 05 59.	18	17.0	GALAXY
RNGC 1542	04 14 35.	+ 04 40		15.0	GALAXY
SHAP415-5234.8	04 14 35.	- 52 42 10.	36	16.5	GALAXY
MCG+01-11-016	04 14 36.	+ 04 39	84	14.5	GALAXY
ZWG 418.017	04 14 36.	+ 04 40		15.1	GALAXY
UGC 03003	04 14 36.	+ 04 40	90	15.1	GALAXY Sa-b
KARA.73B 0145	04 14 36.	+ 04 40	66	15.1	ISOLATED GALAXY S
SHAP415-5334.1	04 14 36.	- 53 41 28.	18	17.0	GALAXY
SHAP415-5320.7	04 14 37.	- 53 28 04.	60	17.0	GALAXY
LB 01295	04 14 38.	+ 21 55 12.		16.1	FAINT BLUE STAR
SHAP416-5016.0	04 14 39.	- 50 23 22.	12	17.5	GALAXY
SHAP416-5017.8	04 14 39.	- 50 25 10.	18	17.5	GALAXY
RNGC 1549	04 14 39.	- 55 42		11.0	GALAXY
SHAP415-5534.6	04 14 41.	- 55 41 58.	180	11.0	GALAXY
ZWG 392.014	04 14 42.	+ 02 17		14.9	GALAXY
UGC 03005	04 14 42.	+ 02 17	72	17.	GALAXY
UGC 03004	04 14 42.	+ 02 17	90	14.9	GALAXY SB
MCG-02-11-032	04 14 42.	- 11 28 30.	48	14.5	GALAXY
SHAP416-5059.5	04 14 43.	- 51 06 52.	36	16.0	GALAXY
SHAP415-5241.6	04 14 43.	- 52 48 58.	12	17.5	GALAXY
SHAP415-5325.4	04 14 43.	- 53 32 46.	24	16.5	GALAXY
LB 01296	04 14 44.	+ 12 06 06.		15.6	FAINT BLUE STAR
PATH 1.151	04 14 44.	- 15 23			NEBULA
SHAP415-5231.0	04 14 44.	- 52 38 22.	18	17.25	GALAXY
SHAP415-5302.9	04 14 44.	- 53 10 16.	36	16.0	GALAXY
MCG+00-11-042	04 14 45.	+ 02 18	48	14.5	GALAXY
MCG+00-11-041	04 14 45.	+ 02 19	60	15.	GALAXY
SHAP415-5259.9	04 14 45.	- 53 07 16.	30	16.25	GALAXY
SHAP416-5001.4	04 14 46.	- 50 08 45.	24	17.0	GALAXY
SHAP416-5257.7	04 14 46.	- 53 05 04.	12	17.0	GALAXY
SHAP415-5321.6	04 14 46.	- 53 28 58.	24	17.0	GALAXY
ZWG 392.015	04 14 48.	+ 00 10		15.4	GALAXY

OBJECT NAME	RIGHT ASCEN.	DECLINATION	DIAM.	MAGN.	TYPE OF OBJECT
MCG+00-11-043	04 14 48.	+ 00 11	15	15.	GALAXY
ZWG 392.016	04 14 48.	+ 02 14		15.0	GALAXY
UGC 03006	04 14 48.	+ 02 14	114	15.0	GALAXY S0
RLWT 099	04 14 48.	+ 56 13		14.	FAINT VERY BLUE STAR
OCL 0393	04 14 48.	+ 58 11	1140	11.2	OPEN STAR CLUSTER
SHAP416-5101.2	04 14 48.	- 51 08 33.	18	16.0	GALAXY
SHAP416-5333.9	04 14 48.	- 53 41 16.	18	17.5	GALAXY
LB 01714	04 14 48.	- 59 37		14.6	FAINT BLUE STAR
LB 03343	04 14 48.	- 79 19		14.4	FAINT BLUE STAR
SHAP416-5326.4	04 14 49.	- 53 33 46.	18	17.0	GALAXY
SHB 082	04 14 49.2	- 06 01 04.		15.	QUASI-STELLAR OBJECT
BC PKS0414-06	04 14 49.3	- 06 01 07.		16.	QUASI-STELLAR OBJECT
SHAP416-5330.4	04 14 50.	- 53 37 46.	24	16.5	GALAXY
MCG+00-11-044	04 14 51.	+ 02 14	78	14.	GALAXY
SG 2.009	04 14 51.	+ 58 09	120		DIFFUSE EMISSION NEBULA
MCG-02-11-033	04 14 51.	- 14 00 30.	36	15.5	GALAXY
MCG-03-11-020	04 14 51.	- 17 59	72	14.5	GALAXY
SHAP416-5333.4	04 14 51.	- 53 40 45.	18	17.0	GALAXY
IC 0361	04 14 52.	+ 58 11		13.	OPEN CLUSTER
SHAP416-5240.2	04 14 52.	- 52 47 33.	18	17.25	GALAXY
SHAP416-5207.0	04 14 53.	- 52 14 21.	12	17.5	GALAXY
SHAP416-5443.6	04 14 53.	- 54 50 57.	18	17.5	GALAXY
MCG+00-11-045	04 14 54.	+ 02 06 30.	30	15.	GALAXY
SHAP416-5017.8	04 14 54.	- 50 25 09.	18	16.75	GALAXY
LB 01715	04 14 54.	- 52 05		13.6	FAINT BLUE STAR
SHAP416-5210.8	04 14 54.	- 52 18 09.	18	17.5	GALAXY
SHAP416-5232.8	04 14 55.	- 52 40 09.	18	17.5	GALAXY
SHAP416-5241.7	04 14 55.	- 52 49 03.	12	16.75	GALAXY
SHAP416-5334.2	04 14 55.	- 53 41 33.	24	16.5	GALAXY
SHAP416-5056.6	04 14 58.	- 51 03 57.	18	17.0	GALAXY
SHAP416-5402.6	04 14 58.	- 54 09 57.	24	17.5	GALAXY
LB 00229	04 14 59.	+ 09 30 18.		15.5	FAINT BLUE STAR
SHAP416-5015.7	04 14 59.	- 50 23 03.	12	17.0	GALAXY
KHAV 054	04 15	+ 25 07	3400		DARK NEBULA
KHAV 053	04 15	+ 37 55	11390		DARK NEBULA
MCG+00-11-046	04 15 00.	+ 01 24	24	15.	GALAXY
LDN 1495	04 15 00.	+ 27 30	6540		DARK NEBULA
LDN 1398	04 15 00.	+ 54 20	2820		DARK NEBULA
SHAP416-5127.0	04 15 00.	- 51 34 21.	12	17.0	GALAXY
SHAP416-5151.2	04 15 00.	- 51 58 33.	54	17.0	GALAXY
SHAP416-5406.4	04 15 00.	- 54 13 45.	18	17.5	GALAXY
SHAP416-5421.9	04 15 00.	- 54 29 15.	12	17.5	GALAXY
HSE 001	04 15 00.	- 68 42			STAR CLUSTER IN LMC
ARC 0485	04 15 01.	+ 04 41		17.7	RICH CLUSTER OF GALAXIES
SHAP416-5320.0	04 15 01.	- 53 27 21.	12	16.5	GALAXY
SHAP416-4946.0	04 15 03.	- 49 53 20.	24	17.0	GALAXY
SHAP416-5320.3	04 15 03.	- 53 27 39.	18	17.0	GALAXY
SHAP416-5238.6	04 15 05.	- 52 45 56.	12	17.5	GALAXY
ZZW 007	04 15 06.	+ 01 25			COMPACT GALAXY
ZWG 392.017	04 15 06.	+ 01 25		14.9	GALAXY
SHAP416-5407.7	04 15 06.	- 54 15 03.	18	16.75	GALAXY
SHAP416-5414.4	04 15 06.	- 54 21 45.	18	17.25	GALAXY
SHAP416-5453.4	04 15 06.	- 55 00 45.	48	16.5	GALAXY
SHAP416-5818.6	04 15 06.	- 58 25 57.	36	17.0	GALAXY
LB 01297	04 15 07.	+ 16 08 36.		15.6	FAINT BLUE STAR
SHAP416-5546.8	04 15 07.	- 55 54 09.	426	10.2	GALAXY
SHAP416-5119.4	04 15 09.	- 51 26 44.	24	16.75	GALAXY
SHAP416-5341.4	04 15 09.	- 53 48 44.	18	17.5	GALAXY
RNGC 1553	04 15 09.	- 55 54		10.5	GALAXY
SHAP416-5049.1	04 15 10.	- 50 56 26.	12	17.25	GALAXY
SHAP416-5056.6	04 15 10.	- 51 03 56.	24	17.0	GALAXY
MCG+00-11-047	04 15 12.	+ 01 25	30	14.5	GALAXY
UGC 03007	04 15 12.	+ 33 39	84	17.	GALAXY Sc
SHAP416-5623.4	04 15 12.	- 56 30 44.	18	17.0	GALAXY
LB 03344	04 15 12.	- 62 50		12.8	FAINT BLUE STAR
LB 03345	04 15 12.	- 69 38		12.6	FAINT BLUE STAR
SHAP416-5250.8	04 15 13.	- 52 58 08.	18	17.5	GALAXY
SHAP416-5335.6	04 15 14.	- 53 42 56.	18	17.0	GALAXY
SHAP416-5412.4	04 15 14.	- 54 19 44.	18	17.5	GALAXY
SHAP416-5739.4	04 15 14.	- 57 46 44.	48	16.0	GALAXY
SHAP416-4937.2	04 15 15.	- 49 44 32.	18	17.0	GALAXY
SHAP416-5107.3	04 15 15.	- 51 14 38.	18	16.5	GALAXY
SHAP416-5218.9	04 15 16.	- 52 26 14.	24	17.0	GALAXY
SHAP416-5413.7	04 15 16.	- 54 21 02.	18	17.0	GALAXY
MCG+00-11-048	04 15 18.	+ 00 05 30.	42	15.	GALAXY
SHAP416-5312.4	04 15 19.	- 53 19 44.	24	17.5	GALAXY
SHAP416-5319.3	04 15 19.	- 53 26 38.	18	16.0	GALAXY
SHAP416-5341.5	04 15 19.	- 53 48 50.	18	16.0	GALAXY
SHAP416-5051.4	04 15 20.	- 50 58 43.	18	17.0	GALAXY
SHAP416-5026.2	04 15 22.	- 50 33 31.	24	17.25	GALAXY
SHAP416-5356.0	04 15 22.	- 54 03 19.	24	17.5	GALAXY
KARA.68 046	04 15 24.	- 21 18	47		DWARF GALAXY
SHAP416-5057.6	04 15 24.	- 51 04 55.	24	17.0	GALAXY
MCG-02-11-034	04 15 27.	- 12 35 30.	54	15.	GALAXY
SHAP416-5012.5	04 15 27.	- 50 19 49.	18	16.75	GALAXY
LB 01298	04 15 28.	+ 08 42 36.		15.6	FAINT BLUE STAR
SHAP416-5439.8	04 15 28.	- 54 47 07.	42	16.5	GALAXY
SHAP416-5009.0	04 15 29.	- 50 16 19.	18	16.75	GALAXY
SHAP416-5042.0	04 15 29.	- 50 49 19.	12	17.25	GALAXY
ZC 0415.5+0227	04 15 30.	+ 02 27	810		CLUSTER OF GALAXIES
SHAP416-5007.8	04 15 30.	- 50 15 07.	24	17.5	GALAXY
SHAP416-5133.2	04 15 30.	- 51 40 31.	36	15.75	GALAXY
SHAP416-5030.2	04 15 31.	- 50 37 31.	24	16.75	GALAXY
LB 01299	04 15 33.	+ 09 50 06.		17.1	FAINT BLUE STAR
SHAP416-5241.4	04 15 33.	- 52 48 43.	18	16.5	GALAXY
B 010	04 15 35.	+ 28 09	480		DARK OBJECT
SG 2.010	04 15 35.	+ 53 01	240		DIFFUSE EMISSION NEBULA
SHAP416-5102.4	04 15 35.	- 51 09 42.	18	17.0	GALAXY
ZC 0415.6+7159	04 15 36.	+ 71 59	740		CLUSTER OF GALAXIES
IC 2056	04 15 36.	- 60 20	96	12.3	GALAXY F
SHAP416-5331.3	04 15 37.	- 53 38 36.	30	16.5	GALAXY
MCG-02-11-035	04 15 39.	- 14 07	36	15.	GALAXY
SHAP417-5002.2	04 15 39.	- 50 09 30.	12	17.5	GALAXY
SHAP416-5415.4	04 15 40.	- 54 22 42.	24	16.75	GALAXY
PK151+02.1	04 15 41.	+ 52 56	250		PLANETARY NEBULA
RNGC 1547	04 15 41.	- 18 00			NON-EXISTENT OBJECT
SHAP417-5031.9	04 15 42.	- 50 39 12.	24	17.25	GALAXY
BOH 1	04 15 43.	+ 28 08			NEBULOUS OBJECT
SHAP417-5056.8	04 15 43.	- 51 04 06.	18	17.5	GALAXY
SHAP416-5315.6	04 15 43.	- 53 22 54.	24	17.0	GALAXY
SHAP416-5513.0	04 15 43.	- 55 20 18.	18	16.5	GALAXY
SHAP416-5345.5	04 15 44.	- 53 52 48.	24	17.0	GALAXY
SHAP416-5443.5	04 15 45.	- 54 50 48.	30	17.0	GALAXY
ZC 0415.8+0747	04 15 48.	+ 07 47	1680		CLUSTER OF GALAXIES
ZC 0415.8+0944	04 15 48.	+ 09 44	1680		CLUSTER OF GALAXIES
PATH 1.152	04 15 48.	- 15 33	8		NEBULA
SHAP417-5002.2	04 15 50.	- 50 09 29.	18	16.5	GALAXY
SHAP417-5118.8	04 15 50.	- 51 26 05.	24	16.5	GALAXY

OBJECT NAME	RIGHT ASCEN.	DECLINATION	DIAM.	MAGN.	TYPE OF OBJECT
SHAP417-5355.0	04 15 50.	- 54 02 18.	24	17.25	GALAXY
SHAP417-5154.8	04 15 51.	- 52 02 05.	12	17.5	GALAXY
SHAP417-5238.7	04 15 53.	- 52 45 59.	18	17.0	GALAXY
CED 030	04 15 54.	+ 28 06	60		DIFFUSE GALACTIC NEBULA
DG 029	04 15 54.	+ 28 07	600		REFLECTION NEBULA
SHAP417-5236.2	04 15 54.	- 52 43 29.	18	16.5	GALAXY
LB 03346	04 15 54.	- 61 33		12.0	FAINT BLUE STAR
SHAP417-5020.0	04 15 55.	- 50 27 17.	42	16.5	GALAXY
SHAP417-5442.9	04 15 56.	- 54 50 11.	18	17.5	GALAXY
SHAP417-5241.4	04 15 57.	- 52 48 41.	18	17.25	GALAXY
RNGC 1539	04 15 58.	+ 26 43		15.5	GALAXY
LBN 0782	04 16	+ 28 05	720		BRIGHT NEBULA
LBN 0708	04 16	+ 53 00	360		BRIGHT NEBULA
LB 09815	04 16	- 85 26		14.4	FAINT BLUE STAR
ZWG 392.018	04 16 00.	+ 02 25		14.8	GALAXY
UGC 03008	04 16 00.	+ 02 25	120	14.8	GALAXY SB?0-a
UGC 03009	04 16 00.	+ 26 05	108	16.0	GALAXY Sc
5ZW 373	04 16 00.	+ 26 43			COMPACT GALAXY
ZWG 488.001	04 16 00.	+ 26 43		15.7	GALAXY
SHAP417-5102.0	04 16 00.	- 51 09 17.	36	16.5	GALAXY
SHAP417-5409.0	04 16 00.	- 54 16 17.	24	17.0	GALAXY
SHAP417-5350.3	04 16 03.	- 53 57 35.	18	17.25	GALAXY
SHAP417-5409.8	04 16 04.	- 54 17 05.	18	16.5	GALAXY
MCG+00-11-049	04 16 06.	+ 02 26	66	14.	GALAXY
ZWG 488.002	04 16 06.	+ 26 04		15.7	GALAXY
RLWT 100	04 16 06.	+ 56 32		14.5	FAINT VERY BLUE STAR
FATH 1.153	04 16 06.	- 14 55	19		NEBULA
SHAP417-5346.6	04 16 06.	- 53 53 53.	18	17.0	GALAXY
LB 03347	04 16 06.	- 62 38		12.2	FAINT BLUE STAR
SHAP417-5026.3	04 16 07.	- 50 33 34.	12	16.25	GALAXY
SHAP417-5026.5	04 16 07.	- 50 33 46.	18	16.25	GALAXY
MCG+00-11-050	04 16 09.	+ 02 14 30.	12	15.5	GALAXY
FATH 1.154	04 16 10.	- 14 58	5		NEBULA
SHAP417-5242.4	04 16 11.	- 52 49 40.	18	16.5	GALAXY
SHAP417-5616.6	04 16 11.	- 56 23 53.	24	17.5	GALAXY
MCG+00-11-051	04 16 12.	+ 02 02 30.	30	15.5	GALAXY
B 212	04 16 12.	+ 25 11			DARK OBJECT
MCG+04-11-001	04 16 12.	+ 26 03	78	15.	GALAXY
MRSL 151+02/1	04 16 12.	+ 53 02	240		HII REGION
SHAP417-5006.7	04 16 13.	- 50 13 58.	18	16.5	GALAXY
FATH 1.155	04 16 16.	- 15 07	95		NEBULA
ZWG 392.019	04 16 18.	+ 02 53		15.4	GALAXY
ZWG 418.018	04 16 18.	+ 05 20		15.4	GALAXY
UGC 03010	04 16 18.	+ 05 20	72	15.4	GALAXY SBc
KARA.73B 0146	04 16 18.	+ 05 20	66	15.4	ISOLATED GALAXY S
SHAP417-5006.9	04 16 18.	- 50 14 09.	18	16.75	GALAXY
LB 03348	04 16 18.	- 66 16		12.6	FAINT BLUE STAR
IC 0363	04 16 20.	+ 02 55			NONSTELLAR OBJECT
SHAP417-5009.4	04 16 23.	- 50 16 39.	84	13.8	GALAXY
SHAP417-5101.9	04 16 23.	- 51 09 09.	18	17.0	GALAXY
SHAP417-5252.2	04 16 23.	- 52 59 27.	18	16.25	GALAXY
MCG-02-11-036	04 16 24.	- 11 38	54	15.	GALAXY
HW 0280	04 16 24.	- 49 03			NEBULA
SHAP417-5252.6	04 16 24.	- 52 59 51.	18	17.5	GALAXY
LB 03349	04 16 24.	- 62 53		13.3	FAINT BLUE STAR
IC 2055	04 16 25.	- 49 03			NONSTELLAR OBJECT
SHAP417-5048.3	04 16 25.	- 50 55 33.	18	17.5	GALAXY
SHAP417-5702.2	04 16 25.	- 57 09 28.	24	16.5	GALAXY
FATH 1.156	04 16 26.	- 14 36	14		NEBULA
MCG-04-11-006	04 16 27.	- 22 37	48	15.	GALAXY
SHAP417-4955.6	04 16 28.	- 50 02 51.	18	17.0	GALAXY
SHAP417-5048.3	04 16 28.	- 50 55 33.	18	17.25	GALAXY
SHAP417-5150.2	04 16 28.	- 51 57 27.	18	17.5	GALAXY
SHAP417-5856.6	04 16 28.	- 59 03 52.	36	17.0	GALAXY
IC 0364	04 16 29.	+ 03 04 18.			NONSTELLAR OBJECT
SHAP417-5100.8	04 16 29.	- 51 08 03.	12	17.0	GALAXY
ZWG 392.020	04 16 30.	+ 03 03		15.2	GALAXY
ZC 0416.5+7719	04 16 30.	+ 77 19	1480		CLUSTER OF GALAXIES
SHAP417-5014.8	04 16 31.	- 50 22 03.	18	17.25	GALAXY
SHAP417-5125.4	04 16 31.	- 51 32 39.	24	17.25	GALAXY
MCG-02-11-037	04 16 33.	- 14 20	42	15.	GALAXY
ZWG 392.021	04 16 36.	+ 03 13		14.8	GALAXY
MCG+01-11-017	04 16 36.	+ 03 13	36	14.	GALAXY
MCG+00-11-052	04 16 36.	- 02 17 30.	48	15.	GALAXY
SHAP417-4933.6	04 16 36.	- 49 40 50.	24	17.25	GALAXY
SHAP417-4955.2	04 16 36.	- 50 02 26.	18	15.75	GALAXY
SHAP417-5409.4	04 16 36.	- 54 16 39.	18	17.5	GALAXY
IC 0365	04 16 37.	+ 03 13 59.			NONSTELLAR OBJECT
SHAP417-5002.3	04 16 37.	- 50 09 32.	30	16.0	GALAXY
SHAP417-5132.0	04 16 37.	- 51 39 14.	12	17.5	GALAXY
SHAP417-5308.3	04 16 37.	- 53 15 33.	24	16.0	GALAXY
SHAP417-5849.8	04 16 37.	- 58 57 03.	42	17.25	GALAXY
SHAP418-5016.0	04 16 39.	- 50 23 14.	24	17.25	GALAXY
SHAP417-5053.6	04 16 39.	- 51 00 50.	12	17.5	GALAXY
SHAP417-5507.8	04 16 39.	- 55 15 03.	18	17.5	GALAXY
SHAP418-5018.8	04 16 41.	- 50 26 02.	18	16.25	GALAXY
ZWG 392.022	04 16 42.	+ 02 28		15.5	GALAXY
UGC 03011	04 16 42.	+ 02 29	66	15.5	GALAXY S
MCG+00-11-053	04 16 42.	+ 02 29 30.	60	15.	GALAXY
LB 01300	04 16 42.	+ 22 23 00.		16.6	FAINT BLUE STAR
ACK 149+04.1	04 16 42.	+ 56 11			PLANETARY NEBULA
MCG-02-11-038	04 16 42.	- 14 20	30	15.	GALAXY
SHAP417-5437.5	04 16 43.	- 54 44 44.	18	17.0	GALAXY
SHAP417-5356.2	04 16 45.	- 54 03 26.	24	17.0	GALAXY
SHAP417-5555.2	04 16 45.	- 56 02 26.	216	14.6	GALAXY
SHAP418-4954.6	04 16 46.	- 50 01 50.	18	16.25	GALAXY
SHAP418-4955.4	04 16 46.	- 50 02 38.	42	16.5	GALAXY
HW 0281	04 16 46.	- 56 03			NEBULA
IC 2058	04 16 46.	- 56 03	180		GALAXY S
SHAP417-5615.5	04 16 46.	- 56 22 44.	30	17.0	GALAXY
SHAP418-5002.4	04 16 47.	- 50 09 38.	24	16.5	GALAXY
SS 15	04 16 48.	+ 28 12			DIFFUSE GALACTIC NEBULA
SHAP417-5636.3	04 16 48.	- 56 43 32.	30	15.0	GALAXY
SHAP418-4955.0	04 16 49.	- 50 02 13.	24	16.75	GALAXY
SHAP418-5021.5	04 16 49.	- 50 28 43.	36	16.0	GALAXY
SHAP418-5055.8	04 16 49.	- 51 03 02.	12	17.0	GALAXY
SHAP418-5240.5	04 16 49.	- 52 47 44.	18	17.5	GALAXY
SHAP418-5247.8	04 16 50.	- 52 55 02.	24	17.25	GALAXY
SHAP418-5416.6	04 16 50.	- 54 23 50.	30	16.25	GALAXY
HW 0282	04 16 50.	- 56 44			NEBULA
IC 2060	04 16 50.	- 56 44			NONSTELLAR OBJECT
SHAP418-5251.3	04 16 51.	- 52 58 32.	24	16.75	GALAXY
SHAP418-5044.0	04 16 52.	- 50 51 13.	12	16.75	GALAXY
SHAP418-5414.7	04 16 52.	- 54 21 56.	18	16.75	GALAXY
ZC 0416.9+2601	04 16 54.	+ 26 01	4840		CLUSTER OF GALAXIES
MCG+00-11-054	04 16 54.	- 01 09	30	16.	GALAXY
MCG-02-11-039	04 16 54.	- 14 15	42	15.	GALAXY
AGU 19	04 16 54.	- 40 58 18.		14.5	2 INTERACTING GALAXIES
SHAP418-5007.5	04 16 54.	- 50 14 43.	24	17.0	GALAXY
SHAP418-4929.5	04 16 55.	- 49 36 43.	24	16.75	GALAXY
SHAP418-5059.0	04 16 55.	- 51 06 13.	12	17.25	GALAXY
SHAP418-5309.1	04 16 55.	- 53 16 19.	18	17.25	GALAXY
SHAP418-4929.0	04 16 56.	- 49 36 13.	24	16.75	GALAXY
MCG+00-11-055	04 16 57.	+ 02 18	66	13.	GALAXY
IC 0366	04 16 58.	+ 02 14 09.			NONSTELLAR OBJECT
LB 01301	04 16 58.	+ 13 10 54.		16.5	FAINT BLUE STAR
FATH 1.157	04 16 58.	- 14 37	14		NEBULA
SHAP418-5339.4	04 16 58.	- 53 46 37.	30	17.0	GALAXY
SHAP418-5410.7	04 16 58.	- 54 17 55.	36	16.75	GALAXY
FATH 1.158	04 16 59.	- 15 34	14		NEBULA
KHAV 055	04 17	+ 27 01	4430		DARK NEBULA
LBN 0731	04 17	+ 44 50	60		BRIGHT NEBULA
RNGC 1551	04 17 00.	+ 01 17			NON-EXISTENT OBJECT
ZWG 393.001	04 17 00.	+ 02 18		14.0	GALAXY
RNGC 1550	04 17 00.	+ 02 18		14.0	GALAXY
UGC 03012	04 17 00.	+ 02 18	114	14.0	GALAXY E
LDN 1511	04 17 00.	+ 24 40	2220		DARK NEBULA
LDN 1506	04 17 00.	+ 25 10	2340		DARK NEBULA
OCL 0404	04 17 00.	+ 44 48	360	15.	OPEN STAR CLUSTER
LDN 1397	04 17 00.	+ 54 45	540		DARK NEBULA
LDN 1396	04 17 00.	+ 55 00	1020		DARK NEBULA
RLWT 101	04 17 00.	+ 55 22		12.5	FAINT VERY BLUE STAR
MCG+13-04-004	04 17 00.	+ 75 11	114	13.	GALAXY
7ZW 012	04 17 00.	+ 75 12			COMPACT GALAXY
ZWG 347.004	04 17 00.	+ 75 12		13.4	GALAXY
ZWG 327.017	04 17 00.	+ 75 12		13.4	GALAXY
UGC 03013	04 17 00.	+ 75 12	312	13.4	GALAXY SBb
KARA.73B 0147	04 17 00.	+ 75 12	264	13.4	ISOLATED GALAXY S
MCG+00-12-001	04 17 00.	- 00 36	30	15.	GALAXY
MCG-06-10-006	04 17 00.	- 37 16	15	15.5	GALAXY
MCG-06-10-005	04 17 00.	- 37 37	18	15.5	GALAXY
SHAP418-5116.3	04 17 00.	- 51 23 31.	18	17.5	GALAXY
LB 01716	04 17 00.	- 52 36		12.5	FAINT BLUE STAR
SHAP418-5245.1	04 17 00.	- 52 52 19.	12	17.25	GALAXY
RNGC 1559	04 17 00.	- 62 55		12.0	GALAXY
SHAP418-5413.9	04 17 01.	- 54 21 07.	18	16.5	GALAXY
SHAP418-5014.2	04 17 02.	- 50 21 25.	18	17.0	GALAXY
SHAP418-5207.2	04 17 02.	- 52 14 25.	18	17.25	GALAXY
SHAP418-5556.0	04 17 02.	- 56 03 13.	30	16.25	GALAXY
SHAP418-5205.2	04 17 03.	- 52 12 25.	24	17.25	GALAXY
SHAP418-5547.0	04 17 04.	- 55 54 13.	18	17.0	GALAXY
LB 00211	04 17 05.	+ 14 18 54.		17.1	FAINT BLUE STAR
SHAP418-5041.0	04 17 05.	- 50 48 12.	30	16.25	GALAXY
SHAP418-5113.2	04 17 05.	- 51 20 24.	36	16.75	GALAXY
SHAP418-5142.7	04 17 05.	- 51 49 55.	18	16.75	GALAXY
ZWG 393.002	04 17 06.	+ 02 14		15.6	GALAXY
MRSL 157-03/2	04 17 06.	+ 44 48	60		HII REGION
ISS 0147	04 17 06.	+ 47 56	219		STELLAR RING
OCL 0399	04 17 06.	+ 50 08	1500	8.0	OPEN STAR CLUSTER
SHAP418-5409.6	04 17 06.	- 54 16 49.	18	17.0	GALAXY
RNGC 1545	04 17 07.	+ 50 08		8.0	OPEN CLUSTER
FATH 2.009	04 17 07.	+ 75 10	150		NEBULA
RNGC 1530	04 17 07.	+ 75 11		13.5	GALAXY
FATH 1.159	04 17 08.	- 14 32	16		NEBULA
SHAP418-5327.9	04 17 08.	- 53 35 06.	18	17.0	GALAXY
SHAP418-5558.0	04 17 08.	- 56 05 13.	18	17.5	GALAXY
SHAP418-5007.9	04 17 10.	- 50 15 06.	18	16.75	GALAXY
SHAP418-5144.4	04 17 10.	- 51 51 36.	18	17.5	GALAXY
SHAP418-5051.3	04 17 11.	- 50 58 30.	18	16.75	GALAXY
SHAP418-5142.2	04 17 11.	- 51 49 24.	18	17.0	GALAXY
SHAP418-5305.3	04 17 11.	- 53 12 30.	18	17.25	GALAXY
MCG+00-12-002	04 17 12.	+ 01 59	54	14.5	GALAXY
RLWT 102	04 17 12.	+ 48 55		17.5	FAINT VERY BLUE STAR
MCG+00-12-003	04 17 12.	- 01 03	30	15.5	GALAXY
MCG-03-12-001	04 17 12.	- 17 44	54	15.	GALAXY
MCG-04-11-007	04 17 12.	- 26 56	72	15.5	GALAXY
TON-S 0408	04 17 12.	- 30 37		15.8	BLUE STAR
SHAP418-5356.2	04 17 13.	- 54 03 24.	12	17.5	GALAXY
SHAP418-5627.2	04 17 13.	- 56 34 24.	18	16.5	GALAXY
LB 01302	04 17 14.	+ 17 48 18.		15.7	FAINT BLUE STAR
SHAP418-5002.6	04 17 15.	- 50 09 48.	18	16.75	GALAXY
ZWG 393.004	04 17 18.	+ 01 59		14.7	GALAXY
UGC 03014	04 17 18.	+ 01 59	84	14.7	GALAXY S
ZWG 393.003	04 17 18.	- 01 02		15.7	GALAXY
SHAP418-4958.8	04 17 18.	- 50 05 59.	18	16.25	GALAXY
SHAP418-5601.8	04 17 18.	- 56 09 00.	12	17.0	GALAXY
SHAP418-5244.4	04 17 20.	- 52 51 36.	18	16.5	GALAXY
MCG+00-12-004	04 17 21.	- 01 56	24	15.5	GALAXY
MCG-04-11-008	04 17 21.	- 26 51 30.	24	15.	GALAXY
SHAP418-5245.2	04 17 21.	- 52 52 24.	12	17.25	GALAXY
ARP 020	04 17 22.	+ 01 58			PECULIAR GALAXY
SHAP418-4933.8	04 17 22.	- 49 40 59.	18	17.0	GALAXY
LB 01303	04 17 23.	+ 09 12 42.		15.7	FAINT BLUE STAR
SHAP418-4936.8	04 17 23.	- 49 43 59.	24	16.5	GALAXY
SHAP418-5202.2	04 17 23.	- 52 09 23.	18	16.75	GALAXY
SHAP418-5403.4	04 17 23.	- 54 10 36.	36	17.5	GALAXY
LB 03350	04 17 24.	- 64 02		11.9	FAINT BLUE STAR
SHAP418-5237.4	04 17 25.	- 52 44 35.	18	17.25	GALAXY
SHAP418-5321.2	04 17 26.	- 53 28 23.	18	17.5	GALAXY
SHAP418-5347.1	04 17 28.	- 53 54 17.	36	16.5	GALAXY
SHAP418-5038.6	04 17 29.	- 50 45 47.	24	17.5	GALAXY
MCG+00-12-005	04 17 30.	+ 00 06	30	15.5	GALAXY
MCG+14-03-004	04 17 30.	+ 82 18	27	17.	GALAXY
SHAP418-5040.0	04 17 30.	- 50 47 11.	18	17.5	GALAXY
SHAP418-4926.2	04 17 31.	- 49 33 23.	24	17.25	GALAXY
SHAP418-5622.6	04 17 31.	- 56 29 47.	36	16.5	GALAXY
SHAP418-5154.6	04 17 32.	- 52 01 47.	18	17.5	GALAXY
MCG+00-12-006	04 17 33.	- 00 07 30.	36	15.	GALAXY
SHAP418-5128.6	04 17 34.	- 51 35 47.	18	17.0	GALAXY
SHAP418-5215.2	04 17 34.	- 52 22 23.	18	17.0	GALAXY
SHAP418-5238.5	04 17 34.	- 52 45 41.	18	15.5	GALAXY
LB 01304	04 17 35.	+ 20 49 06.		16.1	FAINT BLUE STAR
SHAP418-5346.8	04 17 35.	- 53 53 59.	18	17.5	GALAXY
SHAP418-5414.7	04 17 35.	- 54 21 53.	24	17.25	GALAXY
SHAP418-5547.8	04 17 35.	- 55 54 59.	24	17.0	GALAXY
ZC 0417.6+0736	04 17 36.	+ 07 36	1210		CLUSTER OF GALAXIES
OCL 0415	04 17 36.	+ 36 48			OPEN STAR CLUSTER
SHAP418-5041.5	04 17 36.	- 50 48 40.	12	17.5	GALAXY
SHAP418-5241.4	04 17 36.	- 52 48 35.	12	17.0	GALAXY
SHAP418-5412.0	04 17 36.	- 54 19 11.	36	16.75	GALAXY
SHAP418-5110.2	04 17 38.	- 51 17 22.	18	17.25	GALAXY
SHAP418-5531.6	04 17 38.	- 55 38 47.	12	17.25	GALAXY
RNGC 1548	04 17 39.	+ 36 48			OPEN CLUSTER
SHAP419-4927.8	04 17 39.	- 49 34 58.	18	17.5	GALAXY
SHAP418-5649.8	04 17 39.	- 56 56 59.	24	17.25	GALAXY
SHAP418-5449.8	04 17 40.	- 54 56 59.	18	17.0	GALAXY

OBJECT NAME	RIGHT ASCEN.	DECLINATION	DIAM.	MAGN.	TYPE OF OBJECT
SHAP418-5709.8	04 17 40.	- 57 16 59.	30	16.5	GALAXY
ZC 0417.7+3557	04 17 42.	+ 35 57	7530		CLUSTER OF GALAXIES
ZWG 393.005	04 17 42.	- 00 48		14.4	GALAXY
UGC 03015	04 17 42.	- 00 48	126	14.4	GALAXY S0
RNGC 1558	04 17 43.	- 45 10			GALAXY
SHAP418-5218.2	04 17 43.	- 52 25 22.	18	16.5	GALAXY
RNGC 1552	04 17 44.	- 00 48		14.5	GALAXY
SHAP418-5556.8	04 17 44.	- 56 03 58.	18	17.0	GALAXY
MCG+00-12-007	04 17 45.	- 00 50	24	13.	GALAXY
SHAP419-5004.9	04 17 45.	- 50 12 04.	24	17.5	GALAXY
SHAP418-5413.6	04 17 45.	- 54 20 46.	30	17.25	GALAXY
LB 01305	04 17 46.	+ 22 06 36.		16.0	FAINT BLUE STAR
SHAP418-5706.6	04 17 46.	- 57 13 46.	42	16.5	GALAXY
SHAP419-5305.0	04 17 47.	- 53 12 10.	12	17.5	GALAXY
SHAP419-5015.0	04 17 48.	- 50 22 10.	24	17.0	GALAXY
SHAP419-5048.2	04 17 48.	- 50 55 22.	18	17.5	GALAXY
SHAP419-5000.4	04 17 49.	- 50 07 33.	24	16.25	GALAXY
SHAP419-5014.8	04 17 49.	- 50 21 57.	18	17.0	GALAXY
SHAP419-4948.4	04 17 50.	- 49 55 33.	24	16.25	GALAXY
SHAP419-5049.4	04 17 50.	- 50 56 34.	12	17.5	GALAXY
SHAP419-5139.9	04 17 51.	- 51 47 04.	18	16.25	GALAXY
SHAP418-5622.4	04 17 51.	- 56 29 34.	18	17.0	GALAXY
SHAP419-5001.8	04 17 52.	- 50 08 57.	18	17.25	GALAXY
SHAP419-5116.8	04 17 52.	- 51 23 57.	24	17.0	GALAXY
SHAP418-5701.8	04 17 53.	- 57 08 58.	24	17.0	GALAXY
SHAP419-4948.2	04 17 54.	- 49 55 21.	24	16.0	GALAXY
SHAP419-5001.5	04 17 54.	- 50 08 39.	18	17.25	GALAXY
LB 01717	04 17 54.	- 53 22		13.3	FAINT BLUE STAR
SHAP419-5321.2	04 17 54.	- 53 28 21.	24	15.75	GALAXY
SHAP419-5235.7	04 17 55.	- 52 42 51.	12	17.5	GALAXY
MCG+00-12-008	04 17 57.	- 00 52 30.	42	15.5	GALAXY
SHAP419-5217.8	04 17 57.	- 52 24 57.	18	16.5	GALAXY
SHAP419-5145.6	04 17 58.	- 51 52 45.	66	16.5	GALAXY
SHAP419-5142.0	04 17 59.	- 51 49 09.	12	16.75	GALAXY
SHAP419-5222.6	04 17 59.	- 52 29 45.	24	17.25	GALAXY
SHAP419-5433.2	04 17 59.	- 54 40 21.	24	17.0	GALAXY
SHAP419-5541.6	04 17 59.	- 55 48 45.	18	16.5	GALAXY
SHAP419-5625.6	04 17 59.	- 56 32 45.	18	17.5	GALAXY
LBN 0735	04 18	+ 44 15	240		BRIGHT NEBULA
MCG+00-12-009	04 18 00.	+ 02 13	30	15.5	GALAXY
UGC 03016	04 18 00.	+ 36 39	96	17.	GALAXY Sc
MRSL 157-03/1	04 18 00.	+ 44 15	240		HII REGION
ZWG 393.006	04 18 00.	- 02 09		14.9	GALAXY
MCG+00-12-010	04 18 00.	- 02 10	36	14.5	GALAXY
SHAP419-4926.8	04 18 01.	- 49 33 57.	24	16.75	GALAXY
SHAP419-5238.6	04 18 02.	- 52 45 45.	18	17.0	GALAXY
SHAP419-5256.6	04 18 02.	- 53 03 45.	12	17.0	GALAXY
SHAP419-5349.4	04 18 02.	- 53 56 33.	24	17.0	GALAXY
SHAP419-4943.6	04 18 03.	- 49 50 45.	24	17.25	GALAXY
SHAP419-5048.2	04 18 03.	- 50 55 21.	18	16.5	GALAXY
SHAP419-5137.8	04 18 03.	- 51 44 57.	18	17.5	GALAXY
SHAP419-5235.0	04 18 03.	- 52 42 09.	12	17.25	GALAXY
LB 01306	04 18 05.	+ 12 55 54.		15.7	FAINT BLUE STAR
SHAP419-5054.0	04 18 05.	- 51 01 09.	60	15.5	GALAXY
SHAP419-5353.6	04 18 05.	- 54 00 45.	24	16.75	GALAXY
SHAP419-5443.6	04 18 05.	- 54 50 45.	24	17.5	GALAXY
ZWG 393.007	04 18 06.	+ 02 12		15.0	GALAXY
B 213	04 18 06.	+ 26 56			DARK OBJECT
SHAP419-5246.8	04 18 06.	- 52 53 57.	12	17.0	GALAXY
LB 01718	04 18 06.	- 53 37		15.8	FAINT BLUE STAR
SHAP419-5543.2	04 18 06.	- 55 50 21.	12	17.5	GALAXY
SHAP419-5814.9	04 18 06.	- 58 22 03.	72	16.0	GALAXY
LB 01307	04 18 07.	+ 13 23 12.		16.2	FAINT BLUE STAR
SHAP419-5135.6	04 18 07.	- 51 42 44.	18	17.0	GALAXY
SHAP419-5300.1	04 18 07.	- 53 07 15.	18	17.0	GALAXY
SHAP419-5655.4	04 18 07.	- 57 02 33.	18	17.0	GALAXY
SHAP419-4955.6	04 18 09.	- 50 02 44.	24	17.0	GALAXY
SHAP419-5026.8	04 18 10.	- 50 33 56.	24	16.25	GALAXY
SHAP419-5235.9	04 18 11.	- 52 43 02.	12	17.25	GALAXY
SHAP419-5410.0	04 18 11.	- 54 17 08.	18	17.0	GALAXY
SHAP419-5416.8	04 18 11.	- 54 23 56.	24	16.5	GALAXY
MCG+00-12-011	04 18 12.	+ 02 31 30.	48	15.	GALAXY
UGC 03017	04 18 12.	+ 05 33	72	16.5	GALAXY
SHAP419-5552.4	04 18 12.	- 55 59 33.	18	16.75	GALAXY
SHAP419-5128.6	04 18 13.	- 51 35 44.	12	17.5	GALAXY
SHAP419-5243.2	04 18 14.	- 52 50 20.	12	16.75	GALAXY
SHAP419-5134.6	04 18 15.	- 51 41 44.	12	17.0	GALAXY
SHAP419-5408.7	04 18 16.	- 54 15 50.	18	17.5	GALAXY
ZWG 393.008	04 18 18.	+ 02 31		15.2	GALAXY
SHAP419-5000.6	04 18 18.	- 50 07 44.	18	17.0	GALAXY
SHAP419-5012.0	04 18 18.	- 50 19 08.	24	17.0	GALAXY
SHAP419-5240.0	04 18 18.	- 52 47 08.	18	16.0	GALAXY
SHAP419-5357.0	04 18 20.	- 54 04 08.	54	16.25	GALAXY
MCG-02-12-001	04 18 21.	- 14 54	18	15.	GALAXY
SHAP419-5619.6	04 18 21.	- 56 26 44.	30	17.25	GALAXY
SHAP419-5122.9	04 18 23.	- 51 30 01.	18	17.5	GALAXY
UGC 03019	04 18 24.	+ 36 37	72	17.	GALAXY S
MCG+06-10-004	04 18 24.	+ 36 37	48	17.	GALAXY
ISS 0148	04 18 24.	+ 48 52	155		STELLAR RING
ZWG 393.009	04 18 24.	- 00 17		15.3	GALAXY
UGC 03018	04 18 24.	- 00 17	78	15.3	GALAXY S
MCG+00-12-012	04 18 24.	- 00 17 30.	30	15.	GALAXY
IC 0367	04 18 24.	- 14 54 09.			NONSTELLAR OBJECT
SHAP419-5039.2	04 18 24.	- 50 46 19.	12	17.0	GALAXY
SHAP419-5111.5	04 18 24.	- 51 18 37.	18	16.75	GALAXY
SHAP419-5131.0	04 18 24.	- 51 38 07.	24	17.5	GALAXY
SHAP419-5238.6	04 18 25.	- 52 45 43.	12	17.0	GALAXY
SHAP419-5317.8	04 18 25.	- 53 24 55.	18	17.5	GALAXY
LB 01308	04 18 26.	+ 09 13 54.		16.9	FAINT BLUE STAR
SHAP419-5106.0	04 18 26.	- 51 13 07.	18	16.25	GALAXY
IC 2059	04 18 27.	- 31 34 30.			NONSTELLAR OBJECT
SHAP419-5309.0	04 18 28.	- 53 16 07.	18	17.25	GALAXY
SHAP419-5003.2	04 18 29.	- 50 10 19.	24	16.75	GALAXY
UGC 03020	04 18 30.	- 02 25	72	16.5	GALAXY
MCG+00-12-013	04 18 30.	- 02 25	48	15.5	GALAXY
MCG+01-12-001	04 18 30.	- 04 51 30.	66	15.5	GALAXY
MCG-02-12-002	04 18 30.	- 09 52	12	15.5	GALAXY
MCG-05-11-007	04 18 30.	- 31 50	48	15.	GALAXY
SHAP419-5314.3	04 18 31.	- 53 21 25.	12	17.0	GALAXY
SHAP419-5157.5	04 18 32.	- 52 04 37.	18	16.5	GALAXY
SHAP419-5321.2	04 18 32.	- 53 28 19.	30	17.5	GALAXY
HUB C06	04 18 33.	+ 35 14			DIFFUSE NEBULA
SHAP419-5137.6	04 18 33.	- 51 44 43.	18	17.5	GALAXY
SHAP419-5510.2	04 18 33.	- 55 17 19.	18	17.0	GALAXY
SHAP419-5131.0	04 18 34.	- 51 38 07.	18	17.5	GALAXY
SHAP419-5552.6	04 18 34.	- 55 59 43.	24	17.5	GALAXY
SHAP419-5139.6	04 18 35.	- 51 46 43.	30	17.25	GALAXY
ZC 0418.6+0158	04 18 36.	+ 01 58	1080		CLUSTER OF GALAXIES

OBJECT NAME	RIGHT ASCEN.	DECLINATION	DIAM.	MAGN.	TYPE OF OBJECT
UGC 03021	04 18 36.	+ 36 00	138	17.	GALAXY E
MCG+06-10-006	04 18 36.	+ 36 01	6	17.	GALAXY
MCG+06-10-005	04 18 36.	+ 36 01	18	17.	GALAXY
SHAP419-5322.0	04 18 36.	- 53 29 07.	12	17.5	GALAXY
SHAP419-5702.2	04 18 36.	- 57 09 19.	18	17.0	GALAXY
SHAP419-5020.4	04 18 37.	- 50 27 30.	18	17.25	GALAXY
SHAP419-5038.6	04 18 38.	- 50 45 42.	18	17.25	GALAXY
SHAP419-5336.6	04 18 38.	- 53 43 43.	18	17.5	GALAXY
LB 01309	04 18 39.	+ 08 25 18.		16.4	FAINT BLUE STAR
SHAP420-5015.5	04 18 39.	- 50 22 36.	30	17.0	GALAXY
SHAP420-5021.6	04 18 39.	- 50 28 42.	24	16.5	GALAXY
SHAP419-5124.6	04 18 40.	- 51 31 42.	18	16.5	GALAXY
SHAP419-5154.7	04 18 40.	- 52 01 48.	30	16.5	GALAXY
SHAP419-5406.6	04 18 40.	- 54 13 42.	12	17.0	GALAXY
SHAP419-5438.5	04 18 40.	- 54 45 37.	18	17.5	GALAXY
SHAP419-5516.0	04 18 40.	- 55 23 07.	18	17.5	GALAXY
SHAP420-4926.2	04 18 41.	- 49 33 18.	24	16.75	GALAXY
SHAP419-5514.1	04 18 41.	- 55 21 13.	18	17.25	GALAXY
SHAP419-5531.8	04 18 41.	- 55 38 55.	18	17.5	GALAXY
ZWG 419.001	04 18 42.	+ 03 21		15.6	GALAXY
MCG-02-12-003	04 18 42.	- 09 50	48	15.	GALAXY
CED 032A	04 18 43.	+ 19 24			DIFFUSE GALACTIC NEBULA
SHAP420-5100.5	04 18 43.	- 51 07 36.	18	17.25	GALAXY
SHAP420-5109.6	04 18 43.	- 51 16 42.	12	16.0	GALAXY
SHAP420-5147.6	04 18 43.	- 51 54 42.	24	16.5	GALAXY
HUB C05	04 18 44.	+ 23 22			DIFFUSE NEBULA
SHAP420-5049.2	04 18 44.	- 50 56 18.	18	17.25	GALAXY
SHAP420-5207.2	04 18 44.	- 52 14 18.	24	16.25	GALAXY
SHAP419-5703.8	04 18 44.	- 57 10 55.	30	17.0	GALAXY
MCG+01-12-001	04 18 45.	+ 03 19	48	15.	GALAXY
MCG-02-12-004	04 18 45.	- 09 53	24	15.5	GALAXY
MCG-04-11-009	04 18 45.	- 23 52	24	15.	GALAXY
SHAP420-5059.9	04 18 45.	- 51 07 00.	18	17.0	GALAXY
SHAP419-5632.0	04 18 46.	- 56 39 06.	24	17.0	GALAXY
SHAP419-5347.2	04 18 47.	- 53 54 18.	18	17.5	GALAXY
SHAP419-5617.1	04 18 47.	- 56 24 12.	18	17.25	GALAXY
CED 031	04 18 48.	+ 28 20	330		DIFFUSE GALACTIC NEBULA
SG 3.022	04 18 48.	+ 28 20	180		DIFFUSE EMISSION NEBULA
DG 030	04 18 48.	+ 28 21	240		REFLECTION NEBULA
SHAP420-4940.9	04 18 48.	- 49 48 00.	24	17.0	GALAXY
SHAP420-5011.6	04 18 48.	- 50 18 42.	18	17.0	GALAXY
B 214	04 18 49.	+ 28 26	300		DARK OBJECT
RNGC 1566	04 18 50.	- 55 05		10.5	GALAXY
SHAP420-5125.7	04 18 51.	- 51 32 47.	30	16.25	GALAXY
SHAP420-4943.0	04 18 52.	- 49 50 05.	18	17.5	GALAXY
SHAP420-5029.5	04 18 52.	- 50 36 35.	18	16.75	GALAXY
SHAP420-5125.6	04 18 52.	- 51 32 41.	12	16.0	GALAXY
SHAP420-5321.0	04 18 52.	- 53 28 06.	18	16.5	GALAXY
SHAP420-5456.3	04 18 53.	- 55 03 24.	498	10.5	GALAXY
MRSL 176-20/1	04 18 54.	+ 19 25	60		HII REGION
SHAP420-4942.2	04 18 54.	- 49 49 17.	30	17.25	GALAXY
SHAP420-5121.4	04 18 54.	- 51 28 29.	12	16.5	GALAXY
LB 01719	04 18 54.	- 52 41		14.5	FAINT BLUE STAR
SHAP420-5250.9	04 18 54.	- 52 57 59.	18	17.0	GALAXY
SHAP420-5358.2	04 18 54.	- 54 05 18.	18	17.5	GALAXY
VVI 15	04 18 54.	- 55 04	660	10.4	SEYFERT GALAXY
SHAP420-5137.5	04 18 55.	- 51 44 35.	12	16.75	GALAXY
SHAP420-5325.2	04 18 55.	- 53 32 17.	18	17.0	GALAXY
LB 00212	04 18 56.	+ 15 21 54.		16.6	FAINT BLUE STAR
VDB.66N 028	04 18 57.	+ 19 26	744		REFLECTION NEBULA
SHAP420-4927.3	04 18 58.	- 49 34 23.	24	17.25	GALAXY
SHAP420-5121.9	04 18 58.	- 51 28 59.	12	17.25	GALAXY
SHAP420-5324.2	04 18 58.	- 53 31 17.	18	17.0	GALAXY
LBN 0817	04 19	+ 19 30	420		BRIGHT NEBULA
LBN 0785	04 19	+ 28 20	300		BRIGHT NEBULA
DG 031	04 19 00.	+ 19 24	60		REFLECTION NEBULA
MCG-04-11-010	04 19 00.	- 21 57	360	13.	GALAXY
SHAP420-5650.8	04 19 00.	- 56 57 53.	48	16.5	GALAXY
SS 16	04 19 01.	+ 19 24			DIFFUSE GALACTIC NEBULA
CED 032B	04 19 01.	+ 19 24			DIFFUSE GALACTIC NEBULA
SHAP420-5120.5	04 19 01.	- 51 27 35.	12	17.0	GALAXY
MCG+00-12-014	04 19 03.	- 02 28 30.	36	15.	GALAXY
SHAP420-5256.1	04 19 03.	- 53 03 11.	12	17.5	GALAXY
SHAP420-5322.0	04 19 04.	- 53 29 05.	24	16.5	GALAXY
SHAP420-5517.0	04 19 05.	- 55 24 05.	18	17.0	GALAXY
VDB.66N 027	04 19 06.	+ 28 19	516		REFLECTION NEBULA
ZWG 393.010	04 19 06.	- 02 27		15.7	GALAXY
SHAP420-5051.5	04 19 06.	- 50 58 34.	12	16.25	GALAXY
SHAP420-5120.2	04 19 06.	- 51 27 17.	18	16.75	GALAXY
CED 032C	04 19 07.	+ 19 25			DIFFUSE GALACTIC NEBULA
SHAP420-5011.6	04 19 08.	- 50 18 40.	24	17.0	GALAXY
SHAP420-5600.8	04 19 08.	- 56 07 53.	18	16.75	GALAXY
MCG-03-12-002	04 19 09.	- 18 55	15	15.5	GALAXY
SHAP420-5049.8	04 19 09.	- 50 56 52.	18	17.0	GALAXY
SHAP420-5328.1	04 19 10.	- 53 35 10.	18	17.5	GALAXY
SHAP420-5331.3	04 19 10.	- 53 38 22.	24	16.5	GALAXY
SHAP420-5443.2	04 19 10.	- 54 50 17.	24	17.5	GALAXY
MCG+00-12-016	04 19 12.	+ 01 43	24	15.	GALAXY
ZWG 393.011	04 19 12.	- 02 19		15.6	GALAXY
MCG+00-12-015	04 19 12.	- 02 20 30.	24	15.	GALAXY
LB 03351	04 19 12.	- 72 59		14.5	FAINT BLUE STAR
SHAP420-5215.8	04 19 13.	- 52 22 52.	18	17.25	GALAXY
SHAP420-5017.8	04 19 14.	- 50 24 52.	18	16.25	GALAXY
SHAP420-5200.0	04 19 14.	- 52 07 04.	12	16.25	GALAXY
MCG+00-12-016A	04 19 15.	+ 01 43 30.	24	15.	GALAXY
LB 01310	04 19 15.	+ 22 24 42.		15.4	FAINT BLUE STAR
SHAP420-5227.6	04 19 15.	- 52 34 40.	12	16.75	GALAXY
SHAP420-5443.7	04 19 15.	- 54 50 46.	36	16.75	GALAXY
SHAP420-5503.4	04 19 15.	- 55 10 28.	18	17.5	GALAXY
SHAP420-5014.2	04 19 17.	- 50 21 16.	18	17.5	GALAXY
SHAP420-5103.2	04 19 17.	- 51 10 16.	18	17.25	GALAXY
SHAP420-5641.6	04 19 17.	- 56 48 40.	18	17.0	GALAXY
ZWG 419.002	04 19 18.	+ 03 56		15.2	GALAXY
KARA.73B 0148	04 19 18.	+ 03 56	42	15.2	ISOLATED GALAXY S
IC 2057	04 19 18.	+ 03 56 34.			NONSTELLAR OBJECT
MCG+00-12-017	04 19 18.	- 02 19	24	15.5	GALAXY
SHAP420-5344.9	04 19 19.	- 53 51 58.	30	17.0	GALAXY
SHAP420-5534.0	04 19 19.	- 55 41 04.	24	17.0	GALAXY
REIN 4.030	04 19 20.66	- 16 06 44.7			NEBULA
SHAP420-4959.4	04 19 21.	- 50 06 27.	18	16.75	GALAXY
SHAP420-5310.2	04 19 22.	- 53 17 16.	18	17.0	GALAXY
SHAP420-5512.6	04 19 22.	- 55 19 40.	18	17.0	GALAXY
SHAP420-5206.8	04 19 23.	- 52 13 51.	18	17.25	GALAXY
ZWG 393.012	04 19 24.	+ 01 43		14.9	GALAXY
UGC 03023	04 19 24.	+ 01 43	132	14.9	GALAXY SB0
UGC 03024	04 19 24.	+ 27 11	90	16.5	GALAXY E-S0
RLWT 103	04 19 24.	+ 50 01		16.5	FAINT VERY BLUE STAR
UGC 03022	04 19 24.	- 00 56	66	16.0	GALAXY S

OBJECT NAME	RIGHT ASCEN.	DECLINATION	DIAM.	MAGN.	TYPE OF OBJECT
MCG+00-12-018	04 19 24.	- 00 56	24	15.5	GALAXY
MCG-03-12-003	04 19 24.	- 16 06	60	15.	GALAXY
KW 50	04 19 24.	- 55 01			SEYFERT GALAXY
SHAP420-5258.1	04 19 25.	- 53 05 09.	18	17.5	GALAXY
SHAP420-4949.8	04 19 26.	- 49 56 51.	24	17.0	GALAXY
SHAP420-5216.2	04 19 26.	- 52 23 15.	12	16.0	GALAXY
MCG-03-12-004	04 19 27.	- 18 03	66	15.5	GALAXY
SHAP420-5349.8	04 19 27.	- 53 56 51.	24	16.75	GALAXY
RNGC 1562	04 19 28.	- 15 52			GALAXY
SHAP420-5142.8	04 19 28.	- 51 49 51.	12	17.0	GALAXY
SHAP420-5243.0	04 19 28.	- 52 50 03.	18	17.5	GALAXY
SHAP420-5252.6	04 19 28.	- 52 59 39.	24	17.25	GALAXY
SHAP420-5201.8	04 19 29.	- 52 08 51.	18	16.75	GALAXY
ZWG 393.013	04 19 30.	- 02 47		15.4	GALAXY
MCG+01-12-002	04 19 30.	- 06 22	42	15.	GALAXY
SHAP420-5449.4	04 19 30.	- 54 56 27.	18	16.75	GALAXY
MCG+00-12-019	04 19 33.	- 02 47 30.	30	15.	GALAXY
SHAP420-5316.8	04 19 34.	- 53 23 51.	24	16.75	GALAXY
LB 01311	04 19 35.	+ 09 18 54.		16.4	FAINT BLUE STAR
SHAP420-5022.2	04 19 35.	- 50 29 15.	24	16.0	GALAXY
SHAP420-5236.0	04 19 35.	- 52 43 03.	18	17.0	GALAXY
TON-S 0409	04 19 36.	- 30 08		14.2	BLUE STAR
LB 01312	04 19 37.	+ 21 47 00.		16.0	FAINT BLUE STAR
SHAP420-5533.6	04 19 37.	- 55 40 39.	18	17.0	GALAXY
SHAP420-5049.9	04 19 38.	- 50 56 56.	12	16.5	GALAXY
LB 01313	04 19 39.	+ 17 43 36.		15.4	FAINT BLUE STAR
SHAP420-5044.8	04 19 39.	- 50 51 50.	18	16.25	GALAXY
SHAP420-5542.6	04 19 39.	- 55 49 39.	18	17.0	GALAXY
SHAP420-5634.4	04 19 39.	- 56 41 27.	18	17.25	GALAXY
SHAP420-5207.0	04 19 40.	- 52 14 02.	12	16.75	GALAXY
SHAP420-5318.5	04 19 40.	- 53 25 32.	24	17.5	GALAXY
SHAP420-5406.8	04 19 40.	- 54 13 51.	18	17.5	GALAXY
RNGC 1567	04 19 41.	- 48 23			UNVERIFIED SOUTHERN OBJECT
SHAP421-5052.0	04 19 41.	- 50 59 02.	18	16.75	GALAXY
SHAP421-5055.4	04 19 41.	- 51 02 26.	24	16.5	GALAXY
ZWG 419.003	04 19 42.	+ 07 06		15.2	GALAXY
KARA.73B 0149	04 19 42.	+ 07 06	42	15.2	ISOLATED GALAXY S
MCG-02-12-005	04 19 42.	- 10 16 30.	84	14.	GALAXY
MCG-02-12-006	04 19 42.	- 12 53	42	14.5	GALAXY
SHAP421-5044.5	04 19 43.	- 50 51 32.	18	16.75	GALAXY
SHAP420-5455.8	04 19 43.	- 55 02 50.	18	17.5	GALAXY
SHAP420-5504.8	04 19 43.	- 55 11 50.	24	17.0	GALAXY
SHAP420-5619.0	04 19 43.	- 56 26 03.	24	17.0	GALAXY
SHAP420-5312.2	04 19 44.	- 53 19 14.	18	17.0	GALAXY
SHAP420-5345.4	04 19 44.	- 53 52 26.	18	17.5	GALAXY
SHAP420-5402.2	04 19 44.	- 54 09 14.	18	17.0	GALAXY
SHAP420-5702.4	04 19 44.	- 57 09 27.	24	17.5	GALAXY
SHAP421-5010.2	04 19 47.	- 50 17 14.	18	16.75	GALAXY
SHAP420-5334.5	04 19 47.	- 53 41 32.	24	16.5	GALAXY
SHAP420-5345.2	04 19 47.	- 53 52 14.	18	17.5	GALAXY
SHAP420-5609.2	04 19 47.	- 56 16 14.	18	16.5	GALAXY
2ZW 008	04 19 48.	- 03 45			COMPACT GALAXY
LB 01314	04 19 50.	+ 17 02 06.		16.3	FAINT BLUE STAR
SHAP421-5205.8	04 19 50.	- 52 12 50.	24	16.25	GALAXY
SHAP421-5406.8	04 19 50.	- 54 13 50.	18	16.75	GALAXY
MCG-02-12-007	04 19 51.	- 10 37 30.	36	15.	GALAXY
SHAP420-5637.6	04 19 55.	- 56 44 38.	18	16.75	GALAXY
RNGC 1555	04 19 56.	+ 19 25			DIFFUSE NEBULA
RNGC 1554	04 19 56.	+ 19 25			DIFFUSE NEBULA
SHAP421-5027.0	04 19 58.	- 50 34 01.	12	17.25	GALAXY
LB 01315	04 19 59.	+ 23 07 00.		16.4	FAINT BLUE STAR
LBN 0799	04 20	+ 23 30	4500		BRIGHT NEBULA
KHAV 056	04 20	+ 45 25	4860		DARK NEBULA
ZWG 419.004	04 20 00.	+ 03 06		15.4	GALAXY
LDN 1478	04 20 00.	+ 37 00	8520		DARK NEBULA
LDN 1402	04 20 00.	+ 54 30	1680		DARK NEBULA
LDN 1399	04 20 00.	+ 55 00	1620		DARK NEBULA
ARC 0486	04 20 00.	- 05 03		17.5	RICH CLUSTER OF GALAXIES
SHAP421-5023.6	04 20 00.	- 50 30 37.	12	16.5	GALAXY
SHAP421-5050.6	04 20 00.	- 50 57 37.	12	17.0	GALAXY
SHAP421-5049.3	04 20 01.	- 50 56 19.	12	17.25	GALAXY
HOLM 074B	04 20 02.	- 16 02	18	15.3	PART OF MULTIPLE GALAXY
SHAP420-5809.9	04 20 02.	- 58 16 56.	54	17.0	GALAXY
HOLM 074A	04 20 04.	- 16 01	18	14.8	PART OF MULTIPLE GALAXY
SHAP421-5321.3	04 20 04.	- 53 28 19.	18	15.5	GALAXY
REIN 4.031	04 20 04.65	- 16 01 43.9			NEBULA
SHAP421-5607.2	04 20 05.	- 56 14 13.	18	17.0	GALAXY
SS 17	04 20 06.	+ 28 22			DIFFUSE GALACTIC NEBULA
MCG-02-12-008	04 20 06.	- 10 17	24	15.	GALAXY
SHAP421-5321.8	04 20 06.	- 53 28 49.	30	16.25	GALAXY
SHAP421-5408.8	04 20 06.	- 54 15 49.	12	17.0	GALAXY
REIN 4.032	04 20 06.45	- 16 00 54.9			NEBULA
RNGC 1556	04 20 07.	- 49 57			GALAXY
SHAP421-5136.9	04 20 08.	- 51 43 54.	12	17.0	GALAXY
SHAP421-4946.0	04 20 09.	- 49 53 00.	30	16.75	GALAXY
SHAP421-5321.6	04 20 09.	- 53 28 37.	18	16.0	GALAXY
SHAP421-5336.8	04 20 09.	- 53 43 49.	18	17.0	GALAXY
SHAP421-5516.6	04 20 09.	- 55 23 37.	18	17.5	GALAXY
SHAP421-5442.2	04 20 10.	- 54 49 13.	24	17.0	GALAXY
SHAP421-5619.8	04 20 10.	- 56 26 49.	18	17.0	GALAXY
SHAP421-5023.0	04 20 11.	- 50 30 00.	54	16.0	GALAXY
SHAP421-5320.0	04 20 11.	- 53 27 00.	24	16.25	GALAXY
SHAP421-4951.2	04 20 12.	- 49 58 12.	18	17.0	GALAXY
SHAP421-5010.5	04 20 12.	- 50 17 30.	18	17.25	GALAXY
SHAP421-5131.2	04 20 13.	- 51 38 12.	18	17.5	GALAXY
SHAP421-5322.1	04 20 13.	- 53 29 06.	24	17.0	GALAXY
SHAP421-5329.0	04 20 14.	- 53 36 00.	24	16.25	GALAXY
SHAP421-5444.2	04 20 14.	- 54 51 12.	24	17.25	GALAXY
SHAP421-5230.6	04 20 16.	- 52 37 36.	18	16.0	GALAXY
SHAP421-5400.0	04 20 16.	- 54 07 00.	18	17.0	GALAXY
SHAP421-5323.9	04 20 17.	- 53 30 54.	18	17.0	GALAXY
SHAP421-5452.3	04 20 17.	- 54 59 18.	36	17.0	GALAXY
SHAP421-5354.2	04 20 18.	- 54 01 12.	18	16.75	GALAXY
SHAP421-5114.8	04 20 19.	- 51 21 48.	30	17.0	GALAXY
SHAP421-5345.4	04 20 19.	- 53 52 24.	12	17.5	GALAXY
SHAP421-5338.3	04 20 21.	- 53 45 18.	18	17.5	GALAXY
SHAP421-4951.8	04 20 22.	- 49 58 47.	24	16.25	GALAXY
SHAP421-5615.9	04 20 22.	- 56 22 54.	18	17.0	GALAXY
IC 0368	04 20 23.	- 12 43 53.			NONSTELLAR OBJECT
SHAP421-5538.4	04 20 23.	- 55 45 24.	18	16.75	GALAXY
REIN 4.033	04 20 23.63	- 15 46 36.4			NEBULA
MCG+00-12-020	04 20 24.	- 00 11 30.	30	16.	GALAXY
MCG+00-12-021	04 20 24.	- 00 13 30.	18	16.	GALAXY
MCG-02-12-009	04 20 24.	- 12 43 30.	48	14.5	GALAXY
IC 2063	04 20 24.	- 15 46 42.			NONSTELLAR OBJECT
SHAP421-5029.6	04 20 24.	- 50 36 35.	24	16.25	GALAXY
SHAP421-5321.1	04 20 24.	- 53 28 06.	30	16.5	GALAXY
SHAP421-5352.2	04 20 24.	- 53 59 12.	18	17.5	GALAXY
SHAP421-5352.6	04 20 24.	- 53 59 36.	18	17.5	GALAXY
SHAP421-5155.0	04 20 26.	- 52 01 59.	12	16.75	GALAXY
SHAP421-5323.3	04 20 26.	- 53 30 17.	18	17.0	GALAXY
SHAP421-5541.2	04 20 26.	- 55 48 12.	72	17.0	GALAXY
SHAP421-5216.2	04 20 27.	- 52 23 11.	18	17.0	GALAXY
SHAP421-5428.7	04 20 27.	- 54 35 41.	18	17.0	GALAXY
RNGC 1563	04 20 28.	- 15 46		15.0	GALAXY
SHAP421-5551.6	04 20 28.	- 55 58 36.	18	16.75	GALAXY
SHAP421-5300.8	04 20 29.	- 53 07 47.	12	17.5	GALAXY
UGC 03025	04 20 30.	+ 05 28	72	16.0	GALAXY
2ZW 009	04 20 30.	+ 07 01			COMPACT GALAXY
ZC 0420.5+0701	04 20 30.	+ 07 01	610		CLUSTER OF GALAXIES
MCG-03-12-005	04 20 30.	- 15 46	48	15.	GALAXY
SHAP421-5416.8	04 20 31.	- 54 23 47.	18	17.0	GALAXY
B 215	04 20 32.	+ 24 56			DARK OBJECT
SHAP421-5119.9	04 20 32.	- 51 26 53.	18	17.25	GALAXY
SHAP421-5545.4	04 20 32.	- 55 52 23.	18	17.25	GALAXY
SHAP421-5555.9	04 20 32.	- 56 02 53.	78	15.25	GALAXY
REIN 4.034A	04 20 32.16	- 15 54 56.4			NEBULA
REIN 4.034B	04 20 32.17	- 15 54 56.4			NEBULA
MCG+00-12-022	04 20 33.	- 00 12	30	15.5	GALAXY
SHAP421-5047.8	04 20 33.	- 50 54 47.	18	17.25	GALAXY
SHAP421-5247.7	04 20 33.	- 52 54 41.	18	17.5	GALAXY
SHAP421-5153.0	04 20 34.	- 51 59 59.	12	17.25	GALAXY
SHAP421-5335.1	04 20 34.	- 53 42 05.	18	17.0	GALAXY
SHAP421-5355.6	04 20 34.	- 54 02 35.	18	17.5	GALAXY
HN 0284	04 20 34.	- 56 03			NEBULA
IC 2065	04 20 34.	- 56 03			NONSTELLAR OBJECT
REIN 4.035	04 20 34.27	- 15 47 14.1			NEBULA
SHAP421-5056.6	04 20 35.	- 51 03 35.	18	17.25	GALAXY
SHAP421-5153.1	04 20 35.	- 52 00 05.	12	17.5	GALAXY
SHAP421-5226.0	04 20 35.	- 52 32 59.	30	17.0	GALAXY
MCG+12-05-001	04 20 36.	+ 70 02	51	16.	GALAXY
ZWG 393.014	04 20 36.	- 00 12		15.7	GALAXY
RNGC 1570	04 20 36.	- 43 34			NON-EXISTENT OBJECT
RNGC 1571	04 20 36.	- 43 44			GALAXY
SHAP421-5325.9	04 20 36.	- 53 32 53.	24	17.0	GALAXY
SHAP421-5402.6	04 20 36.	- 54 09 35.	18	17.25	GALAXY
SHAP421-5544.7	04 20 36.	- 55 51 41.	24	16.75	GALAXY
LB 01316	04 20 37.	+ 07 44 42.		16.5	FAINT BLUE STAR
SHAP421-5321.3	04 20 37.	- 53 28 17.	24	16.5	GALAXY
SHAP421-5410.5	04 20 37.	- 54 17 29.	12	17.0	GALAXY
REIN 4.036A	04 20 37.45	- 15 50 53.8			NEBULA
REIN 4.036B	04 20 37.49	- 15 50 53.8			NEBULA
SHAP421-5400.8	04 20 38.	- 54 07 47.	36	16.5	GALAXY
SHAP421-5622.6	04 20 38.	- 56 29 35.	18	16.5	GALAXY
RNGC 1564	04 20 40.	- 15 50			GALAXY
SHAP421-5307.0	04 20 40.	- 53 13 58.	24	16.5	GALAXY
SHAP421-5302.8	04 20 41.	- 53 09 46.	18	17.0	GALAXY
ZC 0420.7+1138	04 20 42.	+ 11 38	1480		CLUSTER OF GALAXIES
ZWG 328.001	04 20 42.	+ 70 04		15.6	GALAXY
UGC 03026	04 20 42.	+ 70 04	72	15.6	GALAXY S
ARC 0487	04 20 42.	- 24 22		17.0	RICH CLUSTER OF GALAXIES
SHAP422-5050.4	04 20 42.	- 50 57 22.	12	17.25	GALAXY
LB 01317	04 20 43.	+ 19 08 00.		15.8	FAINT BLUE STAR
BC PKS0420-01	04 20 43.1	- 01 27 29.		18.	QUASI-STELLAR OBJECT
SHB 083	04 20 43.1	- 01 27 29.		18.	QUASI-STELLAR OBJECT
SHAP421-5319.2	04 20 44.	- 53 26 10.	24	17.5	GALAXY
SHAP421-5623.8	04 20 44.	- 56 30 47.	24	17.5	GALAXY
SHAP421-5636.8	04 20 44.	- 56 43 47.	18	17.0	GALAXY
SHAP421-5732.0	04 20 44.	- 57 38 59.	30	16.5	GALAXY
REIN 4.037B	04 20 44.40	- 15 51 17.4			NEBULA
REIN 4.037A	04 20 44.47	- 15 51 17.4			NEBULA
REIN 4.038B	04 20 44.69	- 15 57 40.3			NEBULA
REIN 4.038A	04 20 44.72	- 15 57 40.3			NEBULA
SHAP422-5008.7	04 20 45.	- 50 15 40.	24	16.5	GALAXY
RNGC 1561	04 20 46.	- 15 57		15.0	GALAXY
SHAP422-5013.2	04 20 46.	- 50 20 10.	24	17.0	GALAXY
SHAP421-5315.4	04 20 46.	- 53 22 22.	24	16.75	GALAXY
SHAP422-5013.0	04 20 47.	- 50 19 58.	24	17.0	GALAXY
SHAP421-5354.1	04 20 47.	- 54 01 04.	12	17.25	GALAXY
HOLM 075B	04 20 48.	- 15 55	12	15.0	PART OF MULTIPLE GALAXY
MCG-03-12-006	04 20 48.	- 15 57	36	15.	GALAXY
LB 01720	04 20 48.	- 47 49		14.8	FAINT BLUE STAR
SHAP422-5111.0	04 20 48.	- 51 17 58.	18	16.0	GALAXY
SHAP421-5654.9	04 20 48.	- 57 01 52.	18	17.0	GALAXY
SHAP422-5005.8	04 20 49.	- 50 12 46.	24	17.25	GALAXY
HOLM 075A	04 20 52.	- 15 55	18	14.7	PART OF MULTIPLE GALAXY
SHAP422-5355.2	04 20 52.	- 54 02 10.	18	17.0	GALAXY
SHAP421-5455.0	04 20 52.	- 55 01 58.	30	17.5	GALAXY
REIN 4.039	04 20 52.14	- 15 54 56.1			NEBULA
SHAP422-5127.0	04 20 53.	- 51 33 57.	18	17.25	GALAXY
SHAP422-5131.2	04 20 54.	- 51 38 09.	24	17.5	GALAXY
B 216	04 20 55.	+ 26 31			DARK OBJECT
SHAP422-5330.5	04 20 55.	- 53 37 28.	18	16.75	GALAXY
SHAP422-5336.0	04 20 55.	- 53 42 58.	24	17.5	GALAXY
SHAP422-5458.4	04 20 55.	- 55 05 22.	42	17.0	GALAXY
SHAP422-5543.4	04 20 56.	- 55 50 22.	18	16.75	GALAXY
SHAP421-5618.5	04 20 56.	- 56 25 28.	18	17.5	GALAXY
MCG+00-12-023	04 20 57.	- 00 17 30.	30	15.5	GALAXY
RNGC 1572	04 20 57.	- 40 42			GALAXY
SHAP422-5126.0	04 20 57.	- 51 32 57.	24	16.5	GALAXY
SHAP422-5127.0	04 20 57.	- 51 33 57.	18	16.75	GALAXY
SHAP422-4948.0	04 20 58.	- 49 54 57.	24	17.0	GALAXY
SHAP422-5006.2	04 20 58.	- 50 13 09.	30	16.5	GALAXY
SHAP422-5249.2	04 20 59.	- 52 56 09.	24	17.5	GALAXY
LDN 1400	04 21 00.	+ 54 50	420		DARK NEBULA
7ZW 013	04 21 00.	+ 82 09			COMPACT GALAXY
SHAP422-4948.6	04 21 00.	- 49 55 33.	24	15.5	GALAXY
SHAP422-5006.6	04 21 00.	- 50 13 33.	24	16.5	GALAXY
SHAP422-5009.1	04 21 00.	- 50 16 03.	24	16.0	GALAXY
SHAP422-5105.6	04 21 00.	- 51 12 33.	24	16.5	GALAXY
SHAP422-5619.8	04 21 00.	- 56 26 46.	24	16.5	GALAXY
SHAP422-5658.0	04 21 00.	- 57 04 58.	234	12.2	GALAXY
RNGC 1574	04 21 00.	- 57 05		12.5	GALAXY
SHAP422-5136.1	04 21 01.	- 51 43 03.	72	16.5	GALAXY
SHAP422-5250.4	04 21 01.	- 52 57 21.	18	17.0	GALAXY
SHAP422-5020.3	04 21 02.	- 50 27 15.	12	16.75	GALAXY
SHAP422-5305.4	04 21 02.	- 53 12 21.	30	17.0	GALAXY
HN 0283	04 21 03.	+ 20 58			NEBULA
IC 2061	04 21 03.	+ 20 58			NONSTELLAR OBJECT
SHAP422-5201.6	04 21 03.	- 52 08 33.	18	17.0	GALAXY
SHAP422-5323.8	04 21 03.	- 53 30 45.	30	15.25	GALAXY
REIN 4.041	04 21 03.90	- 15 54 00.0			NEBULA
SHAP422-5620.6	04 21 04.	- 56 27 33.	24	17.0	GALAXY
SHAP422-5340.3	04 21 05.	- 53 47 15.	24	17.5	GALAXY
UGC 03027	04 21 06.	+ 30 48	72	18.	GALAXY Sc
ISS 0115	04 21 06.	+ 51 32	452		STELLAR RING

OBJECT NAME	RIGHT ASCEN.	DECLINATION	DIAM.	MAGN.	TYPE OF OBJECT
SHAP422-5617.6	04 21 07.	- 56 24 33.	18	17.0	GALAXY
REIN 4.042	04 21 07.22	- 15 51 34.1			NEBULA
IC 0369	04 21 08.	- 11 54 23.			NONSTELLAR OBJECT
SHAP422-5017.8	04 21 08.	- 50 24 44.	24	17.25	GALAXY
SHAP422-5322.8	04 21 08.	- 53 29 45.	18	17.0	GALAXY
SHAP422-5356.0	04 21 08.	- 54 02 57.	12	17.5	GALAXY
SHAP422-5616.4	04 21 08.	- 56 23 21.	18	17.5	GALAXY
SHAP422-5125.2	04 21 09.	- 51 32 08.	24	16.75	GALAXY
SHAP422-5401.4	04 21 09.	- 54 08 21.	18	17.25	GALAXY
SHAP422-5403.0	04 21 09.	- 54 09 57.	18	17.0	GALAXY
SHAP422-5614.8	04 21 09.	- 56 21 45.	18	17.5	GALAXY
RNGC 1565	04 21 10.	- 15 51		15.0	GALAXY
SHAP422-5314.6	04 21 10.	- 53 21 33.	12	17.0	GALAXY
SHAP422-5408.2	04 21 10.	- 54 15 09.	30	17.5	GALAXY
REIN 4.043	04 21 10.24	- 15 48 02.1			NEBULA
IC 2064	04 21 11.	- 15 48			NONSTELLAR OBJECT
REIN 4.044	04 21 11.92	- 15 52 10.3			NEBULA
MCG-02-12-010	04 21 12.	- 11 53 30.	30	15.	GALAXY
MCG-03-12-007	04 21 12.	- 15 51	54	15.	GALAXY
SHAP422-5443.4	04 21 12.	- 54 50 21.	24	16.75	GALAXY
SHAP422-5302.4	04 21 13.	- 53 09 20.	24	17.0	GALAXY
SHAP422-5251.8	04 21 14.	- 52 58 44.	18	16.0	GALAXY
SHAP422-5132.0	04 21 15.	- 51 38 56.	18	16.0	GALAXY
SHAP422-5338.3	04 21 15.	- 53 45 14.	24	17.5	GALAXY
SHAP422-5505.3	04 21 16.	- 55 12 14.	18	17.0	GALAXY
SHAP422-5142.4	04 21 17.	- 51 49 20.	12	17.0	GALAXY
UGC 03028	04 21 18.	+ 33 45	102	16.0	GALAXY S
SHAP422-4947.3	04 21 18.	- 49 54 14.	18	17.5	GALAXY
SHAP422-5617.5	04 21 18.	- 56 24 26.	18	17.0	GALAXY
SHAP422-5132.4	04 21 19.	- 51 39 20.	24	17.0	GALAXY
SHAP422-5139.8	04 21 19.	- 51 46 44.	12	17.5	GALAXY
SHAP422-5405.9	04 21 19.	- 54 12 50.	18	17.5	GALAXY
SHAP422-5613.5	04 21 19.	- 56 20 26.	24	16.5	GALAXY
SHAP422-5546.0	04 21 20.	- 55 52 56.	18	16.75	GALAXY
MCG+06-10-007	04 21 21.	+ 33 45	78	16.5	GALAXY
SHAP422-5350.9	04 21 22.	- 53 57 50.	12	17.0	GALAXY
SHAP422-5654.7	04 21 22.	- 57 01 38.	18	17.5	GALAXY
SHAP422-4946.3	04 21 23.	- 49 53 13.	24	16.75	GALAXY
SHAP422-4959.2	04 21 23.	- 50 06 07.	18	17.0	GALAXY
SHAP422-5613.0	04 21 23.	- 56 19 56.	84	16.5	GALAXY
ZWG 393.015	04 21 24.	- 00 57		15.5	GALAXY
UGC 03029	04 21 24.	- 00 57	78	15.5	GALAXY Sb
MCG+00-12-024	04 21 24.	- 00 57	48	14.	GALAXY
LB 01721	04 21 24.	- 54 16		13.9	FAINT BLUE STAR
SHAP422-5611.4	04 21 24.	- 56 18 20.	18	17.0	GALAXY
SHAP422-5657.4	04 21 24.	- 57 04 20.	24	17.5	GALAXY
LB 03352	04 21 24.	- 66 47		13.2	FAINT BLUE STAR
LB 01318	04 21 25.	+ 12 46 48.		16.0	FAINT BLUE STAR
SHAP422-5220.2	04 21 25.	- 52 27 07.	36	16.75	GALAXY
SHAP422-5345.6	04 21 25.	- 53 52 32.	24	17.25	GALAXY
SHAP422-5001.8	04 21 26.	- 50 08 43.	30	16.75	GALAXY
SHAP422-5323.8	04 21 26.	- 53 30 43.	30	16.0	GALAXY
SHAP422-5511.6	04 21 26.	- 55 18 32.	12	17.5	GALAXY
SHAP422-5619.3	04 21 26.	- 56 26 14.	30	16.0	GALAXY
HZ 12	04 21 27.	+ 08 26		12.8	BLUE STAR
SHAP422-5057.6	04 21 27.	- 51 04 31.	30	17.0	GALAXY
SHAP422-5130.4	04 21 27.	- 51 37 19.	12	17.25	GALAXY
SHAP422-5619.4	04 21 28.	- 56 26 20.	12	16.5	GALAXY
SHAP422-5129.3	04 21 29.	- 51 36 13.	18	15.75	GALAXY
MCG-04-11-011	04 21 30.	- 24 13	36	15.	GALAXY
SHAP422-5015.8	04 21 30.	- 50 22 43.	60	17.0	GALAXY
PK146+07.1	04 21 31.	+ 60 00 25.			PLANETARY NEBULA
SHAP422-5055.6	04 21 31.	- 51 02 31.	12	17.0	GALAXY
SHB 084	04 21 32.5	+ 01 57 21.		17.5	QUASI-STELLAR OBJECT
MCG+00-12-025	04 21 33.	- 02 42	48	15.5	GALAXY
BC PKS0421+019	04 21 33.0	+ 01 57 33.		17.5	QUASI-STELLAR OBJECT
SHAP422-4939.6	04 21 34.	- 49 46 31.	24	16.75	GALAXY
SHAP422-5324.2	04 21 34.	- 53 31 07.	18	16.75	GALAXY
MCG-04-11-012	04 21 36.	- 23 13	36	15.	GALAXY
SHAP422-5155.0	04 21 36.	- 52 01 55.	18	17.0	GALAXY
SHAP422-5230.4	04 21 36.	- 52 37 19.	30	16.75	GALAXY
SHAP422-5307.4	04 21 36.	- 53 14 19.	18	17.0	GALAXY
SHAP422-5340.2	04 21 36.	- 53 47 07.	12	16.75	GALAXY
SER 037.02	04 21 36.	- 56 27	720	16.	LOOSE GROUP OF 6 GALAXIES
IC 0370	04 21 38.	- 09 30 50.			NONSTELLAR OBJECT
SHAP423-5006.8	04 21 39.	- 50 13 42.	18	17.25	GALAXY
SHAP422-5607.8	04 21 41.	- 56 14 43.	18	17.5	GALAXY
ZC 0421.7+1556	04 21 42.	+ 15 56	2490		CLUSTER OF GALAXIES
UGC 03030	04 21 42.	+ 66 13	72	18.	GALAXY DWARF?
MCG-02-12-011	04 21 42.	- 09 30	72	14.5	GALAXY
SHAP422-5316.8	04 21 43.	- 53 23 42.	12	16.25	GALAXY
SHAP422-5632.8	04 21 43.	- 56 39 43.	18	17.0	GALAXY
MCG+00-12-026	04 21 45.	- 00 51	12	15.5	GALAXY
SHAP423-5015.4	04 21 46.	- 50 22 18.	24	16.25	GALAXY
SHAP423-5140.0	04 21 46.	- 51 46 54.	24	16.0	GALAXY
SHAP423-5039.5	04 21 47.	- 50 46 24.	18	17.0	GALAXY
SHAP423-5258.3	04 21 47.	- 53 05 12.	18	17.0	GALAXY
SHAP422-5341.5	04 21 47.	- 53 48 24.	18	17.0	GALAXY
UGC 03033	04 21 48.	+ 10 47	72	17.	GALAXY SB
ZWG 393.016	04 21 48.	- 00 51		14.9	GALAXY
UGC 03032	04 21 48.	- 00 51	102	14.9	GALAXY E-S
UGC 03031	04 21 48.	- 00 51	96	14.9	GALAXY
SHAP423-5307.9	04 21 48.	- 53 14 48.	18	17.25	GALAXY
SHAP422-5345.5	04 21 48.	- 53 52 24.	18	17.25	GALAXY
RNGC 1568	04 21 50.	- 00 51		15.0	GALAXY
SHAP423-5034.8	04 21 50.	- 50 41 42.	18	17.25	GALAXY
SHAP423-5357.2	04 21 50.	- 54 04 06.	24	17.5	GALAXY
SHAP422-5625.8	04 21 50.	- 56 32 42.	36	17.5	GALAXY
MCG+00-12-027	04 21 51.	- 00 52	36	14.	GALAXY
MCG-02-12-012	04 21 51.	- 12 23 30.	54	15.	GALAXY
SHAP423-5131.2	04 21 51.	- 51 38 06.	12	16.25	GALAXY
UGC 03034	04 21 54.	+ 09 34	66	17.	GALAXY DWARF?
2ZW 010	04 21 54.	- 00 51			COMPACT GALAXY
SHAP423-5407.4	04 21 55.	- 54 14 18.	12	17.5	GALAXY
MCG+01-12-003	04 21 57.	- 03 46	60	15.	GALAXY
SHAP423-5021.2	04 21 57.	- 50 28 05.	18	17.5	GALAXY
SHAP423-5052.4	04 21 57.	- 50 59 17.	24	17.25	GALAXY
SHAP423-4921.2	04 21 59.	- 49 28 05.	30	17.25	GALAXY
SHAP423-5330.4	04 21 59.	- 53 37 17.	30	17.0	GALAXY
SHAP423-5334.3	04 21 59.	- 53 41 11.	12	17.25	GALAXY
ZWG 419.005	04 22 00.	+ 07 06		14.8	GALAXY
UGC 03035	04 22 00.	+ 07 06	102	14.8	GALAXY SB0
MCG+01-12-002	04 22 00.	+ 07 07	21	14.	GALAXY
MRK 615	04 22 00.	- 00 52	8	16.	GALAXY WITH UV CONTINUUM
SHAP423-5232.0	04 22 00.	- 52 38 53.	18	17.0	GALAXY
SHAP423-5326.6	04 22 00.	- 53 33 29.	18	17.0	GALAXY
SHAP423-5018.4	04 22 02.	- 50 25 17.	18	16.75	GALAXY
SHAP423-5344.4	04 22 03.	- 53 51 17.	18	17.25	GALAXY

OBJECT NAME	RIGHT ASCEN.	DECLINATION	DIAM.	MAGN.	TYPE OF OBJECT
SNO 18	04 22 04.	- 27 51 40.	1200	19.	CLUSTER OF 20 GALAXIES
SHAP423-5347.0	04 22 05.	- 53 53 53.	24	17.25	GALAXY
ZWG 393.017	04 22 06.	- 02 33		15.6	GALAXY
SHAP423-5344.8	04 22 06.	- 53 51 41.	18	17.0	GALAXY
SHAP423-5330.8	04 22 07.	- 53 37 41.	24	17.5	GALAXY
SHAP423-5331.9	04 22 07.	- 53 38 47.	18	17.25	GALAXY
SHAP423-5639.2	04 22 10.	- 56 46 05.	30	17.0	GALAXY
ABC 0488	04 22 11.	- 05 24		17.6	RICH CLUSTER OF GALAXIES
BC PKS0422+00	04 22 12.	+ 00 29 12.		17.	QUASI-STELLAR OBJECT
UGC 03036	04 22 12.	+ 70 47	78	16.5	GALAXY S
LB 01722	04 22 12.	- 45 58		15.3	FAINT BLUE STAR
SHAP423-5319.3	04 22 13.	- 53 26 10.	30	15.75	GALAXY
HSE 002	04 22 13.	- 68 37			STAR CLUSTER IN LMC
SHAP423-5347.8	04 22 14.	- 53 54 40.	18	17.25	GALAXY
SHAP423-5543.4	04 22 14.	- 55 50 17.	24	17.25	GALAXY
SHAP423-5220.4	04 22 15.	- 52 27 16.	36	17.0	GALAXY
SHAP423-5550.8	04 22 16.	- 55 57 40.	24	17.5	GALAXY
SHAP423-5244.2	04 22 17.	- 52 51 04.	24	17.25	GALAXY
SHAP423-5619.6	04 22 17.	- 56 26 28.	12	17.0	GALAXY
MCG+01-12-003	04 22 18.	+ 07 03	72	14.5	GALAXY
ZC 0422.3+2515	04 22 18.	+ 25 15	4770		CLUSTER OF GALAXIES
ZWG 393.018	04 22 18.	- 01 09		15.7	GALAXY
UGC 03037	04 22 18.	- 01 09	84	15.7	GALAXY Sb-c
SHAP423-5549.4	04 22 18.	- 55 56 16.	18	17.0	GALAXY
SHAP423-4923.2	04 22 19.	- 49 30 04.	36	16.0	GALAXY
LB 01319	04 22 21.	+ 21 16 42.		16.8	FAINT BLUE STAR
SHAP423-4927.3	04 22 21.	- 49 34 09.	30	17.25	GALAXY
ZWG 419.006	04 22 24.	+ 07 04		15.5	GALAXY
UGC 03038	04 22 24.	+ 07 04	78	15.5	GALAXY Sb
SHAP423-5408.5	04 22 25.	- 54 15 22.	24	17.5	GALAXY
SHAP423-5444.0	04 22 25.	- 54 50 52.	66	16.0	GALAXY
HN 0285	04 22 25.	- 54 51			NEBULA
IC 2066	04 22 25.	- 54 51			NONSTELLAR OBJECT
SHAP423-5243.2	04 22 26.	- 52 50 03.	18	17.0	GALAXY
SHAP423-5320.2	04 22 27.	- 53 27 03.	12	17.0	GALAXY
LB 00213	04 22 28.	+ 16 51 30.		17.6	FAINT BLUE STAR
SHAP423-5135.6	04 22 29.	- 51 42 27.	60	14.2	GALAXY
SHAP423-5322.6	04 22 29.	- 53 29 27.	18	17.0	GALAXY
SHAP423-5326.7	04 22 29.	- 53 33 33.	18	17.25	GALAXY
ZWG 393.019	04 22 30.	+ 02 55		15.7	GALAXY
UGC 03039	04 22 30.	+ 02 55	66	15.7	GALAXY S0
SHAP423-5331.0	04 22 30.	- 53 37 51.	18	17.5	GALAXY
SHAP423-5333.2	04 22 30.	- 53 40 03.	12	16.75	GALAXY
SHAP423-5400.0	04 22 30.	- 54 06 51.	18	16.5	GALAXY
LB 01723	04 22 30.	- 54 24		14.3	FAINT BLUE STAR
SHAP423-5406.5	04 22 31.	- 54 13 21.	24	17.5	GALAXY
SHAP423-5401.8	04 22 32.	- 54 08 39.	18	16.75	GALAXY
SHAP423-5214.2	04 22 33.	- 52 21 03.	12	17.0	GALAXY
RNGC 1578	04 22 34.	- 51 41			GALAXY
SHAP423-5333.4	04 22 34.	- 53 40 15.	30	17.0	GALAXY
SHAP423-5330.9	04 22 35.	- 53 37 45.	18	17.5	GALAXY
B 011	04 22 36.	+ 54 56			DARK OBJECT
MCG+12-05-002	04 22 36.	+ 70 46	66	16.	GALAXY
SHAP423-5247.6	04 22 36.	- 52 54 27.	12	17.25	GALAXY
SHAP423-5555.8	04 22 36.	- 56 02 39.	24	17.5	GALAXY
SHAP423-5249.0	04 22 37.	- 52 55 51.	18	17.5	GALAXY
SHAP423-5329.5	04 22 37.	- 53 36 21.	12	17.5	GALAXY
SHAP423-5625.4	04 22 37.	- 56 32 15.	24	17.25	GALAXY
LB 00230	04 22 38.	+ 10 42 48.		17.1	FAINT BLUE STAR
SHAP423-5328.2	04 22 39.	- 53 35 03.	24	17.5	GALAXY
SHAP423-5340.4	04 22 40.	- 53 47 15.	18	17.0	GALAXY
SHAP423-5354.6	04 22 40.	- 54 01 27.	24	16.5	GALAXY
SHAP424-5031.9	04 22 41.	- 50 38 44.	24	17.0	GALAXY
MCG-02-12-013	04 22 42.	- 10 00	24	15.	GALAXY
REIN 6.030	04 22 43.43	- 02 54 00.6			NEBULA
SHAP424-5200.4	04 22 45.	- 52 07 14.	18	17.5	GALAXY
SHAP423-5445.2	04 22 45.	- 54 52 02.	12	17.0	GALAXY
SHAP423-5637.8	04 22 45.	- 56 44 39.	24	16.75	GALAXY
SHAP424-5022.4	04 22 46.	- 50 29 14.	18	17.5	GALAXY
SHAP424-5236.2	04 22 46.	- 52 43 02.	24	17.5	GALAXY
SHAP423-5500.0	04 22 46.	- 55 06 50.	24	17.0	GALAXY
SHAP423-5553.0	04 22 46.	- 55 59 50.	18	17.25	GALAXY
SHAP423-5334.5	04 22 47.	- 53 41 20.	18	15.75	GALAXY
SHAP423-5342.4	04 22 47.	- 53 49 14.	18	16.5	GALAXY
SHAP423-5416.8	04 22 47.	- 54 23 38.	24	17.25	GALAXY
UGC 03040	04 22 48.	+ 35 05	84	18.	GALAXY
SHAP424-5152.0	04 22 48.	- 51 58 50.	18	17.5	GALAXY
SHAP424-5332.6	04 22 49.	- 53 39 26.	24	16.25	GALAXY
LB 01724	04 22 54.	- 46 52		14.6	FAINT BLUE STAR
SHAP424-5241.2	04 22 54.	- 52 48 02.	12	17.25	GALAXY
SHAP424-5400.6	04 22 54.	- 54 07 26.	18	17.0	GALAXY
SHAP424-5419.6	04 22 54.	- 54 26 26.	18	17.0	GALAXY
BC PKS0422-380	04 22 55.6	- 38 03 02.		16.5	QUASI-STELLAR OBJECT
SHAP424-5004.0	04 22 56.	- 50 10 49.	24	16.5	GALAXY
SHAP424-5018.2	04 22 56.	- 50 25 01.	18	16.75	GALAXY
SHAP424-5200.0	04 22 56.	- 52 06 49.	24	17.0	GALAXY
LB 00214	04 22 57.	+ 16 48 54.		16.8	FAINT BLUE STAR
MCG+01-12-004	04 22 57.	- 03 36	42	15.5	GALAXY
SHAP424-5318.2	04 22 57.	- 53 25 01.	24	17.3	GALAXY
SHAP424-5530.0	04 22 57.	- 55 36 50.	24	17.0	GALAXY
SHAP424-5309.2	04 22 58.	- 53 16 01.	30	16.7	GALAXY
HSE 003	04 22 58.	- 68 00			STAR CLUSTER IN LMC
ZWG 419.007	04 23 00.	+ 07 12		15.5	GALAXY
MCG+01-12-004	04 23 00.	+ 07 12	42	14.5	GALAXY
SHAP424-5220.2	04 23 01.	- 52 27 01.	24	17.25	GALAXY
REIN 6.031	04 23 01.88	- 03 36 53.8			NEBULA
SHAP424-5022.5	04 23 02.	- 50 29 19.	18	16.5	GALAXY
SHAP424-5301.0	04 23 05.	- 53 07 49.	12	17.0	GALAXY
SHAP424-5337.5	04 23 05.	- 53 44 19.	18	17.5	GALAXY
LB 01320	04 23 06.	+ 12 05 00.		15.0	FAINT BLUE STAR
LB 03353	04 23 06.	- 64 47		12.8	FAINT BLUE STAR
SHAP424-4934.4	04 23 07.	- 49 41 12.	24	16.25	GALAXY
SHAP424-5058.8	04 23 07.	- 51 05 37.	18	17.0	GALAXY
MCG-04-11-013	04 23 09.	- 26 51	54	16.	GALAXY
SHAP424-5022.8	04 23 09.	- 50 29 36.	18	16.75	GALAXY
MCG+12-05-003	04 23 12.	+ 70 14	132	14.	GALAXY
ZWG 328.002	04 23 12.	+ 70 15		14.9	GALAXY
UGC 03042	04 23 12.	+ 70 15	126	15.6	GALAXY Sb/SBc
UGC 03041	04 23 12.	- 00 43	60	16.5	GALAXY Sc
ZC 0423.2-0053	04 23 12.	- 00 53	1010		CLUSTER OF GALAXIES
SHAP424-4928.2	04 23 14.	- 49 35 00.	30	17.0	GALAXY
SHAP424-5000.0	04 23 15.	- 50 06 48.	18	17.0	GALAXY
MCG-03-12-008	04 23 18.	- 16 39	36	15.	GALAXY
SHAP424-5414.6	04 23 18.	- 54 21 24.	18	17.5	GALAXY
SHAP424-5215.5	04 23 19.	- 52 22 18.	12	17.5	GALAXY
SHAP424-5316.6	04 23 19.	- 53 23 24.	24	17.3	GALAXY
SHAP424-5324.7	04 23 19.	- 53 31 30.	48	16.0	GALAXY
SHAP424-5625.2	04 23 19.	- 56 32 00.	18	17.5	GALAXY

OBJECT NAME	RIGHT ASCEN.	DECLINATION	DIAM.	MAGN.	TYPE OF OBJECT
SHAP424-5706.3	04 23 19.	- 57 13 06.	36	16.5	GALAXY
SHAP424-5232.4	04 23 20.	- 52 39 12.	18	17.0	GALAXY
SHAP424-5338.8	04 23 20.	- 53 45 36.	18	17.25	GALAXY
LB 00231	04 23 22.	+ 12 21 18.		16.8	FAINT BLUE STAR
SHAP424-4958.3	04 23 23.	- 50 05 05.	24	17.0	GALAXY
SHAP424-5200.0	04 23 23.	- 52 06 48.	18	17.0	GALAXY
SHAP424-5405.4	04 23 23.	- 54 12 12.	18	15.75	GALAXY
7ZW 014	04 23 24.	+ 69 29			COMPACT GALAXY
ZWG 328.003	04 23 24.	+ 69 29		15.6	GALAXY
UGC 03043	04 23 24.	+ 70 49	78	17.	GALAXY Sc
SHAP424-5057.8	04 23 24.	- 51 04 35.	24	17.0	GALAXY
SHAP424-5347.4	04 23 24.	- 53 54 12.	24	17.0	GALAXY
SHAP424-5605.0	04 23 24.	- 56 11 48.	36	17.0	GALAXY
LB 03354	04 23 24.	- 66 15		13.6	FAINT BLUE STAR
SHAP424-5250.2	04 23 25.	- 52 57 00.	18	17.25	GALAXY
SHAP424-5502.2	04 23 25.	- 55 09 00.	24	17.0	GALAXY
SHAP424-5059.4	04 23 26.	- 51 06 11.	18	17.0	GALAXY
SHAP424-5310.2	04 23 26.	- 53 16 59.	18	17.3	GALAXY
SEY 004	04 23 27.	+ 70 15 00.		14.3	FAINT GALAXY
SHAP424-5044.0	04 23 27.	- 50 50 47.	18	17.0	GALAXY
SHAP424-5225.7	04 23 28.	- 52 32 29.	24	17.0	GALAXY
SHAP424-5314.8	04 23 28.	- 53 21 35.	24	17.4	GALAXY
SHAP424-5026.9	04 23 29.	- 50 33 41.	18	17.0	GALAXY
SHAP424-5042.0	04 23 29.	- 50 48 47.	18	17.0	GALAXY
SHAP424-5341.8	04 23 29.	- 53 48 35.	18	17.25	GALAXY
ISS 0202	04 23 30.	+ 44 14	406		STELLAR RING
MCG+01-12-005	04 23 30.	- 08 42	15	15.5	GALAXY
SHAP424-5308.2	04 23 30.	- 53 14 59.	18	17.4	GALAXY
SHAP424-5315.0	04 23 32.	- 53 21 47.	18	16.6	GALAXY
SHAP424-5358.6	04 23 32.	- 54 05 23.	18	17.0	GALAXY
RNGC 1581	04 23 33.	- 55 02			GALAXY
SHAP424-5456.4	04 23 35.	- 55 03 11.	108	13.9	GALAXY
MCG+01-12-006	04 23 36.	- 08 42	24	15.5	GALAXY
LB 01321	04 23 38.	+ 08 46 48.		15.8	FAINT BLUE STAR
SHAP424-5236.6	04 23 40.	- 52 43 22.	18	17.5	GALAXY
LB 00215	04 23 41.	+ 16 48 12.		14.2	FAINT BLUE STAR
FATH 1.160	04 23 42.	+ 29 50	27		NEBULA
UGC 03044	04 23 42.	+ 29 51	72	18.	GALAXY DWARF?
SHAP425-5103.6	04 23 42.	- 51 10 22.	18	16.0	GALAXY
SHAP424-5317.8	04 23 42.	- 53 24 34.	12	17.4	GALAXY
SHAP425-5028.2	04 23 44.	- 50 34 58.	42	17.0	GALAXY
RNGC 1576	04 23 45.	- 03 43		15.0	GALAXY
SHAP424-5338.2	04 23 47.	- 53 44 58.	12	17.5	GALAXY
UGC 03045	04 23 48.	+ 20 18	84	16.5	GALAXY Sb?
MCG+01-12-007	04 23 48.	- 03 43	42	15.	GALAXY
SHAP425-5033.3	04 23 49.	- 50 40 04.	18	16.5	GALAXY
SHAP425-5246.4	04 23 49.	- 52 53 10.	24	16.5	GALAXY
SHAP424-5622.2	04 23 49.	- 56 28 58.	18	17.0	GALAXY
REIN 6.032	04 23 49.19	- 03 44 00.2			NEBULA
SHAP424-5654.4	04 23 50.	- 57 01 10.	18	17.0	GALAXY
SHAP425-5352.9	04 23 51.	- 53 59 40.	18	16.5	GALAXY
LB 00216	04 23 52.	+ 14 05 36.		17.5	FAINT BLUE STAR
SHAP425-5102.6	04 23 52.	- 51 09 22.	18	16.75	GALAXY
SHAP425-5205.6	04 23 53.	- 52 12 22.	36	17.0	GALAXY
MCG+00-12-028	04 23 54.	+ 02 13 30.	48	16.	GALAXY
ZWG 328.004	04 23 54.	+ 69 26		15.3	GALAXY
UGC 03046	04 23 54.	+ 69 26	102	15.3	GALAXY Sb
ARC 0489	04 23 54.	- 04 43		17.7	RICH CLUSTER OF GALAXIES
SHAP425-5008.8	04 23 54.	- 50 15 33.	18	17.0	GALAXY
RNGC 1577	04 23 55.	- 10 13			NON-EXISTENT OBJECT
SHAP425-5301.9	04 23 55.	- 53 08 40.	12	16.75	GALAXY
SHAP425-5342.0	04 23 55.	- 53 48 46.	18	17.0	GALAXY
HN 0286	04 23 56.	- 58 04			NEBULA
IC 2070	04 23 56.	- 58 04			NONSTELLAR OBJECT
SHAP425-5448.2	04 23 57.	- 54 54 58.	18	17.0	GALAXY
SHAP424-5709.2	04 23 57.	- 57 15 58.	24	17.25	GALAXY
SHAP425-5220.6	04 23 58.	- 52 27 21.	18	17.5	GALAXY
SHAP425-5251.4	04 23 58.	- 52 58 09.	18	17.25	GALAXY
SHAP425-5333.0	04 23 58.	- 53 39 45.	24	17.4	GALAXY
SHAP425-5214.0	04 23 59.	- 52 20 45.	18	17.0	GALAXY
LBN 081[illegible]	04 24	+ 22 50	480		BRIGHT NEBULA
LBN 0788	04 24	+ 25 59	180		BRIGHT NEBULA
KHAV 058	04 24	+ 30 49	4200		DARK NEBULA
MIL 04	04 24	+ 47 00	18000		SUPERNOVA REMNANT
KHAV 057	04 24	+ 47 49	7470		DARK NEBULA
OCL 0456	04 24 00.	+ 15 45	72000	1.4	OPEN STAR CLUSTER
DG 032	04 24 00.	+ 25 58	420		REFLECTION NEBULA
LDN 1497	04 24 00.	+ 28 30	3420		DARK NEBULA
UGC 03047	04 24 00.	+ 32 32	96	17.	GALAXY SBc
LDN 1401	04 24 00.	+ 55 06	480		DARK NEBULA
MCG+12-05-003A	04 24 00.	+ 70 19	51	14.	GALAXY
ZWG 328.005	04 24 00.	+ 70 20		15.3	GALAXY
UGC 03048	04 24 00.	+ 70 20	78	15.3	GALAXY SBb-c
MCG-02-12-014	04 24 00.	- 10 11 30.	84	13.5	GALAXY
SHAP425-5351.1	04 24 00.	- 53 57 51.	18	17.0	GALAXY
RNGC 1575	04 24 01.	- 10 11		13.0	GALAXY
SHAP425-5052.9	04 24 02.	- 50 59 39.	42	17.0	GALAXY
SHAP425-5511.4	04 24 02.	- 55 18 09.	18	16.5	GALAXY
LB 01322	04 24 03.	+ 12 58 00.		17.0	FAINT BLUE STAR
CED 033	04 24 03.	+ 25 58	480		DIFFUSE GALACTIC NEBULA
SEY 005	04 24 03.	+ 70 18 58.		14.5	FAINT GALAXY
SHAP425-5038.9	04 24 03.	- 50 45 39.	18	16.5	GALAXY
SHAP425-5223.8	04 24 04.	- 52 30 33.	12	17.0	GALAXY
LB 01323	04 24 06.	+ 12 58 18.		17.0	FAINT BLUE STAR
MCG+12-05-004	04 24 06.	+ 69 25	78	15.	GALAXY
MCG-06-10-007	04 24 06.	- 37 24	9	15.5	GALAXY
LB 01725	04 24 06.	- 50 32		12.7	FAINT BLUE STAR
SHAP425-5154.6	04 24 06.	- 52 01 21.	12	16.5	GALAXY
SHAP425-5223.5	04 24 06.	- 52 30 15.	12	17.25	GALAXY
SHAP425-5336.2	04 24 07.	- 53 42 57.	18	17.5	GALAXY
SS 18	04 24 09.	+ 26 04			DIFFUSE GALACTIC NEBULA
SHAP425-5345.6	04 24 10.	- 53 52 21.	18	17.5	GALAXY
SHAP425-5645.6	04 24 11.	- 56 52 21.	24	17.0	GALAXY
DG 033	04 24 12.	+ 22 53	1020		REFLECTION NEBULA
MCG-03-12-009	04 24 12.	- 20 50	24	15.	GALAXY
SHAP425-5053.4	04 24 12.	- 51 00 08.	30	16.5	GALAXY
SHAP425-5226.4	04 24 14.	- 52 33 08.	18	17.5	GALAXY
SHAP425-5328.0	04 24 14.	- 53 34 44.	18	17.4	GALAXY
SHAP425-5116.4	04 24 15.	- 51 23 08.	24	17.25	GALAXY
SHAP425-5626.4	04 24 16.	- 56 33 09.	24	16.5	GALAXY
CED 034	04 24 17.	+ 22 53	480		DIFFUSE GALACTIC NEBULA
ZWG 393.020	04 24 18.	+ 01 35		15.4	GALAXY
UGC 03049	04 24 18.	+ 01 35	66	15.4	GALAXY Sb-c
UGC 03050	04 24 18.	+ 30 51	60	18.	GALAXY S
LB 03355	04 24 18.	- 72 13		13.7	FAINT BLUE STAR
SHAP425-5242.0	04 24 19.	- 52 48 44.	24	16.5	GALAXY
SHAP425-5514.2	04 24 19.	- 55 20 56.	24	16.5	GALAXY
SHAP425-5321.5	04 24 20.	- 53 28 14.	24	16.4	GALAXY

OBJECT NAME	RIGHT ASCEN.	DECLINATION	DIAM.	MAGN.	TYPE OF OBJECT
MCG-02-12-015	04 24 21.	- 10 38 30.	24	15.5	GALAXY
SHAP425-5639.2	04 24 21.	- 56 45 56.	18	17.0	GALAXY
SHAP425-5153.8	04 24 22.	- 52 00 32.	18	16.0	GALAXY
SHAP425-4927.1	04 24 24.	- 49 33 49.	36	17.0	GALAXY
LB 01726	04 24 24.	- 51 51		14.5	FAINT BLUE STAR
SHAP425-5312.6	04 24 25.	- 53 19 20.	24	17.5	GALAXY
SHAP425-5157.0	04 24 26.	- 52 03 43.	12	16.75	GALAXY
HSE 004	04 24 27.	- 67 27			STAR CLUSTER IN LMC
SHAP425-5018.0	04 24 28.	- 50 24 43.	18	17.0	GALAXY
SHAP425-5113.9	04 24 28.	- 51 20 37.	24	17.25	GALAXY
SHAP425-5233.9	04 24 28.	- 52 40 37.	18	17.0	GALAXY
SHAP425-5651.0	04 24 28.	- 56 57 44.	66	16.25	GALAXY
SHAP425-5032.2	04 24 29.	- 50 38 55.	24	16.75	GALAXY
SHAP425-5152.8	04 24 29.	- 51 59 31.	24	16.75	GALAXY
SHAP425-5317.9	04 24 29.	- 53 24 37.	18	17.1	GALAXY
LDN 1543	04 24 30.	+ 18 45	1200		DARK NEBULA
LDN 1403	04 24 30.	+ 55 05	180		DARK NEBULA
SHAP425-5138.9	04 24 30.	- 51 45 37.	60	16.0	GALAXY
SHAP425-5351.8	04 24 30.	- 53 58 31.	24	16.0	GALAXY
SHAP425-5616.0	04 24 30.	- 56 22 44.	24	17.5	GALAXY
HN 0287	04 24 31.	- 48 19			NEBULA
IC 2069	04 24 31.	- 48 19			NONSTELLAR OBJECT
SHAP425-5521.8	04 24 31.	- 55 28 31.	30	16.0	GALAXY
SHAP425-4956.0	04 24 33.	- 50 02 43.	36	16.5	GALAXY
SHAP425-5207.8	04 24 33.	- 52 14 31.	12	17.0	GALAXY
SHAP425-5254.0	04 24 33.	- 53 00 43.	12	17.5	GALAXY
SHAP425-5410.3	04 24 33.	- 54 17 01.	18	17.0	GALAXY
SHAP425-5430.4	04 24 33.	- 54 37 07.	18	17.25	GALAXY
SHAP425-5156.7	04 24 34.	- 52 03 25.	30	17.5	GALAXY
SHAP425-5313.9	04 24 34.	- 53 20 37.	18	17.2	GALAXY
B 217	04 24 35.	+ 26 00			DARK OBJECT
MCG+00-12-029	04 24 36.	+ 02 17 30.	48	15.5	GALAXY
MCG-02-12-016	04 24 36.	- 12 04	15	15.	GALAXY
SHAP425-5114.0	04 24 37.	- 51 20 43.	18	15.75	GALAXY
SHAP425-5224.3	04 24 38.	- 52 31 01.	12	17.5	GALAXY
MCG+01-12-008	04 24 39.	- 06 17	72	15.	GALAXY
SHAP425-5250.9	04 24 39.	- 52 57 37.	18	17.25	GALAXY
SHAP425-5511.8	04 24 39.	- 55 18 31.	24	16.5	GALAXY
HUB E05	04 24 40.	+ 50 02			DIFFUSE NEBULA
IC 2068	04 24 41.	- 42 16 29.			NONSTELLAR OBJECT
UGC 03051	04 24 42.	+ 02 15	66	16.0	GALAXY Sc
MCG-02-12-017	04 24 42.	- 12 10	54	15.	GALAXY
MCG-06-10-008	04 24 42.	- 34 42	36	15.	GALAXY
SER 037.03	04 24 42.	- 57 01	60	14.	LOW SURFACE BRIGHT. GALXY
LB 03356	04 24 42.	- 69 25		13.8	FAINT BLUE STAR
SHAP425-5214.4	04 24 43.	- 52 21 06.	24	17.25	GALAXY
LB 01324	04 24 44.	+ 17 32 36.		16.4	FAINT BLUE STAR
ARC 0490	04 24 44.	- 20 50		17.0	RICH CLUSTER OF GALAXIES
SHAP425-5454.2	04 24 44.	- 55 00 54.	12	16.5	GALAXY
MCG+01-12-010	04 24 45.	- 06 14	42	17.	GALAXY
MCG+01-12-009	04 24 45.	- 06 14 30.	54	16.	GALAXY
SHAP426-4943.5	04 24 45.	- 49 50 12.	36	17.5	GALAXY
SHAP426-5034.6	04 24 45.	- 50 41 18.	18	17.0	GALAXY
SHAP425-5401.6	04 24 45.	- 54 08 18.	18	17.5	GALAXY
SHAP426-5036.2	04 24 46.	- 50 42 54.	36	16.75	GALAXY
SHAP426-5036.4	04 24 47.	- 50 43 06.	30	17.0	GALAXY
SHAP426-5251.9	04 24 47.	- 52 58 36.	18	17.25	GALAXY
OCL 0426	04 24 48.	+ 30 49	600		OPEN STAR CLUSTER
SHB 085	04 24 48.	- 13 09 36.		17.5	QUASI-STELLAR OBJECT
BC PKS0424-13	04 24 48.	- 13 10		17.5	QUASI-STELLAR OBJECT
SHAP426-5221.5	04 24 48.	- 52 28 12.	12	17.25	GALAXY
LB 03357	04 24 48.	- 67 08		12.3	FAINT BLUE STAR
SHAP426-5043.8	04 24 50.	- 50 50 30.	12	17.25	GALAXY
SHAP426-5228.2	04 24 50.	- 52 34 54.	12	17.25	GALAXY
SHAP426-5116.5	04 24 51.	- 51 23 12.	18	17.5	GALAXY
SL 002	04 24 51.	- 72 40	35		STAR CLUSTER IN LMC
ZC 0424.9+0832	04 24 54.	+ 08 32	1340		CLUSTER OF GALAXIES
SHAP426-5111.4	04 24 54.	- 51 18 05.	18	16.75	GALAXY
SHAP426-5308.4	04 24 54.	- 53 15 06.	18	17.1	GALAXY
SHAP426-4948.6	04 24 55.	- 49 55 17.	36	17.0	GALAXY
SHAP426-5215.5	04 24 56.	- 52 22 11.	18	16.75	GALAXY
SHAP426-5231.4	04 24 56.	- 52 38 05.	18	17.5	GALAXY
SHAP426-5308.2	04 24 56.	- 53 14 53.	24	16.9	GALAXY
SHAP426-5312.4	04 24 56.	- 53 19 05.	18	17.4	GALAXY
LW 002	04 24 57.	- 72 41			STAR CLUSTER IN LMC
SHAP426-5146.9	04 24 58.	- 51 53 35.	24	16.5	GALAXY
SHAP426-5248.6	04 24 59.	- 52 55 17.	18	17.0	GALAXY
SHAP426-5308.3	04 24 59.	- 53 14 59.	60	16.6	GALAXY
LBN 0819	04 25	+ 18 40	5100		BRIGHT NEBULA
LBN 0800	04 25	+ 24 30	9900		BRIGHT NEBULA
KHAV 059	04 25	+ 54 13	3170		DARK NEBULA
LBN 0896	04 25	- 06 20	1500		BRIGHT NEBULA
LDN 1524	04 25 00.	+ 24 30	2280		DARK NEBULA
LDN 1482	04 25 00.	+ 35 30	1440		DARK NEBULA
UGC 03052	04 25 00.	+ 39 30	96	16.0	GALAXY SBb?
HN 0288	04 25 00.	- 53 15			NEBULA
IC 2071	04 25 00.	- 53 15			NONSTELLAR OBJECT
LB 01727	04 25 00.	- 57 19		15.2	FAINT BLUE STAR
SHAP426-4931.0	04 25 02.	- 49 37 41.	24	16.75	GALAXY
SHAP426-5044.9	04 25 02.	- 50 51 35.	24	16.5	GALAXY
SHAP426-5344.8	04 25 02.	- 53 51 29.	18	17.25	GALAXY
SHAP426-5258.6	04 25 04.	- 53 05 17.	18	16.75	GALAXY
SHAP426-5127.8	04 25 05.	- 51 34 29.	12	17.5	GALAXY
ZWG 393.021	04 25 06.	- 02 40		15.7	GALAXY
KARA.73B 0150	04 25 06.	- 02 40	66	15.7	ISOLATED GALAXY S
LB 01728	04 25 06.	- 49 27		14.6	FAINT BLUE STAR
SHAP426-5244.6	04 25 06.	- 52 51 17.	12	17.25	GALAXY
SHAP426-5310.5	04 25 06.	- 53 17 11.	24	17.1	GALAXY
SHAP426-5311.0	04 25 06.	- 53 17 41.	30	16.2	GALAXY
LB 03358	04 25 06.	- 65 27		15.0	FAINT BLUE STAR
SHAP426-5044.8	04 25 07.	- 50 51 28.	12	17.0	GALAXY
SHAP426-5229.4	04 25 07.	- 52 36 05.	24	17.25	GALAXY
SHAP426-5250.2	04 25 07.	- 52 56 53.	24	17.25	GALAXY
SHAP426-5308.7	04 25 07.	- 53 15 23.	18	17.0	GALAXY
SHAP426-5547.6	04 25 07.	- 55 54 17.	24	17.0	GALAXY
SHAP426-5048.9	04 25 08.	- 50 55 34.	24	17.0	GALAXY
SHAP426-5209.5	04 25 09.	- 52 16 11.	18	17.25	GALAXY
SHAP426-5239.8	04 25 09.	- 52 46 29.	18	17.25	GALAXY
SHAP426-5316.4	04 25 09.	- 53 23 05.	18	17.3	GALAXY
LB 01325	04 25 11.	+ 20 07 18.		15.4	FAINT BLUE STAR
MCG+04-11-002	04 25 12.	+ 21 31	42	14.5	GALAXY
ZWG 488.003	04 25 12.	+ 21 33		14.9	GALAXY
UGC 03053	04 25 12.	+ 21 33	60	14.9	GALAXY Sc
MCG+00-12-030	04 25 12.	- 02 40	60	15.	GALAXY
MCG-02-12-018	04 25 12.	- 10 55	48	15.	GALAXY
SHAP426-4924.0	04 25 12.	- 49 30 40.	42	16.75	GALAXY
SHAP426-5335.6	04 25 12.	- 53 42 16.	18	17.3	GALAXY
LB 01729	04 25 12.	- 54 10		14.4	FAINT BLUE STAR

OBJECT NAME	RIGHT ASCEN.	DECLINATION	DIAM.	MAGN.	TYPE OF OBJECT
SHAP426-5057.0	04 25 13.	- 51 03 40.	24	17.5	GALAXY
SHAP426-5234.5	04 25 13.	- 52 41 10.	18	16.5	GALAXY
SHAP426-5244.4	04 25 13.	- 52 51 04.	30	17.25	GALAXY
HH 31B	04 25 13.8	+ 26 11 35.			HERBIG-HARO OBJECT
LB 01326	04 25 14.	+ 10 18 54.		15.6	FAINT BLUE STAR
HH 31C	04 25 14.3	+ 26 10 24.			HERBIG-HARO OBJECT
HH 31	04 25 14.4	+ 26 11 04.	80		HERBIG-HARO OBJECT
SHAP426-5239.6	04 25 15.	- 52 46 16.	24	16.5	GALAXY
HH 31A	04 25 15.2	+ 26 11 33.			HERBIG-HARO OBJECT
SHAP426-5236.2	04 25 16.	- 52 42 52.	24	16.75	GALAXY
SHAP426-5613.2	04 25 17.	- 56 19 52.	24	17.0	GALAXY
ZWG 419.008	04 25 18.	+ 07 00		15.7	GALAXY
MCG-02-12-019	04 25 18.	- 09 30	54	15.	GALAXY
LB 01730	04 25 18.	- 48 28		14.0	FAINT BLUE STAR
SHAP426-4928.1	04 25 18.	- 49 34 46.	36	16.75	GALAXY
SHAP426-5043.8	04 25 19.	- 50 50 28.	24	16.75	GALAXY
SHAP426-5053.2	04 25 19.	- 50 59 52.	30	17.25	GALAXY
SHAP426-5333.2	04 25 19.	- 53 39 52.	24	16.7	GALAXY
SHAP426-5442.0	04 25 20.	- 54 48 40.	18	17.0	GALAXY
SHAP426-5434.8	04 25 21.	- 54 41 28.	60	16.5	GALAXY
SHAP426-5308.6	04 25 22.	- 53 15 16.	24	17.6	GALAXY
SHAP426-5311.3	04 25 22.	- 53 17 58.	114	15.9	GALAXY
LB 01327	04 25 23.	+ 13 58 42.		15.9	FAINT BLUE STAR
SHAP426-5036.8	04 25 23.	- 50 43 27.	24	17.5	GALAXY
HN 0290	04 25 23.	- 53 18			NEBULA
IC 2073	04 25 23.	- 53 18			NONSTELLAR OBJECT
SHAP426-5405.6	04 25 23.	- 54 12 16.	36	16.0	GALAXY
SHAP426-5600.5	04 25 23.	- 56 07 10.	24	16.75	GALAXY
LB 01328	04 25 24.	+ 17 47 00.		15.8	FAINT BLUE STAR
MCG-02-12-021	04 25 24.	- 12 07 30.	36	14.5	GALAXY
MCG-02-12-020	04 25 24.	- 12 42	30	15.	GALAXY
SHAP426-5200.4	04 25 24.	- 52 07 04.	12	17.25	GALAXY
LB 01329	04 25 25.	+ 13 55 36.		16.0	FAINT BLUE STAR
SHAP426-5034.0	04 25 25.	- 50 40 39.	24	17.25	GALAXY
SHAP426-5347.5	04 25 25.	- 53 54 10.	18	17.5	GALAXY
SHAP426-5545.3	04 25 26.	- 55 51 58.	24	16.25	GALAXY
SHAP426-5514.6	04 25 27.	- 55 21 16.	18	17.5	GALAXY
SHAP426-5035.2	04 25 29.	- 50 41 51.	24	17.5	GALAXY
SHAP426-5046.0	04 25 29.	- 50 52 39.	54	17.25	GALAXY
SHAP426-5513.6	04 25 29.	- 55 20 16.	24	17.0	GALAXY
HN 0289	04 25 30.	- 48 29			NEBULA
IC 2072	04 25 31.	- 48 29			NONSTELLAR OBJECT
SHAP426-5042.4	04 25 31.	- 50 49 03.	18	17.0	GALAXY
SHAP426-5331.2	04 25 31.	- 53 37 51.	24	17.3	GALAXY
SHAP426-5214.6	04 25 32.	- 52 21 15.	24	17.25	GALAXY
SHAP426-5305.4	04 25 32.	- 53 12 03.	18	17.4	GALAXY
SHAP426-5451.4	04 25 32.	- 54 58 03.	24	17.0	GALAXY
SHAP426-5114.2	04 25 33.	- 51 20 51.	30	17.0	GALAXY
SHAP426-5130.2	04 25 33.	- 51 36 51.	12	17.25	GALAXY
SHAP426-5513.0	04 25 33.	- 55 19 39.	18	17.0	GALAXY
PATH 1.161	04 25 35.	+ 30 28	22		NEBULA
SHAP426-4935.8	04 25 35.	- 49 42 27.	36	16.25	GALAXY
SHAP426-5202.6	04 25 35.	- 52 09 15.	12	17.0	GALAXY
SHAP426-5215.3	04 25 35.	- 52 21 57.	24	16.0	GALAXY
SHAP426-5513.0	04 25 35.	- 55 19 39.	18	16.5	GALAXY
UGC 03054	04 25 36.	+ 00 56	96	17.	GALAXY
SHAP426-5045.2	04 25 36.	- 50 51 51.	30	17.25	GALAXY
SHAP426-5514.3	04 25 36.	- 55 20 57.	24	16.5	GALAXY
SHAP426-5233.2	04 25 37.	- 52 39 51.	24	17.0	GALAXY
SHAP426-5145.5	04 25 38.	- 51 52 09.	18	17.0	GALAXY
SHAP426-5225.8	04 25 38.	- 52 32 27.	12	17.5	GALAXY
LB 01330	04 25 39.	+ 14 07 18.		16.0	FAINT BLUE STAR
SHAP426-5030.7	04 25 40.	- 50 37 20.	30	16.0	GALAXY
SHAP426-5423.5	04 25 40.	- 54 30 09.	24	16.5	GALAXY
SHAP426-5104.2	04 25 41.	- 51 10 50.	12	17.5	GALAXY
SHAP426-5342.9	04 25 41.	- 53 49 33.	18	17.0	GALAXY
ZC 0425.7+4120	04 25 42.	+ 41 20	12230		CLUSTER OF GALAXIES
MCG+08-09-001	04 25 42.	+ 48 28	30	16.	GALAXY
MCG-02-12-022	04 25 42.	- 14 53	72	15.	GALAXY
SHAP426-5559.0	04 25 43.	- 56 05 39.	18	17.0	GALAXY
SHAP426-5222.6	04 25 44.	- 52 29 14.	18	17.0	GALAXY
SHAP426-5323.6	04 25 44.	- 53 30 14.	18	17.2	GALAXY
SHAP426-5502.8	04 25 44.	- 55 09 26.	18	17.5	GALAXY
LB 01331	04 25 45.	+ 14 33 30.		16.6	FAINT BLUE STAR
SHAP427-5129.8	04 25 45.	- 51 36 26.	18	17.5	GALAXY
RNGC 1580	04 25 46.	- 05 17		14.5	GALAXY
SHAP427-5025.0	04 25 46.	- 50 31 38.	18	17.25	GALAXY
SHAP426-5647.3	04 25 46.	- 56 53 57.	24	17.0	GALAXY
LB 01332	04 25 47.	+ 08 52 00.		16.5	FAINT BLUE STAR
SHAP427-5104.3	04 25 47.	- 51 10 56.	12	17.5	GALAXY
SHAP426-5411.3	04 25 47.	- 54 17 56.	18	17.0	GALAXY
UGC 03055	04 25 48.	+ 10 57	84	16.0	GALAXY S0?
B 012	04 25 48.	+ 54 08	1440		DARK OBJECT
7ZW 015	04 25 48.	+ 73 10			COMPACT GALAXY
MCG-06-10-009	04 25 48.	- 33 44	30	15.5	GALAXY
SHAP426-5327.2	04 25 48.	- 53 33 50.	18	17.3	GALAXY
SHAP426-5347.0	04 25 48.	- 53 53 38.	18	16.6	GALAXY
SHAP426-5430.0	04 25 49.	- 54 36 38.	24	17.25	GALAXY
SHAP426-5433.1	04 25 50.	- 54 39 44.	24	17.0	GALAXY
REIN 6.033	04 25 50.37	- 05 17 21.2			NEBULA
MCG+01-12-011	04 25 51.	- 05 17 30.	42	14.5	GALAXY
SHAP427-5321.5	04 25 51.	- 53 28 08.	24	16.8	GALAXY
SHAP427-5219.2	04 25 52.	- 52 25 50.	18	17.5	GALAXY
RNGC 1584	04 25 53.	- 17 38			GALAXY
RNGC 1585	04 25 53.	- 42 15			GALAXY
SHAP427-5030.9	04 25 53.	- 50 37 31.	36	16.0	GALAXY
SHAP427-5031.2	04 25 53.	- 50 37 49.	24	17.25	GALAXY
SHAP427-5047.6	04 25 53.	- 50 54 13.	24	17.25	GALAXY
SHAP427-5235.9	04 25 53.	- 52 42 32.	24	17.0	GALAXY
MCG-02-12-023	04 25 54.	- 12 15	54	15.	GALAXY
SHAP427-5106.4	04 25 54.	- 51 13 01.	12	17.25	GALAXY
SHAP426-5530.2	04 25 54.	- 55 36 50.	18	17.5	GALAXY
SHAP427-5034.4	04 25 56.	- 50 41 01.	18	17.0	GALAXY
SHAP427-5055.4	04 25 56.	- 51 02 01.	30	17.0	GALAXY
SHAP427-5415.0	04 25 56.	- 54 21 38.	24	17.0	GALAXY
SHAP427-5326.4	04 25 57.	- 53 33 01.	30	17.5	GALAXY
SHAP427-5333.4	04 25 58.	- 53 40 01.	24	16.9	GALAXY
SHAP427-4946.2	04 25 59.	- 49 52 49.	48	16.5	GALAXY
SHAP427-5100.8	04 25 59.	- 51 07 25.	24	17.0	GALAXY
SHAP427-5431.4	04 25 59.	- 54 38 01.	18	17.25	GALAXY
KHAV 063	04 26	+ 18 19	3940		DARK NEBULA
KHAV 064	04 26	+ 23 13	6270		DARK NEBULA
LBN 0712	04 26	+ 52 30	1200		BRIGHT NEBULA
LDN 1546	04 26 00.	+ 18 20	2460		DARK NEBULA
LDN 1407	04 26 00.	+ 54 10	1440		DARK NEBULA
LDN 1404	04 26 00.	+ 55 10	600		DARK NEBULA
MCG+12-05-005	04 26 00.	+ 71 47 30.	702	12.1	GALAXY
MCG+01-12-012	04 26 00.	- 06 25	36	15.5	GALAXY
LB 01731	04 26 00.	- 48 29		15.1	FAINT BLUE STAR
SHAP427-5156.3	04 26 00.	- 52 02 55.	18	17.0	GALAXY
SHAP427-5515.4	04 26 00.	- 55 22 01.	18	16.5	GALAXY
SHAP427-5133.4	04 26 01.	- 51 40 01.	24	15.0	GALAXY
SHAP427-5324.6	04 26 01.	- 53 31 13.	24	17.0	GALAXY
SHAP427-5313.4	04 26 03.	- 53 20 01.	24	17.3	GALAXY
SHAP427-5400.0	04 26 03.	- 54 06 37.	18	17.5	GALAXY
SHAP427-5125.4	04 26 04.	- 51 32 01.	36	16.25	GALAXY
SHAP427-5318.8	04 26 04.	- 53 25 25.	18	17.4	GALAXY
LB 01333	04 26 05.	+ 20 36 48.		14.9	FAINT BLUE STAR
RNGC 1583	04 26 05.	- 17 43		15.0	GALAXY
SHAP427-5550.8	04 26 05.	- 55 57 25.	18	17.0	GALAXY
MCG+11-06-001	04 26 06.	+ 64 44	156	11.7	GALAXY
7ZW 016	04 26 06.	+ 64 45			COMPACT GALAXY
ZWG 306.001	04 26 06.	+ 64 45		11.8	GALAXY
UGC 03056	04 26 06.	+ 64 45	198	11.8	GALAXY IRR
LB 01732	04 26 06.	- 46 32		14.5	FAINT BLUE STAR
SHAP427-5433.4	04 26 07.	- 54 40 01.	18	17.0	GALAXY
SHAP427-5434.2	04 26 07.	- 54 40 49.	24	16.0	GALAXY
RNGC 1569	04 26 08.	+ 64 44		12.5	GALAXY
SHAP427-5330.6	04 26 08.	- 53 37 13.	24	17.5	GALAXY
MCG-03-12-010	04 26 09.	- 17 43	12	15.	GALAXY
SHAP427-5102.0	04 26 10.	- 51 08 36.	24	16.5	GALAXY
SHAP427-5345.1	04 26 10.	- 53 51 43.	30	17.5	GALAXY
SHAP427-5447.3	04 26 10.	- 54 53 55.	30	17.25	GALAXY
SHAP427-5527.1	04 26 10.	- 55 33 43.	36	16.75	GALAXY
SHAP427-5534.8	04 26 10.	- 55 41 25.	18	17.25	GALAXY
ZWG 419.009	04 26 12.	+ 07 38		15.2	GALAXY
MCG+13-04-005	04 26 12.	+ 76 28	57	16.	GALAXY
LB 00217	04 26 13.	+ 15 13 06.		15.9	FAINT BLUE STAR
SHAP427-5511.4	04 26 13.	- 55 18 01.	18	17.5	GALAXY
SHAP427-5304.8	04 26 14.	- 53 11 24.	24	17.5	GALAXY
SHAP427-5119.5	04 26 15.	- 51 26 06.	42	17.0	GALAXY
SHAP427-5330.2	04 26 15.	- 53 36 48.	18	17.2	GALAXY
SHAP427-5349.5	04 26 17.	- 53 56 06.	18	17.4	GALAXY
SHAP427-5434.7	04 26 17.	- 54 41 18.	18	16.75	GALAXY
ZWG 347.005	04 26 18.	+ 76 28		15.1	GALAXY
UGC 03057	04 26 18.	+ 76 28	72	15.1	GALAXY Sb/SBb
SHAP427-5344.5	04 26 18.	- 53 51 06.	24	16.9	GALAXY
LB 03359	04 26 18.	- 66 02		14.0	FAINT BLUE STAR
IC 2062	04 26 21.	+ 71 49			SAME AS NGC 1560
SHAP427-5355.9	04 26 21.	- 54 02 30.	24	17.25	GALAXY
SHAP427-5128.1	04 26 22.	- 51 34 42.	18	17.5	GALAXY
SHAP427-5249.3	04 26 22.	- 52 55 54.	18	17.0	GALAXY
SHAP427-5323.8	04 26 22.	- 53 30 24.	18	16.7	GALAXY
SHAP427-5410.9	04 26 22.	- 54 17 30.	60	14.5	GALAXY
SHAP427-5510.4	04 26 23.	- 55 17 00.	24	17.5	GALAXY
ZC 0426.4-0002	04 26 24.	- 00 02	11630		CLUSTER OF GALAXIES
2ZW 011	04 26 24.	- 01 58			COMPACT GALAXY
ZWG 393.022	04 26 24.	- 01 59		15.7	GALAXY
MCG-04-11-014	04 26 24.	- 25 24 30.	36	16.	GALAXY
SHAP427-5115.2	04 26 25.	- 51 21 47.	30	15.5	GALAXY
SHAP427-5338.7	04 26 26.	- 53 45 18.	18	17.3	GALAXY
MCG-02-12-024	04 26 27.	- 12 36 30.	72	15.	GALAXY
SHAP427-5237.9	04 26 27.	- 52 44 29.	24	16.75	GALAXY
SHAP427-5347.8	04 26 28.	- 53 54 23.	24	16.9	GALAXY
MCG+04-11-003	04 26 30.	+ 26 24	78	17.	GALAXY
MCG-05-11-008	04 26 30.	- 32 32	60	15.5	GALAXY
SHAP427-5300.5	04 26 30.	- 53 07 05.	24	17.4	GALAXY
SHAP427-5304.0	04 26 30.	- 53 10 35.	42	16.8	GALAXY
SHAP427-5323.0	04 26 31.	- 53 29 35.	18	17.1	GALAXY
SHAP427-5225.0	04 26 32.	- 52 31 35.	24	16.75	GALAXY
SHAP427-5501.4	04 26 32.	- 55 07 59.	216	12.3	GALAXY
RNGC 1596	04 26 33.	- 55 07		12.5	GALAXY
SHAP427-5340.8	04 26 34.	- 53 47 23.	24	17.2	GALAXY
SHAP427-5109.6	04 26 35.	- 51 16 11.	18	17.0	GALAXY
SHAP427-5130.7	04 26 35.	- 51 37 17.	18	17.25	GALAXY
MCG-05-11-009	04 26 36.	- 32 32	18	16.	GALAXY
SHAP427-5255.7	04 26 36.	- 53 02 17.	24	17.4	GALAXY
LB 01334	04 26 37.	+ 10 59 24.		17.3	FAINT BLUE STAR
SHAP427-5311.7	04 26 37.	- 53 18 17.	24	17.3	GALAXY
LB 01335	04 26 38.	+ 22 01 42.		15.7	FAINT BLUE STAR
SHAP427-5108.6	04 26 38.	- 51 15 11.	18	16.25	GALAXY
SHAP427-5331.9	04 26 40.	- 53 38 29.	36	17.3	GALAXY
SS 19	04 26 41.	+ 35 13			DIFFUSE GALACTIC NEBULA
MCG-02-12-025	04 26 42.	- 11 06 30.	72	15.	GALAXY
MCG-02-12-026	04 26 42.	- 12 20	48	14.5	GALAXY
MCG-02-12-027	04 26 42.	- 14 14	54	15.	GALAXY
IC 2076	04 26 42.	- 48 20			NONSTELLAR OBJECT
SHAP427-5241.9	04 26 42.	- 52 48 28.	24	17.5	GALAXY
SHAP427-5558.4	04 26 42.	- 56 04 59.	24	17.5	GALAXY
HN 0291	04 26 43.	- 48 20			NEBULA
SHAP427-5300.4	04 26 43.	- 53 06 58.	18	17.1	GALAXY
SHAP427-5405.3	04 26 43.	- 54 11 52.	24	17.5	GALAXY
SHAP428-5055.9	04 26 44.	- 51 02 28.	12	16.25	GALAXY
SHAP427-5305.4	04 26 44.	- 53 11 58.	24	17.3	GALAXY
SHAP427-5342.9	04 26 44.	- 53 49 28.	24	17.3	GALAXY
SHAP427-5503.1	04 26 44.	- 55 09 40.	144	14.5	GALAXY
SHAP427-5530.9	04 26 44.	- 55 37 29.	18	17.0	GALAXY
SHAP428-5101.2	04 26 45.	- 51 07 46.	18	17.0	GALAXY
SHAP428-5143.2	04 26 45.	- 51 49 46.	12	17.0	GALAXY
SHAP427-5349.0	04 26 45.	- 53 55 34.	42	16.8	GALAXY
RNGC 1602	04 26 45.	- 55 10			GALAXY
LB 01336	04 26 46.	+ 17 53 36.		11.4	FAINT BLUE STAR
SHAP428-5054.6	04 26 46.	- 51 01 10.	42	15.5	GALAXY
SHAP427-5351.8	04 26 47.	- 53 58 22.	24	16.9	GALAXY
RLWT 104	04 26 48.	+ 40 18		14.	FAINT VERY BLUE STAR
ZWG 393.023	04 26 48.	- 00 19		15.6	GALAXY
MCG+00-12-031	04 26 48.	- 00 19	36	15.	GALAXY
MCG-02-12-028	04 26 48.	- 12 40 30.	30	15.5	GALAXY
SHAP427-5543.6	04 26 48.	- 55 50 10.	12	17.5	GALAXY
HZ 11	04 26 49.	+ 07 36		14.7	DECIDEDLY BLUE STAR
SHAP428-5034.9	04 26 49.	- 50 41 28.	24	15.75	GALAXY
SHAP427-5619.0	04 26 49.	- 56 25 34.	24	17.0	GALAXY
RNGC 1579	04 26 52.	+ 35 10			DIFFUSE NEBULA
SHAP428-5032.5	04 26 52.	- 50 39 03.	24	16.75	GALAXY
SHAP428-5344.8	04 26 52.	- 53 51 22.	18	17.4	GALAXY
SG 2.011	04 26 53.	+ 52 35	900		DIFFUSE EMISSION NEBULA
ZC 0426.9+1410	04 26 54.	+ 14 10	1010		CLUSTER OF GALAXIES
LB 01337	04 26 54.	+ 16 51 48.		16.2	FAINT BLUE STAR
MRSL 165-09/1	04 26 54.	+ 35 10	360		HII REGION
DG 034	04 26 54.	+ 35 11	1020		REFLECTION NEBULA
SHAP428-5301.2	04 26 54.	- 53 07 46.	24	16.9	GALAXY
SHAP428-5340.6	04 26 54.	- 53 47 10.	30	17.2	GALAXY
LB 03360	04 26 54.	- 66 58		12.4	FAINT BLUE STAR
SHAP428-5125.8	04 26 55.	- 51 32 21.	24	16.5	GALAXY
SHAP428-5322.9	04 26 55.	- 53 29 28.	48	16.5	GALAXY
SHAP428-5334.0	04 26 55.	- 53 40 34.	24	16.5	GALAXY

OBJECT NAME	RIGHT ASCEN.	DECLINATION	DIAM.	MAGN.	TYPE OF OBJECT
SHAP428-5216.6	04 26 56.	- 52 23 09.	24	17.0	GALAXY
SHAP428-5405.2	04 26 56.	- 54 11 46.	24	17.5	GALAXY
MCG+00-12-032	04 26 57.	+ 00 40	48	15.	GALAXY
SHAP428-5347.0	04 26 58.	- 53 53 33.	18	17.3	GALAXY
CED 035	04 26 59.	+ 35 11	720		DIFFUSE GALACTIC NEBULA
ARC 0491	04 26 59.	- 05 09		17.7	RICH CLUSTER OF GALAXIES
RNGC 1595	04 26 59.	- 47 55			UNVERIFIED SOUTHERN OBJECT
SHAP428-4935.0	04 26 59.	- 49 41 32.	48	16.0	GALAXY
KHAV 065	04 27	+ 28 49	9060		DARK NEBULA
KHAV 061	04 27	+ 35 37	13340		DARK NEBULA
KHAV 062	04 27	+ 46 25	1860		DARK NEBULA
KHAV 060	04 27	+ 51 13	5010		DARK NEBULA
ZWG 393.024	04 27 00.	+ 00 05		15.0	GALAXY
ZWG 393.025	04 27 00.	+ 00 39		15.6	GALAXY
UGC 03058	04 27 00.	+ 00 39	78	15.6	GALAXY SB
MCG+01-12-005	04 27 00.	+ 03 33	144	15.	GALAXY
DG 035	04 27 00.	+ 35 23	120		REFLECTION NEBULA
ARP 210	04 27 00.	+ 64 45			PECULIAR GALAXY
SHAP428-5259.0	04 27 00.	- 53 05 33.	30	17.4	GALAXY
SHAP428-5536.4	04 27 02.	- 55 42 57.	12	17.5	GALAXY
RNGC 1560	04 27 04.	+ 71 46		12.0	GALAXY
SHAP428-5305.5	04 27 04.	- 53 12 03.	24	16.8	GALAXY
LB 01338	04 27 05.	+ 17 06 30.		17.0	FAINT BLUE STAR
CED 036	04 27 05.	+ 35 23	120		DIFFUSE GALACTIC NEBULA
SHAP428-5351.8	04 27 05.	- 53 58 21.	18	17.1	GALAXY
SHAP428-5459.1	04 27 05.	- 55 05 39.	18	17.25	GALAXY
ZWG 419.010	04 27 06.	+ 03 35		15.4	GALAXY
UGC 03059	04 27 06.	+ 03 35	150	15.4	GALAXY
KARA.73B 0151	04 27 06.	+ 03 35	168	15.4	ISOLATED GALAXY S
MRSL 152+02/1	04 27 06.	+ 52 27	1200		HII REGION
ZWG 328.006	04 27 06.	+ 71 47		12.1	GALAXY
UGC 03060	04 27 06.	+ 71 47	600	12.1	GALAXY
MCG-02-12-029	04 27 06.	- 10 36	36	15.5	GALAXY
MCG-06-10-010	04 27 06.	- 37 36	24	15.	GALAXY
SHAP428-5112.9	04 27 09.	- 51 19 26.	18	17.0	GALAXY
SHAP428-5527.5	04 27 10.	- 55 34 03.	30	17.0	GALAXY
SHAP428-5610.2	04 27 10.	- 56 16 45.	24	17.0	GALAXY
RNGC 1598	04 27 11.	- 47 53			UNVERIFIED SOUTHERN OBJECT
SHAP428-5328.4	04 27 11.	- 53 34 56.	18	17.0	GALAXY
SHAP428-5343.0	04 27 15.	- 53 49 32.	54	17.0	GALAXY
SHAP428-5255.8	04 27 16.	- 53 02 20.	24	17.2	GALAXY
SHAP428-5334.6	04 27 16.	- 53 41 08.	18	17.5	GALAXY
SHAP428-5336.4	04 27 16.	- 53 42 56.	30	17.4	GALAXY
SHAP428-5559.8	04 27 16.	- 56 06 20.	18	17.5	GALAXY
B 013	04 27 17.	+ 54 47	660		DARK OBJECT
MCG-05-11-010	04 27 18.	- 29 53	30	15.	GALAXY
SHAP428-5623.0	04 27 19.	- 56 29 32.	30	16.5	GALAXY
SHAP428-5343.7	04 27 20.	- 53 50 14.	90	16.1	GALAXY
SHAP428-5347.6	04 27 20.	- 53 54 08.	18	17.1	GALAXY
MCG+00-12-033	04 27 21.	- 00 51	48	15.	GALAXY
SHAP428-5351.0	04 27 21.	- 53 57 32.	18	17.3	GALAXY
SHAP428-5359.4	04 27 21.	- 54 05 56.	30	17.0	GALAXY
SHAP428-5401.6	04 27 21.	- 54 08 08.	18	17.25	GALAXY
SHAP428-5129.8	04 27 22.	- 51 36 20.	18	17.0	GALAXY
SHAP428-5342.0	04 27 22.	- 53 48 32.	24	16.5	GALAXY
HN 0292	04 27 22.	- 53 50			NEBULA
IC 2079	04 27 22.	- 53 50			NONSTELLAR OBJECT
RNGC 1591	04 27 23.	- 26 50		13.0	GALAXY
SHAP428-5132.3	04 27 23.	- 51 38 50.	18	17.0	GALAXY
7ZW 017	04 27 24.	+ 66 13			COMPACT GALAXY
ZWG 393.026	04 27 24.	- 00 52		15.5	GALAXY
MCG-02-12-030	04 27 24.	- 12 35 30.	54	14.	GALAXY
MCG-04-11-015	04 27 24.	- 26 50	60	13.5	GALAXY
SHAP428-5142.5	04 27 24.	- 51 49 01.	18	16.5	GALAXY
SHAP428-5324.1	04 27 25.	- 53 30 38.	42	16.1	GALAXY
MCG+00-12-034	04 27 27.	- 01 02	36	15.5	GALAXY
SNO 19	04 27 27.	- 27 33 12.		13.	LINEAR GRP OF 6 GALAXIES
SHAP428-5323.0	04 27 27.	- 53 29 31.	18	16.7	GALAXY
SHAP428-5353.8	04 27 27.	- 54 00 19.	18	17.2	GALAXY
SHAP428-5613.4	04 27 27.	- 56 19 56.	18	17.5	GALAXY
SS 20	04 27 29.	+ 35 19			DIFFUSE GALACTIC NEBULA
SHAP428-5350.1	04 27 29.	- 53 56 37.	18	16.7	GALAXY
ZWG 419.011	04 27 30.	+ 06 50		15.5	GALAXY
UGC 03061	04 27 30.	+ 06 50	132	15.5	GALAXY MLT SYS
MCG+01-12-006	04 27 30.	+ 06 50	120	14.	GALAXY
SHAP428-5157.8	04 27 30.	- 52 04 19.	18	17.0	GALAXY
IC 2067	04 27 32.	+ 35 20 16.			NONSTELLAR OBJECT
SHAP428-5304.0	04 27 32.	- 53 10 31.	30	17.3	GALAXY
SHAP428-5340.0	04 27 33.	- 53 46 31.	18	17.3	GALAXY
SHAP428-5357.0	04 27 33.	- 54 03 31.	18	16.7	GALAXY
SHAP428-5502.9	04 27 33.	- 55 09 25.	12	17.25	GALAXY
SHAP428-5443.4	04 27 34.	- 54 49 55.	30	16.5	GALAXY
SG 3.024	04 27 35.	+ 50 22	300		DIFFUSE EMISSION NEBULA
IC 0372	04 27 35.	- 05 07 02.			NONSTELLAR OBJECT
RNGC 1592	04 27 35.	- 27 04			NON-EXISTENT OBJECT
SHAP428-5443.0	04 27 35.	- 54 49 31.	36	16.5	GALAXY
MCG-04-11-017	04 27 36.	- 26 37	54	15.	GALAXY
MCG-04-11-016	04 27 36.	- 26 53 30.	72	15.	GALAXY
SHAP428-5438.6	04 27 36.	- 54 45 07.	42	16.5	GALAXY
REIN 4.045A	04 27 36.02	- 05 07 06.1			NEBULA
REIN 4.045B	04 27 36.03	- 05 07 06.2			NEBULA
SHAP428-5346.8	04 27 37.	- 53 53 19.	30	16.2	GALAXY
SHAP428-5349.6	04 27 37.	- 53 56 07.	30	17.0	GALAXY
IC 0371	04 27 38.	- 00 40			NONSTELLAR OBJECT
SHAP428-5338.2	04 27 38.	- 53 44 43.	24	17.1	GALAXY
SHAP428-5509.2	04 27 38.	- 55 15 43.	24	17.0	GALAXY
SHAP429-5008.2	04 27 40.	- 50 14 42.	18	17.0	GALAXY
SHAP428-5301.2	04 27 40.	- 53 07 42.	42	17.4	GALAXY
SHAP428-5332.3	04 27 40.	- 53 38 49.	18	17.0	GALAXY
SHAP428-5553.1	04 27 40.	- 55 59 37.	18	17.0	GALAXY
SHAP428-5632.4	04 27 40.	- 56 38 55.	72	17.25	GALAXY
SHAP428-5341.0	04 27 41.	- 53 47 30.	24	17.0	GALAXY
SHAP428-5351.0	04 27 41.	- 53 57 31.	36	16.0	GALAXY
RLWT 105	04 27 42.	+ 42 51		15.	FAINT VERY BLUE STAR
SHAP428-5336.0	04 27 44.	- 53 42 30.	24	17.3	GALAXY
SHAP428-5316.0	04 27 45.	- 53 22 30.	18	16.8	GALAXY
SHAP428-5336.8	04 27 45.	- 53 43 18.	24	17.4	GALAXY
SHAP428-5347.1	04 27 46.	- 53 53 36.	18	16.1	GALAXY
MCG+12-05-006	04 27 48.	+ 73 12 30.	39	15.	GALAXY
MCG-04-11-018	04 27 48.	- 26 57	36	15.	GALAXY
MCG-05-11-011	04 27 48.	- 27 30	84	15.	GALAXY
SHAP429-5331.2	04 27 50.	- 53 37 42.	18	17.3	GALAXY
SHAP429-5335.2	04 27 50.	- 53 41 42.	24	17.4	GALAXY
SHAP428-5419.4	04 27 50.	- 54 25 54.	18	17.0	GALAXY
LB 00218	04 27 51.	+ 15 50 36.		16.9	FAINT BLUE STAR
IC 2081	04 27 51.	- 53 43			NONSTELLAR OBJECT
SHAP429-5336.7	04 27 51.	- 53 43 12.	72	16.0	GALAXY
SHAP429-5347.1	04 27 51.	- 53 53 36.	18	17.2	GALAXY
SHAP429-5054.2	04 27 52.	- 51 00 41.	18	17.0	GALAXY
HN 0293	04 27 52.	- 53 43			NEBULA
SHAP428-5429.6	04 27 52.	- 54 36 06.	12	17.5	GALAXY
SHAP429-5156.2	04 27 53.	- 52 02 42.	18	16.75	GALAXY
SHAP429-5341.1	04 27 53.	- 53 47 36.	24	16.3	GALAXY
ARC 0495	04 27 54.	- 26 28		17.0	RICH CLUSTER OF GALAXIES
SHAP429-5205.4	04 27 54.	- 52 11 54.	24	17.0	GALAXY
SHAP429-5338.9	04 27 54.	- 53 45 24.	18	17.1	GALAXY
SHAP429-5344.6	04 27 54.	- 53 51 06.	42	16.8	GALAXY
SHAP429-5226.6	04 27 55.	- 52 33 05.	24	17.25	GALAXY
SHAP429-5400.9	04 27 55.	- 54 07 24.	48	16.9	GALAXY
SHAP429-5335.2	04 27 56.	- 53 41 41.	24	16.8	GALAXY
LB 00219	04 27 57.	+ 16 04 12.		16.7	FAINT BLUE STAR
SHAP429-5202.0	04 27 57.	- 52 08 29.	42	16.75	GALAXY
SHAP429-5230.0	04 27 57.	- 52 36 29.	24	17.0	GALAXY
SHAP429-5300.6	04 27 57.	- 53 07 05.	24	17.0	GALAXY
SHAP429-5349.2	04 27 57.	- 53 55 41.	24	16.7	GALAXY
HN 0294	04 27 57.	- 53 56			NEBULA
SHAP429-5349.7	04 27 57.	- 53 56 11.	30	16.1	GALAXY
HOLM 076A	04 27 58.	+ 00 33	24	13.3	PART OF MULTIPLE GALAXY
SHAP429-5336.9	04 27 59.	- 53 43 23.	24	17.1	GALAXY
LBN 0821	04 28	+ 18 00	240		BRIGHT NEBULA
KHAV 066	04 28	+ 24 13	3540		DARK NEBULA
LBN 0792	04 28	+ 26 30	14400		BRIGHT NEBULA
LBN 0767	04 28	+ 35 20	180		BRIGHT NEBULA
LBN 0766	04 28	+ 35 20	180		BRIGHT NEBULA
KHAV 067	04 28	+ 53 13	3680		DARK NEBULA
MCG+00-12-035	04 28 00.	+ 00 34	36	13.2	GALAXY
LDN 1408	04 28 00.	+ 54 25	1560		DARK NEBULA
SHAP429-5016.4	04 28 00.	- 50 22 53.	30	16.5	GALAXY
SHAP429-5335.2	04 28 00.	- 53 41 41.	24	16.6	GALAXY
SHAP429-5338.3	04 28 00.	- 53 44 47.	18	16.7	GALAXY
IC 2082	04 28 00.	- 53 56			GALAXY
SHAP429-5227.8	04 28 01.	- 52 34 17.	24	17.0	GALAXY
SHAP429-5332.5	04 28 01.	- 53 38 59.	18	17.4	GALAXY
SHAP429-5352.4	04 28 01.	- 53 58 53.	18	16.6	GALAXY
HSE 005	04 28 01.	- 68 23			STAR CLUSTER IN LMC
HOLM 076B	04 28 02.	+ 00 33	18	14.3	PART OF MULTIPLE GALAXY
MCG+00-12-036	04 28 03.	- 00 23 30.	84	13.	GALAXY
SHAP429-5114.8	04 28 03.	- 51 21 17.	18	16.75	GALAXY
SHAP429-5335.3	04 28 04.	- 53 41 47.	24	17.3	GALAXY
REIN 6.034	04 28 04.99	- 00 24 39.4			NEBULA
SHAP429-5117.7	04 28 05.	- 51 24 11.	24	17.5	GALAXY
SHAP429-5316.3	04 28 05.	- 53 22 47.	24	17.2	GALAXY
SHAP429-5325.5	04 28 05.	- 53 31 59.	24	17.5	GALAXY
SHAP429-5329.9	04 28 05.	- 53 36 23.	24	16.9	GALAXY
2ZW 012	04 28 06.	+ 00 33			COMPACT GALAXY
ZWG 393.028	04 28 06.	+ 00 33		13.3	GALAXY
MRK 616	04 28 06.	+ 00 33	20	14.5	GALAXY WITH UV CONTINUUM
UGC 03063	04 28 06.	+ 00 33	120	13.3	GALAXY E
KARA.72 099A	04 28 06.	+ 00 33	114	13.3	PART OF DOUBLE GALAXY
MCG+00-12-037	04 28 06.	+ 00 34	60	14.3	GALAXY
SS 21	04 28 06.	+ 23 18			DIFFUSE GALACTIC NEBULA
ZWG 393.027	04 28 06.	- 00 25		14.3	GALAXY
UGC 03062	04 28 06.	- 00 25	126	14.3	GALAXY Sb
SHAP429-5054.0	04 28 06.	- 51 00 29.	24	16.0	GALAXY
SER 037.01	04 28 06.	- 53 57	1200	14.	RICH CLUSTER OF GALAXIES
SHAP429-5353.8	04 28 06.	- 54 00 17.	30	16.4	GALAXY
SHAP429-5400.8	04 28 06.	- 54 07 17.	30	17.2	GALAXY
LB 03361	04 28 06.	- 61 21		12.8	FAINT BLUE STAR
RNGC 1587	04 28 07.	+ 00 23		13.5	GALAXY
RNGC 1586	04 28 07.	- 00 25		14.5	GALAXY
SHAP429-5326.5	04 28 08.	- 53 32 59.	30	17.2	GALAXY
SHAP429-5558.0	04 28 08.	- 56 04 29.	18	17.5	GALAXY
MCG+00-12-038	04 28 09.	+ 00 44	138	12.	GALAXY
SHAP429-5328.8	04 28 09.	- 53 35 17.	24	16.0	GALAXY
SHAP429-5403.6	04 28 09.	- 54 10 05.	24	16.8	GALAXY
LB 01339	04 28 10.	+ 17 23 48.		15.5	FAINT BLUE STAR
SHAP429-5345.8	04 28 11.	- 53 52 17.	18	17.0	GALAXY
SHAP429-5348.8	04 28 11.	- 53 55 17.	30	16.1	GALAXY
SHAP429-5351.4	04 28 11.	- 53 57 53.	24	16.5	GALAXY
ZWG 393.029	04 28 12.	+ 00 33		14.1	GALAXY
UGC 03064	04 28 12.	+ 00 33	90	14.1	GALAXY COMPACT
KARA.72 099B	04 28 12.	+ 00 33	60	14.1	PART OF DOUBLE GALAXY
ZWG 393.030	04 28 12.	+ 00 45		13.8	GALAXY
UGC 03065	04 28 12.	+ 00 45	210	13.8	GALAXY Sb
DG 036	04 28 12.	+ 17 04	300		REFLECTION NEBULA
B 018	04 28 12.	+ 24 15	3600		DARK OBJECT
MCG+12-05-007	04 28 12.	+ 73 07	33	15.	GALAXY
ZWG 328.007	04 28 12.	+ 73 11		15.5	GALAXY
RNGC 1588	04 28 13.	+ 00 33		14.0	GALAXY
RNGC 1589	04 28 13.	+ 00 45		14.0	GALAXY
SHAP429-5227.2	04 28 13.	- 52 33 40.	18	17.0	GALAXY
SHAP429-5555.0	04 28 13.	- 56 01 29.	12	17.0	GALAXY
HH 28	04 28 13.5	+ 17 57 02.	15		HERBIG-HARO OBJECT
LB 01340	04 28 14.	+ 21 37 12.		15.7	FAINT BLUE STAR
SHAP429-5513.9	04 28 14.	- 55 20 22.	24	17.0	GALAXY
REIN 4.046A	04 28 14.36	- 04 58 38.2			NEBULA
REIN 4.046B	04 28 14.37	- 04 58 37.3			NEBULA
MCG+01-12-013	04 28 15.	- 04 58	60	15.	GALAXY
IC 0373	04 28 15.	- 04 58 55.			NONSTELLAR OBJECT
SS 22	04 28 16.	+ 17 04			DIFFUSE GALACTIC NEBULA
SHAP429-5329.7	04 28 16.	- 53 36 10.	24	16.7	GALAXY
B 015	04 28 17.	+ 46 31	900		DARK OBJECT
SHAP429-5227.4	04 28 17.	- 52 33 52.	18	17.5	GALAXY
SHAP429-5342.0	04 28 17.	- 53 48 28.	30	16.5	GALAXY
SHAP429-5349.2	04 28 17.	- 53 55 40.	18	16.7	GALAXY
ZWG 419.012	04 28 18.	+ 05 26		15.0	GALAXY
UGC 03066	04 28 18.	+ 05 26	120	15.0	GALAXY Sc/SBc
MCG+01-12-007	04 28 18.	+ 05 26	72	14.	GALAXY
ZWG 419.013	04 28 18.	+ 07 59		15.3	GALAXY
UGC 03067	04 28 18.	+ 07 59	66	15.3	GALAXY E?
LB 01734	04 28 18.	- 46 13		14.8	FAINT BLUE STAR
SHAP429-5049.2	04 28 18.	- 50 55 40.	18	17.0	GALAXY
SHAP429-5428.0	04 28 18.	- 54 34 28.	18	16.5	GALAXY
SHAP429-5531.0	04 28 18.	- 55 37 28.	24	17.0	GALAXY
SHAP429-5326.2	04 28 19.	- 53 32 40.	18	17.0	GALAXY
SHAP429-5351.6	04 28 19.	- 53 58 04.	24	16.9	GALAXY
SHAP429-5024.8	04 28 20.	- 50 31 16.	30	17.0	GALAXY
SHAP429-5052.2	04 28 20.	- 50 58 40.	18	17.0	GALAXY
SHAP429-5411.4	04 28 20.	- 54 17 52.	18	17.0	GALAXY
LB 01341	04 28 21.	+ 19 43 06.		12.6	FAINT BLUE STAR
SHAP429-5334.0	04 28 21.	- 53 40 28.	24	17.0	GALAXY
SHAP429-5423.5	04 28 21.	- 54 29 58.	24	17.0	GALAXY
REIN 6.035A	04 28 21.66	- 05 54 18.1			NEBULA
REIN 6.035B	04 26 21.68	- 05 54 18.1			NEBULA
RNGC 1594	04 28 22.	- 05 54		14.0	GALAXY
SHAP429-5318.4	04 28 22.	- 53 24 52.	30	16.1	GALAXY

OBJECT NAME	RIGHT ASCEN.	DECLINATION	DIAM.	MAGN.	TYPE OF OBJECT
SHAP429-5526.2	04 28 22.	- 55 32 40.	24	16.5	GALAXY
SHAP429-5108.0	04 28 23.	- 51 14 27.	24	16.25	GALAXY
SHAP429-5311.4	04 28 23.	- 53 17 52.	36	16.6	GALAXY
SHAP429-5402.4	04 28 23.	- 54 08 52.	18	17.3	GALAXY
UGC 03068	04 28 24.	+ 05 48	102	17.	GALAXY DBL SYS
MRSL 178-20/1	04 28 24.	+ 18 10	300		HII REGION
ZWG 328.008	04 28 24.	+ 73 05		15.1	GALAXY
UGC 03069	04 28 24.	+ 73 05	78	15.1	GALAXY E-S0
IC 2075	04 28 24.	- 05 54			NONSTELLAR OBJECT
ARC 0494	04 28 24.	- 07 51		17.3	RICH CLUSTER OF GALAXIES
SHAP429-5303.0	04 28 24.	- 53 09 28.	24	17.2	GALAXY
SHAP429-5347.9	04 28 24.	- 53 54 22.	24	17.3	GALAXY
REIN 6.037	04 28 24.53	- 05 54 15.0			NEBULA
RNGC 1590	04 28 27.	+ 07 31		14.5	GALAXY
MCG+01-12-014	04 28 27.	- 05 54	96	14.	GALAXY
SHAP429-5130.7	04 28 27.	- 51 37 09.	24	16.0	GALAXY
SS 23	04 28 28.	+ 17 07			DIFFUSE GALACTIC NEBULA
LB 00232	04 28 28.	+ 18 56 12.		15.8	FAINT BLUE STAR
SHAP429-5352.3	04 28 28.	- 53 58 45.	30	16.6	GALAXY
REIN 6.036	04 28 28.29	- 02 06 32.9			NEBULA
SHAP429-5318.1	04 28 29.	- 53 24 33.	30	17.0	GALAXY
SHAP429-5350.0	04 28 29.	- 53 56 27.	36	16.8	GALAXY
2ZW 013	04 28 30.	+ 07 31			COMPACT GALAXY
ZWG 419.014	04 28 30.	+ 07 31		14.6	GALAXY
UGC 03071	04 28 30.	+ 07 31	72	14.6	GALAXY PECULR
MCG+01-12-008	04 28 30.	+ 07 31	36	14.	GALAXY
ZWG 419.015	04 28 30.	+ 08 24		14.8	GALAXY
LDN 1551	04 28 30.	+ 18 00	840		DARK NEBULA
RLWT 106	04 28 30.	+ 41 42		13.	FAINT VERY BLUE STAR
OCL 0407	04 28 30.	+ 43 45	2250	7.3	OPEN STAR CLUSTER
LDN 1445	04 28 30.	+ 46 30	1140		DARK NEBULA
ZWG 393.031	04 28 30.	- 02 07		14.9	GALAXY
UGC 03070	04 28 30.	- 02 07	96	14.9	GALAXY Sb?
MCG+00-12-039	04 28 30.	- 02 07	78	14.5	GALAXY
KARA.73B 0152	04 28 30.	- 02 07	78	14.9	ISOLATED GALAXY S
SHAP429-5602.5	04 28 30.	- 56 08 57.	24	17.5	GALAXY
RNGC 1593	04 28 31.	+ 00 28			NON-EXISTENT OBJECT
SHAP429-5326.9	04 28 32.	- 53 33 21.	18	16.9	GALAXY
SHAP429-5335.4	04 28 32.	- 53 41 51.	24	17.2	GALAXY
SHAP429-5352.8	04 28 32.	- 53 59 15.	24	17.0	GALAXY
HH 29	04 28 33.6	+ 18 00 03.	12		HERBIG-HARO OBJECT
LB 01342	04 28 34.	+ 09 30 18.		16.6	FAINT BLUE STAR
SHAP429-5017.2	04 28 35.	- 50 23 39.	60	16.0	GALAXY
SHAP429-5345.0	04 28 35.	- 53 51 27.	24	16.6	GALAXY
UGC 03072	04 28 36.	+ 00 44	84	18.	GALAXY DWARF
UGC 03073	04 28 36.	+ 02 35	60	16.5	GALAXY S
ZC 0428.6+1647	04 28 36.	+ 16 47	2150		CLUSTER OF GALAXIES
SHAP429-5313.2	04 28 36.	- 53 19 39.	30	17.4	GALAXY
SHAP429-5348.4	04 28 36.	- 53 54 51.	36	16.8	GALAXY
SHAP429-5108.6	04 28 37.	- 51 15 03.	18	17.0	GALAXY
HSE 006	04 28 37.	- 66 56			STAR CLUSTER IN LMC
RNGC 1582	04 28 38.	+ 43 45		7.0	OPEN CLUSTER
SHAP429-5459.3	04 28 38.	- 55 05 45.	24	17.0	GALAXY
IC 2074	04 28 39.	+ 07 36			OPEN CLUSTER
MCG-03-12-011	04 28 39.	- 16 25	60	14.	GALAXY
HUB C07	04 28 41.	+ 25 20			DIFFUSE NEBULA
SHAP429-5312.2	04 28 41.	- 53 18 38.	18	17.3	GALAXY
SHAP429-5312.8	04 28 41.	- 53 19 14.	24	17.2	GALAXY
SHAP429-5348.4	04 28 41.	- 53 54 51.	66	15.9	GALAXY
ZWG 419.016	04 28 42.	+ 06 32		15.6	GALAXY
UGC 03074	04 28 42.	+ 06 32	78	15.6	GALAXY
UGC 03075	04 28 42.	+ 08 03	4266	16.5	GALAXY S
ZWG 393.032	04 28 42.	- 01 25		15.7	GALAXY
SHAP429-5128.8	04 28 43.	- 51 35 14.	18	16.5	GALAXY
HH 30	04 28 43.6	+ 18 06 03.	35		HERBIG-HARO OBJECT
IC 0375	04 28 45.	- 13 04 47.			NONSTELLAR OBJECT
SHAP429-5544.2	04 28 45.	- 55 50 38.	18	17.25	GALAXY
B 016	04 28 47.	+ 46 30			DARK OBJECT
SHAP430-5138.6	04 28 47.	- 51 45 02.	24	16.0	GALAXY
SHAP430-5140.6	04 28 47.	- 51 47 02.	24	15.0	GALAXY
5ZW 374	04 28 48.	+ 44 17			COMPACT GALAXY
MCG-02-12-032	04 28 48.	- 11 23	48	15.	GALAXY
MCG-02-12-031	04 28 48.	- 12 33	48	14.5	GALAXY
SHAP430-5127.8	04 28 48.	- 51 34 14.	24	17.0	GALAXY
RNGC 1597	04 28 49.	- 11 23		15.0	GALAXY
SHAP429-5402.4	04 28 49.	- 54 08 50.	12	16.9	GALAXY
REIN 4.047B	04 28 49.53	- 04 50 42.0			NEBULA
REIN 4.047A	04 28 49.53	- 04 50 42.6			NEBULA
B 017	04 28 52.	+ 46 25			DARK OBJECT
SHAP430-5024.4	04 28 52.	- 50 30 49.	24	17.0	GALAXY
SHAP430-5402.0	04 28 52.	- 54 08 26.	18	16.6	GALAXY
YM 38	04 28 53.	+ 24 22	54		SYMMFTRIC GALACTIC NFBULA
SHAP430-5400.8	04 28 53.	- 54 07 14.	30	16.7	GALAXY
MCG+01-12-015	04 28 54.	- 04 01	66	15.	GALAXY
IC 0376	04 28 54.	- 12 32 06.			NONSTELLAR OBJECT
REIN 6.038	04 28 54.16	- 04 01 57.4			NEBULA
SHAP430-5352.4	04 28 55.	- 53 58 50.	30	17.3	GALAXY
SHAP430-5221.2	04 28 56.	- 52 27 37.	18	16.75	GALAXY
SHAP430-5349.4	04 28 56.	- 53 55 50.	36	16.8	GALAXY
SHAP429-5553.4	04 28 56.	- 55 59 50.	24	17.25	GALAXY
IC 0377	04 28 57.	- 12 33 31.			NONSTELLAR OBJECT
LBN 0812	04 29	+ 24 13	240		BRIGHT NEBULA
KHAV 069	04 29	+ 24 49			DARK NFBULA
KHAV 070	04 29	+ 26 31	5760		DARK NEBULA
KHAV 068	04 29	+ 29 25	7060		DARK NEBULA
KHAV 071	04 29	+ 31 49	10520		DARK NFBULA
LBN 0897	04 29	- 05 50	960		BRIGHT NEBULA
UGC 03076	04 29 00.	+ 08 20	66	16.0	GALAXY Sb
LDN 1531	04 29 00.	+ 24 13	480		DARK NFBULA
LDN 1529	04 29 00.	+ 24 20	1860		DARK NEBULA
LDN 1421	04 29 00.	+ 51 55	360		DARK NEBULA
LDN 1420	04 29 00.	+ 52 10	180		DARK NEBULA
7ZW 018	04 29 00.	+ 73 09			COMPACT GALAXY
ZWG 328.009	04 29 00.	+ 73 09		13.3	GALAXY
UGC 03077	04 29 00.	+ 73 09	138	13.3	GALAXY E
MCG+12-05-008	04 29 00.	+ 73 10	57	14.	GALAXY
RNGC 1573	04 29 01.	+ 73 09		13.5	GALAXY
SHAP430-5337.1	04 29 02.	- 53 43 31.	24	17.2	GALAXY
KARA.73 23	04 29 03.	- 44 19	27		DWARF GALAXY
SHAP430-5316.2	04 29 03.	- 53 22 37.	36	17.3	GALAXY
SHAP430-5610.0	04 29 03.	- 56 16 25.	24	17.0	GALAXY
UGC 03078	04 29 06.	+ 33 08	108	17.	GALAXY Sb-c
MCG-03-12-012	04 29 06.	- 20 38 30.	48	16.	GALAXY
SHAP430-5357.5	04 29 07.	- 54 03 55.	24	16.9	GALAXY
BC 3CR119	04 29 07.96	+ 41 32 08.8			QUASI-STELLAR OBJECT
MCG+00-12-040	04 29 09.	+ 01 40 30.	36	14.5	GALAXY
YM 39	04 29 09.	+ 24 17	108		SYMMETRIC GALACTIC NFBULA
IC 0378	04 29 09.	- 12 24 14.			NONSTELLAR OBJFCT
RNGC 1599	04 29 10.	- 04 41		14.5	GALAXY
RNGC 1601	04 29 10.	- 05 10		15.0	GALAXY
RNGC 1600	04 29 10.	- 05 10		13.0	GALAXY
SHAP430-5329.0	04 29 10.	- 53 35 25.	24	16.8	GALAXY
REIN 6.039	04 29 10.09	- 04 41 40.3			NEBULA
SHAP430-5540.0	04 29 11.	- 55 46 25.	24	17.0	GALAXY
REIN 2.044	04 29 11.72	- 05 11 32.3			NEBULA
ZWG 393.033	04 29 12.	+ 01 40		15.7	GALAXY
UGC 03079	04 29 12.	+ 01 40	84	15.7	GALAXY S
MCG+01-12-016	04 29 12.	- 04 41	42	14.5	GALAXY
MCG+01-12-017	04 29 12.	- 05 11 30.	60	13.	GALAXY
SHAP430-5403.4	04 29 13.	- 54 09 48.	18	16.6	GALAXY
REIN 4.048B	04 29 13.59	- 05 09 59.0			NEBULA
REIN 4.048A	04 29 13.59	- 05 09 59.8			NEBULA
MCG+01-12-018	04 29 15.	- 05 10	30	16.5	GALAXY
SHAP430-5022.8	04 29 15.	- 50 29 12.	30	17.25	GALAXY
RNGC 1603	04 29 16.	- 05 12		15.5	GALAXY
SHAP430-5323.0	04 29 16.	- 53 29 24.	18	17.3	GALAXY
SHAP430-5327.8	04 29 16.	- 53 34 12.	18	17.3	GALAXY
SHAP430-5406.4	04 29 16.	- 54 12 48.	18	16.2	GALAXY
SHAP430-5359.2	04 29 17.	- 54 05 36.	18	17.2	GALAXY
ZWG 393.034	04 29 18.	+ 01 05		14.9	GALAXY
UGC 03080	04 29 18.	+ 01 05	132	14.9	GALAXY Sc
MCG+00-12-042	04 29 18.	+ 01 06 30.	90	13.	GALAXY
MCG+00-12-041	04 29 18.	+ 02 18	36	15.	GALAXY
ZWG 419.017	04 29 18.	+ 06 40		14.9	GALAXY
MCG-02-12-033	04 29 18.	- 11 48	48	15.	GALAXY
LB 03362	04 29 18.	- 62 26		11.6	FAINT BLUE STAR
RNGC 1608	04 29 19.	+ 00 29		15.0	GALAXY
MCG+00-12-043	04 29 21.	+ 01 23	15	14.5	GALAXY
LB 01343	04 29 21.	+ 21 21 00.		16.1	FAINT BLUE STAR
MCG+01-12-019	04 29 21.	- 05 12	18	15.5	GALAXY
MCG-02-12-034	04 29 21.	- 13 02 30.	36	15.	GALAXY
REIN 4.049B	04 29 21.82	- 05 11 59.3			NEBULA
REIN 4.049A	04 29 21.82	- 05 12 00.1			NEBULA
IC 2078	04 29 23.	- 04 47			NONSTELLAR OBJECT
IC 0380	04 29 23.	- 13 01 58.			NONSTELLAR OBJFCT
SHAP430-5310.8	04 29 23.	- 53 17 12.	24	17.4	GALAXY
SHAP430-5347.3	04 29 23.	- 53 53 42.	30	17.2	GALAXY
ZWG 393.035	04 29 24.	+ 01 21		15.6	GALAXY
ZWG 393.036	04 29 24.	+ 02 16		15.7	GALAXY
UGC 03081	04 29 24.	+ 02 16	96	15.7	GALAXY SBa
MCG-02-12-035	04 29 24.	- 12 46 30.	60	15.	GALAXY
SHAP430-5021.6	04 29 24.	- 50 27 59.	60	16.25	GALAXY
SHAP430-5340.0	04 29 24.	- 53 46 24.	24	16.9	GALAXY
IC 0379	04 29 25.	- 07 20 40.			NONSTELLAR OBJFCT
REIN 4.050B	04 29 25.17	- 05 08 56.5			NEBULA
REIN 4.050A	04 29 25.17	- 05 08 57.4			NEBULA
SHAP430-5439.4	04 29 26.	- 54 45 48.	54	17.5	GALAXY
MCG-03-12-013	04 29 27.	- 15 52 30.	30	15.	GALAXY
SHAP430-5338.4	04 29 27.	- 53 44 47.	18	17.3	GALAXY
RNGC 1606	04 29 28.	- 05 09		17.0	GALAXY
RNGC 1604	04 29 28.	- 05 28		14.5	GALAXY
SHAP430-5259.8	04 29 28.	- 53 06 11.	24	17.5	GALAXY
HZ 09	04 29 29.	+ 17 39		14.1	BLUE STAR
ZWG 393.037	04 29 30.	+ 00 27		14.8	GALAXY
UGC 03082	04 29 30.	+ 00 27	102	14.8	GALAXY S0
MCG+00-12-044	04 29 30.	+ 00 29	66	14.	GALAXY
DG 037	04 29 30.	+ 24 19	180		REFLECTION NEBULA
MCG+01-12-020	04 29 30.	- 05 28	42	14.5	GALAXY
MCG+01-12-021	04 29 30.	- 07 21	54	15.	GALAXY
REIN 6.040	04 29 30.53	- 04 35 00.5			NEBULA
REIN 4.051A	04 29 30.72	- 05 28 31.5			NEBULA
REIN 4.051B	04 29 30.73	- 05 28 32.6			NEBULA
SS 24	04 29 31.	+ 24 19			DIFFUSE GALACTIC NEBULA
IC 2077	04 29 32.	+ 00 27 45.			SAME AS NGC 1608
KW 67	04 29 32.	+ 01 55 00.	42		SEYFERT GALAXY
IC 2083	04 29 32.	- 54 04			NONSTELLAR OBJECT
SHAP430-5357.8	04 29 32.	- 54 04 11.	48	16.1	GALAXY
HN 0295	04 29 32.	- 54 05			NEBULA
SHAP430-5345.5	04 29 33.	- 53 51 53.	18	16.9	GALAXY
REIN 6.041B	04 29 33.10	- 04 34 56.4			NEBULA
REIN 6.041A	04 29 33.17	- 04 34 56.4			NEBULA
REIN 4.052B	04 29 33.26	- 05 07 07.2			NEBULA
REIN 4.052A	04 29 33.26	- 05 07 07.9			NEBULA
RNGC 1607	04 29 34.	- 04 33		15.0	GALAXY
SHAP430-5341.4	04 29 34.	- 53 47 47.	18	17.0	GALAXY
SHAP430-5517.2	04 29 34.	- 55 23 35.	24	17.25	GALAXY
REIN 6.042	04 29 34.24	- 04 33 57.5			NEBULA
SHAP430-5342.0	04 29 35.	- 53 48 23.	24	15.9	GALAXY
SHAP430-5349.8	04 29 35.	- 53 56 11.	30	17.4	GALAXY
REIN 4.053B	04 29 35.06	- 05 08 16.4			NEBULA
REIN 4.053A	04 29 35.06	- 05 08 17.1			NEBULA
4ZW 061	04 29 36.	+ 00 44			COMPACT GALAXY
ZWG 467.001	04 29 36.	+ 16 32		15.7	GALAXY
MCG+01-12-022	04 29 36.	- 05 08	15	17.	GALAXY
MCG-05-11-012	04 29 36.	- 29 51	36	15.	GALAXY
SHAP430-5317.0	04 29 37.	- 53 23 23.	24	16.9	GALAXY
LB 00220	04 29 38.	+ 14 18 18.		16.6	FAINT BLUE STAR
MCG+01-12-023	04 29 39.	- 04 33	48	15.	GALAXY
IC 0374	04 29 41.	+ 16 31 51.			NONSTELLAR OBJECT
SHAP430-5347.4	04 29 41.	- 53 53 46.	48	17.2	GALAXY
SHAP430-5347.6	04 29 41.	- 53 53 58.	24	17.3	GALAXY
UGC 03083	04 29 42.	+ 01 37	84	16.5	GALAXY Sc
DG 038	04 29 42.	+ 24 16	240		REFLECTION NEBULA
MCG-05-11-013	04 29 42.	- 29 51	24	15.5	GALAXY
SHAP430-5134.2	04 29 42.	- 51 40 34.	24	17.0	GALAXY
SS 25	04 29 43.	+ 24 16			DIFFUSE GALACTIC NEBULA
SHAP431-5135.3	04 29 45.	- 51 41 40.	24	16.0	GALAXY
SHAP430-5350.1	04 29 47.	- 53 56 28.	18	17.3	GALAXY
MCG+03-12-001	04 29 48.	+ 16 31	9	16.	GALAXY
ZC 0429.8-0121	04 29 48.	- 01 21	940		CLUSTER OF GALAXIES
MCG-02-12-036	04 29 48.	- 12 47	60	15.5	GALAXY
MCG-05-11-015	04 29 48.	- 29 46	30	15.5	GALAXY
MCG-05-11-014	04 29 48.	- 31 38	30	15.5	GALAXY
SHAP430-5330.4	04 29 48.	- 53 36 46.	18	17.3	GALAXY
SHAP430-5531.7	04 29 48.	- 55 38 04.	12	17.25	GALAXY
SHAP430-5335.9	04 29 49.	- 53 42 16.	24	16.8	GALAXY
SHAP431-5132.4	04 29 50.	- 51 38 46.	48	16.25	GALAXY
SHAP431-5336.6	04 29 50.	- 53 42 58.	24	17.4	GALAXY
SHAP430-5521.8	04 29 50.	- 55 28 10.	18	17.0	GALAXY
SHAP430-5620.9	04 29 51.	- 56 27 16.	24	16.5	GALAXY
IC 2080	04 29 52.	- 05 51			NONSTELLAR OBJECT
SHAP431-5331.3	04 29 52.	- 53 37 40.	18	17.4	GALAXY
SHAP431-5332.2	04 29 52.	- 53 38 34.	24	17.5	GALAXY
SHAP430-5531.4	04 29 52.	- 55 37 46.	18	17.0	GALAXY
SHAP431-5354.4	04 29 53.	- 54 00 46.	18	17.3	GALAXY
ZWG 328.010	04 29 54.	+ 70 01		15.5	GALAXY

OBJECT NAME	RIGHT ASCEN.	DECLINATION	DIAM.	MAGN.	TYPE OF OBJECT
RNGC 1629	04 29 54.	- 71 56			OPEN CLUSTER IN LMC
SHAP431-5039.6	04 29 56.	- 50 45 57.	30	16.0	GALAXY
SHAP430-5636.1	04 29 56.	- 56 42 28.	24	16.5	GALAXY
LB 01344	04 29 58.	+ 21 32 00.		16.0	FAINT BLUE STAR
LBN 0849	04 30	+ 05 50	360		BRIGHT NEBULA
LBN 0842	04 30	+ 07 00	11700		BRIGHT NEBULA
LBN 0816	04 30	+ 22 50	4800		BRIGHT NEBULA
B 019	04 30	+ 26 10	3600		DARK OBJECT
LBN 0632	04 30	+ 85 00	4020		BRIGHT NEBULA
LBN 0893	04 30	- 05 00	2700		BRIGHT NEBULA
UGC 03084	04 30 00.	+ 10 17	138	16.0	GALAXY Sc
LDN 1536	04 30 00.	+ 23 00	4920		DARK NEBULA
LDN 1521	04 30 00.	+ 26 00	8220		DARK NEBULA
LDN 1500	04 30 00.	+ 29 20	3960		DARK NEBULA
LDN 1444	04 30 00.	+ 46 50	2940		DARK NEBULA
LDN 1418	04 30 00.	+ 52 30	6540		DARK NEBULA
LDN 1414	04 30 00.	+ 53 20	1800		DARK NEBULA
LDN 1405	04 30 00.	+ 55 10	1680		DARK NEBULA
UGC 03085	04 30 00.	+ 72 50	84	16.5	GALAXY Sc
MCG+01-12-024	04 30 00.	- 07 24 30.	36	16.	GALAXY
LB 01735	04 30 00.	- 53 42		12.6	FAINT BLUE STAR
SL 003	04 30 00.	- 71 56	40		STAR CLUSTER IN LMC
LB 01345	04 30 03.	+ 16 43 36.		15.5	FAINT BLUE STAR
SHAP431-5400.8	04 30 06.	- 54 07 09.	24	17.5	GALAXY
REIN 6.043	04 30 06.73	- 04 31 06.5			NEBULA
SHAP431-5201.5	04 30 07.	- 52 07 51.	30	16.0	GALAXY
SHAP431-5317.8	04 30 07.	- 53 24 09.	24	17.1	GALAXY
SHAP431-5257.9	04 30 08.	- 53 04 15.	24	17.0	GALAXY
SHAP431-5340.6	04 30 08.	- 53 46 57.	24	16.2	GALAXY
HSE 007	04 30 08.	- 67 00			STAR CLUSTER IN LMC
SHAP431-5335.5	04 30 09.	- 53 41 51.	18	16.8	GALAXY
RNGC 1609	04 30 10.	- 04 28		15.0	GALAXY
SHAP431-5159.7	04 30 11.	- 52 06 02.	36	16.5	GALAXY
MRSL 189-27/1	04 30 12.	+ 05 46	2100		HII REGION
ZWG 306.002	04 30 12.	+ 68 48		15.6	GALAXY
ZWG 328.011	04 30 12.	+ 73 18		15.7	GALAXY
MCG+12-05-009	04 30 12.	+ 73 19	12	16.	GALAXY
LW 003	04 30 12.	- 71 57			STAR CLUSTER IN LMC
SHAP431-4948.6	04 30 14.	- 49 54 56.	60	16.75	GALAXY
MCG+01-12-025	04 30 15.	- 04 28	48	15.	GALAXY
SHAP431-5138.4	04 30 16.	- 51 44 44.	42	15.75	GALAXY
REIN 6.044	04 30 16.20	- 04 28 38.4			NEBULA
SHAP431-5347.4	04 30 17.	- 53 53 44.	30	17.5	GALAXY
SHAP431-5538.2	04 30 17.	- 55 44 32.	24	17.5	GALAXY
UGC 03086	04 30 18.	+ 00 26	60	18.	GALAXY DWRF SP
SHAP431-5425.0	04 30 18.	- 54 31 20.	180	14.8	GALAXY
SER 040.06	04 30 18.	- 61 35	420	17.	CLUSTER OF GALAXIES
SHAP431-5149.5	04 30 19.	- 51 55 50.	24	16.5	GALAXY
SHAP431-5244.0	04 30 19.	- 52 50 20.	30	17.0	GALAXY
IC 2085	04 30 19.	- 54 31			NONSTELLAR OBJECT
HN 0296	04 30 19.	- 54 32			NEBULA
SHAP431-5244.2	04 30 20.	- 52 50 32.	36	16.5	GALAXY
REIN 6.045	04 30 21.61	- 04 40 30.3			NEBULA
SHAP431-5311.4	04 30 23.	- 53 17 44.	24	17.1	GALAXY
DG 039	04 30 24.	+ 24 18	120		REFLECTION NEBULA
MCG+12-05-010	04 30 24.	+ 73 20	39	16.	GALAXY
MCG+01-12-026	04 30 24.	- 06 07	24	15.	GALAXY
MCG+01-12-027	04 30 24.	- 07 25	60	16.	GALAXY
SHAP431-5143.7	04 30 24.	- 51 50 01.	48	16.5	GALAXY
SHAP431-5404.4	04 30 24.	- 54 10 44.	18	17.3	GALAXY
SS 26	04 30 25.	+ 24 17			DIFFUSE GALACTIC NEBULA
SHAP431-5338.0	04 30 26.	- 53 44 19.	48	16.6	GALAXY
SHAP431-5401.4	04 30 26.	- 54 07 43.	24	16.7	GALAXY
SHAP431-5317.6	04 30 27.	- 53 23 55.	18	16.8	GALAXY
HN 0298	04 30 27.	- 53 45			NEBULA
SHAP431-5513.6	04 30 27.	- 55 19 56.	24	17.25	GALAXY
SHAP431-5316.8	04 30 28.	- 53 23 07.	42	16.3	GALAXY
IC 2086	04 30 28.	- 53 44			NONSTELLAR OBJECT
2ZW 014	04 30 30.	+ 05 15			COMPACT GALAXY
VVI 16	04 30 30.	+ 05 15	40	14.2	SEYFERT GALAXY
UGC 03087	04 30 30.	+ 05 15	54	14.2	GALAXY S0
MCG+01-12-028	04 30 30.	- 04 17	90	16.	GALAXY
SHAP431-5215.8	04 30 30.	- 52 22 07.	30	16.5	GALAXY
SHAP431-5416.9	04 30 30.	- 54 23 13.	18	16.5	GALAXY
SHAP431-5318.6	04 30 31.	- 53 24 55.	24	16.5	GALAXY
SHAP431-5336.2	04 30 31.	- 53 42 31.	18	17.1	GALAXY
SHAP431-5427.2	04 30 31.	- 54 33 31.	18	17.0	GALAXY
SHAP431-5338.0	04 30 32.	- 53 44 19.	24	17.1	GALAXY
SHAP431-5518.4	04 30 32.	- 55 24 43.	18	17.25	GALAXY
HSE 008	04 30 32.	- 67 04			STAR CLUSTER IN LMC
MCG-02-12-037	04 30 33.	- 12 52	60	15.	GALAXY
RNGC 1617	04 30 33.	- 54 42		12.0	GALAXY
SHAP431-5436.0	04 30 33.	- 54 42 19.	240	11.7	GALAXY
RNGC 1611	04 30 34.	- 04 23		15.0	GALAXY
SHAP431-5139.8	04 30 35.	- 51 46 07.	48	16.25	GALAXY
MCG+01-12-009	04 30 36.	+ 05 15	36	15.	GALAXY
UGC 03088	04 30 36.	+ 07 36	78	18.	GALAXY DWARF
MCG+01-12-029	04 30 36.	- 04 23	48	15.	GALAXY
MCG-03-12-014	04 30 36.	- 15 03	24	16.	GALAXY
SHAP431-5255.0	04 30 36.	- 53 01 19.	24	17.0	GALAXY
REIN 6.047	04 30 37.08	- 04 24 04.6			NEBULA
RNGC 1612	04 30 39.	- 04 16		15.0	GALAXY
SHAP431-5455.5	04 30 39.	- 55 01 49.	36	16.75	GALAXY
SHAP431-5141.6	04 30 41.	- 51 47 54.	12	17.0	GALAXY
SHAP431-5414.8	04 30 41.	- 54 21 06.	36	17.25	GALAXY
SHAP431-5710.0	04 30 41.	- 57 16 19.	24	16.0	GALAXY
MCG-02-12-038	04 30 42.	- 10 55	60	15.	GALAXY
IC 2084	04 30 42.	- 48 23			NONSTELLAR OBJECT
HN 0297	04 30 42.	- 48 24			NEBULA
REIN 6.048	04 30 43.90	- 04 16 36.9			NEBULA
MCG+01-12-030	04 30 45.	- 04 16	60	15.	GALAXY
ZC 0430.8+0354	04 30 48.	+ 03 54	1340		CLUSTER OF GALAXIES
ZC 0430.8-0424	04 30 48.	- 04 24	14720		CLUSTER OF GALAXIES
SHAP431-5557.8	04 30 50.	- 56 04 06.	18	17.0	GALAXY
RNGC 1613	04 30 52.	- 04 21		15.0	GALAXY
MCG+01-12-031	04 30 54.	- 04 21 30.	36	15.	GALAXY
MCG-06-11-001	04 30 54.	- 33 29	36	16.	GALAXY
REIN 6.049	04 30 56.23	- 04 22 09.7			NEBULA
SHAP432-5340.0	04 30 57.	- 53 46 17.	18	17.4	GALAXY
SHAP432-5346.2	04 30 57.	- 53 52 29.	24	16.6	GALAXY
SHAP432-5426.0	04 30 57.	- 54 32 17.	24	17.25	GALAXY
SHAP432-4940.4	04 30 58.	- 49 46 41.	192	14.0	GALAXY
SHAP432-5130.2	04 30 58.	- 51 36 29.	24	17.25	GALAXY
SHAP432-5133.5	04 30 58.	- 51 39 47.	24	16.0	GALAXY
SHAP432-5528.8	04 30 58.	- 55 35 05.	24	17.5	GALAXY
REIN 6.050	04 30 58.99	- 04 23 17.5			NEBULA
SHAP432-5344.6	04 30 59.	- 53 50 53.	18	17.1	GALAXY
LBN 0852	04 31	+ 05 50	600		BRIGHT NEBULA
B 219	04 31	+ 29 30	3300		DARK OBJECT
KHAV 072	04 31	+ 55 12	2760		DARK NEBULA
HZ 07	04 31 00.	+ 12 35		14.5	DECIDEDLY BLUE STAR
ZWG 467.002	04 31 00.	+ 16 48		15.6	GALAXY
UGC 03089	04 31 00.	+ 16 48	78	15.6	GALAXY Sb-c
UGC 03090	04 31 00.	+ 71 27	120	17.	GALAXY DWARF
LB 03363	04 31 00.	- 70 03		11.6	FAINT BLUE STAR
SHAP432-5338.0	04 31 01.	- 53 44 17.	24	16.8	GALAXY
SHAP432-5435.2	04 31 04.	- 54 41 29.	18	17.5	GALAXY
LB 01346	04 31 05.	+ 15 47 18.		16.8	FAINT BLUE STAR
SHAP432-5218.4	04 31 05.	- 52 24 41.	24	16.25	GALAXY
MCG+03-12-002	04 31 06.	+ 16 48	66	15.5	GALAXY
MCG-04-11-019	04 31 06.	- 24 47	60	15.	GALAXY
AGU 20	04 31 06.	- 43 47 24.	66	12.5	PECULIAR GALAXY
RNGC 1616	04 31 07.	- 43 49			GALAXY
SHAP432-5335.0	04 31 08.	- 53 41 17.	18	17.0	GALAXY
MCG-03-12-015	04 31 12.	- 18 48 30.	72	14.5	GALAXY
LB 03364	04 31 12.	- 79 05		14.6	FAINT BLUE STAR
SHAP432-5439.2	04 31 13.	- 54 45 28.	24	16.75	GALAXY
SHAP432-5104.2	04 31 14.	- 51 10 28.	30	16.5	GALAXY
SHAP432-5216.7	04 31 15.	- 52 22 58.	24	16.5	GALAXY
SHAP432-5438.4	04 31 15.	- 54 44 40.	24	17.5	GALAXY
ZWG 393.038	04 31 18.	+ 01 00		15.4	GALAXY
UGC 03091	04 31 18.	+ 01 00	90	15.4	GALAXY Sc
MCG+00-12-045	04 31 18.	+ 01 01 30.	66	14.	GALAXY
ABC 0496	04 31 18.	- 13 22		15.3	RICH CLUSTER OF GALAXIES
MCG-02-12-039	04 31 18.	- 13 22	120	14.	GALAXY
SHAP432-5041.2	04 31 18.	- 50 47 28.	48	17.25	GALAXY
SHAP432-5316.6	04 31 18.	- 53 22 52.	30	16.1	GALAXY
SHAP432-5133.5	04 31 20.	- 51 39 46.	36	16.5	GALAXY
SHAP432-5314.2	04 31 21.	- 53 20 28.	30	17.3	GALAXY
SHAP432-4955.4	04 31 22.	- 50 01 39.	36	17.25	GALAXY
SHAP432-5106.2	04 31 22.	- 51 12 27.	24	15.5	GALAXY
SHAP432-5337.2	04 31 23.	- 53 43 28.	18	17.2	GALAXY
OCL 0406	04 31 24.	+ 45 09	420	11.	OPEN STAR CLUSTER
RNGC 1605	04 31 24.	+ 45 09		11.0	OPEN CLUSTER
MCG-02-12-040	04 31 24.	- 12 38	36	15.5	GALAXY
SHAP432-5432.6	04 31 25.	- 54 38 52.	24	16.0	GALAXY
SHAP432-5347.6	04 31 26.	- 53 53 51.	24	16.7	GALAXY
MCG-02-12-041	04 31 30.	- 11 48	180	14.	GALAXY
SHAP432-5341.2	04 31 31.	- 53 47 27.	30	17.2	GALAXY
2ZW 015	04 31 36.	- 08 41			COMPACT GALAXY
MRK 617	04 31 36.	- 08 41	9	15.	GALAXY WITH UV CONTINUUM
SHAP432-5355.6	04 31 36.	- 54 01 51.	30	17.3	GALAXY
SHAP432-5415.4	04 31 37.	- 54 21 39.	30	17.25	GALAXY
SHAP432-5416.4	04 31 37.	- 54 22 39.	18	17.0	GALAXY
SHAP432-5424.2	04 31 38.	- 54 30 27.	18	17.0	GALAXY
RNGC 1610	04 31 40.	- 04 48			GALAXY
SHAP432-5315.8	04 31 41.	- 53 22 02.	24	16.9	GALAXY
SHAP432-5333.2	04 31 41.	- 53 39 26.	18	17.3	GALAXY
MCG+01-12-032	04 31 42.	- 08 42	60	14.	GALAXY
SHAP432-5207.0	04 31 45.	- 52 13 14.	36	16.75	GALAXY
SHAP432-5311.9	04 31 47.	- 53 18 08.	24	16.9	GALAXY
LB 01347	04 31 50.	+ 10 53 30.		16.2	FAINT BLUE STAR
SHAP433-5331.5	04 31 52.	- 53 37 44.	24	17.0	GALAXY
SHAP433-5319.6	04 31 53.	- 53 25 50.	30	17.2	GALAXY
ZWG 393.039	04 31 54.	+ 01 34		15.6	GALAXY
KARA.72 100A	04 31 54.	+ 08 03	48		PART OF DOUBLE GALAXY
2ZW 016	04 31 54.	- 02 51			COMPACT GALAXY
RNGC 1614	04 31 54.	- 08 41		14.0	GALAXY
ARP 186	04 31 56.	- 08 41			PECULIAR GALAXY
MCG+01-12-010	04 31 57.	+ 08 03	15	15.	GALAXY
SHAP433-5300.2	04 31 57.	- 53 06 25.	24	16.5	GALAXY
LB 01348	04 31 58.	+ 18 27 00.		15.9	FAINT BLUE STAR
LBN 0900	04 32	- 05 56	1020		BRIGHT NEBULA
LB 09816	04 32	- 82 19		14.4	FAINT BLUE STAR
MCG+00-12-046	04 32 00.	+ 01 33	36	14.5	GALAXY
ZWG 419.018	04 32 00.	+ 08 04		15.4	GALAXY
KARA.72 100B	04 32 00.	+ 08 04	54	15.4	PART OF DOUBLE GALAXY
MCG+01-12-011	04 32 00.	+ 08 05	12	15.	GALAXY
LDN 1483	04 32 00.	+ 36 20	1860		DARK NEBULA
LDN 1411	04 32 00.	+ 54 20	360		DARK NEBULA
LDN 1410	04 32 00.	+ 54 45	180		DARK NEBULA
LDN 1409	04 32 00.	+ 54 50	360		DARK NEBULA
MCG+12-05-011	04 32 00.	+ 73 13	27	16.	GALAXY
LB 03365	04 32 00.	- 67 46		13.3	FAINT BLUE STAR
SHAP433-5135.9	04 32 03.	- 51 42 07.	36	16.25	GALAXY
SHAP433-4947.2	04 32 04.	- 49 53 24.	54	17.0	GALAXY
SHAP433-5413.4	04 32 04.	- 54 19 37.	36	16.5	GALAXY
LB 01349	04 32 06.	+ 11 17 12.		16.5	FAINT BLUE STAR
MCG+12-05-012	04 32 06.	+ 73 12 30.	33	16.	GALAXY
SHAP433-5238.4	04 32 06.	- 52 44 37.	24	17.0	GALAXY
SHAP433-5439.2	04 32 10.	- 54 45 25.	24	17.5	GALAXY
LB 01350	04 32 12.	+ 19 18 18.		14.8	FAINT BLUE STAR
UGC 03092	04 32 12.	+ 72 13	96	17.	GALAXY Sc
ZWG 328.012	04 32 12.	+ 73 12		15.5	GALAXY
SHAP433-5440.5	04 32 12.	- 54 46 42.	24	17.0	GALAXY
LB 01351	04 32 15.	+ 12 56 54.		16.1	FAINT BLUE STAR
ZWG 419.019	04 32 18.	+ 07 53		15.6	GALAXY
UGC 03093	04 32 18.	+ 07 53	66	15.6	GALAXY Sc
MCG+01-12-012	04 32 18.	+ 07 54	60	15.	GALAXY
SHAP433-5433.4	04 32 18.	- 54 39 36.	24	16.5	GALAXY
SHAP433-5332.2	04 32 20.	- 53 38 24.	30	17.3	GALAXY
SHAP433-5457.0	04 32 21.	- 55 03 12.	18	17.25	GALAXY
SHAP433-5600.5	04 32 22.	- 56 06 42.	18	17.5	GALAXY
REIN 6.051	04 32 22.86	- 04 28 03.5			NEBULA
ZWG 328.013	04 32 24.	+ 73 12		15.7	GALAXY
MCG-04-11-020	04 32 24.	- 25 02	36	15.	GALAXY
SHAP433-5109.2	04 32 24.	- 51 15 23.	24	16.0	GALAXY
SHAP433-5349.8	04 32 25.	- 53 55 59.	30	16.6	GALAXY
SHAP433-5350.0	04 32 25.	- 53 56 11.	18	17.2	GALAXY
SHAP433-5415.6	04 32 29.	- 54 21 47.	18	17.0	GALAXY
MCG+03-12-003	04 32 30.	+ 17 05	30	17.	GALAXY
LDN 1535	04 32 30.	+ 23 48	1320		DARK NEBULA
LB 01352	04 32 32.	+ 09 44 54.		13.1	FAINT BLUE STAR
RNGC 1623	04 32 33.	- 13 37			GALAXY
SHAP433-5105.2	04 32 33.	- 51 11 23.	30	16.5	GALAXY
SHAP433-5411.8	04 32 35.	- 54 17 59.	24	17.0	GALAXY
UGC 03094	04 32 36.	+ 19 03	72	16.5	GALAXY S
DG 040	04 32 36.	+ 24 19	60		REFLECTION NEBULA
LB 01736	04 32 36.	- 46 12		14.4	FAINT BLUE STAR
SS 28	04 32 37.	+ 24 17			DIFFUSE GALACTIC NEBULA
SS 27	04 32 37.	+ 50 42			DIFFUSE GALACTIC NEBULA
SHAP433-5351.0	04 32 37.	- 53 57 11.	24	16.9	GALAXY
SHAP433-5255.9	04 32 38.	- 53 02 04.	30	17.0	GALAXY
SHAP433-5346.7	04 32 39.	- 53 52 52.	18	16.6	GALAXY
SHAP433-5346.8	04 32 40.	- 53 52 58.	36	16.6	GALAXY
UGC 03095	04 32 42.	+ 18 14	78	16.0	GALAXY E-S0

OBJECT NAME	RIGHT ASCEN.	DECLINATION	DIAM.	MAGN.	TYPE OF OBJECT
MCG+03-12-004	04 32 42.	+ 18 15	12	16.5	GALAXY
SS 29	04 32 42.	+ 22 47			DIFFUSE GALACTIC NEBULA
DG 041	04 32 42.	+ 22 48	120		REFLECTION NEBULA
2ZW 017	04 32 42.	- 01 50			COMPACT GALAXY
MCG-02-12-042	04 32 42.	- 14 20	60	16.	GALAXY
SHAP433-5448.8	04 32 42.	- 54 54 58.	18	16.5	GALAXY
SHAP433-5447.8	04 32 43.	- 54 53 58.	30	17.5	GALAXY
SHAP433-5236.2	04 32 44.	- 52 42 22.	48	16.0	GALAXY
LB 01353	04 32 48.	+ 14 11 18.		16.3	FAINT BLUE STAR
LB 01737	04 32 48.	- 48 48		13.3	FAINT BLUE STAR
SHAP434-5256.0	04 32 50.	- 53 02 10.	24	17.5	GALAXY
SHAP433-5409.6	04 32 50.	- 54 15 46.	18	16.25	GALAXY
MCG-02-12-043	04 32 51.	- 13 20 30.	48	15.	GALAXY
SHAP434-5344.0	04 32 51.	- 53 50 10.	18	16.8	GALAXY
ZC 0432.9+2051	04 32 54.	+ 20 51	1880		CLUSTER OF GALAXIES
YM 21	04 32 54.	+ 50 50	132		SYMMETRIC GALACTIC NEBULA
LB 03366	04 32 54.	- 65 44		12.7	FAINT BLUE STAR
HSE 009	04 32 55.	- 66 48			STAR CLUSTER IN LMC
SHAP434-5514.6	04 32 56.	- 55 20 45.	24	17.25	GALAXY
MCG+01-12-033	04 32 57.	- 07 31	78	15.	GALAXY
KHAV 074	04 33	+ 22 54	5310		DARK NEBULA
LBN 0815	04 33	+ 24 08	60		BRIGHT NEBULA
KHAV 073	04 33	+ 24 42			DARK NEBULA
LBN 0717	04 33	+ 50 45	120		BRIGHT NEBULA
LBN 0981	04 33	- 14 30	4080		BRIGHT NEBULA
ZWG 419.020	04 33 00.	+ 08 09		14.8	GALAXY
ZWG 467.003	04 33 00.	+ 19 50		15.0	GALAXY
UGC 03096	04 33 00.	+ 19 50	102	15.0	GALAXY E-S0
RLWT 107	04 33 00.	+ 40 40		18.	FAINT VERY BLUE STAR
LDN 1413	04 33 00.	+ 54 20	180		DARK NEBULA
LDN 1412	04 33 00.	+ 54 20	180		DARK NEBULA
ZC 0433.0+8437	04 33 00.	+ 84 37	1680		CLUSTER OF GALAXIES
LDN 1642	04 33 00.	- 14 20	2460		DARK NEBULA
SHAP434-5435.2	04 33 00.	- 54 41 21.	24	17.0	GALAXY
LB 01738	04 33 00.	- 55 51		13.2	FAINT BLUE STAR
SHAP434-5401.8	04 33 02.	- 54 07 57.	24	16.25	GALAXY
LB 01354	04 33 03.	+ 16 52 24.		15.8	FAINT BLUE STAR
SHAP434-5531.9	04 33 03.	- 55 38 03.	24	17.5	GALAXY
SHAP434-5356.0	04 33 04.	- 54 02 09.	18	17.5	GALAXY
ZWG 393.040	04 33 06.	+ 02 09		15.6	GALAXY
UGC 03097	04 33 06.	+ 02 09	72	15.6	GALAXY S0
ZC 0433.1+0955	04 33 06.	+ 09 55	1210		CLUSTER OF GALAXIES
MRSL 154+02/1	04 33 06.	+ 50 50	120		HII REGION
MCG+12-05-013	04 33 06.	+ 73 12	27	16.	GALAXY
SHAP434-5258.0	04 33 07.	- 53 04 08.	24	17.0	GALAXY
SHAP434-5344.2	04 33 07.	- 53 50 21.	24	17.2	GALAXY
RNGC 1615	04 33 08.	+ 19 51		15.0	GALAXY
SHAP434-5210.2	04 33 08.	- 52 16 20.	36	16.5	GALAXY
SHAP434-5353.7	04 33 08.	- 53 59 51.	60	15.75	GALAXY
MCG+03-12-005	04 33 09.	+ 19 51 30.	48	14.5	GALAXY
SHAP434-5441.8	04 33 10.	- 54 47 56.	48	17.0	GALAXY
SHAP434-5444.2	04 33 10.	- 54 50 20.	24	17.0	GALAXY
ZC 0433.2+0053	04 33 12.	+ 00 53	1340		CLUSTER OF GALAXIES
ZWG 393.041	04 33 12.	+ 02 42		15.6	GALAXY
MCG+00-12-047	04 33 12.	+ 02 43	36	16.	GALAXY
LDN 1533	04 33 12.	+ 24 49	420		DARK NEBULA
MCG-04-11-021	04 33 12.	- 26 06 30.	72	15.	GALAXY
B 020	04 33 14.	+ 50 53	3600		DARK OBJECT
SHAP434-5109.4	04 33 14.	- 51 15 32.	36	16.0	GALAXY
SHAP434-5354.9	04 33 15.	- 54 01 02.	18	17.0	GALAXY
HSE 010	04 33 15.	- 67 05			STAR CLUSTER IN LMC
LW 004	04 33 15.	- 72 27			STAR CLUSTER IN LMC
LB 01355	04 33 16.	+ 18 11 00.		16.3	FAINT BLUE STAR
ZC 0433.3+0015	04 33 18.	+ 00 15	1340		CLUSTER OF GALAXIES
ISS 0149	04 33 18.	+ 49 52	301		STELLAR RING
MCG-02-12-044	04 33 18.	- 11 48 30.	30	15.	GALAXY
LB 01739	04 33 18.	- 46 52		12.4	FAINT BLUE STAR
SHAP434-5321.6	04 33 18.	- 53 27 44.	24	17.0	GALAXY
SHAP434-5438.8	04 33 18.	- 54 44 56.	18	16.25	GALAXY
LB 03367	04 33 18.	- 61 56		12.0	FAINT BLUE STAR
SHAP434-5440.1	04 33 19.	- 54 46 14.	30	16.25	GALAXY
ACK 167-09.1	04 33 24.	+ 33 33			PLANETARY NEBULA
UGC 03098	04 33 24.	+ 43 51	84	17.	GALAXY Sb-c
MCG+11-06-002	04 33 24.	+ 65 12 30.	39	16.	GALAXY
ZWG 393.043	04 33 24.	- 02 09		15.5	GALAXY
ZWG 393.042	04 33 24.	- 02 17		15.5	GALAXY
SHAP434-5347.9	04 33 24.	- 53 54 01.	24	17.5	GALAXY
REIN 6.052	04 33 24.98	- 02 58 21.8			NEBULA
SHAP434-5453.0	04 33 25.	- 54 59 07.	18	17.25	GALAXY
MCG+00-12-048	04 33 26.	- 02 21	48	15.	GALAXY
UGC 03099	04 33 30.	+ 20 30	96	16.0	GALAXY S
UGC 03100	04 33 30.	+ 65 13	78	16.0	GALAXY Sc
ZWG 328.014	04 33 30.	+ 73 11		15.4	GALAXY
UGC 03101	04 33 30.	+ 79 53	96	16.0	GALAXY
MCG+00-12-049	04 33 30.	- 02 09	36	15.	GALAXY
MCG+00-12-050	04 33 30.	- 02 16 30.	42	15.	GALAXY
MCG-04-11-022	04 33 30.	- 22 06	84	14.	GALAXY
MCG-04-11-023	04 33 30.	- 25 14 30.	66	15.	GALAXY
SHAP434-4919.2	04 33 31.	- 49 25 19.	60	17.0	GALAXY
SHAP434-5340.8	04 33 31.	- 53 46 55.	24	17.0	GALAXY
LB 01356	04 33 33.	+ 10 41 24.		15.4	FAINT BLUE STAR
SL 004	04 33 33.	- 72 27	80		STAR CLUSTER IN LMC
LB 01357	04 33 35.	+ 11 04 42.		14.0	FAINT BLUE STAR
MCG+03-12-006	04 33 36.	+ 20 31	72	17.	GALAXY
MCG+01-12-034	04 33 36.	- 03 13	132	13.5	GALAXY
SHAP434-5512.4	04 33 36.	- 55 18 31.	24	17.5	GALAXY
REIN 6.053	04 33 36.83	- 03 14 39.5			NEBULA
BIGO 485	04 33 37.	- 03 19			NEBULA
REIN 6.054	04 33 37.86	- 03 18 10.4			NEBULA
REIN 6.055	04 33 38.54	- 04 54 02.6			NEBULA
REIN 6.056A	04 33 38.76	- 04 57 37.4			NEBULA
REIN 6.056B	04 33 38.76	- 04 57 37.9			NEBULA
RNGC 1618	04 33 39.	- 03 15		13.5	GALAXY
UGC 03102	04 33 42.	+ 14 14	84	16.0	GALAXY S
MCG+12-05-014	04 33 42.	+ 74 06	39	15.	GALAXY
MCG+13-04-006	04 33 42.	+ 79 52	60	16.	GALAXY
ZWG 393.044	04 33 42.	- 02 57		14.8	GALAXY
SHAP434-5432.9	04 33 44.	- 54 39 00.	24	17.5	GALAXY
RNGC 1619	04 33 46.	- 04 56			NON-EXISTENT OBJECT
SHAP434-5432.2	04 33 46.	- 54 38 18.	24	16.5	GALAXY
SHAP434-5512.4	04 33 46.	- 55 18 30.	18	17.0	GALAXY
REIN 6.057	04 33 47.99	- 02 55 55.6			NEBULA
MCG-04-11-024	04 33 48.	- 23 50	24	15.5	GALAXY
LB 01740	04 33 48.	- 56 45		14.3	FAINT BLUE STAR
SHAP435-5130.9	04 33 49.	- 51 37 00.	42	16.25	GALAXY
LB 01358	04 33 50.	+ 18 40 48.		16.6	FAINT BLUE STAR
SHAP435-5301.2	04 33 50.	- 53 07 18.	18	16.75	GALAXY
MCG+00-12-051	04 33 51.	- 02 56	36	14.	GALAXY
RNGC 1621	04 33 52.	- 05 04		14.5	GALAXY
LB 01359	04 33 53.	+ 21 25 48.		15.2	FAINT BLUE STAR
ZC 0433.9+0827	04 33 54.	+ 08 27	2690		CLUSTER OF GALAXIES
ISS 0116	04 33 54.	+ 51 15	326		STELLAR RING
B 021	04 33 54.	+ 55 16	600		DARK OBJECT
ZWG 328.015	04 33 54.	+ 74 05		14.8	GALAXY
ZWG 393.045	04 33 54.	- 02 59		14.8	GALAXY
HOLM 077B	04 33 54.	- 03 18	18	14.9	PART OF MULTIPLE GALAXY
MCG+01-12-035	04 33 54.	- 05 04	30	14.5	GALAXY
REIN 6.058	04 33 56.87	- 05 05 15.7			NEBULA
SHAP435-5218.5	04 33 59.	- 52 24 35.	36	16.75	GALAXY
REIN 6.059	04 33 59.75	- 02 58 00.3			NEBULA
LDN 1528	04 34 00.	+ 25 40	5700		DARK NEBULA
LDN 1485	04 34 00.	+ 35 35	2280		DARK NEBULA
LDN 1406	04 34 00.	+ 55 18	240		DARK NEBULA
ZWG 393.046	04 34 00.	- 00 15		13.6	GALAXY
UGC 03103	04 34 00.	- 00 15	192	13.6	GALAXY Sc
HOLM 077A	04 34 00.	- 03 18	60	14.0	PART OF MULTIPLE GALAXY
MCG-02-12-046	04 34 00.	- 09 38 30.	78	15.	GALAXY
MRK 618	04 34 00.	- 10 28	10	14.5	GALAXY WITH UV CONTINUUM
MCG-02-12-045	04 34 00.	- 10 28	48	15.	GALAXY
REIN 6.061A	04 34 01.36	- 03 17 44.3			NEBULA
REIN 6.061B	04 34 01.45	- 03 17 44.3			NEBULA
SHAP435-5438.2	04 34 02.	- 54 44 17.	24	16.5	GALAXY
LB 01360	04 34 03.	+ 19 33 48.		16.0	FAINT BLUE STAR
MCG+00-12-052	04 34 03.	- 00 13	138	12.	GALAXY
MCG+00-12-053	04 34 03.	- 02 24 30.	12	15.5	GALAXY
MCG+00-12-054	04 34 03.	- 02 58	27	14.	GALAXY
MCG+01-12-036	04 34 03.	- 03 16	210	13.	GALAXY
RNGC 1622	04 34 03.	- 03 17		13.0	GALAXY
REIN 6.060A	04 34 03.43	- 00 14 48.8			NEBULA
REIN 6.060B	04 34 03.45	- 00 14 47.8			NEBULA
SHAP435-5119.2	04 34 04.	- 51 25 16.	30	16.25	GALAXY
SHAP435-5412.2	04 34 05.	- 54 18 17.	90	16.0	GALAXY
SHAP435-5419.4	04 34 05.	- 54 25 29.	18	16.75	GALAXY
HN 0299	04 34 05.	- 75 39			NEBULA
IC 2089	04 34 05.	- 75 39			NONSTELLAR OBJECT
ZWG 393.047	04 34 06.	- 02 24		15.4	GALAXY
UGC 03104	04 34 06.	- 02 24	90	15.4	GALAXY Sa
SHAP435-5358.8	04 34 06.	- 54 04 53.	24	17.0	GALAXY
SHAP435-5358.8	04 34 06.	- 54 04 53.	24	17.0	GALAXY
REIN 6.062	04 34 06.53	- 03 17 19.3			NEBULA
RNGC 1620	04 34 07.	- 00 15		13.5	GALAXY
ABC 0497	04 34 08.	+ 10 33		17.0	RICH CLUSTER OF GALAXIES
SHAP435-5418.6	04 34 08.	- 54 24 40.	72	15.75	GALAXY
REIN 6.063	04 34 08.75	- 02 23 22.1			NEBULA
MCG+00-12-055	04 34 09.	- 02 23 30.	39	14.5	GALAXY
SHAP435-5303.8	04 34 09.	- 53 09 52.	24	16.75	GALAXY
SHAP435-5341.4	04 34 09.	- 53 47 28.	24	16.5	GALAXY
SHAP435-5512.5	04 34 09.	- 55 18 35.	30	17.5	GALAXY
ARP 061	04 34 12.	- 02 23			PECULIAR GALAXY
SHAP435-5608.6	04 34 14.	- 56 14 40.	24	17.0	GALAXY
SHAP435-5030.8	04 34 16.	- 50 36 52.	18	16.75	GALAXY
SHAP435-5318.0	04 34 16.	- 53 24 04.	24	16.75	GALAXY
ZC 0434.3+1033	04 34 18.	+ 10 33	2080		CLUSTER OF GALAXIES
5ZW 375	04 34 18.	+ 24 56			COMPACT GALAXY
OCL 0400	04 34 18.	+ 50 39	460	15.	OPEN STAR CLUSTER
MCG+01-12-037	04 34 18.	- 03 06 30.	36	16.	GALAXY
MCG-03-12-016	04 34 18.	- 20 58	36	15.	GALAXY
SN 1966N	04 34 19.	- 03 08		15.0	SUPERNOVA
SHAP435-5304.4	04 34 19.	- 53 10 28.	18	17.25	GALAXY
REIN 6.064A	04 34 19.12	- 03 08 42.9			NEBULA
REIN 6.064B	04 34 19.35	- 03 08 44.1			NEBULA
SHAP435-5025.6	04 34 20.	- 50 31 39.	42	16.25	GALAXY
ZC 0434.4+3412	04 34 24.	+ 34 12	3560		CLUSTER OF GALAXIES
SHAP435-5606.6	04 34 24.	- 56 12 40.	24	17.0	GALAXY
LW 005	04 34 24.	- 70 46			STAR CLUSTER IN LMC
SHAP435-5535.0	04 34 25.	- 55 41 03.	24	17.5	GALAXY
SHAP435-5114.0	04 34 26.	- 51 20 03.	30	15.5	GALAXY
ZC 0434.5+0104	04 34 30.	+ 01 04	1140		CLUSTER OF GALAXIES
LDN 1556	04 34 30.	+ 16 50	900		DARK NEBULA
ZC 0434.5+2112	04 34 30.	+ 21 12	3900		CLUSTER OF GALAXIES
ZWG 393.048	04 34 30.	- 02 25		14.5	GALAXY
UGC 03105	04 34 30.	- 02 25	120	14.5	GALAXY S0
RNGC 1625	04 34 33.	- 03 24		13.0	GALAXY
SHAP435-5359.4	04 34 33.	- 54 05 27.	18	17.0	GALAXY
SHAP435-5608.0	04 34 33.	- 56 14 03.	30	17.0	GALAXY
SHAP435-5436.1	04 34 34.	- 54 42 09.	30	16.5	GALAXY
REIN 6.065	04 34 34.27	- 03 23 39.0			NEBULA
UGC 03107	04 34 36.	+ 09 27	66	16.5	GALAXY Sb-c
ZWG 393.049	04 34 36.	- 01 58		15.6	GALAXY
UGC 03106	04 34 36.	- 01 58	96	15.6	GALAXY Sa-b
MCG+00-12-056	04 34 36.	- 02 24	24	14.	GALAXY
MCG+01-12-038	04 34 36.	- 03 22 30.	120	13.	GALAXY
REIN 6.067	04 34 36.28	- 03 24 11.5			NEBULA
REIN 6.066	04 34 36.57	- 02 24 19.5			NEBULA
PK174-14.1	04 34 38.	+ 24 58			PLANETARY NEBULA
SHAP435-5434.1	04 34 38.	- 54 40 08.	24	17.0	GALAXY
MCG+00-12-057	04 34 39.	- 01 56 30.	66	15.5	GALAXY
LB 01361	04 34 42.	+ 19 23 18.		16.8	FAINT BLUE STAR
MCG-02-12-047	04 34 42.	- 09 51 30.	48	15.	GALAXY
SHAP435-5242.4	04 34 42.	- 52 48 26.	30	16.5	GALAXY
LW 006	04 34 42.	- 72 55			STAR CLUSTER IN LMC
SHAP435-5424.4	04 34 44.	- 54 30 26.	30	16.0	GALAXY
RNGC 1626	04 34 46.	- 05 06			NON-EXISTENT OBJECT
UGC 03108	04 34 48.	+ 43 57	120	18.	GALAXY S
ABC 0493	04 34 48.	+ 73 42		17.0	RICH CLUSTER OF GALAXIES
ABC 0498	04 34 49.	+ 21 07		16.7	RICH CLUSTER OF GALAXIES
REIN 6.068	04 34 50.47	- 05 04 22.0			NEBULA
DG 042	04 34 54.	+ 26 11	60		REFLECTION NEBULA
ZC 0434.9+7342	04 34 54.	+ 73 42	2020		CLUSTER OF GALAXIES
ZWG 393.050	04 34 54.	- 00 23		14.8	GALAXY
SHAP436-5117.6	04 34 54.	- 51 23 37.	30	16.25	GALAXY
SHAP436-5357.7	04 34 54.	- 54 03 43.	18	17.0	GALAXY
LB 03368	04 34 54.	- 68 44		12.9	FAINT BLUE STAR
REIN 6.069	04 34 55.75	- 05 06 59.7			NEBULA
MCG+00-12-058	04 34 57.	- 00 21	21	14.5	GALAXY
SS 30	04 34 58.	+ 26 10			DIFFUSE GALACTIC NEBULA
KHAV 078	04 35	+ 15 30			DARK NEBULA
B 022	04 35	+ 25 57	7200		DARK OBJECT
LBN 0805	04 35	+ 26 05	180		BRIGHT NEBULA
KHAV 075	04 35	+ 45 24	9060		DARK NEBULA
LDN 1479	04 35 00.	+ 39 30	4980		DARK NEBULA
ZWG 328.016	04 35 00.	+ 73 34		15.0	GALAXY
UGC 03110	04 35 00.	+ 73 34	150	15.0	GALAXY SBc
MCG+12-05-015	04 35 00.	+ 73 34	102	15.	GALAXY
ZC 0435.0+8701	04 35 00.	+ 87 01	1410		CLUSTER OF GALAXIES
MCG+00-12-059	04 35 00.	- 00 22 30.	42	16.	GALAXY

OBJECT NAME	RIGHT ASCEN.	DECLINATION	DIAM.	MAGN.	TYPE OF OBJECT
UGC 03109	04 35 00.	- 00 24	96	16.5	GALAXY
MCG-04-11-025	04 35 00.	- 24 12 30.	36	15.	GALAXY
AGU 21	04 35 00.	- 41 49 36.	54	13.	PECULIAR GALAXY
SHAP436-5558.2	04 35 00.	- 56 04 13.	24	17.0	GALAXY
ABC 0499	04 35 02.	- 20 31		17.8	RICH CLUSTER OF GALAXIES
LB 01362	04 35 04.	+ 12 31 36.		16.6	FAINT BLUE STAR
RNGC 1628	04 35 04.	- 04 48		14.0	GALAXY
RNGC 1627	04 35 04.	- 04 57		13.0	GALAXY
LB 01363	04 35 06.	+ 18 48 54.		15.2	FAINT BLUE STAR
MCG+01-12-039	04 35 06.	- 04 48	84	14.	GALAXY
RNGC 1630	04 35 06.	- 19 00			GALAXY
LB 01741	04 35 06.	- 53 52		12.0	FAINT BLUE STAR
SHAP436-5424.3	04 35 06.	- 54 30 19.	18	16.75	GALAXY
LB 01742	04 35 06.	- 56 25		15.4	FAINT BLUE STAR
SHAP436-5603.2	04 35 07.	- 56 09 13.	24	17.25	GALAXY
REIN 6.070	04 35 07.60	- 04 48 50.8			NEBULA
HSE 011	04 35 08.	- 66 57			STAR CLUSTER IN LMC
ABC 0492	04 35 09.	+ 76 02		17.7	RICH CLUSTER OF GALAXIES
MCG+01-12-040	04 35 09.	- 04 57 30.	84	13.	GALAXY
REIN 6.071	04 35 09.93	- 04 59 18.3			NEBULA
SHAP436-5106.6	04 35 10.	- 51 12 36.	36	16.25	GALAXY
ZWG 393.051	04 35 12.	- 02 43		15.7	GALAXY
SHAP436-5553.7	04 35 13.	- 55 59 42.	24	17.0	GALAXY
SHAP436-5509.9	04 35 14.	- 55 15 54.	36	16.0	GALAXY
SHAP436-5424.9	04 35 16.	- 54 30 54.	24	16.5	GALAXY
UGC 03111	04 35 18.	+ 08 47	78	16.5	GALAXY Sa
LB 01743	04 35 18.	- 48 01		14.9	FAINT BLUE STAR
SHAP436-5511.7	04 35 18.	- 55 17 42.	24	17.0	GALAXY
LB 01744	04 35 18.	- 58 44		14.4	FAINT BLUE STAR
SHAP436-5401.0	04 35 20.	- 54 07 00.	24	17.5	GALAXY
LB 01364	04 35 22.	+ 08 29 00.		16.6	FAINT BLUE STAR
SHAP436-5536.0	04 35 22.	- 55 42 00.	24	17.0	GALAXY
UGC 03112	04 35 24.	+ 79 55	60	16.5	GALAXY DWRF SP
SHAP436-5353.2	04 35 24.	- 53 59 11.	24	16.0	GALAXY
SER 040.03	04 35 24.	- 58 56	480	18.	DWARF GALAXY
HSE 012	04 35 24.	- 66 09			STAR CLUSTER IN LMC
LB 01365	04 35 26.	+ 18 56 36.		16.1	FAINT BLUE STAR
SHAP436-5122.4	04 35 26.	- 51 28 23.	24	16.75	GALAXY
SHAP436-5126.8	04 35 27.	- 51 32 47.	42	16.75	GALAXY
HSE 013	04 35 29.	- 67 49			STAR CLUSTER IN LMC
ZWG 306.003	04 35 30.	+ 66 32		15.5	GALAXY
UGC 03114	04 35 30.	+ 66 32	168	15.5	GALAXY Sc
MCG+11-06-003	04 35 30.	+ 66 32	78	16.	GALAXY
UGC 03113	04 35 30.	- 01 03	66	16.5	GALAXY Sc
ZWG 393.052	04 35 30.	- 02 10		15.1	GALAXY
RNGC 1632	04 35 30.	- 09 37		13.0	GALAXY
MCG-02-12-049	04 35 30.	- 09 37	120	13.	GALAXY
MCG-02-12-048	04 35 30.	- 12 32 30.	48	15.	GALAXY
IC 0382	04 35 31.	- 09 37 48.			NONSTELLAR OBJECT
LB 01366	04 35 32.	+ 22 05 00.		16.5	FAINT BLUE STAR
SHAP436-5305.2	04 35 32.	- 53 11 11.	24	16.5	GALAXY
MCG+00-12-060	04 35 33.	- 02 08	36	15.5	GALAXY
ZC 0435.6+0022	04 35 36.	+ 00 22	1080		CLUSTER OF GALAXIES
ZWG 393.053	04 35 36.	- 02 15		15.7	GALAXY
LB 01367	04 35 37.	+ 07 37 48.		15.6	FAINT BLUE STAR
SHAP436-5538.0	04 35 38.	- 55 43 59.	24	17.5	GALAXY
SHAP436-5232.8	04 35 40.	- 52 38 46.	42	16.0	GALAXY
7ZW 019	04 35 42.	+ 67 38			COMPACT GALAXY
LB 01368	04 35 43.	+ 12 01 06.		16.5	FAINT BLUE STAR
SG 3.023	04 35 43.	+ 50 28	150		DIFFUSE EMISSION NEBULA
SHAP437-5128.8	04 35 47.	- 51 34 45.	42	16.75	GALAXY
MCG-02-12-051	04 35 48.	- 11 29	24	16.	GALAXY
SHAP436-5549.6	04 35 48.	- 55 55 34.	30	16.75	GALAXY
LW 007	04 35 48.	- 69 28			STAR CLUSTER IN LMC
LB 03369	04 35 48.	- 70 28		12.7	FAINT BLUE STAR
SHAP437-5209.0	04 35 49.	- 52 14 57.	108	16.25	GALAXY
LB 01369	04 35 51.	+ 20 13 18.		15.7	FAINT BLUE STAR
MCG-02-12-050	04 35 51.	- 10 53 30.	54	15.	GALAXY
RNGC 1641	04 35 53.	- 65 52			OPEN CLUSTER IN LMC
2ZW 018	04 35 54.	+ 11 09			COMPACT GALAXY
ZWG 467.004	04 35 54.	+ 18 44		15.7	GALAXY
UGC 03115	04 35 54.	+ 18 44	84	15.7	GALAXY S0
ZC 0435.9+3053	04 35 54.	+ 30 53	2890		CLUSTER OF GALAXIES
LB 03370	04 35 54.	- 66 06		12.9	FAINT BLUE STAR
SHAP436-5657.8	04 35 58.	- 57 03 45.	42	16.5	GALAXY
SL 006	04 35 59.	- 65 52	20		STAR CLUSTER IN LMC
KHAV 077	04 36	+ 38 36	10110		DARK NEBULA
KHAV 076	04 36	+ 48 54	3090		DARK NEBULA
LDN 1527	04 36 00.	+ 26 10	360		DARK NEBULA
LDN 1487	04 36 00.	+ 34 50	3900		DARK NEBULA
LDN 1481	04 36 00.	+ 38 10	3000		DARK NEBULA
LDN 1416	04 36 00.	+ 54 00	300		DARK NEBULA
SHAP437-5436.2	04 36 00.	- 54 42 09.	30	17.0	GALAXY
LB 01745	04 36 00.	- 57 30		15.0	FAINT BLUE STAR
LB 03371	04 36 00.	- 65 20		13.0	FAINT BLUE STAR
HSE 014	04 36 00.	- 67 50			STAR CLUSTER IN LMC
LB 01370	04 36 01.	+ 22 06 42.		15.8	FAINT BLUE STAR
HSE 015	04 36 01.	- 66 38			STAR CLUSTER IN LMC
SHAP437-5612.6	04 36 02.	- 56 18 33.	42	16.5	GALAXY
MCG+03-12-007	04 36 03.	+ 18 44	78	16.	GALAXY
SHAP437-5438.9	04 36 04.	- 54 44 51.	24	17.0	GALAXY
ZWG 393.054	04 36 06.	+ 00 05		15.1	GALAXY
MCG+00-12-061	04 36 06.	+ 02 46 30.	36	15.	GALAXY
MCG-05-12-001	04 36 06.	- 29 43	36	15.5	GALAXY
LB 03372	04 36 06.	- 72 28		13.6	FAINT BLUE STAR
SHAP437-5020.9	04 36 08.	- 50 26 50.	72	16.25	GALAXY
SHAP437-5431.6	04 36 08.	- 54 37 32.	18	17.0	GALAXY
SHAP437-5554.8	04 36 10.	- 56 00 44.	78	16.0	GALAXY
HSE 016	04 36 10.	- 67 27			STAR CLUSTER IN LMC
ZWG 393.055	04 36 12.	+ 00 02		15.7	GALAXY
UGC 03116	04 36 12.	+ 00 02	66	15.7	GALAXY S
ZWG 393.056	04 36 12.	+ 02 44		15.1	GALAXY
UGC 03117	04 36 12.	+ 02 44	102	15.1	GALAXY S
MCG-03-12-017	04 36 12.	- 20 46	42	14.5	GALAXY
RNGC 1631	04 36 13.	- 20 46		14.0	GALAXY
IC 0383	04 36 15.	+ 09 47 39.			NONSTELLAR OBJECT
MCG-04-12-001	04 36 15.	- 21 51	48	15.	GALAXY
SHAP437-5143.4	04 36 16.	- 51 49 20.	36	15.5	GALAXY
SHAP437-5556.1	04 36 17.	- 56 02 02.	30	16.75	GALAXY
UGC 03118	04 36 18.	+ 05 31	78	18.	GALAXY DWARF
UGC 03119	04 36 18.	+ 11 27	72	16.5	GALAXY Sb-c
MCG+00-12-062	04 36 18.	- 00 03 30.	36	15.	GALAXY
SHAP437-5425.9	04 36 19.	- 54 31 50.	30	16.5	GALAXY
SHAP437-5540.8	04 36 19.	- 55 46 44.	24	16.75	GALAXY
HSE 017	04 36 20.	- 68 29			STAR CLUSTER IN LMC
SHAP437-5435.4	04 36 22.	- 54 41 19.	18	17.0	GALAXY
SHAP437-5511.1	04 36 22.	- 55 17 01.	24	17.25	GALAXY
ZWG 393.057	04 36 24.	- 02 27		15.6	GALAXY
SHAP437-5350.6	04 36 24.	- 53 56 31.	24	17.25	GALAXY
SL 005	04 36 25.	- 73 50	50		STAR CLUSTER IN LMC
SHAP437-5321.2	04 36 26.	- 53 27 07.	24	17.0	GALAXY
SHAP437-5430.6	04 36 26.	- 54 36 31.	36	17.25	GALAXY
SHAP437-5435.0	04 36 28.	- 54 40 55.	24	17.25	GALAXY
RNGC 1624	04 36 29.	+ 50 21		12.0	CLUSTER WITH NEBULOSITY
BIGO 486	04 36 29.	- 19 01			NEBULA
ACK 165-06.1	04 36 30.	+ 36 40			PLANETARY NEBULA
SHAP437-5231.1	04 36 32.	- 52 37 01.	30	17.0	GALAXY
SHAP437-5342.1	04 36 32.	- 53 48 01.	36	17.0	GALAXY
SHAP437-5500.3	04 36 32.	- 55 06 13.	24	17.0	GALAXY
ZC 0436.6+0626	04 36 36.	+ 06 26	1810		CLUSTER OF GALAXIES
LB 01371	04 36 36.	+ 22 05 54.		17.1	FAINT BLUE STAR
OCL 0403	04 36 36.	+ 50 21	120	11.8	OPEN STAR CLUSTER
HSE 019	04 36 36.	- 66 13			STAR CLUSTER IN LMC
LW 009	04 36 36.	- 71 47			STAR CLUSTER IN LMC
CED 037	04 36 37.	+ 50 21	210		DIFFUSE GALACTIC NEBULA
SHAP437-5125.0	04 36 37.	- 51 30 54.	60	15.0	GALAXY
HSE 018	04 36 38.	- 68 30			STAR CLUSTER IN LMC
SHAP437-5426.2	04 36 39.	- 54 32 06.	18	17.0	GALAXY
LW 008	04 36 43.	- 73 50			STAR CLUSTER IN LMC
LB 01372	04 36 44.	+ 12 06 42.		17.0	FAINT BLUE STAR
ABC 0500	04 36 45.	- 22 12		15.8	RICH CLUSTER OF GALAXIES
SHAP437-5417.6	04 36 45.	- 54 23 30.	30	17.5	GALAXY
SHAP437-5548.6	04 36 47.	- 55 54 30.	24	17.5	GALAXY
2ZW 019	04 36 48.	+ 01 02			COMPACT GALAXY
UGC 03120	04 36 48.	+ 40 10	66	17.	GALAXY Sc
MRSL 155+02/1	04 36 48.	+ 50 17	300		HII REGION
2ZW 020	04 36 48.	- 02 54			COMPACT GALAXY
SHAP438-5139.8	04 36 48.	- 51 45 41.	42	15.75	GALAXY
SHAP438-5417.2	04 36 49.	- 54 23 06.	18	17.0	GALAXY
LB 01373	04 36 51.	+ 11 32 30.		13.2	FAINT BLUE STAR
MCG+01-12-041	04 36 51.	- 05 00	48	15.	GALAXY
IC 0384	04 36 53.	- 07 56 04.			NONSTELLAR OBJECT
ZC 0436.9+0936	04 36 54.	+ 09 36	1210		CLUSTER OF GALAXIES
DG 043	04 36 54.	+ 25 42	180		REFLECTION NEBULA
B 014	04 36 55.	+ 25 39	180		DARK OBJECT
IC 2087	04 36 57.	+ 25 37 54.			DIFFUSE NEBULA
CED 038	04 36 57.	+ 25 38	1680		DIFFUSE GALACTIC NEBULA
SHAP438-5400.6	04 36 57.	- 54 06 29.	24	16.5	GALAXY
SG 3.025	04 36 58.	+ 25 40	120		DIFFUSE EMISSION NEBULA
SS 31	04 36 58.	+ 25 41			DIFFUSE GALACTIC NEBULA
LBN 0846	04 37	+ 07 10	420		BRIGHT NEBULA
LBN 0813	04 37	+ 25 39	240		BRIGHT NEBULA
KHAV 079	04 37	+ 25 42	6050		DARK NEBULA
LBN 0722	04 37	+ 50 20	300		BRIGHT NEBULA
LB 09817	04 37	- 84 07		14.7	FAINT BLUE STAR
ZWG 393.058	04 37 00.	+ 01 19		15.6	GALAXY
SS 32	04 37 00.	+ 22 55			DIFFUSE GALACTIC NEBULA
LDN 1534	04 37 00.	+ 25 30	3780		DARK NEBULA
LDN 1532	04 37 00.	+ 25 40	1320		DARK NEBULA
LDN 1417	04 37 00.	+ 53 50	240		DARK NEBULA
MCG-04-12-002	04 37 00.	- 21 26 30.	36	15.	GALAXY
MCG-04-12-003	04 37 00.	- 24 17	54	14.	GALAXY
SHAP438-5311.6	04 37 00.	- 53 17 29.	24	17.0	GALAXY
SHAP438-5422.2	04 37 00.	- 54 28 05.	36	17.25	GALAXY
SHAP438-5547.0	04 37 01.	- 55 52 53.	30	17.0	GALAXY
LB 00233	04 37 02.	+ 15 12 36.		15.6	FAINT BLUE STAR
HOLM 078B	04 37 03.	+ 06 58	24	14.9	PART OF MULTIPLE GALAXY
MCG-04-12-004	04 37 03.	- 24 17	78	15.	GALAXY
HOLM 078A	04 37 05.	+ 06 57	36	14.2	PART OF MULTIPLE GALAXY
ZWG 393.059	04 37 06.	+ 02 55		15.6	GALAXY
UGC 03121	04 37 06.	+ 02 55	84	15.6	GALAXY S
MCG+12-05-016	04 37 06.	+ 69 07 30.	51	16.	GALAXY
IC 0385	04 37 06.	- 07 11 24.			NONSTELLAR OBJECT
MCG-04-12-005	04 37 06.	- 22 19	48	15.	GALAXY
LB 01746	04 37 06.	- 46 17		13.9	FAINT BLUE STAR
LB 01375	04 37 07.	+ 09 16 12.		16.2	FAINT BLUE STAR
LB 01374	04 37 07.	+ 21 58 06.		14.4	FAINT BLUE STAR
LB 01376	04 37 08.	+ 22 59 12.		15.6	FAINT BLUE STAR
SHAP438-5415.4	04 37 09.	- 54 21 16.	30	17.0	GALAXY
LB 01377	04 37 10.	+ 09 28 24.		16.5	FAINT BLUE STAR
SHAP438-5651.0	04 37 10.	- 56 56 52.	42	16.5	GALAXY
ZWG 419.021	04 37 12.	+ 06 57		14.8	GALAXY
UGC 03122	04 37 12.	+ 06 57	168	14.8	GALAXY SBc
LDN 1503	04 37 12.	+ 29 50	360		DARK NEBULA
MCG-02-12-052	04 37 12.	- 09 09	36	14.5	GALAXY
MCG+01-12-013	04 37 15.	+ 06 57	120	14.	GALAXY
UGC 03123	04 37 18.	+ 12 25	60	17.	GALAXY DBL SYS
ZWG 328.017	04 37 18.	+ 69 09		15.4	GALAXY
UGC 03124	04 37 18.	+ 69 09	66	15.4	GALAXY SBa-b
SHAP438-5309.1	04 37 18.	- 53 14 57.	30	17.5	GALAXY
SHAP438-5656.4	04 37 18.	- 57 02 16.	48	16.5	GALAXY
RNGC 1634	04 37 21.	+ 07 14		15.0	GALAXY
RNGC 1633	04 37 21.	+ 07 15		14.5	GALAXY
SHAP438-5435.1	04 37 22.	- 54 40 57.	24	17.0	GALAXY
SHAP438-5618.4	04 37 23.	- 56 24 15.	36	16.5	GALAXY
ZWG 419.022	04 37 24.	+ 07 14		15.0	GALAXY
KARA.72 101B	04 37 24.	+ 07 14	42	15.0	PART OF DOUBLE GALAXY
ZWG 419.023	04 37 24.	+ 07 15		14.6	GALAXY
HOLM 079A	04 37 24.	+ 07 15	48	13.9	PART OF MULTIPLE GALAXY
UGC 03125	04 37 24.	+ 07 15	84	14.6	GALAXY Sa/SBb
KARA.72 101A	04 37 24.	+ 07 15	66	14.6	PART OF DOUBLE GALAXY
B 023	04 37 24.	+ 29 47	300		DARK OBJECT
MCG-02-12-053	04 37 24.	- 13 23	96	15.5	GALAXY
MCG-06-11-002	04 37 24.	- 37 22	42	15.5	GALAXY
LW 011	04 37 24.	- 66 18			STAR CLUSTER IN LMC
HOLM 079B	04 37 25.	+ 07 14	18	14.3	PART OF MULTIPLE GALAXY
HSE 020	04 37 27.	- 67 00			STAR CLUSTER IN LMC
MCG+01-12-015	04 37 30.	+ 07 13	24	14.	GALAXY
MCG+01-12-014	04 37 30.	+ 07 14	48	13.	GALAXY
MRSL 189-24/1	04 37 30.	+ 07 16	600		HII REGION
ZWG 393.060	04 37 30.	- 00 40		13.5	GALAXY
UGC 03126	04 37 30.	- 00 40	96	13.5	GALAXY SB0/SBa
LW 010	04 37 30.	- 69 18			STAR CLUSTER IN LMC
REIN 6.072	04 37 34.86	- 00 38 39.9			NEBULA
IC 0386	04 37 35.	- 09 33 09.			NONSTELLAR OBJECT
MCG+00-12-063	04 37 36.	- 00 37	78	13.	GALAXY
MCG-02-12-054	04 37 36.	- 09 01 30.	48	14.5	GALAXY
LB 01378	04 37 38.	+ 10 05 06.		15.6	FAINT BLUE STAR
RNGC 1635	04 37 38.	- 00 39		13.5	GALAXY
SHAP438-5251.6	04 37 39.	- 52 57 26.	36	16.0	GALAXY
SHAP438-5547.9	04 37 39.	- 55 53 44.	24	16.5	GALAXY
SHAP438-5604.2	04 37 40.	- 56 10 02.	36	17.0	GALAXY
SHAP438-5240.5	04 37 41.	- 52 46 20.	66	16.25	GALAXY
LB 01379	04 37 43.	+ 13 06 48.		16.0	FAINT BLUE STAR
SHAP438-5346.2	04 37 43.	- 53 52 02.	30	16.75	GALAXY
SHAP438-5337.4	04 37 45.	- 53 43 14.	30	16.5	GALAXY

OBJECT NAME	RIGHT ASCEN.	DECLINATION	DIAM.	MAGN.	TYPE OF OBJECT
ZWG 419.024	04 37 48.	+ 04 06		15.3	GALAXY
UGC 03128	04 37 48.	+ 04 06	78	15.3	GALAXY S0
UGC 03129	04 37 48.	+ 17 00	150	16.5	GALAXY DWARF
MCG+13-04-007	04 37 48.	+ 75 32 30.	132	13.	GALAXY
ZWG 347.006	04 37 48.	+ 75 34		14.5	GALAXY
UGC 03130	04 37 48.	+ 75 34	180	14.5	GALAXY Sb/SBb
ZWG 393.061	04 37 48.	- 02 08		15.1	GALAXY
UGC 03127	04 37 48.	- 02 08	126	15.1	GALAXY Sc
MCG-02-12-055	04 37 48.	- 13 20	30	15.	GALAXY
LB 01747	04 37 48.	- 53 48		14.0	FAINT BLUE STAR
RNGC 1644	04 37 48.	- 66 18		13.0	GLOBULAR CLUSTER IN LMC
SL 009	04 37 48.	- 66 18	35		STAR CLUSTER IN LMC
IC 0381	04 37 51.	+ 75 32 34.			SAME AS NGC 1530A
RNGC 1530A	04 37 53.	+ 75 33		14.5	GALAXY
SL 008	04 37 53.	- 69 07	50		STAR CLUSTER IN LMC
SG 3.026	04 37 54.	+ 50 00	150		DIFFUSE EMISSION NEBULA
ZWG 393.062	04 37 54.	- 00 14		15.7	GALAXY
MCG+00-12-064	04 37 54.	- 02 07	114	14.	GALAXY
RNGC 1651	04 37 54.	- 70 41			GLOBULAR CLUSTER IN LMC
SL 007	04 37 54.	- 70 41	90		STAR CLUSTER IN LMC
KEEL 107	04 37 54.2	- 02 06 24.		16.	NEBULA
MCG+00-12-065	04 37 57.	+ 00 24	24	15.	GALAXY
SHAP439-5338.8	04 37 57.	- 53 44 37.	30	16.5	GALAXY
HZ 15	04 37 58.	+ 08 36		12.8	BLUE STAR
SHAP439-5411.4	04 37 59.	- 54 17 13.	30	17.0	GALAXY
LBN 0847	04 38	+ 07 20	780		BRIGHT NEBULA
KHAV 080	04 38	+ 32 24	6830		DARK NEBULA
LBN 0909	04 38	- 06 15	1320		BRIGHT NEBULA
ZWG 393.063	04 38 00.	+ 00 20		15.6	GALAXY
ZWG 393.064	04 38 00.	+ 00 23		14.7	GALAXY
MCG+01-12-016	04 38 00.	+ 04 05	36	15.	GALAXY
LDN 1508	04 38 00.	+ 29 00	1680		DARK NEBULA
LDN 1504	04 38 00.	+ 29 50	2340		DARK NEBULA
LDN 1415	04 38 00.	+ 54 20	1020		DARK NEBULA
MCG+00-12-066	04 38 00.	- 00 26 30.	30	14.	GALAXY
LW 012	04 38 00.	- 70 41			STAR CLUSTER IN LMC
LW 013	04 38 05.	- 69 08			STAR CLUSTER IN LMC
HMS 1.08	04 38 06.	+ 04 09			Sa GALAXY
MCG-02-12-056	04 38 06.	- 13 20	54	15.	GALAXY
MCG-04-12-006	04 38 06.	- 24 25 30.	72	14.	GALAXY
SHAP439-5242.4	04 38 07.	- 52 48 12.	36	16.75	GALAXY
SHAP439-5302.4	04 38 08.	- 53 08 12.	36	16.0	GALAXY
SHAP439-5412.4	04 38 09.	- 54 18 12.	66	16.0	GALAXY
SHAP439-5414.1	04 38 09.	- 54 19 54.	36	16.5	GALAXY
KEEN 1530A	04 38 12.	+ 75 37		12.88	GALAXY
HZ 14	04 38 14.	+ 10 53		13.5	DECIDEDLY BLUE STAR
LB 01380	04 38 14.	+ 19 41 06.		15.5	FAINT BLUE STAR
SHAP439-5341.4	04 38 14.	- 53 47 12.	24	17.25	GALAXY
LB 01381	04 38 16.	+ 12 08 54.		17.0	FAINT BLUE STAR
RNGC 1652	04 38 16.	- 68 46			GLOBULAR CLUSTER IN LMC
RNGC 1649	04 38 16.	- 68 55			NON-EXISTENT OBJECT
BIGO 487	04 38 17.	- 20 33			NEBULA
LDN 1461	04 38 18.	+ 44 50	900		DARK NEBULA
RNGC 1636	04 38 18.	- 08 42		14.0	GALAXY
MCG+01-12-042	04 38 18.	- 08 42	54	14.	GALAXY
MCG-04-12-007	04 38 18.	- 24 51	36	16.	GALAXY
SHAP439-5300.4	04 38 18.	- 53 06 11.	96	15.25	GALAXY
SHAP439-5207.4	04 38 19.	- 52 13 11.	90	16.5	GALAXY
SHAP439-5410.6	04 38 19.	- 54 16 23.	24	17.0	GALAXY
SL 010	04 38 22.	- 68 46	35		STAR CLUSTER IN LMC
LW 014	04 38 22.	- 68 47			STAR CLUSTER IN LMC
ZC 0438.4+2238	04 38 24.	+ 22 38	4840		CLUSTER OF GALAXIES
MCG-04-12-008	04 38 24.	- 22 03	24	16.	GALAXY
SHAP439-5428.3	04 38 25.	- 54 34 05.	42	17.5	GALAXY
B 220	04 38 26.	+ 25 54	420		DARK OBJECT
SHAP439-5619.4	04 38 29.	- 56 25 11.	30	17.0	GALAXY
WS 01	04 38 30.	- 70 41 31.		16.05	PLANETARY NEB. IN LMC
KEEL 108	04 38 32.2	- 01 48 43.		18.	NEBULA
MCG-02-12-057	04 38 33.	- 08 49	84	15.	GALAXY
RNGC 1639	04 38 35.	- 17 05			NON-EXISTENT OBJECT
KEEL 109	04 38 35.2	- 01 50 48.		17.	NEBULA
DG 044	04 38 36.	+ 25 04	60		REFLECTION NEBULA
ISS 0150	04 38 36.	+ 50 05	210		STELLAR RING
SHAP439-5452.0	04 38 37.	- 54 57 46.	36	17.5	GALAXY
SS 33	04 38 39.	+ 25 04			DIFFUSE GALACTIC NEBULA
HSE 021	04 38 40.	- 67 21			STAR CLUSTER IN LMC
MCG+01-12-017	04 38 42.	+ 07 20	36	15.5	GALAXY
SHAP439-5645.0	04 38 42.	- 56 50 46.	36	16.5	GALAXY
LB 01382	04 38 43.	+ 16 39 36.		16.9	FAINT BLUE STAR
SHAP439-5332.4	04 38 43.	- 53 38 10.	36	17.0	GALAXY
KEEL 110	04 38 44.5	- 01 40 26.		16.	NEBULA
SHAP439-5415.6	04 38 46.	- 54 21 22.	24	17.0	GALAXY
KEEL 111	04 38 46.5	- 02 07 45.		18.	SPIRAL NEBULA
LB 01383	04 38 48.	+ 16 54 42.		17.0	FAINT BLUE STAR
ZWG 393.065	04 38 48.	- 02 42		15.7	GALAXY
LB 01748	04 38 48.	- 47 05		13.6	FAINT BLUE STAR
SHAP439-5538.6	04 38 48.	- 55 44 22.	24	16.75	GALAXY
RNGC 1637	04 38 51.	- 02 56		11.5	GALAXY
UGC 03131	04 38 54.	+ 72 43	90	16.0	GALAXY S
ZWG 393.067	04 38 54.	- 01 26		15.6	GALAXY
ZWG 393.066	04 38 54.	- 02 58		11.8	GALAXY
SHAP439-5452.2	04 38 54.	- 54 57 57.	24	17.0	GALAXY
REIN 2.045	04 38 54.24	- 02 56 18.0			NEBULA
SHAP440-5333.1	04 38 57.	- 53 38 51.	42	16.5	GALAXY
REIN 2.046	04 38 57.63	- 02 57 10.7			NEBULA
SHAP440-5454.4	04 38 59.	- 55 00 09.	24	17.0	GALAXY
KHAV 081	04 39	+ 44 00	6420		DARK NEBULA
LBN 0906	04 39	- 05 30	2100		BRIGHT NEBULA
LB 09818	04 39	- 81 19		14.2	FAINT BLUE STAR
ZC 0439.0+0906	04 39 00.	+ 09 06	1280		CLUSTER OF GALAXIES
MCG+12-05-017	04 39 00.	+ 72 40	36	16.	GALAXY
ZWG 362.003	04 39 00.	+ 82 34		15.4	GALAXY
UGC 03132	04 39 00.	+ 82 34	96	15.4	GALAXY S0-a
MCG+14-03-005	04 39 00.	+ 82 35	18	16.	GALAXY
MCG+00-12-067	04 39 00.	- 01 25 30.	42	16.	GALAXY
ZWG 393.068	04 39 00.	- 01 55		13.6	GALAXY
UGC 03133	04 39 00.	- 01 55	138	13.6	GALAXY E-S0
MCG+00-12-068	04 39 00.	- 02 56 30.	150	11.2	GALAXY
MCG-06-11-003	04 39 00.	- 37 30	18	15.	GALAXY
LH120-N182	04 39 00.	- 70 41			EMISSION NEBULA IN LMC
SHAP440-5346.8	04 39 02.	- 53 52 32.	24	17.5	GALAXY
REIN 6.073	04 39 04.87	- 01 54 13.8			NEBULA
ZC 0439.1+0946	04 39 06.	+ 09 46	1410		CLUSTER OF GALAXIES
MCG+00-12-069	04 39 09.	- 01 53	24	13.	GALAXY
SHAP440-5453.2	04 39 11.	- 54 58 56.	36	16.25	GALAXY
KEEL 112	04 39 11.8	- 02 04 02.		18.	NEBULA
UGC 03135	04 39 12.	+ 19 57	96	17.	GALAXY SBc
ZWG 328.018	04 39 12.	+ 73 17		15.3	GALAXY

OBJECT NAME	RIGHT ASCEN.	DECLINATION	DIAM.	MAGN.	TYPE OF OBJECT
ZWG 393.069	04 39 12.	- 01 24		14.4	GALAXY
UGC 03134	04 39 12.	- 01 24	78	14.4	GALAXY SBb/SBc
SHAP440-5345.4	04 39 12.	- 53 51 08.	24	16.75	GALAXY
KEEL 113	04 39 12.7	- 01 52 32.		18.	NEBULA
ARC 0503	04 39 15.	- 17 18		17.7	RICH CLUSTER OF GALAXIES
REIN 6.074A	04 39 15.97	- 01 23 45.9			NEBULA
REIN 6.074B	04 39 16.02	- 01 23 45.9			NEBULA
ZWG 393.070	04 39 18.	+ 00 57		15.5	GALAXY
KARA.72 102A	04 39 18.	+ 00 57	24	15.5	PART OF DOUBLE GALAXY
ZWG 393.071	04 39 18.	+ 00 58		15.3	GALAXY
UGC 03136	04 39 18.	+ 00 58	72	15.3	GALAXY Sb
KARA.72 102B	04 39 18.	+ 00 58	42	15.3	PART OF DOUBLE GALAXY
ZC 0439.3+0448	04 39 18.	+ 04 48	1410		CLUSTER OF GALAXIES
ZC 0439.3+0816	04 39 18.	+ 08 16	1340		CLUSTER OF GALAXIES
MCG+00-12-070	04 39 18.	- 01 23	78	14.5	GALAXY
MCG+01-12-044	04 39 18.	- 07 11	66	14.	GALAXY
MCG+01-12-043	04 39 18.	- 08 48	48	15.5	GALAXY
MCG-04-12-009	04 39 18.	- 21 55	24	15.5	GALAXY
IC 0387	04 39 19.	- 07 10 52.			NONSTELLAR OBJECT
ARC 0501	04 39 21.	+ 08 18		17.4	RICH CLUSTER OF GALAXIES
LB 01384	04 39 23.	+ 17 58 06.		16.6	FAINT BLUE STAR
ZWG 419.025	04 39 24.	+ 05 22		15.4	GALAXY
ZWG 347.007	04 39 24.	+ 76 20		15.5	GALAXY
UGC 03137	04 39 24.	+ 76 20	240	15.5	GALAXY Sb?
SER 040.04	04 39 24.	- 58 49	25	15.5	HIGH SURFACE BRIGHT. GLXY
IC 0388	04 39 28.	- 07 23 35.			NONSTELLAR OBJECT
LDN 1514	04 39 30.	+ 29 00	900		DARK NEBULA
MCG+13-04-008	04 39 30.	+ 76 20	192	15.	GALAXY
MCG+01-12-045	04 39 33.	- 07 24	24	15.	GALAXY
SHAP440-5245.9	04 39 33.	- 52 51 36.	84	16.0	GALAXY
SL 011	04 39 33.	- 71 07	55		STAR CLUSTER IN LMC
KEEL 114	04 39 33.8	- 01 59 20.		18.	NEBULA
IC 0389	04 39 34.	- 07 24 06.			NONSTELLAR OBJECT
SHAP440-5405.4	04 39 35.	- 54 11 06.	18	17.0	GALAXY
LW 015	04 39 37.	- 74 34			STAR CLUSTER IN LMC
IC 0390	04 39 38.	- 07 17 48.			NONSTELLAR OBJECT
MCG+01-12-046	04 39 39.	- 07 18 30.	42	15.	GALAXY
LW 016	04 39 39.	- 71 06			STAR CLUSTER IN LMC
SHAP440-5343.2	04 39 41.	- 53 48 54.	30	16.5	GALAXY
BC PKS0439-433	04 39 41.6	- 43 19 09.		16.5	QUASI-STELLAR OBJECT
B 024	04 39 44.	+ 29 39	480		DARK OBJECT
SHAP440-5627.3	04 39 44.	- 56 33 00.	42	16.25	GALAXY
UGC 03138	04 39 54.	+ 18 28	72	16.0	GALAXY Sb
SCHO 0067	04 39 54.	+ 43 12 00.	580		ISOLATED DARK CLOUD
ZWG 393.072	04 39 54.	- 01 52		15.5	GALAXY
MCG-06-11-004	04 39 54.	- 37 04		15.	GALAXY
LB 01385	04 39 57.	+ 22 34 54.		16.0	FAINT BLUE STAR
SHAP441-5402.4	04 39 58.	- 54 08 05.	36	16.5	GALAXY
KEEL 115	04 39 58.4	- 01 51 03.		15.	SPIRAL NEBULA
CED 039	04 40	+ 27			DIFFUSE GALACTIC NEBULA
KHAV 082	04 40	+ 22 36	2760		DARK NEBULA
B 221	04 40	+ 31 39	2700		DARK OBJECT
MCG+03-12-008	04 40 00.	+ 18 28	42	17.	GALAXY
LDN 1507	04 40 00.	+ 29 40	1200		DARK NEBULA
LDN 1496	04 40 00.	+ 32 00	3420		DARK NEBULA
LDN 1428	04 40 00.	+ 52 30	240		DARK NEBULA
LDN 1427	04 40 00.	+ 52 35	240		DARK NEBULA
MCG+00-12-071	04 40 00.	- 01 50	18	15.	GALAXY
MCG-03-12-018	04 40 00.	- 20 32	150	13.	GALAXY
LB 01386	04 40 05.	+ 20 28 12.		16.7	FAINT BLUE STAR
SL 012	04 40 05.	- 69 08	20		STAR CLUSTER IN LMC
SHB 086	04 40 05.4	- 00 23 22.		18.5	QUASI-STELLAR OBJECT
ZC 0440.1+0514	04 40 06.	+ 05 14	1480		CLUSTER OF GALAXIES
LB 01749	04 40 06.	- 49 04		14.4	FAINT BLUE STAR
LB 03373	04 40 06.	- 70 11		13.5	FAINT BLUE STAR
RNGC 1640	04 40 07.	- 20 32		12.5	GALAXY
UGC 03139	04 40 12.	+ 40 02	90	17.	GALAXY S
MCG-03-12-019	04 40 12.	- 17 33 30.	60	15.5	GALAXY
SHAP441-5328.4	04 40 13.	- 53 34 04.	42	16.5	GALAXY
LW 018	04 40 17.	- 69 07			STAR CLUSTER IN LMC
ZWG 393.073	04 40 18.	+ 00 30		13.6	GALAXY
UGC 03140	04 40 18.	+ 00 30	126	13.6	GALAXY Sc
ZC 0440.3+0547	04 40 18.	+ 05 47	1480		CLUSTER OF GALAXIES
ZC 0440.3+0644	04 40 18.	+ 06 44	1280		CLUSTER OF GALAXIES
LB 01750	04 40 18.	- 49 01		15.3	FAINT BLUE STAR
RNGC 1642	04 40 19.	+ 00 31		13.5	GALAXY
MCG+00-12-072	04 40 21.	+ 00 33	90	12.	GALAXY
MCG-04-12-010	04 40 21.	- 21 46	90	14.5	GALAXY
MCG+08-09-002	04 40 24.	+ 47 54	36	18.	GALAXY
SHAP441-5353.8	04 40 27.	- 53 59 27.	48	16.25	GALAXY
LW 019	04 40 28.	- 71 19			STAR CLUSTER IN LMC
ZWG 393.074	04 40 30.	+ 00 38		15.3	GALAXY
UGC 03141	04 40 30.	+ 00 38	60	15.3	GALAXY S
MCG+00-12-073	04 40 30.	+ 00 40	42	15.	GALAXY
ARC 0504	04 40 30.	+ 06 43		17.4	RICH CLUSTER OF GALAXIES
TON-S 0410	04 40 30.	- 32 10		14.9	BLUE STAR
IC 2088	04 40 34.	+ 27 11			NONSTELLAR OBJECT
ZZW 021	04 40 36.	+ 00 40			COMPACT GALAXY
MCG+01-12-047	04 40 36.	- 08 11	90	15.	GALAXY
MCG-02-12-058	04 40 36.	- 12 53	84	14.5	GALAXY
LW 020	04 40 36.	- 71 44			STAR CLUSTER IN LMC
LB 01387	04 40 42.	+ 10 54 18.		15.8	FAINT BLUE STAR
HSE 022	04 40 44.	- 66 49			STAR CLUSTER IN LMC
SL 014	04 40 44.	- 69 45	35		STAR CLUSTER IN LMC
LW 021	04 40 44.	- 69 45			STAR CLUSTER IN LMC
SL 015	04 40 45.	- 68 27	35		STAR CLUSTER IN LMC
SL 013	04 40 45.	- 74 07	60		STAR CLUSTER IN LMC
LW 017	04 40 45.	- 74 07			STAR CLUSTER IN LMC
SHAP441-5537.8	04 40 46.	- 55 43 25.	42	16.75	GALAXY
UGC 03142	04 40 48.	+ 28 54	72	16.5	GALAXY SB0-a
LW 023	04 40 51.	- 68 27			STAR CLUSTER IN LMC
ZC 0440.9+0407	04 40 54.	+ 04 07	810		CLUSTER OF GALAXIES
ARC 0506	04 40 55.	- 09 49		17.5	RICH CLUSTER OF GALAXIES
SHAP441-5533.9	04 40 55.	- 55 39 31.	42	16.0	GALAXY
SHAP441-5539.4	04 40 57.	- 55 45 01.	36	16.0	GALAXY
YM 22	04 40 58.	+ 46 41	3540		SYMMETRIC GALACTIC NEBULA
LBN 0742	04 41	+ 46 40	2520		BRIGHT NEBULA
VDB.66G 033	04 41	+ 74 51	70		DWARF GALAXY
ZC 0441.0+2012	04 41 00.	+ 20 12	2350		CLUSTER OF GALAXIES
LDN 1541	04 41 00.	+ 22 40	2400		DARK NEBULA
LDN 1520	04 41 00.	+ 28 20	2520		DARK NEBULA
OCL 0413	04 41 00.	+ 41 58	720		OPEN STAR CLUSTER
LDN 1424	04 41 00.	+ 53 10	300		DARK NEBULA
LDN 1423	04 41 00.	+ 53 15	120		DARK NEBULA
ZC 0441.0-0145	04 41 00.	- 01 45	940		CLUSTER OF GALAXIES
ZZW 022	04 41 00.	- 08 37			COMPACT GALAXY
ARC 0507	04 41 00.	- 18 35		17.4	RICH CLUSTER OF GALAXIES
LB 01388	04 41 05.	+ 17 20 00.		16.1	FAINT BLUE STAR

OBJECT NAME	RIGHT ASCEN.	DECLINATION	DIAM.	MAGN.	TYPE OF OBJECT
ZC 0441.1+0211	04 41 06.	+ 02 11	1280		CLUSTER OF GALAXIES
LB 01389	04 41 06.	+ 16 11 36.		16.0	FAINT BLUE STAR
OCL 0412	04 41 06.	+ 42 35	420	16.	OPEN STAR CLUSTER
MCG+12-05-018	04 41 06.	+ 74 50	108	16.	GALAXY
LB 01751	04 41 06.	- 52 44		14.1	FAINT BLUE STAR
LB 03374	04 41 06.	- 73 33		13.5	FAINT BLUE STAR
LB 01390	04 41 07.	+ 21 55 00.		15.5	FAINT BLUE STAR
RNGC 1643	04 41 10.	- 05 23		14.0	GALAXY
MCG-01-13-001	04 41 12.	- 05 26	60	14.	GALAXY
ZCG 0441-08	04 41 12.	- 08 34		15.3	COMPACT GALAXY
LW 022	04 41 12.	- 72 43			STAR CLUSTER IN LMC
SHAP442-5543.2	04 41 13.	- 55 48 48.	30	16.5	GALAXY
LB 01391	04 41 16.	+ 20 24 12.		16.3	FAINT BLUE STAR
REIN 6.075A	04 41 16.15	- 05 24 38.2			NEBULA
REIN 6.075B	04 41 16.24	- 05 24 41.7			NEBULA
MRSL 158+00/1	04 41 18.	+ 46 44	4800		HII REGION
UGC 03143	04 41 18.	+ 70 02	114	17.	GALAXY Sc
ZC 0441.4+1036	04 41 24.	+ 10 36	1210		CLUSTER OF GALAXIES
ZWG 328.019	04 41 24.	+ 74 50		15.7	GALAXY
UGC 03144	04 41 24.	+ 74 50	126	15.7	GALAXY DWRF IR
ZC 0441.4-0228	04 41 24.	- 02 28	1550		CLUSTER OF GALAXIES
IC 2103	04 41 24.	- 76 55			NONSTELLAR OBJECT
HN 0300	04 41 25.	- 76 55			NEBULA
LB 01392	04 41 29.	+ 15 49 54.		16.8	FAINT BLUE STAR
TON-S 0411	04 41 30.	- 33 03		15.3	BLUE STAR
SER 040.07	04 41 30.	- 60 22	20	17.	HIGH SURFACE BRIGHT. GLXY
RNGC 1645	04 41 34.	- 05 32		14.0	GALAXY
MCG+00-13-001	04 41 36.	+ 00 17 30.	66	15.	GALAXY
MCG+11-06-004	04 41 36.	+ 63 44 30.	72	16.	GALAXY
MCG-01-13-002	04 41 36.	- 05 35	120	14.	GALAXY
SHAP442-5526.0	04 41 38.	- 55 31 34.	36	16.5	GALAXY
LW 024	04 41 38.	- 70 53			STAR CLUSTER IN LMC
REIN 6.076A	04 41 38.75	- 05 33 26.7			NEBULA
REIN 6.076B	04 41 38.93	- 05 33 23.6			NEBULA
LB 00234	04 41 39.	+ 20 11 30.		17.8	FAINT BLUE STAR
ZWG 394.001	04 41 42.	+ 00 15		15.4	GALAXY
UGC 03145	04 41 42.	+ 00 15	90	15.4	GALAXY SBb-c
ZC 0441.7+0423	04 41 42.	+ 04 23	1340		CLUSTER OF GALAXIES
UGC 03146	04 41 42.	+ 63 46	84	17.	GALAXY S
ZC 0441.7+7555	04 41 42.	+ 75 55	1410		CLUSTER OF GALAXIES
MCG+00-13-002	04 41 45.	- 02 15	30	15.	GALAXY
ZWG 328.020	04 41 48.	+ 72 46		14.8	GALAXY
UGC 03147	04 41 48.	+ 72 46	96	14.8	GALAXY S
LW 025	04 41 50.	- 68 16			STAR CLUSTER IN LMC
MCG+12-05-019	04 41 54.	+ 72 45	66	15.	GALAXY
RNGC 1646	04 41 54.	- 08 39		14.5	GALAXY
SHAP443-5432.6	04 41 55.	- 54 38 09.	36	16.0	GALAXY
MCG-01-13-003	04 41 57.	- 08 39	30	14.5	GALAXY
LB 01394	04 41 58.	+ 21 13 42.		15.8	FAINT BLUE STAR
LB 01393	04 41 58.	+ 22 06 06.		16.5	FAINT BLUE STAR
LB 00235	04 41 59.	+ 12 42 48.		16.5	FAINT BLUE STAR
LBN 0744	04 42	+ 46 40	1920		BRIGHT NEBULA
LB 09819	04 42	- 83 07		15.1	FAINT BLUE STAR
LB 09820	04 42	- 83 17		14.2	FAINT BLUE STAR
UGC 03148	04 42 00.	+ 18 30	72	16.5	GALAXY S
LDN 1425	04 42 00.	+ 53 00	1200		DARK NEBULA
REIN 6.077	04 42 05.04	- 04 55 55.5			NEBULA
RNGC 1648	04 42 06.	- 08 35		16.0	GALAXY
MCG-01-13-004	04 42 06.	- 08 35 30.	24	16.	GALAXY
LB 01752	04 42 06.	- 56 57		13.8	FAINT BLUE STAR
LB 03375	04 42 06.	- 67 31		13.8	FAINT BLUE STAR
SHAP443-5617.0	04 42 08.	- 56 22 32.	48	16.5	GALAXY
LB 01395	04 42 11.	+ 10 07 12.		13.8	FAINT BLUE STAR
LB 03376	04 42 12.	- 67 36		14.0	FAINT BLUE STAR
LB 01396	04 42 14.	+ 17 11 12.		16.8	FAINT BLUE STAR
RNGC 1573A	04 42 15.	+ 73 22		14.5	GALAXY
LW 026	04 42 17.	- 70 17			STAR CLUSTER IN LMC
ZWG 306.004	04 42 18.	+ 65 57		15.1	GALAXY
UGC 03149	04 42 18.	+ 65 57	108	15.1	GALAXY S
ZWG 328.021	04 42 18.	+ 73 22		14.6	GALAXY
UGC 03150	04 42 18.	+ 73 22	108	14.6	GALAXY Sb
MCG+12-05-020	04 42 18.	+ 73 22	84	14.	GALAXY
LH120-N001	04 42 18.	- 66 17			EMISSION NEBULA IN LMC
SHAP443-5419.1	04 42 20.	- 54 24 37.	42	16.75	GALAXY
RNGC 1658	04 42 23.	- 41 33			GALAXY
LW 028	04 42 25.	- 69 28			STAR CLUSTER IN LMC
SHAP443-5409.2	04 42 27.	- 54 14 42.	78	16.25	GALAXY
SHAP443-5648.2	04 42 28.	- 56 53 43.	42	16.75	GALAXY
ZWG 420.001	04 42 30.	+ 05 23		15.5	GALAXY
UGC 03151	04 42 30.	+ 10 59	66	16.5	GALAXY S
TON-S 0413	04 42 30.	- 30 25		15.8	BLUE STAR
TON-S 0412	04 42 30.	- 32 15		15.4	BLUE STAR
LB 01397	04 42 31.	+ 12 57 24.		15.4	FAINT BLUE STAR
LB 01398	04 42 33.	+ 15 23 42.		16.1	FAINT BLUE STAR
SL 016	04 42 33.	- 65 03	15		STAR CLUSTER IN LMC
RNGC 1660	04 42 35.	- 41 35			GALAXY
LW 027	04 42 35.	- 72 29			STAR CLUSTER IN LMC
ZC 0442.6-0013	04 42 36.	- 00 13	940		CLUSTER OF GALAXIES
KEEN 1573A	04 42 42.	+ 73 26		14.0	GALAXY
RNGC 1673	04 42 45.	- 69 57			OPEN CLUSTER IN LMC
TON-S 0414	04 42 46.	- 33 38		15.3	BLUE STAR
RNGC 1669	04 42 48.	- 65 54			GALAXY
LB 03377	04 42 48.	- 67 49		13.0	FAINT BLUE STAR
LW 030	04 42 50.	- 68 10			STAR CLUSTER IN LMC
SHAP443-5622.8	04 42 51.	- 56 28 17.	36	16.25	GALAXY
SL 017	04 42 51.	- 69 57	25		STAR CLUSTER IN LMC
RNGC 1650	04 42 52.	- 15 57		14.0	GALAXY
LB 01399	04 42 53.	+ 09 16 54.		15.5	FAINT BLUE STAR
IC 2090	04 42 53.	- 34 05 04.			NONSTELLAR OBJECT
BA 09	04 42 54.	+ 44 37 48.	468		STELLAR GROUP
OCL 0408	04 42 54.	+ 44 39	180	14.	OPEN STAR CLUSTER
MCG-03-13-001	04 42 54.	- 15 57	30	14.	GALAXY
LW 029	04 42 56.	- 70 48			STAR CLUSTER IN LMC
LB 01400	04 42 57.	+ 09 28 18.		16.0	FAINT BLUE STAR
LW 031	04 42 57.	- 69 55			STAR CLUSTER IN LMC
KHAV 085	04 43	+ 25 06	7370		DARK NEBULA
KHAV 084	04 43	+ 53 00	700		DARK NEBULA
KHAV 083	04 43	+ 54 12	3330		DARK NEBULA
LB 09221	04 43	- 85 44		13.5	FAINT BLUE STAR
LDN 1558	04 43 00.	+ 17 00	3420		DARK NEBULA
LDN 1538	04 43 00.	+ 25 00	7020		DARK NEBULA
LDN 1426	04 43 00.	+ 53 00	540		DARK NEBULA
UGC 03152	04 43 00.	+ 79 19	66	16.0	GALAXY Sc
MCG-03-13-002	04 43 00.	- 15 13	12	15.	GALAXY
SHAP444-5336.9	04 43 00.	- 53 42 22.	54	16.5	GALAXY
LB 01401	04 43 03.	+ 22 33 36.		16.3	FAINT BLUE STAR
MCG+05-12-001	04 43 03.	+ 31 18 30.	54	15.5	GALAXY
ZC 0443.1+1302	04 43 06.	+ 13 02	1010		CLUSTER OF GALAXIES
OCL 0457	04 43 06.	+ 18 59	3180	7.0	OPEN STAR CLUSTER
GCL 007	04 43 06.	+ 31 23	102		GLOBULAR STAR CLUSTER
LH120-N002	04 43 06.	- 68 01	95		EMISSION NEBULA IN LMC
HSE 023	04 43 11.	- 70 14			STAR CLUSTER IN LMC
LW 032	04 43 11.	- 70 14			STAR CLUSTER IN LMC
SL 018	04 43 11.	- 70 15	50		STAR CLUSTER IN LMC
ZC 0443.2+0155	04 43 12.	+ 01 55	1480		CLUSTER OF GALAXIES
MCG+00-13-003	04 43 12.	- 02 29	33	13.	GALAXY
SL 020	04 43 12.	- 67 50	70		STAR CLUSTER IN LMC
LB 03378	04 43 12.	- 70 34		13.6	FAINT BLUE STAR
RNGC 1647	04 43 14.	+ 18 59		6.0	OPEN CLUSTER
SL 019	04 43 15.	- 69 57	10		STAR CLUSTER IN LMC
REIN 6.078	04 43 16.31	- 02 28 59.0			NEBULA
REIN 6.079A	04 43 17.10	- 02 10 27.5			NEBULA
REIN 6.079B	04 43 17.11	- 02 10 27.4			NEBULA
MCG+12-05-021	04 43 18.	+ 71 25	27	16.	GALAXY
ZWG 328.022	04 43 18.	+ 71 27		15.3	GALAXY
UGC 03155	04 43 18.	+ 71 27	90	15.3	GALAXY PECULR?
MCG+12-05-022	04 43 18.	+ 72 17	24	17.	GALAXY
ZWG 394.003	04 43 18.	- 02 11		14.2	GALAXY
UGC 03154	04 43 18.	- 02 11	54	14.4	GALAXY S
SN 1962P	04 43 18.	- 02 11		14.5	SUPERNOVA
ZWG 394.002	04 43 18.	- 02 30		12.9	GALAXY
UGC 03153	04 43 18.	- 02 30	138	12.9	GALAXY E
LB 01753	04 43 18.	- 56 53		12.4	FAINT BLUE STAR
ARC 0508	04 43 20.	+ 01 56		17.4	RICH CLUSTER OF GALAXIES
RNGC 1654	04 43 20.	- 02 11		14.0	GALAXY
RNGC 1653	04 43 21.	- 02 30		13.0	GALAXY
LW 033	04 43 21.	- 69 56			STAR CLUSTER IN LMC
RNGC 1656	04 43 22.	- 05 14		14.0	GALAXY
ZWG 394.004	04 43 24.	- 02 18		15.7	GALAXY
MCG-01-13-005	04 43 24.	- 05 14 30.	36	14.	GALAXY
MCG-03-13-003	04 43 24.	- 16 19	48	15.	GALAXY
LB 01754	04 43 24.	- 46 02		14.0	FAINT BLUE STAR
REIN 6.080A	04 43 25.35	- 05 13 35.7			NEBULA
REIN 6.080B	04 43 25.35	- 05 13 36.2			NEBULA
REIN 6.080C	04 43 25.52	- 05 13 33.6			NEBULA
REIN 6.080D	04 43 25.52	- 05 13 34.3			NEBULA
MCG+00-13-004	04 43 30.	- 02 10	57	14.	GALAXY
LB 01402	04 43 33.	+ 20 52 00.		16.2	FAINT BLUE STAR
REIN 6.081	04 43 35.88	- 02 10 00.4			NEBULA
MCG+01-13-001	04 43 36.	+ 03 25	84	13.5	GALAXY
ZWG 468.001	04 43 36.	+ 18 23		15.7	GALAXY
UGC 03157	04 43 36.	+ 18 23	102	15.7	GALAXY SB
MCG+03-13-001	04 43 36.	+ 18 23	84	15.	GALAXY
ZWG 394.005	04 43 36.	- 02 10		15.0	GALAXY
UGC 03156	04 43 36.	- 02 10	90	15.0	GALAXY Sb/SBb
MCG-04-12-011	04 43 36.	- 21 43 30.	48	15.5	GALAXY
LB 01403	04 43 38.	+ 09 07 54.		15.4	FAINT BLUE STAR
RNGC 1657	04 43 38.	- 02 10		15.0	GALAXY
LW 034	04 43 41.	- 70 15			STAR CLUSTER IN LMC
SL 021	04 43 41.	- 70 16	50		STAR CLUSTER IN LMC
MCG+00-13-005	04 43 42.	+ 00 17 30.	54	15.	GALAXY
MCG-02-13-001	04 43 42.	- 12 32	54	14.5	GALAXY
MCG-03-13-004	04 43 42.	- 17 23	102	14.	GALAXY
ZWG 420.002	04 43 48.	+ 03 25		14.8	GALAXY
UGC 03158	04 43 48.	+ 03 25	90	14.8	GALAXY S0
IC 0392	04 43 48.	+ 03 25 18.			NONSTELLAR OBJECT
ZWG 328.023	04 43 48.	+ 72 53		15.6	GALAXY
UGC 03159	04 43 48.	+ 72 53	66	15.6	GALAXY Sb-c
SL 022	04 43 49.	- 69 29	30		STAR CLUSTER IN LMC
MCG+00-13-006	04 43 51.	- 02 46 30.	30	15.	GALAXY
MCG-04-12-012	04 43 51.	- 22 27	72	14.5	GALAXY
RNGC 1676	04 43 53.	- 68 56			OPEN CLUSTER IN LMC
ZWG 394.006	04 43 54.	+ 00 16		15.2	GALAXY
MCG-03-13-006	04 43 54.	- 15 55	48	15.	GALAXY
MCG-03-13-005	04 43 54.	- 17 10	66	15.	GALAXY
LW 035	04 43 55.	- 69 28			STAR CLUSTER IN LMC
LW 036	04 43 57.	- 69 49			STAR CLUSTER IN LMC
SL 023	04 43 57.	- 69 50	35		STAR CLUSTER IN LMC
RNGC 1659	04 43 58.	- 04 53		13.0	GALAXY
SL 024	04 43 59.	- 68 54	35		STAR CLUSTER IN LMC
SL 025	04 43 59.	- 68 56	50		STAR CLUSTER IN LMC
UGC 03161	04 44 00.	+ 00 31	78	16.5	GALAXY Sc
LDN 1431	04 44 00.	+ 51 55	360		DARK NEBULA
LDN 1429	04 44 00.	+ 52 00	720		DARK NEBULA
MCG+14-03-006	04 44 00.	+ 86 08 30.	54	14.	GALAXY
ZWG 370.001	04 44 00.	+ 86 09		14.2	GALAXY
ZWG 362.004	04 44 00.	+ 86 09		14.2	GALAXY
ZWG 361.011	04 44 00.	+ 86 09		14.2	GALAXY
UGC 03160	04 44 00.	+ 86 09	102	14.2	GALAXY
KARA.73B 0153	04 44 00.	+ 86 09	66	14.2	ISOLATED GALAXY S
ZWG 394.007	04 44 00.	- 02 46		14.8	GALAXY
MCG-01-13-006	04 44 00.	- 04 53	72	13.	GALAXY
MCG-02-13-002	04 44 00.	- 13 27	66	15.	GALAXY
RNGC 1544	04 44 01.	+ 86 08		14.0	GALAXY
REIN 6.082A	04 44 01.54	- 04 52 39.6			NEBULA
REIN 6.082B	04 44 01.54	- 04 52 40.4			NEBULA
REIN 6.082C	04 44 01.59	- 04 52 40.6			NEBULA
REIN 6.082D	04 44 01.60	- 04 52 40.5			NEBULA
MCG-01-13-007	04 44 03.	- 07 20	66	15.	GALAXY
LW 037	04 44 05.	- 68 56			STAR CLUSTER IN LMC
SHAP445-5427.4	04 44 07.	- 54 32 48.	36	16.75	GALAXY
MCG-02-13-003	04 44 09.	- 12 01	36	15.	GALAXY
HSE 024	04 44 09.	- 69 50			STAR CLUSTER IN LMC
IC 2091	04 44 10.	- 04 46 21.			GROUP OF FAINT STARS
LW 038	04 44 10.	- 69 59			STAR CLUSTER IN LMC
LB 01404	04 44 11.	+ 13 18 36.		16.0	FAINT BLUE STAR
SL 027	04 44 11.	- 68 52	15		STAR CLUSTER IN LMC
LW 039	04 44 11.	- 68 53			STAR CLUSTER IN LMC
UGC 03162	04 44 12.	+ 08 13	102	16.0	GALAXY Sc
MCG+01-13-002	04 44 12.	+ 08 13	72	15.	GALAXY
MCG+12-05-023	04 44 12.	+ 69 22 30.	84	15.	GALAXY
ZWG 328.024	04 44 12.	+ 69 24		15.3	GALAXY
UGC 03163	04 44 12.	+ 69 24	96	15.3	GALAXY Sa
RNGC 1655	04 44 13.	+ 20 50			NON-EXISTENT OBJECT
LB 01405	04 44 16.	+ 20 36 06.		15.8	FAINT BLUE STAR
LW 040	04 44 18.	- 70 31			STAR CLUSTER IN LMC
IC 2092	04 44 19.	- 05 02 41.			DOUBLE STAR
SHAP445-5307.2	04 44 19.	- 53 12 35.	60	16.0	GALAXY
SHAP445-5420.8	04 44 23.	- 54 26 10.	30	16.5	GALAXY
LW 041	04 44 23.	- 68 51			STAR CLUSTER IN LMC
LB 01406	04 44 24.	+ 22 39 30.		16.0	FAINT BLUE STAR
MCG+00-13-007	04 44 24.	- 00 01	90	15.	GALAXY
SHAP445-5434.4	04 44 25.	- 54 39 46.	54	16.5	GALAXY
UGC 03165	04 44 30.	+ 23 53	162	16.0	GALAXY IRR
ZWG 394.008	04 44 30.	- 00 01		15.6	GALAXY
UGC 03164	04 44 30.	- 00 01	90	15.6	GALAXY S IV-V

OBJECT NAME	RIGHT ASCEN.	DECLINATION	DIAM.	MAGN.	TYPE OF OBJECT
MCG-02-13-004	04 44 30.	- 11 34	36	15.5	GALAXY
RNGC 1668	04 44 33.	- 44 49			GALAXY
ARC 0510	04 44 35.	- 21 06		17.6	RICH CLUSTER OF GALAXIES
ZC 0444.6+0206	04 44 36.	+ 02 06	810		CLUSTER OF GALAXIES
ZWG 394.009	04 44 36.	- 02 09		14.3	GALAXY
UGC 03166	04 44 36.	- 02 09	108	14.3	GALAXY Sb-c
MCG+00-13-008	04 44 36.	- 02 10	66	14.	GALAXY
MCG-02-13-005	04 44 36.	- 14 58	30	15.	GALAXY
LB 01755	04 44 36.	- 52 33		13.4	FAINT BLUE STAR
REIN 6.083	04 44 36.09	- 02 08 35.4			NEBULA
ARC 0511	04 44 37.	- 25 30		17.0	RICH CLUSTER OF GALAXIES
RNGC 1661	04 44 38.	- 02 09		14.5	GALAXY
LW 042	04 44 40.	- 68 43			STAR CLUSTER IN LMC
ZC 0444.7+0828	04 44 42.	+ 08 28	6920		CLUSTER OF GALAXIES
LW 044	04 44 45.	- 68 20			STAR CLUSTER IN LMC
DG 045	04 44 48.	+ 48 27	60		REFLECTION NEBULA
UGC 03167	04 44 48.	+ 63 51	108	17.	GALAXY Sb-c
MCG-03-13-008	04 44 48.	- 16 34 30.	48	14.5	GALAXY
MCG-03-13-007	04 44 48.	- 20 25	36	15.	GALAXY
GCL 008	04 44 48.	- 84 00	144	12.2	GLOBULAR STAR CLUSTER
LB 01407	04 44 50.	+ 14 47 42.		15.8	FAINT BLUE STAR
SS 34	04 44 51.	+ 48 27			DIFFUSE GALACTIC NEBULA
MCG+00-13-009	04 44 51.	- 01 18 30.	21	15.	GALAXY
LW 043	04 44 53.	- 71 29			STAR CLUSTER IN LMC
MCG+00-13-010	04 44 54.	- 02 24	54	15.	GALAXY
MCG-02-13-006	04 44 54.	- 10 20	48	14.5	GALAXY
RNGC 1672	04 44 54.	- 59 20		11.5	GALAXY
HSE 025	04 44 55.	- 73 39			STAR CLUSTER IN LMC
HZ 13	04 44 56.	+ 07 44		14.6	BLUE STAR
SHAP446-5435.4	04 44 56.	- 54 40 44.	36	16.75	GALAXY
LB 01408	04 44 57.	+ 18 57 00.		14.0	FAINT BLUE STAR
MCG+00-13-011	04 44 57.	- 01 15 30.	54	15.	GALAXY
VDB.66G 034	04 45	+ 00 10	70		DWARF GALAXY
LBN 0793	04 45	+ 29 42	480		BRIGHT NEBULA
KHAV 086	04 45	+ 30 29	7400		DARK NEBULA
LBN 0737	04 45	+ 48 28	60		BRIGHT NEBULA
LBN 0917	04 45	- 06 00	3000		BRIGHT NEBULA
UGC 03169	04 45 00.	+ 09 30	72	17.	GALAXY Sc
LDN 1505	04 45 00.	+ 31 00	4140		DARK NEBULA
LDN 1474	04 45 00.	+ 44 00	8700		DARK NEBULA
LDN 1430	04 45 00.	+ 52 03	360		DARK NEBULA
LDN 1422	04 45 00.	+ 54 00	4260		DARK NEBULA
ZWG 394.011	04 45 00.	- 01 18		15.7	GALAXY
ZWG 394.010	04 45 00.	- 02 24		15.3	GALAXY
UGC 03168	04 45 00.	- 02 24	66	15.3	GALAXY Sc
ARC 0502	04 45 01.	+ 69 43		16.8	RICH CLUSTER OF GALAXIES
IC 2093	04 45 03.	- 02 48			NONSTELLAR OBJECT
SL 026	04 45 05.	- 67 24	35		STAR CLUSTER IN LMC
ZWG 394.013	04 45 06.	+ 01 44		15.2	GALAXY
UGC 03171	04 45 06.	+ 01 44	90	15.2	GALAXY SB:c
KARA.73B 0154	04 45 06.	+ 01 44	48	15.2	ISOLATED GALAXY S
ZC 0445.1+0223	04 45 06.	+ 02 23	1080		CLUSTER OF GALAXIES
ZWG 394.012	04 45 06.	- 01 15		15.5	GALAXY
UGC 03170	04 45 06.	- 01 15	66	15.5	GALAXY Sa
ARC 0509	04 45 08.	+ 02 13		17.4	RICH CLUSTER OF GALAXIES
LW 046	04 45 09.	- 69 45			STAR CLUSTER IN LMC
SS 35	04 45 10.	+ 29 42			DIFFUSE GALACTIC NEBULA
MCG+00-13-012	04 45 12.	+ 01 45	48	13.5	GALAXY
LB 01409	04 45 12.	+ 15 42 00.		15.8	FAINT BLUE STAR
DG 046	04 45 12.	+ 29 43	660		REFLECTION NEBULA
VDB.66N 029	04 45 14.	+ 29 42	816		REFLECTION NEBULA
LB 01410	04 45 17.	+ 21 08 18.		15.5	FAINT BLUE STAR
LW 048	04 45 17.	- 67 24			STAR CLUSTER IN LMC
ISS 0203	04 45 18.	+ 43 00	201		STELLAR RING
UGC 03172	04 45 18.	+ 08 41	102	16.5	GALAXY Sc
SCHO 0068	04 45 18.	+ 44 07 00.	480		ISOLATED DARK CLOUD
ARC 0512	04 45 18.	- 18 23		17.0	RICH CLUSTER OF GALAXIES
MCG-03-13-009	04 45 21.	- 17 41 30.	72	14.5	GALAXY
MCG-03-13-010	04 45 24.	- 17 27	48	14.5	GALAXY
LW 045	04 45 25.	- 73 41			STAR CLUSTER IN LMC
MCG-02-13-007	04 45 27.	- 10 19 30.	48	15.	GALAXY
LW 049	04 45 28.	- 69 55			STAR CLUSTER IN LMC
RLWT 108	04 45 30.	+ 40 25		12.5	FAINT VERY BLUE STAR
MCG-03-13-011	04 45 30.	- 17 32	30	15.	GALAXY
SHAP446-5437.4	04 45 30.	- 54 42 42.	48	16.0	GALAXY
ARC 0514	04 45 31.	- 20 32		15.2	RICH CLUSTER OF GALAXIES
MCG-02-13-008	04 45 33.	- 10 20	27	15.	GALAXY
ZC 0445.6+0539	04 45 36.	+ 05 39	5040		CLUSTER OF GALAXIES
IC 0393	04 45 36.	- 15 36 57.			NONSTELLAR OBJECT
MCG-03-13-012	04 45 36.	- 15 37	15	15.	GALAXY
SER 040.01	04 45 36.	- 57 23	100	14.	IRR. MAGELLANIC GALAXY
LB 03379	04 45 36.	- 68 55		13.1	FAINT BLUE STAR
OCL 0470	04 45 42.	+ 10 51	1800	8.2	OPEN STAR CLUSTER
MCG+12-05-024	04 45 42.	+ 72 12 30.	12	16.	GALAXY
MCG-03-13-013	04 45 42.	- 16 11	42	15.	GALAXY
RNGC 1662	04 45 43.	+ 10 51		8.0	OPEN CLUSTER
RNGC 1665	04 45 46.	- 05 31		14.0	GALAXY
ZWG 394.014	04 45 48.	+ 00 08		15.7	GALAXY
UGC 03173	04 45 48.	+ 07 58	60	16.5	GALAXY
OCL 0461	04 45 48.	+ 13 04		14.1	OPEN STAR CLUSTER
RNGC 1663	04 45 48.	+ 13 04			OPEN CLUSTER
MCG+12-05-025	04 45 48.	+ 72 12	12	16.	GALAXY
MCG-01-13-008	04 45 48.	- 03 27	48	14.	GALAXY
MCG-01-13-009	04 45 48.	- 05 31 30.	84	14.	GALAXY
LW 047	04 45 48.	- 74 21			STAR CLUSTER IN LMC
LB 00236	04 45 49.	+ 16 31 12.		16.3	FAINT BLUE STAR
ARC 0513	04 45 49.	- 09 49		17.5	RICH CLUSTER OF GALAXIES
MCG-02-13-009	04 45 51.	- 13 46	90	14.	GALAXY
ZWG 328.025	04 45 54.	+ 69 06		15.7	GALAXY
MCG+00-13-013	04 45 54.	- 01 37	48	14.	GALAXY
LW 051	04 45 54.	- 67 36			STAR CLUSTER IN LMC
SL 028	04 45 54.	- 74 22	80		STAR CLUSTER IN LMC
LB 01411	04 45 55.	+ 10 15 00.		17.0	FAINT BLUE STAR
SL 032	04 45 56.	- 66 25	15		STAR CLUSTER IN LMC
IC 2094	04 45 57.	- 05 26 47.			NONSTELLAR OBJECT
MCG-04-12-013	04 45 57.	- 25 18 30.	84	14.5	GALAXY
SL 030	04 45 58.	- 68 36	25		STAR CLUSTER IN LMC
LB 01412	04 45 59.	+ 16 48 54.		14.6	FAINT BLUE STAR
RNGC 1666	04 45 59.	- 06 39		13.5	GALAXY
KHAV 087	04 46	+ 32 29	4140		DARK NEBULA
VDB.66G 228	04 46	- 29 17	130		DWARF GALAXY
UGC 03174	04 46 00.	+ 00 10	120	17.	GALAXY DWRF IR
LDN 1492	04 46 00.	+ 35 00	4500		DARK NEBULA
ZWG 394.015	04 46 00.	- 01 37		15.0	GALAXY
MCG-01-13-011	04 46 00.	- 03 57 30.	66	15.	GALAXY
MCG-01-13-010	04 46 00.	- 06 40	48	13.5	GALAXY
LB 03380	04 46 00.	- 79 13		14.8	FAINT BLUE STAR
LW 054	04 46 04.	- 67 01			STAR CLUSTER IN LMC
MCG+00-13-014	04 46 06.	+ 00 10	90	16.	GALAXY
ZWG 328.026	04 46 06.	+ 74 24		15.2	GALAXY
UGC 03175	04 46 06.	+ 74 24	108	15.2	GALAXY Sb-c
SL 031	04 46 10.	- 71 06	20		STAR CLUSTER IN LMC
LB 01413	04 46 11.	+ 14 12 42.		16.1	FAINT BLUE STAR
RNGC 1667	04 46 11.	- 06 24		13.0	GALAXY
ZC 0446.2+0235	04 46 12.	+ 02 35	1210		CLUSTER OF GALAXIES
MCG-01-13-012	04 46 12.	- 05 04 30.	66	15.	GALAXY
MCG-01-13-013	04 46 12.	- 06 26	72	13.	GALAXY
LB 01414	04 46 13.	+ 21 59 36.		16.6	FAINT BLUE STAR
MCG-01-13-014	04 46 15.	- 05 13 30.	78	15.5	GALAXY
LW 057	04 46 15.	- 66 47			STAR CLUSTER IN LMC
LW 053	04 46 15.	- 71 05			STAR CLUSTER IN LMC
IC 2095	04 46 17.	- 05 13 07.			NONSTELLAR OBJECT
LW 052	04 46 19.	- 71 42			STAR CLUSTER IN LMC
HZ 08	04 46 21.	+ 13 45		14.6	BLUE STAR
MCG-04-12-014	04 46 24.	- 24 00	60	16.	GALAXY
LH120-N183	04 46 24.	- 70 55			EMISSION NEBULA IN LMC
SL 035	04 46 25.	- 67 48	20		STAR CLUSTER IN LMC
IC 0394	04 46 26.	- 06 22			MAY NOT EXIST
HSE 026	04 46 26.	- 69 39			STAR CLUSTER IN LMC
LB 01415	04 46 27.	+ 09 36 24.		15.9	FAINT BLUE STAR
MCG-02-13-010	04 46 27.	- 14 29	30	14.5	GALAXY
SL 033	04 46 29.	- 72 27	25		STAR CLUSTER IN LMC
ZWG 306.005	04 46 30.	+ 67 07		14.8	GALAXY
UGC 03176	04 46 30.	+ 67 07	84	14.8	GALAXY E-S0
LB 01756	04 46 30.	- 55 47		10.6	FAINT BLUE STAR
LW 058	04 46 31.	- 67 47			STAR CLUSTER IN LMC
SL 029	04 46 31.	- 75 13	40		STAR CLUSTER IN LMC
LB 01416	04 46 33.	+ 08 10 06.		15.8	FAINT BLUE STAR
ZC 0446.6+0150	04 46 36.	+ 01 50	1550		CLUSTER OF GALAXIES
MCG-03-13-014	04 46 36.	- 20 27	24	15.	GALAXY
SL 034	04 46 36.	- 73 32	60		STAR CLUSTER IN LMC
LW 050	04 46 37.	- 75 12			STAR CLUSTER IN LMC
LW 056	04 46 41.	- 72 27			STAR CLUSTER IN LMC
LW 055	04 46 42.	- 73 33			STAR CLUSTER IN LMC
UGC 03177	04 46 48.	+ 21 35	84	17.	GALAXY DWARF
MCG-04-12-015	04 46 48.	- 23 49	48	15.	GALAXY
LB 03382	04 46 48.	- 68 21		14.3	FAINT BLUE STAR
ARC 0515	04 46 53.	+ 06 05		16.8	RICH CLUSTER OF GALAXIES
LB 01417	04 46 56.	+ 17 52 12.		17.0	FAINT BLUE STAR
MCG+00-13-015	04 46 57.	+ 00 10	27	14.	GALAXY
KHAV 088	04 47	+ 22 47			DARK NEBULA
IC 0395	04 47 00.	+ 00 09 42.			NONSTELLAR OBJECT
ZWG 394.016	04 47 00.	+ 00 10		13.9	GALAXY
UGC 03178	04 47 00.	+ 00 10	96	13.9	GALAXY S0
ZWG 420.003	04 47 00.	+ 03 14		15.7	GALAXY
KARA.72 103A	04 47 00.	+ 03 14	36	15.7	PART OF DOUBLE GALAXY
2ZW 023	04 47 00.	+ 03 15			COMPACT GALAXY
LB 01418	04 47 00.	+ 11 55 42.		15.2	FAINT BLUE STAR
ZC 0447.0+3149	04 47 00.	+ 31 49	2150		CLUSTER OF GALAXIES
LDN 1435	04 47 00.	+ 51 30	1500		DARK NEBULA
LDN 1433	04 47 00.	+ 51 50	540		DARK NEBULA
LDN 1419	04 47 00.	+ 54 40	1140		DARK NEBULA
MCG-02-13-011	04 47 00.	- 12 50	78	15.5	GALAXY
LB 03384	04 47 00.	- 67 11		12.3	FAINT BLUE STAR
LB 03385	04 47 00.	- 78 58		14.3	FAINT BLUE STAR
HOLM 080A	04 47 04.	+ 00 10	18	13.4	PART OF MULTIPLE GALAXY
HOLM 080B	04 47 06.	+ 00 10	18	14.0	PART OF MULTIPLE GALAXY
MCG-05-12-002	04 47 06.	- 32 55	18	16.	GALAXY
IC 2096	04 47 09.	- 05 04 26.			NONSTELLAR OBJECT
SL 036	04 47 11.	- 75 00	20		STAR CLUSTER IN LMC
ZWG 420.004	04 47 12.	+ 03 15		14.8	GALAXY
UGC 03179	04 47 12.	+ 03 15	66	14.8	GALAXY VY CMPT
KARA.72 103B	04 47 12.	+ 03 15	42	14.8	PART OF DOUBLE GALAXY
ISS 0151	04 47 12.	+ 46 40	358		STELLAR RING
MCG+00-13-016	04 47 12.	- 02 50	27	14.	GALAXY
HOLM 081A	04 47 12.	- 02 51	30	13.9	PART OF MULTIPLE GALAXY
MCG-01-13-015	04 47 12.	- 07 58 30.	30	15.5	GALAXY
MCG-05-12-003	04 47 12.	- 29 17	90	14.5	GALAXY
RNGC 1680	04 47 12.	- 47 54			UNVERIFIED SOUTHERN OBJECT
HOLM 081B	04 47 13.	- 02 52	30	13.9	PART OF MULTIPLE GALAXY
SL 038	04 47 13.	- 66 12	15		STAR CLUSTER IN LMC
LW 063	04 47 17.	- 67 11			STAR CLUSTER IN LMC
HZ 01	04 47 18.	+ 17 37		12.7	VERY BLUE STAR
ZWG 394.017	04 47 18.	- 02 50		14.1	GALAXY
MCG-02-13-012	04 47 18.	- 10 48	48	14.5	GALAXY
LB 01757	04 47 18.	- 52 59		13.9	FAINT BLUE STAR
LW 059	04 47 19.	- 72 40			STAR CLUSTER IN LMC
RNGC 1670	04 47 21.	- 02 50		14.0	GALAXY
SL 037	04 47 24.	- 72 32	75		STAR CLUSTER IN LMC
LB 03386	04 47 24.	- 73 31		12.9	FAINT BLUE STAR
RNGC 1664	04 47 25.	+ 43 37		8.0	OPEN CLUSTER
OCL 0411	04 47 30.	+ 43 37	1680	8.7	OPEN STAR CLUSTER
MCG-02-13-013	04 47 30.	- 10 50	60	15.	GALAXY
MCG-03-13-015	04 47 30.	- 17 35	48	15.	GALAXY
SER 040.08	04 47 30.	- 59 32	40	15.5	HIGH SURFACE BRIGHT. GLXY
LB 01419	04 47 35.	+ 18 19 06.		15.8	FAINT BLUE STAR
LW 060	04 47 35.	- 74 59			STAR CLUSTER IN LMC
LW 061	04 47 36.	- 72 29			STAR CLUSTER IN LMC
LB 01420	04 47 38.	+ 19 21 42.		13.5	FAINT BLUE STAR
RNGC 1688	04 47 38.	- 59 53		12.5	GALAXY
ZWG 420.005	04 47 42.	+ 08 37		15.0	GALAXY
UGC 03180	04 47 42.	+ 08 37	66	15.0	GALAXY SB
MCG+01-13-003	04 47 42.	+ 08 40	48	14.	GALAXY
LB 01422	04 47 44.	+ 10 16 36.		15.2	FAINT BLUE STAR
LB 01421	04 47 44.	+ 14 11 48.		15.3	FAINT BLUE STAR
SCHO 0069	04 47 44.	+ 45 50 12.	790		ISOLATED DARK CLOUD
RNGC 1671	04 47 44.	- 00 52			NON-EXISTENT OBJECT
RNGC 1693	04 47 44.	- 69 27			OPEN CLUSTER IN LMC
SL 039	04 47 44.	- 69 27	20		STAR CLUSTER IN LMC
ARC 0516	04 47 45.	- 08 55		17.5	RICH CLUSTER OF GALAXIES
MCG-04-12-016	04 47 48.	- 25 53 30.	60	16.	GALAXY
LB 01758	04 47 48.	- 53 10		13.8	FAINT BLUE STAR
LW 062	04 47 48.	- 74 15			STAR CLUSTER IN LMC
RNGC 1695	04 47 50.	- 69 30			OPEN CLUSTER IN LMC
SL 040	04 47 50.	- 69 30	40		STAR CLUSTER IN LMC
HSE 027	04 47 50.	- 69 33			STAR CLUSTER IN LMC
SL 042	04 47 51.	- 66 52	15		STAR CLUSTER IN LMC
IC 2097	04 47 53.	- 05 10 19.			NONSTELLAR OBJECT
HOLM 082B	04 47 54.	+ 05 55	18	14.6	PART OF MULTIPLE GALAXY
MCG+01-13-004	04 47 54.	+ 05 57	120	13.	GALAXY
FATH 1.162	04 47 54.	- 00 28	27		NEBULA
MCG-01-13-016	04 47 54.	- 03 36	66	14.5	GALAXY
MCG-04-12-017	04 47 54.	- 23 58 30.	48	14.5	GALAXY
HOLM 082A	04 47 58.	+ 05 55	60	14.2	PART OF MULTIPLE GALAXY
LB 01423	04 47 59.	+ 08 59 30.		16.8	FAINT BLUE STAR
BC PKS0448-392	04 47 59.6	- 39 16 13.		16.	QUASI-STELLAR OBJECT

OBJECT NAME	RIGHT ASCEN.	DECLINATION	DIAM.	MAGN.	TYPE OF OBJECT
KHAV 092	04 48	+ 37 41	7960		DARK NEBULA
KHAV 091	04 48	+ 44 53	8180		DARK NEBULA
LBN 0755	04 48	+ 45 40	540		BRIGHT NEBULA
KHAV 090	04 48	+ 45 47	1410		DARK NEBULA
KHAV 089	04 48	+ 51 53	4910		DARK NEBULA
LB 09822	04 48	- 81 03		13.2	FAINT BLUE STAR
ZWG 420.006	04 48 00.	+ 05 55		14.5	GALAXY
UGC 03181	04 48 00.	+ 05 55	144	14.5	GALAXY SBb
LDN 1561	04 48 00.	+ 12 12	2100		DARK NEBULA
LDN 1537	04 48 00.	+ 26 20	2460		DARK NEBULA
LDN 1432	04 48 00.	+ 52 00	540		DARK NEBULA
MCG+13-04-009	04 48 00.	+ 75 30	27	17.	GALAXY
MCG-03-13-016	04 48 00.	- 17 23	108	14.	GALAXY
RNGC 1679	04 48 03.	- 32 04		13.0	GALAXY
RLWT 109	04 48 06.	+ 39 36		13.5	FAINT VERY BLUE STAR
MCG+12-05-026	04 48 06.	+ 72 13	27	14.	GALAXY
MCG-01-13-017	04 48 06.	- 03 13	45	15.	GALAXY
MCG-05-12-004	04 48 06.	- 32 04	138	13.	GALAXY
SL 041	04 48 07.	- 72 43	75		STAR CLUSTER IN LMC
ARC 0517	04 48 08.	- 09 20		17.6	RICH CLUSTER OF GALAXIES
HELW 409	04 48 09.	- 03 13 58.			NEBULA
RNGC 1696	04 48 09.	- 68 22			OPEN CLUSTER IN LMC
LB 01424	04 48 10.	+ 17 26 42.		16.1	FAINT BLUE STAR
ZC 0448.2+0919	04 48 12.	+ 09 19	2020		CLUSTER OF GALAXIES
ZWG 328.027	04 48 12.	+ 72 14		14.5	GALAXY
UGC 03182	04 48 12.	+ 72 14	90	14.5	GALAXY S0
MCG-02-13-014	04 48 12.	- 11 30 30.	60	14.5	GALAXY
LB 01425	04 48 13.	+ 21 59 54.		15.2	FAINT BLUE STAR
FATH 1.163	04 48 13.	- 00 26	14		NEBULA
IC 2098	04 48 15.	- 05 30 40.			NONSTELLAR OBJECT
MCG-01-13-018	04 48 15.	- 05 31	120	14.5	GALAXY
SL 043	04 48 15.	- 68 22	20		STAR CLUSTER IN LMC
HSE 028	04 48 17.	- 68 53			STAR CLUSTER IN LMC
UGC 03183	04 48 18.	+ 23 06	72	16.0	GALAXY Sb
MCG+13-04-010	04 48 18.	+ 75 04	18	16.	GALAXY
ZWG 394.018	04 48 18.	- 00 26		15.7	GALAXY
LW 064	04 48 19.	- 72 41			STAR CLUSTER IN LMC
MCG-01-13-019	04 48 21.	- 04 59	36	15.	GALAXY
IC 2099	04 48 21.	- 04 59 05.			NONSTELLAR OBJECT
RNGC 1677	04 48 22.	- 04 57		15.0	GALAXY
RLWT 110	04 48 24.	+ 39 42		13.5	FAINT VERY BLUE STAR
B 025	04 48 24.	+ 45 56	480		DARK OBJECT
WS 02	04 48 29.	- 72 33 28.		17.05	PLANETARY NEB. IN LMC
MCG+00-13-017	04 48 30.	+ 02 17	48	15.	GALAXY
ZC 0448.5+0551	04 48 30.	+ 05 51	870		CLUSTER OF GALAXIES
ZWG 347.008	04 48 30.	+ 75 35		15.6	GALAXY
LH120-N184	04 48 30.	- 72 30			EMISSION NEBULA IN LMC
LB 01426	04 48 31.	+ 14 41 06.		15.8	FAINT BLUE STAR
SL 044	04 48 35.	- 68 49	75		STAR CLUSTER IN LMC
ZWG 394.019	04 48 36.	+ 02 17		15.7	GALAXY
ISS 0255	04 48 36.	+ 33 41	445		STELLAR RING
MCG-01-13-020	04 48 36.	- 06 04	12	15.5	GALAXY
HSE 029	04 48 36.	- 69 07			STAR CLUSTER IN LMC
LB 01427	04 48 37.	+ 18 56 54.		16.1	FAINT BLUE STAR
LB 01428	04 48 38.	+ 07 43 42.		15.6	FAINT BLUE STAR
FATH 1.164	04 48 41.	+ 00 20	27		NEBULA
LDN 1490	04 48 42.	+ 35 35	1080		DARK NEBULA
LB 01759	04 48 42.	- 52 05		13.6	FAINT BLUE STAR
LH120-N077F	04 48 42.	- 69 13	22		EMISSION NEBULA IN LMC
RNGC 1697	04 48 46.	- 68 38			OPEN CLUSTER IN LMC
IC 2100	04 48 47.	- 04 54 50.			NONSTELLAR OBJECT
LB 00237	04 48 48.	+ 18 09 00.		17.4	FAINT BLUE STAR
DG 047	04 48 48.	+ 32 52	60		REFLECTION NEBULA
MCG-01-13-021	04 48 48.	- 06 04	60	15.	GALAXY
SS 36	04 48 50.	+ 23 52			DIFFUSE GALACTIC NEBULA
ZWG 420.007	04 48 54.	+ 05 38		15.4	GALAXY
MCG-01-13-022	04 48 54.	- 03 12	90	15.	GALAXY
VDB.66N 030	04 48 55.	+ 66 13	2760		REFLECTION NEBULA
HSE 030	04 48 56.	- 69 29			STAR CLUSTER IN LMC
REIN 6.084	04 48 56.81	- 03 12 22.2			NEBULA
HELW 410	04 48 57.	- 03 12 31.			NEBULA
MCG-02-13-015	04 48 57.	- 13 56	72	15.	GALAXY
ARC 0518	04 48 59.	- 10 48		17.5	RICH CLUSTER OF GALAXIES
KHAV 093	04 49	+ 27 59	6990		DARK NEBULA
UGC 03184	04 49 00.	+ 05 23	90	16.0	GALAXY S
LDN 1513	04 49 00.	+ 30 45	960		DARK NEBULA
LDN 1465	04 49 00.	+ 45 55	1080		DARK NEBULA
ZWG 306.006	04 49 00.	+ 63 31		15.7	GALAXY
UGC 03185	04 49 00.	+ 64 24	78	16.5	GALAXY Sc
MCG+00-13-018	04 49 00.	- 02 39	30	14.5	GALAXY
MCG+00-13-019	04 49 00.	- 02 42	24	13.5	GALAXY
REIN 6.085	04 49 04.88	- 02 42 25.2			NEBULA
UGC 03186	04 49 06.	+ 03 35	102	16.0	GALAXY Sc
ZWG 420.008	04 49 06.	+ 06 45		15.7	GALAXY
UGC 03187	04 49 06.	+ 06 45	90	15.7	GALAXY Sc
ZWG 420.009	04 49 06.	+ 08 45		15.0	GALAXY
UGC 03188	04 49 06.	+ 08 45	66	15.0	GALAXY S
MCG+01-13-005	04 49 06.	+ 08 48	60	14.	GALAXY
ZWG 394.021	04 49 06.	- 02 39		15.1	GALAXY
ZWG 394.020	04 49 06.	- 02 43		14.4	GALAXY
HSE 031	04 49 06.	- 69 00			STAR CLUSTER IN LMC
LH120-N077D	04 49 06.	- 69 15	68		EMISSION NEBULA IN LMC
SL 045	04 49 07.	- 69 14	40		STAR CLUSTER IN LMC
HSE 032	04 49 07.	- 69 21			STAR CLUSTER IN LMC
RNGC 1678	04 49 09.	- 02 43		14.5	GALAXY
7ZW 020	04 49 12.	+ 68 18			COMPACT GALAXY
MCG+12-05-027	04 49 12.	+ 69 37 30.	96	15.	GALAXY
ZWG 328.028	04 49 12.	+ 69 39		15.4	GALAXY
UGC 03189	04 49 12.	+ 69 39	132	15.4	GALAXY Sc
MCG-01-13-025	04 49 12.	- 03 53	24	15.	GALAXY
MCG-01-13-024	04 49 12.	- 06 20	90	14.5	GALAXY
MCG-01-13-023	04 49 12.	- 07 23	48	17.	GALAXY
MCG-02-13-016	04 49 12.	- 14 41 30.	30	15.	GALAXY
MCG-03-13-018	04 49 12.	- 15 03	48	15.	GALAXY
MCG-03-13-017	04 49 12.	- 17 36	42	15.	GALAXY
LH120-N076	04 49 12.	- 68 29	111		EMISSION NEBULA IN LMC
LB 01429	04 49 14.	+ 08 01 00.		16.3	FAINT BLUE STAR
IC 2101	04 49 15.	- 06 18 47.			NONSTELLAR OBJECT
RNGC 1681	04 49 16.	- 05 54		13.5	GALAXY
LB 01430	04 49 18.	+ 20 02 36.		14.5	FAINT BLUE STAR
ZC 0449.3-0437	04 49 18.	- 04 37	14920		CLUSTER OF GALAXIES
FATH 1.165	04 49 21.	+ 00 12	8		NEBULA
MCG-01-13-026	04 49 21.	- 05 54	60	13.5	GALAXY
LB 01431	04 49 22.	+ 10 53 12.		15.2	FAINT BLUE STAR
RNGC 1675	04 49 23.	+ 23 49			NON-EXISTENT OBJECT
RNGC 1674	04 49 23.	+ 23 49			NON-EXISTENT OBJECT
ZC 0449.4+7543	04 49 24.	+ 75 43	3560		CLUSTER OF GALAXIES
MCG-01-13-027	04 49 24.	- 05 02 30.	66	15.	GALAXY
IC 2102	04 49 24.	- 05 03 18.			NONSTELLAR OBJECT
LB 01760	04 49 24.	- 48 44		14.4	FAINT BLUE STAR
LB 03387	04 49 24.	- 74 00		14.5	FAINT BLUE STAR
RNGC 1687	04 49 29.	- 34 01			GALAXY
ZWG 347.009	04 49 30.	+ 78 07		12.8	GALAXY
UGC 03190	04 49 30.	+ 78 07	108	12.8	GALAXY S
KARA.73B 0155	04 49 30.	+ 78 07	66	12.8	ISOLATED GALAXY IR
LB 03388	04 49 30.	- 68 45		14.4	FAINT BLUE STAR
SCHO 0070	04 49 32.	+ 43 54 12.	320		ISOLATED DARK CLOUD
HELW 411	04 49 34.	- 03 01 22.			NEBULA
MCG+13-04-011	04 49 36.	+ 78 06	78	12.8	GALAXY
MCG-06-11-005	04 49 36.	- 34 03	72	14.5	GALAXY
RNGC 1698	04 49 37.	- 69 16			DIFFUSE NEBULA IN LMC
HN 0084	04 49 39.	- 69 16			NEBULA
IC 2105	04 49 39.8	- 69 17 20.			HII REGION IN LMC
RNGC 1702	04 49 40.	- 69 57			OPEN CLUSTER IN LMC
SL 046	04 49 40.	- 69 58	40		STAR CLUSTER IN LMC
FATH 1.166	04 49 41.	+ 00 03	11		NEBULA
HSE 033	04 49 41.	- 70 00			STAR CLUSTER IN LMC
ZWG 420.010	04 49 42.	+ 05 04		15.5	GALAXY
LH120-N077A	04 49 42.	- 69 17	32		EMISSION NEBULA IN LMC
SL 047	04 49 42.	- 70 20	25		STAR CLUSTER IN LMC
RNGC 1683	04 49 45.	- 03 05			GALAXY
RNGC 1682	04 49 45.	- 03 10		14.0	GALAXY
MCG-01-13-028	04 49 45.	- 03 10 30.	18	14.	GALAXY
LB 01432	04 49 46.	+ 17 26 48.		16.3	FAINT BLUE STAR
KARA.68 047	04 49 48.	+ 23 06	60		DWARF GALAXY
HELW 412	04 49 48.	- 03 10 23.			NEBULA
MCG-01-13-029	04 49 48.	- 06 15	54	15.	GALAXY
LB 03389	04 49 48.	- 69 16		13.0	FAINT BLUE STAR
REIN 6.086	04 49 48.17	- 03 05 20.1			NEBULA
REIN 6.087	04 49 49.63	- 03 11 19.0			NEBULA
MCG-01-13-030	04 49 51.	- 03 38 30.	24	14.	GALAXY
IC 0391	04 49 52.	+ 78 06	54	12.7	GALAXY SA(s)
FATH 1.167	04 49 55.	+ 00 27	8		NEBULA
RNGC 1684	04 49 57.	- 03 10		13.0	GALAXY
MCG-01-13-031	04 49 57.	- 03 10 30.	48	13.	GALAXY
HSE 034	04 49 57.	- 69 35			STAR CLUSTER IN LMC
HELW 413	04 49 58.	- 03 38 59.			NEBULA
HSE 035	04 49 58.	- 69 50			STAR CLUSTER IN LMC
FATH 1.168	04 49 59.	+ 00 11	41		NEBULA
MCG+01-13-006	04 50 00.	+ 04 19	60	14.	GALAXY
LDN 1515	04 50 00.	+ 30 50	1500		DARK NEBULA
LDN 1488	04 50 00.	+ 36 37	3060		DARK NEBULA
LDN 1436	04 50 00.	+ 51 40	2220		DARK NEBULA
MCG-01-13-032	04 50 00.	- 03 01	60	14.5	GALAXY
MCG-05-12-005	04 50 00.	- 28 40	60	15.	GALAXY
LH120-N003	04 50 00.	- 67 46	573		EMISSION NEBULA IN LMC
LH120-N077E	04 50 00.	- 69 18	370		EMISSION NEBULA IN LMC
LH120-N077B	04 50 00.	- 69 18	34		EMISSION NEBULA IN LMC
REIN 6.088	04 50 00.93	- 03 11 20.8			NEBULA
RNGC 1685	04 50 03.	- 03 01		14.5	GALAXY
RNGC 1704	04 50 04.	- 69 51			OPEN CLUSTER IN LMC
HSE 036	04 50 05.	- 70 09			STAR CLUSTER IN LMC
LW 065	04 50 05.	- 73 17			STAR CLUSTER IN LMC
ZWG 420.011	04 50 06.	+ 04 19		15.3	GALAXY
UGC 03191	04 50 06.	+ 04 19	90	15.3	GALAXY Sb
MCG-06-11-006	04 50 06.	- 33 17	156	14.	GALAXY
SL 049	04 50 06.	- 69 06	35		STAR CLUSTER IN LMC
FATH 1.169	04 50 07.	+ 00 26	33		NEBULA
SL 050	04 50 10.	- 69 52	60		STAR CLUSTER IN LMC
SL 051	04 50 11.	- 67 11	20		STAR CLUSTER IN LMC
ZWG 394.022	04 50 12.	+ 01 10		14.6	GALAXY
UGC 03192	04 50 12.	+ 01 10	96	14.6	GALAXY E-S0
MCG+00-13-020	04 50 12.	+ 01 12	18	14.5	GALAXY
MCG-04-12-018	04 50 12.	- 23 04	36	15.	GALAXY
MCG-05-12-006	04 50 12.	- 33 05	24	15.	GALAXY
REIN 6.089	04 50 13.27	- 02 59 23.1			NEBULA
LB 01433	04 50 14.	+ 15 55 00.		16.8	FAINT BLUE STAR
REIN 6.090	04 50 14.65	- 02 57 23.6			NEBULA
MCG+00-13-021	04 50 15.	+ 02 59	96	14.5	GALAXY
MCG-02-13-017	04 50 15.	- 13 59	36	15.	GALAXY
LW 066	04 50 16.	- 73 56			STAR CLUSTER IN LMC
ZWG 394.023	04 50 18.	+ 02 58		14.7	GALAXY
UGC 03193	04 50 18.	+ 02 58	108	14.7	GALAXY SBb
RLWT 111	04 50 18.	+ 43 32		14.	FAINT VERY BLUE STAR
MCG+11-07-001	04 50 18.	+ 67 40	42	17.	GALAXY
LB 03390	04 50 18.	- 67 44		12.2	FAINT BLUE STAR
LH120-N077C	04 50 16.	- 69 17	25		EMISSION NEBULA IN LMC
FATH 1.170	04 50 21.	+ 00 30	16		NEBULA
SL 048	04 50 21.	- 72 54	20		STAR CLUSTER IN LMC
LW 067	04 50 22.	- 72 08			STAR CLUSTER IN LMC
ZWG 394.024	04 50 24.	+ 01 15		15.4	GALAXY
UGC 03194	04 50 24.	+ 01 15	108	15.4	GALAXY SBb
MCG+00-13-022	04 50 24.	+ 01 16	90	15.	GALAXY
MCG+01-13-007	04 50 24.	+ 04 18	60	15.	GALAXY
ZWG 420.012	04 50 24.	+ 05 43		15.7	GALAXY
FATH 1.171	04 50 25.	- 00 03	8		NEBULA
HELW 414	04 50 26.	- 03 09 19.			NEBULA
LW 068	04 50 26.	- 72 53			STAR CLUSTER IN LMC
FATH 1.172	04 50 27.	- 00 01	8		NEBULA
LW 069	04 50 29.	- 72 20			STAR CLUSTER IN LMC
ZWG 420.013	04 50 30.	+ 04 18		15.6	GALAXY
UGC 03195	04 50 30.	+ 04 18	90	15.6	GALAXY Sc
LH120-N078	04 50 30.	- 69 39			EMISSION NEBULA IN LMC
SL 052	04 50 32.	- 71 45	20		STAR CLUSTER IN LMC
LW 070	04 50 33.	- 71 59			STAR CLUSTER IN LMC
RNGC 1686	04 50 34.	- 15 26		14.0	GALAXY
ZWG 394.025	04 50 36.	+ 02 15		15.1	GALAXY
MCG+00-13-023	04 50 36.	+ 02 17	36	15.	GALAXY
UGC 03196	04 50 36.	+ 67 40	72	16.5	GALAXY Sc
MCG-03-13-019	04 50 39.	- 15 26	102	14.5	GALAXY
HSE 037	04 50 41.	- 68 47			STAR CLUSTER IN LMC
ZWG 420.014	04 50 42.	+ 07 03		14.9	GALAXY
LW 072	04 50 44.	- 71 44			STAR CLUSTER IN LMC
MCG-01-13-033	04 50 45.	- 03 05	72	15.	GALAXY
HELW 415	04 50 47.	- 03 05 15.			NEBULA
LW 071	04 50 47.	- 73 13			STAR CLUSTER IN LMC
ZWG 420.015	04 50 46.	+ 03 59		15.0	GALAXY
MCG-04-12-019	04 50 46.	- 25 18 30.	240	13.5	GALAXY
LB 03391	04 50 48.	- 65 56		14.2	FAINT BLUE STAR
RNGC 1711	04 50 53.	- 70 04		10.0	OPEN CLUSTER IN LMC
LDN 1494	04 50 54.	+ 35 08	420		DARK NEBULA
MCG-01-13-034	04 50 54.	- 04 53	60	16.	GALAXY
SL 057	04 50 54.	- 65 41	25		STAR CLUSTER IN LMC
ARP 180	04 50 55.	- 04 52			PECULIAR GALAXY
LW 074	04 50 55.	- 71 33			STAR CLUSTER IN LMC
SCHO 0071	04 50 58.	+ 43 25 42.	430		ISOLATED DARK CLOUD

OBJECT NAME	RIGHT ASCEN.	DECLINATION	DIAM.	MAGN.	TYPE OF OBJECT
SL 055	04 50 59.	- 70 04	205		STAR CLUSTER IN LMC
SL 056	04 50 59.	- 70 09	35		STAR CLUSTER IN LMC
KHAV 094	04 51	+ 31 17	2110		DARK NEBULA
VDB.66G 229	04 51	- 25 22	170		DWARF GALAXY
ZWG 394.026	04 51 00.	+ 02 22		15.7	GALAXY
ZWG 394.027	04 51 00.	+ 02 25		15.7	GALAXY
LDN 1493	04 51 00.	+ 35 35	2760		DARK NEBULA
ZWG 347.010	04 51 00.	+ 80 05		15.5	GALAXY
UGC 03197	04 51 00.	+ 80 05	84	15.5	GALAXY E?
LW 077	04 51 00.	- 65 41			STAR CLUSTER IN LMC
LH120-N079	04 51 06.	- 69 30	1038		EMISSION NEBULA IN LMC
LB 01434	04 51 07.	+ 18 12 36.		16.8	FAINT BLUE STAR
HELW 416	04 51 07.	- 03 09 58.			NEBULA
LH 001	04 51 08.	- 69 29	450		STELLAR ASSN. IN LMC
SL 054	04 51 08.	- 72 49	35		STAR CLUSTER IN LMC
ARC 0519	04 51 11.	+ 00 37		17.0	RICH CLUSTER OF GALAXIES
RNGC 1689	04 51 11.	- 06 24			NON-EXISTENT OBJECT
ZC 0451.2+0041	04 51 12.	+ 00 41	1950		CLUSTER OF GALAXIES
MCG+01-13-008	04 51 12.	+ 03 30	48	14.5	GALAXY
ZWG 420.016	04 51 12.	+ 03 40		15.2	GALAXY
HSE 038	04 51 13.	- 67 37			STAR CLUSTER IN LMC
SL 053	04 51 13.	- 75 45	35		STAR CLUSTER IN LMC
LW 075	04 51 14.	- 73 44			STAR CLUSTER IN LMC
SL 058	04 51 15.	- 69 43	25		STAR CLUSTER IN LMC
MCG+00-13-025	04 51 18.	+ 01 05	30	15.5	GALAXY
ZWG 394.028	04 51 18.	+ 01 36		15.7	GALAXY
MCG+00-13-024	04 51 18.	+ 01 36	18	15.	GALAXY
ZC 0451.3+0159	04 51 18.	+ 01 59	5650		CLUSTER OF GALAXIES
ZWG 420.017	04 51 18.	+ 03 30		15.3	GALAXY
ZWG 328.029	04 51 18.	+ 72 20		15.2	GALAXY
MCG+13-04-012	04 51 18.	+ 80 07	60	15.	GALAXY
FATH 1.173	04 51 19.	+ 00 12	14		NEBULA
LW 073	04 51 19.	- 75 47			STAR CLUSTER IN LMC
HSE 039	04 51 20.	- 69 19			STAR CLUSTER IN LMC
LB 01435	04 51 21.	+ 17 38 30.		16.8	FAINT BLUE STAR
MCG-03-13-020	04 51 24.	- 18 01 30.	48	15.5	GALAXY
LB 03392	04 51 24.	- 70 22		13.2	FAINT BLUE STAR
B 026	04 51 27.	+ 30 33	300		DARK OBJECT
MCG-03-13-021	04 51 27.	- 18 01 30.	9	15.5	GALAXY
HSE 040	04 51 29.	- 70 04			STAR CLUSTER IN LMC
HSE 041	04 51 30.	- 67 32			STAR CLUSTER IN LMC
LW 076	04 51 32.	- 73 43			STAR CLUSTER IN LMC
SCHO 0072	04 51 33.	+ 37 51 00.	490		ISOLATED DARK CLOUD
RNGC 1712	04 51 33.	- 69 32			CLUSTER/NEBULOSITY IN LMC
SL 060	04 51 33.	- 69 32	270		STAR CLUSTER IN LMC
ZC 0451.6+0251	04 51 36.	+ 02 51	1480		CLUSTER OF GALAXIES
OCL 0402	04 51 36.	+ 52 40	480	15.	OPEN STAR CLUSTER
LB 01761	04 51 36.	- 53 57		13.3	FAINT BLUE STAR
ARC 0505	04 51 38.	+ 79 56		15.2	RICH CLUSTER OF GALAXIES
MCG+00-13-026	04 51 39.	+ 01 35 30.	60	15.	GALAXY
ARC 0520	04 51 40.	+ 02 53		17.4	RICH CLUSTER OF GALAXIES
ZWG 394.029	04 51 42.	+ 01 33		15.0	GALAXY
RNGC 1690	04 51 42.	+ 01 33		15.0	GALAXY
UGC 03198	04 51 42.	+ 01 33	72	15.0	GALAXY E
MCG+00-13-027	04 51 42.	+ 01 34	18	13.5	GALAXY
ZWG 394.030	04 51 42.	+ 01 35		15.5	GALAXY
UGC 03199	04 51 42.	+ 01 35	66	15.5	GALAXY SBb
ZWG 394.031	04 51 42.	+ 02 03		15.2	GALAXY
UGC 03200	04 51 42.	+ 02 03	78	15.2	GALAXY SBb
ZWG 420.018	04 51 42.	+ 03 59		15.2	GALAXY
SL 062	04 51 42.	- 65 41	15		STAR CLUSTER IN LMC
HSE 042	04 51 42.	- 65 50			STAR CLUSTER IN LMC
LH120-N004F	04 51 42.	- 67 00	72		EMISSION NEBULA IN LMC
FATH 1.174	04 51 43.	+ 00 02	41		NEBULA
LW 078	04 51 43.	- 72 40			STAR CLUSTER IN LMC
MCG+00-13-028	04 51 45.	+ 02 05	60	14.5	GALAXY
SL 059	04 51 45.	- 73 45	100		STAR CLUSTER IN LMC
ARC 0521	04 51 48.	- 10 20		17.6	RICH CLUSTER OF GALAXIES
LW 080	04 51 48.	- 65 41			STAR CLUSTER IN LMC
LB 03393	04 51 48.	- 67 48		14.1	FAINT BLUE STAR
SCHO 0073	04 51 49.	+ 37 40 24.	520		ISOLATED DARK CLOUD
LB 01436	04 51 52.	+ 14 48 24.		14.7	FAINT BLUE STAR
MCG+01-13-009	04 51 54.	+ 03 10	96	13.	GALAXY
ZC 0451.9+2711	04 51 54.	+ 27 11	400		CLUSTER OF GALAXIES
MCG-02-13-018	04 51 54.	- 11 51	72	15.	GALAXY
MCG-05-12-007	04 51 54.	- 32 32	36	15.5	GALAXY
HSE 043	04 51 54.	- 68 50			STAR CLUSTER IN LMC
LB 01437	04 51 55.	+ 17 05 48.		16.3	FAINT BLUE STAR
LB 01438	04 51 57.	+ 09 12 12.		15.9	FAINT BLUE STAR
B 027	04 51 57.	+ 30 29	300		DARK OBJECT
RNGC 1691	04 51 59.	+ 03 11		13.0	GALAXY
SL 063	04 51 59.	- 65 29	35		STAR CLUSTER IN LMC
KHAV 095	04 52	+ 32 17	7160		DARK NEBULA
KHAV 096	04 52	+ 35 11	5110		DARK NEBULA
LBN 0754	04 52	+ 46 20	2220		BRIGHT NEBULA
LBN 0747	04 52	+ 47 19	120		BRIGHT NEBULA
ZWG 420.019	04 52 00.	+ 03 11		13.2	GALAXY
UGC 03201	04 52 00.	+ 03 11	108	13.2	GALAXY SB0/SBa
LDN 1517	04 52 00.	+ 30 30	900		DARK NEBULA
LDN 1438	04 52 00.	+ 51 30	1620		DARK NEBULA
MCG+00-13-029	04 52 00.	- 01 27 30.	24	15.	GALAXY
CM 07	04 52 00.	- 69 18			HII REGION IN LMC
MCG+00-13-030	04 52 03.	+ 01 58	36	15.	GALAXY
RNGC 1706	04 52 04.	- 63 04			UNVERIFIED SOUTHERN OBJECT
RNGC 1715	04 52 04.	- 66 59			DIFFUSE NEBULA IN LMC
RNGC 1714	04 52 04.	- 67 00			DIFFUSE NEBULA IN LMC
SL 064	04 52 04.	- 67 00	45		STAR CLUSTER IN LMC
SG 2.013	04 52 05.	+ 26 05	5400		DIFFUSE EMISSION NEBULA
LW 081	04 52 05.	- 65 24			STAR CLUSTER IN LMC
ZWG 394.033	04 52 06.	+ 01 15		15.7	GALAXY
MCG+00-13-031	04 52 06.	+ 01 16 30.	30	16.	GALAXY
ZC 0452.1+0627	04 52 06.	+ 06 27	4370		CLUSTER OF GALAXIES
ZWG 394.032	04 52 06.	- 01 27		15.5	GALAXY
LH120-N004B	04 52 06.	- 66 59	60		EMISSION NEBULA IN LMC
LH120-N004A	04 52 06.	- 67 00	53		EMISSION NEBULA IN LMC
LH120-N079A	04 52 06.	- 69 29	78		EMISSION NEBULA IN LMC
FATH 1.175	04 52 07.	+ 00 09	14		NEBULA
RNGC 1703	04 52 08.	- 59 48			GALAXY
HSE 046	04 52 08.	- 66 12			STAR CLUSTER IN LMC
IC 2111	04 52 08.9	- 69 28 28.			HII REGION IN LMC
HSE 044	04 52 09.	- 70 47			STAR CLUSTER IN LMC
HSE 047	04 52 10.	- 65 02			STAR CLUSTER IN LMC
RNGC 1718	04 52 11.	- 67 08			OPEN CLUSTER IN LMC
SL 065	04 52 11.	- 67 08	55		STAR CLUSTER IN LMC
ZWG 394.034	04 52 12.	+ 01 05		15.7	GALAXY
MCG+00-13-032	04 52 12.	+ 01 05	30	15.	GALAXY
ZC 0452.2+7305	04 52 12.	+ 73 05	12970		CLUSTER OF GALAXIES
MCG-03-13-022	04 52 12.	- 19 01 30.	12	15.5	GALAXY

OBJECT NAME	RIGHT ASCEN.	DECLINATION	DIAM.	MAGN.	TYPE OF OBJECT
MCG-03-13-023	04 52 12.	- 19 02	15	15.	GALAXY
SC 0450-1634.1	04 52 18.	- 16 29 09.	12		NEBULA
LH120-N004C	04 52 18.	- 67 00	31		EMISSION NEBULA IN LMC
LH120-N079C	04 52 18.	- 69 26	51		EMISSION NEBULA IN LMC
LH120-N079B	04 52 18.	- 69 29	21		EMISSION NEBULA IN LMC
SL 061	04 52 18.	- 75 39	100		STAR CLUSTER IN LMC
SG 2.012	04 52 19.	+ 47 19	150		DIFFUSE EMISSION NEBULA
HN 0085	04 52 19.	- 69 28			NEBULA
FATH 1.176	04 52 21.	+ 00 07	14		NEBULA
MCG-02-13-020	04 52 21.	- 10 21	60	15.	GALAXY
MCG-02-13-019	04 52 21.	- 12 17	120	14.5	GALAXY
HSE 048	04 52 22.	- 68 32			STAR CLUSTER IN LMC
LB 01439	04 52 23.	+ 21 27 12.		15.6	FAINT BLUE STAR
SC 0450-1634.3	04 52 23.	- 16 29 21.	12		NEBULA
LW 079	04 52 23.	- 75 36			STAR CLUSTER IN LMC
MFSL 159+02/1	04 52 24.	+ 47 19	180		HII REGION
ZWG 394.035	04 52 24.	- 00 09		15.6	GALAXY
MCG+00-13-033	04 52 24.	- 00 09	24	15.5	GALAXY
SC 0450-1611.9	04 52 25.	- 16 06 58.	18		NEBULA
SL 066	04 52 25.	- 70 28	25		STAR CLUSTER IN LMC
RNGC 1727	04 52 26.	- 69 25			CLUSTER/NEBULOSITY IN LMC
SL 067	04 52 26.	- 69 25	150		STAR CLUSTER IN LMC
LH 002	04 52 26.	- 69 25	180		STELLAR ASSN. IN LMC
RNGC 1722	04 52 26.	- 69 28			CLUSTER/NEBULOSITY IN LMC
FATH 1.177	04 52 27.	- 00 09	24		NEBULA
SC 0450-1628.6	04 52 28.	- 16 23 40.	12		NEBULA
MRSL 193-22/1	04 52 30.	+ 05 35	1320		HII REGION
VDB.66N 031	04 52 30.	+ 30 28	540		REFLECTION NEBULA
LDN 1519	04 52 30.	+ 30 30	960		DARK NEBULA
CS 2	04 52 30.	+ 65 30	27000		FAINT H-ALPHA REGION
MCG-01-13-035	04 52 30.	- 04 11	90	15.	GALAXY
MCG-03-13-024	04 52 30.	- 17 33	30	15.	GALAXY
LH120-N004E	04 52 30.	- 67 00	46		EMISSION NEBULA IN LMC
LB 03394	04 52 30.	- 67 24		13.5	FAINT BLUE STAR
LH120-N079E	04 52 30.	- 69 25	180		EMISSION NEBULA IN LMC
RNGC 1692	04 52 31.	- 20 38			GALAXY
HSE 045	04 52 31.	- 73 31			STAR CLUSTER IN LMC
RNGC 1841	04 52 32.	- 84 05		12.0	GLOBULAR CLUSTER
LB 01440	04 52 35.	+ 18 47 54.		15.9	FAINT BLUE STAR
SS 37	04 52 35.	+ 30 29			DIFFUSE GALACTIC NEBULA
DG 048	04 52 36.	+ 30 29	480		REFLECTION NEBULA
MCG+11-07-002	04 52 36.	+ 68 15	114	14.	GALAXY
MCG+00-13-034	04 52 36.	- 01 17	54	15.	GALAXY
MCG-04-12-020	04 52 36.	- 24 36	36	15.	GALAXY
MCG-05-12-008	04 52 36.	- 28 43	60	17.	GALAXY
LB 03395	04 52 36.	- 61 30		11.2	FAINT BLUE STAR
LH120-N005	04 52 36.	- 67 22	202		EMISSION NEBULA IN LMC
SL 069	04 52 36.	- 67 22	90		STAR CLUSTER IN LMC
SL 070	04 52 36.	- 67 29	35		STAR CLUSTER IN LMC
HSE 050	04 52 39.	- 66 32			STAR CLUSTER IN LMC
HSE 051	04 52 40.	- 66 58			STAR CLUSTER IN LMC
B 028	04 52 41.	+ 30 34	240		DARK OBJECT
HSE 053	04 52 41.	- 65 31			STAR CLUSTER IN LMC
SCHO 0074	04 52 42.	+ 37 38 24.	380		ISOLATED DARK CLOUD
ZC 0452.7-0001	04 52 42.	- 00 01	1410		CLUSTER OF GALAXIES
ZWG 394.036	04 52 42.	- 01 16		15.1	GALAXY
UGC 03202	04 52 42.	- 01 16	78	15.1	GALAXY SBa-b
MCG-02-13-021	04 52 42.	- 10 46	180	14.	GALAXY
MCG-03-13-025	04 52 42.	- 18 13 30.	24	14.	GALAXY
LH 003	04 52 42.	- 67 22	300		STELLAR ASSN. IN LMC
SL 071	04 52 42.	- 67 32	20		STAR CLUSTER IN LMC
LH120-N079D	04 52 42.	- 69 27	75		EMISSION NEBULA IN LMC
HSE 049	04 52 44.	- 69 28			STAR CLUSTER IN LMC
MCG-02-13-022	04 52 45.	- 12 10 30.	21	14.5	GALAXY
RNGC 1694	04 52 46.	- 04 43		15.0	GALAXY
ZWG 307.001	04 52 48.	+ 68 14		13.2	GALAXY
ZWG 306.007	04 52 48.	+ 68 14		13.2	GALAXY
UGC 03203	04 52 48.	+ 68 14	150	13.2	GALAXY S
KARA.72 104B	04 52 48.	+ 68 14	24		PART OF DOUBLE GALAXY
KARA.72 104A	04 52 48.	+ 68 14	24	13.2	PART OF DOUBLE GALAXY
LB 03396	04 52 48.	- 67 00		14.0	FAINT BLUE STAR
HSE 052	04 52 48.	- 68 51			STAR CLUSTER IN LMC
MCG-02-13-023	04 52 51.	- 12 14 30.	72	14.	GALAXY
MCG-03-13-026	04 52 54.	- 16 13 30.	30	16.	GALAXY
MCG-03-13-027	04 52 54.	- 16 14	54	14.5	GALAXY
SL 075	04 52 54.	- 69 00	40		STAR CLUSTER IN LMC
SC 0450-1619.1	04 52 55.	- 16 14 12.	78		NEBULA
MCG-04-12-021	04 52 57.	- 26 06 30.	24	15.	GALAXY
HSE 054	04 52 57.	- 66 30			STAR CLUSTER IN LMC
SL 076	04 52 57.	- 68 17	40		STAR CLUSTER IN LMC
IC 0396	04 52 58.	+ 68 15 51.			NONSTELLAR OBJECT
LBN 0860	04 53	+ 05 20	1320		BRIGHT NEBULA
KHAV 097	04 53	+ 53 47	3010		DARK NEBULA
SIV 03	04 53	+ 65 25	27000		FAINT H EMISSION REGION
LB 09823	04 53	- 83 12		14.5	FAINT BLUE STAR
LDN 1539	04 53 00.	+ 26 30	6060		DARK NEBULA
MCG-06-11-007	04 53 00.	- 37 27	60	15.	GALAXY
CM 01	04 53 00.	- 66 34			HII REGION IN LMC
LH120-N008A	04 53 00.	- 68 09			EMISSION NEBULA IN LMC
LH120-N081A	04 53 00.	- 69 18	100		EMISSION NEBULA IN LMC
CM 06	04 53 00.	- 69 30			HII REGION IN LMC
SL 068	04 53 00.	- 73 19	55		STAR CLUSTER IN LMC
LB 01441	04 53 04.	+ 16 09 12.		16.6	FAINT BLUE STAR
KARA.73 24	04 53 04.	- 37 15	27		DWARF GALAXY
LW 083	04 53 04.	- 73 07			STAR CLUSTER IN LMC
RNGC 1732	04 53 05.	- 68 44			OPEN CLUSTER IN LMC
SL 077	04 53 05.	- 68 44	25		STAR CLUSTER IN LMC
ZWG 306.008	04 53 06.	+ 64 55		15.7	GALAXY
UGC 03204	04 53 06.	+ 64 55	84	15.7	GALAXY Sc
MCG+11-07-003	04 53 06.	+ 64 55	78	16.	GALAXY
SL 078	04 53 06.	- 67 21	20		STAR CLUSTER IN LMC
LH120-N008	04 53 06.	- 68 08	116		EMISSION NEBULA IN LMC
SL 072	04 53 06.	- 72 25	15		STAR CLUSTER IN LMC
LW 084	04 53 06.	- 73 18			STAR CLUSTER IN LMC
LB 01442	04 53 08.	+ 19 17 12.		15.3	FAINT BLUE STAR
RNGC 1705	04 53 08.	- 53 25		13.0	GALAXY
SL 073	04 53 08.	- 72 46	15		STAR CLUSTER IN LMC
MCG+05-12-002	04 53 09.	+ 29 58 30.	84	15.	GALAXY
MCG-04-12-022	04 53 09.	- 22 48 30.	54	15.	GALAXY
RNGC 1736	04 53 09.	- 68 08			DIFFUSE NEBULA IN LMC
UGC 03205	04 53 12.	+ 29 59	138	16.0	GALAXY Sa-b
ZC 0453.2+7955	04 53 12.	+ 79 55	6590		CLUSTER OF GALAXIES
LH120-N004D	04 53 12.	- 66 59	250		EMISSION NEBULA IN LMC
SL 081	04 53 12.	- 67 22	100		STAR CLUSTER IN LMC
LW 082	04 53 12.	- 74 55			STAR CLUSTER IN LMC
SC 0450-1516.2	04 53 13.	- 15 11 19.	12		NEBULA
SC 0450-1404.9	04 53 14.	- 14 00 01.	12		NEBULA
MCG-01-13-035A	04 53 15.	- 04 44	48	15.	GALAXY

OBJECT NAME	RIGHT ASCEN.	DECLINATION	DIAM.	MAGN.	TYPE OF OBJECT
MCG-01-13-036	04 53 15.	- 07 26	36	15.	GALAXY
ZWG 394.037	04 53 18.	+ 02 09		14.7	GALAXY
MCG+00-13-035	04 53 18.	+ 02 11	27	14.5	GALAXY
ZWG 394.038	04 53 18.	+ 02 51		15.3	GALAXY
UGC 03206	04 53 18.	+ 02 51	78	15.3	GALAXY Sb
7ZW 021	04 53 18.	+ 69 02			COMPACT GALAXY
ZWG 328.030	04 53 18.	+ 69 03		15.4	GALAXY
MCG-03-13-028	04 53 18.	- 15 10	36	15.5	GALAXY
LH120-N081B	04 53 18.	- 69 19	101		EMISSION NEBULA IN LMC
SC 0451-1404.3	04 53 19.	- 13 59 25.	12		NEBULA
SC 0451-1515.7	04 53 19.	- 15 10 49.	24		NEBULA
HSE 055	04 53 20.	- 70 42			STAR CLUSTER IN LMC
LW 085	04 53 20.	- 71 44			STAR CLUSTER IN LMC
MCG+00-13-036	04 53 21.	+ 02 52	72	14.	GALAXY
MCG-03-13-029	04 53 21.	- 20 41 30.	36	14.5	GALAXY
RNGC 1731	04 53 23.	- 67 00			CLUSTER/NEBULOSITY IN LMC
SL 082	04 53 23.	- 67 00	70		STAR CLUSTER IN LMC
LH 004	04 53 23.	- 67 00	240		STELLAR ASSN. IN LMC
MCG-03-13-031	04 53 24.	- 15 04	48	15.	GALAXY
MCG-03-13-030	04 53 24.	- 15 10 30.	48	15.	GALAXY
LB 03397	04 53 24.	- 68 34		13.5	FAINT BLUE STAR
SL 074	04 53 24.	- 74 57	25		STAR CLUSTER IN LMC
SL 079	04 53 26.	- 71 45	35		STAR CLUSTER IN LMC
SC 0451-1409.4	04 53 28.	- 14 04 32.	24		NEBULA
ZWG 394.039	04 53 30.	+ 02 05		14.3	GALAXY
UGC 03207	04 53 30.	+ 02 05	168	14.3	GALAXY Sb
LH120-N007	04 53 30.	- 67 28	61		EMISSION NEBULA IN LMC
RNGC 1734	04 53 30.	- 68 51			OPEN CLUSTER IN LMC
SL 083	04 53 30.	- 68 51	60		STAR CLUSTER IN LMC
SC 0451-1408.8	04 53 32.	- 14 03 56.	12		NEBULA
MCG+00-13-037	04 53 36.	+ 02 07	138	13.	GALAXY
MCG-05-12-009	04 53 36.	- 31 53	54	15.	GALAXY
LW 086	04 53 37.	- 72 36			STAR CLUSTER IN LMC
HSE 056	04 53 38.	- 67 57			STAR CLUSTER IN LMC
LB 01443	04 53 41.	+ 08 51 54.		14.8	FAINT BLUE STAR
LW 087	04 53 41.	- 72 15			STAR CLUSTER IN LMC
ZWG 394.040	04 53 42.	+ 01 31		15.6	GALAXY
UGC 03208	04 53 42.	+ 01 31	78	15.6	GALAXY Sc
MCG+00-13-038	04 53 42.	+ 01 32 30.	54	14.	GALAXY
MCG-03-13-032	04 53 42.	- 19 02 30.	36	14.5	GALAXY
MCG+00-13-039	04 53 45.	+ 02 58	72	15.	GALAXY
ZWG 394.041	04 53 48.	+ 02 57		15.3	GALAXY
UGC 03209	04 53 48.	+ 02 57	102	15.3	GALAXY Sb-c
MCG-05-12-010	04 53 48.	- 29 57	60	14.	GALAXY
LH120-N082	04 53 48.	- 69 23			EMISSION NEBULA IN LMC
WS 03	04 53 49.	- 69 39 05.		15.30	PLANETARY NEB. IN LMC
SL 080	04 53 49.	- 75 00	20		STAR CLUSTER IN LMC
RNGC 1701	04 53 50.	- 29 57		14.0	GALAXY
MCG+00-13-040	04 53 51.	+ 00 38	30	14.	GALAXY
MCG-01-13-037	04 53 51.	- 04 31	60	15.	GALAXY
LB 01444	04 53 52.	+ 16 44 18.		16.2	FAINT BLUE STAR
RNGC 1733	04 53 52.	- 66 45			OPEN CLUSTER IN LMC
SC 0451-1559.5	04 53 53.	- 15 54 40.	6		NEBULA
RNGC 1735	04 53 53.	- 67 11			CLUSTER/NEBULOSITY IN LMC
ZWG 394.042	04 53 54.	+ 00 38		15.3	GALAXY
ZC 0453.9+0047	04 53 54.	+ 00 47	680		CLUSTER OF GALAXIES
OCL 0443	04 53 54.	+ 28 42	080		OPEN STAR CLUSTER
ISS 0204	04 53 54.	+ 40 23	236		STELLAR RING
HSE 057	04 53 56.	- 70 41			STAR CLUSTER IN LMC
SL 085	04 53 58.	- 66 45	60		STAR CLUSTER IN LMC
IC 2106	04 53 59.	- 28 36 44.			NONSTELLAR OBJECT
SL 086	04 53 59.	- 67 11	75		STAR CLUSTER IN LMC
UGC 03210	04 54 00.	+ 82 01	72	16.5	GALAXY S
LB 03398	04 54 00.	- 65 40		13.3	FAINT BLUE STAR
LH120-N006	04 54 00.	- 66 50	17		EMISSION NEBULA IN LMC
CM 08	04 54 00.	- 69 18			HII REGION IN LMC
LW 088	04 54 00.	- 74 58			STAR CLUSTER IN LMC
BC PKS0454-22	04 54 02.	- 22 04 06.			QUASI-STELLAR OBJECT
HSE 058	04 54 02.	- 66 18			STAR CLUSTER IN LMC
IC 2104	04 54 04.	- 15 52 25.			NONSTELLAR OBJECT
ZWG 394.043	04 54 06.	+ 00 58		15.7	GALAXY
MCG+00-13-041	04 54 06.	+ 01 00	18	14.5	GALAXY
MCG-03-13-033	04 54 06.	- 17 31	60	15.	GALAXY
LB 01445	04 54 07.	+ 08 57 00.		14.4	FAINT BLUE STAR
SL 084	04 54 08.	- 75 12	70		STAR CLUSTER IN LMC
SHB 087	04 54 08.4	+ 03 56 06.		16.5	QUASI-STELLAR OBJECT
BC PKS0454+039	04 54 08.8	+ 03 56 13.			QUASI-STELLAR OBJECT
MCG-02-13-024	04 54 09.	- 10 18	54	14.5	GALAXY
LW 090	04 54 09.	- 73 47			STAR CLUSTER IN LMC
ZWG 394.044	04 54 12.	+ 02 43		15.5	GALAXY
7ZW 022	04 54 12.	+ 72 15			COMPACT GALAXY
ZWG 328.031	04 54 12.	+ 72 15		15.4	GALAXY
2ZW 024	04 54 12.	- 06 53			COMPACT GALAXY
MCG-03-13-034	04 54 12.	- 15 52 30.	96	14.	GALAXY
LH120-N083C	04 54 12.	- 69 15	21		EMISSION NEBULA IN LMC
LH120-N185	04 54 12.	- 70 05	397		EMISSION NEBULA IN LMC
SC 0451-1548.1	04 54 13.	- 15 43 17.	18		NEBULA
RNGC 1743	04 54 14.	- 69 16			CLUSTER/NEBULOSITY IN LMC
SL 087	04 54 14.	- 69 16	20		STAR CLUSTER IN LMC
LW 089	04 54 14.	- 75 09			STAR CLUSTER IN LMC
MCG+00-13-042	04 54 15.	+ 02 44	30	14.5	GALAXY
ZC 0454.3+0534	04 54 18.	+ 05 34	1080		CLUSTER OF GALAXIES
ZWG 394.045	04 54 18.	- 00 40		15.3	GALAXY
LB 03399	04 54 18.	- 61 24		11.6	FAINT BLUE STAR
LH120-N080	04 54 18.	- 68 27	84		EMISSION NEBULA IN LMC
LH120-N083A	04 54 18.	- 69 17	61		EMISSION NEBULA IN LMC
SC 0452-1420.7	04 54 20.	- 14 15 53.	6		NEBULA
LH 005	04 54 20.	- 69 17	360		STELLAR ASSN. IN LMC
MCG+00-13-043	04 54 21.	- 00 40	42	15.	GALAXY
MCG-02-13-025	04 54 21.	- 10 40	138	14.	GALAXY
SCHO 0075	04 54 22.	+ 23 56 30.	700		ISOLATED DARK CLOUD
LB 01446	04 54 23.	+ 09 02 30.		14.7	FAINT BLUE STAR
MCG-03-13-035	04 54 24.	- 17 45	36	15.	GALAXY
LH120-N083D	04 54 24.	- 69 16	21		EMISSION NEBULA IN LMC
MCG-01-13-038	04 54 27.	- 04 57	60	12.	GALAXY
RNGC 1699	04 54 28.	- 04 50		15.0	GALAXY
RNGC 1700	04 54 28.	- 04 56		12.5	GALAXY
MCG-01-13-039	04 54 30.	- 04 51	48	15.	GALAXY
MCG-03-13-036	04 54 30.	- 18 57	48	15.	GALAXY
MCG-05-12-011	04 54 30.	- 28 33	72	14.5	GALAXY
LH120-N083	04 54 30.	- 69 16	340		EMISSION NEBULA IN LMC
RNGC 1745	04 54 32.	- 69 16			CLUSTER/NEBULOSITY IN LMC
RNGC 1737	04 54 32.	- 69 16			CLUSTER/NEBULOSITY IN LMC
RNGC 1751	04 54 35.	- 69 53			GLOBULAR CLUSTER IN LMC
SL 089	04 54 35.	- 69 53	60		STAR CLUSTER IN LMC
ZWG 394.046	04 54 36.	- 00 55		15.2	GALAXY
LH120-N083B	04 54 36.	- 69 16	44		EMISSION NEBULA IN LMC
SL 088	04 54 36.	- 71 19	20		STAR CLUSTER IN LMC

OBJECT NAME	RIGHT ASCEN.	DECLINATION	DIAM.	MAGN.	TYPE OF OBJECT
RNGC 1748	04 54 38.	- 69 16			CLUSTER/NEBULOSITY IN LMC
ARC 0522	04 54 39.	- 06 14		17.0	RICH CLUSTER OF GALAXIES
ZWG 394.047	04 54 42.	+ 01 49		15.3	GALAXY
UGC 03211	04 54 42.	+ 01 49	84	15.3	GALAXY Sc
LH120-N011L	04 54 42.	- 66 30	67		EMISSION NEBULA IN LMC
LH120-N087	04 54 42.	- 69 35	19		EMISSION NEBULA IN LMC
LB 01447	04 54 44.	+ 08 46 54.		14.6	FAINT BLUE STAR
SL 092	04 54 45.	- 68 13	25		STAR CLUSTER IN LMC
RNGC 1749	04 54 46.	- 68 16			OPEN CLUSTER IN LMC
SL 093	04 54 46.	- 68 16	65		STAR CLUSTER IN LMC
SL 090	04 54 47.	- 70 00	25		STAR CLUSTER IN LMC
MCG+00-13-044	04 54 48.	+ 01 50	84	14.5	GALAXY
LB 01762	04 54 48.	- 45 47		14.8	FAINT BLUE STAR
IC 2114	04 54 49.	- 69 17			SAME AS NGC 1748
HN 0086	04 54 50.	- 69 16			NEBULA
RNGC 1756	04 54 50.	- 69 19			OPEN CLUSTER IN LMC
SL 094	04 54 50.	- 69 19	35		STAR CLUSTER IN LMC
RNGC 1754	04 54 50.	- 70 31			GLOBULAR CLUSTER IN LMC
SL 091	04 54 50.	- 70 31	35		STAR CLUSTER IN LMC
ZWG 394.048	04 54 54.	- 00 56		15.5	GALAXY
MCG+00-13-045	04 54 54.	- 00 56 30.	18	14.5	GALAXY
SC 0452-1521.4	04 54 54.	- 15 16 38.	72		NEBULA
LB 01763	04 54 54.	- 47 22		14.1	FAINT BLUE STAR
RNGC 1747	04 54 54.	- 67 15			CLUSTER/NEBULOSITY IN LMC
SL 097	04 54 55.	- 66 00	20		STAR CLUSTER IN LMC
HSE 060	04 54 55.	- 69 04			STAR CLUSTER IN LMC
HSE 059	04 54 55.	- 70 18			STAR CLUSTER IN LMC
SG 2.014	04 54 56.	+ 47 53	600		DIFFUSE EMISSION NEBULA
SL 096	04 54 56.	- 67 47	25		STAR CLUSTER IN LMC
MCG+00-13-046	04 54 57.	- 01 12 30.	30	15.	GALAXY
SC 0452-1541.5	04 54 57.	- 15 36 44.	42		NEBULA
RNGC 1710	04 54 58.	- 15 22		14.0	GALAXY
LBN 0745	04 55	+ 47 56	540		BRIGHT NEBULA
MRSL 159+03/1	04 55 00.	+ 47 56	540		HII REGION
UGC 03212	04 55 00.	+ 71 08	72	18.	GALAXY DWRF SP
MCG+15-01-002	04 55 00.	+ 88 19	60	15.	GALAXY
UGC 03211A	04 55 00.	+ 89 17	84	16.0	GALAXY Sc
ZWG 394.049	04 55 00.	- 01 12		15.3	GALAXY
MCG-03-13-037	04 55 00.	- 15 22	72	14.	GALAXY
SL 098	04 55 00.	- 67 15	40		STAR CLUSTER IN LMC
LB 03400	04 55 00.	- 70 05		13.0	FAINT BLUE STAR
IC 2108	04 55 01.	- 15 22			NONSTELLAR OBJECT
LB 01448	04 55 03.	+ 18 23 36.		15.0	FAINT BLUE STAR
RNGC 1755	04 55 04.	- 68 16		10.0	OPEN CLUSTER IN LMC
SL 099	04 55 04.	- 68 16	120		STAR CLUSTER IN LMC
SCHO 0076	04 55 05.	+ 22 54 42.	510		ISOLATED DARK CLOUD
SL 101	04 55 05.	- 67 08	25		STAR CLUSTER IN LMC
LH 006	04 55 06.	- 67 15	300		STELLAR ASSN. IN LMC
LH120-N088	04 55 06.	- 69 29	36		EMISSION NEBULA IN LMC
SL 095	04 55 06.	- 71 20	40		STAR CLUSTER IN LMC
LB 01449	04 55 09.	+ 19 16 48.		15.7	FAINT BLUE STAR
ZC 0455.2+0746	04 55 12.	+ 07 46	1140		CLUSTER OF GALAXIES
2ZW 025	04 55 12.	- 04 10			COMPACT GALAXY
LB 03401	04 55 12.	- 65 56		15.1	FAINT BLUE STAR
LH120-N009	04 55 12.	- 67 13	528		EMISSION NEBULA IN LMC
HSE 061	04 55 13.	- 67 35			STAR CLUSTER IN LMC
SL 100	04 55 13.	- 70 27	15		STAR CLUSTER IN LMC
SL 105	04 55 17.	- 68 37	40		STAR CLUSTER IN LMC
MCG+00-13-047	04 55 18.	+ 00 43 30.	21	15.	GALAXY
ZC 0455.3+0155	04 55 18.	+ 01 55	3090		CLUSTER OF GALAXIES
ZWG 394.050	04 55 18.	+ 02 51		15.7	GALAXY
UGC 03213	04 55 18.	+ 02 51	66	15.7	GALAXY Sa
ZWG 420.020	04 55 18.	+ 05 47		15.7	GALAXY
LH120-N089	04 55 18.	- 69 25			EMISSION NEBULA IN LMC
MCG+00-13-048	04 55 21.	- 00 12 30.	222	13.	GALAXY
SL 102	04 55 21.	- 70 42	20		STAR CLUSTER IN LMC
SL 107	04 55 22.	- 68 21	15		STAR CLUSTER IN LMC
SL 106	04 55 22.	- 69 45	70		STAR CLUSTER IN LMC
ZWG 394.052	04 55 24.	+ 00 43		15.0	GALAXY
MCG+00-13-049	04 55 24.	+ 02 52	36	15.	GALAXY
LDN 1437	04 55 24.	+ 52 00	180		DARK NEBULA
ZWG 394.051	04 55 24.	- 00 12		14.4	GALAXY
UGC 03214	04 55 24.	- 00 12	198	14.4	GALAXY Sb
AGU 22	04 55 24.	- 42 46 30.		16.	2 INTERACTING GALAXIES
LH120-N011G	04 55 24.	- 66 27	79		EMISSION NEBULA IN LMC
SL 103	04 55 25.	- 71 28	35		STAR CLUSTER IN LMC
LW 091	04 55 25.	- 73 26			STAR CLUSTER IN LMC
HSE 062	04 55 27.	- 70 41			STAR CLUSTER IN LMC
ZWG 420.021	04 55 30.	+ 03 55		15.1	GALAXY
HSE 063	04 55 34.	- 70 59			STAR CLUSTER IN LMC
HSE 065	04 55 36.	- 68 57			STAR CLUSTER IN LMC
HSE 064	04 55 37.	- 70 16			STAR CLUSTER IN LMC
IC 2107	04 55 38.	+ 08 10			MAY NOT EXIST
ARC 0524	04 55 38.	- 19 48		16.7	RICH CLUSTER OF GALAXIES
SL 109	04 55 39.	- 69 34	40		STAR CLUSTER IN LMC
HSE 066	04 55 41.	- 70 01			STAR CLUSTER IN LMC
SL 108	04 55 41.	- 71 11	15		STAR CLUSTER IN LMC
MCG+00-13-050	04 55 42.	+ 00 39 30.	48	14.	GALAXY
ZWG 394.054	04 55 42.	+ 00 40		14.8	GALAXY
UGC 03215	04 55 42.	+ 00 40	102	14.8	GALAXY S0
ZWG 394.053	04 55 42.	- 00 15		15.4	GALAXY
MCG+00-13-051	04 55 42.	- 00 15	42	15.	GALAXY
LH120-N084	04 55 42.	- 68 31	113		EMISSION NEBULA IN LMC
LH120-N090	04 55 42.	- 69 21	27		EMISSION NEBULA IN LMC
SCHO 0077	04 55 45.	+ 22 56 42.	410		ISOLATED DARK CLOUD
MCG-01-13-040	04 55 45.	- 07 53	66	15.	GALAXY
SC 0453-1521.3	04 55 46.	- 15 16 35.	18		NEBULA
LB 01450	04 55 47.	+ 16 17 06.		15.8	FAINT BLUE STAR
IC 0398	04 55 47.7	- 07 51 18.	54		GALAXY SB(s)
MCG-02-13-026	04 55 48.	- 09 52	72	14.	GALAXY
LH120-N011H	04 55 48.	- 66 33	27		EMISSION NEBULA IN LMC
LH120-N011I	04 55 48.	- 66 39	110		EMISSION NEBULA IN LMC
CM 09	04 55 48.	- 68 41			HII REGION IN LMC
SC 0453-1604.0	04 55 51.	- 15 59 18.	12		NEBULA
MCG-04-12-023	04 55 51.	- 21 41	60	14.	GALAXY
LW 093	04 55 52.	- 72 01			STAR CLUSTER IN LMC
SL 104	04 55 52.	- 75 23	15		STAR CLUSTER IN LMC
ZWG 420.022	04 55 54.	+ 03 13		15.5	GALAXY
MCG+00-13-052	04 55 54.	- 00 39	18	15.5	GALAXY
LH120-N085	04 55 54.	- 68 41	23		EMISSION NEBULA IN LMC
LH120-N086	04 55 54.	- 68 45	223		EMISSION NEBULA IN LMC
LW 092	04 55 54.	- 73 15			STAR CLUSTER IN LMC
SL 110	04 55 56.	- 69 14	20		STAR CLUSTER IN LMC
MCG+00-13-053	04 55 57.	- 00 38	12	15.5	GALAXY
BIGO 488	04 55 57.	- 15 24			NEBULA
HSE 067	04 55 58.	- 70 59			STAR CLUSTER IN LMC
LBN 0798	04 56	+ 30 29	360		BRIGHT NEBULA
LB 09824	04 56	- 81 36		14.5	FAINT BLUE STAR

OBJECT NAME	RIGHT ASCEN.	DECLINATION	DIAM.	MAGN.	TYPE OF OBJECT
MCG+01-13-011	04 56 00.	+ 06 54	60	15.	GALAXY
MCG+01-13-010	04 56 00.	+ 07 02	60	15.5	GALAXY
ZC 0456.0+0837	04 56 00.	+ 08 37	2080		CLUSTER OF GALAXIES
UGC 03217	04 56 00.	+ 53 45	84	17.	GALAXY
UGC 03218	04 56 00.	+ 62 10	108	15.0	GALAXY Sb
MCG+10-08-001	04 56 00.	+ 62 10	90	14.	GALAXY
ZWG 394.056	04 56 00.	- 00 38		15.5	GALAXY
ZWG 394.055	04 56 00.	- 00 39		15.7	GALAXY
UGC 03216	04 56 00.	- 01 32	72	17.	GALAXY DWARF
LB 01764	04 56 00.	- 48 45		14.1	FAINT BLUE STAR
HSE 068	04 56 03.	- 66 25			STAR CLUSTER IN LMC
ZWG 420.023	04 56 06.	+ 06 54		15.7	GALAXY
UGC 03219	04 56 06.	+ 06 54	84	15.7	GALAXY Sc
ZWG 420.024	04 56 06.	+ 07 02		15.7	GALAXY
UGC 03220	04 56 06.	+ 07 02	66	15.7	GALAXY Sc
ZWG 283.001	04 56 06.	+ 62 10		15.0	GALAXY
MCG+00-13-054	04 56 06.	- 00 32 30.	15	15.	GALAXY
ZC 0456.1-0103	04 56 06.	- 01 03	9140		CLUSTER OF GALAXIES
SCHO 0078	04 56 08.	+ 37 35 30.	450		ISOLATED DARK CLOUD
SC 0453-1506.7	04 56 08.	- 15 02 01.	18		NEBULA
RNGC 1707	04 56 09.	+ 08 10			NON-EXISTENT OBJECT
SL 112	04 56 10.	- 69 47	75		STAR CLUSTER IN LMC
HSE 069	04 56 11.	- 65 14			STAR CLUSTER IN LMC
ZWG 445.001	04 56 12.	+ 12 20		15.7	GALAXY
ZWG 394.058	04 56 12.	- 00 33		15.6	GALAXY
MCG+00-13-055	04 56 12.	- 00 57 30.	42	14.	GALAXY
ZWG 394.057	04 56 12.	- 00 58		15.5	GALAXY
UGC 03221	04 56 12.	- 00 58	66	15.5	GALAXY S
MCG-03-13-040	04 56 12.	- 15 02	24	16.	GALAXY
MCG-03-13-039	04 56 12.	- 15 02	30	16.	GALAXY
MCG-03-13-038	04 56 12.	- 20 28	90	14.	GALAXY
MCG-04-12-024	04 56 12.	- 21 37	18	16.	GALAXY
RNGC 1709	04 56 13.	- 00 33		15.5	GALAXY
RNGC 1716	04 56 13.	- 20 28		14.0	GALAXY
RNGC 1766	04 56 13.	- 70 18			OPEN CLUSTER IN LMC
SL 113	04 56 13.	- 70 18	50		STAR CLUSTER IN LMC
SL 111	04 56 13.	- 71 29	90		STAR CLUSTER IN LMC
SCHO 0079	04 56 14.	+ 37 35 48.	280		ISOLATED DARK CLOUD
RNGC 1764	04 56 14.	- 67 45			OPEN CLUSTER IN LMC
SL 115	04 56 14.	- 67 45	25		STAR CLUSTER IN LMC
SL 114	04 56 14.	- 69 19	55		STAR CLUSTER IN LMC
ARC 0523	04 56 16.	+ 08 42		16.7	RICH CLUSTER OF GALAXIES
MCG+00-13-056	04 56 18.	- 00 33	24	13.	GALAXY
ZWG 394.059	04 56 18.	- 00 34		13.9	GALAXY
UGC 03222	04 56 18.	- 00 34	156	13.9	GALAXY E
MCG-04-12-025	04 56 18.	- 21 38	36	15.	GALAXY
RNGC 1713	04 56 19.	- 00 34		14.0	GALAXY
RNGC 1761	04 56 21.	- 66 34			CLUSTER/NEBULOSITY IN LMC
MCG+01-13-012	04 56 24.	+ 04 53	78	13.	GALAXY
MCG+00-13-057	04 56 24.	- 01 34 30.	30	15.	GALAXY
LH120-N011	04 56 24.	- 66 30	1404		EMISSION NEBULA IN LMC
SL 116	04 56 24.	- 68 52	40		STAR CLUSTER IN LMC
SL 117	04 56 25.	- 69 02	110		STAR CLUSTER IN LMC
SCHO 0080	04 56 27.	+ 23 52 12.	350		ISOLATED DARK CLOUD
IC 2109	04 56 27.	- 00 23			NONSTELLAR OBJECT
SL 122	04 56 27.	- 66 34	255		STAR CLUSTER IN LMC
RNGC 1760	04 56 27.	- 66 36			CLUSTER/NEBULOSITY IN LMC
SL 119	04 56 28.	- 68 14	20		STAR CLUSTER IN LMC
ZWG 420.025	04 56 30.	+ 04 54		14.0	GALAXY
UGC 03223	04 56 30.	+ 04 54	96	14.0	GALAXY SBa
IC 2110	04 56 30.	- 00 23			NONSTELLAR OBJECT
ZWG 394.060	04 56 30.	- 01 34		15.5	GALAXY
RNGC 1720	04 56 30.	- 07 55		13.0	GALAXY
HSE 071	04 56 30.	- 65 39			STAR CLUSTER IN LMC
LH120-N011F	04 56 30.	- 66 36	79		EMISSION NEBULA IN LMC
HSE 070	04 56 31.	- 70 23			STAR CLUSTER IN LMC
LH 007	04 56 31.	- 71 29	240		STELLAR ASSN. IN LMC
MCG-04-12-026	04 56 33.	- 21 38	72	14.	GALAXY
RNGC 1767	04 56 33.	- 69 28			CLUSTER/NEBULOSITY IN LMC
SL 120	04 56 33.	- 69 28	55		STAR CLUSTER IN LMC
LB 01451	04 56 34.	+ 21 39 30.		15.5	FAINT BLUE STAR
MCG+01-13-013	04 56 36.	+ 05 32	84	12.	GALAXY
ZWG 420.026	04 56 36.	+ 05 33		14.4	GALAXY
UGC 03224	04 56 36.	+ 05 33	102	14.4	GALAXY Sb
UGC 03225	04 56 36.	+ 12 47	66	16.5	GALAXY S
OCL 0416	04 56 36.	+ 43 24	540	16.	OPEN STAR CLUSTER
MCG+00-13-058	04 56 36.	- 00 40	42	15.	GALAXY
MCG-03-13-041	04 56 36.	- 19 41 30.	18	15.	GALAXY
MCG-05-12-012	04 56 36.	- 27 53	48	15.	GALAXY
LH120-N094A	04 56 36.	- 69 29	90		EMISSION NEBULA IN LMC
SCHO 0081	04 56 37.	+ 36 39 36.	260		ISOLATED DARK CLOUD
RNGC 1717	04 56 37.	- 00 19			NON-EXISTENT OBJECT
SL 126	04 56 39.	- 62 36	20		STAR CLUSTER IN LMC
RNGC 1763	04 56 39.	- 66 29			CLUSTER/NEBULOSITY IN LMC
SL 125	04 56 39.	- 66 29	240		STAR CLUSTER IN LMC
LH 009	04 56 39.	- 66 33	360		STELLAR ASSN. IN LMC
SL 123	04 56 39.	- 69 26	20		STAR CLUSTER IN LMC
LB 01452	04 56 42.	+ 09 39 30.		15.8	FAINT BLUE STAR
ZWG 394.061	04 56 42.	- 00 40		15.7	GALAXY
MCG+00-13-059	04 56 45.	- 01 15	54	14.5	GALAXY
LH 010	04 56 45.	- 66 28	240		STELLAR ASSN. IN LMC
LH 008	04 56 45.	- 69 30	1200		STELLAR ASSN. IN LMC
ZC 0456.8+0801	04 56 48.	+ 08 01	1410		CLUSTER OF GALAXIES
ZWG 394.062	04 56 48.	- 01 14		15.6	GALAXY
LB 03402	04 56 48.	- 65 35		13.6	FAINT BLUE STAR
LH120-N011B	04 56 48.	- 66 28	262		EMISSION NEBULA IN LMC
LH120-N094B	04 56 48.	- 69 31	62		EMISSION NEBULA IN LMC
SL 124	04 56 48.	- 70 03	20		STAR CLUSTER IN LMC
RNGC 1721	04 56 50.	- 11 12		13.0	GALAXY
ARC 0525	04 56 51.	+ 08 05		17.2	RICH CLUSTER OF GALAXIES
MCG-01-13-041	04 56 51.	- 07 57 30.	78	13.	GALAXY
MCG-02-13-027	04 56 51.	- 11 12	84	13.	GALAXY
SCHO 0082	04 56 52.	+ 24 02 18.	450		ISOLATED DARK CLOUD
RNGC 1768	04 56 52.	- 68 19			OPEN CLUSTER IN LMC
LW 094	04 56 53.	- 74 45			STAR CLUSTER IN LMC
HN 0087	04 56 54.	- 66 27			NEBULA
LB 03403	04 56 54.	- 68 51		13.2	FAINT BLUE STAR
HOLM 083B	04 56 56.	- 00 19	12	14.1	PART OF MULTIPLE GALAXY
ARC 0528	04 56 56.	- 09 05		17.5	RICH CLUSTER OF GALAXIES
HOLM 083A	04 56 57.	- 00 19	18	14.0	PART OF MULTIPLE GALAXY
MCG+00-13-060	04 56 57.	- 00 20	60	14.5	GALAXY
IC 2115	04 56 57.4	- 66 28 15.			HII REGION IN LMC
SC 0454-1432.8	04 56 58.	- 14 28 10.	6		NEBULA
SL 127	04 56 58.	- 68 19	40		STAR CLUSTER IN LMC
SL 118	04 56 59.	- 74 46	20		STAR CLUSTER IN LMC
KHAV 098	04 57	+ 25 47	9460		DARK NEBULA
VMT 04	04 57	+ 46 36	7500		SUPERNOVA REMNANT
LBN 0975	04 57	- 10 00	4500		BRIGHT NEBULA

OBJECT NAME	RIGHT ASCEN.	DECLINATION	DIAM.	MAGN.	TYPE OF OBJECT
ZC 0457.0+0511	04 57 00.	+ 05 11	3090		CLUSTER OF GALAXIES
LB 01453	04 57 00.	+ 22 02 30.		14.6	FAINT BLUE STAR
LDN 1439	04 57 00.	+ 52 00	420		DARK NEBULA
ZWG 394.063	04 57 00.	- 00 20		14.5	GALAXY
UGC 03226	04 57 00.	- 00 20	78	14.5	GALAXY Sa
MCG-02-13-029	04 57 00.	- 11 03	180	12.5	GALAXY
MCG-02-13-028	04 57 00.	- 11 12 30.	27	13.	GALAXY
LB 03404	04 57 00.	- 66 38		12.8	FAINT BLUE STAR
RNGC 1719	04 57 01.	- 00 20		14.5	GALAXY
RNGC 1723	04 57 01.	- 11 03		12.0	GALAXY
RNGC 1725	04 57 02.	- 11 12		13.0	GALAXY
RNGC 1777	04 57 02.	- 74 21			OPEN CLUSTER IN LMC
SL 121	04 57 02.	- 74 21	100		STAR CLUSTER IN LMC
HSE 072	04 57 03.	- 68 04			STAR CLUSTER IN LMC
LB 01454	04 57 04.	+ 17 47 12.		15.4	FAINT BLUE STAR
RNGC 1772	04 57 04.	- 69 38			CLUSTER/NEBULOSITY IN LMC
SL 128	04 57 04.	- 69 38	50		STAR CLUSTER IN LMC
LB 01455	04 57 05.	+ 20 02 48.		16.1	FAINT BLUE STAR
SC 0454-1737.9	04 57 05.	- 17 33 17.	6		NEBULA
MCG-02-13-030	04 57 06.	- 11 12	120	13.	GALAXY
CM 10	04 57 06.	- 68 26			HII REGION IN LMC
SL 131	04 57 07.	- 65 56	15		STAR CLUSTER IN LMC
HSE 073	04 57 07.	- 67 30			STAR CLUSTER IN LMC
RNGC 1728	04 57 08.	- 11 12		13.0	GALAXY
LW 096	04 57 08.	- 74 21			STAR CLUSTER IN LMC
ARC 0526	04 57 09.	+ 05 23		16.4	RICH CLUSTER OF GALAXIES
MCG+00-13-061	04 57 09.	- 00 14	36	15.	GALAXY
LW 095	04 57 09.	- 75 12			STAR CLUSTER IN LMC
SL 133	04 57 11.	- 65 21	35		STAR CLUSTER IN LMC
RNGC 1770	04 57 11.	- 68 29		9.0	CLUSTER/NEBULOSITY IN LMC
SL 130	04 57 11.	- 68 29	270		STAR CLUSTER IN LMC
ZC 0457.2+1010	04 57 12.	+ 10 10	1880		CLUSTER OF GALAXIES
ZWG 394.064	04 57 12.	- 00 14		15.5	GALAXY
MCG-01-13-042	04 57 12.	- 07 51	48	13.	GALAXY
SC 0454-1433.8	04 57 12.	- 14 29 11.	12		NEBULA
SC 0454-1642.3	04 57 12.	- 16 37 42.	60		NEBULA
LH120-N011A	04 57 12.	- 66 27			EMISSION NEBULA IN LMC
LH120-N092B	04 57 12.	- 68 50	48		EMISSION NEBULA IN LMC
LH120-N093	04 57 12.	- 69 17	41		EMISSION NEBULA IN LMC
LH120-N094C	04 57 12.	- 69 35	223		EMISSION NEBULA IN LMC
SL 132	04 57 14.	- 67 45	15		STAR CLUSTER IN LMC
RNGC 1775	04 57 14.	- 70 30			OPEN CLUSTER IN LMC
SL 129	04 57 14.	- 70 30	70		STAR CLUSTER IN LMC
IC 2113	04 57 15.	- 15 53 49.			NONSTELLAR OBJECT
MCG-03-13-042	04 57 15.	- 16 38	72	13.5	GALAXY
RNGC 1730	04 57 16.	- 15 54		13.0	GALAXY
IC 2116	04 57 16.1	- 66 29 13.			HII REGION IN LMC
RNGC 1726	04 57 18.	- 07 49		13.5	GALAXY
MCG-02-13-031	04 57 18.	- 11 21 30.	96	14.	GALAXY
MCG-03-13-043	04 57 18.	- 15 54	138	13.	GALAXY
LH120-N091A	04 57 18.	- 68 31	41		EMISSION NEBULA IN LMC
HSE 074	04 57 18.	- 68 49			STAR CLUSTER IN LMC
LH 011	04 57 18.	- 68 49	90		STELLAR ASSN. IN LMC
LH120-N092	04 57 18.	- 68 50	169		EMISSION NEBULA IN LMC
LB 01456	04 57 19.	+ 15 58 36.		16.6	FAINT BLUE STAR
SC 0455-1736.3	04 57 20.	- 17 31 42.	30		NEBULA
HN 0089	04 57 20.	- 68 30			NEBULA
LB 01457	04 57 22.	+ 17 58 24.		16.4	FAINT BLUE STAR
SL 134	04 57 22.	- 68 26	20		STAR CLUSTER IN LMC
SCHO 0083	04 57 23.	+ 23 08 12.	400		ISOLATED DARK CLOUD
SC 0455-1358.3	04 57 23.	- 13 53 42.	6		NEBULA
LW 099	04 57 23.	- 65 21			STAR CLUSTER IN LMC
IC 2117	04 57 23.2	- 68 31 01.			HII REGION IN LMC
UGC 03227	04 57 24.	+ 76 48	84	16.0	GALAXY Sb-c
MCG+13-04-013	04 57 24.	+ 76 48	60	16.	GALAXY
SC 0455-1735.1	04 57 24.	- 17 30 31.	6		NEBULA
MCG-03-13-044	04 57 24.	- 17 32	48	14.5	GALAXY
MCG-04-12-027	04 57 24.	- 22 46 30.	24	16.	GALAXY
LH120-N011J	04 57 24.	- 66 23			EMISSION NEBULA IN LMC
HN 0088	04 57 24.	- 66 28			NEBULA
LH120-N091B	04 57 24.	- 68 31	48		EMISSION NEBULA IN LMC
LW 100	04 57 25.	- 65 45			STAR CLUSTER IN LMC
MCG-03-13-045	04 57 27.	- 19 17	72	14.5	GALAXY
SC 0455-1236.3	04 57 29.	- 12 31 44.	36		NEBULA
MIL 05	04 57 30.	+ 46 30	8400		SUPERNOVA REMNANT
MCG-02-13-032	04 57 30.	- 12 07	54	14.5	GALAXY
MCG-02-13-033	04 57 30.	- 12 31 30.	36	14.5	GALAXY
LH120-N091	04 57 30.	- 68 29	420		EMISSION NEBULA IN LMC
LH120-N092A	04 57 30.	- 68 50	30		EMISSION NEBULA IN LMC
HSE 075	04 57 31.	- 71 30			STAR CLUSTER IN LMC
LH 013	04 57 33.	- 66 32	180		STELLAR ASSN. IN LMC
SCHO 0084	04 57 34.	+ 23 12 48.	340		ISOLATED DARK CLOUD
LH 012	04 57 35.	- 68 29	360		STELLAR ASSN. IN LMC
2ZW 026	04 57 36.	- 01 34			COMPACT GALAXY
LH120-N011K	04 57 36.	- 66 19	61		EMISSION NEBULA IN LMC
SCHO 0085	04 57 37.	+ 45 54 00.	390		ISOLATED DARK CLOUD
SL 136	04 57 37.	- 69 07	35		STAR CLUSTER IN LMC
IC 0397	04 57 38.	+ 40 21 29.			NONSTELLAR OBJECT
RNGC 1729	04 57 39.	- 03 25		13.0	GALAXY
RNGC 1769	04 57 39.	- 66 32			CLUSTER/NEBULOSITY IN LMC
SL 138	04 57 40.	- 66 49	15		STAR CLUSTER IN LMC
2ZW 027	04 57 42.	- 01 33			COMPACT GALAXY
MCG-03-13-046	04 57 42.	- 17 32 30.	84	14.	GALAXY
MCG-03-13-047	04 57 42.	- 18 39	54	15.	GALAXY
MCG-04-12-028	04 57 42.	- 23 59 30.	36	16.	GALAXY
LW 102	04 57 42.	- 65 33			STAR CLUSTER IN LMC
LH120-N011C	04 57 42.	- 66 32	191		EMISSION NEBULA IN LMC
LH120-N011D	04 57 42.	- 66 33			EMISSION NEBULA IN LMC
MCG-01-13-043	04 57 45.	- 03 25	72	13.	GALAXY
SC 0455-1736.3	04 57 46.	- 17 31 44.	36		NEBULA
SL 139	04 57 46.	- 68 16	15		STAR CLUSTER IN LMC
SCHO 0086	04 57 47.	+ 23 18 24.	350		ISOLATED DARK CLOUD
RNGC 1774	04 57 48.	- 67 19		10.5	OPEN CLUSTER IN LMC
SL 141	04 57 48.	- 67 19	70		STAR CLUSTER IN LMC
LW 103	04 57 49.	- 65 54			STAR CLUSTER IN LMC
LW 098	04 57 49.	- 72 32			STAR CLUSTER IN LMC
RNGC 1765	04 57 50.	- 62 06			GALAXY
IC 2112	04 57 51.	+ 04 19 51.			NONSTELLAR OBJECT
MCG-04-12-029	04 57 51.	- 26 05	600	12.	GALAXY
RNGC 1782	04 57 51.	- 69 27			CLUSTER/NEBULOSITY IN LMC
SL 140	04 57 51.	- 69 27	35		STAR CLUSTER IN LMC
SC 0455-1737.4	04 57 53.	- 17 32 51.	6		NEBULA
RNGC 1744	04 57 53.	- 26 06		12.0	GALAXY
SL 135	04 57 53.	- 73 57	25		STAR CLUSTER IN LMC
LW 097	04 57 53.	- 73 57			STAR CLUSTER IN LMC
ZWG 420.027	04 57 54.	+ 04 19		15.3	GALAXY
ZWG 394.065	04 57 54.	- 01 28		15.6	GALAXY
SC 0455-1400.7	04 57 54.	- 13 56 08.	36		NEBULA

OBJECT NAME	RIGHT ASCEN.	DECLINATION	DIAM.	MAGN.	TYPE OF OBJECT
SL 142	04 57 54.	- 65 28	15		STAR CLUSTER IN LMC
MCG-02-13-034	04 57 57.	- 12 47	90	14.5	GALAXY
MCG-02-13-035	04 57 57.	- 13 51	48	15.	GALAXY
SC 0455-1519.8	04 57 57.	- 15 15 15.	12		NEBULA
SL 137	04 57 57.	- 72 48	20		STAR CLUSTER IN LMC
ARC 0529	04 57 59.	+ 06 07		17.0	RICH CLUSTER OF GALAXIES
SL 143	04 57 59.	- 67 08	15		STAR CLUSTER IN LMC
KHAV 101	04 58	+ 33 41	5970		DARK NEBULA
KHAV 100	04 58	+ 35 41	2470		DARK NEBULA
KHAV 099	04 58	+ 44 35	8480		DARK NEBULA
LDN 1542	04 58 00.	+ 25 30	3180		DARK NEBULA
LDN 1540	04 58 00.	+ 26 05	2100		DARK NEBULA
MCG-04-12-030	04 58 00.	- 22 16	48	15.	GALAXY
LB 01765	04 58 00.	- 55 34		12.9	FAINT BLUE STAR
CM 11	04 58 00.	- 69 30			HII REGION IN LMC
ARC 0530	04 58 02.	- 00 56		18.2	RICH CLUSTER OF GALAXIES
LH 014	04 58 03.	- 66 26	90		STELLAR ASSN. IN LMC
LW 101	04 58 03.	- 72 48			STAR CLUSTER IN LMC
KARA.73 25	04 58 04.	- 33 44	27		DWARF GALAXY
LH120-N011E	04 58 06.	- 66 25	127		EMISSION NEBULA IN LMC
LH120-N016	04 58 06.	- 67 46			EMISSION NEBULA IN LMC
LB 03405	04 58 06.	- 72 39		13.6	FAINT BLUE STAR
RNGC 1773	04 58 09.	- 66 25			CLUSTER/NEBULOSITY IN LMC
ZC 0458.2+0137	04 58 12.	+ 01 37	1550		CLUSTER OF GALAXIES
MCG+13-04-014	04 58 12.	+ 75 56	57	17.	GALAXY
LW 105	04 58 12.	- 65 28			STAR CLUSTER IN LMC
LB 01458	04 58 13.	+ 16 08 00.		17.0	FAINT BLUE STAR
RNGC 1776	04 58 15.	- 66 30			OPEN CLUSTER IN LMC
SC 0456-1734.7	04 58 16.	- 17 30 10.	54		NEBULA
ZWG 394.066	04 58 18.	+ 01 17		15.7	GALAXY
MCG-03-13-048	04 58 18.	- 17 31	42	14.5	GALAXY
LB 01766	04 58 18.	- 53 57		12.4	FAINT BLUE STAR
SL 145	04 58 21.	- 66 30	50		STAR CLUSTER IN LMC
LB 01459	04 58 25.	+ 17 48 42.		16.5	FAINT BLUE STAR
HSE 076	04 58 26.	- 69 08			STAR CLUSTER IN LMC
SL 144	04 58 28.	- 71 58	70		STAR CLUSTER IN LMC
RNGC 1771	04 58 29.	- 63 24			GALAXY
ZC 0458.5+0102	04 58 30.	+ 01 02	3490		CLUSTER OF GALAXIES
ZC 0458.5+0536	04 58 30.	+ 05 36	1210		CLUSTER OF GALAXIES
ZWG 420.028	04 58 30.	+ 07 30		15.2	GALAXY
UGC 03228	04 58 30.	+ 75 58	72	16.5	GALAXY Sb-c
CM 02	04 58 30.	- 66 36			HII REGION IN LMC
HSE 077	04 58 31.	- 67 27			STAR CLUSTER IN LMC
RNGC 1789	04 58 34.	- 71 58			OPEN CLUSTER IN LMC
LW 104	04 58 34.	- 71 59			STAR CLUSTER IN LMC
7ZW 023	04 58 36.	+ 65 44			COMPACT GALAXY
MCG-01-13-044	04 58 36.	- 08 41	36	15.5	GALAXY
MCG-02-13-036	04 58 36.	- 09 01	66	14.5	GALAXY
MCG-04-12-031	04 58 36.	- 25 08	33	14.5	GALAXY
LH120-N012	04 58 36.	- 66 16	288		EMISSION NEBULA IN LMC
LB 03406	04 58 36.	- 67 15		13.6	FAINT BLUE STAR
SL 146	04 58 36.	- 70 09	55		STAR CLUSTER IN LMC
LB 03407	04 58 36.	- 70 18		13.3	FAINT BLUE STAR
LB 03408	04 58 36.	- 70 19		13.3	FAINT BLUE STAR
RNGC 1708	04 58 37.	+ 52 49			NON-EXISTENT OBJECT
SC 0456-1334.4	04 58 37.	- 13 29 53.	18		NEBULA
LB 01460	04 58 38.	+ 17 35 12.		16.0	FAINT BLUE STAR
SL 147	04 58 41.	- 66 52	15		STAR CLUSTER IN LMC
SHB 088	04 58 41.3	- 02 03 34.		18.5	QUASI-STELLAR OBJECT
RLWT 112	04 58 42.	+ 40 45		16.5	FAINT VERY BLUE STAR
SCHO 0087	04 58 42.	+ 44 17 30.	760		ISOLATED DARK CLOUD
OCL 0414	04 58 42.	+ 44 23	540	15.	OPEN STAR CLUSTER
ISS 0152	04 58 42.	+ 47 19	186		STELLAR RING
MCG-02-13-037	04 58 42.	- 13 30	24	14.5	GALAXY
LB 03409	04 58 42.	- 65 54		13.1	FAINT BLUE STAR
RNGC 1783	04 58 44.	- 66 04		11.0	GLOBULAR CLUSTER IN LMC
SL 148	04 58 44.	- 66 04	340		STAR CLUSTER IN LMC
SCHO 0088	04 58 46.	+ 23 51 30.	450		ISOLATED DARK CLOUD
VDB.66N 032	04 58 47.	+ 44 11	108		REFLECTION NEBULA
ZWG 394.067	04 58 48.	+ 00 53		15.7	GALAXY
ZWG 420.029	04 58 48.	+ 03 42		15.5	GALAXY
MCG-05-12-013	04 58 48.	- 29 55	48	16.	GALAXY
LH120-N012A	04 58 48.	- 66 18	37		EMISSION NEBULA IN LMC
RNGC 1785	04 58 49.	- 68 56			NON-EXISTENT OBJECT
ARC 0531	04 58 50.	- 03 37		17.0	RICH CLUSTER OF GALAXIES
MCG+13-04-015	04 58 54.	+ 75 35	27	16.	GALAXY
MCG-03-13-049	04 58 54.	- 18 23 30.	48	15.	GALAXY
MCG-04-12-032	04 58 54.	- 21 43 30.	24	15.5	GALAXY
SC 0456-1410.4	04 58 55.	- 14 05 55.	30		NEBULA
RNGC 1787	04 58 55.	- 65 48			OPEN CLUSTER IN LMC
LH 015	04 58 55.	- 65 48	1440		STELLAR ASSN. IN LMC
RNGC 1786	04 58 56.	- 67 49		10.0	GLOBULAR CLUSTER IN LMC
SL 149	04 58 56.	- 67 49	70		STAR CLUSTER IN LMC
LBN 0838	04 59	+ 14 00	3600		BRIGHT NEBULA
KHAV 102	04 59	+ 28 22	4320		DARK NEBULA
LBN 0968	04 59	- 09 00	960		BRIGHT NEBULA
DG 049	04 59	- 09 05	1200		REFLECTION NEBULA
2ZW 028	04 59 00.	+ 03 30			COMPACT GALAXY
LDN 1563	04 59 00.	+ 13 45	900		DARK NEBULA
LDN 1562	04 59 00.	+ 14 00	1560		DARK NEBULA
SC 0456-1353.5	04 59 00.	- 13 49 01.	12		NEBULA
SL 152	04 59 00.	- 65 38	25		STAR CLUSTER IN LMC
LB 03410	04 59 00.	- 70 36		13.5	FAINT BLUE STAR
SL 150	04 59 02.	- 69 17	15		STAR CLUSTER IN LMC
SL 153	04 59 03.	- 66 24	40		STAR CLUSTER IN LMC
RNGC 1741	04 59 04.	- 04 20		15.0	GALAXY
ZWG 394.068	04 59 06.	- 00 35		15.3	GALAXY
MCG+00-13-062	04 59 06.	- 00 35	30	15.	GALAXY
MCG-01-13-045	04 59 06.	- 04 20	90	15.	GALAXY
MCG-03-13-050	04 59 06.	- 20 23	36	15.5	GALAXY
MCG-06-12-002	04 59 06.	- 38 46	12	15.5	GALAXY
MCG-06-12-001	04 59 06.	- 38 47	48	15.	GALAXY
LW 108	04 59 06.	- 65 38			STAR CLUSTER IN LMC
RNGC 1742	04 59 09.	- 03 22			NON-EXISTENT OBJECT
RNGC 1759	04 59 09.	- 38 47			GALAXY
SL 154	04 59 09.	- 67 58	20		STAR CLUSTER IN LMC
ZC 0459.2+0212	04 59 12.	+ 02 12	2290		CLUSTER OF GALAXIES
7ZW 024	04 59 12.	+ 69 02			COMPACT GALAXY
ARP 259	04 59 12.	- 04 20			PECULIAR GALAXY
SL 151	04 59 12.	- 70 01	35		STAR CLUSTER IN LMC
IC 0399	04 59 15.0	- 04 21 39.			GALAXY E0
LB 01461	04 59 17.	+ 18 02 00.		15.8	FAINT BLUE STAR
SL 156	04 59 17.	- 66 54	30		STAR CLUSTER IN LMC
MRSL 187-16/1	04 59 18.	+ 14 02	3900		HII REGION
LB 03411	04 59 18.	- 68 38		13.3	FAINT BLUE STAR
RNGC 1740	04 59 21.	- 03 22		15.0	GALAXY
MCG-03-13-051	04 59 21.	- 16 14	96	14.	GALAXY
MCG-04-12-033	04 59 21.	- 23 11 30.	24	16.	GALAXY
MCG-01-13-046	04 59 24.	- 03 22	36	15.	GALAXY
HOLM 084A	04 59 24.	- 03 23	30	13.9	PART OF MULTIPLE GALAXY
SC 0457-1424.6	04 59 25.	- 14 20 09.	12		NEBULA
ARC 0533	04 59 25.	- 22 41		15.8	RICH CLUSTER OF GALAXIES
RNGC 1791	04 59 25.	- 70 14			DIFFUSE NEBULA IN LMC
SL 155	04 59 25.	- 70 14	25		STAR CLUSTER IN LMC
MCG-03-13-052	04 59 27.	- 15 14	60	14.5	GALAXY
SC 0457-1519.1	04 59 27.	- 15 14 39.	48		NEBULA
SC 0457-1533.2	04 59 27.	- 15 28 45.	6		NEBULA
SC 0457-1618.9	04 59 27.	- 16 14 27.	60		NEBULA
HOLM 084B	04 59 28.	- 03 23	12	14.3	PART OF MULTIPLE GALAXY
SC 0457-1509.0	04 59 29.	- 15 04 33.	48		NEBULA
ZC 0459.5-0235	04 59 30.	- 02 35	3290		CLUSTER OF GALAXIES
MCG-03-13-053	04 59 30.	- 15 31	12	15.	GALAXY
SC 0457-1536.2	04 59 30.	- 15 31 45.	6		NEBULA
RNGC 1738	04 59 30.	- 18 15		13.0	GALAXY
MCG-03-13-054	04 59 30.	- 18 15	72	13.5	GALAXY
RNGC 1739	04 59 30.	- 18 16		14.0	GALAXY
MCG-03-13-055	04 59 30.	- 18 16	72	14.	GALAXY
LW 106	04 59 31.	- 75 36			STAR CLUSTER IN LMC
SCHO 0089	04 59 32.	+ 45 32 12.	280		ISOLATED DARK CLOUD
HUB C08	04 59 33.	- 03 50			DIFFUSE NEBULA
MCG-03-13-056	04 59 33.	- 20 09	36	14.5	GALAXY
ZC 0459.6+0606	04 59 36.	+ 06 06	3560		CLUSTER OF GALAXIES
ZWG 420.030	04 59 36.	+ 07 33		15.7	GALAXY
UGC 03229	04 59 36.	+ 07 33	78	15.7	GALAXY Sc
LB 03412	04 59 36.	- 65 37		13.4	FAINT BLUE STAR
LW 107	04 59 36.	- 74 47			STAR CLUSTER IN LMC
SL 158	04 59 37.	- 70 19	35		STAR CLUSTER IN LMC
SL 161	04 59 38.	- 66 15	35		STAR CLUSTER IN LMC
RNGC 1724	04 59 41.	+ 49 26			NON-EXISTENT OBJECT
OCL 0405	04 59 42.	+ 49 26		10.	OPEN STAR CLUSTER
MCG+13-04-016	04 59 42.	+ 75 31	60	15.	GALAXY
MCG-01-13-047	04 59 42.	- 08 21	138	13.	GALAXY
MCG-04-12-034	04 59 42.	- 21 12	21	14.5	GALAXY
MCG-06-12-003	04 59 42.	- 34 06	60	14.5	GALAXY
SCHO 0090	04 59 43.	+ 23 55 30.	500		ISOLATED DARK CLOUD
SL 157	04 59 43.	- 72 27	15		STAR CLUSTER IN LMC
LW 109	04 59 43.	- 72 27			STAR CLUSTER IN LMC
KLEM 09	04 59 45.	- 34 08	300	13.	CMPT GROUP OF 3 GALAXIES
SL 162	04 59 45.	- 67 59	15		STAR CLUSTER IN LMC
RNGC 1793	04 59 46.	- 69 37			OPEN CLUSTER IN LMC
HSE 079	04 59 47.	- 65 12			STAR CLUSTER IN LMC
ZC 0459.8+0943	04 59 48.	+ 09 43	1340		CLUSTER OF GALAXIES
ZWG 347.011	04 59 48.	+ 75 32		14.1	GALAXY
UGC 03230	04 59 48.	+ 75 32	102	14.1	GALAXY S0-a
RNGC 1752	04 59 48.	- 08 18		13.0	GALAXY
MCG-02-13-038	04 59 48.	- 10 25 30.	96	14.	GALAXY
HSE 078	04 59 49.	- 68 57			STAR CLUSTER IN LMC
SL 159	04 59 50.	- 70 22	50		STAR CLUSTER IN LMC
SL 160	04 59 50.	- 70 29	5		STAR CLUSTER IN LMC
SL 163	04 59 52.	- 69 37	50		STAR CLUSTER IN LMC
RNGC 1795	04 59 53.	- 69 52			GLOBULAR CLUSTER IN LMC
MCG-04-12-035	04 59 54.	- 26 16 30.	60	15.5	GALAXY
LW 112	04 59 54.	- 65 30			STAR CLUSTER IN LMC
RNGC 1753	04 59 57.	- 03 25		15.5	GALAXY
SCHO 0091	04 59 58.	+ 43 51 42.	1160		ISOLATED DARK CLOUD
SL 165	04 59 59.	- 69 52	40		STAR CLUSTER IN LMC
VDB.66G 035	05 00	+ 16 19	70		DWARF GALAXY
KHAV 103	05 00	+ 32 34	1000		DARK NEBULA
LBN 0959	05 00	- 08 00	7200		BRIGHT NEBULA
ZWG 394.069	05 00 00.	+ 00 10		14.9	GALAXY
UGC 03231	05 00 00.	+ 00 10	114	14.9	GALAXY Sc
MCG+00-13-063	05 00 00.	+ 00 11	66	14.	GALAXY
ZC 0500.0+2258	05 00 00.	+ 22 58	2220		CLUSTER OF GALAXIES
LDN 1502	05 00 00.	+ 34 00	2820		DARK NEBULA
LDN 1477	05 00 00.	+ 44 00	2100		DARK NEBULA
LDN 1476	05 00 00.	+ 44 30	1320		DARK NEBULA
LDN 1475	05 00 00.	+ 45 00	2460		DARK NEBULA
LB 01767	05 00 00.	- 51 16		14.1	FAINT BLUE STAR
LH120-N014	05 00 00.	- 66 19	172		EMISSION NEBULA IN LMC
LH120-N016A	05 00 00.	- 68 03	214		EMISSION NEBULA IN LMC
LH120-N186C	05 00 00.	- 70 13	28		EMISSION NEBULA IN LMC
HSE 080	05 00 01.	- 69 02			STAR CLUSTER IN LMC
SL 164	05 00 02.	- 71 37	25		STAR CLUSTER IN LMC
MCG-01-13-048	05 00 03.	- 03 25	60	15.5	GALAXY
MCG-01-13-049	05 00 06.	- 08 24 30.	60	15.	GALAXY
LB 01768	05 00 06.	- 46 49		15.0	FAINT BLUE STAR
LW 113	05 00 06.	- 65 34			STAR CLUSTER IN LMC
LH120-N013	05 00 06.	- 66 09	156		EMISSION NEBULA IN LMC
LH120-N186E	05 00 06.	- 70 15	538		EMISSION NEBULA IN LMC
MCG-03-13-057	05 00 09.	- 20 05 30.	78	14.	GALAXY
RNGC 1757	05 00 10.	- 04 48			NON-EXISTENT OBJECT
MCG-02-13-039	05 00 12.	- 12 28	42	15.	GALAXY
HSE 081	05 00 16.	- 69 35			STAR CLUSTER IN LMC
LW 110	05 00 16.	- 75 19			STAR CLUSTER IN LMC
SCHO 0092	05 00 18.	+ 23 55 48.	500		ISOLATED DARK CLOUD
ZWG 394.070	05 00 18.	- 01 21		15.6	GALAXY
MCG-03-13-058	05 00 18.	- 15 36	84	14.5	GALAXY
MCG-05-12-014	05 00 18.	- 30 30	60	15.	GALAXY
LH120-N186D	05 00 18.	- 70 12	114		EMISSION NEBULA IN LMC
HSE 082	05 00 19.	- 70 10			STAR CLUSTER IN LMC
SL 167	05 00 23.	- 66 57	35		STAR CLUSTER IN LMC
MCG+00-13-064	05 00 24.	+ 00 11	48	15.	GALAXY
ARC 0532	05 00 24.	+ 11 54		17.5	RICH CLUSTER OF GALAXIES
UGC 03232	05 00 24.	+ 18 23	72	17.	GALAXY S
ZC 0500.4+6809	05 00 24.	+ 68 09	1010		CLUSTER OF GALAXIES
ZC 0500.4+7511	05 00 24.	+ 75 11	1340		CLUSTER OF GALAXIES
MCG-04-12-036	05 00 24.	- 21 01	48	15.	GALAXY
SL 168	05 00 24.	- 65 31	20		STAR CLUSTER IN LMC
LH120-N186B	05 00 24.	- 70 08	38		EMISSION NEBULA IN LMC
ZWG 394.071	05 00 30.	+ 00 11		15.3	GALAXY
UGC 03233	05 00 30.	+ 00 11	78	15.3	GALAXY Sc
ZC 0500.5+1151	05 00 30.	+ 11 51	1140		CLUSTER OF GALAXIES
UGC 03234	05 00 30.	+ 16 20	120	16.5	GALAXY DWRF IR
UGC 03235	05 00 30.	+ 66 57	150	16.0	GALAXY Sc
LW 114	05 00 30.	- 65 32			STAR CLUSTER IN LMC
WS 04	05 00 34.	- 70 32 39.		17.32	PLANETARY NEB. IN LMC
RNGC 1746	05 00 35.	+ 23 45		6.0	OPEN CLUSTER
OCL 0452	05 00 36.	+ 23 45	3240	8.4	OPEN STAR CLUSTER
HSE 083	05 00 36.	- 68 50			STAR CLUSTER IN LMC
MCG-04-12-037	05 00 39.	- 25 35 30.	48	15.	GALAXY
SCHO 0093	05 00 41.	+ 32 43 18.	500		ISOLATED DARK CLOUD
SL 166	05 00 41.	- 74 43	35		STAR CLUSTER IN LMC
LW 111	05 00 41.	- 74 43			STAR CLUSTER IN LMC
MCG-03-13-059	05 00 42.	- 18 51 30.	66	15.	GALAXY
LH120-N015	05 00 42.	- 66 27	20		EMISSION NEBULA IN LMC
LB 03413	05 00 42.	- 68 10		13.9	FAINT BLUE STAR

OBJECT NAME	RIGHT ASCEN.	DECLINATION	DIAM.	MAGN.	TYPE OF OBJECT
HSE 084	05 00 43.	- 65 54			STAR CLUSTER IN LMC
LW 115	05 00 44.	- 64 19			STAR CLUSTER IN LMC
MCG+00-13-065	05 00 45.	+ 00 36	78	14.	GALAXY
MCG+00-13-066	05 00 45.	- 03 00	96	14.	GALAXY
MCG-03-13-060	05 00 45.	- 20 27	30	15.	GALAXY
SL 169	05 00 47.	- 71 00	20		STAR CLUSTER IN LMC
ZWG 420.031	05 00 48.	+ 06 35		15.6	GALAXY
UGC 03236	05 00 48.	+ 06 35	66	15.6	GALAXY Sb-c
MCG-01-13-050	05 00 48.	- 03 00	150	15.	GALAXY
HSE 085	05 00 51.	- 67 53			STAR CLUSTER IN LMC
RNGC 1801	05 00 52.	- 69 42			OPEN CLUSTER IN LMC
SL 170	05 00 52.	- 69 42	95		STAR CLUSTER IN LMC
RNGC 1750	05 00 53.	+ 23 35			OPEN CLUSTER
ZWG 394.072	05 00 54.	+ 00 35		15.2	GALAXY
UGC 03237	05 00 54.	+ 00 35	102	15.2	GALAXY Sc
ZWG 420.032	05 00 54.	+ 04 36		15.0	GALAXY
MCG+01-13-014	05 00 54.	+ 06 34	60	15.	GALAXY
OCL 0454	05 00 54.	+ 23 35		8.	OPEN STAR CLUSTER
PK215-30.1	05 00 54.	- 15 40	871	13.2	PLANETARY NEBULA
MCG-04-12-038	05 00 57.	- 22 52	48	15.	GALAXY
SL 171	05 00 59.	- 71 00	20		STAR CLUSTER IN LMC
ZWG 394.073	05 01 00.	+ 01 30		13.5	GALAXY
RNGC 1762	05 01 00.	+ 01 30		13.5	GALAXY
UGC 03238	05 01 00.	+ 01 30	108	13.5	GALAXY Sc
MCG+00-13-067	05 01 00.	+ 01 30	72	13.	GALAXY
LDN 1544	05 01 00.	+ 25 10	1320		DARK NEBULA
ZC 0501.0+8443	05 01 00.	+ 84 43	1080		CLUSTER OF GALAXIES
MCG-04-12-039	05 01 00.	- 22 53 30.	60	14.	GALAXY
HUB C09	05 01 02.	- 07 44			DIFFUSE NEBULA
HSE 086	05 01 03.	- 67 59			STAR CLUSTER IN LMC
ZWG 445.002	05 01 06.	+ 10 57		15.6	GALAXY
SL 173	05 01 07.	- 67 21	15		STAR CLUSTER IN LMC
RNGC 1804	05 01 08.	- 69 09			OPEN CLUSTER IN LMC
SL 172	05 01 08.	- 69 09	25		STAR CLUSTER IN LMC
SL 174	05 01 09.	- 67 53	35		STAR CLUSTER IN LMC
HSE 087	05 01 09.	- 69 25			STAR CLUSTER IN LMC
HSE 088	05 01 10.	- 68 09			STAR CLUSTER IN LMC
ZC 0501.2+5959	05 01 12.	+ 59 59	1480		CLUSTER OF GALAXIES
SCHO 0094	05 01 17.	+ 23 24 12.	510		ISOLATED DARK CLOUD
HSE 089	05 01 17.	- 68 28			STAR CLUSTER IN LMC
OCL 0430	05 01 18.	+ 34 46	1080		OPEN STAR CLUSTER
MCG-03-13-061	05 01 18.	- 20 08	36	15.	GALAXY
MCG-04-12-041	05 01 18.	- 25 29	72	14.	GALAXY
MCG-04-12-040	05 01 18.	- 25 44 30.	72	15.	GALAXY
LB 01769	05 01 18.	- 53 43		13.9	FAINT BLUE STAR
LH120-N186A	05 01 18.	- 70 18			EMISSION NEBULA IN LMC
WS 05	05 01 20.	- 70 17 07.		16.54	PLANETARY NEB. IN LMC
RNGC 1758	05 01 23.	+ 23 42			OPEN CLUSTER
OCL 0453	05 01 24.	+ 23 43			OPEN STAR CLUSTER
LB 03414	05 01 24.	- 62 14		13.6	FAINT BLUE STAR
SL 177	05 01 24.	- 65 37	15		STAR CLUSTER IN LMC
SL 176	05 01 24.	- 68 46	25		STAR CLUSTER IN LMC
SL 175	05 01 24.	- 71 07	20		STAR CLUSTER IN LMC
HSE 090	05 01 25.	- 70 15			STAR CLUSTER IN LMC
SL 178	05 01 26.	- 65 57	25		STAR CLUSTER IN LMC
IC 0400	05 01 28.	- 15 50			NONSTELLAR OBJECT
MCG-04-12-042	05 01 30.	- 23 23	24	15.5	GALAXY
LB 01770	05 01 30.	- 47 39		13.8	FAINT BLUE STAR
SL 179	05 01 30.	- 67 09	20		STAR CLUSTER IN LMC
ZWG 283.002	05 01 36.	+ 62 24		15.7	GALAXY
ZWG 394.074	05 01 36.	- 01 13		15.5	GALAXY
MCG+00-13-068	05 01 36.	- 01 13	48	15.	GALAXY
LW 117	05 01 36.	- 65 34			STAR CLUSTER IN LMC
SL 180	05 01 38.	- 69 06	25		STAR CLUSTER IN LMC
HSE 091	05 01 39.	- 66 28			STAR CLUSTER IN LMC
ZWG 420.033	05 01 42.	+ 04 35		15.7	GALAXY
UGC 03240	05 01 42.	+ 04 35	66	15.7	GALAXY Sb-c
7ZW 025	05 01 42.	+ 64 03			COMPACT GALAXY
ZWG 394.075	05 01 42.	- 00 29		15.7	GALAXY
UGC 03239	05 01 42.	- 00 29	72	15.7	GALAXY Sa-b
SL 182	05 01 42.	- 63 34	5		STAR CLUSTER IN LMC
MCG-02-13-040	05 01 51.	- 10 09	78	14.	GALAXY
MCG-04-12-043	05 01 51.	- 24 02 30.	48	15.	GALAXY
MCG-01-13-051	05 01 54.	- 08 29	78	15.5	GALAXY
SC 0459-1505.4	05 01 54.	- 15 01 07.	6		NEBULA
LW 116	05 01 54.	- 73 07			STAR CLUSTER IN LMC
SL 181	05 01 57.	- 69 17	35		STAR CLUSTER IN LMC
IC 0401	05 01 58.	- 10 08 48.			NONSTELLAR OBJECT
LBN 0759	05 02	+ 46 10	3600		BRIGHT NEBULA
LBN 0923	05 02	- 04 00	5400		BRIGHT NEBULA
ZC 0502.0+0350	05 02 00.	+ 03 50	940		CLUSTER OF GALAXIES
LDN 1480	05 02 00.	+ 43 30	1260		DARK NEBULA
MCG+13-04-017	05 02 00.	+ 75 21	60	15.	GALAXY
RNGC 1805	05 02 02.	- 66 10		10.5	OPEN CLUSTER IN LMC
SL 186	05 02 02.	- 66 10	65		STAR CLUSTER IN LMC
RNGC 1806	05 02 03.	- 68 03		11.5	GLOBULAR CLUSTER IN LMC
SL 184	05 02 03.	- 68 03	135		STAR CLUSTER IN LMC
ZC 0502.1+0201	05 02 06.	+ 02 01	4840		CLUSTER OF GALAXIES
ZWG 347.012	05 02 06.	+ 75 22		14.6	GALAXY
UGC 03241	05 02 06.	+ 75 22	108	14.6	GALAXY S0
RNGC 1796	05 02 06.	- 61 12		13.0	GALAXY
SC 0501-6117.0	05 02 07.	- 61 12 49.	72		NEBULA
LB 01462	05 02 08.	+ 17 33 00.		16.5	FAINT BLUE STAR
ZWG 394.076	05 02 12.	+ 01 45		15.4	GALAXY
ZWG 420.034	05 02 12.	+ 04 25		15.7	GALAXY
UGC 03242	05 02 12.	+ 04 25	84	15.7	GALAXY Sb
UGC 03243	05 02 12.	+ 20 56	60	16.5	GALAXY Sc/SBc
MCG+00-13-069	05 02 15.	+ 01 46	18	15.5	GALAXY
SL 183	05 02 15.	- 71 41	25		STAR CLUSTER IN LMC
SC 0500-1643.0	05 02 17.	- 16 38 45.	6		NEBULA
SC 0500-1643.8	05 02 17.	- 16 39 33.	24		NEBULA
5ZW 376	05 02 18.	+ 23 57			COMPACT GALAXY
UGC 03244	05 02 18.	+ 33 57	84	18.	GALAXY
ZWG 307.002	05 02 18.	+ 67 35		15.7	GALAXY
SER 040.05	05 02 18.	- 61 15	100	13.	HIGH SURFACE BRIGHT. GLXY
LB 03415	05 02 18.	- 71 46		13.6	FAINT BLUE STAR
HSE 093	05 02 20.	- 66 05			STAR CLUSTER IN LMC
RNGC 1809	05 02 22.	- 69 38			GALAXY
SG 3.027	05 02 24.	- 08 12	480		DIFFUSE EMISSION NEBULA
MCG-03-13-062	05 02 24.	- 15 50	36	15.	GALAXY
MCG-03-13-063	05 02 24.	- 16 39	72	14.5	GALAXY
MCG-03-13-064	05 02 24.	- 17 09	84	15.	GALAXY
HSE 092	05 02 24.	- 68 37			STAR CLUSTER IN LMC
LW 118	05 02 24.	- 73 13			STAR CLUSTER IN LMC
LW 120	05 02 29.	- 63 18			STAR CLUSTER IN LMC
LH120-N187	05 02 30.	- 70 46			EMISSION NEBULA IN LMC
SL 185	05 02 30.	- 73 07	50		STAR CLUSTER IN LMC
ARP 187	05 02 33.	- 10 19			PECULIAR GALAXY
HN 0068	05 02 34.	+ 00 04			NEBULA
SCHO 0095	05 02 34.	+ 43 34 48.	770		ISOLATED DARK CLOUD
HUB E06	05 02 35.	+ 33 45			DIFFUSE NEBULA
MCG-02-13-040A	05 02 36.	- 10 19	48	15.	GALAXY
MCG-03-13-065	05 02 36.	- 20 37	36	15.5	GALAXY
MCG-05-13-001	05 02 36.	- 29 09	24	15.5	GALAXY
SL 188	05 02 37.	- 68 53	25		STAR CLUSTER IN LMC
ZC 0502.7-0226	05 02 42.	- 02 26	1280		CLUSTER OF GALAXIES
ZC 0502.7-0245	05 02 42.	- 02 45	1950		CLUSTER OF GALAXIES
PK190-17.1	05 02 47.	+ 10 39	22	12.9	PLANETARY NEBULA
MCG-02-13-041	05 02 48.	- 09 12	150	13.	GALAXY
SER 040.02	05 02 48.	- 58 15	30	17.	3 INTERACTING GALAXIES
LH120-N095	05 02 48.	- 68 33	16		EMISSION NEBULA IN LMC
RNGC 1815	05 02 51.	- 70 41			OPEN CLUSTER IN LMC
SL 189	05 02 51.	- 70 41	55		STAR CLUSTER IN LMC
RNGC 1779	05 02 54.	- 09 13		13.0	GALAXY
SL 187	05 02 54.	- 72 11	70		STAR CLUSTER IN LMC
HSE 094	05 02 56.	- 70 22			STAR CLUSTER IN LMC
MCG-03-13-066	05 02 57.	- 19 04	60	15.	GALAXY
SCHO 0097	05 02 58.	+ 23 25 00.	450		ISOLATED DARK CLOUD
SCHO 0096	05 02 59.	+ 23 05 42.	450		ISOLATED DARK CLOUD
KHAV 104	05 03	+ 38 04	7500		DARK NEBULA
2ZW 029	05 03 00.	+ 10 48			COMPACT GALAXY
LDN 1523	05 03 00.	+ 31 40	480		DARK NEBULA
MCG+12-05-028	05 03 00.	+ 70 25	84	15.	GALAXY
LDN 1615	05 03 00.	- 03 30	1140		DARK NEBULA
MCG-03-13-067	05 03 00.	- 17 25	36	15.	GALAXY
SCHO 0098	05 03 02.	+ 24 25 42.	420		ISOLATED DARK CLOUD
RNGC 1813	05 03 02.	- 70 23			OPEN CLUSTER IN LMC
SL 190	05 03 02.	- 70 23	40		STAR CLUSTER IN LMC
SC 0500-1729.6	05 03 03.	- 17 25 24.	36		NEBULA
RNGC 1810	05 03 03.	- 66 26			OPEN CLUSTER IN LMC
ZC 0503.1+0751	05 03 06.	+ 07 51	1210		CLUSTER OF GALAXIES
ZWG 328.032	05 03 06.	+ 70 26		15.0	GALAXY
UGC 03245	05 03 06.	+ 70 26	96	15.0	GALAXY Sc
MCG-02-13-042	05 03 06.	- 11 57	270	12.	GALAXY
SL 193	05 03 06.	- 65 25	15		STAR CLUSTER IN LMC
LW 119	05 03 06.	- 72 12			STAR CLUSTER IN LMC
SCHO 0099	05 03 07.	+ 22 56 36.	410		ISOLATED DARK CLOUD
HSE 095	05 03 07.	- 68 50			STAR CLUSTER IN LMC
SL 191	05 03 08.	- 69 06	50		STAR CLUSTER IN LMC
SL 194	05 03 09.	- 66 26	40		STAR CLUSTER IN LMC
B 029	05 03 10.	+ 31 31	600		DARK OBJECT
PK167-00.1	05 03 12.	+ 39 04	64	16.6	PLANETARY NEBULA
MCG-05-13-002	05 03 12.	- 31 52	60	15.	GALAXY
SL 195	05 03 12.	- 65 29	15		STAR CLUSTER IN LMC
LH120-N096	05 03 12.	- 69 26			EMISSION NEBULA IN LMC
SL 196	05 03 13.	- 65 52	15		STAR CLUSTER IN LMC
RNGC 1784	05 03 14.	- 11 56		12.5	GALAXY
HSE 096	05 03 14.	- 65 57			STAR CLUSTER IN LMC
LB 01463	05 03 15.	+ 18 48 12.		15.4	FAINT BLUE STAR
ZWG 394.077	05 03 18.	+ 02 25		15.3	GALAXY
SL 197	05 03 20.	- 67 41	15		STAR CLUSTER IN LMC
LH 016	05 03 20.	- 69 06	180		STELLAR ASSN. IN LMC
HSE 098	05 03 21.	- 66 17			STAR CLUSTER IN LMC
ZWG 394.078	05 03 24.	+ 00 29		15.0	GALAXY
UGC 03246	05 03 24.	+ 00 29	66	15.0	GALAXY S
MCG+00-13-070	05 03 24.	+ 00 29	48	14.	GALAXY
LW 122	05 03 24.	- 65 27			STAR CLUSTER IN LMC
LW 123	05 03 24.	- 65 31			STAR CLUSTER IN LMC
RNGC 1792	05 03 27.	- 38 04		11.5	GALAXY
HSE 097	05 03 28.	- 69 33			STAR CLUSTER IN LMC
SCHO 0100	05 03 30.	+ 24 32 00.	490		ISOLATED DARK CLOUD
MCG-06-12-004	05 03 30.	- 38 04	390	10.7	GALAXY
LH120-N019	05 03 30.	- 68 01			EMISSION NEBULA IN LMC
LW 124	05 03 31.	- 65 38			STAR CLUSTER IN LMC
YC 0503-38	05 03 32.	- 38 02 54.			UNUSUAL SOUTHERN GALAXY
ZWG 394.079	05 03 36.	- 00 06		15.7	GALAXY
MCG-03-13-068	05 03 36.	- 19 53 30.	48	14.5	GALAXY
HSE 099	05 03 37.	- 70 10			STAR CLUSTER IN LMC
SCHO 0101	05 03 40.	+ 44 43 24.	380		ISOLATED DARK CLOUD
LH120-N017B	05 03 42.	- 67 22	57		EMISSION NEBULA IN LMC
SL 192	05 03 43.	- 74 54	70		STAR CLUSTER IN LMC
HSE 101	05 03 45.	- 66 20			STAR CLUSTER IN LMC
MCG-02-13-043	05 03 48.	- 09 10 30.	138	13.5	GALAXY
MCG-03-13-069	05 03 48.	- 16 48	72	15.	GALAXY
MCG-05-13-003	05 03 48.	- 28 40	24	15.5	GALAXY
LH120-N017	05 03 48.	- 67 23	110		EMISSION NEBULA IN LMC
SC 0501-1711.7	05 03 49.	- 17 07 34.	12		NEBULA
RNGC 1820	05 03 49.	- 67 21		9.0	OPEN CLUSTER IN LMC
RNGC 1816	05 03 49.	- 67 21		9.0	CLUSTER/NEBULOSITY IN LMC
RNGC 1814	05 03 49.	- 67 21		9.0	CLUSTER/NEBULOSITY IN LMC
SL 199	05 03 49.	- 67 21	410		STAR CLUSTER IN LMC
SL 198	05 03 50.	- 70 24	70		STAR CLUSTER IN LMC
MCG-04-13-001	05 03 51.	- 24 33	30	16.	GALAXY
IC 0402	05 03 53.	- 09 12 02.			NONSTELLAR OBJECT
HSE 100	05 03 53.	- 69 49			STAR CLUSTER IN LMC
ZWG 420.035	05 03 54.	+ 08 35		15.3	GALAXY
UGC 03247	05 03 54.	+ 08 35	60	15.3	GALAXY S
LH120-N017A	05 03 54.	- 67 23	23		EMISSION NEBULA IN LMC
LB 03416	05 03 54.	- 70 16		12.4	FAINT BLUE STAR
LW 121	05 03 55.	- 74 56			STAR CLUSTER IN LMC
HSE 103	05 03 56.	- 66 08			STAR CLUSTER IN LMC
RNGC 1823	05 03 56.	- 70 25			OPEN CLUSTER IN LMC
ARC 0527	05 03 59.	+ 73 38		15.7	RICH CLUSTER OF GALAXIES
CED 041A	05 04	- 07 18	8940		DIFFUSE GALACTIC NEBULA
CED 041B	05 04	- 07 18			DIFFUSE GALACTIC NEBULA
LBN 0868	05 04	+ 06 00	1020		BRIGHT NEBULA
LBN 0809	05 04	+ 30 46	120		BRIGHT NEBULA
LBN 0739	05 04	+ 50 16	360		BRIGHT NEBULA
LBN 0916	05 04	- 03 25	120		BRIGHT NEBULA
DG 052	05 04	- 07 20	9300		REFLECTION NEBULA
LBN 0980	05 04	- 11 00	1860		BRIGHT NEBULA
SS 38	05 04 00.	+ 30 45			DIFFUSE GALACTIC NEBULA
DG 050	05 04 00.	+ 31 45	120		REFLECTION NEBULA
LDN 1522	05 04 00.	+ 32 00	540		DARK NEBULA
LDN 1616	05 04 00.	- 03 30	480		DARK NEBULA
LB 03417	05 04 00.	- 71 09		13.0	FAINT BLUE STAR
LH 019	05 04 01.	- 67 20	420		STELLAR ASSN. IN LMC
LH 017	05 04 02.	- 69 09	120		STELLAR ASSN. IN LMC
RNGC 1818	05 04 04.	- 66 28		10.0	OPEN CLUSTER IN LMC
SL 201	05 04 04.	- 66 28	150		STAR CLUSTER IN LMC
HSE 102	05 04 04.	- 69 28			STAR CLUSTER IN LMC
MRSL 194-19/1	05 04 06.	+ 06 06	1200		HII REGION
ZWG 283.003	05 04 06.	+ 62 22		15.2	GALAXY
7ZW 026	05 04 06.	+ 69 08			COMPACT GALAXY
LH120-N188	05 04 06.	- 70 18			EMISSION NEBULA IN LMC
SC 0501-1458.6	05 04 08.	- 14 54 29.	6		NEBULA

OBJECT NAME	RIGHT ASCEN.	DECLINATION	DIAM.	MAGN.	TYPE OF OBJECT
LH 018	05 04 08.	- 70 24	480		STELLAR ASSN. IN LMC
ZC 0504.2+0251	05 04 12.	+ 02 51	1810		CLUSTER OF GALAXIES
RNGC 1780	05 04 13.	- 19 33		15.0	GALAXY
WS 06	05 04 14.	- 70 17 37.		17.06	PLANETARY NEB. IN LMC
SL 200	05 04 14.	- 70 28	55		STAR CLUSTER IN LMC
MCG-03-13-070	05 04 15.	- 19 33	36	15.	GALAXY
RNGC 1803	05 04 15.	- 49 39			UNVERIFIED SOUTHERN OBJECT
RNGC 1825	05 04 19.	- 68 59			OPEN CLUSTER IN LMC
SL 202	05 04 19.	- 68 59	35		STAR CLUSTER IN LMC
LW 125	05 04 20.	- 66 06			STAR CLUSTER IN LMC
IC 2118	05 04 21.	- 07 17			DIFFUSE NEBULA
VDB.66N 033	05 04 23.	- 03 23	108		REFLECTION NEBULA
RNGC 1781	05 04 23.	- 17 38		14.0	GALAXY
MCG+01-13-015	05 04 24.	+ 03 53	96	15.	GALAXY
DG 051	05 04 24.	- 03 25	420		REFLECTION NEBULA
CED 040	05 04 24.	- 03 25	480		DIFFUSE GALACTIC NEBULA
MCG-03-13-071	05 04 24.	- 17 40	30	14.	GALAXY
DV.56 N1796A	05 04 24.	- 61 15	60		S GALAXY
LB 03418	05 04 24.	- 67 52		13.9	FAINT BLUE STAR
LB 03419	05 04 24.	- 71 19		13.3	FAINT BLUE STAR
SC 0502-1741.7	05 04 26.	- 17 37 36.	54		NEBULA
SL 204	05 04 26.	- 66 05	50		STAR CLUSTER IN LMC
LB 01464	05 04 27.	+ 18 10 18.		16.0	FAINT BLUE STAR
RNGC 1788	05 04 27.	- 03 24			DIFFUSE NEBULA
MCG-03-13-072	05 04 27.	- 17 38	96	14.	GALAXY
RNGC 1828	05 04 28.	- 69 27			OPEN CLUSTER IN LMC
ZWG 420.036	05 04 30.	+ 03 55		15.0	GALAXY
UGC 03248	05 04 30.	+ 03 55	108	15.0	GALAXY Sb
MCG-05-13-005	05 04 30.	- 32 03	48	14.5	GALAXY
MCG-05-13-004	05 04 30.	- 32 03	12	15.	GALAXY
LW 126	05 04 31.	- 65 44			STAR CLUSTER IN LMC
LH 020	05 04 32.	- 69 05	240		STELLAR ASSN. IN LMC
RNGC 1800	05 04 33.	- 32 01		15.0	GALAXY
SL 205	05 04 33.	- 66 25	35		STAR CLUSTER IN LMC
LH 021	05 04 33.	- 67 53	300		STELLAR ASSN. IN LMC
SL 203	05 04 34.	- 69 27	25		STAR CLUSTER IN LMC
HSE 104	05 04 34.	- 70 42			STAR CLUSTER IN LMC
7ZW 027	05 04 36.	+ 66 38			COMPACT GALAXY
RNGC 1796A	05 04 36.	- 61 15			GALAXY
HSE 105	05 04 39.	- 66 24			STAR CLUSTER IN LMC
LW 127	05 04 41.	- 65 12			STAR CLUSTER IN LMC
MCG+00-13-071	05 04 42.	+ 00 51	78	13.	GALAXY
2ZW 030	05 04 42.	+ 09 24			COMPACT GALAXY
OCL 0429	05 04 42.	+ 36 59	1020	8.8	OPEN STAR CLUSTER
MCG-03-13-073	05 04 42.	- 20 25	72	15.	GALAXY
LH120-N018	05 04 42.	- 66 44			EMISSION NEBULA IN LMC
RNGC 1778	05 04 43.	+ 36 59		8.5	OPEN CLUSTER
5ZW 377	05 04 48.	+ 44 16			COMPACT GALAXY
MCG-03-13-074	05 04 48.	- 18 29	24	15.	GALAXY
MCG-05-13-006	05 04 48.	- 27 44	60	15.	GALAXY
LH120-N098	05 04 48.	- 69 51			EMISSION NEBULA IN LMC
LW 129	05 04 50.	- 66 03			STAR CLUSTER IN LMC
LH 022	05 04 50.	- 67 38	300		STELLAR ASSN. IN LMC
HSE 106	05 04 51.	- 66 15			STAR CLUSTER IN LMC
RNGC 1822	05 04 51.	- 66 16			OPEN CLUSTER IN LMC
SL 210	05 04 51.	- 66 16	25		STAR CLUSTER IN LMC
RNGC 1830	05 04 51.	- 69 24			OPEN CLUSTER IN LMC
SL 207	05 04 51.	- 69 24	25		STAR CLUSTER IN LMC
RNGC 1829	05 04 52.	- 68 07		8.5	CLUSTER/NEBULOSITY IN LMC
SL 208	05 04 52.	- 68 07	50		STAR CLUSTER IN LMC
RNGC 1833	05 04 52.	- 70 47			CLUSTER/NEBULOSITY IN LMC
SL 206	05 04 52.	- 70 47	80		STAR CLUSTER IN LMC
ZC 0504.9+0417	05 04 54.	+ 04 17	1080		CLUSTER OF GALAXIES
PK173-05.1	05 04 54.	+ 30 44	132		PLANETARY NEBULA
SL 214	05 04 54.	- 63 21	35		STAR CLUSTER IN LMC
LH120-N021	05 04 54.	- 67 38	221		EMISSION NEBULA IN LMC
CM 32	05 04 54.	- 70 09			HII REGION IN LMC
LH120-N190	05 04 54.	- 70 48	133		EMISSION NEBULA IN LMC
ARC 0535	05 04 58.	- 02 39		17.4	RICH CLUSTER OF GALAXIES
KHAV 105	05 05	+ 31 16	3250		DARK NEBULA
VDB.66G 036	05 05	- 16 22	70		DWARF GALAXY
ZC 0505.0+1510	05 05 00.	+ 15 10	2150		CLUSTER OF GALAXIES
OCL 0424	05 05 00.	+ 39 01	90	14.	OPEN STAR CLUSTER
LB 01771	05 05 00.	- 51 36		13.0	FAINT BLUE STAR
SL 211	05 05 01.	- 68 58	35		STAR CLUSTER IN LMC
IC 2119	05 05 03.	- 10 25 17.			NONSTELLAR OBJECT
SL 209	05 05 05.	- 70 58	70		STAR CLUSTER IN LMC
ZC 0505.1+0655	05 05 06.	+ 06 55	1210		CLUSTER OF GALAXIES
ISS 0205	05 05 06.	+ 41 07	296		STELLAR RING
LW 130	05 05 06.	- 63 20			STAR CLUSTER IN LMC
LH120-N022	05 05 06.	- 67 52			EMISSION NEBULA IN LMC
LH120-N023A	05 05 06.	- 68 08	107		EMISSION NEBULA IN LMC
SL 212	05 05 06.	- 68 37	15		STAR CLUSTER IN LMC
WS 07	05 05 06.	- 68 43 05.		16.60	PLANETARY NEB. IN LMC
LH120-N097	05 05 06.	- 68 44			EMISSION NEBULA IN LMC
LH120-N191B	05 05 06.	- 70 58	54		EMISSION NEBULA IN LMC
B 222	05 05 09.	+ 32 06	600		DARK OBJECT
RNGC 1835	05 05 10.	- 69 28		10.0	GLOBULAR CLUSTER IN LMC
LH 023	05 05 11.	- 70 59	90		STELLAR ASSN. IN LMC
2ZW 031	05 05 12.	+ 00 36			COMPACT GALAXY
ZC 0505.2+1548	05 05 12.	+ 15 48	940		CLUSTER OF GALAXIES
LH120-N020	05 05 12.	- 66 59	106		EMISSION NEBULA IN LMC
LH120-N191A	05 05 12.	- 70 58	53		EMISSION NEBULA IN LMC
RNGC 1826	05 05 15.	- 66 17			OPEN CLUSTER IN LMC
SL 221	05 05 15.	- 66 17	25		STAR CLUSTER IN LMC
SL 222	05 05 16.	- 66 33	50		STAR CLUSTER IN LMC
SL 215	05 05 16.	- 69 28	70		STAR CLUSTER IN LMC
RNGC 1837	05 05 16.	- 70 46			CLUSTER/NEBULOSITY IN LMC
RNGC 1799	05 05 18.	- 08 02		15.0	GALAXY
MCG-01-14-001	05 05 18.	- 08 02 30.	45	15.	GALAXY
RNGC 1797	05 05 18.	- 08 06		15.0	GALAXY
MCG-01-14-002	05 05 18.	- 08 06	36	15.	GALAXY
SL 218	05 05 18.	- 68 34	20		STAR CLUSTER IN LMC
LH120-N189	05 05 18.	- 70 12	90		EMISSION NEBULA IN LMC
SL 216	05 05 19.	- 70 04	20		STAR CLUSTER IN LMC
SL 213	05 05 20.	- 71 28	5		STAR CLUSTER IN LMC
SL 217	05 05 22.	- 70 46	70		STAR CLUSTER IN LMC
ZWG 469.001	05 05 24.	+ 17 18		15.4	GALAXY
ZWG 468.002	05 05 24.	+ 17 18		15.4	GALAXY
ZWG 347.013	05 05 24.	+ 75 52		15.6	GALAXY
UGC 03249	05 05 24.	+ 75 52	90	15.6	GALAXY
LW 128	05 05 24.	- 73 05			STAR CLUSTER IN LMC
ABC 0536	05 05 26.	- 09 18		17.0	RICH CLUSTER OF GALAXIES
LDN 1512	05 05 30.	+ 32 43	480		DARK NEBULA
ZC 0505.5+7331	05 05 30.	+ 73 31	2890		CLUSTER OF GALAXIES
MCG-03-14-001	05 05 30.	- 16 21	132	15.5	GALAXY
RNGC 1836	05 05 30.	- 68 41			OPEN CLUSTER IN LMC
SL 223	05 05 30.	- 68 41	45		STAR CLUSTER IN LMC

OBJECT NAME	RIGHT ASCEN.	DECLINATION	DIAM.	MAGN.	TYPE OF OBJECT
HSE 107	05 05 31.	- 68 49			STAR CLUSTER IN LMC
SC 0503-1625.7	05 05 33.	- 16 21 41.	60		NEBULA
RNGC 1834	05 05 33.	- 69 17			NON-EXISTENT OBJECT
ZC 0505.6-0240	05 05 36.	- 02 40	2350		CLUSTER OF GALAXIES
LW 131	05 05 38.	- 66 06			STAR CLUSTER IN LMC
HSE 108	05 05 40.	- 66 26			STAR CLUSTER IN LMC
LH 024	05 05 40.	- 70 47	960		STELLAR ASSN. IN LMC
RNGC 1794	05 05 42.	- 18 15		14.0	GALAXY
MCG-03-14-002	05 05 42.	- 18 15	42	14.5	GALAXY
MCG-05-13-007	05 05 42.	- 28 08	24	16.	GALAXY
SL 219	05 05 42.	- 73 03	55		STAR CLUSTER IN LMC
HSE 110	05 05 44.	- 66 06			STAR CLUSTER IN LMC
ZWG 307.003	05 05 48.	+ 63 07		14.5	GALAXY
UGC 03250	05 05 48.	+ 63 07	138	14.5	GALAXY SBb
MCG+11-07-004	05 05 48.	+ 67 30	42	16.	GALAXY
LH120-N023	05 05 48.	- 68 12	817		EMISSION NEBULA IN LMC
HSE 109	05 05 49.	- 68 47			STAR CLUSTER IN LMC
RNGC 1831	05 05 53.	- 64 59		11.0	OPEN CLUSTER IN LMC
MCG+11-07-005	05 05 54.	+ 63 06	84	15.	GALAXY
MCG-06-12-005	05 05 54.	- 37 35	300	11.2	GALAXY
HSE 111	05 05 54.	- 68 34			STAR CLUSTER IN LMC
RNGC 1808	05 05 56.	- 37 34		11.0	GALAXY
HSE 112	05 05 57.	- 69 11			STAR CLUSTER IN LMC
SL 227	05 05 59.	- 64 59	220		STAR CLUSTER IN LMC
UGC 03251	05 06 00.	+ 13 13	66	16.0	GALAXY
ZWG 307.004	05 06 00.	+ 67 30		15.7	GALAXY
ZC 0506.0+8558	05 06 00.	+ 85 58	1610		CLUSTER OF GALAXIES
SL 224	05 06 02.	- 70 23	20		STAR CLUSTER IN LMC
LW 133	05 06 05.	- 64 59			STAR CLUSTER IN LMC
HSE 113	05 06 05.	- 68 19			STAR CLUSTER IN LMC
RNGC 1838	05 06 06.	- 68 31			OPEN CLUSTER IN LMC
SL 225	05 06 06.	- 68 31	35		STAR CLUSTER IN LMC
RNGC 1839	05 06 06.	- 68 41			CLUSTER/NEBULOSITY IN LMC
SL 226	05 06 06.	- 68 41	35		STAR CLUSTER IN LMC
RNGC 1824	05 06 09.	- 59 47			GALAXY
SL 220	05 06 09.	- 76 21	25		STAR CLUSTER IN LMC
HSE 114	05 06 10.	- 68 05			STAR CLUSTER IN LMC
SL 228	05 06 11.	- 66 57	45		STAR CLUSTER IN LMC
ZWG 395.001	05 06 12.	+ 01 55		15.6	GALAXY
LW 134	05 06 12.	- 65 15			STAR CLUSTER IN LMC
LH120-N024	05 06 12.	- 67 50			EMISSION NEBULA IN LMC
WS 08	05 06 14.	- 67 49 45.		16.73	PLANETARY NEB. IN LMC
SC 0505-5951.5	05 06 15.	- 59 47 36.	126		NEBULA
HSE 116	05 06 16.	- 68 08			STAR CLUSTER IN LMC
LH 025	05 06 16.	- 68 10	180		STELLAR ASSN. IN LMC
SL 229	05 06 17.	- 68 27	25		STAR CLUSTER IN LMC
ZWG 395.002	05 06 18.	- 00 38		15.1	GALAXY
MCG+00-14-001	05 06 18.	- 00 40	36	15.	GALAXY
MCG-06-12-006	05 06 18.	- 38 24	60	15.5	GALAXY
LH120-N099	05 06 18.	- 69 44			EMISSION NEBULA IN LMC
MCG-03-14-003	05 06 21.	- 17 14	24	15.	GALAXY
SC 0504-1718.8	05 06 21.	- 17 14 50.	18		NEBULA
HSE 115	05 06 21.	- 70 27			STAR CLUSTER IN LMC
SL 230	05 06 23.	- 68 26	35		STAR CLUSTER IN LMC
ZC 0506.4+0016	05 06 24.	+ 00 16	4230		CLUSTER OF GALAXIES
MCG+11-07-006	05 06 24.	+ 67 25	156	14.	GALAXY
MCG-03-14-004	05 06 27.	- 17 00	72	15.	GALAXY
SC 0504-1706.5	05 06 29.	- 17 02 33.	48		NEBULA
ZWG 421.001	05 06 30.	+ 07 18		15.2	GALAXY
LB 03420	05 06 30.	- 70 33		12.3	FAINT BLUE STAR
LB 03421	05 06 30.	- 70 34		13.8	FAINT BLUE STAR
SL 233	05 06 31.	- 63 42	15		STAR CLUSTER IN LMC
LW 135	05 06 31.	- 65 41			STAR CLUSTER IN LMC
ZC 0506.6+1649	05 06 36.	+ 16 49	)010		CLUSTER OF GALAXIES
ZWG 307.005	05 06 36.	+ 67 25		14.1	GALAXY
UGC 03252	05 06 36.	+ 67 25	138	14.1	GALAXY Sc/SBc
MCG+11-07-007	05 06 36.	+ 68 34	39	16.	GALAXY
ZWG 395.004	05 06 36.	- 00 45		15.7	GALAXY
MCG+00-14-002	05 06 36.	- 00 46 30.	48	14.	GALAXY
ZWG 395.003	05 06 36.	- 00 55		15.7	GALAXY
HSE 117	05 06 37.	- 68 47			STAR CLUSTER IN LMC
LW 132	05 06 37.	- 74 48			STAR CLUSTER IN LMC
LH 026	05 06 39.	- 70 36	1020		STELLAR ASSN. IN LMC
SL 231	05 06 40.	- 69 32	25		STAR CLUSTER IN LMC
MCG-05-13-008	05 06 42.	- 29 21	84	15.	GALAXY
RNGC 1811	05 06 43.	- 29 21		15.0	GALAXY
RNGC 1845	05 06 45.	- 70 32			OPEN CLUSTER IN LMC
SL 232	05 06 45.	- 70 32	70		STAR CLUSTER IN LMC
ZWG 307.006	05 06 48.	+ 68 31		15.5	GALAXY
7ZW 028	05 06 48.	+ 68 32			COMPACT GALAXY
MCG-02-14-001	05 06 48.	- 09 18	66	15.5	GALAXY
MCG-03-14-005	05 06 48.	- 20 20	18	15.	GALAXY
SL 238	05 06 51.	- 64 29	30		STAR CLUSTER IN LMC
MCG-05-13-009	05 06 54.	- 29 20	48	14.5	GALAXY
HSE 118	05 06 54.	- 68 40			STAR CLUSTER IN LMC
RNGC 1812	05 06 55.	- 29 20		14.0	GALAXY
SL 234	05 06 55.	- 68 47	35		STAR CLUSTER IN LMC
LW 138	05 06 56.	- 66 00			STAR CLUSTER IN LMC
LW 139	05 06 57.	- 64 30			STAR CLUSTER IN LMC
SL 239	05 06 59.	- 66 43	20		STAR CLUSTER IN LMC
KHAV 107	05 07	+ 19 40			DARK NEBULA
KHAV 106	05 07	+ 33 58	7640		DARK NEBULA
LB 09825	05 07	- 81 06		13.2	FAINT BLUE STAR
ZWG 421.002	05 07 00.	+ 06 29		15.7	GALAXY
UGC 03254	05 07 00.	+ 66 42	90	16.0	GALAXY SB
7ZW 029	05 07 00.	+ 69 46			COMPACT GALAXY
ZWG 362.005	05 07 00.	+ 84 00		13.1	GALAXY
UGC 03253	05 07 00.	+ 84 00	114	13.1	GALAXY SBb
MCG+14-03-007	05 07 00.	+ 84 01	84	15.	GALAXY
SL 237	05 07 03.	- 69 13	40		STAR CLUSTER IN LMC
ZWG 395.005	05 07 06.	- 00 18		15.6	GALAXY
MCG+00-14-003	05 07 06.	- 00 20	36	15.	GALAXY
RNGC 1842	05 07 07.	- 67 20			OPEN CLUSTER IN LMC
SL 241	05 07 07.	- 67 20	20		STAR CLUSTER IN LMC
RNGC 1847	05 07 08.	- 69 03			OPEN CLUSTER IN LMC
SL 240	05 07 08.	- 69 03	80		STAR CLUSTER IN LMC
HOD.72 01	05 07 08.	- 69 04	420		DARK NEBULA IN LMC
RNGC 1840	05 07 10.	- 71 53			OPEN CLUSTER IN LMC
SL 235	05 07 10.	- 71 53	20		STAR CLUSTER IN LMC
RNGC 1802	05 07 11.	+ 24 02			NON-EXISTENT OBJECT
ZWG 421.003	05 07 12.	+ 07 25		15.6	GALAXY
UGC 03255	05 07 12.	+ 07 25	90	15.6	GALAXY SB?b
MCG+01-14-001	05 07 12.	+ 07 25	72	15.	GALAXY
LB 03422	05 07 12.	- 68 35		12.6	FAINT BLUE STAR
RNGC 1844	05 07 13.	- 67 23			OPEN CLUSTER IN LMC
SL 242	05 07 13.	- 67 23	40		STAR CLUSTER IN LMC
LH 027	05 07 15.	- 69 12	240		STELLAR ASSN. IN LMC
SL 236	05 07 15.	- 72 34	45		STAR CLUSTER IN LMC

OBJECT NAME	RIGHT ASCEN.	DECLINATION	DIAM.	MAGN.	TYPE OF OBJECT
LW 136	05 07 15.	- 72 34			STAR CLUSTER IN LMC
SC 0506-6119.3	05 07 18.	- 61 15 29.	72		NEBULA
DV.56 N1796B	05 07 18.	- 61 31	78		GALAXY
CM 12	05 07 18.	- 68 34			HII REGION IN LMC
RNGC 1790	05 07 19.	+ 52 00			NON-EXISTENT OBJECT
RNGC 1796B	05 07 19.	- 61 31			GALAXY
HSE 119	05 07 22.	- 69 24			STAR CLUSTER IN LMC
MCG+00-14-004	05 07 24.	+ 00 51	15	15.	GALAXY
ZWG 258.001	05 07 24.	+ 51 29		15.5	GALAXY
7ZW 030	05 07 24.	+ 65 08			COMPACT GALAXY
SL 243	05 07 26.	- 67 32	170		STAR CLUSTER IN LMC
MCG+00-14-005	05 07 27.	- 00 50	36	15.5	GALAXY
ZWG 395.006	05 07 30.	+ 00 52		15.4	GALAXY
LB 03423	05 07 30.	- 65 23		13.3	FAINT BLUE STAR
LH120-N100	05 07 30.	- 68 37	183		EMISSION NEBULA IN LMC
MCG+00-14-006	05 07 33.	- 00 47	9	16.	GALAXY
HSE 121	05 07 33.	- 67 56			STAR CLUSTER IN LMC
SL 245	05 07 34.	- 66 28	15		STAR CLUSTER IN LMC
HSE 120	05 07 34.	- 70 41			STAR CLUSTER IN LMC
ZWG 395.008	05 07 36.	+ 00 50		14.6	GALAXY
UGC 03256	05 07 36.	+ 00 50	72	14.6	GALAXY Sb/SBb
MCG+00-14-007	05 07 36.	+ 00 50	48	13.	GALAXY
SG 3.029	05 07 36.	+ 34 28	840		DIFFUSE EMISSION NEBULA
SG 3.028	05 07 36.	+ 37 54	60		DIFFUSE EMISSION NEBULA
ZWG 395.007	05 07 36.	- 00 46		15.2	GALAXY
MCG+00-14-008	05 07 36.	- 00 47 30.	42	15.	GALAXY
MCG-06-12-007	05 07 36.	- 37 56	42	15.	GALAXY
SL 244	05 07 36.	- 68 37	25		STAR CLUSTER IN LMC
RNGC 1846	05 07 38.	- 67 32		11.5	GLOBULAR CLUSTER IN LMC
MRSL 168-01/1	05 07 42.	+ 37 56	180		HII REGION
2ZW 032	05 07 42.	- 00 47			COMPACT GALAXY
ZWG 395.009	05 07 42.	- 00 47		15.7	GALAXY
MCG-02-14-002	05 07 42.	- 14 58	72	14.5	GALAXY
LW 137	05 07 43.	- 75 29			STAR CLUSTER IN LMC
MAI 007	05 07 44.	+ 81 37	33		DWARF SPHEROIDAL GALAXY
RNGC 1807	05 07 46.	+ 16 28		8.5	OPEN CLUSTER
SC 0505-1505.1	05 07 47.	- 15 01 14.	72		NEBULA
OCL 0462	05 07 48.	+ 16 28	1260	8.8	OPEN STAR CLUSTER
ZWG 469.002	05 07 48.	+ 17 58		15.2	GALAXY
RLWT 113	05 07 48.	+ 40 49		15.	FAINT VERY BLUE STAR
MCG+00-14-009	05 07 48.	- 00 48 30.	42	15.	GALAXY
MCG-04-13-002	05 07 48.	- 22 22 30.	30	16.	GALAXY
LB 01772	05 07 48.	- 53 23		13.3	FAINT BLUE STAR
LH120-N101	05 07 48.	- 69 13			EMISSION NEBULA IN LMC
RNGC 1798	05 07 49.	+ 47 36			NON-EXISTENT OBJECT
HSE 122	05 07 49.	- 68 51			STAR CLUSTER IN LMC
SL 246	05 07 53.	- 70 52	15		STAR CLUSTER IN LMC
ZWG 469.003	05 07 54.	+ 16 25		15.7	GALAXY
ZWG 395.010	05 07 54.	- 00 47		15.7	GALAXY
ZC 0507.9-0110	05 07 54.	- 01 10	5440		CLUSTER OF GALAXIES
SCHO 0102	05 07 55.	+ 38 06 30.	500		ISOLATED DARK CLOUD
RNGC 1848	05 07 55.	- 71 15			CLUSTER/NEBULOSITY IN LMC
SL 247	05 07 55.	- 71 15	135		STAR CLUSTER IN LMC
KHAV 108	05 08	+ 10 52	3090		DARK NEBULA
LBN 0780	05 08	+ 37 58	120		BRIGHT NEBULA
LBN 0910	05 08	- 02 10	1620		BRIGHT NEBULA
VDB.66G 230	05 08	- 31 39	100		DWARF GALAXY
LDN 1572	05 08 00.	+ 10 30	1140		DARK NEBULA
LDN 1571	05 08 00.	+ 10 30	1860		DARK NEBULA
ZWG 362.006	05 08 00.	+ 84 26		14.9	GALAXY
UGC 03257	05 08 00.	+ 84 26	102	14.9	GALAXY SBa
LB 03424	05 08 00.	- 71 12		13.0	FAINT BLUE STAR
SL 249	05 08 05.	- 70 49	25		STAR CLUSTER IN LMC
ZWG 395.011	05 08 06.	+ 00 21		13.9	GALAXY
UGC 03258	05 08 06.	+ 00 21	48	13.9	GALAXY SB
KARA.73B 0156	05 08 06.	+ 00 21	48	13.9	ISOLATED GALAXY S
OCL 0410	05 08 06.	+ 47 34	360	13.	OPEN STAR CLUSTER
LB 01773	05 08 06.	- 47 50		13.3	FAINT BLUE STAR
LB 03425	05 08 06.	- 71 05		13.3	FAINT BLUE STAR
LH 028	05 08 07.	- 71 15	300		STELLAR ASSN. IN LMC
SL 253	05 08 08.	- 65 55	25		STAR CLUSTER IN LMC
HOD.72 02	05 08 08.	- 69 00	180		DARK NEBULA IN LMC
HSE 124	05 08 09.	- 66 12			STAR CLUSTER IN LMC
SL 254	05 08 09.	- 66 16	20		STAR CLUSTER IN LMC
SL 250	05 08 10.	- 69 31	20		STAR CLUSTER IN LMC
HSE 125	05 08 11.	- 64 56			STAR CLUSTER IN LMC
ZWG 307.007	05 08 12.	+ 66 25		15.4	GALAXY
ZWG 328.033	05 08 12.	+ 72 16		15.6	GALAXY
UGC 03259	05 08 12.	+ 72 16	126	15.6	GALAXY Sc
MCG+12-05-029	05 08 12.	+ 72 16	84	15.	GALAXY
7ZW 031	05 08 12.	+ 79 36			COMPACT GALAXY
SL 258	05 08 17.	- 64 58	25		STAR CLUSTER IN LMC
LW 142	05 08 17.	- 64 59			STAR CLUSTER IN LMC
WS 09	05 08 17.	- 68 43 58.		16.25	PLANETARY NEB. IN LMC
LW 140	05 08 17.	- 72 52			STAR CLUSTER IN LMC
RLWT 114	05 08 18.	+ 41 36		16.	FAINT VERY BLUE STAR
2ZW 033	05 08 18.	- 02 45			COMPACT GALAXY
ZWG 395.012	05 08 18.	- 02 45		14.0	GALAXY
MCG+00-14-010	05 08 18.	- 02 45	30	14.	GALAXY
MCG-05-13-010	05 08 18.	- 29 28	72	15.	GALAXY
MCG-06-12-008	05 08 18.	- 37 02	180	12.	GALAXY
LH120-N102	05 08 18.	- 68 45			EMISSION NEBULA IN LMC
SL 251	05 08 18.	- 69 57	55		STAR CLUSTER IN LMC
LH 029	05 08 19.	- 70 05	300		STELLAR ASSN. IN LMC
SC 0506-1436.6	05 08 20.	- 14 32 47.	6		NEBULA
RNGC 1827	05 08 20.	- 37 02			GALAXY
SL 257	05 08 20.	- 65 56	15		STAR CLUSTER IN LMC
LW 143	05 08 20.	- 65 56			STAR CLUSTER IN LMC
MCG+00-14-011	05 08 21.	- 00 47	36	15.	GALAXY
SL 252	05 08 21.	- 70 27	20		STAR CLUSTER IN LMC
HSE 123	05 08 22.	- 73 38			STAR CLUSTER IN LMC
HSE 126	05 08 23.	- 64 57			STAR CLUSTER IN LMC
ZWG 328.034	05 08 24.	+ 70 50		15.7	GALAXY
LB 03426	05 08 24.	- 63 25		13.2	FAINT BLUE STAR
SL 255	05 08 25.	- 70 07	15		STAR CLUSTER IN LMC
SL 259	05 08 26.	- 64 10	15		STAR CLUSTER IN LMC
LW 145	05 08 26.	- 64 10			STAR CLUSTER IN LMC
LW 144	05 08 26.	- 65 58			STAR CLUSTER IN LMC
LH 030	05 08 27.	- 69 17	120		STELLAR ASSN. IN LMC
SL 248	05 08 27.	- 75 45	15		STAR CLUSTER IN LMC
ZWG 258.002	05 08 30.	+ 51 15		15.2	GALAXY
UGC 03260	05 08 30.	+ 51 15	90	15.2	GALAXY S0?
MCG+14-03-008	05 08 30.	+ 84 25	66	15.	GALAXY
MCG-04-13-003	05 08 30.	- 25 48	48	15.5	GALAXY
CM 33	05 08 30.	- 70 37			HII REGION IN LMC
ZWG 469.004	05 08 36.	+ 16 59		15.5	GALAXY
UGC 03261	05 08 36.	+ 16 59	84	15.5	GALAXY Sc-IRR
ZWG 395.013	05 08 36.	- 01 08		15.7	GALAXY

OBJECT NAME	RIGHT ASCEN.	DECLINATION	DIAM.	MAGN.	TYPE OF OBJECT
SL 256	05 08 37.	- 71 14	25		STAR CLUSTER IN LMC
SL 262	05 08 40.	- 62 26	35		STAR CLUSTER IN LMC
2ZW 034	05 08 42.	+ 01 28			COMPACT GALAXY
SL 260	05 08 43.	- 68 46	15		STAR CLUSTER IN LMC
RNGC 1850	05 08 43.	- 68 50		9.5	CLUSTER/NEBULOSITY IN LMC
SL 261	05 08 43.	- 68 50	205		STAR CLUSTER IN LMC
LW 146	05 08 46.	- 62 26			STAR CLUSTER IN LMC
ZC 0508.8+0241	05 08 48.	+ 02 41	1550		CLUSTER OF GALAXIES
MCG-02-14-003	05 08 48.	- 09 26 30.	144	14.	GALAXY
MCG-03-14-006	05 08 48.	- 18 30	72	16.	GALAXY
MCG-05-13-011	05 08 48.	- 31 41	180	15.	GALAXY
MCG+00-14-012	05 08 51.	- 00 38 30.	42	14.5	GALAXY
SC 0508-6230.1	05 08 53.	- 62 26 24.	12		NEBULA
ZWG 395.014	05 08 54.	- 00 38		15.6	GALAXY
UGC 03262	05 08 54.	- 00 38	60	15.6	GALAXY S
LH120-N103A	05 08 54.	- 68 50	332		EMISSION NEBULA IN LMC
SG 3.030	05 08 55.	+ 34 52	1800		DIFFUSE EMISSION NEBULA
SL 263	05 08 56.	- 65 50	15		STAR CLUSTER IN LMC
MCG+00-14-013	05 08 57.	- 00 39	48	14.5	GALAXY
HSE 127	05 08 57.	- 69 15			STAR CLUSTER IN LMC
HSE 129	05 08 58.	- 66 23			STAR CLUSTER IN LMC
SC 0506-1432.4	05 08 59.	- 14 28 37.	6		NEBULA
LBN 0991	05 09	- 12 30	2220		BRIGHT NEBULA
VDB.66G 231	05 09	- 32 59	170		DWARF GALAXY
ZWG 258.003	05 09 00.	+ 52 11		15.3	GALAXY
ZWG 395.015	05 09 00.	- 00 38		14.7	GALAXY
UGC 03264	05 09 00.	- 00 38	72	14.7	GALAXY S0-a
UGC 03263	05 09 00.	- 00 44	66	16.0	GALAXY SBc
MCG+00-14-014	05 09 00.	- 00 45 30.	48	15.	GALAXY
LW 141	05 09 00.	- 74 42			STAR CLUSTER IN LMC
HSE 128	05 09 03.	- 69 15			STAR CLUSTER IN LMC
RNGC 1819	05 09 04.	+ 05 08		13.5	GALAXY
ZWG 421.004	05 09 06.	+ 05 08		13.7	GALAXY
UGC 03265	05 09 06.	+ 05 08	102	13.7	GALAXY SB0
MCG+01-14-002	05 09 06.	+ 05 08	72	14.	GALAXY
7ZW 032	05 09 06.	+ 67 13			COMPACT GALAXY
RNGC 1817	05 09 09.	+ 16 38		8.0	OPEN CLUSTER
ABC 0534	05 09 09.	+ 73 28		17.0	RICH CLUSTER OF GALAXIES
RNGC 1852	05 09 09.	- 67 51			CLUSTER/NEBULOSITY IN LMC
RNGC 1849	05 09 10.	- 66 22			OPEN CLUSTER IN LMC
OCL 0463	05 09 12.	+ 16 38	1980	7.9	OPEN STAR CLUSTER
MCG-01-14-004	05 09 12.	- 03 04	42	16.	GALAXY
MCG-01-14-003	05 09 12.	- 03 09	60	16.	GALAXY
MCG-04-13-004	05 09 12.	- 22 19 30.	48	15.5	GALAXY
MCG-06-12-009	05 09 12.	- 34 37	66	14.	GALAXY
HSE 130	05 09 15.	- 67 46			STAR CLUSTER IN LMC
SL 264	05 09 15.	- 67 51	70		STAR CLUSTER IN LMC
SL 267	05 09 16.	- 66 22	30		STAR CLUSTER IN LMC
UGC 03266	05 09 18.	+ 13 50	66	16.0	GALAXY S
ZWG 258.004	05 09 18.	+ 51 18		15.4	GALAXY
MCG-02-14-004	05 09 18.	- 14 52	144	13.5	GALAXY
HSE 131	05 09 18.	- 68 29			STAR CLUSTER IN LMC
HSE 132	05 09 20.	- 65 59			STAR CLUSTER IN LMC
RNGC 1854	05 09 20.	- 68 55		10.5	OPEN CLUSTER IN LMC
SL 265	05 09 20.	- 68 55	70		STAR CLUSTER IN LMC
LW 148	05 09 21.	- 64 14			STAR CLUSTER IN LMC
LB 01774	05 09 24.	- 46 59		14.2	FAINT BLUE STAR
LH120-N025	05 09 24.	- 67 52			EMISSION NEBULA IN LMC
SL 266	05 09 24.	- 69 48	15		STAR CLUSTER IN LMC
SC 0507-1454.5	05 09 26.	- 14 50 45.	54		NEBULA
LW 149	05 09 26.	- 65 50			STAR CLUSTER IN LMC
SL 269	05 09 27.	- 67 53	25		STAR CLUSTER IN LMC
RNGC 1821	05 09 28.	- 15 11		14.0	GALAXY
SL 268	05 09 29.	- 69 40	55		STAR CLUSTER IN LMC
ZWG 421.005	05 09 30.	+ 05 24		15.4	GALAXY
MCG+01-14-003	05 09 30.	+ 05 24	36	15.	GALAXY
MCG-03-14-007	05 09 30.	- 15 11	54	14.5	GALAXY
MCG-03-14-008	05 09 30.	- 17 54	48	15.	GALAXY
MCG-03-14-009	05 09 30.	- 20 30	78	14.	GALAXY
SL 273	05 09 31.	- 63 42	15		STAR CLUSTER IN LMC
ZWG 283.004	05 09 36.	+ 62 31		15.3	GALAXY
SC 0507-1438.0	05 09 36.	- 14 34 16.	18		NEBULA
LH120-N103B	05 09 36.	- 68 50	97		EMISSION NEBULA IN LMC
LB 03427	05 09 36.	- 70 37		13.8	FAINT BLUE STAR
RNGC 1855	05 09 38.	- 68 55			NON-EXISTENT OBJECT
RNGC 1856	05 09 39.	- 69 12		10.0	OPEN CLUSTER IN LMC
SL 271	05 09 39.	- 69 12	135		STAR CLUSTER IN LMC
HOD.72 03	05 09 40.	- 69 29	1020		DARK NEBULA IN LMC
SL 270	05 09 40.	- 70 40	20		STAR CLUSTER IN LMC
HSE 133	05 09 41.	- 69 36			STAR CLUSTER IN LMC
MCG-03-14-010	05 09 42.	- 15 44	138	12.5	GALAXY
LH120-N104B	05 09 42.	- 68 33	193		EMISSION NEBULA IN LMC
HSE 134	05 09 42.	- 68 37			STAR CLUSTER IN LMC
HSE 135	05 09 43.	- 65 29			STAR CLUSTER IN LMC
HSE 137	05 09 48.	- 65 22			STAR CLUSTER IN LMC
SG 3.031	05 09 53.	+ 37 25	240		DIFFUSE EMISSION NEBULA
SL 275	05 09 53.	- 64 59	35		STAR CLUSTER IN LMC
RNGC 1858	05 09 56.	- 68 58			CLUSTER/NEBULOSITY IN LMC
SL 274	05 09 56.	- 68 58	205		STAR CLUSTER IN LMC
HSE 136	05 09 57.	- 69 17			STAR CLUSTER IN LMC
RNGC 1832	05 09 58.	- 15 47		12.5	GALAXY
LW 150	05 09 59.	- 64 59			STAR CLUSTER IN LMC
HSE 143	05 09 59.	- 65 02			STAR CLUSTER IN LMC
LBN 0784	05 10	+ 37 20	120		BRIGHT NEBULA
LBN 0783	05 10	+ 37 20	540		BRIGHT NEBULA
SHER 1	05 10	- 33 00	100		DWARF GALAXY
ZC 0510.0+0458	05 10 00.	+ 04 58	21300		CLUSTER OF GALAXIES
SS 39	05 10 00.	+ 37 29			DIFFUSE GALACTIC NEBULA
ZWG 328.035	05 10 00.	+ 71 25		15.2	GALAXY
UGC 03267	05 10 00.	+ 71 25	102	15.2	GALAXY Sc
MCG+12-05-030	05 10 00.	+ 71 25	84	15.	GALAXY
KARA.73 26	05 10 01.	- 33 01	148		DWARF GALAXY
HSE 138	05 10 01.	- 68 39			STAR CLUSTER IN LMC
HSE 139	05 10 01.	- 68 50			STAR CLUSTER IN LMC
LW 147	05 10 01.	- 74 48			STAR CLUSTER IN LMC
SL 272	05 10 02.	- 72 25	5		STAR CLUSTER IN LMC
SL 277	05 10 04.	- 64 32	5		STAR CLUSTER IN LMC
MIN.47 02	05 10 05.	+ 37 24			DIFFUSE NEBULA
MCG+08-10-001	05 10 06.	+ 49 57	16	16.	GALAXY
MCG-02-14-005	05 10 06.	- 14 25	60	15.	GALAXY
MCG-06-12-010	05 10 06.	- 33 02	120		GALAXY
HSE 145	05 10 06.	- 65 22			STAR CLUSTER IN LMC
LH120-N105	05 10 06.	- 68 58	340		EMISSION NEBULA IN LMC
LH120-N192	05 10 06.	- 70 53			EMISSION NEBULA IN LMC
SC 0507-1428.2	05 10 08.	- 14 24 30.	60		NEBULA
LH 031	05 10 08.	- 68 58	240		STELLAR ASSN. IN LMC
SC 0507-1419.0	05 10 09.	- 14 15 18.	54		NEBULA
HSE 141	05 10 09.	- 69 09			STAR CLUSTER IN LMC

OBJECT NAME	RIGHT ASCEN.	DECLINATION	DIAM.	MAGN.	TYPE OF OBJECT
HOD.72 04	05 10 09.	- 69 16	360		DARK NEBULA IN LMC
HSE 140	05 10 09.	- 70 26			STAR CLUSTER IN LMC
WS 10	05 10 09.	- 70 52 54.		17.19	PLANETARY NEB. IN LMC
SL 276	05 10 10.	- 69 26	25		STAR CLUSTER IN LMC
HSE 142	05 10 11.	- 69 39			STAR CLUSTER IN LMC
ZWG 395.016	05 10 12.	+ 02 40		14.6	GALAXY
KARA.73B 0157	05 10 12.	+ 02 40	36	14.6	ISOLATED GALAXY S
MRSL 169-00/1	05 10 12.	+ 37 24	480		HII REGION
ZWG 231.001	05 10 12.	+ 49 57		15.4	GALAXY
LH120-N105A	05 10 12.	- 68 58	270		EMISSION NEBULA IN LMC
HSE 144	05 10 13.	- 68 39			STAR CLUSTER IN LMC
SL 279	05 10 14.	- 65 58	15		STAR CLUSTER IN LMC
ZWG 395.017	05 10 18.	- 02 17		15.6	GALAXY
LH120-N104A	05 10 18.	- 68 34			EMISSION NEBULA IN LMC
SL 278	05 10 18.	- 68 34	30		STAR CLUSTER IN LMC
WS 11	05 10 22.	- 68 33 30.		17.01	PLANETARY NEB. IN LMC
SL 281	05 10 25.	- 67 12	70		STAR CLUSTER IN LMC
LW 151	05 10 26.	- 65 53			STAR CLUSTER IN LMC
HSE 149	05 10 26.	- 65 54			STAR CLUSTER IN LMC
SL 283	05 10 27.	- 66 18	30		STAR CLUSTER IN LMC
HSE 147	05 10 27.	- 69 09			STAR CLUSTER IN LMC
SL 287	05 10 28.	- 64 35	25		STAR CLUSTER IN LMC
LW 152	05 10 28.	- 64 35			STAR CLUSTER IN LMC
SL 280	05 10 28.	- 69 25	20		STAR CLUSTER IN LMC
ZWG 395.018	05 10 30.	+ 00 55		15.7	GALAXY
MCG+00-14-015	05 10 30.	+ 00 55	30	14.5	GALAXY
LH120-N106	05 10 30.	- 68 53			EMISSION NEBULA IN LMC
LB 03428	05 10 30.	- 71 42		13.7	FAINT BLUE STAR
HSE 148	05 10 32.	- 70 10			STAR CLUSTER IN LMC
IC 0404	05 10 35.	+ 09 41 57.			NONSTELLAR OBJECT
ZWG 446.001	05 10 36.	+ 09 42		15.7	GALAXY
WS 12	05 10 37.	- 65 33 05.		17.04	PLANETARY NEB. IN LMC
RNGC 1860	05 10 37.	- 68 50			OPEN CLUSTER IN LMC
SL 284	05 10 37.	- 68 50	30		STAR CLUSTER IN LMC
MCG-02-14-006	05 10 39.	- 09 15	72	15.	GALAXY
SL 282	05 10 39.	- 70 26	15		STAR CLUSTER IN LMC
RNGC 1861	05 10 41.	- 70 50			OPEN CLUSTER IN LMC
MCG-01-14-005	05 10 42.	- 03 24	48	16.	GALAXY
LH120-N026	05 10 42.	- 67 09	30		EMISSION NEBULA IN LMC
HSE 151	05 10 42.	- 68 27			STAR CLUSTER IN LMC
LH 032	05 10 43.	- 67 14	420		STELLAR ASSN. IN LMC
SL 289	05 10 43.	- 67 16	15		STAR CLUSTER IN LMC
HSE 152	05 10 43.	- 68 45			STAR CLUSTER IN LMC
SL 288	05 10 45.	- 69 07	20		STAR CLUSTER IN LMC
SL 285	05 10 45.	- 70 26	55		STAR CLUSTER IN LMC
SL 286	05 10 47.	- 70 50	25		STAR CLUSTER IN LMC
HSE 146	05 10 47.	- 73 44			STAR CLUSTER IN LMC
SG 3.032	05 10 48.	+ 34 02	7200		DIFFUSE EMISSION NEBULA
UGC 03268	05 10 48.	+ 68 12	90	16.5	GALAXY
MCG+11-07-008	05 10 48.	+ 68 12	39	17.	GALAXY
LH120-N108	05 10 48.	- 69 31	44		EMISSION NEBULA IN LMC
SL 292	05 10 49.	- 65 43	20		STAR CLUSTER IN LMC
HSE 153	05 10 50.	- 68 55			STAR CLUSTER IN LMC
WS 13	05 10 51.	- 68 39 23.		17.68	PLANETARY NEB. IN LMC
LB 03429	05 10 54.	- 67 13		13.5	FAINT BLUE STAR
LH120-N107	05 10 54.	- 68 41			EMISSION NEBULA IN LMC
LW 154	05 10 55.	- 65 43			STAR CLUSTER IN LMC
MAI 008	05 10 57.	+ 74 32	40		DWARF SPHEROIDAL GALAXY
HSE 154	05 10 57.	- 67 41			STAR CLUSTER IN LMC
SL 293	05 10 57.	- 67 46	15		STAR CLUSTER IN LMC
LBN 0791	05 11	+ 34 00	3900		BRIGHT NEBULA
LBN 0907	05 11	- 01 30	540		BRIGHT NEBULA
MRSL 172-02/1	05 11 00.	+ 33 45			HII REGION
LH120-N027	05 11 00.	- 67 11	53		EMISSION NEBULA IN LMC
HSE 150	05 11 01.	- 73 07			STAR CLUSTER IN LMC
HSE 155	05 11 04.	- 69 20			STAR CLUSTER IN LMC
SL 290	05 11 04.	- 70 33	25		STAR CLUSTER IN LMC
ZWG 395.019	05 11 06.	+ 01 54		15.7	GALAXY
MCG+05-13-001	05 11 06.	+ 32 43	42	15.	GALAXY
LB 03430	05 11 06.	- 64 03		13.7	FAINT BLUE STAR
LH120-N028	05 11 06.	- 67 52			EMISSION NEBULA IN LMC
SL 291	05 11 06.	- 70 58	20		STAR CLUSTER IN LMC
WS 14	05 11 07.	- 67 51 22.		16.93	PLANETARY NEB. IN LMC
SL 294	05 11 07.	- 70 07	60		STAR CLUSTER IN LMC
MCG+00-14-016	05 11 09.	+ 01 54	42	14.	GALAXY
RNGC 1859	05 11 12.	- 65 17			OPEN CLUSTER IN LMC
SL 297	05 11 12.	- 65 17	35		STAR CLUSTER IN LMC
MCG+05-13-002	05 11 15.	+ 32 46 30.	24	14.	GALAXY
HSE 156	05 11 15.	- 67 41			STAR CLUSTER IN LMC
MCG+05-13-003	05 11 16.	+ 32 46 30.	12	15.5	GALAXY
HSE 157	05 11 17.	- 68 16			STAR CLUSTER IN LMC
SL 296	05 11 17.	- 69 39	15		STAR CLUSTER IN LMC
OCL 0435	05 11 18.	+ 32 40	720		OPEN STAR CLUSTER
LW 156	05 11 18.	- 65 19			STAR CLUSTER IN LMC
SL 298	05 11 18.	- 67 02	65		STAR CLUSTER IN LMC
MCG+05-13-004	05 11 21.	+ 32 46 30.	15	15.	GALAXY
ZC 0511.4+0151	05 11 24.	+ 01 51	1410		CLUSTER OF GALAXIES
ZWG 395.020	05 11 24.	+ 02 15		15.7	GALAXY
MCG-02-14-007	05 11 24.	- 12 49	60	15.	GALAXY
LB 01775	05 11 24.	- 47 18		14.1	FAINT BLUE STAR
LH 033	05 11 27.	- 69 14	360		STELLAR ASSN. IN LMC
RNGC 1853	05 11 28.	- 57 28			UNVERIFIED SOUTHERN OBJECT
ZWG 395.021	05 11 30.	+ 02 28		15.6	GALAXY
ZWG 421.006	05 11 30.	+ 06 27		14.9	GALAXY
UGC 03269	05 11 30.	+ 06 27	60	14.9	GALAXY Sb-c
WS 15	05 11 33.	- 70 05 03.		16.43	PLANETARY NEB. IN LMC
HSE 160	05 11 35.	- 64 54			STAR CLUSTER IN LMC
LW 157	05 11 35.	- 65 03			STAR CLUSTER IN LMC
ZWG 395.022	05 11 36.	+ 01 50		15.5	GALAXY
MCG+00-14-017	05 11 36.	+ 01 50	48	14.	GALAXY
RNGC 1863	05 11 38.	- 68 51			OPEN CLUSTER IN LMC
SL 299	05 11 38.	- 68 51	35		STAR CLUSTER IN LMC
SL 300	05 11 39.	- 67 39	30		STAR CLUSTER IN LMC
HSE 158	05 11 42.	- 68 33			STAR CLUSTER IN LMC
HSE 159	05 11 43.	- 68 39			STAR CLUSTER IN LMC
SL 295	05 11 44.	- 75 36	25		STAR CLUSTER IN LMC
IC 0403	05 11 47.	+ 39 55 05.			NONSTELLAR OBJECT
MCG-02-14-008	05 11 48.	- 10 40	90	13.	GALAXY
LH120-N110	05 11 48.	- 70 06			EMISSION NEBULA IN LMC
RNGC 1843	05 11 49.	- 10 40		13.0	GALAXY
SCHO 0103	05 11 50.	+ 40 55 18.	490		ISOLATED DARK CLOUD
LW 153	05 11 50.	- 75 36			STAR CLUSTER IN LMC
SL 303	05 11 51.	- 64 23	25		STAR CLUSTER IN LMC
SL 301	05 11 51.	- 66 10	15		STAR CLUSTER IN LMC
VDB.66N 036	05 11 52.	- 08 15	49200		REFLECTION NEBULA
MCG+01-14-004	05 11 54.	+ 05 03	24	15.5	GALAXY
YM 32	05 11 57.	- 08 12	5400		SYMMETRIC GALACTIC NEBULA
LW 158	05 11 57.	- 64 22			STAR CLUSTER IN LMC

OBJECT NAME	RIGHT ASCEN.	DECLINATION	DIAM.	MAGN.	TYPE OF OBJECT
HSE 161	05 11 57.	- 69 10			STAR CLUSTER IN LMC
HSE 162	05 11 58.	- 66 22			STAR CLUSTER IN LMC
LBN 0848	05 12	+ 12 58	120		BRIGHT NEBULA
LB 09965	05 12	- 89 04		14.0	FAINT BLUE STAR
ZWG 258.005	05 12 00.	+ 51 45		15.6	GALAXY
LW 159	05 12 02.	- 64 05			STAR CLUSTER IN LMC
SL 305	05 12 02.	- 64 06	100		STAR CLUSTER IN LMC
VDB.66N 035	05 12 05.	+ 12 56	240		REFLECTION NEBULA
HSE 163	05 12 05.	- 68 20			STAR CLUSTER IN LMC
LH120-N109	05 12 06.	- 69 32	29		EMISSION NEBULA IN LMC
SL 302	05 12 08.	- 70 08	40		STAR CLUSTER IN LMC
RNGC 1862	05 12 09.	- 66 12			OPEN CLUSTER IN LMC
HOD.72 05	05 12 09.	- 69 12	240		DARK NEBULA IN LMC
HOD.72 06	05 12 11.	- 69 32	300		DARK NEBULA IN LMC
ZWG 421.007	05 12 12.	+ 05 21		15.5	GALAXY
ZWG 421.008	05 12 12.	+ 06 11		15.5	GALAXY
MCG+01-14-005	05 12 12.	+ 06 11	48	15.	GALAXY
LDN 1545	05 12 12.	+ 26 40	900		DARK NEBULA
MCG-05-13-012	05 12 12.	- 27 49	30	15.5	GALAXY
HSE 164	05 12 14.	- 65 51			STAR CLUSTER IN LMC
MCG+01-14-006	05 12 15.	+ 05 20	36	15.5	GALAXY
SL 306	05 12 15.	- 66 12	25		STAR CLUSTER IN LMC
SL 304	05 12 15.	- 69 17	55		STAR CLUSTER IN LMC
LW 155	05 12 15.	- 75 44			STAR CLUSTER IN LMC
WS 16	05 12 16.	- 66 26 32.		17.95	PLANETARY NEB. IN LMC
SCHO 0104	05 12 19.	+ 32 51 30.	400		ISOLATED DARK CLOUD
SC 0511-6158.8	05 12 19.	- 61 55 21.	12		NEBULA
HSE 168	05 12 19.	- 65 42			STAR CLUSTER IN LMC
RNGC 1851	05 12 23.	- 40 05		7.5	GLOBULAR CLUSTER
IC 0405	05 12 24.	+ 33 48	1080		HII REGION
GCL 009	05 12 24.	- 40 05	792	7.72	GLOBULAR STAR CLUSTER
HSE 165	05 12 25.	- 68 39			STAR CLUSTER IN LMC
RNGC 1865	05 12 26.	- 68 51			OPEN CLUSTER IN LMC
SL 307	05 12 26.	- 68 51	35		STAR CLUSTER IN LMC
HSE 166	05 12 26.	- 69 00			STAR CLUSTER IN LMC
HSE 167	05 12 28.	- 69 17			STAR CLUSTER IN LMC
HSE 170	05 12 29.	- 64 55			STAR CLUSTER IN LMC
SL 308	05 12 30.	- 68 23	30		STAR CLUSTER IN LMC
RNGC 1864	05 12 33.	- 67 42			OPEN CLUSTER IN LMC
SL 309	05 12 33.	- 67 42	25		STAR CLUSTER IN LMC
HSE 169	05 12 34.	- 69 20			STAR CLUSTER IN LMC
ZWG 421.009	05 12 36.	+ 06 12		14.8	GALAXY
MCG+01-14-007	05 12 36.	+ 06 12	48	15.	GALAXY
LH120-N029	05 12 36.	- 66 42			EMISSION NEBULA IN LMC
LH120-N193E	05 12 36.	- 70 32	35		EMISSION NEBULA IN LMC
SL 310	05 12 37.	- 67 19	20		STAR CLUSTER IN LMC
MCG+01-14-008	05 12 39.	+ 06 25	48	15.5	GALAXY
HOD.72 07	05 12 41.	- 69 35	240		DARK NEBULA IN LMC
HOD.72 08	05 12 41.	- 69 42	480		DARK NEBULA IN LMC
ZWG 421.010	05 12 42.	+ 06 25		15.5	GALAXY
UGC 03270	05 12 42.	+ 06 25	60	15.5	GALAXY S
HSE 171	05 12 42.	- 68 26			STAR CLUSTER IN LMC
LH120-N193B	05 12 42.	- 70 32	36		EMISSION NEBULA IN LMC
HSE 173	05 12 43.	- 65 26			STAR CLUSTER IN LMC
HSE 172	05 12 43.	- 68 42			STAR CLUSTER IN LMC
LH 034	05 12 44.	- 67 22	960		STELLAR ASSN. IN LMC
MCG-01-14-006	05 12 45.	- 04 27 30.	11	16.	GALAXY
SC 0512-6159.4	05 12 47.	- 61 55 59.			NEBULA
MCG+01-14-009	05 12 48.	+ 07 08	48	15.	GALAXY
LH120-N193D	05 12 48.	- 70 31	35		EMISSION NEBULA IN LMC
SL 311	05 12 50.	- 68 57	5		STAR CLUSTER IN LMC
VDB.66N 034	05 12 53.	+ 34 13	1200		REFLECTION NEBULA
ZWG 421.011	05 12 54.	+ 07 08		15.4	GALAXY
ZWG 258.006	05 12 54.	+ 53 09		15.6	GALAXY
SL 315	05 12 56.	- 65 47	25		STAR CLUSTER IN LMC
ABC 0538	05 12 59.	- 15 45		17.4	RICH CLUSTER OF GALAXIES
LBN 0915	05 13	- 02 10	2400		BRIGHT NEBULA
LBN 0945	05 13	- 05 00	4500		BRIGHT NEBULA
LDN 1548	05 13 00.	+ 26 10	960		DARK NEBULA
LDN 1547	05 13 00.	+ 26 20	840		DARK NEBULA
CED 042	05 13 00.	+ 34 16	1800		DIFFUSE GALACTIC NEBULA
SG 3.033	05 13 00.	+ 34 16	2400		DIFFUSE EMISSION NEBULA
MRSL 172-02/2	05 13 00.	+ 34 25	3900		HII REGION
LDN 1510	05 13 00.	+ 34 45	1980		DARK NEBULA
HSE 174	05 13 00.	- 69 49			STAR CLUSTER IN LMC
LH120-N193C	05 13 00.	- 70 28	52		EMISSION NEBULA IN LMC
LH120-N193A	05 13 00.	- 70 28	29		EMISSION NEBULA IN LMC
LW 162	05 13 02.	- 65 48			STAR CLUSTER IN LMC
SC 0510-1447.4	05 13 03.	- 14 43 55.	42		NEBULA
MCG+01-14-010	05 13 06.	+ 07 07	36	15.	GALAXY
MCG+13-04-018	05 13 06.	+ 76 09	18	18.	GALAXY
LH120-N111	05 13 06.	- 69 08			EMISSION NEBULA IN LMC
SL 313	05 13 08.	- 70 08	40		STAR CLUSTER IN LMC
HOD.72 09	05 13 10.	- 69 25	420		DARK NEBULA IN LMC
RNGC 1878	05 13 10.	- 70 32			OPEN CLUSTER IN LMC
SL 312	05 13 10.	- 71 45	15		STAR CLUSTER IN LMC
ZWG 421.012	05 13 12.	+ 07 06		15.5	GALAXY
ZWG 307.008	05 13 12.	+ 67 40		15.1	GALAXY
MCG-01-14-007	05 13 12.	- 03 29	12	15.	GALAXY
SC 0510-1409.6	05 13 12.	- 14 06 07.	36		NEBULA
MCG+01-14-011	05 13 15.	+ 06 32 30.	36	16.	GALAXY
MCG-01-14-008	05 13 15.	- 03 25	36	17.	GALAXY
HSE 175	05 13 15.	- 69 08			STAR CLUSTER IN LMC
RNGC 1870	05 13 15.	- 69 12			OPEN CLUSTER IN LMC
SL 316	05 13 16.	- 70 32	20		STAR CLUSTER IN LMC
LW 160	05 13 17.	- 72 55			STAR CLUSTER IN LMC
ZWG 421.013	05 13 18.	+ 06 43		15.7	GALAXY
MCG+13-04-019	05 13 18.	+ 76 15	30	16.	GALAXY
MCG-05-13-013	05 13 18.	- 30 36	150	14.	GALAXY
LH120-N030D	05 13 18.	- 67 32	66		EMISSION NEBULA IN LMC
RNGC 1866	05 13 19.	- 65 31		10.0	OPEN CLUSTER IN LMC
SL 319	05 13 19.	- 65 31	305		STAR CLUSTER IN LMC
SL 317	05 13 21.	- 69 12	25		STAR CLUSTER IN LMC
RNGC 1872	05 13 22.	- 69 24			CLUSTER/NEBULOSITY IN LMC
SL 318	05 13 22.	- 69 24	45		STAR CLUSTER IN LMC
HSE 176	05 13 23.	- 64 58			STAR CLUSTER IN LMC
ZWG 421.014	05 13 24.	+ 05 31		15.7	GALAXY
MCG-03-14-011	05 13 24.	- 16 09	60	14.5	GALAXY
MCG-04-13-005	05 13 24.	- 22 47	66	15.	GALAXY
LW 163	05 13 25.	- 65 31			STAR CLUSTER IN LMC
SL 314	05 13 27.	- 73 27	70		STAR CLUSTER IN LMC
LW 161	05 13 27.	- 73 27			STAR CLUSTER IN LMC
RNGC 1867	05 13 28.	- 66 21			OPEN CLUSTER IN LMC
SL 321	05 13 28.	- 66 21	45		STAR CLUSTER IN LMC
SL 322	05 13 29.	- 66 34	20		STAR CLUSTER IN LMC
LB 01776	05 13 30.	- 47 32		11.3	FAINT BLUE STAR
LH120-N113D	05 13 30.	- 69 27	73		EMISSION NEBULA IN LMC
SC 0511-1612.8	05 13 32.	- 16 09 21.	54		NEBULA

OBJECT NAME	RIGHT ASCEN.	DECLINATION	DIAM.	MAGN.	TYPE OF OBJECT
RNGC 1881	05 13 34.	- 69 25			CLUSTER/NEBULOSITY IN LMC
LH 035	05 13 34.	- 69 25	360		STELLAR ASSN. IN LMC
RNGC 1877	05 13 34.	- 69 27			CLUSTER/NEBULOSITY IN LMC
RNGC 1876	05 13 34.	- 69 27			CLUSTER/NEBULOSITY IN LMC
RNGC 1874	05 13 34.	- 69 27			CLUSTER/NEBULOSITY IN LMC
SL 320	05 13 34.	- 69 27	135		STAR CLUSTER IN LMC
UGC 03272	05 13 36.	+ 06 04	66	16.0	GALAXY S
ZWG 421.015	05 13 36.	+ 06 23		15.5	GALAXY
MCG+09-09-001	05 13 36.	+ 53 30	57	14.	GALAXY
ZWG 395.023	05 13 36.	- 00 12		14.6	GALAXY
UGC 03271	05 13 36.	- 00 12	84	14.6	GALAXY
LH120-N113C	05 13 36.	- 69 26	81		EMISSION NEBULA IN LMC
LH120-N113E	05 13 36.	- 69 27			EMISSION NEBULA IN LMC
MCG+00-14-018	05 13 39.	- 00 12	24	15.	GALAXY
HSE 177	05 13 39.	- 69 07			STAR CLUSTER IN LMC
HSE 178	05 13 41.	- 66 41			STAR CLUSTER IN LMC
MCG+01-14-012	05 13 42.	+ 06 25	30	16.	GALAXY
MCG+01-14-014	05 13 42.	+ 06 26	9	16.	GALAXY
MCG+01-14-013	05 13 42.	+ 06 27	10	16.	GALAXY
ZC 0513.7+2922	05 13 42.	+ 29 22	2550		CLUSTER OF GALAXIES
ISS 0256	05 13 42.	+ 35 29	325		STELLAR RING
ZWG 258.007	05 13 42.	+ 53 30		15.2	GALAXY
UGC 03273	05 13 42.	+ 53 30	198	15.2	GALAXY S IV-V
MCG-04-13-006	05 13 42.	- 23 33 30.	30	15.5	GALAXY
LH120-N112	05 13 42.	- 69 15	48		EMISSION NEBULA IN LMC
LH120-N113B	05 13 42.	- 69 27			EMISSION NEBULA IN LMC
LH120-N113A	05 13 42.	- 69 27	29		EMISSION NEBULA IN LMC
RNGC 1873	05 13 44.	- 67 24			CLUSTER/NEBULOSITY IN LMC
SL 324	05 13 44.	- 67 24	145		STAR CLUSTER IN LMC
RNGC 1871	05 13 44.	- 67 32			CLUSTER/NEBULOSITY IN LMC
SL 325	05 13 44.	- 67 32	135		STAR CLUSTER IN LMC
VV 161F	05 13 45.	+ 06 24	24	18.	INTERACTING GALAXY
VV 161E	05 13 45.	+ 06 24	9	17.5	INTERACTING GALAXY
VV 161D	05 13 45.	+ 06 24	9	17.	INTERACTING GALAXY
VV 161C	05 13 45.	+ 06 24	12	16.	INTERACTING GALAXY
VV 161B	05 13 45.	+ 06 24	12	16.	INTERACTING GALAXY
VV 161A	05 13 45.	+ 06 24	18	16.	INTERACTING GALAXY
VV 161	05 13 45.	+ 06 24	84		INTERACTING GALAXY
SL 323	05 13 46.	- 69 22	20		STAR CLUSTER IN LMC
SC 0513-6054.3	05 13 47.	- 60 50 57.	24		NEBULA
ZWG 421.016	05 13 48.	+ 06 20		15.6	GALAXY
LH120-N030	05 13 48.	- 67 28	564		EMISSION NEBULA IN LMC
LH120-N113	05 13 48.	- 69 24	556		EMISSION NEBULA IN LMC
LH 036	05 13 50.	- 67 24	180		STELLAR ASSN. IN LMC
RNGC 1869	05 13 50.	- 67 27			CLUSTER/NEBULOSITY IN LMC
SL 326	05 13 50.	- 67 27	105		STAR CLUSTER IN LMC
LH 037	05 13 50.	- 67 27	120		STELLAR ASSN. IN LMC
LH 038	05 13 50.	- 67 31	180		STELLAR ASSN. IN LMC
HSE 179	05 13 50.	- 68 53			STAR CLUSTER IN LMC
MCG+01-14-016	05 13 54.	+ 06 22 30.	66	14.	GALAXY
ABC 0539	05 13 54.	+ 06 25		14.4	RICH CLUSTER OF GALAXIES
ZWG 421.017	05 13 54.	+ 06 27		15.6	GALAXY
MCG+01-14-015	05 13 54.	+ 06 27	9	16.	GALAXY
LH120-N030C	05 13 54.	- 67 31	317		EMISSION NEBULA IN LMC
SC 0513-6207.1	05 13 56.	- 62 03 45.	12		NEBULA
RNGC 1880	05 13 58.	- 69 27			CLUSTER/NEBULOSITY IN LMC
LW 166	05 13 59.	- 64 52			STAR CLUSTER IN LMC
SL 329	05 13 59.	- 64 57	30		STAR CLUSTER IN LMC
LBN 0795	05 14	+ 34 20	2400		BRIGHT NEBULA
LBN 0768	05 14	+ 42 00	2400		BRIGHT NEBULA
ZWG 421.018	05 14 00.	+ 06 23		15.1	GALAXY
UGC 03274	05 14 00.	+ 06 23	78	14.5	GALAXY CHAIN
MCG+01-14-018	05 14 00.	+ 06 27	18	16.	GALAXY
MCG+01-14-017	05 14 00.	+ 06 34	96	16.	GALAXY
LDN 1549	05 14 00.	+ 26 10	1140		DARK NEBULA
MCG+13-04-020	05 14 00.	+ 76 18	78	15.	GALAXY
MCG-02-14-009	05 14 00.	- 10 36	42	15.5	GALAXY
MCG-02-14-010	05 14 00.	- 13 32	66	15.	GALAXY
SC 0513-6216.7	05 14 00.	- 62 13 22.	30		NEBULA
LW 165	05 14 00.	- 65 04			STAR CLUSTER IN LMC
LH120-N030A	05 14 00.	- 67 27	41		EMISSION NEBULA IN LMC
LH120-N030B	05 14 00.	- 67 31	25		EMISSION NEBULA IN LMC
CM 14	05 14 00.	- 69 06			HII REGION IN LMC
LH120-N113F	05 14 00.	- 69 27	49		EMISSION NEBULA IN LMC
SL 327	05 14 01.	- 68 46	20		STAR CLUSTER IN LMC
SC 0513-6220.3	05 14 03.	- 62 16 58.	18		NEBULA
SL 328	05 14 05.	- 69 37	135		STAR CLUSTER IN LMC
UGC 03275	05 14 06.	+ 06 34	114	16.0	GALAXY S
ZWG 347.014	05 14 06.	+ 76 18		15.7	GALAXY
UGC 03276	05 14 06.	+ 76 18	102	15.7	GALAXY Sb
MCG-06-12-011	05 14 06.	- 38 48	36	15.5	GALAXY
SC 0513-6135.6	05 14 06.	- 61 32 16.	12		NEBULA
SC 0513-6203.7	05 14 08.	- 62 00 22.	18		NEBULA
RNGC 1868	05 14 08.	- 64 01			GLOBULAR CLUSTER IN LMC
SL 330	05 14 08.	- 64 01	160		STAR CLUSTER IN LMC
SC 0513-6117.9	05 14 09.	- 61 14 34.	18		NEBULA
HSE 180	05 14 09.	- 69 09			STAR CLUSTER IN LMC
MCG+01-14-019	05 14 12.	+ 06 30	18	16.	GALAXY
ZWG 307.009	05 14 12.	+ 65 25		15.4	GALAXY
UGC 03277	05 14 12.	+ 65 25	144	15.4	GALAXY Sc
MCG+11-07-009	05 14 12.	+ 65 25	78	16.	GALAXY
MCG-01-14-009	05 14 12.	- 03 54 30.	36	17.	GALAXY
LW 164	05 14 14.	- 73 15			STAR CLUSTER IN LMC
SC 0512-1659.0	05 14 16.	- 16 55 36.	30		NEBULA
HSE 183	05 14 17.	- 64 56			STAR CLUSTER IN LMC
HSE 181	05 14 17.	- 69 34			STAR CLUSTER IN LMC
LH 039	05 14 17.	- 69 35	360		STELLAR ASSN. IN LMC
MCG+00-14-019	05 14 18.	+ 00 13 30.	30	15.5	GALAXY
ZWG 421.019	05 14 18.	+ 06 30		15.6	GALAXY
MCG-02-14-011	05 14 18.	- 12 24	48	15.	GALAXY
IC 0406	05 14 20.	+ 39 49 57.			OPEN CLUSTER
SC 0513-6131.2	05 14 20.	- 61 27 53.	12		NEBULA
LW 169	05 14 20.	- 64 01			STAR CLUSTER IN LMC
HSE 182	05 14 20.	- 69 00			STAR CLUSTER IN LMC
2ZW 035	05 14 24.	+ 00 52			COMPACT GALAXY
UGC 03278	05 14 24.	+ 06 08	60	16.0	GALAXY S
SL 333	05 14 25.	- 65 23	40		STAR CLUSTER IN LMC
ZWG 421.020	05 14 30.	+ 06 52		15.3	GALAXY
UGC 03279	05 14 30.	+ 06 52	84	15.3	GALAXY Sb
MCG+01-14-020	05 14 30.	+ 06 53	72	15.	GALAXY
LB 03431	05 14 30.	- 63 26		14.2	FAINT BLUE STAR
HSE 185	05 14 30.	- 65 04			STAR CLUSTER IN LMC
LH120-N115	05 14 30.	- 70 12	38		EMISSION NEBULA IN LMC
SL 332	05 14 31.	- 68 37	10		STAR CLUSTER IN LMC
SL 335	05 14 33.	- 66 02	15		STAR CLUSTER IN LMC
HSE 186	05 14 34.	- 66 15			STAR CLUSTER IN LMC
ZWG 421.021	05 14 36.	+ 06 05		15.7	GALAXY
MCG+01-14-021	05 14 36.	+ 06 05	4	15.5	GALAXY
MCG+01-14-022	05 14 36.	+ 06 50	12	15.	GALAXY
LDN 1552	05 14 36.	+ 26 00	360		DARK NEBULA
UGC 03280	05 14 36.	+ 30 25	72	17.	GALAXY
ZWG 329.001	05 14 36.	+ 72 29		14.9	GALAXY
ZWG 328.036	05 14 36.	+ 72 29		14.9	GALAXY
UGC 03281	05 14 36.	+ 72 29	96	14.9	GALAXY S0
LW 170	05 14 37.	- 65 24			STAR CLUSTER IN LMC
LW 167	05 14 37.	- 72 09			STAR CLUSTER IN LMC
LW 171	05 14 38.	- 65 40			STAR CLUSTER IN LMC
HOD.72 10	05 14 40.	- 69 15	240		DARK NEBULA IN LMC
HSE 184	05 14 41.	- 70 46			STAR CLUSTER IN LMC
7ZW 033	05 14 42.	+ 63 29			COMPACT GALAXY
MCG+12-06-001	05 14 42.	+ 72 29	15	16.	GALAXY
RNGC 1890	05 14 43.	- 72 08			OPEN CLUSTER IN LMC
SL 331	05 14 43.	- 72 08	45		STAR CLUSTER IN LMC
LW 168	05 14 43.	- 72 15			STAR CLUSTER IN LMC
HOD.72 11	05 14 44.	- 70 15	540		DARK NEBULA IN LMC
IC 2120	05 14 46.	+ 37 30 21.			PLANETARY NEBULA
PK169-00.1	05 14 46.	+ 37 33	50		PLANETARY NEBULA
SC 0514-6122.7	05 14 46.	- 61 19 25.	12		NEBULA
MCG+01-14-023	05 14 48.	+ 06 44	48	14.5	GALAXY
LW 172	05 14 51.	- 66 03			STAR CLUSTER IN LMC
MCG-06-12-012	05 14 54.	- 37 09	240	12.5	GALAXY
HSE 187	05 14 54.	- 68 24			STAR CLUSTER IN LMC
LH120-N114A	05 14 54.	- 69 34	45		EMISSION NEBULA IN LMC
SL 334	05 14 55.	- 72 15	20		STAR CLUSTER IN LMC
SCHO 0105	05 14 59.	+ 36 57 24.	530		ISOLATED DARK CLOUD
LBN 0883	05 15	+ 04 10	1320		BRIGHT NEBULA
LBN 0872	05 15	+ 06 55	1320		BRIGHT NEBULA
LBN 0851	05 15	+ 13 20	660		BRIGHT NEBULA
LBN 0850	05 15	+ 13 20	240		BRIGHT NEBULA
KHAV 109	05 15	+ 23 45	5760		DARK NEBULA
KHAV 110	05 15	+ 26 21	7130		DARK NEBULA
LBN 0796	05 15	+ 34 00	14400		BRIGHT NEBULA
MIL 07	05 15	+ 41 40	4500		SUPERNOVA REMNANT
ZWG 421.022	05 15 00.	+ 03 05		15.7	GALAXY
SN 1969D	05 15 00.	+ 05 47		17.5	SUPERNOVA
ZWG 421.023	05 15 00.	+ 06 45		15.7	GALAXY
UGC 03282	05 15 00.	+ 06 45	72	15.7	GALAXY SBc
7ZW 024	05 15 00.	+ 82 46			COMPACT GALAXY
CM 13	05 15 00.	- 69 30			HII REGION IN LMC
SC 0514-6206.6	05 15 01.	- 62 03 20.	12		NEBULA
SL 336	05 15 02.	- 71 17	20		STAR CLUSTER IN LMC
MCG+01-39-014A	05 15 06.	+ 05 14		15.7	GALAXY
LH120-N031	05 15 06.	- 66 31	90		EMISSION NEBULA IN LMC
LH120-N114	05 15 06.	- 69 31	528		EMISSION NEBULA IN LMC
RNGC 1882	05 15 09.	- 66 11			OPEN CLUSTER IN LMC
HSE 188	05 15 10.	- 69 17			STAR CLUSTER IN LMC
LW 174	05 15 11.	- 64 46			STAR CLUSTER IN LMC
2ZW 036	05 15 12.	+ 00 10			COMPACT GALAXY
SCHO 0106	05 15 12.	+ 40 53 24.	340		ISOLATED DARK CLOUD
UGC 03284	05 15 12.	+ 72 40	102	16.0	GALAXY Sc
UGC 03283	05 15 12.	- 01 13	96	16.5	GALAXY Sc
SL 340	05 15 15.	- 66 11	65		STAR CLUSTER IN LMC
RNGC 1885	05 15 15.	- 69 02			OPEN CLUSTER IN LMC
SL 338	05 15 15.	- 69 02	35		STAR CLUSTER IN LMC
HSE 189	05 15 16.	- 69 21			STAR CLUSTER IN LMC
HSE 190	05 15 17.	- 69 31			STAR CLUSTER IN LMC
MCG+02-14-001	05 15 18.	+ 13 22	180	12.	GALAXY
MCG-06-12-013	05 15 18.	- 37 08	60	15.	GALAXY
SL 339	05 15 18.	- 68 21	35		STAR CLUSTER IN LMC
MCG-04-13-007	05 15 21.	- 23 50	36	15.	GALAXY
HSE 191	05 15 23.	- 69 29			STAR CLUSTER IN LMC
VDB.66N 037	05 15 24.	+ 13 21	924		REFLECTION NEBULA
MCG-03-14-012	05 15 24.	- 20 56	24	15.5	GALAXY
SL 337	05 15 25.	- 73 10	20		STAR CLUSTER IN LMC
MCG-03-14-013	05 15 27.	- 15 32	78	14.	GALAXY
IC 0407	05 15 27.	- 15 34 14.			NONSTELLAR OBJECT
HSE 192	05 15 28.	- 69 24			STAR CLUSTER IN LMC
HSE 194	05 15 29.	- 66 46			STAR CLUSTER IN LMC
HSE 193	05 15 30.	- 68 29			STAR CLUSTER IN LMC
HSE 196	05 15 34.	- 66 20			STAR CLUSTER IN LMC
LDN 1589	05 15 36.	+ 07 20	660		DARK NEBULA
MCG+12-06-002	05 15 36.	+ 72 17	24	16.	GALAXY
LW 173	05 15 37.	- 73 11			STAR CLUSTER IN LMC
VMT 05	05 15 38.	+ 41 46	4800		SUPERNOVA REMNANT
7ZW 035	05 15 42.	+ 66 12			COMPACT GALAXY
ZWG 329.002	05 15 42.	+ 72 17		14.7	GALAXY
ZWG 328.037	05 15 42.	+ 72 17		14.7	GALAXY
YM 31	05 15 43.	- 05 38	3480		SYMMETRIC GALACTIC NEBULA
HSE 197	05 15 43.	- 68 40			STAR CLUSTER IN LMC
HSE 198	05 15 45.	- 69 07			STAR CLUSTER IN LMC
HSE 195	05 15 45.	- 70 26			STAR CLUSTER IN LMC
RNGC 1887	05 15 46.	- 66 22			OPEN CLUSTER IN LMC
ZWG 307.010	05 15 48.	+ 66 11		14.7	GALAXY
SL 345	05 15 48.	- 63 16	35		STAR CLUSTER IN LMC
IC 0408	05 15 51.	- 25 08 30.			NONSTELLAR OBJECT
RNGC 1884	05 15 52.	- 66 14			NON-EXISTENT OBJECT
SL 343	05 15 52.	- 66 22	35		STAR CLUSTER IN LMC
SL 341	05 15 53.	- 69 32	20		STAR CLUSTER IN LMC
MRSL 195-16/1	05 15 54.	+ 07 23	4200		HII REGION
TON-S 0415	05 15 54.	- 30 52		13.7	BLUE STAR
SC 0515-6101.6	05 15 54.	- 60 58 24.	6		NEBULA
SL 346	05 15 54.	- 63 14	20		STAR CLUSTER IN LMC
LH120-N032	05 15 54.	- 68 02	19		EMISSION NEBULA IN LMC
HSE 200	05 15 57.	- 69 12			STAR CLUSTER IN LMC
SL 342	05 15 57.	- 70 24	25		STAR CLUSTER IN LMC
LBN 0871	05 16	+ 07 05	1020		BRIGHT NEBULA
LBN 0869	05 16	+ 07 45	1320		BRIGHT NEBULA
LBN 0781	05 16	+ 38 50	1500		BRIGHT NEBULA
LBN 0937	05 16	- 03 30	8400		BRIGHT NEBULA
UGC 03285	05 16 00.	+ 19 09	102	16.5	GALAXY
LDN 1553	05 16 00.	+ 26 10	1140		DARK NEBULA
LDN 1634	05 16 00.	- 05 50	2820		DARK NEBULA
MCG-04-13-008	05 16 00.	- 24 10	30	15.5	GALAXY
MCG-05-13-014	05 16 00.	- 29 47	60	15.	GALAXY
MCG-06-12-014	05 16 00.	- 33 57	66	15.5	GALAXY
LW 175	05 16 00.	- 63 16			STAR CLUSTER IN LMC
HSE 199	05 16 00.	- 70 58			STAR CLUSTER IN LMC
LB 03432	05 16 00.	- 71 52		13.2	FAINT BLUE STAR
RNGC 1894	05 16 05.	- 69 31			OPEN CLUSTER IN LMC
SL 344	05 16 05.	- 69 31	50		STAR CLUSTER IN LMC
LW 176	05 16 06.	- 63 14			STAR CLUSTER IN LMC
SCHO 0107	05 16 08.	+ 29 00 48.	430		ISOLATED DARK CLOUD
OCL 0372	05 16 12.	+ 73 14	7200	4.2	OPEN STAR CLUSTER
SL 347	05 16 12.	- 66 52	15		STAR CLUSTER IN LMC
HODG 01	05 16 15.	- 68 59	66		RED GLOBULAR CLSTR IN LMC
SC 0515-6116.2	05 16 17.	- 61 13 01.	24		NEBULA

OBJECT NAME	RIGHT ASCEN.	DECLINATION	DIAM.	MAGN.	TYPE OF OBJECT
MCG-02-14-012	05 16 18.	- 14 12 30.	36	15.5	GALAXY
2ZW 037	05 16 24.	+ 02 34			COMPACT GALAXY
SN 19720	05 16 24.	+ 05 57		18.5	SUPERNOVA
UGC 03286	05 16 24.	+ 16 50	72	16.0	GALAXY SBb-c
MRSL 168+01/1	05 16 24.	+ 38 54	1200		HII REGION
HSE 204	05 16 28.	- 64 36			STAR CLUSTER IN LMC
ZWG 469.005	05 16 30.	+ 17 39		15.4	GALAXY
LH120-N194	05 16 30.	- 71 50	19		EMISSION NEBULA IN LMC
HSE 201	05 16 32.	- 68 49			STAR CLUSTER IN LMC
RNGC 1857	05 16 34.	+ 39 18		8.5	OPEN CLUSTER
ZWG 421.024	05 16 36.	+ 04 45		15.7	GALAXY
ZWG 421.025	05 16 36.	+ 05 49		15.7	GALAXY
OCL 0427	05 16 36.	+ 39 25	1080		OPEN STAR CLUSTER
SCHO 0108	05 16 38.	+ 36 31 42.	430		ISOLATED DARK CLOUD
PATH 1.178	05 16 41.	- 14 46	5		NEBULA
ZWG 395.024	05 16 42.	+ 01 16		14.7	GALAXY
UGC 03287	05 16 42.	+ 01 16	60	14.7	GALAXY S
KARA.72 105B	05 16 42.	+ 01 16	48		PART OF DOUBLE GALAXY
KARA.72 105A	05 16 42.	+ 01 16	36	14.7	PART OF DOUBLE GALAXY
OCL 0428	05 16 42.	+ 39 18	840	8.6	OPEN STAR CLUSTER
ZWG 329.003	05 16 42.	+ 72 55		15.1	GALAXY
ZWG 328.038	05 16 42.	+ 72 55		15.1	GALAXY
PATH 1.179	05 16 43.	- 14 47	3		NEBULA
HSE 203	05 16 43.	- 69 54			STAR CLUSTER IN LMC
HSE 205	05 16 44.	- 68 58			STAR CLUSTER IN LMC
HSE 202	05 16 47.	- 71 54			STAR CLUSTER IN LMC
MCG-04-13-009	05 16 48.	- 21 36	72	18.	GALAXY
UGC 03288	05 16 54.	+ 04 04	66	17.	GALAXY
LDN 1554	05 16 54.	+ 26 10	240		DARK NEBULA
MCG+12-06-003	05 16 54.	+ 72 55	33	16.	GALAXY
RNGC 1892	05 16 54.	- 65 01			GALAXY
LH120-N033	05 16 54.	- 67 23	81		EMISSION NEBULA IN LMC
IC 0409	05 16 56.	+ 03 16 01.			NONSTELLAR OBJECT
RNGC 1895	05 16 56.	- 67 23			DIFFUSE NEBULA IN LMC
ARP 052	05 16 57.	+ 03 41			PECULIAR GALAXY
SC 0516-6124.2	05 16 58.	- 61 21 04.			NEBULA
LBN 0884	05 17	+ 04 10	1620		BRIGHT NEBULA
LBN 0880	05 17	+ 04 43	1260		BRIGHT NEBULA
KHAV 112	05 17	+ 07 57	10400		DARK NEBULA
LBN 0919	05 17	- 01 50	1920		BRIGHT NEBULA
LBN 0960	05 17	- 06 00	1500		BRIGHT NEBULA
VDB.66G 037	05 17	- 21 39	100		DWARF GALAXY
VDB.66G 232	05 17	- 32 10	100		DWARF GALAXY
MCG+01-14-024	05 17 00.	+ 03 15	36	15.	GALAXY
ZWG 421.026	05 17 00.	+ 03 16		15.0	GALAXY
MCG-06-12-016	05 17 00.	- 36 40	24	15.	GALAXY
MCG-06-12-015	05 17 00.	- 36 40	36	15.	GALAXY
LB 01777	05 17 00.	- 47 15		13.8	FAINT BLUE STAR
CM 15	05 17 00.	- 69 06			HII REGION IN LMC
SL 354	05 17 01.	- 63 28	55		STAR CLUSTER IN LMC
SL 348	05 17 01.	- 68 40	35		STAR CLUSTER IN LMC
SL 349	05 17 02.	- 68 55	25		STAR CLUSTER IN LMC
HSE 206	05 17 05.	- 66 34			STAR CLUSTER IN LMC
ZWG 421.027	05 17 06.	+ 03 40		15.6	GALAXY
RNGC 1898	05 17 06.	- 69 40		11.5	GLOBULAR CLUSTER IN LMC
SL 351	05 17 07.	- 68 44	25		STAR CLUSTER IN LMC
HSE 207	05 17 11.	- 66 29			STAR CLUSTER IN LMC
HOD.72 12	05 17 11.	- 69 29	120		DARK NEBULA IN LMC
LW 177	05 17 13.	- 63 29			STAR CLUSTER IN LMC
SL 353	05 17 14.	- 68 55	60		STAR CLUSTER IN LMC
MCG-06-12-017	05 17 15.	- 37 07	72	13.5	GALAXY
SL 352	05 17 16.	- 70 37	20		STAR CLUSTER IN LMC
PATH 1.180	05 17 17.	- 14 41	5		NEBULA
ZC 0517.3+0303	05 17 18.	+ 03 03	1010		CLUSTER OF GALAXIES
ZWG 421.028	05 17 18.	+ 06 37		15.5	GALAXY
LH120-N116	05 17 18.	- 69 57	86		EMISSION NEBULA IN LMC
IC 2122	05 17 19.	- 37 09 50.			NONSTELLAR OBJECT
ZWG 421.029	05 17 24.	+ 06 31		15.6	GALAXY
UGC 03289	05 17 24.	+ 06 31	66	15.6	GALAXY SBb
MCG+01-14-025	05 17 24.	+ 06 32	54	15.5	GALAXY
UGC 03290	05 17 24.	+ 17 40	72	17.	GALAXY S
OCL 0447	05 17 24.	+ 30 33	840	16.	OPEN STAR CLUSTER
MRSL 207-22/1	05 17 24.	- 05 43	3000		HII REGION
MCG-05-13-015	05 17 24.	- 28 46	30	16.	GALAXY
LH120-N117	05 17 24.	- 69 38	28		EMISSION NEBULA IN LMC
LB 03433	05 17 24.	- 70 22		13.3	FAINT BLUE STAR
RNGC 1897	05 17 26.	- 67 30			OPEN CLUSTER IN LMC
SL 355	05 17 26.	- 67 31	25		STAR CLUSTER IN LMC
ZWG 421.030	05 17 30.	+ 05 47		15.7	GALAXY
ZWG 421.031	05 17 30.	+ 05 52		15.2	GALAXY
LH120-N034C	05 17 30.	- 66 47			EMISSION NEBULA IN LMC
LH120-N034B	05 17 30.	- 66 47			EMISSION NEBULA IN LMC
LH120-N034A	05 17 30.	- 66 47			EMISSION NEBULA IN LMC
CM 38	05 17 30.	- 68 12			HII REGION IN LMC
HSE 208	05 17 31.	- 67 05			STAR CLUSTER IN LMC
RNGC 1903	05 17 34.	- 69 22		12.0	OPEN CLUSTER IN LMC
SL 356	05 17 34.	- 69 23	95		STAR CLUSTER IN LMC
MCG-06-12-018	05 17 36.	- 37 11	30	15.	GALAXY
SL 357	05 17 41.	- 69 25	55		STAR CLUSTER IN LMC
ZWG 421.032	05 17 42.	+ 06 31		15.4	GALAXY
UGC 03291	05 17 42.	+ 06 31	66	15.4	GALAXY SB:c
MCG+01-14-026	05 17 42.	+ 06 31	54	15.	GALAXY
IC 2121	05 17 42.	- 25 06 57.			NONSTELLAR OBJECT
HOD.72 13	05 17 43.	- 69 57	240		DARK NEBULA IN LMC
SNO 20	05 17 46.	- 25 20	3000	18.	LOOSE CLSTR OF 20 GLXIES
RNGC 1879	05 17 46.	- 32 13		14.0	GALAXY
SL 358	05 17 47.	- 69 34	35		STAR CLUSTER IN LMC
ZC 0517.8+5108	05 17 48.	+ 51 08	5650		CLUSTER OF GALAXIES
UGC 03292	05 17 48.	+ 52 48	66	17.	GALAXY S
MCG-05-13-016	05 17 48.	- 32 13	150	14.	GALAXY
MCG-05-13-017	05 17 48.	- 32 43	36	15.	GALAXY
LH120-N035	05 17 48.	- 66 04	134		EMISSION NEBULA IN LMC
LB 03434	05 17 48.	- 66 05		12.9	FAINT BLUE STAR
SL 359	05 17 49.	- 68 31	60		STAR CLUSTER IN LMC
SL 360	05 17 52.	- 69 16	135		STAR CLUSTER IN LMC
RNGC 1916	05 17 53.	- 69 26			OPEN CLUSTER IN LMC
SL 361	05 17 53.	- 69 27	60		STAR CLUSTER IN LMC
HSE 210	05 17 53.	- 70 39			STAR CLUSTER IN LMC
LB 03435	05 17 54.	- 67 23		13.6	FAINT BLUE STAR
HSE 209	05 17 54.	- 71 58			STAR CLUSTER IN LMC
HSE 212	05 17 58.	- 66 13			STAR CLUSTER IN LMC
RNGC 1899	05 17 58.	- 67 57			DIFFUSE NEBULA IN LMC
SL 362	05 17 59.	- 69 37	35		STAR CLUSTER IN LMC
LBN 0882	05 18	+ 04 40	1380		BRIGHT NEBULA
KHAV 111	05 18	+ 22 09	7570		DARK NEBULA
LBN 0942	05 18	- 03 50	3300		BRIGHT NEBULA
LBN 0956	05 18	- 05 30	1620		BRIGHT NEBULA
SAP 1	05 18	- 68 30	300		STELLAR GROUP

OBJECT NAME	RIGHT ASCEN.	DECLINATION	DIAM.	MAGN.	TYPE OF OBJECT
ZC 0518.0+0045	05 18 00.	+ 00 45	2550		CLUSTER OF GALAXIES
ZWG 421.033	05 18 00.	+ 05 05		15.6	GALAXY
LDN 1590	05 18 00.	+ 07 20	3720		DARK NEBULA
LDN 1588	05 18 00.	+ 08 20	1560		DARK NEBULA
UGC 03293	05 18 00.	+ 08 45	108	16.0	GALAXY Sc
ISS 0257	05 18 00.	+ 35 16	433		STELLAR RING
LH120-N036	05 18 00.	- 67 57	78		EMISSION NEBULA IN LMC
RNGC 1901	05 18 00.	- 68 30			OPEN CLUSTER
CM 16	05 18 00.	- 69 24			HII REGION IN LMC
CM 17	05 18 00.	- 69 36			HII REGION IN LMC
HSE 214	05 18 00.	- 69 36			STAR CLUSTER IN LMC
SCHO 0109	05 18 02.	+ 33 23 30.	470		ISOLATED DARK CLOUD
SL 366	05 18 03.	- 65 59	25		STAR CLUSTER IN LMC
RNGC 1902	05 18 05.	- 66 41			OPEN CLUSTER IN LMC
SL 367	05 18 05.	- 66 42	50		STAR CLUSTER IN LMC
SL 363	05 18 06.	- 69 41	35		STAR CLUSTER IN LMC
SL 372	05 18 09.	- 64 10	20		STAR CLUSTER IN LMC
LW 178	05 18 09.	- 65 59			STAR CLUSTER IN LMC
ZWG 421.034	05 18 12.	+ 03 13		15.2	GALAXY
KARA.72 106A	05 18 12.	+ 03 13	42	15.2	PART OF DOUBLE GALAXY
MCG-04-13-010	05 18 12.	- 25 46	24	15.5	GALAXY
HSE 213	05 18 12.	- 69 38			STAR CLUSTER IN LMC
LH120-N195A	05 18 12.	- 71 18	34		EMISSION NEBULA IN LMC
IC 0411	05 18 13.	- 25 22 43.			NONSTELLAR OBJECT
HOD.72 14	05 18 13.	- 69 56	240		DARK NEBULA IN LMC
SL 364	05 18 13.	- 71 08	15		STAR CLUSTER IN LMC
RNGC 1905	05 18 14.	- 67 19			OPEN CLUSTER IN LMC
SL 369	05 18 14.	- 67 20	25		STAR CLUSTER IN LMC
RNGC 1914	05 18 14.	- 71 18			CLUSTER/NEBULOSITY IN LMC
SL 365	05 18 14.	- 71 19	60		STAR CLUSTER IN LMC
MCG+01-14-027	05 18 15.	+ 03 12 30.	24	15.	GALAXY
LW 180	05 18 15.	- 64 10			STAR CLUSTER IN LMC
RNGC 1910	05 18 16.	- 69 15			CLUSTER/NEBULOSITY IN LMC
SHB 089	05 18 16.5	+ 16 35 27.		18.8	QUASI-STELLAR OBJECT
BC 3CR138	05 18 16.51	+ 16 35 27.0		18.84	QUASI-STELLAR OBJECT
ABC 0537	05 18 17.	+ 73 51		17.5	RICH CLUSTER OF GALAXIES
HODG 02	05 18 17.	- 69 43	48		RED GLOBULAR CLSTR IN LMC
ZWG 421.035	05 18 18.	+ 03 12		15.3	GALAXY
KARA.72 106B	05 18 18.	+ 03 12	36	15.3	PART OF DOUBLE GALAXY
MCG+01-14-028	05 18 18.	+ 03 12	30	15.	GALAXY
ZWG 421.036	05 18 18.	+ 06 30		15.5	GALAXY
MCG-04-13-011	05 18 18.	- 25 23	30	15.5	GALAXY
LH120-N119A	05 18 18.	- 69 14			EMISSION NEBULA IN LMC
LB 03436	05 18 18.	- 71 03		12.8	FAINT BLUE STAR
LH120-N195B	05 18 18.	- 71 18	46		EMISSION NEBULA IN LMC
SL 370	05 18 19.	- 68 40	40		STAR CLUSTER IN LMC
SL 368	05 18 19.	- 70 00	20		STAR CLUSTER IN LMC
HSE 215	05 18 20.	- 65 45			STAR CLUSTER IN LMC
LW 181	05 18 20.	- 65 45			STAR CLUSTER IN LMC
SL 371	05 18 22.	- 69 16	55		STAR CLUSTER IN LMC
ZC 0518.4+0110	05 18 24.	+ 01 10	1080		CLUSTER OF GALAXIES
MCG+01-14-029	05 18 24.	+ 03 56	156	14.	GALAXY
ZWG 421.037	05 18 24.	+ 03 58		14.5	GALAXY
UGC 03294	05 18 24.	+ 03 58	204	14.5	GALAXY Sb
UGC 03295	05 18 24.	+ 15 12	72	17.	GALAXY S
ASS 49	05 18 24.	+ 33 49	21600		OB ASSOCIATION AUR OB1
LH120-N195	05 18 24.	- 71 18	182		EMISSION NEBULA IN LMC
LH 040	05 18 26.	- 71 18	60		STELLAR ASSN. IN LMC
MCG+00-14-020	05 18 27.	+ 02 35	42	14.5	GALAXY
SL 375	05 18 29.	- 66 34	5		STAR CLUSTER IN LMC
SL 373	05 18 29.	- 69 36	70		STAR CLUSTER IN LMC
OCL 0419	05 18 30.	+ 45 21	1200	16.	OPEN STAR CLUSTER
PATH 1.181	05 18 32.	- 14 44	14		NEBULA
HSE 217	05 18 34.	- 66 17			STAR CLUSTER IN LMC
LH 041	05 18 34.	- 69 17	420		STELLAR ASSN. IN LMC
HSE 216	05 18 34.	- 69 23			STAR CLUSTER IN LMC
LB 03437	05 18 36.	- 61 33		14.4	FAINT BLUE STAR
RNGC 1900	05 18 36.	- 63 04			OPEN CLUSTER IN LMC
SL 376	05 18 36.	- 63 04	70		STAR CLUSTER IN LMC
MEL 1	05 18 37.	+ 08 16			NEBULA
SCHO 0110	05 18 38.	+ 32 34 18.	270		ISOLATED DARK CLOUD
MCG-04-13-012	05 18 39.	- 24 44	36	16.	GALAXY
LW 182	05 18 39.	- 66 07			STAR CLUSTER IN LMC
ZWG 421.038	05 18 42.	+ 04 50		14.3	GALAXY
UGC 03296	05 18 42.	+ 04 50	114	14.3	GALAXY Sa/Sb
IC 0410	05 18 42.	+ 33 26	3300		HII REGION
ZC 0518.7+4832	05 18 42.	+ 48 32	1750		CLUSTER OF GALAXIES
ZWG 258.008	05 18 42.	+ 51 21		15.6	GALAXY
LB 03438	05 18 42.	- 65 45		12.5	FAINT BLUE STAR
LH120-N119	05 18 42.	- 69 15	932		EMISSION NEBULA IN LMC
MCG+01-14-030	05 18 45.	+ 04 50	60	14.	GALAXY
B 223	05 18 45.	+ 08 17	480		DARK OBJECT
SC 0518-6121.8	05 18 47.	- 61 18 48.	18		NEBULA
7ZW 036	05 18 48.	+ 63 02			COMPACT GALAXY
MCG+12-06-004	05 18 48.	+ 71 09	45	17.	GALAXY
SC 0518-6145.4	05 18 48.	- 61 42 24.	60		NEBULA
LW 184	05 18 48.	- 63 04			STAR CLUSTER IN LMC
LW 183	05 18 48.	- 65 08			STAR CLUSTER IN LMC
SL 377	05 18 48.	- 65 09	40		STAR CLUSTER IN LMC
RNGC 1913	05 18 48.	- 69 36			OPEN CLUSTER IN LMC
LH120-N120D	05 18 48.	- 69 43	65		EMISSION NEBULA IN LMC
SEY 006	05 18 52.	+ 72 32 12.		14.8	FAINT GALAXY
HSE 218	05 18 53.	- 69 23			STAR CLUSTER IN LMC
MCG+01-14-032	05 18 54.	+ 06 38	18	15.	GALAXY
MCG+01-14-031	05 18 54.	+ 06 38	72	15.	GALAXY
VV 169D	05 18 54.	+ 06 39	12	18.	INTERACTING GALAXY
VV 169C	05 18 54.	+ 06 39	21	17.	INTERACTING GALAXY
VV 169B	05 18 54.	+ 06 39	30	17.	INTERACTING GALAXY
VV 169A	05 18 54.	+ 06 39	36	15.	INTERACTING GALAXY
VV 169	05 18 54.	+ 06 39	102		INTERACTING GALAXY
UGC 03297	05 18 54.	+ 73 15	72	16.5	GALAXY MLT SYS
LB 01778	05 18 54.	- 52 13		12.5	FAINT BLUE STAR
HSE 219	05 18 54.	- 69 39			STAR CLUSTER IN LMC
SG 3.034	05 18 55.	+ 34 34	8640		DIFFUSE EMISSION NEBULA
HSE 220	05 18 55.	- 69 55			STAR CLUSTER IN LMC
VDB.66N 038	05 18 58.	+ 08 24	1680		REFLECTION NEBULA
SL 374	05 18 58.	- 73 29	55		STAR CLUSTER IN LMC
LW 179	05 18 58.	- 73 30			STAR CLUSTER IN LMC
SL 378	05 18 59.	- 68 14	35		STAR CLUSTER IN LMC
LBN 0867	05 19	+ 08 20	1200		BRIGHT NEBULA
LBN 0866	05 19	+ 08 20	180		BRIGHT NEBULA
LBN 0857	05 19	+ 11 20	2400		BRIGHT NEBULA
LBN 0807	05 19	+ 33 20	2100		BRIGHT NEBULA
LBN 0957	05 19	- 05 30	480		BRIGHT NEBULA
MRSL 194-15/1	05 19 00.	+ 08 21	1320		HII REGION
OCL 0423	05 19 00.	+ 40 57	1800	4.4	OPEN STAR CLUSTER
ZWG 329.004	05 19 00.	+ 72 32		15.2	GALAXY
ZWG 328.039	05 19 00.	+ 72 32		15.2	GALAXY

OBJECT NAME	RIGHT ASCEN.	DECLINATION	DIAM.	MAGN.	TYPE OF OBJECT
HSE 222	05 19 00.	- 66 53			STAR CLUSTER IN LMC
CM 18	05 19 00.	- 69 54			HII REGION IN LMC
SG 3.036	05 19 01.	+ 08 23	960		DIFFUSE EMISSION NEBULA
HSE 221	05 19 01.	- 68 39			STAR CLUSTER IN LMC
RNGC 1875	05 19 03.	+ 06 38		15.0	GALAXY
IC 0415	05 19 05.	- 15 35 24.			NONSTELLAR OBJECT
ZWG 421.039	05 19 06.	+ 06 38		15.2	GALAXY
DG 053	05 19 06.	+ 08 23	1800		REFLECTION NEBULA
ARP 327	05 19 07.	+ 06 38			PECULIAR GALAXY
CED 044	05 19 07.	+ 08 23	2340		DIFFUSE GALACTIC NEBULA
SC 0518-6150.3	05 19 10.	- 61 47 20.	6		NEBULA
MIN.47 03	05 19 11.	+ 33 28			DIFFUSE NEBULA
ISS 0288	05 19 12.	+ 29 25	160		STELLAR RING
MRSL 173-01/2	05 19 12.	+ 34 05	18000		HII REGION
OCL 0420	05 19 12.	+ 45 25	840	15.	OPEN STAR CLUSTER
RNGC 1911	05 19 12.	- 66 48			NON-EXISTENT OBJECT
HOD.72 15	05 19 14.	- 70 06	120		DARK NEBULA IN LMC
RNGC 1917	05 19 15.	- 69 02			GLOBULAR CLUSTER IN LMC
SL 379	05 19 15.	- 69 02	55		STAR CLUSTER IN LMC
IC 2123	05 19 17.	+ 03 27 01.			NONSTELLAR OBJECT
SG 3.035	05 19 17.	+ 33 27	3840		DIFFUSE EMISSION NEBULA
CED 043	05 19 17.	+ 33 28	1380		DIFFUSE GALACTIC NEBULA
SL 382	05 19 17.	- 66 32	15		STAR CLUSTER IN LMC
HSE 223A	05 19 17.	- 69 26			STAR CLUSTER IN LMC
MCG+01-14-033	05 19 18.	+ 03 17	24	15.5	GALAXY
ZWG 421.040	05 19 18.	+ 03 18		14.9	GALAXY
ZWG 421.041	05 19 18.	+ 03 26		14.8	GALAXY
UGC 03298	05 19 18.	+ 03 26	96	14.5	GALAXY
KARA.72 107A	05 19 18.	+ 03 26	54	14.8	PART OF DOUBLE GALAXY
ZWG 469.006	05 19 18.	+ 18 02		15.6	GALAXY
MRSL 183-01/1	05 19 18.	+ 33 19	3300		HII REGION
OCL 0438	05 19 18.	+ 33 28	360	9.	OPEN STAR CLUSTER
MCG+08-10-002	05 19 18.	+ 49 02	30	18.	GALAXY
7ZW 037	05 19 18.	+ 60 35			COMPACT GALAXY
MCG-06-12-019	05 19 18.	- 37 00	132	15.	GALAXY
LH120-N118	05 19 18.	- 68 24	27		EMISSION NEBULA IN LMC
HSE 223	05 19 18.	- 69 42			STAR CLUSTER IN LMC
LH120-N120	05 19 18.	- 69 43	520		EMISSION NEBULA IN LMC
IC 2124	05 19 19.	+ 03 26 55.			NONSTELLAR OBJECT
HSE 224	05 19 19.	- 67 09			STAR CLUSTER IN LMC
IC 0412	05 19 20.	+ 03 26 19.			NONSTELLAR OBJECT
SCHO 0111	05 19 20.	+ 33 19 42.	440		ISOLATED DARK CLOUD
YC 0519-37	05 19 20.	- 37 00 24.			UNUSUAL SOUTHERN OBJECT
WS 17	05 19 20.	- 67 01 05.		18.13	PLANETARY NEB. IN LMC
MAI 009	05 19 21.	+ 72 42	47		DWARF SPHEROIDAL GALAXY
RNGC 1893	05 19 22.	+ 33 21		8.0	OPEN CLUSTER
IC 0414	05 19 23.	+ 03 16 00.			NONSTELLAR OBJECT
IC 0413	05 19 23.	+ 03 26 06.			NONSTELLAR OBJECT
MCG+01-14-035	05 19 24.	+ 03 24 30.	48	15.	GALAXY
MCG+01-14-034	05 19 24.	+ 03 25	48	15.	GALAXY
ZWG 421.042	05 19 24.	+ 03 26		14.9	GALAXY
UGC 03299	05 19 24.	+ 03 26	72	14.5	GALAXY
KARA.72 107B	05 19 24.	+ 03 26	48	14.9	PART OF DOUBLE GALAXY
OCL 0439	05 19 24.	+ 33 21	1200	8.0	OPEN STAR CLUSTER
SL 381	05 19 25.	- 69 50	35		STAR CLUSTER IN LMC
SL 384	05 19 26.	- 65 50	20		STAR CLUSTER IN LMC
HSE 225	05 19 26.	- 67 16			STAR CLUSTER IN LMC
HSE 226	05 19 28.	- 67 46			STAR CLUSTER IN LMC
SL 383	05 19 30.	- 68 18	20		STAR CLUSTER IN LMC
LH120-N120A	05 19 30.	- 69 41	43		EMISSION NEBULA IN LMC
RNGC 1918	05 19 30.	- 69 41			CLUSTER/NEBULOSITY IN LMC
LH120-N120B	05 19 30.	- 69 42	51		EMISSION NEBULA IN LMC
SL 380	05 19 30.	- 71 55	35		STAR CLUSTER IN LMC
SCHO 0112	05 19 31.	+ 33 22 00.	300		ISOLATED DARK CLOUD
RNGC 1891	05 19 31.	- 35 46			NON-EXISTENT OBJECT
RNGC 1886	05 19 34.	- 23 53		14.0	GALAXY
SC 0519-6218.9	05 19 34.	- 62 15 58.			NEBULA
RNGC 1915	05 19 36.	- 66 50			NON-EXISTENT OBJECT
SL 388	05 19 37.	- 63 31	25		STAR CLUSTER IN LMC
SL 386	05 19 37.	- 65 27	15		STAR CLUSTER IN LMC
HSE 227	05 19 37.	- 69 51			STAR CLUSTER IN LMC
HSE 228	05 19 38.	- 68 55			STAR CLUSTER IN LMC
MCG-04-13-013	05 19 39.	- 23 53	180	14.	GALAXY
HSE 229	05 19 40.	- 66 17			STAR CLUSTER IN LMC
SL 385	05 19 41.	- 69 34	35		STAR CLUSTER IN LMC
OCL 0482	05 19 42.	+ 07 04	720		OPEN STAR CLUSTER
LH 042	05 19 42.	- 69 41	120		STELLAR ASSN. IN LMC
LH120-N121	05 19 42.	- 69 50			EMISSION NEBULA IN LMC
RNGC 1921	05 19 43.	- 69 50			DIFFUSE NEBULA IN LMC
SC 0519-6121.3	05 19 44.	- 61 18 22.	18		NEBULA
LW 185	05 19 44.	- 65 49			STAR CLUSTER IN LMC
SG 3.037	05 19 46.	+ 36 13	7200		DIFFUSE EMISSION NEBULA
LH120-N120C	05 19 48.	- 69 42	97		EMISSION NEBULA IN LMC
SC 0519-6123.5	05 19 49.	- 61 20 35.	6		NEBULA
LW 186	05 19 49.	- 63 31			STAR CLUSTER IN LMC
SCHO 0113	05 19 51.	+ 28 58 30.	520		ISOLATED DARK CLOUD
SL 387	05 19 53.	- 69 33	25		STAR CLUSTER IN LMC
UGC 03300	05 19 54.	+ 43 30	66	18.	GALAXY Sc
HSE 230	05 19 56.	- 67 23			STAR CLUSTER IN LMC
LBN 0890	05 20	+ 02 50	4500		BRIGHT NEBULA
LBN 0881	05 20	+ 05 00	1800		BRIGHT NEBULA
LBN 0787	05 20	+ 36 30	6000		BRIGHT NEBULA
LBN 0964	05 20	- 05 50	5100		BRIGHT NEBULA
LDN 1530	05 20 00.	+ 33 00	840		DARK NEBULA
LDN 1516	05 20 00.	+ 35 00	9600		DARK NEBULA
ZWG 307.011	05 20 00.	+ 64 55		15.6	GALAXY
RNGC 1919	05 20 00.	- 66 57			CLUSTER/NEBULOSITY IN LMC
SL 392	05 20 00.	- 66 57	75		STAR CLUSTER IN LMC
SL 390	05 20 03.	- 68 59	25		STAR CLUSTER IN LMC
HSE 233	05 20 05.	- 66 27			STAR CLUSTER IN LMC
RNGC 1922	05 20 05.	- 69 31			OPEN CLUSTER IN LMC
SL 391	05 20 05.	- 69 31	35		STAR CLUSTER IN LMC
ZWG 421.043	05 20 06.	+ 03 39		15.3	GALAXY
MCG-04-13-014	05 20 06.	- 22 24	42	15.5	GALAXY
HSE 231	05 20 06.	- 68 23			STAR CLUSTER IN LMC
SL 389	05 20 08.	- 71 15	50		STAR CLUSTER IN LMC
ARP 123	05 20 10.	- 11 32			PECULIAR GALAXY
HOD.72 16	05 20 10.	- 69 11	240		DARK NEBULA IN LMC
HSE 234	05 20 11.	- 66 29			STAR CLUSTER IN LMC
SL 393	05 20 11.	- 69 28	40		STAR CLUSTER IN LMC
MCG-02-14-014	05 20 12.	- 11 32	15	14.	GALAXY
MCG-02-14-013	05 20 12.	- 11 32 30.	180	13.	GALAXY
HOD.72 17	05 20 12.	- 69 44	180		DARK NEBULA IN LMC
SC 0519-6121.8	05 20 13.	- 61 18 54.			NEBULA
SL 398	05 20 13.	- 65 20	25		STAR CLUSTER IN LMC
RNGC 1889	05 20 14.	- 11 31		14.0	GALAXY
RNGC 1888	05 20 14.	- 11 31		13.0	GALAXY
HSE 237	05 20 17.	- 64 50			STAR CLUSTER IN LMC
HSE 236	05 20 17.	- 66 26			STAR CLUSTER IN LMC
ZWG 421.044	05 20 18.	+ 03 49		15.6	GALAXY
ZC 0520.3+1525	05 20 18.	+ 15 25	1680		CLUSTER OF GALAXIES
LH120-N037	05 20 18.	- 66 56	230		EMISSION NEBULA IN LMC
LH120-N122	05 20 18.	- 69 34			EMISSION NEBULA IN LMC
HSE 235	05 20 19.	- 68 31			STAR CLUSTER IN LMC
HSE 232	05 20 19.	- 71 00			STAR CLUSTER IN LMC
SL 401	05 20 21.	- 64 08	15		STAR CLUSTER IN LMC
SL 397	05 20 21.	- 68 56	25		STAR CLUSTER IN LMC
WS 18	05 20 21.	- 69 33 50.		15.74	PLANETARY NEB. IN LMC
SL 395	05 20 23.	- 70 43	15		STAR CLUSTER IN LMC
ZWG 395.025	05 20 24.	- 00 11		15.5	GALAXY
UGC 03301	05 20 24.	- 00 11	84	15.5	GALAXY SB0-a
MCG+00-14-021	05 20 24.	- 00 11 30.	36	14.5	GALAXY
SC 0519-6129.5	05 20 24.	- 61 26 37.	18		NEBULA
SC 0519-6129.8	05 20 24.	- 61 26 55.	18		NEBULA
SL 394	05 20 24.	- 71 55	35		STAR CLUSTER IN LMC
LW 189	05 20 25.	- 65 19			STAR CLUSTER IN LMC
LW 190	05 20 27.	- 64 02			STAR CLUSTER IN LMC
MCG-06-12-020	05 20 30.	- 36 31	18	16.	GALAXY
SC 0519-6122.1	05 20 31.	- 61 19 14.			NEBULA
LH120-N038	05 20 36.	- 66 50	64		EMISSION NEBULA IN LMC
RNGC 1920	05 20 36.	- 66 50			DIFFUSE NEBULA IN LMC
LH120-N123	05 20 36.	- 69 57			EMISSION NEBULA IN LMC
SL 396	05 20 38.	- 73 09	15		STAR CLUSTER IN LMC
SG 3.038	05 20 41.	+ 33 30	300		DIFFUSE EMISSION NEBULA
WS 19	05 20 41.	- 69 56 20.		15.71	PLANETARY NEB. IN LMC
LH120-N041	05 20 42.	- 68 04	51		EMISSION NEBULA IN LMC
LH120-N196	05 20 42.	- 70 28			EMISSION NEBULA IN LMC
SL 399	05 20 42.	- 70 49	35		STAR CLUSTER IN LMC
HSE 238	05 20 44.	- 70 11			STAR CLUSTER IN LMC
LW 187	05 20 44.	- 73 09			STAR CLUSTER IN LMC
HSE 240	05 20 46.	- 64 33			STAR CLUSTER IN LMC
OCL 0440	05 20 48.	+ 33 15	720		OPEN STAR CLUSTER
SL 402	05 20 48.	- 69 39	15		STAR CLUSTER IN LMC
RNGC 1923	05 20 49.	- 65 32			CLUSTER/NEBULOSITY IN LMC
SL 404	05 20 49.	- 65 32	35		STAR CLUSTER IN LMC
SCHO 0114	05 20 50.	+ 31 11 00.	340		ISOLATED DARK CLOUD
WS 20	05 20 51.	- 67 08 43.		16.69	PLANETARY NEB. IN LMC
SG 3.039	05 20 53.	+ 33 26	420		DIFFUSE EMISSION NEBULA
RNGC 1926	05 20 53.	- 69 32			OPEN CLUSTER IN LMC
SL 403	05 20 53.	- 69 32	40		STAR CLUSTER IN LMC
OCL 0450	05 20 54.	+ 29 33	420	15.	OPEN STAR CLUSTER
MCG+13-04-021	05 20 54.	+ 76 37	57	15.	GALAXY
MCG-02-14-015	05 20 54.	- 11 28	90	14.5	GALAXY
LH120-N039	05 20 54.	- 67 09			EMISSION NEBULA IN LMC
LH 043	05 20 55.	- 65 31	360		STELLAR ASSN. IN LMC
HSE 239	05 20 55.	- 69 54			STAR CLUSTER IN LMC
KHAV 113	05 21	+ 31 39	7770		DARK NEBULA
LBN 0961	05 21	- 05 30	1200		BRIGHT NEBULA
MCG+01-14-036	05 21 00.	+ 05 31	30	15.	GALAXY
OCL 0480	05 21 00.	+ 08 08	1440		OPEN STAR CLUSTER
ZWG 347.015	05 21 00.	+ 76 37		14.9	GALAXY
UGC 03302	05 21 00.	+ 76 37	108	14.9	GALAXY Sc
OCL 0477	05 21 06.	+ 11 25	720		OPEN STAR CLUSTER
HSE 241	05 21 06.	- 69 41			STAR CLUSTER IN LMC
HSE 242	05 21 08.	- 70 03			STAR CLUSTER IN LMC
B 224	05 21 09.	+ 10 34	1200		DARK OBJECT
HOD.72 18	05 21 09.	- 69 03	120		DARK NEBULA IN LMC
RNGC 1928	05 21 11.	- 69 31			OPEN CLUSTER IN LMC
SL 405	05 21 11.	- 69 31	20		STAR CLUSTER IN LMC
ZC 0521.2+6418	05 21 12.	+ 64 18	15320		CLUSTER OF GALAXIES
SL 409	05 21 16.	- 66 08	35		STAR CLUSTER IN LMC
OCL 0444	05 21 18.	+ 32 36	360	14.	OPEN STAR CLUSTER
VDB.66N 039	05 21 18.	+ 32 46	36		REFLECTION NEBULA
MCG-06-12-021	05 21 18.	- 38 55	30	16.	GALAXY
SL 400	05 21 18.	- 75 51	35		STAR CLUSTER IN LMC
HSE 244	05 21 19.	- 65 17			STAR CLUSTER IN LMC
SL 407	05 21 19.	- 68 28	35		STAR CLUSTER IN LMC
LW 191	05 21 19.	- 72 08			STAR CLUSTER IN LMC
SL 408	05 21 21.	- 69 07	205		STAR CLUSTER IN LMC
LH120-N125	05 21 24.	- 70 12			EMISSION NEBULA IN LMC
SL 406	05 21 24.	- 70 55	15		STAR CLUSTER IN LMC
LW 188	05 21 24.	- 75 51			STAR CLUSTER IN LMC
SCHO 0115	05 21 25.	+ 31 09 30.	300		ISOLATED DARK CLOUD
SL 410	05 21 25.	- 65 17	60		STAR CLUSTER IN LMC
LH 044	05 21 27.	- 69 07	300		STELLAR ASSN. IN LMC
HSE 243	05 21 29.	- 69 31			STAR CLUSTER IN LMC
ZC 0521.5+0328	05 21 30.	+ 03 28	1210		CLUSTER OF GALAXIES
SC 0520-6132.5	05 21 30.	- 61 29 42.			NEBULA
LH120-N040	05 21 30.	- 65 30	80		EMISSION NEBULA IN LMC
LW 192	05 21 31.	- 65 16			STAR CLUSTER IN LMC
SL 411	05 21 34.	- 67 43	25		STAR CLUSTER IN LMC
WS 22	05 21 35.	- 67 03 06.		16.52	PLANETARY NEB. IN LMC
LH120-N042	05 21 36.	- 67 03			EMISSION NEBULA IN LMC
LH120-N044A	05 21 36.	- 67 54			EMISSION NEBULA IN LMC
LH120-N124	05 21 36.	- 68 38			EMISSION NEBULA IN LMC
LH120-N197	05 21 36.	- 71 45	21		EMISSION NEBULA IN LMC
WS 21	05 21 38.	- 68 38 12.		16.22	PLANETARY NEB. IN LMC
RNGC 1925	05 21 39.	- 65 50			CLUSTER/NEBULOSITY IN LMC
SL 415	05 21 39.	- 65 50	15		STAR CLUSTER IN LMC
LH 045	05 21 39.	- 65 51	780		STELLAR ASSN. IN LMC
RNGC 1929	05 21 41.	- 67 58			CLUSTER/NEBULOSITY IN LMC
MCG-03-14-014	05 21 42.	- 17 17	48	14.	GALAXY
LH120-N127B	05 21 42.	- 69 44	87		EMISSION NEBULA IN LMC
IC 0416	05 21 44.	- 17 17 51.			NONSTELLAR OBJECT
HSE 245	05 21 47.	- 69 31			STAR CLUSTER IN LMC
MRSL 172-00/1	05 21 48.	+ 35 35			HII REGION
HSE 246	05 21 48.	- 65 04			STAR CLUSTER IN LMC
LH120-N044J	05 21 48.	- 67 49	43		EMISSION NEBULA IN LMC
LH120-N044F	05 21 48.	- 67 58	36		EMISSION NEBULA IN LMC
SL 412	05 21 49.	- 69 51	20		STAR CLUSTER IN LMC
RNGC 1938	05 21 49.	- 69 57			OPEN CLUSTER IN LMC
SL 413	05 21 49.	- 69 57	20		STAR CLUSTER IN LMC
RNGC 1939	05 21 49.	- 69 58			OPEN CLUSTER IN LMC
SL 414	05 21 49.	- 69 58	35		STAR CLUSTER IN LMC
SG 3.040	05 21 50.	+ 35 28	3600		DIFFUSE EMISSION NEBULA
SL 416	05 21 50.	- 67 14	25		STAR CLUSTER IN LMC
SC 0521-6011.7	05 21 52.	- 60 08 55.	12		NEBULA
ZC 0521.9+0818	05 21 54.	+ 08 18	2290		CLUSTER OF GALAXIES
LH120-N126	05 21 54.	- 69 05	48		EMISSION NEBULA IN LMC
RNGC 1934	05 21 59.	- 67 58			CLUSTER/NEBULOSITY IN LMC
SL 417	05 21 59.	- 67 58	270		STAR CLUSTER IN LMC
LBN 0794	05 22	+ 35 30	1920		BRIGHT NEBULA
LBN 0687	05 22	+ 67 30	2400		BRIGHT NEBULA
LBN 0969	05 22	- 06 10	900		BRIGHT NEBULA
LB 09826	05 22	- 80 35		14.2	FAINT BLUE STAR
LB 09966	05 22	- 88 33		14.5	FAINT BLUE STAR

OBJECT NAME	RIGHT ASCEN.	DECLINATION	DIAM.	MAGN.	TYPE OF OBJECT
CM 39	05 22 00.	- 68 54			HII REGION IN LMC
LH120-N127A	05 22 00.	- 69 43	73		EMISSION NEBULA IN LMC
HN 0090	05 22 02.	- 68 00			NEBULA
IC 2126	05 22 02.	- 68 00			SAME AS NGC 1935
IC 2127	05 22 02.	- 68 01			SAME AS NGC 1936
HSE 247	05 22 03.	- 68 57			STAR CLUSTER IN LMC
SC 0521-6122.3	05 22 05.	- 61 19 32.			NEBULA
HOD.72 22	05 22 05.	- 67 58	240		DARK NEBULA IN LMC
RNGC 1935	05 22 05.	- 68 00			CLUSTER/NEBULOSITY IN LMC
ZWG 347.016	05 22 06.	+ 74 52		15.7	GALAXY
ZWG 329.005	05 22 06.	+ 74 52		15.7	GALAXY
SL 418	05 22 06.	- 69 40	25		STAR CLUSTER IN LMC
HSE 248	05 22 06.	- 69 43			STAR CLUSTER IN LMC
LB 03439	05 22 06.	- 71 59		12.9	FAINT BLUE STAR
SC 0521-6146.4	05 22 07.	- 61 43 39.			NEBULA
SC 0521-6158.2	05 22 08.	- 61 55 27.	12		NEBULA
LW 195	05 22 09.	- 61 55			STAR CLUSTER IN LMC
RNGC 1956	05 22 09.	- 77 47			GALAXY
SEY 007	05 22 10.	+ 74 53 17.		15.3	FAINT GALAXY
RNGC 1904	05 22 10.	- 24 34		8.5	GLOBULAR CLUSTER
LH 047	05 22 11.	- 68 00	360		STELLAR ASSN. IN LMC
HSE 249	05 22 11.	- 69 24			STAR CLUSTER IN LMC
LH 046	05 22 11.	- 69 30	240		STELLAR ASSN. IN LMC
OCL 0422	05 22 12.	+ 41 51	720	15.	OPEN STAR CLUSTER
OCL 0417	05 22 12.	+ 46 30	360	12.2	OPEN STAR CLUSTER
MCG+12-06-005	05 22 12.	+ 74 28	39	16.	GALAXY
GCL 010	05 22 12.	- 24 34	468	8.39	GLOBULAR STAR CLUSTER
LH120-N043	05 22 12.	- 65 46	655		EMISSION NEBULA IN LMC
LH120-N044B	05 22 12.	- 68 00	67		EMISSION NEBULA IN LMC
SC 0521-6141.7	05 22 13.	- 61 38 57.			NEBULA
HOD.72 21	05 22 13.	- 69 49	90		DARK NEBULA IN LMC
HOD.72 20	05 22 13.	- 69 53	360		DARK NEBULA IN LMC
RNGC 1883	05 22 14.	+ 46 30		12.0	OPEN CLUSTER
HOD.72 19	05 22 14.	- 70 03	300		DARK NEBULA IN LMC
HSE 250	05 22 14.	- 70 04			STAR CLUSTER IN LMC
RNGC 1932	05 22 16.	- 66 12			OPEN CLUSTER IN LMC
SL 420	05 22 16.	- 66 12	40		STAR CLUSTER IN LMC
SL 419	05 22 16.	- 69 16	25		STAR CLUSTER IN LMC
SL 421	05 22 17.	- 66 34	20		STAR CLUSTER IN LMC
RNGC 1936	05 22 17.	- 68 01			CLUSTER/NEBULOSITY IN LMC
UGC 03303	05 22 18.	+ 04 27	240	16.0	GALAXY DWARF?
LB 05134	05 22 18.	- 03 53		18.7	FAINT BLUE STAR
HSE 251	05 22 18.	- 69 37			STAR CLUSTER IN LMC
HN 0091	05 22 20.	- 68 01			NEBULA
SL 422	05 22 22.	- 67 55	155		STAR CLUSTER IN LMC
RNGC 1906	05 22 23.	- 15 58		14.0	GALAXY
UGC 03304	05 22 24.	+ 21 49	78	17.	GALAXY S
LH120-N044C	05 22 24.	- 68 01	67		EMISSION NEBULA IN LMC
RNGC 1896	05 22 25.	+ 20 08			NON-EXISTENT OBJECT
IC 2125	05 22 25.	- 27 01 46.			NONSTELLAR OBJECT
HSE 252	05 22 25.	- 69 53			STAR CLUSTER IN LMC
LW 193	05 22 26.	- 74 45			STAR CLUSTER IN LMC
MCG-03-14-015	05 22 27.	- 15 58	42	14.	GALAXY
RNGC 1933	05 22 28.	- 66 11			NON-EXISTENT OBJECT
CED 045	05 22 29.	+ 06 19	2400		DIFFUSE GALACTIC NEBULA
RNGC 1937	05 22 29.	- 67 57			CLUSTER/NEBULOSITY IN LMC
SL 424	05 22 29.	- 68 11	30		STAR CLUSTER IN LMC
SL 423	05 22 29.	- 69 32	15		STAR CLUSTER IN LMC
UGC 03463	05 22 30.	+ 59 07	162	13.5	GALAXY Sc
LH120-N044K	05 22 30.	- 68 07	56		EMISSION NEBULA IN LMC
LH120-N044G	05 22 30.	- 68 07			EMISSION NEBULA IN LMC
LH120-N128	05 22 30.	- 68 41	75		EMISSION NEBULA IN LMC
SL 425	05 22 32.	- 68 48	35		STAR CLUSTER IN LMC
SCHO 0116	05 22 34.	+ 34 44 24.	450		ISOLATED DARK CLOUD
LH 048	05 22 34.	- 67 56	120		STELLAR ASSN. IN LMC
USC 03305	05 22 36.	+ 54 23	60	17.	GALAXY DWARF
LB 05101	05 22 36.	- 03 49		17.0	FAINT BLUE STAR
MCG-02-14-016	05 22 36.	- 12 44 30.	84	14.5	GALAXY
LH120-N044	05 22 36.	- 67 59	1163		EMISSION NEBULA IN LMC
LB 03440	05 22 36.	- 70 38		13.6	FAINT BLUE STAR
LB 03441	05 22 36.	- 71 14		13.5	FAINT BLUE STAR
LW 196	05 22 38.	- 65 45			STAR CLUSTER IN LMC
SL 428	05 22 38.	- 65 46	20		STAR CLUSTER IN LMC
RNGC 1940	05 22 38.	- 67 14			OPEN CLUSTER IN LMC
SL 427	05 22 38.	- 67 14	35		STAR CLUSTER IN LMC
HSE 253	05 22 38.	- 70 05			STAR CLUSTER IN LMC
HSE 254	05 22 39.	- 67 38			STAR CLUSTER IN LMC
ZWG 395.026	05 22 42.	+ 00 22		15.0	GALAXY
UGC 03306	05 22 42.	+ 00 22	78	15.0	GALAXY E-S0
MCG+00-14-022	05 22 42.	+ 00 22 30.	18	14.5	GALAXY
LH120-N044I	05 22 42.	- 67 57			EMISSION NEBULA IN LMC
HOD.72 23	05 22 42.	- 69 45	240		DARK NEBULA IN LMC
SG 3.041	05 22 43.	+ 34 49	720		DIFFUSE EMISSION NEBULA
SL 429	05 22 47.	- 68 05	165		STAR CLUSTER IN LMC
MCG+09-09-002	05 22 48.	+ 51 15	30	15.	GALAXY
LB 05135	05 22 48.	- 03 31		18.6	FAINT BLUE STAR
LH120-N045A	05 22 48.	- 66 44			EMISSION NEBULA IN LMC
LH120-N045	05 22 48.	- 66 44	77		EMISSION NEBULA IN LMC
LH120-N129	05 22 48.	- 69 45	27		EMISSION NEBULA IN LMC
RNGC 1944	05 22 52.	- 72 32			OPEN CLUSTER IN LMC
SL 426	05 22 52.	- 72 32	155		STAR CLUSTER IN LMC
LW 194	05 22 52.	- 72 32			STAR CLUSTER IN LMC
MCG+01-14-037	05 22 54.	+ 06 33	48	12.	GALAXY
LB 05136	05 22 54.	- 04 05		18.8	FAINT BLUE STAR
LB 03442	05 22 54.	- 71 02		12.6	FAINT BLUE STAR
IC 2128	05 22 55.21	- 68 05 40.			HII REGION IN LMC
HSE 258	05 22 59.	- 64 44			STAR CLUSTER IN LMC
KHAV 114	05 23	+ 36 27	5270		DARK NEBULA
KHAV 115	05 23	+ 38 03	7990		DARK NEBULA
LBN 0778	05 23	+ 40 30	360		BRIGHT NEBULA
OCL 0474	05 23 00.	+ 16 03	14400	3.0	OPEN STAR CLUSTER
LDN 1559	05 23 00.	+ 23 00	4680		DARK NEBULA
LB 05137	05 23 00.	- 04 06		18.2	FAINT BLUE STAR
LB 03443	05 23 00.	- 64 15		14.6	FAINT BLUE STAR
LH120-N044H	05 23 00.	- 68 04	47		EMISSION NEBULA IN LMC
LH120-N044D	05 23 00.	- 68 07	91		EMISSION NEBULA IN LMC
HSE 255	05 23 00.	- 69 42			STAR CLUSTER IN LMC
LH120-N130	05 23 00.	- 70 13	69		EMISSION NEBULA IN LMC
LB 03444	05 23 00.	- 71 37		13.2	FAINT BLUE STAR
SL 430	05 23 03.	- 70 12	50		STAR CLUSTER IN LMC
VDB.66N 040	05 23 04.	+ 06 33	36		REFLECTION NEBULA
RNGC 1941	05 23 05.	- 66 25			DIFFUSE NEBULA IN LMC
LH 049	05 23 05.	- 68 06	240		STELLAR ASSN. IN LMC
ZC 0523.1+0141	05 23 06.	+ 01 41	1210		CLUSTER OF GALAXIES
OCL 0432	05 23 06.	+ 35 57	540		OPEN STAR CLUSTER
MIL 06	05 23 06.	+ 42 50	2700		SUPERNOVA REMNANT
ZWG 307.012	05 23 06.	+ 63 50		14.8	GALAXY
UGC 03307	05 23 06.	+ 63 50	60	14.8	GALAXY E

OBJECT NAME	RIGHT ASCEN.	DECLINATION	DIAM.	MAGN.	TYPE OF OBJECT
ZWG 307.013	05 23 06.	+ 64 12		15.4	GALAXY
LH120-N046	05 23 06.	- 66 25	64		EMISSION NEBULA IN LMC
RNGC 1943	05 23 09.	- 70 12		12.0	CLUSTER/NEBULOSITY IN LMC
MCG+11-07-010	05 23 12.	+ 63 49	39	16.	GALAXY
7ZW 038	05 23 12.	+ 64 13			COMPACT GALAXY
LB 05138	05 23 12.	- 03 19		18.5	FAINT BLUE STAR
LB 05139	05 23 12.	- 03 48		17.1	FAINT BLUE STAR
LB 05140	05 23 12.	- 03 57		16.4	FAINT BLUE STAR
LB 05141	05 23 12.	- 04 11		18.0	FAINT BLUE STAR
LB 01779	05 23 12.	- 46 49		12.8	FAINT BLUE STAR
LB 03445	05 23 12.	- 67 28		13.5	FAINT BLUE STAR
LH120-N198	05 23 12.	- 71 38	441		EMISSION NEBULA IN LMC
HSE 256	05 23 13.	- 69 47			STAR CLUSTER IN LMC
HOD.72 24	05 23 13.	- 69 49	60		DARK NEBULA IN LMC
HSE 257	05 23 15.	- 70 13			STAR CLUSTER IN LMC
UGC 03308	05 23 18.	+ 08 55	60	16.5	GALAXY Sc
SG 3.042	05 23 18.	+ 34 12	900		DIFFUSE EMISSION NEBULA
ZWG 329.006	05 23 18.	+ 74 28		15.6	GALAXY
ZWG 328.040	05 23 18.	+ 74 28		15.6	GALAXY
LB 05142	05 23 18.	- 03 38		15.5	FAINT BLUE STAR
LH120-N044L	05 23 18.	- 68 03	36		EMISSION NEBULA IN LMC
LH120-N131	05 23 18.	- 69 54	53		EMISSION NEBULA IN LMC
HSE 259	05 23 19.	- 69 54			STAR CLUSTER IN LMC
VMT 06	05 23 21.	+ 43 00	2400		SUPERNOVA REMNANT
HSE 260	05 23 21.	- 67 39			STAR CLUSTER IN LMC
SL 431	05 23 21.	- 70 19	25		STAR CLUSTER IN LMC
MCG+08-10-003	05 23 24.	+ 45 39	24	18.	GALAXY
LB 05102	05 23 24.	- 03 14		19.0	FAINT BLUE STAR
LB 05143	05 23 24.	- 03 48		19.7	FAINT BLUE STAR
TON-S 0416	05 23 24.	- 30 54		15.1	BLUE STAR
LH120-N044E	05 23 24.	- 68 03	36		EMISSION NEBULA IN LMC
LH120-N199	05 23 24.	- 71 21			EMISSION NEBULA IN LMC
LB 03446	05 23 24.	- 71 45		12.4	FAINT BLUE STAR
LB 03447	05 23 24.	- 75 15		13.2	FAINT BLUE STAR
RNGC 1908	05 23 27.	- 02 34			NON-EXISTENT OBJECT
ARC 0540	05 23 27.	- 25 45		17.6	RICH CLUSTER OF GALAXIES
MRSL 168+03/1	05 23 30.	+ 40 35	600		HII REGION
RNGC 1909	05 23 30.	- 08 10			NON-EXISTENT OBJECT
LB 05144	05 23 36.	- 03 21		17.8	FAINT BLUE STAR
LB 05103	05 23 36.	- 03 42		18.6	FAINT BLUE STAR
LB 05145	05 23 36.	- 03 53		18.5	FAINT BLUE STAR
MCG-05-13-018	05 23 36.	- 31 35	60	15.5	GALAXY
HSE 261	05 23 36.	- 69 40			STAR CLUSTER IN LMC
LB 03448	05 23 36.	- 71 23		13.4	FAINT BLUE STAR
SL 432	05 23 36.	- 72 55	15		STAR CLUSTER IN LMC
SL 434	05 23 39.	- 69 02	20		STAR CLUSTER IN LMC
MCG+08-10-004	05 23 42.	+ 45 49	48	17.	GALAXY
SG 3.043	05 23 43.	+ 34 30	480		DIFFUSE EMISSION NEBULA
HSE 263	05 23 43.	- 69 55			STAR CLUSTER IN LMC
LW 197	05 23 43.	- 74 39			STAR CLUSTER IN LMC
MCG+08-10-005	05 23 48.	+ 49 36	24	18.	GALAXY
LB 05146	05 23 48.	- 03 18		19.6	FAINT BLUE STAR
MCG-05-13-020	05 23 48.	- 31 36	6	15.5	GALAXY
MCG-05-13-019	05 23 48.	- 31 39	36	15.5	GALAXY
MCG-06-12-022	05 23 48.	- 36 38	18	16.	GALAXY
LW 199	05 23 48.	- 65 10			STAR CLUSTER IN LMC
LH120-N044M	05 23 48.	- 68 03	49		EMISSION NEBULA IN LMC
LH120-N136	05 23 48.	- 69 07			EMISSION NEBULA IN LMC
HSE 264	05 23 48.	- 70 49			STAR CLUSTER IN LMC
HSE 262	05 23 49.	- 72 01			STAR CLUSTER IN LMC
HSE 265	05 23 51.	- 70 17			STAR CLUSTER IN LMC
ISS 0206	05 23 54.	+ 39 32	4088		STELLAR RING
MCG+11-07-011	05 23 54.	+ 67 20	66	15.	GALAXY
LB 05147	05 23 54.	- 03 02		19.0	FAINT BLUE STAR
MCG-05-13-021	05 23 54.	- 31 35	6	15.5	GALAXY
LB 01780	05 23 54.	- 47 51		12.9	FAINT BLUE STAR
SL 438	05 23 54.	- 66 50	25		STAR CLUSTER IN LMC
LH120-N132F	05 23 54.	- 69 38	32		EMISSION NEBULA IN LMC
HSE 266	05 23 55.	- 69 53			STAR CLUSTER IN LMC
LW 200	05 23 56.	- 65 37			STAR CLUSTER IN LMC
SL 440	05 23 56.	- 65 38	15		STAR CLUSTER IN LMC
HSE 269	05 23 57.	- 65 56			STAR CLUSTER IN LMC
HSE 267	05 23 58.	- 70 24			STAR CLUSTER IN LMC
SL 435	05 23 58.	- 71 30	20		STAR CLUSTER IN LMC
SC 0523-6126.7	05 23 59.	- 61 24 05.	60		NEBULA
KHAV 118	05 24	+ 10 03	3610		DARK NEBULA
ST 42	05 24	+ 10 05	8400		EMISSION NEBULA
LBN 0769	05 24	+ 43 00	660		BRIGHT NEBULA
ZWG 307.014	05 24 00.	+ 67 19		15.3	GALAXY
UGC 03309	05 24 00.	+ 67 19	90	15.3	GALAXY
LB 05148	05 24 00.	- 04 07		18.6	FAINT BLUE STAR
LB 05104	05 24 00.	- 04 15		19.2	FAINT BLUE STAR
LH120-N044N	05 24 00.	- 67 59	25		EMISSION NEBULA IN LMC
SL 436	05 24 00.	- 69 38	20		STAR CLUSTER IN LMC
LH120-N132J	05 24 00.	- 69 41	86		EMISSION NEBULA IN LMC
LH120-N200	05 24 00.	- 71 23	795		EMISSION NEBULA IN LMC
HSE 268	05 24 01.	- 69 48			STAR CLUSTER IN LMC
SL 433	05 24 01.	- 74 39	50		STAR CLUSTER IN LMC
LH120-N132A	05 24 06.	- 69 40	44		EMISSION NEBULA IN LMC
SG 3.044	05 24 07.	+ 34 19	2160		DIFFUSE EMISSION NEBULA
SL 439	05 24 09.	- 70 15	35		STAR CLUSTER IN LMC
2ZW 038	05 24 12.	- 01 15			COMPACT GALAXY
LB 05149	05 24 12.	- 03 08		15.5	FAINT BLUE STAR
LB 05105	05 24 12.	- 03 33		19.2	FAINT BLUE STAR
TON-S 0417	05 24 12.	- 32 10		15.4	BLUE STAR
HUB E07	05 24 15.	+ 22 30			DIFFUSE NEBULA
ZWG 421.045	05 24 18.	+ 04 25		15.7	GALAXY
OCL 0436	05 24 18.	+ 34 23	300	9.	OPEN STAR CLUSTER
LH120-N137A	05 24 18.	- 68 58	27		EMISSION NEBULA IN LMC
SL 441	05 24 20.	- 71 07	20		STAR CLUSTER IN LMC
RNGC 1942	05 24 21.	- 63 58			OPEN CLUSTER IN LMC
SL 445	05 24 21.	- 63 58	50		STAR CLUSTER IN LMC
LW 203	05 24 21.	- 63 59			STAR CLUSTER IN LMC
HSE 271	05 24 21.	- 67 29			STAR CLUSTER IN LMC
ZC 0524.4+0409	05 24 24.	+ 04 09	2020		CLUSTER OF GALAXIES
UGC 03310	05 24 24.	+ 49 56	66	17.	GALAXY S
MCG+08-10-006	05 24 24.	+ 49 56	60	16.	GALAXY
LH120-N138B	05 24 24.	- 68 32	26		EMISSION NEBULA IN LMC
HSE 272	05 24 25.	- 68 32			STAR CLUSTER IN LMC
SG 3.046	05 24 26.	+ 09 06	10800		DIFFUSE EMISSION NEBULA
LH 050	05 24 27.	- 71 25	480		STELLAR ASSN. IN LMC
SL 444	05 24 28.	- 67 43	75		STAR CLUSTER IN LMC
UGC 03311	05 24 30.	+ 79 21	78	16.5	GALAXY
MCG-03-14-016	05 24 30.	- 19 15 30.	84	15.	GALAXY
RNGC 1930	05 24 30.	- 46 46			GALAXY
SL 448	05 24 30.	- 63 05	35		STAR CLUSTER IN LMC
LH120-N047	05 24 30.	- 67 12			EMISSION NEBULA IN LMC
LH120-N138D	05 24 30.	- 68 32	22		EMISSION NEBULA IN LMC

OBJECT NAME	RIGHT ASCEN.	DECLINATION	DIAM.	MAGN.	TYPE OF OBJECT
LH120-N137B	05 24 30.	- 68 58	42		EMISSION NEBULA IN LMC
LH120-N132G	05 24 30.	- 69 41	41		EMISSION NEBULA IN LMC
LH120-N132I	05 24 30.	- 69 43	103		EMISSION NEBULA IN LMC
SL 443	05 24 32.	- 70 02	15		STAR CLUSTER IN LMC
HSE 273	05 24 33.	- 68 54			STAR CLUSTER IN LMC
SL 446	05 24 34.	- 67 46	25		STAR CLUSTER IN LMC
HSE 270	05 24 34.	- 71 37			STAR CLUSTER IN LMC
SL 437	05 24 38.	- 75 27	15		STAR CLUSTER IN LMC
HOD.72 25	05 24 40.	- 69 14	540		DARK NEBULA IN LMC
IC 2134	05 24 41.	- 75 30			NONSTELLAR OBJECT
OCL 0483	05 24 42.	+ 07 02	720		OPEN STAR CLUSTER
OCL 0434	05 24 42.	+ 35 17	660	10.7	OPEN STAR CLUSTER
MCG+08-10-007	05 24 42.	+ 49 51	42	15.	GALAXY
MCG-01-14-010	05 24 42.	- 06 09 30.	48	16.	GALAXY
MCG-04-13-015	05 24 42.	- 21 36	48	15.5	GALAXY
LW 205	05 24 42.	- 63 05			STAR CLUSTER IN LMC
HSE 276	05 24 42.	- 64 58			STAR CLUSTER IN LMC
LH120-N132B	05 24 42.	- 69 42	38		EMISSION NEBULA IN LMC
HN 0301	05 24 42.	- 75 29			NEBULA
RNGC 1907	05 24 44.	+ 35 17		10.5	OPEN CLUSTER
LW 198	05 24 45.	- 75 29			STAR CLUSTER IN LMC
HSE 278	05 24 47.	- 64 43			STAR CLUSTER IN LMC
ZWG 395.027	05 24 48.	+ 01 12		15.7	GALAXY
KARA.73B 0158	05 24 48.	+ 01 12	36	15.7	ISOLATED GALAXY S
MRSL 173-00/1	05 24 48.	+ 34 24	720		HII REGION
IC 0417	05 24 48.	+ 34 25			HII REGION
MRSL 172+00/1	05 24 48.	+ 35 20			HII REGION
ZWG 231.002	05 24 48.	+ 49 50		15.5	GALAXY
LH120-N138	05 24 48.	- 68 33	391		EMISSION NEBULA IN LMC
LH120-N132E	05 24 48.	- 69 42			EMISSION NEBULA IN LMC
LH120-N133	05 24 48.	- 70 07			EMISSION NEBULA IN LMC
SCHO 0117	05 24 49.	+ 30 17 24.	500		ISOLATED DARK CLOUD
CED 046	05 24 49.	+ 34 25	780		DIFFUSE GALACTIC NEBULA
SG 3.045	05 24 49.	+ 34 25	960		DIFFUSE EMISSION NEBULA
HSE 277	05 24 49.	- 67 07			STAR CLUSTER IN LMC
HSE 275	05 24 49.	- 69 49			STAR CLUSTER IN LMC
WS 23	05 24 49.	- 70 07 41.		16.19	PLANETARY NEB. IN LMC
LW 201	05 24 50.	- 74 42			STAR CLUSTER IN LMC
HSE 274	05 24 52.	- 71 37			STAR CLUSTER IN LMC
TON-S 0418	05 24 54.	- 32 03		15.3	BLUE STAR
LH120-N048E	05 24 54.	- 66 30	36		EMISSION NEBULA IN LMC
LH120-N138C	05 24 54.	- 68 31	41		EMISSION NEBULA IN LMC
LH120-N132C	05 24 54.	- 69 43	35		EMISSION NEBULA IN LMC
LB 03449	05 24 54.	- 70 32		13.5	FAINT BLUE STAR
SL 449	05 24 55.	- 69 48	35		STAR CLUSTER IN LMC
SL 447	05 24 55.	- 72 02	5		STAR CLUSTER IN LMC
RNGC 1950	05 24 56.	- 69 59			OPEN CLUSTER IN LMC
SL 450	05 24 56.	- 69 59	70		STAR CLUSTER IN LMC
SL 442	05 24 56.	- 74 42	50		STAR CLUSTER IN LMC
LBN 0804	05 25	+ 34 20	660		BRIGHT NEBULA
KHAV 116	05 25	+ 57 03	4810		DARK NEBULA
KHAV 117	05 25	+ 62 09	7500		DARK NEBULA
LDN 1595	05 25 00.	+ 06 30	4200		DARK NEBULA
ASS 48	05 25 00.	+ 34 52			OB ASSOCIATION AUR OB2
7ZW 039	05 25 00.	+ 63 50			COMPACT GALAXY
ZC 0525.0+6756	05 25 00.	+ 67 56	1210		CLUSTER OF GALAXIES
MRSL 206-20/1	05 25 00.	- 04 00	72000		HII REGION
CM 03	05 25 00.	- 66 19			HII REGION IN LMC
SL 452	05 25 00.	- 66 42	25		STAR CLUSTER IN LMC
RNGC 1946	05 25 05.	- 66 27			OPEN CLUSTER IN LMC
SL 454	05 25 05.	- 66 27	30		STAR CLUSTER IN LMC
LH120-N139	05 25 06.	- 69 22			EMISSION NEBULA IN LMC
PK215-24.1	05 25 08.5	- 12 43 53.	14	12.0	PLANETARY NEBULA
IC 0418	05 25 09.5	- 12 44 15.	14	10.6	PLANETARY NEBULA
SL 453	05 25 11.	- 69 26	50		STAR CLUSTER IN LMC
ISS 0258	05 25 12.	+ 37 12	175		STELLAR RING
ISS 0153	05 25 12.	+ 48 43	186		STELLAR RING
MCG+08-10-008	05 25 12.	+ 49 54	42	17.	GALAXY
LB 03450	05 25 12.	- 61 17		14.5	FAINT BLUE STAR
LB 03451	05 25 12.	- 72 29		12.8	FAINT BLUE STAR
HN 0069	05 25 13.	- 12 43			NEBULA
HSE 279	05 25 15.	- 69 04			STAR CLUSTER IN LMC
OCL 0442	05 25 18.	+ 33 45	720		OPEN STAR CLUSTER
LDN 1526	05 25 18.	+ 34 29	180		DARK NEBULA
OCL 0433	05 25 18.	+ 35 48	2640	7.4	OPEN STAR CLUSTER
LH120-N138A	05 25 18.	- 68 31	47		EMISSION NEBULA IN LMC
LH120-N132H	05 25 18.	- 69 41	30		EMISSION NEBULA IN LMC
RNGC 1912	05 25 19.	+ 35 48		7.0	OPEN CLUSTER
RNGC 1949	05 25 19.	- 68 31			DIFFUSE NEBULA IN LMC
LW 204	05 25 20.	- 73 09			STAR CLUSTER IN LMC
SL 456	05 25 21.	- 67 31	135		STAR CLUSTER IN LMC
LW 208	05 25 23.	- 64 48			STAR CLUSTER IN LMC
RNGC 1945	05 25 23.	- 66 23			DIFFUSE NEBULA IN LMC
LW 202	05 25 23.	- 75 42			STAR CLUSTER IN LMC
LW 209	05 25 24.	- 63 04			STAR CLUSTER IN LMC
LH120-N048	05 25 24.	- 66 23	913		EMISSION NEBULA IN LMC
LH120-N048D	05 25 24.	- 66 24	56		EMISSION NEBULA IN LMC
HSE 280	05 25 25.	- 69 53			STAR CLUSTER IN LMC
RNGC 1924	05 25 28.	- 05 21		13.0	GALAXY
RNGC 1948	05 25 28.	- 66 17			CLUSTER/NEBULOSITY IN LMC
SL 458	05 25 28.	- 66 17	410		STAR CLUSTER IN LMC
LH120-N140	05 25 30.	- 69 14	516		EMISSION NEBULA IN LMC
LH120-N132D	05 25 30.	- 69 40	32		EMISSION NEBULA IN LMC
MIL 97	05 25 32.	- 69 41 00.	22		SUPERNOVA REMNANT
MCG-01-14-011	05 25 33.	- 05 21	78	13.	GALAXY
LH 051	05 25 33.	- 67 32	180		STELLAR ASSN. IN LMC
SL 457	05 25 35.	- 69 27	70		STAR CLUSTER IN LMC
UGC 03312	05 25 36.	+ 22 04	90	16.5	GALAXY S
PK172+00.1	05 25 36.	+ 36 00	40	18.9	PLANETARY NEBULA
UGC 03313	05 25 36.	+ 79 11	72	18.	GALAXY DWRF SP
LH120-N048B	05 25 36.	- 66 20	52		EMISSION NEBULA IN LMC
LH120-N201	05 25 36.	- 71 35			EMISSION NEBULA IN LMC
RNGC 1953	05 25 38.	- 68 50			GLOBULAR CLUSTER IN LMC
SL 459	05 25 38.	- 68 50	40		STAR CLUSTER IN LMC
WS 25	05 25 41.	- 71 35 15.		15.71	PLANETARY NEB. IN LMC
LH120-N141	05 25 42.	- 68 58			EMISSION NEBULA IN LMC
SC 0525-6207.9	05 25 43.	- 62 05 24.	12		NEBULA
SL 460	05 25 43.	- 69 49	20		STAR CLUSTER IN LMC
WS 26	05 25 45.	- 68 59 51.		15.88	PLANETARY NEB. IN LMC
LH 052	05 25 46.	- 66 19	270		STELLAR ASSN. IN LMC
SL 451	05 25 46.	- 75 38	15		STAR CLUSTER IN LMC
HSE 281	05 25 47.	- 69 18			STAR CLUSTER IN LMC
ZC 0525.8-0208	05 25 48.	- 02 08	1340		CLUSTER OF GALAXIES
LH120-N048C	05 25 48.	- 66 17	25		EMISSION NEBULA IN LMC
LH120-N048A	05 25 48.	- 66 18			EMISSION NEBULA IN LMC
LH120-N202	05 25 48.	- 71 30	59		EMISSION NEBULA IN LMC
LH120-N203	05 25 48.	- 73 41			EMISSION NEBULA IN LMC
WS 24	05 25 48.	- 73 43 15.		16.48	PLANETARY NEB. IN LMC
MCG-03-14-017	05 25 51.	- 16 08	90	14.	GALAXY
HSE 282	05 25 51.	- 70 09			STAR CLUSTER IN LMC
SL 463	05 25 52.	- 66 05	25		STAR CLUSTER IN LMC
LW 206	05 25 52.	- 75 36			STAR CLUSTER IN LMC
LW 212	05 25 53.	- 64 36			STAR CLUSTER IN LMC
HSE 283	05 25 53.	- 69 24			STAR CLUSTER IN LMC
UGC 03314	05 25 54.	+ 55 53	108	18.	GALAXY
ZWG 329.007	05 25 54.	+ 70 01		15.7	GALAXY
HUB E08	05 25 54.	- 01 17			DIFFUSE NEBULA
RNGC 1951	05 25 54.	- 66 38		10.5	OPEN CLUSTER IN LMC
SL 464	05 25 54.	- 66 38	80		STAR CLUSTER IN LMC
LH120-N142	05 25 54.	- 69 28	210		EMISSION NEBULA IN LMC
LB 03584	05 25 55.	+ 35 53 42.		19.7	FAINT BLUE STAR
RNGC 1958	05 25 55.	- 69 52			OPEN CLUSTER IN LMC
SL 462	05 25 55.	- 69 52	50		STAR CLUSTER IN LMC
RNGC 1947	05 25 56.	- 63 49		12.5	GALAXY
LH 053	05 25 58.	- 66 16	1140		STELLAR ASSN. IN LMC
SL 465	05 25 59.	- 66 36	50		STAR CLUSTER IN LMC
LBN 0888	05 26	+ 03 50	1920		BRIGHT NEBULA
B 225	05 26	+ 11 34			DARK OBJECT
KHAV 119	05 26	+ 25 03	14250		DARK NEBULA
LDN 1583	05 26 00.	+ 12 00	960		DARK NEBULA
CED 047	05 26 00.	+ 12 32	480		DIFFUSE GALACTIC NEBULA
MCG+09-09-003	05 26 00.	+ 53 18	51	15.	GALAXY
ZWG 362.007	05 26 00.	+ 81 42		15.7	GALAXY
MCG+14-03-009	05 26 00.	+ 81 42	15	16.	GALAXY
LB 05150	05 26 00.	- 03 49		19.7	FAINT BLUE STAR
HUB E09	05 26 00.	- 05 34			DIFFUSE NEBULA
LH120-N049	05 26 00.	- 66 08	82		EMISSION NEBULA IN LMC
MHW 1	05 26 00.	- 66 08	60		SUPERNOVA REMNANT IN LMC
MIL 95	05 26 00.	- 66 08 00.	67		SUPERNOVA REMNANT
LH120-N050	05 26 00.	- 67 12	193		EMISSION NEBULA IN LMC
LH120-N051D	05 26 00.	- 67 32	586		EMISSION NEBULA IN LMC
SL 461	05 26 00.	- 71 52	80		STAR CLUSTER IN LMC
HUB C10	05 26 01.	- 04 30			DIFFUSE NEBULA
SCHO 0118	05 26 02.	+ 37 40 00.	680		ISOLATED DARK CLOUD
LW 207	05 26 04.	- 76 15			STAR CLUSTER IN LMC
DG 054	05 26 06.	+ 23 09	60		REFLECTION NEBULA
ZWG 307.015	05 26 06.	+ 67 41		14.6	GALAXY
RNGC 2012	05 26 06.	- 79 53			UNVERIFIED SOUTHERN OBJECT
HSE 287	05 26 08.	- 67 12			STAR CLUSTER IN LMC
HSE 288	05 26 08.	- 67 18			STAR CLUSTER IN LMC
SS 40	05 26 09.	+ 23 39			DIFFUSE GALACTIC NEBULA
LB 03585	05 26 09.	+ 35 44 00.		18.7	FAINT BLUE STAR
RNGC 1955	05 26 09.	- 67 31		9.0	CLUSTER/NEBULOSITY IN LMC
SL 467	05 26 09.	- 67 31	145		STAR CLUSTER IN LMC
LW 213	05 26 10.	- 66 05			STAR CLUSTER IN LMC
SL 455	05 26 10.	- 76 15	35		STAR CLUSTER IN LMC
SL 470	05 26 11.	- 66 25	40		STAR CLUSTER IN LMC
LH120-N134	05 26 12.	- 69 55	27		EMISSION NEBULA IN LMC
HSE 294	05 26 12.	- 71 55			STAR CLUSTER IN LMC
HSE 285	05 26 13.	- 69 46			STAR CLUSTER IN LMC
RNGC 1959	05 26 14.	- 69 58			OPEN CLUSTER IN LMC
SL 466	05 26 14.	- 69 58	30		STAR CLUSTER IN LMC
LW 210	05 26 14.	- 73 04			STAR CLUSTER IN LMC
VDB.66N 041	05 26 15.	+ 23 38	72		REFLECTION NEBULA
LH 054	05 26 15.	- 67 32	210		STELLAR ASSN. IN LMC
SL 471	05 26 15.	- 67 37	90		STAR CLUSTER IN LMC
HSE 286	05 26 15.	- 70 18			STAR CLUSTER IN LMC
SG 3.047	05 26 16.	+ 10 42	6120		DIFFUSE EMISSION NEBULA
SC 0525-6114.7	05 26 17.	- 61 12 15.			NEBULA
UGC 03315	05 26 18.	+ 53 21	72	17.	GALAXY
LB 05106	05 26 18.	- 03 29		18.0	FAINT BLUE STAR
RNGC 1927	05 26 18.	- 08 25			NON-EXISTENT OBJECT
SL 468	05 26 18.	- 69 31	20		STAR CLUSTER IN LMC
SL 469	05 26 19.	- 69 45	15		STAR CLUSTER IN LMC
HSE 289	05 26 20.	- 68 39			STAR CLUSTER IN LMC
SCHO 0119	05 26 22.	+ 37 09 18.	530		ISOLATED DARK CLOUD
7ZW 040	05 26 24.	+ 67 34			COMPACT GALAXY
LB 03452	05 26 24.	- 64 04		14.2	FAINT BLUE STAR
LH120-N051B	05 26 24.	- 67 40	31		EMISSION NEBULA IN LMC
HSE 290	05 26 24.	- 69 42			STAR CLUSTER IN LMC
LH 055	05 26 28.	- 67 40	300		STELLAR ASSN. IN LMC
SC 0525-6131.5	05 26 29.	- 61 29 03.	6		NEBULA
ZWG 307.016	05 26 30.	+ 67 34		14.9	GALAXY
MCG+12-06-006	05 26 30.	+ 73 42	60	16.	GALAXY
LB 03453	05 26 30.	- 55 36		13.2	FAINT BLUE STAR
SC 0525-5921.3	05 26 30.	- 59 18 51.			NEBULA
SL 473	05 26 31.	- 68 24	35		STAR CLUSTER IN LMC
HSE 291	05 26 32.	- 69 59			STAR CLUSTER IN LMC
LW 214	05 26 33.	- 65 50			STAR CLUSTER IN LMC
SL 472	05 26 34.	- 70 27	35		STAR CLUSTER IN LMC
LW 211	05 26 35.	- 73 32			STAR CLUSTER IN LMC
LDN 1581	05 26 36.	+ 12 10	120		DARK NEBULA
HSE 297	05 26 37.	- 65 22			STAR CLUSTER IN LMC
HSE 294	05 26 38.	- 68 45			STAR CLUSTER IN LMC
HSE 292	05 26 38.	- 70 05			STAR CLUSTER IN LMC
HSE 293	05 26 40.	- 70 23			STAR CLUSTER IN LMC
SL 475	05 26 41.	- 69 19	90		STAR CLUSTER IN LMC
SCHO 0120	05 26 42.	+ 37 04 24.	490		ISOLATED DARK CLOUD
LB 03454	05 26 42.	- 61 26		13.6	FAINT BLUE STAR
LH120-N051E	05 26 42.	- 67 41	422		EMISSION NEBULA IN LMC
RNGC 1970	05 26 44.	- 68 49		8.5	CLUSTER/NEBULOSITY IN LMC
RNGC 1966	05 26 44.	- 68 49		8.5	CLUSTER/NEBULOSITY IN LMC
RNGC 1965	05 26 44.	- 68 49		8.5	CLUSTER/NEBULOSITY IN LMC
RNGC 1962	05 26 44.	- 68 49		8.5	CLUSTER/NEBULOSITY IN LMC
SL 476	05 26 44.	- 68 49	205		STAR CLUSTER IN LMC
HSE 295	05 26 45.	- 70 08			STAR CLUSTER IN LMC
UGC 03316	05 26 48.	+ 55 49	90	18.	GALAXY DWARF?
LB 05107	05 26 48.	- 03 56		17.0	FAINT BLUE STAR
LH120-N144B	05 26 48.	- 68 51			EMISSION NEBULA IN LMC
LH120-N143	05 26 48.	- 69 21	164		EMISSION NEBULA IN LMC
LB 03455	05 26 48.	- 71 36		12.7	FAINT BLUE STAR
SL 474	05 26 48.	- 71 52	20		STAR CLUSTER IN LMC
HSE 296	05 26 49.	- 69 43			STAR CLUSTER IN LMC
HSE 299	05 26 52.	- 67 42			STAR CLUSTER IN LMC
RNGC 1967	05 26 52.	- 69 06			OPEN CLUSTER IN LMC
SL 478	05 26 52.	- 69 06	25		STAR CLUSTER IN LMC
LH 056	05 26 52.	- 71 36	300		STELLAR ASSN. IN LMC
LH 057	05 26 53.	- 69 20	90		STELLAR ASSN. IN LMC
ZWG 421.046	05 26 54.	+ 05 24		15.7	GALAXY
LB 05151	05 26 54.	- 02 58		20.6	FAINT BLUE STAR
MCG-03-14-018	05 26 54.	- 19 58	72	15.	GALAXY
LH120-N051	05 26 54.	- 67 35	1179		EMISSION NEBULA IN LMC
LH120-N144	05 26 54.	- 68 52	528		EMISSION NEBULA IN LMC
HSE 298	05 26 54.	- 69 33			STAR CLUSTER IN LMC
LH120-N205B	05 26 54.	- 71 38	230		EMISSION NEBULA IN LMC
RNGC 1969	05 26 55.	- 69 51			OPEN CLUSTER IN LMC

OBJECT NAME	RIGHT ASCEN.	DECLINATION	DIAM.	MAGN.	TYPE OF OBJECT
SL 479	05 26 55.	- 69 51	50		STAR CLUSTER IN LMC
SCHO 0121	05 26 56.	+ 03 36 42.	370		ISOLATED DARK CLOUD
LH 058	05 26 57.	- 68 51	240		STELLAR ASSN. IN LMC
KHAV 121	05 27	+ 12 50	8020		DARK NEBULA
KHAV 120	05 27	+ 15 20	9980		DARK NEBULA
VDB.66G 038	05 27	+ 73 40	70		DWARF GALAXY
LDN 1577	05 27 00.	+ 12 30	2580		DARK NEBULA
CM 04	05 27 00.	- 67 20			HII REGION IN LMC
LH120-N144A	05 27 00.	- 68 51	46		EMISSION NEBULA IN LMC
HSE 300	05 27 00.	- 69 30			STAR CLUSTER IN LMC
CM 34	05 27 00.	- 70 44			HII REGION IN LMC
SL 477	05 27 00.	- 71 47	35		STAR CLUSTER IN LMC
HOD.72 27	05 27 03.	- 67 33	540		DARK NEBULA IN LMC
SL 482	05 27 05.	- 66 25	25		STAR CLUSTER IN LMC
ZC 0527.1+7202	05 27 06.	+ 72 02	3290		CLUSTER OF GALAXIES
RNGC 1972	05 27 07.	- 69 51			OPEN CLUSTER IN LMC
SL 480	05 27 07.	- 69 51	15		STAR CLUSTER IN LMC
RNGC 1971	05 27 07.	- 69 53			OPEN CLUSTER IN LMC
SL 481	05 27 07.	- 69 53	30		STAR CLUSTER IN LMC
SG 3.048	05 27 08.	+ 34 56	3240		DIFFUSE EMISSION NEBULA
HOD.72 26	05 27 09.	- 68 57	240		DARK NEBULA IN LMC
HSE 301	05 27 12.	- 69 40			STAR CLUSTER IN LMC
HSE 302	05 27 13.	- 66 56			STAR CLUSTER IN LMC
LH 059	05 27 13.	- 69 53	180		STELLAR ASSN. IN LMC
RNGC 1968	05 27 15.	- 67 28		9.0	CLUSTER/NEBULOSITY IN LMC
SL 483	05 27 15.	- 67 28	270		STAR CLUSTER IN LMC
LH 060	05 27 15.	- 67 30	360		STELLAR ASSN. IN LMC
LB 03456	05 27 18.	- 63 21		14.5	FAINT BLUE STAR
SL 484	05 27 23.	- 64 41	35		STAR CLUSTER IN LMC
LB 03586	05 27 24.	+ 35 28 48.		19.4	FAINT BLUE STAR
7ZW 041	05 27 24.	+ 63 48			COMPACT GALAXY
ZWG 307.017	05 27 24.	+ 67 40		15.5	GALAXY
HSE 303	05 27 24.	- 66 49			STAR CLUSTER IN LMC
B 030	05 27 28.	+ 12 44	4020		DARK OBJECT
SL 485	05 27 28.	- 66 03	70		STAR CLUSTER IN LMC
UGC 03317	05 27 30.	+ 73 42	138	17.	GALAXY DWRF IR
SC 0527-6138.5	05 27 31.	- 61 36 08.	12		NEBULA
LW 215	05 27 31.	- 65 20			STAR CLUSTER IN LMC
HSE 301A	05 27 31.	- 72 01			STAR CLUSTER IN LMC
HSE 305	05 27 34.	- 64 13			STAR CLUSTER IN LMC
LW 216	05 27 35.	- 64 41			STAR CLUSTER IN LMC
MCG-03-14-019	05 27 36.	- 18 44 30.	12	15.	GALAXY
MCG-04-13-016	05 27 36.	- 24 45 30.	24	15.	GALAXY
SC 0526-5912.5	05 27 37.	- 59 10 08.			NEBULA
HOD.72 28	05 27 39.	- 68 53	240		DARK NEBULA IN LMC
SCHO 0122	05 27 41.	+ 03 18 12.	630		ISOLATED DARK CLOUD
ZWG 421.047	05 27 42.	+ 03 05		15.7	GALAXY
IC 0419	05 27 45.	+ 30 07			NONSTELLAR OBJECT
MAI 010	05 27 45.	+ 78 44	114		DWARF SPHEROIDAL GALAXY
HUB E10	05 27 47.	- 05 20			DIFFUSE NEBULA
ZWG 421.048	05 27 48.	+ 05 08		15.7	GALAXY
CED 048	05 27 48.	+ 30 06			DIFFUSE GALACTIC NEBULA
UGC 03318	05 27 48.	+ 76 46	102	17.	GALAXY DWRF SP
MCG-01-14-012	05 27 48.	- 08 49	48	16.	GALAXY
LH120-N051C	05 27 48.	- 67 30	92		EMISSION NEBULA IN LMC
RNGC 1987	05 27 48.	- 70 48		12.0	GLOBULAR CLUSTER IN LMC
SL 486	05 27 48.	- 70 48	55		STAR CLUSTER IN LMC
SL 487	05 27 50.	- 71 05	15		STAR CLUSTER IN LMC
MCG-03-14-020	05 27 51.	- 20 23	42	15.	GALAXY
RNGC 1983	05 27 51.	- 68 59		8.5	CLUSTER/NEBULOSITY IN LMC
LH 061	05 27 51.	- 69 01	300		STELLAR ASSN. IN LMC
RNGC 1984	05 27 52.	- 69 08			CLUSTER/NEBULOSITY IN LMC
SL 488	05 27 52.	- 69 08	100		STAR CLUSTER IN LMC
LH120-N145	05 27 54.	- 69 11			EMISSION NEBULA IN LMC
SL 491	05 27 55.	- 68 25	30		STAR CLUSTER IN LMC
RNGC 1986	05 27 56.	- 70 00		11.0	OPEN CLUSTER IN LMC
SL 489	05 27 56.	- 70 00	80		STAR CLUSTER IN LMC
HSE 306	05 27 56.	- 70 03			STAR CLUSTER IN LMC
HSE 304	05 27 56.	- 71 13			STAR CLUSTER IN LMC
RNGC 1974	05 27 57.	- 67 26		9.0	CLUSTER/NEBULOSITY IN LMC
SL 494	05 27 57.	- 67 26	135		STAR CLUSTER IN LMC
SL 492	05 27 57.	- 68 59	35		STAR CLUSTER IN LMC
LBN 0892	05 28	+ 03 30	1140		BRIGHT NEBULA
LBN 0886	05 28	+ 04 38	120		BRIGHT NEBULA
LBN 0885	05 28	+ 04 50	2400		BRIGHT NEBULA
KHAV 124	05 28	+ 09 44	3400		DARK NEBULA
LBN 0810	05 28	+ 34 10	240		BRIGHT NEBULA
KHAV 125	05 28	- 05 22	4080		DARK NEBULA
LBN 1005	05 28	- 12 00	2700		BRIGHT NEBULA
ZC 0528.0+0023	05 28 00.	+ 00 23	1140		CLUSTER OF GALAXIES
LDN 1573	05 28 00.	+ 13 30	4140		DARK NEBULA
LB 03457	05 28 00.	- 61 49		14.3	FAINT BLUE STAR
LH120-N051A	05 28 00.	- 67 28	25		EMISSION NEBULA IN LMC
CM 19	05 28 00.	- 68 33			HII REGION IN LMC
LH120-N204	05 28 00.	- 70 36	242		EMISSION NEBULA IN LMC
LH 063	05 28 03.	- 67 28	240		STELLAR ASSN. IN LMC
OCL 0441	05 28 06.	+ 34 13	180	12.8	OPEN STAR CLUSTER
MRSL 183+00/1	05 28 06.	+ 34 15	420		HII REGION
KEEL 116	05 28 06.6	+ 34 08 54.			NEBULA
CED 049	05 28 07.	+ 34 12	180		DIFFUSE GALACTIC NEBULA
SG 3.049	05 28 07.	+ 34 12	300		DIFFUSE EMISSION NEBULA
SS 41	05 28 07.	+ 34 13			DIFFUSE GALACTIC NEBULA
SCHO 0123	05 28 08.	+ 37 41 36.	710		ISOLATED DARK CLOUD
SL 495	05 28 08.	- 68 49	50		STAR CLUSTER IN LMC
RNGC 1931	05 28 09.	+ 34 13		13.0	CLUSTER WITH NEBULOSITY
LH 062	05 28 11.	- 70 39	90		STELLAR ASSN. IN LMC
SL 496	05 28 13.	- 65 13	15		STAR CLUSTER IN LMC
RNGC 2000	05 28 13.	- 71 57			OPEN CLUSTER IN LMC
SL 493	05 28 13.	- 71 57	70		STAR CLUSTER IN LMC
LH120-N205A	05 28 18.	- 71 26	175		EMISSION NEBULA IN LMC
SL 490	05 28 18.	- 73 44	40		STAR CLUSTER IN LMC
LW 218	05 28 19.	- 65 13			STAR CLUSTER IN LMC
RNGC 1978	05 28 22.	- 66 16		10.5	GLOBULAR CLUSTER IN LMC
SL 497	05 28 22.	- 67 40	25		STAR CLUSTER IN LMC
ACK 178-02.1	05 28 24.	+ 28 57			PLANETARY NEBULA
SCHO 0124	05 28 24.	+ 39 41 48.	650		ISOLATED DARK CLOUD
LB 05152	05 28 24.	- 03 05		20.0	FAINT BLUE STAR
TON-S 0419	05 28 24.	- 30 32		15.4	BLUE STAR
SL 500	05 28 24.	- 64 53	20		STAR CLUSTER IN LMC
LW 219	05 28 24.	- 64 53			STAR CLUSTER IN LMC
SC 0527-6013.3	05 28 25.	- 60 10 59.	12		NEBULA
HSE 309	05 28 25.	- 66 58			STAR CLUSTER IN LMC
SC 0527-6022.0	05 28 26.	- 60 19 42.	6		NEBULA
SL 498	05 28 26.	- 67 15	50		STAR CLUSTER IN LMC
HSE 307	05 28 26.	- 69 57			STAR CLUSTER IN LMC
SC 0527-6011.8	05 28 28.	- 60 09 30.	12		NEBULA
SL 501	05 28 28.	- 66 16	205		STAR CLUSTER IN LMC
SG 3.051	05 28 30.	+ 12 20	2400		DIFFUSE EMISSION NEBULA

OBJECT NAME	RIGHT ASCEN.	DECLINATION	DIAM.	MAGN.	TYPE OF OBJECT
SG 3.050	05 28 30.	+ 12 20	7200		DIFFUSE EMISSION NEBULA
ARC 0543	05 28 30.	- 22 27		16.9	RICH CLUSTER OF GALAXIES
HSE 308	05 28 30.	- 69 36			STAR CLUSTER IN LMC
HSE 312	05 28 32.	- 67 15			STAR CLUSTER IN LMC
RNGC 1994	05 28 34.	- 69 09			OPEN CLUSTER IN LMC
SL 499	05 28 34.	- 69 09	80		STAR CLUSTER IN LMC
ZWG 258.009	05 28 36.	+ 51 34		15.7	GALAXY
HSE 313	05 28 36.	- 68 14			STAR CLUSTER IN LMC
LW 217	05 28 36.	- 73 43			STAR CLUSTER IN LMC
HSE 310	05 28 38.	- 70 02			STAR CLUSTER IN LMC
HSE 314	05 28 39.	- 69 01			STAR CLUSTER IN LMC
HSE 317	05 28 40.	- 64 20			STAR CLUSTER IN LMC
CED 051	05 28 42.	+ 12 07			DIFFUSE GALACTIC NEBULA
SG 3.052	05 28 42.	+ 12 08	60		DIFFUSE EMISSION NEBULA
UGC 03319	05 28 42.	+ 70 09	108	16.5	GALAXY Sc
DG 055	05 28 42.	- 04 43	360		REFLECTION NEBULA
HOD.72 29	05 28 44.	- 69 56	180		DARK NEBULA IN LMC
LDN 1584	05 28 48.	+ 12 15	480		DARK NEBULA
CED 050	05 28 48.	+ 30 24			DIFFUSE GALACTIC NEBULA
LH120-N052	05 28 48.	- 67 36			EMISSION NEBULA IN LMC
HSE 311	05 28 48.	- 71 53			STAR CLUSTER IN LMC
VDB.66N 042	05 28 49.	- 05 42	336		REFLECTION NEBULA
HSE 318	05 28 49.	- 67 06			STAR CLUSTER IN LMC
WS 27	05 28 51.	- 67 35 38.		17.0	PLANETARY NEB. IN LMC
ZWG 421.049	05 28 54.	+ 06 58		15.7	GALAXY
ASS 52	05 28 54.	- 02 43	75600		OB ASSOCIATION ORI OB1
MCG-04-14-001	05 28 54.	- 23 10	48	15.	GALAXY
MCG-06-13-001	05 28 54.	- 33 23	36	15.	GALAXY
HSE 316	05 28 54.	- 69 31			STAR CLUSTER IN LMC
MIN.46 01	05 28 55.	+ 34 12			DIFFUSE NEBULA
ARC 0544	05 28 57.	- 25 59		17.5	RICH CLUSTER OF GALAXIES
HSE 315	05 28 59.	- 71 35			STAR CLUSTER IN LMC
ST 41	05 29	+ 12 25	3600		EMISSION NEBULA
KHAV 123	05 29	+ 28 56	4860		DARK NEBULA
KHAV 122	05 29	+ 31 26	9800		DARK NEBULA
LDN 1582	05 29 00.	+ 12 30	420		DARK NEBULA
ISS 0154	05 29 00.	+ 46 25	382		STELLAR RING
MCG+14-03-010	05 29 00.	+ 81 51 30.	36	16.	GALAXY
SL 502	05 29 00.	- 66 37	40		STAR CLUSTER IN LMC
RNGC 1991	05 29 03.	- 67 28			NON-EXISTENT OBJECT
PK197-14.1	05 29 06.	+ 06 53	36	15.2	PLANETARY NEBULA
SCHO 0127	05 29 06.	+ 40 06 36.	590		ISOLATED DARK CLOUD
MCG-06-13-002	05 29 06.	- 33 24	30	15.	GALAXY
LH120-N146	05 29 06.	- 69 03	53		EMISSION NEBULA IN LMC
SL 503	05 29 07.	- 68 27	45		STAR CLUSTER IN LMC
SCHO 0125	05 29 09.	+ 38 08 12.	580		ISOLATED DARK CLOUD
HSE 319	05 29 09.	- 69 01			STAR CLUSTER IN LMC
SCHO 0126	05 29 10.	+ 39 55 12.	630		ISOLATED DARK CLOUD
IC 2129	05 29 11.	- 23 05 55.			NONSTELLAR OBJECT
ZC 0529.2+0516	05 29 12.	+ 05 16	11420		CLUSTER OF GALAXIES
HSE 320	05 29 12.	- 69 40			STAR CLUSTER IN LMC
B 031	05 29 13.	+ 12 44	1800		DARK OBJECT
ARC 0541	05 29 13.	+ 64 23		17.5	RICH CLUSTER OF GALAXIES
SL 504	05 29 13.	- 69 50	15		STAR CLUSTER IN LMC
HOD.72 30	05 29 14.	- 70 05	180		DARK NEBULA IN LMC
HODG 14	05 29 14.	- 73 39			RED GLOBULAR CLSTR IN LMC
HSE 322	05 29 17.	- 66 33			STAR CLUSTER IN LMC
MCG-02-15-001	05 29 18.	- 10 24	72	14.5	GALAXY
MCG-05-14-001	05 29 18.	- 29 22	30	15.5	GALAXY
B 032	05 29 20.	+ 12 24			DARK OBJECT
SL 509	05 29 20.	- 63 41	60		STAR CLUSTER IN LMC
RNGC 2001	05 29 20.	- 68 46			CLUSTER/NEBULOSITY IN LMC
SL 507	05 29 20.	- 68 46	410		STAR CLUSTER IN LMC
LH 064	05 29 21.	- 68 49	480		STELLAR ASSN. IN LMC
ZWG 258.010	05 29 24.	+ 51 44		15.7	GALAXY
LH120-N053	05 29 24.	- 67 35			EMISSION NEBULA IN LMC
LH120-N147	05 29 24.	- 69 25			EMISSION NEBULA IN LMC
HSE 321	05 29 26.	- 69 53			STAR CLUSTER IN LMC
VDB.66N 043	05 29 27.	+ 06 01	132		REFLECTION NEBULA
HODG 03	05 29 27.	- 68 20	36		RED GLOBULAR CLSTR IN LMC
SL 511	05 29 28.	- 64 28	35		STAR CLUSTER IN LMC
SL 505	05 29 29.	- 71 41	35		STAR CLUSTER IN LMC
LB 03458	05 29 30.	- 64 20		13.4	FAINT BLUE STAR
LH120-N054	05 29 30.	- 67 16			EMISSION NEBULA IN LMC
LW 221	05 29 32.	- 63 41			STAR CLUSTER IN LMC
HSE 323	05 29 33.	- 70 15			STAR CLUSTER IN LMC
VDB.66N 044	05 29 35.	- 04 33	516		REFLECTION NEBULA
IC 2130	05 29 35.	- 23 12 23.			NONSTELLAR OBJECT
LW 222	05 29 35.	- 64 29			STAR CLUSTER IN LMC
SL 508	05 29 36.	- 69 36	35		STAR CLUSTER IN LMC
LH120-N206C	05 29 36.	- 71 11	24		EMISSION NEBULA IN LMC
HSE 324	05 29 38.	- 71 09			STAR CLUSTER IN LMC
DG 056	05 29 42.	- 04 30	480		REFLECTION NEBULA
MCG-04-14-002	05 29 42.	- 23 12	90	14.	GALAXY
HSE 325	05 29 42.	- 66 48			STAR CLUSTER IN LMC
SL 506	05 29 42.	- 73 41	35		STAR CLUSTER IN LMC
SL 515	05 29 43.	- 63 27	70		STAR CLUSTER IN LMC
HSE 326	05 29 44.	- 67 09			STAR CLUSTER IN LMC
HOD.72 31	05 29 44.	- 69 55	240		DARK NEBULA IN LMC
HN 0063	05 29 46.	- 04 32			NEBULA
SL 512	05 29 46.	- 66 08	15		STAR CLUSTER IN LMC
IC 0420	05 29 47.	- 04 32			NONSTELLAR OBJECT
SL 510	05 29 47.	- 70 37	15		STAR CLUSTER IN LMC
MCG+13-05-001	05 29 48.	+ 79 32 30.	54	16.	GALAXY
MCG-01-15-001	05 29 48.	- 07 58	60	15.	GALAXY
SL 513	05 29 48.	- 66 47	60		STAR CLUSTER IN LMC
IC 0421	05 29 49.	- 08 07			NONSTELLAR OBJECT
LW 223	05 29 49.	- 63 27			STAR CLUSTER IN LMC
HN 0066	05 29 54.	- 08 07			NEBULA
LW 220	05 29 54.	- 73 40			STAR CLUSTER IN LMC
LBN 0918	05 30	+ 00 00	64800		BRIGHT NEBULA
LBN 0865	05 30	+ 10 00	15300		BRIGHT NEBULA
LBN 0690	05 30	+ 66 30	3000		BRIGHT NEBULA
LBN 0963	05 30	- 04 30	360		BRIGHT NEBULA
LBN 1001	05 30	- 11 00	3000		BRIGHT NEBULA
ZWG 348.001	05 30 00.	+ 79 33		15.3	GALAXY
ZWG 347.017	05 30 00.	+ 79 33		15.3	GALAXY
UGC 03320	05 30 00.	+ 79 33	78	15.3	GALAXY SBb
SC 0529-6226.3	05 30 00.	- 62 24 07.	30		NEBULA
SL 516	05 30 01.	- 67 03	20		STAR CLUSTER IN LMC
SL 514	05 30 01.	- 69 47	35		STAR CLUSTER IN LMC
ARC 0545	05 30 02.	- 11 35		17.0	RICH CLUSTER OF GALAXIES
SC 0529-6227.3	05 30 02.	- 62 25 07.	30		NEBULA
HSE 327	05 30 02.	- 69 58			STAR CLUSTER IN LMC
IC 2131	05 30 03.	- 17 15 25.			NONSTELLAR OBJECT
IC 0422	05 30 05.	- 17 15 32.			NONSTELLAR OBJECT
HSE 328	05 30 06.	- 69 36			STAR CLUSTER IN LMC
RNGC 1997	05 30 07.	- 63 14			OPEN CLUSTER IN LMC

OBJECT NAME	RIGHT ASCEN.	DECLINATION	DIAM.	MAGN.	TYPE OF OBJECT
SL 520	05 30 07.	- 63 14	20		STAR CLUSTER IN LMC
LH 065	05 30 07.	- 66 59	120		STELLAR ASSN. IN LMC
SC 0529-6140.5	05 30 08.	- 61 38 19.	12		NEBULA
LW 225	05 30 09.	- 64 03			STAR CLUSTER IN LMC
IC 2132	05 30 11.	- 13 57 38.			NONSTELLAR OBJECT
ZWG 307.018	05 30 12.	+ 67 43		15.7	GALAXY
MCG-02-15-002	05 30 12.	- 13 57 30.	78	14.5	GALAXY
MCG-03-15-001	05 30 12.	- 17 15	24	14.5	GALAXY
LW 226	05 30 13.	- 63 14			STAR CLUSTER IN LMC
RNGC 2002	05 30 13.	- 66 54			OPEN CLUSTER IN LMC
SL 517	05 30 13.	- 66 54	80		STAR CLUSTER IN LMC
SL 521	05 30 15.	- 65 56	50		STAR CLUSTER IN LMC
ZC 0530.3-0114	05 30 18.	- 01 14	2690		CLUSTER OF GALAXIES
OCL 0497	05 30 24.	+ 00 11	180	15.	OPEN STAR CLUSTER
RNGC 2005	05 30 25.	- 69 46			OPEN CLUSTER IN LMC
SL 518	05 30 25.	- 69 46	40		STAR CLUSTER IN LMC
RNGC 1963	05 30 26.	- 36 25			NON-EXISTENT OBJECT
SL 519	05 30 26.	- 69 59	60		STAR CLUSTER IN LMC
LW 227	05 30 27.	- 65 57			STAR CLUSTER IN LMC
DG 057	05 30 30.	- 05 33	240		REFLECTION NEBULA
MCG-02-15-003	05 30 30.	- 14 05	228	13.	GALAXY
SCHO 0128	05 30 32.	+ 36 21 36.	490		ISOLATED DARK CLOUD
SL 522	05 30 32.	- 67 12	30		STAR CLUSTER IN LMC
SL 523	05 30 32.	- 67 18	120		STAR CLUSTER IN LMC
HSE 331	05 30 32.	- 69 55			STAR CLUSTER IN LMC
HSE 329	05 30 32.	- 71 03			STAR CLUSTER IN LMC
SCHO 0129	05 30 33.	+ 09 55 00.	470		ISOLATED DARK CLOUD
RNGC 1954	05 30 33.	- 14 06		13.0	GALAXY
SL 525	05 30 33.	- 63 58	20		STAR CLUSTER IN LMC
SL 524	05 30 35.	- 66 21	25		STAR CLUSTER IN LMC
OCL 0459	05 30 36.	+ 26 27	900		OPEN STAR CLUSTER
LB 05153	05 30 36.	- 06 30		18.0	FAINT BLUE STAR
LB 05154	05 30 36.	- 06 34		20.5	FAINT BLUE STAR
LB 05155	05 30 36.	- 06 42		17.1	FAINT BLUE STAR
LB 05108	05 30 36.	- 06 44		18.1	FAINT BLUE STAR
LB 05156	05 30 36.	- 06 52		17.9	FAINT BLUE STAR
LB 05157	05 30 36.	- 07 02		17.0	FAINT BLUE STAR
LB 05158	05 30 36.	- 07 07		18.9	FAINT BLUE STAR
LB 05159	05 30 36.	- 07 11		20.4	FAINT BLUE STAR
LB 05160	05 30 36.	- 07 32		17.0	FAINT BLUE STAR
LB 05161	05 30 36.	- 07 37		20.0	FAINT BLUE STAR
LB 05162	05 30 36.	- 07 49		15.3	FAINT BLUE STAR
MCG-01-15-002	05 30 36.	- 07 49	24	17.	GALAXY
LB 05163	05 30 36.	- 07 58		16.5	FAINT BLUE STAR
SL 529	05 30 38.	- 63 35	20		STAR CLUSTER IN LMC
HSE 330	05 30 38.	- 72 06			STAR CLUSTER IN LMC
RNGC 1957	05 30 40.	- 14 09			UNVERIFIED SOUTHERN OBJECT
HOD.72 34	05 30 40.	- 69 04	240		DARK NEBULA IN LMC
SL 527	05 30 41.	- 66 26	60		STAR CLUSTER IN LMC
RNGC 2003	05 30 41.	- 66 29			OPEN CLUSTER IN LMC
SL 526	05 30 41.	- 66 29	70		STAR CLUSTER IN LMC
LB 05164	05 30 42.	- 06 21		18.7	FAINT BLUE STAR
LB 05165	05 30 42.	- 06 42		18.3	FAINT BLUE STAR
LB 05166	05 30 42.	- 06 50		18.6	FAINT BLUE STAR
LB 05167	05 30 42.	- 07 21		20.4	FAINT BLUE STAR
LB 05109	05 30 42.	- 07 24		17.0	FAINT BLUE STAR
LB 05168	05 30 42.	- 07 31		20.2	FAINT BLUE STAR
LB 05169	05 30 42.	- 08 17		17.0	FAINT BLUE STAR
MCG-05-14-002	05 30 42.	- 29 42	42	15.	GALAXY
HOD.72 33	05 30 43.	- 69 44	120		DARK NEBULA IN LMC
HSE 332	05 30 43.	- 69 49			STAR CLUSTER IN LMC
LW 229	05 30 44.	- 63 34			STAR CLUSTER IN LMC
LW 228	05 30 44.	- 65 37			STAR CLUSTER IN LMC
RNGC 2004	05 30 44.	- 67 19		10.0	OPEN CLUSTER IN LMC
LW 224	05 30 44.	- 72 05			STAR CLUSTER IN LMC
HOD.72 32	05 30 45.	- 70 09	240		DARK NEBULA IN LMC
LDN 1509	05 30 48.	+ 37 15	1320		DARK NEBULA
LB 05171	05 30 48.	- 06 33		19.0	FAINT BLUE STAR
LB 05170	05 30 48.	- 06 33		17.5	FAINT BLUE STAR
LB 05172	05 30 48.	- 06 36		18.3	FAINT BLUE STAR
LB 05173	05 30 48.	- 06 42		18.7	FAINT BLUE STAR
LB 05174	05 30 48.	- 07 03		17.7	FAINT BLUE STAR
LB 05175	05 30 48.	- 07 07		18.5	FAINT BLUE STAR
LB 05176	05 30 48.	- 07 29		18.0	FAINT BLUE STAR
LB 05177	05 30 48.	- 07 32		19.9	FAINT BLUE STAR
LB 05110	05 30 48.	- 07 48		19.7	FAINT BLUE STAR
IC 0423	05 30 50.	- 00 39			NONSTELLAR OBJECT
LH 066	05 30 50.	- 71 07	240		STELLAR ASSN. IN LMC
HN 0058	05 30 51.	- 00 39			NEBULA
CED 052	05 30 51.	- 00 39	360		DIFFUSE GALACTIC NEBULA
UGC 03321	05 30 54.	+ 06 14	90	16.0	GALAXY Sa-b
DG 058	05 30 54.	- 00 38	360		REFLECTION NEBULA
LB 05111	05 30 54.	- 06 53		18.6	FAINT BLUE STAR
LB 05178	05 30 54.	- 07 06		17.8	FAINT BLUE STAR
LB 05179	05 30 54.	- 07 11		19.8	FAINT BLUE STAR
LB 05180	05 30 54.	- 07 13		18.0	FAINT BLUE STAR
LB 05182	05 30 54.	- 07 16		20.6	FAINT BLUE STAR
LB 05181	05 30 54.	- 07 16		20.4	FAINT BLUE STAR
LB 05183	05 30 54.	- 07 23		20.7	FAINT BLUE STAR
LB 05184	05 30 54.	- 07 29		17.6	FAINT BLUE STAR
LB 05185	05 30 54.	- 07 41		18.9	FAINT BLUE STAR
LB 05186	05 30 54.	- 07 53		15.9	FAINT BLUE STAR
LB 05187	05 30 54.	- 08 00		19.1	FAINT BLUE STAR
LB 05188	05 30 54.	- 08 16		18.6	FAINT BLUE STAR
SL 528	05 30 57.	- 70 16	25		STAR CLUSTER IN LMC
LBN 0913	05 31	- 00 42	120		BRIGHT NEBULA
LBN 1006	05 31	- 12 20	1800		BRIGHT NEBULA
MCG+01-15-001	05 31 00.	+ 06 17	36	16.5	GALAXY
LDN 1580	05 31 00.	+ 12 50	1860		DARK NEBULA
ZWG 362.008	05 31 00.	+ 82 26		15.4	GALAXY
LDN 1640	05 31 00.	- 05 50	540		DARK NEBULA
LB 05189	05 31 00.	- 06 29		18.4	FAINT BLUE STAR
LB 05190	05 31 00.	- 07 14		17.8	FAINT BLUE STAR
LB 05191	05 31 00.	- 08 00		18.4	FAINT BLUE STAR
LB 05192	05 31 00.	- 08 19		18.4	FAINT BLUE STAR
LB 05193	05 31 00.	- 08 21		19.0	FAINT BLUE STAR
MCG-06-13-003	05 31 00.	- 32 57	36	16.	GALAXY
LW 230	05 31 01.	- 63 18			STAR CLUSTER IN LMC
SL 540	05 31 03.	- 63 56	15		STAR CLUSTER IN LMC
IC 2135	05 31 04.	- 36 26 19.			NONSTELLAR OBJECT
SL 530	05 31 04.	- 69 04	20		STAR CLUSTER IN LMC
HSE 333	05 31 05.	- 69 16			STAR CLUSTER IN LMC
IC 0424	05 31 06.	- 00 21			NONSTELLAR OBJECT
LB 05194	05 31 06.	- 07 18		17.7	FAINT BLUE STAR
LB 05195	05 31 06.	- 07 43		18.3	FAINT BLUE STAR
HN 0056	05 31 09.	- 00 21			NEBULA
RNGC 1964	05 31 09.	- 21 59		11.5	GALAXY
RNGC 1995	05 31 09.	- 48 44			GALAXY

OBJECT NAME	RIGHT ASCEN.	DECLINATION	DIAM.	MAGN.	TYPE OF OBJECT
WS 28	05 31 09.	- 70 46 51.		16.60	PLANETARY NEB. IN LMC
RNGC 2009	05 31 10.	- 69 11			OPEN CLUSTER IN LMC
HOD.72 35	05 31 10.	- 69 12	540		DARK NEBULA IN LMC
DG 059	05 31 12.	- 00 23	120		REFLECTION NEBULA
LB 05196	05 31 12.	- 06 23		18.6	FAINT BLUE STAR
LB 05197	05 31 12.	- 06 26		18.3	FAINT BLUE STAR
LB 05198	05 31 12.	- 06 52		18.6	FAINT BLUE STAR
LB 05199	05 31 12.	- 07 00		18.4	FAINT BLUE STAR
LB 05200	05 31 12.	- 07 06		17.7	FAINT BLUE STAR
LB 05201	05 31 12.	- 07 08		17.7	FAINT BLUE STAR
LB 05202	05 31 12.	- 07 15		20.0	FAINT BLUE STAR
LB 05204	05 31 12.	- 07 16		17.8	FAINT BLUE STAR
LB 05203	05 31 12.	- 07 16		17.4	FAINT BLUE STAR
LB 05205	05 31 12.	- 07 19		17.5	FAINT BLUE STAR
LB 05206	05 31 12.	- 07 32		17.4	FAINT BLUE STAR
LB 05112	05 31 12.	- 07 36		20.3	FAINT BLUE STAR
LB 05207	05 31 12.	- 07 55		18.6	FAINT BLUE STAR
LH120-N148F	05 31 12.	- 68 36	26		EMISSION NEBULA IN LMC
LH120-N207	05 31 12.	- 70 46			EMISSION NEBULA IN LMC
RNGC 2006	05 31 13.	- 66 59			OPEN CLUSTER IN LMC
SL 538	05 31 13.	- 66 59	45		STAR CLUSTER IN LMC
SL 537	05 31 13.	- 66 59	60		STAR CLUSTER IN LMC
RNGC 2010	05 31 13.	- 70 51			OPEN CLUSTER IN LMC
SL 531	05 31 13.	- 70 51	90		STAR CLUSTER IN LMC
HUB C11	05 31 15.	- 01 39			DIFFUSE NEBULA
MCG-04-14-003	05 31 15.	- 21 59	360	12.	GALAXY
LW 232	05 31 15.	- 63 56			STAR CLUSTER IN LMC
SL 534	05 31 16.	- 69 11	55		STAR CLUSTER IN LMC
ZC 0531.3+0127	05 31 18.	+ 01 27	9680		CLUSTER OF GALAXIES
LB 05208	05 31 18.	- 08 17		16.2	FAINT BLUE STAR
LH120-N206	05 31 18.	- 71 07	1038		EMISSION NEBULA IN LMC
HUB F11	05 31 19.	- 02 43			DIFFUSE NEBULA
SL 532	05 31 19.	- 70 57	35		STAR CLUSTER IN LMC
SL 535	05 31 20.	- 69 59	25		STAR CLUSTER IN LMC
RNGC 2018	05 31 20.	- 71 07			CLUSTER/NEBULOSITY IN LMC
SL 533	05 31 20.	- 71 07	40		STAR CLUSTER IN LMC
LB 05113	05 31 24.	- 06 25		19.6	FAINT BLUE STAR
LB 05114	05 31 24.	- 06 26		19.2	FAINT BLUE STAR
LB 05209	05 31 24.	- 06 31		18.2	FAINT BLUE STAR
LB 05210	05 31 24.	- 07 00		17.4	FAINT BLUE STAR
LB 05211	05 31 24.	- 07 08		18.6	FAINT BLUE STAR
LB 05212	05 31 24.	- 07 13		15.6	FAINT BLUE STAR
LB 05213	05 31 24.	- 07 17		16.7	FAINT BLUE STAR
LB 05214	05 31 24.	- 07 20		15.7	FAINT BLUE STAR
LB 05216	05 31 24.	- 07 21		17.8	FAINT BLUE STAR
LB 05215	05 31 24.	- 07 21		17.0	FAINT BLUE STAR
LB 05217	05 31 24.	- 07 22		18.0	FAINT BLUE STAR
LB 05218	05 31 24.	- 07 24		18.8	FAINT BLUE STAR
LB 05219	05 31 24.	- 07 28		19.4	FAINT BLUE STAR
LB 05220	05 31 24.	- 07 31		18.4	FAINT BLUE STAR
LB 05221	05 31 24.	- 07 36		18.5	FAINT BLUE STAR
LB 05222	05 31 24.	- 07 38		16.4	FAINT BLUE STAR
LB 05223	05 31 24.	- 07 57		18.5	FAINT BLUE STAR
LB 05224	05 31 24.	- 08 01		18.8	FAINT BLUE STAR
LH120-N206B	05 31 24.	- 71 10	38		EMISSION NEBULA IN LMC
SL 541	05 31 27.	- 68 49	25		STAR CLUSTER IN LMC
LH 067	05 31 29.	- 69 20	300		STELLAR ASSN. IN LMC
HSE 335	05 31 29.	- 69 23			STAR CLUSTER IN LMC
RNGC 1952	05 31 30.	+ 21 59			PLANETARY NEBULA
CED 053	05 31 30.	+ 21 59	360		DIFFUSE GALACTIC NEBULA
LB 05225	05 31 30.	- 07 01		18.3	FAINT BLUE STAR
LB 05226	05 31 30.	- 07 09		18.8	FAINT BLUE STAR
LB 05227	05 31 30.	- 07 11		19.0	FAINT BLUE STAR
LB 05115	05 31 30.	- 07 19		17.5	FAINT BLUE STAR
LB 05228	05 31 30.	- 07 48		15.8	FAINT BLUE STAR
LB 05229	05 31 30.	- 07 55		18.6	FAINT BLUE STAR
LB 05230	05 31 30.	- 08 14		20.0	FAINT BLUE STAR
MCG-02-15-005	05 31 30.	- 13 23	54	15.	GALAXY
MCG-02-15-004	05 31 30.	- 13 24	12	15.	GALAXY
MCG-06-13-004	05 31 30.	- 36 25	180	12.5	GALAXY
SL 539	05 31 30.	- 70 44	60		STAR CLUSTER IN LMC
SL 536	05 31 30.	- 71 51	40		STAR CLUSTER IN LMC
VMT 08	05 31 31.	+ 21 58 54.	420		SUPERNOVA REMNANT
ST 40	05 31 31.	+ 21 59	360		EMISSION NEBULA
MIL 09	05 31 31.	+ 21 59 00.	252		SUPERNOVA REMNANT
SL 542	05 31 33.	- 70 15	35		STAR CLUSTER IN LMC
HSE 336	05 31 34.	- 69 08			STAR CLUSTER IN LMC
HSE 334	05 31 35.	- 71 38			STAR CLUSTER IN LMC
UGC 03322	05 31 36.	+ 06 47	66	17.	GALAXY S
MCG+07-12-001	05 31 36.	+ 40 54 30.	30	17.	GALAXY
LB 05231	05 31 36.	- 07 07		16.6	FAINT BLUE STAR
LB 05232	05 31 36.	- 07 18		16.6	FAINT BLUE STAR
LB 05233	05 31 36.	- 07 19		17.9	FAINT BLUE STAR
LB 05234	05 31 36.	- 07 31		18.3	FAINT BLUE STAR
LB 05235	05 31 36.	- 07 34		17.8	FAINT BLUE STAR
LB 05236	05 31 36.	- 07 46		17.3	FAINT BLUE STAR
LB 05237	05 31 36.	- 08 01		19.0	FAINT BLUE STAR
IC 2136	05 31 36.	- 26 28 33.			NONSTELLAR OBJECT
LH120-N148D	05 31 36.	- 68 33	28		EMISSION NEBULA IN LMC
LH120-N148H	05 31 36.	- 68 38	24		EMISSION NEBULA IN LMC
MCG+08-11-001	05 31 39.	+ 49 45	6	16.	GALAXY
HODG 04	05 31 39.	- 64 44	42		RED GLOBULAR CLSTR IN LMC
LH 068	05 31 39.	- 68 52	60		STELLAR ASSN. IN LMC
SL 549	05 31 40.	- 64 17	20		STAR CLUSTER IN LMC
HSE 337	05 31 41.	- 66 18			STAR CLUSTER IN LMC
ZWG 396.001	05 31 42.	+ 02 46		15.5	GALAXY
UGC 03323	05 31 42.	+ 49 43	84	17.	GALAXY
UGC 03324	05 31 42.	+ 63 34	72	17.	GALAXY Sc
LB 05238	05 31 42.	- 07 00		18.2	FAINT BLUE STAR
LB 05239	05 31 42.	- 07 18		16.0	FAINT BLUE STAR
LB 05240	05 31 42.	- 07 36		18.0	FAINT BLUE STAR
LB 05241	05 31 42.	- 07 37		18.7	FAINT BLUE STAR
LB 05242	05 31 42.	- 07 38		18.3	FAINT BLUE STAR
LB 05243	05 31 42.	- 07 41		17.6	FAINT BLUE STAR
LB 05244	05 31 42.	- 07 45		19.0	FAINT BLUE STAR
LB 05245	05 31 42.	- 08 00		18.5	FAINT BLUE STAR
HOD.72 36	05 31 43.	- 69 47	240		DARK NEBULA IN LMC
SL 543	05 31 43.	- 71 56	15		STAR CLUSTER IN LMC
HSE 338	05 31 44.	- 67 13			STAR CLUSTER IN LMC
RNGC 1979	05 31 45.	- 23 21		13.0	GALAXY
SCHO 0130	05 31 46.	+ 31 37 54.	440		ISOLATED DARK CLOUD
SL 550	05 31 46.	- 66 09	70		STAR CLUSTER IN LMC
UGC 03325	05 31 48.	+ 40 53	108	16.5	GALAXY Sb-c
LB 05246	05 31 48.	- 07 00		18.7	FAINT BLUE STAR
LB 05247	05 31 48.	- 07 18		20.7	FAINT BLUE STAR
LB 05248	05 31 48.	- 07 30		17.0	FAINT BLUE STAR
LB 05249	05 31 48.	- 07 34		19.0	FAINT BLUE STAR
LB 05250	05 31 48.	- 07 38		17.8	FAINT BLUE STAR

OBJECT NAME	RIGHT ASCEN.	DECLINATION	DIAM.	MAGN.	TYPE OF OBJECT
LB 05251	05 31 48.	- 07 43		18.8	FAINT BLUE STAR
LB 05252	05 31 48.	- 07 58		18.8	FAINT BLUE STAR
MCG-06-13-005	05 31 48.	- 36 25	30	15.5	GALAXY
MCG-04-14-004	05 31 51.	- 23 21 30.	36	13.5	GALAXY
HSE 339	05 31 51.	- 67 20			STAR CLUSTER IN LMC
SL 544	05 31 51.	- 70 08	25		STAR CLUSTER IN LMC
SL 551	05 31 53.	- 68 00	20		STAR CLUSTER IN LMC
ZC 0531.9+0424	05 31 54.	+ 04 24	1480		CLUSTER OF GALAXIES
LB 05253	05 31 54.	- 07 01		19.4	FAINT BLUE STAR
LB 05254	05 31 54.	- 07 02		19.6	FAINT BLUE STAR
LB 05255	05 31 54.	- 07 09		18.4	FAINT BLUE STAR
LB 05256	05 31 54.	- 07 20		20.4	FAINT BLUE STAR
LB 05257	05 31 54.	- 07 25		17.8	FAINT BLUE STAR
LB 05258	05 31 54.	- 08 00		18.5	FAINT BLUE STAR
LH120-N148E	05 31 54.	- 68 30	33		EMISSION NEBULA IN LMC
LH120-N208	05 31 54.	- 70 42			EMISSION NEBULA IN LMC
RNGC 2016	05 31 56.	- 69 58			OPEN CLUSTER IN LMC
SL 547	05 31 56.	- 69 58	70		STAR CLUSTER IN LMC
WS 29	05 31 56.	- 70 43 02.		15.93	PLANETARY NEB. IN LMC
LH 069	05 31 56.	- 71 06	300		STELLAR ASSN. IN LMC
LB 03587	05 31 57.	+ 34 03 48.		18.5	FAINT BLUE STAR
RNGC 1998	05 31 57.	- 48 43			NON-EXISTENT OBJECT
SC 0531-5922.1	05 31 57.	- 59 20 03.	6		NEBULA
SL 553	05 31 59.	- 66 28	340		STAR CLUSTER IN LMC
LBN 0833	05 32	+ 22 00	360		BRIGHT NEBULA
KHAV 126	05 32	+ 30 26	16650		DARK NEBULA
LBN 0797	05 32	+ 36 10	2100		BRIGHT NEBULA
LBN 0694	05 32	+ 65 00	3720		BRIGHT NEBULA
LBN 0914	05 32	- 00 30	120		BRIGHT NEBULA
LBN 0935	05 32	- 01 10	600		BRIGHT NEBULA
LBN 0940	05 32	- 01 40	1800		BRIGHT NEBULA
LB 05116	05 32 00.	- 07 14		14.7	FAINT BLUE STAR
LB 05259	05 32 00.	- 07 23		18.5	FAINT BLUE STAR
LB 05260	05 32 00.	- 07 38		17.0	FAINT BLUE STAR
LB 05261	05 32 00.	- 07 42		18.8	FAINT BLUE STAR
LB 05117	05 32 00.	- 07 56		19.4	FAINT BLUE STAR
LB 05262	05 32 00.	- 08 03		16.9	FAINT BLUE STAR
LB 05118	05 32 00.	- 08 12		18.8	FAINT BLUE STAR
LH120-N148C	05 32 00.	- 68 34	125		EMISSION NEBULA IN LMC
CM 22	05 32 00.	- 69 00			HII REGION IN LMC
CM 23	05 32 00.	- 69 24			HII REGION IN LMC
LB 03459	05 32 00.	- 69 55		12.2	FAINT BLUE STAR
LH120-N206A	05 32 00.	- 71 06	229		EMISSION NEBULA IN LMC
SC 0531-6046.6	05 32 01.	- 60 44 33.	6		NEBULA
LB 03588	05 32 02.	+ 34 02 24.		17.4	FAINT BLUE STAR
SL 552	05 32 02.	- 68 33	20		STAR CLUSTER IN LMC
SL 545	05 32 02.	- 72 10	20		STAR CLUSTER IN LMC
HSE 341	05 32 03.	- 67 24			STAR CLUSTER IN LMC
LH 070	05 32 03.	- 67 25	420		STELLAR ASSN. IN LMC
LW 233	05 32 05.	- 72 38			STAR CLUSTER IN LMC
SL 546	05 32 05.	- 72 40	15		STAR CLUSTER IN LMC
ZWG 348.002	05 32 06.	+ 77 17		15.6	GALAXY
ZWG 347.018	05 32 06.	+ 77 17		15.6	GALAXY
UGC 03326	05 32 06.	+ 77 17	204	15.6	GALAXY Sc
KARA.73B 0159	05 32 06.	+ 77 17	216	15.6	ISOLATED GALAXY S
LB 05263	05 32 06.	- 07 22		18.9	FAINT BLUE STAR
LB 05264	05 32 06.	- 07 28		16.9	FAINT BLUE STAR
LB 05265	05 32 06.	- 07 31		17.7	FAINT BLUE STAR
LB 05266	05 32 06.	- 07 32		19.2	FAINT BLUE STAR
LB 05267	05 32 06.	- 07 38		17.9	FAINT BLUE STAR
LB 05268	05 32 06.	- 08 16		18.6	FAINT BLUE STAR
SL 556	05 32 06.	- 64 46	65		STAR CLUSTER IN LMC
HSE 343	05 32 06.	- 66 43			STAR CLUSTER IN LMC
LH120-N057E	05 32 06.	- 67 44	377		EMISSION NEBULA IN LMC
LH120-N148I	05 32 06.	- 68 42	229		EMISSION NEBULA IN LMC
HSE 344	05 32 08.	- 67 14			STAR CLUSTER IN LMC
LH 071	05 32 08.	- 68 35	180		STELLAR ASSN. IN LMC
HSE 340	05 32 08.	- 70 03			STAR CLUSTER IN LMC
SL 548	05 32 08.	- 72 05	15		STAR CLUSTER IN LMC
LW 231	05 32 08.	- 75 23			STAR CLUSTER IN LMC
MCG-04-14-005	05 32 09.	- 23 29	48	15.	GALAXY
LH 072	05 32 11.	- 66 29	360		STELLAR ASSN. IN LMC
LB 05269	05 32 12.	- 07 23		18.2	FAINT BLUE STAR
LB 05270	05 32 12.	- 07 40		14.3	FAINT BLUE STAR
LB 05119	05 32 12.	- 07 46		18.5	FAINT BLUE STAR
LB 05271	05 32 12.	- 07 52		19.9	FAINT BLUE STAR
LB 05272	05 32 12.	- 07 58		18.0	FAINT BLUE STAR
LB 05273	05 32 12.	- 08 12		20.2	FAINT BLUE STAR
MCG-04-14-006	05 32 12.	- 23 35	60	14.	GALAXY
MCG-05-14-003	05 32 12.	- 28 32	36	14.	GALAXY
LW 237	05 32 12.	- 64 46			STAR CLUSTER IN LMC
IC 2137	05 32 13.	- 23 22 13.			NONSTELLAR OBJECT
HOD.72 37	05 32 13.	- 69 47	180		DARK NEBULA IN LMC
SCHO 0131	05 32 14.	- 00 21 18.	410		ISOLATED DARK CLOUD
LW 235	05 32 14.	- 72 05			STAR CLUSTER IN LMC
RNGC 2011	05 32 15.	- 67 33		9.5	CLUSTER/NEBULOSITY IN LMC
RNGC 2019	05 32 15.	- 70 12		11.0	GLOBULAR CLUSTER IN LMC
HSE 342	05 32 16.	- 70 20			STAR CLUSTER IN LMC
IC 2138	05 32 17.	- 23 34			NONSTELLAR OBJECT
RNGC 2015	05 32 17.	- 69 15			CLUSTER/NEBULOSITY IN LMC
OCL 0479	05 32 18.	+ 09 54	10800	3.0	OPEN STAR CLUSTER
LB 05274	05 32 18.	- 07 31		17.7	FAINT BLUE STAR
LB 05275	05 32 18.	- 07 32		17.5	FAINT BLUE STAR
LB 05276	05 32 18.	- 07 33		20.4	FAINT BLUE STAR
LB 05278	05 32 18.	- 07 40		18.9	FAINT BLUE STAR
LB 05277	05 32 18.	- 07 40		18.3	FAINT BLUE STAR
LB 05279	05 32 18.	- 07 46		17.8	FAINT BLUE STAR
LB 05280	05 32 18.	- 08 03		19.0	FAINT BLUE STAR
LB 05282	05 32 18.	- 08 04		18.7	FAINT BLUE STAR
LB 05281	05 32 18.	- 08 04		18.7	FAINT BLUE STAR
LB 05283	05 32 18.	- 08 13		16.6	FAINT BLUE STAR
LB 03460	05 32 18.	- 60 48		14.6	FAINT BLUE STAR
LH120-N055	05 32 18.	- 66 28	434		EMISSION NEBULA IN LMC
LH120-N057B	05 32 18.	- 67 48	26		EMISSION NEBULA IN LMC
SNO 21	05 32 19.	- 23 35 25.	840	19.	GROUP OF 8 GALAXIES
LH 073	05 32 20.	- 68 43	180		STELLAR ASSN. IN LMC
CED 054	05 32 21.	+ 09 54	1800		DIFFUSE GALACTIC NEBULA
CED 055A	05 32 21.	- 05 36			DIFFUSE GALACTIC NEBULA
SL 559	05 32 21.	- 67 33	75		STAR CLUSTER IN LMC
SL 554	05 32 21.	- 70 12	70		STAR CLUSTER IN LMC
RNGC 2014	05 32 22.	- 67 42		8.5	CLUSTER/NEBULOSITY IN LMC
SL 560	05 32 22.	- 67 42	340		STAR CLUSTER IN LMC
SL 557	05 32 23.	- 69 15	80		STAR CLUSTER IN LMC
LH 074	05 32 23.	- 69 16	300		STELLAR ASSN. IN LMC
MCG+13-05-002	05 32 24.	+ 77 16	192	14.	GALAXY
LB 05284	05 32 24.	- 06 59		18.5	FAINT BLUE STAR
LB 05285	05 32 24.	- 07 02		19.2	FAINT BLUE STAR
LB 05286	05 32 24.	- 07 16		18.9	FAINT BLUE STAR
LB 05287	05 32 24.	- 08 22		20.4	FAINT BLUE STAR
LH120-N057A	05 32 24.	- 67 44	55		EMISSION NEBULA IN LMC
SL 558	05 32 24.	- 69 30	15		STAR CLUSTER IN LMC
SCHO 0132	05 32 26.	- 05 36 06.	230		ISOLATED DARK CLOUD
LW 234	05 32 26.	- 74 42			STAR CLUSTER IN LMC
SCHO 0133	05 32 28.	+ 10 28 24.	530		ISOLATED DARK CLOUD
HSE 345	05 32 28.	- 70 19			STAR CLUSTER IN LMC
MRSL 195-12/1	05 32 30.	+ 09 54	23400		HII REGION
SCHO 0134	05 32 30.	+ 40 31 18.	440		ISOLATED DARK CLOUD
MRSL 209-19/1	05 32 30.	- 05 30	3600		HII REGION
LB 05120	05 32 30.	- 06 45		16.9	FAINT BLUE STAR
LB 05288	05 32 30.	- 07 20		18.5	FAINT BLUE STAR
LB 05121	05 32 30.	- 07 31		17.7	FAINT BLUE STAR
LB 05289	05 32 30.	- 07 52		19.5	FAINT BLUE STAR
MCG-05-14-004	05 32 30.	- 30 53	30	15.	GALAXY
LH120-N055A	05 32 30.	- 66 29	65		EMISSION NEBULA IN LMC
LH120-N057	05 32 30.	- 67 43	652		EMISSION NEBULA IN LMC
LW 238	05 32 32.	- 63 40			STAR CLUSTER IN LMC
SL 555	05 32 32.	- 72 12	25		STAR CLUSTER IN LMC
RNGC 1989	05 32 33.	- 30 53		15.0	GALAXY
LH 075	05 32 33.	- 67 34	180		STELLAR ASSN. IN LMC
CED 055B	05 32 34.	- 04 46			DIFFUSE GALACTIC NEBULA
LH 076	05 32 34.	- 67 44	360		STELLAR ASSN. IN LMC
HSE 346	05 32 35.	- 69 25			STAR CLUSTER IN LMC
LB 05122	05 32 36.	- 07 29		16.7	FAINT BLUE STAR
LB 05290	05 32 36.	- 07 32		20.2	FAINT BLUE STAR
LB 05291	05 32 36.	- 07 33		17.7	FAINT BLUE STAR
MCG-02-15-006	05 32 36.	- 10 02	60	15.5	GALAXY
MCG-03-15-002	05 32 36.	- 17 53 30.	60	16.	GALAXY
MCG-05-14-005	05 32 36.	- 30 45	12	15.5	GALAXY
REIN 4.054	05 32 36.62	- 04 45 49.0			NEBULA
LW 236	05 32 38.	- 72 11			STAR CLUSTER IN LMC
RNGC 1992	05 32 39.	- 30 58		15.0	GALAXY
HOD.72 38	05 32 39.	- 68 53	600		DARK NEBULA IN LMC
RNGC 1973	05 32 40.	- 04 46			DIFFUSE NEBULA
SCHO 0135	05 32 42.	+ 05 37 42.	820		ISOLATED DARK CLOUD
OCL 0525	05 32 42.	- 04 28	1500	5.2	OPEN STAR CLUSTER
LB 05292	05 32 42.	- 07 52		18.1	FAINT BLUE STAR
MCG-05-14-006	05 32 42.	- 29 17	24	14.5	GALAXY
MCG-05-14-007	05 32 42.	- 30 58	30	15.5	GALAXY
LH120-N148G	05 32 42.	- 68 41			EMISSION NEBULA IN LMC
SL 561	05 32 42.	- 70 44	70		STAR CLUSTER IN LMC
SL 562	05 32 45.	- 71 17	35		STAR CLUSTER IN LMC
RNGC 1981	05 32 46.	- 04 28			OPEN CLUSTER
CED 055C	05 32 46.	- 04 43			DIFFUSE GALACTIC NEBULA
ST 38	05 32 46.	- 05 26			EMISSION NEBULA
RNGC 1980	05 32 47.	- 05 57		2.5	CLUSTER WITH NEBULOSITY
SL 563	05 32 47.	- 69 25	15		STAR CLUSTER IN LMC
OCL 0460	05 32 48.	+ 25 55	1680		OPEN STAR CLUSTER
OCL 0445	05 32 48.	+ 34 06	1500	7.0	OPEN STAR CLUSTER
ZC 0532.8+7046	05 32 48.	+ 70 46	2490		CLUSTER OF GALAXIES
LB 05293	05 32 48.	- 08 07		17.5	FAINT BLUE STAR
LH120-N058	05 32 48.	- 67 31	35		EMISSION NEBULA IN LMC
REIN 4.055A	05 32 49.54	- 04 43 00.6			NEBULA
REIN 4.055B	05 32 50.17	- 04 42 49.9			NEBULA
RNGC 1960	05 32 51.	+ 34 06		6.5	OPEN CLUSTER
CED 055D	05 32 51.	- 05 25	3960		DIFFUSE GALACTIC NEBULA
SG 3.053	05 32 51.	- 05 25	3600		DIFFUSE EMISSION NEBULA
SL 564	05 32 51.	- 67 21	20		STAR CLUSTER IN LMC
HSE 347	05 32 51.	- 70 04			STAR CLUSTER IN LMC
HH 33	05 32 51.5	- 06 19 35.	20		HERBIG-HARO OBJECT
RNGC 1975	05 32 52.	- 04 43			DIFFUSE NEBULA
RNGC 1976	05 32 52.	- 05 25			CLUSTER WITH NEBULOSITY
SCHO 0136	05 32 52.	- 05 35 54.	180		ISOLATED DARK CLOUD
MRSL 208-19/1	05 32 54.	- 04 50	1200		HII REGION
OCL 0528	05 32 54.	- 05 25	3000		OPEN STAR CLUSTER
OCL 0529	05 32 54.	- 05 58	3000	2.5	OPEN STAR CLUSTER
LB 05294	05 32 54.	- 07 33		17.7	FAINT BLUE STAR
HH 40	05 32 54.5	- 06 20 16.			HERBIG-HARO OBJECT
CED 055F	05 32 57.	- 05 56	840		DIFFUSE GALACTIC NEBULA
RNGC 1977	05 32 58.	- 04 50			CLUSTER WITH NEBULOSITY
CED 055E	05 32 58.	- 04 52	2520		DIFFUSE GALACTIC NEBULA
SG 3.054	05 32 58.	- 04 52	1200		DIFFUSE EMISSION NEBULA
HSE 348	05 32 58.	- 70 25			STAR CLUSTER IN LMC
SCHO 0137	05 32 59.	+ 07 19 12.	260		ISOLATED DARK CLOUD
KHAV 127	05 33	+ 08 38	7770		DARK NEBULA
LBN 0699	05 33	+ 63 50	1200		BRIGHT NEBULA
LBN 0974	05 33	- 05 30	4500		BRIGHT NEBULA
LDN 1614	05 33 00.	- 00 40	240		DARK NEBULA
OCL 0503	05 33 00.	- 01 08	11400	0.6	OPEN STAR CLUSTER
LB 05295	05 33 00.	- 07 34		18.5	FAINT BLUE STAR
LB 05296	05 33 00.	- 08 00		18.8	FAINT BLUE STAR
LB 05297	05 33 00.	- 08 01		18.4	FAINT BLUE STAR
LB 05298	05 33 00.	- 08 02		17.3	FAINT BLUE STAR
LH120-N057D	05 33 00.	- 67 43			EMISSION NEBULA IN LMC
LH120-N148A	05 33 00.	- 68 25	25		EMISSION NEBULA IN LMC
LH120-N206D	05 33 00.	- 71 15	27		EMISSION NEBULA IN LMC
SCHO 0138	05 33 01.	+ 07 12 12.	320		ISOLATED DARK CLOUD
ABC 0542	05 33 03.	+ 64 00		17.5	RICH CLUSTER OF GALAXIES
CED 055G	05 33 03.	- 05 18			DIFFUSE GALACTIC NEBULA
RNGC 1982	05 33 04.	- 05 18			DIFFUSE NEBULA
IC 2139	05 33 04.	- 17 58			OPEN CLUSTER
SCHO 0139	05 33 05.	- 02 13 36.	610		ISOLATED DARK CLOUD
HH 34	05 33 05.4	- 06 30 28.	15		HERBIG-HARO OBJECT
LB 05123	05 33 06.	- 07 34		20.5	FAINT BLUE STAR
LB 05299	05 33 06.	- 08 14		17.2	FAINT BLUE STAR
LB 03461	05 33 06.	- 62 49		14.7	FAINT BLUE STAR
SL 565	05 33 11.	- 70 30	40		STAR CLUSTER IN LMC
SCHO 0140	05 33 12.	- 05 45 30.	380		ISOLATED DARK CLOUD
LB 05300	05 33 12.	- 07 45		19.0	FAINT BLUE STAR
ST 39	05 33 13.	+ 09 34	2400		EMISSION NEBULA
RNGC 2020	05 33 16.	- 67 44			DIFFUSE NEBULA IN LMC
B 226	05 33 17.	+ 33 40	1020		DARK OBJECT
LB 05301	05 33 18.	- 08 03		18.6	FAINT BLUE STAR
RNGC 1993	05 33 18.	- 17 51		14.0	GALAXY
MCG-03-15-003	05 33 18.	- 17 51	24	14.	GALAXY
LH120-N057C	05 33 18.	- 67 44	94		EMISSION NEBULA IN LMC
LH120-N149B	05 33 18.	- 69 48			EMISSION NEBULA IN LMC
LH120-N149A	05 33 18.	- 69 48	27		EMISSION NEBULA IN LMC
SL 566	05 33 18.	- 70 49	25		STAR CLUSTER IN LMC
SG 3.055	05 33 19.	- 02 25	3000		DIFFUSE EMISSION NEBULA
LH 077	05 33 19.	- 67 01	3600		STELLAR ASSN. IN LMC
SL 567	05 33 21.	- 67 33	135		STAR CLUSTER IN LMC
VDB.66N 045	05 33 23.	+ 31 49	36		REFLECTION NEBULA
SG 3.057	05 33 26.	- 06 10	5400		DIFFUSE EMISSION NEBULA
MCG+05-14-001	05 33 27.	+ 31 54	48	13.	GALAXY
UGC 03327	05 33 30.	+ 31 54	150	13.	GALAXY E?
7ZW 042	05 33 30.	+ 63 18			COMPACT GALAXY

OBJECT NAME	RIGHT ASCEN.	DECLINATION	DIAM.	MAGN.	TYPE OF OBJECT
SL 573	05 33 30.	- 64 57	35		STAR CLUSTER IN LMC
CM 05	05 33 30.	- 67 54			HII REGION IN LMC
SL 569	05 33 30.	- 68 10	35		STAR CLUSTER IN LMC
RNGC 2021	05 33 33.	- 67 28			CLUSTER/NEBULOSITY IN LMC
SL 570	05 33 33.	- 67 28	25		STAR CLUSTER IN LMC
LH 078	05 33 33.	- 67 33	240		STELLAR ASSN. IN LMC
HSE 349	05 33 33.	- 70 04			STAR CLUSTER IN LMC
HH 41	05 33 34.1	- 05 04 40.	12		HERBIG-HARO OBJECT
SL 568	05 33 35.	- 70 32	15		STAR CLUSTER IN LMC
ZWG 422.001	05 33 36.	+ 07 18		15.6	GALAXY
UGC 03328	05 33 36.	+ 07 18	72	15.6	GALAXY Sb
ZWG 307.019	05 33 36.	+ 63 19		15.6	GALAXY
SCHO 0141	05 33 36.	- 04 35 30.	230		ISOLATED DARK CLOUD
HH 42	05 33 37.3	- 05 06 31.	15		HERBIG-HARO OBJECT
CED 055H	05 33 38.	- 01 14	3000		DIFFUSE GALACTIC NEBULA
MCG-03-15-004	05 33 39.	- 17 48	54	16.	GALAXY
MCG+01-15-002	05 33 42.	+ 07 20	60	16.	GALAXY
UGC 03329	05 33 42.	+ 16 33	96	16.0	GALAXY Sb-c
LB 05302	05 33 42.	- 08 11		19.2	FAINT BLUE STAR
LW 240	05 33 42.	- 64 58			STAR CLUSTER IN LMC
LB 03462	05 33 42.	- 75 40		14.5	FAINT BLUE STAR
RNGC 1990	05 33 44.	- 01 14			DIFFUSE NEBULA
LH 079	05 33 45.	- 67 29	420		STELLAR ASSN. IN LMC
HOD.72 39	05 33 45.	- 70 06	180		DARK NEBULA IN LMC
HH 03	05 33 45.9	- 06 44 55.	6		HERBIG-HARO OBJECT
HSE 350	05 33 46.	- 70 15			STAR CLUSTER IN LMC
SCHO 0142	05 33 47.	- 04 27 06.	310		ISOLATED DARK CLOUD
RNGC 2007	05 33 48.	- 50 58			GALAXY
SL 572	05 33 50.	- 71 03	20		STAR CLUSTER IN LMC
HSE 351	05 33 51.	- 70 04			STAR CLUSTER IN LMC
IC 0425	05 33 52.	+ 32 24			NONSTELLAR OBJECT
LB 03589	05 33 53.	+ 33 47 54.		19.2	FAINT BLUE STAR
UGC 03330	05 33 54.	+ 14 24	72	17.	GALAXY S
SCHO 0143	05 33 54.	- 04 45 36.	320		ISOLATED DARK CLOUD
LB 01781	05 33 54.	- 46 30		13.7	FAINT BLUE STAR
RNGC 2008	05 33 54.	- 51 01			GALAXY
RNGC 2025	05 33 54.	- 71 46			OPEN CLUSTER IN LMC
SL 571	05 33 54.	- 71 46	70		STAR CLUSTER IN LMC
HH 01	05 33 54.9	- 06 47 02.	8		HERBIG-HARO OBJECT
SCHO 0144	05 33 55.	- 10 29 48.	360		ISOLATED DARK CLOUD
SL 574	05 33 56.	- 69 59	30		STAR CLUSTER IN LMC
HH 35	05 33 56.6	- 06 43 40.	5		HERBIG-HARO OBJECT
HSE 352	05 33 57.	- 70 09			STAR CLUSTER IN LMC
HH 02A	05 33 59.4	- 06 49 00.			HERBIG-HARO OBJECT
HH 02D	05 33 59.4	- 06 49 04.			HERBIG-HARO OBJECT
HH 02C	05 33 59.6	- 06 48 53.			HERBIG-HARO OBJECT
HH 02H	05 33 59.7	- 06 49 02.			HERBIG-HARO OBJECT
HH 02I	05 33 59.7	- 06 49 07.			HERBIG-HARO OBJECT
HH 02B	05 33 59.9	- 06 48 56.			HERBIG-HARO OBJECT
CED 056	05 34	+ 32 24	1800		DIFFUSE GALACTIC NEBULA
LBN 0921	05 34	- 00 25	360		BRIGHT NEBULA
LBN 0979	05 34	- 06 46	120		BRIGHT NEBULA
ZC 0534.0+6653	05 34 00.	+ 66 53	1410		CLUSTER OF GALAXIES
LDN 1620	05 34 00.	- 00 10	420		DARK NEBULA
DG 060	05 34 00.	- 06 45	1200		REFLECTION NEBULA
CM 40	05 34 00.	- 68 06			HII REGION IN LMC
LH120-N151	05 34 00.	- 68 38			EMISSION NEBULA IN LMC
LH120-N150	05 34 00.	- 68 47	49		EMISSION NEBULA IN LMC
CM 20	05 34 00.	- 68 54			HII REGION IN LMC
CM 24	05 34 00.	- 69 18			HII REGION IN LMC
HH 02G	05 34 00.1	- 06 48 56.			HERBIG-HARO OBJECT
HH 02E	05 34 00.7	- 06 49 00.			HERBIG-HARO OBJECT
HH 02L	05 34 00.8	- 06 49 15.			HERBIG-HARO OBJECT
HSE 354	05 34 01.	- 66 52			STAR CLUSTER IN LMC
CED 055I	05 34 02.	- 06 45	960		DIFFUSE GALACTIC NEBULA
LH 080	05 34 02.	- 69 59	240		STELLAR ASSN. IN LMC
MCG-03-15-005	05 34 03.	- 16 42	72	15.	GALAXY
VDB.66N 046	05 34 04.	- 06 44	108		REFLECTION NEBULA
RNGC 1999	05 34 05.	- 06 45			DIFFUSE NEBULA
SC 0533-5922.6	05 34 05.	- 59 20 42.	6		NEBULA
LH120-N060	05 34 06.	- 67 55			EMISSION NEBULA IN LMC
IC 0427	05 34 07.	- 06 41			NONSTELLAR OBJECT
SL 578	05 34 07.	- 65 18	35		STAR CLUSTER IN LMC
LW 242	05 34 07.	- 65 19			STAR CLUSTER IN LMC
SG 3.056	05 34 08.	+ 34 42	2400		DIFFUSE EMISSION NEBULA
HN 0067	05 34 08.	- 06 41			NEBULA
WS 31	05 34 08.	- 68 38 45.		16.16	PLANETARY NEB. IN LMC
HSE 353	05 34 08.	- 69 58			STAR CLUSTER IN LMC
RNGC 2028	05 34 08.	- 69 58			OPEN CLUSTER IN LMC
SL 575	05 34 08.	- 69 58	25		STAR CLUSTER IN LMC
SCHO 0145	05 34 09.	- 04 34 54.	240		ISOLATED DARK CLOUD
LB 03501	05 34 10.	- 01 29 42.		16.9	FAINT BLUE STAR
WS 32	05 34 10.	- 67 55 03.		16.28	PLANETARY NEB. IN LMC
IC 0428	05 34 13.	- 06 32			NONSTELLAR OBJECT
SCHO 0146	05 34 14.	- 04 59 54.	360		ISOLATED DARK CLOUD
HN 0064	05 34 14.	- 06 32			NEBULA
HN 0057	05 34 15.	- 00 16			NEBULA
CED 055J	05 34 15.	- 00 16	300		DIFFUSE GALACTIC NEBULA
IC 0426	05 34 16.	- 00 16			NONSTELLAR OBJECT
HOD.72 40	05 34 16.	- 70 16	420		DARK NEBULA IN LMC
LB 03502	05 34 17.	- 00 46 24.		18.8	FAINT BLUE STAR
SL 579	05 34 17.	- 67 52	35		STAR CLUSTER IN LMC
DG 061	05 34 18.	- 00 16	480		REFLECTION NEBULA
LB 05303	05 34 18.	- 08 06		20.5	FAINT BLUE STAR
MCG-05-14-008	05 34 18.	- 33 01	24	15.	GALAXY
LH120-N056	05 34 18.	- 67 36	1940		EMISSION NEBULA IN LMC
LH120-N209	05 34 18.	- 71 53			EMISSION NEBULA IN LMC
LW 244	05 34 19.	- 65 05			STAR CLUSTER IN LMC
SL 582	05 34 20.	- 67 08	25		STAR CLUSTER IN LMC
SL 580	05 34 20.	- 68 36	40		STAR CLUSTER IN LMC
RNGC 2031	05 34 20.	- 71 01			OPEN CLUSTER IN LMC
SL 577	05 34 20.	- 71 01	170		STAR CLUSTER IN LMC
WS 30	05 34 20.	- 71 54 21.		16.57	PLANETARY NEB. IN LMC
HH 36	05 34 20.7	- 06 46 01.	9		HERBIG-HARO OBJECT
SG 3.059	05 34 22.	- 04 26	1500		DIFFUSE EMISSION NEBULA
DG 062	05 34 24.	- 00 23	120		REFLECTION NEBULA
LH120-N062B	05 34 24.	- 66 10	239		EMISSION NEBULA IN LMC
LB 03503	05 34 25.	- 00 52 00.		19.1	FAINT BLUE STAR
SL 583	05 34 26.	- 68 33	25		STAR CLUSTER IN LMC
RNGC 1985	05 34 28.	+ 31 58			PLANETARY NEBULA
ZWG 422.002	05 34 30.	+ 08 56		15.7	GALAXY
RNGC 1988	05 34 30.	+ 21 12			NON-EXISTENT OBJECT
LH120-N152	05 34 30.	- 69 28			EMISSION NEBULA IN LMC
LW 239	05 34 30.	- 74 24			STAR CLUSTER IN LMC
LB 03504	05 34 31.	- 00 40 18.		18.0	FAINT BLUE STAR
PK176+00.1	05 34 32.80	+ 31 57 37.1	70	12.5	PLANETARY NEBULA
CED 057	05 34 33.	+ 31 58	42		DIFFUSE GALACTIC NEBULA
HSE 355	05 34 33.	- 71 19			STAR CLUSTER IN LMC

OBJECT NAME	RIGHT ASCEN.	DECLINATION	DIAM.	MAGN.	TYPE OF OBJECT
HSE 358	05 34 34.	- 66 06			STAR CLUSTER IN LMC
PK196-12.1	05 34 36.	+ 08 14	34	17.1	PLANETARY NEBULA
DG 063	05 34 36.	- 00 21	180		REFLECTION NEBULA
LH120-N062A	05 34 36.	- 66 15	108		EMISSION NEBULA IN LMC
SL 576	05 34 36.	- 74 25	70		STAR CLUSTER IN LMC
SL 586	05 34 37.	- 66 59	40		STAR CLUSTER IN LMC
HSE 357	05 34 38.	- 68 44			STAR CLUSTER IN LMC
SG 3.058	05 34 39.	+ 31 58	30		DIFFUSE EMISSION NEBULA
SS 42	05 34 39.	+ 31 59			DIFFUSE GALACTIC NEBULA
MCG-04-14-007	05 34 39.	- 22 27	54	14.	GALAXY
HOD.72 42	05 34 39.	- 68 51	120		DARK NEBULA IN LMC
SL 585	05 34 39.	- 68 51	30		STAR CLUSTER IN LMC
DG 064	05 34 42.	+ 31 59	60		REFLECTION NEBULA
MCG-03-15-006	05 34 42.	- 15 14	42	14.	GALAXY
LH120-N153	05 34 42.	- 69 00			EMISSION NEBULA IN LMC
HOD.72 41	05 34 42.	- 69 35	480		DARK NEBULA IN LMC
HSE 356	05 34 42.	- 71 48			STAR CLUSTER IN LMC
SCHO 0147	05 34 44.	+ 37 30 30.	520		ISOLATED DARK CLOUD
LB 03505	05 34 44.	- 01 15 42.		19.0	FAINT BLUE STAR
2ZW 039	05 34 48.	+ 13 31			COMPACT GALAXY
LB 05304	05 34 48.	- 08 23		18.7	FAINT BLUE STAR
SCHO 0148	05 34 49.	+ 07 29 48.	450		ISOLATED DARK CLOUD
SL 588	05 34 49.	- 68 19	35		STAR CLUSTER IN LMC
SL 584	05 34 49.	- 71 58	25		STAR CLUSTER IN LMC
HUB E12	05 34 50.	- 02 15			DIFFUSE NEBULA
WS 33	05 34 50.	- 69 00 25.		15.73	PLANETARY NEB. IN LMC
RNGC 2036	05 34 51.	- 70 06			OPEN CLUSTER IN LMC
SL 587	05 34 51.	- 70 06	25		STAR CLUSTER IN LMC
LW 243	05 34 51.	- 73 09			STAR CLUSTER IN LMC
MCG+13-05-003	05 34 54.	+ 77 05	12	16.	GALAXY
DG 065	05 34 54.	- 00 20	60		REFLECTION NEBULA
LH120-N154B	05 34 54.	- 69 48			EMISSION NEBULA IN LMC
HN 0302	05 34 54.	- 75 24			NEBULA
IC 2140	05 34 54.	- 75 24			OPEN CLUSTER
LW 245	05 34 55.	- 65 21			STAR CLUSTER IN LMC
RNGC 2033	05 34 55.	- 69 46			CLUSTER/NEBULOSITY IN LMC
SL 589	05 34 55.	- 69 46	270		STAR CLUSTER IN LMC
LW 241	05 34 57.	- 75 25			STAR CLUSTER IN LMC
SCHO 0150	05 34 58.	+ 09 11 12.	470		ISOLATED DARK CLOUD
SCHO 0149	05 34 59.	- 04 49 12.	370		ISOLATED DARK CLOUD
LBN 0802	05 35	+ 35 50	120		BRIGHT NEBULA
LBN 0789	05 35	+ 37 30	4500		BRIGHT NEBULA
LBN 0950	05 35	- 02 20	3900		BRIGHT NEBULA
LBN 0977	05 35	- 06 00	12600		BRIGHT NEBULA
ZWG 396.002	05 35 00.	+ 01 18		15.7	GALAXY
ZWG 362.009	05 35 00.	+ 82 08		15.7	GALAXY
MCG+14-03-011	05 35 00.	+ 85 55	90	16.	GALAXY
LDN 1641	05 35 00.	- 07 00	10140		DARK NEBULA
CM 21	05 35 00.	- 68 42			HII REGION IN LMC
SL 581	05 35 03.	- 75 24	80		STAR CLUSTER IN LMC
SCHO 0151	05 35 04.	- 05 22 00.	450		ISOLATED DARK CLOUD
RNGC 2038	05 35 05.	- 70 35			OPEN CLUSTER IN LMC
SL 590	05 35 05.	- 70 35	70		STAR CLUSTER IN LMC
LB 05305	05 35 06.	- 08 18		16.7	FAINT BLUE STAR
HSE 359	05 35 06.	- 69 25			STAR CLUSTER IN LMC
LB 03506	05 35 07.	- 01 33 06.		18.0	FAINT BLUE STAR
ARC 0547	05 35 07.	- 14 27		17.0	RICH CLUSTER OF GALAXIES
LW 246	05 35 07.	- 65 15			STAR CLUSTER IN LMC
RNGC 2027	05 35 07.	- 66 57			CLUSTER/NEBULOSITY IN LMC
SL 592	05 35 07.	- 66 57	60		STAR CLUSTER IN LMC
RNGC 2037	05 35 07.	- 69 47			CLUSTER/NEBULOSITY IN LMC
LH 081	05 35 07.	- 69 47	330		STELLAR ASSN. IN LMC
SL 591	05 35 08.	- 69 56	25		STAR CLUSTER IN LMC
RNGC 1996	05 35 09.	+ 25 47			NON-EXISTENT OBJECT
LB 03507	05 35 09.	- 01 07 00.		17.1	FAINT BLUE STAR
SS 43	05 35 10.	+ 35 49			DIFFUSE GALACTIC NEBULA
LB 05306	05 35 12.	- 07 28		20.4	FAINT BLUE STAR
HSE 360	05 35 14.	- 71 07			STAR CLUSTER IN LMC
HOD.72 43	05 35 15.	- 70 04	600		DARK NEBULA IN LMC
SG 3.060	05 35 16.	+ 35 50	60		DIFFUSE EMISSION NEBULA
UGC 03331	05 35 18.	+ 00 05	78	16.0	GALAXY Sb
MCG+00-15-001	05 35 18.	+ 00 05	54	14.	GALAXY
DG 066	05 35 18.	+ 30 39	120		REFLECTION NEBULA
LB 03508	05 35 20.	- 00 51 06.		16.0	FAINT BLUE STAR
RNGC 2030	05 35 22.	- 66 03			CLUSTER/NEBULOSITY IN LMC
MRSL 173+02/1	05 35 24.	+ 35 46	120		HII REGION
SS 44	05 35 25.	+ 30 39			DIFFUSE GALACTIC NEBULA
SCHO 0152	05 35 27.	+ 09 06 42.	210		ISOLATED DARK CLOUD
SL 595	05 35 28.	- 66 03	185		STAR CLUSTER IN LMC
RNGC 2035	05 35 28.	- 67 36			CLUSTER/NEBULOSITY IN LMC
RNGC 2032	05 35 28.	- 67 36			CLUSTER/NEBULOSITY IN LMC
LH120-N059A	05 35 30.	- 67 36	174		EMISSION NEBULA IN LMC
LH120-N154	05 35 30.	- 69 44	1054		EMISSION NEBULA IN LMC
SL 594	05 35 32.	- 68 42	15		STAR CLUSTER IN LMC
RNGC 2029	05 35 34.	- 67 35			CLUSTER/NEBULOSITY IN LMC
LH 082	05 35 34.	- 67 36	360		STELLAR ASSN. IN LMC
SL 593	05 35 34.	- 70 21	5		STAR CLUSTER IN LMC
SCHO 0153	05 35 36.	+ 35 33 00.	470		ISOLATED DARK CLOUD
OCL 0431	05 35 36.	+ 37 54	1500		OPEN STAR CLUSTER
DG 067	05 35 36.	- 00 08	600		REFLECTION NEBULA
LH120-N063	05 35 36.	- 66 01	380		EMISSION NEBULA IN LMC
LH120-N059	05 35 36.	- 67 35	561		EMISSION NEBULA IN LMC
MIL 96	05 35 39.	- 66 03 30.	27		SUPERNOVA REMNANT
LH 083	05 35 40.	- 66 04	240		STELLAR ASSN. IN LMC
HOD.72 44	05 35 40.	- 69 08	780		DARK NEBULA IN LMC
HSE 361	05 35 40.	- 71 30			STAR CLUSTER IN LMC
VDB.66N 048	05 35 41.	- 00 11	1680		REFLECTION NEBULA
SCHO 0155	05 35 41.	- 06 25 18.	490		ISOLATED DARK CLOUD
SCHO 0154	05 35 42.	+ 36 16 12.	530		ISOLATED DARK CLOUD
LH120-N063A	05 35 42.	- 66 02	40		EMISSION NEBULA IN LMC
LH120-N210	05 35 42.	- 74 18			EMISSION NEBULA IN LMC
RNGC 2034	05 35 43.	- 66 58			CLUSTER/NEBULOSITY IN LMC
LH 084	05 35 43.	- 66 58	540		STELLAR ASSN. IN LMC
HN 0303	05 35 43.	- 78 02			NEBULA
SG 3.063	05 35 44.	- 01 47	180		DIFFUSE EMISSION NEBULA
HSE 362	05 35 44.	- 69 57			STAR CLUSTER IN LMC
IC 2142	05 35 44.	- 78 02			NONSTELLAR OBJECT
HH 43	05 35 45.4	- 07 11 04.	18		HERBIG-HARO OBJECT
LW 247	05 35 46.	- 72 25			STAR CLUSTER IN LMC
LH120-N059C	05 35 48.	- 67 38	90		EMISSION NEBULA IN LMC
SL 596	05 35 50.	- 68 41	25		STAR CLUSTER IN LMC
LH 085	05 35 51.	- 68 54	240		STELLAR ASSN. IN LMC
SG 3.061	05 35 52.	+ 35 54	720		DIFFUSE EMISSION NEBULA
SC 0535-6137.8	05 35 52.	- 61 36 02.	12		NEBULA
RNGC 2046	05 35 52.	- 70 16			OPEN CLUSTER IN LMC
IC 0429	05 35 54.	- 07 04 34.			NONSTELLAR OBJECT
HH 38	05 35 56.5	- 07 13 18.	15		HERBIG-HARO OBJECT
SG 3.062	05 35 57.	+ 27 58	10800		DIFFUSE EMISSION NEBULA

OBJECT NAME	RIGHT ASCEN.	DECLINATION	DIAM.	MAGN.	TYPE OF OBJECT
SL 597	05 35 58.	- 70 16	25		STAR CLUSTER IN LMC
LBN 0894	05 36	+ 04 10	720		BRIGHT NEBULA
LBN 0830	05 36	+ 23 20	360		BRIGHT NEBULA
LBN 0822	05 36	+ 28 00	11400		BRIGHT NEBULA
MIL 08	05 36	+ 28 00	10800		SUPERNOVA REMNANT
LBN 0818	05 36	+ 30 40	960		BRIGHT NEBULA
LBN 0803	05 36	+ 35 50	660		BRIGHT NEBULA
KHAV 128	05 36	+ 52 08	4380		DARK NEBULA
LBN 0928	05 36	- 00 10	720		BRIGHT NEBULA
LBN 1002	05 36	- 10 20	1920		BRIGHT NEBULA
MRSL 173+02/2	05 36 00.	+ 35 54	720		HII REGION
LDN 1518	05 36 00.	+ 37 00	8700		DARK NEBULA
ZWG 348.003	05 36 00.	+ 79 34		15.7	GALAXY
ZWG 347.019	05 36 00.	+ 79 34		15.7	GALAXY
LB 01782	05 36 00.	- 48 50		13.8	FAINT BLUE STAR
LB 01783	05 36 00.	- 49 29		12.5	FAINT BLUE STAR
CM 25	05 36 00.	- 69 30			HII REGION IN LMC
VDB.66N 047	05 36 01.	+ 23 18	612		REFLECTION NEBULA
LH 086	05 36 03.	- 67 30	300		STELLAR ASSN. IN LMC
HSE 363	05 36 04.	- 70 23			STAR CLUSTER IN LMC
DG 068	05 36 06.	+ 23 18	300		REFLECTION NEBULA
YM 30	05 36 06.	- 02 38	7680		SYMMETRIC GALACTIC NEBULA
DG 069	05 36 06.	- 07 06	600		REFLECTION NEBULA
HN 0065	05 36 07.	- 07 06			NEBULA
CED 055K	05 36 07.	- 07 06	660		DIFFUSE GALACTIC NEBULA
IC 0430	05 36 07.	- 07 06 35.			NONSTELLAR OBJECT
SS 45	05 36 08.	+ 23 18			DIFFUSE GALACTIC NEBULA
RNGC 2042	05 36 09.	- 68 56			CLUSTER/NEBULOSITY IN LMC
RNGC 2047	05 36 09.	- 70 13			OPEN CLUSTER IN LMC
UGC 03332	05 36 12.	+ 15 33	90	16.5	GALAXY S
UGC 03333	05 36 12.	+ 49 18	66	17.	GALAXY
SL 604	05 36 12.	- 62 57	20		STAR CLUSTER IN LMC
LW 248	05 36 13.	- 65 18			STAR CLUSTER IN LMC
SL 598	05 36 13.	- 69 37	15		STAR CLUSTER IN LMC
SL 599	05 36 14.	- 69 54	250		STAR CLUSTER IN LMC
HSE 364	05 36 14.	- 71 08			STAR CLUSTER IN LMC
SL 601	05 36 15.	- 68 56	70		STAR CLUSTER IN LMC
SL 600	05 36 15.	- 70 13	15		STAR CLUSTER IN LMC
RNGC 2040	05 36 16.	- 67 35			CLUSTER/NEBULOSITY IN LMC
LH 088	05 36 16.	- 67 36	120		STELLAR ASSN. IN LMC
RNGC 2044	05 36 17.	- 69 13			CLUSTER/NEBULOSITY IN LMC
SL 602	05 36 17.	- 69 13	135		STAR CLUSTER IN LMC
MRSL 184-04/1	05 36 18.	+ 23 16	360		HII REGION
LH120-N059B	05 36 18.	- 67 35	118		EMISSION NEBULA IN LMC
RNGC 2048	05 36 19.	- 69 41			CLUSTER/NEBULOSITY IN LMC
LH 087	05 36 19.	- 69 41	420		STELLAR ASSN. IN LMC
LB 03509	05 36 20.	- 00 38 30.		17.3	FAINT BLUE STAR
SG 3.064	05 36 23.	+ 29 05			DIFFUSE EMISSION NEBULA
DG 070	05 36 24.	+ 04 05	1230		REFLECTION NEBULA
LB 03510	05 36 24.	- 00 41 12.		16.9	FAINT BLUE STAR
LW 251	05 36 24.	- 62 57			STAR CLUSTER IN LMC
LH120-N154A	05 36 24.	- 69 40	80		EMISSION NEBULA IN LMC
RNGC 2041	05 36 25.	- 67 00		10.5	OPEN CLUSTER IN LMC
SL 605	05 36 25.	- 67 00	90		STAR CLUSTER IN LMC
LB 03511	05 36 26.	- 01 06 12.		19.2	FAINT BLUE STAR
LB 03512	05 36 26.	- 01 24 36.		18.9	FAINT BLUE STAR
WS 35	05 36 27.	- 67 19 34.		15.76	PLANETARY NEB. IN LMC
LH 089	05 36 28.	- 68 59	540		STELLAR ASSN. IN LMC
LW 250	05 36 29.	- 64 26			STAR CLUSTER IN LMC
LH 090	05 36 29.	- 69 13	240		STELLAR ASSN. IN LMC
LB 01784	05 36 30.	- 49 21		13.0	FAINT BLUE STAR
LH120-N066	05 36 30.	- 67 19			EMISSION NEBULA IN LMC
CM 26	05 36 30.	- 69 33			HII REGION IN LMC
SC 0536-6138.9	05 36 31.	- 61 37 11.	12		NEBULA
VDB.66N 049	05 36 35.	+ 04 08	2520		REFLECTION NEBULA
ZWG 329.008	05 36 36.	+ 69 21		12.2	GALAXY
UGC 03334	05 36 36.	+ 69 21	276	12.2	GALAXY Sb
LH120-N211	05 36 36.	- 73 54			EMISSION NEBULA IN LMC
SEY 008	05 36 38.	+ 74 55 56.		15.3	FAINT GALAXY
LB 03513	05 36 38.	- 00 44 24.		18.6	FAINT BLUE STAR
SL 606	05 36 38.	- 68 35	20		STAR CLUSTER IN LMC
RNGC 1961	05 36 39.	+ 69 21		11.5	GALAXY
SL 607	05 36 39.	- 68 50	60		STAR CLUSTER IN LMC
RNGC 2043	05 36 39.	- 70 07			NON-EXISTENT OBJECT
WS 34	05 36 39.	- 73 57 04.		16.48	PLANETARY NEB. IN LMC
IC 2133	05 36 40.	+ 69 22			MAY NOT EXIST
HOD.72 45	05 36 40.	- 69 01	300		DARK NEBULA IN LMC
MCG+12-06-007	05 36 42.	+ 69 21	234	11.6	GALAXY
ZWG 348.004	05 36 42.	+ 79 13		15.4	GALAXY
ZWG 347.020	05 36 42.	+ 79 13		15.4	GALAXY
UGC 03335	05 36 42.	+ 79 13	72	15.4	GALAXY S0
LH 091	05 36 42.	- 66 29	120		STELLAR ASSN. IN LMC
VMT 07	05 36 45.	+ 27 44 30.	12000		SUPERNOVA REMNANT
LH 092	05 36 45.	- 67 29	150		STELLAR ASSN. IN LMC
MCG+08-11-002	05 36 48.	+ 49 40	48	16.	GALAXY
ARP 184	05 36 48.	+ 69 24			PECULIAR GALAXY
RNGC 2051	05 36 50.	- 71 02			OPEN CLUSTER IN LMC
SL 608	05 36 50.	- 71 02	40		STAR CLUSTER IN LMC
MCG-06-13-006	05 36 54.	- 34 24	60	15.5	GALAXY
RNGC 2050	05 36 54.	- 69 24			CLUSTER/NEBULOSITY IN LMC
SL 609	05 36 54.	- 69 24	135		STAR CLUSTER IN LMC
LW 253	05 36 58.	- 64 18			STAR CLUSTER IN LMC
LH 095	05 36 59.	- 66 23	90		STELLAR ASSN. IN LMC
KHAV 129	05 37	+ 11 14	3680		DARK NEBULA
LBN 0958	05 37	- 03 10	2700		BRIGHT NEBULA
LBN 0965	05 37	- 03 40	300		BRIGHT NEBULA
LBN 0967	05 37	- 04 00	720		BRIGHT NEBULA
KHAV 130	05 37	- 07 58	8050		DARK NEBULA
LBN 1007	05 37	- 11 50	1320		BRIGHT NEBULA
LDN 1579	05 37 00.	+ 13 45	2280		DARK NEBULA
MCG+08-11-003	05 37 00.	+ 50 10	54	16.	GALAXY
ZWG 362.010	05 37 00.	+ 85 54		15.6	GALAXY
ZWG 361.012	05 37 00.	+ 85 54		15.6	GALAXY
UGC 03336	05 37 00.	+ 85 54	102	15.6	GALAXY SBb
DG 071	05 37 00.	- 00 36	120		REFLECTION NEBULA
LB 05124	05 37 00.	- 08 10		19.6	FAINT BLUE STAR
MCG-02-15-007	05 37 00.	- 11 40	36	16.	GALAXY
CM 35	05 37 00.	- 67 54			HII REGION IN LMC
CM 41	05 37 00.	- 68 06			HII REGION IN LMC
SL 610	05 37 00.	- 69 31	70		STAR CLUSTER IN LMC
HSE 365	05 37 00.	- 70 39			STAR CLUSTER IN LMC
LB 03514	05 37 05.	- 02 15 42.		18.9	FAINT BLUE STAR
HSE 366	05 37 05.	- 67 54			STAR CLUSTER IN LMC
RNGC 2045	05 37 06.	+ 12 49			NON-EXISTENT OBJECT
ZWG 329.009	05 37 06.	+ 69 13		15.6	GALAXY
LB 05125	05 37 06.	- 06 07		15.2	FAINT BLUE STAR
MCG-03-15-007	05 37 06.	- 17 03	72	15.	GALAXY
LH120-N064B	05 37 06.	- 66 21	350		EMISSION NEBULA IN LMC
LH120-N064A	05 37 06.	- 66 22	40		EMISSION NEBULA IN LMC
LH 093	05 37 06.	- 69 25	240		STELLAR ASSN. IN LMC
LH 094	05 37 06.	- 69 31	150		STELLAR ASSN. IN LMC
RNGC 2056	05 37 06.	- 70 42			OPEN CLUSTER IN LMC
SL 611	05 37 06.	- 70 42	50		STAR CLUSTER IN LMC
RNGC 2059	05 37 09.	- 70 08			OPEN CLUSTER IN LMC
RNGC 2058	05 37 09.	- 70 10		12.0	OPEN CLUSTER IN LMC
HOD.72 46	05 37 10.	- 69 08	120		DARK NEBULA IN LMC
HSE 367	05 37 11.	- 66 24			STAR CLUSTER IN LMC
UGC 03337	05 37 12.	+ 61 01	78	16.5	GALAXY S
ZWG 329.010	05 37 12.	+ 69 52		15.1	GALAXY
KARA.73B 0160	05 37 12.	+ 69 52	42	15.1	ISOLATED GALAXY S
RNGC 2017	05 37 12.	- 17 53			NON-EXISTENT OBJECT
LH120-N068	05 37 12.	- 68 15	21		EMISSION NEBULA IN LMC
SCHO 0156	05 37 13.	+ 08 58 30.	440		ISOLATED DARK CLOUD
HOD.72 47	05 37 13.	- 69 46	720		DARK NEBULA IN LMC
SL 615	05 37 15.	- 68 45	20		STAR CLUSTER IN LMC
HOD.72 48	05 37 15.	- 70 03	480		DARK NEBULA IN LMC
SL 613	05 37 15.	- 70 08	20		STAR CLUSTER IN LMC
SL 614	05 37 15.	- 70 10	115		STAR CLUSTER IN LMC
SL 603	05 37 15.	- 76 01	15		STAR CLUSTER IN LMC
RNGC 2057	05 37 16.	- 70 17			OPEN CLUSTER IN LMC
LW 252	05 37 16.	- 73 20			STAR CLUSTER IN LMC
LB 03515	05 37 17.	- 00 56 36.		18.2	FAINT BLUE STAR
MRSL 159+11/1	05 37 18.	+ 52 10	4200		HII REGION
LB 01785	05 37 18.	- 46 22		13.7	FAINT BLUE STAR
LH120-N064C	05 37 18.	- 66 17	120		EMISSION NEBULA IN LMC
HSE 368	05 37 18.	- 66 29			STAR CLUSTER IN LMC
LH120-N065	05 37 18.	- 66 38	760		EMISSION NEBULA IN LMC
BC PKS0537-441	05 37 20.5	- 44 06 40.		15.47	QUASI-STELLAR OBJECT
SHB 090	05 37 20.5	- 44 06 40.		15.5	QUASI-STELLAR OBJECT
SL 616	05 37 22.	- 70 17	35		STAR CLUSTER IN LMC
ABC 0546	05 37 23.	+ 66 27		17.3	RICH CLUSTER OF GALAXIES
ZWG 348.005	05 37 24.	+ 75 36		15.3	GALAXY
ZWG 347.021	05 37 24.	+ 75 36		15.3	GALAXY
KARA.73B 0161	05 37 24.	+ 75 36	36	15.3	ISOLATED GALAXY S0
RNGC 2055	05 37 24.	- 69 26			CLUSTER/NEBULOSITY IN LMC
LH 096	05 37 24.	- 69 26	1020		STELLAR ASSN. IN LMC
RNGC 2053	05 37 27.	- 67 26			OPEN CLUSTER IN LMC
SL 617	05 37 27.	- 71 10	40		STAR CLUSTER IN LMC
SL 618	05 37 28.	- 71 23	15		STAR CLUSTER IN LMC
SL 612	05 37 28.	- 73 20	35		STAR CLUSTER IN LMC
MRSL 173+02/3	05 37 30.	+ 35 40	11		HII REGION
7ZW 043	05 37 30.	+ 64 03			COMPACT GALAXY
MCG-05-14-009	05 37 30.	- 29 25	84	15.	GALAXY
CM 36	05 37 30.	- 67 12			HII REGION IN LMC
LB 03516	05 37 31.	- 01 27 30.		19.5	FAINT BLUE STAR
VDB.66N 050	05 37 32.	- 01 28	372		REFLECTION NEBULA
SL 621	05 37 32.	- 69 55	15		STAR CLUSTER IN LMC
LW 249	05 37 32.	- 75 59			STAR CLUSTER IN LMC
SL 623	05 37 33.	- 67 26	60		STAR CLUSTER IN LMC
SL 619	05 37 33.	- 72 18	55		STAR CLUSTER IN LMC
LW 257	05 37 34.	- 64 13			STAR CLUSTER IN LMC
SG 3.066	05 37 35.	+ 33 02	2400		DIFFUSE EMISSION NEBULA
WS 36	05 37 35.	- 71 55 01.		16.72	PLANETARY NEB. IN LMC
MCG+02-15-001	05 37 36.	+ 11 27	36	17.	GALAXY
SL 622	05 37 38.	- 70 00	20		STAR CLUSTER IN LMC
LW 254	05 37 39.	- 72 17			STAR CLUSTER IN LMC
MIN.46 02	05 37 40.	+ 35 50			DIFFUSE NEBULA
SG 3.065	05 37 40.	+ 35 50	360		DIFFUSE EMISSION NEBULA
HOD.72 49	05 37 41.	- 69 19	240		DARK NEBULA IN LMC
MRSL 173+02/4	05 37 42.	+ 35 50	600		HII REGION
HSE 369	05 37 42.	- 69 32			STAR CLUSTER IN LMC
LH120-N155	05 37 42.	- 69 47	51		EMISSION NEBULA IN LMC
LH120-N212	05 37 42.	- 71 54			EMISSION NEBULA IN LMC
IC 0431	05 37 43.	- 01 29			NEBULA
RNGC 2052	05 37 43.	- 69 47			DIFFUSE NEBULA IN LMC
HN 0061	05 37 44.	- 01 28			NEBULA
CED 055L	05 37 44.	- 01 29	300		DIFFUSE GALACTIC NEBULA
UGC 03338	05 37 48.	+ 16 27	66	16.5	GALAXY S
MCG+11-07-012	05 37 48.	+ 67 15	42	16.	GALAXY
ZWG 329.011	05 37 48.	+ 69 25		14.8	GALAXY
DG 072	05 37 48.	- 01 29	360		REFLECTION NEBULA
SL 620	05 37 48.	- 74 26	80		STAR CLUSTER IN LMC
LB 03517	05 37 52.	- 00 39 48.		18.3	FAINT BLUE STAR
RNGC 2065	05 37 52.	- 70 15		11.0	OPEN CLUSTER IN LMC
ACK 170+04.1	05 37 54.	+ 39 14			PLANETARY NEBULA
LB 05307	05 37 54.	- 07 10		20.0	FAINT BLUE STAR
LW 255	05 37 54.	- 74 26			STAR CLUSTER IN LMC
SG 3.067	05 37 56.	+ 27 02			DIFFUSE EMISSION NEBULA
SS 46	05 37 56.	+ 35 05			DIFFUSE GALACTIC NEBULA
SL 624	05 37 56.	- 71 08	25		STAR CLUSTER IN LMC
SL 626	05 37 58.	- 70 15	110		STAR CLUSTER IN LMC
RNGC 2060	05 37 59.	- 69 12			CLUSTER/NEBULOSITY IN LMC
LH 097	05 37 59.	- 69 22	240		STELLAR ASSN. IN LMC
KHAV 133	05 38	+ 09 56	6610		DARK NEBULA
KHAV 132	05 38	+ 14 14	7600		DARK NEBULA
LBN 0837	05 38	+ 20 19	180		BRIGHT NEBULA
LBN 0808	05 38	+ 35 50	420		BRIGHT NEBULA
LBN 0927	05 38	- 00 36	120		BRIGHT NEBULA
LBN 0925	05 38	- 00 40	120		BRIGHT NEBULA
LBN 0924	05 38	- 00 45	720		BRIGHT NEBULA
LBN 0944	05 38	- 01 33	360		BRIGHT NEBULA
LBN 0992	05 38	- 09 25	420		BRIGHT NEBULA
LDN 1585	05 38 00.	+ 13 20	960		DARK NEBULA
OCL 0449	05 38 00.	+ 32 22	360	15.	OPEN STAR CLUSTER
ZWG 307.020	05 38 00.	+ 67 15		15.6	GALAXY
MCG+11-07-013	05 38 00.	+ 68 57	60	16.	GALAXY
MCG+14-03-012	05 38 00.	+ 83 24	30	16.	GALAXY
MCG+14-03-013	05 38 00.	+ 85 51	39	17.	GALAXY
MRSL 206-16/1	05 38 00.	- 02 28	7200		HII REGION
MCG-04-14-008	05 38 00.	- 22 02	66	15.	GALAXY
LH120-N157	05 38 00.	- 69 08	1849		EMISSION NEBULA IN LMC
LH120-N157B	05 38 00.	- 69 12	200		EMISSION NEBULA IN LMC
CM 27	05 38 00.	- 69 24			HII REGION IN LMC
HSE 370	05 38 00.	- 69 29			STAR CLUSTER IN LMC
LH 098	05 38 00.	- 69 29	120		STELLAR ASSN. IN LMC
CM 28	05 38 00.	- 69 42			HII REGION IN LMC
LB 03463	05 38 00.	- 76 33		12.8	FAINT BLUE STAR
PK193-09.1	05 38 01.	+ 12 16			PLANETARY NEBULA
RNGC 2066	05 38 03.	- 70 12			OPEN CLUSTER IN LMC
ZWG 307.021	05 38 06.	+ 68 55		14.9	GALAXY
DG 073	05 38 06.	- 00 30	180		REFLECTION NEBULA
MCG-03-15-008	05 38 06.	- 17 52 30.	30	15.5	GALAXY
HSE 371	05 38 06.	- 69 23			STAR CLUSTER IN LMC
LH120-N156	05 38 06.	- 69 36	36		EMISSION NEBULA IN LMC
MCG+12-06-008	05 38 09.	+ 69 18	102	15.	GALAXY
HOD.72 50	05 38 09.	- 68 51	420		DARK NEBULA IN LMC

OBJECT NAME	RIGHT ASCEN.	DECLINATION	DIAM.	MAGN.	TYPE OF OBJECT
SL 627	05 38 09.	- 70 12	35		STAR CLUSTER IN LMC
SL 625	05 38 09.	- 73 06	20		STAR CLUSTER IN LMC
LH 099	05 38 11.	- 69 12	180		STELLAR ASSN. IN LMC
UGC 03339	05 38 12.	+ 76 53	66	16.5	GALAXY Sc
MCG-02-15-008	05 38 12.	- 11 40	21	14.5	GALAXY
IC 0433	05 38 12.	- 11 40 46.			NONSTELLAR OBJECT
LW 256	05 38 15.	- 73 06			STAR CLUSTER IN LMC
HSE 373	05 38 20.	- 67 08			STAR CLUSTER IN LMC
B 033	05 38 22.	- 02 29	240		DARK OBJECT
IC 0432	05 38 23.	- 01 31			NEBULA
ISS 0207	05 38 24.	+ 44 47	1541		STELLAR RING
7ZW 044	05 38 24.	+ 62 20			COMPACT GALAXY
LB 05126	05 38 24.	- 06 23		16.8	FAINT BLUE STAR
VDB.66N 051	05 38 25.	- 01 27	612		REFLECTION NEBULA
SCHO 0157	05 38 26.	+ 40 15 00.	300		ISOLATED DARK CLOUD
HN 0060	05 38 26.	- 01 30			NEBULA
CED 055M	05 38 26.	- 01 31	480		DIFFUSE GALACTIC NEBULA
SG 3.069	05 38 26.	- 01 31	360		DIFFUSE EMISSION NEBULA
HSE 374	05 38 26.	- 68 36			STAR CLUSTER IN LMC
SL 628	05 38 27.	- 67 21	25		STAR CLUSTER IN LMC
HSE 372	05 38 29.	- 70 25			STAR CLUSTER IN LMC
MCG+12-06-009	05 38 30.	+ 69 04	51	16.	GALAXY
DG 074	05 38 30.	- 01 31	480		REFLECTION NEBULA
LDN 1635	05 38 30.	- 03 10	360		DARK NEBULA
CED 055N	05 38 31.	- 02 26	3600		DIFFUSE GALACTIC NEBULA
SG 3.070	05 38 31.	- 02 26	6600		DIFFUSE EMISSION NEBULA
SL 629	05 38 33.	- 68 48	15		STAR CLUSTER IN LMC
ST 37	05 38 34.	- 02 42	4200		EMISSION NEBULA
LDN 1525	05 38 36.	+ 36 18	300		DARK NEBULA
MCG-02-15-009	05 38 36.	- 13 50	54	15.5	GALAXY
HN 0062	05 38 37.	- 02 25			NEBULA
LB 03518	05 38 39.	- 02 47 24.		17.7	FAINT BLUE STAR
HOD.72 51	05 38 40.	- 69 01	420		DARK NEBULA IN LMC
RNGC 2072	05 38 40.	- 70 16			OPEN CLUSTER IN LMC
MCG+13-05-004	05 38 42.	+ 79 36	78	15.	GALAXY
ZWG 348.006	05 38 42.	+ 79 37		14.7	GALAXY
ZWG 347.022	05 38 42.	+ 79 37		14.7	GALAXY
UGC 03340	05 38 42.	+ 79 37	84	14.7	GALAXY Sa-b
HOD.72 52	05 38 43.	- 69 38	240		DARK NEBULA IN LMC
BC 3CR147	05 38 43.49	+ 49 49 43.3		17.80	QUASI-STELLAR OBJECT
SHB 091	05 38 43.5	+ 49 49 43.		17.8	QUASI-STELLAR OBJECT
LB 03519	05 38 44.	- 01 32 12.		17.5	FAINT BLUE STAR
LW 260	05 38 46.	- 64 19			STAR CLUSTER IN LMC
SL 630	05 38 46.	- 70 16	15		STAR CLUSTER IN LMC
ST 36	05 38 48.	- 01 47	1500		EMISSION NEBULA
MCG-02-15-010	05 38 48.	- 13 01	42	16.	GALAXY
RNGC 2075	05 38 48.	- 70 42			CLUSTER/NEBULOSITY IN LMC
SL 631	05 38 48.	- 70 42	40		STAR CLUSTER IN LMC
MCG-04-14-009	05 38 51.	- 26 21 30.	36	15.5	GALAXY
RNGC 2070	05 38 52.	- 69 07		8.5	CLUSTER/NEBULOSITY IN LMC
RNGC 2069	05 38 52.	- 69 07			DIFFUSE NEBULA IN LMC
SL 633	05 38 52.	- 69 07	340		STAR CLUSTER IN LMC
LB 03520	05 38 53.	- 02 22 18.		17.4	FAINT BLUE STAR
MCG+12-06-010	05 38 54.	+ 69 10	84	15.	GALAXY
LDN 1636	05 38 54.	- 03 11	180		DARK NEBULA
LB 05127	05 38 54.	- 05 45		16.4	FAINT BLUE STAR
LH120-N067	05 38 54.	- 64 41	15		EMISSION NEBULA IN LMC
LH120-N157A	05 38 54.	- 69 06	1063		EMISSION NEBULA IN LMC
LH120-N213A	05 38 54.	- 70 42	48		EMISSION NEBULA IN LMC
SL 635	05 38 56.	- 65 29	15		STAR CLUSTER IN LMC
HSE 375	05 38 56.	- 71 07			STAR CLUSTER IN LMC
VDB.66N 053	05 38 57.	- 10 19	108		REFLECTION NEBULA
LW 261	05 38 58.	- 64 10			STAR CLUSTER IN LMC
KHAV 131	05 39	+ 32 32	1570		DARK NEBULA
LBN 0806	05 39	+ 36 10	2400		BRIGHT NEBULA
LBN 0946	05 39	- 01 35	600		BRIGHT NEBULA
LBN 0954	05 39	- 02 20	540		BRIGHT NEBULA
LBN 0953	05 39	- 02 20	3600		BRIGHT NEBULA
LBN 0962	05 39	- 03 10	1680		BRIGHT NEBULA
ZWG 422.003	05 39 00.	+ 06 40		15.7	GALAXY
UGC 03341	05 39 00.	+ 18 29	120	15.5	GALAXY SBa-b
ZWG 329.012	05 39 00.	+ 69 17		15.4	GALAXY
UGC 03342	05 39 00.	+ 69 17	108	15.4	GALAXY Sc
IC 0434	05 39 00.	- 01 54			HII REGION
LB 05128	05 39 00.	- 05 47		15.7	FAINT BLUE STAR
LH120-N213	05 39 00.	- 70 42	186		EMISSION NEBULA IN LMC
MRSL 173+03/1	05 39 06.	+ 36 11	2400		HII REGION
MCG+12-06-011	05 39 06.	+ 73 35 30.	15	16.	GALAXY
LH120-N158B	05 39 06.	- 69 25	26		EMISSION NEBULA IN LMC
CED 055O	05 39 07.	- 02 15	600		DIFFUSE GALACTIC NEBULA
VDB.66N 052	05 39 09.	- 02 13	516		REFLECTION NEBULA
SG 3.068	05 39 10.	+ 36 10	2400		DIFFUSE EMISSION NEBULA
LH 100	05 39 10.	- 69 07	540		STELLAR ASSN. IN LMC
ZWG 329.013	05 39 12.	+ 72 19		13.9	GALAXY
UGC 03343	05 39 12.	+ 72 19	150	13.9	GALAXY S
KARA.73B 0162	05 39 12.	+ 72 19	120	13.9	ISOLATED GALAXY S
ZWG 329.014	05 39 12.	+ 73 35		15.4	GALAXY
KARA.73B 0163	05 39 12.	+ 73 35	36	15.4	ISOLATED GALAXY S
SEY 009	05 39 12.	+ 73 35 44.		15.4	FAINT GALAXY
MCG-03-15-009	05 39 12.	- 18 20	54	14.5	GALAXY
SL 634	05 39 12.	- 72 44	20		STAR CLUSTER IN LMC
SG 3.072	05 39 13.	- 02 15	180		DIFFUSE EMISSION NEBULA
SEY 010	05 39 14.	+ 72 20 44.		14.1	FAINT GALAXY
RNGC 2023	05 39 14.	- 02 15			DIFFUSE NEBULA
SG 3.071	05 39 15.	+ 10	28800		DIFFUSE EMISSION NEBULA
SL 636	05 39 15.	- 70 01	35		STAR CLUSTER IN LMC
LB 03521	05 39 16.	+ 00 24 30.		17.6	FAINT BLUE STAR
SC 0538-5838.3	05 39 16.	- 58 36 46.	60		NEBULA
HOD.72 53	05 39 16.	- 70 13	1020		DARK NEBULA IN LMC
HN 0304	05 39 16.	- 74 49			NEBULA
LW 258	05 39 16.	- 74 49			STAR CLUSTER IN LMC
IC 2146	05 39 17.	- 74 48			OPEN CLUSTER
MCG-06-13-007	05 39 18.	- 35 44	180	15.	GALAXY
CED 055P	05 39 19.	- 01 52	1860		DIFFUSE GALACTIC NEBULA
SG 3.073	05 39 19.	- 01 52	1800		DIFFUSE EMISSION NEBULA
RNGC 2022	05 39 20.	+ 09 03		13.0	PLANETARY NEBULA
HSE 377	05 39 20.	- 67 05			STAR CLUSTER IN LMC
SL 632	05 39 22.	- 74 49	135		STAR CLUSTER IN LMC
PK196-10.1	05 39 22.19	+ 09 03 48.2	30	12.8	PLANETARY NEBULA
MRSL 206-16/2	05 39 24.	- 01 52	1980		HII REGION
RNGC 2074	05 39 24.	- 69 32		8.5	CLUSTER/NEBULOSITY IN LMC
SL 637	05 39 24.	- 69 32	205		STAR CLUSTER IN LMC
LW 259	05 39 24.	- 72 44			STAR CLUSTER IN LMC
RNGC 2024	05 39 26.	- 01 52			DIFFUSE NEBULA
VDB.66N 054	05 39 26.	- 06 16	72		REFLECTION NEBULA
HSE 378	05 39 26.	- 68 32			STAR CLUSTER IN LMC
LW 262	05 39 27.	- 65 39			STAR CLUSTER IN LMC
SL 638	05 39 27.	- 65 40	40		STAR CLUSTER IN LMC
HSE 376	05 39 29.	- 70 30			STAR CLUSTER IN LMC
UGC 03344	05 39 30.	+ 60 09	162	14.7	GALAXY Sb
ZWG 329.015	05 39 30.	+ 69 09		14.7	GALAXY
LH 101	05 39 30.	- 69 31	300		STELLAR ASSN. IN LMC
LH120-N158C	05 39 30.	- 69 32	217		EMISSION NEBULA IN LMC
LB 03522	05 39 32.	+ 01 06 00.		17.8	FAINT BLUE STAR
MCG-01-15-003	05 39 33.	- 06 17	60	12.	GALAXY
MCG+12-06-013	05 39 36.	+ 69 05	39	15.	GALAXY
MCG+12-06-012	05 39 36.	+ 72 20	108	14.	GALAXY
LB 05129	05 39 36.	- 08 16		18.5	FAINT BLUE STAR
MCG-04-14-010	05 39 36.	- 26 50	48	15.	GALAXY
HSE 379	05 39 39.	- 71 19			STAR CLUSTER IN LMC
DG 075	05 39 42.	- 08 09	480		REFLECTION NEBULA
HOD.72 55	05 39 42.	- 69 28	120		DARK NEBULA IN LMC
LH120-N158D	05 39 42.	- 69 34	40		EMISSION NEBULA IN LMC
HOD.72 54	05 39 44.	- 69 55	1260		DARK NEBULA IN LMC
SG 3.074	05 39 46.	+ 10 35	11880		DIFFUSE EMISSION NEBULA
MCG-04-14-011	05 39 48.	- 22 58	168	13.5	GALAXY
LH120-N171B	05 39 48.	- 70 14	70		EMISSION NEBULA IN LMC
HSE 380	05 39 49.	- 65 02			STAR CLUSTER IN LMC
MCG-04-14-013	05 39 54.	- 26 09	15	15.5	GALAXY
MCG-04-14-012	05 39 54.	- 26 10 30.	18	15.5	GALAXY
LH120-N158	05 39 54.	- 69 28	759		EMISSION NEBULA IN LMC
LH120-N159H	05 39 54.	- 69 48			EMISSION NEBULA IN LMC
LH120-N171A	05 39 54.	- 70 13	21		EMISSION NEBULA IN LMC
SCHO 0158	05 39 57.	+ 40 21 42.	530		ISOLATED DARK CLOUD
VDB.66N 055	05 39 59.	- 08 06	108		REFLECTION NEBULA
SL 639	05 39 59.	- 69 14	20		STAR CLUSTER IN LMC
LBN 0751	05 40	+ 52 00	5400		BRIGHT NEBULA
KHAV 134	05 40	- 05 22	8770		DARK NEBULA
LBN 1010	05 40	- 13 00	6300		BRIGHT NEBULA
LDN 1460	05 40 00.	+ 52 20	1680		DARK NEBULA
7ZW 045	05 40 00.	+ 69 02			COMPACT GALAXY
UGC 03345	05 40 00.	+ 85 49	60	16.5	GALAXY S
LB 05308	05 40 00.	- 06 04		17.0	FAINT BLUE STAR
LDN 1647	05 40 00.	- 10 00	8340		DARK NEBULA
MCG-04-14-014	05 40 00.	- 26 13 30.	15	15.	GALAXY
CM 37	05 40 00.	- 67 51			HII REGION IN LMC
LH120-N160D	05 40 00.	- 69 40	58		EMISSION NEBULA IN LMC
LH120-N159J	05 40 00.	- 69 44			EMISSION NEBULA IN LMC
LH120-N159K	05 40 00.	- 69 47	25		EMISSION NEBULA IN LMC
RNGC 2062	05 40 01.	- 66 55			OPEN CLUSTER IN LMC
SL 640	05 40 01.	- 66 55	35		STAR CLUSTER IN LMC
HUB C12	05 40 02.	- 00 27			DIFFUSE NEBULA
LH 102	05 40 03.	- 67 25	540		STELLAR ASSN. IN LMC
ZWG 259.001	05 40 06.	+ 51 11		15.5	GALAXY
UGC 03346	05 40 06.	+ 51 11	90	15.5	GALAXY S-IRR
SCHO 0159	05 40 06.	- 02 23 18.	590		ISOLATED DARK CLOUD
DG 076	05 40 06.	- 08 07	480		REFLECTION NEBULA
MCG-04-14-015	05 40 06.	- 25 34	30	16.	GALAXY
LH120-N159F	05 40 06.	- 69 46	48		EMISSION NEBULA IN LMC
LH120-N159A	05 40 06.	- 69 47	62		EMISSION NEBULA IN LMC
RNGC 2026	05 40 07.	+ 20 05			NON-EXISTENT OBJECT
RNGC 2079	05 40 07.	- 69 47			CLUSTER/NEBULOSITY IN LMC
LB 03523	05 40 08.	+ 00 56 24.		19.2	FAINT BLUE STAR
SL 643	05 40 08.	- 67 04	35		STAR CLUSTER IN LMC
LW 266	05 40 10.	- 64 20			STAR CLUSTER IN LMC
HOD.72 56	05 40 11.	- 69 14	240		DARK NEBULA IN LMC
RNGC 2013	05 40 12.	+ 55 46			NON-EXISTENT OBJECT
ZWG 329.016	05 40 12.	+ 69 02		15.1	GALAXY
UGC 03347	05 40 12.	+ 79 32	66	16.0	GALAXY SBb
MCG-04-14-016	05 40 12.	- 26 08	12	15.5	GALAXY
LH120-N160A	05 40 12.	- 69 40	94		EMISSION NEBULA IN LMC
LH120-N159I	05 40 12.	- 69 46			EMISSION NEBULA IN LMC
LH 103	05 40 13.	- 69 39	300		STELLAR ASSN. IN LMC
RNGC 2080	05 40 13.	- 69 40			CLUSTER/NEBULOSITY IN LMC
RNGC 2077	05 40 13.	- 69 40			CLUSTER/NEBULOSITY IN LMC
SL 641	05 40 13.	- 69 40	60		STAR CLUSTER IN LMC
B 034	05 40 14.	+ 32 37	1200		DARK OBJECT
LB 03524	05 40 14.	- 01 50 42.		18.6	FAINT BLUE STAR
LB 03525	05 40 15.	+ 00 49 24.		18.0	FAINT BLUE STAR
SG 3.075	05 40 15.	+ 28 15			DIFFUSE EMISSION NEBULA
LB 03526	05 40 17.	- 01 46 12.		16.8	FAINT BLUE STAR
UGC 03348	05 40 18.	+ 16 29	84	16.0	GALAXY S
MCG-04-14-017	05 40 18.	- 25 34	36	15.	GALAXY
RNGC 2084	05 40 19.	- 69 47			CLUSTER/NEBULOSITY IN LMC
RNGC 2083	05 40 19.	- 69 47			CLUSTER/NEBULOSITY IN LMC
RNGC 2078	05 40 19.	- 69 47			CLUSTER/NEBULOSITY IN LMC
SL 644	05 40 19.	- 69 47	205		STAR CLUSTER IN LMC
HSE 381	05 40 21.	- 70 09			STAR CLUSTER IN LMC
LB 03527	05 40 24.	+ 01 08 06.		18.2	FAINT BLUE STAR
MCG+09-10-001	05 40 24.	+ 51 10	72	15.	GALAXY
LB 05130	05 40 24.	- 07 13		18.5	FAINT BLUE STAR
LH120-N159D	05 40 24.	- 69 45	131		EMISSION NEBULA IN LMC
LH120-N159	05 40 24.	- 69 46	288		EMISSION NEBULA IN LMC
LH120-N159C	05 40 24.	- 69 47	140		EMISSION NEBULA IN LMC
CED 055Q	05 40 25.	- 02 20	2430		DIFFUSE GALACTIC NEBULA
IC 0435	05 40 27.	- 02 20	240	8.2	REFLECTION NEBULA
MCG-04-14-018	05 40 27.	- 21 03 30.	36	15.	GALAXY
LB 03528	05 40 30.	+ 00 52 00.		17.3	FAINT BLUE STAR
DG 077	05 40 30.	- 02 20	300		REFLECTION NEBULA
LH120-N069	05 40 30.	- 66 17			EMISSION NEBULA IN LMC
LH120-N161	05 40 30.	- 69 00	65		EMISSION NEBULA IN LMC
RNGC 2081	05 40 30.	- 69 26			CLUSTER/NEBULOSITY IN LMC
LH 104	05 40 30.	- 69 26	360		STELLAR ASSN. IN LMC
LH120-N159B	05 40 30.	- 69 45			EMISSION NEBULA IN LMC
LH120-N159E	05 40 30.	- 69 48	35		EMISSION NEBULA IN LMC
LH120-N159L	05 40 30.	- 69 50	62		EMISSION NEBULA IN LMC
HN 0059	05 40 31.	- 02 19			NEBULA
LH 105	05 40 31.	- 69 46	240		STELLAR ASSN. IN LMC
LW 264	05 40 32.	- 73 01			STAR CLUSTER IN LMC
LB 03529	05 40 33.	+ 00 31 48.		18.5	FAINT BLUE STAR
SL 646	05 40 33.	- 71 12	70		STAR CLUSTER IN LMC
MCG+12-06-014	05 40 36.	+ 69 04	51	15.	GALAXY
LB 03530	05 40 36.	- 02 16 30.		17.7	FAINT BLUE STAR
LH120-N158A	05 40 36.	- 69 24	53		EMISSION NEBULA IN LMC
LH120-N160B	05 40 36.	- 69 41	60		EMISSION NEBULA IN LMC
LH120-N172	05 40 36.	- 69 56	53		EMISSION NEBULA IN LMC
LH120-N214A	05 40 36.	- 71 10	35		EMISSION NEBULA IN LMC
LH120-N214B	05 40 36.	- 71 11	44		EMISSION NEBULA IN LMC
RNGC 2085	05 40 37.	- 69 41			CLUSTER/NEBULOSITY IN LMC
SL 645	05 40 38.	- 73 01	5		STAR CLUSTER IN LMC
SL 649	05 40 39.	- 63 49	15		STAR CLUSTER IN LMC
HSE 382	05 40 39.	- 67 28			STAR CLUSTER IN LMC
LW 263	05 40 40.	- 74 53			STAR CLUSTER IN LMC
SCHO 0160	05 40 41.	+ 40 11 54.	570		ISOLATED DARK CLOUD
PATH 1.182	05 40 41.	+ 45 01	14		NEBULA
HODG 05	05 40 41.	- 72 16	24		RED GLOBULAR CLSTR IN LMC

OBJECT NAME	RIGHT ASCEN.	DECLINATION	DIAM.	MAGN.	TYPE OF OBJECT
ZWG 307.022	05 40 42.	+ 64 00		15.3	GALAXY
LB 03464	05 40 42.	- 60 21		13.2	FAINT BLUE STAR
HOD.72 57	05 40 43.	- 69 43	60		DARK NEBULA IN LMC
HOD.72 58	05 40 44.	- 69 48	60		DARK NEBULA IN LMC
LB 03531	05 40 45.	+ 00 15 30.		17.5	FAINT BLUE STAR
VDB.66N 056	05 40 45.	+ 16 23	72		REFLECTION NEBULA
VDB.66N 057	05 40 45.	- 02 20	312		REFLECTION NEBULA
ZC 0540.8+6657	05 40 48.	+ 66 57	1210		CLUSTER OF GALAXIES
LH120-N160	05 40 48.	- 69 38	787		EMISSION NEBULA IN LMC
LH120-N160C	05 40 48.	- 69 41	44		EMISSION NEBULA IN LMC
LH120-N159G	05 40 48.	- 69 46	50		EMISSION NEBULA IN LMC
LH120-N173	05 40 48.	- 69 54	54		EMISSION NEBULA IN LMC
LH120-N214D	05 40 48.	- 71 12	90		EMISSION NEBULA IN LMC
LH120-N214E	05 40 48.	- 71 13	38		EMISSION NEBULA IN LMC
RNGC 2086	05 40 49.	- 69 41			CLUSTER/NEBULOSITY IN LMC
IC 2145	05 40 49.9	- 69 41 40.			HII REGION IN LMC
LW 269	05 40 51.	- 63 48			STAR CLUSTER IN LMC
SL 642	05 40 53.	- 75 36	20		STAR CLUSTER IN LMC
LW 265	05 40 53.	- 75 36			STAR CLUSTER IN LMC
LH120-N160E	05 40 54.	- 69 42	51		EMISSION NEBULA IN LMC
LH120-N174	05 40 54.	- 69 58	25		EMISSION NEBULA IN LMC
HSE 383	05 40 55.	- 70 48			STAR CLUSTER IN LMC
HN 0305	05 40 56.	- 75 36			NEBULA
IC 2148	05 40 56.	- 75 36			NONSTELLAR OBJECT
HN 0092	05 40 58.	- 69 42			NEBULA
KHAV 135	05 41	+ 06 13	4760		DARK NEBULA
MCG-03-15-010	05 41 00.	- 18 42	48	15.	GALAXY
MCG-05-14-010	05 41 00.	- 29 47	48	14.5	GALAXY
CM 29	05 41 00.	- 69 24			HII REGION IN LMC
SL 650	05 41 03.	- 67 21	25		STAR CLUSTER IN LMC
SL 648	05 41 03.	- 72 13	40		STAR CLUSTER IN LMC
LW 268	05 41 03.	- 72 13			STAR CLUSTER IN LMC
ZWG 329.017	05 41 06.	+ 69 02		14.4	GALAXY
UGC 03349	05 41 06.	+ 69 02	78	14.4	GALAXY Sa-b
SL 651	05 41 06.	- 68 08	60		STAR CLUSTER IN LMC
HSE 384	05 41 07.	- 69 44			STAR CLUSTER IN LMC
IC 2141	05 41 08.	- 51 03			NONSTELLAR OBJECT
HSE 386	05 41 09.	- 67 28			STAR CLUSTER IN LMC
HOD.72 60	05 41 10.	- 69 04	540		DARK NEBULA IN LMC
HOD.72 59	05 41 12.	- 69 27	240		DARK NEBULA IN LMC
RNGC 2088	05 41 12.	- 69 30			OPEN CLUSTER IN LMC
SL 652	05 41 12.	- 69 30	35		STAR CLUSTER IN LMC
LH120-N160F	05 41 12.	- 69 44	58		EMISSION NEBULA IN LMC
LH120-N175	05 41 12.	- 70 03	118		EMISSION NEBULA IN LMC
HSE 385	05 41 13.	- 69 35			STAR CLUSTER IN LMC
LH 106	05 41 13.	- 69 36	1080		STELLAR ASSN. IN LMC
HSE 387	05 41 15.	- 67 19			STAR CLUSTER IN LMC
ZC 0541.3+0101	05 41 18.	+ 01 01	6450		CLUSTER OF GALAXIES
LH120-N162	05 41 18.	- 69 17			EMISSION NEBULA IN LMC
RNGC 2091	05 41 18.	- 69 28			OPEN CLUSTER IN LMC
SL 653	05 41 18.	- 69 28	55		STAR CLUSTER IN LMC
LH120-N176	05 41 18.	- 70 11	41		EMISSION NEBULA IN LMC
LH120-N217	05 41 18.	- 70 29			EMISSION NEBULA IN LMC
LW 267	05 41 19.	- 75 14			STAR CLUSTER IN LMC
RNGC 2039	05 41 20.	+ 08 36			NON-EXISTENT OBJECT
SL 647	05 41 20.	- 75 15	55		STAR CLUSTER IN LMC
LH 107	05 41 21.	- 71 16	720		STELLAR ASSN. IN LMC
MCG-05-14-011	05 41 24.	- 30 08	84	14.	GALAXY
MCG-06-13-008	05 41 24.	- 34 37	60	14.5	GALAXY
RNGC 2049	05 41 26.	- 30 08		14.0	GALAXY
SL 654	05 41 26.	- 69 55	35		STAR CLUSTER IN LMC
ZWG 307.023	05 41 30.	+ 63 20		15.7	GALAXY
MCG-05-14-012	05 41 30.	- 27 41	36	15.	GALAXY
LH120-N218	05 41 30.	- 70 35	28		EMISSION NEBULA IN LMC
RNGC 2082	05 41 34.	- 64 20		13.0	GALAXY
UGC 03350	05 41 36.	+ 58 34	60	16.0	GALAXY Sa-b
MCG+10-09-001	05 41 36.	+ 58 34	57	16.	GALAXY
LB 05309	05 41 36.	- 06 42		19.3	FAINT BLUE STAR
MCG-05-14-013	05 41 36.	- 30 33	72	14.	GALAXY
LB 03465	05 41 36.	- 60 31		14.1	FAINT BLUE STAR
LH120-N177	05 41 36.	- 69 55	175		EMISSION NEBULA IN LMC
LH120-N216	05 41 36.	- 70 55	59		EMISSION NEBULA IN LMC
LH120-N214	05 41 36.	- 71 16	831		EMISSION NEBULA IN LMC
SL 655	05 41 37.	- 71 50	20		STAR CLUSTER IN LMC
HSE 388	05 41 40.	- 70 15			STAR CLUSTER IN LMC
HUB C13	05 41 41.	+ 00 53			DIFFUSE NEBULA
HOD.72 61	05 41 41.	- 69 21	240		DARK NEBULA IN LMC
UGC 03351	05 41 42.	+ 58 40	102	16.0	GALAXY Sa-b
MCG+10-09-002	05 41 42.	+ 58 40	90	15.	GALAXY
LB 03532	05 41 42.	- 02 30 54.		18.4	FAINT BLUE STAR
LB 05131	05 41 42.	- 08 12		17.8	FAINT BLUE STAR
MCG-03-15-011	05 41 42.	- 19 20	120	14.5	GALAXY
MCG-04-14-019	05 41 42.	- 26 29	36	15.	GALAXY
LH120-N219	05 41 42.	- 70 24	40		EMISSION NEBULA IN LMC
LB 03533	05 41 43.	+ 00 29 00.		17.7	FAINT BLUE STAR
LW 271	05 41 43.	- 71 50			STAR CLUSTER IN LMC
SL 656	05 41 46.	- 71 22	20		STAR CLUSTER IN LMC
LB 03534	05 41 48.	+ 00 51 12.		18.6	FAINT BLUE STAR
RNGC 2093	05 41 52.	- 68 57			CLUSTER/NEBULOSITY IN LMC
HSE 389	05 41 52.	- 70 14			STAR CLUSTER IN LMC
HSE 390	05 41 53.	- 69 13			STAR CLUSTER IN LMC
RNGC 2092	05 41 53.	- 69 15			OPEN CLUSTER IN LMC
UGC 03352	05 41 54.	+ 16 45	90	16.0	GALAXY SB
LB 03466	05 41 54.	- 75 17		13.7	FAINT BLUE STAR
LW 270	05 41 56.	- 75 20			STAR CLUSTER IN LMC
MCG+11-07-014	05 41 57.	+ 65 27	18	18.	GALAXY
SL 657	05 41 58.	- 68 57	60		STAR CLUSTER IN LMC
LW 272	05 41 59.	- 62 32			STAR CLUSTER IN LMC
CED 055E	05 42	+ 02			DIFFUSE GALACTIC NEBULA
LBN 0878	05 42	+ 09 10	1020		BRIGHT NEBULA
KHAV 137	05 42	+ 09 13	1410		DARK NEBULA
KHAV 136	05 42	+ 10 37	4380		DARK NEBULA
LBN 0700	05 42	+ 64 20	1800		BRIGHT NEBULA
LBN 1009	05 42	- 12 30	1800		BRIGHT NEBULA
LDN 1603	05 42 00.	+ 06 00	2460		DARK NEBULA
LDN 1602	05 42 00.	+ 06 00	2460		DARK NEBULA
LDN 1594	05 42 00.	+ 09 00	960		DARK NEBULA
7ZW 046	05 42 00.	+ 87 15			COMPACT GALAXY
LH120-N214G	05 42 00.	- 71 15	30		EMISSION NEBULA IN LMC
LH 108	05 42 01.	- 69 41	120		STELLAR ASSN. IN LMC
VDB.66N 058	05 42 04.	- 08 42	36		REFLECTION NEBULA
HSE 391	05 42 04.	- 70 16			STAR CLUSTER IN LMC
RNGC 2061	05 42 06.	- 33 59			NON-EXISTENT OBJECT
LH120-N214F	05 42 06.	- 71 16	47		EMISSION NEBULA IN LMC
LH120-N214H	05 42 06.	- 71 18	59		EMISSION NEBULA IN LMC
LH120-N215	05 42 06.	- 72 42			EMISSION NEBULA IN LMC
WS 37	05 42 08.	- 72 43 22.		17.0	PLANETARY NEB. IN LMC
SL 659	05 42 10.	- 68 57	25		STAR CLUSTER IN LMC
LH 109	05 42 10.	- 68 57	300		STELLAR ASSN. IN LMC
LB 01786	05 42 12.	- 57 14		14.8	FAINT BLUE STAR
CED 058	05 42 13.	+ 12 52			DIFFUSE GALACTIC NEBULA
SL 663	05 42 14.	- 65 23	70		STAR CLUSTER IN LMC
LW 273	05 42 14.	- 65 23			STAR CLUSTER IN LMC
SL 658	05 42 14.	- 71 06	25		STAR CLUSTER IN LMC
HOD.72 62	05 42 15.	- 70 07	300		DARK NEBULA IN LMC
LB 03536	05 42 17.	- 02 09 06.		20.2	FAINT BLUE STAR
LB 03535	05 42 17.	- 02 09 06.		17.6	FAINT BLUE STAR
ISS 0259	05 42 18.	+ 36 40	400		STELLAR RING
MCG+13-05-005	05 42 18.	+ 79 40	60	15.	GALAXY
LH120-N214C	05 42 18.	- 71 20	198		EMISSION NEBULA IN LMC
HSE 392	05 42 19.	- 66 46			STAR CLUSTER IN LMC
RNGC 2094	05 42 19.	- 68 23			OPEN CLUSTER IN LMC
BIGO 489	05 42 20.	- 10 01			NEBULA
RNGC 2103	05 42 22.	- 71 22			CLUSTER/NEBULOSITY IN LMC
LH 110	05 42 22.	- 71 22	180		STELLAR ASSN. IN LMC
LH 111	05 42 23.	- 69 14	360		STELLAR ASSN. IN LMC
RNGC 2100	05 42 23.	- 69 15		9.5	CLUSTER/NEBULOSITY IN LMC
SL 662	05 42 23.	- 69 15	135		STAR CLUSTER IN LMC
ISS 0208	05 42 24.	+ 39 17	271		STELLAR RING
RNGC 2054	05 42 25.	- 10 06			NON-EXISTENT OBJECT
RNGC 2096	05 42 26.	- 68 30			OPEN CLUSTER IN LMC
SL 664	05 42 26.	- 68 30	35		STAR CLUSTER IN LMC
HN 0003	05 42 28.	- 10 06			NEBULA
SL 660	05 42 28.	- 71 22	70		STAR CLUSTER IN LMC
SL 661	05 42 28.	- 71 28	35		STAR CLUSTER IN LMC
LB 03537	05 42 30.	- 02 28 36.		18.3	FAINT BLUE STAR
LB 05132	05 42 30.	- 07 24		16.5	FAINT BLUE STAR
HSE 396	05 42 32.	- 66 59			STAR CLUSTER IN LMC
ISS 0260	05 42 36.	+ 36 45	1676		STELLAR RING
SL 670	05 42 36.	- 62 52	15		STAR CLUSTER IN LMC
RNGC 2102	05 42 36.	- 69 31			OPEN CLUSTER IN LMC
SL 665	05 42 36.	- 69 31	25		STAR CLUSTER IN LMC
HSE 393	05 42 37.	- 70 48			STAR CLUSTER IN LMC
CED 059	05 42 38.	+ 09 03			DIFFUSE GALACTIC NEBULA
SCHO 0161	05 42 39.	+ 09 02 30.	990		ISOLATED DARK CLOUD
LW 275	05 42 39.	- 65 50			STAR CLUSTER IN LMC
LH 112	05 42 39.	- 67 21	300		STELLAR ASSN. IN LMC
HOD.72 63	05 42 40.	- 69 07	360		DARK NEBULA IN LMC
ZWG 348.007	05 42 42.	+ 79 41		14.8	GALAXY
ZWG 347.023	05 42 42.	+ 79 41		14.8	GALAXY
UGC 03353	05 42 42.	+ 79 41	126	14.8	GALAXY SB0
LB 05310	05 42 42.	- 07 03		19.9	FAINT BLUE STAR
MCG-05-14-014	05 42 42.	- 28 35	48	15.	GALAXY
SL 666	05 42 42.	- 68 13	40		STAR CLUSTER IN LMC
LB 03538	05 42 43.	- 01 32 48.		18.7	FAINT BLUE STAR
RNGC 2098	05 42 43.	- 68 18			CLUSTER/NEBULOSITY IN LMC
SL 667	05 42 43.	- 68 18	65		STAR CLUSTER IN LMC
HSE 394	05 42 43.	- 70 51			STAR CLUSTER IN LMC
LW 276	05 42 44.	- 63 39			STAR CLUSTER IN LMC
RNGC 2095	05 42 45.	- 67 21			OPEN CLUSTER IN LMC
SL 669	05 42 45.	- 67 21	30		STAR CLUSTER IN LMC
B 035	05 42 46.	+ 09 02	1200		DARK OBJECT
HSE 395	05 42 46.	- 71 23			STAR CLUSTER IN LMC
LW 278	05 42 47.	- 62 31			STAR CLUSTER IN LMC
ST 35	05 42 48.	+ 09 03	180		EMISSION NEBULA
LB 05133	05 42 48.	- 06 38		16.1	FAINT BLUE STAR
MCG-04-14-020	05 42 48.	- 25 07	36	15.	GALAXY
LB 01787	05 42 48.	- 48 46		14.2	FAINT BLUE STAR
LW 277	05 42 48.	- 62 51			STAR CLUSTER IN LMC
LB 03539	05 42 50.	+ 00 49 18.		18.0	FAINT BLUE STAR
LH120-N164	05 42 54.	- 69 05	310		EMISSION NEBULA IN LMC
SL 671	05 42 58.	- 67 41	40		STAR CLUSTER IN LMC
SL 668	05 42 59.	- 71 37	70		STAR CLUSTER IN LMC
LBN 0879	05 43	+ 09 02	120		BRIGHT NEBULA
LDN 1596	05 43 00.	+ 09 00	780		DARK NEBULA
ZWG 259.002	05 43 00.	+ 56 05		15.3	GALAXY
MCG+09-10-002	05 43 00.	+ 56 07	126	14.	GALAXY
LDN 1630	05 43 00.	- 01 00	9900		DARK NEBULA
LB 05311	05 43 00.	- 06 35		19.8	FAINT BLUE STAR
LB 01788	05 43 00.	- 51 28		13.5	FAINT BLUE STAR
CM 30	05 43 00.	- 69 12			HII REGION IN LMC
HSE 397	05 43 00.	- 70 37			STAR CLUSTER IN LMC
HSE 397A	05 43 04.	- 69 05			STAR CLUSTER IN LMC
LH 113	05 43 04.	- 69 05	120		STELLAR ASSN. IN LMC
LW 274	05 43 05.	- 71 37			STAR CLUSTER IN LMC
UGC 03354	05 43 06.	+ 56 05	132	15.3	GALAXY Sa-b
LH120-N178	05 43 06.	- 70 10			EMISSION NEBULA IN LMC
WS 38	05 43 08.	- 70 10 37.		15.92	PLANETARY NEB. IN LMC
HODG 06	05 43 10.	- 71 40	60		RED GLOBULAR CLSTR IN LMC
SL 674	05 43 11.	- 66 18	50		STAR CLUSTER IN LMC
SL 673	05 43 11.	- 67 54	70		STAR CLUSTER IN LMC
LH120-N165	05 43 12.	- 68 58	103		EMISSION NEBULA IN LMC
SL 675	05 43 16.	- 67 38	20		STAR CLUSTER IN LMC
HH 19	05 43 16.0	- 00 06 20.			HERBIG-HARO OBJECT
HSE 398	05 43 17.	- 70 27			STAR CLUSTER IN LMC
ZWG 232.001	05 43 18.	+ 50 50		15.6	GALAXY
UGC 03355	05 43 18.	+ 50 50	72	15.6	GALAXY (S0)
MCG+08-11-004	05 43 18.	+ 50 51	60	15.	GALAXY
MCG-04-14-021	05 43 18.	- 25 49	36	15.	GALAXY
MCG-05-14-015	05 43 18.	- 28 42	36	16.	GALAXY
RNGC 2087	05 43 20.	- 55 33			UNVERIFIED SOUTHERN OBJECT
SL 677	05 43 20.	- 63 24	20		STAR CLUSTER IN LMC
HH 20	05 43 21.5	- 00 04 14.	3		HERBIG-HARO OBJECT
HH 21	05 43 22.0	- 00 05 36.	4		HERBIG-HARO OBJECT
LH 114	05 43 23.	- 67 53	180		STELLAR ASSN. IN LMC
MCG-04-14-022	05 43 24.	- 25 57	54	15.	GALAXY
SL 672	05 43 26.	- 72 05	20		STAR CLUSTER IN LMC
SL 678	05 43 29.	- 66 14	20		STAR CLUSTER IN LMC
ZCG 127.01	05 43 30.	+ 58 46		18.3	RED COMPACT GALAXY
MCG-04-14-023	05 43 30.	- 25 34	30	15.	GALAXY
MCG-05-14-016	05 43 30.	- 28 17	60	16.	GALAXY
LH120-N070	05 43 30.	- 67 51	420		EMISSION NEBULA IN LMC
HH 26	05 43 31.1	- 00 15 42.			HERBIG-HARO OBJECT
LW 279	05 43 32.	- 72 05			STAR CLUSTER IN LMC
VV 162C	05 43 33.	- 25 56	12	15.5	INTERACTING GALAXY
VV 162B	05 43 33.	- 25 56	12	16.	INTERACTING GALAXY
VV 162A	05 43 33.	- 25 56	12	15.5	INTERACTING GALAXY
VV 162	05 43 33.	- 25 56	72		INTERACTING GALAXY
SL 680	05 43 33.	- 63 57	70		STAR CLUSTER IN LMC
HH 25	05 43 33.4	- 00 14 31.	8		HERBIG-HARO OBJECT
HSE 399	05 43 34.	- 70 19			STAR CLUSTER IN LMC
HH 24C	05 43 34.2	- 00 10 50.			HERBIG-HARO OBJECT
HH 24B	05 43 34.4	- 00 11 12.			HERBIG-HARO OBJECT
HH 24A	05 43 35.6	- 00 11 32.			HERBIG-HARO OBJECT
LB 05312	05 43 36.	- 07 12		18.5	FAINT BLUE STAR
LB 03467	05 43 36.	- 64 37		14.8	FAINT BLUE STAR

OBJECT NAME	RIGHT ASCEN.	DECLINATION	DIAM.	MAGN.	TYPE OF OBJECT
LH120-N072	05 43 36.	- 66 17	32		EMISSION NEBULA IN LMC
LH120-N163	05 43 36.	- 69 46	217		EMISSION NEBULA IN LMC
HH 24D	05 43 36.1	- 00 11 02.			HERBIG-HARO OBJECT
LW 280	05 43 38.	- 63 24			STAR CLUSTER IN LMC
HSE 400	05 43 38.	- 69 48			STAR CLUSTER IN LMC
HH 22	05 43 40.3	- 00 06 36.			HERBIG-HARO OBJECT
HH 23	05 43 40.9	- 00 04 37.	2		HERBIG-HARO OBJECT
ZCG 127.02	05 43 42.	+ 58 48		18.6	RED COMPACT GALAXY
ZWG 308.001	05 43 42.	+ 66 51		15.5	GALAXY
ZWG 307.024	05 43 42.	+ 66 51		15.5	GALAXY
MCG-04-14-024	05 43 42.	- 22 01 30.	24	13.5	GALAXY
RNGC 2097	05 43 42.	- 62 49			OPEN CLUSTER IN LMC
SL 682	05 43 42.	- 62 49	100		STAR CLUSTER IN LMC
SL 676	05 43 42.	- 70 35	25		STAR CLUSTER IN LMC
RNGC 2073	05 43 45.	- 22 01		13.0	GALAXY
CED 055S	05 43 46.	- 00 02	690		DIFFUSE GALACTIC NEBULA
PK184-02.1	05 43 46.1	+ 24 20 59.			PLANETARY NEBULA
HSE 401	05 43 47.	- 70 23			STAR CLUSTER IN LMC
DG 078	05 43 48.	- 00 01	300		REFLECTION NEBULA
LB 01789	05 43 48.	- 47 41		12.6	FAINT BLUE STAR
RNGC 2107	05 43 48.	- 70 39		11.5	OPEN CLUSTER IN LMC
SL 679	05 43 48.	- 70 39	90		STAR CLUSTER IN LMC
RNGC 2064	05 43 49.	- 00 02			DIFFUSE NEBULA
HH 27	05 43 49.4	- 00 14 45.	10		HERBIG-HARO OBJECT
LW 281	05 43 51.	- 63 57			STAR CLUSTER IN LMC
SL 681	05 43 51.	- 67 24	20		STAR CLUSTER IN LMC
7ZW 047	05 43 54.	+ 64 47			COMPACT GALAXY
LW 282	05 43 54.	- 62 49			STAR CLUSTER IN LMC
LH120-N071	05 43 54.	- 67 27	48		EMISSION NEBULA IN LMC
SL 685	05 43 56.	- 65 21	70		STAR CLUSTER IN LMC
CED 055T	05 43 58.	+ 00 05	450		DIFFUSE GALACTIC NEBULA
LBN 0939	05 44	+ 00 00	600		BRIGHT NEBULA
KHAV 138	05 44	- 01 11	15680		DARK NEBULA
KHAV 139	05 44	- 03 59	7570		DARK NEBULA
LDN 1627	05 44 00.	+ 00 00	960		DARK NEBULA
DG 079	05 44 00.	+ 00 06	480		REFLECTION NEBULA
LDN 1466	05 44 00.	+ 52 10	1380		DARK NEBULA
MCG+15-01-003	05 44 00.	+ 87 42 30.	39	16.	GALAXY
RNGC 2067	05 44 01.	+ 00 05			DIFFUSE NEBULA
RNGC 2063	05 44 02.	+ 08 46			NON-EXISTENT OBJECT
LB 03540	05 44 05.	- 01 49 36.		16.6	FAINT BLUE STAR
LH 115	05 44 05.	- 66 20	600		STELLAR ASSN. IN LMC
LW 283	05 44 08.	- 65 19			STAR CLUSTER IN LMC
SL 683	05 44 08.	- 69 49	15		STAR CLUSTER IN LMC
CED 055U	05 44 10.	+ 00 02	480		DIFFUSE GALACTIC NEBULA
VDB.66N 059	05 44 11.	+ 00 06	816		REFLECTION NEBULA
RNGC 2108	05 44 11.	- 69 14			GLOBULAR CLUSTER IN LMC
SL 686	05 44 11.	- 69 14	55		STAR CLUSTER IN LMC
HOD.72 64	05 44 11.	- 69 19	240		DARK NEBULA IN LMC
SL 684	05 44 11.	- 70 31	20		STAR CLUSTER IN LMC
DG 080	05 44 12.	+ 00 03	660		REFLECTION NEBULA
ISS 0363	05 44 12.	+ 16 51	291		STELLAR RING
UGC 03356	05 44 12.	+ 17 33	60	16.0	GALAXY Sb-c
MCG+05-14-002	05 44 12.	+ 30 45	42	15.	GALAXY
MCG-04-14-025	05 44 12.	- 23 30	54	15.	GALAXY
RNGC 2105	05 44 12.	- 66 27			OPEN CLUSTER IN LMC
SL 687	05 44 12.	- 66 27	50		STAR CLUSTER IN LMC
RNGC 2068	05 44 13.	+ 00 02			DIFFUSE NEBULA
HSE 402	05 44 19.	- 68 23			STAR CLUSTER IN LMC
ZC 0544.4+5036	05 44 24.	+ 50 36	11890		CLUSTER OF GALAXIES
LW 285	05 44 24.	- 64 57			STAR CLUSTER IN LMC
LW 284	05 44 25.	- 65 01			STAR CLUSTER IN LMC
SL 689	05 44 25.	- 65 02	25		STAR CLUSTER IN LMC
SL 688	05 44 26.	- 68 36	60		STAR CLUSTER IN LMC
HSE 403	05 44 27.	- 70 02			STAR CLUSTER IN LMC
MCG+08-11-005	05 44 30.	+ 50 17	48	16.	GALAXY
ZCG 127.03	05 44 30.	+ 58 43		18.8	RED COMPACT GALAXY
7ZW 048	05 44 30.	+ 68 47			COMPACT GALAXY
LH120-N073	05 44 30.	- 67 27	40		EMISSION NEBULA IN LMC
HSE 404	05 44 32.	- 67 12			STAR CLUSTER IN LMC
MCG+07-12-002	05 44 33.	+ 42 28	42	17.	GALAXY
SL 694	05 44 33.	- 63 42	15		STAR CLUSTER IN LMC
REIN 2.047	05 44 33.68	+ 00 16 39.3			NEBULA
REIN 2.048	05 44 33.92	+ 00 16 55.6			NEBULA
CED 055V	05 44 34.	+ 00 17	270		DIFFUSE GALACTIC NEBULA
RNGC 2076	05 44 35.	- 16 48		14.0	GALAXY
SL 690	05 44 35.	- 67 50	25		STAR CLUSTER IN LMC
DG 081	05 44 36.	+ 00 18	540		REFLECTION NEBULA
ISS 0364	05 44 36.	+ 20 48	139		STELLAR RING
MCG-03-15-012	05 44 36.	- 16 48 30.	96	14.	GALAXY
RNGC 2071	05 44 37.	+ 00 17			DIFFUSE NEBULA
RNGC 2109	05 44 38.	- 68 34			CLUSTER/NEBULOSITY IN LMC
IC 2143	05 44 41.	- 18 44 32.			NONSTELLAR OBJECT
SL 693	05 44 41.	- 67 46	25		STAR CLUSTER IN LMC
VDB.66N 060	05 44 42.	+ 00 16	516		REFLECTION NEBULA
ZCG 127.04	05 44 42.	+ 58 43		19.1	RED COMPACT GALAXY
MCG+12-06-015	05 44 42.	+ 74 17	84	16.	GALAXY
MCG-03-15-013	05 44 42.	- 18 47	96	14.	GALAXY
LH120-N166	05 44 42.	- 69 25	30		EMISSION NEBULA IN LMC
SEY 011	05 44 44.	+ 74 17 09.		15.0	FAINT GALAXY
DG 082	05 44 48.	+ 00 42	240		REFLECTION NEBULA
ZCG 127.07	05 44 48.	+ 58 40		19.0	RED COMPACT GALAXY
ZCG 127.06	05 44 48.	+ 58 41		18.9	RED COMPACT GALAXY
ZCG 127.05	05 44 48.	+ 58 42		18.7	RED COMPACT GALAXY
ZWG 284.001	05 44 48.	+ 60 39		15.7	GALAXY
LH120-N167	05 44 48.	- 69 23	64		EMISSION NEBULA IN LMC
SL 691	05 44 48.	- 70 40	15		STAR CLUSTER IN LMC
SL 692	05 44 48.	- 70 41	20		STAR CLUSTER IN LMC
MCG-04-14-026	05 44 51.	- 25 39 30.	12	15.5	GALAXY
HSE 405	05 44 51.	- 71 12			STAR CLUSTER IN LMC
SL 695	05 44 53.	- 67 53	20		STAR CLUSTER IN LMC
MCG+10-09-003	05 44 54.	+ 60 39	39	16.	GALAXY
ZWG 329.018	05 44 54.	+ 74 15		15.0	GALAXY
UGC 03357	05 44 54.	+ 74 15	102	15.0	GALAXY S0-a
SL 696	05 44 54.	- 64 49	15		STAR CLUSTER IN LMC
LW 287	05 44 57.	- 63 42			STAR CLUSTER IN LMC
LBN 0938	05 45	+ 00 12	360		BRIGHT NEBULA
LBN 0934	05 45	+ 01 00	4620		BRIGHT NEBULA
LBN 0933	05 45	+ 01 00	13500		BRIGHT NEBULA
KHAV 140	05 45	+ 12 37	7680		DARK NEBULA
LBN 0750	05 45	+ 52 30	3300		BRIGHT NEBULA
LBN 0989	05 45	- 07 20	2100		BRIGHT NEBULA
LDN 1467	05 45 00.	+ 52 10	660		DARK NEBULA
7ZW 049	05 45 00.	+ 58 33			COMPACT GALAXY
7ZW 050	05 45 00.	+ 58 40			COMPACT GALAXY
ZWG 362.011	05 45 00.	+ 82 48		15.6	GALAXY
LW 286	05 45 00.	- 64 48			STAR CLUSTER IN LMC
CM 31	05 45 00.	- 69 12			HII REGION IN LMC

OBJECT NAME	RIGHT ASCEN.	DECLINATION	DIAM.	MAGN.	TYPE OF OBJECT
SL 697	05 45 02.	- 67 10	15		STAR CLUSTER IN LMC
ABC 0548	05 45 03.	- 25 39		13.7	RICH CLUSTER OF GALAXIES
SL 701	05 45 03.	- 63 44	70		STAR CLUSTER IN LMC
LH 116	05 45 03.	- 67 15	540		STELLAR ASSN. IN LMC
HSE 406	05 45 04.	- 70 21			STAR CLUSTER IN LMC
SL 700	05 45 05.	- 66 10	45		STAR CLUSTER IN LMC
ZCG 127.10	05 45 06.	+ 58 38		18.7	RED COMPACT GALAXY
MCG-03-15-014	05 45 06.	- 16 41	30	15.5	GALAXY
SL 698	05 45 06.	- 68 12	55		STAR CLUSTER IN LMC
MCG+08-11-006	05 45 12.	+ 50 48	48	17.	GALAXY
RNGC 2090	05 45 12.	- 34 15		12.0	GALAXY
HOD.72 65	05 45 12.	- 69 27	480		DARK NEBULA IN LMC
RNGC 2101	05 45 14.	- 52 06			UNVERIFIED SOUTHERN OBJECT
HSE 408	05 45 14.	- 67 00			STAR CLUSTER IN LMC
LW 288	05 45 17.	- 66 09			STAR CLUSTER IN LMC
OCL 0455	05 45 18.	+ 30 12	480		OPEN STAR CLUSTER
MCG-06-13-009	05 45 18.	- 34 16	180	12.4	GALAXY
LB 01790	05 45 18.	- 47 12		13.0	FAINT BLUE STAR
HSE 407	05 45 20.	- 70 58			STAR CLUSTER IN LMC
RNGC 2111	05 45 20.	- 70 59			OPEN CLUSTER IN LMC
SL 699	05 45 20.	- 71 00	40		STAR CLUSTER IN LMC
MCG-04-14-027	05 45 21.	- 23 36	66	15.	GALAXY
MCG-04-14-028	05 45 21.	- 25 36	18	15.5	GALAXY
LW 289	05 45 21.	- 63 44			STAR CLUSTER IN LMC
MCG-04-14-029	05 45 24.	- 25 07	36	15.5	GALAXY
MCG-04-14-030	05 45 24.	- 25 09	48	15.	GALAXY
LH120-N074B	05 45 24.	- 67 09	128		EMISSION NEBULA IN LMC
VV 180B	05 45 27.	- 25 14	15	16.	INTERACTING GALAXY
VV 180A	05 45 27.	- 25 14	18	16.	INTERACTING GALAXY
VV 180	05 45 27.	- 25 14	60		INTERACTING GALAXY
ZWG 259.003	05 45 30.	+ 51 19		15.6	GALAXY
MCG-03-15-015	05 45 33.	- 19 54 30.	15	15.5	GALAXY
SL 702	05 45 34.	- 70 15	50		STAR CLUSTER IN LMC
HSE 409	05 45 34.	- 70 17			STAR CLUSTER IN LMC
RNGC 2089	05 45 36.	- 17 37		14.0	GALAXY
MCG-04-14-031	05 45 36.	- 25 56 30.	15	15.5	GALAXY
LB 03468	05 45 36.	- 76 08		12.4	FAINT BLUE STAR
MCG-03-15-016	05 45 39.	- 17 37 30.	30	14.	GALAXY
SL 705	05 45 39.	- 63 44	25		STAR CLUSTER IN LMC
HOD.72 66	05 45 42.	- 69 24	240		DARK NEBULA IN LMC
LH120-N168B	05 45 42.	- 69 47			EMISSION NEBULA IN LMC
OCL 0489	05 45 48.	+ 07 23	180	14.6	OPEN STAR CLUSTER
ISS 0261	05 45 48.	+ 33 43	346		STELLAR RING
MCG-04-14-032	05 45 48.	- 25 10	30	15.	GALAXY
LH120-N074	05 45 48.	- 67 09	795		EMISSION NEBULA IN LMC
LH120-N074A	05 45 48.	- 67 10	32		EMISSION NEBULA IN LMC
LH120-N168	05 45 48.	- 69 46	191		EMISSION NEBULA IN LMC
LW 292	05 45 51.	- 63 44			STAR CLUSTER IN LMC
MCG+08-11-007	05 45 54.	+ 50 22 30.	36	16.	GALAXY
IC 2147	05 45 54.	- 30 30 50.			NONSTELLAR OBJECT
LH120-N168A	05 45 54.	- 69 47	24		EMISSION NEBULA IN LMC
RNGC 2104	05 45 55.	- 51 35			UNVERIFIED SOUTHERN OBJECT
RNGC 2113	05 45 56.	- 69 47			CLUSTER/NEBULOSITY IN LMC
SL 704	05 45 59.	- 70 25	15		STAR CLUSTER IN LMC
KHAV 141	05 46	- 07 11	4810		DARK NEBULA
LB 09827	05 46	- 80 40		13.3	FAINT BLUE STAR
UGC 03358	05 46 00.	+ 50 22	66	16.0	GALAXY S
ZWG 259.004	05 46 00.	+ 51 05		15.7	GALAXY
UGC 03359	05 46 00.	+ 51 05	90	15.7	GALAXY SB:c
MCG-03-15-017	05 46 00.	- 19 37	60	15.	GALAXY
MCG-06-13-010	05 46 00.	- 35 20	36	15.	GALAXY
RNGC 2114	05 46 00.	- 68 05			OPEN CLUSTER IN LMC
SL 706	05 46 00.	- 68 06	40		STAR CLUSTER IN LMC
LB 03469	05 46 00.	- 76 35		15.5	FAINT BLUE STAR
SL 708	05 46 02.	- 65 28	25		STAR CLUSTER IN LMC
LW 291	05 46 04.	- 72 19			STAR CLUSTER IN LMC
OCL 0448	05 46 06.	+ 33 37	480	15.	OPEN STAR CLUSTER
HSE 410	05 46 06.	- 70 43			STAR CLUSTER IN LMC
RNGC 2144	05 46 08.	- 82 08			UNVERIFIED SOUTHERN OBJECT
SL 709	05 46 10.	- 67 37	25		STAR CLUSTER IN LMC
SL 707	05 46 11.	- 69 08	35		STAR CLUSTER IN LMC
HOD.72 67	05 46 13.	- 69 38	600		DARK NEBULA IN LMC
LW 293	05 46 14.	- 65 27			STAR CLUSTER IN LMC
HSE 411	05 46 18.	- 69 25			STAR CLUSTER IN LMC
SL 703	05 46 22.	- 74 53	40		STAR CLUSTER IN LMC
LW 290	05 46 22.	- 74 53			STAR CLUSTER IN LMC
HSE 412	05 46 23.	- 69 17			STAR CLUSTER IN LMC
UGC 03360	05 46 24.	+ 17 50	78	16.0	GALAXY S
MCG+08-11-008	05 46 24.	+ 49 42	42	16.	GALAXY
MCG-03-15-018	05 46 24.	- 18 43 30.	42	14.5	GALAXY
SL 710	05 46 27.	- 63 53	60		STAR CLUSTER IN LMC
OCL 0458	05 46 30.	+ 28 55	11040		OPEN STAR CLUSTER
UGC 03361	05 46 30.	+ 49 42	78	16.5	GALAXY SB:c-Ir
LH120-N169C	05 46 30.	- 69 34	80		EMISSION NEBULA IN LMC
ISS 0365	05 46 36.	+ 15 13	871		STELLAR RING
UGC 03362	05 46 36.	+ 17 39	90	16.0	GALAXY S
UGC 03363	05 46 36.	+ 17 41	102	15.0	GALAXY
MCG-04-14-034	05 46 36.	- 24 24	54	15.	GALAXY
MCG-04-14-033	05 46 36.	- 25 30	12	15.	GALAXY
MCG-04-14-035	05 46 42.	- 25 30	12	15.	GALAXY
MCG-04-14-036	05 46 42.	- 25 34	36	16.	GALAXY
MCG-05-14-017	05 46 42.	- 33 02	24	15.	GALAXY
SL 711	05 46 43.	- 66 41	15		STAR CLUSTER IN LMC
LW 298	05 46 45.	- 63 53			STAR CLUSTER IN LMC
UGC 03364	05 46 48.	+ 76 41	114	16.0	GALAXY Sc
MCG-05-14-018	05 46 48.	- 32 48	36	16.	GALAXY
LH120-N169B	05 46 48.	- 69 35	32		EMISSION NEBULA IN LMC
LW 294	05 46 48.	- 71 45			STAR CLUSTER IN LMC
SL 712	05 46 49.	- 66 47	25		STAR CLUSTER IN LMC
MCG-04-14-037	05 46 51.	- 25 39 30.	30	15.	GALAXY
LH120-N169A	05 46 54.	- 69 34	64		EMISSION NEBULA IN LMC
HSE 413	05 46 58.	- 70 09			STAR CLUSTER IN LMC
B 036	05 47	+ 07 25	7200		DARK OBJECT
LB 09828	05 47	- 81 00		14.5	FAINT BLUE STAR
LDN 1599	05 47 00.	+ 07 30	1680		DARK NEBULA
UGC 03368	05 47 00.	+ 39 49	90	17.	GALAXY Sc
MCG+13-05-006	05 47 00.	+ 76 37 30.	57	15.	GALAXY
SL 714	05 47 01.	- 66 54	30		STAR CLUSTER IN LMC
HSE 415	05 47 08.	- 67 01			STAR CLUSTER IN LMC
RNGC 2116	05 47 08.	- 68 32			OPEN CLUSTER IN LMC
SL 715	05 47 08.	- 68 32	30		STAR CLUSTER IN LMC
IC 2144	05 47 11.	+ 23 51 33.			NONSTELLAR OBJECT
LW 296	05 47 11.	- 71 34			STAR CLUSTER IN LMC
MCG+02-15-002	05 47 12.	+ 13 38	48	15.	GALAXY
MESL 184-01/7	05 47 12.	+ 23 51	60		HII REGION
LW 295	05 47 15.	- 73 07			STAR CLUSTER IN LMC
SL 713	05 47 17.	- 71 33	20		STAR CLUSTER IN LMC
OCL 0464	05 47 18.	+ 22 11	360	15.	OPEN STAR CLUSTER

OBJECT NAME	RIGHT ASCEN.	DECLINATION	DIAM.	MAGN.	TYPE OF OBJECT
MCG-04-14-038	05 47 18.	- 25 22 30.	18	15.	GALAXY
MCG-05-14-019	05 47 18.	- 31 32	36	15.	GALAXY
HSE 414	05 47 19.	- 70 47			STAR CLUSTER IN LMC
MCG-03-15-019	05 47 21.	- 19 36	36	15.	GALAXY
ZWG 308.002	05 47 24.	+ 66 47		15.0	GALAXY
ZWG 307.025	05 47 24.	+ 66 47		15.0	GALAXY
UGC 03365	05 47 24.	+ 66 47	144	15.0	GALAXY Sa
LH120-N170	05 47 24.	- 69 27			EMISSION NEBULA IN LMC
SL 716	05 47 25.	- 70 50	15		STAR CLUSTER IN LMC
RNGC 2117	05 47 28.	- 67 28			OPEN CLUSTER IN LMC
WS 39	05 47 29.	- 69 28 34.		15.80	PLANETARY NEB. IN LMC
LDN 1597	05 47 30.	+ 08 40	720		DARK NEBULA
LW 299	05 47 31.	- 65 01			STAR CLUSTER IN LMC
SL 720	05 47 31.	- 65 03	15		STAR CLUSTER IN LMC
LW 297	05 47 32.	- 74 33			STAR CLUSTER IN LMC
HSE 416	05 47 33.	- 69 58			STAR CLUSTER IN LMC
SL 718	05 47 34.	- 67 28	45		STAR CLUSTER IN LMC
RNGC 2118	05 47 35.	- 69 09			OPEN CLUSTER IN LMC
SL 717	05 47 35.	- 69 09	35		STAR CLUSTER IN LMC
UGC 03366	05 47 36.	+ 48 50	78	16.5	GALAXY Sc
SL 719	05 47 42.	- 69 21	15		STAR CLUSTER IN LMC
LW 300	05 47 43.	- 65 01			STAR CLUSTER IN LMC
HSE 417	05 47 45.	- 69 58			STAR CLUSTER IN LMC
MCG-04-14-039	05 47 54.	- 24 39	36	15.5	GALAXY
ST 34	05 48	+ 00	18000		EMISSION NEBULA
LBN 0905	05 48	+ 04 30	4200		BRIGHT NEBULA
KHAV 142	05 48	+ 06 07	3750		DARK NEBULA
LBN 0826	05 48	+ 27 00	420		BRIGHT NEBULA
VDB.66G 039	05 48	+ 75 17	130		DWARF GALAXY
ST 33	05 48	- 07	24000		EMISSION NEBULA
UGC 03367	05 48 00.	+ 37 52	102	17.	GALAXY S
MCG+07-12-003	05 48 09.	+ 39 50	63	17.	GALAXY
LB 01791	05 48 12.	- 48 34		14.4	FAINT BLUE STAR
DG 083	05 48 18.	+ 27 02	540		REFLECTION NEBULA
ISS 0209	05 48 18.	+ 41 04	287		STELLAR RING
MCG-03-15-020	05 48 18.	- 19 46	144	14.5	GALAXY
LH120-N179C	05 48 18.	- 69 52	32		EMISSION NEBULA IN LMC
SEY 012	05 48 22.	+ 74 43 23.		15.4	FAINT GALAXY
ZWG 329.019	05 48 24.	+ 74 41		15.5	GALAXY
KARA.73B 0164	05 48 24.	+ 74 41	18	15.5	ISOLATED GALAXY E
SL 724	05 48 24.	- 64 38	20		STAR CLUSTER IN LMC
LH120-N179B	05 48 24.	- 69 53	24		EMISSION NEBULA IN LMC
LH120-N179A	05 48 24.	- 69 53	48		EMISSION NEBULA IN LMC
HSE 419	05 48 27.	- 67 15			STAR CLUSTER IN LMC
HSE 418	05 48 27.	- 69 58			STAR CLUSTER IN LMC
LW 301	05 48 29.	- 72 29			STAR CLUSTER IN LMC
MCG+08-11-010	05 48 30.	+ 46 47 30.	48	16.	GALAXY
ZWG 232.002	05 48 30.	+ 46 49		14.8	GALAXY
MCG-04-14-040	05 48 30.	- 21 35	120	13.5	GALAXY
RNGC 2106	05 48 32.	- 21 35		13.0	GALAXY
MCG-03-15-021	05 48 33.	- 18 13	24	15.	GALAXY
SL 722	05 48 33.	- 71 13	15		STAR CLUSTER IN LMC
SL 721	05 48 35.	- 72 29	5		STAR CLUSTER IN LMC
ZWG 308.003	05 48 36.	+ 66 41		15.6	GALAXY
ZWG 307.026	05 48 36.	+ 66 41		15.6	GALAXY
MCG-04-14-041	05 48 36.	- 23 01	30	15.	GALAXY
LH120-N179D	05 48 36.	- 69 52	48		EMISSION NEBULA IN LMC
LW 304	05 48 37.	- 65 09			STAR CLUSTER IN LMC
OCL 0469	05 48 42.	+ 21 46	420	16.	OPEN STAR CLUSTER
MRSL 182+00/1	05 48 42.	+ 27 00	420		HII REGION
UGC 03369	05 48 42.	+ 41 46	102	17.	GALAXY
MCG-02-15-011	05 48 42.	- 14 48	162	13.	GALAXY
LW 305	05 48 42.	- 64 38			STAR CLUSTER IN LMC
LW 306	05 48 42.	- 64 44			STAR CLUSTER IN LMC
SL 726	05 48 42.	- 64 45	5		STAR CLUSTER IN LMC
LH120-N180C	05 48 42.	- 70 02	111		EMISSION NEBULA IN LMC
SL 729	05 48 43.	- 63 08	35		STAR CLUSTER IN LMC
SL 727	05 48 43.	- 65 01	25		STAR CLUSTER IN LMC
YM 43	05 48 44.	- 14 44	162		SYMMETRIC GALACTIC NEBULA
SL 728	05 48 45.	- 65 42	25		STAR CLUSTER IN LMC
HOD.72 68	05 48 46.	- 70 08	480		DARK NEBULA IN LMC
DG 084	05 48 48.	+ 27 01	480		REFLECTION NEBULA
UGC 03370	05 48 48.	+ 78 31	84	17.	GALAXY Sb-c
MCG-02-15-012	05 48 48.	- 14 52	42	16.	GALAXY
LW 307	05 48 49.	- 65 01			STAR CLUSTER IN LMC
SS 47	05 48 50.	+ 27 01			DIFFUSE GALACTIC NEBULA
SG 3.076	05 48 50.	+ 27 02	420		DIFFUSE EMISSION NEBULA
LW 308	05 48 51.	- 65 42			STAR CLUSTER IN LMC
SL 723	05 48 51.	- 73 09	20		STAR CLUSTER IN LMC
RNGC 2121	05 48 53.	- 71 30			GLOBULAR CLUSTER IN LMC
LW 311	05 48 55.	- 63 08			STAR CLUSTER IN LMC
LW 309	05 48 55.	- 65 13			STAR CLUSTER IN LMC
LW 302	05 48 58.	- 73 11			STAR CLUSTER IN LMC
LW 303	05 48 59.	- 71 29			STAR CLUSTER IN LMC
SL 725	05 48 59.	- 71 30	135		STAR CLUSTER IN LMC
KHAV 143	05 49	+ 08 19	4320		DARK NEBULA
LBN 0827	05 49	+ 27 00	600		BRIGHT NEBULA
HSE 421	05 49 02.	- 67 06			STAR CLUSTER IN LMC
RNGC 2099	05 49 04.	+ 32 32		6.0	OPEN CLUSTER
SL 730	05 49 04.	- 67 32	35		STAR CLUSTER IN LMC
OCL 0451	05 49 06.	+ 32 32	2280	7.	OPEN STAR CLUSTER
LH120-N180A	05 49 06.	- 70 06	24		EMISSION NEBULA IN LMC
HSE 420	05 49 06.	- 70 34			STAR CLUSTER IN LMC
MCG-06-13-011	05 49 12.	- 34 48	36	15.	GALAXY
LB 01792	05 49 12.	- 46 22		14.2	FAINT BLUE STAR
RNGC 2122	05 49 15.	- 70 04			CLUSTER/NEBULOSITY IN LMC
MCG+13-05-007	05 49 18.	+ 79 47	39	16.	GALAXY
MCG-06-13-012	05 49 18.	- 34 47	36	15.5	GALAXY
IC 0437	05 49 19.	- 12 34 38.			NONSTELLAR OBJECT
HSE 422	05 49 19.	- 66 47			STAR CLUSTER IN LMC
IC 2150	05 49 21.	- 38 21 54.			NONSTELLAR OBJECT
SL 731	05 49 21.	- 70 04	270		STAR CLUSTER IN LMC
LH 117	05 49 21.	- 70 05	300		STELLAR ASSN. IN LMC
SL 732	05 49 22.	- 67 40	25		STAR CLUSTER IN LMC
LH120-N180B	05 49 24.	- 70 03	318		EMISSION NEBULA IN LMC
LH120-N220	05 49 24.	- 70 17	20		EMISSION NEBULA IN LMC
MCG-05-14-020	05 49 30.	- 31 09	30	15.	GALAXY
MCG-06-13-015	05 49 30.	- 34 55		15.	GALAXY
MCG-06-13-014	05 49 30.	- 34 56		15.	GALAXY
MCG-06-13-013	05 49 30.	- 34 56	8	16.5	GALAXY
MCG-06-13-016	05 49 30.	- 38 22	150	12.5	GALAXY
LH120-N180	05 49 30.	- 70 05	859		EMISSION NEBULA IN LMC
LW 310	05 49 30.	- 71 39			STAR CLUSTER IN LMC
SL 735	05 49 35.	- 67 44	70		STAR CLUSTER IN LMC
MCG-05-14-021	05 49 36.	- 31 04	36	16.	GALAXY
SL 738	05 49 40.	- 64 09	20		STAR CLUSTER IN LMC
LDN 1598	05 49 42.	+ 08 20	1080		DARK NEBULA
SL 733	05 49 42.	- 71 40	35		STAR CLUSTER IN LMC
SL 739	05 49 44.	- 65 31	20		STAR CLUSTER IN LMC
HODG 07	05 49 44.	- 67 45	60		RED GLOBULAR CLSTR IN LMC
SL 734	05 49 44.	- 71 03	50		STAR CLUSTER IN LMC
RNGC 2110	05 49 47.	- 07 18		14.0	GALAXY
UGC 03371	05 49 48.	+ 75 19	300	17.	GALAXY DWARF
MCG-05-14-022	05 49 48.	- 31 34	18	15.	GALAXY
MCG-01-15-004	05 49 51.	- 07 18	42	14.	GALAXY
LW 314	05 49 52.	- 64 09			STAR CLUSTER IN LMC
HSE 423	05 49 52.	- 70 08			STAR CLUSTER IN LMC
MAI 011	05 49 53.	+ 75 25	134		DWARF SPHEROIDAL GALAXY
MCG+08-11-009	05 49 54.	+ 50 52 30.	36	16.	GALAXY
UGC 03372	05 49 54.	+ 76 31	66	16.0	GALAXY Sb
MCG-02-15-013	05 49 54.	- 14 08	60	15.	GALAXY
MCG-03-15-022	05 49 54.	- 18 04	60	15.	GALAXY
LH120-N181	05 49 54.	- 69 09			EMISSION NEBULA IN LMC
SL 736	05 49 55.	- 70 48	15		STAR CLUSTER IN LMC
LW 313	05 49 57.	- 65 31			STAR CLUSTER IN LMC
KHAV 145	05 50	+ 04 43	7160		DARK NEBULA
KHAV 144	05 50	+ 10 19	7770		DARK NEBULA
LBN 0988	05 50	- 06 30	7200		BRIGHT NEBULA
MCG+13-05-007A	05 50 00.	+ 75 17 30.	150		GALAXY
LDN 1648	05 50 00.	- 09 10	5520		DARK NEBULA
LB 03470	05 50 00.	- 77 16		14.9	FAINT BLUE STAR
VDB.66N 061	05 50 01.	+ 05 08	72		REFLECTION NEBULA
SL 741	05 50 01.	- 66 52	15		STAR CLUSTER IN LMC
RNGC 2573	05 50 02.	- 89 52			GALAXY
SL 742	05 50 03.	- 63 40	70		STAR CLUSTER IN LMC
RNGC 2115	05 50 06.	- 50 35			UNVERIFIED SOUTHRN OBJECT
SL 740	05 50 10.	- 70 13	35		STAR CLUSTER IN LMC
IC 0436	05 50 13.	+ 38 37 11.			NONSTELLAR OBJECT
LH 119	05 50 13.	- 68 15	180		STELLAR ASSN. IN LMC
HSE 424	05 50 13.	- 70 43			STAR CLUSTER IN LMC
RNGC 2120	05 50 15.	- 63 41			GLOBULAR CLUSTER IN LMC
LW 316	05 50 15.	- 63 41			STAR CLUSTER IN LMC
SL 743	05 50 15.	- 67 15	15		STAR CLUSTER IN LMC
LH 118	05 50 15.	- 70 05	240		STELLAR ASSN. IN LMC
MCG+13-05-008	05 50 18.	+ 76 29	60	16.	GALAXY
MCG-03-15-023	05 50 18.	- 19 28 30.	24	15.	GALAXY
IC 2151	05 50 24.	- 17 47 41.			NONSTELLAR OBJECT
HSE 426	05 50 25.	- 66 52			STAR CLUSTER IN LMC
MCG-03-15-024	05 50 27.	- 17 49	72	14.5	GALAXY
HSE 425	05 50 27.	- 70 05			STAR CLUSTER IN LMC
MCG+07-12-004	05 50 30.	+ 40 01 30.	18	17.	GALAXY
SL 745	05 50 30.	- 68 01	15		STAR CLUSTER IN LMC
SL 737	05 50 31.	- 75 45	15		STAR CLUSTER IN LMC
LW 318	05 50 32.	- 65 18			STAR CLUSTER IN LMC
ARC 0550	05 50 33.	- 21 05		16.7	RICH CLUSTER OF GALAXIES
SL 746	05 50 34.	- 67 35	20		STAR CLUSTER IN LMC
MCG-06-13-017	05 50 36.	- 34 21	66	14.5	GALAXY
SL 744	05 50 37.	- 70 51	20		STAR CLUSTER IN LMC
LW 312	05 50 37.	- 75 46			STAR CLUSTER IN LMC
IC 0438	05 50 39.	- 17 52 49.			NONSTELLAR OBJECT
MCG-06-13-018	05 50 42.	- 36 58	48	15.	GALAXY
LH 120	05 50 42.	- 68 10	840		STELLAR ASSN. IN LMC
MCG-03-15-025	05 50 45.	- 17 53	150	13.	GALAXY
RNGC 2134	05 50 45.	- 71 07		11.0	OPEN CLUSTER IN LMC
LW 315	05 50 51.	- 74 43			STAR CLUSTER IN LMC
LW 319	05 50 52.	- 64 12			STAR CLUSTER IN LMC
MCG+11-08-001	05 50 54.	+ 68 48	15	16.	GALAXY
HSE 428	05 50 57.	- 67 11			STAR CLUSTER IN LMC
SL 747	05 50 57.	- 71 10	40		STAR CLUSTER IN LMC
SL 748	05 50 59.	- 70 26	40		STAR CLUSTER IN LMC
KHAV 146	05 51	+ 03 01	7770		DARK NEBULA
LBN 0982	05 51	- 05 30	1800		BRIGHT NEBULA
MCG+08-11-011	05 51 00.	+ 46 25	102	15.	GALAXY
ZWG 348.008	05 51 00.	+ 78 30		15.0	GALAXY
ZWG 347.024	05 51 00.	+ 78 30		15.0	GALAXY
UGC 03373	05 51 00.	+ 78 30	96	15.0	GALAXY Sc
MCG+13-05-009	05 51 00.	+ 78 30	78	15.	GALAXY
LB 01793	05 51 00.	- 46 20		13.6	FAINT BLUE STAR
HSE 427	05 51 00.	- 70 38			STAR CLUSTER IN LMC
LW 320	05 51 01.	- 63 00			STAR CLUSTER IN LMC
ZWG 232.003	05 51 06.	+ 46 26		13.9	GALAXY
UGC 03374	05 51 06.	+ 46 26	168	13.9	GALAXY SB
RNGC 2125	05 51 06.	- 69 29			OPEN CLUSTER IN LMC
SL 750	05 51 06.	- 69 29	20		STAR CLUSTER IN LMC
LW 317	05 51 09.	- 73 03			STAR CLUSTER IN LMC
VDB.66N 062	05 51 10.	+ 01 42	168		REFLECTION NEBULA
OCL 0509	05 51 18.	+ 00 23	1320	9.6	OPEN STAR CLUSTER
ZWG 308.004	05 51 18.	+ 68 43		15.2	GALAXY
ZWG 307.027	05 51 18.	+ 68 43		15.2	GALAXY
LB 01794	05 51 18.	- 46 15		13.2	FAINT BLUE STAR
LW 321	05 51 18.	- 66 36			STAR CLUSTER IN LMC
RNGC 2112	05 51 19.	+ 00 23		9.0	OPEN CLUSTER
LW 323	05 51 21.	- 63 52			STAR CLUSTER IN LMC
HSE 431	05 51 21.	- 67 12			STAR CLUSTER IN LMC
SL 749	05 51 21.	- 73 02	70		STAR CLUSTER IN LMC
ZWG 259.005	05 51 24.	+ 51 55		14.6	GALAXY
UGC 03375	05 51 24.	+ 51 55	138	14.6	GALAXY Sc
ZWG 284.002	05 51 24.	+ 60 50		15.6	GALAXY
ZC 0551.4+6546	05 51 24.	+ 65 46	1750		CLUSTER OF GALAXIES
MCG-03-15-026	05 51 24.	- 20 18	24	15.5	GALAXY
HSE 429	05 51 25.	- 69 43			STAR CLUSTER IN LMC
SL 753	05 51 28.	- 65 53	25		STAR CLUSTER IN LMC
MCG+09-10-003	05 51 30.	+ 51 54	96	13.	GALAXY
ZC 0551.5+6454	05 51 30.	+ 64 54	2220		CLUSTER OF GALAXIES
SL 752	05 51 30.	- 68 10	15		STAR CLUSTER IN LMC
RNGC 2127	05 51 30.	- 69 22			OPEN CLUSTER IN LMC
SL 751	05 51 30.	- 69 22	35		STAR CLUSTER IN LMC
RNGC 2123	05 51 32.	- 65 19			OPEN CLUSTER IN LMC
SL 755	05 51 32.	- 65 19	25		STAR CLUSTER IN LMC
LW 324	05 51 32.	- 65 19			STAR CLUSTER IN LMC
HSE 432	05 51 32.	- 69 49			STAR CLUSTER IN LMC
HSE 430	05 51 32.	- 71 04			STAR CLUSTER IN LMC
7ZW 051	05 51 36.	+ 63 13			COMPACT GALAXY
ARC 0549	05 51 38.	+ 65 42		17.3	RICH CLUSTER OF GALAXIES
YM 40	05 51 39.	- 00 59	2460		SYMMETRIC GALACTIC NEBULA
LW 325	05 51 46.	- 65 53			STAR CLUSTER IN LMC
UGC 03376	05 51 48.	+ 15 10	66	15.5	GALAXY SB
HSE 433	05 51 58.	- 67 36			STAR CLUSTER IN LMC
SL 756	05 51 59.	- 70 28	15		STAR CLUSTER IN LMC
KHAV 147	05 52	+ 01 55	700		DARK NEBULA
LBN 0955	05 52	- 01 00	1500		BRIGHT NEBULA
LDN 1622	05 52 00.	+ 02 00	1380		DARK NEBULA
LDN 1617	05 52 00.	+ 03 00	11280		DARK NEBULA
7ZW 052	05 52 00.	+ 84 25			COMPACT GALAXY
ZWG 363.001	05 52 00.	+ 84 25		15.5	GALAXY
ZWG 362.012	05 52 00.	+ 84 25		15.5	GALAXY

OBJECT NAME	RIGHT ASCEN.	DECLINATION	DIAM.	MAGN.	TYPE OF OBJECT
SCHO 0162	05 52 02.	+ 11 08 00.	340		ISOLATED DARK CLOUD
HSE 434	05 52 05.	- 69 11			STAR CLUSTER IN LMC
RNGC 2133	05 52 09.	- 71 11			GLOBULAR CLUSTER IN LMC
SCHO 0163	05 52 12.	+ 11 01 12.	360		ISOLATED DARK CLOUD
OCL 0472	05 52 12.	+ 20 52	360		OPEN STAR CLUSTER
MCG-05-14-023	05 52 12.	- 32 47	48	15.	GALAXY
RNGC 2130	05 52 15.	- 67 20			OPEN CLUSTER IN LMC
HSE 436	05 52 17.	- 69 14			STAR CLUSTER IN LMC
MCG+08-11-012	05 52 18.	+ 48 32	54	16.	GALAXY
HSE 435	05 52 20.	- 71 02			STAR CLUSTER IN LMC
SL 758	05 52 21.	- 67 20	35		STAR CLUSTER IN LMC
SL 757	05 52 21.	- 71 11	55		STAR CLUSTER IN LMC
SL 754	05 52 21.	- 75 23	20		STAR CLUSTER IN LMC
LW 322	05 52 21.	- 75 23			STAR CLUSTER IN LMC
LDN 1621	05 52 24.	+ 02 20	660		DARK NEBULA
ZWG 232.004	05 52 24.	+ 48 31		15.7	GALAXY
MCG-06-13-019	05 52 24.	- 34 44	30	16.	GALAXY
HSE 437	05 52 25.	- 68 14			STAR CLUSTER IN LMC
MCG+07-13-001	05 52 30.	+ 41 03	24	17.	GALAXY
MCG+14-03-014	05 52 30.	+ 82 50	39	16.	GALAXY
LH 121	05 52 31.	- 68 14	840		STELLAR ASSN. IN LMC
SL 759	05 52 31.	- 69 41	35		STAR CLUSTER IN LMC
MCG+11-08-002	05 52 36.	+ 68 29 30.	114	14.	GALAXY
MCG-03-15-027	05 52 36.	- 15 07	36	15.5	GALAXY
SCHO 0164	05 52 37.	+ 02 00 12.	580		ISOLATED DARK CLOUD
HW 0105	05 52 37.	+ 46 07			NEBULA
PK166+10.1	05 52 37.	+ 46 07	15	11.2	PLANETARY NEBULA
HSE 439	05 52 37.	- 69 33			STAR CLUSTER IN LMC
HSE 438	05 52 37.	- 70 52			STAR CLUSTER IN LMC
IC 2149	05 52 40.9	+ 46 05 53.	15	10.5	PLANETARY NEBULA
ABC 0551	05 52 41.	- 17 46		17.5	RICH CLUSTER OF GALAXIES
MCG-06-13-020	05 52 42.	- 35 02	30	15.5	GALAXY
SL 760	05 52 45.	- 71 07	90		STAR CLUSTER IN LMC
HSE 440	05 52 49.	- 68 17			STAR CLUSTER IN LMC
SL 761	05 52 50.	- 70 55	20		STAR CLUSTER IN LMC
UGC 03377	05 52 54.	+ 64 44	66	16.0	GALAXY Sc
LBN 0966	05 53	- 01 40	4200		BRIGHT NEBULA
SCHO 0167	05 53 00.	+ 11 05 00.	300		ISOLATED DARK CLOUD
7ZW 053	05 53 00.	+ 63 49			COMPACT GALAXY
MCG+11-08-003	05 53 00.	+ 64 43	66	16.	GALAXY
ZWG 308.005	05 53 00.	+ 68 26		13.8	GALAXY
ZWG 307.028	05 53 00.	+ 68 26		13.8	GALAXY
UGC 03379	05 53 00.	+ 68 26	102	13.8	GALAXY SBb/Sb
ZWG 362.013	05 53 00.	+ 83 50		14.6	GALAXY
UGC 03378	05 53 00.	+ 83 50	102	14.6	GALAXY
LB 03471	05 53 00.	- 77 10		16.2	FAINT BLUE STAR
HSE 441	05 53 05.	- 69 09			STAR CLUSTER IN LMC
MRSL 203-10/1	05 53 06.	+ 03 23	18		HII REGION
2ZW 040	05 53 06.	+ 03 24			COMPACT GALAXY
MAI 012	05 53 10.	+ 68 41	40		DWARF SPHEROIDAL GALAXY
ISS 0210	05 53 12.	+ 43 18	376		STELLAR RING
LB 01795	05 53 12.	- 45 48		14.0	FAINT BLUE STAR
RNGC 2136	05 53 12.	- 69 30		10.5	OPEN CLUSTER IN LMC
SL 762	05 53 12.	- 69 30	80		STAR CLUSTER IN LMC
SL 763	05 53 14.	- 69 47	35		STAR CLUSTER IN LMC
RNGC 2135	05 53 16.	- 67 26			OPEN CLUSTER IN LMC
HSE 442	05 53 17.	- 67 43			STAR CLUSTER IN LMC
SL 768	05 53 20.	- 63 37	100		STAR CLUSTER IN LMC
CED 060	05 53 21.	+ 32 02	600		DIFFUSE GALACTIC NEBULA
IC 0439	05 53 22.	+ 32 01			NONSTELLAR OBJECT
SL 765	05 53 22.	- 67 26	30		STAR CLUSTER IN LMC
SCHO 0165	05 53 24.	+ 12 22 30.	350		ISOLATED DARK CLOUD
ISS 0366	05 53 24.	+ 17 59	265		STELLAR RING
MCG-05-14-024	05 53 24.	- 31 40	30	15.5	GALAXY
SL 764	05 53 24.	- 69 29	25		STAR CLUSTER IN LMC
LW 326	05 53 32.	- 63 37			STAR CLUSTER IN LMC
HSE 443	05 53 32.	- 70 54			STAR CLUSTER IN LMC
RNGC 2137	05 53 36.	- 69 29			OPEN CLUSTER IN LMC
SL 766	05 53 36.	- 70 33	15		STAR CLUSTER IN LMC
VDB.66N 063	05 53 39.	+ 01 49	36		REFLECTION NEBULA
2ZW 041	05 53 42.	+ 07 50			COMPACT GALAXY
ZWG 203.001	05 53 42.	+ 43 14		15.5	GALAXY
MCG-04-14-042	05 53 42.	- 22 49 30.	66	15.	GALAXY
SL 767	05 53 42.	- 70 31	20		STAR CLUSTER IN LMC
HSE 444	05 53 51.	- 67 19			STAR CLUSTER IN LMC
SL 769	05 53 51.	- 70 05	60		STAR CLUSTER IN LMC
SL 771	05 53 53.	- 67 44	65		STAR CLUSTER IN LMC
LB 01796	05 53 54.	- 46 12		12.2	FAINT BLUE STAR
SL 770	05 53 54.	- 69 19	35		STAR CLUSTER IN LMC
HSE 445	05 53 57.	- 67 23			STAR CLUSTER IN LMC
KHAV 148	05 54	+ 00 42	8950		DARK NEBULA
LBN 0985	05 54	- 05 50	2400		BRIGHT NEBULA
MCG+14-03-015	05 54 00.	+ 82 45	30	16.	GALAXY
7ZW 054	05 54 00.	+ 82 47			COMPACT GALAXY
MCG+14-03-016	05 54 00.	+ 83 45	9	16.	GALAXY
SL 772	05 54 01.	- 69 40	15		STAR CLUSTER IN LMC
MAI 013	05 54 03.	+ 73 23	47		DWARF SPHEROIDAL GALAXY
MCG+08-11-013	05 54 12.	+ 46 50	36	16.	GALAXY
HODG 08	05 54 14.	- 67 48	66		RED GLOBULAR CLSTR IN LMC
RNGC 2140	05 54 14.	- 68 36			OPEN CLUSTER IN LMC
SL 773	05 54 14.	- 68 36	35		STAR CLUSTER IN LMC
SL 774	05 54 16.	- 69 01	20		STAR CLUSTER IN LMC
MCG-03-16-001	05 54 18.	- 18 12	30	14.5	GALAXY
LW 327	05 54 18.	- 71 43			STAR CLUSTER IN LMC
RNGC 2138	05 54 28.	- 65 50			OPEN CLUSTER IN LMC
MCG+14-03-017	05 54 30.	+ 83 50	18	15.	GALAXY
MCG-06-14-001	05 54 30.	- 38 12	48	15.	GALAXY
LW 329	05 54 30.	- 64 54			STAR CLUSTER IN LMC
SCHO 0166	05 54 32.	+ 12 17 18.	370		ISOLATED DARK CLOUD
SL 777	05 54 34.	- 65 50	25		STAR CLUSTER IN LMC
RNGC 2132	05 54 35.	- 59 55			NON-EXISTENT OBJECT
RNGC 2119	05 54 36.	+ 11 57			GALAXY
UGC 03380	05 54 36.	+ 11 58	102	15.0	GALAXY E
SL 775	05 54 36.	- 71 44	25		STAR CLUSTER IN LMC
LW 328	05 54 39.	- 73 04			STAR CLUSTER IN LMC
SL 778	05 54 43.	- 66 42	10		STAR CLUSTER IN LMC
SL 776	05 54 45.	- 73 03	50		STAR CLUSTER IN LMC
LW 330	05 54 46.	- 65 50			STAR CLUSTER IN LMC
ZC 0554.9+5359	05 54 54.	+ 53 59	6520		CLUSTER OF GALAXIES
LW 332	05 54 54.	- 64 39			STAR CLUSTER IN LMC
LW 331	05 54 55.	- 66 43			STAR CLUSTER IN LMC
LBN 0765	05 55	+ 49 00	2100		BRIGHT NEBULA
LBN 0763	05 55	+ 49 40	1500		BRIGHT NEBULA
LBN 0976	05 55	- 03 00	6000		BRIGHT NEBULA
LDN 1624	05 55 00.	+ 02 00			DARK NEBULA
MCG+07-13-002	05 55 00.	+ 40 28 30.	24	18.	GALAXY
MCG+07-13-003	05 55 00.	+ 40 30 30.	30	17.	GALAXY
ZWG 362.014	05 55 00.	+ 82 43		15.5	GALAXY
UGC 03381	05 55 00.	+ 83 43	90	16.5	GALAXY
LDN 1638	05 55 00.	- 02 30	6240		DARK NEBULA
SL 779	05 55 03.	- 71 07	20		STAR CLUSTER IN LMC
MCG+07-13-004	05 55 06.	+ 40 30	12	17.	GALAXY
ZWG 203.002	05 55 06.	+ 40 31		15.6	GALAXY
ZWG 284.003	05 55 06.	+ 62 08		14.8	GALAXY
UGC 03382	05 55 06.	+ 62 08	72	14.8	GALAXY SBa
MCG+10-09-004	05 55 06.	+ 62 09	66	14.	GALAXY
HSE 447	05 55 08.	- 67 07			STAR CLUSTER IN LMC
RNGC 2145	05 55 08.	- 70 55			OPEN CLUSTER IN LMC
SL 780	05 55 08.	- 70 55	50		STAR CLUSTER IN LMC
BA 11	05 55 12.	+ 21 58	600		STELLAR GROUP
UGC 03383	05 55 12.	+ 54 27	66	16.0	GALAXY
7ZW 055	05 55 12.	+ 77 28			COMPACT GALAXY
ZWG 259.006	05 55 18.	+ 54 25		15.2	GALAXY
SL 781	05 55 19.	- 70 43	20		STAR CLUSTER IN LMC
HSE 446	05 55 19.	- 70 47			STAR CLUSTER IN LMC
SL 782	05 55 20.	- 68 29	70		STAR CLUSTER IN LMC
HSE 448	05 55 22.	- 70 11			STAR CLUSTER IN LMC
ASS 50	05 55 24.	+ 21 13			OB ASSOCIATION ORI OB2
UGC 03384	05 55 30.	+ 73 07	114	16.0	GALAXY DWRF SP
MCG-03-16-002	05 55 30.	- 18 36	36	15.	GALAXY
VDB.66N 064	05 55 35.	- 14 03	72		REFLECTION NEBULA
MCG+10-09-005	05 55 36.	+ 60 47	57	16.	GALAXY
MCG-05-15-001	05 55 36.	- 29 56	60	15.5	GALAXY
MCG-06-14-002	05 55 36.	- 37 31	30	15.	GALAXY
OCL 0490	05 55 42.	+ 07 50	120	15.	OPEN STAR CLUSTER
ACK 184+00.1	05 55 42.	+ 25 19			PLANETARY NEBULA
7ZW 056	05 55 42.	+ 62 58			COMPACT GALAXY
MCG-03-16-003	05 55 42.	- 20 04	132	13.	GALAXY
LB 03472	05 55 42.	- 78 18		13.0	FAINT BLUE STAR
RNGC 2124	05 55 43.	- 20 04		13.0	GALAXY
SL 784	05 55 43.	- 68 11	25		STAR CLUSTER IN LMC
LW 334	05 55 47.	- 66 17			STAR CLUSTER IN LMC
IC 2152	05 55 48.	- 23 11 06.			NONSTELLAR OBJECT
MCG-04-15-001	05 55 51.	- 23 11	36	13.5	GALAXY
MCG-03-16-004	05 55 54.	- 16 35	54	14.5	GALAXY
MCG-06-14-003	05 55 54.	- 34 57	30	15.	GALAXY
LH120-N075A	05 55 54.	- 68 07	32		EMISSION NEBULA IN LMC
SL 785	05 55 55.	- 68 12	80		STAR CLUSTER IN LMC
KHAV 149	05 56	- 04 36	9060		DARK NEBULA
LBN 1016	05 56	- 13 40	3060		BRIGHT NEBULA
LBN 1019	05 56	- 14 10	120		BRIGHT NEBULA
ZWG 348.009	05 56 00.	+ 80 08		15.1	GALAXY
ZWG 347.025	05 56 00.	+ 80 08		15.1	GALAXY
UGC 03385	05 56 00.	+ 80 08	90	15.1	GALAXY S0
LW 335	05 56 02.	- 66 57			STAR CLUSTER IN LMC
SL 783	05 56 02.	- 74 36	35		STAR CLUSTER IN LMC
MCG+11-08-004	05 56 06.	+ 66 23 30.	42	16.	GALAXY
LW 333	05 56 09.	- 74 37			STAR CLUSTER IN LMC
LH120-N075B	05 56 12.	- 68 12	334		EMISSION NEBULA IN LMC
RNGC 2147	05 56 13.	- 68 12			CLUSTER/NEBULOSITY IN LMC
LH 122	05 56 13.	- 68 13	120		STELLAR ASSN. IN LMC
RNGC 2150	05 56 19.	- 69 34			GALAXY
MCG+07-13-005	05 56 21.	+ 41 39	30	17.	GALAXY
SL 786	05 56 22.	- 69 01	20		STAR CLUSTER IN LMC
ZWG 284.004	05 56 24.	+ 60 50		15.4	GALAXY
MCG+10-09-006	05 56 24.	+ 60 51	45	15.	GALAXY
SL 787	05 56 27.	- 70 02	15		STAR CLUSTER IN LMC
MCG+10-09-007	05 56 30.	+ 60 46	39	16.	GALAXY
LW 336	05 56 33.	- 71 12			STAR CLUSTER IN LMC
ZWG 308.006	05 56 36.	+ 64 38		15.6	GALAXY
SL 788	05 56 39.	- 71 13	15		STAR CLUSTER IN LMC
RNGC 2151	05 56 40.	- 69 01			OPEN CLUSTER IN LMC
SL 790	05 56 41.	- 70 21	90		STAR CLUSTER IN LMC
MCG+09-10-004	05 56 42.	+ 53 34	45	15.	GALAXY
MCG+11-08-005	05 56 42.	+ 64 37	30	16.	GALAXY
MCG-04-15-002	05 56 42.	- 25 25	72	15.	GALAXY
RNGC 2131	05 56 42.	- 26 38			GALAXY
RLWT 002	05 56 45.	+ 25 35 12.		15.0	FAINT VERY BLUE STAR
RLWT 001	05 56 46.	+ 25 03 12.		14.0	FAINT VERY BLUE STAR
ZWG 259.007	05 56 48.	+ 53 33		15.6	GALAXY
MCG+11-08-006	05 56 48.	+ 66 24	27	16.	GALAXY
RLWT 003	05 56 49.	+ 25 05 54.		15.0	FAINT VERY BLUE STAR
HODG 09	05 56 49.	- 70 18	72		RED GLOBULAR CLSTR IN LMC
RLWT 004	05 56 51.	+ 24 50 00.		14.5	FAINT VERY BLUE STAR
MCG-04-15-003	05 56 51.	- 23 20	78	14.5	GALAXY
SL 791	05 56 56.	- 68 36	35		STAR CLUSTER IN LMC
KHAV 150	05 57	+ 05 18	3010		DARK NEBULA
LDN 1632	05 57 00.	+ 00 30	6180		DARK NEBULA
RLWT 006	05 57 00.	+ 25 28 30.		15.0	FAINT VERY BLUE STAR
LDN 1550	05 57 00.	+ 32 10	780		DARK NEBULA
MCG+11-08-007	05 57 00.	+ 66 25	42	16.	GALAXY
RNGC 2161	05 57 00.	- 74 21			GLOBULAR CLUSTER IN LMC
SL 789	05 57 00.	- 74 21	70		STAR CLUSTER IN LMC
SS 48	05 57 03.	+ 31 57			DIFFUSE GALACTIC NEBULA
LW 337	05 57 07.	- 74 22			STAR CLUSTER IN LMC
RLWT 007	05 57 09.	+ 25 30 36.		15.0	FAINT VERY BLUE STAR
ISS 0289	05 57 12.	+ 32 11	2279		STELLAR RING
LW 338	05 57 15.	- 71 06			STAR CLUSTER IN LMC
RLWT 005	05 57 16.	+ 22 08 12.		15.5	FAINT VERY BLUE STAR
RLWT 009	05 57 17.	+ 25 20 00.		12.0	FAINT VERY BLUE STAR
RLWT 010	05 57 19.	+ 25 30 30.		15.0	FAINT VERY BLUE STAR
RLWT 008	05 57 24.	+ 23 43 24.		16.5	FAINT VERY BLUE STAR
RLWT 012	05 57 24.	+ 25 32 00.		14.5	FAINT VERY BLUE STAR
ZWG 308.007	05 57 30.	+ 65 23		14.4	GALAXY
UGC 03386	05 57 30.	+ 65 23	72	14.4	GALAXY Sa
RNGC 2153	05 57 30.	- 66 24			OPEN CLUSTER IN LMC
SL 792	05 57 30.	- 66 24	25		STAR CLUSTER IN LMC
RLWT 011	05 57 31.	+ 23 35 36.		12.5	FAINT VERY BLUE STAR
RNGC 2154	05 57 33.	- 67 15			GLOBULAR CLUSTER IN LMC
SL 793	05 57 33.	- 67 15	60		STAR CLUSTER IN LMC
SL 795	05 57 34.	- 65 56	25		STAR CLUSTER IN LMC
RLWT 013	05 57 35.	+ 25 28 48.		14.0	FAINT VERY BLUE STAR
MCG+11-08-008	05 57 36.	+ 65 22	78	14.	GALAXY
LW 339	05 57 39.	- 67 16			STAR CLUSTER IN LMC
RLWT 014	05 57 40.	+ 25 33 06.		15.0	FAINT VERY BLUE STAR
RNGC 2157	05 57 41.	- 69 11		10.0	OPEN CLUSTER IN LMC
SL 794	05 57 41.	- 69 11	100		STAR CLUSTER IN LMC
LW 340	05 57 43.	- 65 00			STAR CLUSTER IN LMC
SL 798	05 57 45.	- 63 54	15		STAR CLUSTER IN LMC
HN 0308	05 57 46.	- 76 55			NEBULA
IC 2160	05 57 46.	- 76 55			NONSTELLAR OBJECT
RLWT 015	05 57 48.	+ 25 34 42.		15.0	FAINT VERY BLUE STAR
LW 341	05 57 48.	- 66 24			STAR CLUSTER IN LMC
HUB E13	05 57 49.	+ 20 14			DIFFUSE NEBULA
HODG 10	05 57 49.	- 67 06	60		RED GLOBULAR CLSTR IN LMC
RNGC 2156	05 57 50.	- 68 27			OPEN CLUSTER IN LMC

OBJECT NAME	RIGHT ASCEN.	DECLINATION	DIAM.	MAGN.	TYPE OF OBJECT
SL 796	05 57 50.	- 68 27	70		STAR CLUSTER IN LMC
LW 342	05 57 52.	- 65 57			STAR CLUSTER IN LMC
LBN 0908	05 58	+ 05 00	9900		BRIGHT NEBULA
VDB.66G 233	05 58	- 29 00	70		DWARF GALAXY
LDN 1611	05 58 00.	+ 05 00	1920		DARK NEBULA
OCL 0467	05 58 00.	+ 23 18	840	7.2	OPEN STAR CLUSTER
LDN 1555	05 58 00.	+ 31 50	1200		DARK NEBULA
7ZW 057	05 58 00.	+ 62 04			COMPACT GALAXY
MCG+11-08-009	05 58 00.	+ 64 48 30.	78	15.	GALAXY
UGC 03387	05 58 00.	+ 64 50	90	16.0	GALAXY Sc
RNGC 2148	05 58 03.	- 59 07			GALAXY
LW 344	05 58 03.	- 63 54			STAR CLUSTER IN LMC
HUB C15	05 58 05.	+ 19 07			DIFFUSE NEBULA
RNGC 2129	05 58 05.	+ 23 18		7.0	OPEN CLUSTER
SL 797	05 58 05.	- 70 25	40		STAR CLUSTER IN LMC
SL 800	05 58 08.	- 67 06	35		STAR CLUSTER IN LMC
RNGC 2159	05 58 09.	- 68 38			OPEN CLUSTER IN LMC
SL 799	05 58 09.	- 68 38	80		STAR CLUSTER IN LMC
ZWG 284.005	05 58 12.	+ 60 31		15.7	GALAXY
UGC 03388	05 58 12.	+ 75 23	72	16.0	GALAXY S
MCG-03-16-006	05 58 12.	- 16 10	36	15.	GALAXY
MCG-03-16-005	05 58 12.	- 19 42	48	14.5	GALAXY
RNGC 2160	05 58 13.	- 68 17			OPEN CLUSTER IN LMC
SL 801	05 58 13.	- 68 17	40		STAR CLUSTER IN LMC
RNGC 2155	05 58 14.	- 65 28			GLOBULAR CLUSTER IN LMC
SL 803	05 58 14.	- 65 28	65		STAR CLUSTER IN LMC
MCG+10-09-008	05 58 15.	+ 57 33	27	16.	GALAXY
HN 0306	05 58 19.	- 33 55			NEBULA
IC 2153	05 58 19.	- 33 55			NONSTELLAR OBJECT
LW 347	05 58 20.	- 65 29			STAR CLUSTER IN LMC
RLWT 016	05 58 23.	+ 24 50 24.		15.5	FAINT VERY BLUE STAR
RLWT 017	05 58 23.	+ 25 27 36.		15.5	FAINT VERY BLUE STAR
MCG+13-05-010	05 58 24.	+ 75 22	39	16.	GALAXY
SL 806	05 58 28.	- 66 03	15		STAR CLUSTER IN LMC
LW 343	05 58 29.	- 71 27			STAR CLUSTER IN LMC
ZC 0558.5+5951	05 58 30.	+ 59 51	16460		CLUSTER OF GALAXIES
UGC 03389	05 58 30.	+ 66 18	60	17.	GALAXY Sc
LW 349	05 58 32.	- 67 06			STAR CLUSTER IN LMC
RLWT 018	05 58 33.	+ 24 26 42.		16.5	FAINT VERY BLUE STAR
MCG-06-14-004	05 58 36.	- 38 13	30	16.	GALAXY
ISS 0290	05 58 42.	+ 29 07	200		STELLAR RING
MCG+06-14-001	05 58 42.	+ 36 06	102	17.	GALAXY
MCG-05-15-002	05 58 42.	- 28 57	180	16.	GALAXY
LW 350	05 58 46.	- 66 03			STAR CLUSTER IN LMC
SL 805	05 58 47.	- 71 29	20		STAR CLUSTER IN LMC
UGC 03390	05 58 48.	+ 36 07	138	16.5	GALAXY
HN 0309	05 58 52.	- 75 08			NEBULA
IC 2161	05 58 52.	- 75 08			NONSTELLAR OBJECT
FATH 1.183	05 58 54.	- 00 29	14		NEBULA
RNGC 2164	05 58 56.	- 68 31		10.5	OPEN CLUSTER IN LMC
SL 808	05 58 56.	- 68 31	130		STAR CLUSTER IN LMC
RNGC 2139	05 58 58.	- 23 40		12.5	GALAXY
SL 804	05 58 59.	- 74 11	70		STAR CLUSTER IN LMC
LW 346	05 58 59.	- 74 11			STAR CLUSTER IN LMC
LP 09829	05 59	- 80 25		13.5	FAINT BLUE STAR
IC 2155	05 59 00.	- 34 01			NONSTELLAR OBJECT
LB 01797	05 59 00.	- 46 53		12.9	FAINT BLUE STAR
HN 0307	05 59 01.	- 34 01			NEBULA
SL 802	05 59 01.	- 75 08	35		STAR CLUSTER IN LMC
SL 807	05 59 02.	- 72 59	115		STAR CLUSTER IN LMC
LW 348	05 59 02.	- 72 59			STAR CLUSTER IN LMC
IC 2154	05 59 03.	- 23 40 26.			TWO STARS
RNGC 2126	05 59 04.	+ 49 54		10.0	OPEN CLUSTER
OCL 0418	05 59 06.	+ 49 54	420	10.5	OPEN STAR CLUSTER
LW 345	05 59 07.	- 75 08			STAR CLUSTER IN LMC
RNGC 2173	05 59 08.	- 72 59			GLOBULAR CLUSTER IN LMC
ZWG 232.005	05 59 12.	+ 49 55		15.5	GALAXY
MCG-04-15-004	05 59 12.	- 21 42	120	13.	GALAXY
RNGC 2171	05 59 12.	- 70 41			GALAXY
SL 809	05 59 12.	- 70 41	5		STAR CLUSTER IN LMC
MCG-04-15-005	05 59 15.	- 23 40	120	12.5	GALAXY
SL 810	05 59 15.	- 68 38	20		STAR CLUSTER IN LMC
ZC 0559.3+2935	05 59 18.	+ 29 35	4700		CLUSTER OF GALAXIES
LB 03473	05 59 18.	- 62 00		15.3	FAINT BLUE STAR
UGC 03391	05 59 24.	+ 57 31	72	16.5	GALAXY Sc
MCG+10-09-009	05 59 24.	+ 57 31	66	16.	GALAXY
RNGC 2166	05 59 30.	- 67 56			OPEN CLUSTER IN LMC
SL 811	05 59 30.	- 67 56	25		STAR CLUSTER IN LMC
RNGC 2142	05 59 31.	- 10 36			NON-EXISTENT OBJECT
PK198-06.1	05 59 36.	+ 09 39	37	13.9	PLANETARY NEBULA
MCG-03-16-008	05 59 48.	- 17 56 30.	18	15.	GALAXY
MCG-03-16-007	05 59 48.	- 20 08	30	15.	GALAXY
RNGC 2152	05 59 48.	- 50 44			UNVERIFIED SOUTHERN OBJECT
MCG-03-16-009	05 59 51.	- 17 53	48	15.	GALAXY
LDN 1612	06 00 00.	+ 04 40	1140		DARK NEBULA
ISS 0291	06 00 00.	+ 27 32	430		STELLAR RING
ZWG 259.008	06 00 00.	+ 51 06		15.3	GALAXY
SL 814	06 00 03.	- 63 45	90		STAR CLUSTER IN LMC
LB 01798	06 00 06.	- 45 48		11.9	FAINT BLUE STAR
SL 813	06 00 08.	- 67 07	15		STAR CLUSTER IN LMC
LW 351	06 00 09.	- 63 44			STAR CLUSTER IN LMC
RNGC 2162	06 00 09.	- 63 45			GLOBULAR CLUSTER IN LMC
RNGC 2172	06 00 09.	- 68 38			OPEN CLUSTER IN LMC
SL 812	06 00 09.	- 68 38	70		STAR CLUSTER IN LMC
RLWT 115	06 00 12.	+ 25 25		13.	FAINT VERY BLUE STAR
LB 01799	06 00 12.	- 46 39		11.6	FAINT BLUE STAR
RLWT 019	06 00 15.	+ 24 40 42.		15.0	FAINT VERY BLUE STAR
OCL 0487	06 00 18.	+ 10 26	840	10.8	OPEN STAR CLUSTER
ZWG 284.006	06 00 18.	+ 57 39		13.7	GALAXY
UGC 03392	06 00 18.	+ 57 39	90	13.7	GALAXY S0 OR E
RNGC 2141	06 00 19.	+ 10 26		11.0	OPEN CLUSTER
RNGC 2128	06 00 20.	+ 57 39		13.5	GALAXY
RNGC 2143	06 00 22.	+ 05 43			NON-EXISTENT OBJECT
YM 23	06 00 22.	+ 30 12	600		SYMMETRIC GALACTIC NEBULA
2ZW 042	06 00 24.	+ 07 50			COMPACT GALAXY
UGC 03393	06 00 24.	+ 07 50	36	14.5	GALAXY COMPACT
IC 0441	06 00 24.	- 12 29 53.			NONSTELLAR OBJECT
MCG-02-16-001	06 00 24.	- 12 30	66	14.	GALAXY
ACK 243-25.1	06 00 24.	- 37 25			PLANETARY NEBULA
LW 354	06 00 24.	- 64 45			STAR CLUSTER IN LMC
RLWT 020	06 00 27.	+ 24 52 24.		12.0	FAINT VERY BLUE STAR
ST 32	06 00 28.	+ 10 32	480		EMISSION NEBULA
LW 352	06 00 29.	- 67 46			STAR CLUSTER IN LMC
RLWT 021	06 00 30.	+ 24 53 00.		14.5	FAINT VERY BLUE STAR
MCG+11-08-010	06 00 30.	+ 66 54	42	16.	GALAXY
LW 353	06 00 32.	- 67 08			STAR CLUSTER IN LMC
DG 085	06 00 36.	+ 30 19	240		REFLECTION NEBULA
ZWG 259.009	06 00 36.	+ 56 09		15.4	GALAXY

OBJECT NAME	RIGHT ASCEN.	DECLINATION	DIAM.	MAGN.	TYPE OF OBJECT
UGC 03394	06 00 36.	+ 56 09	120	15.4	GALAXY SB
MCG+10-09-010	06 00 36.	+ 57 40	39	13.	GALAXY
SS 49	06 00 37.	+ 30 19			DIFFUSE GALACTIC NEBULA
RLWT 023	06 00 40.	+ 24 12 36.		15.5	FAINT VERY BLUE STAR
RLWT 022	06 00 43.	+ 23 35 24.		15.5	FAINT VERY BLUE STAR
MCG+13-05-011	06 00 48.	+ 79 56	18	15.	GALAXY
VDB.66N 066	06 00 48.	- 09 44	108		REFLECTION NEBULA
MCG-04-15-006	06 00 51.	- 23 41	18	15.5	GALAXY
SCHO 0168	06 00 54.	+ 11 01 42.	420		ISOLATED DARK CLOUD
RLWT 024	06 00 56.	+ 22 48 24.		14.5	FAINT VERY BLUE STAR
HSE 449	06 00 57.	- 68 41			STAR CLUSTER IN LMC
RLWT 025	06 00 59.	+ 24 15 06.		15.0	FAINT VERY BLUE STAR
KHAV 151	06 01	+ 16 00	12020		DARK NEBULA
LBN 0825	06 01	+ 30 10	600		BRIGHT NEBULA
LBN 0824	06 01	+ 30 15	120		BRIGHT NEBULA
LBN 0823	06 01	+ 30 30	240		BRIGHT NEBULA
YM 26	06 01 01.	+ 09 19	132		SYMMETRIC GALACTIC NEBULA
RLWT 026	06 01 04.	+ 23 31 54.		13.5	FAINT VERY BLUE STAR
RNGC 2149	06 01 07.	- 09 44			DIFFUSE NEBULA
SL 815	06 01 07.	- 66 50	35		STAR CLUSTER IN LMC
VDB.66N 065	06 01 08.	+ 30 30	276		REFLECTION NEBULA
RLWT 028	06 01 11.	+ 24 04 12.		13.5	FAINT VERY BLUE STAR
RNGC 2177	06 01 11.	- 67 43			OPEN CLUSTER IN LMC
SL 816	06 01 11.	- 67 43	45		STAR CLUSTER IN LMC
LW 355	06 01 11.	- 67 44			STAR CLUSTER IN LMC
RLWT 027	06 01 13.	+ 23 08 42.		15.5	FAINT VERY BLUE STAR
RLWT 029	06 01 13.	+ 24 04 24.		15.5	FAINT VERY BLUE STAR
RLWT 030	06 01 17.	+ 24 29 12.		14.0	FAINT VERY BLUE STAR
UGC 03395	06 01 18.	+ 08 40	78	16.5	GALAXY Sb
DG 086	06 01 18.	+ 30 30	600		REFLECTION NEBULA
ISS 0292	06 01 18.	+ 30 33	196		STELLAR RING
ZWG 348.010	06 01 18.	+ 79 56		15.0	GALAXY
UGC 03396	06 01 18.	+ 79 56	72	15.0	GALAXY E
MCG-03-16-010	06 01 18.	- 20 40	120	13.5	GALAXY
CED 061	06 01 19.	+ 30 30	180		DIFFUSE GALACTIC NEBULA
SS 50	06 01 19.	+ 30 34			DIFFUSE GALACTIC NEBULA
SL 817	06 01 21.	- 70 04	20		STAR CLUSTER IN LMC
RNGC 2176	06 01 25.	- 66 51			OPEN CLUSTER IN LMC
LW 356	06 01 25.	- 66 52			STAR CLUSTER IN LMC
MCG+13-05-012	06 01 30.	+ 79 52	54	16.	GALAXY
SL 818	06 01 36.	- 67 59	25		STAR CLUSTER IN LMC
RLWT 031	06 01 41.	+ 24 05 06.		13.0	FAINT VERY BLUE STAR
RLWT 032	06 01 43.	+ 24 04 12.		12.0	FAINT VERY BLUE STAR
IC 2156	06 01 44.	+ 24 09			OPEN CLUSTER
RNGC 2178	06 01 45.	- 63 46			GALAXY
RLWT 033	06 01 46.	+ 24 03 30.		11.5	FAINT VERY BLUE STAR
RLWT 034	06 01 46.	+ 24 05 06.		11.5	FAINT VERY BLUE STAR
RLWT 035	06 01 48.	+ 24 03 48.		11.5	FAINT VERY BLUE STAR
RLWT 116	06 01 48.	+ 25 21		15.	FAINT VERY BLUE STAR
MCG-05-15-003	06 01 48.	- 32 09	36	15.	GALAXY
RLWT 036	06 01 51.	+ 24 10 18.		13.5	FAINT VERY BLUE STAR
RLWT 037	06 01 51.	+ 24 11 30.		14.0	FAINT VERY BLUE STAR
RLWT 039	06 01 53.	+ 24 04 42.		11.5	FAINT VERY BLUE STAR
SL 820	06 01 53.	- 64 21	55		STAR CLUSTER IN LMC
OCL 0465	06 01 54.	+ 24 00	780	9.7	OPEN STAR CLUSTER
IC 2157	06 01 54.	+ 24 00	480		OPEN CLUSTER
RLWT 117	06 01 54.	+ 25 15		14.	FAINT VERY BLUE STAR
ZC 0601.9+7617	06 01 54.	+ 76 17	1010		CLUSTER OF GALAXIES
ZWG 348.011	06 01 54.	+ 79 52		15.6	GALAXY
UGC 03397	06 01 54.	+ 79 52	78	15.6	GALAXY Sa-b
LW 358	06 01 54.	- 68 00			STAR CLUSTER IN LMC
RLWT 038	06 01 56.	+ 22 54 36.		13.5	FAINT VERY BLUE STAR
RLWT 040	06 01 58.	+ 23 29 00.		13.5	FAINT VERY BLUE STAR
KHAV 152	06 02	+ 04 24	4760		DARK NEBULA
LBN 1008	06 02	- 09 45	120		BRIGHT NEBULA
PK204-08.1	06 02 00.	+ 03 57	174	16.0	PLANETARY NEBULA
LDN 1619	06 02 00.	+ 04 10	780		DARK NEBULA
UGC 03398	06 02 00.	+ 60 35	90	16.0	GALAXY Sc
MCG+10-09-011	06 02 00.	+ 60 35	78	15.	GALAXY
ZWG 329.020	06 02 00.	+ 74 50		15.7	GALAXY
SEY 013	06 02 00.	+ 74 51 30.		15.4	FAINT GALAXY
7ZW 058	06 02 00.	+ 76 17			COMPACT GALAXY
LB 03474	06 02 00.	- 60 34		14.4	FAINT BLUE STAR
RLWT 041	06 02 02.	+ 23 40 48.		12.0	FAINT VERY BLUE STAR
LW 361	06 02 03.	- 65 46			STAR CLUSTER IN LMC
SL 823	06 02 05.	- 64 20	35		STAR CLUSTER IN LMC
LW 362	06 02 05.	- 64 20			STAR CLUSTER IN LMC
MCG+11-08-011	06 02 06.	+ 64 40	57	16.	GALAXY
RLWT 042	06 02 07.	+ 23 32 36.		15.0	FAINT VERY BLUE STAR
YM 28	06 02 12.	+ 04 01	174		SYMMETRIC GALACTIC NEBULA
RLWT 043	06 02 12.	+ 24 31 12.		14.0	FAINT VERY BLUE STAR
ZWG 308.008	06 02 12.	+ 64 41		15.3	GALAXY
SL 822	06 02 13.	- 68 19	50		STAR CLUSTER IN LMC
RLWT 044	06 02 14.	+ 24 47 48.		15.5	FAINT VERY BLUE STAR
MCG-02-16-002	06 02 15.	- 12 37	132	14.	GALAXY
ABC 0552	06 02 16.	+ 76 20		17.8	RICH CLUSTER OF GALAXIES
MCG-03-16-011	06 02 18.	- 20 22	72	15.	GALAXY
RLWT 046	06 02 19.	+ 24 45 36.		15.5	FAINT VERY BLUE STAR
SL 824	06 02 19.	- 66 54	15		STAR CLUSTER IN LMC
RLWT 045	06 02 20.	+ 23 25 18.		14.5	FAINT VERY BLUE STAR
RNGC 2190	06 02 21.	- 74 43			GLOBULAR CLUSTER IN LMC
LW 359	06 02 23.	- 72 31			STAR CLUSTER IN LMC
RLWT 047	06 02 24.	+ 24 46 06.		13.5	FAINT VERY BLUE STAR
UGC 03399	06 02 24.	+ 54 44	60	16.5	GALAXY SBc
LW 364	06 02 25.	- 68 20			STAR CLUSTER IN LMC
RNGC 2181	06 02 26.	- 65 15			OPEN CLUSTER IN LMC
SL 825	06 02 26.	- 65 15	50		STAR CLUSTER IN LMC
RLWT 048	06 02 27.	+ 24 49 24.		12.5	FAINT VERY BLUE STAR
LW 357	06 02 28.	- 74 44			STAR CLUSTER IN LMC
SL 821	06 02 29.	- 72 31	25		STAR CLUSTER IN LMC
UGC 03400	06 02 30.	+ 80 38	72	16.5	GALAXY
SL 819	06 02 33.	- 74 43	80		STAR CLUSTER IN LMC
LW 366	06 02 38.	- 65 16			STAR CLUSTER IN LMC
LW 365	06 02 38.	- 66 55			STAR CLUSTER IN LMC
MCG-04-15-007	06 02 39.	- 21 54	48	15.5	GALAXY
MCG-04-15-008	06 02 45.	- 26 07	150	14.	GALAXY
LW 363	06 02 46.	- 72 21			STAR CLUSTER IN LMC
MCG+09-10-005	06 02 48.	+ 54 14	3	14.	GALAXY
7ZW 059	06 02 48.	+ 76 41			COMPACT GALAXY
LW 360	06 02 49.	- 74 23			STAR CLUSTER IN LMC
MCG-03-16-012	06 02 51.	- 18 16	60	15.	GALAXY
SL 827	06 02 51.	- 65 39	40		STAR CLUSTER IN LMC
KHAV 153	06 03	+ 04 06	4430		DARK NEBULA
LBN 0760	06 03	+ 52 10	1200		BRIGHT NEBULA
LBN 1011	06 03	- 11 00	10800		BRIGHT NEBULA
LBN 1025	06 03	- 16 00	6900		BRIGHT NEBULA
LDN 1618	06 03 00.	+ 04 20	1680		DARK NEBULA
LDN 1586	06 03 00.	+ 16 00	10980		DARK NEBULA

OBJECT NAME	RIGHT ASCEN.	DECLINATION	DIAM.	MAGN.	TYPE OF OBJECT
LDN 1557	06 03 00.	+ 30 00	4140		DARK NEBULA
ZC 0603.0+4746	06 03 00.	+ 47 46	2890		CLUSTER OF GALAXIES
7ZW 060	06 03 00.	+ 67 32			COMPACT GALAXY
ZC 0603.0+7922	06 03 00.	+ 79 22	10420		CLUSTER OF GALAXIES
MCG+14-03-018	06 03 00.	+ 81 04	102	16.	GALAXY
LW 368	06 03 03.	- 67 24			STAR CLUSTER IN LMC
SL 826	06 03 04.	- 72 22	25		STAR CLUSTER IN LMC
SL 829	06 03 14.	- 65 20	5		STAR CLUSTER IN LMC
LW 370	06 03 15.	- 65 40			STAR CLUSTER IN LMC
IC 2158	06 03 21.	- 27 51 00.			NONSTELLAR OBJECT
MCG-03-16-013	06 03 24.	- 19 02	78	15.	GALAXY
LW 371	06 03 26.	- 67 03			STAR CLUSTER IN LMC
HSE 450	06 03 26.	- 67 04			STAR CLUSTER IN LMC
MCG+13-05-013	06 03 30.	+ 81 05	114	15.	GALAXY
MCG-05-15-004	06 03 30.	- 27 50	102	14.	GALAXY
LW 372	06 03 32.	- 65 20			STAR CLUSTER IN LMC
SL 828	06 03 35.	- 74 12	40		STAR CLUSTER IN LMC
LW 367	06 03 35.	- 74 12			STAR CLUSTER IN LMC
ZWG 308.009	06 03 36.	+ 64 44		14.7	GALAXY
MCG+11-08-012	06 03 36.	+ 64 44	45	15.	GALAXY
RLWT 049	06 03 40.	+ 22 23 36.		16.5	FAINT VERY BLUE STAR
LW 369	06 03 41.	- 71 28			STAR CLUSTER IN LMC
MCG-03-16-015	06 03 42.	- 15 38	48	15.	GALAXY
MCG-03-16-014	06 03 42.	- 15 38	150	13.5	GALAXY
RLWT 118	06 03 54.	+ 23 52		16.	FAINT VERY BLUE STAR
LDN 1570	06 04 00.	+ 19 20	540		DARK NEBULA
MCG+08-12-001	06 04 00.	+ 47 27	36	16.	GALAXY
ZWG 362.015	06 04 00.	+ 81 05		15.7	GALAXY
UGC 03401	06 04 00.	+ 81 05	114	15.7	GALAXY Sb
LDN 1643	06 04 00.	- 04 40	3720		DARK NEBULA
MRSL 181+04/1	06 04 12.	+ 30 12	23		HII REGION
MCG+10-09-012	06 04 12.	+ 57 50	39	16.	GALAXY
MCG+13-05-014	06 04 12.	+ 80 29 30.	84	15.	GALAXY
MCG-05-15-005	06 04 12.	- 32 51	42	15.5	GALAXY
RNGC 2158	06 04 16.	+ 24 06		12.0	GLOBULAR CLUSTER
RNGC 2187	06 04 19.	- 69 34			GALAXY
SL 830	06 04 23.	- 69 12	25		STAR CLUSTER IN LMC
OCL 0468	06 04 24.	+ 24 06	450	12.5	OPEN STAR CLUSTER
MCG+07-13-006	06 04 24.	+ 39 15	39	16.	GALAXY
SL 831	06 04 24.	- 66 26	25		STAR CLUSTER IN LMC
B 227	06 04 26.	+ 19 40	720		DARK OBJECT
RNGC 2167	06 04 29.	- 06 12			NON-EXISTENT OBJECT
UGC 03402	06 04 30.	+ 67 59	66	16.0	GALAXY Sc
LDN 1645	06 04 30.	- 05 40	900		DARK NEBULA
LW 374	06 04 30.	- 66 26			STAR CLUSTER IN LMC
ISS 0670	06 04 36.	- 12 30	220		STELLAR RING
MCG-04-15-009	06 04 36.	- 24 40	48	14.	GALAXY
LW 373	06 04 40.	- 74 45			STAR CLUSTER IN LMC
LW 375	06 04 41.	- 69 12			STAR CLUSTER IN LMC
DG 087	06 04 42.	+ 18 42	180		REFLECTION NEBULA
ZWG 329.021	06 04 42.	+ 71 23		14.6	GALAXY
UGC 03403	06 04 42.	+ 71 23	162	14.6	GALAXY SB?c
SEY 014	06 04 42.	+ 71 23 52.		14.6	FAINT GALAXY
CED 062	06 04 44.	+ 18 42	120		DIFFUSE GALACTIC NEBULA
LDN 1575	06 04 48.	+ 18 30	600		DARK NEBULA
LDN 1574	06 04 48.	+ 18 30	600		DARK NEBULA
MCG+11-08-013	06 04 48.	+ 67 58	57	15.	GALAXY
MCG+13-05-015	06 04 48.	+ 80 27 30.	96	15.	GALAXY
RNGC 2163	06 04 54.	- 18 40			NON-EXISTENT OBJECT
VDB.66N 067	06 04 58.	- 06 25	132		REFLECTION NEBULA
KHAV 154	06 05	+ 13 06	9190		DARK NEBULA
CED 064	06 05	+ 15 49			DIFFUSE GALACTIC NEBULA
KHAV 155	06 05	+ 19 06	5310		DARK NEBULA
LBN 0994	06 05	- 06 23	120		BRIGHT NEBULA
LDN 1623	06 05 00.	+ 03 30	1440		DARK NEBULA
LDN 1587	06 05 00.	+ 16 00	10980		DARK NEBULA
MCG+06-14-002	06 05 00.	+ 34 16	27	17.	GALAXY
ZWG 348.012	06 05 00.	+ 80 01		15.2	GALAXY
UGC 03404	06 05 00.	+ 80 01	78	15.2	GALAXY S0
MCG+13-05-016	06 05 00.	+ 80 01	18	16.	GALAXY
ZWG 348.013	06 05 00.	+ 80 29		15.5	GALAXY
ZWG 347.026	06 05 00.	+ 80 29		15.5	GALAXY
UGC 03405	06 05 00.	+ 80 29	96	15.5	GALAXY Sb-c
KARA.72 108A	06 05 00.	+ 80 29	90	15.5	PART OF DOUBLE GALAXY
LDN 1646	06 05 00.	- 06 00	7140		DARK NEBULA
SL 833	06 05 05.	- 67 41	15		STAR CLUSTER IN LMC
REIN 2.049	06 05 05.66	- 06 23 30.4			NEBULA
MCG+12-06-016	06 05 06.	+ 71 25	114	15.	GALAXY
DG 088	06 05 06.	- 06 23	180		REFLECTION NEBULA
CED 063	06 05 08.	- 06 23	60		DIFFUSE GALACTIC NEBULA
VDB.66N 069	06 05 09.	- 06 17	372		REFLECTION NEBULA
RNGC 2170	06 05 11.	- 06 23			DIFFUSE NEBULA
LDN 1644	06 05 12.	- 05 30	300		DARK NEBULA
HUB C14	06 05 15.	- 05 58			DIFFUSE NEBULA
LW 378	06 05 16.	- 67 39			STAR CLUSTER IN LMC
MCG-03-16-016	06 05 18.	- 18 55	30	15.	GALAXY
LDN 1576	06 05 24.	+ 18 10	840		DARK NEBULA
MCG+11-08-014	06 05 24.	+ 65 28 30.	30	17.	GALAXY
UGC 03406	06 05 24.	+ 67 16	78	17.	GALAXY
SL 832	06 05 24.	- 73 31	25		STAR CLUSTER IN LMC
SG 3.250	06 05 29.	+ 20 40	60		DIFFUSE EMISSION NEBULA
LDN 1578	06 05 30.	+ 18 00	840		DARK NEBULA
MRSL 188+00/1	06 05 30.	+ 21 38	540		HII REGION
SG 2.015	06 05 30.	+ 21 38	300		DIFFUSE EMISSION NEBULA
ZWG 203.003	06 05 30.	+ 42 05		14.6	GALAXY
UGC 03407	06 05 30.	+ 42 05	78	14.6	GALAXY Sa
MCG+07-13-007	06 05 30.	+ 42 05	72	16.	GALAXY
MCG+13-05-017	06 05 30.	+ 79 42	90	16.	GALAXY
UGC 03408	06 05 30.	+ 79 43	96	16.0	GALAXY TRP SYS
LB 03475	06 05 30.	- 62 38		13.4	FAINT BLUE STAR
VDB.66N 068	06 05 31.	- 06 14	408		REFLECTION NEBULA
LW 377	06 05 33.	- 71 13			STAR CLUSTER IN LMC
RNGC 2169	06 05 35.	+ 13 58		7.0	OPEN CLUSTER
ISS 0423	06 05 36.	+ 12 57	287		STELLAR RING
OCL 0481	06 05 36.	+ 13 58	780	7.5	OPEN STAR CLUSTER
DG 089	06 05 36.	- 06 13	360		REFLECTION NEBULA
DG 090	06 05 36.	- 06 21	180		REFLECTION NEBULA
LW 376	06 05 36.	- 73 31			STAR CLUSTER IN LMC
HELW 054	06 05 37.	- 06 13 07.			NEBULA
CED 065	06 05 38.	- 06 13	360		DIFFUSE GALACTIC NEBULA
CED 066	06 05 38.	- 06 21	180		DIFFUSE GALACTIC NEBULA
SL 837	06 05 38.	- 65 24	5		STAR CLUSTER IN LMC
HSE 451	06 05 38.	- 67 08			STAR CLUSTER IN LMC
HELW 055	06 05 39.	- 06 21 13.			NEBULA
MCG+08-12-002	06 05 42.	+ 50 34	30	16.	GALAXY
LW 382	06 05 44.	- 65 24			STAR CLUSTER IN LMC
MCG-04-15-010	06 05 45.	- 23 28	96	13.5	GALAXY
RNGC 2168	06 05 46.	+ 24 21		5.5	OPEN CLUSTER
SG 2.016	06 05 47.	+ 15 48	1440		DIFFUSE EMISSION NEBULA
OCL 0466	06 05 48.	+ 24 21	3000	6.0	OPEN STAR CLUSTER
ZC 0605.8+6007	06 05 48.	+ 60 07	670		CLUSTER OF GALAXIES
UGC 03409	06 05 48.	+ 64 36	96	17.	GALAXY DWRF SP
LW 381	06 05 49.	- 68 10			STAR CLUSTER IN LMC
RNGC 2179	06 05 50.	- 21 44		13.5	GALAXY
VDB.66N 070	06 05 53.	- 05 19	720		REFLECTION NEBULA
RNGC 2199	06 05 53.	- 73 24			GALAXY
MCG+11-08-015	06 05 54.	+ 64 34	84	17.	GALAXY
RNGC 2197	06 05 56.	- 67 05			OPEN CLUSTER IN LMC
SL 838	06 05 56.	- 67 05	65		STAR CLUSTER IN LMC
KHAV 156	06 06	+ 02 42	5630		DARK NEBULA
LBN 0864	06 06	+ 15 40	1200		BRIGHT NEBULA
LBN 0863	06 06	+ 15 47	2160		BRIGHT NEBULA
LBN 0862	06 06	+ 15 50	1320		BRIGHT NEBULA
MIL 11	06 06	+ 16 40	4800		SUPERNOVA REMNANT
LBN 0843	06 06	+ 21 36	600		BRIGHT NEBULA
LBN 0990	06 06	- 05 18	360		BRIGHT NEBULA
LBN 0993	06 06	- 06 12	180		BRIGHT NEBULA
LBN 0995	06 06	- 06 20	120		BRIGHT NEBULA
LBN 0999	06 06	- 06 30	720		BRIGHT NEBULA
ZWG 348.014	06 06 00.	+ 80 28		15.1	GALAXY
UGC 03410	06 06 00.	+ 80 28	138	15.1	GALAXY Sb
KARA.72 108B	06 06 00.	+ 80 28	132	15.1	PART OF DOUBLE GALAXY
MCG+14-03-019	06 06 00.	+ 81 09	102	14.	GALAXY
DG 091	06 06 00.	- 05 15	360		REFLECTION NEBULA
MCG-04-15-011	06 06 00.	- 21 43	24	14.	GALAXY
LB 03476	06 06 00.	- 76 18		14.2	FAINT BLUE STAR
RNGC 2193	06 06 01.	- 65 05			GLOBULAR CLUSTER IN LMC
SL 839	06 06 01.	- 65 05	25		STAR CLUSTER IN LMC
ST 31	06 06 04.	+ 15 43	600		EMISSION NEBULA
LW 383	06 06 05.	- 69 16			STAR CLUSTER IN LMC
MRSL 194-01/1	06 06 06.	+ 15 49	2700		HII REGION
ZWG 259.010	06 06 06.	+ 53 31		15.7	GALAXY
MCG+10-09-013	06 06 06.	+ 62 02	96	15.	GALAXY
DG 092	06 06 06.	- 06 30	960		REFLECTION NEBULA
LW 387	06 06 07.	- 65 06			STAR CLUSTER IN LMC
YM 41	06 06 08.	- 05 18	540		SYMMETRIC GALACTIC NEBULA
LW 386	06 06 09.	- 65 35			STAR CLUSTER IN LMC
SG 2.017	06 06 11.	+ 15 48	3000		DIFFUSE EMISSION NEBULA
MCG+09-11-001	06 06 12.	+ 53 32	72	16.	GALAXY
ZWG 284.007	06 06 12.	+ 62 01		15.7	GALAXY
UGC 03411	06 06 12.	+ 62 01	102	15.7	GALAXY Sc
LW 385	06 06 14.	- 67 06			STAR CLUSTER IN LMC
RLWT 050	06 06 17.	+ 20 59 36.		16.5	FAINT VERY BLUE STAR
ZC 0606.3+6258	06 06 18.	+ 62 58	940		CLUSTER OF GALAXIES
LW 379	06 06 19.	- 75 06			STAR CLUSTER IN LMC
MCG+07-13-008	06 06 21.	+ 42 47 30.	42	16.	GALAXY
LW 380	06 06 22.	- 75 26			STAR CLUSTER IN LMC
SL 835	06 06 25.	- 75 06	20		STAR CLUSTER IN LMC
RNGC 2203	06 06 28.	- 75 26			OPEN CLUSTER IN LMC
SL 836	06 06 28.	- 75 26	170		STAR CLUSTER IN LMC
SCHO 0169	06 06 30.	+ 07 48 00.	690		ISOLATED DARK CLOUD
LW 388	06 06 31.	- 68 15			STAR CLUSTER IN LMC
LW 384	06 06 31.	- 71 46			STAR CLUSTER IN LMC
LW 390	06 06 35.	- 67 52			STAR CLUSTER IN LMC
SCHO 0170	06 06 36.	+ 08 22 30.	290		ISOLATED DARK CLOUD
MCG+13-05-018	06 06 36.	+ 79 43 30.	57	16.	GALAXY
SCHO 0171	06 06 38.	+ 05 15 06.	280		ISOLATED DARK CLOUD
ST 30	06 06 39.	+ 20 29	1200		EMISSION NEBULA
SCHO 0172	06 06 40.	+ 14 30 36.	700		ISOLATED DARK CLOUD
CED 067A	06 06 41.	+ 20 30	1740		DIFFUSE GALACTIC NEBULA
LW 389	06 06 41.	- 70 28			STAR CLUSTER IN LMC
MRSL 190+00/1	06 06 42.	+ 20 31	2400		HII REGION
SL 840	06 06 42.	- 69 21	20		STAR CLUSTER IN LMC
RNGC 2174	06 06 43.	+ 20 31			DIFFUSE NEBULA
SCHO 0173	06 06 44.	+ 08 28 00.	460		ISOLATED DARK CLOUD
MCG+02-16-001	06 06 45.	+ 13 28	120	14.	GALAXY
SCHO 0174	06 06 46.	+ 10 22 18.	730		ISOLATED DARK CLOUD
SL 834	06 06 47.	- 70 28	20		STAR CLUSTER IN LMC
OCL 0476	06 06 48.	+ 20 20	1080	6.8	OPEN STAR CLUSTER
ASS 51	06 06 48.	+ 21 36	18000		OB ASSOCIATION GEM OB1
ZWG 260.001	06 06 48.	+ 56 46		15.5	GALAXY
ZWG 259.011	06 06 48.	+ 56 46		15.5	GALAXY
UGC 03412	06 06 48.	+ 79 44	96	16.0	GALAXY Sc
RNGC 2175	06 06 49.	+ 20 20		7.0	CLUSTER WITH NEBULOSITY
LW 392	06 06 50.	- 65 19			STAR CLUSTER IN LMC
HSE 452	06 06 50.	- 67 08			STAR CLUSTER IN LMC
CED 067B	06 06 53.	+ 20 26			DIFFUSE GALACTIC NEBULA
IC 2159	06 06 54.	+ 20 25			DIFFUSE NEBULA
MCG-04-15-012	06 06 54.	- 25 07	66	14.5	GALAXY
VDB.66N 072	06 06 57.	- 06 17	276		REFLECTION NEBULA
RNGC 2180	06 06 58.	+ 04 43			NON-EXISTENT OBJECT
SCHO 0175	06 06 58.	+ 10 24 18.	460		ISOLATED DARK CLOUD
SG 1.11	06 06 59.	+ 20 30	1800		DIFFUSE EMISSION NEBULA
LBN 0877	06 07	+ 12 50	60		BRIGHT NEBULA
LBN 0873	06 07	+ 14 10	1260		BRIGHT NEBULA
LBN 0855	06 07	+ 18 42	120		BRIGHT NEBULA
LBN 0854	06 07	+ 20 30	2100		BRIGHT NEBULA
LBN 0998	06 07	- 06 20	120		BRIGHT NEBULA
LDN 1629	06 07 00.	+ 02 30	4380		DARK NEBULA
LDN 1593	06 07 00.	+ 13 40	1020		DARK NEBULA
LDN 1592	06 07 00.	+ 13 40	1020		DARK NEBULA
LDN 1591	06 07 00.	+ 13 40	1020		DARK NEBULA
ZWG 362.016	06 07 00.	+ 81 10		13.8	GALAXY
UGC 03413	06 07 00.	+ 81 10	156	13.8	GALAXY SBc
ZC 0607.0+8613	06 07 00.	+ 86 13	1140		CLUSTER OF GALAXIES
DG 093	06 07 00.	- 06 19	180		REFLECTION NEBULA
LW 391	06 07 00.	- 69 21			STAR CLUSTER IN LMC
CED 068	06 07 02.	- 06 19	180		DIFFUSE GALACTIC NEBULA
RNGC 2182	06 07 05.	- 06 19			DIFFUSE NEBULA
REIN 2.050	06 07 05.05	- 06 18 56.9			NEBULA
SCHO 0176	06 07 06.	+ 10 14 36.	360		ISOLATED DARK CLOUD
MCG+11-08-016	06 07 06.	+ 67 37	27	15.	GALAXY
RNGC 2165	06 07 07.	+ 51 41			NON-EXISTENT OBJECT
REIN 2.051	06 07 07.97	- 06 18 59.6			NEBULA
MCG+11-08-017	06 07 12.	+ 64 17 30.	39	16.	GALAXY
MCG-05-15-006	06 07 12.	- 27 46	60	15.	GALAXY
SCHO 0177	06 07 18.	+ 08 27 06.	620		ISOLATED DARK CLOUD
VDB.66N 071	06 07 18.	+ 14 06	204		REFLECTION NEBULA
RNGC 2191	06 07 21.	- 52 30			UNVERIFIED SOUTHRN OBJECT
MRSL 196-03/1	06 07 24.	+ 12 50	60		HII REGION
MRSL 196-02/1	06 07 24.	+ 13 21	3600		HII REGION
7ZW 061	06 07 24.	+ 64 17			COMPACT GALAXY
ZWG 308.010	06 07 24.	+ 64 18		13.9	GALAXY
UGC 03414	06 07 24.	+ 64 18	42	13.9	GALAXY PECULR
LW 395	06 07 26.	- 65 19			STAR CLUSTER IN LMC
LW 393	06 07 27.	- 72 13			STAR CLUSTER IN LMC

OBJECT NAME	RIGHT ASCEN.	DECLINATION	DIAM.	MAGN.	TYPE OF OBJECT
MCG+06-14-003	06 07 30.	+ 35 39	24	16.	GALAXY
ZWG 174.001	06 07 30.	+ 35 40		15.5	GALAXY
MCG+11-08-018	06 07 30.	+ 64 17	39	14.	GALAXY
UGC 03415	06 07 30.	+ 70 15	102	16.5	GALAXY DWARF
MCG-03-16-017	06 07 36.	- 19 44	24	15.5	GALAXY
SCHO 0178	06 07 38.	+ 04 50 24.	280		ISOLATED DARK CLOUD
SCHO 0179	06 07 39.	+ 10 22 12.	610		ISOLATED DARK CLOUD
7ZW 062	06 07 42.	+ 64 34			COMPACT GALAXY
SL 878	06 07 43.	- 71 46	70		STAR CLUSTER IN LMC
ZWG 308.011	06 07 48.	+ 64 35		15.7	GALAXY
LW 396	06 07 48.	- 64 37			STAR CLUSTER IN LMC
LW 394	06 07 48.	- 71 45			STAR CLUSTER IN LMC
SL 842	06 07 49.	- 62 58	25		STAR CLUSTER IN LMC
SCHO 0180	06 07 52.	+ 10 20 30.	440		ISOLATED DARK CLOUD
CED 067C	06 07 53.	+ 20 38	240		DIFFUSE GALACTIC NEBULA
PIS 1	06 07 53.8	+ 20 37 24.	222		STAR CLUSTER BY NGC 2175
ZC 0607.9+4848	06 07 54.	+ 48 48	2150		CLUSTER OF GALAXIES
ZC 0607.9+5823	06 07 54.	+ 58 23	1210		CLUSTER OF GALAXIES
LW 399	06 07 55.	- 62 59			STAR CLUSTER IN LMC
SL 841	06 07 58.	- 67 30	70		STAR CLUSTER IN LMC
LBN 0874	06 08	+ 13 45	1260		BRIGHT NEBULA
KHAV 157	06 08	+ 14 11	7200		DARK NEBULA
KHAV 158	06 08	+ 17 29	13680		DARK NEBULA
LBN 0996	06 08	- 06 10	120		BRIGHT NEBULA
LBN 0997	06 08	- 06 12	120		BRIGHT NEBULA
ZWG 259.012	06 08 00.	+ 51 53		15.5	GALAXY
MCG+09-11-002	06 08 00.	+ 51 53	30	15.	GALAXY
MCG+11-08-019	06 08 00.	+ 64 34 30.	66	15.	GALAXY
ZWG 329.022	06 08 00.	+ 69 45		15.2	GALAXY
UGC 03416	06 08 00.	+ 69 45	144	15.2	GALAXY Sc
KARA.73B 0165	06 08 00.	+ 69 45	132	15.2	ISOLATED GALAXY S
MCG-04-15-013	06 08 00.	- 22 36 30.	18	17.	GALAXY
MCG-06-14-005	06 08 00.	- 33 34	12	15.5	GALAXY
MCG-06-14-006	06 08 00.	- 33 38	12	15.	GALAXY
RLWT 119	06 08 06.	+ 23 11		16.5	FAINT VERY BLUE STAR
7ZW 063	06 08 06.	+ 61 33			COMPACT GALAXY
MCG+12-06-017	06 08 06.	+ 69 44	114	15.	GALAXY
MCG-02-16-003	06 08 06.	- 09 21 30.	30	15.	GALAXY
MCG-06-14-007	06 08 06.	- 33 37	36	15.	GALAXY
LW 398	06 08 10.	- 67 30			STAR CLUSTER IN LMC
ZWG 233.001	06 08 12.	+ 50 23		15.6	GALAXY
ZWG 308.012	06 08 12.	+ 64 35		14.9	GALAXY
SL 843	06 08 12.	- 67 59	35		STAR CLUSTER IN LMC
DG 094	06 08 18.	- 06 11	120		REFLECTION NEBULA
RNGC 2188	06 08 18.	- 34 05		12.5	GALAXY
LW 400	06 08 18.	- 67 58			STAR CLUSTER IN LMC
CED 069	06 08 20.	- 06 12	60		DIFFUSE GALACTIC NEBULA
YC 0608-34	06 08 22.	- 34 06 24.			UNUSUAL SOUTHERN GALAXY
RNGC 2183	06 08 23.	- 06 12			DIFFUSE NEBULA
PK197-03.1	06 08 24.	+ 11 47	40	18.2	PLANETARY NEBULA
ZWG 259.013	06 08 24.	+ 51 32		15.7	GALAXY
UGC 03417	06 08 24.	+ 51 32	78	15.7	GALAXY Sb
MCG-06-14-008	06 08 24.	- 34 05	180	12.6	GALAXY
LW 401	06 08 24.	- 64 48			STAR CLUSTER IN LMC
HN 0310	06 08 24.	- 75 21			NEBULA
IC 2164	06 08 26.	- 75 20			NONSTELLAR OBJECT
RNGC 2184	06 08 27.	- 03 31			NON-EXISTENT OBJECT
LW 397	06 08 29.	- 72 29			STAR CLUSTER IN LMC
SL 845	06 08 30.	- 64 48	15		STAR CLUSTER IN LMC
SL 844	06 08 33.	- 67 19	25		STAR CLUSTER IN LMC
HSE 453	06 08 33.	- 67 20			STAR CLUSTER IN LMC
VDB.66N 073	06 08 34.	- 06 11	168		REFLECTION NEBULA
MCG+09-11-003	06 08 36.	+ 56 57	24	17.	GALAXY
DG 095	06 08 36.	- 06 11	120		REFLECTION NEBULA
SL 846	06 08 36.	- 66 35	15		STAR CLUSTER IN LMC
CED 070	06 08 38.	- 06 12	150		DIFFUSE GALACTIC NEBULA
LW 402	06 08 39.	- 67 19			STAR CLUSTER IN LMC
RNGC 2185	06 08 41.	- 06 12			DIFFUSE NEBULA
MRSL 192-00/1	06 08 42.	+ 17 28	120		HII REGION
LW 403	06 08 42.	- 66 35			STAR CLUSTER IN LMC
SCHO 0181	06 08 44.	+ 12 18 42.	360		ISOLATED DARK CLOUD
ARC 0553	06 08 46.	+ 48 37		15.3	RICH CLUSTER OF GALAXIES
ZWG 174.002	06 08 48.	+ 37 10		15.6	GALAXY
ZWG 203.004	06 08 48.	+ 44 27		15.0	GALAXY
UGC 03418	06 08 48.	+ 44 27	66	15.0	GALAXY SBb
SCHO 0182	06 08 50.	+ 15 45 00.	500		ISOLATED DARK CLOUD
MCG+07-13-009	06 08 51.	+ 44 27	60	15.	GALAXY
MCG+08-12-003	06 08 54.	+ 47 32	36	17.	GALAXY
KHAV 160	06 09	+ 03 17	9440		DARK NEBULA
KHAV 159	06 09	+ 12 23	1860		DARK NEBULA
LBN 0858	06 09	+ 18 02	600		BRIGHT NEBULA
LBN 1000	06 09	- 06 08	120		BRIGHT NEBULA
7ZW 064	06 09 00.	+ 62 49			COMPACT GALAXY
MCG+11-08-020	06 09 00.	+ 66 09 30.	57	16.	GALAXY
UGC 03419	06 09 00.	+ 66 21	60	16.0	GALAXY Sa-b
ZWG 348.015	06 09 00.	+ 75 58		14.4	GALAXY
UGC 03420	06 09 00.	+ 75 58	204	14.4	GALAXY Sb
KARA.73B 0166	06 09 00.	+ 75 58	162	14.4	ISOLATED GALAXY S
MCG-03-16-018	06 09 00.	- 15 29	48	14.	GALAXY
MCG-06-14-009	06 09 00.	- 33 52	36	15.	GALAXY
LW 406	06 09 01.	- 66 49			STAR CLUSTER IN LMC
LW 404	06 09 05.	- 70 23			STAR CLUSTER IN LMC
ZWG 308.013	06 09 06.	+ 64 43		14.8	GALAXY
MCG+11-08-021	06 09 06.	+ 64 44	18	15.	GALAXY
SL 847	06 09 10.	- 71 18	50		STAR CLUSTER IN LMC
MCG+08-12-004	06 09 12.	+ 47 58	30	17.	GALAXY
UGC 03421	06 09 12.	+ 53 05	66	18.	GALAXY
MCG+11-08-022	06 09 12.	+ 66 19	39	16.	GALAXY
MCG-06-14-010	06 09 12.	- 33 18	60	15.5	GALAXY
MCG-06-14-011	06 09 12.	- 36 13	48	15.	GALAXY
LW 407	06 09 12.	- 68 08			STAR CLUSTER IN LMC
SEY 015	06 09 14.	+ 71 09 32.		15.0	FAINT GALAXY
LW 409	06 09 16.	- 67 34			STAR CLUSTER IN LMC
VDB.66N 074	06 09 17.	- 06 06	72		REFLECTION NEBULA
ZWG 329.023	06 09 18.	+ 71 09		14.5	GALAXY
UGC 03422	06 09 18.	+ 71 09	144	14.5	GALAXY Sb/SBb
MCG+13-05-019	06 09 18.	+ 75 56 30.	138	14.	GALAXY
DG 096	06 09 18.	- 06 08	120		REFLECTION NEBULA
LB 01800	06 09 18.	- 48 44		13.5	FAINT BLUE STAR
CED 071	06 09 20.	- 06 09	60		DIFFUSE GALACTIC NEBULA
RNGC 2186	06 09 22.	+ 05 28		9.0	OPEN CLUSTER
HELW 127	06 09 23.	- 06 07 36.			NEBULA
MRSL 192-00/2	06 09 24.	+ 18 04	660		HII REGION
UGC 03423	06 09 24.	+ 78 52	96	16.5	GALAXY
SG 2.018	06 09 25.	+ 18 02	378		DIFFUSE EMISSION NEBULA
SL 848	06 09 28.	- 74 03	35		STAR CLUSTER IN LMC
OCL 0498	06 09 30.	+ 05 28	600	11.0	OPEN STAR CLUSTER
ZWG 233.002	06 09 30.	+ 46 31		15.1	GALAXY
UGC 03424	06 09 30.	+ 53 05	90	17.	GALAXY DWARF?
LW 405	06 09 34.	- 74 02			STAR CLUSTER IN LMC
SCHO 0183	06 09 36.	+ 12 01 36.	470		ISOLATED DARK CLOUD
MCG+11-08-023	06 09 36.	+ 66 35	156	14.	GALAXY
MCG+12-06-018	06 09 36.	+ 71 09	102	15.	GALAXY
MCG+13-05-020	06 09 36.	+ 78 50 30.	60	16.	GALAXY
SG 2.019	06 09 37.	+ 17 58	18		DIFFUSE EMISSION NEBULA
MRSL 192-00/3	06 09 42.	+ 17 58	60		HII REGION
MCG+07-13-010	06 09 42.	+ 39 03 30.	42	16.	GALAXY
ZWG 203.005	06 09 42.	+ 39 04		15.4	GALAXY
RNGC 2189	06 09 43.	+ 01 08			NON-EXISTENT OBJECT
MIN.47 04	06 09 43.	+ 18 00			DIFFUSE NEBULA
SG 2.020	06 09 43.	+ 18 00	90		DIFFUSE EMISSION NEBULA
LW 411	06 09 45.	- 68 48			STAR CLUSTER IN LMC
HN 0026	06 09 47.	+ 01 08			NEBULA
HN 0025	06 09 47.	+ 01 08			NEBULA
ZWG 308.014	06 09 48.	+ 66 35		15.3	GALAXY
UGC 03425	06 09 48.	+ 66 35	150	15.3	GALAXY Sb
ZWG 329.024	06 09 48.	+ 71 03		13.8	GALAXY
UGC 03426	06 09 48.	+ 71 03	108	13.8	GALAXY S0
SEY 016	06 09 48.	+ 71 03 36.		14.3	FAINT GALAXY
LW 410	06 09 48.	- 71 39			STAR CLUSTER IN LMC
LW 408	06 09 50.	- 73 50			STAR CLUSTER IN LMC
SL 850	06 09 52.	- 65 58	55		STAR CLUSTER IN LMC
MRSL 192-00/4	06 09 54.	+ 17 59	180		HII REGION
RLWT 051	06 09 54.	+ 22 11 00.		16.5	FAINT VERY BLUE STAR
RNGC 2209	06 09 56.	- 73 51			GLOBULAR CLUSTER IN LMC
SL 849	06 09 56.	- 73 51	145		STAR CLUSTER IN LMC
KAZ 1	06 10	+ 12 24	44	12.98	PLANETARY NEBULA
KHAV 162	06 10	+ 17 23	13590		DARK NEBULA
LBN 0859	06 10	+ 18 00	240		BRIGHT NEBULA
KHAV 161	06 10	- 04 43	12350		DARK NEBULA
LDN 1628	06 10 00.	+ 03 00	7440		DARK NEBULA
LDN 1560	06 10 00.	+ 28 00	6120		DARK NEBULA
ZWG 348.016	06 10 00.	+ 80 05		14.3	GALAXY
UGC 03427	06 10 00.	+ 80 05	126	14.3	GALAXY Sb
KARA.72 109B	06 10 00.	+ 80 05	36		PART OF DOUBLE GALAXY
KARA.72 109A	06 10 00.	+ 80 05	30	14.3	PART OF DOUBLE GALAXY
MCG+13-05-021	06 10 00.	+ 80 05 30.	90	14.	GALAXY
LB 03477	06 10 00.	- 63 53		13.9	FAINT BLUE STAR
MIN.47 05	06 10 01.	+ 17 59			DIFFUSE NEBULA
SG 2.021	06 10 01.	+ 18 00	144		DIFFUSE EMISSION NEBULA
RLWT 052	06 10 02.	+ 22 12 06.		15.5	FAINT VERY BLUE STAR
RNGC 2196	06 10 03.	- 21 47		12.5	GALAXY
MCG-04-15-014	06 10 03.	- 21 47	132	12.	GALAXY
UGC 03428	06 10 06.	+ 67 45	96	16.5	GALAXY
MRK 003	06 10 06.	+ 71 03	35	15.	GALAXY WITH UV CONTINUUM
KW 02	06 10 06.	+ 71 03	44		SEYFERT GALAXY
VVI 17	06 10 06.	+ 71 03	35	13.8	SEYFERT GALAXY
MCG+12-06-019	06 10 06.	+ 71 03	33	15.	GALAXY
LW 413	06 10 10.	- 65 57			STAR CLUSTER IN LMC
SCHO 0184	06 10 11.	+ 15 43 36.	460		ISOLATED DARK CLOUD
SCHO 0185	06 10 12.	+ 12 22 30.	1100		ISOLATED DARK CLOUD
MRSL 192-00/5	06 10 12.	+ 17 59	180		HII REGION
MCG+09-11-004	06 10 12.	+ 51 04	57	16.	GALAXY
SL 851	06 10 13.	- 70 46	70		STAR CLUSTER IN LMC
LW 412	06 10 13.	- 70 47			STAR CLUSTER IN LMC
IC 2162	06 10 17.	+ 17 59 18.			HII REGION
MCG+11-08-024	06 10 18.	+ 67 45	78	16.	GALAXY
MCG-02-16-004	06 10 18.	- 14 14 30.	48	16.	GALAXY
MCG-02-16-005	06 10 18.	- 14 21 30.	36	17.	GALAXY
CED 072	06 10 19.	+ 17 59	60		DIFFUSE GALACTIC NEBULA
IC 0440	06 10 19.	+ 80 05			NONSTELLAR OBJECT
LW 414	06 10 19.	- 68 21			STAR CLUSTER IN LMC
OCL 0493	06 10 24.	+ 07 00	420		OPEN STAR CLUSTER
RLWT 053	06 10 24.	+ 20 44 30.		16.5	FAINT VERY BLUE STAR
7ZW 065	06 10 24.	+ 65 39			COMPACT GALAXY
MCG-02-16-006	06 10 27.	- 14 15	36	17.	GALAXY
ZWG 260.002	06 10 30.	+ 51 08		15.7	GALAXY
ZWG 348.017	06 10 30.	+ 78 22		11.1	GALAXY
UGC 03429	06 10 30.	+ 78 22	354	11.1	GALAXY S
KARA.72 110A	06 10 30.	+ 78 22	348	11.1	PART OF DOUBLE GALAXY
SL 852	06 10 32.	- 68 26	40		STAR CLUSTER IN LMC
LW 415	06 10 33.	- 67 14			STAR CLUSTER IN LMC
WS 40	06 10 35.	- 67 55 43.		16.3	PLANETARY NEB. IN LMC
MRSL 192+00/1	06 10 36.	+ 17 56	60		HII REGION
MCG+09-11-005	06 10 36.	+ 51 07	8	16.	GALAXY
ZWG 308.015	06 10 36.	+ 66 51		15.3	GALAXY
MCG+13-05-022	06 10 36.	+ 78 22	306	11.2	GALAXY
LW 416	06 10 38.	- 68 25			STAR CLUSTER IN LMC
SCHO 0186	06 10 39.	+ 12 21 06.	510		ISOLATED DARK CLOUD
MCG+11-08-025	06 10 42.	+ 66 51	60	15.	GALAXY
RNGC 2146	06 10 43.	+ 78 22		11.5	GALAXY
LW 417	06 10 46.	- 70 18			STAR CLUSTER IN LMC
ACK 184+04.1	06 10 48.	+ 26 54			PLANETARY NEBULA
ZWG 329.025	06 10 48.	+ 69 12		15.3	GALAXY
SL 855	06 10 49.	- 65 03	70		STAR CLUSTER IN LMC
SL 854	06 10 50.	- 67 07	20		STAR CLUSTER IN LMC
SL 853	06 10 52.	- 70 17	40		STAR CLUSTER IN LMC
ZWG 308.016	06 10 54.	+ 64 27		15.1	GALAXY
UGC 03430	06 10 54.	+ 64 27	60	15.1	GALAXY
RNGC 2205	06 10 59.	- 62 33			GALAXY
KHAV 163	06 11	+ 07 29	6500		DARK NEBULA
OCL 0485	06 11 00.	+ 12 49	780	10.4	OPEN STAR CLUSTER
RNGC 2194	06 11 00.	+ 12 50		10.5	OPEN CLUSTER
MCG+11-08-026	06 11 00.	+ 64 27	51	15.	GALAXY
ZWG 363.002	06 11 00.	+ 83 14		15.6	GALAXY
ZWG 362.017	06 11 00.	+ 83 14		15.6	GALAXY
LW 420	06 11 01.	- 65 06			STAR CLUSTER IN LMC
LW 418	06 11 02.	- 67 06			STAR CLUSTER IN LMC
MCG+08-12-005	06 11 06.	+ 50 56	30	17.	GALAXY
ZWG 260.003	06 11 12.	+ 51 40		15.6	GALAXY
ZWG 284.008	06 11 12.	+ 61 05		15.6	GALAXY
LW 421	06 11 17.	- 69 08			STAR CLUSTER IN LMC
RLWT 120	06 11 18.	+ 25 55		15.	FAINT VERY BLUE STAR
MCG+08-12-006	06 11 18.	+ 50 57	18	17.	GALAXY
MCG-05-15-007	06 11 18.	- 27 42	84	14.5	GALAXY
RNGC 2198	06 11 19.	+ 01 00			NON-EXISTENT OBJECT
SL 856	06 11 21.	- 70 02	70		STAR CLUSTER IN LMC
MCG+10-09-014	06 11 24.	+ 61 06	42	15.	GALAXY
RNGC 2195	06 11 27.	+ 17 44			DIFFUSE NEBULA
LW 419	06 11 29.	- 71 32			STAR CLUSTER IN LMC
RNGC 2213	06 11 29.	- 71 33			GLOBULAR CLUSTER IN LMC
SL 857	06 11 29.	- 71 33	40		STAR CLUSTER IN LMC
MCG+11-08-027	06 11 30.	+ 66 31	18	15.	GALAXY
MCG-06-14-012	06 11 30.	- 33 39	24	15.	GALAXY
ZC 0611.6+4601	06 11 36.	+ 46 01	4840		CLUSTER OF GALAXIES
MCG+09-11-006	06 11 36.	+ 51 59 30.	30	16.	GALAXY

OBJECT NAME	RIGHT ASCEN.	DECLINATION	DIAM.	MAGN.	TYPE OF OBJECT
RNGC 2192	06 11 39.	+ 39 52		11.0	OPEN CLUSTER
MRSL 196-01/1	06 11 42.	+ 13 50	240		HII REGION
OCL 0437	06 11 42.	+ 39 52	540	10.9	OPEN STAR CLUSTER
ZWG 308.017	06 11 42.	+ 66 31		14.9	GALAXY
MCG+08-12-007	06 11 45.	+ 49 47 30.	54	16.	GALAXY
MCG+09-11-007	06 11 45.	+ 52 19 30.	30	16.	GALAXY
RNGC 2210	06 11 47.	- 69 07		11.0	GLOBULAR CLUSTER IN LMC
SL 858	06 11 47.	- 69 08	100		STAR CLUSTER IN LMC
MIN.46 03	06 11 50.	+ 13 51			DIFFUSE NEBULA
RNGC 2200	06 11 51.	- 43 39			GALAXY
LW 423	06 11 53.	- 69 07			STAR CLUSTER IN LMC
ZWG 233.003	06 11 54.	+ 49 47		15.0	GALAXY
ZWG 348.018	06 11 54.	+ 76 50		13.6	GALAXY
UGC 03431	06 11 54.	+ 76 50	108	13.6	GALAXY E
RNGC 2201	06 11 57.	- 43 41			GALAXY
YM 25	06 11 59.	+ 12 31	132		SYMMETRIC GALACTIC NEBULA
LBN 0876	06 12	+ 13 51	120		BRIGHT NEBULA
LBN 1003	06 12	- 06 15	60		BRIGHT NEBULA
LBN 1018	06 12	- 12 00	1740		BRIGHT NEBULA
ZWG 260.004	06 12 00.	+ 54 33		15.2	GALAXY
MCG+09-11-008	06 12 00.	+ 54 34	41	15.	GALAXY
ZWG 284.009	06 12 00.	+ 57 04		15.5	GALAXY
UGC 03432	06 12 00.	+ 57 04	108	15.5	GALAXY Sc
MCG+11-08-028	06 12 00.	+ 66 32	24	17.	GALAXY
MCG+13-05-023	06 12 00.	+ 76 51	39	14.	GALAXY
MCG+08-12-008	06 12 06.	+ 49 07	42	16.	GALAXY
MCG+11-08-029	06 12 06.	+ 66 34	36	16.	GALAXY
MCG-06-14-013	06 12 06.	- 38 44	30	15.	GALAXY
LW 422	06 12 07.	- 71 54			STAR CLUSTER IN LMC
ACK 197-02.1	06 12 12.	+ 12 21			PLANETARY NEBULA
MCG+08-12-009	06 12 12.	+ 47 07 30.	18	16.	GALAXY
DG 097	06 12 12.	- 06 15	60		REFLECTION NEBULA
MCG-06-14-014	06 12 12.	- 37 40	60	15.	GALAXY
HSE 454	06 12 13.	- 65 12			STAR CLUSTER IN LMC
MCG-02-16-007	06 12 15.	- 12 12	36	16.	GALAXY
ZWG 233.004	06 12 18.	+ 47 07		15.1	GALAXY
ZC 0612.3+5547	06 12 18.	+ 55 47	8060		CLUSTER OF GALAXIES
ZWG 308.018	06 12 18.	+ 66 33		15.5	GALAXY
ZWG 233.005	06 12 24.	+ 47 11		15.2	GALAXY
MCG+11-08-030	06 12 30.	+ 66 37	39	15.	GALAXY
SG 2.022	06 12 31.	+ 22 26	1440		DIFFUSE EMISSION NEBULA
SL 859	06 12 32.	- 67 06	20		STAR CLUSTER IN LMC
UGC 03433	06 12 36.	+ 00 13	162	16.0	GALAXY S
ZWG 308.019	06 12 36.	+ 66 37		15.4	GALAXY
MCG-04-15-015	06 12 36.	- 25 46	18	15.	GALAXY
MCG+00-16-001	06 12 42.	+ 00 12	90	15.5	GALAXY
PK197-02.1	06 12 42.	+ 12 24	135		PLANETARY NEBULA
DG 098	06 12 42.	- 06 17	60		REFLECTION NEBULA
MCG-04-15-016	06 12 42.	- 25 45 30.	24	15.	GALAXY
LW 424	06 12 44.	- 67 04			STAR CLUSTER IN LMC
MCG-03-16-019	06 12 48.	- 19 26	60	14.5	GALAXY
MCG-03-16-020	06 12 51.	- 19 27 30.	42	14.5	GALAXY
MCG-04-15-017	06 12 51.	- 25 51	24	15.5	GALAXY
ZWG 233.006	06 12 54.	+ 50 03		15.6	GALAXY
MAI 014	06 12 59.	+ 74 49	53		DWARF SPHEROIDAL GALAXY
VDB.66G 234	06 13	- 26 31	100		DWARF GALAXY
YM 24	06 13 00.	+ 14 21	324		SYMMETRIC GALACTIC NEBULA
KHAV 164	06 13	+ 13 53	4960		DARK NEBULA
LBN 0875	06 13	+ 14 18	360		BRIGHT NEBULA
LBN 1004	06 13	- 06 17	60		BRIGHT NEBULA
RNGC 2214	06 13 01.	- 68 16		11.0	OPEN CLUSTER IN LMC
SL 860	06 13 01.	- 68 17	185		STAR CLUSTER IN LMC
LW 425	06 13 01.	- 68 25			STAR CLUSTER IN LMC
PK196-01.1	06 13 03.	+ 14 17	350		PLANETARY NEBULA
ZC 0613.1+6720	06 13 06.	+ 67 20	4840		CLUSTER OF GALAXIES
LW 426	06 13 07.	- 68 15			STAR CLUSTER IN LMC
RLWT 121	06 13 12.	+ 25 37		15.5	FAINT VERY BLUE STAR
KARA.68 048	06 13 12.	- 26 34	81		DWARF GALAXY
SG 2.023	06 13 13.	+ 22 18	1800		DIFFUSE EMISSION NEBULA
UGC 03434	06 13 18.	+ 32 14	66	17.	GALAXY SB
ZC 0613.3+7658	06 13 18.	+ 76 58	2290		CLUSTER OF GALAXIES
LW 427	06 13 19.	- 68 21			STAR CLUSTER IN LMC
MCG+08-12-010	06 13 24.	+ 48 07	42	16.	GALAXY
MCG+13-05-024	06 13 24.	+ 80 07 30.	42	17.	GALAXY
IC 0443	06 13 26.	+ 22 29			SUPERNOVA REMNANT
MCG-04-15-018	06 13 27.	- 22 34 30.	48	16.	GALAXY
ABC 0554	06 13 28.	+ 67 28		17.0	RICH CLUSTER OF GALAXIES
MCG+14-03-020	06 13 30.	+ 82 22	48	15.	GALAXY
OCL 0572	06 13 30.	- 18 38	1260	9.6	OPEN STAR CLUSTER
RNGC 2204	06 13 30.	- 18 38		9.5	OPEN CLUSTER
LW 428	06 13 46.	- 70 10			STAR CLUSTER IN LMC
SL 861	06 13 46.	- 70 12	20		STAR CLUSTER IN LMC
7ZW 066	06 13 48.	+ 60 01			COMPACT GALAXY
LW 430	06 13 48.	- 69 25			STAR CLUSTER IN LMC
SCHO 0187	06 13 52.	+ 09 43 00.	310		ISOLATED DARK CLOUD
ISS 0367	06 13 54.	+ 20 33	296		STELLAR RING
ZC 0613.9+4753	06 13 54.	+ 47 53	5380		CLUSTER OF GALAXIES
SL 862	06 13 54.	- 70 42	35		STAR CLUSTER IN LMC
SG 2.024	06 13 55.	+ 22 32	2160		DIFFUSE EMISSION NEBULA
LW 429	06 13 55.	- 70 45			STAR CLUSTER IN LMC
SCHO 0188	06 13 59.	+ 09 33 24.	300		ISOLATED DARK CLOUD
KHAV 165	06 14	+ 03 47	9650		DARK NEBULA
LBN 0844	06 14	+ 22 30	2700		BRIGHT NEBULA
MIL 10	06 14	+ 22 30	2400		SUPERNOVA REMNANT
UGC 03436	06 14 00.	+ 63 28	96	16.0	GALAXY Sc/SBc
MCG+11-08-031	06 14 00.	+ 63 28	78	17.	GALAXY
ZWG 362.018	06 14 00.	+ 82 21		14.6	GALAXY
UGC 03435	06 14 00.	+ 82 21	78	14.6	GALAXY PECULR?
RNGC 2206	06 14 00.	- 26 45		13.0	GALAXY
MCG-04-15-019	06 14 00.	- 26 45	120	13.	GALAXY
SL 863	06 14 00.	- 68 00	15		STAR CLUSTER IN LMC
SL 864	06 14 00.	- 68 06	15		STAR CLUSTER IN LMC
VMT 09	06 14 06.	+ 22 37 12.	2880		SUPERNOVA REMNANT
LW 431	06 14 06.	- 70 40			STAR CLUSTER IN LMC
HOLM 085B	06 14 12.	- 21 22	24	14.1	PART OF MULTIPLE GALAXY
LW 433	06 14 12.	- 67 58			STAR CLUSTER IN LMC
LW 432	06 14 12.	- 68 03			STAR CLUSTER IN LMC
SL 865	06 14 13.	- 68 17	15		STAR CLUSTER IN LMC
HOLM 085A	06 14 14.	- 21 22	24	13.9	PART OF MULTIPLE GALAXY
RNGC 2202	06 14 16.	+ 06 01			NON-EXISTENT OBJECT
SL 866	06 14 16.	- 65 57	20		STAR CLUSTER IN LMC
MCG+08-12-011	06 14 18.	+ 50 53	15	16.	GALAXY
MCG-04-15-020	06 14 18.	- 21 21	210	12.	GALAXY
LW 434	06 14 19.	- 68 15			STAR CLUSTER IN LMC
RNGC 2207	06 14 20.	- 21 20		12.0	GALAXY
IC 2163	06 14 20.5	- 21 21 27.	144		GALAXY SB(rs)
MCG-04-15-021	06 14 21.	- 21 21	180	13.5	GALAXY
MCG+08-12-012	06 14 24.	+ 47 46	36	17.	GALAXY

OBJECT NAME	RIGHT ASCEN.	DECLINATION	DIAM.	MAGN.	TYPE OF OBJECT
ZWG 233.007	06 14 24.	+ 50 54		15.5	GALAXY
ZC 0614.4+6548	06 14 24.	+ 65 48	1010		CLUSTER OF GALAXIES
LW 436	06 14 24.	- 66 25			STAR CLUSTER IN LMC
MCG+08-12-013	06 14 30.	+ 50 55	12	16.	GALAXY
LW 435	06 14 34.	- 70 15			STAR CLUSTER IN LMC
ZWG 233.008	06 14 36.	+ 50 55		15.3	GALAXY
MCG+11-08-032	06 14 36.	+ 63 19	45	17.	GALAXY
HODG 11	06 14 36.	- 69 50	84		RED GLOBULAR CLSTR IN LMC
MCG-02-16-008	06 14 39.	- 13 04	36	15.	GALAXY
LW 438	06 14 40.	- 65 58			STAR CLUSTER IN LMC
MCG+08-12-014	06 14 42.	+ 48 20	36	16.	GALAXY
UGC 03437	06 14 42.	+ 64 21	96	16.5	GALAXY Sc
MCG+11-08-033	06 14 42.	+ 66 46	36	16.	GALAXY
MCG-06-14-015	06 14 42.	- 37 08	36	15.	GALAXY
SL 867	06 14 43.	- 71 52	5		STAR CLUSTER IN LMC
SG 2.025	06 14 44.	+ 22 45	1800		DIFFUSE EMISSION NEBULA
SL 868	06 14 44.	- 69 51	100		STAR CLUSTER IN LMC
MCG-06-14-016	06 14 45.	- 37 08	12	16.5	GALAXY
MCG+11-08-034	06 14 48.	+ 64 18	72	15.	GALAXY
ZWG 308.020	06 14 48.	+ 66 46		15.6	GALAXY
MIN.46 04	06 14 50.	+ 22 48			DIFFUSE NEBULA
LW 437	06 14 50.	- 69 50			STAR CLUSTER IN LMC
ZWG 260.005	06 14 54.	+ 51 57		15.3	GALAXY
CED 073	06 14 56.	+ 22 48	1650		DIFFUSE GALACTIC NEBULA
ISS 0470	06 15 00.	+ 05 20	440		STELLAR RING
LDN 1664	06 15 00.	+ 23 15	720		DARK NEBULA
OCL 0473	06 15 00.	+ 23 39	3600	5.8	OPEN STAR CLUSTER
MCG+11-08-036	06 15 00.	+ 65 47	15	16.	GALAXY
MCG+11-08-035	06 15 00.	+ 66 36 30.	54	14.	GALAXY
MCG+14-03-021	06 15 00.	+ 83 20	30	16.	GALAXY
MCG+14-03-022	06 15 00.	+ 84 57	60	15.	GALAXY
ABC 0555	06 15 01.	- 17 15		17.4	RICH CLUSTER OF GALAXIES
SL 869	06 15 02.	- 69 48	45		STAR CLUSTER IN LMC
ZWG 308.021	06 15 06.	+ 66 36		14.9	GALAXY
UGC 03438	06 15 06.	+ 66 36	60	14.9	GALAXY Sb/SBb
ZC 0615.2+6815	06 15 12.	+ 68 15	1880		CLUSTER OF GALAXIES
MCG-05-15-008	06 15 12.	- 27 22	132	14.5	GALAXY
LW 441	06 15 14.	- 69 47			STAR CLUSTER IN LMC
LW 439	06 15 14.	- 72 03			STAR CLUSTER IN LMC
RLWT 122	06 15 18.	+ 25 41		15.5	FAINT VERY BLUE STAR
SL 870	06 15 23.	- 72 35	15		STAR CLUSTER IN LMC
LW 440	06 15 23.	- 72 35			STAR CLUSTER IN LMC
SL 871	06 15 24.	- 68 00	15		STAR CLUSTER IN LMC
MCG+11-08-037	06 15 30.	+ 66 16	24	16.	GALAXY
MCG+00-16-002	06 15 30.	- 01 38	18	16.	GALAXY
ZWG 174.003	06 15 36.	+ 33 44		15.7	GALAXY
ZWG 203.006	06 15 36.	+ 42 39		15.6	GALAXY
KARA.72 111A	06 15 36.	+ 42 39	42	15.6	PART OF DOUBLE GALAXY
ZWG 284.010	06 15 36.	+ 61 00		15.7	GALAXY
ZWG 348.019	06 15 36.	+ 78 33		14.2	GALAXY
UGC 03439	06 15 36.	+ 78 33	210	14.2	GALAXY Sc
KARA.72 110B	06 15 36.	+ 78 33	192	14.2	PART OF DOUBLE GALAXY
MCG+00-16-003	06 15 36.	- 01 39	24	17.	GALAXY
MCG+10-09-015	06 15 42.	+ 61 02	18	16.	GALAXY
ZWG 308.022	06 15 42.	+ 66 16		15.7	GALAXY
LW 443	06 15 42.	- 67 58			STAR CLUSTER IN LMC
RLWT 054	06 15 43.	+ 21 36 48.		15.0	FAINT VERY BLUE STAR
MCG+08-12-015	06 15 45.	+ 50 58	54	16.	GALAXY
SL 872	06 15 46.	- 71 19	15		STAR CLUSTER IN LMC
LW 442	06 15 46.	- 71 19			STAR CLUSTER IN LMC
RLWT 055	06 15 48.	+ 22 30 54.		16.5	FAINT VERY BLUE STAR
MCG+08-12-016	06 15 48.	+ 46 57	30	17.	GALAXY
LW 444	06 15 49.	- 64 57			STAR CLUSTER IN LMC
SEY 017	06 15 50.	+ 71 15 15.		15.0	FAINT GALAXY
RNGC 2146A	06 15 52.	- 78 33		14.0	GALAXY
ZWG 233.009	06 15 54.	+ 50 58		15.6	GALAXY
ZWG 329.026	06 15 54.	+ 71 15		15.7	GALAXY
UGC 03440	06 15 54.	+ 71 15	66	15.7	GALAXY Sc
PK195-00.1	06 15 58.	+ 15 18	80		PLANETARY NEBULA
LBN 0840	06 16	+ 23 18	360		BRIGHT NEBULA
ZWG 203.007	06 16 00.	+ 42 35		15.7	GALAXY
KARA.72 111B	06 16 00.	+ 42 35	42	15.7	PART OF DOUBLE GALAXY
ZWG 233.010	06 16 00.	+ 50 52		15.7	GALAXY
UGC 03443	06 16 00.	+ 51 24	66	17.	GALAXY DWARF?
MCG+13-05-025	06 16 00.	+ 78 32	156	14.	GALAXY
KEEF 2146A	06 16 00.	+ 78 35	180	13.66	IRR GALAXY
ZWG 363.003	06 16 00.	+ 83 20		15.3	GALAXY
ZWG 362.019	06 16 00.	+ 83 20		15.3	GALAXY
UGC 03441	06 16 00.	+ 83 20	72	15.3	GALAXY Sa
ZWG 363.004	06 16 00.	+ 84 57		14.8	GALAXY
ZWG 362.020	06 16 00.	+ 84 57		14.8	GALAXY
UGC 03442	06 16 00.	+ 84 57	72	14.8	GALAXY S0-a
SG 2.026	06 16 02.	+ 23 07	3600		DIFFUSE EMISSION NEBULA
ZC 0616.1+7433	06 16 06.	+ 74 33	2080		CLUSTER OF GALAXIES
SCHO 0189	06 16 08.	+ 11 08 06.	520		ISOLATED DARK CLOUD
ZWG 144.001	06 16 12.	+ 28 45		15.7	GALAXY
MCG+12-06-020	06 16 12.	+ 71 15 30.	60	16.	GALAXY
RNGC 2211	06 16 12.	- 18 32		14.0	GALAXY
MCG-03-16-021	06 16 12.	- 18 32	36	14.	GALAXY
MCG+07-13-011	06 16 15.	+ 39 03	36	17.	GALAXY
SL 873	06 16 17.	- 69 08	25		STAR CLUSTER IN LMC
ZWG 203.008	06 16 18.	+ 39 03		15.7	GALAXY
RNGC 2212	06 16 18.	- 18 31		14.0	GALAXY
MCG-03-16-022	06 16 18.	- 18 31	48	14.5	GALAXY
VDB.66N 075	06 16 25.	+ 23 16	408		REFLECTION NEBULA
MCG-04-15-022	06 16 27.	- 24 50	15	15.5	GALAXY
SL 874	06 16 27.	- 70 02	25		STAR CLUSTER IN LMC
LW 445	06 16 29.	- 70 25			STAR CLUSTER IN LMC
7ZW 067	06 16 30.	+ 63 06			COMPACT GALAXY
UGC 03444	06 16 30.	+ 73 44	90	17.	GALAXY DWARF
DG 099	06 16 36.	+ 23 22	120		REFLECTION NEBULA
MCG+08-12-017	06 16 36.	+ 46 02	30	18.	GALAXY
LW 446	06 16 39.	- 70 03			STAR CLUSTER IN LMC
LW 447	06 16 40.	- 69 06			STAR CLUSTER IN LMC
ZWG 233.011	06 16 48.	+ 50 07		15.4	GALAXY
MCG-03-16-023	06 16 51.	- 16 56	60	15.	GALAXY
LW 448	06 16 54.	- 68 05			STAR CLUSTER IN LMC
ZWG 348.020	06 17 00.	+ 80 32		15.5	GALAXY
MCG-04-15-023	06 17 00.	- 24 35 30.	54	15.	GALAXY
7ZW 068	06 17 06.	+ 59 08			COMPACT GALAXY
ZWG 284.011	06 17 06.	+ 59 08		13.9	GALAXY
UGC 03445	06 17 06.	+ 59 08	90	13.9	GALAXY S0-a
KARA.72 112A	06 17 06.	+ 59 08	96	13.9	PART OF DOUBLE GALAXY
SL 350	06 17 07.	- 69 40	50		STAR CLUSTER IN LMC
LW 449	06 17 09.	- 68 50			STAR CLUSTER IN LMC
ZWG 284.012	06 17 12.	+ 59 08		14.1	GALAXY
UGC 03446	06 17 12.	+ 59 08	84	14.1	GALAXY S0
KARA.72 112B	06 17 12.	+ 59 08	78	14.1	PART OF DOUBLE GALAXY

OBJECT NAME	RIGHT ASCEN.	DECLINATION	DIAM.	MAGN.	TYPE OF OBJECT
MCG+11-08-038	06 17 12.	+ 66 35	78	15.	GALAXY
MCG+10-09-016	06 17 18.	+ 59 11	78	14.	GALAXY
MCG-04-15-024	06 17 18.	- 24 25 30.	66	14.5	GALAXY
IC 0444	06 17 23.	+ 23 17			NONSTELLAR OBJECT
SL 875	06 17 23.	- 72 33	5		STAR CLUSTER IN LMC
DG 100	06 17 24.	+ 23 19	480		REFLECTION NEBULA
ZWG 144.002	06 17 24.	+ 28 00		15.6	GALAXY
UGC 03447	06 17 24.	+ 28 00	102	15.6	GALAXY E
MCG+10-09-017	06 17 24.	+ 59 11	57	13.	GALAXY
ZWG 308.023	06 17 24.	+ 66 36		14.5	GALAXY
UGC 03448	06 17 24.	+ 66 36	78	14.5	GALAXY S0-a
CED 074	06 17 26.	+ 23 19	330		DIFFUSE GALACTIC NEBULA
MCG+08-12-018	06 17 30.	+ 45 14	60	15.	GALAXY
ZWG 260.006	06 17 30.	+ 51 08		15.7	GALAXY
UGC 03449	06 17 30.	+ 51 08	78	15.7	GALAXY Sc
RLWT 123	06 17 36.	+ 25 39		13.	FAINT VERY BLUE STAR
ZWG 233.012	06 17 36.	+ 45 15		15.7	GALAXY
MCG+08-12-019	06 17 36.	+ 50 24	48	16.	GALAXY
MCG+09-11-009	06 17 36.	+ 51 07	78	16.	GALAXY
MCG+08-12-020	06 17 48.	+ 50 36	18	17.	GALAXY
MRSL 189+04/1	06 17 54.	+ 23 07	4800		HII REGION
SL 876	06 17 57.	- 67 27	10		STAR CLUSTER IN LMC
LBN 0841	06 18	+ 23 30	2700		BRIGHT NEBULA
LDN 1566	06 18 00.	+ 23 10			DARK NEBULA
LDN 1565	06 18 00.	+ 23 20	1980		DARK NEBULA
ZWG 144.003	06 18 00.	+ 27 53		15.6	GALAXY
UGC 03450	06 18 00.	+ 27 53	96	15.6	GALAXY SBb
7ZW 069	06 18 00.	+ 81 03			COMPACT GALAXY
RLWT 124	06 18 06.	+ 26 52		14.	FAINT VERY BLUE STAR
ZWG 233.013	06 18 06.	+ 46 45		15.5	GALAXY
LW 452	06 18 09.	- 67 28			STAR CLUSTER IN LMC
LW 451	06 18 11.	- 69 12			STAR CLUSTER IN LMC
MCG+08-12-021	06 18 12.	+ 50 51	42	16.	GALAXY
MCG+11-08-039	06 18 12.	+ 67 42 30.	39	16.	GALAXY
ZWG 348.021	06 18 12.	+ 77 54		15.7	GALAXY
UGC 03451	06 18 12.	+ 77 54	78	15.7	GALAXY SBb
LW 450	06 18 12.	- 70 33			STAR CLUSTER IN LMC
HUB C19	06 18 15.	+ 09 15			DIFFUSE NEBULA
ZWG 233.014	06 18 18.	+ 50 51		15.7	GALAXY
MCG-04-15-025	06 18 18.	- 22 34	66	16.	GALAXY
RNGC 2215	06 18 23.	- 07 16		8.5	OPEN CLUSTER
MCG+11-08-040	06 18 24.	+ 64 36	18	16.	GALAXY
ZWG 308.024	06 18 24.	+ 64 37		15.5	GALAXY
MCG-04-15-026	06 18 24.	- 22 34	24	16.	GALAXY
MCG-06-14-017	06 18 24.	- 37 10	36	15.	GALAXY
ZWG 260.007	06 18 30.	+ 51 57		14.0	GALAXY
UGC 03452	06 18 30.	+ 51 57	120	14.0	GALAXY S0
ZWG 284.013	06 18 30.	+ 57 35		15.3	GALAXY
LW 454	06 18 30.	- 68 02			STAR CLUSTER IN LMC
SL 877	06 18 30.	- 71 43	55		STAR CLUSTER IN LMC
RNGC 2208	06 18 31.	+ 51 57		14.0	GALAXY
MCG-03-17-001	06 18 33.	- 16 02	96	14.5	GALAXY
YM 42	06 18 34.	- 08 29	210		SYMMETRIC GALACTIC NEBULA
RLWT 125	06 18 36.	+ 26 55		13.	FAINT VERY BLUE STAR
MCG+09-11-010	06 18 36.	+ 51 55 30.	78	14.	GALAXY
ZWG 330.001	06 18 36.	+ 73 08		14.6	GALAXY
ZWG 329.027	06 18 36.	+ 73 08		14.6	GALAXY
UGC 03453	06 18 36.	+ 73 08	78	14.6	GALAXY SB
KARA.73B 0167	06 18 36.	+ 73 08	48	14.6	ISOLATED GALAXY S
OCL 0550	06 18 36.	- 07 16	720	9.1	OPEN STAR CLUSTER
LW 453	06 18 36.	- 71 44			STAR CLUSTER IN LMC
ZWG 144.004	06 18 42.	+ 27 12		15.7	GALAXY
ZC 0618.7+4618	06 18 42.	+ 46 18	2290		CLUSTER OF GALAXIES
SEY 018	06 18 42.	+ 73 09 52.		15.2	FAINT GALAXY
YM 44	06 18 42.	- 20 00	210		SYMMETRIC GALACTIC NEBULA
WS 41	06 18 45.	- 73 11 26.		16.13	PLANETARY NEB. IN LMC
MCG+09-11-011	06 18 48.	+ 54 12	45	17.	GALAXY
UGC 03454	06 18 48.	+ 72 07	72	16.5	GALAXY DWARF
MCG+12-07-001	06 18 48.	+ 73 10	60	15.	GALAXY
MCG-03-17-002	06 18 51.	- 20 01	120	13.	GALAXY
ISS 0424	06 18 54.	+ 08 49	259		STELLAR RING
MCG+10-09-018	06 18 54.	+ 57 37 30.	51	15.	GALAXY
SG 2.027	06 18 56.	+ 23 04	2880		DIFFUSE EMISSION NEBULA
MCG+12-07-002	06 19 00.	+ 72 07	51	17.	GALAXY
ZWG 348.022	06 19 00.	+ 77 26		15.4	GALAXY
UGC 03456	06 19 00.	+ 77 26	78	15.4	GALAXY S0-a
MCG+13-05-026	06 19 00.	+ 77 50	30	16.	GALAXY
UGC 03455	06 19 00.	+ 85 23	66	17.	GALAXY Sc
MCG+15-01-004	06 19 00.	+ 87 35	48	16.	GALAXY
MCG+00-17-001	06 19 06.	+ 00 24	48	15.5	GALAXY
OCL 0519	06 19 06.	+ 02 23	1050	6.9	OPEN STAR CLUSTER
ISS 0368	06 19 06.	+ 14 47	175		STELLAR RING
ZWG 233.015	06 19 06.	+ 46 38		15.6	GALAXY
SL 879	06 19 11.	- 71 34	15		STAR CLUSTER IN LMC
UGC 03457	06 19 12.	+ 00 24	90	15.0	GALAXY S0-a
RLWT 126	06 19 12.	+ 26 22		13.5	FAINT VERY BLUE STAR
7ZW 070	06 19 12.	+ 57 14			COMPACT GALAXY
MCG+13-05-027	06 19 12.	+ 77 25	51	16.	GALAXY
SL 880	06 19 13.	- 70 54	20		STAR CLUSTER IN LMC
LW 455	06 19 13.	- 70 55			STAR CLUSTER IN LMC
LW 458	06 19 15.	- 67 28			STAR CLUSTER IN LMC
HN 0079	06 19 24.	- 12 57			NEBULA
MCG-04-15-027	06 19 24.	- 22 03 30.	60	14.	GALAXY
IC 2165	06 19 24.2	- 12 57 36.	15		PLANETARY NEBULA
PK221-12.1	06 19 25.	- 12 57	9	13.7	PLANETARY NEBULA
RNGC 2216	06 19 27.	- 22 03		14.0	GALAXY
RNGC 2222	06 19 29.	- 57 29			UNVERIFIED SOUTHERN OBJECT
RNGC 2221	06 19 29.	- 57 32			UNVERIFIED SOUTHERN OBJECT
SL 881	06 19 29.	- 71 32	25		STAR CLUSTER IN LMC
LW 456	06 19 29.	- 71 36			STAR CLUSTER IN LMC
LDN 1567	06 19 30.	+ 23 20			DARK NEBULA
MCG+08-12-022	06 19 30.	+ 50 26	36	17.	GALAXY
LH120-N221	06 19 30.	- 71 35			EMISSION NEBULA IN LMC
LW 457	06 19 35.	- 71 34			STAR CLUSTER IN LMC
ISS 0369	06 19 36.	+ 14 48	164		STELLAR RING
OCL 0546	06 19 36.	- 06 19	180	16.	OPEN STAR CLUSTER
MCG-03-17-003	06 19 36.	- 20 12	54	16.	GALAXY
MCG-05-15-009	06 19 36.	- 28 05	36	15.	GALAXY
LW 459	06 19 37.	- 68 18			STAR CLUSTER IN LMC
MCG+11-08-041	06 19 39.	+ 64 48	30	16.	GALAXY
MCG+11-08-042	06 19 42.	+ 65 45	36	15.	GALAXY
MCG-05-15-010	06 19 42.	- 27 13	198	13.	GALAXY
RNGC 2217	06 19 42.	- 27 14		12.0	GALAXY
HELW 128	06 19 43.	- 27 31 52.			NEBULA
WS 42	06 19 47.	- 71 34 53.		16.05	PLANETARY NEB. IN LMC
MCG+08-12-023	06 19 54.	+ 50 55	42	17.	GALAXY
SL 883	06 19 54.	- 68 13	15		STAR CLUSTER IN LMC
HUB C17	06 19 55.	+ 10 36			DIFFUSE NEBULA
MCG+09-11-012	06 19 57.	+ 51 39 30.	36	16.	GALAXY
SL 882	06 19 58.	- 72 20	35		STAR CLUSTER IN LMC
LBN 0845	06 20	+ 23 00	2100		BRIGHT NEBULA
DG 101	06 20	+ 23 00	3000		REFLECTION NEBULA
LB 09830	06 20	- 83 52		14.4	FAINT BLUE STAR
LDN 1568	06 20 00.	+ 23 00			DARK NEBULA
RLWT 127	06 20 00.	+ 26 50		14.5	FAINT VERY BLUE STAR
ZWG 260.008	06 20 00.	+ 51 41		15.2	GALAXY
MCG+11-08-043	06 20 00.	+ 64 35	18	15.	GALAXY
UGC 03457A	06 20 00.	+ 87 33	78	16.5	GALAXY
LDN 1651	06 20 00.	- 09 50	2040		DARK NEBULA
RLWT 056	06 20 01.	+ 20 56 00.		14.5	FAINT VERY BLUE STAR
ZWG 308.025	06 20 06.	+ 64 36		14.6	GALAXY
MCG-06-14-018	06 20 06.	- 34 39	30	15.	GALAXY
LW 461	06 20 06.	- 68 13			STAR CLUSTER IN LMC
CED 075	06 20 10.	+ 05 10	180		DIFFUSE GALACTIC NEBULA
OCL 0504	06 20 12.	+ 05 09	660	8.7	OPEN STAR CLUSTER
MCG-06-14-019	06 20 15.	- 36 31	36	15.	GALAXY
RNGC 2220	06 20 22.	- 44 44			NON-EXISTENT OBJECT
ZC 0620.4+4620	06 20 24.	+ 46 20	6720		CLUSTER OF GALAXIES
LW 460	06 20 24.	- 71 43			STAR CLUSTER IN LMC
MCG+11-08-044	06 20 30.	+ 64 47	36	16.	GALAXY
LW 464	06 20 32.	- 68 35			STAR CLUSTER IN LMC
RNGC 2231	06 20 34.	- 67 30			GLOBULAR CLUSTER IN LMC
SL 884	06 20 34.	- 67 30	75		STAR CLUSTER IN LMC
LW 463	06 20 34.	- 71 18			STAR CLUSTER IN LMC
OCL 0508	06 20 36.	+ 04 41	1080		OPEN STAR CLUSTER
ZWG 260.009	06 20 36.	+ 51 00		15.5	GALAXY
ZWG 233.016	06 20 36.	+ 51 00		15.5	GALAXY
ZWG 260.010	06 20 36.	+ 53 19		15.5	GALAXY
MCG+11-08-045	06 20 36.	+ 65 21	30	15.	GALAXY
ZWG 308.026	06 20 36.	+ 65 22		15.2	GALAXY
MCG-05-16-001	06 20 36.	- 27 54	48	15.	GALAXY
LW 462	06 20 39.	- 72 15			STAR CLUSTER IN LMC
LW 465	06 20 40.	- 69 03			STAR CLUSTER IN LMC
MCG+09-11-013	06 20 42.	+ 51 00	39	16.	GALAXY
MCG+11-08-046	06 20 51.	+ 64 45	42	16.	GALAXY
LW 466	06 20 52.	- 67 30			STAR CLUSTER IN LMC
MCG+12-07-011	06 20 54.	+ 74 27	33	16.	GALAXY
SL 885	06 20 57.	- 67 27	15		STAR CLUSTER IN LMC
RNGC 2219	06 20 58.	- 04 39			NON-EXISTENT OBJECT
RNGC 2228	06 20 59.	- 64 25			GALAXY
KHAV 166	06 21	+ 12 23	4600		DARK NEBULA
LB 09831	06 21	- 81 57		13.8	FAINT BLUE STAR
MCG+08-12-024	06 21 00.	+ 49 32	42	14.	GALAXY
ZWG 308.027	06 21 00.	+ 64 46		15.1	GALAXY
UGC 03458	06 21 00.	+ 64 46	174	15.1	GALAXY Sb
MCG+11-08-047	06 21 00.	+ 64 46	114	14.	GALAXY
MIN.48 01	06 21 04.	- 10 07			DIFFUSE NEBULA
ZWG 233.017	06 21 06.	+ 49 32		14.9	GALAXY
RNGC 2229	06 21 06.	- 64 56			GALAXY
ACK 204-03.1	06 21 12.	+ 05 32			PLANETARY NEBULA
MCG-03-17-004	06 21 12.	- 16 07	84	15.	GALAXY
RNGC 2230	06 21 12.	- 64 58			GALAXY
LW 467	06 21 15.	- 67 27			STAR CLUSTER IN LMC
ISS 0471	06 21 24.	+ 02 49	244		STELLAR RING
UGC 03459	06 21 24.	+ 04 43	72	17.	GALAXY Sc
SEY 019	06 21 29.	+ 74 20 34.		14.6	FAINT GALAXY
ZWG 330.002	06 21 30.	+ 74 18		15.0	GALAXY
ZWG 329.028	06 21 30.	+ 74 18		15.0	GALAXY
UGC 03460	06 21 30.	+ 74 18	102	15.0	GALAXY SBb
RNGC 2233	06 21 31.	- 65 01			GALAXY
ZWG 204.001	06 21 36.	+ 40 20		15.3	GALAXY
ZWG 203.009	06 21 36.	+ 40 20		15.3	GALAXY
MCG+11-08-048	06 21 36.	+ 64 44	27	16.	GALAXY
MRK 004	06 21 36.	+ 74 20	10	16.5	GALAXY WITH UV CONTINUUM
MCG+12-07-003	06 21 36.	+ 74 20	102	16.	GALAXY
SL 886	06 21 41.	- 69 18	25		STAR CLUSTER IN LMC
HUB C16	06 21 42.	+ 10 50			DIFFUSE NEBULA
ZWG 260.011	06 21 42.	+ 51 20		15.7	GALAXY
MCG-06-15-001	06 21 42.	- 37 02	30	15.	GALAXY
RNGC 2218	06 21 43.	+ 19 22			NON-EXISTENT OBJECT
MCG-04-16-001	06 21 48.	- 23 10	66	15.	GALAXY
LW 468	06 21 53.	- 69 16			STAR CLUSTER IN LMC
UGC 03460A	06 21 54.	+ 06 14	60	17.	GALAXY
RLWT 128	06 21 54.	+ 26 53		15.5	FAINT VERY BLUE STAR
RLWT 057	06 21 56.	+ 22 28 12.		16.0	FAINT VERY BLUE STAR
MCG+11-08-049	06 21 57.	+ 63 34	24	16.	GALAXY
LDN 1633	06 22 00.	+ 03 30	1320		DARK NEBULA
UGC 03462	06 22 00.	+ 28 34	72	16.5	GALAXY Sc
UGC 03461	06 22 00.	+ 82 55	72	16.5	GALAXY DWRF SP
MCG-06-15-002	06 22 00.	- 35 36	42	15.	GALAXY
RLWT 058	06 22 01.	+ 21 08 36.		14.0	FAINT VERY BLUE STAR
RLWT 059	06 22 02.	+ 21 15 18.		14.5	FAINT VERY BLUE STAR
RLWT 060	06 22 05.	+ 21 19 42.		14.0	FAINT VERY BLUE STAR
ISS 0472	06 22 06.	+ 07 32	224		STELLAR RING
RNGC 2235	06 22 06.	- 64 55			GALAXY
MCG-06-15-003	06 22 12.	- 34 58	60	14.5	GALAXY
HUB E14	06 22 19.	+ 05 15			DIFFUSE NEBULA
MCG+05-15-001	06 22 27.	+ 30 41	30	15.2	GALAXY
RNGC 2223	06 22 27.	- 22 49		12.0	GALAXY
RLWT 061	06 22 30.	+ 20 44 24.		16.0	FAINT VERY BLUE STAR
ZWG 145.001	06 22 30.	+ 30 40		15.2	GALAXY
ZWG 144.005	06 22 30.	+ 30 40		15.2	GALAXY
MCG+08-12-025	06 22 30.	+ 48 43	60	17.	GALAXY
ZWG 285.001	06 22 30.	+ 59 07		13.5	GALAXY
ZWG 284.014	06 22 30.	+ 59 07		13.5	GALAXY
KARA.73B 0168	06 22 30.	+ 59 07	198	13.5	ISOLATED GALAXY S
SL 887	06 22 30.	- 72 38	15		STAR CLUSTER IN LMC
MCG-04-16-002	06 22 33.	- 22 48	180	13.	GALAXY
RLWT 062	06 22 34.	+ 21 23 06.		12.5	FAINT VERY BLUE STAR
MRSL 203-02/1	06 22 36.	+ 07 33	36		HII REGION
ZC 0622.6+5731	06 22 36.	+ 57 31	940		CLUSTER OF GALAXIES
ZC 0622.6+5923	06 22 36.	+ 59 23	540		CLUSTER OF GALAXIES
ZWG 330.003	06 22 36.	+ 74 30		15.5	GALAXY
ZWG 329.029	06 22 36.	+ 74 30		15.5	GALAXY
UGC 03464	06 22 36.	+ 74 30	84	15.5	GALAXY Sa?
LW 469	06 22 36.	- 72 46			STAR CLUSTER IN LMC
MCG+01-17-001	06 22 42.	+ 07 32	48	14.	GALAXY
MCG+10-10-001	06 22 42.	+ 59 08	156	13.	GALAXY
MRSL 192+03/1	06 22 48.	+ 20 04	300		HII REGION
MCG+09-11-014	06 22 48.	+ 52 28 30.	24	16.	GALAXY
ZWG 260.012	06 22 48.	+ 52 29		15.4	GALAXY
UGC 03465	06 22 48.	+ 52 29	60	15.4	GALAXY SBa-b
MCG+12-07-004	06 22 48.	+ 74 31	84	16.	GALAXY
SCHO 0190	06 22 54.	+ 07 57 00.	480		ISOLATED DARK CLOUD
RLWT 063	06 22 59.	+ 20 53 36.		15.0	FAINT VERY BLUE STAR
LBN 1017	06 23	- 10 32	240		BRIGHT NEBULA

OBJECT NAME	RIGHT ASCEN.	DECLINATION	DIAM.	MAGN.	TYPE OF OBJECT
MCG+14-04-001	06 23 00.	+ 83 33	48	16.	GALAXY
UGC 03466	06 23 00.	+ 83 34	72	16.0	GALAXY Sb-c
SEY 020	06 23 01.	+ 74 32 16.		15.1	FAINT GALAXY
PK194+02.1	06 23 02.27	+ 17 49 14.9	12	12.4	PLANETARY NEBULA
RNGC 2241	06 23 03.	- 68 55			OPEN CLUSTER IN LMC
HSE 455	06 23 05.	- 67 47			STAR CLUSTER IN LMC
SL 888	06 23 09.	- 68 55	20		STAR CLUSTER IN LMC
LW 471	06 23 15.	- 68 54			STAR CLUSTER IN LMC
LW 470	06 23 15.	- 72 14			STAR CLUSTER IN LMC
ARC 0556	06 23 21.	+ 67 04		16.7	RICH CLUSTER OF GALAXIES
LW 472	06 23 25.	- 68 17			STAR CLUSTER IN LMC
SCHO 0191	06 23 27.	+ 15 07 54.	600		ISOLATED DARK CLOUD
ZWG 233.018	06 23 30.	+ 49 38		14.7	GALAXY
UGC 03467	06 23 30.	+ 49 38	108	14.7	GALAXY Sa-b
MRK 071	06 23 30.	+ 69 17	33	14.	GALAXY WITH UV CONTINUUM
MCG-04-16-003	06 23 30.	- 22 54	48	14.5	GALAXY
SL 889	06 23 34.	- 69 03	15		STAR CLUSTER IN LMC
HUB C18	06 23 37.	+ 09 58			DIFFUSE NEBULA
MCG-06-15-004	06 23 42.	- 38 58		15.	GALAXY
SL 890	06 23 47.	- 71 38	25		STAR CLUSTER IN LMC
ACK 247-21.1	06 23 48.	- 39 50			PLANETARY NEBULA
RNGC 2227	06 23 51.	- 21 57		14.0	GALAXY
MCG-04-16-004	06 23 51.	- 21 57	120	14.	GALAXY
LW 474	06 23 52.	- 68 58			STAR CLUSTER IN LMC
ZWG 330.004	06 23 54.	+ 74 28		15.6	GALAXY
ZWG 329.030	06 23 54.	+ 74 28		15.6	GALAXY
LW 473	06 23 54.	- 71 41			STAR CLUSTER IN LMC
SCHO 0192	06 23 56.	+ 11 12 12.	390		ISOLATED DARK CLOUD
SEY 021	06 23 59.	+ 74 29 00.		15.4	FAINT GALAXY
LW 475	06 23 59.	- 70 31			STAR CLUSTER IN LMC
LBN 0887	06 24	+ 12 00	4920		BRIGHT NEBULA
LDN 1652	06 24 00.	- 10 00	4320		DARK NEBULA
OCL 0545	06 24 06.	- 04 43	1200	4.3	OPEN STAR CLUSTER
ZWG 308.028	06 24 12.	+ 67 58		15.6	GALAXY
KARA.73B 0169	06 24 12.	+ 67 58	18	15.6	ISOLATED GALAXY S0
7ZW 071	06 24 12.	+ 68 45			COMPACT GALAXY
RNGC 2225	06 24 13.	- 09 37			NON-EXISTENT OBJECT
LW 476	06 24 14.	- 72 02			STAR CLUSTER IN LMC
RNGC 2232	06 24 16.	- 04 42		4.0	OPEN CLUSTER
UGC 03468	06 24 18.	+ 18 51	84	17.	GALAXY Sc-IRR
RNGC 2226	06 24 19.	- 09 37			NON-EXISTENT OBJECT
ZWG 175.001	06 24 24.	+ 35 33		15.7	GALAXY
ZWG 174.004	06 24 24.	+ 35 33		15.7	GALAXY
UGC 03469	06 24 30.	+ 56 04	78	16.5	GALAXY Sc
ZC 0624.6+5720	06 24 36.	+ 57 20	940		CLUSTER OF GALAXIES
MCG-05-16-002	06 24 36.	- 31 45	48	15.	GALAXY
ZWG 260.013	06 24 42.	+ 52 37		15.5	GALAXY
RNGC 2224	06 24 48.	+ 12 40			NON-EXISTENT OBJECT
7ZW 072	06 24 48.	+ 69 33			COMPACT GALAXY
MCG-05-16-003	06 24 48.	- 28 37	66	14.5	GALAXY
MCG-05-16-004	06 24 48.	- 29 53	60	15.	GALAXY
MCG-06-15-005	06 24 48.	- 34 47	36	15.	GALAXY
ARC 0557	06 24 50.	+ 69 15		17.0	RICH CLUSTER OF GALAXIES
MCG+06-15-001	06 24 54.	+ 35 31	27	15.	GALAXY
MCG+11-08-050	06 24 54.	+ 65 14	39	15.	GALAXY
LDN 1625	06 25 00.	+ 06 00	17880		DARK NEBULA
MCG+11-08-051	06 25 00.	+ 65 00	30	16.	GALAXY
ZWG 330.005	06 25 00.	+ 74 27		14.9	GALAXY
ZWG 329.031	06 25 00.	+ 74 27		14.9	GALAXY
UGC 03471	06 25 00.	+ 74 27	102	14.9	GALAXY Sc
MCG+12-07-005	06 25 00.	+ 74 28	84	15.	GALAXY
ZWG 362.021	06 25 00.	+ 81 37		15.7	GALAXY
ZWG 363.005	06 25 00.	+ 83 00		13.7	GALAXY
ZWG 362.022	06 25 00.	+ 83 00		13.7	GALAXY
UGC 03470	06 25 00.	+ 83 00	84	13.7	GALAXY PECULR
MCG+14-04-002	06 25 00.	+ 83 48	48	16.	GALAXY
PK233-16.1	06 25 00.	- 25 21	34	16.3	PLANETARY NEBULA
SEY 022	06 25 05.	+ 74 28 25.		14.5	FAINT GALAXY
MCG+11-08-052	06 25 18.	+ 65 08	27	16.	GALAXY
SL 891	06 25 29.	- 71 36	30		STAR CLUSTER IN LMC
KARA.72 113A	06 25 30.	+ 83 00	60	13.7	PART OF DOUBLE GALAXY
SCHO 0193	06 25 33.	+ 07 53 06.	440		ISOLATED DARK CLOUD
ZWG 115.001	06 25 36.	+ 21 35		15.4	GALAXY
ZC 0625.6+4608	06 25 36.	+ 46 08	740		CLUSTER OF GALAXIES
KARA.72 113B	06 25 36.	+ 83 00	24		PART OF DOUBLE GALAXY
LW 477	06 25 41.	- 71 39			STAR CLUSTER IN LMC
ISS 0370	06 25 42.	+ 17 36	335		STELLAR RING
7ZW 073	06 25 42.	+ 63 43			COMPACT GALAXY
ZC 0625.7+6640	06 25 42.	+ 66 40	4300		CLUSTER OF GALAXIES
IC 0442	06 25 42.	+ 83 00 18.			NONSTELLAR OBJECT
SL 892	06 25 56.	- 71 02	15		STAR CLUSTER IN LMC
LDN 1604	06 26 00.	+ 11 40	360		DARK NEBULA
MCG+11-08-053	06 26 00.	+ 65 00	30	16.	GALAXY
MCG+13-05-028	06 26 00.	+ 75 54 30.	51	16.	GALAXY
LW 478	06 26 02.	- 71 05			STAR CLUSTER IN LMC
SL 893	06 26 03.	- 68 46	70		STAR CLUSTER IN LMC
LW 479	06 26 03.	- 68 53			STAR CLUSTER IN LMC
7ZW 074	06 26 06.	+ 59 05			COMPACT GALAXY
ZWG 348.023	06 26 06.	+ 75 55		15.5	GALAXY
UGC 03472	06 26 06.	+ 75 55	126	15.5	GALAXY S0
KARA.73B 0170	06 26 06.	+ 75 55	72	15.5	ISOLATED GALAXY S
RNGC 2249	06 26 09.	- 68 46			GLOBULAR CLUSTER IN LMC
UGC 03473	06 26 24.	+ 59 37	78	16.0	GALAXY Sc
MCG+10-10-002	06 26 24.	+ 59 37 30.	78	15.	GALAXY
RNGC 2234	06 26 27.	+ 16 43			NON-EXISTENT OBJECT
MCG+14-04-003	06 26 30.	+ 82 59	48	14.	GALAXY
ZWG 260.014	06 26 42.	+ 51 55		15.2	GALAXY
ZWG 330.006	06 26 42.	+ 71 36		15.6	GALAXY
ZWG 329.032	06 26 42.	+ 71 36		15.6	GALAXY
UGC 03474	06 26 42.	+ 71 36	156	15.6	GALAXY Sc
KARA.73B 0171	06 26 42.	+ 71 36	132	15.6	ISOLATED GALAXY S
MCG+09-11-015	06 26 48.	+ 52 32	45	16.	GALAXY
SEY 023	06 26 53.	+ 71 35 15.		14.8	FAINT GALAXY
MCG+07-14-001	06 26 54.	+ 40 06	42	16.	GALAXY
ZWG 233.019	06 26 54.	+ 46 02		15.0	GALAXY
ZWG 260.015	06 26 54.	+ 52 34		15.6	GALAXY
MCG+10-10-003	06 26 54.	+ 58 49	42	15.	GALAXY
MCG+12-07-006	06 26 54.	+ 71 37	126	15.	GALAXY
RNGC 2236	06 26 57.	+ 06 52		11.5	OPEN CLUSTER
MCG+07-14-002	06 26 57.	+ 39 32	120	15.	GALAXY
SNO 22	06 26 59.	- 35 34 12.		18.	LINEAR GRP OF 7 GALAXIES
KHAV 167	06 27	+ 12 46	5840		DARK NEBULA
OCL 0501	06 27 00.	+ 06 52	720	11.9	OPEN STAR CLUSTER
LDN 1606	06 27 00.	+ 10 15	840		DARK NEBULA
LDN 1601	06 27 00.	+ 12 25	1020		DARK NEBULA
ZWG 204.002	06 27 00.	+ 39 31		15.5	GALAXY
UGC 03475	06 27 00.	+ 39 31	168	15.5	GALAXY DWRF SP
MCG+08-12-026	06 27 00.	+ 45 06	30	17.	GALAXY

OBJECT NAME	RIGHT ASCEN.	DECLINATION	DIAM.	MAGN.	TYPE OF OBJECT
SC 0625-4559.9	06 27 02.	- 46 01 49.	12		NEBULA
SC 0625-4559.8	06 27 04.	- 46 01 43.	12		NEBULA
MCG-03-17-005	06 27 06.	- 17 19	42	14.	GALAXY
ZWG 175.002	06 27 12.	+ 33 20		15.7	GALAXY
UGC 03476	06 27 12.	+ 33 20	66	15.7	GALAXY IRR
MRK 619	06 27 12.	+ 57 15	8	17.	GALAXY WITH UV CONTINUUM
ISS 0425	06 27 18.	+ 14 34	316		STELLAR RING
MCG-04-16-005	06 27 18.	- 26 27	60	15.	GALAXY
MCG-05-16-005	06 27 24.	- 30 17	36	15.	GALAXY
ISS 0473	06 27 30.	+ 06 20	296		STELLAR RING
ZWG 233.020	06 27 30.	+ 50 07		15.7	GALAXY
UGC 03477	06 27 30.	+ 50 07	126	15.7	GALAXY Sc
MCG-05-16-006	06 27 33.	- 30 18	24	16.	GALAXY
ISS 0474	06 27 36.	+ 06 52	263		STELLAR RING
ZWG 260.016	06 27 36.	+ 55 39		15.5	GALAXY
MCG-04-16-006	06 27 36.	- 26 35	18	15.	GALAXY
IC 2166	06 27 37.	+ 59 07			NONSTELLAR OBJECT
RNGC 2237	06 27 40.	+ 05 05			CLUSTER WITH NEBULOSITY
OCL 0523	06 27 42.	+ 02 54	480	9.3	OPEN STAR CLUSTER
OCL 0511	06 27 42.	+ 05 05	3600	7.	OPEN STAR CLUSTER
SCHO 0194	06 27 42.	+ 09 39 24.	470		ISOLATED DARK CLOUD
MCG+09-11-016	06 27 42.	+ 55 38 30.	24	16.	GALAXY
VDB.66N 076	06 27 47.	+ 10 07			REFLECTION NEBULA
MCG+11-08-055	06 27 48.	+ 63 27 30.	15	16.	GALAXY
MCG+11-08-054	06 27 48.	+ 63 42	78	14.	GALAXY
RNGC 2243	06 27 51.	- 31 15		10.5	OPEN CLUSTER
RNGC 2238	06 27 52.	+ 05 05			DIFFUSE NEBULA
ZC 0627.9+7540	06 27 54.	+ 75 40	1550		CLUSTER OF GALAXIES
OCL 0644	06 27 54.	- 31 15	300	10.9	OPEN STAR CLUSTER
CED 076A	06 27 58.	+ 05 05			DIFFUSE GALACTIC NEBULA
LBN 0903	06 28	+ 10 03	1320		BRIGHT NEBULA
KHAV 169	06 28	+ 10 22	6420		DARK NEBULA
LBN 0898	06 28	+ 10 29	240		BRIGHT NEBULA
LBN 0895	06 28	+ 11 00	4800		BRIGHT NEBULA
LBN 0889	06 28	+ 12 20	2220		BRIGHT NEBULA
KHAV 168	06 28	+ 14 58	6380		DARK NEBULA
LBN 1015	06 28	- 09 36	240		BRIGHT NEBULA
LB 09967	06 28	- 87 47		14.2	FAINT BLUE STAR
LDN 1607	06 28 00.	+ 10 20	480		DARK NEBULA
LDN 1600	06 28 00.	+ 13 00	2520		DARK NEBULA
ZWG 115.002	06 28 00.	+ 23 12		15.7	GALAXY
MCG+11-08-056	06 28 00.	+ 63 27	18	17.	GALAXY
ZWG 308.029	06 28 00.	+ 63 28		15.6	GALAXY
ZWG 308.030	06 28 00.	+ 63 43		13.3	GALAXY
UGC 03478	06 28 00.	+ 63 43	114	13.3	GALAXY Sb
DG 102	06 28 06.	+ 10 29	480		REFLECTION NEBULA
ZC 0628.1+2502	06 28 06.	+ 25 02	3490		CLUSTER OF GALAXIES
CED 077	06 28 10.	+ 10 29	300		DIFFUSE GALACTIC NEBULA
MCG-05-16-007	06 28 12.	- 32 53	36	15.	GALAXY
IC 0446	06 28 13.	+ 10 29 14.			SAME AS IC 2167
RNGC 2239	06 28 16.	+ 04 59			OPEN CLUSTER
VDB.66N 080	06 28 16.	- 09 41	684		REFLECTION NEBULA
OCL 0512	06 28 18.	+ 04 59	960		OPEN STAR CLUSTER
MCG+09-11-017	06 28 18.	+ 52 01	15	16.	GALAXY
ZWG 260.017	06 28 18.	+ 52 03		15.3	GALAXY
OCL 0544	06 28 18.	- 04 11	300		OPEN STAR CLUSTER
VDB.66N 078	06 28 21.	+ 09 51	336		REFLECTION NEBULA
VDB.66N 077	06 28 21.	+ 09 51	108		REFLECTION NEBULA
DG 103	06 28 24.	+ 10 03	1620		REFLECTION NEBULA
MCG+07-14-003	06 28 24.	+ 41 04	42	16.	GALAXY
IC 0447	06 28 25.	+ 10 04			SAME AS IC 2167
CED 078	06 28 27.	+ 10 03	1350		DIFFUSE GALACTIC NEBULA
ZC 0628.5+4709	06 28 30.	+ 47 09	2020		CLUSTER OF GALAXIES
DG 104	06 28 30.	- 09 30	360		REFLECTION NEBULA
IC 2167	06 28 31.	+ 10 28 42.			NONSTELLAR OBJECT
OCL 0506	06 28 36.	+ 05 57	1350	5.4	OPEN STAR CLUSTER
ZWG 175.003	06 28 36.	+ 35 35		15.1	GALAXY
MCG+06-15-002	06 28 36.	+ 35 36	48	15.	GALAXY
MCG+08-12-027	06 28 36.	+ 47 33	60	17.	GALAXY
IC 2169	06 28 38.	+ 09 51 23.			NONSTELLAR OBJECT
SCHO 0195	06 28 41.	+ 15 33 54.	450		ISOLATED DARK CLOUD
ZWG 115.003	06 28 42.	+ 26 02		15.7	GALAXY
ZWG 175.004	06 28 42.	+ 33 09		15.5	GALAXY
ZC 0628.9+5232	06 28 54.	+ 52 32	19220		CLUSTER OF GALAXIES
MCG-04-16-007	06 28 54.	- 23 40 30.	60	15.	GALAXY
SL 894	06 28 57.	- 65 56	15		STAR CLUSTER IN LMC
SS 52	06 28 58.	+ 10 22			DIFFUSE GALACTIC NEBULA
SS 51	06 28 58.	+ 10 29			DIFFUSE GALACTIC NEBULA
CED 076B	06 29	+ 05	3840		DIFFUSE GALACTIC NEBULA
LBN 0941	06 29	+ 05 55	780		BRIGHT NEBULA
LBN 0932	06 29	+ 07 00	1920		BRIGHT NEBULA
CED 079	06 29	+ 07 27			DIFFUSE GALACTIC NEBULA
KHAV 170	06 29	+ 15 16	3880		DARK NEBULA
DG 106	06 29 00.	+ 10 22	180		REFLECTION NEBULA
DG 105	06 29 00.	+ 10 26	120		REFLECTION NEBULA
ZWG 175.005	06 29 00.	+ 35 14		15.4	GALAXY
UGC 03479	06 29 00.	+ 35 14	72	15.4	GALAXY Sa-b
MCG+08-12-029	06 29 00.	+ 47 43	30	17.	GALAXY
MCG+08-12-028	06 29 00.	+ 47 43	30	17.	GALAXY
MCG+09-11-018	06 29 00.	+ 53 33	60	15.	GALAXY
ZWG 260.018	06 29 00.	+ 53 35		15.3	GALAXY
UGC 03480	06 29 00.	+ 53 35	72	15.3	GALAXY SBc
MCG+10-10-004	06 29 00.	+ 58 54	45	16.	GALAXY
ZWG 363.006	06 29 00.	+ 82 57		15.6	GALAXY
ZWG 362.023	06 29 00.	+ 82 57		15.6	GALAXY
MCG+14-04-004	06 29 00.	+ 82 59	30	16.	GALAXY
MCG+14-04-005	06 29 00.	+ 83 00	24	17.	GALAXY
VDB.66N 079	06 29 02.	+ 10 23	312		REFLECTION NEBULA
MRSL 209-02/1	06 29 06.	+ 04 58	6000		HII REGION
ZWG 204.003	06 29 06.	+ 40 14		15.7	GALAXY
UGC 03481	06 29 06.	+ 40 14	102	15.7	GALAXY
ZWG 233.021	06 29 06.	+ 47 33		15.3	GALAXY
MCG+12-07-007	06 29 06.	+ 73 54	39	16.	GALAXY
ST 29	06 29 09.	+ 04 56	4800		EMISSION NEBULA
SG 1.12	06 29 09.	+ 04 58	5220		DIFFUSE EMISSION NEBULA
MCG-05-16-008	06 29 12.	- 28 30	72	15.5	GALAXY
UGC 03482	06 29 18.	+ 19 15	72	17.	GALAXY S
UGC 03483	06 29 18.	+ 51 23	72	16.0	GALAXY Sb-c
7ZW 075	06 29 18.	+ 63 54			COMPACT GALAXY
MCG-05-16-009	06 29 18.	- 30 58	18	16.	GALAXY
MCG-05-16-010	06 29 18.	- 30 59	30	16.	GALAXY
SC 0627-2056.5	06 29 22.	- 20 58 33.	18		NEBULA
SEY 024	06 29 28.	+ 73 54 05.		15.3	FAINT GALAXY
MCG+12-07-008	06 29 30.	+ 74 23	72	16.	GALAXY
SCHO 0196	06 29 31.	+ 04 43 48.	300		ISOLATED DARK CLOUD
RNGC 2240	06 29 32.	+ 35 14			NON-EXISTENT OBJECT
SCHO 0197	06 29 35.	+ 15 35 00.	640		ISOLATED DARK CLOUD
RNGC 2244	06 29 40.	+ 04 54		5.0	CLUSTER WITH NEBULOSITY

OBJECT NAME	RIGHT ASCEN.	DECLINATION	DIAM.	MAGN.	TYPE OF OBJECT
RNGC 2257	06 29 40.	- 64 08			GLOBULAR CLUSTER IN LMC
OCL 0515	06 29 42.	+ 04 54	2520	5.8	OPEN STAR CLUSTER
ISS 0426	06 29 42.	+ 10 50	229		STELLAR RING
7ZW 076	06 29 42.	+ 57 43			COMPACT GALAXY
ZWG 285.002	06 29 42.	+ 58 54		15.6	GALAXY
UGC 03484	06 29 42.	+ 58 54	114	15.6	GALAXY Sb-c
UGC 03485	06 29 42.	+ 65 55	60	18.	GALAXY
MCG+10-10-005	06 29 45.	+ 58 53	96	14.	GALAXY
RNGC 2246	06 29 46.	+ 05 09			DIFFUSE NEBULA
SL 895	06 29 46.	- 64 08	205		STAR CLUSTER IN LMC
ZWG 330.007	06 29 48.	+ 74 21		15.5	GALAXY
ZWG 329.033	06 29 48.	+ 74 21		15.5	GALAXY
UGC 03486	06 29 48.	+ 74 21	102	15.5	GALAXY Sa-b
IC 0448	06 29 52.	+ 07 26			NONSTELLAR OBJECT
MCG+11-08-057	06 29 54.	+ 65 31	18	16.	GALAXY
MCG+12-07-009	06 29 54.	+ 73 06	57	17.	GALAXY
MCG-04-16-008	06 29 54.	- 26 43 30.	120	14.5	GALAXY
RNGC 2245	06 29 55.	+ 10 12			DIFFUSE NEBULA
VDB.66N 082	06 29 57.	+ 10 23	372		REFLECTION NEBULA
CED 080	06 29 58.	+ 10 12	300		DIFFUSE GALACTIC NEBULA
LBN 0949	06 30	+ 05 00	1080		BRIGHT NEBULA
LBN 0948	06 30	+ 05 00	4200		BRIGHT NEBULA
DG 107	06 30	+ 07 27	1230		REFLECTION NEBULA
LBN 0904	06 30	+ 10 12	120		BRIGHT NEBULA
LBN 0901	06 30	+ 10 23	120		BRIGHT NEBULA
B 037	06 30	+ 10 31	10800		DARK OBJECT
KHAV 171	06 30	+ 16 46	6350		DARK NEBULA
DG 108	06 30 00.	+ 10 12	300		REFLECTION NEBULA
LDN 1605	06 30 00.	+ 10 40	4500		DARK NEBULA
VDB.66N 081	06 30 01.	+ 07 21	2160		REFLECTION NEBULA
SS 53	06 30 04.	+ 10 12			DIFFUSE GALACTIC NEBULA
LW 481	06 30 04.	- 64 17			STAR CLUSTER IN LMC
ST 28	06 30 06.	+ 06 26	5400		EMISSION NEBULA
ZWG 204.004	06 30 06.	+ 40 43		15.3	GALAXY
UGC 03487	06 30 06.	+ 40 43	108	15.3	GALAXY Sc
MCG+07-14-005	06 30 06.	+ 40 43	42	17.	GALAXY
MCG+07-14-004	06 30 06.	+ 40 44	54	15.	GALAXY
MCG+09-11-019	06 30 06.	+ 52 09	57	15.	GALAXY
ZWG 308.031	06 30 06.	+ 65 31		15.7	GALAXY
IC 2168	06 30 08.	+ 44 43			OPEN CLUSTER
SG 3.077	06 30 09.	+ 05	6000		DIFFUSE EMISSION NEBULA
SG 3.078	06 30 11.	+ 06 36	780		DIFFUSE EMISSION NEBULA
ZWG 260.019	06 30 12.	+ 52 11		15.1	GALAXY
UGC 03488	06 30 12.	+ 52 11	66	15.1	GALAXY S0-a
RNGC 2250	06 30 16.	- 05 00		9.0	OPEN CLUSTER
SL 896	06 30 17.	- 69 18	25		STAR CLUSTER IN LMC
OCL 0547	06 30 18.	- 05 00	480	8.9	OPEN STAR CLUSTER
RNGC 2247	06 30 19.	+ 10 23			DIFFUSE NEBULA
SS 54	06 30 22.	+ 10 22			DIFFUSE GALACTIC NEBULA
CED 081	06 30 22.	+ 10 22	240		DIFFUSE GALACTIC NEBULA
RNGC 2242	06 30 22.	+ 44 48		14.5	GALAXY
LW 480	06 30 23.	- 69 18			STAR CLUSTER IN LMC
ASS 53	06 30 24.	+ 08 52	50400		OB ASSOCIATION MON OB1
DG 109	06 30 24.	+ 10 22	240		REFLECTION NEBULA
ZWG 204.005	06 30 24.	+ 44 48		14.7	GALAXY
ZWG 330.008	06 30 24.	+ 74 18		15.3	GALAXY
ZWG 329.034	06 30 24.	+ 74 18		15.3	GALAXY
MCG+12-07-010	06 30 24.	+ 74 19	30	16.	GALAXY
IC 2170	06 30 27.	+ 44 44			NONSTELLAR OBJECT
OCL 0478	06 30 30.	+ 20 35	240	15.	OPEN STAR CLUSTER
UGC 03489	06 30 30.	+ 21 04	132	17.	GALAXY Sb-c
SEY 025	06 30 32.	+ 74 20 01.		15.0	FAINT GALAXY
SNO 23	06 30 34.	- 37 37	900	19.	GROUP OF 12 GALAXIES
UGC 03490	06 30 36.	+ 12 05	84	15.0	GALAXY S?
ZWG 308.032	06 30 36.	+ 63 45		15.7	GALAXY
ZWG 175.006	06 30 42.	+ 38 16		15.4	GALAXY
MCG-06-15-006	06 30 42.	- 33 28	30	15.	GALAXY
MCG-04-16-009	06 30 51.	- 25 39 30.	54	15.	GALAXY
B 038	06 30 55.	+ 11 07	3600		DARK OBJECT
KHAV 172	06 31	+ 03 58	14350		DARK NEBULA
LBN 0943	06 31	+ 05 40	1140		BRIGHT NEBULA
LBN 0936	06 31	+ 06 40	600		BRIGHT NEBULA
LBN 0931	06 31	+ 07 20	720		BRIGHT NEBULA
LBN 0930	06 31	+ 07 20	1500		BRIGHT NEBULA
HUB C20	06 31 00.	+ 08 40			DIFFUSE NEBULA
ZC 0631.0+4613	06 31 00.	+ 46 13	3090		CLUSTER OF GALAXIES
MCG-06-15-007	06 31 06.	- 34 05	30	15.	GALAXY
SCHO 0198	06 31 20.	+ 04 44 48.	300		ISOLATED DARK CLOUD
MCG-05-16-011	06 31 24.	- 27 57	60	15.	GALAXY
RNGC 2248	06 31 27.	+ 26 21			NON-EXISTENT OBJECT
BA 08	06 31 30.	+ 08 07	1200		STELLAR GROUP
MCG+08-12-030	06 31 30.	+ 48 53	78	14.	GALAXY
UGC 03491	06 31 30.	+ 75 27	60	17.	GALAXY DWARF
UGC 03492	06 31 30.	+ 75 28	60	16.5	GALAXY SBb
ZC 0631.6+2609	06 31 36.	+ 26 09	4370		CLUSTER OF GALAXIES
ZWG 233.022	06 31 36.	+ 48 53		14.8	GALAXY
UGC 03493	06 31 36.	+ 48 53	138	14.8	GALAXY S
UGC 03494	06 31 36.	+ 56 22	60	16.0	GALAXY SBb
SG 3.080	06 31 37.	+ 02 31	1440		DIFFUSE EMISSION NEBULA
SG 3.079	06 31 37.	+ 02 55	3360		DIFFUSE EMISSION NEBULA
MRSL 208-02/1	06 31 42.	+ 02 35	2400		HII REGION
MCG-04-16-010	06 31 42.	- 24 55 30.	90	14.	GALAXY
MCG+09-11-020	06 31 45.	+ 56 24 30.	60	16.	GALAXY
UGC 03495	06 31 48.	+ 76 52	66	16.0	GALAXY Sa-b
SNO 24	06 31 49.	- 36 36 32.	360	18.	GROUP OF 5 GALAXIES
ZC 0631.9+5937	06 31 54.	+ 59 37	5780		CLUSTER OF GALAXIES
ISS 0562	06 31 54.	- 07 53	623		STELLAR RING
MCG-06-15-008	06 31 54.	- 34 13	42	16.	GALAXY
ST 27	06 31 57.	+ 02 29	1200		EMISSION NEBULA
LBN 0973	06 32	+ 02 40	660		BRIGHT NEBULA
LBN 0972	06 32	+ 02 40	360		BRIGHT NEBULA
LBN 0971	06 32	+ 02 40	1320		BRIGHT NEBULA
LBN 0970	06 32	+ 02 40	2400		BRIGHT NEBULA
KHAV 173	06 32	+ 07 46	1720		DARK NEBULA
LBN 0698	06 32	+ 67 00	3600		BRIGHT NEBULA
OCL 0499	06 32 00.	+ 08 24	840	9.0	OPEN STAR CLUSTER
ISS 0371	06 32 00.	+ 15 59	338		STELLAR RING
ZC 0632.0+4704	06 32 00.	+ 47 04	1680		CLUSTER OF GALAXIES
ZWG 308.033	06 32 00.	+ 67 54		14.3	GALAXY
UGC 03497	06 32 00.	+ 67 54	72	14.3	GALAXY S0?
KARA.72 114A	06 32 00.	+ 67 54	66	14.3	PART OF DOUBLE GALAXY
UGC 03496	06 32 00.	+ 85 53	138	17.	GALAXY DWARF
ISS 0563	06 32 00.	- 08 50	536		STELLAR RING
MCG-06-15-009	06 32 00.	- 34 14	48	15.	GALAXY
LB 03478	06 32 00.	- 77 59		15.4	FAINT BLUE STAR
SCHO 0199	06 32 01.	+ 07 38 48.	320		ISOLATED DARK CLOUD
RNGC 2251	06 32 02.	+ 08 24		8.5	OPEN CLUSTER
ZWG 115.004	06 32 06.	+ 26 30		15.5	GALAXY
ZC 0632.1+5521	06 32 06.	+ 55 21	1410		CLUSTER OF GALAXIES
MCG+11-09-001	06 32 06.	+ 67 54	39	15.	GALAXY
MCG-04-16-011	06 32 06.	- 25 37	36	14.	GALAXY
IC 0445	06 32 10.	+ 67 55 01.			NONSTELLAR OBJECT
RNGC 2255	06 32 12.	- 34 45			GALAXY
MCG-06-15-010	06 32 12.	- 34 46	78	15.	GALAXY
MCG-06-15-011	06 32 12.	- 36 23	60	15.5	GALAXY
RNGC 2252	06 32 16.	+ 05 25		7.5	OPEN CLUSTER
OCL 0514	06 32 18.	+ 05 25	1200	8.0	OPEN STAR CLUSTER
ISS 0427	06 32 18.	+ 10 09	278		STELLAR RING
MCG-06-15-012	06 32 18.	- 34 08	30	15.5	GALAXY
SCHO 0200	06 32 23.	+ 07 33 36.	290		ISOLATED DARK CLOUD
ZWG 308.034	06 32 24.	+ 67 53		15.7	GALAXY
KARA.72 114B	06 32 24.	+ 67 53	36	15.7	PART OF DOUBLE GALAXY
ISS 0428	06 32 30.	+ 12 33	175		STELLAR RING
UGC 03498	06 32 30.	+ 15 00	72	17.	GALAXY S
7ZW 077	06 32 30.	+ 69 53			COMPACT GALAXY
ARC 0558	06 32 32.	+ 73 38		17.0	RICH CLUSTER OF GALAXIES
MCG+07-14-006	06 32 36.	+ 39 26 30.	30	15.	GALAXY
ZWG 204.006	06 32 36.	+ 39 27		15.2	GALAXY
MCG+11-09-002	06 32 36.	+ 67 52	18	17.	GALAXY
MCG+01-17-002	06 32 42.	+ 03 54	60	14.	GALAXY
ZC 0632.7+6323	06 32 42.	+ 63 23	4910		CLUSTER OF GALAXIES
MCG-06-15-013	06 32 42.	- 35 43	30	15.	GALAXY
MCG-04-16-012	06 32 48.	- 21 41 30.	78	15.	GALAXY
ZC 0632.9+2244	06 32 54.	+ 22 44	5980		CLUSTER OF GALAXIES
ZC 0633.0+7516	06 33 00.	+ 75 16	1610		CLUSTER OF GALAXIES
MAI 015	06 33 00.	+ 88 56	40		DWARF SPHEROIDAL GALAXY
LB 03479	06 33 00.	- 77 07		15.6	FAINT BLUE STAR
SCHO 0201	06 33 04.	+ 07 36 36.	370		ISOLATED DARK CLOUD
ISS 0372	06 33 06.	+ 15 06	304		STELLAR RING
ZWG 260.020	06 33 06.	+ 56 25		14.8	GALAXY
MCG+09-11-021	06 33 09.	+ 56 25 30.	51	16.	GALAXY
PK211-03.1	06 33 11.	- 00 03 07.	5		PLANETARY NEBULA
SG 3.081	06 33 12.	+ 11 54	2160		DIFFUSE EMISSION NEBULA
ZWG 330.009	06 33 12.	+ 74 21		15.2	GALAXY
ZWG 329.035	06 33 12.	+ 74 21		15.2	GALAXY
LW 482	06 33 16.	- 72 27			STAR CLUSTER IN LMC
OCL 0500	06 33 18.	+ 07 43	480	11.1	OPEN STAR CLUSTER
SCHO 0202	06 33 18.	+ 12 52 54.	350		ISOLATED DARK CLOUD
SCHO 0203	06 33 20.	+ 13 03 06.	450		ISOLATED DARK CLOUD
RNGC 2254	06 33 21.	+ 07 43		11.0	OPEN CLUSTER
SEY 026	06 33 23.	+ 74 22 49.		15.3	FAINT GALAXY
ZWG 175.007	06 33 24.	+ 37 37		14.8	GALAXY
UGC 03499	06 33 24.	+ 37 37	72	14.8	GALAXY S
ZC 0633.4+4353	06 33 24.	+ 43 53	1340		CLUSTER OF GALAXIES
ZWG 233.023	06 33 24.	+ 49 53		15.6	GALAXY
ISS 0430	06 33 30.	+ 09 45	141		STELLAR RING
LDN 1609	06 33 30.	+ 10 30	480		DARK NEBULA
ISS 0429	06 33 30.	+ 11 42	229		STELLAR RING
MCG+06-15-003	06 33 30.	+ 37 38	66	14.5	GALAXY
ST 26	06 33 32.	+ 09 56	2400		EMISSION NEBULA
SL 897	06 33 38.	- 71 04	55		STAR CLUSTER IN LMC
LW 483	06 33 44.	- 71 06			STAR CLUSTER IN LMC
SG 3.082	06 33 46.	+ 10 48	660		DIFFUSE EMISSION NEBULA
RNGC 2273A	06 33 46.	+ 60 08			GALAXY
OCL 0517	06 33 48.	+ 04 52	1320	9.7	OPEN STAR CLUSTER
BA 07	06 33 54.	+ 08 24	600		STELLAR GROUP
SG 3.083	06 33 57.	+ 09 28	960		DIFFUSE EMISSION NEBULA
LBN 0951	06 34	+ 05 10	540		BRIGHT NEBULA
LBN 0902	06 34	+ 10 50	360		BRIGHT NEBULA
LBN 0899	06 34	+ 11 00	1320		BRIGHT NEBULA
MCG+01-17-003	06 34 00.	+ 05 39	96	13.	GALAXY
OCL 0494	06 34 00.	+ 09 29	960	11.0	OPEN STAR CLUSTER
LDN 1608	06 34 00.	+ 10 50	840		DARK NEBULA
7ZW 078	06 34 00.	+ 68 56			COMPACT GALAXY
7ZW 079	06 34 00.	+ 69 55			COMPACT GALAXY
ZWG 363.007	06 34 00.	+ 84 13		14.7	GALAXY
ZWG 362.024	06 34 00.	+ 84 13		14.7	GALAXY
UGC 03500	06 34 00.	+ 84 13	114	14.7	GALAXY PECULR
MCG+01-17-004	06 34 06.	+ 05 40	48	14.	GALAXY
ZC 0634.1+4750	06 34 06.	+ 47 50	1280		CLUSTER OF GALAXIES
OCL 0563	06 34 06.	- 14 08	660	11.	OPEN STAR CLUSTER
7ZW 080	06 34 12.	+ 63 33			COMPACT GALAXY
ZC 0634.2+7002	06 34 12.	+ 70 02	3560		CLUSTER OF GALAXIES
PK189+07.1	06 34 17.8	+ 24 03 13.	38		PLANETARY NEBULA
ZWG 260.021	06 34 18.	+ 56 05		15.4	GALAXY
ZWG 330.010	06 34 18.	+ 73 55		15.7	GALAXY
ZWG 329.036	06 34 18.	+ 73 55		15.7	GALAXY
SG 3.084	06 34 22.	+ 05 08	1200		DIFFUSE EMISSION NEBULA
OCL 0527	06 34 24.	+ 03 07			OPEN STAR CLUSTER
OCL 0510	06 34 24.	+ 06 00	2700	4.6	OPEN STAR CLUSTER
ISS 0431	06 34 24.	+ 11 52	177		STELLAR RING
ASS 54	06 34 30.	+ 04 53	21600		OB ASSOCIATION MON OB2
ZC 0634.5+3457	06 34 30.	+ 34 57	2220		CLUSTER OF GALAXIES
MCG+14-04-006	06 34 30.	+ 84 12	90	15.	GALAXY
ARC 0559	06 34 34.	+ 69 45		15.8	RICH CLUSTER OF GALAXIES
SL 898	06 34 34.	- 06 24	25		STAR CLUSTER IN LMC
ST 25	06 34 35.	+ 05 58	6000		EMISSION NEBULA
UGC 03501	06 34 36.	+ 49 17	72	18.	GALAXY DWARF
MCG+08-12-031	06 34 36.	+ 49 18	60	16.	GALAXY
CED 082	06 34 41.	+ 06 11			DIFFUSE GALACTIC NEBULA
ZWG 260.022	06 34 42.	+ 53 24		15.4	GALAXY
ZWG 308.035	06 34 48.	+ 65 30		14.8	GALAXY
UGC 03502	06 34 48.	+ 65 30	78	14.8	GALAXY SB:b
VMT 10	06 35	+ 06 30	12000		SUPERNOVA REMNANT
MIL 12	06 35	+ 06 30	12600		SUPERNOVA REMNANT
KHAV 174	06 35	+ 07 34	1720		DARK NEBULA
LBN 0929	06 35	+ 08 00	19800		BRIGHT NEBULA
OCL 0518	06 35 00.	+ 04 47	2100	5.4	OPEN STAR CLUSTER
ZWG 115.005	06 35 00.	+ 22 41		15.6	GALAXY
UGC 03503	06 35 00.	+ 22 41	108	15.6	GALAXY Sc-IRR
MCG+10-10-007	06 35 00.	+ 57 46	27	16.	GALAXY
MCG+10-10-006	06 35 00.	+ 57 46	24	16.	GALAXY
MCG+11-09-003	06 35 00.	+ 65 29	84	14.	GALAXY
MRK 005	06 35 00.	+ 75 39	7	17.	GALAXY WITH UV CONTINUUM
ZC 0635.0+7917	06 35 00.	+ 79 17	1480		CLUSTER OF GALAXIES
7ZW 081	06 35 00.	+ 85 06			COMPACT GALAXY
MCG+15-01-005B	06 35 00.	+ 87 22	6	18.	GALAXY
MCG+15-01-005A	06 35 00.	+ 87 22	30	17.	GALAXY
MCG-06-15-014	06 35 00.	- 35 02	18	15.	GALAXY
SCHO 0204	06 35 02.	+ 10 33 18.	400		ISOLATED DARK CLOUD
LDN 1610	06 35 12.	+ 10 30	540		DARK NEBULA
OCL 0539	06 35 12.	- 00 52	600	17.	OPEN STAR CLUSTER
SCHO 0205	06 35 16.	+ 10 17 18.	680		ISOLATED DARK CLOUD
SG 3.085	06 35 17.	+ 01 31	1860		DIFFUSE EMISSION NEBULA
B 039	06 35 17.	+ 10 22			DARK OBJECT
ZWG 175.008	06 35 18.	+ 36 12		15.6	GALAXY

OBJECT NAME	RIGHT ASCEN.	DECLINATION	DIAM.	MAGN.	TYPE OF OBJECT
ZC 0635.3+7547	06 35 18.	+ 75 47	940		CLUSTER OF GALAXIES
MRSL 210-02/2	06 35 24.	+ 01 34	2100		HII REGION
RNGC 2260	06 35 32.	- 01 26			NON-EXISTENT OBJECT
ZWG 115.006	06 35 36.	+ 26 32		15.4	GALAXY
ZWG 285.003	06 35 36.	+ 60 07		13.3	GALAXY
UGC 03504	06 35 36.	+ 60 07	174	13.3	GALAXY Sc
ZWG 308.036	06 35 36.	+ 64 36		15.7	GALAXY
MCG+10-10-009	06 35 42.	+ 60 07 30.	156	13.	GALAXY
MCG+10-10-008	06 35 42.	+ 62 25	30	17.	GALAXY
OCL 0531	06 35 48.	+ 01 14	210	11.3	OPEN STAR CLUSTER
RNGC 2262	06 35 48.	+ 01 14		11.5	OPEN CLUSTER
OCL 0530	06 35 48.	+ 02 04	720	10.6	OPEN STAR CLUSTER
OCL 0492	06 35 48.	+ 10 56	960	10.8	OPEN STAR CLUSTER
ZC 0635.8+4303	06 35 48.	+ 43 03	3700		CLUSTER OF GALAXIES
RNGC 2259	06 35 49.	+ 10 56		11.0	OPEN CLUSTER
MAI 016	06 35 52.	+ 80 38	40		DWARF SPHEROIDAL GALAXY
MRSL 210-02/1	06 35 54.	+ 00 46	180		HII REGION
MCG+09-11-022	06 35 54.	+ 53 25 30.	60	17.	GALAXY
MCG-04-16-013	06 35 54.	- 25 56	60	15.	GALAXY
LBN 0978	06 36	+ 01 40	1620		BRIGHT NEBULA
LBN 0920	06 36	+ 08 46	60		BRIGHT NEBULA
KHAV 175	06 36	+ 11 21	4860		DARK NEBULA
KEEN 2273A	06 36	+ 60 07			GALAXY
OCL 0505	06 36 00.	+ 06 57	192	7.1	OPEN STAR CLUSTER
ZWG 204.007	06 36 00.	+ 43 24		15.6	GALAXY
MCG+07-14-007	06 36 00.	+ 43 25	24	16.	GALAXY
ZWG 363.008	06 36 00.	+ 83 51		15.7	GALAXY
ZWG 362.025	06 36 00.	+ 83 51		15.7	GALAXY
MCG+14-04-007	06 36 00.	+ 83 52	30	17.	GALAXY
ZWG 175.009	06 36 06.	+ 33 37		15.5	GALAXY
7ZW 082	06 36 12.	+ 63 15			COMPACT GALAXY
7ZW 083	06 36 12.	+ 63 49			COMPACT GALAXY
SG 3.086	06 36 15.	+ 09 47	6900		DIFFUSE EMISSION NEBULA
ZWG 145.002	06 36 18.	+ 30 21		15.5	GALAXY
HH 39C	06 36 21.2	+ 08 53 48.			HERBIG-HARO OBJECT
HH 39D	06 36 21.6	+ 08 53 39.			HERBIG-HARO OBJECT
HH 39	06 36 21.6	+ 08 53 58.	45		HERBIG-HARO OBJECT
HH 39E	06 36 21.8	+ 08 53 47.			HERBIG-HARO OBJECT
HH 39A	06 36 21.8	+ 08 54 11.			HERBIG-HARO OBJECT
RNGC 2263	06 36 22.	- 24 47		13.0	GALAXY
HH 39F	06 36 22.8	+ 08 53 35.			HERBIG-HARO OBJECT
ZWG 260.023	06 36 24.	+ 53 16		15.4	GALAXY
MCG-04-16-014	06 36 24.	- 24 47 30.	156	13.	GALAXY
RNGC 2261	06 36 26.	+ 08 46			DIFFUSE NEBULA
CED 083	06 36 26.	+ 08 46			DIFFUSE GALACTIC NEBULA
SG 3.087	06 36 26.	+ 08 47	180		DIFFUSE EMISSION NEBULA
ZC 0636.5+4637	06 36 30.	+ 46 37	3900		CLUSTER OF GALAXIES
ZC 0636.5+5735	06 36 30.	+ 57 35	1550		CLUSTER OF GALAXIES
MCG+11-09-004	06 36 30.	+ 65 06	36	17.	GALAXY
SCHO 0206	06 36 31.	+ 04 01 42.	380		ISOLATED DARK CLOUD
MCG-03-17-006	06 36 33.	- 20 15 30.	72	15.	GALAXY
UGC 03505	06 36 36.	+ 20 49	90	17.	GALAXY Sc
MCG+08-12-032	06 36 36.	+ 50 09 30.	36	14.	GALAXY
7ZW 084	06 36 36.	+ 64 57			COMPACT GALAXY
ZWG 234.001	06 36 42.	+ 50 09		14.4	GALAXY
ZWG 233.024	06 36 42.	+ 50 09		14.4	GALAXY
UGC 03506	06 36 42.	+ 50 09	54	14.4	GALAXY E-SO
ZWG 260.024	06 36 42.	+ 53 17		15.6	GALAXY
SG 3.088	06 36 44.	+ 09 02	15000		DIFFUSE EMISSION NEBULA
MCG+12-07-012	06 36 48.	+ 74 17	51	17.	GALAXY
KHAV 176	06 37	+ 07 51	1570		DARK NEBULA
MCG+11-09-005	06 37 00.	+ 64 01	60	18.	GALAXY
MCG+08-12-033	06 37 06.	+ 50 12	48	15.	GALAXY
ZWG 260.025	06 37 06.	+ 53 05		15.6	GALAXY
MCG+09-11-023	06 37 06.	+ 53 19 30.	27	17.	GALAXY
ZWG 260.026	06 37 06.	+ 53 20		15.6	GALAXY
7ZW 085	06 37 06.	+ 63 17			COMPACT GALAXY
ST 24	06 37 10.	+ 08 15	7200		EMISSION NEBULA
SS 55	06 37 11.	+ 11 44			DIFFUSE GALACTIC NEBULA
ACK 201+02.1	06 37 12.	+ 11 09			PLANETARY NEBULA
DG 110	06 37 12.	+ 11 44	180		REFLECTION NEBULA
ZWG 234.002	06 37 12.	+ 50 12		15.4	GALAXY
ZWG 233.025	06 37 12.	+ 50 12		15.4	GALAXY
ZWG 260.027	06 37 18.	+ 53 47		15.4	GALAXY
UGC 03507	06 37 18.	+ 53 47	66	15.4	GALAXY Sb
MCG+09-11-024	06 37 18.	+ 53 47 30.	18	16.	GALAXY
UGC 03508	06 37 30.	+ 33 40	60	16.0	GALAXY S
MCG+07-14-008	06 37 30.	+ 40 06	42	16.	GALAXY
MCG+08-12-034	06 37 36.	+ 50 18 30.	36	16.	GALAXY
ZC 0637.6+6755	06 37 36.	+ 67 55	670		CLUSTER OF GALAXIES
CED 084A	06 37 37.	+ 08 12			DIFFUSE GALACTIC NEBULA
ZWG 234.003	06 37 42.	+ 50 18		15.2	GALAXY
ZWG 233.026	06 37 42.	+ 50 18		15.2	GALAXY
7ZW 086	06 37 42.	+ 67 55			COMPACT GALAXY
ARC 0560	06 37 44.	+ 67 55		17.6	RICH CLUSTER OF GALAXIES
MCG+09-11-025	06 37 48.	+ 53 44 30.	24	16.	GALAXY
7ZW 087	06 37 48.	+ 65 34			COMPACT GALAXY
ZWG 330.011	06 37 48.	+ 74 28		15.4	GALAXY
SCHO 0207	06 37 52.	+ 07 45 06.	410		ISOLATED DARK CLOUD
MCG+07-14-009	06 37 54.	+ 40 13	36	17.	GALAXY
VDB.66N 083	06 37 54.	- 27 13	36		REFLECTION NEBULA
MCG-06-15-015	06 37 54.	- 37 46	24	15.	GALAXY
SCHO 0208	06 37 56.	+ 09 00 42.	640		ISOLATED DARK CLOUD
MCG+07-14-010	06 37 57.	+ 40 12 30.	78	14.5	GALAXY
VV 134A	06 37 57.	- 27 09	60	10.5	INTERACTING GALAXY
VV 134B	06 37 57.	- 27 12	90	10.	INTERACTING GALAXY
LBN 0952	06 38	+ 05 30	8700		BRIGHT NEBULA
LBN 0922	06 38	+ 09 00	7200		BRIGHT NEBULA
KHAV 177	06 38	+ 09 03	4140		DARK NEBULA
LBN 0912	06 38	+ 09 45	2100		BRIGHT NEBULA
LBN 0911	06 38	+ 09 57	480		BRIGHT NEBULA
LDN 1613	06 38 00.	+ 09 30			DARK NEBULA
ZWG 115.007	06 38 00.	+ 25 25		1[illegible].7	GALAXY
ZWG 145.003	06 38 00.	+ 32 58		15.6	GALAXY
ZWG 204.008	06 38 00.	+ 40 13		15.0	GALAXY
UGC 03510	06 38 00.	+ 40 13	84	15.0	GALAXY S
ZC 0638.0+4155	06 38 00.	+ 41 55	2350		CLUSTER OF GALAXIES
ZC 0638.0+4740	06 38 00.	+ 47 40	2490		CLUSTER OF GALAXIES
ZWG 363.009	06 38 00.	+ 85 42		15.5	GALAXY
ZWG 362.026	06 38 00.	+ 85 42		15.5	GALAXY
UGC 03509	06 38 00.	+ 85 42	84	15.5	GALAXY
VDB.66N 084	06 38 04.	- 27 18	168		REFLECTION NEBULA
MRSL 202+02/1	06 38 06.	+ 09 57	15000		HII REGION
ZWG 145.004	06 38 06.	+ 32 52		15.1	GALAXY
MCG-05-16-012	06 38 06.	- 28 56	30	16.	GALAXY
VV 134D	06 38 09.	- 27 09	12	16.	INTERACTING GALAXY
MCG+05-16-001	06 38 12.	+ 32 53	36	15.1	GALAXY
MCG+11-09-006	06 38 12.	+ 65 34	18	17.	GALAXY

OBJECT NAME	RIGHT ASCEN.	DECLINATION	DIAM.	MAGN.	TYPE OF OBJECT
ZWG 309.001	06 38 12.	+ 65 43		15.0	GALAXY
ZWG 308.037	06 38 12.	+ 65 43		15.0	GALAXY
RNGC 2253	06 38 12.	+ 65 43		15.0	GALAXY
MCG-04-16-015	06 38 12.	- 25 50	72	15.	GALAXY
VV 134C	06 38 12.	- 27 10 30.	36	12.	INTERACTING GALAXY
ST 23	06 38 14.	+ 09 45	3600		EMISSION NEBULA
CED 084B	06 38 15.	+ 09 56	3600		DIFFUSE GALACTIC NEBULA
SG 3.089	06 38 15.	+ 09 56	3900		DIFFUSE EMISSION NEBULA
OCL 0495	06 38 18.	+ 09 56	2280	4.7	OPEN STAR CLUSTER
YM 27	06 38 18.	+ 10 24	480		SYMMETRIC GALACTIC NEBULA
MCG-04-16-016	06 38 21.	- 26 13 30.	36	15.	GALAXY
SG 3.090	06 38 22.	+ 10 15			DIFFUSE EMISSION NEBULA
MCG+11-09-007	06 38 24.	+ 65 42 30.	18	16.	GALAXY
ISS 0500	06 38 24.	- 01 07	249		STELLAR RING
MCG-05-16-013	06 38 24.	- 27 03	72	15.5	GALAXY
MCG-06-15-016	06 38 24.	- 37 44	60	15.	GALAXY
RNGC 2264	06 38 25.	+ 09 56		4.0	CLUSTER WITH NEBULOSITY
SG 3.091	06 38 27.	+ 09 29			DIFFUSE EMISSION NEBULA
ARC 0561	06 38 27.	+ 69 04		17.0	RICH CLUSTER OF GALAXIES
MCG-05-16-014	06 38 36.	- 27 09	12	15.5	GALAXY
MCG+09-11-026	06 38 39.	+ 53 12	30	16.	GALAXY
RNGC 2265	06 38 42.	+ 11 58			NON-EXISTENT OBJECT
MCG+08-13-001	06 38 42.	+ 46 08	30	16.	GALAXY
ZWG 260.028	06 38 42.	+ 53 13		15.4	GALAXY
ZWG 309.002	06 38 42.	+ 65 15		13.0	GALAXY
ZWG 308.038	06 38 42.	+ 65 15		13.0	GALAXY
UGC 03511	06 38 42.	+ 65 15	96	13.0	GALAXY Sc
MCG+12-07-013	06 38 42.	+ 74 21	9	17.	GALAXY
MCG+09-11-027	06 38 45.	+ 52 53	12	16.	GALAXY
ZWG 175.010	06 38 48.	+ 34 31		15.6	GALAXY
KARA.72 115A	06 38 48.	+ 34 31	30	15.6	PART OF DOUBLE GALAXY
ZWG 234.004	06 38 48.	+ 46 08		15.6	GALAXY
ZWG 260.029	06 38 48.	+ 52 54		15.5	GALAXY
OCL 0576	06 38 48.	- 16 28	420	16.	OPEN STAR CLUSTER
SCHO 0210	06 38 49.	+ 09 35 36.	460		ISOLATED DARK CLOUD
ZWG 175.011	06 38 54.	+ 34 30		15.5	GALAXY
KARA.72 115B	06 38 54.	+ 34 30	54	15.5	PART OF DOUBLE GALAXY
ZC 0638.9+6348	06 38 54.	+ 63 48	2550		CLUSTER OF GALAXIES
RNGC 2267	06 38 58.	- 32 27		14.0	GALAXY
LB 09832	06 39	- 84 50		14.0	FAINT BLUE STAR
MCG+08-13-002	06 39 00.	+ 45 50	36	16.	GALAXY
MCG+11-09-008	06 39 00.	+ 65 15	84	13.	GALAXY
MCG-05-16-015	06 39 00.	- 32 27	84	14.5	GALAXY
ST 22	06 39 05.	+ 06 05	7800		EMISSION NEBULA
OCL 0643	06 39 06.	- 29 30	420	12.	OPEN STAR CLUSTER
ZWG 115.008	06 39 12.	+ 26 05		15.4	GALAXY
UGC 03512	06 39 12.	+ 42 28	96	18.	GALAXY DWARF
7ZW 088	06 39 12.	+ 64 17			COMPACT GALAXY
ZWG 309.003	06 39 12.	+ 66 08		15.1	GALAXY
ZWG 308.039	06 39 12.	+ 66 08		15.1	GALAXY
CED 085	06 39 17.	+ 06 24			DIFFUSE GALACTIC NEBULA
SCHO 0211	06 39 18.	+ 09 22 36.	590		ISOLATED DARK CLOUD
ZWG 204.009	06 39 18.	+ 41 28		15.6	GALAXY
UGC 03513	06 39 18.	+ 41 28	66	15.6	GALAXY Sc
PK153+22.1	06 39 18.	+ 61 50	162	15.9	PLANETARY NEBULA
MCG-06-15-017	06 39 18.	- 38 00	72	15.	GALAXY
MCG+08-13-003	06 39 24.	+ 49 08	42	18.	GALAXY
MCG+10-10-010	06 39 24.	+ 57 40	72	15.	GALAXY
ZWG 285.004	06 39 24.	+ 57 41		15.4	GALAXY
UGC 03514	06 39 24.	+ 57 41	78	15.4	GALAXY Sa-b
ZWG 348.024	06 39 24.	+ 79 25		15.7	GALAXY
MCG+07-14-011	06 39 36.	+ 39 29	15	16.	GALAXY
MCG+13-05-029	06 39 36.	+ 75 11	45	16.	GALAXY
ZWG 204.010	06 39 42.	+ 39 30		15.4	GALAXY
7ZW 089	06 39 48.	+ 63 19			COMPACT GALAXY
ZWG 330.012	06 39 48.	+ 71 24		13.7	GALAXY
UGC 03515	06 39 48.	+ 71 24	102	13.7	GALAXY E
MCG+12-07-014	06 39 54.	+ 71 24	33	15.	GALAXY
MCG-06-15-018	06 39 54.	- 34 42	30	15.	GALAXY
IC 0449	06 39 55.	+ 71 24 22.			NONSTELLAR OBJECT
KHAV 179	06 40	+ 03 57	6880		DARK NEBULA
KHAV 178	06 40	+ 08 15	4080		DARK NEBULA
LBN 0702	06 40	+ 66 20	3600		BRIGHT NEBULA
ZC 0640.0+6828	06 40 00.	+ 68 28	1480		CLUSTER OF GALAXIES
UGC 03516	06 40 06.	+ 22 55	90	17.	GALAXY
OCL 0471	06 40 06.	+ 27 01	420	10.	OPEN STAR CLUSTER
ZC 0640.1+6700	06 40 06.	+ 67 00	1080		CLUSTER OF GALAXIES
MCG-05-16-016	06 40 06.	- 30 04	48	15.5	GALAXY
RNGC 2266	06 40 08.	+ 27 01		9.5	OPEN CLUSTER
UGC 03517	06 40 12.	+ 41 11	90	16.0	GALAXY S
MCG+09-11-028	06 40 12.	+ 56 51	42	18.	GALAXY
OCL 0642	06 40 12.	- 29 42	168	11.	OPEN STAR CLUSTER
SCHO 0212	06 40 15.	+ 09 37 30.	780		ISOLATED DARK CLOUD
ZWG 115.009	06 40 18.	+ 21 03		15.3	GALAXY
ZWG 330.013	06 40 18.	+ 74 34		15.4	GALAXY
ZC 0640.4+2559	06 40 24.	+ 25 59	6180		CLUSTER OF GALAXIES
SEY 027	06 40 27.	+ 71 11 16.		15.3	FAINT GALAXY
ZWG 145.005	06 40 30.	+ 28 30		15.5	GALAXY
UGC 03518	06 40 30.	+ 28 30	66	15.5	GALAXY SO-a
OCL 0538	06 40 36.	+ 00 03	720		OPEN STAR CLUSTER
MCG+10-10-011	06 40 36.	+ 57 39	36	16.	GALAXY
7ZW 090	06 40 36.	+ 60 28			COMPACT GALAXY
7ZW 091	06 40 36.	+ 64 17			COMPACT GALAXY
ZC 0640.6+6640	06 40 36.	+ 66 40	1010		CLUSTER OF GALAXIES
ZWG 330.014	06 40 42.	+ 74 17		14.0	GALAXY
UGC 03519	06 40 42.	+ 74 17	138	14.0	GALAXY E?
RNGC 2272	06 40 42.	- 27 24		14.0	GALAXY
MCG-05-16-017	06 40 42.	- 27 24	84	14.	GALAXY
RNGC 2271	06 40 45.	- 23 25		14.0	GALAXY
SCHO 0213	06 40 46.	+ 12 47 00.	440		ISOLATED DARK CLOUD
RNGC 2256	06 40 47.	+ 74 17		14.0	GALAXY
ZWG 115.010	06 40 48.	+ 25 53		15.4	GALAXY
UGC 03520	06 40 48.	+ 62 36	84	17.	GALAXY SB IV
MCG+10-10-012	06 40 48.	+ 62 36	60	15.	GALAXY
MCG+12-07-015	06 40 48.	+ 74 18	27	14.	GALAXY
MCG-04-16-017	06 40 48.	- 23 25	36	14.	GALAXY
LBN 0983	06 41	+ 00 30	3300		BRIGHT NEBULA
KHAV 180	06 41	+ 09 27	2230		DARK NEBULA
LDN 1637	06 41 00.	+ 04 00	3960		DARK NEBULA
MCG+08-13-004	06 41 00.	+ 47 55	42	17.	GALAXY
ZWG 234.005	06 41 00.	+ 47 56		15.7	GALAXY
MCG+12-07-016	06 41 00.	+ 74 32	51	14.	GALAXY
ZWG 363.010	06 41 00.	+ 84 06		15.5	GALAXY
ZWG 362.027	06 41 00.	+ 84 06		15.5	GALAXY
UGC 03521	06 41 00.	+ 84 06	102	15.5	GALAXY S
7ZW 092	06 41 00.	+ 84 58			COMPACT GALAXY
ZWG 363.011	06 41 00.	+ 84 58		15.7	GALAXY
ZWG 362.028	06 41 00.	+ 84 58		15.7	GALAXY

OBJECT NAME	RIGHT ASCEN.	DECLINATION	DIAM.	MAGN.	TYPE OF OBJECT
UGC 03522	06 41 00.	+ 84 58	138	15.7	GALAXY PECULR
MCG+14-04-009	06 41 00.	+ 85 00	108	16.	GALAXY
MCG+14-04-010	06 41 00.	+ 86 38	42	17.	GALAXY
ZWG 330.015	06 41 06.	+ 74 30		13.2	GALAXY
UGC 03523	06 41 06.	+ 74 30	150	13.2	GALAXY S0
MCG-06-15-019	06 41 06.	- 35 32	42	15.	GALAXY
UGC 03524	06 41 12.	+ 12 28	102	16.0	GALAXY S
ZWG 115.011	06 41 12.	+ 25 59		15.5	GALAXY
MIN.47 06	06 41 14.	- 01 05			DIFFUSE NEBULA
MCG+02-18-001	06 41 15.	+ 12 27	60	16.	GALAXY
RNGC 2270	06 41 17.	+ 03 30			NON-EXISTENT OBJECT
RNGC 2269	06 41 17.	+ 04 37		10.0	OPEN CLUSTER
SG 3.092	06 41 17.	+ 06 32	3000		DIFFUSE EMISSION NEBULA
OCL 0524	06 41 18.	+ 04 37	360	10.4	OPEN STAR CLUSTER
ZWG 234.006	06 41 18.	+ 45 33		15.6	GALAXY
ISS 0501	06 41 18.	- 01 23	338		STELLAR RING
ISS 0564	06 41 18.	- 07 42	109		STELLAR RING
RNGC 2258	06 41 21.	+ 74 32		13.0	GALAXY
MRSL 233-12/1	06 41 24.	- 24 05	12000		HII REGION
LDN 1631	06 41 30.	+ 06 25	600		DARK NEBULA
ZWG 175.012	06 41 30.	+ 35 22		15.7	GALAXY
MCG+07-14-012	06 41 30.	+ 40 27	15	15.	GALAXY
ZWG 204.011	06 41 30.	+ 40 28		14.9	GALAXY
UGC 03525	06 41 30.	+ 40 28	90	14.9	GALAXY PECULR
MCG+08-13-005	06 41 30.	+ 45 31	12	16.	GALAXY
7ZW 093	06 41 30.	+ 57 57			COMPACT GALAXY
VV 248B	06 41 30.	+ 86 37	42	16.	INTERACTING GALAXY
VV 248A	06 41 30.	+ 86 37	42	16.	INTERACTING GALAXY
VV 248	06 41 30.	+ 86 37	150		INTERACTING GALAXY
MCG+09-11-029	06 41 33.	+ 53 30	72	16.	GALAXY
OCL 0533	06 41 36.	+ 01 39	1080		OPEN STAR CLUSTER
ZWG 260.030	06 41 36.	+ 53 31		15.7	GALAXY
UGC 03526	06 41 36.	+ 53 31	84	15.7	GALAXY Sb
7ZW 094	06 41 36.	+ 66 33			COMPACT GALAXY
ST 21	06 41 42.	+ 09 15	4200		EMISSION NEBULA
MCG-05-16-018	06 41 42.	- 27 12	48	15.	GALAXY
MCG+11-09-009	06 41 48.	+ 64 06	36	16.	GALAXY
ZWG 330.016	06 41 54.	+ 72 48		15.7	GALAXY
UGC 03527	06 41 54.	+ 72 48	72	15.7	GALAXY S
RNGC 2273B	06 41 57.	+ 60 24		14.0	GALAXY
LBN 0987	06 42	+ 00 10	840		BRIGHT NEBULA
LBN 0984	06 42	+ 00 20	2100		BRIGHT NEBULA
LBN 0947	06 42	+ 06 40	2100		BRIGHT NEBULA
ACK 210-00.2	06 42 00.	+ 01 22			PLANETARY NEBULA
ACK 210-00.1	06 42 00.	+ 01 23			PLANETARY NEBULA
LDN 1626	06 42 00.	+ 08 00	4380		DARK NEBULA
URA 58	06 42 00.	+ 10 45	708		STELLAR RING
ZWG 175.013	06 42 00.	+ 34 32		14.9	GALAXY
UGC 03529	06 42 00.	+ 34 32	60	14.9	GALAXY S
ZWG 204.012	06 42 00.	+ 43 37		15.5	GALAXY
ZWG 260.031	06 42 00.	+ 51 42		15.4	GALAXY
MCG+10-10-013	06 42 00.	+ 60 23	174	13.	GALAXY
ZWG 285.005	06 42 00.	+ 60 24		14.1	GALAXY
UGC 03530	06 42 00.	+ 60 24	150	14.1	GALAXY SBc
7ZW 095	06 42 00.	+ 67 00			COMPACT GALAXY
MCG+12-07-017	06 42 00.	+ 72 49	60	16.	GALAXY
ZC 0642.0+7334	06 42 00.	+ 73 34	8600		CLUSTER OF GALAXIES
MCG+13-05-030	06 42 00.	+ 80 47	39	17.	GALAXY
MCG+14-04-008	06 42 00.	+ 84 05 30.	48	15.	GALAXY
ZWG 363.012	06 42 00.	+ 84 08		14.5	GALAXY
ZWG 362.029	06 42 00.	+ 84 08		14.5	GALAXY
UGC 03528	06 42 00.	+ 84 08	102	14.5	GALAXY SBa-b
MCG+14-04-011	06 42 00.	+ 86 37	42	16.	GALAXY
ZWG 370.002	06 42 00.	+ 86 40		15.6	GALAXY
ZWG 362.030	06 42 00.	+ 86 40		15.6	GALAXY
UGC 03528A	06 42 00.	+ 86 40	138	15.6	GALAXY DBL SYS
KARA.72 116A	06 42 00.	+ 86 40	48	15.6	PART OF DOUBLE GALAXY
MCG+06-15-004	06 42 06.	+ 34 32	48	15.	GALAXY
ZWG 234.007	06 42 06.	+ 46 07		15.5	GALAXY
KARA.73B 0172	06 42 06.	+ 46 07	60	15.5	ISOLATED GALAXY S
MCG+08-13-006	06 42 06.	+ 49 31	48	17.	GALAXY
ZWG 260.032	06 42 06.	+ 51 47		15.5	GALAXY
MCG-05-16-019	06 42 06.	- 27 07	84	16.	GALAXY
LB 03846	06 42 11.	- 20 52 48.		18.5	FAINT BLUE STAR
MCG+01-18-001	06 42 12.	+ 03 01	48	14.	GALAXY
ZC 0642.2+4130	06 42 12.	+ 41 30	22110		CLUSTER OF GALAXIES
ZWG 234.008	06 42 12.	+ 45 31		15.2	GALAXY
MCG+08-13-007	06 42 12.	+ 46 05	66	15.	GALAXY
KEEN 2273B	06 42 12.	+ 60 24	132		SB GALAXY
ZC 0642.2+6653	06 42 12.	+ 66 53	7860		CLUSTER OF GALAXIES
ISS 0565	06 42 12.	- 08 25	188		STELLAR RING
ZWG 145.006	06 42 18.	+ 29 16		15.4	GALAXY
MCG+08-13-008	06 42 18.	+ 45 30	54	16.	GALAXY
MCG+11-09-010	06 42 18.	+ 64 58	27	16.	GALAXY
MCG-03-18-001	06 42 18.	- 17 54	72	15.	GALAXY
MAI 017	06 42 20.	+ 60 15	33		DWARF SPHEROIDAL GALAXY
MCG-04-16-018	06 42 21.	- 26 02 30.	90	14.	GALAXY
URA 55	06 42 24.	+ 09 45	96		STELLAR RING
MCG+05-16-002	06 42 24.	+ 29 15	36	15.4	GALAXY
ZWG 204.013	06 42 24.	+ 43 53		15.7	GALAXY
ARP 096	06 42 25.	+ 86 37			PECULIAR GALAXY
LB 03847	06 42 26.	- 20 34 06.		18.3	FAINT BLUE STAR
MCG+09-11-030	06 42 27.	+ 53 07 30.	39	16.	GALAXY
MRSL 212-01/1	06 42 30.	+ 00 17	4800		HII REGION
OCL 0537	06 42 30.	+ 00 21	1440		OPEN STAR CLUSTER
ZWG 115.012	06 42 30.	+ 25 52		15.4	GALAXY
UGC 03531	06 42 30.	+ 25 52	90	15.4	GALAXY SBb/Sc
ZWG 204.014	06 42 30.	+ 43 50		15.3	GALAXY
UGC 03532	06 42 30.	+ 43 50	108	15.3	GALAXY SBb
KARA.72 117A	06 42 30.	+ 43 50	78	15.3	PART OF DOUBLE GALAXY
ZWG 260.033	06 42 30.	+ 53 08		15.4	GALAXY
UGC 03533	06 42 30.	+ 53 08	66	15.4	GALAXY S
LB 03848	06 42 31.	- 20 54 54.		18.0	FAINT BLUE STAR
LB 03849	06 42 35.	- 20 20 00.		19.0	FAINT BLUE STAR
UGC 03534	06 42 36.	+ 22 28	102	18.	GALAXY S
MCG+07-14-013	06 42 36.	+ 43 51	72	15.	GALAXY
ZC 0642.6+4706	06 42 36.	+ 47 06	1410		CLUSTER OF GALAXIES
MCG+09-11-031	06 42 36.	+ 56 44	39	18.	GALAXY
LB 03850	06 42 40.	- 20 54 12.		18.7	FAINT BLUE STAR
LB 03851	06 42 41.	- 20 40 36.		18.5	FAINT BLUE STAR
ZWG 204.015	06 42 42.	+ 43 52		14.7	GALAXY
UGC 03535	06 42 42.	+ 43 52	72	14.7	GALAXY COMPACT
KARA.72 117B	06 42 42.	+ 43 52	66	14.7	PART OF DOUBLE GALAXY
LB 03852	06 42 44.	- 20 39 00.		15.0	FAINT BLUE STAR
CED 084C	06 42 45.	+ 09 50			DIFFUSE GALACTIC NEBULA
ZWG 145.007	06 42 48.	+ 29 24		14.4	GALAXY
UGC 03536	06 42 48.	+ 29 24	72	14.4	GALAXY S0
MCG+07-14-014	06 42 48.	+ 43 52 30.	72	15.	GALAXY

OBJECT NAME	RIGHT ASCEN.	DECLINATION	DIAM.	MAGN.	TYPE OF OBJECT
RNGC 2280	06 42 48.	- 27 34		12.0	GALAXY
MCG+05-16-003	06 42 51.	+ 29 23	48	14.4	GALAXY
BC OH471	06 42 53.1	+ 44 54 31.		18.	QUASI-STELLAR OBJECT
SHB 092	06 42 53.1	+ 44 54 31.		18.7	QUASI-STELLAR OBJECT
MCG-05-16-020	06 42 54.	- 27 35	390	12.	GALAXY
IC 2171	06 42 59.	- 17 37			NONSTELLAR OBJECT
LBN 0986	06 43	+ 00 20	720		BRIGHT NEBULA
LBN 1049	06 43	- 24 00	6300		BRIGHT NEBULA
LB 09833	06 43	- 80 58		13.0	FAINT BLUE STAR
MCG+14-04-012	06 43 00.	+ 82 32	78	16.	GALAXY
MCG+14-04-013	06 43 00.	+ 84 08	78	15.	GALAXY
ZWG 370.003	06 43 00.	+ 86 38		14.8	GALAXY
ZWG 362.031	06 43 00.	+ 86 38		14.8	GALAXY
UGC 03536A	06 43 00.	+ 86 38	138	14.8	GALAXY DBL SYS
KARA.72 116B	06 43 00.	+ 86 38	48	14.8	PART OF DOUBLE GALAXY
LB 03853	06 43 05.	- 20 18 12.		19.6	FAINT BLUE STAR
URA 57	06 43 12.	+ 10 54	282		STELLAR RING
ZWG 234.009	06 43 12.	+ 46 29		15.4	GALAXY
YM 34	06 43 14.	- 07 14	660		SYMMETRIC GALACTIC NEBULA
MCG+06-15-005	06 43 15.	+ 36 50 30.	30	16.	GALAXY
MCG+08-13-009	06 43 15.	+ 46 27	42	16.	GALAXY
LB 03854	06 43 23.	- 20 45 30.		18.0	FAINT BLUE STAR
ZWG 175.014	06 43 24.	+ 33 41		15.2	GALAXY
UGC 03537	06 43 24.	+ 33 41	60	15.2	GALAXY SBc
SNO 25	06 43 24.	- 36 58 41.	2100	19.	LINEAR GRP OF 6 GALAXIES
MCG+06-15-006	06 43 27.	+ 33 40	54	15.	GALAXY
MCG+07-14-015	06 43 30.	+ 40 03 30.	21	15.	GALAXY
ZWG 204.016	06 43 30.	+ 40 05		15.3	GALAXY
7ZW 096	06 43 30.	+ 64 03			COMPACT GALAXY
MRSL 218-04/1	06 43 30.	- 07 17	6660		HII REGION
ZWG 204.017	06 43 36.	+ 42 23		15.2	GALAXY
ZWG 309.004	06 43 36.	+ 64 51		15.7	GALAXY
YC 0643-46	06 43 39.	- 46 23 42.			UNUSUAL SOUTHERN OBJECT
MCG+08-13-010	06 43 42.	+ 50 15	24	16.	GALAXY
MCG+11-09-011	06 43 42.	+ 64 50	15	16.	GALAXY
DG 111	06 43 42.	- 18 09	120		REFLECTION NEBULA
CED 086	06 43 42.	- 18 09			DIFFUSE GALACTIC NEBULA
RNGC 2283	06 43 42.	- 18 10		13.0	GALAXY
MCG-03-18-002	06 43 42.	- 18 10 30.	138	13.	GALAXY
MCG+02-18-002	06 43 48.	+ 12 58	96	14.	GALAXY
ZWG 234.010	06 43 48.	+ 47 43		14.9	GALAXY
UGC 03538	06 43 48.	+ 47 43	96	14.9	GALAXY SBa
KARA.73B 0173	06 43 48.	+ 47 43	72	14.9	ISOLATED GALAXY S
LB 03855	06 43 48.	- 20 20 12.		17.0	FAINT BLUE STAR
MCG-05-16-021	06 43 48.	- 31 12	60	15.	GALAXY
MCG-06-15-020	06 43 48.	- 36 58		15.	GALAXY
LB 03856	06 43 50.	- 20 38 36.		19.7	FAINT BLUE STAR
LB 03857	06 43 53.	- 20 37 48.		18.4	FAINT BLUE STAR
OCL 0535	06 43 54.	+ 01 49	450	9.2	OPEN STAR CLUSTER
MCG+06-15-008	06 43 54.	+ 33 36	72	14.	GALAXY
MCG+06-15-007	06 43 54.	+ 33 38	66	14.5	GALAXY
MCG+08-13-011	06 43 54.	+ 47 42	72	15.	GALAXY
ZWG 309.005	06 43 54.	+ 66 18		15.4	GALAXY
ZWG 308.040	06 43 54.	+ 66 18		15.4	GALAXY
UGC 03539	06 43 54.	+ 66 18	126	15.4	GALAXY SB?b-c
LB 03858	06 43 54.	- 20 59 36.		17.7	FAINT BLUE STAR
RNGC 2274	06 43 57.	+ 33 38		13.5	GALAXY
RNGC 2275	06 43 57.	+ 33 40		14.5	GALAXY
LB 03859	06 43 59.	- 20 46 36.		19.2	FAINT BLUE STAR
LBN 1014	06 44	- 07 17	360		BRIGHT NEBULA
URA 59	06 44 00.	+ 10 59	324		STELLAR RING
ZWG 175.015	06 44 00.	+ 33 38		13.6	GALAXY
UGC 03541	06 44 00.	+ 33 38	114	13.6	GALAXY E
KARA.72 118A	06 44 00.	+ 33 38	138	13.6	PART OF DOUBLE GALAXY
ZWG 175.016	06 44 00.	+ 33 40		14.5	GALAXY
UGC 03542	06 44 00.	+ 33 40	90	14.5	GALAXY S
KARA.72 118B	06 44 00.	+ 33 40	84	14.5	PART OF DOUBLE GALAXY
ZWG 260.034	06 44 00.	+ 51 13		15.5	GALAXY
ZWG 260.035	06 44 00.	+ 51 21		15.6	GALAXY
MCG+11-09-012	06 44 00.	+ 64 48	15	17.	GALAXY
7ZW 097	06 44 00.	+ 67 09			COMPACT GALAXY
ZWG 363.013	06 44 00.	+ 80 56		15.6	GALAXY
ZWG 348.025	06 44 00.	+ 80 56		15.6	GALAXY
ZWG 363.014	06 44 00.	+ 82 33		15.7	GALAXY
ZWG 362.032	06 44 00.	+ 82 33		15.7	GALAXY
UGC 03540	06 44 00.	+ 82 33	108	15.7	GALAXY Sc
MCG+14-04-014	06 44 00.	+ 86 45 30.	27	16.	GALAXY
RNGC 2297	06 44 01.	- 63 39			UNVERIFIED SOUTHERN OBJECT
LB 03860	06 44 02.	- 20 31 00.		18.1	FAINT BLUE STAR
LB 03861	06 44 03.	- 20 19 36.		17.4	FAINT BLUE STAR
SCHO 0214	06 44 06.	+ 08 32 06.	560		ISOLATED DARK CLOUD
ZWG 145.008	06 44 06.	+ 32 23		15.2	GALAXY
MCG+11-09-013	06 44 06.	+ 66 19	168	14.	GALAXY
7ZW 098	06 44 06.	+ 66 57			COMPACT GALAXY
7ZW 099	06 44 06.	+ 67 17			COMPACT GALAXY
GRA A	06 44 09.	- 74 12	95		PECULIAR RING GALAXY
ZWG 115.013	06 44 12.	+ 26 00		15.6	GALAXY
ZWG 260.036	06 44 12.	+ 51 18		15.7	GALAXY
7ZW 100	06 44 12.	+ 63 22			COMPACT GALAXY
MCG+13-05-031	06 44 12.	+ 80 56	30	16.	GALAXY
IC 2172	06 44 16.	+ 01 22 50.			NONSTELLAR OBJECT
VDB.66N 085	06 44 16.	+ 01 25	132		REFLECTION NEBULA
CED 087	06 44 17.	+ 01 23	180		DIFFUSE GALACTIC NEBULA
DG 112	06 44 18.	+ 01 23	180		REFLECTION NEBULA
RNGC 2282	06 44 18.	+ 01 23			DIFFUSE NEBULA
MRSL 211-00/1	06 44 18.	+ 01 23	180		HII REGION
ISS 0432	06 44 18.	+ 10 52	449		STELLAR RING
URA 52	06 44 18.	+ 10 53	264		STELLAR RING
ZWG 145.009	06 44 18.	+ 29 25		15.2	GALAXY
LB 03862	06 44 19.	- 20 48 30.		17.9	FAINT BLUE STAR
MCG-04-16-019	06 44 21.	- 26 02	102	14.5	GALAXY
ACK 208+01.1	06 44 24.	+ 04 41			PLANETARY NEBULA
MCG+05-16-004	06 44 24.	+ 32 23 30.	42	15.2	GALAXY
LB 03863	06 44 25.	- 20 19 06.		19.3	FAINT BLUE STAR
LB 03864	06 44 25.	- 21 02 42.		17.5	FAINT BLUE STAR
LB 03865	06 44 26.	- 20 25 18.		15.3	FAINT BLUE STAR
ZWG 115.014	06 44 30.	+ 26 47		15.5	GALAXY
ZWG 145.010	06 44 30.	+ 31 22		15.7	GALAXY
MCG+13-05-032	06 44 30.	+ 81 01	27	16.	GALAXY
RNGC 2277	06 44 34.	+ 33 30			NON-EXISTENT OBJECT
ZWG 260.037	06 44 36.	+ 51 54		14.7	GALAXY
7ZW 101	06 44 36.	+ 73 02			COMPACT GALAXY
LB 03866	06 44 38.	- 20 59 06.		18.0	FAINT BLUE STAR
MCG-04-16-020	06 44 42.	- 26 24 30.	108	15.	GALAXY
MCG+01-18-002	06 44 45.	+ 05 46	66	13.	GALAXY
UGC 03543	06 44 48.	+ 05 46	84	15.0	GALAXY
UGC 03544	06 44 48.	+ 33 39	72	16.5	GALAXY Sc
ZC 0644.8+4626	06 44 48.	+ 46 26	2690		CLUSTER OF GALAXIES

OBJECT NAME	RIGHT ASCEN.	DECLINATION	DIAM.	MAGN.	TYPE OF OBJECT
MCG+13-05-033	06 44 48.	+ 81 02	30	16.	GALAXY
MCG-05-16-022	06 44 48.	- 27 12	72	15.	GALAXY
URA 56	06 44 54.	+ 10 55	192		STELLAR RING
OCL 0597	06 44 54.	- 20 41	2640	5.9	OPEN STAR CLUSTER
RNGC 2287	06 44 56.	- 20 42		5.0	OPEN CLUSTER
LB 09834	06 45	- 83 32		14.7	FAINT BLUE STAR
ISS 0433	06 45 00.	+ 08 45	301		STELLAR RING
ZWG 204.018	06 45 00.	+ 43 33		15.3	GALAXY
MCG+08-13-012	06 45 00.	+ 50 28	14	16.	GALAXY
7ZW 102	06 45 00.	+ 67 15			COMPACT GALAXY
RNGC 2278	06 45 04.	+ 33 28			NON-EXISTENT OBJECT
ZWG 234.011	06 45 06.	+ 50 26		15.7	GALAXY
OCL 0548	06 45 06.	- 03 07	900	10.	OPEN STAR CLUSTER
RNGC 2286	06 45 09.	- 03 07		8.5	OPEN CLUSTER
RNGC 2279	06 45 10.	+ 33 28			NON-EXISTENT OBJECT
LB 03867	06 45 10.	- 20 45 48.		17.3	FAINT BLUE STAR
ZWG 115.015	06 45 12.	+ 25 34		15.4	GALAXY
UGC 03545	06 45 18.	+ 61 40	84	17.	GALAXY
MCG+10-10-014	06 45 18.	+ 61 40	57	15.	GALAXY
LB 03868	06 45 18.	- 21 02 48.		17.5	FAINT BLUE STAR
RNGC 2295	06 45 18.	- 26 40		14.0	GALAXY
MCG-04-16-021	06 45 21.	- 26 40	120	14.	GALAXY
ISS 0434	06 45 24.	+ 08 45	136		STELLAR RING
MRK 006	06 45 30.	+ 74 29	20	15.	GALAXY WITH UV CONTINUUM
KW 03	06 45 30.	+ 74 29	52		SEYFERT GALAXY
VVI 18	06 45 30.	+ 74 29	20	15.05	SEYFERT GALAXY
MCG+12-07-018	06 45 30.	+ 74 29	21	16.	GALAXY
ISS 0566	06 45 30.	- 07 02	119		STELLAR RING
VV 178B	06 45 33.	- 26 41	30	15.	INTERACTING GALAXY
VV 178A	06 45 33.	- 26 41	36	14.	INTERACTING GALAXY
VV 178	06 45 33.	- 26 41	120		INTERACTING GALAXY
ZWG 085.001	06 45 36.	+ 15 28		15.7	GALAXY
MCG+10-10-015	06 45 36.	+ 60 53	174	12.	GALAXY
ZWG 285.006	06 45 36.	+ 60 54		12.5	GALAXY
MRK 620	06 45 36.	+ 60 54	53	13.5	GALAXY WITH UV CONTINUUM
UGC 03546	06 45 36.	+ 60 54	216	12.5	GALAXY SBa
7ZW 103	06 45 36.	+ 67 30			COMPACT GALAXY
ZWG 330.017	06 45 36.	+ 74 28		14.8	GALAXY
UGC 03547	06 45 36.	+ 74 28	60	14.8	GALAXY S0-a
ISS 0567	06 45 36.	- 07 49	271		STELLAR RING
RNGC 2292	06 45 36.	- 26 41		14.0	GALAXY
REIN 2.052	06 45 37.45	+ 60 54 09.5			NEBULA
RNGC 2273	06 45 38.	+ 60 54		12.5	GALAXY
MCG-04-16-022	06 45 39.	- 26 41	18	14.	GALAXY
RNGC 2293	06 45 42.	- 26 41		13.0	GALAXY
MCG-04-16-023	06 45 42.	- 26 41 30.	30	13.5	GALAXY
OCL 0446	06 45 48.	+ 41 07	1290	7.7	OPEN STAR CLUSTER
ZWG 204.019	06 45 48.	+ 42 20		15.6	GALAXY
LB 03869	06 45 48.	- 20 46 06.		17.5	FAINT BLUE STAR
LB 03870	06 45 49.	- 20 30 24.		17.1	FAINT BLUE STAR
RNGC 2281	06 45 51.	+ 41 07		7.0	OPEN CLUSTER
RNGC 2284	06 45 52.	+ 33 17			NON-EXISTENT OBJECT
ZWG 085.002	06 45 54.	+ 17 02		15.3	GALAXY
ZWG 085.003	06 45 54.	+ 18 50		15.4	GALAXY
ZWG 204.020	06 45 54.	+ 44 29		14.8	GALAXY
MCG+07-14-016	06 45 54.	+ 44 29 30.	21	15.5	GALAXY
ZWG 348.026	06 45 54.	+ 77 28		14.8	GALAXY
UGC 03548	06 45 54.	+ 77 28	66	14.8	GALAXY Sa
MCG+08-13-013	06 46 00.	+ 48 33	42	16.	GALAXY
ZWG 234.012	06 46 00.	+ 48 34		15.4	GALAXY
MCG+13-05-034	06 46 00.	+ 77 27 30.	60	15.	GALAXY
MCG+14-04-015	06 46 00.	+ 81 00	27	16.	GALAXY
ZWG 363.015	06 46 00.	+ 81 02		14.4	GALAXY
UGC 03549	06 46 00.	+ 81 02	60	14.4	GALAXY E
ZWG 363.016	06 46 00.	+ 81 07		15.7	GALAXY
RNGC 2285	06 46 10.	+ 33 25			NON-EXISTENT OBJECT
MCG+12-07-019	06 46 12.	+ 74 32	66	15.	GALAXY
PK221-04.1	06 46 12.	- 09 29	54	18.5	PLANETARY NEBULA
ZWG 330.018	06 46 18.	+ 74 31		14.9	GALAXY
UGC 03550	06 46 18.	+ 74 31	96	14.9	GALAXY Sb
RNGC 2296	06 46 23.	- 16 51		13.0	GALAXY
ISS 0475	06 46 24.	+ 06 35	271		STELLAR RING
ZC 0646.4+4720	06 46 24.	+ 47 20	1680		CLUSTER OF GALAXIES
ZWG 261.001	06 46 24.	+ 53 31		14.8	GALAXY
ZWG 260.038	06 46 24.	+ 53 31		14.8	GALAXY
MCG+09-12-001	06 46 24.	+ 53 32	24	15.	GALAXY
MCG+11-09-014	06 46 24.	+ 63 22	51	17.	GALAXY
IC 0452	06 46 24.	- 16 51			NONSTELLAR OBJECT
SCHO 0215	06 46 25.	+ 12 38 36.	300		ISOLATED DARK CLOUD
MCG-03-18-003	06 46 27.	- 16 51 30.	36	13.	GALAXY
MCG+05-16-005	06 46 30.	+ 29 33	72	15.7	GALAXY
ZWG 145.011	06 46 30.	+ 29 35		15.7	GALAXY
UGC 03551	06 46 30.	+ 29 35	66	15.7	GALAXY
ARC 0562	06 46 30.	+ 69 20		17.0	RICH CLUSTER OF GALAXIES
MCG+14-04-016	06 46 30.	+ 81 01	24	15.	GALAXY
OCL 0560	06 46 30.	- 10 29	360	15.	OPEN STAR CLUSTER
MCG+05-16-006	06 46 33.	+ 28 24	60	15.3	GALAXY
ZC 0646.6+2340	06 46 36.	+ 23 40	4230		CLUSTER OF GALAXIES
ZWG 145.012	06 46 36.	+ 28 26		15.3	GALAXY
UGC 03552	06 46 36.	+ 28 26	66	15.3	GALAXY SBc
MCG+05-16-007	06 46 39.	+ 29 33 30.	42	15.5	GALAXY
ZWG 115.016	06 46 42.	+ 25 44		15.3	GALAXY
MCG+04-16-001	06 46 42.	+ 25 44	24	15.	GALAXY
ZWG 145.013	06 46 42.	+ 29 35		15.5	GALAXY
IC 0450	06 46 46.	+ 74 24			SEYFERT GALAXY
ZWG 085.004	06 46 48.	+ 20 08		15.6	GALAXY
UGC 03553	06 46 48.	+ 20 08	66	15.6	GALAXY Sc
KARA.72 119A	06 46 48.	+ 25 41	30	14.3	PART OF DOUBLE GALAXY
ZWG 115.017	06 46 48.	+ 25 45		15.1	GALAXY
MCG+04-16-002	06 46 48.	+ 25 45	42	15.1	GALAXY
ZWG 204.021	06 46 48.	+ 43 06		14.7	GALAXY
UGC 03554	06 46 48.	+ 43 06	72	14.7	GALAXY SB?b
MCG+07-14-017	06 46 51.	+ 43 06 30.	66	14.	GALAXY
ZWG 115.018	06 46 54.	+ 25 41		14.3	GALAXY
UGC 03555	06 46 54.	+ 25 41	72	14.3	GALAXY Sb/SBc
KARA.72 119B	06 46 54.	+ 25 41	66		PART OF DOUBLE GALAXY
MCG+04-16-003	06 46 54.	+ 25 41	30	14.3	GALAXY
MCG+11-09-015	06 46 54.	+ 64 42	36	16.	GALAXY
ISS 0568	06 46 54.	- 07 03	362		STELLAR RING
IC 0453	06 46 55.	- 16 59			MAY NOT EXIST
MCG+04-16-004	06 46 57.	+ 25 41	72	14.3	GALAXY
KHAV 181	06 47	- 04 21	7060		DARK NEBULA
ZWG 115.019	06 47 00.	+ 25 43		15.0	GALAXY
MCG+04-16-005	06 47 00.	+ 25 43	24	15.0	GALAXY
ZWG 115.020	06 47 00.	+ 26 26		15.2	GALAXY
MCG+04-16-006	06 47 00.	+ 26 26	30	15.2	GALAXY
MCG+08-13-014	06 47 00.	+ 46 00	36	16.	GALAXY
ZWG 234.013	06 47 00.	+ 46 02		15.3	GALAXY
MCG+08-13-015	06 47 00.	+ 48 40	48	16.	GALAXY
ZWG 261.002	06 47 00.	+ 53 05		15.7	GALAXY
ZWG 260.039	06 47 00.	+ 53 05		15.7	GALAXY
UGC 03556	06 47 00.	+ 53 05	78	15.7	GALAXY Sb-c
MCG+09-12-002	06 47 00.	+ 53 06	78	16.	GALAXY
ZC 0647.0+6921	06 47 00.	+ 69 21	3090		CLUSTER OF GALAXIES
MCG+12-07-020	06 47 00.	+ 71 34	66	17.	GALAXY
7ZW 104	06 47 00.	+ 86 09			COMPACT GALAXY
ISS 0502	06 47 06.	+ 01 57	301		STELLAR RING
UGC 03557	06 47 06.	+ 71 34	78	16.5	GALAXY
ZWG 085.005	06 47 12.	+ 20 17		15.7	GALAXY
ZWG 234.014	06 47 12.	+ 48 47		15.1	GALAXY
OCL 0614	06 47 12.	- 23 54	240	16.	OPEN STAR CLUSTER
GCL 011	06 47 12.	- 35 57	252	10.48	GLOBULAR STAR CLUSTER
RNGC 2298	06 47 13.	- 35 57		10.5	GLOBULAR CLUSTER
MCG+08-13-016	06 47 15.	+ 48 47	42	16.	GALAXY
ZWG 085.006	06 47 18.	+ 20 12		15.4	GALAXY
UGC 03558	06 47 18.	+ 20 12	90	15.4	GALAXY E
ZC 0647.4+3323	06 47 24.	+ 33 23	5040		CLUSTER OF GALAXIES
UGC 03559	06 47 24.	+ 34 07	66	16.0	GALAXY Sc
7ZW 105	06 47 24.	+ 63 10			COMPACT GALAXY
IC 0451	06 47 26.	+ 74 28		14.18	GALAXY E4
MCG+06-15-009	06 47 27.	+ 33 40	24	16.	GALAXY
RNGC 2289	06 47 28.	+ 33 30		14.5	GALAXY
RNGC 2288	06 47 28.	+ 33 31		15.5	GALAXY
URA 61	06 47 30.	+ 08 40	246		STELLAR RING
IC 2173	06 47 30.	+ 33 30			SAME AS NGC 2291
ZWG 175.017	06 47 30.	+ 33 31		15.5	GALAXY
ZWG 175.018	06 47 30.	+ 33 32		14.6	GALAXY
UGC 03560	06 47 30.	+ 33 32	66	14.6	GALAXY S0
ZWG 204.022	06 47 30.	+ 39 44		15.2	GALAXY
ZWG 234.015	06 47 30.	+ 49 56		14.5	GALAXY
UGC 03561	06 47 30.	+ 49 56	48	14.5	GALAXY S?
MCG+08-13-017	06 47 30.	+ 49 56	42	16.	GALAXY
MCG+06-15-011	06 47 33.	+ 33 30	9	15.5	GALAXY
MCG+06-15-010	06 47 33.	+ 33 31	48	14.5	GALAXY
OCL 0520	06 47 36.	+ 05 49	240	19.9	OPEN STAR CLUSTER
MCG+06-15-012	06 47 36.	+ 33 29	66	14.5	GALAXY
ZWG 175.019	06 47 36.	+ 33 30		14.6	GALAXY
UGC 03562	06 47 36.	+ 33 30	78	14.6	GALAXY Sa
ZWG 175.020	06 47 36.	+ 33 35		15.3	GALAXY
MCG+06-15-013	06 47 36.	+ 33 35	48	15.	GALAXY
ZWG 309.006	06 47 36.	+ 63 09		15.6	GALAXY
UGC 03563	06 47 36.	+ 63 09	72	15.6	GALAXY DISTRBD
RNGC 2290	06 47 40.	+ 33 30		14.5	GALAXY
RNGC 2291	06 47 40.	+ 33 35		15.5	GALAXY
UGC 03564	06 47 42.	+ 16 25	66	17.	GALAXY
SCHO 0216	06 47 45.	- 14 01 18.	520		ISOLATED DARK CLOUD
UGC 03565	06 47 48.	+ 09 43	66	17.	GALAXY Sc
MCG+06-15-014	06 47 48.	+ 33 34 30.	36	14.5	GALAXY
ZWG 175.021	06 47 48.	+ 33 35		15.0	GALAXY
ZWG 234.016	06 47 48.	+ 49 12		15.7	GALAXY
MCG+11-09-016	06 47 48.	+ 63 08	45	15.	GALAXY
RNGC 2294	06 47 52.	+ 33 35		15.0	GALAXY
ZWG 175.022	06 48 00.	+ 36 58		15.3	GALAXY
MCG+06-15-015	06 48 00.	+ 36 58 30.	36	15.	GALAXY
UGC 03566	06 48 00.	+ 41 49	60	18.	GALAXY DWRF SP
MCG+08-13-018	06 48 00.	+ 48 32 30.	30	15.	GALAXY
ZWG 234.017	06 48 00.	+ 48 34		14.2	GALAXY
UGC 03567	06 48 00.	+ 48 34	66	14.2	GALAXY S0
ZWG 370.004	06 48 00.	+ 86 45		15.3	GALAXY
ZWG 362.033	06 48 00.	+ 86 45		15.3	GALAXY
MCG+08-13-019	06 48 06.	+ 50 27	84	16.	GALAXY
ZWG 234.018	06 48 06.	+ 50 28		15.7	GALAXY
UGC 03568	06 48 06.	+ 50 28	90	15.7	GALAXY Sb-c
ZC 0648.1+5550	06 48 06.	+ 55 50	740		CLUSTER OF GALAXIES
CED 088	06 48 07.	+ 12 59			DIFFUSE GALACTIC NEBULA
ZWG 085.007	06 48 12.	+ 17 21		15.7	GALAXY
ZWG 115.021	06 48 12.	+ 26 49		15.2	GALAXY
MCG+04-16-007	06 48 12.	+ 26 49	24	15.2	GALAXY
ZC 0648.2+5313	06 48 12.	+ 53 13	870		CLUSTER OF GALAXIES
ZWG 285.007	06 48 12.	+ 57 14		15.7	GALAXY
UGC 03569	06 48 12.	+ 57 14	96	15.7	GALAXY S IV-V
KARA.72 120A	06 48 12.	+ 57 14	54	15.7	PART OF DOUBLE GALAXY
IC 0454	06 48 17.	+ 12 58 02.			NONSTELLAR OBJECT
UGC 03570	06 48 18.	+ 12 58	126	14.5	GALAXY SBa-b
ZWG 204.023	06 48 18.	+ 40 56		15.0	GALAXY
MCG+10-10-016	06 48 18.	+ 57 12 30.	57	14.	GALAXY
ARC 0879	06 48 20.	+ 29 27		17.2	RICH CLUSTER OF GALAXIES
RNGC 2305	06 48 21.	- 64 13			UNVERIFIED SOUTHERN OBJECT
ZC 0648.4+1950	06 48 24.	+ 19 50	3830		CLUSTER OF GALAXIES
MCG+05-16-008	06 48 24.	+ 29 06	72	15.1	GALAXY
MCG+08-13-020	06 48 24.	+ 47 07	21	16.	GALAXY
ZWG 261.003	06 48 24.	+ 53 03		15.5	GALAXY
SCHO 0217	06 48 27.	+ 00 01 54.	650		ISOLATED DARK CLOUD
MCG+09-12-003	06 48 27.	+ 53 02	39	16.	GALAXY
RNGC 2307	06 48 27.	- 64 16			UNVERIFIED SOUTHERN OBJECT
URA 62	06 48 30.	+ 09 13	378		STELLAR RING
MCG+05-16-009	06 48 30.	+ 27 33	114	15.2	GALAXY
ZWG 146.001	06 48 30.	+ 29 08		15.1	GALAXY
ZWG 145.014	06 48 30.	+ 29 08		15.1	GALAXY
UGC 03571	06 48 30.	+ 29 08	84	15.1	GALAXY Sc
ZWG 204.024	06 48 30.	+ 39 30		15.7	GALAXY
UGC 03572	06 48 30.	+ 49 47	66	16.5	GALAXY Sc
OCL 0521	06 48 36.	+ 05 50	180	17.7	OPEN STAR CLUSTER
ISS 0435	06 48 36.	+ 13 00	184		STELLAR RING
ZWG 234.019	06 48 36.	+ 45 31		15.7	GALAXY
ZWG 234.020	06 48 36.	+ 47 03		15.7	GALAXY
ZWG 261.004	06 48 36.	+ 55 40		14.9	GALAXY
ZWG 260.040	06 48 36.	+ 55 40		14.9	GALAXY
MCG+09-12-004	06 48 36.	+ 55 40	45	15.	GALAXY
RNGC 2299	06 48 41.	- 06 56			NON-EXISTENT OBJECT
ZWG 146.002	06 48 42.	+ 27 33		15.2	GALAXY
ZWG 145.015	06 48 42.	+ 27 33		15.2	GALAXY
UGC 03573	06 48 42.	+ 27 33	120	15.2	GALAXY Sb
ZWG 146.003	06 48 42.	+ 32 38		15.7	GALAXY
ZWG 145.016	06 48 42.	+ 32 38		15.7	GALAXY
ZWG 204.025	06 48 42.	+ 39 15		15.3	GALAXY
ZC 0648.7+4601	06 48 42.	+ 46 01	1680		CLUSTER OF GALAXIES
MCG+08-13-021	06 48 42.	+ 47 02	18	16.	GALAXY
ZWG 309.007	06 48 42.	+ 62 59		15.6	GALAXY
ZWG 285.008	06 48 42.	+ 62 59		15.6	GALAXY
ZWG 085.008	06 48 48.	+ 15 26		15.5	GALAXY
MCG+05-16-010	06 48 48.	+ 27 32	24	14.9	GALAXY
ZWG 261.005	06 48 48.	+ 51 37		15.6	GALAXY
ZWG 261.006	06 48 48.	+ 56 18		15.5	GALAXY
ZWG 260.041	06 48 48.	+ 56 18		15.5	GALAXY
ZWG 285.009	06 48 48.	+ 60 02		15.5	GALAXY

OBJECT NAME	RIGHT ASCEN.	DECLINATION	DIAM.	MAGN.	TYPE OF OBJECT
ISS 0503	06 48 48.	- 02 44	325		STELLAR RING
ZWG 115.022	06 48 54.	+ 25 54		15.6	GALAXY
ZWG 146.004	06 48 54.	+ 27 32		14.9	GALAXY
ZWG 145.017	06 48 54.	+ 27 32		14.9	GALAXY
ZWG 285.010	06 48 54.	+ 57 15		13.9	GALAXY
UGC 03574	06 48 54.	+ 57 15	270	13.9	GALAXY Sc
KARA.72 120B	06 48 54.	+ 57 15	258	13.9	PART OF DOUBLE GALAXY
KHAV 182	06 49	+ 04 33	2470		DARK NEBULA
LBN 0926	06 49	+ 10 03	420		BRIGHT NEBULA
LDN 1639	06 49 00.	+ 04 30	3960		DARK NEBULA
MCG+08-13-022	06 49 00.	+ 45 50	48	16.	GALAXY
ZWG 234.021	06 49 00.	+ 45 51		15.3	GALAXY
ZWG 234.022	06 49 00.	+ 47 11		15.7	GALAXY
MCG+10-10-017	06 49 00.	+ 57 14	234	13.	GALAXY
MCG+14-04-017	06 49 00.	+ 83 29	42	17.	GALAXY
ZWG 085.009	06 49 06.	+ 16 21		15.3	GALAXY
ZWG 146.005	06 49 06.	+ 30 54		15.6	GALAXY
ZWG 145.018	06 49 06.	+ 30 54		15.6	GALAXY
MCG+05-17-001	06 49 06.	+ 30 55	48	15.6	GALAXY
UGC 03575	06 49 06.	+ 70 49	108	16.5	GALAXY Sc
MCG-06-16-001	06 49 06.	- 35 18	9	15.5	GALAXY
MCG-06-16-002	06 49 08.	- 35 18	12	15.5	GALAXY
VV 041C	06 49 09.	- 02 11	6	19.	INTERACTING GALAXY
VV 041B	06 49 09.	- 02 11	18	16.5	INTERACTING GALAXY
VV 041A	06 49 09.	- 02 11	42	15.	INTERACTING GALAXY
VV 041	06 49 09.	- 02 11	126		INTERACTING GALAXY
OCL 0540	06 49 12.	+ 00 32	1800	6.4	OPEN STAR CLUSTER
ISS 0436	06 49 12.	+ 08 57	199		STELLAR RING
MCG+08-13-023	06 49 12.	+ 50 05	90	15.	GALAXY
ZC 0649.2+5528	06 49 12.	+ 55 28	1010		CLUSTER OF GALAXIES
ZWG 309.008	06 49 12.	+ 64 43		15.7	GALAXY
MCG+12-07-021	06 49 12.	+ 70 49	102	16.	GALAXY
RNGC 2301	06 49 13.	+ 00 32		6.5	OPEN CLUSTER
ZWG 085.010	06 49 18.	+ 19 35		15.6	GALAXY
ZWG 234.023	06 49 18.	+ 50 05		14.6	GALAXY
UGC 03576	06 49 18.	+ 50 05	102	14.6	GALAXY SBb
ZWG 309.009	06 49 18.	+ 65 16		15.3	GALAXY
UGC 03577	06 49 18.	+ 65 16	96	15.3	GALAXY SBb
ZWG 085.011	06 49 18.	+ 15 18		15.2	GALAXY
MCG-05-17-001	06 49 24.	- 29 28	60	16.	GALAXY
MAI 018	06 49 27.	+ 71 31	40		DWARF SPHEROIDAL GALAXY
RNGC 2302	06 49 29.	- 07 00			OPEN CLUSTER
ZWG 085.012	06 49 30.	+ 15 19		14.1	GALAXY
UGC 03578	06 49 30.	+ 15 19	108	14.1	GALAXY SBa-b
KARA.72 121B	06 49 30.	+ 15 19	90	14.1	PART OF DOUBLE GALAXY
ZWG 146.006	06 49 30.	+ 31 20		15.3	GALAXY
ZWG 145.019	06 49 30.	+ 31 20		15.3	GALAXY
ZWG 204.026	06 49 30.	+ 44 07		15.5	GALAXY
MCG+08-13-024	06 49 30.	+ 47 13	78	16.	GALAXY
ZWG 234.024	06 49 30.	+ 47 15		15.7	GALAXY
UGC 03579	06 49 30.	+ 47 15	72	15.7	GALAXY Sa-b
ZWG 261.007	06 49 30.	+ 54 25		15.1	GALAXY
MCG+09-12-005	06 49 30.	+ 54 29	30	17.	GALAXY
ZWG 261.008	06 49 30.	+ 55 02		15.4	GALAXY
KARA.73B 0174	06 49 30.	+ 55 02	42	15.4	ISOLATED GALAXY E
MCG+11-09-017	06 49 30.	+ 64 42	39	16.	GALAXY
MCG+11-09-018	06 49 30.	+ 65 16	60	15.	GALAXY
7ZW 106	06 49 30.	+ 65 23			COMPACT GALAXY
OCL 0554	06 49 30.	- 07 00	150	12.	OPEN STAR CLUSTER
OCL 0532	06 49 36.	+ 03 00	420	15.2	OPEN STAR CLUSTER
MCG+03-18-001	06 49 36.	+ 15 17	90	14.1	GALAXY
ZWG 175.023	06 49 36.	+ 34 59		15.2	GALAXY
MCG+07-14-018	06 49 36.	+ 44 07	48	15.	GALAXY
KARA.72 121A	06 49 42.	+ 15 18	42	15.2	PART OF DOUBLE GALAXY
MCG+09-12-006	06 49 42.	+ 54 28	39	15.	GALAXY
MCG-03-18-004	06 49 42.	- 20 12	36	15.5	GALAXY
PK204+04.1	06 49 45.	+ 10 02	414		PLANETARY NEBULA
7ZW 107	06 49 48.	+ 66 33			COMPACT GALAXY
ZWG 146.007	06 49 54.	+ 27 42		15.0	GALAXY
MCG+05-17-002	06 49 54.	+ 27 42	36	15.0	GALAXY
ZWG 116.001	06 50 00.	+ 23 12		14.7	GALAXY
MCG+04-17-001	06 50 00.	+ 23 13	30	14.7	GALAXY
MCG+11-09-019	06 50 00.	+ 64 21	60	16.	GALAXY
ZWG 330.019	06 50 00.	+ 69 39		12.9	GALAXY
UGC 03580	06 50 00.	+ 69 39	264	12.9	GALAXY Sa
KARA.73B 0175	06 50 00.	+ 69 39	216	12.9	ISOLATED GALAXY S
ZWG 349.001	06 50 00.	+ 80 04		14.4	GALAXY
ZWG 348.027	06 50 00.	+ 80 04		14.4	GALAXY
UGC 03581	06 50 00.	+ 80 04	108	14.4	GALAXY SBc
MCG+13-05-035	06 50 00.	+ 80 04	54	14.	GALAXY
KARA.73B 0176	06 50 00.	+ 80 04	78	14.4	ISOLATED GALAXY S
ZC 0650.0+8119	06 50 00.	+ 81 19	470		CLUSTER OF GALAXIES
OCL 0613	06 50 06.	- 23 34	336	13.	OPEN STAR CLUSTER
URA 53	06 50 12.	+ 09 59	150		STELLAR RING
UGC 03582	06 50 12.	+ 12 15	66	16.5	GALAXY SBb
ZWG 085.013	06 50 12.	+ 16 59		14.3	GALAXY
UGC 03583	06 50 12.	+ 16 59	240	14.3	GALAXY SYSTEM?
MCG+03-18-002	06 50 12.	+ 16 59	240	14.3	GALAXY
ZWG 116.002	06 50 12.	+ 22 22		14.9	GALAXY
ZWG 146.008	06 50 12.	+ 27 09		15.4	GALAXY
UGC 03584	06 50 12.	+ 27 09	108	15.4	GALAXY Sb-c
ZWG 146.009	06 50 12.	+ 30 21		15.6	GALAXY
MCG+10-10-018	06 50 12.	+ 59 02	27	16.	GALAXY
7ZW 108	06 50 12.	+ 63 26			COMPACT GALAXY
MCG+12-07-022	06 50 12.	+ 69 37	102	14.	GALAXY
ZWG 116.003	06 50 18.	+ 25 21		15.7	GALAXY
URA 60	06 50 24.	+ 08 45	210		STELLAR RING
OCL 0486	06 50 24.	+ 16 59	240	15.	OPEN STAR CLUSTER
ZWG 146.010	06 50 24.	+ 30 23		15.7	GALAXY
ZWG 330.020	06 50 24.	+ 73 45		15.7	GALAXY
MCG+05-17-003	06 50 27.	+ 27 22	48	15.5	GALAXY
ZWG 146.011	06 50 30.	+ 27 23		15.5	GALAXY
UGC 03585	06 50 30.	+ 27 23	66	15.5	GALAXY Sc
ZWG 234.025	06 50 30.	+ 50 26		15.6	GALAXY
ISS 0476	06 50 36.	+ 06 19	323		STELLAR RING
UGC 03586	06 50 36.	+ 14 49	72	17.	GALAXY
MCG+08-13-025	06 50 42.	+ 47 10	24	17.	GALAXY
ZWG 234.026	06 50 42.	+ 50 25		15.1	GALAXY
MRK 373	06 50 42.	+ 50 25	7	16.5	GALAXY WITH UV CONTINUUM
ZWG 261.009	06 50 48.	+ 53 29		15.6	GALAXY
MCG+11-09-020	06 50 48.	+ 64 25	39	18.	GALAXY
PK210+01.1	06 50 54.	+ 03 16	22		PLANETARY NEBULA
ZWG 085.014	06 50 54.	+ 19 22		14.3	GALAXY
UGC 03587	06 50 54.	+ 19 22	180	14.3	GALAXY S
MCG+08-13-026	06 50 54.	+ 45 45	90	15.	GALAXY
ZWG 234.027	06 50 54.	+ 45 47		14.9	GALAXY
UGC 03588	06 50 54.	+ 45 47	102	14.9	GALAXY S0
ZWG 234.028	06 50 54.	+ 50 30		15.7	GALAXY

OBJECT NAME	RIGHT ASCEN.	DECLINATION	DIAM.	MAGN.	TYPE OF OBJECT
MCG+09-12-007	06 50 54.	+ 53 30	30	16.	GALAXY
ISS 0569	06 50 54.	- 08 46	222		STELLAR RING
URA 54	06 51 00.	+ 13 06	228		STELLAR RING
MCG+03-18-003	06 51 00.	+ 19 21	156	14.3	GALAXY
MCG-04-17-001	06 51 00.	- 26 30	24	14.	GALAXY
7ZW 109	06 51 12.	+ 67 10			COMPACT GALAXY
UGC 03589	06 51 18.	+ 26 45	66	16.5	GALAXY S
ZWG 146.012	06 51 24.	+ 30 08		15.3	GALAXY
UGC 03590	06 51 24.	+ 30 08	60	15.3	GALAXY SBa-b
UGC 03591	06 51 24.	+ 34 56	66	16.0	GALAXY S
MCG+08-13-027	06 51 24.	+ 47 37	27	16.	GALAXY
MCG+09-12-008	06 51 24.	+ 53 25	54	17.	GALAXY
ZWG 309.010	06 51 24.	+ 65 01		15.5	GALAXY
7ZW 110	06 51 24.	+ 68 22			COMPACT GALAXY
ISS 0570	06 51 30.	- 07 19	377		STELLAR RING
ZWG 205.001	06 51 36.	+ 41 04		14.6	GALAXY
ZWG 204.027	06 51 36.	+ 41 04		14.6	GALAXY
ZWG 261.010	06 51 36.	+ 53 16		15.7	GALAXY
MCG+11-09-021	06 51 36.	+ 65 00	30	15.	GALAXY
MCG+07-14-020	06 51 39.	+ 40 23	42	15.	GALAXY
MCG+07-14-019	06 51 39.	+ 41 03 30.	15	15.5	GALAXY
ZWG 205.002	06 51 42.	+ 40 24		15.1	GALAXY
ZWG 204.028	06 51 42.	+ 40 24		15.1	GALAXY
UGC 03592	06 51 42.	+ 40 24	60	15.1	GALAXY SBa
MCG+07-14-021	06 51 42.	+ 41 00 30.	90	14.5	GALAXY
ZWG 205.003	06 51 42.	+ 41 01		14.7	GALAXY
ZWG 204.029	06 51 42.	+ 41 01		14.7	GALAXY
UGC 03593	06 51 42.	+ 41 01	108	14.7	GALAXY Sc
MCG+09-12-009	06 51 42.	+ 53 17 30.	42	17.	GALAXY
ZWG 116.004	06 51 48.	+ 23 34		15.2	GALAXY
UGC 03594	06 51 48.	+ 23 34	78	15.2	GALAXY Sb/SBb
MCG+04-17-002	06 51 48.	+ 23 34	48	15.2	GALAXY
ZC 0651.8+5950	06 51 48.	+ 59 50	1480		CLUSTER OF GALAXIES
MCG+08-13-028	06 51 54.	+ 47 40	72	15.	GALAXY
ZWG 261.011	06 51 54.	+ 55 28		15.2	GALAXY
UGC 03595	06 51 54.	+ 55 28	90	15.2	GALAXY SBb
MRSL 233-09/1	06 51 54.	- 22 22	5400		HII REGION
LBN 1051	06 52	- 23 40	720		BRIGHT NEBULA
ZWG 205.004	06 52 00.	+ 39 50		13.5	GALAXY
ZWG 204.030	06 52 00.	+ 39 50		13.5	GALAXY
UGC 03596	06 52 00.	+ 39 50	66	13.5	GALAXY S0?
ZWG 234.029	06 52 00.	+ 48 00		15.2	GALAXY
MCG+09-12-010	06 52 00.	+ 55 30	57	15.	GALAXY
MCG+10-10-019	06 52 00.	+ 60 42 30.	114	14.	GALAXY
ZWG 285.011	06 52 00.	+ 60 43		15.5	GALAXY
UGC 03598	06 52 00.	+ 60 43	132	15.5	GALAXY IRR
UGC 03597	06 52 00.	+ 78 00	126	15.2	GALAXY S
OCL 0484	06 52 06.	+ 18 05	360	11.5	OPEN STAR CLUSTER
ZWG 116.005	06 52 06.	+ 24 17		14.5	GALAXY
UGC 03599	06 52 06.	+ 24 17	66	14.5	GALAXY COMPACT
MCG+07-15-001	06 52 06.	+ 39 48 30.	48	14.5	GALAXY
MRSL 217-01/1	06 52 06.	- 04 27	360		HII REGION
MRSL 234-10/1	06 52 06.	- 23 53	2100		HII REGION
OCL 0619	06 52 06.	- 24 34	3000	6.1	OPEN STAR CLUSTER
RNGC 2304	06 52 09.	+ 18 05		11.0	OPEN CLUSTER
MCG+04-17-003	06 52 09.	+ 24 18	48	14.5	GALAXY
RNGC 2306	06 52 11.	- 07 07			NON-EXISTENT OBJECT
ZWG 116.006	06 52 12.	+ 24 19		15.5	GALAXY
ZWG 176.001	06 52 12.	+ 33 20		14.8	GALAXY
ZWG 175.024	06 52 12.	+ 33 20		14.8	GALAXY
UGC 03600	06 52 12.	+ 39 08	72	18.	GALAXY DWARF
ZWG 205.005	06 52 12.	+ 44 30		15.7	GALAXY
ZWG 204.031	06 52 12.	+ 44 30		15.7	GALAXY
ZC 0652.2+5720	06 52 12.	+ 57 20	1480		CLUSTER OF GALAXIES
MCG+06-16-001	06 52 15.	+ 33 19	36	15.	GALAXY
ZWG 205.006	06 52 18.	+ 40 04		14.5	GALAXY
ZWG 204.032	06 52 18.	+ 40 04		14.5	GALAXY
UGC 03601	06 52 18.	+ 40 04	36	14.5	GALAXY
ZWG 205.007	06 52 18.	+ 40 45		15.1	GALAXY
ZWG 204.033	06 52 18.	+ 40 45		15.1	GALAXY
ZC 0652.3+5736	06 52 18.	+ 57 36	940		CLUSTER OF GALAXIES
MCG+07-15-002	06 52 21.	+ 40 45	36	15.5	GALAXY
RNGC 2310	06 52 23.	- 40 48		12.5	GALAXY
MCG+08-13-029	06 52 30.	+ 46 58	36	16.	GALAXY
ABC 0563	06 52 30.	+ 69 05		17.0	RICH CLUSTER OF GALAXIES
RNGC 2303	06 52 34.	+ 45 34		14.0	GALAXY
SCHO 0218	06 52 34.	- 15 09 48.	520		ISOLATED DARK CLOUD
ZWG 085.015	06 52 36.	+ 16 00		14.8	GALAXY
UGC 03602	06 52 36.	+ 16 00	102	14.8	GALAXY
MCG+03-18-004	06 52 36.	+ 16 00	72	14.8	GALAXY
MCG+08-13-030	06 52 36.	+ 45 14	60	16.	GALAXY
ZWG 234.030	06 52 36.	+ 45 34		13.9	GALAXY
UGC 03603	06 52 36.	+ 45 34	96	13.9	GALAXY E
MCG+08-13-031	06 52 42.	+ 45 32	36	14.	GALAXY
ZC 0652.7+5130	06 52 42.	+ 51 30	610		CLUSTER OF GALAXIES
7ZW 111	06 52 42.	+ 67 21			COMPACT GALAXY
MRSL 213+00/1	06 52 42.	- 00 27	60		HII REGION
LB 00488	06 52 46.	+ 79 56 00.		15.7	FAINT BLUE STAR
MCG+08-13-032	06 52 54.	+ 48 57	42	16.	GALAXY
KHAV 183	06 53	- 04 16	2470		DARK NEBULA
LBN 1020	06 53	- 07 50	540		BRIGHT NEBULA
LBN 1047	06 53	- 22 40	1920		BRIGHT NEBULA
UGC 03605	06 53 00.	+ 14 00	72	16.5	GALAXY Sc
ZWG 085.016	06 53 00.	+ 16 32		15.2	GALAXY
ZWG 234.031	06 53 00.	+ 48 58		14.6	GALAXY
MCG+10-10-020	06 53 00.	+ 59 29	12	17.	GALAXY
ZWG 363.017	06 53 00.	+ 81 01		14.2	GALAXY
UGC 03604	06 53 00.	+ 81 01	48	14.2	GALAXY S
SCHO 0219	06 53 00.	- 15 04 00.	510		ISOLATED DARK CLOUD
ZWG 261.012	06 53 06.	+ 54 22		15.6	GALAXY
ZWG 309.011	06 53 06.	+ 63 45		15.0	GALAXY
UGC 03606	06 53 06.	+ 63 45	96	15.0	GALAXY S IV
MRSL 220-02/1	06 53 06.	- 07 58	480		HII REGION
MCG+06-16-002	06 53 09.	+ 37 00	30	16.	GALAXY
MCG+09-12-011	06 53 12.	+ 54 23	15	16.	GALAXY
OCL 0591	06 53 12.	- 18 40	168	13.	OPEN STAR CLUSTER
LB 03929	06 53 15.	- 19 34 54.		17.3	FAINT BLUE STAR
LB 03930	06 53 16.	- 19 20 30.		17.0	FAINT BLUE STAR
MCG+09-12-012	06 53 18.	+ 53 58	13	17.	GALAXY
MCG+11-09-022	06 53 18.	+ 63 45	90	15.	GALAXY
LB 03931	06 53 22.	- 19 29 00.		17.1	FAINT BLUE STAR
ZWG 146.013	06 53 24.	+ 28 01		15.7	GALAXY
ZWG 234.032	06 53 24.	+ 50 41		15.6	GALAXY
LB 03932	06 53 24.	- 19 12 18.		15.5	FAINT BLUE STAR
LB 03933	06 53 30.	- 19 37 00.		17.2	FAINT BLUE STAR
LB 03934	06 53 30.	- 19 38 54.		17.5	FAINT BLUE STAR
LB 03935	06 53 33.	- 19 37 48.		19.8	FAINT BLUE STAR
LB 03936	06 53 34.	- 19 36 42.		18.6	FAINT BLUE STAR

OBJECT NAME	RIGHT ASCEN.	DECLINATION	DIAM.	MAGN.	TYPE OF OBJECT
RNGC 2309	06 53 35.	- 07 08		10.5	OPEN CLUSTER
ZWG 146.014	06 53 36.	+ 31 38		15.7	GALAXY
ZC 0653.6+6812	06 53 36.	+ 68 12	1280		CLUSTER OF GALAXIES
ISS 0669	06 53 36.	- 13 42	253		STELLAR RING
LB 03937	06 53 37.	- 19 02 30.		16.1	FAINT BLUE STAR
LB 03938	06 53 37.	- 19 08 00.		18.7	FAINT BLUE STAR
LB 03939	06 53 37.	- 19 40 18.		18.0	FAINT BLUE STAR
UGC 03607	06 53 42.	+ 06 20	66	16.5	GALAXY Sc/SBc
ZWG 085.017	06 53 42.	+ 20 49		15.6	GALAXY
TON-N 0267	06 53 42.	+ 28 47		16.0	BLUE STAR
MCG+09-12-013	06 53 42.	+ 54 31	12	18.	GALAXY
PK216-00.1	06 53 42.	- 02 49	80	17.5	PLANETARY NEBULA
OCL 0569	06 53 42.	- 13 13	120		OPEN STAR CLUSTER
OCL 0557	06 53 48.	- 07 08	420	11.7	OPEN STAR CLUSTER
ST 20	06 53 48.	- 07 09	360		EMISSION NEBULA
LB 03940	06 53 48.	- 19 25 06.		16.5	FAINT BLUE STAR
LB 03941	06 53 49.	- 19 09 42.		14.8	FAINT BLUE STAR
LB 03942	06 53 49.	- 19 15 36.		16.7	FAINT BLUE STAR
LB 03943	06 53 50.	- 19 23 42.		16.8	FAINT BLUE STAR
LB 03944	06 53 51.	- 19 23 18.		17.7	FAINT BLUE STAR
ZWG 234.033	06 53 54.	+ 45 21		15.0	GALAXY
MCG+08-13-033	06 53 54.	+ 46 27	108	14.	GALAXY
ZWG 234.034	06 53 54.	+ 46 29		13.7	GALAXY
UGC 03608	06 53 54.	+ 46 29	96	13.7	GALAXY S
ZWG 330.021	06 53 54.	+ 69 21		15.6	GALAXY
UGC 03609	06 53 54.	+ 69 21	66	15.6	GALAXY Sb
LB 03945	06 53 54.	- 19 39 42.		18.6	FAINT BLUE STAR
MCG-06-16-003	06 53 54.	- 36 57	42	15.	GALAXY
LB 03946	06 53 55.	- 19 15 24.		17.0	FAINT BLUE STAR
LB 03947	06 53 55.	- 19 33 18.		18.8	FAINT BLUE STAR
LB 03948	06 53 57.	- 19 33 00.		18.0	FAINT BLUE STAR
LBN 1022	06 54	- 08 10	1080		BRIGHT NEBULA
MCG+08-13-034	06 54 00.	+ 45 25	36	16.	GALAXY
UGC 03610	06 54 00.	+ 58 03	60	16.0	GALAXY SBb
MCG+14-04-018	06 54 00.	+ 81 00	36	15.	GALAXY
MCG+14-04-019	06 54 00.	+ 81 01	51	16.	GALAXY
LB 03949	06 54 02.	- 19 39 18.		16.0	FAINT BLUE STAR
ZC 0654.1+4213	06 54 06.	+ 42 13	2150		CLUSTER OF GALAXIES
LDN 1656	06 54 06.	- 08 20	1080		DARK NEBULA
LB 03950	06 54 11.	- 19 31 30.		17.6	FAINT BLUE STAR
ZWG 085.018	06 54 12.	+ 20 30		15.1	GALAXY
UGC 03611	06 54 12.	+ 20 30	66	15.1	GALAXY S0-a
ZWG 205.008	06 54 12.	+ 39 09		14.2	GALAXY
UGC 03612	06 54 12.	+ 39 09	48	14.2	GALAXY SBb
MCG+09-12-014	06 54 12.	+ 54 35	30	16.	GALAXY
LDN 1655	06 54 12.	- 08 10	480		DARK NEBULA
LB 03951	06 54 12.	- 19 36 54.		16.9	FAINT BLUE STAR
LB 03952	06 54 18.	- 19 21 12.		18.2	FAINT BLUE STAR
UGC 03613	06 54 24.	+ 13 37	96	16.5	GALAXY
MCG+07-15-003	06 54 24.	+ 39 09	54	15.	GALAXY
ZWG 205.009	06 54 24.	+ 44 13		15.4	GALAXY
ZWG 234.035	06 54 24.	+ 47 19		15.6	GALAXY
ZC 0654.4+4823	06 54 24.	+ 48 23	610		CLUSTER OF GALAXIES
LB 03953	06 54 25.	- 19 36 36.		18.6	FAINT BLUE STAR
LB 03954	06 54 25.	- 19 41 42.		18.5	FAINT BLUE STAR
LB 03955	06 54 28.	- 19 16 06.		16.5	FAINT BLUE STAR
MCG+08-13-035	06 54 30.	+ 47 18	33	16.	GALAXY
ZCG 0654+68	06 54 30.	+ 68 41		15.0	COMPACT GALAXY
DG 113	06 54 30.	- 08 10	600		REFLECTION NEBULA
MCG+06-16-003	06 54 36.	+ 35 48	84	15.	GALAXY
ZWG 234.036	06 54 36.	+ 45 08		14.7	GALAXY
UGC 03614	06 54 36.	+ 45 08	84	14.7	GALAXY SBa-b
ZWG 261.013	06 54 36.	+ 53 33		15.7	GALAXY
MCG+09-12-015	06 54 36.	+ 53 34	39	17.	GALAXY
ZWG 309.012	06 54 36.	+ 64 16		15.3	GALAXY
7ZW 112	06 54 36.	+ 68 43			COMPACT GALAXY
LB 03956	06 54 37.	- 19 39 36.		16.0	FAINT BLUE STAR
ZWG 176.002	06 54 42.	+ 35 48		15.4	GALAXY
UGC 03615	06 54 42.	+ 35 48	90	15.4	GALAXY Sa-b
MCG+08-13-036	06 54 42.	+ 45 06	72	14.	GALAXY
LB 03957	06 54 42.	- 19 00 24.		17.5	FAINT BLUE STAR
SCHO 0220	06 54 43.	- 16 54 00.	640		ISOLATED DARK CLOUD
LB 03958	06 54 43.	- 19 02 54.		18.2	FAINT BLUE STAR
VDB.66N 086	06 54 45.	- 10 10	240		REFLECTION NEBULA
ISS 0477	06 54 48.	+ 06 06	353		STELLAR RING
UGC 03616	06 54 48.	+ 22 57	84	17.	GALAXY
ZC 0654.8+2753	06 54 48.	+ 27 53	11690		CLUSTER OF GALAXIES
ZC 0654.8+4911	06 54 48.	+ 49 11	2820		CLUSTER OF GALAXIES
LB 03959	06 54 50.	- 19 02 54.		17.8	FAINT BLUE STAR
LB 03960	06 54 50.	- 19 15 06.		16.6	FAINT BLUE STAR
MCG+04-17-004	06 54 51.	+ 22 56	78		GALAXY
OCL 0513	06 54 54.	+ 08 20	780	17.3	OPEN STAR CLUSTER
TON-N 0268	06 54 54.	+ 28 39		14.7	BLUE STAR
ZC 0654.9+4742	06 54 54.	+ 47 42	3020		CLUSTER OF GALAXIES
MCG+11-09-023	06 54 54.	+ 64 14 30.	36	16.	GALAXY
LB 03961	06 54 56.	- 19 00 48.		17.6	FAINT BLUE STAR
RNGC 2308	06 54 58.	+ 45 17		14.5	GALAXY
KHAV 184	06 55	+ 02 20	4860		DARK NEBULA
MCG+06-16-004	06 55 00.	+ 35 54	24	14.5	GALAXY
MCG+08-13-037	06 55 00.	+ 45 15	90	14.	GALAXY
ZWG 234.037	06 55 00.	+ 45 17		14.4	GALAXY
UGC 03618	06 55 00.	+ 45 17	126	14.4	GALAXY Sa-b
ZC 0655.0+6840	06 55 00.	+ 68 40	2550		CLUSTER OF GALAXIES
UGC 03617A	06 55 00.	+ 87 49	90	16.5	GALAXY DWRF SP
LB 03962	06 55 00.	- 19 23 12.		18.9	FAINT BLUE STAR
LB 03963	06 55 05.	- 19 01 30.		18.4	FAINT BLUE STAR
URA 64	06 55 06.	+ 09 38	156		STELLAR RING
OCL 0534	06 55 06.	+ 03 17	840	16.0	OPEN STAR CLUSTER
ZWG 176.003	06 55 06.	+ 35 55		14.8	GALAXY
ZC 0655.1+4609	06 55 06.	+ 46 09	1140		CLUSTER OF GALAXIES
ZC 0655.1+6020	06 55 06.	+ 60 20	1480		CLUSTER OF GALAXIES
LB 03964	06 55 11.	- 19 39 18.		18.3	FAINT BLUE STAR
ZWG 176.004	06 55 12.	+ 34 17		15.3	GALAXY
UGC 03619	06 55 12.	+ 34 17	60	15.3	GALAXY SBc
ZWG 285.012	06 55 12.	+ 59 36		15.2	GALAXY
KARA.73B 0177	06 55 12.	+ 59 36	48	15.2	ISOLATED GALAXY S
ZWG 309.013	06 55 12.	+ 63 28		15.5	GALAXY
LB 03965	06 55 12.	- 19 16 24.		17.7	FAINT BLUE STAR
ARC 0564	06 55 13.	+ 69 53		16.2	RICH CLUSTER OF GALAXIES
LB 03966	06 55 13.	- 19 01 12.		18.5	FAINT BLUE STAR
LB 03967	06 55 13.	- 19 01 30.		17.0	FAINT BLUE STAR
LB 03968	06 55 13.	- 19 34 24.		19.8	FAINT BLUE STAR
MCG+02-18-003	06 55 15.	+ 12 44	24	14.5	GALAXY
RNGC 2311	06 55 16.	- 04 31		9.5	OPEN CLUSTER
ISS 0478	06 55 18.	+ 08 35	163		STELLAR RING
MCG+06-16-005	06 55 18.	+ 34 16	42	14.5	GALAXY
MCG+08-13-038	06 55 18.	+ 45 05	48	17.	GALAXY
ZWG 234.038	06 55 18.	+ 45 08		15.5	GALAXY
ZC 0655.3+5257	06 55 18.	+ 52 57	340		CLUSTER OF GALAXIES
OCL 0553	06 55 18.	- 04 31	480	9.8	OPEN STAR CLUSTER
LB 03969	06 55 21.	- 19 41 06.		15.0	FAINT BLUE STAR
URA 63	06 55 24.	+ 09 41	186		STELLAR RING
OCL 0522	06 55 24.	+ 06 30	690	15.2	OPEN STAR CLUSTER
MCG+08-13-039	06 55 24.	+ 45 30	24	16.	GALAXY
ZWG 234.039	06 55 24.	+ 45 31		14.8	GALAXY
LDN 1653	06 55 24.	- 07 52	420		DARK NEBULA
OCL 0570	06 55 24.	- 13 09	240	15.	OPEN STAR CLUSTER
LB 03970	06 55 25.	- 19 39 42.		20.0	FAINT BLUE STAR
LB 03971	06 55 28.	- 19 01 18.		16.4	FAINT BLUE STAR
MCG+08-13-040	06 55 30.	+ 45 44	39	16.	GALAXY
ZC 0655.5+5054	06 55 30.	+ 50 54	2420		CLUSTER OF GALAXIES
MCG+11-09-024	06 55 30.	+ 63 27 30.	15	16.	GALAXY
MCG+11-09-025	06 55 30.	+ 64 50	39	16.	GALAXY
LB 03972	06 55 33.	- 18 59 24.		18.6	FAINT BLUE STAR
LB 03973	06 55 34.	- 18 59 00.		17.5	FAINT BLUE STAR
LB 03974	06 55 34.	- 19 43 42.		17.3	FAINT BLUE STAR
ZWG 116.007	06 55 36.	+ 25 16		15.7	GALAXY
MCG+05-17-004	06 55 36.	+ 28 27	30	15.3	GALAXY
ZWG 146.015	06 55 36.	+ 28 28		15.3	GALAXY
MCG+05-17-005	06 55 36.	+ 29 05	48	15.2	GALAXY
LDN 1649	06 55 36.	- 03 53	240		DARK NEBULA
RNGC 2313	06 55 36.	- 07 53			NON-EXISTENT OBJECT
LB 03975	06 55 36.	- 19 40 54.		18.5	FAINT BLUE STAR
ZWG 146.016	06 55 42.	+ 29 05		15.2	GALAXY
MRK 374	06 55 42.	+ 54 17	7	15.5	GALAXY WITH UV CONTINUUM
KW 36	06 55 42.	+ 54 17	41		SEYFERT GALAXY
VVI 19	06 55 42.	+ 54 17	30	15.31	SEYFERT GALAXY
MCG+09-12-016	06 55 42.	+ 54 17 30.	36	17.	GALAXY
MCG+02-18-004	06 55 45.	+ 12 44	36	14.5	GALAXY
LB 03976	06 55 47.	- 19 31 54.		17.0	FAINT BLUE STAR
ZWG 085.019	06 55 48.	+ 18 43		15.5	GALAXY
MCG+09-12-017	06 55 48.	+ 54 17	30	17.	GALAXY
ZC 0655.8+6527	06 55 48.	+ 65 27	940		CLUSTER OF GALAXIES
UGC 03620	06 55 48.	+ 76 40	66	16.5	GALAXY Sa-b
ISS 0504	06 55 48.	- 03 15	220		STELLAR RING
ZWG 116.008	06 55 54.	+ 25 06		15.4	GALAXY
ISS 0571	06 55 54.	- 03 11	214		STELLAR RING
LB 03977	06 55 59.	- 19 25 36.		18.2	FAINT BLUE STAR
SIV 06	06 56	- 04 40	5400		FAINT H EMISSION REGION
KHAV 185	06 56	- 05 34	4380		DARK NEBULA
RNGC 2312	06 56 01.	+ 10 20			NON-EXISTENT OBJECT
LB 03978	06 56 05.	- 19 33 42.		20.0	FAINT BLUE STAR
ZWG 205.010	06 56 06.	+ 41 15		15.4	GALAXY
LB 03979	06 56 09.	- 19 00 54.		19.8	FAINT BLUE STAR
MCG+02-18-005	06 56 15.	+ 14 22	84	17.	GALAXY
UGC 03621	06 56 18.	+ 14 22	96	17.	GALAXY DWRF IR
ZWG 116.009	06 56 18.	+ 25 38		15.6	GALAXY
ZWG 176.005	06 56 18.	+ 33 05		15.7	GALAXY
UGC 03622	06 56 18.	+ 33 05	90	15.7	GALAXY S
ZWG 309.014	06 56 18.	+ 63 20		15.7	GALAXY
LB 03980	06 56 18.	- 19 27 36.		19.2	FAINT BLUE STAR
LB 03981	06 56 20.	- 19 16 12.		18.1	FAINT BLUE STAR
LB 03982	06 56 20.	- 19 27 06.		17.9	FAINT BLUE STAR
IC 2175	06 56 21.	+ 35 21			NONSTELLAR OBJECT
LB 03983	06 56 21.	- 19 32 12.		18.7	FAINT BLUE STAR
LB 03984	06 56 21.	- 19 38 00.		16.7	FAINT BLUE STAR
LB 03985	06 56 22.	- 19 58 24.		19.1	FAINT BLUE STAR
MCG+06-16-006	06 56 24.	+ 33 07	66	14.5	GALAXY
RNGC 2317	06 56 24.	- 07 42			NON-EXISTENT OBJECT
LB 03986	06 56 24.	- 19 16 30.		19.3	FAINT BLUE STAR
LB 03987	06 56 24.	- 19 37 42.		18.8	FAINT BLUE STAR
LB 03988	06 56 25.	- 19 11 24.		17.7	FAINT BLUE STAR
LB 03989	06 56 25.	- 19 32 00.		20.2	FAINT BLUE STAR
LB 03990	06 56 29.	- 19 06 54.		18.6	FAINT BLUE STAR
ZWG 176.006	06 56 30.	+ 35 31		15.0	GALAXY
UGC 03623	06 56 30.	+ 35 31	120	15.0	GALAXY SB:c
MCG+08-13-041	06 56 30.	+ 48 27	24	17.	GALAXY
MCG+06-16-007	06 56 33.	+ 35 31 30.	90	14.	GALAXY
ZWG 205.011	06 56 36.	+ 40 27		15.4	GALAXY
MCG+11-09-026	06 56 36.	+ 63 18	27	16.	GALAXY
UGC 03624	06 56 42.	+ 27 38	78	16.0	GALAXY Sb
ZWG 176.007	06 56 42.	+ 34 09		15.6	GALAXY
SCHO 0221	06 56 44.	- 04 15 30.	590		ISOLATED DARK CLOUD
MCG+07-15-004	06 56 48.	+ 40 27	27	15.5	GALAXY
7ZW 113	06 56 48.	+ 67 25			COMPACT GALAXY
LBN 1012	06 57	- 04 40	720		BRIGHT NEBULA
LBN 1021	06 57	- 07 40	120		BRIGHT NEBULA
UGC 03625	06 57 00.	+ 51 25	66	16.5	GALAXY DBL SYS
ZWG 330.022	06 57 00.	+ 71 02		15.3	GALAXY
UGC 03626	06 57 00.	+ 71 02	78	15.3	GALAXY SB?b-c
MCG+12-07-023	06 57 00.	+ 71 02	114	16.	GALAXY
ZWG 363.018	06 57 00.	+ 86 01		15.7	GALAXY
ZWG 362.034	06 57 00.	+ 86 01		15.7	GALAXY
LDN 1650	06 57 00.	- 04 50	480		DARK NEBULA
LDN 1654	06 57 00.	- 07 40	900		DARK NEBULA
MRSL 220-01/1	06 57 00.	- 07 44	71		HII REGION
PK200+08.1	06 57 06.	+ 14 41		17.0	PLANETARY NEBULA
ZWG 261.014	06 57 06.	+ 51 21		14.7	GALAXY
UGC 03627	06 57 06.	+ 51 21	78	14.7	GALAXY Sc
ZC 0657.1+5945	06 57 06.	+ 59 45	1140		CLUSTER OF GALAXIES
MRSL 218-00/1	06 57 06.	- 04 44	720		HII REGION
MCG+09-12-018	06 57 12.	+ 51 20	66	15.	GALAXY
7ZW 114	06 57 12.	+ 67 39			COMPACT GALAXY
RNGC 2318	06 57 15.	- 13 38			NON-EXISTENT OBJECT
TON-N 0269	06 57 18.	+ 28 48		15.5	BLUE STAR
UGC 03628	06 57 18.	+ 38 42	60	18.	GALAXY Sc-IRR
RNGC 2316	06 57 18.	- 07 40			DIFFUSE NEBULA
ZWG 176.008	06 57 42.	+ 36 18		15.6	GALAXY
ZC 0657.8+6946	06 57 48.	+ 69 46	3290		CLUSTER OF GALAXIES
OCL 0543	06 57 48.	- 00 11	360	20.1	OPEN STAR CLUSTER
ISS 0505	06 57 48.	- 02 52	232		STELLAR RING
UGC 03617	06 57 54.	+ 13 40	60	16.0	GALAXY SBa-b
MRSL 221-02/1	06 57 54.	- 08 47	276		HII REGION
LBN 1023	06 58	- 08 47	180		BRIGHT NEBULA
LB 09835	06 58	- 84 02		13.1	FAINT BLUE STAR
ZWG 116.010	06 58 00.	+ 22 14		15.7	GALAXY
UGC 03629	06 58 00.	+ 42 43	60	16.5	GALAXY Sc
MCG+08-13-042	06 58 00.	+ 47 37	60	16.	GALAXY
ZWG 234.040	06 58 00.	+ 47 38		15.1	GALAXY
ZWG 261.015	06 58 00.	+ 51 12		15.7	GALAXY
MCG+14-04-020	06 58 00.	+ 85 50	15	16.	GALAXY
ZWG 363.019	06 58 00.	+ 85 51		15.4	GALAXY
ZWG 362.035	06 58 00.	+ 85 51		15.4	GALAXY
MCG+09-12-019	06 58 06.	+ 51 10	45	16.	GALAXY
DG 114	06 58 06.	- 08 52	240		REFLECTION NEBULA
VDB.66N 087	06 58 08.	- 08 44	168		REFLECTION NEBULA

OBJECT NAME	RIGHT ASCEN.	DECLINATION	DIAM.	MAGN.	TYPE OF OBJECT
YM 35	06 58 10.	- 08 41	162		SYMMETRIC GALACTIC NEBULA
ZWG 085.020	06 58 12.	+ 19 42		14.9	GALAXY
MCG+03-18-005	06 58 12.	+ 19 42	48	14.9	GALAXY
MCG+08-13-043	06 58 12.	+ 47 44	30	16.	GALAXY
ZWG 285.013	06 58 12.	+ 62 50		14.9	GALAXY
OCL 0603	06 58 12.	- 20 24	360	14.	OPEN STAR CLUSTER
ARC 0567	06 58 18.	+ 32 54		16.9	RICH CLUSTER OF GALAXIES
ZC 0658.3+4908	06 58 18.	+ 49 08	2150		CLUSTER OF GALAXIES
MCG+10-10-021	06 58 18.	+ 62 50	45	14.	GALAXY
ZC 0658.3+6320	06 58 18.	+ 63 20	4100		CLUSTER OF GALAXIES
IC 0456	06 58 19.	- 30 05 57.			NONSTELLAR OBJECT
UGC 03630	06 58 24.	+ 01 59	102	15.5	GALAXY S
MCG+00-18-001	06 58 24.	+ 01 59	78	15.	GALAXY
MCG-05-17-002	06 58 24.	- 30 04	60	15.	GALAXY
MCG+06-16-008	06 58 27.	+ 37 48	42	14.5	GALAXY
RNGC 2319	06 58 29.	+ 03 08			NON-EXISTENT OBJECT
MCG+05-17-006	06 58 30.	+ 29 57	72	15.7	GALAXY
ZWG 146.017	06 58 30.	+ 29 58		15.7	GALAXY
UGC 03631	06 58 30.	+ 29 58	84	15.7	GALAXY SB
ZWG 176.009	06 58 30.	+ 37 48		15.2	GALAXY
MCG+08-13-044	06 58 30.	+ 47 14	24	16.	GALAXY
MCG+15-01-006	06 58 30.	+ 87 52	78	17.	GALAXY
ISS 0506	06 58 30.	- 01 31	360		STELLAR RING
ZC 0658.6+6651	06 58 36.	+ 66 51	610		CLUSTER OF GALAXIES
UGC 03632	06 58 36.	+ 77 58	66	16.5	GALAXY Sc
RNGC 2315	06 58 40.	+ 50 40		14.5	GALAXY
ZWG 116.011	06 58 42.	+ 24 18		15.5	GALAXY
ZC 0658.7+3300	06 58 42.	+ 33 00	3700		CLUSTER OF GALAXIES
ZWG 234.041	06 58 42.	+ 50 40		14.5	GALAXY
UGC 03633	06 58 42.	+ 50 40	102	14.5	GALAXY S0-a
MCG+08-13-045	06 58 42.	+ 50 40	78	14.	GALAXY
7ZW 115	06 58 42.	+ 64 55			COMPACT GALAXY
ZC 0658.7+7031	06 58 42.	+ 70 31	1080		CLUSTER OF GALAXIES
MCG-05-17-003	06 58 42.	- 27 15	96	15.	GALAXY
MCG+02-18-006	06 58 45.	+ 14 12	60	14.5	GALAXY
UGC 03634	06 58 48.	+ 14 12	84	15.5	GALAXY SBa
ZWG 085.021	06 58 48.	+ 17 15		15.4	GALAXY
UGC 03635	06 58 48.	+ 17 15	66	15.4	GALAXY S-IRR
ZWG 116.012	06 58 54.	+ 22 21		15.6	GALAXY
ZWG 146.018	06 58 54.	+ 27 55		15.1	GALAXY
MCG+05-17-007	06 58 54.	+ 27 55	30	15.1	GALAXY
ZC 0658.9+5014	06 58 54.	+ 50 14	2350		CLUSTER OF GALAXIES
MCG+11-09-027	06 58 54.	+ 65 47	42	16.	GALAXY
BC 3CR173	06 58 56.48	+ 38 01 44.8		21.2	QUASI-STELLAR OBJECT
SIV 05	06 59	- 02 39	7200		FAINT H EMISSION REGION
KHAV 186	06 59	- 04 46	4080		DARK NEBULA
ZC 0659.0+4225	06 59 00.	+ 42 25	1140		CLUSTER OF GALAXIES
ZWG 234.042	06 59 00.	+ 50 40		14.9	GALAXY
MCG+08-13-046	06 59 00.	+ 50 40	30	16.	GALAXY
ZWG 309.015	06 59 00.	+ 64 07		15.6	GALAXY
MCG+13-05-036	06 59 00.	+ 77 56	39	16.	GALAXY
ISS 0572	06 59 00.	- 08 42	367		STELLAR RING
ZWG 261.016	06 59 06.	+ 54 42		15.3	GALAXY
ZWG 348.028	06 59 06.	+ 75 32		14.9	GALAXY
UGC 03636	06 59 06.	+ 75 32	72	14.9	GALAXY Sa-b
MCG-06-16-004	06 59 06.	- 37 42	36	14.5	GALAXY
UGC 03637	06 59 12.	+ 05 00	72	16.5	GALAXY Sc-IRR
ZWG 234.043	06 59 12.	+ 49 30		14.4	GALAXY
UGC 03638	06 59 12.	+ 49 30	66	14.4	GALAXY SB:a-b
MCG+08-13-047	06 59 12.	+ 49 30	72	15.	GALAXY
MCG+11-09-028	06 59 12.	+ 64 05	39	16.	GALAXY
SCHO 0222	06 59 14.	+ 02 06 18.	250		ISOLATED DARK CLOUD
MCG+09-12-020	06 59 18.	+ 54 43	15	16.	GALAXY
MCG+09-12-021	06 59 18.	+ 54 57 30.	51	16.	GALAXY
MCG+13-05-037	06 59 18.	+ 75 30 30.	51	16.	GALAXY
ZC 0659.4+5156	06 59 24.	+ 51 56	670		CLUSTER OF GALAXIES
ZWG 309.016	06 59 24.	+ 64 04		15.7	GALAXY
ZC 0659.4+6545	06 59 24.	+ 65 45	810		CLUSTER OF GALAXIES
MRSL 224-02/1	06 59 24.	- 11 14	660		HII REGION
OCL 0573	06 59 24.	- 13 31	240	12.	OPEN STAR CLUSTER
HELW 129	06 59 29.	- 10 48 24.			NEBULA
HELW 130	06 59 29.	- 11 13 42.			NEBULA
ZWG 085.022	06 59 30.	+ 20 02		15.4	GALAXY
UGC 03639	06 59 30.	+ 20 02	60	15.4	GALAXY S
ZWG 176.010	06 59 30.	+ 37 12		15.1	GALAXY
UGC 03640	06 59 30.	+ 37 12	66	15.1	GALAXY S0-a
MCG+06-16-009	06 59 30.	+ 37 12	24	15.5	GALAXY
MCG+11-09-029	06 59 30.	+ 64 03	36	16.	GALAXY
SCHO 0223	06 59 33.	+ 02 01 00.	330		ISOLATED DARK CLOUD
UGC 03641	06 59 36.	+ 11 18	66	16.0	GALAXY Sb
MCG+02-18-007	06 59 36.	+ 11 18	72	15.	GALAXY
TON-N 0270	06 59 36.	+ 28 17		15.6	BLUE STAR
ZWG 205.012	06 59 36.	+ 39 18		15.0	GALAXY
ZC 0659.6+6111	06 59 36.	+ 61 11	2080		CLUSTER OF GALAXIES
ZWG 309.017	06 59 36.	+ 64 06		13.5	GALAXY
UGC 03642	06 59 36.	+ 64 06	90	13.5	GALAXY S0
SG 3.093	06 59 39.	- 11 10	660		DIFFUSE EMISSION NEBULA
7ZW 116	06 59 42.	+ 63 23			COMPACT GALAXY
VDB.66N 088	06 59 43.	- 11 14	852		REFLECTION NEBULA
ARC 0566	06 59 47.	+ 63 22		16.4	RICH CLUSTER OF GALAXIES
ZWG 116.013	06 59 48.	+ 22 55		15.5	GALAXY
ZC 0659.8+3253	06 59 48.	+ 32 53	1080		CLUSTER OF GALAXIES
ZC 0659.8+4627	06 59 48.	+ 46 27	540		CLUSTER OF GALAXIES
UGC 03643	06 59 48.	+ 53 52	90	16.0	GALAXY
ZWG 261.017	06 59 48.	+ 56 33		15.3	GALAXY
MCG+09-12-022	06 59 48.	+ 56 35	57	15.	GALAXY
MCG+11-09-030	06 59 48.	+ 64 05	78	14.	GALAXY
OCL 0610	06 59 48.	- 21 51	84	13.	OPEN STAR CLUSTER
ZWG 261.018	06 59 54.	+ 53 45		15.4	GALAXY
ZWG 330.023	06 59 54.	+ 71 08		14.9	GALAXY
UGC 03644	06 59 54.	+ 71 08	132	14.9	GALAXY SBc
VDB.66G 040	07 00	+ 56 32	70		DWARF GALAXY
KHAV 187	07 00	- 00 04	15100		DARK NEBULA
LBN 1030	07 00	- 11 10	900		BRIGHT NEBULA
ZWG 085.023	07 00 00.	+ 18 36		15.4	GALAXY
ZWG 205.013	07 00 00.	+ 40 23		15.3	GALAXY
MCG+09-12-023	07 00 00.	+ 53 46	39	15.	GALAXY
MCG+09-12-024	07 00 00.	+ 53 50	51	16.	GALAXY
ZC 0700.0+6042	07 00 00.	+ 60 42	270		CLUSTER OF GALAXIES
MCG+11-09-031	07 00 00.	+ 63 24	8	17.	GALAXY
MCG+12-07-024	07 00 00.	+ 71 08	132	15.	GALAXY
ZWG 085.024	07 00 06.	+ 18 33		15.0	GALAXY
MCG+03-18-006	07 00 06.	+ 18 33	21	15.0	GALAXY
ZC 0700.1+3719	07 00 06.	+ 37 19	1010		CLUSTER OF GALAXIES
ZWG 234.044	07 00 06.	+ 50 45		15.5	GALAXY
UGC 03645	07 00 06.	+ 50 45	108	15.5	GALAXY S
ZC 0700.1+7151	07 00 06.	+ 71 51	3490		CLUSTER OF GALAXIES
SCHO 0224	07 00 08.	- 04 01 42.	630		ISOLATED DARK CLOUD
MCG+06-16-010	07 00 09.	+ 36 21	42	16.	GALAXY
MCG+06-16-012	07 00 12.	+ 33 50	36	14.5	GALAXY
MCG+06-16-011	07 00 12.	+ 33 50	36	15.	GALAXY
ZWG 176.011	07 00 12.	+ 33 51		14.7	GALAXY
MCG+08-13-048	07 00 12.	+ 50 45	66	14.	GALAXY
ZWG 261.019	07 00 12.	+ 54 03		15.5	GALAXY
ZWG 261.020	07 00 12.	+ 54 18		15.0	GALAXY
UGC 03646	07 00 12.	+ 54 18	72	15.0	GALAXY E
MCG-05-17-004	07 00 12.	- 28 21	60	15.	GALAXY
SG 3.094	07 00 15.	- 11 23	360		DIFFUSE EMISSION NEBULA
ZWG 085.025	07 00 18.	+ 17 57		15.3	GALAXY
MCG+09-12-025	07 00 18.	+ 54 04	27	16.	GALAXY
MCG+09-12-026	07 00 18.	+ 54 19	2	15.	GALAXY
MRSL 224-02/2	07 00 18.	- 11 23	480		HII REGION
VDB.66N 089	07 00 20.	- 12 10	132		REFLECTION NEBULA
HELW 131	07 00 23.	- 11 22 58.			NEBULA
ZWG 146.019	07 00 24.	+ 27 46		15.6	GALAXY
ZWG 146.020	07 00 24.	+ 32 32		15.7	GALAXY
ZC 0700.4+4801	07 00 24.	+ 48 01	24460		CLUSTER OF GALAXIES
PK226-03.1	07 00 24.	- 13 40			PLANETARY NEBULA
MCG+06-16-013	07 00 27.	+ 38 01 30.	36	15.	GALAXY
VDB.66N 090	07 00 28.	- 11 21	480		REFLECTION NEBULA
OCL 0526	07 00 30.	+ 06 29	300		OPEN STAR CLUSTER
ZWG 176.012	07 00 30.	+ 34 16		15.6	GALAXY
MCG+14-04-021	07 00 30.	+ 85 47	15	15.	GALAXY
RNGC 2323	07 00 30.	- 08 16		7.0	OPEN CLUSTER
SCHO 0225	07 00 35.	- 08 34 06.	390		ISOLATED DARK CLOUD
ZWG 085.026	07 00 36.	+ 19 01		15.1	GALAXY
MCG+03-18-007	07 00 36.	+ 19 01	42	15.1	GALAXY
TON-N 0001	07 00 36.	+ 26 40		14.4	BLUE STAR
MCG+06-16-014	07 00 36.	+ 38 00	48	14.5	GALAXY
UGC 03647	07 00 36.	+ 56 36	84	16.0	GALAXY IRR
ZWG 309.018	07 00 36.	+ 64 08		15.4	GALAXY
MCG-05-17-005	07 00 36.	- 28 35	36	13.	GALAXY
HELW 132	07 00 39.	- 11 22 59.			NEBULA
MCG+06-16-015	07 00 42.	+ 37 46 30.	42	15.5	GALAXY
ZWG 176.013	07 00 42.	+ 37 59		15.0	GALAXY
MCG+09-12-027	07 00 42.	+ 56 36	78	16.	GALAXY
ZWG 309.019	07 00 42.	+ 64 11		15.3	GALAXY
UGC 03648	07 00 42.	+ 64 11	84	15.3	GALAXY Sb-c
RNGC 2325	07 00 43.	- 28 38		13.0	GALAXY
MCG+04-17-005	07 00 48.	+ 22 26	120	14.7	GALAXY
ZWG 146.021	07 00 48.	+ 29 20		14.6	GALAXY
UGC 03649	07 00 48.	+ 29 20	66	14.6	GALAXY Sa
KARA.72 122A	07 00 48.	+ 29 20	66	14.6	PART OF DOUBLE GALAXY
ZWG 176.014	07 00 48.	+ 37 46		15.5	GALAXY
UGC 03650	07 00 48.	+ 37 46	66	15.5	GALAXY S
ZWG 234.045	07 00 48.	+ 45 25		15.6	GALAXY
UGC 03651	07 00 48.	+ 45 25	60	15.6	GALAXY Sa
ZWG 261.021	07 00 48.	+ 54 04		15.5	GALAXY
MCG+09-12-028	07 00 48.	+ 54 04 30.	15	16.	GALAXY
OCL 0559	07 00 48.	- 08 16	1500	7.5	OPEN STAR CLUSTER
MCG-05-17-006	07 00 48.	- 29 20	42	14.5	GALAXY
HELW 133	07 00 49.	- 11 08 12.			NEBULA
MCG+05-17-008	07 00 51.	+ 29 19	42	14.6	GALAXY
ZWG 116.014	07 00 54.	+ 22 27		14.7	GALAXY
UGC 03652	07 00 54.	+ 22 27	126	14.7	GALAXY Sc
ZWG 146.022	07 00 54.	+ 29 20		15.4	GALAXY
KARA.72 122B	07 00 54.	+ 29 20	48	15.4	PART OF DOUBLE GALAXY
MCG+09-12-029	07 00 54.	+ 54 05	24	16.	GALAXY
MCG+11-09-032	07 00 54.	+ 64 09	78	15.	GALAXY
MAI 019	07 00 54.	+ 66 06	40		DWARF SPHEROIDAL GALAXY
RNGC 2328	07 00 54.	- 41 59			GALAXY
PK242-11.1	07 00 56.	- 31 31 00.	29		PLANETARY NEBULA
MCG+05-17-009	07 00 57.	+ 29 19	24	15.4	GALAXY
PK223-02.1	07 00 58.	- 10 30	79		PLANETARY NEBULA
KHAV 188	07 01	- 10 16	8570		DARK NEBULA
LBN 1035	07 01	- 11 25	720		BRIGHT NEBULA
ZC 0701.0+1858	07 01 00.	+ 18 58	5380		CLUSTER OF GALAXIES
MCG+08-13-049	07 01 00.	+ 45 24	54	16.	GALAXY
ZC 0701.0+5622	07 01 00.	+ 56 22	610		CLUSTER OF GALAXIES
MCG+14-04-022	07 01 00.	+ 84 27	192	12.1	GALAXY
ZWG 363.020	07 01 00.	+ 84 28		12.1	GALAXY
ZWG 362.036	07 01 00.	+ 84 28		12.1	GALAXY
UGC 03653	07 01 00.	+ 84 28	222	12.1	GALAXY Sc
ZWG 363.021	07 01 00.	+ 85 48		15.2	GALAXY
ZWG 362.037	07 01 00.	+ 85 48		15.2	GALAXY
UGC 03654	07 01 00.	+ 85 48	78	15.2	GALAXY COMPACT
MCG+06-16-016	07 01 06.	+ 37 36	42	15.	GALAXY
VDB.66N 091	07 01 09.	- 10 36	108		REFLECTION NEBULA
ZWG 234.046	07 01 12.	+ 46 13		14.5	GALAXY
UGC 03655	07 01 12.	+ 46 13	36	14.5	GALAXY PECULR
OCL 0605	07 01 12.	- 20 47	180	16.	OPEN STAR CLUSTER
MCG+04-17-006	07 01 18.	+ 23 58 30.	54	14.8	GALAXY
ZWG 116.015	07 01 18.	+ 23 59		14.8	GALAXY
MCG+08-13-050	07 01 18.	+ 46 12 30.	42	15.	GALAXY
CED 089A	07 01 21.	- 11 28			DIFFUSE GALACTIC NEBULA
ZWG 085.027	07 01 24.	+ 18 37		15.6	GALAXY
UGC 03656	07 01 24.	+ 18 37	66	15.6	GALAXY S0-a
ZWG 176.015	07 01 24.	+ 37 08		15.3	GALAXY
ZC 0701.4+6857	07 01 24.	+ 68 57	340		CLUSTER OF GALAXIES
MRSL 224-02/3	07 01 24.	- 11 24	30		HII REGION
HELW 134	07 01 24.	- 11 28 32.			NEBULA
SCHO 0226	07 01 25.	- 04 28 18.	480		ISOLATED DARK CLOUD
ZWG 085.028	07 01 30.	+ 15 51		15.7	GALAXY
MCG+03-18-008	07 01 30.	+ 18 37	60	15.6	GALAXY
ZWG 085.029	07 01 30.	+ 18 40		15.2	GALAXY
MCG+09-12-030	07 01 30.	+ 54 03	39	17.	GALAXY
ZWG 261.022	07 01 30.	+ 54 09		15.1	GALAXY
MRSL 224-02/4	07 01 30.	- 11 14	90		HII REGION
VDB.66N 092	07 01 34.	- 11 27	240		REFLECTION NEBULA
OCL 0542	07 01 36.	+ 01 08	960	9.6	OPEN STAR CLUSTER
MCG+03-18-009	07 01 36.	+ 18 40	24	15.2	GALAXY
ZC 0701.6+3104	07 01 36.	+ 31 04	1210		CLUSTER OF GALAXIES
ZC 0701.6+5444	07 01 36.	+ 54 44	670		CLUSTER OF GALAXIES
ZWG 261.023	07 01 36.	+ 56 17		15.5	GALAXY
MCG+12-07-025	07 01 36.	+ 71 17	66	16.	GALAXY
OCL 0556	07 01 36.	- 06 03	162	14.	OPEN STAR CLUSTER
RNGC 2324	07 01 37.	+ 01 08		9.0	OPEN CLUSTER
HELW 135	07 01 37.	- 11 30 03.			NEBULA
HUB C21	07 01 37.	- 12 12			DIFFUSE NEBULA
ARC 0565	07 01 39.	+ 71 50		16.5	RICH CLUSTER OF GALAXIES
MCG+13-06-001	07 01 39.	+ 77 56 30.	45	16.	GALAXY
MCG+09-12-031	07 01 42.	+ 54 04	30	18.	GALAXY
UGC 03657	07 01 42.	+ 71 16	72	16.5	GALAXY Sc
SCHO 0227	07 01 42.	- 01 13 36.	430		ISOLATED DARK CLOUD
DG 115	07 01 42.	- 11 14	540		REFLECTION NEBULA
MRSL 224-02/5	07 01 42.	- 11 14	41		HII REGION

OBJECT NAME	RIGHT ASCEN.	DECLINATION	DIAM.	MAGN.	TYPE OF OBJECT
CED 089B	07 01 45.	- 11 14			DIFFUSE GALACTIC NEBULA
RNGC 2320	07 01 46.	+ 50 40		14.0	GALAXY
UGC 03658	07 01 48.	+ 17 41	120	17.	GALAXY DWRF IR
ZC 0701.8+3116	07 01 48.	+ 31 16	4570		CLUSTER OF GALAXIES
ZWG 234.047	07 01 48.	+ 50 40		13.9	GALAXY
UGC 03659	07 01 48.	+ 50 40	102	13.9	GALAXY E
MCG+08-13-051	07 01 48.	+ 50 40	42	13.	GALAXY
ZWG 309.020	07 01 48.	+ 63 57		13.6	GALAXY
UGC 03660	07 01 48.	+ 63 57	108	13.6	GALAXY Sa/SBa
ZC 0701.8+6620	07 01 48.	+ 66 20	1140		CLUSTER OF GALAXIES
MRK 375	07 01 48.	+ 67 48	10	16.	GALAXY WITH UV CONTINUUM
ZWG 085.030	07 01 54.	+ 16 23		15.2	GALAXY
ZWG 116.016	07 01 54.	+ 24 14		15.6	GALAXY
ZWG 234.048	07 01 54.	+ 48 26		15.1	GALAXY
MCG+08-13-052	07 01 54.	+ 48 26	36	16.	GALAXY
ISS 0573	07 01 54.	- 05 29	218		STELLAR RING
RNGC 2327	07 01 56.	- 11 14			DIFFUSE NEBULA
SG 3.095	07 01 57.	- 11 19	90		DIFFUSE EMISSION NEBULA
ST 19	07 01 57.	- 11 28			EMISSION NEBULA
KHAV 189	07 02	- 03 34	1990		DARK NEBULA
LBN 1027	07 02	- 10 20	1200		BRIGHT NEBULA
ZWG 085.031	07 02 00.	+ 18 54		15.7	GALAXY
ZWG 234.049	07 02 00.	+ 45 48		15.6	GALAXY
MCG+08-13-054	07 02 00.	+ 50 35	60	14.	GALAXY
MCG+08-13-053	07 02 00.	+ 50 50	78	15.	GALAXY
MCG+11-09-033	07 02 00.	+ 63 55	90	14.	GALAXY
ZWG 349.002	07 02 00.	+ 77 57		15.7	GALAXY
ZWG 348.029	07 02 00.	+ 77 57		15.7	GALAXY
MCG+14-04-023	07 02 00.	+ 81 02	66	16.	GALAXY
ZWG 363.022	07 02 00.	+ 81 04		15.2	GALAXY
MCG+14-04-024	07 02 00.	+ 85 50	42	16.	GALAXY
ZWG 363.023	07 02 00.	+ 85 51		15.6	GALAXY
ZWG 362.038	07 02 00.	+ 85 51		15.6	GALAXY
UGC 03661	07 02 00.	+ 85 51	108	15.6	GALAXY S
MCG+14-04-025	07 02 00.	+ 85 59	15	17.	GALAXY
SCHO 0228	07 02 00.	- 04 20 48.	430		ISOLATED DARK CLOUD
LDN 1657	07 02 00.	- 10 20	4200		DARK NEBULA
MRSL 223-01/1	07 02 00.	- 10 23	900		HII REGION
VDB.66N 093	07 02 02.	- 10 22	1200		REFLECTION NEBULA
RNGC 2322	07 02 04.	+ 50 35		14.5	GALAXY
RNGC 2321	07 02 04.	+ 50 50		15.0	GALAXY
ZC 0702.1+3215	07 02 06.	+ 32 15	1140		CLUSTER OF GALAXIES
ZWG 234.050	07 02 06.	+ 50 35		14.6	GALAXY
UGC 03662	07 02 06.	+ 50 35	78	14.6	GALAXY SBa
ZWG 234.051	07 02 06.	+ 50 50		14.8	GALAXY
UGC 03663	07 02 06.	+ 50 50	90	14.8	GALAXY SBa
SG 3.096	07 02 10.	- 10 22	1440		DIFFUSE EMISSION NEBULA
HUB E15	07 02 11.	- 17 47			DIFFUSE NEBULA
ZWG 176.016	07 02 12.	+ 33 54		15.4	GALAXY
UGC 03664	07 02 12.	+ 33 54	60	15.4	GALAXY Sc
7ZW 117	07 02 12.	+ 63 00			COMPACT GALAXY
ZC 0702.2+6712	07 02 12.	+ 67 12	3160		CLUSTER OF GALAXIES
ZWG 330.024	07 02 12.	+ 71 38		15.3	GALAXY
UGC 03665	07 02 12.	+ 71 38	72	15.3	GALAXY Sa
ZWG 348.030	07 02 12.	+ 75 27		15.0	GALAXY
UGC 03666	07 02 12.	+ 75 27	66	15.0	GALAXY Sa
ST 18	07 02 12.	- 10 23	1500		EMISSION NEBULA
MCG+06-16-017	07 02 18.	+ 33 54	66	14.5	GALAXY
7ZW 118	07 02 18.	+ 64 42			COMPACT GALAXY
HELW 136	07 02 22.	- 11 27 18.			NEBULA
ZWG 205.014	07 02 24.	+ 41 05		15.1	GALAXY
MCG+12-07-026	07 02 24.	+ 71 38	66	16.	GALAXY
ZC 0702.5+4554	07 02 30.	+ 45 54	4230		CLUSTER OF GALAXIES
IC 2174	07 02 30.	+ 75 20	66		GALAXY SB0
MCG+13-06-002	07 02 30.	+ 75 25 30.	36	15.	GALAXY
ZC 0702.5+7752	07 02 30.	+ 77 52	740		CLUSTER OF GALAXIES
ST 17	07 02 40.	- 11 08			EMISSION NEBULA
PK212+04.1	07 02 42.	+ 02 51 07.			PLANETARY NEBULA
ZWG 085.032	07 02 42.	+ 15 47		15.7	GALAXY
7ZW 119	07 02 42.	+ 64 11			COMPACT GALAXY
KARA.68 049	07 02 42.	+ 71 57	40		DWARF GALAXY
IC 2177	07 02 45.	- 10 38	1080		DIFFUSE NEBULA
CED 089C	07 02 46.	- 10 37	990		DIFFUSE GALACTIC NEBULA
UGC 03667	07 02 48.	+ 39 15	84	16.	GALAXY S
MCG+08-13-055	07 02 48.	+ 46 12	39	16.	GALAXY
ZWG 234.052	07 02 48.	+ 46 13		15.1	GALAXY
ZWG 234.053	07 02 48.	+ 48 34		15.1	GALAXY
SG 3.097	07 02 53.	- 09 39	3600		DIFFUSE EMISSION NEBULA
ZWG 146.023	07 02 54.	+ 28 23		14.9	GALAXY
MCG+05-17-010	07 02 54.	+ 28 23	36	14.9	GALAXY
MCG+08-13-056	07 02 54.	+ 48 34	21	16.	GALAXY
DG 116	07 02 54.	- 12 15	660		REFLECTION NEBULA
MRSL 225-02/1	07 02 54.	- 12 15	420		HII REGION
VDB.66N 094	07 02 54.	- 12 17	672		REFLECTION NEBULA
ST 16	07 02 55.	- 12 20			EMISSION NEBULA
SCHO 0229	07 02 56.	- 01 15 54.	350		ISOLATED DARK CLOUD
CED 090	07 02 56.	- 12 15	630		DIFFUSE GALACTIC NEBULA
LBN 1037	07 03	- 12 00	5100		BRIGHT NEBULA
LBN 1039	07 03	- 12 11	600		BRIGHT NEBULA
LBN 1042	07 03	- 16 20	360		BRIGHT NEBULA
LBN 1056	07 03	- 24 10	2340		BRIGHT NEBULA
ZC 0703.0+4228	07 03 00.	+ 42 28	2820		CLUSTER OF GALAXIES
ZC 0703.0+4935	07 03 00.	+ 49 35	4370		CLUSTER OF GALAXIES
MCG+08-13-057	07 03 00.	+ 50 14	24	17.	GALAXY
7ZW 120	07 03 00.	+ 78 08			COMPACT GALAXY
UGC 03671	07 03 00.	+ 78 30	60	17.	GALAXY DWRF IR
UGC 03668	07 03 00.	+ 81 04	78	16.0	GALAXY Sc
ZWG 363.024	07 03 00.	+ 85 06		15.3	GALAXY
ZWG 362.039	07 03 00.	+ 85 06		15.3	GALAXY
UGC 03669	07 03 00.	+ 85 33	84	17.	GALAXY
ZWG 363.025	07 03 00.	+ 85 41		15.4	GALAXY
ZWG 362.040	07 03 00.	+ 85 41		15.4	GALAXY
UGC 03670	07 03 00.	+ 85 41	78	15.4	GALAXY
ZWG 176.017	07 03 06.	+ 33 42		15.5	GALAXY
RNGC 2348	07 03 06.	- 67 19			UNVERIFIED SOUTHERN OBJECT
MCG+06-16-018	07 03 09.	+ 33 41 30.	15	16.	GALAXY
ZWG 116.017	07 03 12.	+ 23 40		15.4	GALAXY
ZWG 205.015	07 03 12.	+ 42 35		15.6	GALAXY
ZWG 234.054	07 03 12.	+ 50 28		15.3	GALAXY
MCG+08-13-058	07 03 12.	+ 50 29	60	16.	GALAXY
CED 091	07 03 12.	- 08 38			DIFFUSE GALACTIC NEBULA
SCHO 0230	07 03 17.	- 03 26 00.	570		ISOLATED DARK CLOUD
ST 15	07 03 17.	- 10 48			EMISSION NEBULA
UGC 03672	07 03 18.	+ 30 26	78	17.	GALAXY IRR
ZWG 205.016	07 03 18.	+ 44 55		15.7	GALAXY
UGC 03673	07 03 18.	+ 44 55	66	15.7	GALAXY Sc/SBc
ZWG 261.024	07 03 18.	+ 56 22		15.7	GALAXY
ZWG 286.001	07 03 18.	+ 62 48		15.7	GALAXY
ZWG 285.014	07 03 18.	+ 62 48		15.7	GALAXY
ZWG 085.033	07 03 24.	+ 20 52		15.3	GALAXY
MCG+07-15-005	07 03 24.	+ 44 56	48	17.	GALAXY
ZC 0703.4+5925	07 03 24.	+ 59 25	3490		CLUSTER OF GALAXIES
CED 089D	07 03 28.	- 10 34			DIFFUSE GALACTIC NEBULA
MCG+02-18-008	07 03 30.	+ 13 23	42	16.	GALAXY
ZWG 205.017	07 03 30.	+ 44 53		14.7	GALAXY
ZWG 234.055	07 03 30.	+ 48 25		15.6	GALAXY
MCG+09-12-032	07 03 30.	+ 54 31	12	18.	GALAXY
MCG+09-12-033	07 03 30.	+ 56 23	39	16.	GALAXY
MRSL 224-01/1	07 03 30.	- 11 08	12000		HII REGION
ZWG 116.018	07 03 36.	+ 23 50		15.3	GALAXY
ZWG 116.019	07 03 36.	+ 25 32		15.5	GALAXY
UGC 03674	07 03 36.	+ 25 32	66	15.5	GALAXY Sc
MCG+07-15-006	07 03 36.	+ 44 54	36	15.5	GALAXY
ZC 0703.6+6856	07 03 36.	+ 68 56	540		CLUSTER OF GALAXIES
ZWG 348.031	07 03 36.	+ 74 59		15.5	GALAXY
ZWG 330.025	07 03 36.	+ 74 59		15.5	GALAXY
UGC 03675	07 03 36.	+ 74 59	96	15.5	GALAXY Sc
MCG+04-17-007	07 03 39.	+ 23 58	60	15.4	GALAXY
ZWG 085.034	07 03 42.	+ 15 05		15.7	GALAXY
ZWG 116.020	07 03 42.	+ 23 58		15.4	GALAXY
UGC 03676	07 03 42.	+ 23 58	66	15.4	GALAXY Sb
ZWG 146.024	07 03 42.	+ 29 30		15.6	GALAXY
ZC 0703.7+4913	07 03 42.	+ 49 13	1210		CLUSTER OF GALAXIES
MCG+10-11-001	07 03 42.	+ 62 48	36	17.	GALAXY
ZWG 348.032	07 03 42.	+ 75 25		13.1	GALAXY
UGC 03677	07 03 42.	+ 75 25	120	13.1	GALAXY E-S0
ZC 0703.7+7618	07 03 42.	+ 76 18	670		CLUSTER OF GALAXIES
MCG+08-13-059	07 03 45.	+ 48 58	36	16.	GALAXY
ZWG 085.035	07 03 48.	+ 19 27		15.4	GALAXY
UGC 03678	07 03 48.	+ 19 27	60	15.4	GALAXY S
ZWG 205.018	07 03 48.	+ 44 52		15.5	GALAXY
UGC 03679	07 03 48.	+ 44 52	102	15.5	GALAXY S
ZWG 234.056	07 03 48.	+ 48 58		14.8	GALAXY
OCL 0374	07 03 48.	+ 75 19	360		OPEN STAR CLUSTER
SG 3.098	07 03 51.	- 10 45	7200		DIFFUSF EMISSION NEFULA
RNGC 2314	07 03 52.	+ 75 24		13.0	GALAXY
MCG+07-15-007	07 03 54.	+ 44 52	33	17.	GALAXY
ZWG 261.025	07 03 54.	+ 54 21		15.7	GALAXY
ZC 0703.9+6820	07 03 54.	+ 68 20	1750		CLUSTER OF GALAXIES
MCG+12-07-027	07 03 54.	+ 74 58	90	16.	GALAXY
OCL 0583	07 03 54.	- 14 54	144	14.	OPEN STAR CLUSTER
OCL 0648	07 03 54.	- 28 22	150	13.	OPEN STAR CLUSTER
MAI 020	07 03 59.	+ 78 39	40		DWARF SPHEROIDAL GALAXY
LBN 1033	07 04	- 11 00	5100		BRIGHT NEBULA
ZC 0704.0+3821	07 04 00.	+ 38 21	3290		CLUSTER OF GALAXIES
ZWG 234.057	07 04 00.	+ 48 29		15.7	GALAXY
UGC 03680	07 04 00.	+ 53 00	90	17.	GALAXY DWARF
MCG+14-04-026	07 04 00.	+ 85 39	45	16.	GALAXY
OCL 0475	07 04 06.	+ 27 26	1080	8.6	OPEN STAR CLUSTER
ZWG 234.058	07 04 06.	+ 47 06		15.6	GALAXY
MCG+08-13-060	07 04 06.	+ 47 07	36	16.	GALAXY
ZWG 286.002	07 04 06.	+ 61 23		15.3	GALAXY
ZWG 285.015	07 04 06.	+ 61 23		15.3	GALAXY
KARA.73B 0178	07 04 06.	+ 61 23	18	15.3	ISOLATED GALAXY E
OCL 0568	07 04 06.	- 11 32	420	16.	OPEN STAR CLUSTFR
BC 4C38.20	07 04 08.2	+ 38 26 50.		17.5	QUASI-STELLAR OBJECT
SHB 093	07 04 08.4	+ 38 26 57.		17.5	QUASI-STELLAR OBJECT
RNGC 2331	07 04 09.	+ 27 26		8.5	OPEN CLUSTER
ZWG 234.059	07 04 12.	+ 48 44		14.6	GALAXY
MCG+08-13-061	07 04 12.	+ 48 45	60	15.	GALAXY
MCG+08-13-062	07 04 12.	+ 50 46	114	14.	GALAXY
MCG+09-12-034	07 04 12.	+ 52 59	78	17.	GALAXY
ZWG 261.026	07 04 12.	+ 56 08		15.6	GALAXY
ZWG 261.027	07 04 12.	+ 56 16		15.5	GALAXY
ZC 0704.2+6345	07 04 12.	+ 63 45	2420		CLUSTER OF GALAXIES
ZWG 309.021	07 04 12.	+ 65 17		15.6	GALAXY
MCG+13-06-003	07 04 12.	+ 75 23	66	13.3	GALAXY
OCL 0562	07 04 12.	- 10 00	1260	9.5	OPEN STAR CLUSTER
OCL 0604	07 04 12.	- 20 02	270	12.	OPEN STAR CLUSTER
RNGC 2335	07 04 13.	- 10 00		9.5	OPEN CLUSTER
IC 2176	07 04 16.	+ 32 32 27.			NONSTELLAR OBJECT
ST 14	07 04 17.	- 10 33			EMISSION NEBULA
VDB.66N 095	07 04 17.	- 11 12	336		REFLECTION NEBULA
ZWG 116.021	07 04 18.	+ 24 48		15.7	GALAXY
ZWG 146.025	07 04 18.	+ 32 33		15.1	GALAXY
MCG+05-17-011	07 04 18.	+ 32 33	24	15.1	GALAXY
ARC 0568	07 04 18.	+ 35 08		15.4	RICH CLUSTER OF GALAXIES
ZWG 234.060	07 04 18.	+ 50 45		14.3	GALAXY
UGC 03681	07 04 18.	+ 50 45	138	14.3	GALAXY SBb
MCG+09-12-035	07 04 18.	+ 56 09	36	17.	GALAXY
ZC 0704.3+5859	07 04 18.	+ 58 59	870		CLUSTER OF GALAXIES
SG 3.099	07 04 21.	- 11 12	240		DIFFUSE EMISSION NEBULA
IC 2178	07 04 22.	+ 32 35 02.			NONSTELLAR OBJECT
UGC 03682	07 04 24.	+ 14 15	60	17.	GALAXY
ZWG 116.022	07 04 24.	+ 23 12		15.4	GALAXY
ZWG 116.023	07 04 24.	+ 24 35		15.6	GALAXY
ZWG 146.026	07 04 24.	+ 32 36		15.7	GALAXY
MCG+06-16-019	07 04 24.	+ 35 08			GALAXY
ZC 0704.4+4019	07 04 24.	+ 40 19	1210		CLUSTER OF GALAXIES
ZWG 234.061	07 04 24.	+ 49 00		15.5	GALAXY
MCG+08-13-063	07 04 24.	+ 49 00	48	16.	GALAXY
ZWG 234.062	07 04 24.	+ 49 13		15.1	GALAXY
MCG+08-13-064	07 04 24.	+ 49 13	42	16.	GALAXY
ACK 225-01.1	07 04 24.	- 11 41			PLANETARY NEBULA
ZWG 086.001	07 04 30.	+ 18 35		15.4	GALAXY
ZWG 085.036	07 04 30.	+ 18 35		15.4	GALAXY
ZWG 234.063	07 04 30.	+ 46 12		14.1	GALAXY
UGC 03683	07 04 30.	+ 46 12	132	14.1	GALAXY S0
MCG+08-13-065	07 04 30.	+ 46 12	30	15.	GALAXY
ZWG 234.064	07 04 30.	+ 50 12		15.5	GALAXY
MCG+09-12-036	07 04 30.	+ 51 18	57	15.	GALAXY
ZWG 261.028	07 04 30.	+ 51 20		14.8	GALAXY
UGC 03684	07 04 30.	+ 51 20	72	14.8	GALAXY SBb
ZWG 286.003	07 04 30.	+ 61 41		13.1	GALAXY
ZWG 285.016	07 04 30.	+ 61 41		13.1	GALAXY
UGC 03685	07 04 30.	+ 61 41	240	13.1	GALAXY SBb
RNGC 2326	07 04 34.	+ 50 48		14.5	GALAXY
RNGC 2338	07 04 34.	- 05 33			NON-EXISTENT OBJECT
TON-N 0271	07 04 36.	+ 28 26		15.8	BLUE STAR
ZC 0704.6+2947	07 04 36.	+ 29 47	3430		CLUSTER OF GALAXIES
ZC 0704.6+4115	07 04 36.	+ 41 15	1010		CLUSTER OF GALAXIES
MCG+08-13-066	07 04 36.	+ 49 05	30	16.	GALAXY
MCG+08-13-067	07 04 36.	+ 50 42 30.	60	15.	GALAXY
MCG+09-12-037	07 04 36.	+ 56 10	36	17.	GALAXY
MCG+10-11-002	07 04 36.	+ 61 42	156	12.	GALAXY
MCG+11-09-034	07 04 36.	+ 63 04	51	16.	GALAXY

OBJECT NAME	RIGHT ASCEN.	DECLINATION	DIAM.	MAGN.	TYPE OF OBJECT
RLWT 129	07 04 36.	- 03 05		15.	FAINT VERY BLUE STAR
RLWT 130	07 04 36.	- 03 06		14.	FAINT VERY BLUE STAR
ASS 55	07 04 36.	- 10 23	14400		OB ASSOCIATION CMA OB1
RNGC 2326A	07 04 40.	+ 50 43		15.5	GALAXY
ZC 0704.7+3510	07 04 42.	+ 35 10	4570		CLUSTER OF GALAXIES
UGC 03686	07 04 42.	+ 46 44	60	16.0	GALAXY Sc
ZWG 234.065	07 04 42.	+ 48 12		15.6	GALAXY
ZWG 234.066	07 04 42.	+ 50 42		15.5	GALAXY
UGC 03687	07 04 42.	+ 50 42	66	15.5	GALAXY PECULR
MCG+08-13-068	07 04 45.	+ 46 44	42	16.	GALAXY
MCG+08-13-069	07 04 45.	+ 48 12 30.	36	16.	GALAXY
MCG+09-12-038	07 04 45.	+ 54 52	45	17.	GALAXY
PK219+01.1	07 04 46.	- 05 06	60		PLANETARY NEBULA
ZWG 086.002	07 04 48.	+ 18 50		15.3	GALAXY
ZWG 085.037	07 04 48.	+ 18 50		15.3	GALAXY
TON-N 0272	07 04 48.	+ 26 57		14.7	BLUE STAR
TON-N 0273	07 04 48.	+ 27 36		14.6	BLUE STAR
ZWG 146.027	07 04 48.	+ 31 45		14.7	GALAXY
KARA.72 123B	07 04 48.	+ 31 45	36		PART OF DOUBLE GALAXY
KARA.72 123A	07 04 48.	+ 31 45	42	14.7	PART OF DOUBLE GALAXY
MCG+05-17-012	07 04 48.	+ 31 45	24	14.7	GALAXY
ZC 0704.8+5730	07 04 48.	+ 57 30	4970		CLUSTER OF GALAXIES
OCL 0564	07 04 48.	- 10 32	540	10.1	OPEN STAR CLUSTER
PK234-06.1	07 04 50.	- 21 58	76		PLANETARY NEBULA
KEEN 2326A	07 04 54.	+ 50 44	72	14.	E3 GALAXY
MCG+09-12-039	07 04 54.	+ 56 09	27	18.	GALAXY
OCL 0566	07 04 54.	- 10 44	240	11.1	OPEN STAR CLUSTER
ST 13	07 04 58.	- 10 59	3600		EMISSION NEBULA
HMS 0705+3506	07 05	+ 35 06			GEMINI GALAXY CLUSTER
VDB.66G 041	07 05	+ 53 32	70		DWARF GALAXY
ZWG 057.001	07 05 00.	+ 13 10		15.6	GALAXY
UGC 03688	07 05 00.	+ 13 10	66	15.6	GALAXY Sb
ZWG 086.003	07 05 00.	+ 20 35		15.7	GALAXY
ZWG 085.038	07 05 00.	+ 20 35		15.7	GALAXY
ZWG 116.024	07 05 00.	+ 22 17		15.7	GALAXY
MCG+06-16-021	07 05 00.	+ 35 04			GALAXY
ZWG 176.018	07 05 00.	+ 35 15		14.1	GALAXY
UGC 03689	07 05 00.	+ 35 15	78	14.1	GALAXY Sa
MCG+06-16-020	07 05 00.	+ 35 15	48	14.	GALAXY
ZWG 234.067	07 05 00.	+ 49 04		15.1	GALAXY
MCG+08-13-070	07 05 00.	+ 49 05	30	16.	GALAXY
UGC 03690	07 05 00.	+ 53 33	90	17.	GALAXY DWRF IR
ZC 0705.0+6630	07 05 00.	+ 66 30	1010		CLUSTER OF GALAXIES
RNGC 2333	07 05 03.	+ 35 15		14.0	GALAXY
ZWG 086.004	07 05 06.	+ 15 15		13.4	GALAXY
ZWG 085.039	07 05 06.	+ 15 15		13.4	GALAXY
UGC 03691	07 05 06.	+ 15 15	144	13.4	GALAXY Sc
ZWG 234.068	07 05 06.	+ 48 31		15.1	GALAXY
MCG+08-13-071	07 05 06.	+ 48 31	36	16.	GALAXY
ZWG 309.022	07 05 06.	+ 63 08		15.7	GALAXY
ZWG 057.002	07 05 12.	+ 13 14		15.4	GALAXY
MCG+03-19-001	07 05 12.	+ 15 16	120	13.4	GALAXY
MCG+09-12-040	07 05 12.	+ 53 32	84	17.	GALAXY
7ZW 121	07 05 12.	+ 59 37			COMPACT GALAXY
OCL 0607	07 05 12.	- 20 43	174	11.	OPEN STAR CLUSTER
OCL 0647	07 05 12.	- 28 08	300	14.	OPEN STAR CLUSTER
MCG+08-13-072	07 05 15.	+ 48 40	30	15.	GALAXY
ZWG 146.028	07 05 18.	+ 29 58		15.7	GALAXY
UGC 03692	07 05 18.	+ 29 58	96	15.7	GALAXY S0?
ZWG 234.069	07 05 18.	+ 48 39		15.6	GALAXY
MCG+08-13-073	07 05 18.	+ 48 42	30	13.	GALAXY
ZWG 086.005	07 05 24.	+ 18 51		12.3	GALAXY
ZWG 085.040	07 05 24.	+ 18 51		12.3	GALAXY
MCG+05-17-013	07 05 24.	+ 32 46	84	15.3	GALAXY
ZWG 146.029	07 05 24.	+ 32 47		15.3	GALAXY
UGC 03694	07 05 24.	+ 32 47	96	15.3	GALAXY Sc
ZC 0705.4+3642	07 05 24.	+ 36 42	8530		CLUSTER OF GALAXIES
ZWG 234.070	07 05 24.	+ 48 42		13.7	GALAXY
UGC 03695	07 05 24.	+ 48 42	102	13.7	GALAXY E-S0
ARC 0569	07 05 24.	+ 48 43		13.8	RICH CLUSTER OF GALAXIES
ZWG 234.071	07 05 24.	+ 49 54		15.5	GALAXY
MCG-05-17-007	07 05 24.	- 28 06	36	16.	GALAXY
RNGC 2329	07 05 25.	+ 48 42		13.5	GALAXY
RNGC 2339	07 05 26.	+ 18 52		12.5	GALAXY
MCG+03-19-002	07 05 30.	+ 18 52	168	12.3	GALAXY
ZWG 205.019	07 05 30.	+ 43 46		15.2	GALAXY
MCG+08-13-074	07 05 30.	+ 46 10	24	15.	GALAXY
ZWG 234.072	07 05 30.	+ 46 11		14.7	GALAXY
MCG+08-13-075	07 05 30.	+ 49 54	36	16.	GALAXY
MCG+08-13-076	07 05 30.	+ 49 57	39	16.	GALAXY
MCG+10-11-003	07 05 30.	+ 58 50	36	15.	GALAXY
ZWG 286.004	07 05 30.	+ 58 51		15.7	GALAXY
ZWG 285.017	07 05 30.	+ 58 51		15.7	GALAXY
MCG+12-07-028	07 05 30.	+ 71 55	204	15.	GALAXY
LB 00489	07 05 34.	+ 65 25 42.		17.1	FAINT BLUE STAR
ZWG 205.020	07 05 36.	+ 40 25		15.5	GALAXY
ZWG 234.073	07 05 36.	+ 48 43		13.8	GALAXY
UGC 03696	07 05 36.	+ 48 43	66	13.8	GALAXY
MCG+08-13-077	07 05 36.	+ 48 43	36	15.	GALAXY
ZWG 234.074	07 05 36.	+ 50 13		15.7	GALAXY
MCG+08-13-078	07 05 36.	+ 50 14	15	16.	GALAXY
7ZW 122	07 05 36.	+ 64 12			COMPACT GALAXY
ZWG 330.026	07 05 36.	+ 71 55		13.1	GALAXY
UGC 03697	07 05 36.	+ 71 55	198	13.1	GALAXY INTEGRL
RLWT 131	07 05 36.	- 02 48		13.	FAINT VERY BLUE STAR
BIGO 490	07 05 37.	+ 50 13			NEBULA
IC 0457	07 05 38.	+ 50 13 57.			NONSTELLAR OBJECT
RNGC 2332	07 05 41.	+ 50 15		14.0	GALAXY
RNGC 2330	07 05 41.	+ 50 16		14.0	GALAXY
UGC 03693	07 05 42.	+ 18 51	162	12.3	GALAXY Sb/SBc
UGC 03698	07 05 42.	+ 44 28	66	17.	GALAXY DWARF
ZWG 234.075	07 05 42.	+ 50 15		14.0	GALAXY
UGC 03699	07 05 42.	+ 50 15	102	14.0	GALAXY S0
MCG+08-13-079	07 05 42.	+ 50 16	30	14.	GALAXY
SCHO 0231	07 05 46.	- 04 12 24.	660		ISOLATED DARK CLOUD
ZWG 261.029	07 05 48.	+ 55 03		15.4	GALAXY
UGC 03700	07 05 48.	+ 55 03	60	15.4	GALAXY S
ZC 0705.8+6434	07 05 48.	+ 64 34	3900		CLUSTER OF GALAXIES
ZWG 330.027	07 05 48.	+ 72 15		15.2	GALAXY
UGC 03701	07 05 48.	+ 72 15	120	15.2	GALAXY Sc
OCL 0631	07 05 48.	- 25 47	252	13.	OPEN STAR CLUSTER
ZWG 205.021	07 05 54.	+ 39 10		14.8	GALAXY
ZWG 234.076	07 05 54.	+ 46 10		15.3	GALAXY
ZWG 234.077	07 05 54.	+ 48 00		15.2	GALAXY
ZWG 234.078	07 05 54.	+ 48 46		15.2	GALAXY
MCG+12-07-029	07 05 54.	+ 72 14	24	15.	GALAXY
ZWG 330.028	07 05 54.	+ 73 09		15.0	GALAXY
OCL 0565	07 05 54.	- 10 34	720	8.0	OPEN STAR CLUSTER

OBJECT NAME	RIGHT ASCEN.	DECLINATION	DIAM.	MAGN.	TYPE OF OBJECT
RNGC 2343	07 05 55.	- 10 35		8.0	OPEN CLUSTER
LBN 1013	07 06	- 04 10	60		BRIGHT NEBULA
LBN 1038	07 06	- 11 45	900		BRIGHT NEBULA
ISS 0507	07 06 00.	+ 00 27	223		STELLAR RING
ZWG 116.025	07 06 00.	+ 26 10		15.5	GALAXY
ZWG 146.030	07 06 00.	+ 28 47		15.7	GALAXY
UGC 03702	07 06 00.	+ 28 47	66	15.7	GALAXY Sc
ZWG 176.019	07 06 00.	+ 36 22		15.4	GALAXY
UGC 03703	07 06 00.	+ 36 22	72	15.4	GALAXY Sc
MCG+06-16-022	07 06 00.	+ 36 22	72	14.5	GALAXY
RNGC 2337	07 06 00.	+ 44 32		13.0	GALAXY
ZC 0706.0+4600	07 06 00.	+ 46 00	2080		CLUSTER OF GALAXIES
MCG+08-13-080	07 06 00.	+ 46 10	48	16.	GALAXY
MCG+08-13-081	07 06 00.	+ 48 00	30	16.	GALAXY
MCG+08-13-082	07 06 00.	+ 48 47	36	16.	GALAXY
MCG+09-12-041	07 06 00.	+ 54 00	45	17.	GALAXY
MCG+09-12-042	07 06 00.	+ 55 02 30.	66	16.	GALAXY
ZWG 286.005	07 06 00.	+ 61 52		14.5	GALAXY
ZWG 285.018	07 06 00.	+ 61 52		14.5	GALAXY
UGC 03704	07 06 00.	+ 61 52	84	14.5	GALAXY Sa
MCG+10-11-004	07 06 00.	+ 61 52	66	15.	GALAXY
MCG+12-07-030	07 06 00.	+ 73 08	33	16.	GALAXY
ZWG 330.029	07 06 00.	+ 73 32		14.4	GALAXY
UGC 03705	07 06 00.	+ 73 32	90	14.4	GALAXY S
MCG+12-07-031	07 06 00.	+ 73 32	51	15.	GALAXY
MCG+14-04-027	07 06 00.	+ 86 17	57	16.	GALAXY
OCL 0575	07 06 00.	- 13 05	1020	8.3	OPEN STAR CLUSTER
RNGC 2345	07 06 03.	- 13 05		8.0	OPEN CLUSTER
RNGC 2334	07 06 05.	+ 50 16			NON-EXISTENT OBJECT
MIN.47 07	07 06 05.	- 04 14			DIFFUSE NEBULA
ZWG 234.079	07 06 06.	+ 47 59		14.7	GALAXY
UGC 03706	07 06 06.	+ 47 59	66	14.7	GALAXY DBL SYS
KARA.72 124B	07 06 06.	+ 47 59	30		PART OF DOUBLE GALAXY
KARA.72 124A	07 06 06.	+ 48 00	30	14.7	PART OF DOUBLE GALAXY
MCG+08-13-084	07 06 06.	+ 48 00	24	16.	GALAXY
MCG+08-13-083	07 06 06.	+ 48 00	24	16.	GALAXY
7ZW 123	07 06 06.	+ 64 18			COMPACT GALAXY
MRSL 218+01/1	07 06 06.	- 04 14	60		HII REGION
MCG+07-15-008	07 06 10.	+ 39 11	36	15.	GALAXY
IC 0466	07 06 10.	- 04 14 29.			NONSTELLAR OBJECT
CED 092	07 06 11.	- 04 15			DIFFUSE GALACTIC NEBULA
ZWG 176.020	07 06 12.	+ 33 04		15.7	GALAXY
DG 117	07 06 12.	- 04 14	120		REFLECTION NEBULA
HOLM 086B	07 06 15.	+ 20 41	30	14.6	PART OF MULTIPLE GALAXY
ZWG 029.001	07 06 18.	+ 07 01		15.6	GALAXY
UGC 03707	07 06 18.	+ 07 01	60	15.6	GALAXY Sa-b
ZWG 086.006	07 06 18.	+ 20 40		13.7	GALAXY
UGC 03708	07 06 18.	+ 20 40	54	13.7	GALAXY PECULR
KARA.72 125A	07 06 18.	+ 20 40	66	13.7	PART OF DOUBLE GALAXY
MCG+03-19-003	07 06 18.	+ 20 43	48	13.7	GALAXY
ZWG 176.021	07 06 18.	+ 36 27		15.5	GALAXY
ZWG 205.022	07 06 18.	+ 42 33		15.7	GALAXY
RNGC 2341	07 06 19.	+ 20 40		13.5	GALAXY
MCG+07-15-009	07 06 21.	+ 42 33	36	16.	GALAXY
HOLM 086A	07 06 22.	+ 20 43	78	14.0	PART OF MULTIPLE GALAXY
ZWG 086.007	07 06 24.	+ 20 43		12.6	GALAXY
UGC 03709	07 06 24.	+ 20 43	84	12.6	GALAXY S
KARA.72 125B	07 06 24.	+ 20 43	78	12.6	PART OF DOUBLE GALAXY
ZWG 146.031	07 06 24.	+ 28 45		15.6	GALAXY
UGC 03710	07 06 24.	+ 28 45	84	15.6	GALAXY
ZWG 176.022	07 06 24.	+ 33 27		15.4	GALAXY
7ZW 124	07 06 24.	+ 64 09			COMPACT GALAXY
RNGC 2342	07 06 25.	+ 20 43		12.5	GALAXY
MCG+03-19-004	07 06 30.	+ 20 45	78	12.6	GALAXY
ZWG 176.023	07 06 30.	+ 33 07		15.6	GALAXY
ZC 0706.5+5830	07 06 30.	+ 58 30	2020		CLUSTER OF GALAXIES
LB 00490	07 06 35.	+ 81 58 18.		15.8	FAINT BLUE STAR
ZC 0706.6+3221	07 06 36.	+ 32 21	7530		CLUSTER OF GALAXIES
ZWG 205.023	07 06 36.	+ 44 32		13.1	GALAXY
UGC 03711	07 06 36.	+ 44 32	162	13.1	GALAXY IRR
MCG+07-15-010	07 06 36.	+ 44 32	120	14.	GALAXY
UGC 03712	07 06 36.	+ 63 17	66	16.0	GALAXY S
7ZW 125	07 06 36.	+ 63 21			COMPACT GALAXY
ZWG 234.080	07 06 42.	+ 49 56		15.3	GALAXY
ZWG 234.081	07 06 42.	+ 50 11		14.4	GALAXY
UGC 03713	07 06 42.	+ 50 11	60	14.4	GALAXY E-S0
MCG+08-13-085	07 06 42.	+ 50 12	24	15.	GALAXY
ZWG 261.030	07 06 42.	+ 54 35		15.6	GALAXY
BIGO 491	07 06 44.	+ 50 12			NEBULA
MCG+08-13-086	07 06 45.	+ 49 58	36	16.	GALAXY
IC 0458	07 06 45.	+ 50 12 06.			NONSTELLAR OBJECT
IC 0459	07 06 47.	+ 50 15 36.			NONSTELLAR OBJECT
TON-N 0274	07 06 48.	+ 29 30		14.5	BLUE STAR
MCG+08-13-087	07 06 48.	+ 48 43	36	16.	GALAXY
ZWG 234.082	07 06 48.	+ 50 14		15.5	GALAXY
MCG+11-09-035	07 06 48.	+ 63 17 30.	39	16.	GALAXY
ZWG 330.030	07 06 48.	+ 71 50		12.7	GALAXY
UGC 03714	07 06 48.	+ 71 50	120	12.7	GALAXY S?
BIGO 492	07 06 49.	+ 50 15			NEBULA
PK215+03.1	07 06 49.68	- 00 43 29.4	143		PLANETARY NEBULA
RNGC 2346	07 06 50.	- 00 44			PLANETARY NEBULA
SCHO 0232	07 06 51.	- 14 01 12.	510		ISOLATED DARK CLOUD
SCHO 0233	07 06 53.	- 17 05 42.	840		ISOLATED DARK CLOUD
TON-N 0275	07 06 54.	+ 27 33		15.1	BLUE STAR
ZWG 234.083	07 06 54.	+ 50 09		15.7	GALAXY
MCG+08-13-088	07 06 54.	+ 50 10	36	16.	GALAXY
ZWG 234.084	07 06 54.	+ 50 16		15.4	GALAXY
IC 0460	07 06 54.	+ 50 17 05.			NONSTELLAR OBJECT
MCG+08-13-089	07 06 54.	+ 50 18	24	16.	GALAXY
ZC 0706.9+5819	07 06 54.	+ 58 19	400		CLUSTER OF GALAXIES
MCG+12-07-032	07 06 54.	+ 71 50	39	14.	GALAXY
BIGO 493	07 06 55.	+ 50 10			NEBULA
IC 0461	07 06 57.	+ 50 09 53.			NONSTELLAR OBJECT
CED 089E	07 06 58.	- 10 16			DIFFUSE GALACTIC NEBULA
KHAV 190	07 07	- 04 41	3250		DARK NEBULA
KHAV 191	07 07	- 06 29	7270		DARK NEBULA
LBN 1043	07 07	- 18 20	900		BRIGHT NEBULA
ZWG 146.032	07 07 00.	+ 28 23		15.3	GALAXY
TON-N 0276	07 07 00.	+ 28 43		15.1	BLUE STAR
TON-N 0277	07 07 00.	+ 29 01		14.6	BLUE STAR
MCG+06-16-023	07 07 00.	+ 33 44	48	15.	GALAXY
ZWG 176.024	07 07 00.	+ 33 45		15.6	GALAXY
UGC 03716	07 07 00.	+ 39 49	66	16.0	GALAXY Sc
ZWG 234.085	07 07 00.	+ 48 16		15.5	GALAXY
ZWG 234.086	07 07 00.	+ 49 59		15.4	GALAXY
MCG+08-13-090	07 07 00.	+ 50 00	15	16.	GALAXY
ZWG 330.031	07 07 00.	+ 73 55		13.9	GALAXY
UGC 03717	07 07 00.	+ 73 55	138	13.9	GALAXY Sb-c

OBJECT NAME	RIGHT ASCEN.	DECLINATION	DIAM.	MAGN.	TYPE OF OBJECT
MCG+12-07-033	07 07 00.	+ 73 55	114	14.	GALAXY
ZWG 370.005	07 07 00.	+ 86 19		15.5	GALAXY
ZWG 363.026	07 07 00.	+ 86 19		15.5	GALAXY
ZWG 362.041	07 07 00.	+ 86 19		15.5	GALAXY
UGC 03715	07 07 00.	+ 86 19	138	15.5	GALAXY
ST 12	07 07 00.	- 10 04			EMISSION NEBULA
ZWG 205.024	07 07 06.	+ 39 48		15.2	GALAXY
UGC 03718	07 07 06.	+ 39 48	60	15.2	GALAXY S0
ZWG 234.087	07 07 06.	+ 50 13		14.8	GALAXY
ZWG 261.031	07 07 06.	+ 56 13		15.6	GALAXY
ZC 0707.1+6535	07 07 06.	+ 65 35	1750		CLUSTER OF GALAXIES
RLWT 132	07 07 06.	- 02 48		15.	FAINT VERY BLUE STAR
MCG-05-17-008	07 07 06.	- 28 22	30	15.	GALAXY
IC 0462	07 07 07.	+ 50 15 57.			NONSTELLAR OBJECT
IC 0463	07 07 10.	+ 50 11 57.			NONSTELLAR OBJECT
FATH 1.184	07 07 10.	+ 59 47	54		NEBULA
RNGC 2340	07 07 11.	+ 50 15		14.0	GALAXY
ZWG 234.088	07 07 12.	+ 48 35		15.1	GALAXY
UGC 03719	07 07 12.	+ 48 35	78	15.1	GALAXY Sa-b
ZWG 234.089	07 07 12.	+ 48 45		15.3	GALAXY
ZWG 234.090	07 07 12.	+ 49 05		15.2	GALAXY
MCG+08-13-091	07 07 12.	+ 49 07	60	15.	GALAXY
MCG+08-13-092	07 07 12.	+ 50 14	36	15.	GALAXY
ZWG 234.091	07 07 12.	+ 50 15		13.9	GALAXY
UGC 03720	07 07 12.	+ 50 15	138	13.9	GALAXY E
UGC 03721	07 07 12.	+ 59 48	60	16.5	GALAXY Sc
MCG+10-11-005	07 07 12.	+ 59 48	57	17.	GALAXY
7ZW 126	07 07 12.	+ 67 09			COMPACT GALAXY
OCL 0590	07 07 12.	- 16 52	660	13.	OPEN STAR CLUSTER
CED 093	07 07 12.	- 18 24	480		DIFFUSE GALACTIC NEBULA
MCG+07-15-011	07 07 15.	+ 39 49	63	17.	GALAXY
MCG+08-13-093	07 07 15.	+ 48 16	48	15.	GALAXY
BIGO 494	07 07 15.	+ 50 13			NEBULA
IC 0464	07 07 15.	+ 50 13 14.			NONSTELLAR OBJECT
RLWT 133	07 07 18.	+ 02 49		15.	FAINT VERY BLUE STAR
MCG+07-15-012	07 07 18.	+ 39 47 30.	60	16.	GALAXY
MCG+08-13-094	07 07 18.	+ 48 35	48	15.	GALAXY
ZWG 234.092	07 07 18.	+ 49 58		15.7	GALAXY
MCG+08-13-095	07 07 18.	+ 50 00	42	16.	GALAXY
MCG+08-13-096	07 07 18.	+ 50 16	36	13.	GALAXY
7ZW 127	07 07 18.	+ 67 17			COMPACT GALAXY
ZWG 330.032	07 07 18.	+ 71 45		15.7	GALAXY
ZWG 057.003	07 07 24.	+ 13 47		15.7	GALAXY
MCG+06-16-024	07 07 24.	+ 34 30	90	14.	GALAXY
ZWG 261.032	07 07 24.	+ 51 16		15.1	GALAXY
UGC 03722	07 07 24.	+ 54 28	60	17.	GALAXY
ZC 0707.4+5449	07 07 24.	+ 54 49	340		CLUSTER OF GALAXIES
ZWG 176.025	07 07 30.	+ 34 30		14.4	GALAXY
UGC 03723	07 07 30.	+ 34 30	96	14.4	GALAXY SB0
MCG+12-07-034	07 07 30.	+ 73 39	39	16.	GALAXY
RLWT 134	07 07 30.	- 02 50		15.	FAINT VERY BLUE STAR
MCG+08-13-097	07 07 36.	+ 48 57	18	16.	GALAXY
RNGC 2349	07 07 36.	- 08 33			NON-EXISTENT OBJECT
MRSL 231-04/1	07 07 36.	- 18 24	540		HII REGION
ZWG 234.093	07 07 42.	+ 48 19		14.5	GALAXY
UGC 03724	07 07 42.	+ 48 19	102	14.5	GALAXY SBb
ZWG 234.094	07 07 42.	+ 49 10		15.4	GALAXY
ZWG 234.095	07 07 42.	+ 50 19		14.6	GALAXY
MCG+08-13-098	07 07 42.	+ 50 20	48	15.	GALAXY
BIGO 495	07 07 43.	+ 50 19			NEBULA
IC 0465	07 07 43.	+ 50 19 59.			NONSTELLAR OBJECT
MCG+08-13-099	07 07 45.	+ 49 10	42	16.	GALAXY
ZWG 205.025	07 07 48.	+ 39 47		15.6	GALAXY
MCG+08-13-100	07 07 48.	+ 48 20	72	14.	GALAXY
MCG-05-17-009	07 07 48.	- 27 26	60	15.	GALAXY
MCG-05-17-010	07 07 48.	- 27 27	60	15.	GALAXY
ZWG 234.096	07 07 54.	+ 47 01		15.3	GALAXY
KARA.73B 0179	07 07 54.	+ 47 01	24	15.3	ISOLATED GALAXY E
ZWG 234.097	07 07 54.	+ 49 56		14.2	GALAXY
UGC 03725	07 07 54.	+ 49 56	102	14.2	GALAXY E-S0
MCG+08-13-101	07 07 54.	+ 49 57	24	15.	GALAXY
ST 11	07 07 54.	- 18 24	480		EMISSION NEBULA
LBN 1024	07 08	- 07 50	3300		BRIGHT NEBULA
LBN 1034	07 08	- 10 30	5400		BRIGHT NEBULA
LBN 1044	07 08	- 18 23	540		BRIGHT NEBULA
LBN 1055	07 08	- 23 00	5100		BRIGHT NEBULA
ZWG 029.002	07 08 00.	+ 06 32		15.0	GALAXY
MCG+01-19-001	07 08 00.	+ 06 32	36	15.0	GALAXY
ZC 0708.0+2107	07 08 00.	+ 21 07	1480		CLUSTER OF GALAXIES
ZWG 116.026	07 08 00.	+ 26 00		14.8	GALAXY
UGC 03726	07 08 00.	+ 26 00	66	14.8	GALAXY PECULR?
MCG+04-17-008	07 08 00.	+ 26 00	60	14.8	GALAXY
ZWG 146.033	07 08 00.	+ 31 16		15.2	GALAXY
MCG+11-09-036	07 08 00.	+ 65 03	27	16.	GALAXY
ZWG 029.003	07 08 06.	+ 07 58		15.3	GALAXY
ZWG 146.034	07 08 06.	+ 31 14		15.7	GALAXY
MCG+05-17-014	07 08 06.	+ 31 15	48	15.2	GALAXY
ZWG 205.026	07 08 06.	+ 39 45		15.2	GALAXY
MCG+08-13-102	07 08 06.	+ 50 41	36	16.	GALAXY
OCL 0541	07 08 12.	+ 02 49	480	16.	OPEN STAR CLUSTER
ZWG 116.027	07 08 12.	+ 26 28		15.0	GALAXY
UGC 03727	07 08 12.	+ 26 28	72	15.0	GALAXY Sa-b
ZWG 146.035	07 08 12.	+ 30 15		15.3	GALAXY
UGC 03728	07 08 12.	+ 30 15	66	15.3	GALAXY SBb
ZWG 146.036	07 08 12.	+ 31 17		15.4	GALAXY
ZWG 205.027	07 08 12.	+ 39 39		15.5	GALAXY
UGC 03729	07 08 12.	+ 39 39	60	15.5	GALAXY Sc
ZWG 234.098	07 08 12.	+ 50 40		14.9	GALAXY
ZC 0708.2+6958	07 08 12.	+ 69 58	2420		CLUSTER OF GALAXIES
MCG+12-07-035	07 08 12.	+ 73 32	132	15.	GALAXY
ZWG 330.033	07 08 12.	+ 73 33		13.7	GALAXY
UGC 03730	07 08 12.	+ 73 33	162	13.7	GALAXY
IC 2180	07 08 14.	+ 26 27 26.			NONSTELLAR OBJECT
MCG+05-17-016	07 08 18.	+ 29 14	66	14.8	GALAXY
MCG+05-17-015	07 08 18.	+ 31 17	30	15.4	GALAXY
ZC 0708.3+4544	07 08 18.	+ 45 44	810		CLUSTER OF GALAXIES
VV 123C	07 08 18.	+ 73 34	24	17.	INTERACTING GALAXY
VV 123B	07 08 18.	+ 73 34	30	15.5	INTERACTING GALAXY
VV 123A	07 08 18.	+ 73 34	30	15.	INTERACTING GALAXY
VV 123	07 08 18.	+ 73 34	120	15.	INTERACTING GALAXY
ARP 141	07 08 20.	+ 73 34			PECULIAR GALAXY
ZC 0708.4+3849	07 08 24.	+ 38 49	810		CLUSTER OF GALAXIES
MCG+07-15-013	07 08 24.	+ 39 38	60	15.	GALAXY
ZWG 205.028	07 08 24.	+ 43 44		15.5	GALAXY
ZWG 205.029	07 08 24.	+ 44 43		15.3	GALAXY
ZWG 234.099	07 08 24.	+ 46 18		15.7	GALAXY
RLWT 135	07 08 24.	- 00 59		15.	FAINT VERY BLUE STAR
ZWG 146.037	07 08 30.	+ 29 16		14.8	GALAXY

OBJECT NAME	RIGHT ASCEN.	DECLINATION	DIAM.	MAGN.	TYPE OF OBJECT
UGC 03731	07 08 30.	+ 29 16	84	14.8	GALAXY S-IRR
MCG+10-11-006	07 08 30.	+ 62 23	27	16.	GALAXY
7ZW 128	07 08 30.	+ 65 23			COMPACT GALAXY
ZWG 205.030	07 08 36.	+ 41 51		15.2	GALAXY
UGC 03732	07 08 36.	+ 41 51	78	15.2	GALAXY Sa-b
ZWG 205.031	07 08 36.	+ 44 49		15.6	GALAXY
MCG+07-15-014	07 08 36.	+ 44 49	36	17.	GALAXY
UGC 03733	07 08 36.	+ 51 55	66	16.0	GALAXY Sc
7ZW 129	07 08 36.	+ 67 16			COMPACT GALAXY
ZC 0708.7+1936	07 08 42.	+ 19 36	5040		CLUSTER OF GALAXIES
MCG+06-16-025	07 08 42.	+ 33 09 30.	60	15.	GALAXY
ZWG 234.100	07 08 42.	+ 47 15		13.2	GALAXY
UGC 03734	07 08 42.	+ 47 15	132	13.2	GALAXY Sb
KARA.73B 0180	07 08 42.	+ 47 15	102	13.2	ISOLATED GALAXY S
7ZW 130	07 08 42.	+ 67 08			COMPACT GALAXY
ZWG 309.023	07 08 42.	+ 67 08		15.6	GALAXY
RNGC 2268	07 08 47.	+ 84 28		12.5	GALAXY
UGC 03735	07 08 48.	+ 33 10	60	16.0	GALAXY Sc
MCG+08-13-103	07 08 48.	+ 47 15	90	13.	GALAXY
ZWG 261.033	07 08 48.	+ 56 02		15.1	GALAXY
RNGC 2344	07 08 51.	+ 47 15		13.0	GALAXY
ZWG 234.101	07 08 54.	+ 50 08		15.6	GALAXY
LBN 0891	07 09	+ 16 50	7200		BRIGHT NEBULA
KHAV 192	07 09	- 10 47	12280		DARK NEBULA
KHAV 193	07 09	- 31 17	2110		DARK NEBULA
ZWG 029.004	07 09 00.	+ 07 26		15.5	GALAXY
MCG+08-13-104	07 09 00.	+ 49 02	48	16.	GALAXY
MCG+08-13-105	07 09 00.	+ 49 06	60	15.	GALAXY
MRK 085	07 09 00.	+ 55 50	25	14.	GALAXY WITH UV CONTINUUM
SCHO 0234	07 09 04.	- 19 31 00.	610		ISOLATED DARK CLOUD
ZC 0709.1+3714	07 09 06.	+ 37 14	1140		CLUSTER OF GALAXIES
ZWG 234.102	07 09 06.	+ 49 05		15.0	GALAXY
ZWG 234.103	07 09 06.	+ 49 51		15.2	GALAXY
MCG+08-13-106	07 09 06.	+ 49 51	36	15.	GALAXY
PK232-04.1	07 09 06.	- 19 46 00.			PLANETARY NEBULA
ZWG 146.038	07 09 12.	+ 28 40		15.5	GALAXY
KARA.72 126A	07 09 12.	+ 28 40	42	15.5	PART OF DOUBLE GALAXY
ZWG 234.104	07 09 12.	+ 45 53		15.7	GALAXY
MCG+04-17-009	07 09 15.	+ 26 10	48	15.3	GALAXY
ZWG 116.028	07 09 18.	+ 23 28		15.7	GALAXY
UGC 03736	07 09 18.	+ 23 28	78	15.7	GALAXY S
ZWG 116.029	07 09 18.	+ 26 10		15.3	GALAXY
ZWG 146.039	07 09 18.	+ 28 41		15.2	GALAXY
KARA.72 126B	07 09 18.	+ 28 41	42	15.2	PART OF DOUBLE GALAXY
ZC 0709.3+6621	07 09 18.	+ 66 21	940		CLUSTER OF GALAXIES
RLWT 136	07 09 18.	- 01 18		16.	FAINT VERY BLUE STAR
SG 3.100	07 09 22.	- 10 20			DIFFUSE EMISSION NEBULA
ZWG 116.030	07 09 24.	+ 23 49		15.3	GALAXY
UGC 03737	07 09 24.	+ 23 49	66	15.3	GALAXY TRP SYS
ZWG 116.031	07 09 24.	+ 26 49		15.6	GALAXY
ZC 0709.4+3246	07 09 24.	+ 32 46	1340		CLUSTER OF GALAXIES
7ZW 131	07 09 24.	+ 64 04			COMPACT GALAXY
7ZW 132	07 09 24.	+ 64 13			COMPACT GALAXY
ZC 0709.5+0642	07 09 30.	+ 06 42	3360		CLUSTER OF GALAXIES
ZWG 029.005	07 09 30.	+ 07 20		15.2	GALAXY
UGC 03738	07 09 30.	+ 07 20	66	15.2	GALAXY S
ZC 0709.5+5234	07 09 30.	+ 52 34	740		CLUSTER OF GALAXIES
ZC 0709.5+6828	07 09 30.	+ 68 28	810		CLUSTER OF GALAXIES
MRK 086	07 09 36.	+ 46 09	54	12.0	GALAXY WITH UV CONTINUUM
MCG+08-13-107	07 09 39.	+ 50 49	48	16.	GALAXY
ZC 0709.7+6657	07 09 42.	+ 66 57	1010		CLUSTER OF GALAXIES
ZWG 057.004	07 09 48.	+ 12 37		15.3	GALAXY
ZWG 146.040	07 09 48.	+ 27 52		15.1	GALAXY
MCG+05-17-017	07 09 48.	+ 27 52	36	15.1	GALAXY
ZWG 234.105	07 09 48.	+ 50 06		14.9	GALAXY
ZWG 234.106	07 09 48.	+ 50 48		15.1	GALAXY
ZC 0709.8+6033	07 09 48.	+ 60 33	4440		CLUSTER OF GALAXIES
7ZW 133	07 09 48.	+ 65 08			COMPACT GALAXY
ZC 0709.8+7227	07 09 48.	+ 72 27	1210		CLUSTER OF GALAXIES
MCG+13-06-004	07 09 48.	+ 75 49	51	17.	GALAXY
UGC 03739	07 09 48.	+ 75 51	72	16.5	GALAXY DWARF
KHAV 196	07 10	- 16 17	6650		DARK NEBULA
KHAV 195	07 10	- 21 17	6090		DARK NEBULA
KHAV 194	07 10	- 25 35	13060		DARK NEBULA
DV.55 1	07 10	- 78 40	9000		FAINT EMISSION NEBULOSITY
ZWG 029.006	07 10 00.	+ 06 39		15.4	GALAXY
MCG+01-19-002	07 10 00.	+ 06 39	48	15.4	GALAXY
ZWG 234.107	07 10 00.	+ 50 20		15.5	GALAXY
UGC 03741	07 10 00.	+ 50 20	66	15.5	GALAXY Sc
MCG+08-13-108	07 10 00.	+ 50 21	60	15.	GALAXY
ZC 0710.0+5444	07 10 00.	+ 54 44	1810		CLUSTER OF GALAXIES
ZWG 261.034	07 10 00.	+ 55 07		15.3	GALAXY
MCG+09-12-043	07 10 00.	+ 55 08	15	16.	GALAXY
MCG+10-11-007	07 10 00.	+ 61 40	39	17.	GALAXY
MCG+14-04-028	07 10 00.	+ 85 50	132	11.9	GALAXY
ZWG 363.027	07 10 00.	+ 85 51		12.3	GALAXY
ZWG 362.042	07 10 00.	+ 85 51		12.3	GALAXY
UGC 03740	07 10 00.	+ 85 51	186	12.3	GALAXY Sc
KARA.72 127A	07 10 00.	+ 85 51	150	12.3	PART OF DOUBLE GALAXY
MCG+05-17-018	07 10 06.	+ 27 34	78	14.9	GALAXY
ZWG 176.026	07 10 06.	+ 35 12		14.3	GALAXY
UGC 03742	07 10 06.	+ 35 12	90	14.3	GALAXY
ZWG 234.108	07 10 06.	+ 50 28		14.6	GALAXY
MCG+08-13-109	07 10 06.	+ 50 29	30	16.	GALAXY
MCG+09-12-044	07 10 06.	+ 55 07 30.	45	17.	GALAXY
MCG+06-16-026	07 10 09.	+ 35 10 30.	90	14.	GALAXY
ARC 0572	07 10 09.	+ 54 46		17.0	RICH CLUSTER OF GALAXIES
ZWG 029.007	07 10 12.	+ 07 20		15.7	GALAXY
ZWG 116.032	07 10 12.	+ 26 11		15.6	GALAXY
UGC 03743	07 10 12.	+ 35 44	78	16.0	GALAXY Sc
IC 2181	07 10 15.	+ 19 04 45.			NONSTELLAR OBJECT
PK229-02.1	07 10 15.	- 16 03	86		PLANETARY NEBULA
REIN 2.053	07 10 15.05	+ 85 51 03.7			NEBULA
SHB 094	07 10 15.3	+ 11 51 30.		16.6	QUASI-STELLAR OBJECT
BC 3CR175	07 10 15.44	+ 11 51 24.5		16.60	QUASI-STELLAR OBJECT
ZWG 086.008	07 10 18.	+ 19 05		14.8	GALAXY
UGC 03744	07 10 18.	+ 19 05	60	14.8	GALAXY Sa-b
MCG+03-19-005	07 10 18.	+ 19 05 30.	54	14.8	GALAXY
ZWG 146.041	07 10 18.	+ 27 37		14.9	GALAXY
UGC 03745	07 10 18.	+ 27 37	84	14.9	GALAXY SBb
MCG+06-16-027	07 10 18.	+ 35 45	84	15.	GALAXY
ZWG 261.035	07 10 18.	+ 55 14		15.5	GALAXY
UGC 03746	07 10 18.	+ 55 14	72	15.5	GALAXY Sb
ZWG 057.005	07 10 24.	+ 12 21		14.1	GALAXY
RNGC 2350	07 10 24.	+ 12 21		14.0	GALAXY
UGC 03747	07 10 24.	+ 12 21	90	14.1	GALAXY S0-a
MCG+02-19-001	07 10 24.	+ 12 21	72	14.1	GALAXY
ZWG 116.033	07 10 24.	+ 25 11		15.7	GALAXY

OBJECT NAME	RIGHT ASCEN.	DECLINATION	DIAM.	MAGN.	TYPE OF OBJECT
ZWG 234.109	07 10 24.	+ 46 05		15.6	GALAXY
ZWG 261.036	07 10 24.	+ 55 00		15.4	GALAXY
MCG+09-12-045	07 10 24.	+ 55 14	39	16.	GALAXY
ZWG 116.034	07 10 30.	+ 22 53		15.6	GALAXY
MCG+04-17-010	07 10 30.	+ 22 53	30	15.6	GALAXY
ZC 0710.5+4222	07 10 30.	+ 42 22	12500		CLUSTER OF GALAXIES
MRK 376	07 10 30.	+ 45 47	12	16.	GALAXY WITH UV CONTINUUM
KW 37	07 10 30.	+ 45 47	18		SEYFERT GALAXY
VVI 20	07 10 30.	+ 45 47	12	16.	SEYFERT GALAXY
ZWG 261.037	07 10 30.	+ 55 01		15.2	GALAXY
UGC 03748	07 10 30.	+ 65 31	60	16.0	GALAXY DWARF
MCG+11-09-037	07 10 30.	+ 68 05	114	15.	GALAXY
RNGC 2276	07 10 30.	+ 85 51		12.5	GALAXY
SN 1968W	07 10 30.	+ 85 51		16.6	SUPERNOVA
SN 1962Q	07 10 30.	+ 85 51		16.9	SUPERNOVA
SN 1968V	07 10 30.	+ 85 52		15.7	SUPERNOVA
MCG+08-13-110	07 10 36.	+ 45 32	18	18.	GALAXY
ZWG 261.038	07 10 36.	+ 55 04		15.6	GALAXY
MCG+11-09-038	07 10 36.	+ 65 32	78	16.	GALAXY
ZWG 309.024	07 10 36.	+ 68 04		15.2	GALAXY
UGC 03749	07 10 36.	+ 68 04	102	15.2	GALAXY Sc
ZWG 029.008	07 10 42.	+ 06 42		15.5	GALAXY
ZC 0710.7+3855	07 10 42.	+ 38 55	610		CLUSTER OF GALAXIES
ZWG 309.025	07 10 42.	+ 65 02		13.4	GALAXY
UGC 03750	07 10 42.	+ 65 02	60	13.4	GALAXY
KARA.72 128A	07 10 42.	+ 65 02	102	13.4	PART OF DOUBLE GALAXY
SCHO 0235	07 10 42.	- 04 40 24.	640		ISOLATED DARK CLOUD
IC 2179	07 10 43.	+ 65 02			NONSTELLAR OBJECT
HUB E16	07 10 46.	- 12 08			DIFFUSE NEBULA
ZWG 116.035	07 10 48.	+ 23 10		14.8	GALAXY
UGC 03751	07 10 48.	+ 23 10	90	14.8	GALAXY S
ZWG 116.036	07 10 48.	+ 23 36		15.6	GALAXY
ZWG 176.027	07 10 48.	+ 35 22		14.5	GALAXY
UGC 03752	07 10 48.	+ 35 22	27	14.5	GALAXY
MCG+06-16-028	07 10 48.	+ 35 22	30	15.	GALAXY
ZWG 205.032	07 10 48.	+ 42 35		15.4	GALAXY
MCG+04-17-011	07 10 51.	+ 23 09	72	14.8	GALAXY
MCG+04-17-012	07 10 54.	+ 23 19 30.	36	15.1	GALAXY
ZWG 116.037	07 10 54.	+ 23 20		15.1	GALAXY
UGC 03753	07 10 54.	+ 23 20	102	15.1	GALAXY Sa
ZC 0710.9+2959	07 10 54.	+ 29 59	1550		CLUSTER OF GALAXIES
MCG+11-09-038A	07 10 54.	+ 65 01	42	13.1	GALAXY
MRK 377	07 10 54.	+ 74 06	7	16.	GALAXY WITH UV CONTINUUM
MAI 021	07 10 59.	+ 76 04	67		DWARF SPHEROIDAL GALAXY
KHAV 197	07 11	- 04 35	1000		DARK NEBULA
UGC 03754	07 11 00.	+ 12 25	66	16.0	GALAXY Sc/SBc
ZC 0711.0+3311	07 11 00.	+ 33 11	4910		CLUSTER OF GALAXIES
ZWG 234.110	07 11 00.	+ 46 52		15.6	GALAXY
ZWG 261.039	07 11 00.	+ 55 30		15.5	GALAXY
LB 00491	07 11 00.	+ 64 46 48.		17.0	FAINT BLUE STAR
7ZW 134	07 11 00.	+ 85 52			COMPACT GALAXY
MCG+14-04-029	07 11 00.	+ 86 45	66	15.	GALAXY
SCHO 0236	07 11 01.	- 03 36 36.	540		ISOLATED DARK CLOUD
ARP 114	07 11 03.	+ 85 52			PECULIAR GALAXY
ARP 025	07 11 03.	+ 85 52			PECULIAR GALAXY
SHB 095	07 11 05.6	+ 35 39 53.		19.	QUASI-STELLAR OBJECT
ZWG 057.006	07 11 06.	+ 10 35		14.9	GALAXY
UGC 03755	07 11 06.	+ 10 35	108	14.9	GALAXY IRR
MCG+02-19-002	07 11 06.	+ 10 35	90	14.9	GALAXY
ZWG 176.028	07 11 06.	+ 35 29		15.5	GALAXY
UGC 03756	07 11 06.	+ 35 29	72	15.5	GALAXY Sb
RNGC 2351	07 11 08.	- 11 23			NON-EXISTENT OBJECT
ZWG 029.009	07 11 12.	+ 07 50		15.5	GALAXY
ZC 0711.2+3852	07 11 12.	+ 38 52	2890		CLUSTER OF GALAXIES
UGC 03757	07 11 12.	+ 48 49	84	17.	GALAXY Sc
ZWG 234.111	07 11 12.	+ 50 37		14.2	GALAXY
UGC 03758	07 11 12.	+ 50 37	54	14.2	GALAXY COMPACT
ZC 0711.2+5801	07 11 12.	+ 58 01	1210		CLUSTER OF GALAXIES
IC 2182	07 11 15.	+ 19 01 50.			NONSTELLAR OBJECT
MCG+08-13-111	07 11 15.	+ 50 37 30.	36	15.	GALAXY
RNGC 2347	07 11 18.	+ 64 48		13.0	GALAXY
ZWG 309.026	07 11 18.	+ 64 49		13.2	GALAXY
UGC 03759	07 11 18.	+ 64 49	120	13.2	GALAXY Sb
KARA.72 128B	07 11 18.	+ 64 49	120	13.2	PART OF DOUBLE GALAXY
MCG+10-11-008	07 11 24.	+ 57 52	42	16.	GALAXY
MCG+11-09-039	07 11 24.	+ 64 48	114	13.	GALAXY
MCG+08-13-112	07 11 27.	+ 50 39	36	17.	GALAXY
ZWG 234.112	07 11 30.	+ 45 50		15.4	GALAXY
ST 10	07 11 30.	- 10 19	3000		EMISSION NEBULA
MCG+06-16-029	07 11 33.	+ 36 10	48	15.	GALAXY
RNGC 2352	07 11 33.	- 24 00			NON-EXISTENT OBJECT
ZWG 176.029	07 11 36.	+ 33 20		15.2	GALAXY
ZWG 176.030	07 11 36.	+ 35 25		15.7	GALAXY
MCG+06-16-030	07 11 36.	+ 35 25 30.	24	15.	GALAXY
ZWG 176.031	07 11 36.	+ 36 10		15.1	GALAXY
ZWG 261.040	07 11 36.	+ 54 58		15.6	GALAXY
7ZW 136	07 11 36.	+ 60 10			COMPACT GALAXY
MCG+08-13-113	07 11 39.	+ 45 50	24	16.	GALAXY
ZWG 029.010	07 11 42.	+ 05 09		15.3	GALAXY
UGC 03760	07 11 42.	+ 05 09	78	15.3	GALAXY DBL SYS
UGC 03761	07 11 42.	+ 38 13	78	16.0	GALAXY Sc
ZWG 330.034	07 11 42.	+ 73 57		15.7	GALAXY
UGC 03762	07 11 42.	+ 73 57	60	15.7	GALAXY S
UGC 03763	07 11 48.	+ 34 53	60	17.	GALAXY Sc
ZC 0711.8+3641	07 11 48.	+ 36 41	1080		CLUSTER OF GALAXIES
ZWG 309.027	07 11 48.	+ 67 12		14.1	GALAXY
UGC 03764	07 11 48.	+ 67 12	78	14.1	GALAXY SB
KARA.73B 0181	07 11 48.	+ 67 12	78	14.1	ISOLATED GALAXY S
ZWG 029.011	07 11 54.	+ 05 37		15.6	GALAXY
ZWG 176.032	07 11 54.	+ 35 03		15.6	GALAXY
ZWG 235.001	07 11 54.	+ 49 42		15.7	GALAXY
ZWG 234.113	07 11 54.	+ 49 42		15.7	GALAXY
KARA.73B 0182	07 11 54.	+ 49 42	36	15.7	ISOLATED GALAXY S
ZWG 261.041	07 11 54.	+ 56 54		14.3	GALAXY
UGC 03765	07 11 54.	+ 56 54	48	14.3	GALAXY S0
KARA.72 129A	07 11 54.	+ 56 54	66	14.3	PART OF DOUBLE GALAXY
MCG+09-12-046	07 11 54.	+ 56 56	18	15.	GALAXY
MCG+11-09-040	07 11 54.	+ 67 12	78	14.	GALAXY
KHAV 198	07 12	- 19 29	11060		DARK NEBULA
ZWG 029.012	07 12 00.	+ 07 58		15.5	GALAXY
ZWG 086.009	07 12 00.	+ 17 03		15.5	GALAXY
UGC 03766	07 12 00.	+ 17 03	72	15.5	GALAXY Sc
MCG+06-16-031	07 12 00.	+ 36 08	48	15.	GALAXY
ZWG 205.033	07 12 00.	+ 40 35		15.3	GALAXY
MCG+08-14-001	07 12 00.	+ 50 27	18	18.	GALAXY
ZWG 261.042	07 12 00.	+ 56 54		15.7	GALAXY
KARA.72 129B	07 12 00.	+ 56 54	48	15.7	PART OF DOUBLE GALAXY
ZC 0712.0+8415	07 12 00.	+ 84 15	1880		CLUSTER OF GALAXIES
LDN 1658	07 12 00.	- 11 00			DARK NEBULA
ARC 0570	07 12 05.	+ 70 26		18.0	RICH CLUSTER OF GALAXIES
UGC 03767	07 12 06.	+ 06 53	84	16.0	GALAXY
ZC 0712.1+2727	07 12 06.	+ 27 27	3230		CLUSTER OF GALAXIES
ZWG 176.033	07 12 06.	+ 36 07		15.4	GALAXY
MCG+09-12-047	07 12 06.	+ 56 56	27	16.	GALAXY
RNGC 2354	07 12 10.	- 25 38		9.0	OPEN CLUSTER
ZWG 029.013	07 12 12.	+ 06 40		15.3	GALAXY
UGC 03768	07 12 12.	+ 06 40	60	15.3	GALAXY S
ZWG 116.038	07 12 12.	+ 26 10		15.2	GALAXY
OCL 0567	07 12 12.	- 10 13	1800	5.3	OPEN STAR CLUSTER
OCL 0639	07 12 12.	- 25 39	1620	8.9	OPEN STAR CLUSTER
MCG+08-14-002	07 12 15.	+ 47 37	30	16.	GALAXY
ZC 0712.3+6530	07 12 18.	+ 65 30	1410		CLUSTER OF GALAXIES
ARC 0573	07 12 18.	+ 67 38		17.6	RICH CLUSTER OF GALAXIES
RNGC 2353	07 12 19.	- 10 12		5.0	OPEN CLUSTER
ZWG 001.001	07 12 24.	+ 00 51		15.7	GALAXY
UGC 03769	07 12 24.	+ 00 51	60	15.7	GALAXY S
ZWG 116.039	07 12 24.	+ 23 31		15.4	GALAXY
UGC 03770	07 12 24.	+ 23 31	78	15.4	GALAXY IRR
ZC 0712.4+3419	07 12 24.	+ 34 19	1080		CLUSTER OF GALAXIES
ZC 0712.4+4523	07 12 24.	+ 45 23	3430		CLUSTER OF GALAXIES
ZWG 330.035	07 12 24.	+ 70 37		15.0	GALAXY
UGC 03771	07 12 24.	+ 70 37	72	15.0	GALAXY SBb
LB 00492	07 12 25.	+ 62 43 54.		16.6	FAINT BLUE STAR
SCHO 0237	07 12 28.	- 13 49 00.	650		ISOLATED DARK CLOUD
ZWG 086.010	07 12 30.	+ 16 24		15.5	GALAXY
MCG+04-17-013	07 12 30.	+ 23 30 30.	48	15.4	GALAXY
ZWG 116.040	07 12 30.	+ 26 15		15.4	GALAXY
ZWG 235.002	07 12 30.	+ 48 21		15.6	GALAXY
ZWG 234.114	07 12 30.	+ 48 21		15.6	GALAXY
MCG+08-14-003	07 12 30.	+ 48 22	18	16.	GALAXY
MCG+12-07-036	07 12 30.	+ 70 37	57	15.	GALAXY
RLWT 137	07 12 30.	- 00 58		13.	FAINT VERY BLUE STAR
OCL 0671	07 12 30.	- 31 05	5700	3.9	OPEN STAR CLUSTER
LB 00493	07 12 40.	+ 64 19 42.		17.0	FAINT BLUE STAR
ZWG 086.011	07 12 42.	+ 15 13		15.2	GALAXY
UGC 03772	07 12 42.	+ 15 13	84	15.2	GALAXY
ZWG 176.034	07 12 42.	+ 37 30		15.7	GALAXY
MCG+06-16-032	07 12 42.	+ 37 30	48	15.	GALAXY
ZC 0712.7+6740	07 12 42.	+ 67 40	1480		CLUSTER OF GALAXIES
SCHO 0238	07 12 42.	- 14 02 06.	620		ISOLATED DARK CLOUD
PK240-07.1	07 12 43.	- 27 44 12.	10		PLANETARY NEBULA
MCG+08-14-004	07 12 48.	+ 46 11	30	17.	GALAXY
ZWG 309.028	07 12 48.	+ 68 26		14.8	GALAXY
SCHO 0239	07 12 53.	- 19 41 12.	450		ISOLATED DARK CLOUD
ZWG 086.012	07 12 54.	+ 17 33		15.6	GALAXY
UGC 03773	07 12 54.	+ 39 51	78	17.	GALAXY DWARF
ZC 0712.9+5334	07 12 54.	+ 53 34	2620		CLUSTER OF GALAXIES
ZWG 116.041	07 13 00.	+ 25 08		15.6	GALAXY
MCG+06-16-033	07 13 00.	+ 33 00	90	15.	GALAXY
ZWG 176.035	07 13 00.	+ 33 01		15.3	GALAXY
UGC 03774	07 13 00.	+ 33 01	90	15.3	GALAXY Sa-b
ZC 0713.0+3725	07 13 00.	+ 37 25	4500		CLUSTER OF GALAXIES
ZWG 235.003	07 13 00.	+ 49 58		15.7	GALAXY
ZWG 234.115	07 13 00.	+ 49 58		15.7	GALAXY
ZC 0713.0+5440	07 13 00.	+ 54 40	540		CLUSTER OF GALAXIES
ZC 0713.0+5601	07 13 00.	+ 56 01	540		CLUSTER OF GALAXIES
MCG+10-11-009	07 13 00.	+ 59 16	39	15.	GALAXY
ZC 0713.0+6504	07 13 00.	+ 65 04	1810		CLUSTER OF GALAXIES
MCG+14-04-030	07 13 00.	+ 82 44	30	17.	GALAXY
ZWG 370.006	07 13 00.	+ 86 47		15.2	GALAXY
FATH 1.185	07 13 01.	+ 59 41	8		NEBULA
RLWT 138	07 13 06.	+ 00 19		13.	FAINT VERY BLUE STAR
ZWG 029.014	07 13 06.	+ 04 51		15.4	GALAXY
UGC 03775	07 13 06.	+ 12 13	102	17.	GALAXY DWRF SP
ZC 0713.1+4717	07 13 06.	+ 47 17	740		CLUSTER OF GALAXIES
ZC 0713.1+6630	07 13 06.	+ 66 30	1210		CLUSTER OF GALAXIES
OCL 0672	07 13 06.	- 31 16	138	13.	OPEN STAR CLUSTER
ARC 0571	07 13 07.	+ 71 59		17.6	RICH CLUSTER OF GALAXIES
SG 3.101	07 13 10.	- 10 40	3300		DIFFUSE EMISSION NEBULA
LB 00494	07 13 11.	+ 79 01 12.		14.6	FAINT BLUE STAR
ZWG 116.042	07 13 12.	+ 23 05		15.6	GALAXY
MRK 378	07 13 12.	+ 49 47	7	17.	GALAXY WITH UV CONTINUUM
ZWG 309.029	07 13 12.	+ 63 36		14.8	GALAXY
MRK 379	07 13 12.	+ 63 36	30	15.5	GALAXY WITH UV CONTINUUM
ZC 0713.2+6821	07 13 12.	+ 68 21	1280		CLUSTER OF GALAXIES
ZC 0713.2+7208	07 13 12.	+ 72 08	1340		CLUSTER OF GALAXIES
ISS 0822	07 13 12.	- 21 23	134		STELLAR RING
ZWG 146.042	07 13 18.	+ 27 43		15.6	GALAXY
MCG+05-17-019	07 13 18.	+ 27 43	48	15.6	GALAXY
MRK 380	07 13 18.	+ 74 33	6	17.	GALAXY WITH UV CONTINUUM
RLWT 139	07 13 24.	+ 02 15		15.	FAINT VERY BLUE STAR
ZWG 146.043	07 13 24.	+ 29 42		15.6	GALAXY
MCG+05-17-020	07 13 24.	+ 29 56	90	14.5	GALAXY
MCG+06-16-034	07 13 24.	+ 34 04 30.	90	14.	GALAXY
ZWG 176.036	07 13 24.	+ 34 05		14.3	GALAXY
UGC 03776	07 13 24.	+ 34 05	108	14.3	GALAXY S
ZWG 205.034	07 13 24.	+ 42 45		15.6	GALAXY
MCG+11-09-041	07 13 24.	+ 63 34	39	15.	GALAXY
ZCG 0713+66	07 13 24.	+ 66 33		18.0	COMPACT GALAXY
ZC 0713.4+7155	07 13 24.	+ 71 55	670		CLUSTER OF GALAXIES
ISS 0823	07 13 24.	- 24 00	249		STELLAR RING
ACK 258-15.1	07 13 24.	- 46 53			PLANETARY NEBULA
ST 09	07 13 29.	- 10 42	3000		EMISSION NEBULA
ZWG 146.044	07 13 30.	+ 29 57		14.5	GALAXY
UGC 03777	07 13 30.	+ 29 57	120	14.5	GALAXY Sc
ZC 0713.5+5718	07 13 30.	+ 57 18	940		CLUSTER OF GALAXIES
MCG+11-09-042	07 13 30.	+ 65 30	36	17.	GALAXY
MCG-05-18-001	07 13 30.	- 29 16	48	15.	GALAXY
ZWG 116.043	07 13 36.	+ 24 35		15.5	GALAXY
ZWG 286.006	07 13 36.	+ 59 27		15.6	GALAXY
ISS 0671	07 13 36.	- 12 54	98		STELLAR RING
ISS 0717	07 13 36.	- 20 31	771		STELLAR RING
ZWG 235.004	07 13 42.	+ 45 53		15.7	GALAXY
MCG+10-11-010	07 13 42.	+ 59 26	27	16.	GALAXY
7ZW 137	07 13 42.	+ 68 02			COMPACT GALAXY
HN 0311	07 13 46.	- 20 19			NEBULA
ZWG 116.044	07 13 48.	+ 22 42		15.7	GALAXY
UGC 03778	07 13 48.	+ 28 38	66	16.5	GALAXY DWRF SP
MCG+08-14-005	07 13 48.	+ 45 52	12	16.	GALAXY
OCL 0577	07 13 48.	- 13 01	360	17.	OPEN STAR CLUSTER
FATH 1.186	07 13 51.	+ 59 46	8		NEBULA
SG 3.102	07 13 59.	- 09 20	360		DIFFUSE EMISSION NEBULA
LBN 1026	07 14	- 08 40	4500		BRIGHT NEBULA
LBN 1028	07 14	- 09 00	2400		BRIGHT NEBULA
LBN 1032	07 14	- 09 25	420		BRIGHT NEBULA
LBN 1036	07 14	- 10 30	2700		BRIGHT NEBULA

OBJECT NAME	RIGHT ASCEN.	DECLINATION	DIAM.	MAGN.	TYPE OF OBJECT
ZC 0714.0+5515	07 14 00.	+ 55 15	540		CLUSTER OF GALAXIES
MCG+10-11-011	07 14 00.	+ 59 47	42	16.	GALAXY
ZWG 309.030	07 14 00.	+ 63 18		15.5	GALAXY
ZC 0714.0+7101	07 14 00.	+ 71 01	1410		CLUSTER OF GALAXIES
ISS 0824	07 14 00.	- 25 02	176		STELLAR RING
ST 08	07 14 02.	- 09 20	360		EMISSION NEBULA
OCL 0496	07 14 06.	+ 13 52	600	12.2	OPEN STAR CLUSTER
MCG+11-09-043	07 14 06.	+ 66 44 30.	24	17.	GALAXY
7ZW 138	07 14 06.	+ 68 24			COMPACT GALAXY
MRSL 224+01/1	07 14 06.	- 09 20	360		HII REGION
RNGC 2355	07 14 11.	+ 13 52		9.5	OPEN CLUSTER
ZWG 029.015	07 14 12.	+ 07 51		14.9	GALAXY
MCG+01-19-003	07 14 12.	+ 07 51	36	14.9	GALAXY
ZWG 176.037	07 14 12.	+ 34 04		15.5	GALAXY
UGC 03779	07 14 12.	+ 34 04	84	15.5	GALAXY S
MCG+06-16-035	07 14 12.	+ 34 09	60	14.5	GALAXY
ZWG 176.038	07 14 12.	+ 34 10		14.9	GALAXY
UGC 03780	07 14 12.	+ 34 10	66	14.9	GALAXY S
MCG+11-09-044	07 14 12.	+ 63 16	18	16.	GALAXY
MCG+09-12-048	07 14 15.	+ 54 49	42	17.	GALAXY
RNGC 2356	07 14 17.	+ 14 04			NON-EXISTENT OBJECT
MCG+06-16-036	07 14 18.	+ 34 03	66	15.	GALAXY
ZWG 176.039	07 14 18.	+ 36 52		15.6	GALAXY
MCG+10-11-012	07 14 18.	+ 61 21	15	17.	GALAXY
ABC 0574	07 14 23.	+ 71 04		17.4	RICH CLUSTER OF GALAXIES
ZWG 116.045	07 14 24.	+ 25 40		15.6	GALAXY
ZC 0714.4+4015	07 14 24.	+ 40 15	5040		CLUSTER OF GALAXIES
MCG+09-12-049	07 14 24.	+ 52 26	45	15.	GALAXY
ZWG 261.043	07 14 24.	+ 52 27		15.3	GALAXY
ZC 0714.4+5441	07 14 24.	+ 54 41	200		CLUSTER OF GALAXIES
7ZW 13	07 14 24.	+ 59 47			COMPACT GALAXY
ZWG 286.007	07 14 24.	+ 59 47		15.4	GALAXY
PK241-07.1	07 14 24.	- 29 12	6		PLANETARY NEBULA
ZWG 205.035	07 14 30.	+ 41 05		14.9	GALAXY
UGC 03781	07 14 30.	+ 41 05	102	14.9	GALAXY DBL SYS
KARA.72 130A	07 14 30.	+ 41 05	30	14.9	PART OF DOUBLE GALAXY
MCG+10-11-014	07 14 30.	+ 59 46	18	16.	GALAXY
MCG+10-11-013	07 14 30.	+ 59 46	18	16.	GALAXY
RNGC 2357	07 14 35.	+ 23 27		14.5	GALAXY
RNGC 2358	07 14 35.	- 16 57			NON-EXISTENT OBJECT
ZWG 116.046	07 14 36.	+ 23 27		14.6	GALAXY
UGC 03782	07 14 36.	+ 23 27	222	14.6	GALAXY Sc
KARA.72 130B	07 14 36.	+ 41 05	30		PART OF DOUBLE GALAXY
MCG+10-11-015	07 14 36.	+ 58 14	15	17.	GALAXY
MCG+10-11-017	07 14 36.	+ 59 09	27	16.	GALAXY
ZWG 286.008	07 14 36.	+ 59 10		15.4	GALAXY
MCG+10-11-016B	07 14 36.	+ 59 10	8	19.	GALAXY
MCG+10-11-016A	07 14 36.	+ 59 10	8	19.	GALAXY
ZC 0714.6+6245	07 14 36.	+ 62 45	1010		CLUSTER OF GALAXIES
MCG-05-18-002	07 14 36.	- 29 13	66	15.	GALAXY
MCG-06-16-005	07 14 36.	- 35 24	60	15.	GALAXY
MCG+04-17-014	07 14 42.	+ 23 27	210	14.6	GALAXY
ZWG 116.047	07 14 42.	+ 24 36		15.4	GALAXY
ZWG 261.044	07 14 42.	+ 55 35		15.4	GALAXY
ISS 0825	07 14 42.	- 27 21	176		STELLAR RING
SCHO 0240	07 14 45.	- 04 51 00.	440		ISOLATED DARK CLOUD
IC 2183	07 14 46.	- 20 19			NONSTELLAR OBJECT
ZWG 029.016	07 14 48.	+ 03 25		15.7	GALAXY
MCG+04-17-015	07 14 48.	+ 24 36	24	15.4	GALAXY
ZWG 117.001	07 14 48.	+ 25 58		15.7	GALAXY
ZWG 116.048	07 14 48.	+ 25 58		15.7	GALAXY
ZC 0714.8+7140	07 14 48.	+ 71 40	670		CLUSTER OF GALAXIES
ZWG 117.002	07 14 54.	+ 24 50		15.6	GALAXY
ZWG 116.049	07 14 54.	+ 24 50		15.6	GALAXY
UGC 03783	07 14 54.	+ 24 50	84	15.6	GALAXY Sc
ZWG 117.003	07 14 54.	+ 26 44		15.7	GALAXY
ZWG 116.050	07 14 54.	+ 26 44		15.7	GALAXY
UGC 03784	07 14 54.	+ 26 44	66	17.	GALAXY Sc
ZC 0714.9+3149	07 14 54.	+ 31 49	1340		CLUSTER OF GALAXIES
ZC 0714.9+5601	07 14 54.	+ 56 01	940		CLUSTER OF GALAXIES
MCG-06-16-006	07 14 54.	- 35 17	84	14.	GALAXY
MCG+09-12-050	07 14 57.	+ 55 36	30	16.	GALAXY
YC 0714-57	07 14 59.	- 57 15 12.			UNUSUAL SOUTHERN GALAXY
ZWG 029.017	07 15 00.	+ 08 02		14.7	GALAXY
UGC 03785	07 15 00.	+ 08 02	84	14.7	GALAXY Sa
KARA.72 131A	07 15 00.	+ 08 02	66	14.7	PART OF DOUBLE GALAXY
MCG+01-19-004	07 15 00.	+ 08 02	60	14.7	GALAXY
TON-N 0278	07 15 00.	+ 26 33		15.0	BLUE STAR
MCG+04-18-001	07 15 00.	+ 26 37	33	15.5	GALAXY
ZWG 147.001	07 15 00.	+ 30 48		15.6	GALAXY
ZWG 146.045	07 15 00.	+ 30 48		15.6	GALAXY
UGC 03786	07 15 00.	+ 30 48	78	15.6	GALAXY
ZWG 261.045	07 15 00.	+ 52 58		15.4	GALAXY
MCG+09-12-051	07 15 00.	+ 55 34	57	16.	GALAXY
MCG+10-11-018	07 15 00.	+ 59 52	18	17.	GALAXY
MCG+10-11-019	07 15 00.	+ 60 25	39	16.	GALAXY
ZWG 363.028	07 15 00.	+ 82 02		15.6	GALAXY
MCG-06-16-007	07 15 00.	- 38 25	36	15.5	GALAXY
CED 094A	07 15 01.	- 13 04			DIFFUSE GALACTIC NEBULA
IC 0468	07 15 02.	- 13 04			MAY NOT EXIST
KARA.72 131B	07 15 06.	+ 08 02	42		PART OF DOUBLE GALAXY
ZWG 086.013	07 15 06.	+ 18 35		15.7	GALAXY
ZWG 117.004	07 15 06.	+ 26 38		15.5	GALAXY
ZWG 116.051	07 15 06.	+ 26 38		15.5	GALAXY
ZWG 235.005	07 15 06.	+ 49 18		15.5	GALAXY
ZWG 261.046	07 15 06.	+ 56 09		15.7	GALAXY
MCG+10-11-020	07 15 06.	+ 58 15	27	16.	GALAXY
ISS 0826	C7 15 06.	- 23 02	80		STELLAR RING
FATH 1.187	07 15 08.	+ 59 27	54		NEBULA
ZWG 057.007	07 15 12.	+ 09 46		15.6	GALAXY
UGC 03787	07 15 12.	+ 09 46	72	15.6	GALAXY E-SO
ZWG 117.005	07 15 12.	+ 22 05		15.7	GALAXY
MCG+05-18-001	07 15 12.	+ 27 12	66	15.3	GALAXY
ZWG 147.002	07 15 12.	+ 31 39		15.0	GALAXY
ZWG 146.046	07 15 12.	+ 31 39		15.0	GALAXY
UGC 03788	07 15 12.	+ 31 39	84	15.0	GALAXY S
ZWG 147.003	07 15 12.	+ 32 50		15.7	GALAXY
ZWG 146.047	07 15 12.	+ 32 50		15.7	GALAXY
ZC 0715.2+3757	07 15 12.	+ 37 57	1080		CLUSTER OF GALAXIES
MCG+09-12-052	07 15 12.	+ 56 10	18	17.	GALAXY
ZWG 286.009	07 15 12.	+ 57 43		15.5	GALAXY
7ZW 140	07 15 12.	+ 59 28			COMPACT GALAXY
ZWG 286.010	07 15 12.	+ 59 28		13.3	GALAXY
UGC 03789	07 15 12.	+ 59 28	96	13.3	GALAXY Sa
ZWG 349.003	07 15 12.	+ 78 29		15.6	GALAXY
ZWG 348.033	07 15 12.	+ 78 29		15.6	GALAXY
OCL 0700	07 15 12.	- 36 45	3450	3.6	OPEN STAR CLUSTER
CED 094B	07 15 13.	- 13 07	480		DIFFUSE GALACTIC NEBULA
ZWG 147.004	07 15 18.	+ 31 28		15.2	GALAXY
ZWG 146.048	07 15 18.	+ 31 28		15.2	GALAXY
UGC 03790	07 15 18.	+ 31 28	78	15.2	GALAXY DBL SYS
MCG+05-18-002	07 15 18.	+ 31 40 30.	60	15.0	GALAXY
MCG+10-11-021	07 15 18.	+ 59 27	96	14.	GALAXY
ZWG 286.011	07 15 18.	+ 60 24		15.7	GALAXY
RLWT 141	07 15 16.	- 02 32		14.	FAINT VERY BLUE STAR
RLWT 140	07 15 18.	- 02 34		14.	FAINT VERY BLUE STAR
RNGC 2360	07 15 22.	- 15 33		9.0	OPEN CLUSTER
ZWG 117.006	07 15 24.	+ 26 02		15.5	GALAXY
ZWG 147.005	07 15 24.	+ 27 15		15.3	GALAXY
ZWG 146.049	07 15 24.	+ 27 15		15.3	GALAXY
UGC 03791	07 15 24.	+ 27 15	78	15.3	GALAXY Sc-IRR
KARA.73B 0183	07 15 24.	+ 27 15	78	15.3	ISOLATED GALAXY S
MCG+05-18-004	07 15 24.	+ 31 29 30.	36	15.2	GALAXY
MCG+05-18-003	07 15 24.	+ 31 30	42	15.2	GALAXY
ZWG 261.047	07 15 24.	+ 51 24		14.0	GALAXY
UGC 03792	07 15 24.	+ 51 24	114	14.0	GALAXY S0
ZWG 261.048	07 15 24.	+ 55 42		15.7	GALAXY
UGC 03793	07 15 24.	+ 57 15	60	16.5	GALAXY Sc
MCG+10-11-022	07 15 24.	+ 57 30	54	16.	GALAXY
ZC 0715.4+5743	07 15 24.	+ 57 43	1340		CLUSTER OF GALAXIES
MCG+10-11-024	07 15 24.	+ 60 24	18	17.	GALAXY
MCG+10-11-023	07 15 24.	+ 60 24	15	17.	GALAXY
UGC 03794	07 15 24.	+ 77 55	108	16.5	GALAXY DWRF SP
ISS 0827	07 15 24.	- 23 09	167		STELLAR RING
RNGC 2359	07 15 27.	- 13 07			DIFFUSE NEBULA
LB 00495	07 15 29.	+ 64 48 36.		16.2	FAINT BLUE STAR
MCG+09-12-053	07 15 30.	+ 55 43	8	17.	GALAXY
OCL 0589	07 15 30.	- 15 32	1380	9.5	OPEN STAR CLUSTER
LDN 1659	07 15 30.	- 23 15	840		DARK NEBULA
ZWG 117.007	07 15 36.	+ 21 07		15.7	GALAXY
7ZW 141	07 15 36.	+ 63 29			COMPACT GALAXY
UGC 03795	07 15 42.	+ 57 13	60	17.	GALAXY Sc
MCG+10-11-025	07 15 42.	+ 60 20	27	16.	GALAXY
REIN 2.054	07 15 47.34	+ 85 48 33.4			NEBULA
ZC 0715.8+2928	07 15 48.	+ 29 28	1080		CLUSTER OF GALAXIES
ZWG 147.006	07 15 48.	+ 30 38		15.3	GALAXY
ZWG 146.050	07 15 48.	+ 30 38		15.3	GALAXY
UGC 03796	07 15 48.	+ 60 21	66	16.0	GALAXY Sb-c
MCG+10-11-026	07 15 48.	+ 60 21	57	16.	GALAXY
RNGC 2300	07 15 49.	+ 85 49		12.5	GALAXY
MCG+05-18-005	07 15 54.	+ 30 39 30.	30	15.3	GALAXY
MCG+09-12-054	07 15 54.	+ 55 42 30.	8	17.	GALAXY
MCG+10-11-027	07 15 54.	+ 57 27	57	17.	GALAXY
UGC 03797	07 15 54.	+ 59 28	66	16.0	GALAXY SBc
MCG+10-11-028	07 15 54.	+ 59 28	51	16.	GALAXY
7ZW 142	07 15 54.	+ 64 27			COMPACT GALAXY
LBN 1040	07 16	- 13 00	1200		BRIGHT NEBULA
LBN 1041	07 16	- 13 02	420		BRIGHT NEBULA
LBN 1068	07 16	- 30 45	3720		BRIGHT NEBULA
KHAV 199	07 16	- 31 11	1570		DARK NEBULA
ZC 0716.0+4445	07 16 00.	+ 44 45	1480		CLUSTER OF GALAXIES
ZWG 235.006	07 16 00.	+ 49 57		15.4	GALAXY
KARA.73B 0184	07 16 00.	+ 49 57	24	15.4	ISOLATED GALAXY S0
ZWG 261.049	07 16 00.	+ 52 51		15.3	GALAXY
ZWG 261.050	07 16 00.	+ 56 01		15.2	GALAXY
7ZW 144	07 16 00.	+ 64 44			COMPACT GALAXY
ZC 0716.0+6540	07 16 00.	+ 65 40	1550		CLUSTER OF GALAXIES
7ZW 145	07 16 00.	+ 67 30			COMPACT GALAXY
MRK 087	07 16 00.	+ 74 09	20	15.	GALAXY WITH UV CONTINUUM
7ZW 143	07 16 00.	+ 82 02			COMPACT GALAXY
MCG+14-04-031	07 16 00.	+ 85 48	57	12.2	GALAXY
ZWG 363.029	07 16 00.	+ 85 49		12.2	GALAXY
ZWG 362.043	07 16 00.	+ 85 49		12.2	GALAXY
UGC 03798	07 16 00.	+ 85 49	210	12.2	GALAXY E
KARA.72 127B	07 16 00.	+ 85 49	162	12.2	PART OF DOUBLE GALAXY
OCL 0624	07 16 00.	- 22 34	312	15.	OPEN STAR CLUSTER
RNGC 2369	07 16 01.	- 62 15		12.5	GALAXY
RNGC 2361	07 16 03.	- 13 07			NON-EXISTENT OBJECT
ZWG 261.051	07 16 06.	+ 53 06		14.9	GALAXY
UGC 03799	07 16 06.	+ 53 06	66	14.9	GALAXY Sc/SBc
ZC 0716.1+5456	07 16 06.	+ 54 56	2490		CLUSTER OF GALAXIES
ZWG 261.052	07 16 06.	+ 55 37		15.6	GALAXY
MCG+09-12-055	07 16 06.	+ 56 01	39	16.	GALAXY
MRSL 227-00/1	07 16 06.	- 13 09	480		HII REGION
SG 3.103	07 16 07.	- 13 08	720		DIFFUSE EMISSION NEBULA
RLWT 142	07 16 12.	+ 02 26		14.	FAINT VERY BLUE STAR
MCG+09-12-056	07 16 12.	+ 53 06	60	14.	GALAXY
ST 07	07 16 12.	- 13 10	840		EMISSION NEBULA
MCG+09-12-057	07 16 15.	+ 55 37	12	16.	GALAXY
ZWG 261.053	07 16 18.	+ 55 55		15.7	GALAXY
UGC 03800	07 16 18.	+ 56 30	72	16.5	GALAXY DISTRBD
RLWT 143	07 16 24.	+ 02 39		12.	FAINT VERY BLUE STAR
TON-N 0279	07 16 24.	+ 27 48		14.8	BLUE STAR
ZWG 261.054	07 16 24.	+ 50 46		15.5	GALAXY
UGC 03801	07 16 24.	+ 63 44	78	16.5	GALAXY Sc
ACK 247-10.1	07 16 24.	- 34 49			PLANETARY NEBULA
ZWG 117.008	07 16 30.	+ 24 35		15.7	GALAXY
ZWG 205.036	07 16 30.	+ 39 43		15.7	GALAXY
MCG+09-12-058	07 16 30.	+ 56 30	9	17.	GALAXY
ZWG 286.012	07 16 30.	+ 58 19		15.5	GALAXY
SEY 043	07 16 31.	+ 72 04 43.		15.4	FAINT GALAXY
RNGC 2362	07 16 34.	- 24 52		4.0	OPEN CLUSTER
CED 095	07 16 34.	- 24 52			DIFFUSE GALACTIC NEBULA
ZWG 147.007	07 16 36.	+ 31 01		15.1	GALAXY
UGC 03802	07 16 36.	+ 31 01	72	15.1	GALAXY Sa
MCG+05-18-006	07 16 36.	+ 31 01	72	15.1	GALAXY
MCG+09-12-059	07 16 36.	+ 55 45	8	17.	GALAXY
MCG+10-11-029	07 16 36.	+ 58 18	30	15.	GALAXY
MCG+11-09-045	07 16 36.	+ 63 46	39	16.	GALAXY
RLWT 144	07 16 36.	- 03 04		14.	FAINT VERY BLUE STAR
LB 00496	07 16 41.	+ 63 13 00.		16.9	FAINT BLUE STAR
OCL 0633	07 16 42.	- 24 51	1200	3.9	OPEN STAR CLUSTER
MCG+08-14-006	07 16 48.	+ 49 10	42	15.	GALAXY
ZWG 235.007	07 16 48.	+ 49 11		15.0	GALAXY
MCG+11-09-046	07 16 48.	+ 63 45	78	17.	GALAXY
MCG+12-07-037	07 16 48.	+ 71 41	102	14.	GALAXY
ISS 0828	07 16 54.	- 23 48	214		STELLAR RING
SG 3.104	07 16 55.	- 13 05	540		DIFFUSE EMISSION NEBULA
LBN 1054	07 17	- 21 50	2400		BRIGHT NEBULA
ZWG 117.009	07 17 00.	+ 22 11		15.1	GALAXY
UGC 03803	07 17 00.	+ 22 11	60	15.1	GALAXY Sa
MCG+04-18-002	07 17 00.	+ 22 11	36	15.1	GALAXY
ZWG 330.036	07 17 00.	+ 71 42		13.0	GALAXY
UGC 03804	07 17 00.	+ 71 42	114	13.0	GALAXY Sc
MRSL 235-04/1	07 17 00.	- 21 50	360		HII REGION
MAI 022	07 17 03.	+ 66 56	33		DWARF SPHEROIDAL GALAXY

OBJECT NAME	RIGHT ASCEN.	DECLINATION	DIAM.	MAGN.	TYPE OF OBJECT
ZWG 029.018	07 17 06.	+ 06 40		15.5	GALAXY
ZWG 086.014	07 17 06.	+ 18 02		14.9	GALAXY
UGC 03805	07 17 06.	+ 18 02	78	14.9	GALAXY S
ZWG 147.008	07 17 06.	+ 32 28		15.6	GALAXY
ZWG 261.055	07 17 06.	+ 55 08		15.5	GALAXY
7ZW 146	07 17 06.	+ 66 42			COMPACT GALAXY
MCG+03-19-006	07 17 12.	+ 18 02	72	14.9	GALAXY
7ZW 147	07 17 12.	+ 68 37			COMPACT GALAXY
PK235-03.1	07 17 12.	- 21 32			PLANETARY NEBULA
ACK 245-08.1	07 17 12.	- 32 30			PLANETARY NEBULA
ARC 0576	07 17 20.	+ 55 50		14.4	RICH CLUSTER OF GALAXIES
MCG+04-18-003	07 17 24.	+ 22 58	48	15.3	GALAXY
ZWG 117.010	07 17 24.	+ 23 00		15.3	GALAXY
UGC 03806	07 17 24.	+ 23 00	66	15.3	GALAXY SB:c
UGC 03807	07 17 24.	+ 54 47	60	16.5	GALAXY S
ZWG 261.056	07 17 24.	+ 55 50		15.4	GALAXY
MCG+10-11-030	07 17 24.	+ 61 41	42	16.	GALAXY
MCG+13-06-005	07 17 24.	+ 77 54	78	16.	GALAXY
ZC 0717.5+3956	07 17 30.	+ 39 56	2350		CLUSTER OF GALAXIES
MCG+09-12-060	07 17 30.	+ 55 33	45	17.	GALAXY
SCHO 0241	07 17 30.	- 04 44 30.	590		ISOLATED DARK CLOUD
OCL 0609	07 17 30.	- 19 34	102	12.	OPEN STAR CLUSTER
DG 118	07 17 30.	- 23 56	240		REFLECTION NEBULA
CED 096A	07 17 30.	- 23 56	330		DIFFUSE GALACTIC NEBULA
VDB.66N 096	07 17 32.	- 23 52	336		REFLECTION NEBULA
ZWG 117.011	07 17 36.	+ 25 03		15.4	GALAXY
MCG+04-18-004	07 17 36.	+ 26 36	48	15.4	GALAXY
TON-N 0280	07 17 36.	+ 28 35		14.7	BLUE STAR
ZWG 261.057	07 17 36.	+ 55 46		15.2	GALAXY
MCG+09-12-061	07 17 36.	+ 55 52	12	16.	GALAXY
MCG+10-11-031	07 17 36.	+ 58 32	27	16.	GALAXY
ZWG 117.012	07 17 42.	+ 26 37		15.4	GALAXY
ZWG 261.058	07 17 42.	+ 55 28		15.6	GALAXY
DG 119	07 17 42.	- 23 58	180		REFLECTION NEBULA
CED 096B	07 17 42.	- 23 58	300		DIFFUSE GALACTIC NEBULA
MCG-06-17-001	07 17 42.	- 35 34	120	14.	GALAXY
ZWG 117.013	07 17 48.	+ 23 27		15.6	GALAXY
YM 33	07 17 48.	- 00 52	186		SYMMETRIC GALACTIC NEBULA
OCL 0582	07 17 48.	- 13 02	240	16.	OPEN STAR CLUSTER
HSE 211	07 17 53.	- 68 55			STAR CLUSTER IN LMC
ZWG 117.014	07 17 54.	+ 23 24		15.1	GALAXY
MCG+04-18-005	07 17 54.	+ 23 24	48	15.1	GALAXY
MCG+04-18-006	07 17 54.	+ 25 14	90	15.5	GALAXY
ZWG 147.009	07 17 54.	+ 29 27		14.9	GALAXY
KARA.73B 0185	07 17 54.	+ 29 27	48	14.9	ISOLATED GALAXY S
ZWG 147.010	07 17 54.	+ 30 51		15.7	GALAXY
OCL 0551	07 17 54.	- 01 00	480	15.	OPEN STAR CLUSTER
YC 0717-44	07 17 56.	- 44 29 36.			UNUSUAL SOUTHERN NEBULA
ARC 0579	07 17 58.	+ 36 51		17.6	RICH CLUSTER OF GALAXIES
KHAV 202	07 18	- 06 24	5170		DARK NEBULA
KHAV 201	07 18	- 16 48	6380		DARK NEBULA
KHAV 200	07 18	- 21 54	7090		DARK NEBULA
LBN 1058	07 18	- 23 54	420		BRIGHT NEBULA
ZWG 117.015	07 18 00.	+ 25 16		15.5	GALAXY
UGC 03808	07 18 00.	+ 25 16	120	15.5	GALAXY SBc
MCG+05-18-007	07 18 00.	+ 29 27	48	14.9	GALAXY
ZC 0718.0+3650	07 18 00.	+ 36 50	1810		CLUSTER OF GALAXIES
ZC 0718.0+3820	07 18 00.	+ 38 20	3970		CLUSTER OF GALAXIES
ZC 0718.0+4301	07 18 00.	+ 43 01	2150		CLUSTER OF GALAXIES
ZWG 261.059	07 18 00.	+ 55 58		15.3	GALAXY
ZWG 261.060	07 18 00.	+ 56 35		15.0	GALAXY
ZWG 349.004	07 18 00.	+ 80 16		11.3	GALAXY
ZWG 348.034	07 18 00.	+ 80 16		11.3	GALAXY
UGC 03809	07 18 00.	+ 80 16	432	11.3	GALAXY SBc
KARA.72 132A	07 18 00.	+ 80 16	366	11.3	PART OF DOUBLE GALAXY
MCG+13-06-006	07 18 00.	+ 80 17	384	11.0	GALAXY
ZC 0718.0+8304	07 18 00.	+ 83 04	2150		CLUSTER OF GALAXIES
OCL 0621	07 18 00.	- 21 50	420	8.0	OPEN STAR CLUSTER
LDN 1660	07 18 00.	- 23 50	840		DARK NEBULA
DG 120	07 18 00.	- 24 00	300		REFLECTION NEBULA
CED 096C	07 18 00.	- 24 01	330		DIFFUSE GALACTIC NEBULA
RNGC 2367	07 18 02.	- 21 50		8.0	OPEN CLUSTER
ZWG 235.008	07 18 06.	+ 46 56		15.3	GALAXY
UGC 03810	07 18 06.	+ 46 56	60	15.3	GALAXY Sc
ZWG 261.061	07 18 06.	+ 55 51		15.7	GALAXY
ZWG 001.002	07 18 06.	- 01 28		15.5	GALAXY
ZWG 001.002	07 18 06.	- 01 28		15.5	GALAXY
SCHO 0242	07 18 08.	- 04 42 42.	550		ISOLATED DARK CLOUD
RLWT 145	07 18 12.	+ 01 52		16.	FAINT VERY BLUE STAR
MCG+08-14-007	07 18 12.	+ 46 56	60	17.	GALAXY
MCG+09-12-062	07 18 12.	+ 56 36	27	16.	GALAXY
ZWG 001.003	07 18 12.	- 01 36		15.6	GALAXY
ZWG 001.003	07 18 12.	- 01 36		15.6	GALAXY
LB 00497	07 18 13.	+ 81 00 18.		17.1	FAINT BLUE STAR
SCHO 0243	07 18 14.	- 15 32 30.	740		ISOLATED DARK CLOUD
RNGC 2369A	07 18 15.	- 62 49			GALAXY
ZWG 147.011	07 18 18.	+ 31 13		15.7	GALAXY
DV.56 N2369A	07 18 18.	- 62 49			S GALAXY
RNGC 2364	07 18 23.	- 07 28			NON-EXISTENT OBJECT
ZWG 117.016	07 18 24.	+ 22 38		15.5	GALAXY
ZWG 147.012	07 18 24.	+ 32 22		15.2	GALAXY
ZWG 176.040	07 18 24.	+ 35 50		15.4	GALAXY
UGC 03811	07 18 24.	+ 35 50	84		GALAXY Sc
MCG+06-16-037	07 18 24.	+ 35 50	78	14.5	GALAXY
ST 06	07 18 27.	- 11 50	10800		EMISSION NEBULA
RNGC 2336	07 18 29.	+ 80 17		11.5	GALAXY
ZWG 117.017	07 18 30.	+ 23 16		15.2	GALAXY
ZWG 235.009	07 18 30.	+ 49 23		14.4	GALAXY
UGC 03812	07 18 30.	+ 49 23	48	14.4	GALAXY E
ZWG 057.008	07 18 36.	+ 13 21		15.7	GALAXY
UGC 03813	07 18 36.	+ 13 21	84	15.7	GALAXY DWARF
MCG+04-18-007	07 18 36.	+ 23 37	48	14.7	GALAXY
ZWG 117.018	07 18 36.	+ 23 38		14.7	GALAXY
ZC 0718.6+3249	07 18 36.	+ 32 49	9880		CLUSTER OF GALAXIES
ZWG 205.037	07 18 36.	+ 40 34		14.9	GALAXY
UGC 03814	07 18 36.	+ 49 22	60	17.	GALAXY Sc-IRR
MCG+09-12-063	07 18 36.	+ 56 39	39	16.	GALAXY
MCG+10-11-032	07 18 36.	+ 59 57	18	16.	GALAXY
OCL 0571	07 18 36.	- 10 17	480	11.8	OPEN STAR CLUSTER
RNGC 2368	07 18 37.	- 10 17		12.0	OPEN CLUSTER
SCHO 0244	07 18 41.	- 16 49 54.	470		ISOLATED DARK CLOUD
ZWG 029.019	07 18 42.	+ 03 45		15.6	GALAXY
ZWG 086.015	07 18 42.	+ 15 19		15.7	GALAXY
MCG+08-14-008	07 18 42.	+ 49 22	24	14.	GALAXY
FATH 1.188	07 18 44.	+ 15 21	8		NEBULA
ZWG 029.020	07 18 48.	+ 05 04		15.6	GALAXY
ZWG 086.016	07 18 48.	+ 20 20		15.4	GALAXY
MCG+03-19-007	07 18 48.	+ 20 22	45	15.4	GALAXY
ZWG 147.013	07 18 48.	+ 29 36		15.5	GALAXY
KARA.73B 0186	07 18 48.	+ 29 36	24	15.5	ISOLATED GALAXY S
MCG+07-15-015	07 18 48.	+ 40 34	48	16.	GALAXY
7ZW 148	07 18 48.	+ 58 13			COMPACT GALAXY
ZWG 286.013	07 18 48.	+ 58 13		15.2	GALAXY
MCG+06-16-038	07 18 51.	+ 38 27	24	15.	GALAXY
ZWG 176.041	07 18 54.	+ 38 26		15.5	GALAXY
MCG+08-14-009	07 18 54.	+ 48 47	54	16.	GALAXY
MCG+09-12-064	07 18 54.	+ 54 03	57	16.	GALAXY
ZC 0718.9+5412	07 18 54.	+ 54 12	19420		CLUSTER OF GALAXIES
MCG+10-11-033	07 18 54.	+ 58 12	15	15.	GALAXY
CED 097	07 18 54.	- 08 47			DIFFUSE GALACTIC NEBULA
LBN 1029	07 19	- 08 20	1320		BRIGHT NEBULA
MCG+03-19-009	07 19 00.	+ 19 01	54	15.1	GALAXY
MCG+08-14-010	07 19 00.	+ 47 42	48	17.	GALAXY
MCG+08-14-011	07 19 00.	+ 48 58	36	17.	GALAXY
MCG+09-12-065	07 19 00.	+ 54 06	30	17.	GALAXY
ZWG 261.062	07 19 00.	+ 55 07		15.7	GALAXY
MCG+10-11-034	07 19 00.	+ 58 10	57	14.	GALAXY
ZWG 286.014	07 19 00.	+ 58 11		13.4	GALAXY
UGC 03816	07 19 00.	+ 58 11	60	13.4	GALAXY SO
MCG+10-11-035	07 19 00.	+ 59 11	27	16.	GALAXY
7ZW 149	07 19 00.	+ 68 35			COMPACT GALAXY
ZC 0719.0+7910	07 19 00.	+ 79 10	5650		CLUSTER OF GALAXIES
MCG+14-04-032	07 19 00.	+ 81 08	45	17.	GALAXY
MCG+14-04-033	07 19 00.	+ 85 38	30	15.	GALAXY
ZWG 363.030	07 19 00.	+ 85 39		14.3	GALAXY
ZWG 362.044	07 19 00.	+ 85 39		14.3	GALAXY
UGC 03815	07 19 00.	+ 85 39	72	14.3	GALAXY SO
PK232-01.1	07 19 00.	- 18 02	10		PLANETARY NEBULA
UGC 03817	07 19 06.	+ 45 12	120	17.	GALAXY DWARF
OCL 0555	07 19 06.	- 03 14	420	13.	OPEN STAR CLUSTER
MRSL 238-05/1	07 19 06.	- 25 06	25200		HII REGION
IC 0467	07 19 10.	+ 80 01 05.	180		GALAXY SAB(s)
FATH 1.189	07 19 12.	+ 15 05	19		NEBULA
ZWG 086.017	07 19 12.	+ 17 30		15.2	GALAXY
ZWG 205.038	07 19 12.	+ 41 45		15.0	GALAXY
UGC 03818	07 19 12.	+ 41 45	66	15.0	GALAXY S
MCG+09-12-066	07 19 12.	+ 55 43	45	17.	GALAXY
ZC 0719.2+7518	07 19 12.	+ 75 18	1280		CLUSTER OF GALAXIES
MCG+08-14-012	07 19 15.	+ 49 00	15	17.	GALAXY
RLWT 146	07 19 18.	+ 01 38		15.	FAINT VERY BLUE STAR
ZWG 086.018	07 19 18.	+ 18 58		15.6	GALAXY
MCG+07-15-016	07 19 18.	+ 41 46	48	16.	GALAXY
MCG+10-11-036	07 19 18.	+ 58 05	39	16.	GALAXY
ZC 0719.3+6934	07 19 18.	+ 69 34	1080		CLUSTER OF GALAXIES
OCL 0552	07 19 18.	- 00 53	600		OPEN STAR CLUSTER
RNGC 2381	07 19 21.	- 62 59			UNVERIFIED SOUTHERN OBJECT
ZWG 029.021	07 19 24.	+ 05 14		15.7	GALAXY
UGC 03819	07 19 24.	+ 05 14	66	15.7	GALAXY IRR
ZWG 117.019	07 19 24.	+ 22 08		15.4	GALAXY
KEEL 117	07 19 24.1	+ 69 33 51.		17.	NEBULA
ZWG 086.019	07 19 30.	+ 17 22		15.1	GALAXY
UGC 03820	07 19 30.	+ 17 22	96	15.1	GALAXY Sc
MCG+03-19-008	07 19 30.	+ 17 23	90	15.1	GALAXY
MCG+04-18-008	07 19 30.	+ 22 09	96	13.8	GALAXY
ZWG 117.020	07 19 30.	+ 22 10		13.8	GALAXY
RNGC 2365	07 19 30.	+ 22 10		14.0	GALAXY
UGC 03821	07 19 30.	+ 22 10	198	13.8	GALAXY Sa/SBa
ZC 0719.5+3606	07 19 30.	+ 36 06	810		CLUSTER OF GALAXIES
IC 0455	07 19 31.	+ 85 39 03.			NONSTELLAR OBJECT
FATH 1.190	07 19 32.	+ 14 58	19		NEBULA
ZWG 117.021	07 19 36.	+ 26 32		15.4	GALAXY
MCG+04-18-009	07 19 36.	+ 26 36	30	15.1	GALAXY
ZC 0719.6+4216	07 19 36.	+ 42 16	1010		CLUSTER OF GALAXIES
MCG+09-12-067	07 19 36.	+ 54 53	39	17.	GALAXY
RLWT 147	07 19 36.	- 01 20		15.	FAINT VERY BLUE STAR
ZWG 117.022	07 19 42.	+ 21 12		15.7	GALAXY
ZWG 117.023	07 19 42.	+ 26 37		15.1	GALAXY
ZWG 177.001	07 19 42.	+ 32 51		15.7	GALAXY
ZWG 147.014	07 19 42.	+ 32 51		15.7	GALAXY
MCG+10-11-037	07 19 42.	+ 57 23	42	14.	GALAXY
ZWG 286.015	07 19 42.	+ 57 25		15.0	GALAXY
IC 2187	07 19 46.	+ 21 34 23.			NONSTELLAR OBJECT
IC 2188	07 19 46.	+ 21 36 11.			NONSTELLAR OBJECT
IC 2186	07 19 46.	+ 21 37 23.			NONSTELLAR OBJECT
KEEL 118	07 19 46.7	+ 69 26 18.		17.	NEBULA
ZWG 086.020	07 19 48.	+ 20 36		15.6	GALAXY
ZWG 117.024	07 19 48.	+ 21 27		15.6	GALAXY
MCG+04-18-010	07 19 48.	+ 21 34	12	15.3	GALAXY
ZWG 117.025	07 19 48.	+ 21 35		15.3	GALAXY
MCG+04-18-011	07 19 48.	+ 21 36	48	15.1	GALAXY
ZWG 117.026	07 19 48.	+ 21 37		15.1	GALAXY
ZWG 117.027	07 19 48.	+ 25 26		15.6	GALAXY
ZWG 177.002	07 19 48.	+ 37 26		15.5	GALAXY
ZWG 176.042	07 19 48.	+ 37 26		15.5	GALAXY
ZWG 177.003	07 19 48.	+ 37 33		15.0	GALAXY
ZWG 176.043	07 19 48.	+ 37 33		15.0	GALAXY
UGC 03822	07 19 48.	+ 37 33	90	15.0	GALAXY SBc
ZCG 0719+66	07 19 48.	+ 66 58		17.9	COMPACT GALAXY
ZCG 0719+67.1	07 19 48.	+ 67 04		17.6	COMPACT GALAXY
MCG-01-19-001	07 19 48.	- 05 50	78	15.	GALAXY
ISS 0829	07 19 48.	- 25 13	106		STELLAR RING
DV.56 N2369B	07 19 48.	- 61 57			SBc GALAXY
RNGC 2369B	07 19 48.	- 61 57			GALAXY
IC 0470	07 19 53.	+ 46 10 20.			NONSTELLAR OBJECT
ZWG 086.021	07 19 54.	+ 19 00		15.1	GALAXY
UGC 03823	07 19 54.	+ 19 00	78	15.1	GALAXY Sb
ZWG 117.028	07 19 54.	+ 22 40		14.4	GALAXY
UGC 03824	07 19 54.	+ 22 40	54	14.4	GALAXY SO
MCG+04-18-012	07 19 54.	+ 22 40	21	14.4	GALAXY
ZCG 0719+67.0	07 19 54.	+ 67 02		17.4	COMPACT GALAXY
ZC 0719.9+6709	07 19 54.	+ 67 09	3160		CLUSTER OF GALAXIES
LB 00498	07 19 55.	+ 61 48 18.		16.2	FAINT BLUE STAR
KEEL 119	07 19 57.6	+ 69 12 43.		17.	NEBULA
KHAV 203	07 20	- 14 06	6800		DARK NEBULA
ZWG 086.022	07 20 00.	+ 20 32		15.5	GALAXY
ZWG 177.004	07 20 00.	+ 32 36		14.9	GALAXY
ZWG 147.015	07 20 00.	+ 32 36		14.9	GALAXY
MCG+05-18-008	07 20 00.	+ 32 36	30	14.9	GALAXY
MCG+06-17-001	07 20 00.	+ 37 35	66	15.	GALAXY
ZWG 206.001	07 20 00.	+ 41 31		15.3	GALAXY
ZWG 205.039	07 20 00.	+ 41 31		15.3	GALAXY
UGC 03825	07 20 00.	+ 41 31	84	15.3	GALAXY Sb/SBb
KARA.73B 0187	07 20 00.	+ 41 31	54	15.3	ISOLATED GALAXY S
ZWG 286.016	07 20 00.	+ 61 47		15.0	GALAXY
UGC 03826	07 20 00.	+ 61 47	240	15.0	GALAXY Sc
KARA.73B 0188	07 20 00.	+ 61 47	198	15.0	ISOLATED GALAXY S

OBJECT NAME	RIGHT ASCEN.	DECLINATION	DIAM.	MAGN.	TYPE OF OBJECT
MCG+10-11-038	07 20 00.	+ 61 50	192	14.	GALAXY
LDN 1661	07 20 00.	- 24 00	2460		DARK NEBULA
ARC 0578	07 20 02.	+ 67 06		17.0	RICH CLUSTER OF GALAXIES
IC 2185	07 20 03.	+ 32 35 40.			NONSTELLAR OBJECT
MCG+07-15-017	07 20 03.	+ 41 32 30.	78	14.5	GALAXY
ZC 0720.1+2736	07 20 06.	+ 27 36	1550		CLUSTER OF GALAXIES
7ZW 150	07 20 06.	+ 67 03			COMPACT GALAXY
ZCG 0720+67.1	07 20 06.	+ 67 05		17.6	COMPACT GALAXY
MCG+04-18-013	07 20 12.	+ 22 17	54	14.8	GALAXY
ZWG 117.029	07 20 12.	+ 22 18		14.8	GALAXY
UGC 03827	07 20 12.	+ 22 18	66	14.8	GALAXY SB?
MCG+04-18-014	07 20 12.	+ 23 43	42	15.5	GALAXY
ZWG 117.030	07 20 12.	+ 23 45		15.5	GALAXY
MCG+09-12-068	07 20 12.	+ 55 40	9	18.	GALAXY
ZCG 0720+67.0	07 20 12.	+ 67 04		18.4	COMPACT GALAXY
MCG-05-18-003	07 20 12.	- 29 08	72	14.5	GALAXY
RLWT 148	07 20 24.	+ 01 51		16.5	FAINT VERY BLUF STAR
PK214+07.1	07 20 24.	+ 01 52	67	16.3	PLANETARY NEBULA
ZWG 286.017	07 20 24.	+ 58 05		12.7	GALAXY
UGC 03828	07 20 24.	+ 58 05	108	12.7	GALAXY Sb/SBb
MCG+10-11-039	07 20 24.	+ 58 05	96	13.	GALAXY
ZWG 177.005	07 20 30.	+ 33 31		13.7	GALAXY
UGC 03829	07 20 30.	+ 33 31	78	13.7	GALAXY DBL SYS
MCG+06-17-002	07 20 30.	+ 33 32	66	14.	GALAXY
ZC 0720.5+4505	07 20 30.	+ 45 05	2150		CLUSTER OF GALAXIES
ZWG 261.063	07 20 30.	+ 54 40		15.4	GALAXY
MCG+09-12-069	07 20 30.	+ 54 40	12	16.	GALAXY
ISS 0830	07 20 30.	- 25 31	96		STELLAR RING
TON-N 0818	07 20 34.	+ 33 56		15.7	BLUE STAR
ZWG 001.004	07 20 36.	+ 02 52		15.3	GALAXY
ZWG 117.031	07 20 36.	+ 21 12		15.7	GALAXY
ZC 0720.6+2259	07 20 36.	+ 22 59	9540		CLUSTER OF GALAXIES
TON-N 0281	07 20 36.	+ 30 24		14.7	BLUE STAR
ZWG 235.010	07 20 36.	+ 49 23		15.5	GALAXY
MCG+08-14-013	07 20 36.	+ 49 23	12	16.	GALAXY
MCG+09-12-070	07 20 36.	+ 55 42 30.	9	18.	GALAXY
7ZW 151	07 20 36.	+ 63 37			COMPACT GALAXY
ZWG 117.032	07 20 42.	+ 21 57		15.5	GALAXY
ZWG 117.033	07 20 42.	+ 22 44		15.6	GALAXY
TON-N 0282	07 20 42.	+ 29 28		15.5	BLUE STAR
ZWG 177.006	07 20 42.	+ 38 02		15.6	GALAXY
ZWG 176.044	07 20 42.	+ 38 02		15.6	GALAXY
ZC 0720.7+4606	07 20 42.	+ 46 06	1140		CLUSTER OF GALAXIES
ZWG 029.022	07 20 48.	+ 03 00		15.3	GALAXY
ZWG 117.034	07 20 48.	+ 21 10		15.7	GALAXY
MCG+09-12-071	07 20 48.	+ 52 11	30	18.	GALAXY
MCG+09-12-072	07 20 48.	+ 54 40	30	16.	GALAXY
MCG+10-11-040	07 20 48.	+ 58 15	27	16.	GALAXY
ZWG 001.005	07 20 54.	+ 02 43		15.1	GALAXY
ZWG 001.005	07 20 54.	+ 02 43		15.1	GALAXY
UGC 03830	07 20 54.	+ 02 43	90	15.1	GALAXY S
MCG+00-19-001	07 20 54.	+ 02 43	42	15.1	GALAXY
ZWG 086.023	07 20 54.	+ 15 38		15.7	GALAXY
ZWG 286.018	07 20 54.	+ 59 31		15.5	GALAXY
KEEL 129	07 20 59.0	+ 29 36 09.		18.	NEBULA
LBN 1059	07 21	- 24 00	11700		BRIGHT NEBULA
ZWG 029.023	07 21 00.	+ 07 09		15.7	GALAXY
ZWG 177.007	07 21 00.	+ 32 42		15.3	GALAXY
ZWG 147.016	07 21 00.	+ 32 42		15.3	GALAXY
ZWG 206.002	07 21 00.	+ 39 03		15.6	GALAXY
ZWG 205.040	07 21 00.	+ 39 03		15.6	GALAXY
ZWG 206.003	07 21 00.	+ 40 52		15.3	GALAXY
ZWG 205.041	07 21 00.	+ 40 52		15.3	GALAXY
ZC 0721.0+4401	07 21 00.	+ 44 01	2220		CLUSTER OF GALAXIES
MCG+10-11-041	07 21 00.	+ 59 31	18	16.	GALAXY
OCL 0612	07 21 00.	- 19 21	660	13.	OPEN STAR CLUSTFR
KEEL 120	07 21 05.9	+ 69 15 59.		17.	NEBULA
MCG+09-12-073	07 21 06.	+ 52 15	39	18.	GALAXY
ZC 0721.1+6411	07 21 06.	+ 64 11	1480		CLUSTER OF GALAXIES
OCL 0578	07 21 06.	- 12 14	252	12.	OPEN STAR CLUSTER
MCG+05-18-009	07 21 09.	+ 32 43	48	15.3	GALAXY
TON-N 0819	07 21 10.	+ 34 41		15.5	BLUF STAR
ZWG 177.008	07 21 12.	+ 32 45		15.6	GALAXY
ZWG 147.017	07 21 12.	+ 32 45		15.6	GALAXY
ZWG 177.009	07 21 12.	+ 36 33		15.7	GALAXY
ZWG 235.011	07 21 12.	+ 49 35		13.6	GALAXY
UGC 03831	07 21 12.	+ 49 35	72	13.6	GALAXY SBb
MCG+08-14-014	07 21 12.	+ 49 35	72	13.	GALAXY
ZC 0721.2+5043	07 21 12.	+ 50 43	2290		CLUSTER OF GALAXIES
MCG+09-12-074	07 21 12.	+ 53 32	66	15.	GALAXY
MCG+09-12-075	07 21 12.	+ 55 06	39	17.	GALAXY
ZC 0721.2+6602	07 21 12.	+ 66 02	2290		CLUSTER OF GALAXIES
OCL 0661	07 21 12.	- 29 24	174	14.	OPEN STAR CLUSTER
VHA 001	07 21 12.	- 29 24	180		OPEN STAR CLUSTFR
KEEL 121	07 21 14.0	+ 69 35 49.		16.	NEBULA
ZWG 261.064	07 21 18.	+ 53 32		15.2	GALAXY
UGC 03832	07 21 18.	+ 53 32	78	15.2	GALAXY SBa-b
ZWG 261.065	07 21 18.	+ 55 36		14.9	GALAXY
RNGC 2397A	07 21 22.	- 68 45			GALAXY
KEEL 132	07 21 23.5	+ 29 21 59.		18.	NEBULA
7ZW 152	07 21 24.	+ 64 40			COMPACT GALAXY
DV.56 N2397A	07 21 24.	- 68 45	60		SA:c GALAXY
KEEL 122	07 21 26.6	+ 69 31 02.		17.	NEBULA
RNGC 2397	07 21 28.	- 68 54		13.0	GALAXY
KEEL 123	07 21 28.0	+ 69 14 26.		18.	NEBULA
KEEL 133	07 21 29.2	+ 29 22 38.		18.	NEBULA
ZWG 029.024	07 21 30.	+ 06 23		15.6	GALAXY
ZWG 177.010	07 21 30.	+ 32 55		15.3	GALAXY
ZWG 147.018	07 21 30.	+ 32 55		15.3	GALAXY
UGC 03833	07 21 30.	+ 32 55	60	15.3	GALAXY S
MCG+05-18-010	07 21 30.	+ 32 55	54	15.3	GALAXY
ZWG 206.004	07 21 30.	+ 40 12		15.7	GALAXY
ZWG 205.042	07 21 30.	+ 40 12		15.7	GALAXY
MCG+13-06-007	07 21 30.	+ 79 59	174	11.	GALAXY
MRSL 224+03/1	07 21 30.	- 08 30	3600		HII REGION
OCL 0627	07 21 30.	- 23 07	300	12.	OPEN STAR CLUSTER
KEEL 134	07 21 30.7	+ 29 45 36.		18.	NEBULA
TON-N 0283	07 21 42.	+ 28 14		14.8	BLUE STAR
ZWG 206.005	07 21 42.	+ 42 07		15.7	GALAXY
ZWG 205.043	07 21 42.	+ 42 07		15.7	GALAXY
MCG+10-11-042	07 21 42.	+ 60 14	39	15.	GALAXY
ZWG 349.005	07 21 42.	+ 79 58		12.7	GALAXY
ZWG 348.035	07 21 42.	+ 79 58		12.7	GALAXY
UGC 03834	07 21 42.	+ 79 58	210	12.7	GALAXY Sc
KARA.72 132B	07 21 42.	+ 79 58	180	12.7	PART OF DOUBLE GALAXY
RLWT 149	07 21 42.	- 00 27		14.	FAINT VERY BLUF STAR
OCL 0585	07 21 42.	- 13 10	1170	7.4	OPEN STAR CLUSTFR
MCG-05-18-004	07 21 42.	- 29 33	72	15.	GALAXY

OBJECT NAME	RIGHT ASCEN.	DECLINATION	DIAM.	MAGN.	TYPE OF OBJECT
RNGC 2374	07 21 44.	- 13 09		7.5	OPEN CLUSTER
MCG+09-12-076	07 21 45.	+ 55 43	39	18.	GALAXY
OCL 0536	07 21 48.	+ 05 27	480	16.	OPEN STAR CLUSTER
ZC 0721.8+4019	07 21 48.	+ 40 19	3290		CLUSTER OF GALAXIES
MCG+07-16-001	07 21 48.	+ 42 07	18	16.	GALAXY
ZWG 235.012	07 21 48.	+ 46 00		15.6	GALAXY
ZWG 261.066	07 21 48.	+ 53 17		15.6	GALAXY
KEEL 135	07 21 51.2	+ 29 21 39.		18.	NEBULA
RNGC 2380	07 21 53.	- 27 25		13.0	GALAXY
ZWG 057.009	07 21 54.	+ 14 11		15.6	GALAXY
ZWG 147.019	07 21 54.	+ 27 26		14.8	GALAXY
MCG-05-18-005	07 21 54.	- 27 25	30	13.5	GALAXY
KEEL 124	07 21 54.6	+ 69 11 06.		16.	NEBULA
RNGC 2370	07 21 59.	+ 23 53		14.5	GALAXY
KHAV 204	07 22	- 01 00	2470		DARK NEBULA
LBN 1031	07 22	- 08 20	1440		BRIGHT NEBULA
ZWG 086.024	07 22 00.	+ 19 17		15.7	GALAXY
ZWG 117.035	07 22 00.	+ 23 31		15.7	GALAXY
ZWG 117.036	07 22 00.	+ 23 53		14.3	GALAXY
UGC 03835	07 22 00.	+ 23 53	66	14.3	GALAXY S
MCG+04-18-015	07 22 00.	+ 23 54	24	14.3	GALAXY
MCG+05-18-011	07 22 00.	+ 27 25	48	14.8	GALAXY
ZWG 309.031	07 22 00.	+ 64 55		15.5	GALAXY
UGC 03836	07 22 00.	+ 64 55	78	15.5	GALAXY SBc
MCG+11-09-047	07 22 00.	+ 64 55	78	15.	GALAXY
LDN 1664	07 22 00.	- 24 30	1560		DARK NEBULA
OCL 0679	07 22 00.	- 32 06	2550	4.4	OPEN STAR CLUSTFR
VHA 002	07 22 00.	- 32 06	3000		OPEN STAR CLUSTFR
DV.56 N2397B	07 22 00.	- 69 01	57		GALAXY
CED 098	07 22 05.	- 29 12			DIFFUSE GALACTIC NEBULA
KEEL 137	07 22 05.5	+ 29 21 22.		19.	NEBULA
ZWG 029.025	07 22 06.	+ 07 18		15.5	GALAXY
ZWG 235.013	07 22 06.	+ 44 55		15.5	GALAXY
ZWG 205.044	07 22 06.	+ 44 55		15.5	GALAXY
MCG-06-17-002	07 22 06.	- 35 48	90	16.	GALAXY
KEEL 125	07 22 09.8	+ 69 27 35.		17.	NEBULA
TON-N 0820	07 22 10.	+ 34 45		15.4	BLUE STAR
RNGC 2382	07 22 11.	- 27 16			NON-EXISTENT OBJFCT
ZC 0722.2+2231	07 22 12.	+ 22 31	2760		CLUSTER OF GALAXIES
ZWG 177.011	07 22 12.	+ 36 46		15.1	GALAXY
UGC 03837	07 22 12.	+ 36 46	60	15.1	GALAXY S
MCG+06-17-003	07 22 12.	+ 36 48	42	15.	GALAXY
MCG+10-11-043	07 22 12.	+ 58 26	39	16.	GALAXY
7ZW 153	07 22 12.	+ 72 40			COMPACT GALAXY
ISS 0831	07 22 12.	- 24 38	242		STELLAR RING
HN 0075	07 22 14.	+ 09 01			NEBULA
IC 2189	07 22 14.	+ 09 01			NONSTELLAR OBJECT
ARC 0575	07 22 16.	+ 79 18		17.0	RICH CLUSTER OF GALAXIES
ZWG 330.037	07 22 18.	+ 72 40		13.9	GALAXY
UGC 03838	07 22 18.	+ 72 40	54	13.9	GALAXY PECULR
OCL 0600	07 22 18.	- 16 54	126	14.	OPEN STAR CLUSTER
ACK 242-06.1	07 22 18.	- 28 54			PLANETARY NEBULA
KEEL 138	07 22 19.C	+ 29 20 21.		18.	SPIRAL NEBULA
KEEL 139	07 22 20.5	+ 29 16 26.		18.	NEBULA
ZWG 057.010	07 22 24.	+ 09 18		15.5	GALAXY
ZWG 057.011	07 22 24.	+ 09 36		15.2	GALAXY
UGC 03839	07 22 24.	+ 09 36	66	15.2	GALAXY SBb
MCG+02-19-003	07 22 24.	+ 09 36	60	15.2	GALAXY
ZWG 086.025	07 22 24.	+ 19 16		14.4	GALAXY
UGC 03840	07 22 24.	+ 19 16	102	14.4	GALAXY E
ZWG 147.020	07 22 24.	+ 30 04		15.6	GALAXY
ZWG 206.006	07 22 24.	+ 40 47		15.6	GALAXY
ZWG 205.045	07 22 24.	+ 40 47		15.6	GALAXY
MCG+08-14-015	07 22 24.	+ 49 03	36	16.	GALAXY
7ZW 154	07 22 24.	+ 60 19			COMPACT GALAXY
MRK 007	07 22 24.	+ 72 39	40	15.5	GALAXY WITH UV CONTINUUM
MCG+12-07-038	07 22 24.	+ 72 39	51	16.	GALAXY
MCG-06-17-003	07 22 24.	- 36 43	30	15.	GALAXY
PK189+19.1	07 22 24.83	+ 29 35 21.5	129	13.0	PLANETARY NEBULA
RNGC 2372	07 22 25.	+ 29 35		13.0	PLANETARY NEBULA
RNGC 2371	07 22 25.	+ 29 35		13.0	PLANETARY NEBULA
ARC 0580	07 22 26.	+ 41 31		16.8	RICH CLUSTER OF GALAXIES
MCG+03-19-010	07 22 27.	+ 19 16	18	14.4	GALAXY
CED 099	07 22 28.	- 16 06			DIFFUSE GALACTIC NEBULA
KEEL 126	07 22 28.6	+ 69 12 43.		18.	NEBULA
ZWG 147.021	07 22 30.	+ 31 35		15.5	GALAXY
ZWG 235.014	07 22 30.	+ 49 03		15.1	GALAXY
MCG+10-11-044	07 22 30.	+ 60 20	39	17.	GALAXY
ZWG 029.026	07 22 36.	+ 03 19		15.6	GALAXY
MCG+04-18-016	07 22 36.	+ 25 46	42	15.2	GALAXY
ZWG 147.022	07 22 36.	+ 30 04		15.2	GALAXY
UGC 03841	07 22 36.	+ 30 04	60	15.2	GALAXY DBL SYS
MCG+05-18-012	07 22 36.	+ 30 04	18	15.2	GALAXY
ZWG 177.012	07 22 36.	+ 32 33		15.6	GALAXY
ZWG 147.023	07 22 36.	+ 32 33		15.6	GALAXY
ZC 0722.6+3600	07 22 36.	+ 36 00	1010		CLUSTER OF GALAXIES
ZWG 177.013	07 22 36.	+ 37 27		15.5	GALAXY
ZWG 206.007	07 22 36.	+ 40 06		15.7	GALAXY
ZWG 205.046	07 22 36.	+ 40 06		15.7	GALAXY
MCG+10-11-045	07 22 36.	+ 58 01	39	17.	GALAXY
RNGC 2377	07 22 36.	- 09 33			NON-EXISTENT OBJECT
OCL 0616	07 22 36.	- 20 50	600	8.8	OPEN STAR CLUSTER
RNGC 2383	07 22 37.	- 20 50		8.5	OPEN CLUSTER
KEEL 142	07 22 39.2	+ 29 32 35.		18.	NEBULA
ZWG 086.026	07 22 42.	+ 19 13		14.9	GALAXY
UGC 03842	07 22 42.	+ 19 13	66	14.9	GALAXY S0
ZWG 117.037	07 22 42.	+ 25 49		15.2	GALAXY
OCL 0646	07 22 42.	- 26 07	240	12.	OPEN STAR CLUSTER
MCG+03-19-011	07 22 45.	+ 19 13	60	14.9	GALAXY
MCG+09-12-077	07 22 45.	+ 53 17	12	17.	GALAXY
ZWG 057.012	07 22 48.	+ 12 11		15.7	GALAXY
ZWG 086.027	07 22 48.	+ 20 11		15.5	GALAXY
UGC 03843	07 22 48.	+ 20 11	60	15.5	GALAXY Sc
ZC 0722.8+5506	07 22 48.	+ 55 06	810		CLUSTER OF GALAXIES
ARC 0581	07 22 52.	+ 11 12		17.3	RICH CLUSTER OF GALAXIES
RLWT 150	07 22 54.	+ 00 04		13.5	FAINT VERY BLUE STAR
ZWG 057.013	07 22 54.	+ 09 16		15.7	GALAXY
ZWG 117.038	07 22 54.	+ 24 53		15.6	GALAXY
OCL 0618	07 22 54.	- 20 56	420	8.3	OPEN STAR CLUSTER
RNGC 2384	07 22 55.	- 20 56		8.0	OPEN CLUSTFR
KEEL 144	07 22 55.7	+ 29 30 10.		18.	NEBULA
SCHO 0245	07 22 56.	- 15 53 30.	510		ISOLATED DARK CLOUD
KEEL 145	07 22 57.5	+ 29 31 33.		18.	NEBULA
VDB.66G 042	07 23	+ 69 19	340		DWARF GALAXY
LBN 1045	07 23	- 16 30	360		BRIGHT NEBULA
LBN 1064	07 23	- 29 00	6600		BRIGHT NEBULA
ZC 0723.0+1109	07 23 00.	+ 11 09	1010		CLUSTER OF GALAXIES
ZC 0723.0+3156	07 23 00.	+ 31 56	1880		CLUSTER OF GALAXIES

OBJECT NAME	RIGHT ASCEN.	DECLINATION	DIAM.	MAGN.	TYPE OF OBJECT
ZC 0723.0+3347	07 23 00.	+ 33 47	1010		CLUSTER OF GALAXIES
ZWG 206.008	07 23 00.	+ 43 23		14.3	GALAXY
ZWG 205.047	07 23 00.	+ 43 23		14.3	GALAXY
UGC 03844	07 23 00.	+ 43 23	108	14.3	GALAXY E?
KARA.73B 0189	07 23 00.	+ 43 23	72	14.3	ISOLATED GALAXY S0
ZWG 235.015	07 23 00.	+ 47 12		13.6	GALAXY
UGC 03845	07 23 00.	+ 47 12	96	13.6	GALAXY SBb
MCG+10-11-046	07 23 00.	+ 61 05	51	17.	GALAXY
MCG+12-07-039	07 23 00.	+ 69 16	102	15.	GALAXY
UGC 03846	07 23 00.	+ 75 37	84	16.5	GALAXY
LDN 1667	07 23 00.	- 25 40	2580		DARK NEBULA
KEEL 127	07 23 01.9	+ 69 39 45.		17.	NEBULA
ZC 0723.1+4146	07 23 06.	+ 41 46	6250		CLUSTER OF GALAXIES
MCG+07-16-002	07 23 06.	+ 43 24	90	15.	GALAXY
MCG+08-14-016	07 23 06.	+ 47 11	90	13.	GALAXY
ZWG 235.016	07 23 06.	+ 48 48		15.2	GALAXY
UGC 03847	07 23 06.	+ 69 17	102	15.5	GALAXY IRR
MCG-06-17-004	07 23 06.	- 36 49	36	16.	GALAXY
KEEL 128	07 23 07.9	+ 69 34 51.		17.	NEBULA
ZC 0723.2+2935	07 23 12.	+ 29 35	1340		CLUSTER OF GALAXIES
ZWG 262.001	07 23 12.	+ 53 02		14.8	GALAXY
ZWG 261.067	07 23 12.	+ 53 02		14.8	GALAXY
ZC 0723.2+6733	07 23 12.	+ 67 33	1810		CLUSTER OF GALAXIES
KARA.72 133A	07 23 12.	+ 69 17	66	11.6	PART OF DOUBLE GALAXY
KEEL 130	07 23 12.8	+ 69 38 01.		17.	SPIRAL NEBULA
RNGC 2373	07 23 16.	+ 33 54		14.5	GALAXY
SHB 096	07 23 17.9	- 00 48 54.		18.	QUASI-STELLAR OBJECT
ZWG 177.014	07 23 18.	+ 33 54		14.5	GALAXY
UGC 03848	07 23 18.	+ 33 54	54	14.5	GALAXY S
ZWG 177.015	07 23 18.	+ 37 28		14.6	GALAXY
UGC 03849	07 23 18.	+ 37 28	60	14.6	GALAXY DBL SYS
KARA.72 134A	07 23 18.	+ 37 28	36	14.6	PART OF DOUBLE GALAXY
ST 05	07 23 18.	- 16 11			EMISSION NEBULA
MCG+06-17-004	07 23 21.	+ 33 55 30.	24	15.	GALAXY
KARA.72 134B	07 23 24.	+ 37 27	48		PART OF DOUBLE GALAXY
ZC 0723.4+4239	07 23 24.	+ 42 39	1680		CLUSTER OF GALAXIES
MCG+09-13-001A	07 23 24.	+ 53 03	30	17.	GALAXY
MCG+09-13-001	07 23 24.	+ 53 03	24	17.	GALAXY
MCG+10-11-047	07 23 24.	+ 58 08	15	16.	GALAXY
RNGC 2363	07 23 27.	+ 69 10			NON-EXISTENT OBJECT
ZWG 001.006	07 23 30.	+ 02 05		15.4	GALAXY
ZWG 001.006	07 23 30.	+ 02 05		15.4	GALAXY
MCG+08-14-017	07 23 30.	+ 48 23	42	16.	GALAXY
ZC 0723.5+5539	07 23 30.	+ 55 39	670		CLUSTER OF GALAXIES
MCG+12-07-040	07 23 30.	+ 69 17	468	11.4	GALAXY
7ZW 155	07 23 30.	+ 69 36			COMPACT GALAXY
7ZW 156	07 23 30.	+ 72 13			COMPACT GALAXY
OCL 0622	07 23 30.	- 21 25	504	13.	OPEN STAR CLUSTER
KEEL 131	07 23 33.7	+ 69 35 42.		16.	NEBULA
ZWG 117.039	07 23 36.	+ 23 10		14.7	GALAXY
RNGC 2376	07 23 36.	+ 23 10		14.5	GALAXY
MCG+04-18-017	07 23 36.	+ 23 10	30	14.7	GALAXY
ZWG 235.017	07 23 36.	+ 48 23		15.5	GALAXY
MCG+10-11-048	07 23 36.	+ 57 17	36	16.	GALAXY
ZWG 309.032	07 23 36.	+ 63 22		13.9	GALAXY
UGC 03850	07 23 36.	+ 63 22	102	13.9	GALAXY Sa/SBa
MCG+11-09-048	07 23 36.	+ 66 49	15	17.	GALAXY
ZWG 330.038	07 23 36.	+ 69 18		11.6	GALAXY
UGC 03851	07 23 36.	+ 69 18	540	11.6	GALAXY IRR
KARA.72 133B	07 23 36.	+ 69 18	270		PART OF DOUBLE GALAXY
ZWG 330.039	07 23 36.	+ 72 14		13.8	GALAXY
UGC 03852	07 23 36.	+ 72 14	72	13.8	GALAXY DBL SYS
KARA.72 135B	07 23 36.	+ 72 14	42		PART OF DOUBLE GALAXY
KARA.72 135A	07 23 36.	+ 72 14	48	13.8	PART OF DOUBLE GALAXY
RNGC 2366	07 23 39.	+ 69 19		11.5	GALAXY
ZWG 147.024	07 23 42.	+ 30 15		15.3	GALAXY
ZWG 286.019	07 23 42.	+ 57 06		15.6	GALAXY
MCG+11-09-049	07 23 42.	+ 63 21	96	14.	GALAXY
MCG-05-18-006	07 23 42.	- 30 18	78	14.5	GALAXY
ZWG 147.025	07 23 48.	+ 31 34		15.7	GALAXY
ZWG 177.016	07 23 48.	+ 33 34		15.6	GALAXY
MCG+06-17-005	07 23 48.	+ 33 56	78	14.	GALAXY
UGC 03853	07 23 48.	+ 48 33	60	17.	GALAXY Sc
ZWG 262.002	07 23 48.	+ 53 10		15.6	GALAXY
ZWG 261.068	07 23 48.	+ 53 10		15.6	GALAXY
MCG+10-11-049	07 23 48.	+ 58 06	30	17.	GALAXY
MCG+10-11-050	07 23 48.	+ 61 02	39	16.	GALAXY
MRK 008	07 23 48.	+ 72 13	37	15.	GALAXY WITH UV CONTINUUM
MCG+12-07-041	07 23 48.	+ 72 13	39	15.	GALAXY
RNGC 2375	07 23 52.	+ 33 55		14.5	GALAXY
RNGC 2379	07 23 52.	+ 33 56		15.0	GALAXY
MCG+05-18-013	07 23 54.	+ 30 15	30	15.3	GALAXY
ZWG 177.017	07 23 54.	+ 33 55		14.7	GALAXY
UGC 03854	07 23 54.	+ 33 55	84	14.7	GALAXY SBb
MCG+08-14-018	07 23 54.	+ 48 32	60	17.	GALAXY
MCG+10-11-051	07 23 54.	+ 57 06	18	16.	GALAXY
VDB.66G 043	07 24	+ 40 54	70		DWARF GALAXY
KHAV 205	07 24	- 12 36	7640		DARK NEBULA
KHAV 206	07 24	- 16 24	9110		DARK NEBULA
LBN 1063	07 24	- 28 00	2820		BRIGHT NEBULA
LBN 1070	07 24	- 31 10	4620		BRIGHT NEBULA
ZC 0724.0+3020	07 24 00.	+ 30 20	1480		CLUSTER OF GALAXIES
MCG+10-11-052	07 24 00.	+ 58 37	120	14.	GALAXY
ZWG 286.020	07 24 00.	+ 58 38		14.2	GALAXY
UGC 03855	07 24 00.	+ 58 38	132	14.2	GALAXY Sa-b
7ZW 157	07 24 00.	+ 64 03			COMPACT GALAXY
OCL 0638	07 24 00.	- 24 12	480	10.0	OPEN STAR CLUSTER
LB 00499	07 24 03.	+ 64 00 42.		12.8	FAINT BLUE STAR
UGC 03856	07 24 06.	+ 20 28	66	16.5	GALAXY Sc
MCG+03-19-012	07 24 06.	+ 20 29	54	17.	GALAXY
ZWG 117.040	07 24 06.	+ 24 08		15.4	GALAXY
ZC 0724.1+3555	07 24 06.	+ 35 55	1010		CLUSTER OF GALAXIES
ZC 0724.1+6436	07 24 06.	+ 64 36	1280		CLUSTER OF GALAXIES
PK221+05.1	07 24 06.	- 05 16	15		PLANETARY NEBULA
SCHO 0246	07 24 09.	- 17 44 00.	410		ISOLATED DARK CLOUD
RNGC 2378	07 24 10.	+ 33 56			NON-EXISTENT OBJECT
TON-N 0821	07 24 10.	+ 34 52		15.5	BLUE STAR
ZWG 001.007	07 24 12.	+ 01 45		15.6	GALAXY
ZWG 001.007	07 24 12.	+ 01 45		15.6	GALAXY
TON-N 0284	07 24 12.	+ 28 59		16.1	BLUE STAR
ZC 0724.2+3216	07 24 12.	+ 32 16	1480		CLUSTER OF GALAXIES
ZWG 177.018	07 24 12.	+ 33 54		14.9	GALAXY
UGC 03857	07 24 12.	+ 33 54	66	14.9	GALAXY (S0)
MCG+06-17-006	07 24 12.	+ 33 55	24	15.	GALAXY
ZWG 262.003	07 24 12.	+ 52 34		15.0	GALAXY
ZWG 261.069	07 24 12.	+ 52 34		15.0	GALAXY
KEEL 136	07 24 13.0	+ 69 32 43.		17.	SPIRAL NEBULA
MCG+11-09-050	07 24 15.	+ 63 30	42	15.	GALAXY

OBJECT NAME	RIGHT ASCEN.	DECLINATION	DIAM.	MAGN.	TYPE OF OBJECT
OCL 0502	07 24 18.	+ 13 41	1260	9.4	OPEN STAR CLUSTER
RNGC 2395	07 24 18.	+ 13 41		9.5	OPEN CLUSTER
ZC 0724.3+3818	07 24 18.	+ 38 18	5580		CLUSTER OF GALAXIES
MRK 088	07 24 18.	+ 55 52	15	14.5	GALAXY WITH UV CONTINUUM
ZWG 309.033	07 24 18.	+ 63 30		15.2	GALAXY
ZC 0724.3+6510	07 24 18.	+ 65 10	2220		CLUSTER OF GALAXIES
HOLM 087A	07 24 20.	+ 19 45	54	14.4	PART OF MULTIPLE GALAXY
MCG+06-17-007	07 24 21.	+ 35 43 30.	48	17.	GALAXY
HOLM 087B	07 24 22.	+ 19 44	36	14.8	PART OF MULTIPLE GALAXY
ZWG 086.028	07 24 24.	+ 19 44		14.7	GALAXY
KARA.72 136A	07 24 24.	+ 19 44	42	14.7	PART OF DOUBLE GALAXY
ZWG 286.021	07 24 24.	+ 57 52		15.5	GALAXY
MCG+03-19-013	07 24 27.	+ 19 44	36	14.7	GALAXY
BIGO 496	07 24 28.	+ 33 55			NEBULA
SCHO 0247	07 24 28.	- 17 26 12.	710		ISOLATED DARK CLOUD
ZWG 086.029	07 24 30.	+ 19 43		14.9	GALAXY
KARA.72 136B	07 24 30.	+ 19 43	42	14.9	PART OF DOUBLF GALAXY
MCG+03-19-014	07 24 30.	+ 19 43	36	14.9	GALAXY
ZWG 086.030	07 24 30.	+ 20 12		15.3	GALAXY
MCG+03-19-015	07 24 30.	+ 20 12	42	15.3	GALAXY
MCG+10-11-053	07 24 30.	+ 57 51	39	15.	GALAXY
MCG+12-07-042	07 24 30.	+ 69 30	12	17.	GALAXY
MCG-01-19-002	07 24 30.	- 07 26	48	17.	GALAXY
ACK 247-08.1	07 24 30.	- 34 51			PLANETARY NEBULA
KEEL 140	07 24 34.1	+ 69 35 04.		15.	NEBULA
MCG+10-11-054	07 24 36.	+ 57 56	45	17.	GALAXY
MCG+12-07-043	07 24 36.	+ 73 49	57	14.	GALAXY
MCG+12-07-044	07 24 36.	+ 74 34	15	17.	GALAXY
IC 2184	07 24 38.	+ 72 04			NONSTELLAR OBJECT
ARC 0582	07 24 40.	+ 42 05		16.4	RICH CLUSTER OF GALAXIES
MCG+10-11-055	07 24 42.	+ 57 57	39	18.	GALAXY
ZWG 331.001	07 24 42.	+ 73 44		13.8	GALAXY
ZWG 330.040	07 24 42.	+ 73 44		13.8	GALAXY
UGC 03858	07 24 42.	+ 73 44	84	13.8	GALAXY Sa
MCG+12-07-045	07 24 42.	+ 73 44	51	14.	GALAXY
ZWG 331.002	07 24 42.	+ 73 49		13.3	GALAXY
ZWG 330.041	07 24 42.	+ 73 49		13.3	GALAXY
UGC 03859	07 24 42.	+ 73 49	102	13.3	GALAXY Sa
OCL 0659	07 24 42.	- 28 47	600	11.	OPEN STAR CLUSTER
KEEL 141	07 24 47.1	+ 69 32 59.		16.	SPIRAL NEBULA
ZWG 206.009	07 24 48.	+ 40 10		15.7	GALAXY
ZWG 206.010	07 24 48.	+ 40 51		15.5	GALAXY
UGC 03860	07 24 48.	+ 40 51	108	15.5	GALAXY DWRF IR
KARA.73B 0190	07 24 48.	+ 40 51	72	15.5	ISOLATED GALAXY S
MCG+08-14-019	07 24 48.	+ 47 10	18	18.	GALAXY
UGC 03861	07 24 48.	+ 58 19	60	16.0	GALAXY
MCG+07-16-003	07 24 51.	+ 40 53	72	15.	GALAXY
ZWG 086.031	07 24 54.	+ 20 29		15.6	GALAXY
UGC 03862	07 24 54.	+ 20 29	96	15.6	GALAXY S
MCG+03-19-016	07 24 54.	+ 20 30	90	15.6	GALAXY
MCG+04-18-018	07 24 54.	+ 24 28	36	15.5	GALAXY
ZWG 117.041	07 24 54.	+ 26 30		15.6	GALAXY
MAI 023	07 24 54.	+ 58 59	60		DWARF SPHEROIDAL GALAXY
MCG+11-09-051	07 24 54.	+ 63 06	18	16.	GALAXY
ZWG 309.034	07 24 54.	+ 63 07		15.4	GALAXY
VV 141D	07 24 54.	+ 72 37	2	19.	INTERACTING GALAXY
VV 141C	07 24 54.	+ 72 37	6	18.	INTERACTING GALAXY
VV 141B	07 24 54.	+ 72 37	6	18.	INTERACTING GALAXY
VV 141A	07 24 54.	+ 72 37	18	16.	INTERACTING GALAXY
VV 141	07 24 54.	+ 72 37	66		INTERACTING GALAXY
OCL 0739	07 24 54.	- 47 38	2160	10.7	OPEN STAR CLUSTER
KEEL 143	07 24 56.0	+ 69 33 54.		17.	NEBULA
LBN 1060	07 25	- 24 30	6000		BRIGHT NEBULA
ZWG 117.042	07 25 00.	+ 24 29		15.5	GALAXY
ZWG 147.026	07 25 00.	+ 29 02		15.2	GALAXY
ZWG 177.019	07 25 00.	+ 34 05		15.5	GALAXY
MCG+09-13-002	07 25 00.	+ 56 19	24	17.	GALAXY
SHAH 149	07 25 00.	+ 82 04	138		GROUP OF COMPACT GALAXIES
ZWG 029.027	07 25 06.	+ 06 34		15.4	GALAXY
MCG+01-19-005	07 25 06.	+ 06 34	36	15.4	GALAXY
MCG+08-14-020	07 25 06.	+ 49 13	78	14.	GALAXY
ZWG 235.018	07 25 06.	+ 49 14		13.9	GALAXY
UGC 03863	07 25 06.	+ 49 14	84	13.9	GALAXY SBa
KARA.73B 0191	07 25 06.	+ 49 14	78	13.9	ISOLATED GALAXY S
ZC 0725.1+5517	07 25 06.	+ 55 17	940		CLUSTER OF GALAXIES
ZWG 330.042	07 25 06.	+ 72 37		15.0	GALAXY
UGC 03864	07 25 06.	+ 72 37	102	15.0	GALAXY
MCG+12-08-001	07 25 06.	+ 72 37	66	15.	GALAXY
RNGC 2385	07 25 10.	+ 33 56		15.0	GALAXY
ZWG 177.020	07 25 12.	+ 33 56		15.2	GALAXY
MCG+06-17-008	07 25 12.	+ 33 56 30.	36	15.	GALAXY
MCG+12-08-002	07 25 12.	+ 74 34	57	16.	GALAXY
OCL 0635	07 25 12.	- 23 56	480	8.0	OPEN STAR CLUSTER
OCL 0675	07 25 12.	- 31 05	660	11.	OPEN STAR CLUSTER
MCG+09-13-003	07 25 15.	+ 55 37 30.	42	16.	GALAXY
ARC 0577	07 25 16.	+ 79 02		17.5	RICH CLUSTER OF GALAXIES
ZC 0725.3+4520	07 25 18.	+ 45 20	4170		CLUSTER OF GALAXIES
MCG+09-13-004	07 25 18.	+ 51 07	66	16.	GALAXY
UGC 03865	07 25 18.	+ 51 08	60	16.0	GALAXY Sb-c
ZWG 262.004	07 25 18.	+ 56 14		15.7	GALAXY
ZWG 261.070	07 25 18.	+ 56 14		15.7	GALAXY
7ZW 158	07 25 18.	+ 68 10			COMPACT GALAXY
ARC 0583	07 25 20.	+ 43 09		17.6	RICH CLUSTER OF GALAXIES
BC 3CR181	07 25 20.22	+ 14 43 46.6		18.92	QUASI-STELLAR OBJECT
SHB 097	07 25 20.3	+ 14 43 46.		18.9	QUASI-STELLAR OBJECT
RNGC 2386	07 25 22.	+ 33 52			NON-EXISTENT OBJECT
ZWG 117.043	07 25 24.	+ 22 08		15.6	GALAXY
ZWG 147.027	07 25 24.	+ 31 02		15.4	GALAXY
UGC 03866	07 25 24.	+ 31 02	66	15.4	GALAXY S
ZC 0725.4+4310	07 25 24.	+ 43 10	940		CLUSTER OF GALAXIES
MCG+09-13-005B	07 25 24.	+ 56 16	9	16.	GALAXY
MCG+09-13-005A	07 25 24.	+ 56 16	30	16.	GALAXY
MCG+10-11-056	07 25 24.	+ 57 13	36	15.	GALAXY
7ZW 159	07 25 24.	+ 57 15			COMPACT GALAXY
ZWG 286.022	07 25 24.	+ 57 15		15.3	GALAXY
UGC 03867	07 25 24.	+ 57 15	84	15.3	GALAXY PR CMPT
ARC 0584	07 25 28.	+ 26 48		17.3	RICH CLUSTER OF GALAXIES
ZWG 086.032	07 25 30.	+ 18 29		15.2	GALAXY
ZWG 177.021	07 25 30.	+ 35 39		14.9	GALAXY
MCG+06-17-009	07 25 30.	+ 35 39	24	15.	GALAXY
UGC 03868	07 25 30.	+ 40 18	90	17.	GALAXY DWRF IR
MCG+07-16-004	07 25 30.	+ 40 18	60	16.	GALAXY
MCG+09-13-006	07 25 30.	+ 56 15 30.	42	17.	GALAXY
MRK 072	07 25 30.	+ 56 57	8	17.	GALAXY WITH UV CONTINUUM
MCG+10-11-057	07 25 30.	+ 59 10	15	16.	GALAXY
YC 0725-57	07 25 32.	- 57 36 06.			UNUSUAL SOUTHERN NEBULA
RNGC 2388	07 25 34.	+ 33 55		14.5	GALAXY
MCG+04-18-019	07 25 36.	+ 24 33	48	15.1	GALAXY

OBJECT NAME	RIGHT ASCEN.	DECLINATION	DIAM.	MAGN.	TYPE OF OBJECT
ZWG 117.044	07 25 36.	+ 24 35		15.1	GALAXY
TON-N 0285	07 25 36.	+ 28 08		15.3	BLUE STAR
ZWG 177.022	07 25 36.	+ 33 55		14.7	GALAXY
UGC 03870	07 25 36.	+ 33 55	60	14.7	GALAXY
MCG+06-17-010	07 25 36.	+ 33 56	60	14.5	GALAXY
UGC 03869	07 25 36.	+ 34 35	60	15.1	GALAXY S0?
ZWG 177.023	07 25 36.	+ 36 58		15.3	GALAXY
ZC 0725.6+4612	07 25 36.	+ 46 12	1010		CLUSTER OF GALAXIES
RNGC 2387	07 25 38.	+ 36 58		15.5	GALAXY
ZWG 086.033	07 25 42.	+ 17 14		15.5	GALAXY
ZWG 206.011	07 25 42.	+ 42 22		14.9	GALAXY
UGC 03871	07 25 42.	+ 42 22	66	14.9	GALAXY Sc
MCG+10-11-058	07 25 42.	+ 57 26	24	17.	GALAXY
RLWT 151	07 25 42.	- 02 19		13.	FAINT VERY BLUE STAR
MCG+09-13-007	07 25 45.	+ 52 22	42	17.	GALAXY
RNGC 2390	07 25 46.	+ 33 56			NON-EXISTENT OBJECT
RNGC 2389	07 25 46.	+ 33 58		13.5	GALAXY
PK235-01.1	07 25 46.	- 20 07			PLANETARY NEBULA
ZWG 177.024	07 25 48.	+ 33 57		13.5	GALAXY
UGC 03872	07 25 48.	+ 33 57	120	13.5	GALAXY Sc
MCG+06-17-011	07 25 48.	+ 33 58	120	14.	GALAXY
MCG+07-16-005	07 25 48.	+ 42 23	60	15.	GALAXY
MCG+08-14-021	07 25 48.	+ 50 26	42	17.	GALAXY
MRK 089	07 25 48.	+ 52 16	12	15.	GALAXY WITH UV CONTINUUM
MCG+10-11-059	07 25 48.	+ 57 50	15	17.	GALAXY
OCL 0579	07 25 48.	- 11 38	1110	7.6	OPEN STAR CLUSTER
RNGC 2396	07 25 50.	- 11 38		7.5	OPEN CLUSTER
ZWG 086.034	07 25 54.	+ 20 40		14.6	GALAXY
UGC 03873	07 25 54.	+ 20 40	114	14.6	GALAXY Sc
ZWG 147.028	07 25 54.	+ 27 01		15.2	GALAXY
UGC 03874	07 25 54.	+ 27 01	72	15.2	GALAXY PECULR
ZC 0725.9+4400	07 25 54.	+ 44 00	3090		CLUSTER OF GALAXIES
ZWG 262.005	07 25 54.	+ 52 54		15.3	GALAXY
ZWG 261.071	07 25 54.	+ 52 54		15.3	GALAXY
UGC 03875	07 25 54.	+ 52 54	84	15.3	GALAXY Sc/SBc
KARA.73B 0192	07 25 54.	+ 52 54	60	15.3	ISOLATED GALAXY S
MCG+09-13-009	07 25 54.	+ 52 55	72	15.	GALAXY
MCG+09-13-008	07 25 54.	+ 55 29	36	18.	GALAXY
MCG+10-11-060	07 25 54.	+ 57 00	36	16.	GALAXY
MCG+10-11-061	07 25 54.	+ 57 05	24	17.	GALAXY
RNGC 2394	07 25 57.	+ 07 08			NON-EXISTENT OBJECT
RNGC 2391	07 25 58.	+ 33 55			NON-EXISTENT OBJECT
LB 00500	07 25 58.	+ 62 18 06.		15.3	FAINT BLUE STAR
YM 29	07 25 59.	+ 13 27	600		SYMMETRIC GALACTIC NEBULA
KHAV 208	07 26	- 06 24	4910		DARK NEBULA
KHAV 207	07 26	- 24 48	10820		DARK NEBULA
MCG+03-19-017	07 26 00.	+ 20 42	114	14.6	GALAXY
MCG+04-18-020	07 26 00.	+ 26 29	36	15.2	GALAXY
MCG+05-18-014	07 26 00.	+ 27 00	90	15.2	GALAXY
ZWG 262.006	07 26 00.	+ 54 32		15.2	GALAXY
ZWG 261.072	07 26 00.	+ 54 32		15.2	GALAXY
MCG+09-13-010	07 26 00.	+ 54 33	42	16.	GALAXY
MCG+10-11-062	07 26 00.	+ 58 47	27	16.	GALAXY
OCL 0595	07 26 00.	- 15 18	480		OPEN STAR CLUSTER
LDN 1662	07 26 00.	- 23 30	1440		DARK NEBULA
ACK 244-06.1	07 26 00.	- 30 36			PLANETARY NEBULA
ST 04	07 26 02.	- 12 23	1440		EMISSION NEBULA
ZWG 117.045	07 26 06.	+ 26 30		15.2	GALAXY
ZWG 147.029	07 26 06.	+ 28 01		14.4	GALAXY
UGC 03876	07 26 06.	+ 28 01	150	14.4	GALAXY Sc
KARA.73B 0193	07 26 06.	+ 28 01	132	14.4	ISOLATED GALAXY S
ACK 205+14.1	07 26 12.	+ 13 21			PLANETARY NEBULA
MCG+05-18-015	07 26 12.	+ 28 00	126	14.4	GALAXY
MRK 090	07 26 12.	+ 52 52	35	14.	GALAXY WITH UV CONTINUUM
ZC 0726.2+7900	07 26 12.	+ 79 00	610		CLUSTER OF GALAXIES
RNGC 2392	07 26 13.	+ 21 01			PLANETARY NEBULA
PK197+17.1	07 26 13.30	+ 21 00 56.7	47	10.4	PLANETARY NEBULA
ZC 0726.3+4019	07 26 18.	+ 40 19	1480		CLUSTER OF GALAXIES
ZWG 262.007	07 26 18.	+ 52 34		15.7	GALAXY
ZWG 286.023	07 26 18.	+ 62 50		15.3	GALAXY
OCL 0594	07 26 18.	- 15 17	96	15.	OPEN STAR CLUSTER
ZWG 057.014	07 26 24.	+ 14 31		15.2	GALAXY
UGC 03877	07 26 24.	+ 14 31	60	15.2	GALAXY S0
ZC 0726.4+4149	07 26 24.	+ 41 49	470		CLUSTER OF GALAXIES
UGC 03878	07 26 24.	+ 73 02	84	16.0	GALAXY
MCG+06-17-012	07 26 27.	+ 33 47 30.	138	14.5	GALAXY
ZWG 029.028	07 26 30.	+ 06 06		15.5	GALAXY
ZWG 177.025	07 26 30.	+ 33 47		15.3	GALAXY
UGC 03879	07 26 30.	+ 33 47	138	15.3	GALAXY Sc-IRR
ZWG 177.026	07 26 30.	+ 37 33		14.8	GALAXY
UGC 03880	07 26 30.	+ 37 33	60	14.8	GALAXY SB:b-c
MCG+06-17-013	07 26 30.	+ 37 34 30.	48	14.	GALAXY
MCG+08-14-022	07 26 30.	+ 49 36	60	16.	GALAXY
IC 2190	07 26 31.	+ 37 37 17.			NONSTELLAR OBJECT
KEEL 146	07 26 35.7	+ 69 38 52.		17.	SPIRAL NEBULA
ZWG 086.035	07 26 36.	+ 20 08		14.9	GALAXY
MCG+03-19-018	07 26 36.	+ 20 09	45	14.9	GALAXY
ZWG 235.019	07 26 36.	+ 49 37		15.6	GALAXY
UGC 03881	07 26 36.	+ 49 37	90	15.6	GALAXY
UGC 03882	07 26 36.	+ 55 40	102	16.0	GALAXY Sc
MCG+10-11-063	07 26 36.	+ 58 55	27	16.	GALAXY
OCL 0608	07 26 36.	- 18 15	132	16.	OPEN STAR CLUSTER
MCG+04-18-021	07 26 42.	+ 24 31	36		GALAXY
MCG+09-13-011	07 26 42.	+ 55 39 30.	72	16.	GALAXY
MCG+10-11-064	07 26 42.	+ 58 58	39	17.	GALAXY
RNGC 2393	07 26 46.	+ 34 07		15.0	GALAXY
UGC 03883	07 26 48.	+ 07 12	84	18.	GALAXY DWARF
ZWG 177.027	07 26 48.	+ 34 07		14.9	GALAXY
UGC 03884	07 26 48.	+ 34 07	66	14.9	GALAXY Sc
MCG+06-17-014	07 26 48.	+ 34 08	78	14.5	GALAXY
ZWG 235.020	07 26 48.	+ 49 27		15.6	GALAXY
MCG+10-11-065	07 26 48.	+ 58 58	18	16.	GALAXY
ZWG 286.024	07 26 48.	+ 59 36		13.9	GALAXY
UGC 03885	07 26 48.	+ 59 36	60	13.9	GALAXY S
ZWG 206.012	07 26 54.	+ 40 02		15.2	GALAXY
MCG+10-11-066	07 26 54.	+ 59 36	66	14.	GALAXY
UGC 03886	07 26 54.	+ 62 35	72	16.0	GALAXY Sc
MCG+10-11-067	07 26 54.	+ 62 35	72	14.	GALAXY
ZC 0726.9+6634	07 26 54.	+ 66 34	1340		CLUSTER OF GALAXIES
KHAV 209	07 27	- 08 24	5930		DARK NEBULA
LBN 1057	07 27	- 22 00	1140		BRIGHT NEBULA
LB 09836	07 27	- 82 22		14.5	FAINT BLUE STAR
ZWG 177.028	07 27 00.	+ 37 28		15.7	GALAXY
MCG+10-11-068	07 27 00.	+ 58 02	45	16.	GALAXY
7ZW 160	07 27 00.	+ 73 29			COMPACT GALAXY
ZWG 177.029	07 27 06.	+ 36 12		15.3	GALAXY
UGC 03887	07 27 06.	+ 36 12	78	15.3	GALAXY SBc
MCG+06-17-015	07 27 06.	+ 36 13	60	15.	GALAXY
ZC 0727.1+4057	07 27 06.	+ 40 57	1810		CLUSTER OF GALAXIES
MCG+11-09-052	07 27 06.	+ 66 57	42	17.	GALAXY
OCL 0588	07 27 06.	- 13 52	300	12.7	OPEN STAR CLUSTER
PK248-08.1	07 27 06.	- 35 38 39.	7		PLANETARY NEBULA
RNGC 2401	07 27 09.	- 13 52		12.5	OPEN CLUSTER
IC 2195	07 27 10.	- 51 09			NONSTELLAR OBJECT
HN 0312	07 27 11.	- 51 09			NEBULA
MCG+04-18-022	07 27 12.	+ 24 34	30	15.6	GALAXY
ZWG 117.046	07 27 12.	+ 24 35		15.6	GALAXY
MRK 073	07 27 12.	+ 63 20	15	15.	GALAXY WITH UV CONTINUUM
MCG+11-09-053	07 27 12.	+ 63 20	18	16.	GALAXY
ZWG 309.035	07 27 12.	+ 63 22		14.9	GALAXY
OCL 0664	07 27 12.	- 29 07	168	12.	OPEN STAR CLUSTER
MCG+04-18-023	07 27 15.	+ 24 34	24	15.3	GALAXY
MCG+08-14-023	07 27 15.	+ 46 06	30	18.	GALAXY
IC 2191	07 27 16.	+ 24 25 52.			NONSTELLAR OBJECT
HN 0009	07 27 16.	- 00 06			NEBULA
RNGC 2398	07 27 17.	+ 24 35		15.5	GALAXY
OCL 0516	07 27 18.	+ 12 00	1440		OPEN STAR CLUSTER
MCG+04-18-024	07 27 18.	+ 24 24	42	15.3	GALAXY
ZWG 117.047	07 27 18.	+ 24 25		15.3	GALAXY
ZWG 117.048	07 27 18.	+ 24 35		15.3	GALAXY
RNGC 2399	07 27 19.	- 00 15			NON-EXISTENT OBJECT
KEEL 147	07 27 19.9	+ 69 38 40.		17.	SPIRAL NEBULA
HN 0010	07 27 22.	- 00 06			NEBULA
ZWG 029.029	07 27 24.	+ 05 55		15.3	GALAXY
7ZW 161	07 27 24.	+ 64 13			COMPACT GALAXY
RLWT 152	07 27 24.	- 02 00		13.5	FAINT VERY BLUE STAR
DV.56 I2200A	07 27 24.	- 62 10	48		E GALAXY
RNGC 2400	07 27 25.	- 00 15			NON-EXISTENT OBJECT
ARC 0585	07 27 29.	+ 40 58		17.0	RICH CLUSTER OF GALAXIES
MCG+07-16-006	07 27 30.	+ 39 07	72	15.	GALAXY
MCG+14-04-034	07 27 30.	+ 83 54	51	16.	GALAXY
SER 055.01	07 27 30.	- 62 15	150	15.	INTERACTING GALAXIES
ZWG 206.013	07 27 36.	+ 39 07		15.5	GALAXY
UGC 03888	07 27 36.	+ 39 07	84	15.5	GALAXY SBb
MRK 074	07 27 36.	+ 55 21	10	16.	GALAXY WITH UV CONTINUUM
ISS 0718	07 27 36.	- 21 02	305		STELLAR RING
OCL 0491	07 27 42.	+ 09 58	1140	6.9	OPEN STAR CLUSTER
HN 0313	07 27 42.	- 62 14			NEBULA
IC 2200	07 27 42.	- 62 14			NONSTELLAR OBJECT
CED 100	07 27 43.	- 22 55	600		DIFFUSE GALACTIC NEBULA
ZWG 331.003	07 27 48.	+ 73 45		15.1	GALAXY
ZWG 330.043	07 27 48.	+ 73 45		15.1	GALAXY
UGC 03889	07 27 48.	+ 73 45	78	15.1	GALAXY Sa-b
MRSL 233-00/1	07 27 48.	- 18 25	270		HII REGION
ST 03	07 27 49.	- 18 28	180		EMISSION NEBULA
HN 0314	07 27 53.	- 67 27			NEBULA
IC 2202	07 27 53.	- 67 27			NONSTELLAR OBJECT
ZWG 086.036	07 27 54.	+ 18 25		15.3	GALAXY
MCG+03-19-019	07 27 54.	+ 18 26	66	15.3	GALAXY
ZC 0727.9+5608	07 27 54.	+ 56 08	740		CLUSTER OF GALAXIES
ZWG 349.006	07 27 54.	+ 75 36		15.6	GALAXY
LBN 1048	07 28	- 18 29	180		BRIGHT NEBULA
LBN 1050	07 28	- 19 03	1800		BRIGHT NEBULA
KHAV 210	07 28	- 22 54	8630		DARK NEBULA
ZWG 057.015	07 28 00.	+ 09 45		15.4	GALAXY
UGC 03891	07 28 00.	+ 09 45	72	15.4	GALAXY DBL SYS
MCG+02-19-004	07 28 00.	+ 09 45	15	15.4	GALAXY
ZWG 057.016	07 28 00.	+ 13 10		15.4	GALAXY
ZWG 086.037	07 28 00.	+ 19 51		15.7	GALAXY
MCG+12-08-003	07 28 00.	+ 73 45 30.	60	16.	GALAXY
ZWG 363.031	07 28 00.	+ 83 54		14.9	GALAXY
UGC 03890	07 28 00.	+ 83 54	78	14.9	GALAXY
KARA.73B 0194	07 28 00.	+ 83 54	72	14.9	ISOLATED GALAXY S
RNGC 2402	07 28 02.	+ 09 45		15.5	GALAXY
ZWG 057.017	07 28 06.	+ 13 12		15.5	GALAXY
ZWG 147.030	07 28 06.	+ 31 50		15.4	GALAXY
ZWG 086.038	07 28 12.	+ 15 02		15.5	GALAXY
ZC 0728.2+3951	07 28 12.	+ 39 51	4770		CLUSTER OF GALAXIES
MCG+10-11-069	07 28 12.	+ 57 06	36	17.	GALAXY
ZWG 286.025	07 28 12.	+ 58 21		15.7	GALAXY
ZWG 286.026	07 28 12.	+ 59 28		15.6	GALAXY
MCG+11-10-001	07 28 12.	+ 65 01	39	16.	GALAXY
MCG+05-18-016	07 28 15.	+ 31 50 30.	54	15.4	GALAXY
MCG+11-10-002	07 28 15.	+ 65 05	45	15.	GALAXY
ZWG 057.018	07 28 18.	+ 13 10		15.1	GALAXY
UGC 03892	07 28 18.	+ 13 10	84	15.1	GALAXY Sc
MCG+02-19-005	07 28 18.	+ 13 10	84	15.1	GALAXY
ZWG 086.039	07 28 18.	+ 18 20		15.0	GALAXY
MCG+03-19-020	07 28 18.	+ 18 20	60	15.0	GALAXY
MCG+10-11-070	07 28 18.	+ 59 28	39	15.	GALAXY
ZWG 309.036	07 28 18.	+ 65 07		14.7	GALAXY
UGC 03893	07 28 18.	+ 65 07	66	14.7	GALAXY Sb
OCL 0632	07 28 18.	- 23 13	180	13.	OPEN STAR CLUSTER
MCG+11-10-003	07 28 21.	+ 65 10	51	15.	GALAXY
MCG+09-13-013	07 28 24.	+ 56 49	18	17.	GALAXY
ZWG 309.037	07 28 24.	+ 65 12		14.5	GALAXY
UGC 03894	07 28 24.	+ 65 12	102	14.5	GALAXY E
MRSL 231+01/1	07 28 24.	- 15 11	60		HII REGION
MRSL 234-00/1	07 28 24.	- 19 00	1800		HII REGION
MCG-05-18-007	07 28 24.	- 31 30	60	15.	GALAXY
ZWG 087.001	07 28 30.	+ 19 00		15.7	GALAXY
ZWG 086.040	07 28 30.	+ 19 00		15.7	GALAXY
MCG+09-13-012	07 28 30.	+ 52 39	30	17.	GALAXY
MCG+10-11-071	07 28 30.	+ 58 19	27	16.	GALAXY
ZWG 147.031	07 28 36.	+ 30 39		15.7	GALAXY
MCG+08-14-024	07 28 36.	+ 50 14	30	17.	GALAXY
7ZW 162	07 28 36.	+ 60 59			COMPACT GALAXY
7ZW 163	07 28 36.	+ 63 22			COMPACT GALAXY
UGC 03895	07 28 42.	+ 00 09	60	17.	GALAXY DWARF
ZWG 058.001	07 28 42.	+ 13 16		15.3	GALAXY
ZWG 057.019	07 28 42.	+ 13 16		15.3	GALAXY
MCG+04-18-025	07 28 42.	+ 25 42	36	15.3	GALAXY
ZWG 117.049	07 28 42.	+ 25 43		15.3	GALAXY
MRK 091	07 28 42.	+ 52 46	12	15.5	GALAXY WITH UV CONTINUUM
MCG+10-11-072	07 28 42.	+ 57 47	39	16.	GALAXY
ZWG 286.027	07 28 42.	+ 58 25		15.6	GALAXY
ZC 0728.7+6607	07 28 42.	+ 66 07	1280		CLUSTER OF GALAXIES
ZWG 117.050	07 28 48.	+ 21 36		15.6	GALAXY
MCG+10-11-073	07 28 48.	+ 58 22	27	16.	GALAXY
MRSL 231+01/2	07 28 48.	- 15 18	180		HII REGION
MCG+03-19-021	07 28 54.	+ 18 22	12	15.0	GALAXY
ZWG 087.002	07 28 54.	+ 18 24		15.0	GALAXY
ZWG 086.041	07 28 54.	+ 18 24		15.0	GALAXY
OCL 0574	07 28 54.	- 09 52	180		OPEN STAR CLUSTER
RNGC 2406	07 28 57.	+ 18 24		15.0	GALAXY
TON-N 0822	07 28 58.	+ 35 23		16.8	BLUE STAR

OBJECT NAME	RIGHT ASCEN.	DECLINATION	DIAM.	MAGN.	TYPE OF OBJECT
KEEL 148	07 28 58.6	+ 65 33 18.			NEBULA
VDB.66G 044	07 29	+ 66 59	100		DWARF GALAXY
KHAV 212	07 29	- 32 00	1990		DARK NEBULA
ZWG 087.003	07 29 00.	+ 18 27		14.9	GALAXY
ZWG 086.042	07 29 00.	+ 18 27		14.9	GALAXY
UGC 03896	07 29 00.	+ 18 27	78	14.9	GALAXY E-SO
ZWG 147.032	07 29 00.	+ 31 59		15.6	GALAXY
MRK 075	07 29 00.	+ 55 19	10	16.	GALAXY WITH UV CONTINUUM
ZWG 286.028	07 29 00.	+ 59 45		14.2	GALAXY
UGC 03897	07 29 00.	+ 59 45	90	14.2	GALAXY SB0
MCG+10-11-074	07 29 00.	+ 59 45	72	14.	GALAXY
MCG+10-11-075	07 29 00.	+ 61 40	39	16.	GALAXY
MCG+11-10-004	07 29 00.	+ 65 33	30	16.	GALAXY
ZWG 310.001	07 29 00.	+ 65 34		15.3	GALAXY
ZWG 309.038	07 29 00.	+ 65 34		15.3	GALAXY
UGC 03898	07 29 00.	+ 65 34	72	15.3	GALAXY Sb
KARA.73B 0195	07 29 00.	+ 65 34	66	15.3	ISOLATED GALAXY S
RNGC 2407	07 29 03.	+ 18 27		15.0	GALAXY
ARC 0586	07 29 05.	+ 31 44		17.4	RICH CLUSTER OF GALAXIES
MCG+03-20-001	07 29 06.	+ 18 25 30.	54	14.9	GALAXY
MCG+04-18-026	07 29 06.	+ 26 00	36	14.7	GALAXY
ZC 0729.1+3746	07 29 06.	+ 37 46	1080		CLUSTER OF GALAXIES
RNGC 2405	07 29 10.	+ 26 01		14.5	GALAXY
LB 00501	07 29 11.	+ 82 13 30.		16.7	FAINT BLUE STAR
ZWG 087.004	07 29 12.	+ 17 55		15.7	GALAXY
ZWG 086.043	07 29 12.	+ 17 55		15.7	GALAXY
ZWG 087.005	07 29 12.	+ 18 26		15.4	GALAXY
ZWG 086.044	07 29 12.	+ 18 26		15.4	GALAXY
ZWG 117.051	07 29 12.	+ 26 01		14.7	GALAXY
ZWG 147.033	07 29 12.	+ 27 08		15.5	GALAXY
ZWG 177.030	07 29 12.	+ 36 17		15.5	GALAXY
ZC 0729.2+4734	07 29 12.	+ 47 34	4570		CLUSTER OF GALAXIES
ST 02	07 29 16.	- 16 52	1200		EMISSION NEBULA
TON-N 0823	07 29 17.	+ 33 13		14.8	BLUE STAR
MCG+03-20-002	07 29 18.	+ 18 25	36	15.4	GALAXY
ZWG 147.034	07 29 18.	+ 27 13		15.6	GALAXY
ZWG 177.031	07 29 18.	+ 35 43		15.4	GALAXY
UGC 03899	07 29 18.	+ 35 43	66	15.4	GALAXY Sa-b
KARA.73B 0196	07 29 18.	+ 35 43	78	15.4	ISOLATED GALAXY S
MCG+06-17-016	07 29 18.	+ 35 44	66	15.	GALAXY
MCG+06-17-017	07 29 18.	+ 36 19	36	17.	GALAXY
MCG+09-13-014	07 29 18.	+ 54 47 30.	30	16.	GALAXY
MCG+09-13-015	07 29 18.	+ 55 10	48	17.	GALAXY
ZWG 286.029	07 29 18.	+ 57 50		15.7	GALAXY
MCG+10-11-076	07 29 18.	+ 58 00	42	17.	GALAXY
ZWG 002.001	07 29 18.	- 02 07		15.7	GALAXY
ZWG 002.001	07 29 18.	- 02 07		15.7	GALAXY
BC 3C184.1/140	07 29 20.	+ 81 53 00.		17.5	QUASI-STELLAR OBJECT
SHB 098	07 29 20.0	+ 81 52 36.		17.5	QUASI-STELLAR OBJECT
RNGC 2409	07 29 23.	- 17 05			NON-EXISTENT OBJECT
ZC 0729.4+3142	07 29 24.	+ 31 42	1810		CLUSTER OF GALAXIES
ZC 0729.4+3932	07 29 24.	+ 39 32	670		CLUSTER OF GALAXIES
MCG+10-11-077	07 29 24.	+ 57 48	15	16.	GALAXY
MCG+10-11-078	07 29 24.	+ 59 54	18	17.	GALAXY
DG 121	07 29 24.	- 16 52	1230		REFLECTION NEBULA
MRSL 232+00/1	07 29 24.	- 16 52	1260		HII REGION
ZC 0729.5+6750	07 29 30.	+ 67 50	2150		CLUSTER OF GALAXIES
RNGC 2417	07 29 30.	- 62 09			GALAXY
ARC 0587	07 29 36.	+ 39 33		16.6	RICH CLUSTER OF GALAXIES
OCL 0586	07 29 36.	- 12 39	120	11.	OPEN STAR CLUSTER
PK234-00.1	07 29 36.	- 19 21	1		PLANETARY NEBULA
SCHO 0248	07 29 41.	- 14 36 12.	450		ISOLATED DARK CLOUD
ZWG 030.001	07 29 42.	+ 07 12		15.7	GALAXY
ZWG 058.002	07 29 42.	+ 11 00		15.0	GALAXY
UGC 03900	07 29 42.	+ 11 00	60	15.0	GALAXY SBb
MCG+02-20-001	07 29 42.	+ 11 00	60	15.0	GALAXY
ZC 0729.7+3900	07 29 42.	+ 39 00	1210		CLUSTER OF GALAXIES
ZWG 286.030	07 29 42.	+ 60 58		15.7	GALAXY
ZWG 087.006	07 29 48.	+ 17 00		15.4	GALAXY
ZWG 087.007	07 29 48.	+ 18 42		15.5	GALAXY
MCG+03-20-003	07 29 48.	+ 18 42	10	15.5	GALAXY
MCG+10-11-079	07 29 48.	+ 59 29	27	16.	GALAXY
MCG+10-11-080	07 29 48.	+ 61 04	24	16.	GALAXY
7ZW 164	07 29 48.	+ 68 22			COMPACT GALAXY
YM 36	07 29 49.	- 19 18	600		SYMMETRIC GALACTIC NEBULA
ZWG 286.031	07 29 54.	+ 58 50		15.6	GALAXY
MCG+10-11-081	07 29 54.	+ 60 58	24	16.	GALAXY
MRSL 234-00/2	07 29 54.	- 19 19	720		HII REGION
KHAV 211	07 30	- 03 54	9600		DARK NEBULA
LBN 1046	07 30	- 16 54	1200		BRIGHT NEBULA
LBN 1053	07 30	- 19 22	600		BRIGHT NEBULA
MCG+03-20-004	07 30 00.	+ 19 19	90	14.7	GALAXY
ZWG 177.032	07 30 00.	+ 32 40		15.4	GALAXY
ZWG 147.035	07 30 00.	+ 32 40		15.4	GALAXY
MCG+10-11-082	07 30 00.	+ 57 43	42	15.	GALAXY
MCG+10-11-083	07 30 00.	+ 58 49	36	16.	GALAXY
LDN 1663	07 30 00.	- 23 00	2640		DARK NEBULA
ZC 0730.1+1858	07 30 06.	+ 18 58	10480		CLUSTER OF GALAXIES
MCG+08-14-025	07 30 06.	+ 49 23	48	16.	GALAXY
ZWG 235.021	07 30 06.	+ 49 24		15.3	GALAXY
UGC 03901	07 30 06.	+ 49 24	60	15.3	GALAXY SB
IC 2192	07 30 07.	+ 31 28 02.			NONSTELLAR OBJECT
IC 2193	07 30 08.	+ 31 33 14.			NONSTELLAR OBJECT
TON-N 0824	07 30 11.	+ 32 24		16.8	BLUE STAR
ZWG 117.052	07 30 12.	+ 24 10		15.4	GALAXY
ZWG 147.036	07 30 12.	+ 30 37		15.7	GALAXY
ZWG 147.037	07 30 12.	+ 31 36		14.7	GALAXY
UGC 03902	07 30 12.	+ 31 36	96	14.7	GALAXY Sb
MCG+05-18-017	07 30 12.	+ 32 39 30.	48	15.4	GALAXY
VDB.66N 097	07 30 15.	- 16 48	132		REFLECTION NEBULA
ZWG 087.008	07 30 18.	+ 19 18		14.7	GALAXY
UGC 03903	07 30 18.	+ 19 18	102	14.7	GALAXY SBa
ZWG 147.038	07 30 18.	+ 30 40		14.6	GALAXY
UGC 03904	07 30 18.	+ 30 40	84	14.6	GALAXY SB0/SBa
MCG+05-18-018	07 30 18.	+ 31 36	84	14.7	GALAXY
ZWG 286.032	07 30 18.	+ 62 39		14.4	GALAXY
UGC 03905	07 30 18.	+ 62 39	60	14.4	GALAXY Sa-b
ZWG 310.002	07 30 18.	+ 65 08		15.1	GALAXY
ZWG 309.039	07 30 18.	+ 65 08		15.1	GALAXY
ISS 0672	07 30 18.	- 15 20	308		STELLAR RING
ZWG 087.009	07 30 24.	+ 18 24		15.6	GALAXY
MCG+05-18-019	07 30 24.	+ 30 40	78	14.6	GALAXY
ZWG 286.033	07 30 24.	+ 57 39		15.7	GALAXY
MCG+10-11-084	07 30 24.	+ 61 53	42	17.	GALAXY
MCG+10-11-085	07 30 24.	+ 62 40	45	14.	GALAXY
ZWG 349.007	07 30 24.	+ 74 34		14.8	GALAXY
UGC 03906	07 30 24.	+ 74 34	102	14.8	GALAXY DBL SYS
ZWG 330.044	07 30 24.	+ 74 37		14.8	GALAXY

OBJECT NAME	RIGHT ASCEN.	DECLINATION	DIAM.	MAGN.	TYPE OF OBJECT
KARA.72 137A	07 30 24.	+ 74 37	42	14.8	PART OF DOUBLE GALAXY
MCG+11-10-005	07 30 27.	+ 65 07	36	16.	GALAXY
IC 2194	07 30 28.	+ 31 26 17.			NONSTELLAR OBJECT
ZWG 147.039	07 30 30.	+ 31 27		15.1	GALAXY
ZWG 177.033	07 30 30.	+ 37 07		15.7	GALAXY
UGC 03907	07 30 30.	+ 37 07	84	15.7	GALAXY Sc
ZWG 262.008	07 30 30.	+ 54 47		15.7	GALAXY
MCG+09-13-016	07 30 30.	+ 54 48	36	16.	GALAXY
MRK 076	07 30 30.	+ 65 07	25	15.5	GALAXY WITH UV CONTINUUM
UGC 03908	07 30 30.	+ 66 31	66	16.5	GALAXY Sb
ZWG 331.004	07 30 30.	+ 74 06		15.5	GALAXY
ZWG 330.045	07 30 30.	+ 74 06		15.5	GALAXY
KARA.72 137B	07 30 30.	+ 74 37	48		PART OF DOUBLE GALAXY
VHA 013	07 30 30.	- 38 25	900		OPEN STAR CLUSTER
MCG+05-18-020	07 30 33.	+ 31 26	60	15.1	GALAXY
ZWG 002.002	07 30 36.	+ 00 58		15.7	GALAXY
ZWG 002.002	07 30 36.	+ 00 58		15.7	GALAXY
MCG+06-17-018	07 30 36.	+ 37 09	72	15.	GALAXY
MCG+10-11-086A	07 30 36.	+ 57 36	10	17.	GALAXY
MCG+10-11-086	07 30 36.	+ 57 36	24	16.	GALAXY
ZWG 286.034	07 30 36.	+ 59 08		15.6	GALAXY
ZWG 262.009	07 30 42.	+ 54 55		15.2	GALAXY
ZWG 262.010	07 30 42.	+ 56 03		15.1	GALAXY
MCG+10-11-087	07 30 42.	+ 58 01	12	16.	GALAXY
MCG+10-11-088	07 30 42.	+ 59 07	24	16.	GALAXY
MCG+10-11-089	07 30 48.	+ 58 04	12	17.	GALAXY
MCG+11-10-006	07 30 48.	+ 66 31	78	16.	GALAXY
ZWG 331.005	07 30 48.	+ 73 49		14.9	GALAXY
ZWG 330.046	07 30 48.	+ 73 49		14.9	GALAXY
UGC 03909	07 30 48.	+ 73 49	168	14.9	GALAXY SB:c
MCG+12-08-004	07 30 48.	+ 74 07 30.	33	16.	GALAXY
MCG+12-08-005	07 30 48.	+ 74 33 30.	66	16.	GALAXY
TON-N 0825	07 30 52.	+ 34 10		15.4	BLUE STAR
LB 00502	07 30 53.	+ 81 13 18.		16.0	FAINT BLUE STAR
ZWG 147.040	07 30 54.	+ 31 31		14.0	GALAXY
UGC 03910	07 30 54.	+ 31 31	90	14.0	GALAXY E
MCG+09-13-017	07 30 54.	+ 56 05	24	15.	GALAXY
RNGC 2413	07 30 56.	- 13 00			NON-EXISTENT OBJECT
IC 2196	07 30 58.	+ 31 30 38.			NONSTELLAR OBJECT
RNGC 2414	07 30 58.	- 15 20		8.0	OPEN CLUSTER
KHAV 213	07 31	- 16 36	4600		DARK NEBULA
LBN 1061	07 31	- 23 50	840		BRIGHT NEBULA
ZC 0731.0+3905	07 31 00.	+ 39 05	540		CLUSTER OF GALAXIES
ZC 0731.0+4616	07 31 00.	+ 46 16	2020		CLUSTER OF GALAXIES
MCG+12-08-006	07 31 00.	+ 73 49	48	16.	GALAXY
OCL 0598	07 31 00.	- 15 20	600	8.5	OPEN STAR CLUSTER
LB 00503	07 31 02.	+ 61 39 00.		14.4	FAINT BLUE STAR
ZWG 087.010	07 31 06.	+ 15 05		15.3	GALAXY
ZWG 117.053	07 31 06.	+ 23 13		15.2	GALAXY
UGC 03911	07 31 06.	+ 23 13	78	15.2	GALAXY S0
ZWG 117.054	07 31 06.	+ 24 05		15.3	GALAXY
MCG+05-18-021	07 31 06.	+ 31 31	30	14.0	GALAXY
IC 2198	07 31 11.	+ 24 04 47.			NONSTELLAR OBJECT
ZWG 087.011	07 31 12.	+ 18 48		15.5	GALAXY
IC 2197	07 31 12.	+ 31 29 12.			NONSTELLAR OBJECT
MCG+10-11-090	07 31 12.	+ 59 36	24	17.	GALAXY
OCL 0634	07 31 12.	- 20 23	210	13.	OPEN STAR CLUSTER
ZWG 147.041	07 31 18.	+ 30 35		15.7	GALAXY
ZWG 087.012	07 31 24.	+ 20 25		15.6	GALAXY
MCG+09-13-018	07 31 24.	+ 54 57 30.	36	16.	GALAXY
ZC 0731.4+5546	07 31 24.	+ 55 46	610		CLUSTER OF GALAXIES
7ZW 165	07 31 24.	+ 67 51			COMPACT GALAXY
ZC 0731.4+7721	07 31 24.	+ 77 21	740		CLUSTER OF GALAXIES
TON-N 0826	07 31 28.	+ 34 29		14.9	BLUE STAR
ZC 0731.5+3937	07 31 30.	+ 39 37	940		CLUSTER OF GALAXIES
ZWG 262.011	07 31 30.	+ 56 08		15.4	GALAXY
MCG+10-11-091	07 31 30.	+ 59 43	42	16.	GALAXY
ZWG 030.002	07 31 36.	+ 04 39		14.5	GALAXY
UGC 03912	07 31 36.	+ 04 39	126	14.5	GALAXY IRR
MCG+01-20-001	07 31 36.	+ 04 39 30.	96	14.5	GALAXY
ZWG 177.034	07 31 36.	+ 33 38		15.7	GALAXY
UGC 03913	07 31 36.	+ 33 38	96	15.7	GALAXY S
MCG+06-17-019	07 31 36.	+ 33 38	96	15.	GALAXY
ZWG 002.003	07 31 36.	- 01 53		15.6	GALAXY
ISS 0832	07 31 36.	- 21 26	637		STELLAR RING
RNGC 2412	07 31 39.	+ 08 39			NON-EXISTENT OBJECT
UGC 03914	07 31 42.	+ 13 23	66	14.6	GALAXY E:
ZWG 087.013	07 31 42.	+ 18 23		14.6	GALAXY
MCG+03-20-005	07 31 42.	+ 18 23	36	14.6	GALAXY
ZWG 147.042	07 31 42.	+ 31 23		13.6	GALAXY
UGC 03915	07 31 42.	+ 31 23	72	13.6	GALAXY S
ZC 0731.7+4515	07 31 42.	+ 45 15	1140		CLUSTER OF GALAXIES
IC 2199	07 31 44.	+ 31 19 38.			NONSTELLAR OBJECT
RNGC 2411	07 31 45.	+ 18 23		14.5	GALAXY
RNGC 2410	07 31 47.	+ 32 56		14.0	GALAXY
ZWG 117.055	07 31 48.	+ 22 41		15.7	GALAXY
UGC 03916	07 31 48.	+ 22 41	72	15.7	GALAXY Sc
ZWG 177.035	07 31 48.	+ 32 56		13.9	GALAXY
ZWG 147.043	07 31 48.	+ 32 56		13.9	GALAXY
UGC 03917	07 31 48.	+ 32 56	150	13.9	GALAXY SB?b
ZWG 331.006	07 31 48.	+ 71 23		15.6	GALAXY
ZWG 330.047	07 31 48.	+ 71 23		15.6	GALAXY
MCG+05-18-022	07 31 51.	+ 31 23	60	13.6	GALAXY
MCG+04-18-027	07 31 54.	+ 22 40	66	15.7	GALAXY
ZWG 117.056	07 31 54.	+ 24 34		15.7	GALAXY
ZC 0731.9+3125	07 31 54.	+ 31 25	5380		CLUSTER OF GALAXIES
MCG+05-18-023	07 31 54.	+ 32 57	138	13.9	GALAXY
MRK 092	07 31 54.	+ 46 40	10	15.	GALAXY WITH UV CONTINUUM
ZWG 262.012	07 31 54.	+ 51 50		15.6	GALAXY
RNGC 2403	07 31 59.	+ 65 43		9.5	GALAXY
REIN 2.055	07 31 59.95	+ 65 42 24.1			NEBULA
VDB.66G 045	07 32	+ 02 53	130		DWARF GALAXY
KHAV 214	07 32	- 14 12	5630		DARK NEBULA
ZWG 262.013	07 32 00.	+ 51 52		15.1	GALAXY
ZC 0732.0+5603	07 32 00.	+ 56 03	1010		CLUSTER OF GALAXIES
MCG+10-11-092	07 32 00.	+ 62 27	42	15.	GALAXY
ZWG 310.003	07 32 00.	+ 65 43		9.3	GALAXY
ZWG 309.040	07 32 00.	+ 65 43		9.3	GALAXY
UGC 03918	07 32 00.	+ 65 43	1680	9.3	GALAXY Sc
KARA.73B 0197	07 32 00.	+ 65 43	1680	9.3	ISOLATED GALAXY S
ZWG 331.007	07 32 00.	+ 73 51		15.5	GALAXY
ZWG 330.048	07 32 00.	+ 73 51		15.5	GALAXY
ZWG 363.032	07 32 00.	+ 81 15		15.1	GALAXY
MCG+14-04-035	07 32 00.	+ 85 40	27	16.	GALAXY
ZWG 363.033	07 32 00.	+ 85 41		15.5	GALAXY
ZWG 362.045	07 32 00.	+ 85 41		15.5	GALAXY
SN 1954J	07 32 01.	+ 65 45		16.0	SUPERNOVA
MCG+11-10-007	07 32 03.	+ 65 43	1080	8.8	GALAXY

OBJECT NAME	RIGHT ASCEN.	DECLINATION	DIAM.	MAGN.	TYPE OF OBJECT
MCG+10-11-093	07 32 06.	+ 57 33	10	16.	GALAXY
7ZW 166	07 32 06.	+ 65 04			COMPACT GALAXY
MCG+12-08-007	07 32 06.	+ 73 53	150	14.	GALAXY
ZWG 058.003	07 32 12.	+ 11 12		15.7	GALAXY
MCG+07-16-007	07 32 12.	+ 39 52	30	17.5	GALAXY
MCG+08-14-026	07 32 12.	+ 49 02	36	16.	GALAXY
UGC 03919	07 32 12.	+ 64 40	96	16.5	GALAXY Sc
RNGC 2404	07 32 17.	+ 65 48			NON-EXISTENT OBJECT
ZC 0732.3+4442	07 32 18.	+ 44 42	1880		CLUSTER OF GALAXIES
ZWG 286.035	07 32 18.	+ 59 25		15.4	GALAXY
MCG+11-10-008	07 32 21.	+ 64 40	90	16.	GALAXY
ZWG 087.014	07 32 24.	+ 19 10		14.5	GALAXY
UGC 03920	07 32 24.	+ 19 10	84	14.5	GALAXY Sb
ZC 0732.4+3447	07 32 24.	+ 34 47	2550		CLUSTER OF GALAXIES
MCG+10-11-094	07 32 24.	+ 59 23	45	15.	GALAXY
ARP 250	07 32 29.	+ 35 29			PECULIAR GALAXY
MCG+03-20-006	07 32 30.	+ 19 09	72	14.5	GALAXY
UGC 03921	07 32 30.	+ 70 59	72	16.5	GALAXY Sc
ZWG 058.004	07 32 36.	+ 11 50		15.6	GALAXY
UGC 03922	07 32 36.	+ 55 10	102	16.0	GALAXY Sc
MRK 009	07 32 36.	+ 58 52	25	14.5	GALAXY WITH UV CONTINUUM
KW 04	07 32 36.	+ 58 52	22		SEYFERT GALAXY
VVI 21	07 32 36.	+ 58 52	24	14.77	SEYFERT GALAXY
7ZW 167	07 32 36.	+ 67 50			COMPACT GALAXY
ZWG 058.005	07 32 42.	+ 12 25		15.4	GALAXY
ZWG 262.014	07 32 42.	+ 55 25		14.9	GALAXY
UGC 03923	07 32 42.	+ 55 25	60	14.9	GALAXY Sb
ZWG 286.036	07 32 42.	+ 58 54		15.2	GALAXY
7ZW 168	07 32 42.	+ 66 50			COMPACT GALAXY
ZWG 058.006	07 32 48.	+ 09 27		15.7	GALAXY
ZWG 058.007	07 32 48.	+ 11 39		15.7	GALAXY
UGC 03924	07 32 48.	+ 11 39	66	15.7	GALAXY DWRF SP
ZWG 117.057	07 32 48.	+ 24 35		15.4	GALAXY
MCG+09-13-019	07 32 48.	+ 53 10	30	17.	GALAXY
MCG+09-13-020	07 32 48.	+ 55 10	90	15.	GALAXY
MCG+10-11-095	07 32 48.	+ 61 22	24	16.	GALAXY
MCG+11-10-009	07 32 48.	+ 66 27	51	17.	GALAXY
ZWG 002.004	07 32 54.	+ 01 04		15.6	GALAXY
ZWG 002.004	07 32 54.	+ 01 04		15.6	GALAXY
MCG+00-20-001	07 32 54.	+ 01 04	36	15.6	GALAXY
ZWG 058.008	07 32 54.	+ 11 43		14.3	GALAXY
UGC 03925	07 32 54.	+ 11 43	66	14.3	GALAXY Sc
MCG+02-20-002	07 32 54.	+ 11 43	60	14.3	GALAXY
ZWG 058.009	07 32 54.	+ 11 49		15.0	GALAXY
MCG+02-20-003	07 32 54.	+ 11 49	48	15.0	GALAXY
MCG+09-13-021	07 32 54.	+ 55 27 30.	60	15.	GALAXY
ZWG 310.004	07 32 54.	+ 62 26		15.3	GALAXY
ZWG 286.037	07 32 54.	+ 62 27		15.3	GALAXY
RNGC 2416	07 32 55.	+ 11 43		14.5	GALAXY
LBN 1052	07 33	- 16 40	240		BRIGHT NEBULA
ZWG 087.015	07 33 00.	+ 18 10		15.0	GALAXY
MCG+06-17-020	07 33 00.	+ 33 13	84	15.	GALAXY
ZWG 177.036	07 33 00.	+ 33 14		14.9	GALAXY
UGC 03926	07 33 00.	+ 33 14	78	14.9	GALAXY Sa
ZWG 286.038	07 33 00.	+ 57 43		15.7	GALAXY
MCG+10-11-096	07 33 00.	+ 62 27	18	15.	GALAXY
ZC 0733.0+7008	07 33 00.	+ 70 08	1480		CLUSTER OF GALAXIES
LDN 1665	07 33 00.	- 23 30	1680		DARK NEBULA
IC 2201	07 33 03.	+ 33 13 59.			NONSTELLAR OBJECT
TON-N 0827	07 33 04.	+ 34 07		16.	BLUE STAR
ZC 0733.1+1514	07 33 06.	+ 15 14	1750		CLUSTER OF GALAXIES
MCG+03-20-007	07 33 06.	+ 18 09	42	15.0	GALAXY
ZC 0733.1+3729	07 33 06.	+ 37 29	1210		CLUSTER OF GALAXIES
MCG+10-11-097	07 33 06.	+ 57 42	36	16.	GALAXY
ZWG 286.039	07 33 06.	+ 59 49		15.5	GALAXY
UGC 03927	07 33 06.	+ 59 49	78	15.5	GALAXY S0
MRK 093	07 33 06.	+ 66 23	10	16.5	GALAXY WITH UV CONTINUUM
TON-N 0828	07 33 11.	+ 32 12		16.8	BLUE STAR
MCG+10-11-098	07 33 12.	+ 62 24	36	16.	GALAXY
MCG+11-10-010	07 33 12.	+ 65 06	30	16.	GALAXY
TON-N 0829	07 33 16.	+ 35 35		15.3	BLUE STAR
MCG+10-11-099	07 33 18.	+ 59 48	66	15.	GALAXY
ZWG 310.005	07 33 18.	+ 63 38		15.7	GALAXY
MRSL 234+00/1	07 33 18.	- 18 38	360		HII REGION
TON-N 0830	07 33 22.	+ 35 57		15.3	BLUE STAR
MCG+10-11-100	07 33 24.	+ 59 25	57	15.	GALAXY
ZWG 286.040	07 33 24.	+ 59 27		14.9	GALAXY
UGC 03928	07 33 24.	+ 59 27	108	14.9	GALAXY PECULR
ZC 0733.4+6102	07 33 24.	+ 61 02	24390		CLUSTER OF GALAXIES
7ZW 169	07 33 24.	+ 64 20			COMPACT GALAXY
7ZW 170	07 33 24.	+ 75 33			COMPACT GALAXY
ZWG 349.008	07 33 24.	+ 75 33		15.3	GALAXY
UGC 03929	07 33 24.	+ 75 33	96	15.3	GALAXY COMPACT
OCL 0657	07 33 24.	- 27 37	264	16.	OPEN STAR CLUSTER
VHA 003	07 33 24.	- 27 37	300		OPEN STAR CLUSTER
MIN.46 05	07 33 25.	- 18 39			DIFFUSE NEBULA
PK215+11.1	07 33 30.	+ 02 49	261	15.4	PLANETARY NEBULA
ZC 0733.5+2952	07 33 30.	+ 29 52	3630		CLUSTER OF GALAXIES
ZWG 177.037	07 33 30.	+ 33 14		15.5	GALAXY
ZWG 286.041	07 33 30.	+ 60 08		15.7	GALAXY
ZWG 087.016	07 33 36.	+ 17 23		15.5	GALAXY
ZC 0733.6+2956	07 33 36.	+ 29 56	340		CLUSTER OF GALAXIES
ZWG 177.038	07 33 36.	+ 35 21		12.5	GALAXY
UGC 03930	07 33 36.	+ 35 21	60	12.5	GALAXY PECULR
MCG+08-14-027	07 33 36.	+ 49 35	48	16.	GALAXY
MCG+10-11-101	07 33 36.	+ 59 57	27	17.	GALAXY
ARC 0588	07 33 36.	+ 70 04		17.1	RICH CLUSTER OF GALAXIES
RNGC 2418	07 33 39.	+ 18 00		13.5	GALAXY
RNGC 2415	07 33 39.	+ 35 21		12.5	GALAXY
TON-N 0831	07 33 40.	+ 34 35		14.9	BLUE STAR
HARO 01	07 33 40.	+ 35 30			BLUE EMISSION-LINE GALAXY
ZWG 087.017	07 33 42.	+ 18 00		13.7	GALAXY
UGC 03931	07 33 42.	+ 18 00	108	13.7	GALAXY E
MCG+06-17-021	07 33 42.	+ 35 22	42	13.5	GALAXY
ZWG 286.042	07 33 42.	+ 59 40		15.5	GALAXY
MCG+10-11-102	07 33 42.	+ 60 08	39	16.	GALAXY
MCG+13-06-008	07 33 42.	+ 75 32	24	16.	GALAXY
ARP 165	07 33 45.	+ 18 00			PECULIAR GALAXY
ZWG 030.003	07 33 48.	+ 06 28		15.7	GALAXY
MCG+01-20-002	07 33 48.	+ 06 28	48	15.7	GALAXY
ZC 0733.8+2356	07 33 48.	+ 23 56	6250		CLUSTER OF GALAXIES
MCG+03-20-008	07 33 51.	+ 17 59	36	13.7	GALAXY
ZWG 177.039	07 33 54.	+ 35 46		15.3	GALAXY
MCG+06-17-022	07 33 54.	+ 35 47	24	17.	GALAXY
ZC 0733.9+3935	07 33 54.	+ 39 35	2150		CLUSTER OF GALAXIES
MCG+10-11-103	07 33 54.	+ 59 09	30	16.	GALAXY
MCG+10-11-104	07 33 54.	+ 59 42	24	18.	GALAXY
ZWG 310.006	07 33 54.	+ 63 44		15.7	GALAXY

OBJECT NAME	RIGHT ASCEN.	DECLINATION	DIAM.	MAGN.	TYPE OF OBJECT
MCG+11-10-011	07 33 54.	+ 63 44	18	16.	GALAXY
ISS 0833	07 33 54.	- 21 34	371		STELLAR RING
LBN 1062	07 34	- 25 10	600		BRIGHT NEBULA
ZWG 030.004	07 34 00.	+ 04 16		15.4	GALAXY
ZWG 058.010	07 34 00.	+ 10 01		15.3	GALAXY
ZWG 117.058	07 34 00.	+ 22 28		14.9	GALAXY
UGC 03932	07 34 00.	+ 22 28	96	14.9	GALAXY Sa-b
ZC 0734.0+3524	07 34 00.	+ 35 24	940		CLUSTER OF GALAXIES
ZWG 206.014	07 34 00.	+ 42 03		14.5	GALAXY
UGC 03933	07 34 00.	+ 42 03	78	14.5	GALAXY Sb/SBc
ZC 0734.0+5137	07 34 00.	+ 51 37	4230		CLUSTER OF GALAXIES
MRK 094	07 34 00.	+ 51 49	7	16.5	GALAXY WITH UV CONTINUUM
7ZW 171	07 34 00.	+ 59 07			COMPACT GALAXY
MCG+10-11-105	07 34 00.	+ 59 38	36	16.	GALAXY
7ZW 172	07 34 00.	+ 63 53			COMPACT GALAXY
MCG+15-01-007	07 34 00.	+ 88 16	36	17.	GALAXY
LDN 1666	07 34 00.	- 24 00	660		DARK NEBULA
MCG+10-11-106	07 34 06.	+ 57 53	27	17.	GALAXY
MCG+10-11-107	07 34 06.	+ 59 56	27	17.	GALAXY
OCL 0626	07 34 06.	- 20 30	840	9.4	OPEN STAR CLUSTER
RNGC 2421	07 34 07.	- 20 30		9.0	OPEN CLUSTER
ARC 0590	07 34 09.	+ 35 24		17.8	RICH CLUSTER OF GALAXIES
MCG+04-18-028	07 34 12.	+ 22 27	72	14.9	GALAXY
UGC 03934	07 34 12.	+ 35 46	66	17.	GALAXY DWRF SP
MCG+07-16-008	07 34 12.	+ 42 03	72	14.5	GALAXY
MCG+08-14-028	07 34 12.	+ 46 30	36	16.	GALAXY
ZWG 235.022	07 34 12.	+ 46 31		15.7	GALAXY
UGC 03935	07 34 12.	+ 46 31	72	15.7	GALAXY Sc
KARA.73B 0198	07 34 12.	+ 46 31	66	15.7	ISOLATED GALAXY S
ZWG 286.043	07 34 12.	+ 59 32		15.6	GALAXY
MCG+10-11-108	07 34 12.	+ 59 40	27	16.	GALAXY
ISS 0834	07 34 12.	- 24 22	154		STELLAR RING
RNGC 2422	07 34 15.	- 14 22		4.5	OPEN CLUSTER
VDB.66N 098	07 34 16.	- 25 12	960		REFLECTION NEBULA
ZWG 058.011	07 34 18.	+ 13 43		14.1	GALAXY
UGC 03936	07 34 18.	+ 13 43	78	14.1	GALAXY SBb-c
MCG+02-20-004	07 34 18.	+ 13 43	72	14.1	GALAXY
ZWG 177.040	07 34 18.	+ 35 43		14.2	GALAXY
UGC 03937	07 34 18.	+ 35 43	126	14.2	GALAXY SB?
MCG+06-17-023	07 34 18.	+ 35 43 30.	120	14.	GALAXY
MCG+10-11-109	07 34 18.	+ 57 49	36	17.	GALAXY
OCL 0596	07 34 18.	- 14 23	2400	5.2	OPEN STAR CLUSTER
CED 101	07 34 23.	- 25 13	900		DIFFUSE GALACTIC NEBULA
ZWG 030.005	07 34 24.	+ 08 06		15.6	GALAXY
ZWG 058.012	07 34 24.	+ 10 01		15.1	GALAXY
UGC 03938	07 34 24.	+ 10 01	66	15.1	GALAXY Sc
MCG+02-20-005	07 34 24.	+ 10 01	60	15.1	GALAXY
ZWG 058.013	07 34 24.	+ 12 11		15.7	GALAXY
ZWG 087.018	07 34 24.	+ 19 45		15.2	GALAXY
ZWG 087.019	07 34 24.	+ 19 50		15.7	GALAXY
ZWG 087.020	07 34 24.	+ 20 05		14.9	GALAXY
UGC 03939	07 34 24.	+ 20 05	72	14.9	GALAXY S0
MCG+03-20-009	07 34 24.	+ 20 05	30	14.9	GALAXY
MCG+05-18-024	07 34 24.	+ 28 56	48	15.6	GALAXY
ZWG 147.044	07 34 24.	+ 28 57		15.6	GALAXY
ZC 0734.4+3426	07 34 24.	+ 34 26	940		CLUSTER OF GALAXIES
MCG+10-11-110	07 34 24.	+ 59 31	30	17.	GALAXY
UGC 03940	07 34 24.	+ 71 17	90	16.5	GALAXY DWRF IR
OCL 0641	07 34 24.	- 23 15	42	14.	OPEN STAR CLUSTER
ISS 0835	07 34 24.	- 24 58	211		STELLAR RING
DG 122	07 34 24.	- 25 13	780		REFLECTION NEBULA
ZWG 058.014	07 34 30.	+ 10 01		15.7	GALAXY
ZWG 058.015	07 34 30.	+ 14 12		15.4	GALAXY
UGC 03941	07 34 30.	+ 14 12	96	15.4	GALAXY SB:c
ZC 0734.5+5457	07 34 30.	+ 54 57	2620		CLUSTER OF GALAXIES
ZWG 147.045	07 34 36.	+ 27 09		15.3	GALAXY
UGC 03942	07 34 36.	+ 27 09	72	15.3	GALAXY Sb
ZWG 147.046	07 34 36.	+ 28 46		15.3	GALAXY
MCG+10-11-111	07 34 36.	+ 59 56	18	17.	GALAXY
OCL 0625	07 34 36.	- 20 23	540		OPEN STAR CLUSTER
ZWG 058.016	07 34 42.	+ 09 08		15.2	GALAXY
ZWG 117.059	07 34 42.	+ 21 45		15.3	GALAXY
MCG+05-18-025	07 34 42.	+ 28 44 30.	42	15.3	GALAXY
MCG+10-11-112	07 34 42.	+ 57 48	39	17.	GALAXY
REIN 2.056	07 34 45.78	+ 38 59 43.9			NEBULA
ZWG 117.060	07 34 48.	+ 21 46		15.4	GALAXY
GCL 012	07 34 48.	+ 39 00	432	11.51	GLOBULAR STAR CLUSTER
RNGC 2419	07 34 48.	+ 39 00		11.5	GLOBULAR CLUSTER
MRK 077	07 34 48.	+ 60 25	8	16.5	GALAXY WITH UV CONTINUUM
OCL 0592	07 34 48.	- 13 45	2040	7.2	OPEN STAR CLUSTER
RNGC 2423	07 34 51.	- 13 45		7.0	OPEN CLUSTER
RNGC 2408	07 34 53.	+ 71 46			NON-EXISTENT OBJECT
MCG+10-11-113	07 34 54.	+ 57 04	30	16.	GALAXY
MCG+10-11-114	07 34 54.	+ 59 16	90	14.	GALAXY
ZWG 286.044	07 34 54.	+ 59 17		14.9	GALAXY
UGC 03943	07 34 54.	+ 59 17	102	14.9	GALAXY SBc
ZWG 286.045	07 34 54.	+ 59 30		15.4	GALAXY
MCG+10-11-115	07 34 54.	+ 62 21	42	16.	GALAXY
OCL 0602	07 34 54.	- 15 32	630	12.	OPEN STAR CLUSTER
LDN 1668	07 34 54.	- 24 47	540		DARK NEBULA
PK226+05.1	07 34 56.	- 09 32	91		PLANETARY NEBULA
RNGC 2434	07 34 58.	- 69 10		12.5	GALAXY
SCHO 0249	07 34 59.	- 20 03 18.	550		ISOLATED DARK CLOUD
KHAV 216	07 35	- 29 55	7130		DARK NEBULA
TON-N 0002	07 35 00.	+ 26 31		14.5	BLUE STAR
ZC 0735.0+4533	07 35 00.	+ 45 33	4370		CLUSTER OF GALAXIES
MCG+09-13-022	07 35 00.	+ 53 06	36	16.	GALAXY
MCG+10-11-116	07 35 00.	+ 59 30	30	16.	GALAXY
MCG+10-11-117	07 35 00.	+ 59 36	27	16.	GALAXY
ZWG 310.007	07 35 00.	+ 64 36		15.5	GALAXY
ZC 0735.0+8545	07 35 00.	+ 85 45	11360		CLUSTER OF GALAXIES
MCG+11-10-012	07 35 03.	+ 64 36	42	16.	GALAXY
RNGC 2427	07 35 05.	- 47 30		12.0	GALAXY
ZC 0735.1+2920	07 35 06.	+ 29 20	670		CLUSTER OF GALAXIES
MCG+08-14-029	07 35 06.	+ 46 03	36	17.	GALAXY
MCG+09-13-023	07 35 06.	+ 51 00	48	16.	GALAXY
7ZW 173	07 35 06.	+ 73 14			COMPACT GALAXY
TON-N 0832	07 35 11.	+ 32 28		16.8	BLUE STAR
ZWG 058.017	07 35 12.	+ 10 18		15.4	GALAXY
ZWG 177.041	07 35 12.	+ 37 45		15.0	GALAXY
UGC 03944	07 35 12.	+ 37 45	120	15.0	GALAXY Sc
KARA.73B 0199	07 35 12.	+ 37 45	120	15.0	ISOLATED GALAXY S
MCG+10-11-118	07 35 12.	+ 57 25	39	15.	GALAXY
ZWG 286.046	07 35 12.	+ 57 27		15.6	GALAXY
ZWG 331.008	07 35 12.	+ 69 11		15.7	GALAXY
UGC 03945	07 35 12.	+ 69 11	78	15.7	GALAXY Sb
7ZW 174	07 35 12.	+ 69 55			COMPACT GALAXY
ZWG 331.009	07 35 12.	+ 70 03		15.1	GALAXY

OBJECT NAME	RIGHT ASCEN.	DECLINATION	DIAM.	MAGN.	TYPE OF OBJECT
OCL 0587	07 35 12.	- 11 57	900	9.0	OPEN STAR CLUSTER
ZWG 147.047	07 35 18.	+ 29 05		15.6	GALAXY
MCG+06-17-024	07 35 18.	+ 37 45 30.	96	14.	GALAXY
ZWG 331.010	07 35 18.	+ 69 56		15.4	GALAXY
ZWG 030.006	07 35 24.	+ 03 25		14.4	GALAXY
UGC 03946	07 35 24.	+ 03 25	84	14.4	GALAXY IRR
MCG+01-20-003	07 35 24.	+ 03 25	72	14.4	GALAXY
ZWG 147.048	07 35 24.	+ 29 19		15.7	GALAXY
ZWG 262.015	07 35 24.	+ 55 39		15.2	GALAXY
OCL 0654	07 35 24.	- 26 29	1080	12.	OPEN STAR CLUSTER
ACK 244-04.1	07 35 24.	- 30 01			PLANETARY NEBULA
RNGC 2420	07 35 25.	+ 21 41		10.0	OPEN CLUSTER
ST 01	07 35 27.	- 11 58	360		EMISSION NEBULA
KEEL 149	07 35 27.6	+ 65 46 40.			NEBULA
OCL 0488	07 35 30.	+ 21 41	630	10.2	OPEN STAR CLUSTER
7ZW 175	07 35 30.	+ 69 59			COMPACT GALAXY
ZWG 087.021	07 35 36.	+ 16 58		15.7	GALAXY
MCG+09-13-024	07 35 36.	+ 55 40	24	16.	GALAXY
ZWG 030.007	07 35 42.	+ 05 05		15.7	GALAXY
ZC 0735.7+7421	07 35 42.	+ 74 21	1480		CLUSTER OF GALAXIES
ZWG 087.022	07 35 48.	+ 16 49		15.3	GALAXY
ZC 0735.8+3200	07 35 48.	+ 32 00	940		CLUSTER OF GALAXIES
ZWG 147.049	07 35 48.	+ 32 01		15.6	GALAXY
UGC 03947	07 35 48.	+ 34 02	60	17.	GALAXY DWRF:SP
ZWG 235.023	07 35 48.	+ 49 27		14.9	GALAXY
MCG+08-14-030	07 35 48.	+ 49 28	24	15.	GALAXY
ZWG 262.016	07 35 48.	+ 54 26		15.6	GALAXY
UGC 03948	07 35 48.	+ 54 26	66	15.6	GALAXY SBc
ZWG 262.017	07 35 48.	+ 56 00		15.7	GALAXY
ZC 0735.8+7906	07 35 48.	+ 79 06	1210		CLUSTER OF GALAXIES
ZWG 147.050	07 35 54.	+ 29 17		14.9	GALAXY
RNGC 2425	07 35 57.	- 14 45			OPEN CLUSTER
KHAV 217	07 36	- 09 55	10920		DARK NEBULA
KHAV 215	07 36	- 14 31	4910		DARK NEBULA
LB 09837	07 36	- 83 21		14.5	FAINT BLUE STAR
ZWG 087.023	07 36 00.	+ 20 43		15.6	GALAXY
MCG+05-18-026	07 36 00.	+ 29 16	30	14.9	GALAXY
ZWG 235.024	07 36 00.	+ 48 51		14.9	GALAXY
UGC 03949	07 36 00.	+ 48 51	72	14.9	GALAXY SBc
MCG+08-14-031	07 36 00.	+ 48 51	60	14.	GALAXY
MCG+09-13-025	07 36 00.	+ 54 27 30.	54	15.	GALAXY
MCG+09-13-026	07 36 00.	+ 56 00	36	16.	GALAXY
OCL 0581	07 36 00.	- 10 34	540	10.9	OPEN STAR CLUSTER
OCL 0599	07 36 00.	- 14 45	198	14.	OPEN STAR CLUSTER
VHA 004	07 36 00.	- 35 55	150		OPEN STAR CLUSTER
KEEL 150	07 36 00.5	+ 65 40 22.			NEBULA
MCG+11-10-013	07 36 03.	+ 65 40	24	16.	GALAXY
ZWG 002.005	07 36 06.	+ 02 20		14.8	GALAXY
UGC 03950	07 36 06.	+ 02 20	66	14.8	GALAXY S0-a
MCG+00-20-002	07 36 06.	+ 02 20	48	14.8	GALAXY
ZWG 262.018	07 36 06.	+ 52 46		15.3	GALAXY
MCG+11-10-014	07 36 06.	+ 65 40	18	17.	GALAXY
UGC 03951	07 36 06.	+ 67 00	66	16.0	GALAXY Sb
ZWG 058.018	07 36 12.	+ 13 44		15.5	GALAXY
ZWG 087.024	07 36 12.	+ 19 37		15.4	GALAXY
UGC 03952	07 36 12.	+ 19 37	66	15.4	GALAXY Sb-c
MCG+09-13-027	07 36 12.	+ 56 51	42	17.	GALAXY
ZWG 286.047	07 36 12.	+ 58 01		15.6	GALAXY
UGC 03953	07 36 12.	+ 58 01	66	15.6	GALAXY Sb/SBb
ZC 0736.2+6341	07 36 12.	+ 63 41	1610		CLUSTER OF GALAXIES
ZWG 310.008	07 36 12.	+ 64 46		15.6	GALAXY
ZWG 310.009	07 36 12.	+ 65 41		15.2	GALAXY
ZC 0736.3+4805	07 36 18.	+ 48 05	3290		CLUSTER OF GALAXIES
MCG+10-11-119	07 36 18.	+ 58 00	60	16.	GALAXY
MCG+11-10-015	07 36 18.	+ 64 45 30.	18	16.	GALAXY
ZWG 058.019	07 36 24.	+ 09 14		15.3	GALAXY
ZWG 235.025	07 36 24.	+ 49 25		15.5	GALAXY
MCG+09-13-028	07 36 24.	+ 54 58	30	18.	GALAXY
MCG+10-11-120	07 36 24.	+ 61 14	51	16.	GALAXY
RNGC 2442	07 36 28.	- 69 25		11.0	GALAXY
RNGC 2443	07 36 29.	- 69 26			NON-EXISTENT OBJECT
BB 4.09	07 36 29.1	+ 01 47 55.5			GALAXY NEAR QSO PKS0736
UGC 03954	07 36 30.	+ 39 08	66	18.	GALAXY DWRF IR
ZWG 235.026	07 36 30.	+ 49 23		15.0	GALAXY
MCG+08-14-032	07 36 30.	+ 49 23	42	15.	GALAXY
ZWG 286.048	07 36 30.	+ 60 04		15.7	GALAXY
7ZW 176	07 36 30.	+ 67 48			COMPACT GALAXY
BB 4.11	07 36 31.95	+ 01 41 54.5			GALAXY NEAR QSO PKS0736
BB 4.07	07 36 32.6	+ 01 46 40.			GALAXY NEAR QSO PKS0736
BB 4.05	07 36 32.95	+ 01 44 48.			GALAXY NEAR QSO PKS0736
SHB 099	07 36 34.	- 06 18		19.0	QUASI-STELLAR OBJECT
ZWG 087.025	07 36 36.	+ 18 18		15.1	GALAXY
ZWG 147.051	07 36 36.	+ 29 51		15.7	GALAXY
MCG+10-11-121	07 36 36.	+ 60 04	27	16.	GALAXY
BB 4.06	07 36 37.15	+ 01 44 57.5			GALAXY NEAR QSO PKS0736
BB 4.10	07 36 38.4	+ 01 40 47.5			GALAXY NEAR QSO PKS0736
BB 4.08	07 36 41.9	+ 01 46 09.5			GALAXY NEAR QSO PKS0736
ZWG 058.020	07 36 42.	+ 09 01		15.3	GALAXY
UGC 03955	07 36 42.	+ 09 01	96	15.3	GALAXY
ARC 0589	07 36 42.	+ 63 37		17.1	RICH CLUSTER OF GALAXIES
SHB 100	07 36 42.4	+ 01 43 57.		17.5	QUASI-STELLAR OBJECT
BC PKS0736+01	07 36 42.53	+ 01 43 59.9		16.47	QUASI-STELLAR OBJECT
UGC 03956	07 36 48.	+ 24 35	66	16.5	GALAXY
ZWG 147.052	07 36 48.	+ 32 19		15.5	GALAXY
BB 4.02	07 36 48.25	+ 01 43 08.5			GALAXY NEAR QSO PKS0736
BB 4.01	07 36 51.55	+ 01 43 57.5			GALAXY NEAR QSO PKS0736
ZWG 147.053	07 36 54.	+ 32 20		15.6	GALAXY
MCG+09-13-029	07 36 54.	+ 54 33 30.	30	17.	GALAXY
ZWG 262.019	07 36 54.	+ 55 32		14.2	GALAXY
UGC 03957	07 36 54.	+ 55 32	90	14.2	GALAXY E
OCL 0580	07 36 54.	- 10 26	180	11.8	OPEN STAR CLUSTER
ISS 0908	07 36 54.	- 28 46	296		STELLAR RING
RNGC 2428	07 36 58.	- 16 25			NON-EXISTENT OBJECT
BB 4.03	07 36 58.25	+ 01 44 19.5			GALAXY NEAR QSO PKS0736
VDB.66G 046	07 37	+ 40 12	70		DWARF GALAXY
ZWG 147.054	07 37 00.	+ 32 23		15.4	GALAXY
MCG+09-13-030	07 37 00.	+ 55 32 30.	24	15.	GALAXY
MCG+11-10-017	07 37 00.	+ 64 49	27	12.	GALAXY
MCG+11-10-016	07 37 00.	+ 64 50	39	16.	GALAXY
BB 4.04	07 37 01.4	+ 01 46 08.			GALAXY NEAR QSO PKS0736
ZWG 058.021	07 37 06.	+ 13 22		15.6	GALAXY
ZWG 087.026	07 37 06.	+ 17 21		15.3	GALAXY
MCG+10-11-122	07 37 06.	+ 60 03	39	17.	GALAXY
MCG+07-16-009	07 37 09.	+ 39 20	210	13.5	GALAXY
RNGC 2430	07 37 10.	- 16 15			NON-EXISTENT OBJECT
ZWG 058.022	07 37 12.	+ 13 27		15.7	GALAXY
MCG+05-18-027	07 37 12.	+ 32 22	48	15.4	GALAXY
ZWG 002.006	07 37 12.	- 02 31		15.5	GALAXY
ZWG 002.006	07 37 12.	- 02 31		15.5	GALAXY
MCG+00-20-003	07 37 12.	- 02 31	48	15.5	GALAXY
IC 2203	07 37 15.	+ 34 18 58.			NONSTELLAR OBJECT
MCG+06-17-025	07 37 15.	+ 34 21	78	14.	GALAXY
MCG+09-13-031	07 37 15.	+ 54 32 30.	7	17.	GALAXY
ZC 0737.3+2801	07 37 18.	+ 28 01	400		CLUSTER OF GALAXIES
ZWG 177.042	07 37 18.	+ 34 20		14.5	GALAXY
UGC 03958	07 37 18.	+ 34 20	72	14.5	GALAXY SBc
ZWG 206.015	07 37 18.	+ 39 20		13.9	GALAXY
UGC 03959	07 37 18.	+ 39 20	234	13.9	GALAXY Sb
RNGC 2424	07 37 18.	+ 39 21		14.0	GALAXY
ZWG 235.027	07 37 18.	+ 49 30		15.4	GALAXY
ZWG 030.008	07 37 24.	+ 05 41		15.6	GALAXY
ZWG 117.061	07 37 24.	+ 21 48		15.6	GALAXY
ZWG 117.062	07 37 24.	+ 23 23		15.0	GALAXY
UGC 03960	07 37 24.	+ 23 23	90	15.0	GALAXY E
MCG+04-18-030	07 37 24.	+ 23 24	15	15.0	GALAXY
ZWG 117.063	07 37 24.	+ 25 15		15.5	GALAXY
MCG+04-18-029	07 37 24.	+ 26 17	48	15.6	GALAXY
ZC 0737.4+3455	07 37 24.	+ 34 55	1210		CLUSTER OF GALAXIES
ZC 0737.5+0108	07 37 30.	+ 01 08	3630		CLUSTER OF GALAXIES
MCG+04-18-031	07 37 30.	+ 21 47	42	15.6	GALAXY
ZWG 117.064	07 37 30.	+ 26 15		15.6	GALAXY
ZWG 235.028	07 37 30.	+ 47 47		15.6	GALAXY
UGC 03961	07 37 30.	+ 47 47	66	15.6	GALAXY SBc
OCL 0680	07 37 30.	- 30 49	240	14.	OPEN STAR CLUSTER
ZWG 030.009	07 37 36.	+ 05 45		15.3	GALAXY
ZWG 058.023	07 37 36.	+ 14 56		15.5	GALAXY
ZWG 087.027	07 37 36.	+ 20 49		15.2	GALAXY
ZWG 147.055	07 37 36.	+ 29 23		15.7	GALAXY
MCG+09-13-032	07 37 36.	+ 51 48	72	16.	GALAXY
ZWG 058.024	07 37 42.	+ 13 59		14.7	GALAXY
UGC 03962	07 37 42.	+ 13 59	138	14.7	GALAXY Sa/SBb
MCG+02-20-006	07 37 42.	+ 13 59	120	14.7	GALAXY
ZWG 117.065	07 37 42.	+ 23 45		15.3	GALAXY
ZWG 206.016	07 37 42.	+ 42 52		14.6	GALAXY
MCG+07-16-010	07 37 42.	+ 42 52	42	15.5	GALAXY
UGC 03963	07 37 42.	+ 51 49	96	16.5	GALAXY
MCG+10-11-123	07 37 42.	+ 57 20	12	17.	GALAXY
WRAY 19.01	07 37 44.1	- 52 19 53.			DIFFUSE NEBULA
ZWG 058.025	07 37 48.	+ 12 32		15.7	GALAXY
MCG+11-10-018	07 37 48.	+ 65 03	39	16.	GALAXY
MCG+11-10-019	07 37 48.	+ 66 22	45	15.	GALAXY
7ZW 177	07 37 48.	+ 67 13			COMPACT GALAXY
ZWG 002.007	07 37 48.	- 01 28		15.5	GALAXY
ZWG 002.007	07 37 48.	- 01 28		15.5	GALAXY
UGC 03964	07 37 48.	- 01 28	120	15.5	GALAXY
MCG+00-20-004	07 37 48.	- 01 28	90	15.5	GALAXY
ZC 0737.9+2615	07 37 54.	+ 26 15	2620		CLUSTER OF GALAXIES
MCG+09-13-033	07 37 54.	+ 55 24	42	16.	GALAXY
MCG+11-10-020	07 37 54.	+ 64 45	39	17.	GALAXY
PATH 1.191	07 37 57.	+ 45 11	11		NEBULA
KHAV 218	07 38	- 26 13	2840		DARK NEBULA
IC 2204	07 38 00.	+ 34 19 05.			NONSTELLAR OBJECT
ZWG 177.043	07 38 00.	+ 34 20		14.6	GALAXY
UGC 03965	07 38 00.	+ 34 20	66	14.6	GALAXY SBb
MCG+06-17-026	07 38 00.	+ 34 20 30.	60	14.	GALAXY
UGC 03966	07 38 00.	+ 40 13	132	16.0	GALAXY DWRF IR
MCG+07-16-011	07 38 00.	+ 40 14	90	16.	GALAXY
ZC 0738.0+4431	07 38 00.	+ 44 31	5510		CLUSTER OF GALAXIES
MCG+11-10-021	07 38 00.	+ 63 02 30.	45	16.	GALAXY
UGC 03967	07 38 00.	+ 63 03	90	16.5	GALAXY SB?c
ZWG 310.010	07 38 00.	+ 66 22		15.2	GALAXY
UGC 03968	07 38 00.	+ 66 22	90	15.2	GALAXY SBc
KARA.73B 0200	07 38 00.	+ 66 22	102	15.2	ISOLATED GALAXY S
PK228+05.1	07 38 00.	- 11 26	3		PLANETARY NEBULA
BC B20738+31	07 38 00.0	+ 31 19 12.0		17.	QUASI-STELLAR OBJECT
SHB 101	07 38 00.2	+ 31 19 02.		17.	QUASI-STELLAR OBJECT
ZC 0738.1+1634	07 38 06.	+ 16 34	2020		CLUSTER OF GALAXIES
ZWG 147.056	07 38 06.	+ 27 44		15.4	GALAXY
UGC 03969	07 38 06.	+ 27 44	66	15.4	GALAXY Sc
ZWG 147.057	07 38 06.	+ 29 15		15.3	GALAXY
MCG+05-18-028	07 38 12.	+ 27 42 30.	54	15.4	GALAXY
7ZW 178	07 38 12.	+ 60 39			COMPACT GALAXY
MRK 078	07 38 12.	+ 65 17	15	15.	GALAXY WITH UV CONTINUUM
KW 12	07 38 12.	+ 65 17	19		SEYFERT GALAXY
VVI 22	07 38 12.	+ 65 17	15	15.57	SEYFERT GALAXY
VV 349C	07 38 12.	+ 73 57	21	16.	INTERACTING GALAXY
VV 349B	07 38 12.	+ 73 57	12	16.	INTERACTING GALAXY
VV 349A	07 38 12.	+ 73 57	42	15.	INTERACTING GALAXY
VV 349	07 38 12.	+ 73 57	60		INTERACTING GALAXY
ZWG 147.058	07 38 18.	+ 29 47		15.5	GALAXY
ZC 0738.3+4407	07 38 18.	+ 44 07	1140		CLUSTER OF GALAXIES
UGC 03970	07 38 18.	+ 61 40	60	16.0	GALAXY Sc
MCG+10-11-124	07 38 18.	+ 61 40	54	16.	GALAXY
ZWG 286.049	07 38 18.	+ 62 00		15.5	GALAXY
ISS 0909	07 38 18.	- 32 59	691		STELLAR RING
MCG+05-18-029	07 38 24.	+ 29 46	36	15.5	GALAXY
MCG+09-13-034	07 38 24.	+ 52 15 30.	36	16.	GALAXY
ZWG 286.050	07 38 24.	+ 60 19		15.5	GALAXY
ZWG 087.028	07 38 30.	+ 18 50		15.6	GALAXY
MCG+10-11-126	07 38 30.	+ 60 19	24	15.	GALAXY
MCG+10-11-125	07 38 30.	+ 62 00	36	16.	GALAXY
MCG+12-08-008B	07 38 30.	+ 73 57	51	16.	GALAXY
MCG+12-08-008A	07 38 30.	+ 73 57	48	15.	GALAXY
OCL 0678	07 38 30.	- 30 00	900	8.	OPEN STAR CLUSTER
VHA 005	07 38 30.	- 30 00	1800		OPEN STAR CLUSTER
ARC 0591	07 38 33.	+ 44 05		16.8	RICH CLUSTER OF GALAXIES
ZWG 286.051	07 38 36.	+ 60 17		15.6	GALAXY
UGC 03971	07 38 36.	+ 60 17	72	15.6	GALAXY S0-a
ZWG 310.011	07 38 36.	+ 64 23		15.6	GALAXY
ZWG 331.011	07 38 36.	+ 73 56		14.7	GALAXY
ZWG 330.049	07 38 36.	+ 73 56		14.7	GALAXY
UGC 03972	07 38 36.	+ 73 56	78	14.7	GALAXY SB
ZWG 087.029	07 38 42.	+ 20 53		15.6	GALAXY
ZC 0738.7+4449	07 38 42.	+ 44 49	1280		CLUSTER OF GALAXIES
MRK 079	07 38 42.	+ 49 56	40	13.5	GALAXY WITH UV CONTINUUM
KW 13	07 38 42.	+ 49 56	86		SEYFERT GALAXY
VVI 23	07 38 42.	+ 49 56	60	13.3	SEYFERT GALAXY
MCG+08-14-033	07 38 42.	+ 49 56	60	14.	GALAXY
MCG+10-11-127	07 38 42.	+ 60 17 30.	39	15.	GALAXY
MCG+11-10-022	07 38 42.	+ 64 22	15	16.	GALAXY
OCL 0620	07 38 42.	- 18 58	480	10.2	OPEN STAR CLUSTER
RNGC 2432	07 38 42.	- 18 58		10.0	OPEN CLUSTER
ISS 0910	07 38 42.	- 29 47	226		STELLAR RING
MCG+08-14-034	07 38 45.	+ 49 19	48	15.	GALAXY
2ZW 043	07 38 48.	+ 45 33			COMPACT GALAXY
ZWG 235.029	07 38 48.	+ 49 18		14.7	GALAXY

OBJECT NAME	RIGHT ASCEN.	DECLINATION	DIAM.	MAGN.	TYPE OF OBJECT
ZWG 235.030	07 38 48.	+ 49 55		13.3	GALAXY
UGC 03973	07 38 48.	+ 49 55	90	13.3	GALAXY SBb
MCG+09-13-035	07 38 48.	+ 54 08	6	17.	GALAXY
MCG+10-11-128	07 38 48.	+ 59 58	45	17.	GALAXY
MCG+10-11-129	07 38 48.	+ 62 29	30	16.	GALAXY
ARP 017	07 38 53.	+ 73 55			PECULIAR GALAXY
ZWG 117.066	07 38 54.	+ 23 18		15.4	GALAXY
7ZW 179	07 38 54.	+ 73 16			COMPACT GALAXY
OCL 0688	07 38 54.	- 31 32	1200	7.1	OPEN STAR CLUSTER
VHA 006	07 38 54.	- 31 32	360		OPEN STAR CLUSTER
FATH 1.192	07 38 58.	+ 44 21	14		NEBULA
VDB.66G 047	07 39	+ 16 52	200		DWARF GALAXY
KHAV 219	07 39	- 27 49	4140		DARK NEBULA
ZWG 087.030	07 39 00.	+ 16 55		15.4	GALAXY
UGC 03974	07 39 00.	+ 16 55	300	15.4	GALAXY DWRF IR
MCG+03-20-010	07 39 00.	+ 16 55	120	15.4	GALAXY
ZWG 087.031	07 39 00.	+ 18 51		15.5	GALAXY
ZC 0739.0+2028	07 39 00.	+ 20 28	1880		CLUSTER OF GALAXIES
MCG+09-13-036	07 39 00.	+ 52 06	36	16.	GALAXY
ZWG 262.020	07 39 00.	+ 52 07		15.7	GALAXY
ZWG 331.012	07 39 00.	+ 72 55		15.1	GALAXY
ZWG 330.050	07 39 00.	+ 72 55		15.1	GALAXY
UGC 03975	07 39 00.	+ 72 55	138	15.1	GALAXY
MCG+12-08-009	07 39 00.	+ 72 56	54	15.	GALAXY
OCL 0606	07 39 00.	- 16 08	900	12.	OPEN STAR CLUSTER
OCL 0650	07 39 00.	- 24 13	150	13.	OPEN STAR CLUSTER
PK235+01.1	07 39 01.	- 18 52	269		PLANETARY NEBULA
RNGC 2439	07 39 02.	- 31 32		7.0	OPEN CLUSTER
ZWG 087.032	07 39 06.	+ 18 41		15.6	GALAXY
ZWG 262.021	07 39 06.	+ 55 05		15.0	GALAXY
MCG+07-16-012	07 39 09.	+ 41 02	48	15.5	GALAXY
ZWG 206.017	07 39 12.	+ 41 02		15.4	GALAXY
UGC 03976	07 39 12.	+ 41 02	66	15.4	GALAXY S
ZWG 058.026	07 39 18.	+ 12 17		15.5	GALAXY
ZWG 235.031	07 39 18.	+ 50 23		15.6	GALAXY
ZWG 286.052	07 39 18.	+ 58 36		15.7	GALAXY
ZWG 002.008	07 39 18.	- 02 35		15.4	GALAXY
ZWG 002.008	07 39 18.	- 02 35		15.4	GALAXY
RNGC 2426	07 39 21.	+ 52 57		14.5	GALAXY
ZWG 262.022	07 39 24.	+ 52 27		14.4	GALAXY
UGC 03977	07 39 24.	+ 52 27	72	14.4	GALAXY E
KARA.72 138A	07 39 24.	+ 52 27	90	14.4	PART OF DOUBLE GALAXY
MCG+11-10-023	07 39 24.	+ 64 15	39	17.	GALAXY
FATH 1.193	07 39 29.	+ 45 14	8		NEBULA
ZWG 030.010	07 39 30.	+ 06 14		15.6	GALAXY
ZWG 030.011	07 39 30.	+ 06 49		15.5	GALAXY
ZWG 235.032	07 39 30.	+ 45 28		15.6	GALAXY
MCG+09-13-037	07 39 30.	+ 52 19	12	16.	GALAXY
MCG+09-13-038	07 39 30.	+ 52 27	36	13.	GALAXY
7ZW 180	07 39 30.	+ 62 11			COMPACT GALAXY
ZWG 286.053	07 39 30.	+ 62 11		15.0	GALAXY
UGC 03978	07 39 30.	+ 62 11	72	15.0	GALAXY SBb
MCG+11-10-024	07 39 30.	+ 67 23	102	15.	GALAXY
OCL 0601	07 39 30.	- 14 42	2400	9.2	OPEN STAR CLUSTER
FATH 1.194	07 39 33.	+ 44 39	11		NEBULA
RNGC 2438	07 39 33.	- 14 36		11.5	PLANETARY NEBULA
RNGC 2437	07 39 33.	- 14 42		6.5	OPEN CLUSTER
PK231+04.2	07 39 33.27	- 14 37 03.8	130	11.3	PLANETARY NEBULA
ZWG 087.033	07 39 36.	+ 16 40		15.6	GALAXY
IC 0471	07 39 36.	+ 49 48 15.			NONSTELLAR OBJECT
MRK 080	07 39 36.	+ 64 51	8	17.	GALAXY WITH UV CONTINUUM
MCG+13-06-009	07 39 36.	+ 75 39	54	16.	GALAXY
ACK 245-03.1	07 39 36.	- 29 52			PLANETARY NEBULA
IC 0473	07 39 41.	+ 09 22 07.			NONSTELLAR OBJECT
PK234+02.1	07 39 41.43	- 18 05 24.2	74	11.7	PLANETARY NEBULA
ZWG 058.027	07 39 42.	+ 11 22		15.3	GALAXY
KARA.72 139A	07 39 42.	+ 11 22	48	15.3	PART OF DOUBLE GALAXY
MRK 081	07 39 42.	+ 57 07	30	14.5	GALAXY WITH UV CONTINUUM
MCG+10-11-130	07 39 42.	+ 57 07	72	14.	GALAXY
ZWG 310.012	07 39 42.	+ 67 23		14.8	GALAXY
UGC 03979	07 39 42.	+ 67 23	114	14.8	GALAXY Sb-c
KARA.73B 0201	07 39 42.	+ 67 23	120	14.8	ISOLATED GALAXY S
PK231+04.1	07 39 42.	- 14 15	36		PLANETARY NEBULA
ZWG 087.034	07 39 48.	+ 18 27		15.2	GALAXY
UGC 03980	07 39 48.	+ 18 27	84	15.2	GALAXY S
ZWG 117.067	07 39 48.	+ 22 14		15.2	GALAXY
MCG+08-14-035	07 39 48.	+ 49 47	18	14.	GALAXY
ZC 0739.8+4949	07 39 48.	+ 49 49	10280		CLUSTER OF GALAXIES
ZWG 286.054	07 39 48.	+ 57 06		14.3	GALAXY
UGC 03981	07 39 48.	+ 57 06	90	14.3	GALAXY S0
IC 0472	07 39 51.	+ 49 45 43.			NONSTELLAR OBJECT
RNGC 2429B	07 39 52.	+ 52 27		15.0	GALAXY
RNGC 2429A	07 39 52.	+ 52 29		14.0	GALAXY
RNGC 2440	07 39 53.	- 18 05		11.5	PLANETARY NEBULA
ARC 0592	07 39 54.	+ 09 30		15.0	RICH CLUSTER OF GALAXIES
ZWG 058.028	07 39 54.	+ 11 23		15.6	GALAXY
KARA.72 139B	07 39 54.	+ 11 23	42	15.6	PART OF DOUBLE GALAXY
ZWG 235.033	07 39 54.	+ 49 46		14.2	GALAXY
UGC 03982	07 39 54.	+ 49 46	48	14.2	GALAXY E
MCG+09-13-039	07 39 54.	+ 52 28	60	14.	GALAXY
ZWG 262.023	07 39 54.	+ 52 29		14.7	GALAXY
UGC 03983	07 39 54.	+ 52 29	108	14.7	GALAXY S
KARA.72 138B	07 39 54.	+ 52 29	78	14.7	PART OF DOUBLE GALAXY
VV 284B	07 39 54.	+ 52 29 30.	24	16.	INTERACTING GALAXY
VV 284A	07 39 54.	+ 52 29 30.	90	14.	INTERACTING GALAXY
ZWG 331.013	07 39 54.	+ 70 10		14.2	GALAXY
UGC 03984	07 39 54.	+ 70 10	132	14.2	GALAXY SBb
KARA.73B 0202	07 39 54.	+ 70 10	126	14.2	ISOLATED GALAXY S
MCG+09-13-040	07 39 57.	+ 52 27	24	15.	GALAXY
KHAV 220	07 40	- 06 37	3330		DARK NEBULA
MCG+08-14-036	07 40 00.	+ 49 44	78	13.	GALAXY
MCG+12-08-010	07 40 00.	+ 70 09	114	15.	GALAXY
ACK 246-04.1	07 40 00.	- 31 39			PLANETARY NEBULA
SCHO 0250	07 40 04.	- 27 57 06.	410		ISOLATED DARK CLOUD
ZWG 058.029	07 40 06.	+ 09 28		15.7	GALAXY
ZWG 235.034	07 40 06.	+ 49 43		14.5	GALAXY
UGC 03985	07 40 06.	+ 49 43	114	14.5	GALAXY Sb/SBb
ZC 0740.2+0930	07 40 12.	+ 09 30	4030		CLUSTER OF GALAXIES
ZWG 148.001	07 40 12.	+ 31 41		15.7	GALAXY
ZWG 147.059	07 40 12.	+ 31 41		15.7	GALAXY
MCG+10-11-131	07 40 12.	+ 62 10	72	15.	GALAXY
RNGC 2433	07 40 14.	+ 09 27			GALAXY
ZWG 148.002	07 40 18.	+ 31 39		15.7	GALAXY
ZWG 147.060	07 40 18.	+ 31 39		15.7	GALAXY
UGC 03986	07 40 18.	+ 31 39	66	15.7	GALAXY Sc
ZC 0740.3+5940	07 40 18.	+ 59 40	1280		CLUSTER OF GALAXIES
OCL 0687	07 40 18.	- 31 21	240	11.	OPEN STAR CLUSTER
FATH 1.195	07 40 19.	+ 44 50	16		NEBULA

OBJECT NAME	RIGHT ASCEN.	DECLINATION	DIAM.	MAGN.	TYPE OF OBJECT
ZC 0740.4+1740	07 40 24.	+ 17 40	1410		CLUSTER OF GALAXIES
ZC 0740.4+2200	07 40 24.	+ 22 00	810		CLUSTER OF GALAXIES
ZC 0740.4+5323	07 40 24.	+ 53 23	2290		CLUSTER OF GALAXIES
ZWG 030.012	07 40 30.	+ 04 04		15.5	GALAXY
ZWG 117.068	07 40 30.	+ 23 03		15.3	GALAXY
UGC 03987	07 40 30.	+ 23 03	60	15.3	GALAXY S
MCG+14-04-036	07 40 30.	+ 84 45	57	16.	GALAXY
FATH 1.196	07 40 34.	+ 44 41	3		NEBULA
ZWG 058.030	07 40 36.	+ 09 29		15.7	GALAXY
MCG+04-18-032	07 40 36.	+ 23 04	48	15.3	GALAXY
ZC 0740.7+1455	07 40 42.	+ 14 55	1410		CLUSTER OF GALAXIES
ZWG 118.001	07 40 42.	+ 25 59		15.2	GALAXY
ZWG 117.069	07 40 42.	+ 25 59		15.2	GALAXY
UGC 03988	07 40 42.	+ 47 52	66	16.5	GALAXY Sc
UGC 03989	07 40 42.	+ 53 58	66	18.	GALAXY DWRF SP
FATH 1.197	07 40 44.	+ 44 25	11		NEBULA
MCG+09-13-041	07 40 45.	+ 53 56	72	17.	GALAXY
ZWG 177.044	07 40 48.	+ 33 11		15.7	GALAXY
FATH 1.198	07 40 48.	+ 44 33	8		NEBULA
ZC 0740.8+4542	07 40 48.	+ 45 42	1080		CLUSTER OF GALAXIES
MCG+11-10-025	07 40 48.	+ 65 43	60	15.	GALAXY
LB 00182	07 40 49.	+ 44 33 48.		17.3	FAINT BLUE STAR
FATH 1.199	07 40 52.	+ 44 38	16		NEBULA
SCHO 0251	07 40 53.	- 30 34 24.	280		ISOLATED DARK CLOUD
MCG+07-16-013	07 40 54.	+ 40 27 30.	45	16.	GALAXY
ZWG 206.018	07 40 54.	+ 40 28		15.7	GALAXY
UGC 03990	07 40 54.	+ 40 28	66	15.7	GALAXY S
ZWG 310.013	07 40 54.	+ 65 53		14.9	GALAXY
UGC 03991	07 40 54.	+ 65 53	72	14.9	GALAXY Sc
OCL 0711	07 40 54.	- 35 28	120	11.	OPEN STAR CLUSTER
SHB 102	07 40 56.7	+ 38 00 30.		17.6	QUASI-STELLAR OBJECT
BC 3CR186	07 40 56.74	+ 38 00 31.1		17.60	QUASI-STELLAR OBJECT
MCG+06-17-027	07 40 57.	+ 34 11	42	17.	GALAXY
ZWG 148.003	07 41 00.	+ 29 21		13.6	GALAXY
ZWG 147.061	07 41 00.	+ 29 21		13.6	GALAXY
UGC 03995	07 41 00.	+ 29 21	150	13.6	GALAXY S
KARA.72 140A	07 41 00.	+ 29 21	48	13.6	PART OF DOUBLE GALAXY
MCG+05-19-001	07 41 00.	+ 29 21	150	13.6	GALAXY
ZWG 148.004	07 41 00.	+ 31 46		13.5	GALAXY
ZWG 147.062	07 41 00.	+ 31 46		13.5	GALAXY
UGC 03996	07 41 00.	+ 31 46	156	13.5	GALAXY Sa
ZC 0741.0+3417	07 41 00.	+ 34 17	3290		CLUSTER OF GALAXIES
ZWG 206.019	07 41 00.	+ 39 46		15.5	GALAXY
ZC 0741.0+4752	07 41 00.	+ 47 52	610		CLUSTER OF GALAXIES
ZC 0741.0+6552	07 41 00.	+ 65 52	4440		CLUSTER OF GALAXIES
UGC 03992	07 41 00.	+ 84 45	102	16.0	GALAXY
ZWG 363.034	07 41 00.	+ 85 04		14.0	GALAXY
ZWG 362.046	07 41 00.	+ 85 04		14.0	GALAXY
UGC 03993	07 41 00.	+ 85 04	96	14.0	GALAXY S0?
MCG+14-04-037	07 41 00.	+ 85 04	27	15.	GALAXY
MCG+14-04-038	07 41 00.	+ 85 17	126	14.	GALAXY
ZWG 364.001	07 41 00.	+ 85 18		13.6	GALAXY
ZWG 363.035	07 41 00.	+ 85 18		13.6	GALAXY
ZWG 362.047	07 41 00.	+ 85 18		13.6	GALAXY
UGC 03994	07 41 00.	+ 85 18	162	13.6	GALAXY Sa
ZWG 363.036	07 41 00.	+ 85 51		15.3	GALAXY
ZWG 362.048	07 41 00.	+ 85 51		15.3	GALAXY
RNGC 2435	07 41 01.	+ 31 46		13.5	GALAXY
IC 0469	07 41 01.	+ 85 22 04.		13.29	GALAXY S
REIN 2.057	07 41 02.77	+ 31 46 18.9			NEBULA
FATH 1.200	07 41 05.	+ 44 15	19		NEBULA
KARA.72 140B	07 41 06.	+ 29 21	54		PART OF DOUBLE GALAXY
MCG+05-19-002	07 41 06.	+ 31 48	102	13.5	GALAXY
ZWG 206.020	07 41 06.	+ 44 15		15.3	GALAXY
ZC 0741.1+6230	07 41 06.	+ 62 30	810		CLUSTER OF GALAXIES
ZWG 148.005	07 41 12.	+ 29 23		15.5	GALAXY
ZWG 147.063	07 41 12.	+ 29 23		15.5	GALAXY
MCG+05-19-003	07 41 12.	+ 29 23	36	15.5	GALAXY
ZWG 177.045	07 41 12.	+ 38 09		15.7	GALAXY
KARA.73B 0203	07 41 12.	+ 38 09	36	15.7	ISOLATED GALAXY S
UGC 03997	07 41 12.	+ 40 29	90	16.0	GALAXY DWARF
MCG+07-16-014	07 41 12.	+ 40 29	72	15.	GALAXY
UGC 03998	07 41 12.	+ 56 18	78	16.5	GALAXY
ZWG 349.009	07 41 12.	+ 75 05		15.5	GALAXY
TON-N 0833	07 41 16.	+ 34 41		17.	BLUE STAR
TON-N 0003	07 41 18.	+ 27 19		15.6	BLUE STAR
MCG+09-13-042	07 41 18.	+ 53 12	60	15.	GALAXY
SCHO 0252	07 41 23.	- 16 28 48.	540		ISOLATED DARK CLOUD
ZWG 262.024	07 41 24.	+ 53 12		14.3	GALAXY
UGC 03999	07 41 24.	+ 53 12	60	14.3	GALAXY SBa
MCG+09-13-043	07 41 24.	+ 56 15	60	15.	GALAXY
PK249-05.1	07 41 24.	- 34 38	54	13.1	PLANETARY NEBULA
LB 00504	07 41 25.	+ 63 31 12.		17.1	FAINT BLUE STAR
RNGC 2431	07 41 27.	+ 53 12		14.5	GALAXY
ZWG 118.002	07 41 30.	+ 21 07		15.4	GALAXY
ZWG 235.035	07 41 30.	+ 46 11		14.8	GALAXY
KARA.73B 0204	07 41 30.	+ 46 11	90	14.8	ISOLATED GALAXY S
MCG+08-14-037	07 41 30.	+ 46 12	84	15.	GALAXY
ZC 0741.5+5235	07 41 30.	+ 52 35	610		CLUSTER OF GALAXIES
UGC 04000	07 41 30.	+ 56 11	96	14.8	GALAXY SB?b
ZWG 058.031	07 41 36.	+ 10 05		15.4	GALAXY
MCG+09-13-044	07 41 36.	+ 53 05	12	17.	GALAXY
LB 00183	07 41 37.	+ 44 45 06.		15.9	FAINT BLUE STAR
FATH 1.201	07 41 38.	+ 44 31	33		NEBULA
MCG+13-06-010	07 41 42.	+ 75 04	36	16.	GALAXY
MCG+09-13-045	07 41 45.	+ 53 50	60	16.	GALAXY
ZWG 148.006	07 41 48.	+ 29 03		15.3	GALAXY
MCG+09-13-046	07 41 48.	+ 53 08	42	17.	GALAXY
CED 102	07 41 48.	- 28 50			DIFFUSE GALACTIC NEBULA
ZWG 235.036	07 41 54.	+ 45 54		15.5	GALAXY
ZC 0741.9+4824	07 41 54.	+ 48 24	2350		CLUSTER OF GALAXIES
MCG+10-11-132	07 41 54.	+ 57 06	45	14.	GALAXY
MCG+10-11-133	07 41 54.	+ 57 12	24	15.	GALAXY
ZWG 286.055	07 41 54.	+ 57 15		15.5	GALAXY
ZWG 310.014	07 41 54.	+ 62 35		14.6	GALAXY
ZWG 286.056	07 41 54.	+ 62 35		14.6	GALAXY
UGC 04001	07 41 54.	+ 62 35	90	14.6	GALAXY E
FIT.G 02	07 41 54.	- 27 14 00.		19.	Irr GALAXY
MCG+08-14-038	07 42 00.	+ 45 53	36	15.	GALAXY
MCG+08-14-039	07 42 00.	+ 47 16	42	16.	GALAXY
MCG+09-13-047	07 42 00.	+ 51 54	48	16.	GALAXY
ZWG 286.057	07 42 00.	+ 57 07		15.1	GALAXY
MCG+10-11-134	07 42 00.	+ 62 36	39	15.	GALAXY
MCG+14-04-039	07 42 00.	+ 85 50	18	16.	GALAXY
ISS 0836	07 42 00.	- 22 55	282		STELLAR RING
FIT.G 01	07 42 00.	- 27 14 36.		19.	Sb GALAXY
SCHO 0253	07 42 01.	- 21 05 42.	340		ISOLATED DARK CLOUD
ARC 0594	07 42 03.	+ 11 12		17.5	RICH CLUSTER OF GALAXIES

OBJECT NAME	RIGHT ASCEN.	DECLINATION	DIAM.	MAGN.	TYPE OF OBJECT
MCG+09-13-048	07 42 06.	+ 51 06	72	16.	GALAXY
UGC 04002	07 42 06.	+ 51 07	78	17.	GALAXY S IV-V
ZWG 262.025	07 42 06.	+ 51 54		15.7	GALAXY
MCG+09-13-049	07 42 12.	+ 51 52	42	16.	GALAXY
MCG+10-11-135	07 42 12.	+ 57 09	45	15.	GALAXY
ZWG 286.058	07 42 12.	+ 57 10		15.3	GALAXY
ZWG 286.059	07 42 12.	+ 59 05		14.8	GALAXY
UGC 04003	07 42 12.	+ 59 05	84	14.8	GALAXY Sa
MCG+10-11-136	07 42 12.	+ 59 05	84	14.	GALAXY
ZC 0742.2+6817	07 42 12.	+ 68 17	1140		CLUSTER OF GALAXIES
RNGC 2436	07 42 16.	+ 52 12			NON-EXISTENT OBJECT
ZWG 148.007	07 42 18.	+ 27 05		15.4	GALAXY
ZWG 235.037	07 42 18.	+ 48 57		15.5	GALAXY
MCG+09-13-050	07 42 18.	+ 53 53	48	15.	GALAXY
ZWG 262.026	07 42 18.	+ 53 54		14.9	GALAXY
ZWG 002.010	07 42 24.	+ 00 03		15.6	GALAXY
ZWG 002.010	07 42 24.	+ 00 03		15.6	GALAXY
ZWG 148.008	07 42 24.	+ 28 34		15.6	GALAXY
ZC 0742.4+3232	07 42 24.	+ 32 32	2220		CLUSTER OF GALAXIES
ZWG 002.009	07 42 24.	- 00 15		15.6	GALAXY
ZWG 002.009	07 42 24.	- 00 15		15.6	GALAXY
UGC 04004	07 42 24.	- 00 15	78	15.6	GALAXY S
RNGC 2447	07 42 26.	- 23 45		6.5	OPEN CLUSTER
SCHO 0254	07 42 27.	+ 23 49 36.	180		ISOLATED DARK CLOUD
RNGC 2448	07 42 27.	- 24 33			NON-EXISTENT OBJECT
PATH 1.202	07 42 28.	+ 44 59	16		NEBULA
ZWG 030.013	07 42 30.	+ 08 03		15.1	GALAXY
UGC 04005	07 42 30.	+ 08 03	120	15.1	GALAXY S
KARA.72 141A	07 42 30.	+ 08 03	108	15.1	PART OF DOUBLE GALAXY
MCG+01-20-004	07 42 30.	+ 08 03	108	15.1	GALAXY
ZWG 058.032	07 42 30.	+ 11 11		14.8	GALAXY
UGC 04006	07 42 30.	+ 11 11	60	14.8	GALAXY S
MCG+02-20-007	07 42 30.	+ 11 11	48	14.8	GALAXY
ZWG 118.003	07 42 30.	+ 24 06		15.7	GALAXY
ZWG 148.009	07 42 30.	+ 32 13		15.5	GALAXY
ZWG 206.021	07 42 30.	+ 41 15		15.6	GALAXY
OCL 0649	07 42 30.	- 23 45	1920	6.7	OPEN STAR CLUSTER
BC B20742+31	07 42 30.8	+ 31 50 18.0		16.	QUASI-STELLAR OBJECT
SCHO 0255	07 42 34.	- 23 45 48.	200		ISOLATED DARK CLOUD
ZWG 235.038	07 42 36.	+ 46 30		15.7	GALAXY
MCG+08-14-041	07 42 36.	+ 46 30	36	16.	GALAXY
ZWG 235.039	07 42 36.	+ 46 51		15.7	GALAXY
MCG+08-14-040	07 42 36.	+ 46 52	36	16.	GALAXY
ZWG 286.060	07 42 36.	+ 58 13		15.7	GALAXY
SCHO 0256	07 42 36.	- 24 36 30.	240		ISOLATED DARK CLOUD
ZWG 235.040	07 42 42.	+ 48 26		14.6	GALAXY
MCG+08-14-042	07 42 42.	+ 48 27	72	15.	GALAXY
UGC 04007	07 42 42.	+ 48 28	66	14.6	GALAXY SBa
ZWG 030.014	07 42 48.	+ 08 03		14.8	GALAXY
KARA.72 141B	07 42 48.	+ 08 03	42	14.8	PART OF DOUBLE GALAXY
MCG+01-20-005	07 42 48.	+ 08 03	42	14.8	GALAXY
ZWG 148.010	07 42 48.	+ 29 50		15.5	GALAXY
MRK 082	07 42 48.	+ 62 31	10	15.5	GALAXY WITH UV CONTINUUM
OCL 0674	07 42 48.	- 28 15	228	14.	OPEN STAR CLUSTER
VHA 007	07 42 48.	- 28 15	480		OPEN STAR CLUSTER
TON-N 0834	07 42 53.	+ 33 55		15.	BLUE STAR
ZWG 148.011	07 42 54.	+ 28 49		15.6	GALAXY
MCG+07-16-015	07 42 54.	+ 39 11	54	15.	GALAXY
ZC 0742.9+4659	07 42 54.	+ 46 59	870		CLUSTER OF GALAXIES
ISS 0837	07 42 54.	- 22 18	259		STELLAR RING
OCL 0651	07 42 54.	- 25 24	360	10.	OPEN STAR CLUSTER
PATH 1.203	07 42 58.	+ 44 22	8		NEBULA
LBN 1072	07 43	- 32 00	4200		BRIGHT NEBULA
ZC 0743.0+3317	07 43 00.	+ 33 17	2150		CLUSTER OF GALAXIES
ZWG 206.022	07 43 00.	+ 39 10		15.1	GALAXY
ZWG 235.041	07 43 00.	+ 44 54		14.3	GALAXY
UGC 04008	07 43 00.	+ 44 54		14.3	GALAXY S0-a
KARA.72 142B	07 43 00.	+ 44 54	78		PART OF DOUBLE GALAXY
KARA.72 142A	07 43 00.	+ 44 55	42	14.3	PART OF DOUBLE GALAXY
MCG+08-14-043	07 43 00.	+ 44 55	72	14.	GALAXY
MCG+09-13-052	07 43 00.	+ 51 23	60	17.	GALAXY
MCG+09-13-051	07 43 00.	+ 54 10	60	17.	GALAXY
UGC 04009	07 43 00.	+ 54 11	66	16.5	GALAXY PECULR
MCG+10-11-137	07 43 00.	+ 59 08	78	15.	GALAXY
IC 0474	07 43 01.	+ 26 36 04.			NONSTELLAR OBJECT
PATH 1.204	07 43 04.	+ 44 54	35		NEBULA
ZWG 030.015	07 43 06.	+ 05 05		15.1	GALAXY
UGC 04010	07 43 06.	+ 05 05	90	15.1	GALAXY Sa?
ZWG 148.012	07 43 06.	+ 26 38		15.2	GALAXY
MCG+09-13-054	07 43 06.	+ 51 32 30.	42	16.	GALAXY
ZWG 262.027	07 43 06.	+ 51 35		15.5	GALAXY
MCG+09-13-053	07 43 06.	+ 52 20	72	16.	GALAXY
UGC 04011	07 43 06.	+ 52 23	60	16.0	GALAXY Sc
ZWG 286.061	07 43 06.	+ 59 08		15.4	GALAXY
UGC 04012	07 43 06.	+ 59 08	78	15.4	GALAXY S
ZC 0743.1+7002	07 43 06.	+ 70 02	1550		CLUSTER OF GALAXIES
MRSL 241-00/1	07 43 06.	- 25 24	156		HII REGION
MCG+04-19-001	07 43 12.	+ 26 40	54		GALAXY
MCG+10-11-139	07 43 12.	+ 57 10	39	19.	GALAXY
ZWG 286.062	07 43 12.	+ 61 03		14.0	GALAXY
MRK 010	07 43 12.	+ 61 03	20	14.5	GALAXY WITH UV CONTINUUM
KW 05	07 43 12.	+ 61 03	95		SEYFERT GALAXY
VVI 24	07 43 12.	+ 61 03	100	14.95	SEYFERT GALAXY
UGC 04013	07 43 12.	+ 61 03	114	14.0	GALAXY Sb
MCG+10-11-138	07 43 12.	+ 61 03	102	14.	GALAXY
ZWG 331.014	07 43 12.	+ 74 27		14.4	GALAXY
ZWG 330.051	07 43 12.	+ 74 27		14.4	GALAXY
MRK 011	07 43 12.	+ 74 27	26	15.	GALAXY WITH UV CONTINUUM
UGC 04014	07 43 12.	+ 74 27	60	14.4	GALAXY E-S0
MCG+12-08-011	07 43 12.	+ 74 27	27	15.	GALAXY
ASS 56	07 43 12.	- 27 48			OB ASSOCIATION PUP OB2
MCG+09-13-055	07 43 15.	+ 55 04	42	16.	GALAXY
ZWG 118.004	07 43 18.	+ 22 07		15.4	GALAXY
ZWG 118.005	07 43 18.	+ 24 13		15.7	GALAXY
ZWG 206.023	07 43 18.	+ 44 23		15.3	GALAXY
ZWG 235.042	07 43 18.	+ 47 05		15.7	GALAXY
SCHO 0257	07 43 18.	- 20 45 12.	240		ISOLATED DARK CLOUD
PATH 1.205	07 43 21.	+ 44 23	27		NEBULA
SHB 103	07 43 22.1	- 67 19 06.		17.	QUASI-STELLAR OBJECT
ZWG 058.033	07 43 24.	+ 10 47		15.2	GALAXY
ZWG 087.035	07 43 24.	+ 18 30		15.3	GALAXY
MCG+08-14-044	07 43 24.	+ 47 06	24	16.	GALAXY
ZWG 286.063	07 43 24.	+ 62 26		15.4	GALAXY
UGC 04015	07 43 24.	+ 62 26	72	15.4	GALAXY SBc
MCG+10-11-140	07 43 24.	+ 62 27	66	14.	GALAXY
MCG+11-10-026	07 43 24.	+ 67 03	45	16.	GALAXY
OCL 0696	07 43 24.	- 32 40	156	12.	OPEN STAR CLUSTER
VHA 008	07 43 24.	- 32 40	210		OPEN STAR CLUSTER
LB 00505	07 43 27.	+ 64 07 00.		17.2	FAINT BLUE STAR
MCG+03-20-011	07 43 30.	+ 18 30	36	15.3	GALAXY
ZWG 148.013	07 43 30.	+ 26 50		15.7	GALAXY
ZC 0743.5+3110	07 43 30.	+ 31 10	2490		CLUSTER OF GALAXIES
ZWG 206.024	07 43 30.	+ 39 08		13.1	GALAXY
UGC 04017	07 43 30.	+ 39 08	180	13.1	GALAXY DBL SYS
UGC 04016	07 43 30.	+ 39 08	180	13.1	GALAXY DBL SYS
MCG+07-16-016	07 43 30.	+ 39 09	90	14.5	GALAXY
VV 117A	07 43 30.	+ 39 09 30.	27	14.	INTERACTING GALAXY
VV 117	07 43 30.	+ 39 09 30.	180	13.	INTERACTING GALAXY
KARA.72 143A	07 43 30.	+ 41 39	36	14.8	PART OF DOUBLE GALAXY
ZWG 206.025	07 43 30.	+ 41 40		14.8	GALAXY
UGC 04018	07 43 30.	+ 41 40	60	14.8	GALAXY E
KARA.72 143B	07 43 30.	+ 41 40	48		PART OF DOUBLE GALAXY
MCG+10-11-141	07 43 30.	+ 61 03	30	16.	GALAXY
ZWG 310.015	07 43 30.	+ 67 03		15.3	GALAXY
UGC 04019	07 43 30.	+ 67 03	66	15.3	GALAXY S
ZWG 002.011	07 43 30.	- 02 48		15.7	GALAXY
ZWG 002.011	07 43 30.	- 02 48		15.7	GALAXY
VHA 009	07 43 30.	- 37 51	1800		OPEN STAR CLUSTER
RNGC 2445	07 43 31.	+ 39 08		13.0	GALAXY
RNGC 2444	07 43 31.	+ 39 09		14.0	GALAXY
MCG+07-16-017	07 43 32.	+ 39 07 30.	120	13.5	GALAXY
MCG+07-16-018	07 43 33.	+ 41 39	18	15.5	GALAXY
ZC 0743.6+3450	07 43 36.	+ 34 50	1080		CLUSTER OF GALAXIES
VV 117I	07 43 36.	+ 39 08 30.	6	18.	INTERACTING GALAXY
VV 117H	07 43 36.	+ 39 08 30.	9	17.	INTERACTING GALAXY
VV 117G	07 43 36.	+ 39 08 30.	15	15.	INTERACTING GALAXY
VV 117F	07 43 36.	+ 39 08 30.	9	15.	INTERACTING GALAXY
VV 117E	07 43 36.	+ 39 08 30.	15	15.	INTERACTING GALAXY
VV 117D	07 43 36.	+ 39 08 30.	15	15.	INTERACTING GALAXY
VV 117C	07 43 36.	+ 39 08 30.	9	17.	INTERACTING GALAXY
VV 117B	07 43 36.	+ 39 08 30.	15	15.5	INTERACTING GALAXY
ZWG 206.026	07 43 36.	+ 43 25		15.5	GALAXY
ZC 0743.6+6148	07 43 36.	+ 61 48	1340		CLUSTER OF GALAXIES
OCL 0640	07 43 36.	- 21 49	420	13.	OPEN STAR CLUSTER
OCL 0716	07 43 36.	- 37 51	2700	3.8	OPEN STAR CLUSTER
RNGC 2451	07 43 37.	- 37 51		3.5	OPEN CLUSTER
ARP 143	07 43 39.	+ 39 14			PECULIAR GALAXY
ZWG 148.014	07 43 42.	+ 26 52		15.5	GALAXY
ZWG 262.028	07 43 42.	+ 51 20		14.8	GALAXY
MCG+10-11-142	07 43 42.	+ 59 08	114	14.	GALAXY
ZWG 286.064	07 43 42.	+ 60 50		15.5	GALAXY
OCL 0628	07 43 42.	- 20 16	480	13.	OPEN STAR CLUSTER
ISS 0838	07 43 42.	- 26 18	90		STELLAR RING
MCG+09-13-056	07 43 45.	+ 51 18	42	16.	GALAXY
MCG+11-10-027	07 43 45.	+ 65 27 30.	57	15.	GALAXY
ZWG 148.015	07 43 48.	+ 27 00		15.2	GALAXY
ZWG 177.046	07 43 48.	+ 34 25		15.7	GALAXY
ZWG 235.043	07 43 48.	+ 47 59		15.7	GALAXY
MCG+08-14-045	07 43 48.	+ 48 00	24	17.	GALAXY
ZWG 286.065	07 43 48.	+ 59 08		14.7	GALAXY
UGC 04020	07 43 48.	+ 59 08	132	14.7	GALAXY Sb/SBb
ZWG 310.016	07 43 48.	+ 65 28		15.1	GALAXY
UGC 04021	07 43 48.	+ 65 28	72	15.1	GALAXY SBa-b
MCG+11-10-028	07 43 48.	+ 66 20	51	15.	GALAXY
LB 03480	07 43 48.	- 59 06		14.0	FAINT BLUE STAR
IC 2205	07 43 49.	+ 26 59 52.			NONSTELLAR OBJECT
HN 0076	07 43 53.	- 34 15			NEBULA
IC 2206	07 43 53.	- 34 15			NONSTELLAR OBJECT
ZWG 002.012	07 43 54.	+ 02 05		15.4	GALAXY
ZWG 002.012	07 43 54.	+ 02 05		15.4	GALAXY
MCG+00-20-005	07 43 54.	+ 02 05	36	15.4	GALAXY
ZWG 030.016	07 43 54.	+ 06 59		15.7	GALAXY
ZWG 058.034	07 43 54.	+ 09 21		15.1	GALAXY
MCG+02-20-008	07 43 54.	+ 09 21	48	15.1	GALAXY
MCG+05-19-004	07 43 54.	+ 28 03	54	14.7	GALAXY
KHAV 221	07 44	- 22 19	13750		DARK NEBULA
KHAV 222	07 44	- 27 25	6580		DARK NEBULA
ZWG 148.016	07 44 00.	+ 28 04		14.7	GALAXY
IC 0475	07 44 00.	+ 30 36 57.			NONSTELLAR OBJECT
ZWG 148.017	07 44 00.	+ 30 37		14.9	GALAXY
MCG+05-19-005	07 44 00.	+ 30 38	30	14.9	GALAXY
ZWG 235.044	07 44 00.	+ 48 20		15.5	GALAXY
UGC 04022	07 44 00.	+ 48 20	66	15.5	GALAXY Sc
ZWG 310.017	07 44 00.	+ 66 19		14.2	GALAXY
UGC 04023	07 44 00.	+ 66 19	96	14.2	GALAXY S0
MCG+12-08-012	07 44 00.	+ 74 28	27	16.	GALAXY
MCG+14-04-040	07 44 00.	+ 86 51	60	16.	GALAXY
PATH 1.206	07 44 05.	+ 44 35	22		NEBULA
ZWG 087.036	07 44 06.	+ 16 38		15.5	GALAXY
ZWG 118.006	07 44 06.	+ 22 25		15.6	GALAXY
MCG+05-19-006	07 44 06.	+ 27 03	30	15.5	GALAXY
MCG+08-14-046	07 44 06.	+ 48 20	60	15.	GALAXY
ZWG 286.066	07 44 06.	+ 62 19		14.8	GALAXY
UGC 04024	07 44 06.	+ 62 19	102	14.8	GALAXY Sc
MCG+10-11-143	07 44 06.	+ 62 20	84	14.	GALAXY
SCHO 0258	07 44 06.	- 27 45 48.	320		ISOLATED DARK CLOUD
SCHO 0259	07 44 08.	- 20 45 30.	590		ISOLATED DARK CLOUD
IC 0476	07 44 10.	+ 27 04 38.			NONSTELLAR OBJECT
ZWG 030.017	07 44 12.	+ 07 25		14.7	GALAXY
UGC 04025	07 44 12.	+ 07 25	72	14.7	GALAXY SB:b
MCG+01-20-006	07 44 12.	+ 07 25	60	14.7	GALAXY
ZWG 087.037	07 44 12.	+ 17 21		15.7	GALAXY
MCG+05-19-007	07 44 12.	+ 27 02	66	14.3	GALAXY
ZWG 148.018	07 44 12.	+ 27 04		15.5	GALAXY
ZWG 148.019	07 44 12.	+ 29 32		15.5	GALAXY
ZWG 331.015	07 44 12.	+ 74 26		15.6	GALAXY
MCG+13-06-011	07 44 12.	+ 75 05	39	17.	GALAXY
OCL 0561	07 44 12.	- 04 29	720	16.	OPEN STAR CLUSTER
OCL 0689	07 44 12.	- 31 08	48	14.	OPEN STAR CLUSTER
VHA 010	07 44 12.	- 31 08	150		OPEN STAR CLUSTER
RNGC 2449	07 44 16.	+ 27 03		14.5	GALAXY
ZC 0744.3+1839	07 44 18.	+ 18 39	6050		CLUSTER OF GALAXIES
ZWG 148.020	07 44 18.	+ 27 03		14.3	GALAXY
UGC 04026	07 44 18.	+ 27 03	96	14.3	GALAXY Sa-b
ZWG 148.021	07 44 18.	+ 31 20		15.6	GALAXY
MCG+05-19-008	07 44 24.	+ 27 08	48	15.3	GALAXY
ZC 0744.4+4733	07 44 24.	+ 47 33	1410		CLUSTER OF GALAXIES
ZWG 002.013	07 44 24.	- 01 02		15.7	GALAXY
ZWG 002.013	07 44 24.	- 01 02		15.7	GALAXY
MCG+08-14-047	07 44 27.	+ 50 14 30.	36	17.	GALAXY
RNGC 2450	07 44 28.	+ 27 08		15.5	GALAXY
ZWG 148.022	07 44 30.	+ 27 08		15.3	GALAXY
ZWG 206.027	07 44 30.	+ 43 43		15.7	GALAXY
MRK 083	07 44 30.	+ 54 20	7	16.5	GALAXY WITH UV CONTINUUM
LB 00506	07 44 35.	+ 81 49 54.		15.9	FAINT BLUE STAR
ZWG 002.014	07 44 36.	+ 00 55		15.7	GALAXY

OBJECT NAME	RIGHT ASCEN.	DECLINATION	DIAM.	MAGN.	TYPE OF OBJECT
ZWG 002.014	07 44 36.	+ 00 55		15.7	GALAXY
ZWG 262.029	07 44 36.	+ 55 56		15.2	GALAXY
MCG+09-13-057	07 44 36.	+ 55 57	30	16.	GALAXY
ZWG 310.018	07 44 36.	+ 63 11		15.5	GALAXY
MRK 012	07 44 36.	+ 74 30	25	13.5	GALAXY WITH UV CONTINUUM
MCG+12-08-013	07 44 36.	+ 74 30	66	13.	GALAXY
SEY 028	07 44 39.	+ 74 28 38.		15.2	FAINT GALAXY
ZWG 118.007	07 44 42.	+ 22 54		15.3	GALAXY
MCG+08-14-048	07 44 42.	+ 46 18	18	17.	GALAXY
ZWG 262.030	07 44 42.	+ 54 44		13.9	GALAXY
UGC 04027	07 44 42.	+ 54 44	126	13.9	GALAXY Sb
RNGC 2446	07 44 43.	+ 54 44		14.0	GALAXY
MCG+09-13-058	07 44 45.	+ 54 44	108	14.	GALAXY
ZWG 058.035	07 44 48.	+ 10 22		15.5	GALAXY
ZWG 148.023	07 44 48.	+ 28 17		15.5	GALAXY
ZWG 148.024	07 44 48.	+ 28 27		15.6	GALAXY
MRK 095	07 44 48.	+ 70 21	13	15.5	GALAXY WITH UV CONTINUUM
ZWG 331.016	07 44 48.	+ 74 28		12.7	GALAXY
UGC 04028	07 44 48.	+ 74 28	66	12.7	GALAXY S
ZWG 148.025	07 44 54.	+ 28 32		15.6	GALAXY
ZC 0744.9+4551	07 44 54.	+ 45 51	1610		CLUSTER OF GALAXIES
ZWG 236.001	07 44 54.	+ 49 40		15.6	GALAXY
ZWG 235.045	07 44 54.	+ 49 40		15.6	GALAXY
MCG+08-14-049	07 44 54.	+ 49 40	18	16.	GALAXY
MCG+11-10-029	07 44 54.	+ 63 11	18	16.	GALAXY
ZC 0744.9+7543	07 44 54.	+ 75 43	1280		CLUSTER OF GALAXIES
ACK 248-04.1	07 44 54.	- 33 18			PLANETARY NEBULA
KHAV 223	07 45	- 19 49	2570		DARK NEBULA
ZWG 087.038	07 45 00.	+ 19 13		15.7	GALAXY
ZWG 148.026	07 45 00.	+ 28 09		15.6	GALAXY
MCG+06-17-028	07 45 00.	+ 34 27	108	14.	GALAXY
ZWG 177.047	07 45 00.	+ 34 28		14.6	GALAXY
UGC 04029	07 45 00.	+ 34 28	162	14.6	GALAXY SB:b-c
ZC 0745.0+3825	07 45 00.	+ 38 25	1880		CLUSTER OF GALAXIES
ZWG 235.046	07 45 00.	+ 44 48		15.5	GALAXY
MCG+09-13-059	07 45 00.	+ 55 33	60	17.	GALAXY
FATH 1.207	07 45 02.	+ 44 48	14		NEBULA
ARC 0595	07 45 02.	+ 52 12		15.6	RICH CLUSTER OF GALAXIES
SCHO 0260	07 45 03.	- 27 22 06.	320		ISOLATED DARK CLOUD
ZWG 030.018	07 45 06.	+ 06 55		15.0	GALAXY
MCG+01-20-007	07 45 06.	+ 06 55	48	15.0	GALAXY
ZWG 118.008	07 45 06.	+ 23 01		15.7	GALAXY
ZWG 148.027	07 45 06.	+ 28 20		14.4	GALAXY
UGC 04030	07 45 06.	+ 28 20		14.4	GALAXY DBL SYS
KARA.72 144A	07 45 06.	+ 28 20	24	14.4	PART OF DOUBLE GALAXY
MCG+05-19-009	07 45 06.	+ 28 20	24	14.4	GALAXY
ZC 0745.1+5220	07 45 06.	+ 52 20	2820		CLUSTER OF GALAXIES
MCG+09-13-061	07 45 06.	+ 52 42	24	17.	GALAXY
ZWG 262.031	07 45 06.	+ 55 43		15.3	GALAXY
MCG+09-13-060	07 45 06.	+ 55 43	36	16.	GALAXY
7ZW 181	07 45 06.	+ 57 11			COMPACT GALAXY
ZWG 286.067	07 45 06.	+ 57 11		15.6	GALAXY
FIT.G 06	07 45 06.	- 25 30 24.		19.	Sb GALAXY
SEY 029	07 45 07.	+ 74 30 36.		12.7	FAINT GALAXY
ZWG 087.039	07 45 12.	+ 18 42		15.6	GALAXY
MCG+03-20-012	07 45 12.	+ 18 42	9	15.6	GALAXY
MCG+04-19-002	07 45 12.	+ 23 21	48	15.3	GALAXY
ZWG 148.028	07 45 12.	+ 28 09		15.5	GALAXY
KARA.72 144B	07 45 12.	+ 28 19	24		PART OF DOUBLE GALAXY
MCG+10-11-144	07 45 12.	+ 61 34	39	16.	GALAXY
ZWG 087.040	07 45 18.	+ 18 40		15.7	GALAXY
MCG+03-20-013	07 45 18.	+ 18 41	36	15.7	GALAXY
ZWG 148.029	07 45 18.	+ 27 42		15.7	GALAXY
ZWG 286.068	07 45 18.	+ 61 34		15.5	GALAXY
MCG+11-10-030	07 45 18.	+ 64 39	24	17.	GALAXY
ZWG 310.019	07 45 18.	+ 65 12		15.7	GALAXY
MCG+11-10-031	07 45 18.	+ 65 12	42	16.	GALAXY
RNGC 2452	07 45 23.	- 27 13		12.5	PLANETARY NEBULA
PK243-01.1	07 45 23.71	- 27 12 36.6	31	12.6	PLANETARY NEBULA
ZWG 058.036	07 45 24.	+ 10 56		15.7	GALAXY
ZWG 118.009	07 45 24.	+ 23 21		15.3	GALAXY
UGC 04031	07 45 24.	+ 23 21	60	15.3	GALAXY Sc
KARA.73B 0205	07 45 24.	+ 23 21	60	15.3	ISOLATED GALAXY S
ZWG 148.030	07 45 24.	+ 30 16		15.6	GALAXY
UGC 04032	07 45 24.	+ 30 16	72	15.6	GALAXY Sc
ZWG 206.028	07 45 24.	+ 43 55		15.6	GALAXY
ZWG 286.069	07 45 24.	+ 61 28		14.9	GALAXY
UGC 04033	07 45 24.	+ 61 28	66	14.9	GALAXY Sb
MCG+11-10-032	07 45 24.	+ 64 37	24	17.	GALAXY
ZC 0745.4+7252	07 45 24.	+ 72 52	1680		CLUSTER OF GALAXIES
FATH 1.208	07 45 25.	+ 44 29	11		NEBULA
ZC 0745.5+4020	07 45 30.	+ 40 20	11220		CLUSTER OF GALAXIES
MCG+09-13-062	07 45 30.	+ 52 10	9	18.	GALAXY
MCG+09-13-063	07 45 30.	+ 52 47 30.	15	15.	GALAXY
ZWG 262.032	07 45 30.	+ 52 48		15.1	GALAXY
MCG+10-11-145	07 45 30.	+ 61 28	39	14.	GALAXY
7ZW 182	07 45 30.	+ 65 40			COMPACT GALAXY
MCG-03-20-001	07 45 33.	- 18 39	120	14.5	GALAXY
YC 0745-57	07 45 35.	- 57 05 24.			UNUSUAL SOUTHERN NEBULA
TON-N 0836	07 45 36.	+ 21 31		15.5	BLUE STAR
TON-N 0286	07 45 36.	+ 30 41		14.5	BLUE STAR
TON-N 0287	07 45 36.	+ 30 51		14.4	BLUE STAR
MCG+08-14-050	07 45 36.	+ 45 38	24	16.	GALAXY
MRK 096	07 45 36.	+ 46 28	9	15.5	GALAXY WITH UV CONTINUUM
MCG+11-10-033	07 45 36.	+ 65 40	18	17.	GALAXY
RNGC 2466	07 45 36.	- 71 16			GALAXY
MCG+08-14-051	07 45 39.	+ 46 32	42	16.	GALAXY
RNGC 2453	07 45 40.	- 27 07		9.0	OPEN CLUSTER
ZWG 148.031	07 45 42.	+ 29 01		15.2	GALAXY
ZC 0745.7+3644	07 45 42.	+ 36 44	740		CLUSTER OF GALAXIES
MCG+09-13-064	07 45 42.	+ 56 00	30	18.	GALAXY
OCL 0670	07 45 42.	- 27 07	420	9.4	OPEN STAR CLUSTER
SCHO 0261	07 45 46.	- 27 08 00.	290		ISOLATED DARK CLOUD
ZWG 148.032	07 45 48.	+ 27 41		15.5	GALAXY
MCG+05-19-010	07 45 48.	+ 29 01	36	15.2	GALAXY
ZC 0745.8+3550	07 45 48.	+ 35 50	670		CLUSTER OF GALAXIES
ZWG 206.029	07 45 48.	+ 43 54		14.7	GALAXY
MCG+10-11-146	07 45 48.	+ 59 46	39	15.	GALAXY
MCG+10-11-147	07 45 48.	+ 59 48	24	17.	GALAXY
ZWG 002.015	07 45 48.	- 01 21		15.6	GALAXY
ZWG 002.015	07 45 48.	- 01 21		15.6	GALAXY
ZWG 148.033	07 45 54.	+ 30 03		15.6	GALAXY
ZWG 262.033	07 45 54.	+ 56 02		15.4	GALAXY
MCG+10-11-148	07 45 54.	+ 59 45	39	17.	GALAXY
7ZW 183	07 45 54.	+ 65 34			COMPACT GALAXY
SCHO 0262	07 45 59.	- 20 11 12.	270		ISOLATED DARK CLOUD
KHAV 225	07 46	- 26 49	3750		DARK NEBULA
KHAV 224	07 46	- 29 19	5010		DARK NEBULA

OBJECT NAME	RIGHT ASCEN.	DECLINATION	DIAM.	MAGN.	TYPE OF OBJECT
ZWG 087.041	07 46 00.	+ 17 29		15.4	GALAXY
ZWG 118.010	07 46 00.	+ 21 52		14.9	GALAXY
MCG+04-19-003	07 46 00.	+ 21 52	15	14.9	GALAXY
ZWG 148.034	07 46 00.	+ 28 45		15.4	GALAXY
MCG+05-19-011	07 46 00.	+ 28 45	48	15.4	GALAXY
MCG+09-13-065	07 46 00.	+ 50 53	18	16.	GALAXY
MCG+09-13-066	07 46 00.	+ 56 03	24	16.	GALAXY
MCG+09-13-067	07 46 00.	+ 56 53	60	15.	GALAXY
MCG+10-11-149	07 46 00.	+ 62 10	39	16.	GALAXY
MCG+12-08-014	07 46 00.	+ 72 09	21	17.	GALAXY
MCG+12-08-015	07 46 00.	+ 73 09	102	13.0	GALAXY
PK264-12.1	07 46 04.	- 51 08 05.	10		PLANETARY NEBULA
RNGC 2441	07 46 05.	+ 73 09		13.0	GALAXY
ZWG 148.035	07 46 06.	+ 28 44		15.6	GALAXY
MCG+05-19-012	07 46 06.	+ 28 44	36	15.6	GALAXY
ZWG 148.036	07 46 06.	+ 31 01		15.7	GALAXY
UGC 04034	07 46 06.	+ 31 01	72	15.7	GALAXY Sc
ZWG 206.030	07 46 06.	+ 43 05		15.6	GALAXY
MCG+08-15-001	07 46 06.	+ 45 02	9	17.	GALAXY
ZWG 236.002	07 46 06.	+ 48 15		15.4	GALAXY
ZWG 235.047	07 46 06.	+ 48 15		15.4	GALAXY
ZWG 262.034	07 46 06.	+ 55 30		14.1	GALAXY
UGC 04035	07 46 06.	+ 55 30	66	14.1	GALAXY E
ZWG 331.017	07 46 06.	+ 73 08		12.7	GALAXY
UGC 04036	07 46 06.	+ 73 08	138	12.7	GALAXY Sc/SBc
SCHO 0263	07 46 08.	- 20 07 36.	280		ISOLATED DARK CLOUD
MCG+08-15-003	07 46 09.	+ 48 15		16.	GALAXY
MCG+08-15-002	07 46 09.	+ 48 15	36	16.	GALAXY
ARC 0593	07 46 10.	+ 72 58		17.4	RICH CLUSTER OF GALAXIES
ZWG 030.019	07 46 12.	+ 06 20		15.4	GALAXY
MCG+07-16-019	07 46 12.	+ 40 18 30.	27	16.	GALAXY
ZWG 206.031	07 46 12.	+ 40 19		15.4	GALAXY
MCG+09-13-068	07 46 12.	+ 55 30	24	15.	GALAXY
ZWG 262.035	07 46 12.	+ 55 32		14.9	GALAXY
ZWG 030.020	07 46 18.	+ 06 27		15.6	GALAXY
ZWG 058.037	07 46 18.	+ 12 13		15.7	GALAXY
MCG+09-13-069	07 46 18.	+ 55 32 30.	48	16.	GALAXY
ZWG 262.036	07 46 18.	+ 56 01		15.1	GALAXY
ZWG 331.018	07 46 18.	+ 71 10		15.6	GALAXY
KARA.73B 0206	07 46 18.	+ 71 10	42	15.6	ISOLATED GALAXY RING
ZWG 002.016	07 46 18.	- 01 36		15.5	GALAXY
ZWG 002.016	07 46 18.	- 01 36		15.5	GALAXY
TON-N 0837	07 46 22.	+ 35 04		16.3	BLUE STAR
ZWG 118.011	07 46 24.	+ 21 58		15.2	GALAXY
ZWG 177.048	07 46 24.	+ 32 51		15.4	GALAXY
ZWG 236.003	07 46 24.	+ 50 10		15.0	GALAXY
ZWG 235.048	07 46 24.	+ 50 10		15.0	GALAXY
UGC 04037	07 46 24.	+ 73 43	72	16.5	GALAXY Sc
OCL 0660	07 46 24.	- 26 10	240	12.	OPEN STAR CLUSTER
SEY 030	07 46 25.	+ 72 10 42.		15.1	FAINT GALAXY
TON-N 0838	07 46 28.	+ 36 05		16.	BLUE STAR
ZC 0746.5+2315	07 46 30.	+ 23 15	6720		CLUSTER OF GALAXIES
MCG+05-19-013	07 46 30.	+ 27 01	48	15.1	GALAXY
ZWG 148.037	07 46 30.	+ 27 02		15.1	GALAXY
UGC 04038	07 46 30.	+ 27 02	66	15.1	GALAXY S
ZWG 148.038	07 46 30.	+ 30 04		15.2	GALAXY
UGC 04039	07 46 30.	+ 30 04	84	15.2	GALAXY Sb:
MCG+09-13-070	07 46 30.	+ 55 37	36	18.	GALAXY
MCG+09-13-071	07 46 30.	+ 56 02	48	15.	GALAXY
MCG+06-17-029	07 46 33.	+ 34 05	126	14.	GALAXY
ZWG 148.039	07 46 36.	+ 27 58		15.5	GALAXY
MCG+05-19-014	07 46 36.	+ 30 05	72	15.2	GALAXY
ZWG 177.049	07 46 36.	+ 34 05		15.4	GALAXY
UGC 04040	07 46 36.	+ 34 05	120	15.4	GALAXY Sc
MCG+12-08-016	07 46 36.	+ 73 38	51	15.	GALAXY
OCL 0656	07 46 36.	- 25 46	60	14.	OPEN STAR CLUSTER
FIT.G 03	07 46 36.	- 26 07 30.		18.0	Sc GALAXY
IC 2207	07 46 37.	+ 34 05 27.			NONSTELLAR OBJECT
ZC 0746.7+3502	07 46 42.	+ 35 02	1280		CLUSTER OF GALAXIES
ZC 0746.7+5105	07 46 42.	+ 51 05	270		CLUSTER OF GALAXIES
MRK 097	07 46 42.	+ 65 51	20	15.5	GALAXY WITH UV CONTINUUM
ZWG 331.019	07 46 42.	+ 73 38		13.6	GALAXY
UGC 04041	07 46 42.	+ 73 38	54	13.6	GALAXY E
ZWG 148.040	07 46 48.	+ 30 09		15.0	GALAXY
UGC 04042	07 46 48.	+ 30 09	66	15.0	GALAXY SBb
ZWG 148.041	07 46 48.	+ 30 52		15.4	GALAXY
ZWG 177.050	07 46 48.	+ 34 37		15.6	GALAXY
ZWG 262.037	07 46 48.	+ 54 29		14.8	GALAXY
UGC 04043	07 46 48.	+ 54 29	132	14.8	GALAXY Sc
ZWG 287.001	07 46 48.	+ 62 12		14.7	GALAXY
ZWG 286.070	07 46 48.	+ 62 12		14.7	GALAXY
MCG+10-11-150	07 46 48.	+ 62 12	24	15.	GALAXY
OCL 0636	07 46 49.	- 21 10	480	10.3	OPEN STAR CLUSTER
RNGC 2455	07 46 49.	- 21 10		10.0	OPEN CLUSTER
ZWG 087.042	07 46 54.	+ 18 57		15.1	GALAXY
UGC 04044	07 46 54.	+ 18 57	90	15.1	GALAXY Sc
ZWG 118.012	07 46 54.	+ 26 09		15.7	GALAXY
KARA.73B 0207	07 46 54.	+ 26 09	18	15.7	ISOLATED GALAXY E
ZWG 148.042	07 46 54.	+ 27 45		15.7	GALAXY
MCG+05-19-015	07 46 54.	+ 30 10	48	15.0	GALAXY
MCG+05-19-016	07 46 54.	+ 30 52	36	15.4	GALAXY
ZWG 177.051	07 46 54.	+ 34 33		15.5	GALAXY
UGC 04045	07 46 54.	+ 34 33	60	15.5	GALAXY VY CMPT
ZWG 206.032	07 46 54.	+ 41 23		15.6	GALAXY
MCG+09-13-072	07 46 54.	+ 54 30	120	15.	GALAXY
MCG+09-13-073	07 46 54.	+ 55 35	24	18.	GALAXY
ZC 0746.9+7141	07 46 54.	+ 71 41	1950		CLUSTER OF GALAXIES
MRK 098	07 46 54.	+ 72 00	12	15.	GALAXY WITH UV CONTINUUM
MCG+03-20-014	07 46 57.	+ 18 57	78	15.1	GALAXY
ZC 0747.0+2447	07 47 00.	+ 24 47	610		CLUSTER OF GALAXIES
ZWG 148.043	07 47 00.	+ 27 45		15.6	GALAXY
UGC 04046	07 47 00.	+ 30 48	60	17.	GALAXY SB
ZWG 148.044	07 47 00.	+ 30 51		14.5	GALAXY
UGC 04047	07 47 00.	+ 30 51	132	14.5	GALAXY SBb
MCG+05-19-017	07 47 00.	+ 30 52	90	14.5	GALAXY
ZC 0747.0+5233	07 47 00.	+ 52 33	540		CLUSTER OF GALAXIES
MCG+10-11-151	07 47 00.	+ 60 18	39	18.	GALAXY
MCG+10-11-152	07 47 00.	+ 62 00	18	17.	GALAXY
ZWG 370.007	07 47 00.	+ 86 52		15.7	GALAXY
FIT.G 04	07 47 00.	- 26 07 30.		17.5	Irr GALAXY
SEY 031	07 47 03.	+ 73 38 52.		14.1	FAINT GALAXY
ZWG 087.043	07 47 06.	+ 19 42		15.7	GALAXY
UGC 04048	07 47 06.	+ 19 42	60	15.7	GALAXY S
ZWG 148.045	07 47 06.	+ 27 59		15.6	GALAXY
ZC 0747.1+3729	07 47 06.	+ 37 29	3560		CLUSTER OF GALAXIES
ZC 0747.1+5826	07 47 06.	+ 58 26	1210		CLUSTER OF GALAXIES
ZWG 287.002	07 47 06.	+ 62 00		15.4	GALAXY
ZWG 286.071	07 47 06.	+ 62 00		15.4	GALAXY

OBJECT NAME	RIGHT ASCEN.	DECLINATION	DIAM.	MAGN.	TYPE OF OBJECT
PIT.G 05	07 47 06.	- 26 16 12.		17.0	E2 GALAXY
YC 0747-55	07 47 08.	- 55 09 36.			UNUSUAL SOUTHERN NEBULA
TON-N 0288	07 47 12.	+ 27 20		14.8	BLUE STAR
ZWG 148.046	07 47 12.	+ 30 21		15.2	GALAXY
MCG+10-11-153	07 47 12.	+ 57 01	78	15.	GALAXY
MCG+11-10-034	07 47 12.	+ 64 02 30.	36	16.	GALAXY
MCG+12-08-017	07 47 12.	+ 72 10	36	15.	GALAXY
MCG+08-15-004	07 47 15.	+ 48 27	30	16.	GALAXY
TON-N 0839	07 47 18.	+ 21 25		16.	BLUE STAR
MCG+05-19-018	07 47 18.	+ 30 32	30	15.5	GALAXY
ZC 0747.3+4522	07 47 18.	+ 45 22	1080		CLUSTER OF GALAXIES
ZWG 236.004	07 47 18.	+ 48 27		15.7	GALAXY
ZWG 286.072	07 47 18.	+ 57 03		14.6	GALAXY
UGC 04049	07 47 18.	+ 57 03	84	14.6	GALAXY S?
ZC 0747.3+6058	07 47 18.	+ 60 58	1010		CLUSTER OF GALAXIES
ZWG 331.020	07 47 18.	+ 72 10		14.3	GALAXY
UGC 04050	07 47 18.	+ 72 10	42	14.3	GALAXY S?
ISS 1036	07 47 18.	- 43 47	385		STELLAR RING
ZC 0747.4+2241	07 47 24.	+ 22 41	1010		CLUSTER OF GALAXIES
ZC 0747.4+3850	07 47 24.	+ 38 50	1280		CLUSTER OF GALAXIES
MCG+08-15-005	07 47 27.	+ 50 20	24	15.	GALAXY
SEY 032	07 47 29.	+ 72 11 32.		13.7	FAINT GALAXY
ZC 0747.5+2120	07 47 30.	+ 21 20	2620		CLUSTER OF GALAXIES
MCG+04-19-004	07 47 30.	+ 23 06	36	15.5	GALAXY
ZWG 236.005	07 47 30.	+ 50 18		15.4	GALAXY
ZWG 235.049	07 47 30.	+ 50 18		15.4	GALAXY
MCG+08-15-006	07 47 33.	+ 50 19	30	15.	GALAXY
MCG+08-15-007	07 47 33.	+ 50 22	15	17.	GALAXY
ZWG 118.013	07 47 36.	+ 23 07		15.5	GALAXY
ZWG 148.047	07 47 36.	+ 27 56		15.5	GALAXY
ZC 0747.6+3248	07 47 36.	+ 32 48	1010		CLUSTER OF GALAXIES
ZC 0747.6+4349	07 47 36.	+ 43 49	870		CLUSTER OF GALAXIES
ZWG 236.006	07 47 36.	+ 50 17		14.4	GALAXY
ZWG 235.050	07 47 36.	+ 50 17		14.4	GALAXY
UGC 04051	07 47 36.	+ 50 17	42	14.4	GALAXY
ZWG 236.007	07 47 36.	+ 50 21		14.1	GALAXY
ZWG 235.051	07 47 36.	+ 50 21		14.1	GALAXY
UGC 04052	07 47 36.	+ 50 21	48	14.1	GALAXY DBL SYS
KARA.72 145B	07 47 36.	+ 50 21	36		PART OF DOUBLE GALAXY
KARA.72 145A	07 47 36.	+ 50 21	36	14.1	PART OF DOUBLE GALAXY
MCG+08-15-008	07 47 36.	+ 50 25	42	16.	GALAXY
MCG+10-11-154	07 47 36.	+ 57 13	30	16.	GALAXY
ZWG 286.073	07 47 36.	+ 57 15		15.4	GALAXY
OCL 0615	07 47 36.	- 17 09	180	14.	OPEN STAR CLUSTER
ACK 243-00.1	07 47 36.	- 27 28			PLANETARY NEBULA
RNGC 2454	07 47 40.	+ 16 30		14.5	GALAXY
ZWG 087.044	07 47 42.	+ 16 30		14.7	GALAXY
UGC 04053	07 47 42.	+ 16 30	66	14.7	GALAXY
ZWG 118.014	07 47 42.	+ 23 37		15.3	GALAXY
TON-N 0289	07 47 42.	+ 29 23		16.0	BLUE STAR
TON-N 0290	07 47 42.	+ 30 26		15.2	BLUE STAR
ISS 0719	07 47 42.	- 18 26	200		STELLAR RING
ZWG 058.038	07 47 48.	+ 10 27		15.4	GALAXY
MCG+03-20-015	07 47 48.	+ 16 29 30.	39	14.7	GALAXY
MCG+04-19-005	07 47 48.	+ 24 00	120	15.1	GALAXY
ZWG 262.038	07 47 48.	+ 55 13		15.3	GALAXY
ZWG 058.039	07 47 54.	+ 11 13		15.7	GALAXY
TON-N 0840	07 47 54.	+ 21 56		15.8	BLUE STAR
ZWG 118.015	07 47 54.	+ 24 01		15.1	GALAXY
UGC 04054	07 47 54.	+ 24 01	132	15.1	GALAXY Sb
KARA.73B 0208	07 47 54.	+ 24 01	234	15.1	ISOLATED GALAXY S
MCG+06-17-030	07 47 54.	+ 34 10	66	15.	GALAXY
ZWG 178.001	07 47 54.	+ 34 11		15.6	GALAXY
ZWG 177.052	07 47 54.	+ 34 11		15.6	GALAXY
UGC 04055	07 47 54.	+ 34 11	72	15.6	GALAXY Sc
MCG+09-13-074	07 47 54.	+ 55 13	60	15.	GALAXY
MCG+12-08-018	07 47 54.	+ 74 32	150	14.	GALAXY
ABC 0598	07 47 59.	+ 17 48		17.4	RICH CLUSTER OF GALAXIES
KHAV 226	07 48	- 30 20	8110		DARK NEBULA
LB 09838	07 48	- 82 50		15.0	FAINT BLUE STAR
ZWG 206.033	07 48 00.	+ 43 00		15.0	GALAXY
UGC 04056	07 48 00.	+ 43 00	66	15.0	GALAXY Sc
MCG+08-15-009	07 48 00.	+ 49 57	36	16.	GALAXY
ZWG 262.039	07 48 00.	+ 55 25		15.7	GALAXY
MCG+09-13-075	07 48 00.	+ 55 25	12	17.	GALAXY
ZC 0748.0+6117	07 48 00.	+ 61 17	1340		CLUSTER OF GALAXIES
MCG+11-10-035	07 48 00.	+ 68 13	30	16.	GALAXY
ZWG 349.010	07 48 00.	+ 74 32		13.4	GALAXY
UGC 04057	07 48 00.	+ 74 32	162	13.4	GALAXY Sa
PK236+03.1	07 48 01.	- 19 11	38		PLANETARY NEBULA
MCG+07-16-020	07 48 03.	+ 43 00	60	15.	GALAXY
UGC 04058	07 48 06.	+ 18 06	66	16.5	GALAXY S IV
ZWG 287.003	07 48 06.	+ 60 52		15.2	GALAXY
ZWG 286.074	07 48 06.	+ 60 52		15.2	GALAXY
7ZW 184	07 48 06.	+ 66 13			COMPACT GALAXY
ZWG 058.040	07 48 12.	+ 09 42		15.7	GALAXY
ZWG 058.041	07 48 12.	+ 09 43		15.5	GALAXY
ZC 0748.2+1744	07 48 12.	+ 17 44	1140		CLUSTER OF GALAXIES
ZWG 236.008	07 48 12.	+ 49 57		15.1	GALAXY
MCG+10-12-001	07 48 12.	+ 60 52	39	15.	GALAXY
MCG+10-12-002	07 48 12.	+ 62 04	30	16.	GALAXY
ZWG 310.020	07 48 12.	+ 62 40		15.4	GALAXY
ZWG 286.075	07 48 12.	+ 62 40		15.4	GALAXY
UGC 04059	07 48 12.	+ 62 40	90	15.4	GALAXY Sc
MCG+10-12-003	07 48 12.	+ 62 40	57	16.	GALAXY
OCL 0655	07 48 12.	- 25 19	120	12.	OPEN STAR CLUSTER
MCG+08-15-010	07 48 15.	+ 46 11	12	16.	GALAXY
ZWG 310.021	07 48 18.	+ 68 13		15.0	GALAXY
KARA.73B 0209	07 48 18.	+ 68 13	42	15.0	ISOLATED GALAXY S
PIT.G 07	07 48 18.	- 26 38 48.		17.5	SO GALAXY
SEY 033	07 48 23.	+ 74 32 23.		13.4	FAINT GALAXY
ZWG 206.034	07 48 24.	+ 44 08		15.6	GALAXY
MCG+08-15-011	07 48 24.	+ 46 07	36	16.	GALAXY
MCG+09-13-076	07 48 24.	+ 55 46	30	18.	GALAXY
ISS 0720	07 48 24.	- 20 03	105		STELLAR RING
OCL 0629	07 48 24.	- 20 04	420	13.	OPEN STAR CLUSTER
OCL 0683	07 48 24.	- 29 42	300		OPEN STAR CLUSTER
VHA 011	07 48 24.	- 29 42	270		OPEN STAR CLUSTER
ZWG 058.042	07 48 30.	+ 10 16		15.5	GALAXY
ZWG 058.043	07 48 30.	+ 10 58		15.1	GALAXY
ZWG 148.048	07 48 30.	+ 27 30		15.5	GALAXY
ZWG 148.049	07 48 30.	+ 30 34		15.5	GALAXY
ZC 0748.5+3302	07 48 30.	+ 33 02	1010		CLUSTER OF GALAXIES
MRK 099	07 48 30.	+ 61 10	10	16.5	GALAXY WITH UV CONTINUUM
MCG+11-10-036	07 48 30.	+ 62 55	27	17.	GALAXY
ZWG 058.044	07 48 36.	+ 11 06		15.5	GALAXY
ZWG 148.050	07 48 36.	+ 27 44		15.5	GALAXY
ZWG 058.045	07 48 42.	+ 10 39		15.1	GALAXY

OBJECT NAME	RIGHT ASCEN.	DECLINATION	DIAM.	MAGN.	TYPE OF OBJECT
ZWG 058.046	07 48 42.	+ 10 54		15.3	GALAXY
ZWG 058.047	07 48 42.	+ 14 09		14.6	GALAXY
UGC 04060	07 48 42.	+ 14 09	72	14.6	GALAXY S
MCG+02-20-009	07 48 42.	+ 14 09	72	14.6	GALAXY
ZWG 148.051	07 48 42.	+ 27 35		15.6	GALAXY
ZWG 178.002	07 48 42.	+ 32 47		15.7	GALAXY
KARA.73B 0210	07 48 42.	+ 32 47	24	15.7	ISOLATED GALAXY S
ZC 0748.7+3531	07 48 42.	+ 35 31	1950		CLUSTER OF GALAXIES
ABC 0597	07 48 43.	+ 35 30		17.0	RICH CLUSTER OF GALAXIES
ZWG 002.017	07 48 48.	+ 02 27		15.6	GALAXY
ZWG 148.052	07 48 48.	+ 27 26		15.7	GALAXY
UGC 04061	07 48 48.	+ 27 26	78	15.7	GALAXY Sb
MCG+05-19-019	07 48 48.	+ 27 36	36	15.6	GALAXY
7ZW 185	07 48 48.	+ 63 28			COMPACT GALAXY
MCG+13-06-012	07 48 48.	+ 78 09	78	14.	GALAXY
ZWG 058.048	07 48 54.	+ 09 45		15.6	GALAXY
ZWG 058.049	07 48 54.	+ 09 55		15.4	GALAXY
ZWG 058.050	07 48 54.	+ 10 01		15.6	GALAXY
ZWG 058.051	07 48 54.	+ 11 18		15.3	GALAXY
MCG+08-15-012	07 48 54.	+ 48 32	48	16.	GALAXY
UGC 04062	07 48 54.	+ 50 32	78	16.0	GALAXY Sb-c
RNGC 2336A	07 48 59.	+ 78 08		14.0	GALAXY
KHAV 227	07 49	- 32 50	13990		DARK NEBULA
MCG+04-19-006	07 49 00.	+ 23 36	42	15.5	GALAXY
ZWG 148.053	07 49 00.	+ 27 37		15.2	GALAXY
UGC 04064	07 49 00.	+ 30 30	90	17.	GALAXY DWARF
ZC 0749.0+3419	07 49 00.	+ 34 19	3760		CLUSTER OF GALAXIES
MCG+08-15-013	07 49 00.	+ 50 16	30	16.	GALAXY
ZWG 262.040	07 49 00.	+ 55 21		15.5	GALAXY
UGC 04065	07 49 00.	+ 55 21	96	15.5	GALAXY Sc
MCG+10-12-004	07 49 00.	+ 58 22	27	17.	GALAXY
ZC 0749.0+5842	07 49 00.	+ 58 42	1010		CLUSTER OF GALAXIES
ZWG 310.022	07 49 00.	+ 64 01		15.2	GALAXY
MCG+11-10-037	07 49 00.	+ 64 01	18	15.	GALAXY
MCG+11-10-038	07 49 00.	+ 66 29	18	17.	GALAXY
ZWG 349.011	07 49 00.	+ 78 08		14.2	GALAXY
UGC 04066	07 49 00.	+ 78 08	126	14.2	GALAXY Sc
UGC 04063	07 49 00.	+ 85 57	72	17.	GALAXY SB IV-V
PK217+14.1	07 49 06.	+ 03 08	375	13.6	PLANETARY NEBULA
ZWG 030.021	07 49 06.	+ 04 38		15.2	GALAXY
ZWG 058.052	07 49 06.	+ 10 06		15.7	GALAXY
ZWG 118.016	07 49 06.	+ 23 36		15.5	GALAXY
MCG+10-12-005	07 49 06.	+ 58 22	30	17.	GALAXY
FEIG 112	07 49 06.	- 14 35		12.8	FAINT BLUE STAR
IC 0477	07 49 08.	+ 23 36 44.			NONSTELLAR OBJECT
ZWG 002.018	07 49 12.	+ 02 57		14.7	GALAXY
MCG+00-20-006	07 49 12.	+ 02 57	21	14.7	GALAXY
ZWG 030.022	07 49 12.	+ 04 34		15.3	GALAXY
ZWG 118.017	07 49 12.	+ 21 43		15.6	GALAXY
ZC 0749.2+3335	07 49 12.	+ 33 35	1080		CLUSTER OF GALAXIES
ZWG 262.041	07 49 12.	+ 53 10		15.2	GALAXY
MCG+09-13-077	07 49 12.	+ 55 22 30.	96	15.	GALAXY
ZC 0749.2+5542	07 49 12.	+ 55 42	810		CLUSTER OF GALAXIES
KARA.68 050	07 49 12.	+ 61 34	27		DWARF GALAXY
MCG+10-12-006	07 49 12.	+ 62 12	18	17.	GALAXY
ZC 0749.2+6826	07 49 12.	+ 68 26	2290		CLUSTER OF GALAXIES
ZWG 331.021	07 49 12.	+ 72 45		15.4	GALAXY
UGC 04067	07 49 12.	+ 72 45	90	15.4	GALAXY SBc
MCG+12-08-019	07 49 12.	+ 72 45	66	16.	GALAXY
ISS 0839	07 49 12.	- 25 45	575		STELLAR RING
ZWG 058.053	07 49 18.	+ 11 15		15.5	GALAXY
ZWG 178.003	07 49 18.	+ 33 36		15.3	GALAXY
ZC 0749.3+4402	07 49 18.	+ 44 02	6520		CLUSTER OF GALAXIES
ZC 0749.3+5344	07 49 18.	+ 53 44	1010		CLUSTER OF GALAXIES
SEY 034	07 49 22.	+ 72 45 31.		15.1	FAINT GALAXY
IC 2208	07 49 23.	+ 27 38 01.			NONSTELLAR OBJECT
MCG+06-18-001	07 49 24.	+ 33 36 30.	42	15.	GALAXY
ZWG 236.009	07 49 24.	+ 50 15		15.3	GALAXY
MCG+08-15-014	07 49 24.	+ 50 16	60	16.	GALAXY
MCG+10-12-007	07 49 24.	+ 56 52	51	17.	GALAXY
7ZW 186	07 49 24.	+ 60 48			COMPACT GALAXY
7ZW 187	07 49 24.	+ 72 24			COMPACT GALAXY
UGC 04068	07 49 30.	+ 40 17	96	17.	GALAXY Sc
7ZW 188	07 49 30.	+ 59 34			COMPACT GALAXY
MCG+11-10-039	07 49 30.	+ 64 46	27	17.	GALAXY
RNGC 2459	07 49 32.	+ 09 41			NON-EXISTENT OBJECT
MCG+08-15-015	07 49 36.	+ 50 19	36	16.	GALAXY
MCG+10-12-008	07 49 36.	+ 61 17	51	18.	GALAXY
MCG+11-10-040	07 49 36.	+ 63 00	24	17.	GALAXY
MCG+07-16-021	07 49 39.	+ 39 52	54	16.	GALAXY
ZWG 206.035	07 49 42.	+ 39 53		15.7	GALAXY
UGC 04069	07 49 42.	+ 39 53	60	15.7	GALAXY S
ZWG 236.010	07 49 42.	+ 50 21		14.9	GALAXY
UGC 04070	07 49 42.	+ 50 21	60	14.9	GALAXY Sa/SBa
MCG+08-15-016	07 49 42.	+ 50 21	60	15.	GALAXY
MCG+09-13-078	07 49 42.	+ 53 02	30	16.	GALAXY
LB 03481	07 49 42.	- 56 09		15.2	FAINT BLUE STAR
MCG+09-13-079	07 49 45.	+ 51 54	18	16.	GALAXY
MCG+09-13-080	07 49 45.	+ 53 00	54	15.	GALAXY
ZWG 087.045	07 49 48.	+ 18 30		15.6	GALAXY
MCG+07-16-022	07 49 48.	+ 39 10	48	14.5	GALAXY
ZWG 262.042	07 49 48.	+ 51 56		15.0	GALAXY
UGC 04071	07 49 48.	+ 51 56	60	15.0	GALAXY
ZWG 262.043	07 49 48.	+ 53 01		14.6	GALAXY
UGC 04072	07 49 48.	+ 53 01	84	14.6	GALAXY SBb
MRK 381	07 49 48.	+ 58 23	6	17.5	GALAXY WITH UV CONTINUUM
ZC 0749.8+7810	07 49 48.	+ 78 10	2960		CLUSTER OF GALAXIES
OCL 0694	07 49 48.	- 31 43	96	15.	OPEN STAR CLUSTER
VHA 012	07 49 48.	- 31 43	210		OPEN STAR CLUSTER
ZWG 058.054	07 49 54.	+ 14 58		15.4	GALAXY
ZWG 206.036	07 49 54.	+ 39 11		14.7	GALAXY
ZC 0749.9+5121	07 49 54.	+ 51 21	2220		CLUSTER OF GALAXIES
MCG+09-13-081	07 49 54.	+ 52 45	36	17.	GALAXY
ZWG 310.023	07 49 54.	+ 62 59		15.0	GALAXY
KEEN 2336A	07 50	+ 78 19			GALAXY
LBN 1066	07 50	- 26 20	600		BRIGHT NEBULA
LBN 1065	07 50	- 26 20	420		BRIGHT NEBULA
ZC 0750.0+0604	07 50 00.	+ 06 04	470		CLUSTER OF GALAXIES
MCG+08-15-017	07 50 00.	+ 50 21	42	16.	GALAXY
ZC 0750.0+5327	07 50 00.	+ 53 27	4100		CLUSTER OF GALAXIES
OCL 0645	07 50 00.	- 22 20	90	14.	OPEN STAR CLUSTER
ISS 1037	07 50 00.	- 42 42	461		STELLAR RING
ZWG 030.023	07 50 06.	+ 06 37		14.8	GALAXY
MCG+01-20-008	07 50 06.	+ 06 37	12	14.8	GALAXY
ZWG 058.055	07 50 06.	+ 09 06		15.6	GALAXY
ZWG 178.004	07 50 06.	+ 32 40		15.7	GALAXY
KARA.73B 0211	07 50 06.	+ 32 40	48	15.7	ISOLATED GALAXY S
ZC 0750.1+3525	07 50 06.	+ 35 25	340		CLUSTER OF GALAXIES

OBJECT NAME	RIGHT ASCEN.	DECLINATION	DIAM.	MAGN.	TYPE OF OBJECT
ZWG 178.005	07 50 06.	+ 36 24		15.4	GALAXY
ZWG 206.037	07 50 06.	+ 38 52		15.2	GALAXY
ZWG 262.044	07 50 06.	+ 55 37		14.3	GALAXY
RNGC 2456	07 50 06.	+ 55 37		14.5	GALAXY
UGC 04073	07 50 06.	+ 55 37	78	14.3	GALAXY E
MAI 024	07 50 06.	+ 61 35	47		DWARF SPHEROIDAL GALAXY
MCG+11-10-041	07 50 06.	+ 62 59	39	17.	GALAXY
ARC 0601	07 50 10.	+ 34 28		17.6	RICH CLUSTER OF GALAXIES
ZWG 148.054	07 50 12.	+ 27 35		15.4	GALAXY
7ZW 189	07 50 12.	+ 59 30			COMPACT GALAXY
MCG+11-10-042	07 50 12.	+ 62 59	39	17.	GALAXY
ARC 0602	07 50 14.	+ 29 30		15.8	RICH CLUSTER OF GALAXIES
MCG+09-13-082	07 50 15.	+ 55 37 30.	21	15.	GALAXY
WRAY 19.02	07 50 15.8	- 26 18 12.		8.0	STAR-NEBULA ASSOCIATION
ZWG 030.024	07 50 18.	+ 05 37		15.1	GALAXY
ZWG 262.045	07 50 18.	+ 54 23		13.8	GALAXY
UGC 04074	07 50 18.	+ 54 23	66	13.8	GALAXY Sc
ZWG 287.004	07 50 18.	+ 58 25		15.3	GALAXY
ZWG 286.076	07 50 18.	+ 58 25		15.3	GALAXY
UGC 04075	07 50 18.	+ 60 19	72	16.0	GALAXY Sc
MCG+10-12-009	07 50 18.	+ 60 19	66	16.	GALAXY
MCG+10-12-010	07 50 18.	+ 61 19	30	18.	GALAXY
MCG+11-10-043	07 50 18.	+ 63 18	54	16.	GALAXY
7ZW 190	07 50 18.	+ 71 16			COMPACT GALAXY
RNGC 2467	07 50 22.	- 26 16		7.0	CLUSTER WITH NEBULOSITY
CED 103	07 50 22.	- 26 16	270		DIFFUSE GALACTIC NEBULA
ZWG 030.025	07 50 24.	+ 07 24		15.6	GALAXY
ZWG 058.056	07 50 24.	+ 11 45		15.6	GALAXY
ZWG 058.057	07 50 24.	+ 13 02		15.5	GALAXY
MCG+09-13-083	07 50 24.	+ 52 35	48	17.	GALAXY
MCG+09-13-084	07 50 24.	+ 54 23	54	14.	GALAXY
MCG+10-12-011	07 50 24.	+ 58 22	42	16.	GALAXY
ZWG 310.024	07 50 24.	+ 63 08		15.6	GALAXY
UGC 04076	07 50 24.	+ 63 08	60	15.6	GALAXY SB?b-c
ZC 0750.4+6413	07 50 24.	+ 64 13	2290		CLUSTER OF GALAXIES
MCG+12-08-020	07 50 24.	+ 71 16	39	16.	GALAXY
OCL 0667	07 50 24.	- 26 14	60	12.	OPEN STAR CLUSTER
OCL 0666	07 50 24.	- 26 14	60	12.	OPEN STAR CLUSTER
OCL 0665	07 50 24.	- 26 14	66	11.	OPEN STAR CLUSTFR
ZWG 058.058	07 50 30.	+ 09 05		15.4	GALAXY
ZWG 058.059	07 50 30.	+ 09 32		15.3	GALAXY
ZWG 058.060	07 50 30.	+ 12 54		15.7	GALAXY
ZC 0750.5+5143	07 50 30.	+ 51 43	1210		CLUSTER OF GALAXIES
OCL 0668	07 50 30.	- 26 15	960	7.2	OPEN STAR CLUSTER
OCL 0720	07 50 30.	- 38 25	1980	5.92	OPEN STAR CLUSTFR
RNGC 2477	07 50 31.	- 38 25		5.5	OPEN CLUSTER
MIN.47 08	07 50 34.	- 26 17			DIFFUSF NEBULA
ZWG 058.061	07 50 36.	+ 09 03		15.5	GALAXY
ZWG 058.062	07 50 36.	+ 14 45		15.1	GALAXY
UGC 04077	07 50 36.	+ 14 45	72	15.1	GALAXY SBc
MCG+02-20-010	07 50 36.	+ 14 45	48	15.1	GALAXY
MCG+04-19-007	07 50 36.	+ 21 10	45	14.9	GALAXY
ZWG 148.055	07 50 36.	+ 26 37		15.2	GALAXY
OCL 0662	07 50 36.	- 26 07	108	14.	OPEN STAR CLUSTER
IC 0478	07 50 39.	+ 26 37 10.			NONSTELLAR OBJECT
ZWG 030.026	07 50 42.	+ 06 38		15.3	GALAXY
ZWG 118.018	07 50 42.	+ 24 22		15.7	GALAXY
ZWG 262.046	07 50 42.	+ 55 40		15.6	GALAXY
RNGC 2457	07 50 42.	+ 55 40		15.5	GALAXY
MCG+09-13-085	07 50 42.	+ 56 41	30	18.	GALAXY
ZWG 287.005	07 50 42.	+ 57 55		15.7	GALAXY
ZWG 030.027	07 50 48.	+ 06 40		15.4	GALAXY
ZWG 118.019	07 50 48.	+ 21 10		14.9	GALAXY
KARA.73B 0212	07 50 48.	+ 21 10	48	14.9	ISOLATFD GALAXY S
ZWG 148.056	07 50 48.	+ 28 10		15.6	GALAXY
TON-N 0291	07 50 48.	+ 29 59		14.7	BLUE STAR
MCG+09-13-086	07 50 48.	+ 55 41	30	16.	GALAXY
ZWG 262.047	07 50 48.	+ 55 43		15.7	GALAXY
MCG+10-12-012	07 50 48.	+ 57 55	36	15.	GALAXY
ZC 0750.8+7246	07 50 48.	+ 72 46	2350		CLUSTER OF GALAXIES
MCG+12-08-021	07 50 48.	+ 73 57	60	13.	GALAXY
PATH 1.209	07 50 48.	- 00 20	14		NEBULA
ZWG 002.019	07 50 48.	- 00 21		15.3	GALAXY
ZWG 058.063	07 50 54.	+ 09 32		15.6	GALAXY
ZWG 058.064	07 50 54.	+ 13 28		15.4	GALAXY
TON-N 0841	07 50 54.	+ 22 40		16.7	BLUE STAR
ZC 0750.9+2634	07 50 54.	+ 26 34	2020		CLUSTER OF GALAXIES
ZWG 262.048	07 50 54.	+ 54 15		15.1	GALAXY
ACK 259-09.1	07 50 54.	- 45 43			PLANETARY NEBULA
MCG+07-16-023	07 50 57.	+ 39 30	36	15.	GALAXY
LBN 1067	07 51	- 26 20	2400		BRIGHT NEBULA
TON-N 0292	07 51 00.	+ 25 41		15.2	BLUE STAR
ZWG 148.057	07 51 00.	+ 27 05		15.3	GALAXY
ZWG 207.001	07 51 00.	+ 39 30		14.8	GALAXY
ZWG 206.038	07 51 00.	+ 39 30		14.8	GALAXY
KARA.73B 0213	07 51 00.	+ 39 30	36	14.8	ISOLATED GALAXY S
ZWG 207.002	07 51 00.	+ 44 17		15.4	GALAXY
ZWG 206.039	07 51 00.	+ 44 17		15.4	GALAXY
MCG+08-15-018	07 51 00.	+ 50 10	24	14.	GALAXY
MCG+09-13-088	07 51 00.	+ 54 15	30	15.	GALAXY
MCG+09-13-087	07 51 00.	+ 55 40	30	16.	GALAXY
MCG+09-13-089	07 51 00.	+ 55 44	48	16.	GALAXY
ZWG 262.049	07 51 00.	+ 55 50		13.6	GALAXY
UGC 04079	07 51 00.	+ 55 50	60	13.6	GALAXY S
ZWG 331.022	07 51 00.	+ 73 55		13.9	GALAXY
UGC 04080	07 51 00.	+ 73 55	66	13.9	GALAXY Sa
ZWG 363.037	07 51 00.	+ 84 46		15.4	GALAXY
UGC 04078	07 51 00.	+ 84 46	138	15.4	GALAXY Sb-c
ARC 0596	07 51 01.	+ 72 46		17.2	RICH CLUSTER OF GALAXIES
UGC 04081	07 51 06.	+ 38 47	66	17.	GALAXY DWRF IR
ZWG 236.011	07 51 06.	+ 50 10		14.5	GALAXY
UGC 04082	07 51 06.	+ 50 10	66	14.5	GALAXY E
MRSL 243+00/1	07 51 06.	- 26 16	1920		HII REGION
ISS 0840	07 51 06.	- 27 18	169		STELLAR RING
ACK 252-04.1	07 51 06.	- 36 36			PLANETARY NEBULA
ZWG 058.065	07 51 12.	+ 14 05		15.3	GALAXY
ZWG 236.012	07 51 12.	+ 46 21		15.7	GALAXY
MCG+08-15-019	07 51 12.	+ 49 48	18	16.	GALAXY
MCG+09-13-090	07 51 12.	+ 55 50	60	15.	GALAXY
ACK 259-09.2	07 51 12.	- 45 04			PLANETARY NEBULA
SEY 035	07 51 16.	+ 73 56 42.		14.6	FAINT GALAXY
ZWG 118.020	07 51 18.	+ 25 59		15.6	GALAXY
ZWG 148.058	07 51 18.	+ 27 08		15.0	GALAXY
ZWG 148.059	07 51 18.	+ 28 15		15.7	GALAXY
MCG+05-19-020	07 51 18.	+ 30 21	60	15.6	GALAXY
ZWG 207.003	07 51 18.	+ 43 18		15.3	GALAXY
ZWG 206.040	07 51 18.	+ 43 18		15.3	GALAXY
ZWG 236.013	07 51 18.	+ 45 58		14.8	GALAXY

OBJECT NAME	RIGHT ASCEN.	DECLINATION	DIAM.	MAGN.	TYPE OF OBJECT
UGC 04083	07 51 18.	+ 56 21	66	17.	GALAXY Sc
IC 0479	07 51 19.	+ 27 09 17.			NONSTELLAR OBJECT
RNGC 2458	07 51 22.	+ 56 51			GALAXY
ZWG 058.066	07 51 24.	+ 14 24		15.3	GALAXY
ZWG 148.060	07 51 24.	+ 28 38		15.7	GALAXY
ZWG 148.061	07 51 24.	+ 29 50		15.6	GALAXY
UGC 04084	07 51 24.	+ 29 50	72	15.6	GALAXY IRR
ZWG 262.050	07 51 24.	+ 53 28		14.4	GALAXY
UGC 04085	07 51 24.	+ 53 28	60	14.4	GALAXY S
MRK 084	07 51 24.	+ 55 50	30	14.5	GALAXY WITH UV CONTINUUM
MCG+10-12-014	07 51 24.	+ 57 28	45	15.	GALAXY
ZWG 287.006	07 51 24.	+ 57 29		15.4	GALAXY
MCG+10-12-013	07 51 24.	+ 62 42	27	16.	GALAXY
7ZW 191	07 51 24.	+ 66 00			COMPACT GALAXY
ISS 0841	07 51 24.	- 27 08	168		STELLAR RING
UGC 04086	07 51 30.	+ 16 20	60	18.	GALAXY DWARF
ZWG 118.021	07 51 30.	+ 25 56		15.7	GALAXY
ZWG 207.004	07 51 30.	+ 43 37		15.5	GALAXY
ZWG 206.041	07 51 30.	+ 43 37		15.5	GALAXY
UGC 04087	07 51 30.	+ 43 37	66	15.5	GALAXY Sc
MCG+09-13-091	07 51 30.	+ 53 27 30.	66	15.	GALAXY
UGC 04088	07 51 30.	+ 56 20	66	17.	GALAXY Sc
ZWG 310.025	07 51 30.	+ 62 43		15.6	GALAXY
ZWG 286.077	07 51 30.	+ 62 43		15.6	GALAXY
MCG+14-04-041	07 51 30.	+ 81 24	24	17.	GALAXY
ZWG 058.067	07 51 36.	+ 13 52		15.3	GALAXY
UGC 04089	07 51 36.	+ 13 52	72	15.3	GALAXY Sa
ZWG 087.046	07 51 36.	+ 16 56		14.8	GALAXY
UGC 04090	07 51 36.	+ 16 56	60	14.8	GALAXY S?
KARA.72 146B	07 51 36.	+ 16 56	24		PART OF DOUBLE GALAXY
KARA.72 146A	07 51 36.	+ 16 56	18	14.8	PART OF DOUBLE GALAXY
MCG+03-20-016	07 51 36.	+ 16 56	36	14.8	GALAXY
MCG+07-16-024	07 51 36.	+ 43 37	60	15.5	GALAXY
RNGC 2470	07 51 41.	+ 04 35		14.0	GALAXY
ZWG 030.028	07 51 42.	+ 03 30		15.5	GALAXY
ZWG 030.029	07 51 42.	+ 04 35		14.2	GALAXY
UGC 04091	07 51 42.	+ 04 35	132	14.2	GALAXY Sa-b
MCG+01-20-009	07 51 42.	+ 04 35	96	14.2	GALAXY
SHAH 085	07 51 42.	+ 54 55	90	18.0	GROUP OF COMPACT GALAXIES
7ZW 192	07 51 42.	+ 70 59			COMPACT GALAXY
ISS 0911	07 51 42.	- 28 45	362		STELLAR RING
MCG+09-13-092	07 51 45.	+ 56 42	30	16.	GALAXY
ZWG 058.068	07 51 48.	+ 14 32		15.5	GALAXY
TON-N 0842	07 51 48.	+ 22 09		15.3	BLUF STAR
MCG+10-12-015	07 51 48.	+ 58 33	60	14.	GALAXY
OCL 0673	07 51 48.	- 26 51	66	14.	OPEN STAR CLUSTER
ARC 0600	07 51 50.	+ 63 54		16.5	RICH CLUSTER OF GALAXIFS
BC B20751+29	07 51 51.1	+ 29 49 48.0		18.	QUASI-STELLAR OBJECT
MCG+09-13-093	07 51 54.	+ 56 42	15	16.	GALAXY
ZC 0751.9+5819	07 51 54.	+ 58 19	1340		CLUSTER OF GALAXIFS
ZWG 287.007	07 51 54.	+ 58 34		14.3	GALAXY
UGC 04092	07 51 54.	+ 58 34	72	14.3	GALAXY Sa-b
ISS 0842	07 51 54.	- 23 32	256		STELLAR RING
RNGC 2461	07 51 58.	+ 56 49		15.5	GALAXY
IC 2209	07 51 58.	+ 60 26			NONSTELLAR OBJFCT
ZWG 207.005	07 52 00.	+ 39 19		15.5	GALAXY
MRK 382	07 52 00.	+ 39 19	12	15.5	GALAXY WITH UV CONTINUUM
KW 38	07 52 00.	+ 39 19	42		SEYFERT GALAXY
VVI 25	07 52 00.	+ 39 19	42	16.5	SEYFERT GALAXY
KARA.73B 0214	07 52 00.	+ 39 19	54	15.5	ISOLATED GALAXY S
MCG+10-12-016	07 52 00.	+ 56 49	10	16.	GALAXY
MRK 013	07 52 00.	+ 60 25	20	15.5	GALAXY WITH UV CONTINUUM
MCG+10-12-017	07 52 00.	+ 60 25	42	14.	GALAXY
ZWG 287.008	07 52 00.	+ 60 26		14.5	GALAXY
ZWG 286.078	07 52 00.	+ 60 26		14.5	GALAXY
UGC 04093	07 52 00.	+ 60 26	66	14.5	GALAXY S
MCG+11-10-044	07 52 00.	+ 66 45	45	15.	GALAXY
MCG+14-04-042	07 52 00.	+ 84 45	114	16.	GALAXY
ZC 0752.1+1204	07 52 06.	+ 12 04	1010		CLUSTER OF GALAXIES
ZWG 058.069	07 52 06.	+ 14 35		15.1	GALAXY
ZC 0752.1+2445	07 52 06.	+ 24 45	810		CLUSTER OF GALAXIES
MCG+07-17-001	07 52 06.	+ 39 18	42	15.5	GALAXY
MCG+08-15-020	07 52 06.	+ 47 01	48	16.	GALAXY
ZWG 310.026	07 52 06.	+ 63 44		14.9	GALAXY
UGC 04094	07 52 06.	+ 63 44	84	14.9	GALAXY Sa/SBb
MCG+11-10-045	07 52 06.	+ 63 44	39	16.	GALAXY
KARA.73B 0215	07 52 06.	+ 63 44	66	14.9	ISOLATED GALAXY S
ZWG 310.027	07 52 06.	+ 66 44		14.4	GALAXY
UGC 04095	07 52 06.	+ 66 44	54	14.4	GALAXY PECULR
LB 00507	07 52 06.	+ 78 51 42.		17.6	FAINT BLUE STAR
BIGO 497	07 52 10.	+ 56 57			NEBULA
MCG+10-12-018	07 52 12.	+ 61 09	45	17.	GALAXY
ZC 0752.2+6900	07 52 12.	+ 69 00	1010		CLUSTER OF GALAXIFS
IC 0480	07 52 17.	+ 26 54 40.			NONSTELLAR OBJFCT
ZWG 148.062	07 52 18.	+ 26 52		15.2	GALAXY
UGC 04096	07 52 18.	+ 26 52	108	15.2	GALAXY Sb-c
ZWG 148.063	07 52 18.	+ 27 06		15.6	GALAXY
ZC 0752.3+6608	07 52 18.	+ 66 08	4500		CLUSTER OF GALAXIES
ZC 0752.4+0123	07 52 24.	+ 01 23	3160		CLUSTER OF GALAXIES
ZWG 058.070	07 52 24.	+ 14 30		15.4	GALAXY
ZWG 058.071	07 52 24.	+ 14 34		15.4	GALAXY
ZWG 148.064	07 52 24.	+ 27 52		15.7	GALAXY
MCG+10-12-019	07 52 24.	+ 62 04	27	16.	GALAXY
7ZW 193	07 52 24.	+ 63 49			COMPACT GALAXY
MCG+11-10-046	07 52 24.	+ 66 34	57	15.	GALAXY
ARC 0599	07 52 25.	+ 69 00		17.2	RICH CLUSTER OF GALAXIFS
RNGC 2478	07 52 27.	- 15 17			NON-EXISTENT OBJECT
ZWG 087.047	07 52 30.	+ 15 13		15.3	GALAXY
ZWG 118.022	07 52 30.	+ 23 52		15.4	GALAXY
MCG+07-17-002	07 52 30.	+ 41 30	36	17.	GALAXY
ZWG 287.009	07 52 30.	+ 56 49		15.5	GALAXY
MCG+10-12-021	07 52 30.	+ 60 29	90	12.9	GALAXY
MCG+10-12-020	07 52 30.	+ 62 32	15	17.	GALAXY
RNGC 2460	07 52 33.	+ 60 29		12.5	GALAXY
ZWG 088.001	07 52 36.	+ 17 12		15.6	GALAXY
ZWG 087.048	07 52 36.	+ 17 12		15.6	GALAXY
TON-N 0843	07 52 36.	+ 22 09		14.2	BLUE STAR
TON-N 0844	07 52 36.	+ 22 55		15.9	BLUE STAR
MCG+04-19-008	07 52 36.	+ 24 49	78	14.7	GALAXY
ZWG 236.014	07 52 36.	+ 45 37		15.7	GALAXY
MCG+10-12-022	07 52 36.	+ 60 24	39	16.	GALAXY
ZWG 287.010	07 52 36.	+ 60 30		12.5	GALAXY
UGC 04097	07 52 36.	+ 60 30	240	12.5	GALAXY Sb
MCG+10-12-023	07 52 36.	+ 61 47	18	16.	GALAXY
ARC 0603	07 52 38.	+ 33 53		16.8	RICH CLUSTER OF GALAXIES
RNGC 2462	07 52 40.	+ 56 47		14.0	GALAXY
HOLM 088A	07 52 40.	+ 56 48	24	14.1	PART OF MULTIPLF GALAXY
ZWG 118.023	07 52 42.	+ 21 34		15.6	GALAXY

OBJECT NAME	RIGHT ASCEN.	DECLINATION	DIAM.	MAGN.	TYPE OF OBJECT
KARA.73B 0216	07 52 42.	+ 21 34	36	15.6	ISOLATED GALAXY S
ZC 0752.7+3236	07 52 42.	+ 32 36	2550		CLUSTER OF GALAXIES
ZWG 207.006	07 52 42.	+ 44 10		15.7	GALAXY
MCG+08-15-021	07 52 42.	+ 46 01	24	16.	GALAXY
ZC 0752.7+5155	07 52 42.	+ 51 55	810		CLUSTER OF GALAXIES
MCG+10-12-024	07 52 42.	+ 56 47	24	14.1	GALAXY
ZWG 287.011	07 52 42.	+ 59 01		15.5	GALAXY
MCG+10-12-025	07 52 42.	+ 61 01	18	17.	GALAXY
ZWG 310.028	07 52 42.	+ 66 34		14.5	GALAXY
UGC 04098	07 52 42.	+ 66 34	54	14.5	GALAXY S
ASS 57	07 52 42.	- 26 57	14400		OB ASSOCIATION PUP OB1
HOLM 088B	07 52 45.	+ 56 49	24	14.6	PART OF MULTIPLE GALAXY
RNGC 2479	07 52 47.	- 17 35			OPEN CLUSTER
ZWG 118.024	07 52 48.	+ 24 50		14.7	GALAXY
UGC 04099	07 52 48.	+ 24 50	114	14.7	GALAXY SBb
ZWG 148.065	07 52 48.	+ 27 01		15.4	GALAXY
MCG+10-12-026	07 52 48.	+ 59 00	27	15.	GALAXY
MCG+10-12-027	07 52 48.	+ 62 34	10	17.	GALAXY
ZWG 310.029	07 52 48.	+ 66 25		15.7	GALAXY
OCL 0623	07 52 48.	- 17 35	600	9.8	OPEN STAR CLUSTER
OCL 0653	07 52 48.	- 24 10	1080	8.8	OPEN STAR CLUSTER
OCL 0719	07 52 48.	- 38 08	102	16.	OPEN STAR CLUSTFR
VHA 014	07 52 48.	- 38 08	120		OPEN STAR CLUSTER
RNGC 2482	07 52 50.	- 24 10		8.5	OPEN CLUSTER
LB 00508	07 52 53.	+ 79 22 18.		16.3	FAINT BLUE STAR
TON-N 0835	07 52 54.	+ 22 36		13.9	BLUE STAR
TON-N 0845	07 52 54.	+ 22 54		13.9	BLUE STAR
ZC 0752.9+2833	07 52 54.	+ 28 33	15320		CLUSTER OF GALAXIES
ZC 0752.9+3842	07 52 54.	+ 38 42	610		CLUSTER OF GALAXIES
ZWG 287.012	07 52 54.	+ 61 47		15.2	GALAXY
ZWG 286.079	07 52 54.	+ 61 47		15.2	GALAXY
MCG+10-12-028	07 52 54.	+ 62 32	54	15.	GALAXY
IC 2210	07 52 55.	+ 56 49			NONSTELLAR OBJECT
LB 09839	07 53	- 81 38		15.6	FAINT BLUE STAR
ZWG 148.066	07 53 00.	+ 27 00		15.6	GALAXY
ZWG 148.067	07 53 00.	+ 27 34		15.7	GALAXY
ZWG 148.068	07 53 00.	+ 30 20		15.6	GALAXY
MCG+05-19-021	07 53 00.	+ 30 21	60	15.6	GALAXY
ZC 0753.0+3606	07 53 00.	+ 36 06	4440		CLUSTER OF GALAXIES
MCG+08-15-022	07 53 00.	+ 45 40	66	16.	GALAXY
UGC 04101	07 53 00.	+ 45 41	78	16.0	GALAXY S
ZWG 310.030	07 53 00.	+ 62 32		15.1	GALAXY
ZWG 286.080	07 53 00.	+ 62 32		15.1	GALAXY
UGC 04102	07 53 00.	+ 74 02	60	16.5	GALAXY Sc
UGC 04103	07 53 00.	+ 79 30	90	16.0	GALAXY Sc
MCG+13-06-013	07 53 00.	+ 79 30	39	16.	GALAXY
UGC 04100	07 53 00.	+ 84 55	108	16.0	GALAXY IRR
MCG+14-04-043	07 53 00.	+ 84 55	78	16.	GALAXY
PK241+02.1	07 53 00.	- 23 26	14		PLANETARY NEBULA
WRAY 19.03	07 53 03.7	- 23 30 23.			DIFFUSE NEBULA
RNGC 2463	07 53 04.	+ 56 48		15.0	GALAXY
ZWG 059.001	07 53 06.	+ 14 20		15.2	GALAXY
ZWG 058.072	07 53 06.	+ 14 20		15.2	GALAXY
ZC 0753.1+2635	07 53 06.	+ 26 35	870		CLUSTER OF GALAXIES
ZWG 148.069	07 53 06.	+ 27 01		15.7	GALAXY
ZC 0753.1+3331	07 53 06.	+ 33 31	3160		CLUSTER OF GALAXIES
ZWG 207.007	07 53 06.	+ 44 14		15.7	GALAXY
ZWG 236.015	07 53 06.	+ 45 00		15.5	GALAXY
ZWG 236.016	07 53 06.	+ 46 01		15.7	GALAXY
ZWG 287.013	07 53 06.	+ 56 48		14.8	GALAXY
ZWG 287.014	07 53 06.	+ 58 40		15.5	GALAXY
UGC 04104	07 53 06.	+ 58 40	84	15.5	GALAXY Sa
ZC 0753.1-0221	07 53 06.	- 02 21	5240		CLUSTER OF GALAXIES
ZWG 059.002	07 53 12.	+ 13 07		15.3	GALAXY
ZWG 058.073	07 53 12.	+ 13 07		15.3	GALAXY
ZWG 059.003	07 53 12.	+ 14 23		15.6	GALAXY
ZWG 058.074	07 53 12.	+ 14 23		15.6	GALAXY
ZWG 088.002	07 53 12.	+ 19 19		15.7	GALAXY
ZWG 087.049	07 53 12.	+ 19 19		15.7	GALAXY
ZWG 148.070	07 53 12.	+ 29 36		15.2	GALAXY
ZC 0753.[illegible]+4101	07 53 12.	+ 41 01	3020		CLUSTER OF GALAXIES
MCG+08-15-023	07 53 12.	+ 44 59	30	16.	GALAXY
MCG+08-15-024	07 53 12.	+ 46 00	42	15.	GALAXY
ZWG 236.017	07 53 12.	+ 46 02		15.6	GALAXY
MCG+10-12-031	07 53 12.	+ 56 49	24	15.	GALAXY
MCG+10-12-029	07 53 12.	+ 57 13	24	16.	GALAXY
MCG+10-12-030	07 53 12.	+ 58 38	78	16.	GALAXY
OCL 0663	07 53 12.	- 25 48	480	9.1	OPEN STAR CLUSTER
TON-N 0846	07 53 13.	+ 19 59		14.9	BLUE STAR
ARC 0605	07 53 15.	+ 27 32		16.7	RICH CLUSTER OF GALAXIES
MCG+08-15-025	07 53 15.	+ 46 00	9	17.	GALAXY
MCG+08-15-026	07 53 15.	+ 46 01	24	15.	GALAXY
RNGC 2464	07 53 16.	+ 56 48			NON-EXISTENT OBJECT
MCG+05-19-022	07 53 18.	+ 27 08	72	15.0	GALAXY
ZWG 148.071	07 53 18.	+ 27 09		15.0	GALAXY
UGC 04105	07 53 18.	+ 27 09	66	15.0	GALAXY S
ZWG 207.008	07 53 18.	+ 40 04		13.4	GALAXY
ZWG 236.018	07 53 18.	+ 49 42		13.9	GALAXY
UGC 04107	07 53 18.	+ 49 42	102	13.9	GALAXY Sc
MCG+08-15-027	07 53 18.	+ 49 42	78	13.	GALAXY
KARA.73B 0217	07 53 18.	+ 49 42	78	13.9	ISOLATED GALAXY S
ZWG 287.015	07 53 18.	+ 59 14		15.0	GALAXY
UGC 04108	07 53 18.	+ 59 14	72	15.0	GALAXY S
RNGC 2476	07 53 19.	+ 40 04		13.5	GALAXY
ZWG 088.003	07 53 24.	+ 17 09		15.5	GALAXY
ZWG 087.050	07 53 24.	+ 17 09		15.5	GALAXY
MCG+10-12-032	07 53 24.	+ 59 12	60	14.	GALAXY
ZWG 031.001	07 53 30.	+ 07 35		15.5	GALAXY
ZWG 059.004	07 53 30.	+ 11 48		15.2	GALAXY
UGC 04109	07 53 30.	+ 11 48	72	15.2	GALAXY SBb
MCG+07-17-003	07 53 30.	+ 40 03	36	14.	GALAXY
TON-N 0847	07 53 36.	+ 20 54		14.0	BLUE STAR
UGC 04106	07 53 36.	+ 40 04	84	13.4	GALAXY E?
ZC 0753.6+4844	07 53 36.	+ 48 44	940		CLUSTER OF GALAXIES
RNGC 2465	07 53 40.	+ 56 53			NON-EXISTENT OBJECT
ZWG 148.072	07 53 42.	+ 28 31		15.7	GALAXY
MCG+09-13-094	07 53 42.	+ 52 48	15	16.	GALAXY
ZWG 262.051	07 53 42.	+ 52 50		15.4	GALAXY
ZWG 031.002	07 53 48.	+ 02 41		15.7	GALAXY
ZWG 178.006	07 53 48.	+ 33 13		14.7	GALAXY
ZC 0753.8+3850	07 53 48.	+ 38 50	670		CLUSTER OF GALAXIES
MCG+10-12-033	07 53 48.	+ 61 04	27	16.	GALAXY
MCG+11-10-047	07 53 48.	+ 63 28	78	16.	GALAXY
ARC 0606	07 53 50.	+ 36 05		17.7	RICH CLUSTER OF GALAXIES
ZWG 088.004	07 53 54.	+ 17 03		15.0	GALAXY
MCG+03-21-001	07 53 54.	+ 17 03	24	15.0	GALAXY
ZWG 088.005	07 53 54.	+ 18 08		14.7	GALAXY
MCG+03-21-002	07 53 54.	+ 18 08	24	14.7	GALAXY
TON-N 0848	07 53 54.	+ 21 21		16.1	BLUE STAR
PK164+31.1	07 53 54.	+ 53 33 04.	450	14.	PLANETARY NEBULA
MCG+06-18-002	07 53 57.	+ 33 12	33	15.	GALAXY
RNGC 2469	07 53 58.	+ 56 49		13.0	GALAXY
RNGC 2468	07 53 59.	+ 56 30		15.0	GALAXY
RNGC 2483	07 53 59.	- 27 48			OPEN CLUSTER
LBN 0790	07 54	+ 47 20	1200		BRIGHT NEBULA
VDB.66G 048	07 54	+ 58 10	100		DWARF GALAXY
MCG+04-19-009	07 54 00.	+ 23 54 30.	72	15.2	GALAXY
ZC 0754.0+3450	07 54 00.	+ 34 50	810		CLUSTER OF GALAXIES
ZC 0754.0+4619	07 54 00.	+ 46 19	6720		CLUSTER OF GALAXIES
MCG+08-15-028	07 54 00.	+ 46 36	60	16.	GALAXY
ZWG 287.016	07 54 00.	+ 56 30		14.9	GALAXY
UGC 04110	07 54 00.	+ 56 30	60	14.9	GALAXY DBL SYS
MCG+09-13-095	07 54 00.	+ 56 30	60	14.	GALAXY
ZWG 287.017	07 54 00.	+ 56 49		13.2	GALAXY
UGC 04111	07 54 00.	+ 56 49	60	13.2	GALAXY S
7ZW 194	07 54 00.	+ 58 39			COMPACT GALAXY
ZC 0754.0+5844	07 54 00.	+ 58 44	940		CLUSTER OF GALAXIES
MCG+10-12-034	07 54 00.	+ 59 19	36	16.	GALAXY
ZWG 287.018	07 54 00.	+ 59 21		15.6	GALAXY
MCG+11-10-048	07 54 00.	+ 63 28	30	17.	GALAXY
ZWG 331.023	07 54 00.	+ 71 53		15.7	GALAXY
KARA.73B 0218	07 54 00.	+ 71 53	42	15.7	ISOLATED GALAXY S
MCG+12-08-022	07 54 00.	+ 71 54	33	15.	GALAXY
LDN 1669	07 54 00.	- 29 30	3120		DARK NEBULA
TON-N 0849	07 54 01.	+ 19 54		14.0	BLUE STAR
RNGC 2485	07 54 03.	+ 07 37		13.5	GALAXY
RNGC 2474	07 54 04.	+ 53 00		14.0	GALAXY
ZWG 031.003	07 54 06.	+ 07 37		13.3	GALAXY
UGC 04112	07 54 06.	+ 07 37	108	13.3	GALAXY Sa
MCG+01-21-001	07 54 06.	+ 07 37	96	13.3	GALAXY
ZWG 088.006	07 54 06.	+ 15 58		15.2	GALAXY
ZWG 118.025	07 54 06.	+ 22 39		15.5	GALAXY
KARA.73B 0219	07 54 06.	+ 22 39	18	15.5	ISOLATED GALAXY
MCG+04-19-010	07 54 06.	+ 23 54	72	13.4	GALAXY
ZWG 148.073	07 54 06.	+ 27 37		15.5	GALAXY
UGC 04113	07 54 06.	+ 31 37	60	16.5	GALAXY Sc
ZWG 178.007	07 54 06.	+ 33 12		15.2	GALAXY
ZC 0754.1+3506	07 54 06.	+ 35 06	610		CLUSTER OF GALAXIES
SHAH 086	07 54 06.	+ 50 48	30	18.2	GROUP OF COMPACT GALAXIES
MCG+09-13-096	07 54 06.	+ 52 59	18	16.	GALAXY
ZWG 262.052	07 54 06.	+ 53 00		13.9	GALAXY
UGC 04114	07 54 06.	+ 53 00	54	13.9	GALAXY DBL SYS
KARA.72 147B	07 54 06.	+ 53 00	48		PART OF DOUBLE GALAXY
KARA.72 147A	07 54 06.	+ 53 00	36	13.9	PART OF DOUBLE GALAXY
MCG+10-12-035	07 54 06.	+ 56 49	57	14.	GALAXY
MCG+10-12-036	07 54 06.	+ 58 38	18	15.	GALAXY
ZC 0754.1+5928	07 54 06.	+ 59 28	940		CLUSTER OF GALAXIES
ARC 0607	07 54 07.	+ 39 28		16.9	RICH CLUSTER OF GALAXIES
RNGC 2475	07 54 10.	+ 53 00		14.0	GALAXY
ZWG 088.007	07 54 12.	+ 14 31		15.2	GALAXY
UGC 04115	07 54 12.	+ 14 32	120	16.0	GALAXY IRR
ZWG 118.026	07 54 12.	+ 23 55		15.2	GALAXY
RNGC 2480	07 54 12.	+ 23 55		15.0	GALAXY
UGC 04116	07 54 12.	+ 23 55	96	15.2	GALAXY SB
KARA.72 148A	07 54 12.	+ 23 55	72	15.2	PART OF DOUBLE GALAXY
ZWG 178.008	07 54 12.	+ 36 04		15.5	GALAXY
UGC 04117	07 54 12.	+ 36 04	90	15.5	GALAXY IRR
MCG+06-18-003	07 54 12.	+ 36 04	48	15.	GALAXY
MCG+10-12-037	07 54 12.	+ 59 11	45	17.	GALAXY
OCL 0690	07 54 12.	- 29 56	600	9.4	OPEN STAR CLUSTER
VHA 015	07 54 12.	- 29 56	300		OPEN STAR CLUSTER
RNGC 2489	07 54 12.	- 29 56		9.5	OPEN CLUSTER
SEY 036	07 54 13.	+ 71 53 23.		15.4	FAINT GALAXY
HOLM 089B	07 54 15.	+ 23 55	48	14.9	PART OF MULTIPLE GALAXY
RNGC 2471	07 54 16.	+ 56 53			NON-EXISTENT OBJECT
LB 00509	07 54 16.	+ 77 57 42.		17.4	FAINT BLUE STAP
ZWG 118.027	07 54 18.	+ 23 54		13.4	GALAXY
RNGC 2481	07 54 18.	+ 23 54		13.5	GALAXY
HOLM 089A	07 54 18.	+ 23 54	48	13.8	PART OF MULTIPLE GALAXY
UGC 04118	07 54 18.	+ 23 54	96	13.4	GALAXY
KARA.72 148B	07 54 18.	+ 23 54	96	13.4	PART OF DOUBLE GALAXY
ZWG 118.028	07 54 18.	+ 26 00		15.6	GALAXY
MCG+09-13-097	07 54 18.	+ 52 59	30	15.	GALAXY
OCL 0692	07 54 18.	- 30 16	108	14.	OPEN STAR CLUSTER
VHA 016	07 54 18.	- 30 16	180		OPEN STAR CLUSTER
ACK 251-03.1	07 54 18.	- 35 34			PLANETARY NEBULA
RNGC 2473	07 54 22.	+ 56 49			NON-EXISTENT OBJECT
ZWG 059.005	07 54 24.	+ 10 44		15.3	GALAXY
ZWG 178.009	07 54 24.	+ 38 08		15.5	GALAXY
MCG+10-12-038	07 54 24.	+ 57 03	15	17.	GALAXY
ZWG 331.024	07 54 24.	+ 73 12		15.5	GALAXY
ZWG 031.004	07 54 30.	+ 05 55		15.6	GALAXY
ZWG 059.006	07 54 30.	+ 11 11		15.3	GALAXY
ZWG 178.010	07 54 30.	+ 32 42		14.5	GALAXY
UGC 04119	07 54 30.	+ 32 42	48	14.5	GALAXY Sa
MCG+08-15-030	07 54 30.	+ 47 15	12	17.	GALAXY
MCG+08-15-029	07 54 30.	+ 47 16	54	17.	GALAXY
MCG+08-15-031	07 54 30.	+ 47 17	36	16.	GALAXY
ZWG 236.019	07 54 30.	+ 47 27		15.3	GALAXY
MCG+09-13-098	07 54 30.	+ 52 53	15	17.	GALAXY
RNGC 2502	07 54 33.	- 52 09			UNVERIFIED SOUTHERN OBJECT
RNGC 2472	07 54 34.	+ 56 50		15.5	GALAXY
ZWG 059.007	07 54 36.	+ 11 21		15.5	GALAXY
ZWG 287.019	07 54 36.	+ 56 50		15.4	GALAXY
ZC 0754.6+6155	07 54 36.	+ 61 55	3560		CLUSTER OF GALAXIES
IC 2211	07 54 37.	+ 32 41 25.			NONSTELLAR OBJECT
TON-N 0850	07 54 42.	+ 23 09		15.5	BLUE STAR
MCG+05-19-023	07 54 42.	+ 32 42	48	14.5	GALAXY
ZWG 207.009	07 54 42.	+ 42 24		15.6	GALAXY
7ZW 195	07 54 42.	+ 59 07			COMPACT GALAXY
MCG+04-19-011	07 54 48.	+ 25 16	120	14.3	GALAXY
ZWG 207.010	07 54 48.	+ 39 58		15.6	GALAXY
UGC 04120	07 54 48.	+ 39 58	60	15.6	GALAXY Sc
ZWG 207.011	07 54 48.	+ 44 11		15.7	GALAXY
MCG+08-15-032	07 54 48.	+ 50 35	24	15.	GALAXY
MCG+10-12-039	07 54 48.	+ 56 49	30	16.	GALAXY
7ZW 196	07 54 48.	+ 57 38			COMPACT GALAXY
UGC 04121	07 54 48.	+ 58 12	150	16.0	GALAXY DWRF SP
ZWG 287.020	07 54 48.	+ 59 15		14.7	GALAXY
UGC 04122	07 54 48.	+ 59 15	96	14.7	GALAXY E
KARA.72 149A	07 54 48.	+ 59 15	84	14.7	PART OF DOUBLE GALAXY
ZC 0754.8+6949	07 54 48.	+ 69 49	1140		CLUSTER OF GALAXIES
MCG+13-06-014	07 54 48.	+ 75 40	18	17.	GALAXY
RNGC 2486	07 54 53.	+ 25 17		14.5	GALAXY
ZWG 118.029	07 54 54.	+ 25 17		14.3	GALAXY
UGC 04123	07 54 54.	+ 25 17	114	14.3	GALAXY Sa

OBJECT NAME	RIGHT ASCEN.	DECLINATION	DIAM.	MAGN.	TYPE OF OBJECT
KARA.72 150A	07 54 54.	+ 25 17	102	14.3	PART OF DOUBLE GALAXY
HOLM 090B	07 54 54.	+ 25 18	90	14.6	PART OF MULTIPLE GALAXY
ZWG 236.020	07 54 54.	+ 47 47		15.4	GALAXY
MCG+10-12-040	07 54 54.	+ 59 14	36	14.	GALAXY
MCG+10-12-041	07 54 54.	+ 59 15	39	16.	GALAXY
ZWG 287.021	07 54 54.	+ 59 17		15.4	GALAXY
UGC 04124	07 54 54.	+ 59 17	66	15.4	GALAXY S0
KARA.72 149B	07 54 54.	+ 59 17	66	15.4	PART OF DOUBLE GALAXY
ARC 0604	07 54 56.	+ 61 30		17.1	RICH CLUSTER OF GALAXIES
LBN 0811	07 55	+ 45 40	2100		BRIGHT NEBULA
ZWG 088.008	07 55 00.	+ 16 10		15.3	GALAXY
ZWG 148.074	07 55 00.	+ 26 42		15.5	GALAXY
ZWG 262.053	07 55 00.	+ 50 37		15.3	GALAXY
ZWG 349.012	07 55 00.	+ 80 10		15.6	GALAXY
ZWG 003.001	07 55 00.	- 01 25		15.5	GALAXY
ZWG 178.011	07 55 06.	+ 37 55		14.9	GALAXY
UGC 04125	07 55 06.	+ 37 55	102	14.9	GALAXY S0
MCG+07-17-004	07 55 06.	+ 39 58	48	15.5	GALAXY
MCG+10-12-042	07 55 06.	+ 59 33	30	16.	GALAXY
RNGC 2484	07 55 08.	+ 37 55		15.0	GALAXY
MCG+06-18-004	07 55 09.	+ 37 55	24	15.	GALAXY
ZWG 088.009	07 55 12.	+ 14 49		15.1	GALAXY
ZC 0755.2+1450	07 55 12.	+ 14 50	810		CLUSTER OF GALAXIES
TON-N 0851	07 55 12.	+ 21 28		15.6	BLUE STAR
MCG+04-19-012	07 55 12.	+ 25 16	168	14.0	GALAXY
ZC 0755.2+2649	07 55 12.	+ 26 49	740		CLUSTER OF GALAXIES
ZWG 349.013	07 55 12.	+ 77 21		15.6	GALAXY
ISS 0843	07 55 12.	- 21 32	137		STELLAR RING
RNGC 2487	07 55 17.	+ 25 16		14.0	GALAXY
ZWG 059.008	07 55 18.	+ 10 27		15.3	GALAXY
ZWG 118.030	07 55 18.	+ 25 16		14.0	GALAXY
UGC 04126	07 55 18.	+ 25 16	150	14.0	GALAXY SBb
KARA.72 150B	07 55 18.	+ 25 16	138	14.0	PART OF DOUBLE GALAXY
ZWG 148.075	07 55 18.	+ 31 10		15.5	GALAXY
7ZW 197	07 55 18.	+ 56 44			COMPACT GALAXY
MCG+11-10-049	07 55 18.	+ 67 37 30.	39	16.	GALAXY
HOLM 090A	07 55 19.	+ 25 17	108	13.9	PART OF MULTIPLE GALAXY
ZWG 262.054	07 55 24.	+ 55 25		14.9	GALAXY
MCG+10-12-043	07 55 24.	+ 59 18	15	17.	GALAXY
MRK 100	07 55 24.	+ 66 38	28	14.5	GALAXY WITH UV CONTINUUM
7ZW 198	07 55 24.	+ 68 26			COMPACT GALAXY
ZWG 031.005	07 55 30.	+ 08 17		15.5	GALAXY
ZWG 178.012	07 55 30.	+ 33 00		15.6	GALAXY
ZC 0755.5+4915	07 55 30.	+ 49 15	610		CLUSTER OF GALAXIES
7ZW 199	07 55 30.	+ 56 40			COMPACT GALAXY
OCL 0669	07 55 30.	- 25 45	240	15.	OPEN STAR CLUSTER
TON-N 0852	07 55 35.	+ 35 16		15.5	BLUE STAR
MCG+09-13-099	07 55 36.	+ 55 26	42	16.	GALAXY
7ZW 200	07 55 36.	+ 60 46			COMPACT GALAXY
MCG+10-12-044	07 55 36.	+ 61 57	24	16.	GALAXY
ZWG 349.014	07 55 36.	+ 76 05		15.5	GALAXY
ZWG 178.013	07 55 42.	+ 32 45		15.3	GALAXY
MCG+10-12-045	07 55 42.	+ 57 26	30	16.	GALAXY
ZWG 287.022	07 55 42.	+ 57 29		15.7	GALAXY
MCG-02-21-001	07 55 42.	- 14 10	72	15.5	GALAXY
RNGC 2491	07 55 45.	+ 08 07		15.5	GALAXY
ZWG 031.006	07 55 48.	+ 08 00		15.4	GALAXY
ZWG 031.007	07 55 48.	+ 08 07		15.6	GALAXY
ZWG 031.008	07 55 48.	+ 08 17		15.5	GALAXY
ZWG 059.009	07 55 48.	+ 12 18		15.4	GALAXY
MCG+04-19-013	07 55 48.	+ 24 17	60	15.1	GALAXY
MCG+05-19-024	07 55 48.	+ 32 45	27	15.	GALAXY
ZC 0755.8+5408	07 55 48.	+ 54 08	1750		CLUSTER OF GALAXIES
ZWG 287.023	07 55 48.	+ 57 03		15.5	GALAXY
MCG+10-12-046	07 55 48.	+ 58 11	132	15.	GALAXY
MCG+10-12-047	07 55 48.	+ 58 44	27	16.	GALAXY
IC 2212	07 55 49.	+ 32 44 41.			NONSTELLAR OBJECT
ZC 0755.9+0805	07 55 54.	+ 08 05	2690		CLUSTER OF GALAXIES
ZWG 031.009	07 55 54.	+ 08 10		14.8	GALAXY
UGC 04127	07 55 54.	+ 08 10	84	14.8	GALAXY E
MCG+01-21-002	07 55 54.	+ 08 10	72	14.8	GALAXY
ZWG 031.010	07 55 54.	+ 08 23		15.3	GALAXY
MCG+05-19-025	07 55 54.	+ 27 36	42	15.4	GALAXY
MCG+10-12-048	07 55 54.	+ 57 28	24	16.	GALAXY
ZWG 287.024	07 55 54.	+ 60 25		13.9	GALAXY
UGC 04128	07 55 54.	+ 60 25	108	13.9	GALAXY Sc
MCG+10-12-049	07 55 54.	+ 60 26	78	14.	GALAXY
IC 2220	07 55 54.	- 58 59			REFLECTION NEBULA
RNGC 2496	07 55 57.	+ 08 10		15.0	GALAXY
HN 0315	07 55 57.	- 58 59			NEBULA
ZWG 118.031	07 56 00.	+ 24 17		15.1	GALAXY
UGC 04130	07 56 00.	+ 24 17	66	15.1	GALAXY S
MCG+04-19-014	07 56 00.	+ 24 34	60	15.5	GALAXY
ZWG 148.076	07 56 00.	+ 27 36		15.4	GALAXY
ZWG 148.077	07 56 00.	+ 31 56		15.7	GALAXY
UGC 04131	07 56 00.	+ 31 56	84	15.7	GALAXY SBc
MCG+05-19-026	07 56 00.	+ 31 56	54	15.7	GALAXY
ZWG 178.014	07 56 00.	+ 33 03		13.5	GALAXY
UGC 04132	07 56 00.	+ 33 03	120	13.5	GALAXY Sb-c
ZWG 262.055	07 56 00.	+ 55 30		15.4	GALAXY
MCG+09-13-100	07 56 00.	+ 56 26	18	17.	GALAXY
UGC 04133	07 56 00.	+ 56 31	114	16.0	GALAXY Sc
MCG+10-12-050	07 56 00.	+ 57 02	18	15.	GALAXY
MCG+10-12-051	07 56 00.	+ 58 50	27	16.	GALAXY
7ZW 201	07 56 00.	+ 59 30			COMPACT GALAXY
MCG+11-10-050	07 56 00.	+ 63 41	30	17.	GALAXY
UGC 04129	07 56 00.	+ 86 18	90	16.5	GALAXY DBL SYS
IC 2213	07 56 02.	+ 27 35 14.			NONSTELLAR OBJECT
BIGO 498	07 56 03.	+ 07 36			NEBULA
IC 0481	07 56 03.	+ 24 17 56.			NONSTELLAR OBJECT
MCG+06-18-005	07 56 06.	+ 33 02 30.	90	14.	GALAXY
ZC 0756.1+5616	07 56 06.	+ 56 16	11290		CLUSTER OF GALAXIES
ZWG 287.025	07 56 06.	+ 56 31		15.6	GALAXY
KARA.72 151A	07 56 06.	+ 56 31	120	15.6	PART OF DOUBLE GALAXY
RNGC 2499	07 56 09.	+ 07 38		15.0	GALAXY
RNGC 2490	07 56 10.	+ 27 13		15.5	GALAXY
BIGO 499	07 56 12.	+ 07 36			NEBULA
ZWG 031.011	07 56 12.	+ 07 38		15.1	GALAXY
MCG+01-21-003	07 56 12.	+ 07 38	60	15.1	GALAXY
TON-N 0853	07 56 12.	+ 20 14		14.9	BLUE STAR
ZWG 118.032	07 56 12.	+ 22 57		15.7	GALAXY
KARA.73B 0220	07 56 12.	+ 22 57	60	15.7	ISOLATED GALAXY S
MCG+05-19-027	07 56 12.	+ 27 11	24	15.3	GALAXY
ZWG 148.078	07 56 12.	+ 27 13		15.3	GALAXY
ZC 0756.2+2850	07 56 12.	+ 28 50	2290		CLUSTER OF GALAXIES
MCG+08-15-033	07 56 12.	+ 45 29	42	16.	GALAXY
ZWG 236.021	07 56 12.	+ 45 30		15.5	GALAXY
ZWG 287.026	07 56 12.	+ 56 30		15.5	GALAXY
UGC 04134	07 56 12.	+ 56 30	60	15.5	GALAXY SB
KARA.72 151B	07 56 12.	+ 56 30	60	15.5	PART OF DOUBLE GALAXY
MCG-02-21-002	07 56 12.	- 14 14 30.	36	14.5	GALAXY
ARC 0610	07 56 14.	+ 27 16		16.4	RICH CLUSTER OF GALAXIES
MCG+08-15-034	07 56 15.	+ 47 32 30.	78	15.	GALAXY
RNGC 2501	07 56 15.	- 14 14		14.0	GALAXY
ZWG 118.033	07 56 18.	+ 24 35		15.5	GALAXY
UGC 04135	07 56 18.	+ 24 35	60	15.5	GALAXY S
ZC 0756.3+4205	07 56 18.	+ 42 05	1550		CLUSTER OF GALAXIES
ZWG 207.012	07 56 18.	+ 42 23		15.4	GALAXY
ZWG 236.022	07 56 18.	+ 47 33		15.0	GALAXY
UGC 04136	07 56 18.	+ 47 33	96	15.0	GALAXY Sa
MCG+09-13-102	07 56 18.	+ 56 29 30.	120	15.	GALAXY
MCG+09-13-101	07 56 18.	+ 56 29 30.	60	15.	GALAXY
ZWG 331.025	07 56 18.	+ 73 10		15.7	GALAXY
UGC 04137	07 56 18.	+ 73 10	114	15.7	GALAXY S IV-V
MCG+12-08-023	07 56 18.	+ 73 12	66	16.	GALAXY
LB 00510	07 56 21.	+ 78 08 00.		15.4	FAINT BLUE STAR
RNGC 2492	07 56 22.	+ 27 10		14.5	GALAXY
ZWG 031.012	07 56 24.	+ 03 20		15.5	GALAXY
MCG+05-19-029	07 56 24.	+ 27 00	24	15.4	GALAXY
ZWG 148.079	07 56 24.	+ 27 01		15.4	GALAXY
MCG+05-19-028	07 56 24.	+ 27 09	48	14.4	GALAXY
ZWG 148.080	07 56 24.	+ 27 10		14.4	GALAXY
UGC 04138	07 56 24.	+ 27 10	84	14.4	GALAXY E-S0
MCG+07-17-005	07 56 24.	+ 42 24	42	15.5	GALAXY
ZC 0756.4+5607	07 56 24.	+ 56 07	1280		CLUSTER OF GALAXIES
ZC 0756.4+5805	07 56 24.	+ 58 05	1680		CLUSTER OF GALAXIES
LB 00511	07 56 25.	+ 29 32 36.		17.4	FAINT BLUE STAR
ZWG 088.010	07 56 30.	+ 16 34		15.0	GALAXY
UGC 04139	07 56 30.	+ 16 34	84	15.0	GALAXY Sc
ZWG 088.011	07 56 30.	+ 18 15		14.9	GALAXY
UGC 04140	07 56 30.	+ 18 15	90	14.9	GALAXY Sb-c
ZWG 148.081	07 56 30.	+ 26 42		15.6	GALAXY
7ZW 202	07 56 30.	+ 59 16			COMPACT GALAXY
MCG+03-21-003	07 56 33.	+ 18 15	90	14.9	GALAXY
SEY 037	07 56 33.	+ 73 11 45.		15.4	FAINT GALAXY
IC 2215	07 56 34.	+ 25 03			OPEN CLUSTER
ZWG 059.010	07 56 36.	+ 09 08		15.6	GALAXY
MCG+03-21-004	07 56 36.	+ 16 33	78	15.0	GALAXY
ZWG 088.012	07 56 36.	+ 17 37		15.4	GALAXY
ZWG 088.013	07 56 36.	+ 18 27		15.7	GALAXY
TON-N 0855	07 56 36.	+ 23 43		16.0	BLUE STAR
TON-N 0854	07 56 36.	+ 23 59		15.5	BLUE STAR
ZWG 118.034	07 56 36.	+ 25 06		14.4	GALAXY
RNGC 2498	07 56 36.	+ 25 06		14.5	GALAXY
UGC 04142	07 56 36.	+ 25 06	78	14.4	GALAXY SBa
MCG+04-19-015	07 56 36.	+ 25 06	60	14.4	GALAXY
ZWG 178.015	07 56 36.	+ 35 57		14.9	GALAXY
MCG+10-12-052	07 56 36.	+ 61 52	30	16.	GALAXY
MCG+10-12-053	07 56 36.	+ 62 14	15	17.	GALAXY
ZWG 003.002	07 56 36.	- 00 30		14.2	GALAXY
UGC 04141	07 56 36.	- 00 30	66	14.2	GALAXY SBa
MCG+00-21-001	07 56 36.	- 00 30	30	14.2	GALAXY
RNGC 2494	07 56 37.	- 00 30		14.0	GALAXY
IC 0487	07 56 39.	- 00 31 23.			SAME AS NGC 2494
ZWG 031.013	07 56 42.	+ 07 54		15.6	GALAXY
IC 2214	07 56 42.	+ 33 25 29.			NONSTELLAR OBJECT
ZWG 178.016	07 56 42.	+ 33 26		14.4	GALAXY
UGC 04143	07 56 42.	+ 33 26	48	14.4	GALAXY SBa-b
MCG+06-18-006	07 56 42.	+ 35 57	42	15.	GALAXY
MCG+09-13-103	07 56 42.	+ 56 44	12	17.	GALAXY
MCG+13-06-015	07 56 42.	+ 76 05	39	15.	GALAXY
OCL 0684	07 56 42.	- 28 47	840	12.	OPEN STAR CLUSTER
MRSL 258-07/1	07 56 42.	- 43 08	72000		HII REGION
IC 0482	07 56 47.	+ 25 28 57.			NONSTELLAR OBJECT
ZWG 031.014	07 56 48.	+ 07 35		15.6	GALAXY
UGC 04144	07 56 48.	+ 07 35	72	15.6	GALAXY Sc
ZWG 088.014	07 56 48.	+ 15 31		14.0	GALAXY
UGC 04145	07 56 48.	+ 15 31	96	14.0	GALAXY Sa
ZWG 118.035	07 56 48.	+ 25 29		15.0	GALAXY
MCG+04-19-016	07 56 48.	+ 25 29	36	15.0	GALAXY
ZWG 148.082	07 56 48.	+ 26 56		15.7	GALAXY
ZWG 148.083	07 56 48.	+ 27 54		15.6	GALAXY
MCG+06-18-007	07 56 48.	+ 33 26	36	15.	GALAXY
MCG+08-15-035	07 56 48.	+ 45 20	42	15.	GALAXY
ZWG 236.023	07 56 48.	+ 45 22		15.1	GALAXY
UGC 04146	07 56 48.	+ 59 16	78	16.0	GALAXY Sc
MCG+10-12-054	07 56 48.	+ 59 16	51	16.	GALAXY
IC 0483	07 56 50.	+ 26 03 38.			NONSTELLAR OBJECT
HOLM 091A	07 56 53.	+ 39 58	36	13.5	PART OF MULTIPLE GALAXY
MCG+03-21-005	07 56 54.	+ 15 30 30.	78	14.0	GALAXY
ZWG 088.015	07 56 54.	+ 18 33		15.0	GALAXY
UGC 04147	07 56 54.	+ 18 33	78	15.0	GALAXY Sa-b
MCG+03-21-006	07 56 54.	+ 18 33	72	15.0	GALAXY
ZWG 118.036	07 56 54.	+ 23 31		14.8	GALAXY
MCG+04-19-017	07 56 54.	+ 23 31	36	14.8	GALAXY
ZWG 118.037	07 56 54.	+ 24 08		15.5	GALAXY
ZWG 148.084	07 56 54.	+ 26 48		15.3	GALAXY
ZWG 207.013	07 56 54.	+ 42 19		15.5	GALAXY
UGC 04148	07 56 54.	+ 42 19	150	15.5	GALAXY
ZWG 262.056	07 56 54.	+ 55 41		15.2	GALAXY
IC 2216	07 56 55.	+ 08 45			NONSTELLAR OBJECT
TON-N 0856	07 56 55.	+ 20 03		16.0	BLUE STAR
IC 0484	07 56 58.	+ 26 47 56.			NONSTELLAR OBJECT
ZWG 118.038	07 57 00.	+ 23 34		15.4	GALAXY
ZWG 118.039	07 57 00.	+ 25 47		15.6	GALAXY
ZWG 207.014	07 57 00.	+ 39 58		13.1	GALAXY
UGC 04150	07 57 00.	+ 39 58	132	13.1	GALAXY SB0
ZWG 207.015	07 57 00.	+ 41 48		15.4	GALAXY
MCG+07-17-006	07 57 00.	+ 42 20	138	15.	GALAXY
MCG+09-13-104	07 57 00.	+ 55 42	12	16.	GALAXY
ZWG 349.015	07 57 00.	+ 77 58		13.1	GALAXY
UGC 04151	07 57 00.	+ 77 58	102	13.1	GALAXY Sc-IRR
MCG+14-04-045	07 57 00.	+ 86 16	27	18.	GALAXY
MCG+14-04-044	07 57 00.	+ 86 16	8	18.	GALAXY
UGC 04149	07 57 00.	- 01 32	84	16.5	GALAXY IV-V
FIT.G 08	07 57 00.	- 26 10 06.		17.5	Sa GALAXY
OCL 0681	07 57 00.	- 28 27	300	12.	OPEN STAR CLUSTER
RNGC 2493	07 57 01.	+ 39 58		13.0	GALAXY
HOLM 091B	07 57 03.	+ 39 59	18	14.9	PART OF MULTIPLE GALAXY
LB 00512	07 57 04.	+ 30 30 18.		17.3	FAINT BLUE STAR
MCG+04-19-018	07 57 06.	+ 26 44	54		GALAXY
ZWG 148.085	07 57 06.	+ 26 50		15.7	GALAXY
MCG+07-17-007	07 57 06.	+ 39 57 30.	102	14.	GALAXY
ZWG 207.016	07 57 06.	+ 39 59		15.6	GALAXY
MRK 383	07 57 06.	+ 39 59	12	15.5	GALAXY WITH UV CONTINUUM
RNGC 2495	07 57 07.	+ 39 59		15.5	GALAXY

OBJECT NAME	RIGHT ASCEN.	DECLINATION	DIAM.	MAGN.	TYPE OF OBJECT
MCG+07-17-008	07 57 09.	+ 39 58	18	15.5	GALAXY
RNGC 2504	07 57 10.	+ 05 45		14.0	GALAXY
ZWG 031.015	07 57 12.	+ 05 45		14.1	GALAXY
UGC 04152	07 57 12.	+ 05 45	27	14.1	GALAXY PECULR
MCG+01-21-004	07 57 12.	+ 05 45	30	14.1	GALAXY
ZC 0757.2+0828	07 57 12.	+ 08 28	540		CLUSTER OF GALAXIES
ZWG 148.086	07 57 12.	+ 30 48		15.7	GALAXY
ZWG 207.017	07 57 12.	+ 40 00		15.6	GALAXY
ZWG 236.024	07 57 12.	+ 46 50		15.2	GALAXY
ZWG 262.057	07 57 12.	+ 55 53		15.5	GALAXY
UGC 04153	07 57 12.	+ 61 33	60	16.5	GALAXY
MCG+10-12-055	07 57 12.	+ 61 33	54	16.	GALAXY
MCG+13-06-016	07 57 12.	+ 77 57	57	14.	GALAXY
IC 0485	07 57 17.	+ 26 49 47.			NONSTELLAR OBJECT
ZWG 031.016	07 57 18.	+ 08 09		15.2	GALAXY
ZWG 059.011	07 57 18.	+ 13 17		15.1	GALAXY
UGC 04154	07 57 18.	+ 13 17	84	15.1	GALAXY Sc
MCG+02-21-001	07 57 18.	+ 13 17	84	15.1	GALAXY
IC 0486	07 57 18.	+ 26 44 47.			NONSTELLAR OBJECT
ZWG 148.087	07 57 18.	+ 26 45		14.7	GALAXY
UGC 04155	07 57 18.	+ 26 45	60	14.7	GALAXY SBa
ZWG 148.088	07 57 18.	+ 26 50		15.5	GALAXY
UGC 04156	07 57 18.	+ 26 50	84	15.5	GALAXY Sa
ZWG 262.058	07 57 18.	+ 55 52		15.4	GALAXY
MCG+09-13-105	07 57 18.	+ 55 54	60	16.	GALAXY
MCG+09-13-106	07 57 18.	+ 56 30	18	17.	GALAXY
OCL 0617	07 57 18.	- 16 10	660	13.	OPEN STAR CLUSTER
TON-N 0857	07 57 19.	+ 19 45		15.0	BLUE STAR
RNGC 2516	07 57 23.	- 60 44		3.5	OPEN CLUSTER
ZWG 031.017	07 57 24.	+ 06 50		14.7	GALAXY
MCG+01-21-005	07 57 24.	+ 06 50	42	14.7	GALAXY
TON-N 0858	07 57 24.	+ 20 32		15.5	BLUE STAR
ZWG 178.017	07 57 24.	+ 37 20		15.5	GALAXY
UGC 04157	07 57 24.	+ 37 20	60	15.5	GALAXY IRR
MCG+09-13-107	07 57 24.	+ 55 53	36	16.	GALAXY
ISS 0844	07 57 24.	- 22 26	181		STELLAR RING
ZWG 118.040	07 57 30.	+ 21 17		15.6	GALAXY
KARA.73B 0221	07 57 30.	+ 21 17	36	15.6	ISOLATED GALAXY S
MCG+04-19-019	07 57 30.	+ 22 31	66	15.0	GALAXY
ZWG 262.059	07 57 30.	+ 55 43		15.1	GALAXY
ZWG 262.060	07 57 30.	+ 56 03		15.6	GALAXY
ZWG 287.027	07 57 30.	+ 57 03		15.7	GALAXY
MCG+10-12-056	07 57 30.	+ 61 34	57	14.	GALAXY
MCG+11-10-051	07 57 30.	+ 67 15	51	16.	GALAXY
ISS 0845	07 57 30.	- 26 06	264		STELLAR RING
OCL 0776	07 57 30.	- 60 44	6180	3.6	OPEN STAR CLUSTER
TON-N 0859	07 57 35.	+ 34 02		16.	BLUE STAR
ZWG 118.041	07 57 36.	+ 22 32		15.0	GALAXY
UGC 04158	07 57 36.	+ 22 32	66	15.0	GALAXY SBb/Sc
KARA.73B 0222	07 57 36.	+ 22 32	60	15.0	ISOLATED GALAXY S
ZWG 148.089	07 57 36.	+ 28 37		15.7	GALAXY
ZWG 148.090	07 57 36.	+ 29 53		15.0	GALAXY
MCG+05-19-030	07 57 36.	+ 29 53	36	15.0	GALAXY
ZC 0757.6+5214	07 57 36.	+ 52 14	4570		CLUSTER OF GALAXIES
MCG+10-12-057	07 57 36.	+ 57 02	15	17.	GALAXY
ZWG 287.028	07 57 36.	+ 61 32		13.9	GALAXY
UGC 04159	07 57 36.	+ 61 32	54	13.9	GALAXY S
7ZW 203	07 57 36.	+ 68 04			COMPACT GALAXY
RNGC 2503	07 57 37.	+ 22 32		15.0	GALAXY
ABC 0608	07 57 41.	+ 63 55		17.1	RICH CLUSTER OF GALAXIES
ZWG 031.018	07 57 42.	+ 07 57		15.1	GALAXY
MCG+01-21-006	07 57 42.	+ 07 57	24	15.1	GALAXY
ZWG 148.091	07 57 42.	+ 27 38		14.2	GALAXY
UGC 04160	07 57 42.	+ 27 38	36	14.2	GALAXY S?
MCG+05-19-031	07 57 42.	+ 27 38	36	14.2	GALAXY
ZC 0757.7+3610	07 57 42.	+ 36 10	1210		CLUSTER OF GALAXIES
MCG+09-13-108	07 57 42.	+ 55 44	12	16.	GALAXY
RNGC 2506	07 57 43.	- 10 39		8.5	OPEN CLUSTER
ABC 0611	07 57 44.	+ 36 14		17.9	RICH CLUSTER OF GALAXIES
IC 2217	07 57 45.	+ 27 37 32.			NONSTELLAR OBJECT
MCG+09-13-109	07 57 45.	+ 56 42	30	13.	GALAXY
ABC 0612	07 57 46.	+ 34 58		16.5	RICH CLUSTER OF GALAXIES
IC 0488	07 57 47.	+ 26 02 44.			NONSTELLAR OBJECT
RNGC 2488	07 57 47.	+ 56 42		14.0	GALAXY
ZWG 059.012	07 57 48.	+ 13 47		15.7	GALAXY
ZWG 059.013	07 57 48.	+ 13 50		14.9	GALAXY
MCG+02-21-002	07 57 48.	+ 13 50	24	14.9	GALAXY
ZWG 088.016	07 57 48.	+ 15 49		15.0	GALAXY
MCG+03-21-007	07 57 48.	+ 15 49	42	15.0	GALAXY
ZWG 088.017	07 57 48.	+ 16 29		15.4	GALAXY
ZC 0757.8+1829	07 57 48.	+ 18 29	3290		CLUSTER OF GALAXIES
ZWG 118.042	07 57 48.	+ 23 02		15.7	GALAXY
ZWG 148.092	07 57 48.	+ 28 16		15.7	GALAXY
ZC 0757.8+3441	07 57 48.	+ 34 41	1950		CLUSTER OF GALAXIES
ZWG 287.029	07 57 48.	+ 56 42		14.2	GALAXY
UGC 04161	07 57 48.	+ 56 42	96	14.2	GALAXY E-S0
ZC 0757.8+6345	07 57 48.	+ 63 45	2080		CLUSTER OF GALAXIES
OCL 0593	07 57 48.	- 10 39	1020	11.5	OPEN STAR CLUSTER
ABC 0609	07 57 51.	+ 63 41		17.1	RICH CLUSTER OF GALAXIES
ZWG 088.018	07 57 54.	+ 16 40		14.9	GALAXY
UGC 04162	07 57 54.	+ 16 40	60	14.9	GALAXY S
ZC 0757.9+3458	07 57 54.	+ 34 58	540		CLUSTER OF GALAXIES
ZWG 207.018	07 57 54.	+ 44 19		15.5	GALAXY
UGC 04163	07 57 54.	+ 44 19	66	15.5	GALAXY
ACK 259-07.1	07 57 54.	- 44 36			PLANETARY NEBULA
MCG+03-21-008	07 57 57.	+ 16 40 30.	45	14.9	GALAXY
LB 09840	07 58	- 82 41		12.9	FAINT BLUE STAR
ZWG 059.014	07 58 00.	+ 09 48		15.4	GALAXY
ZWG 148.093	07 58 00.	+ 26 56		15.7	GALAXY
ZWG 178.018	07 58 00.	+ 32 59		15.7	GALAXY
ZC 0758.0+4852	07 58 00.	+ 48 52	3830		CLUSTER OF GALAXIES
ZWG 287.030	07 58 00.	+ 56 47		15.2	GALAXY
UGC 04164	07 58 00.	+ 56 47	84	15.2	GALAXY S0-a
MCG+10-12-058	07 58 00.	+ 56 47	78	15.	GALAXY
MCG+10-12-059	07 58 00.	+ 57 02	15	16.	GALAXY
ZWG 287.031	07 58 00.	+ 57 04		15.0	GALAXY
ISS 0846	07 58 00.	- 23 25	421		STELLAR RING
ZWG 207.019	07 58 06.	+ 42 08		15.3	GALAXY
MCG+09-13-110	07 58 06.	+ 50 51	180	12.1	GALAXY
ZWG 262.061	07 58 06.	+ 55 09		15.5	GALAXY
MCG+10-12-060	07 58 06.	+ 57 02	27	16.	GALAXY
MCG+10-12-061	07 58 06.	+ 57 05	18	13.	GALAXY
MCG+10-12-062	07 58 06.	+ 61 32	96	13.	GALAXY
MCG+12-08-024	07 58 06.	+ 74 12	51	15.	GALAXY
RNGC 2500	07 58 08.	+ 50 53		12.5	GALAXY
ABC 0614	07 58 11.	+ 18 07		17.0	RICH CLUSTER OF GALAXIES
ZWG 003.003	07 58 12.	+ 02 26		15.4	GALAXY
ZWG 118.043	07 58 12.	+ 25 41		15.6	GALAXY
KARA.73B 0223	07 58 12.	+ 25 41	54	15.6	ISOLATED GALAXY S
ZWG 118.044	07 58 12.	+ 26 27		15.6	GALAXY
ZWG 148.094	07 58 12.	+ 27 21		15.7	GALAXY
ZWG 148.095	07 58 12.	+ 29 19		15.4	GALAXY
ZWG 262.062	07 58 12.	+ 50 54		12.3	GALAXY
UGC 04165	07 58 12.	+ 50 54	168	12.3	GALAXY S IV
KARA.73B 0224	07 58 12.	+ 50 54	192	12.3	ISOLATFD GALAXY S
ZC 0758.2+6531	07 58 12.	+ 65 31	340		CLUSTER OF GALAXIES
ZWG 331.026	07 58 12.	+ 74 11		14.5	GALAXY
UGC 04166	07 58 12.	+ 74 11	72	14.5	GALAXY SBc
MCG+13-06-017	07 58 12.	+ 80 17	54	14.	GALAXY
OCL 0693	07 58 12.	- 30 07	168	14.	OPEN STAR CLUSTER
RNGC 2523A	07 58 13.	+ 74 11		14.5	GALAXY
RNGC 2497	07 58 16.	+ 57 05		14.5	GALAXY
ZWG 118.045	07 58 18.	+ 25 25		15.5	GALAXY
UGC 04167	07 58 18.	+ 25 25	66	15.5	GALAXY Sc
MCG+04-19-020	07 58 18.	+ 25 25	48	15.5	GALAXY
ZWG 118.046	07 58 18.	+ 26 23		15.7	GALAXY
MCG+05-19-032	07 58 18.	+ 27 24	24	15.5	GALAXY
ZWG 262.063	07 58 18.	+ 56 02		15.6	GALAXY
ZWG 287.032	07 58 18.	+ 57 05		14.5	GALAXY
UGC 04168	07 58 18.	+ 57 05	84	14.5	GALAXY COMPACT
7ZW 204	07 58 18.	+ 60 24			COMPACT GALAXY
ZWG 287.033	07 58 18.	+ 61 31		13.5	GALAXY
UGC 04169	07 58 18.	+ 61 31	108	13.5	GALAXY Sc
MCG+11-10-052	07 58 18.	+ 67 18	57	15.	GALAXY
ZWG 031.019	07 58 24.	+ 04 27		15.6	GALAXY
TON-N 0004	07 58 24.	+ 26 05		14.7	BLUE STAR
ZWG 118.047	07 58 24.	+ 26 21		15.7	GALAXY
ZWG 148.096	07 58 24.	+ 27 25		15.5	GALAXY
MCG+10-12-063	07 58 24.	+ 56 50	27	15.	GALAXY
ZWG 287.034	07 58 24.	+ 56 51		15.6	GALAXY
MCG+10-12-064	07 58 24.	+ 57 03	30	17.	GALAXY
KEEN 2523A	07 58 24.	+ 74 14			GALAXY
RNGC 2509	07 58 29.	- 18 56		9.5	OPEN CLUSTER
ZWG 059.015	07 58 30.	+ 09 24		15.5	GALAXY
OCL 0630	07 58 30.	- 18 56	600	9.3	OPEN STAR CLUSTER
LB 00513	07 58 34.	+ 78 57 36.		17.8	FAINT BLUE STAR
IC 0489	07 58 35.	+ 26 08 14.			NONSTELLAR OBJECT
ZWG 031.020	07 58 36.	+ 04 25		15.4	GALAXY
MCG+03-21-009	07 58 36.	+ 15 29 30.	27	14.5	GALAXY
ZWG 088.019	07 58 36.	+ 15 30		14.5	GALAXY
UGC 04170	07 58 36.	+ 15 30	78	14.5	GALAXY E
ZWG 118.048	07 58 36.	+ 24 34		15.7	GALAXY
ZC 0758.6+3850	07 58 36.	+ 38 50	540		CLUSTER OF GALAXIES
MCG+11-10-053	07 58 36.	+ 63 10	27	16.	GALAXY
ZWG 310.031	07 58 36.	+ 67 17		15.4	GALAXY
7ZW 205	07 58 36.	+ 68 02			COMPACT GALAXY
IC 2218	07 58 38.	+ 24 33 56.			NONSTELLAR OBJECT
MCG+09-13-111	07 58 39.	+ 56 38	9	17.	GALAXY
ZWG 031.021	07 58 42.	+ 06 16		15.3	GALAXY
TON-N 0860	07 58 42.	+ 23 24		15.5	BLUE STAR
ZWG 118.049	07 58 42.	+ 25 22		15.5	GALAXY
TON-N 0293	07 58 42.	+ 30 13		15.2	BLUE STAR
MCG+09-13-112	07 58 42.	+ 56 38	9	17.	GALAXY
HOLM 092A	07 58 44.	+ 15 51	114	13.2	PART OF MULTIPLE GALAXY
BC 3CR190	07 58 45.05	+ 14 23 04.3		21.	QUASI-STELLAR OBJECT
SHB 104	07 58 45.1	+ 14 23 02.		17.5	QUASI-STELLAR OBJECT
HOLM 092B	07 58 46.	+ 15 53	24	14.6	PART OF MULTIPLE GALAXY
RNGC 2507	07 58 47.	+ 15 51		14.0	GALAXY
ZWG 059.016	07 58 48.	+ 09 51		14.6	GALAXY
UGC 04171	07 58 48.	+ 09 51	144	14.6	GALAXY Sb
MCG+02-21-003	07 58 48.	+ 09 51	132	14.6	GALAXY
MCG+03-21-010	07 58 48.	+ 15 50	150	14.0	GALAXY
ZWG 088.020	07 58 48.	+ 15 51		14.0	GALAXY
UGC 04172	07 58 48.	+ 15 51	150	14.0	GALAXY S
TON-N 0861	07 58 48.	+ 21 55		16.2	BLUE STAR
MCG+05-19-033	07 58 48.	+ 27 21	42	15.5	GALAXY
LB 00514	07 58 48.	+ 28 32 30.		17.2	FAINT BLUE STAR
7ZW 206	07 58 48.	+ 57 08			COMPACT GALAXY
ZWG 059.017	07 58 54.	+ 10 52		15.2	GALAXY
ZWG 031.022	07 59 00.	+ 05 19		15.6	GALAXY
ZWG 031.023	07 59 00.	+ 05 57		15.7	GALAXY
ZWG 031.024	07 59 00.	+ 06 46		15.2	GALAXY
ZWG 148.097	07 59 00.	+ 27 22		15.5	GALAXY
ZWG 148.098	07 59 00.	+ 32 27		15.3	GALAXY
KARA.73B 0225	07 59 00.	+ 32 27	42	15.3	ISOLATED GALAXY S
ZC 0759.0+4519	07 59 00.	+ 45 19	540		CLUSTER OF GALAXIES
ZC 0759.0+5359	07 59 00.	+ 53 59	740		CLUSTER OF GALAXIES
ZWG 262.064	07 59 00.	+ 56 13		15.5	GALAXY
ZWG 262.065	07 59 00.	+ 56 21		15.6	GALAXY
ZWG 349.016	07 59 00.	+ 80 15		15.7	GALAXY
UGC 04173	07 59 00.	+ 80 15	198	15.7	GALAXY IRR
MCG+05-19-034	07 59 03.	+ 32 28	36	15.3	GALAXY
ABC 0613	07 59 03.	+ 45 18		17.5	RICH CLUSTER OF GALAXIES
MCG+09-13-113	07 59 03.	+ 56 16	36	16.	GALAXY
TON-N 0294	07 59 06.	+ 30 48		15.9	BLUE STAR
7ZW 207	07 59 06.	+ 65 29			COMPACT GALAXY
ZC 0759.1+7211	07 59 06.	+ 72 11	740		CLUSTER OF GALAXIES
OCL 0677	07 59 06.	- 27 02	66	15.	OPEN STAR CLUSTER
VHE 01	07 59 06.	- 45 19	78		REFLECTION NEBULA
MCG+09-13-114	07 59 09.	+ 56 24	36	17.	GALAXY
ZWG 031.025	07 59 12.	+ 04 25		15.5	GALAXY
ZWG 059.018	07 59 12.	+ 08 42		14.2	GALAXY
UGC 04174	07 59 12.	+ 08 42	120	14.2	GALAXY E?
MCG+02-21-004	07 59 12.	+ 08 42	36	14.2	GALAXY
ZWG 059.019	07 59 12.	+ 09 46		14.9	GALAXY
MCG+02-21-005	07 59 12.	+ 09 46	48	14.9	GALAXY
ZWG 059.020	07 59 12.	+ 11 52		15.1	GALAXY
MCG+02-21-006	07 59 12.	+ 11 52	36	15.1	GALAXY
ZWG 088.021	07 59 12.	+ 15 11		15.7	GALAXY
UGC 04175	07 59 12.	+ 15 11	60	15.2	GALAXY SB?0
ZWG 118.050	07 59 12.	+ 21 24		15.7	GALAXY
KARA.73B 0226	07 59 12.	+ 21 24	30	15.7	ISOLATED GALAXY S
ZWG 148.099	07 59 12.	+ 31 19		15.5	GALAXY
MCG+10-12-065	07 59 12.	+ 56 58	30	17.	GALAXY
TON-N 0862	07 59 13.	+ 19 48		15.5	BLUE STAR
RNGC 2508	07 59 15.	+ 08 42		14.0	GALAXY
ZWG 059.021	07 59 18.	+ 09 49		15.3	GALAXY
ZWG 207.020	07 59 18.	+ 40 49		15.2	GALAXY
UGC 04176	07 59 18.	+ 40 49	108	15.2	GALAXY SBc
ZWG 031.026	07 59 24.	+ 07 49		15.3	GALAXY
UGC 04177	07 59 24.	+ 07 49	72	15.3	GALAXY Sa-b
ZWG 059.022	07 59 24.	+ 09 25		15.6	GALAXY
ZWG 059.023	07 59 24.	+ 09 37		14.7	GALAXY
UGC 04178	07 59 24.	+ 09 37	60	14.7	GALAXY S0
MCG+02-21-007	07 59 24.	+ 09 37	24	14.7	GALAXY
ZC 0759.4+3152	07 59 24.	+ 31 52	870		CLUSTER OF GALAXIES

OBJECT NAME	RIGHT ASCEN.	DECLINATION	DIAM.	MAGN.	TYPE OF OBJECT
ZWG 178.019	07 59 24.	+ 34 55		15.2	GALAXY
ZWG 207.021	07 59 24.	+ 40 36		15.4	GALAXY
MCG+07-17-009	07 59 24.	+ 40 49	90	15.	GALAXY
MCG+10-12-066	07 59 24.	+ 56 57	39	17.	GALAXY
RNGC 2510	07 59 26.	+ 09 37		14.5	GALAXY
ZWG 003.004	07 59 30.	+ 00 57		15.5	GALAXY
UGC 04179	07 59 30.	+ 00 57	78	15.5	GALAXY SBb
MCG+00-21-002	07 59 30.	+ 00 57	66	15.5	GALAXY
ZWG 059.024	07 59 30.	+ 09 32		15.0	GALAXY
MCG+02-21-008	07 59 30.	+ 09 32	48	15.0	GALAXY
MCG+05-19-035	07 59 30.	+ 27 34	66	14.5	GALAXY
ZWG 148.100	07 59 30.	+ 27 35		14.5	GALAXY
UGC 04180	07 59 30.	+ 27 35	90	14.5	GALAXY Sc
ZWG 148.101	07 59 30.	+ 30 23		15.5	GALAXY
MCG+06-18-008	07 59 30.	+ 34 55	18	15.5	GALAXY
MCG+10-12-067	07 59 30.	+ 61 26	54	15.	GALAXY
MCG+10-12-068	07 59 30.	+ 61 30	45	15.	GALAXY
UGC 04181	07 59 30.	+ 65 02	60	16.0	GALAXY SB?b-c
MCG+11-10-054	07 59 30.	+ 65 02 30.	51	15.	GALAXY
ZC 0759.5-0206	07 59 30.	- 02 06	2550		CLUSTER OF GALAXIES
RNGC 2511	07 59 32.	+ 09 32		15.0	GALAXY
IC 2219	07 59 33.	+ 27 34 44.			NONSTELLAR OBJECT
ZWG 031.027	07 59 36.	+ 04 28		15.5	GALAXY
TON-N 0863	07 59 36.	+ 21 29		16.5	BLUE STAR
ZWG 118.051	07 59 36.	+ 23 01		15.7	GALAXY
ZWG 148.102	07 59 36.	+ 29 55		15.6	GALAXY
ZWG 178.020	07 59 36.	+ 35 55		15.6	GALAXY
ZC 0759.6+5427	07 59 36.	+ 54 27	610		CLUSTER OF GALAXIES
ZWG 287.035	07 59 36.	+ 61 25		15.1	GALAXY
ZWG 287.036	07 59 36.	+ 61 29		14.5	GALAXY
UGC 04182	07 59 36.	+ 61 29	54	14.5	GALAXY DISTRBD
ARC 0615	07 59 37.	+ 31 53		17.7	RICH CLUSTER OF GALAXIES
ZWG 031.028	07 59 42.	+ 07 01		14.8	GALAXY
UGC 04183	07 59 42.	+ 07 01	120	14.8	GALAXY SBb
MCG+01-21-007	07 59 42.	+ 07 01	84	14.8	GALAXY
ZWG 059.025	07 59 42.	+ 09 33		13.7	GALAXY
UGC 04184	07 59 42.	+ 09 33	168	13.7	GALAXY E
MCG+02-21-009	07 59 42.	+ 09 33	42	13.7	GALAXY
ZWG 059.026	07 59 42.	+ 09 44		15.0	GALAXY
MCG+02-21-010	07 59 42.	+ 09 44	24	15.0	GALAXY
MCG+11-10-055	07 59 42.	+ 63 06	39	17.	GALAXY
ZWG 310.032	07 59 42.	+ 63 07		15.5	GALAXY
UGC 04186	07 59 42.	+ 63 07	84	15.5	GALAXY S
UGC 04185	07 59 42.	+ 63 07	84	15.5	GALAXY S
MCG+11-10-056	07 59 42.	+ 63 07 30.	78	15.	GALAXY
7ZW 208	07 59 42.	+ 66 39			COMPACT GALAXY
MCG+11-10-057	07 59 42.	+ 66 56	60	15.	GALAXY
RNGC 2513	07 59 44.	+ 09 33		13.5	GALAXY
ZWG 207.022	07 59 48.	+ 40 03		15.4	GALAXY
MCG+11-10-058	07 59 48.	+ 63 07 30.	78	15.	GALAXY
UGC 04187	07 59 48.	+ 66 55	72	16.0	GALAXY Sc
MCG-03-21-001	07 59 48.	- 18 37 30.	36	15.	GALAXY
ZC 0759.9+0927	07 59 54.	+ 09 27	270		CLUSTER OF GALAXIES
ZWG 059.027	07 59 54.	+ 09 39		15.2	GALAXY
ZWG 148.103	07 59 54.	+ 27 53		15.6	GALAXY
ZWG 207.023	07 59 54.	+ 42 03		14.7	GALAXY
UGC 04188	07 59 54.	+ 42 03	90	14.7	GALAXY S0
MCG+11-10-059	07 59 54.	+ 65 32	27	15.	GALAXY
OCL 0637	07 59 54.	- 19 20	120	12.	OPEN STAR CLUSTER
RNGC 2514	07 59 59.	+ 15 57		14.5	GALAXY
DV.55 2	08 00	- 80 30	9000		FAINT EMISSION NEBULOSITY
LB 09841	08 00	- 83 35		13.5	FAINT BLUE STAR
ZC 0800.0+0946	08 00 00.	+ 09 46	5710		CLUSTER OF GALAXIES
MCG+03-21-011	08 00 00.	+ 15 56	72	14.4	GALAXY
ZWG 088.022	08 00 00.	+ 15 57		14.4	GALAXY
UGC 04189	08 00 00.	+ 15 57	72	14.4	GALAXY SBc
LB 09158	08 00 00.	+ 16 49		17.5	FAINT BLUE STAR
LB 09157	08 00 00.	+ 18 15		18.4	FAINT BLUE STAR
LB 09156	08 00 00.	+ 18 44		17.8	FAINT BLUE STAR
LB 09155	08 00 00.	+ 18 56		16.4	FAINT BLUE STAR
MCG+04-19-021	08 00 00.	+ 23 32	90	14.2	GALAXY
ZWG 148.104	08 00 00.	+ 30 55		14.9	GALAXY
ZWG 148.105	08 00 00.	+ 31 55		15.7	GALAXY
MCG+10-12-069	08 00 00.	+ 61 41	24	17.	GALAXY
ZWG 287.037	08 00 00.	+ 61 42		15.6	GALAXY
MCG+10-12-070	08 00 00.	+ 61 42	39	16.	GALAXY
ZWG 310.033	08 00 00.	+ 65 32		15.7	GALAXY
RNGC 2512	08 00 01.	+ 23 32		14.0	GALAXY
KARA.68 051	08 00 03.	+ 20 15	60		DWARF GALAXY
MCG+07-17-010	08 00 03.	+ 42 03	54	15.	GALAXY
LB 09161	08 00 06.	+ 15 44		17.5	FAINT BLUE STAR
MCG+03-21-012	08 00 06.	+ 16 26	66	14.7	GALAXY
ZWG 088.023	08 00 06.	+ 16 27		14.7	GALAXY
UGC 04190	08 00 06.	+ 16 27	72	14.7	GALAXY S0:
LB 09160	08 00 06.	+ 18 06		18.6	FAINT BLUE STAR
LB 09159	08 00 06.	+ 19 34		18.4	FAINT BLUE STAR
ZWG 118.052	08 00 06.	+ 23 32		14.2	GALAXY
MRK 384	08 00 06.	+ 23 32	24	16.	GALAXY WITH UV CONTINUUM
UGC 04191	08 00 06.	+ 23 32	90	14.2	GALAXY SBb
ZWG 118.053	08 00 06.	+ 24 47		15.5	GALAXY
MCG+05-19-036	08 00 06.	+ 30 56	42	14.9	GALAXY
MCG+11-10-060	08 00 06.	+ 66 56	96	13.	GALAXY
MCG+12-08-025	08 00 06.	+ 72 40	57	16.	GALAXY
ARC 0619	08 00 06.	- 02 05		16.8	RICH CLUSTER OF GALAXIES
OCL 0738	08 00 06.	- 44 16	36	13.	OPEN STAR CLUSTER
RNGC 2505	08 00 10.	+ 53 42		14.0	GALAXY
TON-N 0864	08 00 11.	+ 34 30		14.9	BLUE STAR
ZWG 003.005	08 00 12.	+ 00 25		15.5	GALAXY
ZWG 088.024	08 00 12.	+ 16 29		15.6	GALAXY
LB 09165	08 00 12.	+ 17 34		17.8	FAINT BLUE STAR
LB 09164	08 00 12.	+ 19 00		17.0	FAINT BLUE STAR
LB 09163	08 00 12.	+ 20 04		18.3	FAINT BLUE STAR
LB 09162	08 00 12.	+ 20 40		17.0	FAINT BLUE STAR
ZWG 118.054	08 00 12.	+ 24 48		15.7	GALAXY
UGC 04192	08 00 12.	+ 43 29	72	16.0	GALAXY Sc
ZWG 262.066	08 00 12.	+ 53 42		14.0	GALAXY
UGC 04193	08 00 12.	+ 53 42	84	14.0	GALAXY SBa
ZWG 263.001	08 00 12.	+ 55 32		15.0	GALAXY
ZWG 262.067	08 00 12.	+ 55 32		15.0	GALAXY
IC 0490	08 00 17.	+ 25 57 03.			NONSTELLAR OBJECT
ZC 0800.3+0424	08 00 18.	+ 04 24	940		CLUSTER OF GALAXIES
LB 09167	08 00 18.	+ 16 24		18.1	FAINT BLUE STAR
MCG+03-21-013	08 00 18.	+ 16 28	27	15.6	GALAXY
LB 09166	08 00 18.	+ 19 59		16.6	FAINT BLUE STAR
ZWG 118.055	08 00 18.	+ 25 56		15.6	GALAXY
ZWG 178.021	08 00 18.	+ 33 36		14.7	GALAXY
MCG+09-13-115	08 00 18.	+ 53 41	72	15.	GALAXY
MCG+09-13-116	08 00 18.	+ 55 32	48	16.	GALAXY

OBJECT NAME	RIGHT ASCEN.	DECLINATION	DIAM.	MAGN.	TYPE OF OBJECT
MCG+10-12-071	08 00 18.	+ 61 41	57	15.	GALAXY
UGC 04194	08 00 18.	+ 72 41	66	16.5	GALAXY Sc-IRR
ZWG 331.027	08 00 18.	+ 73 28		15.6	GALAXY
KARA.72 152A	08 00 18.	+ 73 28	42	15.6	PART OF DOUBLE GALAXY
OCL 0699	08 00 18.	- 30 58	300	13.	OPEN STAR CLUSTER
MCG+06-18-009	08 00 21.	+ 33 37	24	15.5	GALAXY
LB 09171	08 00 24.	+ 18 40		18.2	FAINT BLUE STAR
LB 09170	08 00 24.	+ 18 57		16.3	FAINT BLUE STAR
LB 09169	08 00 24.	+ 20 25		16.2	FAINT BLUE STAR
LB 09168	08 00 24.	+ 20 35		16.4	FAINT BLUE STAR
ZWG 310.034	08 00 24.	+ 66 55		14.6	GALAXY
UGC 04195	08 00 24.	+ 66 55	126	14.6	GALAXY SBb
ZC 0800.4+6958	08 00 24.	+ 69 58	1880		CLUSTER OF GALAXIES
MCG-02-21-003	08 00 24.	- 12 11	36	14.	GALAXY
RNGC 2517	08 00 25.	- 12 11		14.0	GALAXY
ZWG 031.029	08 00 30.	+ 03 32		15.4	GALAXY
TON-N 0865	08 00 30.	+ 21 43		16.4	BLUE STAR
ZWG 118.056	08 00 30.	+ 25 14		15.2	GALAXY
MRK 385	08 00 30.	+ 25 14	14	15.5	GALAXY WITH UV CONTINUUM
TON-N 0295	08 00 30.	+ 30 54		15.0	BLUE STAR
ZWG 178.022	08 00 30.	+ 35 29		15.4	GALAXY
ZC 0800.5+3646	08 00 30.	+ 36 46	740		CLUSTER OF GALAXIES
ZWG 287.038	08 00 30.	+ 61 41		15.1	GALAXY
UGC 04196	08 00 30.	+ 61 41	66	15.1	GALAXY Sc
PK245+01.1	08 00 30.	- 27 32	25		PLANETARY NEBULA
HN 0005	08 00 31.	+ 20 20			NEBULA
SEY 038	08 00 31.	+ 72 40 06.		15.2	FAINT GALAXY
ZC 0800.6+5232	08 00 36.	+ 52 32	740		CLUSTER OF GALAXIES
ZWG 263.002	08 00 36.	+ 55 40		15.4	GALAXY
ZWG 262.068	08 00 36.	+ 55 40		15.4	GALAXY
ZWG 031.030	08 00 42.	+ 03 10		15.0	GALAXY
MCG+01-21-008	08 00 42.	+ 03 10	60	15.0	GALAXY
ZWG 031.031	08 00 42.	+ 07 08		15.4	GALAXY
ZWG 031.032	08 00 42.	+ 08 20		15.7	GALAXY
ZWG 059.028	08 00 42.	+ 10 11		14.5	GALAXY
UGC 04197	08 00 42.	+ 10 11	114	14.5	GALAXY Sb
MCG+02-21-011	08 00 42.	+ 10 11	108	14.5	GALAXY
MCG+12-08-026	08 00 42.	+ 73 05	33	16.	GALAXY
MCG-03-21-002	08 00 42.	- 18 34 30.	30	15.	GALAXY
ARC 0616	08 00 45.	+ 46 57		17.1	RICH CLUSTER OF GALAXIES
MCG+09-13-117	08 00 45.	+ 55 42	36	16.	GALAXY
ZWG 031.033	08 00 48.	+ 04 34		15.6	GALAXY
ZWG 148.106	08 00 48.	+ 26 40		15.3	GALAXY
ZWG 207.024	08 00 48.	+ 43 43		15.7	GALAXY
ZC 0800.8+5441	08 00 48.	+ 54 41	470		CLUSTER OF GALAXIES
SHAH 087	08 00 48.	+ 54 56	60	17.0	GROUP OF COMPACT GALAXIES
IC 0491	08 00 51.	+ 26 39 17.			NONSTELLAR OBJECT
ZWG 088.025	08 00 54.	+ 15 22		15.5	GALAXY
ZWG 178.023	08 00 54.	+ 35 00		15.4	GALAXY
ZWG 207.025	08 00 54.	+ 40 20		15.0	GALAXY
ZC 0800.9+4653	08 00 54.	+ 46 53	1010		CLUSTER OF GALAXIES
ZC 0800.9+6835	08 00 54.	+ 68 35	1140		CLUSTER OF GALAXIES
OCL 0702	08 00 54.	- 31 53	120	12.	OPEN STAR CLUSTER
ZWG 031.034	08 01 00.	+ 07 42		15.4	GALAXY
ZWG 059.029	08 01 00.	+ 10 06		15.1	GALAXY
UGC 04198	08 01 00.	+ 10 06	90	15.1	GALAXY S0?
MCG+02-21-012	08 01 00.	+ 10 06	18	15.1	GALAXY
ZWG 059.030	08 01 00.	+ 10 12		15.5	GALAXY
ZWG 148.107	08 01 00.	+ 30 19		15.4	GALAXY
ZWG 287.039	08 01 00.	+ 56 37		15.6	GALAXY
MCG+09-14-001	08 01 00.	+ 56 38	24	17.	GALAXY
MCG+10-13-054	08 01 00.	+ 60 08	78	14.	GALAXY
ZWG 331.028	08 01 00.	+ 73 29		14.6	GALAXY
UGC 04199	08 01 00.	+ 73 29	78	14.6	GALAXY S
KARA.72 152B	08 01 00.	+ 73 29	72	14.6	PART OF DOUBLE GALAXY
ZC 0801.0+7355	08 01 00.	+ 73 55	2490		CLUSTER OF GALAXIES
FIT.G 09	08 01 00.	- 26 03 30.		17.0	SO GALAXY
OCL 0676	08 01 00.	- 26 39	150	13.	OPEN STAR CLUSTER
ZWG 031.035	08 01 06.	+ 04 39		15.6	GALAXY
ZWG 059.031	08 01 06.	+ 10 41		15.7	GALAXY
ZWG 088.026	08 01 06.	+ 15 24		15.4	GALAXY
ZWG 088.027	08 01 06.	+ 19 30		15.6	GALAXY
ZWG 207.026	08 01 06.	+ 40 21		14.6	GALAXY
UGC 04200	08 01 06.	+ 40 21	60	14.6	GALAXY SBb
TON-N 0866	08 01 11.	+ 32 34		16.5	BLUE STAR
ZWG 059.032	08 01 12.	+ 08 50		15.1	GALAXY
KARA.72 153B	08 01 12.	+ 29 37	48		PART OF DOUBLE GALAXY
ZWG 148.108	08 01 12.	+ 29 38		14.6	GALAXY
KARA.72 153A	08 01 12.	+ 29 38	42	14.6	PART OF DOUBLE GALAXY
MCG+07-17-011	08 01 12.	+ 40 20	45	14.5	GALAXY
SEY 039	08 01 16.	+ 73 29 45.		14.5	FAINT GALAXY
ZWG 059.033	08 01 18.	+ 08 40		15.2	GALAXY
ZWG 059.034	08 01 18.	+ 10 09		15.2	GALAXY
TON-N 0867	08 01 18.	+ 23 01		15.5	BLUE STAR
MCG+05-19-038	08 01 18.	+ 29 38	42	14.6	GALAXY
MCG+05-19-037	08 01 18.	+ 29 39	36	14.6	GALAXY
ZC 0801.3+3954	08 01 18.	+ 39 54	6380		CLUSTER OF GALAXIES
ISS 1038	08 01 18.	- 43 38	541		STELLAR RING
ZWG 059.035	08 01 24.	+ 09 08		15.6	GALAXY
ZWG 059.036	08 01 24.	+ 09 57		15.2	GALAXY
UGC 04201	08 01 24.	+ 35 34	72	16.5	GALAXY Sc
SHAH 088	08 01 24.	+ 51 14	78	18.7	GROUP OF COMPACT GALAXIES
UGC 04202	08 01 24.	+ 74 44	60	16.0	GALAXY Sc
OCL 0698	08 01 24.	- 30 45	210	12.	OPEN STAR CLUSTER
VHA 017	08 01 24.	- 30 45	180		OPEN STAR CLUSTER
BC PKS0801+17	08 01 28.0	+ 17 30 03.		16.	QUASI-STELLAR OBJECT
ZWG 031.036	08 01 30.	+ 05 15		14.4	GALAXY
UGC 04203	08 01 30.	+ 05 15	54	14.4	GALAXY
MCG+01-21-009	08 01 30.	+ 05 15	24	14.4	GALAXY
ZWG 031.037	08 01 30.	+ 07 31		15.2	GALAXY
ZWG 059.037	08 01 30.	+ 08 47		15.5	GALAXY
ZWG 059.038	08 01 30.	+ 08 52		15.3	GALAXY
UGC 04204	08 01 30.	+ 56 07	72	17.	GALAXY DWARF
BC B20801+30	08 01 35.4	+ 30 20 48.0		18.	QUASI-STELLAR OBJECT
MCG+08-15-036	08 01 36.	+ 47 11	78	15.	GALAXY
ZWG 236.025	08 01 36.	+ 47 12		15.3	GALAXY
UGC 04205	08 01 36.	+ 47 12	90	15.3	GALAXY Sb
7ZW 209	08 01 36.	+ 62 44			COMPACT GALAXY
OCL 0697	08 01 36.	- 30 30	270	14.	OPEN STAR CLUSTER
ZWG 088.028	08 01 42.	+ 15 37		15.7	GALAXY
MCG+04-19-022	08 01 42.	+ 20 50	90	14.8	GALAXY
ZWG 207.027	08 01 42.	+ 40 19		15.1	GALAXY
ZC 0801.7+4908	08 01 42.	+ 49 08	1610		CLUSTER OF GALAXIES
ZC 0801.7+5431	08 01 42.	+ 54 31	610		CLUSTER OF GALAXIES
ZWG 263.003	08 01 42.	+ 55 27		15.7	GALAXY
ZWG 262.069	08 01 42.	+ 55 27		15.7	GALAXY
UGC 04206	08 01 42.	+ 67 12	96	16.0	GALAXY Sb-c
FIT.G 10	08 01 42.	- 25 28 24.		15.5	SBb GALAXY

OBJECT NAME	RIGHT ASCEN.	DECLINATION	DIAM.	MAGN.	TYPE OF OBJECT
MCG+09-14-002	08 01 45.	+ 55 28 30.	36	17.	GALAXY
LB 00515	08 01 46.	+ 30 24 06.		17.1	FAINT BLUE STAR
ZWG 003.006	08 01 48.	+ 01 20		15.6	GALAXY
ZWG 118.057	08 01 48.	+ 20 50		14.8	GALAXY
UGC 04207	08 01 48.	+ 20 50	102	14.8	GALAXY Sb/Sc?
KARA.73B 0227	08 01 48.	+ 20 50	102	14.8	ISOLATED GALAXY S
TON-N 0868	08 01 48.	+ 21 31		17.0	BLUE STAR
UGC 04208	08 01 48.	+ 25 00	60	16.0	GALAXY Sc
ZC 0801.8+2523	08 01 48.	+ 25 23	4840		CLUSTER OF GALAXIES
MCG+07-17-012	08 01 48.	+ 40 18	42	15.	GALAXY
ZWG 207.028	08 01 48.	+ 40 44		15.3	GALAXY
MCG+12-08-027	08 01 48.	+ 74 44	51	17.	GALAXY
IC 2221	08 01 50.	+ 37 35 42.			NONSTELLAR OBJECT
ZWG 088.029	08 01 54.	+ 15 55		15.6	GALAXY
ZWG 148.109	08 01 54.	+ 29 51		15.5	GALAXY
ZWG 178.024	08 01 54.	+ 37 36		15.3	GALAXY
MCG+08-15-037	08 01 54.	+ 46 50	36	15.	GALAXY
ZWG 236.026	08 01 54.	+ 46 51		14.7	GALAXY
UGC 04209	08 01 54.	+ 46 51	72	14.7	GALAXY E-S0
MCG+11-10-061	08 01 54.	+ 67 12	102	15.	GALAXY
IC 2222	08 01 56.	+ 37 37 05.			NONSTELLAR OBJECT
LBN 0801	08 02	+ 46 30	3300		BRIGHT NEBULA
LBN 0727	08 02	+ 60 40	1560		BRIGHT NEBULA
KHAV 228	08 02	- 29 08	5540		DARK NEBULA
ZWG 118.058	08 02 00.	+ 25 12		15.3	GALAXY
UGC 04210	08 02 00.	+ 25 12	66	15.3	GALAXY Sc
MCG+04-19-023	08 02 00.	+ 25 12	54	15.3	GALAXY
ZWG 207.029	08 02 00.	+ 40 19		15.6	GALAXY
ZC 0802.0+7652	08 02 00.	+ 76 52	1010		CLUSTER OF GALAXIES
ZWG 003.007	08 02 00.	- 02 14		15.7	GALAXY
OCL 0741	08 02 00.	- 46 08	22200	0.9	OPEN STAR CLUSTER
BC 3CR191	08 02 03.76	+ 10 23 57.6		18.40	QUASI-STELLAR OBJECT
SHB 105	08 02 03.9	+ 10 23 57.		18.4	QUASI-STELLAR OBJECT
ZWG 059.039	08 02 06.	+ 10 55		14.9	GALAXY
UGC 04211	08 02 06.	+ 10 55	84	14.9	GALAXY S-IRR
MCG+02-21-013	08 02 06.	+ 10 55	60	14.9	GALAXY
ZWG 148.110	08 02 06.	+ 30 53		15.6	GALAXY
MCG+05-19-039	08 02 06.	+ 30 53	24	15.6	GALAXY
TON-N 0296	08 02 06.	+ 33 03		15.4	BLUE STAR
ZWG 178.025	08 02 06.	+ 37 42		15.2	GALAXY
ARC 0620	08 02 10.	+ 45 50		17.4	RICH CLUSTER OF GALAXIES
ZWG 031.038	08 02 12.	+ 02 39		15.6	GALAXY
ZWG 059.040	08 02 12.	+ 10 59		15.1	GALAXY
MCG+02-21-014	08 02 12.	+ 10 59	18	15.1	GALAXY
ZWG 088.030	08 02 12.	+ 14 57		15.6	GALAXY
TON-N 0869	08 02 12.	+ 21 53		16.5	BLUE STAR
ZWG 178.026	08 02 12.	+ 36 05		15.0	GALAXY
MCG+06-18-012	08 02 12.	+ 36 05 30.	51	15.	GALAXY
ZWG 178.027	08 02 12.	+ 37 41		15.7	GALAXY
MCG+06-18-010	08 02 12.	+ 37 41 30.	30	16.	GALAXY
MCG+06-18-011	08 02 12.	+ 37 42 30.	15	15.5	GALAXY
ZC 0802.2+4547	08 02 12.	+ 45 47	1010		CLUSTER OF GALAXIES
MCG+10-12-072	08 02 12.	+ 61 12	24	15.	GALAXY
ISS 0847	08 02 12.	- 22 30	223		STELLAR RING
LB 00516	08 02 15.	+ 30 34 30.		17.4	FAINT BLUE STAR
ZWG 031.039	08 02 18.	+ 05 12		15.3	GALAXY
ZWG 263.004	08 02 18.	+ 53 18		15.5	GALAXY
ZWG 262.070	08 02 18.	+ 53 18		15.5	GALAXY
ZWG 287.040	08 02 18.	+ 61 13		15.4	GALAXY
ACK 251-01.1	08 02 18.	- 34 08			PLANETARY NEBULA
SHAH 089	08 02 24.	+ 50 29	54	18.5	GROUP OF COMPACT GALAXIES
IC 2223	08 02 28.	+ 37 36 31.			NONSTELLAR OBJECT
MCG+04-19-024	08 02 30.	+ 26 20	66	14.6	GALAXY
TON-N 0297	08 02 30.	+ 31 14		14.7	BLUE STAR
MCG+08-15-038	08 02 30.	+ 45 57 30.	42	16.	GALAXY
ZWG 236.027	08 02 30.	+ 46 00		15.7	GALAXY
SHAH 090	08 02 30.	+ 55 19	150	17.5	GROUP OF COMPACT GALAXIES
7ZW 210	08 02 30.	+ 62 36			COMPACT GALAXY
LDN 1670	08 02 30.	- 31 30	780		DARK NEBULA
LB 00517	08 02 31.	+ 30 20 06.		18.2	FAINT BLUE STAR
IC 2224	08 02 32.	+ 37 36 25.			NONSTELLAR OBJECT
MAI 025	08 02 35.	+ 73 19	33		DWARF SPHEROIDAL GALAXY
ZWG 118.059	08 02 36.	+ 26 18		14.6	GALAXY
UGC 04212	08 02 36.	+ 26 18	66	14.6	GALAXY SBb
ZWG 148.111	08 02 36.	+ 30 22		15.5	GALAXY
MCG+09-14-003	08 02 36.	+ 51 05	54	15.	GALAXY
ZWG 263.005	08 02 36.	+ 51 07		15.5	GALAXY
UGC 04213	08 02 36.	+ 51 07	66	15.5	GALAXY Sc
MCG+09-14-004	08 02 36.	+ 56 37	30	17.	GALAXY
MCG+10-12-073	08 02 36.	+ 62 11	27	16.	GALAXY
ZC 0802.6-0104	08 02 36.	- 01 04	3290		CLUSTER OF GALAXIES
FIT.G 12	08 02 36.	- 26 49 24.		18.0	SO GALAXY
IC 0492	08 02 37.	+ 26 18 29.			NONSTELLAR OBJECT
TON-N 0298	08 02 42.	+ 33 54		16.5	BLUE STAR
ZWG 263.006	08 02 42.	+ 55 17		15.4	GALAXY
ZWG 262.071	08 02 42.	+ 55 17		15.4	GALAXY
UGC 04214	08 02 42.	+ 55 17	84	15.4	GALAXY Sb
MCG+10-12-074	08 02 42.	+ 62 14	42	16.	GALAXY
ZWG 310.035	08 02 42.	+ 67 42		15.4	GALAXY
FIT.G 16	08 02 42.	- 25 50 12.		18.5	Sa GALAXY
LB 00518	08 02 44.	+ 31 30 42.		16.8	FAINT BLUE STAR
ZWG 059.041	08 02 48.	+ 10 32		15.1	GALAXY
UGC 04215	08 02 48.	+ 10 32	72	15.1	GALAXY S
MCG+02-21-015	08 02 48.	+ 10 32	60	15.1	GALAXY
ZWG 148.112	08 02 48.	+ 30 30		15.7	GALAXY
ZC 0802.8+3827	08 02 48.	+ 38 27	1140		CLUSTER OF GALAXIES
MCG+09-14-005	08 02 48.	+ 55 17 30.	60	16.	GALAXY
FIT.G 14	08 02 48.	- 25 52 54.		18.5	Sa GALAXY
WRAY 19.04	08 02 53.3	- 27 52 51.			DIFFUSE NEBULA
ZWG 031.040	08 02 54.	+ 07 03		15.6	GALAXY
ZWG 059.042	08 02 54.	+ 10 51		15.2	GALAXY
ZC 0802.9+4714	08 02 54.	+ 47 14	1140		CLUSTER OF GALAXIES
ISS 0848	08 02 54.	- 26 33	379		STELLAR RING
VHE 03A	08 02 54.	- 39 11	78		REFLECTION NEBULA
LB 00519	08 02 59.	+ 28 04 48.		16.2	FAINT BLUE STAR
LBN 0733	08 03	+ 59 50	1260		BRIGHT NEBULA
LBN 0723	08 03	+ 61 30	1980		BRIGHT NEBULA
ZWG 031.041	08 03 00.	+ 07 44		15.2	GALAXY
MCG+01-21-010	08 03 00.	+ 07 44	24	15.2	GALAXY
TON-N 0299	08 03 00.	+ 32 41		14.8	BLUE STAR
ZC 0803.0+4307	08 03 00.	+ 43 07	3230		CLUSTER OF GALAXIES
ZWG 263.007	08 03 00.	+ 56 02		15.6	GALAXY
ZWG 262.072	08 03 00.	+ 56 02		15.6	GALAXY
KARA.73B 0228	08 03 00.	+ 56 02	18	15.6	ISOLATED GALAXY E
MCG+11-10-062	08 03 00.	+ 67 54	39	16.	GALAXY
FIT.G 15	08 03 00.	- 25 53 36.		18.5	E2 GALAXY
OCL 0703	08 03 00.	- 31 39	120	13.	OPEN STAR CLUSTER
VHE 03B	08 03 00.	- 39 12	48		REFLECTION NEBULA
IC 2226	08 03 04.	+ 12 41 36.			NONSTELLAR OBJECT
ZWG 031.042	08 03 06.	+ 05 28		15.7	GALAXY
ZWG 059.043	08 03 06.	+ 12 37		15.3	GALAXY
UGC 04216	08 03 06.	+ 12 37	72	15.3	GALAXY Sa-b
OCL 0705	08 03 06.	- 31 49	150	13.	OPEN STAR CLUSTER
VHE 03C	08 03 06.	- 39 10			REFLECTION NEBULA
ARC 0623	08 03 09.	- 00 49		16.9	RICH CLUSTER OF GALAXIES
RNGC 2527	08 03 10.	- 28 01		8.0	OPEN CLUSTER
IC 2225	08 03 11.	+ 36 05 08.			NONSTELLAR OBJECT
UGC 04217	08 03 12.	+ 06 13	78	16.0	GALAXY S
ZWG 031.043	08 03 12.	+ 06 33		15.6	GALAXY
ZWG 118.060	08 03 12.	+ 21 08		15.5	GALAXY
ZC 0803.2+5012	08 03 12.	+ 50 12	2620		CLUSTER OF GALAXIES
MCG+10-12-075	08 03 12.	+ 61 35	24	16.	GALAXY
ZWG 003.008	08 03 12.	- 03 25		15.5	GALAXY
OCL 0685	08 03 12.	- 28 01	1350	8.5	OPEN STAR CLUSTER
VHE 02	08 03 12.	- 31 22	78		REFLECTION NEBULA
RNGC 2522	08 03 16.	+ 17 51		14.5	GALAXY
ZWG 088.031	08 03 18.	+ 17 51		14.4	GALAXY
UGC 04218	08 03 18.	+ 17 51	72	14.4	GALAXY S0-a
ZWG 207.030	08 03 18.	+ 39 14		15.5	GALAXY
UGC 04219	08 03 18.	+ 39 14	162	15.5	GALAXY Sb
SHAH 091	08 03 18.	+ 51 36	60	17.0	GROUP OF COMPACT GALAXIES
ZWG 263.008	08 03 18.	+ 54 46		15.7	GALAXY
ZWG 262.073	08 03 18.	+ 54 46		15.7	GALAXY
MCG-02-21-004	08 03 18.	- 11 17	144	12.5	GALAXY
RNGC 2525	08 03 19.	- 11 17		12.5	GALAXY
ZWG 059.044	08 03 24.	+ 12 41		14.9	GALAXY
UGC 04220	08 03 24.	+ 12 41	78	14.9	GALAXY Sb
MCG+02-21-016	08 03 24.	+ 12 41	60	14.9	GALAXY
MCG+03-21-014	08 03 24.	+ 17 51 30.	54	14.4	GALAXY
ZWG 148.113	08 03 24.	+ 27 02		15.5	GALAXY
LB 00520	08 03 24.	+ 29 02 42.		17.4	FAINT BLUE STAR
MCG+08-15-039	08 03 24.	+ 46 17	42	16.	GALAXY
MCG+08-15-040	08 03 24.	+ 46 40	48	16.	GALAXY
MCG+09-14-006	08 03 24.	+ 51 15	24	13.	GALAXY
FIT.G 11	08 03 24.	- 26 45 18.		18.5	Sb GALAXY
ZWG 059.045	08 03 30.	+ 11 56		15.2	GALAXY
KARA.73B 0229	08 03 30.	+ 11 56	24	15.2	ISOLATED GALAXY IR
MCG+07-17-013	08 03 30.	+ 39 13	108	15.	GALAXY
MCG+08-15-041	08 03 30.	+ 45 50	15	16.	GALAXY
ZWG 263.009	08 03 30.	+ 51 17		14.2	GALAXY
UGC 04221	08 03 30.	+ 51 17	78	14.2	GALAXY E-S0
RNGC 2518	08 03 32.	+ 51 17		14.0	GALAXY
LB 00521	08 03 34.	+ 31 25 00.		16.7	FAINT BLUE STAR
LB 00522	08 03 39.	+ 78 07 48.		17.3	FAINT BLUE STAR
ZWG 031.044	08 03 42.	+ 05 46		15.4	GALAXY
UGC 04222	08 03 42.	+ 05 46	60	15.4	GALAXY SBb
MCG+09-14-007	08 03 42.	+ 56 35	42	17.	GALAXY
UGC 04223	08 03 42.	+ 62 55	72	16.0	GALAXY Sc
7ZW 211	08 03 42.	+ 64 32			COMPACT GALAXY
ZWG 003.009	08 03 42.	- 03 08		15.1	GALAXY
MCG+00-21-003	08 03 42.	- 03 08	48	15.1	GALAXY
SEY 040	08 03 44.	+ 72 11 17.		15.3	FAINT GALAXY
IC 0494	08 03 46.	+ 01 11			NONSTELLAR OBJECT
TON-N 0870	08 03 47.	+ 32 12		16.8	BLUE STAR
ZWG 003.010	08 03 48.	+ 01 11		14.3	GALAXY
UGC 04224	08 03 48.	+ 01 11	84	14.3	GALAXY S0
MCG+00-21-004	08 03 48.	+ 01 11	36	14.3	GALAXY
MCG+04-19-025	08 03 48.	+ 23 00	63	15.3	GALAXY
TON-N 0300	08 03 48.	+ 33 33		15.3	BLUE STAR
ZWG 178.028	08 03 48.	+ 36 23		14.8	GALAXY
ZWG 310.036	08 03 48.	+ 64 32		15.3	GALAXY
MCG+11-10-063	08 03 48.	+ 64 32	36	15.	GALAXY
FIT.G 17	08 03 48.	- 25 50 12.		16.0	Sa GALAXY
ARC 0622	08 03 49.	+ 48 10		17.4	RICH CLUSTER OF GALAXIES
ZWG 031.045	08 03 54.	+ 04 42		15.4	GALAXY
ZWG 118.061	08 03 54.	+ 22 59		15.3	GALAXY
UGC 04225	08 03 54.	+ 22 59	72	15.3	GALAXY Sa-b
ZWG 207.031	08 03 54.	+ 40 33		15.2	GALAXY
UGC 04226	08 03 54.	+ 40 33	126	15.2	GALAXY Sc
ZC 0803.9+5327	08 03 54.	+ 53 27	3760		CLUSTER OF GALAXIES
ZWG 287.041	08 03 54.	+ 56 50		15.5	GALAXY
FIT.G 13	08 03 54.	- 27 16 06.		14.5	E2 GALAXY
LBN 0732	08 04	+ 60 10	900		BRIGHT NEBULA
ZC 0804.0+1627	08 04 00.	+ 16 27	4910		CLUSTER OF GALAXIES
ZWG 088.032	08 04 00.	+ 18 53		15.0	GALAXY
MCG+03-21-015	08 04 00.	+ 18 53	36	15.0	GALAXY
ZWG 088.033	08 04 00.	+ 19 59		15.4	GALAXY
MCG+03-21-016	08 04 00.	+ 20 00	42	15.4	GALAXY
ZWG 148.114	08 04 00.	+ 27 16		15.2	GALAXY
ZWG 207.032	08 04 00.	+ 39 20		13.5	GALAXY
UGC 04227	08 04 00.	+ 39 20	90	13.5	GALAXY Sb
MCG+10-12-076	08 04 00.	+ 56 50	45	16.	GALAXY
LDN 1671	08 04 00.	- 32 00	1500		DARK NEBULA
RNGC 2528	08 04 02.	+ 39 21		13.5	GALAXY
MCG+07-17-014	08 04 03.	+ 40 32	84	14.5	GALAXY
ARC 0618	08 04 04.	+ 67 42		17.3	RICH CLUSTER OF GALAXIES
IC 2227	08 04 05.	+ 36 10 26.			NONSTELLAR OBJECT
ZWG 031.046	08 04 06.	+ 05 27		13.9	GALAXY
UGC 04228	08 04 06.	+ 05 27	168	13.9	GALAXY S0
MCG+01-21-011	08 04 06.	+ 05 27	60	13.9	GALAXY
MRK 621	08 04 06.	+ 15 45	11	16.5	GALAXY WITH UV CONTINUUM
ZWG 118.062	08 04 06.	+ 24 31		15.3	GALAXY
SHAH 092	08 04 06.	+ 55 10	60	17.5	GROUP OF COMPACT GALAXIES
LB 00523	08 04 07.	+ 28 14 48.		16.7	FAINT BLUE STAR
MCG+07-17-015	08 04 09.	+ 39 19	90	14.	GALAXY
ZWG 088.034	08 04 12.	+ 18 54		15.2	GALAXY
MRK 622	08 04 12.	+ 36 09	14	15.	GALAXY WITH UV CONTINUUM
ZWG 207.033	08 04 12.	+ 39 09		14.4	GALAXY
UGC 04229	08 04 12.	+ 39 09	36	14.4	GALAXY
UGC 04230	08 04 12.	+ 55 32	78	15.5	GALAXY S
ZWG 263.010	08 04 12.	+ 55 39		15.5	GALAXY
ZWG 262.074	08 04 12.	+ 55 39		15.5	GALAXY
MCG+09-14-008	08 04 12.	+ 55 40	90	15.	GALAXY
MCG+11-10-064	08 04 12.	+ 64 43	45	15.	GALAXY
PK224+15.1	08 04 12.	- 02 44	197	15.4	PLANETARY NEBULA
RNGC 2526	08 04 15.	+ 08 09		14.5	GALAXY
MCG+03-21-017	08 04 15.	+ 18 55	30	15.2	GALAXY
ZWG 031.047	08 04 18.	+ 08 09		14.6	GALAXY
UGC 04231	08 04 18.	+ 08 09	60	14.6	GALAXY
KARA.72 154A	08 04 18.	+ 08 09	72	14.6	PART OF DOUBLE GALAXY
MCG+01-21-012	08 04 18.	+ 08 09	48	14.6	GALAXY
TON-N 0871	08 04 18.	+ 21 39		16.7	BLUE STAR
ZWG 118.063	08 04 18.	+ 24 42		15.7	GALAXY
MCG+04-19-026	08 04 18.	+ 25 18	33	14.8	GALAXY
MCG+08-15-042	08 04 18.	+ 46 02	15	16.	GALAXY
ZWG 310.037	08 04 18.	+ 64 43		15.6	GALAXY

OBJFCT NAME	RIGHT ASCEN.	DECLINATION	DIAM.	MAGN.	TYPE OF OBJECT
LB 00524	08 04 19.	+ 30 43 42.		16.3	FAINT BLUE STAP
ZWG 031.048	08 04 24.	+ 08 11		15.3	GALAXY
KARA.72 154B	08 04 24.	+ 08 11	42	15.3	PART OF DOUBLE GALAXY
MCG+01-21-013	08 04 24.	+ 08 11	24	15.3	GALAXY
ZWG 088.035	08 04 24.	+ 15 05		15.3	GALAXY
MCG+03-21-018	08 04 24.	+ 15 05	78	15.3	GALAXY
ZWG 118.064	08 04 24.	+ 25 16		14.8	GALAXY
IC 0493	08 04 25.	+ 25 14 28.			NONSTELLAR OBJFCT
IC 2228	08 04 26.	+ 08 12			NONSTELLAR OBJFCT
ZWG 088.036	08 04 30.	+ 17 54		15.1	GALAXY
UGC 04232	08 04 30.	+ 17 54	60	15.1	GALAXY Sb-c
MCG+03-21-019	08 04 30.	+ 17 54	60	15.1	GALAXY
MCG-05-20-001	08 04 30.	- 27 46	60	16.	GALAXY
ZWG 031.049	08 04 36.	+ 05 49		15.7	GALAXY
ZWG 059.046	08 04 36.	+ 09 00		15.1	GALAXY
UGC 04233	08 04 36.	+ 09 00	72	15.1	GALAXY S
FIT.G 18	08 04 36.	- 25 40 36.		16.0	SO GALAXY
ZWG 031.050	08 04 42.	+ 04 36		15.2	GALAXY
ZWG 207.034	08 04 42.	+ 39 18		13.7	GALAXY
UGC 04234	08 04 42.	+ 39 18	90	13.7	GALAXY S0-a
7ZW 212	08 04 42.	+ 57 55			COMPACT GALAXY
ZWG 287.042	08 04 42.	+ 57 55		14.2	GALAXY
UGC 04235	08 04 42.	+ 57 55	72	14.2	GALAXY E?
7ZW 213	08 04 42.	+ 68 14			COMPACT GALAXY
ZWG 349.017	08 04 42.	+ 79 31		15.1	GALAXY
MCG-04-20-001	08 04 42.	- 25 51 30.	12	15.5	GALAXY
RNGC 2524	08 04 44.	+ 39 18		13.5	GALAXY
RNGC 2521	08 04 46.	+ 57 55		14.0	GALAXY
ZWG 031.051	08 04 48.	+ 07 00		15.4	GALAXY
ZC 0804.8+1959	08 04 48.	+ 19 59	6920		CLUSTER OF GALAXIES
ZWG 178.029	08 04 48.	+ 37 55		15.6	GALAXY
ZC 0804.8+4856	08 04 48.	+ 48 56	1010		CLUSTER OF GALAXIES
ZWG 287.043	08 04 48.	+ 56 44		15.6	GALAXY
MCG+09-14-009	08 04 48.	+ 56 44	36	17.	GALAXY
MCG+10-12-077	08 04 48.	+ 57 55	30	14.	GALAXY
ZWG 031.052	08 04 54.	+ 04 39		14.9	GALAXY
MCG+01-21-014	08 04 54.	+ 04 39	24	14.9	GALAXY
ZWG 031.053	08 04 54.	+ 05 54		15.6	GALAXY
ZWG 088.037	08 04 54.	+ 15 05		15.3	GALAXY
ZWG 118.065	08 04 54.	+ 26 10		15.7	GALAXY
UGC 04236	08 04 54.	+ 26 10	66	15.7	GALAXY S IV
MCG+04-19-027	08 04 54.	+ 26 10	48	15.7	GALAXY
ZWG 148.115	08 04 54.	+ 30 05		15.6	GALAXY
MCG+07-17-016	08 04 54.	+ 39 17	30	14.	GALAXY
7ZW 214	08 04 54.	+ 61 31			COMPACT GALAXY
RNGC 2529	08 04 58.	+ 17 58		14.5	GALAXY
RNGC 2533	08 04 59.	- 29 45		10.0	OPEN CLUSTER
KHAV 229	08 05	- 24 51	6990		DARK NFBULA
ZWG 088.038	08 05 00.	+ 17 58		14.7	GALAXY
UGC 04237	08 05 00.	+ 17 58	96	14.7	GALAXY SBc
ZWG 088.039	08 05 00.	+ 19 03		15.5	GALAXY
ZWG 207.035	08 05 00.	+ 40 07		14.9	GALAXY
ZWG 207.036	08 05 00.	+ 41 44		15.7	GALAXY
ZWG 310.038	08 05 00.	+ 63 44		15.6	GALAXY
MCG+12-08-028	08 05 00.	+ 68 58	33	16.	GALAXY
MCG+12-08-029	08 05 00.	+ 73 29	54	15.	GALAXY
MCG+13-06-018	08 05 00.	+ 76 34	96	13.	GALAXY
ZWG 349.018	08 05 00.	+ 76 35		13.5	GALAXY
UGC 04238	08 05 00.	+ 76 35	168	13.5	GALAXY SBc
OCL 0695	08 05 00.	- 29 45	540	10.1	OPEN STAR CLUSTER
VHA 018	08 05 00.	- 29 45	360		OPEN STAR CLUSTER
MCG+03-21-020	08 05 06.	+ 17 58	84	14.7	GALAXY
ZC 0805.1+2658	08 05 06.	+ 26 58	1010		CLUSTER OF GALAXIES
TON-N 0301	08 05 06.	+ 32 12		13.9	BLUE STAR
MCG+09-14-010	08 05 06.	+ 55 36	24	16.	GALAXY
VHA 019	08 05 06.	- 32 14	150		OPEN STAR CLUSTER
MCG+07-17-017	08 05 09.	+ 40 06	27	15.	GALAXY
MCG+10-12-078	08 05 12.	+ 57 50	51	14.	GALAXY
ZWG 287.044	08 05 12.	+ 57 59		15.3	GALAXY
ZWG 310.039	08 05 12.	+ 63 24		15.7	GALAXY
KARA.73B 0230	08 05 12.	+ 63 24	42	15.7	ISOLATED GALAXY S
ZC 0805.2+6751	08 05 12.	+ 67 51	3290		CLUSTER OF GALAXIES
MCG+11-10-065	08 05 15.	+ 63 24	45	15.	GALAXY
ZWG 031.054	08 05 18.	+ 05 21		15.2	GALAXY
UGC 04239	08 05 18.	+ 05 21	66	15.2	GALAXY S
MCG+03-21-021	08 05 18.	+ 14 57 30.	90	15.4	GALAXY
ZWG 088.040	08 05 18.	+ 14 59		15.4	GALAXY
UGC 04240	08 05 18.	+ 14 59	90	15.4	GALAXY S
ZWG 287.045	08 05 18.	+ 57 54		14.8	GALAXY
UGC 04241	08 05 18.	+ 57 54	78	14.8	GALAXY Sa-b
7ZW 215	08 05 18.	+ 57 59			COMPACT GALAXY
BC 4C05.34	08 05 19.17	+ 04 41 20.5		18.16	QUASI-STELLAR OBJECT
SHB 106	08 05 19.2	+ 04 41 20.		18.2	QUASI-STELLAR OBJECT
MCG+10-12-079	08 05 21.	+ 57 59 30.	18	15.	GALAXY
ZC 0805.4+4654	08 05 24.	+ 46 54	870		CLUSTER OF GALAXIES
MCG+10-12-080	08 05 24.	+ 57 55	57	14.	GALAXY
MCG+11-10-066	08 05 24.	+ 67 34	60	13.	GALAXY
MRK 014	08 05 24.	+ 72 56	20	15.5	GALAXY WITH UV CONTINUUM
ZWG 331.029	08 05 24.	+ 72 57		14.4	GALAXY
UGC 04242	08 05 24.	+ 72 57	72	14.4	GALAXY
ZWG 031.055	08 05 30.	+ 04 30		15.5	GALAXY
TON-N 0872	08 05 30.	+ 21 23		16.9	BLUE STAR
MCG-03-21-003	08 05 30.	- 18 35	36	16.5	GALAXY
VHE 05	08 05 30.	- 35 48	36		REFLECTION NEBULA
ZWG 031.056	08 05 36.	+ 05 21		15.7	GALAXY
ZWG 059.047	08 05 36.	+ 09 09		15.4	GALAXY
IC 0495	08 05 36.	+ 09 09 43.			NONSTELLAR OBJFCT
ZWG 088.041	08 05 36.	+ 19 21		15.7	GALAXY
ZWG 148.116	08 05 36.	+ 28 13		15.4	GALAXY
ZWG 310.040	08 05 36.	+ 67 23		14.5	GALAXY
UGC 04243	08 05 36.	+ 67 23	90	14.5	GALAXY Sa-b
ZWG 059.048	08 05 42.	+ 11 19		15.1	GALAXY
UGC 04244	08 05 42.	+ 11 19	60	15.1	GALAXY Sa-b
ZWG 088.042	08 05 42.	+ 19 35		15.7	GALAXY
TON-N 0873	08 05 42.	+ 21 14		16.9	BLUE STAR
TON-N 0005	08 05 42.	+ 28 25		15.1	BLUE STAR
ZWG 178.030	08 05 42.	+ 36 50		15.3	GALAXY
ZWG 287.046	08 05 42.	+ 58 04		14.9	GALAZY
SEY 041	08 05 42.	+ 72 56 40.		15.3	FAINT GALAXY
ZWG 088.043	08 05 48.	+ 18 14		15.3	GALAXY
MCG+03-21-022	08 05 48.	+ 18 15	45	15.3	GALAXY
ZWG 236.028	08 05 48.	+ 45 38		15.6	GALAXY
MCG+10-12-081	08 05 48.	+ 58 04	18	15.	GALAXY
TON-N 0874	08 05 53.	+ 35 40		15.	BLUE STAR
ZWG 031.057	08 05 54.	+ 02 45		15.2	GALAXY
ZC 0805.9+0728	08 05 54.	+ 07 28	2220		CLUSTER OF GALAXIES
ZC 0805.9+1452	08 05 54.	+ 14 52	1340		CLUSTER OF GALAXIES
ZWG 088.044	08 05 54.	+ 18 20		14.9	GALAXY

OBJECT NAME	RIGHT ASCEN.	DECLINATION	DIAM.	MAGN.	TYPE OF OBJECT
UGC 04245	08 05 54.	+ 18 20	96	14.9	GALAXY SBb
MCG+03-21-023	08 05 54.	+ 18 21	90	14.9	GALAXY
ZWG 207.037	08 05 54.	+ 44 05		15.5	GALAXY
UGC 04246	08 05 54.	+ 44 05	78	15.5	GALAXY Sa-b
KARA.73B 0231	08 05 54.	+ 44 05	90	15.5	ISOLATED GALAXY S
ZC 0805.9+5044	08 05 54.	+ 50 44	540		CLUSTER OF GALAXIES
MCG+11-10-067	08 05 54.	+ 67 37	54	15.	GALAXY
ISS 0912	08 05 54.	- 33 10	710		STELLAR RING
LB 00525	08 05 55.	+ 28 03 36.		17.5	FAINT BLUE STAR
MCG+07-17-018	08 06 00.	+ 44 05	66	16.	GALAXY
MCG+08-15-043	08 06 00.	+ 47 28	42	16.	GALAXY
ZWG 236.029	08 06 00.	+ 47 30		15.5	GALAXY
ZWG 287.047	08 06 00.	+ 57 59		15.1	GALAXY
MCG+10-12-082	08 06 00.	+ 58 00	39	15.	GALAXY
ZC 0806.0+7013	08 06 00.	+ 70 13	740		CLUSTER OF GALAXIES
ARC 0621	08 06 04.	+ 70 11		17.3	RICH CLUSTER OF GALAXIES
TON-N 0875	08 06 06.	+ 22 33		15.3	BLUE STAR
ZWG 148.117	08 06 06.	+ 28 09		15.3	GALAXY
ZWG 207.038	08 06 06.	+ 41 45		15.5	GALAXY
ARC 0627	08 06 08.	+ 34 53		17.4	RICH CLUSTER OF GALAXIES
ZWG 003.011	08 06 12.	+ 01 46		15.4	GALAXY
ZWG 088.045	08 06 12.	+ 16 49		15.5	GALAXY
UGC 04247	08 06 12.	+ 16 49	72	15.5	GALAXY SB?c
ZC 0806.2+3455	08 06 12.	+ 34 55	1340		CLUSTER OF GALAXIES
MCG+07-17-019	08 06 12.	+ 41 44	54	16.	GALAXY
HOLM 093A	08 06 13.	+ 41 44	30	14.9	PART OF MULTIPLE GALAXY
HOLM 093B	08 06 16.	+ 41 44	12	15.1	PART OF MULTIPLE GALAXY
ZWG 003.012	08 06 18.	+ 00 27		14.9	GALAXY
UGC 04248	08 06 18.	+ 00 27	78	14.9	GALAXY Sa/SBa
MCG+00-21-005	08 06 18.	+ 00 27	66	14.9	GALAXY
ZWG 059.049	08 06 18.	+ 11 30		15.1	GALAXY
ZC 0806.3+4415	08 06 18.	+ 44 15	1080		CLUSTER OF GALAXIES
SV 194	08 06 22.16	+ 42 37 45.0		17.72	BLUE STELLAR OBJECT
ZWG 088.046	08 06 24.	+ 17 08		14.9	GALAXY
UGC 04249	08 06 24.	+ 17 08	60	14.9	GALAXY S
MCG+03-21-024	08 06 24.	+ 17 08	54	14.9	GALAXY
ZWG 088.047	08 06 24.	+ 19 38		15.6	GALAXY
MCG-03-21-004	08 06 27.	- 19 05 30.	42	14.5	GALAXY
ZWG 031.058	08 06 30.	+ 05 10		15.5	GALAXY
ZC 0806.5+2822	08 06 30.	+ 28 22	740		CLUSTER OF GALAXIES
ZWG 236.030	08 06 30.	+ 46 20		15.4	GALAXY
UGC 04250	08 06 30.	+ 46 20	66	15.4	GALAXY Sc
ACK 238+07.2	08 06 30.	- 19 05			PLANETARY NEBULA
ACK 246+02.1	08 06 30.	- 27 32			PLANETARY NEBULA
TON-N 0876	08 06 31.	+ 19 35		16.5	BLUE STAR
ZWG 003.013	08 06 36.	+ 00 25		14.7	GALAXY
UGC 04251	08 06 36.	+ 00 25	66	14.7	GALAXY E-S0
MCG+00-21-006	08 06 36.	+ 00 25	54	14.7	GALAXY
ZWG 207.039	08 06 36.	+ 40 15		15.4	GALAXY
UGC 04252	08 06 36.	+ 40 15	126	15.4	GALAXY Sc
ZC 0806.6+4115	08 06 36.	+ 41 15	940		CLUSTER OF GALAXIES
MCG+08-15-044	08 06 36.	+ 46 20	78	15.	GALAXY
TON-N 0877	08 06 42.	+ 21 31		17.	BLUE STAR
ZWG 119.001	08 06 42.	+ 26 01		14.7	GALAXY
ZWG 118.066	08 06 42.	+ 26 01		14.7	GALAXY
KARA.72 155B	08 06 42.	+ 26 01	24		PART OF DOUBLE GALAXY
KARA.72 155A	08 06 42.	+ 26 01	30	14.7	PART OF DOUBLE GALAXY
MCG+04-19-028	08 06 42.	+ 26 01	18	14.7	GALAXY
ZC 0806.7+3831	08 06 42.	+ 38 31	870		CLUSTER OF GALAXIES
MCG+07-17-020	08 06 42.	+ 40 15	90	15.	GALAXY
ZWG 003.014	08 06 42.	- 00 13		14.5	GALAXY
UGC 04253	08 06 42.	- 00 13	84	14.5	GALAXY Sb-c
MCG+00-21-007	08 06 42.	- 00 13	102	14.5	GALAXY
IC 2229	08 06 43.	+ 26 01 31.			NONSTELLAR OBJECT
IC 0496	08 06 43.	+ 26 01 37.			NONSTELLAR OBJFCT
ZWG 003.015	08 06 48.	+ 00 45		14.4	GALAXY
UGC 04254	08 06 48.	+ 00 45	102	14.4	GALAXY S-IRR
MCG+00-21-008	08 06 48.	+ 00 45	42	14.4	GALAXY
ZC 0806.8+0514	08 06 48.	+ 05 14	4370		CLUSTER OF GALAXIES
ZWG 031.059	08 06 48.	+ 05 25		14.6	GALAXY
UGC 04255	08 06 48.	+ 05 25	84	14.6	GALAXY S
MCG+01-21-015	08 06 48.	+ 05 25	60	14.6	GALAXY
TON-N 0878	08 06 48.	+ 22 46		16.3	BLUE STAR
ZWG 178.031	08 06 48.	+ 38 10		15.4	GALAXY
TON-N 0879	08 06 49.	+ 19 20		16.8	BLUE STAR
MCG-03-21-005	08 06 51.	- 18 34	36	15.	GALAXY
ARC 0628	08 06 52.	+ 35 23		15.9	RICH CLUSTER OF GALAXIES
IC 0498	08 06 53.	+ 05 25 46.			NONSTELLAR OBJECT
ZC 0806.9+4223	08 06 54.	+ 42 23	3020		CLUSTFR OF GALAXIES
ACK 238+07.1	08 06 54.	- 18 33			PLANETARY NEBULA
VDB.66G 049	08 07	+ 46 39	70		DWARF GALAXY
KHAV 230	08 07	- 26 51	13020		DARK NFBULA
ZC 0807.0+0504	08 07 00.	+ 05 04	740		CLUSTER OF GALAXIES
ZWG 119.002	08 07 00.	+ 25 03		15.2	GALAXY
ZWG 118.067	08 07 00.	+ 25 03		15.2	GALAXY
MCG+04-20-001	08 07 00.	+ 25 04	42	15.2	GALAXY
ZWG 178.032	08 07 00.	+ 34 06		12.9	GALAXY
RNGC 2532	08 07 00.	+ 34 06		13.0	GALAXY
UGC 04256	08 07 00.	+ 34 06	132	12.9	GALAXY Sc
MCG+06-18-013	08 07 00.	+ 34 06	108	13.5	GALAXY
KARA.73B 0232	08 07 00.	+ 34 06	132	12.9	ISOLATED GALAXY S
ZWG 207.040	08 07 00.	+ 42 25		15.6	GALAXY
ARC 0626	08 07 00.	+ 49 22		17.1	RICH CLUSTER OF GALAXIES
MCG+10-12-083	08 07 00.	+ 56 48	30	16.	GALAXY
MCG+10-12-084	08 07 00.	+ 56 55	45	16.	GALAXY
PK250+00.1	08 07 00.	- 32 31	40	18.1	PLANETARY NEBULA
ARC 0632	08 07 02.	+ 05 05		17.2	RICH CLUSTER OF GALAXIES
ZWG 119.003	08 07 06.	+ 23 00		15.5	GALAXY
ZWG 119.004	08 07 06.	+ 25 00		15.6	GALAXY
ZWG 118.068	08 07 06.	+ 25 00		15.6	GALAXY
ZWG 119.005	08 07 06.	+ 25 01		15.4	GALAXY
ZWG 118.069	08 07 06.	+ 25 01		15.4	GALAXY
UGC 04257	08 07 06.	+ 25 01	138	15.4	GALAXY Sc
MCG+04-20-002	08 07 06.	+ 25 01	120	15.6	GALAXY
MCG+04-20-003	08 07 06.	+ 25 02	27	15.4	GALAXY
IC 0497	08 07 06.	+ 25 02 58.			NONSTELLAR OBJECT
ZWG 207.041	08 07 06.	+ 39 17		15.5	GALAXY
ZC 0807.1+4918	08 07 06.	+ 49 18	2290		CLUSTER OF GALAXIES
MCG+10-12-085	08 07 06.	+ 57 10	36	15.	GALAXY
ZWG 287.048	08 07 06.	+ 57 11		15.2	GALAXY
MCG+12-08-030	08 07 06.	+ 73 42	114	15.	GALAXY
KEEN 2523B	08 07 06.	+ 73 44	138		Sc GALAXY
ZWG 031.060	08 07 12.	+ 05 25		15.5	GALAXY
ZWG 088.048	08 07 12.	+ 17 25		15.3	GALAXY
ZWG 236.031	08 07 12.	+ 47 04		15.7	GALAXY
UGC 04258	08 07 12.	+ 47 04	102	15.7	GALAXY Sc
MCG+08-15-045	08 07 12.	+ 48 25	42	16.	GALAXY
SHAH 093	08 07 12.	+ 54 09	150	17.0	GROUP OF COMPACT GALAXIES

OBJECT NAME	RIGHT ASCEN.	DECLINATION	DIAM.	MAGN.	TYPE OF OBJECT
ZWG 331.030	08 07 12.	+ 73 43		14.8	GALAXY
UGC 04259	08 07 12.	+ 73 43	126	14.8	GALAXY Sb
RNGC 2523B	08 07 13.	+ 73 43		15.0	GALAXY
MCG+03-21-025	08 07 15.	+ 17 25	42	15.3	GALAXY
MCG+08-15-046	08 07 15.	+ 47 03	90	15.	GALAXY
TON-N 0880	08 07 18.	+ 21 28		16.8	BLUE STAR
ABC 0617	08 07 23.	+ 77 28		17.7	RICH CLUSTER OF GALAXIES
ZC 0807.4+0955	08 07 24.	+ 09 55	400		CLUSTER OF GALAXIES
MCG+10-12-086	08 07 24.	+ 58 24	18	16.	GALAXY
MCG-03-21-006	08 07 27.	- 17 37	42	16.	GALAXY
TON-N 0881	08 07 30.	+ 22 35		16.4	BLUE STAR
ZC 0807.5+4344	08 07 30.	+ 43 44	1140		CLUSTER OF GALAXIES
MCG+11-10-068	08 07 30.	+ 63 38	39	16.	GALAXY
ZC 0807.5+7731	08 07 30.	+ 77 31	1410		CLUSTER OF GALAXIES
TON-N 0882	08 07 35.	+ 35 20		16.8	BLUE STAR
ZWG 031.061	08 07 36.	+ 08 05		14.9	GALAXY
MCG+01-21-016	08 07 36.	+ 08 05	36	14.9	GALAXY
LB 00526	08 07 36.	+ 27 10 00.		17.1	FAINT BLUE STAR
ZWG 236.032	08 07 36.	+ 46 37		14.3	GALAXY
UGC 04260	08 07 36.	+ 46 37	108	14.3	GALAXY IRR
ZC 0807.6+5300	08 07 36.	+ 53 00	810		CLUSTER OF GALAXIES
VHE 04	08 07 36.	- 38 43	210		REFLECTION NEBULA
LB 00527	08 07 37.	+ 30 23 48.		18.2	FAINT BLUE STAR
ZC 0807.7+2940	08 07 42.	+ 29 40	940		CLUSTER OF GALAXIES
LB 00528	08 07 42.	+ 30 24 06.		18.2	FAINT BLUE STAR
ZC 0807.7+3520	08 07 42.	+ 35 20	3700		CLUSTER OF GALAXIES
ZWG 178.033	08 07 42.	+ 36 58		14.7	GALAXY
UGC 04261	08 07 42.	+ 36 58	60	14.7	GALAXY PECULR
ZWG 236.033	08 07 42.	+ 45 49		15.6	GALAXY
MCG+08-15-048	08 07 42.	+ 45 49	42	16.	GALAXY
MCG+08-15-047	08 07 42.	+ 46 37	108	14.	GALAXY
ABC 0630	08 07 43.	+ 40 29		16.9	RICH CLUSTER OF GALAXIES
ZWG 178.034	08 07 48.	+ 36 45		15.5	GALAXY
ZWG 287.049	08 07 48.	+ 58 51		15.7	GALAXY
TON-N 0883	08 07 53.	+ 34 57		16.	BLUE STAR
ZWG 119.006	08 07 54.	+ 21 58		15.7	GALAXY
ZWG 119.007	08 07 54.	+ 25 50		15.6	GALAXY
ZC 0807.9+4030	08 07 54.	+ 40 30	1340		CLUSTER OF GALAXIES
MCG+10-12-087	08 07 54.	+ 58 50	30	15.	GALAXY
MCG+10-12-088	08 07 54.	+ 62 23	39	16.	GALAXY
ZC 0807.9+7417	08 07 54.	+ 74 17	1550		CLUSTER OF GALAXIES
IC 2230	08 07 57.	+ 25 49 47.			NONSTELLAR OBJECT
ZWG 031.062	08 08 00.	+ 07 37		15.1	GALAXY
MCG+01-21-017	08 08 00.	+ 07 37	30	15.1	GALAXY
ABC 0631	08 08 00.	+ 36 08		17.1	RICH CLUSTER OF GALAXIES
ZC 0808.0+3612	08 08 00.	+ 36 12	2350		CLUSTER OF GALAXIES
ZWG 287.050	08 08 00.	+ 57 08		15.7	GALAXY
MCG+10-12-089	08 08 00.	+ 57 08	39	16.	GALAXY
MCG+10-12-090	08 08 00.	+ 58 06	27	17.	GALAXY
ZC 0808.0+7020	08 08 00.	+ 70 20	2220		CLUSTER OF GALAXIES
7ZW 216	08 08 00.	+ 79 00			COMPACT GALAXY
ZWG 349.019	08 08 00.	+ 79 00		14.0	GALAXY
UGC 04263	08 08 00.	+ 79 00	54	14.0	GALAXY PECULR
MCG+14-04-046	08 08 00.	+ 83 24	114	14.	GALAXY
ZWG 363.038	08 08 00.	+ 83 25		14.2	GALAXY
UGC 04262	08 08 00.	+ 83 25	144	14.2	GALAXY Sb
CED 104	08 08 03.	- 47 12			DIFFUSE GALACTIC NEBULA
ZWG 031.063	08 08 06.	+ 03 19		15.5	GALAXY
ZWG 031.064	08 08 06.	+ 03 31		15.7	GALAXY
ZC 0808.1+3817	08 08 06.	+ 38 17	1010		CLUSTER OF GALAXIES
ZWG 287.051	08 08 06.	+ 58 43		15.5	GALAXY
ZWG 331.031	08 08 06.	+ 71 35		15.7	GALAXY
ZWG 119.008	08 08 12.	+ 25 20		13.5	GALAXY
UGC 04264	08 08 12.	+ 25 20	216	13.5	GALAXY S(c)
KARA.72 156A	08 08 12.	+ 25 20	126	13.5	PART OF DOUBLE GALAXY
MCG+04-20-005	08 08 12.	+ 25 20 30.	36	14.8	GALAXY
RNGC 2535	08 08 12.	+ 25 21		13.5	GALAXY
VV 009A	08 08 12.	+ 25 21 30.	180	13.0	INTERACTING GALAXY
ARP 082	08 08 12.	+ 25 22			PECULIAR GALAXY
MCG+04-20-004	08 08 12.	+ 25 22	138	13.5	GALAXY
ZWG 149.001	08 08 12.	+ 29 39		15.4	GALAXY
7ZW 217	08 08 12.	+ 58 43			COMPACT GALAXY
7ZW 218	08 08 12.	+ 71 34			COMPACT GALAXY
SN 1901A	08 08 13.	+ 25 20		14.7	SUPERNOVA
HOLM 094A	08 08 13.	+ 25 21	150	13.0	PART OF MULTIPLE GALAXY
MCG+05-20-001	08 08 15.	+ 29 38 30.	24	15.4	GALAXY
HOLM 094B	08 08 16.	+ 25 19	30	14.3	PART OF MULTIPLE GALAXY
ZWG 119.009	08 08 18.	+ 25 19		14.8	GALAXY
KARA.72 156B	08 08 18.	+ 25 19	60	14.8	PART OF DOUBLE GALAXY
RNGC 2536	08 08 18.	+ 25 20		15.0	GALAXY
VV 009B	08 08 18.	+ 25 20	48	14.3	INTERACTING GALAXY
ZWG 207.042	08 08 18.	+ 41 43		15.3	GALAXY
ZWG 263.011	08 08 18.	+ 52 37		15.7	GALAXY
MCG+09-14-011	08 08 18.	+ 53 03	15	16.	GALAXY
ABC 0635	08 08 19.	+ 16 52		17.0	RICH CLUSTER OF GALAXIES
MCG+05-20-002	08 08 21.	+ 29 39	24	15.4	GALAXY
ZWG 031.066	08 08 24.	+ 05 14		15.0	GALAXY
UGC 04265	08 08 24.	+ 05 14	120	15.0	GALAXY E
MCG+01-21-018	08 08 24.	+ 05 14	60	15.0	GALAXY
ZWG 031.065	08 08 24.	+ 06 27		15.2	GALAXY
MCG+09-14-012	08 08 24.	+ 52 36	24	16.	GALAXY
OCL 0611	08 08 24.	- 12 41	1800	8.2	OPEN STAR CLUSTER
TON-N 0884	08 08 25.	+ 19 06		15.4	BLUE STAR
RNGC 2539	08 08 26.	- 12 41		8.0	OPEN CLUSTER
IC 2231	08 08 29.	+ 08 14 11.			NONSTELLAR OBJECT
ZC 0808.5+3847	08 08 30.	+ 38 47	470		CLUSTER OF GALAXIES
ZWG 263.012	08 08 30.	+ 53 38		15.5	GALAXY
MCG+09-14-013	08 08 30.	+ 53 38	36	16.	GALAXY
ZC 0808.5+6536	08 08 30.	+ 65 36	1750		CLUSTER OF GALAXIES
7ZW 219	08 08 30.	+ 66 31			COMPACT GALAXY
ZWG 207.043	08 08 36.	+ 39 10		15.4	GALAXY
ZC 0808.6+5120	08 08 36.	+ 51 20	1280		CLUSTER OF GALAXIES
TON-N 0885	08 08 37.	+ 20 20		15.9	BLUE STAR
LB 00529	08 08 39.	+ 31 02 06.		17.1	FAINT BLUE STAR
LB 00530	08 08 41.	+ 28 14 00.		16.8	FAINT BLUE STAR
ZWG 059.050	08 08 42.	+ 09 02		15.2	GALAXY
ZWG 059.051	08 08 42.	+ 09 05		15.3	GALAXY
ZWG 149.002	08 08 42.	+ 27 41		15.6	GALAXY
KARA.73B 0233	08 08 42.	+ 27 41	24	15.6	ISOLATED GALAXY S0
ACK 240+07.1	08 08 42.	- 20 23			PLANETARY NEBULA
OCL 0682	08 08 42.	- 26 52	1080	10.	OPEN STAR CLUSTER
MCG+07-17-021	08 08 45.	+ 39 09	36	15.5	GALAXY
RNGC 2538	08 08 47.	+ 03 46		14.0	GALAXY
ZC 0808.8+0219	08 08 48.	+ 02 19	1880		CLUSTER OF GALAXIES
ZWG 031.067	08 08 48.	+ 03 46		13.8	GALAXY
UGC 04266	08 08 48.	+ 03 46	102	13.8	GALAXY SBa
MCG+01-21-019	08 08 48.	+ 03 46	72	13.8	GALAXY
ZC 0808.8+5150	08 08 48.	+ 51 50	540		CLUSTER OF GALAXIES

OBJECT NAME	RIGHT ASCEN.	DECLINATION	DIAM.	MAGN.	TYPE OF OBJECT
ZC 0808.8+5252	08 08 48.	+ 52 52	400		CLUSTER OF GALAXIES
ZWG 287.052	08 08 48.	+ 58 05		15.4	GALAXY
ZWG 031.068	08 08 54.	+ 05 14		15.6	GALAXY
MCG+10-12-091	08 08 54.	+ 58 04	39	16.	GALAXY
ISS 0913	08 08 54.	- 29 36	391		STELLAR RING
RNGC 2542	08 08 56.	- 12 47			NON-EXISTENT OBJECT
MCG+09-14-014	08 08 57.	+ 55 50	60	13.	GALAXY
LB 09842	08 09	- 86 25		14.7	FAINT BLUE STAR
ZWG 031.069	08 09 00.	+ 05 05		15.7	GALAXY
ZWG 149.003	08 09 00.	+ 29 33		15.5	GALAXY
ZWG 263.013	08 09 00.	+ 55 07		14.2	GALAXY
UGC 04267	08 09 00.	+ 55 07	60	14.2	GALAXY
MCG+09-14-015	08 09 00.	+ 55 07 30.	36	15.	GALAXY
ZWG 263.014	08 09 00.	+ 55 49		13.8	GALAXY
UGC 04268	08 09 00.	+ 55 49	84	13.8	GALAXY PECULR
MCG+10-12-092	08 09 00.	+ 58 02	27	16.	GALAXY
RNGC 2534	08 09 02.	+ 55 49		14.0	GALAXY
MCG+05-20-003	08 09 03.	+ 29 32	48	15.	GALAXY
ZWG 088.049	08 09 06.	+ 19 31		14.6	GALAXY
UGC 04269	08 09 06.	+ 19 31	72	14.6	GALAXY S
KARA.73B 0234	08 09 06.	+ 19 31	72	14.6	ISOLATED GALAXY S
ZWG 207.044	08 09 06.	+ 38 46		15.7	GALAXY
ZC 0809.1+4355	08 09 06.	+ 43 55	1080		CLUSTER OF GALAXIES
ZWG 263.015	08 09 06.	+ 54 40		15.6	GALAXY
MCG+03-21-026	08 09 09.	+ 19 30 30.	66	14.6	GALAXY
RNGC 2547	08 09 10.	- 49 07		5.0	OPEN CLUSTER
ZWG 119.010	08 09 12.	+ 26 20		15.4	GALAXY
MCG+08-15-049	08 09 12.	+ 46 20	24	16.	GALAXY
ZWG 263.016	08 09 12.	+ 54 46		15.2	GALAXY
MCG+10-12-093	08 09 12.	+ 58 09	42	16.	GALAXY
ZWG 287.053	08 09 12.	+ 58 10		15.7	GALAXY
ZWG 310.041	08 09 12.	+ 64 30		15.1	GALAXY
MCG+12-08-031	08 09 12.	+ 73 43	168	12.6	GALAXY
OCL 0753	08 09 12.	- 49 07	2220	5.1	OPEN STAR CLUSTER
TON-N 0886	08 09 13.	+ 19 21		16.0	BLUE STAR
ARP 009	08 09 17.	+ 73 46			PECULIAR GALAXY
MCG+07-17-022	08 09 18.	+ 44 30 30.	48	15.	GALAXY
ZWG 236.034	08 09 18.	+ 44 31		15.2	GALAXY
VV 138	08 09 18.	+ 46 09	78	12.2	INTERACTING GALAXY
MCG+09-14-016	08 09 18.	+ 54 47 30.	18	17.	GALAXY
ZWG 287.054	08 09 18.	+ 58 00		14.9	GALAXY
UGC 04270	08 09 18.	+ 58 00	102	14.9	GALAXY Sb/SBc
MCG+10-12-094	08 09 18.	+ 61 04	30	15.	GALAXY
ZWG 331.032	08 09 18.	+ 73 44		12.4	GALAXY
UGC 04271	08 09 18.	+ 73 44	180	12.4	GALAXY SBb
OCL 0710	08 09 18.	- 31 48	150	15.	OPEN STAR CLUSTER
VHA 020	08 09 18.	- 31 48	180		OPEN STAR CLUSTER
RNGC 2523	08 09 19.	+ 73 44		12.5	GALAXY
ZWG 031.070	08 09 24.	+ 05 37		15.6	GALAXY
ZWG 059.052	08 09 24.	+ 11 24		15.4	GALAXY
ZWG 059.053	08 09 24.	+ 13 46		15.7	GALAXY
UGC 04272	08 09 24.	+ 44 31	60	16.0	GALAXY S
ZWG 263.017	08 09 24.	+ 52 37		15.0	GALAXY
MCG+10-12-095	08 09 24.	+ 58 00	96	15.	GALAXY
MCG+11-10-069	08 09 24.	+ 64 29	24	15.	GALAXY
MCG+09-14-017	08 09 27.	+ 52 36	42	14.	GALAXY
ZWG 287.055	08 09 30.	+ 58 17		15.3	GALAXY
MCG+11-10-070	08 09 30.	+ 64 29	18	17.	GALAXY
ZC 0809.5+6639	08 09 30.	+ 66 39	2690		CLUSTER OF GALAXIES
LB 00531	08 09 31.	+ 27 24 00.		17.7	FAINT BLUE STAR
MCG+08-15-050	08 09 36.	+ 46 09	90	12.2	GALAXY
MCG+09-14-018	08 09 36.	+ 54 28	42	17.	GALAXY
RNGC 2543	08 09 40.	+ 36 24		12.5	GALAXY
IC 2232	08 09 41.	+ 36 24 30.			SAME AS NGC 2543
ARP 006	08 09 41.	+ 46 09			PECULIAR GALAXY
MAI 026	08 09 41.	+ 73 39	33		DWARF SPHEROIDAL GALAXY
REIN 2.058	08 09 41.36	+ 46 08 43.9			NEBULA
MCG+05-20-004	08 09 42.	+ 26 31	66	14.5	GALAXY
ZWG 178.035	08 09 42.	+ 36 24		12.7	GALAXY
UGC 04273	08 09 42.	+ 36 24	162	12.7	GALAXY SBb
KARA.72 157B	08 09 42.	+ 36 24	30		PART OF DOUBLE GALAXY
KARA.72 157A	08 09 42.	+ 36 24	24	12.7	PART OF DOUBLE GALAXY
MCG+06-18-014	08 09 42.	+ 36 24	120	13.	GALAXY
ZWG 236.035	08 09 42.	+ 46 09		11.7	GALAXY
UGC 04274	08 09 42.	+ 46 09	102	11.7	GALAXY S
KARA.73B 0235	08 09 42.	+ 46 09	402	11.7	ISOLATED GALAXY S
MCG+10-12-096	08 09 42.	+ 58 17	36	16.	GALAXY
REIN 2.059	08 09 42.66	+ 46 08 36.9			NEBULA
REIN 2.060	08 09 42.78	+ 46 08 25.1			NEBULA
RNGC 2537	08 09 44.	+ 46 09		12.5	GALAXY
REIN 2.061	08 09 44.14	+ 46 08 35.6			NEBULA
RNGC 2540	08 09 47.	+ 26 31		14.5	GALAXY
ZWG 149.004	08 09 48.	+ 26 31		14.5	GALAXY
UGC 04275	08 09 48.	+ 26 31	84	14.5	GALAXY SBc
ZC 0809.8+5305	08 09 48.	+ 53 05	1210		CLUSTER OF GALAXIES
MCG+10-12-097	08 09 48.	+ 58 13	24	16.	GALAXY
LB 00532	08 09 52.	+ 79 15 36.		17.5	FAINT BLUE STAR
ZWG 088.050	08 09 54.	+ 17 02		15.4	GALAXY
ZWG 003.016	08 09 54.	- 01 17		15.3	GALAXY
MCG+00-21-009	08 09 54.	- 01 17	30	15.3	GALAXY
TON-N 0887	08 09 55.	+ 19 49		16.6	BLUE STAR
BC 3CR196	08 09 59.39	+ 48 22 07.7		17.79	QUASI-STELLAR OBJECT
SHB 107	08 09 59.4	+ 48 22 07.		17.8	QUASI-STELLAR OBJECT
ZWG 178.036	08 10 00.	+ 35 05		15.1	GALAXY
ZC 0810.0+3905	08 10 00.	+ 39 05	870		CLUSTER OF GALAXIES
MCG+08-15-051	08 10 00.	+ 46 10	30	16.	GALAXY
ZC 0810.0+5554	08 10 00.	+ 55 54	2080		CLUSTER OF GALAXIES
ZC 0810.0-0320	08 10 00.	- 03 20	870		CLUSTER OF GALAXIES
PK264-08.1	08 10 02.6	- 48 34 15.	49	12.4	PLANETARY NEBULA
ZWG 059.054	08 10 06.	+ 09 32		15.1	GALAXY
UGC 04276	08 10 06.	+ 09 32	66	15.1	GALAXY SBa
MCG+02-21-017	08 10 06.	+ 09 32	54	15.1	GALAXY
KARA.73B 0236	08 10 06.	+ 09 32	48	15.1	ISOLATED GALAXY S
ZWG 119.011	08 10 06.	+ 24 42		15.6	GALAXY
ZWG 263.018	08 10 06.	+ 52 48		15.4	GALAXY
UGC 04277	08 10 06.	+ 52 48	228	15.4	GALAXY Sc
MCG+09-14-019	08 10 06.	+ 52 48	210	14.	GALAXY
KARA.73B 0237	08 10 06.	+ 52 48	270	15.4	ISOLATED GALAXY S
ZC 0810.1+5813	08 10 06.	+ 58 13	5780		CLUSTER OF GALAXIES
MCG+10-12-098	08 10 06.	+ 61 20	36	16.	GALAXY
TON-N 0888	08 10 07.	+ 19 24		16.3	BLUE STAR
RNGC 2537A	08 10 08.	+ 46 09			GALAXY
ZC 0810.2+1328	08 10 12.	+ 13 28	2150		CLUSTER OF GALAXIES
ZWG 149.005	08 10 12.	+ 32 06		15.7	GALAXY
ZC 0810.2+3956	08 10 12.	+ 39 56	3090		CLUSTER OF GALAXIES
ZC 0810.2+7703	08 10 12.	+ 77 03	1480		CLUSTER OF GALAXIES
ABC 0629	08 10 16.	+ 66 35		17.1	RICH CLUSTER OF GALAXIES
ABC 0638	08 10 18.	+ 13 35		17.0	RICH CLUSTER OF GALAXIES

OBJECT NAME	RIGHT ASCEN.	DECLINATION	DIAM.	MAGN.	TYPE OF OBJECT
TON-N 0302	08 10 18.	+ 27 31		15.6	BLUE STAR
OCL 0713	08 10 18.	- 32 27	1020	11.	OPEN STAR CLUSTER
MCG-03-21-007	08 10 21.	- 15 54	42	15.	GALAXY
IC 0500	08 10 22.	- 15 54 10.			NONSTELLAR OBJECT
ZWG 119.012	08 10 24.	+ 22 47		15.5	GALAXY
ZWG 119.013	08 10 24.	+ 25 16		15.5	GALAXY
MCG+07-17-023	08 10 24.	+ 41 47 30.	30	15.	GALAXY
ZWG 207.045	08 10 24.	+ 41 48		15.2	GALAXY
ZWG 236.036	08 10 24.	+ 45 54		13.6	GALAXY
UGC 04278	08 10 24.	+ 45 54	282	13.6	GALAXY Sc
UGC 04279	08 10 24.	+ 73 50	66	16.5	GALAXY SB IV
IC 2233	08 10 26.	+ 45 54 08.	270	13.0	GALAXY S0
ARC 0634	08 10 29.	+ 58 12		14.9	RICH CLUSTER OF GALAXIES
ZC 0810.5+1907	08 10 30.	+ 19 07	2960		CLUSTER OF GALAXIES
ZWG 178.037	08 10 30.	+ 37 12		15.5	GALAXY
MCG+08-15-052	08 10 30.	+ 45 54	270	12.9	GALAXY
ZWG 287.056	08 10 30.	+ 58 11		15.3	GALAXY
OCL 0691	08 10 30.	- 27 44	288	15.	OPEN STAR CLUSTER
VHA 021	08 10 30.	- 27 44	360		OPEN STAR CLUSTER
RNGC 2546	08 10 35.	- 37 29		5.0	OPEN CLUSTER
IC 2235	08 10 36.	+ 24 13 47.			NONSTELLAR OBJECT
TON-N 0006	08 10 36.	+ 27 59		15.2	BLUE STAR
ZWG 263.019	08 10 36.	+ 54 56		14.4	GALAXY
UGC 04280	08 10 36.	+ 54 56	90	14.4	GALAXY Sa
ZWG 287.057	08 10 36.	+ 58 07		14.4	GALAXY
UGC 04280A	08 10 36.	+ 58 07	48	14.4	GALAXY DBL SYS
KARA.72 158B	08 10 36.	+ 58 07	18		PART OF DOUBLE GALAXY
KARA.72 158A	08 10 36.	+ 58 07	54	14.4	PART OF DOUBLE GALAXY
ZWG 287.058	08 10 36.	+ 58 23		15.2	GALAXY
UGC 04281	08 10 36.	+ 58 23	66	15.2	GALAXY Sa-b
MCG+10-12-099	08 10 36.	+ 58 28	30	17.	GALAXY
OCL 0726	08 10 36.	- 37 29	3240	5.2	OPEN STAR CLUSTER
VHA 022	08 10 36.	- 37 29	2100		OPEN STAR CLUSTER
ISS 1039	08 10 36.	- 42 42	265		STELLAR RING
IC 2234	08 10 38.	+ 35 38 42.			NONSTELLAR OBJECT
IC 2236	08 10 39.	+ 24 12 04.			NONSTELLAR OBJECT
LB 00533	08 10 41.	+ 31 01 12.		17.1	FAINT BLUE STAR
TON-N 0889	08 10 42.	+ 22 58		16.8	BLUE STAR
TON-N 0303	08 10 42.	+ 28 22		15.8	BLUE STAR
MCG+09-14-021	08 10 42.	+ 54 57	84	15.	GALAXY
ZWG 263.020	08 10 42.	+ 55 17		15.3	GALAXY
MCG+09-14-020	08 10 42.	+ 55 18 30.	15	16.	GALAXY
7ZW 220	08 10 42.	+ 58 06			COMPACT GALAXY
7ZW 221	08 10 42.	+ 58 25			COMPACT GALAXY
TON-N 0890	08 10 47.	+ 33 30		17.	BLUE STAR
ZC 0810.8+1151	08 10 48.	+ 11 51	740		CLUSTER OF GALAXIES
ZWG 119.014	08 10 48.	+ 22 13		15.6	GALAXY
MCG+08-15-053	08 10 48.	+ 46 10	36	16.	GALAXY
ZC 0810.8+4618	08 10 48.	+ 46 18	870		CLUSTER OF GALAXIES
MCG+10-12-100	08 10 48.	+ 58 10	54	15.	GALAXY
MCG+10-12-101	08 10 48.	+ 58 22	72	15.	GALAXY
MCG+10-12-102	08 10 48.	+ 58 24	30	17.	GALAXY
ZWG 349.020	08 10 48.	+ 77 40		15.1	GALAXY
UGC 04282	08 10 48.	+ 77 40	78	15.1	GALAXY Sb-c
MCG-03-21-008	08 10 48.	- 15 58	18	15.5	GALAXY
OCL 0730	08 10 48.	- 40 19	2520	9.	OPEN STAR CLUSTER
ZC 0810.9+1054	08 10 54.	+ 10 54	340		CLUSTER OF GALAXIES
ZWG 088.051	08 10 54.	+ 17 11		15.5	GALAXY
ZC 0810.9+3631	08 10 54.	+ 36 31	1340		CLUSTER OF GALAXIES
MCG+10-12-103	08 10 54.	+ 58 07	18	15.	GALAXY
MCG-04-20-002	08 10 54.	- 27 26	60	14.5	GALAXY
ISS 0914	08 10 54.	- 32 24	220		STELLAR RING
ARC 0633	08 10 58.	+ 63 54		17.1	RICH CLUSTER OF GALAXIES
RNGC 2541	08 10 59.	+ 49 13		12.0	GALAXY
ZWG 207.046	08 11 00.	+ 39 25		15.2	GALAXY
UGC 04283	08 11 00.	+ 39 25	78	15.2	GALAXY SBb
KARA.73B 0238	08 11 00.	+ 39 25	90	15.2	ISOLATED GALAXY S
ZWG 207.047	08 11 00.	+ 43 17		15.1	GALAXY
ZWG 236.037	08 11 00.	+ 49 13		13.0	GALAXY
UGC 04284	08 11 00.	+ 49 13	486	13.0	GALAXY Sc
MCG+13-06-019	08 11 00.	+ 77 37	30	17.	GALAXY
MCG+13-06-020	08 11 00.	+ 77 39	39	15.	GALAXY
ZWG 119.015	08 11 06.	+ 24 00		15.3	GALAXY
ZC 0811.1+4206	08 11 06.	+ 42 06	1140		CLUSTER OF GALAXIES
MCG+07-17-024	08 11 06.	+ 43 18	12	15.	GALAXY
MCG+08-15-054	08 11 06.	+ 49 12 30.	360	11.8	GALAXY
MCG+10-12-104	08 11 06.	+ 61 24	18	17.	GALAXY
MCG+10-12-105	08 11 06.	+ 61 25	18	16.	GALAXY
VHE 06	08 11 06.	- 34 26	66		REFLECTION NEBULA
IC 2239	08 11 08.	+ 24 01 07.			NONSTELLAR OBJECT
IC 2237	08 11 08.	+ 24 49 55.			NONSTELLAR OBJECT
MCG+04-20-006	08 11 09.	+ 24 01	60	15.3	GALAXY
IC 2238	08 11 09.	+ 24 48 49.			NONSTELLAR OBJECT
RNGC 2548	08 11 10.	- 05 38		5.5	OPEN CLUSTER
ZWG 003.017	08 11 12.	+ 00 48		15.2	GALAXY
UGC 04285	08 11 12.	+ 00 48	90	15.2	GALAXY Sc
MCG+00-21-010	08 11 12.	+ 00 48	66	15.2	GALAXY
ZC 0811.2+3838	08 11 12.	+ 38 38	1210		CLUSTER OF GALAXIES
MCG+07-17-025	08 11 12.	+ 39 24	72	15.5	GALAXY
MCG+10-12-106	08 11 12.	+ 60 52	27	16.	GALAXY
ZWG 287.059	08 11 12.	+ 61 25		15.6	GALAXY
TON-N 0891	08 11 13.	+ 19 20		14.8	BLUE STAR
TON-N 0892	08 11 13.	+ 20 30		16.2	BLUE STAR
ARC 0624	08 11 13.	+ 77 02		17.8	RICH CLUSTER OF GALAXIES
ZWG 088.052	08 11 18.	+ 18 36		15.6	GALAXY
KARA.72 159A	08 11 18.	+ 18 36	30	15.6	PART OF DOUBLE GALAXY
ZC 0811.3+1955	08 11 18.	+ 19 55	1950		CLUSTER OF GALAXIES
SCH 01	08 11 18.	+ 21 30		13.2	PECULIAR GALAXY
ZC 0811.3+5227	08 11 18.	+ 52 27	810		CLUSTER OF GALAXIES
LB 00534	08 11 18.	+ 81 56 00.		16.5	FAINT BLUE STAR
OCL 0584	08 11 18.	- 05 39	3240	5.9	OPEN STAR CLUSTER
RNGC 2545	08 11 20.	+ 21 31		13.5	GALAXY
LB 00535	08 11 22.	+ 29 01 36.		17.6	FAINT BLUE STAR
TON-N 0893	08 11 23.	+ 32 45		15.	BLUE STAR
ZWG 031.071	08 11 24.	+ 03 32		15.4	GALAXY
ZWG 088.053	08 11 24.	+ 18 36		14.6	GALAXY
UGC 04286	08 11 24.	+ 18 36	78	14.6	GALAXY S
KARA.72 159B	08 11 24.	+ 18 36	66	14.6	PART OF DOUBLE GALAXY
ZWG 119.016	08 11 24.	+ 21 30		13.2	GALAXY
UGC 04287	08 11 24.	+ 21 30	150	13.2	GALAXY S
MCG+04-20-007	08 11 24.	+ 21 30	108	13.2	GALAXY
ZC 0811.4+2245	08 11 24.	+ 22 45	1010		CLUSTER OF GALAXIES
MCG+08-15-055	08 11 24.	+ 45 57	15	17.	GALAXY
ARC 0637	08 11 24.	+ 48 32		17.5	RICH CLUSTER OF GALAXIES
ZC 0811.4+5615	08 11 24.	+ 56 15	940		CLUSTER OF GALAXIES
ZWG 287.060	08 11 24.	+ 58 20		14.7	GALAXY
MCG+03-21-027	08 11 27.	+ 18 35	90	14.6	GALAXY
LB 00536	08 11 29.	+ 27 10 06.		17.2	FAINT BLUE STAR

OBJECT NAME	RIGHT ASCEN.	DECLINATION	DIAM.	MAGN.	TYPE OF OBJECT
MCG+10-12-107	08 11 30.	+ 58 20	27	15.	GALAXY
ZWG 003.018	08 11 30.	- 00 08		15.3	GALAXY
VHE 07	08 11 30.	- 38 17	432		REFLECTION NEBULA
ZC 0811.6+4828	08 11 36.	+ 48 28	940		CLUSTER OF GALAXIES
ZC 0811.6+4921	08 11 36.	+ 49 21	1680		CLUSTER OF GALAXIES
ZWG 287.061	08 11 36.	+ 58 27		15.3	GALAXY
ZWG 287.062	08 11 36.	+ 60 47		15.5	GALAXY
UGC 04288	08 11 42.	+ 19 30	60	17.	GALAXY
TON-N 0894	08 11 42.	+ 22 37		16.5	BLUE STAR
ZWG 236.038	08 11 42.	+ 46 14		15.2	GALAXY
MCG+10-12-108	08 11 42.	+ 58 27	39	15.	GALAXY
ZWG 287.063	08 11 42.	+ 58 29		14.6	GALAXY
UGC 04289	08 11 42.	+ 58 29	66	14.6	GALAXY E
MCG+10-12-109	08 11 42.	+ 60 47	24	15.	GALAXY
MCG+10-12-110	08 11 42.	+ 60 50	39	16.	GALAXY
ZWG 031.072	08 11 48.	+ 04 29		15.4	GALAXY
IC 2240	08 11 48.	+ 24 37 09.			NONSTELLAR OBJECT
TON-N 0007	08 11 48.	+ 28 21		16.6	BLUE STAR
ZWG 207.048	08 11 48.	+ 41 22		15.3	GALAXY
ZC 0811.8+4414	08 11 48.	+ 44 14	1080		CLUSTER OF GALAXIES
MCG+08-15-056	08 11 48.	+ 46 14	42	15.	GALAXY
MCG+10-12-111	08 11 48.	+ 58 29	30	14.	GALAXY
MCG+13-06-021	08 11 48.	+ 79 23	39	16.	GALAXY
TON-N 0895	08 11 54.	+ 23 08		14.5	BLUE STAR
ARC 0640	08 11 54.	+ 29 44		17.5	RICH CLUSTER OF GALAXIES
MCG+07-17-026	08 11 54.	+ 41 22	45	15.	GALAXY
ZC 0811.9+5344	08 11 54.	+ 53 44	810		CLUSTER OF GALAXIES
ZWG 119.017	08 12 00.	+ 25 45		15.6	GALAXY
LB 00537	08 12 00.	+ 31 10 54.		16.7	FAINT BLUE STAR
ZC 0812.0+3705	08 12 00.	+ 37 05	2420		CLUSTER OF GALAXIES
ZWG 287.064	08 12 00.	+ 58 10		15.4	GALAXY
MCG+10-12-112	08 12 00.	+ 58 28	30	16.	GALAXY
KEEN 2523C	08 12 00.	+ 73 30	54		E4 GALAXY
MCG+14-04-047	08 12 00.	+ 85 07	45	17.	GALAXY
VHE 08	08 12 00.	- 35 59	66		REFLECTION NEBULA
RNGC 2523C	08 12 03.	+ 73 28		14.0	GALAXY
MCG+10-12-113	08 12 06.	+ 58 06	15	16.	GALAXY
MCG+10-12-114	08 12 06.	+ 58 09	15	16.	GALAXY
ZWG 331.033	08 12 06.	+ 73 28		14.1	GALAXY
UGC 04290	08 12 06.	+ 73 28	90	14.1	GALAXY E?
MCG+12-08-032	08 12 06.	+ 73 28	27	15.	GALAXY
IC 2241	08 12 10.	+ 24 17 00.			NONSTELLAR OBJECT
ZWG 003.019	08 12 12.	+ 01 42		15.6	GALAXY
UGC 04291	08 12 12.	+ 01 42	66	15.6	GALAXY Sb-c
ZWG 088.054	08 12 12.	+ 18 02		15.6	GALAXY
ZWG 287.065	08 12 12.	+ 57 52		15.6	GALAXY
ZWG 349.021	08 12 12.	+ 79 23		14.6	GALAXY
UGC 04292	08 12 12.	+ 79 23	72	14.6	GALAXY S0-a
IC 2242	08 12 13.	+ 24 17 12.			NONSTELLAR OBJECT
ZWG 003.020	08 12 18.	+ 00 13		15.4	GALAXY
TON-N 0008	08 12 18.	+ 27 46		16.8	BLUE STAR
ZWG 149.006	08 12 18.	+ 28 54		15.3	GALAXY
TON-N 0304	08 12 18.	+ 32 08		13.8	BLUE STAR
MCG+10-12-115	08 12 18.	+ 57 51	39	16.	GALAXY
MCG-05-20-002	08 12 18.	- 30 32	18	14.5	GALAXY
IC 2243	08 12 20.	+ 24 06 59.			NONSTELLAR OBJECT
TON-N 0896	08 12 24.	+ 22 58		16.9	BLUE STAR
IC 2244	08 12 24.	+ 24 41 53.			NONSTELLAR OBJECT
ZWG 119.018	08 12 24.	+ 26 17		15.6	GALAXY
IC 2245	08 12 29.	+ 24 41 22.			NONSTELLAR OBJECT
ZWG 059.055	08 12 30.	+ 10 25		15.7	GALAXY
ZWG 119.019	08 12 30.	+ 21 43		15.7	GALAXY
ZC 0812.5+2300	08 12 30.	+ 23 00	1140		CLUSTER OF GALAXIES
MCG+09-14-022	08 12 30.	+ 52 34 30.		15.	GALAXY
ZWG 263.021	08 12 30.	+ 52 35		15.4	GALAXY
UGC 04293	08 12 30.	+ 52 35	66	15.4	GALAXY S
MCG+11-10-071	08 12 30.	+ 64 41	39	15.	GALAXY
VHA 023	08 12 30.	- 36 15	780		OPEN STAR CLUSTER
ARC 0641	08 12 32.	+ 22 59		17.5	RICH CLUSTER OF GALAXIES
YC 0812-51	08 12 34.	- 51 22 42.			UNUSUAL SOUTHERN NEBULA
TON-N 0305	08 12 36.	+ 24 59		15.2	BLUE STAR
ZC 0812.6+3902	08 12 36.	+ 39 02	740		CLUSTER OF GALAXIES
UGC 04294	08 12 36.	+ 60 39	66	16.0	GALAXY S
MCG+10-12-116	08 12 36.	+ 60 39	42	16.	GALAXY
MCG+10-12-117	08 12 36.	+ 61 44	42	17.	GALAXY
UGC 04295	08 12 36.	+ 64 40	78	16.0	GALAXY SB:c
ACK 251+00.1	08 12 36.	- 33 38			PLANETARY NEBULA
ZC 0812.7+6341	08 12 42.	+ 63 41	3490		CLUSTER OF GALAXIES
MCG+11-10-072	08 12 42.	+ 64 42 30.	60	15.	GALAXY
OCL 0686	08 12 42.	- 26 50	300	12.	OPEN STAR CLUSTER
OCL 0712	08 12 42.	- 31 48	600	12.	OPEN STAR CLUSTER
TON-N 0897	08 12 43.	+ 20 03		16.5	BLUE STAR
SHB 108	08 12 47.2	+ 02 04 11.		18.5	QUASI-STELLAR OBJECT
BC PKS0812+02	08 12 47.26	+ 02 04 13.1		18.5	QUASI-STELLAR OBJECT
ZWG 031.073	08 12 48.	+ 08 30		15.4	GALAXY
UGC 04296	08 12 48.	+ 08 30	66	15.4	GALAXY Sc
ZWG 207.049	08 12 48.	+ 43 36		15.5	GALAXY
ZC 0812.9+1015	08 12 54.	+ 10 15	400		CLUSTER OF GALAXIES
TON-N 0898	08 12 54.	+ 22 02		16.8	BLUE STAR
MCG+05-20-005	08 12 54.	+ 27 13	72	14.7	GALAXY
ARC 0642	08 12 54.	+ 30 09		17.7	RICH CLUSTER OF GALAXIES
ZWG 207.050	08 12 54.	+ 41 52		15.7	GALAXY
ZWG 207.051	08 12 54.	+ 44 07		15.6	GALAXY
TON-N 0899	08 12 59.	+ 35 26		14.7	BLUE STAR
UGC 04298	08 13 00.	+ 05 25	60	17.	GALAXY DWRF SP
ZC 0813.0+1209	08 13 00.	+ 12 09	3020		CLUSTER OF GALAXIES
ZWG 088.055	08 13 00.	+ 17 10		15.2	GALAXY
ZWG 119.020	08 13 00.	+ 23 20		15.2	GALAXY
UGC 04299	08 13 00.	+ 23 20	120	15.2	GALAXY Sb-c
MCG+04-20-008	08 13 00.	+ 23 21	90	15.2	GALAXY
ZWG 149.007	08 13 00.	+ 27 14		14.7	GALAXY
UGC 04300	08 13 00.	+ 27 14	96	14.7	GALAXY Sc
ZWG 149.008	08 13 00.	+ 28 47		14.6	GALAXY
UGC 04301	08 13 00.	+ 28 47	78	14.6	GALAXY Sb-c
MCG+05-20-006	08 13 00.	+ 28 47	66	14.6	GALAXY
ZWG 149.009	08 13 00.	+ 31 36		15.7	GALAXY
ZC 0813.0+3509	08 13 00.	+ 35 09	3020		CLUSTER OF GALAXIES
ZC 0813.0+4210	08 13 00.	+ 42 10	740		CLUSTER OF GALAXIES
ZC 0813.0+4353	08 13 00.	+ 43 53	1010		CLUSTER OF GALAXIES
ZWG 236.039	08 13 00.	+ 46 13		15.5	GALAXY
ZC 0813.0+5118	08 13 00.	+ 51 18	2890		CLUSTER OF GALAXIES
ZWG 310.042	08 13 00.	+ 64 43		15.5	GALAXY
UGC 04302	08 13 00.	+ 64 43	72	15.5	GALAXY SB:b
ZWG 364.002	08 13 00.	+ 85 06		15.5	GALAXY
ZWG 363.039	08 13 00.	+ 85 06		15.5	GALAXY
MCG+14-04-048	08 13 00.	+ 85 46 30.	66	15.	GALAXY
ZWG 364.003	08 13 00.	+ 85 47		14.5	GALAXY
ZWG 363.040	08 13 00.	+ 85 47		14.5	GALAXY

OBJECT NAME	RIGHT ASCEN.	DECLINATION	DIAM.	MAGN.	TYPE OF OBJECT
ZWG 362.049	08 13 00.	+ 85 47		14.5	GALAXY
UGC 04297	08 13 00.	+ 85 47	126	14.5	GALAXY Sa
SN 1968G	08 13 01.	+ 20 37		17.4	SUPERNOVA
IC 2246	08 13 03.	+ 24 00 12.			NONSTELLAR OBJECT
IC 2247	08 13 05.	+ 24 21 12.			NONSTELLAR OBJECT
ZWG 119.021	08 13 06.	+ 23 16		15.1	GALAXY
MCG+04-20-009	08 13 06.	+ 23 17 30.	24	15.1	GALAXY
7ZW 222	08 13 06.	+ 57 10			COMPACT GALAXY
IC 2248	08 13 07.	+ 23 17 18.			NONSTELLAR OBJECT
MRK 623	08 13 12.	+ 26 08	7	16.5	GALAXY WITH UV CONTINUUM
ZC 0813.3+4848	08 13 18.	+ 48 48	1280		CLUSTER OF GALAXIES
ZC 0813.3+5640	08 13 18.	+ 56 40	870		CLUSTER OF GALAXIES
ZC 0813.4+1437	08 13 24.	+ 14 37	810		CLUSTER OF GALAXIES
ZWG 119.022	08 13 24.	+ 26 09		15.7	GALAXY
UGC 04303	08 13 24.	+ 26 09	60	15.7	GALAXY S
ZWG 287.066	08 13 24.	+ 60 06		15.2	GALAXY
MCG+10-12-118	08 13 24.	+ 60 06	36	14.	GALAXY
OCL 0707	08 13 24.	- 30 41	390	14.	OPEN STAR CLUSTER
LB 00538	08 13 26.	+ 30 21 30.		17.2	FAINT BLUE STAR
MCG+04-20-010	08 13 27.	+ 23 58	90	15.6	GALAXY
ZWG 088.056	08 13 30.	+ 18 03		15.6	GALAXY
ZWG 119.023	08 13 30.	+ 23 58		15.6	GALAXY
UGC 04304	08 13 30.	+ 23 58	90	15.6	GALAXY S0
ZWG 207.052	08 13 30.	+ 38 44		15.6	GALAXY
ZC 0813.5+5449	08 13 30.	+ 54 49	3090		CLUSTER OF GALAXIES
MRSL 253-00/1	08 13 30.	- 35 42	2880		HII REGION
IC 2250	08 13 34.	+ 23 47 03.			NONSTELLAR OBJECT
IC 2249	08 13 35.	+ 24 38 57.			NONSTELLAR OBJECT
ZWG 119.024	08 13 36.	+ 21 34		15.0	GALAXY
TON-N 0900	08 13 37.	+ 20 18		16.9	BLUE STAR
IC 2253	08 13 39.	+ 21 35 20.			NONSTELLAR OBJECT
IC 2251	08 13 41.	+ 24 06 02.			NONSTELLAR OBJECT
MCG+04-20-011	08 13 42.	+ 21 33	24	15.0	GALAXY
ZWG 119.025	08 13 42.	+ 24 56		15.4	GALAXY
ZC 0813.7+4141	08 13 42.	+ 41 41	1280		CLUSTER OF GALAXIES
IC 2252	08 13 43.	+ 24 50 56.			NONSTELLAR OBJECT
IC 2255	08 13 45.	+ 23 36 32.			NONSTELLAR OBJECT
TON-N 0901	08 13 48.	+ 21 47		16.5	BLUE STAR
MCG+04-20-012	08 13 48.	+ 24 20	36	15.2	GALAXY
IC 2254	08 13 48.	+ 24 55 43.			NONSTELLAR OBJECT
ZWG 149.010	08 13 48.	+ 27 45		15.7	GALAXY
MCG+08-15-057	08 13 48.	+ 45 48	42	16.	GALAXY
ARP 268	08 13 48.	+ 70 54			PECULIAR GALAXY
WRAY 19.05	08 13 50.6	- 35 24 34.			DIFFUSE NEBULA
ZC 0813.9+1008	08 13 54.	+ 10 08	1210		CLUSTER OF GALAXIES
ZWG 119.026	08 13 54.	+ 24 19		15.2	GALAXY
MCG+10-12-119	08 13 54.	+ 59 17	45	17.	GALAXY
ACK 251+01.1	08 13 54.	- 33 07			PLANETARY NEBULA
WRAY 19.06	08 13 55.0	- 35 28 52.		7.5	STAR-NEBULA ASSOCIATION
IC 2256	08 13 57.	+ 24 20 00.			NONSTELLAR OBJECT
TON-N 0902	08 13 59.	+ 34 35		17.	BLUE STAR
VDB.66G 050	08 14	+ 70 53	440		DWARF GALAXY
LB 09843	08 14	- 86 15		14.3	FAINT BLUE STAR
ZWG 119.027	08 14 00.	+ 20 40		15.5	GALAXY
MCG+10-12-120	08 14 00.	+ 57 55	42	15.	GALAXY
MCG+10-12-121	08 14 00.	+ 58 14	30	16.	GALAXY
ZC 0814.0+6023	08 14 00.	+ 60 23	1280		CLUSTER OF GALAXIES
MCG+12-08-033	08 14 00.	+ 70 52	438	11.1	GALAXY
VHE 09A	08 14 00.	- 35 59	18		REFLECTION NEBULA
ZC 0814.1+5005	08 14 06.	+ 50 05	740		CLUSTER OF GALAXIES
ZWG 287.067	08 14 06.	+ 57 55		15.3	GALAXY
MCG+10-12-122	08 14 06.	+ 59 16	36	17.	GALAXY
ZWG 331.034	08 14 06.	+ 70 52		11.3	GALAXY
UGC 04305	08 14 06.	+ 70 52	558	11.3	GALAXY DWRF IR
KARA.73B 0239	08 14 06.	+ 70 52	408	11.3	ISOLATED GALAXY IR
7ZW 223	08 14 06.	+ 70 53			COMPACT GALAXY
ZWG 003.022	08 14 06.	- 00 15		15.3	GALAXY
ZWG 003.021	08 14 06.	- 02 50		15.3	GALAXY
VHE 09B	08 14 06.	- 36 00	30		REFLECTION NEBULA
HO 2	08 14 07.	+ 70 52	600	11.14	IRR GALAXY
IC 2257	08 14 12.	+ 23 48 17.			NONSTELLAR OBJECT
ZWG 287.068	08 14 12.	+ 59 17		15.6	GALAXY
TON-N 0903	08 14 17.	+ 34 48		15.5	BLUE STAR
ZWG 059.056	08 14 18.	+ 11 44		15.4	GALAXY
TON-N 0306	08 14 18.	+ 23 01		14.6	BLUE STAR
IC 2258	08 14 18.	+ 23 43 58.			NONSTELLAR OBJECT
MCG+08-15-058	08 14 18.	+ 47 23 30.	18	16.	GALAXY
IC 2259	08 14 20.	+ 23 43 10.			NONSTELLAR OBJECT
ZWG 088.057	08 14 24.	+ 18 14		15.7	GALAXY
ZWG 178.038	08 14 24.	+ 35 36		15.1	GALAXY
UGC 04306	08 14 24.	+ 35 36	78	15.1	GALAXY S
MCG+06-18-015	08 14 24.	+ 35 36	42	15.	GALAXY
MCG+08-15-059	08 14 24.	+ 46 43	42	16.	GALAXY
ZWG 236.040	08 14 24.	+ 48 00		15.5	GALAXY
UGC 04307	08 14 24.	+ 48 00	90	14.4	GALAXY S
MCG+08-15-060	08 14 24.	+ 48 00	72	15.	GALAXY
ZWG 003.023	08 14 24.	- 00 36		15.6	GALAXY
CAR 52	08 14 24.4	+ 20 59 57.	6	16.5	NEBULA
IC 2260	08 14 29.	+ 24 49 45.			NONSTELLAR OBJECT
ZC 0814.5+0802	08 14 30.	+ 08 02	340		CLUSTER OF GALAXIES
ZWG 059.057	08 14 30.	+ 13 09		15.2	GALAXY
ZWG 119.028	08 14 30.	+ 21 19		15.5	GALAXY
ZWG 119.029	08 14 30.	+ 21 50		14.5	GALAXY
UGC 04308	08 14 30.	+ 21 50	132	14.5	GALAXY SBc
MCG+04-20-013	08 14 30.	+ 21 50	132	14.5	GALAXY
ZWG 207.053	08 14 30.	+ 38 57		14.9	GALAXY
MCG+08-15-061	08 14 30.	+ 49 23	60	15.	GALAXY
DG 123	08 14 30.	+ 70 52	480		REFLECTION NEBULA
IC 2262	08 14 31.	+ 18 36 38.			MAY NOT EXIST
SN 1962F	08 14 32.	+ 21 49		16.5	SUPERNOVA
MCG+07-17-027	08 14 33.	+ 38 56 30.	54	14.5	GALAXY
MCG+09-14-023	08 14 33.	+ 56 15	36	16.	GALAXY
ZWG 119.030	08 14 36.	+ 21 20		15.7	GALAXY
IC 2261	08 14 36.	+ 23 40 14.			NONSTELLAR OBJECT
ZWG 236.041	08 14 36.	+ 49 23		15.4	GALAXY
UGC 04309	08 14 36.	+ 49 23	60	15.4	GALAXY S
MCG+10-12-123	08 14 36.	+ 59 07	30	17.	GALAXY
ZC 0814.6+7549	08 14 36.	+ 75 49	1880		CLUSTER OF GALAXIES
ZWG 003.024	08 14 36.	- 00 42		15.6	GALAXY
BC 4C22.20	08 14 38.5	+ 22 46 34.		18.	QUASI-STELLAR OBJECT
CAR 51	08 14 38.8	+ 21 19 14.	38	16.4	NEBULA
SHB 109	08 14 39.	+ 22 46 34.		18.	QUASI-STELLAR OBJECT
CAR 50	08 14 41.2	+ 21 20 20.	16	16.3	NEBULA
ZWG 003.025	08 14 42.	+ 01 22		15.3	GALAXY
UGC 04310	08 14 42.	+ 01 22	78	15.3	GALAXY S IV
MCG+00-21-011	08 14 42.	+ 01 22	54	15.3	GALAXY
ZWG 119.031	08 14 42.	+ 21 04		15.0	GALAXY
TON-N 0307	08 14 42.	+ 30 42		15.3	BLUE STAR
ZC 0814.7+3825	08 14 42.	+ 38 25	1280		CLUSTER OF GALAXIES
MCG+09-14-024	08 14 42.	+ 56 44	6	17.	GALAXY
IC 2263	08 14 43.	+ 23 44 07.			NONSTELLAR OBJECT
MCG+04-20-014	08 14 45.	+ 21 02 30.	48	15.0	GALAXY
RNGC 2553	08 14 45.	+ 21 04		15.0	GALAXY
IC 2266	08 14 47.	+ 18 34 00.			MAY NOT EXIST
IC 2264	08 14 47.	+ 23 52 19.			NONSTELLAR OBJECT
CAR 49	08 14 47.9	+ 21 03 37.	17	15.5	NEBULA
ZWG 119.032	08 14 48.	+ 23 37		15.7	GALAXY
UGC 04311	08 14 48.	+ 37 33	66	16.0	GALAXY Sb
ZC 0814.8+5648	08 14 48.	+ 56 48	3760		CLUSTER OF GALAXIES
MCG+11-10-073	08 14 48.	+ 64 39	18	16.	GALAXY
ZC 0814.8-0055	08 14 48.	- 00 55	2690		CLUSTER OF GALAXIES
IC 2265	08 14 51.	+ 24 21 00.			NONSTELLAR OBJECT
RNGC 2549	08 14 53.	+ 57 58		12.5	GALAXY
ZWG 119.033	08 14 54.	+ 23 37		13.5	GALAXY
UGC 04312	08 14 54.	+ 23 37	210	13.5	GALAXY S0/Sa
MCG+04-20-015	08 14 54.	+ 23 38	48	13.5	GALAXY
TON-N 0308	08 14 54.	+ 32 15		15.4	BLUE STAR
ZC 0814.9+3344	08 14 54.	+ 33 44	540		CLUSTER OF GALAXIES
ZWG 287.069	08 14 54.	+ 57 58		12.1	GALAXY
UGC 04313	08 14 54.	+ 57 58	222	12.1	GALAXY S0
7ZW 224	08 14 54.	+ 58 25			COMPACT GALAXY
ZWG 287.070	08 14 54.	+ 58 26		14.9	GALAXY
UGC 04314	08 14 54.	+ 58 26	60	14.9	GALAXY DBL SYS
MCG-05-20-003	08 14 54.	- 29 24	78	15.	GALAXY
RNGC 2554	08 14 55.	+ 23 37		13.5	GALAXY
ZWG 119.034	08 15 00.	+ 21 16		15.6	GALAXY
ZWG 119.035	08 15 00.	+ 22 35		15.4	GALAXY
MCG+04-20-016	08 15 00.	+ 24 52 30.	132	15.1	GALAXY
ZWG 119.036	08 15 00.	+ 24 53		15.1	GALAXY
UGC 04315	08 15 00.	+ 24 53	138	15.1	GALAXY SB?c
MCG+09-14-025	08 15 00.	+ 56 31	48	17.	GALAXY
MCG+10-12-124	08 15 00.	+ 57 58	210	12.1	GALAXY
MCG+10-12-125	08 15 00.	+ 58 25	36	14.	GALAXY
MCG+10-12-126	08 15 00.	+ 59 54	18	16.	GALAXY
ZC 0815.0+8114	08 15 00.	+ 81 14	2020		CLUSTER OF GALAXIES
ABC 0644	08 15 02.	- 07 26		16.2	RICH CLUSTER OF GALAXIES
IC 2267	08 15 03.	+ 24 53 35.			NONSTELLAR OBJECT
RNGC 2559	08 15 04.	- 27 19		13.0	GALAXY
CAR 48	08 15 04.0	+ 21 16 05.	17	16.3	NEBULA
ZWG 003.026	08 15 06.	+ 00 49		15.7	GALAXY
ZWG 031.074	08 15 06.	+ 04 46		15.4	GALAXY
UGC 04316	08 15 06.	+ 04 46	96	15.4	GALAXY S IV-V
MCG+01-21-020	08 15 06.	+ 04 46	72	15.4	GALAXY
ZWG 059.058	08 15 06.	+ 13 03		15.1	GALAXY
UGC 04317	08 15 06.	+ 13 03	60	15.1	GALAXY SBb
ZWG 119.037	08 15 06.	+ 24 57		15.3	GALAXY
MCG+05-20-007	08 15 06.	+ 26 47	42	15.0	GALAXY
ZWG 149.011	08 15 06.	+ 26 48		15.0	GALAXY
MCG+09-14-026	08 15 06.	+ 56 17 30.	48	16.	GALAXY
IC 2270	08 15 08.	+ 19 15 10.			MAY NOT EXIST
IC 2268	08 15 08.	+ 24 57 11.			NONSTELLAR OBJECT
ABC 0639	08 15 08.	+ 68 05		17.7	RICH CLUSTER OF GALAXIES
CAR 47	08 15 08.6	+ 21 18 04.	8	16.7	NEBULA
MCG-04-20-003	08 15 09.	- 27 19	180	13.	GALAXY
TON-N 0904	08 15 11.	+ 32 38		16.8	BLUE STAR
ZWG 031.075	08 15 12.	+ 04 52		15.6	GALAXY
ZWG 088.058	08 15 12.	+ 20 07		15.4	GALAXY
ZWG 119.038	08 15 12.	+ 23 12		15.6	GALAXY
IC 2269	08 15 12.	+ 23 13 16.			NONSTELLAR OBJECT
TON-N 0309	08 15 12.	+ 27 17		14.8	BLUE STAR
ZC 0815.2+6803	08 15 12.	+ 68 03	610		CLUSTER OF GALAXIES
ZWG 003.027	08 15 12.	- 00 44		15.5	GALAXY
ISS 0915	08 15 12.	- 30 26	201		STELLAR RING
IC 2272	08 15 14.	+ 18 53 33.			MAY NOT EXIST
ABC 0636	08 15 14.	+ 72 55		16.8	RICH CLUSTER OF GALAXIES
ZWG 031.076	08 15 18.	+ 03 32		15.3	GALAXY
UGC 04318	08 15 18.	+ 03 32	60	15.3	GALAXY SB
ZWG 149.012	08 15 18.	+ 30 07		15.7	GALAXY
ZWG 207.054	08 15 18.	+ 39 18		15.0	GALAXY
ZC 0815.3+6637	08 15 18.	+ 66 37	1410		CLUSTER OF GALAXIES
MCG+11-10-074	08 15 18.	+ 67 10	66	14.	GALAXY
MCG+11-10-075	08 15 18.	+ 68 47	78	14.	GALAXY
HOLM 095B	08 15 21.	+ 00 54	18	14.6	PART OF MULTIPLE GALAXY
IC 2273	08 15 21.	+ 18 33 32.			MAY NOT EXIST
IC 2271	08 15 21.	+ 24 40 57.			NONSTELLAR OBJECT
HOLM 095C	08 15 22.	+ 00 55	18	14.8	PART OF MULTIPLE GALAXY
IC 2275	08 15 22.	+ 18 34 08.			MAY NOT EXIST
IC 2274	08 15 22.	+ 18 49 20.			THREE STARS
HOLM 095A	08 15 23.	+ 00 54	42	14.4	PART OF MULTIPLE GALAXY
ZWG 003.028	08 15 24.	+ 00 54		13.5	GALAXY
UGC 04319	08 15 24.	+ 00 54	132	13.5	GALAXY SBa
MCG+00-21-012	08 15 24.	+ 00 54	114	13.5	GALAXY
ZWG 003.029	08 15 24.	+ 02 20		15.5	GALAXY
ZWG 119.039	08 15 24.	+ 24 40		15.3	GALAXY
TON-N 0310	08 15 24.	+ 33 12		15.4	BLUE STAR
MCG+07-17-028	08 15 24.	+ 39 17 30.	24	15.5	GALAXY
ZWG 207.055	08 15 24.	+ 41 00		15.7	GALAXY
MCG+10-12-127	08 15 24.	+ 59 52	30	16.	GALAXY
ZWG 310.043	08 15 24.	+ 65 29		15.7	GALAXY
7ZW 225	08 15 24.	+ 67 35			COMPACT GALAXY
MCG-04-20-004	08 15 24.	- 24 30 30.	48	15.	GALAXY
RNGC 2555	08 15 25.	+ 00 54		13.5	GALAXY
BIGO 500	08 15 25.	+ 50 15			NEBULA
ZC 0815.5+1230	08 15 30.	+ 12 30	270		CLUSTER OF GALAXIES
ZWG 263.022	08 15 30.	+ 55 15		15.5	GALAXY
MCG+09-14-027	08 15 30.	+ 55 15 30.	42	16.	GALAXY
ZWG 287.071	08 15 30.	+ 59 52		15.7	GALAXY
UGC 04320	08 15 30.	+ 59 52	72	15.7	GALAXY S
VHE 10	08 15 30.	- 42 05	48		REFLECTION NEBULA
MCG+04-20-017	08 15 33.	+ 20 56	36	15.3	GALAXY
CAR 46	08 15 35.6	+ 20 56 37.	25	16.4	NEBULA
ZWG 031.077	08 15 36.	+ 04 55		15.7	GALAXY
MCG+01-21-021	08 15 36.	+ 04 55	36	15.7	GALAXY
ZWG 059.059	08 15 36.	+ 11 47		15.3	GALAXY
UGC 04321	08 15 36.	+ 11 47	66	15.7	GALAXY Sa-b
ZWG 119.040	08 15 36.	+ 20 56		15.6	GALAXY
ZWG 149.013	08 15 36.	+ 27 19		15.6	GALAXY
MCG+05-20-008	08 15 36.	+ 27 19	48	15.6	GALAXY
MCG+08-15-062	08 15 36.	+ 50 11	168	12.5	GALAXY
MCG+09-14-028	08 15 36.	+ 54 13	30	16.	GALAXY
ZC 0815.6+5902	08 15 36.	+ 59 02	1140		CLUSTER OF GALAXIES
ZWG 310.044	08 15 36.	+ 62 59		15.0	GALAXY
UGC 04322	08 15 36.	+ 62 59	84	15.0	GALAXY E
MCG+11-10-076	08 15 36.	+ 63 00	39	15.	GALAXY
ZWG 310.045	08 15 36.	+ 67 07		14.4	GALAXY
UGC 04323	08 15 36.	+ 67 07	120	14.4	GALAXY E?

OBJECT NAME	RIGHT ASCEN.	DECLINATION	DIAM.	MAGN.	TYPE OF OBJECT
IC 2276	08 15 37.	+ 18 38 07.			MAY NOT EXIST
CAR 45	08 15 39.2	+ 20 55 07.	33	16.2	NEBULA
IC 2277	08 15 40.	+ 18 48 30.			MAY NOT EXIST
RNGC 2552	08 15 40.	+ 50 10		12.5	GALAXY
ZWG 031.078	08 15 42.	+ 04 36		15.7	GALAXY
IC 2278	08 15 42.	+ 18 37 06.			MAY NOT EXIST
ZWG 119.041	08 15 42.	+ 20 55		15.3	GALAXY
UGC 04324	08 15 42.	+ 20 55	96	15.3	GALAXY S
ZC 0815.7+4137	08 15 42.	+ 41 37	1210		CLUSTER OF GALAXIES
ZC 0815.7+4555	08 15 42.	+ 45 55	740		CLUSTER OF GALAXIES
ZWG 236.042	08 15 42.	+ 50 10		13.5	GALAXY
UGC 04325	08 15 42.	+ 50 10	234	13.5	GALAXY IRR
ZC 0815.7+5240	08 15 42.	+ 52 40	940		CLUSTER OF GALAXIES
ZWG 331.035	08 15 42.	+ 68 47		13.8	GALAXY
UGC 04326	08 15 42.	+ 68 47	90	13.8	GALAXY Sb-c
KARA.73B 0240	08 15 42.	+ 68 47	96	13.8	ISOLATED GALAXY S
IC 2279	08 15 44.	+ 18 43 30.			MAY NOT EXIST
IC 2280	08 15 47.	+ 18 36 30.			MAY NOT EXIST
ZWG 119.042	08 15 48.	+ 24 41		15.5	GALAXY
IC 0501	08 15 48.	+ 24 42 00.			NONSTELLAR OBJECT
TON-N 0311	08 15 48.	+ 30 13		15.5	BLUE STAR
ARC 0643	08 15 48.	+ 52 40		17.1	RICH CLUSTER OF GALAXIES
PATH 2.010	08 15 48.	+ 74 08	14		NEBULA
MCG+13-06-022	08 15 48.	+ 79 22	78	17.	GALAXY
TON-N 0905	08 15 53.	+ 33 26		14.9	BLUF STAR
ZWG 149.014	08 15 54.	+ 28 10		15.6	GALAXY
ZWG 331.036	08 15 54.	+ 74 08		13.4	GALAXY
UGC 04327	08 15 54.	+ 74 08	60	13.4	GALAXY SBa
KARA.72 160A	08 15 54.	+ 74 08	72	13.4	PART OF DOUBLE GALAXY
RNGC 2544	08 15 54.	+ 74 09		13.5	GALAXY
OCL 0727	08 15 54.	- 36 56	276	11.	OPEN STAR CLUSTER
VHA 024	08 15 54.	- 36 56	180		OPEN STAR CLUSTER
CAR 43	08 15 57.3	+ 21 19 35.	16	16.7	NEBULA
CAR 44	08 15 57.4	+ 21 22 35.	35	16.2	NEBULA
CAR 42	08 15 58.7	+ 21 02 22.	16	16.5	NEBULA
TON-N 0906	08 15 59.	+ 32 32		17.	BLUE STAR
ZWG 119.043	08 16 00.	+ 21 22		15.5	GALAXY
ZWG 119.044	08 16 00.	+ 22 16		15.7	GALAXY
ZC 0816.0+3155	08 16 00.	+ 31 55	3760		CLUSTER OF GALAXIES
ZC 0816.0+6445	08 16 00.	+ 64 45	1080		CLUSTER OF GALAXIES
MCG+12-08-034	08 16 00.	+ 74 09	57	14.	GALAXY
ZWG 349.022	08 16 00.	+ 79 20		15.5	GALAXY
UGC 04328	08 16 00.	+ 79 20	102	15.5	GALAXY SB IV
CAR 41	08 16 00.2	+ 20 58 46.	13	16.7	NEBULA
IC 2281	08 16 02.	+ 19 03 58.			MAY NOT EXIST
PATH 1.210	08 16 03.	+ 74 09	33		NEBULA
CAR 40	08 16 04.0	+ 21 13 04.	10	16.6	NEBULA
ZWG 059.060	08 16 06.	+ 11 43		15.6	GALAXY
ZC 0816.1+7256	08 16 06.	+ 72 56	2220		CLUSTER OF GALAXIES
MCG+12-08-035	08 16 06.	+ 74 10	51	16.	GALAXY
IC 2284	08 16 07.	+ 18 45 51.			MAY NOT EXIST
CAR 39	08 16 08.1	+ 21 04 39.	17	16.0	NEBULA
CAR 38	08 16 09.4	+ 21 20 33.	32	16.5	NEBULA
RNGC 2568	08 16 10.	- 36 56			OPEN CLUSTER
IC 2285	08 16 12.	+ 19 04 15.			NONSTELLAR OBJECT
IC 2286	08 16 12.	+ 19 06 45.			NONSTELLAR OBJECT
ZWG 119.045	08 16 12.	+ 21 05		15.5	GALAXY
MCG+04-20-018	08 16 12.	+ 21 19	150	15.0	GALAXY
ZWG 119.046	08 16 12.	+ 21 20		15.0	GALAXY
UGC 04329	08 16 12.	+ 21 20	150	15.0	GALAXY Sc
ZWG 119.047	08 16 12.	+ 21 57		15.2	GALAXY
MCG+04-20-019	08 16 12.	+ 21 57 30.	60	15.2	GALAXY
MCG+04-20-020	08 16 12.	+ 24 57	60	15.3	GALAXY
MCG+06-19-001	08 16 12.	+ 35 10 30.	30	14.5	GALAXY
ZWG 179.001	08 16 12.	+ 35 13		14.6	GALAXY
ZWG 178.039	08 16 12.	+ 35 13		14.6	GALAXY
ZC 0816.2+4853	08 16 12.	+ 48 53	1680		CLUSTER OF GALAXIES
ZC 0816.2+4937	08 16 12.	+ 49 37	3090		CLUSTER OF GALAXIES
7ZW 226	08 16 12.	+ 56 54			COMPACT GALAXY
ZWG 331.037	08 16 12.	+ 74 09		15.5	GALAXY
KARA.72 160B	08 16 12.	+ 74 09	54	15.5	PART OF DOUBLE GALAXY
ASS 58	08 16 12.	- 35 37			OB ASSOCIATION PUP OB3
CAR 37	08 16 12.8	+ 21 24 15.	14	16.5	NEBULA
CAR 36	08 16 13.1	+ 21 18 27.	16	16.6	NEBULA
IC 2287	08 16 15.	+ 19 33 27.			NONSTELLAR OBJECT
RNGC 2556	08 16 15.	+ 21 05		15.5	GALAXY
MCG+04-20-021	08 16 15.	+ 21 34	60	14.6	GALAXY
MCG+08-15-063	08 16 15.	+ 47 08	18	17.	GALAXY
IC 2289	08 16 16.	+ 18 41 20.			MAY NOT EXIST
IC 2283	08 16 17.	+ 24 56 27.			NONSTELLAR OBJECT
IC 2282	08 16 17.	+ 24 56 57.			NONSTELLAR OBJECT
MCG+04-20-022	08 16 18.	+ 20 39	72	14.6	GALAXY
ZWG 119.048	08 16 18.	+ 21 36		14.6	GALAXY
UGC 04330	08 16 18.	+ 21 36	72	14.6	GALAXY SB0
MCG+04-20-023	08 16 18.	+ 23 55	18	15.5	GALAXY
ZWG 119.049	08 16 18.	+ 24 56		15.3	GALAXY
TON-N 0312	08 16 18.	+ 29 07		15.3	BLUE STAR
TON-N 0313	08 16 18.	+ 31 25		14.4	BLUE STAR
ZC 0816.3+5837	08 16 18.	+ 58 37	1550		CLUSTER OF GALAXIES
VHA 025	08 16 18.	- 41 27	240		OPFN STAR CLUSTER
OCL 0733	08 16 18.	- 41 28	258	15.	OPEN STAR CLUSTER
VHE 11	08 16 18.	- 41 57	144		REFLECTION NEBULA
CAR 35	08 16 18.1	+ 21 35 38.	22	15.4	NEBULA
CAR 34	08 16 19.8	+ 20 54 56.	32	16.6	NEBULA
RNGC 2557	08 16 20.	+ 21 36		14.5	GALAXY
CED 105	08 16 22.	- 36 58			DIFFUSE GALACTIC NEBULA
IC 2290	08 16 23.	+ 19 28 20.			MAY NOT EXIST
TON-N 0907	08 16 23.	+ 33 59		16.6	BLUE STAR
SEY 042	08 16 23.	+ 74 09 49.		15.3	FAINT GALAXY
ZWG 031.079	08 16 24.	+ 03 47		15.4	GALAXY
ZWG 089.001	08 16 24.	+ 19 28		15.2	GALAXY
ZWG 088.059	08 16 24.	+ 19 28		15.2	GALAXY
KARA.73B 0241	08 16 24.	+ 19 28	36	15.2	ISOLATFD GALAXY S0
ZWG 119.050	08 16 24.	+ 20 40		14.6	GALAXY
UGC 04331	08 16 24.	+ 20 40	132	14.6	GALAXY Sa-b
ZWG 119.051	08 16 24.	+ 20 55		15.5	GALAXY
T 5	08 16 24.	+ 21 13 00.		15.5	S GALAXY IN CANCER CLSTR
ZWG 119.052	08 16 24.	+ 23 54		15.5	GALAXY
KARA.73B 0242	08 16 24.	+ 23 54	48	15.5	ISOLATED GALAXY S
IC 2288	08 16 24.	+ 23 54 08.			NONSTELLAR OBJECT
ZC 0816.4+5050	08 16 24.	+ 50 50	470		CLUSTER OF GALAXIES
RNGC 2564	08 16 24.	- 21 39			GALAXY
CAR 33	08 16 25.5	+ 21 06 43.	8	16.7	NEBULA
IC 2291	08 16 26.	+ 18 39 55.			MAY NOT EXIST
CAR 32	08 16 26.1	+ 21 12 55.	22	16.2	NEBULA
CAR 31	08 16 26.2	+ 21 14 25.		16.7	NEBULA
RNGC 2558	08 16 27.	+ 20 40		14.5	GALAXY
IC 2292	08 16 29.	+ 19 43 19.			NONSTELLAR OBJECT
ZC 0816.5+0449	08 16 30.	+ 04 49	610		CLUSTER OF GALAXIES
T 6	08 16 30.	+ 21 09 42.		16.2	S GALAXY IN CANCER CLSTR
ZWG 119.053	08 16 30.	+ 21 13		15.5	GALAXY
ZC 0816.5+2324	08 16 30.	+ 23 24	3090		CLUSTER OF GALAXIES
ZWG 263.023	08 16 30.	+ 54 17		15.4	GALAXY
MCG+09-14-029	08 16 30.	+ 54 18	30	15.	GALAXY
7ZW 227	08 16 30.	+ 56 55			COMPACT GALAXY
MCG-04-20-005	08 16 30.	- 25 14	42	16.	GALAXY
CAR 30	08 16 30.4	+ 21 09 43.	13	16.6	NEBULA
IC 2294	08 16 34.	+ 19 08 37.			NONSTELLAR OBJECT
IC 2295	08 16 35.	+ 18 34 18.			NONSTELLAR OBJECT
TON-N 0908	08 16 35.	+ 33 05		15.5	BLUF STAR
KLEM 10	08 16 35.	- 25 17	600	11.	CMPT GROUP OF 4 GALAXIES
RNGC 2567	08 16 35.	- 30 29		8.5	OPEN CLUSTER
CAR 29	08 16 35.7	+ 21 09 48.		16.6	NEBULA
IC 2296	08 16 36.	+ 19 03 24.			NONSTELLAR OBJFCT
IC 2293	08 16 36.	+ 21 32 24.			NONSTELLAR OBJECT
ZC 0816.6+6413	08 16 36.	+ 64 13	2020		CLUSTER OF GALAXIES
ZC 0816.6+6603	08 16 36.	+ 66 03	1750		CLUSTER OF GALAXIES
MCG-04-20-006	08 16 36.	- 25 13	12	17.	GALAXY
OCL 0708	08 16 36.	- 30 29	780	8.6	OPEN STAR CLUSTER
VHA 026	08 16 36.	- 30 29	540		OPEN STAR CLUSTER
CAR 28	08 16 38.1	+ 21 33 06.	35	16.6	NEBULA
IC 2311	08 16 39.	- 25 12 47.			NONSTELLAR OBJECT
TON-N 0909	08 16 41.	+ 35 28		14.9	BLUE STAR
CAR 27	08 16 41.4	+ 21 23 42.		16.7	NEBULA
ZWG 031.080	08 16 42.	+ 03 56		15.7	GALAXY
T 4	08 16 42.	+ 21 16 18.		15.5	PEC GLXY IN CANCER CLSTR
MCG+04-20-024	08 16 42.	+ 21 32	48	15.2	GALAXY
ZWG 119.054	08 16 42.	+ 21 33		15.2	GALAXY
ZWG 207.056	08 16 42.	+ 44 23		15.2	GALAXY
ZC 0816.7+5519	08 16 42.	+ 55 19	2150		CLUSTER OF GALAXIES
MCG-04-20-007	08 16 42.	- 25 12	30	14.	GALAXY
MCG-04-20-008	08 16 42.	- 25 19	180	13.	GALAXY
LB 00184	08 16 43.	+ 14 37 48.		15.1	FAINT BLUE STAR
CAR 25	08 16 43.5	+ 21 16 27.	16	16.4	NEBULA
CAR 26	08 16 43.5	+ 21 26 35.	16	16.6	NEBULA
CAR 24	08 16 43.8	+ 21 13 27.	14	16.6	NEBULA
RNGC 2566	08 16 44.	- 25 19		13.0	GALAXY
MCG+04-20-025	08 16 45.	+ 21 15	84	15.5	GALAXY
TON-N 0910	08 16 47.	+ 32 30		17.	BLUE STAR
ZWG 089.002	08 16 48.	+ 20 04		15.5	GALAXY
ZWG 088.060	08 16 48.	+ 20 04		15.5	GALAXY
MCG+03-21-028	08 16 48.	+ 20 04	42	15.5	GALAXY
KARA.73B 0243	08 16 48.	+ 20 04	42	15.5	ISOLATED GALAXY S
ZL 030	08 16 48.	+ 21 10 42.		21.2	ULTRAFAINT BLUF STAR
T 3	08 16 48.	+ 21 11 42.		16.5	S GALAXY IN CANCER CLSTR
ZWG 119.055	08 16 48.	+ 21 16		15.5	GALAXY
UGC 04332	08 16 48.	+ 21 16	84	15.5	GALAXY PECULR
ZWG 119.056	08 16 49.	+ 22 12		15.5	GALAXY
UGC 04333	08 16 48.	+ 52 42	60	17.	GALAXY DWARF
MCG+09-14-030	08 16 48.	+ 54 17 30.	42	15.	GALAXY
ZWG 287.072	08 16 49.	+ 56 39		14.6	GALAXY
7ZW 228	08 16 48.	+ 60 19			COMPACT GALAXY
MCG+10-12-128	08 16 48.	+ 62 25	18	17.	GALAXY
CAR 23	08 16 48.7	+ 21 11 59.		16.7	NEBULA
ZL 031	08 16 50.	+ 21 15 42.		19.7	ULTRAFAINT BLUF STAR
MCG+09-14-031	08 16 51.	+ 56 39	24	15.	GALAXY
SN 1960M	08 16 53.	+ 22 12		15.7	SUPERNOVA
RNGC 2571	08 16 53.	- 29 35		7.5	OPEN CLUSTER
CAR 22	08 16 53.1	+ 21 23 34.	22	16.6	NEBULA
ZWG 119.057	08 16 54.	+ 22 11		13.8	GALAXY
MRK 386	08 16 54.	+ 22 11	30	15.5	GALAXY WITH UV CONTINUUM
UGC 04334	08 16 54.	+ 22 11	108	13.8	GALAXY SBb
MCG+04-20-026	08 16 54.	+ 22 11	96	13.8	GALAXY
ZWG 263.024	08 16 54.	+ 54 16		14.8	GALAXY
UGC 04335	08 16 54.	+ 54 16	84	14.8	GALAXY SBa
OCL 0701	08 16 54.	- 29 35	810	7.5	OPEN STAR CLUSTER
VHA 027	08 16 54.	- 29 35	300		OPEN STAR CLUSTER
ZL 032	08 16 56.	+ 21 06 12.		20.9	ULTRAFAINT BLUE STAR
RNGC 2565	08 16 56.	+ 22 11		14.0	GALAXY
CAR 21	08 16 56.1	+ 21 08 28.	54	15.1	NEBULA
RNGC 2561	08 16 59.	+ 04 49		14.0	GALAXY
LB 03482	08 17	- 84 20		14.0	FAINT BLUE STAR
ZWG 032.001	08 17 00.	+ 04 49		14.0	GALAXY
ZWG 031.081	08 17 00.	+ 04 49		14.0	GALAXY
UGC 04336	08 17 00.	+ 04 49	72	14.0	GALAXY SB
MCG+01-22-001	08 17 00.	+ 04 49	60	14.0	GALAXY
MCG+04-20-027	08 17 00.	+ 21 07 30.	84	14.9	GALAXY
ZWG 119.058	08 17 00.	+ 21 08		14.9	GALAXY
UGC 04337	08 17 00.	+ 21 08	102	14.9	GALAXY S0-a
ZL 033	08 17 00.	+ 21 18 42.		19.2	ULTRAFAINT BLUE STAR
ZWG 149.015	08 17 00.	+ 26 32		15.7	GALAXY
ZWG 149.016	08 17 00.	+ 27 45		15.4	GALAXY
ZC 0817.0+4007	08 17 00.	+ 40 07	610		CLUSTER OF GALAXIES
ZWG 263.025	08 17 00.	+ 55 25		15.6	GALAXY
UGC 04338	08 17 00.	+ 55 25	60	15.6	GALAXY SBa
ZC 0817.0+5635	08 17 00.	+ 56 35	1340		CLUSTER OF GALAXIES
ZC 0817.0+6207	08 17 00.	+ 62 07	4440		CLUSTER OF GALAXIES
ZWG 310.046	08 17 00.	+ 62 48		15.5	GALAXY
UGC 04339	08 17 00.	+ 62 48	84	15.5	GALAXY S
MCG+10-12-129	08 17 00.	+ 62 49	57	15.	GALAXY
MCG+14-04-049	08 17 00.	+ 86 07	45	15.	GALAXY
ZWG 363.041	08 17 00.	+ 86 09		15.2	GALAXY
LB 03482	08 17 00.	- 84 20		14.0	FAINT BLUE STAR
ZL 034	08 17 02.	+ 21 05 30.		21.5	ULTRAFAINT BLUE STAR
CAR 20	08 17 02.8	+ 21 13 21.	30	16.4	NEBULA
RNGC 2560	08 17 03.	+ 21 08		15.0	GALAXY
MCG+04-20-028	08 17 03.	+ 26 12	72	15.5	GALAXY
MCG+09-14-032	08 17 03.	+ 55 26 30.	15	16.	GALAXY
ZWG 060.001	08 17 06.	+ 14 02		15.6	GALAXY
ZWG 059.061	08 17 06.	+ 14 02		15.6	GALAXY
T 2	08 17 06.	+ 21 13 30.		15.7	S GALAXY IN CANCER CLSTR
ZWG 119.059	08 17 06.	+ 21 14		15.7	GALAXY
TON-N 0314	08 17 06.	+ 22 49		15.0	BLUE STAR
ZWG 119.060	08 17 06.	+ 26 10		15.5	GALAXY
UGC 04340	08 17 06.	+ 26 10	84	15.5	GALAXY Sc
MCG+05-20-009	08 17 06.	+ 27 14 30.	63	15.0	GALAXY
ZWG 149.017	08 17 06.	+ 27 15		15.0	GALAXY
UGC 04341	08 17 06.	+ 27 15	66	15.0	GALAXY S0-a
KARA.73B 0244	08 17 06.	+ 27 15	66	15.0	ISOLATED GALAXY S
ZWG 263.026	08 17 06.	+ 55 20		15.3	GALAXY
MCG+09-14-033	08 17 06.	+ 55 22	15	16.	GALAXY
MCG-01-22-001	08 17 06.	- 06 54	42	15.	GALAXY
ZL 036	08 17 07.	+ 21 13 06.		22.0	ULTRAFAINT BLUE STAR
ZL 035	08 17 07.	+ 21 19 06.		20.7	ULTRAFAINT BLUE STAR
LB 00539	08 17 10.	+ 11 33 36.		16.0	FAINT BLUE STAR
CAR 19	08 17 11.8	+ 20 53 50.	10	16.7	NEBULA

OBJECT NAME	RIGHT ASCEN.	DECLINATION	DIAM.	MAGN.	TYPE OF OBJECT
ZC 0817.2+3802	08 17 12.	+ 38 02	6050		CLUSTER OF GALAXIES
ZWG 207.057	08 17 12.	+ 40 27		15.3	GALAXY
UGC 04342	08 17 12.	+ 40 27	78	15.3	GALAXY S0-a
ZWG 310.047	08 17 12.	+ 66 41		15.5	GALAXY
MCG-04-20-009	08 17 12.	- 25 00 30.	36	15.	GALAXY
ISS 0916	08 17 12.	- 32 17	503		STELLAR RING
OCL 0717	08 17 12.	- 34 18	300	12.	OPEN STAR CLUSTER
IC 2297	08 17 13.	+ 18 32 32.			NONSTELLAR OBJECT
CAR 18	08 17 14.1	+ 21 13 32.	11	16.6	NEBULA
IC 2298	08 17 15.	+ 18 33 44.			NONSTELLAR OBJECT
MCG+04-20-029	08 17 15.	+ 21 12 30.	36	15.5	GALAXY
IC 2299	08 17 18.	+ 19 29 50.			NONSTELLAR OBJECT
T 1	08 17 18.	+ 21 13 42.		15.5	S GALAXY IN CANCER CLSTR
ZWG 119.061	08 17 18.	+ 21 14		15.5	GALAXY
CAR 17	08 17 19.2	+ 21 01 55.	32	16.6	NEBULA
IC 2300	08 17 20.	+ 18 34 43.			NONSTELLAR OBJECT
ZL 037	08 17 20.	+ 21 21 36.		20.6	ULTRAFAINT BLUE STAR
CAR 16	08 17 20.0	+ 20 59 55.	11	16.5	NEBULA
MCG+04-20-030	08 17 21.	+ 21 01	84	15.5	GALAXY
ZL 038	08 17 21.	+ 21 12 06.		21.3	ULTRAFAINT BLUE STAR
IC 2301	08 17 22.	+ 18 35 37.			NONSTELLAR OBJECT
ZL 039	08 17 23.	+ 21 06 54.		22.4	ULTRAFAINT BLUE STAR
ZC 0817.4+0820	08 17 24.	+ 08 20	610		CLUSTER OF GALAXIES
UGC 04343	08 17 24.	+ 17 30	66	17.	GALAXY DWRF IR
IC 2302	08 17 24.	+ 19 30 55.			NONSTELLAR OBJECT
ZWG 119.062	08 17 24.	+ 21 02		15.5	GALAXY
UGC 04344	08 17 24.	+ 21 02	102	15.5	GALAXY
ZWG 179.002	08 17 24.	+ 37 02		15.7	GALAXY
ZC 0817.4+5820	08 17 24.	+ 58 20	400		CLUSTER OF GALAXIES
MCG-04-20-010	08 17 24.	- 24 37	42	15.	GALAXY
SN 1960D	08 17 25.	+ 21 02		16.4	SUPERNOVA
CAR 15	08 17 26.3	+ 21 17 18.	35	14.6	NEBULA
IC 2303	08 17 27.	+ 19 34 37.			NONSTELLAR OBJECT
ZWG 119.063	08 17 30.	+ 21 17		14.0	GALAXY
UGC 04345	08 17 30.	+ 21 17	78	14.3	GALAXY S0-a
MCG+04-20-031	08 17 30.	+ 21 17	36	14.0	GALAXY
MCG+11-10-077	08 17 30.	+ 63 07	54	16.	GALAXY
MCG-04-20-011	08 17 30.	- 22 24	60	16.	GALAXY
LB 00540	08 17 31.	+ 13 11 06.		17.6	FAINT BLUE STAR
FATH 1.213	08 17 32.	+ 14 39	11		NEBULA
FATH 1.212	08 17 32.	+ 15 09	8		NEBULA
RNGC 2562	08 17 32.	+ 21 17		14.0	GALAXY
ZL 040	08 17 33.	+ 21 22 54.		21.4	ULTRAFAINT BLUE STAR
MCG+04-20-032	08 17 33.	+ 26 06	54	15.0	GALAXY
ZL 041	08 17 34.	+ 21 05 24.		20.4	ULTRAFAINT BLUE STAR
CAR 14	08 17 35.3	+ 21 15 47.	10	16.7	NEBULA
ZWG 032.002	08 17 36.	+ 04 52		15.2	GALAXY
ZC 0817.6+0748	08 17 36.	+ 07 48	740		CLUSTER OF GALAXIES
ZL 043	08 17 36.	+ 21 06 24.		17.5	ULTRAFAINT BLUE STAR
ZL 042	08 17 36.	+ 21 17 24.		21.7	ULTRAFAINT BLUE STAR
ZWG 119.064	08 17 36.	+ 26 04		15.0	GALAXY
UGC 04346	08 17 36.	+ 26 04	60	15.0	GALAXY SBa-b
MCG+08-15-064	08 17 36.	+ 45 25	36	16.	GALAXY
ZC 0817.6+5401	08 17 36.	+ 54 01	3230		CLUSTER OF GALAXIES
ZC 0817.6-0236	08 17 36.	- 02 36	3090		CLUSTER OF GALAXIES
MCG-01-22-002	08 17 36.	- 06 57	42	15.	GALAXY
CAR 13	08 17 37.4	+ 21 13 29.	19	14.9	NEBULA
CAR 05	08 17 38.3	+ 21 36 23.	13	16.6	NEBULA
ZL 044	08 17 40.	+ 21 23 24.		22.3	ULTRAFAINT BLUE STAR
CAR 12	08 17 41.0	+ 21 14 11.	10	16.5	NEBULA
ARC 0647	08 17 42.	+ 07 42		17.8	RICH CLUSTER OF GALAXIES
IC 2304	08 17 42.	+ 19 35 59.			NONSTELLAR OBJECT
ZWG 119.065	08 17 42.	+ 21 14		13.7	GALAXY
UGC 04347	08 17 42.	+ 21 14	132	13.7	GALAXY S0
T 7	08 17 42.	+ 21 14 18.		16.2	E GALAXY IN CANCER CLSTR
ZL 045	08 17 42.	+ 21 22 42.		22.2	ULTRAFAINT BLUE STAR
ZC 0817.7+4715	08 17 42.	+ 47 15	2620		CLUSTER OF GALAXIES
ZC 0817.7+7447	08 17 42.	+ 74 47	1140		CLUSTER OF GALAXIES
MCG-02-22-001	08 17 42.	- 10 18 30.	48	15.	GALAXY
CAR 11	08 17 43.7	+ 21 06 40.	6	16.3	NEBULA
MCG+04-20-033	08 17 45.	+ 21 12	30	13.7	GALAXY
RNGC 2563	08 17 45.	+ 21 14		13.5	GALAXY
IC 2306	08 17 47.	+ 19 16 11.			NONSTELLAR OBJECT
IC 2305	08 17 47.	+ 19 36 46.			NONSTELLAR OBJECT
ZWG 032.003	08 17 48.	+ 04 44		15.1	GALAXY
MCG+04-20-034	08 17 48.	+ 22 50	54	14.9	GALAXY
ZWG 207.058	08 17 48.	+ 39 26		14.8	GALAXY
ZC 0817.8+5311	08 17 48.	+ 53 11	1140		CLUSTER OF GALAXIES
MCG+09-14-034	08 17 48.	+ 56 27 30.	42	16.	GALAXY
OCL 0744	08 17 48.	- 45 15	18	15.	OPEN STAR CLUSTER
ZL 046	08 17 49.	+ 21 08 42.		22.3	ULTRAFAINT BLUE STAR
LB 00541	08 17 50.	+ 79 02 48.		15.9	FAINT BLUE STAR
IC 2309	08 17 51.	+ 18 33 22.			NONSTELLAR OBJECT
IC 2307	08 17 51.	+ 19 35 58.			NONSTELLAR OBJECT
MCG+07-17-029	08 17 51.	+ 39 25	15	16.	GALAXY
CAR 10	08 17 51.9	+ 21 12 45.	8	16.1	NEBULA
FATH 1.214	08 17 52.	+ 14 40	11		NEBULA
IC 2308	08 17 52.	+ 19 31 22.			NONSTELLAR OBJECT
ZWG 089.003	08 17 54.	+ 18 33		15.6	GALAXY
IC 2310	08 17 54.	+ 18 37 22.			NONSTELLAR OBJECT
ZWG 089.004	08 17 54.	+ 19 35		15.5	GALAXY
T 8	08 17 54.	+ 21 12 48.		16.0	E GALAXY IN CANCER CLSTR
ZWG 119.066	08 17 54.	+ 22 49		14.9	GALAXY
ZC 0817.9+3935	08 17 54.	+ 39 35	1010		CLUSTER OF GALAXIES
ZWG 263.027	08 17 54.	+ 56 26		15.6	GALAXY
CAR 09	08 17 58.8	+ 21 24 39.		16.4	NEBULA
LBN 0729	08 18	+ 60 25	1140		BRIGHT NEBULA
ZWG 089.005	08 18 00.	+ 16 48		15.0	GALAXY
SCH 02	08 18 00.	+ 16 48		15.0	PECULIAR GALAXY
UGC 04350	08 18 00.	+ 16 48	60	15.0	GALAXY SBb
ZWG 089.006	08 18 00.	+ 19 31		15.6	GALAXY
MCG+03-22-001	08 18 00.	+ 19 31	18	15.6	GALAXY
UGC 04351	08 18 00.	+ 35 58	60	16.0	GALAXY
MCG+10-12-130	08 18 00.	+ 57 28	36	15.	GALAXY
MCG+11-10-078	08 18 00.	+ 67 09	72	14.	GALAXY
ZWG 363.042	08 18 00.	+ 86 07		15.7	GALAXY
UGC 04348	08 18 00.	+ 86 07	108	15.7	GALAXY PECULR
ZWG 004.001	08 18 00.	- 01 13		15.1	GALAXY
UGC 04349	08 18 00.	- 01 13	84	15.1	GALAXY SB?b-c
MCG+00-22-001	08 18 00.	- 01 13	72	15.1	GALAXY
ZL 047	08 18 01.	+ 21 17 24.		22.2	ULTRAFAINT BLUE STAR
IC 2312	08 18 02.	+ 18 40 03.			NONSTELLAR OBJECT
ZL 048	08 18 02.	+ 21 19 06.		22.1	ULTRAFAINT BLUE STAR
MCG+03-22-002	08 18 03.	+ 16 46	60	15.0	GALAXY
IC 2313	08 18 03.	+ 18 40 27.			NONSTELLAR OBJECT
MCG+03-22-003	08 18 06.	+ 16 46	24	16.	GALAXY
ZC 0818.1+3027	08 18 06.	+ 30 27	2490		CLUSTER OF GALAXIES
CAR 08	08 18 06.7	+ 21 19 44.		16.7	NEBULA
CAR 07	08 18 10.8	+ 21 33 31.	13	16.6	NEBULA
IC 2314	08 18 11.	+ 18 55 20.			NONSTELLAR OBJECT
ZWG 311.001	08 18 12.	+ 67 06		15.3	GALAXY
ZWG 310.048	08 18 12.	+ 67 06		15.3	GALAXY
UGC 04353	08 18 12.	+ 67 06	72	15.3	GALAXY Sc
ZC 0818.2+6715	08 18 12.	+ 67 15	1410		CLUSTER OF GALAXIES
ZWG 004.002	08 18 12.	- 01 15		15.0	GALAXY
UGC 04352	08 18 12.	- 01 15	78	15.0	GALAXY TRP SYS
MCG+00-22-003	08 18 12.	- 01 15	48	15.0	GALAXY
MCG+00-22-002	08 18 12.	- 01 15	48	15.0	GALAXY
CAR 06	08 18 15.3	+ 21 37 37.	13	16.6	NEBULA
IC 2315	08 18 18.	+ 19 04 31.			NONSTELLAR OBJECT
ZWG 207.059	08 18 18.	+ 42 23		15.7	GALAXY
ZC 0818.3+5951	08 18 18.	+ 59 51	2220		CLUSTER OF GALAXIES
CAR 04	08 18 18.1	+ 21 36 12.	13	16.5	NEBULA
CAR 03	08 18 20.1	+ 21 01 30.	14	15.7	NEBULA
IC 2316	08 18 23.	+ 19 55 07.			NONSTELLAR OBJECT
RNGC 2574	08 18 23.	- 08 47		14.0	GALAXY
ZWG 207.060	08 18 24.	+ 42 01		15.4	GALAXY
MCG+08-15-065	08 18 24.	+ 45 43	36	16.	GALAXY
ZC 0818.4+5746	08 18 24.	+ 57 46	1410		CLUSTER OF GALAXIES
ZWG 331.038	08 18 24.	+ 70 26		15.7	GALAXY
MCG-01-22-003	08 18 24.	- 08 47	120	14.	GALAXY
CAR 02	08 18 28.9	+ 21 04 05.	17	16.3	NEBULA
IC 2317	08 18 29.	+ 19 00 12.			NONSTELLAR OBJECT
MCG+04-20-035	08 18 30.	+ 21 01	18	15.3	GALAXY
ZWG 119.067	08 18 30.	+ 21 02		15.3	GALAXY
MCG+04-20-036	08 18 30.	+ 21 03 30.	54	15.4	GALAXY
ZWG 119.068	08 18 30.	+ 21 05		15.4	GALAXY
UGC 04354	08 18 30.	+ 21 05	78	15.4	GALAXY Sa-b
ZC 0818.5+4605	08 18 30.	+ 46 05	810		CLUSTER OF GALAXIES
MCG+14-04-050	08 18 30.	+ 86 06	78	15.	GALAXY
VHE 12B	08 18 30.	- 49 58	18		REFLECTION NEBULA
CAR 01	08 18 32.7	+ 21 17 17.	13	16.4	NEBULA
RNGC 2569	08 18 33.	+ 21 02		15.5	GALAXY
RNGC 2570	08 18 33.	+ 21 05		15.5	GALAXY
RNGC 2572	08 18 34.	+ 19 18		15.0	GALAXY
ARC 0645	08 18 34.	+ 56 44		17.7	RICH CLUSTER OF GALAXIES
ARC 0646	08 18 35.	+ 47 16		16.8	RICH CLUSTER OF GALAXIES
MCG+03-22-004	08 18 36.	+ 19 17	33	14.8	GALAXY
ZWG 089.007	08 18 36.	+ 19 18		14.8	GALAXY
UGC 04355	08 18 36.	+ 19 18	102	14.8	GALAXY Sa?
ZWG 119.069	08 18 36.	+ 21 17		15.7	GALAXY
ZC 0818.6+2702	08 18 36.	+ 27 02	3830		CLUSTER OF GALAXIES
ZWG 207.061	08 18 36.	+ 38 32		15.7	GALAXY
ZWG 310.049	08 18 36.	+ 62 31		15.7	GALAXY
IC 2319	08 18 41.	+ 18 38 11.			NONSTELLAR OBJECT
IC 2318	08 18 41.	+ 18 46 59.			NONSTELLAR OBJECT
ZC 0818.7+3241	08 18 42.	+ 32 41	940		CLUSTER OF GALAXIES
ZC 0818.7+6811	08 18 42.	+ 68 11	2820		CLUSTER OF GALAXIES
MCG+12-08-036	08 18 42.	+ 70 25	33	16.	GALAXY
KARA.68 052	08 18 42.	+ 71 13	60		DWARF GALAXY
FATH 2.011	08 18 42.	+ 74 10	41		NEBULA
IC 2320	08 18 43.	+ 18 49 46.			NONSTELLAR OBJECT
MCG+09-14-035	08 18 45.	+ 56 30	15	15.	GALAXY
IC 2321	08 18 47.	+ 18 37 46.			NONSTELLAR OBJECT
IC 2322	08 18 47.	+ 18 38 40.			NONSTELLAR OBJECT
ARC 0625	08 18 47.	+ 82 32		16.7	RICH CLUSTER OF GALAXIES
ZC 0818.8+0124	08 18 48.	+ 01 24	2220		CLUSTER OF GALAXIES
ZWG 032.004	08 18 48.	+ 03 20		13.9	GALAXY
UGC 04356	08 18 48.	+ 03 20	90	13.9	GALAXY Sa?
MCG+01-22-002	08 18 48.	+ 03 20	72	13.9	GALAXY
ZWG 287.073	08 18 48.	+ 56 28		14.9	GALAXY
ZWG 263.028	08 18 48.	+ 56 28		14.9	GALAXY
UGC 04357	08 18 48.	+ 56 28	84	14.9	GALAXY S0
ZC 0818.8+5646	08 18 48.	+ 56 46	340		CLUSTER OF GALAXIES
IC 2323	08 18 49.	+ 18 46 22.			NONSTELLAR OBJECT
RNGC 2550	08 18 49.	+ 74 10		13.0	GALAXY
IC 2327	08 18 52.	+ 03 20 08.			NONSTELLAR OBJECT
ZC 0818.9+4546	08 18 54.	+ 45 46	1010		CLUSTER OF GALAXIES
ZC 0818.9+5021	08 18 54.	+ 50 21	1950		CLUSTER OF GALAXIES
ZWG 331.039	08 18 54.	+ 74 10		13.1	GALAXY
UGC 04359	08 18 54.	+ 74 10	66	13.1	GALAXY S
UGC 04360	08 18 54.	+ 78 24	66	16.5	GALAXY
UGC 04358	08 18 54.	- 00 15	60	16.0	GALAXY
MRSL 254+00/1	08 18 54.	- 36 03	120		HII REGION
WRAY 19.07	08 18 57.1	- 35 07 43.			DIFFUSE NEBULA
VDB.66G 051	08 19	+ 74 38	70		DWARF GALAXY
ZWG 032.005	08 19 00.	+ 04 42		14.9	GALAXY
MCG+01-22-003	08 19 00.	+ 04 42	42	14.9	GALAXY
ZWG 119.070	08 19 00.	+ 22 48		15.6	GALAXY
UGC 04361	08 19 00.	+ 22 48	72	15.6	GALAXY Sc
MCG+04-20-037	08 19 00.	+ 22 49	66	15.6	GALAXY
7ZW 229	08 19 00.	+ 64 06			COMPACT GALAXY
MCG+12-08-037	08 19 00.	+ 74 11	66	15.	GALAXY
ZC 0819.0+8233	08 19 00.	+ 82 33	2290		CLUSTER OF GALAXIES
OCL 0749	08 19 00.	- 36 51	132	13.	OPEN STAR CLUSTER
RNGC 2578	08 19 02.	- 13 09		14.0	GALAXY
MCG-02-22-002	08 19 03.	- 13 09	108	14.	GALAXY
PK263-05.1	08 19 03.	- 46 13	10		PLANETARY NEBULA
TON-N 0911	08 19 05.	+ 34 10		16.5	BLUE STAR
IC 2324	08 19 06.	+ 19 21 14.			NONSTELLAR OBJECT
ZWG 119.071	08 19 06.	+ 21 30		15.7	GALAXY
ARC 0648	08 19 06.	+ 32 42		17.7	RICH CLUSTER OF GALAXIES
ZWG 331.040	08 19 06.	+ 73 35		12.7	GALAXY
UGC 04362	08 19 06.	+ 73 35	114	12.7	GALAXY S
VHE 13B	08 19 06.	- 36 05	36		REFLECTION NEBULA
VHE 13A	08 19 06.	- 36 05	78		REFLECTION NEBULA
VHE 12A	08 19 06.	- 50 09	144		REFLECTION NEBULA
RNGC 2579	08 19 09.	- 36 01		9.0	CLUSTER WITH NEBULOSITY
RNGC 2551	08 19 11.	+ 73 35		13.0	GALAXY
FATH 1.211	08 19 11.	+ 74 34	19		NEBULA
MCG+04-20-038	08 19 12.	+ 21 29	30	15.7	GALAXY
ZWG 310.050	08 19 12.	+ 64 30		15.4	GALAXY
ZC 0819.2+6958	08 19 12.	+ 69 58	1010		CLUSTER OF GALAXIES
OCL 0724	08 19 12.	- 36 01	600	9.3	OPEN STAR CLUSTER
ARC 0653	08 19 13.	+ 01 23		17.0	RICH CLUSTER OF GALAXIES
MCG-02-22-003	08 19 15.	- 13 12	36	15.	GALAXY
IC 2325	08 19 16.	+ 19 04 19.			NONSTELLAR OBJECT
MAI 027	08 19 17.	+ 85 04	40		DWARF SPHEROIDAL GALAXY
ZWG 060.002	08 19 18.	+ 08 55		15.4	GALAXY
KARA.73B 0245	08 19 18.	+ 08 55	24	15.4	ISOLATED GALAXY E
ZC 0819.3+2923	08 19 18.	+ 29 23	3020		CLUSTER OF GALAXIES
MCG+12-08-038	08 19 18.	+ 73 34	78	14.	GALAXY
MCG+12-08-039	08 19 18.	+ 74 35	66	13.2	GALAXY
ZWG 349.023	08 19 18.	+ 74 36		14.9	GALAXY
UGC 04363	08 19 18.	+ 74 36	90	14.9	GALAXY SBc
IC 2326	08 19 19.	+ 19 10 18.			NONSTELLAR OBJECT

OBJECT NAME	RIGHT ASCEN.	DECLINATION	DIAM.	MAGN.	TYPE OF OBJECT
IC 0502	08 19 21.	+ 08 54 53.			NONSTELLAR OBJECT
MCG+04-20-039	08 19 21.	+ 25 40	54	15.4	GALAXY
ZWG 119.072	08 19 24.	+ 21 15		15.5	GALAXY
ZWG 119.073	08 19 24.	+ 25 39		15.4	GALAXY
UGC 04364	08 19 24.	+ 25 39	90	15.4	GALAXY Sa/SBb
MCG+06-19-002	08 19 24.	+ 37 15 30.	42	15.	GALAXY
ZWG 179.003	08 19 24.	+ 37 16		15.5	GALAXY
ZC 0819.4+5420	08 19 24.	+ 54 20	610		CLUSTER OF GALAXIES
ZWG 310.051	08 19 24.	+ 62 26		15.4	GALAXY
ZWG 287.074	08 19 24.	+ 62 26		15.4	GALAXY
MCG+10-12-131	08 19 24.	+ 62 26 30.	15	15.	GALAXY
ZWG 349.024	08 19 24.	+ 74 40		14.7	GALAXY
ISS 0917	08 19 24.	- 29 33	89		STELLAR RING
ACK 261-04.1	08 19 24.	- 44 14			PLANETARY NEBULA
IC 2328	08 19 25.	+ 19 46 36.			NONSTELLAR OBJECT
IC 2329	08 19 28.	+ 19 34 36.			NONSTELLAR OBJECT
ARC 0651	08 19 29.	+ 16 16		17.6	RICH CLUSTER OF GALAXIES
ZWG 089.008	08 19 30.	+ 19 34		15.0	GALAXY
UGC 04365	08 19 30.	+ 19 34	132	15.0	GALAXY Sc-IRR
MCG+03-22-005	08 19 30.	+ 19 34	120	15.0	GALAXY
ZWG 207.062	08 19 30.	+ 38 39		15.5	GALAXY
MCG+12-08-040	08 19 30.	+ 74 39	72	15.	GALAXY
IC 2330	08 19 31.	+ 19 00 53.			NONSTELLAR OBJECT
WRAY 19.08	08 19 33.9	- 42 34 54.			DIFFUSE NEBULA
IC 0503	08 19 34.	+ 03 24 21.			NONSTELLAR OBJECT
ARC 0650	08 19 34.	+ 18 44		17.5	RICH CLUSTER OF GALAXIES
RNGC 2580	08 19 35.	- 30 10		9.5	OPEN CLUSTER
ZWG 032.006	08 19 36.	+ 03 25		14.0	GALAXY
UGC 04366	08 19 36.	+ 03 25	66	14.0	GALAXY SBa
MCG+01-22-004	08 19 36.	+ 03 25	72	14.0	GALAXY
ZWG 089.009	08 19 36.	+ 15 40		15.4	GALAXY
ZC 0819.6+2209	08 19 36.	+ 22 09	14520		CLUSTER OF GALAXIES
ZC 0819.6+5636	08 19 36.	+ 56 36	810		CLUSTER OF GALAXIES
SEY 044	08 19 36.	+ 74 39 37.		14.5	FAINT GALAXY
OCL 0709	08 19 36.	- 30 09	540	9.8	OPEN STAR CLUSTER
VHA 028	08 19 36.	- 30 09	360		OPEN STAR CLUSTER
ZWG 032.007	08 19 42.	+ 05 10		15.5	GALAXY
ZWG 004.003	08 19 42.	- 01 54		15.6	GALAXY
IC 2331	08 19 43.	+ 19 50 34.			NONSTELLAR OBJECT
MCG+04-20-040	08 19 45.	+ 24 27 30.	120	14.3	GALAXY
MCG+11-10-079	08 19 45.	+ 63 49	18	17.	GALAXY
MCG-01-22-004	08 19 45.	- 08 02	72	15.5	GALAXY
IC 2332	08 19 46.	+ 20 05 04.			NONSTELLAR OBJECT
REIN 2.062	08 19 47.05	+ 22 42 50.4			NEBULA
ZWG 119.074	08 19 48.	+ 22 43		13.8	GALAXY
UGC 04367	08 19 48.	+ 22 43	108	13.8	GALAXY E-S0
ZWG 119.075	08 19 48.	+ 24 27		14.3	GALAXY
UGC 04368	08 19 48.	+ 24 27	156	14.3	GALAXY Sc
MCG+04-20-041	08 19 48.	+ 25 55	90	15.4	GALAXY
UGC 04369	08 19 48.	+ 65 47	60	16.0	GALAXY SBc
MCG-02-22-004	08 19 48.	- 13 27	36	15.	GALAXY
RNGC 2575	08 19 49.	+ 24 27		14.5	GALAXY
RNGC 2577	08 19 50.	+ 22 43		14.0	GALAXY
MCG+04-20-042	08 19 51.	+ 22 43	42	13.8	GALAXY
MCG+11-10-080	08 19 51.	+ 65 45	45	15.	GALAXY
ZWG 004.006	08 19 54.	+ 02 29		15.7	GALAXY
MCG+00-22-005	08 19 54.	+ 02 29	36	15.7	GALAXY
IC 0504	08 19 54.	+ 04 24 55.			NONSTELLAR OBJECT
TON-N 0912	08 19 54.	+ 21 34		14.8	BLUE STAR
ZWG 119.076	08 19 54.	+ 25 54		15.4	GALAXY
RNGC 2576	08 19 54.	+ 25 54		15.5	GALAXY
UGC 04371	08 19 54.	+ 25 54	114	15.4	GALAXY Sb
MCG+05-20-010	08 19 54.	+ 27 51	54	15.5	GALAXY
TON-N 0315	08 19 54.	+ 32 13		15.8	BLUE STAR
ZC 0819.9+4424	08 19 54.	+ 44 24	4100		CLUSTER OF GALAXIES
ZWG 004.005	08 19 54.	- 00 53		14.9	GALAXY
UGC 04370	08 19 54.	- 00 53	78	14.9	GALAXY Sc
MCG+00-22-004	08 19 54.	- 00 53	60	14.9	GALAXY
ZWG 004.004	08 19 54.	- 00 57		15.5	GALAXY
SHB 110	08 20	+ 15		19.	QUASI-STELLAR OBJECT
ZWG 032.008	08 20 00.	+ 04 25		14.3	GALAXY
UGC 04372	08 20 00.	+ 04 25	66	14.3	GALAXY S0
MCG+01-22-005	08 20 00.	+ 04 25	60	14.3	GALAXY
ZWG 032.009	08 20 00.	+ 04 29		15.1	GALAXY
ZWG 119.077	08 20 00.	+ 21 15		15.4	GALAXY
ZWG 149.018	08 20 00.	+ 27 52		15.5	GALAXY
UGC 04373	08 20 00.	+ 27 52	78	15.5	GALAXY SB:c
MCG+10-12-132	08 20 00.	+ 58 25	39	16.	GALAXY
ZC 0820.0+5834	08 20 00.	+ 58 34	1010		CLUSTER OF GALAXIES
MCG+11-11-001	08 20 00.	+ 67 02	30	15.	GALAXY
MCG+11-11-002	08 20 00.	+ 67 30	45	17.	GALAXY
MCG-02-22-005	08 20 03.	- 11 15	108	14.5	GALAXY
WRAY 19.09	08 20 04.2	- 42 27 51.			DIFFUSE NEBULA
ZC 0820.1+0647	08 20 06.	+ 06 47	2420		CLUSTER OF GALAXIES
ZWG 060.003	08 20 06.	+ 13 47		15.2	GALAXY
MCG+10-12-133	08 20 06.	+ 62 46	30	16.	GALAXY
ZC 0820.1+6438	08 20 06.	+ 64 38	740		CLUSTER OF GALAXIES
ZC 0820.1-0029	08 20 06.	- 00 29	4370		CLUSTER OF GALAXIES
IC 2334	08 20 08.	+ 18 46 31.			NONSTELLAR OBJECT
IC 2333	08 20 08.	+ 19 14 31.			NONSTELLAR OBJECT
ZWG 032.010	08 20 12.	+ 03 44		14.7	GALAXY
UGC 04374	08 20 12.	+ 03 44	102	14.7	GALAXY SBc
MCG+01-22-006	08 20 12.	+ 03 44	84	14.7	GALAXY
ZWG 119.078	08 20 12.	+ 22 49		14.6	GALAXY
UGC 04375	08 20 12.	+ 22 49	162	14.6	GALAXY Sc
ZWG 149.019	08 20 12.	+ 29 38		15.6	GALAXY
MCG+09-14-036	08 20 12.	+ 53 29	36	16.	GALAXY
ZWG 263.029	08 20 12.	+ 53 30		15.0	GALAXY
MCG+10-12-134	08 20 12.	+ 61 29	30	16.	GALAXY
ZWG 310.052	08 20 12.	+ 62 46		15.6	GALAXY
ZWG 311.002	08 20 12.	+ 67 00		15.1	GALAXY
ZWG 310.053	08 20 12.	+ 67 00		15.1	GALAXY
UGC 04376	08 20 12.	+ 67 00	66	15.1	GALAXY
UGC 04377	08 20 12.	+ 67 29	84	16.5	GALAXY Sc
ZWG 004.007	08 20 12.	- 01 38		15.7	GALAXY
MRSL 260-03/1	08 20 12.	- 42 28	300		HII REGION
IC 2335	08 20 14.	+ 19 34 19.			NONSTELLAR OBJECT
MCG+04-20-043	08 20 15.	+ 22 50	150	14.6	GALAXY
MCG-01-22-005	08 20 15.	- 08 07	42	15.5	GALAXY
ZWG 032.011	08 20 18.	+ 04 17		15.2	GALAXY
ZWG 032.012	08 20 18.	+ 04 26		14.6	GALAXY
MCG+01-22-007	08 20 18.	+ 04 26	15	14.6	GALAXY
ZWG 032.013	08 20 18.	+ 04 33		15.2	GALAXY
ZC 0820.3+1603	08 20 18.	+ 16 03	3160		CLUSTER OF GALAXIES
SHAH 094	08 20 18.	+ 54 03	150	17.5	GROUP OF COMPACT GALAXIES
UGC 04378	08 20 18.	+ 70 04	60	17.	GALAXY DWARF
MCG+05-20-011	08 20 24.	+ 27 17	60	15.4	GALAXY
MCG+10-12-135	08 20 24.	+ 57 26	12	16.	GALAXY
MCG+10-12-136	08 20 24.	+ 61 42 30.	78	15.	GALAXY
ZWG 004.008	08 20 24.	- 00 12		15.2	GALAXY
VHE 15D	08 20 24.	- 41 55			REFLECTION NEBULA
IC 2336	08 20 27.	+ 18 41 53.			NONSTELLAR OBJECT
ARC 0657	08 20 28.	+ 16 07		17.4	RICH CLUSTER OF GALAXIES
IC 2337	08 20 28.	+ 18 41 47.			NONSTELLAR OBJECT
ZWG 089.010	08 20 30.	+ 18 42		15.7	GALAXY
ZWG 149.020	08 20 30.	+ 27 18		15.4	GALAXY
ZWG 287.075	08 20 30.	+ 61 42		15.4	GALAXY
UGC 04379	08 20 30.	+ 61 42	78	15.4	GALAXY SBb
KARA.73B 0246	08 20 30.	+ 61 42	90	15.4	ISOLATED GALAXY S
MCG-01-22-006	08 20 30.	- 04 47	15	16.	GALAXY
VHE 14A	08 20 30.	- 40 17	30		REFLECTION NEBULA
VMT 11	08 20 30.	- 42 50	4800		SUPERNOVA REMNANT
MIL 13	08 20 30.	- 42 50	3300		SUPERNOVA REMNANT
HELW 417	08 20 31.	- 04 47 08.			NEBULA
HELW 418	08 20 32.	- 04 46 20.			NEBULA
IC 0505	08 20 35.	+ 04 32 03.			NONSTELLAR OBJECT
ARC 0649	08 20 35.	+ 49 02		17.3	RICH CLUSTER OF GALAXIES
ZC 0820.6+0436	08 20 36.	+ 04 36	3700		CLUSTER OF GALAXIES
ZWG 032.014	08 20 36.	+ 07 50		15.7	GALAXY
IC 2340	08 20 36.	+ 18 54 10.			NONSTELLAR OBJECT
ZWG 089.011	08 20 36.	+ 18 55		14.6	GALAXY
ZWG 119.079	08 20 36.	+ 25 10		15.7	GALAXY
ZWG 263.030	08 20 36.	+ 55 00		15.1	GALAXY
UGC 04380	08 20 36.	+ 55 00	78	15.1	GALAXY Sc
MCG-01-22-007	08 20 36.	- 04 47 30.	42	16.	GALAXY
OCL 0725	08 20 36.	- 36 00	540	10.1	OPEN STAR CLUSTER
HELW 419	08 20 38.	- 04 46 51.			NEBULA
MCG+09-14-037	08 20 39.	+ 55 00	60	15.	GALAXY
IC 2342	08 20 40.	+ 18 44 34.			NONSTELLAR OBJECT
RNGC 2583	08 20 40.	- 04 51		14.5	GALAXY
ZWG 032.015	08 20 42.	+ 04 32		14.8	GALAXY
UGC 04382	08 20 42.	+ 04 32	84	14.8	GALAXY S
MCG+01-22-008	08 20 42.	+ 04 32	48	14.8	GALAXY
ZWG 060.004	08 20 42.	+ 09 34		15.5	GALAXY
ZWG 089.012	08 20 42.	+ 16 24		15.6	GALAXY
MCG+03-22-006	08 20 42.	+ 18 54	48	14.6	GALAXY
KARA.72 161A	08 20 42.	+ 21 29	42	14.7	PART OF DOUBLE GALAXY
ZWG 119.080	08 20 42.	+ 21 30		14.7	GALAXY
UGC 04383	08 20 42.	+ 21 30	90	14.7	GALAXY DBL SYS
MCG+04-20-044	08 20 42.	+ 21 30	30	14.7	GALAXY
IC 2339	08 20 42.	+ 21 30		14.7	GALAXY IN CANCER CLST
IC 2338	08 20 42.	+ 21 30		14.7	GALAXY IN CANCER CLST
MCG+04-20-045	08 20 42.	+ 21 30 30.	60	14.7	GALAXY
ZC 0820.7+3915	08 20 42.	+ 39 15	1950		CLUSTER OF GALAXIES
ZWG 004.009	08 20 42.	- 00 42		14.8	GALAXY
UGC 04381	08 20 42.	- 00 42	90	14.8	GALAXY Sb-c
MCG+00-22-006	08 20 42.	- 00 42	78	14.8	GALAXY
MCG-01-22-008	08 20 42.	- 04 51	30	14.5	GALAXY
VHE 14B	08 20 42.	- 40 15	18		REFLECTION NEBULA
ARP 247	08 20 44.	+ 21 30			PECULIAR GALAXY
MCG+04-20-046	08 20 45.	+ 21 36	21	14.9	GALAXY
RNGC 2584	08 20 45.	- 04 50		14.5	GALAXY
LB 00542	08 20 47.	+ 11 46 12.		16.7	FAINT BLUE STAR
ZWG 032.016	08 20 48.	+ 04 28		14.7	GALAXY
MCG+01-22-009	08 20 48.	+ 04 28	21	14.7	GALAXY
ZWG 089.013	08 20 48.	+ 18 28		15.2	GALAXY
ZC 0820.8+1915	08 20 48.	+ 19 15	5650		CLUSTER OF GALAXIES
KARA.72 161B	08 20 48.	+ 21 30	54		PART OF DOUBLE GALAXY
IC 2341	08 20 48.	+ 21 35 48.		14.9	GALAXY E
ZWG 119.081	08 20 48.	+ 21 36		14.9	GALAXY
UGC 04384	08 20 48.	+ 21 36	78	14.9	GALAXY E-S0
TON-N 0316	08 20 48.	+ 25 03		15.1	BLUE STAR
ZC 0820.8+3200	08 20 48.	+ 32 00	1140		CLUSTER OF GALAXIES
ZC 0820.8+5106	08 20 48.	+ 51 06	1010		CLUSTER OF GALAXIES
MCG-01-22-009	08 20 48.	- 04 50	60	14.5	GALAXY
IC 0506	08 20 50.	+ 04 28 13.			NONSTELLAR OBJECT
MCG+06-19-003	08 20 51.	+ 38 01	30	15.5	GALAXY
ARC 0654	08 20 53.	+ 39 05		17.1	RICH CLUSTER OF GALAXIES
ZWG 032.017	08 20 54.	+ 03 22		15.4	GALAXY
MCG+03-22-007	08 20 54.	+ 18 27	54	15.2	GALAXY
ZWG 179.004	08 20 54.	+ 38 01		15.5	GALAXY
ZWG 208.001	08 20 54.	+ 42 07		15.7	GALAXY
ZWG 207.063	08 20 54.	+ 42 07		15.7	GALAXY
MCG-02-22-006	08 20 54.	- 14 51 30.	30	15.	GALAXY
VHE 15C	08 20 54.	- 41 56			REFLECTION NEBULA
RNGC 2585	08 20 57.	- 04 46		14.5	GALAXY
ARC 0658	08 20 58.	+ 15 51		17.0	RICH CLUSTER OF GALAXIES
MCG+00-22-008	08 21 00.	+ 00 00	30	15.1	GALAXY
MCG+00-22-007	08 21 00.	+ 00 00	42	15.1	GALAXY
ZWG 004.010	08 21 00.	+ 00 00		15.1	GALAXY
ZWG 060.005	08 21 00.	+ 11 43		15.2	GALAXY
KARA.72 162A	08 21 00.	+ 11 43	54	15.2	PART OF DOUBLE GALAXY
ZWG 089.014	08 21 00.	+ 18 09		15.7	GALAXY
MCG+03-22-008	08 21 00.	+ 18 09	9	15.7	GALAXY
ZWG 119.082	08 21 00.	+ 21 08		15.6	GALAXY
MCG+04-20-047	08 21 00.	+ 21 08	36	15.6	GALAXY
ZWG 363.043	08 21 00.	+ 81 14		15.4	GALAXY
MCG-01-22-010	08 21 00.	- 04 46	102	14.5	GALAXY
TON-N 0913	08 21 01.	+ 20 48		14.8	BLUE STAR
IC 2343	08 21 02.	+ 19 11 13.			NONSTELLAR OBJECT
IC 2344	08 21 04.	+ 18 49 13.			NONSTELLAR OBJECT
ZWG 060.006	08 21 06.	+ 10 00		15.1	GALAXY
MCG+02-22-001	08 21 06.	+ 10 00	48	15.1	GALAXY
FATH 1.215	08 21 06.	+ 14 54	22		NEBULA
ZWG 089.015	08 21 06.	+ 14 55		14.4	GALAXY
UGC 04385	08 21 06.	+ 14 55	54	14.4	GALAXY PECULR
MCG+03-22-009	08 21 06.	+ 14 55	60	14.4	GALAXY
KARA.73B 0247	08 21 06.	+ 14 55	42	14.4	ISOLATED GALAXY IR
ZWG 119.083	08 21 06.	+ 21 12		14.8	GALAXY
UGC 04386	08 21 06.	+ 21 12	114	14.8	GALAXY Sb
ZWG 149.021	08 21 06.	+ 29 11		15.7	GALAXY
ZWG 149.022	08 21 06.	+ 30 39		15.6	GALAXY
MCG+08-16-001	08 21 06.	+ 47 04	72	15.	GALAXY
MCG-02-22-007	08 21 06.	- 15 00	42	16.	GALAXY
ARC 0659	08 21 11.	+ 19 35		16.8	RICH CLUSTER OF GALAXIES
MCG+04-20-048	08 21 12.	+ 21 11	96	14.8	GALAXY
MRK 624	08 21 12.	+ 25 51	10	16.5	GALAXY WITH UV CONTINUUM
OCL 0715	08 21 12.	- 32 49	300	12.2	OPEN STAR CLUSTER
VHA 029	08 21 12.	- 32 49	120		OPEN STAR CLUSTER
VHE 15B	08 21 12.	- 41 55			REFLECTION NEBULA
RNGC 2588	08 21 13.	- 32 49		12.0	OPEN CLUSTER
IC 2345	08 21 15.	+ 20 06 54.			NONSTELLAR OBJECT
RNGC 2587	08 21 16.	- 29 20		9.0	OPEN CLUSTER
ZWG 032.018	08 21 18.	+ 04 19		15.3	GALAXY
ZWG 060.007	08 21 18.	+ 09 52		15.5	GALAXY
ZWG 060.008	08 21 18.	+ 11 40		15.3	GALAXY

OBJECT NAME	RIGHT ASCEN.	DECLINATION	DIAM.	MAGN.	TYPE OF OBJECT
KARA.72 162B	08 21 18.	+ 11 40	48	15.3	PART OF DOUBLE GALAXY
IC 2346	08 21 18.	+ 19 52 12.			NONSTELLAR OBJECT
TON-N 0317	08 21 18.	+ 30 39		15.4	BLUE STAR
UGC 04387	08 21 18.	+ 47 04	90	16.0	GALAXY Sc
ZC 0821.3+6529	08 21 18.	+ 65 29	2550		CLUSTER OF GALAXIES
HELW 420	08 21 18.	- 05 00 41.			NEBULA
MCG-02-22-008	08 21 18.	- 14 52 30.	54	14.5	GALAXY
MCG-04-20-012	08 21 18.	- 26 01	60	14.	GALAXY
IC 2347	08 21 23.	+ 18 56 11.			NONSTELLAR OBJECT
ZWG 060.009	08 21 24.	+ 12 17		15.5	GALAXY
ZWG 089.016	08 21 24.	+ 17 29		15.1	GALAXY
MRK 387	08 21 24.	+ 17 29	18	16.	GALAXY WITH UV CONTINUUM
KARA.73B 0248	08 21 24.	+ 17 29	24	15.1	ISOLATED GALAXY E
IC 2349	08 21 24.	+ 19 10 05.			NONSTELLAR OBJECT
ZWG 119.084	08 21 24.	+ 26 19		15.5	GALAXY
ZWG 004.011	08 21 24.	- 00 08		15.1	GALAXY
VHA 030	08 21 24.	- 28 20	300		OPEN STAR CLUSTER
OCL 0706	08 21 24.	- 29 20	540	9.2	OPEN STAR CLUSTER
VHE 15A	08 21 24.	- 41 56	126		REFLECTION NEBULA
IC 2348	08 21 26.	+ 20 41 47.			NONSTELLAR OBJECT
MCG+04-20-049	08 21 27.	+ 20 41 30.	36	15.6	GALAXY
ZWG 119.085	08 21 30.	+ 20 42		15.6	GALAXY
ZWG 004.012	08 21 30.	- 00 13		15.5	GALAXY
MCG-01-22-011	08 21 30.	- 06 45	60	15.	GALAXY
IC 2350	08 21 35.	+ 19 42 52.			NONSTELLAR OBJECT
ZWG 004.014	08 21 36.	+ 00 08		15.6	GALAXY
ZWG 119.086	08 21 36.	+ 21 13		15.7	GALAXY
ZC 0821.6+3745	08 21 36.	+ 37 45	1080		CLUSTER OF GALAXIES
ZWG 004.013	08 21 36.	- 00 09		15.4	GALAXY
IC 2351	08 21 38.	+ 18 45 03.			NONSTELLAR OBJECT
RNGC 2581	08 21 40.	+ 18 45		14.5	GALAXY
ZWG 004.016	08 21 42.	+ 00 01		15.7	GALAXY
ZWG 032.019	08 21 42.	+ 04 25		14.9	GALAXY
MCG+01-22-010	08 21 42.	+ 04 25	36	14.9	GALAXY
ZWG 089.017	08 21 42.	+ 16 22		15.7	GALAXY
ZWG 089.018	08 21 42.	+ 16 32		15.7	GALAXY
MCG+03-22-010	08 21 42.	+ 18 45	66	14.4	GALAXY
ZWG 089.019	08 21 42.	+ 18 46		14.4	GALAXY
UGC 04388	08 21 42.	+ 18 46	66	14.4	GALAXY S
MCG+07-17-030	08 21 42.	+ 42 28	42	15.	GALAXY
ZWG 004.015	08 21 42.	- 01 56		15.7	GALAXY
IC 2353	08 21 45.	+ 18 49 09.			NONSTELLAR OBJECT
ARC 0655	08 21 46.	+ 47 18		17.1	RICH CLUSTER OF GALAXIES
IC 2352	08 21 47.	+ 19 45 56.			NONSTELLAR OBJECT
ARC 0652	08 21 47.	+ 56 12		17.5	RICH CLUSTER OF GALAXIES
ZC 0821.8+0113	08 21 48.	+ 01 13	2020		CLUSTER OF GALAXIES
IC 2354	08 21 48.	+ 18 49 44.			NONSTELLAR OBJECT
ZWG 119.087	08 21 48.	+ 22 55		15.6	GALAXY
MCG-02-22-009	08 21 51.	- 14 20	72	15.5	GALAXY
IC 2367	08 21 52.	- 18 37 02.			NONSTELLAR OBJECT
ZWG 179.005	08 21 54.	+ 38 09		15.4	GALAXY
ZC 0821.9+5846	08 21 54.	+ 58 46	1010		CLUSTER OF GALAXIES
ZWG 004.017	08 21 54.	- 02 22		15.7	GALAXY
MCG-03-22-001	08 21 54.	- 18 36	150	13.5	GALAXY
ARC 0656	08 21 56.	+ 48 27		17.5	RICH CLUSTER OF GALAXIES
RNGC 2586	08 21 57.	- 04 43		15.5	GALAXY
IC 2355	08 21 58.	+ 20 37 37.			NONSTELLAR OBJECT
LB 05315	08 22 00.	+ 17 58		20.1	FAINT BLUE STAR
LB 05314	08 22 00.	+ 18 46		20.7	FAINT BLUE STAR
LB 05313	08 22 00.	+ 18 48		20.8	FAINT BLUE STAR
ZWG 119.088	08 22 00.	+ 25 57		15.7	GALAXY
MCG+08-16-002	08 22 00.	+ 49 44	42	16.	GALAXY
ZC 0822.0+5605	08 22 00.	+ 56 05	1340		CLUSTER OF GALAXIES
ZWG 349.025	08 22 00.	+ 75 34		14.8	GALAXY
MCG+13-06-023	08 22 00.	+ 75 34	36	16.	GALAXY
MCG+14-04-051	08 22 00.	+ 83 28	78	15.	GALAXY
MCG-01-22-012	08 22 00.	- 04 43	60	15.5	GALAXY
HELW 421	08 22 03.	- 04 43 14.			NEBULA
ARC 0660	08 22 05.	+ 37 01		17.7	RICH CLUSTER OF GALAXIES
RNGC 2589	08 22 05.	- 08 36			NON-EXISTENT OBJECT
ZC 0822.1+0437	08 22 06.	+ 04 37	870		CLUSTER OF GALAXIES
ZWG 089.020	08 22 06.	+ 16 33		15.7	GALAXY
LB 05319	08 22 06.	+ 17 55		18.3	FAINT BLUE STAR
LB 05318	08 22 06.	+ 17 58		20.7	FAINT BLUE STAR
LB 05317	08 22 06.	+ 18 28		19.9	FAINT BLUE STAR
LB 05316	08 22 06.	+ 18 46		18.0	FAINT BLUE STAR
ZWG 119.089	08 22 06.	+ 26 04		15.7	GALAXY
TON-N 0009	08 22 06.	+ 29 21		15.0	BLUE STAR
TON-N 0318	08 22 06.	+ 32 01		14.4	BLUE STAR
ZC 0822.1+3658	08 22 06.	+ 36 58	1140		CLUSTER OF GALAXIES
ZC 0822.1+4824	08 22 06.	+ 48 24	940		CLUSTER OF GALAXIES
VV 157C	08 22 06.	+ 73 48	6	17.	INTERACTING GALAXY
VV 157B	08 22 06.	+ 73 48	6	17.	INTERACTING GALAXY
VV 157A	08 22 06.	+ 73 48	42	16.	INTERACTING GALAXY
VV 157	08 22 06.	+ 73 48	66	16.	INTERACTING GALAXY
MCG+12-08-041	08 22 06.	+ 73 48	51	16.	GALAXY
ZWG 331.041	08 22 06.	+ 73 49		15.5	GALAXY
UGC 04389	08 22 06.	+ 73 49	84	15.5	GALAXY S
OCL 0704	08 22 06.	- 28 59	450	9.6	OPEN STAR CLUSTER
ISS 0918	08 22 06.	- 32 20	168		STELLAR RING
ARC 0662	08 22 08.	+ 08 40		17.4	RICH CLUSTER OF GALAXIES
IC 2356	08 22 08.	+ 19 39 36.			NONSTELLAR OBJECT
IC 2357	08 22 11.	+ 19 40 18.			NONSTELLAR OBJECT
ZWG 032.020	08 22 12.	+ 07 11		15.4	GALAXY
ZC 0822.2+0841	08 22 12.	+ 08 41	940		CLUSTER OF GALAXIES
ZWG 060.010	08 22 12.	+ 13 41		15.1	GALAXY
LB 05321	08 22 12.	+ 17 48		18.2	FAINT BLUE STAR
LB 05320	08 22 12.	+ 17 53		17.4	FAINT BLUE STAR
IC 2358	08 22 12.	+ 19 39 30.			NONSTELLAR OBJECT
ZC 0822.2+6113	08 22 12.	+ 61 13	540		CLUSTER OF GALAXIES
MCG+12-08-042	08 22 12.	+ 73 40	102	16.	GALAXY
ZWG 331.042	08 22 12.	+ 73 41		15.6	GALAXY
UGC 04390	08 22 12.	+ 73 41	132	15.6	GALAXY SBc
ARC 0664	08 22 13.	+ 04 37		17.4	RICH CLUSTER OF GALAXIES
TON-N 0914	08 22 13.	+ 20 17		15.1	BLUE STAR
WRAY 19.10	08 22 15.6	- 42 53 01.			DIFFUSE NEBULA
LB 05324	08 22 18.	+ 18 20		19.6	FAINT BLUE STAR
LB 05323	08 22 18.	+ 18 35		18.3	FAINT BLUE STAR
LB 05322	08 22 18.	+ 18 42		19.3	FAINT BLUE STAR
ZWG 089.021	08 22 18.	+ 18 45		15.6	GALAXY
IC 2359	08 22 18.	+ 20 29 53.			NONSTELLAR OBJECT
MCG+04-20-050	08 22 18.	+ 20 30	54	14.3	GALAXY
ZWG 119.090	08 22 18.	+ 25 10		15.6	GALAXY
KARA.73B 0249	08 22 18.	+ 25 10	18	15.6	ISOLATED GALAXY E
ZWG 179.006	08 22 18.	+ 37 40		15.6	GALAXY
ZWG 287.076	08 22 18.	+ 60 04		15.5	GALAXY
MCG+10-12-137	08 22 18.	+ 60 04	15	15.	GALAXY
ZWG 004.018	08 22 18.	- 02 42		15.5	GALAXY

OBJECT NAME	RIGHT ASCEN.	DECLINATION	DIAM.	MAGN.	TYPE OF OBJECT
RNGC 2582	08 22 21.	+ 20 30		14.5	GALAXY
IC 2360	08 22 22.	+ 19 37 17.			NONSTELLAR OBJECT
SEY 045	08 22 22.	+ 73 40 39.		15.2	FAINT GALAXY
LB 05331	08 22 24.	+ 17 46		19.3	FAINT BLUE STAR
LB 05330	08 22 24.	+ 18 02		18.6	FAINT BLUE STAR
LB 05329	08 22 24.	+ 18 07		19.2	FAINT BLUE STAR
LB 05328	08 22 24.	+ 18 22		19.3	FAINT BLUE STAR
LB 05327	08 22 24.	+ 18 28		20.0	FAINT BLUE STAR
LB 05326	08 22 24.	+ 18 34		19.0	FAINT BLUE STAR
LB 05325	08 22 24.	+ 18 43		20.5	FAINT BLUE STAR
ZWG 119.091	08 22 24.	+ 20 30		14.3	GALAXY
ZWG 089.022	08 22 24.	+ 20 30		14.3	GALAXY
UGC 04391	08 22 24.	+ 20 30	66	14.3	GALAXY SBb
ZC 0822.4+5453	08 22 24.	+ 54 53	9740		CLUSTER OF GALAXIES
ZC 0822.4+6406	08 22 24.	+ 64 06	740		CLUSTER OF GALAXIES
ZWG 032.021	08 22 30.	+ 03 32		15.2	GALAXY
LB 05332	08 22 30.	+ 18 02		19.5	FAINT BLUE STAR
ZC 0822.5+4411	08 22 30.	+ 44 11	940		CLUSTER OF GALAXIES
MCG+11-11-003	08 22 30.	+ 65 23	30	16.	GALAXY
ZWG 004.020	08 22 30.	- 00 26		14.0	GALAXY
UGC 04392	08 22 30.	- 00 26	150	14.0	GALAXY Sb
MCG+00-22-010	08 22 30.	- 00 26	150	14.0	GALAXY
ZWG 004.019	08 22 30.	- 01 03		15.5	GALAXY
MCG+00-22-009	08 22 30.	- 01 03	42	15.5	GALAXY
RNGC 2590	08 22 31.	- 00 26		14.0	GALAXY
ZWG 004.022	08 22 36.	+ 00 40		15.6	GALAXY
ZWG 032.022	08 22 36.	+ 04 05		15.1	GALAXY
ZWG 032.023	08 22 36.	+ 06 56		15.1	GALAXY
ZWG 060.011	08 22 36.	+ 09 52		15.6	GALAXY
LB 05340	08 22 36.	+ 17 50		18.3	FAINT BLUE STAR
LB 05339	08 22 36.	+ 18 00		19.4	FAINT BLUE STAR
LB 05338	08 22 36.	+ 18 16		18.4	FAINT BLUE STAR
LB 05337	08 22 36.	+ 18 17		19.6	FAINT BLUE STAR
LB 05336	08 22 36.	+ 18 17		18.8	FAINT BLUE STAR
LB 05335	08 22 36.	+ 18 25		19.0	FAINT BLUE STAR
LB 05334	08 22 36.	+ 18 36		18.6	FAINT BLUE STAR
LB 05333	08 22 36.	+ 18 42		18.9	FAINT BLUE STAR
MCG+05-20-012	08 22 36.	+ 28 02	60	14.9	GALAXY
TON-N 0319	08 22 36.	+ 31 03		15.1	BLUE STAR
ZWG 208.002	08 22 36.	+ 39 21		15.5	GALAXY
ZWG 237.001	08 22 36.	+ 46 08		13.6	GALAXY
UGC 04393	08 22 36.	+ 46 08	150	13.6	GALAXY SB
MCG+08-16-003	08 22 36.	+ 46 08	132	13.	GALAXY
KARA.73B 0250	08 22 36.	+ 46 08	114	13.6	ISOLATED GALAXY S
MAI 028	08 22 36.	+ 79 06	40		DWARF SPHEROIDAL GALAXY
ZWG 004.021	08 22 36.	- 02 25		15.4	GALAXY
LB 05341	08 22 42.	+ 18 42		17.0	FAINT BLUE STAR
IC 2361	08 22 42.	+ 28 01 51.			NONSTELLAR OBJECT
ZWG 149.023	08 22 42.	+ 28 03		14.9	GALAXY
UGC 04394	08 22 42.	+ 28 03	102	14.9	GALAXY
ZWG 149.024	08 22 42.	+ 28 17		15.5	GALAXY
UGC 04395	08 22 42.	+ 28 17	108	15.5	GALAXY Sc
MCG+05-20-013	08 22 42.	+ 28 17	102	15.5	GALAXY
ZWG 208.003	08 22 42.	+ 41 09		15.4	GALAXY
ZC 0822.7+4454	08 22 42.	+ 44 54	1680		CLUSTER OF GALAXIES
ZC 0822.7+4903	08 22 42.	+ 49 03	3430		CLUSTER OF GALAXIES
ZC 0822.7+5635	08 22 42.	+ 56 35	1550		CLUSTER OF GALAXIES
ZC 0822.7+6226	08 22 42.	+ 62 26	810		CLUSTER OF GALAXIES
MCG+07-18-001	08 22 45.	+ 41 09	48	15.	GALAXY
ZWG 004.023	08 22 48.	+ 02 09		15.4	GALAXY
LB 05348	08 22 48.	+ 17 47		18.7	FAINT BLUE STAR
LB 05347	08 22 48.	+ 17 57		20.3	FAINT BLUE STAR
LB 05346	08 22 48.	+ 18 05		19.4	FAINT BLUE STAR
LB 05345	08 22 48.	+ 18 12		19.8	FAINT BLUE STAR
LB 05344	08 22 48.	+ 18 15		18.2	FAINT BLUE STAR
LB 05343	08 22 48.	+ 18 22		18.5	FAINT BLUE STAR
LB 05342	08 22 48.	+ 18 34		18.7	FAINT BLUE STAR
IC 2362	08 22 48.	+ 20 06 20.			NONSTELLAR OBJECT
ZC 0822.8+4027	08 22 48.	+ 40 27	810		CLUSTER OF GALAXIES
ZC 0822.8+4722	08 22 48.	+ 47 22	2220		CLUSTER OF GALAXIES
IC 2363	08 22 52.6	+ 19 36 49.			GALAXY
ZWG 032.024	08 22 54.	+ 05 17		15.3	GALAXY
LB 05352	08 22 54.	+ 17 57		20.3	FAINT BLUE STAR
LB 05351	08 22 54.	+ 17 58		17.8	FAINT BLUE STAR
LB 05350	08 22 54.	+ 18 22		17.3	FAINT BLUE STAR
LB 05349	08 22 54.	+ 18 28		17.2	FAINT BLUE STAR
ZWG 089.023	08 22 54.	+ 19 36		15.0	GALAXY
MCG+03-22-011	08 22 54.	+ 19 36	48	15.0	GALAXY
OCL 0751	08 22 54.	- 47 02	4200	13.	OPEN STAR CLUSTER
VHA 031	08 22 54.	- 47 02	180		OPEN STAR CLUSTER
SN 1961B	08 22 55.	+ 19 36		18.5	SUPERNOVA
IC 2364	08 22 58.	+ 19 55 25.			NONSTELLAR OBJECT
RNGC 2550A	08 22 58.	+ 73 55			GALAXY
LBN 0728	08 23	+ 60 30	1020		BRIGHT NEBULA
ZWG 004.024	08 23 00.	+ 02 07		15.5	GALAXY
LB 05354	08 23 00.	+ 18 00		20.7	FAINT BLUE STAR
ZWG 089.024	08 23 00.	+ 18 20		15.4	GALAXY
LB 05353	08 23 00.	+ 18 35		19.6	FAINT BLUE STAR
ZC 0823.0+3355	08 23 00.	+ 33 55	2020		CLUSTER OF GALAXIES
ZC 0823.0+3455	08 23 00.	+ 34 55	810		CLUSTER OF GALAXIES
ZC 0823.0+3658	08 23 00.	+ 36 58	3490		CLUSTER OF GALAXIES
ZWG 331.043	08 23 00.	+ 73 55		13.5	GALAXY
UGC 04397	08 23 00.	+ 73 55	102	13.5	GALAXY Sc
ZWG 364.004	08 23 00.	+ 84 29		15.3	GALAXY
ZWG 363.044	08 23 00.	+ 84 29		15.3	GALAXY
UGC 04396	08 23 00.	+ 84 29	132	15.3	GALAXY SBb-c
OCL 0718	08 23 00.	- 34 00	150	14.	OPEN STAR CLUSTER
VHA 032	08 23 00.	- 34 02	180		OPEN STAR CLUSTER
IC 0507	08 23 04.	- 00 16 56.			MAY NOT EXIST
ZC 0823.1+0444	08 23 06.	+ 04 44	1010		CLUSTER OF GALAXIES
LB 05357	08 23 06.	+ 18 09		18.4	FAINT BLUE STAR
LB 05356	08 23 06.	+ 18 28		19.2	FAINT BLUE STAR
LB 05355	08 23 06.	+ 18 29		18.9	FAINT BLUE STAR
MCG+08-16-004	08 23 06.	+ 48 59	72	16.	GALAXY
MCG+11-11-004	08 23 06.	+ 64 22	57	15.	GALAXY
ZWG 311.003	08 23 06.	+ 64 23		15.1	GALAXY
ZWG 310.054	08 23 06.	+ 64 23		15.1	GALAXY
UGC 04398	08 23 06.	+ 64 23	78	15.1	GALAXY Sb/Sc
KARA.73B 0251	08 23 06.	+ 64 23	72	15.1	ISOLATED GALAXY S
MCG+12-08-043	08 23 06.	+ 73 54	84	14.	GALAXY
IC 2368	08 23 08.	+ 20 03 00.			NONSTELLAR OBJECT
ZWG 032.025	08 23 12.	+ 03 20		15.3	GALAXY
ZC 0823.2+0425	08 23 12.	+ 04 25	470		CLUSTER OF GALAXIES
LB 05360	08 23 12.	+ 17 59		18.4	FAINT BLUE STAR
LB 05359	08 23 12.	+ 18 18		17.6	FAINT BLUE STAR
LB 05358	08 23 12.	+ 18 42		19.5	FAINT BLUE STAR
ZWG 119.092	08 23 12.	+ 21 37		15.5	GALAXY
UGC 04399	08 23 12.	+ 21 37	66	15.5	GALAXY S-IRR

OBJECT NAME	RIGHT ASCEN.	DECLINATION	DIAM.	MAGN.	TYPE OF OBJECT
ZWG 119.093	08 23 12.	+ 21 50		15.7	GALAXY
UGC 04400	08 23 12.	+ 21 50	102	15.7	GALAXY Sc
UGC 04401	08 23 12.	+ 48 58	84	17.	GALAXY DWARF
ZC 0823.2+5320	08 23 12.	+ 53 20	1950		CLUSTER OF GALAXIES
ZWG 004.025	08 23 12.	- 02 54		15.6	GALAXY
SN 1960N	08 23 13.	+ 21 37		16.6	SUPERNOVA
SHB 111	08 23 13.1	+ 03 19 24.			QUASI-STELLAR OBJECT
MCG+04-20-051	08 23 15.	+ 21 37 30.	48	15.5	GALAXY
IC 2366	08 23 16.	+ 27 59 48.			NONSTELLAR OBJECT
IC 2365	08 23 16.	+ 28 02 18.			NONSTELLAR OBJECT
ARC 0663	08 23 17.	+ 35 00		17.7	RICH CLUSTER OF GALAXIES
LB 05364	08 23 18.	+ 17 57		20.6	FAINT BLUE STAR
LB 05363	08 23 18.	+ 18 07		16.6	FAINT BLUE STAR
LB 05362	08 23 18.	+ 18 15		19.7	FAINT BLUE STAR
LB 05361	08 23 18.	+ 18 46		19.2	FAINT BLUE STAR
ZWG 149.025	08 23 18.	+ 28 00		14.7	GALAXY
UGC 04402	08 23 18.	+ 28 00	84	14.7	GALAXY S0
MCG+05-20-014	08 23 18.	+ 28 00	42	14.7	GALAXY
KEEN 2550A	08 23 18.	+ 74 04			GALAXY
IC 2369	08 23 22.	+ 20 23 46.			NONSTELLAR OBJECT
LB 05368	08 23 24.	+ 17 46		18.8	FAINT BLUE STAR
LB 05367	08 23 24.	+ 17 55		18.8	FAINT BLUE STAR
LB 05366	08 23 24.	+ 18 03		18.4	FAINT BLUE STAR
LB 05365	08 23 24.	+ 18 40		18.3	FAINT BLUE STAR
ZWG 119.094	08 23 24.	+ 23 02		15.7	GALAXY
ZWG 119.095	08 23 24.	+ 23 04		15.6	GALAXY
MCG+10-12-138	08 23 24.	+ 59 40	57	16.	GALAXY
ZC 0823.4+5958	08 23 24.	+ 59 58	1080		CLUSTER OF GALAXIES
MCG-02-22-010	08 23 24.	- 13 21	48	16.	GALAXY
ZWG 032.026	08 23 30.	+ 04 45		15.5	GALAXY
ZWG 060.012	08 23 30.	+ 11 40		15.4	GALAXY
UGC 04403	08 23 30.	+ 11 40	60	15.4	GALAXY S
KARA.73B 0252	08 23 30.	+ 11 40	60	15.4	ISOLATED GALAXY S
ZWG 060.013	08 23 30.	+ 14 23		15.1	GALAXY
ZWG 089.025	08 23 30.	+ 17 22		15.1	GALAXY
ZWG 089.026	08 23 30.	+ 17 31		15.5	GALAXY
LB 05372	08 23 30.	+ 17 49		18.9	FAINT BLUE STAR
LB 05371	08 23 30.	+ 18 02		19.7	FAINT BLUE STAR
LB 05370	08 23 30.	+ 18 05		20.8	FAINT BLUE STAR
LB 05369	08 23 30.	+ 18 11		19.0	FAINT BLUE STAR
IC 2370	08 23 30.	+ 19 48 09.			NONSTELLAR OBJECT
ZWG 089.027	08 23 30.	+ 20 26		15.2	GALAXY
ZC 0823.5+4649	08 23 30.	+ 46 49	1340		CLUSTER OF GALAXIES
ARC 0661	08 23 30.	+ 53 19		17.4	RICH CLUSTER OF GALAXIES
ZWG 311.004	08 23 30.	+ 63 26		15.2	GALAXY
ZWG 310.055	08 23 30.	+ 63 26		15.2	GALAXY
MCG-02-22-011	08 23 30.	- 13 20	36	15.	GALAXY
MCG+11-11-005	08 23 33.	+ 63 26	39	15.	GALAXY
MCG-02-22-012	08 23 33.	- 11 36	84	15.	GALAXY
ZWG 032.027	08 23 36.	+ 02 45		15.3	GALAXY
ZC 0823.6+1248	08 23 36.	+ 12 48	1610		CLUSTER OF GALAXIES
LB 05376	08 23 36.	+ 18 05		18.9	FAINT BLUE STAR
LB 05375	08 23 36.	+ 18 12		17.7	FAINT BLUE STAR
LB 05374	08 23 36.	+ 18 29		17.6	FAINT BLUE STAR
LB 05373	08 23 36.	+ 18 44		18.3	FAINT BLUE STAR
ZWG 119.096	08 23 36.	+ 22 26		15.5	GALAXY
UGC 04404	08 23 36.	+ 22 26	66	15.5	GALAXY Sc-IRR
ZWG 119.097	08 23 36.	+ 23 21		15.2	GALAXY
UGC 04405	08 23 36.	+ 23 21	96	15.2	GALAXY S0-a
MCG+04-20-052	08 23 36.	+ 23 22 30.	66	15.2	GALAXY
ZC 0823.6+3744	08 23 36.	+ 37 44	1810		CLUSTER OF GALAXIES
MCG+09-14-038	08 23 36.	+ 55 04	18	16.	GALAXY
MCG+12-08-044	08 23 36.	+ 73 47	33	16.	GALAXY
MCG-02-22-013	08 23 36.	- 13 43	48	15.	GALAXY
ZWG 032.028	08 23 42.	+ 03 05		15.1	GALAXY
ZWG 060.014	08 23 42.	+ 08 35		15.3	GALAXY
LB 05378	08 23 42.	+ 17 50		20.6	FAINT BLUE STAR
LB 05377	08 23 42.	+ 18 04		18.2	FAINT BLUE STAR
ZWG 119.098	08 23 42.	+ 22 35		15.7	GALAXY
ZWG 119.099	08 23 42.	+ 23 07		15.5	GALAXY
UGC 04406	08 23 42.	+ 23 07	72	15.5	GALAXY SB0
MCG+04-20-053	08 23 42.	+ 23 07 30.	72	15.5	GALAXY
ZC 0823.7+5404	08 23 42.	+ 54 04	1280		CLUSTER OF GALAXIES
ZWG 263.031	08 23 42.	+ 55 04		14.9	GALAXY
IC 2371	08 23 43.	+ 19 57 50.			NONSTELLAR OBJECT
LB 05382	08 23 48.	+ 17 45		18.7	FAINT BLUE STAR
ZWG 089.028	08 23 48.	+ 18 00		15.6	GALAXY
UGC 04407	08 23 48.	+ 18 00	60	15.6	GALAXY S
LB 05381	08 23 48.	+ 18 18		17.6	FAINT BLUE STAR
LB 05380	08 23 48.	+ 18 22		18.7	FAINT BLUE STAR
LB 05379	08 23 48.	+ 18 47		18.5	FAINT BLUE STAR
IC 2372	08 23 48.	+ 20 02 56.			NONSTELLAR OBJECT
TON-N 0321	08 23 48.	+ 31 28		16.8	BLUE STAR
TON-N 0320	08 23 48.	+ 31 42		14.6	BLUE STAR
MCG+08-16-005	08 23 48.	+ 46 12	18	16.	GALAXY
MCG+12-09-049	08 23 48.	+ 68 38	36	14.	GALAXY
LB 05384	08 23 54.	+ 17 54		18.4	FAINT BLUE STAR
LB 05383	08 23 54.	+ 18 24		18.8	FAINT BLUE STAR
ZC 0823.9+2257	08 23 54.	+ 22 57	670		CLUSTER OF GALAXIES
ZC 0823.9+4211	08 23 54.	+ 42 11	610		CLUSTER OF GALAXIES
MCG+12-09-050	08 23 54.	+ 68 39	84	13.	GALAXY
7ZW 230	08 23 54.	+ 71 13			COMPACT GALAXY
MCG-02-22-014	08 23 54.	- 13 07 30.	96	14.5	GALAXY
IC 2373	08 23 55.	+ 20 31 49.			NONSTELLAR OBJECT
ARC 0666	08 23 57.	+ 38 30		17.9	RICH CLUSTER OF GALAXIES
MCG+09-14-039	08 23 57.	+ 55 20	72	15.	GALAXY
IC 2375	08 23 58.	- 13 08 15.			NONSTELLAR OBJECT
RNGC 2593	08 23 59.	+ 17 32		15.0	GALAXY
ZC 0824.0+0650	08 24 00.	+ 06 50	1480		CLUSTER OF GALAXIES
ZWG 089.029	08 24 00.	+ 17 32		14.9	GALAXY
UGC 04408	08 24 00.	+ 17 32	66	14.9	GALAXY S0-a
LB 05388	08 24 00.	+ 17 51		18.5	FAINT BLUE STAR
LB 05387	08 24 00.	+ 18 23		19.6	FAINT BLUE STAR
LB 05386	08 24 00.	+ 18 34		20.5	FAINT BLUE STAR
LB 05385	08 24 00.	+ 18 38		20.4	FAINT BLUE STAR
ZWG 119.100	08 24 00.	+ 20 32		15.5	GALAXY
UGC 04409	08 24 00.	+ 20 32	84	15.5	GALAXY Sc?
MCG+04-20-054	08 24 00.	+ 20 32	54	15.5	GALAXY
MCG+04-20-055	08 24 00.	+ 26 10	30	13.6	GALAXY
ZC 0824.0+3115	08 24 00.	+ 31 15	1010		CLUSTER OF GALAXIES
MCG+06-19-004	08 24 00.	+ 37 57	42	15.	GALAXY
ZWG 263.032	08 24 00.	+ 55 18		15.1	GALAXY
UGC 04410	08 24 00.	+ 55 18	72	15.1	GALAXY S
MCG+03-22-012	08 24 03.	+ 17 31	42	14.9	GALAXY
MCG-02-22-015	08 24 03.	- 13 08	36	14.5	GALAXY
IC 2377	08 24 05.	- 13 08 28.			NONSTELLAR OBJECT
LB 05390	08 24 06.	+ 18 28		19.0	FAINT BLUE STAR
LB 05389	08 24 06.	+ 18 28		19.7	FAINT BLUE STAR

OBJECT NAME	RIGHT ASCEN.	DECLINATION	DIAM.	MAGN.	TYPE OF OBJECT
ZWG 119.101	08 24 06.	+ 21 53		14.8	GALAXY
ZWG 119.102	08 24 06.	+ 26 08		13.6	GALAXY
RNGC 2592	08 24 06.	+ 26 08		13.5	GALAXY
UGC 04411	08 24 06.	+ 26 08	108	13.6	GALAXY E
ZC 0824.1+3258	08 24 06.	+ 32 58	3830		CLUSTER OF GALAXIES
ZWG 179.007	08 24 06.	+ 37 56		15.2	GALAXY
UGC 04412	08 24 06.	+ 37 56	84	15.2	GALAXY S
ZWG 208.004	08 24 06.	+ 41 40		15.4	GALAXY
ZWG 331.044	08 24 06.	+ 74 01		15.6	GALAXY
UGC 04413	08 24 06.	+ 74 01	102	15.6	GALAXY Sc
ZC 0824.1-0316	08 24 06.	- 03 16	670		CLUSTER OF GALAXIES
MCG-02-22-016	08 24 06.	- 13 07	42	14.5	GALAXY
IC 2379	08 24 06.	- 13 07 40.			NONSTELLAR OBJECT
MCG+09-14-040	08 24 09.	+ 54 36	60	15.	GALAXY
LB 05395	08 24 12.	+ 17 51		18.8	FAINT BLUE STAR
LB 05394	08 24 12.	+ 17 54		18.2	FAINT BLUE STAR
LB 05393	08 24 12.	+ 18 00		19.4	FAINT BLUE STAR
LB 05392	08 24 12.	+ 18 22		18.8	FAINT BLUE STAR
LB 05391	08 24 12.	+ 18 43		17.8	FAINT BLUE STAR
MCG+04-20-058	08 24 12.	+ 21 48	66	15.0	GALAXY
ZWG 119.103	08 24 12.	+ 21 49		15.0	GALAXY
UGC 04414	08 24 12.	+ 21 49	66	15.0	GALAXY SB0/SBa
ZWG 119.104	08 24 12.	+ 23 00		15.4	GALAXY
MCG+04-20-057	08 24 12.	+ 23 01	30	15.4	GALAXY
MCG+04-20-056	08 24 12.	+ 26 04	36	15.0	GALAXY
MCG+07-18-002	08 24 12.	+ 41 40	30	15.	GALAXY
ZWG 263.033	08 24 12.	+ 54 36		15.3	GALAXY
UGC 04415	08 24 12.	+ 54 36	78	15.3	GALAXY SBc
ZWG 263.034	08 24 12.	+ 55 15		14.8	GALAXY
MCG+09-14-041	08 24 12.	+ 55 16	18	16.	GALAXY
ZWG 004.026	08 24 18.	+ 00 12		15.4	GALAXY
LB 05397	08 24 18.	+ 17 42		18.7	FAINT BLUE STAR
LB 05396	08 24 18.	+ 18 24		19.0	FAINT BLUE STAR
ZWG 119.105	08 24 18.	+ 23 02		14.5	GALAXY
UGC 04416	08 24 18.	+ 23 02	150	14.5	GALAXY SBb
MCG+04-20-060	08 24 18.	+ 23 03	132	14.5	GALAXY
MCG+04-20-059	08 24 18.	+ 25 55	48	15.7	GALAXY
ZWG 119.106	08 24 18.	+ 26 02		15.0	GALAXY
RNGC 2594	08 24 18.	+ 26 02		15.0	GALAXY
1ZW 014	08 24 18.	+ 55 52			COMPACT GALAXY
ZWG 263.035	08 24 18.	+ 55 52		14.3	GALAXY
UGC 04417	08 24 18.	+ 55 52	12	14.3	GALAXY EX CMPT
ZC 0824.3+5933	08 24 18.	+ 59 33	1280		CLUSTER OF GALAXIES
LB 05403	08 24 24.	+ 17 48		17.9	FAINT BLUE STAR
LB 05402	08 24 24.	+ 17 49		20.4	FAINT BLUE STAR
LB 05401	08 24 24.	+ 17 53		18.9	FAINT BLUE STAR
LB 05400	08 24 24.	+ 18 02		19.5	FAINT BLUE STAR
LB 05399	08 24 24.	+ 18 05		19.2	FAINT BLUE STAR
LB 05398	08 24 24.	+ 18 11		18.7	FAINT BLUE STAR
TON-N 0915	08 24 24.	+ 22 22		15.5	BLUE STAR
MCG+04-20-061	08 24 24.	+ 23 22	36	15.6	GALAXY
MCG+07-18-003	08 24 24.	+ 40 54	36	16.	GALAXY
ZC 0824.4+5525	08 24 24.	+ 55 25	1550		CLUSTER OF GALAXIES
ARC 0666	08 24 29.	+ 34 57		17.5	RICH CLUSTER OF GALAXIES
LB 05408	08 24 30.	+ 18 07		19.2	FAINT BLUE STAR
LB 05407	08 24 30.	+ 18 10		18.7	FAINT BLUE STAR
LB 05406	08 24 30.	+ 18 12		19.0	FAINT BLUE STAR
LB 05405	08 24 30.	+ 18 30		20.7	FAINT BLUE STAR
LB 05404	08 24 30.	+ 18 50		17.7	FAINT BLUE STAR
ZC 0824.5+2244	08 24 30.	+ 22 44	540		CLUSTER OF GALAXIES
ZWG 119.107	08 24 30.	+ 23 20		15.6	GALAXY
ZWG 119.108	08 24 30.	+ 25 53		15.7	GALAXY
UGC 04418	08 24 30.	+ 25 53	66	15.7	GALAXY Sc
ZWG 237.002	08 24 30.	+ 47 27		15.6	GALAXY
MCG+10-12-139	08 24 30.	+ 59 49	51	16.	GALAXY
7ZW 231	08 24 30.	+ 63 52			COMPACT GALAXY
MCG+11-11-006	08 24 33.	+ 67 21	45	16.	GALAXY
RNGC 2596	08 24 35.	+ 17 27		14.0	GALAXY
TON-N 0916	08 24 35.	+ 32 46		16.	BLUE STAR
ZWG 089.030	08 24 36.	+ 17 27		14.2	GALAXY
UGC 04419	08 24 36.	+ 17 27	90	14.2	GALAXY Sb
LB 05410	08 24 36.	+ 17 54		19.6	FAINT BLUE STAR
LB 05409	08 24 36.	+ 18 09		20.2	FAINT BLUE STAR
ARC 0667	08 24 36.	+ 44 54		17.5	RICH CLUSTER OF GALAXIES
ZWG 311.005	08 24 36.	+ 67 20		15.6	GALAXY
ZWG 310.056	08 24 36.	+ 67 20		15.6	GALAXY
ZWG 004.027	08 24 36.	- 00 55		15.5	GALAXY
MCG+03-22-013	08 24 42.	+ 17 26	84	14.2	GALAXY
LB 05418	08 24 42.	+ 17 57		20.2	FAINT BLUE STAR
LB 05417	08 24 42.	+ 18 06		18.6	FAINT BLUE STAR
LB 05416	08 24 42.	+ 18 08		20.2	FAINT BLUE STAR
LB 05415	08 24 42.	+ 18 08		18.9	FAINT BLUE STAR
LB 05414	08 24 42.	+ 18 23		20.3	FAINT BLUE STAR
LB 05413	08 24 42.	+ 18 24		19.4	FAINT BLUE STAR
LB 05412	08 24 42.	+ 18 31		17.9	FAINT BLUE STAR
LB 05411	08 24 42.	+ 18 42		18.8	FAINT BLUE STAR
ZWG 311.006	08 24 42.	+ 63 30		14.6	GALAXY
UGC 04420	08 24 42.	+ 63 30	96	14.6	GALAXY S0
MCG+04-20-062	08 24 45.	+ 21 38	180	13.9	GALAXY
ZWG 004.028	08 24 48.	+ 02 01		15.3	GALAXY
UGC 04421	08 24 48.	+ 02 01	66	15.3	GALAXY Sb/SBb
KARA.73B 0253	08 24 48.	+ 02 01	66	15.3	ISOLATED GALAXY S
ZC 0824.8+1731	08 24 48.	+ 17 31	6590		CLUSTER OF GALAXIES
LB 05423	08 24 48.	+ 18 04		19.4	FAINT BLUE STAR
LB 05422	08 24 48.	+ 18 10		19.9	FAINT BLUE STAR
LB 05421	08 24 48.	+ 18 18		19.8	FAINT BLUE STAR
LB 05420	08 24 48.	+ 18 19		20.0	FAINT BLUE STAR
LB 05419	08 24 48.	+ 18 44		19.3	FAINT BLUE STAR
3ZW 059	08 24 48.	+ 21 39			COMPACT GALAXY
ZWG 119.109	08 24 48.	+ 21 39		13.9	GALAXY
UGC 04422	08 24 48.	+ 21 39	192	13.9	GALAXY SBb/Sc
ZWG 119.110	08 24 48.	+ 25 01		15.7	GALAXY
ZC 0824.8+5050	08 24 48.	+ 50 50	1480		CLUSTER OF GALAXIES
MCG+10-12-140	08 24 48.	+ 58 19	57	16.	GALAXY
RNGC 2595	08 24 51.	+ 21 39		14.0	GALAXY
ZWG 032.029	08 24 54.	+ 03 02		14.7	GALAXY
MCG+01-22-011	08 24 54.	+ 03 02	24	14.7	GALAXY
LB 05429	08 24 54.	+ 18 08		19.3	FAINT BLUE STAR
LB 05428	08 24 54.	+ 18 14		18.8	FAINT BLUE STAR
LB 05427	08 24 54.	+ 18 30		18.5	FAINT BLUE STAR
LB 05426	08 24 54.	+ 18 38		18.5	FAINT BLUE STAR
LB 05425	08 24 54.	+ 18 41		20.5	FAINT BLUE STAR
LB 05424	08 24 54.	+ 18 44		20.6	FAINT BLUE STAR
ZC 0824.9+3451	08 24 54.	+ 34 51	2760		CLUSTER OF GALAXIES
ZC 0824.9+5544	08 24 54.	+ 55 44	670		CLUSTER OF GALAXIES
MCG+10-12-141	08 24 54.	+ 58 22	18	15.	GALAXY
ZWG 331.045	08 24 54.	+ 71 00		15.4	GALAXY
ZWG 349.026	08 24 54.	+ 74 51		15.4	GALAXY

OBJECT NAME	RIGHT ASCEN.	DECLINATION	DIAM.	MAGN.	TYPE OF OBJECT
UGC 04423	08 24 54.	+ 74 51	84	15.4	GALAXY Sc
VDB.66G 052	08 25	+ 42 02	70		DWARF GALAXY
LB 05431	08 25 00.	+ 17 57		19.3	FAINT BLUE STAR
LB 05430	08 25 00.	+ 18 17		19.5	FAINT BLUE STAR
ZC 0825.0+2236	08 25 00.	+ 22 36	340		CLUSTER OF GALAXIES
ZC 0825.0+2824	08 25 00.	+ 28 24	5240		CLUSTER OF GALAXIES
MCG+07-18-004	08 25 00.	+ 42 01	120	16.	GALAXY
ZC 0825.0+4453	08 25 00.	+ 44 53	1340		CLUSTER OF GALAXIES
MCG+11-11-007	08 25 00.	+ 66 33	30	17.	GALAXY
ZWG 331.046	08 25 00.	+ 70 18		15.4	GALAXY
ZWG 004.030	08 25 00.	- 01 24		15.5	GALAXY
ZWG 004.029	08 25 00.	- 01 28		15.5	GALAXY
TON-N 0917	08 25 01.	+ 20 31		16.0	BLUE STAR
ZWG 032.030	08 25 06.	+ 08 16		15.6	GALAXY
KARA.73B 0254	08 25 06.	+ 08 16	24	15.6	ISOLATED GALAXY E
LB 05434	08 25 06.	+ 17 58		19.8	FAINT BLUE STAR
LB 05433	08 25 06.	+ 18 18		20.8	FAINT BLUE STAR
LB 05432	08 25 06.	+ 18 21		19.3	FAINT BLUE STAR
TON-N 0918	08 25 06.	+ 22 44		14.8	BLUE STAR
ZC 0825.1+3916	08 25 06.	+ 39 16	2760		CLUSTER OF GALAXIES
ZWG 208.005	08 25 06.	+ 40 48		15.0	GALAXY
MCG+11-11-008	08 25 06.	+ 63 27	39	16.	GALAXY
ZWG 032.031	08 25 12.	+ 04 49		15.4	GALAXY
ZWG 032.032	08 25 12.	+ 05 08		15.4	GALAXY
ZWG 060.015	08 25 12.	+ 09 41		15.7	GALAXY
LB 05438	08 25 12.	+ 17 59		20.6	FAINT BLUE STAR
LB 05437	08 25 12.	+ 18 11		19.3	FAINT BLUE STAR
LB 05436	08 25 12.	+ 18 29		19.9	FAINT BLUE STAR
LB 05435	08 25 12.	+ 18 40		17.0	FAINT BLUE STAR
UGC 04424	08 25 12.	+ 20 25	78	16.0	GALAXY S
MCG+05-20-015	08 25 12.	+ 28 13	78	15.2	GALAXY
ZWG 149.026	08 25 12.	+ 28 14		15.2	GALAXY
UGC 04425	08 25 12.	+ 28 14	90	15.2	GALAXY SB
ZWG 149.027	08 25 12.	+ 30 37		15.5	GALAXY
ZWG 149.028	08 25 12.	+ 30 39		15.7	GALAXY
MCG+07-18-005	08 25 12.	+ 40 49	36	15.	GALAXY
UGC 04426	08 25 12.	+ 42 02	120	18.	GALAXY DWARF
SHAH 095	08 25 12.	+ 50 28	66	16.5	GROUP OF COMPACT GALAXIES
MCG+11-11-009	08 25 12.	+ 63 29	36	15.	GALAXY
MCG+12-08-045	08 25 12.	+ 74 49 30.	33	16.	GALAXY
MCG-02-22-017	08 25 12.	- 12 35	120	13.	GALAXY
RNGC 2601	08 25 14.	- 67 57			GALAXY
IC 2374	08 25 16.	+ 30 36 23.			NONSTELLAR OBJECT
LB 05448	08 25 18.	+ 17 46		16.6	FAINT BLUE STAR
LB 05447	08 25 18.	+ 17 48		19.8	FAINT BLUE STAR
LB 05446	08 25 18.	+ 17 54		18.9	FAINT BLUE STAR
LB 05445	08 25 18.	+ 18 04		18.2	FAINT BLUE STAR
LB 05444	08 25 18.	+ 18 11		18.7	FAINT BLUE STAR
LB 05443	08 25 18.	+ 18 12		18.2	FAINT BLUE STAR
LB 05442	08 25 18.	+ 18 19		19.5	FAINT BLUE STAR
LB 05441	08 25 18.	+ 18 31		19.6	FAINT BLUE STAR
LB 05440	08 25 18.	+ 18 32		20.7	FAINT BLUE STAR
LB 05439	08 25 18.	+ 18 42		19.9	FAINT BLUE STAR
MCG+04-20-063	08 25 18.	+ 25 19 30.	45	14.8	GALAXY
MCG+05-20-017	08 25 18.	+ 30 33 30.	90	15.3	GALAXY
ZWG 149.029	08 25 18.	+ 30 35		15.3	GALAXY
MCG+05-20-016	08 25 18.	+ 30 35	24	15.3	GALAXY
ZWG 149.030	08 25 18.	+ 31 00		15.6	GALAXY
MCG+09-14-042	08 25 18.	+ 52 11	30	16.	GALAXY
ZWG 263.036	08 25 18.	+ 55 40		15.2	GALAXY
UGC 04427	08 25 18.	+ 55 40	72	15.2	GALAXY Sb-c
KARA.72 163A	08 25 18.	+ 55 40	66	15.2	PART OF DOUBLE GALAXY
ZWG 331.047	08 25 18.	+ 71 04		15.0	GALAXY
SEY 046	08 25 18.	+ 74 51 17.		15.1	FAINT GALAXY
IC 2376	08 25 21.	+ 30 34 05.			NONSTELLAR OBJECT
ARC 0671	08 25 23.	+ 30 36		14.9	RICH CLUSTER OF GALAXIES
ZWG 060.016	08 25 24.	+ 10 42		15.3	GALAXY
LB 05451	08 25 24.	+ 17 55		18.7	FAINT BLUE STAR
LB 05450	08 25 24.	+ 18 12		20.5	FAINT BLUE STAR
LB 05449	08 25 24.	+ 18 28		18.7	FAINT BLUE STAR
TON-N 0919	08 25 24.	+ 21 49		16.0	BLUE STAR
MCG+04-20-063A	08 25 24.	+ 25 12 30.	30	15.6	GALAXY
ZWG 119.111	08 25 24.	+ 25 17		14.8	GALAXY
IC 0508	08 25 24.	+ 25 17 10.			NONSTELLAR OBJECT
ZWG 149.031	08 25 24.	+ 30 36		15.3	GALAXY
MCG+05-20-018	08 25 24.	+ 30 37	42	15.3	GALAXY
TON-N 0322	08 25 24.	+ 30 38		13.6	BLUE STAR
MCG+07-18-006	08 25 24.	+ 39 07 30.	36	16.	GALAXY
ZWG 263.037	08 25 24.	+ 55 40		14.9	GALAXY
KARA.72 163B	08 25 24.	+ 55 40	48	14.9	PART OF DOUBLE GALAXY
MCG+09-14-043	08 25 24.	+ 55 41	60	15.	GALAXY
UGC 04428	08 25 24.	+ 63 08	84	16.5	GALAXY Sc
ZWG 311.007	08 25 24.	+ 67 11		15.4	GALAXY
ZWG 310.057	08 25 24.	+ 67 11		15.4	GALAXY
IC 2378	08 25 26.	+ 30 35 34.			NONSTELLAR OBJECT
IC 2381	08 25 28.	+ 19 57 27.			NONSTELLAR OBJECT
ZC 0825.5+0005	08 25 30.	+ 00 05	1750		CLUSTER OF GALAXIES
ZWG 004.031	08 25 30.	+ 01 40		15.7	GALAXY
ZWG 060.017	08 25 30.	+ 08 49		15.6	GALAXY
LB 05454	08 25 30.	+ 18 01		18.5	FAINT BLUE STAR
LB 05453	08 25 30.	+ 18 40		18.6	FAINT BLUE STAR
LB 05452	08 25 30.	+ 18 47		20.4	FAINT BLUE STAR
ZWG 119.112	08 25 30.	+ 25 10		15.6	GALAXY
ZWG 208.006	08 25 30.	+ 40 50		15.5	GALAXY
UGC 04429	08 25 30.	+ 40 50	66	15.5	GALAXY Sc
MCG+07-18-007	08 25 30.	+ 40 50	60	14.5	GALAXY
MCG+08-16-006	08 25 30.	+ 47 58	36	16.	GALAXY
MCG+09-14-044	08 25 30.	+ 55 41	36	15.	GALAXY
ZC 0825.5+5711	08 25 30.	+ 57 11	2020		CLUSTER OF GALAXIES
MCG+11-11-010	08 25 30.	+ 63 26 30.	18	16.	GALAXY
ARC 0673	08 25 33.	+ 15 20		17.1	RICH CLUSTER OF GALAXIES
MCG+05-20-019	08 25 33.	+ 30 35	24	15.5	GALAXY
ZWG 004.033	08 25 36.	+ 00 11		15.2	GALAXY
UGC 04431	08 25 36.	+ 00 11	66	15.1	GALAXY Sb-c
ZWG 004.034	08 25 36.	+ 01 10		15.2	GALAXY
UGC 04432	08 25 36.	+ 01 10	102	15.2	GALAXY Sb-c
MCG+00-22-012	08 25 36.	+ 01 10	90	15.2	GALAXY
KARA.73B 0255	08 25 36.	+ 01 10	96	15.2	ISOLATED GALAXY S
LB 05460	08 25 36.	+ 17 55		18.3	FAINT BLUE STAR
LB 05459	08 25 36.	+ 18 02		19.1	FAINT BLUE STAR
LB 05458	08 25 36.	+ 18 05		18.7	FAINT BLUE STAR
LB 05457	08 25 36.	+ 18 16		19.1	FAINT BLUE STAR
LB 05456	08 25 36.	+ 18 20		20.1	FAINT BLUE STAR
LB 05455	08 25 36.	+ 18 42		18.5	FAINT BLUE STAR
ZWG 119.113	08 25 36.	+ 23 14		15.6	GALAXY
ZC 0825.6+2447	08 25 36.	+ 24 47	2220		CLUSTER OF GALAXIES
TON-N 0323	08 25 36.	+ 25 30		14.7	BLUE STAR
MRK 388	08 25 36.	+ 25 30	8	16.	GALAXY WITH UV CONTINUUM

OBJECT NAME	RIGHT ASCEN.	DECLINATION	DIAM.	MAGN.	TYPE OF OBJECT
VVI 26	08 25 36.	+ 25 30	8	16.	SEYFERT GALAXY
ZWG 149.032	08 25 36.	+ 30 35		15.5	GALAXY
ZC 0825.6+3600	08 25 36.	+ 36 00	870		CLUSTER OF GALAXIES
ZC 0825.6+3826	08 25 36.	+ 38 26	2350		CLUSTER OF GALAXIES
ZWG 263.038	08 25 36.	+ 55 10		14.8	GALAXY
MCG+09-14-045	08 25 36.	+ 55 10 30.	36	16.	GALAXY
7ZW 232	08 25 36.	+ 69 13			COMPACT GALAXY
ZWG 331.048	08 25 36.	+ 69 13		15.5	GALAXY
KARA.73B 0256	08 25 36.	+ 69 13	18	15.5	ISOLATED GALAXY E
ZWG 004.032	08 25 36.	- 01 22		15.3	GALAXY
UGC 04430	08 25 36.	- 01 22	66	15.3	GALAXY S0-a
MCG+00-22-011	08 25 36.	- 01 22	24	15.3	GALAXY
IC 2380	08 25 39.	+ 30 33 57.			NONSTELLAR OBJECT
MCG+00-22-013	08 25 42.	+ 00 12	72	14.	GALAXY
ZWG 004.035	08 25 42.	+ 00 27		15.1	GALAXY
ZWG 032.033	08 25 42.	+ 02 36		15.1	GALAXY
MCG+03-22-014	08 25 42.	+ 17 37	60	14.3	GALAXY
ZWG 089.031	08 25 42.	+ 17 38		14.3	GALAXY
SCH 03	08 25 42.	+ 17 38		14.3	PECULIAR GALAXY
UGC 04433	08 25 42.	+ 17 38	66	14.3	GALAXY S
LB 05464	08 25 42.	+ 18 11		20.4	FAINT BLUE STAR
LB 05463	08 25 42.	+ 18 15		18.8	FAINT BLUE STAR
LB 05462	08 25 42.	+ 18 16		19.8	FAINT BLUE STAR
LB 05461	08 25 42.	+ 18 29		18.9	FAINT BLUE STAR
ZWG 149.033	08 25 42.	+ 29 05		15.6	GALAXY
ZWG 179.008	08 25 42.	+ 34 50		14.6	GALAXY
UGC 04434	08 25 42.	+ 34 50	66	14.6	GALAXY PECULR
MCG+07-18-008	08 25 42.	+ 39 11	42	15.	GALAXY
ZWG 208.007	08 25 42.	+ 39 12		15.6	GALAXY
ARC 0672	08 25 46.	+ 32 39		17.4	RICH CLUSTER OF GALAXIES
ZWG 060.018	08 25 48.	+ 08 54		15.5	GALAXY
LB 05466	08 25 48.	+ 18 19		18.5	FAINT BLUE STAR
LB 05465	08 25 48.	+ 18 36		19.8	FAINT BLUE STAR
ZWG 119.114	08 25 48.	+ 22 14		14.9	GALAXY
MCG+04-20-064	08 25 48.	+ 22 14	24	14.9	GALAXY
ZWG 149.034	08 25 48.	+ 27 55		15.6	GALAXY
ZWG 149.035	08 25 48.	+ 30 37		15.5	GALAXY
ZC 0825.8+3240	08 25 48.	+ 32 40	470		CLUSTER OF GALAXIES
MCG+06-19-005	08 25 48.	+ 34 49	30	15.	GALAXY
ZWG 179.009	08 25 48.	+ 38 24		15.1	GALAXY
MCG+11-11-011	08 25 48.	+ 63 06	58	16.	GALAXY
MCG+12-09-051	08 25 48.	+ 74 38	21	17.	GALAXY
IC 2382	08 25 51.	+ 22 13 49.			NONSTELLAR OBJECT
UGC 04435	08 25 54.	+ 04 53	66	16.0	GALAXY S
LB 05469	08 25 54.	+ 18 00		18.8	FAINT BLUE STAR
LB 05468	08 25 54.	+ 18 08		19.3	FAINT BLUE STAR
LB 05467	08 25 54.	+ 18 15		18.9	FAINT BLUE STAR
ZC 0825.9+4512	08 25 54.	+ 45 12	940		CLUSTER OF GALAXIES
MCG-02-22-018	08 25 54.	- 11 58	54	15.	GALAXY
MCG+09-14-046	08 25 57.	+ 55 51 30.	36	16.	GALAXY
LB 05474	08 26 00.	+ 17 49		17.8	FAINT BLUE STAR
LB 05473	08 26 00.	+ 17 53		19.1	FAINT BLUE STAR
LB 05472	08 26 00.	+ 17 56		18.8	FAINT BLUE STAR
LB 05471	08 26 00.	+ 18 04		19.5	FAINT BLUE STAR
LB 05470	08 26 00.	+ 18 35		19.6	FAINT BLUE STAR
TON-N 0920	08 26 00.	+ 23 01		16.0	BLUE STAR
MCG+07-18-009	08 26 00.	+ 43 32 30.	33	16.	GALAXY
ZWG 263.039	08 26 00.	+ 55 50		15.3	GALAXY
MCG+11-11-012	08 26 00.	+ 63 37 30.	27	16.	GALAXY
MCG+12-09-052	08 26 00.	+ 74 39	33	16.	GALAXY
ZC 0826.0-0108	08 26 00.	- 01 08	1080		CLUSTER OF GALAXIES
ZWG 089.032	08 26 06.	+ 17 43		15.3	GALAXY
LB 05476	08 26 06.	+ 18 05		19.8	FAINT BLUE STAR
LB 05475	08 26 06.	+ 18 09		19.0	FAINT BLUE STAR
ZWG 119.115	08 26 06.	+ 25 56		15.7	GALAXY
MCG+09-14-048	08 26 06.	+ 52 07 30.	60	16.	GALAXY
MCG+09-14-047	08 26 06.	+ 52 51	42	14.	GALAXY
ZC 0826.1+5630	08 26 06.	+ 56 30	670		CLUSTER OF GALAXIES
ZC 0826.1+6554	08 26 06.	+ 65 54	2020		CLUSTER OF GALAXIES
TON-N 0921	08 26 07.	+ 19 19		15.3	BLUE STAR
FATH 1.216	08 26 10.	+ 30 21	16		NEBULA
ARC 0665	08 26 11.	+ 66 04		17.5	RICH CLUSTER OF GALAXIES
ZC 0826.2+1044	08 26 12.	+ 10 44	340		CLUSTER OF GALAXIES
LB 05482	08 26 12.	+ 18 03		18.9	FAINT BLUE STAR
LB 05481	08 26 12.	+ 18 03		18.7	FAINT BLUE STAR
LB 05480	08 26 12.	+ 18 06		19.4	FAINT BLUE STAR
LB 05479	08 26 12.	+ 18 11		17.9	FAINT BLUE STAR
LB 05478	08 26 12.	+ 18 20		19.1	FAINT BLUE STAR
LB 05477	08 26 12.	+ 18 24		20.2	FAINT BLUE STAR
ZC 0826.2+3039	08 26 12.	+ 30 39	9070		CLUSTER OF GALAXIES
TON-N 0324	08 26 12.	+ 32 39		14.3	BLUE STAR
MCG+06-19-006	08 26 12.	+ 35 05	18	16.	GALAXY
ZWG 237.003	08 26 12.	+ 48 56		15.5	GALAXY
UGC 04436	08 26 12.	+ 48 56	78	15.5	GALAXY Sb-c
UGC 04437	08 26 12.	+ 52 08	78	16.0	GALAXY Sc
ZWG 263.040	08 26 12.	+ 52 52		13.9	GALAXY
UGC 04438	08 26 12.	+ 52 52	48	13.9	GALAXY S
ZC 0826.2+6805	08 26 12.	+ 68 05	1140		CLUSTER OF GALAXIES
MCG+12-09-053	08 26 12.	+ 74 02 30.	84	14.	GALAXY
VBE 16	08 26 12.	- 50 59	48		REFLECTION NEBULA
MCG+06-19-007	08 26 15.	+ 35 05 30.	15	17.	GALAXY
MCG+08-16-007	08 26 15.	+ 48 57	78	15.	GALAXY
TON-N 0923	08 26 17.	+ 31 43		15.2	BLUE STAR
TON-N 0922	08 26 17.	+ 34 16		16.	BLUE STAR
ARC 0669	08 26 17.	+ 56 30		17.9	RICH CLUSTER OF GALAXIES
LB 05487	08 26 18.	+ 17 49		20.3	FAINT BLUE STAR
LB 05486	08 26 18.	+ 18 24		20.0	FAINT BLUE STAR
LB 05485	08 26 18.	+ 18 32		19.8	FAINT BLUE STAR
LB 05484	08 26 18.	+ 18 33		18.9	FAINT BLUE STAR
LB 05483	08 26 18.	+ 18 33		20.2	FAINT BLUE STAR
ZWG 149.036	08 26 18.	+ 30 35		15.6	GALAXY
ZWG 149.037	08 26 18.	+ 31 50		14.9	GALAXY
MCG+05-20-020	08 26 18.	+ 31 50	36	14.9	GALAXY
ZC 0826.3+6323	08 26 18.	+ 63 23	2890		CLUSTER OF GALAXIES
MCG-01-22-013	08 26 18.	- 06 47 30.	24	15.	GALAXY
FATH 1.217	08 26 23.	+ 30 26	19		NEBULA
ZWG 004.036	08 26 24.	+ 00 47		15.5	GALAXY
LB 05489	08 26 24.	+ 18 10		18.8	FAINT BLUE STAR
LB 05488	08 26 24.	+ 18 42		19.9	FAINT BLUE STAR
MCG+05-20-021	08 26 24.	+ 30 36	45	15.6	GALAXY
TON-N 0325	08 26 24.	+ 31 50		14.2	BLUE STAR
ZC 0826.4+3710	08 26 24.	+ 37 10	540		CLUSTER OF GALAXIES
MCG+08-16-008	08 26 24.	+ 49 50	42	15.	GALAXY
ZWG 237.004	08 26 24.	+ 49 51		15.6	GALAXY
MCG-04-20-013	08 26 24.	- 21 33	60	16.	GALAXY
FATH 1.218	08 26 28.	+ 30 08	14		NEBULA
ZWG 149.038	08 26 30.	+ 30 51		15.7	GALAXY
ZWG 237.005	08 26 30.	+ 45 06		15.4	GALAXY

OBJECT NAME	RIGHT ASCEN.	DECLINATION	DIAM.	MAGN.	TYPE OF OBJECT
MCG+08-16-009	08 26 30.	+ 45 06	12	16.	GALAXY
FATH 1.219	08 26 33.	+ 30 09	8		NEBULA
IC 2383	08 26 37.	+ 30 52 15.			NONSTELLAR OBJECT
PK258-00.1	08 26 39.	- 39 14	10		PLANETARY NEBULA
ZWG 032.034	08 26 42.	+ 02 48		15.7	GALAXY
UGC 04439	08 26 42.	+ 02 48	90	15.7	GALAXY Sc
MCG+01-22-012	08 26 42.	+ 02 48	72	15.7	GALAXY
FATH 1.220	08 26 44.	+ 29 23	22		NEBULA
FATH 1.221	08 26 46.	+ 30 08	14		NEBULA
ZWG 089.033	08 26 48.	+ 15 14		15.2	GALAXY
UGC 04440	08 26 48.	+ 15 14	66	15.2	GALAXY SB
ZC 0826.8+5055	08 26 48.	+ 50 55	1080		CLUSTER OF GALAXIES
MCG+09-14-049	08 26 48.	+ 52 27	66	14.	GALAXY
7ZW 233	08 26 48.	+ 68 45			COMPACT GALAXY
TON-N 0924	08 26 49.	+ 19 01		14.9	BLUE STAR
ZWG 149.039	08 26 54.	+ 29 27		15.6	GALAXY
ZWG 208.008	08 26 54.	+ 40 03		15.6	GALAXY
UGC 04441	08 26 54.	+ 40 03	66	15.6	GALAXY Sb
MCG+07-18-010	08 26 54.	+ 40 03 30.	63	15.	GALAXY
TON-N 0925	08 26 55.	+ 21 08		14.8	BLUE STAR
ZWG 004.037	08 27 00.	+ 01 22		15.7	GALAXY
ZC 0827.0+4856	08 27 00.	+ 48 56	1480		CLUSTER OF GALAXIES
ZWG 263.041	08 27 00.	+ 52 28		14.5	GALAXY
UGC 04442	08 27 00.	+ 52 28	66	14.5	GALAXY Sb/SBc
ZC 0827.0+6619	08 27 00.	+ 66 19	1010		CLUSTER OF GALAXIES
FATH 1.222	08 27 02.	+ 29 27	22		NEBULA
RNGC 2597	08 27 03.	+ 21 40			NON-EXISTENT OBJECT
FATH 1.223	08 27 05.	+ 29 28	19		NEBULA
ZC 0827.1+1526	08 27 06.	+ 15 26	1010		CLUSTER OF GALAXIES
ZWG 119.116	08 27 06.	+ 21 40		15.1	GALAXY
UGC 04443	08 27 06.	+ 21 40	84	15.1	GALAXY S
MCG+04-20-065	08 27 06.	+ 21 40	48	15.1	GALAXY
ZC 0827.1+5421	08 27 06.	+ 54 21	1010		CLUSTER OF GALAXIES
ZWG 004.038	08 27 06.	- 00 59		15.5	GALAXY
RNGC 2598	08 27 09.	+ 21 39		15.0	GALAXY
BC 4C19.30	08 27 10.95	+ 19 20 47.5		17.	QUASI-STELLAR OBJECT
SHB 112	08 27 11.0	+ 19 20 47.		17.	QUASI-STELLAR OBJECT
LB 00543	08 27 12.	+ 11 49 36.		17.0	FAINT BLUE STAR
ZWG 089.034	08 27 12.	+ 17 26		14.4	GALAXY
UGC 04444	08 27 12.	+ 17 26	96	14.4	GALAXY SBc
ZWG 149.040	08 27 12.	+ 31 21		15.7	GALAXY
ZWG 208.009	08 27 12.	+ 43 09		15.5	GALAXY
MCG+12-09-054	08 27 12.	+ 69 59		16.	GALAXY
MCG+10-12-142	08 27 15.	+ 61 10	72	15.	GALAXY
MCG+03-22-015	08 27 18.	+ 17 26	84	14.4	GALAXY
MCG+07-18-011	08 27 18.	+ 43 10	36	15.	GALAXY
ZC 0827.3+4432	08 27 18.	+ 44 32	1340		CLUSTER OF GALAXIES
ZC 0827.3+4628	08 27 18.	+ 46 28	2550		CLUSTER OF GALAXIES
ZWG 287.077	08 27 18.	+ 61 10		15.3	GALAXY
UGC 04445	08 27 18.	+ 61 10	102	15.3	GALAXY Sc
7ZW 234	08 27 18.	+ 63 01			COMPACT GALAXY
ARC 0674	08 27 21.	+ 18 36		17.7	RICH CLUSTER OF GALAXIES
ZC 0827.4+3311	08 27 24.	+ 33 11	810		CLUSTER OF GALAXIES
ZC 0827.4+4045	08 27 24.	+ 40 45	2350		CLUSTER OF GALAXIES
ZC 0827.4+4940	08 27 24.	+ 49 40	1010		CLUSTER OF GALAXIES
7ZW 235	08 27 24.	+ 65 28			COMPACT GALAXY
MCG-03-22-002	08 27 24.	- 17 07	48	15.	GALAXY
CED 106A	08 27 24.	- 44 00	1980		DIFFUSE GALACTIC NEBULA
CED 106B	08 27 29.	- 47 46			DIFFUSE GALACTIC NEBULA
ZWG 004.039	08 27 30.	+ 02 08		15.6	GALAXY
ZC 0827.5+1835	08 27 30.	+ 18 35	740		CLUSTER OF GALAXIES
MCG-01-22-014	08 27 30.	- 07 22	66	15.	GALAXY
OCL 0652	08 27 30.	- 18 56	1020	11.	OPEN STAR CLUSTER
TON-N 0926	08 27 35.	+ 32 56		15.7	BLUE STAR
ZWG 089.035	08 27 36.	+ 14 33		15.6	GALAXY
ZWG 089.036	08 27 36.	+ 19 54		15.6	GALAXY
ZWG 119.117	08 27 36.	+ 21 20		15.7	GALAXY
CED 106C	08 27 41.	- 44 33	1620		DIFFUSE GALACTIC NEBULA
ZWG 089.037	08 27 42.	+ 18 22		15.5	GALAXY
ZC 0827.7+3707	08 27 42.	+ 37 07	2550		CLUSTER OF GALAXIES
ZWG 208.010	08 27 42.	+ 39 47		15.5	GALAXY
MCG+07-18-012	08 27 42.	+ 40 00	30	16.	GALAXY
ZWG 004.040	08 27 42.	- 02 44		15.5	GALAXY
FATH 1.224	08 27 45.	+ 30 02	8		NEBULA
ZWG 119.118	08 27 48.	+ 20 47		15.7	GALAXY
UGC 04446	08 27 48.	+ 20 47	72	15.7	GALAXY Sc
ZWG 119.119	08 27 48.	+ 24 47		15.7	GALAXY
KARA.73B 0257	08 27 48.	+ 24 47	54	15.7	ISOLATED GALAXY S
MCG+07-18-013	08 27 48.	+ 39 59	36	16.	GALAXY
ZC 0827.8+4158	08 27 48.	+ 41 58	870		CLUSTER OF GALAXIES
ZWG 208.011	08 27 48.	+ 43 38		15.7	GALAXY
CED 106D	08 27 53.	- 41 21	870		DIFFUSE GALACTIC NEBULA
MCG+03-22-016	08 27 54.	+ 20 26	42	15.6	GALAXY
ZC 0827.9+3133	08 27 54.	+ 31 33	1080		CLUSTER OF GALAXIES
ZC 0827.9+3236	08 27 54.	+ 32 36	870		CLUSTER OF GALAXIES
MCG+08-16-010	08 27 54.	+ 48 47 30.	9	16.	GALAXY
ZC 0827.9+5731	08 27 54.	+ 57 31	870		CLUSTER OF GALAXIES
ZWG 004.041	08 27 54.	- 00 36		15.6	GALAXY
SHB 113	08 27 54.4	+ 24 21 08.		17.5	QUASI-STELLAR OBJECT
SHB 114	08 27 55.1	+ 37 52 18.		18.1	QUASI-STELLAR OBJECT
BC 4C37.24	08 27 55.3	+ 37 52 16.		18.11	QUASI-STELLAR OBJECT
LB 00544	08 27 59.	+ 13 32 36.		17.3	FAINT BLUE STAR
MAI 029	08 27 59.	+ 59 11	33		DWARF SPHEROIDAL GALAXY
UGC 04447	08 28 00.	+ 20 00	66	16.5	GALAXY
ZWG 089.038	08 28 00.	+ 20 25		15.6	GALAXY
MCG+07-18-014	08 28 00.	+ 39 58	18	16.	GALAXY
ZWG 349.027	08 28 00.	+ 74 35		14.3	GALAXY
UGC 04448	08 28 00.	+ 74 35	54	14.3	GALAXY E
CED 106E	08 28 00.	- 43 56			DIFFUSE GALACTIC NEBULA
ARC 0670	08 28 03.	+ 67 03		17.5	RICH CLUSTER OF GALAXIES
FATH 1.225	08 28 06.	+ 29 57	5		NEBULA
ZWG 208.012	08 28 06.	+ 41 07		15.7	GALAXY
UGC 04449	08 28 06.	+ 41 07	78	15.7	GALAXY PECULR
MCG+12-09-055	08 28 06.	+ 68 40	24	17.	GALAXY
7ZW 236	08 28 06.	+ 70 57			COMPACT GALAXY
MCG+12-08-046	08 28 06.	+ 74 35	27	16.	GALAXY
MCG-01-22-015	08 28 06.	- 07 11	48	15.	GALAXY
MCG+07-18-015	08 28 09.	+ 41 08	54	15.	GALAXY
ZWG 179.010	08 28 12.	+ 36 01		15.7	GALAXY
MCG+06-19-008	08 28 12.	+ 36 02	36	15.5	GALAXY
ZC 0828.3+1300	08 28 18.	+ 13 00	740		CLUSTER OF GALAXIES
ZC 0828.3+3840	08 28 18.	+ 38 40	1010		CLUSTER OF GALAXIES
ZWG 263.042	08 28 18.	+ 56 01		15.5	GALAXY
ZC 0828.3-0008	08 28 18.	- 00 08	1340		CLUSTER OF GALAXIES
SEY 047	08 28 19.	+ 74 34 06.		14.7	FAINT GALAXY
ARC 0675	08 28 20.	+ 38 35		17.7	RICH CLUSTER OF GALAXIES
ARC 0676	08 28 22.	+ 37 20		17.7	RICH CLUSTER OF GALAXIES
ZWG 004.042	08 28 24.	+ 00 04		15.5	GALAXY

OBJECT NAME	RIGHT ASCEN.	DECLINATION	DIAM.	MAGN.	TYPE OF OBJECT
ZWG 060.019	08 28 24.	+ 09 38		15.3	GALAXY
ZWG 149.041	08 28 24.	+ 27 45		15.3	GALAXY
UGC 04450	08 28 24.	+ 27 45	72	15.3	GALAXY Sc
MCG+09-14-050	08 28 24.	+ 56 01 30.	60	15.	GALAXY
1ZW 015	08 28 24.	+ 56 02			COMPACT GALAXY
ZWG 349.028	08 28 24.	+ 75 20		14.5	GALAXY
UGC 04451	08 28 24.	+ 75 20	60	14.5	GALAXY Sa-b
LB 00545	08 28 25.	+ 12 41 24.		18.0	FAINT BLUE STAR
ZWG 060.020	08 28 30.	+ 09 47		15.2	GALAXY
UGC 04452	08 28 30.	+ 09 47	72	15.2	GALAXY S0?
ZC 0828.5+3815	08 28 30.	+ 38 15	940		CLUSTER OF GALAXIES
MCG+09-14-051	08 28 30.	+ 56 01	18	16.	GALAXY
MCG+14-04-052	08 28 30.	+ 84 37	39	17.	GALAXY
FATH 1.226	08 28 31.	+ 29 43	14		NEBULA
RNGC 2609	08 28 31.	- 60 56			NON-EXISTENT OBJECT
MCG-03-22-003	08 28 33.	- 17 31	42	15.	GALAXY
ZWG 119.120	08 28 36.	+ 22 12		15.4	GALAXY
UGC 04453	08 28 36.	+ 22 12	66	15.4	GALAXY S0
FATH 1.227	08 28 36.	+ 29 50	5		NEBULA
MCG+09-14-053	08 28 36.	+ 52 45 30.	48	16.	GALAXY
MCG+09-14-052	08 28 36.	+ 56 38	42	16.	GALAXY
ZWG 004.043	08 28 36.	- 01 57		15.6	GALAXY
MCG-01-22-016	08 28 36.	- 04 01	36	15.	GALAXY
ZWG 032.035	08 28 42.	+ 07 10		15.6	GALAXY
ZWG 149.042	08 28 42.	+ 28 43		15.3	GALAXY
KARA.73B 0258	08 28 42.	+ 28 43	36	15.3	ISOLATED GALAXY S
ZWG 263.043	08 28 42.	+ 52 46		14.7	GALAXY
ZWG 287.078	08 28 42.	+ 61 58		15.0	GALAXY
LB 00546	08 28 47.	+ 13 32 30.		16.8	FAINT BLUE STAR
ZWG 089.039	08 28 48.	+ 19 33		15.7	GALAXY
MCG+10-12-143	08 28 48.	+ 61 59	39	14.	GALAXY
ZWG 004.044	08 28 48.	- 02 50		15.6	GALAXY
MRSL 257+00/1	08 28 48.	- 38 10	60		HII REGION
MCG+09-14-054	08 28 51.	+ 52 36	42	16.	GALAXY
ZC 0828.9+3557	08 28 54.	+ 35 57	1480		CLUSTER OF GALAXIES
MCG-01-22-017	08 28 54.	- 03 51	42	15.5	GALAXY
VDB.66G 053	08 29	+ 66 21	70		DWARF GALAXY
ZWG 119.121	08 29 00.	+ 24 10		14.6	GALAXY
UGC 04456	08 29 00.	+ 24 10	114	14.6	GALAXY Sc
KARA.73B 0260	08 29 00.	+ 24 10	120	14.6	ISOLATED GALAXY S
MCG+07-18-016	08 29 00.	+ 41 31 30.	36	15.	GALAXY
ZWG 208.013	08 29 00.	+ 41 32		15.2	GALAXY
MCG+08-16-011	08 29 00.	+ 50 07	60	16.	GALAXY
ZWG 263.044	08 29 00.	+ 52 37		15.3	GALAXY
MRK 015	08 29 00.	+ 75 19	10	17.	GALAXY WITH UV CONTINUUM
MCG+13-06-025	08 29 00.	+ 75 19	24	17.	GALAXY
MCG+13-06-024	08 29 00.	+ 75 20	51	16.	GALAXY
ZWG 363.045	08 29 00.	+ 80 44		15.3	GALAXY
UGC 04454	08 29 00.	+ 80 44	60	15.3	GALAXY PECULR
ZWG 004.045	08 29 00.	- 01 02		15.0	GALAXY
UGC 04455	08 29 00.	- 01 02	60	15.0	GALAXY SBa
MCG+00-22-014	08 29 00.	- 01 02	48	15.0	GALAXY
KARA.73B 0259	08 29 00.	- 01 02	60	15.0	ISOLATED GALAXY S
ACK 257+00.1	08 29 00.	- 38 10			PLANETARY NEBULA
FATH 1.228	08 29 02.	+ 29 24	19		NEBULA
ARP 058	08 29 03.	+ 19 23			PECULIAR GALAXY
MCG+04-20-066	08 29 03.	+ 24 12 30.	90	14.6	GALAXY
WRAY 19.11	08 29 03.5	- 38 09 44.			DIFFUSE NEBULA
ZC 0829.1+0428	08 29 06.	+ 04 28	870		CLUSTER OF GALAXIES
ZWG 089.040	08 29 06.	+ 19 23		14.9	GALAXY
SCH 04	08 29 06.	+ 19 23		14.9	PECULIAR GALAXY
UGC 04457	08 29 06.	+ 19 23	102	14.9	GALAXY S
MCG+03-22-017	08 29 06.	+ 19 23	102	14.9	GALAXY
IC 0509	08 29 06.	+ 24 10 40.			NONSTELLAR OBJECT
ARC 0677	08 29 06.	+ 35 57		17.9	RICH CLUSTER OF GALAXIES
ZC 0829.1+4200	08 29 06.	+ 42 00	940		CLUSTER OF GALAXIES
ZWG 208.014	08 29 06.	+ 42 10		15.3	GALAXY
MCG+08-16-012	08 29 06.	+ 49 19	54	16.	GALAXY
OCL 0731	08 29 06.	- 38 30	390	13.	OPEN STAR CLUSTER
VHA 033	08 29 06.	- 38 30	360		OPEN STAR CLUSTER
MCG+07-18-017	08 29 09.	+ 42 10	54	15.	GALAXY
ZWG 119.122	08 29 12.	+ 22 44		13.4	GALAXY
MRK 389	08 29 12.	+ 22 44	24	15.	GALAXY WITH UV CONTINUUM
UGC 04458	08 29 12.	+ 22 44	156	13.4	GALAXY Sa
ZWG 149.043	08 29 12.	+ 30 42		15.4	GALAXY
ZWG 149.044	08 29 12.	+ 31 21		15.6	GALAXY
MCG+07-18-018	08 29 12.	+ 44 21 30.	48	15.5	GALAXY
ZC 0829.2+5159	08 29 12.	+ 51 59	2080		CLUSTER OF GALAXIES
ZC 0829.2+6040	08 29 12.	+ 60 40	1080		CLUSTER OF GALAXIES
ZC 0829.2+6656	08 29 12.	+ 66 56	3160		CLUSTER OF GALAXIES
MCG-01-22-018	08 29 12.	- 03 56	18	15.	GALAXY
SN 1965P	08 29 13.	+ 22 44		15.7	SUPERNOVA
LB 01801	08 29 14.	+ 20 44 06.		16.0	FAINT BLUE STAR
RNGC 2599	08 29 14.	+ 22 44		13.5	GALAXY
FATH 1.229	08 29 14.	+ 29 52	8		NEBULA
MCG+04-20-067	08 29 15.	+ 22 45	102	13.4	GALAXY
ISS 0919	08 29 18.	- 31 23	299		STELLAR RING
FATH 1.230	08 29 22.	+ 29 49	27		NEBULA
ZWG 060.021	08 29 24.	+ 12 51		15.6	GALAXY
ZWG 089.041	08 29 24.	+ 19 47		15.4	GALAXY
ZWG 119.123	08 29 24.	+ 24 42		15.7	GALAXY
ZWG 149.045	08 29 24.	+ 30 07		15.6	GALAXY
ZC 0829.4+4552	08 29 24.	+ 45 52	540		CLUSTER OF GALAXIES
MCG+11-11-013	08 29 24.	+ 66 20	78	15.	GALAXY
MCG+09-14-055	08 29 27.	+ 54 00 30.	30	16.	GALAXY
BC B20829+33	08 29 28.0	+ 33 41 54.0		19.	QUASI-STELLAR OBJECT
ZWG 032.036	08 29 30.	+ 03 48		15.1	GALAXY
KARA.73B 0261	08 29 30.	+ 03 48	24	15.1	ISOLATED GALAXY S
FATH 1.231	08 29 30.	+ 30 06	41		NEBULA
ZWG 149.046	08 29 30.	+ 30 32		15.5	GALAXY
7ZW 237	08 29 30.	+ 62 30			COMPACT GALAXY
ZWG 311.008	08 29 30.	+ 66 20		15.4	GALAXY
UGC 04459	08 29 30.	+ 66 20	108	15.4	GALAXY DWRF IR
MCG+09-14-057	08 29 33.	+ 52 40 30.	90	14.	GALAXY
MCG+09-14-056	08 29 33.	+ 53 59 30.	36	17.	GALAXY
MAI 030	08 29 34.	+ 84 36	33		DWARF SPHEROIDAL GALAXY
CED 106F	08 29 35.	- 47 42			DIFFUSE GALACTIC NEBULA
ZWG 149.047	08 29 36.	+ 29 23		15.7	GALAXY
ZC 0829.6+4923	08 29 36.	+ 49 23	1550		CLUSTER OF GALAXIES
ZWG 263.045	08 29 36.	+ 52 42		14.4	GALAXY
UGC 04461	08 29 36.	+ 52 42	114	14.4	GALAXY Sb-c
ZC 0829.6+5245	08 29 36.	+ 52 45	4370		CLUSTER OF GALAXIES
ZWG 263.046	08 29 36.	+ 54 43		15.5	GALAXY
MCG+09-14-059	08 29 36.	+ 54 43	18	16.	GALAXY
MCG+09-14-058	08 29 36.	+ 54 56	36	16.	GALAXY
7ZW 238	08 29 36.	+ 66 21			COMPACT GALAXY
ZWG 004.046	08 29 36.	- 02 00		15.2	GALAXY
UGC 04460	08 29 36.	- 02 00	66	15.2	GALAXY SB+COMP

OBJECT NAME	RIGHT ASCEN.	DECLINATION	DIAM.	MAGN.	TYPE OF OBJECT
MCG+00-22-015	08 29 36.	- 02 00	54	15.2	GALAXY
VHA 034	08 29 36.	- 44 19	780		OPEN STAR CLUSTER
LB 03483	08 29 36.	- 76 05		11.0	FAINT BLUE STAR
FATH 1.232	08 29 37.	+ 29 23	27		NEBULA
LB 00185	08 29 37.	+ 30 09 12.		17.1	FAINT BLUE STAR
IC 0510	08 29 40.	- 01 59 14.			NONSTELLAR OBJECT
ZC 0829.7+5042	08 29 42.	+ 50 42	810		CLUSTER OF GALAXIES
MCG+09-14-060	08 29 45.	+ 55 46	42	17.	GALAXY
ZWG 089.042	08 29 48.	+ 19 41		15.7	GALAXY
MCG+03-22-018	08 29 48.	+ 19 41	30	15.7	GALAXY
ZWG 208.015	08 29 48.	+ 41 16		14.9	GALAXY
MCG+09-14-061	08 29 48.	+ 54 42 30.	48	15.	GALAXY
ZWG 263.047	08 29 48.	+ 55 45		15.7	GALAXY
ZWG 331.049	08 29 48.	+ 70 21		15.7	GALAXY
PK252+04.1	08 29 48.	- 31 55	55	17.9	PLANETARY NEBULA
ZWG 060.022	08 29 54.	+ 09 57		15.7	GALAXY
UGC 04462	08 29 54.	+ 09 57	72	15.7	GALAXY DBL SYS
MCG+04-20-068	08 29 54.	+ 26 14	66	15.5	GALAXY
MCG+07-18-018A	08 29 54.	+ 41 15	60	14.5	GALAXY
ZWG 263.048	08 29 54.	+ 54 42		14.9	GALAXY
ZWG 287.079	08 29 54.	+ 56 59		15.5	GALAXY
MCG+10-12-144	08 29 54.	+ 56 59	27	15.	GALAXY
MCG+10-12-145	08 29 54.	+ 57 42	18	15.	GALAXY
ZWG 287.080	08 29 54.	+ 57 43		15.2	GALAXY
KARA.72 164A	08 29 54.	+ 57 43	42	15.2	PART OF DOUBLE GALAXY
ZWG 004.048	08 29 54.	- 00 44		15.5	GALAXY
ZWG 004.047	08 29 54.	- 03 02		15.3	GALAXY
ZWG 089.043	08 30 00.	+ 20 14		15.7	GALAXY
ZWG 119.124	08 30 00.	+ 26 11		15.5	GALAXY
UGC 04464	08 30 00.	+ 26 11	102	15.5	GALAXY SB?c
MCG+07-18-019	08 30 00.	+ 41 25 30.	72	14.5	GALAXY
ZWG 208.016	08 30 00.	+ 41 26		15.0	GALAXY
UGC 04465	08 30 00.	+ 41 26	72	15.0	GALAXY Sa
ZC 0830.0+5405	08 30 00.	+ 54 05	870		CLUSTER OF GALAXIES
MCG+09-14-062	08 30 00.	+ 55 42	36	16.	GALAXY
MCG+10-12-146	08 30 00.	+ 57 44	18	16.	GALAXY
ZWG 287.081	08 30 00.	+ 60 18		15.7	GALAXY
UGC 04466	08 30 00.	+ 78 00	90	17.	GALAXY DWRF SP
MCG+14-04-053	08 30 00.	+ 84 49	66	16.	GALAXY
ZWG 364.005	08 30 00.	+ 85 55		13.5	GALAXY
ZWG 363.046	08 30 00.	+ 85 55		13.5	GALAXY
UGC 04463	08 30 00.	+ 85 55	138	14.5	GALAXY Sa
MCG+09-14-063	08 30 03.	+ 53 52	24	17.	GALAXY
ZC 0830.1+2126	08 30 06.	+ 21 26	1080		CLUSTER OF GALAXIES
ZWG 263.049	08 30 06.	+ 55 42		15.4	GALAXY
ZC 0830.1+5634	08 30 06.	+ 56 34	1750		CLUSTER OF GALAXIES
ZWG 287.082	08 30 06.	+ 57 45		15.1	GALAXY
KARA.72 164B	08 30 06.	+ 57 45	36	15.1	PART OF DOUBLE GALAXY
KARA.73B 0262	08 30 06.	+ 57 45	36	15.1	ISOLATED GALAXY S
MCG+11-11-014	08 30 06.	+ 66 09	27	17.	GALAXY
MCG-01-22-019	08 30 06.	- 05 28	42	15.	GALAXY
ZWG 004.049	08 30 12.	+ 00 24		15.2	GALAXY
UGC 04467	08 30 12.	+ 00 24	90	15.5	GALAXY S
ZC 0830.2+1547	08 30 12.	+ 15 47	2890		CLUSTER OF GALAXIES
ZWG 089.044	08 30 12.	+ 17 09		15.3	GALAXY
ZWG 208.017	08 30 12.	+ 41 42		14.7	GALAXY
UGC 04468	08 30 12.	+ 41 42	90	14.7	GALAXY S0
ZC 0830.2+4526	08 30 12.	+ 45 26	870		CLUSTER OF GALAXIES
MCG+09-14-064	08 30 12.	+ 52 09	36	16.	GALAXY
7ZW 239	08 30 12.	+ 57 44			COMPACT GALAXY
MCG+05-20-022	08 30 15.	+ 29 43	120	13.5	GALAXY
MCG+07-18-020	08 30 15.	+ 41 42	24	14.	GALAXY
FATH 1.234	08 30 16.	+ 29 37	14		NEBULA
HOLM 096A	08 30 16.	+ 29 42	114	12.7	PART OF MULTIPLE GALAXY
RNGC 2604A	08 30 16.	+ 29 43		13.5	GALAXY
FATH 1.233	08 30 16.	+ 30 01	14		NEBULA
ZC 0830.3+1716	08 30 18.	+ 17 16	810		CLUSTER OF GALAXIES
ZWG 149.048	08 30 18.	+ 29 43		13.5	GALAXY
UGC 04469	08 30 18.	+ 29 43	132	13.5	GALAXY SBc
ZWG 263.050	08 30 18.	+ 55 47		15.7	GALAXY
UGC 04470	08 30 18.	+ 55 47	78	15.7	GALAXY S
MCG+09-14-065	08 30 18.	+ 55 48	60	16.	GALAXY
OCL 0658	08 30 18.	- 19 29	420	11.	OPEN STAR CLUSTER
FATH 2.012	08 30 19.	+ 29 42	68		NEBULA
MCG+07-18-021	08 30 21.	+ 42 19	27	16.	GALAXY
ZWG 089.045	08 30 24.	+ 15 32		15.7	GALAXY
ZWG 089.046	08 30 24.	+ 19 31		15.5	GALAXY
ZC 0830.4+3510	08 30 24.	+ 35 10	4170		CLUSTER OF GALAXIES
MCG+07-18-022	08 30 24.	+ 41 35 30.	60	15.	GALAXY
ZWG 208.018	08 30 24.	+ 41 36		15.6	GALAXY
UGC 04471	08 30 24.	+ 41 36	66	15.6	GALAXY S
ZWG 237.006	08 30 24.	+ 47 18		15.7	GALAXY
ZWG 263.051	08 30 24.	+ 52 11		15.4	GALAXY
ZWG 263.052	08 30 24.	+ 54 10		15.6	GALAXY
MCG+09-14-066	08 30 24.	+ 54 10	24	17.	GALAXY
ZWG 004.050	08 30 24.	- 00 52		15.5	GALAXY
HELW 137	08 30 24.	- 22 47 56.			NEBULA
FATH 1.235	08 30 26.	+ 30 23	14		NEBULA
RNGC 2604B	08 30 28.	+ 29 40		15.5	GALAXY
ZWG 149.049	08 30 30.	+ 29 40		15.6	GALAXY
HOLM 096B	08 30 30.	+ 29 40	48	14.6	PART OF MULTIPLE GALAXY
MCG+05-20-023	08 30 30.	+ 29 40	42	15.6	GALAXY
FATH 1.236	08 30 30.	+ 29 47	14		NEBULA
ZC 0830.5+4746	08 30 30.	+ 47 46	1550		CLUSTER OF GALAXIES
ZC 0830.5+5058	08 30 30.	+ 50 58	1140		CLUSTER OF GALAXIES
ZWG 263.053	08 30 30.	+ 52 00		15.5	GALAXY
7ZW 240	08 30 30.	+ 70 47			COMPACT GALAXY
ZWG 004.051	08 30 30.	- 02 22		15.5	GALAXY
MCG-04-21-001	08 30 30.	- 23 53	24	16.	GALAXY
TON-N 0927	08 30 31.	+ 20 37		11.0	BLUE STAR
SHB 115	08 30 31.2	+ 15 23 48.		19.	QUASI-STELLAR OBJECT
FATH 1.237	08 30 32.	+ 29 40	46		NEBULA
MCG+05-20-024	08 30 36.	+ 27 52 30.	24	15.1	GALAXY
ZWG 149.050	08 30 36.	+ 27 54		15.1	GALAXY
ZWG 208.019	08 30 36.	+ 39 59		15.4	GALAXY
ZWG 331.050	08 30 36.	+ 70 47		15.3	GALAXY
ZWG 350.001	08 30 36.	+ 78 12		12.8	GALAXY
ZWG 349.029	08 30 36.	+ 78 12		12.8	GALAXY
UGC 04472	08 30 36.	+ 78 12	192	12.8	GALAXY Sc
MCG+13-07-001	08 30 36.	+ 78 13	168	12.	GALAXY
MCG-02-22-019	08 30 36.	- 12 10 30.	36	14.5	GALAXY
MCG+07-18-023	08 30 39.	+ 39 58	36	15.	GALAXY
ARC 0678	08 30 40.	+ 50 54		17.1	RICH CLUSTER OF GALAXIES
MCG+07-18-024	08 30 42.	+ 41 34	30	15.	GALAXY
ZWG 208.020	08 30 42.	+ 41 35		15.7	GALAXY
MCG+09-14-067	08 30 42.	+ 54 01	24	17.	GALAXY
IC 0513	08 30 42.	- 12 11 03.			NONSTELLAR OBJECT
MCG-04-21-002	08 30 42.	- 22 42 30.	48	15.5	GALAXY

OBJECT NAME	RIGHT ASCEN.	DECLINATION	DIAM.	MAGN.	TYPE OF OBJECT
RNGC 2591	08 30 44.	+ 78 12		13.0	GALAXY
FATH 1.238	08 30 47.	+ 29 43	16		NEBULA
ZWG 060.023	08 30 48.	+ 13 13		15.3	GALAXY
ZWG 089.047	08 30 48.	+ 19 34		15.6	GALAXY
MCG+05-20-025	08 30 48.	+ 27 08	42	14.9	GALAXY
MCG+08-16-013	08 30 48.	+ 48 15	36	15.	GALAXY
MCG+09-14-068	08 30 48.	+ 52 52 30.	72	15.	GALAXY
ZWG 263.054	08 30 48.	+ 54 01		15.7	GALAXY
ZWG 004.053	08 30 48.	- 02 21		15.4	GALAXY
ZWG 004.052	08 30 48.	- 03 20		14.9	GALAXY
MCG+00-22-016	08 30 48.	- 03 20	30	14.9	GALAXY
ISS 0920	08 30 48.	- 32 16	670		STELLAR RING
ZWG 060.024	08 30 54.	+ 10 07		15.5	GALAXY
TON-N 0928	08 30 54.	+ 23 12		16.0	BLUE STAR
ZWG 149.051	08 30 54.	+ 27 09		14.9	GALAXY
RNGC 2607	08 30 54.	+ 27 09		15.0	GALAXY
UGC 04473	08 30 54.	+ 27 09	96	14.9	GALAXY S
FATH 1.239	08 30 54.	+ 29 37	14		NEBULA
ARC 0679	08 30 54.	+ 36 09		17.9	RICH CLUSTER OF GALAXIES
ZC 0830.9+4814	08 30 54.	+ 48 14	1340		CLUSTER OF GALAXIES
MCG+11-11-015	08 30 54.	+ 63 44	12	16.	GALAXY
MCG-03-22-004	08 30 54.	- 20 44	48	16.5	GALAXY
BIGO 501	08 30 59.	- 16 01			NEBULA
CED 106G	08 31	- 44			DIFFUSE GALACTIC NEBULA
MEL 2	08 31	- 44 10			NEBULA
ZC 0831.0+3324	08 31 00.	+ 33 24	870		CLUSTER OF GALAXIES
ZWG 263.055	08 31 00.	+ 52 54		15.1	GALAXY
UGC 04475	08 31 00.	+ 52 54	78	15.1	GALAXY Sb
ZC 0831.0+6145	08 31 00.	+ 61 45	1610		CLUSTER OF GALAXIES
UGC 04476	08 31 00.	+ 67 30	60	18.	GALAXY DWARF?
MCG+12-09-056	08 31 00.	+ 74 13	33	16.	GALAXY
ZWG 364.006	08 31 00.	+ 84 49		15.7	GALAXY
ZWG 363.047	08 31 00.	+ 84 49		15.7	GALAXY
UGC 04474	08 31 00.	+ 84 49	102	15.7	GALAXY Sb
MCG-03-22-005	08 31 00.	- 17 47	84	15.	GALAXY
RNGC 2600	08 31 02.	+ 52 54		15.0	GALAXY
REIN 2.063	08 31 04.74	- 15 58 42.5			NEBULA
LB 00547	08 31 05.	+ 11 12 00.		16.5	FAINT BLUE STAR
PK239+13.1	08 31 05.25	- 15 58 40.0	60	13.6	PLANETARY NEBULA
REIN 2.064	08 31 05.60	- 15 58 23.0			NEBULA
ZWG 004.054	08 31 06.	+ 02 20		15.6	GALAXY
ZWG 032.037	08 31 06.	+ 06 22		15.6	GALAXY
ZWG 119.125	08 31 06.	+ 22 23		15.7	GALAXY
RNGC 2613	08 31 06.	- 22 48		11.0	GALAXY
OCL 0758	08 31 06.	- 48 08	300	13.	OPEN STAR CLUSTER
VHA 035	08 31 06.	- 48 08	180		OPEN STAR CLUSTER
FATH 1.240	08 31 10.	+ 30 25	16		NEBULA
ZC 0831.2+2100	08 31 12.	+ 21 00	1950		CLUSTER OF GALAXIES
MCG+09-14-069	08 31 15.	+ 52 59 30.	18	16.	GALAXY
MCG-01-22-020	08 31 15.	- 04 28	60	15.	GALAXY
RNGC 2610	08 31 15.	- 15 58		13.5	PLANETARY NEBULA
MCG-04-21-003	08 31 15.	- 22 47 30.	420	11.	GALAXY
IC 2384	08 31 16.	+ 32 36 03.			NONSTELLAR OBJECT
LB 05494	08 31 18.	+ 19 44		19.2	FAINT BLUE STAR
LB 05493	08 31 18.	+ 19 55		20.2	FAINT BLUE STAR
LB 05492	08 31 18.	+ 19 58		19.2	FAINT BLUE STAR
LB 05491	08 31 18.	+ 20 36		17.7	FAINT BLUE STAR
LB 05490	08 31 18.	+ 20 39		20.3	FAINT BLUE STAR
ZWG 179.011	08 31 18.	+ 32 37		15.3	GALAXY
ZC 0831.3+3610	08 31 18.	+ 36 10	810		CLUSTER OF GALAXIES
ZC 0831.3+5745	08 31 18.	+ 57 45	2550		CLUSTER OF GALAXIES
UGC 04477	08 31 18.	+ 63 47	60	16.5	GALAXY SB:b
MCG+12-08-047	08 31 18.	+ 73 46	66	17.	GALAXY
ZWG 060.025	08 31 24.	+ 12 09		15.6	GALAXY
KARA.73B 0263	08 31 24.	+ 12 09	36	15.6	ISOLATED GALAXY S
LB 05500	08 31 24.	+ 18 49		19.5	FAINT BLUE STAR
LB 05499	08 31 24.	+ 18 58		19.2	FAINT BLUE STAR
LB 05498	08 31 24.	+ 19 36		18.4	FAINT BLUE STAR
LB 05497	08 31 24.	+ 19 41		20.2	FAINT BLUE STAR
LB 05496	08 31 24.	+ 19 42		19.8	FAINT BLUE STAR
LB 05495	08 31 24.	+ 19 46		18.5	FAINT BLUE STAR
ZWG 149.052	08 31 24.	+ 30 06		15.5	GALAXY
MCG+06-19-009	08 31 24.	+ 32 36	24	16.	GALAXY
TON-N 0326	08 31 24.	+ 32 39		15.1	BLUE STAR
ZWG 208.021	08 31 24.	+ 38 32		15.2	GALAXY
ZWG 237.007	08 31 24.	+ 44 48		15.7	GALAXY
MCG+08-16-014	08 31 24.	+ 48 05	21	16.	GALAXY
ZC 0831.4+4932	08 31 24.	+ 49 32	610		CLUSTER OF GALAXIES
ZWG 263.056	08 31 24.	+ 53 00		15.4	GALAXY
ZWG 263.057	08 31 24.	+ 55 59		15.2	GALAXY
UGC 04478	08 31 24.	+ 55 59	84	15.2	GALAXY SBa
MCG+11-11-015A	08 31 24.	+ 63 45	51	17.	GALAXY
ZWG 004.056	08 31 24.	- 00 33		15.7	GALAXY
ZWG 004.055	08 31 24.	- 01 56		15.7	GALAXY
RNGC 2612	08 31 25.	- 12 59		13.0	GALAXY
RNGC 2602	08 31 26.	+ 53 00		15.5	GALAXY
MCG+09-14-070	08 31 27.	+ 56 00	36	16.	GALAXY
MCG-02-22-020	08 31 27.	- 12 59	150	13.5	GALAXY
MCG-03-22-006	08 31 27.	- 18 23	36	17.	GALAXY
ZWG 089.048	08 31 30.	+ 14 43		15.6	GALAXY
ZWG 089.049	08 31 30.	+ 15 27		15.5	GALAXY
LB 05508	08 31 30.	+ 18 52		19.3	FAINT BLUE STAR
LB 05506	08 31 30.	+ 18 54		17.9	FAINT BLUE STAR
LB 05507	08 31 30.	+ 18 55		18.7	FAINT BLUE STAR
LB 05505	08 31 30.	+ 20 01		18.7	FAINT BLUE STAR
LB 05504	08 31 30.	+ 20 22		19.4	FAINT BLUE STAR
LB 05503	08 31 30.	+ 20 24		19.8	FAINT BLUE STAR
LB 05502	08 31 30.	+ 20 26		20.6	FAINT BLUE STAR
LB 05501	08 31 30.	+ 20 34		18.0	FAINT BLUE STAR
FATH 1.241	08 31 30.	+ 30 05	22		NEBULA
ZWG 179.012	08 31 30.	+ 35 05		15.7	GALAXY
MCG+07-18-025	08 31 30.	+ 41 22 30.	39	15.	GALAXY
ZWG 208.022	08 31 30.	+ 41 23		15.5	GALAXY
ZC 0831.5+7011	08 31 30.	+ 70 11	1010		CLUSTER OF GALAXIES
MCG-01-22-021	08 31 30.	- 04 12	30	16.	GALAXY
MCG-01-22-022	08 31 30.	- 04 13	30	15.5	GALAXY
FATH 1.242	08 31 31.	+ 30 05	8		NEBULA
LB 05513	08 31 36.	+ 19 10		18.9	FAINT BLUE STAR
LB 05512	08 31 36.	+ 19 28		20.5	FAINT BLUE STAR
LB 05511	08 31 36.	+ 19 38		19.8	FAINT BLUE STAR
LB 05510	08 31 36.	+ 20 29		19.2	FAINT BLUE STAR
LB 05509	08 31 36.	+ 20 33		19.0	FAINT BLUE STAR
ZC 0831.6+3238	08 31 36.	+ 32 38	670		CLUSTER OF GALAXIES
ZWG 208.023	08 31 36.	+ 41 35		15.6	GALAXY
ZC 0831.6+4347	08 31 36.	+ 43 47	810		CLUSTER OF GALAXIES
MCG+11-11-016	08 31 36.	+ 66 08	18	17.	GALAXY
MAI 031	08 31 36.	+ 78 10	47		DWARF SPHEROIDAL GALAXY
MCG-02-22-021	08 31 36.	- 12 15 30.	60	15.	GALAXY

OBJECT NAME	RIGHT ASCEN.	DECLINATION	DIAM.	MAGN.	TYPE OF OBJECT
RNGC 2605	08 31 38.	+ 53 01			NON-EXISTENT OBJECT
MCG+07-18-026	08 31 39.	+ 41 34	15	15.	GALAXY
LB 05522	08 31 42.	+ 19 14		19.2	FAINT BLUE STAR
LB 05520	08 31 42.	+ 19 22		20.4	FAINT BLUE STAR
LB 05521	08 31 42.	+ 19 44		18.7	FAINT BLUE STAR
LB 05519	08 31 42.	+ 19 44		17.8	FAINT BLUE STAR
LB 05518	08 31 42.	+ 20 13		16.8	FAINT BLUE STAR
LB 05517	08 31 42.	+ 20 30		18.5	FAINT BLUE STAR
LB 05516	08 31 42.	+ 20 34		19.0	FAINT BLUE STAR
LB 05515	08 31 42.	+ 20 42		19.1	FAINT BLUE STAR
LB 05514	08 31 42.	+ 20 43		18.9	FAINT BLUE STAR
ZC 0831.7+3935	08 31 42.	+ 39 35	940		CLUSTER OF GALAXIES
ZC 0831.7+5001	08 31 42.	+ 50 01	870		CLUSTER OF GALAXIES
RNGC 2603	08 31 44.	+ 52 57		15.0	GALAXY
MCG+09-14-072	08 31 45.	+ 52 57	36	15.	GALAXY
MCG+09-14-071	08 31 45.	+ 54 42	54	16.	GALAXY
ZWG 004.057	08 31 48.	+ 01 11		15.6	GALAXY
LB 05528	08 31 48.	+ 19 00		19.4	FAINT BLUE STAR
LB 05527	08 31 48.	+ 19 31		20.1	FAINT BLUE STAR
LB 05526	08 31 48.	+ 19 49		16.6	FAINT BLUE STAR
LB 05525	08 31 48.	+ 20 25		19.4	FAINT BLUE STAR
LB 05524	08 31 48.	+ 20 31		18.6	FAINT BLUE STAR
LB 05523	08 31 48.	+ 20 36		19.1	FAINT BLUE STAR
ZWG 208.024	08 31 48.	+ 39 51		15.6	GALAXY
UGC 04479	08 31 48.	+ 39 51	60	15.6	GALAXY SB?c
ZC 0831.8+4010	08 31 48.	+ 40 10	810		CLUSTER OF GALAXIES
ZWG 263.058	08 31 48.	+ 54 42		15.6	GALAXY
MCG+10-12-147	08 31 48.	+ 57 07	18	16.	GALAXY
MCG-04-21-004	08 31 48.	- 21 41 30.	96	15.	GALAXY
IC 2386	08 31 49.	+ 26 00			NONSTELLAR OBJECT
LB 00548	08 31 50.	+ 13 53 36.		16.2	FAINT BLUE STAR
MCG+06-19-010	08 31 51.	+ 37 27	36	15.	GALAXY
MCG+07-18-027	08 31 51.	+ 39 50	54	15.	GALAXY
IC 2385	08 31 53.	+ 37 26 06.			NONSTELLAR OBJECT
ZWG 004.058	08 31 54.	+ 01 50		14.7	GALAXY
UGC 04480	08 31 54.	+ 01 50	84	14.7	GALAXY SBb+CMP
KARA.72 165A	08 31 54.	+ 01 50	42	14.7	PART OF DOUBLE GALAXY
MCG+00-22-018	08 31 54.	+ 01 50	36	14.7	GALAXY
MCG+00-22-017	08 31 54.	+ 01 50	48	14.7	GALAXY
LB 05536	08 31 54.	+ 18 53		20.0	FAINT BLUE STAR
LB 05535	08 31 54.	+ 19 30		19.3	FAINT BLUE STAR
LB 05534	08 31 54.	+ 19 32		18.0	FAINT BLUE STAR
LB 05533	08 31 54.	+ 19 36		19.2	FAINT BLUE STAR
LB 05532	08 31 54.	+ 19 41		18.1	FAINT BLUE STAR
LB 05531	08 31 54.	+ 19 51		20.5	FAINT BLUE STAR
LB 05530	08 31 54.	+ 20 29		19.6	FAINT BLUE STAR
LB 05529	08 31 54.	+ 20 39		20.5	FAINT BLUE STAR
ZWG 179.013	08 31 54.	+ 37 26		15.1	GALAXY
ZWG 263.059	08 31 54.	+ 52 58		15.0	GALAXY
LB 03484	08 31 54.	- 74 05		14.1	FAINT BLUE STAR
LB 00186	08 31 55.	+ 29 49 24.		17.0	FAINT BLUE STAR
RNGC 2606	08 31 56.	+ 52 58		15.0	GALAXY
MCG+09-14-073	08 31 57.	+ 51 44 30.	24	16.	GALAXY
VMT 12	08 32	- 45 00	16200		SUPERNOVA REMNANT
MIL 15	08 32	- 45 00	13200		SUPERNOVA REMNANT
LB 09968	08 32	- 89 23		13.9	FAINT BLUE STAR
ZWG 004.060	08 32 00.	+ 00 22		15.5	GALAXY
KARA.72 165B	08 32 00.	+ 01 50	36		PART OF DOUBLE GALAXY
ZWG 032.038	08 32 00.	+ 05 56		15.4	GALAXY
ZWG 089.050	08 32 00.	+ 14 47		15.7	GALAXY
LB 05547	08 32 00.	+ 18 55		18.9	FAINT BLUE STAR
LB 05546	08 32 00.	+ 19 03		19.3	FAINT BLUE STAR
LB 05545	08 32 00.	+ 19 18		15.4	FAINT BLUE STAR
LB 01802	08 32 00.	+ 19 24		18.4	FAINT BLUE STAR
LB 05544	08 32 00.	+ 19 27		18.9	FAINT BLUE STAR
LB 05543	08 32 00.	+ 19 42		18.6	FAINT BLUE STAR
LB 05542	08 32 00.	+ 19 43		19.1	FAINT BLUE STAR
LB 05541	08 32 00.	+ 20 06		19.6	FAINT BLUE STAR
LB 05540	08 32 00.	+ 20 07		16.5	FAINT BLUE STAR
LB 05539	08 32 00.	+ 20 23		20.5	FAINT BLUE STAR
LB 05538	08 32 00.	+ 20 30		18.5	FAINT BLUE STAR
LB 05537	08 32 00.	+ 20 44		18.6	FAINT BLUE STAR
MCG+05-20-026	08 32 00.	+ 29 52 30.	48	15.5	GALAXY
ZWG 149.053	08 32 00.	+ 29 53		15.5	GALAXY
MCG+08-16-015	08 32 00.	+ 49 45	24	16.	GALAXY
MCG+14-04-054	08 32 00.	+ 85 54	114	14.	GALAXY
ZWG 004.059	08 32 00.	- 02 22		13.5	GALAXY
UGC 04481	08 32 00.	- 02 22	138	13.5	GALAXY SBb
MCG+00-22-019	08 32 00.	- 02 22	102	13.5	GALAXY
TON-N 0929	08 32 01.	+ 20 17		14.9	BLUE STAR
RNGC 2615	08 32 02.	- 02 22		13.5	GALAXY
FATH 1.243	08 32 05.	+ 29 53	41		NEBULA
LB 05552	08 32 06.	+ 19 28		18.8	FAINT BLUE STAR
LB 05551	08 32 06.	+ 19 30		18.0	FAINT BLUE STAR
LB 05550	08 32 06.	+ 20 02		18.7	FAINT BLUE STAR
LB 05548	08 32 06.	+ 20 08		17.8	FAINT BLUE STAR
ZWG 149.054	08 32 06.	+ 28 56		15.6	GALAXY
UGC 04482	08 32 06.	+ 28 56	66	15.6	GALAXY Sc-IRR
ZC 0832.1+3734	08 32 06.	+ 37 34	1140		CLUSTER OF GALAXIES
ZWG 311.009	08 32 06.	+ 66 24		15.7	GALAXY
MCG+12-08-048	08 32 06.	+ 69 56	33	16.	GALAXY
ZWG 331.051	08 32 06.	+ 69 58		15.3	GALAXY
UGC 04483	08 32 06.	+ 69 58	84	15.3	GALAXY DWARF
ARC 0680	08 32 09.	+ 37 03		17.7	RICH CLUSTER OF GALAXIES
SN 1920A	08 32 11.	+ 28 39		11.8	SUPERNOVA
ARP 012	08 32 11.	+ 28 45			PECULIAR GALAXY
ZWG 032.039	08 32 12.	+ 07 55		15.4	GALAXY
ZC 0832.2+1526	08 32 12.	+ 15 26	940		CLUSTER OF GALAXIES
LB 05564	08 32 12.	+ 18 49		19.3	FAINT BLUE STAR
LB 05563	08 32 12.	+ 18 53		19.7	FAINT BLUE STAR
LB 05562	08 32 12.	+ 18 56		19.5	FAINT BLUE STAR
LB 05561	08 32 12.	+ 18 58		17.9	FAINT BLUE STAR
LB 05560	08 32 12.	+ 19 01		19.3	FAINT BLUE STAR
LB 05559	08 32 12.	+ 19 18		20.1	FAINT BLUE STAR
LB 05558	08 32 12.	+ 19 18		19.5	FAINT BLUE STAR
LB 05557	08 32 12.	+ 19 30		20.0	FAINT BLUE STAR
LB 05556	08 32 12.	+ 19 38		18.6	FAINT BLUE STAR
LB 05555	08 32 12.	+ 19 41		20.6	FAINT BLUE STAR
LB 05554	08 32 12.	+ 20 15		20.2	FAINT BLUE STAR
LB 05553	08 32 12.	+ 20 22		18.9	FAINT BLUE STAR
MCG+05-20-027	08 32 12.	+ 28 38 30.	120	13.2	GALAXY
ZWG 149.055	08 32 12.	+ 28 39		13.2	GALAXY
UGC 04484	08 32 12.	+ 28 39	150	13.2	GALAXY SBb/SBc
ZC 0832.2+3811	08 32 12.	+ 38 11	340		CLUSTER OF GALAXIES
ZC 0832.2+3845	08 32 12.	+ 38 45	3700		CLUSTER OF GALAXIES
ZC 0832.2+4050	08 32 12.	+ 40 50	1550		CLUSTER OF GALAXIES
MCG+09-14-074	08 32 12.	+ 54 36	36	17.	GALAXY
UGC 04485	08 32 12.	+ 61 54	60	16.5	GALAXY S

OBJECT NAME	RIGHT ASCEN.	DECLINATION	DIAM.	MAGN.	TYPE OF OBJECT
MCG+10-13-001	08 32 12.	+ 61 54	45	17.	GALAXY
MCG-01-22-023	08 32 12.	- 04 05	48	16.	GALAXY
REIN 2.065	08 32 13.81	+ 28 38 54.4			NEBULA
REIN 2.066A	08 32 15.15	+ 28 38 52.3			NEBULA
REIN 2.066B	08 32 15.18	+ 28 38 52.6			NEBULA
REIN 2.067A	08 32 15.36	+ 28 38 46.6			NEBULA
REIN 2.067B	08 32 15.37	+ 28 38 46.2			NEBULA
RNGC 2608	08 32 17.	+ 28 39		13.0	GALAXY
LB 05571	08 32 18.	+ 19 06		18.5	FAINT BLUE STAR
LB 05570	08 32 18.	+ 19 22		18.6	FAINT BLUE STAR
LB 05569	08 32 18.	+ 19 38		17.7	FAINT BLUE STAR
LB 05567	08 32 18.	+ 19 43		18.8	FAINT BLUE STAR
LB 01803	08 32 18.	+ 19 49		15.5	FAINT BLUE STAR
LB 05566	08 32 18.	+ 20 03		18.3	FAINT BLUE STAR
LB 05565	08 32 18.	+ 20 38		18.7	FAINT BLUE STAR
TON-N 0327	08 32 18.	+ 23 43		15.4	BLUE STAR
ZWG 208.025	08 32 18.	+ 40 00		15.3	GALAXY
MCG+10-13-002	08 32 18.	+ 61 30	27	16.	GALAXY
ZC 0832.3+7935	08 32 18.	+ 79 35	4440		CLUSTER OF GALAXIES
MCG-01-22-024	08 32 18.	- 04 13	24	16.	GALAXY
LB 05576	08 32 24.	+ 18 45		19.0	FAINT BLUE STAR
LB 05575	08 32 24.	+ 18 50		20.3	FAINT BLUE STAR
LB 05574	08 32 24.	+ 19 06		19.7	FAINT BLUE STAR
LB 05573	08 32 24.	+ 19 50		19.2	FAINT BLUE STAR
LB 05572	08 32 24.	+ 20 30		19.4	FAINT BLUE STAR
MCG+04-20-069	08 32 24.	+ 23 44	51	14.7	GALAXY
ZWG 150.001	08 32 24.	+ 30 43		15.0	GALAXY
ZWG 149.056	08 32 24.	+ 30 43		15.0	GALAXY
MRK 390	08 32 24.	+ 30 43	13	15.5	GALAXY WITH UV CONTINUUM
MCG+05-20-028	08 32 24.	+ 30 43	30	15.0	GALAXY
ZWG 288.001	08 32 24.	+ 61 29		15.7	GALAXY
ZWG 287.083	08 32 24.	+ 61 29		15.7	GALAXY
MCG+08-16-016	08 32 27.	+ 50 35	72	14.	GALAXY
UGC 04486	08 32 30.	+ 17 36	60	16.0	GALAXY Sa-b
LB 05584	08 32 30.	+ 18 53		19.5	FAINT BLUE STAR
LB 05583	08 32 30.	+ 19 18		19.1	FAINT BLUE STAR
LB 05582	08 32 30.	+ 19 32		19.4	FAINT BLUE STAR
LB 05581	08 32 30.	+ 19 33		19.2	FAINT BLUE STAR
LB 05580	08 32 30.	+ 19 52		17.4	FAINT BLUE STAR
LB 05579	08 32 30.	+ 20 12		17.0	FAINT BLUE STAR
LB 05578	08 32 30.	+ 20 18		19.5	FAINT BLUE STAR
LB 05577	08 32 30.	+ 20 48		17.6	FAINT BLUE STAR
ZWG 119.126	08 32 30.	+ 23 42		14.7	GALAXY
ZWG 119.127	08 32 30.	+ 25 12		15.3	GALAXY
ZWG 208.026	08 32 30.	+ 38 44		15.5	GALAXY
UGC 04487	08 32 30.	+ 38 44	72	15.5	GALAXY SB?c
ZC 0832.5+4340	08 32 30.	+ 43 40	870		CLUSTER OF GALAXIES
ZWG 263.060	08 32 30.	+ 50 36		15.0	GALAXY
MCG+12-09-057	08 32 30.	+ 73 36	66	16.	GALAXY
ZWG 004.061	08 32 30.	- 03 26		15.6	GALAXY
MRSL 265-04/1	08 32 30.	- 47 06	48		HII REGION
RNGC 2611	08 32 31.	+ 25 12		15.5	GALAXY
LB 05594	08 32 36.	+ 18 54		19.1	FAINT BLUE STAR
LB 05593	08 32 36.	+ 18 56		20.0	FAINT BLUE STAR
LB 01804	08 32 36.	+ 19 10		19.3	FAINT BLUE STAR
LB 05590	08 32 36.	+ 19 15		18.8	FAINT BLUE STAR
LB 05592	08 32 36.	+ 19 31		18.8	FAINT BLUE STAR
LB 05591	08 32 36.	+ 19 32		19.4	FAINT BLUE STAR
LB 05589	08 32 36.	+ 19 59		18.5	FAINT BLUE STAR
LB 05588	08 32 36.	+ 20 03		18.5	FAINT BLUE STAR
LB 05587	08 32 36.	+ 20 24		20.2	FAINT BLUE STAR
LB 05586	08 32 36.	+ 20 29		18.8	FAINT BLUE STAR
LB 05585	08 32 36.	+ 20 32		18.6	FAINT BLUE STAR
ZC 0832.6+2324	08 32 36.	+ 23 24	1140		CLUSTER OF GALAXIES
ZWG 119.128	08 32 36.	+ 23 51		15.6	GALAXY
ZC 0832.6+3424	08 32 36.	+ 34 24	810		CLUSTER OF GALAXIES
ZC 0832.6+4200	08 32 36.	+ 42 00	740		CLUSTER OF GALAXIES
ZC 0832.6-0235	08 32 36.	- 02 35	11690		CLUSTER OF GALAXIES
ZWG 004.063	08 32 36.	- 02 48		15.6	GALAXY
ZWG 004.062	08 32 36.	- 02 56		15.0	GALAXY
MCG+00-22-020	08 32 36.	- 02 56	24	15.0	GALAXY
MCG-01-22-025	08 32 36.	- 05 00	42	15.	GALAXY
LB 05601	08 32 42.	+ 18 48		18.9	FAINT BLUE STAR
LB 05600	08 32 42.	+ 19 22		18.7	FAINT BLUE STAR
LB 05599	08 32 42.	+ 19 27		19.6	FAINT BLUE STAR
LB 05598	08 32 42.	+ 19 54		16.7	FAINT BLUE STAR
LB 05597	08 32 42.	+ 19 56		17.6	FAINT BLUE STAR
LB 05596	08 32 42.	+ 20 10		18.7	FAINT BLUE STAR
LB 05595	08 32 42.	+ 20 10		19.8	FAINT BLUE STAR
LB 01805	08 32 42.	+ 20 32		19.7	FAINT BLUE STAR
ZWG 150.002	08 32 42.	+ 28 56		15.3	GALAXY
ZWG 149.057	08 32 42.	+ 28 56		15.3	GALAXY
ZC 0832.7+4410	08 32 42.	+ 44 10	1880		CLUSTER OF GALAXIES
MCG+09-14-075	08 32 45.	+ 54 12	15	17.	GALAXY
TON-N 0931	08 32 47.	+ 35 18		15.6	BLUE STAR
LB 05608	08 32 48.	+ 18 50		18.7	FAINT BLUE STAR
LB 05607	08 32 48.	+ 18 56		20.7	FAINT BLUE STAR
LB 05606	08 32 48.	+ 19 02		18.9	FAINT BLUE STAR
LB 01806	08 32 48.	+ 19 12		19.9	FAINT BLUE STAR
LB 05605	08 32 48.	+ 19 35		19.1	FAINT BLUE STAR
LB 05604	08 32 48.	+ 19 52		20.4	FAINT BLUE STAR
LB 05603	08 32 48.	+ 20 01		18.6	FAINT BLUE STAR
LB 05602	08 32 48.	+ 20 18		18.5	FAINT BLUE STAR
LB 01807	08 32 48.	+ 20 38		18.6	FAINT BLUE STAR
ARC 0681	08 32 48.	+ 44 02		17.1	RICH CLUSTER OF GALAXIES
MCG+09-14-076	08 32 48.	+ 53 54	42	16.	GALAXY
ZWG 263.061	08 32 48.	+ 54 13		15.5	GALAXY
ZWG 004.064	08 32 48.	- 02 04		15.6	GALAXY
MCG-05-21-001	08 32 48.	- 31 48	144	15.	GALAXY
OCL 0746	08 32 48.	- 44 06	1110	7.7	OPEN STAR CLUSTER
VHA 036	08 32 48.	- 44 06	900		OPEN STAR CLUSTER
TON-N 0930	08 32 49.	+ 19 58		14.6	BLUE STAR
IC 0514	08 32 53.	- 01 52 52.			NONSTELLAR OBJECT
ZWG 060.026	08 32 54.	+ 12 44		15.1	GALAXY
LB 05614	08 32 54.	+ 18 59		18.5	FAINT BLUE STAR
LB 05613	08 32 54.	+ 19 31		19.4	FAINT BLUE STAR
LB 05612	08 32 54.	+ 19 44		18.6	FAINT BLUE STAR
LB 05611	08 32 54.	+ 20 00		19.9	FAINT BLUE STAR
LB 05610	08 32 54.	+ 20 05		19.1	FAINT BLUE STAR
LB 05609	08 32 54.	+ 20 05		20.6	FAINT BLUE STAR
LB 01808	08 32 54.	+ 22 22 36.		16.8	FAINT BLUE STAR
ZWG 150.003	08 32 54.	+ 30 26		15.1	GALAXY
ZWG 149.058	08 32 54.	+ 30 26		15.1	GALAXY
MCG+05-20-029	08 32 54.	+ 30 26	30	15.1	GALAXY
KARA.73B 0264	08 32 54.	+ 30 26	18	15.1	ISOLATED GALAXY E
ZC 0832.9+3119	08 32 54.	+ 31 19	4230		CLUSTER OF GALAXIES
ZWG 263.062	08 32 54.	+ 53 54		15.3	GALAXY
KARA.73B 0265	08 32 54.	+ 53 54	198	15.3	ISOLATED GALAXY S

OBJECT NAME	RIGHT ASCEN.	DECLINATION	DIAM.	MAGN.	TYPE OF OBJECT
ZWG 004.066	08 32 54.	- 01 53		15.3	GALAXY
ZWG 004.065	08 32 54.	- 01 56		15.5	GALAXY
IC 0515	08 32 59.	- 01 43 53.			NONSTELLAR OBJECT
ZWG 089.051	08 33 00.	+ 15 00		15.7	GALAXY
LB 05622	08 33 00.	+ 19 01		20.4	FAINT BLUE STAR
LB 05621	08 33 00.	+ 19 09		18.3	FAINT BLUE STAR
LB 05620	08 33 00.	+ 19 46		18.5	FAINT BLUE STAR
LB 05619	08 33 00.	+ 19 47		18.2	FAINT BLUE STAR
LB 05618	08 33 00.	+ 20 21		18.0	FAINT BLUE STAR
LB 05617	08 33 00.	+ 20 29		20.0	FAINT BLUE STAR
LB 05616	08 33 00.	+ 20 31		20.5	FAINT BLUE STAR
LB 05615	08 33 00.	+ 20 32		20.2	FAINT BLUE STAR
ZC 0833.0+3815	08 33 00.	+ 38 15	540		CLUSTER OF GALAXIES
ZC 0833.0+4650	08 33 00.	+ 46 50	3630		CLUSTER OF GALAXIES
MCG+09-14-077	08 33 00.	+ 55 49 30.	24	17.	GALAXY
ZC 0833.0+6725	08 33 00.	+ 67 25	470		CLUSTER OF GALAXIES
ZC 0833.0+7203	08 33 00.	+ 72 03	400		CLUSTER OF GALAXIES
ZWG 004.069	08 33 00.	- 01 40		15.2	GALAXY
UGC 04489	08 33 00.	- 01 40	90	15.2	GALAXY COMPACT
MCG+00-22-021	08 33 00.	- 01 40	24	15.2	GALAXY
ZWG 004.068	08 33 00.	- 01 44		15.6	GALAXY
UGC 04488	08 33 00.	- 01 44	72	15.6	GALAXY
ZWG 004.067	08 33 00.	- 02 40		15.7	GALAXY
RNGC 2616	08 33 02.	- 01 40		15.0	GALAXY
ZC 0833.1+1545	08 33 06.	+ 15 45	740		CLUSTER OF GALAXIES
LB 05629	08 33 06.	+ 18 43		19.4	FAINT BLUE STAR
LE 05628	08 33 06.	+ 19 01		19.6	FAINT BLUF STAR
LB 01809	08 33 06.	+ 19 12		18.7	FAINT BLUE STAR
LB 05627	08 33 06.	+ 20 11		20.5	FAINT BLUE STAR
LB 05626	08 33 06.	+ 20 11		19.0	FAINT BLUE STAR
LB 05625	08 33 06.	+ 20 14		18.7	FAINT BLUE STAR
LB 05624	08 33 06.	+ 20 42		17.7	FAINT BLUE STAR
LB 05623	08 33 06.	+ 20 46		18.3	FAINT BLUE STAR
ZWG 120.001	08 33 06.	+ 25 47		15.4	GALAXY
ZWG 119.129	08 33 06.	+ 25 47		15.4	GALAXY
ZC 0833.1+4452	08 33 06.	+ 44 52	3160		CLUSTER OF GALAXIFS
MCG+08-16-018	08 33 06.	+ 47 31	36	16.	GALAXY
MCG+08-16-017	08 33 06.	+ 47 31	36	16.	GALAXY
MCG+08-16-019	08 33 06.	+ 48 51	48	15.	GALAXY
UGC 04490	08 33 06.	+ 66 20	66	16.0	GALAXY S
ZWG 004.071	08 33 06.	- 01 35		15.3	GALAXY
ZWG 004.070	08 33 06.	- 02 48		15.7	GALAXY
MCG-01-22-026	08 33 06.	- 03 53	60	14.	GALAXY
ZWG 004.073	08 33 12.	+ 01 54		13.9	GALAXY
UGC 04491	08 33 12.	+ 01 54	162	13.9	GALAXY Sa-b
MCG+00-22-022	08 33 12.	+ 01 54	138	13.9	GALAXY
LB 05638	08 33 12.	+ 19 24		19.9	FAINT BLUE STAR
LB 05637	08 33 12.	+ 19 28		19.2	FAINT BLUE STAR
LB 05636	08 33 12.	+ 19 35		19.0	FAINT BLUE STAR
LB 05635	08 33 12.	+ 19 44		19.1	FAINT BLUE STAR
LB C5634	08 33 12.	+ 19 55		19.9	FAINT BLUE STAR
LB 05633	08 33 12.	+ 20 20		18.6	FAINT BLUE STAR
LB 05632	08 33 12.	+ 20 22		18.5	FAINT BLUE STAR
LB 05631	08 33 12.	+ 20 30		18.2	FAINT BLUE STAR
LB 05630	08 33 12.	+ 20 45		18.2	FAINT BLUE STAR
ZWG 120.002	08 33 12.	+ 25 17		15.4	GALAXY
ZWG 119.130	08 33 12.	+ 25 17		15.4	GALAXY
ZWG 150.004	08 33 12.	+ 28 14		14.9	GALAXY
ZWG 149.059	08 33 12.	+ 28 14		14.9	GALAXY
ZWG 179.014	08 33 12.	+ 38 18		15.5	GALAXY
ZWG 004.072	08 33 12.	- 01 47		15.7	GALAXY
RNGC 2617	08 33 15.	- 03 53		15.0	GALAXY
MCG-01-22-027	08 33 15.	- 03 53	42	15.	GALAXY
ZWG 004.074	08 33 18.	+ 00 52		13.9	GALAXY
UGC 04492	08 33 18.	+ 00 52	180	13.9	GALAXY Sa
MCG+00-22-023	08 33 18.	+ 00 52	132	13.9	GALAXY
LB 05646	08 33 18.	+ 19 13		18.5	FAINT BLUE STAR
LB 05645	08 33 18.	+ 19 27		20.1	FAINT BLUE STAR
LB 00378	08 33 18.	+ 19 28		15.8	FAINT BLUE STAR
LB 05644	08 33 18.	+ 19 41		20.7	FAINT BLUE STAR
LB 05643	08 33 18.	+ 19 41		19.5	FAINT BLUE STAR
LB 05642	08 33 18.	+ 19 44		18.5	FAINT BLUE STAR
LB 05642	08 33 18.	+ 19 44		18.5	FAINT BLUE STAR
LB 01810	08 33 18.	+ 19 48		16.7	FAINT BLUE STAR
LB 05641	08 33 18.	+ 19 58		19.0	FAINT BLUE STAR
LB 05640	08 33 18.	+ 20 16		18.7	FAINT BLUE STAR
LB 05639	08 33 18.	+ 20 22		18.2	FAINT BLUE STAR
MCG+05-21-001	08 33 18.	+ 28 13	42	14.9	GALAXY
ZWG 208.027	08 33 18.	+ 41 39		15.6	GALAXY
ZC 0833.3+4746	08 33 18.	+ 47 46	810		CLUSTER OF GALAXIES
ZWG 004.075	08 33 18.	- 01 42		15.7	GALAXY
BC 3CR204	08 33 18.02	+ 65 24 04.4		18.21	QUASI-STELLAR OBJECT
SHB 116	08 33 18.2	+ 65 24 05.		18.2	QUASI-STELLAR OBJECT
RNGC 2618	08 33 19.	+ 00 52		14.0	GALAXY
IC 0516	08 33 19.	- 01 42 07.			NONSTELLAR OBJECT
LB 01812	08 33 21.	+ 19 17 12.		19.0	FAINT BLUE STAR
LB 05658	08 33 24.	+ 18 56		18.8	FAINT BLUE STAR
LB 05657	08 33 24.	+ 18 58		19.2	FAINT BLUE STAR
LB 05656	08 33 24.	+ 19 12		18.2	FAINT BLUE STAR
LB 05655	08 33 24.	+ 19 13		18.8	FAINT BLUE STAR
LB 05654	08 33 24.	+ 19 21		20.5	FAINT BLUE STAR
LB 05653	08 33 24.	+ 19 38		18.5	FAINT BLUE STAR
LB 05652	08 33 24.	+ 19 50		18.8	FAINT BLUE STAR
LB 05651	08 33 24.	+ 20 00		18.9	FAINT BLUE STAR
LB 05650	08 33 24.	+ 20 04		19.8	FAINT BLUE STAR
LB 05649	08 33 24.	+ 20 07		19.4	FAINT BLUE STAR
LB 05648	08 33 24.	+ 20 18		18.3	FAINT BLUE STAR
LB 01811	08 33 24.	+ 20 29		19.1	FAINT BLUE STAR
LB 01813	08 33 24.	+ 20 42		18.0	FAINT BLUE STAR
LB 05647	08 33 24.	+ 20 46		19.9	FAINT BLUE STAR
ZWG 120.003	08 33 24.	+ 23 57		15.7	GALAXY
ZC 0833.4+6335	08 33 24.	+ 63 35	870		CLUSTER OF GALAXIES
ZWG 004.076	08 33 24.	- 02 48		15.7	GALAXY
MCG+07-18-028	08 33 27.	+ 41 38	36	15.5	GALAXY
LB 01814	08 33 28.	+ 21 08 00.		15.8	FAINT BLUE STAR
ZWG 004.077	08 33 30.	+ 00 47		15.7	GALAXY
UGC 04493	08 33 30.	+ 00 47	60	15.7	GALAXY S
ZWG 004.078	08 33 30.	+ 01 30		15.4	GALAXY
ZWG 004.079	08 33 30.	+ 02 04		15.2	GALAXY
UGC 04494	08 33 30.	+ 02 04	84	15.2	GALAXY Sa-b
LB 05661	08 33 30.	+ 19 03		19.2	FAINT BLUE STAR
LB 05660	08 33 30.	+ 19 22		17.9	FAINT BLUE STAR
LB 05659	08 33 30.	+ 20 17		18.4	FAINT BLUE STAR
ABC 0683	08 33 30.	+ 31 25		17.1	RICH CLUSTER OF GALAXIES
ZC 0833.5+6529	08 33 30.	+ 65 29	1140		CLUSTER OF GALAXIES
ZWG 331.052	08 33 30.	+ 69 13		15.2	GALAXY
UGC 04495	08 33 30.	+ 69 13	60	15.2	GALAXY Sc
KARA.73B 0266	08 33 30.	+ 69 13	60	15.2	ISOLATED GALAXY S
IC 0499	08 33 31.	+ 85 55 55.			NONSTELLAR OBJECT
ZC 0833.6+0917	08 33 36.	+ 09 17	870		CLUSTER OF GALAXIES
ZWG 089.052	08 33 36.	+ 18 32		15.7	GALAXY
LB 05669	08 33 36.	+ 18 44		19.0	FAINT BLUE STAR
LB 05668	08 33 36.	+ 18 58		18.9	FAINT BLUE STAR
LB 05667	08 33 36.	+ 19 15		19.5	FAINT BLUE STAR
LB 05666	08 33 36.	+ 19 31		19.4	FAINT BLUE STAR
ZWG 089.053	08 33 36.	+ 19 36		15.6	GALAXY
LB 05665	08 33 36.	+ 19 39		18.4	FAINT BLUE STAR
LB 05664	08 33 36.	+ 19 42		18.6	FAINT BLUE STAR
LB 05663	08 33 36.	+ 20 08		17.7	FAINT BLUE STAR
LB 05662	08 33 36.	+ 20 22		20.6	FAINT BLUE STAR
ZWG 120.004	08 33 36.	+ 23 01		15.6	GALAXY
ZC 0833.6+5015	08 33 36.	+ 50 15	940		CLUSTER OF GALAXIES
ZWG 311.010	08 33 36.	+ 66 42		15.7	GALAXY
MCG+11-11-017	08 33 36.	+ 66 42	45	16.	GALAXY
MCG+12-08-049	08 33 36.	+ 69 12	54	15.	GALAXY
ZWG 060.027	08 33 42.	+ 10 49		14.9	GALAXY
MCG+02-22-002	08 33 42.	+ 10 49	48	14.9	GALAXY
LB 05674	08 33 42.	+ 19 00		18.5	FAINT BLUE STAR
LB 05673	08 33 42.	+ 19 12		19.2	FAINT BLUE STAR
LB 05672	08 33 42.	+ 19 27		18.5	FAINT BLUE STAR
LB 05671	08 33 42.	+ 19 31		18.6	FAINT BLUE STAR
LB 05670	08 33 42.	+ 20 30		19.0	FAINT BLUE STAR
ZWG 120.005	08 33 42.	+ 25 19		15.7	GALAXY
UGC 04496	08 33 42.	+ 25 19	60	15.7	GALAXY Sc
ZWG 208.028	08 33 42.	+ 41 38		15.1	GALAXY
ZWG 237.008	08 33 42.	+ 45 20		15.5	GALAXY
MCG+08-16-020	08 33 42.	+ 45 20	27	16.	GALAXY
MCG+08-16-021	08 33 42.	+ 47 21	48	16.	GALAXY
ZWG 004.080	08 33 42.	- 03 16		15.7	GALAXY
MCG+00-22-024	08 33 42.	- 03 16	30	15.7	GALAXY
MCG-01-22-028	08 33 42.	- 06 54	24	15.5	GALAXY
MRSL 260-00/1	08 33 42.	- 40 30	281		HII REGION
CED 106H	08 33 43.	- 40 28			DIFFUSE GALACTIC NEBULA
MAI 032	08 33 44.	+ 67 36	40		DWARF SPHEROIDAL GALAXY
RNGC 2626	08 33 47.	- 40 28			DIFFUSE NEBULA
ZWG 060.028	08 33 48.	+ 12 21		15.7	GALAXY
LB 05683	08 33 48.	+ 19 00		20.6	FAINT BLUE STAR
LB 05682	08 33 48.	+ 19 01		20.5	FAINT BLUE STAR
LB 05681	08 33 48.	+ 19 03		20.7	FAINT BLUE STAR
LB 05680	08 33 48.	+ 19 41		20.2	FAINT BLUE STAR
LB 05679	08 33 48.	+ 19 44		20.7	FAINT BLUE STAR
LB 05678	08 33 48.	+ 19 46		18.4	FAINT BLUE STAR
LB 05677	08 33 48.	+ 20 05		19.8	FAINT BLUE STAR
LB 05676	08 33 48.	+ 20 15		19.1	FAINT BLUE STAR
LB 05675	08 33 48.	+ 20 28		19.1	FAINT BLUE STAR
LB 01815	08 33 48.	+ 21 11 36.		14.7	FAINT BLUE STAR
TON-N 0328	08 33 48.	+ 31 27		14.5	BLUE STAR
ZC 0833.8+3806	08 33 48.	+ 38 06	1280		CLUSTER OF GALAXIES
MCG+09-14-078	08 33 48.	+ 51 48 30.	138	14.	GALAXY
ZWG 004.082	08 33 48.	- 01 54		15.4	GALAXY
ZWG 004.081	08 33 48.	- 03 25		15.7	GALAXY
MCG+07-18-029	08 33 51.	+ 41 37	30	14.5	GALAXY
IC 0517	08 33 52.	- 01 53 16.			NONSTELLAR OBJECT
ZWG 060.029	08 33 54.	+ 12 15		15.6	GALAXY
LB 05692	08 33 54.	+ 18 54		17.5	FAINT BLUE STAR
LB 05691	08 33 54.	+ 19 30		18.3	FAINT BLUE STAR
LB 05690	08 33 54.	+ 19 41		18.7	FAINT BLUE STAR
LB 05689	08 33 54.	+ 19 43		19.9	FAINT BLUE STAR
LB 05688	08 33 54.	+ 19 44		19.8	FAINT BLUE STAR
LB 05687	08 33 54.	+ 19 50		19.0	FAINT BLUE STAR
LB 05686	08 33 54.	+ 20 14		18.9	FAINT BLUE STAR
LB 05685	08 33 54.	+ 20 41		16.6	FAINT BLUE STAR
LB 05684	08 33 54.	+ 20 43		18.1	FAINT BLUE STAR
ZWG 150.005	08 33 54.	+ 27 31		15.6	GALAXY
TON-N 0329	08 33 54.	+ 32 31		14.5	BLUE STAR
MCG+10-13-003	08 33 54.	+ 57 32	39	16.	GALAXY
UGC 04497	08 33 54.	+ 67 33	78	16.5	GALAXY DWARF
ZWG 004.083	08 33 54.	- 03 25		15.6	GALAXY
LB 00379	08 33 57.	+ 20 02 18.		17.3	FAINT BLUE STAR
LB 01817	08 33 58.	+ 19 10 48.		18.9	FAINT BLUE STAR
LB 05698	08 34 00.	+ 18 46		18.5	FAINT BLUE STAR
LB 05697	08 34 00.	+ 19 00		20.3	FAINT BLUE STAR
LB 05696	08 34 00.	+ 19 01		20.8	FAINT BLUE STAR
LB 05695	08 34 00.	+ 19 10		20.6	FAINT PLUE STAR
LB 05694	08 34 00.	+ 19 24		20.2	FAINT BLUE STAR
LB 05693	08 34 00.	+ 19 27		19.1	FAINT BLUE STAR
LB 01816	08 34 00.	+ 20 27		19.0	FAINT BLUE STAR
TON-N 0330	08 34 00.	+ 30 53		14.7	BLUE STAR
ZC 0834.0+3743	08 34 00.	+ 37 43	1080		CLUSTER OF GALAXIES
ZWG 208.029	08 34 00.	+ 40 13		14.8	GALAXY
UGC 04498	08 34 00.	+ 40 13	66	14.8	GALAXY SBa
ZWG 208.030	08 34 00.	+ 41 44		15.4	GALAXY
ZWG 263.063	08 34 00.	+ 51 50		14.0	GALAXY
UGC 04499	08 34 00.	+ 51 50	180	14.0	GALAXY S IV
MCG+11-11-018	08 34 00.	+ 67 32	39	17.	GALAXY
MCG+12-09-058	08 34 00.	+ 69 03	51	15.	GALAXY
ZWG 004.084	08 34 00.	- 01 29		15.4	GALAXY
MCG-03-22-007	08 34 00.	- 20 07	60	15.	GALAXY
MCG-04-21-005	08 34 00.	- 26 15	30	14.	GALAXY
VHE 17A	08 34 00.	- 40 30	318		REFLECTION NEBULA
VHA 037	08 34 00.	- 43 25	150		OPEN STAR CLUSTER
LB 03485	08 34 00.	- 78 28		09.7	FAINT BLUE STAR
LB 05704	08 34 06.	+ 19 01		20.3	FAINT BLUE STAR
LB 05703	08 34 06.	+ 19 22		19.6	FAINT BLUE STAR
LB 01819	08 34 06.	+ 19 42		19.8	FAINT BLUE STAR
LB 05702	08 34 06.	+ 20 24		18.8	FAINT BLUE STAR
LB 05701	08 34 06.	+ 20 29		18.1	FAINT BLUE STAR
LB 01818	08 34 06.	+ 20 31		19.0	FAINT BLUE STAR
LB 05700	08 34 06.	+ 20 36		18.9	FAINT BLUE STAR
LB 05699	08 34 06.	+ 20 38		18.8	FAINT BLUE STAR
ZWG 150.006	08 34 06.	+ 28 24		15.7	GALAXY
ZWG 208.031	08 34 06.	+ 41 30		15.5	GALAXY
ZWG 331.053	08 34 06.	+ 71 53		15.4	GALAXY
UGC 04500	08 34 06.	+ 71 53	96	15.4	GALAXY Sc/SBc
KARA.73B 0267	08 34 06.	+ 71 53	96	15.4	ISOLATED GALAXY S
ZWG 004.085	08 34 06.	- 01 22		15.6	GALAXY
MCG-02-22-022	08 34 06.	- 11 38 30.	120	15.	GALAXY
KON HE2-10	08 34 06.	- 26 14		13.	EXTRAGALACTIC OBJECT
PK248+08.1	08 34 08.	- 26 14	10		PLANETARY NEBULA
MCG+07-18-030	08 34 09.	+ 40 12	48	14.5	GALAXY
LB 05711	08 34 12.	+ 19 06		18.1	FAINT BLUE STAR
LB 05710	08 34 12.	+ 19 12		20.3	FAINT BLUE STAR
LB 05709	08 34 12.	+ 19 20		20.6	FAINT BLUF STAR
LB 01820	08 34 12.	+ 19 41		19.6	FAINT BLUE STAR
LB 05708	08 34 12.	+ 19 43		18.9	FAINT BLUE STAR
LB 05707	08 34 12.	+ 20 01		18.5	FAINT BLUE STAR

OBJECT NAME	RIGHT ASCEN.	DECLINATION	DIAM.	MAGN.	TYPE OF OBJECT
LB 05706	08 34 12.	+ 20 05		20.8	FAINT BLUE STAR
LB 05705	08 34 12.	+ 20 20		20.2	FAINT BLUE STAR
ZC 0834.2+3529	08 34 12.	+ 35 29	1880		CLUSTER OF GALAXIES
MCG+12-09-001	08 34 12.	+ 71 52	78	15.	GALAXY
MCG+12-09-002	08 34 12.	+ 73 57	33	15.	GALAXY
ARC 0685	08 34 13.	+ 44 26		17.1	RICH CLUSTER OF GALAXIES
MCG+08-16-022	08 34 15.	+ 47 30	48	15.	GALAXY
LB 05716	08 34 18.	+ 18 50		19.4	FAINT BLUE STAR
LB 05715	08 34 18.	+ 19 04		19.6	FAINT BLUE STAR
LB 05714	08 34 18.	+ 19 33		20.2	FAINT BLUE STAR
LB 01821	08 34 18.	+ 19 41		18.5	FAINT BLUE STAR
LB 05713	08 34 18.	+ 20 02		19.4	FAINT BLUE STAR
LB 05712	08 34 18.	+ 20 22		17.0	FAINT BLUE STAR
TON-N 0331	08 34 18.	+ 29 59		14.2	BLUE STAR
ZWG 237.009	08 34 18.	+ 47 31		15.0	GALAXY
MCG+10-13-004	08 34 18.	+ 61 38	27	16.	GALAXY
MCG-03-22-008	08 34 18.	- 20 19	120	14.	GALAXY
ACK 255+03.1	08 34 18.	- 35 05			PLANETARY NEBULA
ARC 0682	08 34 19.	+ 52 02		17.1	RICH CLUSTER OF GALAXIES
ZWG 004.086	08 34 24.	+ 01 58		15.3	GALAXY
LB 05728	08 34 24.	+ 18 59		19.4	FAINT BLUE STAR
LB 05727	08 34 24.	+ 19 07		20.0	FAINT BLUE STAR
LB 05726	08 34 24.	+ 19 09		19.5	FAINT BLUE STAR
LB 05725	08 34 24.	+ 19 12		18.9	FAINT BLUE STAR
LB 05724	08 34 24.	+ 19 13		19.8	FAINT BLUE STAR
LB 05723	08 34 24.	+ 19 14		19.3	FAINT BLUE STAR
LB 05722	08 34 24.	+ 19 16		18.5	FAINT BLUE STAR
LB 05721	08 34 24.	+ 19 25		18.8	FAINT BLUE STAR
LB 05720	08 34 24.	+ 19 27		18.8	FAINT BLUE STAR
LB 01822	08 34 24.	+ 19 32		19.5	FAINT BLUE STAR
LB 05719	08 34 24.	+ 20 04		20.6	FAINT BLUE STAR
LB 05718	08 34 24.	+ 20 23		19.0	FAINT BLUE STAR
LB 05717	08 34 24.	+ 20 24		18.4	FAINT BLUE STAR
ZC 0834.4+5031	08 34 24.	+ 50 31	940		CLUSTER OF GALAXIES
SHAH 096	08 34 24.	+ 52 46	270	17.5	GROUP OF COMPACT GALAXIES
MCG+09-14-079	08 34 24.	+ 54 55 30.	24	17.	GALAXY
ZWG 288.002	08 34 24.	+ 61 36		15.4	GALAXY
ZWG 287.084	08 34 24.	+ 61 36		15.4	GALAXY
ZC 0834.4+6807	08 34 24.	+ 68 07	1410		CLUSTER OF GALAXIES
TON-N 0932	08 34 25.	+ 22 14		16.0	BLUE STAR
IC 0518	08 34 29.	+ 00 51			OPEN CLUSTER
ZC 0834.5+1555	08 34 30.	+ 15 55	2020		CLUSTER OF GALAXIES
LB 05737	08 34 30.	+ 18 46		19.1	FAINT BLUE STAR
LB 05736	08 34 30.	+ 18 47		19.1	FAINT BLUE STAR
LB 01823	08 34 30.	+ 18 53		18.2	FAINT BLUE STAR
LB 05735	08 34 30.	+ 19 24		19.5	FAINT BLUE STAR
LB 05734	08 34 30.	+ 19 27		18.7	FAINT BLUE STAR
LB 05733	08 34 30.	+ 19 40		19.7	FAINT BLUE STAR
LB 05732	08 34 30.	+ 19 42		18.6	FAINT BLUE STAR
LB 05731	08 34 30.	+ 19 53		18.5	FAINT BLUE STAR
LB 05730	08 34 30.	+ 20 12		19.6	FAINT BLUE STAR
LB 05729	08 34 30.	+ 20 17		17.9	FAINT BLUE STAR
ZC 0834.5+2307	08 34 30.	+ 23 07	3160		CLUSTER OF GALAXIES
ZWG 120.006	08 34 30.	+ 25 07		14.8	GALAXY
UGC 04501	08 34 30.	+ 25 07	132	14.8	GALAXY S
ZWG 150.007	08 34 30.	+ 28 25		15.6	GALAXY
ZC 0834.5+3711	08 34 30.	+ 37 11	870		CLUSTER OF GALAXIES
ZC 0834.5+4610	08 34 30.	+ 46 10	540		CLUSTER OF GALAXIES
ZWG 332.001	08 34 30.	+ 73 55		15.6	GALAXY
ZWG 331.054	08 34 30.	+ 73 55		15.6	GALAXY
UGC 04502	08 34 30.	+ 73 55	102	15.6	GALAXY SBc
RNGC 2620	08 34 31.	+ 25 07		15.0	GALAXY
MCG+04-21-001	08 34 33.	+ 25 08	114	14.8	GALAXY
RNGC 2619	08 34 35.	+ 28 53		13.5	GALAXY
ZWG 032.040	08 34 36.	+ 04 52		15.0	GALAXY
MCG+01-22-013	08 34 36.	+ 04 52	24	15.0	GALAXY
LB 05744	08 34 36.	+ 19 21		19.6	FAINT BLUE STAR
LB 05743	08 34 36.	+ 19 34		19.2	FAINT BLUE STAR
LB 05742	08 34 36.	+ 19 43		19.1	FAINT BLUE STAR
LB 05741	08 34 36.	+ 19 46		18.6	FAINT BLUE STAR
LB 05740	08 34 36.	+ 19 57		18.9	FAINT BLUE STAR
LB 01824	08 34 36.	+ 20 02		18.3	FAINT BLUE STAR
LB 05739	08 34 36.	+ 20 18		19.2	FAINT BLUE STAR
LB 05738	08 34 36.	+ 20 36		19.8	FAINT BLUE STAR
MCG+05-21-002	08 34 36.	+ 28 52	132	13.6	GALAXY
ZWG 150.008	08 34 36.	+ 28 53		13.6	GALAXY
UGC 04503	08 34 36.	+ 28 53	168	13.6	GALAXY Sb/Sc
TON-N 0332	08 34 36.	+ 29 48		14.7	BLUE STAR
ZWG 208.032	08 34 36.	+ 40 05		15.7	GALAXY
ZC 0834.6+6219	08 34 36.	+ 62 19	2420		CLUSTER OF GALAXIES
LB 05750	08 34 42.	+ 18 58		18.6	FAINT BLUE STAR
LB 05749	08 34 42.	+ 19 01		19.7	FAINT BLUE STAR
LB 05748	08 34 42.	+ 19 18		19.8	FAINT BLUE STAR
LB 05747	08 34 42.	+ 20 07		18.5	FAINT BLUE STAR
LB 05746	08 34 42.	+ 20 13		20.5	FAINT BLUE STAR
LB 01825	08 34 42.	+ 20 25		19.1	FAINT BLUE STAR
LB 05745	08 34 42.	+ 20 30		18.9	FAINT BLUE STAR
ZWG 120.007	08 34 42.	+ 25 10		15.4	GALAXY
MCG+07-18-031	08 34 42.	+ 40 04	54	15.5	GALAXY
ZC 0834.7+5854	08 34 42.	+ 58 54	2890		CLUSTER OF GALAXIES
ZWG 004.087	08 34 42.	- 03 12		15.5	GALAXY
ARC 0689	08 34 43.	+ 15 10		17.1	RICH CLUSTER OF GALAXIES
RNGC 2621	08 34 43.	+ 25 10		15.5	GALAXY
ARC 0688	08 34 45.	+ 16 02		17.5	RICH CLUSTER OF GALAXIES
MCG+04-21-002	08 34 45.	+ 20 40 30.	90	15.3	GALAXY
MCG+04-21-003	08 34 45.	+ 25 11	42	15.4	GALAXY
LB 05760	08 34 48.	+ 19 02		18.5	FAINT BLUE STAR
LB 01826	08 34 48.	+ 19 14		19.2	FAINT BLUE STAR
LB 05759	08 34 48.	+ 19 25		18.7	FAINT BLUE STAR
LB 05758	08 34 48.	+ 19 40		19.2	FAINT BLUE STAR
LB 05757	08 34 48.	+ 19 45		18.9	FAINT BLUE STAR
LB 05756	08 34 48.	+ 19 48		18.8	FAINT BLUE STAR
LB 05755	08 34 48.	+ 19 59		19.6	FAINT BLUE STAR
LB 05754	08 34 48.	+ 20 12		18.2	FAINT BLUE STAR
LB 05753	08 34 48.	+ 20 18		18.3	FAINT BLUE STAR
LB 05752	08 34 48.	+ 20 22		19.3	FAINT BLUE STAR
LB 05751	08 34 48.	+ 20 30		19.3	FAINT BLUE STAR
ZWG 120.008	08 34 48.	+ 20 41		15.3	GALAXY
UGC 04504	08 34 48.	+ 20 41	90	15.3	GALAXY Sc
KARA.73B 0268	08 34 48.	+ 20 41	72	15.3	ISOLATED GALAXY S
TON-N 0333	08 34 48.	+ 29 47		15.0	BLUE STAR
ZWG 060.030	08 34 54.	+ 12 13		15.2	GALAXY
KARA.73B 0269	08 34 54.	+ 12 13	48	15.2	ISOLATED GALAXY S
LB 05767	08 34 54.	+ 19 14		18.8	FAINT BLUE STAR
LB 05766	08 34 54.	+ 19 14		20.0	FAINT BLUE STAR
LB 05765	08 34 54.	+ 19 41		19.6	FAINT BLUE STAR
LB 05764	08 34 54.	+ 19 42		18.8	FAINT BLUE STAR
LB 01827	08 34 54.	+ 20 00		17.1	FAINT BLUE STAR
LB 05763	08 34 54.	+ 20 06		18.8	FAINT BLUE STAR
LB 05762	08 34 54.	+ 20 16		18.3	FAINT BLUE STAR
LB 05761	08 34 54.	+ 20 35		17.8	FAINT BLUE STAR
ZC 0834.9+3355	08 34 54.	+ 33 55	670		CLUSTER OF GALAXIES
LB 05774	08 35 00.	+ 18 55		18.7	FAINT BLUE STAR
LB 05773	08 35 00.	+ 18 56		19.6	FAINT BLUE STAR
LB 05772	08 35 00.	+ 19 22		19.4	FAINT BLUE STAR
LB 05771	08 35 00.	+ 19 31		17.7	FAINT BLUE STAR
LB 05770	08 35 00.	+ 19 47		18.7	FAINT BLUE STAR
LB 05769	08 35 00.	+ 20 11		18.6	FAINT BLUE STAR
LB 05768	08 35 00.	+ 20 20		20.3	FAINT BLUE STAR
LB 01828	08 35 00.	+ 20 34		18.2	FAINT BLUE STAR
ZWG 120.009	08 35 00.	+ 25 20		15.5	GALAXY
ZC 0835.0+5208	08 35 00.	+ 52 08	2350		CLUSTER OF GALAXIES
ZC 0835.0+5320	08 35 00.	+ 53 20	1140		CLUSTER OF GALAXIES
MCG+12-09-003	08 35 00.	+ 73 40	66	15.	GALAXY
ZC 0835.0+7812	08 35 00.	+ 78 12	2760		CLUSTER OF GALAXIES
MCG-03-22-009	08 35 00.	- 20 46 30.	102	15.	GALAXY
MCG+04-21-004	08 35 03.	+ 25 21	12	15.5	GALAXY
ZWG 060.031	08 35 06.	+ 12 57		15.3	GALAXY
KARA.73B 0270	08 35 06.	+ 12 57	60	15.3	ISOLATED GALAXY S
LB 05782	08 35 06.	+ 18 45		19.4	FAINT BLUE STAR
LB 01830	08 35 06.	+ 18 51		18.5	FAINT BLUE STAR
LB 05781	08 35 06.	+ 19 00		19.5	FAINT BLUE STAR
LB 01829	08 35 06.	+ 19 41		18.6	FAINT BLUE STAR
LB 05780	08 35 06.	+ 19 48		19.2	FAINT BLUE STAR
LB 05779	08 35 06.	+ 20 04		19.4	FAINT BLUE STAR
LB 05778	08 35 06.	+ 20 05		19.0	FAINT BLUE STAR
LB 05777	08 35 06.	+ 20 24		19.1	FAINT BLUE STAR
LB 05776	08 35 06.	+ 20 25		18.6	FAINT BLUE STAR
LB 05775	08 35 06.	+ 20 26		19.3	FAINT BLUE STAR
ZWG 120.010	08 35 06.	+ 25 30		15.3	GALAXY
ZC 0835.1+3148	08 35 06.	+ 31 48	1080		CLUSTER OF GALAXIES
ZWG 179.015	08 35 06.	+ 33 45		15.6	GALAXY
ZWG 004.089	08 35 06.	- 00 41		15.5	GALAXY
ZWG 004.088	08 35 06.	- 03 02		15.1	GALAXY
MCG+04-21-005	08 35 09.	+ 25 30 30.	30	15.3	GALAXY
BC 3CR205	08 35 09.94	+ 58 04 51.8		17.62	QUASI-STELLAR OBJECT
RNGC 2627	08 35 10.	- 29 46		8.5	OPEN CLUSTER
SHB 117	08 35 10.6	+ 58 04 46.		17.6	QUASI-STELLAR OBJECT
LB 05793	08 35 12.	+ 18 43		18.4	FAINT BLUE STAR
LB 05792	08 35 12.	+ 18 48		18.7	FAINT BLUE STAR
LB 05791	08 35 12.	+ 18 58		20.2	FAINT BLUE STAR
LB 01831	08 35 12.	+ 19 02		18.8	FAINT BLUE STAR
LB 05790	08 35 12.	+ 19 09		18.4	FAINT BLUE STAR
LB 05789	08 35 12.	+ 19 18		19.1	FAINT BLUE STAR
LB 05788	08 35 12.	+ 19 45		19.3	FAINT BLUE STAR
LB 01832	08 35 12.	+ 19 51		18.5	FAINT BLUE STAR
LB 05787	08 35 12.	+ 20 22		17.9	FAINT BLUE STAR
LB 05786	08 35 12.	+ 20 26		20.4	FAINT BLUE STAR
LB 05785	08 35 12.	+ 20 30		20.5	FAINT BLUE STAR
LB 05784	08 35 12.	+ 20 32		18.8	FAINT BLUE STAR
LB 05783	08 35 12.	+ 20 40		19.7	FAINT BLUE STAR
ZWG 120.011	08 35 12.	+ 25 04		15.5	GALAXY
ZWG 120.012	08 35 12.	+ 25 27		15.3	GALAXY
ZWG 150.009	08 35 12.	+ 31 59		15.7	GALAXY
UGC 04505	08 35 12.	+ 50 36	78	15.0	GALAXY S
MCG-01-22-029	08 35 12.	- 06 32 30.	60	14.	GALAXY
OCL 0714	08 35 12.	- 29 46	840	8.5	OPEN STAR CLUSTER
VHA 038	08 35 12.	- 29 46	540		OPEN STAR CLUSTER
VHE 18	08 35 12.	- 39 14	36		REFLECTION NEBULA
PK259+00.1	08 35 14.	- 39 15	65		PLANETARY NEBULA
MCG+04-21-006	08 35 15.	+ 25 17 30.	21	15.5	GALAXY
MCG+04-21-007	08 35 15.	+ 25 27 30.	30	15.3	GALAXY
RNGC 2624	08 35 16.	+ 19 54		14.5	GALAXY
HOLM 097B	08 35 16.	+ 43 47	54	14.6	PART OF MULTIPLE GALAXY
ZWG 004.091	08 35 18.	+ 02 11		15.7	GALAXY
LB 05796	08 35 18.	+ 18 47		19.0	FAINT BLUE STAR
ZWG 089.054	08 35 18.	+ 18 52		15.2	GALAXY
LB 05795	08 35 18.	+ 19 31		18.5	FAINT BLUE STAR
ZWG 089.055	08 35 18.	+ 19 54		14.5	GALAXY
UGC 04506	08 35 18.	+ 19 54	48	14.5	GALAXY
MCG+03-22-019	08 35 18.	+ 19 54	24	14.5	GALAXY
LB 05794	08 35 18.	+ 20 19		19.7	FAINT BLUE STAR
ZWG 120.013	08 35 18.	+ 25 05		14.8	GALAXY
MCG+04-21-008	08 35 18.	+ 25 05	42	14.8	GALAXY
ZWG 120.014	08 35 18.	+ 25 16		15.5	GALAXY
TON-N 0334	08 35 18.	+ 27 58		16.0	BLUE STAR
ZC 0835.3+4230	08 35 18.	+ 42 30	270		CLUSTER OF GALAXIES
ZWG 208.033	08 35 18.	+ 43 43		15.0	GALAXY
UGC 04507	08 35 18.	+ 43 43	114	15.0	GALAXY Sb
HOLM 097A	08 35 18.	+ 43 44	84	14.3	PART OF MULTIPLE GALAXY
ZWG 004.090	08 35 18.	- 00 05		15.6	GALAXY
MCG-03-22-010	08 35 18.	- 16 45	48	15.	GALAXY
RNGC 2622	08 35 19.	+ 25 05		15.0	GALAXY
LB 01833	08 35 20.	+ 21 08 24.		14.2	FAINT BLUE STAR
ZWG 032.041	08 35 24.	+ 08 04		15.0	GALAXY
MCG+01-22-014	08 35 24.	+ 08 04	36	15.0	GALAXY
LB 05805	08 35 24.	+ 18 47		19.4	FAINT BLUE STAR
LB 05804	08 35 24.	+ 18 52		20.6	FAINT BLUE STAR
LB 05803	08 35 24.	+ 19 22		19.5	FAINT BLUE STAR
LB 05802	08 35 24.	+ 19 30		19.0	FAINT BLUE STAR
LB 05801	08 35 24.	+ 19 35		20.7	FAINT BLUE STAR
LB 00380	08 35 24.	+ 19 37		17.3	FAINT BLUE STAR
LB 05800	08 35 24.	+ 19 53		19.0	FAINT BLUE STAR
LB 05799	08 35 24.	+ 20 03		18.4	FAINT BLUE STAR
LB 05798	08 35 24.	+ 20 12		19.2	FAINT BLUE STAR
LB 05797	08 35 24.	+ 20 15		20.3	FAINT BLUE STAR
ZWG 120.015	08 35 24.	+ 25 56		14.4	GALAXY
UGC 04509	08 35 24.	+ 25 56	132	14.4	GALAXY TRP SYS
ARP 243	08 35 24.	+ 25 56			PECULIAR GALAXY
ZWG 150.010	08 35 24.	+ 28 40		15.5	GALAXY
ZWG 150.011	08 35 24.	+ 28 50		15.6	GALAXY
ZWG 150.012	08 35 24.	+ 32 00		15.5	GALAXY
ZC 0835.4+3225	08 35 24.	+ 32 25	2150		CLUSTER OF GALAXIES
MCG+07-18-032	08 35 24.	+ 43 44	96	14.5	GALAXY
MCG+09-14-080	08 35 24.	+ 53 47 30.	36	14.	GALAXY
ZWG 332.002	08 35 24.	+ 73 40		14.5	GALAXY
ZWG 331.055	08 35 24.	+ 73 40		14.5	GALAXY
UGC 04510	08 35 24.	+ 73 40	96	14.5	GALAXY Sa
ZWG 004.092	08 35 24.	- 02 17		14.5	GALAXY
UGC 04508	08 35 24.	- 02 17	21	14.5	GALAXY EX CMPT
MCG+00-22-025	08 35 24.	- 02 17	21	14.5	GALAXY
RNGC 2623	08 35 25.	+ 25 56		14.5	GALAXY
MCG+04-21-009	08 35 27.	+ 25 56 30.	132	14.4	GALAXY
MCG+10-13-005	08 35 27.	+ 61 08	54	15.	GALAXY
IC 0511	08 35 27.	+ 73 40 39.			NONSTELLAR OBJECT
RNGC 2625	08 35 28.	+ 19 53		15.0	GALAXY

OBJECT NAME	RIGHT ASCEN.	DECLINATION	DIAM.	MAGN.	TYPE OF OBJECT
SV 205	08 35 28.22	+ 58 03 04.6		18.13	BLUE STELLAR OBJECT
IC 2387	08 35 29.	+ 30 58 13.			NONSTELLAR OBJECT
ZWG 032.042	08 35 30.	+ 04 01		15.5	GALAXY
ZC 0835.5+1805	08 35 30.	+ 18 05	4910		CLUSTER OF GALAXIES
LB 05815	08 35 30.	+ 18 45		18.8	FAINT BLUE STAR
LB 05814	08 35 30.	+ 18 56		19.9	FAINT BLUE STAR
LB 05813	08 35 30.	+ 19 07		20.7	FAINT BLUE STAR
LB 05812	08 35 30.	+ 19 22		18.9	FAINT BLUE STAR
LB 05810	08 35 30.	+ 19 44		18.8	FAINT BLUE STAR
ZWG 089.056	08 35 30.	+ 19 46		15.2	GALAXY
LB 05809	08 35 30.	+ 19 49		18.1	FAINT BLUE STAR
LB 05808	08 35 30.	+ 19 51		18.7	FAINT BLUE STAR
ZWG 089.057	08 35 30.	+ 19 53		15.1	GALAXY
MRK 625	08 35 30.	+ 19 53	16	15.5	GALAXY WITH UV CONTINUUM
LB 05811	08 35 30.	+ 19 53		19.9	FAINT BLUE STAR
LB 05807	08 35 30.	+ 19 58		17.7	FAINT BLUE STAR
LB 05806	08 35 30.	+ 20 25		19.1	FAINT BLUE STAR
TON-N 0335	08 35 30.	+ 29 46		16.7	BLUE STAR
ZWG 150.013	08 35 30.	+ 30 58		14.8	GALAXY
UGC 04511	08 35 30.	+ 30 58	72	14.8	GALAXY Sc
MCG+05-21-003	08 35 30.	+ 30 58	66	14.8	GALAXY
ZWG 263.064	08 35 30.	+ 53 47		14.6	GALAXY
ZC 0835.5+5418	08 35 30.	+ 54 18	740		CLUSTER OF GALAXIES
ZWG 288.003	08 35 30.	+ 61 08		15.3	GALAXY
ZWG 287.085	08 35 30.	+ 61 08		15.3	GALAXY
UGC 04512	08 35 30.	+ 61 08	90	15.3	GALAXY Sc
KARA.73B 0271	08 35 30.	+ 61 08	66	15.3	ISOLATED GALAXY S
ZWG 004.093	08 35 30.	- 02 32		15.5	GALAXY
MCG-02-22-023	08 35 30.	- 09 38	54	15.	GALAXY
KEEL 151	08 35 31.3	+ 19 46 13.		16.	SPIRAL NEBULA
KEEL 152	08 35 32.7	+ 19 45 36.		17.	NEBULA
ZWG 089.058	08 35 36.	+ 17 48		14.9	GALAXY
KARA.72 166A	08 35 36.	+ 17 48	42	14.9	PART OF DOUBLE GALAXY
MCG+03-22-020	08 35 36.	+ 17 48	24	14.9	GALAXY
LB 05817	08 35 36.	+ 19 08		18.3	FAINT BLUE STAR
LB 00381	08 35 36.	+ 19 58		17.3	FAINT BLUE STAR
LB 05816	08 35 36.	+ 20 10		18.9	FAINT BLUE STAR
LB 01834	08 35 36.	+ 20 28		19.7	FAINT BLUE STAR
LB 01836	08 35 36.	+ 20 34		18.9	FAINT BLUE STAR
ZC 0835.6+4024	08 35 36.	+ 40 24	1010		CLUSTER OF GALAXIES
ZWG 208.034	08 35 36.	+ 40 32		15.0	GALAXY
UGC 04513	08 35 36.	+ 40 32	60	15.0	GALAXY COMPACT
ZC 0835.6+4218	08 35 36.	+ 42 18	810		CLUSTER OF GALAXIES
OCL 0740	08 35 36.	- 39 55	4020	9.	OPEN STAR CLUSTER
VHE 19	08 35 36.	- 47 01	144		REFLECTION NEBULA
LB 01835	08 35 39.	+ 20 37 54.		20.6	FAINT BLUE STAR
ABC 0687	08 35 41.	+ 42 19		17.7	RICH CLUSTER OF GALAXIES
ZWG 004.095	08 35 42.	+ 00 30		15.7	GALAXY
LB 05822	08 35 42.	+ 18 53		20.4	FAINT BLUE STAR
LB 05821	08 35 42.	+ 19 06		18.4	FAINT BLUE STAR
LB 00382	08 35 42.	+ 19 11		16.8	FAINT BLUE STAR
LB 05820	08 35 42.	+ 19 27		18.5	FAINT BLUE STAR
LB 05819	08 35 42.	+ 19 40		19.8	FAINT BLUE STAR
ZC 0835.7+2000	08 35 42.	+ 20 00	1550		CLUSTER OF GALAXIES
LB 05818	08 35 42.	+ 20 12		19.6	FAINT BLUE STAR
MCG+07-18-033	08 35 42.	+ 40 31	54	15.	GALAXY
MCG+08-16-023	08 35 42.	+ 45 17	132	15.	GALAXY
ZWG 237.010	08 35 42.	+ 45 18		15.3	GALAXY
ZWG 004.094	08 35 42.	- 02 56		15.6	GALAXY
OCL 0735	08 35 42.	- 39 29	120	11.	OPEN STAR CLUSTER
BIGO 502	08 35 46.	+ 78 11			NEBULA
ZWG 032.043	08 35 48.	+ 04 11		15.4	GALAXY
ZWG 089.059	08 35 48.	+ 15 54		15.6	GALAXY
KARA.73B 0272	08 35 48.	+ 15 54	18	15.6	ISOLATED GALAXY E
ZWG 089.060	08 35 48.	+ 17 48		15.7	GALAXY
KARA.72 166B	08 35 48.	+ 17 48	24	15.7	PART OF DOUBLE GALAXY
LB 05834	08 35 48.	+ 18 59		19.0	FAINT BLUE STAR
LB 05833	08 35 48.	+ 19 02		20.0	FAINT BLUE STAR
LB 05832	08 35 48.	+ 19 08		17.8	FAINT BLUE STAR
LB 05831	08 35 48.	+ 19 21		19.1	FAINT BLUE STAR
LB 05830	08 35 48.	+ 19 30		20.0	FAINT BLUE STAR
LB 05829	08 35 48.	+ 19 40		19.3	FAINT BLUE STAR
LB 05828	08 35 48.	+ 19 44		19.0	FAINT BLUE STAR
LB 05827	08 35 48.	+ 19 54		19.4	FAINT BLUE STAR
LB 05826	08 35 48.	+ 20 07		20.2	FAINT BLUE STAR
LB 05825	08 35 48.	+ 20 11		20.4	FAINT BLUE STAR
LB 05824	08 35 48.	+ 20 11		18.0	FAINT BLUE STAR
LB 05823	08 35 48.	+ 20 39		17.3	FAINT BLUE STAR
TON-N 0336	08 35 48.	+ 26 24		15.7	BLUE STAR
TON-N 0337	08 35 48.	+ 28 20		16.2	BLUE STAR
MCG+06-19-012	08 35 48.	+ 36 24 30.	12	17.	GALAXY
MCG+06-19-011	08 35 48.	+ 36 24 30.	12	17.	GALAXY
ZC 0835.8+4436	08 35 48.	+ 44 36	470		CLUSTER OF GALAXIES
ZWG 263.065	08 35 48.	+ 52 25		15.6	GALAXY
MCG+09-14-081	08 35 48.	+ 53 57 30.	120	13.	GALAXY
ZWG 032.044	08 35 54.	+ 07 23		15.4	GALAXY
LB 05844	08 35 54.	+ 18 50		19.9	FAINT BLUE STAR
LB 05843	08 35 54.	+ 18 59		18.9	FAINT BLUE STAR
LB 05842	08 35 54.	+ 19 01		19.6	FAINT BLUE STAR
LB 05841	08 35 54.	+ 19 04		18.6	FAINT BLUE STAR
LB 05840	08 35 54.	+ 19 08		19.7	FAINT BLUE STAR
LB 05839	08 35 54.	+ 19 14		19.8	FAINT BLUE STAR
LB 05838	08 35 54.	+ 19 25		17.8	FAINT BLUE STAR
LB 05837	08 35 54.	+ 19 28		19.4	FAINT BLUE STAR
LB 05836	08 35 54.	+ 20 10		20.5	FAINT BLUE STAR
LB 05835	08 35 54.	+ 20 39		19.5	FAINT BLUE STAR
ZWG 263.066	08 35 54.	+ 53 38		14.1	GALAXY
UGC 04514	08 35 54.	+ 53 38	138	14.1	GALAXY SBc
ZWG 263.067	08 35 54.	+ 55 53		15.7	GALAXY
ZWG 004.097	08 35 54.	- 01 02		15.7	GALAXY
ZWG 004.096	08 35 54.	- 01 27		15.6	GALAXY
LB 00383	08 35 56.	+ 19 50 36.		14.9	FAINT BLUE STAR
ZWG 032.045	08 36 00.	+ 07 59		14.8	GALAXY
MCG+01-22-015	08 36 00.	+ 07 59	24	14.8	GALAXY
LB 05849	08 36 00.	+ 18 43		16.7	FAINT BLUE STAR
LB 01838	08 36 00.	+ 18 55		18.7	FAINT BLUE STAR
LB 01837	08 36 00.	+ 18 55		18.6	FAINT BLUE STAR
LB 05848	08 36 00.	+ 19 02		20.1	FAINT BLUE STAR
ZWG 089.061	08 36 00.	+ 19 04		15.6	GALAXY
LB 05847	08 36 00.	+ 19 28		20.2	FAINT BLUE STAR
LB 05846	08 36 00.	+ 20 02		18.8	FAINT BLUE STAR
LB 05845	08 36 00.	+ 20 04		17.3	FAINT BLUE STAR
ZWG 120.016	08 36 00.	+ 26 19		15.6	GALAXY
MCG+04-21-010	08 36 00.	+ 26 20	9	15.6	GALAXY
MCG+10-13-006	08 36 00.	+ 59 04	30	16.	GALAXY
MCG+12-09-059	08 36 00.	+ 71 25	204	13.2	GALAXY
ZWG 032.046	08 36 06.	+ 07 40		15.1	GALAXY
LB 05858	08 36 06.	+ 19 00		18.0	FAINT BLUE STAR
LB 05857	08 36 06.	+ 19 01		19.2	FAINT BLUE STAR
LB 05856	08 36 06.	+ 19 21		20.5	FAINT BLUE STAR
LB 01839	08 36 06.	+ 19 44		17.8	FAINT BLUE STAR
LB 05855	08 36 06.	+ 19 47		20.6	FAINT BLUE STAR
LB 05854	08 36 06.	+ 19 48		18.0	FAINT BLUE STAR
LB 05853	08 36 06.	+ 19 50		18.1	FAINT BLUE STAR
LB 05852	08 36 06.	+ 19 54		18.8	FAINT BLUE STAR
LB 05851	08 36 06.	+ 20 19		19.5	FAINT BLUE STAR
LB 05850	08 36 06.	+ 20 24		16.9	FAINT BLUE STAR
ZC 0836.1+3401	08 36 06.	+ 34 01	1340		CLUSTER OF GALAXIES
ZC 0836.1+4925	08 36 06.	+ 49 25	2350		CLUSTER OF GALAXIES
RNGC 2640	08 36 08.	- 54 57			UNVERIFIED SOUTHRN OBJECT
ZC 0836.2+0832	08 36 12.	+ 08 32	1750		CLUSTER OF GALAXIES
LB 05861	08 36 12.	+ 19 45		19.0	FAINT BLUE STAR
LB 05860	08 36 12.	+ 20 15		19.4	FAINT BLUE STAR
LB 05859	08 36 12.	+ 20 39		19.8	FAINT BLUE STAR
ZWG 120.017	08 36 12.	+ 26 24		15.4	GALAXY
MCG+04-21-011	08 36 12.	+ 26 25	24	15.4	GALAXY
ABC 0690	08 36 12.	+ 29 01		16.9	RICH CLUSTER OF GALAXIES
ZC 0836.2+4145	08 36 12.	+ 41 45	1210		CLUSTER OF GALAXIES
ZC 0836.2+4744	08 36 12.	+ 47 44	1610		CLUSTER OF GALAXIES
SHB 118	08 36 15.0	+ 19 32 25.		17.6	QUASI-STELLAR OBJECT
LB 00385	08 36 16.	+ 19 25 00.		16.0	FAINT BLUE STAR
BC 4C19.31	08 36 16.	+ 19 33		17.73	QUASI-STELLAR OBJECT
ZWG 004.098	08 36 18.	+ 01 34		15.3	GALAXY
ZWG 032.047	08 36 18.	+ 03 34		15.5	GALAXY
LB 00384	08 36 18.	+ 19 32		17.6	FAINT BLUE STAR
ZWG 089.062	08 36 18.	+ 19 39		15.6	GALAXY
LB 01840	08 36 18.	+ 19 58		19.5	FAINT BLUE STAR
LB 05863	08 36 18.	+ 20 28		19.0	FAINT BLUE STAR
LB 05862	08 36 18.	+ 20 46		16.7	FAINT BLUE STAR
ZC 0836.3+4147	08 36 18.	+ 41 47	18550		CLUSTER OF GALAXIES
MCG+09-14-082	08 36 18.	+ 52 37	78	15.	GALAXY
LB 05869	08 36 24.	+ 18 42		18.4	FAINT BLUE STAR
LB 05868	08 36 24.	+ 18 51		18.9	FAINT BLUE STAR
LB 05867	08 36 24.	+ 18 55		18.7	FAINT BLUE STAR
LB 00386	08 36 24.	+ 19 51		17.8	FAINT BLUE STAR
LB 01841	08 36 24.	+ 19 56		19.7	FAINT BLUE STAR
LB 05866	08 36 24.	+ 19 58		18.3	FAINT BLUE STAR
LB 01842	08 36 24.	+ 20 08		20.4	FAINT BLUE STAR
LB 05865	08 36 24.	+ 20 28		19.2	FAINT BLUE STAR
LB 05864	08 36 24.	+ 20 38		19.4	FAINT BLUE STAR
ZC 0836.4+5017	08 36 24.	+ 50 17	1480		CLUSTER OF GALAXIES
ZWG 263.068	08 36 24.	+ 52 38		14.5	GALAXY
UGC 04515	08 36 24.	+ 52 38	102	14.5	GALAXY SB:b
ZWG 263.069	08 36 24.	+ 56 12		15.1	GALAXY
MCG+10-13-007	08 36 24.	+ 59 50	24	16.	GALAXY
MCG-01-22-030	08 36 24.	- 05 34	42	15.	GALAXY
MCG-02-22-024	08 36 24.	- 14 30	24	15.5	GALAXY
TON-N 0933	08 36 29.	+ 34 47		15.8	BLUE STAR
ZWG 032.048	08 36 30.	+ 04 44		15.1	GALAXY
KARA.73B 0273	08 36 30.	+ 04 44	36	15.1	ISOLATED GALAXY S
ZWG 032.049	08 36 30.	+ 07 35		15.3	GALAXY
LB 05874	08 36 30.	+ 18 57		18.5	FAINT BLUE STAR
LB 01843	08 36 30.	+ 19 10		17.9	FAINT BLUE STAR
LB 05873	08 36 30.	+ 19 36		20.3	FAINT BLUE STAR
LB 05872	08 36 30.	+ 19 36		18.5	FAINT BLUE STAR
LB 00387	08 36 30.	+ 19 46		15.5	FAINT BLUE STAR
LB 05871	08 36 30.	+ 19 49		18.8	FAINT BLUE STAR
LB 05870	08 36 30.	+ 20 32		16.2	FAINT BLUE STAR
ZC 0836.5+5600	08 36 30.	+ 56 00	1280		CLUSTER OF GALAXIES
MCG+09-14-083	08 36 30.	+ 56 14	42	15.	GALAXY
MCG+11-11-019	08 36 30.	+ 67 01	96	15.	GALAXY
OCL 0728	08 36 30.	- 34 35	360	10.2	OPEN STAR CLUSTER
VHA 039	08 36 30.	- 34 35	210		OPEN STAR CLUSTER
RNGC 2635	08 36 31.	- 34 35		10.0	OPEN CLUSTER
LB 01845	08 36 34.	+ 21 00 12.		16.3	FAINT BLUE STAR
CED 106I	08 36 34.	- 42 53	1080		DIFFUSE GALACTIC NEBULA
LB 05882	08 36 36.	+ 18 55		20.0	FAINT BLUE STAR
LB 05881	08 36 36.	+ 19 00		18.8	FAINT BLUE STAR
LB 01844	08 36 36.	+ 19 05		18.5	FAINT BLUE STAR
LB 05880	08 36 36.	+ 19 08		18.5	FAINT BLUE STAR
LB 05879	08 36 36.	+ 19 25		20.0	FAINT BLUE STAR
LB 05878	08 36 36.	+ 19 27		20.6	FAINT BLUE STAR
LB 05877	08 36 36.	+ 19 42		19.5	FAINT BLUE STAR
LB 05876	08 36 36.	+ 19 57		20.6	FAINT BLUE STAR
LB 05875	08 36 36.	+ 20 32		18.3	FAINT BLUE STAR
ZC 0836.6+2242	08 36 36.	+ 22 42	810		CLUSTER OF GALAXIES
ZC 0836.6+2915	08 36 36.	+ 29 15	1950		CLUSTER OF GALAXIES
ZWG 311.011	08 36 36.	+ 67 02		15.2	GALAXY
UGC 04516	08 36 36.	+ 67 02	102	15.2	GALAXY Sc-IRR
ZWG 004.100	08 36 36.	- 02 07		15.3	GALAXY
ZWG 004.099	08 36 36.	- 02 12		15.3	GALAXY
LB 00388	08 36 40.	+ 19 49 30.		13.0	FAINT BLUE STAR
LB 00389	08 36 41.	+ 20 06 24.		17.8	FAINT BLUE STAR
ZC 0836.7+1014	08 36 42.	+ 10 14	400		CLUSTER OF GALAXIES
LB 05886	08 36 42.	+ 19 47		18.6	FAINT BLUE STAR
LB 05885	08 36 42.	+ 20 12		17.5	FAINT BLUE STAR
LB 05884	08 36 42.	+ 20 30		18.8	FAINT BLUE STAR
LB 01846	08 36 42.	+ 20 34		18.7	FAINT BLUE STAR
LB 05883	08 36 42.	+ 20 43		20.0	FAINT BLUE STAR
ZC 0836.7+5045	08 36 42.	+ 50 45	610		CLUSTER OF GALAXIES
MCG+12-09-004	08 36 42.	+ 73 52 30.	12	16.	GALAXY
MCG-01-22-031	08 36 42.	- 08 42 30.	60	15.	GALAXY
MCG-02-22-025	08 36 42.	- 14 33	60	15.	GALAXY
BIGO 503	08 36 43.	+ 78 20			NEBULA
LB 05900	08 36 48.	+ 18 58		20.3	FAINT BLUE STAR
LB 05899	08 36 48.	+ 18 59		20.5	FAINT BLUE STAR
LB 05898	08 36 48.	+ 19 11		19.8	FAINT BLUE STAR
LB 05897	08 36 48.	+ 19 15		19.7	FAINT BLUE STAR
LB 05896	08 36 48.	+ 19 17		20.4	FAINT BLUE STAR
LB 05895	08 36 48.	+ 19 23		18.0	FAINT BLUE STAR
LB 05894	08 36 48.	+ 19 38		19.0	FAINT BLUE STAR
LB 05893	08 36 48.	+ 19 42		16.8	FAINT BLUE STAR
LB 05892	08 36 48.	+ 19 47		19.7	FAINT BLUE STAR
LB 05891	08 36 48.	+ 19 53		18.2	FAINT BLUE STAR
LB 05890	08 36 48.	+ 19 56		18.4	FAINT BLUE STAR
LB 05889	08 36 48.	+ 20 02		20.7	FAINT BLUE STAR
LB 05888	08 36 48.	+ 20 02		18.2	FAINT BLUE STAR
LB 05887	08 36 48.	+ 20 28		19.8	FAINT BLUE STAR
ZC 0836.8+4618	08 36 48.	+ 46 18	2290		CLUSTER OF GALAXIES
ZWG 263.070	08 36 48.	+ 53 54		15.0	GALAXY
ZWG 004.102	08 36 48.	- 00 27		15.7	GALAXY
ZWG 004.101	08 36 48.	- 03 10		15.7	GALAXY
LB 05906	08 36 54.	+ 19 24		18.4	FAINT BLUE STAR
LB 05905	08 36 54.	+ 19 52		18.9	FAINT BLUE STAR
LB 01847	08 36 54.	+ 19 57		17.5	FAINT BLUE STAR
LB 05904	08 36 54.	+ 20 02		20.6	FAINT BLUE STAR

OBJECT NAME	RIGHT ASCEN.	DECLINATION	DIAM.	MAGN.	TYPE OF OBJECT
LB 05903	08 36 54.	+ 20 02		19.2	FAINT BLUE STAR
LB 00390	08 36 54.	+ 20 11		17.2	FAINT BLUE STAR
LB 05902	08 36 54.	+ 20 18		20.5	FAINT BLUE STAR
LB 05901	08 36 54.	+ 20 24		18.9	FAINT BLUE STAR
LB 01848	08 36 54.	+ 20 35		18.4	FAINT BLUE STAR
ZWG 120.018	08 36 54.	+ 23 19		15.6	GALAXY
KARA.73B 0274	08 36 54.	+ 23 19	18	15.6	ISOLATED GALAXY
MCG+10-13-008	08 36 54.	+ 57 26	42	16.	GALAXY
ZC 0836.9+6542	08 36 54.	+ 65 42	1680		CLUSTER OF GALAXIES
ZWG 332.003	08 36 54.	+ 73 52		15.5	GALAXY
ZWG 331.056	08 36 54.	+ 73 52		15.5	GALAXY
UGC 04517	08 36 54.	+ 73 52	66	15.5	GALAXY
ZC 0837.0+0113	08 37 00.	+ 01 13	670		CLUSTER OF GALAXIES
LB 05922	08 37 00.	+ 18 54		19.1	FAINT BLUE STAR
LB 05921	08 37 00.	+ 18 56		19.0	FAINT BLUE STAR
LB 05920	08 37 00.	+ 18 56		18.5	FAINT BLUE STAR
LB 05919	08 37 00.	+ 19 13		18.5	FAINT BLUE STAR
LB 05918	08 37 00.	+ 19 14		20.2	FAINT BLUE STAR
LB 05917	08 37 00.	+ 19 14		20.0	FAINT BLUE STAR
LB 05915	08 37 00.	+ 19 18		20.4	FAINT BLUE STAR
LB 05914	08 37 00.	+ 19 25		20.5	FAINT BLUE STAR
LB 05913	08 37 00.	+ 19 29		18.7	FAINT BLUE STAR
LB 05912	08 37 00.	+ 19 41		19.0	FAINT BLUE STAR
LB 05911	08 37 00.	+ 19 42		19.0	FAINT BLUE STAR
LB 05910	08 37 00.	+ 19 55		18.9	FAINT BLUE STAR
LB 05909	08 37 00.	+ 20 17		18.7	FAINT BLUE STAR
LB 05908	08 37 00.	+ 20 29		18.6	FAINT BLUE STAR
LB 05907	08 37 00.	+ 20 33		20.5	FAINT BLUE STAR
ZWG 120.019	08 37 00.	+ 23 35		15.7	GALAXY
ZC 0837.0+2506	08 37 00.	+ 25 06	8530		CLUSTER OF GALAXIES
ZC 0837.0+2815	08 37 00.	+ 28 15	3630		CLUSTER OF GALAXIES
ZWG 150.014	08 37 00.	+ 30 00		15.7	GALAXY
ZWG 288.004	08 37 00.	+ 60 07		15.5	GALAXY
MCG+10-13-009	08 37 00.	+ 60 07	15	15.	GALAXY
ZWG 311.012	08 37 00.	+ 66 37		15.5	GALAXY
ZC 0837.0+8200	08 37 00.	+ 82 00	3160		CLUSTER OF GALAXIES
ZC 0837.0+8308	08 37 00.	+ 83 08	2490		CLUSTER OF GALAXIES
KEEL 153	08 37 04.1	+ 19 49 21.		17.	SPIRAL NEBULA
IC 2388	08 37 05.	+ 19 49 20.			NONSTELLAR OBJECT
ARC 0693	08 37 06.	+ 01 15		17.4	RICH CLUSTER OF GALAXIES
LB 05930	08 37 06.	+ 19 02		19.9	FAINT BLUE STAR
LB 05929	08 37 06.	+ 19 10		20.5	FAINT BLUE STAR
LB 05928	08 37 06.	+ 19 19		19.3	FAINT BLUE STAR
LB 05927	08 37 06.	+ 19 24		16.0	FAINT BLUE STAR
LB 01849	08 37 06.	+ 19 26		17.5	FAINT BLUE STAR
LB 01850	08 37 06.	+ 19 36		17.6	FAINT BLUE STAR
ZWG 089.063	08 37 06.	+ 19 49		15.7	GALAXY
LB 05926	08 37 06.	+ 20 14		18.4	FAINT BLUE STAR
LB 05925	08 37 06.	+ 20 16		19.1	FAINT BLUE STAR
LB 05924	08 37 06.	+ 20 25		18.8	FAINT BLUE STAR
LB 05923	08 37 06.	+ 20 30		19.3	FAINT BLUE STAR
MCG+10-13-010	08 37 06.	+ 60 07 30.	24	16.	GALAXY
ZC 0837.1+6445	08 37 06.	+ 64 45	9810		CLUSTER OF GALAXIES
BB 5.07	08 37 06.2	- 12 05 20.			GALAXY NEAR QSO PKS0837
RNGC 2632	08 37 10.	+ 20 10		4.0	OPEN CLUSTER
LB 05938	08 37 12.	+ 19 38		18.8	FAINT BLUE STAR
LB 05937	08 37 12.	+ 19 41		20.3	FAINT BLUE STAR
LB 05936	08 37 12.	+ 19 47		19.7	FAINT BLUE STAR
LB 05935	08 37 12.	+ 19 50		19.1	FAINT BLUE STAR
LB 05934	08 37 12.	+ 19 57		18.6	FAINT BLUE STAR
OCL 0507	08 37 12.	+ 20 10	9000	5.28	OPEN STAR CLUSTER
LB 05933	08 37 12.	+ 20 19		20.0	FAINT BLUE STAR
LB 05932	08 37 12.	+ 20 20		17.8	FAINT BLUE STAR
LB 05931	08 37 12.	+ 20 28		19.0	FAINT BLUE STAR
7ZW 241	08 37 12.	+ 61 23			COMPACT GALAXY
MCG+12-09-005	08 37 12.	+ 73 10	114	14.	GALAXY
MCG-01-22-032	08 37 15.	- 04 12 30.	24	16.	GALAXY
RNGC 2645	08 37 16.	- 46 03			NON-EXISTENT OBJECT
LB 00391	08 37 17.	+ 19 00 54.		17.5	FAINT BLUE STAR
ZWG 060.032	08 37 18.	+ 11 11		14.9	GALAXY
MCG+02-22-003	08 37 18.	+ 11 11	36	14.9	GALAXY
LB 05944	08 37 18.	+ 18 57		19.4	FAINT BLUE STAR
LB 05943	08 37 18.	+ 19 22		19.1	FAINT BLUE STAR
LB 05948	08 37 18.	+ 19 45		20.7	FAINT BLUE STAR
LB 05947	08 37 18.	+ 19 48		20.7	FAINT BLUE STAR
LB 05946	08 37 18.	+ 20 02		18.7	FAINT BLUE STAR
LB 05945	08 37 18.	+ 20 13		20.6	FAINT BLUE STAR
LB 05942	08 37 18.	+ 20 26		20.5	FAINT BLUE STAR
LB 05941	08 37 18.	+ 20 28		19.2	FAINT BLUE STAR
LB 05940	08 37 18.	+ 20 30		19.8	FAINT BLUE STAR
LB 05939	08 37 18.	+ 20 36		17.2	FAINT BLUE STAR
7ZW 242	08 37 18.	+ 61 24			COMPACT GALAXY
MRSL 260+00/1	08 37 18.	- 40 14	6000		HII REGION
BB 5.02	08 37 20.9	- 12 03 16.			GALAXY NEAR QSO PKS0837
BB 5.08	08 37 21.5	- 12 07 38.			GALAXY NEAR QSO PKS0837
LB 05960	08 37 24.	+ 18 44		18.4	FAINT BLUE STAR
LB 05959	08 37 24.	+ 18 54		17.5	FAINT BLUE STAR
LB 05958	08 37 24.	+ 19 09		18.3	FAINT BLUE STAR
LB 05957	08 37 24.	+ 19 25		19.7	FAINT BLUE STAR
LB 05956	08 37 24.	+ 19 34		18.9	FAINT BLUE STAR
LB 05955	08 37 24.	+ 19 35		20.6	FAINT BLUE STAR
LB 05954	08 37 24.	+ 19 36		18.5	FAINT BLUE STAR
LB 05953	08 37 24.	+ 19 39		18.5	FAINT BLUE STAR
LB 05952	08 37 24.	+ 19 56		19.1	FAINT BLUE STAR
LB 05951	08 37 24.	+ 20 05		20.4	FAINT BLUE STAR
LB 05950	08 37 24.	+ 20 25		17.7	FAINT BLUE STAR
LB 05949	08 37 24.	+ 20 31		20.4	FAINT BLUE STAR
ZC 0837.4+2245	08 37 24.	+ 22 45	540		CLUSTER OF GALAXIES
MCG+06-19-013	08 37 24.	+ 37 06	12	15.5	GALAXY
UGC 04518	08 37 24.	+ 42 07	60	16.0	GALAXY
ZC 0837.4+4431	08 37 24.	+ 44 31	1880		CLUSTER OF GALAXIES
ZWG 237.011	08 37 24.	+ 49 36		15.7	GALAXY
SEY 048	08 37 24.	+ 71 08 33.		15.2	FAINT GALAXY
ZWG 004.103	08 37 24.	- 02 37		15.6	GALAXY
OCL 0750	08 37 24.	- 43 51	660	13.	OPEN STAR CLUSTER
RNGC 2614	08 37 25.	+ 73 09		14.0	GALAXY
BB 5.11	08 37 25.7	- 12 00 01.5			GALAXY NEAR QSO PKS0837
LB 01851	08 37 27.	+ 22 18 42.		16.4	FAINT BLUE STAR
MCG+04-21-012	08 37 27.	+ 23 43	66	14.1	GALAXY
BB 5.03	08 37 27.75	- 12 01 57.			GALAXY NEAR QSO PKS0837
BC PKS0837-12	08 37 27.95	- 12 03 54.2		15.76	QUASI-STELLAR OBJECT
SHB 119	08 37 28.0	- 12 03 54.		15.8	QUASI-STELLAR OBJECT
LB 01853	08 37 30.	+ 18 41		16.1	FAINT BLUE STAR
LB 05964	08 37 30.	+ 18 42		18.6	FAINT BLUE STAR
LB 00392	08 37 30.	+ 19 23		18.0	FAINT BLUE STAR
LB 05963	08 37 30.	+ 19 28		20.6	FAINT BLUE STAR
LB 05962	08 37 30.	+ 20 13		18.9	FAINT BLUE STAR
LB 05961	08 37 30.	+ 20 16		18.5	FAINT BLUE STAR

OBJECT NAME	RIGHT ASCEN.	DECLINATION	DIAM.	MAGN.	TYPE OF OBJECT
LB 01852	08 37 30.	+ 20 36		16.8	FAINT BLUE STAR
ZWG 120.020	08 37 30.	+ 23 43		14.1	GALAXY
UGC 04519	08 37 30.	+ 23 43	72	14.1	GALAXY Sc?
ZWG 179.016	08 37 30.	+ 37 05		15.7	GALAXY
UGC 04520	08 37 30.	+ 37 05	72	15.7	GALAXY S
ZC 0837.5+3845	08 37 30.	+ 38 45	940		CLUSTER OF GALAXIES
ZWG 208.035	08 37 30.	+ 43 01		15.5	GALAXY
UGC 04521	08 37 30.	+ 43 01	72	15.5	GALAXY Sb-c
KARA.73B 0275	08 37 30.	+ 43 01	72	15.5	ISOLATED GALAXY S
ZWG 263.071	08 37 30.	+ 54 06		15.2	GALAXY
MCG+12-09-006	08 37 30.	+ 71 08	60	15.	GALAXY
ZWG 332.004	08 37 30.	+ 71 09		15.0	GALAXY
ZWG 331.057	08 37 30.	+ 71 09		15.0	GALAXY
UGC 04522	08 37 30.	+ 71 09	60	15.0	GALAXY Sc
ZWG 332.005	08 37 30.	+ 73 09		14.0	GALAXY
ZWG 331.058	08 37 30.	+ 73 09		14.0	GALAXY
UGC 04523	08 37 30.	+ 73 09	168	14.0	GALAXY Sc
RNGC 2628	08 37 32.	+ 23 43		14.0	GALAXY
BB 5.01	08 37 32.35	- 12 03 38.5			GALAXY NEAR QSO PKS0837
MCG+06-19-014	08 37 33.	+ 37 06	60	15.5	GALAXY
MCG+07-18-034	08 37 33.	+ 43 00 30.	60	15.	GALAXY
MCG-04-21-006	08 37 33.	- 23 17	30	15.	GALAXY
LB 01854	08 37 34.	+ 23 03 06.		16.1	FAINT BLUE STAR
BB 5.04	08 37 34.85	- 12 01 51.5			GALAXY NEAR QSO PKS0837
BB 5.09	08 37 35.85	- 12 06 30.			GALAXY NEAR QSO PKS0837
ZWG 032.050	08 37 36.	+ 05 49		15.2	GALAXY
UGC 04524	08 37 36.	+ 05 49	84	15.2	GALAXY Sc
KARA.73B 0276	08 37 36.	+ 05 49	90	15.2	ISOLATED GALAXY S
LB 05971	08 37 36.	+ 18 46		18.3	FAINT BLUE STAR
LB 05970	08 37 36.	+ 19 04		18.8	FAINT BLUE STAR
LB 05969	08 37 36.	+ 19 15		19.4	FAINT BLUE STAR
LB 05968	08 37 36.	+ 19 27		18.5	FAINT BLUE STAR
LB 05967	08 37 36.	+ 19 52		18.2	FAINT BLUE STAR
LB 00393	08 37 36.	+ 19 54		17.4	FAINT BLUE STAR
LB 05966	08 37 36.	+ 20 00		20.7	FAINT BLUE STAR
LB 05965	08 37 36.	+ 20 03		18.8	FAINT BLUE STAR
LB 01855	08 37 36.	+ 20 27		17.6	FAINT BLUE STAR
ZWG 150.015	08 37 36.	+ 30 21		15.7	GALAXY
ZC 0837.6+3050	08 37 36.	+ 30 50	1880		CLUSTER OF GALAXIES
ZC 0837.6+3238	08 37 36.	+ 32 38	3360		CLUSTER OF GALAXIES
MCG+07-18-035	08 37 36.	+ 42 39	36	16.	GALAXY
ZC 0837.6+5643	08 37 36.	+ 56 43	940		CLUSTER OF GALAXIES
MCG+11-11-020	08 37 36.	+ 65 27	39	17.	GALAXY
VHE 20	08 37 36.	- 40 17	96		REFLECTION NEBULA
OCL 0754	08 37 36.	- 46 02	90	9.2	OPEN STAR CLUSTER
VHA 040	08 37 36.	- 46 02	90		OPEN STAR CLUSTER
BB 5.05	08 37 41.9	- 11 59 37.			GALAXY NEAR QSO PKS0837
LB 05979	08 37 42.	+ 18 44		19.2	FAINT BLUE STAR
LB 05978	08 37 42.	+ 19 02		18.8	FAINT BLUE STAR
LB 05977	08 37 42.	+ 19 12		18.5	FAINT BLUE STAR
LB 05976	08 37 42.	+ 19 21		18.6	FAINT BLUE STAR
LB 05975	08 37 42.	+ 19 34		18.5	FAINT BLUE STAR
LB 05974	08 37 42.	+ 19 54		20.1	FAINT BLUE STAR
LB 05973	08 37 42.	+ 20 08		19.3	FAINT BLUE STAR
LB 05972	08 37 42.	+ 20 37		18.8	FAINT BLUE STAR
ZWG 150.016	08 37 42.	+ 27 26		15.7	GALAXY
MCG+09-14-084	08 37 42.	+ 54 08	42	16.	GALAXY
PK158+37.1	08 37 42.	+ 58 24	300	14.6	PLANETARY NEBULA
LB 01856	08 37 44.	+ 22 03 24.		16.7	FAINT BLUE STAR
MCG+09-14-085	08 37 45.	+ 51 24	48	15.	GALAXY
BB 5.06	08 37 45.4	- 11 59 15.5			GALAXY NEAR QSO PKS0837
ARC 0691	08 37 46.	+ 42 13		17.1	RICH CLUSTER OF GALAXIES
ZWG 060.033	08 37 48.	+ 11 03		15.7	GALAXY
LB 05989	08 37 48.	+ 18 50		19.0	FAINT BLUE STAR
LB 05988	08 37 48.	+ 18 56		17.9	FAINT BLUE STAR
LB 05987	08 37 48.	+ 19 02		17.8	FAINT BLUE STAR
LB 05986	08 37 48.	+ 19 06		18.7	FAINT BLUE STAR
LB 05985	08 37 48.	+ 19 08		18.8	FAINT BLUE STAR
LB 05984	08 37 48.	+ 19 38		18.3	FAINT BLUE STAR
LB 05983	08 37 48.	+ 19 40		19.9	FAINT BLUE STAR
LB 05982	08 37 48.	+ 19 41		19.7	FAINT BLUE STAR
LB 05981	08 37 48.	+ 19 43		18.5	FAINT BLUE STAR
LB 05980	08 37 48.	+ 20 37		19.3	FAINT BLUE STAR
ZC 0837.8+3130	08 37 48.	+ 31 30	2080		CLUSTER OF GALAXIES
MCG+12-09-007	08 37 48.	+ 70 42	33	17.	GALAXY
ZWG 004.104	08 37 48.	- 00 58		15.7	GALAXY
BB 5.10	08 37 49.05	- 12 02 23.5			GALAXY NEAR QSO PKS0837
LB 05992	08 37 54.	+ 19 56		19.5	FAINT BLUE STAR
LB 05991	08 37 54.	+ 20 05		19.8	FAINT BLUE STAR
LB 05990	08 37 54.	+ 20 36		18.4	FAINT BLUE STAR
ZWG 263.072	08 37 54.	+ 51 26		15.1	GALAXY
UGC 04525	08 37 54.	+ 51 26	78	15.1	GALAXY SBb
ZWG 350.002	08 37 54.	+ 79 47		15.5	GALAXY
ZWG 349.030	08 37 54.	+ 79 47		15.5	GALAXY
ARC 0692	08 37 55.	+ 26 55		16.2	RICH CLUSTER OF GALAXIES
IC 0519	08 37 58.	+ 02 47 19.			NONSTELLAR OBJECT
LBN 1071	08 38	- 22 10	2520		BRIGHT NEBULA
LBN 1073	08 38	- 24 30	1740		BRIGHT NEBULA
ZWG 004.105	08 38 00.	+ 02 21		15.5	GALAXY
ZWG 032.051	08 38 00.	+ 02 47		15.4	GALAXY
LB 06004	08 38 00.	+ 19 14		18.7	FAINT BLUE STAR
LB 06003	08 38 00.	+ 19 21		18.6	FAINT BLUE STAR
LB 06002	08 38 00.	+ 19 23		19.0	FAINT BLUE STAR
LB 06001	08 38 00.	+ 19 24		16.7	FAINT BLUE STAR
LB 06000	08 38 00.	+ 19 27		18.6	FAINT BLUE STAR
LB 05999	08 38 00.	+ 19 31		19.8	FAINT BLUE STAR
ZWG 089.064	08 38 00.	+ 19 32		14.8	GALAXY
UGC 04526	08 38 00.	+ 19 32	84	14.8	GALAXY Sa-b
MCG+03-22-021	08 38 00.	+ 19 32	84	14.8	GALAXY
LB 05998	08 38 00.	+ 19 35		19.9	FAINT BLUE STAR
LB 05997	08 38 00.	+ 19 37		19.2	FAINT BLUE STAR
LB 05996	08 38 00.	+ 19 43		18.3	FAINT BLUE STAR
LB 05995	08 38 00.	+ 19 48		19.1	FAINT BLUE STAR
LB 05994	08 38 00.	+ 19 56		19.2	FAINT BLUE STAR
LB 01857	08 38 00.	+ 20 39		19.9	FAINT BLUE STAR
LB 05993	08 38 00.	+ 20 48		18.7	FAINT BLUE STAR
ZC 0838.0+5331	08 38 00.	+ 53 31	1210		CLUSTER OF GALAXIES
7ZW 243	08 38 00.	+ 62 30			COMPACT GALAXY
UGC 04527	08 38 00.	+ 77 06	90	17.	GALAXY DWRF IR
PK244+12.1	08 38 00.	- 20 43	482	14.3	PLANETARY NEBULA
SHB 120	08 38 01.7	+ 13 23 05.		18.2	QUASI-STELLAR OBJECT
BC 3CR207	08 38 01.73	+ 13 23 05.4		18.15	QUASI-STELLAR OBJECT
ARC 0684	08 38 03.	+ 72 59		17.1	RICH CLUSTER OF GALAXIES
LB 06017	08 38 06.	+ 18 44		18.5	FAINT BLUE STAR
LB 06016	08 38 06.	+ 18 45		18.5	FAINT BLUE STAR
LB 06015	08 38 06.	+ 19 17		18.5	FAINT BLUE STAR
LB 06014	08 38 06.	+ 19 18		18.7	FAINT BLUE STAR
LB 06013	08 38 06.	+ 19 23		19.1	FAINT BLUE STAR

OBJECT NAME	RIGHT ASCEN.	DECLINATION	DIAM.	MAGN.	TYPE OF OBJECT
LB 06012	08 38 06.	+ 19 38		18.2	FAINT BLUE STAR
LB 06011	08 38 06.	+ 19 43		18.1	FAINT BLUE STAR
LB 06010	08 38 06.	+ 19 48		20.6	FAINT BLUE STAR
LB 06009	08 38 06.	+ 19 48		17.7	FAINT BLUE STAR
LB 06008	08 38 06.	+ 20 14		18.6	FAINT BLUE STAR
LB 06007	08 38 06.	+ 20 31		19.2	FAINT BLUE STAR
LB 06006	08 38 06.	+ 20 32		19.4	FAINT BLUE STAR
LB 06005	08 38 06.	+ 20 43		18.7	FAINT BLUE STAR
ZC 0838.1+5406	08 38 06.	+ 54 06	1340		CLUSTER OF GALAXIES
MCG+10-13-011	08 38 06.	+ 60 57	30	17.	GALAXY
MCG+12-09-060	08 38 06.	+ 68 39	51	16.	GALAXY
UGC 04528	08 38 12.	+ 16 22	60	16.0	GALAXY
MCG+03-22-022	08 38 12.	+ 16 22	72	17.	GALAXY
LB 06024	08 38 12.	+ 18 52		19.1	FAINT BLUE STAR
LB 06023	08 38 12.	+ 19 27		19.0	FAINT BLUE STAR
LB 06022	08 38 12.	+ 19 46		18.8	FAINT BLUE STAR
LB 06021	08 38 12.	+ 19 52		19.0	FAINT BLUE STAR
LB 01859	08 38 12.	+ 20 01		18.6	FAINT BLUE STAR
LB 06020	08 38 12.	+ 20 10		20.0	FAINT BLUE STAR
LB 06019	08 38 12.	+ 20 11		19.1	FAINT BLUE STAR
LB 06018	08 38 12.	+ 20 17		18.4	FAINT BLUE STAR
LB 01858	08 38 12.	+ 20 32		18.3	FAINT BLUE STAR
MCG+04-21-013	08 38 12.	+ 21 04	36	15.4	GALAXY
UGC 04529	08 38 12.	+ 46 58	60	16.5	GALAXY Sb-c
ZWG 263.073	08 38 12.	+ 50 58		15.7	GALAXY
UGC 04530	08 38 12.	+ 57 46	66	16.0	GALAXY Sb/Sc
MCG+10-13-012	08 38 12.	+ 57 46	60	15.	GALAXY
ARC 0694	08 38 13.	+ 32 14		17.5	RICH CLUSTER OF GALAXIES
RNGC 2642	08 38 15.	- 03 57		12.5	GALAXY
RNGC 2637	08 38 16.	+ 19 52			GALAXY
ZC 0838.3+1410	08 38 18.	+ 14 10	1550		CLUSTER OF GALAXIES
LB 06031	08 38 18.	+ 18 52		18.5	FAINT BLUE STAR
LB 06030	08 38 18.	+ 19 19		17.4	FAINT BLUE STAR
LB 06029	08 38 18.	+ 19 49		18.8	FAINT BLUE STAR
LB 06028	08 38 18.	+ 19 51		18.8	FAINT BLUE STAR
ZWG 089.065	08 38 18.	+ 19 52		15.4	GALAXY
LB 06027	08 38 18.	+ 20 15		19.2	FAINT BLUE STAR
LB 06026	08 38 18.	+ 20 40		19.6	FAINT BLUE STAR
LB 06025	08 38 18.	+ 20 41		18.2	FAINT BLUE STAR
ZWG 120.021	08 38 18.	+ 21 05		15.4	GALAXY
KARA.73B 0277	08 38 18.	+ 21 05	60	15.4	ISOLATED GALAXY S
ZC 0838.3+2305	08 38 18.	+ 23 05	2220		CLUSTER OF GALAXIES
KARA.72 167A	08 38 18.	+ 40 49	24	14.9	PART OF DOUBLE GALAXY
ZWG 208.036	08 38 18.	+ 40 50		14.9	GALAXY
KARA.72 167B	08 38 18.	+ 40 50	42		PART OF DOUBLE GALAXY
ZC 0838.3+4214	08 38 18.	+ 42 14	2420		CLUSTER OF GALAXIES
MCG+13-07-002	08 38 18.	+ 77 06	33	14.	GALAXY
MCG-01-22-033	08 38 18.	- 03 56	120	13.	GALAXY
LB 00394	08 38 19.	+ 19 22 48.		17.8	FAINT BLUE STAR
ARC 0695	08 38 19.	+ 32 28		17.1	RICH CLUSTER OF GALAXIES
KEEL 154	08 38 21.3	+ 19 52 13.		17.	NEBULA
LB 01860	08 38 23.	+ 21 13 18.		16.8	FAINT BLUE STAR
LB 06035	08 38 24.	+ 19 11		20.1	FAINT BLUE STAR
LB 06034	08 38 24.	+ 19 26		18.6	FAINT BLUE STAR
LB 06033	08 38 24.	+ 19 42		20.5	FAINT BLUE STAR
LB 06032	08 38 24.	+ 20 01		19.1	FAINT BLUE STAR
ZC 0838.4+4030	08 38 24.	+ 40 30	1080		CLUSTER OF GALAXIES
MCG+07-18-037	08 38 24.	+ 40 49 30.	15	15.5	GALAXY
MCG+07-18-036	08 38 24.	+ 40 50	30	15.	GALAXY
ZC 0838.4+7301	08 38 24.	+ 73 01	1410		CLUSTER OF GALAXIES
MCG+13-07-003	08 38 24.	+ 76 48	30	16.	GALAXY
TON-N 0934	08 38 29.	+ 34 36		16.9	BLUE STAR
TON-N 0935	08 38 29.	+ 35 04		16.	BLUE STAR
LB 01861	08 38 30.	+ 19 22		18.4	FAINT BLUE STAR
LB 06040	08 38 30.	+ 19 34		18.7	FAINT BLUE STAR
LB 06039	08 38 30.	+ 19 42		18.3	FAINT BLUE STAR
LB 06038	08 38 30.	+ 19 57		18.9	FAINT BLUE STAR
LB 00395	08 38 30.	+ 20 13		17.0	FAINT BLUE STAR
LB 06037	08 38 30.	+ 20 14		18.7	FAINT BLUE STAR
LB 06036	08 38 30.	+ 20 15		17.8	FAINT BLUE STAR
ZWG 120.022	08 38 30.	+ 25 25		15.4	GALAXY
ZWG 263.074	08 38 30.	+ 52 56		15.1	GALAXY
MCG+14-04-055	08 38 30.	+ 84 27 30.	51	17.	GALAXY
ZC 0838.6+0810	08 38 36.	+ 08 10	470		CLUSTER OF GALAXIES
LB 06048	08 38 36.	+ 19 02		20.0	FAINT BLUE STAR
LB 06047	08 38 36.	+ 19 15		18.9	FAINT BLUE STAR
LB 06046	08 38 36.	+ 19 16		17.5	FAINT BLUE STAR
LB 06045	08 38 36.	+ 19 37		19.1	FAINT BLUE STAR
LB 06044	08 38 36.	+ 19 45		19.4	FAINT BLUE STAR
LB 06043	08 38 36.	+ 20 08		19.8	FAINT BLUE STAR
LB 01864	08 38 36.	+ 20 12		19.9	FAINT BLUE STAR
LB 01863	08 38 36.	+ 20 12		20.5	FAINT BLUE STAR
LB 06042	08 38 36.	+ 20 13		18.1	FAINT BLUE STAR
LB 01862	08 38 36.	+ 20 24		15.5	FAINT BLUE STAR
LB 06041	08 38 36.	+ 20 49		18.7	FAINT BLUE STAR
ZC 0838.6+3823	08 38 36.	+ 38 23	400		CLUSTER OF GALAXIES
ZC 0838.6+5013	08 38 36.	+ 50 13	810		CLUSTER OF GALAXIES
MCG+09-14-086	08 38 36.	+ 52 57 30.	18	16.	GALAXY
7ZW 244	08 38 36.	+ 77 05			COMPACT GALAXY
OCL 0732	08 38 36.	- 37 53	120	15.	OPEN STAR CLUSTER
VHA 041	08 38 36.	- 37 53	150		OPEN STAR CLUSTER
LB 00396	08 38 38.	+ 19 39 36.		17.1	FAINT BLUE STAR
LB 06053	08 38 42.	+ 18 56		19.9	FAINT BLUE STAR
LB 06052	08 38 42.	+ 18 56		19.7	FAINT BLUE STAR
LB 06051	08 38 42.	+ 20 01		18.4	FAINT BLUE STAR
LB 06050	08 38 42.	+ 20 09		19.1	FAINT BLUE STAR
LB 01865	08 38 42.	+ 20 12		18.9	FAINT BLUE STAR
LB 06049	08 38 42.	+ 20 21		18.9	FAINT BLUE STAR
ZWG 179.017	08 38 42.	+ 33 03		14.5	GALAXY
UGC 04531	08 38 42.	+ 33 03	72	14.5	GALAXY SBb
KARA.73B 0278	08 38 42.	+ 33 03	66	14.5	ISOLATED GALAXY S
ZWG 311.013	08 38 42.	+ 68 09		15.4	GALAXY
MCG-01-22-034	08 38 42.	- 04 32	78	14.5	GALAXY
MCG-04-21-007	08 38 42.	- 23 44	48	15.	GALAXY
ARC 0696	08 38 45.	+ 16 21		17.4	RICH CLUSTER OF GALAXIES
ZC 0838.8+0119	08 38 48.	+ 01 19	2350		CLUSTER OF GALAXIES
LB 06061	08 38 48.	+ 18 52		19.0	FAINT BLUE STAR
ZWG 089.066	08 38 48.	+ 19 02		15.4	GALAXY
UGC 04532	08 38 48.	+ 19 02	96	15.4	GALAXY Sb-c
MCG+03-22-023	08 38 48.	+ 19 03	90	15.5	GALAXY
LB 06060	08 38 48.	+ 19 05		20.2	FAINT BLUE STAR
LB 06059	08 38 48.	+ 19 11		17.4	FAINT BLUE STAR
LB 06058	08 38 48.	+ 19 13		19.5	FAINT BLUE STAR
LB 06057	08 38 48.	+ 19 19		19.8	FAINT BLUE STAR
LB 01866	08 38 48.	+ 19 21		17.8	FAINT BLUE STAR
LB 06056	08 38 48.	+ 19 26		19.6	FAINT BLUE STAR
LB 01867	08 38 48.	+ 19 50		18.7	FAINT BLUE STAR
LB 06055	08 38 48.	+ 20 13		19.2	FAINT BLUE STAR
LB 06054	08 38 48.	+ 20 31		18.5	FAINT BLUE STAR
LB 01868	08 38 48.	+ 20 33		19.7	FAINT BLUE STAR
MCG+06-19-015	08 38 48.	+ 33 02	66	14.5	GALAXY
7ZW 245	08 38 48.	+ 70 13			COMPACT GALAXY
OCL 0767	08 38 48.	- 52 53	3000	2.6	OPEN STAR CLUSTER
VHA 042	08 38 48.	- 52 53	1800		OPEN STAR CLUSTER
IC 2391	08 38 48.	- 52 53	2700		OPEN CLUSTER
RNGC 2643	08 38 52.	+ 19 53			GALAXY
RNGC 2644	08 38 53.	+ 05 09		13.5	GALAXY
ZWG 032.052	08 38 54.	+ 05 09		13.4	GALAXY
UGC 04533	08 38 54.	+ 05 09	114	13.4	GALAXY PECULR
MCG+01-22-016	08 38 54.	+ 05 09	84	13.4	GALAXY
KARA.73B 0279	08 38 54.	+ 05 09	168	13.4	ISOLATED GALAXY S
ZC 0838.9+1620	08 38 54.	+ 16 20	1340		CLUSTER OF GALAXIES
LB 01869	08 38 54.	+ 19 03		17.5	FAINT BLUE STAR
LB 06068	08 38 54.	+ 19 04		18.2	FAINT BLUE STAR
LB 06067	08 38 54.	+ 19 26		18.6	FAINT BLUE STAR
LB 06066	08 38 54.	+ 19 42		19.6	FAINT BLUE STAR
LB 06065	08 38 54.	+ 19 45		20.4	FAINT BLUE STAR
LB 06064	08 38 54.	+ 19 52		20.5	FAINT BLUE STAR
ZWG 089.067	08 38 54.	+ 19 53		15.6	GALAXY
LB 06063	08 38 54.	+ 20 04		20.7	FAINT BLUE STAR
LB 06062	08 38 54.	+ 20 43		17.8	FAINT BLUE STAR
MCG+12-09-008	08 38 54.	+ 70 43	39	17.	GALAXY
CED 106J	08 38 58.	- 46 29	690		DIFFUSE GALACTIC NEBULA
TON-N 0936	08 38 59.	+ 35 27		17.	BLUE STAR
TON-N 0937	08 38 59.	+ 35 28		15.8	BLUE STAR
KEEL 155	08 38 59.8	+ 19 52 57.		17.	NEBULA
LB 03486	08 39	- 80 52		14.5	FAINT BLUE STAR
LB 06083	08 39 00.	+ 18 50		18.9	FAINT BLUE STAR
LB 06082	08 39 00.	+ 18 55		19.7	FAINT BLUE STAR
LB 06081	08 39 00.	+ 19 06		19.8	FAINT BLUE STAR
LB 06080	08 39 00.	+ 19 09		20.0	FAINT BLUE STAR
LB 06079	08 39 00.	+ 19 13		18.2	FAINT BLUE STAR
LB 06078	08 39 00.	+ 19 15		20.7	FAINT BLUE STAR
LB 06077	08 39 00.	+ 19 34		18.1	FAINT BLUE STAR
LB 06076	08 39 00.	+ 19 36		18.6	FAINT BLUE STAR
LB 06075	08 39 00.	+ 19 40		17.8	FAINT BLUE STAR
LB 06074	08 39 00.	+ 19 52		20.8	FAINT BLUE STAR
IC 2390	08 39 00.	+ 19 52 50.			NONSTELLAR OBJECT
LB 06073	08 39 00.	+ 19 54		18.3	FAINT BLUE STAR
LB 06072	08 39 00.	+ 19 55		18.0	FAINT BLUE STAR
LB 06071	08 39 00.	+ 20 15		18.2	FAINT BLUE STAR
LB 06070	08 39 00.	+ 20 18		19.1	FAINT BLUE STAR
LB 06069	08 39 00.	+ 20 46		20.4	FAINT BLUE STAR
ZWG 150.017	08 39 00.	+ 31 00		15.5	GALAXY
ZC 0839.0+3824	08 39 00.	+ 38 24	1950		CLUSTER OF GALAXIES
ZWG 208.037	08 39 00.	+ 42 05		15.0	GALAXY
ZC 0839.0+4800	08 39 00.	+ 48 00	470		CLUSTER OF GALAXIES
MCG+09-15-001	08 39 00.	+ 52 56	36	16.	GALAXY
ZWG 264.001	08 39 00.	+ 54 12		15.7	GALAXY
ZWG 263.075	08 39 00.	+ 54 12		15.7	GALAXY
SHAH 150	08 39 00.	+ 82 57	180		GROUP OF COMPACT GALAXIES
MCG-03-22-011	08 39 00.	- 20 06	84	17.	GALAXY
LB 03486	08 39 00.	- 80 52		14.5	FAINT BLUE STAR
MCG+07-18-038	08 39 03.	+ 42 05	30	15.	GALAXY
LB 06095	08 39 06.	+ 18 44		19.5	FAINT BLUE STAR
LB 06094	08 39 06.	+ 19 02		18.8	FAINT BLUE STAR
LB 06093	08 39 06.	+ 19 15		18.5	FAINT BLUE STAR
LB 06092	08 39 06.	+ 19 19		20.6	FAINT BLUE STAR
LB 06091	08 39 06.	+ 19 21		18.5	FAINT BLUE STAR
LB 06090	08 39 06.	+ 19 24		18.5	FAINT BLUE STAR
LB 06089	08 39 06.	+ 19 57		18.1	FAINT BLUE STAR
LB 06088	08 39 06.	+ 20 00		19.2	FAINT BLUE STAR
LB 06087	08 39 06.	+ 20 07		19.6	FAINT BLUE STAR
LB 06086	08 39 06.	+ 20 27		19.3	FAINT BLUE STAR
LB 06085	08 39 06.	+ 20 32		18.2	FAINT BLUE STAR
LB 06084	08 39 06.	+ 20 44		20.6	FAINT BLUE STAR
ZC 0839.1+4952	08 39 06.	+ 49 52	1140		CLUSTER OF GALAXIES
ZC 0839.1+5316	08 39 06.	+ 53 16	540		CLUSTER OF GALAXIES
MCG+11-11-021	08 39 06.	+ 65 21	78	15.	GALAXY
MCG+11-11-022	08 39 06.	+ 65 58	39	17.	GALAXY
MAI 033	08 39 06.	+ 77 21	40		DWARF SPHEROIDAL GALAXY
TON-N 0938	08 39 11.	+ 32 58		15.7	BLUE STAR
LB 06102	08 39 12.	+ 19 02		18.9	FAINT BLUE STAR
LB 06101	08 39 12.	+ 19 04		18.2	FAINT BLUE STAR
LB 00397	08 39 12.	+ 19 18		17.1	FAINT BLUE STAR
LB 06100	08 39 12.	+ 19 20		19.0	FAINT BLUE STAR
LB 06099	08 39 12.	+ 19 47		17.9	FAINT BLUE STAR
LB 06098	08 39 12.	+ 20 04		19.7	FAINT BLUE STAR
LB 06097	08 39 12.	+ 20 06		19.8	FAINT BLUE STAR
LB 06096	08 39 12.	+ 20 48		18.3	FAINT BLUE STAR
ZC 0839.2+3616	08 39 12.	+ 36 16	1950		CLUSTER OF GALAXIES
ZWG 179.018	08 39 12.	+ 37 23		13.7	GALAXY
RNGC 2638	08 39 12.	+ 37 23		13.5	GALAXY
UGC 04534	08 39 12.	+ 37 23	114	13.7	GALAXY S0-a
MCG+06-19-016	08 39 12.	+ 37 25	90	14.	GALAXY
MCG+09-15-002	08 39 12.	+ 54 13	36	16.	GALAXY
ZWG 311.014	08 39 12.	+ 64 26		15.7	GALAXY
UGC 04535	08 39 12.	+ 64 26	84	15.7	GALAXY Sb
MCG+11-11-023	08 39 12.	+ 65 12	57	16.	GALAXY
ZWG 350.003	08 39 12.	+ 75 21		15.7	GALAXY
ZWG 349.031	08 39 12.	+ 75 21		15.7	GALAXY
ACK 256+03.1	08 39 12.	- 35 52			PLANETARY NEBULA
OCL 0734	08 39 12.	- 38 30	150	13.	OPEN STAR CLUSTER
VHA 043	08 39 12.	- 38 30	180		OPEN STAR CLUSTER
LB 01870	08 39 14.	+ 22 01 36.		17.0	FAINT BLUE STAR
LB 00549	08 39 16.	+ 80 36 24.		16.4	FAINT BLUE STAR
TON-N 0939	08 39 17.	+ 32 31		16.9	BLUE STAR
LB 06106	08 39 18.	+ 18 46		18.7	FAINT BLUE STAR
LB 06105	08 39 18.	+ 19 31		18.9	FAINT BLUE STAR
LB 06104	08 39 18.	+ 20 27		16.2	FAINT BLUE STAR
LB 06103	08 39 18.	+ 20 34		18.1	FAINT BLUE STAR
ZWG 208.038	08 39 18.	+ 41 00		15.5	GALAXY
ZWG 208.039	08 39 18.	+ 43 00		15.2	GALAXY
ZWG 311.015	08 39 18.	+ 65 22		15.5	GALAXY
UGC 04536	08 39 18.	+ 65 22	138	15.5	GALAXY SBc
MCG+11-11-024	08 39 18.	+ 67 08	51	15.	GALAXY
ZC 0839.3-0300	08 39 18.	- 03 00	1010		CLUSTER OF GALAXIES
MCG-03-22-012	08 39 18.	- 20 10	84	15.	GALAXY
PK254+05.1	08 39 19.	- 32 08 30.	11		PLANETARY NEBULA
LB 06118	08 39 24.	+ 18 42		18.9	FAINT BLUE STAR
LB 06117	08 39 24.	+ 18 43		15.6	FAINT BLUE STAR
LB 06116	08 39 24.	+ 18 46		19.2	FAINT BLUE STAR
LB 06115	08 39 24.	+ 18 49		19.4	FAINT BLUE STAR
LB 06114	08 39 24.	+ 18 51		20.5	FAINT BLUE STAR
LB 01871	08 39 24.	+ 18 59		18.4	FAINT BLUE STAR
LB 06113	08 39 24.	+ 19 04		19.3	FAINT BLUE STAR

OBJECT NAME	RIGHT ASCEN.	DECLINATION	DIAM.	MAGN.	TYPE OF OBJECT
LB 00398	08 39 24.	+ 19 16		16.7	FAINT BLUE STAR
LB 06112	08 39 24.	+ 19 25		18.3	FAINT BLUE STAR
LB 06111	08 39 24.	+ 19 30		17.4	FAINT BLUE STAR
LB 06110	08 39 24.	+ 19 44		18.0	FAINT BLUE STAR
LB 06109	08 39 24.	+ 20 07		19.1	FAINT BLUE STAR
LB 06108	08 39 24.	+ 20 19		18.7	FAINT BLUE STAR
LB 06107	08 39 24.	+ 20 22		18.3	FAINT BLUE STAR
MCG+07-18-039	08 39 24.	+ 43 00	42	15.	GALAXY
ZWG 237.012	08 39 24.	+ 45 50		15.2	GALAXY
ZC 0839.4+5704	08 39 24.	+ 57 04	1410		CLUSTER OF GALAXIES
MCG-03-22-013	08 39 24.	- 20 36	72	15.	GALAXY
ZWG 004.106	08 39 30.	+ 01 05		15.4	GALAXY
LB 06125	08 39 30.	+ 19 19		19.5	FAINT BLUE STAR
LB 06124	08 39 30.	+ 19 38		18.4	FAINT BLUE STAR
LB 06123	08 39 30.	+ 19 41		18.0	FAINT BLUE STAR
LB 06122	08 39 30.	+ 19 42		19.3	FAINT BLUE STAR
LB 06121	08 39 30.	+ 19 48		19.4	FAINT BLUE STAR
LB 06120	08 39 30.	+ 20 12		19.4	FAINT BLUE STAR
LB 06119	08 39 30.	+ 20 34		17.8	FAINT BLUE STAR
UGC 04537	08 39 30.	+ 35 58	60	16.5	GALAXY Sc
ZC 0839.5+3631	08 39 30.	+ 36 31	1140		CLUSTER OF GALAXIES
MCG+11-11-025	08 39 30.	+ 64 25	54	16.	GALAXY
UGC 04538	08 39 30.	+ 65 14	90	16.0	GALAXY Sc
ZWG 311.016	08 39 30.	+ 67 08		14.5	GALAXY
UGC 04539	08 39 30.	+ 67 08	60	14.5	GALAXY Sc
ZWG 332.006	08 39 30.	+ 69 01		15.3	GALAXY
ZWG 331.059	08 39 30.	+ 69 01		15.3	GALAXY
7ZW 246	08 39 30.	+ 69 02			COMPACT GALAXY
OCL 0763	08 39 30.	- 48 01	1680	4.7	OPEN STAR CLUSTER
IC 2395	08 39 30.	- 48 01	1200		OPEN CLUSTER
CED 106K	08 39 32.	- 47 08			DIFFUSE GALACTIC NEBULA
LB 01872	08 39 36.	+ 18 58		18.4	FAINT BLUE STAR
LB 06135	08 39 36.	+ 19 10		20.1	FAINT BLUE STAR
LB 06134	08 39 36.	+ 19 10		19.9	FAINT BLUE STAR
LB 06133	08 39 36.	+ 19 22		19.3	FAINT BLUE STAR
LB 06132	08 39 36.	+ 19 26		18.3	FAINT BLUE STAR
LB 06131	08 39 36.	+ 19 28		20.2	FAINT BLUE STAR
LB 06130	08 39 36.	+ 19 34		19.3	FAINT BLUE STAR
LB 06129	08 39 36.	+ 19 39		19.2	FAINT BLUE STAR
LB 06128	08 39 36.	+ 20 15		17.8	FAINT BLUE STAR
LB 06127	08 39 36.	+ 20 41		18.0	FAINT BLUE STAR
LB 06126	08 39 36.	+ 20 43		20.0	FAINT BLUE STAR
SHAH 097	08 39 36.	+ 51 49	180	17.5	GROUP OF COMPACT GALAXIES
ZWG 264.002	08 39 36.	+ 52 56		15.6	GALAXY
ZWG 263.076	08 39 36.	+ 52 56		15.6	GALAXY
MCG+10-13-013	08 39 36.	+ 59 13	39	17.	GALAXY
ZWG 311.017	08 39 36.	+ 65 10		15.4	GALAXY
TON-N 0940	08 39 41.	+ 34 34		15.9	BLUE STAR
TON-N 0941	08 39 41.	+ 34 35		15.6	BLUE STAR
ARC 0697	08 39 41.	+ 36 32		17.9	RICH CLUSTER OF GALAXIES
LB 06142	08 39 42.	+ 18 51		18.7	FAINT BLUE STAR
LB 06141	08 39 42.	+ 19 15		18.8	FAINT BLUE STAR
LB 06140	08 39 42.	+ 19 24		17.7	FAINT BLUE STAR
LB 06139	08 39 42.	+ 19 37		18.7	FAINT BLUE STAR
LB 06138	08 39 42.	+ 20 23		18.9	FAINT BLUE STAR
LB 06137	08 39 42.	+ 20 24		20.6	FAINT BLUE STAR
LB 06136	08 39 42.	+ 20 35		17.9	FAINT BLUE STAR
ZWG 120.023	08 39 42.	+ 22 27		15.6	GALAXY
ZWG 150.018	08 39 42.	+ 27 28		15.4	GALAXY
ZWG 179.019	08 39 42.	+ 37 43		15.2	GALAXY
MCG+11-11-026	08 39 42.	+ 65 08	18	16.	GALAXY
ZWG 311.018	08 39 42.	+ 65 10		15.5	GALAXY
RNGC 2647	08 39 46.	+ 19 50		15.0	GALAXY
TON-N 0942	08 39 47.	+ 32 46		15.4	BLUE STAR
ZC 0839.8+0828	08 39 48.	+ 08 28	540		CLUSTER OF GALAXIES
ZC 0839.8+1325	08 39 48.	+ 13 25	1680		CLUSTER OF GALAXIES
LB 06147	08 39 48.	+ 18 54		18.5	FAINT BLUE STAR
LB 06146	08 39 48.	+ 19 24		19.1	FAINT BLUE STAR
LB 06145	08 39 48.	+ 19 37		19.4	FAINT BLUE STAR
LB 06144	08 39 48.	+ 19 41		20.6	FAINT BLUE STAR
ZWG 089.068	08 39 48.	+ 19 50		15.2	GALAXY
LB 01874	08 39 48.	+ 20 01		20.5	FAINT BLUE STAR
LB 01873	08 39 48.	+ 20 18		19.1	FAINT BLUE STAR
LB 06143	08 39 48.	+ 20 20		18.8	FAINT BLUE STAR
ZWG 150.019	08 39 48.	+ 29 35		15.7	GALAXY
MCG+06-19-017	08 39 48.	+ 37 45	30	15.	GALAXY
MCG+11-11-027	08 39 48.	+ 65 08	27	16.	GALAXY
ZC 0839.8+6837	08 39 48.	+ 68 37	1280		CLUSTER OF GALAXIES
MCG-02-22-026	08 39 48.	- 13 40	36	15.5	GALAXY
OCL 0755	08 39 48.	- 46 06	120	10.5	OPEN STAR CLUSTER
VHA 044	08 39 48.	- 46 06	150		OPEN STAR CLUSTER
ARP 089	08 39 53.	+ 14 28			PECULIAR GALAXY
ZWG 060.034	08 39 54.	+ 10 46		14.6	GALAXY
UGC 04540	08 39 54.	+ 10 46	102	14.6	GALAXY
MCG+02-22-004	08 39 54.	+ 10 46	84	14.6	GALAXY
ZWG 060.035	08 39 54.	+ 14 28		13.0	GALAXY
UGC 04541	08 39 54.	+ 14 28	216	13.0	GALAXY Sa
KARA.72 168A	08 39 54.	+ 14 28	228	13.0	PART OF DOUBLE GALAXY
MCG+02-22-005	08 39 54.	+ 14 28	168	13.0	GALAXY
LB 08586	08 39 54.	+ 16 56		18.0	FAINT BLUE STAR
LB 06153	08 39 54.	+ 18 58		18.3	FAINT BLUE STAR
LB 06152	08 39 54.	+ 19 11		19.3	FAINT BLUE STAR
LB 06151	08 39 54.	+ 19 30		17.5	FAINT BLUE STAR
LB 06150	08 39 54.	+ 19 46		19.8	FAINT BLUE STAR
LB 01875	08 39 54.	+ 19 59		19.4	FAINT BLUE STAR
LB 06149	08 39 54.	+ 20 09		18.6	FAINT BLUE STAR
LB 06148	08 39 54.	+ 20 23		18.7	FAINT BLUE STAR
ZWG 120.024	08 39 54.	+ 25 15		15.5	GALAXY
UGC 04542	08 39 54.	+ 25 15	90	15.5	GALAXY
MCG+04-21-014	08 39 54.	+ 25 15	60	15.5	GALAXY
KARA.73B 0280	08 39 54.	+ 25 15	66	15.5	ISOLATED GALAXY S
ZC 0839.9+2937	08 39 54.	+ 29 37	1080		CLUSTER OF GALAXIES
ZWG 179.020	08 39 54.	+ 37 42		15.7	GALAXY
ZWG 237.013	08 39 54.	+ 45 55		15.1	GALAXY
UGC 04543	08 39 54.	+ 45 55	216	15.1	GALAXY
ZWG 332.007	08 39 54.	+ 69 16		15.4	GALAXY
ZWG 331.060	08 39 54.	+ 69 16		15.4	GALAXY
7ZW 247	08 39 54.	+ 69 17			COMPACT GALAXY
RNGC 2648	08 39 55.	+ 14 28		13.0	GALAXY
LB 09969	08 40	- 88 57		13.5	FAINT BLUE STAR
ZC 0840.0+0915	08 40 00.	+ 09 15	1140		CLUSTER OF GALAXIES
ZWG 060.036	08 40 00.	+ 14 27		15.0	GALAXY
KARA.72 168B	08 40 00.	+ 14 27	84	15.0	PART OF DOUBLE GALAXY
MCG+02-22-006	08 40 00.	+ 14 27	48	15.0	GALAXY
LB 08588	08 40 00.	+ 18 34		17.7	FAINT BLUE STAR
LB 06158	08 40 00.	+ 18 41		17.3	FAINT BLUE STAR
LB 08587	08 40 00.	+ 20 01		15.3	FAINT BLUE STAR
LB 01876	08 40 00.	+ 20 02		17.1	FAINT BLUE STAR

OBJECT NAME	RIGHT ASCEN.	DECLINATION	DIAM.	MAGN.	TYPE OF OBJECT
LB 06157	08 40 00.	+ 20 16		20.0	FAINT BLUE STAR
LB 06156	08 40 00.	+ 20 18		18.5	FAINT BLUE STAR
LB 06155	08 40 00.	+ 20 21		19.0	FAINT BLUE STAR
LB 06154	08 40 00.	+ 20 42		18.0	FAINT BLUE STAR
TON-N 0338	08 40 00.	+ 26 54		15.0	BLUE STAR
TON-N 0339	08 40 00.	+ 27 04		16.5	BLUE STAR
TON-N 0340	08 40 00.	+ 30 44		15.2	BLUE STAR
ZWG 237.014	08 40 00.	+ 50 22		12.4	GALAXY
UGC 04544	08 40 00.	+ 50 22	120	12.4	GALAXY Sa?
MCG+08-16-024	08 40 00.	+ 50 23	66	12.6	GALAXY
ZC 0840.0+6557	08 40 00.	+ 65 57	1010		CLUSTER OF GALAXIES
7ZW 248	08 40 00.	+ 71 07			COMPACT GALAXY
MCG+12-09-009	08 40 00.	+ 73 12	84	16.	GALAXY
OCL 0748	08 40 00.	- 43 12	360	12.	OPEN STAR CLUSTER
RNGC 2630	08 40 01.	+ 73 12		16.0	GALAXY
REIN 2.068	08 40 03.26	+ 50 23 07.2			NEBULA
TON-N 0943	08 40 05.	+ 33 39		14.9	BLUE STAR
LB 06161	08 40 06.	+ 18 52		18.9	FAINT BLUE STAR
LB 01877	08 40 06.	+ 19 06		17.1	FAINT BLUE STAR
LB 06160	08 40 06.	+ 20 07		17.7	FAINT BLUE STAR
LB 06159	08 40 06.	+ 20 32		19.6	FAINT BLUE STAR
ZWG 120.025	08 40 06.	+ 25 43		15.6	GALAXY
RNGC 2639	08 40 06.	+ 50 23		13.0	GALAXY
ZC 0840.1+5128	08 40 06.	+ 51 28	2960		CLUSTER OF GALAXIES
MCG+09-15-003	08 40 06.	+ 52 10	78	15.	GALAXY
ZC 0840.1+7457	08 40 06.	+ 74 57	3090		CLUSTER OF GALAXIES
ZWG 004.107	08 40 06.	- 00 23		15.3	GALAXY
MCG-01-22-035	08 40 06.	- 05 29	42	15.	GALAXY
ARC 0698	08 40 08.	+ 41 44		17.9	RICH CLUSTER OF GALAXIES
ZWG 060.037	08 40 12.	+ 13 50		14.5	GALAXY
UGC 04545	08 40 12.	+ 13 50	78	14.5	GALAXY S
MCG+02-22-007	08 40 12.	+ 13 50	60	14.5	GALAXY
LB 08591	08 40 12.	+ 17 12		16.7	FAINT BLUE STAR
LB 08590	08 40 12.	+ 17 44		15.5	FAINT BLUE STAR
LB 06168	08 40 12.	+ 18 46		19.0	FAINT BLUE STAR
LB 06167	08 40 12.	+ 19 07		16.8	FAINT BLUE STAR
LB 06166	08 40 12.	+ 19 14		19.0	FAINT BLUE STAR
LB 06165	08 40 12.	+ 19 17		18.4	FAINT BLUE STAR
LB 06164	08 40 12.	+ 19 21		18.6	FAINT BLUE STAR
LB 00399	08 40 12.	+ 19 26		17.7	FAINT BLUE STAR
LB 08589	08 40 12.	+ 19 36		16.9	FAINT BLUE STAR
LB 06163	08 40 12.	+ 19 59		18.7	FAINT BLUE STAR
LB 06162	08 40 12.	+ 20 14		18.5	FAINT BLUE STAR
LB 01878	08 40 12.	+ 20 36		19.0	FAINT BLUE STAR
ZWG 150.020	08 40 12.	+ 27 00		15.3	GALAXY
ZWG 179.021	08 40 12.	+ 36 22		15.5	GALAXY
ZWG 264.003	08 40 12.	+ 52 11		14.9	GALAXY
ZWG 263.077	08 40 12.	+ 52 11		14.9	GALAXY
UGC 04546	08 40 12.	+ 52 11	84	14.9	GALAXY Sa
ZC 0840.2+5359	08 40 12.	+ 53 59	400		CLUSTER OF GALAXIES
7ZW 249	08 40 12.	+ 63 53			COMPACT GALAXY
ZWG 332.008	08 40 12.	+ 73 11		15.2	GALAXY
ZWG 331.061	08 40 12.	+ 73 11		15.2	GALAXY
UGC 04547	08 40 12.	+ 73 11	84	15.2	GALAXY Sb-c
CED 106L	08 40 12.	- 45 14	1080		DIFFUSE GALACTIC NEBULA
MCG+03-22-024	08 40 15.	+ 18 25	60	15.1	GALAXY
MCG+09-15-004	08 40 15.	+ 50 54	36	16.	GALAXY
SN 1950D	08 40 17.	+ 18 21		16.6	SUPERNOVA
LB 01878	08 40 17.	+ 20 35 54.		19.4	FAINT BLUE STAR
LB 08592	08 40 18.	+ 17 23		16.8	FAINT BLUE STAR
ZWG 089.069	08 40 18.	+ 18 24		15.1	GALAXY
UGC 04548	08 40 18.	+ 18 24	72	15.1	GALAXY S
LB 06176	08 40 18.	+ 18 48		18.7	FAINT BLUE STAR
LB 06175	08 40 18.	+ 18 53		19.6	FAINT BLUE STAR
LB 06174	08 40 18.	+ 18 53		18.4	FAINT BLUE STAR
LB 06173	08 40 18.	+ 19 01		19.5	FAINT BLUE STAR
LB 06172	08 40 18.	+ 19 25		19.0	FAINT BLUE STAR
LB 01879	08 40 18.	+ 19 49		16.6	FAINT BLUE STAR
LB 06171	08 40 18.	+ 20 07		18.1	FAINT BLUE STAR
LB 06170	08 40 18.	+ 20 21		19.4	FAINT BLUE STAR
LB 06169	08 40 18.	+ 20 27		18.6	FAINT BLUE STAR
ZC 0840.3+3323	08 40 18.	+ 33 23	1340		CLUSTER OF GALAXIES
ZC 0840.3+3717	08 40 18.	+ 37 17	2490		CLUSTER OF GALAXIES
ZC 0840.3+4142	08 40 18.	+ 41 42	540		CLUSTER OF GALAXIES
MCG+09-15-005	08 40 18.	+ 55 26	36	16.	GALAXY
MCG+07-18-040	08 40 21.	+ 41 28 30.	27	16.	GALAXY
MCG-03-22-014	08 40 21.	- 19 41	84	16.	GALAXY
TON-N 0944	08 40 23.	+ 34 50		15.3	BLUE STAR
LB 08593	08 40 24.	+ 15 08		15.8	FAINT BLUE STAR
LB 06182	08 40 24.	+ 18 42		19.2	FAINT BLUE STAR
LB 06181	08 40 24.	+ 18 56		18.5	FAINT BLUE STAR
LB 06180	08 40 24.	+ 19 05		19.1	FAINT BLUE STAR
LB 06179	08 40 24.	+ 19 11		19.2	FAINT BLUE STAR
LB 06178	08 40 24.	+ 19 46		18.8	FAINT BLUE STAR
LB 06177	08 40 24.	+ 20 13		18.5	FAINT BLUE STAR
ZC 0840.4+3914	08 40 24.	+ 39 14	3020		CLUSTER OF GALAXIES
MCG+09-15-006	08 40 24.	+ 56 27	30	16.	GALAXY
ZWG 288.005	08 40 24.	+ 59 00		14.2	GALAXY
UGC 04549	08 40 24.	+ 59 00	102	14.2	GALAXY Sc?
ZWG 060.038	08 40 30.	+ 13 16		14.8	GALAXY
UGC 04550	08 40 30.	+ 13 16	168	14.8	GALAXY Sb
MCG+02-22-008	08 40 30.	+ 13 16	156	14.8	GALAXY
LB 08596	08 40 30.	+ 15 18		15.4	FAINT BLUE STAR
LB 06186	08 40 30.	+ 18 45		20.2	FAINT BLUE STAR
LB 06185	08 40 30.	+ 19 13		20.3	FAINT BLUE STAR
LB 08595	08 40 30.	+ 20 01		16.4	FAINT BLUE STAR
LB 06184	08 40 30.	+ 20 02		18.8	FAINT BLUE STAR
LB 08594	08 40 30.	+ 20 12		15.3	FAINT BLUE STAR
LB 01880	08 40 30.	+ 20 14		17.4	FAINT BLUE STAR
LB 06183	08 40 30.	+ 20 43		18.5	FAINT BLUE STAR
ZC 0840.5+2917	08 40 30.	+ 29 17	870		CLUSTER OF GALAXIES
ZWG 237.015	08 40 30.	+ 49 58		13.1	GALAXY
UGC 04551	08 40 30.	+ 49 58	108	13.1	GALAXY S0?
MCG+10-13-014	08 40 30.	+ 59 00	90	13.	GALAXY
MCG+08-16-025	08 40 33.	+ 49 58	120	14.	GALAXY
MCG-02-22-027	08 40 33.	- 11 17	72	15.	GALAXY
RNGC 2660	08 40 35.	- 46 58		11.0	OPEN CLUSTER
ZWG 060.039	08 40 36.	+ 10 55		15.0	GALAXY
UGC 04552	08 40 36.	+ 10 55	66	15.0	GALAXY Sb
MCG+02-22-009	08 40 36.	+ 10 55	60	15.0	GALAXY
LB 08597	08 40 36.	+ 17 22		16.3	FAINT BLUE STAR
LB 06195	08 40 36.	+ 18 41		20.1	FAINT BLUE STAR
LB 06194	08 40 36.	+ 18 46		20.0	FAINT BLUE STAR
LB 06193	08 40 36.	+ 19 09		18.0	FAINT BLUE STAR
LB 06192	08 40 36.	+ 19 13		19.7	FAINT BLUE STAR
LB 06191	08 40 36.	+ 19 22		18.5	FAINT BLUE STAR
LB 06189	08 40 36.	+ 19 57		19.0	FAINT BLUE STAR
LB 06188	08 40 36.	+ 20 15		17.1	FAINT BLUE STAR

OBJECT NAME	RIGHT ASCEN.	DECLINATION	DIAM.	MAGN.	TYPE OF OBJECT
LB 01881	08 40 36.	+ 22 10 00.		17.0	FAINT BLUE STAR
ZC 0840.6+3105	08 40 36.	+ 31 05	1280		CLUSTER OF GALAXIES
MCG+09-15-007	08 40 36.	+ 53 37 30.	12	17.	GALAXY
ZWG 264.004	08 40 36.	+ 55 11		15.4	GALAXY
ZWG 263.078	08 40 36.	+ 55 11		15.4	GALAXY
MCG+09-15-008	08 40 36.	+ 55 12	30	16.	GALAXY
ZWG 264.005	08 40 36.	+ 55 48		15.6	GALAXY
ZWG 263.079	08 40 36.	+ 55 48		15.6	GALAXY
MCG-03-22-015	08 40 36.	- 19 53 30.	144	14.	GALAXY
OCL 0759	08 40 36.	- 46 58	240	10.8	OPEN STAR CLUSTER
VHA 045	08 40 36.	- 46 58	180		OPEN STAR CLUSTER
ZWG 032.053	08 40 42.	+ 08 16		15.3	GALAXY
LB 08599	08 40 42.	+ 18 38		16.6	FAINT BLUE STAR
LB 06196	08 40 42.	+ 18 41		19.9	FAINT BLUE STAR
LB 08598	08 40 42.	+ 19 02		17.3	FAINT BLUE STAR
LB 06190	08 40 42.	+ 19 46		16.4	FAINT BLUE STAR
LB 01882	08 40 42.	+ 19 59		18.4	FAINT BLUE STAR
LB 06187	08 40 42.	+ 20 22		16.7	FAINT BLUE STAR
MCG+09-15-009	08 40 42.	+ 53 38	15	17.	GALAXY
MCG+09-15-010	08 40 45.	+ 53 41	12	16.	GALAXY
RNGC 2652	08 40 45.	- 03 27			NON-EXISTENT OBJECT
TON-N 0945	08 40 47.	+ 34 44		16.8	BLUE STAR
ZWG 032.054	08 40 48.	+ 03 48		14.8	GALAXY
UGC 04553	08 40 48.	+ 03 48	78	14.8	GALAXY S
MCG+01-22-017	08 40 48.	+ 03 48	72	14.8	GALAXY
ZWG 032.055	08 40 48.	+ 04 37		15.6	GALAXY
LB 08601	08 40 48.	+ 16 03		18.2	FAINT BLUE STAR
LB 08600	08 40 48.	+ 18 32		16.5	FAINT BLUE STAR
LB 06204	08 40 48.	+ 18 43		18.1	FAINT BLUE STAR
LB 06203	08 40 48.	+ 18 52		16.8	FAINT BLUE STAR
LB 06202	08 40 48.	+ 19 01		19.5	FAINT BLUE STAR
LB 01883	08 40 48.	+ 19 17		19.5	FAINT BLUE STAR
LB 06201	08 40 48.	+ 19 19		18.6	FAINT BLUE STAR
LB 06200	08 40 48.	+ 19 34		19.3	FAINT BLUE STAR
LB 06199	08 40 48.	+ 19 55		16.8	FAINT BLUE STAR
LB 06198	08 40 48.	+ 20 08		16.9	FAINT BLUE STAR
LB 06197	08 40 48.	+ 20 24		17.6	FAINT BLUE STAR
ZWG 120.026	08 40 48.	+ 22 17		15.4	GALAXY
UGC 04554	08 40 48.	+ 22 17	84	15.4	GALAXY Sc
MCG+04-21-015	08 40 48.	+ 22 17	72	15.4	GALAXY
TON-N 0341	08 40 48.	+ 29 08		15.4	BLUE STAR
ZWG 264.006	08 40 48.	+ 55 11		14.6	GALAXY
ZWG 263.080	08 40 48.	+ 55 11		14.6	GALAXY
LB 08605	08 40 54.	+ 14 51		15.5	FAINT BLUE STAR
LB 08604	08 40 54.	+ 16 39		18.2	FAINT BLUE STAR
LB 08603	08 40 54.	+ 16 58		18.1	FAINT BLUE STAR
LB 06215	09 40 54.	+ 18 40		20.6	FAINT BLUE STAR
LB 06214	08 40 54.	+ 18 47		20.5	FAINT BLUE STAR
LB 06213	08 40 54.	+ 18 51		18.6	FAINT BLUE STAR
LB 06212	08 40 54.	+ 18 55		19.0	FAINT BLUE STAR
LB 06211	08 40 54.	+ 19 13		18.2	FAINT BLUE STAR
LB 08602	08 40 54.	+ 19 14		16.3	FAINT BLUE STAR
LB 06210	08 40 54.	+ 19 18		18.5	FAINT BLUE STAR
LB 06209	08 40 54.	+ 19 43		18.6	FAINT BLUE STAR
LB 01884	08 40 54.	+ 19 48		18.5	FAINT BLUE STAR
LB 06208	08 40 54.	+ 19 54		19.0	FAINT BLUE STAR
LB 06207	08 40 54.	+ 19 58		19.6	FAINT BLUE STAR
LB 06206	08 40 54.	+ 20 18		17.8	FAINT BLUE STAR
LB 06205	08 40 54.	+ 20 21		18.2	FAINT BLUE STAR
ZWG 179.022	08 40 54.	+ 34 54		13.1	GALAXY
UGC 04555	08 40 54.	+ 34 54	102	13.1	GALAXY SBb/Sc
KARA.73B 0281	08 40 54.	+ 34 54	78	13.1	ISOLATED GALAXY S
MCG+07-18-041	08 40 54.	+ 41 52	45	15.	GALAXY
ZWG 208.040	08 40 54.	+ 41 55		14.1	GALAXY
UGC 04556	08 40 54.	+ 41 55	48	14.1	GALAXY PECULR
ZWG 264.007	08 40 54.	+ 54 07		15.2	GALAXY
ZWG 263.081	08 40 54.	+ 54 07		15.2	GALAXY
OCL 0752	08 40 54.	- 44 46	840	9.8	OPEN STAR CLUSTER
VHA 046	08 40 54.	- 44 46	600		OPEN STAR CLUSTER
RNGC 2659	08 40 56.	- 44 46		9.5	OPEN CLUSTER
LB 06223	08 41 00.	+ 18 48		19.8	FAINT BLUE STAR
LB 01885	08 41 00.	+ 19 08		18.5	FAINT BLUE STAR
LB 06222	08 41 00.	+ 19 23		17.7	FAINT BLUE STAR
LB 06221	08 41 00.	+ 19 44		19.4	FAINT BLUE STAR
LB 06220	08 41 00.	+ 19 46		18.8	FAINT BLUE STAR
LB 06219	08 41 00.	+ 19 54		18.4	FAINT BLUE STAR
LB 08606	08 41 00.	+ 20 06		16.1	FAINT BLUE STAR
LB 06217	08 41 00.	+ 20 37		19.8	FAINT BLUE STAR
LB 06216	08 41 00.	+ 20 39		18.8	FAINT BLUE STAR
TON-N 0010	08 41 00.	+ 26 14		14.5	BLUE STAR
TON-N 0342	08 41 00.	+ 31 14		14.0	BLUE STAR
ZWG 179.023	08 41 00.	+ 33 42		15.2	GALAXY
UGC 04558	08 41 00.	+ 33 42	108	15.2	GALAXY Sb-c
MCG+06-19-018	08 41 00.	+ 34 54 30.	90	13.	GALAXY
MCG+09-15-012	08 41 00.	+ 54 09	36	16.	GALAXY
MCG+09-15-011	08 41 00.	+ 55 12 30.	60	15.	GALAXY
MCG+12-09-061	08 41 00.	+ 72 14	33	16.	GALAXY
UGC 04557	08 41 00.	+ 84 29	66	16.5	GALAXY IRR
MCG-03-23-001	08 41 00.	- 20 29	150	14.5	GALAXY
VHA 047	08 41 00.	- 47 56	780		OPEN STAR CLUSTER
RNGC 2649	08 41 01.	+ 34 54		13.0	GALAXY
MAI 034	08 41 01.	+ 78 51	40		DWARF SPHEROIDAL GALAXY
LB 08608	08 41 06.	+ 17 46		18.3	FAINT BLUE STAR
LB 08607	08 41 06.	+ 18 35		17.8	FAINT BLUE STAR
LB 06227	08 41 06.	+ 18 43		19.1	FAINT BLUE STAR
LB 06226	08 41 06.	+ 18 51		19.0	FAINT BLUE STAR
LB 06225	08 41 06.	+ 18 59		19.7	FAINT BLUE STAR
LB 06224	08 41 06.	+ 19 11		20.4	FAINT BLUE STAR
LB 01887	08 41 06.	+ 19 22		18.1	FAINT BLUE STAR
ZWG 090.001	08 41 06.	+ 19 37		15.7	GALAXY
ZWG 089.070	08 41 06.	+ 19 37		15.7	GALAXY
LB 06218	08 41 06.	+ 20 18		18.3	FAINT BLUE STAR
ZWG 150.021	08 41 06.	+ 30 18		14.7	GALAXY
UGC 04559	08 41 06.	+ 30 18	180	14.7	GALAXY Sa-b
MCG+06-19-019	08 41 06.	+ 33 42	90	14.5	GALAXY
ZC 0841.1+3758	08 41 06.	+ 37 58	1140		CLUSTER OF GALAXIES
MCG+09-15-013	08 41 06.	+ 55 15	15	17.	GALAXY
MCG+12-09-062	08 41 06.	+ 68 49	66	14.	GALAXY
ZWG 033.001	08 41 12.	+ 02 54		15.0	GALAXY
MCG+01-23-001	08 41 12.	+ 02 54	48	15.0	GALAXY
ZWG 061.001	08 41 12.	+ 11 57		15.5	GALAXY
KARA.73B 0282	08 41 12.	+ 11 57	36	15.5	ISOLATED GALAXY S
LB 06233	08 41 12.	+ 18 51		19.4	FAINT BLUE STAR
LB 06232	08 41 12.	+ 19 11		17.9	FAINT BLUE STAR
LB 01888	08 41 12.	+ 19 24		19.8	FAINT BLUE STAR
LB 06231	08 41 12.	+ 19 46		19.0	FAINT BLUE STAR
LB 01886	08 41 12.	+ 19 54		16.6	FAINT BLUE STAR
LB 06230	08 41 12.	+ 20 03		18.9	FAINT BLUE STAR

OBJECT NAME	RIGHT ASCEN.	DECLINATION	DIAM.	MAGN.	TYPE OF OBJECT
LB 06229	08 41 12.	+ 20 14		18.2	FAINT BLUE STAR
LB 06228	08 41 12.	+ 20 15		18.8	FAINT BLUE STAR
LB 08609	08 41 12.	+ 20 32		15.8	FAINT BLUE STAR
MCG+05-21-004	08 41 12.	+ 30 18	210	14.7	GALAXY
ZWG 264.008	08 41 12.	+ 52 40		15.7	GALAXY
UGC 04560	08 41 12.	+ 52 40	72	15.7	GALAXY Sa-b
MCG+09-15-014	08 41 12.	+ 52 40	72	16.	GALAXY
ZWG 005.001	08 41 12.	- 00 21		15.6	GALAXY
MCG-01-23-001	08 41 12.	- 06 32	30	15.5	GALAXY
RNGC 2651	08 41 14.	+ 11 57		15.5	GALAXY
TON-N 0946	08 41 17.	+ 33 41		15.7	BLUE STAR
ZWG 005.002	08 41 18.	+ 01 01		15.0	GALAXY
UGC 04561	08 41 18.	+ 01 01	66	15.0	GALAXY S
MCG+00-23-001	08 41 18.	+ 01 01	66	15.0	GALAXY
LB 08610	08 41 18.	+ 16 56		18.6	FAINT BLUE STAR
LB 06239	08 41 18.	+ 18 42		18.1	FAINT BLUE STAR
LB 06238	08 41 18.	+ 18 57		19.2	FAINT BLUE STAR
LB 06237	08 41 18.	+ 20 15		18.3	FAINT BLUE STAR
LB 06236	08 41 18.	+ 20 19		19.0	FAINT BLUE STAR
LB 06235	08 41 18.	+ 20 21		18.7	FAINT BLUE STAR
LB 06234	08 41 18.	+ 20 45		18.5	FAINT BLUE STAR
ZWG 264.009	08 41 18.	+ 55 09		15.3	GALAXY
ZWG 263.082	08 41 18.	+ 55 09		15.3	GALAXY
MCG+09-15-015	08 41 18.	+ 55 09 30.	15	16.	GALAXY
ZC 0841.3+5855	08 41 18.	+ 58 55	2020		CLUSTER OF GALAXIES
ZC 0841.3+6106	08 41 18.	+ 61 06	4030		CLUSTER OF GALAXIES
MCG+12-09-063	08 41 18.	+ 68 50	45	16.	GALAXY
FATH 1.244	08 41 20.	+ 44 33	22		NEBULA
RNGC 2658	08 41 23.	- 32 29		9.0	OPEN CLUSTER
LB 08612	08 41 24.	+ 16 21		18.2	FAINT BLUE STAR
LB 08611	08 41 24.	+ 18 32		18.3	FAINT BLUE STAR
LB 06243	08 41 24.	+ 19 07		17.2	FAINT BLUE STAR
LB 06242	08 41 24.	+ 19 23		18.2	FAINT BLUE STAR
LB 06241	08 41 24.	+ 19 40		19.0	FAINT BLUE STAR
LB 01889	08 41 24.	+ 19 59		17.1	FAINT BLUE STAR
LB 06240	08 41 24.	+ 20 14		18.5	FAINT BLUE STAR
TON-N 0343	08 41 24.	+ 31 43		14.8	BLUE STAR
UGC 04562	08 41 24.	+ 41 55	78	15.4	GALAXY SBc
ZWG 208.041	08 41 24.	+ 43 14		15.6	GALAXY
ZWG 237.016	08 41 24.	+ 47 55		15.4	GALAXY
OCL 0723	08 41 24.	- 32 28	840	9.2	OPEN STAR CLUSTER
VHA 048	08 41 24.	- 32 28	600		OPEN STAR CLUSTER
LB 01890	08 41 26.	+ 21 10 36.		16.6	FAINT BLUE STAR
ABC 0686	08 41 28.	+ 77 54		17.7	RICH CLUSTER OF GALAXIES
LB 08613	08 41 30.	+ 16 51		17.9	FAINT BLUE STAR
LB 06249	08 41 30.	+ 19 10		18.9	FAINT BLUE STAR
LB 06248	08 41 30.	+ 19 34		19.1	FAINT BLUE STAR
LB 06247	08 41 30.	+ 19 46		17.4	FAINT BLUE STAR
LB 06246	08 41 30.	+ 20 02		19.0	FAINT BLUE STAR
LB 06245	08 41 30.	+ 20 16		19.7	FAINT BLUE STAR
LB 06244	08 41 30.	+ 20 27		18.6	FAINT BLUE STAR
ZWG 179.024	08 41 30.	+ 33 05		15.6	GALAXY
ZWG 208.042	08 41 30.	+ 43 10		15.4	GALAXY
ZC 0841.5+5619	08 41 30.	+ 56 19	1610		CLUSTER OF GALAXIES
7ZW 250	08 41 30.	+ 60 22			COMPACT GALAXY
SN 1954K	08 41 30.	+ 61 19		15.0	SUPERNOVA
ZC 0841.5+6638	08 41 30.	+ 66 38	1010		CLUSTER OF GALAXIES
UGC 04563	08 41 30.	+ 78 45	66	16.5	GALAXY DWARF
ZWG 350.004	08 41 30.	+ 78 50		15.7	GALAXY
ZWG 349.032	08 41 30.	+ 78 50		15.7	GALAXY
MCG+07-18-042	08 41 33.	+ 43 14	45	16.	GALAXY
MCG+09-15-016	08 41 33.	+ 55 18 30.	18	17.	GALAXY
ZWG 005.003	08 41 36.	+ 00 05		15.7	GALAXY
LB 06255	08 41 36.	+ 19 02		17.9	FAINT BLUE STAR
LB 06254	08 41 36.	+ 19 04		20.2	FAINT BLUE STAR
LB 06253	08 41 36.	+ 19 20		19.9	FAINT BLUE STAR
LB 06252	08 41 36.	+ 19 22		19.1	FAINT BLUE STAR
LB 01891	08 41 36.	+ 19 26		18.8	FAINT BLUE STAR
LB 06251	08 41 36.	+ 19 51		19.1	FAINT BLUE STAR
LB 06250	08 41 36.	+ 20 03		19.2	FAINT BLUE STAR
ZWG 264.010	08 41 36.	+ 55 17		15.0	GALAXY
ZWG 263.083	08 41 36.	+ 55 17		15.0	GALAXY
UGC 04564	08 41 36.	+ 55 17	60	15.0	GALAXY S
MCG+09-15-017	08 41 36.	+ 55 19	60	15.	GALAXY
MCG-02-23-001	08 41 36.	- 12 40	90	14.5	GALAXY
CED 106M	08 41 37.	- 48 00			DIFFUSE GALACTIC NEBULA
ABC 0699	08 41 38.	+ 27 59		16.5	RICH CLUSTER OF GALAXIES
MCG+07-18-043	08 41 39.	+ 43 10	48	16.	GALAXY
IC 2392	08 41 40.3	+ 18 28 03.			GALAXY SA
UGC 04565	08 41 42.	+ 09 43	60	16.0	GALAXY Sc
ZWG 061.002	08 41 42.	+ 11 22		15.5	GALAXY
LB 08615	08 41 42.	+ 17 49		17.7	FAINT BLUE STAR
LB 08614	08 41 42.	+ 17 58		18.1	FAINT BLUE STAR
ZWG 090.002	08 41 42.	+ 18 28		14.6	GALAXY
MCG+03-23-001	08 41 42.	+ 18 28	48	14.6	GALAXY
LB 06259	08 41 42.	+ 18 40		18.4	FAINT BLUE STAR
LB 06258	08 41 42.	+ 18 42		20.5	FAINT BLUE STAR
LB 06257	08 41 42.	+ 19 45		20.4	FAINT BLUE STAR
LB 06256	08 41 42.	+ 20 40		16.7	FAINT BLUE STAR
ZWG 208.043	08 41 42.	+ 40 45		15.2	GALAXY
ZC 0841.7+4131	08 41 42.	+ 41 31	1010		CLUSTER OF GALAXIES
MCG+08-16-026	08 41 42.	+ 44 45	9	16.	GALAXY
MCG+08-16-027	08 41 42.	+ 44 46	30	16.	GALAXY
ZC 0841.7+4838	08 41 42.	+ 48 38	540		CLUSTER OF GALAXIES
ZWG 264.011	08 41 42.	+ 53 20		15.6	GALAXY
MCG+09-15-018	08 41 42.	+ 53 20	30	17.	GALAXY
FATH 1.245	08 41 44.	+ 44 47	14		NEBULA
ZWG 061.003	08 41 48.	+ 09 21		15.6	GALAXY
LB 06263	08 41 48.	+ 19 07		19.6	FAINT BLUE STAR
LB 06262	08 41 48.	+ 19 09		18.4	FAINT BLUE STAR
LB 06261	08 41 48.	+ 19 50		17.7	FAINT BLUE STAR
LB 06260	08 41 48.	+ 19 52		18.9	FAINT BLUE STAR
LB 01892	08 41 48.	+ 20 23		19.0	FAINT BLUE STAR
MCG+09-15-019	08 41 48.	+ 53 20	30	16.	GALAXY
MCG+12-09-010	08 41 48.	+ 73 11	45	14.	GALAXY
UGC 04566	08 41 48.	+ 78 44	66	16.5	GALAXY DWARF
RNGC 2631	08 41 50.	+ 73 11			NON-EXISTENT OBJECT
MCG+07-18-044	08 41 51.	+ 40 44	36	15.	GALAXY
ZWG 005.004	08 41 54.	+ 02 15		15.7	GALAXY
LB 08617	08 41 54.	+ 17 55		17.3	FAINT BLUE STAR
LB 08616	08 41 54.	+ 18 40		17.8	FAINT BLUE STAR
LB 06269	08 41 54.	+ 18 41		19.0	FAINT BLUE STAR
LB 06268	08 41 54.	+ 18 46		18.5	FAINT BLUE STAR
LB 06267	08 41 54.	+ 19 37		19.7	FAINT BLUE STAR
LB 06265	08 41 54.	+ 20 20		18.4	FAINT BLUE STAR
LB 06264	08 41 54.	+ 20 24		20.4	FAINT BLUE STAR
ZWG 150.022	08 41 54.	+ 30 44		15.7	GALAXY
ZC 0841.9+4149	08 41 54.	+ 41 49	1140		CLUSTER OF GALAXIES

OBJECT NAME	RIGHT ASCEN.	DECLINATION	DIAM.	MAGN.	TYPE OF OBJECT
MCG+08-16-028	08 41 54.	+ 44 41	18		GALAXY
ZC 0841.9+5341	08 41 54.	+ 53 41	1480		CLUSTER OF GALAXIES
MCG+11-11-028	08 41 54.	+ 64 17 30.	39	16.	GALAXY
PK265-02.1	08 41 55.6	- 45 55 12.	5		PLANETARY NEBULA
RNGC 2629	08 41 56.	+ 73 10		13.0	GALAXY
FATH 1.246	08 41 57.	+ 44 42	5		NEBULA
FATH 1.247	08 41 57.	+ 44 43	19		NEBULA
FATH 1.248	08 41 58.	+ 44 42	19		NEBULA
LBN 1074	08 42	- 24 00	2400		BRIGHT NEBULA
ZWG 061.004	08 42 00.	+ 09 59		15.3	GALAXY
UGC 04567	08 42 00.	+ 09 59	60	15.3	GALAXY S
ZWG 061.005	08 42 00.	+ 10 40		14.9	GALAXY
UGC 04568	08 42 00.	+ 10 40	102	14.9	GALAXY IRR
MCG+02-23-001	08 42 00.	+ 10 40	84	14.9	GALAXY
KARA.73B 0283	08 42 00.	+ 10 40	102	14.9	ISOLATED GALAXY S
LB 08622	08 42 00.	+ 16 25		18.1	FAINT BLUE STAR
LB 08621	08 42 00.	+ 17 29		17.9	FAINT BLUE STAR
LB 08620	08 42 00.	+ 18 42		16.9	FAINT BLUE STAR
LB 06275	08 42 00.	+ 18 48		18.8	FAINT BLUE STAR
LB 08619	08 42 00.	+ 19 07		18.0	FAINT BLUE STAR
LB 06274	08 42 00.	+ 19 09		18.5	FAINT BLUE STAR
LB 06273	08 42 00.	+ 19 23		19.8	FAINT BLUE STAR
LB 06272	08 42 00.	+ 19 30		19.9	FAINT BLUE STAR
LB 06271	08 42 00.	+ 19 54		17.3	FAINT BLUE STAR
LB 08618	08 42 00.	+ 19 56		18.1	FAINT BLUE STAR
LB 06266	08 42 00.	+ 20 01		17.6	FAINT BLUE STAR
LB 06270	08 42 00.	+ 20 04		19.2	FAINT BLUE STAR
LB 01893	08 42 00.	+ 20 34		20.3	FAINT BLUE STAR
ZC 0842.0+4332	08 42 00.	+ 43 32	870		CLUSTER OF GALAXIES
ZWG 237.017	08 42 00.	+ 46 08		15.6	GALAXY
ZWG 264.012	08 42 00.	+ 54 10		15.1	GALAXY
MCG+10-13-015	08 42 00.	+ 60 40	27	16.	GALAXY
ZWG 332.009	08 42 00.	+ 73 10		12.8	GALAXY
ZWG 331.062	08 42 00.	+ 73 10		12.8	GALAXY
UGC 04569	08 42 00.	+ 73 10	138	12.8	GALAXY E
7ZW 251	08 42 00.	+ 78 43			COMPACT GALAXY
LB 06279	08 42 06.	+ 19 23		18.5	FAINT BLUE STAR
LB 06278	08 42 06.	+ 19 38		19.3	FAINT BLUE STAR
LB 06277	08 42 06.	+ 19 51		18.9	FAINT BLUE STAR
LB 06276	08 42 06.	+ 20 09		20.4	FAINT BLUE STAR
ZWG 120.027	08 42 06.	+ 22 47		15.5	GALAXY
MCG+09-15-020	08 42 06.	+ 54 11	18	16.	GALAXY
LB 08624	08 42 12.	+ 14 28		15.2	FAINT BLUE STAR
LB 08623	08 42 12.	+ 17 03		18.1	FAINT BLUE STAR
LB 06285	08 42 12.	+ 19 08		20.2	FAINT BLUE STAR
LB 06284	08 42 12.	+ 19 18		18.4	FAINT BLUE STAR
LB 06283	08 42 12.	+ 19 26		18.2	FAINT BLUE STAR
LB 06282	08 42 12.	+ 19 31		18.6	FAINT BLUE STAR
LB 06281	08 42 12.	+ 20 04		19.4	FAINT BLUE STAR
LB 06280	08 42 12.	+ 20 26		18.0	FAINT BLUE STAR
LB 01895	08 42 12.	+ 20 36		16.5	FAINT BLUE STAR
ZWG 120.028	08 42 12.	+ 26 09		15.7	GALAXY
UGC 04570	08 42 12.	+ 28 00	66	16.5	GALAXY
ZWG 179.025	08 42 12.	+ 34 37		14.7	GALAXY
ZC 0842.2+6651	08 42 12.	+ 66 51	470		CLUSTER OF GALAXIES
MCG+04-21-016	08 42 15.	+ 26 08 30.	36	15.7	GALAXY
ZWG 005.005	08 42 18.	+ 01 20		15.7	GALAXY
UGC 04571	08 42 18.	+ 01 20	60	15.7	GALAXY Sc
ZC 0842.3+0943	08 42 18.	+ 09 43	270		CLUSTER OF GALAXIES
LB 06294	08 42 18.	+ 18 56		19.4	FAINT BLUE STAR
LB 06293	08 42 18.	+ 19 02		19.1	FAINT BLUE STAR
LB 06292	08 42 18.	+ 19 12		19.6	FAINT BLUE STAR
LB 06291	08 42 18.	+ 19 16		18.7	FAINT BLUE STAR
LB 06290	08 42 18.	+ 19 30		18.5	FAINT BLUE STAR
LB 06289	08 42 18.	+ 19 42		19.4	FAINT BLUE STAR
LB 06288	08 42 18.	+ 19 55		18.5	FAINT BLUE STAR
LB 06287	08 42 18.	+ 20 01		20.5	FAINT BLUE STAR
LB 01894	08 42 18.	+ 20 06		17.9	FAINT BLUE STAR
LB 08625	08 42 18.	+ 20 19		18.4	FAINT BLUE STAR
LB 06286	08 42 18.	+ 20 22		18.0	FAINT BLUE STAR
ZC 0842.3+2820	08 42 18.	+ 28 20	5110		CLUSTER OF GALAXIES
ZWG 208.044	08 42 18.	+ 41 28		15.1	GALAXY
ZC 0842.3+4805	08 42 18.	+ 48 05	740		CLUSTER OF GALAXIES
ZWG 264.013	08 42 18.	+ 52 43		14.9	GALAXY
MCG+09-15-021	08 42 18.	+ 52 43	36	16.	GALAXY
MCG+12-09-064	08 42 18.	+ 69 11	144	15.	GALAXY
MCG+12-09-011	08 42 18.	+ 73 44	78	14.	GALAXY
MCG-03-23-002	08 42 21.	- 20 11	84	14.	GALAXY
MCG-04-21-008	08 42 21.	- 25 04	36	16.	GALAXY
TON-N 0947	08 42 23.	+ 34 11		16.9	BLUE STAR
LB 08626	08 42 24.	+ 16 06		18.4	FAINT BLUE STAR
LB 01896	08 42 24.	+ 18 49		19.0	FAINT BLUE STAR
LB 06299	08 42 24.	+ 19 09		18.5	FAINT BLUE STAR
LB 06298	08 42 24.	+ 19 29		18.8	FAINT BLUE STAR
LB 06297	08 42 24.	+ 19 53		18.1	FAINT BLUE STAR
LB 06296	08 42 24.	+ 20 05		17.4	FAINT BLUE STAR
LB 06295	08 42 24.	+ 20 10		20.4	FAINT BLUE STAR
MCG+06-19-020	08 42 24.	+ 34 36 30.	36	14.5	GALAXY
ZWG 179.026	08 42 24.	+ 37 07		13.8	GALAXY
MRK 626	08 42 24.	+ 37 07	22	14.5	GALAXY WITH UV CONTINUUM
UGC 04572	08 42 24.	+ 37 07	42	13.8	GALAXY COMPACT
MCG+06-19-021	08 42 24.	+ 37 08	39	14.5	GALAXY
MCG+07-18-045	08 42 24.	+ 41 27 30.	36	15.	GALAXY
7ZW 252	08 42 24.	+ 60 59			COMPACT GALAXY
MCG+12-09-012	08 42 24.	+ 73 06	66	15.	GALAXY
LB 08630	08 42 30.	+ 16 04		18.1	FAINT BLUE STAR
LB 08629	08 42 30.	+ 17 04		18.6	FAINT BLUE STAR
LB 06302	08 42 30.	+ 18 42		20.6	FAINT BLUE STAR
LB 08628	08 42 30.	+ 18 55		17.1	FAINT BLUE STAR
LB 06301	08 42 30.	+ 19 11		20.5	FAINT BLUE STAR
LB 06300	08 42 30.	+ 19 28		18.7	FAINT BLUE STAR
LB 08627	08 42 30.	+ 20 40		17.7	FAINT BLUE STAR
MCG+12-09-013	08 42 30.	+ 74 18	102	12.8	GALAXY
MCG+13-07-004	08 42 30.	+ 76 26	27	17.	GALAXY
OCL 0729	08 42 30.	- 35 45	600	14.	OPEN STAR CLUSTER
VHA 049	08 42 30.	- 35 45	540		OPEN STAR CLUSTER
IC 2389	08 42 31.	+ 73 42	78	14.4	GALAXY SBO
ZWG 061.006	08 42 36.	+ 09 50		14.0	GALAXY
UGC 04573	08 42 36.	+ 09 50	96	14.0	GALAXY Sc
MCG+02-23-002	08 42 36.	+ 09 50	72	14.0	GALAXY
LB 08634	08 42 36.	+ 14 49		17.5	FAINT BLUE STAR
LB 08633	08 42 36.	+ 18 12		17.5	FAINT BLUE STAR
LB 08632	08 42 36.	+ 18 49		17.0	FAINT BLUE STAR
LB 06304	08 42 36.	+ 19 28		18.7	FAINT BLUE STAR
LB 08631	08 42 36.	+ 19 38		18.1	FAINT BLUE STAR
LB 06303	08 42 36.	+ 20 22		20.3	FAINT BLUE STAR
ZWG 120.029	08 42 36.	+ 23 47		15.5	GALAXY
MCG+04-21-017	08 42 36.	+ 23 47	60	15.5	GALAXY
ZWG 150.023	08 42 36.	+ 26 30		15.5	GALAXY
ZWG 120.030	08 42 36.	+ 26 30		15.5	GALAXY
TON-N 0344	08 42 36.	+ 32 14		15.4	BLUE STAR
FATH 1.249	08 42 36.	+ 44 57	14		NEBULA
ZWG 350.005	08 42 36.	+ 74 17		12.4	GALAXY
ZWG 332.010	08 42 36.	+ 74 17		12.4	GALAXY
ZWG 331.063	08 42 36.	+ 74 17		12.4	GALAXY
UGC 04574	08 42 36.	+ 74 17	168	12.4	GALAXY SBb
KARA.72 169B	08 42 36.	+ 74 17	24		PART OF DOUBLE GALAXY
KARA.72 169A	08 42 36.	+ 74 17	42	12.4	PART OF DOUBLE GALAXY
RNGC 2633	08 42 38.	+ 74 17		12.5	GALAXY
RNGC 2657	08 42 39.	+ 09 50		14.0	GALAXY
LB 08638	08 42 42.	+ 16 58		17.6	FAINT BLUE STAR
LB 08637	08 42 42.	+ 17 06		18.0	FAINT BLUE STAR
LB 08636	08 42 42.	+ 19 11		17.6	FAINT BLUE STAR
LB 00400	08 42 42.	+ 19 53		13.1	FAINT BLUE STAR
LB 06306	08 42 42.	+ 20 03		19.1	FAINT BLUE STAR
LB 08635	08 42 42.	+ 20 04		18.0	FAINT BLUE STAR
LB 06305	08 42 42.	+ 20 05		18.5	FAINT BLUE STAR
LB 01897	08 42 42.	+ 20 24		17.5	FAINT BLUE STAR
ZWG 120.031	08 42 42.	+ 24 03		15.6	GALAXY
UGC 04575	08 42 42.	+ 24 03	72	15.6	GALAXY Sb
ZC 0842.7+6756	08 42 42.	+ 67 56	2080		CLUSTER OF GALAXIES
ZWG 332.011	08 42 42.	+ 73 43		13.2	GALAXY
ZWG 331.064	08 42 42.	+ 73 43		13.2	GALAXY
UGC 04576	08 42 42.	+ 73 43	96	13.2	GALAXY S
MCG+04-21-019	08 42 45.	+ 23 53	48		GALAXY
MCG+04-21-018	08 42 45.	+ 24 03	78	15.6	GALAXY
TON-N 0948	08 42 47.	+ 34 34		14.9	BLUE STAR
ARP 080	08 42 47.	+ 74 18			PECULIAR GALAXY
ZC 0842.8+0336	08 42 48.	+ 03 36	400		CLUSTER OF GALAXIES
ZWG 090.003	08 42 48.	+ 16 16		15.7	GALAXY
LB 08642	08 42 48.	+ 16 33		18.7	FAINT BLUE STAR
ZC 0842.8+1802	08 42 48.	+ 18 02	1280		CLUSTER OF GALAXIES
LB 08641	08 42 48.	+ 18 46		16.3	FAINT BLUE STAR
LB 08640	08 42 48.	+ 19 44		17.0	FAINT BLUE STAR
LB 06308	08 42 48.	+ 20 02		18.6	FAINT BLUE STAR
LB 06307	08 42 48.	+ 20 21		18.2	FAINT BLUE STAR
LB 08639	08 42 48.	+ 20 42		17.8	FAINT BLUE STAR
TON-N 0345	08 42 48.	+ 23 10		15.2	BLUE STAR
ARC 0700	08 42 48.	+ 37 09		18.3	RICH CLUSTER OF GALAXIES
ZWG 237.018	08 42 48.	+ 47 20		14.7	GALAXY
KARA.72 170B	08 42 48.	+ 47 20	24		PART OF DOUBLE GALAXY
KARA.72 170A	08 42 48.	+ 47 20	24	14.7	PART OF DOUBLE GALAXY
MCG+09-15-022	08 42 48.	+ 53 14	36	17.	GALAXY
MCG+12-09-014	08 42 48.	+ 70 56	9	18.	GALAXY
ZWG 332.012	08 42 48.	+ 73 05		15.0	GALAXY
ZWG 331.065	08 42 48.	+ 73 05		15.0	GALAXY
UGC 04577	08 42 48.	+ 73 05	84	15.0	GALAXY (S0)
MCG+12-09-015	08 42 48.	+ 74 10	33	14.	GALAXY
CED 106N	08 42 49.	- 41 06	960		DIFFUSE GALACTIC NEBULA
WRAY 19.12	08 42 50.2	- 41 05 35.		7.5	STAR-NEBULA ASSOCIATION
RNGC 2641	08 42 51.	+ 73 05		15.0	GALAXY
ZWG 033.002	08 42 54.	+ 07 38		15.4	GALAXY
KARA.73B 0284	08 42 54.	+ 07 38	18	15.4	ISOLATED GALAXY E
LB 01898	08 42 54.	+ 20 14		17.7	FAINT BLUE STAR
LB 08643	08 42 54.	+ 20 22		18.1	FAINT BLUE STAR
ZC 0842.9+3708	08 42 54.	+ 37 08	470		CLUSTER OF GALAXIES
MCG+07-18-046	08 42 54.	+ 41 45	60	14.	GALAXY
ZWG 208.045	08 42 54.	+ 41 46		15.1	GALAXY
UGC 04578	08 42 54.	+ 41 46	78	15.1	GALAXY Sb
ZWG 005.006	08 42 54.	- 02 18		15.6	GALAXY
OCL 0742	08 42 54.	- 41 11	1050	7.7	OPEN STAR CLUSTER
RNGC 2634	08 42 57.	+ 74 09		12.5	GALAXY
RNGC 2636	08 42 58.	+ 73 51		14.5	GALAXY
LB 02888	08 42 59.	- 00 11 24.		15.9	FAINT BLUE STAR
LB 08644	08 43 00.	+ 18 11		17.8	FAINT BLUE STAR
UGC 04579	08 43 00.	+ 35 53	60	16.5	GALAXY Sc
ZWG 180.001	08 43 00.	+ 37 52		15.4	GALAXY
ZWG 179.027	08 43 00.	+ 37 52		15.4	GALAXY
MCG+06-19-022	08 43 00.	+ 37 54 30.	30	15.	GALAXY
ZWG 237.019	08 43 00.	+ 48 37		14.5	GALAXY
UGC 04580	08 43 00.	+ 48 37	54	14.5	GALAXY Sb-c
MCG+12-09-016	08 43 00.	+ 74 08	102	15.	GALAXY
ZWG 332.013	08 43 00.	+ 74 09		12.6	GALAXY
ZWG 331.066	08 43 00.	+ 74 09		12.6	GALAXY
UGC 04581	08 43 00.	+ 74 09	120	12.6	GALAXY E
OCL 0721	08 43 00.	- 31 27	300	10.5	OPEN STAR CLUSTER
MRSL 261+00/1	08 43 00.	- 41 09	1620		HII REGION
OCL 0762	08 43 00.	- 47 25	132	14.	OPEN STAR CLUSTER
BC 4C13.39	08 43 01.22	+ 13 39 55.9		17.80	QUASI-STELLAR OBJECT
SHB 121	08 43 01.3	+ 13 39 56.		17.8	QUASI-STELLAR OBJECT
ZWG 061.007	08 43 06.	+ 12 58		15.0	GALAXY
UGC 04582	08 43 06.	+ 12 58	78	15.0	GALAXY S
KARA.72 171A	08 43 06.	+ 12 58	66	15.0	PART OF DOUBLE GALAXY
MCG+02-23-003	08 43 06.	+ 12 58	48	15.0	GALAXY
LB 08646	08 43 06.	+ 16 01		18.5	FAINT BLUE STAR
LB 08645	08 43 06.	+ 17 14		17.7	FAINT BLUE STAR
ZWG 332.014	08 43 06.	+ 73 50		14.4	GALAXY
ZWG 331.067	08 43 06.	+ 73 50		14.4	GALAXY
UGC 04583	08 43 06.	+ 73 50	21	14.4	GALAXY COMPACT
ZC 0843.1-0017	08 43 06.	- 00 17	940		CLUSTER OF GALAXIES
RNGC 2663	08 43 06.	- 33 39			GALAXY
RNGC 2634A	08 43 09.	+ 74 07		14.5	GALAXY
ZWG 061.008	08 43 12.	+ 12 49		13.9	GALAXY
UGC 04584	08 43 12.	+ 12 49	96	13.9	GALAXY Sc
MCG+02-23-004	08 43 12.	+ 12 49	72	13.9	GALAXY
LB 08648	08 43 12.	+ 18 42		18.0	FAINT BLUE STAR
LB 08647	08 43 12.	+ 20 17		18.4	FAINT BLUE STAR
TON-N 0346	08 43 12.	+ 24 38		14.6	BLUE STAR
MCG-06-20-001	08 43 12.	- 33 34	90	12.	GALAXY
RNGC 2661	08 43 13.	+ 12 49		14.0	GALAXY
ZWG 061.009	08 43 18.	+ 12 59		15.0	GALAXY
KARA.72 171B	08 43 18.	+ 12 59	54	15.0	PART OF DOUBLE GALAXY
MCG+02-23-005	08 43 18.	+ 12 59	48	15.0	GALAXY
LB 08650	08 43 18.	+ 19 18		18.2	FAINT BLUE STAR
ZWG 090.004	08 43 18.	+ 19 32		15.4	GALAXY
LB 08649	08 43 18.	+ 20 34		18.0	FAINT BLUE STAR
ZWG 150.024	08 43 18.	+ 27 33		15.1	GALAXY
MCG+05-21-005	08 43 18.	+ 27 33	48	15.1	GALAXY
TON-N 0347	08 43 18.	+ 31 25		15.1	BLUE STAR
ZWG 180.002	08 43 18.	+ 36 37		15.6	GALAXY
ZWG 179.028	08 43 18.	+ 36 37		15.6	GALAXY
MRK 627	08 43 18.	+ 36 37	13	16.	GALAXY WITH UV CONTINUUM
ZWG 332.015	08 43 18.	+ 74 07		14.3	GALAXY
ZWG 331.068	08 43 18.	+ 74 07		14.3	GALAXY
UGC 04585	08 43 18.	+ 74 07	102	14.3	GALAXY S
MCG-02-23-002	08 43 18.	- 14 57	30	15.	GALAXY

OBJECT NAME	RIGHT ASCEN.	DECLINATION	DIAM.	MAGN.	TYPE OF OBJECT
OCL 0722	08 43 18.	- 31 35	360	11.2	OPEN STAR CLUSTER
RNGC 2662	08 43 20.	- 14 57		15.0	GALAXY
RNGC 2669	08 43 22.	- 52 47			OPEN CLUSTER
LB 01899	08 43 23.	+ 20 51 54.		16.6	FAINT BLUE STAR
LB 08652	08 43 24.	+ 19 46		17.9	FAINT BLUE STAR
LB 08651	08 43 24.	+ 20 20		17.8	FAINT BLUE STAR
TON-N 0348	08 43 24.	+ 31 00		15.2	BLUE STAR
OCL 0768	08 43 24.	- 52 47	1380	6.2	OPEN STAR CLUSTER
LB 08655	08 43 30.	+ 19 02		17.7	FAINT BLUE STAR
LB 08654	08 43 30.	+ 20 18		16.6	FAINT BLUE STAR
LB 08653	08 43 30.	+ 20 30		18.1	FAINT BLUE STAR
ZC 0843.5+6203	08 43 30.	+ 62 03	1080		CLUSTER OF GALAXIES
ZC 0843.5+6612	08 43 30.	+ 66 12	1810		CLUSTER OF GALAXIES
UGC 04586	08 43 30.	+ 75 29	78	16.5	GALAXY Sc
MCG+13-07-005	08 43 30.	+ 76 59	10	16.	GALAXY
ABC 0703	08 43 31.	+ 05 20		17.8	RICH CLUSTER OF GALAXIES
MCG+05-21-006	08 43 33.	+ 29 47	30	16.	GALAXY
TON-N 0949	08 43 35.	+ 35 53		15.	BLUE STAR
LB 08656	08 43 36.	+ 18 57		17.3	FAINT BLUE STAR
ZWG 120.032	08 43 36.	+ 21 02		15.3	GALAXY
ZWG 150.025	08 43 36.	+ 30 04		15.6	GALAXY
MCG+13-07-006	08 43 36.	+ 76 58	6	17.	GALAXY
MCG-03-23-003	08 43 36.	- 17 21	18	15.	GALAXY
ZC 0843.7+0516	08 43 42.	+ 05 16	670		CLUSTER OF GALAXIES
LB 08658	08 43 42.	+ 19 56		17.0	FAINT BLUE STAR
LB 08657	08 43 42.	+ 20 03		17.9	FAINT BLUE STAR
ZWG 150.026	08 43 42.	+ 31 15		15.6	GALAXY
ZWG 237.020	08 43 42.	+ 49 44		13.8	GALAXY
UGC 04587	08 43 42.	+ 49 44	102	13.8	GALAXY S0?
MCG+09-15-023	08 43 42.	+ 53 50	60	16.	GALAXY
LB 02889	08 43 44.	- 00 25 00.		16.2	FAINT BLUE STAR
MCG+12-09-017	08 43 45.	+ 70 55	24	16.	GALAXY
RNGC 2665	08 43 46.	- 19 06		13.0	GALAXY
TON-N 0950	08 43 47.	+ 36 28		15.	BLUE STAR
ZC 0843.8+0215	08 43 48.	+ 02 15	4370		CLUSTER OF GALAXIES
LB 08661	08 43 48.	+ 16 22		18.2	FAINT BLUE STAR
LB 08660	08 43 48.	+ 17 12		18.1	FAINT BLUE STAR
LB 08659	08 43 48.	+ 17 49		18.0	FAINT BLUE STAR
ZWG 090.005	08 43 48.	+ 19 12		15.0	GALAXY
UGC 04588	08 43 48.	+ 19 12	102	15.0	GALAXY Sc
MCG+03-23-002	08 43 48.	+ 19 12	102	15.0	GALAXY
MCG+05-21-007	08 43 48.	+ 28 21	24	14.6	GALAXY
ZWG 150.027	08 43 48.	+ 28 22		14.6	GALAXY
UGC 04589	08 43 48.	+ 28 22	84	14.6	GALAXY E
ZWG 150.028	08 43 48.	+ 32 07		15.7	GALAXY
ZC 0843.8+3842	08 43 48.	+ 38 42	1080		CLUSTER OF GALAXIES
MCG+08-16-029	08 43 48.	+ 49 44	54	14.	GALAXY
ZWG 264.014	08 43 48.	+ 54 05		15.7	GALAXY
ZC 0843.8+5940	08 43 48.	+ 59 40	1410		CLUSTER OF GALAXIES
MCG-03-23-004	08 43 48.	- 19 06	120	13.5	GALAXY
IC 2393	08 43 49.	+ 28 21 28.			NONSTELLAR OBJECT
IC 2403	08 43 50.	- 15 10 18.			NONSTELLAR OBJECT
IC 2396	08 43 51.	+ 17 49 57.			TWO STARS
LB 01900	08 43 51.	+ 20 59 54.		16.5	FAINT BLUE STAR
IC 2397	08 43 52.	+ 17 50 38.			MAY NOT EXIST
WRAY 19.13	08 43 52.7	- 43 39 44.			DIFFUSE NEBULA
ZWG 061.010	08 43 54.	+ 13 24		15.0	GALAXY
UGC 04590	08 43 54.	+ 13 24	84	15.0	GALAXY S
MCG+02-23-006	08 43 54.	+ 13 24	48	15.0	GALAXY
ZWG 090.006	08 43 54.	+ 17 56		14.8	GALAXY
MCG+03-23-003	08 43 54.	+ 17 56	30	14.8	GALAXY
ZWG 150.029	08 43 54.	+ 28 27		15.6	GALAXY
UGC 04591	08 43 54.	+ 28 27	84	15.6	GALAXY Sc
OCL 0764	08 43 54.	- 48 36	900	9.3	OPEN STAR CLUSTER
VHA 050	08 43 54.	- 48 36	360		OPEN STAR CLUSTER
RNGC 2670	08 43 54.	- 48 36		9.5	OPEN CLUSTER
IC 2398	08 43 54.4	+ 17 56 20.			GALAXY SAb
ZWG 033.003	08 44 00.	+ 02 35		15.4	GALAXY
ZWG 033.004	06 44 00.	+ 02 43		14.8	GALAXY
MCG+01-23-002	08 44 00.	+ 02 43	12	14.8	GALAXY
ZWG 061.011	08 44 00.	+ 13 54		15.5	GALAXY
LB 08665	08 44 00.	+ 14 44		15.4	FAINT BLUE STAR
LB 08664	08 44 00.	+ 17 15		17.7	FAINT BLUE STAR
PK208+33.1	08 44 00.	+ 18 04	127	15.6	PLANETARY NEBULA
LB 08663	08 44 00.	+ 18 54		17.6	FAINT BLUE STAR
LB 08662	08 44 00.	+ 20 23		18.4	FAINT BLUE STAR
MCG+04-21-020	08 44 00.	+ 21 52 30.	90	15.5	GALAXY
ZWG 120.033	08 44 00.	+ 21 55		15.5	GALAXY
UGC 04592	08 44 00.	+ 21 55	96	15.5	GALAXY SB
ZC 0844.0+2508	08 44 00.	+ 25 08	940		CLUSTER OF GALAXIES
MCG+05-21-008	08 44 00.	+ 28 25	84	15.6	GALAXY
ZWG 180.003	08 44 00.	+ 35 39		15.7	GALAXY
ZWG 179.029	08 44 00.	+ 35 39		15.7	GALAXY
MCG+06-20-001	08 44 00.	+ 35 41	15	16.	GALAXY
ZWG 332.016	08 44 00.	+ 70 18		13.4	GALAXY
UGC 04593	08 44 00.	+ 70 18	42	13.4	GALAXY PECULR
KARA.72 172B	08 44 00.	+ 70 18	36		PART OF DOUBLE GALAXY
KARA.72 172A	08 44 00.	+ 70 18	36	13.4	PART OF DOUBLE GALAXY
ABC 0702	08 44 02.	+ 25 10		17.1	RICH CLUSTER OF GALAXIES
IC 2394	08 44 05.	+ 28 25 20.			NONSTELLAR OBJECT
ZWG 033.005	08 44 06.	+ 07 09		14.8	GALAXY
UGC 04594	08 44 06.	+ 07 09	66	14.8	GALAXY Sa-b
MCG+01-23-003	08 44 06.	+ 07 09	48	14.8	GALAXY
LB 08667	08 44 06.	+ 14 35		17.8	FAINT BLUE STAR
LB 08666	08 44 06.	+ 20 32		18.1	FAINT BLUE STAR
ZWG 150.030	08 44 06.	+ 28 05		15.6	GALAXY
ZWG 150.031	08 44 06.	+ 28 27		15.2	GALAXY
UGC 04595	08 44 06.	+ 28 27	102	15.2	GALAXY SBb
ZC 0844.1+3718	08 44 06.	+ 37 18	1280		CLUSTER OF GALAXIES
ZWG 180.004	08 44 06.	+ 37 49		15.7	GALAXY
ZWG 179.030	08 44 06.	+ 37 49		15.7	GALAXY
ZWG 264.015	08 44 06.	+ 54 03		15.0	GALAXY
MCG+09-15-025	08 44 06.	+ 54 04	60	15.	GALAXY
MCG+09-15-024	08 44 06.	+ 54 13	48	16.	GALAXY
MCG-02-23-003	08 44 06.	- 11 47	42	15.	GALAXY
FATH 1.250	08 44 07.	+ 44 22	19		NEBULA
IC 0521	08 44 09.	+ 02 43 35.			NONSTELLAR OBJECT
RNGC 2656	08 44 09.	+ 54 03		15.0	GALAXY
ZWG 033.006	08 44 12.	+ 02 40		15.6	GALAXY
LB 08670	08 44 12.	+ 15 18		16.0	FAINT BLUE STAR
LB 08669	08 44 12.	+ 17 00		18.1	FAINT BLUE STAR
LB 08668	08 44 12.	+ 17 13		17.5	FAINT BLUE STAR
MCG+05-21-009	08 44 12.	+ 28 25	78	15.2	GALAXY
ZWG 150.032	08 44 12.	+ 31 22		15.7	GALAXY
ZWG 264.016	08 44 12.	+ 54 37		15.7	GALAXY
MCG+12-09-018	08 44 12.	+ 70 57	33	17.	GALAXY
ZWG 005.007	08 44 18.	+ 00 29		15.6	GALAXY
ZWG 090.007	08 44 18.	+ 19 49		15.0	GALAXY
UGC 04596	08 44 18.	+ 19 49	72	15.0	GALAXY S0
MCG+03-23-004	08 44 18.	+ 19 49	66	15.0	GALAXY
MCG+09-15-026	08 44 18.	+ 54 37	42	16.	GALAXY
MCG+09-15-027	08 44 18.	+ 54 54	30	16.	GALAXY
MCG+10-13-016	08 44 18.	+ 61 49	51	17.	GALAXY
VHE 21A	08 44 18.	- 43 29	66		REFLECTION NEBULA
CED 1060	08 44 18.	- 45 51	1320		DIFFUSE GALACTIC NEBULA
FATH 1.251	08 44 23.	+ 44 46	22		NEBULA
ZWG 090.008	08 44 24.	+ 16 31		15.5	GALAXY
LB 08671	08 44 24.	+ 20 16		18.2	FAINT BLUE STAR
ZC 0844.4+3542	08 44 24.	+ 35 42	4370		CLUSTER OF GALAXIES
ZWG 180.005	08 44 24.	+ 37 47		15.7	GALAXY
ZWG 179.031	08 44 24.	+ 37 47		15.7	GALAXY
ZC 0844.4+4820	08 44 24.	+ 48 20	810		CLUSTER OF GALAXIES
ZC 0844.4+6328	08 44 24.	+ 63 28	1480		CLUSTER OF GALAXIES
ZWG 332.017	08 44 24.	+ 70 21		15.5	GALAXY
OCL 0745	08 44 24.	- 41 42	1800	11.9	OPEN STAR CLUSTER
VHA 051	08 44 24.	- 41 42	360		OPEN STAR CLUSTER
RNGC 2671	08 44 24.	- 41 42		11.5	OPEN CLUSTER
RNGC 2664	08 44 25.	+ 12 48			NON-EXISTENT OBJECT
ABC 0701	08 44 27.	+ 38 19		17.7	RICH CLUSTER OF GALAXIES
LB 08673	08 44 30.	+ 15 41		18.4	FAINT BLUE STAR
LB 08672	08 44 30.	+ 20 40		18.2	FAINT BLUE STAR
TON-N 0349	08 44 30.	+ 23 14		14.6	BLUE STAR
ZC 0844.5+3208	08 44 30.	+ 32 08	2220		CLUSTER OF GALAXIES
MCG+07-18-047	08 44 30.	+ 39 42 30.	45	15.5	GALAXY
ZWG 208.046	08 44 30.	+ 39 43		15.7	GALAXY
TON-N 0951	08 44 35.	+ 34 59		14.8	BLUE STAR
LB 08674	08 44 36.	+ 18 50		18.0	FAINT BLUE STAR
ZWG 090.009	08 44 36.	+ 19 44		15.6	GALAXY
ABC 0705	08 44 37.	+ 30 11		17.1	RICH CLUSTER OF GALAXIES
FATH 1.252	08 44 37.	+ 44 44	5		NEBULA
ZWG 033.007	08 44 42.	+ 02 57		15.2	GALAXY
LB 08678	08 44 42.	+ 15 16		18.2	FAINT BLUE STAR
LB 08677	08 44 42.	+ 16 09		18.2	FAINT BLUE STAR
LB 08676	08 44 42.	+ 17 12		18.3	FAINT BLUE STAR
LB 08675	08 44 42.	+ 19 18		16.3	FAINT BLUE STAR
ZC 0844.7+2306	08 44 42.	+ 23 06	4370		CLUSTER OF GALAXIES
ZWG 120.034	08 44 42.	+ 26 05		14.9	GALAXY
UGC 04597	08 44 42.	+ 26 05	66	14.9	GALAXY S
ZWG 180.006	08 44 42.	+ 38 15		15.0	GALAXY
ZWG 179.032	08 44 42.	+ 38 15		15.0	GALAXY
ZC 0844.7+3819	08 44 42.	+ 38 19	870		CLUSTER OF GALAXIES
MCG+12-09-019	08 44 42.	+ 73 40	45	13.1	GALAXY
IC 2400	08 44 47.	+ 38 15 29.			NONSTELLAR OBJECT
LB 08682	08 44 48.	+ 17 52		17.5	FAINT BLUE STAR
LB 08681	08 44 48.	+ 18 22		17.0	FAINT BLUE STAR
LB 08680	08 44 48.	+ 19 37		16.1	FAINT BLUE STAR
LB 08679	08 44 48.	+ 20 10		18.0	FAINT BLUE STAR
ZWG 120.035	08 44 48.	+ 24 04		15.6	GALAXY
MCG+04-21-021	08 44 48.	+ 26 04	60	14.9	GALAXY
TON-N 0350	08 44 48.	+ 31 44		14.6	BLUE STAR
ZWG 180.007	08 44 48.	+ 37 50		15.6	GALAXY
ZWG 179.033	08 44 48.	+ 37 50		15.6	GALAXY
MCG+06-20-002	08 44 48.	+ 38 17	36	15.	GALAXY
MCG+07-18-048	08 44 48.	+ 42 02 30.	42	15.5	GALAXY
UGC 04598	08 44 48.	+ 42 03	60	16.5	GALAXY Sc
LB 02890	08 44 50.	- 00 47 00.		13.9	FAINT BLUE STAR
MCG+06-20-003	08 44 51.	+ 36 59	24	15.	GALAXY
MCG+06-20-004	08 44 51.	+ 37 52	24	15.5	GALAXY
SBB 122	08 44 53.7	+ 31 58 45.			QUASI-STELLAR OBJECT
ZWG 033.008	08 44 54.	+ 02 52		15.6	GALAXY
ZWG 061.012	08 44 54.	+ 13 36		14.9	GALAXY
UGC 04599	08 44 54.	+ 13 36	126	14.9	GALAXY S0
MCG+02-23-007	08 44 54.	+ 13 36	84	14.9	GALAXY
ZC 0844.9+1434	08 44 54.	+ 14 34	1210		CLUSTER OF GALAXIES
LB 08683	08 44 54.	+ 17 25		17.3	FAINT BLUE STAR
ZC 0844.9+2856	08 44 54.	+ 28 56	1080		CLUSTER OF GALAXIES
ZWG 150.033	08 44 54.	+ 31 58		15.5	GALAXY
ZC 0844.9+3407	08 44 54.	+ 34 07	3490		CLUSTER OF GALAXIES
ZWG 180.008	08 44 54.	+ 37 56		15.0	GALAXY
ZWG 179.034	08 44 54.	+ 37 56		15.0	GALAXY
UGC 04600	08 44 54.	+ 37 56	78	15.0	GALAXY (S0)
ZC 0844.9+4322	08 44 54.	+ 43 22	3700		CLUSTER OF GALAXIES
ZWG 237.021	08 44 54.	+ 47 28		14.8	GALAXY
MCG+08-16-030	08 44 54.	+ 47 28	24	16.	GALAXY
IC 2402	08 44 55.	+ 31 58 22.			NONSTELLAR OBJECT
IC 2401	08 44 58.	+ 37 55 22.			NONSTELLAR OBJECT
IC 2399	08 44 58.6	+ 19 05 04.			GALAXY SB(r)
LBN 1075	08 45	- 23 50	2100		BRIGHT NEBULA
LB 08686	08 45 00.	+ 17 02		17.8	FAINT BLUE STAR
LB 08685	08 45 00.	+ 18 08		17.8	FAINT BLUE STAR
LB 08684	08 45 00.	+ 18 42		17.1	FAINT BLUE STAR
ZWG 090.010	08 45 00.	+ 19 05		15.6	GALAXY
ZWG 120.036	08 45 00.	+ 26 00		15.1	GALAXY
UGC 04602	08 45 00.	+ 26 00	84	15.1	GALAXY Sa-b
ZC 0845.0+3030	08 45 00.	+ 30 30	7330		CLUSTER OF GALAXIES
MCG+05-21-010	08 45 00.	+ 32 00	24	15.5	GALAXY
ZWG 180.009	08 45 00.	+ 35 25		15.7	GALAXY
ZWG 180.010	08 45 00.	+ 36 57		15.2	GALAXY
MCG+06-20-005	08 45 00.	+ 37 58	24	15.	GALAXY
MCG+12-09-020	08 45 00.	+ 70 29	84	12.8	GALAXY
ZWG 332.018	08 45 00.	+ 70 30		14.3	GALAXY
UGC 04603	08 45 00.	+ 70 30	108	14.3	GALAXY SBb
RNGC 2646	08 45 00.	+ 73 39		13.0	GALAXY
ZC 0845.0+7631	08 45 00.	+ 76 31	5440		CLUSTER OF GALAXIES
ZWG 364.007	08 45 00.	+ 85 00		14.9	GALAXY
ZWG 363.048	08 45 00.	+ 85 00		14.9	GALAXY
UGC 04601	08 45 00.	+ 85 00	96	14.9	GALAXY
VHA 052	08 45 00.	- 52 44	420		OPEN STAR CLUSTER
MCG+04-21-022	08 45 03.	+ 26 00	72	15.1	GALAXY
MCG-03-23-005	08 45 03.	- 19 52	150	14.	GALAXY
ZC 0845.1+2445	08 45 06.	+ 24 45	5380		CLUSTER OF GALAXIES
ZWG 150.034	08 45 06.	+ 29 41		15.5	GALAXY
MCG+09-15-028	08 45 06.	+ 52 10	36	16.	GALAXY
ZWG 332.019	08 45 06.	+ 73 39		13.0	GALAXY
ZWG 331.069	08 45 06.	+ 73 39		13.0	GALAXY
UGC 04604	08 45 06.	+ 73 39	102	13.0	GALAXY SB0
MCG-02-23-004	08 45 06.	- 11 11	60	15.	GALAXY
IC 2404	08 45 09.	+ 29 41 02.			NONSTELLAR OBJECT
RNGC 2650	08 45 09.	+ 70 29		13.0	GALAXY
RNGC 2654	08 45 11.	+ 60 24		13.0	GALAXY
LB 08689	08 45 12.	+ 16 48		17.7	FAINT BLUE STAR
LB 08688	08 45 12.	+ 18 37		18.4	FAINT BLUE STAR
LB 08687	08 45 12.	+ 19 22		16.6	FAINT BLUE STAR
ZWG 150.035	08 45 12.	+ 31 27		15.6	GALAXY
ZC 0845.2+4929	08 45 12.	+ 49 29	200		CLUSTER OF GALAXIES
ZWG 288.006	08 45 12.	+ 60 25		12.8	GALAXY

OBJECT NAME	RIGHT ASCEN.	DECLINATION	DIAM.	MAGN.	TYPE OF OBJECT
UGC 04605	08 45 12.	+ 60 25	258	12.8	GALAXY Sa-b
KARA.73B 0285	08 45 12.	+ 60 25	270	12.8	ISOLATED GALAXY S
MCG-06-20-002	08 45 12.	- 33 32	60	15.	GALAXY
MAI 035	08 45 14.	+ 74 12	40		DWARF SPHEROIDAL GALAXY
IC 2406	08 45 14.5	+ 17 53 17.			GALAXY S0
MCG+03-23-005	08 45 15.	+ 17 53	72	14.4	GALAXY
ARC 0706	08 45 16.	+ 28 55		17.5	RICH CLUSTER OF GALAXIES
LB 08691	08 45 18.	+ 14 50		18.1	FAINT BLUE STAR
MCG+03-23-006	08 45 18.	+ 17 48	66	15.1	GALAXY
ZWG 090.011	08 45 18.	+ 17 54		14.4	GALAXY
UGC 04606	08 45 18.	+ 17 54	96	14.4	GALAXY S0-a
LB 08690	08 45 18.	+ 19 25		18.2	FAINT BLUE STAR
ZC 0845.3+4635	08 45 18.	+ 46 35	2820		CLUSTER OF GALAXIES
MCG+09-15-029	08 45 18.	+ 52 27	36	17.	GALAXY
MCG+10-13-017	08 45 18.	+ 60 26	228	12.8	GALAXY
IC 2407	08 45 19.1	+ 17 47 52.			GALAXY Sb
LB 08693	08 45 24.	+ 15 27		18.3	FAINT BLUE STAR
ZWG 090.012	08 45 24.	+ 17 48		15.1	GALAXY
UGC 04607	08 45 24.	+ 17 48	78	15.1	GALAXY Sb-c
LB 08692	08 45 24.	+ 19 50		18.3	FAINT BLUE STAR
VHE 21B	08 45 24.	- 43 32	18		REFLECTION NEBULA
HOLM 098C	08 45 26.	+ 19 13	18	15.7	PART OF MULTIPLE GALAXY
RNGC 2667A	08 45 28.	+ 19 13		15.0	GALAXY
ZWG 090.013	08 45 30.	+ 17 09		15.5	GALAXY
LB 08696	08 45 30.	+ 18 06		16.5	FAINT BLUE STAR
LB 08695	08 45 30.	+ 18 48		18.0	FAINT BLUE STAR
LB 08694	08 45 30.	+ 19 05		16.2	FAINT BLUE STAR
ZWG 090.014	08 45 30.	+ 19 08		15.7	GALAXY
IC 2408	08 45 30.	+ 19 13 23.			SINGLE STAR
ZWG 120.037	08 45 30.	+ 23 55		15.5	GALAXY
ZWG 120.038	08 45 30.	+ 24 01		15.6	GALAXY
ZWG 180.011	08 45 30.	+ 37 23		14.8	GALAXY
IC 2405	08 45 30.	+ 37 24 31.			NONSTELLAR OBJECT
MCG+06-20-006	08 45 30.	+ 37 25 30.	24	15.	GALAXY
ZC 0845.5+3740	08 45 30.	+ 37 40	1280		CLUSTER OF GALAXIES
HOLM 098A	08 45 33.	+ 19 12	18	14.7	PART OF MULTIPLE GALAXY
MCG+03-23-007	08 45 33.	+ 19 13 30.	42	14.8	GALAXY
BIGO 504	08 45 34.	+ 19 13			NEBULA
RNGC 2667B	08 45 34.	+ 19 14		15.5	GALAXY
IC 2409	08 45 34.3	+ 18 31 00.			GALAXY Sab
LB 08699	08 45 36.	+ 15 45		17.4	FAINT BLUE STAR
ZWG 090.015	08 45 36.	+ 18 31		14.4	GALAXY
UGC 04608	08 45 36.	+ 18 31	60	14.4	GALAXY SBa
MCG+03-23-008	08 45 36.	+ 18 32	60	14.4	GALAXY
LB 08698	08 45 36.	+ 19 00		17.7	FAINT BLUE STAR
ZWG 090.016	08 45 36.	+ 19 12		14.8	GALAXY
HOLM 098B	08 45 36.	+ 19 14	30	15.0	PART OF MULTIPLE GALAXY
MCG+03-23-009	08 45 36.	+ 19 14	42	14.8	GALAXY
LB 08697	08 45 36.	+ 20 34		18.2	FAINT BLUE STAR
ZC 0845.6+4045	08 45 36.	+ 40 45	1410		CLUSTER OF GALAXIES
MCG+09-15-030	08 45 36.	+ 51 05	30	16.	GALAXY
IC 2410	08 45 36.4	+ 19 12 16.			GALAXY S0
IC 2411	08 45 38.9	+ 19 13 41.			GALAXY Sb
ZWG 033.009	08 45 42.	+ 04 34		15.7	GALAXY
ZWG 090.017	08 45 42.	+ 14 46		15.6	GALAXY
LB 08702	08 45 42.	+ 14 48		17.7	FAINT BLUE STAR
LB 08701	08 45 42.	+ 16 20		17.8	FAINT BLUE STAR
LB 08700	08 45 42.	+ 18 56		17.5	FAINT BLUE STAR
ZWG 090.018	08 45 42.	+ 19 14		15.4	GALAXY
TON-N 0351	08 45 42.	+ 32 56		15.6	BLUE STAR
MCG+07-18-050	08 45 42.	+ 41 10	45	15.	GALAXY
ZWG 208.047	08 45 42.	+ 41 12		14.9	GALAXY
MCG+07-18-049	08 45 42.	+ 41 22 30.	96	15.5	GALAXY
ZWG 208.048	08 45 42.	+ 41 23		15.4	GALAXY
UGC 04609	08 45 42.	+ 41 23	120	15.4	GALAXY PECULR
ZWG 005.010	08 45 48.	+ 01 15		15.7	GALAXY
UGC 04610	08 45 48.	+ 01 15	66	15.7	GALAXY Sb-c
KARA.72 174A	08 45 48.	+ 01 15	60	15.7	PART OF DOUBLE GALAXY
ZWG 005.011	08 45 48.	+ 01 35		15.7	GALAXY
ZWG 061.013	08 45 48.	+ 13 00		15.6	GALAXY
ZWG 150.036	08 45 48.	+ 30 04		15.6	GALAXY
UGC 04611	08 45 48.	+ 30 04	108	15.6	GALAXY Sc
MCG+05-21-011	08 45 48.	+ 30 04	108	15.6	GALAXY
ZC 0845.8+6435	08 45 48.	+ 64 35	1140		CLUSTER OF GALAXIES
ZC 0845.8+7858	08 45 48.	+ 78 58	1410		CLUSTER OF GALAXIES
ZWG 005.009	08 45 48.	- 02 47		15.6	GALAXY
KARA.72 173B	08 45 48.	- 02 47	48	15.6	PART OF DOUBLE GALAXY
ZWG 005.008	08 45 48.	- 02 50		14.7	GALAXY
KARA.72 173A	08 45 48.	- 02 50	102	14.7	PART OF DOUBLE GALAXY
MCG+00-23-002	08 45 48.	- 02 50	90	14.7	GALAXY
MCG-03-23-006	08 45 48.	- 19 29	36	17.	GALAXY
ARC 0708	08 45 49.	+ 37 43		17.9	RICH CLUSTER OF GALAXIES
RNGC 2674	08 45 49.	- 14 05			NON-EXISTENT OBJECT
LB 08704	08 45 54.	+ 18 36		18.4	FAINT BLUE STAR
LB 08703	08 45 54.	+ 19 12		18.3	FAINT BLUE STAR
TON-N 0352	08 45 54.	+ 32 17		16.1	BLUE STAR
ZC 0845.9+4745	08 45 54.	+ 47 45	810		CLUSTER OF GALAXIES
MCG+09-15-031	08 45 54.	+ 51 03	36	17.	GALAXY
ZWG 264.017	08 45 54.	+ 51 04		15.3	GALAXY
MCG+11-11-029	08 45 54.	+ 66 39	39	17.	GALAXY
ZWG 061.014	08 46 00.	+ 10 16		15.4	GALAXY
LB 08707	08 46 00.	+ 14 32		18.1	FAINT BLUE STAR
LB 08706	08 46 00.	+ 19 42		18.0	FAINT BLUE STAR
LB 08705	08 46 00.	+ 20 34		17.4	FAINT BLUE STAR
TON-N 0011	08 46 00.	+ 30 13		15.5	BLUE STAR
MCG+10-13-018	08 46 00.	+ 57 20	27	16.	GALAXY
MCG+12-09-021	08 46 00.	+ 72 01	18	17.	GALAXY
ZWG 332.020	08 46 00.	+ 72 02		15.5	GALAXY
MCG+12-09-022	08 46 00.	+ 72 02	27	16.	GALAXY
UGC 04612	08 46 00.	+ 85 44	90	17.	GALAXY DWRF SP
MCG-02-23-005	08 46 00.	- 13 02	36	16.	GALAXY
OCL 0747	08 46 00.	- 42 18	1800	5.0	OPEN STAR CLUSTER
VHA 053	08 46 00.	- 42 18	1200		OPEN STAR CLUSTER
BC LB8707	08 46 01.	+ 14 32 00.			QUASI-STELLAR OBJECT
SHB 123	08 46 01.	+ 14 32 00.		18.1	QUASI-STELLAR OBJECT
TON-N 0952	08 46 05.	+ 32 58		15.9	BLUE STAR
TON-N 0953	08 46 05.	+ 34 45		15.7	BLUE STAR
ZWG 005.012	08 46 06.	+ 01 13		14.9	GALAXY
UGC 04613	08 46 06.	+ 01 13	102	14.9	GALAXY SBb
KARA.72 174B	08 46 06.	+ 01 13	108	14.9	PART OF DOUBLE GALAXY
MCG+00-23-003	08 46 06.	+ 01 13	144	14.9	GALAXY
ZWG 061.015	08 46 06.	+ 09 13		15.7	GALAXY
LB 08710	08 46 06.	+ 17 14		17.1	FAINT BLUE STAR
LB 08709	08 46 06.	+ 18 34		18.5	FAINT BLUE STAR
LB 08708	08 46 06.	+ 18 48		17.9	FAINT BLUE STAR
ZWG 180.012	08 46 06.	+ 36 17		14.3	GALAXY
UGC 04614	08 46 06.	+ 36 17	72	14.3	GALAXY PECULR
ZWG 208.049	08 46 06.	+ 41 03		15.7	GALAXY

OBJECT NAME	RIGHT ASCEN.	DECLINATION	DIAM.	MAGN.	TYPE OF OBJECT
MCG+07-18-051	08 46 06.	+ 41 57	60	15.	GALAXY
ZWG 208.050	08 46 06.	+ 41 58		15.6	GALAXY
UGC 04615	08 46 06.	+ 41 58	66	15.6	GALAXY SBc
ZWG 288.007	08 46 06.	+ 57 19		15.7	GALAXY
MCG+10-13-019	08 46 06.	+ 57 19	24	16.	GALAXY
MRSL 264-00/1	08 46 06.	- 44 41	3180		HII REGION
MCG+10-13-020	08 46 09.	+ 57 19	39	15.	GALAXY
LB 08712	08 46 12.	+ 15 30		16.0	FAINT BLUE STAR
LB 08711	08 46 12.	+ 17 41		17.9	FAINT BLUE STAR
TON-N 0353	08 46 12.	+ 24 59		15.1	BLUE STAR
ZWG 180.013	08 46 12.	+ 36 53		14.9	GALAXY
UGC 04616	08 46 12.	+ 36 53	96	14.9	GALAXY Sa-b
MCG+06-20-007	08 46 12.	+ 36 55	60	15.	GALAXY
ZC 0846.2+3809	08 46 12.	+ 38 09	1080		CLUSTER OF GALAXIES
MCG+07-18-053	08 46 12.	+ 39 22 30.	48	14.5	GALAXY
ZWG 208.051	08 46 12.	+ 39 24		15.3	GALAXY
ZC 0846.2+3948	08 46 12.	+ 39 48	1340		CLUSTER OF GALAXIES
MCG+07-18-052	08 46 12.	+ 42 27 30.	42	15.5	GALAXY
ZWG 208.052	08 46 12.	+ 42 28		15.7	GALAXY
ZWG 264.018	08 46 12.	+ 53 42		15.5	GALAXY
RNGC 2668	08 46 13.	+ 36 53		15.0	GALAXY
ZWG 033.010	08 46 18.	+ 04 31		15.6	GALAXY
ZC 0846.3+1910	08 46 18.	+ 19 10	6990		CLUSTER OF GALAXIES
LB 08713	08 46 18.	+ 19 42		18.3	FAINT BLUE STAR
ZWG 120.039	08 46 18.	+ 23 34		15.5	GALAXY
MCG+09-15-032	08 46 18.	+ 53 42 30.	18	17.	GALAXY
MCG+10-13-021	08 46 18.	+ 60 16	39	15.	GALAXY
ZWG 033.011	08 46 24.	+ 07 34		15.3	GALAXY
KARA.73B 0286	08 46 24.	+ 07 34	24	15.3	ISOLATED GALAXY E
LB 08716	08 46 24.	+ 15 00		18.1	FAINT BLUE STAR
LB 08715	08 46 24.	+ 18 12		17.7	FAINT BLUE STAR
LB 08714	08 46 24.	+ 19 21		18.2	FAINT BLUE STAR
MCG+05-21-012	08 46 24.	+ 29 42 30.	78	15.5	GALAXY
ZWG 150.037	08 46 24.	+ 29 43		15.5	GALAXY
UGC 04617	08 46 24.	+ 29 43	96	15.5	GALAXY Sc
ZC 0846.4+4021	08 46 24.	+ 40 21	270		CLUSTER OF GALAXIES
ZWG 208.053	08 46 24.	+ 40 26		15.5	GALAXY
MCG+10-13-022	08 46 24.	+ 61 49	24	17.	GALAXY
UGC 04618	08 46 24.	+ 66 29	60	16.5	GALAXY S:IV-V
HOLM 099A	08 46 27.	+ 19 15	36	13.2	PART OF MULTIPLE GALAXY
MCG+07-18-054	08 46 27.	+ 40 25	42	15.	GALAXY
RNGC 2673	08 46 28.	+ 19 16		14.5	GALAXY
RNGC 2672	08 46 28.	+ 19 16		13.0	GALAXY
LB 08718	08 46 30.	+ 15 12		18.4	FAINT BLUE STAR
LB 08717	08 46 30.	+ 16 20		18.2	FAINT BLUE STAR
HOLM 099B	08 46 30.	+ 19 15	24	14.5	PART OF MULTIPLE GALAXY
ZWG 090.019	08 46 30.	+ 19 16		13.1	GALAXY
UGC 04620	08 46 30.	+ 19 16	168	14.4	GALAXY E+COMP
UGC 04619	08 46 30.	+ 19 16	168	13.4	GALAXY E+COMP
KARA.72 175A	08 46 30.	+ 19 16	108	13.4	PART OF DOUBLE GALAXY
ZC 0846.5+4923	08 46 30.	+ 49 23	1950		CLUSTER OF GALAXIES
MCG+08-16-031	08 46 30.	+ 49 26	72	16.	GALAXY
ZWG 264.019	08 46 30.	+ 55 33		15.2	GALAXY
ZWG 288.008	08 46 30.	+ 60 16		15.4	GALAXY
ZWG 311.019	08 46 30.	+ 65 49		15.3	GALAXY
MCG+14-05-001	08 46 30.	+ 85 00	78	15.	GALAXY
IC 2412	08 46 34.	+ 18 43 46.			TWO STARS
ARP 167	08 46 34.	+ 19 16			PECULIAR GALAXY
RNGC 2666	08 46 34.	+ 47 15			NON-EXISTENT OBJECT
ZC 0846.6+1413	08 46 36.	+ 14 13	1210		CLUSTER OF GALAXIES
LB 08723	08 46 36.	+ 14 57		18.4	FAINT BLUE STAR
LB 08722	08 46 36.	+ 18 44		17.7	FAINT BLUE STAR
LB 08721	08 46 36.	+ 18 44		17.8	FAINT BLUE STAR
LB 08720	08 46 36.	+ 18 47		18.0	FAINT BLUE STAR
LB 08719	08 46 36.	+ 18 56		18.0	FAINT BLUE STAR
KARA.72 175B	08 46 36.	+ 19 16	54	14.4	PART OF DOUBLE GALAXY
MCG+03-23-010	08 46 36.	+ 19 16	42	13.4	GALAXY
ZC 0846.6+3150	08 46 36.	+ 31 50	810		CLUSTER OF GALAXIES
ZC 0846.6+3512	08 46 36.	+ 35 12	1080		CLUSTER OF GALAXIES
MCG-01-23-002	08 46 36.	- 07 30	78	13.5	GALAXY
MCG+03-23-011	08 46 39.	+ 19 16	42	14.4	GALAXY
TON-N 0954	08 46 41.	+ 35 01		16.2	BLUE STAR
LB 08724	08 46 42.	+ 16 41		17.8	FAINT BLUE STAR
IC 2413	08 46 42.	+ 18 55 46.			TWO STARS
MCG+09-15-033	08 46 42.	+ 55 34	30	15.	GALAXY
ZWG 288.009	08 46 42.	+ 57 33		15.5	GALAXY
MCG+10-13-023	08 46 42.	+ 61 50	30	16.	GALAXY
TON-N 0955	08 46 47.	+ 33 59		16.2	BLUE STAR
ZC 0846.8+0554	08 46 48.	+ 05 54	1010		CLUSTER OF GALAXIES
MCG+07-18-055	08 46 48.	+ 40 25	36	15.	GALAXY
MCG+09-15-034	08 46 48.	+ 55 43	36	16.	GALAXY
ZWG 332.021	08 46 48.	+ 72 00		15.0	GALAXY
SEY 049	08 46 52.	+ 71 59 25.		15.4	FAINT GALAXY
TON-N 0956	08 46 53.	+ 33 29		16.	BLUE STAR
LB 08727	08 46 54.	+ 18 37		16.8	FAINT BLUE STAR
LB 08726	08 46 54.	+ 18 58		16.6	FAINT BLUE STAR
LB 08725	08 46 54.	+ 20 07		17.1	FAINT BLUE STAR
ZWG 208.054	08 46 54.	+ 40 27		15.1	GALAXY
7ZW 253	08 46 54.	+ 65 45			COMPACT GALAXY
MCG+12-09-023	08 46 54.	+ 72 00	24	16.	GALAXY
MRSL 263+00/1	08 46 54.	- 42 41	168		HII REGION
ACK 263+00.1	08 46 54.	- 42 43			PLANETARY NEBULA
MCG-04-21-009	08 46 57.	- 26 08 30.	24	15.	GALAXY
BC 4C09.31	08 46 57.3	+ 10 00 42.		19.2	QUASI-STELLAR OBJECT
SHB 124	08 46 57.7	+ 10 00 32.		19.2	QUASI-STELLAR OBJECT
IC 2414	08 46 59.5	+ 18 58 49.			GALAXY S0
LB 08728	08 47 00.	+ 17 36		17.1	FAINT BLUE STAR
ZWG 090.020	08 47 00.	+ 18 58		15.7	GALAXY
ZWG 150.038	08 47 00.	+ 30 40		15.7	GALAXY
ZWG 180.014	08 47 00.	+ 35 15		13.6	GALAXY
UGC 04621	08 47 00.	+ 35 15	78	13.6	GALAXY PECULR
MCG+06-20-008	08 47 00.	+ 35 17	42	14.	GALAXY
ZWG 208.055	08 47 00.	+ 40 11		15.7	GALAXY
MCG+07-18-056	08 47 00.	+ 41 28	78	14.5	GALAXY
ZWG 208.056	08 47 00.	+ 41 29		15.3	GALAXY
UGC 04622	08 47 00.	+ 41 29	84	15.3	GALAXY Sc
ZC 0847.0+4707	08 47 00.	+ 47 07	270		CLUSTER OF GALAXIES
MCG+10-13-024	08 47 00.	+ 57 32	39	15.	GALAXY
ZC 0847.0+7137	08 47 00.	+ 71 37	1140		CLUSTER OF GALAXIES
SHAH 018	08 47 00.	+ 79 21	42	16.	GROUP OF COMPACT GALAXIES
MCG-01-23-003	08 47 00.	- 06 22	30	15.5	GALAXY
MCG-03-23-007	08 47 00.	- 18 50	12	15.	GALAXY
TON-N 0957	08 47 05.	+ 35 25		16.8	BLUE STAR
LB 08730	08 47 06.	+ 15 17		18.1	FAINT BLUE STAR
LB 08729	08 47 06.	+ 17 31		17.3	FAINT BLUE STAR
ZWG 350.006	08 47 06.	+ 76 41		13.5	GALAXY
UGC 04623	08 47 06.	+ 76 41	210	13.5	GALAXY Sc
RNGC 2677	08 47 11.	+ 19 12		15.0	GALAXY

OBJECT NAME	RIGHT ASCEN.	DECLINATION	DIAM.	MAGN.	TYPE OF OBJECT
ZWG 005.013	08 47 12.	+ 01 11		15.3	GALAXY
LB 08732	08 47 12.	+ 14 50		18.3	FAINT BLUE STAR
LB 08731	08 47 12.	+ 16 00		17.2	FAINT BLUE STAR
IC 2415	08 47 12.	+ 18 50 19.			THREE STARS
ZWG 090.021	08 47 12.	+ 19 12		15.2	GALAXY
MCG+03-23-012	08 47 12.	+ 19 12	18	15.2	GALAXY
ZWG 120.040	08 47 12.	+ 21 53		15.4	GALAXY
ZC 0847.2+3617	08 47 12.	+ 36 17	810		CLUSTER OF GALAXIES
ZC 0847.2+4522	08 47 12.	+ 45 22	940		CLUSTER OF GALAXIES
ZC 0847.2+5029	08 47 12.	+ 50 29	610		CLUSTER OF GALAXIES
MCG+13-07-007	08 47 12.	+ 76 41 30.	216	12.	GALAXY
MCG+13-07-008	08 47 12.	+ 77 08	12	17.	GALAXY
ZC 0847.3+1258	08 47 18.	+ 12 58	610		CLUSTER OF GALAXIES
ZWG 208.057	08 47 18.	+ 40 27		15.4	GALAXY
MCG+12-09-024	08 47 18.	+ 73 07 30.	27	16.	GALAXY
MCG+07-18-057	08 47 21.	+ 40 25	39	15.	GALAXY
ZWG 061.016	08 47 24.	+ 12 32		15.4	GALAXY
LB 08734	08 47 24.	+ 15 51		17.9	FAINT BLUE STAR
LB 08733	08 47 24.	+ 15 52		16.6	FAINT BLUE STAR
ZWG 120.041	08 47 24.	+ 26 08		15.4	GALAXY
UGC 04624	08 47 24.	+ 26 08	66	15.4	GALAXY Sa-b
MCG+04-21-023	08 47 24.	+ 26 08	51	15.4	GALAXY
KARA.73B 0287	08 47 24.	+ 26 08	60	15.4	ISOLATED GALAXY S
TON-N 0012	08 47 24.	+ 27 31		15.9	BLUE STAR
ZC 0847.4+3928	08 47 24.	+ 39 28	870		CLUSTER OF GALAXIES
ZC 0847.4+5440	08 47 24.	+ 54 40	810		CLUSTER OF GALAXIES
MCG+10-13-025	08 47 24.	+ 61 12	27	16.	GALAXY
OCL 0760	08 47 24.	- 46 39	336	13.	OPEN STAR CLUSTER
RNGC 2678	08 47 26.	+ 11 32			NON-EXISTENT OBJECT
ABC 0709	08 47 26.	+ 12 56		17.5	RICH CLUSTER OF GALAXIES
LB 03590	08 47 28.	+ 11 47 42.		18.2	FAINT BLUE STAR
LB 08737	08 47 30.	+ 16 30		17.0	FAINT BLUE STAR
LB 08736	08 47 30.	+ 18 56		16.3	FAINT BLUE STAR
LB 08735	08 47 30.	+ 19 30		16.8	FAINT BLUE STAR
ZWG 180.015	08 47 30.	+ 36 39		15.4	GALAXY
MCG+06-20-009	08 47 30.	+ 36 41	24	16.	GALAXY
TON-N 0958	08 47 35.	+ 34 11		16.7	BLUE STAR
ZWG 005.014	08 47 36.	+ 00 50		15.7	GALAXY
ZWG 033.012	08 47 36.	+ 03 40		15.4	GALAXY
UGC 04625	08 47 36.	+ 03 40	102	15.4	GALAXY Sc
LB 08739	08 47 36.	+ 18 01		18.2	FAINT BLUE STAR
LB 08738	08 47 36.	+ 19 28		17.3	FAINT BLUE STAR
ZC 0847.6+4132	08 47 36.	+ 41 32	2690		CLUSTER OF GALAXIES
ZWG 208.058	08 47 36.	+ 42 55		15.4	GALAXY
ZC 0847.6+5648	08 47 36.	+ 56 48	740		CLUSTER OF GALAXIES
ZWG 311.020	08 47 36.	+ 64 37		15.7	GALAXY
ZWG 332.022	08 47 36.	+ 73 06		15.7	GALAXY
ZWG 331.070	08 47 36.	+ 73 06		15.7	GALAXY
LB 03591	08 47 37.	+ 11 56 48.		17.6	FAINT BLUE STAR
LB 03592	08 47 38.	+ 11 45 24.		19.7	FAINT BLUE STAR
IC 2416	08 47 41.	+ 18 44 46.			TWO STARS
OCL 0549	08 47 42.	+ 12 00	4620	7.5	OPEN STAR CLUSTER
LB 08742	08 47 42.	+ 18 08		15.9	FAINT BLUE STAR
LB 08741	08 47 42.	+ 19 06		16.6	FAINT BLUE STAR
ZWG 090.022	08 47 42.	+ 19 25		15.6	GALAXY
ZWG 090.023	08 47 42.	+ 19 34		14.9	GALAXY
MCG+03-23-013	08 47 42.	+ 19 34	42	14.9	GALAXY
LB 08740	08 47 42.	+ 20 24		16.8	FAINT BLUE STAR
TON-N 0354	08 47 42.	+ 31 37		15.1	BLUE STAR
MCG+06-20-010	08 47 42.	+ 36 32	36	16.	GALAXY
ZC 0847.7+3839	08 47 42.	+ 38 39	1140		CLUSTER OF GALAXIES
ZWG 208.059	08 47 42.	+ 40 31		15.7	GALAXY
MRK 016	08 47 42.	+ 73 23	20	15.	GALAXY WITH UV CONTINUUM
ABC 0711	08 47 47.	+ 00 30		17.3	RICH CLUSTER OF GALAXIES
ZC 0847.8+0029	08 47 48.	+ 00 29	870		CLUSTER OF GALAXIES
LB 08745	08 47 48.	+ 14 52		17.9	FAINT BLUE STAR
LB 08744	08 47 48.	+ 18 16		17.6	FAINT BLUE STAR
LB 08743	08 47 48.	+ 19 39		17.4	FAINT BLUE STAR
ZC 0847.8+2348	08 47 48.	+ 23 48	870		CLUSTER OF GALAXIES
TON-N 0355	08 47 48.	+ 28 07		15.7	BLUE STAR
ZWG 180.016	08 47 48.	+ 36 30		15.5	GALAXY
ZC 0847.8+5020	08 47 48.	+ 50 20	2960		CLUSTER OF GALAXIES
MCG-03-23-008	08 47 48.	- 19 33 30.	54	15.	GALAXY
OCL 0761	08 47 48.	- 46 37	420	11.	OPEN STAR CLUSTER
KEEL 156	08 47 48.0	+ 34 02 16.			NEBULA
LB 02891	08 47 53.	+ 00 14 24.		16.3	FAINT BLUE STAR
LB 03593	08 47 54.	+ 11 50 30.		17.8	FAINT BLUE STAR
ZWG 090.024	08 47 54.	+ 17 17		15.3	GALAXY
LB 08747	08 47 54.	+ 20 15		18.2	FAINT BLUE STAR
LB 08746	08 47 54.	+ 20 26		17.9	FAINT BLUE STAR
ZWG 120.042	08 47 54.	+ 22 49		15.7	GALAXY
ZWG 150.039	08 47 54.	+ 29 25		15.3	GALAXY
MRK 628	08 47 54.	+ 29 25	12	15.5	GALAXY WITH UV CONTINUUM
MRK 017	08 47 54.	+ 57 16	7	17.	GALAXY WITH UV CONTINUUM
ZC 0847.9+5754	08 47 54.	+ 57 54	1340		CLUSTER OF GALAXIES
MCG+11-11-030	08 47 54.	+ 64 25	39	16.	GALAXY
ZC 0847.9+7317	08 47 54.	+ 73 17	1480		CLUSTER OF GALAXIES
RNGC 2676	08 47 58.	+ 47 44		14.5	GALAXY
LBN 1076	08 48	- 24 40	2400		BRIGHT NEBULA
LBN 1079	08 48	- 28 20	720		BRIGHT NEBULA
LB 08750	08 48 00.	+ 15 53		15.5	FAINT BLUE STAR
LB 08749	08 48 00.	+ 16 42		15.7	FAINT BLUE STAR
LB 08748	08 48 00.	+ 17 38		15.0	FAINT BLUE STAR
UGC 04626	08 48 00.	+ 24 30	60	18.	GALAXY DWARF
MCG+05-21-013	08 48 00.	+ 29 22	42	15.5	GALAXY
ZWG 150.040	08 48 00.	+ 29 23		15.5	GALAXY
ZWG 208.060	08 48 00.	+ 43 00		15.7	GALAXY
ZWG 237.022	08 48 00.	+ 47 44		14.3	GALAXY
UGC 04627	08 48 00.	+ 47 44	72	14.3	GALAXY S0
MCG+09-15-035	08 48 00.	+ 51 18	90	15.	GALAXY
ZWG 264.020	08 48 00.	+ 51 19		15.5	GALAXY
UGC 04628	08 48 00.	+ 51 19	108	15.5	GALAXY Sc
MCG+10-13-026	08 48 00.	+ 57 40	42	16.	GALAXY
ZWG 332.023	08 48 00.	+ 73 23		14.6	GALAXY
ZWG 331.071	08 48 00.	+ 73 23		14.6	GALAXY
MCG+13-07-009	08 48 00.	+ 76 32 30.	24	16.	GALAXY
ZWG 005.015	08 48 00.	- 02 55		15.0	GALAXY
VV 028B	08 48 00.	- 16 23	6	18.	INTERACTING GALAXY
VV 028A	08 48 00.	- 16 23	72	15.	INTERACTING GALAXY
VHA 054	08 48 00.	- 44 14	180		OPEN STAR CLUSTER
LB 06309	08 48 01.	+ 11 57 00.		17.9	FAINT BLUE STAR
MCG-03-23-009	08 48 03.	- 16 23 30.	72	14.5	GALAXY
LB 03594	08 48 05.	+ 12 14 12.		20.0	FAINT BLUE STAR
SHB 125	08 48 05.	+ 15 33 30.		17.7	QUASI-STELLAR OBJECT
ZWG 005.016	08 48 06.	+ 01 32		15.0	GALAXY
MCG+00-23-004	08 48 06.	+ 01 32	36	15.0	GALAXY
LB 08755	08 48 06.	+ 15 33		17.7	FAINT BLUE STAR
BC LB8755	08 48 06.	+ 15 33		17.7	QUASI-STELLAR OBJECT

OBJECT NAME	RIGHT ASCEN.	DECLINATION	DIAM.	MAGN.	TYPE OF OBJECT
LB 08754	08 48 06.	+ 18 34		17.1	FAINT BLUE STAR
LB 08753	08 48 06.	+ 19 18		17.7	FAINT BLUE STAR
LB 08752	08 48 06.	+ 19 48		17.8	FAINT BLUE STAR
LB 08751	08 48 06.	+ 20 27		18.0	FAINT BLUE STAR
TON-N 0356	08 48 06.	+ 30 05		15.0	BLUE STAR
TON-N 0959	08 48 07.	+ 20 44		15.6	BLUE STAR
LB 06310	08 48 08.	+ 11 56 00.		21.0	FAINT BLUE STAR
LB 06312	08 48 09.	+ 11 58 24.		20.2	FAINT BLUE STAR
LB 06311	08 48 09.	+ 12 03 18.		21.2	FAINT BLUE STAR
ARP 007	08 48 10.	- 16 26			PECULIAR GALAXY
TON-N 0960	08 48 11.	+ 34 17		17.	BLUE STAR
LB 03595	08 48 12.	+ 11 47 54.		19.8	FAINT BLUE STAR
LB 06314	08 48 12.	+ 11 55 12.		19.6	FAINT BLUE STAR
LB 06313	08 48 12.	+ 12 01 00.		21.5	FAINT BLUE STAR
LB 08759	08 48 12.	+ 16 37		18.0	FAINT BLUE STAR
LB 08758	08 48 12.	+ 17 31		15.8	FAINT BLUE STAR
LB 08757	08 48 12.	+ 18 16		18.0	FAINT BLUE STAR
LB 08756	08 48 12.	+ 19 39		16.3	FAINT BLUE STAR
ZC 0848.2+3610	08 48 12.	+ 36 10	400		CLUSTER OF GALAXIES
MCG+12-09-025	08 48 12.	+ 73 08	39	16.	GALAXY
MCG+12-09-026	08 48 12.	+ 73 41	102	13.	GALAXY
MCG-01-23-004	08 48 12.	- 04 42	30	16.	GALAXY
ASS 59	08 48 12.	- 44 49	21600		OB ASSOCIATION VELA OB1
LB 06316	08 48 13.	+ 11 56 36.		20.0	FAINT BLUE STAR
LB 06315	08 48 13.	+ 11 57 24.		21.3	FAINT BLUE STAR
LB 03596	08 48 14.	+ 12 01 54.		18.7	FAINT BLUE STAR
LB 03597	08 48 15.	+ 12 02 42.		19.7	FAINT BLUE STAR
RNGC 2675	08 48 15.	+ 53 48		14.5	GALAXY
LB 06317	08 48 16.	+ 11 59 00.		20.8	FAINT BLUE STAR
IC 0520	08 48 17.	+ 73 40 50.	720	12.8	GALAXY SAB(rs)
LB 03598	08 48 18.	+ 11 48 54.		20.2	FAINT BLUE STAR
LB 06320	08 48 18.	+ 11 55 24.		20.5	FAINT BLUE STAR
LB 06319	08 48 18.	+ 11 59 42.		21.0	FAINT BLUE STAR
LB 06318	08 48 18.	+ 12 03 48.		21.5	FAINT BLUE STAR
LB 08762	08 48 18.	+ 14 50		16.5	FAINT BLUE STAR
LB 08761	08 48 18.	+ 15 44		18.2	FAINT BLUE STAR
LB 08760	08 48 18.	+ 20 22		17.8	FAINT BLUE STAR
ZWG 120.043	08 48 18.	+ 21 55		15.7	GALAXY
ZWG 264.021	08 48 18.	+ 53 48		14.4	GALAXY
UGC 04629	08 48 18.	+ 53 48	96	14.4	GALAXY E
LB 06321	08 48 19.	+ 12 05 18.		21.6	FAINT BLUE STAR
IC 2417	08 48 19.3	+ 18 59 22.			GALAXY S0
LB 06322	08 48 20.	+ 11 57 48.		21.2	FAINT BLUE STAR
RNGC 2682	08 48 20.	+ 12 00		7.5	OPEN CLUSTER
LB 06324	08 48 21.	+ 11 55 18.		21.2	FAINT BLUE STAR
LB 06323	08 48 21.	+ 12 00 36.		19.6	FAINT BLUE STAR
LB 06325	08 48 22.	+ 11 55 42.		20.3	FAINT BLUE STAR
LB 08764	08 48 24.	+ 14 50		17.8	FAINT BLUE STAR
LB 08763	08 48 24.	+ 19 46		15.6	FAINT BLUE STAR
ZWG 264.022	08 48 24.	+ 53 07		15.3	GALAXY
MCG+09-15-036	08 48 24.	+ 53 07	66	15.	GALAXY
ZWG 332.024	08 48 24.	+ 73 08		15.6	GALAXY
ZWG 331.072	08 48 24.	+ 73 08		15.6	GALAXY
ZWG 332.025	08 48 24.	+ 73 41		11.9	GALAXY
ZWG 331.073	08 48 24.	+ 73 41		11.9	GALAXY
UGC 04630	08 48 24.	+ 73 41	156	11.9	GALAXY Sa
LB 06328	08 48 25.	+ 11 55 12.		20.9	FAINT BLUE STAR
LB 06327	08 48 25.	+ 11 56 36.		21.2	FAINT BLUE STAR
LB 06326	08 48 25.	+ 11 57 00.		21.1	FAINT BLUE STAR
MCG+03-23-014	08 48 27.	+ 19 32	66	15.3	GALAXY
LB 06330	08 48 28.	+ 12 02 06.		19.9	FAINT BLUE STAR
LB 06329	08 48 28.	+ 12 03 48.		20.9	FAINT BLUE STAR
RNGC 2680	08 48 28.	+ 31 04		14.5	GALAXY
RNGC 2679	08 48 28.	+ 31 04		14.5	GALAXY
LB 02892	08 48 28.	- 01 46 42.		16.5	FAINT BLUE STAR
LB 06331	08 48 29.	+ 12 05 18.		20.4	FAINT BLUE STAR
ZWG 033.013	08 48 30.	+ 05 15		15.3	GALAXY
ZWG 090.025	08 48 30.	+ 19 32		15.3	GALAXY
UGC 04631	08 48 30.	+ 19 32	72	15.3	GALAXY S0
LB 08765	08 48 30.	+ 20 11		15.5	FAINT BLUE STAR
MCG+05-21-014	08 48 30.	+ 31 03	30	14.3	GALAXY
ZWG 150.041	08 48 30.	+ 31 04		14.3	GALAXY
UGC 04632	08 48 30.	+ 31 04	120	14.3	GALAXY SB:0
KARA.72 176B	08 48 30.	+ 31 04	30		PART OF DOUBLE GALAXY
KARA.72 176A	08 48 30.	+ 31 04	60	14.3	PART OF DOUBLE GALAXY
ZC 0848.5+3341	08 48 30.	+ 33 41	400		CLUSTER OF GALAXIES
ZC 0848.5+3909	08 48 30.	+ 39 09	1340		CLUSTER OF GALAXIES
ZC 0848.5+4114	08 48 30.	+ 41 14	400		CLUSTER OF GALAXIES
ZWG 264.023	08 48 30.	+ 53 04		15.5	GALAXY
UGC 04633	08 48 30.	+ 53 04	66	15.5	GALAXY SBb
SHAH 098	08 48 30.	+ 53 46	180	16.7	GROUP OF COMPACT GALAXIES
MCG+09-15-037	08 48 30.	+ 53 48	36	15.	GALAXY
UGC 04634	08 48 30.	+ 73 44	78	16.5	GALAXY DWRF IR
MCG-04-21-010	08 48 30.	- 21 46 30.	90	14.	GALAXY
LB 06332	08 48 31.	+ 12 02 12.		19.5	FAINT BLUE STAR
LB 02893	08 48 33.	+ 00 41 36.		16.3	FAINT BLUE STAR
LB 06337	08 48 34.	+ 11 55 24.		20.5	FAINT BLUE STAR
LB 06336	08 48 34.	+ 11 58 30.		22.0	FAINT BLUE STAR
LB 06335	08 48 34.	+ 11 59 48.		19.3	FAINT BLUE STAR
LB 06334	08 48 34.	+ 12 03 36.		18.4	FAINT BLUE STAR
LB 06333	08 48 34.	+ 12 05 00.		21.0	FAINT BLUE STAR
LB 06338	08 48 35.	+ 12 02 48.		19.4	FAINT BLUE STAR
IC 2418	08 48 35.3	+ 18 08 02.			GALAXY S
ZWG 005.017	08 48 36.	+ 00 26		15.5	GALAXY
LB 06341	08 48 36.	+ 11 55 48.		21.4	FAINT BLUE STAR
LB 06340	08 48 36.	+ 11 59 18.		20.6	FAINT BLUE STAR
LB 06339	08 48 36.	+ 11 59 54.		18.2	FAINT BLUE STAR
LB 08771	08 48 36.	+ 14 50		18.0	FAINT BLUE STAR
LB 08770	08 48 36.	+ 15 48		16.0	FAINT BLUE STAR
LB 08769	08 48 36.	+ 16 56		18.0	FAINT BLUE STAR
LB 08768	08 48 36.	+ 17 36		16.3	FAINT BLUE STAR
LB 08766	08 48 36.	+ 18 16		17.7	FAINT BLUE STAR
LB 08767	08 48 36.	+ 18 23		17.5	FAINT BLUE STAR
MCG+07-18-058	08 48 36.	+ 41 00 30.	78	15.	GALAXY
ZWG 208.061	08 48 36.	+ 41 02		15.2	GALAXY
UGC 04635	08 48 36.	+ 41 02	108	15.2	GALAXY PECULR
ZC 0848.6+4655	08 48 36.	+ 46 55	810		CLUSTER OF GALAXIES
MCG+08-16-032	08 48 36.	+ 47 44	18	14.	GALAXY
MCG+09-15-038	08 48 36.	+ 53 04	15	16.	GALAXY
ZWG 264.024	08 48 36.	+ 56 17		15.6	GALAXY
LB 06343	08 48 37.	+ 12 01 54.		21.2	FAINT BLUE STAR
LB 06342	08 48 37.	+ 12 03 18.		21.6	FAINT BLUE STAR
LB 06345	08 48 38.	+ 11 57 54.		21.9	FAINT BLUE STAR
LB 06344	08 48 38.	+ 12 04 42.		21.4	FAINT BLUE STAR
LB 02894	08 48 38.	- 00 52 06.		15.8	FAINT BLUE STAR
LB 06347	08 48 39.	+ 11 57 24.		21.8	FAINT BLUE STAR
LB 06346	08 48 39.	+ 12 07 54.		22.2	FAINT BLUE STAR
LB 06349	08 48 40.	+ 12 05 06.		20.6	FAINT BLUE STAR

OBJECT NAME	RIGHT ASCEN.	DECLINATION	DIAM.	MAGN.	TYPE OF OBJECT
LB 06348	08 48 40.	+ 12 07 54.		18.6	FAINT BLUE STAR
LB 06351	08 48 41.	+ 11 53 00.		22.0	FAINT BLUE STAR
LB 06350	08 48 41.	+ 12 08 00.		22.0	FAINT BLUE STAR
LB 06352	08 48 42.	+ 12 06 36.		21.1	FAINT BLUE STAR
LB 08772	08 48 42.	+ 16 30		16.6	FAINT BLUE STAR
ZC 0848.7+2552	08 48 42.	+ 25 52	1340		CLUSTER OF GALAXIES
ZC 0848.7+3727	08 48 42.	+ 37 27	940		CLUSTER OF GALAXIES
ZWG 264.025	08 48 42.	+ 53 11		15.6	GALAXY
ZWG 332.026	08 48 42.	+ 68 40		14.8	GALAXY
LB 06356	08 48 43.	+ 11 56 18.		22.2	FAINT BLUE STAR
LB 06355	08 48 43.	+ 11 56 48.		20.5	FAINT BLUE STAR
LB 06354	08 48 43.	+ 11 59 00.		21.7	FAINT BLUE STAR
LB 06353	08 48 43.	+ 11 59 42.		21.6	FAINT BLUE STAR
LB 06358	08 48 44.	+ 11 52 42.		21.3	FAINT BLUE STAR
LB 06357	08 48 44.	+ 11 54 18.		21.6	FAINT BLUE STAR
LB 03599	08 48 44.	+ 12 17 36.		18.5	FAINT BLUE STAR
LB 06361	08 48 45.	+ 11 51 42.		21.4	FAINT BLUE STAR
LB 06360	08 48 45.	+ 11 52 00.		19.4	FAINT BLUE STAR
LB 06359	08 48 45.	+ 11 54 12.		18.0	FAINT BLUE STAR
LB 02895	08 48 45.	- 00 03 06.		15.2	FAINT BLUE STAR
LB 06362	08 48 46.	+ 12 05 24.		22.2	FAINT BLUE STAR
LB 08773	08 48 48.	+ 19 02		15.4	FAINT BLUE STAR
TON-N 0357	08 48 48.	+ 24 56		13.7	BLUE STAR
ZWG 150.042	08 48 48.	+ 29 29		15.6	GALAXY
UGC 04636	08 48 48.	+ 29 29	60	15.6	GALAXY S
ZC 0848.8+6836	08 48 48.	+ 68 36	1080		CLUSTER OF GALAXIES
MCG-03-23-010	08 48 48.	- 17 22 30.	240	14.	GALAXY
LB 06364	08 48 49.	+ 11 54 48.		21.6	FAINT BLUE STAR
LB 06363	08 48 49.	+ 12 02 00.		21.7	FAINT BLUE STAR
LB 03600	08 48 50.	+ 11 53 42.		18.6	FAINT BLUE STAR
LB 06365	08 48 50.	+ 12 00 18.		22.3	FAINT BLUE STAR
LB 06366	08 48 51.	+ 12 06 06.		20.2	FAINT BLUE STAR
LB 06367	08 48 52.	+ 12 07 54.		19.3	FAINT BLUE STAR
ZWG 033.014	08 48 54.	+ 03 16		15.1	GALAXY
LB 08775	08 48 54.	+ 16 24		16.9	FAINT BLUE STAR
LB 08774	08 48 54.	+ 17 06		16.0	FAINT BLUE STAR
ZC 0848.9+3645	08 48 54.	+ 36 45	1280		CLUSTER OF GALAXIES
MCG+10-13-027	08 48 54.	+ 57 00	18	17.	GALAXY
LB 06368	08 48 55.	+ 12 05 42.		20.7	FAINT BLUE STAR
TON-N 0961	08 48 55.	+ 21 59		16.0	BLUE STAR
IC 2420	08 48 56.	+ 03 17 07.			NONSTELLAR OBJECT
LB 06370	08 48 56.	+ 11 56 06.		19.5	FAINT BLUE STAR
LB 03601	08 48 56.	+ 11 57 18.		19.9	FAINT BLUE STAR
LB 06369	08 48 56.	+ 12 01 48.		21.4	FAINT BLUE STAR
LB 06371	08 48 57.	+ 11 59 18.		19.9	FAINT BLUE STAR
ARC 0710	08 48 57.	+ 36 46		17.9	RICH CLUSTER OF GALAXIES
MCG+10-13-028	08 48 57.	+ 56 59	42	17.	GALAXY
LB 06372	08 49 00.	+ 11 58 00.		21.0	FAINT BLUE STAR
LB 08778	08 49 00.	+ 16 08		17.8	FAINT BLUE STAR
LB 08777	08 49 00.	+ 17 18		16.7	FAINT BLUE STAR
LB 08776	08 49 00.	+ 19 46		16.5	FAINT BLUE STAR
ZWG 150.043	08 49 00.	+ 29 54		15.7	GALAXY
ZC 0849.0+3800	08 49 00.	+ 38 00	1810		CLUSTER OF GALAXIES
ZWG 350.007	08 49 00.	+ 78 25		10.8	GALAXY
ZWG 349.033	08 49 00.	+ 78 25		10.8	GALAXY
UGC 04637	08 49 00.	+ 78 25	408	10.8	GALAXY S0/Sa
MCG+13-07-010	08 49 00.	+ 78 27	234	10.9	GALAXY
RNGC 2653	08 49 00.	+ 78 37			NON-EXISTENT OBJECT
MCG-06-20-003	08 49 00.	- 34 20	90	14.5	GALAXY
LB 03487	08 49 00.	- 78 04		12.8	FAINT BLUE STAR
LB 06373	08 49 01.	+ 11 59 48.		20.8	FAINT BLUE STAR
LB 03602	08 49 01.	+ 12 21 12.		20.4	FAINT BLUE STAR
LB 06375	08 49 02.	+ 11 56 48.		21.4	FAINT BLUE STAR
LB 06374	08 49 02.	+ 12 03 18.		20.0	FAINT BLUE STAR
LB 06377	08 49 04.	+ 11 57 30.		21.8	FAINT BLUE STAR
LB 06376	08 49 04.	+ 12 00 00.		21.5	FAINT BLUE STAR
LB 06378	08 49 05.	+ 12 01 06.		20.1	FAINT BLUE STAR
ARP 257	08 49 05.	- 02 10			PECULIAR GALAXY
LB 08783	08 49 06.	+ 14 42		17.9	FAINT BLUE STAR
LB 08782	08 49 06.	+ 15 08		17.2	FAINT BLUE STAR
LB 08781	08 49 06.	+ 15 30		15.4	FAINT BLUE STAR
ZWG 090.026	08 49 06.	+ 17 08		14.5	GALAXY
UGC 04639	08 49 06.	+ 17 08	84	14.5	GALAXY S0?
MCG+03-23-015	08 49 06.	+ 17 08	72	14.5	GALAXY
LB 08780	08 49 06.	+ 19 24		17.6	FAINT BLUE STAR
LB 08779	08 49 06.	+ 20 07		15.3	FAINT BLUE STAR
KARA.72 177B	08 49 06.	- 02 09	48		PART OF DOUBLE GALAXY
ZWG 005.018	08 49 06.	- 02 10		14.5	GALAXY
UGC 04638	08 49 06.	- 02 10	120	14.5	GALAXY DBL SYS
KARA.72 177A	08 49 06.	- 02 10	90	14.5	PART OF DOUBLE GALAXY
MCG+00-23-006	08 49 06.	- 02 10	30	17.	GALAXY
MCG+00-23-005	08 49 06.	- 02 10	66	14.5	GALAXY
MRSL 265-01/1	08 49 06.	- 45 37	2400		HII REGION
TON-N 0962	08 49 07.	+ 20 36		15.8	BLUE STAR
LB 06379	08 49 08.	+ 12 03 42.		20.5	FAINT BLUE STAR
ARC 0712	08 49 08.	+ 25 50		17.7	RICH CLUSTER OF GALAXIES
RNGC 2655	08 49 08.	+ 78 25		11.5	GALAXY
KEEL 157	08 49 08.8	+ 33 39 48.			NEBULA
LB 06380	08 49 09.	+ 12 02 42.		21.6	FAINT BLUE STAR
LB 06382	08 49 10.	+ 11 59 12.		19.3	FAINT BLUE STAR
LB 06381	08 49 10.	+ 12 01 30.		21.0	FAINT BLUE STAR
LB 06383	08 49 11.	+ 12 00 24.		20.4	FAINT BLUE STAR
LB 08785	08 49 12.	+ 17 03		17.8	FAINT BLUE STAR
LB 08784	08 49 12.	+ 20 37		15.4	FAINT BLUE STAR
ZWG 005.019	08 49 12.	- 01 58		14.2	GALAXY
UGC 04640	08 49 12.	- 01 58	204	14.2	GALAXY Sc
MCG+00-23-007	08 49 12.	- 01 58	210	14.2	GALAXY
MCG-01-23-005	08 49 12.	- 06 57	96	14.	GALAXY
MCG+09-15-039	08 49 15.	+ 51 34	18	16.	GALAXY
LB 08788	08 49 18.	+ 18 46		16.0	FAINT BLUE STAR
LB 08787	08 49 18.	+ 19 33		16.6	FAINT BLUE STAR
LB 08786	08 49 18.	+ 19 52		16.2	FAINT BLUE STAR
ZWG 120.044	08 49 18.	+ 25 20		15.6	GALAXY
MCG+11-11-031	08 49 18.	+ 68 06	39	15.	GALAXY
MRSL 262+01/1	08 49 18.	- 41 44	4020		HII REGION
IC 2419	08 49 20.	+ 18 17 24.			SINGLE STAR
ZC 0849.4+1702	08 49 24.	+ 17 02	3560		CLUSTER OF GALAXIES
LB 08790	08 49 24.	+ 18 26		16.3	FAINT BLUE STAR
ZWG 090.027	08 49 24.	+ 19 15		15.5	GALAXY
LB 08789	08 49 24.	+ 20 25		16.0	FAINT BLUE STAR
ZC 0849.4+3133	08 49 24.	+ 31 33	2550		CLUSTER OF GALAXIES
MCG+07-18-059	08 49 24.	+ 40 52	48	14.5	GALAXY
ZWG 208.062	08 49 24.	+ 40 53		15.0	GALAXY
MCG-02-23-006	08 49 24.	- 14 43 30.	24	15.5	GALAXY
MCG+06-20-011	08 49 27.	+ 33 36	480	09.	GALAXY
TON-N 0963	08 49 29.	+ 33 14		15.9	BLUE STAR
HOLM 100B	08 49 29.	+ 42 37	30	14.4	PART OF MULTIPLE GALAXY
MAI 036	08 49 29.	+ 73 58	33		DWARF SPHEROIDAL GALAXY
ARP 225	08 49 29.	+ 78 26			PECULIAR GALAXY
LB 03603	08 49 30.	+ 11 50 30.		17.6	FAINT BLUE STAR
LB 08792	08 49 30.	+ 16 16		16.8	FAINT BLUE STAR
LB 08791	08 49 30.	+ 16 21		17.6	FAINT BLUE STAR
ZWG 311.021	08 49 30.	+ 68 06		15.2	GALAXY
VHE 22E	08 49 30.	- 48 54			REFLECTION NEBULA
KEEL 158	08 49 33.5	+ 33 33 15.			SPIRAL NEBULA
LB 08795	08 49 36.	+ 16 05		17.1	FAINT BLUE STAR
LB 08794	08 49 36.	+ 16 55		16.9	FAINT BLUE STAR
LB 08793	08 49 36.	+ 17 38		16.0	FAINT BLUE STAR
ZWG 180.017	08 49 36.	+ 33 36		9.7	GALAXY
UGC 04641	08 49 36.	+ 33 36	552	9.7	GALAXY Sb
ZWG 209.001	08 49 36.	+ 42 36		14.1	GALAXY
ZWG 208.063	08 49 36.	+ 42 36		14.1	GALAXY
UGC 04642	08 49 36.	+ 42 36	42	14.1	GALAXY S
BIGO 505	08 49 36.	- 02 31			NEBULA
HOLM 100A	08 49 37.	+ 42 37	30	13.3	PART OF MULTIPLE GALAXY
MCG+04-21-024	08 49 39.	+ 21 35	90	14.6	GALAXY
RNGC 2683	08 49 39.	+ 33 37		11.0	GALAXY
MCG+07-18-060	08 49 39.	+ 42 36 30.	36	14.5	GALAXY
ZWG 033.015	08 49 42.	+ 06 07		15.5	GALAXY
KARA.73B 0288	08 49 42.	+ 06 07	24	15.5	ISOLATED GALAXY S
LB 08796	08 49 42.	+ 15 30		18.0	FAINT BLUE STAR
LB 02896	08 49 42.	+ 21 09 30.		16.4	FAINT BLUE STAR
ZC 0849.7+4015	08 49 42.	+ 40 15	1280		CLUSTER OF GALAXIES
MCG+09-15-040	08 49 42.	+ 55 06	15	16.	GALAXY
MCG+10-13-029	08 49 42.	+ 62 07	18	16.	GALAXY
ZC 0849.7+6357	08 49 42.	+ 63 57	1480		CLUSTER OF GALAXIES
MCG+13-07-011	08 49 42.	+ 77 03	18	17.	GALAXY
VHE 22D	08 49 42.	- 48 54			REFLECTION NEBULA
VHE 22C	08 49 42.	- 48 56			REFLECTION NEBULA
ARC 0713	08 49 48.	+ 18 25		16.8	RICH CLUSTER OF GALAXIES
ZWG 090.028	08 49 48.	+ 19 31		15.7	GALAXY
LB 08798	08 49 48.	+ 20 16		17.0	FAINT BLUE STAR
LB 08797	08 49 48.	+ 20 41		15.6	FAINT BLUE STAR
ZWG 120.045	08 49 48.	+ 21 37		14.6	GALAXY
UGC 04643	08 49 48.	+ 21 37	102	14.6	GALAXY Sb/Sc
TON-N 0358	08 49 48.	+ 31 56		14.0	BLUE STAR
ZC 0849.8+5937	08 49 48.	+ 59 37	2350		CLUSTER OF GALAXIES
ZWG 350.008	08 49 48.	+ 79 53		15.4	GALAXY
ZWG 349.034	08 49 48.	+ 79 53		15.4	GALAXY
UGC 04644	08 49 48.	+ 79 53	72	15.4	GALAXY SBb
LB 02897	08 49 49.	- 01 49 06.		16.2	FAINT BLUE STAR
LB 08799	08 49 54.	+ 14 38		17.7	FAINT BLUE STAR
ZC 0849.9+1816	08 49 54.	+ 18 16	1880		CLUSTER OF GALAXIES
MCG+09-15-041	08 49 54.	+ 51 30	180	11.3	GALAXY
ZWG 264.026	08 49 54.	+ 51 31		10.4	GALAXY
UGC 04645	08 49 54.	+ 51 31	240	10.4	GALAXY Sa
MRSL 264+00/1	08 49 54.	- 43 51	1920		HII REGION
LB 03604	08 49 55.	+ 11 51 06.		18.8	FAINT BLUE STAR
TON-N 0964	08 49 55.	+ 22 36		15.7	BLUE STAR
LB 03605	08 49 56.	+ 12 51 06.		17.0	FAINT BLUE STAR
KEEL 159	08 49 59.3	+ 33 34 07.			NEBULA
LB 09844	08 50	- 86 22		13.5	FAINT BLUE STAR
LB 08802	08 50 00.	+ 15 45		17.4	FAINT BLUE STAR
ZWG 090.029	08 50 00.	+ 19 12		15.7	GALAXY
LB 08801	08 50 00.	+ 19 49		15.5	FAINT BLUE STAR
LB 08800	08 50 00.	+ 20 36		18.0	FAINT BLUE STAR
ZC 0850.0+4143	08 50 00.	+ 41 43	1010		CLUSTER OF GALAXIES
MCG+08-16-033	08 50 00.	+ 45 30	120	16.	GALAXY
RNGC 2681	08 50 00.	+ 51 30		11.5	GALAXY
7ZW 254	08 50 00.	+ 57 22			COMPACT GALAXY
MCG+13-07-012	08 50 00.	+ 79 54	12	15.	GALAXY
ZWG 364.008	08 50 00.	+ 85 42		13.2	GALAXY
ZWG 363.049	08 50 00.	+ 85 42		13.2	GALAXY
UGC 04646	08 50 00.	+ 85 42	216	13.2	GALAXY Sc
MCG-03-23-011	08 50 03.	- 17 33	18	14.	GALAXY
LB 03606	08 50 05.	+ 12 18 54.		16.8	FAINT BLUE STAR
LB 02898	08 50 05.	- 00 30 06.		16.0	FAINT BLUE STAR
LB 08804	08 50 06.	+ 18 46		18.1	FAINT BLUE STAR
ZWG 090.030	08 50 06.	+ 19 32		15.7	GALAXY
LB 08803	08 50 06.	+ 19 47		18.3	FAINT BLUE STAR
ZC 0850.1+2113	08 50 06.	+ 21 13	1410		CLUSTER OF GALAXIES
ZC 0850.1+2239	08 50 06.	+ 22 39	1610		CLUSTER OF GALAXIES
ZWG 005.020	08 50 06.	- 02 25		14.1	GALAXY
UGC 04647	08 50 06.	- 02 25	120	14.1	GALAXY Sa-b
MCG+00-23-008	08 50 06.	- 02 25	120	14.1	GALAXY
RNGC 2690	08 50 08.	- 02 25		14.0	GALAXY
ZC 0850.2+1327	08 50 12.	+ 13 27	1140		CLUSTER OF GALAXIES
LB 08811	08 50 12.	+ 15 06		16.5	FAINT BLUE STAR
LB 08810	08 50 12.	+ 16 43		18.2	FAINT BLUE STAR
LB 08809	08 50 12.	+ 17 22		17.9	FAINT BLUE STAR
LB 08808	08 50 12.	+ 18 00		17.0	FAINT BLUE STAR
LB 08807	08 50 12.	+ 18 12		17.2	FAINT BLUE STAR
LB 08806	08 50 12.	+ 19 38		18.1	FAINT BLUE STAR
LB 08805	08 50 12.	+ 20 21		17.9	FAINT BLUE STAR
ZC 0850.2+2650	08 50 12.	+ 26 50	2420		CLUSTER OF GALAXIES
UGC 04648	08 50 12.	+ 45 31	72	17.	GALAXY DWARF
ZC 0850.2+4829	08 50 12.	+ 48 29	1210		CLUSTER OF GALAXIES
ZWG 264.027	08 50 12.	+ 52 35		15.2	GALAXY
ZC 0850.2+5835	08 50 12.	+ 58 35	2550		CLUSTER OF GALAXIES
MCG+12-09-027	08 50 12.	+ 72 54	21	16.	GALAXY
ZWG 005.021	08 50 12.	- 01 38		15.5	GALAXY
MCG+00-23-009	08 50 12.	- 01 38	54	15.5	GALAXY
OCL 0737	08 50 12.	- 37 25	84	13.	OPEN STAR CLUSTER
MCG+09-15-042	08 50 15.	+ 52 35	18	16.	GALAXY
ZWG 061.017	08 50 18.	+ 09 30		15.1	GALAXY
MCG+02-23-008	08 50 18.	+ 09 30	36	15.1	GALAXY
LB 08813	08 50 18.	+ 17 51		18.1	FAINT BLUE STAR
LB 08812	08 50 18.	+ 19 33		18.0	FAINT BLUE STAR
ZWG 209.002	08 50 18.	+ 40 55		14.9	GALAXY
ZWG 208.064	08 50 18.	+ 40 55		14.9	GALAXY
UGC 04649	08 50 18.	+ 40 55	78	14.9	GALAXY
ZC 0850.3+6747	08 50 18.	+ 67 47	1410		CLUSTER OF GALAXIES
ZWG 332.027	08 50 18.	+ 72 54		15.3	GALAXY
VV 168C	08 50 18.	- 01 36	15	16.	INTERACTING GALAXY
VV 168B	08 50 18.	- 01 36	30	16.5	INTERACTING GALAXY
VV 168B	08 50 18.	- 01 36	30	17.5	INTERACTING GALAXY
VV 168A	08 50 18.	- 01 36	24	17.	INTERACTING GALAXY
VV 168	08 50 18.	- 01 36	60		INTERACTING GALAXY
PK266-01.1	08 50 19.6	- 46 05 56.	10		PLANETARY NEBULA
BC 3CR208	08 50 22.64	+ 14 04 16.5		17.42	QUASI-STELLAR OBJECT
SHB 126	08 50 22.7	+ 14 04 16.		17.4	QUASI-STELLAR OBJECT
LB 08819	08 50 24.	+ 16 20		18.0	FAINT BLUE STAR
LB 08818	08 50 24.	+ 17 34		17.2	FAINT BLUE STAR
LB 08817	08 50 24.	+ 17 44		17.7	FAINT BLUE STAR
LB 08816	08 50 24.	+ 18 06		18.5	FAINT BLUE STAR
LB 08815	08 50 24.	+ 18 19		17.0	FAINT BLUE STAR

OBJECT NAME	RIGHT ASCEN.	DECLINATION	DIAM.	MAGN.	TYPE OF OBJECT
LB 08814	08 50 24.	+ 18 59		18.5	FAINT BLUE STAR
LB 08820	08 50 24.	+ 19 05		17.5	FAINT BLUE STAR
MCG+07-18-061	08 50 24.	+ 39 18 30.	108	14.5	GALAXY
ZWG 209.003	08 50 24.	+ 39 20		14.6	GALAXY
ZWG 208.065	08 50 24.	+ 39 20		14.6	GALAXY
UGC 04650	08 50 24.	+ 39 20	114	14.6	GALAXY Sa-b
ZC 0850.4+5437	08 50 24.	+ 54 37	470		CLUSTER OF GALAXIES
OCL 0736	08 50 24.	- 37 23	150	12.	OPEN STAR CLUSTER
SEY 050	08 50 26.	+ 72 53 50.		15.4	FAINT GALAXY
MCG+07-18-062	08 50 27.	+ 40 54 30.	42	15.	GALAXY
KEEL 160	08 50 27.8	+ 34 03 30.			NEBULA
LB 02899	08 50 30.	+ 00 16 24.		16.0	FAINT BLUE STAR
ZWG 061.018	08 50 30.	+ 09 20		15.0	GALAXY
UGC 04652	08 50 30.	+ 09 20	84	15.0	GALAXY Sa-b
MCG+02-23-009	08 50 30.	+ 09 20	96	15.0	GALAXY
IC 0523	08 50 30.	+ 09 20 40.			NONSTELLAR OBJECT
LB 08821	08 50 30.	+ 16 10		18.0	FAINT BLUE STAR
ZWG 120.046	08 50 30.	+ 20 47		15.6	GALAXY
TON-N 0359	08 50 30.	+ 28 53		16.6	BLUE STAR
ZC 0850.5+3237	08 50 30.	+ 32 37	6590		CLUSTER OF GALAXIES
MCG+10-13-030	08 50 30.	+ 59 08	30	17.	GALAXY
UGC 04651	08 50 30.	- 00 51	60	17.	GALAXY
LB 03488	08 50 30.	- 70 47		15.2	FAINT BLUE STAR
LB 08823	08 50 36.	+ 19 28		17.8	FAINT BLUE STAR
LB 08822	08 50 36.	+ 19 32		18.0	FAINT BLUE STAR
ZC 0850.6+3536	08 50 36.	+ 35 36	5380		CLUSTER OF GALAXIES
ZWG 332.028	08 50 36.	+ 73 59		15.7	GALAXY
ZWG 331.074	08 50 36.	+ 73 59		15.7	GALAXY
ZWG 005.022	08 50 36.	- 01 28		15.5	GALAXY
TON-N 0965	08 50 37.	+ 21 56		17.0	BLUE STAR
LB 08825	08 50 42.	+ 15 06		18.3	FAINT BLUE STAR
LB 08824	08 50 42.	+ 16 11		18.3	FAINT BLUE STAR
VV 243C	08 50 42.	+ 35 19	15	17.5	INTERACTING GALAXY
VV 243B	08 50 42.	+ 35 19	21	16.5	INTERACTING GALAXY
VV 243A	08 50 42.	+ 35 19	30	16.	INTERACTING GALAXY
VV 243	08 50 42.	+ 35 19	90		INTERACTING GALAXY
ZWG 180.018	08 50 42.	+ 35 20		15.0	GALAXY
UGC 04653	08 50 42.	+ 35 20	108	15.0	GALAXY TRP SYS
ZWG 209.004	08 50 42.	+ 40 33		15.6	GALAXY
ZWG 208.066	08 50 42.	+ 40 33		15.6	GALAXY
MCG+09-15-043	08 50 42.	+ 54 39	60	17.	GALAXY
ZWG 288.010	08 50 42.	+ 57 21		13.9	GALAXY
UGC 04654	08 50 42.	+ 57 21	60	13.9	GALAXY S0
MCG-04-21-011	08 50 42.	- 25 08	48	14.	GALAXY
IC 0522	08 50 44.	+ 57 21 33.			NONSTELLAR OBJECT
ARP 195	08 50 45.	+ 35 20			PECULIAR GALAXY
SEY 051	08 50 46.	+ 73 59 13.		15.3	FAINT GALAXY
LB 08830	08 50 48.	+ 16 50		17.8	FAINT BLUE STAR
LB 08829	08 50 48.	+ 17 10		16.6	FAINT BLUE STAR
LB 08828	08 50 43.	+ 18 31		16.7	FAINT BLUE STAR
LB 08827	08 50 48.	+ 19 13		16.1	FAINT BLUE STAR
LB 08826	08 50 48.	+ 19 57		17.9	FAINT BLUE STAR
ZWG 120.047	08 50 48.	+ 20 48		15.5	GALAXY
ZWG 150.044	08 50 48.	+ 30 09		15.5	GALAXY
MCG+06-20-012	08 50 48.	+ 35 20 30.	102	15.	GALAXY
ZC 0850.8+5416	08 50 48.	+ 54 16	1280		CLUSTER OF GALAXIES
ZC 0850.8+5605	08 50 48.	+ 56 05	1680		CLUSTER OF GALAXIES
MCG+10-13-031	08 50 48.	+ 57 20	51	14.	GALAXY
KARA.68 053	08 50 48.	+ 72 12	60		DWARF GALAXY
VHE 22A	08 50 48.	- 48 33			REFLECTION NEBULA
LB 02900	08 50 54.	+ 02 03 18.		16.2	FAINT BLUE STAR
ZWG 033.016	08 50 54.	+ 04 58		15.6	GALAXY
UGC 04655	08 50 54.	+ 04 58	72	15.6	GALAXY Sc
ZC 0850.9+1852	08 50 54.	+ 18 52	1010		CLUSTER OF GALAXIES
ZWG 120.048	08 50 54.	+ 22 10		15.6	GALAXY
ZC 0850.9+2416	08 50 54.	+ 24 16	5110		CLUSTER OF GALAXIES
ZC 0850.9+2903	08 50 54.	+ 29 03	3230		CLUSTER OF GALAXIES
ZWG 264.028	08 50 54.	+ 53 58		15.7	GALAXY
MCG+09-15-044	08 50 54.	+ 54 30	36	18.	GALAXY
ZC 0850.9+5514	08 50 54.	+ 55 14	540		CLUSTER OF GALAXIES
ZC 0850.9+5535	08 50 54.	+ 55 35	740		CLUSTER OF GALAXIES
TON-N 0966	08 50 55.	+ 19 52		16.5	BLUE STAR
LBN 1080	08 51	- 28 10	420		BRIGHT NEBULA
LB 08833	08 51 00.	+ 17 51		18.6	FAINT BLUE STAR
ZWG 090.031	08 51 00.	+ 18 22		15.3	GALAXY
UGC 04656	08 51 00.	+ 18 22	90	15.3	GALAXY Sb
LB 08832	08 51 00.	+ 18 23		18.2	FAINT BLUE STAR
LB 08831	08 51 00.	+ 18 45		17.5	FAINT BLUE STAR
TON-N 0360	08 51 00.	+ 28 10		16.0	BLUE STAR
ZWG 288.011	08 51 00.	+ 57 51		15.7	GALAXY
MCG+10-13-032	08 51 00.	+ 57 51	39	15.	GALAXY
MCG+10-13-033	08 51 00.	+ 61 34	18	16.	GALAXY
MCG+11-11-032	08 51 00.	+ 64 35	42	17.	GALAXY
KEEL 161	08 51 04.1	+ 34 07 26.			NEBULA
ZWG 005.023	08 51 06.	+ 01 44		15.2	GALAXY
KARA.73B 0289	08 51 06.	+ 01 44	36	15.2	ISOLATED GALAXY S0
ZWG 090.032	08 51 06.	+ 18 52		15.6	GALAXY
UGC 04657	08 51 06.	+ 18 52	90	15.6	GALAXY Sc
LB 08834	08 51 06.	+ 19 54		17.5	FAINT BLUE STAR
TON-N 0361	08 51 06.	+ 31 16		16.5	BLUE STAR
MCG+08-16-034	08 51 06.	+ 47 17	72	14.	GALAXY
ZC 0851.1+4925	08 51 06.	+ 49 25	4030		CLUSTER OF GALAXIES
MCG+09-15-046	08 51 06.	+ 52 40	30	16.	GALAXY
ZWG 264.029	08 51 06.	+ 53 53		15.5	GALAXY
MCG+09-15-045	08 51 06.	+ 54 46	36	16.	GALAXY
MCG+10-13-034	08 51 06.	+ 61 37	18	16.	GALAXY
VHE 22B	08 51 06.	- 48 35			REFLECTION NEBULA
TON-N 0967	08 51 07.	+ 20 26		16.6	BLUE STAR
TON-N 0968	08 51 07.	+ 22 27		16.3	BLUE STAR
LB 02901	08 51 09.	+ 02 28 42.		16.2	FAINT BLUE STAR
MCG+06-20-013	08 51 09.	+ 32 51 30.	150	14.	GALAXY
ZC 0851.2+0234	08 51 12.	+ 02 34	610		CLUSTER OF GALAXIES
ZWG 033.017	08 51 12.	+ 05 26		15.5	GALAXY
LB 08837	08 51 12.	+ 17 27		18.1	FAINT BLUE STAR
LB 08836	08 51 12.	+ 18 54		14.8	FAINT BLUE STAR
LB 08835	08 51 12.	+ 20 17		18.1	FAINT BLUE STAR
ZC 0851.2+2723	08 51 12.	+ 27 23	1550		CLUSTER OF GALAXIES
ZWG 180.019	08 51 12.	+ 32 52		14.9	GALAXY
UGC 04658	08 51 12.	+ 32 52	144	14.9	GALAXY Sc
KARA.72 178A	08 51 12.	+ 32 52	114	14.9	PART OF DOUBLE GALAXY
ZWG 237.023	08 51 12.	+ 47 17		15.5	GALAXY
UGC 04659	08 51 12.	+ 47 17	144	15.5	GALAXY
KARA.73B 0290	08 51 12.	+ 47 17	126	15.5	ISOLATED GALAXY S
ZC 0851.2+4932	08 51 12.	+ 49 32	670		CLUSTER OF GALAXIES
MCG+10-13-035	08 51 12.	+ 58 48	30	16.	GALAXY
MCG+06-20-013A	08 51 15.	+ 32 48	36	15.5	GALAXY
MCG+06-20-014	08 51 15.	+ 34 45	54	16.	GALAXY
LB 02902	08 51 16.	+ 21 12 48.		16.1	FAINT BLUE STAR
ZWG 061.019	08 51 18.	+ 09 49		15.7	GALAXY
LB 08840	08 51 18.	+ 16 08		17.9	FAINT BLUE STAR
ZWG 090.033	08 51 18.	+ 17 37		15.2	GALAXY
LB 08839	08 51 18.	+ 19 34		18.2	FAINT BLUE STAR
LB 08838	08 51 18.	+ 20 28		16.8	FAINT BLUE STAR
ZWG 180.020	08 51 18.	+ 32 49		15.3	GALAXY
KARA.72 178B	08 51 18.	+ 32 49	36	15.3	PART OF DOUBLE GALAXY
IC 2421	08 51 18.	+ 32 52 38.			NONSTELLAR OBJECT
HOLM 101A	08 51 18.	+ 32 53	120	13.9	PART OF MULTIPLE GALAXY
UGC 04660	08 51 18.	+ 34 44	102	17.	GALAXY DWRF SP
ZWG 180.021	08 51 18.	+ 36 38		15.1	GALAXY
UGC 04661	08 51 18.	+ 36 38	60	15.1	GALAXY S
MCG+06-20-015	08 51 18.	+ 36 39	54	15.	GALAXY
ZWG 180.022	08 51 18.	+ 36 41		15.7	GALAXY
ZWG 209.005	08 51 18.	+ 42 43		15.4	GALAXY
ZWG 208.067	08 51 18.	+ 42 43		15.4	GALAXY
ZWG 237.024	08 51 18.	+ 49 20		13.4	GALAXY
UGC 04662	08 51 18.	+ 49 20	60	13.4	GALAXY S
MCG+08-16-035	08 51 18.	+ 49 21	60	13.	GALAXY
ZC 0851.3+6521	08 51 18.	+ 65 21	1880		CLUSTER OF GALAXIES
MCG-01-23-006	08 51 18.	- 06 51	36	15.	GALAXY
TON-N 0969	08 51 19.	+ 22 34		16.7	BLUE STAR
RNGC 2684	08 51 21.	+ 49 20		13.5	GALAXY
LB 08843	08 51 24.	+ 18 44		17.6	FAINT BLUE STAR
LB 08842	08 51 24.	+ 19 04		17.6	FAINT BLUE STAR
LB 08841	08 51 24.	+ 19 25		18.1	FAINT BLUE STAR
ZWG 120.049	08 51 24.	+ 23 21		15.5	GALAXY
MCG+08-16-037	08 51 24.	+ 49 20	6	16.	GALAXY
MCG+08-16-036	08 51 24.	+ 49 20	12	16.	GALAXY
MCG+09-15-047	08 51 24.	+ 52 14	42	16.	GALAXY
ZWG 264.030	08 51 24.	+ 52 15		15.7	GALAXY
MCG+10-13-036	08 51 24.	+ 57 32	30	16.	GALAXY
MCG+10-13-037	08 51 24.	+ 58 43	39	16.	GALAXY
MCG+11-11-033	08 51 24.	+ 68 26	51	16.	GALAXY
TON-N 0970	08 51 25.	+ 20 19		16.6	BLUE STAR
IC 0512	08 51 25.	+ 85 41 18.			NONSTELLAR OBJECT
HOLM 101B	08 51 26.	+ 32 50	30	14.6	PART OF MULTIPLE GALAXY
MCG+07-18-063	08 51 27.	+ 42 42 30.	60	15.	GALAXY
RNGC 2686B	08 51 27.	+ 49 20		16.0	GALAXY
RNGC 2686A	08 51 27.	+ 49 20		16.0	GALAXY
RNGC 2691	08 51 29.	+ 39 44		14.0	GALAXY
ZWG 033.018	08 51 30.	+ 07 10		15.4	GALAXY
PK219+31.1	08 51 30.	+ 09 06	1040	12.2	PLANETARY NEBULA
LB 08846	08 51 30.	+ 15 32		17.7	FAINT BLUE STAR
LB 08845	08 51 30.	+ 19 34		17.9	FAINT BLUE STAR
ZWG 090.034	08 51 30.	+ 20 25		15.6	GALAXY
MCG+03-23-016	08 51 30.	+ 20 25	24	15.6	GALAXY
UGC 04663	08 51 30.	+ 20 34	108	18.	GALAXY DWARF
LB 08844	08 51 30.	+ 20 39		17.7	FAINT BLUE STAR
ZC 0851.5+3358	08 51 30.	+ 33 58	3020		CLUSTER OF GALAXIES
MCG+07-18-064	08 51 30.	+ 39 42 30.	66	14.5	GALAXY
ZWG 209.006	08 51 30.	+ 39 44		13.9	GALAXY
ZWG 208.068	08 51 30.	+ 39 44		13.9	GALAXY
MRK 391	08 51 30.	+ 39 44	24	14.5	GALAXY WITH UV CONTINUUM
KW 39	08 51 30.	+ 39 44	47		SEYFERT GALAXY
VVI 27	08 51 30.	+ 39 44	24	14.5	SEYFERT GALAXY
UGC 04664	08 51 30.	+ 39 44	96	13.9	GALAXY Sa?
ZC 0851.5+4207	08 51 30.	+ 42 07	3090		CLUSTER OF GALAXIES
ARC 0714	08 51 30.	+ 42 07		16.8	RICH CLUSTER OF GALAXIES
MCG+08-16-038	08 51 30.	+ 49 21	12	17.	GALAXY
MCG+08-16-039	08 51 30.	+ 49 21 30.	21	16.	GALAXY
MCG+09-15-048	08 51 30.	+ 52 40	42	16.	GALAXY
MCG-01-23-007	08 51 30.	- 06 40	30	17.	GALAXY
MRSL 263+01/1	08 51 30.	- 42 24	2400		HII REGION
IC 2422	08 51 33.	+ 20 24 47.			NONSTELLAR OBJECT
RNGC 2687B	08 51 33.	+ 49 21		16.0	GALAXY
RNGC 2687A	08 51 33.	+ 49 21		17.0	GALAXY
ZWG 005.024	08 51 36.	+ 00 41		15.5	GALAXY
ZWG 033.019	08 51 36.	+ 06 15		15.7	GALAXY
ZWG 061.020	08 51 36.	+ 09 32		15.6	GALAXY
LB 08848	08 51 36.	+ 15 05		17.9	FAINT BLUE STAR
LB 08847	08 51 36.	+ 18 50		18.1	FAINT BLUE STAR
ARC 0715	08 51 36.	+ 35 36		17.9	RICH CLUSTER OF GALAXIES
MCG+09-15-049	08 51 36.	+ 52 13	60	17.	GALAXY
7ZW 255	08 51 36.	+ 59 43			COMPACT GALAXY
ZWG 332.029	08 51 36.	+ 68 28		15.5	GALAXY
ZWG 311.022	08 51 36.	+ 68 28		15.5	GALAXY
UGC 04665	08 51 36.	+ 68 28	72	15.5	GALAXY S
LB 02903	08 51 36.	- 00 59 18.		16.2	FAINT BLUE STAR
MCG-01-23-008	08 51 36.	- 06 53	42	15.	GALAXY
MCG-05-21-002	08 51 36.	- 32 45	30	15.	GALAXY
PK261+02.1	08 51 37.	- 39 53	20		PLANETARY NEBULA
LB 08850	08 51 42.	+ 16 11		18.4	FAINT BLUE STAR
LB 08849	08 51 42.	+ 19 27		17.9	FAINT BLUE STAR
MCG+04-21-025	08 51 42.	+ 20 45	42	15.0	GALAXY
ZWG 120.050	08 51 42.	+ 20 47		15.0	GALAXY
ZC 0851.7+4820	08 51 42.	+ 48 20	740		CLUSTER OF GALAXIES
ZWG 288.012	08 51 42.	+ 58 55		12.1	GALAXY
UGC 04666	08 51 42.	+ 58 55	282	12.1	GALAXY PECULR
TON-N 0971	08 51 43.	+ 22 30		17.1	BLUE STAR
RNGC 2689	08 51 45.	+ 49 21			GALAXY
RNGC 2685	08 51 45.	+ 58 56		12.0	GALAXY
LB 02904	08 51 48.	+ 01 49 00.		15.8	FAINT BLUE STAR
LB 08859	08 51 48.	+ 15 00		18.0	FAINT BLUE STAR
LB 08858	08 51 48.	+ 15 43		18.8	FAINT BLUE STAR
LB 08857	08 51 48.	+ 17 32		18.0	FAINT BLUE STAR
LB 08856	08 51 48.	+ 18 17		18.0	FAINT BLUE STAR
LB 08855	08 51 48.	+ 18 30		18.0	FAINT BLUE STAR
LB 08854	08 51 48.	+ 18 32		18.3	FAINT BLUE STAR
LB 08853	08 51 48.	+ 18 44		17.8	FAINT BLUE STAR
LB 08852	08 51 48.	+ 19 19		18.0	FAINT BLUE STAR
LB 08851	08 51 48.	+ 20 04		18.2	FAINT BLUE STAR
ZC 0851.8+2036	08 51 48.	+ 20 36	940		CLUSTER OF GALAXIES
ZC 0851.8+3036	08 51 48.	+ 30 36	870		CLUSTER OF GALAXIES
ZC 0851.8+3534	08 51 48.	+ 35 34	870		CLUSTER OF GALAXIES
MCG+08-16-040	08 51 48.	+ 49 18	18	16.	GALAXY
ZC 0851.8+5101	08 51 48.	+ 51 01	540		CLUSTER OF GALAXIES
ZWG 288.013	08 51 48.	+ 57 45		15.6	GALAXY
MCG-01-23-009	08 51 48.	- 06 52	18	17.	GALAXY
RNGC 2688	08 51 51.	+ 49 18		16.0	GALAXY
LB 08862	08 51 54.	+ 15 33		18.4	FAINT BLUE STAR
LB 08861	08 51 54.	+ 16 20		17.9	FAINT BLUE STAR
LB 08860	08 51 54.	+ 20 06		16.8	FAINT BLUE STAR
ZWG 090.035	08 51 54.	+ 20 25		14.7	GALAXY
UGC 04667	08 51 54.	+ 20 25	66	14.7	GALAXY Sb/SBb
MCG+10-13-038	08 51 54.	+ 57 45	57	17.	GALAXY
MCG+10-13-039	08 51 54.	+ 58 55	300	12.0	GALAXY
ZWG 005.025	08 51 54.	- 02 53		13.3	GALAXY

OBJECT NAME	RIGHT ASCEN.	DECLINATION	DIAM.	MAGN.	TYPE OF OBJECT
MCG+00-23-010	08 51 54.	- 02 53	48	13.3	GALAXY
REIN 2.069	08 51 55.89	- 02 52 34.2			NEBULA
RNGC 2695	08 51 56.	- 02 53		13.5	GALAXY
LB 02905	08 51 57.	+ 02 05 42.		16.4	FAINT BLUE STAR
RNGC 2696	08 51 57.	- 04 47			NON-EXISTENT OBJECT
BC 0J287	08 51 57.3	+ 20 17 58.		14.5	QUASI-STELLAR OBJECT
REIN 4.056	08 51 57.53	- 02 52 35.4			NEBULA
LBN 0692	08 52	+ 67 15	900		BRIGHT NEBULA
LBN 1081	08 52	- 28 00	2700		BRIGHT NEBULA
LB 08865	08 52 00.	+ 15 13		18.4	FAINT BLUE STAR
LB 08864	08 52 00.	+ 15 22		17.7	FAINT BLUE STAR
LB 08863	08 52 00.	+ 19 42		18.0	FAINT BLUE STAR
MCG+03-23-017	08 52 00.	+ 20 26	60	14.7	GALAXY
ZWG 150.045	08 52 00.	+ 26 53		15.5	GALAXY
ZC 0852.0+3935	08 52 00.	+ 39 35	1140		CLUSTER OF GALAXIES
MCG+10-13-040	08 52 00.	+ 62 32	39	16.	GALAXY
ZWG 311.023	08 52 00.	+ 62 33		15.3	GALAXY
7ZW 256	08 52 00.	+ 82 53			COMPACT GALAXY
MCG+14-05-002	08 52 00.	+ 85 41	120	14.	GALAXY
ZWG 033.020	08 52 06.	+ 03 16		15.6	GALAXY
ZC 0852.1+1340	08 52 06.	+ 13 40	1480		CLUSTER OF GALAXIES
ZC 0852.1+1606	08 52 06.	+ 16 06	670		CLUSTER OF GALAXIES
LB 08869	08 52 06.	+ 16 33		18.1	FAINT BLUE STAR
LB 08868	08 52 06.	+ 18 56		18.1	FAINT BLUE STAR
LB 08867	08 52 06.	+ 19 22		18.1	FAINT BLUE STAR
LB 08866	08 52 06.	+ 20 32		17.6	FAINT BLUE STAR
ZWG 209.007	08 52 06.	+ 42 13		15.3	GALAXY
ZWG 208.069	08 52 06.	+ 42 13		15.3	GALAXY
ZC 0852.1+5430	08 52 06.	+ 54 30	870		CLUSTER OF GALAXIES
MCG+10-13-041	08 52 06.	+ 57 31	27	16.	GALAXY
ZWG 288.014	08 52 06.	+ 57 52		15.6	GALAXY
UGC 04668	08 52 06.	+ 57 52	66	15.6	GALAXY Sb-c
MCG+10-13-042	08 52 06.	+ 62 28	51	16.	GALAXY
ZWG 311.024	08 52 06.	+ 66 31		15.6	GALAXY
LB 02906	08 52 08.	+ 21 27 30.		16.2	FAINT BLUE STAR
ZC 0852.2+0858	08 52 12.	+ 08 58	470		CLUSTER OF GALAXIES
ZC 0852.2+1551	08 52 12.	+ 15 51	670		CLUSTER OF GALAXIES
LB 08871	08 52 12.	+ 17 08		17.5	FAINT BLUE STAR
LB 08870	08 52 12.	+ 17 35		16.9	FAINT BLUE STAR
ZC 0852.2+4840	08 52 12.	+ 48 40	940		CLUSTER OF GALAXIES
ZC 0852.2+5403	08 52 12.	+ 54 03	400		CLUSTER OF GALAXIES
MCG+10-13-043	08 52 12.	+ 57 35	36	16.	GALAXY
MCG+10-13-044	08 52 12.	+ 57 51	66	15.	GALAXY
ZWG 005.026	08 52 12.	- 01 42		15.6	GALAXY
ARP 336	08 52 14.	+ 59 00			PECULIAR GALAXY
MCG+09-15-050	08 52 15.	+ 56 13	42	18.	GALAXY
LB 08875	08 52 18.	+ 16 45		17.9	FAINT BLUE STAR
LB 08874	08 52 18.	+ 18 04		18.6	FAINT BLUE STAR
LB 08873	08 52 18.	+ 18 22		18.2	FAINT BLUE STAR
LB 08872	08 52 18.	+ 18 25		16.7	FAINT BLUE STAR
ZC 0852.3+1840	08 52 18.	+ 18 40	540		CLUSTER OF GALAXIES
ZWG 090.036	08 52 18.	+ 19 07		15.5	GALAXY
UGC 04669	08 52 18.	+ 19 07	96	15.5	GALAXY Sc/IRR
MCG+03-23-018	08 52 18.	+ 19 07	72	15.5	GALAXY
7ZW 257	08 52 18.	+ 67 04			COMPACT GALAXY
RNGC 2714	08 52 18.	- 59 01			UNVERIFIED SOUTHERN OBJECT
ARC 0720	08 52 19.	+ 15 50		17.7	RICH CLUSTER OF GALAXIES
ARC 0704	08 52 23.	+ 79 30		17.5	RICH CLUSTER OF GALAXIES
LB 08880	08 52 24.	+ 15 17		18.4	FAINT BLUE STAR
LB 08879	08 52 24.	+ 16 16		18.1	FAINT BLUE STAR
LB 08878	08 52 24.	+ 19 51		16.8	FAINT BLUE STAR
LB 08877	08 52 24.	+ 19 59		18.1	FAINT BLUE STAR
LB 08876	08 52 24.	+ 20 04		18.4	FAINT BLUE STAR
TON-N 0972	08 52 25.	+ 19 52		16.6	BLUE STAR
ARC 0716	08 52 27.	+ 48 40		17.1	RICH CLUSTER OF GALAXIES
HOLM 102B	08 52 28.	+ 42 28	30	14.6	PART OF MULTIPLE GALAXY
REIN 4.057	08 52 28.27	- 02 47 48.0			NEBULA
LB 08882	08 52 30.	+ 14 46		18.5	FAINT BLUE STAR
LB 08881	08 52 30.	+ 19 25		17.5	FAINT BLUE STAR
ZWG 005.027	08 52 30.	- 02 48		13.6	GALAXY
MCG+00-23-011	08 52 30.	- 02 48	48	13.6	GALAXY
HELW 138	08 52 31.	- 02 57 03.			NEBULA
RNGC 2697	08 52 32.	- 02 48		13.5	GALAXY
ZWG 033.021	08 52 36.	+ 03 13		15.6	GALAXY
ZWG 061.021	08 52 36.	+ 12 15		15.3	GALAXY
LB 08887	08 52 36.	+ 16 52		18.4	FAINT BLUE STAR
LB 08886	08 52 36.	+ 18 40		17.1	FAINT BLUE STAR
LB 08885	08 52 36.	+ 19 02		18.3	FAINT BLUE STAR
LB 08884	08 52 36.	+ 19 45		18.0	FAINT BLUE STAR
LB 08883	08 52 36.	+ 19 58		17.4	FAINT BLUE STAR
ZWG 090.037	08 52 36.	+ 20 14		15.6	GALAXY
MCG+07-19-001	08 52 36.	+ 42 28	42	16.	GALAXY
LB 03489	08 52 36.	- 72 42		16.5	FAINT BLUE STAR
ZC 0852.7+0054	08 52 42.	+ 00 54	2960		CLUSTER OF GALAXIES
LB 08889	08 52 42.	+ 17 57		16.5	FAINT BLUE STAR
LB 08888	08 52 42.	+ 19 16		16.4	FAINT BLUE STAR
ZWG 090.038	08 52 42.	+ 19 34		15.7	GALAXY
ZC 0852.7+3739	08 52 42.	+ 37 39	740		CLUSTER OF GALAXIES
ZC 0852.7+4745	08 52 42.	+ 47 45	1080		CLUSTER OF GALAXIES
MCG+09-15-051	08 52 42.	+ 52 03	24	15.	GALAXY
ZWG 264.031	08 52 42.	+ 52 05		15.2	GALAXY
ZWG 005.028	08 52 42.	- 02 05		15.4	GALAXY
BC PKS0852-07	08 52 42.4	- 07 03 37.		18.	QUASI-STELLAR OBJECT
HOLM 102A	08 52 43.	+ 42 28	42	14.1	PART OF MULTIPLE GALAXY
LB 02907	08 52 43.	- 00 07 30.		16.8	FAINT BLUE STAR
PK269-03.1	08 52 44.	- 50 21	25		PLANETARY NEBULA
LB 02908	08 52 45.	+ 21 35 48.		16.7	FAINT BLUE STAR
ZWG 005.029	08 52 48.	+ 01 02		15.6	GALAXY
ZWG 033.022	08 52 48.	+ 07 47		14.9	GALAXY
MCG+01-23-004	08 52 48.	+ 07 47	12	14.9	GALAXY
LB 08891	08 52 48.	+ 18 07		18.2	FAINT BLUE STAR
LB 08890	08 52 48.	+ 20 23		18.1	FAINT BLUE STAR
ZWG 332.030	08 52 48.	+ 68 27		15.7	GALAXY
ZWG 311.025	08 52 48.	+ 68 27		15.7	GALAXY
MCG+07-19-003	08 52 51.	+ 42 28	30	16.	GALAXY
MCG+07-19-002	08 52 51.	+ 42 28	36	16.	GALAXY
MCG+08-16-041	08 52 51.	+ 48 49	42	16.	GALAXY
ZWG 033.023	08 52 54.	+ 06 51		15.5	GALAXY
LB 08896	08 52 54.	+ 15 23		18.1	FAINT BLUE STAR
LB 08895	08 52 54.	+ 15 50		18.3	FAINT BLUE STAR
LB 08894	08 52 54.	+ 16 32		17.8	FAINT BLUE STAR
LB 08893	08 52 54.	+ 18 57		17.9	FAINT BLUE STAR
LB 08892	08 52 54.	+ 18 58		17.0	FAINT BLUE STAR
ZC 0852.9+4009	08 52 54.	+ 40 09	1610		CLUSTER OF GALAXIES
ZC 0852.9-0015	08 52 54.	- 00 15	1480		CLUSTER OF GALAXIES
VHE 23	08 52 54.	- 46 44			REFLECTION NEBULA
TON-N 0973	08 52 55.	+ 20 10		16.7	BLUE STAR
ZWG 033.024	08 53 00.	+ 05 10		15.4	GALAXY

OBJECT NAME	RIGHT ASCEN.	DECLINATION	DIAM.	MAGN.	TYPE OF OBJECT
LB 08898	08 53 00.	+ 17 14		18.0	FAINT BLUE STAR
ZWG 090.039	08 53 00.	+ 18 21		14.9	GALAXY
MCG+03-23-019	08 53 00.	+ 18 21	30	14.9	GALAXY
LB 08897	08 53 00.	+ 20 27		18.5	FAINT BLUE STAR
ZWG 120.051	08 53 00.	+ 20 48		15.7	GALAXY
TON-N 0362	08 53 00.	+ 26 03		15.1	BLUE STAR
ZWG 150.046	08 53 00.	+ 31 20		15.7	GALAXY
MCG+09-15-054	08 53 00.	+ 51 40	42	16.	GALAXY
MCG+09-15-052	08 53 00.	+ 52 06	30	15.	GALAXY
ZWG 264.032	08 53 00.	+ 52 07		15.4	GALAXY
MCG+09-15-053	08 53 00.	+ 52 18	54	14.	GALAXY
ZWG 288.015	08 53 00.	+ 56 59		15.7	GALAXY
ZC 0853.0+8003	08 53 00.	+ 80 03	870		CLUSTER OF GALAXIES
MCG+14-05-003	08 53 00.	+ 86 17	42	17.	GALAXY
BC B20853+29	08 53 00.3	+ 29 10 09.6		19.2	QUASI-STELLAR OBJECT
RNGC 2698	08 53 02.	- 02 59		13.0	GALAXY
REIN 4.058	08 53 04.99	- 02 59 45.8			NEBULA
REIN 2.072	08 53 05.38	- 02 59 32.0			NEBULA
ZWG 061.022	08 53 06.	+ 13 25		14.2	GALAXY
UGC 04670	08 53 06.	+ 13 25	102	14.2	GALAXY S0
MCG+02-23-010	08 53 06.	+ 13 25	72	14.2	GALAXY
ZWG 061.023	08 53 06.	+ 13 45		15.5	GALAXY
LB 08904	08 53 06.	+ 17 26		17.4	FAINT BLUE STAR
LB 08903	08 53 06.	+ 17 47		18.3	FAINT BLUE STAR
LB 08902	08 53 06.	+ 18 27		18.4	FAINT BLUE STAR
LB 08901	08 53 06.	+ 18 53		17.8	FAINT BLUE STAR
LB 08900	08 53 06.	+ 19 28		18.3	FAINT BLUE STAR
LB 08899	08 53 06.	+ 20 10		17.9	FAINT BLUE STAR
ZC 0853.1+3728	08 53 06.	+ 37 28	4840		CLUSTER OF GALAXIES
ZWG 264.033	08 53 06.	+ 52 18		13.6	GALAXY
UGC 04671	08 53 06.	+ 52 18	96	13.6	GALAXY S
KARA.72 179A	08 53 06.	+ 52 18	108	13.6	PART OF DOUBLE GALAXY
MCG+10-13-045	08 53 06.	+ 56 58	18	16.	GALAXY
ZWG 005.030	08 53 06.	- 03 00		13.2	GALAXY
MCG+00-23-012	08 53 06.	- 03 00	60	13.2	GALAXY
MCG-04-21-012	08 53 06.	- 24 50	54	15.5	GALAXY
MCG-04-21-013	08 53 06.	- 24 54	30	17.	GALAXY
LB 02909	08 53 11.	+ 20 58 42.		16.6	FAINT BLUE STAR
REIN 4.059	08 53 11.16	- 02 59 06.1			NEBULA
ZWG 005.032	08 53 12.	+ 00 58		15.2	GALAXY
ZWG 033.025	08 53 12.	+ 04 33		15.7	GALAXY
LB 08908	08 53 12.	+ 15 03		17.9	FAINT BLUE STAR
LB 08907	08 53 12.	+ 15 14		16.1	FAINT BLUE STAR
LB 08906	08 53 12.	+ 18 32		18.0	FAINT BLUE STAR
LB 08905	08 53 12.	+ 19 25		18.6	FAINT BLUE STAR
ZWG 150.047	08 53 12.	+ 30 53		15.6	GALAXY
ZWG 005.031	08 53 12.	- 02 06		14.8	GALAXY
UGC 04672	08 53 12.	- 02 06	60	14.8	GALAXY S
MCG+00-23-013	08 53 12.	- 02 06	48	14.8	GALAXY
RNGC 2702	08 53 14.	- 02 52			GALAXY
RNGC 2700	08 53 15.	- 04 54			NON-EXISTENT OBJECT
REIN 4.060	08 53 15.68	- 02 56 32.9			NEBULA
REIN 4.061	08 53 15.83	- 03 07 00.5			NEBULA
REIN 4.062	08 53 16.35	- 03 06 50.6			NEBULA
REIN 2.073	08 53 17.56	- 02 56 09.6			NEBULA
ZWG 033.026	08 53 18.	+ 02 43		14.9	GALAXY
UGC 04673	08 53 18.	+ 02 43	90	14.9	GALAXY SBc
MCG+01-23-005	08 53 18.	+ 02 43	90	14.9	GALAXY
ZWG 033.027	08 53 18.	+ 03 56		15.5	GALAXY
ZC 0853.3+0900	08 53 18.	+ 09 00	270		CLUSTER OF GALAXIES
LB 08910	08 53 18.	+ 17 48		18.4	FAINT BLUE STAR
LB 08909	08 53 18.	+ 18 24		18.1	FAINT BLUE STAR
ZWG 209.008	08 53 18.	+ 39 20		14.9	GALAXY
ZC 0853.3+4245	08 53 18.	+ 42 45	1080		CLUSTER OF GALAXIES
ZWG 264.034	08 53 18.	+ 51 02		15.2	GALAXY
ZWG 264.035	08 53 18.	+ 51 03		13.1	GALAXY
MCG+09-15-056	08 53 18.	+ 51 30	18	15.5	GALAXY
MCG+09-15-055	08 53 18.	+ 51 31	90	13.3	GALAXY
UGC 04674	08 53 18.	+ 51 33	132	13.1	GALAXY E
MCG+09-15-057	08 53 18.	+ 52 15	78	14.	GALAXY
ZWG 264.036	08 53 18.	+ 52 16		14.1	GALAXY
UGC 04675	08 53 18.	+ 52 16	84	14.1	GALAXY SB:a-b
KARA.72 179B	08 53 18.	+ 52 16	90	14.1	PART OF DOUBLE GALAXY
ZC 0853.3+5710	08 53 18.	+ 57 10	2620		CLUSTER OF GALAXIES
ZWG 005.033	08 53 18.	- 02 57		13.6	GALAXY
MCG+00-23-014	08 53 18.	- 02 57	36	13.6	GALAXY
IC 2425	08 53 18.	- 03 12			MAY NOT EXIST
VHE 24	08 53 18.	- 43 16	96		REFLECTION NEBULA
MRSL 267-01/1	08 53 18.	- 47 24	3000		HII REGION
REIN 4.063	08 53 19.44	- 02 55 27.7			NEBULA
RNGC 2699	08 53 20.	- 02 57		13.5	GALAXY
RNGC 2703	08 53 20.	- 03 05			NON-EXISTENT OBJECT
HOLM 103B	08 53 22.	+ 39 35	18	15.2	PART OF MULTIPLE GALAXY
REIN 4.064	08 53 23.54	- 02 52 23.9			NEBULA
LB 08915	08 53 24.	+ 16 23		15.7	FAINT BLUE STAR
LB 08914	08 53 24.	+ 17 40		17.8	FAINT BLUE STAR
LB 08913	08 53 24.	+ 17 42		18.0	FAINT BLUE STAR
LB 08912	08 53 24.	+ 18 24		17.8	FAINT BLUE STAR
ZC 0853.4+1945	08 53 24.	+ 19 45	1410		CLUSTER OF GALAXIES
LB 08911	08 53 24.	+ 19 49		17.2	FAINT BLUE STAR
ZWG 150.048	08 53 24.	+ 29 52		15.6	GALAXY
ZWG 150.049	08 53 24.	+ 32 01		15.6	GALAXY
MCG+07-19-004	08 53 24.	+ 39 19	48	15.	GALAXY
ZC 0853.4+4321	08 53 24.	+ 43 21	3020		CLUSTER OF GALAXIES
MCG+09-15-058	08 53 24.	+ 52 00	60	15.	GALAXY
RNGC 2692	08 53 24.	+ 52 16		14.0	GALAXY
ZC 0853.4+5528	08 53 24.	+ 55 28	610		CLUSTER OF GALAXIES
REIN 2.070	08 53 24.85	+ 51 31 29.9			NEBULA
RNGC 2694	08 53 25.	+ 51 31		15.0	GALAXY
RNGC 2693	08 53 25.	+ 51 32		13.0	GALAXY
REIN 2.071	08 53 25.00	+ 51 32 26.5			NEBULA
RNGC 2705	08 53 26.	- 02 48			NON-EXISTENT OBJECT
LB 02910	08 53 28.	- 00 09 36.		16.2	FAINT BLUE STAR
REIN 4.065	08 53 28.37	- 02 53 58.4			NEBULA
REIN 4.066	08 53 28.76	- 02 49 32.5			NEBULA
LB 03626	08 53 30.	+ 09 29 48.		18.7	FAINT BLUE STAR
LB 08922	08 53 30.	+ 15 48		17.7	FAINT BLUE STAR
LB 08921	08 53 30.	+ 16 11		18.2	FAINT BLUE STAR
LB 08920	08 53 30.	+ 17 02		15.8	FAINT BLUE STAR
LB 08919	08 53 30.	+ 19 06		16.8	FAINT BLUE STAR
LB 08918	08 53 30.	+ 19 20		18.1	FAINT BLUE STAR
LB 08917	08 53 30.	+ 19 30		17.5	FAINT BLUE STAR
LB 08916	08 53 30.	+ 20 02		18.1	FAINT BLUE STAR
MCG+09-15-059	08 53 30.	+ 51 40	78	16.	GALAXY
UGC 04676	08 53 30.	+ 52 00	78	16.0	GALAXY Sc
MCG+09-15-060	08 53 30.	+ 55 16	12	16.	GALAXY
ZC 0853.5-0312	08 53 30.	- 03 12	6990		CLUSTER OF GALAXIES
REIN 4.067	08 53 31.87	- 02 58 06.0			NEBULA

OBJECT NAME	RIGHT ASCEN.	DECLINATION	DIAM.	MAGN.	TYPE OF OBJECT
HOLM 103A	08 53 32.	+ 39 35	54	13.8	PART OF MULTIPLE GALAXY
RNGC 2708	08 53 33.	- 03 09		13.5	GALAXY
REIN 4.068	08 53 34.45	- 02 52 28.3			NEBULA
LB 02911	08 53 35.	+ 20 26 06.		16.0	FAINT BLUE STAR
IC 2424	08 53 35.	+ 39 35			SAME AS NGC 2704
ZWG 061.024	08 53 36.	+ 12 37		15.1	GALAXY
MCG+02-23-011	08 53 36.	+ 12 37	48	15.1	GALAXY
ZWG 061.025	08 53 36.	+ 13 22		15.1	GALAXY
UGC 04677	08 53 36.	+ 13 22	90	15.1	GALAXY Sc
MCG+02-23-012	08 53 36.	+ 13 22	72	15.1	GALAXY
LB 08925	08 53 36.	+ 16 49		16.0	FAINT BLUE STAR
LB 08924	08 53 36.	+ 18 07		18.2	FAINT BLUE STAR
LB 08923	08 53 36.	+ 20 22		17.1	FAINT BLUE STAR
ZWG 120.052	08 53 36.	+ 21 35		15.7	GALAXY
ZWG 209.009	08 53 36.	+ 39 34		14.4	GALAXY
UGC 04678	08 53 36.	+ 39 34	66	14.4	GALAXY SBb
UGC 04679	08 53 36.	+ 51 40	84	16.5	GALAXY Sc
ZC 0853.6+5913	08 53 36.	+ 59 13	2350		CLUSTER OF GALAXIES
ZWG 005.034	08 53 36.	- 03 10		13.6	GALAXY
MCG+00-23-015	08 53 36.	- 03 10	150	13.6	GALAXY
LB 02912	08 53 37.	+ 00 16 06.		16.9	FAINT BLUE STAR
REIN 4.069	08 53 37.06	- 03 10 02.4			NEBULA
RNGC 2707	08 53 38.	- 02 52			NON-EXISTENT OBJECT
REIN 4.070	08 53 40.61	- 02 22 09.5			NEBULA
RNGC 2704	08 53 41.	+ 39 34		14.0	GALAXY
MAI 037	08 53 41.	+ 59 17	60		DWARF SPHEROIDAL GALAXY
REIN 4.071	08 53 41.41	- 03 03 05.7			NEBULA
ZC 0853.7+0537	08 53 42.	+ 05 37	870		CLUSTER OF GALAXIES
ZWG 061.026	08 53 42.	+ 11 45		15.6	GALAXY
UGC 04681	08 53 42.	+ 11 45	60	15.6	GALAXY S
MCG+02-23-013	08 53 42.	+ 11 45	36	15.6	GALAXY
LB 08926	08 53 42.	+ 16 46		15.5	FAINT BLUE STAR
ZWG 150.050	08 53 42.	+ 30 07		15.7	GALAXY
TON-N 0974	08 53 42.	+ 32 50		16.9	BLUE STAR
MCG+07-19-005	08 53 42.	+ 39 34	60	14.5	GALAXY
ZC 0853.7+5942	08 53 42.	+ 59 42	940		CLUSTER OF GALAXIES
ZWG 005.036	08 53 42.	- 02 23		13.8	GALAXY
UGC 04680	08 53 42.	- 02 23	108	13.8	GALAXY Sa?
MCG+00-23-017	08 53 42.	- 02 23	114	13.8	GALAXY
ZWG 005.035	08 53 42.	- 03 04		14.8	GALAXY
MCG+00-23-016	08 53 42.	- 03 04	36	14.8	GALAXY
RNGC 2706	08 53 44.	- 02 23		14.0	GALAXY
RNGC 2709	08 53 44.	- 03 04		15.0	GALAXY
LB 02913	08 53 46.	+ 21 36 12.		16.8	FAINT BLUE STAR
ABC 0707	08 53 47.	+ 80 06		17.5	RICH CLUSTER OF GALAXIES
LB 02914	08 53 48.	+ 01 42 30.		16.5	FAINT BLUE STAR
ZWG 061.027	08 53 48.	+ 11 13		15.4	GALAXY
LB 08928	08 53 48.	+ 18 16		17.7	FAINT BLUE STAR
ZWG 090.040	08 53 48.	+ 19 11		15.7	GALAXY
LB 08927	08 53 48.	+ 20 13		18.0	FAINT BLUE STAR
ZC 0853.8+4048	08 53 48.	+ 40 48	3760		CLUSTER OF GALAXIES
ZWG 005.038	08 53 48.	- 00 15		15.3	GALAXY
ZWG 005.037	08 53 48.	- 01 27		15.5	GALAXY
ZC 0853.9+0319	08 53 54.	+ 03 19	670		CLUSTER OF GALAXIES
LB 08933	08 53 54.	+ 16 12		17.9	FAINT BLUE STAR
LB 08932	08 53 54.	+ 16 43		18.5	FAINT BLUE STAR
LB 08931	08 53 54.	+ 17 13		18.4	FAINT BLUE STAR
LB 08930	08 53 54.	+ 18 34		18.7	FAINT BLUE STAR
LB 08929	08 53 54.	+ 20 40		17.7	FAINT BLUE STAR
ZC 0854.0+0602	08 54 00.	+ 06 02	1340		CLUSTER OF GALAXIES
LB 08938	08 54 00.	+ 14 25		17.2	FAINT BLUE STAR
LB 08937	08 54 00.	+ 15 00		18.0	FAINT BLUE STAR
LB 08936	08 54 00.	+ 15 48		18.7	FAINT BLUE STAR
LB 08935	08 54 00.	+ 18 05		17.0	FAINT BLUE STAR
LB 08934	08 54 00.	+ 18 08		15.6	FAINT BLUE STAR
ZWG 090.041	08 54 00.	+ 20 12		15.7	GALAXY
ZC 0854.0+3017	08 54 00.	+ 30 17	2490		CLUSTER OF GALAXIES
TON-N 0363	08 54 00.	+ 30 40		15.1	BLUE STAR
ZC 0854.0+4217	08 54 00.	+ 42 17	1080		CLUSTER OF GALAXIES
ZWG 264.037	08 54 00.	+ 55 15		15.4	GALAXY
UGC 04683	08 54 00.	+ 59 16	120	16.5	GALAXY DWRF IR
MCG+10-13-046	08 54 00.	+ 59 16	57	16.	GALAXY
ZC 0854.0+8209	08 54 00.	+ 82 09	540		CLUSTER OF GALAXIES
UGC 04682	08 54 00.	+ 86 20	78	16.5	GALAXY Sc
REIN 4.072	08 54 01.33	- 03 08 56.7			NEBULA
LB 03627	08 54 02.	+ 09 08 48.		18.0	FAINT BLUE STAR
ZWG 005.039	08 54 06.	+ 00 33		14.7	GALAXY
UGC 04684	08 54 06.	+ 00 33	84	14.7	GALAXY Sc
MCG+00-23-018	08 54 06.	+ 00 33	84	14.7	GALAXY
KARA.73B 0291	08 54 06.	+ 00 33	90	14.7	ISOLATED GALAXY S
ZWG 061.028	08 54 06.	+ 10 34		15.7	GALAXY
LB 08941	08 54 06.	+ 16 04		18.0	FAINT BLUE STAR
LB 08940	08 54 06.	+ 17 38		17.5	FAINT BLUE STAR
LB 08939	08 54 06.	+ 19 52		18.3	FAINT BLUE STAR
ZC 0854.1+2824	08 54 06.	+ 28 24	1010		CLUSTER OF GALAXIES
ZWG 150.051	08 54 06.	+ 31 53		15.7	GALAXY
ZC 0854.1+4819	08 54 06.	+ 48 19	1880		CLUSTER OF GALAXIES
LB 08945	08 54 12.	+ 14 47		16.0	FAINT BLUE STAR
ZC 0854.2+1600	08 54 12.	+ 16 00	4030		CLUSTER OF GALAXIES
LB 08944	08 54 12.	+ 16 55		18.3	FAINT BLUE STAR
LB 08943	08 54 12.	+ 17 00		16.1	FAINT BLUE STAR
LB 08942	08 54 12.	+ 18 54		17.9	FAINT BLUE STAR
ZWG 120.053	08 54 12.	+ 23 19		15.7	GALAXY
TON-N 0364	08 54 12.	+ 29 53		15.9	BLUE STAR
ZWG 209.010	08 54 12.	+ 42 22		15.7	GALAXY
MCG+10-13-047	08 54 12.	+ 59 22 30.	54	15.	GALAXY
ZWG 288.016	08 54 12.	+ 59 24		15.2	GALAXY
VHA 055	08 54 12.	- 39 20	240		OPEN STAR CLUSTER
LB 02915	08 54 17.	- 01 13 12.		16.5	FAINT BLUE STAR
LB 02916	08 54 18.	+ 00 02 18.		15.8	FAINT BLUE STAR
ZWG 061.029	08 54 18.	+ 13 24		14.2	GALAXY
UGC 04685	08 54 18.	+ 13 24	84	14.2	GALAXY S
MCG+02-23-014	08 54 18.	+ 13 24	72	14.2	GALAXY
LB 08950	08 54 18.	+ 14 55		17.8	FAINT BLUE STAR
LB 08949	08 54 18.	+ 18 38		17.0	FAINT BLUE STAR
LB 08948	08 54 18.	+ 19 20		16.9	FAINT BLUE STAR
LB 08947	08 54 18.	+ 19 26		17.9	FAINT BLUE STAR
LB 08946	08 54 18.	+ 20 10		18.2	FAINT BLUE STAR
ABC 0722	08 54 18.	+ 30 57		16.9	RICH CLUSTER OF GALAXIES
LB 02917	08 54 22.	+ 22 06 24.		17.2	FAINT BLUE STAR
LB 08953	08 54 24.	+ 15 08		17.6	FAINT BLUE STAR
LB 08952	08 54 24.	+ 15 31		5.0	FAINT BLUE STAR
LB 08951	08 54 24.	+ 15 56		17.8	FAINT BLUE STAR
ZWG 120.054	08 54 24.	+ 22 38		15.4	GALAXY
VV 079	08 54 24.	+ 25 56	150	15.	INTERACTING GALAXY
TON-N 0365	08 54 24.	+ 27 40		15.7	BLUE STAR
ZWG 150.052	08 54 24.	+ 30 04		15.6	GALAXY
ZWG 209.011	08 54 24.	+ 43 18		14.5	GALAXY
UGC 04686	08 54 24.	+ 43 18	48	14.5	GALAXY S
ZWG 311.026	08 54 24.	+ 66 40		14.2	GALAXY
UGC 04687	08 54 24.	+ 66 40	42	14.2	GALAXY PECULR
MRSL 264+01/1	08 54 24.	- 42 56	120		HII REGION
ZL 049	08 54 28.	+ 03 21 00.		18.1	ULTRAFAINT BLUE STAR
LB 02918	08 54 29.	+ 20 07 06.		15.4	FAINT BLUE STAR
LB 08954	08 54 30.	+ 16 26		17.4	FAINT BLUE STAR
ZWG 090.042	08 54 30.	+ 17 29		14.6	GALAXY
RNGC 2711	08 54 30.	+ 17 29		14.5	GALAXY
UGC 04688	08 54 30.	+ 17 29	60	14.6	GALAXY SB
MCG+07-19-006	08 54 30.	+ 39 42	48	17.	GALAXY
UGC 04689	08 54 30.	+ 39 43	60	16.0	GALAXY S
MCG+07-19-007	08 54 30.	+ 43 20	42	15.	GALAXY
ZC 0854.5+5106	08 54 30.	+ 51 06	540		CLUSTER OF GALAXIES
MCG+09-15-061	08 54 30.	+ 52 22	72	16.	GALAXY
ZWG 264.038	08 54 30.	+ 52 23		14.9	GALAXY
UGC 04690	08 54 30.	+ 52 23	72	14.9	GALAXY Sa-b
ZL 050	08 54 34.	+ 03 31 00.		18.6	ULTRAFAINT BLUE STAR
ZL 051	08 54 35.	+ 03 19 24.		21.6	ULTRAFAINT BLUE STAR
LB 02919	08 54 35.	- 01 26 24.		15.5	FAINT BLUE STAR
ZL 052	08 54 36.	+ 03 23 30.		21.4	ULTRAFAINT BLUE STAR
LB 08959	08 54 36.	+ 16 03		17.9	FAINT BLUE STAR
MCG+03-23-020	08 54 36.	+ 17 29	48	14.6	GALAXY
LB 08958	08 54 36.	+ 18 02		16.3	FAINT BLUE STAR
LB 08957	08 54 36.	+ 18 36		18.1	FAINT BLUE STAR
LB 08956	08 54 36.	+ 19 07		17.9	FAINT BLUE STAR
ZWG 090.043	08 54 36.	+ 20 19		15.7	GALAXY
LB 08955	08 54 36.	+ 20 20		18.4	FAINT BLUE STAR
ZC 0854.6+4841	08 54 36.	+ 48 41	870		CLUSTER OF GALAXIES
MCG-01-23-010	08 54 36.	- 05 35	36	15.5	GALAXY
MCG-03-23-012	08 54 36.	- 20 22	60	14.5	GALAXY
ZL 053	08 54 39.	+ 03 27 24.		21.5	ULTRAFAINT BLUE STAR
MCG-04-21-014	08 54 39.	- 24 39	60	16.	GALAXY
WRAY 19.14	08 54 39.2	- 42 54 14.		10.7	STAR-NEBULA ASSOCIATION
ZWG 033.028	08 54 42.	+ 03 06		12.9	GALAXY
UGC 04691	08 54 42.	+ 03 06	246	12.9	GALAXY SBb
SN 1968E	08 54 42.	+ 03 06		13.5	SUPERNOVA
MCG+01-23-006	08 54 42.	+ 03 06	180	12.9	GALAXY
RNGC 2713	08 54 42.	+ 03 07		12.5	GALAXY
LB 08962	08 54 42.	+ 14 34		17.8	FAINT BLUE STAR
LB 08961	08 54 42.	+ 15 12		17.6	FAINT BLUE STAR
LB 08960	08 54 42.	+ 16 12		17.6	FAINT BLUE STAR
ZC 0854.7+4843	08 54 42.	+ 48 43	270		CLUSTER OF GALAXIES
ZWG 264.039	08 54 42.	+ 55 40		15.6	GALAXY
ZWG 005.040	08 54 42.	- 02 10		15.7	GALAXY
MCG-01-23-011	08 54 42.	- 08 25	96	15.	GALAXY
MCG-03-23-013	08 54 42.	- 19 52	36	15.5	GALAXY
VHE 25A	08 54 42.	- 42 54	30		REFLECTION NEBULA
ZL 054	08 54 45.	+ 03 15 06.		20.9	ULTRAFAINT BLUE STAR
LB 08967	08 54 48.	+ 15 53		17.4	FAINT BLUE STAR
LB 08966	08 54 48.	+ 18 02		18.6	FAINT BLUE STAR
LB 08965	08 54 48.	+ 18 46		18.3	FAINT BLUE STAR
LB 08964	08 54 48.	+ 19 43		17.6	FAINT BLUE STAR
LB 08963	08 54 48.	+ 20 32		17.8	FAINT BLUE STAR
ZC 0854.8+5434	08 54 48.	+ 54 34	2020		CLUSTER OF GALAXIES
IZW 016	08 54 48.	+ 55 41			COMPACT GALAXY
ZWG 005.041	08 54 48.	- 00 48		15.6	GALAXY
ZC 0854.8-0323	08 54 48.	- 03 23	1140		CLUSTER OF GALAXIES
RNGC 2717	08 54 48.	- 24 28		14.0	GALAXY
MCG-04-21-015	08 54 48.	- 24 28 30.	24	14.	GALAXY
VHE 25B	08 54 48.	- 43 01	48		REFLECTION NEBULA
ZL 055	08 54 50.	+ 03 15 18.		18.3	ULTRAFAINT BLUE STAR
MCG-03-23-014	08 54 51.	- 20 24 30.	72	14.5	GALAXY
MCG-04-21-016	08 54 51.	- 24 34	78	15.	GALAXY
ZL 056	08 54 52.	+ 03 23 06.		22.6	ULTRAFAINT BLUE STAR
ABC 0731	08 54 53.	- 03 30		17.6	RICH CLUSTER OF GALAXIES
ZWG 033.029	08 54 54.	+ 03 16		13.7	GALAXY
UGC 04692	08 54 54.	+ 03 16	96	13.7	GALAXY S0
MCG+01-23-007	08 54 54.	+ 03 16	72	13.7	GALAXY
HOLM 104B	08 54 54.	+ 03 17	18	14.8	PART OF MULTIPLE GALAXY
ZC 0854.9+1011	08 54 54.	+ 10 11	610		CLUSTER OF GALAXIES
ZC 0854.9+1147	08 54 54.	+ 11 47	1950		CLUSTER OF GALAXIES
LB 08969	08 54 54.	+ 19 34		18.2	FAINT BLUE STAR
LB 08968	08 54 54.	+ 20 09		17.6	FAINT BLUE STAR
ZWG 264.040	08 54 54.	+ 54 14		15.6	GALAXY
MCG-01-23-012	08 54 54.	- 09 07	24	16.	GALAXY
ABC 0728	08 54 57.	+ 10 10		17.5	RICH CLUSTER OF GALAXIES
HMS 0855+0321	08 55	+ 03 21			HYDRA GALAXY CLUSTER
LBN 1077	08 55	- 25 00	27900		BRIGHT NEBULA
RNGC 2716	08 55 00.	+ 03 17		13.5	GALAXY
LB 08975	08 55 00.	+ 15 35		18.3	FAINT BLUE STAR
LB 08974	08 55 00.	+ 16 14		17.1	FAINT BLUE STAR
LB 08973	08 55 00.	+ 16 54		18.3	FAINT BLUE STAR
LB 08972	08 55 00.	+ 17 44		17.7	FAINT BLUE STAR
LB 08971	08 55 00.	+ 19 56		17.3	FAINT BLUE STAR
ZWG 090.044	08 55 00.	+ 20 19		14.9	GALAXY
LB 08970	08 55 00.	+ 20 21		17.5	FAINT BLUE STAR
ZC 0855.0+2340	08 55 00.	+ 23 40	870		CLUSTER OF GALAXIES
ZWG 264.041	08 55 00.	+ 52 05		15.3	GALAXY
ZC 0855.0+5248	08 55 00.	+ 52 48	12100		CLUSTER OF GALAXIES
MCG+09-15-062	08 55 00.	+ 54 30	30	17.	GALAXY
MCG+11-11-034	08 55 00.	+ 66 24	39	17.	GALAXY
UGC 04693	08 55 00.	+ 66 25	78	16.5	GALAXY S IV
HOLM 104A	08 55 01.	+ 03 18	48	13.5	PART OF MULTIPLE GALAXY
MCG+03-23-021	08 55 03.	+ 17 17	36	14.7	GALAXY
MCG+03-23-023	08 55 03.	+ 20 19	27	15.7	GALAXY
MCG+03-23-022	08 55 03.	+ 20 19	30	14.9	GALAXY
MCG+01-23-009	08 55 06.	+ 03 23	1	20.	GALAXY
MCG+01-23-008	08 55 06.	+ 03 23	1	20.	GALAXY
ZWG 061.030	08 55 06.	+ 12 41		15.1	GALAXY
UGC 04694	08 55 06.	+ 12 41	60	15.1	GALAXY SBc
MCG+02-23-015	08 55 06.	+ 12 41	48	15.1	GALAXY
LB 08980	08 55 06.	+ 15 58		18.6	FAINT BLUE STAR
LB 08979	08 55 06.	+ 16 01		16.0	FAINT BLUE STAR
ZWG 090.045	08 55 06.	+ 17 17		14.7	GALAXY
LB 08978	08 55 06.	+ 18 33		15.2	FAINT BLUE STAR
LB 08977	08 55 06.	+ 18 36		18.0	FAINT BLUE STAR
LB 08976	08 55 06.	+ 18 36		17.2	FAINT BLUE STAR
ZWG 120.055	08 55 06.	+ 22 58		15.6	GALAXY
ZC 0855.1+3918	08 55 06.	+ 39 18	3700		CLUSTER OF GALAXIES
ZC 0855.1+4922	08 55 06.	+ 49 22	270		CLUSTER OF GALAXIES
LB 02920	08 55 07.	+ 22 19 30.		16.7	FAINT BLUE STAR
ABC 0724	08 55 07.	+ 38 46		16.7	RICH CLUSTER OF GALAXIES
MCG+08-17-001	08 55 09.	+ 45 29	18	17.	GALAXY
TON-N 0975	08 55 11.	+ 35 14		17.	BLUE STAR
ZC 0855.2+0324	08 55 12.	+ 03 24	810		CLUSTER OF GALAXIES
ZWG 033.030	08 55 12.	+ 07 56		15.5	GALAXY
ZC 0855.2+1154	08 55 12.	+ 11 54	270		CLUSTER OF GALAXIES

OBJECT NAME	RIGHT ASCEN.	DECLINATION	DIAM.	MAGN.	TYPE OF OBJECT
LB 08981	08 55 12.	+ 15 30		17.7	FAINT BLUE STAR
ZC 0855.2+7034	08 55 12.	+ 70 34	1210		CLUSTER OF GALAXIES
ZWG 005.042	08 55 12.	- 00 01		15.1	GALAXY
LB 02921	08 55 13.	- 01 05 06.		16.3	FAINT BLUE STAR
ZL 057	08 55 14.	+ 03 29 42.		22.5	ULTRAFAINT BLUE STAR
ARC 0732	08 55 16.	+ 03 22		17.7	RICH CLUSTER OF GALAXIES
TON-N 0978	08 55 17.	+ 34 36		16.8	BLUE STAR
MCG+01-23-010	08 55 18.	+ 03 22	1	19.	GALAXY
HMS 1.09	08 55 18.	+ 03 23			Sb GALAXY
MCG+01-23-011	08 55 18.	+ 03 23	1	19.2	GALAXY
ZC 0855.3+0905	08 55 18.	+ 09 05	1610		CLUSTER OF GALAXIES
ZWG 061.031	08 55 18.	+ 13 11		15.2	GALAXY
LB 08985	08 55 18.	+ 14 39		16.4	FAINT BLUE STAR
LB 08984	08 55 18.	+ 15 32		17.6	FAINT BLUE STAR
LB 08983	08 55 18.	+ 15 53		15.6	FAINT BLUE STAR
LB 08982	08 55 18.	+ 16 28		17.7	FAINT BLUE STAR
TON-N 0977	08 55 18.	+ 32 14		16.8	BLUE STAR
ZWG 180.023	08 55 18.	+ 37 16		15.7	GALAXY
ZWG 180.024	08 55 18.	+ 37 20		15.6	GALAXY
ZWG 264.042	08 55 18.	+ 52 46		15.7	GALAXY
ZWG 264.043	08 55 18.	+ 53 57		12.3	GALAXY
UGC 04695	08 55 18.	+ 53 57	120	12.3	GALAXY Sc
LB 03628	08 55 19.	+ 09 34 54.		18.7	FAINT BLUE STAR
TON-N 0976	08 55 19.	+ 19 17		16.0	BLUE STAR
ARC 0726	08 55 19.	+ 31 19		16.7	RICH CLUSTER OF GALAXIES
RNGC 2701	08 55 22.	+ 53 58		12.5	GALAXY
ZL 058	08 55 23.	+ 03 25 00.		19.2	ULTRAFAINT BLUE STAR
TON-N 0979	08 55 23.	+ 36 35		16.8	BLUE STAR
HMS 1.10	08 55 24.	+ 03 21			Sa GALAXY
MCG+01-23-012	08 55 24.	+ 03 21		20.3	GALAXY
LB 08989	08 55 24.	+ 16 25		17.8	FAINT BLUE STAR
LB 08988	08 55 24.	+ 16 36		16.3	FAINT BLUE STAR
LB 08987	08 55 24.	+ 18 05		17.8	FAINT BLUE STAR
LB 08986	08 55 24.	+ 18 24		17.6	FAINT BLUE STAR
ZWG 264.044	08 55 24.	+ 53 49		15.7	GALAXY
UGC 04696	08 55 24.	+ 53 49	66	15.7	GALAXY Sc
MCG+09-15-063	08 55 24.	+ 53 58	108	12.5	GALAXY
UGC 04697	08 55 24.	+ 70 13	102	16.0	GALAXY SBc
REIN 2.074	08 55 24.20	+ 53 58 03.7			NEBULA
ZL 059	08 55 25.	+ 03 19 12.		20.5	ULTRAFAINT BLUE STAR
REIN 2.075	08 55 26.92	+ 53 58 03.7			NEBULA
MCG+08-17-002	08 55 27.	+ 46 15	15	16.	GALAXY
ZC 0855.5+0125	08 55 30.	+ 01 25	1950		CLUSTER OF GALAXIES
ZWG 033.031	08 55 30.	+ 03 56		15.7	GALAXY
LB 08990	08 55 30.	+ 16 12		16.0	FAINT BLUE STAR
ZWG 090.046	08 55 30.	+ 17 44		15.7	GALAXY
ZWG 150.053	08 55 30.	+ 28 29		14.6	GALAXY
UGC 04698	08 55 30.	+ 28 29	102	14.6	GALAXY SBa-b
MCG+09-15-064	08 55 30.	+ 53 50	72	15.	GALAXY
MCG+09-15-065	08 55 30.	+ 54 14	9	16.	GALAXY
ZC 0855.5+5628	08 55 30.	+ 56 28	670		CLUSTER OF GALAXIES
ZC 0855.5+6130	08 55 30.	+ 61 30	1210		CLUSTER OF GALAXIES
ZC 0855.5+6658	08 55 30.	+ 66 58	3290		CLUSTER OF GALAXIES
MCG+12-09-028	08 55 30.	+ 70 12 30.	84	16.	GALAXY
MCG-06-20-004	08 55 30.	- 39 06	72	13.	GALAXY
VHA 056	08 55 30.	- 43 01	720		OPEN STAR CLUSTER
ZL 060	08 55 32.	+ 03 27 18.		22.7	ULTRAFAINT BLUE STAR
ARC 0721	08 55 32.	+ 61 29		17.7	RICH CLUSTER OF GALAXIES
ZL 061	08 55 33.	+ 03 27 00.		22.4	ULTRAFAINT BLUE STAR
ZL 062	08 55 35.	+ 03 22 18.		22.8	ULTRAFAINT BLUE STAR
LB 08993	08 55 36.	+ 14 52		17.3	FAINT BLUE STAR
LB 08992	08 55 36.	+ 18 40		17.6	FAINT BLUE STAR
LB 08991	08 55 36.	+ 18 49		17.3	FAINT BLUE STAR
MCG+05-21-015	08 55 36.	+ 28 28	90	14.6	GALAXY
TON-N 0366	08 55 36.	+ 33 00		16.8	BLUE STAR
ZWG 209.012	08 55 36.	+ 39 42		14.7	GALAXY
UGC 04699	08 55 36.	+ 39 42	78	14.7	GALAXY PECULR
KARA.72 180B	08 55 36.	+ 39 42	24		PART OF DOUBLE GALAXY
KARA.72 180A	08 55 36.	+ 39 42	36	14.7	PART OF DOUBLE GALAXY
MCG+07-19-008	08 55 36.	+ 39 42	24	15.	GALAXY
ZWG 209.013	08 55 36.	+ 41 46		14.6	GALAXY
UGC 04700	08 55 36.	+ 41 46	72	14.6	GALAXY Sb
ZWG 350.009	08 55 36.	+ 78 29		15.7	GALAXY
UGC 04701	08 55 36.	+ 78 29	132	15.7	GALAXY
MCG+13-07-013	08 55 36.	+ 78 29	114	15.	GALAXY
LB 03629	08 55 37.	+ 13 25 24.		16.6	FAINT BLUE STAR
LB 02922	08 55 41.	+ 19 46 30.		16.5	FAINT BLUE STAR
LB 08996	08 55 42.	+ 14 45		18.1	FAINT BLUE STAR
LB 08995	08 55 42.	+ 15 55		16.4	FAINT BLUE STAR
LB 08994	08 55 42.	+ 17 46		16.4	FAINT BLUE STAR
LB 02922	08 55 42.	+ 15 46		15.7	FAINT BLUE STAR
ZWG 209.014	08 55 42.	+ 39 00		15.0	GALAXY
UGC 04702	08 55 42.	+ 39 00	90	15.0	GALAXY S0?
MCG+07-19-009	08 55 42.	+ 39 00	18	15.	GALAXY
MCG+07-19-010	08 55 42.	+ 41 46	66	15.	GALAXY
MCG+10-13-048	08 55 42.	+ 61 57	27	16.	GALAXY
MCG-01-23-013	08 55 42.	- 05 52	108	15.5	GALAXY
ZL 063	08 55 43.	+ 03 30 42.		20.7	ULTRAFAINT BLUE STAR
PK253+10.1	08 55 43.	- 28 46	65		PLANETARY NEBULA
LB 03631	08 55 44.	+ 09 43 00.		16.2	FAINT BLUE STAR
ZWG 033.032	08 55 48.	+ 06 31		15.5	GALAXY
UGC 04703	08 55 48.	+ 06 31	114	15.5	GALAXY DBL SYS
MCG+01-23-013	08 55 48.	+ 06 31	12	15.5	GALAXY
LB 08999	08 55 48.	+ 17 13		18.0	FAINT BLUE STAR
LB 08998	08 55 48.	+ 18 26		17.5	FAINT BLUE STAR
LB 08997	08 55 48.	+ 19 05		18.0	FAINT BLUE STAR
ZWG 209.015	08 55 48.	+ 39 24		15.5	GALAXY
UGC 04704	08 55 48.	+ 39 24	246	15.5	GALAXY
MCG+10-13-049	08 55 48.	+ 57 32	18	16.	GALAXY
MCG+11-11-035	08 55 48.	+ 65 06	78	15.	GALAXY
MCG+07-19-011	08 55 51.	+ 39 23	240	14.5	GALAXY
ZL 064	08 55 52.	+ 03 19 12.		20.3	ULTRAFAINT BLUE STAR
ZWG 005.043	08 55 54.	+ 00 12		15.3	GALAXY
ZWG 033.033	08 55 54.	+ 03 06		15.1	GALAXY
MCG+01-23-014	08 55 54.	+ 03 06	18	15.1	GALAXY
IC 2426	08 55 54.	+ 03 06 03.			NONSTELLAR OBJECT
ZWG 061.032	08 55 54.	+ 12 52		15.2	GALAXY
LB 09006	08 55 54.	+ 15 35		18.0	FAINT BLUE STAR
LB 09005	08 55 54.	+ 15 58		18.6	FAINT BLUE STAR
LB 09004	08 55 54.	+ 16 54		16.8	FAINT BLUE STAR
LB 09003	08 55 54.	+ 17 30		17.7	FAINT BLUE STAR
LB 09002	08 55 54.	+ 20 10		15.7	FAINT BLUE STAR
LB 09001	08 55 54.	+ 20 28		17.6	FAINT BLUE STAR
LB 09000	08 55 54.	+ 20 35		17.7	FAINT BLUE STAR
ZWG 150.054	08 55 54.	+ 29 06		15.4	GALAXY
ZWG 264.045	08 55 54.	+ 53 07		15.6	GALAXY
LB 02923	08 55 56.	+ 19 56 06.		16.2	FAINT BLUE STAR
IC 0524	08 55 56.	- 18 59 29.			NONSTELLAR OBJECT

OBJECT NAME	RIGHT ASCEN.	DECLINATION	DIAM.	MAGN.	TYPE OF OBJECT
ARC 0727	08 55 57.	+ 39 37		16.7	RICH CLUSTER OF GALAXIES
LB 02924	08 55 57.	- 01 30 54.		14.4	FAINT BLUE STAR
LB 03630	08 55 58.	+ 13 01 00.		16.8	FAINT BLUE STAR
LB 02925	08 55 58.	+ 21 34 00.		16.4	FAINT BLUE STAR
ZL 065	08 55 59.	+ 03 25 48.		21.8	ULTRAFAINT BLUE STAR
TON-N 0981	08 55 59.	+ 33 40		16.5	BLUE STAR
LB 09970	08 56	- 88 31		13.3	FAINT BLUE STAR
LB 09015	08 56 00.	+ 15 15		14.2	FAINT BLUE STAR
LB 09014	08 56 00.	+ 16 32		18.0	FAINT BLUE STAR
LB 09013	08 56 00.	+ 17 02		17.4	FAINT BLUE STAR
LB 09012	08 56 00.	+ 17 44		18.1	FAINT BLUE STAR
LB 09011	08 56 00.	+ 18 06		18.1	FAINT BLUE STAR
LB 09010	08 56 00.	+ 18 38		18.3	FAINT BLUE STAR
LB 09009	08 56 00.	+ 18 44		17.0	FAINT BLUE STAR
LB 09008	08 56 00.	+ 19 00		16.5	FAINT BLUE STAR
LB 09007	08 56 00.	+ 20 26		17.5	FAINT BLUE STAR
LB 02926	08 56 00.	+ 22 28 30.		16.1	FAINT BLUE STAR
ZC 0856.0+2959	08 56 00.	+ 29 59	1280		CLUSTER OF GALAXIES
TON-N 0367	08 56 00.	+ 33 05		16.0	BLUE STAR
TON-N 0368	08 56 00.	+ 33 12		14.7	BLUE STAR
ZWG 264.046	08 56 00.	+ 55 53		13.8	GALAXY
UGC 04705	08 56 00.	+ 55 53	132	13.8	GALAXY SBb
ZWG 311.027	08 56 00.	+ 65 06		15.5	GALAXY
UGC 04706	08 56 00.	+ 65 06	84	15.5	GALAXY SBb
LB 02927	08 56 01.	+ 01 19 30.		16.3	FAINT BLUE STAR
TON-N 0980	08 56 01.	+ 21 49		17.0	BLUE STAR
RNGC 2710	08 56 02.	+ 55 53		14.0	GALAXY
BC 4C17.46	08 56 03.	+ 17 05		17.90	QUASI-STELLAR OBJECT
SHB 127	08 56 04.3	+ 17 03 07.		17.4	QUASI-STELLAR OBJECT
TON-N 0982	08 56 05.	+ 33 42		16.9	BLUE STAR
LB 09017	08 56 06.	+ 20 00		16.7	FAINT BLUE STAR
LB 09016	08 56 06.	+ 20 07		17.9	FAINT BLUE STAR
MCG+09-15-066	08 56 06.	+ 55 54	108	13.	GALAXY
ZC 0856.1+5818	08 56 06.	+ 58 18	3490		CLUSTER OF GALAXIES
MCG+10-13-050	08 56 06.	+ 60 17	36	16.	GALAXY
7ZW 258	08 56 06.	+ 66 48			COMPACT GALAXY
RNGC 2712	08 56 07.	+ 45 07		12.5	GALAXY
RNGC 2718	08 56 10.	+ 06 30		13.5	GALAXY
ZWG 033.034	08 56 12.	+ 06 30		13.3	GALAXY
UGC 04707	08 56 12.	+ 06 30	138	13.3	GALAXY Sa/SBb
MCG+01-23-015	08 56 12.	+ 06 30	120	13.3	GALAXY
ZWG 061.033	08 56 12.	+ 11 03		15.3	GALAXY
LB 09019	08 56 12.	+ 16 46		18.1	FAINT BLUE STAR
LB 09018	08 56 12.	+ 16 52		17.6	FAINT BLUE STAR
ZC 0856.2+3103	08 56 12.	+ 31 03	4970		CLUSTER OF GALAXIES
MRK 392	08 56 12.	+ 33 08	7	15.	GALAXY WITH UV CONTINUUM
ZWG 238.001	08 56 12.	+ 45 06		12.3	GALAXY
UGC 04708	08 56 12.	+ 45 06	210	12.3	GALAXY SBb
MCG+08-17-003	08 56 12.	+ 45 06	108	12.8	GALAXY
KARA.73B 0292	08 56 12.	+ 45 06	186	12.3	ISOLATED GALAXY S
MCG+08-17-004	08 56 12.	+ 45 56	15	17.	GALAXY
ZWG 238.002	08 56 12.	+ 46 08		14.5	GALAXY
UGC 04709	08 56 12.	+ 46 08	33	14.5	GALAXY S
MCG+08-17-005	08 56 12.	+ 46 08	30	15.	GALAXY
VHE 27B	08 56 12.	- 42 30			REFLECTION NEBULA
VHE 27A	08 56 12.	- 42 38	48		REFLECTION NEBULA
TON-N 0983	08 56 13.	+ 22 05		16.7	BLUE STAR
LB 03632	08 56 14.	+ 09 48 54.		16.0	FAINT BLUE STAR
MCG+08-17-006	08 56 15.	+ 45 54	12	16.	GALAXY
LB 02928	08 56 16.	+ 20 48 30.		16.2	FAINT BLUE STAR
TON-N 0984	08 56 17.	+ 35 01		16.8	BLUE STAR
LB 09021	08 56 18.	+ 16 32		17.9	FAINT BLUE STAR
LB 09020	08 56 18.	+ 18 30		17.7	FAINT BLUE STAR
ZC 0856.3+4554	08 56 18.	+ 45 54	4570		CLUSTER OF GALAXIES
MCG+09-15-067	08 56 18.	+ 55 00	12	17.	GALAXY
MCG-01-23-014	08 56 18.	- 03 22	120	13.	GALAXY
RNGC 2721	08 56 21.	- 04 24		12.5	GALAXY
LB 03633	08 56 24.	+ 09 30 12.		19.0	FAINT BLUE STAR
ZWG 061.034	08 56 24.	+ 11 22		14.2	GALAXY
UGC 04710	08 56 24.	+ 11 22	84	14.2	GALAXY E-S0
MCG+02-23-016	08 56 24.	+ 11 22	23	14.2	GALAXY
LB 09024	08 56 24.	+ 15 39		17.9	FAINT BLUE STAR
LB 09023	08 56 24.	+ 18 24		17.5	FAINT BLUE STAR
LB 09022	08 56 24.	+ 18 27		18.0	FAINT BLUE STAR
ZC 0856.4+3520	08 56 24.	+ 35 20	5040		CLUSTER OF GALAXIES
MCG-01-23-015	08 56 24.	- 04 34	120	12.5	GALAXY
VHE 26	08 56 24.	- 47 11	102		REFLECTION NEBULA
LB 03634	08 56 25.	+ 09 22 24.		18.4	FAINT BLUE STAR
PNGC 2720	08 56 26.	+ 11 22		14.0	GALAXY
LB 02929	08 56 27.	+ 00 50 24.		15.5	FAINT BLUE STAR
MCG+08-17-007	08 56 27.	+ 50 08 30.	42	16.	GALAXY
TON-N 0986	08 56 29.	+ 35 36		15.6	BLUE STAR
LB 09027	08 56 30.	+ 17 44		18.6	FAINT BLUE STAR
LB 09026	08 56 30.	+ 18 37		17.9	FAINT BLUE STAR
LB 09025	08 56 30.	+ 19 27		18.0	FAINT BLUE STAR
ZC 0856.5+2521	08 56 30.	+ 25 21	1010		CLUSTER OF GALAXIES
ZC 0856.5+4310	08 56 30.	+ 43 10	1140		CLUSTER OF GALAXIES
MCG+08-17-008	08 56 30.	+ 46 08	18	16.	GALAXY
MCG+08-17-009	08 56 30.	+ 49 58	36	15.	GALAXY
MCG+09-15-068	08 56 30.	+ 54 07	36	17.	GALAXY
ZWG 005.044	08 56 30.	- 01 36		15.3	GALAXY
TON-N 0985	08 56 31.	+ 22 52		15.3	BLUE STAR
MCG+08-17-010	08 56 33.	+ 49 23	30	17.	GALAXY
MCG+09-15-069	08 56 33.	+ 54 08	48	18.	GALAXY
LB 09030	08 56 36.	+ 15 35		16.3	FAINT BLUE STAR
LB 09029	08 56 36.	+ 18 56		17.7	FAINT BLUE STAR
LB 09028	08 56 36.	+ 19 06		18.1	FAINT BLUE STAR
LB 02930	08 56 36.	+ 21 39 48.		16.7	FAINT BLUE STAR
TON-N 0369	08 56 36.	+ 31 36		15.7	BLUE STAR
ZC 0856.6+3710	08 56 36.	+ 37 10	1140		CLUSTER OF GALAXIES
ZC 0856.6+5202	08 56 36.	+ 52 02	940		CLUSTER OF GALAXIES
ZCG 0856+62.1	08 56 36.	+ 62 48		18.0	COMPACT GALAXY
ZCG 0856+62.6	08 56 36.	+ 62 52		18.4	COMPACT GALAXY
ZWG 350.010	08 56 36.	+ 78 57		15.4	GALAXY
ZWG 349.035	08 56 36.	+ 78 57		15.4	GALAXY
UGC 04711	08 56 36.	+ 78 57	66	15.4	GALAXY
VHE 28	08 56 36.	- 43 14	60		REFLECTION NEBULA
TON-N 0987	08 56 37.	+ 21 43		16.0	BLUE STAR
ZC 0856.7+0320	08 56 42.	+ 03 20	810		CLUSTER OF GALAXIES
ZWG 061.035	08 56 42.	+ 11 20		15.2	GALAXY
UGC 04712	08 56 42.	+ 11 20	66	15.2	GALAXY S
LB 09036	08 56 42.	+ 15 12		18.8	FAINT BLUE STAR
LB 09035	08 56 42.	+ 15 19		17.5	FAINT BLUE STAR
LB 09034	08 56 42.	+ 19 00		18.0	FAINT BLUE STAR
LB 09033	08 56 42.	+ 19 53		18.0	FAINT BLUE STAR
LB 09032	08 56 42.	+ 19 59		17.6	FAINT BLUE STAR
LB 09031	08 56 42.	+ 20 21		17.6	FAINT BLUE STAR
ZC 0856.7+3832	08 56 42.	+ 38 32	540		CLUSTER OF GALAXIES

OBJECT NAME	RIGHT ASCEN.	DECLINATION	DIAM.	MAGN.	TYPE OF OBJECT
ZC 0856.7+4404	08 56 42.	+ 44 04	1280		CLUSTER OF GALAXIES
ZC 0856.7+5133	08 56 42.	+ 51 33	870		CLUSTER OF GALAXIES
MCG+09-15-070	08 56 42.	+ 52 40	72	13.	GALAXY
ZWG 264.047	08 56 42.	+ 52 42		13.7	GALAXY
UGC 04713	08 56 42.	+ 52 42	114	13.7	GALAXY Sb
ZWG 350.011	08 56 42.	+ 78 46		13.7	GALAXY
UGC 04714	08 56 42.	+ 78 46	84	13.7	GALAXY Sb
TON-N 0988	08 56 43.	+ 23 13		16.7	BLUE STAR
LB 02931	08 56 47.	+ 19 25 18.		16.4	FAINT BLUE STAR
ZWG 005.045	08 56 48.	+ 00 59		15.4	GALAXY
LB 09039	08 56 48.	+ 14 42		17.9	FAINT BLUE STAR
LB 09038	08 56 48.	+ 16 08		18.0	FAINT BLUE STAR
LB 09037	08 56 48.	+ 20 25		16.5	FAINT BLUE STAR
MCG+09-15-071	08 56 48.	+ 53 15	60	16.	GALAXY
UGC 04715	08 56 48.	+ 53 16	60	16.0	GALAXY PECULR
ZCG 0856+62.3	08 56 48.	+ 62 50		17.6	COMPACT GALAXY
ZCG 0856+62.4	08 56 48.	+ 62 51		17.8	COMPACT GALAXY
ARC 0730	08 56 49.	+ 51 32		17.5	RICH CLUSTER OF GALAXIES
LB 02932	08 56 49.	- 01 25 06.		16.5	FAINT BLUE STAR
LB 09043	08 56 54.	+ 14 57		18.0	FAINT BLUE STAR
LB 09042	08 56 54.	+ 15 04		15.3	FAINT BLUE STAR
LB 09041	08 56 54.	+ 15 14		18.0	FAINT BLUE STAR
LB 09040	08 56 54.	+ 17 58		17.4	FAINT BLUE STAR
LB 02933	08 56 54.	+ 22 38 12.		16.6	FAINT BLUE STAR
ZCG 0856+62.2	08 56 54.	+ 62 49		18.2	COMPACT GALAXY
ZCG 0856+62.5	08 56 54.	+ 62 51		18.4	COMPACT GALAXY
MCG+13-07-014	08 56 54.	+ 78 46	66	14.	GALAXY
TON-N 0989	08 56 55.	+ 19 37		16.5	BLUE STAR
LB 00550	08 56 55.	- 01 36 12.		17.8	FAINT BLUE STAR
RNGC 2722	08 56 57.	- 03 32			GALAXY
ZWG 033.035	08 57 00.	+ 03 41		15.7	GALAXY
ZWG 033.036	08 57 00.	+ 08 01		15.6	GALAXY
LB 09044	08 57 00.	+ 19 40		15.4	FAINT BLUE STAR
ZWG 150.055	08 57 00.	+ 29 50		15.6	GALAXY
TON-N 0990	08 57 00.	+ 32 32		17.	BLUE STAR
TON-N 0370	08 57 00.	+ 33 37		15.4	BLUE STAR
MCG+06-20-016	08 57 00.	+ 34 52	48	15.	GALAXY
MCG+06-20-018	08 57 00.	+ 35 56	24	14.	GALAXY
MCG+06-20-017	08 57 00.	+ 35 56	48	14.	GALAXY
ZWG 209.016	08 57 00.	+ 40 30		15.5	GALAXY
UGC 04716	08 57 00.	+ 40 30	84	15.5	GALAXY SBb
MCG+09-15-072	08 57 00.	+ 50 51	138	14.	GALAXY
MCG+09-15-073	08 57 00.	+ 51 22 30.	60	15.	GALAXY
ZWG 264.048	08 57 00.	+ 51 25		15.1	GALAXY
UGC 04717	08 57 00.	+ 51 25	66	15.1	GALAXY Sa-b
7ZW 259	08 57 00.	+ 62 50			COMPACT GALAXY
ZC 0857.0+6253	08 57 00.	+ 62 53	2760		CLUSTER OF GALAXIES
MCG-02-23-007	08 57 00.	- 11 38 30.	54	15.	GALAXY
MRSL 251+13/1	08 57 00.	- 25 30	43200		HII REGION
MCG+07-19-012	08 57 03.	+ 40 29	66	15.	GALAXY
ARC 0725	08 57 05.	+ 62 48		17.4	RICH CLUSTER OF GALAXIES
ZWG 033.037	08 57 06.	+ 05 15		14.8	GALAXY
MCG+01-23-016	08 57 06.	+ 05 15	24	14.8	GALAXY
ZWG 061.036	08 57 06.	+ 13 23		15.3	GALAXY
LB 09046	08 57 06.	+ 16 02		16.9	FAINT BLUE STAR
ZWG 090.047	08 57 06.	+ 18 01		15.5	GALAXY
LB 09045	08 57 06.	+ 18 32		17.8	FAINT BLUE STAR
ZWG 150.056	08 57 06.	+ 32 12		15.2	GALAXY
ZWG 180.025	08 57 06.	+ 35 55		13.7	GALAXY
UGC 04718	08 57 06.	+ 35 55	84	13.7	GALAXY PECULR
KARA.72 181B	08 57 06.	+ 35 55	30		PART OF DOUBLE GALAXY
KARA.72 181A	08 57 06.	+ 35 56	90	13.7	PART OF DOUBLE GALAXY
ZWG 264.049	08 57 06.	+ 50 53		15.0	GALAXY
UGC 04719	08 57 06.	+ 50 53	138	15.0	GALAXY Sc
ZCG 0857+62.9	08 57 06.	+ 62 52		18.6	COMPACT GALAXY
MCG+12-09-029	08 57 06.	+ 73 31	84	16.	GALAXY
RNGC 2719A	08 57 08.	+ 35 55		13.5	GALAXY
RNGC 2719	08 57 08.	+ 35 55		13.5	GALAXY
HOLM 105B	08 57 09.	+ 35 55	36	14.5	PART OF MULTIPLE GALAXY
HOLM 105A	08 57 09.	+ 35 55	60	14.0	PART OF MULTIPLE GALAXY
ARP 202	08 57 09.	+ 35 56			PECULIAR GALAXY
LB 09053	08 57 12.	+ 14 27		16.6	FAINT BLUE STAR
LB 09052	08 57 12.	+ 15 27		17.7	FAINT BLUE STAR
LB 09051	08 57 12.	+ 15 41		16.8	FAINT BLUE STAR
LB 09050	08 57 12.	+ 16 32		17.5	FAINT BLUE STAR
LB 09049	08 57 12.	+ 18 04		17.1	FAINT BLUE STAR
LB 09048	08 57 12.	+ 18 11		14.6	FAINT BLUE STAR
LB 09047	08 57 12.	+ 19 11		17.6	FAINT BLUE STAR
ZWG 180.026	08 57 12.	+ 34 51		15.5	GALAXY
UGC 04720	08 57 12.	+ 34 51	66	15.5	GALAXY S
MCG+08-17-011	08 57 12.	+ 46 53 30.	18	16.	GALAXY
ZCG 0857+62.8	08 57 12.	+ 62 48		18.2	COMPACT GALAXY
LB 02934	08 57 15.	+ 21 24 24.		16.6	FAINT BLUE STAR
MCG+05-21-016	08 57 15.	+ 28 51	24	17.	GALAXY
MCG+08-17-012	08 57 15.	+ 45 52	30	16.	GALAXY
ARC 0729	08 57 17.	+ 58 27		17.1	RICH CLUSTER OF GALAXIES
ZC 0857.3+1510	08 57 18.	+ 15 10	2890		CLUSTER OF GALAXIES
LB 09055	08 57 18.	+ 15 20		17.7	FAINT BLUE STAR
LB 09054	08 57 18.	+ 15 35		17.0	FAINT BLUE STAR
ZWG 090.048	08 57 18.	+ 17 07		15.5	GALAXY
UGC 04721	08 57 18.	+ 17 07	66	15.5	GALAXY SBc
ZWG 150.057	08 57 18.	+ 30 00		15.5	GALAXY
ZWG 238.003	08 57 18.	+ 46 53		15.3	GALAXY
MCG+09-15-074	08 57 18.	+ 52 27	12	17.	GALAXY
WRAY 19.15	08 57 19.8	- 47 18 57.			DIFFUSE NEBULA
LB 09058	08 57 24.	+ 16 08		17.0	FAINT BLUE STAR
LB 09057	08 57 24.	+ 17 11		18.0	FAINT BLUE STAR
MCG+03-23-024	08 57 24.	+ 17 35	18	16.	GALAXY
ZWG 090.049	08 57 24.	+ 17 36		15.2	GALAXY
MCG+03-23-025	08 57 24.	+ 17 36	39	15.2	GALAXY
LB 09056	08 57 24.	+ 19 07		17.1	FAINT BLUE STAR
ZC 0857.4+4116	08 57 24.	+ 41 16	2690		CLUSTER OF GALAXIES
MCG+09-15-075	08 57 24.	+ 53 24	42	16.	GALAXY
ZWG 264.050	08 57 24.	+ 53 25		15.4	GALAXY
KARA.72 182A	08 57 24.	+ 53 25	42	15.4	PART OF DOUBLE GALAXY
ZC 0857.4+5552	08 57 24.	+ 55 52	670		CLUSTER OF GALAXIES
MRSL 267-01/3	08 57 24.	- 47 18	120		HII REGION
LB 03490	08 57 24.	- 73 27		14.9	FAINT BLUE STAR
LB 02935	08 57 25.	+ 21 33 12.		16.4	FAINT BLUE STAR
LB 02936	08 57 26.	+ 19 42 30.		16.4	FAINT BLUE STAR
MCG+08-17-013	08 57 27.	+ 46 50	30	17.	GALAXY
ZWG 033.038	08 57 30.	+ 07 55		15.4	GALAXY
LB 09062	08 57 30.	+ 14 35		16.8	FAINT BLUE STAR
LB 09061	08 57 30.	+ 16 08		16.5	FAINT BLUE STAR
LB 09060	08 57 30.	+ 17 25		15.4	FAINT BLUE STAR
LB 09059	08 57 30.	+ 17 58		17.7	FAINT BLUE STAR
ZWG 120.056	08 57 30.	+ 25 47		15.2	GALAXY
UGC 04722	08 57 30.	+ 25 47	96	15.2	GALAXY S-IRR
KARA.73B 0293	08 57 30.	+ 25 47	96	15.2	ISOLATED GALAXY S
MCG+04-21-026	08 57 30.	+ 25 48	30	15.2	GALAXY
MCG+09-15-076	08 57 30.	+ 52 50	30	16.	GALAXY
MCG+09-15-077	08 57 30.	+ 53 24	48	16.	GALAXY
ZWG 264.051	08 57 30.	+ 53 25		15.5	GALAXY
KARA.72 182B	08 57 30.	+ 53 25	48	15.5	PART OF DOUBLE GALAXY
ZC 0857.5+5606	08 57 30.	+ 56 06	740		CLUSTER OF GALAXIES
ZWG 332.031	08 57 30.	+ 73 30		15.4	GALAXY
MCG-01-23-016	08 57 30.	- 07 05	84	15.	GALAXY
MRSL 267-01/2	08 57 30.	- 47 16	2400		HII REGION
LB 02937	08 57 32.	- 00 40 42.		15.4	FAINT BLUE STAR
PK266+00.1	08 57 33.	- 45 24			PLANETARY NEBULA
WRAY 19.16	08 57 34.7	- 43 33 29.			DIFFUSE NEBULA
ZWG 033.039	08 57 36.	+ 03 22		14.5	GALAXY
RNGC 2723	08 57 36.	+ 03 22		14.5	GALAXY
UGC 04723	08 57 36.	+ 03 22	66	14.5	GALAXY S0
MCG+01-23-017	08 57 36.	+ 03 22	36	14.5	GALAXY
ZWG 033.040	08 57 36.	+ 07 57		15.7	GALAXY
ZWG 090.050	08 57 36.	+ 17 22		15.7	GALAXY
LB 09064	08 57 36.	+ 17 42		17.6	FAINT BLUE STAR
LB 09063	08 57 36.	+ 20 24		16.9	FAINT BLUE STAR
ZC 0857.6+3837	08 57 36.	+ 38 37	610		CLUSTER OF GALAXIES
ZWG 238.004	08 57 36.	+ 48 00		15.6	GALAXY
MCG+08-17-014	08 57 36.	+ 48 00	54	15.	GALAXY
ZC 0857.6+6332	08 57 36.	+ 63 32	1680		CLUSTER OF GALAXIES
ARC 0733	08 57 38.	+ 55 49		17.7	RICH CLUSTER OF GALAXIES
ZC 0857.7+1630	08 57 42.	+ 16 30	870		CLUSTER OF GALAXIES
ZWG 090.051	08 57 42.	+ 17 34		15.3	GALAXY
UGC 04724	08 57 42.	+ 17 34	72	15.3	GALAXY IRR
MCG+03-23-026	08 57 42.	+ 17 34	66	15.3	GALAXY
LB 09067	08 57 42.	+ 18 10		16.0	FAINT BLUE STAR
LB 09066	08 57 42.	+ 18 31		16.6	FAINT BLUE STAR
LB 09065	08 57 42.	+ 19 11		15.2	FAINT BLUE STAR
ZWG 150.058	08 57 42.	+ 31 21		15.7	GALAXY
MRSL 265+01/1	08 57 42.	- 43 33	420		HII REGION
TON-N 0991	08 57 43.	+ 21 58		16.6	BLUE STAR
LB 09071	08 57 48.	+ 15 45		15.3	FAINT BLUE STAR
LB 09070	08 57 48.	+ 18 42		14.8	FAINT BLUE STAR
LB 09069	08 57 48.	+ 19 46		17.4	FAINT BLUE STAR
LB 09068	08 57 48.	+ 20 39		16.5	FAINT BLUE STAR
ZWG 150.059	08 57 48.	+ 32 12		15.7	GALAXY
UGC 04725	08 57 48.	+ 32 12	90	15.7	GALAXY Sc
ZC 0857.8+3220	08 57 48.	+ 32 20	1010		CLUSTER OF GALAXIES
ZWG 180.027	08 57 48.	+ 35 57		14.8	GALAXY
UGC 04726	08 57 48.	+ 35 57	108	14.8	GALAXY SBc/Sc
MCG+06-20-019	08 57 48.	+ 35 58	102	14.	GALAXY
ZWG 180.028	08 57 48.	+ 38 04		15.7	GALAXY
ZWG 264.052	08 57 48.	+ 55 14		15.4	GALAXY
MCG+09-15-078	08 57 48.	+ 55 14	15	16.	GALAXY
MCG+10-13-051	08 57 48.	+ 60 20	57	15.	GALAXY
UGC 04727	08 57 48.	+ 60 21	78	16.0	GALAXY Sc
ARC 0734	08 57 49.	+ 16 28		17.7	RICH CLUSTER OF GALAXIES
RNGC 2724	08 57 50.	+ 35 58		15.0	GALAXY
IC 2427	08 57 51.	+ 38 04 19.			NONSTELLAR OBJECT
ZWG 033.041	08 57 54.	+ 04 33		15.6	GALAXY
LB 09076	08 57 54.	+ 16 13		17.4	FAINT BLUE STAR
LB 09075	08 57 54.	+ 17 14		17.5	FAINT BLUE STAR
UGC 04728	08 57 54.	+ 17 23	66	16.0	GALAXY Sa-b
ZWG 090.052	08 57 54.	+ 17 49		14.7	GALAXY
UGC 04729	08 57 54.	+ 17 49	66	14.7	GALAXY SBc
MCG+03-23-027	08 57 54.	+ 17 49	60	14.7	GALAXY
LB 09074	08 57 54.	+ 19 19		17.1	FAINT BLUE STAR
LB 09073	08 57 54.	+ 19 37		18.3	FAINT BLUE STAR
LB 09072	08 57 54.	+ 19 56		17.8	FAINT BLUE STAR
ZC 0857.9+2107	08 57 54.	+ 21 07	1280		CLUSTER OF GALAXIES
LB 02938	08 57 54.	+ 22 52 24.		17.1	FAINT BLUE STAR
MCG+09-15-079	08 57 54.	+ 50 47	18	18.	GALAXY
TON-N 0992	08 57 55.	+ 19 34		16.6	BLUE STAR
YM 45	08 57 59.	- 47 07	360		SYMMETRIC GALACTIC NEBULA
LBN 0695	08 58	+ 66 50	4500		BRIGHT NEBULA
LBN 1078	08 58	- 26 00	6900		BRIGHT NEBULA
LB 09082	08 58 00.	+ 15 05		18.0	FAINT BLUE STAR
LB 09081	08 58 00.	+ 16 11		16.1	FAINT BLUE STAR
LB 09080	08 58 00.	+ 17 02		18.1	FAINT BLUE STAR
LB 09079	08 58 00.	+ 17 24		15.3	FAINT BLUE STAR
LB 09078	08 58 00.	+ 17 52		16.6	FAINT BLUE STAR
LB 09077	08 58 00.	+ 18 19		15.5	FAINT BLUE STAR
ZC 0858.0+2318	08 58 00.	+ 23 18	1080		CLUSTER OF GALAXIES
TON-N 0993	08 58 00.	+ 33 12		17.	BLUE STAR
TON-N 0371	08 58 00.	+ 33 32		14.7	BLUE STAR
MCG+06-20-020	08 58 00.	+ 35 00	42	15.5	GALAXY
ZWG 180.029	08 58 00.	+ 38 08		15.7	GALAXY
MCG+08-17-015	08 58 00.	+ 49 31	42	16.	GALAXY
MCG+10-13-052	08 58 00.	+ 60 20	51	15.	GALAXY
ZWG 288.017	08 58 00.	+ 60 21		14.3	GALAXY
UGC 04730	08 58 00.	+ 60 21	54	14.3	GALAXY
ZWG 311.028	08 58 00.	+ 64 00		15.0	GALAXY
MRSL 268-00/1	08 58 00.	- 47 15	1200		HII REGION
LB 02939	08 58 04.	+ 20 32 42.		16.6	FAINT BLUE STAR
ZWG 033.042	08 58 06.	+ 06 05		15.7	GALAXY
LB 09087	08 58 06.	+ 14 41		16.7	FAINT BLUE STAR
LB 09086	08 58 06.	+ 15 14		18.7	FAINT BLUE STAR
LB 09085	08 58 06.	+ 16 18		16.8	FAINT BLUE STAR
LB 09084	08 58 06.	+ 18 22		16.1	FAINT BLUE STAR
LB 09083	08 58 06.	+ 18 44		16.0	FAINT BLUE STAR
ZC 0858.1+2503	08 58 06.	+ 25 03	1550		CLUSTER OF GALAXIES
ZC 0858.1+5943	08 58 06.	+ 59 43	2890		CLUSTER OF GALAXIES
MCG+11-11-036	08 58 06.	+ 65 58	39	16.	GALAXY
LB 02940	08 58 11.	+ 19 05 12.		16.6	FAINT BLUE STAR
LB 09090	08 58 12.	+ 14 48		16.0	FAINT BLUE STAR
LB 09089	08 58 12.	+ 18 06		12.3	FAINT BLUE STAR
LB 09088	08 58 12.	+ 18 48		17.2	FAINT BLUE STAR
ZWG 120.057	08 58 12.	+ 20 49		15.5	GALAXY
TON-N 0372	08 58 12.	+ 30 33		15.8	BLUE STAR
LB 02941	08 58 13.	+ 22 13 48.		15.6	FAINT BLUE STAR
SEY 052	08 58 13.	+ 71 30 54.		15.3	FAINT GALAXY
ZWG 061.037	08 58 18.	+ 10 49		14.7	GALAXY
UGC 04731	08 58 18.	+ 10 49	90	14.7	GALAXY S0-a
MCG+02-23-017	08 58 18.	+ 10 49	72	14.7	GALAXY
ZWG 061.038	08 58 18.	+ 11 17		14.1	GALAXY
UGC 04732	08 58 18.	+ 11 17	48	14.1	GALAXY PECULR
MCG+02-23-018	08 58 18.	+ 11 17	42	14.1	GALAXY
LB 09092	08 58 18.	+ 15 44		17.1	FAINT BLUE STAR
LB 09091	08 58 18.	+ 15 46		18.0	FAINT BLUE STAR
TON-N 0994	08 58 19.	+ 22 37		16.0	BLUE STAR
RNGC 2725	08 58 20.	+ 11 17		14.0	GALAXY
ZC 0858.4+0310	08 58 24.	+ 03 10	1340		CLUSTER OF GALAXIES
ZWG 061.039	08 58 24.	+ 13 15		15.7	GALAXY

OBJECT NAME	RIGHT ASCEN.	DECLINATION	DIAM.	MAGN.	TYPE OF OBJECT
ZWG 061.040	08 58 24.	+ 13 57		15.7	GALAXY
LB 09094	08 58 24.	+ 14 58		15.5	FAINT BLUE STAR
LB 09093	08 58 24.	+ 19 44		17.6	FAINT BLUE STAR
TON-N 0995	08 58 25.	+ 21 59		16.9	BLUE STAR
ZWG 033.043	08 58 30.	+ 06 18		15.3	GALAXY
LB 09096	08 58 30.	+ 15 34		18.5	FAINT BLUE STAR
LB 09095	08 58 30.	+ 16 25		16.7	FAINT BLUE STAR
ZC 0858.5+6028	08 58 30.	+ 60 28	540		CLUSTER OF GALAXIES
ZC 0858.5+6740	08 58 30.	+ 67 40	610		CLUSTER OF GALAXIES
MRSL 266+00/1	08 58 30.	- 45 45	780		HII REGION
RNGC 2727	08 58 33.	- 03 11			NON-EXISTENT OBJECT
ZWG 033.044	08 58 36.	+ 04 18		15.7	GALAXY
UGC 04733	08 58 36.	+ 04 18	78	15.7	GALAXY Sc
KARA.73B 0294	08 58 36.	+ 04 18	72	15.7	ISOLATED GALAXY S
LB 09104	08 58 36.	+ 14 43		18.1	FAINT BLUE STAR
LB 09103	08 58 36.	+ 15 20		16.5	FAINT BLUE STAR
LB 09102	08 58 36.	+ 18 24		17.0	FAINT BLUE STAR
LB 09101	08 58 36.	+ 18 36		15.8	FAINT BLUE STAR
ZWG 090.053	08 58 36.	+ 19 10		15.4	GALAXY
UGC 04734	08 58 36.	+ 19 10	60	15.4	GALAXY S
LB 09100	08 58 36.	+ 19 12		16.2	FAINT BLUE STAR
LB 09099	08 58 36.	+ 19 15		16.8	FAINT BLUE STAR
LB 09098	08 58 36.	+ 19 20		18.2	FAINT BLUE STAR
LB 09097	08 58 36.	+ 20 28		16.9	FAINT BLUE STAR
ZC 0858.6+3836	08 58 36.	+ 38 36	3090		CLUSTER OF GALAXIES
ZWG 264.053	08 58 36.	+ 55 35		15.7	GALAXY
MCG+10-13-053	08 58 36.	+ 58 12	39	16.	GALAXY
TON-N 0996	08 58 37.	+ 21 29		17.0	BLUE STAR
RNGC 2736	08 58 38.	- 45 42			GALAXY
ZWG 061.041	08 58 42.	+ 10 24		15.0	GALAXY
MCG+02-23-019	08 58 42.	+ 10 24	36	15.0	GALAXY
LB 09109	08 58 42.	+ 15 06		15.0	FAINT BLUE STAR
LB 09108	08 58 42.	+ 16 10		18.0	FAINT BLUE STAR
LB 09107	08 58 42.	+ 16 53		18.4	FAINT BLUE STAR
LB 09106	08 58 42.	+ 19 01		16.9	FAINT BLUE STAR
LB 09105	08 58 42.	+ 20 04		17.3	FAINT BLUE STAR
ZC 0858.7+4937	08 58 42.	+ 49 37	870		CLUSTER OF GALAXIES
ZC 0858.7-0215	08 58 42.	- 02 15	870		CLUSTER OF GALAXIES
MCG+09-15-080	08 58 45.	+ 55 35	30	16.	GALAXY
LB 02942	08 58 46.	+ 22 56 24.		16.0	FAINT BLUE STAR
ZWG 033.045	08 58 48.	+ 07 13		15.3	GALAXY
LB 09111	08 58 48.	+ 16 06		18.2	FAINT BLUE STAR
ZWG 090.054	08 58 48.	+ 16 40		15.6	GALAXY
LB 09110	08 58 48.	+ 17 09		16.6	FAINT BLUE STAR
ZWG 151.001	08 58 48.	+ 29 16		15.6	GALAXY
ZWG 150.060	08 58 48.	+ 29 16		15.6	GALAXY
ZC 0858.8+3732	08 58 48.	+ 37 32	2350		CLUSTER OF GALAXIES
UGC 04736	08 58 48.	+ 75 07	96	16.5	GALAXY
ZWG 005.046	08 58 48.	- 01 40		14.9	GALAXY
UGC 04735	08 58 48.	- 01 40	60	14.9	GALAXY S
MCG+00-23-019	08 58 48.	- 01 40	54	14.9	GALAXY
KARA.73B 0295	08 58 48.	- 01 40	60	14.9	ISOLATED GALAXY S0
HOLM 106A	08 58 49.	+ 03 54	48	14.0	PART OF MULTIPLE GALAXY
LB 02943	08 58 49.	+ 21 36 30.		17.0	FAINT BLUE STAR
IC 0525	08 58 50.	- 01 39 02.			NONSTELLAR OBJECT
HOLM 106B	08 58 51.	+ 03 54	18	14.4	PART OF MULTIPLE GALAXY
MCG+08-17-016	08 58 51.	+ 46 37	36	16.	GALAXY
ZWG 033.046	08 58 54.	+ 03 55		14.0	GALAXY
RNGC 2729	08 58 54.	+ 03 55		14.0	GALAXY
UGC 04737	08 58 54.	+ 03 55	48	14.0	GALAXY S0?
MCG+01-23-018	08 58 54.	+ 03 55	36	14.0	GALAXY
LB 09113	08 58 54.	+ 16 38		17.7	FAINT BLUE STAR
LB 09112	08 58 54.	+ 17 41		16.8	FAINT BLUE STAR
ZWG 151.002	08 58 54.	+ 29 12		15.7	GALAXY
ZWG 150.061	08 58 54.	+ 29 12		15.7	GALAXY
TON-N 0373	08 58 54.	+ 33 38		16.2	BLUE STAR
ZWG 238.005	08 58 54.	+ 46 37		15.7	GALAXY
ZC 0858.9+7314	08 58 54.	+ 73 14	1140		CLUSTER OF GALAXIES
ZWG 005.047	08 58 54.	- 00 14		15.6	GALAXY
OCL 0766	08 58 54.	- 48 47	180	8.4	OPEN STAR CLUSTER
VHA 057	08 58 54.	- 48 47	180		OPEN STAR CLUSTER
LB 02944	08 58 56.	+ 22 02 30.		16.8	FAINT BLUE STAR
ZWG 061.042	08 59 00.	+ 11 16		14.9	GALAXY
UGC 04738	08 59 00.	+ 11 16	72	14.9	GALAXY Sb
MCG+02-23-020	08 59 00.	+ 11 16	60	14.9	GALAXY
ZWG 061.043	08 59 00.	+ 13 11		15.3	GALAXY
LB 09116	08 59 00.	+ 17 24		18.3	FAINT BLUE STAR
LB 09115	08 59 00.	+ 17 58		18.3	FAINT BLUE STAR
LB 09114	08 59 00.	+ 19 21		18.2	FAINT BLUE STAR
TON-N 0374	08 59 00.	+ 33 49		16.2	BLUE STAR
RNGC 2728	08 59 02.	+ 11 16		15.0	GALAXY
LB 09122	08 59 06.	+ 14 34		17.8	FAINT BLUE STAR
LB 09121	08 59 06.	+ 16 09		17.1	FAINT BLUE STAR
LB 09120	08 59 06.	+ 18 53		17.9	FAINT BLUE STAR
LB 09119	08 59 06.	+ 19 01		18.1	FAINT BLUE STAR
LB 09118	08 59 06.	+ 19 13		18.0	FAINT BLUE STAR
LB 09117	08 59 06.	+ 20 15		16.9	FAINT BLUE STAR
LB 02945	08 59 06.	+ 23 00 48.		16.0	FAINT BLUE STAR
ZC 0859.1+5641	08 59 06.	+ 56 41	540		CLUSTER OF GALAXIES
ZWG 332.032	08 59 06.	+ 69 42		15.2	GALAXY
UGC 04739	08 59 06.	+ 69 42	84	15.2	GALAXY PECULR
ZWG 005.048	08 59 12.	+ 00 42		15.6	GALAXY
LB 09126	08 59 12.	+ 14 43		17.9	FAINT BLUE STAR
LB 09125	08 59 12.	+ 17 03		17.1	FAINT BLUE STAR
LB 09124	08 59 12.	+ 18 51		17.7	FAINT BLUE STAR
LB 09123	08 59 12.	+ 19 21		18.3	FAINT BLUE STAR
ZWG 121.001	08 59 12.	+ 23 35		14.9	GALAXY
UGC 04740	08 59 12.	+ 23 35	78	14.9	GALAXY S
ZC 0859.2+7945	08 59 12.	+ 79 45	3700		CLUSTER OF GALAXIES
ZC 0859.2-0130	08 59 12.	- 01 30	740		CLUSTER OF GALAXIES
SEY 053	08 59 14.	+ 69 40 09.		15.4	FAINT GALAXY
LB 02946	08 59 15.	+ 10 31 42.		16.5	FAINT BLUE STAR
MCG+04-22-001	08 59 15.	+ 23 36	66	14.9	GALAXY
MCG+08-17-017	08 59 15.	+ 46 35 30.	24	17.	GALAXY
ZWG 061.044	08 59 18.	+ 11 50		15.2	GALAXY
LB 09130	08 59 18.	+ 15 41		18.4	FAINT BLUE STAR
LB 09129	08 59 18.	+ 16 57		16.7	FAINT BLUE STAR
LB 09128	08 59 18.	+ 19 30		17.0	FAINT BLUE STAR
LB 09127	08 59 18.	+ 20 12		18.0	FAINT BLUE STAR
TON-N 0375	08 59 18.	+ 30 48		15.1	BLUE STAR
MCG+09-15-081	08 59 18.	+ 55 35	36	16.	GALAXY
TON-N 0997	08 59 19.	+ 22 38		16.3	BLUE STAR
HOLM 107B	08 59 19.	+ 44 43	30	14.1	PART OF MULTIPLE GALAXY
ARC 0719	08 59 22.	+ 78 12		17.6	RICH CLUSTER OF GALAXIES
ZWG 090.055	08 59 24.	+ 14 43		15.6	GALAXY
LB 09135	08 59 24.	+ 16 30		15.7	FAINT BLUE STAR
LB 09134	08 59 24.	+ 16 51		18.0	FAINT BLUE STAR
LB 09133	08 59 24.	+ 16 56		18.4	FAINT BLUE STAR

OBJECT NAME	RIGHT ASCEN.	DECLINATION	DIAM.	MAGN.	TYPE OF OBJECT
MCG+03-23-028	08 59 24.	+ 17 01	102	13.7	GALAXY
LB 09132	08 59 24.	+ 17 21		16.4	FAINT BLUE STAR
LB 09131	08 59 24.	+ 18 18		17.6	FAINT BLUE STAR
ZC 0859.4+5224	08 59 24.	+ 52 24	1080		CLUSTER OF GALAXIES
RNGC 2731	08 59 28.	+ 08 30		14.0	GALAXY
LB 02947	08 59 28.	+ 09 07 36.		16.5	FAINT BLUE STAR
LB 02948	08 59 29.	+ 10 41 24.		16.2	FAINT BLUE STAR
ZWG 033.047	08 59 30.	+ 03 40		15.4	GALAXY
ZWG 061.045	08 59 30.	+ 08 30		14.2	GALAXY
ZWG 033.048	08 59 30.	+ 08 30		14.2	GALAXY
UGC 04741	08 59 30.	+ 08 30	48	14.2	GALAXY S
MCG+02-23-021	08 59 30.	+ 08 30	48	14.2	GALAXY
ZWG 090.056	08 59 30.	+ 14 44		15.5	GALAXY
UGC 04742	08 59 30.	+ 14 44	78	15.5	GALAXY Sa-b
LB 09137	08 59 30.	+ 16 26		18.4	FAINT BLUE STAR
ZWG 090.057	08 59 30.	+ 17 02		13.7	GALAXY
RNGC 2730	08 59 30.	+ 17 02		13.5	GALAXY
UGC 04743	08 59 30.	+ 17 02	102	13.7	GALAXY SBc-IRR
LB 09136	08 59 30.	+ 17 41		18.6	FAINT BLUE STAR
ZC 0859.5+3217	08 59 30.	+ 32 17	1080		CLUSTER OF GALAXIES
MCG-02-23-008	08 59 30.	- 13 20	30	15.5	GALAXY
MCG-04-22-001	08 59 30.	- 21 52 30.	36	17.	GALAXY
HOLM 107A	08 59 33.	+ 44 42	60	13.3	PART OF MULTIPLE GALAXY
ARC 0736	08 59 33.	+ 52 25		17.5	RICH CLUSTER OF GALAXIES
RNGC 2733	08 59 33.	- 03 32			NON-EXISTENT OBJECT
LB 02950	08 59 34.	+ 20 49 36.		16.5	FAINT BLUE STAR
LB 02949	08 59 34.	+ 21 01 30.		17.1	FAINT BLUE STAR
LB 02951	08 59 35.	+ 23 18 18.		16.6	FAINT BLUE STAR
ZWG 033.049	08 59 36.	+ 03 21		15.5	GALAXY
ZC 0859.6+0850	08 59 36.	+ 08 50	1750		CLUSTER OF GALAXIES
ZWG 090.058	08 59 36.	+ 14 40		15.7	GALAXY
LB 09142	08 59 36.	+ 16 08		17.9	FAINT BLUE STAR
LB 09141	08 59 36.	+ 16 31		16.1	FAINT BLUE STAR
LB 09140	08 59 36.	+ 17 16		17.0	FAINT BLUE STAR
LB 09139	08 59 36.	+ 18 18		18.0	FAINT BLUE STAR
LB 09138	08 59 36.	+ 20 09		18.5	FAINT BLUE STAR
ZWG 121.002	08 59 36.	+ 22 51		15.5	GALAXY
ZWG 151.003	08 59 36.	+ 31 28		15.4	GALAXY
ZWG 150.062	08 59 36.	+ 31 28		15.4	GALAXY
TON-N 0376	08 59 36.	+ 31 36		15.3	BLUE STAR
ZC 0859.6+4315	08 59 36.	+ 43 15	870		CLUSTER OF GALAXIES
ZC 0859.6+6156	08 59 36.	+ 61 56	3360		CLUSTER OF GALAXIES
ZC 0859.7+1438	08 59 42.	+ 14 38	810		CLUSTER OF GALAXIES
LB 09148	08 59 42.	+ 17 12		17.9	FAINT BLUE STAR
LB 09147	08 59 42.	+ 17 40		18.6	FAINT BLUE STAR
LB 09146	08 59 42.	+ 18 05		18.1	FAINT BLUE STAR
LB 09145	08 59 42.	+ 19 11		16.7	FAINT BLUE STAR
LB 09144	08 59 42.	+ 19 19		18.2	FAINT BLUE STAR
LB 09143	08 59 42.	+ 19 42		18.0	FAINT BLUE STAR
ZWG 121.003	08 59 42.	+ 26 08		14.2	GALAXY
ZCG 0859+26	08 59 42.	+ 26 08		14.2	COMPACT GALAXY
UGC 04744	08 59 42.	+ 26 08	66	14.2	GALAXY
SHAH 017	08 59 42.	+ 77 51	36	18.5	GROUP OF COMPACT GALAXIES
ZC 0859.7-0120	08 59 42.	- 01 20	2820		CLUSTER OF GALAXIES
OCL 0769	08 59 42.	- 50 44	216	14.	OPEN STAR CLUSTER
RNGC 2735	08 59 44.	+ 26 08		14.0	GALAXY
HOLM 108A	08 59 45.	+ 26 08	48	13.4	PART OF MULTIPLE GALAXY
VV 040B	08 59 45.	+ 26 09	18	16.	INTERACTING GALAXY
VV 040A	08 59 45.	+ 26 09	18	15.	INTERACTING GALAXY
MCG+07-19-013	08 59 45.	+ 40 30	30	15.	GALAXY
ZWG 033.050	08 59 48.	+ 03 34		15.2	GALAXY
LB 09151	08 59 48.	+ 14 51		18.1	FAINT BLUE STAR
LB 09150	08 59 48.	+ 17 34		17.1	FAINT BLUE STAR
LB 09149	08 59 48.	+ 19 25		17.3	FAINT BLUE STAR
ZWG 121.004	08 59 48.	+ 25 17		15.7	GALAXY
UGC 04745	08 59 48.	+ 25 17	60	15.7	GALAXY Sc
ZWG 121.005	08 59 48.	+ 25 36		15.3	GALAXY
UGC 04746	08 59 48.	+ 25 36	84	15.3	GALAXY Sb-c
ZWG 209.017	08 59 48.	+ 40 38		14.9	GALAXY
MCG+09-15-082	08 59 48.	+ 52 10	30	17.	GALAXY
ZC 0859.8+7042	08 59 48.	+ 70 42	870		CLUSTER OF GALAXIES
ZWG 005.049	08 59 48.	- 00 54		15.5	GALAXY
TON-N 0998	08 59 49.	+ 21 58		16.3	BLUE STAR
HOLM 108B	08 59 49.	+ 26 08	12	14.9	PART OF MULTIPLE GALAXY
ARP 287	08 59 49.	+ 26 08			PECULIAR GALAXY
LB 02952	08 59 51.	+ 19 17 06.		16.2	FAINT BLUE STAR
MCG+04-22-002	08 59 51.	+ 26 08 30.	48	14.2	GALAXY
MCG+07-19-014	08 59 51.	+ 40 37	36	15.	GALAXY
ZWG 061.046	08 59 54.	+ 11 02		14.7	GALAXY
MCG+02-23-022	08 59 54.	+ 11 02	42	14.7	GALAXY
ZWG 061.047	08 59 54.	+ 13 18		14.8	GALAXY
MCG+02-23-023	08 59 54.	+ 13 18	48	14.8	GALAXY
LB 09154	08 59 54.	+ 15 24		18.4	FAINT BLUE STAR
LB 09153	08 59 54.	+ 16 05		18.1	FAINT BLUE STAR
LB 09152	08 59 54.	+ 16 28		17.2	FAINT BLUE STAR
ZWG 090.059	08 59 54.	+ 17 55		15.5	GALAXY
MCG+04-22-004	08 59 54.	+ 25 38	84	15.3	GALAXY
MCG+04-22-003	08 59 54.	+ 26 08 30.	12	14.2	GALAXY
MCG-06-20-021	08 59 54.	+ 34 17	30	15.	GALAXY
ZWG 209.018	08 59 54.	+ 42 01		15.7	GALAXY
ZWG 264.054	08 59 54.	+ 51 07		15.6	GALAXY
ZWG 264.055	08 59 54.	+ 52 11		15.3	GALAXY
BC PKS0859-14	08 59 54.83	- 14 03 38.2		16.59	QUASI-STELLAR OBJECT
SHB 128	08 59 55.	- 14 03 37.		17.8	QUASI-STELLAR OBJECT
IC 0526	08 59 58.	+ 11 02 34.			NONSTELLAR OBJECT
LB 09845	09 00	- 85 21		13.5	FAINT BLUE STAR
LB 09846	09 00	- 86 18		15.5	FAINT BLUE STAR
ZWG 033.051	09 00 00.	+ 04 09		15.7	GALAXY
ZC 0900.0+1343	09 00 00.	+ 13 43	4370		CLUSTER OF GALAXIES
ZC 0900.0+3347	09 00 00.	+ 33 47	810		CLUSTER OF GALAXIES
ZWG 180.030	09 00 00.	+ 34 17		15.5	GALAXY
ZC 0900.0+5121	09 00 00.	+ 51 21	670		CLUSTER OF GALAXIES
ZC 0900.0+5419	09 00 00.	+ 54 19	740		CLUSTER OF GALAXIES
ZWG 264.056	09 00 00.	+ 55 47		15.5	GALAXY
ZWG 090.060	09 00 06.	+ 16 15		15.6	GALAXY
MCG+07-19-015	09 00 06.	+ 40 39	48	17.	GALAXY
MCG+07-19-016	09 00 06.	+ 42 01	39	16.	GALAXY
LB 02953	09 00 07.	+ 23 19 24.		17.0	FAINT BLUE STAR
ARC 0718	09 00 09.	+ 79 25		17.1	RICH CLUSTER OF GALAXIES
TON-N 0999	09 00 11.	+ 34 40		17.	BLUE STAR
ZC 0900.2+1142	09 00 12.	+ 11 42	400		CLUSTER OF GALAXIES
RNGC 2734	09 00 12.	+ 17 02			GALAXY
ZWG 151.004	09 00 12.	+ 30 47		14.7	GALAXY
UGC 04747	09 00 12.	+ 30 47	114	14.7	GALAXY Sc
KARA.73B 0296	09 00 12.	+ 30 47	120	14.7	ISOLATED GALAXY S
ZC 0900.2+4739	09 00 12.	+ 47 39	610		CLUSTER OF GALAXIES
ZC 0900.2+5525	09 00 12.	+ 55 25	3970		CLUSTER OF GALAXIES
IC 2428	09 00 13.	+ 30 47 04.			NONSTELLAR OBJECT

OBJECT NAME	RIGHT ASCEN.	DECLINATION	DIAM.	MAGN.	TYPE OF OBJECT
TON-N 1000	09 00 17.	+ 34 45		16.8	BLUE STAR
LB 02954	09 00 18.	+ 22 41 18.		15.5	FAINT BLUE STAR
ZWG 151.005	09 00 18.	+ 29 19		15.6	GALAXY
MCG+05-22-001	09 00 18.	+ 30 46	102	14.7	GALAXY
LB 02955	09 00 20.	+ 21 19 54.		15.5	FAINT BLUE STAR
ZWG 061.048	09 00 24.	+ 13 50		14.9	GALAXY
MCG+02-23-024	09 00 24.	+ 13 50	18	14.9	GALAXY
LB 09174	09 00 24.	+ 14 34		17.9	FAINT BLUE STAR
LB 09173	09 00 24.	+ 14 46		17.5	FAINT BLUE STAR
LB 09172	09 00 24.	+ 16 54		16.3	FAINT BLUE STAR
MCG+05-22-002	09 00 24.	+ 29 19 30.	30	15.6	GALAXY
ZC 0900.4+4152	09 00 24.	+ 41 52	810		CLUSTER OF GALAXIES
ZC 0900.4+6639	09 00 24.	+ 66 39	740		CLUSTER OF GALAXIES
MCG-03-23-015	09 00 24.	- 20 33 30.	24	15.	GALAXY
MCG-03-23-016	09 00 24.	- 20 34	36	14.5	GALAXY
ARC 0735	09 00 26.	+ 61 58		17.5	RICH CLUSTER OF GALAXIES
ZWG 061.049	09 00 30.	+ 09 49		15.5	GALAXY
ZWG 061.050	09 00 30.	+ 10 21		15.3	GALAXY
ZWG 061.051	09 00 30.	+ 13 54		15.5	GALAXY
LB 09175	09 00 30.	+ 19 14		18.2	FAINT BLUE STAR
ZWG 121.006	09 00 30.	+ 25 50		15.3	GALAXY
LB 00551	09 00 32.	- 01 22 24.		17.2	FAINT BLUE STAR
ZWG 061.052	09 00 36.	+ 10 13		15.4	GALAXY
UGC 04748	09 00 36.	+ 10 13	78	15.4	GALAXY Sc+COMP
ZWG 061.053	09 00 36.	+ 13 35		15.6	GALAXY
ZWG 061.054	09 00 36.	+ 13 43		15.1	GALAXY
MCG+02-23-025	09 00 36.	+ 13 43	48	15.1	GALAXY
LB 09179	09 00 36.	+ 15 25		18.1	FAINT BLUE STAR
LB 09178	09 00 36.	+ 19 37		17.0	FAINT BLUE STAR
LB 09177	09 00 36.	+ 19 50		17.5	FAINT BLUE STAR
LB 09176	09 00 36.	+ 20 36		16.5	FAINT BLUE STAR
ZWG 121.007	09 00 36.	+ 20 51		15.4	GALAXY
ZC 0900.6+3940	09 00 36.	+ 39 40	1010		CLUSTER OF GALAXIES
ZC 0900.6+4521	09 00 36.	+ 45 21	740		CLUSTER OF GALAXIES
ZC 0900.6+5153	09 00 36.	+ 51 53	1410		CLUSTER OF GALAXIES
WRAY 19.17	09 00 41.5	- 48 30 24.		8.6	STAR-NEBULA ASSOCIATION
LB 09182	09 00 42.	+ 16 26		17.6	FAINT BLUE STAR
LB 09181	09 00 42.	+ 16 38		16.0	FAINT BLUE STAR
LB 09180	09 00 42.	+ 17 02		18.1	FAINT BLUE STAR
LB 02956	09 00 42.	+ 21 53 12.		16.1	FAINT BLUE STAR
ZWG 151.006	09 00 42.	+ 29 29		15.2	GALAXY
TON-N 0377	09 00 42.	+ 33 02		15.1	BLUE STAR
MBSL 269-01/1	09 00 42.	- 48 30	420		HII REGION
IC 2429	09 00 44.	+ 29 30 31.			NONSTELLAR OBJECT
MCG+05-22-003	09 00 45.	+ 29 30	24	15.2	GALAXY
ZWG 033.052	09 00 48.	+ 03 34		15.2	GALAXY
KARA.72 183A	09 00 48.	+ 03 34	48	15.2	PART OF DOUBLE GALAXY
MCG+01-23-019	09 00 48.	+ 03 34	48	15.2	GALAXY
LB 09188	09 00 48.	+ 15 18		17.2	FAINT BLUE STAR
LB 09187	09 00 48.	+ 16 10		18.4	FAINT BLUE STAR
LB 09186	09 00 48.	+ 16 47		17.9	FAINT BLUE STAR
LB 09185	09 00 48.	+ 19 26		17.1	FAINT BLUE STAR
LB 09184	09 00 48.	+ 19 35		16.0	FAINT BLUE STAR
LB 09183	09 00 48.	+ 20 20		17.6	FAINT BLUE STAR
MCG+09-15-083	09 00 48.	+ 51 57 30.	48	15.	GALAXY
OCL 0757	09 00 48.	- 43 26	150	10.	OPEN STAR CLUSTER
LB 02957	09 00 51.	+ 22 36 48.		16.0	FAINT BLUE STAR
ZWG 061.055	09 00 54.	+ 13 53		15.3	GALAXY
LB 09195	09 00 54.	+ 16 06		18.4	FAINT BLUE STAR
LB 09194	09 00 54.	+ 16 16		16.9	FAINT BLUE STAR
LB 09193	09 00 54.	+ 16 30		18.7	FAINT BLUE STAR
LB 09192	09 00 54.	+ 16 51		18.2	FAINT BLUE STAR
LB 09191	09 00 54.	+ 17 22		18.1	FAINT BLUE STAR
LB 09190	09 00 54.	+ 17 54		18.0	FAINT BLUE STAR
LB 02958	09 00 54.	+ 18 23		17.7	FAINT BLUE STAR
LB 02958	09 00 54.	+ 18 24 18.		16.7	FAINT BLUE STAR
LB 09189	09 00 54.	+ 19 04		18.0	FAINT BLUE STAR
ZWG 005.050	09 00 54.	- 02 23		14.8	GALAXY
MCG+00-23-020	09 00 54.	- 02 23	30	14.8	GALAXY
ZWG 033.053	09 01 00.	+ 03 34		15.1	GALAXY
KARA.72 183B	09 01 00.	+ 03 34	78	15.1	PART OF DOUBLE GALAXY
MCG+01-23-020	09 01 00.	+ 03 34	36	15.1	GALAXY
LB 09201	09 01 00.	+ 14 49		18.0	FAINT BLUE STAR
LB 09200	09 01 00.	+ 15 30		17.8	FAINT BLUE STAR
LB 09199	09 01 00.	+ 17 08		15.8	FAINT BLUE STAR
LB 09198	09 01 00.	+ 17 32		16.2	FAINT BLUE STAR
LB 09197	09 01 00.	+ 17 44		17.9	FAINT BLUE STAR
LB 09196	09 01 00.	+ 19 49		15.6	FAINT BLUE STAR
ZWG 121.008	09 01 00.	+ 20 45		15.6	GALAXY
ZC 0901.0+4300	09 01 00.	+ 43 00	940		CLUSTER OF GALAXIES
ZWG 264.057	09 01 00.	+ 51 50		13.6	GALAXY
MRK 101	09 01 00.	+ 51 50	30	13.5	GALAXY WITH UV CONTINUUM
UGC 04749	09 01 00.	+ 51 50	42	13.6	GALAXY S
ZWG 264.058	09 01 00.	+ 52 09		15.5	GALAXY
1ZW 017	09 01 00.	+ 55 43			COMPACT GALAXY
ZWG 288.018	09 01 00.	+ 60 08		13.1	GALAXY
MRK 018	09 01 00.	+ 60 08	50	15.	GALAXY WITH UV CONTINUUM
UGC 04750	09 01 00.	+ 60 08	108	13.1	GALAXY Sa?
ZWG 005.051	09 01 00.	- 01 11		15.2	GALAXY
RNGC 2726	09 01 02.	+ 60 08		13.0	GALAXY
RNGC 2737	09 01 04.	+ 22 06		15.0	GALAXY
RNGC 2738	09 01 04.	+ 22 10		14.0	GALAXY
LB 09204	09 01 06.	+ 16 46		18.0	FAINT BLUE STAR
LB 09203	09 01 06.	+ 16 48		18.1	FAINT BLUE STAR
LB 09202	09 01 06.	+ 19 24		17.5	FAINT BLUE STAR
LB 02959	09 01 06.	+ 21 02 36.		16.6	FAINT BLUE STAR
ZWG 121.009	09 01 06.	+ 22 06		14.8	GALAXY
UGC 04751	09 01 06.	+ 22 06	72	14.8	GALAXY Sa-b
MCG+04-22-005	09 01 06.	+ 22 08	42	14.8	GALAXY
ZWG 121.010	09 01 06.	+ 22 10		13.8	GALAXY
UGC 04752	09 01 06.	+ 22 10	90	13.8	GALAXY S
MCG+04-22-006	09 01 06.	+ 22 11	84	13.8	GALAXY
ZWG 151.007	09 01 06.	+ 30 00		15.4	GALAXY
TON-N 0378	09 01 06.	+ 32 24		15.1	BLUE STAR
ZWG 209.019	09 01 06.	+ 41 41		15.6	GALAXY
MCG+08-17-018	09 01 06.	+ 45 29	84	15.	GALAXY
ZWG 005.052	09 01 06.	- 00 57		15.7	GALAXY
LB 02960	09 01 07.	+ 22 11 12.		15.8	FAINT BLUE STAR
ARC 0739	09 01 07.	+ 47 27		17.5	RICH CLUSTER OF GALAXIES
LB 00552	09 01 09.	+ 01 43 00.		17.9	FAINT BLUE STAR
MCG+07-19-017	09 01 09.	+ 41 40	42	16.	GALAXY
LB 09209	09 01 12.	+ 14 47		15.5	FAINT BLUE STAR
LB 09208	09 01 12.	+ 14 54		17.9	FAINT BLUE STAR
LB 09207	09 01 12.	+ 15 23		17.8	FAINT BLUE STAR
LB 09206	09 01 12.	+ 17 12		17.8	FAINT BLUE STAR
LB 09205	09 01 12.	+ 19 19		18.3	FAINT BLUE STAR
ZWG 121.011	09 01 12.	+ 22 15		15.4	GALAXY
MCG+04-22-007	09 01 12.	+ 22 17	36	15.4	GALAXY
MCG+05-22-004	09 01 12.	+ 30 00	36	15.4	GALAXY
TON-N 0379	09 01 12.	+ 32 13		15.4	BLUE STAR
MCG+07-19-018	09 01 12.	+ 41 39 30.	36	17.	GALAXY
ZWG 238.006	09 01 12.	+ 45 29		15.7	GALAXY
UGC 04753	09 01 12.	+ 45 29	90	15.7	GALAXY Sc
ZC 0901.2+4722	09 01 12.	+ 47 22	870		CLUSTER OF GALAXIES
ZC 0901.2+6640	09 01 12.	+ 66 40	11220		CLUSTER OF GALAXIES
ZWG 033.054	09 01 18.	+ 04 46		15.6	GALAXY
LB 09213	09 01 18.	+ 17 04		16.0	FAINT BLUE STAR
LB 09212	09 01 18.	+ 19 03		16.3	FAINT BLUE STAR
LB 09211	09 01 18.	+ 19 12		16.9	FAINT BLUE STAR
LB 09210	09 01 18.	+ 20 11		16.8	FAINT BLUE STAR
ZWG 209.020	09 01 18.	+ 41 36		15.6	GALAXY
MCG+08-17-019	09 01 18.	+ 47 36	36	16.	GALAXY
MCG+08-17-020	09 01 18.	+ 48 57	9	16.	GALAXY
UGC 04754	09 01 18.	- 00 18	66	16.0	GALAXY Sb-c
LB 02961	09 01 23.	+ 21 28 00.		16.3	FAINT BLUE STAR
ZWG 033.055	09 01 24.	+ 03 47		14.9	GALAXY
MCG+01-23-021	09 01 24.	+ 03 47	36	14.9	GALAXY
LB 09221	09 01 24.	+ 14 36		17.5	FAINT BLUE STAR
LB 09220	09 01 24.	+ 15 05		18.0	FAINT BLUE STAR
LB 09219	09 01 24.	+ 15 44		17.0	FAINT BLUE STAR
LB 09218	09 01 24.	+ 16 17		17.2	FAINT BLUE STAR
LB 09217	09 01 24.	+ 16 37		16.9	FAINT BLUE STAR
ZWG 090.061	09 01 24.	+ 17 37		15.4	GALAXY
MCG+03-23-029	09 01 24.	+ 17 37	45	15.4	GALAXY
LB 09216	09 01 24.	+ 19 21		15.2	FAINT BLUE STAR
LB 09215	09 01 24.	+ 19 52		17.3	FAINT BLUE STAR
LB 09214	09 01 24.	+ 20 04		17.7	FAINT BLUE STAR
MCG+07-19-019	09 01 24.	+ 41 35	36	16.	GALAXY
ZWG 238.007	09 01 24.	+ 46 47		15.7	GALAXY
MCG+08-17-021	09 01 24.	+ 46 47	30	16.	GALAXY
IC 2430	09 01 26.	+ 28 08 45.			NONSTELLAR OBJECT
LB 02962	09 01 28.	+ 20 38 48.		17.1	FAINT BLUE STAR
RNGC 2741	09 01 29.	+ 18 28			NON-EXISTENT OBJECT
ZWG 005.053	09 01 30.	+ 00 06		15.7	GALAXY
LB 09223	09 01 30.	+ 16 28		18.1	FAINT BLUE STAR
ZWG 090.062	09 01 30.	+ 17 29		15.6	GALAXY
LB 09222	09 01 30.	+ 19 06		17.4	FAINT BLUE STAR
ZWG 151.008	09 01 30.	+ 28 09		14.4	GALAXY
UGC 04755	09 01 30.	+ 28 09	66	14.4	GALAXY S0-a
MCG+05-22-005	09 01 30.	+ 28 09	66	14.4	GALAXY
ZWG 209.021	09 01 30.	+ 42 31		15.5	GALAXY
LB 02964	09 01 34.	+ 09 38 24.		16.6	FAINT BLUE STAR
LB 02963	09 01 34.	+ 22 00 00.		15.7	FAINT BLUE STAR
LB 09229	09 01 36.	+ 14 45		17.8	FAINT BLUE STAR
LB 09228	09 01 36.	+ 15 52		17.9	FAINT BLUE STAR
LB 09227	09 01 36.	+ 19 13		17.1	FAINT BLUE STAR
LB 09226	09 01 36.	+ 19 18		17.1	FAINT BLUE STAR
LB 09225	09 01 36.	+ 19 19		17.8	FAINT BLUE STAR
LB 09224	09 01 36.	+ 20 26		17.4	FAINT BLUE STAR
MCG+05-22-006	09 01 36.	+ 28 32	36	15.5	GALAXY
TON-N 0380	09 01 36.	+ 31 47		14.7	BLUE STAR
ZWG 209.022	09 01 36.	+ 40 48		15.6	GALAXY
ZWG 209.023	09 01 36.	+ 42 27		15.6	GALAXY
MCG+07-19-020	09 01 36.	+ 42 31	54	15.	GALAXY
WRAY 19.18	09 01 40.9	- 48 13 49.			DIFFUSE NEBULA
ZWG 090.063	09 01 42.	+ 14 48		14.3	GALAXY
UGC 04756	09 01 42.	+ 14 48	36	14.3	GALAXY QUD SYS
MCG+03-23-030	09 01 42.	+ 14 48	30	14.3	GALAXY
LB 09231	09 01 42.	+ 17 25		18.1	FAINT BLUE STAR
LB 02965	09 01 42.	+ 19 26		16.8	FAINT BLUE STAR
LB 09230	09 01 42.	+ 19 39		16.3	FAINT BLUE STAR
ZC 0901.7+2854	09 01 42.	+ 28 54	540		CLUSTER OF GALAXIES
MCG+07-19-021	09 01 42.	+ 40 46 30.	42	15.	GALAXY
MCG+07-19-022	09 01 42.	+ 42 27	42	15.	GALAXY
MCG+09-15-084	09 01 42.	+ 56 36	18	18.	GALAXY
MRSL 269-01/2	09 01 42.	- 48 12	180		HII REGION
LB 02965	09 01 44.	+ 19 26 42.		16.8	FAINT BLUE STAR
RNGC 2715	09 01 45.	+ 78 17		12.0	GALAXY
RNGC 2745	09 01 47.	+ 18 27		15.5	GALAXY
RNGC 2744	09 01 47.	+ 18 40		13.5	GALAXY
LB 02966	09 01 47.	+ 20 52 12.		16.3	FAINT BLUE STAR
ZWG 061.056	09 01 48.	+ 13 22		15.3	GALAXY
ZWG 061.057	09 01 48.	+ 13 45		14.9	GALAXY
KARA.72 184A	09 01 48.	+ 13 45	60	14.9	PART OF DOUBLE GALAXY
MCG+02-23-026	09 01 48.	+ 13 45	36	14.9	GALAXY
LB 09234	09 01 48.	+ 14 50		17.3	FAINT BLUE STAR
LB 09233	09 01 48.	+ 17 10		16.0	FAINT BLUE STAR
ZWG 090.064	09 01 48.	+ 18 27		15.5	GALAXY
ZWG 090.065	09 01 48.	+ 18 39		13.7	GALAXY
UGC 04757	09 01 48.	+ 18 39	96	13.7	GALAXY DBL SYS
LB 09232	09 01 48.	+ 18 41		18.4	FAINT BLUE STAR
LB 02967	09 01 48.	+ 21 40 06.		16.3	FAINT BLUE STAR
ZC 0901.8+4229	09 01 48.	+ 42 29	1280		CLUSTER OF GALAXIES
IC 2431	09 01 49.	+ 14 48 00.			NONSTELLAR OBJECT
TON-N 1001	09 01 49.	+ 21 42		14.8	BLUE STAR
MCG+03-23-031	09 01 51.	+ 18 39	90	13.7	GALAXY
MCG+04-22-008	09 01 51.	+ 22 15	60	15.4	GALAXY
LB 02968	09 01 54.	+ 11 43 30.		15.5	FAINT BLUE STAR
LB 09237	09 01 54.	+ 16 08		17.3	FAINT BLUE STAR
ZWG 090.066	09 01 54.	+ 19 00		15.1	GALAXY
MCG+03-23-032	09 01 54.	+ 19 00	36	15.1	GALAXY
LB 09236	09 01 54.	+ 20 00		16.9	FAINT BLUE STAR
LB 09235	09 01 54.	+ 20 09		18.2	FAINT BLUE STAR
LB 02971	09 01 54.	+ 20 14		16.0	FAINT BLUE STAR
ZWG 121.012	09 01 54.	+ 22 13		15.4	GALAXY
UGC 04758	09 01 54.	+ 22 13	96	15.4	GALAXY SBb
ZWG 151.009	09 01 54.	+ 28 35		15.3	GALAXY
ZWG 350.012	09 01 54.	+ 78 17		11.9	GALAXY
UGC 04759	09 01 54.	+ 78 17	300	11.9	GALAXY Sc
LB 09971	09 02	- 87 29		13.0	FAINT BLUE STAR
ZWG 033.056	09 02 00.	+ 05 43		15.5	GALAXY
LB 09242	09 02 00.	+ 15 19		18.1	FAINT BLUE STAR
LB 09241	09 02 00.	+ 17 14		17.6	FAINT BLUE STAR
LB 09240	09 02 00.	+ 17 16		16.7	FAINT BLUE STAR
MCG+03-23-033	09 02 00.	+ 17 38	42	16.5	GALAXY
LB 09239	09 02 00.	+ 19 00		16.0	FAINT BLUE STAR
LB 09238	09 02 00.	+ 19 20		17.4	FAINT BLUE STAR
ZWG 121.013	09 02 00.	+ 25 12		14.3	GALAXY
UGC 04760	09 02 00.	+ 25 12	78	14.3	GALAXY Sc-IRR
MCG+05-22-007	09 02 00.	+ 28 35	33	15.3	GALAXY
MCG+08-17-022	09 02 00.	+ 45 58	54	17.	GALAXY
MCG+13-07-015	09 02 00.	+ 78 18	300	11.8	GALAXY
ZC 0902.0+8307	09 02 00.	+ 83 07	3020		CLUSTER OF GALAXIES
IC 2432	09 02 01.	+ 05 41 40.			NONSTELLAR OBJECT
RNGC 2743	09 02 02.	+ 25 12		14.5	GALAXY
MCG+04-22-009	09 02 03.	+ 25 12	66	14.3	GALAXY

OBJECT NAME	RIGHT ASCEN.	DECLINATION	DIAM.	MAGN.	TYPE OF OBJECT
RNGC 2788A	09 02 05.	- 68 02			GALAXY
ZWG 061.058	09 02 06.	+ 13 46		14.8	GALAXY
KARA.72 184B	09 02 06.	+ 13 46	60	14.8	PART OF DOUBLE GALAXY
MCG+02-23-027	09 02 06.	+ 13 46	48	14.8	GALAXY
ZWG 090.067	09 02 06.	+ 16 02		15.7	GALAXY
KARA.73B 0297	09 02 06.	+ 16 02	36	15.7	ISOLATED GALAXY S
LB 09246	09 02 06.	+ 16 32		17.0	FAINT BLUE STAR
LB 09245	09 02 06.	+ 16 49		16.9	FAINT BLUE STAR
MCG+03-23-034	09 02 06.	+ 17 38	60	14.8	GALAXY
ZWG 090.068	09 02 06.	+ 17 39		14.8	GALAXY
UGC 04761	09 02 06.	+ 17 39	66	14.8	GALAXY E
LB 09244	09 02 06.	+ 17 47		16.0	FAINT BLUE STAR
BIGO 506	09 02 06.	+ 18 34			NEBULA
LB 09243	09 02 06.	+ 18 57		15.8	FAINT BLUE STAR
DV.56 N2788A	09 02 06.	- 68 02			GALAXY
LB 02969	09 02 07.	+ 18 59 54.		15.3	FAINT BLUE STAR
LB 02970	09 02 09.	+ 22 13 48.		15.9	FAINT BLUE STAR
ZWG 033.057	09 02 12.	+ 08 14		15.3	GALAXY
LB 09251	09 02 12.	+ 14 51		16.2	FAINT BLUE STAR
LB 09250	09 02 12.	+ 16 04		17.7	FAINT BLUE STAR
LB 09249	09 02 12.	+ 16 23		17.7	FAINT BLUE STAR
LB 09248	09 02 12.	+ 17 02		17.1	FAINT BLUE STAR
LB 09247	09 02 12.	+ 17 43		17.5	FAINT BLUE STAR
ZWG 121.014	09 02 12.	+ 20 59		15.7	GALAXY
ZC 0902.2+6253	09 02 12.	+ 62 53	1340		CLUSTER OF GALAXIES
LB 09258	09 02 18.	+ 15 17		17.4	FAINT BLUE STAR
LB 09257	09 02 18.	+ 15 34		18.0	FAINT BLUE STAR
ZC 0902.3+1540	09 02 18.	+ 15 40	1340		CLUSTER OF GALAXIES
LB 09256	09 02 18.	+ 16 11		17.9	FAINT BLUE STAR
LB 09255	09 02 18.	+ 16 17		17.1	FAINT BLUE STAR
LB 09254	09 02 18.	+ 17 10		18.1	FAINT BLUE STAR
LB 09253	09 02 18.	+ 19 18		16.9	FAINT BLUE STAR
LB 09252	09 02 18.	+ 19 27		17.6	FAINT BLUE STAR
UGC 04762	09 02 18.	+ 45 31	66	18.	GALAXY DWRF IR
ABC 0740	09 02 21.	+ 42 32		17.1	RICH CLUSTER OF GALAXIES
LB 02971	09 02 22.	+ 20 14 48.		16.1	FAINT BLUE STAR
MIL 14	09 02 22.	- 38 29	2100		SUPERNOVA REMNANT
PK266+01.1	09 02 23.	- 44 21 00.			PLANETARY NEBULA
LB 09264	09 02 24.	+ 15 09		17.1	FAINT BLUE STAR
LB 09263	09 02 24.	+ 15 57		17.4	FAINT BLUE STAR
LB 09262	09 02 24.	+ 16 59		17.6	FAINT BLUE STAR
LB 09261	09 02 24.	+ 17 34		17.1	FAINT BLUE STAR
LB 09260	09 02 24.	+ 18 35		18.2	FAINT BLUE STAR
LB 09259	09 02 24.	+ 18 53		17.1	FAINT BLUE STAR
ZWG 151.010	09 02 24.	+ 28 27		15.5	GALAXY
ZC 0902.4+3324	09 02 24.	+ 33 24	2690		CLUSTER OF GALAXIES
ZWG 209.024	09 02 24.	+ 41 17		15.7	GALAXY
MCG+09-15-085	09 02 24.	+ 51 55	48	16.	GALAXY
MCG+09-15-086	09 02 24.	+ 51 56	42	15.	GALAXY
ZWG 264.059	09 02 24.	+ 51 57		15.5	GALAXY
KARA.72 185A	09 02 24.	+ 51 57	42	15.5	PART OF DOUBLE GALAXY
ISS 1040	09 02 24.	- 43 02	188		STELLAR RING
LB 02972	09 02 25.	+ 19 46 42.		16.7	FAINT BLUE STAR
RNGC 2739	09 02 25.	+ 51 57		15.5	GALAXY
MCG+03-23-035	09 02 27.	+ 18 31	6	19.5	GALAXY
RNGC 2749	09 02 29.	+ 18 31		13.5	GALAXY
RNGC 2747	09 02 29.	+ 18 38		15.5	GALAXY
LB 09268	09 02 30.	+ 14 39		17.6	FAINT BLUE STAR
LB 09267	09 02 30.	+ 15 26		16.6	FAINT BLUE STAR
ZWG 090.069	09 02 30.	+ 18 30		13.3	GALAXY
UGC 04763	09 02 30.	+ 18 30	132	13.3	GALAXY E
MCG+03-23-036	09 02 30.	+ 18 30	36	13.3	GALAXY
ZWG 090.070	09 02 30.	+ 18 38		15.5	GALAXY
LB 09266	09 02 30.	+ 19 16		16.8	FAINT BLUE STAR
LB 09265	09 02 30.	+ 19 45		16.8	FAINT BLUE STAR
LB 02972	09 02 30.	+ 19 48		16.6	FAINT BLUE STAR
ZWG 121.015	09 02 30.	+ 25 45		15.4	GALAXY
UGC 04764	09 02 30.	+ 25 45	60	15.4	GALAXY PECULR?
TON-N 0381	09 02 30.	+ 29 22		17.0	BLUE STAR
MCG+08-17-024	09 02 30.	+ 47 22	18	16.	GALAXY
ZWG 238.008	09 02 30.	+ 47 23		14.5	GALAXY
UGC 04765	09 02 30.	+ 47 23	66	14.5	GALAXY PECULR
MCG+08-17-023	09 02 30.	+ 47 23	36	14.	GALAXY
ZWG 264.060	09 02 30.	+ 51 56		15.1	GALAXY
KARA.72 185B	09 02 30.	+ 51 56	54	15.1	PART OF DOUBLE GALAXY
UGC 04766	09 02 30.	+ 72 03	60	16.0	GALAXY Sc
RNGC 2740	09 02 32.	+ 51 56		15.0	GALAXY
LB 09271	09 02 36.	+ 17 06		17.8	FAINT BLUE STAR
LB 09270	09 02 36.	+ 17 27		17.9	FAINT BLUE STAR
LB 09269	09 02 36.	+ 18 27		16.4	FAINT BLUE STAR
ZWG 121.016	09 02 36.	+ 22 48		15.4	GALAXY
MCG+04-22-010	09 02 36.	+ 22 49	48	15.4	GALAXY
MCG+04-22-011	09 02 36.	+ 25 46	66	15.4	GALAXY
ZC 0902.6+2814	09 02 36.	+ 28 14	740		CLUSTER OF GALAXIES
ZWG 180.031	09 02 36.	+ 36 32		14.0	GALAXY
UGC 04767	09 02 36.	+ 36 32	78	14.0	GALAXY S0
MCG+06-20-022	09 02 36.	+ 36 33	24	14.	GALAXY
MCG+12-09-030	09 02 36.	+ 72 02	45	16.	GALAXY
IC 2433	09 02 38.	+ 22 48 26.			NONSTELLAR OBJECT
LB 02973	09 02 40.	+ 09 59 36.		16.8	FAINT BLUE STAR
RNGC 2751	09 02 41.	+ 18 27		15.0	GALAXY
LB 09277	09 02 42.	+ 15 16		17.1	FAINT BLUE STAR
LB 09276	09 02 42.	+ 15 48		16.0	FAINT BLUE STAR
LB 09275	09 02 42.	+ 15 58		17.0	FAINT BLUE STAR
LB 09274	09 02 42.	+ 16 10		16.4	FAINT BLUE STAR
LB 09273	09 02 42.	+ 16 59		18.3	FAINT BLUE STAR
LB 09272	09 02 42.	+ 17 42		18.0	FAINT BLUE STAR
ZWG 090.071	09 02 42.	+ 18 27		15.1	GALAXY
MCG+03-23-037	09 02 42.	+ 18 27	45	15.1	GALAXY
ZC 0902.7+4120	09 02 42.	+ 41 20	940		CLUSTER OF GALAXIES
ZC 0902.7+5246	09 02 42.	+ 52 46	2350		CLUSTER OF GALAXIES
ZWG 264.061	09 02 42.	+ 55 25		15.7	GALAXY
ZC 0902.7+6039	09 02 42.	+ 60 39	1480		CLUSTER OF GALAXIES
UGC 04768	09 02 42.	- 00 18	60	16.5	GALAXY Sb-c
RNGC 2754	09 02 45.	- 18 53			GALAXY
LB 02975	09 02 47.	+ 08 37 06.		16.8	FAINT BLUE STAR
LB 02974	09 02 47.	+ 22 00 54.		15.8	FAINT BLUE STAR
LB 09281	09 02 48.	+ 16 03		16.0	FAINT BLUE STAR
LB 09280	09 02 48.	+ 16 53		17.2	FAINT BLUE STAR
LB 09279	09 02 48.	+ 18 22		18.2	FAINT BLUE STAR
LB 09278	09 02 48.	+ 19 38		17.2	FAINT BLUE STAR
ZWG 121.017	09 02 48.	+ 25 38		12.7	GALAXY
UGC 04769	09 02 48.	+ 25 38	144	12.7	GALAXY Sc
KARA.72 186A	09 02 48.	+ 25 38	54	12.7	PART OF DOUBLE GALAXY
ZWG 180.032	09 02 48.	+ 35 34		14.4	GALAXY
UGC 04770	09 02 48.	+ 35 34	108	14.4	GALAXY SBa
KARA.73B 0298	09 02 48.	+ 35 34	96	14.4	ISOLATED GALAXY S
MCG+06-20-023	09 02 48.	+ 35 34 30.	90	14.	GALAXY
MCG+07-19-023	09 02 48.	+ 40 01	27	16.5	GALAXY
ZC 0902.8+4802	09 02 48.	+ 48 02	740		CLUSTER OF GALAXIES
ZWG 238.009	09 02 48.	+ 50 17		15.6	GALAXY
UGC 04771	09 02 48.	+ 50 17	84	15.6	GALAXY SBb
MCG+08-17-025	09 02 48.	+ 50 17	66	15.	GALAXY
MCG-03-23-017	09 02 48.	- 18 20	60	15.	GALAXY
LB 02976	09 02 49.	+ 23 28 18.		16.8	FAINT BLUE STAR
RNGC 2750	09 02 50.	+ 25 38		12.5	GALAXY
MCG+03-23-038	09 02 51.	+ 18 32	108	14.8	GALAXY
RNGC 2746	09 02 51.	+ 35 34		14.5	GALAXY
RNGC 2752	09 02 53.	+ 18 32		15.0	GALAXY
LB 02977	09 02 54.	+ 09 16 48.		16.7	FAINT BLUE STAR
LB 09283	09 02 54.	+ 16 02		16.3	FAINT BLUE STAR
LB 09282	09 02 54.	+ 16 54		16.1	FAINT BLUE STAR
ZWG 090.072	09 02 54.	+ 18 32		14.8	GALAXY
UGC 04772	09 02 54.	+ 18 32	126	14.8	GALAXY Sb
KARA.72 186B	09 02 54.	+ 25 38	114		PART OF DOUBLE GALAXY
MCG+08-17-026	09 02 54.	+ 46 24	30	16.	GALAXY
MCG+10-13-055	09 02 54.	+ 56 59	39	16.	GALAXY
LB 02978	09 02 58.	+ 21 35 06.		16.4	FAINT BLUE STAR
ZWG 005.054	09 03 00.	+ 02 14		15.6	GALAXY
ZC 0903.0+0445	09 03 00.	+ 04 45	610		CLUSTER OF GALAXIES
LB 09287	09 03 00.	+ 15 35		18.1	FAINT BLUE STAR
LB 09286	09 03 00.	+ 18 34		18.1	FAINT BLUE STAR
LB 09285	09 03 00.	+ 18 56		16.5	FAINT BLUE STAR
LB 09284	09 03 00.	+ 19 03		18.0	FAINT BLUE STAR
LB 02979	09 03 00.	+ 22 25 06.		16.5	FAINT BLUE STAR
MCG+04-22-012	09 03 00.	+ 25 39	132	12.7	GALAXY
MCG+11-11-037	09 03 00.	+ 66 46	39	15.	GALAXY
LB 02980	09 03 01.	+ 22 15 18.		15.7	FAINT BLUE STAR
RNGC 2788B	09 03 05.	- 68 03			GALAXY
ZWG 005.055	09 03 06.	+ 02 13		15.7	GALAXY
LB 09291	09 03 06.	+ 15 39		14.7	FAINT BLUE STAR
LB 09290	09 03 06.	+ 17 04		17.4	FAINT BLUE STAR
LB 09289	09 03 06.	+ 17 32		15.3	FAINT BLUE STAR
ZWG 090.073	09 03 06.	+ 18 58		14.8	GALAXY
UGC 04773	09 03 06.	+ 18 58	90	14.8	GALAXY S
LB 09288	09 03 06.	+ 19 00		17.1	FAINT BLUE STAR
MCG+09-15-087	09 03 06.	+ 53 58	36	17.	GALAXY
HN 0316	09 03 06.	- 18 58			NEBULA
IC 2436	09 03 06.	- 18 58			NONSTELLAR OBJECT
DV.56 N2788B	09 03 06.	- 68 03			GALAXY
MCG+03-23-039	09 03 09.	+ 18 57	90	14.8	GALAXY
MCG-03-23-018	09 03 09.	- 18 20	36	14.5	GALAXY
RNGC 2758	09 03 09.	- 18 51		14.0	GALAXY
LB 09298	09 03 12.	+ 14 57		18.0	FAINT BLUE STAR
LB 09297	09 03 12.	+ 15 52		17.9	FAINT BLUE STAR
LB 09296	09 03 12.	+ 17 02		16.7	FAINT BLUE STAR
LB 09295	09 03 12.	+ 19 13		16.5	FAINT BLUE STAR
LB 09294	09 03 12.	+ 19 18		17.6	FAINT BLUE STAR
LB 09293	09 03 12.	+ 19 30		17.9	FAINT BLUE STAR
LB 09292	09 03 12.	+ 19 54		16.3	FAINT BLUE STAR
LB 02981	09 03 12.	+ 23 14 30.		16.9	FAINT BLUE STAR
ZC 0903.2+3538	09 03 12.	+ 35 38	1340		CLUSTER OF GALAXIES
MCG+09-15-088	09 03 12.	+ 56 30	12	19.	GALAXY
MCG-03-23-019	09 03 12.	- 18 51 30.	120	14.	GALAXY
MCG+09-15-089	09 03 15.	+ 53 03	42	18.	GALAXY
RNGC 2757	09 03 15.	- 18 50			NON-EXISTENT OBJECT
MCG-03-23-020	09 03 15.	- 19 01	84	14.	GALAXY
IC 2437	09 03 16.	- 19 00 16.			NONSTELLAR OBJECT
ZWG 033.058	09 03 18.	+ 07 24		15.3	GALAXY
LB 09301	09 03 18.	+ 15 20		17.7	FAINT BLUE STAR
LB 09300	09 03 18.	+ 17 36		18.2	FAINT BLUE STAR
LB 09299	09 03 18.	+ 20 43		15.8	FAINT BLUE STAR
ZWG 121.018	09 03 18.	+ 25 47		15.7	GALAXY
UGC 04774	09 03 18.	+ 25 47	60	15.7	GALAXY
TON-N 0382	09 03 18.	+ 28 50		16.8	BLUE STAR
ZC 0903.3+3731	09 03 18.	+ 37 31	740		CLUSTER OF GALAXIES
MCG+10-13-056	09 03 18.	+ 57 08	27	16.	GALAXY
ZC 0903.3+5802	09 03 18.	+ 58 02	1480		CLUSTER OF GALAXIES
ZWG 311.029	09 03 18.	+ 66 46		14.3	GALAXY
UGC 04775	09 03 18.	+ 66 46	90	14.3	GALAXY E-S0
SHAH 109	09 03 18.	+ 67 41	66	18.5	GROUP OF COMPACT GALAXIES
UGC 04776	09 03 18.	+ 79 34	84	16.5	GALAXY DWRF SP
ABC 0741	09 03 19.	+ 37 31		17.8	RICH CLUSTER OF GALAXIES
LB 09306	09 03 24.	+ 14 52		18.2	FAINT BLUE STAR
LB 09305	09 03 24.	+ 16 04		18.5	FAINT BLUE STAR
LB 09304	09 03 24.	+ 18 14		17.7	FAINT BLUE STAR
LB 09303	09 03 24.	+ 18 24		16.4	FAINT BLUE STAR
LB 09302	09 03 24.	+ 20 26		16.2	FAINT BLUE STAR
ZWG 209.025	09 03 24.	+ 41 35		15.4	GALAXY
ZWG 209.026	09 03 24.	+ 42 00		15.6	GALAXY
TON-N 0383	09 03 30.	+ 29 07		16.6	BLUE STAR
MCG+06-20-024	09 03 30.	+ 34 48	120	14.5	GALAXY
ZC 0903.5+4122	09 03 30.	+ 41 22	6120		CLUSTER OF GALAXIES
MCG+07-19-024	09 03 30.	+ 41 33	18	16.	GALAXY
MCG+07-19-026	09 03 30.	+ 42 00	15	17.	GALAXY
MCG+07-19-025	09 03 30.	+ 42 00	21	17.	GALAXY
MCG+08-17-027	09 03 30.	+ 48 58	30	15.	GALAXY
ZWG 238.010	09 03 30.	+ 49 58		15.4	GALAXY
MCG+08-17-028	09 03 30.	+ 49 58	24	16.	GALAXY
MCG-01-23-017	09 03 30.	- 04 00	84	14.5	GALAXY
MCG+04-22-013	09 03 33.	+ 25 48	54	15.7	GALAXY
LB 09309	09 03 36.	+ 16 04		16.0	FAINT BLUE STAR
LB 09308	09 03 36.	+ 16 58		18.3	FAINT BLUE STAR
LB 09307	09 03 36.	+ 17 19		16.8	FAINT BLUE STAR
ZWG 180.033	09 03 36.	+ 34 49		15.2	GALAXY
UGC 04777	09 03 36.	+ 34 49	144	15.2	GALAXY IRR
KARA.73B 0299	09 03 36.	+ 34 49	132	15.2	ISOLATED GALAXY S
ZWG 180.034	09 03 36.	+ 37 50		15.6	GALAXY
ZWG 209.027	09 03 36.	+ 41 37		15.1	GALAXY
MCG+08-17-029	09 03 36.	+ 46 24 30.	15	16.	GALAXY
ZC 0903.6+4625	09 03 36.	+ 46 25	810		CLUSTER OF GALAXIES
ZWG 264.062	09 03 36.	+ 50 55		14.2	GALAXY
UGC 04778	09 03 36.	+ 50 55	54	14.2	GALAXY Sb
ZWG 288.019	09 03 36.	+ 60 40		12.0	GALAXY
UGC 04779	09 03 36.	+ 60 40	198	12.0	GALAXY Sc
LB 02982	09 03 38.	+ 21 38 00.		16.2	FAINT BLUE STAR
RNGC 2742	09 03 38.	+ 60 41		12.5	GALAXY
ZC 0903.7+0011	09 03 42.	+ 00 11	1610		CLUSTER OF GALAXIES
ZC 0903.7+1033	09 03 42.	+ 10 33	1610		CLUSTER OF GALAXIES
LB 09312	09 03 42.	+ 17 21		17.3	FAINT BLUE STAR
LB 09311	09 03 42.	+ 18 01		18.3	FAINT BLUE STAR
LB 09310	09 03 42.	+ 18 52		16.4	FAINT BLUE STAR
MCG+09-15-091	09 03 42.	+ 51 45	30	16.	GALAXY
MCG+09-15-090	09 03 42.	+ 56 13	18	17.	GALAXY
ZC 0903.7+5700	09 03 42.	+ 57 00	3560		CLUSTER OF GALAXIES
MCG+10-13-057	09 03 42.	+ 60 40	162	12.1	GALAXY

OBJECT NAME	RIGHT ASCEN.	DECLINATION	DIAM.	MAGN.	TYPE OF OBJECT
BC 3CR215	09 03 44.15	+ 16 58 15.7		18.27	QUASI-STELLAR OBJECT
SHB 129	09 03 44.2	+ 16 58 16.		18.3	QUASI-STELLAR OBJECT
MCG+07-19-027	09 03 45.	+ 41 36	42	16.	GALAXY
ARC 0717	09 03 47.	+ 83 04		16.7	RICH CLUSTER OF GALAXIES
ZC 0903.8+1512	09 03 48.	+ 15 12	940		CLUSTER OF GALAXIES
LB 09314	09 03 48.	+ 16 29		14.8	FAINT BLUE STAR
LB 09313	09 03 48.	+ 16 46		17.9	FAINT BLUE STAR
ZWG 090.074	09 03 48.	+ 19 32		15.0	GALAXY
UGC 04780	09 03 48.	+ 19 32	102	15.0	GALAXY
TON-N 0384	09 03 48.	+ 30 19		14.9	BLUE STAR
TON-N 1003	09 03 48.	+ 33 35		16.5	BLUE STAR
ZWG 209.028	09 03 48.	+ 41 29		15.5	GALAXY
MCG+08-17-030	09 03 48.	+ 46 24 30.	15	16.	GALAXY
MCG-02-23-009	09 03 48.	- 15 06 30.	36	16.	GALAXY
TON-N 1002	09 03 49.	+ 21 29		16.2	BLUE STAR
LB 02984	09 03 50.	+ 11 18 24.		16.8	FAINT BLUE STAR
LB 02983	09 03 50.	+ 21 14 00.		17.1	FAINT BLUE STAR
MCG+03-23-040	09 03 51.	+ 19 31	108	15.0	GALAXY
MCG+07-19-028	09 03 51.	+ 41 37	30	17.	GALAXY
LB 00553	09 03 51.	- 00 11 06.		17.6	FAINT BLUE STAR
IC 2435	09 03 52.	+ 26 28 44.			NONSTELLAR OBJECT
ZWG 033.059	09 03 54.	+ 06 30		14.9	GALAXY
UGC 04781	09 03 54.	+ 06 30	114	14.9	GALAXY Sc
MCG+01-23-022	09 03 54.	+ 06 30	120	14.9	GALAXY
KARA.73B 0300	09 03 54.	+ 06 30	114	14.9	ISOLATED GALAXY S
LB 09322	09 03 54.	+ 14 54		17.8	FAINT BLUE STAR
LB 09321	09 03 54.	+ 15 04		17.9	FAINT BLUE STAR
LB 09320	09 03 54.	+ 15 23		18.3	FAINT BLUE STAR
LB 09319	09 03 54.	+ 16 25		17.6	FAINT BLUE STAR
LB 09318	09 03 54.	+ 18 41		17.7	FAINT BLUE STAR
LB 09317	09 03 54.	+ 18 46		18.1	FAINT BLUE STAR
LB 09316	09 03 54.	+ 19 34		17.5	FAINT BLUE STAR
LB 09315	09 03 54.	+ 20 17		17.6	FAINT BLUE STAR
ZWG 121.019	09 03 54.	+ 26 28		15.2	GALAXY
UGC 04782	09 03 54.	+ 26 28	60	15.2	GALAXY E
ZC 0903.9+2907	09 03 54.	+ 29 07	1080		CLUSTER OF GALAXIES
ZC 0903.9+3716	09 03 54.	+ 37 16	6320		CLUSTER OF GALAXIES
MCG+07-19-029	09 03 54.	+ 40 50	36	17.	GALAXY
MCG+07-19-030	09 03 54.	+ 41 28	48	15.	GALAXY
ZWG 238.011	09 03 54.	+ 46 25		15.6	GALAXY
MCG+08-17-031	09 03 54.	+ 46 25	90	16.	GALAXY
LB 09324	09 04 00.	+ 14 49		16.7	FAINT BLUE STAR
LB 09323	09 04 00.	+ 17 53		16.3	FAINT BLUE STAR
MCG+04-22-014	09 04 00.	+ 26 28	24	15.2	GALAXY
ZWG 209.029	09 04 00.	+ 39 09		15.2	GALAXY
UGC 04783	09 04 00.	+ 39 09	60	15.2	GALAXY S
MCG+07-19-031	09 04 00.	+ 42 47 30.	30	15.	GALAXY
MCG+08-17-032	09 04 00.	+ 45 59	36	16.	GALAXY
MCG+08-17-033	09 04 00.	+ 46 00	42	16.	GALAXY
UGC 04784	09 04 00.	+ 46 26	96	16.0	GALAXY S
MCG+09-15-092	09 04 00.	+ 52 17 30.	9	18.	GALAXY
ZC 0904.0+8140	09 04 00.	+ 81 40	610		CLUSTER OF GALAXIES
MCG-01-23-018	09 04 00.	- 04 46	54	15.	GALAXY
HOLM 109B	09 04 02.	+ 37 22	18	14.4	PART OF MULTIPLE GALAXY
ARC 0743	09 04 03.	+ 10 29		17.5	RICH CLUSTER OF GALAXIES
LB 02985	09 04 05.	+ 11 36 00.		15.7	FAINT BLUE STAR
LB 09330	09 04 06.	+ 15 25		17.4	FAINT BLUE STAR
LB 09329	09 04 06.	+ 17 10		17.4	FAINT BLUE STAR
ZWG 090.075	09 04 06.	+ 18 10		15.6	GALAXY
LB 09328	09 04 06.	+ 18 22		18.6	FAINT BLUE STAR
LB 09327	09 04 06.	+ 18 58		18.0	FAINT BLUE STAR
ZWG 090.076	09 04 06.	+ 19 10		15.4	GALAXY
LB 09326	09 04 06.	+ 19 14		18.3	FAINT BLUE STAR
LB 09325	09 04 06.	+ 19 29		18.1	FAINT BLUE STAR
ZWG 151.011	09 04 06.	+ 29 40		15.7	GALAXY
ZWG 180.035	09 04 06.	+ 37 24		14.5	GALAXY
UGC 04785	09 04 06.	+ 37 24	114	14.5	GALAXY SB
MCG+06-20-025	09 04 06.	+ 37 25	66	14.	GALAXY
ZWG 180.036	09 04 06.	+ 37 42		15.1	GALAXY
MCG+07-19-032	09 04 06.	+ 39 07	48	16.	GALAXY
ARC 0723	09 04 06.	+ 81 41		17.1	RICH CLUSTER OF GALAXIES
ZWG 005.056	09 04 06.	- 00 40		15.5	GALAXY
KARA.73B 0301	09 04 06.	- 00 40	24	15.5	ISOLATED GALAXY S0
MCG-01-23-019	09 04 06.	- 06 57	42	15.	GALAXY
IC 2434	09 04 08.	+ 37 25 55.			NONSTELLAR OBJECT
LB 02986	09 04 09.	+ 19 49 30.		16.5	FAINT BLUE STAR
HOLM 109A	09 04 10.	+ 37 25	36	13.7	PART OF MULTIPLE GALAXY
HOLM 109C	09 04 11.	+ 37 24	18	14.5	PART OF MULTIPLE GALAXY
ZC 0904.2+0459	09 04 12.	+ 04 59	940		CLUSTER OF GALAXIES
ARC 0745	09 04 12.	+ 04 59		17.4	RICH CLUSTER OF GALAXIES
LB 09333	09 04 12.	+ 15 40		17.6	FAINT BLUE STAR
LB 09332	09 04 12.	+ 16 12		18.6	FAINT BLUE STAR
LB 09331	09 04 12.	+ 17 25		18.1	FAINT BLUE STAR
ZWG 121.020	09 04 12.	+ 25 32		14.8	GALAXY
KARA.72 187B	09 04 12.	+ 25 32	18		PART OF DOUBLE GALAXY
KARA.72 187A	09 04 12.	+ 25 32	24	14.8	PART OF DOUBLE GALAXY
UGC 04786	09 04 12.	+ 28 30	60	16.5	GALAXY Sc
MCG+06-20-027	09 04 12.	+ 37 42	30	15.5	GALAXY
MCG+06-20-026	09 04 12.	+ 37 42	42	15.	GALAXY
MCG+07-19-033	09 04 12.	+ 40 37 30.	42	17.	GALAXY
MCG+09-15-093	09 04 12.	+ 52 15	36	17.	GALAXY
MCG+09-15-094	09 04 12.	+ 52 17	18	17.	GALAXY
MCG+09-15-095	09 04 12.	+ 52 22 30.	30	17.	GALAXY
MCG-01-23-020	09 04 12.	- 06 13	90	16.	GALAXY
RNGC 2753	09 04 14.	+ 25 32		15.0	GALAXY
LB 02987	09 04 16.	+ 19 40 06.		16.4	FAINT BLUE STAR
ZWG 033.060	09 04 18.	+ 03 23		15.1	GALAXY
ZC 0904.3+0626	09 04 18.	+ 06 26	740		CLUSTER OF GALAXIES
LB 09335	09 04 18.	+ 14 34		17.5	FAINT BLUE STAR
LB 09334	09 04 18.	+ 20 04		18.0	FAINT BLUE STAR
MCG+06-20-028	09 04 18.	+ 33 27	120	14.	GALAXY
TON-N 1004	09 04 19.	+ 21 12		16.6	BLUE STAR
LB 02988	09 04 20.	+ 12 41 54.		15.8	FAINT BLUE STAR
MCG+04-22-015	09 04 21.	+ 25 32	30	14.8	GALAXY
ZWG 005.057	09 04 24.	+ 00 10		15.6	GALAXY
LB 09340	09 04 24.	+ 15 36		17.2	FAINT BLUE STAR
LB 09339	09 04 24.	+ 17 44		14.8	FAINT BLUE STAR
LB 09338	09 04 24.	+ 18 40		17.7	FAINT BLUE STAR
LB 09337	09 04 24.	+ 18 51		17.8	FAINT BLUE STAR
LB 09336	09 04 24.	+ 19 54		16.6	FAINT BLUE STAR
LB 02989	09 04 24.	+ 22 52 18.		16.2	FAINT BLUE STAR
ZWG 121.021	09 04 24.	+ 23 09		15.5	GALAXY
MCG+10-13-058	09 04 24.	+ 62 20	24	16.	GALAXY
LB 09342	09 04 30.	+ 14 55		15.8	FAINT BLUE STAR
LB 09341	09 04 30.	+ 15 16		18.1	FAINT BLUE STAR
ZWG 180.037	09 04 30.	+ 33 28		14.3	GALAXY
UGC 04787	09 04 30.	+ 33 28	138	14.3	GALAXY
ZC 0904.5+3501	09 04 30.	+ 35 01	1140		CLUSTER OF GALAXIES
VV 196B	09 04 30.	+ 49 47 30.	15	16.	INTERACTING GALAXY
VV 196A	09 04 30.	+ 49 47 30.	18	16.	INTERACTING GALAXY
VV 196	09 04 30.	+ 49 47 30.	48		INTERACTING GALAXY
ZWG 238.012	09 04 30.	+ 49 48		15.0	GALAXY
MCG+08-17-035	09 04 30.	+ 49 48	12	15.	GALAXY
MCG+08-17-034	09 04 30.	+ 49 48	12	15.	GALAXY
ZC 0904.5+5650	09 04 30.	+ 56 50	670		CLUSTER OF GALAXIES
ZWG 288.020	09 04 30.	+ 59 16		15.5	GALAXY
UGC 04788	09 04 30.	+ 59 16	102	15.5	GALAXY SB
ZWG 288.021	09 04 30.	+ 62 20		15.5	GALAXY
ARC 0744	09 04 31.	+ 16 52		16.6	RICH CLUSTER OF GALAXIES
RNGC 2763	09 04 32.	- 15 17		12.5	GALAXY
MCG-02-23-010	09 04 33.	- 15 18	120	13.	GALAXY
ZC 0904.6+0613	09 04 36.	+ 06 13	670		CLUSTER OF GALAXIES
LB 09345	09 04 36.	+ 16 39		18.0	FAINT BLUE STAR
LB 09344	09 04 36.	+ 17 20		18.6	FAINT BLUE STAR
LB 09343	09 04 36.	+ 17 54		17.2	FAINT BLUE STAR
LB 02991	09 04 36.	+ 20 33		16.8	FAINT BLUE STAR
MCG+09-15-096	09 04 36.	+ 51 50	72	15.	GALAXY
MCG+10-13-059	09 04 36.	+ 59 17	57	15.	GALAXY
PATH 1.253	09 04 37.	+ 59 16	24		NEBULA
LB 02990	09 04 39.	+ 11 58 00.		16.4	FAINT BLUE STAR
MCG+08-17-036	09 04 39.	+ 50 06	12	16.	GALAXY
RNGC 2761	09 04 41.	+ 18 38		15.0	GALAXY
LB 02991	09 04 41.	+ 20 33 30.		16.7	FAINT BLUE STAR
RNGC 2755	09 04 41.	+ 41 55		14.0	GALAXY
LB 09347	09 04 42.	+ 15 34		16.7	FAINT BLUE STAR
LB 09346	09 04 42.	+ 15 44		15.4	FAINT BLUE STAR
ZWG 091.001	09 04 42.	+ 18 38		14.8	GALAXY
ZWG 090.077	09 04 42.	+ 18 38		14.8	GALAXY
MCG+03-23-041	09 04 42.	+ 18 38	30	14.8	GALAXY
KARA.73B 0302	09 04 42.	+ 18 38	36	14.8	ISOLATED GALAXY IR
ZWG 180.038	09 04 42.	+ 33 40		15.3	GALAXY
ZWG 209.030	09 04 42.	+ 41 55		14.2	GALAXY
UGC 04789	09 04 42.	+ 41 55	72	14.2	GALAXY S
ZWG 264.063	09 04 42.	+ 51 51		15.3	GALAXY
UGC 04790	09 04 42.	+ 51 51	66	15.3	GALAXY SB:a-b
SHAH 099	09 04 42.	+ 60 29	108	18.0	GROUP OF COMPACT GALAXIES
MAI 038	09 04 47.	+ 60 18	586		DWARF SPHEROIDAL GALAXY
LB 09350	09 04 48.	+ 18 58		17.6	FAINT BLUE STAR
LB 09349	09 04 48.	+ 19 12		17.0	FAINT BLUE STAR
LB 09348	09 04 48.	+ 20 19		17.7	FAINT BLUE STAR
7ZW 260	09 04 48.	+ 77 23			COMPACT GALAXY
MCG-01-23-021	09 04 48.	- 07 06	90	14.5	GALAXY
LB 02992	09 04 51.	+ 21 24 36.		16.9	FAINT BLUE STAR
MCG+07-19-034	09 04 51.	+ 41 53 30.	72	14.	GALAXY
LB 09353	09 04 54.	+ 15 18		17.2	FAINT BLUE STAR
LB 09352	09 04 54.	+ 15 58		17.2	FAINT BLUE STAR
ZC 0904.9+1606	09 04 54.	+ 16 06	1810		CLUSTER OF GALAXIES
LB 09351	09 04 54.	+ 18 13		18.3	FAINT BLUE STAR
ZC 0904.9+3037	09 04 54.	+ 30 37	670		CLUSTER OF GALAXIES
ZC 0904.9+4543	09 04 54.	+ 45 43	1080		CLUSTER OF GALAXIES
ZC 0904.9+5555	09 04 54.	+ 55 55	1210		CLUSTER OF GALAXIES
VDB.66G 054	09 05	+ 06 07	100		DWARF GALAXY
LBN 1082	09 05	- 28 30	12600		BRIGHT NEBULA
LBN 1084	09 05	- 30 30	5400		BRIGHT NEBULA
ZWG 033.061	09 05 00.	+ 03 35		13.3	GALAXY
RNGC 2765	09 05 00.	+ 03 35		13.5	GALAXY
UGC 04791	09 05 00.	+ 03 35	120	13.3	GALAXY S0
MCG+01-24-001	09 05 00.	+ 03 35	120	13.3	GALAXY
KARA.73B 0303	09 05 00.	+ 03 35	120	13.3	ISOLATED GALAXY S0
LB 09354	09 05 00.	+ 16 11		18.3	FAINT BLUE STAR
ZC 0905.0+4351	09 05 00.	+ 43 51	3970		CLUSTER OF GALAXIES
MCG+08-17-037	09 05 00.	+ 47 31	15	16.	GALAXY
ZWG 264.064	09 05 00.	+ 53 27		15.4	GALAXY
ZWG 264.065	09 05 00.	+ 55 42		15.6	GALAXY
ZC 0905.0+5958	09 05 00.	+ 59 58	470		CLUSTER OF GALAXIES
ZC 0905.0+6701	09 05 00.	+ 67 01	1480		CLUSTER OF GALAXIES
LB 02993	09 05 05.	+ 21 36 12.		16.7	FAINT BLUE STAR
LB 09355	09 05 06.	+ 14 51		17.0	FAINT BLUE STAR
ZWG 151.012	09 05 06.	+ 27 27		15.3	GALAXY
TON-N 1005	09 05 06.	+ 33 59		15.1	BLUE STAR
ZWG 180.039	09 05 06.	+ 37 44		15.6	GALAXY
LB 02994	09 05 11.	+ 21 18 54.		15.6	FAINT BLUE STAR
LB 09356	09 05 12.	+ 20 17		17.9	FAINT BLUE STAR
ZWG 121.022	09 05 12.	+ 20 42		15.6	GALAXY
UGC 04792	09 05 12.	+ 20 42	66	15.6	GALAXY Sb-c
LB 02996	09 05 12.	+ 21 58 36.		15.4	FAINT BLUE STAR
LB 02995	09 05 12.	+ 22 33 24.		15.7	FAINT BLUE STAR
MCG+05-22-008	09 05 12.	+ 27 26	36	15.3	GALAXY
ZC 0905.2+3309	09 05 12.	+ 33 09	1010		CLUSTER OF GALAXIES
MCG+06-20-029	09 05 12.	+ 37 45	48	17.	GALAXY
ZC 0905.2+5846	09 05 12.	+ 58 46	3090		CLUSTER OF GALAXIES
ZC 0905.2+6413	09 05 12.	+ 64 13	400		CLUSTER OF GALAXIES
ZWG 006.001	09 05 12.	- 01 27		15.6	GALAXY
LB 02997	09 05 15.	+ 20 23 06.		15.2	FAINT BLUE STAR
ZWG 034.001	09 05 18.	+ 03 29		14.8	GALAXY
MCG+01-24-002	09 05 18.	+ 03 29	36	14.8	GALAXY
LB 09358	09 05 18.	+ 17 02		17.7	FAINT BLUE STAR
ZWG 091.002	09 05 18.	+ 19 02		15.7	GALAXY
ZWG 090.078	09 05 18.	+ 19 02		15.7	GALAXY
UGC 04793	09 05 18.	+ 19 02	60	15.7	GALAXY S
LB 09357	09 05 18.	+ 20 23		17.0	FAINT BLUE STAR
ZWG 121.023	09 05 18.	+ 22 11		15.4	GALAXY
ZWG 180.040	09 05 18.	+ 35 50		15.6	GALAXY
KARA.72 188A	09 05 18.	+ 35 50	36	15.6	PART OF DOUBLE GALAXY
MCG+09-15-097	09 05 18.	+ 52 27	36	16.	GALAXY
MCG+09-15-098	09 05 18.	+ 54 02	108	13.	GALAXY
ZC 0905.3+5519	09 05 18.	+ 55 19	1140		CLUSTER OF GALAXIES
TON-N 1006	09 05 19.	+ 20 33		15.3	BLUE STAR
MCG+04-22-016	09 05 21.	+ 22 12 30.	48	15.4	GALAXY
RNGC 2764	09 05 22.	+ 21 39		14.0	GALAXY
HOLM 110B	09 05 22.	+ 35 52	24	14.8	PART OF MULTIPLE GALAXY
LB 09361	09 05 24.	+ 15 49		16.5	FAINT BLUE STAR
LB 09360	09 05 24.	+ 18 04		17.5	FAINT BLUE STAR
LB 09359	09 05 24.	+ 18 49		17.5	FAINT BLUE STAR
MCG+04-22-017	09 05 24.	+ 21 38	48	13.9	GALAXY
ZWG 121.024	09 05 24.	+ 21 39		13.9	GALAXY
UGC 04794	09 05 24.	+ 21 39	102	13.9	GALAXY S?
MCG+06-20-032	09 05 24.	+ 32 45		15.	GALAXY
ZWG 180.041	09 05 24.	+ 35 49		15.4	GALAXY
KARA.72 188B	09 05 24.	+ 35 49	36	15.4	PART OF DOUBLE GALAXY
MCG+06-20-031	09 05 24.	+ 35 51	18	15.	GALAXY
MCG+06-20-030	09 05 24.	+ 35 52	24	15.	GALAXY
ZWG 180.042	09 05 24.	+ 37 49		14.2	GALAXY
UGC 04795	09 05 24.	+ 37 49	78	14.2	GALAXY E-S0
MCG+08-17-038	09 05 24.	+ 45 00	66	15.	GALAXY
MCG+08-17-039	09 05 24.	+ 49 57	36	16.	GALAXY

OBJECT NAME	RIGHT ASCEN.	DECLINATION	DIAM.	MAGN.	TYPE OF OBJECT
ZWG 264.066	09 05 24.	+ 52 28		15.7	GALAXY
ZWG 264.067	09 05 24.	+ 54 03		13.2	GALAXY
RNGC 2756	09 05 24.	+ 54 03		13.0	GALAXY
UGC 04796	09 05 24.	+ 54 03	102	13.2	GALAXY Sb
ZWG 264.068	09 05 24.	+ 54 06		15.5	GALAXY
ISS 1041	09 05 24.	- 42 57	103		STELLAR RING
TON-N 1007	09 05 25.	+ 21 17		16.6	BLUE STAR
HOLM 110A	09 05 25.	+ 35 51	24	14.8	PART OF MULTIPLE GALAXY
RNGC 2759	09 05 25.	+ 37 49		14.0	GALAXY
RNGC 2772	09 05 29.	- 23 25		15.0	GALAXY
ZC 0905.5+0437	09 05 30.	+ 04 37	470		CLUSTER OF GALAXIES
ZWG 034.002	09 05 30.	+ 06 08		15.7	GALAXY
UGC 04797	09 05 30.	+ 06 08	150	15.7	GALAXY DWRF SP
MCG+01-24-003	09 05 30.	+ 06 08	96	15.7	GALAXY
8ZW 0905+10.7	09 05 30.	+ 10 41		16.2	COMPACT GALAXY
LB 09364	09 05 30.	+ 14 40		15.5	FAINT BLUE STAR
LB 09363	09 05 30.	+ 15 16		16.7	FAINT BLUE STAR
LB 09362	09 05 30.	+ 16 02		17.5	FAINT BLUE STAR
ZWG 151.013	09 05 30.	+ 29 06		15.6	GALAXY
ZWG 180.043	09 05 30.	+ 32 47		14.7	GALAXY
ZC 0905.5+3740	09 05 30.	+ 37 40	1750		CLUSTER OF GALAXIES
MCG+06-20-033	09 05 30.	+ 37 49	24	14.5	GALAXY
ZWG 238.013	09 05 30.	+ 45 00		15.2	GALAXY
UGC 04798	09 05 30.	+ 45 00	66	15.2	GALAXY Sc
MCG+08-17-040	09 05 30.	+ 49 56	42	15.	GALAXY
MCG-04-22-002	09 05 30.	- 23 25 30.	84	15.	GALAXY
LB 09369	09 05 36.	+ 14 24		16.9	FAINT BLUE STAR
LB 09368	09 05 36.	+ 16 33		17.8	FAINT BLUE STAR
LB 09367	09 05 36.	+ 17 04		17.6	FAINT BLUE STAR
LB 09366	09 05 36.	+ 18 37		16.0	FAINT BLUE STAR
LB 09365	09 05 36.	+ 19 34		17.6	FAINT BLUE STAR
ZWG 091.003	09 05 36.	+ 20 10		15.7	GALAXY
ZWG 090.079	09 05 36.	+ 20 10		15.7	GALAXY
ZWG 180.044	09 05 36.	+ 37 42		15.2	GALAXY
UGC 04799	09 05 36.	+ 37 42	90	15.2	GALAXY SB
ZWG 238.014	09 05 36.	+ 50 10		15.6	GALAXY
MCG+08-17-041	09 05 36.	+ 50 10	24	16.	GALAXY
ZWG 264.069	09 05 36.	+ 51 03		15.7	GALAXY
ZC 0905.6+5322	09 05 36.	+ 53 22	200		CLUSTER OF GALAXIES
IC 2439	09 05 37.	+ 32 48 59.			NONSTELLAR OBJECT
LB 02998	09 05 39.	+ 10 42 24.		16.2	FAINT BLUE STAR
MCG+08-17-042	09 05 39.	+ 49 57	24	16.	GALAXY
SHB 130	09 05 41.0	+ 38 00 27.		18.5	QUASI-STELLAR OBJECT
LB 09373	09 05 42.	+ 16 54		17.9	FAINT BLUE STAR
LB 09372	09 05 42.	+ 17 30		17.5	FAINT BLUE STAR
LB 09371	09 05 42.	+ 18 32		17.9	FAINT BLUE STAR
LB 09370	09 05 42.	+ 19 48		17.0	FAINT BLUE STAR
MCG+06-20-034	09 05 42.	+ 37 42	66	15.	GALAXY
ZWG 264.070	09 05 42.	+ 55 06		14.8	GALAXY
UGC 04800	09 05 42.	+ 55 06	102	14.8	GALAXY SBc
MCG+09-15-099	09 05 42.	+ 55 06	96	14.	GALAXY
MRK 393	09 05 42.	+ 60 30	6	17.	GALAXY WITH UV CONTINUUM
7ZW 261	09 05 42.	+ 64 37			COMPACT GALAXY
ZWG 006.002	09 05 42.	- 01 36		15.6	GALAXY
LB 09377	09 05 48.	+ 15 00		15.6	FAINT BLUE STAR
LB 09376	09 05 48.	+ 15 15		17.8	FAINT BLUE STAR
LB 09375	09 05 48.	+ 20 00		17.1	FAINT BLUE STAR
LB 09374	09 05 48.	+ 20 14		17.9	FAINT BLUE STAR
ZWG 151.014	09 05 48.	+ 30 04		14.6	GALAXY
RNGC 2766	09 05 48.	+ 30 04		14.5	GALAXY
UGC 04801	09 05 48.	+ 30 04	90	14.6	GALAXY Sa-b
MCG+05-22-009	09 05 48.	+ 30 04	72	14.6	GALAXY
ZC 0905.8+6030	09 05 48.	+ 60 30	1680		CLUSTER OF GALAXIES
MCG+10-13-060	09 05 48.	+ 62 26	78	14.	GALAXY
RNGC 2742A	09 05 53.	+ 62 27		14.0	GALAXY
ZWG 034.003	09 05 54.	+ 03 25		15.6	GALAXY
LB 09378	09 05 54.	+ 16 02		16.6	FAINT BLUE STAR
MCG+05-22-010	09 05 54.	+ 32 28	36	15.5	GALAXY
MCG+06-20-035	09 05 54.	+ 37 41 30.	36	17.	GALAXY
ZC 0905.9+5143	09 05 54.	+ 51 43	1080		CLUSTER OF GALAXIES
ZWG 006.003	09 05 54.	- 01 25		15.3	GALAXY
MCG+00-24-001	09 05 54.	- 01 25	48	15.3	GALAXY
TON-N 1008	09 05 55.	+ 19 49		17.0	BLUE STAR
ARC 0749	09 05 57.	+ 07 11		17.7	RICH CLUSTER OF GALAXIES
KEEN 2742A	09 06	+ 62 22		13.7	GALAXY
ZC 0906.0+0706	09 06 00.	+ 07 06	1410		CLUSTER OF GALAXIES
ZWG 062.001	09 06 00.	+ 13 27		15.4	GALAXY
KARA.73B 0304	09 06 00.	+ 13 27	42	15.4	ISOLATED GALAXY S
ZC 0906.0+1540	09 06 00.	+ 15 40	670		CLUSTER OF GALAXIES
LB 09380	09 06 00.	+ 18 24		18.0	FAINT BLUE STAR
LB 09379	09 06 00.	+ 19 06		16.0	FAINT BLUE STAR
ZC 0906.0+2902	09 06 00.	+ 29 02	1210		CLUSTER OF GALAXIES
ZWG 151.015	09 06 00.	+ 32 29		15.5	GALAXY
MCG+08-17-043	09 06 00.	+ 50 29	36	16.	GALAXY
ZWG 311.030	09 06 00.	+ 62 27		14.0	GALAXY
ZWG 288.022	09 06 00.	+ 62 27		14.0	GALAXY
UGC 04803	09 06 00.	+ 62 27	108	14.0	GALAXY S
MCG+08-17-044	09 06 03.	+ 46 09	66	15.	GALAXY
ARC 0746	09 06 03.	+ 51 45		17.7	RICH CLUSTER OF GALAXIES
ZC 0906.1+1447	09 06 06.	+ 14 47	2890		CLUSTER OF GALAXIES
LB 09381	09 06 06.	+ 16 21		16.4	FAINT BLUE STAR
ZWG 091.004	09 06 06.	+ 20 12		15.7	GALAXY
TON-N 1009	09 06 06.	+ 32 50		16.9	BLUE STAR
MCG+06-20-036	09 06 06.	+ 33 17	42	15.	GALAXY
ZC 0906.1+4656	09 06 06.	+ 46 56	1950		CLUSTER OF GALAXIES
ZWG 006.004	09 06 06.	- 01 33		15.2	GALAXY
UGC 04804	09 06 06.	- 01 33	60	15.2	GALAXY SBb-c
MCG+00-24-002	09 06 06.	- 01 33	60	15.2	GALAXY
ARC 0753	09 06 06.	- 06 42		17.3	RICH CLUSTER OF GALAXIES
MCG+04-22-018	09 06 09.	+ 20 47	48		GALAXY
RNGC 2770B	09 06 10.	+ 33 20			GALAXY
ARC 0738	09 06 10.	+ 78 15		17.5	RICH CLUSTER OF GALAXIES
ZWG 006.005	09 06 12.	+ 00 42		15.3	GALAXY
KARA.73B 0305	09 06 12.	+ 00 42	24	15.3	ISOLATED GALAXY S
ZWG 034.004	09 06 12.	+ 03 39		15.1	GALAXY
ZC 0906.2+1112	09 06 12.	+ 11 12	2020		CLUSTER OF GALAXIES
LB 09384	09 06 12.	+ 15 40		16.6	FAINT BLUE STAR
LB 09383	09 06 12.	+ 17 44		17.1	FAINT BLUE STAR
LB 09382	09 06 12.	+ 19 50		17.0	FAINT BLUE STAR
ZWG 151.016	09 06 12.	+ 26 48		15.6	GALAXY
ZWG 180.045	09 06 12.	+ 37 53		15.7	GALAXY
ZWG 238.015	09 06 12.	+ 46 10		14.6	GALAXY
UGC 04805	09 06 12.	+ 46 10	60	14.6	GALAXY S
TON-N 1010	09 06 13.	+ 20 50		16.5	BLUE STAR
LB 02999	09 06 13.	+ 21 49 06.		17.1	FAINT BLUE STAR
BC 3CR216	09 06 17.26	+ 43 05 59.0		18.48	QUASI-STELLAR OBJECT
SHB 131	09 06 17.3	+ 43 05 59.		18.5	QUASI-STELLAR OBJECT
LB 09386	09 06 18.	+ 18 34		17.8	FAINT BLUE STAR
LB 09385	09 06 18.	+ 19 48		16.9	FAINT BLUE STAR
ZWG 091.005	09 06 18.	+ 20 18		15.7	GALAXY
TON-N 0385	09 06 18.	+ 33 15		15.2	BLUE STAR
MCG+08-17-045	09 06 18.	+ 50 37	9	16.	GALAXY
ZWG 264.071	09 06 18.	+ 53 26		15.7	GALAXY
MCG+09-15-100	09 06 18.	+ 55 50	24	18.	GALAXY
MCG+10-13-061	09 06 18.	+ 60 09	57	15.	GALAXY
ARC 0742	09 06 18.	+ 60 30		17.7	RICH CLUSTER OF GALAXIES
HOLM 111B	09 06 19.	+ 33 21	30	15.3	PART OF MULTIPLE GALAXY
LB 03001	09 06 20.	+ 10 35 36.		15.9	FAINT BLUE STAR
LB 03000	09 06 20.	+ 12 20 30.		16.5	FAINT BLUE STAR
HOLM 111C	09 06 20.	+ 33 20	30	15.0	PART OF MULTIPLE GALAXY
LB 03002	09 06 21.	+ 21 49 24.		16.2	FAINT BLUE STAR
MCG+06-20-037	09 06 21.	+ 32 40	48	15.	GALAXY
RNGC 2762	09 06 21.	+ 50 38		15.5	GALAXY
ARC 0750	09 06 22.	+ 11 14		17.1	RICH CLUSTER OF GALAXIES
KARA.73B 0306	09 06 24.	+ 01 38	24	15.7	ISOLATED GALAXY IR
ZWG 006.006	09 06 24.	+ 01 38		15.7	GALAXY
LB 09388	09 06 24.	+ 16 48		17.2	FAINT BLUE STAR
LB 09387	09 06 24.	+ 19 30		16.8	FAINT BLUE STAR
ZWG 091.006	09 06 24.	+ 19 37		15.3	GALAXY
ZWG 121.025	09 06 24.	+ 24 27		15.5	GALAXY
ZWG 180.046	09 06 24.	+ 32 43		15.4	GALAXY
MCG+06-20-038	09 06 24.	+ 33 17	210	12.	GALAXY
ZWG 180.047	09 06 24.	+ 33 19		12.1	GALAXY
UGC 04806	09 06 24.	+ 33 19	216	12.1	GALAXY Sc
ZWG 180.048	09 06 24.	+ 38 14		15.5	GALAXY
ZWG 264.072	09 06 24.	+ 50 38		15.7	GALAXY
MCG+09-15-102	09 06 24.	+ 52 14	12	17.	GALAXY
ZWG 264.073	09 06 24.	+ 54 46		14.5	GALAXY
UGC 04807	09 06 24.	+ 54 46	66	14.5	GALAXY Sc
MCG+09-15-101	09 06 24.	+ 54 46	60	14.	GALAXY
ZWG 288.023	09 06 24.	+ 60 10		15.7	GALAXY
UGC 04808	09 06 24.	+ 60 10	84	15.7	GALAXY Sc
7ZW 262	09 06 24.	+ 60 29			COMPACT GALAXY
ZC 0906.4+6416	09 06 24.	+ 64 16	870		CLUSTER OF GALAXIES
ARC 0754	09 06 26.	- 09 27		15.2	RICH CLUSTER OF GALAXIES
MCG+04-22-019	09 06 27.	+ 20 54	90	15.1	GALAXY
RNGC 2770	09 06 28.	+ 33 20		12.0	GALAXY
ZWG 121.026	09 06 30.	+ 20 55		15.1	GALAXY
UGC 04809	09 06 30.	+ 20 55	90	15.1	GALAXY SBc
ZWG 151.017	09 06 30.	+ 32 15		15.7	GALAXY
ZC 0906.5+3542	09 06 30.	+ 35 42	2760		CLUSTER OF GALAXIES
ZWG 180.049	09 06 30.	+ 37 47		14.6	GALAXY
UGC 04810	09 06 30.	+ 37 47	114	14.6	GALAXY S
ZWG 264.074	09 06 30.	+ 52 07		15.5	GALAXY
MCG+09-15-103	09 06 30.	+ 55 52 30.	36	18.	GALAXY
HOLM 111A	09 06 34.	+ 33 20	192	11.7	PART OF MULTIPLE GALAXY
SHB 132	09 06 35.3	+ 01 33 40.		18.	QUASI-STELLAR OBJECT
ZWG 006.007	09 06 36.	+ 01 02		15.6	GALAXY
KARA.73B 0307	09 06 36.	+ 01 02	24	15.6	ISOLATED GALAXY E
ZC 0906.6+1012	09 06 36.	+ 10 12	1210		CLUSTER OF GALAXIES
ZWG 091.007	09 06 36.	+ 15 49		15.6	GALAXY
MCG+03-24-001	09 06 36.	+ 15 59	78	14.6	GALAXY
IC 0528	09 06 36.	+ 15 59 34.			NONSTELLAR OBJECT
ZWG 091.008	09 06 36.	+ 16 00		14.6	GALAXY
UGC 04811	09 06 36.	+ 16 00	102	14.6	GALAXY Sb
ZC 0906.6+1611	09 06 36.	+ 16 11	610		CLUSTER OF GALAXIES
ZWG 121.027	09 06 36.	+ 25 25		15.4	GALAXY
MCG+06-20-039	09 06 36.	+ 37 48	90	14.5	GALAXY
ZWG 180.050	09 06 36.	+ 37 54		15.1	GALAXY
MCG+09-15-104	09 06 36.	+ 52 54	48	18.	GALAXY
HN 0080	09 06 36.	- 69 44			NEBULA
PK285-14.1	09 06 37.	- 69 44	9	11.5	PLANETARY NEBULA
IC 2448	09 06 37.	- 69 44	8	11.5	PLANETARY NEBULA
IC 0527	09 06 38.	+ 37 48 47.			NONSTELLAR OBJECT
RNGC 2767	09 06 39.	+ 50 37		14.5	GALAXY
ZWG 121.028	09 06 42.	+ 20 37		15.5	GALAXY
ZWG 121.029	09 06 42.	+ 20 41		15.6	GALAXY
ZC 0906.7+2546	09 06 42.	+ 25 46	2420		CLUSTER OF GALAXIES
MCG+08-17-046	09 06 42.	+ 44 49	24	14.	GALAXY
ZC 0906.7+4603	09 06 42.	+ 46 03	1950		CLUSTER OF GALAXIES
ZWG 238.016	09 06 42.	+ 50 15		14.4	GALAXY
UGC 04812	09 06 42.	+ 50 15	78	14.4	GALAXY Sc+COMP
ZWG 264.075	09 06 42.	+ 50 37		14.4	GALAXY
UGC 04813	09 06 42.	+ 50 37	42	14.4	GALAXY
MCG+08-17-047	09 06 42.	+ 50 40	24	16.	GALAXY
TON-N 1011	09 06 43.	+ 20 55		17.0	BLUE STAR
ARC 0737	09 06 44.	+ 80 08		16.9	RICH CLUSTER OF GALAXIES
ZWG 091.009	09 06 48.	+ 18 49		14.8	GALAXY
MCG+03-24-002	09 06 48.	+ 18 49	30	14.8	GALAXY
KARA.73B 0308	09 06 48.	+ 18 49	24	14.8	ISOLATED GALAXY E
ZWG 091.010	09 06 48.	+ 19 43		15.7	GALAXY
ZWG 091.011	09 06 48.	+ 20 40		15.2	GALAXY
UGC 04814	09 06 48.	+ 44 49	66	16.5	GALAXY Sb
MCG+08-17-049	09 06 48.	+ 49 21	30	17.	GALAXY
MCG+08-17-048	09 06 48.	+ 50 36 30.	42	14.	GALAXY
ZWG 288.024	09 06 48.	+ 62 10		15.7	GALAXY
MCG+10-13-062	09 06 48.	+ 62 10	18	16.	GALAXY
MCG+10-13-063	09 06 48.	+ 62 11	30	16.	GALAXY
ZWG 006.008	09 06 48.	- 00 58		15.4	GALAXY
KARA.72 189A	09 06 48.	- 00 58	54	15.4	PART OF DOUBLE GALAXY
ZC 0906.9+6651	09 06 54.	+ 66 51	670		CLUSTER OF GALAXIES
RNGC 2732	09 06 54.	+ 79 24		13.5	GALAXY
ZWG 006.009	09 06 54.	- 01 32		15.1	GALAXY
RNGC 2769	09 06 57.	+ 50 39		14.0	GALAXY
RNGC 2773	09 06 58.	+ 07 23		14.5	GALAXY
ARC 0747	09 06 59.	+ 61 20		17.5	RICH CLUSTER OF GALAXIES
VDB.66G 056	09 07	- 22 48	70		DWARF GALAXY
ZWG 034.005	09 07 00.	+ 07 23		14.5	GALAXY
UGC 04815	09 07 00.	+ 07 23	48	14.5	GALAXY
MCG+01-24-004	09 07 00.	+ 07 23	30	14.5	GALAXY
ZWG 091.012	09 07 00.	+ 20 04		15.7	GALAXY
ZC 0907.0+4948	09 07 00.	+ 49 48	1550		CLUSTER OF GALAXIES
ZWG 264.076	09 07 00.	+ 50 39		13.8	GALAXY
UGC 04816	09 07 00.	+ 50 39	120	13.8	GALAXY Sa
KARA.72 190A	09 07 00.	+ 50 39	102	13.8	PART OF DOUBLE GALAXY
ZWG 332.033	09 07 00.	+ 73 47		14.8	GALAXY
ZC 0907.0+8009	09 07 00.	+ 80 09	870		CLUSTER OF GALAXIES
ZC 0907.0+8158	09 07 00.	+ 81 58	1410		CLUSTER OF GALAXIES
ZWG 006.010	09 07 00.	- 00 59		15.5	GALAXY
KARA.72 189B	09 07 00.	- 00 59	48	15.5	PART OF DOUBLE GALAXY
RNGC 2771	09 07 03.	+ 50 36		14.0	GALAXY
8ZW 0907+10.0	09 07 06.	+ 10 00		16.6	COMPACT GALAXY
MCG+03-24-003	09 07 06.	+ 20 04	36	15.7	GALAXY
ZWG 121.030	09 07 06.	+ 23 03		15.3	GALAXY
ZWG 264.077	09 07 06.	+ 50 36		14.0	GALAXY
UGC 04817	09 07 06.	+ 50 36	120	14.0	GALAXY SBa

OBJECT NAME	RIGHT ASCEN.	DECLINATION	DIAM.	MAGN.	TYPE OF OBJECT
KARA.72 190B	09 07 06.	+ 50 36	114	14.0	PART OF DOUBLE GALAXY
MCG+08-17-050	09 07 06.	+ 50 38	90	13.	GALAXY
ZC 0907.1+7151	09 07 06.	+ 71 51	1210		CLUSTER OF GALAXIES
ZWG 350.013	09 07 06.	+ 79 24		12.6	GALAXY
UGC 04818	09 07 06.	+ 79 24	114	12.6	GALAXY S0
TON-N 1012	09 07 07.	+ 19 31		15.1	BLUE STAR
IC 2441	09 07 08.	+ 23 03 55.			NONSTELLAR OBJECT
PK273-03.1	09 07 09.	- 53 07 11.	25		PLANETARY NEBULA
IC 2442	09 07 11.	+ 23 02 55.			NONSTELLAR OBJECT
ZWG 121.031	09 07 12.	+ 23 02		15.2	GALAXY
MCG+04-22-020	09 07 12.	+ 23 03	42	15.3	GALAXY
ZC 0907.2+5422	09 07 12.	+ 54 22	610		CLUSTER OF GALAXIES
ZC 0907.2+6120	09 07 12.	+ 61 20	470		CLUSTER OF GALAXIES
UGC 04819	09 07 12.	+ 71 58	84	16.5	GALAXY Sc
MCG+04-22-021	09 07 15.	+ 23 02	45	15.2	GALAXY
PK266+02.2	09 07 15.	- 44 05 28.			PLANETARY NEBULA
ZWG 121.032	09 07 18.	+ 23 19		15.7	GALAXY
TON-N 0386	09 07 18.	+ 27 44		17.0	BLUE STAR
ZC 0907.3+3403	09 07 18.	+ 34 03	1480		CLUSTER OF GALAXIES
MCG+08-17-051	09 07 18.	+ 50 34	150	13.	GALAXY
MCG+10-13-064	09 07 18.	+ 62 12	30	15.	GALAXY
MCG+13-07-016	09 07 18.	+ 79 24	114	12.7	GALAXY
ABC 0752	09 07 20.	+ 35 39		17.8	RICH CLUSTER OF GALAXIES
ZC 0907.4+2849	09 07 24.	+ 28 49	4030		CLUSTER OF GALAXIES
TON-N 0387	09 07 24.	+ 32 14		15.4	BLUE STAR
ZWG 288.025	09 07 24.	+ 62 13		15.3	GALAXY
LB 03003	09 07 27.	+ 21 17 42.		15.8	FAINT BLUE STAR
ZC 0907.5+3214	09 07 30.	+ 32 14	1950		CLUSTER OF GALAXIES
MCG+07-19-035	09 07 30.	+ 40 47	30	16.	GALAXY
7ZW 263	09 07 30.	+ 64 33			COMPACT GALAXY
MCG+11-11-038	09 07 30.	+ 64 35	30	17.	GALAXY
MCG+11-11-039	09 07 30.	+ 67 11	57	16.	GALAXY
7ZW 264	09 07 30.	+ 68 00			COMPACT GALAXY
TON-N 1013	09 07 31.	+ 19 18		15.2	BLUE STAR
TON-N 1014	09 07 31.	+ 19 53		17.0	BLUE STAR
LB 03004	09 07 35.	+ 21 21 54.		16.7	FAINT BLUE STAR
ZWG 091.013	09 07 36.	+ 19 40		15.6	GALAXY
TON-N 1015	09 07 36.	+ 33 44		16.	BLUE STAR
ZC 0907.6+4149	09 07 36.	+ 41 49	1210		CLUSTER OF GALAXIES
MCG-04-22-003	09 07 36.	- 22 47	60	17.	GALAXY
RNGC 2775	09 07 40.	+ 07 15		11.5	GALAXY
RNGC 2768	09 07 40.	+ 60 15		12.0	GALAXY
ZWG 034.006	09 07 42.	+ 07 15		11.4	GALAXY
UGC 04820	09 07 42.	+ 07 15	330	11.4	GALAXY Sa
MCG+01-24-005	09 07 42.	+ 07 15	270	11.4	GALAXY
KARA.73B 0309	09 07 42.	+ 07 15	270	11.4	ISOLATED GALAXY S
ZWG 121.033	09 07 42.	+ 20 46		15.4	GALAXY
MCG+10-13-065	09 07 42.	+ 60 14	108	11.0	GALAXY
ZWG 288.026	09 07 42.	+ 60 15		11.1	GALAXY
UGC 04821	09 07 42.	+ 60 15	420	11.1	GALAXY E-S0
MCG+11-11-040	09 07 42.	+ 65 04	39	17.	GALAXY
MCG+11-11-041	09 07 42.	+ 67 06	24	16.	GALAXY
MCG+11-11-041A	09 07 45.	+ 67 06	39	16.	GALAXY
REIN 2.076	09 07 45.30	+ 60 14 33.4			NEBULA
RNGC 2774	09 07 47.	+ 18 54		15.0	GALAXY
ZWG 034.007	09 07 48.	+ 03 35		15.5	GALAXY
ZWG 091.014	09 07 48.	+ 15 45		15.7	GALAXY
ZWG 091.015	09 07 48.	+ 18 54		14.8	GALAXY
UGC 04822	09 07 48.	+ 19 41	66	16.5	GALAXY
ZC 0907.8+2500	09 07 48.	+ 25 00	340		CLUSTER OF GALAXIES
TON-N 1016	09 07 48.	+ 32 17		16.9	BLUE STAR
ZWG 238.017	09 07 48.	+ 50 20		15.7	GALAXY
MCG+08-17-052	09 07 48.	+ 50 20	30	16.	GALAXY
MCG+11-11-042	09 07 48.	+ 64 30	36	17.	GALAXY
ZWG 311.031	09 07 48.	+ 67 06		15.6	GALAXY
KARA.72 191A	09 07 48.	+ 67 06	30	15.6	PART OF DOUBLE GALAXY
REIN 2.077	09 07 50.98	+ 18 54 04.1			NEBULA
ZC 0907.9+2604	09 07 54.	+ 26 04	470		CLUSTER OF GALAXIES
TON-N 1018	09 07 54.	+ 31 56		16.8	BLUE STAR
MCG+09-15-105	09 07 54.	+ 51 26	144	14.	GALAXY
ZWG 311.032	09 07 54.	+ 67 06		15.6	GALAXY
KARA.72 191B	09 07 54.	+ 67 06	42	15.6	PART OF DOUBLE GALAXY
TON-N 1017	09 07 55.	+ 21 54		17.0	BLUE STAR
MCG+03-24-004	09 07 57.	+ 18 54	48	14.8	GALAXY
RNGC 2777	09 07 58.	+ 07 25		14.0	GALAXY
VDB.66G 055	09 08	+ 35 42	70		DWARF GALAXY
VDB.66G 057	09 08	- 14 48	70		DWARF GALAXY
ZWG 034.008	09 08 00.	+ 07 25		13.9	GALAXY
UGC 04823	09 08 00.	+ 07 25	48	13.9	GALAXY
MCG+01-24-006	09 08 00.	+ 07 25	48	13.9	GALAXY
ZWG 091.016	09 08 00.	+ 19 49		15.6	GALAXY
ZC 0908.0+2755	09 08 00.	+ 27 55	1010		CLUSTER OF GALAXIES
ZWG 151.018	09 08 00.	+ 32 18		15.5	GALAXY
ZWG 264.078	09 08 00.	+ 51 28		14.5	GALAXY
UGC 04824	09 08 00.	+ 51 28	150	14.5	GALAXY Sc
MCG+11-11-043	09 08 00.	+ 67 30	39	17.	GALAXY
MCG+12-09-031	09 08 00.	+ 70 47 30.	72	15.	GALAXY
ZWG 350.014	09 08 00.	+ 76 41		11.7	GALAXY
UGC 04825	09 08 00.	+ 76 41	186	11.7	GALAXY Sb-c
ZC 0908.0+7812	09 08 00.	+ 78 12	870		CLUSTER OF GALAXIES
RNGC 2748	09 08 02.	+ 76 41		12.5	GALAXY
ZC 0908.1+3939	09 08 06.	+ 39 39	810		CLUSTER OF GALAXIES
MCG-05-22-001	09 08 06.	- 32 57	120	15.	GALAXY
TON-N 1019	09 08 07.	+ 19 16		15.2	BLUE STAR
ABC 0760	09 08 08.	- 05 16		17.6	RICH CLUSTER OF GALAXIES
ZWG 034.009	09 08 12.	+ 03 10		15.3	GALAXY
KARA.73B 0310	09 08 12.	+ 03 10	60	15.3	ISOLATED GALAXY S
UGC 04826	09 08 12.	+ 03 10	60	15.3	GALAXY Sa-b
TON-N 0388	09 08 12.	+ 31 08		17.0	BLUE STAR
MCG-04-22-004	09 08 12.	- 23 18	60	16.	GALAXY
MCG+08-17-053	09 08 15.	+ 46 50	18	15.	GALAXY
RNGC 2788	09 08 15.	- 67 44			GALAXY
TON-N 1020	09 08 17.	+ 34 42		14.9	BLUE STAR
ZWG 034.010	09 08 18.	+ 08 05		15.6	GALAXY
KARA.73B 0311	09 08 18.	+ 08 05	24	15.6	ISOLATED GALAXY S
8ZW 0908+09.5	09 08 18.	+ 09 33		16.4	COMPACT GALAXY
ZWG 062.002	09 08 18.	+ 13 38		15.0	GALAXY
UGC 04827	09 08 18.	+ 13 38	84	15.0	GALAXY S
MCG+02-24-001	09 08 18.	+ 13 38	72	15.0	GALAXY
KARA.73B 0312	09 08 18.	+ 13 38	66	15.0	ISOLATED GALAXY S
ZWG 091.017	09 08 18.	+ 19 52		15.7	GALAXY
UGC 04828	09 08 18.	+ 19 52	96	15.7	GALAXY Sc
TON-N 0389	09 08 18.	+ 28 48		15.6	BLUE STAR
ZC 0908.3+4847	09 08 18.	+ 48 47	3490		CLUSTER OF GALAXIES
ABC 0761	09 08 18.	- 10 23		17.0	RICH CLUSTER OF GALAXIES
MCG+06-20-040	09 08 21.	+ 33 00	60	15.5	GALAXY
ZC 0908.4+1309	09 08 24.	+ 13 09	1480		CLUSTER OF GALAXIES
ZWG 091.018	09 08 24.	+ 19 36		15.4	GALAXY

OBJECT NAME	RIGHT ASCEN.	DECLINATION	DIAM.	MAGN.	TYPE OF OBJECT
ZWG 238.018	09 08 24.	+ 46 51		14.3	GALAXY
MRK 102	09 08 24.	+ 46 51	17	14.5	GALAXY WITH UV CONTINUUM
UGC 04829	09 08 24.	+ 46 51	48	14.3	GALAXY COMPACT
MCG-01-24-001	09 08 24.	- 08 42	144	12.	GALAXY
TON-N 1021	09 08 25.	+ 22 34		16.8	BLUE STAR
PK266+02.1	09 08 26.	- 45 12 21.			PLANETARY NEBULA
ZWG 091.019	09 08 30.	+ 19 30		15.3	GALAXY
UGC 04830	09 08 30.	+ 19 30	72	15.3	GALAXY Sa-b
MCG+03-24-005	09 08 30.	+ 19 37	36	15.4	GALAXY
ZC 0908.5+2716	09 08 30.	+ 27 16	810		CLUSTER OF GALAXIES
TON-N 0390	09 08 30.	+ 28 08		16.9	BLUE STAR
ZWG 151.019	09 08 30.	+ 29 01		15.1	GALAXY
MCG+05-22-011	09 08 30.	+ 29 01	36	15.1	GALAXY
ZWG 180.051	09 08 30.	+ 33 02		15.4	GALAXY
UGC 04831	09 08 30.	+ 33 02	66	15.4	GALAXY Sc
ZWG 264.079	09 08 30.	+ 55 53		15.5	GALAXY
MCG+11-11-043A	09 08 30.	+ 67 29	39	17.	GALAXY
ZWG 332.034	09 08 30.	+ 69 20		15.6	GALAXY
7ZW 265	09 08 30.	+ 69 22			COMPACT GALAXY
MCG+13-07-017	09 08 30.	+ 79 24	51	15.	GALAXY
ZWG 350.015	09 08 30.	+ 79 25		15.0	GALAXY
UGC 04832	09 08 30.	+ 79 25	90	15.0	GALAXY PECULR
IC 2443	09 08 33.	+ 29 02 06.			NONSTELLAR OBJECT
ZC 0908.6+0314	09 08 36.	+ 03 14	670		CLUSTER OF GALAXIES
MCG+03-24-006	09 08 36.	+ 19 31	63	15.3	GALAXY
ZWG 121.034	09 08 36.	+ 23 15		15.6	GALAXY
MCG+04-22-022	09 08 36.	+ 23 15	24	15.6	GALAXY
ZWG 151.020	09 08 36.	+ 30 16		15.4	GALAXY
UGC 04833	09 08 36.	+ 30 16	60	15.4	GALAXY Sa
ZWG 350.016	09 08 36.	+ 76 09		15.7	GALAXY
TON-N 1022	09 08 37.	+ 19 41		15.5	BLUE STAR
TON-N 0391	09 08 42.	+ 28 03		16.5	BLUE STAR
MCG+06-20-041	09 08 42.	+ 35 08 30.	48	17.	GALAXY
UGC 04834	09 08 42.	+ 35 10	78	16.5	GALAXY
MCG+09-15-106	09 08 42.	+ 51 47 30.	60	16.	GALAXY
MCG+13-07-018	09 08 42.	+ 75 48	60	16.	GALAXY
UGC 04835	09 08 42.	+ 75 49	60	17.	GALAXY S
TON-N 1023	09 08 43.	+ 18 50		16.8	BLUE STAR
MCG+08-17-054	09 08 45.	+ 46 14	30	16.	GALAXY
ZC 0908.8+0126	09 08 48.	+ 01 26	1080		CLUSTER OF GALAXIES
VHA 058	09 08 48.	- 56 05	180		OPEN STAR CLUSTER
MCG+08-17-055	09 08 48.	+ 49 57	30	15.	GALAXY
ZWG 238.019	09 08 48.	+ 49 58		15.5	GALAXY
LB 03005	09 08 50.	+ 20 39 18.		15.3	FAINT BLUE STAR
IC 2438	09 08 52.	+ 73 38			OPEN CLUSTER
ZWG 091.020	09 08 54.	+ 17 58		15.3	GALAXY
KARA.73B 0313	09 08 54.	+ 17 58	36	15.3	ISOLATED GALAXY S
MCG+08-17-056	09 08 54.	+ 45 10	108	11.9	GALAXY
MCG+08-17-057	09 08 54.	+ 46 18	30	16.	GALAXY
UGC 04836	09 08 54.	+ 70 48	90	15.3	GALAXY SBb
7ZW 266	09 08 54.	+ 70 49			COMPACT GALAXY
UGC 04802	09 08 54.	- 01 25	66	15.3	GALAXY Sc
ZWG 006.011	09 08 54.	- 02 41		15.3	GALAXY
TON-N 1024	09 08 55.	+ 19 33		16.0	BLUE STAR
RNGC 2776	09 08 56.	+ 45 10		12.0	GALAXY
TON-N 1025	09 08 59.	+ 35 54		16.9	BLUE STAR
8ZW 0909+08.7	09 09 00.	+ 08 42		17.4	COMPACT GALAXY
ZWG 091.021	09 09 00.	+ 17 21		15.4	GALAXY
MCG+03-24-007	09 09 00.	+ 17 58	42	15.3	GALAXY
ZWG 091.022	09 09 00.	+ 20 12		15.5	GALAXY
MCG+06-20-042	09 09 00.	+ 35 42	120	15.	GALAXY
ZWG 180.052	09 09 00.	+ 35 44		15.5	GALAXY
UGC 04837	09 09 00.	+ 35 44	126	15.5	GALAXY DWRF SP
ZWG 238.020	09 09 00.	+ 45 10		12.1	GALAXY
UGC 04838	09 09 00.	+ 45 10	198	12.1	GALAXY Sc
KARA.73B 0314	09 09 00.	+ 45 10	180	12.1	ISOLATED GALAXY S
ZC 0909.0+4842	09 09 00.	+ 48 42	670		CLUSTER OF GALAXIES
MCG+13-07-019	09 09 00.	+ 76 41	192	12.3	GALAXY
MCG-02-24-001	09 09 00.	- 14 50	96	15.	GALAXY
LB 03006	09 09 02.	+ 21 19 54.		15.8	FAINT BLUE STAR
ABC 0756	09 09 02.	+ 48 42		17.5	RICH CLUSTER OF GALAXIES
ABC 0755	09 09 04.	+ 49 11		17.1	RICH CLUSTER OF GALAXIES
REIN 2.078	09 09 05.72	- 14 36 41.1			NEBULA
MCG+03-24-010	09 09 06.	+ 16 28	66	15.0	GALAXY
ZWG 091.023	09 09 06.	+ 16 29		15.0	GALAXY
UGC 04839	09 09 06.	+ 16 29	84	15.0	GALAXY Sb-c
MCG+03-24-009	09 09 06.	+ 17 20	36	15.4	GALAXY
MCG+03-24-008	09 09 06.	+ 20 14	60	15.5	GALAXY
ZWG 180.053	09 09 06.	+ 37 52		15.5	GALAXY
KARA.73B 0316	09 09 06.	+ 37 52	48	15.5	ISOLATED GALAXY S
ZC 0909.1+4248	09 09 06.	+ 42 48	870		CLUSTER OF GALAXIES
ZWG 006.012	09 09 06.	- 03 20		15.5	GALAXY
KARA.73B 0315	09 09 06.	- 03 20	54	15.5	ISOLATED GALAXY S
RNGC 2781	09 09 07.	- 14 36		12.5	GALAXY
RNGC 2778	09 09 09.	+ 35 13		13.0	GALAXY
MCG-02-24-002	09 09 09.	- 14 37	210	12.5	GALAXY
ZC 0909.2+1110	09 09 12.	+ 11 10	470		CLUSTER OF GALAXIES
ZWG 180.054	09 09 12.	+ 35 13		13.1	GALAXY
UGC 04840	09 09 12.	+ 35 13	90	13.1	GALAXY E
ZC 0909.2+3529	09 09 12.	+ 35 29	1080		CLUSTER OF GALAXIES
7ZW 267	09 09 12.	+ 60 16			COMPACT GALAXY
ZWG 288.027	09 09 12.	+ 60 27		15.4	GALAXY
MCG+10-13-066	09 09 12.	+ 60 27	12	16.	GALAXY
TON-N 1026	09 09 13.	+ 21 53		16.9	BLUE STAR
MCG+06-20-043	09 09 15.	+ 35 12	36	14.	GALAXY
RNGC 2779	09 09 15.	+ 35 15		15.5	GALAXY
ZC 0909.3+1352	09 09 18.	+ 13 52	670		CLUSTER OF GALAXIES
ZC 0909.3+1450	09 09 18.	+ 14 50	670		CLUSTER OF GALAXIES
TON-N 0392	09 09 18.	+ 25 04		15.1	BLUE STAR
HOLM 112A	09 09 18.	+ 35 14	30	13.8	PART OF MULTIPLE GALAXY
MCG+06-20-044	09 09 18.	+ 35 14	24	15.	GALAXY
ZWG 180.055	09 09 18.	+ 35 15		15.5	GALAXY
MCG+10-13-067	09 09 18.	+ 60 26	18	15.	GALAXY
MCG-05-22-002	09 09 18.	- 30 39	60	16.	GALAXY
TON-N 1027	09 09 19.	+ 23 08		16.5	BLUE STAR
ABC 0758	09 09 19.	+ 42 46		17.9	RICH CLUSTER OF GALAXIES
HOLM 112B	09 09 22.	+ 35 16	30	15.0	PART OF MULTIPLE GALAXY
ABC 0757	09 09 23.	+ 47 56		15.6	RICH CLUSTER OF GALAXIES
ZWG 006.013	09 09 24.	+ 02 17		15.7	GALAXY
ZWG 034.011	09 09 24.	+ 03 04		15.7	GALAXY
ZC 0909.4+1947	09 09 24.	+ 19 47	670		CLUSTER OF GALAXIES
ZWG 332.035	09 09 24.	+ 74 26		14.4	GALAXY
UGC 04841	09 09 24.	+ 74 26	258	14.4	GALAXY SBc
MCG+12-09-032	09 09 24.	+ 74 27	150	12.9	GALAXY
LB 03007	09 09 27.	+ 20 45 48.		16.8	FAINT BLUE STAR
MCG+06-20-046	09 09 30.	+ 35 58	14	16.	GALAXY
MCG+06-20-045	09 09 30.	+ 35 58 30.	12	16.	GALAXY
ZWG 180.056	09 09 30.	+ 35 59		14.9	GALAXY

OBJECT NAME	RIGHT ASCEN.	DECLINATION	DIAM.	MAGN.	TYPE OF OBJECT
ZC 0909.5+3738	09 09 30.	+ 37 38	2960		CLUSTER OF GALAXIES
ZC 0909.5+4205	09 09 30.	+ 42 05	2020		CLUSTER OF GALAXIES
ZC 0909.5+5221	09 09 30.	+ 52 21	1340		CLUSTER OF GALAXIES
HELW 139	09 09 30.	- 14 24 26.			NEBULA
VHE 29C	09 09 30.	- 45 21	78		REFLECTION NEBULA
VHE 29B	09 09 30.	- 45 23	96		REFLECTION NEBULA
HO 3	09 09 31.	+ 74 28	240	12.94	Sc GALAXY
RNGC 2780	09 09 33.	+ 35 07		14.0	GALAXY
ARC 0759	09 09 35.	+ 42 07		17.5	RICH CLUSTER OF GALAXIES
ZWG 062.003	09 09 36.	+ 12 43		15.2	GALAXY
UGC 04842	09 09 36.	+ 12 43	72	15.2	GALAXY Sa-b
ZWG 062.004	09 09 36.	+ 13 53		15.5	GALAXY
TON-N 0393	09 09 36.	+ 27 13		15.2	BLUE STAR
MCG+06-20-047	09 09 36.	+ 35 06	48	14.	GALAXY
ZWG 180.057	09 09 36.	+ 35 07		14.2	GALAXY
UGC 04843	09 09 36.	+ 35 07	60	14.2	GALAXY SB
ZWG 180.058	09 09 36.	+ 36 08		15.4	GALAXY
ZWG 238.021	09 09 36.	+ 49 50		14.1	GALAXY
UGC 04844	09 09 36.	+ 49 50	102	14.1	GALAXY Sc
MCG+08-17-058	09 09 36.	+ 49 50	66	14.	GALAXY
ZWG 006.014	09 09 42.	+ 01 08		15.7	GALAXY
ZWG 062.005	09 09 42.	+ 10 10		15.1	GALAXY
UGC 04845	09 09 42.	+ 10 10	96	15.1	GALAXY SB:c
KARA.73B 0317	09 09 42.	+ 10 10	102	15.1	ISOLATFD GALAXY S
ZWG 062.006	09 09 42.	+ 14 13		15.7	GALAXY
ARC 0763	09 09 42.	+ 16 13		16.5	RICH CLUSTER OF GALAXIES
ZC 0909.7+1814	09 09 42.	+ 18 14	21300		CLUSTER OF GALAXIES
ZC 0909.7+2705	09 09 42.	+ 27 05	1480		CLUSTER OF GALAXIES
ZWG 151.021	09 09 42.	+ 29 38		15.6	GALAXY
MCG+09-15-107	09 09 42.	+ 54 50	42	18.	GALAXY
ZC 0909.7+5612	09 09 42.	+ 56 12	400		CLUSTER OF GALAXIES
ZWG 311.033	09 09 42.	+ 62 32		15.1	GALAXY
UGC 04846	09 09 42.	+ 62 32	60	15.1	GALAXY SBa
MCG+10-13-068	09 09 42.	+ 62 32	57	15.	GALAXY
ZWG 350.017	09 09 42.	+ 77 13		15.2	GALAXY
UGC 04847	09 09 42.	+ 79 29	72	16.0	GALAXY Sc
MCG-03-24-001	09 09 42.	- 19 55	240	15.	GALAXY
SEY 054	09 09 44.	+ 74 26 50.		13.5	FAINT GALAXY
ZWG 062.007	09 09 48.	+ 14 01		15.3	GALAXY
ZWG 151.022	09 09 48.	+ 30 25		15.1	GALAXY
MCG+05-22-012	09 09 48.	+ 30 26	54	15.1	GALAXY
UGC 04848	09 09 48.	+ 34 15	60	16.5	GALAXY Sc
UGC 04849	09 09 48.	+ 43 14	90	17.	GALAXY
ZWG 264.080	09 09 48.	+ 50 57		15.7	GALAXY
MCG+09-15-108	09 09 48.	+ 53 10	66	15.	GALAXY
ZC 0909.8+5328	09 09 48.	+ 53 28	270		CLUSTEF OF GALAXIES
ZWG 332.036	09 09 48.	+ 74 20		15.7	GALAXY
7ZW 268	09 09 48.	+ 77 13			COMPACT GALAXY
MCG-02-24-003	09 09 48.	- 15 13 30.	72	14.5	GALAXY
IC 2444	09 09 52.	+ 30 25 23.			NONSTELLAR OBJECT
ZC 0909.9+1611	09 09 54.	+ 16 11	870		CLUSTER OF GALAXIES
UGC 04850	09 09 54.	+ 33 37	60	17.	GALAXY
ZC 0909.9+4752	09 09 54.	+ 47 52	3020		CLUSTER OF GALAXIES
ZWG 264.081	09 09 54.	+ 53 11		14.2	GALAXY
UGC 04851	09 09 54.	+ 53 11	66	14.2	GALAXY E-SO
RNGC 2770A	09 09 57.	+ 35 16			NON-EXISTENT OBJECT
VDB.66G 058	09 10	+ 19 37	70		DWARF GALAXY
LBN 0683	09 10	+ 71 00	9900		BRIGHT NEBULA
LBN 1083	09 10	- 28 00	9600		BRIGHT NEBULA
ZC 0910.0+1523	09 10 00.	+ 15 23	2890		CLUSTER OF GALAXIES
ZC 0910.0+2745	09 10 00.	+ 27 45	2490		CLUSTER OF GALAXIES
MCG+08-17-059	09 10 00.	+ 45 32 30.	24	16.	GALAXY
7ZW 269	09 10 00.	+ 66 23			COMPACT GALAXY
MCG+13-07-020	09 10 00.	+ 79 29	60	15.	GALAXY
ZWG 364.009	09 10 00.	+ 80 15		15.3	GALAXY
ZWG 350.018	09 10 00.	+ 80 15		15.3	GALAXY
UGC 04852	09 10 00.	+ 80 15	90	15.3	GALAXY IRR
VHE 29A	09 10 00.	- 45 26	18		REFLECTION NEBULA
LB 03008	09 10 03.	+ 21 16 12.		15.9	FAINT BLUE STAR
RNGC 2760	09 10 04.	+ 76 35			NON-EXISTENT OBJECT
RNGC 2784	09 10 05.	- 23 58		12.0	GALAXY
MCG+04-22-023	09 10 06.	+ 20 33	66	15.3	GALAXY
MCG+05-22-013	09 10 06.	+ 32 01	36	15.2	GALAXY
ZWG 238.022	09 10 06.	+ 45 34		15.6	GALAXY
MCG+09-15-109	09 10 06.	+ 52 40	36	18.	GALAXY
MCG-04-22-005	09 10 06.	- 23 58 30.	72	12.	GALAXY
IC 2445	09 10 11.	+ 32 00 52.			NONSTELLAR OBJECT
ZC 0910.2+0037	09 10 12.	+ 00 37	400		CLUSTER OF GALAXIES
ZWG 006.015	09 10 12.	+ 02 11		15.7	GALAXY
ZWG 121.035	09 10 12.	+ 20 35		15.3	GALAXY
UGC 04853	09 10 12.	+ 20 35	72	15.3	GALAXY SBc
ZWG 151.023	09 10 12.	+ 31 33		15.5	GALAXY
MCG+05-22-014	09 10 12.	+ 31 33	60	15.5	GALAXY
ZWG 151.024	09 10 12.	+ 32 00		15.2	GALAXY
UGC 04854	09 10 12.	+ 32 00	60	15.2	GALAXY SB
MCG+09-15-110	09 10 12.	+ 52 28	42	15.	GALAXY
MCG-05-22-003	09 10 12.	- 30 42	96	14.5	GALAXY
ZWG 264.082	09 10 18.	+ 51 03		15.6	GALAXY
ZWG 264.083	09 10 18.	+ 52 30		15.1	GALAXY
MCG+04-22-024	09 10 24.	+ 23 17 30.	24	15.5	GALAXY
ZWG 121.036	09 10 24.	+ 23 18		15.5	GALAXY
TON-N 1028	09 10 24.	+ 33 34		17.	BLUE STAR
MCG+08-17-060	09 10 24.	+ 47 54	48	16.	GALAXY
ZWG 238.023	09 10 24.	+ 47 55		15.7	GALAXY
ZWG 264.084	09 10 24.	+ 52 26		15.5	GALAXY
ZWG 264.085	09 10 24.	+ 52 50		15.4	GALAXY
IC 2440	09 10 24.	+ 73 40			NONSTELLAR OBJFCT
LB 03009	09 10 25.	+ 09 48 00.		17.0	FAINT BLUE STAP
ZWG 006.016	09 10 30.	+ 02 15		15.4	GALAXY
ZWG 091.024	09 10 30.	+ 17 51		15.1	GALAXY
MCG+03-24-011	09 10 30.	+ 17 51	15	15.1	GALAXY
MCG+04-22-027	09 10 30.	+ 23 03	39	15.7	GALAXY
ZWG 121.037	09 10 30.	+ 23 04		15.7	GALAXY
MCG+04-22-025	09 10 30.	+ 23 16 30.	30	15.3	GALAXY
ZWG 121.038	09 10 30.	+ 23 17		15.3	GALAXY
MCG+04-22-026	09 10 30.	+ 23 17	27	15.3	GALAXY
ZWG 151.025	09 10 30.	+ 29 10		15.0	GALAXY
UGC 04855	09 10 30.	+ 29 10	102	15.0	GALAXY Sa
MCG+05-22-015	09 10 30.	+ 29 10	78	15.0	GALAXY
SHAH 110	09 10 30.	+ 29 57	66	17.7	GROUP OF COMPACT GALAXIES
ZWG 151.026	09 10 30.	+ 30 12		15.2	GALAXY
RNGC 2783B	09 10 30.	+ 30 12		15.0	GALAXY
UGC 04856	09 10 30.	+ 30 12	102	15.2	GALAXY Sb
KARA.72 192A	09 10 30.	+ 30 12	102	15.2	PART OF DOUBLE GALAXY
MCG+05-22-017	09 10 30.	+ 30 12	120	15.2	GALAXY
MCG+05-22-016	09 10 30.	+ 30 13	12	16.	GALAXY
HOLM 113B	09 10 30.	+ 30 14	96	15.0	PART OF MULTIPLE GALAXY
MCG+05-22-018	09 10 30.	+ 30 14 30.	6	16.	GALAXY
ZC 0910.5+5105	09 10 30.	+ 51 05	610		CLUSTER OF GALAXIES
IC 2447	09 10 33.	+ 28 56 55.			NONSTELLAR OBJECT
IC 2446	09 10 33.	+ 29 09 37.			NONSTELLAR OBJECT
MCG+06-20-048	09 10 33.	+ 35 00 30.	24	15.5	GALAXY
ARC 0766	09 10 33.	- 04 31		17.1	RICH CLUSTER OF GALAXIES
RNGC 2792	09 10 34.	- 42 14		13.5	PLANETARY NEBULA
TON-N 1029	09 10 35.	+ 35 40		16.9	BLUE STAR
ZWG 034.012	09 10 36.	+ 03 26		14.7	GALAXY
UGC 04857	09 10 36.	+ 03 26	90	14.7	GALAXY SBc
MCG+01-24-007	09 10 36.	+ 03 26	90	14.7	GALAXY
ZWG 034.013	09 10 36.	+ 05 49		15.5	GALAXY
KARA.73B 0318	09 10 36.	+ 05 49	36	15.5	ISOLATED GALAXY S
ZWG 091.025	09 10 36.	+ 19 35		15.5	GALAXY
UGC 04858	09 10 36.	+ 19 35	96	15.5	GALAXY DWRF IR
MCG+03-24-012	09 10 36.	+ 19 35	90	15.5	GALAXY
ZWG 151.027	09 10 36.	+ 30 11		13.9	GALAXY
RNGC 2783A	09 10 36.	+ 30 11		14.0	GALAXY
UGC 04859	09 10 36.	+ 30 11	126	13.9	GALAXY E
KARA.72 192B	09 10 36.	+ 30 11	126	13.9	PART OF DOUBLE GALAXY
MCG+05-22-020	09 10 36.	+ 30 12	18	15.5	GALAXY
MCG+05-22-019	09 10 36.	+ 30 12	33	13.9	GALAXY
HOLM 113A	09 10 36.	+ 30 14	30	13.7	PART OF MULTIPLE GALAXY
ZWG 180.059	09 10 36.	+ 33 31		15.7	GALAXY
ZWG 180.060	09 10 36.	+ 35 02		15.4	GALAXY
HELW 140	09 10 36.	- 14 22 41.			NEBULA
PK265+04.1	09 10 36.	- 42 14	13	13.5	PLANETARY NEBULA
LB 03491	09 10 36.	- 78 36		11.8	FAINT BLUE STAR
ZWG 006.017	09 10 42.	+ 02 04		15.5	GALAXY
ZC 0910.7+3714	09 10 42.	+ 37 14	1210		CLUSTER OF GALAXIES
ZC 0910.7+5421	09 10 42.	+ 54 21	2220		CLUSTER OF GALAXIES
7ZW 270	09 10 42.	+ 68 34			COMPACT GALAXY
MCG+04-22-028	09 10 48.	+ 20 45	36	15.4	GALAXY
ZC 0910.8+2639	09 10 48.	+ 26 39	1550		CLUSTER OF GALAXIES
ZWG 151.028	09 10 48.	+ 29 23		15.0	GALAXY
UGC 04860	09 10 48.	+ 29 23	60	15.0	GALAXY DBL SYS
MCG+05-22-021	09 10 48.	+ 29 23	36	15.0	GALAXY
MCG+13-07-021	09 10 48.	+ 80 15	66	14.	GALAXY
TON-N 1030	09 10 49.	+ 20 13		15.8	BLUE STAR
TON-N 1031	09 10 49.	+ 21 15		15.1	BLUE STAR
HELW 141	09 10 49.	- 14 35 06.			NEBULA
MCG+07-19-036	09 10 51.	+ 40 17 30.	222	12.5	GALAXY
LB 03010	09 10 53.	+ 20 42 18.		16.2	FAINT BLUE STAR
REIN 2.079	09 10 53.93	+ 40 19 15.8			NEBULA
ZWG 062.008	09 10 54.	+ 12 39		13.8	GALAXY
UGC 04861	09 10 54.	+ 12 39	60	13.8	GALAXY Sa?
MCG+02-24-002	09 10 54.	+ 12 39	48	13.8	GALAXY
ZWG 091.026	09 10 54.	+ 17 04		14.9	GALAXY
ZWG 121.039	09 10 54.	+ 20 46		15.4	GALAXY
RNGC 2782	09 10 54.	+ 40 19		12.5	GALAXY
ZWG 209.031	09 10 54.	+ 40 20		12.3	GALAXY
UGC 04862	09 10 54.	+ 40 20	252	12.3	GALAXY S
ZWG 209.032	09 10 54.	+ 43 20		14.8	GALAXY
ARC 0748	09 10 54.	+ 75 59		17.4	RICH CLUSTER OF GALAXIES
ARP 215	09 10 55.	+ 40 19			PECULIAR GALAXY
RNGC 2808	09 10 55.	- 64 39		8.0	GLOBULAR CLUSTFR
ZWG 062.009	09 11 00.	+ 13 06		15.7	GALAXY
MCG+03-24-013	09 11 00.	+ 17 04	39	14.9	GALAXY
ZC 0911.0+3025	09 11 00.	+ 30 25	6590		CLUSTER OF GALAXIES
GCL 013	09 11 00.	- 64 39	1128	7.8	GLOBULAR STAR CLUSTEP
MCG+07-19-037	09 11 03.	+ 43 20	39	15.	GALAXY
ZC 0911.1+0332	09 11 06.	+ 03 32	1680		CLUSTER OF GALAXIES
ZWG 091.027	09 11 06.	+ 19 09		15.4	GALAXY
TON-N 1032	09 11 07.	+ 19 14		17.	BLUE STAR
ZWG 181.001	09 11 12.	+ 36 31		15.6	GALAXY
ZWG 180.061	09 11 12.	+ 36 31		15.6	GALAXY
ZC 0911.2+3846	09 11 12.	+ 38 46	4100		CLUSTER OF GALAXIES
UGC 04863	09 11 12.	+ 41 00	66	16.5	GALAXY Sb-c
ZC 0911.2+4359	09 11 12.	+ 43 59	1010		CLUSTER OF GALAXIES
ZC 0911.2+5955	09 11 12.	+ 59 55	3900		CLUSTER OF GALAXIES
ZC 0911.2-0035	09 11 12.	- 00 35	610		CLUSTER OF GALAXIES
TON-N 1033	09 11 13.	+ 19 58		17.0	BLUE STAR
TON-N 1034	09 11 13.	+ 20 40		14.8	BLUE STAR
RNGC 2786	09 11 14.	+ 12 21			NON-EXISTENT OBJECT
MCG+07-19-038	09 11 15.	+ 40 57 30.	60	16.	GALAXY
HN 0027	09 11 16.	+ 00 49			NEBULA
MCG+06-20-049	09 11 18.	+ 36 00	54	15.5	GALAXY
ZWG 181.002	09 11 18.	+ 36 18		15.0	GALAXY
ZWG 180.062	09 11 18.	+ 36 18		15.0	GALAXY
KARA.72 193A	09 11 18.	+ 36 18	48	15.0	PART OF DOUBLE GALAXY
MCG+06-20-050	09 11 18.	+ 36 19	48	15.	GALAXY
MCG+09-15-111	09 11 18.	+ 51 32	60	15.	GALAXY
ZWG 264.086	09 11 18.	+ 53 35		14.8	GALAXY
SHAH 100	09 11 18.	+ 53 48	300	17.5	GROUP OF COMPACT GALAXIES
ZWG 006.018	09 11 18.	- 01 39		15.6	GALAXY
TON-N 1035	09 11 19.	+ 19 03		17.0	BLUE STAR
TON-N 1036	09 11 19.	+ 23 08		16.1	BLUE STAR
ZC 0911.4+0015	09 11 24.	+ 00 15	870		CLUSTER OF GALAXIES
ZWG 034.014	09 11 24.	+ 07 15		15.6	GALAXY
ZWG 091.028	09 11 24.	+ 16 57		14.7	GALAXY
UGC 04864	09 11 24.	+ 16 57	114	14.7	GALAXY Sa
ZWG 121.040	09 11 24.	+ 22 29		15.5	GALAXY
ZC 0911.4+2413	09 11 24.	+ 24 13	470		CLUSTER OF GALAXIES
ZWG 151.029	09 11 24.	+ 30 50		15.6	GALAXY
ZWG 181.003	09 11 24.	+ 35 55		15.6	GALAXY
ZWG 180.063	09 11 24.	+ 35 55		15.6	GALAXY
UGC 04865	09 11 24.	+ 35 55	66	15.6	GALAXY S
MCG+08-17-061	09 11 24.	+ 49 58	30	16.	GALAXY
ZWG 264.087	09 11 24.	+ 51 34		14.6	GALAXY
7ZW 271	09 11 24.	+ 58 25			COMPACT GALAXY
MRK 103	09 11 24.	+ 68 00	18	16.5	GALAXY WITH UV CONTINUUM
SHB 133	09 11 24.1	+ 05 23 00.		17.4	QUASI-STELLAR OBJECT
BC 4C05.38	09 11 24.3	+ 05 19 30.		17.43	QUASI-STELLAR OBJECT
FATH 1.254	09 11 25.	+ 14 28	11		NEBULA
MCG+06-20-051	09 11 27.	+ 35 53 30.	60	16.	GALAXY
MCG+03-24-014	09 11 30.	+ 16 56	114	14.7	GALAXY
ZWG 151.030	09 11 30.	+ 30 02		15.6	GALAXY
ZWG 181.004	09 11 30.	+ 36 17		15.5	GALAXY
ZWG 180.064	09 11 30.	+ 36 17		15.5	GALAXY
UGC 04866	09 11 30.	+ 36 17	72	15.5	GALAXY Sb
KARA.72 193B	09 11 30.	+ 36 18	60	15.5	PART OF DOUBLE GALAXY
MCG+06-20-052	09 11 30.	+ 36 18	60	16.	GALAXY
MCG+07-19-039	09 11 30.	+ 41 04	90	14.5	GALAXY
ZWG 209.033	09 11 30.	+ 41 05		15.2	GALAXY
UGC 04867	09 11 30.	+ 41 05	120	15.2	GALAXY SBc/Sc
UGC 04868	09 11 30.	+ 48 48	60	16.5	GALAXY
MCG+08-17-062	09 11 30.	+ 48 48	60	14.	GALAXY
SC 0909-1246.4	09 11 31.	- 12 58 43.	12		NEBULA
MCG+05-22-022	09 11 33.	+ 30 21	120	14.5	GALAXY

OBJECT NAME	RIGHT ASCEN.	DECLINATION	DIAM.	MAGN.	TYPE OF OBJECT
MCG+08-17-063	09 11 33.	+ 47 05 30.	48	16.	GALAXY
ZWG 151.031	09 11 36.	+ 30 12		15.7	GALAXY
ZWG 151.032	09 11 36.	+ 30 20		14.5	GALAXY
UGC 04869	09 11 36.	+ 30 20	120	14.5	GALAXY S0?
ZWG 238.024	09 11 36.	+ 47 07		14.0	GALAXY
UGC 04870	09 11 36.	+ 47 07	66	14.0	GALAXY S
SN 1966A	09 11 36.	+ 47 07		15.5	SUPERNOVA
ARC 0769	09 11 40.	+ 03 32		16.5	RICH CLUSTER OF GALAXIES
ZWG 151.033	09 11 42.	+ 26 55		15.7	GALAXY
MCG+05-22-024	09 11 42.	+ 30 11	48	15.7	GALAXY
ZWG 151.034	09 11 42.	+ 30 23		15.7	GALAXY
MCG+05-22-023	09 11 42.	+ 30 23	36	15.7	GALAXY
UGC 04871	09 11 42.	+ 39 28	108	16.0	GALAXY DWRF SP
ZWG 209.034	09 11 42.	+ 42 14		15.4	GALAXY
MCG+09-15-112	09 11 42.	+ 51 14	24	17.	GALAXY
MCG+11-11-044	09 11 42.	+ 64 57	36	17.	GALAXY
ZWG 312.001	09 11 42.	+ 67 58		15.6	GALAXY
ZWG 311.034	09 11 42.	+ 67 58		15.6	GALAXY
ZWG 006.019	09 11 42.	- 02 15		15.4	GALAXY
MCG+07-19-040	09 11 45.	+ 39 26 30.	96	15.	GALAXY
ZC 0911.8+0031	09 11 48.	+ 00 31	1080		CLUSTER OF GALAXIES
ZWG 034.015	09 11 48.	+ 07 54		15.5	GALAXY
UGC 04872	09 11 48.	+ 40 16	114	16.0	GALAXY SBb
VV 131B	09 11 48.	+ 48 51	18	17.	INTERACTING GALAXY
VV 131A	09 11 48.	+ 48 51	15	17.	INTERACTING GALAXY
VV 131	09 11 48.	+ 48 51	60	16.	INTERACTING GALAXY
MCG+08-17-064	09 11 48.	+ 48 51	66	15.	GALAXY
ZWG 332.037	09 11 48.	+ 70 05		15.7	GALAXY
ZC 0911.8+7602	09 11 48.	+ 76 02	2620		CLUSTER OF GALAXIES
ZWG 006.020	09 11 48.	- 02 35		15.7	GALAXY
MCG+00-24-003	09 11 48.	- 02 35	72	15.7	GALAXY
REIN 2.080	09 11 50.58	+ 40 14 39.2			NEBULA
MCG+07-19-041	09 11 51.	+ 40 13	120	15.	GALAXY
ZWG 034.016	09 11 54.	+ 08 19		14.7	GALAXY
MCG+01-24-008	09 11 54.	+ 08 19	42	14.7	GALAXY
8ZW 0911+13.0	09 11 54.	+ 12 58		17.0	COMPACT GALAXY
MCG+03-24-015	09 11 54.	+ 15 39	72	14.2	GALAXY
ZWG 091.029	09 11 54.	+ 15 40		14.2	GALAXY
UGC 04873	09 11 54.	+ 15 40	90	14.2	GALAXY S?
ZWG 091.030	09 11 54.	+ 19 07		15.5	GALAXY
ZWG 121.041	09 11 54.	+ 22 47		15.7	GALAXY
TON-N 0394	09 11 54.	+ 28 37		16.6	BLUE STAR
UGC 04874	09 11 54.	+ 48 53	66	16.5	GALAXY
MCG+12-09-033	09 11 54.	+ 70 06	39	15.	GALAXY
VDB.66G 059	09 12	+ 39 27	70		DWARF GALAXY
LB 03011	09 12 00.	+ 20 16 42.		16.8	FAINT BLUE STAR
ZWG 091.031	09 12 00.	+ 20 20		15.7	GALAXY
ZWG 121.042	09 12 00.	+ 23 40		15.6	GALAXY
MCG+05-22-025	09 12 00.	+ 29 28	48	14.5	GALAXY
ZWG 151.035	09 12 00.	+ 29 56		13.8	GALAXY
RNGC 2789	09 12 00.	+ 29 56		14.0	GALAXY
UGC 04875	09 12 00.	+ 29 56	102	13.8	GALAXY S0-a
MCG+05-22-026	09 12 00.	+ 29 56	114	13.8	GALAXY
ZWG 209.035	09 12 00.	+ 41 07		14.9	GALAXY
RNGC 2785	09 12 00.	+ 41 07		15.0	GALAXY
UGC 04876	09 12 00.	+ 41 07	102	14.9	GALAXY IRR
UGC 04877	09 12 00.	+ 43 15	72	16.0	GALAXY S
MCG+10-13-069	09 12 00.	+ 59 10	18	16.	GALAXY
MCG+07-19-042	09 12 03.	+ 41 06	90	14.	GALAXY
ZWG 091.032	09 12 06.	+ 15 54		15.6	GALAXY
ZWG 121.043	09 12 06.	+ 22 48		15.6	GALAXY
ZC 0912.1+2542	09 12 06.	+ 25 42	2550		CLUSTER OF GALAXIES
MCG+12-09-034	09 12 06.	+ 74 32	66	15.	GALAXY
ZC 0912.1+7937	09 12 06.	+ 79 37	1010		CLUSTER OF GALAXIES
MCG+07-19-043	09 12 09.	+ 43 13	72	16.5	GALAXY
FATH 1.255	09 12 10.	+ 14 14	8		NEBULA
RNGC 2790	09 12 11.	+ 19 55		14.5	GALAXY
ZWG 034.017	09 12 12.	+ 04 50		15.6	GALAXY
ZWG 091.033	09 12 12.	+ 17 48		15.6	GALAXY
RNGC 2791	09 12 12.	+ 17 48		15.5	GALAXY
ZWG 091.034	09 12 12.	+ 19 55		14.7	GALAXY
MCG+03-24-016	09 12 12.	+ 19 55	24	14.7	GALAXY
ZC 0912.2+2448	09 12 12.	+ 24 48	1410		CLUSTER OF GALAXIES
MCG+10-13-070	09 12 12.	+ 59 11	18	17.	GALAXY
ZC 0912.2+7045	09 12 12.	+ 70 45	1410		CLUSTER OF GALAXIES
IC 2449	09 12 15.	+ 30 12 16.			NONSTELLAR OBJECT
ZWG 034.018	09 12 18.	+ 04 54		15.7	GALAXY
ZWG 121.044	09 12 18.	+ 22 51		15.7	GALAXY
TON-N 0395	09 12 18.	+ 25 47		17.0	BLUE STAR
MCG+05-22-027	09 12 18.	+ 32 01	48	15.0	GALAXY
TON-N 1037	09 12 19.	+ 20 18		16.0	BLUE STAR
FATH 1.256	09 12 23.	+ 13 55	14		NEBULA
PK275-04.2	09 12 23.	- 55 16	10		PLANETARY NEBULA
ZC 0912.4+1637	09 12 24.	+ 16 37	940		CLUSTER OF GALAXIES
ZWG 151.036	09 12 24.	+ 27 30		15.6	GALAXY
UGC 04878	09 12 24.	+ 31 36	114	16.0	GALAXY
ZWG 151.037	09 12 24.	+ 32 00		15.0	GALAXY
ZWG 209.036	09 12 24.	+ 44 26		15.4	GALAXY
TON-N 1038	09 12 25.	+ 20 29		15.9	BLUE STAR
VV 155	09 12 27.	+ 44 24 30.	60	15.	INTERACTING GALAXY
MCG-01-24-002	09 12 27.	- 07 18	42	15.5	GALAXY
MCG-03-24-002	09 12 27.	- 19 21	108	14.	GALAXY
TON-N 1039	09 12 29.	+ 35 46		17.	BLUE STAR
ZWG 091.035	09 12 30.	+ 17 59		15.3	GALAXY
ZWG 209.037	09 12 30.	+ 42 28		14.8	GALAXY
VV 124	09 12 30.	+ 53 02 30.	72	13.	INTERACTING GALAXY
MCG+09-15-113	09 12 30.	+ 53 02 30.	72	14.	GALAXY
ZWG 264.088	09 12 30.	+ 53 03		14.0	GALAXY
UGC 04879	09 12 30.	+ 53 03	150	14.0	GALAXY IRR
IC 0530	09 12 33.	+ 12 04 49.			NONSTELLAR OBJECT
MCG+07-19-044	09 12 33.	+ 42 27 30.	48	15.	GALAXY
MCG+08-17-065	09 12 33.	+ 44 31	54	14.	GALAXY
LB 03012	09 12 34.	+ 10 09 36.		16.0	FAINT BLUE STAR
HOLM 114B	09 12 35.	+ 12 05	30	15.5	PART OF MULTIPLE GALAXY
ARP 055	09 12 35.	+ 44 32			PECULIAR GALAXY
ZWG 062.010	09 12 36.	+ 12 06		14.3	GALAXY
UGC 04880	09 12 36.	+ 12 06	126	14.3	GALAXY Sa-b
MCG+02-24-003	09 12 36.	+ 12 06	72	14.3	GALAXY
KARA.73B 0319	09 12 36.	+ 12 06	132	14.3	ISOLATED GALAXY S
ZWG 151.038	09 12 36.	+ 30 20		15.6	GALAXY
ZC 0912.6+6800	09 12 36.	+ 68 00	1410		CLUSTER OF GALAXIES
ZWG 006.021	09 12 36.	- 00 30		15.7	GALAXY
MCG+00-24-004	09 12 36.	- 00 30	24	15.7	GALAXY
HOLM 114A	09 12 37.	+ 12 06	78	14.2	PART OF MULTIPLE GALAXY
ZC 0912.7+0540	09 12 42.	+ 05 40	1610		CLUSTER OF GALAXIES
ZWG 238.025	09 12 42.	+ 44 33		14.9	GALAXY
UGC 04881	09 12 42.	+ 44 33	60	14.9	GALAXY DBL SYS
MCG+11-12-001	09 12 42.	+ 65 01	39	17.	GALAXY
ZWG 121.045	09 12 48.	+ 23 42		15.7	GALAXY
IC 2450	09 12 48.	+ 25 39 43.			NONSTELLAR OBJECT
MCG-05-22-004	09 12 48.	- 28 02	84	15.	GALAXY
TON-N 1040	09 12 49.	+ 19 51		17.	BLUE STAR
ZWG 091.036	09 12 54.	+ 15 26		15.1	GALAXY
TON-N 0396	09 12 54.	+ 29 47		17.0	BLUE STAR
TON-N 0397	09 12 54.	+ 29 49		15.2	BLUE STAR
MCG+06-21-001	09 12 54.	+ 34 15	24	16.	GALAXY
ZC 0912.9+4343	09 12 54.	+ 43 43	1010		CLUSTER OF GALAXIES
ZWG 288.028	09 12 54.	+ 59 59		15.6	GALAXY
MRK 019	09 12 54.	+ 59 59	15	16.	GALAXY WITH UV CONTINUUM
MCG+10-13-071	09 12 54.	+ 59 59	42	16.	GALAXY
FATH 1.257	09 12 55.	+ 14 46	14		NEBULA
IC 2451	09 12 55.	+ 23 43 06.			NONSTELLAR OBJECT
ZL 066	09 12 59.	- 00 11 24.		18.5	ULTRAFAINT BLUE STAR
MCG+03-24-017	09 13 00.	+ 15 24	30	15.1	GALAXY
ZC 0913.0+1934	09 13 00.	+ 19 34	670		CLUSTER OF GALAXIES
ZWG 121.046	09 13 00.	+ 21 09		15.5	GALAXY
ZWG 121.047	09 13 00.	+ 23 40		15.6	GALAXY
TON-N 0398	09 13 00.	+ 29 04		15.3	BLUE STAR
MCG+10-13-072	09 13 00.	+ 58 17	18	16.	GALAXY
ARC 0774	09 13 01.	+ 05 45		16.9	RICH CLUSTER OF GALAXIES
TON-N 1041	09 13 01.	+ 19 49		16.5	BLUE STAR
TON-N 1042	09 13 01.	+ 21 29		17.	BLUE STAR
FATH 1.258	09 13 03.	+ 13 57	11		NEBULA
FATH 1.259	09 13 05.	+ 13 47	27		NEBULA
FATH 1.260	09 13 05.	+ 14 02	5		NEBULA
IC 2453	09 13 05.	+ 21 07 53.			NONSTELLAR OBJECT
IC 2452	09 13 05.	+ 23 41 41.			NONSTELLAR OBJECT
TON-N 1043	09 13 05.	+ 35 33		15.8	BLUE STAR
ZC 0913.1+3651	09 13 06.	+ 36 51	1410		CLUSTER OF GALAXIES
ZWG 209.038	09 13 06.	+ 41 45		15.7	GALAXY
ZC 0913.1+4205	09 13 06.	+ 42 05	2960		CLUSTER OF GALAXIES
ZWG 209.039	09 13 06.	+ 43 12		15.3	GALAXY
UGC 04882	09 13 06.	+ 43 12	66	15.3	GALAXY
ZWG 264.089	09 13 06.	+ 53 43		15.5	GALAXY
ZC 0913.1+6358	09 13 06.	+ 63 58	2890		CLUSTER OF GALAXIES
ZWG 350.019	09 13 06.	+ 74 33		12.8	GALAXY
UGC 04883	09 13 06.	+ 74 33	84	12.8	GALAXY S
ZWG 006.022	09 13 06.	- 01 08		15.6	GALAXY
ARC 0764	09 13 08.	+ 64 03		17.4	RICH CLUSTER OF GALAXIES
ZL 067	09 13 08.	- 00 08 48.		17.3	ULTRAFAINT BLUE STAR
MCG+07-19-045	09 13 09.	+ 41 44	9	16.	GALAXY
FATH 1.261	09 13 11.	+ 13 49	14		NEBULA
ZWG 062.011	09 13 12.	+ 10 20		14.7	GALAXY
UGC 04884	09 13 12.	+ 10 20	60	14.7	GALAXY Sa-b
MCG+02-24-004	09 13 12.	+ 10 20	60	14.7	GALAXY
ZWG 091.037	09 13 12.	+ 17 48		14.0	GALAXY
RNGC 2794	09 13 12.	+ 17 48		14.0	GALAXY
UGC 04885	09 13 12.	+ 17 48	96	14.0	GALAXY S
MCG+03-24-018	09 13 12.	+ 17 48	36	14.0	GALAXY
ZWG 091.038	09 13 12.	+ 18 02		14.4	GALAXY
UGC 04886	09 13 12.	+ 18 02	72	14.4	GALAXY Sa?
ZWG 151.039	09 13 12.	+ 31 01		15.6	GALAXY
ZWG 209.040	09 13 12.	+ 42 20		15.5	GALAXY
ZWG 264.090	09 13 12.	+ 53 39		14.7	GALAXY
MRK 104	09 13 12.	+ 53 39	13	15.	GALAXY WITH UV CONTINUUM
MCG+12-09-035	09 13 12.	+ 73 59	204	12.5	GALAXY
ZL 068	09 13 12.	- 00 03 12.		17.4	ULTRAFAINT BLUE STAR
ZL 069	09 13 12.	- 00 10 42.		19.6	ULTRAFAINT BLUE STAR
ZL 070	09 13 12.	- 00 13 42.		20.0	ULTRAFAINT BLUE STAR
MCG-01-24-003	09 13 12.	- 06 15	78	15.5	GALAXY
MCG-05-22-005	09 13 12.	- 32 43	48	16.	GALAXY
RNGC 2836	09 13 12.	- 69 08			GALAXY
IC 2454	09 13 13.	+ 18 00 46.			NONSTELLAR OBJECT
TON-N 1044	09 13 13.	+ 23 01		16.5	BLUE STAR
MCG+03-24-019	09 13 15.	+ 18 01	66	14.4	GALAXY
MCG+07-19-046	09 13 15.	+ 42 20	9	15.	GALAXY
MCG+07-19-047	09 13 15.	+ 43 11	63	15.	GALAXY
MCG+07-19-048	09 13 15.	+ 43 12	24	16.5	GALAXY
IC 0529	09 13 17.	+ 73 57			NONSTELLAR OBJECT
ZWG 091.039	09 13 18.	+ 17 50		14.1	GALAXY
RNGC 2795	09 13 18.	+ 17 50		14.0	GALAXY
UGC 04887	09 13 18.	+ 17 50	102	14.1	GALAXY E
MCG+03-24-020	09 13 18.	+ 17 50	30	14.1	GALAXY
MCG+03-24-021	09 13 18.	+ 17 56	48	15.7	GALAXY
ZWG 091.040	09 13 18.	+ 17 57		15.7	GALAXY
MCG+05-22-028	09 13 18.	+ 29 06	30	16.	GALAXY
ZWG 181.005	09 13 18.	+ 33 02		15.7	GALAXY
ZWG 332.038	09 13 18.	+ 73 57		12.0	GALAXY
UGC 04888	09 13 18.	+ 73 57	234	12.0	GALAXY Sc
MCG-02-24-004	09 13 18.	- 13 25	36	15.5	GALAXY
LB 03013	09 13 21.	+ 10 23 42.		16.4	FAINT BLUE STAR
ZWG 091.041	09 13 24.	+ 17 52		15.5	GALAXY
MCG+03-24-022	09 13 24.	+ 17 52	24	15.5	GALAXY
TON-N 0399	09 13 24.	+ 26 58		15.2	BLUE STAR
TON-N 0400	09 13 24.	+ 31 18		14.8	BLUE STAR
MCG+07-19-049	09 13 24.	+ 40 03	78	14.5	GALAXY
ZWG 209.041	09 13 24.	+ 40 05		15.0	GALAXY
UGC 04889	09 13 24.	+ 40 05	96	15.0	GALAXY S
MCG+07-19-050	09 13 24.	+ 41 54	36	15.	GALAXY
ZWG 209.042	09 13 24.	+ 41 55		15.3	GALAXY
MCG+08-17-066	09 13 24.	+ 46 00 30.	18	17.	GALAXY
ZWG 238.026	09 13 24.	+ 49 59		14.9	GALAXY
ZC 0913.4+5258	09 13 24.	+ 52 58	1340		CLUSTER OF GALAXIES
RNGC 2822	09 13 24.	- 69 26			GALAXY
FATH 1.263	09 13 26.	- 15 09	5		NEBULA
ARC 0772	09 13 27.	+ 36 51		17.7	RICH CLUSTER OF GALAXIES
MCG+08-17-067	09 13 27.	+ 46 00	24	16.	GALAXY
FATH 1.264	09 13 27.	- 15 08	8		NEBULA
ZWG 034.019	09 13 30.	+ 06 33		14.9	GALAXY
UGC 04890	09 13 30.	+ 06 33	72	14.9	GALAXY SBc
MCG+01-24-009	09 13 30.	+ 06 33	48	14.9	GALAXY
SHAH 111	09 13 30.	+ 27 54	210	16.5	GROUP OF COMPACT GALAXIES
ZWG 209.043	09 13 30.	+ 43 55		15.3	GALAXY
MCG+08-17-068	09 13 30.	+ 45 54	42	15.	GALAXY
ZWG 238.027	09 13 30.	+ 46 02		15.4	GALAXY
MCG+08-17-069	09 13 30.	+ 49 58	36	15.	GALAXY
MRSL 258+12/1	09 13 30.	- 31 10	2700		HII REGION
FATH 1.265	09 13 32.	- 14 51	8		NEBULA
ZWG 006.023	09 13 36.	+ 00 55		15.6	GALAXY
FATH 1.262	09 13 36.	+ 14 58	5		NEBULA
ZWG 091.042	09 13 36.	+ 17 56		14.3	GALAXY
RNGC 2797	09 13 36.	+ 17 56		14.5	GALAXY
UGC 04891	09 13 36.	+ 17 56	33	14.3	GALAXY PECULR
MCG+03-24-023	09 13 36.	+ 17 56	36	14.3	GALAXY
ZWG 121.048	09 13 36.	+ 25 30		15.7	GALAXY
ZWG 238.028	09 13 36.	+ 45 55		14.8	GALAXY

OBJECT NAME	RIGHT ASCEN.	DECLINATION	DIAM.	MAGN.	TYPE OF OBJECT
UGC 04892	09 13 36.	+ 45 55	72	14.8	GALAXY PECULR
ZWG 238.029	09 13 36.	+ 50 15		14.9	GALAXY
MCG+10-14-001	09 13 36.	+ 62 24	30	17.	GALAXY
ZC 0913.6-0010	09 13 36.	- 00 10	540		CLUSTER OF GALAXIES
MCG-02-24-005	09 13 36.	- 13 46	48	15.	GALAXY
MCG-06-21-001	09 13 36.	- 35 26	72	14.	GALAXY
MCG+07-19-051	09 13 37.	+ 43 55	30	15.5	GALAXY
MCG+06-21-002	09 13 39.	+ 34 40	60	13.5	GALAXY
MCG+08-17-070	09 13 39.	+ 50 14 30.	36	15.	GALAXY
PK275-04.1	09 13 39.	- 54 40 10.	10		PLANETARY NEBULA
RNGC 2793	09 13 40.	+ 34 38		13.5	GALAXY
ARC 0776	09 13 40.	- 00 12		17.9	RICH CLUSTER OF GALAXIES
MCG+07-19-052	09 13 41.	+ 43 55 30.	9	16.	GALAXY
ZWG 034.020	09 13 42.	+ 07 18		15.5	GALAXY
ZWG 062.012	09 13 42.	+ 10 26		15.6	GALAXY
ZC 0913.7+1620	09 13 42.	+ 16 20	1010		CLUSTER OF GALAXIES
ZWG 151.040	09 13 42.	+ 27 34		15.7	GALAXY
ZWG 151.041	09 13 42.	+ 29 33		15.6	GALAXY
ZWG 151.042	09 13 42.	+ 31 07		14.6	GALAXY
RNGC 2796	09 13 42.	+ 31 07		14.5	GALAXY
UGC 04893	09 13 42.	+ 31 07	84	14.6	GALAXY Sa?
MCG+05-22-029	09 13 42.	+ 31 08	60	14.6	GALAXY
ZWG 181.006	09 13 42.	+ 34 39		13.9	GALAXY
UGC 04894	09 13 42.	+ 34 39	78	13.9	GALAXY SB
MCG-01-24-005	09 13 42.	- 06 22	24	16.	GALAXY
MCG-01-24-004	09 13 42.	- 08 58	90	15.	GALAXY
ARC 0775	09 13 43.	+ 06 05		17.4	RICH CLUSTER OF GALAXIES
HOLM 115A	09 13 43.	+ 31 09	48	14.7	PART OF MULTIPLE GALAXY
MCG+07-19-053	09 13 43.	+ 43 55	18	16.	GALAXY
MCG+03-24-024	09 13 45.	+ 18 08	54	14.8	GALAXY
HOLM 116A	09 13 45.	+ 27 34	36	14.0	PART OF MULTIPLE GALAXY
ZL 071	09 13 45.	- 00 17 12.		21.4	ULTRAFAINT BLUE STAR
HOLM 115B	09 13 47.	+ 31 09	18	15.5	PART OF MULTIPLE GALAXY
PK227+33.1	09 13 48.	+ 04 06	134	16.1	PLANETARY NEBULA
ZWG 062.013	09 13 48.	+ 13 19		15.5	GALAXY
ZWG 091.043	09 13 48.	+ 18 08		14.8	GALAXY
MCG+05-22-030	09 13 48.	+ 27 33	36	15.7	GALAXY
HOLM 116B	09 13 48.	+ 27 35	18	14.9	PART OF MULTIPLE GALAXY
ZWG 151.043	09 13 48.	+ 27 42		14.9	GALAXY
UGC 04895	09 13 48.	+ 27 42	78	14.9	GALAXY Sb
ZC 0913.8+2839	09 13 48.	+ 28 39	1080		CLUSTER OF GALAXIES
ZC 0913.8+3750	09 13 48.	+ 37 50	540		CLUSTER OF GALAXIES
MCG+10-14-002	09 13 48.	+ 62 24	24	16.	GALAXY
ZWG 332.039	09 13 48.	+ 70 01		14.8	GALAXY
UGC 04896	09 13 48.	+ 70 01	126	14.8	GALAXY Sa-b
MCG-03-24-003	09 13 48.	- 16 06	120	12.5	GALAXY
MCG-05-22-006	09 13 48.	- 31 58	36	16.	GALAXY
ZL 072	09 13 50.	- 00 11 54.		20.5	ULTRAFAINT BLUE STAR
RNGC 2801	09 13 53.	+ 20 08		15.5	GALAXY
ZC 0913.9+0610	09 13 54.	+ 06 10	1140		CLUSTER OF GALAXIES
HOLM 118B	09 13 54.	+ 07 27	18	14.7	PART OF MULTIPLE GALAXY
ZC 0913.9+1652	09 13 54.	+ 16 52	740		CLUSTER OF GALAXIES
ZWG 091.044	09 13 54.	+ 19 10		14.3	GALAXY
RNGC 2803	09 13 54.	+ 19 10		14.5	GALAXY
RNGC 2802	09 13 54.	+ 19 10		14.5	GALAXY
UGC 04898	09 13 54.	+ 19 10	102	14.3	GALAXY DBL SYS
UGC 04897	09 13 54.	+ 19 10	102	14.3	GALAXY DBL SYS
KARA.72 194A	09 13 54.	+ 19 11	42	14.3	PART OF DOUBLE GALAXY
MCG+03-24-026	09 13 54.	+ 19 11	48	14.3	GALAXY
ZWG 091.045	09 13 54.	+ 20 03		15.6	GALAXY
ZWG 091.046	09 13 54.	+ 20 08		15.4	GALAXY
UGC 04899	09 13 54.	+ 20 08	72	15.4	GALAXY Sc
MCG+03-24-025	09 13 54.	+ 20 10	66	15.4	GALAXY
MCG+05-22-031	09 13 54.	+ 27 41	66	14.9	GALAXY
ZWG 151.044	09 13 54.	+ 28 16		15.7	GALAXY
ZC 0913.9+3731	09 13 54.	+ 37 31	1080		CLUSTER OF GALAXIES
MCG+08-17-071	09 13 54.	+ 44 40	42	16.	GALAXY
FATH 1.266	09 13 54.	- 15 40	8		NEBULA
ARC 0762	09 13 55.	+ 74 31		16.2	RICH CLUSTER OF GALAXIES
RNGC 2811	09 13 56.	- 16 06		13.0	GALAXY
MCG+03-24-027	09 13 57.	+ 19 10 30.	66	14.3	GALAXY
HOLM 118A	09 13 58.	+ 07 29	54	13.9	PART OF MULTIPLE GALAXY
RNGC 2804	09 13 59.	+ 20 24		14.0	GALAXY
RNGC 2818A	09 13 59.	- 36 24		10.0	PLANETARY NEBULA
RNGC 2818	09 13 59.	- 36 24		10.0	OPEN CLUSTER
LBN 1086	09 14	- 31 00	3120		BRIGHT NEBULA
ZWG 034.021	09 14 00.	+ 07 29		14.8	GALAXY
UGC 04900	09 14 00.	+ 07 29	84	14.8	GALAXY Sb-c
MCG+01-24-010	09 14 00.	+ 07 29	72	14.8	GALAXY
KARA.72 194B	09 14 00.	+ 19 10	54		PART OF DOUBLE GALAXY
IC 2455	09 14 00.	+ 20 19 36.			NONSTELLAR OBJECT
ZWG 091.047	09 14 00.	+ 20 24		14.0	GALAXY
UGC 04901	09 14 00.	+ 20 24	102	14.0	GALAXY S0
MCG+03-24-028	09 14 00.	+ 20 26	114	14.0	GALAXY
ZWG 151.045	09 14 00.	+ 29 34		15.7	GALAXY
ZC 0914.0+3814	09 14 00.	+ 38 14	870		CLUSTER OF GALAXIES
ZWG 238.030	09 14 00.	+ 44 42		15.1	GALAXY
MCG+09-15-114	09 14 00.	+ 53 10	108	14.	GALAXY
MCG+12-09-036	09 14 00.	+ 70 02	84	15.	GALAXY
ZWG 332.040	09 14 00.	+ 71 56		15.1	GALAXY
MCG+12-09-037	09 14 00.	+ 71 57	33	15.	GALAXY
ZWG 364.010	09 14 00.	+ 81 47		15.6	GALAXY
ZWG 363.050	09 14 00.	+ 81 47		15.6	GALAXY
KARA.73B 0320	09 14 00.	+ 81 47	36	15.6	ISOLATED GALAXY S
OCL 0743	09 14 00.	- 36 24	840	10.0	OPEN STAR CLUSTER
VHA 059	09 14 00.	- 36 24	360		OPEN STAR CLUSTER
PK261+08.1	09 14 00.	- 36 24 16.	40	13.0	PLANETARY NEBULA
ZL 073	09 14 04.	- 00 12 30.		21.8	ULTRAFAINT BLUE STAR
FATH 1.267	09 14 04.	- 15 39	14		NEBULA
RNGC 2815	09 14 05.	- 23 24		13.0	GALAXY
ZWG 006.024	09 14 06.	+ 00 25		15.4	GALAXY
ZWG 062.014	09 14 06.	+ 10 06		15.3	GALAXY
ZWG 062.015	09 14 06.	+ 11 40		15.4	GALAXY
ZWG 091.048	09 14 06.	+ 17 17		15.4	GALAXY
MCG+03-24-029	09 14 06.	+ 17 40	90	15.0	GALAXY
ZWG 091.049	09 14 06.	+ 17 41		15.0	GALAXY
ZWG 121.049	09 14 06.	+ 25 38		14.0	GALAXY
UGC 04902	09 14 06.	+ 25 38	90	14.0	GALAXY S0?
MCG+05-22-032	09 14 06.	+ 26 58	12	15.5	GALAXY
ZWG 151.046	09 14 06.	+ 26 59		15.5	GALAXY
SHAH 112	09 14 06.	+ 28 13	48	17.4	GROUP OF COMPACT GALAXIES
ZWG 151.047	09 14 06.	+ 29 35		15.7	GALAXY
ZC 0914.1+3010	09 14 06.	+ 30 10	1550		CLUSTER OF GALAXIES
UGC 04903	09 14 06.	+ 37 10	90	16.5	GALAXY
ZWG 209.044	09 14 06.	+ 42 07		15.0	GALAXY
UGC 04904	09 14 06.	+ 42 07	66	15.0	GALAXY SB
ZWG 209.045	09 14 06.	+ 42 12		12.9	GALAXY
UGC 04905	09 14 06.	+ 42 12	168	12.9	GALAXY SBa
KARA.72 195A	09 14 06.	+ 42 12	156	12.9	PART OF DOUBLE GALAXY
ZWG 264.091	09 14 06.	+ 53 12		13.4	GALAXY
UGC 04906	09 14 06.	+ 53 12	132	13.4	GALAXY Sa
ZC 0914.1+6533	09 14 06.	+ 65 33	2690		CLUSTER OF GALAXIES
MCG-04-22-006	09 14 06.	- 23 26	210	13.	GALAXY
VHA 060	09 14 06.	- 49 48	150		OPEN STAR CLUSTER
HOLM 117A	09 14 08.	+ 42 13	36	13.2	PART OF MULTIPLE GALAXY
ZL 074	09 14 08.	- 00 15 24.		21.0	ULTRAFAINT BLUE STAR
MCG+07-19-054	09 14 09.	+ 42 06 30.	54	14.5	GALAXY
RNGC 2807	09 14 11.	+ 20 14		15.0	GALAXY
RNGC 2806	09 14 11.	+ 20 15		15.0	GALAXY
RNGC 2798	09 14 11.	+ 42 13		13.0	GALAXY
ZWG 091.050	09 14 12.	+ 16 31		15.5	GALAXY
UGC 04907	09 14 12.	+ 16 31	66	15.5	GALAXY Sc
ZWG 091.051	09 14 12.	+ 20 14		15.1	GALAXY
MCG+03-24-030	09 14 12.	+ 20 15 30.	24	15.1	GALAXY
MCG+05-22-034	09 14 12.	+ 27 33	15	14.7	GALAXY
ZWG 151.048	09 14 12.	+ 32 13		15.7	GALAXY
UGC 04908	09 14 12.	+ 32 13	60	15.7	GALAXY Sb
MCG+05-22-033	09 14 12.	+ 32 14	36	15.7	GALAXY
MCG+07-19-055	09 14 12.	+ 42 11 30.	150	14.	GALAXY
ZWG 209.046	09 14 12.	+ 42 12		14.4	GALAXY
UGC 04909	09 14 12.	+ 42 12	126	14.4	GALAXY S
KARA.72 195B	09 14 12.	+ 42 12	108	14.4	PART OF DOUBLE GALAXY
MCG+09-15-115	09 14 12.	+ 53 30	36	16.	GALAXY
ZC 0914.2+6124	09 14 12.	+ 61 24	1810		CLUSTER OF GALAXIES
MCG+11-12-002	09 14 12.	+ 65 18	10	18.	GALAXY
ARC 0770	09 14 13.	+ 60 39		17.7	RICH CLUSTER OF GALAXIES
ZL 075	09 14 13.	- 00 17 30.		19.6	ULTRAFAINT BLUE STAR
IC 2457	09 14 14.	+ 20 18 17.			NONSTELLAR OBJECT
MCG+03-24-032	09 14 15.	+ 16 30	60	15.5	GALAXY
MCG+03-24-031	09 14 15.	+ 20 16	21	15.1	GALAXY
HOLM 119A	09 14 15.	+ 27 34	18	14.7	PART OF MULTIPLE GALAXY
HOLM 119B	09 14 16.	+ 27 35	12	14.8	PART OF MULTIPLE GALAXY
RNGC 2809	09 14 17.	+ 20 16		14.0	GALAXY
RNGC 2799	09 14 17.	+ 42 12		14.5	GALAXY
HOLM 117B	09 14 17.	+ 42 13	54	14.1	PART OF MULTIPLE GALAXY
ZWG 006.025	09 14 18.	+ 00 47		15.5	GALAXY
MCG+03-24-034	09 14 18.	+ 17 37	36	15.4	GALAXY
ZWG 091.052	09 14 18.	+ 17 38		15.4	GALAXY
ZWG 091.053	09 14 18.	+ 20 05		15.6	GALAXY
ZWG 091.054	09 14 18.	+ 20 16		13.9	GALAXY
UGC 04910	09 14 18.	+ 20 16	102	13.9	GALAXY S0
MCG+03-24-033	09 14 18.	+ 20 18	72	13.9	GALAXY
MCG+04-22-029	09 14 18.	+ 25 37	78	14.0	GALAXY
ZWG 151.049	09 14 18.	+ 29 14		15.7	GALAXY
ZWG 209.047	09 14 18.	+ 40 08		15.4	GALAXY
ZC 0914.3+4035	09 14 18.	+ 40 35	1410		CLUSTER OF GALAXIES
MCG+07-19-056	09 14 18.	+ 42 11	90	14.	GALAXY
MCG+09-15-116	09 14 18.	+ 52 56	66	15.	GALAXY
ZC 0914.3+5743	09 14 18.	+ 57 43	2760		CLUSTER OF GALAXIES
MCG-03-24-004	09 14 18.	- 17 24	102	15.	GALAXY
PK268+02.1	09 14 21.	- 45 16 18.	5		PLANETARY NEBULA
HELW 422	09 14 22.	- 15 47 28.			NEBULA
ZC 0914.4+0014	09 14 24.	+ 00 14	740		CLUSTER OF GALAXIES
ZWG 091.055	09 14 24.	+ 20 21		15.0	GALAXY
ZWG 121.050	09 14 24.	+ 25 05		15.4	GALAXY
ZC 0914.4+4104	09 14 24.	+ 41 04	1010		CLUSTER OF GALAXIES
VV 050A	09 14 24.	+ 42 10 30.	60	13.0	INTERACTING GALAXY
ZWG 209.048	09 14 24.	+ 44 14		15.5	GALAXY
ZWG 264.092	09 14 24.	+ 52 56		15.3	GALAXY
UGC 04911	09 14 24.	+ 54 26	60	16.5	GALAXY Sc
MCG+13-07-022	09 14 24.	+ 75 19	51	17.	GALAXY
ZL 076	09 14 24.	- 00 21 12.		18.3	ULTRAFAINT BLUE STAR
ARP 283	09 14 25.	+ 42 10			PECULIAR GALAXY
FATH 1.268	09 14 25.	- 15 48	19		NEBULA
IC 2456	09 14 27.	+ 34 59 41.			NONSTELLAR OBJECT
ZWG 121.051	09 14 30.	+ 26 10		14.4	GALAXY
UGC 04912	09 14 30.	+ 26 10	48	14.4	GALAXY
ZC 0914.5+3122	09 14 30.	+ 31 22	470		CLUSTER OF GALAXIES
MCG+06-21-003	09 14 30.	+ 37 12	66	16.	GALAXY
ZWG 209.049	09 14 30.	+ 40 18		15.7	GALAXY
VV 050B	09 14 30.	+ 42 10	72	14.1	INTERACTING GALAXY
MCG+07-19-057	09 14 30.	+ 44 14	33	15.5	GALAXY
ZC 0914.5+5156	09 14 30.	+ 51 56	1480		CLUSTER OF GALAXIES
ARC 0773	09 14 30.	+ 51 56		17.5	RICH CLUSTER OF GALAXIES
7ZW 272	09 14 30.	+ 58 12			COMPACT GALAXY
SHAH 101	09 14 30.	+ 62 04	54	18.3	GROUP OF COMPACT GALAXIES
RNGC 2842	09 14 32.	- 62 52			UNVERIFIED SOUTHRN OBJECT
ZWG 264.093	09 14 36.	+ 55 52		15.7	GALAXY
UGC 04913	09 14 36.	+ 55 52	72	15.7	GALAXY Sc
RNGC 2821	09 14 36.	- 26 37		14.0	GALAXY
MCG-04-22-007	09 14 36.	- 26 37	72	14.5	GALAXY
ZWG 091.056	09 14 42.	+ 19 10		15.4	GALAXY
ZWG 091.057	09 14 42.	+ 19 16		15.5	GALAXY
ZWG 091.058	09 14 42.	+ 20 04		15.7	GALAXY
ZWG 151.050	09 14 42.	+ 27 08		15.6	GALAXY
ZWG 151.051	09 14 42.	+ 28 25		15.6	GALAXY
ARC 0771	09 14 42.	+ 61 24		17.7	RICH CLUSTER OF GALAXIES
MCG-01-24-006	09 14 42.	- 04 32	108	13.	GALAXY
LB 03014	09 14 44.	+ 21 03 18.		16.4	FAINT BLUE STAR
FATH 1.269	09 14 45.	+ 14 04	14		NEBULA
MCG+04-22-030	09 14 45.	+ 26 09	36	14.4	GALAXY
RNGC 2817	09 14 45.	- 04 32		13.0	GALAXY
RNGC 2812	09 14 47.	+ 20 07		15.5	GALAXY
ZWG 091.059	09 14 48.	+ 16 40		14.8	GALAXY
MCG+03-24-035	09 14 48.	+ 19 10	60	15.4	GALAXY
ZWG 091.060	09 14 48.	+ 20 07		15.7	GALAXY
MCG+08-17-072	09 14 48.	+ 45 50 30.	72	14.	GALAXY
ZC 0914.8+4827	09 14 48.	+ 48 27	740		CLUSTER OF GALAXIES
ZWG 332.041	09 14 48.	+ 69 25		11.7	GALAXY
UGC 04914	09 14 48.	+ 69 25	222	11.7	GALAXY SB0
MCG+12-09-038	09 14 48.	+ 70 01	21	16.	GALAXY
MRSL 270+00/1	09 14 48.	- 47 45	480		HII REGION
FATH 1.270	09 14 49.	- 15 41	8		NEBULA
REIN 2.081	09 14 49.59	+ 69 24 52.9			NEBULA
RNGC 2787	09 14 51.	+ 69 25		12.0	GALAXY
RNGC 2813	09 14 53.	+ 20 06		15.5	GALAXY
FATH 1.271	09 14 53.	- 15 07	14		NEBULA
MCG+03-24-036	09 14 54.	+ 16 39	66	14.8	GALAXY
ZWG 091.061	09 14 54.	+ 20 06		15.4	GALAXY
UGC 04916	09 14 54.	+ 20 06	66	15.4	GALAXY S0
ZWG 151.052	09 14 54.	+ 27 56		15.3	GALAXY
ZWG 181.007	09 14 54.	+ 34 43		15.7	GALAXY
MCG+08-17-073	09 14 54.	+ 48 09	24	15.	GALAXY
ZWG 238.031	09 14 54.	+ 48 10		15.5	GALAXY
UGC 04917	09 14 54.	+ 48 10	60	15.5	GALAXY Sc
MCG+12-09-039	09 14 54.	+ 69 24	132	11.7	GALAXY

OBJECT NAME	RIGHT ASCEN.	DECLINATION	DIAM.	MAGN.	TYPE OF OBJECT
UGC 04918	09 14 54.	+ 70 00	60	16.0	GALAXY S
ZWG 006.026	09 14 54.	- 00 25		15.0	GALAXY
UGC 04915	09 14 54.	- 00 25	66	15.0	GALAXY Sb-c
MCG+00-24-005	09 14 54.	- 00 25	60	15.0	GALAXY
PATH 1.272	09 14 57.	- 15 31	8		NEBULA
PATH 1.273	09 14 57.	- 15 33			NEBULA
MCG+03-24-037	09 15 00.	+ 20 08	66	15.4	GALAXY
ZC 0915.0+2744	09 15 00.	+ 27 44	6180		CLUSTER OF GALAXIES
MCG+05-22-035	09 15 00.	+ 27 56	12	15.3	GALAXY
ZC 0915.0+3037	09 15 00.	+ 30 37	2420		CLUSTER OF GALAXIES
ZWG 238.032	09 15 00.	+ 45 52		14.4	GALAXY
UGC 04919	09 15 00.	+ 45 52	90	14.4	GALAXY Sc
MCG+09-15-117	09 15 00.	+ 52 42	30	14.	GALAXY
ZWG 264.094	09 15 00.	+ 52 43		14.0	GALAXY
UGC 04920	09 15 00.	+ 52 43	90	14.0	GALAXY E
ZC 0915.0+5521	09 15 00.	+ 55 21	4500		CLUSTER OF GALAXIES
OCL 0772	09 15 00.	- 49 30	120	11.5	OPEN STAR CLUSTER
TON-N 1045	09 15 01.	+ 21 35		17.	BLUE STAR
RNGC 2800	09 15 02.	+ 52 43		14.0	GALAXY
YM 46	09 15 05.	- 47 39	360		SYMMETRIC GALACTIC NEBULA
ZWG 062.016	09 15 06.	+ 09 55		15.5	GALAXY
ZC 0915.1+1200	09 15 06.	+ 12 00	1810		CLUSTER OF GALAXIES
ZC 0915.1+1515	09 15 06.	+ 15 15	2020		CLUSTER OF GALAXIES
ZWG 121.052	09 15 06.	+ 20 41		15.6	GALAXY
ZC 0915.1+3903	09 15 06.	+ 39 03	1210		CLUSTER OF GALAXIES
UGC 04921	09 15 06.	+ 49 46	66	16.5	GALAXY Sc
TON-N 1046	09 15 07.	+ 21 17		17.	BLUE STAR
LB 03015	09 15 11.	+ 19 27 06.		14.2	FAINT BLUE STAR
ZWG 006.027	09 15 12.	+ 01 16		15.7	GALAXY
KARA.73B 0321	09 15 12.	+ 01 16	24	15.7	ISOLATED GALAXY S
ZC 0915.2+1005	09 15 12.	+ 10 05	670		CLUSTER OF GALAXIES
TON-N 0401	09 15 12.	+ 30 48		15.4	BLUE STAR
MCG+06-21-005	09 15 12.	+ 33 47	30	15.	GALAXY
MCG+06-21-004	09 15 12.	+ 34 06	30	16.	GALAXY
MCG+08-17-074	09 15 12.	+ 48 04 30.	180	14.	GALAXY
ZWG 238.033	09 15 12.	+ 48 05		15.1	GALAXY
UGC 04922	09 15 12.	+ 48 05	240	15.1	GALAXY S IV-V
MCG+08-17-075	09 15 12.	+ 49 44 30.	72	16.	GALAXY
MCG-01-24-007	09 15 15.	- 07 43	78	15.	GALAXY
IC 0531	09 15 17.	- 00 02 26.			NONSTELLAR OBJECT
ZC 0915.3+0059	09 15 18.	+ 00 59	540		CLUSTER OF GALAXIES
ZWG 181.008	09 15 18.	+ 33 46		15.6	GALAXY
ZWG 181.009	09 15 18.	+ 34 04		15.6	GALAXY
MCG+09-15-119	09 15 18.	+ 51 30	72	16.	GALAXY
MCG+09-15-118	09 15 18.	+ 51 33	60	15.	GALAXY
ZWG 006.028	09 15 18.	- 00 04		14.9	GALAXY
UGC 04923	09 15 18.	- 00 04	108	14.9	GALAXY SB:a-b
MCG+00-24-006	09 15 18.	- 00 04	114	14.9	GALAXY
PATH 1.274	09 15 21.	+ 13 41	27		NEBULA
MCG-02-24-006	09 15 21.	- 14 17	48	15.	GALAXY
ZC 0915.4+0740	09 15 24.	+ 07 40	1880		CLUSTER OF GALAXIES
ZWG 062.017	09 15 24.	+ 09 59		15.6	GALAXY
ZWG 062.018	09 15 24.	+ 13 41		14.8	GALAXY
MCG+02-24-005	09 15 24.	+ 13 41	36	14.8	GALAXY
MCG+03-24-040	09 15 24.	+ 16 23	24	14.3	GALAXY
ZWG 091.062	09 15 24.	+ 16 25		14.3	GALAXY
UGC 04924	09 15 24.	+ 16 25	90	14.3	GALAXY E
MCG+03-24-039	09 15 24.	+ 17 52	60	16.	GALAXY
MCG+03-24-038	09 15 24.	+ 18 59	42	15.	GALAXY
ZWG 121.053	09 15 24.	+ 20 35		15.7	GALAXY
MCG+06-21-006	09 15 24.	+ 34 06	24	16.	GALAXY
ZWG 264.095	09 15 24.	+ 51 35		15.2	GALAXY
RNGC 2819	09 15 25.	+ 16 25		14.5	GALAXY
MCG+06-21-007	09 15 27.	+ 34 47	66	15.	GALAXY
HOLM 120B	09 15 28.	+ 34 45	42	15.1	PART OF MULTIPLE GALAXY
ARC 0778	09 15 28.	- 08 07		17.8	RICH CLUSTER OF GALAXIES
HOLM 120A	09 15 29.	+ 34 46	78	14.1	PART OF MULTIPLE GALAXY
KEEL 162	09 15 29.7	+ 51 34 50.			NEBULA
ZWG 034.022	09 15 30.	+ 07 31		15.5	GALAXY
ZWG 091.063	09 15 30.	+ 17 58		15.4	GALAXY
UGC 04925	09 15 30.	+ 17 58	108	15.4	GALAXY Sc
ZWG 091.064	09 15 30.	+ 20 28		15.5	GALAXY
MCG+03-24-041	09 15 30.	+ 20 29	30	15.5	GALAXY
ZWG 181.010	09 15 30.	+ 32 28		15.6	GALAXY
ZWG 151.053	09 15 30.	+ 32 28		15.6	GALAXY
ZWG 181.011	09 15 30.	+ 34 04		15.6	GALAXY
ZWG 181.012	09 15 30.	+ 34 30		15.5	GALAXY
ZWG 181.013	09 15 30.	+ 34 46		15.4	GALAXY
UGC 04926	09 15 30.	+ 34 46	102	15.4	GALAXY Sb?
ZWG 238.034	09 15 30.	+ 50 13		14.7	GALAXY
UGC 04927	09 15 30.	+ 50 13	78	14.7	GALAXY E-S0
MCG+08-17-076	09 15 30.	+ 50 13	36	15.	GALAXY
MCG+08-17-077	09 15 30.	+ 50 23	30	16.	GALAXY
ZWG 264.096	09 15 30.	+ 51 32		15.3	GALAXY
UGC 04928	09 15 30.	+ 51 32	60	15.3	GALAXY SBb
KEEL 163	09 15 31.7	+ 51 32 02.		16.	NEBULA
MCG+03-24-042	09 15 33.	+ 17 58	102	15.4	GALAXY
MCG+03-24-043	09 15 36.	+ 16 30	36	15.2	GALAXY
ZWG 091.065	09 15 36.	+ 16 31		15.2	GALAXY
ZWG 121.054	09 15 36.	+ 20 37		15.7	GALAXY
ZC 0915.6+3409	09 15 36.	+ 34 09	5650		CLUSTER OF GALAXIES
ZWG 209.050	09 15 36.	+ 41 19		15.4	GALAXY
ZC 0915.6+5258	09 15 36.	+ 52 58	3760		CLUSTER OF GALAXIES
ZWG 264.097	09 15 36.	+ 55 03		15.4	GALAXY
MCG+09-15-120	09 15 36.	+ 55 03 30.	36	16.	GALAXY
7ZW 273	09 15 36.	+ 59 14			COMPACT GALAXY
UGC 04929	09 15 36.	+ 67 22	78	17.	GALAXY Sb-c
MRK 105	09 15 36.	+ 71 37	8	16.	GALAXY WITH UV CONTINUUM
VVI 28	09 15 36.	+ 71 37	8	16.49	SEYFERT GALAXY
ARC 0765	09 15 37.	+ 74 03		16.9	RICH CLUSTER OF GALAXIES
RNGC 2835	09 15 40.	- 22 08		11.5	GALAXY
SVEN 228	09 15 40.	- 22 09	390	11.2	GALAXY
KEEL 164	09 15 41.8	+ 51 24 24.		17.	NEBULA
TON-N 1047	09 15 42.	+ 33 20		16.7	BLUE STAR
ZWG 181.014	09 15 42.	+ 34 35		15.7	GALAXY
MCG+07-19-059	09 15 42.	+ 41 17 30.	9	17.	GALAXY
MCG+07-19-058	09 15 42.	+ 41 17 30.	15	17.	GALAXY
ZWG 209.051	09 15 42.	+ 42 13		15.6	GALAXY
ZC 0915.7+5815	09 15 42.	+ 58 15	810		CLUSTER OF GALAXIES
HMS 1.11	09 15 42.	- 11 53			S0 GALAXY
MCG-02-24-007	09 15 42.	- 11 53	36	15.5	GALAXY
MCG-04-22-008	09 15 42.	- 22 09 30.	360	12.	GALAXY
LB 03016	09 15 43.	+ 20 06 54.		16.0	FAINT BLUE STAR
PATH 1.275	09 15 46.	- 15 37	22		NEBULA
ZWG 151.054	09 15 48.	+ 28 10		15.7	GALAXY
ZWG 238.035	09 15 48.	+ 49 10		15.4	GALAXY
UGC 04930	09 15 48.	+ 49 10	60	15.4	GALAXY Sc
MCG+08-17-078	09 15 48.	+ 49 10	36	15.	GALAXY
UGC 04966	09 15 48.	+ 51 12	456	9.9	GALAXY Sb
PATH 1.276	09 15 51.	- 15 43	14		NEBULA
MCG-03-24-005	09 15 51.	- 18 14 30.	96	14.5	GALAXY
ZC 0915.9+0225	09 15 54.	+ 02 25	1140		CLUSTER OF GALAXIES
8ZW 0915+09.9	09 15 54.	+ 09 52		17.7	COMPACT GALAXY
MCG+03-24-044	09 15 54.	+ 16 41	36	15.0	GALAXY
ZWG 091.066	09 15 54.	+ 16 42		15.0	GALAXY
ZWG 121.055	09 15 54.	+ 20 57		15.6	GALAXY
ZWG 151.055	09 15 54.	+ 32 03		15.5	GALAXY
ZC 0915.9+4030	09 15 54.	+ 40 30	740		CLUSTER OF GALAXIES
RNGC 2837	09 15 56.	- 16 16			NON-EXISTENT OBJECT
RNGC 2520	09 15 58.	- 34 38			NON-EXISTENT OBJECT
ZWG 062.019	09 16 00.	+ 14 02		15.0	GALAXY
UGC 04931	09 16 00.	+ 14 02	66	15.0	GALAXY S0-a
MCG+02-24-006	09 16 00.	+ 14 02	48	15.0	GALAXY
ZWG 091.067	09 16 00.	+ 20 22		15.7	GALAXY
ZWG 121.056	09 16 00.	+ 20 56		15.6	GALAXY
ZWG 151.056	09 16 00.	+ 28 36		15.7	GALAXY
ZC 0916.0+4050	09 16 00.	+ 40 50	540		CLUSTER OF GALAXIES
ZWG 238.036	09 16 00.	+ 49 22		15.6	GALAXY
MCG+08-17-079	09 16 00.	+ 49 22	9	16.	GALAXY
ZWG 264.098	09 16 00.	+ 51 19		15.4	GALAXY
UGC 04932	09 16 00.	+ 51 19	108	15.4	GALAXY
ZC 0916.0+5634	09 16 00.	+ 56 34	940		CLUSTER OF GALAXIES
MCG+14-05-004	09 16 00.	+ 85 34	36	17.	GALAXY
HOLM 122B	09 16 02.	- 16 17	18	14.1	PART OF MULTIPLE GALAXY
IC 2459	09 16 03.	+ 35 05 27.			NONSTELLAR OBJECT
HOLM 122A	09 16 03.	- 16 17	18	14.1	PART OF MULTIPLE GALAXY
ARC 0780	09 16 04.	- 12 04		16.6	RICH CLUSTER OF GALAXIES
ZWG 091.068	09 16 06.	+ 14 31		15.6	GALAXY
ZWG 121.057	09 16 06.	+ 26 29		14.3	GALAXY
MRK 394	09 16 06.	+ 26 29	20	16.	GALAXY WITH UV CONTINUUM
UGC 04933	09 16 06.	+ 26 29	72	14.3	GALAXY S0
ZWG 151.057	09 16 06.	+ 28 03		15.6	GALAXY
MCG+06-21-008	09 16 06.	+ 34 15	42	15.	GALAXY
ZC 0916.1+3657	09 16 06.	+ 36 57	3700		CLUSTER OF GALAXIES
MCG+09-15-121	09 16 06.	+ 51 03	66	16.	GALAXY
ZWG 264.099	09 16 06.	+ 51 05		15.7	GALAXY
UGC 04934	09 16 06.	+ 51 05	60	15.7	GALAXY Sb
MCG+09-15-122	09 16 06.	+ 51 17 30.	78	15.	GALAXY
ZC 0916.1+6448	09 16 06.	+ 64 48	1410		CLUSTER OF GALAXIES
MCG+13-07-023	09 16 06.	+ 75 05	21	16.	GALAXY
MCG+13-07-024	09 16 06.	+ 75 20	72	15.	GALAXY
KEEL 165	09 16 07.1	+ 51 19 11.		16.	NEBULA
RNGC 2824	09 16 08.	+ 26 29		14.5	GALAXY
MCG+09-15-123	09 16 09.	+ 51 02	30	16.	GALAXY
KEEL 166	09 16 09.0	+ 51 05 28.		16.	NEBULA
RNGC 2827	09 16 10.	+ 34 04		15.5	GALAXY
RNGC 2823	09 16 10.	+ 34 13		15.5	GALAXY
ZWG 121.058	09 16 12.	+ 21 44		15.7	GALAXY
MCG+06-21-010	09 16 12.	+ 33 58 30.	48	15.	GALAXY
ZWG 181.015	09 16 12.	+ 34 04		15.6	GALAXY
MCG+06-21-009	09 16 12.	+ 34 07	42	15.	GALAXY
ZWG 181.016	09 16 12.	+ 34 13		15.7	GALAXY
UGC 04935	09 16 12.	+ 34 13	66	15.7	GALAXY SBa
ZWG 264.100	09 16 12.	+ 51 04		15.7	GALAXY
MCG-02-24-008	09 16 12.	- 12 47	48	15.5	GALAXY
LB 03017	09 16 13.	+ 19 47 24.		16.1	FAINT BLUE STAR
KEEL 167	09 16 13.9	+ 51 04 02.		16.	SPIRAL NEBULA
HOLM 121A	09 16 14.	+ 06 05	30	14.4	PART OF MULTIPLE GALAXY
MCG+04-22-031	09 16 15.	+ 26 27	30	14.3	GALAXY
MCG+06-21-011	09 16 15.	+ 33 51 30.	78	15.	GALAXY
MCG-02-24-009	09 16 15.	- 11 31	48	15.	GALAXY
RNGC 2805	09 16 16.	+ 64 19		12.0	GALAXY
RNGC 2825	09 16 17.	+ 33 57		15.5	GALAXY
IC 2460	09 16 17.	+ 34 03 55.			MAY NOT EXIST
ZWG 034.023	09 16 18.	+ 03 00		15.7	GALAXY
ZWG 034.024	09 16 18.	+ 06 06		14.9	GALAXY
MCG+01-24-011	09 16 18.	+ 06 06	24	14.9	GALAXY
ZWG 181.017	09 16 18.	+ 33 57		15.3	GALAXY
MCG+08-17-080	09 16 18.	+ 48 27	24	16.	GALAXY
ZWG 312.002	09 16 18.	+ 64 19		11.9	GALAXY
UGC 04936	09 16 18.	+ 64 19	450	11.9	GALAXY Sc
MCG+13-07-025	09 16 18.	+ 75 06	9	16.	GALAXY
ZWG 350.020	09 16 18.	+ 75 21		15.2	GALAXY
UGC 04937	09 16 18.	+ 75 21	84	15.2	GALAXY
HOLM 121B	09 16 20.	+ 06 05	24	15.1	PART OF MULTIPLE GALAXY
MAI 039	09 16 22.	+ 75 57	74		DWARF SPHEROIDAL GALAXY
RNGC 2826	09 16 23.	+ 33 50		14.5	GALAXY
HOLM 124B	09 16 23.	+ 64 17	270	12.0	PART OF MULTIPLE GALAXY
ZWG 091.069	09 16 24.	+ 20 27		15.4	GALAXY
UGC 04938	09 16 24.	+ 20 27	78	15.4	GALAXY Sb
MCG+03-24-045	09 16 24.	+ 20 28 30.	72	15.4	GALAXY
TON-N 0402	09 16 24.	+ 27 50		16.9	BLUE STAR
ZWG 181.018	09 16 24.	+ 33 50		14.6	GALAXY
UGC 04939	09 16 24.	+ 33 50	90	14.6	GALAXY S0-a
ZWG 181.019	09 16 24.	+ 34 31		15.6	GALAXY
ZWG 209.052	09 16 24.	+ 39 11		15.7	GALAXY
MRK 106	09 16 24.	+ 55 33	5	16.	GALAXY WITH UV CONTINUUM
VVI 29	09 16 24.	+ 55 33	5	16.44	SEYFERT GALAXY
MCG+06-21-012	09 16 27.	+ 33 39 30.	36	15.	GALAXY
PK275-03.1	09 16 27.	- 54 27	5		PLANETARY NEBULA
RNGC 2828	09 16 28.	+ 34 05		15.5	GALAXY
RNGC 2829	09 16 29.	+ 33 52			GALAXY
ZWG 034.025	09 16 30.	+ 05 39		15.7	GALAXY
ZWG 034.026	09 16 30.	+ 06 06		15.5	GALAXY
ZWG 151.058	09 16 30.	+ 27 08		15.4	GALAXY
TON-N 0403	09 16 30.	+ 28 48		16.2	BLUE STAR
ZWG 181.020	09 16 30.	+ 32 52		15.7	GALAXY
ZWG 181.021	09 16 30.	+ 34 05		15.7	GALAXY
MCG+09-15-124	09 16 30.	+ 54 16	30	18.	GALAXY
SHAH 102	09 16 30.	+ 61 29	132	17.7	GROUP OF COMPACT GALAXIES
MCG+11-12-003	09 16 30.	+ 64 19	348	11.6	GALAXY
7ZW 274	09 16 30.	+ 70 57			COMPACT GALAXY
MCG+13-07-026	09 16 30.	+ 75 05	66	16.	GALAXY
BIGC 507	09 16 32.	+ 34 08			NEBULA
LB 03018	09 16 33.	+ 19 00 42.		15.0	FAINT BLUE STAR
MCG+08-17-081	09 16 33.	+ 48 17	60	16.	GALAXY
ZC 0916.6+0121	09 16 36.	+ 01 21	1950		CLUSTER OF GALAXIES
ZWG 151.059	09 16 36.	+ 27 40		14.8	GALAXY
UGC 04940	09 16 36.	+ 27 40	78	14.8	GALAXY Sa/SBb
ZWG 181.022	09 16 36.	+ 33 38		15.6	GALAXY
MCG+06-21-015	09 16 36.	+ 33 57 30.	30	14.	GALAXY
MCG+06-21-014	09 16 36.	+ 33 57 30.	60	15.	GALAXY
MCG+06-21-013	09 16 36.	+ 33 58 30.	42	15.	GALAXY
ZC 0916.6+3605	09 16 36.	+ 36 05	1280		CLUSTER OF GALAXIES
ZWG 265.001	09 16 36.	+ 54 19		15.6	GALAXY
ZWG 264.101	09 16 36.	+ 54 19		15.6	GALAXY

OBJECT NAME	RIGHT ASCEN.	DECLINATION	DIAM.	MAGN.	TYPE OF OBJECT
RNGC 2845	09 16 36.	- 37 48			GALAXY
REIN 2.082	09 16 38.80	+ 33 57 01.6			NEBULA
HOLM 123B	09 16 39.	+ 33 56	54	14.2	PART OF MULTIPLE GALAXY
RNGC 2831	09 16 41.	+ 33 57		14.5	GALAXY
RNGC 2830	09 16 41.	+ 33 57		15.5	GALAXY
RNGC 2832	09 16 41.	+ 33 58		13.0	GALAXY
ZWG 034.027	09 16 42.	+ 08 14		15.4	GALAXY
ZWG 091.070	09 16 42.	+ 19 50		15.5	GALAXY
MCG+05-22-036	09 16 42.	+ 27 40	60	14.8	GALAXY
MCG+06-21-016	09 16 42.	+ 33 09	66	15.	GALAXY
ZWG 181.023	09 16 42.	+ 33 57		15.4	GALAXY
UGC 04941	09 16 42.	+ 33 57	72	15.4	GALAXY S
ZC 0916.7+4952	09 16 42.	+ 49 52	10550		CLUSTER OF GALAXIES
MCG-01-24-008	09 16 42.	- 06 30	24	15.	GALAXY
IC 0532	09 16 42.	- 16 33			MAY NOT EXIST
MCG-06-21-002	09 16 42.	- 37 49	108	13.5	GALAXY
REIN 2.083	09 16 42.85	+ 33 57 27.9			NEBULA
TON-N 1048	09 16 43.	+ 20 44		16.3	BLUE STAR
HOLM 123C	09 16 43.	+ 33 57	18	14.4	PART OF MULTIPLE GALAXY
REIN 2.084	09 16 44.22	+ 33 57 43.5			NEBULA
MCG+06-21-017	09 16 45.	+ 33 20	24	16.	GALAXY
HOLM 123A	09 16 45.	+ 33 57	24	13.8	PART OF MULTIPLE GALAXY
ARP 315	09 16 47.	+ 33 59			PECULIAR GALAXY
ZWG 181.024	09 16 48.	+ 33 58		13.3	GALAXY
UGC 04942	09 16 48.	+ 33 58	210	13.6	GALAXY E
ZWG 181.025	09 16 48.	+ 37 23		15.1	GALAXY
UGC 04943	09 16 48.	+ 37 23	162	15.1	GALAXY Sb
ZWG 265.002	09 16 48.	+ 55 09		15.7	GALAXY
ZWG 264.102	09 16 48.	+ 55 09		15.7	GALAXY
MCG+09-15-125	09 16 48.	+ 55 09	30	16.	GALAXY
ZWG 332.042	09 16 48.	+ 69 37		15.7	GALAXY
UGC 04944	09 16 48.	+ 69 37	84	15.7	GALAXY Sc
UGC 04945	09 16 48.	+ 76 00	90	17.	GALAXY DWRF IR
ARC 0779	09 16 49.	+ 33 59		13.8	RICH CLUSTER OF GALAXIES
MCG+06-21-018	09 16 51.	+ 33 13	24	15.5	GALAXY
MCG+06-21-019	09 16 51.	+ 37 26	120	14.	GALAXY
RNGC 2833	09 16 52.	+ 34 08		15.5	GALAXY
IC 2461	09 16 53.	+ 37 24 40.			NONSTELLAR OBJECT
ZWG 034.028	09 16 54.	+ 06 06		15.7	GALAXY
UGC 04946	09 16 54.	+ 06 06	90	15.7	GALAXY SBb
MCG+01-24-012	09 16 54.	+ 06 06	72	15.7	GALAXY
ZWG 181.026	09 16 54.	+ 33 08		15.3	GALAXY
UGC 04947	09 16 54.	+ 33 08	66	15.3	GALAXY SB
MCG+06-21-020	09 16 54.	+ 33 20	66	16.	GALAXY
MCG+06-21-021	09 16 54.	+ 33 57	12	16.	GALAXY
ZWG 181.027	09 16 54.	+ 34 08		15.6	GALAXY
MCG+12-09-040	09 16 54.	+ 69 37	84	16.	GALAXY
VHE 30	09 16 54.	- 48 11			REFLECTION NEBULA
TON-N 1049	09 16 55.	+ 20 29		16.6	BLUE STAR
MCG+07-19-060	09 16 57.	+ 39 20	48	14.5	GALAXY
RNGC 2834	09 16 59.	+ 33 55		15.5	GALAXY
VDB.66G 060	09 17	- 11 59	70		DWARF GALAXY
ZWG 151.060	09 17 00.	+ 26 46		15.7	GALAXY
ZWG 151.061	09 17 00.	+ 26 58		15.3	GALAXY
MCG+05-22-037	09 17 00.	+ 26 58	30	15.3	GALAXY
ZWG 151.062	09 17 00.	+ 27 00		15.7	GALAXY
ZWG 181.028	09 17 00.	+ 33 13		15.7	GALAXY
UGC 04949	09 17 00.	+ 33 18	66	16.0	GALAXY Sb-c
ZWG 181.029	09 17 00.	+ 33 55		15.6	GALAXY
ZWG 209.053	09 17 00.	+ 39 22		15.1	GALAXY
UGC 04950	09 17 00.	+ 39 22	60	15.1	GALAXY Sa
ZWG 209.054	09 17 00.	+ 40 10		15.7	GALAXY
ZWG 332.043	09 17 00.	+ 71 45		14.7	GALAXY
UGC 04951	09 17 00.	+ 71 45	72	14.7	GALAXY
ZWG 332.044	09 17 00.	+ 71 54		15.4	GALAXY
7ZW 275	09 17 00.	+ 74 39			COMPACT GALAXY
UGC 04948	09 17 00.	+ 85 32	72	16.5	GALAXY DWRF SP
MCG+15-01-008	09 17 00.	+ 89 15	48	16.	GALAXY
MCG-06-21-003	09 17 00.	- 38 34	60	15.	GALAXY
RNGC 2814	09 17 04.	+ 64 28		14.0	GALAXY
ZWG 034.029	09 17 06.	+ 04 41		15.6	GALAXY
MCG+05-22-039	09 17 06.	+ 27 08	12	16.	GALAXY
ZWG 151.063	09 17 06.	+ 27 16		15.7	GALAXY
MCG+05-22-038	09 17 06.	+ 27 17	42	15.7	GALAXY
ZC 0917.1+4026	09 17 06.	+ 40 26	610		CLUSTER OF GALAXIES
ZC 0917.1+4218	09 17 06.	+ 42 18	1550		CLUSTER OF GALAXIES
ZWG 312.003	09 17 06.	+ 64 28		14.0	GALAXY
UGC 04952	09 17 06.	+ 64 28	78	14.0	GALAXY
MCG+11-12-004	09 17 06.	+ 64 28	57	13.6	GALAXY
MRK 107	09 17 06.	+ 71 45	21	15.	GALAXY WITH UV CONTINUUM
MRK 020	09 17 06.	+ 71 45	23	16.	GALAXY WITH UV CONTINUUM
MCG+12-09-041	09 17 06.	+ 71 45	27	15.	GALAXY
MCG-01-24-009	09 17 06.	- 04 06 30.	36	15.5	GALAXY
MCG-02-24-010	09 17 09.	- 10 17	30	15.	GALAXY
ZC 0917.2+0849	09 17 12.	+ 08 49	1080		CLUSTER OF GALAXIES
ZWG 091.071	09 17 12.	+ 17 29		15.5	GALAXY
TON-N 1051	09 17 12.	+ 34 10		14.I	BLUE STAR
ZC 0917.2+3848	09 17 12.	+ 38 48	1340		CLUSTER OF GALAXIES
ZC 0917.2+4053	09 17 12.	+ 40 53	1550		CLUSTER OF GALAXIES
HOLM 124C	09 17 12.	+ 64 26	48	13.6	PART OF MULTIPLE GALAXY
LB 03019	09 17 13.	+ 21 07 36.		16.5	FAINT BLUE STAR
TON-N 1050	09 17 13.	+ 22 14		15.0	BLUE STAR
SVEN 229	09 17 16.	- 22 35 07.	18	15.0	GALAXY
RNGC 2816	09 17 17.	+ 60 39			NON-EXISTENT OBJECT
ZWG 006.029	09 17 18.	+ 01 09		15.4	GALAXY
UGC 04953	09 17 18.	+ 36 27	96	16.0	GALAXY
MCG+06-21-022	09 17 18.	+ 36 28 30.	90	15.	GALAXY
ZWG 332.045	09 17 18.	+ 72 03		13.4	GALAXY
UGC 04954	09 17 18.	+ 72 03	108	13.4	GALAXY E
MCG+12-09-042	09 17 18.	+ 72 03	66	14.	GALAXY
MCG-03-24-006	09 17 18.	- 16 54 30.	54	15.	GALAXY
RNGC 2810A	09 17 19.	+ 72 03		13.5	GALAXY
RNGC 2810B	09 17 19.	+ 72 04		15.5	GALAXY
RNGC 2846	09 17 19.	- 14 38			NON-EXISTENT OBJECT
RNGC 2849	09 17 20.	- 40 19			OPEN CLUSTER
RNGC 2820A	09 17 22.	+ 64 27		15.0	GALAXY
KEEL 168	09 17 22.7	+ 51 10 10.		18.	NEBULA
ZWG 006.030	09 17 24.	+ 01 08		15.5	GALAXY
ZWG 034.030	09 17 24.	+ 05 56		15.4	GALAXY
ZWG 091.072	09 17 24.	+ 16 45		15.6	GALAXY
UGC 04955	09 17 24.	+ 25 30	60	16.5	GALAXY
ARC 0781	09 17 24.	+ 30 39		17.6	RICH CLUSTER OF GALAXIES
SHAH 103	09 17 24.	+ 53 34	240	17.5	GROUP OF COMPACT GALAXIES
MCG+10-14-003	09 17 24.	+ 58 44	27	16.	GALAXY
7ZW 276	09 17 24.	+ 64 27			COMPACT GALAXY
MRK 108	09 17 24.	+ 64 27	13	15.	GALAXY WITH UV CONTINUUM
MCG+11-12-005	09 17 24.	+ 64 27	27	14.9	GALAXY
OCL 0756	09 17 24.	- 40 20	138	12.8	OPEN STAR CLUSTER

OBJECT NAME	RIGHT ASCEN.	DECLINATION	DIAM.	MAGN.	TYPE OF OBJECT
VHA 061	09 17 24.	- 40 20	120		OPEN STAR CLUSTER
MCG-02-24-011	09 17 27.	- 12 01 30.	72	15.	GALAXY
SVEN 230	09 17 28.	- 22 33 07.	30	14.8	GALAXY
KEEL 169	09 17 29.0	+ 51 14 17.		18.	NEBULA
ZWG 006.031	09 17 30.	+ 01 15		14.1	GALAXY
UGC 04956	09 17 30.	+ 01 15	120	14.1	GALAXY E
MKW 01S	09 17 30.	+ 01 15		14.1	POOR GALAXY CLUSTER
MCG+00-24-007	09 17 30.	+ 01 15	33	14.1	GALAXY
ZWG 062.020	09 17 30.	+ 09 00		15.6	GALAXY
UGC 04957	09 17 30.	+ 09 00	114	15.6	GALAXY
ZWG 151.064	09 17 30.	+ 30 23		15.6	GALAXY
MCG+06-21-023	09 17 30.	+ 33 52	18	15.	GALAXY
ZC 0917.5+3545	09 17 30.	+ 35 45	1010		CLUSTER OF GALAXIES
MCG+07-19-061	09 17 30.	+ 39 30	21	14.	GALAXY
ZWG 209.055	09 17 30.	+ 39 31		14.7	GALAXY
ZWG 312.004	09 17 30.	+ 64 27		15.1	GALAXY
RNGC 2838	09 17 31.	+ 39 31		14.5	GALAXY
HOLM 124D	09 17 32.	+ 64 26	24	14.9	PART OF MULTIPLE GALAXY
RNGC 2839	09 17 35.	+ 33 52		15.5	GALAXY
RNGC 2843	09 17 36.	+ 19 09			GALAXY
ZWG 151.065	09 17 36.	+ 26 45		15.6	GALAXY
ZWG 181.030	09 17 36.	+ 33 17		15.5	GALAXY
ZWG 181.031	09 17 36.	+ 33 52		15.3	GALAXY
MCG+06-21-024	09 17 36.	+ 33 56	15	15.	GALAXY
ZWG 238.037	09 17 36.	+ 49 48		15.0	GALAXY
UGC 04958	09 17 36.	+ 49 48	90	15.0	GALAXY SB
MCG+08-17-082	09 17 36.	+ 49 48	66	14.	GALAXY
SHAH 113	09 17 36.	+ 61 27	90	17.0	GROUP OF COMPACT GALAXIES
MCG+11-12-006	09 17 36.	+ 64 28	282	13.1	GALAXY
ZC 0917.6+7356	09 17 36.	+ 73 56	1810		CLUSTER OF GALAXIES
KARA.68 054	09 17 36.	+ 75 57	60		DWARF GALAXY
LB 03492	09 17 36.	- 73 17		14.6	FAINT BLUE STAR
RNGC 2820	09 17 40.	+ 64 28		13.0	GALAXY
ZC 0917.7+0520	09 17 42.	+ 05 20	270		CLUSTER OF GALAXIES
ZWG 091.073	09 17 42.	+ 17 55		15.1	GALAXY
MCG+03-24-046	09 17 42.	+ 17 55	48	15.1	GALAXY
LB 03020	09 17 42.	+ 20 23 06.		16.6	FAINT BLUE STAR
ZC 0917.7+3041	09 17 42.	+ 30 41	1340		CLUSTER OF GALAXIES
RNGC 2847	09 17 44.	- 16 18			NON-EXISTENT OBJECT
MCG+06-21-025	09 17 45.	+ 35 36	66	14.	GALAXY
IC 2458	09 17 45.	+ 64 28			SAME AS NGC 2820A
HOLM 128C	09 17 45.	- 16 18	24	14.8	PART OF MULTIPLE GALAXY
HOLM 128A	09 17 45.	- 16 19	42	13.9	PART OF MULTIPLE GALAXY
RNGC 2840	09 17 46.	+ 35 35		15.0	GALAXY
IC 0533	09 17 47.	- 07 46 45.			NONSTELLAR OBJECT
ZC 0917.8+0704	09 17 48.	+ 07 04	2420		CLUSTER OF GALAXIES
ZWG 034.031	09 17 48.	+ 07 17		15.0	GALAXY
UGC 04959	09 17 48.	+ 07 17	60	15.0	GALAXY Sb-c
MCG+01-24-013	09 17 48.	+ 07 17	36	15.0	GALAXY
ZWG 062.021	09 17 48.	+ 10 49		15.6	GALAXY
ZWG 091.074	09 17 48.	+ 18 20		15.3	GALAXY
MCG+03-24-047	09 17 48.	+ 18 21	30	15.3	GALAXY
MCG+04-22-032	09 17 48.	+ 22 15	18	15.1	GALAXY
ZWG 121.059	09 17 48.	+ 22 17		15.1	GALAXY
ZWG 151.066	09 17 48.	+ 26 52		15.7	GALAXY
TON-N 0404	09 17 48.	+ 30 49		15.2	BLUE STAR
ZWG 181.032	09 17 48.	+ 35 35		14.8	GALAXY
UGC 04960	09 17 48.	+ 35 35	60	14.8	GALAXY SBb
MCG+07-19-062	09 17 48.	+ 39 20	42	15.	GALAXY
ZWG 209.056	09 17 48.	+ 39 22		15.5	GALAXY
MCG+08-17-083	09 17 48.	+ 45 09	15	16.	GALAXY
MCG+08-17-084	09 17 48.	+ 46 05	30	14.	GALAXY
ZWG 238.038	09 17 48.	+ 46 06		14.8	GALAXY
KARA.73B 0322	09 17 48.	+ 46 06	24	14.8	ISOLATED GALAXY
ZWG 265.003	09 17 48.	+ 52 39		15.6	GALAXY
ZWG 264.103	09 17 48.	+ 52 39		15.6	GALAXY
ZWG 265.004	09 17 48.	+ 55 04		15.3	GALAXY
ZWG 264.104	09 17 48.	+ 55 04		15.3	GALAXY
HOLM 124A	09 17 48.	+ 64 27	180	13.2	PART OF MULTIPLF GALAXY
ZWG 312.005	09 17 48.	+ 64 29		13.1	GALAXY
UGC 04961	09 17 48.	+ 64 29	264	13.1	GALAXY Sc
HOLM 128B	09 17 48.	- 16 18	18	14.8	PART OF MULTIPLE GALAXY
MCG-03-24-007	09 17 48.	- 16 18	132	12.5	GALAXY
RNGC 2848	09 17 50.	- 16 18		12.5	GALAXY
HOLM 127A	09 17 52.	- 07 40	36	13.4	PART OF MULTIPLE GALAXY
MCG+03-24-048	09 17 54.	+ 15 18	72	15.2	GALAXY
ZWG 091.075	09 17 54.	+ 15 19		15.2	GALAXY
UGC 04962	09 17 54.	+ 15 19	84	15.2	GALAXY SBa
MCG+03-24-049	09 17 54.	+ 18 27	54	15.6	GALAXY
UGC 04963	09 17 54.	+ 18 29	72	16.0	GALAXY S0
ZWG 151.067	09 17 54.	+ 28 21		15.4	GALAXY
ZWG 151.068	09 17 54.	+ 29 05		15.7	GALAXY
UGC 04964	09 17 54.	+ 29 05	66	15.7	GALAXY S-IRR
KARA.72 196A	09 17 54.	+ 29 05	54	15.7	PART OF DOUBLE GALAXY
ZWG 151.069	09 17 54.	+ 30 20		15.7	GALAXY
MCG+05-22-040	09 17 54.	+ 30 20	36	15.7	GALAXY
ZWG 238.039	09 17 54.	+ 45 10		15.5	GALAXY
ZC 0917.9+4548	09 17 54.	+ 45 48	1080		CLUSTER OF GALAXIES
MCG+09-16-001	09 17 54.	+ 52 37 30.	60	16.	GALAXY
ZC 0917.9+5508	09 17 54.	+ 55 08	2080		CLUSTER OF GALAXIES
ZWG 006.032	09 17 54.	- 00 48		15.4	GALAXY
ZC 0917.9-0100	09 17 54.	- 01 00	1010		CLUSTER OF GALAXIES
TON-N 1052	09 17 55.	+ 20 21		15.	BLUE STAR
HOLM 127B	09 17 56.	- 07 40	42	14.2	PART OF MULTIPLE GALAXY
MCG+05-22-041	09 17 57.	+ 28 21	36	15.4	GALAXY
HOLM 125A	09 17 57.	+ 28 22	48	14.9	PART OF MULTIPLE GALAXY
MCG-01-24-010	09 17 57.	- 07 41	84	13.5	GALAXY
VDB.66G 061	09 18	- 12 24	70		DWARF GALAXY
LB 09972	09 18	- 87 26		12.5	FAINT BLUE STAR
ZC 0918.0+0300	09 18 00.	+ 03 00	610		CLUSTER OF GALAXIES
8ZW 0918+13.8	09 18 00.	+ 13 48		15.7	COMPACT GALAXY
ZWG 091.076	09 18 00.	+ 15 20		15.6	GALAXY
TON-N 0405	09 18 00.	+ 25 49		14.6	BLUE STAR
ZC 0918.0+3912	09 18 00.	+ 39 12	810		CLUSTER OF GALAXIES
PK278-06.1	09 18 00.	- 58 59 22.	10		PLANETARY NEBULA
HOLM 125B	09 18 04.	+ 28 24	18	15.3	PART OF MULTIPLE GALAXY
ZWG 062.022	09 18 06.	+ 13 24		15.1	GALAXY
ZWG 121.060	09 18 06.	+ 24 31		15.2	GALAXY
UGC 04965	09 18 06.	+ 24 31	66	15.2	GALAXY S
KARA.73B 0323	09 18 06.	+ 24 31	60	15.2	ISOLATED GALAXY S
ZWG 151.070	09 18 06.	+ 29 09		15.6	GALAXY
KARA.72 196B	09 18 06.	+ 29 09	54	15.6	PART OF DOUBLE GALAXY
ZC 0918.1+3505	09 18 06.	+ 35 05	940		CLUSTER OF GALAXIES
ZC 0918.1+5209	09 18 06.	+ 52 09	1880		CLUSTER OF GALAXIES
ZWG 006.033	09 18 06.	- 00 28		14.8	GALAXY
MCG+00-24-008	09 18 06.	- 00 28	30	14.8	GALAXY
OCL 0765	09 18 06.	- 44 55	270	12.5	OPEN STAR CLUSTER
VHA 062	09 18 06.	- 44 55	360		OPEN STAR CLUSTER

OBJECT NAME	RIGHT ASCEN.	DECLINATION	DIAM.	MAGN.	TYPE OF OBJECT
LB 03021	09 18 08.	+ 19 46 24.		16.9	FAINT BLUE STAR
ZC 0918.2+1141	09 18 12.	+ 11 41	610		CLUSTER OF GALAXIES
ZWG 091.077	09 18 12.	+ 18 57		15.5	GALAXY
MCG+03-24-050	09 18 12.	+ 18 57	36	15.5	GALAXY
ZWG 121.061	09 18 12.	+ 21 51		15.7	GALAXY
ZWG 265.005	09 18 12.	+ 54 16		15.6	GALAXY
ZWG 264.105	09 18 12.	+ 54 16		15.6	GALAXY
MCG-03-24-008	09 18 12.	- 16 16	60	15.	GALAXY
RNGC 2851	09 18 13.	- 16 16		15.0	GALAXY
ZWG 062.023	09 18 18.	+ 10 55		14.9	GALAXY
MCG+02-24-007	09 18 18.	+ 10 55	36	14.9	GALAXY
ZWG 062.024	09 18 18.	+ 11 16		15.5	GALAXY
ZWG 151.071	09 18 18.	+ 26 55		15.6	GALAXY
MCG+09-16-002	09 18 18.	+ 54 16	36	16.	GALAXY
ZC 0918.3+6048	09 18 18.	+ 60 48	1480		CLUSTER OF GALAXIES
MCG+12-09-043	09 18 18.	+ 70 39	51	16.	GALAXY
MCG-02-24-012	09 18 21.	- 12 23 30.	60	16.	GALAXY
ZWG 034.032	09 18 24.	+ 06 25		15.6	GALAXY
ZWG 121.062	09 18 24.	+ 24 44		15.4	GALAXY
MCG+09-16-003	09 18 24.	+ 52 36	15	16.	GALAXY
ZC 0918.4+6611	09 18 24.	+ 66 11	1480		CLUSTER OF GALAXIES
MCG-01-24-011	09 18 24.	- 07 50 30.	36	16.	GALAXY
MCG-01-24-012	09 18 24.	- 07 52	78	15.	GALAXY
MCG-01-24-013	09 18 24.	- 07 55	48	15.	GALAXY
VHA 063	09 18 24.	- 49 02	90		OPEN STAR CLUSTER
LB 03022	09 18 25.	+ 19 30 48.		16.4	FAINT BLUE STAR
SN 1957A	09 18 26.	+ 51 13		14.0	SUPERNOVA
RNGC 2850	09 18 27.	- 04 43			GALAXY
SN 1972R	09 18 28.	+ 51 11		16.0	SUPERNOVA
SN 1912A	09 18 28.	+ 51 12		13.0	SUPERNOVA
ARC 0782	09 18 28.	+ 52 11		17.6	RICH CLUSTER OF GALAXIES
BZW 0918+10.5	09 18 30.	+ 10 30		15.4	COMPACT GALAXY
ZWG 091.078	09 18 30.	+ 18 26		14.9	GALAXY
KARA.72 197A	09 18 30.	+ 18 26	60	14.9	PART OF DOUBLE GALAXY
MCG+03-24-051	09 18 30.	+ 18 26	60	14.9	GALAXY
ZWG 121.063	09 18 30.	+ 25 17		15.7	GALAXY
ZC 0918.5+3140	09 18 30.	+ 31 40	3490		CLUSTER OF GALAXIES
ZWG 265.006	09 18 30.	+ 51 12		9.9	GALAXY
KARA.73B 0324	09 18 30.	+ 51 12	438	9.9	ISOLATED GALAXY S
MCG+09-16-004	09 18 30.	+ 52 37	24	16.	GALAXY
ZC 0918.5+5805	09 18 30.	+ 58 05	940		CLUSTER OF GALAXIES
ZWG 332.046	09 18 30.	+ 72 16		14.9	GALAXY
UGC 04967	09 18 30.	+ 72 16	72	14.9	GALAXY Sa
MCG+12-09-044	09 18 30.	+ 72 17	51	15.	GALAXY
MCG-02-24-013	09 18 30.	- 09 31	48	15.	GALAXY
MCG-02-24-014	09 18 30.	- 12 40 30.	42	15.	GALAXY
RNGC 2841	09 18 34.	+ 51 11		10.5	GALAXY
ZWG 034.033	09 18 36.	+ 03 22		15.1	GALAXY
UGC 04968	09 18 36.	+ 03 22	114	15.1	GALAXY Sb
MCG+03-24-053	09 18 36.	+ 16 58	30	15.5	GALAXY
ZWG 091.079	09 18 36.	+ 16 59		15.5	GALAXY
ZWG 091.080	09 18 36.	+ 18 27		15.5	GALAXY
KARA.72 197B	09 18 36.	+ 18 27	42	15.5	PART OF DOUBLE GALAXY
MCG+03-24-052	09 18 36.	+ 18 27	12	15.5	GALAXY
ZWG 091.081	09 18 36.	+ 19 46		15.7	GALAXY
UGC 04969	09 18 36.	+ 19 46	84	15.7	GALAXY Sb
ZC 0918.6+3625	09 18 36.	+ 36 25	1010		CLUSTER OF GALAXIES
MCG+07-19-063	09 18 36.	+ 39 42	96	15.	GALAXY
UGC 04970	09 18 36.	+ 39 45	96	16.0	GALAXY Sc
MCG+07-19-064	09 18 36.	+ 40 20	96	13.5	GALAXY
ZWG 209.057	09 18 36.	+ 40 22		13.6	GALAXY
UGC 04971	09 18 36.	+ 40 22	114	13.6	GALAXY S0-a
ZC 0918.6+4703	09 18 36.	+ 47 03	1480		CLUSTER OF GALAXIES
MCG+09-16-005	09 18 36.	+ 51 10	420	10.	GALAXY
LB 03023	09 18 37.	+ 20 48 06.		16.0	FAINT BLUE STAR
RNGC 2844	09 18 37.	+ 40 22		13.5	GALAXY
IC 0534	09 18 40.	+ 03 21 47.			NONSTELLAR OBJECT
MRK 395	09 18 42.	+ 33 31	8	17.	GALAXY WITH UV CONTINUUM
MCG+06-21-026	09 18 42.	+ 33 38	42	14.5	GALAXY
TON-N 1053	09 18 42.	+ 34 12		15.2	BLUE STAR
KARA.68 055	09 18 42.	+ 48 47	47		DWARF GALAXY
7ZW 277	09 18 42.	+ 76 45			COMPACT GALAXY
ZWG 350.021	09 18 42.	+ 76 45		15.1	GALAXY
ARC 0790	09 18 44.	- 13 26		17.3	RICH CLUSTER OF GALAXIES
ZWG 181.033	09 18 48.	+ 33 37		14.6	GALAXY
UGC 04972	09 18 48.	+ 33 37	102	14.6	GALAXY (S0)
ZC 0918.8+4637	09 18 48.	+ 46 37	340		CLUSTER OF GALAXIES
ZWG 265.007	09 18 48.	+ 50 59		15.7	GALAXY
MCG+10-14-004	09 18 48.	+ 61 04	45	15.	GALAXY
ZWG 289.001	09 18 48.	+ 61 05		14.9	GALAXY
ZWG 288.029	09 18 48.	+ 61 05		14.9	GALAXY
UGC 04973	09 18 48.	+ 61 05	60	14.9	GALAXY Sb-c
KARA.73B 0325	09 18 48.	+ 61 05	54	14.9	ISOLATED GALAXY S
MCG+10-14-005	09 18 48.	+ 62 14	30	16.	GALAXY
KEEL 170	09 18 51.5	+ 50 59 06.		17.	NEBULA
KEEL 171	09 18 52.2	+ 50 57 32.		17.	SPIRAL NEBULA
ARC 0784	09 18 53.	+ 55 11		17.5	RICH CLUSTER OF GALAXIES
ZWG 062.025	09 18 54.	+ 14 11		15.4	GALAXY
ZWG 181.034	09 18 54.	+ 33 13		15.6	GALAXY
ZWG 238.040	09 18 54.	+ 49 22		15.7	GALAXY
MCG+13-07-027	09 18 54.	+ 76 45	24	15.	GALAXY
LB 03493	09 18 54.	- 70 29		14.6	FAINT BLUE STAR
KEEL 172	09 18 56.9	+ 50 59 19.		17.	NEBULA
HOLM 126A	09 18 57.	+ 62 14	36	14.8	PART OF MULTIPLE GALAXY
SVEN 231	09 18 58.	- 22 16 11.	120	14.4	GALAXY
VDB.66G 062	09 19	- 22 14	130		DWARF GALAXY
TON-N 0406	09 19 00.	+ 29 10		16.9	BLUE STAR
ZC 0919.0+3010	09 19 00.	+ 30 10	2020		CLUSTER OF GALAXIES
MCG+06-21-027	09 19 00.	+ 34 05	24	14.	GALAXY
ZC 0919.0+4140	09 19 00.	+ 41 40	610		CLUSTER OF GALAXIES
ZWG 238.041	09 19 00.	+ 47 27		15.6	GALAXY
KARA.72 198A	09 19 00.	+ 47 27	18	15.6	PART OF DOUBLE GALAXY
MCG+08-17-085	09 19 00.	+ 49 21	36	16.	GALAXY
ZWG 265.008	09 19 00.	+ 50 40		14.9	GALAXY
MCG+09-16-006	09 19 00.	+ 50 56	24	16.	GALAXY
ZWG 289.002	09 19 00.	+ 57 44		15.7	GALAXY
ZWG 289.003	09 19 00.	+ 60 42		15.6	GALAXY
ZWG 288.030	09 19 00.	+ 60 42		15.6	GALAXY
OCL 0795	09 19 00.	- 60 11	42	15.	OPEN STAR CLUSTER
HOLM 126B	09 19 03.	+ 62 14	18	15.9	PART OF MULTIPLE GALAXY
ZWG 121.064	09 19 06.	+ 25 03		15.7	GALAXY
TON-N 0407	09 19 06.	+ 29 11		16.9	BLUE STAR
TON-N 0408	09 19 06.	+ 31 18		15.6	BLUE STAR
ZWG 181.035	09 19 06.	+ 34 03		14.5	GALAXY
UGC 04974	09 19 06.	+ 34 03	90	14.5	GALAXY (S0)
MRK 109	09 19 06.	+ 47 27	15	16.	GALAXY WITH UV CONTINUUM
MCG+08-17-086	09 19 06.	+ 47 27	24	16.	GALAXY
ZWG 238.042	09 19 06.	+ 47 28		15.7	GALAXY

OBJECT NAME	RIGHT ASCEN.	DECLINATION	DIAM.	MAGN.	TYPE OF OBJECT
KARA.72 198B	09 19 06.	+ 47 28	30	15.7	PART OF DOUBLE GALAXY
MCG+08-17-087	09 19 06.	+ 50 38	120	16.	GALAXY
RNGC 2855	09 19 06.	- 11 41		12.5	GALAXY
MCG-05-22-007	09 19 06.	- 31 41	60	15.	GALAXY
MCG+08-17-088	09 19 09.	+ 46 51	54	15.	GALAXY
MCG-02-24-015	09 19 09.	- 11 41 30.	54	12.	GALAXY
ZWG 091.082	09 19 12.	+ 17 25		15.3	GALAXY
TON-N 0480	09 19 12.	+ 31 51		15.7	BLUE STAR
ZWG 181.036	09 19 12.	+ 34 08		15.7	GALAXY
MCG+08-17-089	09 19 12.	+ 50 24	72	16.	GALAXY
UGC 04975	09 19 12.	+ 50 27	72	16.0	GALAXY Sc
ZC 0919.2+5431	09 19 12.	+ 54 31	2080		CLUSTER OF GALAXIES
KEEL 173	09 19 12.1	+ 50 58 45.		15.	NEBULA
MCG-04-22-009	09 19 15.	- 22 17	90	16.	GALAXY
ZWG 034.034	09 19 18.	+ 06 13		15.7	GALAXY
ZWG 091.083	09 19 18.	+ 18 33		15.2	GALAXY
ZWG 238.043	09 19 18.	+ 46 52		15.5	GALAXY
UGC 04976	09 19 18.	+ 46 52	60	15.5	GALAXY
MCG+09-16-007	09 19 18.	+ 50 57	30	17.	GALAXY
ARC 0783	09 19 21.	+ 61 27		17.7	RICH CLUSTER OF GALAXIES
ZC 0919.4+0255	09 19 24.	+ 02 55	1480		CLUSTER OF GALAXIES
ZWG 034.035	09 19 24.	+ 03 35		15.3	GALAXY
UGC 04977	09 19 24.	+ 03 55	66	15.3	GALAXY S
MCG+03-24-054	09 19 24.	+ 17 25	30	15.3	GALAXY
ZC 0919.4+2748	09 19 24.	+ 27 48	400		CLUSTER OF GALAXIES
ARC 0768	09 19 24.	+ 79 36		17.7	RICH CLUSTER OF GALAXIES
ARC 0751	09 19 29.	+ 83 39		17.8	RICH CLUSTER OF GALAXIES
ZWG 034.036	09 19 30.	+ 04 06		15.6	GALAXY
UGC 04978	09 19 30.	+ 04 06	96	15.6	GALAXY Sc
MCG+01-24-014	09 19 30.	+ 04 06	90	15.6	GALAXY
ZC 0919.5+1238	09 19 30.	+ 12 38	610		CLUSTER OF GALAXIES
ZWG 091.084	09 19 30.	+ 17 38		15.5	GALAXY
UGC 04979	09 19 30.	+ 36 53	66	17.	GALAXY DWARF
ZC 0919.5+3856	09 19 30.	+ 38 56	1810		CLUSTER OF GALAXIES
MCG+08-17-090	09 19 30.	+ 46 44	36	16.	GALAXY
ZC 0919.5+5251	09 19 30.	+ 52 51	1340		CLUSTER OF GALAXIES
KEEL 174	09 19 34.7	+ 51 29 29.		17.	NEBULA
ARC 0791	09 19 35.	+ 12 39		17.5	RICH CLUSTER OF GALAXIES
ZC 0919.6+2529	09 19 36.	+ 25 29	610		CLUSTER OF GALAXIES
ZC 0919.6+4246	09 19 36.	+ 42 46	1140		CLUSTER OF GALAXIES
ZWG 238.044	09 19 36.	+ 46 45		15.0	GALAXY
ZWG 265.009	09 19 36.	+ 51 04		15.5	GALAXY
MCG-02-24-016	09 19 36.	- 11 26	54	16.	GALAXY
ZWG 034.037	09 19 42.	+ 04 16		14.9	GALAXY
MCG+01-24-015	09 19 42.	+ 04 16	30	14.9	GALAXY
ZWG 034.038	09 19 42.	+ 04 55		14.8	GALAXY
UGC 04980	09 19 42.	+ 04 55	60	14.8	GALAXY SB:b
MCG+01-24-016	09 19 42.	+ 04 55	60	14.8	GALAXY
TON-N 0013	09 19 42.	+ 27 19		12.8	BLUE STAR
ZWG 209.058	09 19 42.	+ 43 17		15.7	GALAXY
MCG+08-17-091	09 19 42.	+ 44 45	60	14.	GALAXY
ZWG 332.047	09 19 42.	+ 68 47		14.6	GALAXY
UGC 04981	09 19 42.	+ 68 47	72	14.6	GALAXY Sb
ZC 0919.7+7935	09 19 42.	+ 79 35	940		CLUSTER OF GALAXIES
ZC 0919.7-0016	09 19 42.	- 00 16	3560		CLUSTER OF GALAXIES
ZWG 006.034	09 19 42.	- 00 50		15.6	GALAXY
MCG-02-24-017	09 19 42.	- 09 31 30.	60	15.5	GALAXY
KEEL 175	09 19 42.1	+ 51 03 00.		17.	NEBULA
IC 0535	09 19 44.	- 00 49 31.			NONSTELLAR OBJECT
ZC 0919.8+3326	09 19 48.	+ 33 26	1680		CLUSTER OF GALAXIES
ZC 0919.8+3755	09 19 48.	+ 37 55	1140		CLUSTER OF GALAXIES
ZC 0919.8+4419	09 19 48.	+ 44 19	1140		CLUSTER OF GALAXIES
ZWG 238.045	09 19 48.	+ 44 46		14.0	GALAXY
UGC 04982	09 19 48.	+ 44 46	78	14.0	GALAXY S-IRR
MCG+09-16-008	09 19 48.	+ 51 02	60	15.	GALAXY
MCG+12-09-045	09 19 48.	+ 68 46	66	15.	GALAXY
MCG+09-16-009	09 19 51.	+ 55 02	48	15.	GALAXY
ZWG 034.039	09 19 54.	+ 05 21		15.6	GALAXY
ZWG 062.026	09 19 54.	+ 08 58		15.4	GALAXY
ZWG 151.072	09 19 54.	+ 29 32		15.7	GALAXY
UGC 04983	09 19 54.	+ 29 32	60	15.7	GALAXY Sc/SBc
MCG+05-22-042	09 19 54.	+ 29 33	42	15.7	GALAXY
ZC 0919.9+3526	09 19 54.	+ 35 26	1410		CLUSTER OF GALAXIES
UGC 04984	09 19 54.	+ 54 43	66	16.5	GALAXY DWRF SP
ZWG 265.010	09 19 54.	+ 55 02		15.3	GALAXY
ZWG 264.106	09 19 54.	+ 55 02		15.3	GALAXY
MCG+04-22-033	09 19 57.	+ 22 11 30.	96	15.0	GALAXY
MCG+09-16-010	09 19 57.	+ 52 10	42	16.	GALAXY
ZWG 121.065	09 20 00.	+ 22 12		15.0	GALAXY
UGC 04985	09 20 00.	+ 22 12	102	15.0	GALAXY Sb/SBb
ZC 0920.0+3646	09 20 00.	+ 36 46	870		CLUSTER OF GALAXIES
ZC 0920.0+3948	09 20 00.	+ 39 48	1480		CLUSTER OF GALAXIES
MCG+07-19-065	09 20 00.	+ 40 20	48	14.	GALAXY
ZWG 209.059	09 20 00.	+ 40 23		14.0	GALAXY
UGC 04986	09 20 00.	+ 40 23	66	14.0	GALAXY
KARA.72 199A	09 20 00.	+ 40 23	102	14.0	PART OF DOUBLE GALAXY
ZC 0920.0+5348	09 20 00.	+ 53 48	810		CLUSTER OF GALAXIES
ZWG 265.011	09 20 00.	+ 53 49		15.3	GALAXY
ZWG 350.022	09 20 00.	+ 78 01		15.3	GALAXY
ZC 0920.0+8207	09 20 00.	+ 82 07	2020		CLUSTER OF GALAXIES
RNGC 2867	09 20 00.	- 58 06		9.5	PLANETARY NEBULA
KEEL 176	09 20 00.1	+ 51 33 49.		17.	NEBULA
PK278-05.1	09 20 01.	- 58 06	13	9.7	PLANETARY NEBULA
MAI 040	09 20 04.	+ 85 50	33		DWARF SPHEROIDAL GALAXY
IC 2462	09 20 05.	+ 22 53 53.			NONSTELLAR OBJECT
ZWG 121.066	09 20 06.	+ 22 50		15.7	GALAXY
MCG+07-19-066	09 20 06.	+ 40 23	63	14.	GALAXY
ZWG 209.060	09 20 06.	+ 40 25		14.6	GALAXY
UGC 04987	09 20 06.	+ 40 25	114	14.6	GALAXY Sa
KARA.72 199B	09 20 06.	+ 40 25	114	14.6	PART OF DOUBLE GALAXY
MCG+09-16-011	09 20 06.	+ 53 48	42	16.	GALAXY
MCG+09-16-012	09 20 06.	+ 54 41 30.	60	16.	GALAXY
ZC 0920.1+6126	09 20 06.	+ 61 26	1410		CLUSTER OF GALAXIES
RNGC 2852	09 20 07.	+ 40 23		14.0	GALAXY
RNGC 2853	09 20 07.	+ 40 25		14.5	GALAXY
SHB 134	09 20 08.4	+ 31 11 54.		18.	QUASI-STELLAR OBJECT
BC B20920+31A	09 20 08.4	+ 31 12 14.0		19.6	QUASI-STELLAR OBJECT
MCG+06-21-028	09 20 09.	+ 34 58 30.	66	15.	GALAXY
IC 2463	09 20 10.	+ 22 49 16.			NONSTELLAR OBJECT
REIN 2.085	09 20 11.28	+ 34 56 56.5			NEBULA
ZWG 151.073	09 20 12.	+ 27 01		15.5	GALAXY
ZWG 181.037	09 20 12.	+ 34 56		15.7	GALAXY
UGC 04988	09 20 12.	+ 34 56	72	15.7	GALAXY S(B)
ZC 0920.2+4613	09 20 12.	+ 46 13	940		CLUSTER OF GALAXIES
MCG+10-14-006	09 20 12.	+ 62 14	57	15.	GALAXY
IC 2469	09 20 17.	- 32 15 42.			NONSTELLAR OBJECT
ZWG 034.040	09 20 18.	+ 03 21		13.8	GALAXY
RNGC 2858	09 20 18.	+ 03 21		14.0	GALAXY

OBJECT NAME	RIGHT ASCEN.	DECLINATION	DIAM.	MAGN.	TYPE OF OBJECT
UGC 04989	09 20 18.	+ 03 21	114	13.8	GALAXY S0-a
MCG+01-24-017	09 20 18.	+ 03 21	96	13.8	GALAXY
MCG+05-22-043	09 20 18.	+ 27 01	42	15.5	GALAXY
TON-N 0014	09 20 18.	+ 29 43		14.6	BLUE STAR
ZC 0920.3+4940	09 20 18.	+ 49 40	610		CLUSTER OF GALAXIES
ZWG 289.004	09 20 18.	+ 62 14		15.2	GALAXY
ZWG 288.031	09 20 18.	+ 62 14		15.2	GALAXY
UGC 04990	09 20 18.	+ 62 14	66	15.2	GALAXY SBc
KARA.73B 0326	09 20 18.	+ 62 14	54	15.2	ISOLATED GALAXY S
LB 03024	09 20 19.	+ 21 51 00.		17.0	FAINT BLUE STAR
MCG+04-22-034	09 20 21.	+ 22 28 30.	24		GALAXY
RNGC 2866	09 20 22.	- 50 53			NON-EXISTENT OBJECT
ZWG 006.035	09 20 24.	+ 01 38		15.4	GALAXY
MCG+04-22-035	09 20 24.	+ 22 27	30		GALAXY
VV 171B	09 20 24.	+ 22 30 30.	15	16.	INTERACTING GALAXY
VV 171A	09 20 24.	+ 22 30 30.	18	16.	INTERACTING GALAXY
ZWG 121.067	09 20 24.	+ 25 15		15.5	GALAXY
ZC 0920.4+3058	09 20 24.	+ 30 58	340		CLUSTER OF GALAXIES
MCG+06-21-029	09 20 24.	+ 33 59	18	15.	GALAXY
ZC 0920.4+4037	09 20 24.	+ 40 37	3490		CLUSTER OF GALAXIES
7ZW 278	09 20 24.	+ 73 29			COMPACT GALAXY
VV 171C	09 20 27.	+ 22 31 30.	24	16.	INTERACTING GALAXY
MCG+07-19-067	09 20 27.	+ 41 01	48	15.	GALAXY
ARC 0785	09 20 27.	+ 59 42		17.6	RICH CLUSTER OF GALAXIES
ZWG 034.041	09 20 30.	+ 02 42		15.2	GALAXY
MCG+01-24-018	09 20 30.	+ 02 42	48	15.2	GALAXY
ZWG 091.085	09 20 30.	+ 16 50		15.5	GALAXY
MCG+04-22-037	09 20 30.	+ 22 29 30.	36	15.2	GALAXY
ZWG 121.068	09 20 30.	+ 22 32		15.2	GALAXY
UGC 04991	09 20 30.	+ 22 32	72	15.2	GALAXY DBL SYS
ZWG 121.069	09 20 30.	+ 22 40		15.7	GALAXY
MCG+04-22-036	09 20 30.	+ 22 48	36	15.4	GALAXY
ZWG 121.070	09 20 30.	+ 22 50		15.4	GALAXY
TON-N 0409	09 20 30.	+ 24 47		14.9	BLUE STAR
ZWG 121.071	09 20 30.	+ 25 59		15.5	GALAXY
ZWG 181.038	09 20 30.	+ 33 58		15.4	GALAXY
ZC 0920.5+3410	09 20 30.	+ 34 10	740		CLUSTER OF GALAXIES
ZC 0920.5+3610	09 20 30.	+ 36 10	540		CLUSTER OF GALAXIES
ZWG 210.001	09 20 30.	+ 41 03		15.5	GALAXY
ZWG 209.061	09 20 30.	+ 41 03		15.5	GALAXY
ZC 0920.5+4150	09 20 30.	+ 41 50	3090		CLUSTER OF GALAXIES
ZWG 210.002	09 20 30.	+ 42 24		15.6	GALAXY
ZWG 209.062	09 20 30.	+ 42 24		15.6	GALAXY
UGC 04992	09 20 30.	+ 42 24	72	15.6	GALAXY Sc
ZWG 332.048	09 20 30.	+ 74 07		15.3	GALAXY
MCG+12-09-046	09 20 30.	+ 74 07	33	16.	GALAXY
IC 2464	09 20 32.	+ 22 49 56.			NONSTELLAR OBJECT
PK275-02.1	09 20 32.	- 53 56 33.	25		PLANETARY NEBULA
MCG+04-22-038	09 20 33.	+ 22 29 30.	36	15.2	GALAXY
ZWG 034.042	09 20 36.	+ 02 49		15.1	GALAXY
UGC 04993	09 20 36.	+ 02 49	66	15.1	GALAXY S
MCG+04-22-039	09 20 36.	+ 22 30	15	15.1	GALAXY
ZWG 121.072	09 20 36.	+ 22 33		15.1	GALAXY
ZWG 121.073	09 20 36.	+ 24 58		14.9	GALAXY
UGC 04994	09 20 36.	+ 24 58	66	14.9	GALAXY Sa-b
MCG+07-19-068	09 20 36.	+ 42 22	66	15.5	GALAXY
ZWG 238.046	09 20 36.	+ 49 25		13.8	GALAXY
RNGC 2854	09 20 36.	+ 49 25		14.0	GALAXY
UGC 04995	09 20 36.	+ 49 25	114	13.8	GALAXY SBb
ZWG 265.012	09 20 36.	+ 51 45		15.4	GALAXY
KARA.73B 0327	09 20 36.	+ 51 45	42	15.4	ISOLATED GALAXY S
ZWG 312.006	09 20 36.	+ 63 43		15.5	GALAXY
ZC 0920.6-0226	09 20 36.	- 02 26	2550		CLUSTER OF GALAXIES
OCL 0774	09 20 36.	- 50 55	120	12.	OPEN STAR CLUSTER
VHA 064	09 20 36.	- 50 55	120		OPEN STAR CLUSTER
MCG+08-17-092	09 20 39.	+ 49 26	84	14.	GALAXY
IC 2465	09 20 41.	+ 24 39 32.			NONSTELLAR OBJECT
8ZW 0920+10.4	09 20 42.	+ 10 22		16.8	COMPACT GALAXY
MCG+04-22-040	09 20 42.	+ 22 29	24		GALAXY
ZWG 181.039	09 20 42.	+ 35 42		15.5	GALAXY
MCG+11-12-007	09 20 42.	+ 63 44	42	16.	GALAXY
ZWG 006.036	09 20 42.	- 00 31		14.2	GALAXY
UGC 04996	09 20 42.	- 00 31	90	14.2	GALAXY Sb-c
MCG+00-24-009	09 20 42.	- 00 31	90	14.2	GALAXY
OCL 0786	09 20 42.	- 56 04	240	12.	OPEN STAR CLUSTER
VHA 065	09 20 42.	- 56 04	120		OPEN STAR CLUSTER
IC 2467	09 20 45.	+ 38 33 09.			NONSTELLAR OBJECT
ARP 285	09 20 45.	+ 49 28			PECULIAR GALAXY
ZWG 006.037	09 20 48.	+ 01 47		15.6	GALAXY
ZC 0920.8+0855	09 20 48.	+ 08 55	1280		CLUSTER OF GALAXIES
ZWG 121.074	09 20 48.	+ 22 34		15.7	GALAXY
ZWG 121.075	09 20 48.	+ 23 08		15.5	GALAXY
MCG+04-22-042	09 20 48.	+ 23 08	42	15.5	GALAXY
ZWG 121.076	09 20 48.	+ 24 44		15.5	GALAXY
MCG+04-22-041	09 20 48.	+ 24 57	60	14.9	GALAXY
ZWG 238.047	09 20 48.	+ 49 27		13.9	GALAXY
RNGC 2856	09 20 48.	+ 49 27		14.0	GALAXY
UGC 04997	09 20 48.	+ 49 27	66	13.9	GALAXY S
ZWG 332.049	09 20 48.	+ 68 37		15.4	GALAXY
UGC 04998	09 20 48.	+ 68 37	120	15.4	GALAXY DWARF
BC 0920+313	09 20 48.46	+ 31 20 45.2		18.	QUASI-STELLAR OBJECT
HOLM 129B	09 20 50.	+ 65 41	18	14.9	PART OF MULTIPLE GALAXY
LB 03025	09 20 51.	+ 21 41 30.		15.6	FAINT BLUE STAR
MCG+04-22-043	09 20 51.	+ 22 30 30.	45	15.7	GALAXY
HOLM 129A	09 20 52.	+ 65 41	42	14.8	PART OF MULTIPLE GALAXY
WRAY 19.19	09 20 53.7	- 32 13 54.			DIFFUSE NEBULA
ZC 0920.9+1425	09 20 54.	+ 14 25	1680		CLUSTER OF GALAXIES
IC 2466	09 20 54.	+ 24 43 55.			NONSTELLAR OBJECT
ZWG 121.077	09 20 54.	+ 25 01		15.6	GALAXY
TON-N 1054	09 20 54.	+ 36 20		15.	BLUE STAR
ZC 0920.9+4644	09 20 54.	+ 46 44	1140		CLUSTER OF GALAXIES
MCG+08-17-093	09 20 54.	+ 49 28	66	14.	GALAXY
ZWG 312.007	09 20 54.	+ 65 41		15.4	GALAXY
MCG+11-12-008A	09 20 54.	+ 65 42	36	14.9	GALAXY
MCG+11-12-008B	09 20 54.	+ 65 42 30.	24	14.8	GALAXY
MCG-05-22-008	09 20 54.	- 32 14	210	12.	GALAXY
LB 03026	09 20 59.	+ 21 38 48.		16.8	FAINT BLUE STAR
ZWG 006.038	09 21 00.	+ 02 20		14.0	GALAXY
RNGC 2861	09 21 00.	+ 02 20		14.0	GALAXY
UGC 04999	09 21 00.	+ 02 20	102	14.0	GALAXY SBb
MCG+00-24-010	09 21 00.	+ 02 20	90	14.0	GALAXY
ZWG 091.086	09 21 00.	+ 17 22		15.2	GALAXY
ZWG 121.078	09 21 00.	+ 23 18		15.7	GALAXY
MCG+04-22-044	09 21 00.	+ 24 41 30.	48	15.5	GALAXY
ZC 0921.0+3722	09 21 00.	+ 37 22	740		CLUSTER OF GALAXIES
ZWG 238.048	09 21 00.	+ 48 23		15.7	GALAXY
ZWG 265.013	09 21 00.	+ 56 20		15.2	GALAXY
MCG+11-12-009	09 21 00.	+ 64 45	27	16.	GALAXY
ZWG 312.008	09 21 00.	+ 64 47		14.7	GALAXY
KARA.72 200A	09 21 00.	+ 64 47	48	14.7	PART OF DOUBLE GALAXY
MCG+11-12-010	09 21 00.	+ 64 47	36	15.	GALAXY
MCG+12-09-047	09 21 00.	+ 68 35	84	15.	GALAXY
MCG-04-22-010	09 21 00.	- 26 45 30.	60	17.	GALAXY
MCG+09-16-013	09 21 03.	+ 56 20	36	16.	GALAXY
ARC 0794	09 21 05.	+ 08 53		17.5	RICH CLUSTER OF GALAXIES
KARA.72 201B	09 21 06.	+ 01 32	48		PART OF DOUBLE GALAXY
ZWG 006.039	09 21 06.	+ 01 33		14.8	GALAXY
KARA.72 201A	09 21 06.	+ 01 33	60	14.8	PART OF DOUBLE GALAXY
MCG+00-24-012	09 21 06.	+ 01 33	36	15.5	GALAXY
MCG+00-24-011	09 21 06.	+ 01 33	42	14.8	GALAXY
ZWG 062.027	09 21 06.	+ 13 12		15.5	GALAXY
ZC 0921.1+2846	09 21 06.	+ 28 46	740		CLUSTER OF GALAXIES
MCG+07-20-001	09 21 06.	+ 40 49 30.	30	15.	GALAXY
UGC 05000	09 21 06.	+ 49 34	150	14.3	GALAXY Sc
ZWG 312.009	09 21 06.	+ 64 45		15.3	GALAXY
KARA.72 200B	09 21 06.	+ 64 45	36	15.3	PART OF DOUBLE GALAXY
RNGC 2865	09 21 10.	- 22 58		13.0	GALAXY
RNGC 2863	09 21 11.	- 10 12		13.0	GALAXY
ZWG 006.040	09 21 12.	+ 02 19		15.4	GALAXY
MCG+00-24-013	09 21 12.	+ 02 19	48	15.4	GALAXY
ZWG 121.079	09 21 12.	+ 22 33		15.5	GALAXY
ZWG 121.080	09 21 12.	+ 23 18		15.7	GALAXY
MCG+06-21-030	09 21 12.	+ 34 45 30.	240	12.	GALAXY
MRK 396	09 21 12.	+ 35 30	7	15.5	GALAXY WITH UV CONTINUUM
ZWG 210.003	09 21 12.	+ 40 50		15.4	GALAXY
ZWG 209.063	09 21 12.	+ 40 50		15.4	GALAXY
ZWG 238.049	09 21 12.	+ 49 34		14.3	GALAXY
RNGC 2857	09 21 12.	+ 49 34		14.5	GALAXY
MCG-02-24-018	09 21 12.	- 10 12 30.	48	13.5	GALAXY
MCG+07-20-002	09 21 15.	+ 40 49	48	15.	GALAXY
REIN 2.086	09 21 15.62	+ 34 43 44.7			NEBULA
RNGC 2859	09 21 16.	+ 34 44		12.0	GALAXY
ZWG 121.081	09 21 18.	+ 24 53		15.7	GALAXY
TON-N 0410	09 21 18.	+ 27 46		16.7	BLUE STAR
ZC 0921.3+2932	09 21 18.	+ 29 32	940		CLUSTER OF GALAXIES
ZWG 181.040	09 21 18.	+ 34 44		11.8	GALAXY
UGC 05001	09 21 18.	+ 34 44	282	11.8	GALAXY SB0
ZWG 238.050	09 21 18.	+ 48 30		15.2	GALAXY
ZWG 265.014	09 21 18.	+ 56 06		14.8	GALAXY
MCG+09-16-014	09 21 18.	+ 56 06 30.	42	15.	GALAXY
MCG-04-22-011	09 21 18.	- 22 58 30.	30	14.	GALAXY
ARC 0795	09 21 19.	+ 14 24		17.5	RICH CLUSTER OF GALAXIES
TON-N 1055	09 21 19.	+ 20 52		15.5	BLUE STAR
MCG-02-24-019	09 21 21.	- 15 10	48	13.	GALAXY
LB 00554	09 21 23.	+ 60 24 00.		17.8	FAINT BLUE STAR
ZC 0921.4+0358	09 21 24.	+ 03 58	1950		CLUSTER OF GALAXIES
ZWG 034.043	09 21 24.	+ 05 25		14.9	GALAXY
MCG+01-24-019	09 21 24.	+ 05 25	48	14.9	GALAXY
MRK 397	09 21 24.	+ 18 02	18	16.5	GALAXY WITH UV CONTINUUM
TON-N 0411	09 21 24.	+ 27 50		16.0	BLUE STAR
ZWG 151.074	09 21 24.	+ 28 30		14.8	GALAXY
UGC 05002	09 21 24.	+ 28 30	60	14.8	GALAXY SB
MCG+05-22-044	09 21 24.	+ 28 30	36	14.8	GALAXY
KARA.73B 0328	09 21 24.	+ 28 30	42	14.8	ISOLATED GALAXY S
TON-N 0412	09 21 24.	+ 29 00		15.0	BLUE STAR
MCG+08-17-094	09 21 24.	+ 48 29	30	16.	GALAXY
MCG-04-22-012	09 21 24.	- 26 42	72	14.5	GALAXY
MCG+06-21-031	09 21 27.	+ 34 54	72	15.	GALAXY
ZWG 006.041	09 21 30.	+ 02 20		15.5	GALAXY
ZWG 062.028	09 21 30.	+ 11 20		15.1	GALAXY
UGC 05003	09 21 30.	+ 11 20	66	15.1	GALAXY Sc
ZC 0921.5+1325	09 21 30.	+ 13 25	1550		CLUSTER OF GALAXIES
TON-N 0413	09 21 30.	+ 29 28		15.2	BLUE STAR
UGC 05004	09 21 30.	+ 34 52	84	16.5	GALAXY IV-V
MCG+08-17-095	09 21 30.	+ 49 33	126	13.	GALAXY
ARC 0789	09 21 30.	+ 61 15		17.5	RICH CLUSTER OF GALAXIES
VHE 31A	09 21 30.	- 48 00			REFLECTION NEBULA
LB 03027	09 21 33.	+ 21 05 24.		15.5	FAINT BLUE STAR
ARP 001	09 21 33.	+ 49 33			PECULIAR GALAXY
RNGC 2864	09 21 35.	+ 06 09		15.0	GALAXY
ZWG 034.044	09 21 36.	+ 06 09		14.8	GALAXY
MCG+01-24-020	09 21 36.	+ 06 09	42	14.8	GALAXY
UGC 05005	09 21 36.	+ 22 29	96	18.	GALAXY DWRF IR
ZWG 121.082	09 21 36.	+ 22 39		15.6	GALAXY
ZC 0921.6+2354	09 21 36.	+ 23 54	10890		CLUSTER OF GALAXIES
ZC 0921.6+3054	09 21 36.	+ 30 54	1340		CLUSTER OF GALAXIES
ZWG 210.004	09 21 36.	+ 39 37		15.7	GALAXY
ZWG 209.064	09 21 36.	+ 39 37		15.7	GALAXY
ZWG 238.051	09 21 36.	+ 49 35		15.6	GALAXY
MRK 110	09 21 36.	+ 52 29	10	16.	GALAXY WITH UV CONTINUUM
KW 14	09 21 36.	+ 52 29	47		SEYFERT GALAXY
VVI 30	09 21 36.	+ 52 29	10	16.13	SEYFERT GALAXY
MCG-04-22-013	09 21 36.	- 25 27	12	15.	GALAXY
ZWG 091.087	09 21 42.	+ 18 13		15.5	GALAXY
ZWG 121.083	09 21 42.	+ 23 25		15.6	GALAXY
ZWG 121.084	09 21 42.	+ 25 20		15.3	GALAXY
UGC 05006	09 21 42.	+ 25 20	72	15.3	GALAXY Sa?
TON-N 0414	09 21 42.	+ 31 07		14.6	BLUE STAR
ZWG 181.041	09 21 42.	+ 34 00		15.5	GALAXY
MCG+06-21-032	09 21 42.	+ 34 01	30	15.	GALAXY
TON-N 1056	09 21 42.	+ 35 22		17.	BLUE STAR
ZC 0921.7+3642	09 21 42.	+ 36 42	740		CLUSTER OF GALAXIES
ZWG 210.005	09 21 42.	+ 41 16		14.8	GALAXY
ZWG 209.065	09 21 42.	+ 41 16		14.8	GALAXY
UGC 05007	09 21 42.	+ 41 16	84	14.8	GALAXY SBa
ZC 0921.7+4520	09 21 42.	+ 45 20	610		CLUSTER OF GALAXIES
UGC 05008	09 21 42.	+ 47 12	60	18.	GALAXY DWRF IR
RNGC 2860	09 21 43.	+ 41 16		15.0	GALAXY
MCG+07-20-005	09 21 45.	+ 38 34	36	14.5	GALAXY
MCG+07-20-003	09 21 45.	+ 41 15	90	14.	GALAXY
MCG+07-20-004	09 21 45.	+ 42 21 30.	24	15.5	GALAXY
ARC 0792	09 21 45.	+ 43 02		17.5	RICH CLUSTER OF GALAXIES
ZWG 091.088	09 21 48.	+ 17 53		14.9	GALAXY
MRK 398	09 21 48.	+ 17 53	18	15.5	GALAXY WITH UV CONTINUUM
MCG+03-24-055	09 21 48.	+ 17 53	24	14.9	GALAXY
MCG+04-22-046	09 21 48.	+ 23 43	48	15.2	GALAXY
ZWG 121.085	09 21 48.	+ 23 45		15.2	GALAXY
MCG+04-22-045	09 21 48.	+ 25 18	72	15.3	GALAXY
ZWG 151.075	09 21 48.	+ 29 01		15.6	GALAXY
MCG+06-21-033	09 21 48.	+ 33 58	30	15.	GALAXY
ZWG 210.006	09 21 48.	+ 38 34		14.8	GALAXY
MCG+07-20-006	09 21 48.	+ 40 35 30.	36	15.	GALAXY
ZWG 210.007	09 21 48.	+ 40 37		15.2	GALAXY
ZWG 209.066	09 21 48.	+ 40 37		15.2	GALAXY
ZC 0921.8+4302	09 21 48.	+ 43 02	2020		CLUSTER OF GALAXIES
ZC 0921.8+6244	09 21 48.	+ 62 44	3090		CLUSTER OF GALAXIES

OBJECT NAME	RIGHT ASCEN.	DECLINATION	DIAM.	MAGN.	TYPE OF OBJECT
MCG-05-22-009	09 21 48.	- 27 56	72	15.	GALAXY
IC 0536	09 21 49.	+ 25 20 32.			NONSTELLAR OBJECT
MCG+07-20-007	09 21 51.	+ 38 34 30.	24	15.5	GALAXY
LB 03029	09 21 53.	+ 18 59 30.		16.4	FAINT BLUE STAR
LB 03028	09 21 53.	+ 19 32 54.		16.3	FAINT BLUE STAR
RNGC 2869	09 21 53.	- 10 12			NON-EXISTENT OBJECT
RNGC 2868	09 21 53.	- 10 12			NON-EXISTENT OBJECT
ZWG 091.089	09 21 54.	+ 20 14		15.5	GALAXY
UGC 05009	09 21 54.	+ 20 14	66	15.5	GALAXY Sc
MCG+03-24-056	09 21 54.	+ 20 15	66	15.5	GALAXY
MCG+04-22-047	09 21 54.	+ 21 44	36		GALAXY
VV 154C	09 21 54.	+ 21 45 30.	6	17.	INTERACTING GALAXY
VV 154B	09 21 54.	+ 21 45 30.	9	16.	INTERACTING GALAXY
VV 154A	09 21 54.	+ 21 45 30.	12	17.	INTERACTING GALAXY
VV 154	09 21 54.	+ 21 45 30.	30	16.	INTERACTING GALAXY
ZWG 151.076	09 21 54.	+ 26 59		13.8	GALAXY
UGC 05010	09 21 54.	+ 26 59	150	13.8	GALAXY S
KARA.73B 0329	09 21 54.	+ 26 59	204	13.8	ISOLATED GALAXY S
ZWG 181.042	09 21 54.	+ 33 57		15.6	GALAXY
IC 2468	09 21 54.	+ 38 32 33.			NONSTELLAR OBJECT
ZWG 265.015	09 21 54.	+ 50 47		15.7	GALAXY
RNGC 2862	09 21 56.	+ 26 59		14.0	GALAXY
ZWG 091.090	09 22 00.	+ 18 09		15.6	GALAXY
MCG+05-22-045	09 22 00.	+ 26 59	138	13.8	GALAXY
ZC 0922.0+3116	09 22 00.	+ 31 16	810		CLUSTER OF GALAXIES
MCG+07-20-008	09 22 00.	+ 38 35 30.	33	15.	GALAXY
ZC 0922.0+4111	09 22 00.	+ 41 11	1010		CLUSTER OF GALAXIES
MCG+08-17-096	09 22 00.	+ 50 36	60	16.	GALAXY
ZWG 265.016	09 22 00.	+ 50 38		15.1	GALAXY
MCG+12-09-048B	09 22 00.	+ 74 40		18.	GALAXY
MCG+12-09-048A	09 22 00.	+ 74 40	36	17.	GALAXY
MCG-01-24-014	09 22 00.	- 06 22 30.	42	15.	GALAXY
ZC 0922.1+1847	09 22 06.	+ 18 47	340		CLUSTER OF GALAXIES
ZWG 091.091	09 22 06.	+ 19 22		15.6	GALAXY
TON-N 0415	09 22 06.	+ 32 07		15.2	BLUE STAR
MCG+09-16-015	09 22 09.	+ 50 45	42	17.	GALAXY
MCG+05-22-046	09 22 12.	+ 29 43	27	15.7	GALAXY
UGC 05011	09 22 12.	+ 34 20	72	16.0	GALAXY SBc
MCG+06-21-034	09 22 12.	+ 34 21	54	15.	GALAXY
ZWG 238.052	09 22 12.	+ 45 45		15.3	GALAXY
KARA.73P 0330	09 22 12.	+ 45 45	42	15.3	ISOLATED GALAXY S
ARC 0793	09 22 14.	+ 50 32		17.1	RICH CLUSTER OF GALAXIES
RNGC 2887	09 22 14.	- 63 36			UNVERIFIED SOUTHERN OBJECT
ZWG 062.029	09 22 18.	+ 11 45		15.2	GALAXY
ZWG 091.092	09 22 18.	+ 15 14		15.7	GALAXY
ZWG 151.077	09 22 18.	+ 29 41		15.7	GALAXY
IC 2470	09 22 22.	+ 03 35 28.			NONSTELLAR OBJECT
SHB 135	09 22 22.03	+ 14 57 26.		18.0	QUASI-STELLAR OBJECT
BC PKS0922+14	09 22 22.41	+ 14 57 23.2		17.96	QUASI-STELLAR OBJECT
ARC 0777	09 22 23.	+ 78 28		17.5	RICH CLUSTER OF GALAXIES
MCG+03-24-057	09 22 24.	+ 17 48	48	14.6	GALAXY
ZWG 091.093	09 22 24.	+ 17 49		14.6	GALAXY
UGC 05012	09 22 24.	+ 17 49	66	14.6	GALAXY Sa-b
ZWG 238.053	09 22 24.	+ 49 32		15.6	GALAXY
MCG+09-16-016	09 22 24.	+ 52 36	42	17.	GALAXY
ZWG 289.005	09 22 24.	+ 61 35		15.5	GALAXY
UGC 05013	09 22 24.	+ 61 35	108	15.5	GALAXY Sb
KARA.73B 0331	09 22 24.	+ 61 35	96	15.5	ISOLATED GALAXY S
MCG+10-14-007	09 22 24.	+ 61 37	78	15.	GALAXY
ARC 0767	09 22 24.	+ 82 38		17.1	RICH CLUSTER OF GALAXIES
MCG-04-23-001	09 22 27.	- 24 54 30.	48	14.	GALAXY
LB 03030	09 22 28.	+ 20 44 24.		16.7	FAINT BLUE STAR
ZWG 121.086	09 22 30.	+ 23 50		15.6	GALAXY
ZC 0922.5+3715	09 22 30.	+ 37 15	870		CLUSTER OF GALAXIES
ZWG 210.008	09 22 30.	+ 40 52		15.6	GALAXY
ZWG 238.054	09 22 30.	+ 49 38		15.5	GALAXY
MCG-06-21-004	09 22 30.	- 37 33	60	15.	GALAXY
OCL 0775	09 22 30.	- 51 31	360	13.	OPEN STAR CLUSTER
BC PKS0922+005	09 22 34.1	+ 00 32 09.		18.07	QUASI-STELLAR OBJECT
SHB 136	09 22 36.	+ 00 32 06.		18.1	QUASI-STELLAR OBJECT
ZWG 062.030	09 22 36.	+ 11 06		15.2	GALAXY
ZWG 062.031	09 22 36.	+ 12 21		15.7	GALAXY
MCG+04-22-048	09 22 36.	+ 23 47	42	15.6	GALAXY
ZWG 151.078	09 22 36.	+ 27 34		15.6	GALAXY
UGC 05014	09 22 36.	+ 35 04	66	17.	GALAXY
MCG+07-20-009	09 22 36.	+ 40 51	36	15.5	GALAXY
ZWG 265.017	09 22 36.	+ 50 40		14.8	GALAXY
MCG+09-16-017	09 22 36.	+ 55 54	30	17.	GALAXY
ZWG 312.010	09 22 36.	+ 62 38		15.0	GALAXY
ZC 0922.6+7828	09 22 36.	+ 78 28	1210		CLUSTER OF GALAXIES
7ZW 279	09 22 36.	+ 80 33			COMPACT GALAXY
MRSL 274-01/1	09 22 36.	- 51 46	360		HII REGION
LB 03031	09 22 39.	+ 21 59 24.		15.0	FAINT BLUE STAR
MCG-04-23-002	09 22 39.	- 25 36	48	15.	GALAXY
ZWG 181.043	09 22 42.	+ 34 30		15.7	GALAXY
UGC 05015	09 22 42.	+ 34 30	120	15.7	GALAXY S IV
MCG+06-21-035	09 22 42.	+ 34 31	102	15.	GALAXY
ZWG 238.055	09 22 42.	+ 49 30		15.4	GALAXY
UGC 05016	09 22 42.	+ 49 30	66	15.4	GALAXY SBa
MCG+10-14-008	09 22 42.	+ 62 37	18	15.	GALAXY
IC 2471	09 22 44.	- 06 36			NONSTELLAR OBJECT
MCG-01-24-015	09 22 45.	- 06 37	42	14.5	GALAXY
RNGC 2876	09 22 46.	- 06 30		14.5	GALAXY
ZC 0922.8+0745	09 22 48.	+ 07 45	1340		CLUSTER OF GALAXIES
ZWG 062.032	09 22 48.	+ 11 17		15.7	GALAXY
UGC 05017	09 22 48.	+ 11 17	66	15.7	GALAXY Sc
MCG+04-22-050	09 22 48.	+ 22 30	18	15.1	GALAXY
ZWG 121.087	09 22 48.	+ 22 33		15.1	GALAXY
MCG+04-22-049	09 22 48.	+ 23 30	15	15.6	GALAXY
ZWG 121.088	09 22 48.	+ 23 35		15.6	GALAXY
MCG+08-17-097	09 22 48.	+ 49 30	60	15.	GALAXY
ZC 0922.8+7321	09 22 48.	+ 73 21	1610		CLUSTER OF GALAXIES
MCG-01-24-016	09 22 48.	- 06 30 30.	60	14.5	GALAXY
LB 03032	09 22 53.	+ 21 48 00.		15.8	FAINT BLUE STAR
VV 063B	09 22 54.	+ 22 35	24	16.	INTERACTING GALAXY
VV 063A	09 22 54.	+ 22 35	18	16.	INTERACTING GALAXY
TON-N 0416	09 22 54.	+ 24 12		14.7	BLUE STAR
TON-N 0015	09 22 54.	+ 25 55		15.0	BLUE STAR
ZC 0922.9+2928	09 22 54.	+ 29 28	470		CLUSTER OF GALAXIES
MCG+06-21-036	09 22 54.	+ 34 52	24	15.	GALAXY
MCG+06-21-037	09 22 54.	+ 34 53 30.	120	14.	GALAXY
ZC 0922.9+3640	09 22 54.	+ 36 40	740		CLUSTER OF GALAXIES
MCG+07-20-010	09 22 54.	+ 42 37	24	16.	GALAXY
MCG-01-24-017	09 22 54.	- 06 28	24	16.5	GALAXY
WRAY 19.20	09 22 56.3	- 51 44 38.		10.7	STAR-NEBULA ASSOCIATION
RNGC 2871	09 22 57.	+ 11 39			NON-EXISTENT OBJECT
HOLM 130C	09 22 58.	+ 11 39	12	15.1	PART OF MULTIPLE GALAXY
RNGC 2879	09 22 59.	- 11 27			NON-EXISTENT OBJECT
IC 0537	09 22 59.	- 12 10 24.			NONSTELLAR OBJECT
LBN 1085	09 23	- 27 50	780		BRIGHT NEBULA
LBN 1087	09 23	- 29 30	3120		BRIGHT NEBULA
ZWG 062.033	09 23 00.	+ 11 39		13.0	GALAXY
UGC 05018	09 23 00.	+ 11 39	120	13.0	GALAXY E
KARA.72 202A	09 23 00.	+ 11 39	96	13.0	PART OF DOUBLE GALAXY
MCG+02-24-008	09 23 00.	+ 11 39	30	13.0	GALAXY
MCG+04-22-051	09 23 00.	+ 22 32	36	15.6	GALAXY
ZWG 121.089	09 23 00.	+ 22 36		15.6	GALAXY
ZWG 121.090	09 23 00.	+ 24 21		15.7	GALAXY
UGC 05019	09 23 00.	+ 24 21	60	15.7	GALAXY S
ZWG 181.044	09 23 00.	+ 34 52		15.3	GALAXY
UGC 05020	09 23 00.	+ 34 52	138	15.3	GALAXY Sc
ZC 0923.0+4628	09 23 00.	+ 46 28	610		CLUSTER OF GALAXIES
MCG+08-17-098	09 23 00.	+ 48 21	30	16.	GALAXY
MCG-02-24-020	09 23 00.	- 12 10	60	14.	GALAXY
HOLM 130A	09 23 01.	+ 11 38	36	13.6	PART OF MULTIPLE GALAXY
TON-N 1057	09 23 01.	+ 20 09		15.0	BLUE STAR
SN 1952B	09 23 01.	+ 29 40		17.8	SUPERNOVA
RNGC 2872	09 23 03.	+ 11 39		13.0	GALAXY
HOLM 130B	09 23 05.	+ 11 38	150	13.8	PART OF MULTIPLE GALAXY
ZWG 062.034	09 23 06.	+ 11 39		13.5	GALAXY
UGC 05021	09 23 06.	+ 11 39	144	13.5	GALAXY Sc
KARA.72 202B	09 23 06.	+ 11 39	144	13.5	PART OF DOUBLE GALAXY
MCG+02-24-010	09 23 06.	+ 11 39	120	13.5	GALAXY
MCG+02-24-009	09 23 06.	+ 11 40	36	15.3	GALAXY
MRK 399	09 23 06.	+ 35 07	18	16.5	GALAXY WITH UV CONTINUUM
ZC 0923.1+5925	09 23 06.	+ 59 25	3760		CLUSTER OF GALAXIES
ZC 0923.1-0130	09 23 06.	- 01 30	810		CLUSTER OF GALAXIES
MCG-06-21-005	09 23 06.	- 33 52		16.	GALAXY
SER 069.01	09 23 06.	- 74 47	70		LOW SURFACE BRIGHT. GALXY
HOLM 130D	09 23 07.	+ 11 39	30	15.3	PART OF MULTIPLE GALAXY
ARP 307	09 23 07.	+ 11 39			PECULIAR GALAXY
MCG-06-21-006	09 23 07.	- 35 52		17.	GALAXY
RNGC 2875	09 23 09.	+ 11 38			NON-EXISTENT OBJECT
RNGC 2874	09 23 09.	+ 11 39		13.5	GALAXY
RNGC 2873	09 23 09.	+ 11 40		15.5	GALAXY
RNGC 2883	09 23 09.	- 33 53			GALAXY
PK275-02.2	09 23 10.	- 54 23 26.	20		PLANETARY NEBULA
ZC 0923.2+0057	09 23 12.	+ 00 57	670		CLUSTER OF GALAXIES
ZWG 006.042	09 23 12.	+ 02 18		14.9	GALAXY
RNGC 2878	09 23 12.	+ 02 18		15.0	GALAXY
UGC 05022	09 23 12.	+ 02 18	66	14.9	GALAXY Sa-b
MCG+00-24-014	09 23 12.	+ 02 18	42	14.9	GALAXY
ZWG 006.043	09 23 12.	+ 02 26		14.7	GALAXY
RNGC 2877	09 23 12.	+ 02 26		14.5	GALAXY
MCG+00-24-015	09 23 12.	+ 02 26	24	14.7	GALAXY
ZWG 091.094	09 23 12.	+ 19 36		14.4	GALAXY
MRK 400	09 23 12.	+ 19 36	24	15.5	GALAXY WITH UV CONTINUUM
UGC 05023	09 23 12.	+ 19 36	48	14.4	GALAXY
MCG+03-24-058	09 23 12.	+ 19 36	42	14.4	GALAXY
LB 10748	09 23 12.	+ 26 02		16.8	FAINT BLUE STAR
TON-N 1058	09 23 12.	+ 32 29		17.	BLUE STAR
ZC 0923.2+3934	09 23 12.	+ 39 34	2760		CLUSTER OF GALAXIES
UGC 05024	09 23 12.	+ 70 27	66	16.5	GALAXY Sc
ARC 0788	09 23 16.	+ 72 32		17.0	RICH CLUSTER OF GALAXIES
ZWG 062.035	09 23 18.	+ 12 57		13.9	GALAXY
8ZW 0923+12.9	09 23 18.	+ 12 57		13.9	COMPACT GALAXY
UGC 05025	09 23 18.	+ 12 57	42	13.9	GALAXY S0?
MCG+02-24-011	09 23 18.	+ 12 57	36	13.9	GALAXY
ZC 0923.3+1926	09 23 18.	+ 19 26	2760		CLUSTER OF GALAXIES
MCG+08-17-099	09 23 18.	+ 45 53	36	15.	GALAXY
MCG+08-17-100	09 23 18.	+ 46 03	36	15.	GALAXY
ZWG 238.056	09 23 18.	+ 46 05		14.2	GALAXY
UGC 05026	09 23 18.	+ 46 05	66	14.2	GALAXY S0
KARA.73B 0332	09 23 18.	+ 46 05	66	14.2	ISOLATED GALAXY S0
LB 10749	09 23 24.	+ 26 22		17.0	FAINT BLUE STAR
ZC 0923.4+3315	09 23 24.	+ 33 15	670		CLUSTER OF GALAXIES
ZC 0923.4+7459	09 23 24.	+ 74 59	1550		CLUSTER OF GALAXIES
ARP 275	09 23 26.	- 11 47			PECULIAR GALAXY
MCG+04-22-052	09 23 27.	+ 22 18	36	15.0	GALAXY
ZWG 091.095	09 23 30.	+ 20 01		15.2	GALAXY
ZWG 121.091	09 23 30.	+ 22 21		15.0	GALAXY
LB 10750	09 23 30.	+ 23 44		15.9	FAINT BLUE STAR
ZC 0923.5+2416	09 23 30.	+ 24 16	1810		CLUSTER OF GALAXIES
ZWG 151.079	09 23 30.	+ 26 47		15.7	GALAXY
MCG+10-14-009	09 23 30.	+ 60 05	27	16.	GALAXY
RNGC 2881	09 23 30.	- 11 46		14.0	GALAXY
VV 293B	09 23 30.	- 11 47	18	15.	INTERACTING GALAXY
VV 293A	09 23 30.	- 11 47	21	14.5	INTERACTING GALAXY
VV 293	09 23 30.	- 11 47	60		INTERACTING GALAXY
ARC 0787	09 23 31.	+ 74 38		15.9	RICH CLUSTER OF GALAXIES
MCG-02-24-021	09 23 33.	- 11 46 30.	60	14.	GALAXY
ZWG 034.045	09 23 36.	+ 03 20		15.4	GALAXY
UGC 05027	09 23 36.	+ 03 20	66	15.4	GALAXY Sb-c
ZC 0923.6+0830	09 23 36.	+ 08 30	1340		CLUSTER OF GALAXIES
ZWG 121.092	09 23 36.	+ 22 18		15.7	GALAXY
LB 10751	09 23 36.	+ 26 28		17.6	FAINT BLUE STAR
ZC 0923.6+2946	09 23 36.	+ 29 46	610		CLUSTER OF GALAXIES
ZC 0923.6+3254	09 23 36.	+ 32 54	1680		CLUSTER OF GALAXIES
ZC 0923.6+5340	09 23 36.	+ 53 40	940		CLUSTER OF GALAXIES
ZWG 332.050	09 23 36.	+ 68 39		13.9	GALAXY
UGC 05028	09 23 36.	+ 68 39	36	13.9	GALAXY PECULR
KARA.72 203A	09 23 36.	+ 68 39	42	13.9	PART OF DOUBLE GALAXY
7ZW 280	09 23 36.	+ 68 40			COMPACT GALAXY
VHA 066	09 23 36.	- 54 34	210		OPEN STAR CLUSTER
PRA 6	09 23 36.0	+ 32 28 54.		17.	NONSTELLAR BLUE OBJECT
LB 03033	09 23 37.	+ 21 10 00.		17.3	FAINT BLUE STAR
ARP 300	09 23 40.	+ 68 37			PECULIAR GALAXY
ARC 0786	09 23 40.	+ 75 02		15.9	RICH CLUSTER OF GALAXIES
LB 03034	09 23 42.	+ 19 25 24.		16.7	FAINT BLUE STAR
ZWG 121.093	09 23 42.	+ 21 36		15.5	GALAXY
ZWG 151.080	09 23 42.	+ 26 48		15.7	GALAXY
SHAH 104	09 23 42.	+ 53 12	72	17.8	GROUP OF COMPACT GALAXIES
ZC 0923.7+5453	09 23 42.	+ 54 53	3020		CLUSTER OF GALAXIES
ZC 0923.7+5557	09 23 42.	+ 55 57	1340		CLUSTER OF GALAXIES
ZC 0923.7+6407	09 23 42.	+ 64 07	3630		CLUSTER OF GALAXIES
IC 2472	09 23 43.	+ 21 37 22.			NONSTELLAR OBJECT
ARC 0797	09 23 48.	+ 17 54		16.5	RICH CLUSTER OF GALAXIES
LB 10753	09 23 48.	+ 20 30		15.5	FAINT BLUE STAR
LB 10752	09 23 48.	+ 21 04		16.4	FAINT BLUE STAR
ZWG 121.094	09 23 48.	+ 22 18		15.5	GALAXY
TON-N 1059	09 23 48.	+ 32 59		14.9	BLUE STAR
ZC 0923.8+3656	09 23 48.	+ 36 56	940		CLUSTER OF GALAXIES
MRK 111	09 23 48.	+ 68 38	30	14.	GALAXY WITH UV CONTINUUM
VV 106B	09 23 48.	+ 68 39	30	14.	INTERACTING GALAXY
ZWG 332.051	09 23 48.	+ 68 40		14.3	GALAXY
UGC 05029	09 23 48.	+ 68 40	108	14.3	GALAXY Sc

OBJECT NAME	RIGHT ASCEN.	DECLINATION	DIAM.	MAGN.	TYPE OF OBJECT
KARA.72 203B	09 23 48.	+ 68 40	96	14.3	PART OF DOUBLE GALAXY
MCG+13-07-028	09 23 48.	+ 77 51	33	15.	GALAXY
ZWG 350.023	09 23 48.	+ 77 52		15.1	GALAXY
RNGC 2882	09 23 52.	+ 08 10		13.5	GALAXY
ZWG 034.046	09 23 54.	+ 08 10		13.5	GALAXY
UGC 05030	09 23 54.	+ 08 10	102	13.5	GALAXY S
MCG+01-24-021	09 23 54.	+ 08 10	84	13.5	GALAXY
ZC 0923.9+1113	09 23 54.	+ 11 13	340		CLUSTER OF GALAXIES
ZC 0923.9+1758	09 23 54.	+ 17 58	1750		CLUSTER OF GALAXIES
ZWG 121.095	09 23 54.	+ 22 37		15.7	GALAXY
TON-N 1060	09 23 54.	+ 35 44		17.	BLUE STAR
ZC 0923.9+3732	09 23 54.	+ 37 32	610		CLUSTER OF GALAXIES
ZWG 210.009	09 23 54.	+ 42 47		15.5	GALAXY
KARA.73B 0333	09 23 54.	+ 42 47	48	15.5	ISOLATED GALAXY S
ZC 0923.9+4410	09 23 54.	+ 44 10	1810		CLUSTER OF GALAXIES
ZC 0923.9+7353	09 23 54.	+ 73 53	7860		CLUSTER OF GALAXIES
BC 4C39.25	09 23 55.	+ 39 15 20.		17.86	QUASI-STELLAR OBJECT
SHB 137	09 23 55.4	+ 39 15 24.		17.9	QUASI-STELLAR OBJECT
LB 03035	09 23 59.	+ 22 49 36.		15.4	FAINT BLUE STAR
RNGC 2884	09 23 59.	- 11 19		13.0	GALAXY
MCG+03-24-059	09 24 00.	+ 17 55	60	15.6	GALAXY
ZWG 091.096	09 24 00.	+ 17 56		15.6	GALAXY
MCG+04-22-053	09 24 00.	+ 23 44	48		GALAXY
LB 10754	09 24 00.	+ 25 24		17.0	FAINT BLUE STAR
MCG+07-20-011	09 24 00.	+ 42 47	42	15.	GALAXY
ZWG 265.018	09 24 00.	+ 52 26		15.0	GALAXY
ZWG 265.019	09 24 00.	+ 55 34		15.1	GALAXY
MCG+10-14-010	09 24 00.	+ 59 21	24	16.	GALAXY
ZWG 289.006	09 24 00.	+ 59 22		15.7	GALAXY
MCG+11-12-011	09 24 00.	+ 65 10	90	16.	GALAXY
VV 106A	09 24 00.	+ 68 39 30.	60	14.	INTERACTING GALAXY
ZWG 350.024	09 24 00.	+ 77 25		15.4	GALAXY
UGC 05031	09 24 00.	+ 77 25	66	15.4	GALAXY S
MCG+13-07-029	09 24 00.	+ 77 25	66	15.	GALAXY
MCG-02-24-022	09 24 00.	- 11 20	120	13.	GALAXY
ZWG 006.044	09 24 06.	+ 01 22		15.7	GALAXY
UGC 05032	09 24 06.	+ 01 22	66	15.7	GALAXY Sa-b
KARA.73B 0334	09 24 06.	+ 01 22	78	15.7	ISOLATED GALAXY S
ZC 0924.1+0411	09 24 06.	+ 04 11	340		CLUSTER OF GALAXIES
MCG+04-22-054	09 24 06.	+ 24 52	42		GALAXY
ZWG 151.081	09 24 06.	+ 29 36		15.7	GALAXY
ZWG 265.020	09 24 06.	+ 56 07		15.7	GALAXY
ZWG 289.007	09 24 06.	+ 58 45		15.7	GALAXY
MCG+10-14-011	09 24 06.	+ 59 16	36	17.	GALAXY
UGC 05033	09 24 06.	+ 65 09	96	16.5	GALAXY Sc
MCG-05-23-001	09 24 06.	- 27 47	24	13.	GALAXY
RNGC 2870	09 24 10.	+ 57 36		14.0	GALAXY
LB 03036	09 24 11.	+ 21 19 46.		15.7	FAINT BLUE STAR
FATH 1.277	09 24 11.	+ 29 37	27		NEBULA
ZWG 121.096	09 24 12.	+ 23 12		15.5	GALAXY
MCG+04-22-055	09 24 12.	+ 23 12	24	15.5	GALAXY
LB 10755	09 24 12.	+ 26 38		16.8	FAINT BLUE STAR
ZWG 181.045	09 24 12.	+ 34 39		15.7	GALAXY
ZWG 289.008	09 24 12.	+ 57 36		13.9	GALAXY
UGC 05034	09 24 12.	+ 57 36	162	13.9	GALAXY Sb-c
KARA.73B 0335	09 24 12.	+ 57 36	162	13.9	ISOLATED GALAXY S
MCG+10-14-012	09 24 12.	+ 58 46	39	15.	GALAXY
ARC 0796	09 24 12.	+ 60 38		17.4	RICH CLUSTER OF GALAXIES
MCG-02-24-024	09 24 12.	- 14 18 30.	36	15.	GALAXY
MCG-02-24-023	09 24 12.	- 15 30	78	15.	GALAXY
RNGC 2888	09 24 12.	- 27 47		14.0	GALAXY
RNGC 2890	09 24 13.	- 14 18		15.0	GALAXY
MCG+09-16-018	09 24 15.	+ 52 25	36	16.	GALAXY
RNGC 2886	09 24 15.	- 21 32			NON-EXISTENT OBJECT
MCG+04-22-056	09 24 18.	+ 21 48	66	15.6	GALAXY
ZWG 121.097	09 24 18.	+ 21 49		15.6	GALAXY
UGC 05035	09 24 18.	+ 21 49	66	15.6	GALAXY SBa
LB 10756	09 24 18.	+ 25 04		17.4	FAINT BLUE STAR
TON-N 0418	09 24 18.	+ 32 42		16.0	BLUE STAR
TON-N 0417	09 24 18.	+ 32 44		15.8	BLUE STAR
MCG+06-21-038	09 24 18.	+ 33 56	42	15.5	GALAXY
UGC 05036	09 24 18.	+ 41 03	72	16.5	GALAXY Sc-IRR
ZC 0924.3+4220	09 24 18.	+ 42 20	1080		CLUSTER OF GALAXIES
MCG+10-14-013	09 24 18.	+ 57 36	120	14.	GALAXY
ZC 0924.3+6041	09 24 18.	+ 60 41	2220		CLUSTER OF GALAXIES
IC 2474	09 24 21.	+ 23 15			NONSTELLAR OBJECT
MCG+04-22-057	09 24 21.	+ 23 16	27		GALAXY
MCG+09-16-019	09 24 21.	+ 54 45	60	16.	GALAXY
ZC 0924.4+0511	09 24 24.	+ 05 11	940		CLUSTER OF GALAXIES
LB 10757	09 24 24.	+ 20 37		17.6	FAINT BLUE STAR
ZWG 121.098	09 24 24.	+ 23 14		14.8	GALAXY
UGC 05037	09 24 24.	+ 23 14	96	14.8	GALAXY S0
LB 10758	09 24 24.	+ 25 30		16.7	FAINT BLUE STAR
ZWG 152.001	09 24 24.	+ 30 39		14.6	GALAXY
ZWG 151.082	09 24 24.	+ 30 39		14.6	GALAXY
UGC 05038	09 24 24.	+ 30 39	96	14.6	GALAXY SBb
KARA.73B 0336	09 24 24.	+ 30 39	108	14.6	ISOLATED GALAXY S
IC 2473	09 24 24.	+ 30 39 31.			NONSTELLAR OBJECT
MCG+05-22-047	09 24 24.	+ 30 40	120	14.6	GALAXY
MCG+07-20-012	09 24 24.	+ 41 02	66	15.	GALAXY
MCG+04-22-058	09 24 27.	+ 23 15	30	14.8	GALAXY
IC 0538	09 24 28.	+ 23 14			NONSTELLAR OBJECT
LB 10759	09 24 30.	+ 22 14		16.9	FAINT BLUE STAR
UGC 05039	09 24 30.	+ 22 28	66	16.0	GALAXY S
MCG+04-22-059	09 24 30.	+ 23 14	27		GALAXY
ZWG 121.099	09 24 30.	+ 23 15		15.6	GALAXY
TON-N 0419	09 24 30.	+ 25 20		15.1	BLUE STAR
ZWG 152.002	09 24 30.	+ 30 12		15.6	GALAXY
ZWG 151.083	09 24 30.	+ 30 12		15.6	GALAXY
ZWG 152.003	09 24 30.	+ 30 15		15.7	GALAXY
ZWG 151.084	09 24 30.	+ 30 15		15.7	GALAXY
ZWG 238.057	09 24 30.	+ 48 44		15.3	GALAXY
MCG+08-17-101	09 24 30.	+ 48 44	42	16.	GALAXY
ZC 0924.5+6121	09 24 30.	+ 61 21	1210		CLUSTER OF GALAXIES
IC 2482	09 24 34.	- 11 53 26.			NONSTELLAR OBJECT
RNGC 2891	09 24 35.	- 24 35		14.0	GALAXY
LB 10761	09 24 36.	+ 21 34		16.4	FAINT BLUE STAR
MCG+04-22-060	09 24 36.	+ 23 15	24	15.6	GALAXY
LB 10760	09 24 36.	+ 26 08		16.3	FAINT BLUE STAR
MCG+08-17-102	09 24 36.	+ 44 42	36	16.	GALAXY
MCG+08-17-103	09 24 36.	+ 48 05	60	14.	GALAXY
MCG+09-16-020	09 24 36.	+ 55 42 30.	48	16.	GALAXY
MCG-02-24-025	09 24 36.	- 11 53	120	13.	GALAXY
LB 03494	09 24 36.	- 71 33		13.0	FAINT BLUE STAR
MCG-04-23-003	09 24 39.	- 24 35 30.	24	14.	GALAXY
ZWG 152.004	09 24 42.	+ 29 01		15.7	GALAXY
ZWG 151.085	09 24 42.	+ 29 01		15.7	GALAXY
UGC 05040	09 24 42.	+ 29 01	132	15.7	GALAXY IV-V
MCG+05-22-048	09 24 42.	+ 29 01	132	15.7	GALAXY
MCG+06-21-039	09 24 42.	+ 32 53	42	15.	GALAXY
ZWG 238.058	09 24 42.	+ 44 44		14.7	GALAXY
MRK 112	09 24 42.	+ 48 18	8	16.5	GALAXY WITH UV CONTINUUM
MCG+09-16-021	09 24 42.	+ 55 42 30.	36	17.	GALAXY
TON-N 1061	09 24 43.	+ 19 59		16.4	BLUE STAR
FATH 1.278	09 24 43.	+ 29 46	14		NEBULA
MAI 041	09 24 44.	+ 74 38	80		DWARF SPHEROIDAL GALAXY
HOLM 131B	09 24 45.	+ 12 30	30	15.2	PART OF MULTIPLE GALAXY
MCG+08-17-104	09 24 45.	+ 44 52	72	14.	GALAXY
RNGC 2889	09 24 47.	- 11 24		12.5	GALAXY
ZWG 034.047	09 24 48.	+ 04 09		14.5	GALAXY
UGC 05040A	09 24 48.	+ 04 09	54	14.5	GALAXY S(a-b)
MCG+01-24-022	09 24 48.	+ 04 09	60	14.5	GALAXY
ZWG 181.046	09 24 48.	+ 32 52		15.4	GALAXY
UGC 05042	09 24 48.	+ 66 42	72	16.0	GALAXY DISTRBD
MCG-02-24-026	09 24 48.	- 11 26	108	12.	GALAXY
FATH 1.279	09 24 49.	+ 29 38	16		NEBULA
MCG-03-24-009	09 24 51.	- 19 18 30.	60	15.	GALAXY
IC 2481	09 24 52.	+ 04 09 57.			NONSTELLAR OBJECT
ZWG 062.036	09 24 54.	+ 12 29		15.6	GALAXY
MCG+02-24-012	09 24 54.	+ 12 29	42	15.6	GALAXY
ZWG 121.100	09 24 54.	+ 24 09		15.4	GALAXY
MCG+05-22-049	09 24 54.	+ 30 01	48	15.7	GALAXY
ZWG 152.005	09 24 54.	+ 30 13		14.5	GALAXY
ZWG 151.086	09 24 54.	+ 30 13		14.5	GALAXY
UGC 05043	09 24 54.	+ 30 13	78	14.5	GALAXY E-S0
MCG+08-17-105	09 24 54.	+ 48 42	36	16.	GALAXY
ZWG 238.059	09 24 54.	+ 48 43		15.2	GALAXY
HOLM 131A	09 24 55.	+ 12 31	42	14.4	PART OF MULTIPLE GALAXY
IC 2484	09 24 55.	- 42 37 30.			NONSTELLAR OBJECT
IC 2476	09 24 56.	+ 30 12	19		BRIGHT NEBULA
FATH 2.013	09 24 56.	+ 30 12	19		NEBULA
IC 2475	09 24 57.	+ 30 01	27		BRIGHT NEBULA
FATH 2.014	09 24 57.	+ 30 01	27		NEBULA
IC 2478	09 24 58.	+ 30 14	8		FAINT NEBULA
IC 2477	09 24 59.	+ 29 55 28.			NONSTELLAR OBJECT
HMS 0925+2044	09 25	+ 20 44			CLUSTER OF GALAXIES
ZWG 034.048	09 25 00.	+ 07 17		15.7	GALAXY
ZWG 062.037	09 25 00.	+ 12 30		14.7	GALAXY
UGC 05044	09 25 00.	+ 12 30	66	14.7	GALAXY DBL SYS
KARA.72 204B	09 25 00.	+ 12 30	24		PART OF DOUBLE GALAXY
KARA.72 204A	09 25 00.	+ 12 30	24	14.7	PART OF DOUBLE GALAXY
MCG+02-24-014	09 25 00.	+ 12 30	48	14.7	GALAXY
MCG+02-24-013	09 25 00.	+ 12 30	48	14.7	GALAXY
ZC 0925.0+2042	09 25 00.	+ 20 42	2220		CLUSTER OF GALAXIES
LB 10763	09 25 00.	+ 25 43		16.4	FAINT BLUE STAR
LB 10762	09 25 00.	+ 26 20		17.3	FAINT BLUE STAR
ZWG 152.006	09 25 00.	+ 30 01		15.7	GALAXY
ZWG 151.087	09 25 00.	+ 30 01		15.7	GALAXY
MCG+05-23-001	09 25 00.	+ 30 13	72	14.5	GALAXY
ZWG 152.007	09 25 00.	+ 30 16		15.6	GALAXY
ZWG 151.088	09 25 00.	+ 30 16		15.6	GALAXY
ZWG 238.060	09 25 00.	+ 44 53		14.3	GALAXY
UGC 05045	09 25 00.	+ 44 53	78	14.3	GALAXY Sc/SBc
ZWG 265.021	09 25 00.	+ 51 02		15.2	GALAXY
MCG+11-12-012	09 25 00.	+ 66 41	66	16.	GALAXY
ZC 0925.0+8441	09 25 00.	+ 84 41	4840		CLUSTER OF GALAXIES
VHA 067	09 25 00.	- 51 06	240		OPEN STAR CLUSTER
ARP 237	09 25 01.	+ 12 30			PECULIAR GALAXY
FATH 2.015	09 25 02.	+ 30 14	8		NEBULA
LB 03037	09 25 05.	+ 19 56 06.		15.5	FAINT BLUE STAR
ZC 0925.1+0246	09 25 06.	+ 02 46	340		CLUSTER OF GALAXIES
LB 10766	09 25 06.	+ 21 37		16.6	FAINT BLUE STAR
LB 10765	09 25 06.	+ 22 13		15.8	FAINT BLUE STAR
LB 10764	09 25 06.	+ 25 34		16.9	FAINT BLUE STAR
TON-N 0420	09 25 06.	+ 25 48		16.5	BLUE STAR
MCG+05-23-002	09 25 06.	+ 30 13	24	15.5	GALAXY
ZWG 152.008	09 25 06.	+ 30 14		15.5	GALAXY
ZWG 151.089	09 25 06.	+ 30 14		15.5	GALAXY
MCG+05-23-003	09 25 06.	+ 30 16	30	15.6	GALAXY
ZC 0925.1+6518	09 25 06.	+ 65 18	1880		CLUSTER OF GALAXIES
IC 2479	09 25 08.	+ 30 13	19		BRIGHT NEBULA
FATH 2.016	09 25 08.	+ 30 13	19		NEBULA
ARC 0801	09 25 11.	+ 20 47		17.7	RICH CLUSTER OF GALAXIES
IC 2485	09 25 11.	- 39 04			NONSTELLAR OBJECT
ZWG 152.009	09 25 12.	+ 26 37		15.6	GALAXY
ZWG 151.090	09 25 12.	+ 26 37		15.6	GALAXY
IC 2480	09 25 12.	+ 29 56	14		BRIGHT NEBULA
ZWG 152.010	09 25 12.	+ 30 28		15.7	GALAXY
ZWG 151.091	09 25 12.	+ 30 28		15.7	GALAXY
HN 0317	09 25 12.	- 39 04			NEBULA
FATH 1.280	09 25 13.	+ 29 51	8		NEBULA
MCG+03-24-060	09 25 18.	+ 17 24	45	14.5	GALAXY
ZWG 091.097	09 25 18.	+ 17 25		14.5	GALAXY
UGC 05046	09 25 18.	+ 17 25	66	14.5	GALAXY S?
ZC 0925.3+1848	09 25 18.	+ 18 48	270		CLUSTER OF GALAXIES
LB 10767	09 25 18.	+ 20 51		16.9	FAINT BLUE STAR
TON-N 0421	09 25 18.	+ 26 48		16.4	BLUE STAR
ZWG 152.011	09 25 18.	+ 30 25		15.6	GALAXY
ZWG 151.092	09 25 18.	+ 30 25		15.6	GALAXY
TON-N 1062	09 25 18.	+ 36 36		17.	BLUE STAR
UGC 05047	09 25 18.	+ 51 47	96	16.5	GALAXY
MCG-03-24-010	09 25 18.	- 19 29	42	15.	GALAXY
FATH 2.017	09 25 21.	+ 29 56	14		NEBULA
MCG+09-16-022	09 25 21.	+ 50 59	54	15.	GALAXY
ZWG 034.049	09 25 24.	+ 02 47		15.5	GALAXY
ZWG 152.012	09 25 24.	+ 26 38		15.5	GALAXY
ZWG 151.093	09 25 24.	+ 26 38		15.5	GALAXY
ZWG 152.013	09 25 24.	+ 29 56		15.5	GALAXY
ZWG 151.094	09 25 24.	+ 29 56		15.5	GALAXY
TON-N 1063	09 25 24.	+ 35 24		16.	BLUE STAR
ARC 0800	09 25 25.	+ 38 01		17.7	RICH CLUSTER OF GALAXIES
ARP 207	09 25 25.	+ 76 41			PECULIAR GALAXY
MCG-04-23-004	09 25 27.	- 24 01	36	15.5	GALAXY
PK277-03.1	09 25 27.	- 55 53 43.	120		PLANETARY NEBULA
ZWG 006.046	09 25 30.	+ 01 53		15.7	GALAXY
KARA.73B 0337	09 25 30.	+ 01 53	42	15.7	ISOLATED GALAXY S
LB 10769	09 25 30.	+ 23 36		16.7	FAINT BLUE STAR
LB 10768	09 25 30.	+ 24 40		18.7	FAINT BLUE STAR
TON-N 0422	09 25 30.	+ 26 53		16.2	BLUE STAR
ZC 0925.5+3517	09 25 30.	+ 35 17	1140		CLUSTER OF GALAXIES
ZC 0925.5+3800	09 25 30.	+ 38 00	870		CLUSTER OF GALAXIES
ZWG 210.010	09 25 30.	+ 38 45		15.2	GALAXY
UGC 05048	09 25 30.	+ 38 45	78	15.2	GALAXY IRR
ZWG 238.061	09 25 30.	+ 49 26		15.7	GALAXY
UGC 05049	09 25 30.	+ 49 26	72	15.7	GALAXY Sc
MCG+10-14-014	09 25 30.	+ 58 31	15	16.	GALAXY

OBJECT NAME	RIGHT ASCEN.	DECLINATION	DIAM.	MAGN.	TYPE OF OBJECT
MRK 113	09 25 30.	+ 62 53	13	14.5	GALAXY WITH UV CONTINUUM
VV 058B	09 25 30.	+ 76 41 30.	3	19.	INTERACTING GALAXY
VV 058A	09 25 30.	+ 76 41 30.	48	14.	INTERACTING GALAXY
VV 058	09 25 30.	+ 76 41 30.	72		INTERACTING GALAXY
ZWG 006.045	09 25 30.	- 02 13		15.7	GALAXY
MCG+00-24-016	09 25 30.	- 02 13	36	15.7	GALAXY
OCL 0784	09 25 30.	- 54 54	120	14.	OPEN STAR CLUSTER
VHA 068	09 25 30.	- 54 54	300		OPEN STAR CLUSTER
RNGC 2899	09 25 32.	- 55 54			PLANETARY NEBULA
FATH 1.281	09 25 35.	+ 29 45	5		NEBULA
ZWG 034.050	09 25 36.	+ 03 38		14.6	GALAXY
MCG+01-24-023	09 25 36.	+ 03 38	36	14.6	GALAXY
KARA.73B 0338	09 25 36.	+ 03 38	60	14.6	ISOLATED GALAXY S
LB 10771	09 25 36.	+ 24 30		17.5	FAINT BLUE STAR
LB 10770	09 25 36.	+ 24 40		17.3	FAINT BLUE STAR
TON-N 0423	09 25 36.	+ 27 24		15.3	BLUE STAR
MCG+07-20-013	09 25 36.	+ 38 44	72	15.	GALAXY
MCG+08-17-106	09 25 36.	+ 49 26	66	15.	GALAXY
MCG+09-16-023	09 25 36.	+ 51 45	72	16.	GALAXY
ZWG 350.025	09 25 36.	+ 76 41		15.4	GALAXY
UGC 05050	09 25 36.	+ 76 41	72	15.4	GALAXY S?
RNGC 2880	09 25 39.	+ 62 43		13.0	GALAXY
MCG+04-23-001	09 25 42.	+ 20 45	7		GALAXY
ZWG 122.001	09 25 42.	+ 21 35		15.6	GALAXY
LB 10772	09 25 42.	+ 26 18		17.1	FAINT BLUE STAR
ZC 0925.7+5626	09 25 42.	+ 56 26	810		CLUSTER OF GALAXIES
ZC 0925.7+6132	09 25 42.	+ 61 32	540		CLUSTER OF GALAXIES
ZWG 312.011	09 25 42.	+ 62 43		12.6	GALAXY
UGC 05051	09 25 42.	+ 62 43	144	12.6	GALAXY SB0
ZWG 312.012	09 25 42.	+ 62 46		15.1	GALAXY
MCG+13-07-030	09 25 42.	+ 76 42	66	15.	GALAXY
MCG+10-14-015	09 25 45.	+ 62 42	39	12.6	GALAXY
MCG-03-24-011	09 25 45.	- 16 40	42	15.	GALAXY
LB 10774	09 25 48.	+ 21 00		16.8	FAINT BLUE STAR
LB 10773	09 25 48.	+ 22 38		16.3	FAINT BLUE STAR
ZC 0925.8+4305	09 25 48.	+ 43 05	1340		CLUSTER OF GALAXIES
LB 03038	09 25 54.	+ 18 54 42.		16.6	FAINT BLUE STAR
LB 10776	09 25 54.	+ 23 24		18.5	FAINT BLUE STAR
LB 10775	09 25 54.	+ 25 19		17.6	FAINT BLUE STAR
ZWG 062.038	09 26 00.	+ 12 50		15.1	GALAXY
MCG+02-24-015	09 26 00.	+ 12 50	48	15.1	GALAXY
ZC 0926.0+4813	09 26 00.	+ 48 13	670		CLUSTER OF GALAXIES
ZC 0926.0+5135	09 26 00.	+ 51 35	1210		CLUSTER OF GALAXIES
ZC 0926.0+5700	09 26 00.	+ 57 00	1280		CLUSTER OF GALAXIES
ZC 0926.0+5730	09 26 00.	+ 57 30	940		CLUSTER OF GALAXIES
MCG+10-14-016	09 26 00.	+ 60 04	18	16.	GALAXY
SHB 138	09 26 01.1	+ 11 47 32.		19.1	QUASI-STELLAR OBJECT
BC 4C11.32	09 26 02.3	+ 11 47 16.		19.06	QUASI-STELLAR OBJECT
FATH 1.282	09 26 04.	+ 30 18	19		NEBULA
ZC 0926.1+0013	09 26 06.	+ 00 13	270		CLUSTER OF GALAXIES
ZC 0926.1+2115	09 26 06.	+ 21 15	870		CLUSTER OF GALAXIES
ZC 0926.1+3809	09 26 06.	+ 38 09	1210		CLUSTER OF GALAXIES
MCG+08-17-107	09 26 06.	+ 44 37	42	16.	GALAXY
IC 2488	09 26 06.	- 56 45			OPEN CLUSTER
OCL 0789	09 26 06.	- 56 46	2580	7.5	OPEN STAR CLUSTER
VHA 069	09 26 06.	- 56 46	600		OPEN STAR CLUSTER
FATH 1.283	09 26 07.	+ 29 41	16		NEBULA
MAI 042	09 26 08.	+ 63 24	47		DWARF SPHEROIDAL GALAXY
ZC 0926.2+0105	09 26 12.	+ 01 05	340		CLUSTER OF GALAXIES
LB 10777	09 26 12.	+ 24 57		16.5	FAINT BLUE STAR
FATH 1.284	09 26 12.	+ 30 21	22		NEBULA
ZWG 152.014	09 26 12.	+ 30 22		15.4	GALAXY
TON-N 1064	09 26 12.	+ 35 21		17.	BLUE STAR
MCG+08-17-108	09 26 12.	+ 45 29	30	16.	GALAXY
ZWG 332.052	09 26 12.	+ 74 01		14.2	GALAXY
UGC 05052	09 26 12.	+ 74 01	102	14.2	GALAXY SBa
MCG+05-23-004	09 26 15.	+ 30 23	42	15.4	GALAXY
ZC 0926.3+0418	09 26 18.	+ 04 18	670		CLUSTER OF GALAXIES
ZWG 062.039	09 26 18.	+ 12 43		15.4	GALAXY
LB 03039	09 26 18.	+ 21 00 06.		16.3	FAINT BLUE STAR
ZWG 152.015	09 26 18.	+ 26 40		15.5	GALAXY
ZWG 152.016	09 26 18.	+ 30 37		15.7	GALAXY
TON-N 1065	09 26 18.	+ 33 53		16.5	BLUE STAR
ZWG 238.062	09 26 18.	+ 44 39		15.1	GALAXY
ZWG 238.063	09 26 18.	+ 45 30		15.7	GALAXY
UGC 05053	09 26 18.	+ 60 22	72	17.	GALAXY
LB 00555	09 26 18.	+ 60 40 12.		16.1	FAINT BLUE STAR
ZWG 350.026	09 26 18.	+ 76 46		15.6	GALAXY
MCG-04-23-005	09 26 18.	- 24 10	48	16.	GALAXY
MCG-06-21-007	09 26 18.	- 35 56 30.	60	13.5	GALAXY
ARC 0805	09 26 22.	+ 04 20		17.5	RICH CLUSTER OF GALAXIES
LB 10783	09 26 24.	+ 23 22		18.6	FAINT BLUE STAR
LB 10782	09 26 24.	+ 24 55		17.2	FAINT BLUE STAR
LB 10781	09 26 24.	+ 24 55		18.3	FAINT BLUE STAR
LB 10780	09 26 24.	+ 25 04		17.2	FAINT BLUE STAR
LB 10779	09 26 24.	+ 25 58		17.0	FAINT BLUE STAR
LB 10778	09 26 24.	+ 26 12		17.0	FAINT BLUE STAR
MCG+06-21-040	09 26 24.	+ 35 36 30.	36	15.5	GALAXY
ZWG 181.047	09 26 24.	+ 36 25		15.2	GALAXY
MCG-06-21-009	09 26 24.	- 35 54	24	16.5	GALAXY
MCG-06-21-008	09 26 24.	- 35 54	30	16.	GALAXY
IC 2483	09 26 26.	+ 31 12 33.			NONSTELLAR OBJECT
RNGC 2915	09 26 27.	- 76 24			GALAXY
ARC 0803	09 26 28.	+ 12 21		17.5	RICH CLUSTER OF GALAXIES
ZC 0926.5+1222	09 26 30.	+ 12 22	1810		CLUSTER OF GALAXIES
LB 10787	09 26 30.	+ 22 43		17.2	FAINT BLUE STAR
LB 10786	09 26 30.	+ 23 02		16.1	FAINT BLUE STAR
LB 10785	09 26 30.	+ 23 24		17.4	FAINT BLUE STAR
LB 10784	09 26 30.	+ 24 44		17.5	FAINT BLUE STAR
MCG+04-23-002	09 26 30.	+ 25 18 30.	48	15.6	GALAXY
TON-N 0424	09 26 30.	+ 29 18		16.6	BLUE STAR
ZC 0926.5+3026	09 26 30.	+ 30 26	7060		CLUSTER OF GALAXIES
ZWG 152.017	09 26 30.	+ 31 13		15.7	GALAXY
MCG+07-20-014	09 26 30.	+ 39 33 30.	42	16.	GALAXY
ARC 0807	09 26 32.	- 06 14		17.9	RICH CLUSTER OF GALAXIES
MCG+06-21-041	09 26 33.	+ 35 35	42	15.	GALAXY
HOLM 132B	09 26 34.	+ 25 47	18	15.0	PART OF MULTIPLE GALAXY
IC 0539	09 26 35.	- 02 20 12.			NONSTELLAR OBJECT
LB 10789	09 26 36.	+ 22 52		16.0	FAINT BLUE STAR
ZWG 122.002	09 26 36.	+ 25 18		15.7	GALAXY
LB 10788	09 26 36.	+ 25 41		17.4	FAINT BLUE STAR
ZWG 122.003	09 26 36.	+ 25 46		15.2	GALAXY
HOLM 132A	09 26 36.	+ 25 46	24	14.6	PART OF MULTIPLE GALAXY
MCG+04-23-003	09 26 36.	+ 25 46	30	15.2	GALAXY
TON-N 1066	09 26 36.	+ 35 46		17.	BLUE STAR
ZWG 265.022	09 26 36.	+ 56 04		14.5	GALAXY
MRK 114	09 26 36.	+ 56 04	30	14.	GALAXY WITH UV CONTINUUM
UGC 05055	09 26 36.	+ 56 04	90	14.5	GALAXY SBb
KARA.73B 0339	09 26 36.	+ 56 04	96	14.5	ISOLATED GALAXY S
MCG+09-16-024	09 26 36.	+ 56 04 30.	90	14.	GALAXY
MCG+13-07-031	09 26 36.	+ 74 50	24	16.	GALAXY
ZWG 006.047	09 26 36.	- 02 20		14.3	GALAXY
UGC 05054	09 26 36.	- 02 20	72	14.3	GALAXY Sc
MCG+00-24-017	09 26 36.	- 02 20	48	14.3	GALAXY
ARC 0799	09 26 37.	+ 58 58		17.7	RICH CLUSTER OF GALAXIES
MCG+04-23-004	09 26 39.	+ 24 06	42		GALAXY
MCG-02-24-027	09 26 39.	- 14 36	72	14.	GALAXY
ZC 0926.7+4851	09 26 42.	+ 48 51	540		CLUSTER OF GALAXIES
RNGC 2894	09 26 46.	+ 07 57		13.5	GALAXY
RNGC 2885	09 26 46.	+ 23 11			GALAXY
SEY 055	09 26 47.	+ 67 59 47.		15.2	FAINT GALAXY
ZWG 034.051	09 26 48.	+ 07 57		13.4	GALAXY
UGC 05056	09 26 48.	+ 07 57	138	13.4	GALAXY Sa
MCG+01-24-024	09 26 48.	+ 07 57	96	13.4	GALAXY
ZWG 062.040	09 26 48.	+ 11 52		15.5	GALAXY
ZC 0926.8+1520	09 26 48.	+ 15 20	5170		CLUSTER OF GALAXIES
LB 10791	09 26 48.	+ 22 48		17.6	FAINT BLUE STAR
LB 10790	09 26 48.	+ 23 11		17.7	FAINT BLUE STAR
ZWG 122.004	09 26 48.	+ 24 49		15.6	GALAXY
ZWG 239.001	09 26 48.	+ 49 42		14.6	GALAXY
ZWG 238.064	09 26 48.	+ 49 42		14.6	GALAXY
HOLM 133B	09 26 50.	+ 07 56	24	14.8	PART OF MULTIPLE GALAXY
MCG+06-21-042	09 26 51.	+ 34 28	18	15.	GALAXY
MCG-03-24-012	09 26 51.	- 20 10	78	14.5	GALAXY
HOLM 133A	09 26 52.	+ 07 56	24	14.0	PART OF MULTIPLE GALAXY
ZWG 122.005	09 26 54.	+ 21 39		15.5	GALAXY
MCG+04-23-005	09 26 54.	+ 21 39	30	15.5	GALAXY
LB 10793	09 26 54.	+ 22 58		17.2	FAINT BLUE STAR
ZWG 122.006	09 26 54.	+ 23 10		15.5	GALAXY
ZC 0926.9+2506	09 26 54.	+ 25 06	2690		CLUSTER OF GALAXIES
LB 10792	09 26 54.	+ 26 18		16.6	FAINT BLUE STAR
MCG+08-17-109	09 26 54.	+ 49 41 30.	18	15.	GALAXY
UGC 05057	09 26 54.	- 01 07	66	16.5	GALAXY CHAIN
SEY 056	09 26 59.	+ 67 50 22.		15.4	FAINT GALAXY
ZWG 122.007	09 27 00.	+ 21 40		15.4	GALAXY
MCG+04-23-006	09 27 00.	+ 21 40	42	15.4	GALAXY
LB 10794	09 27 00.	+ 25 06		17.1	FAINT BLUE STAR
TON-N 0425	09 27 00.	+ 31 56		16.4	BLUE STAR
ZWG 181.048	09 27 00.	+ 34 27		15.7	GALAXY
ZC 0927.0+4721	09 27 00.	+ 47 21	670		CLUSTER OF GALAXIES
FATH 1.285	09 27 03.	+ 30 21	8		NEBULA
FATH 1.286	09 27 05.	+ 30 13	14		NEBULA
LB 10796	09 27 06.	+ 22 37		16.6	FAINT BLUE STAR
LB 10795	09 27 06.	+ 25 10		16.8	FAINT BLUE STAR
ZWG 181.049	09 27 06.	+ 34 53		15.5	GALAXY
ZWG 181.050	09 27 06.	+ 36 54		15.2	GALAXY
ZC 0927.1+3845	09 27 06.	+ 38 45	940		CLUSTER OF GALAXIES
ZC 0927.1+4408	09 27 06.	+ 44 08	1140		CLUSTER OF GALAXIES
ZWG 312.013	09 27 06.	+ 68 00		15.4	GALAXY
UGC 05058	09 27 06.	+ 68 00	72	15.4	GALAXY Sc
ZWG 332.053	09 27 06.	+ 69 59		15.5	GALAXY
MCG+06-21-043	09 27 09.	+ 36 56	42	15.	GALAXY
KEEL 177	09 27 10.4	+ 21 36 46.			NEBULA
ZWG 006.048	09 27 12.	+ 02 17		14.8	GALAXY
RNGC 2898	09 27 12.	+ 02 17		15.0	GALAXY
MCG+00-24-018	09 27 12.	+ 02 17	12	14.8	GALAXY
RNGC 2897	09 27 12.	+ 02 26			GALAXY
LB 10797	09 27 12.	+ 25 01		17.3	FAINT BLUE STAR
TON-N 0426	09 27 12.	+ 25 55		15.3	BLUE STAR
TON-N 0427	09 27 12.	+ 31 04		14.2	BLUE STAR
ZC 0927.2+3446	09 27 12.	+ 34 46	7330		CLUSTER OF GALAXIES
MCG+11-12-013	09 27 12.	+ 68 00	72	15.	GALAXY
ZWG 034.052	09 27 18.	+ 03 52		15.7	GALAXY
ZWG 091.098	09 27 18.	+ 20 17		14.2	GALAXY
UGC 05059	09 27 18.	+ 20 17	114	14.2	GALAXY Sb
KARA.73B 0340	09 27 18.	+ 20 17	102	14.2	ISOLATED GALAXY S
MCG+03-24-061	09 27 18.	+ 20 18	102	14.2	GALAXY
LB 10799	09 27 18.	+ 23 12		16.6	FAINT BLUE STAR
LB 10798	09 27 18.	+ 24 43		17.0	FAINT BLUE STAR
ZWG 152.018	09 27 18.	+ 29 46		13.6	GALAXY
MRK 401	09 27 18.	+ 29 46	30	14.	GALAXY WITH UV CONTINUUM
UGC 05060	09 27 18.	+ 29 46	72	13.6	GALAXY SBa
ZWG 152.019	09 27 18.	+ 30 40		15.6	GALAXY
ZWG 312.014	09 27 18.	+ 67 50		15.3	GALAXY
UGC 05061	09 27 18.	+ 67 50	72	15.3	GALAXY SBa
ZWG 006.049	09 27 18.	- 02 08		15.6	GALAXY
RNGC 2893	09 27 19.	+ 29 46		13.5	GALAXY
FATH 2.018	09 27 20.	+ 29 45	27		NEBULA
IC 2487	09 27 20.3	+ 20 18 39.			GALAXY
IC 2486	09 27 23.	+ 26 51 46.			NONSTELLAR OBJECT
ZWG 062.041	09 27 24.	+ 09 00		15.3	GALAXY
8ZW 0927+09.0	09 27 24.	+ 09 00		15.3	COMPACT GALAXY
ZWG 062.042	09 27 24.	+ 12 01		15.7	GALAXY
ZWG 062.043	09 27 24.	+ 14 11		15.6	GALAXY
KARA.73B 0341	09 27 24.	+ 14 11	12	15.6	ISOLATED GALAXY E
KARA.68 056	09 27 24.	+ 20 12	67		DWARF GALAXY
LB 10802	09 27 24.	+ 24 06		15.8	FAINT BLUE STAR
LB 10801	09 27 24.	+ 24 11		16.9	FAINT BLUE STAR
LB 10800	09 27 24.	+ 26 32		16.4	FAINT BLUE STAR
MCG+05-23-006	09 27 24.	+ 26 51 30.	42	15.1	GALAXY
ZWG 152.020	09 27 24.	+ 26 52		15.1	GALAXY
UGC 05062	09 27 24.	+ 26 52	66	15.1	GALAXY SBb-c
MCG+05-23-005	09 27 24.	+ 29 47	60	13.6	GALAXY
ZWG 239.002	09 27 24.	+ 49 28		15.4	GALAXY
ZWG 238.065	09 27 24.	+ 49 28		15.4	GALAXY
MRK 115	09 27 24.	+ 49 28	10	16.	GALAXY WITH UV CONTINUUM
UGC 05063	09 27 24.	+ 49 28	78	15.4	GALAXY PECULR
ARC 0808	09 27 27.	+ 07 55		17.4	RICH CLUSTER OF GALAXIES
KEEL 178	09 27 27.0	+ 21 35 02.			NEBULA
RNGC 2896	09 27 28.	+ 23 53		15.0	GALAXY
SHB 139	09 27 29.9	+ 36 14 37.		18.2	QUASI-STELLAR OBJECT
BC 3CR220.2	09 27 29.96	+ 36 14 36.7		18.4	QUASI-STELLAR OBJECT
ZWG 034.053	09 27 30.	+ 05 41		15.6	GALAXY
ZWG 034.054	09 27 30.	+ 08 07		14.8	GALAXY
UGC 05064	09 27 30.	+ 08 07	84	14.8	GALAXY S
MCG+01-24-025	09 27 30.	+ 08 07	60	14.8	GALAXY
ZWG 091.099	09 27 30.	+ 19 40		15.3	GALAXY
KARA.73B 0342	09 27 30.	+ 19 40	42	15.3	ISOLATED GALAXY S
ZWG 122.008	09 27 30.	+ 21 35		15.6	GALAXY
ZWG 122.009	09 27 30.	+ 23 53		14.8	GALAXY
MCG+04-23-007	09 27 30.	+ 23 53	48	14.8	GALAXY
LB 10805	09 27 30.	+ 24 28		17.6	FAINT BLUE STAR
LB 10804	09 27 30.	+ 25 10		17.8	FAINT BLUE STAR
LB 10803	09 27 30.	+ 25 39		16.7	FAINT BLUE STAR
MCG+11-12-014	09 27 30.	+ 67 52	42	16.	GALAXY
MCG-01-24-018	09 27 30.	- 06 26	30	15.5	GALAXY

OBJECT NAME	RIGHT ASCEN.	DECLINATION	DIAM.	MAGN.	TYPE OF OBJECT
IC 0540	09 27 31.	+ 08 07 20.			NONSTELLAR OBJECT
MCG+03-24-064	09 27 36.	+ 16 33	27	15.5	GALAXY
ZWG 091.100	09 27 36.	+ 16 34		15.7	GALAXY
KARA.72 205A	09 27 36.	+ 16 34	42	15.7	PART OF DOUBLE GALAXY
MCG+03-24-063	09 27 36.	+ 16 34	42	15.7	GALAXY
ZWG 091.101	09 27 36.	+ 16 35		15.5	GALAXY
KARA.72 205B	09 27 36.	+ 16 35	36	15.5	PART OF DOUBLE GALAXY
MCG+03-24-062	09 27 36.	+ 19 40	54	15.3	GALAXY
LB 10807	09 27 36.	+ 22 57		15.5	FAINT BLUE STAR
LB 10806	09 27 36.	+ 23 46		18.0	FAINT BLUE STAR
ZWG 152.021	09 27 36.	+ 28 00		15.2	GALAXY
TON-N 1067	09 27 36.	+ 34 40		17.	BLUE STAR
ZC 0927.6+5752	09 27 36.	+ 57 52	1680		CLUSTER OF GALAXIES
OCL 0780	09 27 36.	- 52 30	36	14.	OPEN STAR CLUSTER
OCL 0782	09 27 36.	- 53 27	36	15.	OPEN STAR CLUSTER
VHA 070	09 27 36.	- 53 27	210		OPEN STAR CLUSTER
ZWG 034.055	09 27 42.	+ 04 22		14.6	GALAXY
RNGC 2900	09 27 42.	+ 04 22		14.5	GALAXY
UGC 05065	09 27 42.	+ 04 22	96	14.6	GALAXY SBc
MCG+01-24-026	09 27 42.	+ 04 22	72	14.6	GALAXY
KARA.73B 0343	09 27 42.	+ 04 22	126	14.6	ISOLATED GALAXY S
LB 10808	09 27 42.	+ 20 58		16.0	FAINT BLUE STAR
ZWG 239.003	09 27 42.	+ 49 17		15.3	GALAXY
ZWG 238.066	09 27 42.	+ 49 17		15.3	GALAXY
ZWG 265.023	09 27 42.	+ 52 51		14.9	GALAXY
TON-N 1068	09 27 43.	+ 21 01		17.0	BLUE STAR
LB 10814	09 27 48.	+ 21 23		16.7	FAINT BLUE STAR
LB 10813	09 27 48.	+ 22 32		15.5	FAINT BLUE STAR
TON-N 0428	09 27 48.	+ 22 56		16.6	BLUE STAR
LB 10812	09 27 48.	+ 23 32		16.7	FAINT BLUE STAR
LB 10811	09 27 48.	+ 24 08		16.6	FAINT BLUE STAR
LB 10810	09 27 48.	+ 25 43		16.7	FAINT BLUE STAR
LB 10809	09 27 48.	+ 26 24		17.0	FAINT BLUE STAR
UGC 05041	09 27 48.	+ 32 52	78	15.4	GALAXY Sc
MCG+08-17-111	09 27 48.	+ 46 35	60	15.	GALAXY
ZWG 239.004	09 27 48.	+ 46 37		15.5	GALAXY
UGC 05066	09 27 48.	+ 46 37	72	15.5	GALAXY SBb-c
KARA.73B 0344	09 27 48.	+ 46 37	60	15.5	ISOLATED GALAXY S
ZWG 239.005	09 27 48.	+ 48 15		15.7	GALAXY
ZWG 238.067	09 27 48.	+ 48 15		15.7	GALAXY
MCG+08-17-110	09 27 48.	+ 48 16	90	13.	GALAXY
MCG+09-16-025	09 27 48.	+ 55 48	48	16.	GALAXY
LB 03040	09 27 49.	+ 19 02 06.		15.4	FAINT BLUE STAR
LB 10815	09 27 54.	+ 22 44		17.0	FAINT BLUE STAR
ZC 0927.9+4119	09 27 54.	+ 41 19	3760		CLUSTER OF GALAXIES
ZC 0927.9+4252	09 27 54.	+ 42 52	1080		CLUSTER OF GALAXIES
REIZ 0001	09 27 54.	+ 46 37	60	14.2	GALAXY
UGC 05067	09 27 54.	+ 68 40	84	16.5	GALAXY DWRF SP
ZWG 006.050	09 27 54.	- 01 38		15.7	GALAXY
KARA.73B 0345	09 27 54.	- 01 38	48	15.7	ISOLATED GALAXY S
MCG-03-24-013	09 27 54.	- 16 25	48	15.	GALAXY
LBN 1088	09 28	- 31 00	6000		BRIGHT NEBULA
UGC 05069	09 28 00.	+ 06 20	72	15.0	GALAXY S0
LB 10816	09 28 00.	+ 23 58		16.3	FAINT BLUE STAR
ZWG 122.010	09 28 00.	+ 25 58		15.6	GALAXY
MCG+04-23-008	09 28 00.	+ 25 58	36	15.6	GALAXY
ZC 0928.0+2904	09 28 00.	+ 29 04	670		CLUSTER OF GALAXIES
ZWG 364.011	09 28 00.	+ 81 22		14.3	GALAXY
UGC 05068	09 28 00.	+ 81 22	54	14.3	GALAXY S
KARA.73B 0346	09 28 00.	+ 81 22	54	14.3	ISOLATED GALAXY S
ZC 0928.0-0203	09 28 00.	- 02 03	2890		CLUSTER OF GALAXIES
MCG-05-23-003	09 28 00.	- 30 09	24	13.5	GALAXY
MCG-05-23-002	09 28 00.	- 30 23	48	16.	GALAXY
IC 0541	09 28 01.	- 04 01 43.			NONSTELLAR OBJECT
RNGC 2904	09 28 01.	- 30 09		13.0	GALAXY
LB 03041	09 28 04.	+ 19 37 18.		16.0	FAINT BLUE STAR
LB 10817	09 28 06.	+ 23 35		16.5	FAINT BLUE STAR
ZC 0928.1+3943	09 28 06.	+ 39 43	470		CLUSTER OF GALAXIES
ARC 0813	09 28 08.	- 14 21		17.5	RICH CLUSTER OF GALAXIES
TON-N 0429	09 28 12.	+ 24 20		15.4	BLUE STAR
TON-N 1069	09 28 12.	+ 33 56		16.8	BLUE STAR
UGC 05166	09 28 12.	+ 36 07	90	13.9	GALAXY Sb
FATH 1.287	09 28 13.	+ 29 44	19		NEBULA
ARC 0810	09 28 14.	- 01 57		17.4	RICH CLUSTER OF GALAXIES
MCG-02-24-028	09 28 15.	- 11 48 30.	36	15.	GALAXY
FATH 1.288	09 28 17.	+ 30 15	54		NEBULA
ZWG 122.011	09 28 18.	+ 22 58		15.7	GALAXY
ZWG 152.022	09 28 18.	+ 30 15		14.5	GALAXY
UGC 05070	09 28 18.	+ 30 15	66	14.5	GALAXY S
UGC 05071	09 28 18.	+ 42 34		18.	GALAXY DWARF?
MCG+08-18-001	09 28 18.	+ 48 49	36	16.	GALAXY
MCG+05-23-007	09 28 21.	+ 30 16	48	14.5	GALAXY
RNGC 2910	09 28 22.	- 52 41		8.5	OPEN CLUSTER
ZWG 034.056	09 28 24.	+ 06 47		15.4	GALAXY
ZWG 122.012	09 28 24.	+ 21 08		15.6	GALAXY
LB 10823	09 28 24.	+ 21 24		17.6	FAINT BLUE STAR
LB 10822	09 28 24.	+ 22 08		16.4	FAINT BLUE STAR
LB 10821	09 28 24.	+ 22 22		16.5	FAINT BLUE STAR
LB 10820	09 28 24.	+ 23 18		15.0	FAINT BLUE STAR
ZWG 122.013	09 28 24.	+ 25 28		15.1	GALAXY
LB 10819	09 28 24.	+ 25 30		16.2	FAINT BLUE STAR
LB 10818	09 28 24.	+ 26 38		16.4	FAINT BLUE STAR
ZWG 210.011	09 28 24.	+ 41 32		15.7	GALAXY
UGC 05072	09 28 24.	+ 41 32	60	15.7	GALAXY SBb
ZC 0928.5+0929	09 28 30.	+ 09 29	540		CLUSTER OF GALAXIES
MCG+07-20-015	09 28 30.	+ 41 31	54	14.5	GALAXY
MCG+10-14-017	09 28 30.	+ 57 47	39	16.	GALAXY
RNGC 2902	09 28 30.	- 14 29		13.5	GALAXY
ARC 0806	09 28 34.	+ 56 15		17.5	RICH CLUSTER OF GALAXIES
ARC 0816	09 28 34.	- 13 11		17.4	RICH CLUSTER OF GALAXIES
LB 10825	09 28 36.	+ 23 02		17.1	FAINT BLUE STAR
LB 10824	09 28 36.	+ 25 56		17.0	FAINT BLUE STAR
ZWG 152.023	09 28 36.	+ 26 42		15.7	GALAXY
ZC 0928.6+3617	09 28 36.	+ 36 17	740		CLUSTER OF GALAXIES
ZC 0928.6+4932	09 28 36.	+ 49 32	670		CLUSTER OF GALAXIES
ZC 0928.6+5610	09 28 36.	+ 56 10	1480		CLUSTER OF GALAXIES
ZWG 312.015	09 28 36.	+ 67 50		14.4	GALAXY
UGC 05073	09 28 36.	+ 67 50	90	14.4	GALAXY E
MCG-02-24-029	09 28 36.	- 12 28 30.	24	15.	GALAXY
MCG-02-24-030	09 28 36.	- 14 31 30.	36	13.	GALAXY
LB 03042	09 28 41.	+ 20 58 54.		16.7	FAINT BLUE STAR
FATH 1.289	09 28 41.	+ 29 38	5		NEBULA
FATH 1.290	09 28 41.	+ 30 11	11		NEBULA
LB 00556	09 28 41.	+ 59 28 06.		17.0	FAINT BLUE STAR
RNGC 2892	09 28 41.	+ 67 50		14.5	GALAXY
LB 10826	09 28 42.	+ 25 00		17.5	FAINT BLUE STAR
ZC 0928.7+3456	09 28 42.	+ 34 56	1140		CLUSTER OF GALAXIES
MCG+08-18-002	09 28 42.	+ 48 34 30.	60	16.	GALAXY
IC 2489	09 28 42.	- 05 39 46.			NONSTELLAR OBJECT
OCL 0781	09 28 42.	- 52 41	780	8.4	OPEN STAR CLUSTER
VHA 071	09 28 42.	- 52 41	360		OPEN STAR CLUSTER
FATH 1.291	09 28 43.	+ 29 21	8		NEBULA
IC 0542	09 28 43.	- 12 57 47.			NONSTELLAR OBJECT
MCG+11-12-015	09 28 45.	+ 67 51	30	14.	GALAXY
IC 0543	09 28 46.	- 14 34			NONSTELLAR OBJECT
RNGC 2895	09 28 47.	+ 57 43		14.5	GALAXY
ZWG 034.057	09 28 48.	+ 05 30		15.6	GALAXY
LB 03043	09 28 48.	+ 20 05 42.		15.8	FAINT BLUE STAR
ZC 0928.8+2034	09 28 48.	+ 20 34	940		CLUSTER OF GALAXIES
TON-N 1070	09 28 48.	+ 33 26		16.9	BLUE STAR
ZWG 181.051	09 28 48.	+ 35 46		15.5	GALAXY
MCG+06-21-044	09 28 48.	+ 35 48	21	15.	GALAXY
ZWG 289.009	09 28 48.	+ 57 43		14.7	GALAXY
7ZW 281	09 28 48.	+ 61 41			COMPACT GALAXY
MCG-02-24-031	09 28 48.	- 12 57	60	15.	GALAXY
MCG-05-23-004	09 28 48.	- 30 07	42	14.	GALAXY
KEEL 179	09 28 48.5	+ 21 32 17.			NEBULA
LB 03044	09 28 52.	+ 19 37 42.		15.2	FAINT BLUE STAR
ARC 0804	09 28 52.	+ 62 47		17.4	RICH CLUSTER OF GALAXIES
ZWG 034.058	09 28 54.	+ 07 10		15.6	GALAXY
ZWG 152.024	09 28 54.	+ 30 01		15.0	GALAXY
UGC 05074	09 28 54.	+ 30 01	60	15.0	GALAXY Sa?
MCG+06-21-045	09 28 54.	+ 35 47	36	15.	GALAXY
MCG+10-14-018	09 28 54.	+ 57 43	45	14.	GALAXY
MCG+10-14-019	09 28 54.	+ 62 33	39	15.	GALAXY
ZWG 312.016	09 28 54.	+ 62 34		15.6	GALAXY
ZC 0928.9+6248	09 28 54.	+ 62 48	1810		CLUSTER OF GALAXIES
MCG-06-21-010	09 28 54.	- 35 27	120	13.5	GALAXY
FATH 1.292	09 28 56.	+ 30 00	41		NEBULA
MCG+05-23-008	09 28 57.	+ 30 02	66	15.0	GALAXY
ZWG 035.001	09 29 00.	+ 04 43		14.6	GALAXY
UGC 05075	09 29 00.	+ 04 43	66	14.6	GALAXY S0-a?
MCG+01-25-001	09 29 00.	+ 04 43	60	14.6	GALAXY
LB 10829	09 29 00.	+ 21 55		17.0	FAINT BLUE STAR
LB 10828	09 29 00.	+ 24 43		17.3	FAINT BLUE STAR
LB 10827	09 29 00.	+ 24 54		17.5	FAINT BLUE STAR
ZWG 181.052	09 29 00.	+ 35 45		15.4	GALAXY
ZC 0929.0+3805	09 29 00.	+ 38 05	870		CLUSTER OF GALAXIES
MCG+10-14-020	09 29 00.	+ 59 58	102	15.	GALAXY
ZC 0929.0+6542	09 29 00.	+ 65 42	610		CLUSTER OF GALAXIES
MCG-03-25-001	09 29 00.	- 16 22	72	15.	GALAXY
ZC 0929.1+0047	09 29 06.	+ 00 47	1550		CLUSTER OF GALAXIES
ZWG 035.002	09 29 06.	+ 07 38		15.6	GALAXY
ZWG 035.003	09 29 06.	+ 07 48		15.6	GALAXY
ZC 0929.1+2240	09 29 06.	+ 22 40	740		CLUSTER OF GALAXIES
TON-N 0430	09 29 06.	+ 28 24		16.0	BLUE STAR
ZWG 265.024	09 29 06.	+ 52 06		15.5	GALAXY
UGC 05076	09 29 06.	+ 52 06	90	15.5	GALAXY DWARF
MCG+10-14-021	09 29 06.	+ 58 58	18	16.	GALAXY
ZWG 289.010	09 29 06.	+ 59 58		15.2	GALAXY
UGC 05077	09 29 06.	+ 59 58	102	15.2	GALAXY SBb
MCG+08-18-003	09 29 09.	+ 49 00	42	16.	GALAXY
ARC 0802	09 29 09.	+ 67 16		17.3	RICH CLUSTER OF GALAXIES
PK275-01.1	09 29 11.	- 52 56 45.	10		PLANETARY NEBULA
UGC 05078	09 29 12.	+ 03 57	60	16.0	GALAXY Sc
LB 10831	09 29 12.	+ 20 30		16.0	FAINT BLUE STAR
LB 10830	09 29 12.	+ 21 50		16.2	FAINT BLUE STAR
TON-N 0016	09 29 12.	+ 29 04		16.2	BLUE STAR
MCG+07-20-016	09 29 12.	+ 39 43	33	17.	GALAXY
KEEL 180	09 29 12.5	+ 21 35 41.			NEBULA
LB 03045	09 29 13.	+ 20 29 18.		16.8	FAINT BLUE STAR
MCG+07-20-017	09 29 15.	+ 39 46	30	16.	GALAXY
MCG+09-16-026	09 29 15.	+ 52 04	42	15.	GALAXY
MCG-02-25-001	09 29 15.	- 12 48 30.	21	15.	GALAXY
RNGC 2903	09 29 17.	+ 21 43		10.0	GALAXY
8ZW 0929+10.6	09 29 18.	+ 10 36		16.4	COMPACT GALAXY
TON-N 0431	09 29 18.	+ 27 03		15.6	BLUE STAR
ZC 0929.3+6729	09 29 18.	+ 67 29	2760		CLUSTER OF GALAXIES
MCG-03-25-002	09 29 18.	- 16 31 30.	90	13.	GALAXY
VHA 072	09 29 18.	- 52 46	180		OPEN STAR CLUSTER
RNGC 2907	09 29 19.	- 16 31		13.5	GALAXY
MCG-03-25-003	09 29 21.	- 15 50	108	15.	GALAXY
ARC 0817	09 29 23.	+ 17 35		17.1	RICH CLUSTER OF GALAXIES
RNGC 2905	09 29 23.	+ 21 45			NON-EXISTENT OBJECT
PK278-04.1	09 29 23.	- 57 23 16.	40		PLANETARY NEBULA
LB 10832	09 29 24.	+ 21 06		16.8	FAINT BLUE STAR
ZWG 122.014	09 29 24.	+ 21 44		9.8	GALAXY
UGC 05079	09 29 24.	+ 21 44	798	9.8	GALAXY Sb/Sc
MCG+04-23-009	09 29 24.	+ 21 44	780	09.8	GALAXY
KARA.73B 0347	09 29 24.	+ 21 44	732	9.8	ISOLATED GALAXY S
ARC 0815	09 29 24.	+ 29 17		17.7	RICH CLUSTER OF GALAXIES
TON-N 1071	09 29 24.	+ 34 57		17.	BLUE STAR
UGC 05080	09 29 24.	+ 42 00	60	16.5	GALAXY
MCG-05-23-005	09 29 24.	- 30 04	54	14.	GALAXY
RNGC 2901	09 29 25.	+ 31 21			NON-EXISTENT OBJECT
RNGC 2906	09 29 28.	+ 08 40		13.0	GALAXY
ZWG 063.001	09 29 30.	+ 08 40		13.1	GALAXY
UGC 05081	09 29 30.	+ 08 40	90	13.1	GALAXY Sc
MCG+02-25-001	09 29 30.	+ 08 40	84	13.1	GALAXY
ZC 0929.5+1734	09 29 30.	+ 17 34	1010		CLUSTER OF GALAXIES
ZC 0929.5+2407	09 29 30.	+ 24 07	1340		CLUSTER OF GALAXIES
ZC 0929.5+2925	09 29 30.	+ 29 25	1550		CLUSTER OF GALAXIES
ZWG 181.053	09 29 30.	+ 32 45		15.6	GALAXY
ARC 0812	09 29 30.	+ 38 07		17.8	RICH CLUSTER OF GALAXIES
MCG+07-20-018	09 29 30.	+ 44 30 30.	90	14.5	GALAXY
MCG+10-14-022	09 29 30.	+ 59 09	24	15.	GALAXY
7ZW 282	09 29 30.	+ 66 25			COMPACT GALAXY
MCG+07-20-019	09 29 33.	+ 41 59 30.	48	15.	GALAXY
MCG-01-25-001	09 29 33.	- 08 30 30.	84	15.	GALAXY
ZC 0929.6+0825	09 29 36.	+ 08 25	610		CLUSTER OF GALAXIES
ARC 0819	09 29 36.	+ 09 53		16.5	RICH CLUSTER OF GALAXIES
ZC 0929.6+0958	09 29 36.	+ 09 58	2290		CLUSTER OF GALAXIES
ZWG 063.002	09 29 36.	+ 12 19		15.3	GALAXY
LB 10834	09 29 36.	+ 23 46		17.3	FAINT BLUE STAR
TON-N 0432	09 29 36.	+ 23 49		15.3	BLUE STAR
LB 10833	09 29 36.	+ 24 28		17.0	FAINT BLUE STAR
ZC 0929.6+3257	09 29 36.	+ 32 57	610		CLUSTER OF GALAXIES
MCG+09-16-027	09 29 36.	+ 51 20	30	16.	GALAXY
MRSL 277-03/1	09 29 36.	- 56 05	120		HII REGION
ZWG 063.003	09 29 42.	+ 10 59		15.5	GALAXY
ZWG 063.004	09 29 42.	+ 11 02		14.8	GALAXY
UGC 05082	09 29 42.	+ 11 02	72	14.8	GALAXY S
MCG+02-25-002	09 29 42.	+ 11 02	72	14.8	GALAXY
LB 10835	09 29 42.	+ 23 31		17.5	FAINT BLUE STAR
TON-N 0433	09 29 42.	+ 28 27		15.2	BLUE STAR
ZWG 210.012	09 29 42.	+ 42 17		15.7	GALAXY

OBJECT NAME	RIGHT ASCEN.	DECLINATION	DIAM.	MAGN.	TYPE OF OBJECT
ZC 0929.7+6156	09 29 42.	+ 61 56	2620		CLUSTER OF GALAXIES
ACK 277-03.1	09 29 42.	- 56 00			PLANETARY NEBULA
MCG+07-20-020	09 29 45.	+ 42 16 30.	48	15.5	GALAXY
ZWG 007.001	09 29 48.	+ 00 33		15.3	GALAXY
UGC 05083	09 29 48.	+ 00 33	60	15.3	GALAXY S
LB 10837	09 29 48.	+ 21 38		16.6	FAINT BLUE STAR
LB 10836	09 29 48.	+ 26 04		17.2	FAINT BLUE STAR
7ZW 283	09 29 48.	+ 75 13			COMPACT GALAXY
MCG+08-18-004	09 29 51.	+ 49 15 30.	12	16.	GALAXY
ZWG 063.005	09 29 54.	+ 10 03		15.6	GALAXY
LB 10838	09 29 54.	+ 24 30		16.0	FAINT BLUE STAR
ZC 0929.9+2605	09 29 54.	+ 26 05	940		CLUSTER OF GALAXIES
TON-N 0434	09 29 54.	+ 28 54		16.6	BLUE STAR
ZC 0929.9+3601	09 29 54.	+ 36 01	1480		CLUSTER OF GALAXIES
LB 10842	09 30 00.	+ 22 04		17.0	FAINT BLUE STAR
LB 10841	09 30 00.	+ 23 35		17.0	FAINT BLUE STAR
LB 10840	09 30 00.	+ 23 36		17.2	FAINT BLUE STAR
LB 10839	09 30 00.	+ 25 57		16.9	FAINT BLUE STAR
ZWG 152.025	09 30 00.	+ 27 44		15.0	GALAXY
UGC 05084	09 30 00.	+ 27 44	66	15.0	GALAXY S
TON-N 1072	09 30 00.	+ 35 01		16.5	BLUE STAR
ZC 0930.0+3927	09 30 00.	+ 39 27	1280		CLUSTER OF GALAXIES
UGC 05085	09 30 00.	+ 42 35	60	18.	GALAXY DWARF
MCG+08-18-005	09 30 00.	+ 46 21	24	16.	GALAXY
MCG+14-05-005	09 30 00.	+ 81 21	45	14.	GALAXY
ZC 0930.0+8348	09 30 00.	+ 83 48	1140		CLUSTER OF GALAXIES
UGC 05083A	09 30 00.	+ 88 23	144	16.0	GALAXY
UGC 05086	09 30 06.	+ 21 43	60	18.	GALAXY DWARF
MCG+05-23-009	09 30 06.	+ 27 44	60	15.0	GALAXY
ZWG 152.026	09 30 06.	+ 30 10		14.7	GALAXY
UGC 05087	09 30 06.	+ 30 10	90	14.7	GALAXY Sb
ZWG 239.006	09 30 06.	+ 46 23		15.2	GALAXY
FATH 1.293	09 30 07.	+ 29 22	5		NEBULA
IC 2490	09 30 07.	+ 30 08	16		BRIGHT NEBULA
FATH 2.019	09 30 07.	+ 30 08	16		NEBULA
MCG+05-23-010	09 30 09.	+ 30 10	90	14.7	GALAXY
ZWG 063.006	09 30 12.	+ 12 31		15.2	GALAXY
LB 10843	09 30 12.	+ 20 23		15.8	FAINT BLUE STAR
LB 10844	09 30 12.	+ 21 21		16.2	FAINT BLUE STAR
ZWG 181.054	09 30 12.	+ 34 08		15.3	GALAXY
MCG+06-21-046	09 30 12.	+ 34 09	30	14.5	GALAXY
ZWG 265.025	09 30 12.	+ 51 54		15.5	GALAXY
LB 10846	09 30 18.	+ 24 33		17.0	FAINT BLUE STAR
LB 10845	09 30 18.	+ 24 55		17.5	FAINT BLUE STAR
ZWG 152.027	09 30 18.	+ 27 50		15.5	GALAXY
MCG+06-21-047	09 30 18.	+ 34 17	60	14.5	GALAXY
MCG-01-25-002	09 30 18.	- 05 30	84	14.5	GALAXY
MCG+05-23-011	09 30 21.	+ 27 50	48	15.5	GALAXY
LB 10851	09 30 24.	+ 21 06		16.7	FAINT BLUE STAR
LB 10850	09 30 24.	+ 21 10		17.3	FAINT BLUE STAR
LB 10849	09 30 24.	+ 22 05		16.4	FAINT BLUE STAR
ZWG 122.015	09 30 24.	+ 23 21		15.3	GALAXY
MCG+04-23-010	09 30 24.	+ 23 21 30.	42	15.3	GALAXY
LB 10848	09 30 24.	+ 24 03		16.5	FAINT BLUE STAR
LB 10847	09 30 24.	+ 25 56		16.9	FAINT BLUE STAR
MCG+06-21-048	09 30 24.	+ 33 53 30.	15	15.5	GALAXY
ZWG 181.055	09 30 24.	+ 34 16		15.2	GALAXY
UGC 05088	09 30 24.	+ 34 16	78	15.2	GALAXY
ZC 0930.4+3620	09 30 24.	+ 36 20	610		CLUSTER OF GALAXIES
ZC 0930.4+4030	09 30 24.	+ 40 30	870		CLUSTER OF GALAXIES
MCG+09-16-028	09 30 24.	+ 51 52 30.	30	16.	GALAXY
MRK 116	09 30 24.	+ 55 30	6	17.	GALAXY WITH UV CONTINUUM
VHA 073	09 30 24.	- 49 59	90		OPEN STAR CLUSTER
MCG+09-16-029	09 30 27.	+ 52 07	42	15.	GALAXY
ZWG 181.056	09 30 30.	+ 33 53		15.3	GALAXY
ZWG 265.026	09 30 30.	+ 51 46		14.7	GALAXY
1ZW 018	09 30 30.	+ 55 27			COMPACT GALAXY
LB 03046	09 30 32.	+ 20 57 12.		16.8	FAINT BLUE STAR
ZWG 035.004	09 30 36.	+ 05 46		15.5	GALAXY
KARA.73B 0348	09 30 36.	+ 05 46	24	15.5	ISOLATED GALAXY S
ZWG 092.001	09 30 36.	+ 15 39		15.5	GALAXY
ZWG 122.016	09 30 36.	+ 21 10		15.6	GALAXY
LB 10852	09 30 36.	+ 26 36		17.0	FAINT BLUE STAR
ZWG 181.057	09 30 36.	+ 33 50		15.6	GALAXY
MCG+06-21-049	09 30 36.	+ 33 50	24	16.	GALAXY
ZWG 210.013	09 30 36.	+ 40 13		15.5	GALAXY
UGC 05089	09 30 36.	+ 40 13	66	15.5	GALAXY SBb-c
ZWG 239.007	09 30 36.	+ 47 05		15.6	GALAXY
UGC 05090	09 30 36.	+ 47 05	60	15.6	GALAXY Sc
MCG+08-18-006	09 30 36.	+ 47 05	36	15.	GALAXY
MCG+10-14-023	09 30 36.	+ 59 25	15	16.	GALAXY
ZC 0930.6+6100	09 30 36.	+ 61 00	1210		CLUSTER OF GALAXIES
REIZ 0002	09 30 38.	+ 40 12	60	15.1	GALAXY
MCG+09-16-030	09 30 39.	+ 51 45	60	15.	GALAXY
LB 03047	09 30 41.	+ 19 50 42.		16.4	FAINT BLUE STAR
LB 10853	09 30 42.	+ 21 24		16.8	FAINT BLUE STAR
ZC 0930.7+2459	09 30 42.	+ 24 59	670		CLUSTER OF GALAXIES
TON-N 0017	09 30 42.	+ 29 23		15.5	BLUE STAR
MCG+06-21-050	09 30 42.	+ 34 18	36	15.	GALAXY
MCG+07-20-021	09 30 42.	+ 40 11 30.	54	14.5	GALAXY
ZWG 239.008	09 30 42.	+ 44 29		15.1	GALAXY
ZWG 210.014	09 30 42.	+ 44 29		15.1	GALAXY
UGC 05091	09 30 42.	+ 44 29	66	15.1	GALAXY Sc-IRR
ZWG 239.009	09 30 42.	+ 50 14		15.5	GALAXY
MRK 117	09 30 42.	+ 50 14	12	16.	GALAXY WITH UV CONTINUUM
REIZ 0003	09 30 44.	+ 44 29	42	14.1	GALAXY
ZWG 092.002	09 30 48.	+ 14 48		15.0	GALAXY
MCG+03-25-001	09 30 48.	+ 14 48	15	15.0	GALAXY
LB 10854	09 30 48.	+ 20 54		16.5	FAINT BLUE STAR
TON-N 0435	09 30 48.	+ 29 21		15.0	BLUE STAR
ZC 0930.8+5130	09 30 48.	+ 51 30	740		CLUSTER OF GALAXIES
SHAH 114	09 30 48.	+ 54 38	108	17.5	GROUP OF COMPACT GALAXIES
KEEL 181	09 30 50.1	+ 21 39 22.			NEBULA
LB 03048	09 30 51.	+ 20 46 48.		16.7	FAINT BLUE STAR
MCG-03-25-004	09 30 51.	- 16 33 30.	84	14.	GALAXY
REIZ 0004	09 30 53.	+ 46 54	24	14.6	GALAXY
LB 10856	09 30 54.	+ 21 28		16.5	FAINT BLUE STAR
LB 10855	09 30 54.	+ 25 49		16.3	FAINT BLUE STAR
ZWG 152.028	09 30 54.	+ 30 38		15.5	GALAXY
MCG+06-21-051	09 30 54.	+ 34 13	42	15.	GALAXY
ZWG 181.058	09 30 54.	+ 34 17		15.7	GALAXY
REIZ 0005	09 30 54.	+ 46 56	36	13.8	GALAXY
VVI 31	09 31 00.	+ 10 22	27	13.38	SEYFERT GALAXY
LB 10858	09 31 00.	+ 20 58		14.9	FAINT BLUE STAR
LB 10857	09 31 00.	+ 24 22		16.6	FAINT BLUE STAR
MCG+05-23-012	09 31 00.	+ 30 39	24	15.5	GALAXY
ZWG 181.059	09 31 00.	+ 34 13		15.1	GALAXY
MCG+08-18-007	09 31 00.	+ 46 55	24	14.	GALAXY
ZWG 239.010	09 31 00.	+ 46 56		14.7	GALAXY
ZWG 332.054	09 31 00.	+ 74 13		15.7	GALAXY
ABC 0823	09 31 03.	- 25 38		17.3	RICH CLUSTER OF GALAXIES
RNGC 2912	09 31 04.	+ 10 23			GALAXY
RNGC 2911	09 31 04.	+ 10 23		13.5	GALAXY
ZWG 063.007	09 31 06.	+ 10 22		13.6	GALAXY
UGC 05092	09 31 06.	+ 10 22	240	13.6	GALAXY
MCG+02-25-003	09 31 06.	+ 10 22	27	13.6	GALAXY
ZC 0931.1+1055	09 31 06.	+ 10 55	1480		CLUSTER OF GALAXIES
LB 10860	09 31 06.	+ 24 59		17.7	FAINT BLUE STAR
ZWG 122.017	09 31 06.	+ 25 07		15.7	GALAXY
LB 10859	09 31 06.	+ 25 49		17.8	FAINT BLUE STAR
ZWG 152.029	09 31 06.	+ 27 36		15.6	GALAXY
ZC 0931.1+4947	09 31 06.	+ 49 47	670		CLUSTER OF GALAXIES
MCG+10-14-024	09 31 06.	+ 59 26	18	16.	GALAXY
MCG-05-23-006	09 31 06.	- 32 48	360	13.	GALAXY
ARP 232	09 31 08.	+ 10 21			PECULIAR GALAXY
HN 0318	09 31 08.	- 37 41			NEBULA
IC 2492	09 31 08.	- 37 41			NONSTELLAR OBJECT
LB 10864	09 31 12.	+ 22 14		16.8	FAINT BLUE STAR
LB 10863	09 31 12.	+ 23 12		16.8	FAINT BLUE STAR
LB 10862	09 31 12.	+ 23 14		17.4	FAINT BLUE STAR
LB 10861	09 31 12.	+ 25 08		18.0	FAINT BLUE STAR
MCG+05-23-014	09 31 12.	+ 27 35	42	16.	GALAXY
MCG+05-23-013	09 31 12.	+ 27 36	30	15.6	GALAXY
TON-N 1073	09 31 12.	+ 34 04		16.9	BLUE STAR
TON-N 1074	09 31 12.	+ 34 44		17.	BLUE STAR
ZC 0931.2-0241	09 31 12.	- 02 41	670		CLUSTER OF GALAXIES
ABC 0822	09 31 16.	- 13 12		17.5	RICH CLUSTER OF GALAXIES
ZWG 063.008	09 31 18.	+ 10 15		14.9	GALAXY
UGC 05093	09 31 18.	+ 10 15	66	14.9	GALAXY S
MCG+02-25-004	09 31 18.	+ 10 15	60	14.9	GALAXY
ZWG 092.003	09 31 18.	+ 18 03		15.6	GALAXY
LB 10865	09 31 18.	+ 24 10		17.8	FAINT BLUE STAR
ZWG 122.018	09 31 18.	+ 24 26		15.5	GALAXY
UGC 05094	09 31 18.	+ 24 26	60	15.5	GALAXY S0
ZWG 181.060	09 31 18.	+ 34 55		15.7	GALAXY
MCG+08-18-008	09 31 18.	+ 46 38	42	16.	GALAXY
MCG+08-18-009	09 31 18.	+ 46 40	24	16.	GALAXY
ABC 0820	09 31 19.	- 02 42		17.5	RICH CLUSTER OF GALAXIES
MCG+05-23-016	09 31 21.	+ 27 33	24	16.	GALAXY
MCG+05-23-015	09 31 21.	+ 27 34	18	16.	GALAXY
MCG-04-23-006	09 31 21.	- 24 54 30.	18	15.	GALAXY
RNGC 2913	09 31 22.	+ 09 42		14.0	GALAXY
RNGC 2914	09 31 22.	+ 10 20		13.5	GALAXY
LB 03049	09 31 22.	+ 20 30 06.		17.3	FAINT BLUE STAR
ZWG 035.005	09 31 24.	+ 02 41		15.5	GALAXY
ZWG 063.009	09 31 24.	+ 09 42		14.1	GALAXY
UGC 05095	09 31 24.	+ 09 42	72	14.1	GALAXY S
MCG+02-25-005	09 31 24.	+ 09 42	66	14.1	GALAXY
ZWG 063.010	09 31 24.	+ 10 20		13.7	GALAXY
UGC 05096	09 31 24.	+ 10 20	66	13.7	GALAXY Sa?
MCG+02-25-006	09 31 24.	+ 10 20	48	13.7	GALAXY
ZWG 063.011	09 31 24.	+ 11 14		15.2	GALAXY
MCG+03-25-002	09 31 24.	+ 18 05	33	15.6	GALAXY
LB 10870	09 31 24.	+ 22 39		16.9	FAINT BLUE STAR
LB 10869	09 31 24.	+ 22 42		17.4	FAINT BLUE STAR
LB 10868	09 31 24.	+ 25 12		17.5	FAINT BLUE STAR
LB 10867	09 31 24.	+ 25 28		16.3	FAINT BLUE STAR
LB 10866	09 31 24.	+ 25 56		15.9	FAINT BLUE STAR
TON-N 0436	09 31 24.	+ 30 04		14.5	BLUE STAR
7ZW 284	09 31 24.	+ 65 37			COMPACT GALAXY
MCG-02-25-002	09 31 24.	- 11 05	90	14.	GALAXY
ARP 137	09 31 26.	+ 10 19			PECULIAR GALAXY
ABC 0821	09 31 28.	- 04 28		17.7	RICH CLUSTER OF GALAXIES
LB 10874	09 31 30.	+ 22 06		16.3	FAINT BLUE STAR
LB 10873	09 31 30.	+ 23 08		17.3	FAINT BLUE STAR
LB 10872	09 31 30.	+ 25 45		16.0	FAINT BLUE STAR
LB 10871	09 31 30.	+ 26 18		16.9	FAINT BLUE STAR
ZC 0931.5-0002	09 31 30.	- 00 02	1140		CLUSTER OF GALAXIES
MCG-03-25-005	09 31 30.	- 16 46	60	15.5	GALAXY
ZWG 007.002	09 31 36.	+ 00 29		13.9	GALAXY
UGC 05097	09 31 36.	+ 00 29	36	13.9	GALAXY PECULR
KARA.72 206B	09 31 36.	+ 00 29	42		PART OF DOUBLE GALAXY
MCG+00-25-001	09 31 36.	+ 00 29	33	13.9	GALAXY
KARA.72 206A	09 31 36.	+ 00 30	30	13.9	PART OF DOUBLE GALAXY
LB 10875	09 31 36.	+ 23 24		17.2	FAINT BLUE STAR
ZWG 122.019	09 31 36.	+ 24 26		15.5	GALAXY
MCG+09-16-031	09 31 36.	+ 52 34	42	16.	GALAXY
MCG+03-25-003	09 31 42.	+ 16 47	36	15.6	GALAXY
LB 10876	09 31 42.	+ 24 53		17.2	FAINT BLUE STAR
TON-N 0437	09 31 42.	+ 30 03		15.9	BLUE STAR
ZWG 152.030	09 31 42.	+ 32 17		15.0	GALAXY
ZC 0931.7-0238	09 31 42.	- 02 38	3230		CLUSTER OF GALAXIES
MCG-02-25-003	09 31 42.	- 15 04	36	15.5	GALAXY
ZWG 063.012	09 31 48.	+ 10 08		15.6	GALAXY
ZWG 092.004	09 31 48.	+ 16 47		15.6	GALAXY
MCG+05-23-017	09 31 48.	+ 32 18 30.	42	15.0	GALAXY
TON-N 1075	09 31 48.	+ 35 49		16.8	BLUE STAR
RNGC 2925	09 31 52.	- 53 13		8.5	OPEN CLUSTER
ZWG 092.005	09 31 54.	+ 19 46		15.2	GALAXY
LB 10879	09 31 54.	+ 20 48		16.4	FAINT BLUE STAR
LB 10878	09 31 54.	+ 22 58		16.4	FAINT BLUE STAR
LB 10877	09 31 54.	+ 23 11		16.8	FAINT BLUE STAR
ZC 0931.9+4321	09 31 54.	+ 43 21	1080		CLUSTER OF GALAXIES
ZC 0931.9+5329	09 31 54.	+ 53 29	1210		CLUSTER OF GALAXIES
ZWG 007.003	09 31 54.	- 02 16		14.5	GALAXY
UGC 05098	09 31 54.	- 02 16	90	14.5	GALAXY S0-a
MCG+00-25-002	09 31 54.	- 02 16	66	14.5	GALAXY
RNGC 2917	09 31 56.	- 02 16		14.5	GALAXY
RNGC 2920	09 31 57.	- 20 37			GALAXY
ZWG 007.004	09 32 00.	+ 00 19		14.7	GALAXY
UGC 05099	09 32 00.	+ 00 19	72	14.7	GALAXY S
MCG+00-25-003	09 32 00.	+ 00 19	60	14.7	GALAXY
ZC 0932.0+0356	09 32 00.	+ 03 56	870		CLUSTER OF GALAXIES
ZWG 035.006	09 32 00.	+ 06 04		14.9	GALAXY
UGC 05100	09 32 00.	+ 06 04	96	14.9	GALAXY SBb
MCG+01-25-002	09 32 00.	+ 06 04	60	14.9	GALAXY
LB 10861	09 32 00.	+ 21 18		17.5	FAINT BLUE STAR
LB 10880	09 32 00.	+ 22 21		16.6	FAINT BLUE STAR
ZWG 122.020	09 32 00.	+ 26 09		15.5	GALAXY
ZWG 289.011	09 32 00.	+ 61 34		15.5	GALAXY
UGC 05101	09 32 00.	+ 61 34	72	15.5	GALAXY PECULR
KARA.73B 0349	09 32 00.	+ 61 34	18	15.5	ISOLATED GALAXY PEC
MCG+10-14-025	09 32 00.	+ 61 35	66	15.	GALAXY
ZC 0932.0+6343	09 32 00.	+ 63 43	2150		CLUSTER OF GALAXIES
OCL 0783	09 32 00.	- 53 13	1020	8.4	OPEN STAR CLUSTER
RNGC 2919	09 32 03.	+ 10 30		13.5	GALAXY

OBJECT NAME	RIGHT ASCEN.	DECLINATION	DIAM.	MAGN.	TYPE OF OBJECT
MCG+09-16-032	09 32 03.	+ 55 09	42	16.	GALAXY
RNGC 2916	09 32 05.	+ 21 56		12.5	GALAXY
ZWG 007.005	09 32 06.	+ 00 56		15.6	GALAXY
KARA.73B 0350	09 32 06.	+ 00 56	78	15.6	ISOLATED GALAXY S
ZWG 035.007	09 32 06.	+ 06 39		15.0	GALAXY
MCG+01-25-003	09 32 06.	+ 06 39	48	15.0	GALAXY
ZWG 063.013	09 32 06.	+ 10 30		13.6	GALAXY
UGC 05102	09 32 06.	+ 10 30	102	13.6	GALAXY Sb-c
MCG+02-25-007	09 32 06.	+ 10 30	96	13.6	GALAXY
ZWG 122.021	09 32 06.	+ 21 56		12.3	GALAXY
UGC 05103	09 32 06.	+ 21 56	150	12.3	GALAXY S
TON-N 0438	09 32 06.	+ 31 24		14.6	BLUE STAR
MCG+06-21-052	09 32 06.	+ 34 12	30	14.5	GALAXY
MCG+06-21-053	09 32 06.	+ 34 58	24	15.	GALAXY
ZC 0932.1+5708	09 32 06.	+ 57 08	3020		CLUSTER OF GALAXIES
ARC 0827	09 32 07.	- 02 44		17.5	RICH CLUSTER OF GALAXIES
ZC 0932.2+0834	09 32 12.	+ 08 34	1750		CLUSTER OF GALAXIES
ZWG 063.014	09 32 12.	+ 11 55		14.9	GALAXY
MCG+02-25-008	09 32 12.	+ 11 55	12	14.5	GALAXY
ZC 0932.2+1508	09 32 12.	+ 15 08	870		CLUSTER OF GALAXIES
ZWG 092.006	09 32 12.	+ 18 23		15.7	GALAXY
MCG+04-23-011	09 32 12.	+ 21 55	150	12.3	GALAXY
LB 10884	09 32 12.	+ 24 44		15.7	FAINT BLUE STAR
LB 10883	09 32 12.	+ 24 57		16.8	FAINT BLUE STAR
LB 10882	09 32 12.	+ 25 48		15.1	FAINT BLUE STAR
ZC 0932.2+3147	09 32 12.	+ 31 47	6450		CLUSTER OF GALAXIES
ZWG 181.061	09 32 12.	+ 34 13		15.2	GALAXY
ZWG 181.062	09 32 12.	+ 34 57		15.0	GALAXY
UGC 05104	09 32 12.	+ 34 57	66	15.0	GALAXY S0
ZC 0932.2+6828	09 32 12.	+ 68 28	1010		CLUSTER OF GALAXIES
ZC 0932.2-0239	09 32 12.	- 02 39	870		CLUSTER OF GALAXIES
IC 2491	09 32 13.	+ 34 56 29.			NONSTELLAR OBJECT
ZWG 035.008	09 32 18.	+ 05 00		15.6	GALAXY
ZC 0932.3+1942	09 32 18.	+ 19 42	1280		CLUSTER OF GALAXIES
ZWG 122.022	09 32 18.	+ 21 52		15.6	GALAXY
LB 10865	09 32 18.	+ 26 38		17.0	FAINT BLUE STAR
MRK 402	09 32 18.	+ 30 39	12	16.	GALAXY WITH UV CONTINUUM
UGC 05105	09 32 18.	+ 36 08	72	16.5	GALAXY DWARF
MCG+06-21-054	09 32 18.	+ 36 08	42	15.	GALAXY
ZC 0932.3+4034	09 32 18.	+ 40 34	1080		CLUSTER OF GALAXIES
MCG-03-25-006	09 32 18.	- 20 43	156	13.	GALAXY
MCG-04-23-007	09 32 18.	- 21 41 30.	18	15.5	GALAXY
RNGC 2921	09 32 21.	- 20 43		13.0	GALAXY
MCG-04-23-008	09 32 21.	- 21 41	36	14.	GALAXY
LB 00557	09 32 22.	+ 62 41 00.		15.3	FAINT BLUE STAR
ZC 0932.4+1126	09 32 24.	+ 11 26	1680		CLUSTER OF GALAXIES
LB 10891	09 32 24.	+ 21 16		15.8	FAINT BLUE STAR
LB 10890	09 32 24.	+ 22 04		16.9	FAINT BLUE STAR
LB 10889	09 32 24.	+ 23 03		15.2	FAINT BLUE STAR
LB 10888	09 32 24.	+ 23 21		16.9	FAINT BLUE STAR
LB 10887	09 32 24.	+ 23 21		17.3	FAINT BLUE STAR
ZWG 122.023	09 32 24.	+ 24 49		15.6	GALAXY
LB 10886	09 32 24.	+ 25 36		14.9	FAINT BLUE STAR
ZWG 181.063	09 32 24.	+ 35 13		15.1	GALAXY
MCG+06-21-055	09 32 24.	+ 35 13	18	16.	GALAXY
ZWG 210.015	09 32 24.	+ 39 21		15.7	GALAXY
ZC 0932.4+3945	09 32 24.	+ 39 45	940		CLUSTER OF GALAXIES
ZWG 265.027	09 32 24.	+ 54 49		14.8	GALAXY
UGC 05106	09 32 24.	+ 54 49	84	14.8	GALAXY
ARC 0824	09 32 27.	+ 24 09		17.5	RICH CLUSTER OF GALAXIES
MCG+09-16-033	09 32 27.	+ 54 50	48	15.	GALAXY
IC 0546	09 32 28.	- 16 09 53.			NONSTELLAR OBJECT
ZWG 035.009	09 32 30.	+ 05 20		15.2	GALAXY
UGC 05107	09 32 30.	+ 05 20	132	15.2	GALAXY SBc
MCG+01-25-004	09 32 30.	+ 05 20	78	15.2	GALAXY
LB 10892	09 32 30.	+ 26 37		16.8	FAINT BLUE STAR
ZWG 152.031	09 32 30.	+ 30 03		15.0	GALAXY
UGC 05108	09 32 30.	+ 30 03	102	15.0	GALAXY SBa-b
ZC 0932.5+3249	09 32 30.	+ 32 49	740		CLUSTER OF GALAXIES
ZC 0932.5+3400	09 32 30.	+ 34 00	940		CLUSTER OF GALAXIES
UGC 05109	09 32 30.	+ 48 23	84	16.5	GALAXY
ZWG 332.055	09 32 30.	+ 73 36		15.6	GALAXY
UGC 05110	09 32 30.	+ 73 36	108	15.6	GALAXY IRR
MCG-03-25-007	09 32 30.	- 16 10	36	14.5	GALAXY
BIGO 508	09 32 33.	+ 10 30			NEBULA
MCG+05-23-018	09 32 33.	+ 30 03 30.	72	15.0	GALAXY
REIZ 0006	09 32 34.	+ 46 08	36	15.0	GALAXY
ZC 0932.6+0103	09 32 36.	+ 01 03	340		CLUSTER OF GALAXIES
ZWG 063.015	09 32 36.	+ 13 47		15.0	GALAXY
MCG+02-25-009	09 32 36.	+ 13 47	36	15.0	GALAXY
KARA.73B 0351	09 32 36.	+ 13 47	42	15.0	ISOLATED GALAXY S
LB 10895	09 32 36.	+ 21 08		16.5	FAINT BLUE STAR
ZC 0932.6+2410	09 32 36.	+ 24 10	340		CLUSTER OF GALAXIES
LB 10894	09 32 36.	+ 24 38		16.8	FAINT BLUE STAR
LB 10893	09 32 36.	+ 26 30		16.1	FAINT BLUE STAR
SEY 057	09 32 36.	+ 67 01 01.		15.0	FAINT GALAXY
ZC 0932.6+7506	09 32 36.	+ 75 06	2350		CLUSTER OF GALAXIES
SEY 058	09 32 39.	+ 66 47 19.		15.1	FAINT GALAXY
LB 10898	09 32 42.	+ 22 37		16.9	FAINT BLUE STAR
LB 10897	09 32 42.	+ 22 44		17.3	FAINT BLUE STAR
LB 10896	09 32 42.	+ 24 30		17.0	FAINT BLUE STAR
TON-N 1076	09 32 42.	+ 35 08		15.1	BLUE STAR
MCG+08-18-010	09 32 42.	+ 46 08	42	16.	GALAXY
MCG+08-18-011	09 32 42.	+ 48 22	60	16.	GALAXY
ZWG 312.017	09 32 42.	+ 67 01		15.5	GALAXY
UGC 05111	09 32 42.	+ 67 01	96	15.5	GALAXY Sb-c
ARC 0831	09 32 42.	- 02 49		17.5	RICH CLUSTER OF GALAXIES
SHB 140	09 32 42.6	+ 02 16 18.		17.4	QUASI-STELLAR OBJECT
BC PKS0932+02	09 32 42.9	+ 02 17 35.		17.39	QUASI-STELLAR OBJECT
ZWG 035.010	09 32 48.	+ 03 53		15.6	GALAXY
ZC 0932.8+1231	09 32 48.	+ 12 31	1210		CLUSTER OF GALAXIES
LB 10900	09 32 48.	+ 23 38		17.9	FAINT BLUE STAR
LB 10899	09 32 48.	+ 25 22		17.0	FAINT BLUE STAR
ZWG 152.032	09 32 48.	+ 31 56		13.6	GALAXY
UGC 05112	09 32 48.	+ 31 56	96	13.6	GALAXY E
MCG+05-23-019	09 32 48.	+ 31 57 30.	36	13.6	GALAXY
TON-N 1077	09 32 48.	+ 35 34		17.	BLUE STAR
ZCG 0932+59	09 32 48.	+ 59 36		17.6	COMPACT GALAXY
ZWG 289.012	09 32 48.	+ 59 37		15.3	GALAXY
ZWG 312.018	09 32 48.	+ 66 47		15.5	GALAXY
UGC 05113	09 32 48.	+ 66 47	72	15.5	GALAXY Sa-b
MCG+11-12-016	09 32 48.	+ 67 01	96	15.	GALAXY
MCG-01-25-003	09 32 48.	- 08 35	66	15.	GALAXY
MCG-03-25-008	09 32 48.	- 16 10 30.	54	13.5	GALAXY
VHA 074	09 32 48.	- 47 48	210		OPEN STAR CLUSTER
RNGC 2918	09 32 49.	+ 31 56		13.5	GALAXY
RNGC 2924	09 32 49.	- 16 11		13.0	GALAXY
ARC 0830	09 32 51.	+ 07 45		17.7	RICH CLUSTER OF GALAXIES

OBJECT NAME	RIGHT ASCEN.	DECLINATION	DIAM.	MAGN.	TYPE OF OBJECT
ARC 0828	09 32 51.	+ 12 26		17.7	RICH CLUSTER OF GALAXIES
ARC 0798	09 32 52.	+ 81 11		17.9	RICH CLUSTER OF GALAXIES
ZC 0932.9+0745	09 32 54.	+ 07 45	810		CLUSTER OF GALAXIES
LB 10903	09 32 54.	+ 22 13		17.7	FAINT BLUE STAR
LB 10902	09 32 54.	+ 23 41		16.2	FAINT BLUE STAR
LB 10901	09 32 54.	+ 24 08		17.1	FAINT BLUE STAR
ZWG 122.024	09 32 54.	+ 24 36		15.6	GALAXY
ZC 0932.9+2719	09 32 54.	+ 27 19	1280		CLUSTER OF GALAXIES
ZC 0932.9+3430	09 32 54.	+ 34 30	740		CLUSTER OF GALAXIES
ZC 0932.9+3825	09 32 54.	+ 38 25	3830		CLUSTER OF GALAXIES
7ZW 285	09 32 54.	+ 59 36			COMPACT GALAXY
MCG+11-12-017	09 32 54.	+ 66 47	66	15.	GALAXY
OCL 0773	09 32 54.	- 47 54	270	13.5	OPEN STAR CLUSTER
LBN 0691	09 33	+ 66 00	9000		BRIGHT NEBULA
ZC 0933.0+2406	09 33 00.	+ 24 06	1810		CLUSTER OF GALAXIES
ZWG 122.025	09 33 00.	+ 25 07		15.3	GALAXY
KARA.72 207A	09 33 00.	+ 25 07	42	15.3	PART OF DOUBLE GALAXY
MCG+04-23-012	09 33 00.	+ 25 07	42	15.3	GALAXY
LB 10904	09 33 00.	+ 25 43		16.0	FAINT BLUE STAR
ZWG 350.027	09 33 00.	+ 76 33		14.4	GALAXY
UGC 05115	09 33 00.	+ 76 33	144	14.4	GALAXY SBc
ZWG 364.012	09 33 00.	+ 82 20		15.7	GALAXY
UGC 05114	09 33 00.	+ 82 21	114	16.0	GALAXY IRR
IC 0544	09 33 03.	+ 25 06 55.			NONSTELLAR OBJECT
RNGC 2938	09 33 03.	+ 76 33		14.5	GALAXY
ZWG 007.006	09 33 06.	+ 01 28		15.7	GALAXY
ZC 0933.1+0939	09 33 06.	+ 09 39	2350		CLUSTER OF GALAXIES
ZWG 092.007	09 33 06.	+ 20 25		15.7	GALAXY
LB 10906	09 33 06.	+ 21 47		16.8	FAINT BLUE STAR
ZWG 122.026	09 33 06.	+ 25 55		15.5	GALAXY
LB 10905	09 33 06.	+ 26 37		16.9	FAINT BLUE STAR
ZWG 152.033	09 33 06.	+ 26 54		15.7	GALAXY
ZC 0933.1+7207	09 33 06.	+ 72 07	2550		CLUSTER OF GALAXIES
AR 01	09 33 07.06	+ 76 32 39.0			NEBULA
ZC 0933.2+1540	09 33 12.	+ 15 40	940		CLUSTER OF GALAXIES
LB 10908	09 33 12.	+ 20 20		16.6	FAINT BLUE STAR
ZC 0933.2+2022	09 33 12.	+ 20 22	2020		CLUSTER OF GALAXIES
LB 10907	09 33 12.	+ 23 34		17.0	FAINT BLUE STAR
ZWG 122.027	09 33 12.	+ 25 10		14.8	GALAXY
KARA.72 207B	09 33 12.	+ 25 10	42	14.8	PART OF DOUBLE GALAXY
MCG+04-23-013	09 33 12.	+ 25 10	24	14.8	GALAXY
ZWG 152.034	09 33 12.	+ 29 20		15.4	GALAXY
KARA.73B 0352	09 33 12.	+ 29 20	48	15.4	ISOLATED GALAXY S
ZWG 181.064	09 33 12.	+ 37 28		15.4	GALAXY
ZWG 181.065	09 33 12.	+ 37 35		15.0	GALAXY
MCG+06-21-056	09 33 12.	+ 37 36	30	15.	GALAXY
IC 2493	09 33 12.	+ 37 36 19.			NONSTELLAR OBJECT
ZWG 265.028	09 33 12.	+ 53 20		15.4	GALAXY
ZWG 289.013	09 33 12.	+ 58 14		15.7	GALAXY
MCG+10-14-026	09 33 12.	+ 59 25	18	16.	GALAXY
MCG+13-07-032	09 33 12.	+ 76 32	84	14.	GALAXY
ZC 0933.2+7731	09 33 12.	+ 77 31	1880		CLUSTER OF GALAXIES
VHA 075	09 33 12.	- 54 30	300		OPEN STAR CLUSTER
RNGC 2923	09 33 13.	+ 16 59		15.0	GALAXY
ARC 0832	09 33 14.	+ 16 06		17.1	RICH CLUSTER OF GALAXIES
IC 0545	09 33 14.	+ 25 10 18.			NONSTELLAR OBJECT
MCG+08-18-012	09 33 15.	+ 48 41	48	15.	GALAXY
MCG-01-25-004	09 33 15.	- 04 30 30.	15	15.5	GALAXY
ZWG 035.011	09 33 18.	+ 04 14		15.3	GALAXY
KARA.73B 0353	09 33 18.	+ 04 14	36	15.3	ISOLATED GALAXY S
ZWG 092.008	09 33 18.	+ 16 59		15.2	GALAXY
LB 10910	09 33 18.	+ 23 39		17.9	FAINT BLUE STAR
LB 10909	09 33 18.	+ 24 44		17.1	FAINT BLUE STAR
ZWG 239.011	09 33 18.	+ 48 42		15.0	GALAXY
MCG-03-25-009	09 33 18.	- 16 40	60	15.	GALAXY
MCG-03-25-010	09 33 18.	- 17 10	48	15.	GALAXY
LB 10919	09 33 24.	+ 21 07		16.4	FAINT BLUE STAR
LB 10918	09 33 24.	+ 21 44		16.7	FAINT BLUE STAR
LB 10917	09 33 24.	+ 22 24		16.8	FAINT BLUE STAR
LB 10916	09 33 24.	+ 24 10		17.4	FAINT BLUE STAR
LB 10915	09 33 24.	+ 24 39		16.5	FAINT BLUE STAR
LB 10914	09 33 24.	+ 24 45		16.6	FAINT BLUE STAR
LB 10913	09 33 24.	+ 24 58		16.7	FAINT BLUE STAR
LB 10912	09 33 24.	+ 26 16		16.1	FAINT BLUE STAR
LB 10911	09 33 24.	+ 26 21		16.0	FAINT BLUE STAR
MCG+08-18-013	09 33 24.	+ 48 41 30.	42	15.	GALAXY
ARC 0809	09 33 28.	+ 77 32		16.9	RICH CLUSTER OF GALAXIES
RNGC 2932	09 33 28.	- 46 43			NON-EXISTENT OBJECT
ZWG 035.012	09 33 30.	+ 03 14		15.5	GALAXY
UGC 05116	09 33 30.	+ 03 14	72	15.5	GALAXY S
ZWG 035.013	09 33 30.	+ 06 57		15.7	GALAXY
ZWG 092.009	09 33 30.	+ 17 43		15.0	GALAXY
MCG+03-25-004	09 33 30.	+ 17 43	36	15.0	GALAXY
LB 10920	09 33 30.	+ 23 16		17.5	FAINT BLUE STAR
ZWG 122.028	09 33 30.	+ 23 58		15.4	GALAXY
TON-N 0439	09 33 30.	+ 31 36		15.0	BLUE STAR
ZWG 152.035	09 33 30.	+ 32 02		15.5	GALAXY
ZC 0933.5+3451	09 33 30.	+ 34 51	1010		CLUSTER OF GALAXIES
MCG+13-07-033	09 33 30.	+ 77 11	9	16.	GALAXY
MCG+15-01-009	09 33 30.	+ 88 23	120	15.	GALAXY
MCG-01-25-005	09 33 30.	- 07 29 30.	102	15.	GALAXY
ARC 0833	09 33 31.	+ 11 06		16.7	RICH CLUSTER OF GALAXIES
LB 10922	09 33 36.	+ 23 06		16.8	FAINT BLUE STAR
LB 10921	09 33 36.	+ 24 52		15.5	FAINT BLUE STAR
ZWG 152.036	09 33 36.	+ 28 21		15.7	GALAXY
MCG+05-23-020	09 33 36.	+ 28 41	21	15.6	GALAXY
ZWG 152.037	09 33 36.	+ 28 44		15.6	GALAXY
ZC 0933.6+3250	09 33 36.	+ 32 50	810		CLUSTER OF GALAXIES
ZC 0933.6+5343	09 33 36.	+ 53 43	1010		CLUSTER OF GALAXIES
ZC 0933.6+6127	09 33 36.	+ 61 27	670		CLUSTER OF GALAXIES
MCG+11-12-018	09 33 36.	+ 67 06	36	16.	GALAXY
REIZ 0007	09 33 40.	+ 46 23	48	15.5	GALAXY
IC 2494	09 33 41.	- 12 12 47.			NONSTELLAR OBJECT
UGC 05117	09 33 42.	+ 04 49	60	16.5	GALAXY DBL SYS
ZWG 063.016	09 33 42.	+ 11 33		15.0	GALAXY
MCG+02-25-010	09 33 42.	+ 11 33	60	15.0	GALAXY
LB 10925	09 33 42.	+ 21 37		17.0	FAINT BLUE STAR
ZWG 122.029	09 33 42.	+ 23 40		15.6	GALAXY
LB 10924	09 33 42.	+ 23 42		17.0	FAINT BLUE STAR
LB 10923	09 33 42.	+ 24 22		17.6	FAINT BLUE STAR
MCG+06-21-057	09 33 42.	+ 37 56	60	14.	GALAXY
IC 0547	09 33 42.	- 12 12 59.			NONSTELLAR OBJECT
MCG-04-23-009	09 33 42.	- 24 45	54	15.	GALAXY
MCG-01-25-006	09 33 45.	- 08 13	42	14.5	GALAXY
MCG-02-25-004	09 33 45.	- 12 12	72	13.	GALAXY
RNGC 2922	09 33 46.	+ 37 55		14.5	GALAXY
LB 10926	09 33 48.	+ 25 46		15.7	FAINT BLUE STAR
ZWG 152.038	09 33 48.	+ 31 18		15.4	GALAXY

OBJECT NAME	RIGHT ASCEN.	DECLINATION	DIAM.	MAGN.	TYPE OF OBJECT
ZWG 181.066	09 33 48.	+ 37 55		14.6	GALAXY
UGC 05118	09 33 48.	+ 37 55	66	14.6	GALAXY IRR
KARA.73B 0355	09 33 48.	+ 37 55	72	14.6	ISOLATED GALAXY S
ZC 0933.8+6017	09 33 48.	+ 60 17	1750		CLUSTER OF GALAXIES
ZC 0933.8+7810	09 33 48.	+ 78 10	1550		CLUSTER OF GALAXIES
ZWG 007.007	09 33 48.	- 00 20		15.6	GALAXY
MCG+00-25-004	09 33 48.	- 00 20	48	15.6	GALAXY
KARA.73B 0354	09 33 48.	- 00 20	54	15.6	ISOLATED GALAXY S
ABC 0826	09 33 50.	+ 53 45		17.7	RICH CLUSTER OF GALAXIES
MCG-02-25-005	09 33 51.	- 10 44	72	15.	GALAXY
ZC 0933.9+1102	09 33 54.	+ 11 02	1340		CLUSTER OF GALAXIES
LB 10927	09 33 54.	+ 25 49		16.0	FAINT BLUE STAR
MCG+06-21-058	09 33 54.	+ 35 56	42	15.	GALAXY
MCG+07-20-023	09 33 54.	+ 41 35 30.	24	16.	GALAXY
MCG+07-20-022	09 33 54.	+ 41 35 30.	18	16.	GALAXY
ZWG 332.056	09 33 54.	+ 69 04		15.7	GALAXY
ZWG 007.009	09 34 00.	+ 01 21		14.9	GALAXY
MCG+00-25-005	09 34 00.	+ 01 21	36	14.9	GALAXY
ZC 0934.0+0825	09 34 00.	+ 08 25	400		CLUSTER OF GALAXIES
LB 10929	09 34 00.	+ 22 37		17.6	FAINT BLUE STAR
LB 10928	09 34 00.	+ 25 14		16.8	FAINT BLUE STAR
ZWG 152.039	09 34 00.	+ 31 34		15.5	GALAXY
ZWG 181.067	09 34 00.	+ 38 18		14.5	GALAXY
MCG+07-20-024	09 34 00.	+ 41 35 30.	24	16.	GALAXY
MCG+07-20-025	09 34 00.	+ 41 50	24	15.5	GALAXY
ZWG 210.016	09 34 00.	+ 41 52		15.7	GALAXY
MCG+07-20-026	09 34 00.	+ 43 12	42	15.5	GALAXY
ZC 0934.0+6505	09 34 00.	+ 65 05	1550		CLUSTER OF GALAXIES
ABC 0818	09 34 00.	+ 74 13		16.5	RICH CLUSTER OF GALAXIES
ZC 0934.0+8115	09 34 00.	+ 81 15	1340		CLUSTER OF GALAXIES
MCG+14-05-006	09 34 00.	+ 82 19	18	16.	GALAXY
ZC 0934.0+8317	09 34 00.	+ 83 17	2080		CLUSTER OF GALAXIES
ZWG 007.008	09 34 00.	- 02 43		15.7	GALAXY
ABC 0814	09 34 01.	+ 76 10		17.1	RICH CLUSTER OF GALAXIES
ARP 221	09 34 02.	- 11 06			PECULIAR GALAXY
ZWG 007.010	09 34 06.	+ 00 38		15.6	GALAXY
ZWG 007.011	09 34 06.	+ 01 29		15.5	GALAXY
ZC 0934.1+1040	09 34 06.	+ 10 40	870		CLUSTER OF GALAXIES
ZWG 063.017	09 34 06.	+ 11 58		15.5	GALAXY
ZC 0934.1+1245	09 34 06.	+ 12 45	540		CLUSTER OF GALAXIES
ZWG 122.030	09 34 06.	+ 21 42		15.4	GALAXY
MCG+05-23-021	09 34 06.	+ 31 35	36	15.5	GALAXY
TON-N 1078	09 34 06.	+ 34 41		16.6	BLUE STAR
TON-N 1079	09 34 06.	+ 35 54		15.8	BLUE STAR
MCG+06-21-059	09 34 06.	+ 38 20	24	15.	GALAXY
MCG+08-18-014	09 34 06.	+ 48 44	48	15.	GALAXY
ZWG 239.012	09 34 06.	+ 48 45		15.7	GALAXY
MCG+04-23-014	09 34 09.	+ 21 40	30	15.4	GALAXY
MCG-02-25-006	09 34 09.	- 11 05 30.	42	14.	GALAXY
LB 10935	09 34 12.	+ 23 38		17.3	FAINT BLUE STAR
LB 10934	09 34 12.	+ 23 38		16.8	FAINT BLUE STAR
LB 10933	09 34 12.	+ 24 40		17.5	FAINT BLUE STAR
LB 10932	09 34 12.	+ 26 14		18.6	FAINT BLUE STAR
LB 10931	09 34 12.	+ 26 18		16.7	FAINT BLUE STAR
TON-N 1080	09 34 12.	+ 33 49		16.5	BLUE STAR
ZWG 181.068	09 34 12.	+ 34 02		15.7	GALAXY
UGC 05120	09 34 12.	+ 36 50	60	16.5	GALAXY Sa-b
ZWG 035.014	09 34 18.	+ 03 21		15.6	GALAXY
ZWG 063.018	09 34 18.	+ 14 07		15.7	GALAXY
UGC 05121	09 34 18.	+ 14 07	90	15.7	GALAXY Sc
UGC 05122	09 34 18.	+ 23 49	108	14.1	GALAXY Sb/SBb
LB 10936	09 34 18.	+ 24 34		15.8	FAINT BLUE STAR
ZWG 181.069	09 34 18.	+ 35 27		15.7	GALAXY
1ZW 019	09 34 18.	+ 48 51			COMPACT GALAXY
ZC 0934.3+5926	09 34 18.	+ 59 26	3360		CLUSTER OF GALAXIES
ZC 0934.3+7608	09 34 18.	+ 76 08	2020		CLUSTER OF GALAXIES
ZWG 007.012	09 34 18.	- 02 20		15.5	GALAXY
RNGC 2927	09 34 22.	+ 23 49		14.0	GALAXY
LB 00558	09 34 23.	+ 60 48 42.		16.3	FAINT BLUE STAR
MCG+03-25-005	09 34 24.	+ 17 13	60	15.2	GALAXY
ZWG 092.010	09 34 24.	+ 20 03		14.3	GALAXY
UGC 05123	09 34 24.	+ 20 03	42	14.3	GALAXY
MCG+03-25-006	09 34 24.	+ 20 03	36	14.3	GALAXY
MCG+04-23-015	09 34 24.	+ 21 53	60	15.7	GALAXY
ZWG 122.031	09 34 24.	+ 21 54		15.7	GALAXY
UGC 05124	09 34 24.	+ 21 54	66	15.7	GALAXY S
LB 10938	09 34 24.	+ 23 11		17.0	FAINT BLUE STAR
LB 10937	09 34 24.	+ 23 16		16.6	FAINT BLUE STAR
MCG+04-23-016	09 34 24.	+ 23 48	120	14.1	GALAXY
ZWG 122.032	09 34 24.	+ 23 49		14.1	GALAXY
MCG+06-21-060	09 34 24.	+ 33 03 30.	48	14.5	GALAXY
ZWG 181.070	09 34 24.	+ 35 49		15.2	GALAXY
ZC 0934.4+5556	09 34 24.	+ 55 56	1010		CLUSTER OF GALAXIES
MCG+10-14-027	09 34 24.	+ 62 06	36	16.	GALAXY
MCG-02-25-007	09 34 24.	- 09 40	48	15.5	GALAXY
MCG-03-25-011	09 34 24.	- 20 55 30.	192	12.	GALAXY
ABC 0811	09 34 27.	+ 77 46		17.4	RICH CLUSTER OF GALAXIES
RNGC 2935	09 34 27.	- 20 54		12.0	GALAXY
ZWG 007.013	09 34 30.	+ 01 20		15.4	GALAXY
ZC 0934.5+0842	09 34 30.	+ 08 42	1080		CLUSTER OF GALAXIES
ZWG 092.011	09 34 30.	+ 17 12		15.2	GALAXY
LB 10940	09 34 30.	+ 24 08		16.5	FAINT BLUE STAR
LB 10939	09 34 30.	+ 25 13		16.3	FAINT BLUE STAR
ZWG 122.033	09 34 30.	+ 26 07		15.5	GALAXY
ZWG 181.071	09 34 30.	+ 33 04		14.4	GALAXY
RNGC 2926	09 34 30.	+ 33 04		14.5	GALAXY
UGC 05125	09 34 30.	+ 33 04	54	14.4	GALAXY S
RNGC 2928	09 34 31.	+ 17 12		15.0	GALAXY
MCG+06-21-061	09 34 33.	+ 35 27 30.	33	15.	GALAXY
ABC 0838	09 34 34.	- 04 47		15.3	RICH CLUSTER OF GALAXIES
RNGC 2929	09 34 35.	+ 23 23		14.5	GALAXY
SHAH 151	09 34 36.	+ 00 55	102		GROUP OF COMPACT GALAXIES
ZC 0934.6+1309	09 34 36.	+ 13 09	940		CLUSTER OF GALAXIES
ZWG 063.019	09 34 36.	+ 14 01		15.6	GALAXY
LB 10943	09 34 36.	+ 20 35		15.5	FAINT BLUE STAR
LB 10942	09 34 36.	+ 21 10		16.0	FAINT BLUE STAR
LB 10941	09 34 36.	+ 22 49		17.2	FAINT BLUE STAR
ZWG 122.034	09 34 36.	+ 23 23		14.4	GALAXY
UGC 05126	09 34 36.	+ 23 23	78	14.4	GALAXY S
ZWG 181.072	09 34 36.	+ 35 26		15.5	GALAXY
UGC 05127	09 34 36.	+ 37 18	90	16.0	GALAXY Sc
ZWG 210.017	09 34 36.	+ 43 19		15.0	GALAXY
MCG+07-20-030	09 34 36.	+ 43 19	39	15.	GALAXY
ZC 0934.6+6244	09 34 36.	+ 62 44	1410		CLUSTER OF GALAXIES
MCG-01-25-007	09 34 39.	- 04 48	60	16.	GALAXY
ABC 0835	09 34 40.	+ 13 06		17.7	RICH CLUSTER OF GALAXIES
RNGC 2930	09 34 41.	+ 23 26		14.5	GALAXY
ZWG 092.012	09 34 42.	+ 19 49		14.6	GALAXY
LB 10945	09 34 42.	+ 22 52		17.5	FAINT BLUE STAR
MCG+04-23-017	09 34 42.	+ 23 22	72	14.4	GALAXY
ZWG 122.035	09 34 42.	+ 23 26		14.7	GALAXY
LB 10944	09 34 42.	+ 23 32		16.6	FAINT BLUE STAR
TON-N 1081	09 34 42.	+ 35 21		17.	BLUE STAR
MCG+06-21-062	09 34 42.	+ 37 20	78	15.	GALAXY
HOLM 134B	09 34 43.	+ 23 23	72	13.3	PART OF MULTIPLE GALAXY
MCG+04-23-018	09 34 45.	+ 23 25	36	14.7	GALAXY
HOLM 134A	09 34 45.	+ 23 26	36	13.9	PART OF MULTIPLE GALAXY
ABC 0837	09 34 46.	+ 08 46		17.5	RICH CLUSTER OF GALAXIES
RNGC 2931	09 34 47.	+ 23 28		15.0	GALAXY
ZWG 063.020	09 34 48.	+ 14 26		15.7	GALAXY
ZWG 092.013	09 34 48.	+ 17 20		15.5	GALAXY
MCG+03-25-007	09 34 48.	+ 19 50	48	14.6	GALAXY
ZWG 092.014	09 34 48.	+ 19 56		15.3	GALAXY
LB 10949	09 34 48.	+ 20 35		15.8	FAINT BLUE STAR
LB 10948	09 34 48.	+ 20 46		15.4	FAINT BLUE STAR
ZWG 122.036	09 34 48.	+ 23 28		14.9	GALAXY
MCG+04-23-019	09 34 48.	+ 23 28	42	14.9	GALAXY
LB 10947	09 34 48.	+ 24 52		17.4	FAINT BLUE STAR
LB 10946	09 34 48.	+ 26 24		11.2	FAINT BLUE STAR
ZC 0934.8+3335	09 34 48.	+ 33 35	340		CLUSTER OF GALAXIES
TON-N 1082	09 34 48.	+ 38 14		16.9	BLUE STAR
ZC 0934.8+5216	09 34 48.	+ 52 16	940		CLUSTER OF GALAXIES
ABC 0825	09 34 50.	+ 65 40		18.2	RICH CLUSTER OF GALAXIES
HOLM 134C	09 34 51.	+ 23 28	42	13.7	PART OF MULTIPLE GALAXY
ZWG 063.021	09 34 54.	+ 12 51		15.1	GALAXY
ZWG 122.037	09 34 54.	+ 21 31		15.7	GALAXY
LB 10951	09 34 54.	+ 22 01		16.4	FAINT BLUE STAR
LB 10950	09 34 54.	+ 23 58		16.2	FAINT BLUE STAR
ZWG 210.018	09 34 54.	+ 42 57		15.6	GALAXY
KARA.73B 0356	09 34 54.	+ 42 57	24	15.6	ISOLATED GALAXY S0
ZWG 210.019	09 34 54.	+ 43 12		15.7	GALAXY
MCG+07-20-027	09 34 54.	+ 43 12	48	15.5	GALAXY
ZWG 239.013	09 34 54.	+ 48 37		15.1	GALAXY
ZC 0934.9+7436	09 34 54.	+ 74 36	810		CLUSTER OF GALAXIES
ZC 0935.0+1835	09 35 00.	+ 18 35	940		CLUSTER OF GALAXIES
LB 10953	09 35 00.	+ 22 24		16.8	FAINT BLUE STAR
LB 10952	09 35 00.	+ 25 04		16.5	FAINT BLUE STAR
ZWG 122.038	09 35 00.	+ 25 43		14.3	GALAXY
UGC 05129	09 35 00.	+ 25 43	108	14.3	GALAXY Sa
ZWG 152.040	09 35 00.	+ 30 10		15.5	GALAXY
TON-N 1083	09 35 00.	+ 35 43		16.8	BLUE STAR
MCG+08-18-015	09 35 00.	+ 48 37	18	16.	GALAXY
ZWG 364.013	09 35 00.	+ 84 02		15.4	GALAXY
ZWG 363.051	09 35 00.	+ 84 02		15.4	GALAXY
UGC 05128	09 35 00.	+ 84 02	72	15.4	GALAXY Sc
KARA.73B 0357	09 35 00.	+ 84 02	60	15.4	ISOLATED GALAXY S
MCG+14-05-007	09 35 00.	+ 84 03	51	15.	GALAXY
ZWG 035.015	09 35 06.	+ 02 58		14.4	GALAXY
RNGC 2937	09 35 06.	+ 02 58		15.0	GALAXY
RNGC 2936	09 35 06.	+ 02 58		14.5	GALAXY
UGC 05131	09 35 06.	+ 02 58	96	14.4	GALAXY E+SYS
UGC 05130	09 35 06.	+ 02 58	96	14.4	GALAXY E+SYS
MCG+01-25-006	09 35 06.	+ 02 58	96	14.8	GALAXY
MCG+01-25-005	09 35 06.	+ 02 58	72	14.4	GALAXY
VV 316B	09 35 06.	+ 03 48	42	14.8	INTERACTING GALAXY
VV 316A	09 35 06.	+ 03 48	72	14.4	INTERACTING GALAXY
MCG+04-23-022	09 35 06.	+ 23 21 30.	42	15.5	GALAXY
ZWG 122.039	09 35 06.	+ 23 22		15.5	GALAXY
MCG+04-23-021	09 35 06.	+ 23 22	36	15.5	GALAXY
LB 10956	09 35 06.	+ 24 42		16.9	FAINT BLUE STAR
LB 10955	09 35 06.	+ 24 43		16.0	FAINT BLUE STAR
LB 10954	09 35 06.	+ 25 24		16.3	FAINT BLUE STAR
MCG+04-23-020	09 35 06.	+ 25 42	84	14.3	GALAXY
ZWG 152.041	09 35 06.	+ 27 54		15.5	GALAXY
TON-N 0018	09 35 06.	+ 28 11		15.5	BLUE STAR
ABC 0829	09 35 08.	+ 62 23		17.1	RICH CLUSTER OF GALAXIES
MCG+05-23-022	09 35 09.	+ 30 10 30.	36	15.5	GALAXY
ARP 142	09 35 10.	+ 02 58			PECULIAR GALAXY
HOLM 135A	09 35 12.	+ 03 00	54	14.4	PART OF MULTIPLE GALAXY
ZWG 092.015	09 35 12.	+ 17 15		14.9	GALAXY
UGC 05132	09 35 12.	+ 17 15	60	14.9	GALAXY S
MCG+03-25-008	09 35 12.	+ 17 15	66	14.9	GALAXY
LB 10959	09 35 12.	+ 21 16		17.0	FAINT BLUE STAR
LB 10958	09 35 12.	+ 21 31		17.3	FAINT BLUE STAR
LB 10957	09 35 12.	+ 24 04		16.8	FAINT BLUE STAR
ZWG 152.042	09 35 12.	+ 28 17		14.8	GALAXY
ZWG 210.020	09 35 12.	+ 43 44		14.7	GALAXY
UGC 05133	09 35 12.	+ 43 44	78	14.7	GALAXY S0
MCG+07-20-028	09 35 12.	+ 43 44	60	15.	GALAXY
ZC 0935.2-0118	09 35 12.	- 01 18	1950		CLUSTER OF GALAXIES
HOLM 135B	09 35 13.	+ 02 59	12	14.8	PART OF MULTIPLE GALAXY
RNGC 2934	09 35 13.	+ 17 15		15.0	GALAXY
RNGC 2933	09 35 13.	+ 17 15		15.0	GALAXY
IC 2495	09 35 15.	+ 28 17 44.			NONSTELLAR OBJECT
MCG+09-16-034	09 35 15.	+ 52 31	12	16.	GALAXY
RNGC 2945	09 35 15.	- 21 48		14.0	GALAXY
ZWG 035.016	09 35 18.	+ 03 41		15.7	GALAXY
ZC 0935.3+1701	09 35 18.	+ 17 01	5240		CLUSTER OF GALAXIES
LB 10961	09 35 18.	+ 25 02		16.6	FAINT BLUE STAR
LB 10960	09 35 18.	+ 26 06		16.4	FAINT BLUE STAR
MCG+05-23-023	09 35 18.	+ 28 17	42	14.8	GALAXY
REIZ 0008	09 35 18.	+ 43 45	36	14.1	GALAXY
ZWG 265.029	09 35 18.	+ 52 04		15.4	GALAXY
MCG-04-23-010	09 35 21.	- 21 48	21	14.5	GALAXY
RNGC 2939	09 35 22.	+ 09 45		13.5	GALAXY
RNGC 2940	09 35 22.	+ 09 50		15.0	GALAXY
ZC 0935.4+0453	09 35 24.	+ 04 53	1750		CLUSTER OF GALAXIES
ZWG 063.022	09 35 24.	+ 09 45		13.5	GALAXY
UGC 05134	09 35 24.	+ 09 45	156	13.5	GALAXY Sb-c
MCG+02-25-011	09 35 24.	+ 09 45	144	13.5	GALAXY
ZWG 063.023	09 35 24.	+ 09 50		14.8	GALAXY
MCG+02-25-012	09 35 24.	+ 09 50	12	14.8	GALAXY
LB 10965	09 35 24.	+ 23 22		17.6	FAINT BLUE STAR
LB 10964	09 35 24.	+ 23 39		16.9	FAINT BLUE STAR
LB 10963	09 35 24.	+ 24 18		17.0	FAINT BLUE STAR
LB 10962	09 35 24.	+ 24 50		16.7	FAINT BLUE STAR
MCG+07-20-029	09 35 24.	+ 43 24	60	15.	GALAXY
ZWG 239.014	09 35 24.	+ 44 52		15.7	GALAXY
ZC 0935.4+5353	09 35 24.	+ 53 53	810		CLUSTER OF GALAXIES
ZWG 007.014	09 35 24.	- 03 02		15.7	GALAXY
MCG-02-25-008	09 35 27.	- 10 44	78	14.5	GALAXY
ZWG 092.016	09 35 30.	+ 16 17		15.6	GALAXY
LB 10968	09 35 30.	+ 23 09		16.7	FAINT BLUE STAR
LB 10967	09 35 30.	+ 24 01		15.6	FAINT BLUE STAR
LB 10966	09 35 30.	+ 24 50		17.4	FAINT BLUE STAR
ZC 0935.5+2945	09 35 30.	+ 29 45	4170		CLUSTER OF GALAXIES

OBJECT NAME	RIGHT ASCEN.	DECLINATION	DIAM.	MAGN.	TYPE OF OBJECT
MCG+06-21-063	09 35 30.	+ 37 26	24	16.	GALAXY
ZWG 210.021	09 35 30.	+ 43 24		15.3	GALAXY
UGC 05135	09 35 30.	+ 43 24	72	15.3	GALAXY S
ZWG 239.015	09 35 30.	+ 48 47		14.7	GALAXY
MCG+08-18-016A	09 35 30.	+ 48 48	30	16.	GALAXY
MCG+08-18-016	09 35 30.	+ 48 48	24	15.	GALAXY
MCG-04-23-011	09 35 30.	- 22 10	90	15.	GALAXY
ZWG 035.017	09 35 36.	+ 07 58		15.7	GALAXY
ZWG 063.024	09 35 36.	+ 09 40		15.3	GALAXY
MCG+03-25-009	09 35 36.	+ 17 16	30	15.1	GALAXY
ZWG 092.017	09 35 36.	+ 17 17		15.1	GALAXY
LB 10971	09 35 36.	+ 20 54		16.8	FAINT BLUE STAR
LB 10970	09 35 36.	+ 25 22		17.3	FAINT BLUE STAR
LB 10969	09 35 36.	+ 26 26		16.0	FAINT BLUE STAR
MCG+08-18-017	09 35 36.	+ 48 34	42	16.	GALAXY
ZC 0935.6-0251	09 35 36.	- 02 51	940		CLUSTER OF GALAXIES
MCG-02-25-009	09 35 36.	- 11 26	60	15.5	GALAXY
MCG-05-23-007	09 35 36.	- 29 55	66	14.	GALAXY
RNGC 2941	09 35 37.	+ 17 17		15.0	GALAXY
HOLM 136B	09 35 37.	+ 17 17	42	14.5	PART OF MULTIPLE GALAXY
IC 0548	09 35 39.	+ 09 40 29.			NONSTELLAR OBJECT
MCG+03-25-010	09 35 39.	+ 17 40	36	15.4	GALAXY
MCG+09-16-035	09 35 39.	+ 52 31 30.	24	18.	GALAXY
ZWG 092.018	09 35 42.	+ 17 38		15.4	GALAXY
MCG+04-23-023	09 35 42.	+ 20 52	48	15.7	GALAXY
ZC 0935.7+2251	09 35 42.	+ 22 51	1480		CLUSTER OF GALAXIES
LB 10974	09 35 42.	+ 23 04		16.8	FAINT BLUE STAR
ZWG 122.040	09 35 42.	+ 24 07		15.3	GALAXY
LB 10973	09 35 42.	+ 24 22		16.5	FAINT BLUE STAR
LB 10972	09 35 42.	+ 24 42		16.4	FAINT BLUE STAR
ZWG 181.073	09 35 42.	+ 34 57		15.5	GALAXY
ZWG 210.022	09 35 42.	+ 41 04		15.7	GALAXY
MCG-03-25-013	09 35 42.	- 20 08	48	15.	GALAXY
MCG-03-25-012	09 35 42.	- 20 39	42	15.	GALAXY
IC 2496	09 35 43.	+ 34 55 19.			NONSTELLAR OBJECT
MCG+03-25-011	09 35 45.	+ 17 15	24	14.0	GALAXY
HOLM 136A	09 35 46.	+ 17 16	42	13.5	PART OF MULTIPLE GALAXY
ZWG 035.018	09 35 48.	+ 05 26		15.6	GALAXY
ZWG 063.025	09 35 48.	+ 13 46		15.5	GALAXY
ZC 0935.8+1442	09 35 48.	+ 14 42	870		CLUSTER OF GALAXIES
ZWG 092.019	09 35 48.	+ 17 16		14.0	GALAXY
UGC 05136	09 35 48.	+ 17 16	138	14.0	GALAXY E
ZWG 122.041	09 35 48.	+ 20 53		15.7	GALAXY
UGC 05137	09 35 48.	+ 20 53	60	15.7	GALAXY SB
LB 10988	09 35 48.	+ 22 22		14.8	FAINT BLUE STAR
LB 10975	09 35 48.	+ 26 26		15.8	FAINT BLUE STAR
TON-N 1084	09 35 48.	+ 32 47		17.	BLUE STAR
MCG+06-21-064	09 35 48.	+ 34 31 30.	30	15.	GALAXY
ZWG 181.074	09 35 48.	+ 35 09		15.7	GALAXY
RNGC 2973	09 35 48.	- 29 54			GALAXY
RNGC 2943	09 35 49.	+ 17 16		14.0	GALAXY
ZWG 035.019	09 35 54.	+ 02 47		15.4	GALAXY
UGC 05138	09 35 54.	+ 02 47	72	15.4	GALAXY Sb
ZWG 063.026	09 35 54.	+ 10 34		15.7	GALAXY
8ZW 0935+10.6	09 35 54.	+ 10 34		15.7	COMPACT GALAXY
TON-N 0440	09 35 54.	+ 31 20		14.8	BLUE STAR
ZWG 181.075	09 35 54.	+ 34 32		15.1	GALAXY
ZC 0935.9+4301	09 35 54.	+ 43 01	1010		CLUSTER OF GALAXIES
ARC 0842	09 35 54.	- 20 42		16.7	RICH CLUSTER OF GALAXIES
MCG+03-25-012	09 35 57.	+ 17 16	15	15.	GALAXY
REIZ 0009	09 35 59.	+ 47 44	48	14.8	GALAXY
RNGC 2947	09 35 59.	- 12 12			NON-EXISTENT OBJECT
VDB.66G 063	09 36	+ 71 25	170		DWARF GALAXY
LB 10979	09 36 00.	+ 21 11		16.5	FAINT BLUE STAR
TON-N 0441	09 36 00.	+ 24 46		15.4	BLUE STAR
LB 10978	09 36 00.	+ 25 52		16.7	FAINT BLUE STAR
LB 10977	09 36 00.	+ 25 52		16.6	FAINT BLUE STAR
ZC 0936.0+2609	09 36 00.	+ 26 09	1550		CLUSTER OF GALAXIES
LB 10976	09 36 00.	+ 26 39		17.0	FAINT BLUE STAR
ZC 0936.0+3329	09 36 00.	+ 33 29	1140		CLUSTER OF GALAXIES
SN 1961T	09 36 00.	+ 33 40		17.5	SUPERNOVA
MCG+06-21-066	09 36 00.	+ 33 40	9	18.5	GALAXY
MCG+06-21-065	09 36 00.	+ 34 14	120	13.	GALAXY
ZC 0936.0+3752	09 36 00.	+ 37 52	870		CLUSTER OF GALAXIES
ZWG 239.016	09 36 00.	+ 47 43		15.5	GALAXY
MCG+10-14-028	09 36 00.	+ 60 01	36	16.	GALAXY
ZWG 332.057	09 36 00.	+ 71 25		15.5	GALAXY
KARA.68 057	09 36 00.	+ 71 25	215		DWARF GALAXY
UGC 05139	09 36 00.	+ 71 25	240	15.5	GALAXY DWRF IR
7ZW 286	09 36 00.	+ 71 27			COMPACT GALAXY
ZWG 007.015	09 36 00.	- 03 20		15.7	GALAXY
ARC 0841	09 36 05.	- 03 58		16.5	RICH CLUSTER OF GALAXIES
ZC 0936.1+1755	09 36 06.	+ 17 55	610		CLUSTER OF GALAXIES
ZWG 181.076	09 36 06.	+ 34 14		14.1	GALAXY
RNGC 2942	09 36 06.	+ 34 14		13.0	GALAXY
UGC 05140	09 36 06.	+ 34 14	126	14.1	GALAXY Sc
TON-N 1085	09 36 06.	+ 36 16		17.	BLUE STAR
ZWG 239.017	09 36 06.	+ 47 21		15.3	GALAXY
MCG+08-18-018	09 36 06.	+ 47 21	24	16.	GALAXY
ZWG 350.028	09 36 06.	+ 79 44		15.6	GALAXY
MCG+06-21-067	09 36 09.	+ 32 31	66	14.5	GALAXY
HO 1	09 36 11.	+ 71 26	300	13.27	IRR GALAXY
ZWG 063.027	09 36 12.	+ 09 58		15.2	GALAXY
ZWG 181.077	09 36 12.	+ 36 47		15.3	GALAXY
KARA.72 208A	09 36 12.	+ 36 47	60	15.3	PART OF DOUBLE GALAXY
ARP 063	09 36 13.	+ 32 32			PECULIAR GALAXY
REIZ 0012	09 36 13.	+ 34 15	120	13.0	GALAXY
MCG+03-25-013	09 36 15.	+ 17 15	72	14.8	GALAXY
MCG+06-21-068	09 36 15.	+ 36 47	30	15.	GALAXY
MCG+08-18-019	09 36 15.	+ 48 39	90	14.	GALAXY
MAI 043	09 36 15.	+ 58 46	33		DWARF SPHEROIDAL GALAXY
RNGC 2948	09 36 17.	+ 07 11		14.0	GALAXY
ZWG 035.020	09 36 18.	+ 07 11		13.8	GALAXY
UGC 05141	09 36 18.	+ 07 11	102	13.8	GALAXY SBb/SBc
MCG+01-25-007	09 36 18.	+ 07 11	60	13.8	GALAXY
UGC 05142	09 36 18.	+ 08 00	60	16.0	GALAXY
ZWG 092.020	09 36 18.	+ 17 15		14.8	GALAXY
UGC 05143	09 36 18.	+ 17 15	72	14.8	GALAXY SB
TON-N 0442	09 36 18.	+ 25 15		15.2	BLUE STAR
LB 10980	09 36 18.	+ 26 16		17.2	FAINT BLUE STAR
ZWG 181.078	09 36 18.	+ 32 33		14.7	GALAXY
ZCG 0936+32	09 36 18.	+ 32 33		14.7	COMPACT GALAXY
UGC 05144	09 36 18.	+ 32 33	66	14.7	GALAXY DBL SYS
ZWG 181.079	09 36 18.	+ 36 48		15.4	GALAXY
KARA.72 208B	09 36 18.	+ 36 48	48	15.4	PART OF DOUBLE GALAXY
ZWG 239.018	09 36 18.	+ 48 39		15.1	GALAXY
UGC 05145	09 36 18.	+ 48 39	108	15.1	GALAXY S
RNGC 2946	09 36 19.	+ 17 15		15.0	GALAXY
RNGC 2944	09 36 19.	+ 32 32		14.5	GALAXY
REIZ 0010	09 36 19.	+ 48 38	60	13.9	GALAXY
MCG+06-21-072	09 36 21.	+ 32 35	42	14.	GALAXY
MCG+06-21-071	09 36 21.	+ 32 35	24	15.	GALAXY
MCG+06-21-070	09 36 21.	+ 34 05	24	15.	GALAXY
MCG+06-21-069	09 36 21.	+ 36 48	30	15.	GALAXY
VV 116E	09 36 21.	- 04 37	24	15.	INTERACTING GALAXY
VV 116D	09 36 21.	- 04 37	30	16.	INTERACTING GALAXY
VV 116C	09 36 21.	- 04 37	24	15.	INTERACTING GALAXY
VV 116B	09 36 21.	- 04 37	30	16.	INTERACTING GALAXY
VV 116A	09 36 21.	- 04 37	36	15.	INTERACTING GALAXY
VV 116	09 36 21.	- 04 37	150		INTERACTING GALAXY
LB 10981	09 36 24.	+ 21 38		16.7	FAINT BLUE STAR
ZWG 181.080	09 36 24.	+ 32 36		14.3	GALAXY
UGC 05146	09 36 24.	+ 32 36	66	14.3	GALAXY DBL SYS
KARA.72 209B	09 36 24.	+ 32 36	36		PART OF DOUBLE GALAXY
KARA.72 209A	09 36 24.	+ 32 36	42	14.3	PART OF DOUBLE GALAXY
TON-N 1086	09 36 24.	+ 34 10		16.	BLUE STAR
UGC 05147	09 36 24.	+ 38 40	72	16.5	GALAXY Sc
ZC 0936.4+4100	09 36 24.	+ 41 00	810		CLUSTER OF GALAXIES
ZC 0936.4+6655	09 36 24.	+ 66 55	2690		CLUSTER OF GALAXIES
MCG-01-25-009	09 36 24.	- 04 36	30	14.5	GALAXY
ARP 321	09 36 24.	- 04 37			PECULIAR GALAXY
MCG-01-25-008	09 36 24.	- 04 37 30.	42	15.	GALAXY
MCG-06-21-011	09 36 24.	- 38 48	42	15.	GALAXY
REIZ 0011	09 36 25.	+ 48 39	48	15.0	GALAXY
MCG-01-25-012	09 36 27.	- 04 35	24	15.5	GALAXY
MCG-01-25-011	09 36 27.	- 04 36 30.	24	11.	GALAXY
MCG-01-25-010	09 36 27.	- 04 37 30.	36	15.	GALAXY
SHAH 152	09 36 30.	+ 02 10	162		GROUP OF COMPACT GALAXIES
ZWG 122.042	09 36 30.	+ 23 48		15.6	GALAXY
3ZW 060	09 36 30.	+ 32 34			COMPACT GALAXY
VV 083B	09 36 30.	+ 32 36	36	16.	INTERACTING GALAXY
VV 083A	09 36 30.	+ 32 36	33	14.	INTERACTING GALAXY
ZWG 181.081	09 36 30.	+ 34 05		14.8	GALAXY
MCG+10-14-029	09 36 30.	+ 58 46	39	17.	GALAXY
ARP 129	09 36 31.	+ 32 34			PECULIAR GALAXY
VV 082B	09 36 33.	+ 32 32	15	17.	INTERACTING GALAXY
VV 082A	09 36 33.	+ 32 32	54	15.	INTERACTING GALAXY
VV 082	09 36 33.	+ 32 32	66		INTERACTING GALAXY
ARC 0834	09 36 34.	+ 66 55		16.3	RICH CLUSTER OF GALAXIES
LB 10982	09 36 36.	+ 25 38		17.0	FAINT BLUE STAR
ZC 0936.6+2847	09 36 36.	+ 28 47	1080		CLUSTER OF GALAXIES
ZC 0936.6+5939	09 36 36.	+ 59 39	1750		CLUSTER OF GALAXIES
PK238+34.1	09 36 36.	- 02 34	275	13.4	PLANETARY NEBULA
ZWG 035.021	09 36 42.	+ 06 38		15.0	GALAXY
MCG+01-25-008	09 36 42.	+ 06 38	48	15.0	GALAXY
ZC 0936.7+1425	09 36 42.	+ 14 25	1750		CLUSTER OF GALAXIES
MCG+03-25-015	09 36 42.	+ 17 00	45	15.5	GALAXY
ZWG 092.021	09 36 42.	+ 19 20		15.4	GALAXY
MCG+03-25-014	09 36 42.	+ 19 20	24	15.4	GALAXY
LB 10984	09 36 42.	+ 23 00		17.5	FAINT BLUE STAR
LB 10983	09 36 42.	+ 24 44		17.4	FAINT BLUE STAR
ZWG 122.043	09 36 42.	+ 25 56		15.6	GALAXY
ZC 0936.7+3316	09 36 42.	+ 33 16	610		CLUSTER OF GALAXIES
MCG-03-25-014	09 36 42.	- 17 36	36	15.5	GALAXY
MCG+03-25-016	09 36 45.	+ 17 00	9	15.5	GALAXY
ZWG 035.022	09 36 48.	+ 06 27		15.5	GALAXY
ZWG 035.023	09 36 48.	+ 06 40		15.2	GALAXY
ZWG 063.028	09 36 48.	+ 11 44		15.0	GALAXY
UGC 05148	09 36 48.	+ 11 44	66	15.0	GALAXY Sc
MCG+02-25-013	09 36 48.	+ 11 44	72	15.0	GALAXY
ZWG 092.022	09 36 48.	+ 16 59		15.5	GALAXY
LB 10987	09 36 48.	+ 22 30		16.0	FAINT BLUE STAR
LB 10986	09 36 48.	+ 24 25		17.4	FAINT BLUE STAR
LB 10985	09 36 48.	+ 25 53		16.4	FAINT BLUE STAR
MCG+05-23-024	09 36 48.	+ 28 26	48	15.4	GALAXY
ZWG 152.043	09 36 48.	+ 28 27		15.4	GALAXY
TON-N 0019	09 36 48.	+ 29 09		15.0	BLUE STAR
TON-N 1087	09 36 48.	+ 34 57		16.8	BLUE STAR
7ZW 287	09 36 48.	+ 69 12			COMPACT GALAXY
HOLM 137B	09 36 49.	+ 06 39	18	14.7	PART OF MULTIPLE GALAXY
LB 00559	09 36 49.	+ 62 21 48.		16.9	FAINT BLUE STAR
RNGC 2952	09 36 53.	- 09 55			NON-EXISTENT OBJECT
ZWG 035.024	09 36 54.	+ 03 22		15.6	GALAXY
ZWG 035.025	09 36 54.	+ 06 39		15.3	GALAXY
ZC 0936.9+3812	09 36 54.	+ 38 12	1480		CLUSTER OF GALAXIES
ZC 0936.9+4135	09 36 54.	+ 41 35	870		CLUSTER OF GALAXIES
DG 124	09 36 54.	- 02 40	300		REFLECTION NEBULA
HOLM 137A	09 36 56.	+ 06 41	48	14.3	PART OF MULTIPLE GALAXY
RNGC 2956	09 36 56.	- 18 52			GALAXY
MCG-01-25-013	09 36 57.	- 07 23	48	16.	GALAXY
UGC 05149	09 37 00.	+ 21 14	66	16.0	GALAXY Sc
LB 10990	09 37 00.	+ 21 44		17.7	FAINT BLUE STAR
LB 10989	09 37 00.	+ 22 02		17.0	FAINT BLUE STAR
ZWG 122.044	09 37 00.	+ 25 10		15.3	GALAXY
UGC 05119	09 37 00.	+ 38 18	48	14.5	GALAXY PECULR
ZC 0937.0+5117	09 37 00.	+ 51 17	740		CLUSTER OF GALAXIES
ZWG 007.016	09 37 00.	- 02 57		15.6	GALAXY
HOLM 137C	09 37 01.	+ 06 39	18	15.1	PART OF MULTIPLE GALAXY
MAI 044	09 37 02.	+ 71 29	182		DWARF SPHEROIDAL GALAXY
ZWG 007.017	09 37 06.	+ 00 00		15.1	GALAXY
MCG+00-25-006	09 37 06.	+ 00 00	30	15.1	GALAXY
ZC 0937.1+1015	09 37 06.	+ 10 15	670		CLUSTER OF GALAXIES
MCG+03-25-018	09 37 06.	+ 16 13	48	15.4	GALAXY
ZWG 092.023	09 37 06.	+ 16 14		15.4	GALAXY
ZWG 092.024	09 37 06.	+ 17 08		15.2	GALAXY
MCG+03-25-017	09 37 06.	+ 17 09	39	15.2	GALAXY
LB 10992	09 37 06.	+ 21 09		12.2	FAINT BLUE STAR
LB 10991	09 37 06.	+ 23 24		17.3	FAINT BLUE STAR
RNGC 2951	09 37 07.	- 00 00		15.0	GALAXY
ZWG 063.029	09 37 12.	+ 11 15		15.6	GALAXY
ZC 0937.2+1516	09 37 12.	+ 15 16	2820		CLUSTER OF GALAXIES
ZWG 092.025	09 37 12.	+ 17 01		15.5	GALAXY
LB 10993	09 37 12.	+ 22 51		16.0	FAINT BLUE STAR
ZWG 152.044	09 37 12.	+ 31 54		15.7	GALAXY
UGC 05150	09 37 12.	+ 47 44	78	16.0	GALAXY
MCG+08-18-020	09 37 12.	+ 47 44	36	16.	GALAXY
MCG+08-18-021	09 37 12.	+ 48 33	30	14.	GALAXY
ZWG 239.019	09 37 12.	+ 48 34		13.5	GALAXY
UGC 05151	09 37 12.	+ 48 34	48	13.5	GALAXY
ZWG 289.014	09 37 12.	+ 56 53		15.4	GALAXY
RNGC 2949	09 37 13.	+ 17 01		15.5	GALAXY
ZC 0937.3+1452	09 37 18.	+ 14 52	810		CLUSTER OF GALAXIES
LB 10997	09 37 18.	+ 20 29		16.5	FAINT BLUE STAR
LB 10996	09 37 18.	+ 23 36		16.5	FAINT BLUE STAR
LB 10995	09 37 18.	+ 23 37		15.7	FAINT BLUE STAR
LB 10994	09 37 18.	+ 25 54		17.6	FAINT BLUE STAR

OBJECT NAME	RIGHT ASCEN.	DECLINATION	DIAM.	MAGN.	TYPE OF OBJECT
MCG-02-25-010	09 37 18.	- 13 11 30.	15	15.	GALAXY
PK281-05.1	09 37 21.	- 59 51	25	11.3	PLANETARY NEBULA
IC 2501	09 37 21.	- 59 51	2	11.3	PLANETARY NEBULA
HN 0101	09 37 21.	- 59 52			NEBULA
LB 10999	09 37 24.	+ 24 08		17.0	FAINT BLUE STAR
LB 10998	09 37 24.	+ 24 31		16.7	FAINT BLUE STAR
ZWG 210.023	09 37 24.	+ 40 19		15.6	GALAXY
ZWG 350.029	09 37 24.	+ 79 55		14.2	GALAXY
UGC 05152	09 37 24.	+ 79 55	60	14.2	GALAXY S
ZC 0937.4-0043	09 37 24.	- 00 43	870		CLUSTER OF GALAXIES
MCG-02-25-011	09 37 24.	- 11 06 30.	54	14.5	GALAXY
REIZ 0014	09 37 25.	+ 40 18	36	15.2	GALAXY
RNGC 2908	09 37 25.	+ 79 55		14.0	GALAXY
MCG+07-20-031	09 37 27.	+ 40 18	45	15.	GALAXY
ARC 0844	09 37 27.	- 00 44		17.6	RICH CLUSTER OF GALAXIES
REIZ 0013	09 37 29.	+ 47 28	48	15.4	GALAXY
ZC 0937.5+0850	09 37 30.	+ 08 50	670		CLUSTER OF GALAXIES
8ZW 0937+13.4	09 37 30.	+ 13 27		17.6	COMPACT GALAXY
LB 11001	09 37 30.	+ 21 34		17.2	FAINT BLUE STAR
LB 11002	09 37 30.	+ 22 42		17.4	FAINT BLUE STAR
LB 11000	09 37 30.	+ 26 19		18.0	FAINT BLUE STAR
ZWG 289.015	09 37 30.	+ 61 17		15.6	GALAXY
UGC 05153	09 37 30.	+ 61 17	66	15.6	GALAXY Sb-c
MCG+10-14-030	09 37 30.	+ 61 17 30.	45	15.	GALAXY
ZWG 332.058	09 37 30.	+ 68 37		15.2	GALAXY
ZWG 063.030	09 37 36.	+ 12 48		15.6	GALAXY
MCG+03-25-019	09 37 36.	+ 15 09	27	15.3	GALAXY
LB 11005	09 37 36.	+ 23 16		17.3	FAINT BLUE STAR
LB 11004	09 37 36.	+ 23 50		16.0	FAINT BLUE STAR
LB 11003	09 37 36.	+ 25 44		16.8	FAINT BLUE STAR
ZWG 152.045	09 37 36.	+ 29 13		15.6	GALAXY
UGC 05154	09 37 36.	+ 29 13	60	15.6	GALAXY Sa-b
TON-N 1088	09 37 36.	+ 32 02		17.	BLUE STAR
ZWG 007.018	09 37 36.	- 03 21		15.6	GALAXY
RNGC 2953	09 37 38.	+ 15 04			NON-EXISTENT OBJECT
REIZ 0015	09 37 41.	+ 47 51	90	13.4	GALAXY
HN 0319	09 37 41.	- 68 52			NEBULA
ZWG 092.026	09 37 42.	+ 15 09		13.5	GALAXY
UGC 05155	09 37 42.	+ 15 09	102	13.5	GALAXY E
KARA.73B 0358	09 37 42.	+ 15 09	102	13.5	ISOLATED GALAXY S
LB 11006	09 37 42.	+ 24 41		17.6	FAINT BLUE STAR
ZWG 122.045	09 37 42.	+ 25 43		15.4	GALAXY
UGC 05156	09 37 42.	+ 25 43	72	15.4	GALAXY SB
ZWG 239.020	09 37 42.	+ 47 50		14.5	GALAXY
UGC 05157	09 37 42.	+ 47 50	114	14.5	GALAXY S
MCG+08-18-022	09 37 42.	+ 47 50	84	13.7	GALAXY
IC 2504	09 37 42.	- 68 51			NONSTELLAR OBJECT
RNGC 2954	09 37 44.	+ 15 09		13.5	GALAXY
HOLM 138A	09 37 47.	+ 47 51	60	13.7	PART OF MULTIPLE GALAXY
ZWG 063.031	09 37 48.	+ 11 43		14.7	GALAXY
MCG+02-25-014	09 37 48.	+ 11 43	36	14.7	GALAXY
ZC 0937.8+1247	09 37 48.	+ 12 47	1550		CLUSTER OF GALAXIES
ZWG 063.032	09 37 48.	+ 13 46		15.2	GALAXY
8ZW 0937+13.8	09 37 48.	+ 13 46		15.2	COMPACT GALAXY
ZWG 092.027	09 37 48.	+ 19 48		15.7	GALAXY
LB 11008	09 37 48.	+ 22 17		17.0	FAINT BLUE STAR
LB 11007	09 37 48.	+ 23 18		16.0	FAINT BLUE STAR
ZWG 239.021	09 37 48.	+ 48 48		15.3	GALAXY
ZC 0937.8+5550	09 37 48.	+ 55 50	1210		CLUSTER OF GALAXIES
SHAH 105	09 37 48.	+ 62 16	156	16.5	GROUP OF COMPACT GALAXIES
MCG+13-07-034	09 37 48.	+ 79 55	39	14.	GALAXY
MCG-04-23-012	09 37 48.	- 24 50	48	15.	GALAXY
ARC 0847	09 37 50.	+ 02 41		17.5	RICH CLUSTER OF GALAXIES
HOLM 138B	09 37 51.	+ 47 50	36	14.5	PART OF MULTIPLE GALAXY
KARA.68 058	09 37 54.	+ 00 16	54		DWARF GALAXY
ZC 0937.9+0241	09 37 54.	+ 02 41	870		CLUSTER OF GALAXIES
LB 11010	09 37 54.	+ 22 06		16.9	FAINT BLUE STAR
LB 11009	09 37 54.	+ 22 36		16.0	FAINT BLUE STAR
UGC 05158	09 37 54.	+ 53 27	60	16.0	GALAXY S
MCG+11-12-019	09 37 54.	+ 64 18	30	17.	GALAXY
RNGC 2958	09 37 57.	+ 12 07		14.0	GALAXY
IC 0550	09 37 59.	- 06 43 07.			NONSTELLAR OBJECT
SHB 141	09 37 59.4	+ 39 07 29.		18.	QUASI-STELLAR OBJECT
SHB 142	09 38	+ 11		19.	QUASI-STELLAR OBJECT
ZWG 035.026	09 38 00.	+ 03 48		13.6	GALAXY
RNGC 2960	09 38 00.	+ 03 48		13.5	GALAXY
UGC 05159	09 38 00.	+ 03 48	162	13.6	GALAXY Sa?
MCG+01-25-009	09 38 00.	+ 03 48	24	13.6	GALAXY
KARA.73B 0359	09 38 00.	+ 03 48	168	13.6	ISOLATED GALAXY S
ZWG 063.033	09 38 00.	+ 12 07		13.9	GALAXY
UGC 05160	09 38 00.	+ 12 07	66	13.9	GALAXY SB:b-c
MCG+02-25-015	09 38 00.	+ 12 07	60	13.9	GALAXY
ZWG 063.034	09 38 00.	+ 13 33		15.1	GALAXY
ZWG 122.046	09 38 00.	+ 21 28		15.4	GALAXY
MRK 403	09 38 00.	+ 21 28	14	16.5	GALAXY WITH UV CONTINUUM
LB 11012	09 38 00.	+ 21 42		16.6	FAINT BLUE STAR
LB 11011	09 38 00.	+ 24 35		15.8	FAINT BLUE STAR
ZWG 152.046	09 38 00.	+ 27 10		15.6	GALAXY
ZWG 152.047	09 38 00.	+ 28 00		15.5	GALAXY
UGC 05162	09 38 00.	+ 28 00	66	15.5	GALAXY S
UGC 05161	09 38 00.	+ 28 00	60	15.5	GALAXY S
MCG+05-23-025	09 38 00.	+ 28 00	60	15.5	GALAXY
ZWG 312.019	09 38 00.	+ 64 18		15.7	GALAXY
MCG-01-25-014	09 38 00.	- 06 42	36	14.5	GALAXY
IC 2497	09 38 03.	+ 34 55 44.			NONSTELLAR OBJECT
MCG+09-16-036	09 38 03.	+ 53 25	66	16.	GALAXY
ZWG 035.027	09 38 06.	+ 04 10		14.8	GALAXY
MCG+01-25-010	09 38 06.	+ 04 10	30	14.8	GALAXY
IC 0549	09 38 06.	+ 04 12 54.			NONSTELLAR OBJECT
ZC 0938.1+0800	09 38 06.	+ 08 00	670		CLUSTER OF GALAXIES
ZWG 122.047	09 38 06.	+ 25 41		15.4	GALAXY
ZWG 122.048	09 38 06.	+ 26 19		15.7	GALAXY
TON-N 1089	09 38 06.	+ 36 05		16.9	BLUE STAR
ZC 0938.1+4119	09 38 06.	+ 41 19	810		CLUSTER OF GALAXIES
ZC 0938.1+5645	09 38 06.	+ 56 45	1810		CLUSTER OF GALAXIES
ZWG 332.059	09 38 06.	+ 68 38		15.6	GALAXY
UGC 05163	09 38 06.	+ 68 38	60	15.6	GALAXY Sc-IRR
MCG-01-25-015	09 38 06.	- 08 43 30.	120	13.5	GALAXY
RNGC 2955	09 38 11.	+ 36 07		13.0	GALAXY
ZC 0938.2+1130	09 38 12.	+ 11 30	5040		CLUSTER OF GALAXIES
ZWG 063.035	09 38 12.	+ 11 47		14.9	GALAXY
8ZW 0938+11.8	09 38 12.	+ 11 47		14.9	COMPACT GALAXY
UGC 05164	09 38 12.	+ 11 47	66	14.9	GALAXY S-IRR
MCG+02-25-016	09 38 12.	+ 11 47	72	14.9	GALAXY
ZWG 092.028	09 38 12.	+ 20 07		15.7	GALAXY
KARA.73B 0360	09 38 12.	+ 20 07	42	15.7	ISOLATED GALAXY S
ZWG 122.049	09 38 12.	+ 21 25		15.7	GALAXY
UGC 05165	09 38 12.	+ 21 25	72	15.7	GALAXY S
MCG+04-23-024	09 38 12.	+ 21 25	72	15.7	GALAXY
ZC 0938.2+2246	09 38 12.	+ 22 46	1810		CLUSTER OF GALAXIES
LB 11014	09 38 12.	+ 25 22		16.8	FAINT BLUE STAR
LB 11013	09 38 12.	+ 26 01		16.2	FAINT BLUE STAR
ZWG 181.082	09 38 12.	+ 36 07		13.9	GALAXY
KARA.73B 0361	09 38 12.	+ 36 07	102	13.9	ISOLATED GALAXY S
REIZ 0016	09 38 14.	+ 36 06	42	13.2	GALAXY
IC 0553	09 38 14.5	- 05 12 30.			GALAXY SB(rs)
MCG+06-21-073	09 38 15.	+ 36 07	90	14.	GALAXY
MCG-01-25-016	09 38 15.	- 05 12 30.	60	15.	GALAXY
RNGC 2962	09 38 17.	+ 05 24		12.5	GALAXY
ZWG 035.028	09 38 18.	+ 05 24		13.1	GALAXY
UGC 05167	09 38 18.	+ 05 24	180	13.1	GALAXY S0/Sa
MCG+01-25-011	09 38 18.	+ 05 24	156	13.1	GALAXY
ZWG 035.029	09 38 18.	+ 07 10		14.5	GALAXY
UGC 05168	09 38 18.	+ 07 10	54	14.5	GALAXY
MCG+01-25-012	09 38 18.	+ 07 10	36	14.5	GALAXY
ZWG 122.050	09 38 18.	+ 21 27		15.2	GALAXY
MCG+04-23-025	09 38 18.	+ 21 27	30	15.2	GALAXY
ARC 0846	09 38 18.	+ 22 43		17.0	RICH CLUSTER OF GALAXIES
LB 11017	09 38 18.	+ 23 40		16.7	FAINT BLUE STAR
LB 11016	09 38 18.	+ 24 05		16.8	FAINT BLUE STAR
LB 11015	09 38 18.	+ 24 36		16.0	FAINT BLUE STAR
ZWG 122.051	09 38 18.	+ 26 18		15.6	GALAXY
TON-N 0020	09 38 18.	+ 28 37		15.0	BLUE STAR
ZWG 152.048	09 38 18.	+ 32 23		15.5	GALAXY
TON-N 1090	09 38 18.	+ 35 05		15.7	BLUE STAR
MCG+08-18-023	09 38 18.	+ 48 53	30	15.	GALAXY
MCG+11-12-020	09 38 18.	+ 63 24	51	16.	GALAXY
ARC 0850	09 38 21.	+ 12 48		17.5	RICH CLUSTER OF GALAXIES
IC 0551	09 38 22.	+ 07 09 23.			NONSTELLAR OBJECT
ZWG 063.036	09 38 24.	+ 10 32		15.7	GALAXY
ZWG 063.037	09 38 24.	+ 11 44		15.6	GALAXY
LB 11023	09 38 24.	+ 21 16		15.4	FAINT BLUE STAR
LB 11022	09 38 24.	+ 21 37		15.7	FAINT BLUE STAR
LB 11021	09 38 24.	+ 22 23		15.8	FAINT BLUE STAR
LB 11020	09 38 24.	+ 22 31		16.6	FAINT BLUE STAR
LB 11019	09 38 24.	+ 23 32		16.6	FAINT BLUE STAR
LB 11018	09 38 24.	+ 24 08		17.0	FAINT BLUE STAR
ZWG 152.049	09 38 24.	+ 28 20		15.3	GALAXY
MCG+07-20-032	09 38 24.	+ 41 36 30.	48	15.	GALAXY
ZWG 210.024	09 38 24.	+ 41 37		15.6	GALAXY
ARC 0843	09 38 24.	+ 56 48		17.5	RICH CLUSTER OF GALAXIES
IC 2498	09 38 27.	+ 28 20 05.			NONSTELLAR OBJECT
LB 11024	09 38 30.	+ 26 26		16.5	FAINT BLUE STAR
TON-N 0443	09 38 30.	+ 30 01		15.0	BLUE STAR
MCG+08-18-024	09 38 30.	+ 47 11	54	15.	GALAXY
ZWG 239.022	09 38 30.	+ 47 12		15.5	GALAXY
UGC 05169	09 38 30.	+ 47 12	66	15.5	GALAXY
MCG+08-18-025	09 38 30.	+ 48 20	18	15.	GALAXY
MCG+08-18-026	09 38 30.	+ 48 51 30.	30	17.	GALAXY
ZWG 265.030	09 38 30.	+ 54 31		15.7	GALAXY
MCG+10-14-031	09 38 30.	+ 57 18	24	16.	GALAXY
7ZW 288	09 38 30.	+ 64 38			COMPACT GALAXY
OCL 0778	09 38 30.	- 50 06	420	10.2	OPEN STAR CLUSTER
VHA 076	09 38 30.	- 50 06	180		OPEN STAR CLUSTER
RNGC 2972	09 38 30.	- 50 06		10.0	OPEN CLUSTER
IC 2499	09 38 32.	+ 28 07 29.			NONSTELLAR OBJECT
IC 0552	09 38 34.	+ 10 52 34.			NONSTELLAR OBJECT
ARC 0849	09 38 34.	+ 21 35		17.2	RICH CLUSTER OF GALAXIES
REIZ 0017	09 38 34.	+ 47 11	54	14.3	GALAXY
FATE 1.294	09 38 35.	+ 44 24	19		NEBULA
UGC 05170	09 38 36.	+ 03 02	60	16.0	GALAXY S
ZWG 063.038	09 38 36.	+ 10 52		14.5	GALAXY
UGC 05171	09 38 36.	+ 10 52	66	14.5	GALAXY S0
MCG+02-25-017	09 38 36.	+ 10 52	48	14.5	GALAXY
ZWG 063.039	09 38 36.	+ 11 26		15.6	GALAXY
LB 11025	09 38 36.	+ 21 17		17.8	FAINT BLUE STAR
ZWG 122.052	09 38 36.	+ 21 31		15.5	GALAXY
ZWG 122.053	09 38 36.	+ 26 20		15.6	GALAXY
ZC 0938.6+3934	09 38 36.	+ 39 34	1140		CLUSTER OF GALAXIES
MCG+08-18-027	09 38 36.	+ 48 53	54	15.	GALAXY
ZWG 239.023	09 38 36.	+ 48 54		15.6	GALAXY
UGC 05172	09 38 36.	+ 48 54	120	15.6	GALAXY
ZC 0938.6+6228	09 38 36.	+ 62 28	1080		CLUSTER OF GALAXIES
ZWG 312.020	09 38 36.	+ 63 24		15.6	GALAXY
KARA.73B 0362	09 38 36.	+ 63 24	60	15.6	ISOLATED GALAXY S
ZWG 092.029	09 38 42.	+ 16 56		15.7	GALAXY
LB 11026	09 38 42.	+ 24 42		16.6	FAINT BLUE STAR
TON-N 0444	09 38 42.	+ 24 53		15.1	BLUE STAR
ZC 0938.7+3620	09 38 42.	+ 36 20	940		CLUSTER OF GALAXIES
TON-N 1091	09 38 42.	+ 36 23		17.	BLUE STAR
ZC 0938.7+3957	09 38 42.	+ 39 57	5240		CLUSTER OF GALAXIES
MCG+11-12-021	09 38 42.	+ 65 40	39	16.	GALAXY
MCG+13-07-035	09 38 42.	+ 75 06	84	13.2	GALAXY
MCG-01-25-017	09 38 42.	- 07 33 30.	42	16.	GALAXY
ZWG 063.040	09 38 48.	+ 11 38		15.1	GALAXY
UGC 05173	09 38 48.	+ 11 38	132	15.1	GALAXY Sb
MCG+02-25-018	09 38 48.	+ 11 38	144	15.1	GALAXY
ZWG 063.041	09 38 48.	+ 13 35		15.6	GALAXY
LB 11027	09 38 48.	+ 23 22		17.4	FAINT BLUE STAR
ZWG 152.050	09 38 48.	+ 28 50		15.7	GALAXY
UGC 05174	09 38 48.	+ 33 44	60	16.5	GALAXY Sc
ZC 0938.8+3425	09 38 48.	+ 34 25	670		CLUSTER OF GALAXIES
ZWG 350.030	09 38 48.	+ 75 05		12.7	GALAXY
UGC 05175	09 38 48.	+ 75 05	96	12.7	GALAXY S
KARA.73B 0363	09 38 48.	+ 75 05	96	12.7	ISOLATED GALAXY S
ZC 0938.8-0027	09 38 48.	- 00 27	1750		CLUSTER OF GALAXIES
RNGC 2975	09 38 49.	- 16 25			GALAXY
RNGC 2977	09 38 50.	+ 75 05		12.5	GALAXY
AR 03	09 38 50.82	+ 75 05 21.6			NEBULA
ZC 0938.9+0547	09 38 54.	+ 05 47	1680		CLUSTER OF GALAXIES
ZWG 063.042	09 38 54.	+ 10 52		15.4	GALAXY
ZWG 063.043	09 38 54.	+ 11 41		15.5	GALAXY
LB 11028	09 38 54.	+ 24 18		16.3	FAINT BLUE STAR
ZWG 122.054	09 38 54.	+ 26 01		15.6	GALAXY
MCG+05-23-026	09 38 54.	+ 27 52	60	15.3	GALAXY
ZWG 152.051	09 38 54.	+ 27 53		15.3	GALAXY
TON-N 0445	09 38 54.	+ 29 49		15.2	BLUE STAR
ZWG 152.052	09 38 54.	+ 31 44		15.7	GALAXY
ZWG 239.024	09 38 54.	+ 48 22		15.4	GALAXY
OCL 0779	09 38 54.	- 50 23	360		OPEN STAR CLUSTER
ARC 0855	09 38 58.	- 09 04		17.4	RICH CLUSTER OF GALAXIES
RNGC 2950	09 38 59.	+ 59 05		12.5	GALAXY
LB 11030	09 39 00.	+ 21 28		15.9	FAINT BLUE STAR
LB 11029	09 39 00.	+ 25 36		16.2	FAINT BLUE STAR
MCG+10-14-032	09 39 00.	+ 59 04	132	11.8	GALAXY
ZWG 289.016	09 39 00.	+ 59 05		11.8	GALAXY

OBJECT NAME	RIGHT ASCEN.	DECLINATION	DIAM.	MAGN.	TYPE OF OBJECT
UGC 05176	09 39 00.	+ 59 05	210	11.8	GALAXY SB0
ZWG 350.031	09 39 00.	+ 76 35		15.2	GALAXY
MRK 118	09 39 00.	+ 76 35	22	15.5	GALAXY WITH UV CONTINUUM
MCG+13-07-036	09 39 00.	+ 76 35	51	15.	GALAXY
ARC 0852	09 39 05.	+ 29 39		17.8	RICH CLUSTER OF GALAXIES
ZC 0939.1+0733	09 39 06.	+ 07 33	470		CLUSTER OF GALAXIES
8ZW 0939+10.9	09 39 06.	+ 10 57		17.3	COMPACT GALAXY
ZWG 063.044	09 39 06.	+ 11 42		15.0	GALAXY
MCG+02-25-019	09 39 06.	+ 11 42	36	15.0	GALAXY
IC 0554	09 39 06.	+ 12 39 43.			NONSTELLAR OBJECT
ZWG 063.045	09 39 06.	+ 13 43		15.3	GALAXY
ZWG 092.030	09 39 06.	+ 15 12		15.2	GALAXY
ZWG 092.031	09 39 06.	+ 19 38		15.7	GALAXY
LB 11031	09 39 06.	+ 21 34		16.1	FAINT BLUE STAR
ZC 0939.1+4142	09 39 06.	+ 41 42	870		CLUSTER OF GALAXIES
AR 02	09 39 06.68	+ 76 34 53.0			NEBULA
ZWG 063.046	09 39 12.	+ 11 52		15.2	GALAXY
UGC 05177	09 39 12.	+ 11 52	66	15.2	GALAXY S?
ZWG 063.047	09 39 12.	+ 12 31		14.4	GALAXY
UGC 05178	09 39 12.	+ 12 31	78	14.4	GALAXY S0
MCG+02-25-020	09 39 12.	+ 12 31	60	14.4	GALAXY
IC 0555	09 39 12.	+ 12 31 19.			NONSTELLAR OBJECT
LB 11034	09 39 12.	+ 22 38		15.6	FAINT BLUE STAR
LB 11033	09 39 12.	+ 23 51		15.8	FAINT BLUE STAR
ZWG 122.055	09 39 12.	+ 23 55		15.3	GALAXY
ZWG 122.056	09 39 12.	+ 26 16		15.7	GALAXY
LB 11032	09 39 12.	+ 26 35		16.0	FAINT BLUE STAR
MCG-01-25-018	09 39 12.	- 08 48	60	15.	GALAXY
ACK 279-03.2	09 39 12.	- 56 45			PLANETARY NEBULA
REIZ 0018	09 39 15.	+ 42 07	36	15.4	GALAXY
ZWG 035.030	09 39 18.	+ 03 21		15.4	GALAXY
ZC 0939.3+0908	09 39 18.	+ 09 08	1080		CLUSTER OF GALAXIES
LB 11035	09 39 18.	+ 23 43		16.0	FAINT BLUE STAR
ZWG 152.053	09 39 18.	+ 27 04		15.7	GALAXY
IC 2500	09 39 18.	+ 36 34			MAY NOT EXIST
ZWG 182.001	09 39 18.	+ 36 35		15.2	GALAXY
ZWG 181.083	09 39 18.	+ 36 35		15.2	GALAXY
MCG-01-25-019	09 39 18.	- 04 04	18	15.	GALAXY
OCL 0787	09 39 18.	- 53 36	660	11.	OPEN STAR CLUSTER
VHA 077	09 39 18.	- 53 36	420		OPEN STAR CLUSTER
MCG+06-21-074	09 39 21.	+ 36 35	60	15.	GALAXY
MCG-01-25-020	09 39 21.	- 07 06	54	15.	GALAXY
ARC 0854	09 39 22.	+ 09 10		17.7	RICH CLUSTER OF GALAXIES
RNGC 2969	09 39 22.	- 08 23		13.0	GALAXY
ZWG 007.019	09 39 24.	+ 01 33		15.7	GALAXY
KARA.73B 0364	09 39 24.	+ 01 33	54	15.7	ISOLATED GALAXY S
LB 11036	09 39 24.	+ 24 28		16.8	FAINT BLUE STAR
ZWG 152.054	09 39 24.	+ 31 12		15.6	GALAXY
ZC 0939.4+4403	09 39 24.	+ 44 03	1550		CLUSTER OF GALAXIES
ZWG 289.017	09 39 24.	+ 59 12		15.1	GALAXY
UGC 05179	09 39 24.	+ 59 12	78	15.1	GALAXY S
PK274+02.1	09 39 24.	- 49 09 24.	10		PLANETARY NEBULA
MCG-01-25-021	09 39 27.	- 08 23 30.	66	13.	GALAXY
ZWG 007.020	09 39 30.	+ 00 34		12.2	GALAXY
UGC 05180	09 39 30.	+ 00 34	168	12.2	GALAXY Sc
MCG+00-25-007	09 39 30.	+ 00 34	150	12.2	GALAXY
ZWG 035.031	09 39 30.	+ 02 41		15.7	GALAXY
ZC 0939.5+1535	09 39 30.	+ 15 35	870		CLUSTER OF GALAXIES
LB 11038	09 39 30.	+ 21 57		17.5	FAINT BLUE STAR
LB 11037	09 39 30.	+ 25 15		16.4	FAINT BLUE STAR
ZWG 152.055	09 39 30.	+ 29 11		15.6	GALAXY
TON-N 0446	09 39 30.	+ 29 35		16.8	BLUE STAR
ZC 0939.5+3245	09 39 30.	+ 32 45	1680		CLUSTER OF GALAXIES
MCG+07-20-033	09 39 30.	+ 42 04	36	15.5	GALAXY
ZC 0939.5+6102	09 39 30.	+ 61 02	6450		CLUSTER OF GALAXIES
RNGC 2967	09 39 31.	+ 00 34		12.0	GALAXY
ARC 0853	09 39 31.	+ 15 37		17.5	RICH CLUSTER OF GALAXIES
ZWG 035.032	09 39 36.	+ 04 30		15.7	GALAXY
ZWG 035.033	09 39 36.	+ 04 54		14.0	GALAXY
RNGC 2966	09 39 36.	+ 04 54		14.0	GALAXY
UGC 05181	09 39 36.	+ 04 54	138	14.0	GALAXY SB?
MCG+01-25-013	09 39 36.	+ 04 54	132	14.0	GALAXY
ZWG 092.032	09 39 36.	+ 14 51		15.6	GALAXY
LB 11039	09 39 36.	+ 25 01		15.9	FAINT BLUE STAR
ZWG 122.057	09 39 36.	+ 25 28		15.6	GALAXY
ZC 0939.6+2541	09 39 36.	+ 25 41	3760		CLUSTER OF GALAXIES
MCG+10-14-033	09 39 36.	+ 59 11	45	15.	GALAXY
ARC 0851	09 39 37.	+ 47 14		18.4	RICH CLUSTER OF GALAXIES
REIZ 0019	09 39 40.	+ 24 43	48	13.9	GALAXY
LB 11041	09 39 42.	+ 23 24		17.5	FAINT BLUE STAR
LB 11040	09 39 42.	+ 24 30		16.4	FAINT BLUE STAR
7ZW 289	09 39 42.	+ 77 41			COMPACT GALAXY
MCG-01-25-022	09 39 42.	- 04 28	102	14.	GALAXY
BIGO 509	09 39 43.	+ 73 06			NEBULA
MCG-01-25-023	09 39 45.	- 04 29	42	17.	GALAXY
ARC 0857	09 39 46.	- 22 24		16.9	RICH CLUSTER OF GALAXIES
ZWG 035.034	09 39 48.	+ 03 35		15.5	GALAXY
ZWG 035.035	09 39 48.	+ 04 30		14.1	GALAXY
UGC 05182	09 39 48.	+ 04 30	72	14.1	GALAXY S0
MCG+01-25-014	09 39 48.	+ 04 30	60	14.1	GALAXY
ZWG 035.036	09 39 48.	+ 04 35		15.5	GALAXY
ZWG 035.037	09 39 48.	+ 07 20		14.7	GALAXY
MCG+01-25-015	09 39 48.	+ 07 20	36	14.7	GALAXY
KARA.73B 0365	09 39 48.	+ 07 20	42	14.7	ISOLATED GALAXY IR
LB 11044	09 39 48.	+ 21 06		17.4	FAINT BLUE STAR
LB 11043	09 39 48.	+ 21 15		14.6	FAINT BLUE STAR
LB 11042	09 39 48.	+ 25 34		15.9	FAINT BLUE STAR
TON-N 0447	09 39 48.	+ 31 54		15.3	BLUE STAR
ZWG 182.002	09 39 48.	+ 36 26		15.7	GALAXY
ZWG 181.084	09 39 48.	+ 36 26		15.7	GALAXY
ZC 0939.8+4714	09 39 48.	+ 47 14	470		CLUSTER OF GALAXIES
ZWG 332.060	09 39 48.	+ 72 29		15.7	GALAXY
ZWG 007.021	09 39 48.	- 01 46		15.7	GALAXY
PK274+02.2	09 39 48.	- 49 44 38.	5		PLANETARY NEBULA
LB 11046	09 39 54.	+ 24 19		16.1	FAINT BLUE STAR
LB 11045	09 39 54.	+ 26 14		17.1	FAINT BLUE STAR
ZWG 122.058	09 39 54.	+ 26 18		15.7	GALAXY
ZWG 152.056	09 39 54.	+ 32 05		12.0	GALAXY
MRK 404	09 39 54.	+ 32 05		15.5	GALAXY WITH UV CONTINUUM
UGC 05183	09 39 54.	+ 32 05	216	12.0	GALAXY Sb/Sc
KARA.72 210A	09 39 54.	+ 32 05	174	12.0	PART OF DOUBLE GALAXY
ZWG 182.003	09 39 54.	+ 38 03		14.9	GALAXY
ZWG 181.085	09 39 54.	+ 38 03		14.9	GALAXY
UGC 05184	09 39 54.	+ 38 03	72	14.9	GALAXY SBb
KARA.73B 0366	09 39 54.	+ 38 03	48	14.9	ISOLATED GALAXY S
MCG+10-14-034	09 39 54.	+ 58 04 30.	27	16.	GALAXY
MCG-01-25-024	09 39 54.	- 06 01	96	14.	GALAXY
RNGC 2964	09 39 55.	+ 32 05		12.5	GALAXY
REIN 2.087	09 39 57.33	+ 32 04 32.8			NEBULA
REIZ 0020	09 39 58.	+ 32 04	60	12.1	GALAXY
ZWG 092.033	09 40 00.	+ 14 54		15.5	GALAXY
ZC 0940.0+1915	09 40 00.	+ 19 15	3020		CLUSTER OF GALAXIES
LB 11047	09 40 00.	+ 23 29		17.0	FAINT BLUE STAR
TON-N 0021	09 40 00.	+ 26 14		14.4	BLUE STAR
MCG+05-23-028	09 40 00.	+ 29 12 30.	72	14.5	GALAXY
ZWG 152.057	09 40 00.	+ 29 13		14.5	GALAXY
UGC 05185	09 40 00.	+ 29 13	114	14.5	GALAXY Sc
ZC 0940.0+2958	09 40 00.	+ 29 58	1550		CLUSTER OF GALAXIES
MCG+05-23-027	09 40 00.	+ 32 06	180	12.0	GALAXY
UGC 05186	09 40 00.	+ 33 28	84	16.5	GALAXY IRR
MCG+06-22-001	09 40 00.	+ 38 04	48	14.5	GALAXY
MCG+07-20-034	09 40 00.	+ 41 19	48	14.	GALAXY
ZWG 210.025	09 40 00.	+ 41 20		14.1	GALAXY
UGC 05167	09 40 00.	+ 41 20	66	14.1	GALAXY SB
MRK 119	09 40 00.	+ 66 12	20	14.	GALAXY WITH UV CONTINUUM
MCG+11-12-022	09 40 00.	+ 66 12 30.	24	15.	GALAXY
ZWG 312.021	09 40 00.	+ 66 13		14.1	GALAXY
RNGC 2909	09 40 00.	+ 66 13		14.0	GALAXY
UGC 05188	09 40 00.	+ 66 13	27	14.1	GALAXY PECULR
ZC 0940.0+7835	09 40 00.	+ 78 35	940		CLUSTER OF GALAXIES
ZWG 364.014	09 40 00.	+ 82 04		15.1	GALAXY
ZWG 007.022	09 40 00.	- 03 28		12.3	GALAXY
MCG+00-25-008	09 40 00.	- 03 28	66	12.3	GALAXY
SV 225	09 40 00.36	+ 14 02 41.6		19.25	BLUE STELLAR OBJECT
HOLM 139A	09 40 02.	+ 41 19	60	13.3	PART OF MULTIPLE GALAXY
RNGC 2974	09 40 02.	- 03 29		12.5	GALAXY
REIN 2.089	09 40 02.04	- 03 28 12.2			NEBULA
MCG-01-25-025	09 40 03.	- 06 19	30	17.	GALAXY
ZWG 035.038	09 40 06.	+ 04 31		15.4	GALAXY
ZWG 122.059	09 40 06.	+ 26 13		15.7	GALAXY
ZC 0940.1+3606	09 40 06.	+ 36 06	1080		CLUSTER OF GALAXIES
MCG-03-25-015	09 40 06.	- 16 45	84	16.	GALAXY
IC 2502	09 40 11.	+ 33 22 10.			NONSTELLAR OBJECT
RNGC 2965	09 40 11.	+ 36 28		14.5	GALAXY
MCG+02-25-022	09 40 12.	+ 09 42	96		GALAXY
ZWG 063.048	09 40 12.	+ 09 43		13.7	GALAXY
UGC 05189	09 40 12.	+ 09 43	198	13.8	GALAXY IRR SYS
MCG+02-25-021	09 40 12.	+ 09 43	48	13.7	GALAXY
ZWG 063.049	09 40 12.	+ 12 36		14.6	GALAXY
MCG+02-25-023	09 40 12.	+ 12 36	42	14.6	GALAXY
ZWG 122.060	09 40 12.	+ 23 00		15.5	GALAXY
LB 11048	09 40 12.	+ 23 38		18.3	FAINT BLUE STAR
TON-N 0448	09 40 12.	+ 31 25		14.5	BLUE STAR
ZWG 152.058	09 40 12.	+ 32 10		13.1	GALAXY
UGC 05190	09 40 12.	+ 32 10	198	13.1	GALAXY
KARA.72 210B	09 40 12.	+ 32 10	138	13.1	PART OF DOUBLE GALAXY
UGC 05191	09 40 12.	+ 36 25	72	14.7	GALAXY S0
ZWG 182.004	09 40 12.	+ 36 28		14.7	GALAXY
ZWG 181.086	09 40 12.	+ 36 28		14.7	GALAXY
ZWG 210.026	09 40 12.	+ 39 07		15.5	GALAXY
ARC 0845	09 40 12.	+ 64 38		17.4	RICH CLUSTER OF GALAXIES
MCG-06-22-001	09 40 12.	- 38 40	36	15.5	GALAXY
RNGC 2968	09 40 13.	+ 32 10		13.0	GALAXY
HOLM 139B	09 40 13.	+ 41 18	12	15.1	PART OF MULTIPLE GALAXY
IC 2503	09 40 14.	+ 33 25 03.			NONSTELLAR OBJECT
BIGO 510	09 40 14.	+ 36 29			NEBULA
REIN 2.088	09 40 14.89	+ 32 09 35.2			NEBULA
REIZ 0021	09 40 16.	+ 32 09	60	13.0	GALAXY
ZWG 063.050	09 40 18.	+ 14 22		15.4	GALAXY
MCG+04-23-027	09 40 18.	+ 21 23 30.	90	14.8	GALAXY
ZWG 122.061	09 40 18.	+ 21 25		14.8	GALAXY
UGC 05192	09 40 18.	+ 21 25	126	14.8	GALAXY SB:b
LB 11050	09 40 18.	+ 22 58		17.5	FAINT BLUE STAR
MCG+04-23-026	09 40 18.	+ 22 59	42	15.5	GALAXY
LB 11049	09 40 18.	+ 25 21		16.8	FAINT BLUE STAR
MCG+06-22-002	09 40 18.	+ 35 22	24	16.	GALAXY
MCG+06-22-003	09 40 18.	+ 36 28	18	14.5	GALAXY
MCG-04-23-013	09 40 18.	- 26 34 30.	36	16.	GALAXY
OCL 0770	09 40 18.	- 43 48	756	13.	OPEN STAR CLUSTER
SN 1970L	09 40 19.	+ 32 11		13.0	SUPERNOVA
MCG+05-23-029	09 40 21.	+ 32 10 30.	90	13.1	GALAXY
MCG+07-20-035	09 40 21.	+ 39 06 30.	36	15.	GALAXY
MCG-01-25-026	09 40 21.	- 06 30	108	15.	GALAXY
ZC 0940.4+0636	09 40 24.	+ 06 36	540		CLUSTER OF GALAXIES
ZC 0940.4+1840	09 40 24.	+ 18 40	1010		CLUSTER OF GALAXIES
LB 11051	09 40 24.	+ 23 40		17.4	FAINT BLUE STAR
TON-N 1092	09 40 24.	+ 36 20		16.	BLUE STAR
MCG+06-22-004	09 40 24.	+ 36 27 30.	12	15.	GALAXY
ZWG 210.027	09 40 24.	+ 39 39		15.5	GALAXY
UGC 05193	09 40 24.	+ 39 39	60	15.5	GALAXY E
ZC 0940.4-0226	09 40 24.	- 02 26	2290		CLUSTER OF GALAXIES
RNGC 2982	09 40 26.	- 43 58			NON-EXISTENT OBJECT
MCG-03-25-016	09 40 27.	- 15 53 30.	72	15.	GALAXY
ZC 0940.5+0331	09 40 30.	+ 03 31	940		CLUSTER OF GALAXIES
ZC 0940.5+1729	09 40 30.	+ 17 29	1010		CLUSTER OF GALAXIES
LB 11053	09 40 30.	+ 20 52		13.0	FAINT BLUE STAR
ZWG 122.062	09 40 30.	+ 23 03		15.6	GALAXY
LB 11052	09 40 30.	+ 24 20		16.9	FAINT BLUE STAR
UGC 05194	09 40 30.	+ 34 16	66	16.0	GALAXY Sc
ZC 0940.5+3722	09 40 30.	+ 37 22	470		CLUSTER OF GALAXIES
REIZ 0022	09 40 30.	+ 39 38	18	15.3	GALAXY
MCG+07-20-036	09 40 30.	+ 39 38	78	15.	GALAXY
MCG+08-18-028	09 40 30.	+ 48 30	48	16.	GALAXY
ZC 0940.5+5415	09 40 30.	+ 54 15	610		CLUSTER OF GALAXIES
ZWG 007.023	09 40 30.	- 02 01		14.7	GALAXY
MCG+00-25-009	09 40 30.	- 02 01	24	14.7	GALAXY
RNGC 2970	09 40 31.	+ 32 13		14.5	GALAXY
REIZ 0023	09 40 33.	+ 42 19	30	14.7	GALAXY
REIN 7.051	09 40 33.44	+ 32 12 17.5			NEBULA
REIZ 0024	09 40 34.	+ 32 12	60	14.0	GALAXY
ZWG 007.024	09 40 36.	+ 00 39		15.4	GALAXY
UGC 05195	09 40 36.	+ 00 39	66	15.4	GALAXY Sb-c
ZWG 035.039	09 40 36.	+ 03 39		15.3	GALAXY
LB 11054	09 40 36.	+ 22 42		17.5	FAINT BLUE STAR
MCG+04-23-028	09 40 36.	+ 23 02	36	15.6	GALAXY
ZWG 152.059	09 40 36.	+ 32 13		14.7	GALAXY
MRK 405	09 40 36.	+ 32 13	22	15.5	GALAXY WITH UV CONTINUUM
MCG+05-23-030	09 40 36.	+ 32 14	24	14.7	GALAXY
ZC 0940.6+3334	09 40 36.	+ 33 34	870		CLUSTER OF GALAXIES
ZWG 210.028	09 40 36.	+ 42 42		15.3	GALAXY
ZC 0940.6+7900	09 40 36.	+ 79 00	1010		CLUSTER OF GALAXIES
PATH 1.295	09 40 37.	+ 44 32	14		NEBULA
MCG+07-20-037	09 40 39.	+ 42 42 30.	12	15.	GALAXY
RNGC 2971	09 40 41.	+ 36 24		15.0	GALAXY
RNGC 2979	09 40 41.	- 10 09		14.0	GALAXY
ZWG 063.051	09 40 42.	+ 10 19		15.7	GALAXY

OBJECT NAME	RIGHT ASCEN.	DECLINATION	DIAM.	MAGN.	TYPE OF OBJECT
MCG+02-25-024	09 40 42.	+ 10 19	36	15.7	GALAXY
ZWG 092.034	09 40 42.	+ 17 10		15.7	GALAXY
UGC 05196	09 40 42.	+ 17 10	60	15.7	GALAXY ?
MCG+03-25-020	09 40 42.	+ 17 10	54	15.7	GALAXY
LB 11055	09 40 42.	+ 22 40		17.3	FAINT BLUE STAR
ZWG 182.005	09 40 42.	+ 36 24		15.0	GALAXY
ZWG 181.087	09 40 42.	+ 36 24		15.0	GALAXY
UGC 05197	09 40 42.	+ 36 24	72	15.0	GALAXY SBb
MCG+06-22-005	09 40 42.	+ 36 24 30.	66	15.5	GALAXY
MCG+09-16-037	09 40 42.	+ 56 19	30	16.	GALAXY
ZC 0940.7+6605	09 40 42.	+ 66 05	540		CLUSTER OF GALAXIES
7ZW 290	09 40 42.	+ 78 18			COMPACT GALAXY
MCG+07-20-038	09 40 45.	+ 42 53 30.	48	16.	GALAXY
MCG-01-25-027	09 40 45.	- 04 14	48	14.5	GALAXY
MCG-01-25-028	09 40 45.	- 09 23 30.	84	13.	GALAXY
MCG-02-25-012	09 40 45.	- 10 09 30.	72	14.	GALAXY
REIZ 0025	09 40 46.	+ 42 53	18	15.1	GALAXY
RNGC 2980	09 40 46.	- 09 23		13.0	GALAXY
ABC 0858	09 40 48.	+ 06 07		16.6	RICH CLUSTER OF GALAXIES
ZWG 063.052	09 40 48.	+ 09 39		15.4	GALAXY
ZWG 122.063	09 40 48.	+ 22 01		15.7	GALAXY
ZWG 122.064	09 40 48.	+ 23 01		15.6	GALAXY
MCG+04-23-029	09 40 48.	+ 23 01	36	15.6	GALAXY
LB 11057	09 40 48.	+ 23 02		17.3	FAINT BLUE STAR
LB 11056	09 40 48.	+ 25 42		17.6	FAINT BLUE STAR
MCG+05-23-031	09 40 48.	+ 26 34	48	15.6	GALAXY
ZWG 152.060	09 40 48.	+ 26 35		15.6	GALAXY
ZWG 210.029	09 40 48.	+ 41 55		15.0	GALAXY
UGC 05198	09 40 48.	+ 41 55	66	15.0	GALAXY SBa
ZWG 210.030	09 40 48.	+ 42 40		15.0	GALAXY
MCG+07-20-040	09 40 48.	+ 42 40	24	14.5	GALAXY
MCG+07-20-041	09 40 48.	+ 42 52	30	17.	GALAXY
HOLM 141B	09 40 48.	+ 42 53	24	14.8	PART OF MULTIPLE GALAXY
ZWG 210.031	09 40 48.	+ 42 54		15.0	GALAXY
UGC 05199	09 40 48.	+ 42 54	66	15.0	GALAXY Sa-b
MCG+07-20-039	09 40 48.	+ 42 54 30.	84	14.	GALAXY
MCG-01-25-030	09 40 48.	- 04 15	42	16.	GALAXY
MCG-01-25-029	09 40 48.	- 09 30	48	13.5	GALAXY
MCG+07-20-042	09 40 51.	+ 41 55 30.	66	14.	GALAXY
HOLM 140A	09 40 51.	+ 42 39	18	14.3	PART OF MULTIPLE GALAXY
VV 052B	09 40 51.	- 05 02 30.	66	14.	INTERACTING GALAXY
VV 052A	09 40 51.	- 05 02 30.	72	14.	INTERACTING GALAXY
HELW 423	09 40 51.	- 09 42 52.			NEBULA
REIZ 0026	09 40 52.	+ 42 54	60	13.6	GALAXY
HOLM 141A	09 40 52.	+ 42 54	60	13.3	PART OF MULTIPLE GALAXY
RNGC 2978	09 40 52.	- 09 33		13.5	GALAXY
ZC 0940.9+0220	09 40 54.	+ 02 20	1210		CLUSTER OF GALAXIES
ZWG 035.040	09 40 54.	+ 07 41		15.5	GALAXY
ZC 0940.9+0911	09 40 54.	+ 09 11	1080		CLUSTER OF GALAXIES
LB 11058	09 40 54.	+ 23 04		16.0	FAINT BLUE STAR
ZWG 152.061	09 40 54.	+ 30 47		15.7	GALAXY
MCG+07-20-043	09 40 54.	+ 42 40	9	16.	GALAXY
MCG+10-14-035	09 40 54.	+ 62 01	36	16.	GALAXY
MCG-01-25-031	09 40 54.	- 05 03	72	15.	GALAXY
ARP 253	09 40 54.	- 05 04			PECULIAR GALAXY
MCG-02-25-013	09 40 54.	- 09 42 30.	132	14.	GALAXY
ABC 0836	09 40 55.	+ 78 40		17.3	RICH CLUSTER OF GALAXIES
REIZ 0027	09 40 56.	+ 41 55	42	13.9	GALAXY
RNGC 2984	09 40 57.	+ 11 18		14.5	GALAXY
MCG-01-25-032	09 40 57.	- 05 03	60	15.	GALAXY
ABC 0859	09 40 58.	+ 09 05		17.1	RICH CLUSTER OF GALAXIES
REIZ 0028	09 40 58.	+ 42 52	36	15.6	GALAXY
ABC 0861	09 40 59.	+ 00 17		17.4	RICH CLUSTER OF GALAXIES
IC 0556	09 41 00.	+ 11 17 35.			NONSTELLAR OBJECT
ZWG 063.053	09 41 00.	+ 11 18		14.3	GALAXY
UGC 05200	09 41 00.	+ 11 18	54	14.3	GALAXY
MCG+02-25-025	09 41 00.	+ 11 18	36	14.3	GALAXY
LB 11061	09 41 00.	+ 22 34		16.4	FAINT BLUE STAR
LB 11060	09 41 00.	+ 23 33		16.9	FAINT BLUE STAR
LB 11059	09 41 00.	+ 25 12		17.3	FAINT BLUE STAR
HOLM 140B	09 41 00.	+ 42 39	18	15.2	PART OF MULTIPLE GALAXY
ZWG 265.031	09 41 00.	+ 56 00		15.3	GALAXY
UGC 05201	09 41 00.	+ 56 00	90	15.3	GALAXY Sc
MCG+09-16-038	09 41 00.	+ 56 00	78	15.	GALAXY
KARA.73B 0367	09 41 00.	+ 56 00	96	15.3	ISOLATED GALAXY S
ZC 0941.0+6426	09 41 00.	+ 64 26	2490		CLUSTER OF GALAXIES
MCG+11-12-023	09 41 00.	+ 64 30	36	16.	GALAXY
ZWG 332.061	09 41 00.	+ 68 49		13.7	GALAXY
UGC 05202	09 41 00.	+ 68 49	90	13.7	GALAXY Sa
KARA.72 211A	09 41 00.	+ 68 49	84	13.7	PART OF DOUBLE GALAXY
RNGC 2959	09 41 00.	+ 68 50		13.5	GALAXY
UGC 05203	09 41 00.	+ 80 02	156	15.5	GALAXY Sc
MCG+13-07-037	09 41 00.	+ 80 02	138	14.	GALAXY
MCG-01-25-033	09 41 00.	- 05 08	72	14.	GALAXY
ABC 0860	09 41 01.	+ 02 20		17.4	RICH CLUSTER OF GALAXIES
HELW 424	09 41 01.	- 09 26 41.			NEBULA
HELW 425	09 41 02.	- 09 27 05.			NEBULA
ZC 0941.1+0019	09 41 06.	+ 00 19	1140		CLUSTER OF GALAXIES
ZC 0941.1+0611	09 41 06.	+ 06 11	2690		CLUSTER OF GALAXIES
LB 11062	09 41 06.	+ 24 14		16.7	FAINT BLUE STAR
TON-N 0022	09 41 06.	+ 28 03		13.7	BLUE STAR
MRK 406	09 41 06.	+ 29 51	15	15.5	GALAXY WITH UV CONTINUUM
ZWG 182.006	09 41 06.	+ 34 55		15.2	GALAXY
ZWG 333.001	09 41 06.	+ 72 16		15.7	GALAXY
ZWG 332.062	09 41 06.	+ 72 16		15.7	GALAXY
MCG-01-25-034	09 41 06.	- 05 41	120	13.5	GALAXY
HELW 426	09 41 07.	- 09 27 11.			NEBULA
LB 00560	09 41 08.	+ 58 22 48.		17.6	FAINT BLUE STAR
ZWG 063.054	09 41 12.	+ 09 53		14.8	GALAXY
UGC 05204	09 41 12.	+ 09 53	66	14.8	GALAXY
MCG+02-25-026	09 41 12.	+ 09 53	36	14.8	GALAXY
ZWG 092.035	09 41 12.	+ 15 02		15.7	GALAXY
LB 11066	09 41 12.	+ 20 34		14.5	FAINT BLUE STAR
LB 11065	09 41 12.	+ 21 58		16.5	FAINT BLUE STAR
LB 11064	09 41 12.	+ 23 11		16.2	FAINT BLUE STAR
LB 11063	09 41 12.	+ 26 02		16.7	FAINT BLUE STAR
ZC 0941.2+3051	09 41 12.	+ 30 51	740		CLUSTER OF GALAXIES
ZWG 312.022	09 41 12.	+ 64 30		15.7	GALAXY
ZWG 332.063	09 41 12.	+ 68 50		15.6	GALAXY
KARA.72 211B	09 41 12.	+ 68 50	42	15.6	PART OF DOUBLE GALAXY
KEEN 2959A	09 41 12.	+ 68 51	48		Sb GALAXY
RNGC 2959A	09 41 12.	+ 68 51			GALAXY
RNGC 2961	09 41 13.	+ 68 50		16.0	GALAXY
ZWG 092.036	09 41 18.	+ 18 21		15.3	GALAXY
LB 11067	09 41 18.	+ 23 28		17.5	FAINT BLUE STAR
ZC 0941.3+4419	09 41 18.	+ 44 19	1410		CLUSTER OF GALAXIES
ZWG 265.032	09 41 18.	+ 51 55		15.0	GALAXY
KARA.73B 0368	09 41 18.	+ 51 55	42	15.0	ISOLATED GALAXY S

OBJECT NAME	RIGHT ASCEN.	DECLINATION	DIAM.	MAGN.	TYPE OF OBJECT
ZC 0941.3+6633	09 41 18.	+ 66 33	940		CLUSTER OF GALAXIES
MCG-03-25-017	09 41 18.	- 20 16	120	13.	GALAXY
RNGC 2983	09 41 20.	- 20 15		13.0	GALAXY
IC 0557	09 41 21.	+ 11 13 09.			NONSTELLAR OBJECT
ZWG 035.041	09 41 24.	+ 03 05		15.4	GALAXY
SN 1950G	09 41 24.	+ 09 35		18.3	SUPERNOVA
ZWG 063.055	09 41 24.	+ 11 13		14.7	GALAXY
MCG+02-25-027	09 41 24.	+ 11 13	30	14.7	GALAXY
ZC 0941.4+1658	09 41 24.	+ 16 58	2150		CLUSTER OF GALAXIES
MCG+03-25-021	09 41 24.	+ 18 22	18	15.3	GALAXY
LB 11069	09 41 24.	+ 20 34		15.8	FAINT BLUE STAR
LB 11068	09 41 24.	+ 23 44		17.3	FAINT BLUE STAR
ZC 0941.4+4502	09 41 24.	+ 45 02	870		CLUSTER OF GALAXIES
ZC 0941.4+5120	09 41 24.	+ 51 20	810		CLUSTER OF GALAXIES
MCG+09-16-039	09 41 24.	+ 54 52	18	16.	GALAXY
ZWG 035.042	09 41 30.	+ 03 37		15.6	GALAXY
ZWG 007.025	09 41 30.	- 00 25		15.5	GALAXY
KARA.72 212A	09 41 30.	- 00 25	42	15.5	PART OF DOUBLE GALAXY
MCG+00-25-010	09 41 30.	- 00 25	30	15.5	GALAXY
ABC 0863	09 41 30.	- 12 24		17.7	RICH CLUSTER OF GALAXIES
MAI 045	09 41 33.	+ 69 45	74		DWARF SPHEROIDAL GALAXY
ZWG 035.043	09 41 36.	+ 02 38		15.2	GALAXY
ZC 0941.6+1540	09 41 36.	+ 15 40	4700		CLUSTER OF GALAXIES
LB 11073	09 41 36.	+ 20 56		16.0	FAINT BLUE STAR
ZC 0941.6+2233	09 41 36.	+ 22 33	940		CLUSTER OF GALAXIES
LB 11072	09 41 36.	+ 24 37		17.9	FAINT BLUE STAR
LB 11071	09 41 36.	+ 25 36		16.9	FAINT BLUE STAR
LB 11070	09 41 36.	+ 25 52		17.2	FAINT BLUE STAR
ZWG 182.007	09 41 36.	+ 33 02		15.2	GALAXY
ZC 0941.6+4046	09 41 36.	+ 40 46	1410		CLUSTER OF GALAXIES
ZWG 007.026	09 41 36.	- 00 26		15.4	GALAXY
UGC 05205	09 41 36.	- 00 26	72	15.4	GALAXY IRR
KARA.72 212B	09 41 36.	- 00 26	66	15.4	PART OF DOUBLE GALAXY
MCG+00-25-011	09 41 36.	- 00 26	42	15.4	GALAXY
MCG+06-22-006	09 41 39.	+ 33 01	18	15.	GALAXY
ZWG 063.056	09 41 42.	+ 11 27		15.0	GALAXY
MCG+02-25-028	09 41 42.	+ 11 27	48	15.0	GALAXY
UGC 05206	09 41 42.	+ 14 21	78	14.3	GALAXY F
ZC 0941.7+2430	09 41 42.	+ 24 30	22580		CLUSTER OF GALAXIES
ZWG 122.065	09 41 42.	+ 24 44		15.5	GALAXY
LB 11074	09 41 42.	+ 25 35		17.7	FAINT BLUE STAR
MCG+07-20-044	09 41 42.	+ 39 36 30.	30	17.	GALAXY
ZC 0941.7+5756	09 41 42.	+ 57 56	1680		CLUSTER OF GALAXIES
ZC 0941.7+7307	09 41 42.	+ 73 07	1340		CLUSTER OF GALAXIES
MCG-03-25-018	09 41 42.	- 21 05	60	14.5	GALAXY
MCG+06-22-007	09 41 45.	+ 37 50 30.	60	15.	GALAXY
ABC 0862	09 41 47.	+ 09 48		17.7	RICH CLUSTER OF GALAXIES
ZC 0941.8+0944	09 41 48.	+ 09 44	1210		CLUSTER OF GALAXIES
MCG+03-25-022	09 41 48.	+ 15 16	42	15.1	GALAXY
ZC 0941.8+2023	09 41 48.	+ 20 23	610		CLUSTER OF GALAXIES
LB 11076	09 41 48.	+ 21 24		15.6	FAINT BLUE STAR
LB 11075	09 41 48.	+ 21 36		17.4	FAINT BLUE STAR
TON-N 0449	09 41 48.	+ 26 21		14.6	BLUE STAR
ZWG 122.066	09 41 48.	+ 26 25		15.6	GALAXY
ZC 0941.8+2854	09 41 48.	+ 28 54	5380		CLUSTER OF GALAXIES
ZC 0941.8+3018	09 41 48.	+ 30 18	810		CLUSTER OF GALAXIES
ZWG 182.008	09 41 48.	+ 37 50		15.5	GALAXY
ZWG 265.033	09 41 48.	+ 53 55		15.1	GALAXY
UGC 05207	09 41 48.	+ 53 55	66	15.0	GALAXY SBa-b
MCG-05-23-008	09 41 48.	- 28 37	36	14.	GALAXY
MCG-03-25-019	09 41 51.	- 21 04	48	12.5	GALAXY
PK279-03.1	09 41 51.	- 57 03 10.	25	10.4	PLANETARY NEBULA
ZWG 092.037	09 41 54.	+ 15 17		15.1	GALAXY
LB 11078	09 41 54.	+ 25 12		17.2	FAINT BLUE STAR
LB 11077	09 41 54.	+ 25 24		16.6	FAINT BLUE STAR
ZC 0941.9+5653	09 41 54.	+ 56 53	2550		CLUSTER OF GALAXIES
RNGC 2986	09 41 56.	- 21 03		12.5	GALAXY
ABC 0856	09 41 57.	+ 56 47		17.1	RICH CLUSTER OF GALAXIES
ABC 0840	09 41 57.	+ 78 56		17.3	RICH CLUSTER OF GALAXIES
ZWG 063.057	09 42 00.	+ 09 51		15.0	GALAXY
MCG+02-25-029	09 42 00.	+ 09 51	48	15.0	GALAXY
ZWG 092.038	09 42 00.	+ 15 07		15.0	GALAXY
MCG+03-25-023	09 42 00.	+ 15 07	36	15.0	GALAXY
ZC 0942.0+1526	09 42 00.	+ 15 26	810		CLUSTER OF GALAXIES
LB 11081	09 42 00.	+ 21 48		15.7	FAINT BLUE STAR
ZWG 122.067	09 42 00.	+ 23 08		15.5	GALAXY
LB 11080	09 42 00.	+ 24 02		16.0	FAINT BLUE STAR
LB 11079	09 42 00.	+ 26 14		15.7	FAINT BLUE STAR
TON-N 0450	09 42 00.	+ 30 01		16.1	BLUE STAR
ZWG 152.062	09 42 00.	+ 31 20		15.0	GALAXY
UGC 05208	09 42 00.	+ 31 20	72	15.0	GALAXY Sb/Sc
MCG+05-23-032	09 42 00.	+ 31 20	72	15.0	GALAXY
UGC 05209	09 42 00.	+ 32 29	60	19.	GALAXY DWRF IR
ZC 0942.0+3425	09 42 00.	+ 34 25	2150		CLUSTER OF GALAXIES
MCG+07-20-045	09 42 00.	+ 39 01	42	15.	GALAXY
UGC 05210	09 42 00.	+ 69 12	126	16.0	GALAXY Sc
ZC 0942.0+8007	09 42 00.	+ 80 07	610		CLUSTER OF GALAXIES
IC 0559	09 42 02.	+ 09 50 48.			NONSTELLAR OBJECT
RNGC 2981	09 42 02.	+ 31 20		15.0	GALAXY
IC 0558	09 42 03.	+ 29 41 19.			NONSTELLAR OBJECT
ZWG 007.027	09 42 06.	+ 00 01		15.5	GALAXY
UGC 05211	09 42 06.	+ 00 01	60	15.5	GALAXY
MCG+00-25-012	09 42 06.	+ 00 01	42	15.5	GALAXY
8ZW 0942+13.8	09 42 06.	+ 13 45		16.0	COMPACT GALAXY
MCG+03-25-024	09 42 06.	+ 16 57	72	14.3	GALAXY
MCG+04-23-030	09 42 06.	+ 23 07	48	15.5	GALAXY
LB 11082	09 42 06.	+ 23 44		15.8	FAINT BLUE STAR
ZWG 152.063	09 42 06.	+ 29 41		14.9	GALAXY
MCG+05-23-033	09 42 06.	+ 29 41	54	14.9	GALAXY
ZWG 182.009	09 42 06.	+ 36 45		15.3	GALAXY
UGC 05212	09 42 06.	+ 39 41	60	16.0	GALAXY Sc
ZC 0942.1+4325	09 42 06.	+ 43 25	1010		CLUSTER OF GALAXIES
MCG+10-14-036	09 42 06.	+ 57 28	18	16.	GALAXY
VHA 078	09 42 06.	- 56 20	90		OPEN STAR CLUSTER
REIZ 0029	09 42 11.	+ 39 39	72	15.2	GALAXY
ZC 0942.2+0257	09 42 12.	+ 02 57	1140		CLUSTER OF GALAXIES
ZWG 092.039	09 42 12.	+ 16 56		14.3	GALAXY
UGC 05213	09 42 12.	+ 16 56	72	14.3	GALAXY SBa
ZWG 092.040	09 42 12.	+ 18 02		15.2	GALAXY
LB 11084	09 42 12.	+ 22 08		16.0	FAINT BLUE STAR
LB 11083	09 42 12.	+ 23 30		17.3	FAINT BLUE STAR
ZWG 152.064	09 42 12.	+ 27 30		15.7	GALAXY
KARA.72 213A	09 42 12.	+ 27 30	66	15.7	PART OF DOUBLE GALAXY
MCG+05-23-034	09 42 12.	+ 27 30	48	15.7	GALAXY
ZWG 182.010	09 42 12.	+ 34 56		15.4	GALAXY
MCG+07-20-046	09 42 12.	+ 39 40	54	15.	GALAXY
IC 2505	09 42 14.	+ 27 29 42.			NONSTELLAR OBJECT
ZWG 007.028	09 42 18.	+ 00 18		15.7	GALAXY

OBJECT NAME	RIGHT ASCEN.	DECLINATION	DIAM.	MAGN.	TYPE OF OBJECT
LB 11087	09 42 18.	+ 23 41		16.9	FAINT BLUE STAR
LB 11086	09 42 18.	+ 24 17		17.2	FAINT BLUE STAR
LB 11085	09 42 18.	+ 25 06		16.2	FAINT BLUE STAR
ZWG 152.065	09 42 18.	+ 27 29		15.6	GALAXY
KARA.72 213B	09 42 18.	+ 27 29	54	15.6	PART OF DOUBLE GALAXY
ZC 0942.3+3457	09 42 18.	+ 34 57	2080		CLUSTER OF GALAXIES
ZWG 182.011	09 42 18.	+ 36 23		15.0	GALAXY
ZWG 210.032	09 42 18.	+ 39 30		15.6	GALAXY
1ZW 020	09 42 18.	+ 51 43			COMPACT GALAXY
MCG+09-16-040	09 42 18.	+ 55 00	54	16.	GALAXY
MCG-05-23-009	09 42 18.	- 31 34	84	13.	GALAXY
IC 2506	09 42 20.	+ 27 28 35.			NONSTELLAR OBJECT
MCG+05-23-035	09 42 21.	+ 27 29	21	15.6	GALAXY
LB 11091	09 42 24.	+ 21 40		16.5	FAINT BLUE STAR
LB 11090	09 42 24.	+ 22 20		16.6	FAINT BLUE STAR
LB 11089	09 42 24.	+ 23 41		16.8	FAINT BLUE STAR
LB 11088	09 42 24.	+ 24 51		16.3	FAINT BLUE STAR
ZWG 152.066	09 42 24.	+ 30 55		15.7	GALAXY
ZC 0942.4+3610	09 42 24.	+ 36 10	1010		CLUSTER OF GALAXIES
MCG+09-16-041	09 42 24.	+ 55 08	18	16.	GALAXY
MCG+10-14-037	09 42 24.	+ 58 45	27	16.	GALAXY
7ZW 291	09 42 24.	+ 69 39			COMPACT GALAXY
VHA 079	09 42 24.	- 53 05	150		OPEN STAR CLUSTER
IC 2507	09 42 25.	- 31 33 40.			NONSTELLAR OBJECT
ARC 0839	09 42 27.	+ 80 07		17.9	RICH CLUSTER OF GALAXIES
ARP 252	09 42 28.	- 19 27			PECULIAR GALAXY
RNGC 2995	09 42 28.	- 54 33			NON-EXISTENT OBJECT
ZWG 035.044	09 42 30.	+ 07 21		15.7	GALAXY
ZWG 122.068	09 42 30.	+ 21 08		15.5	GALAXY
UGC 05214	09 42 30.	+ 23 19	66	17.	GALAXY DWARF
LB 11092	09 42 30.	+ 23 40		17.1	FAINT BLUE STAR
ZC 0942.5+3915	09 42 30.	+ 39 15	1550		CLUSTER OF GALAXIES
MRK 120	09 42 30.	+ 72 04	6	17.	GALAXY WITH UV CONTINUUM
MCG-03-25-019B	09 42 30.	- 19 27	48	16.	GALAXY
MCG-03-25-019A	09 42 30.	- 19 27 30.	72	16.5	GALAXY
MCG-05-23-010	09 42 30.	- 31 36	120	14.	GALAXY
ARC 0867	09 42 35.	+ 00 47		17.5	RICH CLUSTER OF GALAXIES
SN 1954Z	09 42 35.	+ 09 20		16.0	SUPERNOVA
ZWG 035.045	09 42 36.	+ 07 38		15.6	GALAXY
ZWG 063.058	09 42 36.	+ 09 20		13.8	GALAXY
UGC 05215	09 42 36.	+ 09 20	102	13.8	GALAXY Sb-c
MCG+02-25-030	09 42 36.	+ 09 20	96	13.8	GALAXY
ZC 0942.6+2517	09 42 36.	+ 25 17	810		CLUSTER OF GALAXIES
TON-N 0023	09 42 36.	+ 28 12		15.4	BLUE STAR
MRK 121	09 42 36.	+ 73 13	10	16.	GALAXY WITH UV CONTINUUM
MCG+12-10-002	09 42 36.	+ 73 13	33	16.	GALAXY
MCG+12-10-001	09 42 36.	+ 73 13	36	16.	GALAXY
ARC 0848	09 42 40.	+ 75 06		16.9	RICH CLUSTER OF GALAXIES
ZWG 063.059	09 42 42.	+ 09 59		14.9	GALAXY
UGC 05216	09 42 42.	+ 09 59	108	14.8	GALAXY
MCG+02-25-031	09 42 42.	+ 09 59	72	14.8	GALAXY
ZC 0942.7+1010	09 42 42.	+ 10 10	1680		CLUSTER OF GALAXIES
ZC 0942.7+2030	09 42 42.	+ 20 30	670		CLUSTER OF GALAXIES
LB 11094	09 42 42.	+ 21 34		15.8	FAINT BLUE STAR
ZWG 122.069	09 42 42.	+ 22 08		15.6	GALAXY
LB 11093	09 42 42.	+ 24 26		17.3	FAINT BLUE STAR
ZWG 265.034	09 42 42.	+ 54 50		15.0	GALAXY
MCG+09-16-042	09 42 42.	+ 54 51	36	15.	GALAXY
MCG+10-14-038	09 42 42.	+ 58 46	18	16.	GALAXY
MCG+11-12-024	09 42 42.	+ 65 17	39	16.	GALAXY
ZC 0942.7+6559	09 42 42.	+ 65 59	1080		CLUSTER OF GALAXIES
ZWG 333.002	09 42 42.	+ 73 13		15.3	GALAXY
ZWG 332.064	09 42 42.	+ 73 13		15.3	GALAXY
ZC 0942.8+0051	09 42 48.	+ 00 51	610		CLUSTER OF GALAXIES
ZWG 063.060	09 42 48.	+ 09 50		15.5	GALAXY
ZWG 092.041	09 42 48.	+ 15 29		15.6	GALAXY
LB 11099	09 42 48.	+ 21 14		16.4	FAINT BLUE STAR
LB 11098	09 42 48.	+ 23 17		17.2	FAINT BLUE STAR
ZWG 122.070	09 42 48.	+ 23 55		15.5	GALAXY
LB 11097	09 42 48.	+ 24 58		16.9	FAINT BLUE STAR
LB 11096	09 42 48.	+ 25 02		16.3	FAINT BLUE STAR
LB 11095	09 42 48.	+ 25 28		15.0	FAINT BLUE STAR
ZWG 122.071	09 42 48.	+ 26 19		15.7	GALAXY
ZWG 152.067	09 42 48.	+ 30 29		15.6	GALAXY
ZWG 182.012	09 42 48.	+ 34 05		15.5	GALAXY
UGC 05217	09 42 48.	+ 34 05	60	15.5	GALAXY SBc
ZWG 182.013	09 42 48.	+ 34 55		14.7	GALAXY
MCG+06-22-008	09 42 51.	+ 34 05 30.	60	15.	GALAXY
MCG+06-22-009	09 42 51.	+ 34 55 30.	42	15.	GALAXY
ZWG 035.046	09 42 54.	+ 06 37		15.0	GALAXY
UGC 05218	09 42 54.	+ 06 37	120	15.0	GALAXY Sc
MCG+01-25-016	09 42 54.	+ 06 37	90	15.0	GALAXY
ZWG 092.042	09 42 54.	+ 16 25		15.5	GALAXY
UGC 05219	09 42 54.	+ 28 43	72	16.5	GALAXY Sc/SBc
MCG+05-23-036	09 42 54.	+ 28 43	66	15.	GALAXY
TON-N 1093	09 42 54.	+ 35 00		16.9	BLUE STAR
ZC 0942.9+4902	09 42 54.	+ 49 02	1210		CLUSTER OF GALAXIES
ZWG 007.029	09 42 54.	- 02 48		15.5	GALAXY
KARA.73B 0369	09 42 54.	- 02 48	36	15.5	ISOLATED GALAXY S
MCG-01-25-035	09 42 57.	- 05 27	60	15.5	GALAXY
RNGC 2987	09 42 59.	+ 05 10		14.0	GALAXY
ARC 0868	09 42 59.	- 08 25		17.6	RICH CLUSTER OF GALAXIES
VDB.66G 235	09 43	- 31 36	70		DWARF GALAXY
ZWG 035.047	09 43 00.	+ 05 10		13.9	GALAXY
UGC 05220	09 43 00.	+ 05 10	90	13.9	GALAXY Sa-b
MCG+01-25-017	09 43 00.	+ 05 10	90	13.9	GALAXY
ZWG 092.043	09 43 00.	+ 18 02		15.6	GALAXY
ZWG 092.044	09 43 00.	+ 18 47		15.6	GALAXY
LB 11103	09 43 00.	+ 21 44		16.4	FAINT BLUE STAR
LB 11102	09 43 00.	+ 21 45		15.7	FAINT BLUE STAR
LB 11101	09 43 00.	+ 23 44		17.0	FAINT BLUE STAR
LB 11100	09 43 00.	+ 24 16		16.8	FAINT BLUE STAR
ZC 0943.0+2955	09 43 00.	+ 29 55	2290		CLUSTER OF GALAXIES
MCG+08-18-029	09 43 00.	+ 47 20	45	15.	GALAXY
ZWG 239.025	09 43 00.	+ 47 22		15.4	GALAXY
MCG+11-12-025	09 43 00.	+ 68 10	306	10.7	GALAXY
MCG-03-25-020	09 43 00.	- 18 10	60	14.	GALAXY
REIN 2.090	09 43 01.53	+ 68 08 20.1			NEBULA
REIZ 0030	09 43 03.	+ 47 22	36	14.4	GALAXY
ZWG 122.072	09 43 06.	+ 21 52		15.7	GALAXY
ZWG 312.023	09 43 06.	+ 68 09		10.9	GALAXY
UGC 05221	09 43 06.	+ 68 09	360	10.9	GALAXY Sc
MRK 122	09 43 06.	+ 73 12	36	14.	GALAXY WITH UV CONTINUUM
MCG+12-10-003	09 43 06.	+ 73 13	78	15.	GALAXY
MCG-04-23-015	09 43 06.	- 26 51 30.	36	15.5	GALAXY
MCG-04-23-014	09 43 06.	- 27 11	66	15.	GALAXY
RNGC 2989	09 43 07.	- 18 09		13.5	GALAXY
SHB 143	09 43 08.1	+ 12 19 23.		19.5	QUASI-STELLAR OBJECT

OBJECT NAME	RIGHT ASCEN.	DECLINATION	DIAM.	MAGN.	TYPE OF OBJECT
REIN 2.091	09 43 08.91	+ 68 08 54.1			NEBULA
ZC 0943.2+2047	09 43 12.	+ 20 47	810		CLUSTER OF GALAXIES
ZWG 122.073	09 43 12.	+ 21 49		15.7	GALAXY
ZC 0943.2+4307	09 43 12.	+ 43 07	3430		CLUSTER OF GALAXIES
ZWG 333.003	09 43 12.	+ 73 12		14.3	GALAXY
ZWG 332.065	09 43 12.	+ 73 12		14.3	GALAXY
RNGC 2963	09 43 12.	+ 73 12		14.5	GALAXY
UGC 05222	09 43 12.	+ 73 12	72	14.3	GALAXY SBa-b
PK248+29.1	09 43 12.	- 12 56	330	14.5	PLANETARY NEBULA
MCG-05-23-011	09 43 12.	- 30 07	180	15.	GALAXY
LB 00561	09 43 14.	+ 60 20 48.		17.4	FAINT BLUE STAR
RNGC 2976	09 43 15.	+ 68 09		11.5	GALAXY
ZWG 035.048	09 43 18.	+ 03 12		15.3	GALAXY
UGC 05224	09 43 18.	+ 03 12	102	15.3	GALAXY
MCG+01-25-018	09 43 18.	+ 03 12	96	15.3	GALAXY
ZWG 092.045	09 43 18.	+ 16 43		15.1	GALAXY
ZWG 092.046	09 43 18.	+ 19 40		15.4	GALAXY
MCG+03-25-025	09 43 18.	+ 19 41	42	15.4	GALAXY
LB 11104	09 43 18.	+ 21 23		16.6	FAINT BLUE STAR
ZWG 182.014	09 43 18.	+ 35 01		15.7	GALAXY
MCG+08-18-030	09 43 18.	+ 45 57	36	15.	GALAXY
1ZW 021	09 43 18.	+ 45 59			COMPACT GALAXY
ZWG 239.026	09 43 18.	+ 46 00		14.9	GALAXY
UGC 05225	09 43 18.	+ 46 00	66	14.9	GALAXY COMPACT
ZWG 265.035	09 43 18.	+ 54 40		15.2	GALAXY
MCG+09-16-043	09 43 18.	+ 54 40	36	15.	GALAXY
RNGC 2957	09 43 18.	+ 73 12		15.5	GALAXY
ZWG 007.030	09 43 18.	- 00 02		14.6	GALAXY
UGC 05223	09 43 18.	- 00 02	90	14.6	GALAXY S0-a
MCG+00-25-013	09 43 18.	- 00 02	36	14.6	GALAXY
RNGC 2992	09 43 18.	- 14 06		13.0	GALAXY
WEED 2	09 43 18.	- 14 07		17.5	VERY BLUE STELLAR OBJECT
MCG-02-25-014	09 43 18.	- 14 07	240	13.	GALAXY
MCG-05-23-012	09 43 18.	- 30 57	480	10.	GALAXY
IC 0560	09 43 19.	- 00 03 24.			NONSTELLAR OBJECT
ARP 245	09 43 20.	- 14 05			PECULIAR GALAXY
MCG+06-22-010	09 43 21.	+ 35 01	42	15.5	GALAXY
ZWG 035.049	09 43 24.	+ 03 22		15.0	GALAXY
MCG+01-25-019	09 43 24.	+ 03 22	24	15.0	GALAXY
IC 0561	09 43 24.	+ 03 22 41.			NONSTELLAR OBJECT
ZWG 035.050	09 43 24.	+ 04 38		14.0	GALAXY
UGC 05226	09 43 24.	+ 04 38	72	14.0	GALAXY S0
MCG+01-25-020	09 43 24.	+ 04 38	24	14.0	GALAXY
ZC 0943.4+0905	09 43 24.	+ 09 05	1480		CLUSTER OF GALAXIES
LB 11108	09 43 24.	+ 21 14		16.1	FAINT BLUE STAR
LB 11107	09 43 24.	+ 22 22		17.4	FAINT BLUE STAR
LB 11106	09 43 24.	+ 22 51		16.6	FAINT BLUE STAR
LB 11105	09 43 24.	+ 24 48		15.7	FAINT BLUE STAR
UGC 05227	09 43 24.	+ 42 26	72	16.0	GALAXY Sc
ZWG 265.036	09 43 24.	+ 54 41		15.1	GALAXY
MCG+09-16-044	09 43 24.	+ 54 42	24	15.	GALAXY
RNGC 2993	09 43 24.	- 14 08		13.5	GALAXY
MCG-02-25-015	09 43 24.	- 14 09	72	13.	GALAXY
RNGC 2999	09 43 24.	- 50 12			NON-EXISTENT OBJECT
HOLM 143C	09 43 25.	+ 03 22	60	14.3	PART OF MULTIPLE GALAXY
REIZ 0031	09 43 26.	+ 42 26	60	14.6	GALAXY
MCG+07-20-047	09 43 27.	+ 42 26	60	15.	GALAXY
MCG-03-25-021	09 43 27.	- 17 22 30.	72	16.	GALAXY
SVEN 232	09 43 29.	- 30 58	384	10.7	GALAXY
ZWG 007.031	09 43 30.	+ 01 54		14.1	GALAXY
UGC 05228	09 43 30.	+ 01 54	156	14.1	GALAXY SB
MCG+00-25-014	09 43 30.	+ 01 54	156	14.1	GALAXY
ZWG 122.074	09 43 30.	+ 23 27		15.7	GALAXY
ZWG 182.015	09 43 30.	+ 35 33		15.5	GALAXY
MRK 123	09 43 30.	+ 56 22	15	15.5	GALAXY WITH UV CONTINUUM
BC WEEDMAN2	09 43 30.	- 13 59 30.		17.5	QUASI-STELLAR OBJECT
RNGC 2997	09 43 30.	- 30 58		10.5	GALAXY
OCL 0771	09 43 30.	- 43 54	300	12.	OPEN STAR CLUSTER
ARC 0865	09 43 32.	+ 43 44		16.6	RICH CLUSTER OF GALAXIES
IC 0562	09 43 32.1	- 03 44 25.			GALAXY
MCG-01-25-036	09 43 33.	- 03 43	66	14.5	GALAXY
ZC 0943.6+0240	09 43 36.	+ 02 40	870		CLUSTER OF GALAXIES
ZWG 035.051	09 43 36.	+ 05 56		12.5	GALAXY
UGC 05229	09 43 36.	+ 05 56	66	12.5	GALAXY S?
MCG+01-25-021	09 43 36.	+ 05 56	66	12.5	GALAXY
ZWG 092.047	09 43 36.	+ 19 38		15.5	GALAXY
LB 11109	09 43 36.	+ 22 28		17.2	FAINT BLUE STAR
ZWG 122.075	09 43 36.	+ 25 46		15.6	GALAXY
ZWG 152.068	09 43 36.	+ 30 53		14.9	GALAXY
ZC 0943.6+3718	09 43 36.	+ 37 18	610		CLUSTER OF GALAXIES
ZWG 265.037	09 43 36.	+ 54 32		15.4	GALAXY
MCG+09-16-045	09 43 36.	+ 54 40	12	17.	GALAXY
ARC 0869	09 43 37.	+ 02 36		17.4	RICH CLUSTER OF GALAXIES
REIZ 0032	09 43 39.	+ 47 11	24	15.0	GALAXY
ARP 303	09 43 40.	+ 03 18			PECULIAR GALAXY
RNGC 2990	09 43 41.	+ 05 56		13.5	GALAXY
ZWG 035.052	09 43 42.	+ 05 55		15.3	GALAXY
ZWG 063.061	09 43 42.	+ 09 23		15.1	GALAXY
ZWG 092.048	09 43 42.	+ 17 46		15.7	GALAXY
MCG+05-23-037	09 43 42.	+ 30 54	42	14.9	GALAXY
TON-N 1094	09 43 42.	+ 35 32		13.8	BLUE STAR
ZC 0943.7+5454	09 43 42.	+ 54 54	5910		CLUSTER OF GALAXIES
ZWG 265.038	09 43 42.	+ 56 20		15.2	GALAXY
ZC 0943.7+5835	09 43 42.	+ 58 35	2490		CLUSTER OF GALAXIES
IC 0563	09 43 45.	+ 03 16 40.			SAME AS "A0944B"
PATH 1.296	09 43 45.	+ 44 12	27		NEBULA
REIZ 0033	09 43 45.	+ 47 10	9	15.9	GALAXY
IC 0564	09 43 46.	+ 03 18 16.			SAME AS "A0994A"
HOLM 143B	09 43 47.	+ 03 16	60	13.9	PART OF MULTIPLE GALAXY
ZWG 035.053	09 43 48.	+ 03 16		14.7	GALAXY
MCG+01-25-022	09 43 48.	+ 03 16	48	14.7	GALAXY
ZWG 035.054	09 43 48.	+ 03 18		14.1	GALAXY
HOLM 143A	09 43 48.	+ 03 18	120	13.7	PART OF MULTIPLE GALAXY
UGC 05230	09 43 48.	+ 03 18	108	14.1	GALAXY IRR
MCG+01-25-023	09 43 48.	+ 03 18	108	14.1	GALAXY
ZC 0943.8+1126	09 43 48.	+ 11 26	340		CLUSTER OF GALAXIES
ZWG 092.049	09 43 48.	+ 16 08		15.3	GALAXY
ZWG 122.076	09 43 48.	+ 22 00		15.5	GALAXY
LB 11111	09 43 48.	+ 23 51		16.0	FAINT BLUE STAR
MCG+04-23-031	09 43 48.	+ 24 21	36		GALAXY
LB 11110	09 43 48.	+ 25 41		17.6	FAINT BLUE STAR
ZC 0943.8+4148	09 43 48.	+ 41 48	940		CLUSTER OF GALAXIES
MCG+09-16-046	09 43 48.	+ 52 56	42	16.	GALAXY
ZWG 265.039	09 43 48.	+ 54 03		15.7	GALAXY
MCG+12-10-004	09 43 48.	+ 73 12	21	16.	GALAXY
REIZ 0034	09 43 51.	+ 42 43	18	15.4	GALAXY
REIZ 0035	09 43 51.	+ 42 44	24	15.2	GALAXY
MCG+09-16-047C	09 43 51.	+ 54 02 30.	15	18.	GALAXY

OBJECT NAME	RIGHT ASCEN.	DECLINATION	DIAM.	MAGN.	TYPE OF OBJECT
MCG+09-16-047B	09 43 51.	+ 54 02 30.	30	17.	GALAXY
MCG+09-16-047A	09 43 51.	+ 54 02 30.	24	17.	GALAXY
MCG+09-16-048	09 43 51.	+ 54 45	15	15.	GALAXY
MCG-01-25-037	09 43 51.	- 09 30 30.	54	16.	GALAXY
MCG-02-25-016	09 43 51.	- 11 51	42	16.	GALAXY
ZWG 063.062	09 43 54.	+ 13 54		15.2	GALAXY
ZWG 122.077	09 43 54.	+ 26 10		15.7	GALAXY
ZWG 182.016	09 43 54.	+ 34 14		15.4	GALAXY
MCG+07-20-048	09 43 54.	+ 42 44	18	15.2	GALAXY
ZWG 210.033	09 43 54.	+ 42 45		15.4	GALAXY
UGC 05231	09 43 54.	+ 42 45	60	15.4	GALAXY SB:a-b
MCG+07-20-049	09 43 54.	+ 42 46	60	14.0	GALAXY
ZWG 265.040	09 43 54.	+ 54 43		15.5	GALAXY
OCL 0788	09 43 54.	- 53 45	216	12.	OPEN STAR CLUSTER
HOLM 142B	09 43 55.	+ 42 43	18	15.2	PART OF MULTIPLE GALAXY
REIZ 0036	09 43 57.	+ 42 45	30	14.4	GALAXY
HOLM 142A	09 43 57.	+ 42 45	30	14.0	PART OF MULTIPLE GALAXY
ZWG 035.055	09 44 00.	+ 06 55		15.3	GALAXY
ZC 0944.0+0948	09 44 00.	+ 09 48	1340		CLUSTER OF GALAXIES
ZWG 063.063	09 44 00.	+ 13 46		14.9	GALAXY
MCG+02-25-032	09 44 00.	+ 13 46	39	14.9	GALAXY
ZWG 063.064	09 44 00.	+ 14 00		14.5	GALAXY
UGC 05232	09 44 00.	+ 14 00	66	14.5	GALAXY S
MCG+02-25-033	09 44 00.	+ 14 00	24	14.5	GALAXY
MCG+03-25-026	09 44 00.	+ 16 16	96	14.9	GALAXY
LB 11115	09 44 00.	+ 22 12		15.7	FAINT BLUE STAR
MCG+04-23-032	09 44 00.	+ 22 13	48	14.3	GALAXY
MCG+04-23-033	09 44 00.	+ 22 14	24	14.3	GALAXY
ZWG 122.078	09 44 00.	+ 22 15		14.3	GALAXY
RNGC 2991	09 44 00.	+ 22 15		14.5	GALAXY
RNGC 2988	09 44 00.	+ 22 15		14.5	GALAXY
UGC 05233	09 44 00.	+ 22 15	102	14.3	GALAXY S0
KARA.72 214A	09 44 00.	+ 22 15	54	14.3	PART OF DOUBLE GALAXY
LB 11114	09 44 00.	+ 22 40		16.8	FAINT BLUE STAR
LB 11113	09 44 00.	+ 23 05		15.9	FAINT BLUE STAR
ZWG 122.079	09 44 00.	+ 25 38		15.7	GALAXY
LB 11112	09 44 00.	+ 26 04		16.4	FAINT BLUE STAR
ZWG 122.080	09 44 00.	+ 26 15		15.6	GALAXY
MCG-05-23-013	09 44 00.	- 28 53	66	16.	GALAXY
MCG-05-23-014	09 44 00.	- 30 12	138	13.	GALAXY
REIZ 0038	09 44 01.	+ 22 14	42	13.9	GALAXY
KEEL 182	09 44 02.6	+ 33 44 37.		16.	NEBULA
BIGO 511	09 44 05.	+ 05 56			NEBULA
ABC 0870	09 44 05.	+ 09 51		17.7	RICH CLUSTER OF GALAXIES
ZWG 063.065	09 44 06.	+ 09 29		15.3	GALAXY
ZWG 092.050	09 44 06.	+ 16 17		14.9	GALAXY
UGC 05234	09 44 06.	+ 16 17	102	14.9	GALAXY Sc
KARA.72 214B	09 44 06.	+ 22 15	66		PART OF DOUBLE GALAXY
UGC 05235	09 44 06.	+ 23 16	84	16.0	GALAXY Sc
ZWG 122.081	09 44 06.	+ 24 56		15.4	GALAXY
KARA.73B 0370	09 44 06.	+ 24 56	24	15.4	ISOLATED GALAXY E
ZWG 182.017	09 44 06.	+ 33 44		15.7	GALAXY
ZC 0944.1+4400	09 44 06.	+ 44 00	1080		CLUSTER OF GALAXIES
MCG+08-18-031	09 44 06.	+ 46 50	84	14.	GALAXY
RNGC 3001	09 44 06.	- 30 13		13.0	GALAXY
REIZ 0037	09 44 08.	+ 46 50	60	14.1	GALAXY
RNGC 2996	09 44 08.	- 21 22		14.0	GALAXY
KEEL 183	09 44 08.8	+ 33 44 24.		15.	SPIRAL NEBULA
IC 2508	09 44 09.	+ 33 45 34.			NONSTELLAR OBJECT
MCG-03-25-022	09 44 09.	- 21 22	30	14.5	GALAXY
ZWG 063.066	09 44 12.	+ 09 58		14.6	GALAXY
MCG+02-25-034	09 44 12.	+ 09 58	36	14.6	GALAXY
ZC 0944.2+1526	09 44 12.	+ 15 26	810		CLUSTER OF GALAXIES
MCG+04-23-034	09 44 12.	+ 21 57	72		GALAXY
UGC 05236	09 44 12.	+ 21 58	78	16.5	GALAXY DWRF SP
LB 11118	09 44 12.	+ 22 00		17.1	FAINT BLUE STAR
LB 11117	09 44 12.	+ 22 17		17.3	FAINT BLUE STAR
MCG+04-23-035	09 44 12.	+ 22 19	42	14.4	GALAXY
LB 11116	09 44 12.	+ 25 10		15.5	FAINT BLUE STAR
SHAH 115	09 44 12.	+ 33 11	198	18.1	GROUP OF COMPACT GALAXIES
ZWG 239.027	09 44 12.	+ 46 51		15.0	GALAXY
UGC 05237	09 44 12.	+ 46 51	96	15.0	GALAXY Sc
MCG+09-16-049	09 44 12.	+ 54 39	48	17.	GALAXY
MCG+09-16-050	09 44 15.	+ 54 36	18	17.	GALAXY
ZWG 007.033	09 44 18.	+ 00 44		15.3	GALAXY
UGC 05238	09 44 18.	+ 00 44	150	15.3	GALAXY Sc
MCG+00-25-015	09 44 18.	+ 00 44	120	15.3	GALAXY
ZC 0944.3+1040	09 44 18.	+ 10 40	340		CLUSTER OF GALAXIES
ZC 0944.3+2138	09 44 18.	+ 21 38	540		CLUSTER OF GALAXIES
LB 11120	09 44 18.	+ 23 52		15.8	FAINT BLUE STAR
LB 11119	09 44 18.	+ 26 36		15.5	FAINT BLUE STAR
ZWG 182.018	09 44 18.	+ 35 00		15.2	GALAXY
ZC 0944.3+6230	09 44 18.	+ 62 30	1280		CLUSTER OF GALAXIES
ZWG 007.032	09 44 18.	- 00 48		15.6	GALAXY
MCG-02-25-017	09 44 18.	- 14 39 30.	48	15.	GALAXY
MCG-04-23-016	09 44 18.	- 23 41	60	15.	GALAXY
IC 2509	09 44 19.	+ 05 57			SINGLE STAR
ZWG 035.056	09 44 24.	+ 02 47		15.1	GALAXY
ZWG 063.067	09 44 24.	+ 12 47		15.4	GALAXY
ZWG 122.082	09 44 24.	+ 22 20		14.4	GALAXY
RNGC 2994	09 44 24.	+ 22 20		14.5	GALAXY
REIZ 0039	09 44 25.	+ 22 19	42	14.0	GALAXY
ABC 0866	09 44 26.	+ 58 23		17.5	RICH CLUSTER OF GALAXIES
MCG+06-22-011	09 44 27.	+ 35 01	48	14.5	GALAXY
ZC 0944.5+0002	09 44 30.	+ 00 02	1810		CLUSTER OF GALAXIES
UGC 05239	09 44 30.	+ 22 20	78	14.4	GALAXY S0
ZWG 122.083	09 44 30.	+ 23 15		15.7	GALAXY
LB 11121	09 44 30.	+ 24 52		16.0	FAINT BLUE STAR
ZWG 122.084	09 44 30.	+ 25 59		15.6	GALAXY
UGC 05240	09 44 30.	+ 25 59	72	15.6	GALAXY Sc
TON-N 0024	09 44 30.	+ 27 33		15.5	BLUE STAR
TON-N 1095	09 44 30.	+ 35 23		16.9	BLUE STAR
ZC 0944.5+3959	09 44 30.	+ 39 59	1140		CLUSTER OF GALAXIES
ZWG 265.041	09 44 30.	+ 54 02		15.6	GALAXY
ZWG 265.042	09 44 30.	+ 54 15		14.0	GALAXY
UGC 05241	09 44 30.	+ 54 15	48	14.0	GALAXY S
MCG+09-16-051	09 44 30.	+ 54 16	60	14.	GALAXY
MCG-05-23-015	09 44 30.	- 33 23	72	15.	GALAXY
RNGC 2515	09 44 33.	+ 13 16			NON-EXISTENT OBJECT
ZWG 007.034	09 44 36.	+ 00 56		15.5	GALAXY
UGC 05242	09 44 36.	+ 01 12	90	16.0	GALAXY DWARF
MCG+00-25-016	09 44 36.	+ 01 12	78	16.	GALAXY
ZWG 063.068	09 44 36.	+ 12 23		15.7	GALAXY
ZWG 092.051	09 44 36.	+ 14 58		15.0	GALAXY
MCG+03-25-027	09 44 36.	+ 14 58	30	15.0	GALAXY
ZC 0944.6+1715	09 44 36.	+ 17 15	2350		CLUSTER OF GALAXIES
LB 11123	09 44 36.	+ 21 34		16.0	FAINT BLUE STAR
LB 11122	09 44 36.	+ 26 30		15.5	FAINT BLUE STAR
MCG+09-16-052	09 44 39.	+ 52 06	42	16.	GALAXY
ZWG 007.035	09 44 42.	+ 00 54		15.6	GALAXY
ZWG 063.069	09 44 42.	+ 08 48		14.9	GALAXY
MCG+02-25-035	09 44 42.	+ 08 48	24	14.9	GALAXY
LB 11124	09 44 42.	+ 21 17		16.6	FAINT BLUE STAR
TON-N 1096	09 44 42.	+ 33 30		15.	BLUE STAR
ZWG 182.019	09 44 42.	+ 35 17		14.9	GALAXY
MCG+06-22-012	09 44 42.	+ 35 18	48	15.	GALAXY
ZWG 210.034	09 44 42.	+ 39 20		15.1	GALAXY
MRK 407	09 44 42.	+ 39 20	12	15.	GALAXY WITH UV CONTINUUM
ZWG 210.035	09 44 42.	+ 39 46		15.7	GALAXY
MCG+11-12-026	09 44 42.	+ 64 24	96	15.	GALAXY
REIZ 0041	09 44 47.	+ 27 40	48	15.4	GALAXY
REIZ 0040	09 44 47.	+ 39 46	24	14.9	GALAXY
LB 11127	09 44 48.	+ 20 47		17.3	FAINT BLUE STAR
ZWG 122.085	09 44 48.	+ 21 53		15.7	GALAXY
LB 11126	09 44 48.	+ 25 17		16.4	FAINT BLUE STAR
LB 11125	09 44 48.	+ 25 54		17.6	FAINT BLUE STAR
TON-N 0451	09 44 48.	+ 26 18		15.7	BLUE STAR
MCG+07-20-050	09 44 48.	+ 39 46	30	15.	GALAXY
SHAH 048	09 44 48.	+ 47 07	78	17.5	GROUP OF COMPACT GALAXIES
AR 04	09 44 51.78	+ 76 24 29.5			NEBULA
ZWG 035.057	09 44 54.	+ 07 03		15.6	GALAXY
MCG+01-25-024	09 44 54.	+ 07 03	48	15.6	GALAXY
ZWG 122.086	09 44 54.	+ 21 55		15.6	GALAXY
ZWG 265.043	09 44 54.	+ 54 14		15.4	GALAXY
ZWG 289.018	09 44 54.	+ 58 13		15.1	GALAXY
UGC 05243	09 44 54.	+ 58 13	66	15.1	GALAXY SB
ZWG 312.024	09 44 54.	+ 64 25		15.2	GALAXY
UGC 05244	09 44 54.	+ 64 25	108	15.2	GALAXY Sc
MCG-04-23-017	09 44 54.	- 24 35 30.	144	14.	GALAXY
MCG+03-25-028	09 45 00.	+ 16 05	96	15.1	GALAXY
LB 11128	09 45 00.	+ 21 00		15.5	FAINT BLUE STAR
ZWG 122.087	09 45 00.	+ 23 57		15.1	GALAXY
UGC 05246	09 45 00.	+ 23 57	66	15.1	GALAXY S
MCG+04-23-036	09 45 00.	+ 23 57 30.	63	15.1	GALAXY
ZC 0945.0+3251	09 45 00.	+ 32 51	1340		CLUSTER OF GALAXIES
ZC 0945.0+3441	09 45 00.	+ 34 41	13310		CLUSTER OF GALAXIES
MCG+09-16-053	09 45 00.	+ 54 17	36	16.	GALAXY
ZWG 289.019	09 45 00.	+ 59 30		15.4	GALAXY
MCG+10-14-039	09 45 00.	+ 59 30	57	15.	GALAXY
ZWG 332.066	09 45 00.	+ 69 39		15.6	GALAXY
UGC 05247	09 45 00.	+ 69 39	108	15.6	GALAXY
MCG+13-07-038	09 45 00.	+ 76 00	39	14.	GALAXY
ZWG 007.036	09 45 00.	- 01 48		15.5	GALAXY
UGC 05245	09 45 00.	- 01 48	168	15.5	GALAXY
MCG+00-25-017	09 45 00.	- 01 48	150	15.5	GALAXY
KARA.73B 0371	09 45 00.	- 01 48	174	15.5	ISOLATED GALAXY S
REIZ 0042	09 45 02.	+ 23 57	42	14.7	GALAXY
ZC 0945.1+0739	09 45 06.	+ 07 39	1280		CLUSTER OF GALAXIES
VVI 33	09 45 06.	+ 07 39		17.3	SEYFERT GALAXY
ZWG 063.070	09 45 06.	+ 09 22		15.4	GALAXY
ZWG 092.052	09 45 06.	+ 16 05		15.1	GALAXY
UGC 05248	09 45 06.	+ 16 05	102	15.1	GALAXY Sc
IC 0565	09 45 06.	+ 16 05 10.			NONSTELLAR OBJECT
ZC 0945.1+1810	09 45 06.	+ 18 10	810		CLUSTER OF GALAXIES
ZC 0945.1+1956	09 45 06.	+ 19 56	540		CLUSTER OF GALAXIES
LB 11129	09 45 06.	+ 22 51		17.6	FAINT BLUE STAR
ZWG 182.020	09 45 06.	+ 33 06		14.8	GALAXY
MRK 408	09 45 06.	+ 33 06	14	15.	GALAXY WITH UV CONTINUUM
MRK 124	09 45 06.	+ 50 45	7	16.	GALAXY WITH UV CONTINUUM
KW 15	09 45 06.	+ 50 45	18		SEYFERT GALAXY
VVI 32	09 45 06.	+ 50 45	7	15.94	SEYFERT GALAXY
MCG+10-14-040	09 45 06.	+ 57 25	39	16.	GALAXY
MRK 021	09 45 06.	+ 58 12	19	15.5	GALAXY WITH UV CONTINUUM
MCG+10-14-041	09 45 06.	+ 58 12	42	15.	GALAXY
MCG+12-10-005	09 45 06.	+ 69 39	102	14.	GALAXY
7ZW 292	09 45 06.	+ 73 28			COMPACT GALAXY
KW 68	09 45 07.	+ 07 39 25.	8		SEYFERT GALAXY
KEEL 184	09 45 07.6	+ 33 31 59.		17.	SPIRAL NEBULA
SEY 059	09 45 09.	+ 69 40 05.		15.4	FAINT GALAXY
RNGC 3007	09 45 09.	- 06 12		15.0	GALAXY
ZWG 035.058	09 45 12.	+ 02 51		14.4	GALAXY
UGC 05249	09 45 12.	+ 02 51	144	14.4	GALAXY SB:c
MCG+01-25-025	09 45 12.	+ 02 51	144	14.4	GALAXY
ZWG 063.071	09 45 12.	+ 09 03		14.7	GALAXY
MCG+02-25-036	09 45 12.	+ 09 03	42	14.7	GALAXY
ZWG 122.088	09 45 12.	+ 20 33		15.2	GALAXY
ZWG 122.089	09 45 12.	+ 21 53		15.6	GALAXY
LB 11132	09 45 12.	+ 22 04		17.3	FAINT BLUE STAR
LB 11131	09 45 12.	+ 25 11		16.0	FAINT BLUE STAR
LB 11130	09 45 12.	+ 25 22		17.4	FAINT BLUE STAR
MCG+11-12-027	09 45 12.	+ 64 25	36	16.	GALAXY
MCG-01-25-038	09 45 15.	- 06 12	60	15.	GALAXY
HOLM 145A	09 45 18.	+ 20 32	36	14.7	PART OF MULTIPLE GALAXY
ZWG 122.090	09 45 18.	+ 20 33		15.6	GALAXY
TON-N 1097	09 45 18.	+ 33 31		16.	BLUE STAR
MCG+08-18-032	09 45 18.	+ 47 35	30	15.	GALAXY
ZWG 239.028	09 45 18.	+ 47 36		15.7	GALAXY
KARA.72 215A	09 45 18.	+ 47 36	42	15.7	PART OF DOUBLE GALAXY
SHAH 106	09 45 18.	+ 52 00	120	18.5	GROUP OF COMPACT GALAXIES
ZWG 350.032	09 45 18.	+ 75 59		15.6	GALAXY
ZWG 007.037	09 45 18.	- 02 18		15.7	GALAXY
SVEN 233	09 45 18.	- 30 43 22.	24	14.0	GALAXY
HOLM 145B	09 45 23.	+ 20 32	12	15.1	PART OF MULTIPLE GALAXY
LB 11134	09 45 24.	+ 23 48		16.0	FAINT BLUE STAR
LB 11133	09 45 24.	+ 24 54		16.5	FAINT BLUE STAR
ZC 0945.4+4151	09 45 24.	+ 41 51	4770		CLUSTER OF GALAXIES
ZWG 312.025	09 45 24.	+ 64 27		15.7	GALAXY
MCG-02-25-018	09 45 24.	- 13 31 30.	24	15.	GALAXY
MCG-05-23-016	09 45 24.	- 30 43	48	14.5	GALAXY
TON-N 1098	09 45 25.	+ 20 50		16.5	BLUE STAR
FATE 2.020	09 45 29.	+ 44 19	54		NEBULA
ZWG 035.059	09 45 30.	+ 02 46		15.6	GALAXY
LB 11137	09 45 30.	+ 21 08		16.6	FAINT BLUE STAR
LB 11136	09 45 30.	+ 22 47		16.3	FAINT BLUE STAR
LB 11135	09 45 30.	+ 24 42		17.2	FAINT BLUE STAR
TON-N 0452	09 45 30.	+ 32 40		15.2	BLUE STAR
TON-N 1099	09 45 30.	+ 35 08		17.	BLUE STAR
ZWG 210.036	09 45 30.	+ 44 18		13.3	GALAXY
UGC 05250	09 45 30.	+ 44 18	156	13.3	GALAXY Sc
MCG+07-20-051	09 45 30.	+ 44 19	180	13.	GALAXY
MCG+08-18-033	09 45 30.	+ 47 34	42	15.	GALAXY
ZWG 239.029	09 45 30.	+ 47 35		15.6	GALAXY
KARA.72 215B	09 45 30.	+ 47 35	42	15.6	PART OF DOUBLE GALAXY
MCG-05-23-017	09 45 30.	- 32 37	54	14.	GALAXY
HOLM 144A	09 45 34.	+ 44 20	120	12.3	PART OF MULTIPLE GALAXY
HN 0320	09 45 34.	- 32 37			NEBULA

OBJECT NAME	RIGHT ASCEN.	DECLINATION	DIAM.	MAGN.	TYPE OF OBJECT
IC 2510	09 45 34.	- 32 37			NONSTELLAR OBJECT
PK274+03.1	09 45 34.	- 48 44 17.	30		PLANETARY NEBULA
ZWG 035.060	09 45 36.	+ 04 08		15.4	GALAXY
ZC 0945.6+0810	09 45 36.	+ 08 10	1550		CLUSTER OF GALAXIES
LB 11142	09 45 36.	+ 20 50		15.0	FAINT BLUE STAR
ZWG 122.091	09 45 36.	+ 22 36		15.7	GALAXY
LB 11141	09 45 36.	+ 23 06		17.4	FAINT BLUE STAR
LB 11140	09 45 36.	+ 24 20		18.4	FAINT BLUE STAR
LB 11139	09 45 36.	+ 24 26		17.3	FAINT BLUE STAR
LB 11138	09 45 36.	+ 25 43		16.7	FAINT BLUE STAR
MCG+06-22-013	09 45 36.	+ 33 38	360	12.	GALAXY
ZWG 182.021	09 45 36.	+ 33 39		12.3	GALAXY
UGC 05251	09 45 36.	+ 33 39	342	12.3	GALAXY SB?c
ZC 0945.6+3856	09 45 36.	+ 38 56	1210		CLUSTER OF GALAXIES
ZC 0945.6+4301	09 45 36.	+ 43 01	940		CLUSTER OF GALAXIES
ARC 0864	09 45 36.	+ 71 27		17.4	RICH CLUSTER OF GALAXIES
RNGC 3003	09 45 37.	+ 33 39		12.0	GALAXY
RNGC 2998	09 45 37.	+ 44 19		12.5	GALAXY
REIN 7.052	09 45 37.37	+ 33 39 22.2			NEBULA
SN 1961F	09 45 38.	+ 33 39		13.1	SUPERNOVA
LB 03050	09 45 40.	- 00 26 36.		16.1	FAINT BLUE STAR
REIZ 0044	09 45 41.	+ 33 39	300	12.5	GALAXY
HOLM 144E	09 45 41.	+ 44 23	24	14.7	PART OF MULTIPLE GALAXY
MCG+11-12-028	09 45 42.	+ 66 12 30.	27	18.	GALAXY
REIZ 0043	09 45 46.	+ 44 20	150	12.7	GALAXY
ZC 0945.8+2009	09 45 48.	+ 20 09	670		CLUSTER OF GALAXIES
LB 11145	09 45 48.	+ 20 36		16.4	FAINT BLUE STAR
LB 11144	09 45 48.	+ 21 54		16.0	FAINT BLUE STAR
LB 11143	09 45 48.	+ 24 14		16.1	FAINT BLUE STAR
UGC 05252	09 45 48.	+ 41 42	66	16.0	GALAXY DBL SYS
MCG+07-20-052	09 45 48.	+ 44 17	60	16.	GALAXY
MCG+08-18-034	09 45 48.	+ 47 15	15	15.	GALAXY
MRK 022	09 45 48.	+ 55 50	13	16.5	GALAXY WITH UV CONTINUUM
ZWG 333.004	09 45 48.	+ 72 31		11.1	GALAXY
ZWG 332.067	09 45 48.	+ 72 31		11.1	GALAXY
UGC 05253	09 45 48.	+ 72 31	270	11.1	GALAXY Sb
MCG+12-10-006	09 45 48.	+ 72 32	270	11.1	GALAXY
ZWG 007.038	09 45 48.	- 03 30		14.5	GALAXY
MCG+00-25-018	09 45 48.	- 03 30	42	14.5	GALAXY
KEEL 185	09 45 48.3	+ 33 52 19.		16.	NEBULA
RNGC 3002	09 45 49.	+ 44 17		16.0	GALAXY
RNGC 2985	09 45 51.	+ 72 31		11.5	GALAXY
REIZ 0045	09 45 52.	+ 44 23	24	15.2	GALAXY
ZC 0945.9+1458	09 45 54.	+ 14 58	2220		CLUSTER OF GALAXIES
LB 11146	09 45 54.	+ 24 35		13.4	FAINT BLUE STAR
ZWG 122.092	09 45 54.	+ 25 43		15.5	GALAXY
MCG+04-23-037	09 45 54.	+ 25 43	48	15.5	GALAXY
MCG+07-20-053	09 45 54.	+ 41 41	42	15.	GALAXY
REIZ 0046	09 45 54.	+ 45 41	24	14.8	GALAXY
ZWG 239.030	09 45 54.	+ 47 16		15.7	GALAXY
REIN 2.092	09 45 54.59	+ 72 30 44.4			NEBULA
RNGC 3000	09 45 55.	+ 44 22			NON-EXISTENT OBJECT
LB 03495	09 46	- 81 15		13.5	FAINT BLUE STAR
ZWG 035.061	09 46 00.	+ 03 36		15.3	GALAXY
ZWG 063.072	09 46 00.	+ 09 14		15.0	GALAXY
MCG+02-25-037	09 46 00.	+ 09 14	24	15.0	GALAXY
TON-N 0453	09 46 00.	+ 24 57		13.7	BLUE STAR
ZWG 122.093	09 46 00.	+ 25 06		15.4	GALAXY
KARA.73B 0372	09 46 00.	+ 25 06	54	15.4	ISOLATED GALAXY S
MCG+07-20-056	09 46 00.	+ 43 36	66	15.5	GALAXY
MCG+07-20-055	09 46 00.	+ 44 16	39	15.	GALAXY
MCG+07-20-054	09 46 00.	+ 44 22	60	16.	GALAXY
MCG+08-18-035	09 46 00.	+ 45 49	42	15.	GALAXY
ZWG 265.044	09 46 00.	+ 55 48		15.7	GALAXY
ZWG 312.026	09 46 00.	+ 67 25		15.5	GALAXY
LB 03495	09 46 00.	- 81 15		13.5	FAINT BLUE STAR
RNGC 3004	09 46 01.	+ 44 21			NON-EXISTENT OBJECT
RNGC 3005	09 46 01.	+ 44 22		16.0	GALAXY
REIZ 0049	09 46 03.	+ 25 06	18	15.2	GALAXY
MCG+04-23-038	09 46 03.	+ 25 08	48	15.4	GALAXY
LB 11148	09 46 06.	+ 23 44		17.7	FAINT BLUE STAR
LB 11147	09 46 06.	+ 24 09		16.9	FAINT BLUE STAR
ZC 0946.1+3139	09 46 06.	+ 31 39	2020		CLUSTER OF GALAXIES
ZC 0946.1+3345	09 46 06.	+ 33 45	940		CLUSTER OF GALAXIES
TON-N 1100	09 46 06.	+ 33 58		17.	BLUE STAR
UGC 05254	09 46 06.	+ 43 35	60	16.5	GALAXY S
MCG+07-20-058	09 46 06.	+ 43 41	48	15.	GALAXY
MCG+07-20-057	09 46 06.	+ 44 10	39	15.	GALAXY
ZWG 210.037	09 46 06.	+ 44 15		15.6	GALAXY
HOLM 144C	09 46 06.	+ 44 23	48	14.2	PART OF MULTIPLE GALAXY
ZWG 239.031	09 46 06.	+ 45 51		15.5	GALAXY
7ZW 293	09 46 06.	+ 67 24			COMPACT GALAXY
SEY 060	09 46 06.	+ 67 25 08.		15.4	FAINT GALAXY
ZC 0946.1+7730	09 46 06.	+ 77 30	1080		CLUSTER OF GALAXIES
RNGC 3006	09 46 07.	+ 44 15		15.5	GALAXY
HOLM 144D	09 46 09.	+ 44 17	48	14.4	PART OF MULTIPLE GALAXY
KEEL 186	09 46 11.1	+ 33 50 50.		15.	SPIRAL NEBULA
ZWG 007.039	09 46 12.	+ 01 17		15.7	GALAXY
LB 11149	09 46 12.	+ 25 14		17.6	FAINT BLUE STAR
TON-N 0454	09 46 12.	+ 32 41		14.8	BLUE STAR
ZWG 182.022	09 46 12.	+ 35 16		15.7	GALAXY
UGC 05255	09 46 12.	+ 35 16	60	15.7	GALAXY S
ZWG 210.038	09 46 12.	+ 44 08		15.2	GALAXY
REIZ 0047	09 46 12.	+ 45 50	42	14.4	GALAXY
MCG+08-18-036	09 46 12.	+ 47 12	18	16.	GALAXY
ZC 0946.2+4835	09 46 12.	+ 48 35	1480		CLUSTER OF GALAXIES
REIZ 0048	09 46 16.	+ 44 22	42	14.9	GALAXY
RNGC 3029	09 46 16.	- 07 49			GALAXY
ZWG 063.073	09 46 18.	+ 14 07		15.7	GALAXY
MCG+02-25-038	09 46 18.	+ 14 07	60	15.7	GALAXY
LB 11150	09 46 18.	+ 22 52		17.7	FAINT BLUE STAR
MCG+06-22-014	09 46 18.	+ 35 16	48	15.	GALAXY
UGC 05256	09 46 18.	+ 35 24	60	16.0	GALAXY SB
MCG-01-25-039	09 46 18.	- 04 51 30.	36	15.	GALAXY
MCG-01-25-040	09 46 18.	- 06 22	60	15.	GALAXY
MCG+06-22-015	09 46 21.	+ 35 25	48	15.	GALAXY
MCG+07-20-059	09 46 21.	+ 44 20	27	16.	GALAXY
REIZ 0050	09 46 22.	+ 44 16	36	14.5	GALAXY
ZC 0946.4+0212	09 46 24.	+ 02 12	670		CLUSTER OF GALAXIES
ZWG 035.062	09 46 24.	+ 02 42		15.3	GALAXY
ZWG 035.063	09 46 24.	+ 04 32		15.1	GALAXY
LB 11151	09 46 24.	+ 25 55		16.4	FAINT BLUE STAR
ZWG 210.039	09 46 24.	+ 44 19		15.4	GALAXY
ZWG 312.027	09 46 24.	+ 66 59		15.4	GALAXY
RNGC 3008	09 46 25.	+ 44 19		15.5	GALAXY
HOLM 144B	09 46 26.	+ 44 22	30	13.9	PART OF MULTIPLE GALAXY
KEEL 187	09 46 27.8	+ 33 48 33.		16.	NEBULA
REIZ 0051	09 46 28.	+ 44 09	30	14.1	GALAXY
REIZ 0052	09 46 28.	+ 44 17	60	15.2	GALAXY
LB 03051	09 46 28.	- 00 05 30.		16.1	FAINT BLUE STAR
ZWG 092.053	09 46 30.	+ 18 05		15.7	GALAXY
ZC 0946.5+2010	09 46 30.	+ 20 10	740		CLUSTER OF GALAXIES
LB 11154	09 46 30.	+ 21 54		14.6	FAINT BLUE STAR
LB 11153	09 46 30.	+ 22 47		18.0	FAINT BLUE STAR
LB 11152	09 46 30.	+ 25 26		11.0	FAINT BLUE STAR
ZWG 007.040	09 46 30.	- 02 36		14.4	GALAXY
MCG+00-25-019	09 46 30.	- 02 36	21	14.4	GALAXY
MCG-01-25-041	09 46 30.	- 06 54 30.	78	14.5	GALAXY
RNGC 3017	09 46 32.	- 02 36		14.5	GALAXY
MCG+07-20-060	09 46 33.	+ 42 19	24	16.	GALAXY
MCG-01-25-042	09 46 33.	- 06 55 30.	42	15.	GALAXY
ZWG 035.064	09 46 36.	+ 02 43		14.6	GALAXY
MCG+01-25-026	09 46 36.	+ 02 43	42	14.6	GALAXY
MCG+03-25-029	09 46 36.	+ 14 52	90	15.4	GALAXY
ZWG 122.094	09 46 36.	+ 21 55		15.5	GALAXY
UGC 05257	09 46 36.	+ 21 55	60	15.5	GALAXY S
LB 11157	09 46 36.	+ 22 44		16.1	FAINT BLUE STAR
LB 11156	09 46 36.	+ 23 28		16.9	FAINT BLUE STAR
LB 11155	09 46 36.	+ 24 07		15.4	FAINT BLUE STAR
ZC 0946.6+4047	09 46 36.	+ 40 47	810		CLUSTER OF GALAXIES
MCG+09-16-054	09 46 36.	+ 51 07	60	16.	GALAXY
MCG+11-12-029	09 46 36.	+ 64 43	66	16.	GALAXY
7ZW 294	09 46 36.	+ 77 29			COMPACT GALAXY
MCG-03-25-023	09 46 36.	- 16 30	30	15.	GALAXY
MCG+06-22-016	09 46 39.	+ 34 40	48	15.	GALAXY
RNGC 3014	09 46 39.	- 04 30		14.0	GALAXY
MCG-01-25-043	09 46 39.	- 04 30	48	14.	GALAXY
REIZ 0053	09 46 40.	+ 44 19	24	14.2	GALAXY
REIZ 0054	09 46 40.	+ 44 20	12	16.2	GALAXY
REIZ 0055	09 46 40.	+ 44 26	12	15.4	GALAXY
ZC 0946.7+0605	09 46 42.	+ 06 05	2150		CLUSTER OF GALAXIES
ZWG 092.054	09 46 42.	+ 14 54		15.4	GALAXY
UGC 05258	09 46 42.	+ 14 54	84	15.4	GALAXY SC
MCG+04-23-039	09 46 42.	+ 21 53 30.	54	15.5	GALAXY
LB 11159	09 46 42.	+ 22 50		17.1	FAINT BLUE STAR
TON-N 0455	09 46 42.	+ 26 19		15.6	BLUE STAR
LB 11158	09 46 42.	+ 26 26		16.7	FAINT BLUE STAR
TON-N 0456	09 46 42.	+ 30 31		15.4	BLUE STAR
ZWG 152.069	09 46 42.	+ 32 27		14.2	GALAXY
MRK 409	09 46 42.	+ 32 27	18	15.	GALAXY WITH UV CONTINUUM
UGC 05259	09 46 42.	+ 32 27	54	14.2	GALAXY S0
ZWG 210.040	09 46 42.	+ 39 05		15.5	GALAXY
MCG+09-16-055	09 46 42.	+ 54 43	30	17.	GALAXY
ZC 0946.7+5816	09 46 42.	+ 58 16	470		CLUSTER OF GALAXIES
UGC 05260	09 46 42.	+ 64 44	72	16.0	GALAXY Sb-c
LB 03052	09 46 42.	- 01 03 18.		16.5	FAINT BLUE STAR
RNGC 3011	09 46 44.	+ 32 27		14.0	GALAXY
ZWG 007.041	09 46 48.	+ 01 22		14.2	GALAXY
UGC 05261	09 46 48.	+ 01 22	30	14.2	GALAXY
MCG+00-25-020	09 46 48.	+ 01 22	30	14.2	GALAXY
ZWG 035.065	09 46 48.	+ 04 14		15.5	GALAXY
ZC 0946.8+0656	09 46 48.	+ 06 56	400		CLUSTER OF GALAXIES
ZC 0946.8+1130	09 46 48.	+ 11 30	540		CLUSTER OF GALAXIES
LB 11161	09 46 48.	+ 22 13		17.2	FAINT BLUE STAR
LB 11160	09 46 48.	+ 25 52		15.3	FAINT BLUE STAR
ZWG 182.023	09 46 48.	+ 34 56		14.9	GALAXY
RNGC 3012	09 46 48.	+ 34 56		15.0	GALAXY
UGC 05262	09 46 48.	+ 34 56	66	14.9	GALAXY (E)
STOCK 04	09 46 48.	+ 37 47			BLUE KNOT NEAR ELLIP GLXY
ZWG 210.041	09 46 48.	+ 39 09		15.3	GALAXY
MCG+08-18-037	09 46 48.	+ 46 22	48	15.	GALAXY
ZWG 239.032	09 46 48.	+ 46 35		15.2	GALAXY
UGC 05263	09 46 48.	+ 46 35	60	15.2	GALAXY PECULR
RNGC 3015	09 46 49.	+ 01 22		14.0	GALAXY
REIZ 0056	09 46 49.	+ 46 34	54	14.0	GALAXY
MCG+05-23-038	09 46 51.	+ 32 28	48	14.2	GALAXY
MCG+07-20-061	09 46 51.	+ 39 08	15	15.	GALAXY
ZC 0946.9+2936	09 46 54.	+ 29 36	1340		CLUSTER OF GALAXIES
REIZ 0057	09 46 54.	+ 34 56	54	13.5	GALAXY
MCG+06-22-017	09 46 54.	+ 34 57	48	14.5	GALAXY
MCG+07-20-063	09 46 54.	+ 42 17 30.	36	16.	GALAXY
ZC 0946.9+4320	09 46 54.	+ 43 20	1080		CLUSTER OF GALAXIES
MCG+07-20-062	09 46 54.	+ 44 33	45	14.	GALAXY
MCG-01-25-044	09 46 54.	- 04 56 30.	48	15.	GALAXY
REIZ 0059	09 46 58.	+ 26 01	48	15.1	GALAXY
VDB.66G 064	09 47	+ 31 44	100		DWARF GALAXY
LBN 1069	09 47	- 07 00	1920		BRIGHT NEBULA
ZWG 063.074	09 47 00.	+ 09 14		14.7	GALAXY
MCG+02-25-039	09 47 00.	+ 09 14	48	14.7	GALAXY
ZWG 063.075	09 47 00.	+ 12 54		15.6	GALAXY
ZWG 063.076	09 47 00.	+ 13 53		15.7	GALAXY
LB 11162	09 47 00.	+ 24 20		16.0	FAINT BLUE STAR
ZWG 239.033	09 47 00.	+ 44 32		14.5	GALAXY
HOLM 146B	09 47 00.	+ 44 32	48	13.3	PART OF MULTIPLE GALAXY
UGC 05264	09 47 00.	+ 44 32	48	14.5	GALAXY S
ZC 0947.0+5030	09 47 00.	+ 50 30	810		CLUSTER OF GALAXIES
ZC 0947.0+5854	09 47 00.	+ 58 54	1680		CLUSTER OF GALAXIES
7ZW 295	09 47 00.	+ 59 04			COMPACT GALAXY
SHAH 116	09 47 00.	+ 59 54	162	18.2	GROUP OF COMPACT GALAXIES
RNGC 3009	09 47 01.	+ 44 32		14.5	GALAXY
ARC 0876	09 47 02.	+ 29 32		17.6	RICH CLUSTER OF GALAXIES
RNGC 3016	09 47 03.	+ 12 56		13.5	GALAXY
RNGC 3033	09 47 04.	- 56 11		8.5	OPEN CLUSTER
ZWG 007.042	09 47 06.	+ 00 51		14.2	GALAXY
UGC 05265	09 47 06.	+ 00 51	78	14.2	GALAXY S
KARA.72 216A	09 47 06.	+ 00 51	66	14.2	PART OF DOUBLE GALAXY
MCG+00-25-021	09 47 06.	+ 00 51	60	14.2	GALAXY
ARC 0878	09 47 06.	+ 06 01		16.8	RICH CLUSTER OF GALAXIES
ZWG 063.077	09 47 06.	+ 12 56		13.7	GALAXY
UGC 05266	09 47 06.	+ 12 56	78	13.7	GALAXY Sb
MCG+02-25-040	09 47 06.	+ 12 56	72	13.7	GALAXY
ZWG 152.070	09 47 06.	+ 31 58		15.3	GALAXY
ZWG 182.024	09 47 06.	+ 33 48		15.6	GALAXY
TON-N 1101	09 47 06.	+ 34 04		16.1	BLUE STAR
ZC 0947.1+3853	09 47 06.	+ 38 53	200		CLUSTER OF GALAXIES
MRK 125	09 47 06.	+ 46 12	12	15.	GALAXY WITH UV CONTINUUM
ARC 0874	09 47 06.	+ 58 17		17.7	RICH CLUSTER OF GALAXIES
KARA.68 059	09 47 06.	+ 72 19	40		DWARF GALAXY
MCG-04-23-018	09 47 06.	- 21 31	24	15.	GALAXY
MRSL 268+11/1	09 47 06.	- 38 14	25200		HII REGION
OCL 0796	09 47 06.	- 56 11	720	8.4	OPEN STAR CLUSTER
RNGC 3018	09 47 07.	+ 00 51		14.0	GALAXY
RNGC 3013	09 47 07.	+ 33 48		15.5	GALAXY
HOLM 147C	09 47 08.	+ 12 56	72	13.9	PART OF MULTIPLE GALAXY
RNGC 3025	09 47 08.	- 21 31		15.0	GALAXY
MCG+05-23-039	09 47 09.	+ 28 14	30	15.3	GALAXY

OBJECT NAME	RIGHT ASCEN.	DECLINATION	DIAM.	MAGN.	TYPE OF OBJECT
LB 03053	09 47 10.	+ 01 34 30.		15.4	FAINT BLUE STAR
SN 1965E	09 47 10.	+ 34 39		16.4	SUPERNOVA
IC 2511	09 47 10.	- 32 37 01.			NONSTELLAR OBJECT
ARC 0880	09 47 11.	- 03 56		17.4	RICH CLUSTER OF GALAXIES
KEEL 188	09 47 11.5	+ 33 48 12.		14.	SPIRAL NEBULA
ZWG 063.078	09 47 12.	+ 09 19		14.9	GALAXY
UGC 05267	09 47 12.	+ 09 19	96	14.9	GALAXY Sb
MCG+02-25-041	09 47 12.	+ 09 19	96	14.9	GALAXY
ZWG 063.079	09 47 12.	+ 10 40		15.2	GALAXY
REIZ 0063	09 47 12.	+ 12 55	72	13.7	GALAXY
ZC 0947.2+1723	09 47 12.	+ 17 23	1480		CLUSTER OF GALAXIES
ZC 0947.2+1821	09 47 12.	+ 18 21	2350		CLUSTER OF GALAXIES
LB 11165	09 47 12.	+ 20 34		16.4	FAINT BLUE STAR
LB 11164	09 47 12.	+ 20 37		16.0	FAINT BLUE STAR
LB 11163	09 47 12.	+ 25 12		17.3	FAINT BLUE STAR
ZWG 152.071	09 47 12.	+ 28 15		15.3	GALAXY
MCG+07-20-064	09 47 12.	+ 43 24	36	16.	GALAXY
MCG+08-18-038	09 47 12.	+ 45 33	24	16.	GALAXY
REIZ 0058	09 47 12.	+ 45 34	18	14.8	GALAXY
ZWG 265.045	09 47 12.	+ 52 31		15.4	GALAXY
KARA.73B 0373	09 47 12.	+ 52 31	36	15.4	ISOLATED GALAXY S
MCG+10-14-042	09 47 12.	+ 61 14	36	17.	GALAXY
ZWG 289.020	09 47 12.	+ 62 25		15.2	GALAXY
UGC 05268	09 47 12.	+ 62 25	90	15.2	GALAXY Sa-b
MCG+10-14-043	09 47 12.	+ 62 25	72	15.	GALAXY
KARA.73B 0374	09 47 12.	+ 62 25	78	15.2	ISOLATED GALAXY S
MCG-01-25-045	09 47 12.	- 04 41	36	15.	GALAXY
MCG-01-25-046	09 47 12.	- 04 55	36	14.	GALAXY
MCG-05-23-018	09 47 12.	- 32 37	96	14.	GALAXY
RNGC 3022	09 47 15.	- 04 55		14.0	GALAXY
IC 2512	09 47 15.	- 32 41 37.			NONSTELLAR OBJECT
REIZ 0060	09 47 16.	+ 44 33	48	13.9	GALAXY
ARC 0871	09 47 16.	+ 66 02		17.1	RICH CLUSTER OF GALAXIES
ZWG 007.043	09 47 18.	+ 00 51		13.5	GALAXY
UGC 05269	09 47 18.	+ 00 51	192	13.5	GALAXY S
KARA.72 216B	09 47 18.	+ 00 51	138	13.5	PART OF DOUBLE GALAXY
MCG+00-25-022	09 47 18.	+ 00 51	180	13.5	GALAXY
ZWG 063.080	09 47 18.	+ 09 23		14.6	GALAXY
UGC 05270	09 47 18.	+ 09 23	66	14.6	GALAXY S0
MCG+02-25-043	09 47 18.	+ 09 23	36		GALAXY
MCG+02-25-042	09 47 18.	+ 09 23	27	14.6	GALAXY
LB 11166	09 47 18.	+ 23 36		15.0	FAINT BLUE STAR
MCG+07-20-066	09 47 18.	+ 44 34	21	15.5	GALAXY
MCG+07-20-065	09 47 18.	+ 44 34	36	15.5	GALAXY
MCG+08-18-039	09 47 18.	+ 48 42	18	16.	GALAXY
ZWG 239.034	09 47 18.	+ 48 43		15.4	GALAXY
KARA.73B 0375	09 47 18.	+ 48 43	24	15.4	ISOLATED GALAXY E
RNGC 3023	09 47 19.	+ 00 51		13.5	GALAXY
LB 03054	09 47 21.	+ 01 21 30.		16.6	FAINT BLUE STAR
RNGC 3019	09 47 21.	+ 12 58		15.0	GALAXY
MCG+05-23-040	09 47 21.	+ 28 15	18	15.3	GALAXY
MCG+06-22-018	09 47 21.	+ 33 47	24	15.5	GALAXY
MCG+07-20-068	09 47 21.	+ 44 00	42	15.	GALAXY
MCG+07-20-067	09 47 21.	+ 44 34 30.	30	15.5	GALAXY
HOLM 146A	09 47 22.	+ 44 33	24	14.1	PART OF MULTIPLE GALAXY
KEEL 189	09 47 23.1	+ 33 49 06.		16.	SPIRAL NEBULA
ZWG 007.044	09 47 24.	+ 00 00		15.5	GALAXY
ZC 0947.4+0536	09 47 24.	+ 05 36	610		CLUSTER OF GALAXIES
ZWG 063.081	09 47 24.	+ 12 58		15.0	GALAXY
MCG+02-25-044	09 47 24.	+ 12 58	42	15.0	GALAXY
REIZ 0066	09 47 24.	+ 12 59	36	14.8	GALAXY
ZWG 063.082	09 47 24.	+ 13 03		13.2	GALAXY
REIZ 0067	09 47 24.	+ 13 03	180	13.2	GALAXY
HOLM 147A	09 47 24.	+ 13 03	192	13.1	PART OF MULTIPLE GALAXY
UGC 05271	09 47 24.	+ 13 03	192	13.2	GALAXY SBc
MCG+02-25-045	09 47 24.	+ 13 03	168	13.2	GALAXY
MCG+03-25-030	09 47 24.	+ 16 30	78	14.9	GALAXY
LB 11170	09 47 24.	+ 22 04		15.6	FAINT BLUE STAR
LB 11169	09 47 24.	+ 22 12		16.8	FAINT BLUE STAR
LB 11168	09 47 24.	+ 23 41		14.8	FAINT BLUE STAR
LB 11167	09 47 24.	+ 24 15		16.8	FAINT BLUE STAR
HARO 22	09 47 24.	+ 28 13			BLUE EMISSION-LINE GALAXY
ZWG 152.072	09 47 24.	+ 31 43		14.7	GALAXY
UGC 05272	09 47 24.	+ 31 43	132	14.7	GALAXY IRR
BFGS 1	09 47 24.	+ 31 43	3	16.5	GALAXY WITH UV KNOTS
ZWG 210.042	09 47 24.	+ 44 00		14.9	GALAXY
ZWG 239.035	09 47 24.	+ 44 34		14.3	GALAXY
HOLM 146C	09 47 24.	+ 44 34	24	14.1	PART OF MULTIPLE GALAXY
UGC 05273	09 47 24.	+ 44 34	114	14.3	GALAXY TRP SYS
IC 0566	09 47 24.	- 00 00 13.			NONSTELLAR OBJECT
MCG-01-25-047	09 47 24.	- 07 50	78	14.5	GALAXY
HOLM 147D	09 47 25.	+ 12 59	54	14.7	PART OF MULTIPLE GALAXY
RNGC 3010	09 47 25.	+ 44 34		14.5	GALAXY
RNGC 3020	09 47 27.	+ 13 03		13.0	GALAXY
MCG+05-23-041	09 47 27.	+ 31 44	126	14.7	GALAXY
REIZ 0061	09 47 28.	+ 44 35	15	14.6	GALAXY
HOLM 146D	09 47 29.	+ 44 34	30	14.3	PART OF MULTIPLE GALAXY
ZC 0947.5+1355	09 47 30.	+ 13 55	540		CLUSTER OF GALAXIES
ZWG 092.055	09 47 30.	+ 16 32		14.9	GALAXY
UGC 05274	09 47 30.	+ 16 32	66	14.9	GALAXY Sc
LB 11173	09 47 30.	+ 21 02		16.2	FAINT BLUE STAR
LB 11172	09 47 30.	+ 24 14		17.8	FAINT BLUE STAR
LB 11171	09 47 30.	+ 24 48		17.3	FAINT BLUE STAR
ZWG 122.095	09 47 30.	+ 26 01		15.6	GALAXY
ZC 0947.5+2908	09 47 30.	+ 29 08	1750		CLUSTER OF GALAXIES
TON-N 1102	09 47 30.	+ 34 39		17.	BLUE STAR
MCG-02-25-019	09 47 30.	- 11 53	42	15.	GALAXY
OCL 0791	09 47 30.	- 54 20	204	12.	OPEN STAR CLUSTER
VHA 080	09 47 30.	- 54 20	240		OPEN STAR CLUSTER
RNGC 3028	09 47 31.	- 18 56			GALAXY
REIZ 0069	09 47 34.	+ 26 02	18	15.7	GALAXY
REIZ 0068	09 47 34.	+ 26 02	18	15.5	GALAXY
REIZ 0062	09 47 34.	+ 44 35	18	14.1	GALAXY
ZWG 063.083	09 47 36.	+ 13 36		15.1	GALAXY
LB 11175	09 47 36.	+ 22 16		17.7	FAINT BLUE STAR
LB 11174	09 47 36.	+ 23 38		14.8	FAINT BLUE STAR
MCG+09-16-056	09 47 36.	+ 54 17	42	17.	GALAXY
MCG+11-12-030	09 47 36.	+ 65 43	78	14.	GALAXY
RNGC 3036	09 47 36.	- 62 27			NON-EXISTENT OBJECT
TON-N 1103	09 47 37.	+ 21 46		17.	BLUE STAR
REIZ 0064	09 47 39.	+ 43 59	12	14.0	GALAXY
RNGC 2531	09 47 40.	+ 10 46			NON-EXISTENT OBJECT
RNGC 2530	09 47 40.	+ 10 46			NON-EXISTENT OBJECT
REIZ 0070	09 47 40.	+ 26 02	6	15.9	GALAXY
REIZ 0065	09 47 40.	+ 44 36	18	15.1	GALAXY
IC 2513	09 47 40.	- 32 41 45.			NONSTELLAR OBJECT
LB 11176	09 47 42.	+ 25 20		17.4	FAINT BLUE STAR
ZWG 122.096	09 47 42.	+ 26 01		15.6	GALAXY

OBJECT NAME	RIGHT ASCEN.	DECLINATION	DIAM.	MAGN.	TYPE OF OBJECT
REIZ 0071	09 47 42.	+ 28 49	120	13.8	GALAXY
TON-N 0457	09 47 42.	+ 30 26		16.7	BLUE STAR
TON-N 0458	09 47 42.	+ 32 37		15.1	BLUE STAR
MCG+07-20-069	09 47 42.	+ 43 56	54	16.	GALAXY
MCG-04-23-019	09 47 42.	- 24 45 30.	120	15.	GALAXY
KEEL 190	09 47 42.4	+ 33 23 31.		17.	NEBULA
HOLM 147B	09 47 45.	+ 13 00	90	13.7	PART OF MULTIPLE GALAXY
LB 03055	09 47 45.	- 00 50 42.		16.3	FAINT BLUE STAR
IC 2514	09 47 45.	- 32 39 04.			NONSTELLAR OBJECT
REIZ 0072	09 47 46.	+ 26 02	12	15.7	GALAXY
SEY 061	09 47 46.	+ 65 43 04.		14.3	FAINT GALAXY
LB 03056	09 47 47.	+ 01 09 24.		16.7	FAINT BLUE STAR
RNGC 3030	09 47 47.	- 12 00		15.0	GALAXY
ZWG 007.045	09 47 48.	+ 01 48		14.6	GALAXY
MCG+00-25-023	09 47 48.	+ 01 48	60	14.6	GALAXY
REIZ 0075	09 47 48.	+ 12 59	108	13.6	GALAXY
ZWG 063.084	09 47 48.	+ 13 00		13.7	GALAXY
UGC 05275	09 47 48.	+ 13 00	120	13.7	GALAXY S
MCG+02-25-046	09 47 48.	+ 13 00	120	13.7	GALAXY
8ZW 0947+14.4	09 47 48.	+ 14 27		17.9	COMPACT GALAXY
ZC 0947.8+1612	09 47 48.	+ 16 12	1210		CLUSTER OF GALAXIES
LB 11178	09 47 48.	+ 22 47		14.7	FAINT BLUE STAR
LB 11177	09 47 48.	+ 24 22		16.1	FAINT BLUE STAR
ZWG 152.073	09 47 48.	+ 30 44		15.1	GALAXY
UGC 05276	09 47 48.	+ 30 44	102	15.1	GALAXY
MCG+05-23-042	09 47 48.	+ 30 44	72	15.1	GALAXY
ZWG 239.036	09 47 48.	+ 47 25		15.0	GALAXY
ZWG 289.021	09 47 48.	+ 58 10		15.7	GALAXY
ZC 0947.8+6017	09 47 48.	+ 60 17	2350		CLUSTER OF GALAXIES
ZWG 312.028	09 47 48.	+ 65 44		15.3	GALAXY
UGC 05277	09 47 48.	+ 65 44	108	15.3	GALAXY SBb
KARA.73B 0376	09 47 48.	+ 65 44	78	15.3	ISOLATED GALAXY S
MCG-02-25-020	09 47 48.	- 11 51	90	14.5	GALAXY
MCG-02-25-021	09 47 48.	- 12 00 30.	30	15.	GALAXY
MCG-05-23-019	09 47 48.	- 32 40	84	14.	GALAXY
IC 0567	09 47 49.	+ 13 03			NONSTELLAR OBJECT
TON-N 1104	09 47 49.	+ 20 56		16.8	BLUE STAR
RNGC 3024	09 47 51.	+ 13 00		13.5	GALAXY
LB 03057	09 47 53.	+ 01 10 30.		15.8	FAINT BLUE STAR
ZWG 122.097	09 47 54.	+ 22 59		15.0	GALAXY
UGC 05278	09 47 54.	+ 22 59	66	15.0	GALAXY SBb
MCG+08-18-040	09 47 54.	+ 47 23	30	15.	GALAXY
MCG+10-14-044	09 47 54.	+ 58 10	18	15.	GALAXY
ZC 0947.9+7338	09 47 54.	+ 73 38	1480		CLUSTER OF GALAXIES
MCG-04-23-020	09 47 54.	- 22 47	48	14.	GALAXY
OCL 0814	09 47 54.	- 65 02	216	11.	OPEN STAR CLUSTER
MCG+04-23-040	09 47 57.	+ 22 59	51	15.0	GALAXY
RNGC 3026	09 47 57.	+ 28 47		14.0	GALAXY
REIZ 0073	09 47 58.	+ 33 47	54	12.4	GALAXY
REIN 2.093	09 47 59.77	+ 33 47 20.1			NEBULA
VDB.66G 065	09 48	+ 01 38	70		DWARF GALAXY
ZWG 035.066	09 48 00.	+ 05 01		15.2	GALAXY
REIZ 0077	09 48 00.	+ 13 03	9	15.2	GALAXY
ZWG 092.056	09 48 00.	+ 17 22		15.7	GALAXY
ZC 0948.0+2036	09 48 00.	+ 20 36	2550		CLUSTER OF GALAXIES
ZC 0948.0+2235	09 48 00.	+ 22 35	940		CLUSTER OF GALAXIES
LB 11179	09 48 00.	+ 25 41		17.4	FAINT BLUE STAR
ZWG 152.074	09 48 00.	+ 28 47		13.8	GALAXY
UGC 05279	09 48 00.	+ 28 47	156	13.8	GALAXY IRR
MCG+05-23-043	09 48 00.	+ 28 47	168	13.8	GALAXY
KARA.73B 0377	09 48 00.	+ 28 47	162	13.8	ISOLATED GALAXY S
ZWG 182.025	09 48 00.	+ 33 47		12.6	GALAXY
UGC 05280	09 48 00.	+ 33 47	90	12.6	GALAXY S
MCG+06-22-019	09 48 00.	+ 33 47	90	13.	GALAXY
ZC 0948.0+4017	09 48 00.	+ 40 17	2890		CLUSTER OF GALAXIES
ZC 0948.0+5142	09 48 00.	+ 51 42	1140		CLUSTER OF GALAXIES
7ZW 296	09 48 00.	+ 58 17			COMPACT GALAXY
ZC 0948.0+6312	09 48 00.	+ 63 12	1880		CLUSTER OF GALAXIES
RNGC 3021	09 48 01.	+ 33 47		13.0	GALAXY
REIN 2.094	09 48 03.97	+ 33 46 41.8			NEBULA
LB 03058	09 48 05.	- 01 24 42.		16.4	FAINT BLUE STAR
LB 11181	09 48 06.	+ 23 54		15.7	FAINT BLUE STAR
LB 11180	09 48 06.	+ 24 16		16.1	FAINT BLUE STAR
ZWG 152.075	09 48 06.	+ 30 49		15.1	GALAXY
UGC 05281	09 48 06.	+ 30 49	126	15.1	GALAXY DBL SYS
MCG+05-23-044	09 48 06.	+ 30 50	24	15.1	GALAXY
MCG+05-23-045	09 48 06.	+ 30 51	54	15.1	GALAXY
ZC 0948.1+6107	09 48 06.	+ 61 07	540		CLUSTER OF GALAXIES
REIZ 0074	09 48 09.	+ 43 55	48	15.1	GALAXY
LB 03059	09 48 12.	+ 00 40 12.		15.9	FAINT BLUE STAR
LB 11182	09 48 12.	+ 24 38		16.5	FAINT BLUE STAR
ZWG 152.076	09 48 12.	+ 26 32		15.5	GALAXY
UGC 05282	09 48 12.	+ 33 22	60	16.0	GALAXY
MCG+08-18-041	09 48 12.	+ 45 08	72	15.	GALAXY
ZWG 239.037	09 48 12.	+ 45 10		15.5	GALAXY
UGC 05283	09 48 12.	+ 45 10	90	15.5	GALAXY Sc
ZC 0948.2+5820	09 48 12.	+ 58 20	1080		CLUSTER OF GALAXIES
ZWG 350.033	09 48 12.	+ 77 07		15.6	GALAXY
MCG-02-25-022	09 48 12.	- 10 47 30.	36	16.	GALAXY
MCG-04-24-001	09 48 12.	- 21 33	72	14.5	GALAXY
REIZ 0076	09 48 17.	+ 45 10	36	14.5	GALAXY
UGC 05284	09 48 18.	+ 04 31	66	16.5	GALAXY
MCG+03-25-031	09 48 18.	+ 15 57	78	14.8	GALAXY
LB 11183	09 48 18.	+ 22 42		17.5	FAINT BLUE STAR
MCG+06-22-020	09 48 21.	+ 33 21	84	15.5	GALAXY
MCG-01-25-048	09 48 21.	- 09 06 30.	90	14.5	GALAXY
AR 05	09 48 22.09	+ 76 16 45.1			NEBULA
ZWG 092.057	09 48 24.	+ 15 58		14.8	GALAXY
UGC 05285	09 48 24.	+ 15 58	90	14.8	GALAXY SBb
LB 11184	09 48 24.	+ 24 32		12.0	FAINT BLUE STAR
TON-N 1105	09 48 24.	+ 33 49		16.8	BLUE STAR
MCG+06-22-022	09 48 24.	+ 36 00	30	15.5	GALAXY
ZC 0948.4+5858	09 48 24.	+ 58 58	610		CLUSTER OF GALAXIES
IC 0568	09 48 26.	+ 15 57 56.			NONSTELLAR OBJECT
ARC 0873	09 48 27.	+ 71 33		17.4	RICH CLUSTER OF GALAXIES
MCG-01-25-049	09 48 27.	- 04 45	120	14.	GALAXY
REIZ 0078	09 48 29.	+ 45 11	36	14.5	GALAXY
ZC 0948.5+0515	09 48 30.	+ 05 15	1410		CLUSTER OF GALAXIES
ZWG 063.085	09 48 30.	+ 09 14		14.4	GALAXY
UGC 05286	09 48 30.	+ 09 14	126	14.4	GALAXY Sc
MCG+02-25-047	09 48 30.	+ 09 14	120	14.4	GALAXY
TON-N 0459	09 48 30.	+ 26 52		16.4	BLUE STAR
ZWG 182.026	09 48 30.	+ 33 10		14.7	GALAXY
UGC 05287	09 48 30.	+ 33 10	90	14.7	GALAXY SBc
TON-N 1106	09 48 30.	+ 33 58		16.	BLUE STAR
ZC 0948.5+3834	09 48 30.	+ 38 34	400		CLUSTER OF GALAXIES
ZC 0948.5-0035	09 48 30.	- 00 35	940		CLUSTER OF GALAXIES
MCG+00-25-024	09 48 36.	+ 01 41	48	16.	GALAXY

OBJECT NAME	RIGHT ASCEN.	DECLINATION	DIAM.	MAGN.	TYPE OF OBJECT
ZWG 035.067	09 48 36.	+ 04 20		15.5	GALAXY
MCG+01-25-027	09 48 36.	+ 04 20	54	15.5	GALAXY
ZWG 035.068	09 48 36.	+ 05 24		15.5	GALAXY
LB 11186	09 48 36.	+ 25 53		16.2	FAINT BLUE STAR
LB 11185	09 48 36.	+ 26 16		17.0	FAINT BLUE STAR
ZWG 007.046	09 48 36.	- 01 37		15.7	GALAXY
ARC 0882	09 48 39.	+ 08 30		17.1	RICH CLUSTER OF GALAXIES
MCG+06-22-021	09 48 39.	+ 33 10	96	14.	GALAXY
ARC 0883	09 48 41.	+ 05 45		16.8	RICH CLUSTER OF GALAXIES
ZWG 035.069	09 48 42.	+ 08 04		14.4	GALAXY
UGC 05288	09 48 42.	+ 08 04	90	14.4	GALAXY SB-IRR
KARA.72 217B	09 48 42.	+ 08 04	30		PART OF DOUBLE GALAXY
KARA.72 217A	09 48 42.	+ 08 04	30	14.4	PART OF DOUBLE GALAXY
MCG+01-25-028	09 48 42.	+ 08 04	60	14.4	GALAXY
ZC 0948.7+0831	09 48 42.	+ 08 31	610		CLUSTER OF GALAXIES
ZWG 092.058	09 48 42.	+ 17 30		15.7	GALAXY
ZWG 122.098	09 48 42.	+ 24 02		15.5	GALAXY
LB 11187	09 48 42.	+ 25 59		16.3	FAINT BLUE STAR
MCG-03-25-025	09 48 42.	- 17 16	42	14.5	GALAXY
MCG-03-25-024	09 48 42.	- 18 15 30.	96	15.	GALAXY
VV 110B	09 48 45.	- 04 44	24		INTERACTING GALAXY
VV 110A	09 48 45.	- 04 44	24		INTERACTING GALAXY
VV 110	09 48 45.	- 04 44	102	14.5	INTERACTING GALAXY
ZC 0948.8+0456	09 48 48.	+ 04 56	740		CLUSTER OF GALAXIES
ZWG 035.070	09 48 48.	+ 07 30		15.3	GALAXY
ZWG 063.086	09 48 48.	+ 11 09		15.1	GALAXY
MCG+02-25-048	09 48 48.	+ 11 09	12	15.1	GALAXY
KARA.73B 0378	09 48 48.	+ 11 09	36	15.1	ISOLATED GALAXY E
IC 0569	09 48 48.	+ 11 09 54.			NONSTELLAR OBJECT
LB 11189	09 48 48.	+ 22 28		15.3	FAINT BLUE STAR
LB 11188	09 48 48.	+ 22 39		17.7	FAINT BLUE STAR
SHAH 117	09 48 48.	+ 50 37	42	18.4	GROUP OF COMPACT GALAXIES
MCG+10-14-045	09 48 48.	+ 59 10	18	16.	GALAXY
ARC 0884	09 48 52.	+ 04 59		17.1	RICH CLUSTER OF GALAXIES
ZWG 063.087	09 48 54.	+ 14 16		15.5	GALAXY
ZC 0948.9+6123	09 48 54.	+ 61 23	670		CLUSTER OF GALAXIES
MCG-01-25-050	09 48 57.	- 04 51 30.	90	14.	GALAXY
ZWG 007.047	09 49 00.	+ 01 20		15.7	GALAXY
ZC 0949.0+0745	09 49 00.	+ 07 45	340		CLUSTER OF GALAXIES
ZWG 063.088	09 49 00.	+ 12 40		15.0	GALAXY
MCG+02-25-049	09 49 00.	+ 12 40	12	15.0	GALAXY
MCG+03-25-032	09 49 00.	+ 15 58	30	15.6	GALAXY
LB 11190	09 49 00.	+ 23 47		13.2	FAINT BLUE STAR
ZC 0949.0+3017	09 49 00.	+ 30 17	540		CLUSTER OF GALAXIES
TON-N 1107	09 49 00.	+ 34 23		16.8	BLUE STAR
MCG+07-20-070	09 49 00.	+ 41 05 30.	60	14.	GALAXY
ZWG 210.043	09 49 00.	+ 41 06		14.9	GALAXY
UGC 05290	09 49 00.	+ 41 06	66	14.9	GALAXY S0-a?
ZC 0949.0+4356	09 49 00.	+ 43 56	670		CLUSTER OF GALAXIES
ZC 0949.0+5725	09 49 00.	+ 57 25	3560		CLUSTER OF GALAXIES
ZC 0949.0+6700	09 49 00.	+ 67 00	1340		CLUSTER OF GALAXIES
UGC 05289	09 49 00.	- 01 32	84	16.0	GALAXY DBL SYS
MCG-01-25-051	09 49 00.	- 04 47	48	15.	GALAXY
MCG-02-25-023	09 49 00.	- 12 25	54	15.	GALAXY
MCG-05-24-001	09 49 00.	- 32 33	120	13.5	GALAXY
KEEL 191	09 49 03.7	+ 69 14 15.			NEBULA
ZWG 035.071	09 49 06.	+ 03 30		15.5	GALAXY
ZC 0949.1+0759	09 49 06.	+ 07 59	1610		CLUSTER OF GALAXIES
ZC 0949.1+1359	09 49 06.	+ 13 59	610		CLUSTER OF GALAXIES
ZWG 092.059	09 49 06.	+ 15 10		15.3	GALAXY
MCG+03-25-033	09 49 06.	+ 15 10	42	15.3	GALAXY
KARA.73B 0379	09 49 06.	+ 15 10	36	15.3	ISOLATED GALAXY S
ZWG 092.060	09 49 06.	+ 16 00		15.6	GALAXY
ZC 0949.1+2205	09 49 06.	+ 22 05	1680		CLUSTER OF GALAXIES
LB 11191	09 49 06.	+ 22 23		16.7	FAINT BLUE STAR
TON-N 0460	09 49 06.	+ 30 52		16.0	BLUE STAR
7ZW 297	09 49 06.	+ 67 23			COMPACT GALAXY
ZC 0949.1+7845	09 49 06.	+ 78 45	1680		CLUSTER OF GALAXIES
TON-N 1108	09 49 07.	+ 20 16		16.9	BLUE STAR
TON-N 1109	09 49 07.	+ 21 27		16.3	BLUE STAR
LB 03060	09 49 08.	- 00 14 24.		17.4	FAINT BLUE STAR
IC 0570	09 49 09.	+ 15 59 29.			NONSTELLAR OBJECT
RNGC 3032	09 49 09.	+ 29 28		13.0	GALAXY
MCG-02-25-024	09 49 09.	- 12 47	48	15.	GALAXY
RNGC 3037	09 49 10.	- 26 48		14.0	GALAXY
ZWG 063.089	09 49 12.	+ 13 11		14.8	GALAXY
UGC 05291	09 49 12.	+ 13 11	66	14.8	GALAXY E-S0
MCG+02-25-050	09 49 12.	+ 13 11	15	14.8	GALAXY
LB 11192	09 49 12.	+ 22 22		16.8	FAINT BLUE STAR
ZWG 152.077	09 49 12.	+ 29 28		13.0	GALAXY
UGC 05292	09 49 12.	+ 29 28	138	13.0	GALAXY S0
ZC 0949.2+2951	09 49 12.	+ 29 51	670		CLUSTER OF GALAXIES
TON-N 1110	09 49 12.	+ 32 50		14.9	BLUE STAR
UGC 05293	09 49 12.	+ 41 30	60	16.5	GALAXY
MCG+07-20-071	09 49 12.	+ 41 30	63	17.	GALAXY
RNGC 3038	09 49 13.	- 32 32		13.0	GALAXY
MCG+05-23-046	09 49 15.	+ 29 28	120	13.0	GALAXY
MCG-04-24-002	09 49 15.	- 26 48	60	14.	GALAXY
ZC 0949.3+1204	09 49 18.	+ 12 04	1810		CLUSTER OF GALAXIES
ZWG 092.061	09 49 18.	+ 19 53		15.1	GALAXY
LB 11194	09 49 18.	+ 24 06		16.0	FAINT BLUE STAR
LB 11193	09 49 18.	+ 24 44		15.8	FAINT BLUE STAR
REIZ 0080	09 49 18.	+ 29 28	9	14.8	GALAXY
ZWG 182.027	09 49 18.	+ 33 17		15.2	GALAXY
UGC 05294	09 49 18.	+ 33 17	90	15.2	GALAXY JET?
MCG+07-20-072	09 49 18.	+ 43 04	48	15.	GALAXY
MRK 126	09 49 18.	+ 52 26	15	15.5	GALAXY WITH UV CONTINUUM
7ZW 298	09 49 18.	+ 63 05			COMPACT GALAXY
MCG-02-25-025	09 49 18.	- 13 30	72	14.5	GALAXY
OCL 0790	09 49 18.	- 52 57	90	8.7	OPEN STAR CLUSTER
VHA 081	09 49 18.	- 52 57	90		OPEN STAR CLUSTER
TON-N 1111	09 49 19.	+ 21 12		16.9	BLUE STAR
ARC 0875	09 49 21.	+ 71 11		17.4	RICH CLUSTER OF GALAXIES
RNGC 3035	09 49 21.	- 06 35		13.5	GALAXY
ZWG 035.072	09 49 24.	+ 04 23		15.0	GALAXY
MCG+01-25-029	09 49 24.	+ 04 23	48	15.0	GALAXY
ZWG 063.090	09 49 24.	+ 12 52		15.3	GALAXY
MCG+03-25-034	09 49 24.	+ 17 49	60	15.6	GALAXY
ZWG 092.062	09 49 24.	+ 17 50		15.6	GALAXY
LB 11196	09 49 24.	+ 20 58		16.9	FAINT BLUE STAR
LB 11195	09 49 24.	+ 21 48		16.6	FAINT BLUE STAR
ZC 0949.4+2413	09 49 24.	+ 24 13	670		CLUSTER OF GALAXIES
MCG+12-10-007	09 49 24.	+ 69 15 30.	8	17.	GALAXY
ZC 0949.4+7125	09 49 24.	+ 71 25	1140		CLUSTER OF GALAXIES
MCG-01-25-052	09 49 24.	- 06 35	90	13.5	GALAXY
LB 03062	09 49 25.	+ 01 34 54.		16.6	FAINT BLUE STAR
LB 03061	09 49 25.	+ 01 46 30.		16.4	FAINT BLUE STAR
REIZ 0079	09 49 26.	+ 43 04	48	14.7	GALAXY
MCG+06-22-023	09 49 27.	+ 33 18	18	15.	GALAXY
RNGC 3059	09 49 27.	- 73 41		12.0	GALAXY
ZWG 035.073	09 49 30.	+ 04 29		15.4	GALAXY
LB 11198	09 49 30.	+ 24 29		16.0	FAINT BLUE STAR
LB 11197	09 49 30.	+ 25 12		12.5	FAINT BLUE STAR
ZC 0949.5+3947	09 49 30.	+ 39 47	540		CLUSTER OF GALAXIES
MAI 046	09 49 33.	+ 58 44	47		DWARF SPHEROIDAL GALAXY
ZWG 007.048	09 49 36.	+ 01 19		15.0	GALAXY
MCG+00-25-025	09 49 36.	+ 01 19	54	15.0	GALAXY
ZC 0949.6+0425	09 49 36.	+ 04 25	670		CLUSTER OF GALAXIES
ZC 0949.6+1010	09 49 36.	+ 10 10	810		CLUSTER OF GALAXIES
ZC 0949.6+2513	09 49 36.	+ 25 13	740		CLUSTER OF GALAXIES
ZC 0949.6+4311	09 49 36.	+ 43 11	670		CLUSTER OF GALAXIES
ZC 0949.6+5207	09 49 36.	+ 52 07	1340		CLUSTER OF GALAXIES
LB 03063	09 49 42.	+ 00 31 54.		16.1	FAINT BLUE STAR
TON-N 0461	09 49 42.	+ 31 23		16.1	BLUE STAR
ZWG 211.001	09 49 42.	+ 43 05		14.5	GALAXY
ZWG 210.044	09 49 42.	+ 43 05		14.5	GALAXY
UGC 05295	09 49 42.	+ 43 05	138	14.5	GALAXY SBb/Sb
ZWG 239.038	09 49 42.	+ 47 10		15.3	GALAXY
MCG+10-14-046	09 49 42.	+ 58 42	39	16.	GALAXY
UGC 05296	09 49 42.	+ 58 43	78	16.5	GALAXY DWRF SP
ZC 0949.7+7107	09 49 42.	+ 71 07	740		CLUSTER OF GALAXIES
MCG-01-25-053	09 49 42.	- 09 17	60	15.	GALAXY
MCG-03-25-026	09 49 42.	- 20 37	18	15.	GALAXY
MCG-05-24-002	09 49 42.	- 32 52	96	14.	GALAXY
LB 03064	09 49 43.	- 00 30 36.		16.2	FAINT BLUE STAR
MCG+03-25-035	09 49 45.	+ 16 00	15	15.3	GALAXY
ZWG 063.091	09 49 48.	+ 11 27		15.7	GALAXY
ZWG 092.063	09 49 48.	+ 16 01		15.3	GALAXY
ZWG 092.064	09 49 48.	+ 16 04		14.8	GALAXY
MCG+03-25-036	09 49 48.	+ 16 04	24	14.8	GALAXY
LB 11199	09 49 48.	+ 21 06		17.2	FAINT BLUE STAR
ZC 0949.8+2355	09 49 48.	+ 23 55	940		CLUSTER OF GALAXIES
ZC 0949.8+4048	09 49 48.	+ 40 48	740		CLUSTER OF GALAXIES
MCG+07-20-073	09 49 48.	+ 43 05 30.	120	13.5	GALAXY
SHAH 107	09 49 48.	+ 50 25	240	18.2	GROUP OF COMPACT GALAXIES
ZWG 007.049	09 49 48.	- 01 25		15.5	GALAXY
MCG+00-25-026	09 49 48.	- 01 25	24	15.5	GALAXY
IC 0571	09 49 49.	+ 16 00 38.			NONSTELLAR OBJECT
IC 0572	09 49 50.	+ 16 03 44.			NONSTELLAR OBJECT
REIZ 0081	09 49 50.	+ 43 05	36	13.7	GALAXY
ZL 077	09 49 53.	+ 69 23 54.		17.0	ULTRAFAINT BLUE STAR
ZWG 007.051	09 49 54.	+ 02 22		14.4	GALAXY
UGC 05297	09 49 54.	+ 02 22	66	14.4	GALAXY Sa-b
MCG+00-25-027	09 49 54.	+ 02 22	66	14.4	GALAXY
RNGC 3039	09 49 54.	+ 02 23		14.5	GALAXY
ZWG 035.074	09 49 54.	+ 05 05		15.4	GALAXY
ZWG 063.092	09 49 54.	+ 10 07		14.9	GALAXY
MCG+02-25-051	09 49 54.	+ 10 07	24	14.9	GALAXY
TON-N 0462	09 49 54.	+ 25 38		15.4	BLUE STAR
ZC 0949.9+4409	09 49 54.	+ 44 09	740		CLUSTER OF GALAXIES
ZWG 007.050	09 49 54.	- 02 11		15.6	GALAXY
OCL 0797	09 49 54.	- 56 04	360		OPEN STAR CLUSTER
ZL 078	09 49 55.	+ 69 16 30.		20.0	ULTRAFAINT BLUE STAR
ZWG 063.093	09 50 00.	+ 14 26		15.3	GALAXY
LB 11201	09 50 00.	+ 21 33		16.4	FAINT BLUE STAR
LB 11200	09 50 00.	+ 25 56		16.4	FAINT BLUE STAR
UGC 05298	09 50 00.	+ 82 32	60	16.5	GALAXY Sc-IRR
ZWG 007.052	09 50 00.	- 02 13		15.5	GALAXY
MCG+00-25-029	09 50 00.	- 02 13	36	15.5	GALAXY
MCG+00-25-028	09 50 00.	- 02 13	60	15.5	GALAXY
TON-N 1112	09 50 01.	+ 20 11		15.5	BLUE STAR
UGC 05299	09 50 06.	+ 00 02	60	17.	GALAXY
ZWG 092.065	09 50 06.	+ 19 51		15.2	GALAXY
ZWG 123.001	09 50 06.	+ 23 17		15.5	GALAXY
ZWG 122.099	09 50 06.	+ 23 17		15.5	GALAXY
LB 11202	09 50 06.	+ 25 02		17.3	FAINT BLUE STAR
MRK 410	09 50 06.	+ 37 59	12	16.	GALAXY WITH UV CONTINUUM
ZL 079	09 50 06.	+ 69 23 36.		19.8	ULTRAFAINT BLUE STAR
ZL 080	09 50 08.	+ 69 10 24.		18.6	ULTRAFAINT BLUE STAR
ZL 082	09 50 09.	+ 69 16 54.		20.2	ULTRAFAINT BLUE STAR
ZL 081	09 50 09.	+ 69 22 24.		18.5	ULTRAFAINT BLUE STAR
MCG-01-25-054	09 50 09.	- 07 24	30	14.	GALAXY
ZWG 007.053	09 50 12.	+ 00 10		15.4	GALAXY
KARA.68 060	09 50 12.	+ 11 23	27		DWARF GALAXY
ZWG 092.066	09 50 12.	+ 19 36		15.4	GALAXY
ZWG 092.067	09 50 12.	+ 19 40		14.2	GALAXY
UGC 05300	09 50 12.	+ 19 40	96	14.2	GALAXY DBL SYS
ZWG 153.001	09 50 12.	+ 27 51		15.7	GALAXY
ZWG 152.078	09 50 12.	+ 27 51		15.7	GALAXY
KARA.73B 0380	09 50 12.	+ 27 51	24	15.7	ISOLATED GALAXY E
UGC 05301	09 50 12.	+ 43 05	84	16.0	GALAXY Sc
ZWG 265.046	09 50 12.	+ 52 39		15.3	GALAXY
RNGC 3040	09 50 13.	+ 19 40		14.0	GALAXY
LB 03065	09 50 13.	- 02 05 42.		15.7	FAINT BLUE STAR
LB 03066	09 50 14.	+ 01 13 48.		16.1	FAINT BLUE STAR
HOLM 148B	09 50 15.	+ 19 41	18	15.3	PART OF MULTIPLE GALAXY
MCG+07-20-074	09 50 15.	+ 43 05	72	15.	GALAXY
MCG+09-16-057	09 50 15.	+ 52 40	18	16.	GALAXY
ARC 0872	09 50 16.	+ 77 31		17.6	RICH CLUSTER OF GALAXIES
HOLM 148A	09 50 17.	+ 19 40	18	14.5	PART OF MULTIPLE GALAXY
SEY 062	09 50 17.	+ 68 34 29.		15.0	FAINT GALAXY
MCG+03-25-039	09 50 18.	+ 16 54	240	13.1	GALAXY
MCG+03-25-038	09 50 18.	+ 19 35	39	15.4	GALAXY
MCG+03-25-037	09 50 18.	+ 19 40	36	14.2	GALAXY
LB 11204	09 50 18.	+ 24 49		16.7	FAINT BLUE STAR
LB 11203	09 50 18.	+ 25 24		16.5	FAINT BLUE STAR
TON-N 1113	09 50 18.	+ 35 18		17.	BLUE STAR
ZWG 182.028	09 50 18.	+ 36 18		14.9	GALAXY
SN 1963U	09 50 18.	+ 36 18		15.0	SUPERNOVA
ZWG 333.005	09 50 18.	+ 68 34		15.1	GALAXY
UGC 05302	09 50 18.	+ 68 34	132	15.1	GALAXY
MCG+12-10-008	09 50 18.	+ 68 35	102	15.	GALAXY
KN 15.001	09 50 19.4	+ 36 19 18.			NEBULA
REIZ 0082	09 50 20.	+ 43 05	72	14.8	GALAXY
ZL 083	09 50 20.	+ 69 23 24.		19.5	ULTRAFAINT BLUE STAR
MCG+06-22-024	09 50 21.	+ 36 20	36	14.5	GALAXY
ZWG 035.075	09 50 24.	+ 08 05		15.2	GALAXY
ZWG 092.068	09 50 24.	+ 16 55		13.1	GALAXY
UGC 05303	09 50 24.	+ 16 55	228	13.1	GALAXY Sc
LB 11206	09 50 24.	+ 24 26		16.1	FAINT BLUE STAR
LB 11205	09 50 24.	+ 25 20		17.0	FAINT BLUE STAR
ZL 084	09 50 24.	+ 69 17 24.		20.0	ULTRAFAINT BLUE STAR
RNGC 3041	09 50 26.	+ 16 55		12.5	GALAXY
ARP 255	09 50 27.	+ 08 07			PECULIAR GALAXY
ARC 0890	09 50 29.	- 04 37		17.2	RICH CLUSTER OF GALAXIES
ZC 0950.5+0649	09 50 30.	+ 06 49	1880		CLUSTER OF GALAXIES

OBJECT NAME	RIGHT ASCEN.	DECLINATION	DIAM.	MAGN.	TYPE OF OBJECT
ZWG 035.076	09 50 30.	+ 08 07		14.8	GALAXY
UGC 05304	09 50 30.	+ 08 07	72	14.8	GALAXY S+COMP
MCG+01-25-030	09 50 30.	+ 08 07	60	14.8	GALAXY
VV 342C	09 50 30.	+ 08 08	48	16.	INTERACTING GALAXY
VV 342B	09 50 30.	+ 08 08	42	15.	INTERACTING GALAXY
VV 342A	09 50 30.	+ 08 08	60	14.	INTERACTING GALAXY
UGC 05305	09 50 30.	+ 10 40	66	16.0	GALAXY S
MCG+08-18-042	09 50 30.	+ 47 09	30	15.	GALAXY
ZC 0950.5+6820	09 50 30.	+ 68 20	1480		CLUSTER OF GALAXIES
ZL 085	09 50 31.	+ 69 20 54.		18.6	ULTRAFAINT BLUE STAR
ARC 0887	09 50 34.	+ 40 35		17.6	RICH CLUSTER OF GALAXIES
LB 00562	09 50 34.	+ 59 15 24.		15.8	FAINT BLUE STAR
HN 0321	09 50 34.	- 33 30			NEBULA
IC 2517	09 50 34.	- 33 30			NONSTELLAR OBJECT
ZC 0950.6+0120	09 50 36.	+ 01 20	270		CLUSTER OF GALAXIES
ZWG 063.094	09 50 36.	+ 11 29		15.0	GALAXY
MCG+02-25-052	09 50 36.	+ 11 29	36	15.0	GALAXY
ZC 0950.6+3250	09 50 36.	+ 32 50	2290		CLUSTER OF GALAXIES
ZWG 289.022	09 50 36.	+ 58 35		15.5	GALAXY
UGC 05306	09 50 36.	+ 58 35	72	15.5	GALAXY SBb
MCG+10-14-047	09 50 36.	+ 58 38	39	15.	GALAXY
ZC 0950.6+6045	09 50 36.	+ 60 45	2080		CLUSTER OF GALAXIES
ZL 086	09 50 38.	+ 69 10 24.		18.4	ULTRAFAINT BLUE STAR
ARC 0889	09 50 40.	+ 23 01		17.6	RICH CLUSTER OF GALAXIES
ZWG 123.002	09 50 42.	+ 24 28		15.4	GALAXY
ZWG 122.100	09 50 42.	+ 24 28		15.4	GALAXY
MCG+05-24-001	09 50 42.	+ 30 21	36	15.	GALAXY
MCG+10-14-048	09 50 42.	+ 58 34	45	15.	GALAXY
LB 03067	09 50 43.	+ 02 06 00.		15.6	FAINT BLUE STAR
MCG-03-25-027	09 50 45.	- 15 44	36	15.5	GALAXY
LB 03068	09 50 46.	+ 01 31 24.		16.3	FAINT BLUE STAR
ZWG 007.054	09 50 48.	+ 00 56		13.8	GALAXY
UGC 05307	09 50 48.	+ 00 56	66	13.8	GALAXY S0
MCG+00-25-030	09 50 48.	+ 00 56	54	13.8	GALAXY
ZWG 007.055	09 50 48.	+ 02 12		15.6	GALAXY
ZWG 035.077	09 50 48.	+ 07 18		15.5	GALAXY
ZWG 035.078	09 50 48.	+ 08 07		15.0	GALAXY
UGC 05308	09 50 48.	+ 08 07	72	15.0	GALAXY S
MCG+01-25-031	09 50 48.	+ 08 07	48	15.0	GALAXY
ZC 0950.8+2258	09 50 48.	+ 22 58	1610		CLUSTER OF GALAXIES
LB 11207	09 50 48.	+ 25 49		16.3	FAINT BLUE STAR
ZC 0950.8+4833	09 50 48.	+ 48 33	2020		CLUSTER OF GALAXIES
MRK 127	09 50 48.	+ 51 29	8	16.5	GALAXY WITH UV CONTINUUM
MCG+13-07-039	09 50 48.	+ 79 05	21	15.	GALAXY
MCG-03-25-028	09 50 48.	- 18 25 30.	66	14.	GALAXY
RNGC 3042	09 50 49.	+ 00 56		14.0	GALAXY
RNGC 3045	09 50 49.	- 18 25		14.0	GALAXY
ARC 0886	09 50 51.	+ 58 08		17.5	RICH CLUSTER OF GALAXIES
REIZ 0083	09 50 53.	+ 28 42	6	16.0	GALAXY
ZWG 063.095	09 50 54.	+ 12 21		15.5	GALAXY
ZC 0950.9+1843	09 50 54.	+ 18 43	1480		CLUSTER OF GALAXIES
MCG+10-14-049	09 50 54.	+ 57 07	30	16.	GALAXY
ZC 0950.9+5811	09 50 54.	+ 58 11	1410		CLUSTER OF GALAXIES
MCG+10-14-050	09 50 54.	+ 59 20	24	16.	GALAXY
ARC 0892	09 50 59.	+ 00 48		17.1	RICH CLUSTER OF GALAXIES
MCG+01-25-032	09 51 00.	+ 02 38	60	15.	GALAXY
ZWG 035.079	09 51 00.	+ 03 43		15.6	GALAXY
ZWG 063.096	09 51 00.	+ 09 47		15.4	GALAXY
ZWG 063.097	09 51 00.	+ 11 26		15.5	GALAXY
ZC 0951.0+1229	09 51 00.	+ 12 29	740		CLUSTER OF GALAXIES
ZWG 092.069	09 51 00.	+ 17 20		15.6	GALAXY
ZC 0951.0+3303	09 51 00.	+ 33 03	340		CLUSTER OF GALAXIES
ZC 0951.0+3407	09 51 00.	+ 34 07	2420		CLUSTER OF GALAXIES
ZC 0951.0+3617	09 51 00.	+ 36 17	470		CLUSTER OF GALAXIES
SHAH 108	09 51 00.	+ 50 21	90	17.5	GROUP OF COMPACT GALAXIES
ZWG 265.047	09 51 00.	+ 53 53		15.6	GALAXY
ZWG 351.001	09 51 00.	+ 79 32		15.7	GALAXY
ZWG 350.034	09 51 00.	+ 79 32		15.7	GALAXY
UGC 05310	09 51 00.	+ 79 32	72	15.7	GALAXY SBb
UGC 05309	09 51 00.	+ 80 59	84	16.0	GALAXY
MCG-04-24-003	09 51 00.	- 25 41 30.	18	15.	GALAXY
TON-N 1114	09 51 01.	+ 20 16		15.8	BLUE STAR
ZL 087	09 51 02.	+ 69 05 42.		17.5	ULTRAFAINT BLUE STAR
RNGC 3046	09 51 04.	- 27 05			NON-EXISTENT OBJECT
ARC 0877	09 51 05.	+ 75 38		17.7	RICH CLUSTER OF GALAXIES
RNGC 3058	09 51 05.	- 12 16		15.0	GALAXY
ZWG 007.056	09 51 06.	+ 01 48		12.4	GALAXY
UGC 05311	09 51 06.	+ 01 48	282	12.4	GALAXY Sc
MCG+00-25-031	09 51 06.	+ 01 48	270	12.4	GALAXY
ZWG 035.080	09 51 06.	+ 02 37		15.5	GALAXY
UGC 05312	09 51 06.	+ 02 37	60	15.5	GALAXY Sc
ZWG 063.098	09 51 06.	+ 09 25		15.6	GALAXY
KARA.73B 0381	09 51 06.	+ 09 25	24	15.6	ISOLATED GALAXY S
ZWG 123.003	09 51 06.	+ 23 37		14.5	GALAXY
UGC 05313	09 51 06.	+ 23 37	48	14.5	GALAXY
MCG+04-24-001	09 51 06.	+ 23 37	36	14.5	GALAXY
7ZW 299	09 51 06.	+ 67 56			COMPACT GALAXY
MCG+12-10-009	09 51 06.	+ 72 28	270	12.3	GALAXY
MCG+13-07-040	09 51 06.	+ 76 06	84	13.	GALAXY
RNGC 3044	09 51 07.	+ 01 49		12.5	GALAXY
ZL 088	09 51 07.	+ 69 14 06.		18.5	ULTRAFAINT BLUE STAR
ARC 0885	09 51 08.	+ 62 43		17.7	RICH CLUSTER OF GALAXIES
IC 0573	09 51 08.	- 12 14 53.			NONSTELLAR OBJECT
MCG-02-25-026	09 51 09.	- 12 16 30.	72	15.	GALAXY
MCG-03-25-029	09 51 09.	- 19 23	78	13.5	GALAXY
ZWG 063.099	09 51 12.	+ 09 07		15.7	GALAXY
UGC 05314	09 51 12.	+ 09 07	66	15.7	GALAXY Sc
ZWG 063.100	09 51 12.	+ 13 52		15.0	GALAXY
MCG+02-25-053	09 51 12.	+ 13 52	24	15.0	GALAXY
ZC 0951.2+2842	09 51 12.	+ 28 42	940		CLUSTER OF GALAXIES
ZWG 153.002	09 51 12.	+ 31 51		15.4	GALAXY
ZWG 152.079	09 51 12.	+ 31 51		15.4	GALAXY
ZWG 182.029	09 51 12.	+ 33 37		15.7	GALAXY
ZWG 182.030	09 51 12.	+ 34 22		15.4	GALAXY
UGC 05315	09 51 12.	+ 37 32	72	15.6	GALAXY Sc
ZWG 182.031	09 51 12.	+ 37 46		15.6	GALAXY
BEM 1	09 51 12.	+ 68 50	270	14.7	SPHEROIDAL GALAXY
ZWG 333.006	09 51 12.	+ 72 27		12.3	GALAXY
ZWG 332.068	09 51 12.	+ 72 27		12.3	GALAXY
UGC 05316	09 51 12.	+ 72 27	300	12.3	GALAXY SBc
ZC 0951.2+7726	09 51 12.	+ 77 26	1010		CLUSTER OF GALAXIES
REIZ 0085	09 51 13.	+ 13 51	24	14.6	GALAXY
REIZ 0084	09 51 13.	+ 23 38	24	14.2	GALAXY
KEEL 192	09 51 13.3	+ 68 51 24.			NEBULA
KN 15.002	09 51 14.8	+ 34 22 36.			NEBULA
MCG+06-22-025	09 51 15.	+ 37 35	66	15.	GALAXY
ZL 089	09 51 15.	+ 69 10 18.		20.0	ULTRAFAINT BLUE STAR
AR 06B	09 51 17.48	+ 76 06 11.8			NEBULA
ZC 0951.3+0052	09 51 18.	+ 00 52	1480		CLUSTER OF GALAXIES
LB 11208	09 51 18.	+ 23 05		16.9	FAINT BLUE STAR
MCG+05-24-002	09 51 18.	+ 31 52	42	15.4	GALAXY
TON-N 1116	09 51 18.	+ 33 50		16.9	BLUE STAR
ZC 0951.3+7155	09 51 18.	+ 71 55	870		CLUSTER OF GALAXIES
RNGC 3027	09 51 18.	+ 72 26		12.5	GALAXY
ZWG 350.035	09 51 18.	+ 75 15		15.5	GALAXY
UGC 05317	09 51 18.	+ 75 15	84	15.5	GALAXY S
MCG-01-25-055	09 51 18.	- 09 12	36	15.	GALAXY
AR 06A	09 51 18.56	+ 76 06 10.7			NEBULA
TON-N 1115	09 51 19.	+ 19 28		15.9	BLUE STAR
REIZ 0086	09 51 20.	+ 24 34	6	16.0	GALAXY
RNGC 3061	09 51 20.	+ 76 06		14.0	GALAXY
LB 03069	09 51 20.	- 01 38 18.		16.0	FAINT BLUE STAR
MCG+06-22-026	09 51 21.	+ 34 23	30	15.	GALAXY
ZWG 063.101	09 51 24.	+ 10 51		15.0	GALAXY
MCG+02-25-054	09 51 24.	+ 10 51	12	15.0	GALAXY
8ZW 0951+14.1	09 51 24.	+ 14 05		18.9	COMPACT GALAXY
ZWG 092.070	09 51 24.	+ 19 58		15.6	GALAXY
LB 11209	09 51 24.	+ 24 30		16.5	FAINT BLUE STAR
ARC 0891	09 51 24.	+ 28 39		17.6	RICH CLUSTER OF GALAXIES
ZWG 333.007	09 51 24.	+ 69 18		8.1	GALAXY
UGC 05318	09 51 24.	+ 69 18	1560	8.1	GALAXY Sb
KARA.72 218A	09 51 24.	+ 69 18	1368	8.1	PART OF DOUBLE GALAXY
ZWG 350.036	09 51 24.	+ 76 06		13.9	GALAXY
UGC 05319	09 51 24.	+ 76 06	120	13.9	GALAXY SBc
KARA.73B 0382	09 51 24.	+ 76 06	96	13.9	ISOLATED GALAXY S
REIN 2.095	09 51 25.97	+ 69 54 52.8			NEBULA
KEEL 193	09 51 29.6	+ 68 51 21.			NEBULA
ZWG 035.081	09 51 30.	+ 03 25		15.6	GALAXY
ZC 0951.5+2037	09 51 30.	+ 20 37	2220		CLUSTER OF GALAXIES
ZWG 123.004	09 51 30.	+ 23 31		14.6	GALAXY
UGC 05320	09 51 30.	+ 23 31	126	14.6	GALAXY SBc
MCG+04-24-002	09 51 30.	+ 23 31	96	14.6	GALAXY
ZWG 239.039	09 51 30.	+ 46 37		15.6	GALAXY
ZC 0951.5+6243	09 51 30.	+ 62 43	1950		CLUSTER OF GALAXIES
MCG+12-10-010	09 51 30.	+ 69 18	1200	7.8	GALAXY
ZC 0951.5-0048	09 51 30.	- 00 48	2290		CLUSTER OF GALAXIES
HOLM 149A	09 51 31.	+ 23 31	108	13.1	PART OF MULTIPLE GALAXY
REIZ 0087	09 51 31.	+ 23 32	90	14.6	GALAXY
LB 03070	09 51 33.	+ 01 51 00.		16.0	FAINT BLUE STAR
RNGC 3031	09 51 33.	+ 69 18		8.5	GALAXY
HOLM 149B	09 51 34.	+ 23 29	30	14.4	PART OF MULTIPLE GALAXY
RNGC 3051	09 51 34.	- 27 03		14.0	GALAXY
ZWG 007.057	09 51 36.	+ 00 53		15.7	GALAXY
ZWG 035.082	09 51 36.	+ 07 23		15.2	GALAXY
ZC 0951.6+3154	09 51 36.	+ 31 54	1410		CLUSTER OF GALAXIES
ZWG 182.032	09 51 36.	+ 36 54		15.6	GALAXY
ZWG 182.033	09 51 36.	+ 37 38		15.1	GALAXY
UGC 05321	09 51 36.	+ 37 38	66	15.1	GALAXY Sb
MCG+06-22-027	09 51 36.	+ 37 40	66	14.5	GALAXY
MCG-04-24-004	09 51 36.	- 27 03 30.	30	14.	GALAXY
TON-N 1117	09 51 37.	+ 20 02		17.	BLUE STAR
TON-N 1118	09 51 37.	+ 21 22		17.	BLUE STAR
TON-N 1119	09 51 37.	+ 22 17		16.8	BLUE STAR
REIZ 0089	09 51 37.	+ 23 30	12	15.8	GALAXY
FATH 1.297	09 51 38.	+ 00 11	16		NEBULA
REIN 2.096	09 51 38.45	+ 69 55 01.7			NEBULA
IC 2515	09 51 39.	+ 37 39 43.			NONSTELLAR OBJECT
LB 03071	09 51 39.	- 00 53 30.		16.3	FAINT BLUE STAR
KN 15.003	09 51 39.3	+ 37 38 48.			NEBULA
ARC 0881	09 51 40.	+ 71 57		17.7	RICH CLUSTER OF GALAXIES
ZWG 007.058	09 51 42.	+ 00 52		15.3	GALAXY
ZWG 035.083	09 51 42.	+ 02 31		14.9	GALAXY
MCG+01-25-033	09 51 42.	+ 02 31	72	14.9	GALAXY
ZC 0951.7+3126	09 51 42.	+ 31 26	540		CLUSTER OF GALAXIES
MCG+06-22-028	09 51 42.	+ 37 56 30.	24	14.5	GALAXY
ZWG 211.002	09 51 42.	+ 42 03		15.7	GALAXY
ZWG 210.045	09 51 42.	+ 42 03		15.7	GALAXY
ZC 0951.7+4742	09 51 42.	+ 47 42	610		CLUSTER OF GALAXIES
ZWG 333.008	09 51 42.	+ 69 55		9.2	GALAXY
UGC 05322	09 51 42.	+ 69 55	780	9.2	GALAXY IRR
KARA.72 218B	09 51 42.	+ 69 55	714	9.2	PART OF DOUBLE GALAXY
REIZ 0088	09 51 43.	+ 37 39	48	14.8	GALAXY
RNGC 3034	09 51 43.	+ 69 55		9.5	GALAXY
REIN 2.097	09 51 44.23	+ 69 55 02.3			NEBULA
KN 15.004	09 51 47.5	+ 37 55 32.			NEBULA
ZWG 182.034	09 51 48.	+ 37 55		15.0	GALAXY
IC 2516	09 51 50.	+ 37 54 24.			NONSTELLAR OBJECT
FATH 1.298	09 51 51.	- 00 29	14		NEBULA
REIN 2.098	09 51 51.02	+ 69 55 26.1			NEBULA
RNGC 3050	09 51 52.	- 10 08			NON-EXISTENT OBJECT
KN 15.005	09 51 52.8	+ 34 46 29.			NEBULA
ZC 0951.9+4047	09 51 54.	+ 40 47	810		CLUSTER OF GALAXIES
MCG+12-10-011	09 51 54.	+ 69 56	504	9.2	GALAXY
MCG-01-25-056	09 51 54.	- 06 43 30.	30	15.	GALAXY
ARP 337	09 51 56.	+ 69 57			PECULIAR GALAXY
MCG-01-25-057	09 51 57.	- 06 38 30.	24	17.	GALAXY
IC 0574	09 51 57.	- 06 43 08.			NONSTELLAR OBJECT
ZWG 035.084	09 52 00.	+ 06 52		15.6	GALAXY
ZC 0952.0+4622	09 52 00.	+ 46 22	610		CLUSTER OF GALAXIES
ZC 0952.0+7312	09 52 00.	+ 73 12	740		CLUSTER OF GALAXIES
ZWG 007.059	09 52 00.	- 01 04		14.2	GALAXY
UGC 05323	09 52 00.	- 01 04	54	14.2	GALAXY DBL SYS
MCG+00-25-033	09 52 00.	- 01 04	24	14.2	GALAXY
MCG+00-25-032	09 52 00.	- 01 04	12	15.5	GALAXY
MCG-01-25-058	09 52 00.	- 06 38	60	14.	GALAXY
MCG-03-25-030	09 52 00.	- 18 25 30.	108	12.5	GALAXY
RNGC 3052	09 52 01.	- 18 24		13.0	GALAXY
RNGC 3047A	09 52 02.	- 01 04			GALAXY
RNGC 3047	09 52 02.	- 01 04		14.0	GALAXY
LB 03072	09 52 04.	+ 00 16 24.		16.5	FAINT BLUE STAR
IC 0575	09 52 04.	- 06 37 21.			NONSTELLAR OBJECT
RNGC 3054	09 52 04.	- 25 28		12.5	GALAXY
KN 15.006	09 52 04.2	+ 37 54 02.			NEBULA
REIN 2.099	09 52 04.65	+ 69 55 53.0			NEBULA
ZWG 035.085	09 52 06.	+ 07 15		15.7	GALAXY
ZWG 063.102	09 52 06.	+ 13 37		15.3	GALAXY
UGC 05324	09 52 06.	+ 13 37	72	15.3	GALAXY S
VV 111B	09 52 06.	- 06 37	30	16.	INTERACTING GALAXY
VV 111A	09 52 06.	- 06 37	30	10.	INTERACTING GALAXY
VV 111	09 52 06.	- 06 37	60		INTERACTING GALAXY
ARP 292	09 52 06.	- 06 37			PECULIAR GALAXY
RNGC 3049	09 52 10.	+ 09 30		13.5	GALAXY
BC A00952+17	09 52 11.84	+ 17 57 44.5		17.23	QUASI-STELLAR OBJECT
SHB 184	09 52 11.9	+ 17 57 44.		17.2	QUASI-STELLAR OBJECT
ZWG 063.103	09 52 12.	+ 09 30		13.5	GALAXY
UGC 05325	09 52 12.	+ 09 30	150	13.5	GALAXY SBb

OBJECT NAME	RIGHT ASCEN.	DECLINATION	DIAM.	MAGN.	TYPE OF OBJECT
MCG+02-25-055	09 52 12.	+ 09 30	120	13.5	GALAXY
KARA.73B 0383	09 52 12.	+ 09 30	126	13.5	ISOLATED GALAXY S
ZC 0952.2+1625	09 52 12.	+ 16 25	810		CLUSTER OF GALAXIES
ZWG 092.071	09 52 12.	+ 16 42		15.2	GALAXY
ZWG 153.003	09 52 12.	+ 31 44		15.3	GALAXY
MCG+07-21-001	09 52 12.	+ 41 23	27	17.	GALAXY
MCG+09-16-058	09 52 12.	+ 54 50	42	16.	GALAXY
ZC 0952.2+5958	09 52 12.	+ 59 58	2490		CLUSTER OF GALAXIES
MCG-03-25-031	09 52 12.	- 19 40 30.	48	15.	GALAXY
MCG-04-24-005	09 52 12.	- 25 27 30.	240	12.5	GALAXY
LB 03073	09 52 13.	+ 02 14 12.		16.8	FAINT BLUE STAR
TON-N 1120	09 52 13.	+ 20 05		14.9	BLUE STAR
PNGC 3048	09 52 14.	+ 16 42		15.0	GALAXY
LB 03074	09 52 14.	- 00 19 42.		15.7	FAINT BLUE STAR
MCG-01-25-059	09 52 15.	- 06 34	60	15.5	GALAXY
LB 03075	09 52 16.	+ 01 48 12.		16.6	FAINT BLUE STAR
HOLM 150A	09 52 17.	+ 41 22	48	14.3	PART OF MULTIPLE GALAXY
RNGC 3056	09 52 17.	- 28 04		13.5	GALAXY
SHB 145	09 52 17.2	+ 09 44 08.		17.2	QUASI-STELLAR OBJECT
BC 4C09.35	09 52 17.6	+ 09 44 19.		17.24	QUASI-STELLAR OBJECT
HOLM 150B	09 52 18.	+ 41 20	18	15.1	PART OF MULTIPLE GALAXY
ZWG 211.003	09 52 18.	+ 41 23		15.7	GALAXY
ZC 0952.3+5136	09 52 18.	+ 51 36	1680		CLUSTER OF GALAXIES
TON-N 1121	09 52 19.	+ 21 21		17.	BLUE STAR
LB 03076	09 52 22.	- 00 41 06.		16.2	FAINT BLUE STAR
SVEN 234	09 52 23.	- 32 53	54	14.8	GALAXY
ZC 0952.4+0946	09 52 24.	+ 09 46	940		CLUSTER OF GALAXIES
ZWG 063.104	09 52 24.	+ 11 16		15.4	GALAXY
ZWG 092.072	09 52 24.	+ 14 32		15.1	GALAXY
KARA.72 219A	09 52 24.	+ 14 32	36	15.1	PART OF DOUBLE GALAXY
ZC 0952.4+1829	09 52 24.	+ 18 29	610		CLUSTER OF GALAXIES
ZWG 182.035	09 52 24.	+ 33 30		14.5	GALAXY
UGC 05326	09 52 24.	+ 33 30	66	14.5	GALAXY IRR
TON-N 1123	09 52 24.	+ 34 28		16.3	BLUE STAR
MCG+07-21-002	09 52 24.	+ 41 23	45	15.	GALAXY
ZC 0952.4+4328	09 52 24.	+ 43 28	670		CLUSTER OF GALAXIES
PATH 1.299	09 52 24.	- 00 31	8		NEBULA
MCG-05-24-003	09 52 24.	- 28 03	96	13.	GALAXY
REIZ 0091	09 52 25.	+ 14 33	36	14.6	GALAXY
TON-N 1122	09 52 25.	+ 21 41		17.1	BLUE STAR
LB 03077	09 52 25.	- 00 01 42.		15.4	FAINT BLUE STAR
IC 0576	09 52 26.	+ 11 16 32.			NONSTELLAR OBJECT
LB 03078	09 52 26.	- 00 29 36.		15.6	FAINT BLUE STAR
KN 15.007	09 52 27.9	+ 33 29 54.			NEBULA
ZWG 035.086	09 52 30.	+ 06 47		15.5	GALAXY
ZWG 092.073	09 52 30.	+ 14 33		15.4	GALAXY
KARA.72 219B	09 52 30.	+ 14 33	30	15.4	PART OF DOUBLE GALAXY
ZC 0952.5+3841	09 52 30.	+ 38 41	400		CLUSTER OF GALAXIES
ZC 0952.5+4232	09 52 30.	+ 42 32	1080		CLUSTER OF GALAXIES
ZWG 239.040	09 52 30.	+ 47 37		15.6	GALAXY
ZWG 265.048	09 52 30.	+ 53 55		15.3	GALAXY
KARA.73B 0384	09 52 30.	+ 53 55	36	15.3	ISOLATED GALAXY S
MCG-06-22-002	09 52 30.	- 33 47	15	15.	GALAXY
LB 03079	09 52 31.	+ 02 29 00.		16.8	FAINT BLUE STAR
REIZ 0092	09 52 31.	+ 14 34	30	14.6	GALAXY
MCG+06-22-029	09 52 33.	+ 33 30	60	14.	GALAXY
REIZ 0090	09 52 33.	+ 33 31	54	14.2	GALAXY
ZC 0952.6+3622	09 52 36.	+ 36 22	2350		CLUSTER OF GALAXIES
MCG+09-16-059	09 52 36.	+ 53 56 30.	48	16.	GALAXY
ZWG 289.023	09 52 36.	+ 59 32		13.3	GALAXY
UGC 05327	09 52 36.	+ 59 32	114	13.3	GALAXY S
KARA.73B 0385	09 52 36.	+ 59 32	102	13.3	ISOLATED GALAXY S
KARA.68 061	09 52 36.	+ 68 48	107		DWARF GALAXY
ZL 090	09 52 38.	+ 69 11 48.		18.0	ULTRAFAINT BLUE STAR
RNGC 3055	09 52 42.	+ 04 30		13.0	GALAXY
ZWG 035.087	09 52 42.	+ 04 31		12.3	GALAXY
UGC 05328	09 52 42.	+ 04 31	132	12.3	GALAXY Sc
MCG+01-25-034	09 52 42.	+ 04 31	120	12.3	GALAXY
ARC 0893	09 52 42.	+ 36 13		17.8	RICH CLUSTER OF GALAXIES
ZWG 182.036	09 52 42.	+ 37 15		15.5	GALAXY
MCG+06-22-030	09 52 42.	+ 37 17	42	15.	GALAXY
ZC 0952.7+4413	09 52 42.	+ 44 13	2760		CLUSTER OF GALAXIES
MCG+10-14-051	09 52 42.	+ 58 28	18	16.	GALAXY
MCG+10-14-052	09 52 42.	+ 59 32	102	14.	GALAXY
LB 03080	09 52 43.	+ 02 32 42.		16.7	FAINT BLUE STAR
RNGC 3043	09 52 43.	+ 59 33		13.5	GALAXY
KN 15.008	09 52 43.7	+ 37 14 55.			NEBULA
ZWG 035.088	09 52 48.	+ 02 50		15.6	GALAXY
ZWG 035.089	09 52 48.	+ 03 25		15.7	GALAXY
HMS 1.12	09 52 48.	+ 08 37			PEC GALAXY
ZWG 063.105	09 52 48.	+ 08 37		15.3	GALAXY
MCG+02-25-056	09 52 48.	+ 08 37	48	15.3	GALAXY
ZWG 093.001	09 52 48.	+ 16 40		13.7	GALAXY
ZWG 092.074	09 52 48.	+ 16 40		13.7	GALAXY
UGC 05329	09 52 48.	+ 16 40	108	13.7	GALAXY SB?a
ZWG 093.002	09 52 48.	+ 19 33		14.8	GALAXY
ZWG 092.075	09 52 48.	+ 19 33		14.8	GALAXY
UGC 05330	09 52 48.	+ 19 33	120	14.8	GALAXY PECULR
ZWG 182.037	09 52 48.	+ 32 40		15.6	GALAXY
UGC 05331	09 52 48.	+ 32 40	60	15.6	GALAXY Sc
TON-N 1125	09 52 48.	+ 33 51		17.	BLUE STAR
ZC 0952.8+4102	09 52 48.	+ 41 02	4030		CLUSTER OF GALAXIES
OCL 0785	09 52 49.	- 50 29	1020	9.4	OPEN STAR CLUSTER
TON-N 1124	09 52 49.	+ 22 15		16.9	BLUE STAR
RNGC 3053	09 52 50.	+ 16 40		13.5	GALAXY
KN 15.010	09 52 50.2	+ 32 39 48.			NEBULA
MCG+03-25-040	09 52 51.	+ 16 39	108	13.7	GALAXY
MCG+06-22-031	09 52 51.	+ 37 30	18	16.	GALAXY
LB 03081	09 52 51.	- 01 11 30.		15.8	FAINT BLUE STAR
KN 15.009	09 52 51.7	+ 37 23 36.			NEBULA
MCG+03-25-041	09 52 54.	+ 19 33	33	14.8	GALAXY
MCG+06-22-033	09 52 54.	+ 32 40	60	14.5	GALAXY
ZWG 182.038	09 52 54.	+ 36 12		15.7	GALAXY
MCG+06-22-032	09 52 54.	+ 36 13	48	15.	GALAXY
TON-N 1126	09 52 55.	+ 22 27		16.1	BLUE STAR
ARC 0894	09 52 55.	+ 36 24		17.8	RICH CLUSTER OF GALAXIES
ZL 091	09 52 55.	+ 69 23 42.		19.8	ULTRAFAINT BLUE STAR
PATH 1.300	09 52 56.	- 00 03	8		NEBULA
IC 2518	09 52 58.	+ 37 23 49.			NONSTELLAR OBJECT
IC 2522	09 52 58.0	- 32 54 00.	150	12.9	GALAXY SB
KN 15.011	09 52 58.7	+ 37 23 33.			NEBULA
IC 2523	09 52 58.7	- 32 58 24.			GALAXY
SVEN 235	09 52 59.	- 32 54	96	12.2	GALAXY
HN 0322	09 52 59.	- 32 54			NEBULA
SVEN 236	09 52 59.	- 32 58	54	13.1	GALAXY
HN 0323	09 52 59.	- 32 59			NEBULA
VDB.66G 068	09 53	+ 29 03	130		DWARF GALAXY
BEM 2	09 53	+ 69 18	270	14.3	MAGELLENIC GALAXY

OBJECT NAME	RIGHT ASCEN.	DECLINATION	DIAM.	MAGN.	TYPE OF OBJECT
VDB.66G 066	09 53	+ 69 18	100		DWARF GALAXY
ZWG 036.001	09 53 00.	+ 05 16		15.3	GALAXY
ZWG 035.090	09 53 00.	+ 05 16		15.3	GALAXY
UGC 05332	09 53 00.	+ 16 39	72	17.	GALAXY DWARF
ZWG 182.039	09 53 00.	+ 34 16		15.6	GALAXY
ZWG 182.040	09 53 00.	+ 37 23		15.6	GALAXY
ZC 0953.0+6651	09 53 00.	+ 66 51	610		CLUSTER OF GALAXIES
MCG+14-05-008	09 53 00.	+ 80 58	54	16.	GALAXY
7ZW 300	09 53 00.	+ 82 07			COMPACT GALAXY
MCG-02-26-001	09 53 00.	- 13 35	30	15.	GALAXY
MCG-04-24-006	09 53 00.	- 22 13 30.	60	15.	GALAXY
MCG-05-24-004	09 53 00.	- 32 55	120	13.5	GALAXY
KN 15.012	09 53 01.7	+ 34 16 31.			NEBULA
IC 2519	09 53 02.	+ 34 17 01.			NONSTELLAR OBJECT
LB 03082	09 53 03.	+ 00 09 18.		18.2	FAINT BLUE STAR
PK280-02.1	09 53 05.	- 57 05	10		PLANETARY NEBULA
ZC 0953.1+0028	09 53 06.	+ 00 28	4230		CLUSTER OF GALAXIES
ZWG 036.002	09 53 06.	+ 03 37		15.7	GALAXY
8ZW 0953+12.4	09 53 06.	+ 12 23		16.6	COMPACT GALAXY
MCG+03-26-001	09 53 06.	+ 16 39	66	17.	GALAXY
ZC 0953.1+7406	09 53 06.	+ 74 06	740		CLUSTER OF GALAXIES
MCG-05-24-005	09 53 06.	- 33 00	60	14.	GALAXY
LB 03083	09 53 08.	+ 01 01 12.		15.7	FAINT BLUE STAR
LB 03084	09 53 09.	- 01 45 00.		16.2	FAINT BLUE STAR
RNGC 3064	09 53 09.	- 06 08		14.5	GALAXY
ZWG 036.003	09 53 12.	+ 06 48		15.5	GALAXY
STOCK 05	09 53 12.	+ 37 49			BLUE KNOT NEAR ELLIP GLXY
ZWG 182.041	09 53 12.	+ 37 49		15.4	GALAXY
ZWG 182.042	09 53 12.	+ 37 53		15.7	GALAXY
ZWG 239.041	09 53 12.	+ 46 25		15.7	GALAXY
UGC 05333	09 53 12.	+ 46 25	60	15.7	GALAXY
MCG+08-18-043	09 53 12.	+ 46 25	60	15.	GALAXY
ZC 0953.2+4942	09 53 12.	+ 49 42	610		CLUSTER OF GALAXIES
MRK 128	09 53 12.	+ 60 20	20	16.	GALAXY WITH UV CONTINUUM
MCG+10-14-053	09 53 12.	+ 60 20	24	16.	GALAXY
MCG-01-26-001	09 53 12.	- 06 08	60	14.5	GALAXY
TON-N 1127	09 53 13.	+ 19 21		16.3	BLUE STAR
KN 15.013	09 53 15.2	+ 36 39 41.			NEBULA
PATH 1.301	09 53 16.	+ 00 09	5		NEBULA
REIZ 0094	09 53 17.	+ 46 25	15	15.6	GALAXY
REIZ 0093	09 53 17.	+ 46 25	36	14.9	GALAXY
REIZ 0097	09 53 18.	+ 13 46	12	15.5	GALAXY
ZWG 064.001	09 53 18.	+ 13 47		15.3	GALAXY
ZWG 063.106	09 53 18.	+ 13 47		15.3	GALAXY
TON-N 0463	09 53 18.	+ 26 03		15.6	BLUE STAR
MCG+06-22-034	09 53 18.	+ 37 17	36	15.5	GALAXY
KARA.68 062	09 53 18.	+ 69 16	87		DWARF GALAXY
MCG+12-10-012	09 53 18.	+ 69 16	132	16.	GALAXY
ZWG 008.001	09 53 18.	- 01 37		15.7	GALAXY
REIZ 0095	09 53 19.	+ 37 15	48	14.7	GALAXY
PATH 1.302	09 53 22.	+ 00 03	22		NEBULA
ARC 0896	09 53 22.	+ 41 15		17.0	RICH CLUSTER OF GALAXIES
MAI 047	09 53 22.	+ 68 47	53		DWARF SPHEROIDAL GALAXY
KN 15.014	09 53 22.5	+ 37 15 33.			NEBULA
IC 0577	09 53 23.	+ 10 44 28.			NONSTELLAR OBJECT
ARC 0897	09 53 23.	+ 28 26		17.6	RICH CLUSTER OF GALAXIES
ZWG 064.002	09 53 24.	+ 10 44		14.4	GALAXY
UGC 05334	09 53 24.	+ 10 44	42	14.4	GALAXY S
KARA.72 220A	09 53 24.	+ 10 44	36	14.4	PART OF DOUBLE GALAXY
MCG+02-26-001	09 53 24.	+ 10 44	30	14.4	GALAXY
ZWG 064.003	09 53 24.	+ 13 39		15.4	GALAXY
ZC 0953.4+1845	09 53 24.	+ 18 45	2490		CLUSTER OF GALAXIES
ZC 0953.4+3017	09 53 24.	+ 30 17	540		CLUSTER OF GALAXIES
TON-N 1128	09 53 24.	+ 34 42		14.9	BLUE STAR
ZWG 182.043	09 53 24.	+ 37 15		15.3	GALAXY
ZWG 265.049	09 53 24.	+ 53 23		15.0	GALAXY
7ZW 301	09 53 24.	+ 60 12			COMPACT GALAXY
ZWG 289.024	09 53 24.	+ 60 12		15.6	GALAXY
MRK 023	09 53 24.	+ 60 12	12	16.	GALAXY WITH UV CONTINUUM
MCG+10-14-054	09 53 24.	+ 60 12	18	16.	GALAXY
LB 03086	09 53 27.	+ 02 05 24.		16.9	FAINT BLUE STAR
LB 03085	09 53 27.	+ 02 18 18.		16.1	FAINT BLUE STAR
MCG+09-16-061	09 53 27.	+ 53 25	30	16.	GALAXY
MCG+09-16-060	09 53 27.	+ 53 25	36	16.	GALAXY
REIZ 0098	09 53 28.	+ 27 28	48	13.7	GALAXY
MCG+04-24-003	09 53 30.	+ 20 43 30.	36	15.5	GALAXY
ZWG 123.005	09 53 30.	+ 20 44		15.5	GALAXY
ZWG 153.004	09 53 30.	+ 27 28		14.3	GALAXY
UGC 05335	09 53 30.	+ 27 28	42	14.3	GALAXY PECULR
MCG+05-24-003	09 53 30.	+ 27 28	36	14.3	GALAXY
IC 2520	09 53 30.	+ 27 29 29.			NONSTELLAR OBJECT
TON-N 0025	09 53 30.	+ 27 43		16.5	BLUE STAR
ZC 0953.5+2800	09 53 30.	+ 28 00	2350		CLUSTER OF GALAXIES
ZC 0953.5+2829	09 53 30.	+ 28 29	810		CLUSTER OF GALAXIES
TON-N 0464	09 53 30.	+ 31 25		16.2	BLUE STAR
TON-N 1129	09 53 30.	+ 32 50		17.	BLUE STAR
ZWG 182.044	09 53 30.	+ 36 09		15.6	GALAXY
MCG+08-18-044	09 53 30.	+ 46 41	36	16.	GALAXY
UGC 05336	09 53 30.	+ 69 17	180	16.5	GALAXY DWRF IR
MCG-02-26-002	09 53 30.	- 13 33	72	15.	GALAXY
REIZ 0096	09 53 31.	+ 42 57	30	15.4	GALAXY
ARC 0895	09 53 31.	+ 49 44		18.0	RICH CLUSTER OF GALAXIES
KN 15.015	09 53 32.1	+ 36 09 12.			NEBULA
HOLM 151C	09 53 33.	+ 20 45	24	15.1	PART OF MULTIPLE GALAXY
KN 15.016	09 53 34.8	+ 37 27 30.			NEBULA
IC 0578	09 53 35.	+ 10 43 52.			NONSTELLAR OBJECT
HOLM 151D	09 53 35.	+ 20 48	24	15.1	PART OF MULTIPLE GALAXY
ZWG 064.004	09 53 36.	+ 10 43		14.7	GALAXY
UGC 05337	09 53 36.	+ 10 43	78	14.7	GALAXY SB:a
KARA.72 220B	09 53 36.	+ 10 43	66	14.7	PART OF DOUBLE GALAXY
MCG+02-26-002	09 53 36.	+ 10 43	48	14.7	GALAXY
ZWG 093.003	09 53 36.	+ 17 05		13.8	GALAXY
UGC 05338	09 53 36.	+ 17 05	144	13.8	GALAXY Sb
MCG+03-26-002	09 53 36.	+ 17 05	126	13.8	GALAXY
ZWG 123.006	09 53 36.	+ 20 45		15.7	GALAXY
ZWG 239.042	09 53 36.	+ 46 42		15.1	GALAXY
MRK 129	09 53 36.	+ 46 42	15	15.5	GALAXY WITH UV CONTINUUM
OCL 0777	09 53 36.	- 46 47	102	14.	OPEN STAR CLUSTER
HOLM 151E	09 53 37.	+ 20 47	24	15.4	PART OF MULTIPLE GALAXY
RNGC 3060	09 53 38.	+ 17 05		14.0	GALAXY
REIZ 0099	09 53 39.	+ 17 05	108	13.5	GALAXY
MAI 048	09 53 40.	+ 69 24	94		DWARF SPHEROIDAL GALAXY
HOLM 151A	09 53 41.	+ 20 44	24	14.8	PART OF MULTIPLE GALAXY
ARC 0901	09 53 41.	- 09 43		17.7	RICH CLUSTER OF GALAXIES
PATH 1.303	09 53 42.	+ 00 21	27		NEBULA
ZWG 064.005	09 53 42.	+ 10 04		15.3	GALAXY
MCG+04-24-005	09 53 42.	+ 20 42 30.	15	14.4	GALAXY
ZWG 123.007	09 53 42.	+ 20 43		14.4	GALAXY

OBJECT NAME	RIGHT ASCEN.	DECLINATION	DIAM.	MAGN.	TYPE OF OBJECT
UGC 05339	09 53 42.	+ 20 43	54	14.4	GALAXY DBL SYS
KARA.72 221B	09 53 42.	+ 20 43	30		PART OF DOUBLE GALAXY
KARA.72 221A	09 53 42.	+ 20 43	36	14.4	PART OF DOUBLE GALAXY
MCG+04-24-004	09 53 42.	+ 20 43	36	14.4	GALAXY
HOLM 151B	09 53 42.	+ 20 44	18	14.8	PART OF MULTIPLE GALAXY
ZC 0953.7+3850	09 53 42.	+ 38 50	1080		CLUSTER OF GALAXIES
ZWG 239.043	09 53 42.	+ 47 39		15.4	GALAXY
ZC 0953.7+5803	09 53 42.	+ 58 03	130		CLUSTER OF GALAXIES
MCG-02-26-003	09 53 42.	- 12 45	12	16.	GALAXY
HOLM 151F	09 53 43.	+ 20 45	12	15.4	PART OF MULTIPLE GALAXY
FATH 1.304	09 53 44.	- 00 18	8		NEBULA
KN 15.017	09 53 44.5	+ 36 35 57.			NEBULA
ARC 0900	09 53 45.	+ 18 50		17.5	RICH CLUSTER OF GALAXIES
ZWG 153.005	09 53 48.	+ 29 04		15.2	GALAXY
UGC 05340	09 53 48.	+ 29 04	156	15.2	GALAXY DWF PEC
ZC 0953.8+6102	09 53 48.	+ 61 02	540		CLUSTER OF GALAXIES
MCG-02-26-004	09 53 48.	- 12 45	36	15.	GALAXY
TON-N 1130	09 53 49.	+ 19 01		14.7	BLUE STAR
ZWG 064.006	09 53 54.	+ 10 20		15.0	GALAXY
MCG+02-26-003	09 53 54.	+ 10 20	36	15.0	GALAXY
ZWG 064.007	09 53 54.	+ 11 24		15.0	GALAXY
MCG+02-26-004	09 53 54.	+ 11 24	36	15.0	GALAXY
MCG+04-24-006	09 53 54.	+ 20 52	180	15.5	GALAXY
ZWG 123.008	09 53 54.	+ 20 53		15.5	GALAXY
UGC 05341	09 53 54.	+ 20 53	174	15.5	GALAXY Sc
MCG+05-24-004	09 53 54.	+ 29 04	156	15.2	GALAXY
ZC 0953.9+4022	09 53 54.	+ 40 22	1340		CLUSTER OF GALAXIES
ZWG 239.044	09 53 54.	+ 47 13		15.5	GALAXY
FATH 1.305	09 53 54.	- 00 23	11		NEBULA
REIZ 0100	09 53 57.	+ 17 07	12	15.6	GALAXY
ARC 0902	09 53 58.	- 09 57		17.7	RICH CLUSTER OF GALAXIES
SHB 146	09 53 59.8	+ 25 29 33.		16.5	QUASI-STELLAR OBJECT
ZWG 008.002	09 54 00.	+ 01 40		15.1	GALAXY
ZWG 008.003	09 54 00.	+ 02 06		15.7	GALAXY
MCG+03-26-003	09 54 00.	+ 15 52	66	14.3	GALAXY
ZWG 093.004	09 54 00.	+ 15 53		14.3	GALAXY
UGC 05342	09 54 00.	+ 15 53	66	14.3	GALAXY S
ZWG 093.005	09 54 00.	+ 17 03		15.5	GALAXY
UGC 05343	09 54 00.	+ 17 03	72	15.5	GALAXY
MCG+03-26-004	09 54 00.	+ 17 03	84	15.5	GALAXY
TON-N 0465	09 54 00.	+ 30 04		16.2	BLUE STAR
RNGC 3062	09 54 01.	+ 01 40		15.0	GALAXY
REIZ 0102	09 54 02.	+ 15 53	48	15.2	GALAXY
REIZ 0103	09 54 03.	+ 17 03	90	15.2	GALAXY
BIGO 512	09 54 06.	+ 01 34			NEBULA
MCG-06-22-003	09 54 06.	- 37 32	60	15.	GALAXY
LB 03087	09 54 09.	+ 00 56 48.		16.5	FAINT BLUE STAR
IC 0579	09 54 09.	- 13 54 42.			NONSTELLAR OBJECT
MCG-04-24-007	09 54 09.	- 25 50 30.	36	14.	GALAXY
REIZ 0101	09 54 10.	+ 34 13	30	14.7	GALAXY
ZWG 064.008	09 54 12.	+ 14 14		15.7	GALAXY
ZWG 182.045	09 54 12.	+ 34 13		15.3	GALAXY
ZC 0954.2+4042	09 54 12.	+ 40 42	610		CLUSTER OF GALAXIES
SHB 147	09 54 14.9	+ 55 37 16.		16.5	QUASI-STELLAR OBJECT
MCG-02-26-005	09 54 15.	- 13 31	60	14.5	GALAXY
ZWG 036.004	09 54 18.	+ 03 35		15.6	GALAXY
ZC 0954.3+0523	09 54 18.	+ 05 23	1010		CLUSTER OF GALAXIES
MCG+03-26-005	09 54 18.	+ 15 48	66	15.1	GALAXY
IC 2521	09 54 18.	+ 34 13 20.			NONSTELLAR OBJECT
MCG+06-22-035	09 54 18.	+ 34 15	27	15.	GALAXY
ZWG 182.046	09 54 18.	+ 36 18		15.7	GALAXY
HOLM 152C	09 54 18.	+ 36 21	30	14.3	PART OF MULTIPLE GALAXY
MCG+08-18-045	09 54 18.	+ 45 26	30	17.	GALAXY
MCG+08-18-046	09 54 18.	+ 45 28	90	16.	GALAXY
ZC 0954.3+4931	09 54 18.	+ 49 31	610		CLUSTER OF GALAXIES
ZC 0954.3+5915	09 54 18.	+ 59 15	610		CLUSTER OF GALAXIES
ZWG 312.029	09 54 18.	+ 68 20		15.7	GALAXY
MCG-02-26-006	09 54 18.	- 15 20	36	15.	GALAXY
MCG-05-24-006	09 54 18.	- 31 04	120	15.	GALAXY
KN 15.018	09 54 18.3	+ 37 03 59.			NEBULA
KN 15.020	09 54 19.0	+ 34 12 54.			NEBULA
REIZ 0104	09 54 20.	+ 15 50	48	14.2	GALAXY
FATH 1.306	09 54 21.	- 00 18	5		NEBULA
MCG-01-26-002	09 54 21.	- 06 56	102	13.5	GALAXY
KN 15.019	09 54 22.6	+ 36 48 20.			NEBULA
KN 15.021	09 54 23.5	+ 36 19 14.			NEBULA
ZWG 036.005	09 54 24.	+ 04 00		14.8	GALAXY
MCG+01-26-001	09 54 24.	+ 04 00	18	14.8	GALAXY
ZWG 036.006	09 54 24.	+ 04 03		15.2	GALAXY
MCG+01-26-002	09 54 24.	+ 04 03	36	15.2	GALAXY
ZC 0954.4+1338	09 54 24.	+ 13 38	470		CLUSTER OF GALAXIES
ZWG 093.006	09 54 24.	+ 15 48		15.1	GALAXY
UGC 05344	09 54 24.	+ 15 48	66	15.1	GALAXY Sc
MCG+06-22-036	09 54 24.	+ 36 20	33	15.	GALAXY
ZC 0954.4+4057	09 54 24.	+ 40 57	470		CLUSTER OF GALAXIES
MCG+08-18-047	09 54 24.	+ 45 27	24	14.	GALAXY
ZWG 239.045	09 54 24.	+ 45 29		14.4	GALAXY
UGC 05345	09 54 24.	+ 45 29	90	14.4	GALAXY S
ZC 0954.4+4701	09 54 24.	+ 47 01	810		CLUSTER OF GALAXIES
MCG+10-14-055	09 54 24.	+ 57 49	18	16.	GALAXY
7ZW 302	09 54 24.	+ 65 58			COMPACT GALAXY
KEEL 194	09 54 27.0	+ 69 16 29.			NEBULA
LB 03088	09 54 28.	+ 02 23 54.		17.2	FAINT BLUE STAR
HOLM 152B	09 54 28.	+ 36 22	24	14.2	PART OF MULTIPLE GALAXY
FATH 1.307	09 54 28.	- 00 02	5		NEBULA
KN 15.022	09 54 28.0	+ 37 03 50.			NEBULA
ZWG 036.007	09 54 30.	+ 05 25		15.5	GALAXY
ZC 0954.5+2906	09 54 30.	+ 29 06	2490		CLUSTER OF GALAXIES
TON-N 0466	09 54 30.	+ 31 44		15.5	BLUE STAR
ZWG 182.047	09 54 30.	+ 33 51		14.8	GALAXY
MRK 411	09 54 30.	+ 33 51	18	15.5	GALAXY WITH UV CONTINUUM
ZWG 182.048	09 54 30.	+ 36 19		15.5	GALAXY
HOLM 152A	09 54 30.	+ 36 20	72	14.5	PART OF MULTIPLE GALAXY
MCG+08-18-049	09 54 30.	+ 45 07	45	16.	GALAXY
ZWG 239.046	09 54 30.	+ 45 08		15.5	GALAXY
MCG+08-18-048	09 54 30.	+ 45 29	54	15.	GALAXY
ZWG 239.047	09 54 30.	+ 45 30		15.2	GALAXY
UGC 05346	09 54 30.	+ 45 30	60	15.2	GALAXY S
ARC 0898	09 54 30.	+ 49 30		18.0	RICH CLUSTER OF GALAXIES
ZWG 266.001	09 54 30.	+ 52 05		15.1	GALAXY
ZWG 265.050	09 54 30.	+ 52 05		15.1	GALAXY
KARA.73B 0386	09 54 30.	+ 52 05	36	15.1	ISOLATED GALAXY S
MCG+14-05-009	09 54 30.	+ 86 30	27	16.	GALAXY
LB 03089	09 54 32.	- 01 18 54.		15.4	FAINT BLUE STAR
KN 15.023	09 54 32.7	+ 36 19 52.			NEBULA
KN 15.024	09 54 35.1	+ 36 18 16.			NEBULA
8ZW 0954+03.9	09 54 36.	+ 03 52		17.1	COMPACT GALAXY
TON-N 0467	09 54 36.	+ 30 14		14.6	BLUE STAR
IC 2524	09 54 36.	+ 33 51 24.			NONSTELLAR OBJECT
MCG+06-22-037	09 54 36.	+ 36 20	24	15.5	GALAXY
ZC 0954.6+5802	09 54 36.	+ 58 02	3360		CLUSTER OF GALAXIES
MCG-05-24-007	09 54 36.	- 31 38	42	15.	GALAXY
MCG-06-22-004	09 54 36.	- 36 53	30	14.	GALAXY
KN 15.026	09 54 36.2	+ 33 51 31.			NEBULA
FATH 1.308	09 54 38.	- 00 08	11		NEBULA
FATH 1.309	09 54 38.	- 00 11	14		NEBULA
MCG+06-22-039	09 54 39.	+ 33 51 30.	36	15.	GALAXY
MCG+06-22-038	09 54 39.	+ 36 19 30.	42	15.	GALAXY
MCG-01-26-003	09 54 39.	- 07 39	54	15.	GALAXY
KN 15.025	09 54 39.8	+ 37 33 57.			NEBULA
HOLM 152D	09 54 40.	+ 36 20	72	14.9	PART OF MULTIPLE GALAXY
ZWG 036.008	09 54 42.	+ 04 46		15.0	GALAXY
UGC 05347	09 54 42.	+ 04 46	90	15.0	GALAXY Sc
MCG+01-26-003	09 54 42.	+ 04 46	78	15.0	GALAXY
ZWG 036.009	09 54 42.	+ 07 26		15.0	GALAXY
MCG+01-26-004	09 54 42.	+ 07 26	18	15.0	GALAXY
ZC 0954.7+1636	09 54 42.	+ 16 36	610		CLUSTER OF GALAXIES
ZWG 123.009	09 54 42.	+ 23 50		15.7	GALAXY
ZWG 123.010	09 54 42.	+ 25 45		15.4	GALAXY
MCG+04-24-007	09 54 42.	+ 25 45	36	15.4	GALAXY
MCG+06-22-040	09 54 42.	+ 36 58 30.	36	15.	GALAXY
IC 2526	09 54 44.	- 32 02 15.			NONSTELLAR OBJECT
KN 15.027	09 54 44.1	+ 36 27 14.			NEBULA
HOLM 154C	09 54 45.	+ 37 36	36	15.2	PART OF MULTIPLE GALAXY
KN 15.029	09 54 45.0	+ 36 18 31.			NEBULA
KN 15.028	09 54 45.4	+ 36 57 04.			NEBULA
ARC 0888	09 54 46.	+ 77 11		17.7	RICH CLUSTER OF GALAXIES
FATH 1.310	09 54 46.	- 00 13	8		NEBULA
ZWG 036.010	09 54 48.	+ 08 21		15.3	GALAXY
ZC 0954.8+2016	09 54 48.	+ 20 16	1950		CLUSTER OF GALAXIES
TON-N 0468	09 54 48.	+ 24 46		14.8	BLUE STAR
MCG+06-22-041	09 54 48.	+ 36 19 30.	48	16.	GALAXY
MCG+08-18-050	09 54 48.	+ 47 32	66	15.	GALAXY
UGC 05348	09 54 48.	+ 47 35	66	16.0	GALAXY
ZWG 008.004	09 54 48.	- 02 10		15.7	GALAXY
MCG+06-22-042	09 54 51.	+ 36 37	24	14.5	GALAXY
KN 15.030	09 54 51.3	+ 37 30 20.			NEBULA
ZWG 182.049	09 54 54.	+ 36 35		15.0	GALAXY
REIZ 0105	09 54 54.	+ 36 37	42	14.4	GALAXY
ZC 0954.9+4620	09 54 54.	+ 46 20	740		CLUSTER OF GALAXIES
MCG-05-24-008	09 54 54.	- 32 02	30	14.5	GALAXY
KN 15.031	09 54 54.1	+ 36 35 15.			NEBULA
TON-N 1131	09 54 55.	+ 22 02		17.	BLUE STAR
ARC 0899	09 54 55.	+ 55 31		17.2	RICH CLUSTER OF GALAXIES
KN 15.032	09 54 55.7	+ 36 50 54.			NEBULA
HOLM 154B	09 54 56.	+ 37 32	36	15.1	PART OF MULTIPLE GALAXY
ARC 0903	09 54 58.	+ 19 52		17.2	RICH CLUSTER OF GALAXIES
LBN 1089	09 55	- 29 00	5220		BRIGHT NEBULA
ZC 0955.0+2413	09 55 00.	+ 24 13	2960		CLUSTER OF GALAXIES
MRK 412	09 55 00.	+ 32 28	14	15.5	GALAXY WITH UV CONTINUUM
TON-N 1132	09 55 00.	+ 34 14		16.8	BLUE STAR
MCG+06-22-043	09 55 00.	+ 37 34	138	13.5	GALAXY
ZC 0955.0+3815	09 55 00.	+ 38 15	2420		CLUSTER OF GALAXIES
ZC 0955.0+5530	09 55 00.	+ 55 30	940		CLUSTER OF GALAXIES
MCG-02-26-007	09 55 00.	- 13 39	48	16.	GALAXY
KEEL 195	09 55 00.4	+ 69 21 14.			NEBULA
KN 15.033	09 55 00.8	+ 38 13 59.			NEBULA
KN 15.034	09 55 02.2	+ 36 48 31.			NEBULA
KN 15.035	09 55 04.6	+ 36 51 11.			NEBULA
ZWG 036.011	09 55 06.	+ 08 04		15.6	GALAXY
KARA.73B 0387	09 55 06.	+ 08 04	18	15.6	ISOLATED GALAXY E
ZWG 064.009	09 55 06.	+ 10 16		15.1	GALAXY
MCG+03-26-006	09 55 06.	+ 15 21	72	14.8	GALAXY
ZWG 093.007	09 55 06.	+ 15 22		14.8	GALAXY
ZC 0955.1+1950	09 55 06.	+ 19 50	1010		CLUSTER OF GALAXIES
HOLM 153A	09 55 06.	+ 36 23	36	14.5	PART OF MULTIPLE GALAXY
ZWG 182.050	09 55 06.	+ 37 31		14.5	GALAXY
UGC 05349	09 55 06.	+ 37 31	156	14.5	GALAXY
MCG-02-26-008	09 55 06.	- 13 39	48	14.5	GALAXY
REIZ 0106	09 55 07.	+ 37 31	168	13.7	GALAXY
KN 15.036	09 55 07.2	+ 37 31 57.			NEBULA
KEEL 196	09 55 07.3	+ 69 29 48.			SPIRAL NEBULA
FATH 1.311	09 55 08.	+ 00 14	41		NEBULA
MCG+05-24-005	09 55 09.	+ 32 28	36	15.	GALAXY
MCG+06-22-044	09 55 09.	+ 36 22	42	15.	GALAXY
FATH 1.312	09 55 09.	- 00 04	11		NEBULA
KN 15.037	09 55 10.8	+ 36 21 49.			NEBULA
ZWG 093.008	09 55 12.	+ 15 10		15.0	GALAXY
MCG+03-26-007	09 55 12.	+ 15 10	24	15.0	GALAXY
ZC 0955.2+1924	09 55 12.	+ 19 24	2020		CLUSTER OF GALAXIES
MCG-03-26-001	09 55 12.	- 19 07	90	13.5	GALAXY
HOLM 154A	09 55 13.	+ 37 34	138	12.6	PART OF MULTIPLE GALAXY
RNGC 3072	09 55 13.	- 19 07		13.0	GALAXY
FATH 1.313	09 55 14.	- 00 03	5		NEBULA
FATH 1.314	09 55 14.	- 00 11	8		NEBULA
KN 15.038	09 55 14.0	+ 34 14 30.			NEBULA
RNGC 3069	09 55 16.	+ 10 41		15.0	GALAXY
REIZ 0107	09 55 16.	+ 34 14	30	14.9	GALAXY
FATH 1.315	09 55 16.	- 00 01	8		NEBULA
IC 0580	09 55 17.	+ 10 40 44.			NONSTELLAR OBJECT
HOLM 153B	09 55 17.	+ 36 24	30	15.3	PART OF MULTIPLE GALAXY
KN 15.039	09 55 17.7	+ 36 29 55.			NEBULA
ZWG 064.010	09 55 18.	+ 10 41		15.0	GALAXY
MCG+02-26-005	09 55 18.	+ 10 41	36	15.0	GALAXY
ZWG 093.009	09 55 18.	+ 15 42		15.5	GALAXY
ZC 0955.3+1608	09 55 18.	+ 16 08	810		CLUSTER OF GALAXIES
TON-N 0469	09 55 18.	+ 32 40		15.6	BLUE STAR
ZC 0955.3+4308	09 55 18.	+ 43 08	810		CLUSTER OF GALAXIES
MCG-03-26-002	09 55 18.	- 17 57	48	14.	GALAXY
RNGC 3076	09 55 19.	- 17 57		14.0	GALAXY
REIZ 0110	09 55 20.	+ 15 43	30	15.4	GALAXY
KN 15.041	09 55 21.4	+ 36 22 27.			NEBULA
RNGC 3070	09 55 22.	+ 10 36		13.0	GALAXY
KN 15.040	09 55 22.5	+ 37 29 57.			NEBULA
ZWG 064.011	09 55 24.	+ 10 36		13.2	GALAXY
UGC 05350	09 55 24.	+ 10 36	126	13.2	GALAXY E
MCG+02-26-006	09 55 24.	+ 10 36	36	13.2	GALAXY
ZWG 182.051	09 55 24.	+ 32 37		12.7	GALAXY
UGC 05351	09 55 24.	+ 32 37	132	12.7	GALAXY Sa-b
TON-N 0470	09 55 24.	+ 32 38		16.1	BLUE STAR
ZWG 182.052	09 55 24.	+ 37 20		15.7	GALAXY
ZC 0955.4+4848	09 55 24.	+ 48 48	540		CLUSTER OF GALAXIES
ZC 0955.4+5605	09 55 24.	+ 56 05	1880		CLUSTER OF GALAXIES
MCG+13-07-041	09 55 24.	+ 77 02	51	16.	GALAXY
MCG-02-26-009	09 55 24.	- 13 23	36	15.	GALAXY
IC 2525	09 55 25.	+ 37 18 51.			NONSTELLAR OBJECT

OBJECT NAME	RIGHT ASCEN.	DECLINATION	DIAM.	MAGN.	TYPE OF OBJECT
BC 3C232	09 55 25.44	+ 32 38 23.2		15.78	QUASI-STELLAR OBJECT
KN 15.042	09 55 25.6	+ 37 20 31.			NEBULA
RNGC 3067	09 55 26.	+ 32 36		13.0	GALAXY
REIZ 0108	09 55 26.	+ 32 37	120	12.7	GALAXY
KN 15.043	09 55 26.2	+ 32 36 28.			NEBULA
REIZ 0112	09 55 28.	+ 10 36	24	13.7	GALAXY
REIZ 0111	09 55 28.	+ 10 40	24	14.5	GALAXY
IC 0581	09 55 29.	+ 16 11 26.			NONSTELLAR OBJECT
ZWG 064.012	09 55 30.	+ 10 14		15.7	GALAXY
MCG+02-26-007	09 55 30.	+ 10 14	30	15.7	GALAXY
ZWG 093.010	09 55 30.	+ 16 10		15.0	GALAXY
UGC 05352	09 55 30.	+ 16 10	66	15.0	GALAXY SB:a-b
MCG+03-26-008	09 55 30.	+ 16 10	60	15.0	GALAXY
TON-N 0471	09 55 30.	+ 22 52		14.6	BLUE STAR
ZC 0955.5+3112	09 55 30.	+ 31 12	810		CLUSTER OF GALAXIES
MCG+06-22-046	09 55 30.	+ 32 36 30.	120	12.5	GALAXY
ZWG 182.053	09 55 30.	+ 36 30		15.6	GALAXY
MCG+06-22-045	09 55 30.	+ 36 30	24	15.	GALAXY
KN 15.044	09 55 31.5	+ 36 29 42.			NEBULA
ARP 174	09 55 33.	+ 29 05			PECULIAR GALAXY
KN 15.045	09 55 33.2	+ 36 29 42.			NEBULA
TON-N 0472	09 55 36.	+ 22 59		15.6	BLUE STAR
LB 03090	09 55 36.	- 00 50 24.		16.1	FAINT BLUE STAR
RNGC 3068	09 55 40.	+ 29 07		15.0	GALAXY
FATH 1.136	09 55 41.	- 00 26	14		NEBULA
ZWG 153.006	09 55 42.	+ 29 07		15.1	GALAXY
UGC 05353	09 55 42.	+ 29 07	66	15.1	GALAXY E-SO
STOCK 06	09 55 42.	+ 39 52			BLUE KNOT NEAR ELLIP GLXY
MCG+08-18-051	09 55 42.	+ 47 57	120	14.	GALAXY
ZWG 239.048	09 55 42.	+ 47 58		14.4	GALAXY
UGC 05354	09 55 42.	+ 47 58	138	14.4	GALAXY S
ZWG 289.025	09 55 42.	+ 58 12		15.3	GALAXY
MCG+10-14-056	09 55 42.	+ 58 12	45	15.	GALAXY
MCG-05-24-009	09 55 42.	- 28 15	144	14.5	GALAXY
REIZ 0109	09 55 43.	+ 47 59	72	13.5	GALAXY
REIZ 0113	09 55 47.	+ 29 13	24	14.9	GALAXY
MCG+05-24-006	09 55 48.	+ 29 07	36	15.1	GALAXY
ZWG 153.007	09 55 48.	+ 32 18		15.5	GALAXY
ZC 0955.8+3401	09 55 48.	+ 34 01	340		CLUSTER OF GALAXIES
FATH 1.317	09 55 49.	- 00 36	8		NEBULA
ABC 0907	09 55 51.	- 10 49		17.5	RICH CLUSTER OF GALAXIES
UGC 05355	09 55 54.	+ 01 18	78	16.0	GALAXY Sc
ZWG 064.013	09 55 54.	+ 13 29		14.7	GALAXY
MCG+02-26-008	09 55 54.	+ 13 29	24	14.7	GALAXY
ZC 0955.9+3038	09 55 54.	+ 30 38	1480		CLUSTER OF GALAXIES
ZWG 153.008	09 55 54.	+ 31 51		15.4	GALAXY
TON-N 1133	09 55 54.	+ 36 19		14.	BLUE STAR
ZWG 266.002	09 55 54.	+ 51 15		15.2	GALAXY
ZWG 265.051	09 55 54.	+ 51 15		15.2	GALAXY
UGC 05356	09 55 54.	+ 51 15	72	15.2	GALAXY SBb-c
IZW 022	09 55 54.	+ 51 45			COMPACT GALAXY
FATH 1.318	09 55 54.	- 00 11	19		NEBULA
FATH 1.319	09 55 54.	- 00 37	11		NEBULA
FATH 1.320	09 55 56.	- 00 36	8		NEBULA
RNGC 3071	09 55 57.	+ 31 51		15.5	GALAXY
KN 15.046	09 55 58.3	+ 31 51 30.			NEBULA
VDB.66G 070	09 56	+ 05 33	200		DWARF GALAXY
VDB.66G 069	09 56	+ 30 58	230		DWARF GALAXY
UGC 05357	09 56 00.	+ 05 30	66	16.0	GALAXY Sc
MCG+05-24-007	09 56 00.	+ 32 20	48	15.5	GALAXY
MCG+09-17-001	09 56 00.	+ 51 11	72	15.	GALAXY
IZW 023	09 56 00.	+ 52 29			COMPACT GALAXY
ZWG 266.003	09 56 00.	+ 52 30		14.9	GALAXY
ZWG 265.052	09 56 00.	+ 52 30		14.9	GALAXY
MCG+09-17-002	09 56 00.	+ 52 30	36	16.	GALAXY
ZC 0956.0+5429	09 56 00.	+ 54 29	610		CLUSTER OF GALAXIES
MRK 024	09 56 00.	+ 54 44	13	17.	GALAXY WITH UV CONTINUUM
ZC 0956.0+6430	09 56 00.	+ 64 30	2420		CLUSTER OF GALAXIES
MCG+11-12-031	09 56 00.	+ 65 36	36	16.	GALAXY
ZWG 008.005	09 56 00.	- 02 35		15.4	GALAXY
REIZ 0114	09 56 01.	+ 14 48	18	15.1	GALAXY
FATH 1.322	09 56 02.	+ 00 19	19		NEBULA
FATH 1.321	09 56 02.	- 00 36	8		NEBULA
FATH 1.323	09 56 04.	- 00 41	11		NEBULA
ZWG 008.006	09 56 06.	+ 01 18		15.2	GALAXY
ZWG 064.014	09 56 06.	+ 11 38		14.9	GALAXY
UGC 05358	09 56 06.	+ 11 38	102	14.9	GALAXY SBb
MCG+02-26-009	09 56 06.	+ 11 38	84	14.9	GALAXY
ZWG 093.011	09 56 06.	+ 19 27		15.0	GALAXY
UGC 05359	09 56 06.	+ 19 27	78	15.0	GALAXY Sb-c
ZC 0956.1+4248	09 56 06.	+ 42 48	1140		CLUSTER OF GALAXIES
ZC 0956.1-0155	09 56 06.	- 01 55	2020		CLUSTER OF GALAXIES
MCG-04-24-008	09 56 06.	- 24 55 30.	150	14.	GALAXY
OCL 0794	09 56 06.	- 54 25	240		OPEN STAR CLUSTER
RNGC 3078	09 56 10.	- 26 41		12.5	GALAXY
REIZ 0116	09 56 11.	+ 11 37	72	14.7	GALAXY
ZWG 093.012	09 56 12.	+ 14 39		14.5	GALAXY
UGC 05360	09 56 12.	+ 14 39	72	14.5	GALAXY Sc
MCG+03-26-009	09 56 12.	+ 14 39	72	14.5	GALAXY
ZWG 093.013	09 56 12.	+ 15 37		15.1	GALAXY
ZWG 093.014	09 56 12.	+ 18 52		15.7	GALAXY
MCG+03-26-010	09 56 12.	+ 19 29	72	15.0	GALAXY
MCG+04-24-008	09 56 12.	+ 25 27 30.	42	16.	GALAXY
UGC 05361	09 56 12.	+ 25 28	66	16.0	GALAXY
ZC 0956.2+4618	09 56 12.	+ 46 18	540		CLUSTER OF GALAXIES
ZC 0956.2+6015	09 56 12.	+ 60 15	1680		CLUSTER OF GALAXIES
ZWG 008.007	09 56 12.	- 02 50		15.7	GALAXY
MCG-04-24-009	09 56 12.	- 26 41 30.	48	12.	GALAXY
OCL 0793	09 56 12.	- 54 21	240		OPEN STAR CLUSTER
KN 15.047	09 56 12.4	+ 37 37 59.			NEBULA
REIZ 0117	09 56 13.	+ 14 40	54	14.5	GALAXY
REIZ 0118	09 56 13.	+ 15 38	24	15.1	GALAXY
KN 15.048	09 56 13.2	+ 36 51 43.			NEBULA
RNGC 3075	09 56 15.	+ 14 39		14.5	GALAXY
IC 0582	09 56 15.	+ 18 02 47.			NONSTELLAR OBJECT
KN 15.049	09 56 15.1	+ 33 28 43.			NEBULA
REIZ 0115	09 56 16.	+ 34 29	18	15.1	GALAXY
ZC 0956.3+1516	09 56 18.	+ 15 16	870		CLUSTER OF GALAXIES
ZWG 093.015	09 56 18.	+ 15 55		15.2	GALAXY
ZWG 093.016	09 56 18.	+ 18 03		14.7	GALAXY
UGC 05362	09 56 18.	+ 18 03	66	14.7	GALAXY S
MCG+03-26-011	09 56 18.	+ 18 04	54	14.7	GALAXY
ZC 0956.3+2602	09 56 18.	+ 26 02	1880		CLUSTER OF GALAXIES
ZWG 153.009	09 56 18.	+ 31 57		15.5	GALAXY
MRK 413	09 56 18.	+ 31 57	9	15.5	GALAXY WITH UV CONTINUUM
ZC 0956.3+5933	09 56 18.	+ 59 33	2080		CLUSTER OF GALAXIES
REIZ 0120	09 56 19.	+ 14 41	72	13.7	GALAXY
TON-N 1134	09 56 19.	+ 20 49		16.	BLUE STAR

OBJECT NAME	RIGHT ASCEN.	DECLINATION	DIAM.	MAGN.	TYPE OF OBJECT
IC 0583	09 56 20.	+ 18 03 04.			NONSTELLAR OBJECT
REIZ 0121	09 56 21.	+ 18 04	36	14.2	GALAXY
HOLM 155A	09 56 21.	+ 18 04	54	13.9	PART OF MULTIPLE GALAXY
IC 2528	09 56 21.	- 26 56 52.			SINGLE STAR
KN 15.050	09 56 23.4	+ 36 22 02.			NEBULA
ZWG 064.015	09 56 24.	+ 10 36		14.7	GALAXY
MCG+02-26-010	09 56 24.	+ 10 36	18	14.7	GALAXY
ZWG 093.017	09 56 24.	+ 18 04		15.2	GALAXY
UGC 05363	09 56 24.	+ 18 04	60	15.2	GALAXY S
MCG+03-26-012	09 56 24.	+ 18 04	54	15.2	GALAXY
TON-N 0473	09 56 24.	+ 24 21		16.6	BLUE STAR
ZWG 153.010	09 56 24.	+ 30 59		13.6	GALAXY
UGC 05364	09 56 24.	+ 30 59	330	14.6	GALAXY DWRF IR
ZC 0956.4+3730	09 56 24.	+ 37 30	9680		CLUSTER OF GALAXIES
LB 03091	09 56 26.	+ 01 11 12.		16.2	FAINT BLUE STAR
IC 0584	09 56 26.	+ 10 36 22.			NONSTELLAR OBJECT
HOLM 155B	09 56 26.	+ 18 04	60	14.9	PART OF MULTIPLE GALAXY
REIZ 0123	09 56 27.	+ 18 04	54	14.9	GALAXY
ZC 0956.5+0806	09 56 30.	+ 08 06	1140		CLUSTER OF GALAXIES
ZWG 064.016	09 56 30.	+ 13 18		15.0	GALAXY
MCG+02-26-012	09 56 30.	+ 13 18	24	15.0	GALAXY
MCG+02-26-011	09 56 30.	+ 13 18	60	15.0	GALAXY
TON-N 0474	09 56 30.	+ 31 02		15.8	BLUE STAR
TON-N 1135	09 56 30.	+ 33 54		16.	BLUE STAR
ZC 0956.5+3951	09 56 30.	+ 39 51	2080		CLUSTER OF GALAXIES
MCG+07-21-003	09 56 30.	+ 42 00	33	17.	GALAXY
KEEL 197	09 56 33.4	+ 68 52 35.			NEBULA
LB 03092	09 56 34.	+ 01 37 30.		15.9	FAINT BLUE STAR
FATH 1.324	09 56 36.	+ 00 15	22		NEBULA
UGC 05365	09 56 36.	+ 03 32	60	17.	GALAXY DWARF
ZWG 064.017	09 56 36.	+ 13 13		15.6	GALAXY
ZWG 093.018	09 56 36.	+ 16 09		15.5	GALAXY
MCG+05-24-008	09 56 36.	+ 31 00	300	14.6	GALAXY
ZC 0956.6+4041	09 56 36.	+ 40 41	810		CLUSTER OF GALAXIES
VV 321B	09 56 36.	+ 45 30	48	15.	INTERACTING GALAXY
VV 321A	09 56 36.	+ 45 31	66	15.	INTERACTING GALAXY
MCG+09-17-003	09 56 36.	+ 52 28 30.	30	18.	GALAXY
MCG-05-24-010	09 56 36.	- 28 22	90	14.5	GALAXY
ABC 0904	09 56 38.	+ 60 20		17.4	RICH CLUSTER OF GALAXIES
REIZ 0119	09 56 39.	+ 45 33	42	14.6	GALAXY
KN 15.051	09 56 39.5	+ 34 33 39.			NEBULA
RNGC 3082	09 56 41.	- 30 06		14.0	GALAXY
ZWG 182.054	09 56 42.	+ 35 38		14.8	GALAXY
UGC 05366	09 56 42.	+ 35 38	156	14.8	GALAXY Sc
SN 1965N	09 56 42.	+ 35 38		15.8	SUPERNOVA
MCG+06-22-047	09 56 42.	+ 35 39	138	13.	GALAXY
MCG+08-18-052	09 56 42.	+ 45 30	60	15.	GALAXY
MCG+08-18-053	09 56 42.	+ 45 30 30.	42	15.	GALAXY
ZWG 239.049	09 56 42.	+ 45 31		15.6	GALAXY
UGC 05367	09 56 42.	+ 45 31	66	15.6	GALAXY DBL SYS
ZC 0956.7+5716	09 56 42.	+ 57 16	1880		CLUSTER OF GALAXIES
MCG-05-24-011	09 56 42.	- 30 06	96	14.5	GALAXY
REIZ 0124	09 56 43.	+ 30 59	18	15.7	GALAXY
RNGC 3074	09 56 43.	+ 35 38		15.0	GALAXY
KN 15.052	09 56 43.5	+ 35 38 00.			NEBULA
RNGC 3081	09 56 44.	- 22 33		12.5	GALAXY
MCG+07-21-004	09 56 45.	+ 41 55	24	17.	GALAXY
REIZ 0122	09 56 45.	+ 45 32	36	14.7	GALAXY
RNGC 3084	09 56 46.	- 26 54		14.0	GALAXY
REIZ 0125	09 56 47.	+ 35 39	120	14.0	GALAXY
ZWG 064.018	09 56 48.	+ 11 26		15.1	GALAXY
TON-N 0475	09 56 48.	+ 26 11		15.4	BLUE STAR
MCG+06-22-048	09 56 48.	+ 38 11	36	15.	GALAXY
ZWG 239.050	09 56 48.	+ 46 29		14.8	GALAXY
UGC 05368	09 56 48.	+ 46 29	66	14.8	GALAXY
MCG+08-18-054	09 56 48.	+ 46 38	48	15.	GALAXY
MRK 130	09 56 48.	+ 47 32	8	16.5	GALAXY WITH UV CONTINUUM
SHAH 118	09 56 48.	+ 52 29	72	18.6	GROUP OF COMPACT GALAXIES
MCG+12-10-013	09 56 48.	+ 71 12 30.	33	15.	GALAXY
MCG-04-24-010	09 56 48.	- 26 54	54	14.	GALAXY
TON-N 1136	09 56 49.	+ 21 44		16.8	BLUE STAR
ABC 0908	09 56 50.	+ 22 40		18.4	RICH CLUSTER OF GALAXIES
REIZ 0127	09 56 52.	+ 34 34	36	15.0	GALAXY
ZWG 008.008	09 56 54.	+ 00 00		14.7	GALAXY
MCG+00-26-001	09 56 54.	+ 00 00	54	14.7	GALAXY
ZC 0956.9+4411	09 56 54.	+ 44 11	1550		CLUSTER OF GALAXIES
ZWG 239.051	09 56 54.	+ 45 45		15.6	GALAXY
MCG+08-18-055	09 56 54.	+ 45 45	36	16.	GALAXY
ZC 0956.9+5332	09 56 54.	+ 53 32	740		CLUSTER OF GALAXIES
MCG+09-17-004	09 56 54.	+ 54 46 30.	102	15.	GALAXY
ZWG 266.004	09 56 54.	+ 54 47		14.9	GALAXY
ZWG 265.053	09 56 54.	+ 54 47		14.9	GALAXY
UGC 05369	09 56 54.	+ 54 47	90	14.9	GALAXY S
ZC 0956.9+6630	09 56 54.	+ 66 30	200		CLUSTER OF GALAXIES
FATH 1.325	09 56 56.	- 00 01	22		NEBULA
LB 03093	09 56 57.	+ 01 47 00.		16.2	FAINT BLUE STAR
REIZ 0126	09 56 57.	+ 45 47	36	13.9	GALAXY
MCG+09-17-005	09 56 57.	+ 55 31	18	17.	GALAXY
UGC 05370	09 57 00.	+ 00 48	60	16.0	GALAXY Sb
ZC 0957.0+0406	09 57 00.	+ 04 06	1880		CLUSTER OF GALAXIES
ZWG 064.019	09 57 00.	+ 10 12		15.2	GALAXY
ZWG 064.020	09 57 00.	+ 10 34		15.2	GALAXY
ZWG 064.021	09 57 00.	+ 11 54		14.6	GALAXY
MCG+02-26-013	09 57 00.	+ 11 54	60	14.6	GALAXY
ZWG 064.022	09 57 00.	+ 13 13		14.8	GALAXY
MCG+02-26-014	09 57 00.	+ 13 13	36	14.8	GALAXY
REIZ 0128	09 57 00.	+ 13 16	12	14.7	GALAXY
ZC 0957.0+1359	09 57 00.	+ 13 59	1140		CLUSTER OF GALAXIES
ZWG 064.023	09 57 00.	+ 14 18		15.7	GALAXY
ZWG 153.011	09 57 00.	+ 29 32		15.7	GALAXY
TON-N 0476	09 57 00.	+ 31 42		15.4	BLUE STAR
TON-N 1137	09 57 00.	+ 35 58		14.9	BLUE STAR
MCG+09-17-006	09 57 00.	+ 55 32	36	17.	GALAXY
ABC 0905	09 57 00.	+ 57 15		17.8	RICH CLUSTER OF GALAXIES
MCG+10-14-057	09 57 00.	+ 58 20	24	16.	GALAXY
MCG+12-10-014	09 57 00.	+ 72 24	84	12.9	GALAXY
MCG-02-26-010	09 57 00.	- 12 08	48	15.	GALAXY
MCG-02-26-011	09 57 00.	- 13 17	42	15.	GALAXY
MCG-05-24-012	09 57 00.	- 30 01	168	14.	GALAXY
MCG-06-22-005	09 57 00.	- 33 59	90	12.5	GALAXY
RNGC 3087	09 57 01.	- 33 59			GALAXY
KN 15.054	09 57 01.7	+ 33 39 20.			NEBULA
IC 0585	09 57 03.	+ 13 15			NONSTELLAR OBJECT
IC 2527	09 57 04.	+ 38 24 38.			NONSTELLAR OBJECT
KN 15.055	09 57 04.9	+ 32 29 58.			NEBULA
ZC 0957.1+0309	09 57 06.	+ 03 09	1880		CLUSTER OF GALAXIES
ZWG 064.024	09 57 06.	+ 11 44		15.7	GALAXY
UGC 05371	09 57 06.	+ 13 13	84	14.8	GALAXY E-SO

OBJECT NAME	RIGHT ASCEN.	DECLINATION	DIAM.	MAGN.	TYPE OF OBJECT
ZWG 182.055	09 57 06.	+ 38 24		15.2	GALAXY
ZC 0957.1+4607	09 57 06.	+ 46 07	740		CLUSTER OF GALAXIES
ZC 0957.1+5628	09 57 06.	+ 56 28	470		CLUSTER OF GALAXIES
ZWG 333.009	09 57 06.	+ 71 14		15.6	GALAXY
KN 15.053	09 57 06.3	+ 38 24 46.			NEBULA
IC 2529	09 57 09.	- 22 36 13.			NONSTELLAR OBJECT
ZWG 008.010	09 57 12.	- 01 55		15.5	GALAXY
ZWG 008.009	09 57 12.	- 02 56		15.6	GALAXY
MCG-04-24-012	09 57 12.	- 22 35	84	12.5	GALAXY
MCG-04-24-011	09 57 12.	- 26 37	90	14.	GALAXY
RNGC 3085	09 57 13.	- 19 15		14.0	GALAXY
RNGC 3080	09 57 15.	+ 13 17		14.5	GALAXY
MCG-03-26-003	09 57 15.	- 19 15	54	14.5	GALAXY
RNGC 3089	09 57 16.	- 28 04		13.0	GALAXY
ZWG 064.025	09 57 18.	+ 13 17		14.5	GALAXY
UGC 05372	09 57 18.	+ 13 17	60	14.5	GALAXY Sa
MCG+02-26-015	09 57 18.	+ 13 17	48	14.5	GALAXY
REIZ 0129	09 57 18.	+ 13 20	12	14.5	GALAXY
ZC 0957.3+4335	09 57 18.	+ 43 35	740		CLUSTER OF GALAXIES
ZWG 266.005	09 57 18.	+ 53 16		15.4	GALAXY
ZC 0957.3+6329	09 57 18.	+ 63 29	470		CLUSTER OF GALAXIES
ZWG 008.011	09 57 18.	- 02 38		14.2	GALAXY
MCG+00-26-002	09 57 18.	- 02 38	48	14.2	GALAXY
MCG-03-26-004	09 57 18.	- 20 33	48	14.5	GALAXY
MCG-05-24-013	09 57 18.	- 30 30	72	15.5	GALAXY
KN 15.056	09 57 18.6	+ 36 22 30.			NEBULA
RNGC 3063	09 57 20.	+ 72 22			NON-EXISTENT OBJECT
RNGC 3083	09 57 20.	- 02 38		14.0	GALAXY
IC 0586	09 57 20.	- 06 40 49.			NONSTELLAR OBJECT
KN 15.057	09 57 20.1	+ 36 20 24.			NEBULA
ZWG 036.012	09 57 24.	+ 05 34		12.2	GALAXY
UGC 05373	09 57 24.	+ 05 34	360	12.2	GALAXY DWRF IR
MCG+01-26-005	09 57 24.	+ 05 34	210	12.2	GALAXY
KARA.73B 0388	09 57 24.	+ 05 34	324	12.2	ISOLATED GALAXY IR
ZWG 064.026	09 57 24.	+ 13 47		15.1	GALAXY
ZC 0957.4+1749	09 57 24.	+ 17 49	610		CLUSTER OF GALAXIES
ZC 0957.4+2002	09 57 24.	+ 20 02	810		CLUSTER OF GALAXIES
ZC 0957.4+3757	09 57 24.	+ 37 57	870		CLUSTER OF GALAXIES
ZC 0957.4+4105	09 57 24.	+ 41 05	540		CLUSTER OF GALAXIES
ZWG 266.006	09 57 24.	+ 55 51		13.8	GALAXY
ZWG 265.054	09 57 24.	+ 55 51		13.8	GALAXY
UGC 05374	09 57 24.	+ 55 51	72	13.8	GALAXY S0
HOLM 156B	09 57 24.	+ 55 52	48	13.5	PART OF MULTIPLE GALAXY
MRK 131	09 57 24.	+ 55 54	25	14.	GALAXY WITH UV CONTINUUM
MCG+09-17-007	09 57 24.	+ 55 54	36	13.5	GALAXY
ZC 0957.4+7450	09 57 24.	+ 74 50	4500		CLUSTER OF GALAXIES
ZWG 351.002	09 57 24.	+ 79 43		15.5	GALAXY
ZWG 350.037	09 57 24.	+ 79 43		15.5	GALAXY
MCG-01-26-004	09 57 24.	- 06 41	24	15.	GALAXY
MCG-05-24-014	09 57 24.	- 28 05	84	13.	GALAXY
MCG-03-26-005	09 57 27.	- 20 34	36	14.5	GALAXY
REIN 2.102	09 57 28.69	+ 55 51 39.7			NEBULA
KEEL 198	09 57 28.8	+ 55 51 34.		11.	NEBULA
ZC 0957.5+1246	09 57 30.	+ 12 46	1950		CLUSTER OF GALAXIES
ZC 0957.5+3850	09 57 30.	+ 38 50	1480		CLUSTER OF GALAXIES
MCG+09-17-008	09 57 30.	+ 54 01	30	17.	GALAXY
RNGC 3073	09 57 30.	+ 55 52		14.0	GALAXY
IC 0587	09 57 34.	- 02 13 32.			SAME AS BD-01 2334
ZWG 008.013	09 57 36.	+ 02 24		14.6	GALAXY
MCG+00-26-004	09 57 36.	+ 02 24	24	14.6	GALAXY
ZC 0957.6+2241	09 57 36.	+ 22 41	1950		CLUSTER OF GALAXIES
TON-N 0477	09 57 36.	+ 28 14		16.3	BLUE STAR
MCG+10-14-058	09 57 36.	+ 60 04	78	16.	GALAXY
ZWG 333.010	09 57 36.	+ 72 25		12.9	GALAXY
UGC 05375	09 57 36.	+ 72 25	114	12.9	GALAXY S0
ZWG 008.012	09 57 36.	- 02 44		14.5	GALAXY
MCG+00-26-003	09 57 36.	- 02 44	60	14.5	GALAXY
REIN 2.100	09 57 36.77	+ 72 24 39.0			NEBULA
RNGC 3065	09 57 38.	+ 72 25		13.0	GALAXY
RNGC 3086	09 57 38.	- 02 44		14.5	GALAXY
MCG-01-26-005	09 57 39.	- 06 11	48	14.	GALAXY
IC 2531	09 57 39.	- 29 24 51.			NONSTELLAR OBJECT
LB 00563	09 57 40.	+ 00 29 00.		16.5	FAINT BLUE STAR
ZWG 008.014	09 57 42.	+ 00 45		15.7	GALAXY
ZWG 064.027	09 57 42.	+ 11 35		15.1	GALAXY
ZC 0957.7+5059	09 57 42.	+ 50 59	2150		CLUSTER OF GALAXIES
ZC 0957.7+5525	09 57 42.	+ 55 25	870		CLUSTER OF GALAXIES
7ZW 303	09 57 42.	+ 72 25			COMPACT GALAXY
SHB 149	09 57 43.8	+ 00 19 50.		17.6	QUASI-STELLAR OBJECT
BC PKS0957+00	09 57 43.84	+ 00 19 50.0		17.57	QUASI-STELLAR OBJECT
MCG+09-17-009	09 57 45.	+ 55 59	42	15.	GALAXY
MCG-03-26-006	09 57 45.	- 19 23	18	15.	GALAXY
ZWG 008.015	09 57 48.	+ 00 57		15.1	GALAXY
ZWG 036.013	09 57 48.	+ 03 31		15.6	GALAXY
ZWG 036.014	09 57 48.	+ 03 37		14.5	GALAXY
UGC 05376	09 57 48.	+ 03 37	126	14.5	GALAXY S-IRR
MCG+01-26-006	09 57 48.	+ 03 37	108	14.5	GALAXY
ZWG 064.028	09 57 48.	+ 13 07		15.1	GALAXY
ZWG 123.011	09 57 48.	+ 21 14		15.7	GALAXY
ZC 0957.8+3146	09 57 48.	+ 31 46	870		CLUSTER OF GALAXIES
ZC 0957.8+3345	09 57 48.	+ 33 45	340		CLUSTER OF GALAXIES
ZC 0957.8+4838	09 57 48.	+ 48 38	810		CLUSTER OF GALAXIES
ZWG 266.007	09 57 48.	+ 55 57		14.6	GALAXY
ZWG 265.055	09 57 48.	+ 55 57		14.6	GALAXY
MCG-03-26-007	09 57 48.	- 19 24	48	12.5	GALAXY
MCG-05-24-015	09 57 48.	- 29 23	420	13.	GALAXY
RNGC 3091	09 57 49.	- 19 23		13.0	GALAXY
KEEL 199	09 57 49.7	+ 56 04 19.		18.	NEBULA
ARC 0906	09 57 50.	+ 65 38		17.4	RICH CLUSTER OF GALAXIES
KEEL 200	09 57 51.3	+ 55 57 34.		15.	NEBULA
RNGC 3095	09 57 53.	- 31 18		12.5	GALAXY
HN 0324	09 57 53.	- 33 59			NEBULA
IC 2532	09 57 53.	- 34 00			NONSTELLAR OBJECT
REIN 2.101	09 57 53.14	+ 72 21 58.8			NEBULA
ZWG 036.015	09 57 54.	+ 03 27		14.8	GALAXY
UGC 05377	09 57 54.	+ 03 27	96	14.8	GALAXY
MCG+01-26-007	09 57 54.	+ 03 27	72	14.8	GALAXY
ZWG 036.016	09 57 54.	+ 04 39		14.2	GALAXY
UGC 05378	09 57 54.	+ 04 39	108	14.2	GALAXY Sb
MCG+01-26-008	09 57 54.	+ 04 39	48	14.2	GALAXY
ZC 0957.9+1111	09 57 54.	+ 11 11	6050		CLUSTER OF GALAXIES
ZWG 289.026	09 57 54.	+ 56 54		15.6	GALAXY
MCG+10-14-059	09 57 54.	+ 56 55	36	16.	GALAXY
ZC 0957.9+6538	09 57 54.	+ 65 38	3020		CLUSTER OF GALAXIES
ZWG 333.011	09 57 54.	+ 72 22		12.8	GALAXY
ZCG 0957+72	09 57 54.	+ 72 22		12.8	COMPACT GALAXY
UGC 05379	09 57 54.	+ 72 22	84	12.8	GALAXY S
MCG-06-22-006	09 57 54.	- 38 57	72	14.5	GALAXY

OBJECT NAME	RIGHT ASCEN.	DECLINATION	DIAM.	MAGN.	TYPE OF OBJECT
RNGC 3066	09 57 56.	+ 72 22		13.0	GALAXY
KN 15.058	09 57 56.4	+ 37 09 17.			NEBULA
MCG+06-22-049	09 57 57.	+ 38 23	27	16.	GALAXY
ARC 0911	09 57 58.	- 15 09		17.5	RICH CLUSTER OF GALAXIES
ZC 0958.0+0654	09 58 00.	+ 06 54	400		CLUSTER OF GALAXIES
ZWG 064.029	09 58 00.	+ 12 11		15.7	GALAXY
REIZ 0131	09 58 00.	+ 13 04	18	14.8	GALAXY
ZC 0958.0+1340	09 58 00.	+ 13 40	1750		CLUSTER OF GALAXIES
ZWG 093.019	09 58 00.	+ 15 31		15.5	GALAXY
ZWG 123.012	09 58 00.	+ 22 33		15.4	GALAXY
UGC 05381	09 58 00.	+ 22 33	66	15.4	GALAXY S
TON-N 0478	09 58 00.	+ 27 47		15.2	BLUE STAR
TON-N 1138	09 58 00.	+ 35 29		17.	BLUE STAR
ZWG 182.056	09 58 00.	+ 36 51		14.7	GALAXY
UGC 05382	09 58 00.	+ 36 51	60	14.7	GALAXY Sa-b
MCG+06-22-050	09 58 00.	+ 36 53	54	14.5	GALAXY
ZWG 008.017	09 58 00.	- 01 55		14.3	GALAXY
UGC 05380	09 58 00.	- 01 55	66	14.3	GALAXY SBa-b
MCG+00-26-006	09 58 00.	- 01 55	54	14.3	GALAXY
ZWG 008.016	09 58 00.	- 02 43		14.2	GALAXY
MKW 01	09 58 00.	- 02 43		14.2	POOR GALAXY CLUSTER
MCG+00-26-005	09 58 00.	- 02 43	24	14.2	GALAXY
MCG-01-26-006	09 58 00.	- 07 40	12	15.5	GALAXY
MCG-05-24-016	09 58 00.	- 31 20	180	12.5	GALAXY
MCG-06-22-007	09 58 00.	- 33 59	90	15.	GALAXY
KN 15.059	09 58 00.7	+ 37 29 11.			NEBULA
RNGC 3090	09 58 02.	- 02 43		14.0	GALAXY
MCG-01-26-007	09 58 03.	- 07 39	30	15.5	GALAXY
KN 15.060	09 58 04.6	+ 36 51 30.			NEBULA
REIZ 0130	09 58 05.	+ 36 52	36	14.1	GALAXY
KN 15.061	09 58 05.8	+ 37 40 15.			NEBULA
ZWG 036.017	09 58 06.	+ 04 58		14.6	GALAXY
UGC 05383	09 58 06.	+ 04 58	78	14.6	GALAXY S0
MCG+01-26-009	09 58 06.	+ 04 58	60	14.6	GALAXY
ZWG 093.020	09 58 06.	+ 17 18		15.6	GALAXY
MCG+03-26-013	09 58 06.	+ 17 40	54	17.	GALAXY
MCG+04-24-009	09 58 06.	+ 22 32 30.	60	15.4	GALAXY
MRK 132	09 58 06.	+ 55 02	6	15.	GALAXY WITH UV CONTINUUM
MCG+12-10-016	09 58 06.	+ 70 58	24	15.	GALAXY
MRK 133	09 58 06.	+ 72 22	40	14.5	GALAXY WITH UV CONTINUUM
VVI 34	09 58 06.	+ 72 22	45	14.03	SEYFERT GALAXY
MCG+12-10-015	09 58 06.	+ 72 22	45	13.5	GALAXY
SHB 150	09 58 08.0	+ 55 09 10.		16.0	QUASI-STELLAR OBJECT
KN 15.062	09 58 08.4	+ 37 09 41.			NEBULA
MCG-02-26-012	09 58 09.	- 14 44	90	14.	GALAXY
KN 15.063	09 58 09.6	+ 37 40 25.			NEBULA
REIZ 0132	09 58 11.	+ 11 27	30	14.6	GALAXY
ZWG 036.018	09 58 12.	+ 07 51		15.7	GALAXY
ZWG 182.057	09 58 12.	+ 37 05		15.5	GALAXY
MCG+06-22-051	09 58 12.	+ 37 07	24	15.	GALAXY
HOLM 157A	09 58 12.	+ 37 41	48	15.0	PART OF MULTIPLE GALAXY
ZC 0958.2+3744	09 58 12.	+ 37 44	400		CLUSTER OF GALAXIES
ZWG 008.018	09 58 12.	- 02 42		15.5	GALAXY
KN 15.064	09 58 12.2	+ 36 20 00.			NEBULA
KN 15.065	09 58 13.8	+ 37 05 31.			NEBULA
LB 03094	09 58 15.	+ 00 15 18.		14.6	FAINT BLUE STAR
KN 15.066	09 58 15.5	+ 37 18 40.			NEBULA
HOLM 157B	09 58 16.	+ 37 41	48	15.6	PART OF MULTIPLE GALAXY
ZWG 008.020	09 58 18.	+ 01 30		15.4	GALAXY
MCG+00-26-009	09 58 18.	+ 01 30	48	15.4	GALAXY
ZC 0958.3+0507	09 58 18.	+ 05 07	1550		CLUSTER OF GALAXIES
ZWG 064.030	09 58 18.	+ 11 25		15.0	GALAXY
MCG+02-26-016	09 58 18.	+ 11 25	36	15.0	GALAXY
ZWG 123.013	09 58 18.	+ 22 38		14.7	GALAXY
RNGC 3088B	09 58 18.	+ 22 38		16.0	GALAXY
RNGC 3088A	09 58 18.	+ 22 38		15.0	GALAXY
UGC 05384	09 58 18.	+ 22 38	84	14.7	GALAXY DBL SYS
ZC 0958.3+2813	09 58 18.	+ 28 13	2420		CLUSTER OF GALAXIES
MCG+00-26-007	09 58 18.	- 02 43	36	15.1	GALAXY
ZWG 008.019	09 58 18.	- 02 46		14.5	GALAXY
MCG+00-26-008	09 58 18.	- 02 46	66	14.5	GALAXY
MCG-03-26-008	09 58 18.	- 19 25	48	15.	GALAXY
TON-N 1139	09 58 19.	+ 21 09		16.5	BLUE STAR
RNGC 3096	09 58 19.	- 19 25		15.0	GALAXY
HN 0325	09 58 19.	- 31 00			NEBULA
RNGC 3092	09 58 20.	- 02 46		14.5	GALAXY
IC 2533	09 58 20.	- 31 01			NONSTELLAR OBJECT
RNGC 3100	09 58 23.	- 31 25		12.0	GALAXY
ZWG 036.019	09 58 24.	+ 03 23		15.1	GALAXY
ZWG 036.020	09 58 24.	+ 04 50		15.5	GALAXY
ZWG 064.031	09 58 24.	+ 11 42		14.9	GALAXY
MCG+02-26-017	09 58 24.	+ 11 42	36	14.9	GALAXY
ZWG 093.021	09 58 24.	+ 15 34		15.6	GALAXY
ZWG 093.022	09 58 24.	+ 17 10		15.3	GALAXY
UGC 05385	09 58 24.	+ 17 10	84	15.3	GALAXY Sc
MCG+03-26-014	09 58 24.	+ 17 10	72	15.3	GALAXY
MCG+04-24-011	09 58 24.	+ 22 38	30	14.7	GALAXY
MCG+04-24-010	09 58 24.	+ 22 38	30	14.7	GALAXY
REIZ 0134	09 58 24.	+ 22 39	18	14.5	GALAXY
ZWG 182.058	09 58 24.	+ 36 54		15.3	GALAXY
MCG+06-22-052	09 58 24.	+ 36 56	36	15.	GALAXY
ZC 0958.4+4101	09 58 24.	+ 41 01	1080		CLUSTER OF GALAXIES
ZC 0958.4+4132	09 58 24.	+ 41 32	2020		CLUSTER OF GALAXIES
ZC 0958.4+6635	09 58 24.	+ 66 35	810		CLUSTER OF GALAXIES
ZWG 333.012	09 58 24.	+ 70 59		15.2	GALAXY
UGC 05386	09 58 24.	+ 70 59	72	15.2	GALAXY Sb
ZWG 008.021	09 58 24.	- 02 43		15.1	GALAXY
MCG-02-26-013	09 58 24.	- 13 10	36	15.	GALAXY
MCG-05-24-017	09 58 24.	- 31 00	36	13.5	GALAXY
LB 03095	09 58 25.	+ 00 43 54.		16.5	FAINT BLUE STAR
RNGC 3093	09 58 26.	- 02 43		15.0	GALAXY
LB 03096	09 58 28.	+ 00 34 36.		16.2	FAINT BLUE STAR
KN 15.067	09 58 28.2	+ 36 54 42.			NEBULA
REIZ 0133	09 58 29.	+ 36 55	18	14.7	GALAXY
REIZ 0135	09 58 30.	+ 23 16	36	15.3	GALAXY
ZWG 182.059	09 58 30.	+ 37 26		15.0	GALAXY
MCG+06-22-053	09 58 30.	+ 37 28 30.	24	15.	GALAXY
ZWG 266.008	09 58 30.	+ 55 55		11.2	GALAXY
UGC 05387	09 58 30.	+ 55 55	522	11.2	GALAXY
HOLM 156A	09 58 30.	+ 55 56	270	11.8	PART OF MULTIPLE GALAXY
MCG+09-17-010	09 58 30.	+ 55 56	450	11.1	GALAXY
ZWG 290.001	09 58 30.	+ 59 30		15.0	GALAXY
ZWG 289.027	09 58 30.	+ 59 30		15.0	GALAXY
MCG-03-26-009	09 58 30.	- 20 08	54	14.5	GALAXY
IC 2530	09 58 31.	+ 37 25 14.			NONSTELLAR OBJECT
KN 15.068	09 58 32.5	+ 37 26 41.			NEBULA
ARC 0912	09 58 34.	+ 00 09		15.9	RICH CLUSTER OF GALAXIES
REIN 2.103	09 58 35.55	+ 55 55 05.8			NEBULA

OBJECT NAME	RIGHT ASCEN.	DECLINATION	DIAM.	MAGN.	TYPE OF OBJECT
ZWG 008.022	09 58 36.	+ 00 28		15.0	GALAXY
UGC 05388	09 58 36.	+ 00 28	66	15.0	GALAXY S?
MCG+00-26-010	09 58 36.	+ 00 28	54	15.0	GALAXY
ZWG 008.023	09 58 36.	+ 00 35		15.4	GALAXY
SHAH 119	09 58 36.	+ 37 57	72	17.4	GROUP OF COMPACT GALAXIES
ZWG 211.004	09 58 36.	+ 39 52		15.7	GALAXY
UGC 05389	09 58 36.	+ 39 52	120	15.7	GALAXY Sc
ZWG 266.009	09 58 36.	+ 53 30		14.8	GALAXY
RNGC 3079	09 58 36.	+ 55 55		12.0	GALAXY
MCG+10-15-001	09 58 36.	+ 59 30	18	16.	GALAXY
MCG+11-12-032	09 58 36.	+ 67 58	66	16.	GALAXY
MCG-02-26-014	09 58 36.	- 15 08 30.	36	15.	GALAXY
MCG-05-24-018	09 58 36.	- 31 26	30	12.5	GALAXY
MCG-06-22-008	09 58 36.	- 36 00	30	16.	GALAXY
MCG-06-22-009	09 58 37.	- 36 00	36	15.	GALAXY
SNO 26	09 58 39.	- 30		18.	LOOSE CLSTR OF 30 GLXIES
KEEL 201	09 58 39.0	+ 56 12 52.		17.	NEBULA
ZWG 093.023	09 58 42.	+ 16 01		13.5	GALAXY
UGC 05390	09 58 42.	+ 16 01	96	13.5	GALAXY SBa
ZC 0958.7+2000	09 58 42.	+ 20 00	810		CLUSTER OF GALAXIES
ZC 0958.7+3033	09 58 42.	+ 30 33	1880		CLUSTER OF GALAXIES
TON-N 1140	09 58 42.	+ 35 57		17.	BLUE STAR
ZWG 182.060	09 58 42.	+ 37 29		14.9	GALAXY
UGC 05391	09 58 42.	+ 37 29	138	14.9	GALAXY
HOLM 158A	09 58 42.	+ 37 30	60	14.1	PART OF MULTIPLE GALAXY
MCG+06-22-054	09 58 42.	+ 37 31	132	14.	GALAXY
MCG+07-21-005	09 58 42.	+ 39 51	108	15.	GALAXY
KN 15.069	09 58 42.6	+ 37 29 37.			NEBULA
RNGC 3094	09 58 44.	+ 16 01		13.5	GALAXY
REIZ 0136	09 58 44.	+ 33 23	72	14.1	GALAXY
SEY 063	09 58 44.	+ 67 58 39.		14.3	FAINT GALAXY
KN 15.070	09 58 44.5	+ 37 30 30.			NEBULA
MCG+03-26-015	09 58 45.	+ 16 00	120	13.5	GALAXY
HOLM 158B	09 58 45.	+ 37 31	24	15.2	PART OF MULTIPLE GALAXY
KN 15.071	09 58 45.3	+ 37 16 21.			NEBULA
KN 15.073	09 58 46.4	+ 33 22 42.			NEBULA
HOLM 159A	09 58 47.	+ 36 45	66	14.7	PART OF MULTIPLE GALAXY
ZWG 064.032	09 58 48.	+ 14 15		15.3	GALAXY
ZWG 093.024	09 58 48.	+ 15 22		15.5	GALAXY
ZC 0958.8+1906	09 58 48.	+ 19 06	2820		CLUSTER OF GALAXIES
UGC 05392	09 58 48.	+ 21 51	72	16.0	GALAXY Sc
ZWG 182.061	09 58 48.	+ 33 22		14.9	GALAXY
UGC 05393	09 58 48.	+ 33 22	138	14.9	GALAXY
MCG+06-22-055	09 58 48.	+ 33 23	120	14.	GALAXY
UGC 05394	09 58 48.	+ 36 45	96	16.5	GALAXY Sc
HOLM 159B	09 58 48.	+ 36 46	24	15.4	PART OF MULTIPLE GALAXY
ZC 0958.8+3755	09 58 48.	+ 37 55	740		CLUSTER OF GALAXIES
ZC 0958.8+4015	09 58 48.	+ 40 15	740		CLUSTER OF GALAXIES
ZWG 312.030	09 58 48.	+ 67 58		15.0	GALAXY
MCG-03-26-010	09 58 48.	- 19 13	36	15.	GALAXY
KN 15.072	09 58 49.6	+ 36 44 12.			NEBULA
ZC 0958.9+0038	09 58 54.	+ 00 38	6250		CLUSTER OF GALAXIES
ZC 0958.9+0702	09 58 54.	+ 07 02	940		CLUSTER OF GALAXIES
ZWG 036.021	09 58 54.	+ 07 10		15.6	GALAXY
ZWG 064.033	09 58 54.	+ 13 00		14.6	GALAXY
UGC 05395	09 58 54.	+ 13 00	72	14.6	GALAXY E-SO
MCG+02-26-018	09 58 54.	+ 13 00	24	14.6	GALAXY
ZWG 093.025	09 58 54.	+ 20 02		15.6	GALAXY
MCG+03-26-016	09 58 54.	+ 20 02	30	15.6	GALAXY
ZC 0958.9+2950	09 58 54.	+ 29 50	2550		CLUSTER OF GALAXIES
ZC 0958.9+4228	09 58 54.	+ 42 28	870		CLUSTER OF GALAXIES
MCG+13-07-042	09 58 54.	+ 76 01	9	17.	GALAXY
RNGC 3105	09 58 55.	- 54 32		11.0	OPEN CLUSTER
REIZ 0137	09 58 58.	+ 11 01	90	14.1	GALAXY
KN 15.074	09 58 58.8	+ 37 23 52.			NEBULA
BC PKS0959-443	09 58 58.91	- 44 23 25.1			QUASI-STELLAR OBJECT
KN 15.075	09 58 59.9	+ 37 55 27.			NEBULA
VDB.66G 067	09 59	+ 80 33	100		DWARF GALAXY
LBN 1090	09 59	- 32 00	5100		BRIGHT NEBULA
ZWG 036.022	09 59 00.	+ 03 50		15.4	GALAXY
ZC 0959.0+0438	09 59 00.	+ 04 38	1280		CLUSTER OF GALAXIES
ZWG 064.034	09 59 00.	+ 11 00		14.6	GALAXY
UGC 05396	09 59 00.	+ 11 00	102	14.6	GALAXY Sc
MCG+02-26-019	09 59 00.	+ 11 00	84	14.6	GALAXY
ZWG 093.026	09 59 00.	+ 15 18		15.6	GALAXY
TON-N 0479	09 59 00.	+ 30 34		15.2	BLUE STAR
ZC 0959.0+3354	09 59 00.	+ 33 54	810		CLUSTER OF GALAXIES
ZC 0959.0+5217	09 59 00.	+ 52 17	1210		CLUSTER OF GALAXIES
7ZW 304	09 59 00.	+ 63 30			COMPACT GALAXY
ZC 0959.0+6722	09 59 00.	+ 67 22	2420		CLUSTER OF GALAXIES
OCL 0798	09 59 00.	- 54 32	480	11.1	OPEN STAR CLUSTER
VHA 082	09 59 00.	- 54 32	150		OPEN STAR CLUSTEP
TON-N 1141	09 59 01.	+ 21 27		16.8	BLUE STAR
KN 15.076	09 59 02.6	+ 36 47 13.			NEBULA
MCG+09-17-011	09 59 03.	+ 54 06	36	15.	GALAXY
ZWG 064.035	09 59 06.	+ 14 01		15.7	GALAXY
TON-N 0026	09 59 06.	+ 28 15		15.5	BLUE STAR
ZWG 008.024	09 59 06.	- 02 45		15.6	GALAXY
MCG+00-26-011	09 59 06.	- 02 45	60	15.6	GALAXY
MCG-02-26-016	09 59 06.	- 14 58	54	15.5	GALAXY
MCG-02-26-015	09 59 06.	- 14 58	15	15.5	GALAXY
TON-N 1142	09 59 07.	+ 20 04		16.1	BLUE STAR
ARC 0910	09 59 08.	+ 67 25		17.5	RICH CLUSTER OF GALAXIES
RNGC 3101	09 59 08.	- 02 45		15.5	GALAXY
MCG-01-26-008	09 59 09.	- 06 26	60	15.5	GALAXY
MCG-02-26-017	09 59 09.	- 13 18 30.	60	15.	GALAXY
LB 03098	09 59 10.	+ 00 19 42.		15.8	FAINT BLUE STAR
LB 03097	09 59 10.	+ 00 50 06.		16.4	FAINT BLUE STAR
ZC 0959.2+1327	09 59 12.	+ 13 27	1010		CLUSTER OF GALAXIES
ZC 0959.2+1520	09 59 12.	+ 15 20	3020		CLUSTER OF GALAXIES
ZC 0959.2+2656	09 59 12.	+ 26 56	2150		CLUSTER OF GALAXIES
MCG+09-17-012	09 59 12.	+ 52 07 30.	30	16.	GALAXY
7ZW 305	09 59 12.	+ 62 08			COMPACT GALAXY
MCG-01-26-009	09 59 12.	- 06 17	30	17.	GALAXY
MCG-01-26-011	09 59 12.	- 08 01	48	15.5	GALAXY
MCG-02-26-018	09 59 12.	- 13 27	36	15.	GALAXY
MCG-03-26-011	09 59 15.	- 19 18	60	15.	GALAXY
RNGC 3103	09 59 17.	- 31 27			NON-EXISTENT OBJECT
HN 0326	09 59 17.	- 33 52			NEBULA
IC 2534	09 59 17.	- 33 53			NONSTELLAR OBJECT
ZWG 182.062	09 59 18.	+ 34 05		15.3	GALAXY
MCG+08-18-056	09 59 18.	+ 45 00 30.	42	15.	GALAXY
ZC 0959.3+6810	09 59 18.	+ 68 10	2150		CLUSTER OF GALAXIES
MCG-06-22-011	09 59 18.	- 33 51	60	14.	GALAXY
MCG-06-22-010	09 59 18.	- 34 31	36	15.	GALAXY
KN 15.077	09 59 18.6	+ 34 34 10.			NEBULA
KN 15.078	09 59 19.3	+ 34 06 00.			NEBULA
REIZ 0138	09 59 21.	+ 34 06	24	14.4	GALAXY
MCG+06-22-056	09 59 21.	+ 34 07	36	16.	GALAXY
REIZ 0139	09 59 21.	+ 34 35	12	15.4	GALAXY
ZC 0959.4+0844	09 59 24.	+ 08 44	1010		CLUSTER OF GALAXIES
ZWG 093.027	09 59 24.	+ 18 04		15.6	GALAXY
ZWG 123.014	09 59 24.	+ 24 57		13.0	GALAXY
UGC 05397	09 59 24.	+ 24 57	138	13.0	GALAXY S0-a
KARA.73B 0389	09 59 24.	+ 24 57	132	13.0	ISOLATED GALAXY S
TON-N 0027	09 59 24.	+ 29 42		15.0	BLUE STAR
TON-N 1143	09 59 24.	+ 34 07		14.8	BLUE STAR
ZWG 239.052	09 59 24.	+ 45 02		15.0	GALAXY
MCG+09-17-013	09 59 24.	+ 55 10	18	16.	GALAXY
ZWG 333.013	09 59 24.	+ 68 58		10.7	GALAXY
UGC 05398	09 59 24.	+ 68 58	390	10.7	GALAXY IRR
MCG+12-10-017	09 59 24.	+ 68 58	204	10.5	GALAXY
RNGC 3077	09 59 24.	+ 68 59		11.5	GALAXY
REIZ 0141	09 59 26.	+ 24 58	78	13.0	GALAXY
RNGC 3098	09 59 29.	+ 24 57		13.5	GALAXY
SVEN 237	09 59 29.	- 25 47	42	15.2	GALAXY
ZWG 064.036	09 59 30.	+ 11 44		15.5	GALAXY
ZWG 093.028	09 59 30.	+ 19 27		15.7	GALAXY
MCG+04-24-012	09 59 30.	+ 24 57 30.	108	13.0	GALAXY
ZWG 266.010	09 59 30.	+ 51 46		15.3	GALAXY
MCG+09-17-014	09 59 30.	+ 55 04	30	16.	GALAXY
ZWG 008.025	09 59 30.	- 02 01		15.7	GALAXY
IC 0588	09 59 31.	+ 03 17 57.			NONSTELLAR OBJECT
REIZ 0140	09 59 32.	+ 45 03	48	13.8	GALAXY
REIZ 0142	09 59 33.	+ 26 47	24	14.7	GALAXY
KN 15.079	09 59 35.5	+ 37 20 10.			NEBULA
KN 15.081	09 59 35.7	+ 32 57 26.			NEBULA
ZWG 036.023	09 59 36.	+ 03 17		14.9	GALAXY
UGC 05399	09 59 36.	+ 03 17	66	14.9	GALAXY SBa
MCG+01-26-010	09 59 36.	+ 03 17	30	14.9	GALAXY
ZWG 064.037	09 59 36.	+ 11 56		15.7	GALAXY
ZWG 064.038	09 59 36.	+ 13 56		14.3	GALAXY
UGC 05400	09 59 36.	+ 13 56	72	14.3	GALAXY S0
MCG+02-26-020	09 59 36.	+ 13 56	23	14.3	GALAXY
ZC 0959.6+3257	09 59 36.	+ 32 57	2290		CLUSTER OF GALAXIES
ZWG 182.063	09 59 36.	+ 35 10		15.6	GALAXY
ZWG 266.011	09 59 36.	+ 55 04		15.7	GALAXY
MCG-02-26-019	09 59 36.	- 15 01	30	15.	GALAXY
MCG-02-26-020	09 59 36.	- 15 04	60	15.	GALAXY
MCG-02-26-021	09 59 36.	- 15 04 30.	15	15.5	GALAXY
HOLM 160B	09 59 37.	+ 32 58	30	14.9	PART OF MULTIPLE GALAXY
KN 15.080	09 59 37.6	+ 35 10 37.			NEBULA
REIZ 0143	09 59 38.	+ 32 58	18	15.3	GALAXY
MCG+06-22-057	09 59 39.	+ 35 11	42	15.	GALAXY
KN 15.082	09 59 41.7	+ 32 56 52.			NEBULA
ZWG 064.039	09 59 42.	+ 13 50		15.5	GALAXY
ZWG 182.064	09 59 42.	+ 32 56		15.4	GALAXY
MCG+06-22-058	09 59 42.	+ 32 57	9	16.	GALAXY
ZC 0959.7+3752	09 59 42.	+ 37 52	270		CLUSTER OF GALAXIES
MRK 134	09 59 42.	+ 43 25	11	17.	GALAXY WITH UV CONTINUUM
MCG+10-15-002	09 59 42.	+ 57 59	18	16.	GALAXY
7ZW 306	09 59 42.	+ 65 20			COMPACT GALAXY
MCG+13-07-043	09 59 42.	+ 79 30	39	14.	GALAXY
MCG-03-26-012	09 59 42.	- 19 43	72	15.5	GALAXY
REIZ 0146	09 59 43.	+ 15 50	18	15.2	GALAXY
RNGC 3099A	09 59 44.	+ 32 56		15.5	GALAXY
RNGC 3099B	09 59 44.	+ 32 57		16.0	GALAXY
HOLM 160A	09 59 44.	+ 32 57	36	14.5	PART OF MULTIPLE GALAXY
REIZ 0145	09 59 44.	+ 32 58	36	14.6	GALAXY
MCG+06-22-059	09 59 45.	+ 32 56 30.	24	15.	GALAXY
ARC 0913	09 59 47.	+ 20 44		18.0	RICH CLUSTER OF GALAXIFS
ZWG 064.040	09 59 48.	+ 14 11		15.5	GALAXY
UGC 05401	09 59 48.	+ 19 15	84	17.	GALAXY DWARF
MCG+03-26-017	09 59 48.	+ 19 18	66	17.	GALAXY
ZC 0959.8+2045	09 59 48.	+ 20 45	810		CLUSTER OF GALAXIES
ZWG 211.005	09 59 48.	+ 42 39		15.6	GALAXY
ZWG 351.003	09 59 48.	+ 79 30		14.8	GALAXY
ZWG 350.038	09 59 48.	+ 79 30		14.8	GALAXY
UGC 05402	09 59 48.	+ 79 30	96	14.8	GALAXY Sb
KARA.73B 0390	09 59 48.	+ 79 30	66	14.8	ISOLATED GALAXY S
REIZ 0144	09 59 51.	+ 45 53	42	14.2	GALAXY
MCG-02-26-022	09 59 51.	- 10 36	30	15.5	GALAXY
REIZ 0147	09 59 52.	+ 19 26	54	14.3	GALAXY
ZWG 093.029	09 59 54.	+ 19 25		14.6	GALAXY
UGC 05403	09 59 54.	+ 19 25	78	14.6	GALAXY S0-a
MCG+03-26-018	09 59 54.	+ 19 26	78	14.6	GALAXY
ZC 0959.9+3512	09 59 54.	+ 35 12	870		CLUSTER OF GALAXIES
MCG+08-18-057	09 59 54.	+ 45 50	48	15.	GALAXY
ZWG 239.053	09 59 54.	+ 45 51		15.7	GALAXY
MCG+09-17-015	09 59 54.	+ 54 10	60	15.	GALAXY
MCG+10-15-003	09 59 54.	+ 59 59	39	14.	GALAXY
ZWG 008.026	09 59 54.	- 02 47		15.6	GALAXY
OCL 0799	09 59 54.	- 54 48	102	15.	OPEN STAR CLUSTEP
VHA 083	09 59 54.	- 54 48	360		OPEN STAR CLUSTER
VHA 084	09 59 54.	- 57 56	270		OPEN STAR CLUSTER
KN 15.083	09 59 56.4	+ 35 28 40.			NEBULA
RNGC 3057	09 59 59.	+ 80 32		14.0	GALAXY
VDB.66G 236	10 00	- 25 52	810		DWARF GALAXY
ZWG 008.027	10 00 00.	+ 00 33		15.2	GALAXY
ZWG 093.030	10 00 00.	+ 14 52		15.5	GALAXY
MCG+03-26-019	10 00 00.	+ 14 52	48	15.5	GALAXY
MCG+03-26-020	10 00 00.	+ 16 41	15	15.4	GALAXY
ZWG 123.015	10 00 00.	+ 21 55		15.6	GALAXY
ZWG 266.012	10 00 00.	+ 54 10		15.6	GALAXY
UGC 05405	10 00 00.	+ 54 10	72	15.6	GALAXY S0-a
MCG+11-12-033	10 00 00.	+ 63 50	45	15.	GALAXY
7ZW 307	10 00 00.	+ 77 53			COMPACT GALAXY
ZWG 364.015	10 00 00.	+ 80 32		14.2	GALAXY
UGC 05404	10 00 00.	+ 80 32	168	14.2	GALAXY S IV
VHA 085	10 00 00.	- 49 24	300		OPEN STAR CLUSTER
OCL 0801	10 00 00.	- 59 14	738	12.	OPEN STAR CLUSTER
REIZ 0148	10 00 01.	+ 14 52	12	15.3	GALAXY
TON-N 1144	10 00 01.	+ 22 12		16.9	BLUE STAR
LB 03099	10 00 03.	- 00 12 24.		15.2	FAINT BLUE STAR
SVEN 238	10 00 05.	- 27 00	18	15.2	GALAXY
ZWG 064.041	10 00 06.	+ 13 24		15.4	GALAXY
ZWG 093.031	10 00 06.	+ 16 42		15.4	GALAXY
ZWG 093.032	10 00 06.	+ 20 27		15.7	GALAXY
KARA.73B 0391	10 00 06.	+ 20 27	54	15.7	ISOLATFD GALAXY S
ZC 1000.1+3946	10 00 06.	+ 39 46	2690		CLUSTER OF GALAXIES
MCG+07-21-006	10 00 06.	+ 39 54	48	16.	GALAXY
ZC 1000.1+4901	10 00 06.	+ 49 01	3560		CLUSTER OF GALAXIES
ZC 1000.1+4911	10 00 06.	+ 49 11	3560		CLUSTER OF GALAXIES
UGC 05406	10 00 06.	+ 63 50	78	16.5	GALAXY Sc
MCG-01-26-012	10 00 06.	- 05 47	126	14.	GALAXY
LB 00116	10 00 06.	- 33 47		16.0	FAINT BLUE STAR

OBJECT NAME	RIGHT ASCEN.	DECLINATION	DIAM.	MAGN.	TYPE OF OBJECT
LB 00117	10 00 06.	- 34 27		16.9	FAINT BLUE STAR
KEEL 202	10 00 09.7	+ 55 45 53.		17.	NEBULA
KN 15.084	10 00 09.8	+ 33 17 23.			NEBULA
8ZW 1000+04.2	10 00 12.	+ 04 12		19.6	COMPACT GALAXY
ZWG 064.042	10 00 12.	+ 11 58		15.7	GALAXY
ZWG 064.043	10 00 12.	+ 12 24		15.7	GALAXY
UGC 05407	10 00 12.	+ 33 17	66	16.0	GALAXY S
MCG+06-22-060	10 00 12.	+ 33 18	60	15.	GALAXY
MCG+08-18-058	10 00 12.	+ 45 59	24	16.	GALAXY
ZWG 240.001	10 00 12.	+ 48 37		14.9	GALAXY
ZWG 239.054	10 00 12.	+ 48 37		14.9	GALAXY
KARA.73B 0392	10 00 12.	+ 48 37	36	14.9	ISOLATED GALAXY S
ZWG 351.004	10 00 12.	+ 77 53		15.7	GALAXY
ZWG 350.039	10 00 12.	+ 77 53		15.7	GALAXY
LB 00118	10 00 12.	- 27 44		16.0	FAINT BLUE STAR
LB 00119	10 00 12.	- 28 48		15.5	FAINT BLUE STAR
TON-N 1145	10 00 13.	+ 22 08		16.8	BLUE STAR
REIZ 0149	10 00 14.	+ 33 18	18	14.9	GALAXY
SVEN 239	10 00 17.	- 25 49 56.	12	15.4	GALAXY
ZWG 239.055	10 00 18.	+ 46 52		14.6	GALAXY
ZC 1000.3+4820	10 00 18.	+ 48 20	870		CLUSTER OF GALAXIES
ZC 1000.3+5644	10 00 18.	+ 56 44	810		CLUSTER OF GALAXIES
ZWG 290.002	10 00 18.	+ 59 40		14.2	GALAXY
ZWG 289.028	10 00 18.	+ 59 40		14.2	GALAXY
UGC 05408	10 00 18.	+ 59 40	33	14.2	GALAXY VY CMPT
KARA.73B 0393	10 00 18.	+ 59 40	24	14.2	ISOLATED GALAXY E
7ZW 308	10 00 18.	+ 59 41			COMPACT GALAXY
MCG+12-10-018	10 00 18.	+ 72 26	90	15.	GALAXY
MAI 049	10 00 20.	+ 66 58	74		DWARF SPHEROIDAL GALAXY
REIZ 0150	10 00 22.	+ 11 01	54	14.4	GALAXY
RNGC 3108	10 00 23.	- 31 27		13.0	GALAXY
ZWG 036.024	10 00 24.	+ 02 35		15.0	GALAXY
MCG+01-26-011	10 00 24.	+ 02 35	48	15.0	GALAXY
ZWG 064.044	10 00 24.	+ 10 59		15.0	GALAXY
UGC 05409	10 00 24.	+ 10 59	78	15.0	GALAXY S
MCG+02-26-021	10 00 24.	+ 10 59	72	15.0	GALAXY
ZC 1000.4+1555	10 00 24.	+ 15 55	810		CLUSTER OF GALAXIES
MCG+08-18-059	10 00 24.	+ 46 52 30.	30	14.	GALAXY
MCG+08-18-060	10 00 24.	+ 48 35	42	15.	GALAXY
MRK 025	10 00 24.	+ 59 41	18	15.	GALAXY WITH UV CONTINUUM
VVI 35	10 00 24.	+ 59 41	60	14.02	SEYFERT GALAXY
MCG+10-15-004	10 00 24.	+ 59 41	18	16.	GALAXY
ZC 1000.4+6024	10 00 24.	+ 60 24	1810		CLUSTER OF GALAXIES
MCG-03-26-013	10 00 24.	- 15 35	30	15.	GALAXY
MCG-05-24-019	10 00 24.	- 31 27	30	13.5	GALAXY
VHE 32A	10 00 24.	- 57 26	36		REFLECTION NEBULA
TON-N 1146	10 00 25.	+ 21 52		17.	BLUE STAR
KN 15.085	10 00 27.2	+ 37 16 48.			NEBULA
TON-N 0481	10 00 30.	+ 33 12		14.7	BLUE STAR
ZWG 182.065	10 00 30.	+ 37 42		15.7	GALAXY
UGC 05410	10 00 30.	+ 50 38	72	16.0	GALAXY Sc
MCG-02-26-023	10 00 30.	- 11 30 30.	36	15.	GALAXY
MCG-03-26-014	10 00 30.	- 19 50	36	15.	GALAXY
ACK 284-05.1	10 00 30.	- 61 43			PLANETARY NEBULA
KN 15.086	10 00 31.2	+ 37 42 03.			NEBULA
VV 119	10 00 33.	+ 40 58 30.	90	14.	INTERACTING GALAXY
KN 15.087	10 00 35.8	+ 37 41 17.			NEBULA
ZWG 064.045	10 00 36.	+ 12 22		15.6	GALAXY
ZWG 182.066	10 00 36.	+ 34 00		15.4	GALAXY
MCG+06-22-061	10 00 36.	+ 34 00	30	15.	GALAXY
TON-N 1147	10 00 36.	+ 36 11		17.	BLUE STAR
ZWG 182.067	10 00 36.	+ 37 41		15.5	GALAXY
HOLM 161B	10 00 36.	+ 37 43	24	14.6	PART OF MULTIPLE GALAXY
MCG+06-22-062	10 00 36.	+ 37 43	24	15.5	GALAXY
UGC 05412	10 00 36.	+ 50 36	72	16.0	GALAXY Sc
ZWG 266.013	10 00 36.	+ 54 57		15.6	GALAXY
MCG+09-17-016	10 00 36.	+ 54 57	54	15.	GALAXY
MCG+10-15-005	10 00 36.	+ 58 29	15	16.	GALAXY
ZWG 290.003	10 00 36.	+ 58 30		15.5	GALAXY
ZWG 289.029	10 00 36.	+ 58 30		15.5	GALAXY
MCG+10-15-006	10 00 36.	+ 61 59	39	16.	GALAXY
ZWG 008.028	10 00 36.	- 02 08		14.3	GALAXY
UGC 05411	10 00 36.	- 02 08	90	14.3	GALAXY S
MCG+00-26-012	10 00 36.	- 02 08	54	14.3	GALAXY
MCG-02-26-024	10 00 36.	- 15 08 30.	90	15.	GALAXY
KN 15.088	10 00 36.1	+ 34 00 00.			NEBULA
ABC 0909	10 00 37.	+ 75 05		17.7	RICH CLUSTER OF GALAXIES
HELW 014	10 00 37.	- 15 07			NEBULA
KN 15.089	10 00 40.9	+ 37 14 27.			NEBULA
LB 00564	10 00 41.	- 00 10 12.		16.2	FAINT BLUE STAR
ZWG 153.012	10 00 42.	+ 31 24		15.7	GALAXY
ZC 1000.7+3210	10 00 42.	+ 32 10	470		CLUSTER OF GALAXIES
HOLM 161A	10 00 42.	+ 37 42	30	14.2	PART OF MULTIPLE GALAXY
MCG+06-22-063	10 00 42.	+ 37 42	24	15.	GALAXY
ZC 1000.7+7503	10 00 42.	+ 75 03	810		CLUSTER OF GALAXIES
MCG-02-26-025	10 00 42.	- 15 15	30	15.	GALAXY
MCG-02-26-026	10 00 42.	- 15 16	60	16.	GALAXY
REIZ 0152	10 00 44.	+ 16 29	6	16.0	GALAXY
RNGC 3097	10 00 44.	+ 60 22			NON-EXISTENT OBJECT
MCG+02-03-027	10 00 45.	+ 13 46	36	15.	GALAXY
MCG+08-18-061	10 00 45.	+ 50 34	72	16.	GALAXY
HN 0039	10 00 45.	+ 60 22			NEBULA
RNGC 3109	10 00 45.	- 25 55		10.5	GALAXY
ARP 264	10 00 47.	+ 40 58			PECULIAR GALAXY
SVEN 240	10 00 47.	- 25 55	1080	11.5	GALAXY
ZC 1000.8+0639	10 00 48.	+ 06 39	940		CLUSTER OF GALAXIES
ZWG 064.046	10 00 48.	+ 12 20		15.5	GALAXY
UGC 05413	10 00 48.	+ 13 20	66	17.	GALAXY Sc
MCG+08-18-062	10 00 48.	+ 45 15	42	15.	GALAXY
ZC 1000.8+5725	10 00 48.	+ 57 25	1280		CLUSTER OF GALAXIES
MCG+12-10-019	10 00 48.	+ 70 47	30	16.	GALAXY
MCG-02-26-027	10 00 48.	- 14 43	90	14.	GALAXY
MCG-04-24-013	10 00 48.	- 25 55	1020	12.	GALAXY
REIZ 0151	10 00 50.	+ 33 39	42	14.4	GALAXY
RNGC 3104	10 00 53.	+ 41 00		14.0	GALAXY
KEEL 203	10 00 53.4	+ 56 06 07.		15.	NEBULA
ZC 1000.9+0922	10 00 54.	+ 09 22	540		CLUSTER OF GALAXIES
REIZ 0153	10 00 54.	+ 23 41	24	15.8	GALAXY
ZWG 123.016	10 00 54.	+ 23 46		15.7	GALAXY
ZWG 182.068	10 00 54.	+ 33 39		15.2	GALAXY
ZC 1000.9+3745	10 00 54.	+ 37 45	540		CLUSTER OF GALAXIES
ZWG 211.006	10 00 54.	+ 41 00		14.2	GALAXY
UGC 05414	10 00 54.	+ 41 00	216	14.2	GALAXY IRR
ZWG 266.014	10 00 54.	+ 56 06		15.7	GALAXY
MCG+12-10-020	10 00 54.	+ 70 34	24	15.	GALAXY
ZWG 333.014	10 00 54.	+ 72 26		14.5	GALAXY
UGC 05415	10 00 54.	+ 72 26	72	14.5	GALAXY S
ZWG 008.029	10 00 54.	- 01 18		15.6	GALAXY
MCG+00-26-013	10 00 54.	- 01 18	30	15.6	GALAXY
HELW 015	10 00 55.	- 14 42			NEBULA
KN 15.090	10 00 55.2	+ 33 38 39.			NEBULA
ZC 1001.0+0828	10 01 00.	+ 08 28	340		CLUSTER OF GALAXIES
ZC 1001.0+1225	10 01 00.	+ 12 25	1750		CLUSTER OF GALAXIES
MCG+03-26-021	10 01 00.	+ 16 40	60	15.4	GALAXY
ZWG 093.033	10 01 00.	+ 16 41		15.4	GALAXY
ZWG 093.034	10 01 00.	+ 17 10		15.7	GALAXY
KARA.73B 0394	10 01 00.	+ 17 10	42	15.7	ISOLATED GALAXY S
ZC 1001.0+2131	10 01 00.	+ 21 31	1410		CLUSTER OF GALAXIES
ZC 1001.0+2720	10 01 00.	+ 27 20	1010		CLUSTER OF GALAXIES
ZWG 211.007	10 01 00.	+ 38 55		15.7	GALAXY
KARA.73B 0395	10 01 00.	+ 38 55	36	15.7	ISOLATED GALAXY S
ZWG 211.008	10 01 00.	+ 39 23		15.7	GALAXY
UGC 05416	10 01 00.	+ 39 23	60	15.7	GALAXY SB?
MCG+07-21-008	10 01 00.	+ 39 23	66	15.	GALAXY
MCG+07-21-007	10 01 00.	+ 41 00	180	13.	GALAXY
ZWG 240.002	10 01 00.	+ 45 17		14.7	GALAXY
ZWG 266.015	10 01 00.	+ 53 38		15.7	GALAXY
UGC 05417	10 01 00.	+ 53 38	90	15.7	GALAXY SBb
MCG+09-17-017	10 01 00.	+ 53 38	60	16.	GALAXY
ZWG 290.004	10 01 00.	+ 60 21		14.3	GALAXY
ZWG 289.030	10 01 00.	+ 60 21		14.3	GALAXY
UGC 05418	10 01 00.	+ 60 21	54	14.3	GALAXY E-S0
MCG+10-15-007	10 01 00.	+ 60 21	420	14.	GALAXY
KARA.73B 0396	10 01 00.	+ 60 21	48	14.3	ISOLATED GALAXY E
MCG+10-15-008	10 01 00.	+ 60 42	39	16.	GALAXY
ZWG 333.015	10 01 00.	+ 70 48		15.0	GALAXY
MCG-06-22-012	10 01 00.	- 37 11	30	15.	GALAXY
RNGC 3102	10 01 02.	+ 60 21		14.5	GALAXY
RNGC 3106	10 01 03.	+ 31 25		14.0	GALAXY
ZWG 036.025	10 01 06.	+ 03 55		15.4	GALAXY
ZWG 153.013	10 01 06.	+ 31 25		14.0	GALAXY
UGC 05419	10 01 06.	+ 31 25	108	14.0	GALAXY S0
REIZ 0154	10 01 06.	+ 31 26	54	14.2	GALAXY
TON-N 0482	10 01 06.	+ 31 52		15.1	BLUE STAR
TON-N 1148	10 01 06.	+ 33 19		16.	BLUE STAR
TON-N 1149	10 01 06.	+ 34 05		14.7	BLUE STAR
MCG+07-21-009	10 01 06.	+ 38 54	30	16.	GALAXY
MCG+12-10-021	10 01 06.	+ 70 35	39	15.	GALAXY
MCG-03-26-015	10 01 06.	- 21 12	72	15.	GALAXY
OCL 0802	10 01 06.	- 59 52	3000	4.6	OPEN STAR CLUSTER
VHA 086	10 01 06.	- 59 52	180		OPEN STAR CLUSTER
RNGC 3114	10 01 06.	- 59 53		4.5	OPEN CLUSTER
KN 15.091	10 01 07.1	+ 33 19 58.			NEBULA
HELW 142	10 01 10.	- 19 01 52.			NEBULA
ZWG 093.035	10 01 12.	+ 14 40		15.3	GALAXY
ZWG 123.017	10 01 12.	+ 22 31		14.5	GALAXY
UGC 05420	10 01 12.	+ 22 31	66	14.5	GALAXY S0
TON-N 0028	10 01 12.	+ 29 13		15.5	BLUE STAR
MCG+05-24-009	10 01 12.	+ 31 26	78	14.0	GALAXY
ZWG 211.009	10 01 12.	+ 43 25		15.6	GALAXY
ZWG 266.016	10 01 12.	+ 51 10		15.6	GALAXY
ZWG 266.017	10 01 12.	+ 53 41		15.3	GALAXY
ZWG 333.016	10 01 12.	+ 70 35		15.7	GALAXY
ZC 1001.2+7831	10 01 12.	+ 78 31	870		CLUSTER OF GALAXIES
MCG-02-26-028	10 01 12.	- 14 54	42	15.	GALAXY
MCG-06-22-013	10 01 12.	- 38 37	36	15.	GALAXY
VHE 33D	10 01 12.	- 59 02	30		REFLECTION NEBULA
TON-N 1150	10 01 13.	+ 20 27		16.5	BLUE STAR
REIZ 0157	10 01 14.	+ 16 32	18	15.4	GALAXY
REIZ 0156	10 01 14.	+ 33 42	36	15.5	GALAXY
MCG+04-24-013	10 01 15.	+ 22 31	48	14.5	GALAXY
HN 0327	10 01 17.	- 33 42			NEBULA
KN 15.092	10 01 17.3	+ 33 41 57.			NEBULA
ZC 1001.3+0305	10 01 18.	+ 03 05	1410		CLUSTER OF GALAXIES
ZWG 064.047	10 01 18.	+ 14 27		15.2	GALAXY
UGC 05421	10 01 18.	+ 55 34	120	16.5	GALAXY Sc-IRR
MCG+09-17-018	10 01 18.	+ 55 34	90	16.	GALAXY
IC 2536	10 01 18.	- 33 42			NONSTELLAR OBJECT
MCG-06-22-014	10 01 18.	- 37 09	36	15.	GALAXY
TON-N 1151	10 01 19.	+ 20 26		16.	BLUE STAR
REIZ 0155	10 01 20.	+ 45 20	24	14.0	GALAXY
MCG+06-22-064	10 01 21.	+ 38 06	24	15.5	GALAXY
ZWG 008.030	10 01 24.	+ 01 45		15.4	GALAXY
ZWG 036.026	10 01 24.	+ 02 39		14.9	GALAXY
MCG+01-26-012	10 01 24.	+ 02 39	48	14.9	GALAXY
ZC 1001.4+3107	10 01 24.	+ 31 07	740		CLUSTER OF GALAXIES
TON-N 1152	10 01 24.	+ 34 24		15.9	BLUE STAR
ZWG 182.069	10 01 24.	+ 38 03		15.6	GALAXY
ZC 1001.4+4031	10 01 24.	+ 40 31	540		CLUSTER OF GALAXIES
ZC 1001.4+4912	10 01 24.	+ 49 12	810		CLUSTER OF GALAXIES
ZWG 290.005	10 01 24.	+ 57 50		15.6	GALAXY
ZWG 289.031	10 01 24.	+ 57 50		15.6	GALAXY
UGC 05422	10 01 24.	+ 57 50	66	15.6	GALAXY SBb
MCG+10-15-009	10 01 24.	+ 57 51	57	15.	GALAXY
ZWG 333.017	10 01 24.	+ 70 37		15.3	GALAXY
UGC 05423	10 01 24.	+ 70 37	78	15.3	GALAXY IRR
SEY 064	10 01 24.	+ 70 37 15.		14.7	FAINT GALAXY
LB 00120	10 01 24.	- 32 26		16.6	FAINT BLUE STAR
LB 00121	10 01 24.	- 34 40		16.4	FAINT BLUE STAR
KN 15.093	10 01 24.6	+ 38 03 32.			NEBULA
KN 15.094	10 01 25.1	+ 36 37 31.			NEBULA
MCG-01-26-013	10 01 27.	- 06 15	30	15.	GALAXY
MCG-04-24-014	10 01 27.	- 26 46 30.	150	14.	GALAXY
KN 15.095	10 01 27.9	+ 36 36 09.			NEBULA
SVEN 241	10 01 29.	- 26 22	30	15.4	GALAXY
ZWG 036.027	10 01 30.	+ 06 45		15.2	GALAXY
TON-N 0483	10 01 30.	+ 24 57		15.3	BLUE STAR
MCG+06-22-066	10 01 30.	+ 37 20	24	15.	GALAXY
ZWG 182.070	10 01 30.	+ 38 14		14.6	GALAXY
MCG+06-22-065	10 01 30.	+ 38 17 30.	42	14.	GALAXY
ZWG 266.018	10 01 30.	+ 54 48		15.4	GALAXY
MCG+09-17-019	10 01 30.	+ 54 48	36	15.	GALAXY
7ZW 309	10 01 30.	+ 76 13			COMPACT GALAXY
ABC 0916	10 01 30.	- 19 08		17.3	RICH CLUSTER OF GALAXIES
HOLM 162B	10 01 32.	+ 37 19	30	15.2	PART OF MULTIPLE GALAXY
KN 15.097	10 01 32.1	+ 37 18 32.			NEBULA
KN 15.096	10 01 32.7	+ 38 14 49.			NEBULA
RNGC 3110	10 01 33.	- 06 14		13.5	GALAXY
MCG-01-26-014	10 01 33.	- 06 14	66	13.5	GALAXY
KN 15.098	10 01 33.8	+ 37 36 51.			NEBULA
IC 2535	10 01 34.	+ 38 17 02.			NONSTELLAR OBJECT
IC 2537	10 01 35.5	- 27 19 32.	180	12.8	GALAXY SAB(rs)
MCG+06-22-067	10 01 36.	+ 37 20 30.	14	15.	GALAXY
MCG+08-18-063	10 01 36.	+ 45 56	48	16.	GALAXY
ZWG 240.003	10 01 36.	+ 46 58		15.4	GALAXY
KARA.68 063	10 01 36.	+ 66 46	74		DWARF GALAXY

OBJECT NAME	RIGHT ASCEN.	DECLINATION	DIAM.	MAGN.	TYPE OF OBJECT
RNGC 3112	10 01 37.	- 20 32			GALAXY
RNGC 3107	10 01 39.	+ 13 52		13.5	GALAXY
LB 03100	10 01 39.	- 00 18 42.		16.3	FAINT BLUE STAR
MCG-04-24-015	10 01 39.	- 27 19	138	13.	GALAXY
KN 15.099	10 01 40.2	+ 37 19 12.			NEBULA
REIN 2.104	10 01 40.70	+ 13 51 52.1			NEBULA
HOLM 162A	10 01 41.	+ 37 20	24	14.3	PART OF MULTIPLE GALAXY
ZWG 036.028	10 01 42.	+ 03 55		15.3	GALAXY
UGC 05424	10 01 42.	+ 03 55	66	15.3	GALAXY Sc
ZWG 064.048	10 01 42.	+ 13 52		13.6	GALAXY
UGC 05425	10 01 42.	+ 13 52	54	13.6	GALAXY
MCG+02-26-022	10 01 42.	+ 13 52	36	13.6	GALAXY
KARA.73B 0397	10 01 42.	+ 13 52	60	13.6	ISOLATED GALAXY E
REIZ 0158	10 01 42.	+ 13 54	36	13.6	GALAXY
ZWG 093.036	10 01 42.	+ 15 00		15.4	GALAXY
UGC 05426	10 01 42.	+ 15 00	90	15.4	GALAXY Sc
MCG+05-24-010	10 01 42.	+ 29 36	66	15.2	GALAXY
ZWG 153.014	10 01 42.	+ 29 37		15.2	GALAXY
UGC 05427	10 01 42.	+ 29 37	84	15.2	GALAXY
ZC 1001.7+4211	10 01 42.	+ 42 11	1210		CLUSTER OF GALAXIES
ZWG 240.004	10 01 42.	+ 46 56		15.7	GALAXY
UGC 05428	10 01 42.	+ 66 47	60	18.	GALAXY DWARF
ZC 1001.7+7351	10 01 42.	+ 73 51	670		CLUSTER OF GALAXIES
KN 15.100	10 01 43.8	+ 37 31 12.			NEBULA
KN 15.103	10 01 45.4	+ 32 31 53.			NEBULA
HN 0328	10 01 47.	- 34 34			NEBULA
IC 2538	10 01 47.	- 34 34			NONSTELLAR OBJECT
ZC 1001.8+1947	10 01 48.	+ 19 47	1550		CLUSTER OF GALAXIES
HOLM 163B	10 01 48.	+ 37 36	18	15.1	PART OF MULTIPLE GALAXY
MCG+08-18-064	10 01 48.	+ 44 52	21	15.	GALAXY
ZWG 240.005	10 01 48.	+ 44 54		15.6	GALAXY
UGC 05429	10 01 48.	+ 44 57	66	17.	GALAXY Sc-IRR
LB 00122	10 01 48.	- 30 39		16.0	FAINT BLUE STAR
MCG-06-22-015	10 01 48.	- 34 32	66	14.5	GALAXY
KN 15.101	10 01 48.9	+ 37 36 06.			NEBULA
MCG+06-22-068	10 01 51.	+ 37 38 30.	42	14.5	GALAXY
MCG+08-18-065	10 01 51.	+ 44 42	30	17.	GALAXY
KN 15.102	10 01 51.5	+ 37 36 15.			NEBULA
KN 15.104	10 01 52.7	+ 37 41 34.			NEBULA
REIZ 0159	10 01 53.	+ 29 36	60	15.8	GALAXY
HOLM 163A	10 01 53.	+ 37 36	48	13.8	PART OF MULTIPLE GALAXY
KN 15.106	10 01 53.7	+ 35 21 32.			NEBULA
ZWG 064.049	10 01 54.	+ 10 39		15.6	GALAXY
MCG+06-22-069	10 01 54.	+ 35 22 30.	36	15.	GALAXY
ZWG 182.071	10 01 54.	+ 37 35		15.1	GALAXY
ZC 1001.9-0139	10 01 54.	- 01 39	2080		CLUSTER OF GALAXIES
ZWG 008.031	10 01 54.	- 02 10		14.6	GALAXY
MCG+00-26-014	10 01 54.	- 02 10	54	14.6	GALAXY
IC 0589	10 01 54.	- 05 26 19.			NONSTELLAR OBJECT
KN 15.105	10 01 55.9	+ 37 36 07.			NEBULA
MCG+08-18-066	10 01 57.	+ 44 44	45	16.	GALAXY
KN 15.108	10 01 58.4	+ 32 34 45.			NEBULA
VDB.66G 071	10 02	+ 66 48	70		DWARF GALAXY
ZWG 036.029	10 02 00.	+ 05 03		15.4	GALAXY
ZWG 093.037	10 02 00.	+ 15 01		15.6	GALAXY
ZC 1002.0+2919	10 02 00.	+ 29 19	6650		CLUSTER OF GALAXIES
MCG+06-22-070	10 02 00.	+ 35 24	30	15.	GALAXY
ZWG 240.006	10 02 00.	+ 46 54		15.5	GALAXY
ZWG 266.019	10 02 00.	+ 51 52		15.7	GALAXY
MCG+09-17-020	10 02 00.	+ 53 23	24	16.	GALAXY
ZWG 266.020	10 02 00.	+ 53 24		15.3	GALAXY
ZC 1002.0+5807	10 02 00.	+ 58 07	1140		CLUSTER OF GALAXIES
ZC 1002.0+6347	10 02 00.	+ 63 47	1340		CLUSTER OF GALAXIES
MCG+13-08-001	10 02 00.	+ 78 28	6	18.	GALAXY
MCG+14-05-010	10 02 00.	+ 80 30	120	14.	GALAXY
LB 00123	10 02 00.	- 29 43		17.2	FAINT BLUE STAR
REIZ 0160	10 02 01.	+ 32 35	12	14.8	GALAXY
KN 15.107	10 02 01.7	+ 38 13 54.			NEBULA
HN 0329	10 02 02.	- 31 07			NEBULA
IC 2539	10 02 02.	- 31 07			NONSTELLAR OBJECT
MCG-01-26-015	10 02 03.	- 06 26	36	15.5	GALAXY
RNGC 3136A	10 02 04.	- 67 13			GALAXY
ZWG 008.033	10 02 06.	+ 02 08		15.6	GALAXY
ZWG 036.030	10 02 06.	+ 05 36		15.7	GALAXY
ZC 1002.1+3830	10 02 06.	+ 38 30	870		CLUSTER OF GALAXIES
UGC 05430	10 02 06.	+ 44 45	84	17.	GALAXY Sc
MCG+08-19-001	10 02 06.	+ 46 53	30	16.	GALAXY
MCG+10-15-010	10 02 06.	+ 57 45	39	17.	GALAXY
ZWG 008.032	10 02 06.	- 01 30		15.6	GALAXY
MCG+00-26-015	10 02 06.	- 01 30	24	15.6	GALAXY
DV.56 N3136A	10 02 06.	- 67 13	108		S GALAXY
RNGC 3113	10 02 10.	- 28 12		14.0	GALAXY
ZWG 064.050	10 02 12.	+ 12 09		15.2	GALAXY
MCG+02-26-023	10 02 12.	+ 12 09	36	15.2	GALAXY
UGC 05431	10 02 12.	+ 21 45	78	16.0	GALAXY Sc
MCG+04-24-014	10 02 12.	+ 21 45	84	15.	GALAXY
ZC 1002.2+2606	10 02 12.	+ 26 06	4230		CLUSTER OF GALAXIES
ZC 1002.2+5110	10 02 12.	+ 51 10	1550		CLUSTER OF GALAXIES
KARA.68 064	10 02 12.	+ 68 03	74		DWARF GALAXY
LB 00124	10 02 12.	- 29 46		17.3	FAINT BLUE STAR
MCG-05-24-020	10 02 12.	- 31 07	120	14.5	GALAXY
PK283-04.1	10 02 15.	- 60 30	25		PLANETARY NEBULA
ZWG 036.031	10 02 18.	+ 05 18		14.7	GALAXY
UGC 05432	10 02 18.	+ 05 18	72	14.7	GALAXY E
MCG+01-26-013	10 02 18.	+ 05 18	21	14.7	GALAXY
ZWG 064.051	10 02 18.	+ 10 32		15.7	GALAXY
UGC 05433	10 02 18.	+ 22 22	84	17.	GALAXY DWARF
TON-N 1153	10 02 18.	+ 35 28		16.2	BLUE STAR
ZWG 290.006	10 02 18.	+ 57 38		15.3	GALAXY
MCG-05-24-021	10 02 18.	- 28 12	180	14.	GALAXY
LB 00125	10 02 18.	- 31 15		15.7	FAINT BLUE STAR
LB 00126	10 02 18.	- 35 29		16.5	FAINT BLUE STAR
OCL 0844	10 02 18.	- 61 06	240	11.	OPEN STAR CLUSTER
KN 15.109	10 02 20.2	+ 36 37 49.			NEBULA
KN 15.110	10 02 21.1	+ 36 41 05.			NEBULA
ARC 0919	10 02 22.	- 00 27		17.1	RICH CLUSTER OF GALAXIES
KN 15.111	10 02 22.5	+ 36 40 40.			NEBULA
REIZ 0161	10 02 24.	+ 23 00	12	15.7	GALAXY
TON-N 1154	10 02 24.	+ 35 18		17.	BLUE STAR
MCG+09-17-021	10 02 24.	+ 55 30	42	17.	GALAXY
MCG+10-15-011	10 02 24.	+ 57 39	18	16.	GALAXY
ZWG 290.007	10 02 24.	+ 58 41		15.1	GALAXY
MCG+10-15-012	10 02 24.	+ 58 42	36	16.	GALAXY
ZC 1002.4-0026	10 02 24.	- 00 26	670		CLUSTER OF GALAXIES
MCG-01-26-016	10 02 24.	- 06 29	18	15.	GALAXY
MCG-02-26-029	10 02 24.	- 13 24 30.	48	15.	GALAXY
MRSL 282-02/2	10 02 24.	- 57 44	2880		HII REGION
KN 15.112	10 02 29.4	+ 37 17 50.			NEBULA
MCG+04-24-015	10 02 30.	+ 21 41	72	14.8	GALAXY
ZWG 123.018	10 02 30.	+ 21 42		14.8	GALAXY
UGC 05434	10 02 30.	+ 21 42	84	14.8	GALAXY Sb/SBb
TON-N 0029	10 02 30.	+ 28 12		15.5	BLUE STAR
STOCK 07	10 02 30.	+ 29 19			BLUE KNOT NEAR ELLIP GLXY
ZWG 211.010	10 02 30.	+ 39 56		15.7	GALAXY
ZWG 290.008	10 02 30.	+ 59 03		13.6	GALAXY
UGC 05435	10 02 30.	+ 59 03	48	13.6	GALAXY
MCG+10-15-013	10 02 30.	+ 59 03	39	15.	GALAXY
LB 00128	10 02 30.	- 28 16		16.9	FAINT BLUE STAR
LB 00127	10 02 30.	- 33 04		16.2	FAINT BLUE STAR
MCG-06-22-016	10 02 30.	- 37 05	72	14.5	GALAXY
VHA 087	10 02 30.	- 55 12	180		OPEN STAR CLUSTER
KN 15.113	10 02 32.9	+ 37 17 07.			NEBULA
REIZ 0162	10 02 34.	+ 29 12	54	15.8	GALAXY
REIZ 0163	10 02 35.	+ 29 36	60	14.8	GALAXY
ZWG 036.032	10 02 36.	+ 05 49		15.5	GALAXY
ZWG 064.052	10 02 36.	+ 14 14		15.7	GALAXY
ZWG 093.038	10 02 36.	+ 19 30		15.2	GALAXY
KARA.72 222B	10 02 36.	+ 19 30	36	15.2	PART OF DOUBLE GALAXY
ZWG 093.039	10 02 36.	+ 19 31		14.3	GALAXY
UGC 05436	10 02 36.	+ 19 31	90	14.3	GALAXY S
KARA.72 222A	10 02 36.	+ 19 31	84	14.3	PART OF DOUBLE GALAXY
TON-N 1156	10 02 36.	+ 33 45		16.8	BLUE STAR
ARC 0915	10 02 36.	+ 51 11		17.2	RICH CLUSTER OF GALAXIES
ZWG 266.021	10 02 36.	+ 55 30		15.7	GALAXY
MCG+09-17-022	10 02 36.	+ 55 30	24	16.	GALAXY
ZC 1002.6+6458	10 02 36.	+ 64 58	1410		CLUSTER OF GALAXIES
MCG+11-12-034	10 02 36.	+ 67 05	39	16.	GALAXY
MCG-02-26-030	10 02 36.	- 09 45	36	15.	GALAXY
MCG-02-26-031	10 02 36.	- 11 00	60	14.5	GALAXY
LB 00129	10 02 36.	- 29 40		17.0	FAINT BLUE STAR
TON-N 1155	10 02 37.	+ 20 02		15.5	BLUE STAR
LB 00565	10 02 37.	- 02 08 42.		15.3	FAINT BLUE STAR
KN 15.114	10 02 38.3	+ 36 59 06.			NEBULA
REIZ 0166	10 02 39.	+ 19 31	24	14.8	GALAXY
REIZ 0165	10 02 39.	+ 19 31	42	14.2	GALAXY
ZWG 036.033	10 02 42.	+ 02 44		15.2	GALAXY
MCG+03-26-022	10 02 42.	+ 19 32	24	15.2	GALAXY
MCG+03-26-023	10 02 42.	+ 19 33	48	14.3	GALAXY
ZC 1002.7+2432	10 02 42.	+ 24 32	1340		CLUSTER OF GALAXIES
UGC 05437	10 02 42.	+ 27 45	78	17.	GALAXY
MRK 135	10 02 42.	+ 53 57	7	16.	GALAXY WITH UV CONTINUUM
ZC 1002.7+5939	10 02 42.	+ 59 39	1210		CLUSTER OF GALAXIES
ZC 1002.7+6848	10 02 42.	+ 68 48	610		CLUSTER OF GALAXIES
LB 00130	10 02 42.	- 29 04		16.4	FAINT BLUE STAR
KEEL 204	10 02 44.5	- 07 48 05.		17.	SPIRAL NEBULA
MCG-01-26-017	10 02 45.	- 06 23 30.	12	15.	GALAXY
RNGC 3115	10 02 45.	- 07 28		10.5	GALAXY
SVEN 242	10 02 47.	- 07 29	150	9.4	GALAXY
ZWG 008.034	10 02 48.	+ 00 47		15.0	GALAXY
MCG+00-26-016	10 02 48.	+ 00 47	60	15.0	GALAXY
ZWG 036.034	10 02 48.	+ 05 55		15.6	GALAXY
ZC 1002.8+1105	10 02 48.	+ 11 05	1480		CLUSTER OF GALAXIES
ZWG 064.053	10 02 48.	+ 11 58		15.3	GALAXY
ZWG 093.040	10 02 48.	+ 15 06		15.7	GALAXY
ZWG 093.041	10 02 48.	+ 15 31		15.2	GALAXY
ZC 1002.8+2259	10 02 48.	+ 22 59	1950		CLUSTER OF GALAXIES
TON-N 0484	10 02 48.	+ 30 47		15.2	BLUE STAR
ZC 1002.8+5050	10 02 48.	+ 50 50	940		CLUSTER OF GALAXIES
ZWG 266.022	10 02 48.	+ 52 25		15.0	GALAXY
MCG-01-26-018	10 02 48.	- 07 29	180	11.	GALAXY
HELW 016	10 02 49.	- 14 59			NEBULA
ZC 1002.9+3952	10 02 54.	+ 39 52	2550		CLUSTER OF GALAXIES
ZC 1002.9+4655	10 02 54.	+ 46 55	1210		CLUSTER OF GALAXIES
ZWG 266.023	10 02 54.	+ 50 30		15.7	GALAXY
ARC 0921	10 02 55.	+ 07 40		16.7	RICH CLUSTER OF GALAXIES
MCG+09-17-023	10 02 57.	+ 52 23	72	15.	GALAXY
REIZ 0164	10 02 58.	+ 47 30	18	14.0	GALAXY
GCL 014	10 03 00.	+ 00 18	132	14.7	GLOBULAR STAR CLUSTER
ZWG 008.035	10 03 00.	+ 00 19		15.5	GALAXY
UGC 05439	10 03 00.	+ 00 19	240	15.5	GALAXY DWARF
MCG+00-26-017	10 03 00.	+ 00 19	180	15.5	GALAXY
ZWG 036.035	10 03 00.	+ 03 17		15.3	GALAXY
UGC 05440	10 03 00.	+ 04 31	78	16.5	GALAXY Sc
ZC 1003.0+0445	10 03 00.	+ 04 45	740		CLUSTER OF GALAXIES
MCG+03-26-024	10 03 00.	+ 14 34	48	15.3	GALAXY
ZWG 093.042	10 03 00.	+ 14 35		15.3	GALAXY
TON-N 1157	10 03 00.	+ 36 29		17.	BLUE STAR
ZWG 240.007	10 03 00.	+ 47 30		14.0	GALAXY
UGC 05441	10 03 00.	+ 47 30	60	14.0	GALAXY E-SO
UGC 05442	10 03 00.	+ 68 05	120	18.	GALAXY DWARF
7ZW 310	10 03 00.	+ 69 48			COMPACT GALAXY
ZWG 351.005	10 03 00.	+ 77 15		15.5	GALAXY
ZWG 350.040	10 03 00.	+ 77 15		15.5	GALAXY
ZC 1003.0+8259	10 03 00.	+ 82 59	1610		CLUSTER OF GALAXIES
UGC 05438	10 03 00.	+ 85 04	78	16.5	GALAXY Sb-c
MCG+14-05-011	10 03 00.	+ 85 07	39	17.	GALAXY
RNGC 3111	10 03 02.	+ 47 30		14.0	GALAXY
VV 240B	10 03 03.	+ 14 42 30.	18	15.	INTERACTING GALAXY
MCG-03-26-016	10 03 03.	- 17 33 30.	60	14.	GALAXY
ZWG 064.054	10 03 06.	+ 10 31		15.5	GALAXY
REIZ 0167	10 03 06.	+ 14 34	12	15.2	GALAXY
VV 240A	10 03 06.	+ 14 43	66	14.	INTERACTING GALAXY
ZC 1003.1+3943	10 03 06.	+ 39 43	540		CLUSTER OF GALAXIES
MCG+08-19-002	10 03 06.	+ 47 30	24	14.0	GALAXY
MCG+09-17-024	10 03 06.	+ 53 03	15	16.	GALAXY
KEEL 205	10 03 09.9	- 07 13 57.		17.	NEBULA
ZWG 008.036	10 03 12.	+ 02 29		15.5	GALAXY
8ZW 1003+02.5	10 03 12.	+ 02 29		16.0	COMPACT GALAXY
ZWG 064.055	10 03 12.	+ 13 12		15.3	GALAXY
ZWG 064.056	10 03 12.	+ 13 23		15.6	GALAXY
MCG+08-19-003	10 03 12.	+ 50 28	30	16.	GALAXY
MCG+09-17-025	10 03 12.	+ 53 11	36	16.	GALAXY
ZC 1003.2+5846	10 03 12.	+ 58 46	1750		CLUSTER OF GALAXIES
ZC 1003.2+7759	10 03 12.	+ 77 59	1550		CLUSTER OF GALAXIES
KARA.68 065	10 03 12.	- 07 30	40		DWARF GALAXY
MCG-04-24-016	10 03 12.	- 21 32 30.	36	15.	GALAXY
MCG-06-22-017	10 03 12.	- 33 57	72	14.	GALAXY
RNGC 3120	10 03 12.	- 33 59			GALAXY
KEEL 206	10 03 12.0	- 07 44 22.		11.	NEBULA
TON-N 1158	10 03 13.	+ 22 20		16.8	BLUE STAR
KN 15.115	10 03 13.4	+ 36 47 03.			NEBULA
MCG-01-26-019	10 03 15.	- 04 35	36	15.	GALAXY
IC 0590	10 03 16.	+ 00 52 42.			NONSTELLAR OBJECT
MAI 050	10 03 16.	+ 68 16	94		DWARF SPHEROIDAL GALAXY
ZWG 008.037	10 03 18.	+ 00 53		14.2	GALAXY
UGC 05443	10 03 18.	+ 00 53	66	14.2	GALAXY DBL SYS

OBJECT NAME	RIGHT ASCEN.	DECLINATION	DIAM.	MAGN.	TYPE OF OBJECT
KARA.72 223B	10 03 18.	+ 00 53	54		PART OF DOUBLE GALAXY
KARA.72 223A	10 03 18.	+ 00 53	48	14.2	PART OF DOUBLE GALAXY
MCG+00-26-018	10 03 18.	+ 00 53	30	14.2	GALAXY
ZC 1003.3+4739	10 03 18.	+ 47 39	870		CLUSTER OF GALAXIES
MCG+13-08-002	10 03 18.	+ 77 14	39	16.	GALAXY
MCG-01-26-020	10 03 18.	- 04 39 30.	42	15.5	GALAXY
MCG-01-26-021	10 03 18.	- 07 45	60	13.5	GALAXY
ZWG 036.036	10 03 24.	+ 04 45		15.7	GALAXY
ZC 1003.4+0734	10 03 24.	+ 07 34	1480		CLUSTER OF GALAXIES
ZC 1003.4+1209	10 03 24.	+ 12 09	940		CLUSTER OF GALAXIES
MCG+05-24-011	10 03 24.	+ 29 11	30	14.6	GALAXY
ZWG 153.015	10 03 24.	+ 29 12		14.6	GALAXY
MCG-03-26-017	10 03 24.	- 17 12	72	14.	GALAXY
LB 00131	10 03 24.	- 32 29		17.0	FAINT BLUE STAR
OCL 0808	10 03 24.	- 61 22	600	10.9	OPEN STAR CLUSTER
TON-N 1159	10 03 25.	+ 21 34		17.	BLUE STAR
REIZ 0168	10 03 26.	+ 26 08	21	15.4	GALAXY
HARO 23	10 03 27.	+ 29 10			BLUE EMISSION-LINE GALAXY
MCG-03-26-018	10 03 27.	- 18 00	60	15.	GALAXY
IC 2541	10 03 29.	- 17 11			NONSTELLAR OBJECT
ZWG 036.037	10 03 30.	+ 03 20		15.6	GALAXY
UGC 05444	10 03 30.	+ 14 53	60	16.0	GALAXY Sb
RNGC 3128	10 03 30.	- 15 53		14.0	GALAXY
HN 0330	10 03 30.	- 17 12			NEBULA
MRSL 282-02/1	10 03 30.	- 58 42	1500		HII REGION
KN 15.116	10 03 30.5	+ 36 28 03.			NEBULA
TON-N 1160	10 03 31.	+ 20 47		16.7	BLUE STAR
REIZ 0169	10 03 31.	+ 33 11	90	14.2	GALAXY
LB 00566	10 03 32.	+ 01 07 42.		12.4	FAINT BLUE STAR
REIZ 0170	10 03 32.	+ 26 08	12	15.6	GALAXY
REIZ 0171	10 03 32.	+ 26 11	12	15.5	GALAXY
MCG-03-26-019	10 03 33.	- 21 02	36	15.	GALAXY
ZWG 036.038	10 03 36.	+ 03 08		14.6	GALAXY
RNGC 3117	10 03 36.	+ 03 08		14.5	GALAXY
UGC 05445	10 03 36.	+ 03 08	60	14.6	GALAXY E?
MCG+01-26-014	10 03 36.	+ 03 08	36	14.6	GALAXY
ZWG 064.057	10 03 36.	+ 13 58		15.6	GALAXY
ZWG 064.058	10 03 36.	+ 14 13		15.1	GALAXY
ZWG 064.059	10 03 36.	+ 14 19		15.5	GALAXY
ZC 1003.6+1443	10 03 36.	+ 14 43	5650		CLUSTER OF GALAXIES
ZWG 182.072	10 03 36.	+ 33 11		15.3	GALAXY
UGC 05446	10 03 36.	+ 33 11	90	15.3	GALAXY Sc
MCG+06-22-071	10 03 36.	+ 33 11	78	15.	GALAXY
KN 15.118	10 03 36.	+ 36 31 14.			NEBULA
ZC 1003.6+3829	10 03 36.	+ 38 29	670		CLUSTER OF GALAXIES
ZC 1003.6+4011	10 03 36.	+ 40 11	670		CLUSTER OF GALAXIES
ZWG 240.008	10 03 36.	+ 48 52		15.7	GALAXY
MCG+13-08-003	10 03 36.	+ 77 19	51	17.	GALAXY
MCG-03-26-020	10 03 36.	- 15 53	78	14.	GALAXY
KN 15.121	10 03 36.2	+ 33 11 26.			NEBULA
KN 15.117	10 03 36.8	+ 38 25 37.			NEBULA
TON-N 1161	10 03 37.	+ 21 54		17.	BLUE STAR
MCG-04-24-017	10 03 39.	- 22 48 30.	21	15.	GALAXY
KN 15.120	10 03 40.0	+ 36 29 52.			NEBULA
ARC 0923	10 03 41.	+ 26 09		17.2	RICH CLUSTER OF GALAXIES
KN 15.119	10 03 41.1	+ 38 24 40.			NEBULA
ZWG 036.039	10 03 42.	+ 04 04		14.8	GALAXY
MCG+01-26-015	10 03 42.	+ 04 04	48	14.8	GALAXY
ZC 1003.7+1930	10 03 42.	+ 19 30	740		CLUSTER OF GALAXIES
ZC 1003.7+2725	10 03 42.	+ 27 25	2350		CLUSTER OF GALAXIES
ZC 1003.7+2815	10 03 42.	+ 28 15	1140		CLUSTER OF GALAXIES
TON-N 1162	10 03 42.	+ 33 42		14.6	BLUE STAR
TON-N 1163	10 03 42.	+ 35 48		16.	BLUE STAR
ZWG 182.073	10 03 42.	+ 38 24		15.5	GALAXY
MCG+06-22-072	10 03 42.	+ 38 26	48	15.	GALAXY
ZC 1003.7+6246	10 03 42.	+ 62 46	1480		CLUSTER OF GALAXIES
ZWG 008.038	10 03 42.	- 01 18		15.2	GALAXY
UGC 05447	10 03 42.	- 01 18	60	15.2	GALAXY DBL SYS
MCG+00-26-019	10 03 42.	- 01 18	54	15.2	GALAXY
REIZ 0172	10 03 45.	+ 28 24	24	14.8	GALAXY
RNGC 3116	10 03 45.	+ 31 20		15.5	GALAXY
ZWG 064.060	10 03 48.	+ 13 07		15.2	GALAXY
MCG+03-26-025	10 03 48.	+ 14 40	9	15.1	GALAXY
MCG+03-26-026	10 03 48.	+ 14 40 30.	60	15.1	GALAXY
ZWG 153.016	10 03 48.	+ 28 25		15.4	GALAXY
ZWG 153.017	10 03 48.	+ 31 20		15.3	GALAXY
MCG+05-24-012	10 03 48.	+ 31 21	18	15.3	GALAXY
ZWG 153.018	10 03 48.	+ 31 43		15.2	GALAXY
ZWG 153.019	10 03 48.	+ 32 29		15.5	GALAXY
ZWG 008.039	10 03 48.	- 01 34		15.6	GALAXY
MCG-03-26-021	10 03 48.	- 15 46	96	14.5	GALAXY
LB 00132	10 03 48.	- 28 01		15.8	FAINT BLUE STAR
KN 15.123	10 03 50.9	+ 32 28 33.			NEBULA
KN 15.125	10 03 53.3	+ 31 43 06.			NEBULA
KN 15.122	10 03 53.8	+ 37 29 30.			NEBULA
ZC 1003.9+0608	10 03 54.	+ 06 08	1810		CLUSTER OF GALAXIES
ZWG 064.061	10 03 54.	+ 13 59		15.7	GALAXY
ZWG 093.043	10 03 54.	+ 14 41		15.1	GALAXY
REIZ 0175	10 03 54.	+ 14 41	24	14.8	GALAXY
REIZ 0174	10 03 54.	+ 14 41	30	14.7	GALAXY
UGC 05448	10 03 54.	+ 14 41	66	15.1	GALAXY Sb
ZWG 093.044	10 03 54.	+ 16 11		15.4	GALAXY
IC 2540	10 03 54.	+ 31 43 05.			NONSTELLAR OBJECT
MCG+05-24-013	10 03 54.	+ 32 28	30	15.5	GALAXY
ZC 1003.9+3557	10 03 54.	+ 35 57	1480		CLUSTER OF GALAXIES
HOLM 164B	10 03 54.	+ 37 29	36	15.2	PART OF MULTIPLE GALAXY
MCG+10-15-014	10 03 54.	+ 57 45	39	16.	GALAXY
RNGC 3127	10 03 54.	- 15 53		14.0	GALAXY
LB 00133	10 03 54.	- 29 28		16.2	FAINT BLUE STAR
REIZ 0173	10 03 55.	+ 32 27	30	15.2	GALAXY
RNGC 3119	10 03 57.	+ 14 33		15.5	GALAXY
RNGC 3122	10 03 57.	- 06 18			GALAXY
KN 15.124	10 03 58.5	+ 37 31 24.			NEBULA
HOLM 164A	10 03 59.	+ 37 31	36	15.0	PART OF MULTIPLE GALAXY
ZC 1004.0+1145	10 04 00.	+ 11 45	2150		CLUSTER OF GALAXIES
ZWG 093.045	10 04 00.	+ 14 33		15.3	GALAXY
ARC 0926	10 04 00.	+ 21 56		17.8	RICH CLUSTER OF GALAXIES
ZC 1004.0+3130	10 04 00.	+ 31 30	740		CLUSTER OF GALAXIES
MCG+05-24-014	10 04 00.	+ 31 43	48	15.2	GALAXY
MCG+10-15-015	10 04 00.	+ 57 22	12	16.	GALAXY
MCG+11-12-035	10 04 00.	+ 64 13	27	17.	GALAXY
ZWG 333.018	10 04 00.	+ 68 36		15.3	GALAXY
UGC 05449	10 04 00.	+ 68 36	60	15.3	GALAXY SB?c
MCG+13-08-004	10 04 00.	+ 77 11	30	17.	GALAXY
MCG+13-08-005	10 04 00.	+ 77 19	30	17.	GALAXY
MCG-03-26-022	10 04 00.	- 15 53	72	14.5	GALAXY
MCG-03-26-023	10 04 00.	- 18 03	42	14.5	GALAXY
ARC 0925	10 04 01.	+ 27 22		17.7	RICH CLUSTER OF GALAXIES

OBJECT NAME	RIGHT ASCEN.	DECLINATION	DIAM.	MAGN.	TYPE OF OBJECT
ARC 0920	10 04 01.	+ 55 31		17.2	RICH CLUSTER OF GALAXIES
SEY 065	10 04 02.	+ 68 37 27.		14.9	FAINT GALAXY
KN 15.126	10 04 02.0	+ 38 08 32.			NEBULA
ZC 1004.1+0914	10 04 06.	+ 09 14	1480		CLUSTER OF GALAXIES
ZWG 064.062	10 04 06.	+ 11 55		15.7	GALAXY
REIZ 0177	10 04 06.	+ 14 36	24	14.5	GALAXY
MCG+03-26-027	10 04 06.	+ 14 36	90	14.2	GALAXY
ZWG 093.046	10 04 06.	+ 14 37		14.2	GALAXY
UGC 05450	10 04 06.	+ 14 37	66	14.2	GALAXY E
KARA.72 224B	10 04 06.	+ 14 37	36		PART OF DOUBLE GALAXY
KARA.72 224A	10 04 06.	+ 14 37	60	14.2	PART OF DOUBLE GALAXY
MRK 136	10 04 06.	+ 77 08	8	17.	GALAXY WITH UV CONTINUUM
LB 00134	10 04 06.	- 35 15		16.5	FAINT BLUE STAR
ARC 0924	10 04 08.	+ 35 54		17.2	RICH CLUSTER OF GALAXIES
KN 15.127	10 04 08.9	+ 35 21 02.			NEBULA
RNGC 3121	10 04 09.	+ 14 37		14.0	GALAXY
MCG-01-26-022	10 04 09.	- 07 07	30	17.	GALAXY
RNGC 3125	10 04 10.	- 29 41		14.0	GALAXY
ARC 0928	10 04 11.	+ 11 45		17.2	RICH CLUSTER OF GALAXIES
LB 00567	10 04 12.	+ 00 46 24.		16.4	FAINT BLUE STAR
ZC 1004.2+0704	10 04 12.	+ 07 04	940		CLUSTER OF GALAXIES
ZWG 064.063	10 04 12.	+ 14 19		15.7	GALAXY
REIZ 0179	10 04 12.	+ 14 37	30	13.9	GALAXY
ZC 1004.2+2157	10 04 12.	+ 21 57	2420		CLUSTER OF GALAXIES
ZC 1004.2+3250	10 04 12.	+ 32 50	670		CLUSTER OF GALAXIES
ZWG 182.074	10 04 12.	+ 35 14		15.4	GALAXY
MCG+06-22-073	10 04 12.	+ 35 15	42	15.	GALAXY
ZC 1004.2+4002	10 04 12.	+ 40 02	340		CLUSTER OF GALAXIES
ZWG 240.009	10 04 12.	+ 46 48		15.1	GALAXY
ZWG 240.010	10 04 12.	+ 47 16		14.1	GALAXY
UGC 05451	10 04 12.	+ 47 16	96	14.1	GALAXY IRR
KN 15.128	10 04 12.1	+ 35 13 56.			NEBULA
TON-N 1164	10 04 13.	+ 22 32		15.5	BLUE STAR
REIZ 0176	10 04 13.	+ 33 16	120	13.7	GALAXY
RNGC 3124	10 04 13.	- 19 00		12.5	GALAXY
ARC 0917	10 04 14.	+ 62 45		17.4	RICH CLUSTER OF GALAXIES
MCG+08-19-004	10 04 15.	+ 47 16	66	14.	GALAXY
ARC 0914	10 04 16.	+ 71 30		17.7	RICH CLUSTER OF GALAXIES
KN 15.129	10 04 17.3	+ 33 16 18.			NEBULA
ZWG 064.064	10 04 18.	+ 14 13		15.6	GALAXY
ZWG 182.075	10 04 18.	+ 33 16		14.4	GALAXY
UGC 05452	10 04 18.	+ 33 16	150	14.4	GALAXY Sb-c
MCG+06-22-074	10 04 18.	+ 33 16	144	14.	GALAXY
ZWG 211.011	10 04 18.	+ 39 13		15.5	GALAXY
MCG+09-17-026	10 04 18.	+ 53 07 30.	48	15.	GALAXY
MCG-03-26-024	10 04 18.	- 19 00	198	12.	GALAXY
KEEL 207	10 04 19.2	- 07 26 46.		17.	NEBULA
RNGC 3118	10 04 21.	+ 33 16		14.5	GALAXY
ARC 0930	10 04 22.	- 05 24		16.5	RICH CLUSTER OF GALAXIES
RNGC 3149	10 04 23.	- 80 11			UNVERIFIED SOUTHERN OBJECT
ZWG 064.065	10 04 24.	+ 14 10		15.7	GALAXY
MCG+03-26-028	10 04 24.	+ 16 13	60	15.1	GALAXY
ZWG 093.047	10 04 24.	+ 16 14		15.1	GALAXY
UGC 05453	10 04 24.	+ 16 14	66	15.1	GALAXY IRR
ZWG 093.048	10 04 24.	+ 19 11		15.4	GALAXY
TON-N 1166	10 04 24.	+ 33 58		16.	BLUE STAR
STOCK 08	10 04 24.	+ 34 41			BLUE KNOT NEAR ELLIP GLXY
STOCK 09	10 04 24.	+ 38 07			BLUE KNOT NEAR ELLIP GLXY
ZWG 008.040	10 04 24.	- 01 00		15.4	GALAXY
MCG-01-26-023	10 04 24.	- 05 35	42	16.	GALAXY
MCG-05-24-022	10 04 24.	- 29 42	48	14.	GALAXY
IC 2545	10 04 24.	- 33 36			NONSTELLAR OBJECT
HN 0331	10 04 24.	- 33 37			NEBULA
KN 15.130	10 04 24.7	+ 37 35 44.			NEBULA
RNGC 3123	10 04 25.	+ 00 18			NON-EXISTENT OBJECT
TON-N 1165	10 04 25.	+ 21 56		15.8	BLUE STAR
ARC 0931	10 04 26.	- 13 11		17.4	RICH CLUSTER OF GALAXIES
MCG+07-21-010	10 04 27.	+ 39 12	54	15.	GALAXY
RNGC 3136	10 04 27.	- 67 08		12.5	GALAXY
HN 0015	10 04 28.	+ 00 18			NEBULA
REIZ 0178	10 04 28.	+ 42 32	60	14.0	GALAXY
ZC 1004.5+0805	10 04 30.	+ 08 05	1210		CLUSTER OF GALAXIES
ZWG 064.066	10 04 30.	+ 12 54		15.3	GALAXY
UGC 05454	10 04 30.	+ 12 54	72	15.3	GALAXY
ZWG 093.049	10 04 30.	+ 18 37		15.7	GALAXY
ZWG 123.019	10 04 30.	+ 21 15		15.7	GALAXY
MCG+06-22-075	10 04 30.	+ 38 09	30	16.	GALAXY
ZWG 240.011	10 04 30.	+ 46 21		15.3	GALAXY
MCG+08-19-005	10 04 30.	+ 46 24	30	16.	GALAXY
ZC 1004.5+5527	10 04 30.	+ 55 27	1680		CLUSTER OF GALAXIES
VHA 088	10 04 30.	- 51 20	180		OPEN STAR CLUSTER
KN 15.131	10 04 32.5	+ 38 06 57.			NEBULA
REIZ 0180	10 04 34.	+ 10 36	72	13.6	GALAXY
ZWG 064.067	10 04 36.	+ 09 06		15.7	GALAXY
KARA.73B 0398	10 04 36.	+ 09 06	36	15.7	ISOLATED GALAXY S
ZWG 093.050	10 04 36.	+ 15 07		15.7	GALAXY
ZWG 240.012	10 04 36.	+ 46 25		15.2	GALAXY
UGC 05455	10 04 36.	+ 70 53	120	17.	GALAXY DWRF IR
MCG-01-26-024	10 04 36.	- 05 38 30.	66	15.	GALAXY
OCL 0804	10 04 36.	- 60 08	180		OPEN STAR CLUSTER
TON-N 1168	10 04 37.	+ 20 24		16.5	BLUE STAR
TON-N 1167	10 04 37.	+ 21 19		15.5	BLUE STAR
KN 15.132	10 04 41.5	+ 38 14 30.			NEBULA
KN 15.133	10 04 41.7	+ 36 53 52.			NEBULA
ZWG 064.068	10 04 42.	+ 10 36		13.5	GALAXY
UGC 05456	10 04 42.	+ 10 36	102	13.5	GALAXY PECULR
MCG+02-26-024	10 04 42.	+ 10 36	96	13.5	GALAXY
ZC 1004.7+3814	10 04 42.	+ 38 14	810		CLUSTER OF GALAXIES
ZC 1004.7+5038	10 04 42.	+ 50 38	1550		CLUSTER OF GALAXIES
MCG+11-12-036	10 04 42.	+ 67 20	60	16.	GALAXY
UGC 05457	10 04 42.	+ 67 22	60	16.0	GALAXY S
KARA.68 066	10 04 42.	+ 70 50	107		DWARF GALAXY
ZC 1004.7+7128	10 04 42.	+ 71 28	870		CLUSTER OF GALAXIES
ACK 289-11.1	10 04 42.	- 69 47			PLANETARY NEBULA
BC PKS1004+13	10 04 44.	+ 13 03 30.		15.15	QUASI-STELLAR OBJECT
SHB 151	10 04 44.0	+ 13 05 18.		15.2	QUASI-STELLAR OBJECT
REIZ 0182	10 04 47.	+ 12 30	30	13.9	GALAXY
RNGC 3133	10 04 47.	- 11 43			GALAXY
ZWG 036.040	10 04 48.	+ 03 55		15.7	GALAXY
ZWG 036.041	10 04 48.	+ 04 19		15.1	GALAXY
IC 0591	10 04 48.	+ 12 30 54.			NONSTELLAR OBJECT
ZWG 064.069	10 04 48.	+ 12 31		14.0	GALAXY
UGC 05458	10 04 48.	+ 12 31	78	14.0	GALAXY S
MCG+02-26-025	10 04 48.	+ 12 31	36	14.0	GALAXY
ZWG 093.051	10 04 48.	+ 14 51		15.5	GALAXY
ZWG 093.052	10 04 48.	+ 15 13		15.3	GALAXY
TON-N 1169	10 04 48.	+ 36 54		16.8	BLUE STAR
ZC 1004.8+4108	10 04 48.	+ 41 08	1680		CLUSTER OF GALAXIES

OBJECT NAME	RIGHT ASCEN.	DECLINATION	DIAM.	MAGN.	TYPE OF OBJECT
ZWG 266.024	10 04 48.	+ 53 20		13.8	GALAXY
UGC 05459	10 04 48.	+ 53 20	264	13.8	GALAXY Sc
LB 00135	10 04 48.	- 33 07		14.8	FAINT BLUE STAR
OCL 0803	10 04 48.	- 60 04	540	10.6	OPEN STAR CLUSTER
VHA 089	10 04 48.	- 60 04	300		OPEN STAR CLUSTER
OCL 0805	10 04 48.	- 60 15	180		OPEN STAR CLUSTER
REIZ 0181	10 04 50.	+ 34 33	42	14.1	GALAXY
ZWG 093.053	10 04 54.	+ 17 20		15.1	GALAXY
MCG+03-26-029	10 04 54.	+ 17 20	12	15.1	GALAXY
TON-N 1170	10 04 54.	+ 34 14		14.9	BLUE STAR
ZWG 182.076	10 04 54.	+ 34 33		14.6	GALAXY
KARA.73B 0399	10 04 54.	+ 34 33	60	14.6	ISOLATED GALAXY S
ARC 0929	10 04 54.	+ 38 15		17.6	RICH CLUSTER OF GALAXIES
ZWG 266.025	10 04 54.	+ 52 06		13.9	GALAXY
UGC 05460	10 04 54.	+ 52 06	156	13.9	GALAXY SBc
ZC 1004.9+5743	10 04 54.	+ 57 43	940		CLUSTER OF GALAXIES
IC 2546	10 04 55.	- 33 00			NONSTELLAR OBJECT
HN 0332	10 04 55.	- 33 01			NEBULA
KN 15.134	10 04 55.3	+ 34 33 34.			NEBULA
IC 2542	10 04 56.	+ 34 25 01.			NONSTELLAR OBJECT
MCG+06-22-076	10 04 57.	+ 34 34 30.	42	14.	GALAXY
MCG+09-17-027	10 04 57.	+ 53 19	312	12.	GALAXY
RNGC 3132	10 04 57.	- 40 11		8.0	PLANETARY NEBULA
REIZ 0183	10 04 58.	+ 29 43	42	15.4	GALAXY
MAI 051	10 04 58.	+ 70 56	53		DWARF SPHEROIDAL GALAXY
VDB.66G 072	10 05	+ 29 47	70		DWARF GALAXY
ZWG 008.042	10 05 00.	+ 00 31		15.7	GALAXY
ZC 1005.0+1750	10 05 00.	+ 17 50	470		CLUSTER OF GALAXIES
MCG+05-24-015	10 05 00.	+ 29 42	54	15.4	GALAXY
ZWG 153.020	10 05 00.	+ 29 43		15.4	GALAXY
UGC 05461	10 05 00.	+ 29 43	66	15.4	GALAXY Sb-c
ZWG 153.021	10 05 00.	+ 30 47		15.4	GALAXY
TON-N 0485	10 05 00.	+ 31 21		15.6	BLUE STAR
MCG+09-17-028	10 05 00.	+ 52 04	120	14.	GALAXY
MCG+10-15-016	10 05 00.	+ 60 27	24	17.	GALAXY
ZC 1005.0-0022	10 05 00.	- 00 22	1210		CLUSTER OF GALAXIES
ZWG 008.041	10 05 00.	- 02 17		15.5	GALAXY
MCG-01-26-025	10 05 00.	- 04 31	42	16.	GALAXY
ARC 0927	10 05 03.	+ 50 34		17.4	RICH CLUSTER OF GALAXIES
ARC 0933	10 05 05.	+ 00 46		15.9	RICH CLUSTER OF GALAXIES
MCG+05-24-016	10 05 06.	+ 30 48	36	15.4	GALAXY
ZWG 211.012	10 05 06.	+ 42 40		14.9	GALAXY
UGC 05462	10 05 06.	+ 42 40	60	14.9	GALAXY Sa
MCG+11-13-001	10 05 06.	+ 67 22	51	16.	GALAXY
REIZ 0184	10 05 10.	+ 29 47	42	15.8	GALAXY
REIZ 0185	10 05 11.	+ 30 47	30	14.9	GALAXY
ZWG 064.070	10 05 12.	+ 13 38		15.3	GALAXY
TON-N 1171	10 05 12.	+ 35 11		16.4	BLUE STAR
MCG+07-21-011	10 05 12.	+ 42 39	51	15.	GALAXY
ZWG 351.006	10 05 12.	+ 78 25		15.6	GALAXY
ZWG 350.041	10 05 12.	+ 78 25		15.6	GALAXY
UGC 05463	10 05 12.	+ 78 25	72	15.6	GALAXY
MCG-03-26-025	10 05 12.	- 21 14	48	15.	GALAXY
KN 15.135	10 05 14.8	+ 38 06 54.			NEBULA
ZWG 008.044	10 05 18.	+ 00 33		15.4	GALAXY
ZC 1005.3+0053	10 05 18.	+ 00 53	1480		CLUSTER OF GALAXIES
ZWG 064.071	10 05 18.	+ 13 28		15.6	GALAXY
MCG+03-26-030	10 05 18.	+ 15 01	60	15.1	GALAXY
ZC 1005.3+1840	10 05 18.	+ 18 40	3630		CLUSTER OF GALAXIES
UGC 05464	10 05 18.	+ 29 48	90	16.5	GALAXY DWRF SP
MCG+05-24-017	10 05 18.	+ 29 48	78	16.	GALAXY
ZWG 153.022	10 05 18.	+ 30 45		15.7	GALAXY
ZC 1005.3+4034	10 05 18.	+ 40 34	1550		CLUSTER OF GALAXIES
ZC 1005.3+4310	10 05 18.	+ 43 10	1140		CLUSTER OF GALAXIES
ZWG 008.043	10 05 18.	- 03 14		15.3	GALAXY
ARC 0932	10 05 21.	+ 19 50		17.5	RICH CLUSTER OF GALAXIES
RNGC 3126	10 05 21.	+ 32 06		13.5	GALAXY
KN 15.136	10 05 21.2	+ 37 43 39.			NEBULA
HOLM 166A	10 05 22.	+ 00 32	30	14.3	PART OF MULTIPLE GALAXY
HOLM 166B	10 05 22.	+ 00 33	12	15.7	PART OF MULTIPLE GALAXY
REIZ 0186	10 05 23.	+ 30 44	48	15.3	GALAXY
KN 15.137	10 05 23.8	+ 38 27 47.			NEBULA
ZWG 008.046	10 05 24.	+ 00 31		15.5	GALAXY
ZWG 093.054	10 05 24.	+ 15 03		15.1	GALAXY
TON-N 0486	10 05 24.	+ 26 08		15.3	BLUE STAR
MCG+05-24-018	10 05 24.	+ 30 44	42	15.7	GALAXY
ZWG 153.023	10 05 24.	+ 32 06		13.5	GALAXY
UGC 05466	10 05 24.	+ 32 06	174	13.5	GALAXY Sb
KARA.73B 0400	10 05 24.	+ 32 06	204	13.5	ISOLATED GALAXY S
ZWG 182.077	10 05 24.	+ 38 05		15.7	GALAXY
MCG+13-08-006	10 05 24.	+ 78 24	51	16.	GALAXY
MCG+13-08-007	10 05 24.	+ 79 04	33	17.	GALAXY
ZWG 008.045	10 05 24.	- 02 15		14.0	GALAXY
UGC 05465	10 05 24.	- 02 15	60	14.0	GALAXY Sb-c
MCG+00-26-020	10 05 24.	- 02 15	48	14.0	GALAXY
LB 00136	10 05 24.	- 30 52		17.1	FAINT BLUE STAR
LB 00137	10 05 24.	- 34 02		15.7	FAINT BLUE STAR
TON-N 1172	10 05 25.	+ 22 38		16.	BLUE STAR
IC 2543	10 05 25.	+ 38 04 59.			NONSTELLAR OBJECT
KN 15.138	10 05 25.6	+ 38 05 06.			NEBULA
REIZ 0188	10 05 27.	+ 18 57	24	14.1	GALAXY
KN 15.139	10 05 27.1	+ 37 43 38.			NEBULA
KN 15.141	10 05 27.2	+ 32 06 27.			NEBULA
IC 0592	10 05 28.	- 02 16 03.			NONSTELLAR OBJECT
HOLM 165A	10 05 29.	+ 38 05	36	14.7	PART OF MULTIPLE GALAXY
KN 15.140	10 05 29.8	+ 38 04 57.			NEBULA
ZWG 093.055	10 05 30.	+ 16 05		15.6	GALAXY
ZWG 093.056	10 05 30.	+ 18 57		14.3	GALAXY
UGC 05467	10 05 30.	+ 18 57	48	14.3	GALAXY S0?
MCG+03-26-031	10 05 30.	+ 18 58	60	14.3	GALAXY
ZC 1005.5+2025	10 05 30.	+ 20 25	1610		CLUSTER OF GALAXIES
TON-N 0487	10 05 30.	+ 24 17		14.7	BLUE STAR
REIZ 0187	10 05 30.	+ 32 05	180	13.6	GALAXY
ZWG 182.078	10 05 30.	+ 33 36		15.1	GALAXY
ZWG 240.013	10 05 30.	+ 50 21		15.4	GALAXY
ZWG 351.007	10 05 30.	+ 77 32		15.5	GALAXY
ZWG 350.042	10 05 30.	+ 77 32		15.5	GALAXY
RNGC 3130	10 05 34.	+ 10 13		14.5	GALAXY
REIZ 0189	10 05 34.	+ 10 13	30	14.4	GALAXY
HOLM 165B	10 05 34.	+ 38 05	42	15.3	PART OF MULTIPLE GALAXY
IC 2544	10 05 35.	+ 33 35 40.			NONSTELLAR OBJECT
KN 15.143	10 05 35.0	+ 33 35 32.			NEBULA
ZWG 036.042	10 05 36.	+ 02 42		15.5	GALAXY
ZWG 036.043	10 05 36.	+ 05 09		15.6	GALAXY
ZWG 064.072	10 05 36.	+ 10 13		14.3	GALAXY
UGC 05468	10 05 36.	+ 10 13	66	14.3	GALAXY S0-a
MCG+02-26-026	10 05 36.	+ 10 13	30	14.3	GALAXY
MCG+03-26-032	10 05 36.	+ 14 46	36	15.4	GALAXY

OBJECT NAME	RIGHT ASCEN.	DECLINATION	DIAM.	MAGN.	TYPE OF OBJECT
ZWG 093.057	10 05 36.	+ 17 09		15.1	GALAXY
ZC 1005.6+1953	10 05 36.	+ 19 53	1280		CLUSTER OF GALAXIES
MCG+05-24-019	10 05 36.	+ 32 07	150	13.5	GALAXY
PK272+12.1	10 05 36.	- 40 11 23.	84	8.2	PLANETARY NEBULA
REIZ 0190	10 05 38.	+ 17 08	15	15.3	GALAXY
RNGC 3129	10 05 38.	+ 18 39			NON-EXISTENT OBJECT
KN 15.142	10 05 38.2	+ 36 34 22.			NEBULA
ZWG 093.058	10 05 42.	+ 14 48		15.4	GALAXY
7ZW 311	10 05 42.	+ 61 15			COMPACT GALAXY
MCG+13-08-008	10 05 42.	+ 77 32	36	15.	GALAXY
LB 00138	10 05 42.	- 28 29		16.0	FAINT BLUE STAR
MCG-06-23-001	10 05 42.	- 34 58	66	15.	GALAXY
LB 00568	10 05 45.	+ 01 04 00.		17.0	FAINT BLUE STAR
IC 0593	10 05 47.	- 02 17 40.			NONSTELLAR OBJECT
IC 2548	10 05 47.	- 34 59			NONSTELLAR OBJECT
HN 0333	10 05 47.	- 35 00			NEBULA
ZWG 064.073	10 05 48.	+ 12 33		11.3	GALAXY
HAW 1	10 05 48.	+ 12 33	17	20.0	GALAXY
KARA.68 067	10 05 48.	+ 12 33	470		DWARF GALAXY
UGC 05470	10 05 48.	+ 12 33	900	11.3	GALAXY DWRF EL
MCG+02-26-027	10 05 48.	+ 12 33	480	11.3	GALAXY
ZC 1005.8+1739	10 05 48.	+ 17 39	1280		CLUSTER OF GALAXIES
STOCK 10	10 05 48.	+ 30 08			BLUE KNOT NEAR ELLIP GLXY
ZC 1005.8+3715	10 05 48.	+ 37 15	870		CLUSTER OF GALAXIES
MCG+10-15-017	10 05 48.	+ 58 19	27	16.	GALAXY
MCG+11-13-002	10 05 48.	+ 65 52	39	17.	GALAXY
ZWG 008.047	10 05 48.	- 02 17		14.2	GALAXY
UGC 05469	10 05 48.	- 02 17	48	14.2	GALAXY S
MCG+00-26-021	10 05 48.	- 02 17	42	14.2	GALAXY
LB 00139	10 05 48.	- 33 05		16.2	FAINT BLUE STAR
ZWG 008.048	10 05 54.	+ 00 37		15.0	GALAXY
MCG+00-26-022	10 05 54.	+ 00 37	15	15.0	GALAXY
ZWG 064.074	10 05 54.	+ 12 47		15.4	GALAXY
ZWG 093.059	10 05 54.	+ 17 02		15.0	GALAXY
ARC 0934	10 05 54.	+ 17 30		17.0	RICH CLUSTER OF GALAXIES
ZWG 093.060	10 05 54.	+ 18 29		14.0	GALAXY
UGC 05471	10 05 54.	+ 18 29	150	14.0	GALAXY SB:b
MCG+03-26-033	10 05 54.	+ 18 29	150	14.0	GALAXY
MCG+09-17-029	10 05 54.	+ 54 28	24	16.	GALAXY
MCG+10-15-018	10 05 54.	+ 58 21	18	16.	GALAXY
REIZ 0191	10 05 56.	+ 17 02	36	15.1	GALAXY
REIZ 0192	10 05 56.	+ 18 28	72	13.7	GALAXY
RNGC 3131	10 05 56.	+ 18 29		14.0	GALAXY
PK286-06.1	10 05 56.	- 63 40	10		PLANETARY NEBULA
IC 0594	10 05 57.	- 00 25 59.			NONSTELLAR OBJECT
MCG-03-26-026	10 05 57.	- 19 56	60	15.	GALAXY
VDB.66G 074	10 06	+ 12 32	570		DWARF GALAXY
VDB.66G 073	10 06	+ 30 22	70		DWARF GALAXY
ZC 1006.0+0014	10 06 00.	+ 00 14	2690		CLUSTER OF GALAXIES
ZC 1006.0+2149	10 06 00.	+ 21 49	1480		CLUSTER OF GALAXIES
ZC 1006.0+4255	10 06 00.	+ 42 55	4230		CLUSTER OF GALAXIES
MCG+09-17-030	10 06 00.	+ 54 27	24	16.	GALAXY
MCG+10-15-019	10 06 00.	+ 58 41	114	15.	GALAXY
ZWG 008.049	10 06 00.	- 00 25		14.7	GALAXY
UGC 05472	10 06 00.	- 00 25	66	14.7	GALAXY SBb-c
MCG+00-26-023	10 06 00.	- 00 25	60	14.7	GALAXY
KARA.73B 0401	10 06 00.	- 00 25	66	14.7	ISOLATED GALAXY S
MRSL 282-01/1	10 06 00.	- 57 15	900		HII REGION
ARC 0918	10 06 04.	+ 73 59		17.1	RICH CLUSTER OF GALAXIES
ZWG 064.075	10 06 06.	+ 09 42		14.9	GALAXY
UGC 05473	10 06 06.	+ 09 42	66	14.9	GALAXY S
MCG+02-26-028	10 06 06.	+ 09 42	60	14.9	GALAXY
ZC 1006.1+1201	10 06 06.	+ 12 01	670		CLUSTER OF GALAXIES
ZWG 064.076	10 06 06.	+ 12 56		15.6	GALAXY
REIZ 0193	10 06 06.	+ 32 44	60	14.9	GALAXY
ZWG 182.079	10 06 06.	+ 32 45		15.0	GALAXY
UGC 05474	10 06 06.	+ 32 45	90	15.0	GALAXY Sc/SBc
ZWG 290.009	10 06 06.	+ 58 42		15.5	GALAXY
UGC 05475	10 06 06.	+ 58 42	102	15.5	GALAXY SBc
MCG-01-26-026	10 06 06.	- 04 13	12	16.	GALAXY
LB 00140	10 06 06.	- 30 53		15.8	FAINT BLUE STAR
KN 15.144	10 06 07.7	+ 32 44 23.			NEBULA
ZWG 064.077	10 06 12.	+ 13 28		15.6	GALAXY
MCG+06-22-077	10 06 12.	+ 32 44 30.	72	14.5	GALAXY
MCG+07-21-012	10 06 12.	+ 40 35	78	14.5	GALAXY
ZWG 211.013	10 06 12.	+ 40 37		14.6	GALAXY
UGC 05476	10 06 12.	+ 40 37	84	14.6	GALAXY Sb/SBb
ZWG 351.008	10 06 12.	+ 76 00		15.4	GALAXY
ZWG 350.043	10 06 12.	+ 76 00		15.4	GALAXY
ARC 0937	10 06 14.	+ 14 13		17.2	RICH CLUSTER OF GALAXIES
ARC 0922	10 06 16.	+ 71 16		17.9	RICH CLUSTER OF GALAXIES
ARC 0939	10 06 16.	- 11 05		17.7	RICH CLUSTER OF GALAXIES
ZWG 064.078	10 06 18.	+ 11 53		15.7	GALAXY
ZC 1006.3+1415	10 06 18.	+ 14 15	1280		CLUSTER OF GALAXIES
ZC 1006.3+2320	10 06 18.	+ 23 20	3760		CLUSTER OF GALAXIES
ZC 1006.3+3349	10 06 18.	+ 33 49	1480		CLUSTER OF GALAXIES
ZWG 008.050	10 06 18.	- 00 02		15.5	GALAXY
REIZ 0194	10 06 19.	+ 40 33	42	14.2	GALAXY
ARC 0936	10 06 22.	+ 29 48		17.2	RICH CLUSTER OF GALAXIES
MCG+03-26-034	10 06 24.	+ 15 14	84	14.5	GALAXY
ZC 1006.4+2625	10 06 24.	+ 26 25	810		CLUSTER OF GALAXIES
REIZ 0195	10 06 24.	+ 32 50	42	15.3	GALAXY
MCG+09-17-031	10 06 24.	+ 54 33	36	17.	GALAXY
ZC 1006.4+6802	10 06 24.	+ 68 02	1480		CLUSTER OF GALAXIES
REIZ 0196	10 06 25.	+ 26 06	30	14.5	GALAXY
REIZ 0197	10 06 26.	+ 27 11	24	15.2	GALAXY
KN 15.145	10 06 28.2	+ 32 50 15.			NEBULA
ZWG 093.061	10 06 30.	+ 15 15		14.5	GALAXY
UGC 05477	10 06 30.	+ 15 15	60	14.5	GALAXY E
ZC 1006.5+3054	10 06 30.	+ 30 54	940		CLUSTER OF GALAXIES
TON-N 1173	10 06 30.	+ 35 15		16.8	BLUE STAR
ZC 1006.5+3541	10 06 30.	+ 35 41	810		CLUSTER OF GALAXIES
ZC 1006.5+4459	10 06 30.	+ 44 59	470		CLUSTER OF GALAXIES
MCG+09-17-032	10 06 30.	+ 53 12 30.	24	17.	GALAXY
MCG+10-15-020	10 06 30.	+ 57 08	30	16.	GALAXY
ZC 1006.5+6328	10 06 30.	+ 63 28	2550		CLUSTER OF GALAXIES
LB 00141	10 06 30.	- 30 48		16.6	FAINT BLUE STAR
ZWG 153.024	10 06 36.	+ 30 24		15.5	GALAXY
UGC 05478	10 06 36.	+ 30 24	108	15.5	GALAXY DWRF IR
MCG+05-24-020	10 06 36.	+ 30 25	90	15.5	GALAXY
MCG+10-15-021	10 06 36.	+ 58 23	39	16.	GALAXY
ZC 1006.6+6015	10 06 36.	+ 60 15	1340		CLUSTER OF GALAXIES
SEY 066	10 06 36.	+ 66 46 27.		15.0	FAINT GALAXY
ZWG 313.001	10 06 36.	+ 66 47		15.7	GALAXY
ZWG 312.031	10 06 36.	+ 66 47		15.7	GALAXY
ZC 1006.6+7115	10 06 36.	+ 71 15	1140		CLUSTER OF GALAXIES
RNGC 3138	10 06 40.	- 11 44		15.0	GALAXY
ARC 0940	10 06 40.	- 16 24		17.5	RICH CLUSTER OF GALAXIES

OBJECT NAME	RIGHT ASCEN.	DECLINATION	DIAM.	MAGN.	TYPE OF OBJECT
REIZ 0198	10 06 41.	+ 30 23	90	15.2	GALAXY
ZWG 064.079	10 06 42.	+ 11 01		15.5	GALAXY
ZWG 064.080	10 06 42.	+ 12 51		15.7	GALAXY
TON-N 1174	10 06 42.	+ 36 27		15.7	BLUE STAR
ZC 1006.7+4013	10 06 42.	+ 40 13	1410		CLUSTER OF GALAXIES
ZWG 266.026	10 06 42.	+ 54 55		15.6	GALAXY
ZWG 313.002	10 06 42.	+ 63 10		15.2	GALAXY
ZWG 312.032	10 06 42.	+ 63 10		15.2	GALAXY
KARA.73E 0402	10 06 42.	+ 63 10	66	15.2	ISOLATED GALAXY S
MCG-05-24-023	10 06 42.	- 30 09	36	16.5	GALAXY
ARC 0938	10 06 43.	+ 18 39		16.6	RICH CLUSTER OF GALAXIES
REIZ 0199	10 06 44.	+ 27 11	36	15.8	GALAXY
MCG+09-17-033	10 06 45.	+ 54 54	54	15.	GALAXY
MCG-02-26-032	10 06 45.	- 11 44	66	15.	GALAXY
ZWG 266.027	10 06 48.	+ 54 33		15.5	GALAXY
ZWG 266.028	10 06 48.	+ 54 44		14.8	GALAXY
UGC 05479	10 06 48.	+ 54 44	72	14.8	GALAXY S
ZWG 290.010	10 06 48.	+ 58 44		15.6	GALAXY
UGC 05480	10 06 48.	+ 58 44	60	15.6	GALAXY
RNGC 3134	10 06 52.	+ 12 34			NON-EXISTENT OBJECT
RNGC 3137	10 06 52.	- 28 48		13.0	GALAXY
ZC 1006.9+1715	10 06 54.	+ 17 15	610		CLUSTER OF GALAXIES
TON-N 0488	10 06 54.	+ 30 19		16.7	BLUE STAR
ZWG 153.025	10 06 54.	+ 30 34		15.1	GALAXY
UGC 05481	10 06 54.	+ 30 34	102	15.1	GALAXY Sa
MCG+09-17-035	10 06 54.	+ 54 34	30	16.	GALAXY
ZWG 266.029	10 06 54.	+ 54 44		15.7	GALAXY
MCG+09-17-034	10 06 54.	+ 54 45	90	14.	GALAXY
MCG+10-15-022	10 06 54.	+ 58 44	54	15.	GALAXY
ZWG 351.009	10 06 54.	+ 77 55		15.7	GALAXY
ZWG 350.044	10 06 54.	+ 77 55		15.7	GALAXY
RNGC 3141	10 06 54.	- 16 24			NON-EXISTENT OBJECT
MCG-05-24-024	10 06 54.	- 28 48	360	13.	GALAXY
LB 00569	10 06 57.	+ 01 42 48.		16.9	FAINT BLUE STAR
IC 0595	10 06 58.	+ 11 15 04.			NONSTELLAR OBJECT
REIZ 0201	10 06 59.	+ 30 34	72	14.4	GALAXY
ZC 1007.0+0351	10 07 00.	+ 03 51	810		CLUSTER OF GALAXIES
ZWG 064.081	10 07 00.	+ 11 15		15.1	GALAXY
ZWG 064.082	10 07 00.	+ 14 28		15.5	GALAXY
ZC 1007.0+3024	10 07 00.	+ 30 24	1680		CLUSTER OF GALAXIES
MCG+05-24-021	10 07 00.	+ 30 35	90	15.1	GALAXY
TON-N 1175	10 07 00.	+ 35 54		16.8	BLUE STAR
ZC 1007.0+4413	10 07 00.	+ 44 13	3090		CLUSTER OF GALAXIES
MCG+08-19-006	10 07 00.	+ 47 12	36	16.	GALAXY
MCG+09-17-036	10 07 00.	+ 54 44	15	16.	GALAXY
ZWG 290.011	10 07 00.	+ 58 43		15.7	GALAXY
MCG-03-26-027	10 07 00.	- 19 46	36	15.	GALAXY
MCG-06-23-002	10 07 00.	- 38 10 30.	90	14.5	GALAXY
REIZ 0200	10 07 01.	+ 40 35	42	14.5	GALAXY
MCG-06-23-003	10 07 03.	- 38 10	60	15.	GALAXY
ZC 1007.1+0046	10 07 06.	+ 00 46	940		CLUSTER OF GALAXIES
ZWG 153.026	10 07 06.	+ 32 19		14.9	GALAXY
MCG+05-24-022	10 07 06.	+ 32 19	48	14.9	GALAXY
TON-N 1176	10 07 06.	+ 35 47		16.7	BLUE STAR
ZWG 183.001	10 07 06.	+ 36 45		15.7	GALAXY
ZWG 182.080	10 07 06.	+ 36 45		15.7	GALAXY
ZC 1007.1+3834	10 07 06.	+ 38 34	670		CLUSTER OF GALAXIES
ZWG 240.014	10 07 06.	+ 47 13		15.2	GALAXY
ZC 1007.1+5613	10 07 06.	+ 56 13	1480		CLUSTER OF GALAXIES
RNGC 3140	10 07 06.	- 16 22		14.0	GALAXY
MCG-03-26-028	10 07 06.	- 16 22	60	14.5	GALAXY
IC 2547	10 07 07.	+ 36 45 28.			NONSTELLAR OBJECT
KN 15.146	10 07 07.8	+ 36 44 58.			NEBULA
ARC 0941	10 07 08.	+ 03 56		17.1	RICH CLUSTER OF GALAXIES
ZWG 036.044	10 07 12.	+ 03 47		15.6	GALAXY
ZWG 036.045	10 07 12.	+ 05 25		14.6	GALAXY
MCG+01-26-016	10 07 12.	+ 05 25	48	14.6	GALAXY
ZC 1007.2+3619	10 07 12.	+ 36 19	870		CLUSTER OF GALAXIES
ZWG 183.002	10 07 12.	+ 36 42		15.6	GALAXY
ZWG 182.081	10 07 12.	+ 36 42		15.6	GALAXY
OCL 0810	10 07 12.	- 60 57	1980	11.	OPEN STAR CLUSTER
IC 2549	10 07 13.	+ 36 43 04.			NONSTELLAR OBJECT
KN 15.147	10 07 13.4	+ 36 42 45.			NEBULA
ZWG 008.051	10 07 18.	+ 01 28		15.3	GALAXY
ZWG 064.083	10 07 18.	+ 12 10		15.2	GALAXY
MCG+02-26-029	10 07 18.	+ 12 10	48	15.2	GALAXY
UGC 05482	10 07 18.	+ 32 32	78	16.5	GALAXY Sc
MCG+07-21-013	10 07 18.	+ 43 18	27	17.	GALAXY
LB 00570	10 07 18.	- 00 43 00.		15.1	FAINT BLUE STAR
TON-N 1177	10 07 19.	+ 20 10		16.7	BLUE STAR
ARC 0935	10 07 19.	+ 56 14		17.2	RICH CLUSTER OF GALAXIES
KN 15.148	10 07 22.9	+ 32 31 27.			NEBULA
MCG+03-26-035	10 07 24.	+ 16 36	36	15.5	GALAXY
ZWG 093.062	10 07 24.	+ 16 37		15.5	GALAXY
ZWG 093.063	10 07 24.	+ 17 56		15.5	GALAXY
KARA.72 225A	10 07 24.	+ 17 56	18	15.5	PART OF DOUBLE GALAXY
TON-N 0030	10 07 24.	+ 30 37		14.6	BLUE STAR
REIZ 0202	10 07 24.	+ 32 30	48	15.2	GALAXY
ZC 1007.4+3724	10 07 24.	+ 37 24	740		CLUSTER OF GALAXIES
MCG+07-21-014	10 07 24.	+ 39 59	48	15.5	GALAXY
ZWG 211.014	10 07 24.	+ 40 11		15.5	GALAXY
MCG+09-17-037	10 07 24.	+ 54 27	24	16.	GALAXY
ZWG 008.052	10 07 24.	- 02 35		15.5	GALAXY
LB 00571	10 07 25.	+ 01 44 48.		16.3	FAINT BLUE STAR
REIZ 0203	10 07 26.	+ 17 55	12	15.3	GALAXY
REIZ 0204	10 07 26.	+ 17 57	36	15.3	GALAXY
ZWG 093.064	10 07 30.	+ 17 58		15.6	GALAXY
KARA.72 225B	10 07 30.	+ 17 58	30	15.6	PART OF DOUBLE GALAXY
MCG+06-22-078	10 07 30.	+ 32 31	72	15.5	GALAXY
ZWG 266.030	10 07 30.	+ 54 28		15.4	GALAXY
ZWG 008.053	10 07 30.	- 02 13		15.5	GALAXY
UGC 05483	10 07 30.	- 02 13	60	15.5	GALAXY S
MCG+00-26-024	10 07 30.	- 02 13	48	15.5	GALAXY
LB 00142	10 07 30.	- 28 49		16.6	FAINT BLUE STAR
IC 2554	10 07 30.	- 66 48	186		GALAXY SB(s)
TON-N 1178	10 07 31.	+ 22 31		15.5	BLUE STAR
KN 15.149	10 07 31.4	+ 37 50 09.			NEBULA
HN 0334	10 07 32.	- 66 47			NEBULA
RNGC 3143	10 07 35.	- 12 20		14.0	GALAXY
ZWG 036.046	10 07 36.	+ 08 21		15.6	GALAXY
MCG+05-24-023	10 07 36.	+ 28 11	60	14.6	GALAXY
ZWG 153.027	10 07 36.	+ 28 12		14.6	GALAXY
UGC 05484	10 07 36.	+ 28 12	66	14.6	GALAXY Sb
MCG+10-15-023	10 07 36.	+ 58 08	39	17.	GALAXY
MCG-02-26-034	10 07 36.	- 11 32 30.	24	15.	GALAXY
MCG-02-26-033	10 07 36.	- 12 21	30	14.	GALAXY
TON-N 1179	10 07 37.	+ 22 50		16.5	BLUE STAR
IC 2550	10 07 37.	+ 28 11 38.			NONSTELLAR OBJECT
KN 15.150	10 07 38.4	+ 36 15 27.			NEBULA
RNGC 3142	10 07 39.	- 08 15		14.0	GALAXY
MCG-02-26-035	10 07 39.	- 10 27 30.	48	15.	GALAXY
PK285-05.1	10 07 39.	- 62 19	4	13.0	PLANETARY NEBULA
IC 2553	10 07 39.	- 62 19	4	13.0	PLANETARY NEBULA
RNGC 3139	10 07 40.	- 11 33		15.0	GALAXY
HN 0070	10 07 40.	- 62 20			NEBULA
RNGC 3145	10 07 41.	- 12 10		12.5	GALAXY
REIZ 0206	10 07 42.	+ 32 33	24	15.0	GALAXY
ZWG 313.003	10 07 42.	+ 65 31		15.1	GALAXY
ZWG 312.033	10 07 42.	+ 65 31		15.1	GALAXY
UGC 05485	10 07 42.	+ 65 31	72	15.1	GALAXY PECULR
MCG-01-26-027	10 07 42.	- 04 44 30.	10	16.	GALAXY
MCG-01-26-028	10 07 42.	- 08 15	36	14.	GALAXY
MCG-02-26-036	10 07 42.	- 12 12 30.	192	12.	GALAXY
MCG-04-24-018	10 07 42.	- 24 04	36	15.	GALAXY
IC 0597	10 07 43.	- 06 39 00.			NONSTELLAR OBJECT
MCG+11-13-003	10 07 45.	+ 65 32	72	16.	GALAXY
KN 15.151	10 07 47.3	+ 32 33 26.			NEBULA
ZWG 093.065	10 07 48.	+ 19 09		15.5	GALAXY
TON-N 0031	10 07 48.	+ 26 16		16.1	BLUE STAR
ZWG 183.003	10 07 48.	+ 32 33		15.0	GALAXY
ZWG 182.082	10 07 48.	+ 32 33		15.0	GALAXY
ZWG 240.015	10 07 48.	+ 46 12		14.3	GALAXY
UGC 05486	10 07 48.	+ 46 12	60	14.3	GALAXY S
REIZ 0205	10 07 49.	+ 46 12	60	14.2	GALAXY
HOLM 167B	10 07 50.	+ 02 25	36	14.8	PART OF MULTIPLE GALAXY
REIZ 0207	10 07 51.	+ 20 19	60	14.0	GALAXY
RNGC 3135	10 07 51.	+ 46 12		14.5	GALAXY
MCG+11-13-005	10 07 51.	+ 65 29	12	18.	GALAXY
MCG+11-13-004	10 07 51.	+ 65 30	18	18.	GALAXY
MCG-02-26-037	10 07 51.	- 11 42	30	15.	GALAXY
IC 2551	10 07 52.	+ 24 39 37.			NONSTELLAR OBJECT
HOLM 167A	10 07 53.	+ 02 28	36	14.7	PART OF MULTIPLE GALAXY
IC 0596	10 07 53.	+ 10 17 18.			NONSTELLAR OBJFCT
ZWG 008.054	10 07 54.	+ 02 29		15.2	GALAXY
UGC 05487	10 07 54.	+ 02 29	66	15.2	GALAXY SBb-c
MCG+00-26-025	10 07 54.	+ 02 29	66	15.2	GALAXY
ZWG 036.047	10 07 54.	+ 06 25		15.4	GALAXY
ZWG 064.084	10 07 54.	+ 10 17		15.0	GALAXY
MCG+02-26-030	10 07 54.	+ 10 17	48	15.0	GALAXY
ZWG 093.066	10 07 54.	+ 16 56		14.9	GALAXY
MCG+03-26-036	10 07 54.	+ 16 56	15	14.9	GALAXY
ZC 1007.9+1755	10 07 54.	+ 17 55	2290		CLUSTER OF GALAXIES
ZWG 123.020	10 07 54.	+ 24 40		14.6	GALAXY
UGC 05488	10 07 54.	+ 24 40	78	14.6	GALAXY
MCG+04-24-016	10 07 54.	+ 24 40	36	14.6	GALAXY
MCG+08-19-007	10 07 54.	+ 46 11	48	14.2	GALAXY
MCG+10-15-024	10 07 54.	+ 58 28	24	16.	GALAXY
VDB.66G 075	10 08	- 04 28	270		DWARF GALAXY
VDB.66G 076	10 08	- 13 33	100		DWARF GALAXY
VDB.66G 237	10 08	- 25 37	70		DWARF GALAXY
ZWG 036.048	10 08 00.	+ 05 23		14.9	GALAXY
MCG+01-26-017	10 08 00.	+ 05 23	36	14.9	GALAXY
ZWG 036.049	10 08 00.	+ 07 49		15.5	GALAXY
ZWG 093.067	10 08 00.	+ 16 15		15.6	GALAXY
MCG+03-26-038	10 08 00.	+ 16 15	54	15.6	GALAXY
ZWG 093.068	10 08 00.	+ 20 18		14.0	GALAXY
UGC 05489	10 08 00.	+ 20 18	132	14.0	GALAXY Sa
MCG+03-26-037	10 08 00.	+ 20 20	120	14.0	GALAXY
ZC 1008.0+3847	10 08 00.	+ 38 47	540		CLUSTER OF GALAXIES
ZWG 240.016	10 08 00.	+ 47 10		14.6	GALAXY
MCG+08-19-008	10 08 00.	+ 47 10	24	14.	GALAXY
MCG+10-15-025	10 08 00.	+ 58 32	18	16.	GALAXY
ZWG 008.055	10 08 00.	- 01 33		15.3	GALAXY
MCG-06-23-004	10 08 00.	- 37 53	48	15.	GALAXY
ZC 1008.1+6600	10 08 06.	+ 66 00	1550		CLUSTER OF GALAXIES
ZWG 008.057	10 08 06.	- 01 48		15.3	GALAXY
ZWG 008.056	10 08 06.	- 02 11		15.1	GALAXY
LB 00143	10 08 06.	- 30 40		16.8	FAINT BLUE STAR
REIZ 0208	10 08 07.	+ 16 14	36	15.1	GALAXY
ZC 1008.2+0032	10 08 12.	+ 00 32	740		CLUSTER OF GALAXIES
ZWG 123.021	10 08 12.	+ 23 22		15.5	GALAXY
MCG-06-23-005	10 08 12.	- 37 47	42	15.	GALAXY
REIZ 0209	10 08 17.	+ 23 23	30	14.7	GALAXY
ZC 1008.3+1620	10 08 18.	+ 16 20	870		CLUSTER OF GALAXIES
ZC 1008.3+4002	10 08 18.	+ 40 02	540		CLUSTER OF GALAXIES
ZWG 008.058	10 08 18.	- 00 24		15.7	GALAXY
MCG-04-24-019	10 08 18.	- 25 34 30.	120	15.	GALAXY
KN 15.152	10 08 20.5	+ 36 55 26.			NEBULA
MCG-02-26-038	10 08 21.	- 15 16 30.	36	15.	GALAXY
ZWG 036.050	10 08 24.	+ 03 28		15.6	GALAXY
ZWG 064.085	10 08 24.	+ 10 06		15.1	GALAXY
ZC 1008.4+6455	10 08 24.	+ 64 55	1480		CLUSTER OF GALAXIES
MCG-01-26-029	10 08 24.	- 06 40	60	14.5	GALAXY
LB 00144	10 08 24.	- 27 58		14.5	FAINT BLUE STAR
MCG-06-23-006	10 08 24.	- 38 54	42	15.	GALAXY
SER 072.01	10 08 24.	- 66 48	180		INTERACTING GALAXIES
TON-N 1180	10 08 25.	+ 23 12		17.	BLUE STAR
ZWG 036.051	10 08 30.	+ 08 01		15.2	GALAXY
ZWG 064.086	10 08 30.	+ 10 00		15.4	GALAXY
ZC 1008.5+3354	10 08 30.	+ 33 54	1480		CLUSTER OF GALAXIES
ZC 1008.5+3730	10 08 30.	+ 37 30	610		CLUSTER OF GALAXIES
ZC 1008.5+5517	10 08 30.	+ 55 17	740		CLUSTER OF GALAXIES
ZWG 290.012	10 08 30.	+ 59 08		15.7	GALAXY
MCG-01-26-030	10 08 30.	- 04 26	300	13.	GALAXY
ARP 338	10 08 30.	- 07 40			PECULIAR GALAXY
ZWG 008.059	10 08 36.	+ 00 12		15.5	GALAXY
ZWG 008.060	10 08 36.	+ 01 28		15.2	GALAXY
ZC 1008.6+1926	10 08 36.	+ 19 26	1480		CLUSTER OF GALAXIES
ZC 1008.6+2612	10 08 36.	+ 26 12	1480		CLUSTER OF GALAXIES
ZWG 153.028	10 08 36.	+ 31 02		15.0	GALAXY
UGC 05490	10 08 36.	+ 31 02	84	15.0	GALAXY Sc
ZWG 290.013	10 08 36.	+ 59 07		15.5	GALAXY
UGC 05491	10 08 36.	+ 59 07	60	15.5	GALAXY Sb
MRK 026	10 08 36.	+ 59 08	18	16.	GALAXY WITH UV CONTINUUM
MCG+10-15-026	10 08 36.	+ 59 08	18	16.	GALAXY
UGC 05492	10 08 36.	+ 59 58	84	16.0	GALAXY Sa-b
MCG+10-15-027	10 08 36.	+ 59 58	60	15.	GALAXY
ZC 1008.6-0148	10 08 36.	- 01 48	810		CLUSTER OF GALAXIES
MCG-05-24-025	10 08 36.	- 28 38	12	16.	GALAXY
MCG-06-23-007	10 08 36.	- 34 34	72	13.5	GALAXY
IC 2552	10 08 36.	- 34 35			NONSTELLAR OBJECT
HN 0335	10 08 36.	- 34 36			NEBULA
ARC 0944	10 08 38.	- 01 48		17.1	RICH CLUSTER OF GALAXIES
ZWG 008.061	10 08 42.	+ 00 10		15.6	GALAXY
ZWG 036.052	10 08 42.	+ 06 08		15.1	GALAXY
ZC 1008.7+3905	10 08 42.	+ 39 05	1080		CLUSTER OF GALAXIES

OBJECT NAME	RIGHT ASCEN.	DECLINATION	DIAM.	MAGN.	TYPE OF OBJECT
ZWG 290.014	10 08 42.	+ 58 19		15.6	GALAXY
MRK 027	10 08 42.	+ 58 58	10	17.	GALAXY WITH UV CONTINUUM
MCG+10-15-028	10 08 42.	+ 59 06	57	15.	GALAXY
HMS 1.14	10 08 42.	- 04 28			IRR GALAXY
HMS 1.13	10 08 42.	- 04 28			IRR GALAXY
MCG-02-26-039	10 08 42.	- 13 33	120	15.	GALAXY
TON-N 1181	10 08 44.	+ 18 31		15.3	BLUE STAR
ARC 0942	10 08 44.	+ 19 32		17.0	RICH CLUSTER OF GALAXIES
MCG+05-24-024	10 08 45.	+ 31 02 30.	72	15.0	GALAXY
KN 15.153	10 08 46.7	+ 37 27 57.			NEBULA
ZWG 008.062	10 08 48.	+ 00 13		15.4	GALAXY
ZWG 008.063	10 08 48.	+ 00 41		14.0	GALAXY
UGC 05493	10 08 48.	+ 00 41	102	14.0	GALAXY Sb/SBc
MCG+00-26-026	10 08 48.	+ 00 41	90	14.0	GALAXY
ZWG 211.015	10 08 48.	+ 43 40		15.7	GALAXY
MCG+10-15-029	10 08 48.	+ 58 18	39	16.	GALAXY
MCG+10-15-030	10 08 48.	+ 62 20	39	16.	GALAXY
MCG-03-26-029	10 08 48.	- 20 38	42	13.5	GALAXY
RNGC 3146	10 08 49.	- 20 38		13.0	GALAXY
ZC 1008.9+3811	10 08 54.	+ 38 11	870		CLUSTER OF GALAXIES
MCG-03-26-030	10 08 54.	- 16 57	48	14.	GALAXY
MCG-06-23-008	10 08 54.	- 39 28	72	14.5	GALAXY
TON-N 1182	10 08 55.	+ 20 20		16.	BLUE STAR
ZWG 036.053	10 09 00.	+ 03 31		15.7	GALAXY
ZC 1009.0+0643	10 09 00.	+ 06 43	2420		CLUSTER OF GALAXIES
ZWG 064.087	10 09 00.	+ 08 51		15.5	GALAXY
ZC 1009.0+0855	10 09 00.	+ 08 55	1950		CLUSTER OF GALAXIES
ZC 1009.0+5407	10 09 00.	+ 54 07	740		CLUSTER OF GALAXIES
MCG+10-15-031	10 09 00.	+ 59 54	39	16.	GALAXY
ZWG 313.004	10 09 00.	+ 65 31		15.6	GALAXY
ZWG 312.034	10 09 00.	+ 65 31		15.6	GALAXY
UGC 05494	10 09 00.	+ 67 40	66	16.0	GALAXY PAIR?
MCG-06-23-009	10 09 00.	- 35 26		15.	GALAXY
MCG+03-26-039	10 09 06.	+ 16 40	180	14.7	GALAXY
ZWG 093.069	10 09 06.	+ 16 41		14.7	GALAXY
UGC 05495	10 09 06.	+ 16 41	180	14.7	GALAXY Sc
ZWG 240.017	10 09 06.	+ 46 33		15.5	GALAXY
UGC 05496	10 09 06.	+ 46 33	84	15.5	GALAXY Sc-IRR
MCG+08-19-009	10 09 06.	+ 46 33	66	15.	GALAXY
ZWG 313.005	10 09 06.	+ 64 22		15.7	GALAXY
UGC 05497	10 09 06.	+ 64 22	66	15.7	GALAXY
MRK 137	10 09 06.	+ 65 32	12	16.5	GALAXY WITH UV CONTINUUM
LB 00572	10 09 09.	- 01 04 30.		16.4	FAINT BLUE STAR
MCG-04-24-020	10 09 09.	- 25 02 30.	48	15.5	GALAXY
ZC 1009.2+3612	10 09 12.	+ 36 12	670		CLUSTER OF GALAXIES
ZC 1009.2+4125	10 09 12.	+ 41 25	940		CLUSTER OF GALAXIES
MRK 028	10 09 12.	+ 58 39	8	17.	GALAXY WITH UV CONTINUUM
ZC 1009.2+6115	10 09 12.	+ 61 15	1010		CLUSTER OF GALAXIES
MCG+11-13-007	10 09 12.	+ 64 22	27	15.	GALAXY
MCG+11-13-006	10 09 12.	+ 65 32	18	17.	GALAXY
MRK 138	10 09 12.	+ 67 38	13	16.	GALAXY WITH UV CONTINUUM
ZWG 008.064	10 09 12.	- 02 30		15.1	GALAXY
MCG+00-26-027	10 09 12.	- 02 30	48	15.1	GALAXY
TON-N 1183	10 09 13.	+ 22 58		17.	BLUE STAR
REIZ 0211	10 09 17.	+ 23 20	12	14.0	GALAXY
ZWG 008.065	10 09 18.	+ 00 10		15.4	GALAXY
ZWG 123.022	10 09 18.	+ 23 20		15.3	GALAXY
UGC 05498	10 09 18.	+ 23 20	96	15.3	GALAXY Sa+COMP
MCG+04-24-017	10 09 18.	+ 23 20	84	15.3	GALAXY
ZC 1009.3+6512	10 09 18.	+ 65 12	670		CLUSTER OF GALAXIES
TON-N 1184	10 09 19.	+ 22 32		15.5	BLUE STAR
ARC 0943	10 09 21.	+ 33 52		17.2	RICH CLUSTER OF GALAXIES
REIZ 0210	10 09 23.	+ 39 05	72	14.1	GALAXY
ZWG 036.054	10 09 24.	+ 03 35		15.3	GALAXY
MCG+05-24-025	10 09 24.	+ 28 06	156	14.3	GALAXY
ZWG 153.029	10 09 24.	+ 28 07		14.3	GALAXY
UGC 05499	10 09 24.	+ 28 07	180	14.3	GALAXY SBb
MCG+10-15-033	10 09 24.	+ 58 28			GALAXY
MCG+10-15-032	10 09 24.	+ 58 29	15	16.	GALAXY
ZC 1009.4-0011	10 09 24.	- 00 11	810		CLUSTER OF GALAXIES
KEEL 208	10 09 25.8	+ 03 36 12.		14.	NEBULA
IC 2555	10 09 26.	- 31 24			NONSTELLAR OBJECT
KN 15.154	10 09 26.1	+ 36 49 04.			NEBULA
HN 0336	10 09 27.	- 31 24			NEBULA
RNGC 2519	10 09 28.	+ 43 44			NON-EXISTENT OBJECT
HOLM 168A	10 09 29.	+ 38 21	54	15.2	PART OF MULTIPLE GALAXY
HOLM 168B	10 09 29.	+ 38 22	42	15.6	PART OF MULTIPLE GALAXY
ZWG 036.055	10 09 30.	+ 06 19		15.6	GALAXY
ZWG 290.015	10 09 30.	+ 59 07		15.7	GALAXY
MCG+11-13-008	10 09 30.	+ 65 46	24	16.	GALAXY
ZWG 351.010	10 09 30.	+ 78 03		14.5	GALAXY
ZWG 350.045	10 09 30.	+ 78 03		14.5	GALAXY
UGC 05500	10 09 30.	+ 78 03	84	14.5	GALAXY Sb
ZWG 008.066	10 09 30.	- 01 54		15.0	GALAXY
MCG+00-26-028	10 09 30.	- 01 54	30	15.0	GALAXY
LB 00145	10 09 30.	- 30 33		16.6	FAINT BLUE STAR
MCG-06-23-010	10 09 30.	- 37 40	36	14.5	GALAXY
MRSL 282-01/2	10 09 30.	- 57 25	720		HII REGION
KN 15.155	10 09 30.6	+ 38 21 47.			NEBULA
RNGC 3197	10 09 31.	+ 78 03		14.5	GALAXY
KN 15.156	10 09 31.0	+ 38 22 24.			NEBULA
AR 07	10 09 31.48	+ 78 04 01.1			NEBULA
REIZ 0212	10 09 32.	+ 28 07	126	14.2	GALAXY
SEY 067	10 09 34.	+ 65 45 27.		14.9	FAINT GALAXY
RNGC 3157	10 09 34.	- 30 43			NON-EXISTENT OBJECT
KN 15.157	10 09 34.1	+ 36 51 50.			NEBULA
ZC 1009.6+0340	10 09 36.	+ 03 40	340		CLUSTER OF GALAXIES
ZWG 036.056	10 09 36.	+ 05 10		14.5	GALAXY
UGC 05501	10 09 36.	+ 05 10	36	14.5	GALAXY
MCG+01-26-018	10 09 36.	+ 05 10	30	14.5	GALAXY
ZC 1009.6+4052	10 09 36.	+ 40 52	2080		CLUSTER OF GALAXIES
ZWG 313.006	10 09 36.	+ 65 46		15.3	GALAXY
MCG-05-24-026	10 09 36.	- 31 24	144	14.5	GALAXY
SEY 068	10 09 38.	+ 65 59 33.		14.5	FAINT GALAXY
LB 00573	10 09 40.	- 00 06 00.		16.0	FAINT BLUE STAR
LB 00574	10 09 42.	+ 00 26 24.		17.3	FAINT BLUE STAR
ZC 1009.7+3630	10 09 42.	+ 36 30	1480		CLUSTER OF GALAXIES
ZWG 211.016	10 09 42.	+ 39 38		15.5	GALAXY
ZC 1009.7+4316	10 09 42.	+ 43 16	540		CLUSTER OF GALAXIES
ZWG 313.007	10 09 42.	+ 66 00		15.6	GALAXY
ZWG 312.035	10 09 42.	+ 66 00		15.6	GALAXY
MCG+11-13-009	10 09 42.	+ 66 10	12	18.	GALAXY
7ZW 312	10 09 42.	+ 74 36			COMPACT GALAXY
MCG+13-08-009	10 09 42.	+ 78 03	72	14.	GALAXY
REIZ 0214	10 09 43.	+ 16 45	24	15.8	GALAXY
TON-N 1186	10 09 43.	+ 22 54		17.	BLUE STAR
ARC 0946	10 09 43.	+ 24 06		17.6	RICH CLUSTER OF GALAXIES
TON-N 1185	10 09 44.	+ 19 21		15.8	BLUE STAR
REIZ 0213	10 09 45.	+ 29 32	18	15.2	GALAXY
MCG+07-21-015	10 09 45.	+ 39 36	45	15.	GALAXY
MCG+11-13-010	10 09 45.	+ 66 00	54	16.	GALAXY
ARC 0949	10 09 47.	+ 06 40		16.8	RICH CLUSTER OF GALAXIES
ZC 1009.8+0254	10 09 48.	+ 02 54	610		CLUSTER OF GALAXIES
ZWG 064.088	10 09 48.	+ 12 37		14.8	GALAXY
MCG+02-26-031	10 09 48.	+ 12 37	48	14.8	GALAXY
ZWG 093.070	10 09 48.	+ 20 14		15.6	GALAXY
ZWG 123.023	10 09 48.	+ 22 32		15.6	GALAXY
ZWG 153.030	10 09 48.	+ 29 34		15.7	GALAXY
MCG+07-21-016	10 09 48.	+ 43 23	78	14.	GALAXY
MCG+08-19-010	10 09 48.	+ 46 39	18	16.	GALAXY
MCG+09-17-038	10 09 48.	+ 56 20	18	16.	GALAXY
MCG+10-15-034	10 09 48.	+ 59 07	27	15.	GALAXY
MCG+13-08-010	10 09 48.	+ 76 47	12	16.	GALAXY
KEEL 209	10 09 53.9	+ 03 50 44.		16.	NEBULA
ZWG 064.089	10 09 54.	+ 11 42		15.4	GALAXY
ZC 1009.9+2409	10 09 54.	+ 24 09	2350		CLUSTER OF GALAXIES
MCG+05-24-026	10 09 54.	+ 29 33	42	15.7	GALAXY
ZWG 153.031	10 09 54.	+ 30 50		15.3	GALAXY
ZWG 211.017	10 09 54.	+ 43 24		13.8	GALAXY
UGC 05502	10 09 54.	+ 43 24	96	13.8	GALAXY S0-a
MCG-06-23-011	10 09 54.	- 38 39	36	15.	GALAXY
DV.56 N3136B	10 09 54.	- 66 44	63		E5 GALAXY
KEEL 210	10 09 54.3	+ 03 49 39.		16.	SPIRAL NEBULA
RNGC 3136B	10 09 55.	- 66 44			GALAXY
ZC 1010.0+4331	10 10 00.	+ 43 31	940		CLUSTER OF GALAXIES
ZC 1010.0+8030	10 10 00.	+ 80 30	1210		CLUSTER OF GALAXIES
MRSL 282-00/1	10 10 00.	- 56 09	960		HII REGION
HOLM 169C	10 10 01.	+ 35 32	18	15.7	PART OF MULTIPLE GALAXY
SEY 069	10 10 03.	+ 69 22 15.		15.2	FAINT GALAXY
PK296-20.1	10 10 03.	- 80 36 56.	44		PLANETARY NEBULA
REIZ 0216	10 10 05.	+ 22 58	36	14.5	GALAXY
IC 0598	10 10 05.	+ 43 29 18.			NONSTELLAR OBJECT
RNGC 3195	10 10 05.	- 80 37			PLANETARY NEBULA
ZWG 036.057	10 10 06.	+ 03 22		12.8	GALAXY
UGC 05503	10 10 06.	+ 03 22	120	12.8	GALAXY S0
MCG+01-26-019	10 10 06.	+ 03 22	42	12.8	GALAXY
RNGC 3156	10 10 06.	+ 03 23		13.0	GALAXY
ZWG 036.058	10 10 06.	+ 05 04		14.9	GALAXY
MCG+01-26-020	10 10 06.	+ 05 04	48	14.9	GALAXY
ZWG 036.059	10 10 06.	+ 07 21		15.6	GALAXY
UGC 05504	10 10 06.	+ 07 21	90	15.6	GALAXY
ZWG 123.024	10 10 06.	+ 22 58		15.7	GALAXY
ZWG 008.067	10 10 06.	- 02 36		15.6	GALAXY
ACK 280+03.1	10 10 06.	- 52 24			PLANETARY NEBULA
HOLM 169A	10 10 08.	+ 35 31	18	14.1	PART OF MULTIPLE GALAXY
REIZ 0215	10 10 08.	+ 35 32	12	16.0	GALAXY
HOLM 169B	10 10 09.	+ 35 33	30	14.6	PART OF MULTIPLE GALAXY
RNGC 3153	10 10 10.	+ 12 55		13.5	GALAXY
KN 15.158	10 10 10.8	+ 35 31 45.			NEBULA
ZWG 064.090	10 10 12.	+ 12 55		13.6	GALAXY
SCH 05	10 10 12.	+ 12 55		13.6	PECULIAR GALAXY
UGC 05505	10 10 12.	+ 12 55	138	13.6	GALAXY Sc
MCG+02-26-032	10 10 12.	+ 12 55	120	13.6	GALAXY
MCG+04-24-018	10 10 12.	+ 23 00	36	15.7	GALAXY
ZWG 183.004	10 10 12.	+ 35 31		15.5	GALAXY
MRK 414	10 10 12.	+ 35 31	18	16.5	GALAXY WITH UV CONTINUUM
KARA.72 226A	10 10 12.	+ 35 31	36	15.5	PART OF DOUBLE GALAXY
ZWG 183.005	10 10 12.	+ 35 33		15.7	GALAXY
KARA.72 226B	10 10 12.	+ 35 33	42	15.7	PART OF DOUBLE GALAXY
TON-N 1187	10 10 12.	+ 36 08		16.3	BLUE STAR
MCG+08-19-011	10 10 12.	+ 50 42	30	16.	GALAXY
MCG+09-17-039	10 10 12.	+ 54 55	60	16.	GALAXY
ZC 1010.2+5724	10 10 12.	+ 57 24	1610		CLUSTER OF GALAXIES
ZWG 290.016	10 10 12.	+ 59 41		15.6	GALAXY
MCG+12-10-022	10 10 12.	+ 69 02	66	16.	GALAXY
ZWG 008.069	10 10 12.	- 02 24		15.4	GALAXY
ZWG 008.068	10 10 12.	- 03 15		15.7	GALAXY
VHA 090	10 10 12.	- 57 50	240		OPEN STAR CLUSTER
KN 15.159	10 10 12.0	+ 35 33 37.			NEBULA
LB 00575	10 10 13.	+ 01 16 12.		16.6	FAINT BLUE STAR
RNGC 3148	10 10 13.	+ 50 42		16.0	GALAXY
REIZ 0217	10 10 14.	+ 35 31	18	14.4	GALAXY
REIZ 0218	10 10 14.	+ 35 32	30	14.8	GALAXY
ZWG 036.060	10 10 18.	+ 05 02		14.9	GALAXY
UGC 05506	10 10 18.	+ 05 02	84	14.9	GALAXY S
MCG+01-26-021	10 10 18.	+ 05 02	48	14.9	GALAXY
MCG+03-26-040	10 10 18.	+ 17 16	48	14.3	GALAXY
ZWG 093.071	10 10 18.	+ 17 17		14.3	GALAXY
UGC 05507	10 10 18.	+ 17 17	60	14.3	GALAXY S
MCG+06-23-002	10 10 18.	+ 35 31	42	15.5	GALAXY
MCG+06-23-001	10 10 18.	+ 35 31	30	15.	GALAXY
ZWG 211.018	10 10 18.	+ 39 56		15.7	GALAXY
ZWG 333.019	10 10 18.	+ 69 22		15.4	GALAXY
UGC 05508	10 10 18.	+ 69 22	78	15.4	GALAXY Sc
ZWG 008.070	10 10 18.	- 02 27		15.0	GALAXY
MCG+00-26-029	10 10 18.	- 02 27	36	15.0	GALAXY
MCG-01-26-031	10 10 18.	- 08 21 30.	54	15.	GALAXY
REIZ 0219	10 10 19.	+ 17 16	30	15.8	GALAXY
REIZ 0220	10 10 19.	+ 17 17	60	13.7	GALAXY
RNGC 3154	10 10 20.	+ 17 17		14.5	GALAXY
ZWG 036.061	10 10 24.	+ 06 25		15.1	GALAXY
MCG+01-26-022	10 10 24.	+ 06 25	36	15.1	GALAXY
ZC 1010.4+2540	10 10 24.	+ 25 40	2350		CLUSTER OF GALAXIES
ZWG 211.019	10 10 24.	+ 38 55		15.4	GALAXY
LB 00146	10 10 24.	- 30 10		16.4	FAINT BLUE STAR
MCG-06-23-012	10 10 24.	- 34 28	90	15.5	GALAXY
HN 0337	10 10 24.	- 34 29			NEBULA
IC 2556	10 10 24.	- 34 29			NONSTELLAR OBJECT
REIZ 0223	10 10 25.	+ 17 23	12	16.0	GALAXY
RNGC 3150	10 10 25.	+ 38 55		15.5	GALAXY
HOLM 170B	10 10 26.	+ 38 54	54	14.9	PART OF MULTIPLE GALAXY
HOLM 170A	10 10 29.	+ 38 54	54	14.3	PART OF MULTIPLE GALAXY
ZWG 064.091	10 10 30.	+ 12 54		15.5	GALAXY
ZC 1010.5+1956	10 10 30.	+ 19 56	2420		CLUSTER OF GALAXIES
ZC 1010.5+3458	10 10 30.	+ 34 58	1280		CLUSTER OF GALAXIES
ZWG 183.006	10 10 30.	+ 36 12		15.5	GALAXY
KARA.73B 0403	10 10 30.	+ 36 12	36	15.5	ISOLATED GALAXY S
MCG+07-21-018	10 10 30.	+ 38 51	36	15.	GALAXY
ZWG 211.020	10 10 30.	+ 38 53		15.1	GALAXY
MCG+07-21-017	10 10 30.	+ 38 53 30.	27	15.	GALAXY
ZC 1010.5+3922	10 10 30.	+ 39 22	4030		CLUSTER OF GALAXIES
MCG+10-15-035	10 10 30.	+ 59 41	18	16.	GALAXY
ZWG 008.071	10 10 30.	- 02 25		15.6	GALAXY
RNGC 3151	10 10 31.	+ 38 52		15.0	GALAXY
TON-N 1188	10 10 32.	+ 19 03		14.9	BLUE STAR
KN 15.160	10 10 32.1	+ 36 12 12.			NEBULA

OBJECT NAME	RIGHT ASCEN.	DECLINATION	DIAM.	MAGN.	TYPE OF OBJECT
BC PKS1010-427	10 10 32.32	- 42 43 38.7		17.5	QUASI-STELLAR OBJECT
KEEL 211	10 10 34.1	+ 03 43 57.		16.	SPIRAL NEBULA
HOLM 171A	10 10 35.	+ 03 05	36	14.5	PART OF MULTIPLE GALAXY
REIZ 0225	10 10 35.	+ 23 31	36	14.4	GALAXY
REIZ 0222	10 10 35.	+ 38 52	42	14.2	GALAXY
REIZ 0221	10 10 35.	+ 38 54	48	15.0	GALAXY
ZWG 123.025	10 10 36.	+ 23 30		15.6	GALAXY
TON-N 1189	10 10 36.	+ 36 14		17.	BLUE STAR
ZWG 211.021	10 10 36.	+ 39 06		15.5	GALAXY
MCG-05-24-027	10 10 36.	- 27 35	108	15.	GALAXY
LB 00147	10 10 36.	- 30 18		17.0	FAINT BLUE STAR
KN 15.161	10 10 36.1	+ 36 35 56.			NEBULA
RNGC 3152	10 10 37.	+ 39 05		15.5	GALAXY
ABC 0955	10 10 38.	- 24 12		17.4	RICH CLUSTER OF GALAXIES
REIZ 0224	10 10 41.	+ 39 06	30	14.8	GALAXY
IC 0599	10 10 41.	- 05 23 05.			NONSTELLAR OBJECT
ABC 0953	10 10 41.	- 15 42		17.9	RICH CLUSTER OF GALAXIES
MCG+07-21-018A	10 10 42.	+ 39 06	24	15.	GALAXY
TON-N 1190	10 10 43.	+ 22 45		17.	BLUE STAR
HOLM 171B	10 10 45.	+ 03 05	18	15.3	PART OF MULTIPLE GALAXY
REIZ 0227	10 10 45.	+ 20 25	78	14.2	GALAXY
MCG-01-26-032	10 10 45.	- 05 23	60	14.5	GALAXY
REIZ 0231	10 10 47.	+ 22 59	72	12.5	GALAXY
ZWG 036.062	10 10 48.	+ 08 09		15.7	GALAXY
ZWG 093.072	10 10 48.	+ 20 25		14.9	GALAXY
UGC 05509	10 10 48.	+ 20 25	96	14.9	GALAXY Sb
MCG+03-26-041	10 10 48.	+ 20 26	84	14.9	GALAXY
ZWG 123.026	10 10 48.	+ 22 59		12.2	GALAXY
UGC 05510	10 10 48.	+ 22 59	204	12.2	GALAXY Sc
MCG+04-24-019	10 10 48.	+ 23 00	180	12.2	GALAXY
MCG+07-21-019	10 10 48.	+ 38 54 30.	27	15.	GALAXY
ZWG 211.022	10 10 48.	+ 39 01		13.4	GALAXY
UGC 05511	10 10 48.	+ 39 01	150	13.4	GALAXY E
ZC 1010.8+3915	10 10 48.	+ 39 15	1410		CLUSTER OF GALAXIES
MCG+11-13-011	10 10 48.	+ 66 11	18	16.	GALAXY
7ZW 313	10 10 48.	+ 68 02			COMPACT GALAXY
ZC 1010.8-0121	10 10 48.	- 01 21	740		CLUSTER OF GALAXIES
ZWG 008.072	10 10 48.	- 02 17		15.3	GALAXY
MCG-01-26-033	10 10 48.	- 05 47	36	15.5	GALAXY
MCG-01-26-034	10 10 48.	- 07 49 30.	12	16.	GALAXY
LB 00148	10 10 48.	- 28 56		15.8	FAINT BLUE STAR
RNGC 3162	10 10 49.	+ 22 59		12.5	GALAXY
TON-N 1191	10 10 49.	+ 23 20		16.	BLUE STAR
REIZ 0226	10 10 53.	+ 39 01	48	12.9	GALAXY
ZC 1010.9+0011	10 10 54.	+ 00 11	1550		CLUSTER OF GALAXIES
UGC 05512	10 10 54.	+ 03 28	96	14.5	GALAXY IRR
ZWG 036.063	10 10 54.	+ 03 38		14.5	GALAXY
RNGC 3165	10 10 54.	+ 03 38		14.5	GALAXY
MCG+01-26-023	10 10 54.	+ 03 38	66	14.5	GALAXY
ZC 1010.9+1335	10 10 54.	+ 13 35	1210		CLUSTER OF GALAXIES
ZC 1010.9+1549	10 10 54.	+ 15 49	3020		CLUSTER OF GALAXIES
TON-N 0489	10 10 54.	+ 26 22		15.5	BLUE STAR
TON-N 0032	10 10 54.	+ 29 01		14.9	BLUE STAR
MCG+07-21-021	10 10 54.	+ 38 53	48	14.5	GALAXY
ZWG 211.023	10 10 54.	+ 38 55		14.9	GALAXY
MCG+07-21-020	10 10 54.	+ 39 00	30	13.	GALAXY
ZWG 211.024	10 10 54.	+ 39 06		15.2	GALAXY
UGC 05513	10 10 54.	+ 39 06	84	15.2	GALAXY S
ZC 1010.9+4759	10 10 54.	+ 47 59	1010		CLUSTER OF GALAXIES
MCG-01-26-035	10 10 54.	- 07 53	15	15.5	GALAXY
MCG-01-26-036	10 10 54.	- 08 46	36	15.	GALAXY
LB 00149	10 10 54.	- 29 26		16.8	FAINT BLUE STAR
RNGC 3159	10 10 55.	+ 38 54		15.0	GALAXY
RNGC 3158	10 10 55.	+ 39 01		13.0	GALAXY
RNGC 3160	10 10 55.	+ 39 06		15.0	GALAXY
REIZ 0232	10 10 56.	+ 18 23	42	14.6	GALAXY
KEEL 212	10 10 56.3	+ 03 37 25.		15.	NEBULA
HOLM 173C	10 10 58.	+ 03 37	60	14.2	PART OF MULTIPLE GALAXY
ABC 0951	10 10 58.	+ 34 59		17.6	RICH CLUSTER OF GALAXIES
HOLM 172C	10 10 58.	+ 38 56	60	14.0	PART OF MULTIPLE GALAXY
REIZ 0230	10 10 59.	+ 38 54	30	14.3	GALAXY
REIZ 0228	10 10 59.	+ 38 54	36	13.9	GALAXY
REIZ 0229	10 10 59.	+ 39 06	60	14.2	GALAXY
ZWG 093.073	10 11 00.	+ 18 22		15.5	GALAXY
UGC 05514	10 11 00.	+ 18 22	72	15.5	GALAXY Sc
KARA.72 227A	10 11 00.	+ 18 22	78	15.5	PART OF DOUBLE GALAXY
MCG+03-26-042	10 11 00.	+ 18 22	84	15.5	GALAXY
ZC 1011.0+2759	10 11 00.	+ 27 59	2220		CLUSTER OF GALAXIES
TON-N 0033	10 11 00.	+ 29 25		13.6	BLUE STAR
ZC 1011.0+3009	10 11 00.	+ 30 09	810		CLUSTER OF GALAXIES
MCG+07-21-022	10 11 00.	+ 38 53 30.	18	15.5	GALAXY
ZWG 211.025	10 11 00.	+ 38 55		15.3	GALAXY
MCG+07-21-023	10 11 00.	+ 39 05	72	14.5	GALAXY
MCG+07-21-024	10 11 00.	+ 40 02	30	16.	GALAXY
ZWG 211.026	10 11 00.	+ 40 04		15.5	GALAXY
ZWG 240.018	10 11 00.	+ 50 23		15.5	GALAXY
ZC 1011.0+6302	10 11 00.	+ 63 02	1210		CLUSTER OF GALAXIES
ZWG 008.073	10 11 00.	- 02 56		15.6	GALAXY
KARA.73B 0404	10 11 00.	- 02 56	66	15.6	ISOLATED GALAXY S
RNGC 3161	10 11 01.	+ 38 54		15.5	GALAXY
ABC 0952	10 11 02.	+ 20 02		16.9	RICH CLUSTER OF GALAXIES
ABC 0954	10 11 04.	+ 00 08		16.5	RICH CLUSTER OF GALAXIES
REIZ 0234	10 11 04.	+ 29 52	24	16.0	GALAXY
HOLM 172A	10 11 05.	+ 38 56	48	14.3	PART OF MULTIPLE GALAXY
MCG+07-21-025	10 11 05.	+ 38 59	30	16.	GALAXY
ZWG 036.064	10 11 06.	+ 03 40		11.2	GALAXY
UGC 05516	10 11 06.	+ 03 40	300	11.1	GALAXY S0/Sa
KARA.72 228A	10 11 06.	+ 03 40	306	11.2	PART OF DOUBLE GALAXY
MCG+01-26-024	10 11 06.	+ 03 40	270	11.2	GALAXY
ZWG 093.074	10 11 06.	+ 18 22		15.3	GALAXY
KARA.72 227B	10 11 06.	+ 18 22	42	15.3	PART OF DOUBLE GALAXY
TON-N 0490	10 11 06.	+ 25 06		15.4	BLUE STAR
BC TON490	10 11 06.	+ 25 06		15.4	QUASI-STELLAR OBJECT
SHB 152	10 11 06.	+ 25 06 00.		15.4	QUASI-STELLAR OBJECT
ZWG 211.027	10 11 06.	+ 38 55		14.4	GALAXY
UGC 05517	10 11 06.	+ 38 55	78	14.4	GALAXY E-S0
ZC 1011.1+4600	10 11 06.	+ 46 00	1010		CLUSTER OF GALAXIES
ZWG 008.075	10 11 06.	- 00 41		14.4	GALAXY
UGC 05515	10 11 06.	- 00 41	96	14.4	GALAXY E OR D
MCG+00-26-030	10 11 06.	- 00 41	60	14.4	GALAXY
ZWG 008.074	10 11 06.	- 03 08		15.7	GALAXY
MCG+07-21-026	10 11 09.	+ 38 53	42	14.5	GALAXY
REIZ 0233	10 11 11.	+ 38 53	36	13.5	GALAXY
AR 08	10 11 11.14	+ 74 28 03.9			NEBULA
RNGC 3166	10 11 12.	+ 03 40		11.5	GALAXY
HOLM 173A	10 11 12.	+ 03 40	66	12.2	PART OF MULTIPLE GALAXY
MCG+03-26-044	10 11 12.	+ 14 43	60	14.7	GALAXY
ZWG 093.075	10 11 12.	+ 14 45		14.7	GALAXY
MCG+03-26-045	10 11 12.	+ 14 45	18	15.2	GALAXY
MCG+03-26-043	10 11 12.	+ 18 22	36	15.3	GALAXY
TON-N 0491	10 11 12.	+ 25 54		15.0	BLUE STAR
MCG+07-21-027	10 11 12.	+ 39 00	15	17.	GALAXY
ZWG 211.028	10 11 12.	+ 39 18		15.5	GALAXY
MCG+07-21-028	10 11 12.	+ 39 41	66	16.	GALAXY
UGC 05518	10 11 12.	+ 39 43	126	17.	GALAXY DWRF IR
ZWG 240.019	10 11 12.	+ 49 46		15.7	GALAXY
7ZW 314	10 11 12.	+ 58 27			COMPACT GALAXY
ZWG 313.008	10 11 12.	+ 64 37		15.7	GALAXY
ZWG 351.011	10 11 12.	+ 74 29		14.3	GALAXY
ZWG 333.020	10 11 12.	+ 74 29		14.3	GALAXY
UGC 05519	10 11 12.	+ 74 29	84	14.3	GALAXY SBa-b
ZWG 008.077	10 11 12.	- 00 40		15.5	GALAXY
ZWG 008.076	10 11 12.	- 02 35		15.7	GALAXY
MCG+00-26-031	10 11 12.	- 02 35	36	15.7	GALAXY
RNGC 3163	10 11 13.	+ 38 54		14.5	GALAXY
HOLM 172E	10 11 13.	+ 38 56	54	13.3	PART OF MULTIPLE GALAXY
RNGC 3144	10 11 13.	+ 74 28		14.5	GALAXY
MCG+03-26-046	10 11 15.	+ 14 44 30.	24	15.2	GALAXY
REIZ 0235	10 11 16.	+ 30 24	48	15.6	GALAXY
KN 15.162	10 11 16.4	+ 34 35 36.			NEBULA
ZWG 093.076	10 11 18.	+ 14 47		15.2	GALAXY
ZWG 240.020	10 11 18.	+ 50 23		15.7	GALAXY
ZC 1011.3+5035	10 11 18.	+ 50 35	340		CLUSTER OF GALAXIES
MCG+09-17-040	10 11 18.	+ 56 35	24	16.	GALAXY
ZWG 313.009	10 11 18.	+ 65 24		14.4	GALAXY
UGC 05520	10 11 18.	+ 65 24	132	14.4	GALAXY Sc
MCG-06-23-013	10 11 18.	- 37 57	60	15.	GALAXY
TON-N 1192	10 11 19.	+ 22 39		16.9	BLUE STAR
PK278+05.1	10 11 20.	- 50 04 43.	25		PLANETARY NEBULA
ABC 0957	10 11 21.	- 00 40		15.9	RICH CLUSTER OF GALAXIES
ZWG 008.078	10 11 24.	+ 00 47		15.0	GALAXY
UGC 05521	10 11 24.	+ 00 47	66	15.0	GALAXY Sc
MCG+00-26-032	10 11 24.	+ 00 47	54	15.0	GALAXY
KARA.73B 0405	10 11 24.	+ 00 47	48	15.0	ISOLATED GALAXY S
ZWG 036.065	10 11 24.	+ 07 16		14.4	GALAXY
UGC 05522	10 11 24.	+ 07 16	180	14.4	GALAXY Sc
MCG+01-26-025	10 11 24.	+ 07 16	180	14.4	GALAXY
ZWG 093.077	10 11 24.	+ 14 38		15.6	GALAXY
ZWG 153.032	10 11 24.	+ 30 26		15.6	GALAXY
UGC 05523	10 11 24.	+ 30 26	60	15.6	GALAXY S
ZC 1011.4+3052	10 11 24.	+ 30 52	4230		CLUSTER OF GALAXIES
ZWG 211.029	10 11 24.	+ 38 49		15.6	GALAXY
ZC 1011.4+5007	10 11 24.	+ 50 07	740		CLUSTER OF GALAXIES
MCG+09-17-041	10 11 24.	+ 56 26	36	17.	GALAXY
MCG+11-13-012	10 11 24.	+ 64 37 30.	39	16.	GALAXY
MCG+11-13-013	10 11 24.	+ 65 25	114	15.	GALAXY
MCG+12-10-023	10 11 24.	+ 74 27 30.	51	15.	GALAXY
TON-N 1193	10 11 25.	+ 22 25		16.9	BLUE STAR
TON-N 1194	10 11 26.	+ 19 39		17.	BLUE STAR
ABC 0950	10 11 28.	+ 50 05		17.6	RICH CLUSTER OF GALAXIES
ZWG 064.092	10 11 30.	+ 10 23		15.3	GALAXY
KARA.73B 0406	10 11 30.	+ 10 23	42	15.3	ISOLATED GALAXY S
ZC 1011.5+1914	10 11 30.	+ 19 14	1680		CLUSTER OF GALAXIES
MCG+07-21-029	10 11 30.	+ 39 44	9	15.	GALAXY
ZWG 211.030	10 11 30.	+ 39 45		15.4	GALAXY
ZC 1011.5+4246	10 11 30.	+ 42 46	810		CLUSTER OF GALAXIES
7ZW 315	10 11 30.	+ 58 28			COMPACT GALAXY
MCG-06-23-015	10 11 30.	- 34 35	15	15.5	GALAXY
MCG-06-23-014	10 11 30.	- 34 35	24	15.5	GALAXY
REIZ 0236	10 11 34.	+ 22 22	102	14.0	GALAXY
ABC 0947	10 11 34.	+ 63 20		16.8	RICH CLUSTER OF GALAXIES
KN 15.163	10 11 35.5	+ 36 05 29.			NEBULA
RNGC 3169	10 11 36.	+ 03 43		11.5	GALAXY
ZWG 123.027	10 11 36.	+ 22 22		15.7	GALAXY
UGC 05524	10 11 36.	+ 22 22	108	15.7	GALAXY Sc
ZC 1011.6+3119	10 11 36.	+ 31 19	400		CLUSTER OF GALAXIES
ZWG 290.017	10 11 36.	+ 56 33		15.4	GALAXY
MCG+09-17-042	10 11 36.	+ 56 33	24	15.	GALAXY
MCG+12-10-024	10 11 36.	+ 74 07	21	17.	GALAXY
LB 00576	10 11 38.	- 00 43 18.		15.9	FAINT BLUE STAR
ZWG 036.066	10 11 42.	+ 03 43		11.9	GALAXY
REIZ 0237	10 11 42.	+ 03 43	180	11.8	GALAXY
HOLM 173B	10 11 42.	+ 03 43	180	12.2	PART OF MULTIPLE GALAXY
UGC 05525	10 11 42.	+ 03 43	330	11.9	GALAXY Sa
KARA.72 228B	10 11 42.	+ 03 43	282	11.9	PART OF DOUBLE GALAXY
MCG+01-26-026	10 11 42.	+ 03 43	210	11.9	GALAXY
ZWG 036.067	10 11 42.	+ 06 45		15.2	GALAXY
MCG+03-26-047	10 11 42.	+ 16 08	60	14.8	GALAXY
ZWG 093.078	10 11 42.	+ 16 09		14.8	GALAXY
UGC 05526	10 11 42.	+ 16 09	60	14.8	GALAXY S
MCG+04-24-020	10 11 42.	+ 23 37	108	15.	GALAXY
MCG+09-17-043	10 11 42.	+ 56 36	24	17.	GALAXY
MCG-04-24-021	10 11 42.	- 21 43	96	14.	GALAXY
MCG-06-23-017	10 11 42.	- 34 52 30.	24	15.5	GALAXY
MCG-06-23-016	10 11 42.	- 34 53	36	15.	GALAXY
BC B21011+28	10 11 46.6	+ 28 04 00.4		18.9	QUASI-STELLAR OBJECT
RNGC 3167	10 11 47.	+ 29 50			NON-EXISTENT OBJECT
MCG+07-21-030	10 11 48.	+ 39 13	36	15.	GALAXY
ZWG 211.031	10 11 48.	+ 39 14		15.2	GALAXY
ZWG 290.018	10 11 48.	+ 56 55		14.5	GALAXY
UGC 05527	10 11 48.	+ 56 55	54	14.5	GALAXY S
ZC 1011.8+6320	10 11 48.	+ 63 20	870		CLUSTER OF GALAXIES
REIZ 0238	10 11 49.	+ 16 09	54	14.9	GALAXY
RNGC 3164	10 11 50.	+ 56 55		14.5	GALAXY
ABC 0945	10 11 50.	+ 69 21		17.2	RICH CLUSTER OF GALAXIES
KARA.68 068	10 11 54.	+ 03 37	34		DWARF GALAXY
ZWG 064.093	10 11 54.	+ 12 07		14.9	GALAXY
MCG+02-26-033	10 11 54.	+ 12 07	24	14.9	GALAXY
ZC 1011.9+2203	10 11 54.	+ 22 03	740		CLUSTER OF GALAXIES
TON-N 0492	10 11 54.	+ 31 13		14.7	BLUE STAR
ZC 1011.9+4035	10 11 54.	+ 40 35	1480		CLUSTER OF GALAXIES
ZC 1011.9+4428	10 11 54.	+ 44 28	740		CLUSTER OF GALAXIES
MCG+09-17-044	10 11 54.	+ 54 48	60	16.	GALAXY
MCG+10-15-036	10 11 54.	+ 56 55	51	15.	GALAXY
MCG+10-15-037	10 11 54.	+ 57 54	27	16.	GALAXY
SHB 153	10 11 59.5	+ 23 16 17.		17.5	QUASI-STELLAR OBJECT
ZWG 036.068	10 12 00.	+ 05 12		15.6	GALAXY
ZC 1012.0+3522	10 12 00.	+ 35 22	2150		CLUSTER OF GALAXIES
ZWG 211.032	10 12 00.	+ 39 55		15.6	GALAXY
MCG+07-21-031	10 12 00.	+ 41 06 30.	33	16.	GALAXY
ZC 1012.0+5334	10 12 00.	+ 53 34	1550		CLUSTER OF GALAXIES
MCG+10-15-038	10 12 00.	+ 58 09	39	16.	GALAXY
ZWG 290.019	10 12 00.	+ 59 09		15.1	GALAXY
MCG+10-15-039	10 12 00.	+ 60 22	24	17.	GALAXY
7ZW 316	10 12 00.	+ 60 23			COMPACT GALAXY
MCG+10-15-040	10 12 00.	+ 60 25	36	16.	GALAXY

OBJPCT NAME	RIGHT ASCEN.	DECLINATION	DIAM.	MAGN.	TYPE OF OBJECT
ZWG 333.021	10 12 00.	+ 72 48		15.7	GALAXY
MCG+13-08-011	10 12 00.	+ 77 08	9	16.	GALAXY
ZC 1012.0-0047	10 12 00.	- 00 47	3490		CLUSTER OF GALAXIES
MCG+07-21-032	10 12 03.	+ 39 54	33	15.5	GALAXY
REIZ 0239	10 12 05.	+ 38 57	120	13.8	GALAXY
ZC 1012.1+0849	10 12 06.	+ 08 49	470		CLUSTER OF GALAXIES
ZC 1012.1+1056	10 12 06.	+ 10 56	870		CLUSTER OF GALAXIES
ZWG 064.094	10 12 06.	+ 12 51		15.4	GALAXY
ZC 1012.1+2236	10 12 06.	+ 22 36	870		CLUSTER OF GALAXIES
MCG+10-15-041	10 12 06.	+ 59 08	57	16.	GALAXY
MCG+10-15-042	10 12 06.	+ 60 23	6	17.	GALAXY
ZC 1012.1+6336	10 12 06.	+ 63 36	1080		CLUSTER OF GALAXIES
ZWG 008.079	10 12 06.	- 00 35		14.8	GALAXY
UGC 05528	10 12 06.	- 00 35	66	14.8	GALAXY
MCG+00-26-033	10 12 06.	- 00 35	54	14.8	GALAXY
MCG-01-26-037	10 12 06.	- 06 23	12	15.5	GALAXY
LB 00150	10 12 06.	- 28 27		16.5	FAINT BLUE STAR
MCG-02-26-040	10 12 09.	- 09 42	48	14.5	GALAXY
RNGC 3173	10 12 09.	- 27 27		14.0	GALAXY
ZWG 064.095	10 12 12.	+ 13 35		15.5	GALAXY
ZWG 093.079	10 12 12.	+ 17 07		15.0	GALAXY
MCG+03-26-048	10 12 12.	+ 17 07	48	15.0	GALAXY
ZWG 153.033	10 12 12.	+ 29 06		15.6	GALAXY
ZC 1012.2+4125	10 12 12.	+ 41 25	1340		CLUSTER OF GALAXIES
ZWG 008.081	10 12 12.	- 00 06		15.6	GALAXY
ZWG 008.080	10 12 12.	- 01 57		15.7	GALAXY
MCG-04-24-022	10 12 12.	- 27 27	42	14.5	GALAXY
REIZ 0240	10 12 15.	+ 29 06	15	15.4	GALAXY
KEEL 213	10 12 16.5	+ 03 53 43.		18.	NEBULA
ZWG 036.069	10 12 18.	+ 08 03		15.7	GALAXY
ZWG 093.080	10 12 18.	+ 14 51		15.7	GALAXY
ZWG 123.028	10 12 18.	+ 21 25		15.4	GALAXY
UGC 05529	10 12 18.	+ 21 25	84	15.4	GALAXY Sc-IRR
MCG+04-24-021	10 12 18.	+ 21 26	66	15.4	GALAXY
ZC 1012.3+3246	10 12 18.	+ 32 46	670		CLUSTER OF GALAXIES
SHAH 049	10 12 18.	+ 39 11	72	16.8	GROUP OF COMPACT GALAXIES
ZWG 008.082	10 12 18.	- 01 08		15.3	GALAXY
KEEL 214	10 12 20.6	+ 03 54 56.		17.	SPIRAL NEBULA
ARC 0956	10 12 21.	+ 47 26		17.2	RICH CLUSTER OF GALAXIES
RNGC 3175	10 12 21.	- 28 38		12.0	GALAXY
REIZ 0244	10 12 22.	+ 11 08	24	15.7	GALAXY
REIZ 0242	10 12 22.	+ 21 25	78	14.3	GALAXY
SEY 070	10 12 22.	+ 66 09 21.		14.4	FAINT GALAXY
ZWG 153.034	10 12 24.	+ 28 43		15.7	GALAXY
ZC 1012.4+4725	10 12 24.	+ 47 25	1010		CLUSTER OF GALAXIES
ZWG 266.031	10 12 24.	+ 55 55		14.6	GALAXY
MCG-03-26-031	10 12 24.	- 20 34 30.	42	15.	GALAXY
REIZ 0243	10 12 26.	+ 28 43	24	15.0	GALAXY
KEEL 216	10 12 26.0	+ 03 55 59.		17.	NEBULA
ZWG 036.070	10 12 30.	+ 05 29		15.4	GALAXY
ZWG 036.071	10 12 30.	+ 05 55		15.1	GALAXY
ZWG 123.029	10 12 30.	+ 21 18		15.7	GALAXY
2ZW 044	10 12 30.	+ 21 20			COMPACT GALAXY
ZWG 123.030	10 12 30.	+ 21 21		15.5	GALAXY
ZC 1012.5+3350	10 12 30.	+ 33 50	3230		CLUSTER OF GALAXIES
ZC 1012.5+5732	10 12 30.	+ 57 32	470		CLUSTER OF GALAXIES
MCG+10-15-043	10 12 30.	+ 60 12	24	16.	GALAXY
ZWG 313.010	10 12 30.	+ 66 10		15.1	GALAXY
UGC 05530	10 12 30.	+ 66 10	84	15.1	GALAXY S
MCG-05-24-028	10 12 30.	- 28 37	270	12.	GALAXY
MCG-06-23-019	10 12 30.	- 33 47	66	15.	GALAXY
MCG-06-23-018	10 12 30.	- 34 04	36	15.	GALAXY
HN 0338	10 12 31.	- 34 05			NEBULA
IC 2558	10 12 31.	- 34 05			NONSTELLAR OBJECT
REIZ 0246	10 12 34.	+ 11 05	30	15.4	GALAXY
REIZ 0245	10 12 34.	+ 21 18	30	14.6	GALAXY
ZWG 064.096	10 12 36.	+ 11 04		15.6	GALAXY
MCG+08-19-012	10 12 36.	+ 49 41	36	16.	GALAXY
MCG+10-15-044	10 12 36.	+ 56 57	18	16.	GALAXY
ZC 1012.6+6649	10 12 36.	+ 66 49	1210		CLUSTER OF GALAXIES
MCG-04-24-023	10 12 36.	- 22 48	18	14.5	GALAXY
MCG-05-24-029	10 12 36.	- 28 42	72	15.5	GALAXY
HN 0339	10 12 37.	- 33 49			NEBULA
IC 2559	10 12 38.	- 33 49			NONSTELLAR OBJECT
REIN 2.105	10 12 39.88	+ 73 39 00.1			NEBULA
RNGC 3147	10 12 40.	+ 73 39		12.0	GALAXY
ZWG 036.072	10 12 42.	+ 05 38		15.1	GALAXY
ZWG 211.033	10 12 42.	+ 44 14		15.4	GALAXY
UGC 05531	10 12 42.	+ 44 14	72	15.4	GALAXY SB
MCG+07-21-033	10 12 42.	+ 44 14 30.	66	15.	GALAXY
MCG+10-15-045	10 12 42.	+ 58 43	18	16.	GALAXY
MCG+11-13-014	10 12 42.	+ 66 11	78	15.	GALAXY
ZWG 333.022	10 12 42.	+ 73 38		11.3	GALAXY
UGC 05532	10 12 42.	+ 73 38	282	11.3	GALAXY Sb/Sc
MCG-06-23-020	10 12 42.	- 37 56	78	14.5	GALAXY
SN 1972H	10 12 43.	+ 73 39		14.9	SUPERNOVA
ZWG 064.097	10 12 48.	+ 14 17		15.4	GALAXY
UGC 05533	10 12 48.	+ 14 17	72	15.4	GALAXY Sb
ZWG 211.034	10 12 48.	+ 44 02		14.7	GALAXY
MRK 139	10 12 48.	+ 44 02	18	14.5	GALAXY WITH UV CONTINUUM
MCG+07-21-034	10 12 48.	+ 44 02	36	16.	GALAXY
MCG+08-19-013	10 12 48.	+ 49 12	54	16.	GALAXY
ZC 1012.8+5337	10 12 48.	+ 53 37	19020		CLUSTER OF GALAXIES
MCG+10-15-046	10 12 48.	+ 57 02	10	16.	GALAXY
ZWG 290.020	10 12 48.	+ 58 40		14.9	GALAXY
UGC 05534	10 12 48.	+ 58 40	96	14.9	GALAXY Sb/SBb
MCG+10-15-047	10 12 48.	+ 60 40	30	16.	GALAXY
MCG+12-10-025	10 12 48.	+ 73 39	234	11.4	GALAXY
MCG-02-26-041	10 12 48.	- 14 52	72	14.	GALAXY
BC 4C48.28	10 12 49.	+ 48 53 12.		19.	QUASI-STELLAR OBJECT
SHB 154	10 12 49.	+ 48 53 12.		19.0	QUASI-STELLAR OBJECT
KEEL 215	10 12 49.3	+ 41 45 28.		16.	NEBULA
ARC 0948	10 12 51.	+ 72 34		17.1	RICH CLUSTER OF GALAXIES
ZWG 240.021	10 12 54.	+ 49 13		15.4	GALAXY
ZWG 290.021	10 12 54.	+ 60 34		15.7	GALAXY
RNGC 3176	10 12 54.	- 18 47			NON-EXISTENT OBJECT
TON-N 1195	10 12 56.	+ 19 40		17.	BLUE STAR
MCG+10-15-048	10 12 57.	+ 58 40	78	14.	GALAXY
KEEL 222	10 12 57.4	+ 03 37 00.		17.	SPIRAL NEBULA
REIZ 0241	10 12 58.	+ 60 29	18	14.7	GALAXY
RNGC 3168	10 12 59.	+ 60 29		14.5	GALAXY
KEEL 217	10 12 59.9	+ 41 49 14.		17.	NEBULA
ZWG 093.081	10 13 00.	+ 19 11		15.2	GALAXY
UGC 05535	10 13 00.	+ 19 11	72	15.2	GALAXY S0-a
MCG+03-26-049	10 13 00.	+ 19 11	60	15.2	GALAXY
ZC 1013.0+2502	10 13 00.	+ 25 02	3020		CLUSTER OF GALAXIES
ZC 1013.0+5634	10 13 00.	+ 56 34	1410		CLUSTER OF GALAXIES
MCG+10-15-049	10 13 00.	+ 57 02	15	16.	GALAXY

OBJECT NAME	RIGHT ASCEN.	DECLINATION	DIAM.	MAGN.	TYPE OF OBJECT
ZWG 290.022	10 13 00.	+ 57 03		15.7	GALAXY
MCG+10-15-050	10 13 00.	+ 57 03	15	16.	GALAXY
ZWG 290.023	10 13 00.	+ 60 29		14.6	GALAXY
UGC 05536	10 13 00.	+ 60 29	72	14.6	GALAXY E
MCG+10-15-052	10 13 00.	+ 60 29	18	14.7	GALAXY
MCG+10-15-051	10 13 00.	+ 60 34	27	15.	GALAXY
ZC 1013.0+6613	10 13 00.	+ 66 13	740		CLUSTER OF GALAXIES
REIZ 0249	10 13 01.	+ 16 39	120	14.3	GALAXY
REIZ 0250	10 13 02.	+ 19 12	48	14.6	GALAXY
RNGC 3170	10 13 03.	+ 46 50			NON-EXISTENT OBJECT
KEEL 218	10 13 04.1	+ 41 57 23.		17.	NEBULA
ZWG 036.073	10 13 06.	+ 07 34		15.1	GALAXY
UGC 05537	10 13 06.	+ 07 34	150	15.1	GALAXY Sc
MCG+01-26-027	10 13 06.	+ 07 34	150	15.1	GALAXY
ZWG 093.082	10 13 06.	+ 18 37		15.7	GALAXY
ZWG 123.031	10 13 06.	+ 20 55		15.6	GALAXY
ZWG 183.007	10 13 06.	+ 38 21		15.7	GALAXY
ZWG 266.032	10 13 06.	+ 54 44		15.5	GALAXY
MCG+10-15-053	10 13 06.	+ 57 02	15	16.	GALAXY
ZC 1013.1+6244	10 13 06.	+ 62 44	1010		CLUSTER OF GALAXIES
KN 15.165	10 13 06.4	+ 34 23 10.			NEBULA
IC 2557	10 13 07.	+ 38 21 00.			NONSTELLAR OBJECT
KN 15.164	10 13 08.8	+ 38 21 19.			NEBULA
REIZ 0251	10 13 09.	+ 20 54	18	15.0	GALAXY
MCG+06-23-003	10 13 09.	+ 38 20	18	15.	GALAXY
ZWG 036.074	10 13 12.	+ 07 12		15.0	GALAXY
MCG+01-26-028	10 13 12.	+ 07 12	24	15.0	GALAXY
ZC 1013.2+2219	10 13 12.	+ 22 19	1480		CLUSTER OF GALAXIES
ZWG 211.035	10 13 12.	+ 39 48		15.5	GALAXY
REIZ 0248	10 13 12.	+ 46 51	9	14.2	GALAXY
ZWG 351.012	10 13 12.	+ 74 36		13.9	GALAXY
UGC 05538	10 13 12.	+ 74 36	96	13.9	GALAXY S
TON-N 1196	10 13 13.	+ 21 53		17.	BLUE STAR
RNGC 3155	10 13 13.	+ 74 36		14.0	GALAXY
RNGC 3171	10 13 13.	- 20 33			GALAXY
ARC 0958	10 13 14.	+ 41 16		18.0	RICH CLUSTER OF GALAXIES
KEEL 219	10 13 16.1	+ 41 25 00.		15.	NEBULA
KEEL 221	10 13 17.9	+ 41 59 14.		18.	SPIRAL NEBULA
KEEL 220	10 13 17.9	+ 42 01 03.			NEBULA
ZWG 036.075	10 13 18.	+ 02 56		15.3	GALAXY
UGC 05539	10 13 18.	+ 02 56	132	15.3	GALAXY DWARF
MCG+01-26-029	10 13 18.	+ 02 56	36	15.3	GALAXY
ZWG 036.076	10 13 18.	+ 06 30		15.4	GALAXY
MCG+07-21-035	10 13 18.	+ 41 24	27	15.	GALAXY
ZWG 211.036	10 13 18.	+ 41 25		15.3	GALAXY
MCG-03-26-032	10 13 18.	- 20 24	54	14.	GALAXY
AR 09	10 13 19.93	+ 74 35 49.8			NEBULA
KEEL 223	10 13 23.5	+ 41 23 31.		16.	SPIRAL NEBULA
ZWG 036.077	10 13 24.	+ 05 12		14.7	GALAXY
MCG+01-26-030	10 13 24.	+ 05 12	18	14.7	GALAXY
ZC 1013.4+1955	10 13 24.	+ 19 55	2350		CLUSTER OF GALAXIES
ZWG 183.008	10 13 24.	+ 38 01		14.6	GALAXY
UGC 05540	10 13 24.	+ 38 01	102	14.6	GALAXY Sc
KARA.73B 0407	10 13 24.	+ 38 01	96	14.6	ISOLATED GALAXY S
ZWG 240.022	10 13 24.	+ 45 35		15.0	GALAXY
MRK 140	10 13 24.	+ 45 35	15	15.	GALAXY WITH UV CONTINUUM
MCG-03-26-033	10 13 24.	- 20 03	72	14.5	GALAXY
KEEL 224	10 13 24.5	+ 41 27 54.		18.	SPIRAL NEBULA
TON-N 1197	10 13 25.	+ 21 46		17.	BLUE STAR
KEEL 225	10 13 25.1	+ 41 27 48.		18.	NEBULA
KN 15.166	10 13 25.6	+ 38 01 39.			NEBULA
REIZ 0252	10 13 27.	+ 37 59	72	14.2	GALAXY
REIZ 0247	10 13 28.	+ 60 36	48	14.5	GALAXY
ZWG 093.083	10 13 30.	+ 18 50		15.4	GALAXY
TON-N 0493	10 13 30.	+ 24 18		16.2	BLUE STAR
TON-N 0494	10 13 30.	+ 25 43		15.4	BLUE STAR
ZC 1013.5+2800	10 13 30.	+ 28 00	1410		CLUSTER OF GALAXIES
MCG+06-23-004	10 13 30.	+ 38 02	90	14.5	GALAXY
ZC 1013.5+4120	10 13 30.	+ 41 20	810		CLUSTER OF GALAXIES
ZWG 290.024	10 13 30.	+ 58 39		15.6	GALAXY
UGC 05541	10 13 30.	+ 58 39	90	15.6	GALAXY IRR
ZWG 290.025	10 13 30.	+ 60 32		14.9	GALAXY
UGC 05542	10 13 30.	+ 60 32	60	14.9	GALAXY (E)
MCG+10-15-054	10 13 30.	+ 60 32	18	14.	GALAXY
ZWG 351.013	10 13 30.	+ 74 53		15.7	GALAXY
MCG-06-23-021	10 13 30.	- 34 08	30	15.5	GALAXY
OCL 0792	10 13 30.	- 50 28	132	11.	OPEN STAR CLUSTER
KN 15.167	10 13 35.8	+ 35 02 36.			NEBULA
ZWG 036.078	10 13 36.	+ 03 54		15.4	GALAXY
ZWG 036.079	10 13 36.	+ 06 30		15.7	GALAXY
STOCK 11	10 13 36.	+ 36 00			BLUE KNOT NEAR ELLIP GLXY
MCG+09-17-045	10 13 36.	+ 56 33	18	16.	GALAXY
LB 00577	10 13 36.	- 01 04 12.		16.0	FAINT BLUE STAR
TON-N 1198	10 13 37.	+ 22 37		16.5	BLUE STAR
ARC 0961	10 13 37.	+ 33 53		17.2	RICH CLUSTER OF GALAXIES
MCG+10-15-055	10 13 39.	+ 58 39	78	13.	GALAXY
RNGC 3178	10 13 41.	- 15 32		13.0	GALAXY
ZWG 036.080	10 13 42.	+ 05 04		14.6	GALAXY
UGC 05543	10 13 42.	+ 05 04	84	14.6	GALAXY Sc
MCG+01-26-031	10 13 42.	+ 05 04	60	14.6	GALAXY
ZC 1013.7+1422	10 13 42.	+ 14 22	1680		CLUSTER OF GALAXIES
ZC 1013.7+2416	10 13 42.	+ 24 16	1480		CLUSTER OF GALAXIES
ZC 1013.7+3915	10 13 42.	+ 39 15	2020		CLUSTER OF GALAXIES
MCG+09-17-046	10 13 42.	+ 56 08	54	16.	GALAXY
MCG+10-15-056	10 13 42.	+ 61 48	15	16.	GALAXY
MCG-03-26-034	10 13 42.	- 15 32	66	13.	GALAXY
HOLM 174B	10 13 44.	+ 05 05	48	15.4	PART OF MULTIPLE GALAXY
REIZ 0254	10 13 48.	+ 05 04	54	14.0	GALAXY
HOLM 174A	10 13 48.	+ 05 06	18	13.6	PART OF MULTIPLE GALAXY
ZWG 036.081	10 13 48.	+ 05 56		15.6	GALAXY
ZWG 123.032	10 13 48.	+ 21 22		12.8	GALAXY
UGC 05544	10 13 48.	+ 21 22	96	12.8	GALAXY Sb
ZWG 123.033	10 13 48.	+ 22 20		15.0	GALAXY
MCG+04-24-022	10 13 48.	+ 22 21	30	15.0	GALAXY
ZWG 240.023	10 13 48.	+ 49 53		14.4	GALAXY
UGC 05545	10 13 48.	+ 49 53	60	14.4	GALAXY S
ZWG 266.033	10 13 48.	+ 55 04		15.2	GALAXY
UGC 05546	10 13 48.	+ 55 04	72	15.2	GALAXY SBb
MCG+09-17-047	10 13 48.	+ 55 04	60	15.	GALAXY
MRSL 283-00/2	10 13 48.	- 57 06	360		HII REGION
SN 1947A	10 13 49.	+ 21 21		16.5	SUPERNOVA
RNGC 3177	10 13 49.	+ 21 22		13.5	GALAXY
ARC 0964	10 13 49.	+ 25 04		17.2	RICH CLUSTER OF GALAXIES
KEEL 226	10 13 51.6	+ 41 41 51.		18.	NEBULA
REIZ 0253	10 13 52.	+ 21 23	30	13.0	GALAXY
ZWG 064.098	10 13 54.	+ 12 54		15.1	GALAXY
KARA.72 229A	10 13 54.	+ 12 54	60	15.1	PART OF DOUBLE GALAXY
MCG+03-26-050	10 13 54.	+ 17 12	54	14.9	GALAXY

OBJECT NAME	RIGHT ASCEN.	DECLINATION	DIAM.	MAGN.	TYPE OF OBJECT
ZWG 093.084	10 13 54.	+ 17 13		14.9	GALAXY
UGC 05547	10 13 54.	+ 17 13	66	14.9	GALAXY S0
MCG+04-24-023	10 13 54.	+ 21 23	90	12.8	GALAXY
MCG+08-19-014	10 13 54.	+ 49 53	54	15.	GALAXY
MCG+11-13-015	10 13 54.	+ 64 59	51	16.	GALAXY
MCG+12-10-026	10 13 54.	+ 74 36	33	15.	GALAXY
ARC 0966	10 13 55.	- 25 08		17.4	RICH CLUSTER OF GALAXIES
KEEL 227	10 13 55.2	+ 41 47 29.		17.	NEBULA
MCG-05-24-030	10 13 57.	- 28 56	36	15.	GALAXY
PK283-01.1	10 13 57.	- 58 36	30		PLANETARY NEBULA
ZWG 064.099	10 14 00.	+ 12 50		15.1	GALAXY
UGC 05548	10 14 00.	+ 12 50	72	15.1	GALAXY S
KARA.72 229B	10 14 00.	+ 12 50	78	15.1	PART OF DOUBLE GALAXY
ZC 1014.0+3155	10 14 00.	+ 31 55	1880		CLUSTER OF GALAXIES
ZWG 266.034	10 14 00.	+ 53 43		15.3	GALAXY
UGC 05549	10 14 00.	+ 53 43	72	15.3	GALAXY Sb
ZC 1014.0+5654	10 14 00.	+ 56 54	740		CLUSTER OF GALAXIES
VHE 34	10 14 00.	- 60 03	144		REFLECTION NEBULA
IC 2560	10 14 05.	- 33 17 42.			NONSTELLAR OBJECT
ZC 1014.1+0027	10 14 06.	+ 00 27	3760		CLUSTER OF GALAXIES
ZC 1014.1+2215	10 14 06.	+ 22 15	6650		CLUSTER OF GALAXIES
ZC 1014.1+4016	10 14 06.	+ 40 16	810		CLUSTER OF GALAXIES
MCG+09-17-048	10 14 06.	+ 53 42	60	14.	GALAXY
ZWG 266.035	10 14 06.	+ 54 03		15.7	GALAXY
ZC 1014.1+5949	10 14 06.	+ 59 49	810		CLUSTER OF GALAXIES
UGC 05550	10 14 06.	+ 64 40	60	17.	GALAXY Sc
MCG-05-25-001	10 14 06.	- 33 20	96	14.	GALAXY
MCG+09-17-049	10 14 09.	+ 54 02	36	15.	GALAXY
ARC 0959	10 14 09.	+ 59 49		17.8	RICH CLUSTER OF GALAXIES
ARC 0963	10 14 10.	+ 39 17		17.2	RICH CLUSTER OF GALAXIES
ZWG 093.085	10 14 12.	+ 15 15		15.6	GALAXY
ZC 1014.2+5721	10 14 12.	+ 57 21	1610		CLUSTER OF GALAXIES
ZWG 036.082	10 14 18.	+ 04 00		15.5	GALAXY
ZWG 036.083	10 14 18.	+ 05 55		15.6	GALAXY
ZC 1014.3+0902	10 14 18.	+ 09 02	2350		CLUSTER OF GALAXIES
ZC 1014.3+1244	10 14 18.	+ 12 44	1210		CLUSTER OF GALAXIES
FATH 1.326	10 14 18.	+ 15 16	16		NEBULA
TON-N 0495	10 14 18.	+ 25 22		16.9	BLUE STAR
ZWG 008.083	10 14 18.	- 03 26		15.7	GALAXY
ZWG 064.100	10 14 24.	+ 08 55		15.5	GALAXY
ZWG 093.086	10 14 24.	+ 15 55		15.4	GALAXY
ZC 1014.4+1925	10 14 24.	+ 19 25	610		CLUSTER OF GALAXIES
SHAH 050	10 14 24.	+ 45 37	138	16.5	GROUP OF COMPACT GALAXIES
ZC 1014.4+6807	10 14 24.	+ 68 07	1340		CLUSTER OF GALAXIES
ZWG 008.084	10 14 24.	- 02 15		15.7	GALAXY
MCG-01-26-038	10 14 24.	- 05 37 30.	18	15.	GALAXY
MCG-03-26-035	10 14 24.	- 18 27	42	15.	GALAXY
TON-N 1199	10 14 25.	+ 22 13		16.6	BLUE STAR
TON-N 0496	10 14 30.	+ 23 54		17.0	BLUE STAR
ZC 1014.5+4120	10 14 30.	+ 41 20	940		CLUSTER OF GALAXIES
MCG+09-17-050	10 14 30.	+ 51 42	36	17.	GALAXY
MRK 029	10 14 30.	+ 60 19	13	16.5	GALAXY WITH UV CONTINUUM
ZC 1014.5+6058	10 14 30.	+ 60 58	1810		CLUSTER OF GALAXIES
ZC 1014.5+6904	10 14 30.	+ 69 04	870		CLUSTER OF GALAXIES
MCG-03-26-036	10 14 30.	- 20 23	30	15.	GALAXY
REIZ 0256	10 14 34.	+ 21 29	24	15.0	GALAXY
REIZ 0255	10 14 35.	+ 58 14	18	14.7	GALAXY
UGC 05551	10 14 36.	+ 04 35	60	17.	GALAXY DWARF
ZWG 064.101	10 14 36.	+ 13 26		15.2	GALAXY
MRK 629	10 14 36.	+ 15 45	13	16.	GALAXY WITH UV CONTINUUM
ZC 1014.6+7604	10 14 36.	+ 76 04	1080		CLUSTER OF GALAXIES
MRSL 284-01/1	10 14 36.	- 58 44	6120		HII REGION
IC 0600	10 14 40.	- 03 14 49.			NONSTELLAR OBJECT
ZWG 093.087	10 14 42.	+ 17 20		15.0	GALAXY
UGC 05552	10 14 42.	+ 17 20	96	15.0	GALAXY Sb-c
MCG+03-26-051	10 14 42.	+ 17 20	90	15.0	GALAXY
ZC 1014.7+3555	10 14 42.	+ 35 55	870		CLUSTER OF GALAXIES
UGC 05553	10 14 42.	+ 60 36	72	16.0	GALAXY Sa-b
MCG+10-15-057	10 14 42.	+ 60 36	60	16.	GALAXY
7ZW 317	10 14 42.	+ 66 54			COMPACT GALAXY
ZWG 008.085	10 14 42.	- 03 15		13.3	GALAXY
MCG+00-26-034	10 14 42.	- 03 15	132	13.3	GALAXY
REIZ 0257	10 14 43.	+ 17 20	108	14.8	GALAXY
KEEL 229	10 14 44.7	+ 41 52 08.		18.	SPIRAL NEBULA
KEEL 228	10 14 44.7	+ 41 52 36.		18.	NEBULA
ZC 1014.8+0352	10 14 48.	+ 03 52	2890		CLUSTER OF GALAXIES
ZC 1014.8+1259	10 14 48.	+ 12 59	810		CLUSTER OF GALAXIES
MCG+13-08-012	10 14 48.	+ 79 42	39	15.	GALAXY
MRSL 283-01/1	10 14 48.	- 57 39	900		HII REGION
KEEL 230	10 14 48.1	+ 41 49 32.		17.	SPIRAL NEBULA
REIZ 0258	10 14 52.	+ 21 57	96	12.5	GALAXY
REIZ 0259	10 14 52.	+ 23 03	30	14.5	GALAXY
KEEL 232	10 14 52.8	+ 41 52 00.		18.	NEBULA
ZWG 123.034	10 14 54.	+ 21 56		12.9	GALAXY
UGC 05554	10 14 54.	+ 21 56	168	12.9	GALAXY SBa
ZWG 123.035	10 14 54.	+ 23 03		15.3	GALAXY
ZC 1014.9+4341	10 14 54.	+ 43 41	1880		CLUSTER OF GALAXIES
7ZW 318	10 14 54.	+ 73 24			COMPACT GALAXY
RNGC 3185	10 14 55.	+ 21 57		12.5	GALAXY
KEEL 231	10 14 56.3	+ 45 25 56.			NEBULA
KN 15.168	10 14 58.6	+ 37 47 51.			NEBULA
RNGC 3186	10 14 59.	+ 07 13		15.0	GALAXY
CED 107	10 14 59.	- 57 43			DIFFUSE GALACTIC NEBULA
ZWG 036.084	10 15 00.	+ 06 38		15.1	GALAXY
ZWG 036.085	10 15 00.	+ 07 13		15.0	GALAXY
MCG+01-26-032	10 15 00.	+ 07 13	48	15.0	GALAXY
MCG+04-24-024	10 15 00.	+ 21 57 30.	102	12.9	GALAXY
VV 307B	10 15 00.	+ 21 58	150	13.	INTERACTING GALAXY
ZC 1015.0+2357	10 15 00.	+ 23 57	1210		CLUSTER OF GALAXIES
TON-N 1200	10 15 00.	+ 36 10		15.8	BLUE STAR
ZWG 183.009	10 15 00.	+ 37 25		15.5	GALAXY
MCG+07-21-036	10 15 00.	+ 41 21 30.	66	14.5	GALAXY
ZWG 211.037	10 15 00.	+ 41 22		14.2	GALAXY
RNGC 3179	10 15 00.	+ 41 22		14.0	GALAXY
UGC 05555	10 15 00.	+ 41 22	120	14.2	GALAXY S0
ZWG 351.014	10 15 00.	+ 79 42		15.6	GALAXY
ZWG 350.046	10 15 00.	+ 79 42		15.6	GALAXY
RNGC 3187	10 15 01.	+ 22 07		14.0	GALAXY
BC B21015+27	10 15 01.3	+ 27 47 05.6		19.2	QUASI-STELLAR OBJECT
KN 15.169	10 15 02.2	+ 37 25 49.			NEBULA
REIZ 0261	10 15 04.	+ 22 08	108	13.5	GALAXY
ZWG 123.036	10 15 06.	+ 22 07		13.8	GALAXY
UGC 05556	10 15 06.	+ 22 07	210	13.8	GALAXY S
ZC 1015.1+2301	10 15 06.	+ 23 01	1340		CLUSTER OF GALAXIES
RNGC 3180	10 15 06.	+ 41 41			NON-EXISTENT OBJECT
ARC 0960	10 15 06.	+ 66 29		17.2	RICH CLUSTER OF GALAXIES
ARC 0970	10 15 06.	- 10 27		16.5	RICH CLUSTER OF GALAXIES
RNGC 3199	10 15 06.	- 57 43			DIFFUSE NEBULA
KEEL 233	10 15 06.3	+ 41 21 54.		10.	NEBULA
REIZ 0262	10 15 07.	+ 17 11	72	14.4	GALAXY
ZWG 008.087	10 15 12.	+ 02 03		15.7	GALAXY
ZWG 064.102	10 15 12.	+ 09 23		15.7	GALAXY
MCG+04-24-025	10 15 12.	+ 22 09	204	13.8	GALAXY
RNGC 3181	10 15 12.	+ 41 40			NON-EXISTENT OBJECT
REIZ 0260	10 15 12.	+ 41 41	360	11.7	GALAXY
ZWG 008.086	10 15 12.	- 02 08		15.5	GALAXY
REIN 2.106	10 15 17.13	+ 41 42 19.0			NEBULA
REIN 2.107	10 15 17.62	+ 41 40 29.5			NEBULA
ZC 1015.3+1246	10 15 18.	+ 12 46	470		CLUSTER OF GALAXIES
ZWG 093.088	10 15 18.	+ 16 12		15.6	GALAXY
SN 1937F	10 15 18.	+ 41 38		13.5	SUPERNOVA
ZWG 211.038	10 15 18.	+ 41 40		10.4	GALAXY
UGC 05557	10 15 18.	+ 41 40	510	10.4	GALAXY Sc
MCG+07-21-037	10 15 18.	+ 41 40	420	10.	GALAXY
RNGC 3184	10 15 18.	+ 41 41		11.0	GALAXY
MCG+08-19-015	10 15 18.	+ 46 12 30.	60	16.	GALAXY
ZWG 240.024	10 15 18.	+ 46 13		15.3	GALAXY
UGC 05558	10 15 18.	+ 46 13	72	15.3	GALAXY Sa
MCG+10-15-058	10 15 18.	+ 59 54	36	16.	GALAXY
KN 15.170	10 15 18.2	+ 34 55 26.			NEBULA
RNGC 3190	10 15 19.	+ 22 05		12.0	GALAXY
RNGC 3189	10 15 19.	+ 22 06			NON-EXISTENT OBJECT
FATH 1.327	10 15 20.	+ 15 22	16		NEBULA
SN 1921B	10 15 20.	+ 41 37		13.5	SUPERNOVA
ARC 0962	10 15 20.	+ 63 44		17.1	RICH CLUSTER OF GALAXIES
REIZ 0264	10 15 22.	+ 22 05	198	12.0	GALAXY
HOLM 175A	10 15 22.	+ 22 05	210	12.4	PART OF MULTIPLE GALAXY
SN 1921C	10 15 22.	+ 41 36		11.0	SUPERNOVA
MAI 052	10 15 22.	+ 65 53	67		DWARF SPHEROIDAL GALAXY
KEEL 234	10 15 22.5	+ 42 04 58.		17.	SPIRAL NEBULA
ARC 0967	10 15 23.	+ 43 42		17.0	RICH CLUSTER OF GALAXIES
ZC 1015.4+0712	10 15 24.	+ 07 12	810		CLUSTER OF GALAXIES
ZC 1015.4+1254	10 15 24.	+ 12 54	270		CLUSTER OF GALAXIES
ZC 1015.4+2025	10 15 24.	+ 20 25	740		CLUSTER OF GALAXIES
ZWG 123.037	10 15 24.	+ 22 05		11.9	GALAXY
UGC 05559	10 15 24.	+ 22 05	270	11.9	GALAXY Sa
VV 307A	10 15 24.	+ 22 05	180	12.1	INTERACTING GALAXY
ARP 316	10 15 24.	+ 22 05			PECULIAR GALAXY
ZC 1015.4+5007	10 15 24.	+ 50 07	740		CLUSTER OF GALAXIES
MCG+09-17-051	10 15 24.	+ 51 43	24	17.	GALAXY
ZWG 313.011	10 15 24.	+ 64 14		15.2	GALAXY
ZWG 313.012	10 15 24.	+ 66 58		15.7	GALAXY
ZC 1015.4+7405	10 15 24.	+ 74 05	610		CLUSTER OF GALAXIES
ZWG 351.015	10 15 24.	+ 78 42		15.5	GALAXY
ZWG 350.047	10 15 24.	+ 78 42		15.5	GALAXY
FATH 1.328	10 15 24.	- 15 13			NEBULA
WRAY 19.21	10 15 25.7	- 57 36 53.		9.3	STAR-NEBULA ASSOCIATION
REIZ 0265	10 15 26.	+ 28 16	24	15.6	GALAXY
RNGC 3201	10 15 28.	- 46 09		9.0	GLOBULAR CLUSTER
KEEL 235	10 15 29.6	+ 41 53 57.		17.	SPIRAL NEBULA
MCG+04-24-026	10 15 30.	+ 22 06	210	11.9	GALAXY
ZWG 183.010	10 15 30.	+ 34 52		15.6	GALAXY
UGC 05560	10 15 30.	+ 34 52	60	15.6	GALAXY S
MCG+09-17-052	10 15 30.	+ 53 57 30.	36	17.	GALAXY
MCG+10-15-059	10 15 30.	+ 57 32	18	16.	GALAXY
ZC 1015.5+6624	10 15 30.	+ 66 24	940		CLUSTER OF GALAXIES
GCL 015	10 15 30.	- 46 09	1758	8.8	GLOBULAR STAR CLUSTER
VHA 091	10 15 30.	- 58 28	300		OPEN STAR CLUSTER
ARC 0965	10 15 31.	+ 50 09		17.5	RICH CLUSTER OF GALAXIES
KEEL 236	10 15 32.1	+ 41 58 19.		16.	NEBULA
KEEL 237	10 15 33.9	+ 41 52 49.		18.	SPIRAL NEBULA
KN 15.171	10 15 34.1	+ 34 52 35.			NEBULA
ZWG 036.086	10 15 36.	+ 06 41		15.5	GALAXY
ZWG 036.087	10 15 36.	+ 07 17		15.0	GALAXY
SCH 06	10 15 36.	+ 07 17		15.0	PECULIAR GALAXY
KARA.72 230A	10 15 36.	+ 07 17	42	15.0	PART OF DOUBLE GALAXY
MCG+01-26-033	10 15 36.	+ 07 17	36	15.0	GALAXY
ZC 1015.6+1121	10 15 36.	+ 11 21	1080		CLUSTER OF GALAXIES
ZWG 240.025	10 15 36.	+ 46 18		15.6	GALAXY
MCG+09-17-053	10 15 36.	+ 51 42 30.	24	17.	GALAXY
MCG+11-13-016	10 15 36.	+ 64 08	27	17.	GALAXY
REIZ 0266	10 15 37.	+ 34 53	30	14.8	GALAXY
HOLM 176B	10 15 39.	+ 07 17	36	14.9	PART OF MULTIPLE GALAXY
IC 0601	10 15 39.	+ 07 17 14.			NONSTELLAR OBJECT
REIZ 0267	10 15 40.	+ 22 09	36	12.6	GALAXY
VMT 13	10 15 40.	- 58 40 30.	300		SUPERNOVA REMNANT
MIL 16	10 15 40.	- 58 40 30.	1380		SUPERNOVA REMNANT
ZWG 036.088	10 15 42.	+ 03 36		15.5	GALAXY
ZWG 036.089	10 15 42.	+ 07 18		13.4	GALAXY
UGC 05561	10 15 42.	+ 07 18	54	13.4	GALAXY S
KARA.72 230B	10 15 42.	+ 07 18	54	13.4	PART OF DOUBLE GALAXY
MCG+01-26-034	10 15 42.	+ 07 18	48	13.4	GALAXY
ZWG 064.103	10 15 42.	+ 13 32		15.3	GALAXY
ZWG 123.038	10 15 42.	+ 22 09		12.4	GALAXY
HOLM 175B	10 15 42.	+ 22 09	60	12.6	PART OF MULTIPLE GALAXY
UGC 05562	10 15 42.	+ 22 09	168	12.4	GALAXY E
ZWG 123.039	10 15 42.	+ 23 08		15.5	GALAXY
ZWG 123.040	10 15 42.	+ 24 42		15.7	GALAXY
TON-N 0497	10 15 42.	+ 33 36		14.9	BLUE STAR
MCG+06-23-005	10 15 42.	+ 34 52 30.	48	15.	GALAXY
ZC 1015.7+3953	10 15 42.	+ 39 53	810		CLUSTER OF GALAXIES
MCG+08-19-016B	10 15 42.	+ 46 18	54	17.	GALAXY
MCG+08-19-016A	10 15 42.	+ 46 18	54	17.	GALAXY
MCG+11-13-017	10 15 42.	+ 64 13	39	17.	GALAXY
KN 15.172	10 15 42.3	+ 37 34 26.			NEBULA
IC 0602	10 15 43.	+ 07 18 14.			NONSTELLAR OBJECT
RNGC 3193	10 15 43.	+ 22 09		12.5	GALAXY
KEEL 239	10 15 43.5	+ 41 44 11.		17.	NEBULA
KEEL 240	10 15 43.9	+ 41 50 19.		16.	NEBULA
HOLM 176A	10 15 44.	+ 07 18	48	13.7	PART OF MULTIPLE GALAXY
ARC 0969	10 15 44.	+ 30 38		17.6	RICH CLUSTER OF GALAXIES
KEEL 238	10 15 45.7	+ 45 36 37.			NEBULA
REIZ 0270	10 15 46.	+ 23 08	24	14.6	GALAXY
ZC 1015.8+0603	10 15 48.	+ 06 03	1680		CLUSTER OF GALAXIES
LB 10281	10 15 48.	+ 18 32		16.3	FAINT BLUE STAR
LB 10280	10 15 48.	+ 20 34		15.3	FAINT BLUE STAR
MCG+04-24-027	10 15 48.	+ 22 10	42	12.4	GALAXY
MRK 141	10 15 48.	+ 64 12	14	14.5	GALAXY WITH UV CONTINUUM
KW 16	10 15 48.	+ 64 12	19		SEYFERT GALAXY
WVI 36	10 15 48.	+ 64 12	18	15.69	SEYFERT GALAXY
MCG+11-13-018	10 15 48.	+ 64 12 30.	18	16.	GALAXY
ZWG 008.088	10 15 48.	- 03 22		15.7	GALAXY
TON-N 1201	10 15 49.	+ 21 56		15.8	BLUE STAR
FATH 1.329	10 15 50.	+ 14 28	14		NEBULA
MCG+07-21-038	10 15 51.	+ 38 42	66	15.	GALAXY
KEEL 243	10 15 51.7	+ 41 50 25.		18.	NEBULA

OBJECT NAME	RIGHT ASCEN.	DECLINATION	DIAM.	MAGN.	TYPE OF OBJECT
REIZ 0268	10 15 52.	+ 38 43	42	13.9	GALAXY
RNGC 3192	10 15 52.	+ 46 42		16.0	GALAXY
KN 15.173	10 15 52.1	+ 35 05 04.			NEBULA
KEEL 241	10 15 52.3	+ 45 38 43.			NEBULA
ZWG 183.011	10 15 54.	+ 35 05		15.2	GALAXY
ZWG 211.039	10 15 54.	+ 38 44		14.6	GALAXY
UGC 05563	10 15 54.	+ 38 44	72	14.6	GALAXY S
MCG+08-19-017	10 15 54.	+ 46 42	15	16.	GALAXY
UGC 05564	10 15 54.	+ 59 23	84	16.0	GALAXY Sb
ZC 1015.9+7035	10 15 54.	+ 70 35	1140		CLUSTER OF GALAXIES
MCG-02-26-042	10 15 54.	- 12 51 30.	48	15.5	GALAXY
KEEL 242	10 15 54.8	+ 45 39 13.			NEBULA
MCG-01-26-039	10 15 57.	- 06 04	42	15.	GALAXY
RNGC 3191	10 15 58.	+ 46 43		14.0	GALAXY
REIZ 0263	10 15 58.	+ 58 28	21	14.0	GALAXY
KEEL 244	10 15 58.0	+ 41 43 41.		18.	NEBULA
RNGC 3196	10 15 59.	+ 27 56		15.5	GALAXY
ZWG 036.090	10 16 00.	+ 03 30		15.6	GALAXY
ZWG 064.104	10 16 00.	+ 11 02		15.7	GALAXY
ZC 1016.0+1220	10 16 00.	+ 12 20	2020		CLUSTER OF GALAXIES
ZWG 064.105	10 16 00.	+ 13 29		15.5	GALAXY
LB 10282	10 16 00.	+ 17 21		16.5	FAINT BLUE STAR
ZWG 154.001	10 16 00.	+ 27 56		15.7	GALAXY
ZWG 153.035	10 16 00.	+ 27 56		15.7	GALAXY
ZC 1016.0+3015	10 16 00.	+ 30 15	4970		CLUSTER OF GALAXIES
ZC 1016.0+3936	10 16 00.	+ 39 36	1950		CLUSTER OF GALAXIES
ZWG 211.040	10 16 00.	+ 40 52		15.4	GALAXY
MCG+08-19-018	10 16 00.	+ 46 42	42	14.1	GALAXY
ZWG 240.026	10 16 00.	+ 46 43		13.9	GALAXY
UGC 05565	10 16 00.	+ 46 43	42	13.9	GALAXY S
ZWG 290.026	10 16 00.	+ 58 45		15.5	GALAXY
UGC 05566	10 16 00.	+ 58 45	78	15.5	GALAXY
MCG+10-15-060	10 16 00.	+ 58 45	24	15.	GALAXY
MCG+10-15-061	10 16 00.	+ 59 22	72	16.	GALAXY
MCG+11-13-019	10 16 00.	+ 64 15	18	16.	GALAXY
7ZW 319	10 16 00.	+ 78 58			COMPACT GALAXY
ZWG 008.089	10 16 00.	- 03 20		15.6	GALAXY
KN 15.174	10 16 00.3	+ 35 20 48.			NEBULA
REIZ 0273	10 16 01.	+ 35 21	36	15.3	GALAXY
KEEL 245	10 16 01.0	+ 42 08 07.		16.	SPIRAL NEBULA
REIZ 0274	10 16 03.	+ 38 38	24	14.5	GALAXY
REIZ 0272	10 16 05.	+ 46 42	24	14.1	GALAXY
MCG+03-26-053	10 16 06.	+ 16 23	36	15.2	GALAXY
ZWG 093.089	10 16 06.	+ 16 24		15.2	GALAXY
ZWG 093.090	10 16 06.	+ 18 25		15.6	GALAXY
MCG+03-26-052	10 16 06.	+ 18 25	30	15.6	GALAXY
LB 10283	10 16 06.	+ 19 27		17.2	FAINT BLUE STAR
STOCK 12	10 16 06.	+ 38 37			BLUE KNOT NEAR ELLIP GLXY
ZWG 211.041	10 16 06.	+ 38 38		15.6	GALAXY
TON-N 1202	10 16 07.	+ 21 58		15.3	BLUE STAR
KEEL 246	10 16 08.6	+ 45 59 19.			SPIRAL NEBULA
BIGO 513	10 16 09.	- 17 43			NEBULA
KN 15.175	10 16 09.5	+ 37 07 27.			NEBULA
RNGC 3211	10 16 11.	- 62 26		12.0	PLANETARY NEBULA
LB 10289	10 16 12.	+ 16 26		15.8	FAINT BLUE STAR
LB 10288	10 16 12.	+ 16 48		17.2	FAINT BLUE STAR
LB 10287	10 16 12.	+ 17 43		15.3	FAINT BLUE STAR
LB 10286	10 16 12.	+ 18 27		15.2	FAINT BLUE STAR
LB 10285	10 16 12.	+ 19 54		16.4	FAINT BLUE STAR
LB 10284	10 16 12.	+ 20 20		15.5	FAINT BLUE STAR
TON-N 0498	10 16 12.	+ 31 46		14.8	BLUE STAR
ZWG 183.012	10 16 12.	+ 34 49		15.6	GALAXY
ZWG 183.013	10 16 12.	+ 34 55		14.9	GALAXY
UGC 05567	10 16 12.	+ 34 55	66	14.9	GALAXY S
ZWG 240.027	10 16 12.	+ 46 05		15.5	GALAXY
MCG+08-19-019	10 16 12.	+ 49 22	30	16.	GALAXY
MCG+10-15-062	10 16 12.	+ 58 27	42	14.0	GALAXY
ZWG 290.027	10 16 12.	+ 58 28		13.0	GALAXY
UGC 05568	10 16 12.	+ 58 28	138	13.0	GALAXY S0-a
MCG+10-15-063	10 16 12.	+ 58 44	36	16.	GALAXY
ZC 1016.2+6347	10 16 12.	+ 63 47	3360		CLUSTER OF GALAXIES
ZWG 313.013	10 16 12.	+ 64 13		15.5	GALAXY
RNGC 3200	10 16 12.	- 17 43		12.5	GALAXY
MCG-03-26-037	10 16 12.	- 17 43	258	12.	GALAXY
REIZ 0275	10 16 13.	+ 27 55	42	15.8	GALAXY
SN 1953D	10 16 13.	- 17 43		19.5	SUPERNOVA
PK286-04.1	10 16 13.	- 62 25 33.	14	11.8	PLANETARY NEBULA
IC 2562	10 16 14.	+ 16 24 24.			NONSTELLAR OBJECT
RNGC 3182	10 16 14.	+ 58 27		13.0	GALAXY
RNGC 3203	10 16 14.	- 26 26		13.0	GALAXY
KEEL 247	10 16 14.6	+ 46 03 52.			NEBULA
KN 15.176	10 16 14.9	+ 34 55 35.			NEBULA
REIZ 0277	10 16 15.	+ 21 32	42	14.3	GALAXY
IC 2561	10 16 15.	+ 34 56 13.			NONSTELLAR OBJECT
REIZ 0269	10 16 16.	+ 58 15	18	14.6	GALAXY
ZWG 064.106	10 16 18.	+ 13 24		15.2	GALAXY
ZC 1016.3+1831	10 16 18.	+ 18 31	2020		CLUSTER OF GALAXIES
ZWG 124.001	10 16 18.	+ 21 32		15.3	GALAXY
REIZ 0276	10 16 18.	+ 34 58	54	14.2	GALAXY
ZWG 240.028	10 16 18.	+ 49 22		15.0	GALAXY
MRK 030	10 16 18.	+ 57 40	15	17.	GALAXY WITH UV CONTINUUM
MCG+10-15-064	10 16 18.	+ 57 40	15	16.	GALAXY
7ZW 320	10 16 18.	+ 77 20			COMPACT GALAXY
ZWG 008.090	10 16 18.	- 01 00		15.7	GALAXY
KN 15.177	10 16 18.5	+ 33 41 25.			NEBULA
REIZ 0271	10 16 21.	+ 57 40	60	15.5	GALAXY
ZWG 008.091	10 16 24.	+ 01 54		15.1	GALAXY
LB 10294	10 16 24.	+ 15 46		15.3	FAINT BLUE STAR
ZWG 094.001	10 16 24.	+ 16 21		15.6	GALAXY
ZWG 093.091	10 16 24.	+ 16 21		15.6	GALAXY
LB 10293	10 16 24.	+ 17 33		15.4	FAINT BLUE STAR
LB 10292	10 16 24.	+ 18 24		15.2	FAINT BLUE STAR
LB 10291	10 16 24.	+ 18 38		16.2	FAINT BLUE STAR
ZWG 094.002	10 16 24.	+ 19 35		15.7	GALAXY
ZWG 093.092	10 16 24.	+ 19 35		15.7	GALAXY
LB 10290	10 16 24.	+ 19 54		16.0	FAINT BLUE STAR
MCG+06-23-006	10 16 24.	+ 34 55	36	16.	GALAXY
MCG+06-23-007	10 16 24.	+ 34 56	42	14.5	GALAXY
ZC 1016.4+4257	10 16 24.	+ 42 57	1880		CLUSTER OF GALAXIES
ZWG 240.029	10 16 24.	+ 44 40		15.2	GALAXY
ZWG 290.028	10 16 24.	+ 57 40		14.7	GALAXY
MRK 031	10 16 24.	+ 57 40	10	15.5	GALAXY WITH UV CONTINUUM
UGC 05569	10 16 24.	+ 57 40	72	14.7	GALAXY SBb
MCG+10-15-065	10 16 24.	+ 57 40	42	15.5	GALAXY
7ZW 321	10 16 24.	+ 61 56			COMPACT GALAXY
MCG+11-13-020	10 16 24.	+ 64 12	18	16.	GALAXY
UGC 05570	10 16 24.	+ 73 32	66	16.5	GALAXY Sb-c
RNGC 3174	10 16 24.	+ 75 25			NON-EXISTENT OBJECT

OBJECT NAME	RIGHT ASCEN.	DECLINATION	DIAM.	MAGN.	TYPE OF OBJECT
KEEL 248	10 16 24.0	+ 45 52 16.			NEBULA
REIZ 0278	10 16 26.	+ 28 17	12	15.2	GALAXY
RNGC 3188	10 16 26.	+ 57 40		14.5	GALAXY
SN 1951C	10 16 29.	+ 08 21		18.2	SUPERNOVA
ZC 1016.5+4022	10 16 30.	+ 40 22	1280		CLUSTER OF GALAXIES
ZC 1016.5+4200	10 16 30.	+ 42 00	1950		CLUSTER OF GALAXIES
ZC 1016.5+4243	10 16 30.	+ 42 43	470		CLUSTER OF GALAXIES
ZWG 266.036	10 16 30.	+ 54 28		15.6	GALAXY
MCG+12-10-027	10 16 30.	+ 73 32	66	16.	GALAXY
ZWG 351.016	10 16 30.	+ 79 56		15.7	GALAXY
ZWG 350.048	10 16 30.	+ 79 56		15.7	GALAXY
KEEL 261	10 16 31.1	- 42 24 42.		19.	NEBULA
FATH 1.330	10 16 32.	- 15 04	14		NEBULA
KEEL 249	10 16 34.2	+ 42 02 29.		18.	NEBULA
LB 10297	10 16 36.	+ 18 06		18.5	FAINT BLUE STAR
LB 10296	10 16 36.	+ 18 28		17.5	FAINT BLUE STAR
LB 10295	10 16 36.	+ 19 56		13.3	FAINT BLUE STAR
ZC 1016.6+4116	10 16 36.	+ 41 16	1750		CLUSTER OF GALAXIES
UGC 05571	10 16 36.	+ 52 20	78	17.	GALAXY DWRF SP
ZWG 266.037	10 16 36.	+ 55 40		15.3	GALAXY
ZWG 008.092	10 16 36.	- 01 00		15.6	GALAXY
VHE 35	10 16 36.	- 60 39	66		REFLECTION NEBULA
KEEL 250	10 16 36.9	+ 41 27 40.		17.	SPIRAL NEBULA
TON-N 1203	10 16 37.	+ 35 17		16.8	BLUE STAR
FATH 1.332	10 16 37.	- 15 34			NEBULA
REIZ 0280	10 16 38.	+ 28 08	18	14.9	GALAXY
HN 0340	10 16 39.	- 32 21			NEBULA
IC 2563	10 16 39.	- 32 21			NONSTELLAR OBJECT
KN 15.178	10 16 39.9	+ 35 55 33.			NEBULA
KN 15.180	10 16 41.9	+ 35 55 54.			NEBULA
KN 15.179	10 16 41.9	+ 35 56 26.			NEBULA
ZWG 036.091	10 16 42.	+ 05 01		15.1	GALAXY
ZC 1016.7+0820	10 16 42.	+ 08 20	870		CLUSTER OF GALAXIES
LB 10303	10 16 42.	+ 14 38		17.1	FAINT BLUE STAR
ZWG 094.003	10 16 42.	+ 16 05		15.7	GALAXY
ZWG 093.093	10 16 42.	+ 16 05		15.7	GALAXY
REIZ 0281	10 16 42.	+ 16 15	36	14.6	GALAXY
LB 10302	10 16 42.	+ 16 48		17.6	FAINT BLUE STAR
LB 10301	10 16 42.	+ 17 01		15.5	FAINT BLUE STAR
LB 10300	10 16 42.	+ 17 32		15.6	FAINT BLUE STAR
LB 10299	10 16 42.	+ 19 20		15.4	FAINT BLUE STAR
LB 10298	10 16 42.	+ 20 28		16.3	FAINT BLUE STAR
ZWG 154.002	10 16 42.	+ 28 09		15.7	GALAXY
MCG+09-17-054	10 16 42.	+ 54 12 30.	18	16.	GALAXY
MCG-01-26-040	10 16 42.	- 05 22	42	15.	GALAXY
MCG-03-26-038	10 16 42.	- 17 40	36	15.	GALAXY
KEEL 251	10 16 44.5	+ 42 01 18.		17.	NEBULA
REIZ 0279	10 16 46.	+ 45 49	360	12.1	GALAXY
KEEL 252	10 16 46.1	+ 41 46 16.		17.	SPIRAL NEBULA
REIZ 0283	10 16 47.	+ 02 59	36	15.0	GALAXY
FATH 1.331	10 16 47.	+ 14 38	27		NEBULA
ZWG 036.092	10 16 48.	+ 03 00		15.6	GALAXY
ARC 0973	10 16 48.	+ 08 20		17.6	RICH CLUSTER OF GALAXIES
ZC 1016.8+1255	10 16 48.	+ 12 55	740		CLUSTER OF GALAXIES
LB 10305	10 16 48.	+ 16 18		16.5	FAINT BLUE STAR
ZC 1016.8+1649	10 16 48.	+ 16 49	810		CLUSTER OF GALAXIES
LB 10304	10 16 48.	+ 17 51		15.8	FAINT BLUE STAR
ZWG 094.004	10 16 48.	+ 19 38		15.3	GALAXY
ZWG 093.094	10 16 48.	+ 19 38		15.3	GALAXY
MCG+03-26-054	10 16 48.	+ 19 39	48	15.3	GALAXY
ZWG 124.002	10 16 48.	+ 20 53		15.7	GALAXY
ARC 0971	10 16 48.	+ 41 14		16.6	RICH CLUSTER OF GALAXIES
ZC 1016.8+4716	10 16 48.	+ 47 16	340		CLUSTER OF GALAXIES
ZWG 266.038	10 16 48.	+ 54 13		15.5	GALAXY
ZC 1016.8+6027	10 16 48.	+ 60 27	1010		CLUSTER OF GALAXIES
MCG-04-25-001	10 16 48.	- 26 46	48	15.	GALAXY
MRSL 283-00/1	10 16 48.	- 57 31	360		HII REGION
SN 1966J	10 16 49.	+ 45 46		13.0	SUPERNOVA
RNGC 3198	10 16 52.	+ 45 48		11.0	GALAXY
KEEL 253	10 16 52.8	+ 46 02 58.			SPIRAL NEBULA
KEEL 254	10 16 53.7	+ 41 26 01.		18.	NEBULA
LB 10306	10 16 54.	+ 15 14		15.1	FAINT BLUE STAR
MCG+08-19-020	10 16 54.	+ 45 47 30.	360	10.8	GALAXY
ZWG 240.030	10 16 54.	+ 45 49		10.7	GALAXY
UGC 05572	10 16 54.	+ 45 49	600	10.7	GALAXY Sc
ZWG 009.001	10 16 54.	- 03 05		15.3	GALAXY
ZWG 008.093	10 16 54.	- 03 05		15.3	GALAXY
MCG-01-26-041	10 16 54.	- 05 24	24	14.5	GALAXY
IC 0603	10 16 54.	- 05 24 27.			NONSTELLAR OBJECT
KEEL 255	10 16 54.6	+ 41 48 31.		17.	NEBULA
REIZ 0286	10 16 55.	+ 06 35	48	14.1	GALAXY
ARC 0972	10 16 57.	+ 39 49		17.2	RICH CLUSTER OF GALAXIES
REIZ 0284	10 16 58.	+ 22 42	60	14.3	GALAXY
KEEL 256	10 16 58.0	+ 41 48 05.		17.	NEBULA
KN 15.181	10 16 59.3	+ 35 49 43.			NEBULA
KEEL 257	10 16 59.7	+ 41 26 09.		18.	NEBULA
ZWG 036.093	10 17 00.	+ 06 35		14.6	GALAXY
UGC 05573	10 17 00.	+ 06 35	96	14.6	GALAXY Sc
MCG+01-27-001	10 17 00.	+ 06 35	90	14.6	GALAXY
ZC 1017.0+1414	10 17 00.	+ 14 14	1810		CLUSTER OF GALAXIES
REIZ 0287	10 17 00.	+ 16 12	60	14.8	GALAXY
LB 10307	10 17 00.	+ 19 12		16.0	FAINT BLUE STAR
UGC 05574	10 17 00.	+ 22 43	84	16.0	GALAXY Sc
MCG+04-25-001	10 17 00.	+ 22 43	72	15.	GALAXY
UGC 05575	10 17 00.	+ 22 51	72	16.5	GALAXY DWARF
ZC 1017.0+2419	10 17 00.	+ 24 19	1880		CLUSTER OF GALAXIES
ZWG 183.014	10 17 00.	+ 33 37		15.6	GALAXY
ZC 1017.0+3848	10 17 00.	+ 38 48	1010		CLUSTER OF GALAXIES
ZWG 313.014	10 17 00.	+ 65 26		14.1	GALAXY
UGC 05576	10 17 00.	+ 65 26	90	14.1	GALAXY S?
MCG+11-13-021	10 17 00.	+ 68 27	51	17.	GALAXY
ZWG 364.016	10 17 00.	+ 81 55		15.5	GALAXY
ZWG 009.002	10 17 00.	- 02 14		15.7	GALAXY
ARC 0974	10 17 01.	+ 14 19		17.4	RICH CLUSTER OF GALAXIES
TON-N 1204	10 17 01.	+ 34 56		16.9	BLUE STAR
ARC 0976	10 17 02.	- 13 40		17.7	RICH CLUSTER OF GALAXIES
FATH 1.333	10 17 02.	- 14 52	14		NEBULA
KN 15.182	10 17 02.4	+ 35 50 35.			NEBULA
KN 15.183	10 17 03.7	+ 35 51 24.			NEBULA
REIZ 0288	10 17 04.	+ 22 51	60	14.5	GALAXY
KN 15.184	10 17 04.7	+ 34 01 01.			NEBULA
LB 10308	10 17 06.	+ 20 28		15.3	FAINT BLUE STAR
TON-N 0034	10 17 06.	+ 28 01		16.0	BLUE STAR
MCG+07-21-039	10 17 06.	+ 38 51	48	14.	GALAXY
ZWG 211.042	10 17 06.	+ 38 53		14.3	GALAXY
UGC 05577	10 17 06.	+ 38 53	66	14.3	GALAXY PECULR
VHA 092	10 17 06.	- 56 11	90		OPEN STAR CLUSTER
REIZ 0285	10 17 09.	+ 38 53	48	13.6	GALAXY

OBJECT NAME	RIGHT ASCEN.	DECLINATION	DIAM.	MAGN.	TYPE OF OBJECT
HOLM 177B	10 17 09.	+ 74 23	18	15.3	PART OF MULTIPLE GALAXY
KEEL 258	10 17 09.0	+ 45 22 01.			NEBULA
KEEL 259	10 17 11.6	+ 41 54 10.		18.	SPIRAL NEBULA
LB 10310	10 17 12.	+ 19 42		18.4	FAINT BLUE STAR
LB 10309	10 17 12.	+ 19 56		15.5	FAINT BLUE STAR
ZC 1017.2+2307	10 17 12.	+ 23 07	3560		CLUSTER OF GALAXIES
ZWG 266.039	10 17 12.	+ 54 07		15.2	GALAXY
MCG+09-17-055	10 17 12.	+ 54 07	36	15.	GALAXY
MCG+11-13-022	10 17 12.	+ 65 26	72	14.	GALAXY
ZC 1017.2+6829	10 17 12.	+ 68 29	1410		CLUSTER OF GALAXIES
MCG+11-13-022A	10 17 12.	+ 68 29	45	16.	GALAXY
ZWG 351.017	10 17 12.	+ 78 04		15.4	GALAXY
ZWG 350.049	10 17 12.	+ 78 04		15.4	GALAXY
KEEL 260	10 17 12.6	+ 41 54 44.		17.	SPIRAL NEBULA
REIZ 0282	10 17 14.	+ 54 07	36	14.9	GALAXY
MCG-04-25-002	10 17 15.	- 26 28 30.	120	13.	GALAXY
AR 10	10 17 15.90	+ 78 05 25.3			NEBULA
LB 10312	10 17 18.	+ 16 16		15.7	FAINT BLUE STAR
LB 10311	10 17 18.	+ 19 36		15.7	FAINT BLUE STAR
MCG+07-21-040	10 17 18.	+ 43 04	78	14.5	GALAXY
ZWG 211.043	10 17 18.	+ 43 05		15.6	GALAXY
UGC 05578	10 17 18.	+ 43 05	84	15.6	GALAXY Sc
ZWG 290.029	10 17 18.	+ 57 28		15.5	GALAXY
UGC 05579	10 17 18.	+ 57 28	60	15.5	GALAXY Sa-b
MCG+13-08-013	10 17 18.	+ 80 04	12	16.	GALAXY
ZWG 009.003	10 17 18.	- 00 46		15.7	GALAXY
REIZ 0290	10 17 19.	+ 28 04	48	14.4	GALAXY
8ZW 1017+08.4	10 17 24.	+ 08 22		18.0	COMPACT GALAXY
ZC 1017.4+1805	10 17 24.	+ 18 05	940		CLUSTER OF GALAXIES
LB 10314	10 17 24.	+ 18 22		15.2	FAINT BLUE STAR
LB 10313	10 17 24.	+ 19 58		15.5	FAINT BLUE STAR
MCG+05-25-001	10 17 24.	+ 28 03	66	14.8	GALAXY
ZWG 154.003	10 17 24.	+ 28 05		14.8	GALAXY
RNGC 3204	10 17 24.	+ 28 05		15.0	GALAXY
UGC 05580	10 17 24.	+ 28 05	90	14.8	GALAXY SBb
MCG+10-15-066	10 17 24.	+ 57 27	45	15.	GALAXY
ARC 0968	10 17 24.	+ 68 32		17.5	RICH CLUSTER OF GALAXIES
MCG-04-25-003	10 17 24.	- 25 36	90	14.	GALAXY
LB 03496	10 17 24.	- 77 29		13.6	FAINT BLUE STAR
SN 1950E	10 17 25.	+ 13 35		17.8	SUPERNOVA
RNGC 3208	10 17 26.	- 25 36		14.0	GALAXY
RNGC 3183	10 17 28.	+ 74 26		12.5	GALAXY
ZWG 037.001	10 17 30.	+ 07 13		15.5	GALAXY
ZWG 037.002	10 17 30.	+ 08 01		15.4	GALAXY
ZWG 065.001	10 17 30.	+ 13 35		15.7	GALAXY
LB 10315	10 17 30.	+ 20 30		15.2	FAINT BLUE STAR
TON-N 0499	10 17 30.	+ 31 32		15.4	BLUE STAR
ZWG 211.044	10 17 30.	+ 43 16		14.2	GALAXY
RNGC 3202	10 17 30.	+ 43 16		14.0	GALAXY
UGC 05581	10 17 30.	+ 43 16	78	14.2	GALAXY SBa
MCG+07-21-041	10 17 30.	+ 43 16	66	14.5	GALAXY
ZC 1017.5+4410	10 17 30.	+ 44 10	2220		CLUSTER OF GALAXIES
ZC 1017.5+6540	10 17 30.	+ 65 40	740		CLUSTER OF GALAXIES
ZWG 351.018	10 17 30.	+ 74 26		12.5	GALAXY
ZWG 333.023	10 17 30.	+ 74 26		12.5	GALAXY
UGC 05582	10 17 30.	+ 74 26	150	12.5	GALAXY SBb
MCG+13-08-014	10 17 30.	+ 78 04	27	15.	GALAXY
MCG-03-27-001	10 17 30.	- 19 41	30	15.	GALAXY
KN 15.185	10 17 30.4	+ 37 48 22.			NEBULA
KEEL 262	10 17 30.6	+ 45 42 26.			NEBULA
TON-N 1205	10 17 31.	+ 21 42		16.	BLUE STAR
REIZ 0289	10 17 31.	+ 43 15	60	14.3	GALAXY
HOLM 177A	10 17 32.	+ 74 26	84	13.5	PART OF MULTIPLE GALAXY
MCG+09-17-056	10 17 33.	+ 54 14	60	14.	GALAXY
AR 11	10 17 34.24	+ 74 25 42.1			NEBULA
KEEL 264	10 17 35.6	+ 41 45 30.		17.	SPIRAL NEBULA
LB 10318	10 17 36.	+ 15 40		17.0	FAINT BLUE STAR
LB 10317	10 17 36.	+ 17 58		16.6	FAINT BLUE STAR
LB 10316	10 17 36.	+ 18 28		16.0	FAINT BLUE STAR
UGC 05583	10 17 36.	+ 25 38	60	16.0	GALAXY Sb
ZC 1017.6+3310	10 17 36.	+ 33 10	2690		CLUSTER OF GALAXIES
ZWG 240.031	10 17 36.	+ 48 14		15.6	GALAXY
FATH 1.334	10 17 36.	- 15 05	19		NEBULA
KEEL 263	10 17 36.4	+ 45 21 38.			NEBULA
ZWG 009.004	10 17 42.	+ 01 28		15.7	GALAXY
MCG+00-27-001	10 17 42.	+ 01 28	30	15.7	GALAXY
LB 10322	10 17 42.	+ 14 48		16.0	FAINT BLUE STAR
LB 10321	10 17 42.	+ 17 30		17.0	FAINT BLUE STAR
LB 10320	10 17 42.	+ 17 54		15.8	FAINT BLUE STAR
ZC 1017.7+1909	10 17 42.	+ 19 09	1950		CLUSTER OF GALAXIES
LB 10319	10 17 42.	+ 20 16		15.7	FAINT BLUE STAR
VV 097	10 17 42.	- 03 15	90	14.	INTERACTING GALAXY
MCG-03-27-002	10 17 42.	- 21 15	30	14.5	GALAXY
ZWG 037.003	10 17 48.	+ 07 57		15.5	GALAXY
8ZW 1017+08.6	10 17 48.	+ 08 37		17.4	COMPACT GALAXY
REIZ 0293	10 17 48.	+ 25 37	48	15.1	GALAXY
ZWG 124.003	10 17 48.	+ 25 45		13.9	GALAXY
RNGC 3209	10 17 48.	+ 25 45		14.0	GALAXY
UGC 05584	10 17 48.	+ 25 45	108	13.9	GALAXY E
MCG+04-25-002	10 17 48.	+ 25 45	24	13.9	GALAXY
REIZ 0295	10 17 48.	+ 25 46	18	14.3	GALAXY
TON-N 0500	10 17 48.	+ 31 47		14.7	BLUE STAR
ZWG 211.045	10 17 48.	+ 38 47		15.6	GALAXY
ZWG 211.046	10 17 48.	+ 43 13		14.4	GALAXY
RNGC 3205	10 17 48.	+ 43 13		14.5	GALAXY
UGC 05585	10 17 48.	+ 43 13	96	14.4	GALAXY PECULR
ZC 1017.8+6452	10 17 48.	+ 64 52	2690		CLUSTER OF GALAXIES
FATH 1.336	10 17 48.	- 15 12	8		NEBULA
MCG-03-27-004	10 17 48.	- 21 25 30.	24	15.	GALAXY
MCG-03-27-003	10 17 48.	- 21 26 30.	24	15.	GALAXY
OCL 0822	10 17 48.	- 62 50	360	12.	OPEN STAR CLUSTER
BC B21017+31	10 17 48.0	+ 31 53 13.1		20.	QUASI-STELLAR OBJECT
REIZ 0291	10 17 49.	+ 43 13	24	14.3	GALAXY
HOLM 179A	10 17 49.	+ 43 13	30	13.8	PART OF MULTIPLE GALAXY
KEEL 265	10 17 50.4	+ 46 01 39.			NEBULA
FATH 1.335	10 17 51.	+ 14 33	16		NEBULA
REIZ 0292	10 17 51.	+ 38 46	12	15.0	GALAXY
MCG+07-21-042	10 17 51.	+ 43 13 30.	78	14.	GALAXY
ZWG 037.004	10 17 54.	+ 03 29		15.6	GALAXY
ZWG 037.005	10 17 54.	+ 08 17		15.5	GALAXY
LB 10324	10 17 54.	+ 14 29		15.3	FAINT BLUE STAR
LB 10323	10 17 54.	+ 20 30		16.6	FAINT BLUE STAR
REIZ 0296	10 17 54.	+ 25 54	18	16.0	GALAXY
MCG-01-27-002	10 17 54.	- 06 18	12	15.	GALAXY
MCG-01-27-001	10 17 54.	- 07 39	12	15.	GALAXY
KEEL 266	10 17 55.7	+ 45 24 57.			NEBULA
ARC 0979	10 17 57.	- 07 38		15.3	RICH CLUSTER OF GALAXIES
HOLM 179B	10 17 58.	+ 43 14	18	13.8	PART OF MULTIPLE GALAXY
ARC 0978	10 17 58.	- 06 16		15.6	RICH CLUSTER OF GALAXIES
ARC 0977	10 17 59.	+ 33 30		17.2	RICH CLUSTER OF GALAXIES
ZWG 037.006	10 18 00.	+ 08 07		15.3	GALAXY
ZWG 065.002	10 18 00.	+ 14 14		15.6	GALAXY
LB 10325	10 18 00.	+ 18 08		14.4	FAINT BLUE STAR
ZC 1018.0+3140	10 18 00.	+ 31 40	3090		CLUSTER OF GALAXIES
ZWG 211.047	10 18 00.	+ 43 14		14.3	GALAXY
RNGC 3207	10 18 00.	+ 43 14		14.5	GALAXY
UGC 05587	10 18 00.	+ 43 14	90	14.3	GALAXY PECULR
MCG+07-21-043	10 18 00.	+ 43 14	60	14.5	GALAXY
MCG+10-15-067	10 18 00.	+ 57 14 30.	45	15.	GALAXY
ZC 1018.0+5908	10 18 00.	+ 59 08	1010		CLUSTER OF GALAXIES
MCG+12-10-028	10 18 00.	+ 74 25	132	12.5	GALAXY
UGC 05586	10 18 00.	- 02 15	102	16.0	GALAXY DBL SYS
TON-N 1206	10 18 01.	+ 36 43		15.8	BLUE STAR
REIZ 0294	10 18 01.	+ 43 15	18	14.3	GALAXY
LB 10326	10 18 06.	+ 17 26		15.2	FAINT BLUE STAR
REIZ 0299	10 18 06.	+ 25 36	36	14.0	GALAXY
ZWG 124.004	10 18 06.	+ 25 37		14.0	GALAXY
UGC 05588	10 18 06.	+ 25 37	33	14.0	GALAXY PECULR
MCG+04-25-003	10 18 06.	+ 25 37	30	14.0	GALAXY
REIZ 0300	10 18 06.	+ 25 48	24	15.6	GALAXY
TON-N 1207	10 18 07.	+ 36 03		16.	BLUE STAR
ZWG 037.007	10 18 12.	+ 06 25		15.3	GALAXY
ZC 1018.2+0658	10 18 12.	+ 06 58	1950		CLUSTER OF GALAXIES
8ZW 1018+09.6	10 18 12.	+ 09 38		17.9	COMPACT GALAXY
ZWG 065.003	10 18 12.	+ 11 38		15.7	GALAXY
MCG+02-27-001	10 18 12.	+ 11 38	36	15.7	GALAXY
LB 10329	10 18 12.	+ 15 12		17.0	FAINT BLUE STAR
LB 10328	10 18 12.	+ 16 39		15.1	FAINT BLUE STAR
LB 10327	10 18 12.	+ 19 06		16.8	FAINT BLUE STAR
REIZ 0301	10 18 12.	+ 25 45	12	15.0	GALAXY
ZWG 124.005	10 18 12.	+ 25 46		15.2	GALAXY
MCG+04-25-004	10 18 12.	+ 25 46	18	15.2	GALAXY
ZWG 037.008	10 18 18.	+ 06 55		15.6	GALAXY
ZWG 037.009	10 18 18.	+ 07 03		15.6	GALAXY
ZC 1018.3+1227	10 18 18.	+ 12 27	1410		CLUSTER OF GALAXIES
LB 10332	10 18 18.	+ 19 16		16.0	FAINT BLUE STAR
LB 10331	10 18 18.	+ 20 09		16.0	FAINT BLUE STAR
LB 10330	10 18 18.	+ 20 17		16.6	FAINT BLUE STAR
ZWG 094.005	10 18 18.	+ 20 24		15.7	GALAXY
MCG+03-27-001	10 18 18.	+ 20 26	48	15.7	GALAXY
MCG+04-25-005	10 18 18.	+ 21 58	9	16.5	GALAXY
MCG+10-15-068	10 18 18.	+ 61 29	18	16.	GALAXY
MRSL 284-00/1	10 18 18.	- 57 49	1800		HII REGION
SN 1961S	10 18 19.	+ 21 58		18.3	SUPERNOVA
REIZ 0302	10 18 19.	+ 28 11	24	14.4	GALAXY
ZWG 065.004	10 18 24.	+ 08 48		15.6	GALAXY
LB 10336	10 18 24.	+ 17 20		16.0	FAINT BLUE STAR
LB 10335	10 18 24.	+ 18 20		15.8	FAINT BLUE STAR
LB 10334	10 18 24.	+ 18 45		17.7	FAINT BLUE STAR
LB 10333	10 18 24.	+ 20 26		15.7	FAINT BLUE STAR
ZC 1018.4+4044	10 18 24.	+ 40 44	870		CLUSTER OF GALAXIES
MCG+08-19-021	10 18 24.	+ 48 14	24	16.	GALAXY
ZC 1018.4+6245	10 18 24.	+ 62 45	1010		CLUSTER OF GALAXIES
REIZ 0297	10 18 25.	+ 57 11	180	13.8	GALAXY
KEEL 267	10 18 26.3	+ 45 56 38.			NEBULA
HOLM 180A	10 18 27.	+ 38 33	42	14.3	PART OF MULTIPLE GALAXY
RNGC 3206	10 18 27.	+ 57 11		12.5	GALAXY
MCG-01-27-003	10 18 27.	- 07 41	36	15.	GALAXY
IC 2565	10 18 29.	+ 28 10 53.			NONSTELLAR OBJECT
8ZW 1018+02.9	10 18 30.	+ 02 57		17.8	COMPACT GALAXY
ZC 1018.5+1359	10 18 30.	+ 13 59	1080		CLUSTER OF GALAXIES
ZC 1018.5+1425	10 18 30.	+ 14 25	670		CLUSTER OF GALAXIES
ZC 1018.5+1501	10 18 30.	+ 15 01	670		CLUSTER OF GALAXIES
ZWG 094.006	10 18 30.	+ 16 33		14.8	GALAXY
MCG+03-27-002	10 18 30.	+ 16 33	48	14.8	GALAXY
1ZW 024	10 18 30.	+ 28 11			COMPACT GALAXY
ZWG 154.004	10 18 30.	+ 28 11		15.2	GALAXY
ZWG 183.015	10 18 30.	+ 38 00		15.4	GALAXY
MCG+07-21-044	10 18 30.	+ 38 32	33	15.5	GALAXY
ZWG 211.048	10 18 30.	+ 38 34		15.7	GALAXY
MCG+10-15-069	10 18 30.	+ 57 11	168	13.3	GALAXY
ZWG 290.030	10 18 30.	+ 57 12		12.7	GALAXY
UGC 05589	10 18 30.	+ 57 12	180	12.7	GALAXY SBc
TON-N 1208	10 18 31.	+ 34 56		17.	BLUE STAR
IC 2564	10 18 31.	+ 36 42 11.			NONSTELLAR OBJECT
HOLM 180B	10 18 31.	+ 38 34	42	14.6	PART OF MULTIPLE GALAXY
REIZ 0298	10 18 31.	+ 57 14	60	15.1	GALAXY
REIZ 0306	10 18 32.	+ 19 55	48	13.7	GALAXY
FATH 1.337	10 18 33.	+ 14 59	27		NEBULA
MCG+07-21-045	10 18 33.	+ 38 32 30.	36	16.5	GALAXY
REIZ 0303	10 18 33.	+ 38 33	36	14.4	GALAXY
REIZ 0304	10 18 33.	+ 38 34	42	14.8	GALAXY
KN 15.186	10 18 33.2	+ 36 42 05.			NEBULA
ZWG 037.010	10 18 36.	+ 08 22		15.0	GALAXY
SCH 07	10 18 36.	+ 08 22		15.0	PECULIAR GALAXY
MCG+01-27-002	10 18 36.	+ 08 22	36	15.0	GALAXY
8ZW 1018+12.4	10 18 36.	+ 12 25		17.5	COMPACT GALAXY
LB 10342	10 18 36.	+ 15 46		15.9	FAINT BLUE STAR
ZWG 094.007	10 18 36.	+ 16 22		15.0	GALAXY
MCG+03-27-003	10 18 36.	+ 16 22	36	15.0	GALAXY
LB 10341	10 18 36.	+ 18 32		15.6	FAINT BLUE STAR
LB 10340	10 18 36.	+ 19 08		16.0	FAINT BLUE STAR
LB 10339	10 18 36.	+ 19 24		18.0	FAINT BLUE STAR
LB 10338	10 18 36.	+ 19 38		15.8	FAINT BLUE STAR
ZWG 094.008	10 18 36.	+ 19 53		14.3	GALAXY
UGC 05590	10 18 36.	+ 19 53	72	14.3	GALAXY Sb-c
LB 10337	10 18 36.	+ 20 27		14.9	FAINT BLUE STAR
ZWG 124.006	10 18 36.	+ 24 36		15.3	GALAXY
UGC 05591	10 18 36.	+ 24 36	60	15.3	GALAXY S0-a
REIZ 0305	10 18 36.	+ 35 16	12	15.1	GALAXY
ZWG 183.016	10 18 36.	+ 35 17		15.7	GALAXY
ZC 1018.6+5119	10 18 36.	+ 51 19	870		CLUSTER OF GALAXIES
MCG-03-27-005	10 18 36.	- 17 02 30.	36	15.	GALAXY
MCG-06-23-022	10 18 36.	- 37 32	72	15.	GALAXY
TON-N 1209	10 18 37.	+ 22 17		16.4	BLUE STAR
TON-N 1210	10 18 37.	+ 35 47		16.8	BLUE STAR
RNGC 3213	10 18 38.	+ 19 54		14.5	GALAXY
REIZ 0308	10 18 40.	+ 12 21	18	15.9	GALAXY
REIZ 0309	10 18 40.	+ 12 26	30	15.8	GALAXY
ARC 0984	10 18 40.	+ 12 27		17.8	RICH CLUSTER OF GALAXIES
8ZW 1018+05.6	10 18 42.	+ 05 38		17.3	COMPACT GALAXY
8ZW 1018+09.2	10 18 42.	+ 09 10		19.5	COMPACT GALAXY
LB 10343	10 18 42.	+ 19 32		15.8	FAINT BLUE STAR
MCG+03-27-004	10 18 42.	+ 19 56	54	14.3	GALAXY
MCG+09-17-057	10 18 42.	+ 52 50	12	16.	GALAXY
HOLM 178A	10 18 43.	+ 77 08	30	14.6	PART OF MULTIPLE GALAXY

OBJECT NAME	RIGHT ASCEN.	DECLINATION	DIAM.	MAGN.	TYPE OF OBJECT
RNGC 3194	10 18 44.	+ 75 03			NON-EXISTENT OBJECT
RNGC 3217	10 18 46.	+ 11 09			NON-EXISTENT OBJECT
REIZ 0310	10 18 46.	+ 22 48	60	14.3	GALAXY
FATH 1.338	10 18 46.	- 15 21	16		NEBULA
8ZW 1018+08.9	10 18 48.	+ 08 54		17.2	COMPACT GALAXY
ZWG 065.005	10 18 48.	+ 09 38		15.7	GALAXY
LB 10348	10 18 48.	+ 18 12		17.7	FAINT BLUE STAR
LB 10347	10 18 48.	+ 18 52		16.0	FAINT BLUE STAR
LB 10346	10 18 48.	+ 19 09		17.8	FAINT BLUE STAR
LB 10345	10 18 48.	+ 19 22		16.9	FAINT BLUE STAR
LB 10344	10 18 48.	+ 20 35		16.0	FAINT BLUE STAR
ZWG 124.007	10 18 48.	+ 22 48		15.7	GALAXY
UGC 05592	10 18 48.	+ 22 48	66	15.7	GALAXY Sc
MCG+04-25-006	10 18 48.	+ 22 48	60	15.7	GALAXY
ZC 1018.8+4115	10 18 48.	+ 41 15	1080		CLUSTER OF GALAXIES
ZC 1018.8+5020	10 18 48.	+ 50 20	3090		CLUSTER OF GALAXIES
MRK 142	10 18 48.	+ 51 56	9	16.5	GALAXY WITH UV CONTINUUM
KW 17	10 18 48.	+ 51 56	24		SEYFERT GALAXY
VVI 37	10 18 48.	+ 51 56	9	16.21	SEYFERT GALAXY
7ZW 322	10 18 48.	+ 69 56			COMPACT GALAXY
ZWG 009.005	10 18 48.	- 00 18		15.3	GALAXY
REIZ 0307	10 18 51.	+ 38 23	18	14.5	GALAXY
REIZ 0311	10 18 52.	+ 12 13	30	15.1	GALAXY
ZWG 037.011	10 18 54.	+ 06 12		15.1	GALAXY
ZC 1018.9+0739	10 18 54.	+ 07 39	1340		CLUSTER OF GALAXIES
ZC 1018.9+0828	10 18 54.	+ 08 28	1280		CLUSTER OF GALAXIES
ZWG 065.006	10 18 54.	+ 12 12		15.5	GALAXY
KARA.73B 0408	10 18 54.	+ 12 12	42	15.5	ISOLATED GALAXY S
LB 10351	10 18 54.	+ 14 18		16.0	FAINT BLUE STAR
ARC 0986	10 18 54.	+ 14 23		17.7	RICH CLUSTER OF GALAXIES
LB 10350	10 18 54.	+ 19 48		17.7	FAINT BLUE STAR
LB 10349	10 18 54.	+ 20 16		15.3	FAINT BLUE STAR
ZWG 124.008	10 18 54.	+ 24 10		15.1	GALAXY
UGC 05593	10 18 54.	+ 24 10	96	15.1	GALAXY E
MCG+04-25-007	10 18 54.	+ 24 10	60	15.1	GALAXY
TON-N 0501	10 18 54.	+ 31 49		13.7	BLUE STAR
TON-N 0502	10 18 54.	+ 32 12		13.9	BLUE STAR
ZWG 183.017	10 18 54.	+ 38 22		15.5	GALAXY
ZWG 240.032	10 18 54.	+ 48 17		15.4	GALAXY
UGC 05594	10 18 54.	+ 48 17	72	15.4	GALAXY Sb
MCG+08-19-022	10 18 54.	+ 48 17	60	15.	GALAXY
MCG+13-08-015	10 18 54.	+ 79 07	21	15.	GALAXY
MCG-01-27-004	10 18 54.	- 04 35	15	15.5	GALAXY
REIZ 0312	10 18 55.	+ 18 25	36	14.6	GALAXY
RNGC 3216	10 18 55.	+ 24 10		15.0	GALAXY
HOLM 178B	10 18 56.	+ 77 09	12	15.4	PART OF MULTIPLE GALAXY
MCG+07-21-046	10 18 57.	+ 41 56 30.	24	17.	GALAXY
REIZ 0315	10 18 58.	+ 12 50	54	15.0	GALAXY
REIZ 0313	10 18 58.	+ 22 57	18	14.6	GALAXY
REIZ 0314	10 18 59.	+ 24 12	30	14.5	GALAXY
ZWG 037.012	10 19 00.	+ 06 17		15.0	GALAXY
MCG+01-27-003	10 19 00.	+ 06 17	24	15.0	GALAXY
ZWG 065.007	10 19 00.	+ 12 49		15.3	GALAXY
UGC 05595	10 19 00.	+ 12 49	66	15.3	GALAXY Sb-c
MCG+02-27-002	10 19 00.	+ 12 49	78	15.3	GALAXY
ZWG 094.009	10 19 00.	+ 15 10		15.5	GALAXY
ZWG 094.010	10 19 00.	+ 16 01		15.6	GALAXY
LB 10353	10 19 00.	+ 16 50		16.0	FAINT BLUE STAR
LB 10352	10 19 00.	+ 17 12		17.6	FAINT BLUE STAR
ZWG 124.009	10 19 00.	+ 22 56		15.7	GALAXY
ZC 1019.0+2901	10 19 00.	+ 29 01	1080		CLUSTER OF GALAXIES
ZC 1019.0+3455	10 19 00.	+ 34 55	1410		CLUSTER OF GALAXIES
MCG+12-10-029	10 19 00.	+ 73 32	27	16.	GALAXY
VV 330A	10 19 00.	+ 78 53 30.	72	14.5	INTERACTING GALAXY
ZWG 351.019	10 19 00.	+ 79 07		14.7	GALAXY
ZWG 350.050	10 19 00.	+ 79 07		14.7	GALAXY
UGC 05596	10 19 00.	+ 79 07	78	14.7	GALAXY E?
ZWG 009.006	10 19 00.	- 02 24		15.7	GALAXY
KN 15.187	10 19 00.0	+ 37 02 22.			NEBULA
ARC 0982	10 19 01.	+ 34 53		17.6	RICH CLUSTER OF GALAXIES
FATH 1.339	10 19 03.	+ 15 09	14		NEBULA
ARC 0975	10 19 04.	+ 64 54		16.8	RICH CLUSTER OF GALAXIES
FATH 1.340	10 19 04.	- 15 03			NEBULA
ARC 0987	10 19 05.	+ 06 39		17.2	RICH CLUSTER OF GALAXIES
ZWG 065.008	10 19 06.	+ 13 42		15.6	GALAXY
REIZ 0318	10 19 06.	+ 15 10	48	15.3	GALAXY
LB 10356	10 19 06.	+ 15 47		10.7	FAINT BLUE STAR
LB 10355	10 19 06.	+ 19 40		15.9	FAINT BLUE STAR
LB 10354	10 19 06.	+ 20 14		17.2	FAINT BLUE STAR
ZWG 124.010	10 19 06.	+ 23 55		15.7	GALAXY
MCG+04-25-009	10 19 06.	+ 23 58	48	15.7	GALAXY
ZWG 124.011	10 19 06.	+ 24 07		15.5	GALAXY
UGC 05597	10 19 06.	+ 24 07	78	15.5	GALAXY Sa-b
MCG+04-25-008	10 19 06.	+ 24 07	66	15.5	GALAXY
ZC 1019.1+3934	10 19 06.	+ 39 34	1950		CLUSTER OF GALAXIES
ZC 1019.1+4131	10 19 06.	+ 41 31	870		CLUSTER OF GALAXIES
ZC 1019.1+4325	10 19 06.	+ 43 25	3230		CLUSTER OF GALAXIES
ZWG 290.031	10 19 06.	+ 57 18		15.6	GALAXY
ZWG 333.024	10 19 06.	+ 70 52		15.7	GALAXY
REIZ 0319	10 19 11.	+ 23 56	36	14.6	GALAXY
REIZ 0320	10 19 11.	+ 24 07	72	14.7	GALAXY
REIZ 0321	10 19 11.	+ 24 27	24	14.7	GALAXY
ZWG 065.009	10 19 12.	+ 13 12		15.4	GALAXY
ZC 1019.2+1418	10 19 12.	+ 14 18	810		CLUSTER OF GALAXIES
LB 10358	10 19 12.	+ 15 14		07.0	FAINT BLUE STAR
ZWG 094.011	10 19 12.	+ 17 14		15.6	GALAXY
LB 10357	10 19 12.	+ 18 35		17.4	FAINT BLUE STAR
ZC 1019.2+2341	10 19 12.	+ 23 41	740		CLUSTER OF GALAXIES
ZWG 124.012	10 19 12.	+ 24 27		15.6	GALAXY
MCG+04-25-010	10 19 12.	+ 24 54	42	15.7	GALAXY
IC 2567	10 19 12.	+ 24 54 02.			NONSTELLAR OBJECT
ZWG 124.013	10 19 12.	+ 24 55		15.7	GALAXY
ZWG 124.014	10 19 12.	+ 26 08		15.4	GALAXY
KARA.73B 0409	10 19 12.	+ 26 08	24	15.4	ISOLATED GALAXY IR
ZC 1019.2+3246	10 19 12.	+ 32 46	670		CLUSTER OF GALAXIES
MCG+07-21-047	10 19 12.	+ 38 45	12	16.5	GALAXY
MCG+07-21-048	10 19 12.	+ 38 45 30.	7	16.5	GALAXY
MCG+10-15-070	10 19 12.	+ 57 17	36	15.	GALAXY
ZC 1019.2+6007	10 19 12.	+ 60 07	2080		CLUSTER OF GALAXIES
VV 330B	10 19 12.	+ 78 52	60	15.	INTERACTING GALAXY
ZWG 009.007	10 19 12.	- 03 12		14.8	GALAXY
MCG+00-27-002	10 19 12.	- 03 12	12	14.8	GALAXY
RNGC 3229	10 19 13.	+ 00 32			GALAXY
REIZ 0316	10 19 15.	+ 38 46	12	15.1	GALAXY
REIZ 0317	10 19 15.	+ 38 47	12	15.1	GALAXY
HOLM 181A	10 19 15.	+ 38 47	18	14.7	PART OF MULTIPLE GALAXY
HOLM 181B	10 19 16.	+ 38 47	18	14.7	PART OF MULTIPLE GALAXY
ZWG 009.008	10 19 18.	+ 00 33		15.1	GALAXY

OBJECT NAME	RIGHT ASCEN.	DECLINATION	DIAM.	MAGN.	TYPE OF OBJECT
ZWG 009.009	10 19 18.	+ 00 40		15.4	GALAXY
ZC 1019.3+1527	10 19 18.	+ 15 27	3090		CLUSTER OF GALAXIES
ZWG 094.012	10 19 18.	+ 15 57		15.6	GALAXY
LB 10360	10 19 18.	+ 16 20		16.0	FAINT BLUE STAR
LB 10359	10 19 18.	+ 18 46		16.0	FAINT BLUE STAR
ZC 1019.3+4033	10 19 18.	+ 40 33	610		CLUSTER OF GALAXIES
ARC 0980	10 19 18.	+ 50 22		17.5	RICH CLUSTER OF GALAXIES
FATH 1.341	10 19 18.	- 15 01	14		NEBULA
MCG-06-23-023	10 19 18.	- 34 00	180	12.1	GALAXY
KEEL 268	10 19 20.7	+ 45 42 05.			NEBULA
MCG+04-25-011	10 19 21.	+ 21 18	24	15.7	GALAXY
HN 0341	10 19 21.	- 33 22			NEBULA
IC 2570	10 19 21.	- 33 22			NONSTELLAR OBJECT
IC 2571	10 19 21.	- 34 00 54.			SAME AS NGC 3223
REIZ 0323	10 19 22.	+ 22 37	24	14.7	GALAXY
IC 2566	10 19 23.	+ 36 50 08.			NONSTELLAR OBJECT
RNGC 3223	10 19 23.	- 34 00		12.0	GALAXY
RNGC 3224	10 19 23.	- 34 26			GALAXY
ZWG 037.013	10 19 24.	+ 04 11		15.7	GALAXY
ZC 1019.4+0925	10 19 24.	+ 09 25	870		CLUSTER OF GALAXIES
LB 10362	10 19 24.	+ 17 54		16.1	FAINT BLUE STAR
LB 10361	10 19 24.	+ 19 18		15.4	FAINT BLUE STAR
ZWG 124.015	10 19 24.	+ 21 20		15.7	GALAXY
ZC 1019.4+2501	10 19 24.	+ 25 01	5980		CLUSTER OF GALAXIES
ZWG 183.018	10 19 24.	+ 36 50		14.9	GALAXY
KARA.72 231A	10 19 24.	+ 36 50	60	14.9	PART OF DOUBLE GALAXY
MCG+12-10-030	10 19 24.	+ 70 06	39	15.	GALAXY
MCG-01-27-005	10 19 24.	- 04 38	15	15.5	GALAXY
ARC 0993	10 19 24.	- 04 43		14.9	RICH CLUSTER OF GALAXIES
MCG-06-23-024	10 19 24.	- 34 26	90	13.	GALAXY
KN 15.188	10 19 24.8	+ 36 50 01.			NEBULA
ARC 0989	10 19 25.	+ 09 27		17.2	RICH CLUSTER OF GALAXIES
REIZ 0325	10 19 25.	+ 18 05	42	15.3	GALAXY
REIZ 0326	10 19 25.	+ 18 11	42	14.4	GALAXY
HOLM 183B	10 19 26.	+ 36 51	48	14.3	PART OF MULTIPLE GALAXY
REIZ 0327	10 19 27.	+ 20 51	78	14.2	GALAXY
REIZ 0328	10 19 27.	+ 21 20	12	14.9	GALAXY
REIZ 0329	10 19 29.	+ 24 49	60	14.8	GALAXY
ZC 1019.5+0041	10 19 30.	+ 00 41	4370		CLUSTER OF GALAXIES
REIZ 0331	10 19 30.	+ 04 11	12	15.7	GALAXY
ZWG 094.013	10 19 30.	+ 17 03		15.5	GALAXY
LB 10363	10 19 30.	+ 17 36		15.6	FAINT BLUE STAR
ZWG 094.014	10 19 30.	+ 18 06		15.2	GALAXY
MCG+03-27-005	10 19 30.	+ 18 06	24	15.2	GALAXY
ZWG 124.016	10 19 30.	+ 20 51		15.5	GALAXY
UGC 05598	10 19 30.	+ 20 51	96	15.5	GALAXY S
TON-N 0035	10 19 30.	+ 26 32		15.6	BLUE STAR
MCG+06-23-008	10 19 30.	+ 36 51	42	15.	GALAXY
UGC 05599	10 19 30.	+ 44 07	90	16.0	GALAXY Sc
ZWG 351.020	10 19 30.	+ 78 52		14.4	GALAXY
ZWG 350.051	10 19 30.	+ 78 52		14.4	GALAXY
UGC 05600	10 19 30.	+ 78 52	84	14.4	GALAXY S0?
KARA.72 232A	10 19 30.	+ 78 52	72	14.4	PART OF DOUBLE GALAXY
MCG+13-08-016B	10 19 30.	+ 78 52	84	15.	GALAXY
MCG+13-08-016A	10 19 30.	+ 78 52	66	14.	GALAXY
MCG+13-08-017	10 19 30.	+ 79 10	30	16.	GALAXY
MCG-03-27-006	10 19 30.	- 19 13	42	15.	GALAXY
TON-N 1211	10 19 31.	+ 22 18		17.	BLUE STAR
HOLM 184B	10 19 31.	+ 38 51	60	14.7	PART OF MULTIPLE GALAXY
MCG+07-21-049	10 19 33.	+ 38 50	42	16.	GALAXY
REIZ 0324	10 19 33.	+ 38 51	48	14.9	GALAXY
IC 2568	10 19 34.	+ 36 53 31.			NONSTELLAR OBJFCT
SN 1961G	10 19 35.	+ 21 29		18.2	SUPERNOVA
SN 1961L	10 19 35.	+ 21 51		17.5	SUPERNOVA
KN 15.189	10 19 35.0	+ 36 51 03.			NEBULA
LB 10364	10 19 36.	+ 14 22		13.3	FAINT BLUE STAR
ZWG 094.015	10 19 36.	+ 15 09		15.4	GALAXY
MCG+03-27-006	10 19 36.	+ 17 04	42	15.5	GALAXY
MCG+03-27-008	10 19 36.	+ 18 07	18	15.2	GALAXY
MCG+03-27-007	10 19 36.	+ 18 07	8	15.2	GALAXY
MCG+04-25-012	10 19 36.	+ 21 29	24	17.	GALAXY
MCG+04-25-013	10 19 36.	+ 21 48	180	14.3	GALAXY
ZWG 124.017	10 19 36.	+ 21 50		14.3	GALAXY
UGC 05601	10 19 36.	+ 21 50	198	14.3	GALAXY SBc-IRR
UGC 05602	10 19 36.	+ 22 41	66	16.0	GALAXY SB?
ZWG 183.019	10 19 36.	+ 36 51		15.0	GALAXY
UGC 05603	10 19 36.	+ 36 51	72	15.0	GALAXY SBa
KARA.72 231B	10 19 36.	+ 36 51	66	15.0	PART OF DOUBLE GALAXY
MCG+07-21-050	10 19 36.	+ 44 06	60	15.	GALAXY
ZC 1019.6+5600	10 19 36.	+ 56 00	870		CLUSTER OF GALAXIES
MCG+09-17-058	10 19 36.	+ 56 11	60	16.	GALAXY
ZWG 313.015	10 19 36.	+ 65 00		15.7	GALAXY
ZC 1019.6-0100	10 19 36.	- 01 00	1880		CLUSTER OF GALAXIES
MCG-04-25-004	10 19 36.	- 22 00 30.	78	14.	GALAXY
HOLM 183A	10 19 37.	+ 36 52	60	14.1	PART OF MULTIPLE GALAXY
RNGC 3233	10 19 37.	- 22 00		14.0	GALAXY
RNGC 3221	10 19 38.	+ 21 49		14.5	GALAXY
REIZ 0332	10 19 39.	+ 21 51	180	13.2	GALAXY
REIZ 0330	10 19 39.	+ 38 50	30	14.7	GALAXY
HOLM 184A	10 19 39.	+ 38 50	42	14.3	PART OF MULTIPLE GALAXY
BC OL333	10 19 39.86	+ 30 56 14.9		17.	QUASI-STELLAR OBJECT
SHB 155	10 19 39.9	+ 30 56 14.		17.	QUASI-STELLAR OBJECT
KEEL 270	10 19 40.4	+ 45 45 56.			NEBULA
ARC 0991	10 19 41.	+ 19 08		17.2	RICH CLUSTER OF GALAXIES
MCG+03-27-010	10 19 42.	+ 14 42	27	15.4	GALAXY
MCG+03-27-009	10 19 42.	+ 18 09	24	16.5	GALAXY
2ZW 045	10 19 42.	+ 21 46			COMPACT GALAXY
MCG+06-23-009	10 19 42.	+ 36 51 30.	78	14.5	GALAXY
MCG+07-21-051	10 19 42.	+ 38 49	12	16.	GALAXY
ZWG 211.049	10 19 42.	+ 38 50		15.4	GALAXY
MCG+08-19-023	10 19 42.	+ 46 29	120	14.	GALAXY
ZWG 240.033	10 19 42.	+ 46 30		14.8	GALAXY
UGC 05604	10 19 42.	+ 46 30	156	14.8	GALAXY Sc
ZWG 240.034	10 19 42.	+ 48 53		15.5	GALAXY
UGC 05605	10 19 42.	+ 48 53	66	15.5	GALAXY S
MCG+09-17-059	10 19 42.	+ 54 58	36	18.	GALAXY
ZWG 266.040	10 19 42.	+ 56 09		15.7	GALAXY
MCG+12-10-031	10 19 42.	+ 70 05	51	16.	GALAXY
HOLM 182A	10 19 43.	+ 57 18	60	14.0	PART OF MULTIPLF GALAXY
RNGC 3228	10 19 43.	- 51 28		6.5	OPEN CLUSTER
RNGC 3219	10 19 44.	+ 38 50		15.5	GALAXY
REIZ 0333	10 19 45.	+ 21 52	30	16.3	GALAXY
ARC 0988	10 19 45.	+ 32 34		18.0	RICH CLUSTER OF GALAXIES
MCG+07-21-052	10 19 45.	+ 38 49	9	17.	GALAXY
MCG+08-19-024	10 19 45.	+ 48 53	48	15.	GALAXY
ARC 0985	10 19 46.	+ 52 18		17.0	RICH CLUSTER OF GALAXIES
ZWG 009.010	10 19 48.	+ 01 27		14.5	GALAXY
UGC 05606	10 19 48.	+ 01 27	42	14.5	GALAXY S

OBJFCT NAME	RIGHT ASCEN.	DECLINATION	DIAM.	MAGN.	TYPE OF OBJECT
MCG+00-27-003	10 19 48.	+ 01 27	36	14.5	GALAXY
KARA.73B 0410	10 19 48.	+ 01 27	36	14.5	ISOLATED GALAXY S
ZWG 037.014	10 19 48.	+ 04 15		14.2	GALAXY
REIZ 0337	10 19 48.	+ 04 15	60	14.3	GALAXY
UGC 05607	10 19 48.	+ 04 15	114	14.2	GALAXY Sb
MCG+01-27-004	10 19 48.	+ 04 15	84	14.2	GALAXY
ZWG 094.016	10 19 48.	+ 14 42		15.4	GALAXY
ZC 1019.8+1905	10 19 48.	+ 19 05	870		CLUSTER OF GALAXIES
MCG+05-25-002	10 19 48.	+ 27 36 30.	60	15.	GALAXY
UGC 05608	10 19 48.	+ 27 37	66	16.0	GALAXY Sb
ZC 1019.8+3233	10 19 48.	+ 32 33	740		CLUSTER OF GALAXIES
ZC 1019.8+3341	10 19 48.	+ 33 41	1340		CLUSTER OF GALAXIES
ZC 1019.8+3653	10 19 48.	+ 36 53	12300		CLUSTER OF GALAXIES
ZWG 290.032	10 19 48.	+ 57 18		15.2	GALAXY
MCG+10-15-071	10 19 48.	+ 57 18	39	14.0	GALAXY
MCG+12-10-032	10 19 48.	+ 71 08	204	14.	GALAXY
ZWG 351.021	10 19 48.	+ 78 51		14.5	GALAXY
ZWG 350.052	10 19 48.	+ 78 51		14.5	GALAXY
UGC 05609	10 19 48.	+ 78 51	72	14.5	GALAXY
KARA.72 232B	10 19 48.	+ 78 51	66	14.5	PART OF DOUBLE GALAXY
7ZW 323	10 19 48.	+ 78 55			COMPACT GALAXY
OCL 0800	10 19 48.	- 51 28	1800	6.5	OPEN STAR CLUSTER
VHA 093	10 19 48.	- 51 28	360		OPEN STAR CLUSTER
ARC 0992	10 19 49.	+ 20 45		17.8	RICH CLUSTER OF GALAXIES
TON-N 1212	10 19 49.	+ 34 44		15.9	BLUE STAR
REIZ 0322	10 19 49.	+ 57 18	72	14.3	GALAXY
IC 0605	10 19 50.	+ 01 27 47.			NONSTELLAR OBJECT
RNGC 3222	10 19 50.	+ 20 08		14.5	GALAXY
REIZ 0335	10 19 50.	+ 20 09	30	13.7	GALAXY
RNGC 3214	10 19 51.	+ 57 17		15.0	GALAXY
KEEL 271	10 19 52.2	+ 20 04 41.		13.	SPIRAL NEBULA
ZWG 009.011	10 19 54.	+ 00 56		15.2	GALAXY
REIZ 0339	10 19 54.	+ 04 09	12	15.8	GALAXY
ZWG 037.015	10 19 54.	+ 04 29		15.6	GALAXY
ZWG 037.016	10 19 54.	+ 04 50		14.6	GALAXY
MCG+01-27-005	10 19 54.	+ 04 50	36	14.6	GALAXY
ZWG 065.010	10 19 54.	+ 08 38		15.6	GALAXY
8ZW 1019+14.0	10 19 54.	+ 13 58		10.2	COMPACT GALAXY
8ZW 1019+14.2	10 19 54.	+ 14 11		17.8	COMPACT GALAXY
ZWG 094.017	10 19 54.	+ 16 00		15.7	GALAXY
LB 10366	10 19 54.	+ 17 40		15.1	FAINT BLUE STAR
LB 10365	10 19 54.	+ 18 01		15.8	FAINT BLUE STAR
ZWG 094.018	10 19 54.	+ 20 07		14.5	GALAXY
UGC 05610	10 19 54.	+ 20 07	72	14.5	GALAXY E-SO
ZWG 212.001	10 19 54.	+ 43 06		15.7	GALAXY
ZWG 211.050	10 19 54.	+ 43 06		15.7	GALAXY
MCG+11-13-023	10 19 54.	+ 65 00	30	16.	GALAXY
REIZ 0338	10 19 56.	+ 19 41	24	14.4	GALAXY
KEEL 272	10 19 56.9	+ 20 03 10.		13.	SPIRAL NEBULA
MCG+03-27-011	10 19 57.	+ 20 10	72	14.5	GALAXY
VDB.66G 077	10 20	+ 71 11	130		DWARF GALAXY
ZWG 094.019	10 20 00.	+ 16 11		15.5	GALAXY
ZWG 094.020	10 20 00.	+ 19 38		15.4	GALAXY
MCG+03-27-012	10 20 00.	+ 19 41	48	15.4	GALAXY
LB 10367	10 20 00.	+ 20 20		15.8	FAINT BLUE STAR
ZWG 124.018	10 20 00.	+ 21 07		15.2	GALAXY
MCG+10-15-072	10 20 00.	+ 58 42	24	16.	GALAXY
ZC 1020.0+7323	10 20 00.	+ 73 23	540		CLUSTER OF GALAXIES
MCG+13-08-018	10 20 00.	+ 78 55	12	16.	GALAXY
KEEL 273	10 20 03.7	+ 20 02 41.		14.	NEBULA
ARC 0983	10 20 05.	+ 60 04		17.7	RICH CLUSTER OF GALAXIES
ZC 1020.1+1306	10 20 06.	+ 13 06	4030		CLUSTER OF GALAXIES
ZWG 094.021	10 20 06.	+ 15 10		15.3	GALAXY
LB 10368	10 20 06.	+ 18 33		17.3	FAINT BLUE STAR
ARC 0994	10 20 06.	+ 19 35		17.5	RICH CLUSTER OF GALAXIES
ZC 1020.1+1939	10 20 06.	+ 19 39	2150		CLUSTER OF GALAXIES
ZWG 094.022	10 20 06.	+ 20 02		15.4	GALAXY
ZC 1020.1+2046	10 20 06.	+ 20 46	5440		CLUSTER OF GALAXIES
MCG+07-21-053	10 20 06.	+ 42 05	48	15.	GALAXY
ZWG 212.002	10 20 06.	+ 42 06		15.7	GALAXY
ZWG 211.051	10 20 06.	+ 42 06		15.7	GALAXY
UGC 05611	10 20 06.	+ 42 06	66	15.7	GALAXY SBc
ZWG 333.025	10 20 06.	+ 71 08		14.8	GALAXY
UGC 05612	10 20 06.	+ 71 08	210	14.8	GALAXY SB IV-V
ZWG 009.012	10 20 06.	- 03 21		15.5	GALAXY
FATH 1.342	10 20 06.	- 15 01	14		NEBULA
IC 2569	10 20 08.	+ 24 51 05.			NONSTELLAR OBJECT
ZWG 037.017	10 20 12.	+ 04 00		15.1	GALAXY
REIZ 0340	10 20 12.	+ 04 01	24	14.8	GALAXY
LB 10374	10 20 12.	+ 14 34		15.8	FAINT BLUE STAR
LB 10373	10 20 12.	+ 16 26		16.1	FAINT BLUE STAR
LB 10372	10 20 12.	+ 18 14		17.5	FAINT BLUE STAR
ZWG 094.023	10 20 12.	+ 18 36		15.5	GALAXY
LB 10371	10 20 12.	+ 19 21		15.6	FAINT BLUE STAR
LB 10370	10 20 12.	+ 19 56		15.8	FAINT BLUE STAR
LB 10369	10 20 12.	+ 20 30		16.0	FAINT BLUE STAR
ZC 1020.2+3733	10 20 12.	+ 37 33	3090		CLUSTER OF GALAXIES
MCG+11-13-024	10 20 12.	+ 68 34	36	17.	GALAXY
ZWG 333.026	10 20 12.	+ 68 38		15.5	GALAXY
ZWG 009.013	10 20 12.	- 03 01		15.6	GALAXY
ARC 0996	10 20 13.	+ 15 24		17.6	RICH CLUSTER OF GALAXIES
TON-N 1213	10 20 14.	+ 22 08		17.	BLUE STAR
KEEL 269	10 20 14.5	+ 68 38 05.			NEBULA
REIZ 0336	10 20 17.	+ 52 36	30	14.5	GALAXY
ZWG 037.018	10 20 18.	+ 05 01		15.1	GALAXY
MCG+01-27-006	10 20 18.	+ 05 01	60	15.1	GALAXY
ZC 1020.3+1526	10 20 18.	+ 15 26	870		CLUSTER OF GALAXIES
MCG+03-27-013	10 20 18.	+ 18 37	42	15.5	GALAXY
TON-N 0503	10 20 18.	+ 25 37		15.2	BLUE STAR
ZC 1020.3+4922	10 20 18.	+ 49 22	1080		CLUSTER OF GALAXIES
ZC 1020.3+5246	10 20 18.	+ 52 46	2550		CLUSTER OF GALAXIES
ZWG 290.033	10 20 18.	+ 61 00		15.0	GALAXY
HOLM 182B	10 20 20.	+ 57 17	90	14.0	PART OF MULTIPLE GALAXY
ZWG 065.011	10 20 24.	+ 13 30		15.6	GALAXY
LB 10375	10 20 24.	+ 17 22		15.0	FAINT BLUE STAR
ZWG 094.024	10 20 24.	+ 19 06		15.7	GALAXY
TON-N 0505	10 20 24.	+ 26 57		14.9	BLUE STAR
ZWG 154.005	10 20 24.	+ 30 07		15.6	GALAXY
TON-N 0504	10 20 24.	+ 31 26		15.4	BLUE STAR
ZC 1020.4+4210	10 20 24.	+ 42 10	4300		CLUSTER OF GALAXIES
ZC 1020.4+5216	10 20 24.	+ 52 16	1340		CLUSTER OF GALAXIES
MCG+09-17-060	10 20 24.	+ 52 35	72	15.	GALAXY
ZWG 266.041	10 20 24.	+ 52 36		15.0	GALAXY
UGC 05613	10 20 24.	+ 52 36	60	15.0	GALAXY S-IRR
ZWG 290.034	10 20 24.	+ 57 17		13.7	GALAXY
UGC 05614	10 20 24.	+ 57 17	78	13.7	GALAXY
MCG+10-15-073	10 20 24.	+ 57 17	78	14.0	GALAXY
REIZ 0334	10 20 24.	+ 57 18	66	14.3	GALAXY

OBJECT NAME	RIGHT ASCEF.	DECLINATION	DIAM.	MAGN.	TYPE OF OBJECT
ZC 1020.4-0316	10 20 24.	- 03 16	14650		CLUSTER OF GALAXIES
MCG-04-25-005	10 20 24.	- 22 21	30	15.	GALAXY
TON-N 1214	10 20 26.	+ 21 58		17.	BLUE STAR
ARC 0990	10 20 27.	+ 49 25		17.4	RICH CLUSTER OF GALAXIES
RNGC 3220	10 20 28.	+ 57 17		13.5	GALAXY
REIZ 0347	10 20 30.	+ 04 23	12	15.7	GALAXY
ZWG 037.019	10 20 30.	+ 04 24		15.6	GALAXY
ZC 1020.5+0724	10 20 30.	+ 07 24	470		CLUSTER OF GALAXIES
8ZW 1020+09.1	10 20 30.	+ 09 08		17.0	COMPACT GALAXY
LB 10378	10 20 30.	+ 16 15		15.2	FAINT BLUE STAR
ZWG 094.025	10 20 30.	+ 18 12		14.6	GALAXY
MRK 630	10 20 30.	+ 18 12	20	15.	GALAXY WITH UV CONTINUUM
LB 10377	10 20 30.	+ 18 20		16.5	FAINT BLUE STAR
LB 10376	10 20 30.	+ 19 08		14.4	FAINT BLUE STAR
MCG+04-25-014	10 20 30.	+ 22 39	42	15.5	GALAXY
TON-N 0506	10 20 30.	+ 30 19		15.0	BLUE STAR
VV 312B	10 20 30.	+ 53 21	30	15.	INTERACTING GALAXY
VV 312A	10 20 30.	+ 53 21	36	15.	INTERACTING GALAXY
IC 0604	10 20 30.	+ 57 16 53.			NONSTELLAR OBJECT
REIZ 0344	10 20 31.	+ 18 23	24	15.6	GALAXY
KEEL 274	10 20 31.4	+ 20 25 51.		15.	NEBULA
REIZ 0345	10 20 34.	+ 22 40	30	14.6	GALAXY
ZWG 065.012	10 20 36.	+ 08 40		15.7	GALAXY
ZWG 065.013	10 20 36.	+ 13 21		15.5	GALAXY
MCG+03-27-014	10 20 36.	+ 18 14	18	14.6	GALAXY
ZC 1020.6+4043	10 20 36.	+ 40 43	740		CLUSTER OF GALAXIES
ZWG 212.003	10 20 36.	+ 42 36		15.5	GALAXY
ZWG 211.052	10 20 36.	+ 42 36		15.5	GALAXY
KARA.72 233B	10 20 36.	+ 53 20	36		PART OF DOUBLE GALAXY
ZWG 266.042	10 20 36.	+ 53 21		14.2	GALAXY
UGC 05615	10 20 36.	+ 53 21	72	14.2	GALAXY DBL SYS
KARA.72 233A	10 20 36.	+ 53 21	42	14.2	PART OF DOUBLE GALAXY
MCG+09-17-062	10 20 36.	+ 53 21	36	13.5	GALAXY
MCG+09-17-061	10 20 36.	+ 53 21	36	13.8	GALAXY
ZC 1020.6+5829	10 20 36.	+ 58 29	1480		CLUSTER OF GALAXIES
MCG+10-15-074	10 20 36.	+ 61 00	27	16.	GALAXY
ZWG 009.014	10 20 36.	- 03 21		15.5	GALAXY
YM 47	10 20 37.	- 57 38	540		SYMMETRIC GALACTIC NEBULA
KEEL 275	10 20 37.8	+ 20 16 50.		14.	SPIRAL NEBULA
REIZ 0343	10 20 38.	+ 37 53	36	14.5	GALAXY
HOLM 186A	10 20 38.	+ 37 53	48	14.2	PART OF MULTIPLE GALAXY
REIZ 0348	10 20 39.	+ 21 23	18	15.5	GALAXY
REIZ 0349	10 20 39.	+ 21 27	12	16.2	GALAXY
HOLM 186B	10 20 39.	+ 37 53	42	14.4	PART OF MULTIPLF GALAXY
HOLM 187B	10 20 40.	+ 20 10	36	13.4	PART OF MULTIPLE GALAXY
ARC 0995	10 20 40.	+ 37 32		17.2	RICH CLUSTER OF GALAXIES
ARC 1001	10 20 40.	- 06 22		17.7	RICH CLUSTER OF GALAXIES
ARC 0999	10 20 41.	+ 13 06		15.6	RICH CLUSTER OF GALAXIES
ZWG 065.014	10 20 42.	+ 10 12		15.7	GALAXY
UGC 05616	10 20 42.	+ 10 12	78	15.7	GALAXY Sc
MCG+02-27-003	10 20 42.	+ 10 12	84	15.7	GALAXY
MCG+02-27-005	10 20 42.	+ 13 04	15	15.3	GALAXY
ZWG 065.015	10 20 42.	+ 13 05		15.3	GALAXY
MCG+02-27-004	10 20 42.	+ 13 05	72	15.3	GALAXY
ZWG 065.016	10 20 42.	+ 13 20		15.7	GALAXY
LB 10381	10 20 42.	+ 16 22		15.4	FAINT BLUE STAR
LB 10380	10 20 42.	+ 16 38		18.5	FAINT BLUE STAR
LB 10379	10 20 42.	+ 18 35		15.8	FAINT BLUE STAR
VVI 38	10 20 42.	+ 20 07	360	12.54	SEYFERT GALAXY
ZWG 094.026	10 20 42.	+ 20 08		13.3	GALAXY
UGC 05617	10 20 42.	+ 20 08	180	13.3	GALAXY E
KARA.72 234A	10 20 42.	+ 20 08	150	13.3	PART OF DOUBLE GALAXY
ARP 094	10 20 42.	+ 20 09			PECULIAR GALAXY
UGC 05618	10 20 42.	+ 21 22	60	16.5	GALAXY S
ZWG 290.035	10 20 42.	+ 58 15		15.5	GALAXY
ARC 0981	10 20 43.	+ 68 22		17.9	RICH CLUSTER OF GALAXIES
HOLM 187A	10 20 44.	+ 20 08	138	13.0	PART OF MULTIPLE GALAXY
RNGC 3226	10 20 44.	+ 20 09		12.5	GALAXY
REIZ 0351	10 20 44.	+ 20 32	18	14.8	GALAXY
REIZ 0346	10 20 44.	+ 37 53	30	14.7	GALAXY
HOLM 188A	10 20 45.	+ 10 11	36	15.1	PART OF MULTIPLE GALAXY
REIZ 0352	10 20 45.	+ 21 59	24	15.6	GALAXY
MCG+06-23-010	10 20 45.	+ 37 54 30.	42	15.	GALAXY
HOLM 188B	10 20 46.	+ 10 11	24	15.5	PART OF MULTIPLE GALAXY
REIZ 0355	10 20 46.	+ 13 05	60	14.4	GALAXY
REIZ 0356	10 20 46.	+ 13 06	48	14.5	GALAXY
VV 209A	10 20 46.	+ 20 07	162	11.3	INTERACTING GALAXY
VV 209B	10 20 46.	+ 20 09	102	12.6	INTERACTING GALAXY
REIZ 0353	10 20 47.	+ 24 36	30	15.2	GALAXY
REIZ 0350	10 20 47.	+ 34 02	30	14.6	GALAXY
8ZW 1020+09.1	10 20 48.	+ 09 07		18.2	COMPACT GALAXY
ZC 1020.8+1031	10 20 48.	+ 10 31	2020		CLUSTER OF GALAXIES
ZWG 065.017	10 20 48.	+ 11 13		15.2	GALAXY
8ZW 1020+11.2	10 20 48.	+ 11 13		15.2	COMPACT GALAXY
MCG+02-27-006	10 20 48.	+ 11 13	24	15.2	GALAXY
LB 10383	10 20 48.	+ 14 16		14.7	FAINT BLUE STAR
ZWG 094.027	10 20 48.	+ 18 22		15.1	GALAXY
UGC 05619	10 20 48.	+ 18 22	60	15.1	GALAXY S0-a
LB 10382	10 20 48.	+ 19 40		18.6	FAINT BLUE STAR
ZWG 094.028	10 20 48.	+ 20 06		12.2	GALAXY
UGC 05620	10 20 48.	+ 20 06	390	12.2	GALAXY Sb
KARA.72 234B	10 20 48.	+ 20 06	222	12.2	PART OF DOUBLE GALAXY
MCG+05-25-003	10 20 48.	+ 28 34	66	15.7	GALAXY
ZWG 154.006	10 20 48.	+ 28 35		15.7	GALAXY
UGC 05621	10 20 48.	+ 28 35	66	15.7	GALAXY Sb/Sc
ZWG 183.020	10 20 48.	+ 34 01		15.4	GALAXY
UGC 05622	10 20 48.	+ 34 01	78	15.4	GALAXY Sb/SBc
REIZ 0342	10 20 48.	+ 53 23	18	14.6	GALAXY
REIZ 0341	10 20 48.	+ 53 23	30	14.0	GALAXY
HOLM 185B	10 20 48.	+ 53 23	24	13.8	PART OF MULTIPLE GALAXY
HOLM 185A	10 20 48.	+ 53 23	24	13.5	PART OF MULTIPLE GALAXY
VV 022B	10 20 48.	+ 54 07	24	16.	INTERACTING GALAXY
VV 022A	10 20 48.	+ 54 07	66	14.	INTERACTING GALAXY
VV 022	10 20 48.	+ 54 07	84		INTERACTING GALAXY
ZC 1020.8+7021	10 20 48.	+ 70 21	670		CLUSTER OF GALAXIES
ZWG 009.015	10 20 48.	- 03 00		14.6	GALAXY
MCG+00-27-004	10 20 48.	- 03 00	30	14.6	GALAXY
REIZ 0357	10 20 49.	+ 18 22	30	14.8	GALAXY
REIZ 0354	10 20 49.	+ 28 33	48	15.1	GALAXY
RNGC 3227	10 20 50.	+ 20 07		12.0	GALAXY
REIZ 0359	10 20 50.	+ 20 07	210	12.2	GALAXY
REIZ 0358	10 20 50.	+ 20 10	60	12.9	GALAXY
TON-N 1215	10 20 50.	+ 21 17		14.8	BLUE STAR
MCG+03-27-015	10 20 51.	+ 20 10 30.	42	13.6	GALAXY
KEEL 276	10 20 51.5	+ 20 32 18.		13.	SPIRAL NEBULA
HN 0016	10 20 52.	+ 00 19			NEBULA
ZC 1020.9+0759	10 20 54.	+ 07 59	1550		CLUSTER OF GALAXIES
ZWG 065.018	10 20 54.	+ 14 13		15.2	GALAXY

OBJECT NAME	RIGHT ASCEN.	DECLINATION	DIAM.	MAGN.	TYPE OF OBJECT
MCG+03-27-016	10 20 54.	+ 20 09	360	12.2	GALAXY
MCG+06-23-011	10 20 54.	+ 34 01	66	14.5	GALAXY
UGC 05623	10 20 54.	+ 34 04	60	16.5	GALAXY Sc
MCG+06-23-012	10 20 54.	+ 37 54 30.	36	15.	GALAXY
ZC 1020.9+3843	10 20 54.	+ 38 43	1210		CLUSTER OF GALAXIES
ZWG 009.017	10 20 54.	- 01 55		15.6	GALAXY
ZWG 009.016	10 20 54.	- 03 08		15.5	GALAXY
MCG-06-23-025	10 20 54.	- 39 23	60	14.	GALAXY
IC 0606	10 20 55.	+ 11 12 32.			NONSTELLAR OBJECT
ARC 0997	10 20 59.	+ 37 46		17.2	RICH CLUSTER OF GALAXIES
ZC 1021.0+0426	10 21 00.	+ 04 26	470		CLUSTER OF GALAXIES
ZWG 037.020	10 21 00.	+ 08 07		15.7	GALAXY
ZWG 065.019	10 21 00.	+ 12 58		15.4	GALAXY
LB 10385	10 21 00.	+ 16 40		16.1	FAINT BLUE STAR
LB 10384	10 21 00.	+ 18 04		15.9	FAINT BLUE STAR
MCG+03-27-017	10 21 00.	+ 18 23	60	15.1	GALAXY
TON-N 0507	10 21 00.	+ 29 28		14.9	BLUE STAR
ZC 1021.0+2938	10 21 00.	+ 29 38	670		CLUSTER OF GALAXIES
ZWG 154.007	10 21 00.	+ 31 54		15.7	GALAXY
KARA.73B 0411	10 21 00.	+ 31 54	42	15.7	ISOLATED GALAXY S
MCG+06-23-013	10 21 00.	+ 34 04	60	15.	GALAXY
ZC 1021.0+4211	10 21 00.	+ 42 11	2020		CLUSTER OF GALAXIES
ZWG 212.004	10 21 00.	+ 44 07		15.4	GALAXY
ZWG 211.053	10 21 00.	+ 44 07		15.4	GALAXY
ZWG 240.035	10 21 00.	+ 45 06		15.4	GALAXY
ZC 1021.0+7728	10 21 00.	+ 77 28	21640		CLUSTER OF GALAXIES
RNGC 3230	10 21 04.	+ 12 49		14.0	GALAXY
REIZ 0360	10 21 04.	+ 12 49	48	13.1	GALAXY
ZWG 065.020	10 21 06.	+ 12 49		14.9	GALAXY
UGC 05624	10 21 06.	+ 12 49	138	14.9	GALAXY S0
MCG+02-27-007	10 21 06.	+ 12 49	48	14.9	GALAXY
ZWG 065.021	10 21 06.	+ 12 53		15.2	GALAXY
UGC 05625	10 21 06.	+ 12 53	60	15.2	GALAXY S
LB 10387	10 21 06.	+ 17 34		15.3	FAINT BLUE STAR
LB 10386	10 21 06.	+ 19 17		15.2	FAINT BLUE STAR
ZC 1021.1+5153	10 21 06.	+ 51 53	870		CLUSTER OF GALAXIES
REIZ 0362	10 21 10.	+ 12 53	30	14.6	GALAXY
ZWG 037.021	10 21 12.	+ 07 07		15.6	GALAXY
LB 10389	10 21 12.	+ 19 01		16.3	FAINT BLUE STAR
LB 10388	10 21 12.	+ 19 50		17.7	FAINT BLUE STAR
TON-N 0508	10 21 12.	+ 29 10		14.9	BLUE STAR
TON-N 0509	10 21 12.	+ 30 09		15.6	BLUE STAR
MCG+09-17-063	10 21 12.	+ 53 34	36	16.	GALAXY
ZWG 290.036	10 21 12.	+ 57 39		15.2	GALAXY
UGC 05626	10 21 12.	+ 57 39	102	15.2	GALAXY IRR
ARP 043	10 21 13.	+ 16 59			PECULIAR GALAXY
TON-N 1216	10 21 14.	+ 22 16		14.8	BLUE STAR
MCG+10-15-075	10 21 18.	+ 57 38 30.	96	14.	GALAXY
HN 0342	10 21 20.	- 35 12			NEBULA
IC 2573	10 21 20.	- 35 12			NONSTELLAR OBJECT
REIZ 0363	10 21 21.	+ 21 22	15	15.5	GALAXY
REIZ 0364	10 21 22.	+ 12 56	18	15.4	GALAXY
PK285-02.1	10 21 23.	- 60 17 39.	5		PLANETARY NEBULA
ZWG 037.022	10 21 24.	+ 06 44		15.3	GALAXY
LB 10397	10 21 24.	+ 14 30		15.5	FAINT BLUE STAR
LB 10396	10 21 24.	+ 14 46		17.2	FAINT BLUE STAR
LB 10395	10 21 24.	+ 15 46		16.3	FAINT BLUE STAR
ZWG 094.029	10 21 24.	+ 17 18		15.2	GALAXY
LB 10394	10 21 24.	+ 17 22		15.8	FAINT BLUE STAR
LB 10393	10 21 24.	+ 17 38		15.0	FAINT BLUE STAR
LB 10392	10 21 24.	+ 18 36		14.7	FAINT BLUE STAR
LB 10391	10 21 24.	+ 19 34		16.3	FAINT BLUE STAR
LB 10390	10 21 24.	+ 19 54		13.2	FAINT BLUE STAR
MCG+07-22-001	10 21 24.	+ 41 57 30.	15	15.	GALAXY
ZWG 212.005	10 21 24.	+ 41 58		15.2	GALAXY
ZWG 211.054	10 21 24.	+ 41 58		15.2	GALAXY
ZWG 009.018	10 21 24.	- 02 56		13.4	GALAXY
MCG+00-27-005	10 21 24.	- 02 56	108	13.4	GALAXY
REIZ 0365	10 21 29.	+ 13 49	42	14.9	GALAXY
REIZ 0366	10 21 29.	+ 13 50	18	15.6	GALAXY
REIZ 0367	10 21 29.	+ 14 05	24	15.2	GALAXY
ZC 1021.5+0406	10 21 30.	+ 04 06	1610		CLUSTER OF GALAXIES
ZWG 037.023	10 21 30.	+ 06 45		15.5	GALAXY
ZWG 065.022	10 21 30.	+ 13 49		14.8	GALAXY
UGC 05627	10 21 30.	+ 13 49	60	14.8	GALAXY S
MCG+02-27-008	10 21 30.	+ 13 49	72	14.8	GALAXY
ZWG 065.023	10 21 30.	+ 13 51		15.2	GALAXY
ZWG 094.030	10 21 30.	+ 17 00		14.9	GALAXY
UGC 05628	10 21 30.	+ 17 00	90	14.9	GALAXY SBb
MCG+03-27-018	10 21 30.	+ 17 00	120	14.9	GALAXY
IC 0607	10 21 30.	+ 17 01 42.			NONSTELLAR OBJECT
LB 10398	10 21 30.	+ 19 54		15.8	FAINT BLUE STAR
UGC 05629	10 21 30.	+ 21 19	84	17.	GALAXY DWRF SP
ZWG 154.008	10 21 30.	+ 28 17		15.4	GALAXY
RNGC 3232	10 21 30.	+ 28 17		15.5	GALAXY
ZWG 240.036	10 21 30.	+ 47 05		15.5	GALAXY
MCG+08-19-025	10 21 30.	+ 47 05	48	16.	GALAXY
ZWG 313.016	10 21 30.	+ 67 32		15.7	GALAXY
MCG-03-27-007	10 21 30.	- 20 43	18	15.5	GALAXY
REIZ 0368	10 21 31.	+ 18 36	54	14.7	GALAXY
MCG-01-27-006	10 21 33.	- 09 00	30	15.	GALAXY
ZWG 009.019	10 21 36.	+ 00 24		15.2	GALAXY
ZWG 037.024	10 21 36.	+ 07 28		15.7	GALAXY
ZC 1021.6+1554	10 21 36.	+ 15 54	1750		CLUSTER OF GALAXIES
LB 10400	10 21 36.	+ 18 40		16.5	FAINT BLUE STAR
LB 10399	10 21 36.	+ 19 15		17.6	FAINT BLUE STAR
MCG+05-25-004	10 21 36.	+ 28 16	36	15.4	GALAXY
MCG+05-25-005	10 21 36.	+ 28 17	12	17.	GALAXY
ZWG 240.037	10 21 36.	+ 48 14		15.7	GALAXY
MCG+10-15-076	10 21 36.	+ 60 40	18	16.	GALAXY
ZC 1021.6+6522	10 21 36.	+ 65 22	670		CLUSTER OF GALAXIES
SEY 072	10 21 36.	+ 67 33 10.		15.4	FAINT GALAXY
7ZW 324	10 21 36.	+ 68 08			COMPACT GALAXY
MCG-01-27-007	10 21 36.	- 05 23	102	15.	GALAXY
MCG-03-27-009	10 21 36.	- 17 01	24	15.	GALAXY
MCG-03-27-008	10 21 36.	- 17 01	36	15.	GALAXY
REIZ 0369	10 21 37.	+ 28 17	18	15.0	GALAXY
REIZ 0371	10 21 41.	+ 14 01	18	15.1	GALAXY
REIZ 0370	10 21 41.	+ 24 31	42	14.2	GALAXY
SEY 071	10 21 41.	+ 70 20 04.		15.1	FAINT GALAXY
ZWG 065.024	10 21 42.	+ 13 55		15.1	GALAXY
MCG+03-27-019	10 21 42.	+ 16 00	36	15.4	GALAXY
ZWG 094.031	10 21 42.	+ 16 01		15.4	GALAXY
LB 10402	10 21 42.	+ 16 11		15.6	FAINT BLUE STAR
REIZ 0372	10 21 42.	+ 16 58	72	14.6	GALAXY
LB 10401	10 21 42.	+ 17 44		16.1	FAINT BLUE STAR
MCG+04-25-015	10 21 42.	+ 24 29	30	15.4	GALAXY
ZWG 124.019	10 21 42.	+ 24 30		15.4	GALAXY

OBJECT NAME	RIGHT ASCEN.	DECLINATION	DIAM.	MAGN.	TYPE OF OBJECT
ZC 1021.7+2916	10 21 42.	+ 29 16	2350		CLUSTER OF GALAXIES
ZWG 240.038	10 21 42.	+ 47 15		15.6	GALAXY
ZWG 240.039	10 21 42.	+ 48 42		15.6	GALAXY
MCG+10-15-077	10 21 42.	+ 58 23 30.	108	13.7	GALAXY
ZWG 333.027	10 21 42.	+ 70 20		14.7	GALAXY
UGC 05630	10 21 42.	+ 70 20	84	14.7	GALAXY DBL SYS
MCG+12-10-033	10 21 42.	+ 70 20	51	15.	GALAXY
TON-N 1217	10 21 44.	+ 22 09		17.	BLUE STAR
REIZ 0361	10 21 44.	+ 58 24	132	13.7	GALAXY
REIZ 0373	10 21 45.	+ 11 21	30	15.5	GALAXY
ARC 1000	10 21 46.	+ 50 26		17.6	RICH CLUSTER OF GALAXIES
ZWG 037.025	10 21 48.	+ 04 56		15.7	GALAXY
ZC 1021.8+1725	10 21 48.	+ 17 25	5240		CLUSTER OF GALAXIES
LB 10404	10 21 48.	+ 18 17		15.5	FAINT BLUE STAR
ZWG 094.032	10 21 48.	+ 19 09		15.6	GALAXY
LB 10403	10 21 48.	+ 19 09		16.7	FAINT BLUE STAR
ZC 1021.8+2147	10 21 48.	+ 21 47	470		CLUSTER OF GALAXIES
ZWG 240.040	10 21 48.	+ 46 43		15.7	GALAXY
KARA.73B 0412	10 21 48.	+ 46 43	24	15.7	ISOLATED GALAXY E
ZC 1021.8+5022	10 21 48.	+ 50 22	610		CLUSTER OF GALAXIES
ZWG 290.037	10 21 48.	+ 58 25		13.3	GALAXY
UGC 05631	10 21 48.	+ 58 25	138	13.3	GALAXY Sc
MCG+12-10-034	10 21 48.	+ 71 26	54	15.	GALAXY
7ZW 325	10 21 48.	+ 72 07			COMPACT GALAXY
MCG-01-27-008	10 21 48.	- 05 47	42	15.	GALAXY
MRSL 284-00/2	10 21 48.	- 57 28	540		HII REGION
KEEL 277	10 21 49.4	+ 20 23 33.		13.	SPIRAL NEBULA
RNGC 3225	10 21 51.	+ 58 25		13.5	GALAXY
IC 0608	10 21 51.	- 05 47 38.			NONSTELLAR OBJECT
KEEL 278	10 21 51.0	+ 20 17 12.		16.	NEBULA
REIZ 0374	10 21 53.	+ 24 30	36	14.9	GALAXY
ZWG 037.026	10 21 54.	+ 08 24		15.5	GALAXY
LB 10405	10 21 54.	+ 15 22		15.5	FAINT BLUE STAR
ZWG 094.033	10 21 54.	+ 16 57		15.7	GALAXY
ZWG 094.034	10 21 54.	+ 20 22		15.5	GALAXY
UGC 05632	10 21 54.	+ 20 22	60	15.5	GALAXY SBb
ZWG 183.021	10 21 54.	+ 34 54		15.3	GALAXY
ZC 1021.9+6811	10 21 54.	+ 68 11	1140		CLUSTER OF GALAXIES
KEEL 279	10 21 54.5	+ 20 32 29.		14.	NEBULA
RNGC 3210	10 21 55.	+ 80 06			NON-EXISTENT OBJECT
TON-N 1218	10 21 56.	+ 21 15		16.8	BLUE STAR
SHB 156	10 21 56.	- 00 36 48.		18.5	QUASI-STELLAR OBJECT
ARC 1002	10 21 57.	+ 50 07		17.6	RICH CLUSTER OF GALAXIES
ARC 1003	10 21 59.	+ 48 03		16.6	RICH CLUSTER OF GALAXIES
VDB.66G 079	10 22	+ 15 01	70		DWARF GALAXY
ZWG 009.021	10 22 00.	+ 01 21		15.3	GALAXY
KARA.73B 0413	10 22 00.	+ 01 21	18	15.3	ISOLATED GALAXY E
ZWG 065.025	10 22 00.	+ 11 21		15.6	GALAXY
KARA.73B 0414	10 22 00.	+ 11 21	24	15.6	ISOLATED GALAXY S
ZWG 094.035	10 22 00.	+ 15 00		15.4	GALAXY
UGC 05633	10 22 00.	+ 15 00	156	15.4	GALAXY SB IV-V
MCG+03-27-020	10 22 00.	+ 15 00	180	15.4	GALAXY
MCG+03-27-021	10 22 00.	+ 15 59	30	15.3	GALAXY
ZWG 094.036	10 22 00.	+ 16 00		15.3	GALAXY
LB 10407	10 22 00.	+ 16 59		17.2	FAINT BLUE STAR
LB 10406	10 22 00.	+ 19 28		17.5	FAINT BLUE STAR
MCG+03-27-022	10 22 00.	+ 20 25	66	15.5	GALAXY
ZWG 154.009	10 22 00.	+ 32 17		15.7	GALAXY
MCG+05-25-006	10 22 00.	+ 32 18	24	15.7	GALAXY
ZWG 212.006	10 22 00.	+ 44 15		15.6	GALAXY
ZWG 211.055	10 22 00.	+ 44 15		15.6	GALAXY
MCG+08-19-026	10 22 00.	+ 48 05	6	17.	GALAXY
ZC 1022.0+5120	10 22 00.	+ 51 20	2020		CLUSTER OF GALAXIES
ZWG 266.043	10 22 00.	+ 52 19		15.3	GALAXY
ZWG 333.028	10 22 00.	+ 71 27		15.6	GALAXY
UGC 05634	10 22 00.	+ 71 27	78	15.6	GALAXY Sb/SBb
MCG+13-08-019	10 22 00.	+ 79 31	12	16.	GALAXY
ZWG 009.020	10 22 00.	- 00 36		15.5	GALAXY
MCG+00-27-006	10 22 00.	- 00 36	42	15.5	GALAXY
MCG-05-25-002	10 22 00.	- 32 14	60	13.	GALAXY
OCL 0815	10 22 00.	- 59 50	780	11.3	OPEN STAR CLUSTER
VHA 094	10 22 00.	- 59 50	330		OPEN STAR CLUSTER
TON-N 1220	10 22 01.	+ 32 58		15.8	BLUE STAR
BC 4C19.34	10 22 01.2	+ 19 27 40.		17.49	QUASI-STELLAR OBJECT
SHB 157	10 22 01.4	+ 19 27 35.		17.5	QUASI-STELLAR OBJECT
TON-N 1219	10 22 02.	+ 20 26		16.	BLUE STAR
HOLM 189B	10 22 03.	- 00 37	30	14.4	PART OF MULTIPLE GALAXY
SEY 073	10 22 04.	+ 70 32 57.		15.4	FAINT GALAXY
RNGC 3241	10 22 04.	- 32 12		13.5	GALAXY
KARA.73 27	10 22 05.	- 35 44	27		DWARF GALAXY
ZWG 037.027	10 22 06.	+ 06 40		15.5	GALAXY
KARA.72 235B	10 22 06.	+ 06 40	24	15.5	PART OF DOUBLE GALAXY
ZWG 037.028	10 22 06.	+ 06 41		15.1	GALAXY
KARA.72 235A	10 22 06.	+ 06 41	48	15.1	PART OF DOUBLE GALAXY
ZWG 094.037	10 22 06.	+ 15 14		15.5	GALAXY
LB 10409	10 22 06.	+ 18 50		17.4	FAINT BLUE STAR
LB 10408	10 22 06.	+ 20 24		15.6	FAINT BLUE STAR
RNGC 3234	10 22 06.	+ 27 17			NON-EXISTENT OBJECT
ZWG 154.010	10 22 06.	+ 28 17		14.7	GALAXY
RNGC 3235	10 22 06.	+ 28 17		14.5	GALAXY
UGC 05635	10 22 06.	+ 28 17	78	14.7	GALAXY E-S0
ZC 1022.1+3310	10 22 06.	+ 33 10	2550		CLUSTER OF GALAXIES
ZC 1022.1+4813	10 22 06.	+ 48 13	5980		CLUSTER OF GALAXIES
ZWG 266.044	10 22 06.	+ 55 46		15.1	GALAXY
ZWG 313.017	10 22 06.	+ 67 12		15.0	GALAXY
ZWG 333.029	10 22 06.	+ 70 33		15.7	GALAXY
ZWG 009.023	10 22 06.	- 00 37		15.6	GALAXY
MCG+00-27-007	10 22 06.	- 00 37	36	15.6	GALAXY
ZWG 009.022	10 22 06.	- 02 04		15.2	GALAXY
OCL 0807	10 22 06.	- 57 30		11.26	OPEN STAR CLUSTER
VHA 095	10 22 06.	- 57 30	90		STAR CLSTR IN NEBULOSITY
REIZ 0375	10 22 07.	+ 28 17	30	14.7	GALAXY
RNGC 3240	10 22 07.	- 21 31		14.0	GALAXY
TON-N 1221	10 22 08.	+ 19 47		16.7	BLUE STAR
ARP 181	10 22 08.	+ 80 06			PECULIAR GALAXY
MCG+09-17-064	10 22 09.	+ 55 47 30.	54	15.	GALAXY
HOLM 189A	10 22 09.	- 00 38	30	14.1	PART OF MULTIPLE GALAXY
MCG-04-25-007	10 22 09.	- 21 31 30.	60	14.	GALAXY
MCG-04-25-006	10 22 09.	- 23 18 30.	60	14.5	GALAXY
SEY 074	10 22 10.	+ 67 12 09.		14.5	FAINT GALAXY
ARC 1008	10 22 11.	- 05 07		17.7	RICH CLUSTER OF GALAXIES
MCG+02-27-010	10 22 12.	+ 09 49	12	15.1	GALAXY
MCG+02-27-009	10 22 12.	+ 09 51	36	14.2	GALAXY
ZWG 065.026	10 22 12.	+ 09 52		15.4	GALAXY
8ZW 1022+09.9	10 22 12.	+ 09 52		17.0	COMPACT GALAXY
LB 10413	10 22 12.	+ 15 44		11.0	FAINT BLUE STAR
LB 10412	10 22 12.	+ 16 06		16.0	FAINT BLUE STAR
LB 10411	10 22 12.	+ 19 46		15.4	FAINT BLUE STAR

OBJECT NAME	RIGHT ASCEN.	DECLINATION	DIAM.	MAGN.	TYPE OF OBJECT
KW 51	10 22 12.	+ 19 59	330		SEYFERT GALAXY
LB 10410	10 22 12.	+ 20 27		15.3	FAINT BLUE STAR
MCG+05-25-007	10 22 12.	+ 28 16	24	14.7	GALAXY
ZWG 154.011	10 22 12.	+ 28 21		15.6	GALAXY
UGC 05636	10 22 12.	+ 28 21	60	15.6	GALAXY Sa
TON-N 0510	10 22 12.	+ 30 28		16.8	BLUE STAR
ZC 1022.2+4123	10 22 12.	+ 41 23	2490		CLUSTER OF GALAXIES
MCG+11-13-025	10 22 12.	+ 67 13	42	15.	GALAXY
MCG-06-23-026	10 22 12.	- 36 01	36	15.	GALAXY
HOLM 190B	10 22 15.	+ 09 49	18	15.1	PART OF MULTIPLE GALAXY
HOLM 190A	10 22 15.	+ 09 51	36	14.2	PART OF MULTIPLE GALAXY
REIZ 0377	10 22 15.	+ 09 52	18	15.0	GALAXY
MCG+09-17-065	10 22 15.	+ 54 08	60	14.	GALAXY
HOLM 189C	10 22 16.	- 00 39	18	14.7	PART OF MULTIPLE GALAXY
RNGC 3218	10 22 17.	+ 74 55			NON-EXISTENT OBJECT
ZWG 037.029	10 22 18.	+ 04 06		15.0	GALAXY
MCG+01-27-007	10 22 18.	+ 04 06	60	15.0	GALAXY
LB 10414	10 22 18.	+ 20 26		15.6	FAINT BLUE STAR
MCG+05-25-008	10 22 18.	+ 28 20	54	15.6	GALAXY
MCG+08-19-027	10 22 18.	+ 45 48	48	17.	GALAXY
ZWG 290.038	10 22 18.	+ 57 24		15.3	GALAXY
ZWG 009.024	10 22 18.	- 00 38		15.5	GALAXY
MCG+00-27-008	10 22 18.	- 00 38	24	15.5	GALAXY
ARP 263	10 22 19.	+ 17 25			PECULIAR GALAXY
IC 2572	10 22 19.	+ 28 20 58.			NONSTELLAR OBJECT
REIZ 0376	10 22 19.	+ 28 21	30	15.2	GALAXY
PK261+32.1	10 22 21.44	- 18 23 16.4	45	8.8	PLANETARY NEBULA
REIN 2.108	10 22 22.68	+ 17 24 27.8			NEBULA
REIN 2.109	10 22 22.86	+ 17 25 04.5			NEBULA
ARC 1009	10 22 23.	- 05 33		17.7	RICH CLUSTER OF GALAXIES
ZWG 037.030	10 22 24.	+ 03 28		15.5	GALAXY
VV 095B	10 22 24.	+ 17 24	30	16.	INTERACTING GALAXY
VV 095A	10 22 24.	+ 17 24	120	14.	INTERACTING GALAXY
VV 095	10 22 24.	+ 17 24	240		INTERACTING GALAXY
ZWG 094.038	10 22 24.	+ 17 25		13.5	GALAXY
REIZ 0378	10 22 24.	+ 17 25	150	13.4	GALAXY
UGC 05637	10 22 24.	+ 17 25	360	13.5	GALAXY PECULR
KARA.72 236A	10 22 24.	+ 17 25	162	13.5	PART OF DOUBLE GALAXY
LB 10416	10 22 24.	+ 17 26		16.2	FAINT BLUE STAR
ZWG 094.039	10 22 24.	+ 17 32		15.3	GALAXY
MCG+03-27-023	10 22 24.	+ 17 33	39	15.3	GALAXY
LB 10415	10 22 24.	+ 17 55		15.2	FAINT BLUE STAR
ZWG 183.022	10 22 24.	+ 37 42		15.6	GALAXY
ZC 1022.4+4130	10 22 24.	+ 41 30	400		CLUSTER OF GALAXIES
MCG+08-19-028	10 22 24.	+ 47 44	24	16.	GALAXY
ZWG 240.041	10 22 24.	+ 47 45		15.6	GALAXY
ZC 1022.4+5006	10 22 24.	+ 50 06	940		CLUSTER OF GALAXIES
ARC 1004	10 22 24.	+ 51 19		17.2	RICH CLUSTER OF GALAXIES
7ZW 326	10 22 24.	+ 56 42			COMPACT GALAXY
ZC 1022.4+5651	10 22 24.	+ 56 51	1280		CLUSTER OF GALAXIES
MCG+13-08-020	10 22 24.	+ 79 34	15	16.	GALAXY
RNGC 3242	10 22 24.	- 18 23			PLANETARY NEBULA
REIZ 0379	10 22 25.	+ 17 35	60	13.2	GALAXY
REIN 2.110	10 22 26.04	+ 17 24 30.7			NEBULA
RNGC 3239	10 22 27.	+ 17 25		13.5	GALAXY
MCG+03-27-025	10 22 27.	+ 17 25	330	13.5	GALAXY
MCG+03-27-024	10 22 27.	+ 17 29	36	15.7	GALAXY
WRAY 19.22	10 22 27.7	- 57 30 46.			DIFFUSE NEBULA
8ZW 1022+08.9	10 22 30.	+ 08 54		18.2	COMPACT GALAXY
ZWG 065.027	10 22 30.	+ 13 52		15.3	GALAXY
LB 10417	10 22 30.	+ 16 18		15.9	FAINT BLUE STAR
KARA.72 236B	10 22 30.	+ 17 24	54		PART OF DOUBLE GALAXY
ZWG 094.040	10 22 30.	+ 17 28		15.7	GALAXY
ZWG 094.041	10 22 30.	+ 18 25		15.6	GALAXY
ZC 1022.5+3439	10 22 30.	+ 34 39	610		CLUSTER OF GALAXIES
MCG+07-22-002	10 22 30.	+ 40 42	30	16.	GALAXY
ZC 1022.5+4146	10 22 30.	+ 41 46	940		CLUSTER OF GALAXIES
MCG+10-15-078	10 22 30.	+ 57 24	30	16.	GALAXY
MCG-03-27-010	10 22 30.	- 17 10	60	14.5	GALAXY
MCG-06-23-027	10 22 30.	- 39 04	90	13.	GALAXY
VHE 36	10 22 30.	- 57 19	60		REFLECTION NEBULA
REIZ 0380	10 22 32.	+ 20 17	30	14.3	GALAXY
MCG+03-27-026	10 22 33.	+ 17 31	66	15.3	GALAXY
ARC 1007	10 22 33.	+ 33 11		17.5	RICH CLUSTER OF GALAXIES
MCG+06-23-014	10 22 33.	+ 37 42 30.	12	15.	GALAXY
LB 10422	10 22 36.	+ 16 07		16.0	FAINT BLUE STAR
ZWG 094.042	10 22 36.	+ 17 24		15.2	GALAXY
ZWG 094.043	10 22 36.	+ 17 30		15.1	GALAXY
LB 10421	10 22 36.	+ 18 20		17.9	FAINT BLUE STAR
LB 10420	10 22 36.	+ 19 34		13.7	FAINT BLUE STAR
LB 10419	10 22 36.	+ 19 58		15.5	FAINT BLUE STAR
ZWG 094.044	10 22 36.	+ 20 15		15.0	GALAXY
LB 10418	10 22 36.	+ 20 18		15.6	FAINT BLUE STAR
ZWG 154.012	10 22 36.	+ 26 43		15.5	GALAXY
UGC 05638	10 22 36.	+ 26 43	84	15.5	GALAXY S0
STOCK 13	10 22 36.	+ 37 37			BLUE KNOT NEAR ELLIP GLXY
ZC 1022.6+6334	10 22 36.	+ 63 34	3700		CLUSTER OF GALAXIES
TON-N 1222	10 22 37.	+ 23 13		17.	BLUE STAR
REIZ 0382	10 22 39.	+ 10 45	36	15.7	GALAXY
ARC 0998	10 22 39.	+ 68 13		17.5	RICH CLUSTER OF GALAXIES
REIZ 0383	10 22 41.	+ 13 46	24	15.0	GALAXY
ZWG 037.031	10 22 42.	+ 05 12		15.5	GALAXY
8ZW 1022+08.9	10 22 42.	+ 08 54		18.2	COMPACT GALAXY
ZC 1022.7+1232	10 22 42.	+ 12 32	670		CLUSTER OF GALAXIES
ZWG 065.028	10 22 42.	+ 13 47		15.5	GALAXY
MCG+03-27-027	10 22 42.	+ 17 24	54	15.2	GALAXY
LB 10423	10 22 42.	+ 17 56		15.5	FAINT BLUE STAR
MCG+03-27-028	10 22 42.	+ 20 18	48	15.0	GALAXY
ZC 1022.7+2236	10 22 42.	+ 22 36	2420		CLUSTER OF GALAXIES
ZC 1022.7+3218	10 22 42.	+ 32 18	2620		CLUSTER OF GALAXIES
ZWG 351.022	10 22 42.	+ 79 33		15.3	GALAXY
ZWG 350.053	10 22 42.	+ 79 33		15.3	GALAXY
REIZ 0384	10 22 43.	+ 17 31	48	15.5	GALAXY
ARC 1011	10 22 46.	+ 12 31		17.8	RICH CLUSTER OF GALAXIES
ZWG 009.025	10 22 48.	+ 00 54		15.3	GALAXY
8ZW 1022+10.9	10 22 48.	+ 10 52		17.1	COMPACT GALAXY
LB 10429	10 22 48.	+ 14 30		15.3	FAINT BLUE STAR
LB 10428	10 22 48.	+ 14 51		16.7	FAINT BLUE STAR
LB 10427	10 22 48.	+ 16 29		18.6	FAINT BLUE STAR
LB 10426	10 22 48.	+ 17 12		16.3	FAINT BLUE STAR
ZWG 094.045	10 22 48.	+ 17 30		15.2	GALAXY
UGC 05639	10 22 48.	+ 17 30	72	15.2	GALAXY Sc
MCG+03-27-029	10 22 48.	+ 17 31	90	15.2	GALAXY
LB 10425	10 22 48.	+ 18 16		17.7	FAINT BLUE STAR
LB 10424	10 22 48.	+ 19 15		16.5	FAINT BLUE STAR
ZC 1022.8+2209	10 22 48.	+ 22 09	1080		CLUSTER OF GALAXIES
TON-N 0511	10 22 48.	+ 31 37		14.7	BLUE STAR
TON-N 0512	10 22 48.	+ 31 53		16.0	BLUE STAR

OBJECT NAME	RIGHT ASCEN.	DECLINATION	DIAM.	MAGN.	TYPE OF OBJECT
ZWG 212.007	10 22 48.	+ 39 55		14.2	GALAXY
UGC 05640	10 22 48.	+ 39 55	66	14.2	GALAXY S0
ZC 1022.8+6745	10 22 48.	+ 67 45	1080		CLUSTER OF GALAXIES
RNGC 3237	10 22 50.	+ 39 55		14.0	GALAXY
REIZ 0381	10 22 51.	+ 39 53	30	14.0	GALAXY
MCG+07-22-003	10 22 51.	+ 39 54	72	13.5	GALAXY
ZWG 037.032	10 22 54.	+ 07 43		15.5	GALAXY
KARA.73B 0415	10 22 54.	+ 07 43	18	15.5	ISOLATED GALAXY E
ZWG 094.046	10 22 54.	+ 17 20		15.4	GALAXY
MCG+07-22-005	10 22 54.	+ 40 19	24	16.	GALAXY
MCG+07-22-004	10 22 54.	+ 40 20	27	16.	GALAXY
ZC 1022.9+4301	10 22 54.	+ 43 01	1410		CLUSTER OF GALAXIES
MCG+07-22-006	10 22 54.	+ 44 15	42	16.	GALAXY
VDB.66G 078	10 23	+ 67 51	70		DWARF GALAXY
HOFF L01	10 23	- 59 09	220		DARK HOLE
ZWG 065.029	10 23 00.	+ 11 59		14.8	GALAXY
UGC 05642	10 23 00.	+ 11 59	102	14.8	GALAXY Sb-c
MCG+02-27-011	10 23 00.	+ 11 59	102	14.8	GALAXY
KARA.73B 0416	10 23 00.	+ 11 59	120	14.8	ISOLATED GALAXY S
LB 10430	10 23 00.	+ 15 22		15.2	FAINT BLUE STAR
ZWG 094.047	10 23 00.	+ 15 27		15.4	GALAXY
MCG+03-27-030	10 23 00.	+ 17 20	12	15.4	GALAXY
MCG+05-25-009	10 23 00.	+ 26 49	48	14.6	GALAXY
ZWG 154.013	10 23 00.	+ 26 50		14.6	GALAXY
ZWG 183.023	10 23 00.	+ 36 12		15.7	GALAXY
MCG+06-23-015	10 23 00.	+ 36 13	12	15.	GALAXY
ZWG 290.039	10 23 00.	+ 57 33		15.1	GALAXY
MCG+10-15-079	10 23 00.	+ 57 33	18	15.	GALAXY
MCG+12-10-035	10 23 00.	+ 71 40	78	16.	GALAXY
7ZW 327	10 23 00.	+ 79 57			COMPACT GALAXY
ZWG 351.023	10 23 00.	+ 80 03		14.3	GALAXY
ZWG 350.054	10 23 00.	+ 80 03		14.3	GALAXY
UGC 05643	10 23 00.	+ 80 03	102	14.3	GALAXY SB
KARA.72 237A	10 23 00.	+ 80 03	54	14.3	PART OF DOUBLE GALAXY
ZWG 009.026	10 23 00.	- 01 58		14.4	GALAXY
UGC 05641	10 23 00.	- 01 58	96	14.4	GALAXY SBb
MCG+00-27-009	10 23 00.	- 01 58	90	14.4	GALAXY
MCG-02-27-001	10 23 00.	- 15 06 30.	84	15.	GALAXY
TON-N 1224	10 23 01.	+ 22 51		16.	BLUE STAR
TON-N 1223	10 23 02.	+ 20 03		16.9	BLUE STAR
RNGC 3212	10 23 03.	+ 80 03		14.5	GALAXY
IC 0609	10 23 03.	- 01 57 42.			NONSTELLAR OBJECT
REIZ 0388	10 23 04.	+ 12 00	60	14.4	GALAXY
REIZ 0389	10 23 05.	+ 13 59	24	15.0	GALAXY
HOLM 191A	10 23 05.	+ 13 59	42	13.7	PART OF MULTIPLE GALAXY
REIZ 0390	10 23 05.	+ 14 00	30	15.1	GALAXY
RNGC 3231	10 23 05.	+ 67 04			NON-EXISTENT OBJECT
ARP 044	10 23 05.	- 01 57			PECULIAR GALAXY
ZWG 009.028	10 23 06.	+ 01 00		15.0	GALAXY
MCG+00-27-010	10 23 06.	+ 01 00	36	15.0	GALAXY
ZWG 009.029	10 23 06.	+ 01 41		15.4	GALAXY
ZWG 037.033	10 23 06.	+ 05 54		15.7	GALAXY
ZWG 065.030	10 23 06.	+ 13 58		14.6	GALAXY
SCH 08	10 23 06.	+ 13 58		14.6	PECULIAR GALAXY
UGC 05644	10 23 06.	+ 13 58	66	14.6	GALAXY S
MCG+02-27-012	10 23 06.	+ 13 58	60	14.6	GALAXY
MCG+02-27-013	10 23 06.	+ 13 59	36	14.1	GALAXY
MCG+02-27-015	10 23 06.	+ 14 00	12	17.	GALAXY
MCG+02-27-014	10 23 06.	+ 14 00	30	15.	GALAXY
LB 10431	10 23 06.	+ 16 42		16.4	FAINT BLUE STAR
ZC 1023.1+6715	10 23 06.	+ 67 15	1680		CLUSTER OF GALAXIES
MCG+12-10-036	10 23 06.	+ 68 54	51	17.	GALAXY
UGC 05645	10 23 06.	+ 71 41	72	16.0	GALAXY SBb
VV 354B	10 23 06.	- 01 58	18	18.	INTERACTING GALAXY
VV 354A	10 23 06.	- 01 58	96	14.	INTERACTING GALAXY
ZWG 009.027	10 23 06.	- 03 20		15.6	GALAXY
HOLM 191B	10 23 07.	+ 14 00	30	14.1	PART OF MULTIPLE GALAXY
KEEL 280	10 23 07.1	+ 68 54 40.			NEBULA
REIZ 0391	10 23 09.	+ 10 39	18	16.0	GALAXY
HN 0343	10 23 10.	- 32 23			NEBULA
IC 2575	10 23 10.	- 32 23			NONSTELLAR OBJECT
HOLM 192B	10 23 11.	+ 14 36	18	14.1	PART OF MULTIPLE GALAXY
REIZ 0392	10 23 11.	+ 14 38	72	14.0	GALAXY
HOLM 192A	10 23 11.	+ 14 38	120	13.1	PART OF MULTIPLE GALAXY
ZWG 065.031	10 23 12.	+ 12 58		15.3	GALAXY
ZWG 094.048	10 23 12.	+ 14 37		14.2	GALAXY
UGC 05646	10 23 12.	+ 14 37	150	14.2	GALAXY S
LB 10435	10 23 12.	+ 16 54		15.7	FAINT BLUE STAR
LB 10434	10 23 12.	+ 17 12		16.5	FAINT BLUE STAR
LB 10433	10 23 12.	+ 17 56		15.6	FAINT BLUE STAR
LB 10432	10 23 12.	+ 20 18		16.4	FAINT BLUE STAR
ZC 1023.2+3921	10 23 12.	+ 39 21	740		CLUSTER OF GALAXIES
MCG+08-19-029	10 23 12.	+ 45 49	18	16.	GALAXY
ZC 1023.2+5543	10 23 12.	+ 55 43	2220		CLUSTER OF GALAXIES
ZWG 009.030	10 23 12.	- 03 22		15.5	GALAXY
REIZ 0387	10 23 13.	+ 37 43	48	14.3	GALAXY
REIZ 0393	10 23 14.	+ 20 15	12	14.6	GALAXY
MCG+03-27-031	10 23 15.	+ 14 35	180	14.2	GALAXY
8ZW 1023+08.9	10 23 18.	+ 08 56		18.1	COMPACT GALAXY
ZC 1023.3+1257	10 23 18.	+ 12 57	940		CLUSTER OF GALAXIES
ZWG 094.049	10 23 18.	+ 19 35		15.3	GALAXY
ZWG 094.050	10 23 18.	+ 20 13		15.6	GALAXY
ZC 1023.3+3135	10 23 18.	+ 31 35	1340		CLUSTER OF GALAXIES
ZWG 183.024	10 23 18.	+ 37 43		15.1	GALAXY
ARC 1010	10 23 18.	+ 39 19		17.2	RICH CLUSTER OF GALAXIES
ZWG 240.042	10 23 18.	+ 45 50		15.1	GALAXY
MCG+13-08-022	10 23 18.	+ 80 05	66	14.	GALAXY
MCG+13-08-021	10 23 18.	+ 80 05	84	15.	GALAXY
VV 319A	10 23 18.	+ 80 06	66	13.5	INTERACTING GALAXY
ZC 1023.3-0020	10 23 18.	- 00 20	1280		CLUSTER OF GALAXIES
RNGC 3244	10 23 18.	- 39 35			GALAXY
DV.56 N3256A	10 23 18.	- 43 32	84		S GALAXY
ARC 1012	10 23 19.	+ 31 32		17.8	RICH CLUSTER OF GALAXIES
REIZ 0394	10 23 20.	+ 19 34	48	14.5	GALAXY
MCG-02-27-002	10 23 21.	- 11 11 30.	24	15.	GALAXY
ARC 1013	10 23 23.	- 05 59		17.7	RICH CLUSTER OF GALAXIES
ZWG 037.034	10 23 24.	+ 05 44		15.3	GALAXY
REIZ 0396	10 23 24.	+ 05 44	24	15.2	GALAXY
MCG+01-27-007A	10 23 24.	+ 05 44	36	15.3	GALAXY
LB 10439	10 23 24.	+ 16 28		15.0	FAINT BLUE STAR
LB 10438	10 23 24.	+ 16 44		15.5	FAINT BLUE STAR
LB 10437	10 23 24.	+ 18 29		16.3	FAINT BLUE STAR
MCG+03-27-032	10 23 24.	+ 19 36	45	15.3	GALAXY
LB 10436	10 23 24.	+ 19 46		16.0	FAINT BLUE STAR
MCG+05-25-011	10 23 24.	+ 28 08	72	15.7	GALAXY
ZWG 154.014	10 23 24.	+ 28 09		15.7	GALAXY
UGC 05647	10 23 24.	+ 28 09	72	15.7	GALAXY S
MCG+05-25-010	10 23 24.	+ 32 21	15	15.2	GALAXY

OBJECT NAME	RIGHT ASCEN.	DECLINATION	DIAM.	MAGN.	TYPE OF OBJECT
ZWG 154.015	10 23 24.	+ 32 22		15.2	GALAXY
MCG+06-23-016	10 23 24.	+ 37 43 30.	30	15.	GALAXY
ZWG 290.040	10 23 24.	+ 61 32		15.3	GALAXY
RNGC 3236	10 23 24.	+ 61 32		15.5	GALAXY
ZWG 313.018	10 23 24.	+ 62 37		15.6	GALAXY
KARA.73B 0417	10 23 24.	+ 62 37	24	15.6	ISOLATED GALAXY S0
MCG-02-27-003	10 23 24.	- 11 11 30.	15	15.	GALAXY
REIZ 0385	10 23 25.	+ 61 31	12	15.3	GALAXY
RNGC 3238	10 23 28.	+ 57 29		14.0	GALAXY
REIZ 0386	10 23 29.	+ 57 29	18	14.7	GALAXY
ZWG 037.035	10 23 30.	+ 04 37		15.0	GALAXY
UGC 05648	10 23 30.	+ 04 37	84	15.0	GALAXY S
MCG+01-27-008	10 23 30.	+ 04 37	60	15.0	GALAXY
LB 10442	10 23 30.	+ 14 48		11.7	FAINT BLUE STAR
ZC 1023.5+1835	10 23 30.	+ 18 35	1950		CLUSTER OF GALAXIES
LB 10441	10 23 30.	+ 19 36		19.0	FAINT BLUE STAR
LB 10440	10 23 30.	+ 19 57		15.6	FAINT BLUE STAR
ZC 1023.5+2803	10 23 30.	+ 28 03	1480		CLUSTER OF GALAXIES
TON-N 0513	10 23 30.	+ 31 46		16.7	BLUE STAR
ZC 1023.5+4005	10 23 30.	+ 40 05	1080		CLUSTER OF GALAXIES
MCG+08-19-030	10 23 30.	+ 47 24 30.	18	16.	GALAXY
MCG+10-15-080	10 23 30.	+ 57 28	27	14.7	GALAXY
ZWG 290.041	10 23 30.	+ 57 29		14.1	GALAXY
UGC 05649	10 23 30.	+ 57 29	78	14.1	GALAXY E-S0
REIZ 0395	10 23 31.	+ 28 08	54	14.8	GALAXY
REIZ 0397	10 23 32.	+ 19 33	12	15.4	GALAXY
REIZ 0398	10 23 33.	+ 21 41	12	14.9	GALAXY
ZC 1023.6+0133	10 23 36.	+ 01 33	4300		CLUSTER OF GALAXIES
REIZ 0400	10 23 36.	+ 04 38	24	15.2	GALAXY
LB 10444	10 23 36.	+ 17 30		17.7	FAINT BLUE STAR
LB 10443	10 23 36.	+ 19 26		15.8	FAINT BLUE STAR
ZC 1023.6+2624	10 23 36.	+ 26 24	1010		CLUSTER OF GALAXIES
MCG+08-19-031	10 23 36.	+ 45 29	66	17.	GALAXY
UGC 05650	10 23 36.	+ 45 30	66	16.5	GALAXY Sb
MCG+10-15-081	10 23 36.	+ 61 31	18	15.3	GALAXY
ZWG 009.031	10 23 36.	- 03 09		15.0	GALAXY
MCG+00-27-011	10 23 36.	- 03 09	48	15.0	GALAXY
MCG-05-25-003	10 23 36.	- 32 38	42	15.	GALAXY
TON-N 1225	10 23 38.	+ 20 02		17.	BLUE STAR
REIZ 0399	10 23 38.	+ 30 41	30	15.5	GALAXY
REIZ 0401	10 23 39.	+ 11 11	24	15.4	GALAXY
ZWG 065.032	10 23 42.	+ 11 10		15.6	GALAXY
LB 10448	10 23 42.	+ 14 55		16.0	FAINT BLUE STAR
ZWG 094.051	10 23 42.	+ 17 46		14.4	GALAXY
UGC 05651	10 23 42.	+ 17 46	90	14.4	GALAXY Sc
LB 10447	10 23 42.	+ 18 58		15.3	FAINT BLUE STAR
LB 10446	10 23 42.	+ 19 36		16.6	FAINT BLUE STAR
LB 10445	10 23 42.	+ 20 08		15.4	FAINT BLUE STAR
ZC 1023.7+3014	10 23 42.	+ 30 14	3090		CLUSTER OF GALAXIES
MCG+10-15-082	10 23 42.	+ 58 12	18	16.	GALAXY
MRK 143	10 23 42.	+ 62 38	12	16.	GALAXY WITH UV CONTINUUM
ZC 1023.7+6826	10 23 42.	+ 68 26	810		CLUSTER OF GALAXIES
VV 319B	10 23 42.	+ 80 05	42	14.5	INTERACTING GALAXY
SER 077.04	10 23 42.	- 43 30	40	16.	LOW SURFACE BRIGHT. GALXY
REIZ 0402	10 23 43.	+ 17 46	42	14.4	GALAXY
TON-N 1227	10 23 43.	+ 23 22		16.8	BLUE STAR
TON-N 1226	10 23 44.	+ 19 02		16.9	BLUE STAR
REIZ 0403	10 23 44.	+ 20 30	120	13.6	GALAXY
RNGC 3256A	10 23 44.	- 43 29			GALAXY
IC 0610	10 23 44.6	+ 20 28 57.			SAME AS IC 611
MCG+03-27-033	10 23 45.	+ 17 46	96	14.4	GALAXY
MCG+03-27-034	10 23 45.	+ 20 28	108	14.8	GALAXY
IC 0611	10 23 45.0	+ 20 28 54.			GALAXY SAbc
ARC 1005	10 23 46.	+ 68 29		17.5	RICH CLUSTER OF GALAXIES
HN 0344	10 23 46.	- 32 39			NEBULA
IC 2576	10 23 46.	- 32 39			NONSTELLAR OBJECT
ZWG 009.034	10 23 48.	+ 00 56		15.3	GALAXY
ZC 1023.8+1056	10 23 48.	+ 10 56	4030		CLUSTER OF GALAXIES
ZWG 065.033	10 23 48.	+ 12 48		15.6	GALAXY
LB 10450	10 23 48.	+ 14 29		15.3	FAINT BLUE STAR
MCG+03-27-035	10 23 48.	+ 15 35	66	15.5	GALAXY
LB 10449	10 23 48.	+ 18 06		16.8	FAINT BLUE STAR
ZWG 124.020	10 23 48.	+ 20 29		14.8	GALAXY
ZWG 094.052	10 23 48.	+ 20 29		14.8	GALAXY
UGC 05653	10 23 48.	+ 20 29	114	14.8	GALAXY Sb
ZWG 240.043	10 23 48.	+ 47 20		15.6	GALAXY
MRK 032	10 23 48.	+ 56 31	13	16.	GALAXY WITH UV CONTINUUM
7ZW 328	10 23 48.	+ 72 32			COMPACT GALAXY
ZWG 009.033	10 23 48.	- 02 22		14.0	GALAXY
UGC 05652	10 23 48.	- 02 22	102	14.0	GALAXY E-S0
MCG+00-27-012	10 23 48.	- 02 22	27	14.0	GALAXY
ZWG 009.032	10 23 48.	- 03 14		15.7	GALAXY
MCG-03-27-011	10 23 48.	- 19 58	24	15.5	GALAXY
TON-N 1228	10 23 50.	+ 19 30		17.	BLUE STAR
RNGC 3243	10 23 50.	- 02 22		14.0	GALAXY
ZWG 037.036	10 23 54.	+ 04 31		15.3	GALAXY
REIZ 0406	10 23 54.	+ 04 32	18	14.9	GALAXY
ZWG 065.034	10 23 54.	+ 11 12		15.6	GALAXY
ZWG 065.035	10 23 54.	+ 12 18		15.4	GALAXY
MCG+02-27-016	10 23 54.	+ 12 18	42	15.4	GALAXY
LB 10453	10 23 54.	+ 14 59		15.7	FAINT BLUE STAR
ZWG 094.053	10 23 54.	+ 15 36		15.5	GALAXY
UGC 05654	10 23 54.	+ 15 36	72	15.5	GALAXY Sb-c
ZC 1023.9+1559	10 23 54.	+ 15 59	470		CLUSTER OF GALAXIES
MCG+03-27-036	10 23 54.	+ 16 23	60	15.1	GALAXY
ZWG 094.054	10 23 54.	+ 16 24		15.1	GALAXY
UGC 05655	10 23 54.	+ 16 24	60	15.1	GALAXY S
LB 10452	10 23 54.	+ 17 41		13.4	FAINT BLUE STAR
LB 10451	10 23 54.	+ 19 04		15.8	FAINT BLUE STAR
ZWG 183.025	10 23 54.	+ 35 11		15.1	GALAXY
UGC 05656	10 23 54.	+ 35 11	90	15.1	GALAXY Sb/SBc
MCG+07-22-007	10 23 54.	+ 42 08 30.	72	15.	GALAXY
ZWG 212.008	10 23 54.	+ 42 09		15.4	GALAXY
UGC 05657	10 23 54.	+ 42 09	102	15.4	GALAXY S
ZWG 212.009	10 23 54.	+ 44 15		15.0	GALAXY
MRK 144	10 23 54.	+ 44 15	12	15.	GALAXY WITH UV CONTINUUM
ZWG 240.044	10 23 54.	+ 47 47		15.7	GALAXY
7ZW 329	10 23 54.	+ 67 13			COMPACT GALAXY
UGC 05658	10 23 54.	+ 71 29	60	17.	GALAXY DWARF
SHB 158	10 23 55.1	+ 06 42 50.		18.3	QUASI-STELLAR OBJECT
BC 4C06.40	10 23 55.2	+ 06 43 32.		18.30	QUASI-STELLAR OBJECT
TON-N 1229	10 23 56.	+ 21 19		16.	BLUE STAR
REIZ 0405	10 23 57.	+ 21 59	18	15.2	GALAXY
RNGC 3215	10 23 57.	+ 80 02		14.0	GALAXY
REIZ 0407	10 23 58.	+ 12 18	18	15.0	GALAXY
REIZ 0404	10 23 59.	+ 35 11	60	14.1	GALAXY
HMS 1024+1039	10 24	+ 10 39			LEO GALAXY CLUSTER
ZC 1024.0+1038	10 24 00.	+ 10 38	400		CLUSTER OF GALAXIES
LB 10455	10 24 00.	+ 17 54		15.4	FAINT BLUE STAR
LB 10454	10 24 00.	+ 18 50		16.4	FAINT BLUE STAR
ZC 1024.0+1858	10 24 00.	+ 18 58	540		CLUSTER OF GALAXIES
MCG+04-25-016	10 24 00.	+ 21 58	24	15.5	GALAXY
ZWG 124.021	10 24 00.	+ 21 59		15.5	GALAXY
MCG+06-23-017	10 24 00.	+ 35 10	66	14.5	GALAXY
MCG+08-19-032	10 24 00.	+ 46 18	30	15.	GALAXY
ZWG 240.045	10 24 00.	+ 46 19		15.5	GALAXY
ZWG 266.045	10 24 00.	+ 52 14		15.4	GALAXY
ZWG 351.024	10 24 00.	+ 80 02		14.0	GALAXY
ZWG 350.055	10 24 00.	+ 80 02		14.0	GALAXY
UGC 05659	10 24 00.	+ 80 02	66	14.0	GALAXY S
KARA.72 237B	10 24 00.	+ 80 02	66	14.0	PART OF DOUBLE GALAXY
ZWG 009.036	10 24 00.	- 02 35		14.9	GALAXY
MCG+00-27-013	10 24 00.	- 02 35	78	14.9	GALAXY
ZWG 009.035	10 24 00.	- 03 25		15.7	GALAXY
ARC 1006	10 24 01.	+ 67 18		17.7	RICH CLUSTER OF GALAXIES
SEY 075	10 24 02.	+ 67 12 29.		15.3	FAINT GALAXY
MAI 053	10 24 04.	+ 71 31	47		DWARF SPHEROIDAL GALAXY
RNGC 3249	10 24 05.	- 34 42			GALAXY
RNGC 3247	10 24 05.	- 57 41		9.5	OPEN CLUSTER
ZWG 009.037	10 24 06.	+ 01 15		15.3	GALAXY
UGC 05660	10 24 06.	+ 01 15	66	15.3	GALAXY S
ZWG 037.037	10 24 06.	+ 04 07		13.8	GALAXY
RNGC 3246	10 24 06.	+ 04 07		14.0	GALAXY
REIZ 0410	10 24 06.	+ 04 07	72	13.3	GALAXY
UGC 05661	10 24 06.	+ 04 07	150	13.8	GALAXY
MCG+01-27-009	10 24 06.	+ 04 07	156	13.8	GALAXY
ZWG 037.038	10 24 06.	+ 08 26		15.2	GALAXY
ZWG 094.055	10 24 06.	+ 17 06		15.5	GALAXY
MCG+08-19-033	10 24 06.	+ 48 29	30	16.	GALAXY
MCG+12-10-037	10 24 06.	+ 71 21	27	16.	GALAXY
MCG-06-23-028	10 24 06.	- 34 41	90	13.5	GALAXY
OCL 0809	10 24 06.	- 57 41	480	10.4	OPEN STAR CLUSTER
REIZ 0411	10 24 09.	+ 10 11	18	16.0	GALAXY
REIZ 0412	10 24 09.	+ 11 11	42	15.5	GALAXY
REIZ 0413	10 24 09.	+ 11 12	18	15.8	GALAXY
REIZ 0414	10 24 09.	+ 11 16	18	15.0	GALAXY
REIZ 0408	10 24 09.	+ 21 58	36	14.5	GALAXY
REIZ 0415	10 24 10.	+ 12 18	30	16.0	GALAXY
ZWG 065.036	10 24 12.	+ 11 55		15.6	GALAXY
8ZW 1024+13.0	10 24 12.	+ 13 00		18.5	COMPACT GALAXY
LB 10460	10 24 12.	+ 14 41		15.3	FAINT BLUE STAR
LB 10459	10 24 12.	+ 15 57		15.5	FAINT BLUE STAR
ZWG 094.056	10 24 12.	+ 16 23		15.1	GALAXY
LB 10458	10 24 12.	+ 16 41		18.6	FAINT BLUE STAR
LB 10457	10 24 12.	+ 16 56		15.8	FAINT BLUE STAR
ZWG 094.057	10 24 12.	+ 19 05		15.3	GALAXY
LB 10456	10 24 12.	+ 19 44		15.0	FAINT BLUE STAR
ZWG 154.016	10 24 12.	+ 28 54		15.4	GALAXY
RNGC 3245A	10 24 12.	+ 28 54		15.5	GALAXY
UGC 05662	10 24 12.	+ 28 54	216	15.4	GALAXY SBb
TON-N 0514	10 24 12.	+ 31 47		17.1	BLUE STAR
ZC 1024.2+4228	10 24 12.	+ 42 28	870		CLUSTER OF GALAXIES
ZC 1024.2+6035	10 24 12.	+ 60 35	1210		CLUSTER OF GALAXIES
MCG-03-27-012	10 24 12.	- 19 47	42	15.5	GALAXY
REIZ 0409	10 24 13.	+ 28 54	120	15.0	GALAXY
REIZ 0416	10 24 15.	+ 10 10	12	15.1	GALAXY
MCG+05-25-012	10 24 15.	+ 28 52 30.	210	15.4	GALAXY
MCG+08-19-034	10 24 15.	+ 50 14	15	16.	GALAXY
ZWG 037.039	10 24 18.	+ 08 24		15.4	GALAXY
MCG+03-27-037	10 24 18.	+ 16 22	60	15.1	GALAXY
LB 10463	10 24 18.	+ 17 20		17.9	FAINT BLUE STAR
LB 10462	10 24 18.	+ 18 55		17.8	FAINT BLUE STAR
LB 10461	10 24 18.	+ 20 20		15.8	FAINT BLUE STAR
RNGC 3250	10 24 18.	- 39 40		12.5	GALAXY
IC 0614	10 24 20.	- 03 12 28.			NONSTELLAR OBJECT
REIZ 0417	10 24 21.	+ 11 20	48	15.4	GALAXY
MCG+02-27-017	10 24 24.	+ 10 39	24	17.	GALAXY
CHR 5	10 24 24.	+ 10 53 19.		17.7	GALAXY IN LEO CLUSTER
LB 10467	10 24 24.	+ 14 28		17.2	FAINT BLUE STAR
LB 10466	10 24 24.	+ 15 57		15.7	FAINT BLUE STAR
ZWG 094.058	10 24 24.	+ 16 16		15.2	GALAXY
MCG+03-27-038	10 24 24.	+ 16 16	45	15.2	GALAXY
ZWG 094.059	10 24 24.	+ 16 20		15.4	GALAXY
ZWG 094.060	10 24 24.	+ 18 23		15.6	GALAXY
ZWG 094.061	10 24 24.	+ 18 57		15.5	GALAXY
LB 10465	10 24 24.	+ 19 03		17.5	FAINT BLUE STAR
LB 10464	10 24 24.	+ 19 50		15.8	FAINT BLUE STAR
TON-N 0515	10 24 24.	+ 30 58		16.0	BLUE STAR
ZWG 333.030	10 24 24.	+ 71 21		15.7	GALAXY
ZWG 009.041	10 24 24.	- 00 17		15.5	GALAXY
ZWG 009.040	10 24 24.	- 02 59		15.7	GALAXY
ZWG 009.039	10 24 24.	- 03 12		14.8	GALAXY
MCG+00-27-015	10 24 24.	- 03 12	42	14.8	GALAXY
ZWG 009.038	10 24 24.	- 03 27		15.6	GALAXY
MCG+00-27-014	10 24 24.	- 03 27	30	15.6	GALAXY
REIZ 0418	10 24 26.	+ 20 46	60	14.7	GALAXY
REIZ 0420	10 24 27.	+ 10 18	12	15.9	GALAXY
ARC 1016	10 24 27.	+ 11 14		15.4	RICH CLUSTER OF GALAXIES
IC 0612	10 24 27.	+ 11 18 20.			NONSTELLAR OBJECT
MCG-03-27-013	10 24 27.	- 18 48	90	14.5	GALAXY
SN 1964M	10 24 29.	+ 20 42		18.0	SUPERNOVA
ZWG 065.037	10 24 30.	+ 10 17		15.7	GALAXY
IC 0613	10 24 30.	+ 11 15 38.			NONSTELLAR OBJECT
ZWG 065.038	10 24 30.	+ 11 16		15.1	GALAXY
MCG+02-27-018	10 24 30.	+ 11 16	15	15.1	GALAXY
ZWG 065.039	10 24 30.	+ 11 18		15.3	GALAXY
MCG+02-27-019	10 24 30.	+ 11 18	18	15.3	GALAXY
LB 10468	10 24 30.	+ 18 41		18.0	FAINT BLUE STAR
MCG+03-27-039	10 24 30.	+ 18 58	54	15.5	GALAXY
MCG+04-25-017	10 24 30.	+ 22 04	24	15.6	GALAXY
MCG+04-25-018	10 24 30.	+ 22 04 30.	30	15.6	GALAXY
ZWG 124.022	10 24 30.	+ 22 05		15.6	GALAXY
MCG+05-25-013	10 24 30.	+ 28 45	78	11.6	GALAXY
ZWG 154.017	10 24 30.	+ 28 46		11.6	GALAXY
RNGC 3245	10 24 30.	+ 28 46		12.0	GALAXY
UGC 05663	10 24 30.	+ 28 46	174	11.6	GALAXY S0
REIZ 0419	10 24 31.	+ 28 46	108	12.2	GALAXY
REIZ 0421	10 24 33.	+ 10 18	18	15.8	GALAXY
REIZ 0422	10 24 35.	+ 01 45	30	14.8	GALAXY
ZC 1024.6+0016	10 24 36.	+ 00 16	340		CLUSTER OF GALAXIES
ZWG 009.043	10 24 36.	+ 01 13		15.4	GALAXY
UGC 05664	10 24 36.	+ 01 13	60	15.4	GALAXY S
ZWG 009.044	10 24 36.	+ 01 44		15.1	GALAXY
ZWG 065.040	10 24 36.	+ 08 48		15.6	GALAXY
LB 10472	10 24 36.	+ 14 20		14.5	FAINT BLUE STAR
LB 10471	10 24 36.	+ 15 19		15.7	FAINT BLUE STAR

OBJECT NAME	RIGHT ASCEN.	DECLINATION	DIAM.	MAGN.	TYPE OF OBJECT
ZWG 094.062	10 24 36.	+ 16 18		15.1	GALAXY
MCG+03-27-041	10 24 36.	+ 16 18	84	15.1	GALAXY
ZWG 094.063	10 24 36.	+ 16 43		15.6	GALAXY
ZC 1024.6+1703	10 24 36.	+ 17 03	1480		CLUSTER OF GALAXIES
LB 10470	10 24 36.	+ 17 08		17.7	FAINT BLUE STAR
LB 10469	10 24 36.	+ 17 08		16.5	FAINT BLUE STAR
ZWG 094.064	10 24 36.	+ 18 20		15.3	GALAXY
MCG+03-27-040	10 24 36.	+ 18 21	42	15.3	GALAXY
ZWG 009.042	10 24 36.	- 03 04		15.2	GALAXY
MKW 02S	10 24 36.	- 03 04		15.2	POOR GALAXY CLUSTER
MCG-04-25-008	10 24 36.	- 23 50	48	14.5	GALAXY
CHR 3	10 24 42.	+ 10 39 00.		17.2	GALAXY IN LEO CLUSTER
ZWG 065.041	10 24 42.	+ 11 20		15.1	GALAXY
UGC 05665	10 24 42.	+ 11 20	84	15.1	GALAXY S
MCG+02-27-020	10 24 42.	+ 11 20	72	15.1	GALAXY
LB 10474	10 24 42.	+ 16 46		16.0	FAINT BLUE STAR
LB 10473	10 24 42.	+ 20 10		14.7	FAINT BLUE STAR
MCG+10-15-083	10 24 42.	+ 60 53	15	16.	GALAXY
ZWG 009.046	10 24 42.	- 02 19		15.7	GALAXY
MCG+00-27-016	10 24 42.	- 02 19	48	15.7	GALAXY
ZWG 009.045	10 24 42.	- 03 05		15.6	GALAXY
MCG-06-23-029	10 24 42.	- 35 58	120	14.5	GALAXY
OCL 0817	10 24 42.	- 60 25	480	13.4	OPEN STAR CLUSTER
VHA 096	10 24 42.	- 60 25	180		OPEN STAR CLUSTER
RNGC 3255	10 24 43.	- 60 25		12.5	OPEN CLUSTER
IC 0615	10 24 44.	+ 11 19 49.			NONSTELLAR OBJECT
HOLM 193C	10 24 47.	+ 01 30	24	14.8	PART OF MULTIPLE GALAXY
ZWG 009.048	10 24 48.	+ 01 30		15.4	GALAXY
MCG+00-27-017	10 24 48.	+ 01 30	18	15.4	GALAXY
LB 10480	10 24 48.	+ 15 23		15.1	FAINT BLUE STAR
LB 10479	10 24 48.	+ 16 02		15.8	FAINT BLUE STAR
ZWG 094.065	10 24 48.	+ 17 12		15.3	GALAXY
LB 10478	10 24 48.	+ 18 08		15.5	FAINT BLUE STAR
LB 10477	10 24 48.	+ 18 50		17.9	FAINT BLUE STAR
LB 10476	10 24 48.	+ 19 18		16.0	FAINT BLUE STAR
LB 10475	10 24 48.	+ 20 14		15.8	FAINT BLUE STAR
MCG+04-25-019	10 24 48.	+ 24 33	24	15.0	GALAXY
ZWG 124.023	10 24 48.	+ 24 43		15.0	GALAXY
ZC 1024.8+3100	10 24 48.	+ 31 00	2220		CLUSTER OF GALAXIES
ZWG 240.046	10 24 48.	+ 46 16		15.4	GALAXY
ZWG 333.031	10 24 48.	+ 68 40		11.2	GALAXY
UGC 05666	10 24 48.	+ 68 40	900	11.2	GALAXY DWRF SP
ZWG 009.047	10 24 48.	- 02 56		15.3	GALAXY
REIZ 0424	10 24 52.	+ 01 30	30	15.5	GALAXY
HOLM 193B	10 24 52.	+ 01 30	18	14.8	PART OF MULTIPLE GALAXY
HOLM 193A	10 24 52.	+ 01 31	12	14.4	PART OF MULTIPLE GALAXY
MAI 054	10 24 52.	+ 67 08	67		DWARF SPHEROIDAL GALAXY
ARC 1015	10 24 53.	+ 34 48		17.5	RICH CLUSTER OF GALAXIES
MCG+00-27-019	10 24 54.	+ 01 30	48	14.8	GALAXY
ZWG 009.049	10 24 54.	+ 01 31		14.9	GALAXY
UGC 05667	10 24 54.	+ 01 31	90	14.9	GALAXY DBL SYS
MCG+00-27-018	10 24 54.	+ 01 31	36	14.9	GALAXY
ZWG 037.040	10 24 54.	+ 03 09		15.5	GALAXY
ZWG 065.042	10 24 54.	+ 10 09		15.4	GALAXY
LB 10481	10 24 54.	+ 17 52		16.4	FAINT BLUE STAR
ZC 1024.9+2254	10 24 54.	+ 22 54	470		CLUSTER OF GALAXIES
FATH 1.343	10 24 54.	+ 30 07	8		NEBULA
ZC 1024.9+3450	10 24 54.	+ 34 50	1010		CLUSTER OF GALAXIES
MCG+08-19-035	10 24 54.	+ 46 16	30	15.	GALAXY
UGC 05668	10 24 54.	+ 50 15	78	16.5	GALAXY Sc
IC 2574	10 24 54.	+ 68 42 46.	840	10.9	GALAXY SAB(s)
MRSL 284-00/1	10 24 54.	- 56 56	900		HII REGION
REIZ 0426	10 24 58.	+ 01 30	36	15.2	GALAXY
REIZ 0427	10 24 58.	+ 01 31	30	15.0	GALAXY
ZWG 037.041	10 25 00.	+ 04 43		15.3	GALAXY
REIZ 0428	10 25 00.	+ 04 43	12	15.3	GALAXY
MCG+01-27-010	10 25 00.	+ 04 43	36	15.3	GALAXY
ZC 1025.0+1141	10 25 00.	+ 11 41	740		CLUSTER OF GALAXIES
ZC 1025.0+1542	10 25 00.	+ 15 42	1550		CLUSTER OF GALAXIES
ZC 1025.0+1910	10 25 00.	+ 19 10	1950		CLUSTER OF GALAXIES
LB 10483	10 25 00.	+ 19 44		16.5	FAINT BLUE STAR
MCG+03-27-042	10 25 00.	+ 19 44	18	16.	GALAXY
LB 10482	10 25 00.	+ 19 56		15.9	FAINT BLUE STAR
ZWG 124.024	10 25 00.	+ 23 06		13.9	GALAXY
UGC 05669	10 25 00.	+ 23 06	150	13.9	GALAXY S0
MCG+04-25-020	10 25 00.	+ 23 06	30	13.9	GALAXY
UGC 05670	10 25 00.	+ 27 23	84	17.	GALAXY Sb-c
MCG+07-22-008	10 25 00.	+ 43 53	27	17.	GALAXY
ZWG 290.042	10 25 00.	+ 60 01		15.2	GALAXY
MCG+12-10-038	10 25 00.	+ 68 41	720	13.	GALAXY
7ZW 330	10 25 00.	+ 68 43			COMPACT GALAXY
ZWG 009.051	10 25 00.	- 01 04		15.4	GALAXY
ZWG 009.050	10 25 00.	- 03 26		15.7	GALAXY
MCG-02-27-004	10 25 00.	- 10 28 30.	36	15.5	GALAXY
MCG-04-25-009	10 25 00.	- 25 17 30.	36	15.	GALAXY
RNGC 3248	10 25 01.	+ 23 06		14.0	GALAXY
REIZ 0423	10 25 01.	+ 37 23	24	15.5	GALAXY
TON-N 1230	10 25 02.	+ 21 49		16.9	BLUE STAR
REIZ 0429	10 25 03.	+ 11 07	24	15.7	GALAXY
REIZ 0425	10 25 03.	+ 23 06	36	13.3	GALAXY
MAI 055	10 25 04.	+ 63 38	47		DWARF SPHEROIDAL GALAXY
LB 10486	10 25 06.	+ 16 58		17.6	FAINT BLUE STAR
ZWG 094.066	10 25 06.	+ 17 48		15.3	GALAXY
REIZ 0430	10 25 06.	+ 17 49	54	15.0	GALAXY
ZC 1025.1+1750	10 25 06.	+ 17 50	1080		CLUSTER OF GALAXIES
LB 10485	10 25 06.	+ 19 06		16.5	FAINT BLUE STAR
LB 10484	10 25 06.	+ 19 32		17.6	FAINT BLUE STAR
2ZW 046	10 25 06.	+ 22 00			COMPACT GALAXY
TON-N 0516	10 25 06.	+ 24 26		14.7	BLUE STAR
ZWG 183.026	10 25 06.	+ 33 01		15.3	GALAXY
KARA.73B 0418	10 25 06.	+ 33 01	36	15.3	ISOLATED GALAXY S
ZC 1025.1+4709	10 25 06.	+ 47 09	2150		CLUSTER OF GALAXIES
MCG+10-15-084	10 25 06.	+ 60 01	30	16.	GALAXY
TON-N 1231	10 25 08.	+ 22 07		17.	BLUE STAR
MCG+03-27-044	10 25 09.	+ 17 49	54	15.3	GALAXY
MCG+03-27-043	10 25 09.	+ 19 36	42	17.	GALAXY
HN 0345	10 25 10.	- 33 37			NEBULA
IC 2578	10 25 10.	- 33 37			NONSTELLAR OBJECT
CHR 1	10 25 11.	+ 10 41 47.		16.8	GALAXY IN LEO CLUSTER
HARO 24	10 25 11.	+ 19 45			BLUE EMISSION-LINE GALAXY
IC 2577	10 25 11.	+ 33 02 06.			NONSTELLAR OBJECT
ZC 1025.2+1050	10 25 12.	+ 10 50	2620		CLUSTER OF GALAXIES
LB 10489	10 25 12.	+ 15 42		15.8	FAINT BLUE STAR
LB 10488	10 25 12.	+ 16 10		16.1	FAINT BLUE STAR
LB 10487	10 25 12.	+ 16 16		16.1	FAINT BLUE STAR
MCG+03-27-045	10 25 12.	+ 16 24	48	15.0	GALAXY
ZWG 094.067	10 25 12.	+ 16 25		15.0	GALAXY
3ZW 061	10 25 12.	+ 19 45			COMPACT GALAXY

OBJECT NAME	RIGHT ASCEN.	DECLINATION	DIAM.	MAGN.	TYPE OF OBJECT
2ZW 047	10 25 12.	+ 19 45			COMPACT GALAXY
ZWG 094.068	10 25 12.	+ 19 45		15.5	GALAXY
MCG+06-23-018	10 25 12.	+ 33 01 30.	30	15.	GALAXY
ZC 1025.2+4537	10 25 12.	+ 45 37	1210		CLUSTER OF GALAXIES
KARA.68 069	10 25 12.	+ 67 04	74		DWARF GALAXY
UGC 05671	10 25 12.	+ 67 05	90	17.	GALAXY DWARF
MCG-03-27-014	10 25 12.	- 18 33	36	15.	GALAXY
CHR 4	10 25 13.	+ 10 48 41.		17.5	GALAXY IN LEO CLUSTER
TON-N 1232	10 25 13.	+ 23 39		15.8	BLUE STAR
ARC 1020	10 25 14.	+ 10 40		16.0	RICH CLUSTER OF GALAXIES
CHR 2	10 25 14.	+ 10 41 35.		17.0	GALAXY IN LEO CLUSTER
ARC 1018	10 25 15.	+ 17 49		17.6	RICH CLUSTER OF GALAXIES
ZWG 266.046	10 25 18.	+ 54 07		14.9	GALAXY
TON-N 1233	10 25 20.	+ 19 14		17.	BLUE STAR
ZC 1025.4+0558	10 25 24.	+ 05 58	2890		CLUSTER OF GALAXIES
LB 10491	10 25 24.	+ 15 28		15.3	FAINT BLUE STAR
ZWG 094.069	10 25 24.	+ 17 43		15.4	GALAXY
LB 10490	10 25 24.	+ 18 45		17.3	FAINT BLUE STAR
ZWG 124.025	10 25 24.	+ 21 02		15.7	GALAXY
KARA.73B 0419	10 25 24.	+ 21 02	24	15.7	ISOLATED GALAXY S
TON-N 0517	10 25 24.	+ 29 35		13.6	BLUE STAR
ZC 1025.4+3312	10 25 24.	+ 33 12	1010		CLUSTER OF GALAXIES
MCG+12-10-039	10 25 24.	+ 69 52	33	17.	GALAXY
MCG+13-08-023	10 25 24.	+ 79 03	6	17.	GALAXY
REIZ 0431	10 25 27.	+ 11 05	24	15.5	GALAXY
ARC 1023	10 25 28.	- 06 31		17.1	RICH CLUSTER OF GALAXIES
LB 10492	10 25 30.	+ 17 24		15.5	FAINT BLUE STAR
ZWG 094.070	10 25 30.	+ 18 52		15.4	GALAXY
ZC 1025.5+2500	10 25 30.	+ 25 00	670		CLUSTER OF GALAXIES
ZC 1025.5+3803	10 25 30.	+ 38 03	4910		CLUSTER OF GALAXIES
RNGC 3250C	10 25 30.	- 39 45			GALAXY
DV.56 N3250A	10 25 30.	- 39 49	84		S GALAXY
DV.56 N3250B	10 25 30.	- 40 10	144		S0 GALAXY
RNGC 3250B	10 25 30.	- 40 10			GALAXY
SER 077.05	10 25 30.	- 43 42			NGC 3256
OCL 0811	10 25 30.	- 57 23	960	5.4	OPEN STAR CLUSTER
VHA 097	10 25 30.	- 57 23	360		OPEN STAR CLUSTER
IC 2581	10 25 30.	- 57 23	600		OPEN CLUSTER
ARC 1014	10 25 32.	+ 65 42		17.1	RICH CLUSTER OF GALAXIES
MCG+03-27-046	10 25 33.	+ 18 52 30.	108	15.4	GALAXY
LB 10500	10 25 36.	+ 14 21		16.3	FAINT BLUE STAR
ZC 1025.6+1446	10 25 36.	+ 14 46	1810		CLUSTER OF GALAXIES
LB 10499	10 25 36.	+ 16 39		16.0	FAINT BLUE STAR
LB 10498	10 25 36.	+ 16 52		15.8	FAINT BLUE STAR
LB 10497	10 25 36.	+ 17 30		17.3	FAINT BLUE STAR
LB 10496	10 25 36.	+ 17 54		16.4	FAINT BLUE STAR
LB 10495	10 25 36.	+ 19 23		16.4	FAINT BLUE STAR
LB 10494	10 25 36.	+ 20 05		15.6	FAINT BLUE STAR
LB 10493	10 25 36.	+ 20 28		17.0	FAINT BLUE STAR
ZWG 124.026	10 25 36.	+ 22 50		14.9	GALAXY
UGC 05672	10 25 36.	+ 22 50	138	14.9	GALAXY S
MCG+04-25-021	10 25 36.	+ 22 50	72	14.9	GALAXY
ARC 1019	10 25 36.	+ 31 04		17.8	RICH CLUSTER OF GALAXIES
ZWG 240.047	10 25 36.	+ 49 20		15.3	GALAXY
ZWG 290.043	10 25 36.	+ 59 14		14.8	GALAXY
UGC 05673	10 25 36.	+ 59 14	60	14.8	GALAXY SB
ZC 1025.6+6618	10 25 36.	+ 66 18	1410		CLUSTER OF GALAXIES
MCG+12-10-040	10 25 36.	+ 69 52	21	17.	GALAXY
MCG+00-27-020	10 25 36.	- 03 05	48	14.4	GALAXY
MCG-04-25-010	10 25 36.	- 23 00	54	14.	GALAXY
REIZ 0433	10 25 39.	+ 22 50	72	13.7	GALAXY
LB 10501	10 25 42.	+ 18 34		15.9	FAINT BLUE STAR
ZC 1025.7+6542	10 25 42.	+ 65 42	1480		CLUSTER OF GALAXIES
RNGC 3250D	10 25 42.	- 39 33			GALAXY
DV.56 N3250C	10 25 42.	- 39 45	102		S0 GALAXY
RNGC 3250A	10 25 42.	- 39 49			GALAXY
VV 065	10 25 42.	- 43 39	450	12.0	INTERACTING GALAXY
ARC 1022	10 25 43.	+ 09 56		17.2	RICH CLUSTER OF GALAXIES
ARC 1024	10 25 44.	+ 04 01		17.0	RICH CLUSTER OF GALAXIES
RNGC 3256	10 25 44.	- 43 37			GALAXY
RNGC 3253	10 25 46.	+ 12 57		14.5	GALAXY
REIZ 0435	10 25 46.	+ 12 58	42	13.9	GALAXY
ZWG 037.042	10 25 48.	+ 04 01		15.5	GALAXY
MCG+01-27-011	10 25 48.	+ 04 01		15.5	GALAXY
ZWG 037.043	10 25 48.	+ 06 29		15.7	GALAXY
ZC 1025.8+0959	10 25 48.	+ 09 59	1550		CLUSTER OF GALAXIES
ZWG 065.043	10 25 48.	+ 12 57		14.4	GALAXY
SCH 09	10 25 48.	+ 12 57		14.4	PECULIAR GALAXY
UGC 05674	10 25 48.	+ 12 57	72	14.4	GALAXY Sb/SBc
MCG+02-27-021	10 25 48.	+ 12 57	72	14.4	GALAXY
UGC 05675	10 25 48.	+ 19 50	120	18.	GALAXY DWRF SP
REIZ 0434	10 25 48.	+ 37 21	30	15.2	GALAXY
ZWG 212.010	10 25 48.	+ 40 02		15.7	GALAXY
ZWG 212.011	10 25 48.	+ 40 06		15.5	GALAXY
MRK 415	10 25 48.	+ 40 06	12	16.	GALAXY WITH UV CONTINUUM
ZC 1025.8+5140	10 25 48.	+ 51 40	2690		CLUSTER OF GALAXIES
ZC 1025.8-0025	10 25 48.	- 00 25	1080		CLUSTER OF GALAXIES
DV.56 N3250D	10 25 48.	- 39 33	72		S0 GALAXY
FATH 1.344	10 25 49.	+ 29 24	8		NEBULA
MCG+03-27-047	10 25 51.	+ 19 50	102	17.	GALAXY
ARC 1021	10 25 51.	+ 37 55		16.6	RICH CLUSTER OF GALAXIES
REIZ 0437	10 25 53.	+ 02 08	15	15.7	GALAXY
ZC 1025.9+1313	10 25 54.	+ 13 13	740		CLUSTER OF GALAXIES
LB 10504	10 25 54.	+ 15 24		16.0	FAINT BLUE STAR
LB 10503	10 25 54.	+ 16 02		16.1	FAINT BLUE STAR
LB 10502	10 25 54.	+ 16 10		15.3	FAINT BLUE STAR
ZC 1025.9+3128	10 25 54.	+ 31 28	1410		CLUSTER OF GALAXIES
ZWG 212.012	10 25 54.	+ 40 05		15.6	GALAXY
ZWG 266.047	10 25 54.	+ 54 58		14.8	GALAXY
UGC 05676	10 25 54.	+ 54 58	96	14.8	GALAXY SBc/IRR
MCG+09-17-066	10 25 54.	+ 54 59	72	15.	GALAXY
MCG+10-15-085	10 25 54.	+ 59 12	39	15.	GALAXY
ZWG 313.019	10 25 54.	+ 64 34		15.4	GALAXY
MCG+12-10-041	10 25 54.	+ 69 06	51	16.	GALAXY
MCG+12-10-042	10 25 54.	+ 74 09	102	15.	GALAXY
ZWG 009.052	10 25 54.	- 02 59		15.5	GALAXY
REIZ 0439	10 25 59.	+ 02 07	12	15.7	GALAXY
HN 0346	10 25 59.	- 31 15			NEBULA
IC 2580	10 25 59.	- 31 15			NONSTELLAR OBJECT
LYNG 01	10 25 59.	- 57 26 00.	1440		OB CONCENTRATION
VDB.66G 081	10 26	+ 68 41	770		DWARF GALAXY
VDB.66G 080	10 26	+ 70 16	130		DWARF GALAXY
ZWG 037.044	10 26 00.	+ 03 49		15.2	GALAXY
UGC 05677	10 26 00.	+ 03 49	96	15.2	GALAXY
ZWG 037.045	10 26 00.	+ 04 11		15.6	GALAXY
ZWG 094.071	10 26 00.	+ 14 35		15.6	GALAXY
ZC 1026.0+2022	10 26 00.	+ 20 22	1210		CLUSTER OF GALAXIES
TON-N 0518	10 26 00.	+ 25 48		15.4	BLUE STAR

OBJECT NAME	RIGHT ASCEN.	DECLINATION	DIAM.	MAGN.	TYPE OF OBJECT
MCG+05-25-015	10 26 00.	+ 26 34	96	15.4	GALAXY
MCG+05-25-014	10 26 00.	+ 27 02	48	15.	GALAXY
ZC 1026.0+2738	10 26 00.	+ 27 38	2690		CLUSTER OF GALAXIES
MCG+07-22-009	10 26 00.	+ 39 16	72	14.5	GALAXY
ZC 1026.0+4102	10 26 00.	+ 41 02	1080		CLUSTER OF GALAXIES
ZC 1026.0+5400	10 26 00.	+ 54 00	870		CLUSTER OF GALAXIES
REIZ 0432	10 26 00.	+ 54 59	60	14.7	GALAXY
MCG+10-15-086	10 26 00.	+ 59 21	39	16.	GALAXY
MCG+10-15-087	10 26 00.	+ 60 32	39	15.	GALAXY
MCG+11-13-026	10 26 00.	+ 64 34	57	16.	GALAXY
MCG+12-10-043	10 26 00.	+ 69 20	51	15.	GALAXY
ZWG 351.025	10 26 00.	+ 75 16		15.2	GALAXY
MCG+13-08-024	10 26 00.	+ 79 11	33	16.	GALAXY
MCG-05-25-004	10 26 00.	- 31 15	96	14.	GALAXY
MCG-06-23-030	10 26 00.	- 35 11	60	15.	GALAXY
KARA.73 28	10 26 01.	- 35 23	27		DWARF GALAXY
CHR 6	10 26 04.	+ 10 41 52.		18.1	GALAXY IN LEO CLUSTER
REIZ 0438	10 26 05.	+ 26 35	54	15.1	GALAXY
RNGC 3258A	10 26 05.	- 35 12			GALAXY
ZWG 009.054	10 26 06.	+ 01 36		15.6	GALAXY
ZWG 037.046	10 26 06.	+ 03 56		15.0	GALAXY
UGC 05678	10 26 06.	+ 03 56	66	15.0	GALAXY Sb-c
MCG+01-27-012	10 26 06.	+ 03 56	66	15.0	GALAXY
ZC 1026.1+0412	10 26 06.	+ 04 12	4230		CLUSTER OF GALAXIES
LB 10506	10 26 06.	+ 17 33		14.7	FAINT BLUE STAR
LB 10505	10 26 06.	+ 20 02		15.6	FAINT BLUE STAR
ZWG 154.018	10 26 06.	+ 26 36		15.4	GALAXY
UGC 05679	10 26 06.	+ 26 36	102	15.4	GALAXY S
ZWG 266.048	10 26 06.	+ 55 26		14.6	GALAXY
ZWG 290.044	10 26 06.	+ 60 33		14.9	GALAXY
UGC 05680	10 26 06.	+ 60 33	72	14.9	GALAXY S
KARA.73B 0420	10 26 06.	+ 60 33	66	14.9	ISOLATED GALAXY S
ZWG 333.032	10 26 06.	+ 69 20		15.3	GALAXY
ZWG 351.026	10 26 06.	+ 79 10		15.6	GALAXY
ZWG 350.056	10 26 06.	+ 79 10		15.5	GALAXY
ZWG 009.053	10 26 06.	- 01 42		15.7	GALAXY
DV.56 N3258A	10 26 06.	- 35 13	66		SO GALAXY
TON-N 1235	10 26 07.	+ 34 52		16.9	BLUE STAR
TON-N 1234	10 26 08.	+ 23 14		16.6	BLUE STAR
AR 12	10 26 11.93	+ 75 15 13.2			NEBULA
ZWG 037.047	10 26 12.	+ 04 28		15.4	GALAXY
8ZW 1026+04.5	10 26 12.	+ 04 28		15.4	COMPACT GALAXY
LB 10509	10 26 12.	+ 15 35		15.2	FAINT BLUE STAR
LB 10508	10 26 12.	+ 18 38		16.3	FAINT BLUE STAR
LB 10507	10 26 12.	+ 19 38		18.3	FAINT BLUE STAR
ZWG 094.072	10 26 12.	+ 20 00		15.6	GALAXY
UGC 05681	10 26 12.	+ 20 00	78	15.6	GALAXY SBb
ZC 1026.2+2215	10 26 12.	+ 22 15	5850		CLUSTER OF GALAXIES
MCG+05-25-016	10 26 12.	+ 27 53 30.	54	15.	GALAXY
MCG+13-08-025	10 26 12.	+ 79 09	60	15.	GALAXY
REIZ 0441	10 26 14.	+ 20 01	48	14.9	GALAXY
REIZ 0442	10 26 14.	+ 20 07	18	15.1	GALAXY
MCG-03-27-015	10 26 15.	- 20 41	30	15.5	GALAXY
REIZ 0443	10 26 16.	+ 12 43	24	14.9	GALAXY
REIZ 0444	10 26 16.	+ 12 51	12	15.6	GALAXY
HOLM 194A	10 26 17.	+ 02 41	12	15.5	PART OF MULTIPLE GALAXY
HOLM 194B	10 26 18.	+ 02 40	30	15.7	PART OF MULTIPLE GALAXY
ZWG 065.044	10 26 18.	+ 12 42		15.3	GALAXY
ZWG 065.045	10 26 18.	+ 12 50		15.7	GALAXY
SCH 10	10 26 18.	+ 12 50		15.7	PECULIAR GALAXY
ZC 1026.3+1534	10 26 18.	+ 15 34	1010		CLUSTER OF GALAXIES
LB 10510	10 26 18.	+ 19 17		18.5	FAINT BLUE STAR
MCG+03-27-048	10 26 18.	+ 20 02 30.	72	15.6	GALAXY
ZWG 124.027	10 26 18.	+ 22 28		15.4	GALAXY
MCG+04-25-022	10 26 18.	+ 22 28	36	15.4	GALAXY
MCG+05-25-017	10 26 18.	+ 27 56	42	15.7	GALAXY
ZWG 154.019	10 26 18.	+ 27 58		15.7	GALAXY
MCG+10-15-088	10 26 18.	+ 57 35	18	16.	GALAXY
ZWG 351.027	10 26 18.	+ 79 07		15.5	GALAXY
ZWG 350.057	10 26 18.	+ 79 07		15.5	GALAXY
UGC 05682	10 26 18.	+ 79 07	72	15.5	GALAXY SBc
REIZ 0440	10 26 19.	+ 37 52	24	15.2	GALAXY
REIZ 0436	10 26 19.	+ 55 27	18	14.7	GALAXY
REIZ 0445	10 26 20.	+ 20 23	30	15.0	GALAXY
REIZ 0446	10 26 21.	+ 22 29	36	14.0	GALAXY
LB 10511	10 26 24.	+ 14 45		16.7	FAINT BLUE STAR
ZWG 094.073	10 26 24.	+ 20 00		15.6	GALAXY
TON-N 0519	10 26 24.	+ 25 58		15.5	BLUE STAR
MCG+12-10-044	10 26 24.	+ 70 18	234	15.	GALAXY
REIZ 0447	10 26 29.	+ 26 21	150	13.6	GALAXY
HOLM 195A	10 26 29.	+ 26 21	180	13.1	PART OF MULTIPLE GALAXY
SN 1941B	10 26 29.	+ 29 44		15.1	SUPERNOVA
RNGC 3257	10 26 29.	- 35 25			GALAXY
LB 10512	10 26 30.	+ 17 44		16.0	FAINT BLUE STAR
MCG+03-27-049	10 26 30.	+ 20 02	36	15.6	GALAXY
ZWG 124.028	10 26 30.	+ 23 59		15.5	GALAXY
UGC 05683	10 26 30.	+ 23 59	96	15.5	GALAXY DBL SYS
MCG+04-25-024	10 26 30.	+ 23 59	84	15.5	GALAXY
ZWG 124.029	10 26 30.	+ 26 21		14.2	GALAXY
UGC 05684	10 26 30.	+ 26 21	120	14.2	GALAXY SB
MCG+04-25-023	10 26 30.	+ 26 21	120	14.2	GALAXY
IC 2579	10 26 30.	+ 26 21 56.			SAME AS NGC 3251
ZWG 154.020	10 26 30.	+ 29 45		12.4	GALAXY
RNGC 3254	10 26 30.	+ 29 45		12.0	GALAXY
UGC 05685	10 26 30.	+ 29 45	318	12.4	GALAXY Sb
TON-N 0520	10 26 30.	+ 33 40		15.7	BLUE STAR
ZWG 240.048	10 26 30.	+ 50 02		15.1	GALAXY
ZC 1026.5+6006	10 26 30.	+ 60 06	1340		CLUSTER OF GALAXIES
VV 294B	10 26 30.	+ 70 18	24	17.	INTERACTING GALAXY
VV 294A	10 26 30.	+ 70 18	138	15.	INTERACTING GALAXY
VV 294	10 26 30.	+ 70 18	180		INTERACTING GALAXY
UGC 05686	10 26 30.	+ 74 29	108	17.	GALAXY DWARF
MCG+13-08-026	10 26 30.	+ 75 14	45	15.	GALAXY
MCG-06-23-031	10 26 30.	- 35 23	30	15.	GALAXY
KLEM 12	10 26 30.	- 43 58	3000		LOW LAT GRP OF 10 GLXIES
RNGC 3251	10 26 31.	+ 26 21		14.0	GALAXY
REIZ 0448	10 26 31.	+ 29 45	270	12.8	GALAXY
PATH 2.021	10 26 32.	+ 29 45	272		NEBULA
KLEM 11	10 26 32.	- 31 20	900	12.	GROUP OF 7 GALAXIES
REIZ 0449	10 26 35.	+ 26 19	18	15.3	GALAXY
ARC 1017	10 26 35.	+ 65 23		17.9	RICH CLUSTER OF GALAXIES
RNGC 3258	10 26 35.	- 35 20		13.0	GALAXY
ZWG 037.048	10 26 36.	+ 05 23		15.3	GALAXY
REIZ 0452	10 26 36.	+ 05 24	24	15.6	GALAXY
ZWG 037.049	10 26 36.	+ 06 23		14.9	GALAXY
UGC 05687	10 26 36.	+ 06 23	126	14.9	GALAXY Sc
MCG+01-27-013	10 26 36.	+ 06 23	120	14.9	GALAXY
LB 10514	10 26 36.	+ 16 57		15.6	FAINT BLUE STAR

OBJECT NAME	RIGHT ASCEN.	DECLINATION	DIAM.	MAGN.	TYPE OF OBJECT
LB 10513	10 26 36.	+ 18 14		16.1	FAINT BLUE STAR
HOLM 195B	10 26 36.	+ 26 19	18	14.7	PART OF MULTIPLE GALAXY
MCG+05-25-018	10 26 36.	+ 29 45	276	12.4	GALAXY
ZWG 333.033	10 26 36.	+ 70 19		14.6	GALAXY
UGC 05688	10 26 36.	+ 70 19	216	14.6	GALAXY
MCG+12-10-045	10 26 36.	+ 70 52	168	15.	GALAXY
ZWG 333.034	10 26 36.	+ 74 08		15.4	GALAXY
UGC 05689	10 26 36.	+ 74 08	102	15.4	GALAXY Sc
MCG-05-25-005	10 26 36.	- 31 21	72	13.5	GALAXY
MCG-06-23-032	10 26 36.	- 35 20	120	13.0	GALAXY
LB 10518	10 26 42.	+ 14 47		15.4	FAINT BLUE STAR
LB 10517	10 26 42.	+ 15 50		15.3	FAINT BLUE STAR
LB 10516	10 26 42.	+ 17 50		16.4	FAINT BLUE STAR
LB 10515	10 26 42.	+ 20 16		15.7	FAINT BLUE STAR
MCG+04-25-025	10 26 42.	+ 22 15	66	15.6	GALAXY
ZC 1026.7+3045	10 26 42.	+ 30 45	470		CLUSTER OF GALAXIES
REIZ 0450	10 26 42.	+ 36 26	30	14.5	GALAXY
HOLM 196B	10 26 42.	+ 36 26	30	14.1	PART OF MULTIPLE GALAXY
REIZ 0451	10 26 42.	+ 36 27	18	14.3	GALAXY
HOLM 196A	10 26 42.	+ 36 27	30	13.8	PART OF MULTIPLE GALAXY
ZWG 266.049	10 26 42.	+ 50 42		15.4	GALAXY
ZWG 266.050	10 26 42.	+ 52 04		15.3	GALAXY
ZWG 009.056	10 26 42.	- 01 46		15.6	GALAXY
ZWG 009.055	10 26 42.	- 02 15		15.4	GALAXY
MCG+00-27-021	10 26 42.	- 02 15	30	15.4	GALAXY
OCL 0812	10 26 42.	- 57 55	120	13.	OPEN STAR CLUSTER
REIZ 0453	10 26 43.	+ 19 54	36	14.1	GALAXY
PK285-01.1	10 26 44.4	- 58 48 03.			PLANETARY NEBULA
REIZ 0454	10 26 45.	+ 22 17	24	14.4	GALAXY
REIZ 0456	10 26 46.	+ 12 29	30	14.3	GALAXY
LB 10521	10 26 48.	+ 19 00		18.6	FAINT BLUE STAR
LB 10520	10 26 48.	+ 19 01		18.5	FAINT BLUE STAR
ZWG 094.074	10 26 48.	+ 19 52		14.5	GALAXY
UGC 05690	10 26 48.	+ 19 52	54	14.5	GALAXY SB
LB 10519	10 26 48.	+ 20 25		16.1	FAINT BLUE STAR
ZWG 124.030	10 26 48.	+ 21 15		15.6	GALAXY
UGC 05691	10 26 48.	+ 22 15	66	15.6	GALAXY S
ZWG 212.013	10 26 48.	+ 40 45		15.3	GALAXY
MCG+07-22-010	10 26 48.	+ 40 45	51	15.5	GALAXY
ZWG 240.049	10 26 48.	+ 47 33		15.7	GALAXY
ZC 1026.8+6522	10 26 48.	+ 65 22	810		CLUSTER OF GALAXIES
7ZW 331	10 26 48.	+ 70 18			COMPACT GALAXY
ZWG 333.035	10 26 48.	+ 70 53		15.1	GALAXY
UGC 05692	10 26 48.	+ 70 53	240	15.1	GALAXY DWRF SP
MCG+12-10-046	10 26 48.	+ 72 23	84	15.	GALAXY
MCG-06-23-033	10 26 48.	- 35 19	60	14.	GALAXY
RNGC 3250E	10 26 48.	- 39 50			GALAXY
SER 077.07	10 26 48.	- 44 25	138		GROUP OF THREE GALAXIES
RNGC 3261	10 26 50.	- 44 22		12.5	GALAXY
RNGC 3260	10 26 52.	- 35 18			GALAXY
LB 10522	10 26 54.	+ 17 32		16.0	FAINT BLUE STAR
MCG+03-27-050	10 26 54.	+ 19 54	63	14.5	GALAXY
ZWG 154.021	10 26 54.	+ 27 31		15.6	GALAXY
TON-N 0521	10 26 54.	+ 33 39		14.8	BLUE STAR
ZWG 212.014	10 26 54.	+ 39 16		15.4	GALAXY
UGC 05693	10 26 54.	+ 39 16	90	15.4	GALAXY Sb-c
ZC 1026.9+4023	10 26 54.	+ 40 23	8740		CLUSTER OF GALAXIES
ZC 1026.9+4625	10 26 54.	+ 46 25	1610		CLUSTER OF GALAXIES
ZWG 240.050	10 26 54.	+ 48 37		15.2	GALAXY
ZWG 266.051	10 26 54.	+ 51 06		15.5	GALAXY
ZWG 290.045	10 26 54.	+ 61 31		15.4	GALAXY
7ZW 332	10 26 54.	+ 65 34			COMPACT GALAXY
ZWG 333.036	10 26 54.	+ 73 04		15.7	GALAXY
ZWG 009.057	10 26 54.	- 03 29		15.5	GALAXY
MCG-05-25-006	10 26 54.	- 30 04	72	14.	GALAXY
HN 0347	10 26 54.	- 30 05			NEBULA
MCG-06-23-034	10 26 54.	- 35 24	24	15.	GALAXY
DV.56 N3250E	10 26 54.	- 39 50	84		Sc GALAXY
DV.56 N3256C	10 26 54.	- 43 36	72		Sd GALAXY
DV.56 N3256B	10 26 54.	- 44 08	96		S GALAXY
IC 2582	10 26 55.	- 30 06			NONSTELLAR OBJECT
REIZ 0455	10 26 56.	+ 39 17	60	14.4	GALAXY
RNGC 3256C	10 26 56.	- 43 35			GALAXY
RNGC 3256B	10 26 56.	- 44 07			GALAXY
VDB.66G 082	10 27	+ 70 51	100		DWARF GALAXY
SER 077.06	10 27	- 43 53	168		INTERACTING PAIR OF GLXYS
8ZW 1027+10.3	10 27 00.	+ 10 15		18.2	COMPACT GALAXY
LB 10523	10 27 00.	+ 19 25		16.1	FAINT BLUE STAR
ZC 1027.0+2041	10 27 00.	+ 20 41	810		CLUSTER OF GALAXIES
ZWG 124.031	10 27 00.	+ 22 37		15.3	GALAXY
MCG+04-25-026	10 27 00.	+ 22 37	36	15.3	GALAXY
ZC 1027.0+3030	10 27 00.	+ 30 30	810		CLUSTER OF GALAXIES
ZWG 212.015	10 27 00.	+ 42 23		15.5	GALAXY
MCG+12-10-047	10 27 00.	+ 74 28	51	17.	GALAXY
ZWG 009.058	10 27 00.	- 03 30		15.0	GALAXY
MCG+00-27-022	10 27 00.	- 03 30	18	15.0	GALAXY
GO 023	10 27 01.	- 59 31	31	19.0	STAR CHAIN
REIZ 0458	10 27 04.	+ 13 18	60	14.1	GALAXY
REIZ 0461	10 27 05.	+ 02 11	12	15.6	GALAXY
REIZ 0462	10 27 05.	+ 02 12	24	15.5	GALAXY
ZWG 009.059	10 27 06.	+ 02 10		15.3	GALAXY
UGC 05694	10 27 06.	+ 02 10	66	15.3	GALAXY DBL SYS
ZWG 065.046	10 27 06.	+ 13 16		14.9	GALAXY
UGC 05695	10 27 06.	+ 13 16	90	14.9	GALAXY S
MCG+02-27-022	10 27 06.	+ 13 16	66	14.9	GALAXY
LB 10525	10 27 06.	+ 15 30		12.8	FAINT BLUE STAR
ZWG 094.075	10 27 06.	+ 16 26		15.2	GALAXY
MRK 631	10 27 06.	+ 16 26	12	16.5	GALAXY WITH UV CONTINUUM
LB 10524	10 27 06.	+ 17 08		15.9	FAINT BLUE STAR
ZWG 094.076	10 27 06.	+ 20 05		14.8	GALAXY
UGC 05696	10 27 06.	+ 20 05	78	14.8	GALAXY Sb
MCG+07-22-011	10 27 06.	+ 38 43	66	15.	GALAXY
ZWG 240.051	10 27 06.	+ 47 32		15.7	GALAXY
ZWG 266.052	10 27 06.	+ 52 49		15.6	GALAXY
MCG+10-15-089	10 27 06.	+ 61 31	27	16.	GALAXY
MCG-01-27-009	10 27 06.	- 03 34	54	14.5	GALAXY
MCG-03-27-016	10 27 06.	- 17 09	60	15.	GALAXY
REIZ 0459	10 27 08.	+ 20 13	48	14.3	GALAXY
RNGC 3262	10 27 08.	- 43 55			GALAXY
8ZW 1027+04.9	10 27 12.	+ 04 52		18.9	COMPACT GALAXY
LB 10528	10 27 12.	+ 15 10		16.0	FAINT BLUE STAR
LB 10527	10 27 12.	+ 18 04		15.8	FAINT BLUE STAR
MCG+03-27-051	10 27 12.	+ 20 03	54	17.	GALAXY
LB 10526	10 27 12.	+ 20 12		15.7	FAINT BLUE STAR
ARC 1026	10 27 12.	+ 40 21		17.2	RICH CLUSTER OF GALAXIES
ZWG 212.016	10 27 12.	+ 43 37		15.4	GALAXY
MCG+07-22-012	10 27 12.	+ 43 37 30.	66	14.5	GALAXY
MCG+07-22-013	10 27 12.	+ 44 23	114	15.	GALAXY

OBJECT NAME	RIGHT ASCEN.	DECLINATION	DIAM.	MAGN.	TYPE OF OBJECT
ZWG 240.052	10 27 12.	+ 47 25		15.3	GALAXY
MCG+13-08-027	10 27 12.	+ 76 11	33	16.	GALAXY
ZWG 009.060	10 27 12.	- 01 23		15.7	GALAXY
MCG-04-25-011	10 27 12.	- 23 52	60	15.	GALAXY
MCG-06-23-035	10 27 12.	- 34 59	60	14.5	GALAXY
REIZ 0457	10 27 13.	+ 38 43	36	15.0	GALAXY
HOLM 197B	10 27 14.	+ 36 38	66	15.2	PART OF MULTIPLE GALAXY
RNGC 3263	10 27 14.	- 43 53			GALAXY
MCG+03-27-052	10 27 15.	+ 20 07	84	14.8	GALAXY
MCG+10-15-090	10 27 15.	+ 57 54	57	15.	GALAXY
TON-N 0522	10 27 18.	+ 31 27		15.3	BLUE STAR
REIZ 0460	10 27 18.	+ 36 38	30	15.4	GALAXY
HOLM 197A	10 27 18.	+ 36 38	30	14.8	PART OF MULTIPLE GALAXY
UGC 05697	10 27 18.	+ 38 43	72	16.5	GALAXY S
UGC 05698	10 27 18.	+ 44 23	114	16.0	GALAXY Sb-c
UGC 05699	10 27 18.	+ 60 16	60	16.5	GALAXY Sb-c
ZWG 333.037	10 27 18.	+ 72 22		15.2	GALAXY
UGC 05700	10 27 18.	+ 72 22	102	15.2	GALAXY Sb/SBb?
KARA.73B 0421	10 27 18.	+ 72 22	90	15.2	ISOLATED GALAXY S
OCL 0820	10 27 18.	- 60 28	240		OPEN STAR CLUSTER
LB 10529	10 27 24.	+ 16 54		17.6	FAINT BLUE STAR
ZC 1027.4+1816	10 27 24.	+ 18 16	3360		CLUSTER OF GALAXIES
ZWG 212.017	10 27 24.	+ 39 59		15.7	GALAXY
ZC 1027.4+4123	10 27 24.	+ 41 23	940		CLUSTER OF GALAXIES
MCG+07-22-014	10 27 24.	+ 43 58 30.	48	15.5	GALAXY
MCG+09-17-067	10 27 24.	+ 53 47	72	16.	GALAXY
UGC 05701	10 27 24.	+ 78 04	102	16.0	GALAXY DWARF
ZWG 009.061	10 27 24.	- 02 59		15.3	GALAXY
MCG-01-27-010	10 27 24.	- 04 43	42	15.	GALAXY
MCG-03-27-017	10 27 24.	- 16 48 30.	54	15.	GALAXY
REIZ 0463	10 27 25.	+ 38 40	30	15.4	GALAXY
REIZ 0464	10 27 26.	+ 31 10	30	15.5	GALAXY
RNGC 3267	10 27 28.	- 35 02			GALAXY
ZWG 037.050	10 27 30.	+ 04 15		15.5	GALAXY
ZWG 037.051	10 27 30.	+ 05 18		15.2	GALAXY
REIZ 0466	10 27 30.	+ 05 18	30	14.9	GALAXY
UGC 05702	10 27 30.	+ 05 18	60	15.2	GALAXY S
ZWG 037.052	10 27 30.	+ 07 29		15.7	GALAXY
ZC 1027.5+1416	10 27 30.	+ 14 16	1010		CLUSTER OF GALAXIES
ZC 1027.5+1634	10 27 30.	+ 16 34	870		CLUSTER OF GALAXIES
ZWG 266.053	10 27 30.	+ 53 46		15.6	GALAXY
UGC 05703	10 27 30.	+ 53 46	66	15.6	GALAXY Sb
MCG-02-27-005	10 27 30.	- 11 01 30.	24	16.	GALAXY
MCG-06-23-037	10 27 30.	- 34 39	60	14.	GALAXY
MCG-06-23-036	10 27 30.	- 35 03	138	13.	GALAXY
OCL 0821	10 27 30.	- 60 39	180		OPEN STAR CLUSTER
HN 0348	10 27 33.	- 34 39			NEBULA
IC 2584	10 27 33.	- 34 40	186		GALAXY S0
8ZW 1027+12.7	10 27 36.	+ 12 43		17.9	COMPACT GALAXY
MCG+03-27-053	10 27 36.	+ 15 26	24	15.0	GALAXY
LB 10536	10 27 36.	+ 16 10		15.7	FAINT BLUE STAR
LB 10535	10 27 36.	+ 17 24		18.2	FAINT BLUE STAR
LB 10534	10 27 36.	+ 17 46		15.7	FAINT BLUE STAR
LB 10533	10 27 36.	+ 18 40		16.0	FAINT BLUE STAR
LB 10532	10 27 36.	+ 19 05		16.0	FAINT BLUE STAR
LB 10531	10 27 36.	+ 19 32		18.0	FAINT BLUE STAR
LB 10530	10 27 36.	+ 19 37		15.9	FAINT BLUE STAR
UGC 05704	10 27 36.	+ 23 00	90	16.5	GALAXY Sc
ZC 1027.6+4022	10 27 36.	+ 40 22	1140		CLUSTER OF GALAXIES
ZWG 009.063	10 27 36.	- 01 25		15.6	GALAXY
ZWG 009.062	10 27 36.	- 02 55		14.4	GALAXY
MKW 02	10 27 36.	- 02 55		14.4	POOR GALAXY CLUSTER
MCG+00-27-023	10 27 36.	- 02 55	18	14.4	GALAXY
MCG-06-23-039	10 27 36.	- 34 51	24	15.	GALAXY
MCG-06-23-040	10 27 36.	- 34 56	120	13.	GALAXY
MCG-06-23-038	10 27 36.	- 38 05	66	15.	GALAXY
REIZ 0465	10 27 37.	+ 30 11	42	16.0	GALAXY
REIZ 0467	10 27 38.	+ 21 30	24	14.5	GALAXY
REIZ 0468	10 27 39.	+ 11 20	12	15.1	GALAXY
RNGC 3268	10 27 40.	- 35 04		13.0	GALAXY
REIZ 0469	10 27 41.	+ 15 28	12	14.6	GALAXY
ZWG 037.053	10 27 42.	+ 04 25		15.7	GALAXY
ZWG 037.054	10 27 42.	+ 07 28		15.3	GALAXY
ZWG 037.055	10 27 42.	+ 08 19		15.1	GALAXY
ZWG 065.047	10 27 42.	+ 12 52		15.7	GALAXY
ZWG 094.077	10 27 42.	+ 15 27		15.0	GALAXY
LB 10537	10 27 42.	+ 15 40		16.0	FAINT BLUE STAR
ZWG 154.022	10 27 42.	+ 26 49		15.7	GALAXY
ZC 1027.7+2836	10 27 42.	+ 28 36	1140		CLUSTER OF GALAXIES
SHAH 051	10 27 42.	+ 39 28	300	16.1	GROUP OF COMPACT GALAXIES
ZWG 212.018	10 27 42.	+ 44 15		15.2	GALAXY
MCG+07-22-015	10 27 42.	+ 44 16	48	15.	GALAXY
7ZW 333	10 27 42.	+ 78 04			COMPACT GALAXY
ZWG 009.064	10 27 42.	- 02 05		15.6	GALAXY
MCG-06-23-041	10 27 42.	- 35 03	90	13.0	GALAXY
ARC 1028	10 27 43.	+ 41 24		17.0	RICH CLUSTER OF GALAXIES
ARC 1032	10 27 44.	+ 04 17		15.7	RICH CLUSTER OF GALAXIES
RNGC 3269	10 27 46.	- 34 57			GALAXY
ARC 1030	10 27 47.	+ 31 16		17.8	RICH CLUSTER OF GALAXIES
RNGC 3275A	10 27 47.	- 36 26			GALAXY
LB 10539	10 27 48.	+ 16 13		15.3	FAINT BLUE STAR
LB 10538	10 27 48.	+ 19 36		17.3	FAINT BLUE STAR
ZC 1027.8+3117	10 27 48.	+ 31 17	1010		CLUSTER OF GALAXIES
MCG+09-17-068	10 27 48.	+ 51 35	36	16.	GALAXY
ZC 1027.8+5334	10 27 48.	+ 53 34	610		CLUSTER OF GALAXIES
ZWG 009.065	10 27 48.	- 02 50		15.6	GALAXY
DV.56 N3275A	10 27 48.	- 36 26			S GALAXY
KEEL 281	10 27 48.8	+ 68 44 12.			NEBULA
ARC 1027	10 27 50.	+ 53 40		17.2	RICH CLUSTER OF GALAXIES
REIZ 0470	10 27 51.	+ 22 09	36	14.4	GALAXY
ZWG 037.056	10 27 54.	+ 07 26		15.5	GALAXY
LB 10542	10 27 54.	+ 18 42		15.6	FAINT BLUE STAR
LB 10541	10 27 54.	+ 19 10		15.8	FAINT BLUE STAR
LB 10540	10 27 54.	+ 19 30		16.8	FAINT BLUE STAR
ZWG 183.027	10 27 54.	+ 37 13		15.5	GALAXY
ZC 1027.9+5826	10 27 54.	+ 58 26	1680		CLUSTER OF GALAXIES
ZC 1027.9+6150	10 27 54.	+ 61 50	5170		CLUSTER OF GALAXIES
MAI 056	10 27 58.	+ 78 08	53		DWARF SPHEROIDAL GALAXY
REIZ 0471	10 27 59.	+ 15 48	9	16.0	GALAXY
ZWG 037.057	10 28 00.	+ 07 58		15.7	GALAXY
LB 10544	10 28 00.	+ 16 53		16.4	FAINT BLUE STAR
ZC 1028.0+1720	10 28 00.	+ 17 20	540		CLUSTER OF GALAXIES
LB 10543	10 28 00.	+ 19 30		16.5	FAINT BLUE STAR
ZC 1028.0+4223	10 28 00.	+ 42 23	1680		CLUSTER OF GALAXIES
MCG+10-15-091	10 28 00.	+ 61 11	42	16.	GALAXY
MCG+13-08-028	10 28 00.	+ 78 04 30.	66	16.	GALAXY
ZWG 009.067	10 28 00.	- 02 34		15.7	GALAXY
MCG+00-27-024	10 28 00.	- 02 34	36	15.7	GALAXY
ZWG 009.066	10 28 00.	- 02 57		15.7	GALAXY
MCG-06-23-042	10 28 00.	- 34 06	60	14.5	GALAXY
TON-N 1236	10 28 01.	+ 34 03		16.5	BLUE STAR
RNGC 3258B	10 28 04.	- 35 17			GALAXY
KARA.73 29	10 28 05.	- 36 35	94		DWARF GALAXY
ZWG 037.058	10 28 06.	+ 04 28		15.6	GALAXY
ZWG 037.059	10 28 06.	+ 04 58		15.2	GALAXY
ZC 1028.1+1429	10 28 06.	+ 14 29	1010		CLUSTER OF GALAXIES
TON-N 0523	10 28 06.	+ 32 53		15.4	BLUE STAR
MCG+10-15-092	10 28 06.	+ 57 39	15	16.	GALAXY
ZWG 009.068	10 28 06.	- 01 35		15.0	GALAXY
MCG+00-27-025	10 28 06.	- 01 35	60	15.0	GALAXY
MCG-06-23-043	10 28 06.	- 34 48	36	15.	GALAXY
MCG-06-23-044	10 28 06.	- 35 05	96	12.9	GALAXY
DV.56 N3258B	10 28 06.	- 35 18	78		S GALAXY
HN 0349	10 28 09.	- 35 06			NEBULA
IC 2585	10 28 09.	- 35 07			NONSTELLAR OBJECT
SHB 159	10 28 09.8	+ 31 18 20.		16.7	QUASI-STELLAR OBJECT
ARC 1031	10 28 10.	+ 39 00		17.2	RICH CLUSTER OF GALAXIES
ARC 1025	10 28 10.	+ 63 07		16.9	RICH CLUSTER OF GALAXIES
RNGC 3271	10 28 10.	- 35 05		13.0	GALAXY
RNGC 3273	10 28 10.	- 35 20			GALAXY
REIZ 0472	10 28 12.	+ 04 59	30	15.7	GALAXY
ZC 1028.2+1540	10 28 12.	+ 15 40	740		CLUSTER OF GALAXIES
LB 10545	10 28 12.	+ 18 38		18.6	FAINT BLUE STAR
ZWG 154.023	10 28 12.	+ 29 04		14.1	GALAXY
UGC 05705	10 28 12.	+ 29 04	54	14.1	GALAXY E
ZC 1028.2+4357	10 28 12.	+ 43 57	6920		CLUSTER OF GALAXIES
MCG-05-25-007	10 28 12.	- 30 07	90	14.	GALAXY
MCG-06-23-045	10 28 12.	- 35 20	90	13.	GALAXY
TON-N 1237	10 28 14.	+ 21 27		15.9	BLUE STAR
MCG+07-22-016	10 28 15.	+ 43 24	138	13.	GALAXY
MCG+05-25-019	10 28 18.	+ 29 02	42	14.1	GALAXY
RNGC 3265	10 28 18.	+ 29 03		14.0	GALAXY
UGC 05706	10 28 18.	+ 34 46	90	18.	GALAXY DWRF IR
ZWG 212.019	10 28 18.	+ 43 23		14.7	GALAXY
UGC 05707	10 28 18.	+ 43 23	156	14.7	GALAXY SBc
ZC 1028.3+5340	10 28 18.	+ 53 40	2350		CLUSTER OF GALAXIES
7ZW 334	10 28 18.	+ 60 58			COMPACT GALAXY
ZC 1028.3+6306	10 28 18.	+ 63 06	2220		CLUSTER OF GALAXIES
MCG+12-10-048	10 28 18.	+ 69 22	33	16.	GALAXY
ZC 1028.3-0149	10 28 18.	- 01 49	740		CLUSTER OF GALAXIES
ZWG 009.069	10 28 18.	- 02 30		15.3	GALAXY
IC 2586	10 28 23.	- 28 27 47.			NONSTELLAR OBJECT
ZC 1028.4+0855	10 28 24.	+ 08 55	1610		CLUSTER OF GALAXIES
LB 10548	10 28 24.	+ 15 40		15.5	FAINT BLUE STAR
LB 10547	10 28 24.	+ 17 48		17.4	FAINT BLUE STAR
LB 10546	10 28 24.	+ 18 40		17.4	FAINT BLUE STAR
ZWG 124.032	10 28 24.	+ 26 18		15.1	GALAXY
MCG+04-25-027	10 28 24.	+ 26 18 30.	18	15.1	GALAXY
IC 2583	10 28 24.	+ 26 19 14.			NONSTELLAR OBJECT
REIZ 0473	10 28 24.	+ 29 03	18	14.6	GALAXY
MCG+07-22-017	10 28 24.	+ 43 13	30	16.	GALAXY
ZC 1028.4+5228	10 28 24.	+ 52 28	1140		CLUSTER OF GALAXIES
ZWG 009.070	10 28 24.	- 02 28		15.1	GALAXY
TON-N 1238	10 28 25.	+ 34 15		16.9	BLUE STAR
ZWG 037.060	10 28 30.	+ 04 21		15.6	GALAXY
LB 10550	10 28 30.	+ 17 16		16.5	FAINT BLUE STAR
LB 10549	10 28 30.	+ 20 14		16.0	FAINT BLUE STAR
ZWG 124.033	10 28 30.	+ 26 06		15.0	GALAXY
MCG+04-25-028	10 28 30.	+ 26 06	36	15.0	GALAXY
KARA.68 070	10 28 30.	+ 34 43	47		DWARF GALAXY
ZWG 313.020	10 28 30.	+ 62 41		15.5	GALAXY
ARC 1034	10 28 34.	+ 18 59		17.5	RICH CLUSTER OF GALAXIES
RNGC 3275	10 28 35.	- 36 27		12.5	GALAXY
ZC 1028.6+0045	10 28 36.	+ 00 45	400		CLUSTER OF GALAXIES
ZWG 037.061	10 28 36.	+ 04 44		14.6	GALAXY
UGC 05708	10 28 36.	+ 04 44	204	14.6	GALAXY
MCG+01-27-014	10 28 36.	+ 04 44	204	14.6	GALAXY
KARA.73B 0422	10 28 36.	+ 04 44	222	14.6	ISOLATED GALAXY S
ZWG 037.062	10 28 36.	+ 07 07		15.6	GALAXY
MCG+01-27-015	10 28 36.	+ 07 07	12	15.6	GALAXY
LB 10553	10 28 36.	+ 16 24		15.5	FAINT BLUE STAR
LB 10552	10 28 36.	+ 17 56		16.2	FAINT BLUE STAR
ZWG 094.078	10 28 36.	+ 19 38		15.7	GALAXY
UGC 05709	10 28 36.	+ 19 38	78	15.7	GALAXY
MCG+03-27-054	10 28 36.	+ 19 39	84	15.7	GALAXY
LB 10551	10 28 36.	+ 20 30		15.3	FAINT BLUE STAR
ZC 1028.6+3903	10 28 36.	+ 39 03	1340		CLUSTER OF GALAXIES
MCG+10-15-093	10 28 36.	+ 62 40	36	15.	GALAXY
ZWG 009.071	10 28 36.	- 03 10		15.7	GALAXY
MCG-06-23-046	10 28 36.	- 36 28	90	12.8	GALAXY
TON-N 1239	10 28 38.	+ 20 43		16.6	BLUE STAR
TON-N 1240	10 28 38.	+ 21 06		17.	BLUE STAR
ARC 1033	10 28 40.	+ 35 21		17.2	RICH CLUSTER OF GALAXIES
HN 0350	10 28 40.	- 34 18			NEBULA
REIZ 0476	10 28 42.	+ 04 44	138	14.0	GALAXY
ZC 1028.7+0742	10 28 42.	+ 07 42	1680		CLUSTER OF GALAXIES
ZC 1028.7+1125	10 28 42.	+ 11 25	2150		CLUSTER OF GALAXIES
ZC 1028.7+1858	10 28 42.	+ 18 58	1280		CLUSTER OF GALAXIES
ZWG 094.079	10 28 42.	+ 20 00		15.6	GALAXY
ZWG 094.080	10 28 42.	+ 20 19		15.7	GALAXY
UGC 05710	10 28 42.	+ 24 24	108	16.0	GALAXY S
ZWG 124.034	10 28 42.	+ 25 07		14.1	GALAXY
UGC 05711	10 28 42.	+ 25 07	186	14.1	GALAXY Sb
KARA.73B 0423	10 28 42.	+ 25 07	204	14.1	ISOLATED GALAXY S
UGC 05712	10 28 42.	+ 32 28	66	16.0	GALAXY Sc
ZC 1028.7+3822	10 28 42.	+ 38 22	1210		CLUSTER OF GALAXIES
MCG+10-15-094	10 28 42.	+ 59 16	24	16.	GALAXY
MCG+10-15-095	10 28 42.	+ 60 14	36	16.	GALAXY
MCG-05-25-008	10 28 42.	- 28 26	24	15.	GALAXY
MCG-06-23-047	10 28 42.	- 34 17	72	12.5	GALAXY
OCL 0818	10 28 42.	- 59 34	540	9.6	OPEN STAR CLUSTER
RNGC 3270	10 28 43.	+ 25 07		14.0	GALAXY
IC 2587	10 28 43.	- 34 19			NONSTELLAR OBJECT
MCG+04-25-029	10 28 45.	+ 25 07 30.	192	14.1	GALAXY
MCG+08-19-036	10 28 45.	+ 46 56	48	15.	GALAXY
REIZ 0475	10 28 46.	+ 25 08	240	14.8	GALAXY
ZWG 037.063	10 28 48.	+ 05 16		15.7	GALAXY
ZWG 037.064	10 28 48.	+ 07 44		15.6	GALAXY
ZC 1028.8+1419	10 28 48.	+ 14 19	540		CLUSTER OF GALAXIES
LB 10558	10 28 48.	+ 14 26		15.4	FAINT BLUE STAR
LB 10557	10 28 48.	+ 18 50		17.5	FAINT BLUE STAR
LB 10556	10 28 48.	+ 19 46		16.3	FAINT BLUE STAR
ZWG 094.081	10 28 48.	+ 19 57		15.6	GALAXY
MCG+03-27-055	10 28 48.	+ 20 02	36	15.6	GALAXY
LB 10555	10 28 48.	+ 20 15		15.5	FAINT BLUE STAR
LB 10554	10 28 48.	+ 20 22		15.5	FAINT BLUE STAR

OBJECT NAME	RIGHT ASCEN.	DECLINATION	DIAM.	MAGN.	TYPE OF OBJECT
ZWG 124.035	10 28 48.	+ 26 14		15.0	GALAXY
UGC 05713	10 28 48.	+ 26 14	108	15.0	GALAXY Sb-0
MCG+04-25-030	10 28 48.	+ 26 14	90	15.0	GALAXY
TON-N 0524	10 28 48.	+ 29 06		16.4	BLUE STAR
VVI 40	10 28 48.	+ 29 06	2	18.4	SEYFERT GALAXY
VVI 39	10 28 48.	+ 29 06	4	16.6	SEYFERT GALAXY
ZC 1028.8+3306	10 28 48.	+ 33 06	940		CLUSTER OF GALAXIES
ZWG 240.053	10 28 48.	+ 46 55		14.7	GALAXY
UGC 05714	10 28 48.	+ 46 55	66	14.7	GALAXY SBc
PRA 1	10 28 48.0	+ 29 07 06.		17.	NONSTELLAR BLUE OBJECT
PRA 2	10 28 48.7	+ 29 07 00.		17.	NONSTELLAR BLUE OBJECT
ZWG 037.065	10 28 54.	+ 05 47		15.6	GALAXY
ZWG 065.048	10 28 54.	+ 12 45		14.9	GALAXY
MCG+02-27-023	10 28 54.	+ 12 45	48	14.9	GALAXY
ZC 1028.9+1553	10 28 54.	+ 15 53	670		CLUSTER OF GALAXIES
MCG+03-27-056	10 28 54.	+ 19 59	66	15.6	GALAXY
ZC 1028.9+3521	10 28 54.	+ 35 21	2420		CLUSTER OF GALAXIES
MCG+07-22-018	10 28 54.	+ 43 07	12	16.	GALAXY
ZC 1028.9+5042	10 28 54.	+ 50 42	1210		CLUSTER OF GALAXIES
MCG-01-27-011	10 28 54.	- 08 21	48	14.	GALAXY
RNGC 3276	10 28 54.	- 39 41			GALAXY
SER 077.08	10 28 54.	- 45 56			PECULIAR GALAXIES
SC 1027-5743.9	10 28 55.	- 57 59 17.	12		NEBULA
ZWG 009.072	10 29 00.	+ 00 43		14.3	GALAXY
UGC 05715	10 29 00.	+ 00 43	66	14.3	GALAXY Sb-c
MCG+00-27-026	10 29 00.	+ 00 43	66	14.3	GALAXY
ZWG 037.066	10 29 00.	+ 04 16		15.5	GALAXY
ZC 1029.0+1815	10 29 00.	+ 18 15	870		CLUSTER OF GALAXIES
ZWG 094.082	10 29 00.	+ 18 27		15.7	GALAXY
MCG+03-27-057	10 29 00.	+ 18 27	24	15.7	GALAXY
ZWG 094.083	10 29 00.	+ 19 42		15.5	GALAXY
UGC 05716	10 29 00.	+ 25 35	84	17.	GALAXY DWRF SP
REIZ 0477	10 29 00.	+ 28 45	18	15.8	GALAXY
ZWG 313.021	10 29 00.	+ 65 19		12.9	GALAXY
UGC 05717	10 29 00.	+ 65 19	138	12.9	GALAXY Sb
MCG+11-13-027	10 29 00.	+ 65 19	126	12.9	GALAXY
GO 073	10 29 01.	- 59 56			STAR CHAIN
RNGC 3259	10 29 03.	+ 65 18		13.0	GALAXY
MCG-01-27-012	10 29 03.	- 04 33	36	15.5	GALAXY
RNGC 3258C	10 29 04.	- 34 57			GALAXY
LB 10559	10 29 06.	+ 18 22		17.6	FAINT BLUE STAR
ZC 1029.1+2436	10 29 06.	+ 24 36	740		CLUSTER OF GALAXIES
ZC 1029.1+2744	10 29 06.	+ 27 44	670		CLUSTER OF GALAXIES
MCG-06-23-048	10 29 06.	- 34 56	60	15.	GALAXY
DV.56 N3258C	10 29 06.	- 34 57	72		S GALAXY
OCL 0813	10 29 06.	- 57 59	540	12.	OPEN STAR CLUSTER
REIZ 0474	10 29 07.	+ 56 21	108	14.3	GALAXY
ARP 233	10 29 08.	+ 54 38			PECULIAR GALAXY
MCG+09-17-069	10 29 09.	+ 56 21	150	14.3	GALAXY
ARC 1035	10 29 11.	+ 40 29		15.4	RICH CLUSTER OF GALAXIES
ZWG 037.067	10 29 12.	+ 08 14		15.1	GALAXY
LB 10560	10 29 12.	+ 19 26		18.4	FAINT BLUE STAR
UGC 05718	10 29 12.	+ 21 10	60	16.5	GALAXY Sa-b
ZC 1029.2+5420	10 29 12.	+ 54 20	1410		CLUSTER OF GALAXIES
ZWG 266.054	10 29 12.	+ 56 19		14.3	GALAXY
RNGC 3264	10 29 12.	+ 56 19		14.5	GALAXY
UGC 05719	10 29 12.	+ 56 19	210	14.3	GALAXY SBc/IRR
ZWG 009.073	10 29 12.	- 02 28		15.6	GALAXY
ZC 1029.2-0255	10 29 12.	- 02 55	940		CLUSTER OF GALAXIES
MCG-03-27-018	10 29 12.	- 17 34	42	15.	GALAXY
REIZ 0478	10 29 13.	+ 08 14	24	14.7	GALAXY
MCG+07-22-019	10 29 15.	+ 40 31	36	16.5	GALAXY
HARO 02	10 29 16.	+ 54 40			BLUE EMISSION-LINE GALAXY
ZWG 037.068	10 29 18.	+ 07 02		15.7	GALAXY
8ZW 1029+11.6	10 29 18.	+ 11 34		18.2	COMPACT GALAXY
LB 10563	10 29 18.	+ 14 33		14.9	FAINT BLUE STAR
ZWG 094.084	10 29 18.	+ 14 50		15.5	GALAXY
LB 10562	10 29 18.	+ 15 56		15.6	FAINT BLUE STAR
LB 10561	10 29 18.	+ 16 41		16.1	FAINT BLUE STAR
ARC 1036	10 29 18.	+ 32 08		17.8	RICH CLUSTER OF GALAXIES
ZC 1029.3+3416	10 29 18.	+ 34 16	1480		CLUSTER OF GALAXIES
MCG+07-22-020	10 29 18.	+ 43 15	48	16.	GALAXY
LB 00578	10 29 18.	+ 45 34 30.		17.5	FAINT BLUE STAR
MRK 033	10 29 18.	+ 54 40	30	13.5	GALAXY WITH UV CONTINUUM
VVI 41	10 29 18.	+ 54 40	60	14.29	SEYFERT GALAXY
MCG+09-17-070	10 29 18.	+ 54 40	60	14.	GALAXY
ZWG 290.046	10 29 18.	+ 56 58		15.4	GALAXY
ZC 1029.3+5736	10 29 18.	+ 57 36	8060		CLUSTER OF GALAXIES
ZC 1029.3+7739	10 29 18.	+ 77 39	2220		CLUSTER OF GALAXIES
REIZ 0479	10 29 21.	+ 10 30	60	16.0	GALAXY
MCG+10-15-096	10 29 21.	+ 56 57	27	15.	GALAXY
HOLM 198A	10 29 21.	- 02 26	42	14.6	PART OF MULTIPLE GALAXY
HOLM 198B	10 29 23.	- 02 27	36	14.6	PART OF MULTIPLE GALAXY
ZWG 037.069	10 29 24.	+ 03 19		15.4	GALAXY
ZWG 037.070	10 29 24.	+ 07 55		15.6	GALAXY
LB 10564	10 29 24.	+ 15 32		15.1	FAINT BLUE STAR
MCG+05-25-020	10 29 24.	+ 27 53	120	13.3	GALAXY
RNGC 3272	10 29 24.	+ 28 43			GALAXY
ZC 1029.4+3143	10 29 24.	+ 31 43	1550		CLUSTER OF GALAXIES
ZWG 266.055	10 29 24.	+ 54 38		13.2	GALAXY
UGC 05720	10 29 24.	+ 54 38	60	13.2	GALAXY PECULR
RNGC 3278	10 29 24.	- 39 41			GALAXY
REIZ 0480	10 29 28.	+ 25 47	30	15.4	GALAXY
IC 2588	10 29 28.	- 30 07 57.			NONSTELLAR OBJECT
ZC 1029.5+1349	10 29 30.	+ 13 49	1410		CLUSTER OF GALAXIES
LB 10565	10 29 30.	+ 15 42		15.9	FAINT BLUE STAR
ZWG 154.024	10 29 30.	+ 27 56		13.3	GALAXY
REIZ 0481	10 29 30.	+ 27 56	90	13.1	GALAXY
UGC 05721	10 29 30.	+ 27 56	126	13.3	GALAXY IRR?
ZWG 290.047	10 29 30.	+ 59 07		15.1	GALAXY
UGC 05722	10 29 30.	+ 59 07	84	15.1	GALAXY S
ZWG 009.074	10 29 30.	- 02 11		15.1	GALAXY
MCG+00-27-027	10 29 30.	- 02 11	42	15.1	GALAXY
MCG-05-25-009	10 29 30.	- 30 07	60	14.	GALAXY
PK282+03.1	10 29 32.	- 53 17 34.	30		PLANETARY NEBULA
MCG+10-15-097	10 29 33.	+ 59 06	72	15.	GALAXY
MCG-04-25-012	10 29 33.	- 26 18 30.	18	14.5	GALAXY
REIZ 0482	10 29 34.	+ 24 37	24	15.5	GALAXY
RNGC 3281A	10 29 34.	- 34 57			GALAXY
RNGC 3258D	10 29 34.	- 35 08			GALAXY
HOLM 199C	10 29 35.	- 01 15	12	15.1	PART OF MULTIPLE GALAXY
ZWG 037.071	10 29 36.	+ 03 48		15.6	GALAXY
ZWG 065.049	10 29 36.	+ 12 19		15.0	GALAXY
MCG+02-27-024	10 29 36.	+ 12 19	36	15.0	GALAXY
LB 10566	10 29 36.	+ 18 12		16.1	FAINT BLUE STAR
RNGC 3274	10 29 36.	+ 27 56		13.0	GALAXY
TON-N 0525	10 29 36.	+ 32 41		15.0	BLUE STAR
UGC 05724	10 29 36.	+ 67 53	150	16.0	GALAXY DBL SYS
MCG+12-10-049	10 29 36.	+ 74 02	114	14.	GALAXY
MCG+12-10-050	10 29 36.	+ 74 31	33	16.	GALAXY
ZWG 009.075	10 29 36.	- 01 15		15.1	GALAXY
UGC 05723	10 29 36.	- 01 15	108	15.1	GALAXY S
MCG-06-23-050	10 29 36.	- 34 35	150	12.9	GALAXY
MCG-06-23-049	10 29 36.	- 34 55 30.	18	15.	GALAXY
DV.56 N3258D	10 29 36.	- 35 09	84		S GALAXY
PK283+02.1	10 29 36.	- 55 05 18.	10		PLANETARY NEBULA
HOLM 199D	10 29 38.	- 01 19	30	15.1	PART OF MULTIPLE GALAXY
HOLM 199A	10 29 39.	- 01 14	36	14.1	PART OF MULTIPLE GALAXY
RNGC 3281	10 29 40.	- 34 35		12.5	GALAXY
RNGC 3281B	10 29 40.	- 34 57			GALAXY
MCG+05-25-021	10 29 42.	+ 27 25	42	15.5	GALAXY
ZWG 154.025	10 29 42.	+ 27 26		15.5	GALAXY
ZC 1029.7+2953	10 29 42.	+ 29 53	2220		CLUSTER OF GALAXIES
TON-N 0526	10 29 42.	+ 31 22		15.1	BLUE STAR
ZWG 212.020	10 29 42.	+ 44 14		15.2	GALAXY
MCG+07-22-021	10 29 42.	+ 44 15 30.	72	14.5	GALAXY
ZWG 290.048	10 29 42.	+ 57 00		15.1	GALAXY
ZWG 313.022	10 29 42.	+ 65 01		13.5	GALAXY
UGC 05725	10 29 42.	+ 65 01	84	13.5	GALAXY SB0
ZWG 009.076	10 29 42.	- 01 18		15.4	GALAXY
MCG+00-27-028	10 29 42.	- 01 18	48	15.4	GALAXY
MCG-06-23-051	10 29 42.	- 35 08	60	14.	GALAXY
HOLM 199B	10 29 43.	- 01 18	36	14.4	PART OF MULTIPLE GALAXY
MCG-01-27-014	10 29 45.	- 04 46	18	15.	GALAXY
MCG-01-27-013	10 29 45.	- 09 00	30	14.	GALAXY
MCG-06-23-051A	10 29 45.	- 34 55	36	15.	GALAXY
RNGC 3266	10 29 46.	+ 65 00		13.5	GALAXY
UGC 05726	10 29 48.	+ 02 49	66	17.	GALAXY DWARF?
LB 10569	10 29 48.	+ 17 22		16.3	FAINT BLUE STAR
LB 10568	10 29 48.	+ 17 25		17.9	FAINT BLUE STAR
ZC 1029.8+2023	10 29 48.	+ 20 23	4970		CLUSTER OF GALAXIES
LB 10567	10 29 48.	+ 20 24		15.2	FAINT BLUE STAR
ZWG 124.036	10 29 48.	+ 20 38		15.6	GALAXY
KARA.73B 0424	10 29 48.	+ 20 38	36	15.6	ISOLATED GALAXY E
MCG+10-15-098	10 29 48.	+ 56 50	24	16.	GALAXY
MCG+10-15-099	10 29 48.	+ 56 59 30.	36	15.	GALAXY
ZWG 313.023	10 29 48.	+ 64 46		15.0	GALAXY
UGC 05727	10 29 48.	+ 64 46	90	15.0	GALAXY Sb-c
ZC 1029.8+6635	10 29 48.	+ 66 35	400		CLUSTER OF GALAXIES
7ZW 335	10 29 48.	+ 67 51			COMPACT GALAXY
MCG+11-13-028	10 29 48.	+ 67 52	14	17.	GALAXY
MCG+12-10-051	10 29 48.	+ 72 40	24	16.	GALAXY
UGC 05728	10 29 48.	+ 79 25	90	17.	GALAXY DWRF SP
ZWG 009.077	10 29 48.	- 02 30		15.7	GALAXY
DV.56 N3281A	10 29 48.	- 34 56			S0 GALAXY
MCG+11-13-029	10 29 51.	+ 67 53	12	17.	GALAXY
IC 2589	10 29 52.	- 23 47 40.			NONSTELLAR OBJECT
LB 10571	10 29 54.	+ 16 38		15.6	FAINT BLUE STAR
LB 10570	10 29 54.	+ 20 00		15.8	FAINT BLUE STAR
ZWG 094.085	10 29 54.	+ 20 01		15.3	GALAXY
ZWG 212.021	10 29 54.	+ 43 25		15.7	GALAXY
MCG+07-22-022	10 29 54.	+ 43 26	36	15.	GALAXY
MCG+08-19-037	10 29 54.	+ 49 08	24	16.	GALAXY
MCG+11-13-030	10 29 54.	+ 65 01	57	14.	GALAXY
MCG+11-13-031	10 29 54.	+ 67 52	24	16.	GALAXY
RNGC 3282	10 29 54.	- 22 02		14.0	GALAXY
MCG-04-25-013	10 29 57.	- 22 02 30.	96	14.	GALAXY
ZWG 037.072	10 30 00.	+ 03 23		15.6	GALAXY
ZWG 065.050	10 30 00.	+ 12 26		15.3	GALAXY
ZC 1030.0+1515	10 30 00.	+ 15 15	810		CLUSTER OF GALAXIES
LB 10574	10 30 00.	+ 18 28		16.0	FAINT BLUE STAR
LB 10573	10 30 00.	+ 19 26		17.5	FAINT BLUE STAR
LB 10572	10 30 00.	+ 19 52		16.6	FAINT BLUE STAR
MCG+03-27-058	10 30 00.	+ 20 03	36	15.3	GALAXY
ZWG 094.086	10 30 00.	+ 20 11		15.2	GALAXY
UGC 05729	10 30 00.	+ 20 11	66	15.2	GALAXY SBa-b
ZC 1030.0+5300	10 30 00.	+ 53 00	1480		CLUSTER OF GALAXIES
MCG+11-13-032	10 30 00.	+ 64 46	78	16.	GALAXY
LDN 1321	10 30 00.	+ 84 30	5340		DARK NEBULA
DV.56 N3258E	10 30 00.	- 34 45	84		S GALAXY
MCG-06-23-051B	10 30 00.	- 34 55	90	15.	GALAXY
DV.56 N3281B	10 30 00.	- 34 56			E GALAXY
SEY 076	10 30 02.	+ 57 17 37.		14.9	FAINT GALAXY
MCG+07-22-023	10 30 03.	+ 43 11	30	15.	GALAXY
HOLM 200B	10 30 05.	+ 16 07	12	15.0	PART OF MULTIPLE GALAXY
REIZ 0484	10 30 05.	+ 16 08	72	14.5	GALAXY
HOLM 200A	10 30 05.	+ 16 08	54	13.6	PART OF MULTIPLE GALAXY
IC 0616	10 30 06.	+ 16 05 14.			NONSTELLAR OBJECT
ZWG 094.087	10 30 06.	+ 16 07		14.6	GALAXY
UGC 05730	10 30 06.	+ 16 07	66	14.6	GALAXY Sc
MCG+03-27-060	10 30 06.	+ 16 07	78	14.6	GALAXY
MCG+03-27-059	10 30 06.	+ 20 14	60	15.2	GALAXY
MCG+05-25-022	10 30 06.	+ 28 45	84	12.3	GALAXY
RNGC 3277	10 30 06.	+ 28 46		13.0	GALAXY
ZWG 154.026	10 30 06.	+ 28 47		12.3	GALAXY
UGC 05731	10 30 06.	+ 28 47	156	12.3	GALAXY Sa/Sb
ZC 1030.1+3717	10 30 06.	+ 37 17	1140		CLUSTER OF GALAXIES
RNGC 3280B	10 30 10.	- 12 23		16.0	GALAXY
RNGC 3280A	10 30 10.	- 12 23		16.0	GALAXY
RNGC 3258E	10 30 10.	- 34 43			GALAXY
ZC 1030.2+0225	10 30 12.	+ 02 25	1410		CLUSTER OF GALAXIES
ZC 1030.2+0516	10 30 12.	+ 05 16	940		CLUSTER OF GALAXIES
ZC 1030.2+1408	10 30 12.	+ 14 08	340		CLUSTER OF GALAXIES
LB 10577	10 30 12.	+ 18 21		16.6	FAINT BLUE STAR
LB 10576	10 30 12.	+ 18 48		17.6	FAINT BLUE STAR
LB 10575	10 30 12.	+ 19 06		17.8	FAINT BLUE STAR
ZC 1030.2+2314	10 30 12.	+ 23 14	4230		CLUSTER OF GALAXIES
REIZ 0485	10 30 12.	+ 28 47	36	12.6	GALAXY
ZWG 290.049	10 30 12.	+ 57 19		14.7	GALAXY
MCG-02-27-007	10 30 15.	- 12 23	8	16.	GALAXY
MCG-02-27-006	10 30 15.	- 12 23	10	16.	GALAXY
RNGC 3280C	10 30 16.	- 12 23		16.0	GALAXY
RNGC 3296	10 30 16.	- 12 27			GALAXY
IC 0617	10 30 17.	- 12 23 35.			NONSTELLAR OBJECT
IC 0618	10 30 17.	- 12 28 17.			NONSTELLAR OBJECT
ZWG 065.051	10 30 18.	+ 12 18		15.3	GALAXY
LB 10578	10 30 18.	+ 17 50		15.5	FAINT BLUE STAR
ZWG 266.056	10 30 18.	+ 50 58		15.5	GALAXY
ZWG 290.050	10 30 18.	+ 56 32		15.6	GALAXY
MCG+10-15-100	10 30 18.	+ 57 18	39	15.	GALAXY
ZWG 313.024	10 30 18.	+ 68 01		15.5	GALAXY
MCG+11-13-033	10 30 18.	+ 68 02	15	16.	GALAXY
ARC 1039	10 30 18.	- 04 32		17.5	RICH CLUSTER OF GALAXIES
MCG-02-27-008	10 30 18.	- 12 23	8	16.	GALAXY
RNGC 3252	10 30 23.	+ 74 01		15.5	GALAXY
AR 13	10 30 23.22	+ 74 01 18.6			NEBULA

OBJECT NAME	RIGHT ASCEN.	DECLINATION	DIAM.	MAGN.	TYPE OF OBJECT
ARC 1038	10 30 24.	+ 02 31		17.6	RICH CLUSTER OF GALAXIES
ZWG 037.073	10 30 24.	+ 03 27		15.7	GALAXY
8ZW 1030+13.6	10 30 24.	+ 13 37		18.5	COMPACT GALAXY
LB 10582	10 30 24.	+ 15 38		15.9	FAINT BLUE STAR
LB 10581	10 30 24.	+ 19 32		16.0	FAINT BLUE STAR
LB 10580	10 30 24.	+ 19 58		17.8	FAINT BLUE STAR
LB 10579	10 30 24.	+ 20 06		17.5	FAINT BLUE STAR
ZWG 240.054	10 30 24.	+ 49 35		15.6	GALAXY
REIZ 0483	10 30 24.	+ 50 57	36	15.3	GALAXY
ZWG 313.025	10 30 24.	+ 62 31		15.7	GALAXY
KARA.73B 0425	10 30 24.	+ 62 31	24	15.7	ISOLATED GALAXY E
ZWG 333.038	10 30 24.	+ 73 16		15.6	GALAXY
ZWG 333.039	10 30 24.	+ 74 01		14.2	GALAXY
UGC 05732	10 30 24.	+ 74 01	138	14.2	GALAXY Sc-IRR
MCG-01-27-015	10 30 24.	- 06 14	60	14.	GALAXY
MCG-04-25-014	10 30 24.	- 27 16 30.	48	16.	GALAXY
MCG-05-25-010	10 30 24.	- 28 19	96	15.	GALAXY
ARC 1041	10 30 27.	- 08 38		17.1	RICH CLUSTER OF GALAXIES
ZWG 065.052	10 30 30.	+ 12 09		15.6	GALAXY
ZC 1030.5+1944	10 30 30.	+ 19 44	2020		CLUSTER OF GALAXIES
ZWG 212.022	10 30 30.	+ 44 19		15.3	GALAXY
ZWG 266.057	10 30 30.	+ 53 52		15.5	GALAXY
MCG+09-17-071	10 30 30.	+ 56 30	24	16.	GALAXY
MCG+10-15-101	10 30 30.	+ 62 29	15	16.	GALAXY
ZWG 009.078	10 30 30.	- 01 50		14.9	GALAXY
MCG-04-25-015	10 30 30.	- 27 16 30.	72	14.5	GALAXY
MCG-05-25-011	10 30 30.	- 30 00	72	16.	GALAXY
MCG-06-23-052	10 30 30.	- 34 07	36	14.5	GALAXY
SEY 077	10 30 35.	+ 59 39 36.		15.1	FAINT GALAXY
ZWG 065.053	10 30 36.	+ 12 06		15.6	GALAXY
MCG+07-22-024	10 30 36.	+ 42 01	30	17.	GALAXY
ZWG 266.058	10 30 36.	+ 52 37		14.9	GALAXY
UGC 05733	10 30 36.	+ 52 37	60	14.9	GALAXY SB
MCG+09-17-072	10 30 36.	+ 53 53	36	16.	GALAXY
ZWG 290.051	10 30 36.	+ 57 19		15.5	GALAXY
MCG-01-27-016	10 30 38.	- 07 06	42	16.	GALAXY
RNGC 3285A	10 30 38.	- 27 15		14.0	GALAXY
MCG-01-27-017	10 30 40.	- 07 05	48	16.	GALAXY
RNGC 3281C	10 30 40.	- 34 38			GALAXY
ZWG 037.074	10 30 42.	+ 02 52		15.1	GALAXY
ZWG 037.075	10 30 42.	+ 02 55		15.3	GALAXY
LB 10583	10 30 42.	+ 18 20		15.8	FAINT BLUE STAR
ZC 1030.7+4231	10 30 42.	+ 42 31	870		CLUSTER OF GALAXIES
MCG+09-17-073	10 30 42.	+ 52 38	36	15.	GALAXY
ZWG 290.052	10 30 42.	+ 59 40		15.4	GALAXY
ZWG 290.053	10 30 42.	+ 61 52		15.6	GALAXY
MCG-06-23-053	10 30 42.	- 34 36	66	14.5	GALAXY
DV.56 N3281C	10 30 42.	- 34 39	78		SO GALAXY
SER 077.01	10 30 42.	- 43 07	20	15.	PECULIAR GALAXY
RNGC 3283	10 30 44.	- 45 49			UNVERIFIED SOUTHERN OBJECT
GO 077	10 30 44.	- 60 00	50	17.8	STAR CHAIN
MCG-01-27-018	10 30 45.	- 07 12	84	14.	GALAXY
IC 0621	10 30 48.	+ 02 54 48.			NONSTELLAR OBJECT
ZWG 037.076	10 30 48.	+ 07 23		15.2	GALAXY
MCG+01-27-016	10 30 48.	+ 07 23	18	15.2	GALAXY
ZWG 065.054	10 30 48.	+ 12 05		15.7	GALAXY
ZC 1030.8+1426	10 30 48.	+ 14 26	940		CLUSTER OF GALAXIES
LB 10586	10 30 48.	+ 14 44		16.6	FAINT BLUE STAR
LB 10585	10 30 48.	+ 15 02		16.5	FAINT BLUE STAR
ZWG 094.088	10 30 48.	+ 16 38		15.7	GALAXY
LB 10584	10 30 48.	+ 17 42		17.6	FAINT BLUE STAR
MCG+10-15-102	10 30 48.	+ 57 18	18	16.	GALAXY
ZWG 290.054	10 30 48.	+ 61 54		14.6	GALAXY
MCG+11-13-034	10 30 48.	+ 68 05	18	17.	GALAXY
ZC 1030.8-0235	10 30 48.	- 02 35	1480		CLUSTER OF GALAXIES
MCG-04-25-017	10 30 48.	- 24 17	90	15.5	GALAXY
MCG-04-25-016	10 30 48.	- 26 50 30.	66	15.	GALAXY
SN 1962D	10 30 48.	- 27 39		16.0	SUPERNOVA
MCG-05-25-012	10 30 48.	- 27 39	18	18.	GALAXY
MRSL 286-01/1	10 30 48.	- 60 03	240		HII REGION
MRSL 287-02/1	10 30 48.	- 60 53	3600		HII REGION
REIZ 0486	10 30 50.	+ 52 38	24	14.8	GALAXY
IC 0619	10 30 51.	+ 12 48 12.			NONSTELLAR OBJECT
ZWG 065.055	10 30 54.	+ 12 08		15.2	GALAXY
MCG+07-22-025	10 30 54.	+ 40 32 30.	54	15.	GALAXY
ZWG 212.023	10 30 54.	+ 40 33		15.1	GALAXY
MCG+10-15-103	10 30 54.	+ 59 39	15	15.	GALAXY
MRK 034	10 30 54.	+ 60 17	25	16.	GALAXY WITH UV CONTINUUM
KW 06	10 30 54.	+ 60 17	39		SEYFERT GALAXY
VVI 42	10 30 54.	+ 60 17	25	15.79	SEYFERT GALAXY
MCG+10-15-104	10 30 54.	+ 60 17	27	16.	GALAXY
MCG+10-15-105	10 30 54.	+ 61 52	15	15.	GALAXY
MRSL 285-00/3	10 30 54.	- 57 53	180		HII REGION
VHE 37	10 30 54.	- 58 24	192		REFLECTION NEBULA
IC 0620	10 30 55.	+ 12 07 30.			NONSTELLAR OBJECT
MCG+10-15-106	10 30 57.	+ 61 53	18	15.	GALAXY
ARC 1029	10 30 58.	+ 77 36		17.1	RICH CLUSTER OF GALAXIES
ZWG 037.077	10 31 00.	+ 04 31		15.2	GALAXY
ZWG 037.078	10 31 00.	+ 08 04		15.2	GALAXY
LB 10588	10 31 00.	+ 14 48		17.6	FAINT BLUE STAR
LB 10587	10 31 00.	+ 16 44		15.9	FAINT BLUE STAR
SHAH 052	10 31 00.	+ 48 09	108	17.2	GROUP OF COMPACT GALAXIES
MCG+12-10-052	10 31 00.	+ 72 57	51	16.	GALAXY
REIZ 0487	10 31 01.	+ 08 04	24	14.6	GALAXY
LB 00579	10 31 02.	+ 46 50 30.		17.7	FAINT BLUE STAR
ZC 1031.1+0434	10 31 06.	+ 04 34	3360		CLUSTER OF GALAXIES
8ZW 1031+08.4	10 31 06.	+ 08 22		18.4	COMPACT GALAXY
LB 10589	10 31 06.	+ 18 00		17.0	FAINT BLUE STAR
TON-N 0527	10 31 06.	+ 23 28		15.1	BLUE STAR
MCG+04-25-031	10 31 06.	+ 24 16	30	15.6	GALAXY
ZWG 124.037	10 31 06.	+ 24 17		15.6	GALAXY
ZC 1031.1+4635	10 31 06.	+ 46 35	3830		CLUSTER OF GALAXIES
ZWG 266.059	10 31 06.	+ 53 07		14.0	GALAXY
UGC 05734	10 31 06.	+ 53 07	102	14.0	GALAXY SO-a
MCG+10-15-107	10 31 06.	+ 57 19 30.	39	16.	GALAXY
ZWG 290.055	10 31 06.	+ 59 39		15.6	GALAXY
ZC 1031.1+6021	10 31 06.	+ 60 21	610		CLUSTER OF GALAXIES
ZWG 009.079	10 31 06.	- 02 00		15.4	GALAXY
MCG+09-17-074	10 31 09.	+ 53 08 30.	84	14.	GALAXY
MCG+10-15-108	10 31 09.	+ 59 38	12	15.	GALAXY
ZWG 065.056	10 31 12.	+ 13 08		14.8	GALAXY
UGC 05735	10 31 12.	+ 13 08	60	14.8	GALAXY Sb-c
MCG+02-27-025	10 31 12.	+ 13 08	48	14.8	GALAXY
ZWG 240.055	10 31 12.	+ 44 46		15.6	GALAXY
MCG-04-25-018	10 31 12.	- 26 38 30.	48	14.5	GALAXY
REIZ 0488	10 31 15.	+ 11 28	60	15.0	GALAXY
ZWG 065.057	10 31 18.	+ 11 28		15.2	GALAXY
UGC 05737	10 31 18.	+ 11 28	66	15.2	GALAXY Sc
TON-N 0528	10 31 18.	+ 24 46		15.0	BLUE STAR
ZWG 009.080	10 31 18.	- 00 18		15.6	GALAXY
UGC 05736	10 31 18.	- 00 18	66	15.6	GALAXY Sb-c
MCG+00-27-029	10 31 18.	- 00 18	66	15.6	GALAXY
TON-N 1241	10 31 20.	+ 21 49		16.4	BLUE STAR
RNGC 3285	10 31 20.	- 27 11		13.5	GALAXY
MCG-04-25-019	10 31 21.	- 27 12 30.	120	13.	GALAXY
LB 10592	10 31 24.	+ 16 56		17.4	FAINT BLUE STAR
LB 10591	10 31 24.	+ 17 57		17.2	FAINT BLUE STAR
LB 10590	10 31 24.	+ 19 26		15.9	FAINT BLUE STAR
MRSL 285-00/2	10 31 24.	- 58 10	900		HII REGION
ARC 1040	10 31 25.	+ 45 45		17.2	RICH CLUSTER OF GALAXIES
MCG-01-27-019	10 31 27.	- 08 00	48	14.5	GALAXY
ZC 1031.5+0555	10 31 30.	+ 05 55	1480		CLUSTER OF GALAXIES
ZC 1031.5+1202	10 31 30.	+ 12 02	1080		CLUSTER OF GALAXIES
MCG+03-27-062	10 31 30.	+ 15 12	54	15.5	GALAXY
ZC 1031.5+1524	10 31 30.	+ 15 24	870		CLUSTER OF GALAXIES
ZC 1031.5+1634	10 31 30.	+ 16 34	1140		CLUSTER OF GALAXIES
LB 10593	10 31 30.	+ 18 41		15.8	FAINT BLUE STAR
ZWG 094.089	10 31 30.	+ 19 57		15.1	GALAXY
MCG+03-27-061	10 31 30.	+ 19 58	60	15.1	GALAXY
ZC 1031.5+3056	10 31 30.	+ 30 56	1080		CLUSTER OF GALAXIES
ZC 1031.5+5335	10 31 30.	+ 53 35	940		CLUSTER OF GALAXIES
ZWG 267.001	10 31 30.	+ 55 05		15.7	GALAXY
ZWG 266.060	10 31 30.	+ 55 05		15.7	GALAXY
MCG+12-10-053	10 31 30.	+ 72 19	36	16.	GALAXY
MCG+12-10-054	10 31 30.	+ 72 53	21	15.	GALAXY
ZWG 009.081	10 31 30.	- 02 07		15.4	GALAXY
ARC 1044	10 31 33.	+ 05 56		17.6	RICH CLUSTER OF GALAXIES
ARC 1042	10 31 33.	+ 12 04		17.2	RICH CLUSTER OF GALAXIES
REIZ 0489	10 31 34.	+ 35 32	30	13.7	GALAXY
ZWG 037.079	10 31 36.	+ 08 11		15.3	GALAXY
ZWG 094.090	10 31 36.	+ 15 13		15.5	GALAXY
LB 10594	10 31 36.	+ 18 54		16.7	FAINT BLUE STAR
ZWG 183.028	10 31 36.	+ 35 30		14.1	GALAXY
UGC 05738	10 31 36.	+ 35 30	60	14.1	GALAXY S
ZWG 212.024	10 31 36.	+ 39 53		15.6	GALAXY
KARA.72 238A	10 31 36.	+ 39 53	30	15.6	PART OF DOUBLE GALAXY
MCG+08-19-038	10 31 36.	+ 44 52	24	16.	GALAXY
ZWG 313.026	10 31 36.	+ 64 43		15.4	GALAXY
ZWG 333.040	10 31 36.	+ 72 55		15.4	GALAXY
MCG-05-25-013	10 31 36.	- 27 33	24	17.	GALAXY
MCG-05-25-014	10 31 36.	- 29 54	72	15.5	GALAXY
VHE 39	10 31 36.	- 59 25	30		REFLECTION NEBULA
TON-N 1242	10 31 37.	+ 34 37		16.8	BLUE STAR
MCG+10-15-109	10 31 39.	+ 61 57 30.	36	16.	GALAXY
REIZ 0490	10 31 40.	+ 14 01	54	13.8	GALAXY
ZWG 065.058	10 31 42.	+ 14 00		13.9	GALAXY
UGC 05739	10 31 42.	+ 14 00	72	13.9	GALAXY IRR?
MCG+02-27-026	10 31 42.	+ 14 00	60	13.9	GALAXY
LB 10596	10 31 42.	+ 15 36		15.8	FAINT BLUE STAR
LB 10595	10 31 42.	+ 16 11		16.0	FAINT BLUE STAR
ZWG 212.025	10 31 42.	+ 39 54		15.6	GALAXY
KARA.72 238B	10 31 42.	+ 39 54	30	15.6	PART OF DOUBLE GALAXY
ZWG 240.056	10 31 42.	+ 44 53		15.3	GALAXY
ZWG 266.061	10 31 42.	+ 51 01		15.7	GALAXY
UGC 05740	10 31 42.	+ 51 01	132	15.7	GALAXY DWRF SP
MCG+09-17-075	10 31 42.	+ 51 01	72	15.	GALAXY
ARC 1043	10 31 43.	+ 16 33		17.2	RICH CLUSTER OF GALAXIES
MCG+06-23-019	10 31 45.	+ 35 31	48	14.	GALAXY
ZC 1031.8+0528	10 31 48.	+ 05 28	810		CLUSTER OF GALAXIES
ZC 1031.8+1430	10 31 48.	+ 14 30	670		CLUSTER OF GALAXIES
LB 10599	10 31 48.	+ 14 45		16.6	FAINT BLUE STAR
ZC 1031.8+1547	10 31 48.	+ 15 47	810		CLUSTER OF GALAXIES
LB 10598	10 31 48.	+ 17 24		14.5	FAINT BLUE STAR
LB 10597	10 31 48.	+ 18 32		16.0	FAINT BLUE STAR
ZC 1031.8+4145	10 31 48.	+ 41 45	1750		CLUSTER OF GALAXIES
MCG+08-19-039	10 31 48.	+ 45 31	18	15.	GALAXY
ZWG 240.057	10 31 48.	+ 47 14		15.6	GALAXY
MRK 145	10 31 48.	+ 64 45	10	16.	GALAXY WITH UV CONTINUUM
MCG-06-23-054	10 31 48.	- 35 02	90	14.	GALAXY
MCG-04-25-020	10 31 51.	- 27 05	48	15.5	GALAXY
RNGC 3289	10 31 52.	- 35 04			GALAXY
ZWG 037.080	10 31 54.	+ 04 00		15.7	GALAXY
HOLM 201C	10 31 54.	+ 11 27	18	15.2	PART OF MULTIPLE GALAXY
LB 10600	10 31 54.	+ 16 20		17.2	FAINT BLUE STAR
ZWG 240.058	10 31 54.	+ 45 32		14.8	GALAXY
MCG+11-13-035	10 31 54.	+ 64 27 30.	36	16.	GALAXY
MRSL 285+00/1	10 31 54.	- 57 53	900		HII REGION
VHE 38	10 31 54.	- 61 27	18		REFLECTION NEBULA
TON-N 1243	10 31 55.	+ 33 28		16.9	BLUE STAR
IC 0622	10 31 56.	+ 11 27 33.			NONSTELLAR OBJECT
RNGC 3281D	10 31 58.	- 34 08			GALAXY
RNGC 3279	10 31 59.	+ 11 28		14.0	GALAXY
ZWG 037.081	10 32 00.	+ 06 53		15.2	GALAXY
ZWG 037.082	10 32 00.	+ 06 55		15.3	GALAXY
ZWG 065.059	10 32 00.	+ 11 27		14.1	GALAXY
UGC 05741	10 32 00.	+ 11 27	174	14.1	GALAXY Sc
MCG+02-27-027	10 32 00.	+ 11 27	180	14.1	GALAXY
ZC 1032.0+1250	10 32 00.	+ 12 50	740		CLUSTER OF GALAXIES
LB 10602	10 32 00.	+ 16 42		17.0	FAINT BLUE STAR
LB 10601	10 32 00.	+ 18 30		15.8	FAINT BLUE STAR
MCG+04-25-032	10 32 00.	+ 21 53	120	12.9	GALAXY
ZWG 124.038	10 32 00.	+ 21 55		12.9	GALAXY
UGC 05742	10 32 00.	+ 21 55	126	12.9	GALAXY IRR
UGC 05743	10 32 00.	+ 25 47	66	16.0	GALAXY Sc
MCG+11-13-035A	10 32 00.	+ 64 26	12	16.	GALAXY
ZWG 333.041	10 32 00.	+ 72 51		14.9	GALAXY
MCG-05-25-015	10 32 00.	- 31 56	72	15.5	GALAXY
DV.56 N3281D	10 32 00.	- 34 08	114		Sc GALAXY
MCG-06-23-055	10 32 00.	- 34 08	102	14.	GALAXY
REIZ 0491	10 32 03.	+ 11 27	150	13.4	GALAXY
HOLM 201A	10 32 03.	+ 11 27	138	13.2	PART OF MULTIPLE GALAXY
KARA.68 071	10 32 03.	+ 16 29	40		DWARF GALAXY
MCG+08-19-040	10 32 03.	+ 45 20	72	14.	GALAXY
HOLM 201B	10 32 04.	+ 11 29	18	15.2	PART OF MULTIPLE GALAXY
REIZ 0493	10 32 05.	+ 02 48	24	15.2	GALAXY
ZC 1032.1+1301	10 32 06.	+ 13 01	540		CLUSTER OF GALAXIES
8ZW 1032+13.1	10 32 06.	+ 13 08		17.3	COMPACT GALAXY
ZC 1032.1+4046	10 32 06.	+ 40 46	270		CLUSTER OF GALAXIES
MCG+08-19-041	10 32 06.	+ 44 50	18	16.	GALAXY
ZWG 240.059	10 32 06.	+ 45 00		15.3	GALAXY
ZWG 240.060	10 32 06.	+ 46 49		14.1	GALAXY
MRK 146	10 32 06.	+ 46 49	20	14.	GALAXY WITH UV CONTINUUM
UGC 05744	10 32 06.	+ 46 49	33	14.1	GALAXY COMPACT
MCG+10-15-110	10 32 06.	+ 61 55	24	16.	GALAXY
MCG-04-25-021	10 32 06.	- 26 14	36	15.5	GALAXY
RNGC 3287	10 32 08.	+ 21 55		13.0	GALAXY

OBJECT NAME	RIGHT ASCEN.	DECLINATION	DIAM.	MAGN.	TYPE OF OBJECT
REIZ 0492	10 32 08.	+ 21 56	132	12.8	GALAXY
ABC 1037	10 32 08.	+ 69 03		17.7	RICH CLUSTER OF GALAXIES
ABC 1045	10 32 10.	+ 30 58		17.2	RICH CLUSTER OF GALAXIES
ZWG 009.083	10 32 12.	+ 00 55		15.6	GALAXY
ZC 1032.2+0350	10 32 12.	+ 03 50	2350		CLUSTER OF GALAXIES
ZWG 037.083	10 32 12.	+ 05 50		15.6	GALAXY
ZC 1032.2+0625	10 32 12.	+ 06 25	540		CLUSTER OF GALAXIES
LB 10604	10 32 12.	+ 14 35		16.3	FAINT BLUE STAR
LB 10603	10 32 12.	+ 16 59		15.9	FAINT BLUE STAR
ZWG 094.091	10 32 12.	+ 18 01		15.7	GALAXY
MCG+03-27-063	10 32 12.	+ 18 01	30	15.7	GALAXY
MCG+07-22-026	10 32 12.	+ 38 46	12	17.	GALAXY
ZWG 212.026	10 32 12.	+ 43 01		15.5	GALAXY
ZWG 240.061	10 32 12.	+ 44 52		15.5	GALAXY
ZWG 240.062	10 32 12.	+ 45 21		14.2	GALAXY
UGC 05746	10 32 12.	+ 45 21	102	14.2	GALAXY SBa-b
ZWG 009.082	10 32 12.	- 01 43		13.8	GALAXY
UGC 05745	10 32 12.	- 01 43	66	13.8	GALAXY SB0-a
MCG+00-27-030	10 32 12.	- 01 43	36	13.8	GALAXY
ABC 1047	10 32 14.	+ 04 41		17.2	RICH CLUSTER OF GALAXIES
MCG+07-22-027	10 32 15.	+ 43 02	42	15.	GALAXY
REIZ 0494	10 32 17.	+ 15 46	24	15.2	GALAXY
ZWG 037.084	10 32 18.	+ 05 04		15.1	GALAXY
ZC 1032.3+1231	10 32 18.	+ 12 31	610		CLUSTER OF GALAXIES
ZWG 094.092	10 32 18.	+ 15 44		15.2	GALAXY
LB 10605	10 32 18.	+ 18 24		16.5	FAINT BLUE STAR
ZWG 212.027	10 32 18.	+ 38 46		15.5	GALAXY
MCG+07-22-028	10 32 18.	+ 38 46	30	16.	GALAXY
ZWG 313.027	10 32 18.	+ 63 48		15.5	GALAXY
7ZW 336	10 32 18.	+ 79 50			COMPACT GALAXY
ZWG 351.028	10 32 18.	+ 79 50		15.6	GALAXY
SER 077.03	10 32 18.	- 42 15	60	16.	DWARF GALAXY
PK283+03.1	10 32 18.	- 53 25 29.	25		PLANETARY NEBULA
TON-N 1245	10 32 19.	+ 35 27		17.	BLUE STAR
TON-N 1244	10 32 20.	+ 20 27		17.	BLUE STAR
RNGC 3285B	10 32 20.	- 27 24		14.0	GALAXY
MCG-04-25-022	10 32 21.	- 27 24 30.	48	14.	GALAXY
LB 10608	10 32 24.	+ 17 29		15.8	FAINT BLUE STAR
LB 10607	10 32 24.	+ 18 42		17.6	FAINT BLUE STAR
LB 10606	10 32 24.	+ 20 02		16.9	FAINT BLUE STAR
ZWG 124.039	10 32 24.	+ 25 18		15.7	GALAXY
MCG+07-22-029	10 32 24.	+ 41 07	45	15.	GALAXY
MRK 147	10 32 24.	+ 63 50	12	15.5	GALAXY WITH UV CONTINUUM
ZC 1032.4+6759	10 32 24.	+ 67 59	670		CLUSTER OF GALAXIES
MCG-01-27-020	10 32 24.	- 06 12	60	14.5	GALAXY
MCG-05-25-016	10 32 24.	- 28 12	48	15.	GALAXY
REIZ 0495	10 32 29.	+ 37 56	18	14.5	GALAXY
LB 10609	10 32 30.	+ 18 28		15.9	FAINT BLUE STAR
ZWG 183.029	10 32 30.	+ 37 55		15.0	GALAXY
MCG-03-27-019	10 32 30.	- 20 19	36	15.	GALAXY
MCG-04-25-023	10 32 30.	- 26 58	36	15.	GALAXY
KLEM 13	10 32 30.	- 28 13	1800		LOOSE GRP OF 8 GALAXIES
MCG-05-25-017	10 32 30.	- 28 17	120	15.	GALAXY
SER 077.02	10 32 30.	- 43 02	50	16.	DWARF GALAXY
REIZ 0496	10 32 31.	+ 31 47	24	15.0	GALAXY
TON-N 1246	10 32 32.	+ 20 26		16.5	BLUE STAR
MCG+06-23-020	10 32 33.	+ 37 56 30.	18	16.	GALAXY
REIZ 0497	10 32 35.	+ 37 55	21	15.3	GALAXY
ZWG 065.060	10 32 36.	+ 12 11		15.5	GALAXY
LB 10611	10 32 36.	+ 19 46		17.6	FAINT BLUE STAR
LB 10610	10 32 36.	+ 20 26		16.0	FAINT BLUE STAR
MRK 148	10 32 36.	+ 44 33	13	16.	GALAXY WITH UV CONTINUUM
ZC 1032.6+5015	10 32 36.	+ 50 15	1140		CLUSTER OF GALAXIES
MCG+10-15-111	10 32 36.	+ 60 52	30	16.	GALAXY
MCG-05-25-018	10 32 36.	- 27 47	48	15.	GALAXY
ARP 053	10 32 38.	- 16 54			PECULIAR GALAXY
RNGC 3290	10 32 41.	- 17 00		14.0	GALAXY
ZC 1032.7+1210	10 32 42.	+ 12 10	1140		CLUSTER OF GALAXIES
LB 10615	10 32 42.	+ 16 26		17.9	FAINT BLUE STAR
LB 10614	10 32 42.	+ 16 48		17.3	FAINT BLUE STAR
LB 10613	10 32 42.	+ 18 52		15.6	FAINT BLUE STAR
LB 10612	10 32 42.	+ 20 24		15.4	FAINT BLUE STAR
ZWG 240.063	10 32 42.	+ 44 35		15.4	GALAXY
UGC 05747	10 32 42.	+ 44 35	78	15.4	GALAXY PECULR?
MCG+08-19-042	10 32 42.	+ 46 09	30	16.	GALAXY
ZWG 009.084	10 32 42.	- 01 45		15.7	GALAXY
MCG-03-27-020	10 32 42.	- 17 00	60	14.5	GALAXY
IC 0623	10 32 44.	+ 03 48 48.			NONSTELLAR OBJECT
REIZ 0498	10 32 44.	+ 41 08	36	14.6	GALAXY
MCG-01-27-021	10 32 45.	- 06 59	54	14.5	GALAXY
RNGC 3284	10 32 47.	+ 58 48			NON-EXISTENT OBJECT
ZWG 037.085	10 32 48.	+ 03 49		15.0	GALAXY
UGC 05748	10 32 48.	+ 03 49	66	15.0	GALAXY S
MCG+01-27-017	10 32 48.	+ 03 49	60	15.0	GALAXY
ZWG 037.086	10 32 48.	+ 07 48		15.7	GALAXY
LB 10617	10 32 48.	+ 17 30		15.6	FAINT BLUE STAR
LB 10616	10 32 48.	+ 18 20		15.9	FAINT BLUE STAR
ZC 1032.8+2142	10 32 48.	+ 21 42	400		CLUSTER OF GALAXIES
ZWG 154.027	10 32 48.	+ 26 32		15.6	GALAXY
ZWG 154.028	10 32 48.	+ 28 50		14.2	GALAXY
UGC 05749	10 32 48.	+ 28 50	27	14.2	GALAXY COMPACT
MCG+05-25-022A	10 32 48.	+ 28 50	27	14.2	GALAXY
MCG+12-10-055	10 32 48.	+ 72 17	51	16.	GALAXY
GO 072	10 32 50.	- 60 35	114	18.6	STAR CHAIN
HN 0351	10 32 52.	- 43 26			NEBULA
IC 2592	10 32 52.	- 43 27			NONSTELLAR OBJECT
8ZW 1032+09.2	10 32 54.	+ 09 12		17.5	COMPACT GALAXY
LB 10618	10 32 54.	+ 18 20		15.7	FAINT BLUE STAR
TON-N 0529	10 32 54.	+ 27 21		15.0	BLUE STAR
ZWG 154.029	10 32 54.	+ 29 52		15.7	GALAXY
ZC 1032.9+3131	10 32 54.	+ 31 31	610		CLUSTER OF GALAXIES
MCG-01-27-022	10 32 54.	- 05 55	24	15.	GALAXY
RNGC 3295	10 32 58.	- 12 24			NON-EXISTENT OBJECT
RNGC 3297	10 32 58.	- 12 26			NON-EXISTENT OBJECT
VDB.66G 083	10 33	+ 31 46	70		DWARF GALAXY
VDB.66G 238	10 33	- 24 28	170		DWARF GALAXY
HOFF L02	10 33	- 60 54	180		DARK HOLE
ZC 1033.0+1604	10 33 00.	+ 16 04	4440		CLUSTER OF GALAXIES
LB 10621	10 33 00.	+ 18 57		17.9	FAINT BLUE STAR
LB 10620	10 33 00.	+ 19 20		17.8	FAINT BLUE STAR
LB 10619	10 33 00.	+ 19 36		15.5	FAINT BLUE STAR
UGC 05750	10 33 00.	+ 21 16	72	17.	GALAXY
MCG+04-25-033	10 33 00.	+ 21 17 30.	120	14.9	GALAXY
ZWG 124.040	10 33 00.	+ 21 19		14.9	GALAXY
UGC 05751	10 33 00.	+ 21 19	126	14.9	GALAXY S
ZWG 124.041	10 33 00.	+ 26 23		15.3	GALAXY
ZC 1033.0+3100	10 33 00.	+ 31 00	3160		CLUSTER OF GALAXIES
MCG-01-27-023	10 33 00.	- 05 55	36	14.5	GALAXY
MCG-04-25-024	10 33 00.	- 24 30	180	14.	GALAXY
MCG-04-25-025	10 33 00.	- 27 07 30.	60	15.5	GALAXY
MCG-05-25-019	10 33 00.	- 28 02	66	14.5	GALAXY
RNGC 3366	10 33 01.	- 43 25			GALAXY
HN 0787	10 33 01.	- 72 59	30	15.	NEBULA
IC 2596	10 33 01.	- 73 00			NONSTELLAR OBJECT
TON-N 1247	10 33 02.	+ 21 31		16.8	BLUE STAR
RNGC 3292	10 33 02.	- 05 55		14.5	GALAXY
ABC 1048	10 33 03.	+ 44 13		17.2	RICH CLUSTER OF GALAXIES
MCG-02-27-009	10 33 03.	- 13 52	102	14.	GALAXY
REIZ 0499	10 33 04.	+ 58 53	15	14.3	GALAXY
RNGC 3286	10 33 05.	+ 58 53		14.5	GALAXY
KARA.73 30	10 33 05.	- 36 58	94		DWARF GALAXY
ZWG 037.087	10 33 06.	+ 05 52		15.1	GALAXY
LB 10624	10 33 06.	+ 15 56		16.2	FAINT BLUE STAR
LB 10623	10 33 06.	+ 17 02		15.5	FAINT BLUE STAR
LB 10622	10 33 06.	+ 20 02		15.8	FAINT BLUE STAR
ZWG 094.093	10 33 06.	+ 20 18		15.5	GALAXY
MCG+03-27-064	10 33 06.	+ 20 18	30	15.5	GALAXY
ZWG 124.042	10 33 06.	+ 20 55		15.5	GALAXY
ZWG 124.043	10 33 06.	+ 21 21		15.6	GALAXY
MCG+10-15-114	10 33 06.	+ 58 49	51	14.8	GALAXY
MCG+10-15-112	10 33 06.	+ 58 52	24	14.3	GALAXY
ZWG 290.056	10 33 06.	+ 58 53		14.6	GALAXY
KARA.72 239A	10 33 06.	+ 58 53	84	14.6	PART OF DOUBLE GALAXY
MCG+10-15-113	10 33 06.	+ 60 25	51	17.	GALAXY
7ZW 337	10 33 06.	+ 62 58			COMPACT GALAXY
MCG-04-25-027	10 33 06.	- 26 24	36	14.5	GALAXY
MCG-04-25-026	10 33 06.	- 27 14	36	15.5	GALAXY
VHE 40C	10 33 06.	- 58 45	96		REFLECTION NEBULA
REIZ 0503	10 33 08.	+ 21 19	42	14.2	GALAXY
REIZ 0501	10 33 09.	+ 35 14	12	15.8	GALAXY
REIZ 0500	10 33 09.	+ 58 49	42	14.8	GALAXY
RNGC 3291	10 33 10.	+ 37 31			NON-EXISTENT OBJECT
REIZ 0502	10 33 11.	+ 37 32	60	13.9	GALAXY
RNGC 3288	10 33 11.	+ 58 49		15.0	GALAXY
LYNG 02	10 33 11.	- 57 46 06.	2520		OB CONCENTRATION
ZWG 065.061	10 33 12.	+ 12 27		15.7	GALAXY
LB 10625	10 33 12.	+ 18 08		15.9	FAINT BLUE STAR
ZWG 290.057	10 33 12.	+ 58 49		15.0	GALAXY
UGC 05752	10 33 12.	+ 58 49	66	15.0	GALAXY Sb
KARA.72 239B	10 33 12.	+ 58 49	54	15.0	PART OF DOUBLE GALAXY
MCG+12-10-056	10 33 12.	+ 72 07	33	16.	GALAXY
MCG-01-27-024	10 33 12.	- 03 35	60	15.	GALAXY
HOLM 202B	10 33 13.	+ 37 31	60	13.7	PART OF MULTIPLE GALAXY
MCG+07-22-030	10 33 15.	+ 42 55	51	15.5	GALAXY
ZWG 037.088	10 33 18.	+ 04 09		15.0	GALAXY
MCG+01-27-018	10 33 18.	+ 04 09	18	15.0	GALAXY
LB 10626	10 33 18.	+ 16 52		15.7	FAINT BLUE STAR
MCG+05-25-023	10 33 18.	+ 27 11	36	16.	GALAXY
ZWG 183.030	10 33 18.	+ 37 35		11.5	GALAXY
UGC 05753	10 33 18.	+ 37 35	228	11.5	GALAXY Sc
ZC 1033.3+3804	10 33 18.	+ 38 04	4700		CLUSTER OF GALAXIES
UGC 05754	10 33 18.	+ 42 56	66	16.5	GALAXY Sc
ZWG 240.064	10 33 18.	+ 49 38		15.7	GALAXY
KARA.73B 0426	10 33 18.	+ 49 38	24	15.7	ISOLATED GALAXY S0
MCG+12-10-057	10 33 18.	+ 72 06	21	16.	GALAXY
DV.56 N3318A	10 33 18.	- 41 29	96		S GALAXY
RNGC 3318A	10 33 18.	- 41 29			GALAXY
REIZ 0504	10 33 21.	+ 35 11	36	15.2	GALAXY
REIN 2.111	10 33 21.27	+ 37 35 00.2			NEBULA
RNGC 3294	10 33 22.	+ 37 35		12.0	GALAXY
HOLM 202A	10 33 23.	+ 37 35	180	11.4	PART OF MULTIPLE GALAXY
REIZ 0505	10 33 23.	+ 37 36	180	11.5	GALAXY
REIN 2.112	10 33 23.90	+ 37 35 03.6			NEBULA
ZWG 094.094	10 33 24.	+ 16 45		15.6	GALAXY
LB 10629	10 33 24.	+ 18 07		16.4	FAINT BLUE STAR
LB 10628	10 33 24.	+ 18 18		16.0	FAINT BLUE STAR
LB 10627	10 33 24.	+ 19 54		17.6	FAINT BLUE STAR
MCG+05-25-024	10 33 24.	+ 27 12	18	14.7	GALAXY
STOCK 14	10 33 24.	+ 30 05			BLUE KNOT NEAR ELLIP GLXY
MCG+13-08-029	10 33 24.	+ 79 37 30.	30	15.	GALAXY
MCG-05-25-020	10 33 24.	- 32 06	48	14.5	GALAXY
MCG+03-27-065	10 33 27.	+ 16 28	48	15.4	GALAXY
REIZ 0506	10 33 27.	+ 35 14	30	15.5	GALAXY
MCG+06-23-021	10 33 27.	+ 37 36 30.	150	12.	GALAXY
RNGC 3302	10 33 27.	- 32 06		14.0	GALAXY
LB 00580	10 33 28.	+ 45 36 18.		16.4	FAINT BLUE STAR
ZWG 094.095	10 33 30.	+ 16 29		15.4	GALAXY
LB 10630	10 33 30.	+ 18 30		17.3	FAINT BLUE STAR
UGC 05755	10 33 30.	+ 22 22	60	16.5	GALAXY Sa-b
ZC 1033.5+2632	10 33 30.	+ 26 32	540		CLUSTER OF GALAXIES
ZWG 154.030	10 33 30.	+ 27 14		14.7	GALAXY
UGC 05756	10 33 30.	+ 27 14	66	14.7	GALAXY S0
ZC 1033.5+6823	10 33 30.	+ 68 23	2080		CLUSTER OF GALAXIES
MCG+12-10-058	10 33 30.	+ 71 55	36	16.	GALAXY
ZWG 351.029	10 33 30.	+ 79 38		14.9	GALAXY
UGC 05757	10 33 30.	+ 79 38	138	14.9	GALAXY Sa?
MCG+14-05-012	10 33 30.	+ 85 34	39	17.	GALAXY
IC 2590	10 33 31.	+ 27 12 28.			NONSTELLAR OBJECT
REIZ 0507	10 33 35.	+ 37 58	12	15.5	GALAXY
UGC 05758	10 33 36.	+ 13 42	60	19.	GALAXY DWARF
LB 10634	10 33 36.	+ 15 54		16.8	FAINT BLUE STAR
LB 10633	10 33 36.	+ 16 23		15.7	FAINT BLUE STAR
LB 10632	10 33 36.	+ 18 50		15.5	FAINT BLUE STAR
LB 10631	10 33 36.	+ 19 50		17.2	FAINT BLUE STAR
MCG+04-25-034	10 33 36.	+ 21 16	24	14.9	GALAXY
ZWG 124.044	10 33 36.	+ 21 17		14.9	GALAXY
ZC 1033.6+2815	10 33 36.	+ 28 15	1340		CLUSTER OF GALAXIES
ABC 1052	10 33 36.	+ 28 20		17.8	RICH CLUSTER OF GALAXIES
MCG+07-22-031	10 33 36.	+ 38 41	60	14.5	GALAXY
ZWG 212.028	10 33 36.	+ 38 42		14.5	GALAXY
UGC 05759	10 33 36.	+ 38 42	60	14.5	GALAXY E-S0
ZC 1033.6+3849	10 33 36.	+ 38 49	540		CLUSTER OF GALAXIES
ZWG 267.002	10 33 36.	+ 55 02		15.5	GALAXY
ZWG 266.062	10 33 36.	+ 55 02		15.5	GALAXY
MCG+09-18-002	10 33 36.	+ 55 04	24	17.	GALAXY
MCG+09-18-001	10 33 36.	+ 55 04	18	16.	GALAXY
ZWG 313.028	10 33 36.	+ 62 48		15.5	GALAXY
KARA.73B 0427	10 33 36.	+ 62 48	24	15.5	ISOLATED GALAXY E
MCG-01-27-025	10 33 36.	- 06 41	42	15.	GALAXY
REIZ 0508	10 33 37.	+ 21 18	24	14.3	GALAXY
HOLM 203B	10 33 38.	+ 14 00	24	15.2	PART OF MULTIPLE GALAXY
RNGC 3299	10 33 40.	+ 12 53		14.0	GALAXY
REIZ 0510	10 33 40.	+ 13 58	72	14.0	GALAXY
HOLM 203A	10 33 40.	+ 13 59	78	13.3	PART OF MULTIPLE GALAXY
ZWG 065.062	10 33 42.	+ 13 42		15.2	GALAXY
ZWG 065.063	10 33 42.	+ 13 58		14.2	GALAXY

OBJECT NAME	RIGHT ASCEN.	DECLINATION	DIAM.	MAGN.	TYPE OF OBJECT
UGC 05760	10 33 42.	+ 13 58	90	14.2	GALAXY S
MCG+02-27-028	10 33 42.	+ 13 58	84	14.2	GALAXY
LB 10636	10 33 42.	+ 15 43		16.2	FAINT BLUE STAR
LB 10635	10 33 42.	+ 19 24		16.2	FAINT BLUE STAR
ZWG 290.058	10 33 42.	+ 60 16		15.7	GALAXY
7ZW 338	10 33 42.	+ 71 27			COMPACT GALAXY
ZWG 333.042	10 33 42.	+ 72 16		15.4	GALAXY
IC 2594	10 33 42.	- 24 03 46.			NONSTELLAR OBJECT
MCG-04-25-028	10 33 42.	- 24 04	24	13.5	GALAXY
IC 0624	10 33 44.	- 08 04 51.			NONSTELLAR OBJECT
REIZ 0513	10 33 45.	+ 12 58	108	13.6	GALAXY
REIZ 0509	10 33 45.	+ 35 20	60	14.3	GALAXY
ZWG 065.064	10 33 48.	+ 12 57		14.1	GALAXY
UGC 05761	10 33 48.	+ 12 57	132	14.1	GALAXY
MCG+02-27-029	10 33 48.	+ 12 57	108	14.1	GALAXY
LB 10640	10 33 48.	+ 17 28		15.6	FAINT BLUE STAR
LB 10639	10 33 48.	+ 17 36		15.4	FAINT BLUE STAR
LB 10638	10 33 48.	+ 18 14		15.7	FAINT BLUE STAR
LB 10637	10 33 48.	+ 19 07		15.1	FAINT BLUE STAR
ZC 1033.8+1952	10 33 48.	+ 19 52	740		CLUSTER OF GALAXIES
UGC 05762	10 33 48.	+ 23 28	60	16.5	GALAXY S?
ZWG 183.031	10 33 48.	+ 35 18		14.5	GALAXY
UGC 05763	10 33 48.	+ 35 18	90	14.5	GALAXY S
IC 2591	10 33 48.	+ 35 18 40.			NONSTELLAR OBJECT
MCG+12-10-059	10 33 48.	+ 69 05	102	15.	GALAXY
MCG-01-27-026	10 33 48.	- 08 05	156	13.5	GALAXY
REIZ 0511	10 33 49.	+ 31 50	72	15.1	GALAXY
IC 2593	10 33 49.	- 12 27 52.			NONSTELLAR OBJECT
REIZ 0512	10 33 50.	+ 32 33	36	15.5	GALAXY
ARP 267	10 33 53.	+ 31 49			PECULIAR GALAXY
ARC 1050	10 33 53.	+ 45 05		17.2	RICH CLUSTER OF GALAXIES
ZWG 037.089	10 33 54.	+ 02 37		15.6	GALAXY
8ZW 1033+02.6	10 33 54.	+ 02 37		15.6	COMPACT GALAXY
MCG+01-27-019	10 33 54.	+ 02 37	24	15.6	GALAXY
ZWG 037.090	10 33 54.	+ 05 48		15.3	GALAXY
ZC 1033.9+0824	10 33 54.	+ 08 24	940		CLUSTER OF GALAXIES
ZWG 065.065	10 33 54.	+ 13 35		15.4	GALAXY
ZC 1033.9+1552	10 33 54.	+ 15 52	940		CLUSTER OF GALAXIES
ZWG 154.031	10 33 54.	+ 31 49		15.6	GALAXY
UGC 05764	10 33 54.	+ 31 49	120	15.6	GALAXY DWRF IR
TON-N 0530	10 33 54.	+ 32 42		14.6	BLUE STAR
MCG+06-23-022	10 33 54.	+ 35 19	60	14.	GALAXY
ZC 1033.9+3624	10 33 54.	+ 36 24	3430		CLUSTER OF GALAXIES
ZWG 333.043	10 33 54.	+ 69 05		15.1	GALAXY
UGC 05765	10 33 54.	+ 69 05	108	15.1	GALAXY Sb-c
MRSL 285+00/2	10 33 54.	- 57 48	900		HII REGION
OCL 0816	10 33 54.	- 57 58	1620	7.5	OPEN STAR CLUSTER
VHA 098	10 33 54.	- 57 58	480		STAR CLSTR IN NEBULOSITY
VHE 40B	10 33 54.	- 58 59	102		REFLECTION NEBULA
TON-N 1250	10 33 55.	+ 34 52		15.8	BLUE STAR
SEY 078	10 33 55.	+ 69 04 54.		14.7	FAINT GALAXY
RNGC 3307	10 33 55.	- 27 16		16.0	GALAXY
TON-N 1248	10 33 56.	+ 20 09		15.8	BLUE STAR
TON-N 1249	10 33 56.	+ 21 35		16.9	BLUE STAR
HOLM 204B	10 33 56.	+ 38 18	24	15.1	PART OF MULTIPLE GALAXY
ARC 1053	10 33 57.	+ 31 02		17.2	RICH CLUSTER OF GALAXIES
MCG-04-25-029	10 33 57.	- 27 17	36	16.	GALAXY
RNGC 3293	10 33 57.	- 57 58		7.5	OPEN CLUSTER
RNGC 3300	10 33 58.	+ 14 26		13.0	GALAXY
REIZ 0517	10 33 58.	+ 14 26	36	13.7	GALAXY
REIZ 0514	10 33 59.	+ 38 17	48	14.8	GALAXY
HOLM 204A	10 33 59.	+ 38 17	60	14.3	PART OF MULTIPLE GALAXY
REIZ 0515	10 33 59.	+ 38 18	12	15.5	GALAXY
ZWG 037.091	10 34 00.	+ 03 29		15.6	GALAXY
ZWG 037.092	10 34 00.	+ 06 09		14.7	GALAXY
MCG+01-27-020	10 34 00.	+ 06 09	36	14.7	GALAXY
ZC 1034.0+0732	10 34 00.	+ 07 32	940		CLUSTER OF GALAXIES
ZWG 065.066	10 34 00.	+ 14 26		13.4	GALAXY
UGC 05766	10 34 00.	+ 14 26	108	13.4	GALAXY SB0
MCG+02-27-030	10 34 00.	+ 14 26	78	13.4	GALAXY
LB 10643	10 34 00.	+ 16 54		16.0	FAINT BLUE STAR
LB 10642	10 34 00.	+ 19 01		18.0	FAINT BLUE STAR
LB 10641	10 34 00.	+ 20 08		15.3	FAINT BLUE STAR
MCG+05-25-025	10 34 00.	+ 31 48	120	15.6	GALAXY
ZC 1034.0+4236	10 34 00.	+ 42 36	870		CLUSTER OF GALAXIES
SHAH 053	10 34 00.	+ 45 10		17.0	GROUP OF COMPACT GALAXIES
ZC 1034.0+4626	10 34 00.	+ 46 26	810		CLUSTER OF GALAXIES
ZC 1034.0+5743	10 34 00.	+ 57 43	1280		CLUSTER OF GALAXIES
ZC 1034.0+6030	10 34 00.	+ 60 30	1480		CLUSTER OF GALAXIES
ZWG 313.029	10 34 00.	+ 65 17		15.7	GALAXY
MCG+11-13-036	10 34 00.	+ 65 17	36	15.	GALAXY
ARC 1046	10 34 00.	+ 68 14		17.5	RICH CLUSTER OF GALAXIES
ZWG 009.085	10 34 00.	- 01 00		15.3	GALAXY
MCG-04-25-030	10 34 00.	- 25 07	36	14.	GALAXY
MCG-04-25-031	10 34 00.	- 26 56	21	14.	GALAXY
REIZ 0518	10 34 01.	+ 20 41	12	15.1	GALAXY
ARC 1051	10 34 01.	+ 46 29		17.5	RICH CLUSTER OF GALAXIES
RNGC 3305	10 34 01.	- 26 56		14.0	GALAXY
RNGC 3308	10 34 01.	- 27 11		13.0	GALAXY
PK288-05.1	10 34 02.3	- 64 03 37.	12	14.2	PLANETARY NEBULA
ZC 1034.1+3135	10 34 06.	+ 31 35	870		CLUSTER OF GALAXIES
MCG+06-23-023	10 34 06.	+ 38 20	30	16.5	GALAXY
MCG+06-23-024	10 34 06.	+ 38 20 30.	24	17.5	GALAXY
ZWG 212.029	10 34 06.	+ 40 13		15.7	GALAXY
MCG-01-27-027	10 34 06.	- 06 51	60	15.	GALAXY
MCG-04-25-032	10 34 06.	- 27 11 30.	18	13.	GALAXY
IC 0625	10 34 07.	- 23 39 41.			NONSTELLAR OBJECT
MCG+08-19-043	10 34 09.	+ 50 23	15	15.2	GALAXY
MCG-04-25-033	10 34 09.	- 26 44 30.	42	15.	GALAXY
HELW 017	10 34 09.	- 27 04			NEBULA
RNGC 3298	10 34 11.	+ 50 22		15.0	GALAXY
ZC 1034.2+1240	10 34 12.	+ 12 40	1750		CLUSTER OF GALAXIES
LB 10645	10 34 12.	+ 17 44		17.1	FAINT BLUE STAR
LB 10644	10 34 12.	+ 20 22		15.7	FAINT BLUE STAR
MCG+04-25-035	10 34 12.	+ 22 08	162	12.2	GALAXY
ZWG 124.045	10 34 12.	+ 22 09		12.2	GALAXY
UGC 05767	10 34 12.	+ 22 09	204	12.2	GALAXY SB0
MCG+05-25-026	10 34 12.	+ 29 04	36	15.7	GALAXY
ZWG 154.032	10 34 12.	+ 29 06		15.7	GALAXY
MCG+07-22-032	10 34 12.	+ 40 01	36	16.	GALAXY
ZC 1034.2+4154	10 34 12.	+ 41 54	2490		CLUSTER OF GALAXIES
ZC 1034.2+4510	10 34 12.	+ 45 10	2220		CLUSTER OF GALAXIES
ZWG 241.001	10 34 12.	+ 50 22		15.0	GALAXY
ZWG 240.065	10 34 12.	+ 50 22		15.0	GALAXY
ZWG 290.059	10 34 12.	+ 60 15		15.7	GALAXY
UGC 05768	10 34 12.	+ 60 15	78	15.7	GALAXY S
MCG+10-15-115	10 34 12.	+ 60 15	60	16.	GALAXY
MCG+12-10-060	10 34 12.	+ 72 28	33	16.	GALAXY
RNGC 3301	10 34 14.	+ 22 08		12.5	GALAXY
REIZ 0520	10 34 14.	+ 22 08	192	12.2	GALAXY
GO 014	10 34 14.	- 60 50	162		STAR CHAIN
MCG-05-25-021	10 34 15.	- 27 46	54	15.5	GALAXY
REIZ 0516	10 34 16.	+ 50 22	30	15.2	GALAXY
REIZ 0519	10 34 17.	+ 37 43	60	15.0	GALAXY
ZWG 065.067	10 34 18.	+ 09 55		15.7	GALAXY
REIZ 0522	10 34 18.	+ 18 23	30	14.0	GALAXY
MCG+03-27-066	10 34 18.	+ 18 24	180	14.5	GALAXY
ZWG 124.046	10 34 18.	+ 20 43		15.4	GALAXY
UGC 05770	10 34 18.	+ 20 43	60	15.4	GALAXY
UGC 05769	10 34 18.	+ 20 43	66	15.4	GALAXY
REIZ 0521	10 34 18.	+ 29 05	24	15.3	GALAXY
ZC 1034.3+4251	10 34 18.	+ 42 51	740		CLUSTER OF GALAXIES
ZWG 212.030	10 34 18.	+ 43 51		14.6	GALAXY
UGC 05771	10 34 18.	+ 43 51	102	14.6	GALAXY S0-a
MCG-04-25-034	10 34 18.	- 27 16 30.	24	12.5	GALAXY
ARP 192	10 34 19.	+ 18 23			PECULIAR GALAXY
TON-N 1251	10 34 19.	+ 36 06		16.8	BLUE STAR
RNGC 3309	10 34 19.	- 27 16		13.0	GALAXY
RNGC 3303	10 34 21.	+ 18 24		14.5	GALAXY
MCG-03-27-021	10 34 21.	- 17 51	84	14.	GALAXY
SEY 079	10 34 23.	+ 55 17 00.		15.4	FAINT GALAXY
ZWG 009.086	10 34 24.	+ 00 29		14.7	GALAXY
UGC 05772	10 34 24.	+ 00 29	60	14.7	GALAXY S
MCG+00-27-031	10 34 24.	+ 00 29	60	14.7	GALAXY
ZWG 094.096	10 34 24.	+ 18 24		14.5	GALAXY
UGC 05773	10 34 24.	+ 18 24	162	14.5	GALAXY PECULR
VV 071	10 34 24.	+ 18 24	108		INTERACTING GALAXY
KARA.72 240B	10 34 24.	+ 18 24	36		PART OF DOUBLE GALAXY
KARA.72 240A	10 34 24.	+ 18 24	36	14.5	PART OF DOUBLE GALAXY
ZC 1034.4+1932	10 34 24.	+ 19 32	1340		CLUSTER OF GALAXIES
LB 10646	10 34 24.	+ 20 33		16.8	FAINT BLUE STAR
ZC 1034.4+3435	10 34 24.	+ 34 35	3700		CLUSTER OF GALAXIES
MCG+07-22-033	10 34 24.	+ 43 51	24	14.5	GALAXY
ZC 1034.4+5416	10 34 24.	+ 54 16	1810		CLUSTER OF GALAXIES
ZWG 333.044	10 34 24.	+ 72 28		15.5	GALAXY
MCG-04-25-035	10 34 24.	- 25 00	48	16.	GALAXY
MCG-04-25-036	10 34 24.	- 27 17	18	13.	GALAXY
MRSL 287-02/2	10 34 24.	- 60 34	4500		HII REGION
REIZ 0524	10 34 25.	+ 20 42	18	15.1	GALAXY
ARC 1054	10 34 25.	+ 42 54		17.2	RICH CLUSTER OF GALAXIES
RNGC 3311	10 34 25.	- 27 17		13.0	GALAXY
REIZ 0525	10 34 27.	+ 12 53	42	13.3	GALAXY
MCG-01-27-028	10 34 27.	- 06 45		14.	GALAXY
IC 0626	10 34 27.	- 06 45 59.			NONSTELLAR OBJECT
MCG-05-25-022	10 34 27.	- 27 37	36	15.5	GALAXY
RNGC 3306	10 34 28.	+ 12 55		13.5	GALAXY
ARC 1060	10 34 28.	- 27 16		12.7	RICH CLUSTER OF GALAXIES
REIZ 0523	10 34 29.	+ 37 20	90	15.0	GALAXY
ARC 1055	10 34 29.	+ 37 26		17.2	RICH CLUSTER OF GALAXIES
ZWG 065.068	10 34 30.	+ 12 54		13.7	GALAXY
UGC 05774	10 34 30.	+ 12 54	84	13.7	GALAXY Sc?
MCG+02-27-032	10 34 30.	+ 12 54	84	13.7	GALAXY
ZWG 065.069	10 34 30.	+ 13 02		15.2	GALAXY
MCG+02-27-031	10 34 30.	+ 13 02	60	15.2	GALAXY
LB 10647	10 34 30.	+ 17 20		16.0	FAINT BLUE STAR
MCG+05-25-027	10 34 30.	+ 27 36	42	15.5	GALAXY
ZC 1034.5+3329	10 34 30.	+ 33 29	3090		CLUSTER OF GALAXIES
UGC 05775	10 34 30.	+ 37 20	90	16.0	GALAXY Sc
ZWG 212.031	10 34 30.	+ 43 55		15.7	GALAXY
ZC 1034.5+5254	10 34 30.	+ 52 54	2890		CLUSTER OF GALAXIES
MCG+09-18-003	10 34 30.	+ 55 07 30.	36	18.	GALAXY
MCG+09-18-004	10 34 30.	+ 55 18 30.	54	15.	GALAXY
ZWG 313.030	10 34 30.	+ 64 32		14.4	GALAXY
UGC 05776	10 34 30.	+ 64 32	30	14.4	GALAXY VY CMPT
ZC 1034.5-0015	10 34 30.	- 00 15	2350		CLUSTER OF GALAXIES
MCG-05-25-023	10 34 30.	- 32 06	84	14.	GALAXY
VHE 40A	10 34 30.	- 58 42	96		REFLECTION NEBULA
ARC 1057	10 34 33.	+ 12 41		17.2	RICH CLUSTER OF GALAXIES
MCG-04-25-037	10 34 33.	- 26 48	24	15.5	GALAXY
MCG-05-25-024	10 34 33.	- 27 37	42	15.5	GALAXY
ZWG 154.033	10 34 36.	+ 27 37		15.5	GALAXY
TON-N 0036	10 34 36.	+ 28 38		16.7	BLUE STAR
MCG+06-23-025	10 34 36.	+ 37 20	66	16.	GALAXY
MCG+07-22-034	10 34 36.	+ 43 55	48	15.	GALAXY
MCG+08-19-044	10 34 36.	+ 46 18	24	16.	GALAXY
ZWG 267.003	10 34 36.	+ 55 16		15.4	GALAXY
ZWG 266.063	10 34 36.	+ 55 16		15.4	GALAXY
ZC 1034.6+6320	10 34 36.	+ 63 20	2690		CLUSTER OF GALAXIES
7ZW 339	10 34 36.	+ 64 33			COMPACT GALAXY
MCG-04-25-038	10 34 36.	- 25 56 30.	60	15.5	GALAXY
MCG-05-25-025	10 34 36.	- 27 53	60	14.5	GALAXY
ARC 1059	10 34 41.	- 05 43		17.7	RICH CLUSTER OF GALAXIES
KLEM 14	10 34 41.	- 42 38	600	14.	LINEAR GROUP OF 30 GLXIES
ZWG 037.093	10 34 42.	+ 05 41		15.6	GALAXY
ZWG 065.070	10 34 42.	+ 13 32		15.3	GALAXY
MCG+03-27-067	10 34 42.	+ 17 17	48	15.3	GALAXY
ZWG 094.097	10 34 42.	+ 17 18		15.3	GALAXY
LB 10648	10 34 42.	+ 20 22		15.5	FAINT BLUE STAR
MCG+05-25-028	10 34 42.	+ 27 35	24	15.5	GALAXY
ZWG 184.001	10 34 42.	+ 37 43		14.4	GALAXY
ZWG 183.032	10 34 42.	+ 37 43		14.4	GALAXY
UGC 05777	10 34 42.	+ 37 43	90	14.4	GALAXY Sa
ZC 1034.7+4357	10 34 42.	+ 43 57	1080		CLUSTER OF GALAXIES
MCG+08-19-045	10 34 42.	+ 46 18	60	16.	GALAXY
UGC 05778	10 34 42.	+ 46 20	66	17.	GALAXY Sc
ZC 1034.7+6246	10 34 42.	+ 62 46	1340		CLUSTER OF GALAXIES
MRK 149	10 34 42.	+ 64 34	15	14.5	GALAXY WITH UV CONTINUUM
ZC 1034.7+7116	10 34 42.	+ 71 16	1010		CLUSTER OF GALAXIES
IC 0629	10 34 42.	- 27 18			SAME AS NGC 3312
TON-N 1253	10 34 43.	+ 32 51		13.8	BLUE STAR
TON-N 1252	10 34 44.	+ 22 06		16.2	BLUE STAR
MCG-04-25-039	10 34 45.	- 27 18 30.	192	13.	GALAXY
RNGC 3304	10 34 46.	+ 37 43		14.5	GALAXY
REIZ 0526	10 34 47.	+ 37 42	60	14.5	GALAXY
IC 0627	10 34 47.	- 03 06 12.			NONSTELLAR OBJECT
ZWG 037.094	10 34 48.	+ 02 34		15.4	GALAXY
8ZW 1034+02.6	10 34 48.	+ 02 34		15.4	COMPACT GALAXY
ZWG 065.071	10 34 48.	+ 10 01		15.7	GALAXY
ZWG 065.072	10 34 48.	+ 12 24		15.5	GALAXY
LB 10652	10 34 48.	+ 17 18		18.4	FAINT BLUE STAR
ZWG 094.098	10 34 48.	+ 18 19		15.1	GALAXY
ZC 1034.8+1820	10 34 48.	+ 18 20	870		CLUSTER OF GALAXIES
LB 10651	10 34 48.	+ 18 45		15.8	FAINT BLUE STAR
LB 10650	10 34 48.	+ 18 56		16.2	FAINT BLUE STAR
LB 10649	10 34 48.	+ 19 22		18.3	FAINT BLUE STAR
ZC 1034.8+4057	10 34 48.	+ 40 57	1210		CLUSTER OF GALAXIES

OBJECT NAME	RIGHT ASCEN.	DECLINATION	DIAM.	MAGN.	TYPE OF OBJECT
ZC 1034.8+4800	10 34 48.	+ 48 00	1340		CLUSTER OF GALAXIES
ZWG 009.087	10 34 48.	- 03 06		14.1	GALAXY
MCG+00-27-032	10 34 48.	- 03 06	21	14.1	GALAXY
MCG-04-25-040	10 34 48.	- 26 25	30	16.	GALAXY
RNGC 3312	10 34 49.	- 27 20		13.0	GALAXY
MCG+06-23-026	10 34 51.	+ 37 44	78	14.	GALAXY
ZWG 037.095	10 34 54.	+ 05 53		15.1	GALAXY
UGC 05779	10 34 54.	+ 05 53	66	15.1	GALAXY S
KARA.72 241A	10 34 54.	+ 05 53	66	15.1	PART OF DOUBLE GALAXY
MCG+01-27-021	10 34 54.	+ 05 53	60	15.1	GALAXY
ZC 1034.9+1412	10 34 54.	+ 14 12	1080		CLUSTER OF GALAXIES
LB 10653	10 34 54.	+ 14 52		14.4	FAINT BLUE STAR
ZWG 124.047	10 34 54.	+ 23 56		15.6	GALAXY
ZC 1034.9+2912	10 34 54.	+ 29 12	6180		CLUSTER OF GALAXIES
ZWG 009.088	10 34 54.	- 02 29		15.2	GALAXY
MCG-04-25-041	10 34 54.	- 27 26	84	14.	GALAXY
MRSL 285-00/1	10 34 54.	- 58 34	180		HII REGION
RNGC 3315	10 34 55.	- 27 13			GALAXY
RNGC 3314	10 34 55.	- 27 25		14.0	GALAXY
HELW 018	10 34 57.	- 26 56			NEBULA
HN 0040	10 34 57.	- 27 30			NEBULA
BIGO 514	10 34 58.	- 27 24			NEBULA
GO 024	10 34 58.	- 60 17	42	18.5	STAR CHAIN
IC 0628	10 34 59.	+ 05 51 23.			NONSTELLAR OBJECT
ARC 1056	10 34 59.	+ 42 05		17.0	RICH CLUSTER OF GALAXIES
REIZ 0527	10 34 59.	+ 46 02	24	14.5	GALAXY
HOFF L03	10 35	- 59 57	180		DARK HOLE
HOFF L04	10 35	- 60 22	110		DARK HOLE
8ZW 1035+04.5	10 35 00.	+ 04 31		16.3	COMPACT GALAXY
ZWG 037.096	10 35 00.	+ 05 51		14.8	GALAXY
UGC 05780	10 35 00.	+ 05 51	60	14.8	GALAXY Sa-b
KARA.72 241B	10 35 00.	+ 05 51	66	14.8	PART OF DOUBLE GALAXY
MCG+01-27-022	10 35 00.	+ 05 51	48	14.8	GALAXY
8ZW 1035+11.9	10 35 00.	+ 11 54		17.8	COMPACT GALAXY
LB 10655	10 35 00.	+ 17 32		8.0	FAINT BLUE STAR
LB 10654	10 35 00.	+ 20 19		15.3	FAINT BLUE STAR
ZWG 184.002	10 35 00.	+ 34 05		15.7	GALAXY
ZWG 183.033	10 35 00.	+ 34 05		15.7	GALAXY
MCG+10-15-116	10 35 00.	+ 58 15	51	15.	GALAXY
ZC 1035.0+6156	10 35 00.	+ 61 56	470		CLUSTER OF GALAXIES
MCG-04-25-042	10 35 00.	- 26 56 30.	36	14.5	GALAXY
MCG-04-25-043	10 35 00.	- 27 13 30.	36	14.5	GALAXY
ARC 1058	10 35 01.	+ 34 39		17.2	RICH CLUSTER OF GALAXIES
TON-N 1254	10 35 02.	+ 19 51		16.	BLUE STAR
MCG+08-20-001	10 35 03.	+ 45 05	42	16.	GALAXY
MCG+09-18-005	10 35 03.	+ 54 05	36	16.	GALAXY
ARC 1049	10 35 04.	+ 68 01		17.7	RICH CLUSTER OF GALAXIES
IC 2595	10 35 04.	- 10 51 25.			NONSTELLAR OBJECT
ZC 1035.1+2149	10 35 06.	+ 21 49	2290		CLUSTER OF GALAXIES
ZWG 212.032	10 35 06.	+ 39 26		14.9	GALAXY
ZC 1035.1+7226	10 35 06.	+ 72 26	11220		CLUSTER OF GALAXIES
MCG-04-25-044	10 35 06.	- 25 03 30.	192	13.	GALAXY
RNGC 3318	10 35 06.	- 41 22		13.0	GALAXY
RNGC 3313	10 35 07.	- 25 03		13.0	GALAXY
MCG+07-22-035	10 35 09.	+ 39 25 30.	42	15.	GALAXY
MCG-04-25-045	10 35 09.	- 26 04	144	14.	GALAXY
ZWG 065.073	10 35 12.	+ 11 01		15.1	GALAXY
SN 1970D	10 35 12.	+ 11 01		17.	SUPERNOVA
ZWG 154.034	10 35 12.	+ 28 25		15.7	GALAXY
ZWG 241.002	10 35 12.	+ 45 06		14.9	GALAXY
MCG+09-18-006	10 35 12.	+ 55 13	36	18.	GALAXY
ZC 1035.2+6800	10 35 12.	+ 68 00	540		CLUSTER OF GALAXIES
VHE 41B	10 35 12.	- 58 24	18		REFLECTION NEBULA
RNGC 3317	10 35 13.	- 27 16			NON-EXISTENT OBJECT
RNGC 3316	10 35 13.	- 27 21		15.0	GALAXY
REIZ 0528	10 35 14.	+ 11 01	36	14.6	GALAXY
HN 0041	10 35 15.	- 27 16			NEBULA
UGC 05781	10 35 18.	+ 13 59	66	18.	GALAXY DWARF
ZC 1035.3+5013	10 35 18.	+ 50 13	3700		CLUSTER OF GALAXIES
ZWG 290.060	10 35 18.	+ 59 03		15.4	GALAXY
TON-N 1255	10 35 19.	+ 32 25		17.	BLUE STAR
MCG-04-25-046	10 35 21.	- 27 21	42	15.	GALAXY
LB 10658	10 35 24.	+ 14 41		13.6	FAINT BLUE STAR
LB 10657	10 35 24.	+ 16 52		16.0	FAINT BLUE STAR
LB 10656	10 35 24.	+ 20 08		16.7	FAINT BLUE STAR
MCG+10-15-117	10 35 24.	+ 59 01	15	16.	GALAXY
UGC 05782	10 35 24.	+ 78 53	66	16.0	GALAXY Sb-c
MCG-04-25-048	10 35 24.	- 26 03	42	15.	GALAXY
MCG-04-25-047	10 35 24.	- 27 09	54	15.5	GALAXY
VHE 41C	10 35 24.	- 58 13	18		REFLECTION NEBULA
OCL 0819	10 35 24.	- 58 22	360		OPEN STAR CLUSTER
MCG-04-25-049	10 35 27.	- 26 23	66	14.	GALAXY
HELW 019	10 35 27.	- 26 50			NEBULA
HELW 020	10 35 27.	- 26 52			NEBULA
IC 2599	10 35 27.	- 58 21 28.			NONSTELLAR OBJECT
ZWG 009.089	10 35 30.	+ 02 00		15.0	GALAXY
UGC 05783	10 35 30.	+ 02 00	72	15.0	GALAXY SBb
MCG+00-27-033	10 35 30.	+ 02 00	66	15.0	GALAXY
ZWG 037.097	10 35 30.	+ 05 09		15.0	GALAXY
REIZ 0530	10 35 30.	+ 05 09	54	14.4	GALAXY
UGC 05784	10 35 30.	+ 05 09	84	15.0	GALAXY S
MCG+01-27-023	10 35 30.	+ 05 09	60	15.0	GALAXY
ZWG 065.074	10 35 30.	+ 10 38		15.1	GALAXY
MCG+02-27-033	10 35 30.	+ 10 38	42	15.1	GALAXY
LB 10659	10 35 30.	+ 19 58		15.8	FAINT BLUE STAR
TON-N 0531	10 35 30.	+ 25 48		15.6	BLUE STAR
ZWG 154.035	10 35 30.	+ 27 57		15.6	GALAXY
SHAH 054	10 35 30.	+ 40 30	330	16.0	GROUP OF COMPACT GALAXIES
ZWG 290.061	10 35 30.	+ 56 49		15.2	GALAXY
MCG-04-25-051	10 35 30.	- 26 50	30	13.5	GALAXY
MCG-04-25-050	10 35 30.	- 26 52 30.	36	15.	GALAXY
DV.56 N3318B	10 35 30.	- 41 12	72		SB(s)c GALAXY
RNGC 3318B	10 35 30.	- 41 12			GALAXY
CED 108	10 35 30.	- 58 22	930		DIFFUSE GALACTIC NEBULA
VHE 41A	10 35 30.	- 58 28	144		REFLECTION NEBULA
TON-N 1256	10 35 31.	+ 32 27		16.8	BLUE STAR
RNGC 3324	10 35 33.	- 58 22			DIFFUSE NEBULA
GO 027	10 35 35.	- 60 17	107	19.0	STAR CHAIN
ZWG 037.098	10 35 36.	+ 06 12		15.5	GALAXY
8ZW 1035+07.1	10 35 36.	+ 07 04		17.7	COMPACT GALAXY
ZC 1035.6+1901	10 35 36.	+ 19 01	1410		CLUSTER OF GALAXIES
LB 10660	10 35 36.	+ 20 04		17.0	FAINT BLUE STAR
ZWG 154.036	10 35 36.	+ 30 25		15.6	GALAXY
UGC 05785	10 35 36.	+ 30 25	60	15.6	GALAXY S
ZWG 009.090	10 35 36.	- 02 24		15.3	GALAXY
MCG-05-25-026	10 35 36.	- 28 37	48	15.	GALAXY
TON-N 1257	10 35 38.	+ 20 02		17.	BLUE STAR
TON-N 1258	10 35 38.	+ 20 41		14.7	BLUE STAR
GO 013	10 35 38.	- 61 09	61	19.9	STAR CHAIN
MCG+05-25-029	10 35 39.	+ 30 23	36	15.6	GALAXY
ARP 217	10 35 40.	+ 53 46			PECULIAR GALAXY
ZWG 037.099	10 35 42.	+ 04 11		15.5	GALAXY
LB 10661	10 35 42.	+ 18 18		18.3	FAINT BLUE STAR
REIZ 0531	10 35 42.	+ 30 25	36	15.2	GALAXY
ZC 1035.7+3157	10 35 42.	+ 31 57	1210		CLUSTER OF GALAXIES
MCG+08-20-002	10 35 42.	+ 44 46 30.	24	17.	GALAXY
ZWG 241.003	10 35 42.	+ 44 47		15.0	GALAXY
MRK 150	10 35 42.	+ 44 47	12	15.5	GALAXY WITH UV CONTINUUM
ZWG 267.004	10 35 42.	+ 53 45		11.0	GALAXY
UGC 05786	10 35 42.	+ 53 45	228	11.0	GALAXY S(b)
SN 1974C	10 35 42.	+ 53 45		16.5	SUPERNOVA
MCG+09-18-008	10 35 42.	+ 53 46 30.	198	10.8	GALAXY
MCG+09-18-007	10 35 42.	+ 54 14	60	16.	GALAXY
MCG+10-15-118	10 35 42.	+ 56 48	30	15.	GALAXY
ZWG 351.030	10 35 42.	+ 79 10		15.7	GALAXY
MCG-05-25-027	10 35 42.	- 28 30	18	15.	GALAXY
VHE 42B	10 35 42.	- 57 45	96		REFLECTION NEBULA
REIZ 0529	10 35 43.	+ 53 45	54	10.8	GALAXY
GO 038	10 35 44.	- 59 04	74	18.9	STAR CHAIN
RNGC 3310	10 35 45.	+ 53 46		11.5	GALAXY
IC 2597	10 35 46.	- 26 47 52.			NONSTELLAR OBJECT
LB 10663	10 35 48.	+ 18 34		16.8	FAINT BLUE STAR
LB 10662	10 35 48.	+ 19 24		17.5	FAINT BLUE STAR
ZC 1035.8+2551	10 35 48.	+ 25 51	5650		CLUSTER OF GALAXIES
MCG+09-18-009	10 35 48.	+ 54 57 30.	42	17.	GALAXY
REIZ 0532	10 35 53.	+ 29 27	60	15.9	GALAXY
8ZW 1035+07.0	10 35 54.	+ 06 59		17.3	COMPACT GALAXY
ZC 1035.9+1220	10 35 54.	+ 12 20	610		CLUSTER OF GALAXIES
ZC 1035.9+1429	10 35 54.	+ 14 29	670		CLUSTER OF GALAXIES
ZC 1035.9+1605	10 35 54.	+ 16 05	1550		CLUSTER OF GALAXIES
STOCK 15	10 35 54.	+ 28 55			BLUE KNOT NEAR ELLIP GLXY
ZWG 009.091	10 35 54.	- 02 19		14.7	GALAXY
UGC 05787	10 35 54.	- 02 19	120	14.7	GALAXY SBb
MCG+00-27-034	10 35 54.	- 02 19	114	14.7	GALAXY
MCG-05-25-028	10 35 54.	- 28 36	72	15.5	GALAXY
MRSL 285+00/3	10 35 54.	- 57 47	240		HII REGION
MRSL 286-00/1	10 35 54.	- 58 24	900		HII REGION
REIZ 0534	10 35 57.	+ 11 29	60	15.1	GALAXY
GO 018	10 35 58.	- 60 50	59	19.3	STAR CHAIN
REIZ 0533	10 35 59.	+ 29 21	24	15.4	GALAXY
ZC 1036.0+0140	10 36 00.	+ 01 40	3160		CLUSTER OF GALAXIES
ZWG 065.075	10 36 00.	+ 11 28		15.5	GALAXY
MCG+02-27-034	10 36 00.	+ 11 28	60	15.5	GALAXY
ARC 1062	10 36 00.	+ 16 09		17.2	RICH CLUSTER OF GALAXIES
ZC 1036.0+1759	10 36 00.	+ 17 59	540		CLUSTER OF GALAXIES
LB 10664	10 36 00.	+ 19 06		16.4	FAINT BLUE STAR
ZC 1036.0+2659	10 36 00.	+ 26 59	3560		CLUSTER OF GALAXIES
MCG+09-18-010	10 36 00.	+ 54 36	30	16.	GALAXY
MCG-01-27-029	10 36 00.	- 06 54 30.		13.5	GALAXY
VHA 099	10 36 00.	- 58 56	900		STAR CLSTR IN NEBULOSITY
TON-N 1259	10 36 01.	+ 37 00		17.	BLUE STAR
IC 0630	10 36 03.	- 06 54 46.			NONSTELLAR OBJECT
LB 10666	10 36 06.	+ 16 50		15.9	FAINT BLUE STAR
LB 10665	10 36 06.	+ 20 00		15.2	FAINT BLUE STAR
ZC 1036.1-0158	10 36 06.	- 01 58	1280		CLUSTER OF GALAXIES
ARC 1063	10 36 08.	+ 18 56		17.2	RICH CLUSTER OF GALAXIES
TON-N 1260	10 36 08.	+ 20 42		17.	BLUE STAR
REIZ 0536	10 36 09.	+ 11 38	66	14.6	GALAXY
MCG+07-22-036	10 36 09.	+ 41 56	390	11.	GALAXY
HELW 021	10 36 09.	- 27 29			NEBULA
ARC 1064	10 36 11.	+ 01 32		17.1	RICH CLUSTER OF GALAXIES
ZWG 037.100	10 36 12.	+ 05 57		14.6	GALAXY
UGC 05788	10 36 12.	+ 05 57	78	14.6	GALAXY Sb
MCG+01-27-024	10 36 12.	+ 05 57	48	14.6	GALAXY
ZWG 065.076	10 36 12.	+ 12 23		15.4	GALAXY
8ZW 1036+12.4	10 36 12.	+ 12 23		15.4	COMPACT GALAXY
SCH 11	10 36 12.	+ 12 23		15.4	PECULIAR GALAXY
MCG+02-27-036	10 36 12.	+ 12 23	24	15.4	GALAXY
MCG+02-27-035	10 36 12.	+ 12 23	18	15.4	GALAXY
ZC 1036.2+1534	10 36 12.	+ 15 34	810		CLUSTER OF GALAXIES
LB 10668	10 36 12.	+ 19 47		15.6	FAINT BLUE STAR
LB 10667	10 36 12.	+ 20 20		17.4	FAINT BLUE STAR
ZWG 212.033	10 36 12.	+ 41 57		12.0	GALAXY
UGC 05789	10 36 12.	+ 41 57	450	12.0	GALAXY SBc
KARA.73B 0428	10 36 12.	+ 41 57	402	12.0	ISOLATED GALAXY S
MCG+13-08-030	10 36 12.	+ 78 52	39	15.	GALAXY
RNGC 3319	10 36 15.	+ 41 57		11.5	GALAXY
REIZ 0535	10 36 16.	+ 37 30	24	15.9	GALAXY
RNGC 3322	10 36 16.	- 11 07			NON-EXISTENT OBJECT
ZWG 037.101	10 36 18.	+ 05 55		15.5	GALAXY
MCG+08-20-003	10 36 18.	+ 49 08	18	17.	GALAXY
MCG-04-25-052	10 36 18.	- 27 28	60	14.5	GALAXY
MCG-05-25-029	10 36 18.	- 28 17	60	15.	GALAXY
MCG-03-27-022	10 36 21.	- 16 31	18	15.5	GALAXY
RNGC 3321	10 36 22.	- 11 23		14.0	GALAXY
ZWG 037.102	10 36 24.	+ 08 11		15.7	GALAXY
ZC 1036.4+1101	10 36 24.	+ 11 01	540		CLUSTER OF GALAXIES
ZC 1036.4+1632	10 36 24.	+ 16 32	810		CLUSTER OF GALAXIES
LB 10672	10 36 24.	+ 17 34		17.8	FAINT BLUE STAR
LB 10671	10 36 24.	+ 17 48		15.5	FAINT BLUE STAR
LB 10670	10 36 24.	+ 18 00		15.2	FAINT BLUE STAR
LB 10669	10 36 24.	+ 19 34		16.4	FAINT BLUE STAR
TON-N 0037	10 36 24.	+ 29 31		14.5	BLUE STAR
MCG+07-22-038	10 36 24.	+ 39 20	24	16.	GALAXY
MCG+07-22-037	10 36 24.	+ 39 20	24	16.	GALAXY
MCG-02-27-010	10 36 24.	- 11 23	132	14.	GALAXY
MCG-05-25-030	10 36 24.	- 31 11	54	17.	GALAXY
TON-N 1261	10 36 26.	+ 20 39		15.9	BLUE STAR
TON-N 1262	10 36 26.	+ 21 16		17.	BLUE STAR
IC 0631	10 36 29.	- 06 47 41.			NONSTELLAR OBJECT
ZWG 037.103	10 36 30.	+ 04 54		15.6	GALAXY
UGC 05790	10 36 30.	+ 04 54	90	15.6	GALAXY S?
LB 10673	10 36 30.	+ 14 34		15.5	FAINT BLUE STAR
ZWG 124.048	10 36 30.	+ 20 34		15.6	GALAXY
MCG+05-25-030	10 36 30.	+ 31 22	42	15.	GALAXY
TON-N 0532	10 36 30.	+ 31 35		15.3	BLUE STAR
ZWG 241.004	10 36 30.	+ 48 12		14.4	GALAXY
UGC 05791	10 36 30.	+ 48 12	102	14.4	GALAXY
KARA.72 242A	10 36 30.	+ 48 12	78	14.4	PART OF DOUBLE GALAXY
MCG+08-20-004	10 36 30.	+ 48 12 30.	72	14.	GALAXY
MCG+08-20-005	10 36 30.	+ 48 24	42	16.	GALAXY
GO 031	10 36 30.	- 59 56			STAR CHAIN
REIZ 0537	10 36 31.	+ 20 33	12	15.0	GALAXY
PK285+01.1	10 36 34.5	- 56 30 55.	5	8.6	PLANETARY NEBULA
RNGC 3330	10 36 35.	- 53 53		8.5	OPEN CLUSTER
8ZW 1036+08.1	10 36 36.	+ 08 08		17.7	COMPACT GALAXY

OBJECT NAME	RIGHT ASCEN.	DECLINATION	DIAM.	MAGN.	TYPE OF OBJECT
ZWG 094.099	10 36 36.	+ 15 06		15.2	GALAXY
UGC 05793	10 36 36.	+ 15 06	60	15.2	GALAXY Sa-b
LB 10676	10 36 36.	+ 16 19		16.1	FAINT BLUE STAR
LB 10675	10 36 36.	+ 18 43		17.1	FAINT BLUE STAR
LB 10674	10 36 36.	+ 19 42		17.7	FAINT BLUE STAR
MCG+08-20-006	10 36 36.	+ 44 57	18	17.	GALAXY
MCG+08-20-007	10 36 36.	+ 45 00	9	17.	GALAXY
MCG+08-20-008	10 36 36.	+ 45 02	36	17.	GALAXY
ZWG 241.005	10 36 36.	+ 47 39		13.1	GALAXY
RNGC 3320	10 36 36.	+ 47 39		13.0	GALAXY
UGC 05794	10 36 36.	+ 47 39	132	13.1	GALAXY Sc
MCG+08-20-010	10 36 36.	+ 47 40	120	12.9	GALAXY
MCG+08-20-009	10 36 36.	+ 49 11	42	16.	GALAXY
ZWG 009.092	10 36 36.	- 00 09		14.8	GALAXY
UGC 05792	10 36 36.	- 00 09	60	14.8	GALAXY Sa?
MCG+00-27-035	10 36 36.	- 00 09	54	14.8	GALAXY
OCL 0806	10 36 36.	- 53 53	840	9.0	OPEN STAR CLUSTER
VHA 100	10 36 36.	- 53 53	360		OPEN STAR CLUSTER
HOLM 206B	10 36 39.	- 00 09	48	13.8	PART OF MULTIPLE GALAXY
IC 0632	10 36 39.	- 00 09 12.			NONSTELLAR OBJECT
ZWG 009.093	10 36 42.	+ 00 03		14.0	GALAXY
UGC 05795	10 36 42.	+ 00 03	66	14.0	GALAXY E
MCG+00-27-036	10 36 42.	+ 00 03	30	14.0	GALAXY
ZC 1036.7+0236	10 36 42.	+ 02 36	1280		CLUSTER OF GALAXIES
LB 10678	10 36 42.	+ 18 46		16.0	FAINT BLUE STAR
LB 10677	10 36 42.	+ 20 10		15.9	FAINT BLUE STAR
ZC 1036.7+4300	10 36 42.	+ 43 00	1280		CLUSTER OF GALAXIES
FEIG 034	10 36 42.	+ 43 21		10.9	FAINT BLUE STAR
MCG+08-20-011	10 36 42.	+ 48 11	60	14.	GALAXY
VHE 42A	10 36 42.	- 57 47	240		REFLECTION NEBULA
RNGC 3325	10 36 43.	+ 00 03		14.0	GALAXY
TON-N 1263	10 36 44.	+ 20 09		16.	BLUE STAR
ZWG 009.095	10 36 48.	+ 01 58		14.7	GALAXY
UGC 05797	10 36 48.	+ 01 58	60	14.7	GALAXY IRR
MCG+00-27-038	10 36 48.	+ 01 58	60	14.7	GALAXY
ZC 1036.8+0514	10 36 48.	+ 05 14	4440		CLUSTER OF GALAXIES
ZWG 065.077	10 36 48.	+ 13 15		15.3	GALAXY
SCH 12	10 36 48.	+ 13 15		15.3	PECULIAR GALAXY
MCG+02-27-037	10 36 48.	+ 13 15	24	15.3	GALAXY
LB 10682	10 36 48.	+ 15 07		15.3	FAINT BLUE STAR
LB 10681	10 36 48.	+ 17 11		16.3	FAINT BLUE STAR
LB 10680	10 36 48.	+ 18 00		16.0	FAINT BLUE STAR
ZWG 094.100	10 36 48.	+ 18 05		15.3	GALAXY
LB 10679	10 36 48.	+ 18 22		17.2	FAINT BLUE STAR
ZC 1036.8+2434	10 36 48.	+ 24 34	670		CLUSTER OF GALAXIES
LB 00581	10 36 49.	+ 47 22 54.		17.8	FAINT BLUE STAR
ZWG 241.006	10 36 48.	+ 48 11		14.5	GALAXY
UGC 05798	10 36 48.	+ 48 11	60	14.5	GALAXY
KARA.72 242B	10 36 48.	+ 48 11	66	14.5	PART OF DOUBLE GALAXY
ZWG 313.031	10 36 48.	+ 65 45		15.6	GALAXY
ZWG 351.031	10 36 48.	+ 79 13		15.7	GALAXY
ZWG 009.094	10 36 48.	- 00 08		14.5	GALAXY
UGC 05796	10 36 48.	- 00 08	33	14.5	GALAXY
MCG+00-27-037	10 36 48.	- 00 08	36	14.5	GALAXY
HOLM 205A	10 36 50.	+ 24 06	60	13.7	PART OF MULTIPLE GALAXY
ABC 1066	10 36 51.	+ 05 26		16.6	RICH CLUSTER OF GALAXIES
IC 0633	10 36 51.	- 00 07 48.			NONSTELLAR OBJECT
HOLM 206A	10 36 52.	- 00 08	48	13.6	PART OF MULTIPLE GALAXY
ZWG 037.104	10 36 54.	+ 05 22		14.2	GALAXY
RNGC 3326	10 36 54.	+ 05 22		14.0	GALAXY
UGC 05799	10 36 54.	+ 05 22	60	14.2	GALAXY Sa
MCG+01-27-025	10 36 54.	+ 05 22	30	14.2	GALAXY
ZWG 037.105	10 36 54.	+ 06 03		15.3	GALAXY
8ZW 1036+12.1	10 36 54.	+ 12 04		17.3	COMPACT GALAXY
LB 10683	10 36 54.	+ 14 56		16.3	FAINT BLUE STAR
ZWG 124.049	10 36 54.	+ 25 35		14.3	GALAXY
UGC 05800	10 36 54.	+ 25 35	84	14.3	GALAXY SB
MCG+04-25-036	10 36 54.	+ 25 35	66	14.3	GALAXY
ZWG 154.037	10 36 54.	+ 27 00		15.1	GALAXY
TON-N 0533	10 36 54.	+ 31 44		15.4	BLUE STAR
MCG+12-10-061	10 36 54.	+ 70 12	27	16.	GALAXY
MCG-04-25-053	10 36 54.	- 26 35 30.	60	14.5	GALAXY
MCG-05-25-031	10 36 54.	- 30 02	120	14.	GALAXY
RNGC 3323	10 36 56.	+ 25 35		14.5	GALAXY
IC 2598	10 36 57.	+ 26 59 36.			NONSTELLAR OBJECT
LB 10684	10 37 00.	+ 20 02		17.8	FAINT BLUE STAR
ZWG 212.034	10 37 00.	+ 39 23		14.7	GALAXY
MCG+08-20-012	10 37 00.	+ 49 48	42	15.	GALAXY
ZWG 333.045	10 37 00.	+ 70 12		15.5	GALAXY
ZWG 333.046	10 37 00.	+ 70 58		15.6	GALAXY
REIZ 0538	10 37 01.	+ 21 13	42	14.9	GALAXY
HOLM 205B	10 37 01.	+ 24 06	60	14.7	PART OF MULTIPLE GALAXY
MCG+07-22-039	10 37 03.	+ 39 22	45	14.5	GALAXY
GO 078	10 37 05.	- 60 32	57	17.9	STAR CHAIN
ZWG 065.078	10 37 06.	+ 11 55		15.2	GALAXY
KARA.73B 0429	10 37 06.	+ 11 55	42	15.2	ISOLATED GALAXY S
MCG+04-25-037	10 37 06.	+ 22 06	30	15.7	GALAXY
ZWG 124.050	10 37 06.	+ 22 07		15.7	GALAXY
UGC 05801	10 37 06.	+ 22 07	66	15.7	GALAXY
TON-N 0534	10 37 06.	+ 31 24		16.3	BLUE STAR
ZC 1037.1+3952	10 37 06.	+ 39 52	940		CLUSTER OF GALAXIES
ZC 1037.1+5848	10 37 06.	+ 58 48	3160		CLUSTER OF GALAXIES
ZC 1037.1+6731	10 37 06.	+ 67 31	610		CLUSTER OF GALAXIES
MCG-04-25-054	10 37 06.	- 23 30 30.	72	15.	GALAXY
MCG-05-25-032	10 37 06.	- 27 37	36	15.	GALAXY
MCG-05-25-033	10 37 06.	- 29 19	72	15.5	GALAXY
REIZ 0539	10 37 07.	+ 22 06	36	14.9	GALAXY
SN 1965D	10 37 07.	- 27 39		14.0	SUPERNOVA
HOLM 207A	10 37 09.	+ 24 22	60	13.2	PART OF MULTIPLE GALAXY
UGC 05802	10 37 12.	+ 15 52	60	16.0	GALAXY Sc
LB 10688	10 37 12.	+ 17 12		17.7	FAINT BLUE STAR
LB 10687	10 37 12.	+ 17 20		16.3	FAINT BLUE STAR
LB 10686	10 37 12.	+ 19 04		16.1	FAINT BLUE STAR
LB 10685	10 37 12.	+ 20 26		15.9	FAINT BLUE STAR
ZWG 124.051	10 37 12.	+ 24 21		14.2	GALAXY
UGC 05803	10 37 12.	+ 24 21	84	14.2	GALAXY Sb
MCG+04-25-038	10 37 12.	+ 24 21 30.	36	14.2	GALAXY
TON-N 0038	10 37 12.	+ 32 02		14.7	BLUE STAR
MCG+07-22-040	10 37 12.	+ 39 21	36	17.	GALAXY
ZC 1037.2+4117	10 37 12.	+ 41 17	2690		CLUSTER OF GALAXIES
MCG-01-27-031	10 37 12.	- 05 12 30.	48	14.	GALAXY
MCG-01-27-030	10 37 12.	- 08 26	36	15.	GALAXY
RNGC 3335	10 37 12.	- 23 40		14.0	GALAXY
MCG-04-25-055	10 37 12.	- 23 40 30.	48	14.5	GALAXY
RNGC 3327	10 37 14.	+ 24 21		14.0	GALAXY
HOLM 207B	10 37 16.	+ 24 23	24	14.4	PART OF MULTIPLE GALAXY
ABC 1061	10 37 16.	+ 67 28		17.7	RICH CLUSTER OF GALAXIES
RNGC 3328	10 37 17.	+ 09 34			NON-EXISTENT OBJECT

OBJECT NAME	RIGHT ASCEN.	DECLINATION	DIAM.	MAGN.	TYPE OF OBJECT
REIZ 0540	10 37 17.	+ 38 44	72	14.3	GALAXY
LB 10689	10 37 18.	+ 19 33		15.5	FAINT BLUE STAR
ZC 1037.3+3134	10 37 18.	+ 31 34	4230		CLUSTER OF GALAXIES
FATH 1.345	10 37 18.	+ 44 52	11		NEBULA
ABC 1072	10 37 18.	+ 57 40		17.2	RICH CLUSTER OF GALAXIES
MCG+12-10-062	10 37 18.	+ 70 10	9	16.	GALAXY
TON-N 1264	10 37 20.	+ 19 36		16.	BLUE STAR
REIZ 0541	10 37 21.	+ 24 22	72	13.7	GALAXY
ABC 1069	10 37 21.	- 08 22		15.1	RICH CLUSTER OF GALAXIES
ZWG 037.106	10 37 24.	+ 03 22		15.7	GALAXY
LB 10691	10 37 24.	+ 15 25		15.0	FAINT BLUE STAR
LB 10690	10 37 24.	+ 20 09		16.3	FAINT BLUE STAR
ZC 1037.4+2156	10 37 24.	+ 21 56	6320		CLUSTER OF GALAXIES
MCG+07-22-041	10 37 24.	+ 38 44	84	14.5	GALAXY
ZWG 212.035	10 37 24.	+ 38 45		15.2	GALAXY
UGC 05804	10 37 24.	+ 38 45	96	15.2	GALAXY Sb
GO 028	10 37 24.	- 60 25	93	18.3	STAR CHAIN
GO 025	10 37 29.	- 60 35	56	18.2	STAR CHAIN
ZWG 037.107	10 37 30.	+ 07 50		15.3	GALAXY
LB 10692	10 37 30.	+ 16 54		16.4	FAINT BLUE STAR
ZWG 094.101	10 37 30.	+ 19 28		15.5	GALAXY
KARA.73B 0430	10 37 30.	+ 19 28	18	15.5	ISOLATED GALAXY E
GO 032	10 37 31.	- 59 51	80	17.1	STAR CHAIN
TON-N 1265	10 37 32.	+ 20 10		17.5	BLUE STAR
RNGC 3333	10 37 34.	- 35 47			GALAXY
PK285+01.2	10 37 34.0	- 56 50 47.	10		PLANETARY NEBULA
ZWG 037.108	10 37 36.	+ 05 37		15.4	GALAXY
ZWG 065.079	10 37 36.	+ 10 35		15.4	GALAXY
8ZW 1037+10.6	10 37 36.	+ 10 35		15.4	COMPACT GALAXY
KARA.73B 0431	10 37 36.	+ 10 35	24	15.4	ISOLATED GALAXY
LB 10693	10 37 36.	+ 14 52		16.5	FAINT BLUE STAR
MCG+04-25-039	10 37 36.	+ 21 52	72	15.4	GALAXY
ZWG 124.052	10 37 36.	+ 21 53		15.4	GALAXY
UGC 05805	10 37 36.	+ 21 53	102	15.4	GALAXY
ZC 1037.6+4013	10 37 36.	+ 40 13	1140		CLUSTER OF GALAXIES
MCG-03-27-023	10 37 36.	- 17 24	60	15.	GALAXY
MCG-05-25-034	10 37 36.	- 29 55	84	14.5	GALAXY
MCG-06-24-001	10 37 36.	- 35 47	108	14.	GALAXY
LB 10694	10 37 42.	+ 14 22		15.1	FAINT BLUE STAR
ZC 1037.7+3322	10 37 42.	+ 33 22	540		CLUSTER OF GALAXIES
MCG+06-24-001	10 37 42.	+ 36 09 30.	24	16.	GALAXY
UGC 05806	10 37 42.	+ 37 36	66	16.0	GALAXY Sc
MCG+07-22-042	10 37 42.	+ 39 18 30.	18	18.	GALAXY
KARA.72 243A	10 37 42.	+ 39 19	36	14.6	PART OF DOUBLE GALAXY
RNGC 3331	10 37 42.	- 23 34		14.0	GALAXY
MCG-04-25-056	10 37 42.	- 23 34	42	14.	GALAXY
MCG-05-25-035	10 37 42.	- 30 00	72	14.	GALAXY
REIZ 0544	10 37 43.	+ 21 52	72	14.7	GALAXY
TON-N 1266	10 37 43.	+ 34 51		16.8	BLUE STAR
ABC 1067	10 37 44.	+ 40 30		16.6	RICH CLUSTER OF GALAXIES
REIZ 0542	10 37 45.	+ 36 32	60	14.8	GALAXY
REIZ 0543	10 37 46.	+ 37 34	42	14.8	GALAXY
ABC 1065	10 37 46.	+ 57 09		17.2	RICH CLUSTER OF GALAXIES
RNGC 3332	10 37 47.	+ 09 27		13.5	GALAXY
ZWG 037.109	10 37 48.	+ 07 06		15.7	GALAXY
8ZW 1037+07.4	10 37 49.	+ 07 23		17.5	COMPACT GALAXY
ZWG 065.080	10 37 48.	+ 09 27		13.7	GALAXY
UGC 05807	10 37 48.	+ 09 27	180	13.7	GALAXY S0
MCG+02-27-038	10 37 48.	+ 09 27	24	13.7	GALAXY
ZWG 065.081	10 37 48.	+ 12 33		14.6	GALAXY
UGC 05808	10 37 48.	+ 12 33	72	14.6	GALAXY Sb
MCG+02-27-039	10 37 48.	+ 12 33	72	14.6	GALAXY
ZWG 065.082	10 37 48.	+ 13 53		15.2	GALAXY
LB 10698	10 37 48.	+ 15 41		17.3	FAINT BLUE STAR
ZC 1037.8+1550	10 37 48.	+ 15 50	1480		CLUSTER OF GALAXIES
LB 10697	10 37 48.	+ 17 18		15.5	FAINT BLUE STAR
LB 10696	10 37 48.	+ 17 39		15.8	FAINT BLUE STAR
LB 10695	10 37 48.	+ 19 52		16.0	FAINT BLUE STAR
MCG+07-22-043	10 37 48.	+ 39 19	45	15.	GALAXY
ZWG 212.036	10 37 48.	+ 39 20		14.6	GALAXY
KARA.72 243B	10 37 48.	+ 39 20	48		PART OF DOUBLE GALAXY
FATH 1.346	10 37 48.	+ 45 26	14		NEBULA
UGC 05809	10 37 48.	+ 69 58	60	17.	GALAXY Sc
REIN 2.113	10 37 50.97	+ 09 26 37.3			NEBULA
MCG+06-24-002	10 37 51.	+ 37 36 30.	42	15.	GALAXY
ZWG 037.110	10 37 54.	+ 07 07		15.6	GALAXY
LB 10699	10 37 54.	+ 17 48		15.8	FAINT BLUE STAR
TON-N 0039	10 37 54.	+ 26 45		16.2	BLUE STAR
UGC 05810	10 37 54.	+ 40 04	72	16.5	GALAXY Sb
7ZW 340	10 37 54.	+ 65 39			COMPACT GALAXY
MCG-05-25-036	10 37 54.	- 27 31	108	13.	GALAXY
ABC 1068	10 37 55.	+ 40 13		17.0	RICH CLUSTER OF GALAXIES
RNGC 3336	10 37 55.	- 27 30		13.0	GALAXY
VDB.66G 085	10 38	- 23 10	70		DWARF GALAXY
SER 077.09	10 38	- 46 03	2700	15.5	CLOUD, ABOUT 50 GALAXIES
MCG+14-05-013	10 38 00.	+ 81 08	54	15.	GALAXY
KLEM 15	10 38 02.	- 45 53		12.	LW LAT CLSTR OF 30 GLXIES
MCG-01-27-032	10 38 03.	- 03 50	42	15.5	GALAXY
RNGC 3347A	10 38 04.	- 36 10			GALAXY
HELW 143	10 38 05.	- 36 09 07.			NEBULA
LB 10702	10 38 06.	+ 15 27		18.0	FAINT BLUE STAR
LB 10701	10 38 06.	+ 19 54		15.8	FAINT BLUE STAR
LB 10700	10 38 06.	+ 20 12		15.5	FAINT BLUE STAR
TON-N 0040	10 38 06.	+ 29 05		15.0	BLUE STAR
ZC 1038.1+4330	10 38 06.	+ 43 30	810		CLUSTER OF GALAXIES
MCG-06-24-002	10 38 06.	- 36 09	120	13.	GALAXY
DV.56 N3347A	10 38 06.	- 36 10	120		S GALAXY
ZC 1038.2+1948	10 38 12.	+ 19 48	940		CLUSTER OF GALAXIES
MCG+05-25-031	10 38 12.	+ 28 52	48	14.5	GALAXY
GO 069	10 38 17.	- 60 44	44	17.0	STAR CHAIN
ZWG 037.111	10 38 18.	+ 06 15		15.1	GALAXY
UGC 05811	10 38 18.	+ 06 15	78	15.1	GALAXY PECULR?
IC 0634	10 38 18.	+ 06 15 14.			NONSTELLAR OBJECT
ZWG 065.083	10 38 18.	+ 12 44		15.6	GALAXY
UGC 05812	10 38 18.	+ 12 44	60	15.6	GALAXY IRR
MCG+02-27-040	10 38 18.	+ 12 44	66	15.6	GALAXY
LB 10704	10 38 18.	+ 17 10		14.5	FAINT BLUE STAR
LB 10703	10 38 18.	+ 19 32		17.7	FAINT BLUE STAR
ZWG 184.003	10 38 18.	+ 34 59		15.7	GALAXY
ZWG 184.004	10 38 18.	+ 36 38		15.3	GALAXY
UGC 05813	10 38 18.	+ 36 38	108	15.3	GALAXY Sc
KARA.73B 0432	10 38 18.	+ 36 38	102	15.3	ISOLATED GALAXY S
MCG+06-24-003	10 38 18.	+ 36 38 30.	60	15.	GALAXY
ZWG 351.032	10 38 18.	+ 77 46		15.0	GALAXY
UGC 05814	10 38 18.	+ 77 46	96	15.0	GALAXY PECULR
ZWG 009.097	10 38 18.	- 00 40		15.7	GALAXY
ZWG 009.096	10 38 18.	- 02 03		15.4	GALAXY
GO 047	10 38 18.	- 57 54	37	14.6	STAR CHAIN

OBJECT NAME	RIGHT ASCEN.	DECLINATION	DIAM.	MAGN.	TYPE OF OBJECT
REIZ 0545	10 38 19.	+ 32 31	24	15.4	GALAXY
TON-N 1268	10 38 19.	+ 34 12		15.7	BLUE STAR
TON-N 1267	10 38 20.	+ 22 10		17.	BLUE STAR
AR 14	10 38 20.96	+ 77 45 24.8			NEBULA
VV 113	10 38 21.	+ 01 02	78	15.	INTERACTING GALAXY
REIZ 0546	10 38 21.	+ 36 37	90	14.3	GALAXY
REIZ 0547	10 38 21.	+ 36 38	18	15.0	GALAXY
8ZW 1038+11.4	10 38 24.	+ 11 22		17.5	COMPACT GALAXY
ZC 1038.4+2408	10 38 24.	+ 24 08	1140		CLUSTER OF GALAXIES
ZWG 212.037	10 38 24.	+ 42 38		15.5	GALAXY
MCG+08-20-013	10 38 24.	+ 47 11 30.	30	16.	GALAXY
MCG+09-18-011	10 38 24.	+ 51 04	60	16.	GALAXY
REIZ 0548	10 38 25.	+ 21 33	24	14.3	GALAXY
RNGC 3347C	10 38 28.	- 36 02			GALAXY
GO 019	10 38 29.	- 61 04	50	18.7	STAR CHAIN
ZWG 037.112	10 38 30.	+ 07 04		15.7	GALAXY
8ZW 1038+07.1	10 38 30.	+ 07 04		15.7	COMPACT GALAXY
ZWG 065.084	10 38 30.	+ 10 50		15.6	GALAXY
LB 10705	10 38 30.	+ 20 27		18.5	FAINT BLUE STAR
TON-N 0041	10 38 30.	+ 26 39		15.0	BLUE STAR
ZC 1038.5+3802	10 38 30.	+ 38 02	3490		CLUSTER OF GALAXIES
UGC 05815	10 38 30.	+ 51 05	66	17.	GALAXY Sc
ZC 1038.5+6206	10 38 30.	+ 62 06	3230		CLUSTER OF GALAXIES
MCG-05-25-037	10 38 30.	- 27 40	36	15.	GALAXY
DV.56 N3347C	10 38 30.	- 36 02			SAd GALAXY
TON-N 1269	10 38 32.	+ 21 17		15.5	BLUE STAR
MCG+08-20-014	10 38 33.	+ 47 03 30.	18	16.	GALAXY
RNGC 3334	10 38 34.	+ 37 34		14.0	GALAXY
ARP 156	10 38 34.	+ 77 43			PECULIAR GALAXY
ZWG 037.113	10 38 36.	+ 06 32		15.6	GALAXY
UGC 05816	10 38 36.	+ 06 32	60	15.6	GALAXY SB
ZWG 037.114	10 38 36.	+ 06 38		15.2	GALAXY
KARA.72 244A	10 38 36.	+ 06 38	60	15.2	PART OF DOUBLE GALAXY
MCG+01-27-026	10 38 36.	+ 06 38	12	12.5	GALAXY
ZWG 065.085	10 38 36.	+ 14 10		15.5	GALAXY
ZC 1038.6+1505	10 38 36.	+ 15 05	810		CLUSTER OF GALAXIES
ZC 1038.6+3406	10 38 36.	+ 34 06	2820		CLUSTER OF GALAXIES
ZWG 184.005	10 38 36.	+ 37 34		14.1	GALAXY
UGC 05817	10 38 36.	+ 37 34	84	14.1	GALAXY S0?
MCG+08-20-015	10 38 36.	+ 47 05 30.	24	16.	GALAXY
7ZW 341	10 38 36.	+ 77 45			COMPACT GALAXY
MCG+13-08-031	10 38 36.	+ 77 45 30.	66	14.	GALAXY
ZC 1038.6-0033	10 38 36.	- 00 33	3490		CLUSTER OF GALAXIES
ZWG 009.098	10 38 36.	- 02 42		15.6	GALAXY
MCG+00-27-039	10 38 36.	- 02 42	24	15.6	GALAXY
GO 030	10 38 37.	- 60 19	80	18.2	STAR CHAIN
TON-N 1270	10 38 38.	+ 20 25		14.8	BLUE STAR
MCG-04-25-057	10 38 39.	- 26 50	120	14.	GALAXY
REIZ 0549	10 38 40.	+ 37 33	24	14.1	GALAXY
SHB 160	10 38 40.9	+ 06 25 58.		16.8	QUASI-STELLAR OBJECT
BC 4C06.41	10 38 41.4	+ 06 25 53.		16.81	QUASI-STELLAR OBJECT
ZWG 037.115	10 38 42.	+ 06 37		15.4	GALAXY
UGC 05818	10 38 42.	+ 06 37	90	15.4	GALAXY Sc
KARA.72 244B	10 38 42.	+ 06 37	72	15.4	PART OF DOUBLE GALAXY
MCG+01-27-027	10 38 42.	+ 06 37	60	15.4	GALAXY
MCG+06-24-004	10 38 42.	+ 37 35 30.	24	14.5	GALAXY
ZWG 241.007	10 38 42.	+ 47 05		15.6	GALAXY
ZWG 267.005	10 38 42.	+ 56 12		15.5	GALAXY
ZWG 009.099	10 38 42.	- 02 34		14.8	GALAXY
MCG+00-27-040	10 38 42.	- 02 34	24	14.8	GALAXY
MCG-03-27-024	10 38 42.	- 17 15 30.	66	14.	GALAXY
MCG-06-24-003	10 38 42.	- 36 01	78	16.	GALAXY
TON-N 1271	10 38 43.	+ 36 39		16.9	BLUE STAR
ZWG 037.116	10 38 48.	+ 06 31		15.6	GALAXY
LB 10710	10 38 48.	+ 18 05		15.8	FAINT BLUE STAR
LB 10709	10 38 48.	+ 18 12		15.8	FAINT BLUE STAR
LB 10708	10 38 48.	+ 18 12		15.5	FAINT BLUE STAR
LB 10707	10 38 48.	+ 18 48		18.4	FAINT BLUE STAR
LB 10706	10 38 48.	+ 19 19		17.2	FAINT BLUE STAR
TON-N 0535	10 38 48.	+ 31 35		14.1	BLUE STAR
MCG+07-22-044	10 38 48.	+ 40 17 30.	45	15.	GALAXY
ZWG 212.038	10 38 48.	+ 40 18		15.1	GALAXY
ZWG 313.032	10 38 48.	+ 63 16		15.6	GALAXY
ZC 1038.8-0301	10 38 48.	- 03 01	3160		CLUSTER OF GALAXIES
MCG-03-27-025	10 38 48.	- 20 47	48	14.5	GALAXY
MCG+06-24-005	10 38 51.	+ 33 58 30.	42	16.	GALAXY
LB 10713	10 38 54.	+ 14 27		15.4	FAINT BLUE STAR
LB 10712	10 38 54.	+ 17 04		14.8	FAINT BLUE STAR
LB 10711	10 38 54.	+ 17 30		17.3	FAINT BLUE STAR
ZWG 124.053	10 38 54.	+ 21 28		15.6	GALAXY
MCG+07-22-045	10 38 54.	+ 38 58	84	14.	GALAXY
ZWG 212.039	10 38 54.	+ 38 59		14.5	GALAXY
UGC 05819	10 38 54.	+ 38 59	96	14.5	GALAXY Sb-0
7ZW 342	10 38 54.	+ 63 16			COMPACT GALAXY
ZC 1038.9+6704	10 38 54.	+ 67 04	670		CLUSTER OF GALAXIES
TON-N 1272	10 38 56.	+ 18 59		12.6	BLUE STAR
GO 037	10 38 57.	- 59 20	141	15.5	STAR CHAIN
VDB.66G 084	10 39	+ 34 45	270		DWARF GALAXY
ZWG 037.117	10 39 00.	+ 06 30		15.3	GALAXY
SN 1972D	10 39 00.	+ 12 26		18.5	SUPERNOVA
LB 10714	10 39 00.	+ 15 34		14.9	FAINT BLUE STAR
MCG+03-27-069	10 39 00.	+ 15 53	96	15.2	GALAXY
ZWG 094.102	10 39 00.	+ 15 55		15.2	GALAXY
UGC 05821	10 39 00.	+ 15 55	78	15.2	GALAXY S
ZWG 094.103	10 39 00.	+ 18 36		15.6	GALAXY
MCG+03-27-068	10 39 00.	+ 18 36	42	15.6	GALAXY
ZWG 124.054	10 39 00.	+ 21 35		15.1	GALAXY
ZC 1039.0+2434	10 39 00.	+ 24 34	400		CLUSTER OF GALAXIES
REIZ 0550	10 39 00.	+ 30 59	30	15.1	GALAXY
ZC 1039.0+3258	10 39 00.	+ 32 58	1810		CLUSTER OF GALAXIES
MCG+10-15-119	10 39 00.	+ 57 33	27	17.	GALAXY
ZWG 364.017	10 39 00.	+ 81 10		15.2	GALAXY
UGC 05820	10 39 00.	+ 81 10	108	15.2	GALAXY Sb-c
KARA.73B 0433	10 39 00.	+ 81 10	72	15.2	ISOLATED GALAXY S
MCG-04-25-058	10 39 00.	- 23 08	72	15.	GALAXY
MCG-06-24-004	10 39 00.	- 36 53	60	15.	GALAXY
HELW 144	10 39 03.	- 36 05 21.			NEBULA
ZC 1039.1+0145	10 39 06.	+ 01 45	340		CLUSTER OF GALAXIES
ZWG 037.118	10 39 06.	+ 06 27		15.7	GALAXY
ZC 1039.1+1245	10 39 06.	+ 12 45	1680		CLUSTER OF GALAXIES
IC 0635	10 39 06.	+ 15 53 42.			NONSTELLAR OBJECT
ZWG 124.055	10 39 06.	+ 21 31		15.5	GALAXY
UGC 05822	10 39 06.	+ 21 31	84	15.5	GALAXY SBa
REIZ 0551	10 39 07.	+ 21 30	9	15.0	GALAXY
MCG+04-25-040	10 39 09.	+ 21 33	42	15.1	GALAXY
HOLM 208A	10 39 10.	+ 14 09	24	15.2	PART OF MULTIPLE GALAXY
GO 016	10 39 11.	- 61 16	55	18.8	STAR CHAIN
ZWG 037.119	10 39 12.	+ 05 15		15.3	GALAXY

OBJECT NAME	RIGHT ASCEN.	DECLINATION	DIAM.	MAGN.	TYPE OF OBJECT
RNGC 3337	10 39 12.	+ 05 15		15.5	GALAXY
LB 10717	10 39 12.	+ 15 08		15.4	FAINT BLUE STAR
LB 10716	10 39 12.	+ 18 02		16.0	FAINT BLUE STAR
LB 10715	10 39 12.	+ 18 16		18.5	FAINT BLUE STAR
ZC 1039.2+2020	10 39 12.	+ 20 20	810		CLUSTER OF GALAXIES
MCG+04-25-041	10 39 12.	+ 21 30	48	15.5	GALAXY
MCG+08-20-016	10 39 12.	+ 48 01 30.	30	16.	GALAXY
ZC 1039.2-0221	10 39 12.	- 02 21	540		CLUSTER OF GALAXIES
TON-N 1273	10 39 14.	+ 21 57		12.6	BLUE STAR
IC 0636	10 39 15.	+ 04 35 23.			NONSTELLAR OBJECT
HOLM 208B	10 39 15.	+ 14 09	18	15.4	PART OF MULTIPLE GALAXY
HUB E17	10 39 15.	- 59 39			DIFFUSE NEBULA
GO 017	10 39 17.	- 61 12	44	19.4	STAR CHAIN
ZWG 009.100	10 39 18.	+ 01 03		14.9	GALAXY
UGC 05823	10 39 18.	+ 01 03	60	14.9	GALAXY IRR
MCG+00-27-041	10 39 18.	+ 01 03	60	14.9	GALAXY
ZWG 037.120	10 39 18.	+ 04 05		15.5	GALAXY
ZWG 037.121	10 39 18.	+ 04 36		14.9	GALAXY
UGC 05824	10 39 18.	+ 04 36	66	14.9	GALAXY SB
MCG+01-27-028	10 39 18.	+ 04 36	42	14.9	GALAXY
ZC 1039.3+1109	10 39 18.	+ 11 09	5310		CLUSTER OF GALAXIES
LB 10719	10 39 18.	+ 18 30		17.1	FAINT BLUE STAR
ZWG 094.104	10 39 18.	+ 18 35		15.4	GALAXY
MCG+03-27-070	10 39 18.	+ 18 36	30	15.4	GALAXY
LB 10718	10 39 18.	+ 19 38		18.5	FAINT BLUE STAR
MCG+05-25-032	10 39 19.	+ 31 02	48	15.5	GALAXY
TON-N 0536	10 39 18.	+ 31 05		15.7	BLUE STAR
ZWG 241.008	10 39 18.	+ 48 01		15.1	GALAXY
MRK 151	10 39 18.	+ 48 01	15	14.5	GALAXY WITH UV CONTINUUM
PK285+02.1	10 39 19.7	- 55 53 36.			PLANETARY NEBULA
HOLM 209B	10 39 21.	+ 18 31	12	14.8	PART OF MULTIPLE GALAXY
ZWG 065.086	10 39 24.	+ 12 35		15.7	GALAXY
ZC 1039.4+1649	10 39 24.	+ 16 49	3830		CLUSTER OF GALAXIES
LB 10721	10 39 24.	+ 17 36		15.5	FAINT BLUE STAR
ZWG 094.105	10 39 24.	+ 18 30		15.1	GALAXY
MCG+03-27-071	10 39 24.	+ 18 30	48	15.1	GALAXY
HOLM 209A	10 39 24.	+ 18 31	18	14.2	PART OF MULTIPLE GALAXY
ZC 1039.4+1853	10 39 24.	+ 18 53	4840		CLUSTER OF GALAXIES
LB 10720	10 39 24.	+ 20 18		19.0	FAINT BLUE STAR
ZWG 124.056	10 39 24.	+ 24 00		15.1	GALAXY
UGC 05825	10 39 24.	+ 24 00	72	15.1	GALAXY Sa?
MCG+04-25-042	10 39 24.	+ 24 01	66	15.1	GALAXY
ZWG 154.038	10 39 24.	+ 31 03		15.5	GALAXY
REIZ 0553	10 39 24.	+ 31 03	54	14.6	GALAXY
ZC 1039.4+3941	10 39 24.	+ 39 41	1410		CLUSTER OF GALAXIES
MCG+09-18-012	10 39 24.	+ 55 28	24	17.	GALAXY
ZC 1039.4+5722	10 39 24.	+ 57 22	2350		CLUSTER OF GALAXIES
MCG-01-27-033	10 39 24.	- 07 57	18	15.	GALAXY
KLEM 16	10 39 25.	- 36 03	2400	11.	BRIGHT GRP OF 5 GALAXIES
TON-N 1274	10 39 26.	+ 21 59		17.	BLUE STAR
REIZ 0554	10 39 28.	+ 14 00	150	12.2	GALAXY
RNGC 3338	10 39 28.	+ 14 01		11.5	GALAXY
REIN 2.114	10 39 28.30	+ 14 00 33.5			NEBULA
ZWG 065.087	10 39 30.	+ 14 00		12.1	GALAXY
UGC 05826	10 39 30.	+ 14 00	312	12.1	GALAXY Sc
MCG+02-27-041	10 39 30.	+ 14 00	300	12.1	GALAXY
LB 10722	10 39 30.	+ 18 06		16.0	FAINT BLUE STAR
ZWG 290.062	10 39 30.	+ 59 06		15.6	GALAXY
MCG+12-10-063	10 39 30.	+ 70 06	45	16.	GALAXY
TON-N 1275	10 39 31.	+ 33 59		17.	BLUE STAR
RNGC 3347B	10 39 34.	- 36 40			GALAXY
ZWG 065.088	10 39 36.	+ 12 45		15.6	GALAXY
MCG+03-27-072	10 39 36.	+ 16 32	42	15.5	GALAXY
LB 10725	10 39 36.	+ 18 32		16.1	FAINT BLUE STAR
LB 10724	10 39 36.	+ 19 00		17.2	FAINT BLUE STAR
ZWG 094.106	10 39 36.	+ 19 22		15.6	GALAXY
LB 10723	10 39 36.	+ 19 58		16.0	FAINT BLUE STAR
MCG+05-25-033	10 39 36.	+ 28 16	48	15.5	GALAXY
ZWG 154.039	10 39 36.	+ 28 17		15.5	GALAXY
MCG+10-15-120	10 39 36.	+ 59 06	36	16.	GALAXY
ZWG 290.063	10 39 36.	+ 60 35		15.2	GALAXY
MCG+12-10-064	10 39 36.	+ 70 37	27	17.	GALAXY
MCG-05-25-038	10 39 36.	- 28 30	30	15.	GALAXY
DV.56 N3347B	10 39 36.	- 36 40	210		S GALAXY
ARC 1073	10 39 38.	+ 36 54		17.2	RICH CLUSTER OF GALAXIES
HOLM 210B	10 39 38.	- 00 07	12	14.8	PART OF MULTIPLE GALAXY
REIZ 0552	10 39 39.	+ 60 36	54	14.7	GALAXY
IC 0637	10 39 41.	+ 15 37 10.			NONSTELLAR OBJECT
ARC 1071	10 39 41.	+ 43 21		17.2	RICH CLUSTER OF GALAXIES
ZC 1039.7+0102	10 39 42.	+ 01 02	4910		CLUSTER OF GALAXIES
ZWG 037.122	10 39 42.	+ 06 28		15.0	GALAXY
MCG+01-27-029	10 39 42.	+ 06 28	36	15.0	GALAXY
8ZW 1039+09.4	10 39 42.	+ 09 27		19.1	COMPACT GALAXY
LB 10726	10 39 42.	+ 15 06		15.5	FAINT BLUE STAR
ZWG 094.107	10 39 42.	+ 15 36		15.3	GALAXY
ZWG 094.108	10 39 42.	+ 16 33		15.5	GALAXY
ZC 1039.7+3650	10 39 42.	+ 36 50	2150		CLUSTER OF GALAXIES
ZC 1039.7+6828	10 39 42.	+ 68 28	540		CLUSTER OF GALAXIES
ZWG 351.033	10 39 42.	+ 79 23		15.5	GALAXY
ZWG 009.101	10 39 42.	- 00 07		13.6	GALAXY
UGC 05827	10 39 42.	- 00 07	60	13.6	GALAXY S
MCG+00-27-042	10 39 42.	- 00 07	48	13.6	GALAXY
MCG-05-25-039	10 39 42.	- 33 00	60	15.	GALAXY
RNGC 3339	10 39 43.	- 00 07		13.5	GALAXY
RNGC 3340	10 39 43.	- 00 49			NON-EXISTENT OBJECT
REIZ 0555	10 39 45.	+ 36 11	120	15.2	GALAXY
REIZ 0556	10 39 46.	- 00 07	60	13.5	GALAXY
HOLM 210A	10 39 46.	- 00 07	60	12.5	PART OF MULTIPLE GALAXY
RNGC 3436	10 39 48.	+ 08 12			GALAXY
MCG+03-27-073	10 39 48.	+ 15 59 30.	78	17.	GALAXY
ZWG 094.109	10 39 48.	+ 16 01		15.7	GALAXY
UGC 05828	10 39 48.	+ 16 01	72	15.7	GALAXY S
KARA.72 245A	10 39 48.	+ 16 01	66	15.7	PART OF DOUBLE GALAXY
LB 10728	10 39 48.	+ 18 02		17.6	FAINT BLUE STAR
ZWG 094.110	10 39 48.	+ 18 51		15.6	GALAXY
LB 10727	10 39 48.	+ 19 58		17.7	FAINT BLUE STAR
ZC 1039.8+2324	10 39 48.	+ 23 24	810		CLUSTER OF GALAXIES
MCG+04-25-043	10 39 48.	+ 24 13	72	15.2	GALAXY
ZWG 184.006	10 39 48.	+ 34 43		15.1	GALAXY
UGC 05829	10 39 48.	+ 34 43	318	15.1	GALAXY DWRF IR
KARA.73B 0434	10 39 48.	+ 34 43	282	15.1	ISOLATED GALAXY S
MCG+10-15-121	10 39 48.	+ 60 35	36	16.	GALAXY
MCG-06-24-005	10 39 48.	- 36 41	180	13.	GALAXY
ZWG 037.123	10 39 54.	+ 03 01		14.9	GALAXY
MCG+01-27-030	10 39 54.	+ 03 01	24	14.9	GALAXY
ZC 1039.9+0913	10 39 54.	+ 09 13	400		CLUSTER OF GALAXIES
MCG+03-27-075	10 39 54.	+ 15 59	48	15.0	GALAXY
ZWG 094.111	10 39 54.	+ 16 00		15.0	GALAXY

OBJECT NAME	RIGHT ASCEN.	DECLINATION	DIAM.	MAGN.	TYPE OF OBJECT
KARA.72 245B	10 39 54.	+ 16 00	48	15.0	PART OF DOUBLE GALAXY
MCG+03-27-074	10 39 54.	+ 16 32	42	15.6	GALAXY
ZC 1039.9+1740	10 39 54.	+ 17 40	740		CLUSTER OF GALAXIES
ZWG 124.057	10 39 54.	+ 24 13		15.2	GALAXY
UGC 05830	10 39 54.	+ 24 13	78	15.2	GALAXY Sb
MCG+05-25-034	10 39 54.	+ 27 03 30.	33	15.3	GALAXY
MCG+06-24-006	10 39 54.	+ 34 42 30.	240	14.	GALAXY
ZC 1039.9+3545	10 39 54.	+ 35 45	1480		CLUSTER OF GALAXIES
ZC 1039.9+7819	10 39 54.	+ 78 19	1550		CLUSTER OF GALAXIES
TON-N 1276	10 39 56.	+ 22 27		16.8	BLUE STAR
MCG+07-22-046	10 39 57.	+ 40 23	36	15.5	GALAXY
HOFF L05	10 40	- 60 29	180		DARK HOLE
ZWG 037.124	10 40 00.	+ 05 18		14.9	GALAXY
RNGC 3341	10 40 00.	+ 05 18		15.0	GALAXY
UGC 05831	10 40 00.	+ 05 18	84	14.9	GALAXY PECULR
MCG+01-27-031	10 40 00.	+ 05 18	60	14.9	GALAXY
ZWG 094.112	10 40 00.	+ 16 34		15.6	GALAXY
LB 10730	10 40 00.	+ 17 02		15.5	FAINT BLUE STAR
LB 10729	10 40 00.	+ 18 26		17.0	FAINT BLUE STAR
ZWG 154.040	10 40 00.	+ 27 06		15.3	GALAXY
MCG+05-25-035	10 40 00.	+ 28 40	18	16.	GALAXY
ZWG 212.040	10 40 00.	+ 40 24		15.2	GALAXY
MCG+09-18-013	10 40 00.	+ 51 44	36	17.	GALAXY
MCG+12-10-065	10 40 00.	+ 69 57	60	16.	GALAXY
MCG+12-10-066	10 40 00.	+ 70 02	39	15.	GALAXY
7ZW 343	10 40 00.	+ 79 16			COMPACT GALAXY
MCG-03-27-026	10 40 00.	- 17 24	54	14.5	GALAXY
TON-N 1277	10 40 02.	+ 19 32		17.	BLUE STAR
ABC 1075	10 40 03.	- 09 29		17.1	RICH CLUSTER OF GALAXIES
RNGC 3342	10 40 05.	+ 09 42			NON-EXISTENT OBJECT
LB 10731	10 40 06.	+ 18 54		15.0	FAINT BLUE STAR
MCG+03-27-076	10 40 06.	+ 19 08	36	16.	GALAXY
ZC 1040.1+3206	10 40 06.	+ 32 06	1480		CLUSTER OF GALAXIES
ZC 1040.1+3335	10 40 06.	+ 33 35	540		CLUSTER OF GALAXIES
FATH 1.347	10 40 06.	+ 44 47	14		NEBULA
MCG+13-08-032	10 40 06.	+ 79 04 30.	12	15.	GALAXY
MCG-06-24-006	10 40 06.	- 35 54	48	15.	GALAXY
BC 3CR245	10 40 06.03	+ 12 19 15.0		17.29	QUASI-STELLAR OBJECT
SHB 161	10 40 06.1	+ 12 19 15.		17.3	QUASI-STELLAR OBJECT
FATH 1.348	10 40 07.	+ 44 46	5		NEBULA
HELW 145	10 40 07.	- 35 54 28.			NEBULA
ARP 291	10 40 08.	+ 13 43			PECULIAR GALAXY
REIZ 0558	10 40 09.	+ 13 43	48	13.5	GALAXY
ZWG 065.089	10 40 12.	+ 13 43		13.8	GALAXY
UGC 05832	10 40 12.	+ 13 43	72	13.8	GALAXY SB?
VV 112	10 40 12.	+ 13 43	66	14.	INTERACTING GALAXY
MCG+02-27-042	10 40 12.	+ 13 43	72	13.8	GALAXY
LB 10732	10 40 12.	+ 19 34		17.4	FAINT BLUE STAR
ZWG 212.041	10 40 12.	+ 39 18		15.1	GALAXY
ZC 1040.2+4213	10 40 12.	+ 42 13	1610		CLUSTER OF GALAXIES
MCG+12-10-067	10 40 12.	+ 70 41	54	15.	GALAXY
MCG-04-26-001	10 40 12.	- 23 40	120	14.	GALAXY
MCG+04-25-044	10 40 15.	+ 20 40 30.	63	15.1	GALAXY
MCG+07-22-047	10 40 15.	+ 39 17	27	15.	GALAXY
ABC 1074	10 40 16.	+ 46 54		17.3	RICH CLUSTER OF GALAXIES
ZWG 037.125	10 40 18.	+ 06 07		15.5	GALAXY
ZWG 065.090	10 40 18.	+ 13 46		15.7	GALAXY
ZWG 124.058	10 40 18.	+ 20 41		15.1	GALAXY
MRK 416	10 40 18.	+ 20 41	15	15.5	GALAXY WITH UV CONTINUUM
UGC 05833	10 40 18.	+ 20 41	96	15.1	GALAXY S0
ZWG 184.007	10 40 18.	+ 37 31		15.6	GALAXY
SHAH 055	10 40 18.	+ 48 38	90	17.2	GROUP OF COMPACT GALAXIES
ZC 1040.3+6058	10 40 18.	+ 60 58	2350		CLUSTER OF GALAXIES
UGC 05834	10 40 18.	+ 69 58	84	16.0	GALAXY SBb
ZWG 333.047	10 40 18.	+ 70 01		14.6	GALAXY
UGC 05835	10 40 18.	+ 70 01	66	14.6	GALAXY Sa?
ZWG 333.048	10 40 18.	+ 70 38		15.4	GALAXY
OCL 0841	10 40 18.	- 64 50	2220	8.4	OPEN STAR CLUSTER
VHA 101	10 40 18.	- 64 50	660		OPEN STAR CLUSTER
TON-N 1278	10 40 19.	+ 34 10		14.9	BLUE STAR
ZWG 094.113	10 40 24.	+ 15 49		15.7	GALAXY
LB 10734	10 40 24.	+ 19 18		16.4	FAINT BLUE STAR
LB 10733	10 40 24.	+ 19 30		17.0	FAINT BLUE STAR
REIZ 0559	10 40 24.	+ 20 40	42	14.3	GALAXY
ZWG 124.059	10 40 24.	+ 21 55		15.6	GALAXY
UGC 05836	10 40 24.	+ 21 55	60	15.6	GALAXY Sb
MCG+04-25-045	10 40 24.	+ 21 55	60	15.6	GALAXY
MCG+06-24-007	10 40 24.	+ 37 32 30.	36	15.	GALAXY
ZWG 267.006	10 40 24.	+ 56 02		15.3	GALAXY
ZWG 290.064	10 40 24.	+ 57 56		15.6	GALAXY
ZC 1040.4+6745	10 40 24.	+ 67 45	1810		CLUSTER OF GALAXIES
MCG+12-10-068	10 40 24.	+ 70 39	10	16.	GALAXY
ZWG 351.034	10 40 24.	+ 77 05		12.9	GALAXY
UGC 05837	10 40 24.	+ 77 05	138	12.9	GALAXY Sa?
GO 026	10 40 24.	- 61 02	63	18.8	STAR CHAIN
TON-N 1279	10 40 25.	+ 35 53		17.	BLUE STAR
RNGC 3347	10 40 27.	- 36 06		12.0	GALAXY
LB 10735	10 40 30.	+ 18 16		15.2	FAINT BLUE STAR
ZWG 154.041	10 40 30.	+ 31 47		15.5	GALAXY
REIZ 0560	10 40 30.	+ 31 47	24	14.6	GALAXY
ZC 1040.5+3620	10 40 30.	+ 36 20	1410		CLUSTER OF GALAXIES
MCG+07-22-048	10 40 30.	+ 41 02	60	14.	GALAXY
ZWG 212.042	10 40 30.	+ 41 03		14.0	GALAXY
UGC 05838	10 40 30.	+ 41 03	78	14.0	GALAXY SBb
MCG+10-15-122	10 40 30.	+ 57 55	18	16.	GALAXY
MCG+10-15-123	10 40 30.	+ 58 20	10	16.	GALAXY
MCG+12-10-069	10 40 30.	+ 70 39	9	17.	GALAXY
ZWG 333.049	10 40 30.	+ 70 40		15.5	GALAXY
MCG+12-10-070	10 40 30.	+ 70 40	21	17.	GALAXY
GO 042	10 40 30.	- 58 42	86	17.6	STAR CHAIN
AR 15	10 40 30.16	+ 77 04 19.2			NEBULA
RNGC 3329	10 40 31.	+ 77 04		13.5	GALAXY
REIZ 0557	10 40 34.	+ 57 42	48	15.1	GALAXY
LB 10736	10 40 36.	+ 19 44		15.3	FAINT BLUE STAR
ZC 1040.6+2709	10 40 36.	+ 27 09	1280		CLUSTER OF GALAXIES
MCG+07-22-049	10 40 36.	+ 39 56	60	15.	GALAXY
ZWG 212.043	10 40 36.	+ 39 57		14.8	GALAXY
UGC 05839	10 40 36.	+ 39 57	78	14.8	GALAXY S0-a
SHAH 056	10 40 36.	+ 42 17	108	17.6	GROUP OF COMPACT GALAXIES
ZC 1040.6+4318	10 40 36.	+ 43 18	1340		CLUSTER OF GALAXIES
MCG+12-10-071	10 40 36.	+ 70 22	12	17.	GALAXY
MCG+12-10-072	10 40 36.	+ 72 50	24	16.	GALAXY
MCG-06-24-007	10 40 36.	- 36 05	180	12.	GALAXY
RNGC 3354	10 40 39.	- 36 06			GALAXY
GO 033	10 40 39.	- 59 51	39	14.8	STAR CHAIN
ZC 1040.7+0530	10 40 42.	+ 05 30	610		CLUSTER OF GALAXIES
ZWG 124.060	10 40 42.	+ 25 10		11.1	GALAXY
UGC 05840	10 40 42.	+ 25 10	450	11.1	GALAXY Sc
KARA.73B 0435	10 40 42.	+ 25 10	438	11.1	ISOLATED GALAXY S
ZWG 154.042	10 40 42.	+ 31 48		15.7	GALAXY
REIZ 0561	10 40 42.	+ 31 48	18	15.2	GALAXY
ZWG 267.007	10 40 42.	+ 56 00		15.2	GALAXY
ZWG 351.035	10 40 42.	+ 76 28		15.3	GALAXY
MCG-05-26-001	10 40 42.	- 29 47	72	15.5	GALAXY
GO 045	10 40 43.	- 58 17	114	18.7	STAR CHAIN
MCG+03-27-077	10 40 45.	+ 19 16	48	15.6	GALAXY
MCG+04-25-046	10 40 45.	+ 25 12	420	11.1	GALAXY
VV 057C	10 40 45.	+ 58 27 30.	12	18.	INTERACTING GALAXY
VV 057B	10 40 45.	+ 58 27 30.	21	16.5	INTERACTING GALAXY
VV 057A	10 40 45.	+ 58 27 30.	21	16.5	INTERACTING GALAXY
VV 057	10 40 45.	+ 58 27 30.	108	16.	INTERACTING GALAXY
LB 10738	10 40 48.	+ 17 25		16.0	FAINT BLUE STAR
ZWG 095.001	10 40 48.	+ 19 16		15.6	GALAXY
ZWG 094.114	10 40 48.	+ 19 16		15.6	GALAXY
ZWG 095.002	10 40 48.	+ 19 19		15.4	GALAXY
ZWG 094.115	10 40 48.	+ 19 19		15.4	GALAXY
LB 10737	10 40 48.	+ 20 09		16.8	FAINT BLUE STAR
ZC 1040.8+3504	10 40 48.	+ 35 04	670		CLUSTER OF GALAXIES
ZWG 351.036	10 40 48.	+ 76 58		15.1	GALAXY
UGC 05841	10 40 48.	+ 76 58	102	15.1	GALAXY SBc/Sc
MCG+13-08-033	10 40 48.	+ 77 05	102	12.9	GALAXY
MCG-04-26-002	10 40 48.	- 26 00	72	14.5	GALAXY
MCG-06-24-008	10 40 48.	- 36 05	30	14.5	GALAXY
GO 020	10 40 48.	- 61 19	159	18.7	STAR CHAIN
AR 16	10 40 48.16	+ 76 57 01.9			NEBULA
TON-N 1280	10 40 50.	+ 19 21		13.5	BLUE STAR
TON-N 1281	10 40 50.	+ 23 25		13.5	BLUE STAR
RNGC 3344	10 40 50.	+ 25 11		11.0	GALAXY
REIZ 0562	10 40 51.	+ 25 11	300	12.0	GALAXY
REIN 2.115	10 40 51.47	+ 25 11 07.3			NEBULA
AR 17	10 40 52.14	+ 76 27 47.4			NEBULA
ABC 1079	10 40 53.	- 07 08		17.1	RICH CLUSTER OF GALAXIES
REIN 2.116	10 40 53.97	+ 25 10 54.3			NEBULA
ZWG 066.001	10 40 54.	+ 12 22		15.2	GALAXY
ZWG 065.091	10 40 54.	+ 12 22		15.2	GALAXY
KARA.72 246A	10 40 54.	+ 12 22	42	15.2	PART OF DOUBLE GALAXY
TON-N 1282	10 40 55.	+ 35 06		15.7	BLUE STAR
GO 079	10 40 55.	- 61 01			STAR CHAIN
MCG+10-15-124	10 40 57.	+ 60 37	90	16.	GALAXY
RNGC 3346	10 40 58.	+ 15 08		12.5	GALAXY
REIZ 0564	10 40 58.	+ 15 08	120	12.1	GALAXY
ABC 1078	10 40 59.	+ 00 54		17.0	RICH CLUSTER OF GALAXIES
RNGC 3345	10 40 59.	+ 12 15			NON-EXISTENT OBJECT
VDB.66G 086	10 41	+ 60 35	100		DWARF GALAXY
HOFF L06	10 41	- 59 59	430		DARK HOLE
ZWG 038.001	10 41 00.	+ 05 14		15.0	GALAXY
ZWG 037.126	10 41 00.	+ 05 14		15.0	GALAXY
MCG+01-28-001	10 41 00.	+ 05 14	54	15.0	GALAXY
ZWG 066.002	10 41 00.	+ 12 21		15.4	GALAXY
ZWG 065.092	10 41 00.	+ 12 21		15.4	GALAXY
KARA.72 246B	10 41 00.	+ 12 21	36	15.4	PART OF DOUBLE GALAXY
ZWG 095.003	10 41 00.	+ 15 08		12.8	GALAXY
ZWG 094.116	10 41 00.	+ 15 08		12.8	GALAXY
UGC 05842	10 41 00.	+ 15 08	180	12.8	GALAXY SBc
MCG+03-28-001	10 41 00.	+ 15 08	168	12.8	GALAXY
KARA.73B 0436	10 41 00.	+ 15 08	168	12.8	ISOLATED GALAXY S
ZC 1041.0+2423	10 41 00.	+ 24 23	610		CLUSTER OF GALAXIES
REIZ 0563	10 41 00.	+ 31 50	18	15.4	GALAXY
ZWG 333.050	10 41 00.	+ 69 20		15.5	GALAXY
UGC 05843	10 41 00.	+ 69 20	66	15.5	GALAXY S
MCG+12-10-072A	10 41 00.	+ 69 20	66	15.	GALAXY
SEY 080	10 41 05.	+ 56 41 19.		15.0	FAINT GALAXY
ZC 1041.1+0924	10 41 06.	+ 09 24	1610		CLUSTER OF GALAXIES
ZWG 066.003	10 41 06.	+ 11 46		15.6	GALAXY
ZWG 065.093	10 41 06.	+ 11 46		15.6	GALAXY
ZWG 095.004	10 41 06.	+ 16 09		15.4	GALAXY
ZWG 094.117	10 41 06.	+ 16 09		15.4	GALAXY
MRK 632	10 41 06.	+ 16 09	25	16.	GALAXY WITH UV CONTINUUM
LB 10739	10 41 06.	+ 16 27		17.9	FAINT BLUE STAR
MCG+04-25-047	10 41 06.	+ 21 43	36	15.5	GALAXY
MCG+05-25-036	10 41 06.	+ 28 23	72	15.5	GALAXY
UGC 05844	10 41 06.	+ 28 25	96	16.0	GALAXY Sc
ZWG 155.001	10 41 06.	+ 31 23		15.6	GALAXY
ZWG 154.043	10 41 06.	+ 31 23		15.6	GALAXY
MCG+07-22-050	10 41 06.	+ 38 54	48	15.	GALAXY
MCG+07-22-051	10 41 06.	+ 44 09 30.	72	14.5	GALAXY
MCG+10-16-001	10 41 06.	+ 60 46	27	16.	GALAXY
MCG-02-28-001	10 41 06.	- 09 36	84	14.	GALAXY
MCG-02-28-002	10 41 06.	- 09 54	54	14.5	GALAXY
RNGC 3355	10 41 06.	- 22 56			NON-EXISTENT OBJECT
HN 0029	10 41 06.	- 22 57			NEBULA
MCG-04-26-003	10 41 06.	- 25 36 30.	30	15.	GALAXY
OCL 0828	10 41 06.	- 59 45	1350	5.0	OPEN STAR CLUSTER
IC 0638	10 41 07.	+ 16 09 00.			NONSTELLAR OBJECT
HOLM 211C	10 41 11.	- 01 04	18	15.3	PART OF MULTIPLE GALAXY
ZWG 038.002	10 41 12.	+ 07 01		15.2	GALAXY
RNGC 3349	10 41 12.	+ 07 01		15.0	GALAXY
MCG+01-28-002	10 41 12.	+ 07 01	30	15.2	GALAXY
SN 1974F	10 41 12.	+ 14 17		18.0	SUPERNOVA
ZWG 095.005	10 41 12.	+ 18 54		15.7	GALAXY
ZWG 094.118	10 41 12.	+ 18 54		15.7	GALAXY
ZC 1041.2+2818	10 41 12.	+ 28 18	810		CLUSTER OF GALAXIES
ZWG 212.044	10 41 12.	+ 44 10		14.8	GALAXY
UGC 05845	10 41 12.	+ 44 10	66	14.8	GALAXY
ZC 1041.2+5253	10 41 12.	+ 52 53	810		CLUSTER OF GALAXIES
ZWG 267.008	10 41 12.	+ 53 01		14.6	GALAXY
KARA.73B 0437	10 41 12.	+ 53 01	36	14.6	ISOLATED GALAXY E
ZWG 290.065	10 41 12.	+ 60 38		15.4	GALAXY
UGC 05846	10 41 12.	+ 60 38	120	15.4	GALAXY DWRF IR
MCG+10-16-002	10 41 12.	+ 60 38	90	16.	GALAXY
ZC 1041.2+6350	10 41 12.	+ 63 50	3560		CLUSTER OF GALAXIES
ZC 1041.2+7137	10 41 12.	+ 71 37	1140		CLUSTER OF GALAXIES
MCG+13-08-034	10 41 12.	+ 76 57	84	14.	GALAXY
MCG-01-28-001	10 41 12.	- 09 23	96	14.	GALAXY
MCG-05-26-002	10 41 12.	- 30 31	54	14.5	GALAXY
IC 2602	10 41 12.	- 64 08	5400		OPEN CLUSTER
REIZ 0565	10 41 13.	+ 07 02	18	14.3	GALAXY
MCG-05-26-003	10 41 15.	- 30 33	42	15.	GALAXY
RNGC 3358	10 41 15.	- 36 07		12.5	GALAXY
MAI 057	10 41 16.	+ 60 45	101		DWARF SPHEROIDAL GALAXY
GO 036	10 41 16.	- 59 35	60	13.0	STAR CHAIN
RNGC 3351	10 41 17.	+ 11 58		11.0	GALAXY
HOLM 211B	10 41 17.	- 01 02	18	15.2	PART OF MULTIPLE GALAXY
ZWG 010.002	10 41 18.	+ 02 13		15.5	GALAXY
MCG+00-28-002	10 41 18.	+ 02 13	48	15.5	GALAXY
ZWG 095.006	10 41 18.	+ 18 38		15.5	GALAXY

OBJECT NAME	RIGHT ASCEN.	DECLINATION	DIAM.	MAGN.	TYPE OF OBJECT
ZWG 094.119	10 41 18.	+ 18 38		15.5	GALAXY
TON-N 0537	10 41 18.	+ 33 24		15.2	BLUE STAR
ZC 1041.3+4035	10 41 18.	+ 40 35	610		CLUSTER OF GALAXIES
MCG+09-18-014	10 41 18.	+ 53 02	24	14.	GALAXY
ZWG 290.066	10 41 18.	+ 56 41		15.3	GALAXY
UGC 05848	10 41 18.	+ 56 41	126	15.3	GALAXY DWRF SP
KARA.72 247A	10 41 18.	+ 56 41	150	15.3	PART OF DOUBLE GALAXY
ZWG 010.001	10 41 18.	- 02 15		15.5	GALAXY
UGC 05847	10 41 18.	- 02 15	78	15.5	GALAXY SBb
MCG+00-28-001	10 41 18.	- 02 15	60	15.5	GALAXY
HN 1244	10 41 19.0	- 01 01 30.	30		NEBULA
REIZ 0566	10 41 21.	+ 11 58	480	11.6	GALAXY
HOLM 211A	10 41 21.	- 01 02	42	14.1	PART OF MULTIPLE GALAXY
ARC 1080	10 41 23.	+ 01 21		17.0	RICH CLUSTER OF GALAXIES
ZWG 038.003	10 41 24.	+ 04 56		15.0	GALAXY
MCG+01-28-003	10 41 24.	+ 04 56	36	15.0	GALAXY
ZC 1041.4+0843	10 41 24.	+ 08 43	2490		CLUSTER OF GALAXIES
ZWG 066.004	10 41 24.	+ 11 58		11.2	GALAXY
SCH 13	10 41 24.	+ 11 58		11.2	PECULIAR GALAXY
UGC 05850	10 41 24.	+ 11 58	540	11.2	GALAXY SBb
MCG+02-28-001	10 41 24.	+ 11 58	420	11.2	GALAXY
ZC 1041.4+1257	10 41 24.	+ 12 57	2290		CLUSTER OF GALAXIES
ZC 1041.4+1759	10 41 24.	+ 17 59	1210		CLUSTER OF GALAXIES
MCG+09-18-015	10 41 24.	+ 56 39	120	14.	GALAXY
ZWG 010.003	10 41 24.	- 01 01		14.1	GALAXY
MCG+00-28-003	10 41 24.	- 01 01	30	14.1	GALAXY
UGC 05849	10 41 24.	- 04 04	96	14.1	GALAXY
MCG-04-26-005	10 41 24.	- 22 33	24	15.	GALAXY
MCG-04-26-004	10 41 24.	- 24 06 30.	96	14.	GALAXY
MCG-06-24-009	10 41 24.	- 36 09	240	11.5	GALAXY
GO 022	10 41 24.	- 61 25	87	18.3	STAR CHAIN
OCL 0838	10 41 24.	- 64 08	9300	1.6	OPEN STAR CLUSTER
MCG+09-18-016	10 41 27.	+ 56 39	48	14.	GALAXY
SEY 081	10 41 29.	+ 56 37 37.		15.0	FAINT GALAXY
MCG+04-25-048	10 41 30.	+ 22 37 30.	78	14.1	GALAXY
ZWG 124.061	10 41 30.	+ 22 38		14.1	GALAXY
UGC 05851	10 41 30.	+ 22 38	96	14.1	GALAXY S0
ZWG 124.062	10 41 30.	+ 22 59		15.2	GALAXY
KARA.73B 0438	10 41 30.	+ 22 59	36	15.2	ISOLATED GALAXY S
ZC 1041.5+4647	10 41 30.	+ 46 47	1410		CLUSTER OF GALAXIES
GO 080	10 41 30.	- 61 32	91	18.0	STAR CHAIN
REIZ 0568	10 41 31.	+ 22 38	24	13.8	GALAXY
GO 043	10 41 31.	- 58 36	92	15.7	STAR CHAIN
FATH 1.349	10 41 32.	+ 44 28	5		NEBULA
RNGC 3352	10 41 33.	+ 22 38		14.0	GALAXY
REIZ 0569	10 41 35.	+ 30 59	15	15.2	GALAXY
ZWG 038.004	10 41 36.	+ 03 03		15.3	GALAXY
ZWG 038.005	10 41 36.	+ 07 01		13.3	GALAXY
RNGC 3356	10 41 36.	+ 07 01		13.5	GALAXY
UGC 05852	10 41 36.	+ 07 01	126	13.3	GALAXY Sb
MCG+01-28-004	10 41 36.	+ 07 01	96	13.3	GALAXY
ZWG 066.005	10 41 36.	+ 11 10		15.7	GALAXY
LB 10740	10 41 36.	+ 18 40		17.5	FAINT BLUE STAR
ZWG 155.002	10 41 36.	+ 30 59		15.4	GALAXY
ZWG 154.044	10 41 36.	+ 30 59		15.4	GALAXY
ZWG 184.008	10 41 36.	+ 37 15		15.6	GALAXY
MCG+09-18-017	10 41 36.	+ 52 57	9	18.	GALAXY
ZWG 290.067	10 41 36.	+ 56 38		15.1	GALAXY
KARA.72 247B	10 41 36.	+ 56 38	60	15.1	PART OF DOUBLE GALAXY
MCG+10-16-003	10 41 36.	+ 58 42	57	15.	GALAXY
ZWG 290.068	10 41 36.	+ 58 43		15.5	GALAXY
UGC 05853	10 41 36.	+ 58 43	90	15.5	GALAXY Sc
ZWG 351.037	10 41 36.	+ 77 22		15.2	GALAXY
UGC 05854	10 41 36.	+ 77 22	66	15.2	GALAXY Sc-IRR
MCG-05-26-004	10 41 36.	- 30 51	36	15.5	GALAXY
REIZ 0571	10 41 37.	+ 07 01	72	13.5	GALAXY
RNGC 3350	10 41 37.	+ 30 59		15.5	GALAXY
MCG+07-22-052	10 41 39.	+ 39 24	48	15.	GALAXY
ARC 1077	10 41 39.	+ 46 46		17.5	RICH CLUSTER OF GALAXIES
RNGC 3360	10 41 39.	- 10 59		14.0	GALAXY
RNGC 3357	10 41 40.	+ 14 21		14.5	GALAXY
REIZ 0572	10 41 40.	+ 14 21	18	14.6	GALAXY
REIZ 0570	10 41 41.	+ 30 55	6	15.9	GALAXY
ZWG 038.006	10 41 42.	+ 03 03		15.2	GALAXY
ZWG 066.006	10 41 42.	+ 14 21		14.3	GALAXY
UGC 05854A	10 41 42.	+ 14 21	60	14.3	GALAXY S0
MCG+02-28-002	10 41 42.	+ 14 21	72	14.3	GALAXY
LB 10742	10 41 42.	+ 17 49		16.8	FAINT BLUE STAR
LB 10741	10 41 42.	+ 18 14		14.8	FAINT BLUE STAR
ZWG 095.007	10 41 42.	+ 18 45		14.6	GALAXY
ZWG 095.008	10 41 42.	+ 18 48		15.6	GALAXY
MCG+04-25-049	10 41 42.	+ 21 58 30.	42	14.9	GALAXY
ZWG 212.045	10 41 42.	+ 39 01		15.0	GALAXY
MCG+07-22-053	10 41 42.	+ 39 01	48	14.5	GALAXY
ZC 1041.7+4731	10 41 42.	+ 47 31	1480		CLUSTER OF GALAXIES
MCG-03-28-001	10 41 42.	- 16 13	36	14.5	GALAXY
MCG-06-24-010	10 41 42.	- 38 00	180	13.	GALAXY
MCG-02-28-003	10 41 45.	- 10 59	48	14.5	GALAXY
REIZ 0567	10 41 46.	+ 52 44	24	14.9	GALAXY
AR 18	10 41 46.23	+ 77 21 56.1			NEBULA
REIZ 0573	10 41 47.	+ 18 44	18	14.4	GALAXY
LB 10744	10 41 48.	+ 16 19		16.4	FAINT BLUE STAR
LB 10743	10 41 48.	+ 16 29		15.4	FAINT BLUE STAR
ZWG 095.009	10 41 48.	+ 18 36		15.5	GALAXY
MCG+03-28-002	10 41 48.	+ 18 46	48	14.6	GALAXY
MCG+04-25-050	10 41 48.	+ 21 55 30.	36	15.5	GALAXY
ZWG 124.063	10 41 48.	+ 21 59		14.9	GALAXY
MCG+07-22-054	10 41 48.	+ 38 33 30.	36	15.	GALAXY
ZWG 212.046	10 41 48.	+ 38 34		15.3	GALAXY
ZWG 212.047	10 41 48.	+ 44 19		15.7	GALAXY
MCG+09-18-019	10 41 48.	+ 51 45	42	16.	GALAXY
MCG+09-18-020	10 41 48.	+ 52 14	6	18.	GALAXY
MCG+09-18-021	10 41 48.	+ 52 14 30.	42	17.	GALAXY
MCG+09-18-018	10 41 48.	+ 54 02	30	18.	GALAXY
ZWG 351.038	10 41 48.	+ 75 22		15.6	GALAXY
ARC 1076	10 41 52.	+ 58 26		17.2	RICH CLUSTER OF GALAXIES
ZC 1041.9+2622	10 41 54.	+ 26 22	670		CLUSTER OF GALAXIES
ZWG 125.001	10 41 54.	+ 26 27		15.0	GALAXY
ZWG 124.064	10 41 54.	+ 26 27		15.0	GALAXY
UGC 05855	10 41 54.	+ 26 27	90	15.0	GALAXY S
ZC 1041.9+4406	10 41 54.	+ 44 06	540		CLUSTER OF GALAXIES
MCG+12-10-073	10 41 54.	+ 73 37	24	13.	GALAXY
MCG-04-26-006	10 41 54.	- 22 33	72	14.	GALAXY
ARC 1082	10 41 56.	+ 32 38		17.6	RICH CLUSTER OF GALAXIES
REIZ 0574	10 41 57.	+ 26 27	54	14.8	GALAXY
RNGC 3361	10 41 57.	- 10 57		13.0	GALAXY
ZWG 125.002	10 42 00.	+ 24 06		15.7	GALAXY
ZWG 124.065	10 42 00.	+ 24 06		15.7	GALAXY

OBJECT NAME	RIGHT ASCEN.	DECLINATION	DIAM.	MAGN.	TYPE OF OBJECT
MCG+04-26-001	10 42 00.	+ 26 26	90	15.0	GALAXY
TON-N 0538	10 42 00.	+ 31 09		16.0	BLUE STAR
REIZ 0575	10 42 00.	+ 31 45	18	15.0	GALAXY
ARC 1081	10 42 00.	+ 35 50		17.2	RICH CLUSTER OF GALAXIES
ZWG 184.009	10 42 00.	+ 38 26		15.1	GALAXY
UGC 05856	10 42 00.	+ 38 26	60	15.1	GALAXY S
MCG+06-24-008	10 42 00.	+ 38 28	36	15.	GALAXY
ZC 1042.0+6127	10 42 00.	+ 61 27	870		CLUSTER OF GALAXIES
ZWG 351.039	10 42 00.	+ 75 00		15.4	GALAXY
MCG+13-08-035	10 42 00.	+ 77 22	54	14.	GALAXY
MCG-02-28-004	10 42 00.	- 10 57	120	13.	GALAXY
MCG-05-26-005	10 42 00.	- 31 57	54	14.5	GALAXY
OCL 0826	10 42 00.	- 59 18	780	7.1	OPEN STAR CLUSTER
VHA 102	10 42 00.	- 59 18	90		STAR CLSTR IN NEBULOSITY
GO 021	10 42 00.	- 61 29	119	19.5	STAR CHAIN
TON-N 1283	10 42 02.	+ 22 39		17.	BLUE STAR
MCG-04-26-007	10 42 03.	- 25 07	21	15.	GALAXY
ARC 1084	10 42 05.	- 06 50		17.4	RICH CLUSTER OF GALAXIES
ZC 1042.1+0004	10 42 06.	+ 00 04	1680		CLUSTER OF GALAXIES
TON-N 0539	10 42 06.	+ 27 27		14.5	BLUE STAR
ZWG 155.003	10 42 06.	+ 27 44		15.4	GALAXY
ZWG 154.045	10 42 06.	+ 27 44		15.4	GALAXY
ZC 1042.1+3547	10 42 06.	+ 35 47	1950		CLUSTER OF GALAXIES
ZWG 184.010	10 42 06.	+ 37 23		15.7	GALAXY
MCG+07-22-055	10 42 06.	+ 43 57 30.	72	14.5	GALAXY
ZC 1042.1+5426	10 42 06.	+ 54 26	400		CLUSTER OF GALAXIES
MCG+12-10-074	10 42 06.	+ 72 50	24	17.	GALAXY
MCG+13-08-036	10 42 06.	+ 74 54	39	16.	GALAXY
MCG-02-28-005	10 42 06.	- 14 04	24	15.	GALAXY
REIZ 0577	10 42 08.	+ 25 03	12	15.2	GALAXY
GO 074	10 42 08.	- 51 23	90	18.9	STAR CHAIN
ARC 1086	10 42 09.	- 16 15		17.5	RICH CLUSTER OF GALAXIES
ZWG 038.007	10 42 12.	+ 06 52		13.6	GALAXY
RNGC 3362	10 42 12.	+ 06 52		13.5	GALAXY
UGC 05857	10 42 12.	+ 06 52	84	13.6	GALAXY Sc
MCG+01-28-005	10 42 12.	+ 06 52	84	13.6	GALAXY
LB 10745	10 42 12.	+ 15 34		15.4	FAINT BLUE STAR
ZWG 095.010	10 42 12.	+ 16 13		15.6	GALAXY
UGC 05858	10 42 12.	+ 16 13	96	15.6	GALAXY
MCG+03-28-003	10 42 12.	+ 16 13	90	15.6	GALAXY
KARA.73B 0439	10 42 12.	+ 16 13	60	15.6	ISOLATED GALAXY S
TON-N 0540	10 42 12.	+ 28 46		15.5	BLUE STAR
ZC 1042.2+3236	10 42 12.	+ 32 36	810		CLUSTER OF GALAXIES
MCG+07-22-056	10 42 12.	+ 39 14	90	15.	GALAXY
ZWG 212.048	10 42 12.	+ 43 58		14.9	GALAXY
UGC 05859	10 42 12.	+ 43 58	96	14.9	GALAXY SBa
ZWG 241.009	10 42 12.	+ 50 08		15.5	GALAXY
ZWG 267.009	10 42 12.	+ 56 13		12.9	GALAXY
UGC 05860	10 42 12.	+ 56 13	90	12.9	GALAXY PECULR
SEY 082	10 42 14.	+ 58 51 18.		15.3	FAINT GALAXY
LB 00582	10 42 15.	+ 46 16 36.		17.6	FAINT BLUE STAR
MCG+08-20-017	10 42 15.	+ 50 09	30	15.	GALAXY
HARO 03	10 42 16.	+ 56 13			BLUE EMISSION-LINE GALAXY
REIZ 0579	10 42 18.	+ 06 51	36	13.7	GALAXY
ZC 1042.3+1614	10 42 18.	+ 16 14	2690		CLUSTER OF GALAXIES
UGC 05861	10 42 18.	+ 39 15	90	16.5	GALAXY Sc
MRK 035	10 42 18.	+ 56 14	45	14.	GALAXY WITH UV CONTINUUM
VVI 43	10 42 18.	+ 56 14	60	13.53	SEYFERT GALAXY
MCG+09-18-022	10 42 18.	+ 56 14	60	13.0	GALAXY
ZWG 290.069	10 42 18.	+ 58 50		15.3	GALAXY
UGC 05862	10 42 18.	+ 58 50	84	15.3	GALAXY S
MCG+10-16-004	10 42 18.	+ 58 50	72	14.	GALAXY
ZWG 333.051	10 42 18.	+ 73 37		14.7	GALAXY
RNGC 3343	10 42 18.	+ 73 37		14.5	GALAXY
UGC 05863	10 42 18.	+ 73 37	96	14.7	GALAXY E
7ZW 344	10 42 18.	+ 78 43			COMPACT GALAXY
LB 01901	10 42 19.	+ 57 05 12.		15.8	FAINT BLUE STAR
ARC 1088	10 42 19.	- 19 14		17.0	RICH CLUSTER OF GALAXIES
GO 076	10 42 19.	- 61 17	33	19.1	STAR CHAIN
REIZ 0576	10 42 20.	+ 56 13	30	13.0	GALAXY
RNGC 3353	10 42 21.	+ 56 13		13.5	GALAXY
ZWG 066.007	10 42 24.	+ 10 27		15.5	GALAXY
UGC 05864	10 42 24.	+ 10 27	60	15.5	GALAXY S
ZC 1042.4+1315	10 42 24.	+ 13 15	470		CLUSTER OF GALAXIES
ZWG 155.004	10 42 24.	+ 28 15		15.7	GALAXY
ZWG 154.046	10 42 24.	+ 28 15		15.7	GALAXY
TON-N 0541	10 42 24.	+ 29 43		16.5	BLUE STAR
ZC 1042.4+3132	10 42 24.	+ 31 32	1080		CLUSTER OF GALAXIES
ZC 1042.4+3910	10 42 24.	+ 39 10	8530		CLUSTER OF GALAXIES
MCG+13-08-037	10 42 24.	+ 74 58	33	16.	GALAXY
REIZ 0580	10 42 25.	+ 22 20	36	13.8	GALAXY
MCG+04-26-002	10 42 27.	+ 22 20	60	14.7	GALAXY
ARC 1070	10 42 27.	+ 78 24		17.5	RICH CLUSTER OF GALAXIES
HN 1245	10 42 28.9	+ 00 41 46.			NEBULA
REIZ 0578	10 42 29.	+ 54 09	36	14.7	GALAXY
ZWG 010.004	10 42 30.	+ 00 43		15.1	GALAXY
ZWG 038.008	10 42 30.	+ 05 12		15.7	GALAXY
UGC 05865	10 42 30.	+ 05 12	60	15.7	GALAXY Sc?
MCG+01-28-006	10 42 30.	+ 05 12	60	15.7	GALAXY
KARA.73B 0440	10 42 30.	+ 05 12	54	15.7	ISOLATED GALAXY S
ZWG 095.011	10 42 30.	+ 17 33		15.7	GALAXY
LB 10746	10 42 30.	+ 18 48		18.0	FAINT BLUE STAR
ZWG 095.012	10 42 30.	+ 19 14		15.5	GALAXY
ZWG 125.003	10 42 30.	+ 22 20		14.7	GALAXY
UGC 05866	10 42 30.	+ 22 20	96	14.7	GALAXY S
MCG+05-26-001	10 42 30.	+ 28 15	48	15.7	GALAXY
MCG+07-22-057	10 42 30.	+ 39 25	36	15.	GALAXY
MCG+07-22-058	10 42 30.	+ 39 25 30.	42	15.	GALAXY
ZWG 212.049	10 42 30.	+ 39 26		14.9	GALAXY
MCG+12-10-075	10 42 30.	+ 72 35	33	15.	GALAXY
VHA 103	10 42 30.	- 64 04	3600		OPEN STAR CLUSTER
HN 1246	10 42 31.0	+ 00 22 10.	24		NEBULA
RNGC 3363	10 42 33.	+ 22 20		14.5	GALAXY
REIZ 0582	10 42 34.	+ 00 22	36	14.7	GALAXY
ZWG 010.005	10 42 36.	+ 00 23		15.0	GALAXY
UGC 05867	10 42 36.	+ 00 23	60	15.0	GALAXY SBb
MCG+00-28-004	10 42 36.	+ 00 23	60	15.0	GALAXY
MCG+03-28-004	10 42 36.	+ 19 16	51	15.5	GALAXY
FEIG 035	10 42 36.	+ 25 08		12.7	FAINT BLUE STAR
ZWG 155.005	10 42 36.	+ 29 28		14.9	GALAXY
ZWG 154.047	10 42 36.	+ 29 28		14.9	GALAXY
MCG+05-26-002	10 42 36.	+ 29 28	24	14.9	GALAXY
ZC 1042.6+2930	10 42 36.	+ 29 30	1080		CLUSTER OF GALAXIES
TON-N 0042	10 42 36.	+ 32 15		16.1	BLUE STAR
ZC 1042.6+4600	10 42 36.	+ 46 00	5310		CLUSTER OF GALAXIES
ARC 1085	10 42 38.	+ 20 32		16.6	RICH CLUSTER OF GALAXIES
REIZ 0581	10 42 38.	+ 25 17	18	15.2	GALAXY
PK288-02.1	10 42 38.1	- 61 23 53.	25		PLANETARY NEBULA

OBJECT NAME	RIGHT ASCEN.	DECLINATION	DIAM.	MAGN.	TYPE OF OBJECT
MCG+00-28-005	10 42 42.	+ 00 20 30.	24	17.	GALAXY
ZC 1042.7+2041	10 42 42.	+ 20 41	2420		CLUSTER OF GALAXIES
ZC 1042.7+3836	10 42 42.	+ 38 36	400		CLUSTER OF GALAXIES
MCG-02-28-006	10 42 42.	- 09 49	72	14.	GALAXY
GO 075	10 42 43.	- 61 20	77	18.2	STAR CHAIN
WRAY 19.23	10 42 44.1	- 60 32 16.			DIFFUSE NEBULA
ZWG 095.013	10 42 48.	+ 16 35		14.8	GALAXY
MCG+03-28-005	10 42 48.	+ 16 35	36	14.8	GALAXY
KARA.73B 0441	10 42 48.	+ 16 35	48	14.8	ISOLATED GALAXY S
ZWG 125.004	10 42 48.	+ 24 25		15.2	GALAXY
MCG+04-26-003	10 42 48.	+ 24 25	42	15.2	GALAXY
ZC 1042.8+3353	10 42 48.	+ 33 53	2020		CLUSTER OF GALAXIES
ZC 1042.8+5521	10 42 48.	+ 55 21	870		CLUSTER OF GALAXIES
MCG+10-16-005	10 42 48.	+ 56 45	27	17.	GALAXY
OCL 0825	10 42 48.	- 59 06	660	9.2	OPEN STAR CLUSTER
VHA 104	10 42 48.	- 59 06	180		STAR CLSTR IN NEBULOSITY
REIZ 0583	10 42 51.	+ 37 30	42	14.8	GALAXY
ZWG 095.014	10 42 54.	+ 17 52		15.3	GALAXY
MCG+03-28-006	10 42 54.	+ 17 54	72	15.3	GALAXY
LB 10747	10 42 54.	+ 19 50		15.6	FAINT BLUE STAR
ZWG 184.011	10 42 54.	+ 37 28		15.0	GALAXY
UGC 05868	10 42 54.	+ 37 28	84	15.0	GALAXY Sb
MCG+07-22-059	10 42 54.	+ 44 10	30	15.5	GALAXY
ZC 1042.9+5810	10 42 54.	+ 58 10	4100		CLUSTER OF GALAXIES
7ZW 345	10 42 54.	+ 68 38			COMPACT GALAXY
MCG+12-10-076	10 42 54.	+ 72 26	18	17.	GALAXY
ZWG 333.052	10 42 54.	+ 72 34		15.6	GALAXY
MCG+12-10-077	10 42 54.	+ 73 07	60	12.0	GALAXY
OCL 0827	10 42 54.	- 59 13	240	7.0	OPEN STAR CLUSTER
TON-N 1284	10 42 56.	+ 21 54		17.	BLUE STAR
IC 2600	10 42 58.	+ 72 35			NONSTELLAR OBJECT
ZWG 095.015	10 43 00.	+ 18 05		15.7	GALAXY
ZWG 155.006	10 43 00.	+ 27 53		15.0	GALAXY
MCG+05-26-003	10 43 00.	+ 27 53	36	15.0	GALAXY
MCG+06-24-009	10 43 00.	+ 37 30	60	14.	GALAXY
MCG+08-20-018	10 43 00.	+ 46 11	42	15.	GALAXY
MCG+09-18-023	10 43 00.	+ 52 15	36	17.	GALAXY
MCG+12-10-078	10 43 00.	+ 72 35	21	16.	GALAXY
ARC 1083	10 43 02.	+ 59 55		17.6	RICH CLUSTER OF GALAXIES
ZWG 038.009	10 43 06.	+ 06 28		15.6	GALAXY
ZWG 066.008	10 43 06.	+ 11 37		14.9	GALAXY
UGC 05869	10 43 06.	+ 11 37	84	14.9	GALAXY Sb
MCG+02-28-003	10 43 06.	+ 11 37	84	14.9	GALAXY
ZC 1043.1+1854	10 43 06.	+ 18 54	1950		CLUSTER OF GALAXIFS
ZWG 125.005	10 43 06.	+ 26 13		15.2	GALAXY
STOCK 16	10 43 06.	+ 29 31			BLUE KNOT NEAR ELLIP GLXY
ZWG 184.012	10 43 06.	+ 35 14		14.3	GALAXY
UGC 05870	10 43 06.	+ 35 14	66	14.3	GALAXY S0?
UGC 05871	10 43 06.	+ 37 26	78	16.0	GALAXY S
ZWG 212.050	10 43 06.	+ 39 25		15.7	GALAXY
MCG+08-20-019	10 43 06.	+ 44 59	24	16.	GALAXY
ZWG 241.010	10 43 06.	+ 46 11		15.4	GALAXY
ZWG 241.011	10 43 06.	+ 49 48		15.1	GALAXY
UGC 05872	10 43 06.	+ 49 48	96	15.1	GALAXY Sb-c
MCG+08-20-020	10 43 06.	+ 49 49	48	15.	GALAXY
MCG+09-18-024	10 43 06.	+ 51 20	42	15.	GALAXY
IC 0639	10 43 07.	+ 17 11 13.			NONSTELLAR OBJFCT
CED 109A	10 43 08.	- 59 24	7200		DIFFUSE GALACTIC NEBULA
RNGC 3372	10 43 08.	- 59 25			DIFFUSE NEBULA
REIZ 0584	10 43 09.	+ 26 13	12	15.4	GALAXY
MCG+04-26-004	10 43 09.	+ 26 13	48	15.2	GALAXY
ZWG 010.006	10 43 12.	+ 02 17		15.4	GALAXY
ZC 1043.2+0436	10 43 12.	+ 04 36	940		CLUSTER OF GALAXIES
ZWG 066.009	10 43 12.	+ 10 00		15.0	GALAXY
MCG+02-28-004	10 43 12.	+ 10 00	30	15.0	GALAXY
ZWG 095.016	10 43 12.	+ 17 11		14.8	GALAXY
REIZ 0585	10 43 12.	+ 21 03	12	15.0	GALAXY
MCG+06-24-011	10 43 12.	+ 35 13 30.		16.	GALAXY
MCG+06-24-010	10 43 12.	+ 35 13 30.	24	14.	GALAXY
OCL 0829	10 43 12.	- 59 27	1920	6.8	OPEN STAR CLUSTER
VHA 105	10 43 12.	- 59 27	360		STAR CLSTR IN NEBULOSITY
MCG+03-28-007	10 43 15.	+ 17 12	42	14.8	GALAXY
RNGC 3359	10 43 15.	+ 63 29		11.5	GALAXY
ARC 1087	10 43 16.	+ 44 10		17.8	RICH CLUSTER OF GALAXIES
REIZ 0586	10 43 17.	+ 17 11	60	14.3	GALAXY
ZWG 125.006	10 43 18.	+ 21 04		15.5	GALAXY
ZC 1043.3+4944	10 43 18.	+ 49 44	1680		CLUSTER OF GALAXIES
SHAH 057	10 43 18.	+ 49 44		17.5	GROUP OF COMPACT GALAXIES
ZWG 267.010	10 43 18.	+ 51 21		15.1	GALAXY
MCG+11-13-037	10 43 18.	+ 63 29	420	10.8	GALAXY
ZWG 313.033	10 43 18.	+ 63 30		11.0	GALAXY
UGC 05873	10 43 18.	+ 63 30	480	11.0	GALAXY SBc
KARA.73B 0442	10 43 18.	+ 63 30	522	11.0	ISOLATED GALAXY S
ARC 1090	10 43 20.	- 18 05		17.1	RICH CLUSTER OF GALAXIES
REIZ 0587	10 43 21.	+ 26 10	30	15.0	GALAXY
PATH 1.350	10 43 23.	+ 45 01	5		NEBULA
ZC 1043.4+0141	10 43 24.	+ 01 41	2420		CLUSTER OF GALAXIES
ZWG 125.007	10 43 24.	+ 26 10		15.5	GALAXY
UGC 05874	10 43 24.	+ 26 10	66	15.5	GALAXY S
MCG+04-26-005	10 43 24.	+ 26 10	36	15.5	GALAXY
ZWG 212.051	10 43 24.	+ 38 50		15.5	GALAXY
ZC 1043.4+4848	10 43 24.	+ 48 48	1010		CLUSTER OF GALAXIES
ZC 1043.4+5332	10 43 24.	+ 53 32	540		CLUSTER OF GALAXIES
ZC 1043.4+5956	10 43 24.	+ 59 56	1340		CLUSTER OF GALAXIES
ZWG 333.053	10 43 24.	+ 72 34		15.5	GALAXY
ZWG 333.054	10 43 24.	+ 73 06		12.0	GALAXY
UGC 05875	10 43 24.	+ 73 06	150	12.0	GALAXY E
MCG+12-10-079	10 43 24.	+ 73 06	33	16.	GALAXY
MCG-02-28-007	10 43 24.	- 12 07	54	15.5	GALAXY
OCL 0830	10 43 24.	- 59 29	240	7.5	OPEN STAR CLUSTER
ARC 1091	10 43 27.	- 16 45		17.7	RICH CLUSTER OF GALAXIES
REIN 2.117	10 43 27.21	+ 73 06 14.5			NEBULA
REIZ 0589	10 43 29.	+ 19 12	24	15.3	GALAXY
ZWG 066.010	10 43 30.	+ 13 34		15.3	GALAXY
ZWG 095.017	10 43 30.	+ 19 12		15.4	GALAXY
MCG+06-24-012	10 43 30.	+ 37 37	48	16.	GALAXY
MCG+09-18-025	10 43 30.	+ 52 24	78	15.	GALAXY
MCG+10-16-006	10 43 30.	+ 57 07	24	17.	GALAXY
ZC 1043.5+6708	10 43 30.	+ 67 08	2020		CLUSTER OF GALAXIES
MCG+12-10-080	10 43 30.	+ 72 42	30	17.	GALAXY
TON-N 1286	10 43 31.	+ 35 21		15.	BLUE STAR
TON-N 1285	10 43 32.	+ 20 40		16.	BLUE STAR
IC 2601	10 43 32.	+ 72 35			NONSTELLAR OBJECT
RNGC 3348	10 43 32.	+ 73 06		12.5	GALAXY
GO 102	10 43 32.	- 58 27	60	17.7	STAR CHAIN
REIZ 0590	10 43 34.	+ 28 44	30	15.1	GALAXY
REIZ 0591	10 43 34.	+ 29 37	15	15.3	GALAXY
ZWG 267.011	10 43 36.	+ 52 25		15.7	GALAXY

OBJECT NAME	RIGHT ASCEN.	DECLINATION	DIAM.	MAGN.	TYPE OF OBJECT
UGC 05876	10 43 36.	+ 52 25	84	15.7	GALAXY Sc
MCG+10-16-007	10 43 36.	+ 60 10	66	16.	GALAXY
ZWG 333.055	10 43 36.	+ 69 07		15.7	GALAXY
ZWG 351.040	10 43 36.	+ 77 07		15.3	GALAXY
UGC 05877	10 43 36.	+ 77 07	60	15.3	GALAXY S
MCG-03-28-002	10 43 36.	- 17 26	42	14.5	GALAXY
TON-N 1287	10 43 37.	+ 34 46		17.	BLUE STAR
MCG+07-22-060	10 43 39.	+ 40 14 30.	24	15.5	GALAXY
REIZ 0592	10 43 40.	+ 02 03	240	13.8	GALAXY
REIZ 0588	10 43 40.	+ 53 36	24	14.9	GALAXY
HN 1247	10 43 40.3	- 01 27 33.	18		NEBULA
ZWG 010.007	10 43 42.	+ 00 19		15.5	GALAXY
ZWG 010.008	10 43 42.	+ 02 05		13.6	GALAXY
UGC 05878	10 43 42.	+ 02 05	276	13.6	GALAXY Sc
MCG+00-28-006	10 43 42.	+ 02 05	300	13.6	GALAXY
ZWG 212.052	10 43 42.	+ 40 15		15.7	GALAXY
RNGC 3365	10 43 43.	+ 02 05		13.5	GALAXY
ARC 1089	10 43 43.	+ 18 54		17.7	RICH CLUSTER OF GALAXIES
MCG+05-26-004	10 43 45.	+ 28 43	45	15.	GALAXY
ARC 1092	10 43 47.	+ 01 37		17.6	RICH CLUSTER OF GALAXIES
REIZ 0593	10 43 47.	+ 18 47	18	15.7	GALAXY
AR 19	10 43 47.89	+ 77 06 39.8			NEBULA
ZWG 038.010	10 43 48.	+ 07 51		15.7	GALAXY
ZWG 095.018	10 43 48.	+ 18 47		15.5	GALAXY
ZWG 291.001	10 43 48.	+ 60 11		15.7	GALAXY
ZWG 290.070	10 43 48.	+ 60 11		15.7	GALAXY
UGC 05879	10 43 48.	+ 60 11	90	15.7	GALAXY Sc
ZWG 010.009	10 43 48.	- 01 27		15.6	GALAXY
MCG-04-26-008	10 43 48.	- 21 33 30.	24	15.5	GALAXY
HN 1248	10 43 48.4	- 02 01 16.			NEBULA
GO 081	10 43 50.	- 61 31	76	17.7	STAR CHAIN
RNGC 3367	10 43 52.	+ 14 01		12.5	GALAXY
GO 035	10 43 53.	- 60 05	40	17.2	STAR CHAIN
ZWG 066.011	10 43 54.	+ 14 02		12.0	GALAXY
SCH 14	10 43 54.	+ 14 02		12.0	PECULIAR GALAXY
UGC 05880	10 43 54.	+ 14 02	138	12.0	GALAXY SBc
MCG+02-28-005	10 43 54.	+ 14 02	120	12.0	GALAXY
LB 01902	10 43 54.	+ 50 53 18.		17.1	FAINT BLUE STAR
ZC 1043.9+5635	10 43 54.	+ 56 35	3700		CLUSTER OF GALAXIES
ZWG 010.010	10 43 54.	- 02 01		14.8	GALAXY
MCG+00-28-007	10 43 54.	- 02 01	24	14.8	GALAXY
REIZ 0594	10 43 57.	+ 14 00	96	12.3	GALAXY
MCG+04-26-006	10 43 57.	+ 26 11 30.	63	15.0	GALAXY
GO 029	10 43 57.	- 61 04	60	19.6	STAR CHAIN
ZWG 066.012	10 44 00.	+ 09 10		15.6	GALAXY
ZWG 125.008	10 44 00.	+ 26 12		15.0	GALAXY
UGC 05881	10 44 00.	+ 26 12	72	15.0	GALAXY Sa
ZC 1044.0+3300	10 44 00.	+ 33 00	1280		CLUSTER OF GALAXIES
ZC 1044.0+4403	10 44 00.	+ 44 03	1750		CLUSTER OF GALAXIES
MCG+10-16-008	10 44 00.	+ 58 20	24	16.	GALAXY
ZWG 291.002	10 44 00.	+ 58 21		15.7	GALAXY
ZWG 290.071	10 44 00.	+ 58 21		15.7	GALAXY
ZC 1044.0+5915	10 44 00.	+ 59 15	1810		CLUSTER OF GALAXIES
MCG+13-08-038	10 44 00.	+ 77 07	57	15.	GALAXY
ZWG 010.011	10 44 00.	- 02 27		14.8	GALAXY
MCG+00-28-008	10 44 00.	- 02 27	18	14.8	GALAXY
MRSL 287+00/1	10 44 00.	- 58 24	360		HII REGION
IC 0640	10 44 01.	+ 35 01			NONSTELLAR OBJECT
WRAY 19.24	10 44 03.3	- 58 22 06.		10.0	DIFFUSE NEBULA
RNGC 3368	10 44 05.	+ 12 05		10.5	GALAXY
ZWG 066.013	10 44 06.	+ 12 05		10.0	GALAXY
SCH 15	10 44 06.	+ 12 05		10.0	PECULIAR GALAXY
UGC 05882	10 44 06.	+ 12 05	480	10.0	GALAXY Sa/Sb
MCG+02-28-006	10 44 06.	+ 12 05	450	10.0	GALAXY
ZC 1044.1+1701	10 44 06.	+ 17 01	1340		CLUSTER OF GALAXIES
ZWG 184.013	10 44 06.	+ 37 45		15.6	GALAXY
MCG+12-10-081	10 44 06.	+ 72 27 30.	24	17.	GALAXY
MCG+12-10-082	10 44 06.	+ 72 41 30.	84	13.4	GALAXY
REIZ 0595	10 44 08.	+ 12 05	180	10.8	GALAXY
MCG+08-20-021	10 44 09.	+ 46 58	30	16.	GALAXY
ZC 1044.2+0319	10 44 12.	+ 03 19	1210		CLUSTER OF GALAXIES
ZC 1044.2+0919	10 44 12.	+ 09 19	740		CLUSTER OF GALAXIES
ZWG 267.012	10 44 12.	+ 54 18		15.6	GALAXY
UGC 05883	10 44 12.	+ 54 18	66	15.6	GALAXY DWRF IR
MCG+09-18-026	10 44 12.	+ 54 18	60	14.	GALAXY
ZC 1044.2+5525	10 44 12.	+ 55 25	270		CLUSTER OF GALAXIES
ZC 1044.2+6013	10 44 12.	+ 60 13	3630		CLUSTER OF GALAXIES
TON-N 1288	10 44 14.	+ 22 10		13.5	BLUE STAR
HN 1249	10 44 14.8	+ 00 10 08.	18		NEBULA
MCG+08-20-022	10 44 15.	+ 47 41	60	15.	GALAXY
HN 1250	10 44 16.7	- 01 07 16.	24		NEBULA
ARC 1093	10 44 17.	+ 09 20		17.8	RICH CLUSTER OF GALAXIES
ZWG 155.007	10 44 18.	+ 26 48		14.4	GALAXY
UGC 05884	10 44 18.	+ 26 48	72	14.4	GALAXY Sb
MCG+05-26-005	10 44 18.	+ 30 18	84	15.5	GALAXY
ZWG 155.008	10 44 18.	+ 30 19		15.5	GALAXY
UGC 05885	10 44 18.	+ 30 19	90	15.5	GALAXY Sb
MCG+06-24-013	10 44 18.	+ 37 50	30	15.5	GALAXY
MCG+10-16-010	10 44 18.	+ 58 27	36	16.	GALAXY
RNGC 3369	10 44 18.	- 24 59		15.0	GALAXY
TON-N 1289	10 44 19.	+ 35 10		16.8	BLUE STAR
REIZ 0596	10 44 21.	+ 26 48	30	13.8	GALAXY
MCG-04-26-009	10 44 21.	- 24 59 30.	60	15.	GALAXY
RNGC 3371	10 44 22.	+ 14 04			NON-EXISTENT OBJFCT
RNGC 3370	10 44 22.	+ 17 32		12.5	GALAXY
RNGC 3378	10 44 22.	- 39 45			GALAXY
REIZ 0600	10 44 23.	+ 17 32	108	12.6	GALAXY
REIZ 0597	10 44 23.	+ 31 00	18	15.1	GALAXY
REIZ 0598	10 44 23.	+ 31 26	18	14.9	GALAXY
ZWG 095.019	10 44 24.	+ 17 32		12.4	GALAXY
UGC 05887	10 44 24.	+ 17 32	174	12.4	GALAXY Sc
MCG+05-26-006	10 44 24.	+ 26 47	66	14.4	GALAXY
ZWG 155.009	10 44 24.	+ 31 00		15.6	GALAXY
ZWG 184.014	10 44 24.	+ 35 36		15.2	GALAXY
ZWG 241.012	10 44 24.	+ 47 41		15.4	GALAXY
ZWG 351.041	10 44 24.	+ 77 05		15.0	GALAXY
ZWG 010.012	10 44 24.	- 01 07		14.4	GALAXY
UGC 05886	10 44 24.	- 01 07	72	14.4	GALAXY S
MCG+00-28-009	10 44 24.	- 01 07	48	14.4	GALAXY
MCG-03-28-003	10 44 24.	- 15 53 30.	138	14.5	GALAXY
VHE 43	10 44 24.	- 59 45	96		REFLECTION NEBULA
SEY 083	10 44 26.	+ 59 40 15.		15.2	FAINT GALAXY
RNGC 3375	10 44 27.	- 09 41		13.0	GALAXY
ZWG 038.011	10 44 30.	+ 07 30		15.5	GALAXY
ZWG 038.012	10 44 30.	+ 07 32		15.1	GALAXY
MCG+03-28-008	10 44 30.	+ 17 33	180	12.4	GALAXY
MCG+07-22-061	10 44 30.	+ 39 11 30.	30	14.5	GALAXY
ZWG 212.053	10 44 30.	+ 39 12		15.1	GALAXY

OBJECT NAME	RIGHT ASCEN.	DECLINATION	DIAM.	MAGN.	TYPE OF OBJECT
MCG+09-18-027	10 44 30.	+ 52 06	48	16.	GALAXY
ZWG 267.013	10 44 30.	+ 52 08		15.5	GALAXY
MCG-01-28-002	10 44 30.	- 09 30	36	14.	GALAXY
MCG-02-28-008	10 44 30.	- 09 41	30	13.5	GALAXY
HN 1251	10 44 30.9	- 02 04 29.	12		NEBULA
REIZ 0601	10 44 31.	+ 07 30	12	15.4	GALAXY
TON-N 1290	10 44 32.	+ 21 52		16.9	BLUE STAR
MCG+08-20-023	10 44 33.	+ 45 50	36	16.	GALAXY
AR 20	10 44 34.43	+ 77 03 57.5			NEBULA
RNGC 3373	10 44 35.	+ 13 56			NON-EXISTENT OBJECT
HN 1252	10 44 35.2	- 02 05 29.	18		NEBULA
TON-N 0542	10 44 36.	+ 30 40		15.6	BLUE STAR
ZC 1044.6+4213	10 44 36.	+ 42 13	1010		CLUSTER OF GALAXIES
ZWG 267.014	10 44 36.	+ 56 20		15.2	GALAXY
UGC 05888	10 44 36.	+ 56 20	96	15.2	GALAXY IRR
ZWG 291.003	10 44 36.	+ 59 40		15.0	GALAXY
ZWG 290.072	10 44 36.	+ 59 40		15.0	GALAXY
MCG+10-16-009	10 44 36.	+ 59 40	18	15.	GALAXY
GO 059	10 44 36.	- 57 12	76	18.7	STAR CHAIN
HN 1253	10 44 39.2	- 02 03 59.	12		NEBULA
RNGC 3377A	10 44 40.	+ 14 20		15.0	GALAXY
ZWG 066.014	10 44 42.	+ 14 20		15.0	GALAXY
UGC 05889	10 44 42.	+ 14 20	102	15.0	GALAXY DWRF SP
MCG+02-28-007	10 44 42.	+ 14 20	120	15.0	GALAXY
ZWG 095.020	10 44 42.	+ 17 52		15.5	GALAXY
ZC 1044.7+2745	10 44 42.	+ 27 45	870		CLUSTER OF GALAXIES
MCG+07-22-062	10 44 42.	+ 39 11	30	15.	GALAXY
MCG+09-18-028	10 44 42.	+ 56 21	60	15.	GALAXY
ZWG 333.056	10 44 42.	+ 72 40		13.8	GALAXY
UGC 05890	10 44 42.	+ 72 40	108	13.8	GALAXY Sc
ZWG 010.013	10 44 42.	- 02 05		15.5	GALAXY
ASS 60	10 44 42.	- 58 49	32400		OB ASSOCIATION CAR OB1
REIZ 0599	10 44 43.	+ 56 21	60	14.8	GALAXY
MCG+07-22-063	10 44 44.	+ 39 11 30.	60	14.5	GALAXY
GO 095	10 44 44.	- 61 46	121	18.0	STAR CHAIN
MCG+07-22-064	10 44 45.	+ 39 12 30.	15	17.	GALAXY
RNGC 3364	10 44 45.	+ 72 41		14.0	GALAXY
ARC 1095	10 44 46.	+ 15 29		17.6	RICH CLUSTER OF GALAXIES
GO 053	10 44 46.	- 57 58	84	18.2	STAR CHAIN
ZWG 038.013	10 44 48.	+ 06 19		14.4	GALAXY
RNGC 3376	10 44 48.	+ 06 19		14.5	GALAXY
REIZ 0603	10 44 48.	+ 06 19	12	14.3	GALAXY
UGC 05891	10 44 48.	+ 06 19	60	14.4	GALAXY S
MCG+01-28-007	10 44 48.	+ 06 19	48	14.4	GALAXY
KARA.73B 0443	10 44 48.	+ 06 19	48	14.4	ISOLATED GALAXY E
ZWG 038.014	10 44 48.	+ 07 31		14.5	GALAXY
UGC 05892	10 44 48.	+ 07 31	66	14.5	GALAXY SBb
MCG+01-28-008	10 44 48.	+ 07 31	48	14.5	GALAXY
TON-N 0543	10 44 48.	+ 29 42		16.8	BLUE STAR
ZC 1044.8+3350	10 44 48.	+ 33 50	810		CLUSTER OF GALAXIES
ZWG 212.054	10 44 48.	+ 39 12		14.4	GALAXY
UGC 05893	10 44 48.	+ 39 12	66	14.4	GALAXY E?
MCG+13-08-039	10 44 48.	+ 74 53	51	16.	GALAXY
MCG+13-08-040	10 44 48.	+ 80 26	66	15.	GALAXY
MCG-05-26-006	10 44 48.	- 28 38	24	15.5	GALAXY
ARC 1094	10 44 50.	+ 27 47		18.0	RICH CLUSTER OF GALAXIES
REIZ 0602	10 44 53.	+ 30 54	18	15.2	GALAXY
ZC 1044.9+1527	10 44 54.	+ 15 27	1280		CLUSTER OF GALAXIES
ZWG 155.010	10 44 54.	+ 26 33		14.7	GALAXY
UGC 05894	10 44 54.	+ 26 33	96	14.7	GALAXY SBa
ZWG 155.011	10 44 54.	+ 30 02		15.7	GALAXY
UGC 05895	10 44 54.	+ 30 02	72	15.7	GALAXY SB?a-b
ZWG 155.012	10 44 54.	+ 30 53		15.7	GALAXY
MCG+05-26-007	10 44 54.	+ 30 53	15	15.5	GALAXY
MCG+07-22-065	10 44 54.	+ 39 11	8	17.	GALAXY
RNGC 3383	10 44 54.	- 24 11		13.0	GALAXY
MCG-04-26-010	10 44 54.	- 24 11	72	13.	GALAXY
MRSL 287-00/1	10 44 54.	- 59 16	7200		HII REGION
REIZ 0607	10 44 55.	+ 07 31	18	14.7	GALAXY
REIZ 0604	10 44 57.	+ 26 33	54	14.7	GALAXY
HN 1254	10 44 57.4	- 01 13 53.	30		NEBULA
REIZ 0605	10 44 58.	+ 28 38	30	15.8	GALAXY
REIZ 0606	10 44 58.	+ 30 02	30	15.2	GALAXY
REIZ 0608	10 44 59.	+ 18 09	12	15.4	GALAXY
VDB.66G 088	10 45	+ 14 20	70		DWARF GALAXY
ZWG 066.015	10 45 00.	+ 11 22		14.0	GALAXY
UGC 05897	10 45 00.	+ 11 22	168	14.0	GALAXY Sc
MCG+02-28-008	10 45 00.	+ 11 22	150	14.0	GALAXY
ZWG 095.021	10 45 00.	+ 18 10		15.7	GALAXY
ZC 1045.0+2146	10 45 00.	+ 21 46	1410		CLUSTER OF GALAXIES
MCG+05-26-008	10 45 00.	+ 26 31	90	14.7	GALAXY
MCG+05-26-009	10 45 00.	+ 30 02	60	15.7	GALAXY
UGC 05898	10 45 00.	+ 33 59	96	16.0	GALAXY
STOCK 17	10 45 00.	+ 39 12			BLUE KNOT NEAR ELLIP GLXY
ZWG 212.055	10 45 00.	+ 43 24		15.6	GALAXY
MCG+13-08-041	10 45 00.	+ 77 05	57	14.	GALAXY
ZWG 364.018	10 45 00.	+ 80 26		15.6	GALAXY
ZWG 351.042	10 45 00.	+ 80 26		15.6	GALAXY
ZWG 010.014	10 45 00.	- 01 13		15.2	GALAXY
UGC 05896	10 45 00.	- 01 13	90	15.3	GALAXY Sa-b
MCG+00-28-010	10 45 00.	- 01 13	72	15.2	GALAXY
REIZ 0611	10 45 02.	+ 11 20	132	13.2	GALAXY
IC 0641	10 45 02.	+ 34 56			NONSTELLAR OBJECT
REIZ 0612	10 45 03.	+ 14 15	90	11.5	GALAXY
MCG+06-24-014	10 45 03.	+ 33 59 30.	66	16.	GALAXY
RNGC 3374	10 45 03.	+ 43 26		14.5	GALAXY
MCG+07-22-066	10 45 03.	+ 43 27	66	14.	GALAXY
HN 1255	10 45 03.0	- 01 16 17.	24		NEBULA
RNGC 3377	10 45 04.	+ 14 15		11.5	GALAXY
REIZ 0609	10 45 05.	+ 30 55	12	15.4	GALAXY
ZWG 066.016	10 45 06.	+ 14 15		10.7	GALAXY
UGC 05899	10 45 06.	+ 14 15	252	10.7	GALAXY E
MCG+02-28-009	10 45 06.	+ 14 15	90	10.7	GALAXY
ZWG 155.013	10 45 06.	+ 26 31		15.4	GALAXY
MCG+05-26-010	10 45 06.	+ 30 53 30.	18	15.7	GALAXY
REIZ 0610	10 45 06.	+ 33 59	66	15.2	GALAXY
ZWG 212.056	10 45 06.	+ 41 02		15.6	GALAXY
UGC 05900	10 45 06.	+ 41 02	84	15.6	GALAXY S
MCG+07-22-067	10 45 06.	+ 41 02	72	15.	GALAXY
ZWG 212.057	10 45 06.	+ 43 26		14.6	GALAXY
UGC 05901	10 45 06.	+ 43 26	84	14.6	GALAXY SBc
MCG+08-20-024	10 45 06.	+ 48 35	30	15.	GALAXY
MCG+11-13-038A	10 45 06.	+ 66 37 30.	108	15.	GALAXY
KARA.72 248A	10 45 06.	+ 66 38	48	14.5	PART OF DOUBLE GALAXY
MCG+11-13-038B	10 45 06.	+ 66 38	27	16.	GALAXY
ZWG 010.015	10 45 06.	- 02 50		15.7	GALAXY
LYNG 03	10 45 06.	- 59 50 06.	1800		OB CONCENTRATION
REIZ 0613	10 45 10.	+ 28 30	30	14.5	GALAXY
RNGC 3379	10 45 11.	+ 12 51		11.0	GALAXY
ZWG 066.017	10 45 12.	+ 11 11		14.7	GALAXY
MCG+02-28-010	10 45 12.	+ 11 11	60	14.7	GALAXY
ZWG 066.018	10 45 12.	+ 12 51		9.6	GALAXY
SCH 16	10 45 12.	+ 12 51		9.6	PECULIAR GALAXY
UGC 05902	10 45 12.	+ 12 51	270	9.6	GALAXY E
MCG+02-28-011	10 45 12.	+ 12 51	276	09.6	GALAXY
ZWG 155.014	10 45 12.	+ 28 31		15.1	GALAXY
UGC 05903	10 45 12.	+ 28 31	72	15.1	GALAXY SBa-b
ZWG 313.034	10 45 12.	+ 66 37		14.5	GALAXY
UGC 05904	10 45 12.	+ 66 37	132	14.5	GALAXY Sb
KARA.72 248B	10 45 12.	+ 66 37	90		PART OF DOUBLE GALAXY
REIZ 0614	10 45 14.	+ 11 09	42	14.5	GALAXY
REIZ 0615	10 45 15.	+ 12 51	120	11.0	GALAXY
HOLM 212A	10 45 15.	+ 12 51	120	11.3	PART OF MULTIPLE GALAXY
IC 0642	10 45 18.	+ 18 27 07.			NONSTELLAR OBJECT
MCG+05-26-011	10 45 18.	+ 28 30	72	15.1	GALAXY
ZWG 184.015	10 45 18.	+ 34 00		15.6	GALAXY
MCG+08-20-025	10 45 18.	+ 45 27	42	16.	GALAXY
LB 00583	10 45 18.	+ 46 31 00.		18.0	FAINT BLUE STAR
ZWG 241.013	10 45 18.	+ 50 17		14.9	GALAXY
MCG+08-20-026	10 45 18.	+ 50 17	42	16.	GALAXY
7ZW 346	10 45 18.	+ 66 37			COMPACT GALAXY
ZC 1045.3-0228	10 45 18.	- 02 28	2220		CLUSTER OF GALAXIES
HN 1256	10 45 19.9	+ 01 04 54.	18		NEBULA
SHB 162	10 45 22.8	+ 60 22 36.		17.5	QUASI-STELLAR OBJECT
BC 4C60.15	10 45 23.07	+ 60 24 39.7		17.	QUASI-STELLAR OBJECT
ZWG 010.016	10 45 24.	+ 01 05		15.1	GALAXY
MCG+03-28-009	10 45 24.	+ 15 40	60		GALAXY
ZWG 095.022	10 45 24.	+ 18 27		14.0	GALAXY
UGC 05905	10 45 24.	+ 18 27	96	14.0	GALAXY
TON-N 0544	10 45 24.	+ 28 59		15.5	BLUE STAR
ZWG 241.014	10 45 24.	+ 45 28		15.6	GALAXY
ZC 1045.4+4737	10 45 24.	+ 47 37	4640		CLUSTER OF GALAXIES
ARC 1098	10 45 25.	- 03 41		16.9	RICH CLUSTER OF GALAXIES
ARC 1096	10 45 27.	+ 28 19		17.8	RICH CLUSTER OF GALAXIES
REIZ 0616	10 45 28.	+ 28 52	72	13.7	GALAXY
REIZ 0617	10 45 29.	+ 18 27	24	14.3	GALAXY
ZWG 066.019	10 45 30.	+ 09 59		15.3	GALAXY
ZWG 095.023	10 45 30.	+ 14 57		15.5	GALAXY
ZWG 095.024	10 45 30.	+ 18 53		15.6	GALAXY
MCG+05-26-012	10 45 30.	+ 28 51	102	13.6	GALAXY
ZWG 155.015	10 45 30.	+ 28 53		13.6	GALAXY
UGC 05906	10 45 30.	+ 28 53	102	13.6	GALAXY S
TON-N 0043	10 45 30.	+ 30 05		14.6	BLUE STAR
MCG+07-22-068	10 45 30.	+ 38 39	108	14.	GALAXY
ZC 1045.5+3911	10 45 30.	+ 39 11	670		CLUSTER OF GALAXIES
MCG+09-18-029	10 45 30.	+ 54 21	42	17.	GALAXY
UGC 05907	10 45 30.	+ 66 27	108	17.	GALAXY DWARF
RNGC 3380	10 45 31.	+ 28 52		13.5	GALAXY
HN 1257	10 45 31.5	+ 00 26 54.	18		NEBULA
HN 1258	10 45 31.6	- 01 40 12.	24		NEBULA
TON-N 1291	10 45 32.	+ 20 35		16.8	BLUE STAR
MCG+05-26-013	10 45 33.	+ 26 50	36	15.5	GALAXY
MCG+07-22-069	10 45 33.	+ 38 32 30.	30	15.5	GALAXY
MCG+09-18-030	10 45 33.	+ 54 20	9	17.	GALAXY
REIZ 0619	10 45 33.	- 01 40	18	14.9	GALAXY
RNGC 3384	10 45 35.	+ 12 54		11.5	GALAXY
RNGC 3450	10 45 35.	- 20 35		12.0	GALAXY
ZWG 038.015	10 45 36.	+ 05 11		13.7	GALAXY
RNGC 3385	10 45 36.	+ 05 11		13.5	GALAXY
REIZ 0621	10 45 36.	+ 05 11	18	14.4	GALAXY
UGC 05908	10 45 36.	+ 05 11	102	13.7	GALAXY S0
MCG+01-28-009	10 45 36.	+ 05 11	72	13.7	GALAXY
ZWG 038.016	10 45 36.	+ 05 16		14.8	GALAXY
RNGC 3387	10 45 36.	+ 05 16		15.0	GALAXY
RNGC 3386	10 45 36.	+ 05 16		15.0	GALAXY
REIZ 0622	10 45 36.	+ 05 16	12	14.8	GALAXY
MCG+01-28-010	10 45 36.	+ 05 16	24	14.8	GALAXY
ZWG 066.020	10 45 36.	+ 13 30		15.6	GALAXY
MCG+03-28-010	10 45 36.	+ 18 29	18	14.0	GALAXY
ZWG 184.016	10 45 36.	+ 34 58		12.8	GALAXY
RNGC 3381	10 45 36.	+ 34 58		13.0	GALAXY
UGC 05909	10 45 36.	+ 34 58	138	12.8	GALAXY SB
MCG+06-24-015	10 45 36.	+ 34 58 30.	120	13.	GALAXY
ZC 1045.6+3504	10 45 36.	+ 35 04	2350		CLUSTER OF GALAXIES
ZWG 212.058	10 45 36.	+ 38 33		15.6	GALAXY
ZWG 212.059	10 45 36.	+ 38 40		14.7	GALAXY
UGC 05910	10 45 36.	+ 38 40	90	14.7	GALAXY SBb
ZC 1045.6+4008	10 45 36.	+ 40 08	3160		CLUSTER OF GALAXIES
MCG+09-18-031	10 45 36.	+ 56 00	30	17.	GALAXY
ZC 1045.6+6105	10 45 36.	+ 61 05	540		CLUSTER OF GALAXIES
ZWG 010.017	10 45 36.	- 01 39		15.4	GALAXY
MCG-05-26-007	10 45 36.	- 31 16	210	13.	GALAXY
SN 1970M	10 45 37.	+ 14 19		16.5	SUPERNOVA
REIZ 0618	10 45 37.	+ 34 58	60	13.7	GALAXY
HN 1259	10 45 37.7	- 02 24 36.			NEBULA
IC 2603	10 45 38.	+ 33 12			SINGLE STAR
REIZ 0620	10 45 39.	+ 12 53	180	11.2	GALAXY
RNGC 3382	10 45 41.	+ 37 00			NON-EXISTENT OBJECT
RNGC 3388	10 45 42.	+ 08 52			NON-EXISTENT OBJECT
ZWG 066.021	10 45 42.	+ 12 54		10.0	GALAXY
UGC 05911	10 45 42.	+ 12 54	336	10.0	GALAXY S0
MCG+02-28-012	10 45 42.	+ 12 54	96	10.0	GALAXY
ZWG 095.025	10 45 42.	+ 18 55		15.5	GALAXY
ZC 1045.7+2010	10 45 42.	+ 20 10	2220		CLUSTER OF GALAXIES
ZWG 155.016	10 45 42.	+ 26 51		14.5	GALAXY
UGC 05912	10 45 42.	+ 26 51	84	14.5	GALAXY Sc
ZC 1045.7+2820	10 45 42.	+ 28 20	940		CLUSTER OF GALAXIES
ARC 1097	10 45 42.	+ 31 44		16.0	RICH CLUSTER OF GALAXIES
BELW 146	10 45 42.	- 20 31 24.			NEBULA
OCL 0824	10 45 42.	- 57 13	102	10.	OPEN STAR CLUSTER
HOLM 212B	10 45 43.	+ 12 54	210	11.4	PART OF MULTIPLE GALAXY
SN 1971O	10 45 43.	+ 26 51		15.5	SUPERNOVA
HN 1260	10 45 44.5	+ 01 06 24.	24		NEBULA
SN 1967C	10 45 45.	+ 12 49		13.0	SUPERNOVA
REIZ 0623	10 45 45.	+ 26 51	60	14.9	GALAXY
MCG-03-28-004	10 45 45.	- 20 37	138	12.	GALAXY
REIZ 0625	10 45 46.	+ 00 15	36	15.0	GALAXY
REIZ 0626	10 45 46.	+ 01 07	48	14.1	GALAXY
HN 1261	10 45 46.8	+ 00 14 48.	18		NEBULA
RNGC 3389	10 45 47.	+ 12 48		12.5	GALAXY
ZWG 010.019	10 45 48.	+ 01 07		14.3	GALAXY
UGC 05913	10 45 48.	+ 01 07	48	14.3	GALAXY
MCG+00-28-011A	10 45 48.	+ 01 07	42	14.3	GALAXY
ZWG 038.017	10 45 48.	+ 05 05		15.5	GALAXY
ZC 1045.8+0510	10 45 48.	+ 05 10	2690		CLUSTER OF GALAXIES
ZWG 038.018	10 45 48.	+ 07 14		15.4	GALAXY

OBJECT NAME	RIGHT ASCEN.	DECLINATION	DIAM.	MAGN.	TYPE OF OBJECT
ZWG 066.022	10 45 48.	+ 12 48		12.0	GALAXY
UGC 05914	10 45 48.	+ 12 48	174	12.0	GALAXY Sc
MCG+02-28-013	10 45 48.	+ 12 48	156	12.0	GALAXY
ZC 1045.8+1251	10 45 48.	+ 12 51	2690		CLUSTER OF GALAXIES
MCG+05-26-014	10 45 48.	+ 26 49	84	14.5	GALAXY
ZC 1045.8+3155	10 45 48.	+ 31 55	5110		CLUSTER OF GALAXIES
TON-N 0545	10 45 48.	+ 32 12		15.3	BLUE STAR
ZC 1045.8+3548	10 45 48.	+ 35 48	740		CLUSTER OF GALAXIES
UGC 05915	10 45 48.	+ 48 18	66	16.0	GALAXY Sa-b
ZC 1045.8+5508	10 45 48.	+ 55 08	1950		CLUSTER OF GALAXIES
KARA.68 072	10 45 48.	+ 65 47	101		DWARF GALAXY
MCG+11-13-039	10 45 48.	+ 65 47	90	16.	GALAXY
ZWG 010.018	10 45 48.	- 03 18		15.6	GALAXY
RNGC 3390	10 45 50.	- 31 17		13.0	GALAXY
WRAY 19.25	10 45 50.1	- 59 56 45.			DIFFUSE NEBULA
REIZ 0627	10 45 51.	+ 12 47	90	11.8	GALAXY
MCG+04-26-007	10 45 51.	+ 22 31	42	15.	GALAXY
MCG+08-20-027	10 45 51.	+ 48 19	42	16.	GALAXY
MCG+08-20-028	10 45 51.	+ 50 18	42	16.	GALAXY
MCG-03-28-005	10 45 51.	- 20 53	60	17.	GALAXY
REIZ 0624	10 45 52.	+ 29 58	12	15.6	GALAXY
HN 1262	10 45 52.4	- 03 03 13.	18		NEBULA
ZWG 038.019	10 45 54.	+ 05 04		15.3	GALAXY
ZWG 066.023	10 45 54.	+ 13 15		15.2	GALAXY
REIZ 0628	10 45 54.	+ 21 51	12	15.5	GALAXY
TON-N 0546	10 45 54.	+ 31 26		16.0	BLUE STAR
ZC 1045.9+3615	10 45 54.	+ 36 15	1480		CLUSTER OF GALAXIES
MCG+08-20-029	10 45 54.	+ 46 59	60	15.	GALAXY
ZWG 241.015	10 45 54.	+ 50 18		14.9	GALAXY
MRK 152	10 45 54.	+ 50 18	25	15.	GALAXY WITH UV CONTINUUM
ZWG 010.020	10 45 54.	- 03 02		15.7	GALAXY
MCG+00-28-011B	10 45 54.	- 03 02	66	15.7	GALAXY
RNGC 3393	10 45 54.	- 24 54		13.0	GALAXY
MCG-04-26-011	10 45 54.	- 24 54	108	13.	GALAXY
HOLM 212C	10 45 55.	+ 12 48	120	12.0	PART OF MULTIPLF GALAXY
HARO 25	10 45 55.	+ 26 18			BLUE EMISSION-LINE GALAXY
VDB.66G 087	10 46	+ 65 45	70		DWARF GALAXY
HOFF L07	10 46	- 62 04	180		DARK HOLE
UGC 05916	10 46 00.	+ 21 59	66	16.0	GALAXY Sc
REIZ 0629	10 46 00.	+ 22 00	24	15.3	GALAXY
MCG+04-26-008	10 46 00.	+ 22 00	60	15.	GALAXY
ZWG 125.009	10 46 00.	+ 22 29		15.7	GALAXY
MCG+04-26-010	10 46 00.	+ 22 29 30.	27	15.7	GALAXY
MCG+04-26-009	10 46 00.	+ 26 20	15	15.5	GALAXY
ZWG 155.017	10 46 00.	+ 27 02		15.4	GALAXY
ZC 1046.0+3320	10 46 00.	+ 33 20	470		CLUSTER OF GALAXIES
ZWG 241.016	10 46 00.	+ 46 59		15.0	GALAXY
UGC 05917	10 46 00.	+ 46 59	72	15.0	GALAXY IRR
ZWG 241.017	10 46 00.	+ 48 18		15.5	GALAXY
MCG+09-18-032	10 46 00.	+ 52 36	54	16.	GALAXY
MCG+10-16-011	10 46 00.	+ 58 05	24	16.	GALAXY
ZC 1046.0+6354	10 46 00.	+ 63 54	1140		CLUSTER OF GALAXIES
LB 03497	10 46 00.	- 79 13		13.9	FAINT BLUE STAR
KARA.73 31	10 46 01.	- 38 07	47		DWARF GALAXY
ARC 1099	10 46 04.	+ 35 14		17.0	RICH CLUSTER OF GALAYIES
ZWG 038.020	10 46 06.	+ 04 34		15.6	GALAXY
ZC 1046.1+0801	10 46 06.	+ 08 01	870		CLUSTER OF GALAXIES
ZWG 095.026	10 46 06.	+ 18 56		15.7	GALAXY
ZWG 155.018	10 46 06.	+ 27 13		15.7	GALAXY
ZC 1046.1+5140	10 46 06.	+ 51 40	870		CLUSTER OF GALAXIES
ZWG 267.015	10 46 06.	+ 52 36		14.6	GALAXY
MRK 153	10 46 06.	+ 52 36	20	15.5	GALAXY WITH UV CONTINUUM
UGC 05918	10 46 06.	+ 56 50	180	17.	GALAXY DWARF
MCG-03-28-006	10 46 06.	- 21 24	90	14.	GALAXY
ARC 1100	10 46 09.	+ 22 30		15.7	RICH CLUSTER OF GALAXIES
MAI 058	10 46 10.	+ 65 05	101		DWARF SPHEROIDAL GALAXY
ARC 1104	10 46 10.	- 16 53		17.4	RICH CLUSTER OF GALAXIES
GO 082	10 46 10.	- 61 22	115	18.1	STAR CHAIN
8ZW 1046+12.1	10 46 12.	+ 12 04		16.8	COMPACT GALAXY
ZWG 095.027	10 46 12.	+ 17 23		15.6	GALAXY
ZC 1046.2+5243	10 46 12.	+ 52 43	340		CLUSTER OF GALAXIES
MCG+10-16-012	10 46 12.	+ 58 35	60	15.	GALAXY
ZWG 291.004	10 46 12.	+ 58 37		15.5	GALAXY
UGC 05919	10 46 12.	+ 58 37	60	15.5	GALAXY Sa
ARC 1102	10 46 16.	+ 07 27		17.5	RICH CLUSTER OF GALAXIES
RNGC 3391	10 46 16.	+ 14 29		13.5	GALAXY
ZC 1046.3+0038	10 46 18.	+ 00 38	3090		CLUSTER OF GALAXIES
ZWG 066.024	10 46 18.	+ 11 06		15.7	GALAXY
ZWG 066.025	10 46 18.	+ 12 28		15.4	GALAXY
ZWG 066.026	10 46 18.	+ 14 24		15.5	GALAXY
ZWG 066.027	10 46 18.	+ 14 29		13.5	GALAXY
UGC 05920	10 46 18.	+ 14 29	60	13.1	GALAXY
MCG+02-28-014	10 46 18.	+ 14 29	60	13.5	GALAXY
HOLM 213C	10 46 18.	+ 30 05	18	15.7	PART OF MULTIPLE GALAXY
ZWG 010.021	10 46 18.	- 03 19		15.5	GALAXY
GO 063	10 46 18.	- 57 37	86	19.1	STAR CHAIN
HOLM 213B	10 46 19.	+ 30 04	18	15.5	PART OF MULTIPLE GALAXY
REIZ 0632	10 46 21.	+ 14 29	12	14.7	GALAXY
REIZ 0631	10 46 21.	+ 14 29	24	14.6	GALAXY
REIZ 0630	10 46 22.	+ 30 05	18	15.1	GALAXY
HOLM 213A	10 46 22.	+ 30 05	18	14.8	PART OF MULTIPLE GALAXY
ZC 1046.4+0725	10 46 24.	+ 07 25	1410		CLUSTER OF GALAXIES
ZWG 066.028	10 46 24.	+ 13 17		15.1	GALAXY
ZWG 095.028	10 46 24.	+ 18 39		15.6	GALAXY
ZWG 095.029	10 46 24.	+ 19 45		15.4	GALAXY
ZWG 155.019	10 46 24.	+ 28 01		15.6	GALAXY
ZWG 155.020	10 46 24.	+ 28 11		15.3	GALAXY
UGC 05921	10 46 24.	+ 28 11	102	15.3	GALAXY Sc-IRR
MCG+05-26-015	10 46 24.	+ 30 04	36	15.6	GALAXY
ZWG 155.021	10 46 24.	+ 30 05		15.6	GALAXY
MCG+07-22-070	10 46 24.	+ 43 34 30.	69	15.	GALAXY
MCG-03-28-007	10 46 24.	- 19 26	72	15.	GALAXY
HN 1263	10 46 26.6	- 00 23 01.	36		NEBULA
MCG+04-26-011	10 46 27.	+ 22 17	84	14.9	GALAXY
LB 01903	10 46 28.	+ 57 25 42.		14.8	FAINT BLUE STAR
REIZ 0637	10 46 28.	- 00 23	42	13.8	GALAXY
ZWG 038.021	10 46 30.	+ 02 30		15.3	GALAXY
ZWG 038.022	10 46 30.	+ 07 11		14.2	GALAXY
UGC 05923	10 46 30.	+ 07 11	60	14.4	GALAXY S0-a?
MCG+01-28-011	10 46 30.	+ 07 11	60	14.2	GALAXY
ZC 1046.5+1859	10 46 30.	+ 18 59	3490		CLUSTER OF GALAXIES
ZWG 125.010	10 46 30.	+ 22 17		14.9	GALAXY
REIZ 0633	10 46 30.	+ 22 17	60	14.2	GALAXY
UGC 05924	10 46 30.	+ 22 17	102	14.9	GALAXY Sa
ZWG 212.060	10 46 30.	+ 43 34		15.6	GALAXY
UGC 05925	10 46 30.	+ 43 34	84	15.6	GALAXY Sc
7ZW 347	10 46 30.	+ 65 47			COMPACT GALAXY
ZWG 351.043	10 46 30.	+ 77 13		14.4	GALAXY

OBJECT NAME	RIGHT ASCEN.	DECLINATION	DIAM.	MAGN.	TYPE OF OBJECT
UGC 05926	10 46 30.	+ 77 13	72	14.4	GALAXY S
ZWG 010.022	10 46 30.	- 00 22		14.7	GALAXY
UGC 05922	10 46 30.	- 00 22	66	14.7	GALAXY SB
MCG+00-28-012	10 46 30.	- 00 22	60	14.7	GALAXY
HOLM 214B	10 46 30.	- 00 23	48	13.7	PART OF MULTIPLE GALAXY
ARC 1103	10 46 32.	+ 13 43		17.6	RICH CLUSTER OF GALAXIES
REIZ 0634	10 46 33.	+ 28 00	54	14.7	GALAXY
REIZ 0635	10 46 33.	+ 28 10	84	14.1	GALAXY
REIZ 0638	10 46 34.	- 00 25	18	15.5	GALAXY
HOLM 214A	10 46 34.	- 00 25	30	15.1	PART OF MULTIPLE GALAXY
HN 1264	10 46 34.7	+ 00 38 53.	24		NEBULA
ARC 1105	10 46 35.	+ 09 43		16.8	RICH CLUSTER OF GALAXIES
ZWG 010.023	10 46 36.	+ 02 30		15.3	GALAXY
ZWG 066.029	10 46 36.	+ 12 41		15.6	GALAXY
ZWG 095.030	10 46 36.	+ 16 04		15.0	GALAXY
MCG+03-28-011	10 46 36.	+ 16 04	24	15.0	GALAXY
ZC 1046.6+1745	10 46 36.	+ 17 45	2890		CLUSTER OF GALAXIES
MCG+05-26-017	10 46 36.	+ 28 00	48	15.6	GALAXY
MCG+05-26-016	10 46 36.	+ 28 10	90	15.3	GALAXY
ZWG 184.017	10 46 36.	+ 33 02		15.0	GALAXY
UGC 05927	10 46 36.	+ 33 02	84	15.0	GALAXY IRR
REIZ 0636	10 46 36.	+ 33 03	48	14.2	GALAXY
MCG+09-18-033	10 46 36.	+ 56 13	18	18.	GALAXY
ZWG 291.005	10 46 36.	+ 56 54		15.4	GALAXY
MCG+10-16-013	10 46 36.	+ 56 54	13	16.	GALAXY
MCG+10-16-014	10 46 36.	+ 60 32	36	16.	GALAXY
IC 2604	10 46 37.6	+ 33 02 18.			GALAXY SB(s)
HN 1265	10 46 40.3	+ 00 35 52.	18		NEBULA
PK286+02.1	10 46 41.	- 55 47 08.	25		PLANETARY NEBULA
ZWG 010.024	10 46 42.	+ 00 39		15.7	GALAXY
ZC 1046.7+0116	10 46 42.	+ 01 16	1610		CLUSTER OF GALAXIES
8ZW 1046+13.5	10 46 42.	+ 13 29		17.7	COMPACT GALAXY
MCG+06-24-016	10 46 42.	+ 33 01 30.	72	14.	GALAXY
ZC 1046.7+4430	10 46 42.	+ 44 30	2490		CLUSTER OF GALAXIES
ZWG 241.018	10 46 42.	+ 48 11		15.1	GALAXY
MCG+09-18-034	10 46 42.	+ 52 09	30	15.	GALAXY
AR 21	10 46 42.62	+ 77 12 08.0			NEBULA
ARC 1101	10 46 45.	+ 44 38		17.8	RICH CLUSTER OF GALAXIES
MCG-03-28-008	10 46 45.	- 19 23 30.	60	14.	GALAXY
HN 1266	10 46 45.9	- 00 24 38.			NEBULA
RNGC 3399	10 46 46.	+ 16 29		14.5	GALAXY
REIZ 0639	10 46 46.	+ 16 29	12	15.3	GALAXY
REIZ 0641	10 46 46.	- 00 25	36	14.5	GALAXY
ZWG 010.026	10 46 48.	+ 00 36		15.2	GALAXY
MCG+00-28-014	10 46 48.	+ 00 36	48	15.2	GALAXY
ZWG 066.030	10 46 48.	+ 12 29		15.3	GALAXY
ZWG 066.031	10 46 48.	+ 12 36		15.5	GALAXY
ZWG 095.031	10 46 48.	+ 16 29		14.7	GALAXY
MCG+03-28-012	10 46 48.	+ 16 29	78	14.7	GALAXY
REIZ 0640	10 46 48.	+ 22 03	18	14.2	GALAXY
MCG+04-26-012	10 46 48.	+ 22 03	54	15.	GALAXY
MRK 417	10 46 48.	+ 23 13	10	16.	GALAXY WITH UV CONTINUUM
ZC 1046.8+2747	10 46 48.	+ 27 47	10350		CLUSTER OF GALAXIES
TON-N 0547	10 46 48.	+ 28 10		14.7	BLUE STAR
ZC 1046.8+4253	10 46 48.	+ 42 53	340		CLUSTER OF GALAXIES
ZC 1046.8+4436	10 46 48.	+ 44 36	470		CLUSTER OF GALAXIES
ZWG 267.016	10 46 48.	+ 52 10		15.0	GALAXY
UGC 05928	10 46 48.	+ 52 10	60	15.0	GALAXY E-S0
ZWG 010.025	10 46 48.	- 00 24		14.9	GALAXY
MCG+00-28-013	10 46 48.	- 00 24	36	14.9	GALAXY
MCG-05-26-008	10 46 48.	- 31 03	60	15.	GALAXY
GO 039	10 46 49.	- 60 11	65	17.1	STAR CHAIN
HN 1267	10 46 50.2	+ 00 30 10.	18		NEBULA
IC 0643	10 46 51.	+ 12 28 09.			NONSTELLAR OBJECT
HOLM 214C	10 46 51.	- 00 25	36	14.3	PART OF MULTIPLE GALAXY
LB 01904	10 46 53.	+ 51 17 18.		17.0	FAINT BLUE STAR
ZWG 038.023	10 46 54.	+ 05 04		15.1	GALAXY
REIZ 0642	10 46 54.	+ 05 04	48	14.0	GALAXY
UGC 05929	10 46 54.	+ 05 04	72	15.1	GALAXY Sb-c
8ZW 1046+09.6	10 46 54.	+ 09 38		19.2	COMPACT GALAXY
UGC 05930	10 46 54.	+ 22 15	66	16.5	GALAXY S+COMP
ZC 1046.9+2744	10 46 54.	+ 27 44	400		CLUSTER OF GALAXIES
TON-N 0548	10 46 54.	+ 30 11		16.2	BLUE STAR
ZC 1046.9+3559	10 46 54.	+ 35 59	870		CLUSTER OF GALAXIES
MCG+09-18-036	10 46 54.	+ 52 07	24	17.	GALAXY
MCG+09-18-035	10 46 54.	+ 53 08	48	17.	GALAXY
ZC 1046.9+6441	10 46 54.	+ 64 41	2150		CLUSTER OF GALAXIES
MCG+11-13-040	10 46 54.	+ 65 04	72	15.	GALAXY
ARC 1107	10 46 55.	+ 03 55		17.0	RICH CLUSTER OF GALAXIES
GO 062	10 46 55.	- 57 19	172	18.6	STAR CHAIN
SHB 163	10 46 56.6	+ 05 21 25.		18.9	QUASI-STELLAR OBJECT
BC 4C05.46	10 46 57.3	+ 05 21 36.		18.94	QUASI-STELLAR OBJECT
ZC 1047.0+1349	10 47 00.	+ 13 49	2220		CLUSTER OF GALAXIES
ZWG 095.032	10 47 00.	+ 17 46		15.7	GALAXY
ZWG 184.018	10 47 00.	+ 33 14		12.1	GALAXY
UGC 05931	10 47 00.	+ 33 14	108	12.1	GALAXY S(c)
KARA.72 249A	10 47 00.	+ 33 14	108	12.1	PART OF DOUBLE GALAXY
ZWG 267.017	10 47 00.	+ 52 09		15.6	GALAXY
ZWG 313.035	10 47 00.	+ 65 05		15.2	GALAXY
UGC 05932	10 47 00.	+ 65 05	102	15.2	GALAXY Sb-c
ZC 1047.0+6750	10 47 00.	+ 67 50	1280		CLUSTER OF GALAXIES
MCG+12-10-083	10 47 00.	+ 68 59	27	17.	GALAXY
MCG+12-10-084	10 47 00.	+ 71 57	12	16.	GALAXY
ZWG 351.044	10 47 00.	+ 77 04		15.1	GALAXY
MCG+13-08-042	10 47 00.	+ 77 13	33	14.	GALAXY
IC 2605	10 47 03.	+ 33 14			MAY NOT EXIST
GO 046	10 47 03.	- 59 05	74	17.5	STAR CHAIN
RNGC 3405	10 47 04.	+ 16 30		14.5	GALAXY
REIZ 0645	10 47 04.	+ 16 30	12	15.3	GALAXY
ARP 270	10 47 05.	+ 33 15			PECULIAR GALAXY
AR 22	10 47 05.52	+ 77 03 39.3			NEBULA
ZWG 010.027	10 47 06.	+ 00 38		14.9	GALAXY
MCG+00-28-015	10 47 06.	+ 00 38	15	14.9	GALAXY
ZC 1047.1+0354	10 47 06.	+ 03 54	1340		CLUSTER OF GALAXIES
ZWG 095.033	10 47 06.	+ 16 30		14.4	GALAXY
UGC 05933	10 47 06.	+ 16 30	54	14.4	GALAXY DBL SYS
KARA.72 250B	10 47 06.	+ 16 30	30		PART OF DOUBLE GALAXY
KARA.72 250A	10 47 06.	+ 16 30	42	14.4	PART OF DOUBLE GALAXY
MCG+03-28-015	10 47 06.	+ 16 30	12	15.6	GALAXY
MCG+03-28-014	10 47 06.	+ 16 30	54	14.4	GALAXY
ZWG 095.034	10 47 06.	+ 16 32		15.6	GALAXY
MCG+03-28-013	10 47 06.	+ 19 38	60	17.	GALAXY
ZWG 155.022	10 47 06.	+ 32 10		15.7	GALAXY
UGC 05934	10 47 06.	+ 32 10	108	15.7	GALAXY
ZC 1047.1+3256	10 47 06.	+ 32 56	870		CLUSTER OF GALAXIES
MCG+06-24-017	10 47 06.	+ 33 14	90	13.5	GALAXY
ZWG 184.019	10 47 06.	+ 33 15		12.6	GALAXY
RNGC 3395	10 47 06.	+ 33 15		12.5	GALAXY

OBJECT NAME	RIGHT ASCEN.	DECLINATION	DIAM.	MAGN.	TYPE OF OBJECT
REIZ 0643	10 47 06.	+ 33 15	120	12.6	GALAXY
HOLM 215A	10 47 06.	+ 33 15	78	12.1	PART OF MULTIPLE GALAXY
UGC 05935	10 47 06.	+ 33 15	222	12.6	GALAXY IRR
VV 246B	10 47 06.	+ 33 15	102	12.1	INTERACTING GALAXY
KARA.72 249B	10 47 06.	+ 33 15	132	12.6	PART OF DOUBLE GALAXY
MCG+09-18-037	10 47 06.	+ 52 11	18	16.	GALAXY
MCG+11-13-041	10 47 06.	+ 65 59	108	14.	GALAXY
ZWG 351.045	10 47 06.	+ 78 55		15.6	GALAXY
HN 1268	10 47 07.2	- 01 30 50.	12		NEBULA
ARC 1108	10 47 10.	+ 16 30		17.2	RICH CLUSTER OF GALAXIES
HOLM 215B	10 47 11.	+ 33 15	90	13.1	PART OF MULTIPLE GALAXY
8ZW 1047+12.1	10 47 12.	+ 12 04		19.5	COMPACT GALAXY
MCG+05-26-018	10 47 12.	+ 32 11	60	15.7	GALAXY
RNGC 3396	10 47 12.	+ 33 15		12.5	GALAXY
REIZ 0644	10 47 12.	+ 33 15	108	12.9	GALAXY
MCG+06-24-018	10 47 12.	+ 33 15	144	14.	GALAXY
VV 246A	10 47 12.	+ 33 16	180	12.7	INTERACTING GALAXY
ZWG 184.020	10 47 12.	+ 36 36		14.4	GALAXY
UGC 05936	10 47 12.	+ 36 36	72	14.4	GALAXY S0
ZC 1047.2+4339	10 47 12.	+ 43 39	610		CLUSTER OF GALAXIES
ZWG 313.036	10 47 12.	+ 65 59		13.1	GALAXY
UGC 05937	10 47 12.	+ 65 59	120	13.1	GALAXY Sc
MCG+12-10-085	10 47 12.	+ 71 59	18	17.	GALAXY
RNGC 3394	10 47 14.	+ 65 59		13.0	GALAXY
HN 1269	10 47 14.2	+ 01 27 58.	18		NEBULA
ZWG 010.028	10 47 18.	+ 01 29		15.5	GALAXY
ZC 1047.3+2227	10 47 18.	+ 22 27	1950		CLUSTER OF GALAXIES
TON-N 0549	10 47 18.	+ 30 07		16.6	BLUE STAR
ZC 1047.3+3725	10 47 18.	+ 37 25	1080		CLUSTER OF GALAXIES
ZWG 184.021	10 47 18.	+ 38 25		15.4	GALAXY
ZC 1047.3+5549	10 47 18.	+ 55 49	610		CLUSTER OF GALAXIES
MCG+13-08-043	10 47 18.	+ 77 04	51	14.	GALAXY
UGC 05938	10 47 18.	+ 77 52	60	16.0	GALAXY PECULR
MCG+06-24-019	10 47 21.	+ 36 37	66	14.	GALAXY
HN 1270	10 47 21.7	- 01 36 45.	12		NEBULA
HOLM 216A	10 47 22.	+ 00 35	30	14.3	PART OF MULTIPLE GALAXY
HOLM 216B	10 47 22.	+ 00 37	24	15.3	PART OF MULTIPLE GALAXY
ARC 1109	10 47 23.	+ 18 02		17.2	RICH CLUSTER OF GALAXIES
ZWG 066.032	10 47 24.	+ 11 18		15.7	GALAXY
ZC 1047.4+2346	10 47 24.	+ 23 46	740		CLUSTER OF GALAXIES
ZC 1047.4+3055	10 47 24.	+ 30 55	810		CLUSTER OF GALAXIES
ZWG 184.022	10 47 24.	+ 33 01		15.6	GALAXY
REIZ 0646	10 47 24.	+ 33 02	30	14.7	GALAXY
ZWG 184.023	10 47 24.	+ 35 17		15.6	GALAXY
MCG+06-24-020	10 47 24.	+ 38 27	42	14.5	GALAXY
ZWG 334.001	10 47 24.	+ 71 55		15.6	GALAXY
ZWG 333.057	10 47 24.	+ 71 55		15.6	GALAXY
ZWG 351.046	10 47 24.	+ 77 12		14.9	GALAXY
UGC 05939	10 47 24.	+ 77 12	72	14.9	GALAXY PECULR
ZWG 010.029	10 47 24.	- 01 20		15.7	GALAXY
MCG+00-28-016	10 47 24.	- 01 20	42	15.7	GALAXY
PHL 6717	10 47 24.	- 15 00		18.6	BLUE STELLAR OBJECT
GO 057	10 47 24.	- 57 46	33	15.0	STAR CHAIN
GO 061	10 47 25.	- 57 26	44	17.8	STAR CHAIN
HN 1271	10 47 25.2	+ 00 35 21.			NEBULA
HN 1272	10 47 25.7	+ 00 37 21.	24		NEBULA
TON-N 1292	10 47 26.	+ 21 10		16.2	BLUE STAR
ARC 1106	10 47 26.	+ 44 16		17.8	RICH CLUSTER OF GALAXIES
IC 2606	10 47 27.	+ 38 12 50.			NONSTELLAR OBJECT
MCG+07-22-071	10 47 27.	+ 41 43	60	14.5	GALAXY
HN 1273	10 47 27.2	- 01 35 57.	18		NEBULA
REIZ 0648	10 47 28.	+ 00 34	18	14.4	GALAXY
IC 2607	10 47 28.	+ 38 14 56.			NONSTELLAR OBJECT
IC 2608	10 47 28.4	+ 33 02 00.			GALAXY Sa
ZWG 010.030	10 47 30.	+ 00 36		14.8	GALAXY
MCG+00-28-017	10 47 30.	+ 00 36	36	14.8	GALAXY
ZWG 010.031	10 47 30.	+ 00 38		15.6	GALAXY
ZWG 038.024	10 47 30.	+ 05 35		15.2	GALAXY
UGC 05940	10 47 30.	+ 05 35	72	15.2	GALAXY SBb
ZC 1047.5+1534	10 47 30.	+ 15 34	2820		CLUSTER OF GALAXIES
ZWG 095.035	10 47 30.	+ 16 34		15.7	GALAXY
ZWG 184.024	10 47 30.	+ 33 53		15.5	GALAXY
REIZ 0647	10 47 30.	+ 33 54	36	14.7	GALAXY
ZWG 184.025	10 47 30.	+ 38 13		14.8	GALAXY
ZWG 184.026	10 47 30.	+ 38 15		15.3	GALAXY
MCG+06-24-021	10 47 30.	+ 38 16	42	15.	GALAXY
ZWG 212.061	10 47 30.	+ 41 44		14.5	GALAXY
UGC 05941	10 47 30.	+ 41 55	48	14.5	GALAXY DBL SYS
MCG+08-20-030	10 47 30.	+ 48 12	66	16.	GALAXY
MRK 154	10 47 30.	+ 50 20	13	16.	GALAXY WITH UV CONTINUUM
MCG+13-08-044	10 47 30.	+ 79 22	30	16.	GALAXY
MCG-02-28-009	10 47 30.	- 14 30	24	15.5	GALAXY
AR 23	10 47 31.95	+ 77 11 02.8			NEBULA
MCG+04-26-013	10 47 33.	+ 21 24	36	15.5	GALAXY
MCG+06-24-022	10 47 33.	+ 33 53	15	15.	GALAXY
MCG+06-24-023	10 47 33.	+ 38 18	36	15.	GALAXY
MCG+08-20-031	10 47 33.	+ 48 57	54	15.	GALAXY
RNGC 3402	10 47 33.	- 12 24			GALAXY
MCG-02-28-010	10 47 33.	- 14 30	42	15.5	GALAXY
MCG-03-28-009	10 47 33.	- 19 01	18	15.	GALAXY
ZWG 038.025	10 47 36.	+ 03 46		15.3	GALAXY
ZWG 038.026	10 47 36.	+ 06 23		15.5	GALAXY
ZC 1047.6+1623	10 47 36.	+ 16 23	3020		CLUSTER OF GALAXIES
REIZ 0649	10 47 36.	+ 21 24	18	15.5	GALAXY
ZC 1047.6+2524	10 47 36.	+ 25 24	810		CLUSTER OF GALAXIES
ZWG 155.023	10 47 36.	+ 27 00		15.3	GALAXY
ZC 1047.6+3809	10 47 36.	+ 38 09	540		CLUSTER OF GALAXIES
MCG+08-20-032	10 47 36.	+ 47 06	48	15.	GALAXY
ZWG 241.019	10 47 36.	+ 48 57		15.1	GALAXY
ZWG 313.037	10 47 36.	+ 66 02		14.8	GALAXY
MCG+11-13-042	10 47 36.	+ 66 02	30	15.	GALAXY
UGC 05942	10 47 36.	+ 77 50	72	16.5	GALAXY PECULR
REIZ 0650	10 47 38.	+ 27 00	24	14.9	GALAXY
RNGC 3392	10 47 38.	+ 66 02		15.0	GALAXY
IC 0645	10 47 38.	- 05 46 53.			NONSTELLAR OBJECT
HN 1274	10 47 38.0	- 01 28 45.	30		NEBULA
MCG-03-28-010	10 47 39.	- 16 57	36	15.5	GALAXY
REIZ 0651	10 47 40.	+ 17 41	60	14.7	GALAXY
RNGC 3420	10 47 40.	- 16 59		14.0	GALAXY
HN 1275	10 47 40.1	- 01 01 39.			NEBULA
ZWG 066.033	10 47 42.	+ 13 32		15.7	GALAXY
UGC 05944	10 47 42.	+ 13 32	66	15.7	GALAXY DWARF
ZWG 095.036	10 47 42.	+ 17 50		15.1	GALAXY
UGC 05945	10 47 42.	+ 17 50	114	15.1	GALAXY IRR
ZWG 155.024	10 47 42.	+ 26 37		15.6	GALAXY
ZC 1047.7+6145	10 47 42.	+ 61 45	940		CLUSTER OF GALAXIES
UGC 05946	10 47 42.	+ 76 25	72	14.3	GALAXY DBL SYS
ZWG 351.047	10 47 42.	+ 79 25		14.3	GALAXY

OBJECT NAME	RIGHT ASCEN.	DECLINATION	DIAM.	MAGN.	TYPE OF OBJECT
KARA.72 251A	10 47 42.	+ 79 25	36	14.3	PART OF DOUBLE GALAXY
ZWG 010.032	10 47 42.	- 01 00		14.7	GALAXY
UGC 05943	10 47 42.	- 01 00	66	14.8	GALAXY Sb/SBc
MCG+00-28-018	10 47 42.	- 01 00	72	14.7	GALAXY
GO 083	10 47 42.	- 60 55	66	18.1	STAR CHAIN
MCG+05-26-019	10 47 45.	+ 26 58	42	15.3	GALAXY
RNGC 3404	10 47 45.	- 11 51		14.0	GALAXY
MCG-03-28-011	10 47 45.	- 16 59	60	14.5	GALAXY
IC 2609	10 47 46.	- 11 50			SAME AS NGC 3404
RNGC 3409	10 47 46.	- 16 47		15.0	GALAXY
BC 4C09.37	10 47 47.0	+ 09 41 54.		17.86	QUASI-STELLAR OBJECT
RNGC 3401	10 47 48.	+ 06 04			NON-EXISTENT OBJECT
ZWG 066.034	10 47 48.	+ 09 01		15.4	GALAXY
ZWG 095.037	10 47 48.	+ 15 54		15.7	GALAXY
MCG+03-28-016	10 47 48.	+ 17 50	90	15.1	GALAXY
ZWG 095.038	10 47 48.	+ 19 55		15.0	GALAXY
UGC 05947	10 47 48.	+ 19 55	84	15.0	GALAXY DWF PEC
ZWG 125.011	10 47 48.	+ 20 37		15.4	GALAXY
ZC 1047.8+3747	10 47 48.	+ 37 47	1550		CLUSTER OF GALAXIES
ZWG 241.020	10 47 48.	+ 47 07		15.0	GALAXY
ZWG 241.021	10 47 48.	+ 50 20		15.5	GALAXY
7ZW 348	10 47 48.	+ 66 16			COMPACT GALAXY
7ZW 349	10 47 48.	+ 77 51			COMPACT GALAXY
KARA.72 251B	10 47 48.	+ 79 24	42		PART OF DOUBLE GALAXY
MCG+13-08-045	10 47 48.	+ 79 25 30.	12	15.	GALAXY
ZC 1047.8-0315	10 47 48.	- 03 15	940		CLUSTER OF GALAXIES
MCG-02-28-011	10 47 48.	- 11 51	120	14.	GALAXY
MCG-06-24-011	10 47 48.	- 35 24	30	15.	GALAXY
HN 1276	10 47 48.5	+ 00 29 33.	12		NEBULA
SHB 164	10 47 49.0	+ 09 41 47.		17.9	QUASI-STELLAR OBJECT
REIZ 0652	10 47 50.	+ 12 46	24	15.0	GALAXY
TON-N 1293	10 47 50.	+ 22 33		17.	BLUE STAR
RNGC 3411	10 47 51.	- 12 35		13.0	GALAXY
MCG-03-28-012	10 47 51.	- 16 47 30.	72	15.	GALAXY
ZWG 066.035	10 47 54.	+ 09 02		15.5	GALAXY
ZC 1047.9+1146	10 47 54.	+ 11 46	1480		CLUSTER OF GALAXIES
ZWG 066.036	10 47 54.	+ 12 46		15.1	GALAXY
MCG+02-28-015	10 47 54.	+ 12 46	54	15.1	GALAXY
ZWG 095.039	10 47 54.	+ 16 10		15.4	GALAXY
MCG+03-28-017	10 47 54.	+ 19 56	72	15.0	GALAXY
ZC 1047.9+4042	10 47 54.	+ 40 42	670		CLUSTER OF GALAXIES
MCG+13-08-046	10 47 54.	+ 77 12	45	15.	GALAXY
MCG+13-08-047	10 47 54.	+ 79 25	15	15.	GALAXY
ZWG 010.033	10 47 54.	- 03 10		15.7	GALAXY
MCG-02-28-012	10 47 54.	- 12 35	30	13.	GALAXY
HN 1277	10 47 57.8	- 00 31 45.			NEBULA
REIZ 0653	10 47 59.	+ 19 54	72	14.5	GALAXY
VDB.66G 089	10 48	+ 19 55	70		DWARF GALAXY
UGC 05948	10 48 00.	+ 16 00	84	17.	GALAXY DWARF?
ZWG 155.025	10 48 00.	+ 28 44		14.3	GALAXY
UGC 05949	10 48 00.	+ 28 44	84	14.3	GALAXY SB0-a
ZWG 155.026	10 48 00.	+ 32 28		15.7	GALAXY
ZC 1048.0+3343	10 48 00.	+ 33 43	3430		CLUSTER OF GALAXIES
ZC 1048.0+5329	10 48 00.	+ 53 29	340		CLUSTER OF GALAXIES
ZC 1048.0+6116	10 48 00.	+ 61 16	610		CLUSTER OF GALAXIES
MCG+12-10-086	10 48 00.	+ 68 58	54	16.	GALAXY
ZWG 333.058	10 48 00.	+ 69 45		15.6	GALAXY
MCG+12-10-087	10 48 00.	+ 72 03 30.	21	14.	GALAXY
MCG+14-05-014	10 48 00.	+ 80 42	30	16.	GALAXY
ZC 1048.0-0215	10 48 00.	- 02 15	1010		CLUSTER OF GALAXIES
ZC 1048.0-0250	10 48 00.	- 02 50	870		CLUSTER OF GALAXIES
MRSL 288+00/1	10 48 00.	- 58 59	900		HII REGION
RNGC 3400	10 48 02.	+ 28 44		14.5	GALAXY
REIZ 0654	10 48 03.	+ 28 44	48	14.2	GALAXY
HN 1278	10 48 03.4	- 00 33 40.			NEBULA
IC 0647	10 48 05.	- 12 35 37.			NONSTELLAR OBJECT
UGC 05950	10 48 06.	+ 15 35	120	18.	GALAXY PR DWFS
MCG+05-26-020	10 48 06.	+ 28 43	66	14.3	GALAXY
ZC 1048.1+3844	10 48 06.	+ 38 44	940		CLUSTER OF GALAXIES
ZWG 333.059	10 48 06.	+ 68 58		14.8	GALAXY
MCG-03-28-013	10 48 06.	- 16 48 30.	54	15.	GALAXY
MCG-06-24-012	10 48 06.	- 39 36	72	15.5	GALAXY
ARC 1111	10 48 08.	- 02 18		17.8	RICH CLUSTER OF GALAXIES
RNGC 3412	10 48 11.	+ 13 40		12.0	GALAXY
REIZ 0657	10 48 11.	+ 18 21	12	15.3	GALAXY
REIZ 0655	10 48 11.	+ 32 24	30	14.7	GALAXY
REIZ 0656	10 48 11.	+ 32 27	36	14.8	GALAXY
ZWG 038.027	10 48 12.	+ 06 21		15.3	GALAXY
ZWG 095.040	10 48 12.	+ 16 42		15.3	GALAXY
ZWG 095.041	10 48 12.	+ 18 05		15.7	GALAXY
ZWG 125.012	10 48 12.	+ 25 30		15.4	GALAXY
UGC 05951	10 48 12.	+ 36 28	72	16.0	GALAXY Sc
MCG+12-10-088	10 48 12.	+ 71 52	30	16.	GALAXY
IC 0650	10 48 12.	- 13 10 37.			NONSTELLAR OBJECT
RNGC 3429	10 48 17.	+ 09 31			GALAXY
REIZ 0658	10 48 17.	+ 33 06	18	14.5	GALAXY
IC 0649	10 48 18.	+ 01 25 53.			NONSTELLAR OBJECT
ZWG 010.034	10 48 18.	+ 01 27		14.9	GALAXY
KARA.72 252B	10 48 18.	+ 01 27	18		PART OF DOUBLE GALAXY
KARA.72 252A	10 48 18.	+ 01 27	24	14.9	PART OF DOUBLE GALAXY
MCG+00-28-019	10 48 18.	+ 01 27	36	14.9	GALAXY
ZWG 066.037	10 48 18.	+ 09 02		15.5	GALAXY
ZWG 066.038	10 48 18.	+ 13 41		10.8	GALAXY
UGC 05952	10 48 18.	+ 13 41	198	10.8	GALAXY S0
MCG+02-28-016	10 48 18.	+ 13 41	78	10.8	GALAXY
ZC 1048.3+3010	10 48 18.	+ 30 10	2080		CLUSTER OF GALAXIES
MCG+08-20-033	10 48 18.	+ 44 49 30.	30	15.	GALAXY
LB 01905	10 48 19.	+ 56 26 00.		13.0	FAINT BLUE STAR
REIZ 0662	10 48 21.	+ 13 41	120	11.8	GALAXY
RNGC 3421	10 48 21.	- 12 11		14.0	GALAXY
ARC 1113	10 48 22.	+ 08 55		17.6	RICH CLUSTER OF GALAXIES
HN 1279	10 48 22.6	- 00 31 46.	12		NEBULA
IC 0648	10 48 23.	+ 12 33 35.			NONSTELLAR OBJECT
ZWG 066.039	10 48 24.	+ 08 45		15.3	GALAXY
RNGC 3417	10 48 24.	+ 08 45		15.5	GALAXY
ZWG 066.040	10 48 24.	+ 12 33		14.9	GALAXY
MCG+02-28-017	10 48 24.	+ 12 33	48	14.9	GALAXY
ZWG 095.042	10 48 24.	+ 18 30		15.4	GALAXY
ZC 1048.4+2025	10 48 24.	+ 20 25	1480		CLUSTER OF GALAXIES
ZWG 155.027	10 48 24.	+ 31 52		15.7	GALAXY
ZWG 241.022	10 48 24.	+ 44 50		13.2	GALAXY
MRK 155	10 48 24.	+ 44 50	25	13.5	GALAXY WITH UV CONTINUUM
UGC 05953	10 48 24.	+ 44 50	27	13.2	GALAXY PECULR
MCG+08-20-034	10 48 24.	+ 46 59 30.	30	16.	GALAXY
ZWG 267.018	10 48 24.	+ 55 39		14.4	GALAXY
UGC 05954	10 48 24.	+ 55 39	60	14.4	GALAXY S
ZWG 334.002	10 48 24.	+ 72 02		14.4	GALAXY
ZWG 333.060	10 48 24.	+ 72 02		14.4	GALAXY

OBJECT NAME	RIGHT ASCEN.	DECLINATION	DIAM.	MAGN.	TYPE OF OBJECT
UGC 05955	10 48 24.	+ 72 02	72	14.4	GALAXY E
MCG-02-28-013	10 48 24.	- 12 11	108	14.5	GALAXY
REIZ 0667	10 48 25.	+ 08 43	24	15.2	GALAXY
LB 01906	10 48 25.	+ 58 16 18.		16.1	FAINT BLUE STAR
REIZ 0669	10 48 27.	- 01 53	30	13.3	GALAXY
RNGC 3398	10 48 28.	+ 55 39		15.5	GALAXY
HN 1280	10 48 28.0	- 02 06 28.	12		NEBULA
REIZ 0668	10 48 29.	+ 20 26	54	14.0	GALAXY
IC 0644	10 48 29.0	+ 55 39 25.			GALAXY Sa
ZC 1048.5+0852	10 48 30.	+ 08 52	1280		CLUSTER OF GALAXIES
ZC 1048.5+1056	10 48 30.	+ 10 56	2890		CLUSTER OF GALAXIES
ZWG 095.043	10 48 30.	+ 20 26		14.7	GALAXY
UGC 05957	10 48 30.	+ 20 26	72	14.7	GALAXY S
ZWG 155.028	10 48 30.	+ 28 07		15.6	GALAXY
UGC 05958	10 48 30.	+ 28 07	96	15.6	GALAXY Sb-c
ZWG 155.029	10 48 30.	+ 28 15		12.1	GALAXY
UGC 05959	10 48 30.	+ 28 15	192	12.1	GALAXY (SB0)?
ZWG 184.027	10 48 30.	+ 33 01		13.1	GALAXY
UGC 05960	10 48 30.	+ 33 01	114	13.1	GALAXY Sa?
ZC 1048.5+3417	10 48 30.	+ 34 17	870		CLUSTER OF GALAXIES
UGC 05961	10 48 30.	+ 48 13	90	16.0	GALAXY Sa-b
MCG+09-18-038	10 48 30.	+ 55 39	54	14.6	GALAXY
ZWG 267.019	10 48 30.	+ 55 44		15.7	GALAXY
MCG+09-18-039	10 48 30.	+ 55 44	48	15.	GALAXY
ZWG 334.003	10 48 30.	+ 71 50		15.5	GALAXY
ZWG 333.061	10 48 30.	+ 71 50		15.5	GALAXY
ZWG 010.035	10 48 30.	- 01 52		12.9	GALAXY
UGC 05956	10 48 30.	- 01 52	42	12.9	GALAXY PECULR
MCG+00-28-020	10 48 30.	- 01 52	42	12.9	GALAXY
KARA.73B 0444	10 48 30.	- 01 52	48	12.9	ISOLATED GALAXY S
MCG-02-28-014	10 48 30.	- 10 09	36	15.	GALAXY
GO 054	10 48 30.	- 58 14	96	18.7	STAR CHAIN
IC 0646	10 48 31.8	+ 55 43 51.			GALAXY SA0
RNGC 3414	10 48 32.	+ 28 14		12.0	GALAXY
IC 0652	10 48 32.	- 12 22 20.			NONSTELLAR OBJECT
REIZ 0665	10 48 33.	+ 28 14	132	12.3	GALAXY
REIZ 0666	10 48 33.	+ 28 16	18	15.5	GALAXY
REIZ 0659	10 48 34.	+ 55 41	60	14.3	GALAXY
IC 0651	10 48 34.	- 01 53 02.			NONSTELLAR OBJECT
RNGC 3419	10 48 35.	+ 14 13		13.5	GALAXY
REIZ 0664	10 48 35.	+ 33 02	90	13.4	GALAXY
HOLM 218C	10 48 35.	+ 33 02	90	13.5	PART OF MULTIPLE GALAXY
REIZ 0660	10 48 35.	+ 55 46	30	15.2	GALAXY
ZWG 038.028	10 48 36.	+ 02 42		15.6	GALAXY
ZWG 038.029	10 48 36.	+ 06 06		12.1	GALAXY
RNGC 3423	10 48 36.	+ 06 06		12.0	GALAXY
UGC 05962	10 48 36.	+ 06 06	258	12.1	GALAXY Sc
MCG+01-28-012	10 48 36.	+ 06 06	240	12.1	GALAXY
ZWG 038.030	10 48 36.	+ 07 08		15.3	GALAXY
ZC 1048.6+0722	10 48 36.	+ 07 22	1410		CLUSTER OF GALAXIES
MCG+03-28-018	10 48 36.	+ 20 28	66	14.7	GALAXY
ZC 1048.6+2358	10 48 36.	+ 23 58	11760		CLUSTER OF GALAXIES
MCG+05-26-022	10 48 36.	+ 28 06	72	15.6	GALAXY
MCG+05-26-021	10 48 36.	+ 28 13	180	12.1	GALAXY
ARP 162	10 48 36.	+ 28 15			PECULIAR GALAXY
ZWG 155.030	10 48 36.	+ 28 23		14.5	GALAXY
UGC 05963	10 48 36.	+ 28 23	78	14.5	GALAXY S0/Sa
MCG+06-24-024	10 48 36.	+ 33 01	60	14.	GALAXY
ZC 1048.6+4239	10 48 36.	+ 42 39	810		CLUSTER OF GALAXIES
MCG+09-18-040	10 48 36.	+ 51 16	48	13.4	GALAXY
HOLM 217A	10 48 36.	+ 51 17	30	13.4	PART OF MULTIPLE GALAXY
REIZ 0661	10 48 36.	+ 51 18	42	13.8	GALAXY
MCG+10-16-015	10 48 36.	+ 58 38	18	16.	GALAXY
ZC 1048.6+6925	10 48 36.	+ 69 25	1140		CLUSTER OF GALAXIES
MCG-04-26-012	10 48 36.	- 23 23 30.	72	13.5	GALAXY
RNGC 3413	10 48 37.	+ 33 02		13.0	GALAXY
REIZ 0670	10 48 39.	+ 28 06	90	14.3	GALAXY
REIZ 0675	10 48 39.	+ 28 23	60	14.1	GALAXY
RNGC 3422	10 48 39.	- 12 08		15.0	GALAXY
RNGC 3419A	10 48 41.	+ 14 17		15.0	GALAXY
ARC 1110	10 48 41.	+ 42 40		17.8	RICH CLUSTER OF GALAXIES
REIZ 0678	10 48 42.	+ 06 06	180	12.0	GALAXY
ZWG 066.041	10 48 42.	+ 14 14		13.4	GALAXY
UGC 05964	10 48 42.	+ 14 14	42	13.4	GALAXY S0
MCG+02-28-018	10 48 42.	+ 14 14	42	13.4	GALAXY
ZWG 066.042	10 48 42.	+ 14 18		14.9	GALAXY
UGC 05965	10 48 42.	+ 14 18	102	14.9	GALAXY SBb-c
MCG+02-28-019	10 48 42.	+ 14 18	96	14.9	GALAXY
ZWG 095.044	10 48 42.	+ 19 16		15.3	GALAXY
KARA.73B 0445	10 48 42.	+ 19 16	36	15.3	ISOLATED GALAXY E
GO 096	10 48 42.	- 61 07	101	18.7	STAR CHAIN
REIZ 0680	10 48 43.	+ 08 13	18	14.0	GALAXY
REIZ 0679	10 48 43.	+ 08 48	18	14.3	GALAXY
ARC 1114	10 48 43.	+ 20 26		17.1	RICH CLUSTER OF GALAXIES
GO 084	10 48 43.	- 60 38	98	16.7	STAR CHAIN
RNGC 3418	10 48 44.	+ 28 23		14.5	GALAXY
REIZ 0677	10 48 45.	+ 14 13	12	14.4	GALAXY
MCG+03-28-019	10 48 45.	+ 19 18	48	15.3	GALAXY
MCG+05-26-023	10 48 45.	+ 28 22	66	14.5	GALAXY
RNGC 3415	10 48 45.	+ 43 59		13.0	GALAXY
MCG+09-18-041	10 48 45.	+ 55 39	15	16.	GALAXY
LB 01907	10 48 45.	+ 60 20 00.		16.6	FAINT BLUE STAR
MCG-02-28-015	10 48 45.	- 12 09	66	15.	GALAXY
HOLM 217B	10 48 46.	+ 51 16	48	14.2	PART OF MULTIPLE GALAXY
REIZ 0663	10 48 46.	+ 55 41	6	15.5	GALAXY
RNGC 3431	10 48 46.	- 16 45		14.0	GALAXY
BC 4C24.23	10 48 46.8	+ 24 04 00.		18.5	QUASI-STELLAR OBJECT
SHB 165	10 48 46.8	+ 24 04 00.		18.5	QUASI-STELLAR OBJECT
RNGC 3428	10 48 47.	+ 09 33		14.0	GALAXY
ZWG 066.043	10 48 48.	+ 08 34		14.0	GALAXY
RNGC 3427	10 48 48.	+ 08 34		14.0	GALAXY
UGC 05966	10 48 48.	+ 08 34	78	14.0	GALAXY S0-a
MCG+02-28-020	10 48 48.	+ 08 34	60	14.0	GALAXY
ZWG 066.044	10 48 48.	+ 08 50		14.5	GALAXY
RNGC 3425	10 48 48.	+ 08 50		14.5	GALAXY
UGC 05967	10 48 48.	+ 08 50	66	14.5	GALAXY S0
MCG+02-28-021	10 48 48.	+ 08 50	36	14.5	GALAXY
ZWG 066.045	10 48 48.	+ 09 33		14.1	GALAXY
UGC 05968	10 48 48.	+ 09 33	96	14.1	GALAXY Sb/SBb
MCG+02-28-022	10 48 48.	+ 09 53	90	14.1	GALAXY
8ZW 1048+13.6	10 48 48.	+ 13 38		17.3	COMPACT GALAXY
ZC 1048.8+3535	10 48 48.	+ 35 35	870		CLUSTER OF GALAXIES
ZC 1048.8+3813	10 48 48.	+ 38 13	1080		CLUSTER OF GALAXIES
MCG+07-22-072	10 48 48.	+ 43 58 30.	102	13.5	GALAXY
ZWG 213.001	10 48 48.	+ 43 59		13.2	GALAXY
ZWG 212.062	10 48 48.	+ 43 59		13.2	GALAXY
REIZ 0674	10 48 48.	+ 43 59	42	13.1	GALAXY
UGC 05969	10 48 48.	+ 43 59	126	13.2	GALAXY S0-a
MCG+09-18-042	10 48 48.	+ 51 15	24	14.2	GALAXY
ZWG 267.020	10 48 48.	+ 51 18		13.7	GALAXY
UGC 05970	10 48 48.	+ 51 18	66	13.7	GALAXY DBL SYS
KARA.72 253B	10 48 48.	+ 51 18	36		PART OF DOUBLE GALAXY
KARA.72 253A	10 48 48.	+ 51 18	48	13.7	PART OF DOUBLE GALAXY
ZWG 313.038	10 48 48.	+ 67 02		15.4	GALAXY
UGC 05971	10 48 48.	+ 67 02	66	15.4	GALAXY Sa-b
KARA.68 073	10 48 48.	+ 69 56	40		DWARF GALAXY
ZWG 010.036	10 48 48.	- 03 20		15.1	GALAXY
MCG-01-28-003	10 48 48.	- 09 20	60	15.	GALAXY
MCG-03-28-014	10 48 48.	- 16 45	72	14.	GALAXY
RNGC 3406	10 48 49.	+ 51 18		13.5	GALAXY
RNGC 3416	10 48 51.	+ 44 02		15.0	GALAXY
ARC 1115	10 48 53.	+ 09 19		17.6	RICH CLUSTER OF GALAXIES
REIZ 0681	10 48 53.	+ 18 36	18	14.1	GALAXY
ZWG 095.045	10 48 54.	+ 17 06		15.7	GALAXY
ZWG 184.028	10 48 54.	+ 33 10		13.2	GALAXY
UGC 05972	10 48 54.	+ 33 10	174	13.2	GALAXY S
ZWG 213.002	10 48 54.	+ 44 02		15.2	GALAXY
ZWG 212.063	10 48 54.	+ 44 02		15.2	GALAXY
MCG+07-22-073	10 48 54.	+ 44 02	24	15.	GALAXY
LB 00584	10 48 54.	+ 46 54 18.		18.1	FAINT BLUE STAR
REIZ 0673	10 48 54.	+ 51 16	60	14.7	GALAXY
ZWG 267.021	10 48 54.	+ 51 17		14.8	GALAXY
MCG-02-28-016	10 48 54.	- 09 52	96	15.	GALAXY
MCG-03-28-015	10 48 54.	- 19 38	216	13.	GALAXY
REIZ 0684	10 48 55.	+ 08 31	36	13.7	GALAXY
REIZ 0686	10 48 55.	+ 09 31	48	13.8	GALAXY
REIZ 0685	10 48 55.	+ 09 32	9	15.3	GALAXY
RNGC 3410	10 48 55.	+ 51 17		15.0	GALAXY
REIZ 0682	10 48 58.	+ 17 53	18	15.7	GALAXY
RNGC 3426	10 48 58.	+ 18 44		14.0	GALAXY
REIZ 0687	10 48 59.	+ 04 51	48	14.5	GALAXY
HOLM 218A	10 48 59.	+ 33 10	90	13.1	PART OF MULTIPLE GALAXY
SEY 084	10 48 59.	+ 55 51 08.		14.7	FAINT GALAXY
REIZ 0676	10 48 59.	+ 55 55	48	14.9	GALAXY
HN 1281	10 48 59.1	+ 00 31 19.	18		NEBULA
BC PKS1049-09	10 48 59.42	- 09 02 13.2		16.79	QUASI-STELLAR OBJECT
SHB 166	10 48 59.5	- 09 02 12.		16.8	QUASI-STELLAR OBJECT
HN 1282	10 48 59.6	- 00 43 17.	12		NEBULA
VDB.66G 090	10 49	+ 07 55	70		DWARF GALAXY
LB 09973	10 49	- 88 31		14.0	FAINT BLUE STAR
UGC 05973	10 49 00.	+ 00 32	66	16.0	GALAXY SB?a-b
ZWG 038.031	10 49 00.	+ 03 29		15.6	GALAXY
ZWG 038.032	10 49 00.	+ 04 51		14.7	GALAXY
UGC 05974	10 49 00.	+ 04 51	126	14.7	GALAXY Sc
MCG+01-28-013	10 49 00.	+ 04 51	120	14.7	GALAXY
ZWG 038.033	10 49 00.	+ 04 55		15.7	GALAXY
ZC 1049.0+0520	10 49 00.	+ 05 20	1550		CLUSTER OF GALAXIES
ZWG 038.034	10 49 00.	+ 08 26		15.3	GALAXY
MCG+01-28-014	10 49 00.	+ 08 26	120	15.3	GALAXY
ZWG 095.046	10 49 00.	+ 18 44		13.9	GALAXY
UGC 05975	10 49 00.	+ 18 44	108	13.9	GALAXY
ZC 1049.0+1905	10 49 00.	+ 19 05	1080		CLUSTER OF GALAXIES
MCG+06-24-025	10 49 00.	+ 33 09	180	13.5	GALAXY
ZC 1049.0+4019	10 49 00.	+ 40 19	1410		CLUSTER OF GALAXIES
ZC 1049.0+4941	10 49 00.	+ 49 41	1340		CLUSTER OF GALAXIES
MCG+09-18-043	10 49 00.	+ 55 51	60	14.	GALAXY
ZWG 267.022	10 49 00.	+ 55 52		15.3	GALAXY
UGC 05976	10 49 00.	+ 55 52	90	15.3	GALAXY Sc
MCG+10-16-016	10 49 00.	+ 58 42	45	14.3	GALAXY
MCG+10-16-017	10 49 00.	+ 61 39	24	14.2	GALAXY
7ZW 350	10 49 00.	+ 67 02			COMPACT GALAXY
ZC 1049.0-0226	10 49 00.	- 02 26	740		CLUSTER OF GALAXIES
ZWG 010.037	10 49 00.	- 03 27		14.5	GALAXY
MCG+00-28-021	10 49 00.	- 03 27	72	14.5	GALAXY
RNGC 3424	10 49 01.	+ 33 10		13.0	GALAXY
REIZ 0672	10 49 02.	+ 58 42	42	14.4	GALAXY
HN 1283	10 49 02.3	- 00 48 53.			NEBULA
REIZ 0688	10 49 03.	+ 14 17	72	14.8	GALAXY
MCG+03-28-020	10 49 03.	+ 18 46	30	13.9	GALAXY
MCG-06-24-013	10 49 03.	- 34 10	8	13.5	GALAXY
REIZ 0689	10 49 04.	+ 17 58	12	15.5	GALAXY
HN 1284	10 49 04.4	- 00 51 11.	18		NEBULA
REIZ 0683	10 49 05.	+ 33 21	90	13.2	GALAXY
ZC 1049.1+0914	10 49 06.	+ 09 14	2020		CLUSTER OF GALAXIES
ZWG 095.047	10 49 06.	+ 16 14		15.4	GALAXY
MCG+03-28-021	10 49 06.	+ 16 14	48	15.4	GALAXY
ZWG 095.048	10 49 06.	+ 18 01		15.6	GALAXY
MCG+09-18-044	10 49 06.	+ 50 48	18	16.	GALAXY
ZC 1049.1+5547	10 49 06.	+ 55 47	1210		CLUSTER OF GALAXIES
ZWG 291.006	10 49 06.	+ 58 43		14.1	GALAXY
UGC 05977	10 49 06.	+ 58 43	54	14.1	GALAXY S
MCG+10-16-018	10 49 06.	+ 59 57	27	15.	GALAXY
REIZ 0671	10 49 06.	+ 61 38	30	14.2	GALAXY
ZWG 291.007	10 49 06.	+ 61 39		14.8	GALAXY
UGC 05978	10 49 06.	+ 61 39	84	14.8	GALAXY E-S0
ZWG 010.038	10 49 06.	- 00 48		15.6	GALAXY
MCG-06-24-014	10 49 06.	- 35 25	36	15.	GALAXY
RNGC 3407	10 49 07.	+ 61 39		15.0	GALAXY
RNGC 3408	10 49 09.	+ 58 42		14.0	GALAXY
ZWG 038.035	10 49 12.	+ 04 03		15.6	GALAXY
ZC 1049.2+0902	10 49 12.	+ 09 02	3700		CLUSTER OF GALAXIES
ZWG 095.049	10 49 12.	+ 16 16		15.6	GALAXY
MCG+03-28-022	10 49 12.	+ 16 16	60	15.6	GALAXY
ZWG 095.050	10 49 12.	+ 18 06		15.5	GALAXY
ZWG 095.051	10 49 12.	+ 19 00		15.1	GALAXY
REIZ 0690	10 49 12.	+ 22 30	42	14.6	GALAXY
ZC 1049.2+2422	10 49 12.	+ 24 22	740		CLUSTER OF GALAXIES
ZC 1049.2+2925	10 49 12.	+ 29 25	1080		CLUSTER OF GALAXIES
ZC 1049.2+3501	10 49 12.	+ 35 01	670		CLUSTER OF GALAXIES
MCG+08-20-035	10 49 12.	+ 49 57	36	16.	GALAXY
BIGO 515	10 49 12.	+ 61 39			NEBULA
ZWG 313.039	10 49 12.	+ 68 14		15.6	GALAXY
UGC 05979	10 49 12.	+ 68 14	120	15.6	GALAXY DWRF IR
MCG+11-13-043	10 49 12.	+ 68 16	96	14.	GALAXY
SEY 085	10 49 13.	+ 59 57 08.		15.1	FAINT GALAXY
REIZ 0691	10 49 14.	+ 11 07	18	15.3	GALAXY
ARC 1112	10 49 16.	+ 55 50		17.5	RICH CLUSTER OF GALAXIES
ZC 1049.3+1234	10 49 18.	+ 12 34	1880		CLUSTER OF GALAXIES
ZWG 241.023	10 49 18.	+ 49 57		15.4	GALAXY
ZWG 291.008	10 49 18.	+ 59 57		14.9	GALAXY
KARA.73B 0446	10 49 18.	+ 59 57	48	14.9	ISOLATED GALAXY S
ZC 1049.3+6805	10 49 18.	+ 68 05	740		CLUSTER OF GALAXIES
SHB 167	10 49 22.4	+ 61 41 18.		15.	QUASI-STELLAR OBJECT
BC 4C61.20	10 49 22.43	+ 61 41 18.2		15.	QUASI-STELLAR OBJECT
REIZ 0696	10 49 23.	+ 04 03	36	13.8	GALAXY
RNGC 3433	10 49 23.	+ 10 25		12.5	GALAXY

OBJECT NAME	RIGHT ASCEN.	DECLINATION	DIAM.	MAGN.	TYPE OF OBJECT
IC 2610	10 49 23.	+ 33 20			MAY NOT EXIST
LB 01908	10 49 23.	+ 56 24 48.		15.0	FAINT BLUE STAR
ZWG 038.036	10 49 24.	+ 04 04		13.4	GALAXY
RNGC 3434	10 49 24.	+ 04 04		13.5	GALAXY
UGC 05980	10 49 24.	+ 04 04	150	13.4	GALAXY Sb
MCG+01-28-015	10 49 24.	+ 04 04	120	13.4	GALAXY
ZWG 038.037	10 49 24.	+ 04 05		15.6	GALAXY
ZWG 066.046	10 49 24.	+ 09 03		15.5	GALAXY
ZWG 066.047	10 49 24.	+ 09 13		15.2	GALAXY
ZWG 066.048	10 49 24.	+ 10 25		13.6	GALAXY
UGC 05981	10 49 24.	+ 10 25	240	13.6	GALAXY Sc
MCG+02-28-023	10 49 24.	+ 10 25	210	13.6	GALAXY
8ZW 1049+12.8	10 49 24.	+ 12 48		19.8	COMPACT GALAXY
MCG+04-26-015	10 49 24.	+ 21 11 30.	9	16.	GALAXY
MCG+04-26-014	10 49 24.	+ 21 13	9	16.	GALAXY
TON-N 0550	10 49 24.	+ 31 28		15.3	BLUE STAR
ZWG 184.029	10 49 24.	+ 33 13		12.2	GALAXY
UGC 05982	10 49 24.	+ 33 13	270	12.2	GALAXY Sc
MCG+06-24-026	10 49 24.	+ 33 13 30.	222	12.	GALAXY
UGC 05983	10 49 24.	+ 36 52	72	17.	GALAXY DWARF
ZC 1049.4+5004	10 49 24.	+ 50 04	540		CLUSTER OF GALAXIES
ZC 1049.4+5525	10 49 24.	+ 55 25	3020		CLUSTER OF GALAXIES
REIZ 0695	10 49 25.	+ 10 25	120	12.7	GALAXY
RNGC 3430	10 49 25.	+ 33 13		12.0	GALAXY
HOLM 218B	10 49 25.	+ 33 13	192	12.3	PART OF MULTIPLE GALAXY
REIZ 0692	10 49 28.	+ 30 50	72	14.9	GALAXY
REIZ 0693	10 49 29.	+ 33 14	210	12.4	GALAXY
REIZ 0694	10 49 29.	+ 33 19	12	16.0	GALAXY
ZWG 038.038	10 49 30.	+ 03 38		15.2	GALAXY
ZWG 066.049	10 49 30.	+ 13 04		15.0	GALAXY
MCG+02-28-024	10 49 30.	+ 13 04	36	15.0	GALAXY
ZC 1049.5+1316	10 49 30.	+ 13 16	470		CLUSTER OF GALAXIES
ZWG 155.031	10 49 30.	+ 30 20		14.6	GALAXY
UGC 05984	10 49 30.	+ 30 20	144	14.6	GALAXY DBL SYS
VV 233B	10 49 30.	+ 30 20	21		INTERACTING GALAXY
VV 233A	10 49 30.	+ 30 20	114	15.	INTERACTING GALAXY
KARA.72 254A	10 49 30.	+ 30 20	60	14.6	PART OF DOUBLE GALAXY
ARP 107	10 49 30.	+ 30 21			PECULIAR GALAXY
MCG+06-24-027	10 49 30.	+ 36 52	48	15.5	GALAXY
ZWG 241.024	10 49 30.	+ 48 46		15.7	GALAXY
ARC 1116	10 49 31.	+ 12 30		17.2	RICH CLUSTER OF GALAXIES
TON-N 1294	10 49 32.	+ 20 58		16.4	PLUE STAR
IC 0653	10 49 32.	- 00 18 16.			NONSTELLAR OBJECT
MCG+05-26-024	10 49 33.	+ 30 20	102	14.6	GALAXY
REIZ 0697	10 49 34.	+ 18 04	21	15.0	GALAXY
REIZ 0699	10 49 34.	- 00 18	72	13.9	GALAXY
HOLM 220A	10 49 34.	- 00 18	72	14.2	PART OF MULTIPLF GALAXY
ZC 1049.6+0549	10 49 36.	+ 05 49	540		CLUSTER OF GALAXIES
TON-N 0551	10 49 36.	+ 26 44		16.0	BLUE STAR
KARA.72 254B	10 49 36.	+ 30 21	60		PART OF DOUBLE GALAXY
MCG+05-26-025	10 49 36.	+ 30 21	15	14.6	GALAXY
ZC 1049.6+3242	10 49 36.	+ 32 42	940		CLUSTER OF GALAXIES
ZC 1049.6+3700	10 49 36.	+ 37 00	2080		CLUSTER OF GALAXIES
ZWG 241.025	10 49 36.	+ 46 10		15.4	GALAXY
ZC 1049.6+4723	10 49 36.	+ 47 23	1210		CLUSTER OF GALAXIES
ZC 1049.6+6652	10 49 36.	+ 66 52	1410		CLUSTER OF GALAXIES
ZWG 010.039	10 49 36.	- 00 17		14.2	GALAXY
UGC 05985	10 49 36.	- 00 17	138	14.2	GALAXY Sa
MCG+00-28-022	10 49 36.	- 00 17	60	14.2	GALAXY
GO 056	10 49 37.	- 58 11	43	18.4	STAR CHAIN
HN 1285	10 49 38.6	- 02 08 42.			NEBULA
HOLM 220B	10 49 39.	- 00 16	18	15.0	PART OF MULTIPLE GALAXY
ARP 206	10 49 41.	+ 36 54			PECULIAR GALAXY
BC 5C2.10	10 49 41.	+ 48 55 53.			QUASI-STELLAR OBJECT
SHB 168	10 49 41.0	+ 48 55 53.		18.0	QUASI-STELLAR OBJECT
ZC 1049.7+0610	10 49 42.	+ 06 10	1480		CLUSTER OF GALAXIES
ZWG 095.052	10 49 42.	+ 18 12		15.2	GALAXY
MCG+03-28-023	10 49 42.	+ 18 14	42	15.2	GALAXY
ZC 1049.7+3620	10 49 42.	+ 36 20	940		CLUSTER OF GALAXIES
ZWG 184.030	10 49 42.	+ 36 53		11.7	GALAXY
RNGC 3432	10 49 42.	+ 36 53		12.0	GALAXY
UGC 05986	10 49 42.	+ 36 53	450	11.7	GALAXY
VV 011B	10 49 42.	+ 36 53	54	15.	INTERACTING GALAXY
VV 011A	10 49 42.	+ 36 53	360	12.2	INTERACTING GALAXY
KARA.73B 0447	10 49 42.	+ 36 53	402	11.7	ISOLATED GALAXY S
MCG+06-24-028	10 49 42.	+ 36 54	390	11.	GALAXY
ZWG 213.003	10 49 42.	+ 39 32		15.7	GALAXY
ZWG 212.064	10 49 42.	+ 39 32		15.7	GALAXY
UGC 05987	10 49 42.	+ 39 32	60	15.7	GALAXY SBb-c
MCG+07-22-074	10 49 42.	+ 39 32	42	16.	GALAXY
ZC 1049.7+6043	10 49 42.	+ 60 43	1080		CLUSTER OF GALAXIES
ZWG 010.040	10 49 42.	- 02 07		14.8	GALAXY
MCG+00-28-023	10 49 42.	- 02 07	12	14.8	GALAXY
MCG-05-26-009	10 49 42.	- 32 26	84	14.	GALAXY
REIZ 0698	10 49 43.	+ 36 53	360	12.7	GALAXY
TON-N 1295	10 49 44.	+ 21 58		16.5	BLUE STAR
BIGO 516	10 49 45.	+ 08 10			NEBULA
REIZ 0700	10 49 46.	+ 30 52	15	15.5	GALAXY
RNGC 3438	10 49 47.	+ 10 50		14.5	GALAXY
ZWG 038.039	10 49 48.	+ 08 21		15.0	GALAXY
MCG+01-28-016	10 49 48.	+ 08 21	36	15.0	GALAXY
ZWG 066.050	10 49 48.	+ 08 50		15.2	GALAXY
RNGC 3439	10 49 48.	+ 08 50		15.0	GALAXY
ZC 1049.8+1017	10 49 48.	+ 10 17	610		CLUSTER OF GALAXIES
ZWG 066.051	10 49 48.	+ 10 46		15.5	GALAXY
ZWG 066.052	10 49 48.	+ 10 50		14.3	GALAXY
UGC 05988	10 49 48.	+ 10 50	54	14.3	GALAXY S?
MCG+02-28-025	10 49 48.	+ 10 50	48	14.3	GALAXY
ZC 1049.8+1054	10 49 48.	+ 10 54	1280		CLUSTER OF GALAXIES
ZWG 095.053	10 49 48.	+ 20 03		14.2	GALAXY
UGC 05989	10 49 48.	+ 20 03	90	14.2	GALAXY IRR
ZWG 184.031	10 49 48.	+ 34 45		15.1	GALAXY
UGC 05990	10 49 48.	+ 34 45	90	15.1	GALAXY Sa-b
MCG+06-24-029	10 49 48.	+ 34 45	66	14.5	GALAXY
ZC 1049.8+4234	10 49 48.	+ 42 34	610		CLUSTER OF GALAXIES
ZWG 241.026	10 49 48.	+ 49 52		14.7	GALAXY
UGC 05991	10 49 48.	+ 49 52	78	14.7	GALAXY E-S0
MCG+08-20-036	10 49 48.	+ 49 52	24	15.	GALAXY
ZC 1049.8+6321	10 49 48.	+ 63 21	940		CLUSTER OF GALAXIES
UGC 05992	10 49 48.	+ 66 00	60	16.5	GALAXY
MCG-03-28-016	10 49 48.	- 16 53	24	14.5	GALAXY
DVDV 1	10 49 48.	- 46 18	66		GALAXY
REIZ 0704	10 49 49.	+ 08 48	12	15.0	GALAXY
REIZ 0703	10 49 50.	+ 10 48	18	14.3	GALAXY
MCG+04-26-016	10 49 51.	+ 23 11 30.	132	12.6	GALAXY
REIZ 0701	10 49 51.	+ 28 51	18	15.4	GALAXY
REIZ 0705	10 49 53.	+ 20 04	60	13.7	GALAXY
RNGC 3446	10 49 53.	- 44 53			NON-EXISTENT OBJECT

OBJECT NAME	RIGHT ASCEN.	DECLINATION	DIAM.	MAGN.	TYPE OF OBJECT
ZWG 038.040	10 49 54.	+ 07 30		13.9	GALAXY
RNGC 3441	10 49 54.	+ 07 30		14.0	GALAXY
UGC 05993	10 49 54.	+ 07 30	48	13.9	GALAXY S
MCG+01-28-017	10 49 54.	+ 07 30	42	13.9	GALAXY
ZWG 038.041	10 49 54.	+ 08 10		15.2	GALAXY
ZWG 066.053	10 49 54.	+ 10 17		15.6	GALAXY
UGC 05994	10 49 54.	+ 10 17	102	15.6	GALAXY Sc
ZWG 125.013	10 49 54.	+ 23 12		12.6	GALAXY
REIZ 0702	10 49 54.	+ 23 12	90	12.3	GALAXY
UGC 05995	10 49 54.	+ 23 12	168	12.6	GALAXY Sc
KARA.73B 0448	10 49 54.	+ 23 12	168	12.6	ISOLATED GALAXY S
STOCK 18	10 49 54.	+ 28 49			BLUE KNOT NEAR ELLIP GLXY
ZC 1049.9+4923	10 49 54.	+ 49 23	940		CLUSTER OF GALAXIES
SC 1047-5801.3	10 49 56.	- 58 17 13.	18		NEBULA
HN 1286	10 49 56.7	- 00 23 18.	18		NEBULA
MCG+03-28-024	10 49 57.	+ 20 05	90	14.2	GALAXY
RNGC 3437	10 49 57.	+ 23 12		13.0	GALAXY
HOLM 219B	10 49 58.	+ 70 58	30	14.2	PART OF MULTIPLE GALAXY
HOLM 219A	10 49 58.	+ 70 59	36	13.7	PART OF MULTIPLE GALAXY
REIZ 0706	10 49 58.	- 00 04	18	15.6	GALAXY
LBN 0629	10 50	+ 84 00	3120		BRIGHT NEBULA
LB 09847	10 50	- 84 56		13.5	FAINT BLUE STAR
REIZ 0707	10 50 00.	+ 07 29	30	13.5	GALAXY
ZC 1050.0+1030	10 50 00.	+ 10 30	1010		CLUSTER OF GALAXIES
ZC 1050.0+1143	10 50 00.	+ 11 43	1480		CLUSTER OF GALAXIES
ZWG 095.054	10 50 00.	+ 17 10		15.6	GALAXY
TON-N 0044	10 50 00.	+ 25 16		15.9	BLUE STAR
TON-N 0552	10 50 00.	+ 27 02		16.0	BLUE STAR
TON-N 0553	10 50 00.	+ 31 06		17.1	BLUE STAR
ZWG 184.032	10 50 00.	+ 38 00		15.7	GALAXY
UGC 05996	10 50 00.	+ 40 39	66	16.5	GALAXY
ZWG 334.004	10 50 00.	+ 73 57		13.3	GALAXY
ZWG 333.062	10 50 00.	+ 73 57		13.3	GALAXY
UGC 05997	10 50 00.	+ 73 57	210	13.3	GALAXY Sc
KARA.73B 0449	10 50 00.	+ 73 57	204	13.3	ISOLATED GALAXY S
ZC 1050.0+8542	10 50 00.	+ 85 42	1680		CLUSTER OF GALAXIES
IC 2611	10 50 02.	+ 10 23			SINGLE STAR
AR 24	10 50 02.20	+ 75 04 25.8			NEBULA
BIGO 517	10 50 06.	+ 10 24			NEBULA
ARC 1119	10 50 06.	+ 10 59		17.5	RICH CLUSTER OF GALAXIES
ZC 1050.1+2027	10 50 06.	+ 20 27	1080		CLUSTER OF GALAXIES
ZWG 184.033	10 50 06.	+ 37 52		15.1	GALAXY
MCG+07-22-075	10 50 06.	+ 40 38	42	15.	GALAXY
MCG+08-20-037	10 50 06.	+ 50 32 30.	60	15.	GALAXY
MCG+12-10-089	10 50 06.	+ 73 57	150	12.9	GALAXY
VHE 44	10 50 06.	- 55 54	96		REFLECTION NEBULA
ARC 1121	10 50 11.	+ 09 18		17.7	RICH CLUSTER OF GALAXIES
RNGC 3397	10 50 11.	+ 77 33			NON-EXISTENT OBJFCT
ZWG 095.055	10 50 12.	+ 17 15		15.7	GALAXY
MCG+06-24-030	10 50 12.	+ 37 54	36	15.	GALAXY
ZC 1050.2+3810	10 50 12.	+ 38 10	1340		CLUSTER OF GALAXIES
ZWG 267.023	10 50 12.	+ 50 34		14.5	GALAXY
MRK 156	10 50 12.	+ 50 34	40	15.	GALAXY WITH UV CONTINUUM
UGC 05998	10 50 12.	+ 50 34	54	14.5	GALAXY PECULR
ZC 1050.2+5618	10 50 12.	+ 56 18	2550		CLUSTER OF GALAXIES
HN 1287	10 50 12.9	+ 02 08 06.	18		NEBULA
ARC 1117	10 50 13.	+ 40 01		17.2	RICH CLUSTER OF GALAXIES
AR 25	10 50 14.22	+ 73 57 22.9			NEBULA
HN 1288	10 50 14.5	+ 00 27 06.			NEBULA
RNGC 3403	10 50 15.	+ 73 57		13.0	GALAXY
HN 1289	10 50 15.8	+ 00 18 24.			NEBULA
REIZ 0712	10 50 16.	+ 17 42	120	14.4	GALAXY
RNGC 3443	10 50 16.	+ 17 50		14.5	GALAXY
REIZ 0709	10 50 17.	+ 33 08	18	15.3	GALAXY
ZWG 010.041	10 50 18.	+ 00 19		15.4	GALAXY
UGC 05999	10 50 18.	+ 07 54	96	18.	GALAXY DWRF IR
ZWG 066.054	10 50 18.	+ 11 06		15.7	GALAXY
ZC 1050.3+1745	10 50 18.	+ 17 45	2420		CLUSTER OF GALAXIES
ZWG 095.056	10 50 18.	+ 17 50		14.7	GALAXY
UGC 06000	10 50 18.	+ 17 50	156	14.3	GALAXY Sc
ZWG 184.034	10 50 18.	+ 34 10		13.2	GALAXY
MRK 418	10 50 18.	+ 34 10	24	14.5	GALAXY WITH UV CONTINUUM
RNGC 3442	10 50 18.	+ 34 10		13.0	GALAXY
UGC 06001	10 50 18.	+ 34 10	39	13.2	GALAXY PECULR
ZWG 184.035	10 50 18.	+ 37 41		15.5	GALAXY
UGC 06002	10 50 18.	+ 37 41	78	15.5	GALAXY SBb
MCG+09-18-046	10 50 18.	+ 56 32 30.	9	16.	GALAXY
MCG+09-18-045	10 50 18.	+ 56 34 30.	42	16.	GALAXY
HN 1290	10 50 18.1	+ 02 08 42.	18		NEBULA
HN 1291	10 50 19.8	- 00 24 19.	18		NEBULA
REIZ 0713	10 50 20.	+ 11 06	30	15.3	GALAXY
PATH 2.022	10 50 20.	- 00 25	19		NEBULA
HN 1292	10 50 20.0	- 01 46 01.			NEBULA
HN 1293	10 50 20.6	- 01 36 55.	18		NEBULA
REIZ 0714	10 50 22.	+ 17 35	18	16.0	GALAXY
RNGC 3444	10 50 23.	+ 10 29		15.5	GALAXY
ARC 1118	10 50 23.	+ 37 50		17.6	RICH CLUSTER OF GALAXIES
ZC 1050.4+0451	10 50 24.	+ 04 51	1750		CLUSTER OF GALAXIES
ZWG 038.042	10 50 24.	+ 04 54		14.1	GALAXY
UGC 06003	10 50 24.	+ 04 54	42	14.1	GALAXY
MCG+01-28-018	10 50 24.	+ 04 54	18	14.1	GALAXY
ZWG 066.055	10 50 24.	+ 10 29		15.4	GALAXY
UGC 06004	10 50 24.	+ 10 29	66	15.4	GALAXY Sb-c
SN 1970B	10 50 24.	+ 14 21		15.	SUPERNOVA
MCG+03-28-025	10 50 24.	+ 17 51	180	14.7	GALAXY
UGC 06003A	10 50 24.	+ 25 57	66	16.5	GALAXY S
ZWG 155.032	10 50 24.	+ 27 09		15.1	GALAXY
REIZ 0710	10 50 24.	+ 34 10	18	14.0	GALAXY
MCG+06-24-031	10 50 24.	+ 37 42 30.	60	14.5	GALAXY
ZWG 213.004	10 50 24.	+ 43 43		15.3	GALAXY
ZWG 212.065	10 50 24.	+ 43 43		15.3	GALAXY
MCG-05-26-010	10 50 24.	- 32 41	120	13.5	GALAXY
REIZ 0716	10 50 25.	+ 10 27	42	14.3	GALAXY
HN 1294	10 50 25.9	- 01 40 01.	12		NEBULA
MCG+06-24-032	10 50 27.	+ 37 34 30.	42	15.	GALAXY
ARC 1120	10 50 28.	+ 31 05		18.0	RICH CLUSTER OF GALAXIES
REIZ 0708	10 50 28.	+ 55 50	18	15.5	GALAXY
ZWG 125.014	10 50 30.	+ 26 10		15.6	GALAXY
MCG+04-26-017	10 50 30.	+ 26 10	24	15.6	GALAXY
UGC 06005	10 50 30.	+ 46 16	60	17.	GALAXY
MCG+09-18-047	10 50 30.	+ 51 01	102	14.	GALAXY
MCG+14-05-015	10 50 30.	+ 80 34	27	17.	GALAXY
ZC 1050.6+1431	10 50 36.	+ 14 31	3230		CLUSTER OF GALAXIES
ZWG 095.057	10 50 36.	+ 16 30		14.6	GALAXY
MCG+03-28-026	10 50 36.	+ 16 30	21	14.6	GALAXY
ZWG 155.033	10 50 36.	+ 32 05		15.5	GALAXY
ZC 1050.6+4316	10 50 36.	+ 43 16	3560		CLUSTER OF GALAXIES
ZC 1050.6+4530	10 50 36.	+ 45 30	870		CLUSTER OF GALAXIES

OBJECT NAME	RIGHT ASCEN.	DECLINATION	DIAM.	MAGN.	TYPE OF OBJECT
MCG-01-28-004	10 50 36.	- 06 59	66	14.5	GALAXY
MCG-02-28-017	10 50 36.	- 10 15	36	15.	GALAXY
VHA 106	10 50 36.	- 53 59	300		OPEN STAR CLUSTER
RNGC 3449	10 50 38.	- 32 40		13.0	GALAXY
RNGC 3447	10 50 40.	+ 17 02		14.5	GALAXY
ARC 1122	10 50 41.	+ 38 11		17.7	RICH CLUSTER OF GALAXIES
ZWG 038.043	10 50 42.	+ 02 53		15.4	GALAXY
ZWG 066.056	10 50 42.	+ 11 27		15.7	GALAXY
ZWG 066.057	10 50 42.	+ 12 21		15.4	GALAXY
ZWG 095.058	10 50 42.	+ 17 02		14.3	GALAXY
UGC 06007	10 50 42.	+ 17 02	102	14.3	GALAXY IRR
UGC 06006	10 50 42.	+ 17 02	252	14.3	GALAXY
KARA.72 255A	10 50 42.	+ 17 02	210	14.3	PART OF DOUBLE GALAXY
MCG+03-28-027	10 50 42.	+ 17 02 30.	210	14.3	GALAXY
ZC 1050.7+1930	10 50 42.	+ 19 30	870		CLUSTER OF GALAXIES
ZWG 267.024	10 50 42.	+ 51 03		15.5	GALAXY
UGC 06008	10 50 42.	+ 51 03	114	15.5	GALAXY SBb
ZWG 291.009	10 50 42.	+ 57 24		14.0	GALAXY
UGC 06009	10 50 42.	+ 57 24	138	14.0	GALAXY S
UGC 06010	10 50 42.	+ 62 47	66	17.	GALAXY DWRF SP
HN 1295	10 50 43.1	+ 02 08 29.	18		NEBULA
HN 1296	10 50 44.8	+ 01 18 29.	18		NEBULA
MCG+07-22-076	10 50 45.	+ 39 01	36	15.	GALAXY
RNGC 3447A	10 50 46.	+ 17 03		15.5	GALAXY
RNGC 3440	10 50 46.	+ 57 23		14.0	GALAXY
HN 1297	10 50 46.4	- 00 20 55.	24		NEBULA
HN 1298	10 50 46.7	- 00 25 43.	18		NEBULA
REIZ 0717	10 50 47.	+ 33 03	24	15.0	GALAXY
REIZ 0711	10 50 47.	+ 61 33	60	14.1	GALAXY
VV 252B	10 50 48.	+ 17 02	90	15.	INTERACTING GALAXY
VV 252A	10 50 48.	+ 17 02	240	14.	INTERACTING GALAXY
KARA.72 255B	10 50 48.	+ 17 03	102		PART OF DOUBLE GALAXY
MCG+03-28-028	10 50 48.	+ 17 03	84	14.3	GALAXY
ZWG 095.059	10 50 48.	+ 19 11		14.9	GALAXY
MCG+03-28-029	10 50 48.	+ 19 13	48	14.9	GALAXY
ZC 1050.8+3101	10 50 48.	+ 31 01	940		CLUSTER OF GALAXIES
MCG+06-24-033	10 50 48.	+ 33 01 30.	12	16.	GALAXY
ZWG 184.036	10 50 48.	+ 34 17		14.7	GALAXY
REIZ 0718	10 50 48.	+ 34 18	21	14.7	GALAXY
MCG+08-20-038	10 50 48.	+ 46 28 30.	36	15.	GALAXY
MCG+09-18-048	10 50 48.	+ 55 05	30	17.	GALAXY
MCG+10-16-019	10 50 48.	+ 57 22 30.	114	14.4	GALAXY
REIZ 0715	10 50 48.	+ 57 23	60	14.2	GALAXY
FATH 2.023	10 50 48.	- 00 20	35		NEBULA
FATH 2.024	10 50 48.	- 00 25	16		NEBULA
TON-N 1296	10 50 50.	+ 20 14		15.8	BLUE STAR
IC 2612	10 50 50.6	+ 33 02 05.			GALAXY E1
MCG+06-24-034	10 50 51.	+ 34 18	48	14.5	GALAXY
REIZ 0719	10 50 52.	+ 17 02	120	15.7	GALAXY
ZC 1050.9+1421	10 50 54.	+ 14 21	940		CLUSTER OF GALAXIES
ZC 1050.9+3451	10 50 54.	+ 34 51	2080		CLUSTER OF GALAXIES
ZWG 241.027	10 50 54.	+ 46 29		15.3	GALAXY
MCG+09-18-049	10 50 54.	+ 56 22 30.	24	17.	GALAXY
UGC 06011	10 50 54.	- 00 20	90	16.5	GALAXY GROUP
TON-N 1297	10 50 56.	+ 21 59		15.9	BLUE STAR
PK288-00.1	10 51	- 60 11	72		PLANETARY NEBULA
ZC 1051.0+2104	10 51 00.	+ 21 04	1810		CLUSTER OF GALAXIES
UGC 06012	10 51 00.	+ 27 10	72	16.0	GALAXY Sb-c
MCG+05-26-026	10 51 00.	+ 27 10	78	15.	GALAXY
MCG+08-20-039	10 51 00.	+ 46 17	48	14.	GALAXY
MCG+08-20-040	10 51 0C.	+ 49 51	48	15.	GALAXY
ZWG 241.028	10 51 00.	+ 49 52		14.8	GALAXY
ZWG 241.029	10 51 00.	+ 49 55		13.9	GALAXY
UGC 06013	10 51 00.	+ 49 55	54	13.9	GALAXY S0
MCG+08-20-041	10 51 00.	+ 49 55	42	14.	GALAXY
ZC 1051.0+5920	10 51 00.	+ 59 20	1880		CLUSTER OF GALAXIES
ZWG 351.048	10 51 00.	+ 78 07		15.7	GALAXY
LB 00585	10 51 02.	+ 46 50 06.		17.9	FAINT BLUE STAR
ZWG 066.058	10 51 06.	+ 10 00		15.2	GALAXY
UGC 06014	10 51 06.	+ 10 00	96	15.2	GALAXY
ZC 1051.1+1258	10 51 06.	+ 12 58	1880		CLUSTER OF GALAXIES
ZWG 241.030	10 51 06.	+ 46 17		14.9	GALAXY
UGC 06015	10 51 06.	+ 46 17	60	14.9	GALAXY Sc
ZWG 267.025	10 51 06.	+ 53 18		15.4	GALAXY
OCL 0840	10 51 06.	- 62 01	270	12.	OPEN STAR CLUSTER
ARC 1125	10 51 11.	+ 10 31		17.6	RICH CLUSTER OF GALAXIES
ZC 1051.2+0520	10 51 12.	+ 05 20	340		CLUSTER OF GALAXIES
ZC 1051.2+3838	10 51 12.	+ 38 38	1480		CLUSTER OF GALAXIES
MCG+08-20-042	10 51 12.	+ 44 51 30.	72	16.	GALAXY
MCG+08-20-043	10 51 12.	+ 45 10	42	16.	GALAXY
UGC 06016	10 51 12.	+ 54 34	132	17.	GALAXY DWRF IR
MCG+09-18-050	10 51 12.	+ 55 06	12	17.	GALAXY
MCG+10-16-020	10 51 12.	+ 57 04	18	16.	GALAXY
MCG+10-16-021	10 51 12.	+ 59 18	57	15.	GALAXY
MCG-02-28-018	10 51 12.	- 11 27 30.	54	15.	GALAXY
MCG-04-26-013	10 51 12.	- 21 31	48	13.5	GALAXY
MCG-05-26-011	10 51 12.	- 29 07	36	15.	GALAXY
HN 1299	10 51 12.4	- 02 50 32.	18		NEBULA
TON-N 1298	10 51 14.	+ 20 09		17.	BLUE STAR
MCG+05-26-027	10 51 15.	+ 27 10	24	15.5	GALAXY
ARC 1126	10 51 16.	+ 17 08		16.0	RICH CLUSTER OF GALAXIES
RNGC 3453	10 51 17.	- 21 31			GALAXY
ZC 1051.3+1034	10 51 18.	+ 10 34	1340		CLUSTER OF GALAXIES
ZC 1051.3+1950	10 51 18.	+ 19 50	870		CLUSTER OF GALAXIES
UGC 06017	10 51 18.	+ 44 53	84	16.0	GALAXY Sc
ZWG 241.031	10 51 18.	+ 45 12		15.7	GALAXY
ZWG 241.032	10 51 18.	+ 49 18		15.4	GALAXY
FATH 1.351	10 51 19.	- 00 52	11		NEBULA
FATH 1.352	10 51 20.	- 00 33	5		NEBULA
IC 0654	10 51 21.	- 11 27 39.			NONSTELLAR OBJECT
GO 097	10 51 21.	- 60 46	101	17.8	STAR CHAIN
FATH 1.353	10 51 22.	- 00 33	11		NEBULA
ZWG 038.044	10 51 24.	+ 07 00		15.4	GALAXY
ZC 1051.4+1656	10 51 24.	+ 16 56	4030		CLUSTER OF GALAXIES
MCG+04-26-018	10 51 24.	+ 20 54 30.	72	15.	GALAXY
UGC 06018	10 51 24.	+ 20 55	72	17.	GALAXY DWRF IR
MCG+04-26-019	10 51 24.	+ 21 23	72	15.	GALAXY
ZC 1051.4+2231	10 51 24.	+ 22 31	610		CLUSTER OF GALAXIES
TON-N 0554	10 51 24.	+ 29 35		16.2	BLUE STAR
ZC 1051.4+4041	10 51 24.	+ 40 41	1610		CLUSTER OF GALAXIES
MCG+09-18-051	10 51 24.	+ 51 02	60	16.	GALAXY
MCG+09-18-052	10 51 24.	+ 54 33	120	15.	GALAXY
ZC 1051.4+5440	10 51 24.	+ 54 40	610		CLUSTER OF GALAXIES
ZWG 291.010	10 51 24.	+ 59 18		15.6	GALAXY
UGC 06019	10 51 24.	+ 59 18	66	15.6	GALAXY Sa
REIZ 0720	10 51 29.	+ 57 15	60	12.2	GALAXY
REIZ 0721	10 51 29.	+ 57 22	12	15.6	GALAXY
UGC 06020	10 51 30.	+ 21 24	84	16.0	GALAXY Sb
TON-N 0555	10 51 30.	+ 28 07		15.2	BLUE STAR
ZWG 155.034	10 51 30.	+ 29 58		15.7	GALAXY
IC 2613	10 51 30.	+ 33 14 16.			SAME AS NGC 3430
ZWG 267.026	10 51 30.	+ 56 13		15.7	GALAXY
MCG+09-18-053	10 51 30.	+ 56 14	48	16.	GALAXY
ZWG 291.011	10 51 30.	+ 57 15		12.8	GALAXY
UGC 06021	10 51 30.	+ 57 15	102	12.8	GALAXY IRR
KARA.72 256A	10 51 30.	+ 57 15	84	12.8	PART OF DOUBLE GALAXY
VV 014B	10 51 30.	+ 57 16	24	15.5	INTERACTING GALAXY
VV 014A	10 51 30.	+ 57 16	72	13.	INTERACTING GALAXY
MCG+10-16-022	10 51 30.	+ 61 33	108	14.	GALAXY
MCG-03-28-017	10 51 30.	- 15 53	30	14.5	GALAXY
HELW 427	10 51 31.	- 15 50 56.			NEBULA
ARC 1127	10 51 32.	+ 14 57		17.8	RICH CLUSTER OF GALAXIES
REIZ 0722	10 51 32.	+ 59 20	36	14.9	GALAXY
REIZ 0724	10 51 33.	+ 29 58	30	15.0	GALAXY
ARC 1128	10 51 34.	+ 09 18		17.0	RICH CLUSTER OF GALAXIES
RNGC 3445	10 51 34.	+ 57 15		13.0	GALAXY
ARP 024	10 51 34.	+ 57 15			PECULIAR GALAXY
UGC 06022	10 51 36.	+ 18 04	78	17.	GALAXY DWARF
REIZ 0725	10 51 36.	+ 21 24	60	14.7	GALAXY
ZWG 155.035	10 51 36.	+ 27 30		13.5	GALAXY
UGC 06023	10 51 36.	+ 27 30	114	13.5	GALAXY Sc
ZC 1051.6+3519	10 51 36.	+ 35 19	670		CLUSTER OF GALAXIES
ZWG 241.033	10 51 36.	+ 49 28		15.6	GALAXY
ZWG 267.027	10 51 36.	+ 54 35		12.2	GALAXY
UGC 06024	10 51 36.	+ 54 35	318	12.2	GALAXY
KARA.72 256B	10 51 36.	+ 57 14	36		PART OF DOUBLE GALAXY
MCG+10-16-023	10 51 36.	+ 57 15	72	12.9	GALAXY
ZWG 291.012	10 51 36.	+ 61 34		14.2	GALAXY
UGC 06025	10 51 36.	+ 61 34	120	14.2	GALAXY Sb
MCG-03-28-018	10 51 36.	- 15 47	96	13.	GALAXY
LYNG 04	10 51 36.	- 58 27 06.	1440		OB CONCENTRATION
RNGC 3435	10 51 37.	+ 61 34		14.0	GALAXY
RNGC 3451	10 51 38.	+ 27 30		13.5	GALAXY
REIZ 0726	10 51 38.	+ 27 30	78	13.3	GALAXY
REIZ 0723	10 51 38.	+ 54 34	150	12.3	GALAXY
RNGC 3452	10 51 39.	- 11 08		15.0	GALAXY
RNGC 3456	10 51 40.	- 15 46		13.0	GALAXY
HN 1300	10 51 41.0	- 01 57 08.	24		NEBULA
ZWG 038.045	10 51 42.	+ 06 41		15.6	GALAXY
ZC 1051.7+1454	10 51 42.	+ 14 54	670		CLUSTER OF GALAXIES
ZC 1051.7+1849	10 51 42.	+ 18 49	2020		CLUSTER OF GALAXIES
TON-N 0045	10 51 42.	+ 30 25		15.8	BLUE STAR
REIZ 0727	10 51 42.	+ 36 13	12	15.7	GALAXY
MCG+06-24-035	10 51 42.	+ 36 14	48	16.	GALAXY
MCG+08-20-044	10 51 42.	+ 46 49	60	16.	GALAXY
ZWG 241.034	10 51 42.	+ 47 44		15.0	GALAXY
MCG+08-20-045	10 51 42.	+ 47 44	36	15.	GALAXY
MCG+09-18-054	10 51 42.	+ 51 00	24	18.	GALAXY
ARP 205	10 51 42.	+ 54 34			PECULIAR GALAXY
MCG+09-18-055	10 51 42.	+ 54 34	330	12.6	GALAXY
RNGC 3448	10 51 42.	+ 54 35		12.5	GALAXY
MCG+10-16-024	10 51 42.	+ 57 14	27	15.	GALAXY
MCG-01-28-005	10 51 42.	- 03 53	48	15.	GALAXY
MCG-05-26-012	10 51 42.	- 32 53	66	14.5	GALAXY
HOLM 221B	10 51 45.	+ 17 38	138	13.4	PART OF MULTIPLE GALAXY
MCG+05-26-028	10 51 45.	+ 27 30	90	13.5	GALAXY
MCG-02-28-019	10 51 45.	- 11 08	54	15.	GALAXY
HOLM 221A	10 51 46.	+ 17 34	72	12.6	PART OF MULTIPLE GALAXY
RNGC 3454	10 51 46.	+ 17 37		14.0	GALAXY
ZWG 038.046	10 51 48.	+ 07 25		15.7	GALAXY
ZWG 095.060	10 51 48.	+ 17 36		14.1	GALAXY
UGC 06026	10 51 48.	+ 17 36	144	14.1	GALAXY
KARA.72 257A	10 51 48.	+ 17 36	144	14.1	PART OF DOUBLE GALAXY
ZWG 095.061	10 51 48.	+ 18 01		15.4	GALAXY
MCG+04-26-020	10 51 48.	+ 21 25	24	15.0	GALAXY
ZWG 125.015	10 51 48.	+ 21 26		15.0	GALAXY
REIZ 0728	10 51 48.	+ 21 26	18	14.3	GALAXY
ZC 1051.8+3531	10 51 48.	+ 35 31	610		CLUSTER OF GALAXIES
ZWG 241.035	10 51 48.	+ 48 07		15.4	GALAXY
UGC 06027	10 51 48.	+ 64 17	84	17.	GALAXY DWRF IR
MCG+12-11-001	10 51 48.	+ 71 47	27	16.	GALAXY
ZWG 010.042	10 51 48.	- 00 05		15.3	GALAXY
MCG-02-28-020	10 51 48.	- 11 00	48	15.	GALAXY
MCG-03-28-019	10 51 48.	- 17 57	48	16.	GALAXY
IC 0655	10 51 50.	- 00 06	49		FAINT NEBULA
FATH 2.025	10 51 50.	- 00 06	49		NEBULA
REIN 2.118	10 51 51.47	+ 17 33 07.6			NEBULA
RNGC 3455	10 51 52.	+ 17 33		13.0	GALAXY
REIZ 0730	10 51 52.	+ 17 33	60	13.1	GALAXY
REIZ 0729	10 51 52.	+ 17 36	90	13.5	GALAXY
REIZ 0732	10 51 52.	- 00 06	48	14.9	GALAXY
ZWG 066.059	10 51 54.	+ 10 13		14.9	GALAXY
MCG+02-28-026	10 51 54.	+ 10 13	24	14.9	GALAXY
ZWG 066.060	10 51 54.	+ 11 43		15.6	GALAXY
ARC 1129	10 51 54.	+ 12 06		17.8	RICH CLUSTER OF GALAXIFS
ZC 1051.9+1207	10 51 54.	+ 12 07	740		CLUSTER OF GALAXIES
ZWG 095.062	10 51 54.	+ 17 33		13.1	GALAXY
UGC 06028	10 51 54.	+ 17 33	168	13.1	GALAXY Sb
KARA.72 257B	10 51 54.	+ 17 33	162	13.1	PART OF DOUBLE GALAXY
MCG+03-28-031	10 51 54.	+ 17 34	162	13.1	GALAXY
MCG+03-28-030	10 51 54.	+ 17 37	90	14.1	GALAXY
ZC 1051.9+3258	10 51 54.	+ 32 58	1010		CLUSTER OF GALAXIES
ZC 1051.9+3505	10 51 54.	+ 35 05	740		CLUSTER OF GALAXIES
ZWG 241.036	10 51 54.	+ 46 49		14.9	GALAXY
LB 00586	10 51 54.	+ 46 49 36.		17.5	FAINT BLUE STAR
MCG+12-11-002	10 51 54.	+ 73 12	66	16.	GALAXY
FATH 1.354	10 51 54.	- 00 53	8		NEBULA
MCG-03-28-020	10 51 54.	- 17 12	30	15.	GALAXY
OCL 0834	10 51 54.	- 61 28	132	12.	OPEN STAR CLUSTER
REIZ 0731	10 51 56.	+ 27 42	54	15.6	GALAXY
GO 085	10 51 59.	- 60 00	95	17.0	STAR CHAIN
ZC 1052.0+0629	10 52 00.	+ 06 29	1340		CLUSTER OF GALAXIES
ZC 1052.0+0921	10 52 00.	+ 09 21	1280		CLUSTER OF GALAXIES
ZWG 066.061	10 52 00.	+ 11 17		15.7	GALAXY
TON-N 0046	10 52 00.	+ 30 47		16.3	BLUE STAR
SHAH 001	10 52 00.	+ 40 44	84	17.	GROUP OF COMPACT GALAXIES
ZWG 213.005	10 52 00.	+ 41 46		15.7	GALAXY
ZWG 241.037	10 52 00.	+ 49 59		14.0	GALAXY
MRK 157	10 52 00.	+ 49 59	23	14.	GALAXY WITH UV CONTINUUM
UGC 06029	10 52 00.	+ 49 59	60	14.0	GALAXY PECULR
MCG+08-20-046	10 52 00.	+ 49 59	60	14.	GALAXY
MCG+09-18-056	10 52 00.	+ 53 20	15	16.	GALAXY
TON-N 1299	10 52 01.	+ 34 15		16.8	BLUE STAR
MCG+07-23-001	10 52 03.	+ 41 45	48	15.	GALAXY
MCG+08-20-047	10 52 03.	+ 49 51	30	16.	GALAXY
RNGC 3457	10 52 04.	+ 17 53		13.0	GALAXY

OBJECT NAME	RIGHT ASCEN.	DECLINATION	DIAM.	MAGN.	TYPE OF OBJECT
ZWG 095.063	10 52 06.	+ 16 20		15.5	GALAXY
ZWG 095.064	10 52 06.	+ 16 42		15.7	GALAXY
ZWG 095.065	10 52 06.	+ 17 53		13.0	GALAXY
UGC 06030	10 52 06.	+ 17 53	60	13.0	GALAXY
ZC 1052.1+2411	10 52 06.	+ 24 11	1680		CLUSTER OF GALAXIES
TON-N 0556	10 52 06.	+ 27 25		14.6	BLUE STAR
GCL 016	10 52 06.	+ 40 44	72	15.0	GLOBULAR STAR CLUSTER
ZWG 241.038	10 52 06.	+ 49 52		15.3	GALAXY
7ZW 351	10 52 06.	+ 65 43			COMPACT GALAXY
ZWG 334.005	10 52 06.	+ 71 45		15.4	GALAXY
ZWG 333.063	10 52 06.	+ 71 45		15.4	GALAXY
PATH 1.355	10 52 06.	- 00 27	8		NEBULA
HELW 428	10 52 07.	- 15 34 21.			NEBULA
TON-N 1300	10 52 08.	+ 20 14		16.	BLUE STAR
GO 058	10 52 08.	- 58 20	86	17.9	STAR CHAIN
REIZ 0733	10 52 10.	+ 17 52	24	13.4	GALAXY
RNGC 3464	10 52 11.	- 20 49		13.0	GALAXY
ZWG 095.066	10 52 12.	+ 14 41		15.4	GALAXY
MCG+03-28-032	10 52 12.	+ 17 55	36	13.0	GALAXY
MCG+05-26-029	10 52 12.	+ 28 18	42	15.5	GALAXY
ZWG 155.036	10 52 12.	+ 29 48		15.7	GALAXY
UGC 06031	10 52 12.	+ 29 48	66	15.7	GALAXY S
MCG+08-20-048	10 52 12.	+ 48 59	42	16.	GALAXY
MCG+03-28-033	10 52 15.	+ 14 41	42	15.4	GALAXY
MCG+05-26-030	10 52 15.	+ 29 48 30.	90	15.7	GALAXY
REIZ 0734	10 52 15.	+ 29 49	24	15.2	GALAXY
ARC 1130	10 52 15.	- 10 24		17.1	RICH CLUSTER OF GALAXIES
MCG-03-28-021	10 52 15.	- 20 51	150	13.	GALAXY
RNGC 3461	10 52 16.	+ 17 55			GALAXY
RNGC 3459	10 52 16.	- 16 48		14.0	GALAXY
ZWG 066.062	10 52 18.	+ 10 19		15.1	GALAXY
ZWG 241.039	10 52 18.	+ 48 59		15.7	GALAXY
ZC 1052.3+4945	10 52 18.	+ 49 45	2020		CLUSTER OF GALAXIES
ZC 1052.3+5941	10 52 18.	+ 59 41	470		CLUSTER OF GALAXIES
MCG-03-28-022	10 52 18.	- 16 48	90	14.	GALAXY
AR 26	10 52 18.78	+ 75 06 17.9			NEBULA
MCG+03-28-034	10 52 21.	+ 14 33	48	15.5	GALAXY
ZWG 095.067	10 52 24.	+ 14 32		15.5	GALAXY
MCG+10-16-025	10 52 24.	+ 58 09	27	15.	GALAXY
ZWG 291.013	10 52 24.	+ 58 10		15.7	GALAXY
ZC 1052.4+7201	10 52 24.	+ 72 01	2420		CLUSTER OF GALAXIES
UGC 06032	10 52 24.	+ 73 10	102	16.5	GALAXY Sc
TON-N 1301	10 52 25.	+ 34 49		16.8	BLUE STAR
RNGC 3460	10 52 28.	+ 17 52			NON-EXISTENT OBJECT
IC 0656	10 52 28.	+ 17 52			OPEN CLUSTER
PATH 1.356	10 52 28.	- 00 07	8		NEBULA
PATH 1.357	10 52 28.	- 00 53	11		NEBULA
ZWG 038.047	10 52 30.	+ 03 22		15.7	GALAXY
KARA.73B 0445	10 52 30.	+ 03 22	36	15.7	ISOLATED GALAXY S
ZC 1052.5+1359	10 52 30.	+ 13 59	670		CLUSTER OF GALAXIES
ZC 1052.5+3320	10 52 30.	+ 33 20	740		CLUSTER OF GALAXIES
ZC 1052.5+3544	10 52 30.	+ 35 44	740		CLUSTER OF GALAXIES
ZC 1052.5+4741	10 52 30.	+ 47 41	1950		CLUSTER OF GALAXIES
LB 01909	10 52 31.	+ 56 04 36.		16.4	FAINT BLUE STAR
HN 1301	10 52 31.5	- 00 48 52.	12		NEBULA
PATH 1.358	10 52 32.	- 00 50	8		NEBULA
PK288+00.1	10 52 33.7	- 58 53 44.	35		PLANETARY NEBULA
ARC 1124	10 52 35.	+ 72 02		17.3	RICH CLUSTER OF GALAXIES
ZWG 038.048	10 52 36.	+ 06 07		15.3	GALAXY
ZWG 213.006	10 52 36.	+ 42 34		15.5	GALAXY
UGC 06033	10 52 36.	+ 42 34	78	15.5	GALAXY Sb
SC 1050-5821.6	10 52 36.	- 58 37 34.	12		NEBULA
HN 1302	10 52 36.1	- 02 37 40.			NEBULA
MCG+08-20-049	10 52 39.	+ 47 50	15	16.	GALAXY
GO 101	10 52 40.	- 57 45	94	18.3	STAR CHAIN
REIZ 0735	10 52 41.	+ 21 02	24	14.5	GALAXY
ZWG 038.049	10 52 42.	+ 07 58		13.4	GALAXY
RNGC 3462	10 52 42.	+ 07 58		13.5	GALAXY
UGC 06034	10 52 42.	+ 07 58	156	13.4	GALAXY S0
MCG+01-28-019	10 52 42.	+ 07 58	24	13.4	GALAXY
MCG+04-26-021	10 52 42.	+ 21 01	30	15.5	GALAXY
ZWG 125.016	10 52 42.	+ 21 02		15.5	GALAXY
MCG+07-23-002	10 52 42.	+ 42 33 30.	48	15.	GALAXY
ZWG 241.040	10 52 42.	+ 47 51		15.6	GALAXY
MCG+09-18-057	10 52 42.	+ 54 07	9	16.	GALAXY
ZC 1052.7+5453	10 52 42.	+ 54 53	610		CLUSTER OF GALAXIES
ZC 1052.7+7550	10 52 42.	+ 75 50	1750		CLUSTER OF GALAXIES
ZWG 351.049	10 52 42.	+ 77 59		15.5	GALAXY
PATH 1.359	10 52 42.	- 00 40	22		NEBULA
RNGC 3463	10 52 42.	- 25 51		14.0	GALAXY
HN 1303	10 52 42.3	- 01 42 40.			NEBULA
TON-N 1302	10 52 44.	+ 19 48		16.2	BLUE STAR
PATH 1.360	10 52 45.	- 00 18	16		NEBULA
MCG-04-26-014	10 52 45.	- 25 51	72	14.	GALAXY
REIZ 0737	10 52 46.	+ 17 55	18	15.0	GALAXY
ZWG 095.068	10 52 48.	+ 17 24		15.2	GALAXY
UGC 06035	10 52 48.	+ 17 24	78	15.2	GALAXY DWRF IR
MCG+03-28-035	10 52 48.	+ 17 25	96	15.2	GALAXY
MCG+05-26-031	10 52 48.	+ 27 20	36	15.5	GALAXY
PATH 1.361	10 52 48.	- 00 41	11		NEBULA
ZC 1052.8-0345	10 52 48.	- 03 45	2550		CLUSTER OF GALAXIES
MCG-01-28-006	10 52 48.	- 07 12	36	14.5	GALAXY
MCG-05-26-013	10 52 48.	- 30 12	48	15.	GALAXY
TON-N 1303	10 52 50.	+ 22 38		16.	BLUE STAR
GO 055	10 52 50.	- 58 36	50	16.6	STAR CHAIN
ARC 1123	10 52 51.	+ 75 48		16.9	RICH CLUSTER OF GALAXIES
LB 01910	10 52 53.	+ 55 02 06.		16.9	FAINT BLUE STAR
ZWG 038.050	10 52 54.	+ 04 43		15.6	GALAXY
ZWG 066.063	10 52 54.	+ 09 07		15.0	GALAXY
MCG+02-28-027	10 52 54.	+ 09 07	36	15.0	GALAXY
ZWG 213.007	10 52 54.	+ 44 19		15.2	GALAXY
ZC 1052.9+4834	10 52 54.	+ 48 34	1010		CLUSTER OF GALAXIES
RNGC 3458	10 52 58.	+ 57 23		13.5	GALAXY
ZWG 038.051	10 53 00.	+ 02 39		15.5	GALAXY
ZWG 038.052	10 53 00.	+ 04 38		14.9	GALAXY
MCG+01-28-020	10 53 00.	+ 04 38	36	14.9	GALAXY
ARC 1131	10 53 00.	+ 11 16		17.2	RICH CLUSTER OF GALAXIES
ZC 1053.0+1118	10 53 00.	+ 11 18	1750		CLUSTER OF GALAXIES
ZC 1053.0+1536	10 53 00.	+ 15 36	870		CLUSTER OF GALAXIES
ZWG 155.037	10 53 00.	+ 30 06		15.7	GALAXY
ZWG 155.038	10 53 00.	+ 31 40		15.7	GALAXY
ZWG 184.037	10 53 00.	+ 37 07		14.6	GALAXY
UGC 06036	10 53 00.	+ 37 07	102	14.6	GALAXY Sa-b
MCG+10-16-026	10 53 00.	+ 57 23	42	13.0	GALAXY
ZWG 291.014	10 53 00.	+ 57 24		13.2	GALAXY
UGC 06037	10 53 00.	+ 57 24	84	13.2	GALAXY S0
ZC 1053.0+6740	10 53 00.	+ 67 40	1410		CLUSTER OF GALAXIES
TON-N 1304	10 53 01.	+ 34 57		16.8	BLUE STAR
GO 060	10 53 03.	- 58 17	144	18.6	STAR CHAIN
REIZ 0736	10 53 05.	+ 57 23	12	13.3	GALAXY
ZC 1053.1+1010	10 53 06.	+ 10 10	870		CLUSTER OF GALAXIES
MCG+03-28-036	10 53 06.	+ 15 07 30.	36	15.1	GALAXY
ZWG 095.069	10 53 06.	+ 15 08		15.1	GALAXY
ZWG 095.070	10 53 06.	+ 17 16		15.1	GALAXY
MCG+08-20-050	10 53 06.	+ 47 09	60	16.	GALAXY
MCG+08-20-051	10 53 06.	+ 47 39	60	15.	GALAXY
MCG+09-18-058	10 53 06.	+ 54 06	18	16.	GALAXY
MCG+10-16-027	10 53 06.	+ 57 48	51	16.	GALAXY
MCG+10-16-028	10 53 06.	+ 60 57	12	16.	GALAXY
REIZ 0738	10 53 07.	+ 37 07	36	14.5	GALAXY
REIZ 0739	10 53 07.	+ 38 17	24	14.6	GALAXY
GO 052	10 53 08.	- 59 02	104	18.2	STAR CHAIN
MCG+06-24-036	10 53 09.	+ 37 08	90	14.5	GALAXY
MCG+08-20-053	10 53 09.	+ 47 08	24	16.	GALAXY
MCG+08-20-052	10 53 09.	+ 50 12	21	15.	GALAXY
REIZ 0740	10 53 10.	+ 31 39	36	14.5	GALAXY
ZC 1053.2+0805	10 53 12.	+ 08 05	1880		CLUSTER OF GALAXIES
ZWG 066.064	10 53 12.	+ 10 00		15.7	GALAXY
ZC 1053.2+1027	10 53 12.	+ 10 27	270		CLUSTER OF GALAXIES
ZC 1053.2+3208	10 53 12.	+ 32 08	2220		CLUSTER OF GALAXIES
ZWG 241.041	10 53 12.	+ 44 40		15.7	GALAXY
ZWG 241.042	10 53 12.	+ 47 40		15.5	GALAXY
UGC 06038	10 53 12.	+ 47 40	72	15.5	GALAXY Sb
ZWG 241.043	10 53 12.	+ 50 12		15.0	GALAXY
MCG+10-16-029	10 53 12.	+ 57 00	66	15.	GALAXY
ZC 1053.2-0144	10 53 12.	- 01 44	740		CLUSTER OF GALAXIES
HN 1304	10 53 13.3	+ 01 16 20.	18		NEBULA
LB 01911	10 53 14.	+ 50 52 48.		15.7	FAINT BLUE STAR
REIZ 0741	10 53 15.	+ 30 04	24	15.5	GALAXY
REIZ 0742	10 53 15.	+ 30 55	24	15.5	GALAXY
MCG-02-28-021	10 53 15.	- 09 35	60	14.	GALAXY
ZC 1053.3+3548	10 53 18.	+ 35 48	670		CLUSTER OF GALAXIES
ZWG 213.008	10 53 18.	+ 42 36		15.4	GALAXY
STOCK 19	10 53 18.	+ 44 42			BLUE KNOT NEAR ELLIP GLXY
MCG+08-20-054	10 53 18.	+ 47 07	18	16.	GALAXY
ZWG 241.044	10 53 18.	+ 47 09		15.7	GALAXY
UGC 06039	10 53 18.	+ 57 02	102	16.5	GALAXY
ZWG 291.015	10 53 18.	+ 60 57		15.7	GALAXY
HELW 429	10 53 19.	- 16 00 53.			NEBULA
LB 01912	10 53 23.	+ 57 20 36.		15.1	FAINT BLUE STAR
HN 1305	10 53 23.2	+ 01 47 19.	12		NEBULA
HN 1306	10 53 23.8	+ 00 15 43.	18		NEBULA
ZWG 038.053	10 53 24.	+ 02 46		15.0	GALAXY
UGC 06040	10 53 24.	+ 02 46	60	15.0	GALAXY S
MCG+01-28-021	10 53 24.	+ 02 46	48	15.0	GALAXY
UGC 06041	10 53 24.	+ 41 53	66	17.	GALAXY
MCG+07-23-003	10 53 24.	+ 42 35	27	15.	GALAXY
ZWG 241.045	10 53 24.	+ 47 08		15.7	GALAXY
ZC 1053.4+5427	10 53 24.	+ 54 27	1010		CLUSTER OF GALAXIES
ZC 1053.4+5618	10 53 24.	+ 56 18	740		CLUSTER OF GALAXIES
MCG+10-16-030	10 53 24.	+ 57 10	39	16.	GALAXY
MCG+10-16-031	10 53 24.	+ 57 37	30	16.	GALAXY
7ZW 352	10 53 24.	+ 60 58			COMPACT GALAXY
ZWG 314.001	10 53 24.	+ 67 28		15.4	GALAXY
ZWG 313.040	10 53 24.	+ 67 28		15.4	GALAXY
SC 1051-5754.8	10 53 25.	- 58 10 47.	24		NEBULA
GO 087	10 53 25.	- 59 34	83	18.1	STAR CHAIN
LB 01913	10 53 26.	+ 57 35 24.		14.9	FAINT BLUE STAR
SC 1051-5742.0	10 53 26.	- 57 57 59.	18		NEBULA
PATH 1.362	10 53 28.	- 00 42	3		NEBULA
ZWG 038.054	10 53 30.	+ 06 26		14.8	GALAXY
MCG+01-28-022	10 53 30.	+ 06 26	18	14.8	GALAXY
8ZW 1053+09.6	10 53 30.	+ 09 37		18.5	COMPACT GALAXY
ZWG 095.071	10 53 30.	+ 14 45		15.7	GALAXY
ZC 1053.5+2238	10 53 30.	+ 22 38	810		CLUSTER OF GALAXIES
ZWG 213.009	10 53 30.	+ 44 01		15.2	GALAXY
ZWG 066.065	10 53 36.	+ 10 01		14.6	GALAXY
RNGC 3466	10 53 36.	+ 10 01		14.5	GALAXY
UGC 06042	10 53 36.	+ 10 01	72	14.6	GALAXY Sa-b
MCG+02-28-028	10 53 36.	+ 10 01	60	14.6	GALAXY
ZWG 095.072	10 53 36.	+ 15 29		15.5	GALAXY
UGC 06043	10 53 36.	+ 15 29	78	15.5	GALAXY Sc
MCG+03-28-037	10 53 36.	+ 15 30	54	15.5	GALAXY
ZC 1053.6+6027	10 53 36.	+ 60 27	1280		CLUSTER OF GALAXIES
MCG+12-11-003	10 53 36.	+ 68 41	33	17.	GALAXY
REIZ 0745	10 53 37.	+ 10 01	18	14.1	GALAXY
REIZ 0743	10 53 39.	+ 30 09	18	15.7	GALAXY
REIZ 0744	10 53 40.	+ 31 32	12	15.4	GALAXY
GO 065	10 53 40.	- 57 47			STAR CHAIN
ZC 1053.7+4155	10 53 42.	+ 41 55	1950		CLUSTER OF GALAXIES
MCG+10-16-032	10 53 42.	+ 59 48	18	16.	GALAXY
MCG-01-28-007	10 53 42.	- 09 26	60	15.	GALAXY
MCG-05-26-014	10 53 42.	- 31 42	66	15.	GALAXY
HN 1307	10 53 46.0	+ 01 47 43.	18		NEBULA
ZC 1053.8+0350	10 53 48.	+ 03 50	340		CLUSTER OF GALAXIES
ZC 1053.8+3605	10 53 48.	+ 36 05	540		CLUSTER OF GALAXIES
MCG-03-28-023	10 53 48.	- 20 38	72	14.5	GALAXY
LB 01914	10 53 50.	+ 51 50 30.		15.5	FAINT BLUE STAR
PATH 1.363	10 53 50.	- 00 38	5		NEBULA
HELW 430	10 53 52.	- 15 36 05.			NEBULA
ZWG 038.055	10 53 54.	+ 03 40		15.6	GALAXY
ZWG 038.056	10 53 54.	+ 05 51		15.5	GALAXY
ZWG 066.066	10 53 54.	+ 10 12		14.7	GALAXY
MCG+02-28-029	10 53 54.	+ 10 12	54	14.7	GALAXY
ZWG 095.073	10 53 54.	+ 17 03		14.8	GALAXY
MCG+03-28-038	10 53 54.	+ 17 03	24	14.8	GALAXY
UGC 06044	10 53 54.	+ 20 40	60	17.	GALAXY Sc
ZWG 155.039	10 53 54.	+ 30 08		15.7	GALAXY
MCG+11-14-001	10 53 54.	+ 67 00	54	16.	GALAXY
MCG-03-28-024	10 53 54.	- 15 37	36	14.	GALAXY
MCG-02-28-022	10 53 54.	- 15 37	60	14.5	GALAXY
HELW 431	10 53 54.	- 15 39 35.			NEBULA
HN 1308	10 53 57.0	+ 00 49 07.	18		NEBULA
PK289-01.1	10 53 59.	- 61 11	25		PLANETARY NEBULA
ZC 1054.0+2721	10 54 00.	+ 27 21	1080		CLUSTER OF GALAXIES
ZWG 241.046	10 54 00.	+ 44 35		15.6	GALAXY
ZC 1054.0+4612	10 54 00.	+ 46 12	1950		CLUSTER OF GALAXIES
ZWG 241.047	10 54 00.	+ 50 24		15.7	GALAXY
7ZW 353	10 54 00.	+ 58 21			COMPACT GALAXY
TON-N 1305	10 54 02.	+ 22 57		17.	BLUE STAR
PATH 1.364	10 54 02.	- 00 44	14		NEBULA
PATH 1.365	10 54 04.	- 00 48	16		NEBULA
HN 1309	10 54 05.5	+ 00 20 31.	12		NEBULA
ZWG 010.043	10 54 06.	+ 00 21		15.7	GALAXY
ZWG 038.057	10 54 06.	+ 03 59		15.7	GALAXY
KARA.73B 0451	10 54 06.	+ 03 59	48	15.7	ISOLATED GALAXY S

OBJECT NAME	RIGHT ASCEN.	DECLINATION	DIAM.	MAGN.	TYPE OF OBJECT
ZWG 038.058	10 54 06.	+ 05 55		15.2	GALAXY
ZWG 038.059	10 54 06.	+ 07 07		15.2	GALAXY
ZWG 066.067	10 54 06.	+ 10 02		14.2	GALAXY
RNGC 3467	10 54 06.	+ 10 02		14.0	GALAXY
UGC 06045	10 54 06.	+ 10 02	54	14.2	GALAXY S0
MCG+02-28-030	10 54 06.	+ 10 02	18	14.2	GALAXY
ZC 1054.1+3059	10 54 06.	+ 30 59	870		CLUSTER OF GALAXIES
ZC 1054.1+3753	10 54 06.	+ 37 53	810		CLUSTER OF GALAXIES
ZC 1054.1+4900	10 54 06.	+ 49 00	1140		CLUSTER OF GALAXIES
MCG-03-28-025	10 54 06.	- 18 33	30	16.	GALAXY
REIZ 0746	10 54 07.	+ 10 02	18	13.7	GALAXY
GO 088	10 54 08.	- 59 30			STAR CHAIN
ZWG 038.060	10 54 12.	+ 07 10		14.9	GALAXY
UGC 06046	10 54 12.	+ 07 10	60	14.9	GALAXY SBc
MCG+01-28-023	10 54 12.	+ 07 10	54	14.9	GALAXY
ZC 1054.2+2505	10 54 12.	+ 25 05	810		CLUSTER OF GALAXIES
ZC 1054.2+3517	10 54 12.	+ 35 17	1550		CLUSTER OF GALAXIES
7ZW 354	10 54 12.	+ 58 02			COMPACT GALAXY
MCG-02-28-023	10 54 12.	- 12 37	54	15.	GALAXY
OCL 0831	10 54 12.	- 58 57	840	9.4	OPEN STAR CLUSTER
VHA 107	10 54 12.	- 58 57	360		OPEN STAR CLUSTER
PK289-00.1	10 54 12.	- 60 08 12.	39		PLANETARY NEBULA
MRSL 290-03/1	10 54 12.	- 62 45	480		HII REGION
GO 051	10 54 14.	- 59 11	85	17.4	STAR CHAIN
VV 149	10 54 15.	+ 07 12	60	15.	INTERACTING GALAXY
REIZ 0747	10 54 16.	+ 18 29	48	14.6	GALAXY
LB 01915	10 54 16.	+ 60 26 24.		14.4	FAINT BLUE STAR
REIZ 0749	10 54 18.	+ 07 11	54	14.8	GALAXY
ZWG 095.074	10 54 18.	+ 18 29		15.3	GALAXY
MCG+03-28-039	10 54 18.	+ 18 31	66	15.3	GALAXY
REIZ 0748	10 54 18.	+ 24 07	18	15.9	GALAXY
ZWG 184.038	10 54 18.	+ 37 30		15.7	GALAXY
UGC 06047	10 54 18.	+ 37 30	72	15.7	GALAXY SB
MCG+06-24-037	10 54 21.	+ 37 31	60	15.	GALAXY
8ZW 1054+09.4	10 54 24.	+ 09 26		19.2	COMPACT GALAXY
ZWG 095.075	10 54 24.	+ 14 40		15.6	GALAXY
ZC 1054.4-0010	10 54 24.	- 00 10	270		CLUSTER OF GALAXIES
ZC 1054.4-0044	10 54 24.	- 00 44	340		CLUSTER OF GALAXIES
ZC 1054.4-0157	10 54 24.	- 01 57	740		CLUSTER OF GALAXIES
VHE 45B	10 54 24.	- 62 46			REFLECTION NEBULA
VHE 45A	10 54 24.	- 62 46			REFLECTION NEBULA
RNGC 3469	10 54 28.	- 14 02		14.0	GALAXY
MIL 17	10 54 28.	- 59 49 36.	180		SUPERNOVA REMNANT
ZWG 095.076	10 54 30.	+ 15 11		15.5	GALAXY
ZWG 155.040	10 54 30.	+ 28 49		15.3	GALAXY
MRK 633	10 54 30.	+ 37 50	10	17.	GALAXY WITH UV CONTINUUM
MCG+07-23-004	10 54 30.	+ 44 17	48	16.	GALAXY
MCG-02-28-024	10 54 30.	- 14 02	84	14.	GALAXY
MCG-03-28-026	10 54 30.	- 19 56	24	15.5	GALAXY
MCG-05-26-015	10 54 30.	- 32 55	48	15.	GALAXY
ARC 1134	10 54 33.	- 01 52		17.6	RICH CLUSTER OF GALAXIES
RNGC 3468	10 54 35.	+ 41 13		14.0	GALAXY
ZWG 038.061	10 54 36.	+ 06 28		15.7	GALAXY
ZWG 066.068	10 54 36.	+ 08 43		15.1	GALAXY
8ZW 1054+08.9	10 54 36.	+ 08 52		20.1	COMPACT GALAXY
ZC 1054.6+2017	10 54 36.	+ 20 17	7530		CLUSTER OF GALAXIES
ZC 1054.6+3029	10 54 36.	+ 30 29	740		CLUSTER OF GALAXIES
ZWG 213.010	10 54 36.	+ 41 13		14.2	GALAXY
UGC 06048	10 54 36.	+ 41 13	108	14.2	GALAXY S0
MCG+12-11-004	10 54 36.	+ 72 35 30.	51	16.	GALAXY
REIZ 0750	10 54 38.	+ 28 48	30	14.5	GALAXY
8ZW 1054+09.9	10 54 42.	+ 09 52		19.6	COMPACT GALAXY
TON-N 0557	10 54 42.	+ 22 40		14.6	BLUE STAR
MCG+08-20-055	10 54 42.	+ 46 59	60	16.	GALAXY
7ZW 355	10 54 42.	+ 59 31			COMPACT GALAXY
TON-N 1306	10 54 43.	+ 35 18		15.8	BLUE STAR
ZC 1054.8+0911	10 54 48.	+ 09 11	810		CLUSTER OF GALAXIES
ZWG 066.069	10 54 48.	+ 09 49		15.5	GALAXY
ZC 1054.8+3419	10 54 48.	+ 34 19	1810		CLUSTER OF GALAXIES
MCG+07-23-006	10 54 48.	+ 41 12	24	14.5	GALAXY
MCG+07-23-005	10 54 48.	+ 44 20	27	16.	GALAXY
7ZW 356	10 54 48.	+ 67 16			COMPACT GALAXY
TON-N 1307	10 54 50.	+ 20 18		16.7	BLUE STAR
HN 1310	10 54 50.9	- 01 06 00.	12		NEBULA
ARC 1136	10 54 52.	+ 08 55		17.6	RICH CLUSTER OF GALAXIES
RNGC 3472	10 54 52.	- 19 22			NON-EXISTENT OBJECT
ZWG 038.062	10 54 54.	+ 05 58		15.2	GALAXY
UGC 06049	10 54 54.	+ 05 58	78	15.2	GALAXY Sa-b
ZWG 066.070	10 54 54.	+ 08 34		14.7	GALAXY
MCG+02-28-031	10 54 54.	+ 08 34	48	14.7	GALAXY
ZWG 184.039	10 54 54.	+ 37 55		15.2	GALAXY
ZC 1054.9+4104	10 54 54.	+ 41 04	2150		CLUSTER OF GALAXIES
ZC 1054.9+6120	10 54 54.	+ 61 20	1410		CLUSTER OF GALAXIES
MCG+12-11-005	10 54 54.	+ 73 20	33	16.	GALAXY
LB 01916	10 54 55.	+ 62 02 06.		15.7	FAINT BLUE STAR
ARC 1137	10 54 59.	+ 09 54		17.2	RICH CLUSTER OF GALAXIES
HMS 1055+5702	10 55	+ 57 02			URSA MAJOR GLXY CLSTR 1
ZWG 066.071	10 55 00.	+ 08 37		15.2	GALAXY
ZC 1055.0+1725	10 55 00.	+ 17 25	3830		CLUSTER OF GALAXIES
ZC 1055.0+3259	10 55 00.	+ 32 59	3560		CLUSTER OF GALAXIES
ZWG 184.040	10 55 00.	+ 36 31		15.5	GALAXY
MCG+06-24-038	10 55 00.	+ 36 31 30.	48	15.	GALAXY
MCG+06-24-039	10 55 00.	+ 37 56	18	15.	GALAXY
OCL 0835	10 55 00.	- 60 46	660	7.9	OPEN STAR CLUSTER
ZWG 038.063	10 55 06.	+ 06 20		15.6	GALAXY
ZWG 095.077	10 55 06.	+ 15 34		15.4	GALAXY
MCG+03-28-040	10 55 06.	+ 15 34	42	15.4	GALAXY
ZWG 213.011	10 55 06.	+ 41 06		15.6	GALAXY
ZC 1055.1-0245	10 55 06.	- 02 45	670		CLUSTER OF GALAXIES
MCG+06-24-040	10 55 09.	+ 37 47 30.	42	16.	GALAXY
MCG+06-24-041	10 55 09.	+ 37 49	12	16.	GALAXY
HN 1311	10 55 11.9	+ 01 30 47.	18		NEBULA
ZWG 038.064	10 55 12.	+ 08 18		15.1	GALAXY
ZC 1055.2+0855	10 55 12.	+ 08 55	940		CLUSTER OF GALAXIES
8ZW 1055+09.9	10 55 12.	+ 09 56		18.5	COMPACT GALAXY
ZC 1055.2+1530	10 55 12.	+ 15 30	940		CLUSTER OF GALAXIES
UGC 06050	10 55 12.	+ 40 28	66	16.0	GALAXY Sc
ZC 1055.2+4340	10 55 12.	+ 43 40	1340		CLUSTER OF GALAXIES
ZWG 267.028	10 55 12.	+ 50 29		15.6	GALAXY
ZWG 241.048	10 55 12.	+ 50 29		15.6	GALAXY
ZC 1055.2+5802	10 55 12.	+ 58 02	2550		CLUSTER OF GALAXIES
UGC 06051	10 55 12.	+ 76 43	66	16.0	GALAXY Sc
MCG-04-26-015	10 55 12.	- 25 09	24	14.5	GALAXY
ARC 1133	10 55 13.	+ 50 08		17.4	RICH CLUSTER OF GALAXIES
TON-N 1308	10 55 14.	+ 20 05		17.	BLUE STAR
MCG-04-26-016	10 55 15.	- 25 13 30.	48	14.5	GALAXY
HN 1312	10 55 17.1	+ 01 56 23.	18		NEBULA
HN 1313	10 55 17.9	- 01 59 19.	12		NEBULA
ZWG 010.045	10 55 18.	+ 01 58		15.4	GALAXY
ZWG 038.065	10 55 18.	+ 04 49		15.7	GALAXY
ZWG 095.078	10 55 18.	+ 19 46		15.6	GALAXY
MCG+07-23-007	10 55 18.	+ 40 27	63	15.	GALAXY
BC 5C2.56	10 55 18.	+ 49 55 36.			QUASI-STELLAR OBJECT
ZC 1055.3+5010	10 55 18.	+ 50 10	540		CLUSTER OF GALAXIES
ZWG 267.029	10 55 18.	+ 55 53		15.7	GALAXY
ZWG 334.006	10 55 18.	+ 72 54		14.6	GALAXY
ZWG 333.064	10 55 18.	+ 72 54		14.6	GALAXY
KARA.72 258A	10 55 18.	+ 72 54	18	14.6	PART OF DOUBLE GALAXY
ZWG 334.007	10 55 18.	+ 73 18		15.6	GALAXY
ZWG 333.065	10 55 18.	+ 73 18		15.6	GALAXY
ZWG 010.044	10 55 18.	- 01 59		15.7	GALAXY
MCG-01-28-008	10 55 18.	- 03 54	42	15.5	GALAXY
MCG-03-28-027	10 55 18.	- 19 46	54	14.	GALAXY
SHB 169	10 55 18.3	+ 49 55 36.		19.5	QUASI-STELLAR OBJECT
ARC 1135	10 55 19.	+ 41 20		16.9	RICH CLUSTER OF GALAXIES
ARC 1132	10 55 19.	+ 57 04		17.0	RICH CLUSTER OF GALAXIES
LB 01917	10 55 21.	+ 50 58 36.		16.8	FAINT BLUE STAR
IC 0657	10 55 21.	- 04 38 12.			NONSTELLAR OBJECT
RNGC 3473	10 55 22.	+ 17 23		15.0	GALAXY
REIZ 0752	10 55 23.	+ 19 46	36	14.9	GALAXY
ZC 1055.4+0142	10 55 24.	+ 01 42	4370		CLUSTER OF GALAXIES
ZWG 066.072	10 55 24.	+ 12 16		15.6	GALAXY
ZWG 095.079	10 55 24.	+ 17 23		14.8	GALAXY
UGC 06052	10 55 24.	+ 17 23	72	14.8	GALAXY SBb
MCG+03-28-041	10 55 24.	+ 17 25	72	14.8	GALAXY
ZWG 095.080	10 55 24.	+ 20 04		15.7	GALAXY
MRK 634	10 55 24.	+ 20 45	13	16.	GALAXY WITH UV CONTINUUM
ZC 1055.4+2308	10 55 24.	+ 23 08	1680		CLUSTER OF GALAXIES
MCG+09-18-059	10 55 24.	+ 55 52 30.	42	15.	GALAXY
MCG+10-16-033	10 55 24.	+ 57 02	6	18.	GALAXY
HMS 1.15	10 55 24.	+ 57 03			E1 GALAXY
MCG+10-16-034	10 55 24.	+ 57 03	9	17.6	GALAXY
MCG+10-16-035	10 55 24.	+ 57 15	24	16.	GALAXY
ZWG 334.008	10 55 24.	+ 72 53		15.7	GALAXY
ZWG 333.066	10 55 24.	+ 72 53		15.7	GALAXY
KARA.72 258B	10 55 24.	+ 72 53	18	15.7	PART OF DOUBLE GALAXY
MCG-01-28-009	10 55 24.	- 04 27	48	15.	GALAXY
HN 1314	10 55 26.7	+ 01 53 53.	12		NEBULA
HN 1315	10 55 27.5	- 02 49 19.			NEBULA
RNGC 3474	10 55 28.	+ 17 21		15.0	GALAXY
REIZ 0753	10 55 28.	+ 17 22	24	15.0	GALAXY
ARC 1139	10 55 29.	+ 01 47		15.0	RICH CLUSTER OF GALAXIES
ZWG 038.066	10 55 30.	+ 06 18		15.1	GALAXY
UGC 06053	10 55 30.	+ 06 18	66	15.1	GALAXY Sc
MCG+01-28-024	10 55 30.	+ 06 18	60	15.1	GALAXY
ZWG 066.073	10 55 30.	+ 09 33		15.0	GALAXY
RNGC 3476	10 55 30.	+ 09 33		15.0	GALAXY
MCG+02-28-032	10 55 30.	+ 09 33	12	15.0	GALAXY
ZWG 095.081	10 55 30.	+ 17 21		14.9	GALAXY
MCG+03-28-042	10 55 30.	+ 17 22 30.	42	14.9	GALAXY
ZWG 095.082	10 55 30.	+ 20 21		15.6	GALAXY
UGC 06054	10 55 30.	+ 20 21	72	15.6	GALAXY Sc
ZC 1055.5+2606	10 55 30.	+ 26 06	3290		CLUSTER OF GALAXIES
ZWG 213.012	10 55 30.	+ 41 55		15.7	GALAXY
UGC 06055	10 55 30.	+ 49 03	72	17.	GALAXY IRR
7ZW 357	10 55 30.	+ 66 17			COMPACT GALAXY
MCG-01-28-011	10 55 30.	- 04 18	78	15.	GALAXY
MCG-01-28-010	10 55 30.	- 05 48	18	15.5	GALAXY
MCG-03-28-028	10 55 30.	- 18 12 30.	36	15.	GALAXY
IC 0659	10 55 32.	- 05 59 36.			NONSTELLAR OBJECT
GO 049	10 55 33.	- 59 28	50	16.8	STAR CHAIN
REIZ 0754	10 55 34.	+ 17 22	60	14.3	GALAXY
HN 1316	10 55 34.9	- 00 29 55.	36		NEBULA
ZWG 038.067	10 55 36.	+ 02 54		15.4	GALAXY
ZWG 038.068	10 55 36.	+ 07 50		15.2	GALAXY
ZWG 066.074	10 55 36.	+ 09 29		15.7	GALAXY
RNGC 3477	10 55 36.	+ 09 29		15.5	GALAXY
TON-N 0047	10 55 36.	+ 22 54		15.6	BLUE STAR
ZWG 213.013	10 55 36.	+ 39 30		15.7	GALAXY
ZC 1055.6+4432	10 55 36.	+ 44 32	3760		CLUSTER OF GALAXIES
MCG+08-20-056	10 55 36.	+ 46 31	42	15.	GALAXY
MCG+08-20-057	10 55 36.	+ 49 02	36	17.	GALAXY
MCG+09-18-060	10 55 36.	+ 53 22	30	16.	GALAXY
ZC 1055.6+5655	10 55 36.	+ 56 55	2890		CLUSTER OF GALAXIES
ZC 1055.6+6053	10 55 36.	+ 60 53	1140		CLUSTER OF GALAXIES
ZC 1055.6+6757	10 55 36.	+ 67 57	610		CLUSTER OF GALAXIES
ZWG 351.050	10 55 36.	+ 75 28		14.6	GALAXY
UGC 06056	10 55 36.	+ 75 28	90	14.6	GALAXY Sb
ZWG 010.046	10 55 36.	- 00 30		15.2	GALAXY
MCG-06-24-015	10 55 36.	- 39 12	120	15.5	GALAXY
SHB 170	10 55 36.9	+ 20 08 18.		17.1	QUASI-STELLAR OBJECT
BC PKS1055+20	10 55 37.	+ 20 08 14.		17.07	QUASI-STELLAR OBJECT
HN 1317	10 55 37.3	+ 01 52 53.	18		NEBULA
RNGC 3465	10 55 38.	+ 75 28		14.5	GALAXY
FATH 1.366	10 55 38.	- 00 31	33		NEBULA
IC 0658	10 55 39.	+ 08 32 18.			NONSTELLAR OBJECT
RNGC 3475	10 55 39.	+ 24 30		14.5	GALAXY
RNGC 3470	10 55 40.	+ 59 47		14.5	GALAXY
FATH 1.367	10 55 40.	- 00 45	14		NEBULA
FATH 1.368	10 55 41.	- 00 48	14		NEBULA
HN 1318	10 55 41.5	+ 01 53 29.	18		NEBULA
ZWG 010.047	10 55 42.	+ 01 54		14.9	GALAXY
UGC 06057	10 55 42.	+ 01 54	66	14.9	GALAXY TRP SYS
MCG+00-28-024	10 55 42.	+ 01 54	12	14.9	GALAXY
ZWG 038.069	10 55 42.	+ 04 51		15.1	GALAXY
ZWG 038.070	10 55 42.	+ 05 03		15.7	GALAXY
ZWG 066.075	10 55 42.	+ 08 31		14.6	GALAXY
MCG+02-28-033	10 55 42.	+ 08 31	15	14.6	GALAXY
MCG+04-26-022	10 55 42.	+ 24 28	102	14.6	GALAXY
ZWG 125.017	10 55 42.	+ 24 30		14.6	GALAXY
REIZ 0755	10 55 42.	+ 24 30	18	14.8	GALAXY
UGC 06058	10 55 42.	+ 24 30	102	14.6	GALAXY Sa
ZC 1055.7+3317	10 55 42.	+ 33 17	810		CLUSTER OF GALAXIES
ZWG 241.049	10 55 42.	+ 46 05		15.7	GALAXY
ZWG 241.050	10 55 42.	+ 46 32		15.1	GALAXY
MCG+09-18-061	10 55 42.	+ 55 30	60	14.	GALAXY
ZWG 267.030	10 55 42.	+ 55 52		15.7	GALAXY
UGC 06059	10 55 42.	+ 55 52	102	15.7	GALAXY Sb/Sc
MCG+09-18-062	10 55 42.	+ 55 52	108	15.	GALAXY
MCG+10-16-036	10 55 42.	+ 57 02	6	19.1	GALAXY
MCG+10-16-037	10 55 42.	+ 59 44	18	16.	GALAXY
ZWG 291.016	10 55 42.	+ 59 47		14.3	GALAXY
REIZ 0751	10 55 42.	+ 59 47	36	14.3	GALAXY
UGC 06060	10 55 42.	+ 59 47	102	14.3	GALAXY Sa
KARA.72 259A	10 55 42.	+ 59 47	78	14.3	PART OF DOUBLE GALAXY
ZWG 334.009	10 55 42.	+ 72 06		15.6	GALAXY

OBJECT NAME	RIGHT ASCEN.	DECLINATION	DIAM.	MAGN.	TYPE OF OBJECT
ZWG 333.067	10 55 42.	+ 72 06		15.6	GALAXY
KARA.73B 0452	10 55 42.	+ 72 06	18	15.6	ISOLATED GALAXY E
MCG+07-23-008	10 55 45.	+ 39 29	36	15.	GALAXY
ZWG 038.071	10 55 48.	+ 06 59		15.0	GALAXY
MCG+01-28-025	10 55 48.	+ 06 59	12	15.0	GALAXY
ZC 1055.8+1850	10 55 48.	+ 18 50	810		CLUSTER OF GALAXIES
ZWG 125.018	10 55 48.	+ 24 28		15.7	GALAXY
MRK 419	10 55 48.	+ 24 28	8	16.5	GALAXY WITH UV CONTINUUM
ZWG 125.019	10 55 48.	+ 24 38		15.1	GALAXY
MCG+04-26-023	10 55 48.	+ 24 38	42	15.1	GALAXY
UGC 06061	10 55 48.	+ 55 32	72	16.0	GALAXY SBc
ZWG 267.031	10 55 48.	+ 55 54		15.7	GALAXY
MCG+09-18-063	10 55 48.	+ 55 54	30	15.	GALAXY
7ZW 358	10 55 48.	+ 57 46			COMPACT GALAXY
ZWG 291.017	10 55 48.	+ 59 45		15.6	GALAXY
KARA.72 259B	10 55 48.	+ 59 45	30	15.6	PART OF DOUBLE GALAXY
MCG+10-16-038	10 55 48.	+ 59 46	78	14.5	GALAXY
AR 27	10 55 49.14	+ 75 27 36.8			NEBULA
REIZ 0756	10 55 51.	+ 32 00	48	15.4	GALAXY
ARC 1138	10 55 52.	+ 33 16		17.5	RICH CLUSTER OF GALAXIES
IC 0660	10 55 53.	+ 01 40 05.			NONSTELLAR OBJECT
GO 064	10 55 53.	- 58 08	96	17.1	STAR CHAIN
ZWG 010.048	10 55 54.	+ 01 39		15.7	GALAXY
ZWG 038.072	10 55 54.	+ 04 54		15.3	GALAXY
TON-N 0558	10 55 54.	+ 23 07		16.7	BLUE STAR
ZWG 213.014	10 55 54.	+ 43 35		15.5	GALAXY
MCG+07-23-009	10 55 54.	+ 43 35	30	15.	GALAXY
MRK 158	10 55 54.	+ 61 48	45	14.	GALAXY WITH UV CONTINUUM
MCG+10-16-039	10 55 54.	+ 61 48	96	14.	GALAXY
ZC 1055.9-0154	10 55 54.	- 01 54	740		CLUSTER OF GALAXIES
MCG-03-28-029	10 55 54.	- 18 56	48	15.	GALAXY
SHB 171	10 55 55.2	+ 01 50 08.		18.0	QUASI-STELLAR OBJECT
HN 1319	10 55 56.5	- 02 17 50.	18		NEBULA
ARC 1140	10 55 59.	+ 34 13		17.8	RICH CLUSTER OF GALAXIES
ZC 1056.0+0252	10 56 00.	+ 02 52	340		CLUSTER OF GALAXIES
ZWG 066.076	10 56 00.	+ 09 19		13.7	GALAXY
UGC 06062	10 56 00.	+ 09 19	90	13.7	GALAXY SB0
MCG+02-28-034	10 56 00.	+ 09 19	48	13.7	GALAXY
MCG+04-26-024	10 56 00.	+ 25 24	78	15.4	GALAXY
ZWG 125.020	10 56 00.	+ 25 25		15.4	GALAXY
UGC 06063	10 56 00.	+ 25 25	90	15.4	GALAXY S
KARA.73B 0453	10 56 00.	+ 25 25	96	15.4	ISOLATED GALAXY S
ZC 1056.0+4305	10 56 00.	+ 43 05	1210		CLUSTER OF GALAXIES
ZWG 291.018	10 56 00.	+ 61 48		13.0	GALAXY
UGC 06064	10 56 00.	+ 61 48	120	13.0	GALAXY Sa
MCG+10-16-040	10 56 00.	+ 61 49	30	16.	GALAXY
ZC 1056.0+6717	10 56 00.	+ 67 17	810		CLUSTER OF GALAXIES
MCG+13-08-048	10 56 00.	+ 75 28	66	14.	GALAXY
MCG+13-08-049	10 56 00.	+ 76 43	51	16.	GALAXY
ZWG 010.049	10 56 00.	- 02 17		15.7	GALAXY
MRSL 289+00/2	10 56 00.	- 59 26	900		HII REGION
RNGC 3471	10 56 02.	+ 61 48		13.0	GALAXY
GO 050	10 56 03.	- 59 20	45	16.9	STAR CHAIN
HN 1320	10 56 03.2	- 02 20 14.	18		NEBULA
GO 090	10 56 04.	- 59 04	68	17.2	STAR CHAIN
GO 066	10 56 05.	- 58 31	243	19.6	STAR CHAIN
ZC 1056.1+0644	10 56 06.	+ 06 44	3760		CLUSTER OF GALAXIES
TON-N 0559	10 56 06.	+ 33 05		15.2	BLUE STAR
MRK 159	10 56 06.	+ 72 56	10	14.5	GALAXY WITH UV CONTINUUM
REIZ 0757	10 56 09.	- 00 43	60	14.9	GALAXY
ZWG 038.073	10 56 12.	+ 05 47		15.6	GALAXY
ZC 1056.2+2219	10 56 12.	+ 22 19	2420		CLUSTER OF GALAXIES
ZC 1056.2+4228	10 56 12.	+ 42 28	1340		CLUSTER OF GALAXIES
ZC 1056.2+5439	10 56 12.	+ 54 39	1080		CLUSTER OF GALAXIES
MCG+12-11-006	10 56 12.	+ 72 25	54	17.	GALAXY
MCG-02-28-025	10 56 12.	- 09 34	30	15.	GALAXY
LB 01918	10 56 14.	+ 48 57 48.		17.1	FAINT BLUE STAR
RNGC 3482	10 56 17.	- 46 18			UNVERIFIED SOUTHERN OBJECT
IC 0661	10 56 18.	+ 01 55 04.			NONSTELLAR OBJECT
ZWG 010.050	10 56 18.	+ 01 56		15.7	GALAXY
RNGC 3480	10 56 18.	+ 09 37			NON-EXISTENT OBJECT
ZWG 066.077	10 56 18.	+ 13 26		15.6	GALAXY
ZC 1056.3+2759	10 56 18.	+ 27 59	1950		CLUSTER OF GALAXIES
ZC 1056.3+3529	10 56 18.	+ 35 29	340		CLUSTER OF GALAXIES
SN 1953G	10 56 18.	+ 50 32		16.5	SUPERNOVA
ZWG 351.051	10 56 18.	+ 77 12		15.6	GALAXY
UGC 06065	10 56 18.	+ 77 12	78	15.6	GALAXY Sb
MCG-02-28-026	10 56 18.	- 15 16	108	14.5	GALAXY
OCL 0832	10 56 18.	- 58 47	90		OPEN STAR CLUSTER
HN 1321	10 56 19.5	- 02 31 14.	18		NEBULA
TON-N 1309	10 56 20.	+ 21 09		16.9	BLUE STAR
LB 01919	10 56 20.	+ 51 41 30.		16.9	FAINT BLUE STAR
LB 01920	10 56 22.	+ 52 00 18.		16.1	FAINT BLUE STAR
HN 1322	10 56 22.4	+ 01 41 52.	24		NEBULA
ZWG 038.074	10 56 24.	+ 06 47		15.0	GALAXY
UGC 06066	10 56 24.	+ 06 47	96	15.0	GALAXY Sa?
MCG+01-28-026	10 56 24.	+ 06 47	72	15.0	GALAXY
8ZW 1056+09.3	10 56 24.	+ 09 16		17.8	COMPACT GALAXY
ZC 1056.4+2235	10 56 24.	+ 22 35	1080		CLUSTER OF GALAXIES
ZC 1056.4+7125	10 56 24.	+ 71 25	940		CLUSTER OF GALAXIES
MCG+13-08-050	10 56 24.	+ 77 13		18.	GALAXY
TON-N 1310	10 56 26.	+ 22 05		17.	BLUE STAR
LB 01921	10 56 26.	+ 55 18 12.		17.0	FAINT BLUE STAR
GO 098	10 56 26.	- 60 10	35	17.8	STAR CHAIN
MCG+08-20-058	10 56 27.	+ 46 21	36	17.	GALAXY
RNGC 3479	10 56 28.	- 14 42		13.0	GALAXY
ZWG 010.051	10 56 30.	+ 01 43		15.1	GALAXY
ZWG 038.075	10 56 30.	+ 05 16		15.5	GALAXY
ZC 1056.5+1240	10 56 30.	+ 12 40	4500		CLUSTER OF GALAXIES
MCG+08-20-059	10 56 30.	+ 46 23	162	13.2	GALAXY
ZC 1056.5+5141	10 56 30.	+ 51 41	870		CLUSTER OF GALAXIES
ZC 1056.5+6218	10 56 30.	+ 62 18	1210		CLUSTER OF GALAXIES
ZC 1056.5+6437	10 56 30.	+ 64 37	1210		CLUSTER OF GALAXIES
MCG+13-08-051	10 56 30.	+ 77 12	84	15.	GALAXY
MCG-02-28-027	10 56 30.	- 14 42	90	13.5	GALAXY
MRSL 289-01/1	10 56 30.	- 60 59	600		HII REGION
HN 1323	10 56 30.7	+ 01 26 46.	24		NEBULA
HN 1324	10 56 31.4	+ 02 09 58.	24		NEBULA
MCG+08-20-060	10 56 33.	+ 50 15	15	16.	GALAXY
RNGC 3478	10 56 34.	+ 46 23		13.0	GALAXY
HN 1325	10 56 34.6	+ 01 38 46.	12		NEBULA
AR 28	10 56 35.21	+ 77 11 51.2			NEBULA
HN 1326	10 56 35.4	+ 02 05 39.	24		NEBULA
ZWG 010.052	10 56 36.	+ 01 28		15.1	GALAXY
ZWG 010.053	10 56 36.	+ 02 10		15.2	GALAXY
UGC 06067	10 56 36.	+ 02 10	84	15.2	GALAXY TRP SYS
MCG+00-28-025	10 56 36.	+ 02 10	24	15.2	GALAXY
ZWG 038.076	10 56 36.	+ 05 33		15.4	GALAXY
UGC 06068	10 56 36.	+ 05 33	90	15.4	GALAXY S0?
TON-N 0048	10 56 36.	+ 29 43		13.3	BLUE STAR
ZWG 241.051	10 56 36.	+ 46 23		13.7	GALAXY
UGC 06069	10 56 36.	+ 46 23	156	13.7	GALAXY SBb
ZC 1056.6+5010	10 56 36.	+ 50 10	1010		CLUSTER OF GALAXIES
ZWG 241.052	10 56 36.	+ 50 15		15.5	GALAXY
MCG-02-28-028	10 56 36.	- 12 45 30.	48	15.	GALAXY
RNGC 3483	10 56 36.	- 28 12		13.0	GALAXY
MCG-05-26-016	10 56 36.	- 28 12	72	13.5	GALAXY
TON-N 1311	10 56 38.	+ 22 27		17.0	BLUE STAR
REIZ 0759	10 56 40.	- 00 34	36	15.3	GALAXY
HN 1327	10 56 41.6	- 00 25 03.	12		NEBULA
ZWG 010.054	10 56 42.	+ 02 05		15.6	GALAXY
ZC 1056.7+1721	10 56 42.	+ 17 21	740		CLUSTER OF GALAXIES
MCG+09-18-064	10 56 42.	+ 54 03	30	16.	GALAXY
MCG-01-28-013	10 56 42.	- 09 23	48	14.	GALAXY
MCG-01-28-012	10 56 42.	- 09 25	60	15.	GALAXY
MCG-02-28-029	10 56 42.	- 09 31	60	15.	GALAXY
MCG-02-28-030	10 56 42.	- 09 32 30.	48	14.	GALAXY
MCG-03-28-030	10 56 42.	- 18 02	54	15.	GALAXY
HN 1328	10 56 42.8	+ 01 25 09.	18		NEBULA
TON-N 1312	10 56 43.	+ 34 33		15.	BLUE STAR
REIZ 0758	10 56 44.	+ 56 16	78	15.2	GALAXY
LB 01922	10 56 46.	+ 60 35 36.		15.1	FAINT BLUE STAR
IC 0662	10 56 47.	+ 01 51 57.			NONSTELLAR OBJECT
ZWG 010.055	10 56 48.	+ 01 27		15.4	GALAXY
ZWG 010.056	10 56 48.	+ 01 52		15.6	GALAXY
ZC 1056.8+0557	10 56 48.	+ 05 57	470		CLUSTER OF GALAXIES
ZWG 038.077	10 56 48.	+ 07 15		15.5	GALAXY
ZWG 095.083	10 56 48.	+ 19 06		15.7	GALAXY
ZC 1056.8+4353	10 56 48.	+ 43 53	340		CLUSTER OF GALAXIES
ZC 1056.8+4402	10 56 48.	+ 44 02	670		CLUSTER OF GALAXIES
ZC 1056.8+4753	10 56 48.	+ 47 53	470		CLUSTER OF GALAXIES
MCG-01-28-015	10 56 48.	- 06 32	12	18.	GALAXY
MCG-01-28-014	10 56 48.	- 06 32	30	18.	GALAXY
ZWG 038.078	10 56 54.	+ 07 57		15.6	GALAXY
ZWG 066.078	10 56 54.	+ 09 12		15.5	GALAXY
ZC 1056.9+0922	10 56 54.	+ 09 22	6720		CLUSTER OF GALAXIES
ARC 1141	10 56 54.	+ 12 31		17.2	RICH CLUSTER OF GALAXIES
ZWG 184.041	10 56 54.	+ 33 39		13.3	GALAXY
UGC 06070	10 56 54.	+ 33 39	36	13.3	GALAXY PECULR
ZWG 241.053	10 56 54.	+ 50 17		14.8	GALAXY
KARA.72 260A	10 56 54.	+ 50 17	78	14.8	PART OF DOUBLE GALAXY
MCG+08-20-061	10 56 54.	+ 50 17 30.	42	14.	GALAXY
MCG+09-18-065	10 56 54.	+ 51 09	90	14.	GALAXY
MCG-01-28-016	10 56 54.	- 07 05	42	14.	GALAXY
RNGC 3481	10 56 56.	- 07 15		14.0	GALAXY
REIZ 0760	10 56 57.	+ 32 22	30	15.2	GALAXY
LB 01923	10 56 58.	+ 54 36 18.		16.6	FAINT BLUE STAR
ZC 1057.0+2432	10 57 00.	+ 24 32	340		CLUSTER OF GALAXIES
ZWG 213.015	10 57 00.	+ 43 25		15.7	GALAXY
ZWG 213.016	10 57 00.	+ 43 38		15.7	GALAXY
MCG+08-20-062	10 57 00.	+ 46 20	30	16.	GALAXY
ZWG 241.054	10 57 00.	+ 50 19		14.4	GALAXY
UGC 06071	10 57 00.	+ 50 19	84	14.4	GALAXY E
KARA.72 260B	10 57 00.	+ 50 19	78	14.4	PART OF DOUBLE GALAXY
MCG+08-20-063	10 57 00.	+ 50 20	36	14.	GALAXY
MCG+09-18-067	10 57 00.	+ 51 06	60	16.	GALAXY
MCG+09-18-066	10 57 00.	+ 54 40	72	16.	GALAXY
MCG+10-16-041	10 57 00.	+ 57 41	36	15.	GALAXY
ZC 1057.0+6551	10 57 00.	+ 65 51	870		CLUSTER OF GALAXIES
TON-N 1313	10 57 02.	+ 21 58		16.5	BLUE STAR
MCG+03-28-043	10 57 03.	+ 17 56	72	15.3	GALAXY
ZWG 066.079	10 57 06.	+ 10 20		14.9	GALAXY
UGC 06072	10 57 06.	+ 10 20	60	14.9	GALAXY Sa-b
MCG+02-28-035	10 57 06.	+ 10 20	60	14.9	GALAXY
ZC 1057.1+1752	10 57 06.	+ 17 52	1610		CLUSTER OF GALAXIES
ZWG 095.084	10 57 06.	+ 17 55		15.3	GALAXY
UGC 06073	10 57 06.	+ 17 55	78	15.3	GALAXY DBL SYS
VV 267B	10 57 06.	+ 17 56 30.	42	16.	INTERACTING GALAXY
VV 267A	10 57 06.	+ 17 56 30.	54	16.	INTERACTING GALAXY
VV 267	10 57 06.	+ 17 56 30.	90		INTERACTING GALAXY
ZC 1057.1+3411	10 57 06.	+ 34 11	1080		CLUSTER OF GALAXIES
MCG+07-23-010	10 57 06.	+ 43 25	33	15.	GALAXY
ZWG 241.055	10 57 06.	+ 46 20		15.1	GALAXY
ZWG 267.032	10 57 06.	+ 51 12		14.3	GALAXY
UGC 06074	10 57 06.	+ 51 12	54	14.3	GALAXY PECULR
LB 01924	10 57 06.	+ 53 51 06.		16.8	FAINT BLUE STAR
MCG+09-18-068	10 57 06.	+ 54 19	36	16.	GALAXY
ZWG 291.019	10 57 06.	+ 57 43		15.5	GALAXY
ARP 198	10 57 08.	+ 17 55			PECULIAR GALAXY
TON-N 1314	10 57 08.	+ 20 30		15.7	BLUE STAR
ZWG 038.079	10 57 12.	+ 04 15		15.2	GALAXY
TON-N 0049	10 57 12.	+ 27 56		15.6	BLUE STAR
ZC 1057.2+3804	10 57 12.	+ 38 04	8870		CLUSTER OF GALAXIES
MCG+08-20-065	10 57 12.	+ 46 00	84	14.	GALAXY
MCG+08-20-064	10 57 12.	+ 46 11	24	15.	GALAXY
MCG-02-28-031	10 57 15.	- 15 16	108	14.5	GALAXY
HN 1329	10 57 15.8	- 02 23 21.			NEBULA
ZWG 066.080	10 57 18.	+ 09 38		15.0	GALAXY
RNGC 3490	10 57 18.	+ 09 38		15.0	GALAXY
MCG+02-28-036	10 57 18.	+ 09 38	15	15.0	GALAXY
ZC 1057.3+1240	10 57 18.	+ 12 40	1610		CLUSTER OF GALAXIES
MCG+03-28-044	10 57 18.	+ 15 07	132	12.8	GALAXY
ZWG 184.042	10 57 18.	+ 35 35		15.1	GALAXY
ZWG 241.056	10 57 18.	+ 46 00		14.4	GALAXY
UGC 06075	10 57 18.	+ 46 00	108	14.4	GALAXY Sb
ZWG 241.057	10 57 18.	+ 46 11		14.5	GALAXY
UGC 06076	10 57 18.	+ 46 11	54	14.5	GALAXY E?
MCG-01-28-017	10 57 18.	- 05 41	15	15.	GALAXY
RNGC 3485	10 57 23.	+ 15 07		13.0	GALAXY
ZWG 095.085	10 57 24.	+ 15 06		12.8	GALAXY
UGC 06077	10 57 24.	+ 15 06	156	12.8	GALAXY SBb
ZWG 095.086	10 57 24.	+ 17 42		14.8	GALAXY
MCG+03-28-045	10 57 24.	+ 17 42	54	14.8	GALAXY
MCG+06-24-042	10 57 24.	+ 35 35	48	15.	GALAXY
MCG-01-28-018	10 57 24.	- 05 40	30	17.	GALAXY
MCG-01-28-019	10 57 24.	- 05 42	24	17.	GALAXY
MCG-04-26-017	10 57 24.	- 25 13 30.	36	14.5	GALAXY
TON-N 1315	10 57 26.	+ 20 37		17.	BLUE STAR
HN 1331	10 57 26.9	+ 01 22 32.	18		NEBULA
REIZ 0761	10 57 27.	+ 15 06	84	13.0	GALAXY
HN 1330	10 57 27.1	+ 02 02 56.	18		NEBULA
ZWG 010.057	10 57 30.	+ 01 24		15.7	GALAXY
ZWG 010.058	10 57 30.	+ 02 03		15.6	GALAXY
MCG+00-28-026	10 57 30.	+ 02 03	12	15.6	GALAXY
ZC 1057.5+1001	10 57 30.	+ 10 01	670		CLUSTER OF GALAXIES
ZWG 066.081	10 57 30.	+ 10 39		15.5	GALAXY

OBJECT NAME	RIGHT ASCEN.	DECLINATION	DIAM.	MAGN.	TYPE OF OBJECT
ZWG 066.082	10 57 30.	+ 12 30		15.1	GALAXY
UGC 06078	10 57 30.	+ 12 30	78	15.1	GALAXY Sb-c
MCG+02-28-037	10 57 30.	+ 12 30	60	15.1	GALAXY
SN 1968C	10 57 30.	+ 26 59		17.8	SUPERNOVA
MCG+10-16-042	10 57 30.	+ 61 35	132	15.	GALAXY
REIZ 0763	10 57 32.	+ 12 30	30	15.1	GALAXY
ZC 1057.6+0655	10 57 36.	+ 06 55	1010		CLUSTER OF GALAXIES
ZWG 155.041	10 57 36.	+ 29 15		11.2	GALAXY
UGC 06079	10 57 36.	+ 29 15	432	11.2	GALAXY Sc
ZWG 291.020	10 57 36.	+ 61 36		15.6	GALAXY
UGC 06080	10 57 36.	+ 61 36	126	15.6	GALAXY Sc
MRK 160	10 57 36.	+ 70 21	10	16.5	GALAXY WITH UV CONTINUUM
MCG-01-28-020	10 57 36.	- 09 15	42	15.	GALAXY
RNGC 3489	10 57 41.	+ 14 10		11.5	GALAXY
ZWG 038.080	10 57 42.	+ 03 11		15.5	GALAXY
ZWG 038.081	10 57 42.	+ 06 29		15.3	GALAXY
ZWG 038.082	10 57 42.	+ 07 51		15.6	GALAXY
ZWG 066.083	10 57 42.	+ 10 19		14.8	GALAXY
UGC 06081	10 57 42.	+ 10 19	66	14.8	GALAXY DBL SYS
MCG+02-28-038	10 57 42.	+ 10 19	60	14.8	GALAXY
ZWG 066.084	10 57 42.	+ 14 10		10.9	GALAXY
UGC 06082	10 57 42.	+ 14 10	210	10.9	GALAXY S0
MCG+02-28-039	10 57 42.	+ 14 10	192	10.9	GALAXY
ZWG 095.087	10 57 42.	+ 16 57		15.2	GALAXY
UGC 06083	10 57 42.	+ 16 57	84	15.2	GALAXY Sb-c
MCG+03-28-046	10 57 42.	+ 16 58	69	15.2	GALAXY
ZC 1057.7+2418	10 57 42.	+ 24 18	1410		CLUSTER OF GALAXIES
MCG+05-26-032	10 57 42.	+ 29 14	420	11.2	GALAXY
ZWG 155.042	10 57 42.	+ 29 58		15.2	GALAXY
UGC 06084	10 57 42.	+ 29 58	60	15.2	GALAXY Sa-b
UGC 06085	10 57 42.	+ 35 55	60	16.5	GALAXY Sb-c
LB 01925	10 57 43.	+ 62 00 48.		16.0	FAINT BLUE STAR
REIZ 0766	10 57 44.	+ 14 10	120	10.8	GALAXY
RNGC 3486	10 57 44.	+ 29 15		11.0	GALAXY
REIZ 0764	10 57 44.	+ 29 15	420	12.3	GALAXY
REIZ 0767	10 57 45.	+ 16 57	54	14.5	GALAXY
RNGC 3496	10 57 47.	- 60 04		9.0	OPEN CLUSTER
UGC 06098	10 57 48.	+ 03 54	288	13.1	GALAXY Sc
ZWG 038.083	10 57 48.	+ 04 27		15.1	GALAXY
ZWG 038.084	10 57 48.	+ 06 40		15.4	GALAXY
ZWG 066.085	10 57 48.	+ 10 36		15.6	GALAXY
ZWG 095.088	10 57 48.	+ 15 07		15.5	GALAXY
MCG+05-26-033	10 57 48.	+ 29 58	30	15.2	GALAXY
ZWG 213.017	10 57 48.	+ 44 21		15.6	GALAXY
REIZ 0762	10 57 48.	+ 55 23	60	14.5	GALAXY
MCG+10-16-043	10 57 48.	+ 58 02	18	17.	GALAXY
ZC 1057.8+6623	10 57 48.	+ 66 23	1010		CLUSTER OF GALAXIES
ZC 1057.8+6700	10 57 48.	+ 67 00	1480		CLUSTER OF GALAXIES
MCG-02-28-032	10 57 48.	- 09 42 30.	84	14.	GALAXY
OCL 0836	10 57 48.	- 60 04	1440	9.3	OPEN STAR CLUSTER
VHA 108	10 57 48.	- 60 04	480		OPEN STAR CLUSTER
REIZ 0768	10 57 50.	+ 29 58	36	13.9	GALAXY
LB 01926	10 57 51.	+ 50 45 48.		15.5	FAINT BLUE STAR
LB 01927	10 57 53.	+ 55 22 48.		16.2	FAINT BLUE STAR
ZWG 066.086	10 57 54.	+ 10 13		15.5	GALAXY
UGC 06086	10 57 54.	+ 10 13	60	15.5	GALAXY Sc
ZC 1057.9+2256	10 57 54.	+ 22 56	540		CLUSTER OF GALAXIES
TON-N 0050	10 57 54.	+ 24 52		14.5	BLUE STAR
ZC 1057.9+5442	10 57 54.	+ 54 42	400		CLUSTER OF GALAXIES
MCG+10-16-044	10 57 54.	+ 58 03	36	15.	GALAXY
7ZW 359	10 57 54.	+ 65 13			COMPACT GALAXY
MCG+05-26-034	10 57 57.	+ 29 22	30	16.	GALAXY
IC 0663	10 57 59.	+ 10 42 01.			NONSTELLAR OBJECT
RNGC 3491	10 57 59.	+ 12 26		14.0	GALAXY
LB 09974	10 58	- 87 28		15.0	FAINT BLUE STAR
ZWG 010.059	10 58 00.	+ 02 24		15.0	GALAXY
UGC 06087	10 58 00.	+ 02 24	72	15.0	GALAXY SBb
ZWG 066.087	10 58 00.	+ 10 43		15.6	GALAXY
ZWG 066.088	10 58 00.	+ 11 41		15.0	GALAXY
MCG+02-28-040	10 58 00.	+ 11 41	36	15.0	GALAXY
ZWG 066.089	10 58 00.	+ 12 26		14.1	GALAXY
UGC 06088	10 58 00.	+ 12 26	60	14.1	GALAXY E-S0
MCG+02-28-041	10 58 00.	+ 12 26	18	14.1	GALAXY
ZWG 184.043	10 58 00.	+ 38 21		15.4	GALAXY
UGC 06089	10 58 00.	+ 38 21	66	15.4	GALAXY Sc
ZC 1058.0+4016	10 58 00.	+ 40 16	2420		CLUSTER OF GALAXIES
7ZW 360	10 58 00.	+ 58 03			COMPACT GALAXY
ZWG 291.021	10 58 00.	+ 58 04		15.2	GALAXY
MCG+11-14-002	10 58 00.	+ 67 22	30	16.	GALAXY
ZWG 314.002	10 58 00.	+ 67 23		15.6	GALAXY
ZWG 313.041	10 58 00.	+ 67 23		15.6	GALAXY
KARA.73B 0454	10 58 00.	+ 67 23	36	15.6	ISOLATED GALAXY S
ZWG 351.052	10 58 00.	+ 75 29		14.8	GALAXY
UGC 06090	10 58 00.	+ 75 29	90	14.8	GALAXY Sa-b
OCL 0837	10 58 00.	- 60 06			OPEN STAR CLUSTER
IC 0665	10 58 01.	- 13 35 48.			NONSTELLAR OBJECT
HN 1332	10 58 01.8	+ 00 15 08.	12		NEBULA
REIZ 0770	10 58 02.	+ 12 26	12	14.4	GALAXY
LB 01928	10 58 02.	+ 54 51 54.		16.4	FAINT BLUE STAR
REIZ 0765	10 58 03.	+ 58 03	12	15.0	GALAXY
RNGC 3487	10 58 04.	+ 17 51		14.5	GALAXY
ZWG 038.085	10 58 06.	+ 04 23		15.7	GALAXY
ZWG 066.090	10 58 06.	+ 10 09		15.1	GALAXY
UGC 06091	10 58 06.	+ 10 09	66	15.1	GALAXY SBb
MCG+02-28-043	10 58 06.	+ 10 09	60	15.1	GALAXY
ZWG 066.091	10 58 06.	+ 10 49		14.8	GALAXY
MCG+02-28-042	10 58 06.	+ 10 49	60	14.8	GALAXY
ZWG 095.089	10 58 06.	+ 17 51		14.6	GALAXY
UGC 06092	10 58 06.	+ 17 51	60	14.6	GALAXY S
MCG+03-28-047	10 58 06.	+ 17 52	54	14.6	GALAXY
ZC 1058.1+2638	10 58 06.	+ 26 38	740		CLUSTER OF GALAXIES
TON-N 0560	10 58 06.	+ 28 40		16.7	BLUE STAR
MCG+06-24-043	10 58 06.	+ 38 23	54	15.	GALAXY
ZWG 241.058	10 58 06.	+ 45 10		15.2	GALAXY
ZWG 314.003	10 58 06.	+ 65 02		15.3	GALAXY
ZWG 313.042	10 58 06.	+ 65 02		15.3	GALAXY
ZWG 010.060	10 58 06.	- 02 37		15.4	GALAXY
MCG-02-28-033	10 58 06.	- 13 49	36	17.	GALAXY
TON-N 1317	10 58 07.	+ 35 12		17.	BLUE STAR
TON-N 1316	10 58 08.	+ 21 30		17.	BLUE STAR
IC 0664	10 58 09.	+ 10 48 54.			NONSTELLAR OBJECT
MCG-02-28-034	10 58 09.	- 13 49	12	16.	GALAXY
BC 4C10.30	10 58 10.5	+ 11 02 23.		17.10	QUASI-STELLAR OBJECT
SHB 172	10 58 10.8	+ 11 02 20.		17.1	QUASI-STELLAR OBJECT
ZWG 066.092	10 58 12.	+ 11 00		14.8	GALAXY
UGC 06093	10 58 12.	+ 11 00	84	14.8	GALAXY Sb/SBc
MCG+02-28-044	10 58 12.	+ 11 00	84	14.8	GALAXY
ZWG 095.090	10 58 12.	+ 15 52		15.5	GALAXY
AR 29	10 58 12.16	+ 75 28 14.3			NEBULA
REIZ 0771	10 58 13.	+ 10 59	36	14.6	GALAXY
LB 01929	10 58 13.	+ 56 29 00.		15.0	FAINT BLUE STAR
MCG-02-28-035	10 58 15.	- 13 47 30.	30	15.	GALAXY
ARC 1142	10 58 17.	+ 10 50		15.4	RICH CLUSTER OF GALAXIES
LB 01930	10 58 17.	+ 50 14 12.		15.9	FAINT BLUE STAR
ZWG 066.093	10 58 18.	+ 10 46		14.0	GALAXY
RNGC 3492	10 58 18.	+ 10 46		14.0	GALAXY
8ZW 1058+10.8	10 58 18.	+ 10 46		14.0	COMPACT GALAXY
UGC 06094	10 58 18.	+ 10 46	66	14.0	GALAXY DBL SYS
MCG+02-28-045	10 58 18.	+ 10 46	27	14.0	GALAXY
ZWG 066.094	10 58 18.	+ 11 17		15.1	GALAXY
UGC 06095	10 58 18.	+ 19 21	90	17.	GALAXY DWARF
STOCK 20	10 58 18.	+ 28 29			BLUE KNOT NEAR ELLIP GLXY
ZC 1058.3+6603	10 58 18.	+ 66 03	400		CLUSTER OF GALAXIES
MCG+13-08-052	10 58 18.	+ 75 29	72	15.	GALAXY
REIZ 0772	10 58 20.	+ 30 00	18	14.6	GALAXY
LB 01931	10 58 21.	+ 57 37 36.		16.3	FAINT BLUE STAR
MCG-02-28-036	10 58 21.	- 13 55	42	15.	GALAXY
RNGC 3488	10 58 23.	+ 57 57		13.5	GALAXY
GO 092	10 58 23.	- 59 23	72	17.1	STAR CHAIN
PK291-04.1	10 58 23.5	- 64 58 47.	5		PLANETARY NEBULA
MCG+10-16-045	10 58 24.	+ 57 56	84	12.8	GALAXY
ZWG 291.022	10 58 24.	+ 57 57		13.7	GALAXY
UGC 06096	10 58 24.	+ 57 57	126	13.7	GALAXY Sc
ZC 1058.4+5953	10 58 24.	+ 59 53	870		CLUSTER OF GALAXIES
MCG-01-28-021	10 58 24.	- 06 14	36	15.	GALAXY
REIZ 0774	10 58 25.	+ 10 46	36	14.2	GALAXY
HN 0106	10 58 25.	- 64 58			NEBULA
IC 2621	10 58 25.	- 64 58 26.	2	9.5	PLANETARY NEBULA
TON-N 1318	10 58 26.	+ 23 00		16.5	BLUE STAR
REIZ 0773	10 58 26.	+ 30 00	24	14.6	GALAXY
REIZ 0769	10 58 27.	+ 57 56	90	13.6	GALAXY
SN 1954L	10 58 29.	+ 12 36		19.3	SUPERNOVA
ZWG 038.086	10 58 30.	+ 04 54		15.5	GALAXY
ZWG 066.095	10 58 30.	+ 12 36		15.7	GALAXY
MCG+03-28-048	10 58 30.	+ 17 00	48	15.3	GALAXY
TON-N 0561	10 58 30.	+ 27 23		16.6	BLUE STAR
TON-N 0562	10 58 30.	+ 27 28		16.8	BLUE STAR
MCG+09-18-069	10 58 30.	+ 51 04	60	15.	GALAXY
ZWG 010.061	10 58 30.	- 02 07		15.7	GALAXY
MCG-02-28-037	10 58 30.	- 13 47	60	15.	GALAXY
GO 091	10 58 30.	- 58 57	110	17.0	STAR CHAIN
REIZ 0775	10 58 32.	+ 12 45	12	14.3	GALAXY
8ZW 1058+10.0	10 58 36.	+ 10 02		17.3	COMPACT GALAXY
ZC 1058.6+1049	10 58 36.	+ 10 49	3360		CLUSTER OF GALAXIES
ZWG 066.096	10 58 36.	+ 12 45		14.6	GALAXY
MCG+02-28-046	10 58 36.	+ 12 45	60	14.6	GALAXY
ZWG 095.091	10 58 36.	+ 16 59		15.3	GALAXY
ZC 1058.6+1916	10 58 36.	+ 19 16	1280		CLUSTER OF GALAXIES
MCG+05-26-035	10 58 36.	+ 30 03	72	15.6	GALAXY
ZWG 155.043	10 58 36.	+ 30 04		15.6	GALAXY
UGC 06097	10 58 36.	+ 30 04	66	15.6	GALAXY S0-a
MCG+08-20-066	10 58 36.	+ 45 55	48	14.	GALAXY
ZC 1058.6+4611	10 58 36.	+ 46 11	7530		CLUSTER OF GALAXIES
ZWG 351.053	10 58 36.	+ 79 39		15.6	GALAXY
RNGC 3494	10 58 37.	+ 03 59			NON-EXISTENT OBJECT
IC 0666	10 58 37.	+ 10 44 41.			NONSTELLAR OBJECT
REIZ 0776	10 58 38.	+ 30 03	30	14.3	GALAXY
ZWG 038.087	10 58 42.	+ 03 49		15.5	GALAXY
ZWG 038.088	10 58 42.	+ 03 54		13.1	GALAXY
MCG+01-28-027	10 58 42.	+ 03 54	270	13.1	GALAXY
KARA.73B 0455	10 58 42.	+ 03 54	336	13.1	ISOLATED GALAXY S
ZWG 066.097	10 58 42.	+ 10 45		15.3	GALAXY
ZC 1058.7+2156	10 58 42.	+ 21 56	540		CLUSTER OF GALAXIES
ZWG 155.044	10 58 42.	+ 28 00		15.3	GALAXY
UGC 06099	10 58 42.	+ 28 00	66	15.3	GALAXY S
ZWG 241.059	10 58 42.	+ 45 55		14.1	GALAXY
UGC 06100	10 58 42.	+ 45 55	54	14.1	GALAXY Sa?
MCG+08-20-067	10 58 42.	+ 46 34	42	16.	GALAXY
MCG+08-20-067A	10 58 42.	+ 50 38	42	17.	GALAXY
MCG+10-16-046	10 58 42.	+ 58 55	27	16.	GALAXY
RNGC 3495	10 58 43.	+ 03 54		12.5	GALAXY
REIZ 0777	10 58 43.	+ 27 59	72	14.8	GALAXY
RNGC 3493	10 58 45.	+ 28 00		15.5	GALAXY
MCG+05-26-036	10 58 45.	+ 28 00	72	15.3	GALAXY
MCG-02-28-038	10 58 45.	- 13 17	54	14.5	GALAXY
REIZ 0778	10 58 47.	+ 03 18	36	14.8	GALAXY
IC 2614	10 58 47.	+ 39 04 23.			NONSTELLAR OBJECT
MCG+00-28-027	10 58 48.	+ 02 24	42	15.0	GALAXY
ZWG 038.089	10 58 48.	+ 03 18		15.1	GALAXY
ZC 1058.8+0716	10 58 48.	+ 07 16	1010		CLUSTER OF GALAXIES
ZWG 066.098	10 58 48.	+ 10 30		15.6	GALAXY
ZC 1058.8+4149	10 58 48.	+ 41 49	3900		CLUSTER OF GALAXIES
ZWG 267.033	10 58 48.	+ 51 07		15.6	GALAXY
HN 1333	10 58 49.9	- 00 51 59.	12		NEBULA
REIZ 0779	10 58 50.	+ 12 24	18	14.6	GALAXY
LB 01932	10 58 50.	+ 55 54 00.		16.2	FAINT BLUE STAR
MCG+05-26-036A	10 58 51.	+ 27 58 30.	18	15.	GALAXY
HN 1334	10 58 52.3	- 01 56 35.	18		NEBULA
ZWG 066.099	10 58 54.	+ 12 24		15.6	GALAXY
MCG+08-20-068	10 58 54.	+ 47 21	78	16.	GALAXY
UGC 06101	10 58 54.	+ 47 23	96	16.0	GALAXY Sb-c
ZWG 010.062	10 58 54.	- 01 56		15.3	GALAXY
HN 1335	10 58 54.7	- 00 26 23.	12		NEBULA
ARC 1146	10 58 56.	- 22 28		17.0	RICH CLUSTER OF GALAXIES
ARC 1145	10 58 57.	+ 17 00		15.7	RICH CLUSTER OF GALAXIES
LB 01933	10 58 57.	+ 55 45 48.		14.8	FAINT BLUE STAR
RNGC 3497	10 58 58.	- 19 12			NON-EXISTENT OBJECT
ARC 1143	10 58 59.	+ 50 37		17.2	RICH CLUSTER OF GALAXIES
ZWG 095.092	10 59 00.	+ 18 59		15.5	GALAXY
ZWG 155.045	10 59 00.	+ 28 58		15.6	GALAXY
UGC 06102	10 59 00.	+ 28 58	72	15.6	GALAXY DWRF IR
REIZ 0780	10 59 00.	+ 37 37	30	14.8	GALAXY
ZWG 241.060	10 59 00.	+ 44 43		14.7	GALAXY
MCG+08-20-069	10 59 00.	+ 45 30	42	15.	GALAXY
ZC 1059.0+4614	10 59 00.	+ 46 14	1610		CLUSTER OF GALAXIES
ZC 1059.0+5044	10 59 00.	+ 50 44	1080		CLUSTER OF GALAXIES
ZC 1059.0+6125	10 59 00.	+ 61 25	870		CLUSTER OF GALAXIES
MCG-02-28-039	10 59 00.	- 12 10	36	15.	GALAXY
OCL 0833	10 59 00.	- 59 33	36	9.2	OPEN STAR CLUSTER
MRSL 289+00/1	10 59 00.	- 59 39	1800		HII REGION
HN 1336	10 59 00.0	- 01 37 48.			NEBULA
TON-N 1319	10 59 02.	+ 21 53		15.7	BLUE STAR
MIL 18	10 59 02.	- 60 33 48.	312		SUPERNOVA REMNANT
MCG+03-28-049	10 59 06.	+ 16 53	90	14.4	GALAXY
REIZ 0781	10 59 06.	+ 24 09	24	14.6	GALAXY
ZWG 125.021	10 59 06.	+ 24 11		15.7	GALAXY

OBJECT NAME	RIGHT ASCEN.	DECLINATION	DIAM.	MAGN.	TYPE OF OBJECT
ZWG 241.061	10 59 06.	+ 45 30		13.4	GALAXY
MRK 161	10 59 06.	+ 45 30	30	13.5	GALAXY WITH UV CONTINUUM
UGC 06103	10 59 06.	+ 45 30	48	13.4	GALAXY PECULR
MCG+08-20-069A	10 59 06.	+ 50 35	60		GALAXY
SHAH 026	10 59 06.	+ 50 36	60	16.5	GROUP OF COMPACT GALAXIES
MCG+09-18-070	10 59 06.	+ 54 42	24	15.	GALAXY
MRSL 289+00/3	10 59 06.	- 59 34	120		HII REGION
MCG+04-26-025	10 59 09.	+ 22 37	30	15.2	GALAXY
RNGC 3523	10 59 09.	+ 75 24		14.0	GALAXY
REIZ 0783	10 59 11.	+ 22 37	36	14.5	GALAXY
ZWG 066.100	10 59 12.	+ 10 36		15.7	GALAXY
ZWG 095.093	10 59 12.	+ 16 52		14.4	GALAXY
UGC 06104	10 59 12.	+ 16 52	90	14.4	GALAXY Sb-c
ZC 1059.2+2014	10 59 12.	+ 20 14	940		CLUSTER OF GALAXIES
ZWG 125.022	10 59 12.	+ 22 37		15.2	GALAXY
REIZ 0784	10 59 12.	+ 23 44	30	15.6	GALAXY
ZWG 184.044	10 59 12.	+ 38 12		15.5	GALAXY
ZWG 213.018	10 59 12.	+ 39 03		15.6	GALAXY
MCG+08-20-070	10 59 12.	+ 46 09	72	14.	GALAXY
MCG+09-18-071	10 59 12.	+ 55 59	30	17.	GALAXY
ZWG 351.054	10 59 12.	+ 75 24		13.8	GALAXY
UGC 06105	10 59 12.	+ 75 24	96	13.8	GALAXY Sb
REIZ 0785	10 59 13.	+ 28 02	30	15.4	GALAXY
CED 109B	10 59 14.	- 60 27			DIFFUSE GALACTIC NEBULA
REIZ 0786	10 59 15.	+ 16 52	54	14.6	GALAXY
IC 2618	10 59 16.	+ 28 02			NONSTELLAR OBJECT
IC 2615	10 59 16.	+ 38 12 28.			NONSTELLAR OBJECT
RNGC 3498	10 59 17.	+ 14 39			NON-EXISTENT OBJECT
RNGC 3503	10 59 17.	- 60 27			DIFFUSE NEBULA
ZWG 066.101	10 59 18.	+ 10 34		15.5	GALAXY
ZWG 095.094	10 59 18.	+ 18 51		15.7	GALAXY
TON-N 0563	10 59 18.	+ 25 46		14.3	BLUE STAR
ZC 1059.3+2634	10 59 18.	+ 26 34	810		CLUSTER OF GALAXIES
ZC 1059.3+3401	10 59 18.	+ 34 01	1950		CLUSTER OF GALAXIES
ZWG 213.019	10 59 18.	+ 38 56		15.0	GALAXY
ZWG 241.062	10 59 18.	+ 46 09		14.5	GALAXY
UGC 06106	10 59 18.	+ 46 09	102	14.5	GALAXY SB?b
MCG-02-28-040	10 59 18.	- 12 13	24	15.	GALAXY
IC 2616	10 59 19.	+ 39 03 10.			NONSTELLAR OBJECT
IC 2617	10 59 20.	+ 38 54 34.			NONSTELLAR OBJECT
RNGC 3484	10 59 21.	+ 76 05			NON-EXISTENT OBJECT
REIZ 0782	10 59 23.	+ 54 43	42	15.4	GALAXY
ZWG 038.090	10 59 24.	+ 03 51		15.1	GALAXY
ZWG 066.102	10 59 24.	+ 12 07		15.5	GALAXY
ZC 1059.4+3606	10 59 24.	+ 36 06	470		CLUSTER OF GALAXIES
ZWG 184.045	10 59 24.	+ 38 13		15.7	GALAXY
MCG+07-23-011	10 59 24.	+ 38 55	48	15.	GALAXY
MCG+07-23-012	10 59 24.	+ 39 02	12	15.5	GALAXY
ZWG 241.063	10 59 24.	+ 44 33		15.2	GALAXY
ZWG 241.064	10 59 24.	+ 50 21		15.5	GALAXY
MCG+10-16-047	10 59 24.	+ 59 24	57	14.	GALAXY
LB 01934	10 59 24.	+ 61 59 54.		16.9	FAINT BLUE STAR
KARA.68 074	10 59 24.	+ 70 33	34		DWARF GALAXY
VHE 46	10 59 24.	- 59 35	192		REFLECTION NEBULA
ARC 1144	10 59 26.	+ 59 02		17.2	RICH CLUSTER OF GALAXIES
AR 30	10 59 28.59	+ 75 23 07.0			NEBULA
IC 2619	10 59 29.	+ 38 13 46.			NONSTELLAR OBJECT
SEY 086	10 59 29.	+ 59 24 19.		15.1	FAINT GALAXY
ZC 1059.5+2148	10 59 30.	+ 21 48	540		CLUSTER OF GALAXIES
ZC 1059.5+2257	10 59 30.	+ 22 57	810		CLUSTER OF GALAXIES
ZWG 213.020	10 59 30.	+ 38 46		14.9	GALAXY
UGC 06107	10 59 30.	+ 38 46	66	14.9	GALAXY COMPACT
ZC 1059.5+4250	10 59 30.	+ 42 50	740		CLUSTER OF GALAXIES
MCG+08-20-071	10 59 30.	+ 50 22 30.	36	15.	GALAXY
MCG+09-18-073	10 59 30.	+ 50 50	60	13.	GALAXY
MCG+09-18-072	10 59 30.	+ 51 25	60	14.	GALAXY
7ZW 361	10 59 30.	+ 65 44			COMPACT GALAXY
MCG+13-08-053	10 59 30.	+ 75 23	84	13.	GALAXY
HN 1337	10 59 31.1	- 00 21 18.	12		NEBULA
HN 1338	10 59 33.1	+ 02 02 24.	18		NEBULA
IC 2620	10 59 35.	+ 38 44 52.			NONSTELLAR OBJECT
8ZW 1059+09.1	10 59 36.	+ 09 07		16.9	COMPACT GALAXY
ZC 1059.6+1219	10 59 36.	+ 12 19	870		CLUSTER OF GALAXIES
ZC 1059.6+3315	10 59 36.	+ 33 15	610		CLUSTER OF GALAXIES
UGC 06108	10 59 36.	+ 37 18	60	16.5	GALAXY S
ZWG 267.034	10 59 36.	+ 50 52		15.3	GALAXY
UGC 06109	10 59 36.	+ 50 52	84	15.3	GALAXY Sc
ZWG 291.023	10 59 36.	+ 59 24		15.0	GALAXY
UGC 06110	10 59 36.	+ 59 24	102	15.0	GALAXY
MCG+13-08-054	10 59 36.	+ 79 12	33	15.	GALAXY
RNGC 3502	10 59 39.	- 13 52		14.0	GALAXY
LB 01935	10 59 40.	+ 54 48 24.		16.7	FAINT BLUE STAR
ZWG 066.103	10 59 42.	+ 10 30		15.5	GALAXY
ZC 1059.7+1030	10 59 42.	+ 10 30	340		CLUSTER OF GALAXIES
ZC 1059.7+1204	10 59 42.	+ 12 04	940		CLUSTER OF GALAXIES
ARC 1147	10 59 42.	+ 12 16		17.6	RICH CLUSTER OF GALAXIES
MCG+04-26-026	10 59 42.	+ 22 31	18	15.5	GALAXY
ZWG 155.046	10 59 42.	+ 27 11		15.7	GALAXY
MCG+07-23-013	10 59 42.	+ 38 45	36	14.5	GALAXY
ZWG 267.035	10 59 42.	+ 51 28		15.6	GALAXY
SC 1057-5955.0	10 59 42.	- 60 11 07.	12		NEBULA
LB 03498	10 59 42.	- 79 00		13.3	FAINT BLUE STAR
HN 1339	10 59 42.9	- 01 42 42.	18		NEBULA
REIZ 0787	10 59 43.	+ 27 11	18	15.2	GALAXY
REIZ 0788	10 59 47.	+ 22 33	12	15.0	GALAXY
HN 1340	10 59 47.3	+ 01 05 12.	24		NEBULA
ZC 1059.8+0611	10 59 48.	+ 06 11	670		CLUSTER OF GALAXIES
ZWG 038.091	10 59 48.	+ 06 20		15.0	GALAXY
MCG+01-28-028	10 59 48.	+ 06 20	36	15.0	GALAXY
REIZ 0789	10 59 48.	+ 06 21	36	14.9	GALAXY
ZWG 066.104	10 59 48.	+ 08 45		15.2	GALAXY
ZC 1059.8+1428	10 59 48.	+ 14 28	1810		CLUSTER OF GALAXIES
ZC 1059.8+1710	10 59 48.	+ 17 10	3360		CLUSTER OF GALAXIES
ZWG 095.095	10 59 48.	+ 19 02		15.4	GALAXY
MCG+07-23-014	10 59 48.	+ 38 48	36	16.	GALAXY
MCG+09-18-075	10 59 48.	+ 50 54 30.	18	15.	GALAXY
MCG+09-18-074	10 59 48.	+ 50 55	15	16.	GALAXY
7ZW 362	10 59 48.	+ 65 46			COMPACT GALAXY
MCG-02-28-041	10 59 48.	- 13 52	60	14.	GALAXY
ARC 1148	10 59 52.	- 00 48		17.6	RICH CLUSTER OF GALAXIES
HN 1341	10 59 53.3	+ 00 06 36.	18		NEBULA
ZWG 038.092	10 59 54.	+ 02 54		15.1	GALAXY
KARA.72 261A	10 59 54.	+ 02 54	48	15.1	PART OF DOUBLE GALAXY
MCG+01-28-029	10 59 54.	+ 02 54	36	15.1	GALAXY
ZWG 038.093	10 59 54.	+ 03 22		15.2	GALAXY
UGC 06111	10 59 54.	+ 03 22	60	15.2	GALAXY S
ZWG 038.094	10 59 54.	+ 04 19		15.3	GALAXY
ZWG 095.096	10 59 54.	+ 17 00		14.5	GALAXY
UGC 06112	10 59 54.	+ 17 00	138	14.5	GALAXY Sc-IRR
MCG+03-28-050	10 59 54.	+ 17 00	162	14.5	GALAXY
REIZ 0790	10 59 54.	+ 24 10	30	15.2	GALAXY
UGC 06113	10 59 54.	+ 52 26	60	19.	GALAXY DWARF
MCG+09-18-076	10 59 54.	+ 53 02	30	17.	GALAXY
ZWG 314.004	10 59 54.	+ 65 03		14.8	GALAXY
TON-N 1320	10 59 55.	+ 35 12		16.8	BLUE STAR
REIZ 0792	10 59 57.	+ 17 01	90	14.5	GALAXY
HN 1342	10 59 59.8	- 00 00 49.	12		NEBULA
LBN 0627	11 00	+ 84 30	12900		BRIGHT NEBULA
LB 09848	11 00	- 85 48		14.6	FAINT BLUE STAR
ZWG 038.095	11 00 00.	+ 02 53		15.0	GALAXY
KARA.72 261B	11 00 00.	+ 02 53	42	15.0	PART OF DOUBLE GALAXY
MCG+01-28-030	11 00 00.	+ 02 53	24	15.0	GALAXY
TON-N 0564	11 00 00.	+ 25 38		14.2	BLUE STAR
ZWG 184.046	11 00 00.	+ 37 02		15.5	GALAXY
ZC 1100.0+5018	11 00 00.	+ 50 18	940		CLUSTER OF GALAXIES
MCG+08-20-072	11 00 00.	+ 50 35	30	16.	GALAXY
KARA.72 262A	11 00 00.	+ 50 56	60	14.5	PART OF DOUBLE GALAXY
ZWG 267.036	11 00 00.	+ 50 57		14.5	GALAXY
UGC 06114	11 00 00.	+ 50 57	90	14.5	GALAXY DBL SYS
KARA.72 262B	11 00 00.	+ 50 57	42		PART OF DOUBLE GALAXY
MCG+09-18-078	11 00 00.	+ 53 17	60	16.	GALAXY
ZC 1100.0+5541	11 00 00.	+ 55 41	470		CLUSTER OF GALAXIES
MCG+09-18-077	11 00 00.	+ 55 57	36	16.	GALAXY
UGC 06115	11 00 00.	+ 56 29	48	14.3	GALAXY
ZC 1100.0+5957	11 00 00.	+ 59 57	610		CLUSTER OF GALAXIES
7ZW 363	11 00 00.	+ 65 45			COMPACT GALAXY
ZWG 010.063	11 00 00.	- 03 22		15.4	GALAXY
MCG+00-28-028	11 00 00.	- 03 22	48	15.4	GALAXY
HN 1343	11 00 00.3	+ 00 27 53.	18		NEBULA
REIZ 0793	11 00 05.	+ 37 04	12	15.1	GALAXY
LB 01936	11 00 05.	+ 51 26 18.		16.8	FAINT BLUE STAR
ZWG 038.096	11 00 06.	+ 05 39		15.5	GALAXY
MCG+03-28-051	11 00 06.	+ 18 16	240	13.8	GALAXY
ZWG 125.023	11 00 06.	+ 25 20		15.6	GALAXY
ZWG 184.047	11 00 06.	+ 33 08		15.4	GALAXY
KARA.73B 0456	11 00 06.	+ 33 08	24	15.4	ISOLATED GALAXY S0
MCG+06-24-044	11 00 06.	+ 37 03	27	15.	GALAXY
MCG+08-20-073	11 00 06.	+ 50 29 30.	60	14.	GALAXY
MCG-01-28-022	11 00 06.	- 07 03	60	15.	GALAXY
MCG-04-26-018	11 00 06.	- 25 53 30.	48	14.5	GALAXY
HOLM 224B	11 00 08.	+ 18 16	210	13.5	PART OF MULTIPLE GALAXY
MCG+05-26-037	11 00 09.	+ 31 41	24	15.0	GALAXY
MCG+06-24-045	11 00 09.	+ 35 30	42	16.	GALAXY
RNGC 3501	11 00 10.	+ 18 15		14.0	GALAXY
UGC 06116	11 00 12.	+ 18 15	228	13.8	GALAXY Sc
KARA.72 263A	11 00 12.	+ 18 15	216	13.8	PART OF DOUBLE GALAXY
MCG+05-26-038	11 00 12.	+ 32 16	48	15.2	GALAXY
UGC 06117	11 00 12.	+ 50 29	60	14.5	GALAXY SBa
ZWG 267.037	11 00 12.	+ 56 29		14.3	GALAXY
ZWG 038.097	11 00 12.	+ 04 15		15.6	GALAXY
ZC 1100.2+0753	11 00 12.	+ 07 53	2490		CLUSTER OF GALAXIES
ZWG 095.097	11 00 12.	+ 18 15		13.8	GALAXY
ZWG 267.038	11 00 12.	+ 50 29		14.5	GALAXY
ZWG 241.065	11 00 12.	+ 50 29		14.5	GALAXY
ZWG 291.024	11 00 12.	+ 56 29		14.3	GALAXY
REIZ 0791	11 00 12.	+ 56 29	24	14.3	GALAXY
MCG+10-16-048	11 00 12.	+ 57 25	18	17.	GALAXY
7ZW 364	11 00 12.	+ 68 29			COMPACT GALAXY
HOLM 222A	11 00 15.	+ 31 41	24	15.0	PART OF MULTIPLE GALAXY
MCG+09-18-079	11 00 15.	+ 54 22	36	16.	GALAXY
MCG+09-18-080	11 00 15.	+ 56 29 30.	42	14.4	GALAXY
HOLM 222C	11 00 16.	+ 31 41	60	16.0	PART OF MULTIPLE GALAXY
RNGC 3505	11 00 16.	- 15 13			NON-EXISTENT OBJECT
REIZ 0794	11 00 17.	+ 37 10	36	15.5	GALAXY
ZWG 095.098	11 00 18.	+ 17 36		14.9	GALAXY
MCG+03-28-052	11 00 18.	+ 17 36	36	14.9	GALAXY
ZWG 155.047	11 00 18.	+ 32 14		15.2	GALAXY
ZWG 213.021	11 00 18.	+ 41 58		15.1	GALAXY
ZC 1100.3-0053	11 00 18.	- 00 53	1410		CLUSTER OF GALAXIES
KLEM 17	11 00 20.	- 25 31	900	15.	CLUSTER OF 10 GALAXIES
HOLM 222B	11 00 21.	+ 31 39	18	15.4	PART OF MULTIPLE GALAXY
HOLM 223A	11 00 21.	+ 32 16	24	15.0	PART OF MULTIPLE GALAXY
HOLM 223B	11 00 23.	+ 32 16	24	15.1	PART OF MULTIPLE GALAXY
ZWG 155.048	11 00 24.	+ 26 31		15.6	GALAXY
ZWG 155.049	11 00 24.	+ 28 15		11.5	GALAXY
UGC 06118	11 00 24.	+ 28 15	156	11.5	GALAXY Sa/SBb
MCG+07-23-015	11 00 24.	+ 41 58	12	15.	GALAXY
ZC 1100.4+4423	11 00 24.	+ 44 23	1080		CLUSTER OF GALAXIES
MCG+09-18-081	11 00 24.	+ 55 55	12	16.	GALAXY
HELW 432	11 00 24.	- 23 19 19.			NEBULA
MCG-04-26-019	11 00 24.	- 23 20	60	14.	GALAXY
HN 1344	11 00 24.1	- 00 41 25.	18		NEBULA
SVEN 243	11 00 26.	- 23 19	42	14.0	GALAXY
ARC 1149	11 00 27.	+ 07 54		16.0	RICH CLUSTER OF GALAXIES
BC 3CR249.1	11 00 27.32	+ 77 15 08.6		15.72	QUASI-STELLAR OBJECT
SHB 173	11 00 27.6	+ 77 15 08.		15.7	QUASI-STELLAR OBJECT
SHB 173	11 00 27.6	+ 77 15 08.		15.7	QUASI-STELLAR OBJECT
RNGC 3508	11 00 28.	- 16 02		14.0	GALAXY
ZWG 038.098	11 00 30.	+ 03 44		15.3	GALAXY
ZWG 038.099	11 00 30.	+ 05 22		15.3	GALAXY
ZWG 095.099	11 00 30.	+ 15 01		15.5	GALAXY
TON-N 0565	11 00 30.	+ 25 38		14.5	BLUE STAR
MCG+05-26-039	11 00 30.	+ 28 15	144	11.5	GALAXY
MCG-03-28-031	11 00 30.	- 16 02	66	14.	GALAXY
MRSL 290-00/1	11 00 30.	- 60 30	7200		HII REGION
REIZ 0795	11 00 31.	+ 28 15	120	12.2	GALAXY
ZC 1100.6+0005	11 00 36.	+ 00 05	3430		CLUSTER OF GALAXIES
ZWG 038.100	11 00 36.	+ 03 36		14.4	GALAXY
UGC 06119	11 00 36.	+ 03 36	36	14.4	GALAXY
MCG+01-28-031	11 00 36.	+ 03 36	30	14.4	GALAXY
ZC 1100.6+0431	11 00 36.	+ 04 31	1480		CLUSTER OF GALAXIES
ZWG 066.105	11 00 36.	+ 11 21		12.9	GALAXY
UGC 06120	11 00 36.	+ 11 21	78	12.9	GALAXY Sc?
MCG+02-28-047	11 00 36.	+ 11 21	60	12.9	GALAXY
ZC 1100.6+2010	11 00 36.	+ 20 10	340		CLUSTER OF GALAXIES
ZWG 213.022	11 00 36.	+ 39 31		15.5	GALAXY
UGC 06121	11 00 36.	+ 39 31	102	15.5	GALAXY
KARA.73B 0457	11 00 36.	+ 39 31	108	15.5	ISOLATED GALAXY S
REIZ 0796	11 00 37.	+ 11 21	24	13.1	GALAXY
AR 31	11 00 38.12	+ 74 13 44.3			NEBULA
ZWG 038.101	11 00 42.	+ 04 00		15.5	GALAXY
ZC 1100.7+0510	11 00 42.	+ 05 10	1550		CLUSTER OF GALAXIES
ZWG 038.102	11 00 42.	+ 05 20		15.6	GALAXY
ZC 1100.7+2308	11 00 42.	+ 23 08	3230		CLUSTER OF GALAXIES
MCG+07-23-016	11 00 42.	+ 39 30	72	15.	GALAXY
MCG+08-20-074	11 00 42.	+ 45 27	72	15.	GALAXY

OBJECT NAME	RIGHT ASCEN.	DECLINATION	DIAM.	MAGN.	TYPE OF OBJECT
MCG-02-28-042	11 00 42.	- 15 04	78	15.	GALAXY
MCG-03-28-032	11 00 42.	- 16 31	72	15.5	GALAXY
VV 075	11 00 45.	+ 05 03	126	13.	INTERACTING GALAXY
MCG+03-28-053	11 00 45.	+ 18 25	210	11.4	GALAXY
LB 01937	11 00 45.	+ 49 27 18.		17.6	FAINT BLUE STAR
MIL 19	11 00 45.	- 60 38 00.	780		SUPERNOVA REMNANT
ARP 335	11 00 46.	+ 05 02			PECULIAR GALAXY
RNGC 3507	11 00 46.	+ 18 24		11.5	GALAXY
HOLM 224A	11 00 46.	+ 18 25	180	12.5	PART OF MULTIPLE GALAXY
LB 01938	11 00 46.	+ 59 06 36.		13.2	FAINT BLUE STAR
HN 1345	11 00 46.3	- 02 24 14.	12		NEBULA
RNGC 3511	11 00 47.	- 22 50		12.0	GALAXY
ZWG 038.103	11 00 48.	+ 07 06		15.5	GALAXY
ZWG 038.104	11 00 48.	+ 07 11		15.2	GALAXY
UGC 06122	11 00 48.	+ 11 24	78	16.0	GALAXY DWRF SP
ZC 1100.8+1156	11 00 48.	+ 11 56	470		CLUSTER OF GALAXIES
ZWG 066.106	11 00 48.	+ 13 05		15.5	GALAXY
KARA.73B 0458	11 00 48.	+ 13 05	24	15.5	ISOLATED GALAXY S
ZWG 095.100	11 00 48.	+ 18 24		11.4	GALAXY
UGC 06123	11 00 48.	+ 18 24	204	11.4	GALAXY SBb
KARA.72 263B	11 00 48.	+ 18 24	180	11.4	PART OF DOUBLE GALAXY
UGC 06124	11 00 48.	+ 32 08	90	16.5	GALAXY S
ZWG 184.048	11 00 48.	+ 38 10		15.6	GALAXY
MRK 420	11 00 48.	+ 38 10	14	15.5	GALAXY WITH UV CONTINUUM
ZWG 241.066	11 00 48.	+ 45 27		14.8	GALAXY
UGC 06125	11 00 48.	+ 45 27	84	14.8	GALAXY SBa-b
SN 1953H	11 00 48.	+ 50 07		17.0	SUPERNOVA
ZC 1100.8+5841	11 00 48.	+ 58 41	1080		CLUSTER OF GALAXIES
7ZW 365	11 00 48.	+ 70 42			COMPACT GALAXY
ZC 1100.8+7042	11 00 48.	+ 70 42	340		CLUSTER OF GALAXIES
LB 01939	11 00 49.	+ 53 05 30.		16.8	FAINT BLUE STAR
MCG-04-26-020	11 00 51.	- 22 50	360	11.	GALAXY
ZWG 038.105	11 00 54.	+ 07 08		15.4	GALAXY
ZWG 095.101	11 00 54.	+ 16 38		15.6	GALAXY
ZWG 155.050	11 00 54.	+ 29 10		13.6	GALAXY
UGC 06126	11 00 54.	+ 29 10	258	13.6	GALAXY
MCG+07-23-017	11 00 54.	+ 44 07	48	16.	GALAXY
MCG+08-20-075	11 00 54.	+ 50 05	18	18.	GALAXY
SVEN 244	11 00 56.	- 22 49	276	11.5	GALAXY
MCG+03-28-054	11 00 57.	+ 19 36	48	14.9	GALAXY
LB 01940	11 00 58.	+ 49 10 06.		15.7	FAINT BLUE STAR
HARO 26	11 00 59.	+ 29 09			BLUE EMISSION-LINE GALAXY
IC 2622	11 00 59.	- 15 57 48.			NONSTELLAR OBJECT
ZC 1101.0+1245	11 01 00.	+ 12 45	1550		CLUSTER OF GALAXIES
ZWG 095.102	11 01 00.	+ 17 35		15.4	GALAXY
ZWG 095.103	11 01 00.	+ 19 35		14.9	GALAXY
TON-N 0051	11 01 00.	+ 27 36		16.0	BLUE STAR
MCG+05-26-040	11 01 00.	+ 29 10	240	13.6	GALAXY
ZC 1101.0+2912	11 01 00.	+ 29 12	1410		CLUSTER OF GALAXIES
HMS 1.16	11 01 00.	+ 41 05			PEC GALAXY
ZC 1101.0+4858	11 01 00.	+ 48 58	2550		CLUSTER OF GALAXIES
MCG+08-20-076	11 01 00.	+ 50 05 30.	45	14.	GALAXY
ZWG 010.064	11 01 00.	- 01 14		15.4	GALAXY
REIZ 0797	11 01 01.	+ 29 10	192	13.0	GALAXY
RNGC 3510	11 01 02.	+ 29 09		13.0	GALAXY
HN 1346	11 01 02.9	- 01 07 38.			NEBULA
MCG+07-23-018	11 01 03.	+ 43 45	30	15.	GALAXY
ARP 148	11 01 05.	+ 41 06			PECULIAR GALAXY
RNGC 3513	11 01 05.	- 22 58		12.0	GALAXY
ZC 1101.1+0329	11 01 06.	+ 03 29	400		CLUSTER OF GALAXIES
ZWG 038.106	11 01 06.	+ 03 36		15.0	GALAXY
MCG+01-28-032	11 01 06.	+ 03 36	60	15.0	GALAXY
ZC 1101.1+3158	11 01 06.	+ 31 58	610		CLUSTER OF GALAXIES
MCG+07-23-019	11 01 06.	+ 41 06	30	15.	GALAXY
ZWG 213.023	11 01 06.	+ 41 20		15.6	GALAXY
ZWG 213.024	11 01 06.	+ 43 45		14.6	GALAXY
KARA.72 264B	11 01 06.	+ 43 45	24		PART OF DOUBLE GALAXY
KARA.72 264A	11 01 06.	+ 43 45	30	14.6	PART OF DOUBLE GALAXY
MCG+08-20-077	11 01 06.	+ 45 31	18	15.	GALAXY
MCG+08-20-078	11 01 06.	+ 45 35 30.	30	16.	GALAXY
ZWG 241.067	11 01 06.	+ 50 05		14.4	GALAXY
UGC 06127	11 01 06.	+ 50 05	54	14.4	GALAXY Sc
LB 01941	11 01 06.	+ 52 54 18.		15.0	FAINT BLUE STAR
ZWG 010.065	11 01 06.	- 01 06		15.1	GALAXY
MCG-03-28-033	11 01 06.	- 17 14	84	15.	GALAXY
8ZW 1101+09.4	11 01 12.	+ 09 23		20.0	COMPACT GALAXY
8ZW 1101+11.3	11 01 12.	+ 11 20		18.1	COMPACT GALAXY
ZC 1101.2+2253	11 01 12.	+ 22 53	540		CLUSTER OF GALAXIES
ZWG 184.049	11 01 12.	+ 35 51		15.7	GALAXY
VV 032B	11 01 12.	+ 41 07	21	17.	INTERACTING GALAXY
VV 032A	11 01 12.	+ 41 07	24	15.	INTERACTING GALAXY
VV 032	11 01 12.	+ 41 07	36		INTERACTING GALAXY
ZWG 241.068	11 01 12.	+ 45 32		14.8	GALAXY
MCG+10-16-049	11 01 12.	+ 57 39	45	15.	GALAXY
REIZ 0798	11 01 13.	+ 28 18	66	12.8	GALAXY
RNGC 3499	11 01 13.	+ 56 29		14.5	GALAXY
SEY 087	11 01 16.	+ 57 39 59.		14.7	FAINT GALAXY
HN 1347	11 01 16.1	+ 00 04 22.	18		NEBULA
ARC 1153	11 01 17.	+ 01 36		17.5	RICH CLUSTER OF GALAXIES
ZWG 010.066	11 01 18.	+ 00 04		15.6	GALAXY
ZWG 155.051	11 01 18.	+ 28 18		12.9	GALAXY
UGC 06128	11 01 18.	+ 28 18	102	12.9	GALAXY Sc
MCG+05-26-041	11 01 18.	+ 28 19	90	12.9	GALAXY
ZC 1101.3+4459	11 01 18.	+ 44 59	340		CLUSTER OF GALAXIES
ZWG 241.069	11 01 18.	+ 50 17		15.2	GALAXY
UGC 06129	11 01 18.	+ 50 17	60	15.2	GALAXY Sb-c
ZWG 291.025	11 01 18.	+ 57 41		15.1	GALAXY
SVEN 245	11 01 20.	- 22 58 38.	114	12.2	GALAXY
RNGC 3512	11 01 21.	+ 28 18		13.0	GALAXY
MCG+08-20-079	11 01 21.	+ 50 19	36	15.	GALAXY
MCG-03-28-034	11 01 21.	- 19 52	30	15.	GALAXY
MCG-04-26-021	11 01 21.	- 22 58 30.	138	11.5	GALAXY
REIZ 0799	11 01 23.	+ 22 56	18	15.2	GALAXY
IC 2623	11 01 23.	- 19 49 25.			NONSTELLAR OBJECT
ZC 1101.4+0556	11 01 24.	+ 05 56	1550		CLUSTER OF GALAXIES
ARC 1152	11 01 24.	+ 12 49		17.6	RICH CLUSTER OF GALAXIES
ZC 1101.4+1350	11 01 24.	+ 13 50	1480		CLUSTER OF GALAXIES
ZWG 066.107	11 01 24.	+ 14 08		15.1	GALAXY
ZC 1101.4+1800	11 01 24.	+ 18 00	2150		CLUSTER OF GALAXIES
ZC 1101.4+2521	11 01 24.	+ 25 21	1810		CLUSTER OF GALAXIES
TON-N 0052	11 01 24.	+ 31 58		16.5	BLUE STAR
RNGC 3514	11 01 28.	- 18 33		13.0	GALAXY
LB 01942	11 01 29.	+ 55 19 00.		15.4	FAINT BLUE STAR
SC 1059-6019.9	11 01 29.	- 60 36 03.	12		NEBULA
ZWG 066.108	11 01 30.	+ 08 38		15.0	GALAXY
UGC 06130	11 01 30.	+ 08 38	66	15.0	GALAXY Sc
MCG+02-28-048	11 01 30.	+ 08 38	60	15.0	GALAXY
MCG+08-20-080	11 01 30.	+ 45 24	24	16.	GALAXY
ZWG 241.070	11 01 30.	+ 46 23		15.5	GALAXY
MCG-03-28-035	11 01 30.	- 18 33	60	13.5	GALAXY
RNGC 3504	11 01 33.	+ 28 15		12.0	GALAXY
ARC 1151	11 01 35.	+ 36 13		17.5	RICH CLUSTER OF GALAXIES
ZWG 038.107	11 01 36.	+ 07 20		15.2	GALAXY
RNGC 3506	11 01 36.	+ 11 21		13.5	GALAXY
ZWG 213.025	11 01 36.	+ 41 07		15.5	GALAXY
ZWG 213.026	11 01 36.	+ 44 18		14.7	GALAXY
UGC 06131	11 01 36.	+ 44 18	78	14.7	GALAXY S0
ZWG 241.071	11 01 36.	+ 45 24		15.3	GALAXY
MRSL 289+00/4	11 01 36.	- 59 06	2700		HII REGION
ZWG 038.108	11 01 42.	+ 02 34		15.6	GALAXY
ZWG 155.052	11 01 42.	+ 29 20		15.7	GALAXY
ZWG 184.050	11 01 42.	+ 38 28		13.1	GALAXY
MRK 421	11 01 42.	+ 38 28	18	13.5	GALAXY WITH UV CONTINUUM
UGC 06132	11 01 42.	+ 38 28	48	13.1	GALAXY EX CMPT
MCG+07-23-020	11 01 42.	+ 44 17	72	14.5	GALAXY
MCG+08-20-081	11 01 42.	+ 45 24	60	14.	GALAXY
MCG+08-20-082	11 01 42.	+ 46 15	120	15.	GALAXY
MCG+10-16-050	11 01 42.	+ 61 52	18	16.	GALAXY
UGC 06133	11 01 42.	+ 64 16	108	17.	GALAXY DWARF
REIZ 0800	11 01 43.	+ 28 30	60	13.6	GALAXY
REIZ 0801	11 01 43.	+ 29 22	42	14.5	GALAXY
RNGC 3518	11 01 44.	- 06 12			GALAXY
ARC 1155	11 01 46.	+ 35 28		16.6	RICH CLUSTER OF GALAXIES
KARA.72 265A	11 01 48.	+ 05 05	54	14.0	PART OF DOUBLE GALAXY
ZWG 038.109	11 01 48.	+ 05 06		14.0	GALAXY
RNGC 3509	11 01 48.	+ 05 06		14.0	GALAXY
UGC 06134	11 01 48.	+ 05 06	126	14.0	GALAXY S
KARA.72 265B	11 01 48.	+ 05 06	78		PART OF DOUBLE GALAXY
MCG+01-28-033	11 01 48.	+ 05 06	132	14.0	GALAXY
ZWG 066.109	11 01 48.	+ 12 02		15.7	GALAXY
ZWG 095.104	11 01 48.	+ 17 24		15.1	GALAXY
FEIG 036	11 01 48.	+ 24 56		12.5	FAINT BLUE STAR
MCG+05-26-042	11 01 48.	+ 29 22	36	15.7	GALAXY
ZWG 155.053	11 01 48.	+ 29 47		15.5	GALAXY
SHAH 120	11 01 48.	+ 36 09	96	16.8	GROUP OF COMPACT GALAXIES
ZWG 241.072	11 01 48.	+ 45 24		13.0	GALAXY
UGC 06135	11 01 48.	+ 45 24	60	13.0	GALAXY S
ZWG 241.073	11 01 48.	+ 46 15		14.7	GALAXY
UGC 06136	11 01 48.	+ 46 15	114	14.7	GALAXY S?
ZC 1101.8+4741	11 01 48.	+ 47 41	1140		CLUSTER OF GALAXIES
MCG-01-28-023	11 01 48.	- 09 20	72	15.	GALAXY
MCG-02-28-043	11 01 48.	- 09 31	72	15.	GALAXY
LB 01943	11 01 51.	+ 60 06 12.		11.8	FAINT BLUE STAR
HN 1348	11 01 51.8	- 00 16 33.	18		NEBULA
REIZ 0803	11 01 52.	+ 01 39	60	14.7	GALAXY
PK290-00.1	11 01 52.	- 60 19	36		PLANETARY NEBULA
REIZ 0804	11 01 53.	+ 05 06	120	13.3	GALAXY
ZWG 038.110	11 01 54.	+ 02 56		15.3	GALAXY
ZWG 038.111	11 01 54.	+ 03 59		15.1	GALAXY
ARC 1157	11 01 54.	+ 13 54		17.2	RICH CLUSTER OF GALAXIES
ZWG 095.105	11 01 54.	+ 16 20		14.2	GALAXY
UGC 06137	11 01 54.	+ 16 20	78	14.2	GALAXY S0?
MCG+03-28-055	11 01 54.	+ 16 20	78	14.2	GALAXY
TON-N 0053	11 01 54.	+ 24 17		15.5	BLUE STAR
ZWG 155.054	11 01 54.	+ 28 00		15.7	GALAXY
UGC 06138	11 01 54.	+ 28 00	120	15.7	GALAXY
ZWG 155.055	11 01 54.	+ 28 30		14.8	GALAXY
UGC 06139	11 01 54.	+ 28 30	60	14.8	GALAXY Sb-c
MCG+05-26-044	11 01 54.	+ 28 30	45	14.8	GALAXY
MCG+05-26-043	11 01 54.	+ 29 48	33	15.5	GALAXY
ZC 1101.9+3446	11 01 54.	+ 34 46	2550		CLUSTER OF GALAXIES
UGC 06140	11 01 54.	+ 38 29	84	16.0	GALAXY S
MCG+11-14-002A	11 01 54.	+ 64 16	18	16.	GALAXY
ZWG 010.067	11 01 54.	- 00 16		15.7	GALAXY
TON-N 1322	11 01 55.	+ 35 11		17.	BLUE STAR
TON-N 1323	11 01 55.	+ 36 26		14.4	BLUE STAR
TON-N 1321	11 01 56.	+ 21 42		16.6	BLUE STAR
REIZ 0802	11 01 56.	+ 29 47	42	13.7	GALAXY
RNGC 3515	11 01 57.	+ 28 30		15.0	GALAXY
RNGC 3500	11 01 58.	+ 76 04			NON-EXISTENT OBJECT
RNGC 3519	11 01 58.	- 61 06			NON-EXISTENT OBJECT
REIZ 0805	11 01 59.	+ 04 34	24	13.8	GALAXY
SEY 088	11 01 59.	+ 59 58 22.		15.0	FAINT GALAXY
ZC 1102.0+0126	11 02 00.	+ 01 26	1810		CLUSTER OF GALAXIES
ZWG 038.112	11 02 00.	+ 05 28		15.2	GALAXY
UGC 06141	11 02 00.	+ 05 28	72	15.2	GALAXY Sb/SBb
ZWG 038.113	11 02 00.	+ 06 40		15.3	GALAXY
REIZ 0806	11 02 00.	+ 06 40	24	14.9	GALAXY
REIZ 0807	11 02 00.	+ 06 42	18	15.5	GALAXY
MCG+05-26-045	11 02 00.	+ 28 00	132	15.7	GALAXY
ZWG 184.051	11 02 00.	+ 33 19		15.7	GALAXY
KARA.73B 0459	11 02 00.	+ 33 19	36	15.7	ISOLATED GALAXY S
MCG+07-23-021	11 02 00.	+ 38 30	30	15.5	GALAXY
ZC 1102.0+4307	11 02 00.	+ 43 07	1880		CLUSTER OF GALAXIES
ZC 1102.0+5004	11 02 00.	+ 50 04	740		CLUSTER OF GALAXIES
MCG+10-16-051	11 02 00.	+ 59 56	27	16.	GALAXY
MCG+10-16-052	11 02 00.	+ 59 57	36	15.	GALAXY
ZWG 291.026	11 02 00.	+ 59 58		15.2	GALAXY
ZWG 010.068	11 02 00.	- 00 27		15.6	GALAXY
HN 1349	11 02 00.4	- 01 15 03.			NEBULA
ARC 1154	11 02 02.	+ 50 06		17.8	RICH CLUSTER OF GALAXIES
KARA.73 32	11 02 03.	- 37 02	27		DWARF GALAXY
HOLM 225A	11 02 05.	+ 04 34	42	14.1	PART OF MULTIPLE GALAXY
ARC 1156	11 02 05.	+ 47 41		17.8	RICH CLUSTER OF GALAXIES
FATH 1.369	11 02 05.	+ 59 58	14		NEBULA
AR 32	11 02 05.70	+ 75 10 10.5			NEBULA
ZWG 038.114	11 02 06.	+ 04 33		14.5	GALAXY
UGC 06142	11 02 06.	+ 04 33	54	14.5	GALAXY S
MCG+01-28-034	11 02 06.	+ 04 33	36	14.5	GALAXY
ARC 1158	11 02 06.	+ 22 30		17.6	RICH CLUSTER OF GALAXIES
ZC 1102.1+2233	11 02 06.	+ 22 33	810		CLUSTER OF GALAXIES
TON-N 0566	11 02 06.	+ 28 18		15.4	BLUE STAR
ZC 1102.1+3136	11 02 06.	+ 31 36	5910		CLUSTER OF GALAXIES
MCG+09-18-082	11 02 06.	+ 54 57	18	16.	GALAXY
ARP 021	11 02 07.	+ 30 21			PECULIAR GALAXY
TON-N 1324	11 02 07.	+ 36 38		17.	BLUE STAR
HN 1350	11 02 07.0	- 01 16 51.			NEBULA
HOLM 225B	11 02 09.	+ 04 34	24	14.6	PART OF MULTIPLE GALAXY
MCG+08-20-083	11 02 09.	+ 45 00	30	16.	GALAXY
FATH 1.370	11 02 09.	+ 59 57	16		NEBULA
HN 1351	11 02 09.7	- 02 13 09.	12		NEBULA
ZWG 038.115	11 02 12.	+ 04 33		15.2	GALAXY
MCG+01-28-035	11 02 12.	+ 04 33	24	15.2	GALAXY
REIZ 0808	11 02 12.	+ 25 44	18	15.3	GALAXY
MRK 036	11 02 12.	+ 29 24	15	15.5	GALAXY WITH UV CONTINUUM
MCG+05-26-046	11 02 12.	+ 29 24	18	15.	GALAXY

OBJECT NAME	RIGHT ASCEN.	DECLINATION	DIAM.	MAGN.	TYPE OF OBJECT
ZWG 155.056	11 02 12.	+ 30 18		14.7	GALAXY
SN 1955L	11 02 12.	+ 30 18		16.5	SUPERNOVA
ZWG 185.001	11 02 12.	+ 35 38		15.3	GALAXY
ZWG 184.052	11 02 12.	+ 35 38		15.3	GALAXY
UGC 06143	11 02 12.	+ 35 38	84	15.3	GALAXY Sb
ZWG 241.074	11 02 12.	+ 45 00		14.6	GALAXY
MRK 162	11 02 12.	+ 45 00	22	15.	GALAXY WITH UV CONTINUUM
ZC 1102.2+4710	11 02 12.	+ 47 10	1950		CLUSTER OF GALAXIES
MCG+10-16-053	11 02 12.	+ 59 55	36	16.	GALAXY
MCG+10-16-054	11 02 12.	+ 60 01	12	16.	GALAXY
MCG+12-11-007	11 02 12.	+ 71 59	39	16.	GALAXY
ZWG 010.070	11 02 12.	- 01 15		14.8	GALAXY
MCG+00-28-029	11 02 12.	- 01 15	18	14.8	GALAXY
ZWG 010.069	11 02 12.	- 02 13		15.7	GALAXY
MCG+04-26-027	11 02 15.	+ 25 05	42	15.5	GALAXY
MCG+05-26-047	11 02 15.	+ 30 19	48	14.7	GALAXY
REIZ 0809	11 02 16.	+ 35 38	36	15.2	GALAXY
HARO 04	11 02 17.	+ 29 24			BLUE EMISSION-LINE GALAXY
ZWG 155.057	11 02 18.	+ 30 13		15.6	GALAXY
MCG+05-26-048	11 02 18.	+ 30 13	48	15.6	GALAXY
ZC 1102.3+3155	11 02 18.	+ 31 55	940		CLUSTER OF GALAXIES
MCG+06-24-046	11 02 18.	+ 35 37 30.	84	14.5	GALAXY
ZWG 185.002	11 02 18.	+ 38 20		15.4	GALAXY
ZWG 184.053	11 02 18.	+ 38 20		15.4	GALAXY
STOCK 21	11 02 18.	+ 47 15			BLUE KNOT NEAR ELLIP GLXY
REIZ 0810	11 02 19.	+ 28 53	54	14.4	GALAXY
REIZ 0811	11 02 20.	+ 30 14	18	14.4	GALAXY
REIZ 0812	11 02 20.	+ 30 18	30	13.6	GALAXY
REIZ 0814	11 02 20.	+ 31 27	12	15.9	GALAXY
REIZ 0813	11 02 20.	+ 31 27	6	16.0	GALAXY
REIZ 0815	11 02 20.	+ 31 28	12	15.7	GALAXY
TON-N 1325	11 02 20.	+ 33 57		17.	BLUE STAR
FATH 1.371	11 02 21.	+ 60 02	8		NEBULA
FATH 1.372	11 02 22.	+ 59 56	16		NEBULA
REIZ 0817	11 02 23.	+ 04 34	18	14.8	GALAXY
ZWG 038.116	11 02 24.	+ 04 33		15.1	GALAXY
MCG+01-28-036	11 02 24.	+ 04 33	36	15.1	GALAXY
ZWG 095.106	11 02 24.	+ 15 09		15.1	GALAXY
KARA.73B 0460	11 02 24.	+ 15 09	36	15.1	ISOLATED GALAXY S
GO 099	11 02 25.	- 59 30	37	18.4	STAR CHAIN
TON-N 1326	11 02 26.	+ 32 10		17.	BLUE STAR
TON-N 1327	11 02 26.	+ 33 21		15.9	BLUE STAR
MCG+06-24-047	11 02 27.	+ 38 21	36	15.	GALAXY
SVEN 246	11 02 27.	- 22 52	24	14.9	GALAXY
LB 00252	11 02 28.	+ 59 54 06.		14.1	FAINT BLUE STAR
HELW 433	11 02 28.	- 22 52 10.			NEBULA
ZWG 038.117	11 02 30.	+ 04 51		15.4	GALAXY
ZWG 038.118	11 02 30.	+ 08 13		15.2	GALAXY
ZWG 155.058	11 02 30.	+ 30 08		15.7	GALAXY
ZWG 213.027	11 02 30.	+ 38 30		15.4	GALAXY
ZWG 334.010	11 02 30.	+ 72 46		15.4	GALAXY
REIZ 0818	11 02 32.	+ 30 09	30	14.6	GALAXY
REIZ 0819	11 02 32.	+ 30 15	12	14.3	GALAXY
REIZ 0820	11 02 32.	+ 30 21	24	14.8	GALAXY
HN 1352	11 02 34.7	- 00 30 52.	30		NEBULA
ZC 1102.6+0709	11 02 36.	+ 07 09	940		CLUSTER OF GALAXIES
8ZW 1102+11.6	11 02 36.	+ 11 38		19.5	COMPACT GALAXY
ZWG 155.059	11 02 36.	+ 30 25		15.7	GALAXY
MCG+05-26-049	11 02 36.	+ 30 26	18	15.7	GALAXY
ZWG 185.003	11 02 36.	+ 38 17		15.5	GALAXY
ZWG 184.054	11 02 36.	+ 38 17		15.5	GALAXY
KARA.72 266A	11 02 36.	+ 56 47	66	13.8	PART OF DOUBLE GALAXY
MCG+10-16-055	11 02 36.	+ 56 47	24	17.	GALAXY
ZWG 291.027	11 02 36.	+ 56 48		13.8	GALAXY
UGC 06144	11 02 36.	+ 56 48	60	13.8	GALAXY Sa-b
KARA.72 266B	11 02 36.	+ 56 48	36		PART OF DOUBLE GALAXY
LB 01944	11 02 36.	+ 57 40 06.		15.0	FAINT BLUE STAR
MCG+10-16-056	11 02 36.	+ 60 38	57	16.	GALAXY
ZC 1102.6+6110	11 02 36.	+ 61 10	1080		CLUSTER OF GALAXIES
RNGC 3517	11 02 37.	+ 56 48		14.0	GALAXY
MCG+10-16-057	11 02 39.	+ 56 47	60	14.1	GALAXY
ARC 1160	11 02 41.	- 18 42		17.4	RICH CLUSTER OF GALAXIES
ZWG 155.060	11 02 42.	+ 26 39		15.3	GALAXY
STOCK 22	11 02 42.	+ 30 25			BLUE KNOT NEAR ELLIP GLXY
ZC 1102.7+3325	11 02 42.	+ 33 25	1140		CLUSTER OF GALAXIES
MCG+08-20-084	11 02 42.	+ 46 26	48	16.	GALAXY
REIZ 0816	11 02 42.	+ 56 48	36	14.3	GALAXY
MCG+10-16-058	11 02 42.	+ 61 01	27	16.	GALAXY
ZWG 010.071	11 02 42.	- 00 31		15.1	GALAXY
REIZ 0821	11 02 44.	+ 31 25	18	15.1	GALAXY
MCG+05-26-050	11 02 45.	+ 26 38	30	15.3	GALAXY
LB 01945	11 02 46.	+ 51 41 06.		16.7	FAINT BLUE STAR
ZWG 038.119	11 02 48.	+ 02 49		15.7	GALAXY
ZC 1102.8+1436	11 02 48.	+ 14 36	400		CLUSTER OF GALAXIES
ZWG 095.107	11 02 48.	+ 17 55		14.9	GALAXY
MCG+03-28-056	11 02 48.	+ 17 55	30	14.9	GALAXY
ZWG 095.108	11 02 48.	+ 19 06		15.5	GALAXY
ZC 1102.8+3437	11 02 48.	+ 34 37	400		CLUSTER OF GALAXIES
MCG+06-24-048	11 02 48.	+ 38 01	24	16.	GALAXY
MCG+10-16-059	11 02 48.	+ 60 38	30	16.	GALAXY
MCG-04-26-022	11 02 48.	- 26 22	84	15.	GALAXY
ARC 1150	11 02 49.	+ 73 58		16.5	RICH CLUSTER OF GALAXIES
TON-N 1328	11 02 50.	+ 22 00		16.2	BLUE STAR
ZC 1102.9+0340	11 02 54.	+ 03 40	870		CLUSTER OF GALAXIES
ARC 1159	11 02 54.	+ 12 49		17.2	RICH CLUSTER OF GALAXIES
ZC 1102.9+1913	11 02 54.	+ 19 13	7190		CLUSTER OF GALAXIES
ZWG 185.004	11 02 54.	+ 35 23		15.5	GALAXY
ZWG 184.055	11 02 54.	+ 35 23		15.5	GALAXY
ZC 1102.9+4230	11 02 54.	+ 42 30	810		CLUSTER OF GALAXIES
MCG-02-28-044	11 02 54.	- 09 38	24	15.5	GALAXY
OCL 0857	11 02 54.	- 67 40	102	12.	OPEN STAR CLUSTER
REIZ 0822	11 02 56.	+ 31 26	12	14.9	GALAXY
TON-N 1329	11 02 56.	+ 34 43		15.3	BLUE STAR
REIZ 0823	11 02 58.	+ 35 23	24	15.0	GALAXY
REIZ 0824	11 02 58.	+ 35 46	18	15.5	GALAXY
VDB.66G 091	11 03	+ 20 04	70		DWARF GALAXY
UGC 06145	11 03 00.	+ 04 26	90	17.	GALAXY DWRF IR
UGC 06146	11 03 00.	+ 04 29	78	17.	GALAXY IRR
ZWG 095.109	11 03 00.	+ 20 04		15.5	GALAXY
ZC 1103.0+2550	11 03 00.	+ 25 50	810		CLUSTER OF GALAXIES
MCG+05-26-051	11 03 00.	+ 30 49	27	15.5	GALAXY
ZC 1103.0+3239	11 03 00.	+ 32 39	2220		CLUSTER OF GALAXIES
STOCK 23	11 03 00.	+ 34 03			BLUE KNOT NEAR ELLIP GLXY
MCG+06-25-001	11 03 00.	+ 35 22	36	15.	GALAXY
ZC 1103.0+5115	11 03 00.	+ 51 15	740		CLUSTER OF GALAXIES
ZWG 291.028	11 03 00.	+ 59 13		15.7	GALAXY
FATH 1.373	11 03 00.	+ 59 13			NEBULA
ZWG 365.001	11 03 00.	+ 83 08		15.4	GALAXY
ZWG 364.019	11 03 00.	+ 83 08		15.4	GALAXY
MCG-02-28-045	11 03 00.	- 09 39	48	14.5	GALAXY
HN 1353	11 03 00.3	- 01 53 04.	24		NEBULA
REIN 1.001	11 03 01.61	+ 00 06 25.5			NEBULA
ARC 1161	11 03 03.	- 21 51		17.3	RICH CLUSTER OF GALAXIES
REIZ 0825	11 03 04.	+ 35 45	30	15.4	GALAXY
RNGC 3520	11 03 04.	- 17 40			NON-EXISTENT OBJECT
REIN 1.002	11 03 04.39	+ 00 03 09.3			NEBULA
ZWG 010.072	11 03 06.	+ 00 07		15.7	GALAXY
ZC 1103.1+0756	11 03 06.	+ 07 56	1080		CLUSTER OF GALAXIES
8ZW 1103+10.5	11 03 06.	+ 10 33		19.2	COMPACT GALAXY
UGC 06147	11 03 06.	+ 29 05	84	16.0	GALAXY S
MCG+05-26-052	11 03 06.	+ 31 40	24	16.	GALAXY
ZC 1103.1+3300	11 03 06.	+ 33 00	610		CLUSTER OF GALAXIES
ZC 1103.1+3517	11 03 06.	+ 35 17	400		CLUSTER OF GALAXIES
ZWG 185.005	11 03 06.	+ 36 40		15.3	GALAXY
ZWG 184.056	11 03 06.	+ 36 40		15.3	GALAXY
UGC 06148	11 03 06.	+ 36 40	90	15.3	GALAXY Sb
SHAH 007	11 03 06.	+ 40 03	48	17.5	GROUP OF COMPACT GALAXIES
MCG+07-23-022	11 03 06.	+ 42 47 30.	36	16.	GALAXY
MCG-02-28-046	11 03 06.	- 09 38	9	16.	GALAXY
REIZ 0826	11 03 08.	+ 31 40	12	15.0	GALAXY
ZWG 038.120	11 03 12.	+ 06 58		15.5	GALAXY
ZC 1103.2+1253	11 03 12.	+ 12 53	1680		CLUSTER OF GALAXIES
MCG+03-28-057	11 03 12.	+ 20 06 30.	108	15.5	GALAXY
MCG+05-26-053	11 03 12.	+ 29 05	66	15.5	GALAXY
ZWG 213.028	11 03 12.	+ 43 36		14.6	GALAXY
UGC 06149	11 03 12.	+ 43 36	66	14.6	GALAXY S
ZWG 241.075	11 03 12.	+ 48 32		15.7	GALAXY
ZC 1103.2+5835	11 03 12.	+ 58 35	740		CLUSTER OF GALAXIES
LB 01946	11 03 12.	+ 61 02 30.		15.0	FAINT BLUE STAR
ZWG 010.073	11 03 12.	- 01 13		15.6	GALAXY
MCG-01-28-024	11 03 12.	- 09 15	60	15.	GALAXY
MCG-02-28-048	11 03 12.	- 09 39	54	15.	GALAXY
MCG-02-28-047	11 03 12.	- 15 16 30.	48	15.5	GALAXY
REIN 1.003	11 03 12.84	+ 00 14 27.9			NEBULA
REIN 1.004	11 03 12.93	+ 00 15 46.2			NEBULA
REIN 1.005	11 03 13.04	+ 00 07 56.8			NEBULA
REIN 1.006	11 03 13.16	+ 00 13 44.0			NEBULA
REIN 1.007	11 03 13.19	+ 00 15 12.5			NEBULA
REIN 1.008	11 03 13.55	+ 00 14 43.8			NEBULA
REIN 1.009	11 03 14.31	+ 00 12 09.0			NEBULA
REIN 1.010	11 03 14.58	+ 00 15 27.8			NEBULA
REIN 1.011	11 03 14.97	+ 00 14 04.3			NEBULA
MCG+06-25-002	11 03 15.	+ 36 40	72	15.	GALAXY
REIN 1.012	11 03 15.87	+ 00 15 40.5			NEBULA
REIZ 0827	11 03 16.	+ 00 14	420	10.6	GALAXY
REIN 1.013	11 03 17.10	+ 00 13 24.4			NEBULA
ZWG 010.074	11 03 18.	+ 00 15		10.1	GALAXY
UGC 06150	11 03 18.	+ 00 15	810	10.1	GALAXY Sb
MCG+00-28-030	11 03 18.	+ 00 15	420	10.1	GALAXY
KARA.73B 0461	11 03 18.	+ 00 15	942	10.1	ISOLATED GALAXY S
ZWG 038.121	11 03 18.	+ 02 56		15.4	GALAXY
UGC 06151	11 03 18.	+ 20 06	108	17.	GALAXY DWRF SP
ZWG 155.061	11 03 18.	+ 30 12		15.3	GALAXY
UGC 06152	11 03 18.	+ 30 12	78	15.3	GALAXY SBb
MCG+05-26-054	11 03 18.	+ 30 13	66	15.3	GALAXY
ZWG 185.006	11 03 18.	+ 34 58		15.7	GALAXY
ZWG 184.057	11 03 18.	+ 34 58		15.7	GALAXY
MCG+07-23-023	11 03 18.	+ 43 35	60	14.5	GALAXY
MCG+10-16-060	11 03 18.	+ 58 24	30	16.	GALAXY
ZWG 334.011	11 03 18.	+ 72 49		12.3	GALAXY
UGC 06153	11 03 18.	+ 72 49	138	12.3	GALAXY SB0
MCG-03-28-036	11 03 18.	- 20 34	60	14.	GALAXY
RNGC 3521	11 03 19.	+ 00 14		10.5	GALAXY
REIN 2.119A	11 03 22.61	+ 72 50 22.9			NEBULA
REIN 2.119B	11 03 22.89	+ 72 50 21.4			NEBULA
RNGC 3516	11 03 23.	+ 72 50		12.5	GALAXY
ZWG 066.110	11 03 24.	+ 08 38		15.0	GALAXY
MCG+02-28-049	11 03 24.	+ 08 38	48	15.0	GALAXY
ZWG 125.024	11 03 24.	+ 21 00		15.5	GALAXY
KARA.73B 0462	11 03 24.	+ 21 00	18	15.5	ISOLATED GALAXY E
ZWG 155.062	11 03 24.	+ 31 53		15.6	GALAXY
MCG+06-25-003	11 03 24.	+ 34 57	30	15.5	GALAXY
ZC 1103.4+5405	11 03 24.	+ 54 05	3430		CLUSTER OF GALAXIES
MCG+09-18-083	11 03 24.	+ 54 32	30	16.	GALAXY
7ZW 366	11 03 24.	+ 61 46			COMPACT GALAXY
VVI 44	11 03 24.	+ 72 50	100	12.6	SEYFERT GALAXY
ZWG 351.055	11 03 24.	+ 76 58		14.8	GALAXY
UGC 06154	11 03 24.	+ 76 58	60	14.8	GALAXY SBa
REIZ 0828	11 03 25.	+ 30 12	42	13.5	GALAXY
REIZ 0829	11 03 26.	+ 31 54	24	14.5	GALAXY
ARC 1163	11 03 28.	- 21 17		17.0	RICH CLUSTER OF GALAXIES
REIZ 0830	11 03 29.	+ 04 56	18	14.2	GALAXY
ZWG 038.122	11 03 30.	+ 03 35		15.3	GALAXY
ZWG 038.123	11 03 30.	+ 04 42		14.4	GALAXY
UGC 06155	11 03 30.	+ 04 42	84	14.4	GALAXY Sb/SBc
MCG+01-28-037	11 03 30.	+ 04 42	84	14.4	GALAXY
MCG+04-26-027A	11 03 30.	+ 24 38	24	15.5	GALAXY
TON-N 0567	11 03 30.	+ 28 31		16.8	BLUE STAR
MCG+06-25-004	11 03 30.	+ 32 30	24	16.	GALAXY
ZC 1103.5+3500	11 03 30.	+ 35 00	810		CLUSTER OF GALAXIES
ZC 1103.5+4100	11 03 30.	+ 41 00	4030		CLUSTER OF GALAXIES
ZC 1103.5+4515	11 03 30.	+ 45 15	3020		CLUSTER OF GALAXIES
MCG+08-20-085	11 03 30.	+ 45 46	30	16.	GALAXY
ZWG 241.076	11 03 30.	+ 45 47		15.6	GALAXY
MRK 163	11 03 30.	+ 48 54	6	17.5	GALAXY WITH UV CONTINUUM
ZC 1103.5+5029	11 03 30.	+ 50 29	4370		CLUSTER OF GALAXIES
7ZW 367	11 03 30.	+ 76 58			COMPACT GALAXY
TON-N 1330	11 03 32.	+ 33 38		15.6	BLUE STAR
MCG+08-20-086	11 03 33.	+ 48 55	60	15.	GALAXY
REIZ 0831	11 03 35.	+ 04 36	24	14.4	GALAXY
ZWG 038.124	11 03 36.	+ 03 35		15.1	GALAXY
ZWG 038.125	11 03 36.	+ 04 36		14.7	GALAXY
MCG+01-28-038	11 03 36.	+ 04 36	60	14.7	GALAXY
ZC 1103.6+0529	11 03 36.	+ 05 29	1480		CLUSTER OF GALAXIES
ZC 1103.6+0838	11 03 36.	+ 08 38	1610		CLUSTER OF GALAXIES
TON-N 0568	11 03 36.	+ 31 59		15.1	BLUE STAR
ZWG 185.007	11 03 36.	+ 38 00		15.6	GALAXY
ZWG 184.058	11 03 36.	+ 38 00		15.6	GALAXY
ZWG 241.077	11 03 36.	+ 48 55		15.1	GALAXY
UGC 06156	11 03 36.	+ 48 55	78	15.1	GALAXY S
MCG+09-18-084	11 03 36.	+ 53 37 30.	60	15.	GALAXY
MCG+12-11-008	11 03 36.	+ 71 55	21	16.	GALAXY
TON-N 1332	11 03 37.	+ 35 12		15.4	BLUE STAR
AR 33B	11 03 37.96	+ 76 57 56.5			NEBULA
TON-N 1331	11 03 38.	+ 22 00		16.9	BLUE STAR
AR 33A	11 03 38.20	+ 76 57 58.3			NEBULA

OBJECT NAME	RIGHT ASCEN.	DECLINATION	DIAM.	MAGN.	TYPE OF OBJECT
ARC 1162	11 03 42.	+ 04 13		17.6	RICH CLUSTER OF GALAXIES
ZWG 095.110	11 03 42.	+ 17 46		15.0	GALAXY
UGC 06157	11 03 42.	+ 17 46	108	15.0	GALAXY
MCG+03-28-058	11 03 42.	+ 17 47	120	15.0	GALAXY
ZC 1103.7+2700	11 03 42.	+ 27 00	1140		CLUSTER OF GALAXIES
ZC 1103.7+4932	11 03 42.	+ 49 32	870		CLUSTER OF GALAXIES
MCG+09-18-085	11 03 42.	+ 53 29	36	16.	GALAXY
ZWG 267.039	11 03 42.	+ 53 39		15.3	GALAXY
ZWG 314.005	11 03 42.	+ 67 08		15.5	GALAXY
MCG+13-08-055	11 03 42.	+ 76 59	54	14.	GALAXY
PATH 1.374	11 03 45.	+ 59 28	8		NEBULA
RNGC 6298	11 03 46.	+ 62 08			GALAXY
HN 1354	11 03 46.5	- 02 02 41.			NEBULA
ZC 1103.8+0411	11 03 48.	+ 04 11	1340		CLUSTER OF GALAXIES
8ZW 1103+08.6	11 03 48.	+ 08 38		19.6	COMPACT GALAXY
ZWG 066.111	11 03 48.	+ 11 41		15.7	GALAXY
KARA.72 267A	11 03 48.	+ 11 41	30	15.7	PART OF DOUBLE GALAXY
ZWG 241.078	11 03 48.	+ 46 18		15.6	GALAXY
MCG+09-18-086	11 03 48.	+ 51 27	144	13.	GALAXY
ZWG 267.040	11 03 48.	+ 53 30		15.7	GALAXY
MCG+10-16-061	11 03 48.	+ 57 57	57	14.	GALAXY
ZWG 291.029	11 03 48.	+ 57 58		15.0	GALAXY
MCG+12-11-009	11 03 48.	+ 72 50	102	12.7	GALAXY
ZWG 010.075	11 03 48.	- 02 02		15.7	GALAXY
TON-N 1333	11 03 50.	+ 32 26		17.	BLUE STAR
TON-N 1334	11 03 50.	+ 34 01		17.	BLUE STAR
ARC 1165	11 03 50.	- 24 28		17.5	RICH CLUSTER OF GALAXIES
SEY 089	11 03 51.	+ 57 57 50.		14.8	FAINT GALAXY
ARC 1164	11 03 53.	+ 02 20		17.7	RICH CLUSTER OF GALAXIES
ZWG 038.126	11 03 54.	+ 06 50		15.6	GALAXY
ZWG 066.112	11 03 54.	+ 11 40		13.4	GALAXY
RNGC 3524	11 03 54.	+ 11 40		13.5	GALAXY
UGC 06158	11 03 54.	+ 11 40	102	13.4	GALAXY S0-a
KARA.72 267B	11 03 54.	+ 11 40	90	13.4	PART OF DOUBLE GALAXY
MCG+02-28-050	11 03 54.	+ 11 40	96	13.4	GALAXY
ZWG 095.111	11 03 54.	+ 15 21		15.5	GALAXY
ZC 1103.9+2201	11 03 54.	+ 22 01	1750		CLUSTER OF GALAXIES
ZWG 241.079	11 03 54.	+ 46 14		14.6	GALAXY
MCG+08-20-087	11 03 54.	+ 46 14	18	14.	GALAXY
ZWG 241.080	11 03 54.	+ 48 07		15.6	GALAXY
MCG+10-16-062	11 03 54.	+ 60 31	39	16.	GALAXY
OCL 0842	11 03 54.	- 59 33			OPEN STAR CLUSTER
ASS 61	11 03 54.	- 59 35			OB ASSOCIATION CAR OB2
REIZ 0832	11 03 55.	+ 11 40	60	13.6	GALAXY
RNGC 3522	11 03 58.	+ 20 20		14.0	GALAXY
RNGC 3525	11 03 58.	- 19 11			NON-EXISTENT OBJECT
SHB 174	11 03 58.2	- 00 36 38.		16.	QUASI-STELLAR OBJECT
CED 110	11 04	- 77 04			DIFFUSE GALACTIC NEBULA
ZWG 010.076	11 04 00.	+ 00 34		15.6	GALAXY
ZC 1104.0+0211	11 04 00.	+ 02 11	1550		CLUSTER OF GALAXIES
ZWG 066.113	11 04 00.	+ 13 40		15.5	GALAXY
ZC 1104.0+1417	11 04 00.	+ 14 17	1210		CLUSTER OF GALAXIES
MCG+03-28-059	11 04 00.	+ 15 18	36	15.4	GALAXY
ZWG 095.112	11 04 00.	+ 15 19		15.4	GALAXY
ZWG 095.113	11 04 00.	+ 20 20		14.2	GALAXY
UGC 06159	11 04 00.	+ 20 20	72	14.2	GALAXY E
MCG+03-28-060	11 04 00.	+ 20 22	30	14.2	GALAXY
ZC 1104.0+2549	11 04 00.	+ 25 49	1010		CLUSTER OF GALAXIES
ZWG 155.063	11 04 00.	+ 28 59		15.1	GALAXY
UGC 06160	11 04 00.	+ 28 59	60	15.1	GALAXY Sb-c
ZWG 155.064	11 04 00.	+ 32 00		15.6	GALAXY
MCG+07-23-024	11 04 00.	+ 43 59	138	14.	GALAXY
ZWG 213.029	11 04 00.	+ 44 00		14.4	GALAXY
UGC 06161	11 04 00.	+ 44 00	180	14.4	GALAXY
ZWG 267.041	11 04 00.	+ 51 30		14.2	GALAXY
UGC 06162	11 04 00.	+ 51 30	150	14.2	GALAXY Sc
KARA.73B 0463	11 04 00.	+ 51 30	156	14.2	ISOLATED GALAXY S
LB 01947	11 04 00.	+ 62 58 24.		14.4	FAINT BLUE STAR
ZC 1104.0+6507	11 04 00.	+ 65 07	1950		CLUSTER OF GALAXIES
7ZW 368	11 04 00.	+ 67 08			COMPACT GALAXY
MCG-01-28-025	11 04 00.	- 05 27	30	15.	GALAXY
IC 0667	11 04 01.	+ 15 21 24.			NONSTELLAR OBJECT
REIZ 0833	11 04 01.	+ 28 59	30	13.9	GALAXY
TON-N 1335	11 04 02.	+ 21 46		16.2	BLUE STAR
REIZ 0834	11 04 02.	+ 32 01	18	14.5	GALAXY
TON-N 1336	11 04 02.	+ 33 12		17.	BLUE STAR
IC 0668	11 04 03.	+ 15 18 30.			NONSTELLAR OBJECT
MCG+05-26-055	11 04 03.	+ 29 00	48	15.1	GALAXY
LB 01948	11 04 04.	+ 56 17 48.		15.7	FAINT BLUE STAR
MCG+04-26-028	11 04 06.	+ 23 17	72	14.9	GALAXY
ZWG 125.025	11 04 06.	+ 23 18		14.9	GALAXY
UGC 06163	11 04 06.	+ 23 18	66	14.9	GALAXY Sa
MCG+05-26-056	11 04 06.	+ 28 59	60	15.1	GALAXY
MCG+08-20-088	11 04 06.	+ 46 05 30.	96	15.	GALAXY
LB 01949	11 04 06.	+ 51 38 18.		16.4	FAINT BLUE STAR
MCG+09-18-087	11 04 06.	+ 55 27	12	17.	GALAXY
ZWG 314.006	11 04 06.	+ 65 22		15.0	GALAXY
ZC 1104.1-0226	11 04 06.	- 02 26	1610		CLUSTER OF GALAXIES
REIZ 0836	11 04 11.	+ 23 17	36	14.0	GALAXY
HOLM 226A	11 04 11.	+ 23 17	48	14.2	PART OF MULTIPLE GALAXY
MCG+04-26-029	11 04 12.	+ 23 16	60	15.7	GALAXY
ZWG 125.026	11 04 12.	+ 23 17		15.7	GALAXY
UGC 06164	11 04 12.	+ 23 17	60	15.7	GALAXY S
TON-N 0054	11 04 12.	+ 29 57		14.8	BLUE STAR
ZWG 241.081	11 04 12.	+ 46 06		15.3	GALAXY
UGC 06165	11 04 12.	+ 46 06	90	15.3	GALAXY Sa-b
MCG+10-16-063	11 04 12.	+ 57 20	27	16.	GALAXY
7ZW 369	11 04 12.	+ 61 44			COMPACT GALAXY
MCG+11-14-003A	11 04 12.	+ 65 22		16.	GALAXY
MCG+11-14-003B	11 04 12.	+ 65 22 30.	39	16.	GALAXY
ZC 1104.2+6857	11 04 12.	+ 68 57	870		CLUSTER OF GALAXIES
7ZW 370	11 04 12.	+ 69 48			COMPACT GALAXY
MCG-06-25-001	11 04 12.	- 37 23	120	13.	GALAXY
LB 01950	11 04 14.	+ 60 51 00.		13.2	FAINT BLUE STAR
MCG+06-25-005	11 04 15.	+ 35 49	10	15.5	GALAXY
LB 01951	11 04 17.	+ 55 11 54.		16.0	FAINT BLUE STAR
PK288+05.1	11 04 17.	- 54 32	10		PLANETARY NEBULA
ZWG 038.127	11 04 18.	+ 06 18		15.7	GALAXY
REIZ 0837	11 04 18.	+ 07 27	120	13.4	GALAXY
ZWG 066.114	11 04 18.	+ 14 28		14.9	GALAXY
MCG+02-28-051	11 04 18.	+ 14 28	9	14.9	GALAXY
ZWG 095.114	11 04 18.	+ 18 38		15.4	GALAXY
ZWG 155.065	11 04 18.	+ 28 52		15.5	GALAXY
UGC 06166	11 04 18.	+ 28 52	66	15.5	GALAXY Sb-c
ZC 1104.3+2911	11 04 18.	+ 29 11	1140		CLUSTER OF GALAXIES
ZWG 010.077	11 04 18.	- 02 14		15.7	GALAXY
MCG-01-28-026	11 04 18.	- 05 08	30	15.	GALAXY
OCL 0839	11 04 18.	- 58 24	4560	3.6	OPEN STAR CLUSTER

OBJECT NAME	RIGHT ASCEN.	DECLINATION	DIAM.	MAGN.	TYPE OF OBJECT
VHA 109	11 04 18.	- 58 24	1500		OPEN STAR CLUSTER
MRSL 290+00/2	11 04 18.	- 59 48	2400		HII REGION
MRSL 292-04/1	11 04 18.	- 65 18	420		HII REGION
REIZ 0838	11 04 19.	+ 12 22	42	14.5	GALAXY
HOLM 226B	11 04 19.	+ 23 16	36	14.8	PART OF MULTIPLE GALAXY
RNGC 3532	11 04 20.	- 58 24		3.5	OPEN CLUSTER
ZWG 038.128	11 04 24.	+ 02 53		15.7	GALAXY
ZWG 038.129	11 04 24.	+ 07 26		13.7	GALAXY
RNGC 3526	11 04 24.	+ 07 26		13.5	GALAXY
UGC 06167	11 04 24.	+ 07 26	150	13.7	GALAXY Sc
MCG+01-28-039	11 04 24.	+ 07 26	102	13.7	GALAXY
KARA.73B 0464	11 04 24.	+ 07 26	132	13.7	ISOLATED GALAXY S
TON-N 0569	11 04 24.	+ 28 07		14.9	BLUE STAR
MCG+05-26-058	11 04 24.	+ 28 42	42	15.	GALAXY
MCG+05-26-057	11 04 24.	+ 28 52	54	15.2	GALAXY
ZC 1104.4+3339	11 04 24.	+ 33 39	1140		CLUSTER OF GALAXIES
ZC 1104.4+3708	11 04 24.	+ 37 08	610		CLUSTER OF GALAXIES
ZWG 241.082	11 04 24.	+ 46 39		14.8	GALAXY
MCG+08-20-089	11 04 24.	+ 46 39	18	15.	GALAXY
MCG+09-18-088	11 04 24.	+ 54 47	36	16.	GALAXY
REIZ 0835	11 04 24.	+ 57 57	48	14.5	GALAXY
REIZ 0839	11 04 25.	+ 28 51	36	14.4	GALAXY
TON-N 1337	11 04 26.	+ 32 23		17.1	BLUE STAR
ZWG 038.130	11 04 30.	+ 08 04		15.1	GALAXY
UGC 06168	11 04 30.	+ 08 04	84	15.1	GALAXY Sb
ZWG 066.115	11 04 30.	+ 12 20		14.6	GALAXY
UGC 06169	11 04 30.	+ 12 20	114	14.6	GALAXY Sb
MCG+02-28-052	11 04 30.	+ 12 20	108	14.6	GALAXY
MCG+03-28-061	11 04 30.	+ 18 51	150	15.1	GALAXY
ZWG 155.066	11 04 30.	+ 28 48		15.1	GALAXY
UGC 06170	11 04 30.	+ 28 48	72	15.1	GALAXY SBa
ZWG 155.067	11 04 30.	+ 29 03		15.6	GALAXY
SHAH 058	11 04 30.	+ 34 20	210	18.5	GROUP OF COMPACT GALAXIES
ZC 1104.5+3857	11 04 30.	+ 38 57	2290		CLUSTER OF GALAXIES
MCG+07-23-025	11 04 30.	+ 42 42	27	17.	GALAXY
ZC 1104.5+4638	11 04 30.	+ 46 38	1550		CLUSTER OF GALAXIES
ZWG 267.042	11 04 30.	+ 54 47		15.5	GALAXY
LB 03499	11 04 30.	- 79 06		12.8	FAINT BLUE STAR
RNGC 3527	11 04 33.	+ 28 48		15.0	GALAXY
BC 4C16.30	11 04 35.2	+ 16 44 06.		15.70	QUASI-STELLAR OBJECT
ZWG 038.131	11 04 36.	+ 06 40		15.5	GALAXY
8ZW 1104+08.3	11 04 36.	+ 08 21		18.3	COMPACT GALAXY
ZC 1104.6+1606	11 04 36.	+ 16 06	2550		CLUSTER OF GALAXIES
ZWG 096.001	11 04 36.	+ 18 50		15.1	GALAXY
ZWG 095.115	11 04 36.	+ 18 50		15.1	GALAXY
UGC 06171	11 04 36.	+ 18 50	150	15.1	GALAXY IRR
ZWG 125.027	11 04 36.	+ 23 13		15.4	GALAXY
UGC 06172	11 04 36.	+ 23 13	66	15.4	GALAXY DBL SYS
ZWG 125.028	11 04 36.	+ 23 45		14.4	GALAXY
UGC 06173	11 04 36.	+ 23 45	60	14.4	GALAXY S
MCG+04-26-030	11 04 36.	+ 23 45	60	14.4	GALAXY
MCG+05-26-059	11 04 36.	+ 28 48	42	15.1	GALAXY
8ZW 1104-19.5	11 04 36.	- 19 32		15.5	COMPACT GALAXY
SHB 175	11 04 36.7	+ 16 44 17.		15.7	QUASI-STELLAR OBJECT
REIZ 0840	11 04 37.	+ 28 48	36	14.4	GALAXY
ARP 191	11 04 38.	+ 18 42			PECULIAR GALAXY
REIZ 0841	11 04 38.	+ 31 37	12	16.0	GALAXY
RNGC 3528	11 04 40.	- 19 14		13.0	GALAXY
REIZ 0842	11 04 41.	+ 23 46	30	13.8	GALAXY
ZWG 038.132	11 04 42.	+ 06 34		14.3	GALAXY
UGC 06174	11 04 42.	+ 06 34	96	14.3	GALAXY S0-a
MCG+01-28-040	11 04 42.	+ 06 34	36	14.3	GALAXY
IC 0669	11 04 42.	+ 06 34 23.			NONSTELLAR OBJECT
ZWG 096.002	11 04 42.	+ 18 42		14.7	GALAXY
ZWG 095.116	11 04 42.	+ 18 42		14.7	GALAXY
UGC 06175	11 04 42.	+ 18 42	114	14.7	GALAXY DBL SYS
KARA.72 268A	11 04 42.	+ 18 42	42	15.7	PART OF DOUBLE GALAXY
MCG+03-28-062	11 04 42.	+ 18 42	12	16.	GALAXY
VV 239B	11 04 42.	+ 18 42 30.	15	17.	INTERACTING GALAXY
VV 239A	11 04 42.	+ 18 42 30.	27	16.	INTERACTING GALAXY
MCG+04-26-031	11 04 42.	+ 21 55	72	14.8	GALAXY
ZWG 125.029	11 04 42.	+ 21 56		14.8	GALAXY
UGC 06176	11 04 42.	+ 21 56	72	14.8	GALAXY SB0
TON-N 0055	11 04 42.	+ 28 21		14.9	BLUE STAR
ZC 1104.7+3610	11 04 42.	+ 36 10	1010		CLUSTER OF GALAXIES
MCG+07-23-026	11 04 42.	+ 40 48	42	16.	GALAXY
ZWG 314.007	11 04 42.	+ 66 53		14.6	GALAXY
UGC 06177	11 04 42.	+ 72 20	84	16.0	GALAXY S
MCG+03-28-063	11 04 45.	+ 18 42	60	14.7	GALAXY
LB 00253	11 04 45.	+ 60 13 18.		12.9	FAINT BLUE STAR
MCG-02-28-049	11 04 45.	- 10 57	60	15.	GALAXY
MCG-03-28-037	11 04 45.	- 19 14	96	13.	GALAXY
LB 00253	11 04 46.	+ 60 14 36.		13.5	FAINT BLUE STAR
RNGC 3529	11 04 46.	- 19 19		14.0	GALAXY
ZWG 038.133	11 04 48.	+ 03 10		15.4	GALAXY
ZWG 038.134	11 04 48.	+ 06 59		14.7	GALAXY
UGC 06178	11 04 48.	+ 06 59	78	14.7	GALAXY S0
MCG+01-28-041	11 04 48.	+ 06 59	36	14.7	GALAXY
ZC 1104.8+0914	11 04 48.	+ 09 14	3970		CLUSTER OF GALAXIES
KARA.72 268B	11 04 48.	+ 18 42	30	15.1	PART OF DOUBLE GALAXY
MCG+04-26-032	11 04 48.	+ 23 38 30.	36	15.5	GALAXY
ZC 1104.8+5852	11 04 48.	+ 58 52	2350		CLUSTER OF GALAXIES
ZWG 314.008	11 04 48.	+ 64 12		15.7	GALAXY
UGC 06179	11 04 48.	+ 64 12	108	15.7	GALAXY S
ZC 1104.8-0206	11 04 48.	- 02 06	1010		CLUSTER OF GALAXIES
ZWG 010.078	11 04 48.	- 02 16		15.6	GALAXY
IC 2624	11 04 48.	- 19 17 20.			NONSTELLAR OBJECT
MCG-03-28-038	11 04 48.	- 19 19	42	14.	GALAXY
MCG-06-25-002	11 04 48.	- 36 54	138	13.	GALAXY
DV.56 N3557A	11 04 48.	- 36 56	120		S GALAXY
ARC 1168	11 04 49.	+ 16 11		16.8	RICH CLUSTER OF GALAXIES
RNGC 3557A	11 04 49.	- 36 56			NON-EXISTENT OBJECT
RNGC 3533	11 04 49.	- 36 56			GALAXY
TON-N 1338	11 04 50.	+ 21 11		16.9	BLUE STAR
KEEL 282	11 04 50.4	+ 55 42 48.		16.	NEBULA
BC PKS1104-445	11 04 50.5	- 44 32 54.		18.	QUASI-STELLAR OBJECT
IC 0670	11 04 53.	+ 06 59 04.			NONSTELLAR OBJECT
REIZ 0843	11 04 53.	+ 23 39	42	14.8	GALAXY
IC 2625	11 04 53.	- 19 18 02.			NONSTELLAR OBJECT
MCG+00-28-031	11 04 54.	+ 01 03	66	14.8	GALAXY
ARC 1171	11 04 54.	+ 03 13		16.2	RICH CLUSTER OF GALAXIES
ZWG 038.135	11 04 54.	+ 04 04		15.3	GALAXY
ZWG 067.001	11 04 54.	+ 13 44		15.7	GALAXY
ZWG 066.116	11 04 54.	+ 13 44		15.7	GALAXY
MCG+04-26-033	11 04 54.	+ 23 07 30.	36	15.1	GALAXY
ZWG 125.030	11 04 54.	+ 23 08		15.1	GALAXY
ZC 1104.9+6007	11 04 54.	+ 60 07	1280		CLUSTER OF GALAXIES
ZC 1104.9+7102	11 04 54.	+ 71 02	1480		CLUSTER OF GALAXIES

OBJECT NAME	RIGHT ASCEN.	DECLINATION	DIAM.	MAGN.	TYPE OF OBJECT
MCG+12-11-010	11 04 54.	+ 72 20	78	16.	GALAXY
IC 0671	11 04 55.	+ 01 02 34.			NONSTELLAR OBJECT
ARC 1170	11 04 57.	+ 08 17		17.6	RICH CLUSTER OF GALAXIES
REIZ 0845	11 04 58.	+ 01 03	60	14.3	GALAXY
ZWG 010.079	11 05 00.	+ 01 03		14.8	GALAXY
UGC 06180	11 05 00.	+ 01 03	96	14.8	GALAXY Sa
MCG+01-29-001	11 05 00.	+ 02 59	18	18.	GALAXY
ZWG 039.001	11 05 00.	+ 05 32		15.3	GALAXY
ZWG 038.136	11 05 00.	+ 05 32		15.3	GALAXY
ZC 1105.0+0815	11 05 00.	+ 08 15	1340		CLUSTER OF GALAXIES
MCG+08-20-090	11 05 00.	+ 47 07	42	16.	GALAXY
MCG+09-18-090	11 05 00.	+ 53 52 30.	54	14.	GALAXY
MCG+09-18-089	11 05 00.	+ 56 10	15	16.	GALAXY
MCG+11-14-004	11 05 00.	+ 64 12	72	16.	GALAXY
MCG+11-14-005	11 05 00.	+ 66 53	69	15.	GALAXY
TON-N 1339	11 05 02.	+ 34 38		15.4	BLUE STAR
KEEL 283	11 05 03.9	+ 56 09 04.		15.	NEBULA
ZWG 039.002	11 05 06.	+ 03 32		15.5	GALAXY
RNGC 3531	11 05 06.	+ 06 59			NON-EXISTENT OBJECT
ZWG 067.002	11 05 06.	+ 13 18		14.8	GALAXY
ZWG 066.117	11 05 06.	+ 13 18		14.8	GALAXY
MCG+02-29-001	11 05 06.	+ 13 18	36	14.8	GALAXY
MCG+03-29-001	11 05 06.	+ 18 45	36	16.	GALAXY
ZWG 096.003	11 05 06.	+ 19 49		15.5	GALAXY
ZWG 095.117	11 05 06.	+ 19 49		15.5	GALAXY
UGC 06181	11 05 06.	+ 19 49	60	15.5	GALAXY DWARF
KARA.73B 0465	11 05 06.	+ 19 49	42	15.5	ISOLATED GALAXY S
ZWG 155.068	11 05 06.	+ 28 46		15.6	GALAXY
ZWG 185.008	11 05 06.	+ 35 58		15.7	GALAXY
ZWG 241.083	11 05 06.	+ 47 07		15.2	GALAXY
ZWG 267.043	11 05 06.	+ 53 54		14.1	GALAXY
UGC 06182	11 05 06.	+ 53 54	60	14.1	GALAXY PECULR
7ZW 371	11 05 06.	+ 58 48			COMPACT GALAXY
KW 52	11 05 06.	+ 72 42	80		SEYFERT GALAXY
ZC 1105.1-0055	11 05 06.	- 00 55	810		CLUSTER OF GALAXIES
ZWG 011.001	11 05 06.	- 02 40		15.7	GALAXY
KEEL 284	11 05 07.0	+ 55 37 19.		16.	NEBULA
REIZ 0844	11 05 08.	+ 53 52	18	14.7	GALAXY
ZWG 039.003	11 05 12.	+ 02 38		15.1	GALAXY
ZWG 067.003	11 05 12.	+ 11 07		15.6	GALAXY
ZC 1105.2+1342	11 05 12.	+ 13 42	4300		CLUSTER OF GALAXIES
MCG+03-29-002	11 05 12.	+ 19 50	78	15.5	GALAXY
TON-N 0570	11 05 12.	+ 28 02		14.5	BLUE STAR
ZC 1105.2+3455	11 05 12.	+ 34 55	610		CLUSTER OF GALAXIES
ZWG 185.009	11 05 12.	+ 35 44		14.9	GALAXY
UGC 06183	11 05 12.	+ 35 44	90	14.9	GALAXY Sb
ZWG 185.010	11 05 12.	+ 37 08		14.6	GALAXY
ZWG 241.084	11 05 12.	+ 46 15		15.2	GALAXY
UGC 06184	11 05 12.	+ 46 15	66	15.2	GALAXY Sc
MCG+08-20-091	11 05 12.	+ 46 15	60	15.	GALAXY
VHA 110	11 05 12.	- 61 12	120		OPEN STAR CLUSTER
MCG+06-25-006	11 05 15.	+ 37 08	48	14.	GALAXY
REIZ 0847	11 05 16.	+ 02 20	48	15.1	GALAXY
ARC 1169	11 05 17.	+ 44 13		16.6	RICH CLUSTER OF GALAXIES
ARC 1167	11 05 17.	+ 49 09		17.3	RICH CLUSTER OF GALAXIES
ZWG 039.004	11 05 18.	+ 07 43		15.3	GALAXY
ZWG 039.005	11 05 18.	+ 07 45		15.5	GALAXY
ZWG 039.006	11 05 18.	+ 08 16		14.6	GALAXY
UGC 06185	11 05 18.	+ 08 16	102	14.6	GALAXY
MCG+01-29-002	11 05 18.	+ 08 16	84	14.6	GALAXY
ZWG 067.004	11 05 18.	+ 11 18		15.6	GALAXY
ZWG 067.005	11 05 18.	+ 13 24		15.6	GALAXY
ZWG 066.118	11 05 18.	+ 13 24		15.6	GALAXY
ZC 1105.3+2835	11 05 18.	+ 28 35	14110		CLUSTER OF GALAXIES
MCG+06-25-007	11 05 18.	+ 35 43 30.	66	14.5	GALAXY
ZC 1105.3+4907	11 05 18.	+ 49 07	400		CLUSTER OF GALAXIES
MCG+09-18-091	11 05 18.	+ 54 05	72	14.	GALAXY
7ZW 372	11 05 18.	+ 58 48			COMPACT GALAXY
ZC 1105.3+7130	11 05 18.	+ 71 30	1340		CLUSTER OF GALAXIES
REIZ 0846	11 05 20.	+ 54 05	60	14.5	GALAXY
KEEL 285	11 05 20.6	+ 56 12 17.		15.	NEBULA
REIZ 0848	11 05 22.	+ 02 17	54	14.7	GALAXY
ZWG 011.002	11 05 24.	+ 02 17		15.3	GALAXY
ZWG 011.003	11 05 24.	+ 02 19		15.7	GALAXY
ZWG 039.007	11 05 24.	+ 03 00		15.6	GALAXY
ZWG 067.006	11 05 24.	+ 14 23		15.0	GALAXY
ZWG 066.119	11 05 24.	+ 14 23		15.0	GALAXY
MCG+02-29-002	11 05 24.	+ 14 23	18	15.0	GALAXY
ZWG 267.044	11 05 24.	+ 54 06		14.7	GALAXY
UGC 06186	11 05 24.	+ 54 06	78	14.7	GALAXY S
ZC 1105.4+5921	11 05 24.	+ 59 21	940		CLUSTER OF GALAXIES
MCG-02-29-001	11 05 24.	- 12 10	48	15.	GALAXY
REIZ 0852	11 05 26.	- 06 34	48	14.7	GALAXY
HOLM 227B	11 05 28.	+ 24 15	36	14.9	PART OF MULTIPLE GALAXY
HOLM 227A	11 05 29.	+ 24 14	30	14.8	PART OF MULTIPLE GALAXY
ZC 1105.5+0304	11 05 30.	+ 03 04	2420		CLUSTER OF GALAXIES
ZWG 067.007	11 05 30.	+ 13 19		15.2	GALAXY
ZC 1105.5+1551	11 05 30.	+ 15 51	670		CLUSTER OF GALAXIES
ZWG 155.069	11 05 30.	+ 28 48		15.7	GALAXY
ZC 1105.5+3257	11 05 30.	+ 32 57	7330		CLUSTER OF GALAXIES
MCG-01-29-001	11 05 30.	- 06 34 30.	48	15.	GALAXY
ARC 1172	11 05 30.	- 06 54		17.2	RICH CLUSTER OF GALAXIES
MCG-02-29-002	11 05 30.	- 12 12 30.	30	15.	GALAXY
MRSL 290-00/2	11 05 30.	- 60 28	1440		HII REGION
REIZ 0849	11 05 31.	+ 28 41	18	15.0	GALAXY
IC 0672	11 05 31.	- 12 12 39.			NONSTELLAR OBJECT
REIZ 0850	11 05 32.	+ 31 37	24	14.9	GALAXY
MCG+08-20-092	11 05 33.	+ 47 36	54	15.	GALAXY
ZC 1105.6+2323	11 05 36.	+ 23 23	7860		CLUSTER OF GALAXIES
ZWG 155.070	11 05 36.	+ 32 17		15.5	GALAXY
MCG+05-26-060	11 05 36.	+ 32 17	36	15.5	GALAXY
ZWG 241.085	11 05 36.	+ 45 24		14.0	GALAXY
UGC 06187	11 05 36.	+ 45 24	42	14.0	GALAXY S?
ZWG 241.086	11 05 36.	+ 47 36		15.5	GALAXY
TON-N 1340	11 05 38.	+ 34 12		17.	BLUE STAR
LB 01952	11 05 39.	+ 58 35 48.		16.3	FAINT BLUE STAR
ZWG 067.008	11 05 42.	+ 10 15		15.7	GALAXY
ZWG 067.009	11 05 42.	+ 10 19		15.1	GALAXY
ZWG 067.010	11 05 42.	+ 13 30		15.0	GALAXY
MCG+02-29-003	11 05 42.	+ 13 30	24	15.0	GALAXY
ZWG 067.011	11 05 42.	+ 14 08		14.7	GALAXY
MCG+02-29-004	11 05 42.	+ 14 08	36	14.7	GALAXY
ZWG 155.071	11 05 42.	+ 30 06		15.4	GALAXY
MCG+10-16-064	11 05 42.	+ 57 29		14.4	GALAXY
ZWG 291.030	11 05 42.	+ 57 30		14.4	GALAXY
UGC 06188	11 05 42.	+ 57 30	42	14.4	GALAXY
MCG+10-16-065	11 05 42.	+ 59 01	42	16.	GALAXY
7ZW 373	11 05 42.	+ 79 16			COMPACT GALAXY

OBJECT NAME	RIGHT ASCEN.	DECLINATION	DIAM.	MAGN.	TYPE OF OBJECT
RNGC 3530	11 05 43.	+ 57 30		14.5	GALAXY
FATH 1.375	11 05 43.	+ 59 02	22		NEBULA
REIZ 0851	11 05 47.	+ 57 30	18	14.5	GALAXY
ZWG 039.008	11 05 48.	+ 03 46		14.7	GALAXY
MCG+01-29-003	11 05 48.	+ 03 46	36	14.7	GALAXY
ZWG 039.009	11 05 48.	+ 06 06		15.4	GALAXY
TON-N 0056	11 05 48.	+ 29 56		14.9	BLUE STAR
MCG+06-25-008	11 05 48.	+ 33 53 30.	42	15.	GALAXY
ZC 1105.8+5750	11 05 48.	+ 57 50	1140		CLUSTER OF GALAXIES
MCG+10-16-068	11 05 48.	+ 60 24	24	17.	GALAXY
MCG+10-16-067	11 05 48.	+ 60 24	15	17.	GALAXY
MCG+10-16-066	11 05 48.	+ 60 24	27	16.	GALAXY
ZWG 291.031	11 05 48.	+ 60 25		15.6	GALAXY
LB 01953	11 05 48.	+ 61 56 42.		16.8	FAINT BLUE STAR
MCG-01-29-002	11 05 48.	- 04 51	15	15.5	GALAXY
REIZ 0853	11 05 49.	+ 30 06	24	13.9	GALAXY
REIZ 0854	11 05 50.	+ 31 36	24	14.8	GALAXY
KEEL 286	11 05 52.1	+ 55 55 30.		15.	NEBULA
ZWG 039.010	11 05 54.	+ 05 06		14.3	GALAXY
RNGC 3535	11 05 54.	+ 05 06		14.5	GALAXY
UGC 06189	11 05 54.	+ 05 06	72	14.3	GALAXY Sa
MCG+01-29-004	11 05 54.	+ 05 06	60	14.3	GALAXY
ZC 1105.9+2034	11 05 54.	+ 20 34	810		CLUSTER OF GALAXIES
TON-N 0571	11 05 54.	+ 28 25		15.1	BLUE STAR
TON-N 0058	11 05 54.	+ 31 43		16.7	BLUE STAR
TON-N 0057	11 05 54.	+ 32 12		16.7	BLUE STAR
ZWG 351.056	11 05 54.	+ 79 16		15.4	GALAXY
TON-N 1341	11 05 56.	+ 22 10		16.8	BLUE STAR
ARC 1166	11 05 57.	+ 69 01		17.1	RICH CLUSTER OF GALAXIES
ZWG 039.011	11 06 00.	+ 06 16		15.7	GALAXY
ZC 1106.0+2334	11 06 00.	+ 23 34	1410		CLUSTER OF GALAXIES
REIZ 0855	11 06 00.	+ 26 53	48	15.0	GALAXY
REIZ 0856	11 06 00.	+ 26 54	60	14.7	GALAXY
MCG+06-25-010	11 06 00.	+ 35 20	36	16.	GALAXY
MCG+06-25-009	11 06 00.	+ 36 25	15	15.	GALAXY
MCG+06-25-009	11 06 00.	+ 36 25	15	15.	GALAXY
MCG+10-16-069	11 06 00.	+ 61 40	42	15.	GALAXY
MCG-02-29-003	11 06 00.	- 10 10	72	14.5	GALAXY
MCG-06-25-003	11 06 00.	- 37 22	60	13.5	GALAXY
REIZ 0857	11 06 02.	+ 31 39	6	15.6	GALAXY
TON-N 1342	11 06 02.	+ 35 17		16.5	BLUE STAR
RNGC 3534B	11 06 03.	+ 26 51		15.5	GALAXY
RNGC 3534A	11 06 03.	+ 26 52		15.5	GALAXY
RNGC 3537	11 06 03.	- 10 10		14.0	GALAXY
REIZ 0858	11 06 04.	+ 02 09	24	14.8	GALAXY
KEEL 287	11 06 05.3	+ 55 50 13.		17.	NEBULA
ZWG 011.004	11 06 06.	+ 00 33		14.8	GALAXY
KARA.72 269B	11 06 06.	+ 00 33	42		PART OF DOUBLE GALAXY
MCG+00-29-001	11 06 06.	+ 00 33	36	14.8	GALAXY
KARA.72 269A	11 06 06.	+ 00 34	24	14.8	PART OF DOUBLE GALAXY
ZWG 011.005	11 06 06.	+ 00 40		15.0	GALAXY
MCG+00-29-002	11 06 06.	+ 00 40	15	15.0	GALAXY
ZWG 011.006	11 06 06.	+ 02 09		15.3	GALAXY
ZC 1106.1+2551	11 06 06.	+ 25 51	3090		CLUSTER OF GALAXIES
ZWG 155.072	11 06 06.	+ 26 54		15.4	GALAXY
UGC 06190	11 06 06.	+ 26 54	90	15.4	GALAXY S
ZWG 155.073	11 06 06.	+ 28 45		15.2	GALAXY
UGC 06191	11 06 06.	+ 28 45	108	15.2	GALAXY SB
MCG+09-18-092	11 06 06.	+ 53 37	42	16.	GALAXY
ZC 1106.1+6028	11 06 06.	+ 60 28	1210		CLUSTER OF GALAXIES
ZWG 291.032	11 06 06.	+ 61 41		15.7	GALAXY
UGC 06192	11 06 06.	+ 61 41	66	15.7	GALAXY
LB 01954	11 06 08.	+ 57 02 06.		16.8	FAINT BLUE STAR
RNGC 3536	11 06 09.	+ 28 45		15.0	GALAXY
MCG+05-26-061	11 06 09.	+ 28 45	42	15.2	GALAXY
KEEL 288	11 06 09.7	+ 55 55 04.		16.	SPIRAL NEBULA
ZC 1106.2+0516	11 06 12.	+ 05 16	7120		CLUSTER OF GALAXIES
ZWG 039.012	11 06 12.	+ 05 40		15.5	GALAXY
ZWG 155.074	11 06 12.	+ 26 53		15.6	GALAXY
UGC 06193	11 06 12.	+ 26 53	66	15.6	GALAXY Sb
REIZ 0859	11 06 12.	+ 28 44	42	14.1	GALAXY
ZC 1106.2+4617	11 06 12.	+ 46 17	940		CLUSTER OF GALAXIES
MCG-02-29-004	11 06 12.	- 10 40	24	15.	GALAXY
REIZ 0860	11 06 13.	+ 30 04	12	14.4	GALAXY
LYNG 06	11 06 14.	- 60 15 18.	6840		OB CONCENTRATION
MCG+05-26-063	11 06 15.	+ 26 51	48	15.6	GALAXY
MCG+05-26-062	11 06 15.	+ 26 52	72	15.4	GALAXY
RNGC 3541	11 06 15.	- 10 40		15.0	GALAXY
HOLM 228A	11 06 17.	+ 23 12	36	14.2	PART OF MULTIPLE GALAXY
LB 01955	11 06 17.	+ 58 00 36.		14.7	FAINT BLUE STAR
ZWG 039.013	11 06 18.	+ 02 57		15.3	GALAXY
8ZW 1106+02.9	11 06 18.	+ 02 57		15.3	COMPACT GALAXY
ZWG 039.014	11 06 18.	+ 06 35		15.2	GALAXY
KARA.72 270A	11 06 18.	+ 06 35	36	15.2	PART OF DOUBLE GALAXY
MCG+04-26-034	11 06 18.	+ 23 11	90	14.7	GALAXY
ZWG 125.031	11 06 18.	+ 23 12		14.7	GALAXY
UGC 06194	11 06 18.	+ 23 12	96	14.7	GALAXY S
KARA.73B 0466	11 06 18.	+ 23 12	90	14.7	ISOLATED GALAXY S
HOLM 229A	11 06 18.	+ 26 53	84	14.1	PART OF MULTIPLE GALAXY
ZWG 155.075	11 06 18.	+ 27 11		15.7	GALAXY
MCG+09-18-093	11 06 18.	+ 55 42 30.	48	17.	GALAXY
ZWG 314.009	11 06 18.	+ 62 31		15.7	GALAXY
ZC 1106.3+6752	11 06 18.	+ 67 52	1340		CLUSTER OF GALAXIES
KEEL 289	11 06 19.8	+ 55 42 30.		14.	NEBULA
HOLM 228B	11 06 20.	+ 23 12	18	15.3	PART OF MULTIPLE GALAXY
HOLM 229B	11 06 20.	+ 26 52	24	14.5	PART OF MULTIPLE GALAXY
MCG-02-29-005	11 06 21.	- 14 20	42	15.	GALAXY
REIZ 0861	11 06 23.	+ 23 13	60	14.1	GALAXY
ZWG 039.015	11 06 24.	+ 05 18		15.6	GALAXY
ZWG 039.016	11 06 24.	+ 06 37		15.1	GALAXY
KARA.72 270B	11 06 24.	+ 06 37	30	15.1	PART OF DOUBLE GALAXY
MCG+05-26-064	11 06 24.	+ 26 53	60	15.6	GALAXY
ZWG 155.076	11 06 24.	+ 26 54		15.6	GALAXY
ZWG 155.077	11 06 24.	+ 28 56		15.4	GALAXY
REIZ 0862	11 06 24.	+ 28 56	18	14.5	GALAXY
MCG+05-26-065	11 06 24.	+ 29 57	60	15.4	GALAXY
ZC 1106.4+3326	11 06 24.	+ 33 26	1210		CLUSTER OF GALAXIES
UGC 06195	11 06 24.	+ 34 13	60	16.0	GALAXY Sc
ZC 1106.4+4150	11 06 24.	+ 41 50	1810		CLUSTER OF GALAXIES
MCG+11-14-006	11 06 24.	+ 62 31	57	16.	GALAXY
ZC 1106.4-0149	11 06 24.	- 01 49	670		CLUSTER OF GALAXIES
IC 2626	11 06 25.	+ 27 10 26.			NONSTELLAR OBJECT
FATH 1.376	11 06 25.	+ 59 30	5		NEBULA
ARC 1176	11 06 26.	+ 06 54		17.7	RICH CLUSTER OF GALAXIES
ARC 1173	11 06 26.	+ 41 51		16.8	RICH CLUSTER OF GALAXIES
RNGC 3539	11 06 27.	+ 28 56		15.5	GALAXY
MCG+05-26-066	11 06 27.	+ 31 48	42	15.7	GALAXY
LB 00254	11 06 29.	+ 60 51 00.		14.0	FAINT BLUE STAR

OBJECT NAME	RIGHT ASCEN.	DECLINATION	DIAM.	MAGN.	TYPE OF OBJECT
ZWG 039.017	11 06 30.	+ 05 03		15.6	GALAXY
TON-N 0059	11 06 30.	+ 27 06		14.9	BLUE STAR
ZC 1106.5+3105	11 06 30.	+ 31 05	810		CLUSTER OF GALAXIES
ZWG 155.078	11 06 30.	+ 31 49		15.7	GALAXY
ARC 1175	11 06 30.	+ 33 27		17.8	RICH CLUSTER OF GALAXIES
ZC 1106.5+3605	11 06 30.	+ 36 05	540		CLUSTER OF GALAXIES
ZWG 185.011	11 06 30.	+ 36 17		14.6	GALAXY
UGC 06196	11 06 30.	+ 36 17	90	14.6	GALAXY SB0
KARA.73B 0467	11 06 30.	+ 36 17	90	14.6	ISOLATED GALAXY S
ZC 1106.5+3650	11 06 30.	+ 36 50	610		CLUSTER OF GALAXIES
7ZW 374	11 06 30.	+ 60 39			COMPACT GALAXY
RNGC 3540	11 06 32.	+ 36 17		14.5	GALAXY
LB 00254	11 06 32.	+ 60 50 36.		14.7	FAINT BLUE STAR
ARC 1174	11 06 33.	+ 43 33		17.0	RICH CLUSTER OF GALAXIES
ZWG 011.007	11 06 36.	+ 00 59		15.6	GALAXY
ZWG 039.018	11 06 36.	+ 08 28		15.2	GALAXY
8ZW 1106+08.5	11 06 36.	+ 08 28		15.2	COMPACT GALAXY
UGC 06197	11 06 36.	+ 08 28	84	15.2	GALAXY S+COMP
MCG+01-29-005	11 06 36.	+ 08 28	48	15.2	GALAXY
ZWG 155.079	11 06 36.	+ 29 50		15.0	GALAXY
UGC 06198	11 06 36.	+ 29 50	60	15.0	GALAXY S0?
LB 10004	11 06 36.	+ 31 26		17.3	FAINT BLUE STAR
MCG+06-25-011	11 06 36.	+ 36 17	48	14.5	GALAXY
MCG+09-18-094	11 06 36.	+ 51 10	60	16.	GALAXY
CED 111	11 06 37.	- 77 23			DIFFUSE GALACTIC NEBULA
REIZ 0863	11 06 38.	+ 31 49	24	15.1	GALAXY
REIZ 0864	11 06 38.	+ 32 10	18	15.3	GALAXY
LB 01956	11 06 40.	+ 57 12 24.		15.0	FAINT BLUE STAR
FATH 1.377	11 06 41.	+ 59 32	8		NEBULA
ZWG 039.019	11 06 42.	+ 04 14		15.7	GALAXY
ZWG 067.012	11 06 42.	+ 13 33		15.4	GALAXY
ZC 1106.7+4414	11 06 42.	+ 44 14	6180		CLUSTER OF GALAXIES
ZC 1106.7+5128	11 06 42.	+ 51 28	340		CLUSTER OF GALAXIES
ZWG 314.010	11 06 42.	+ 62 34		14.9	GALAXY
UGC 06199	11 06 42.	+ 62 34	84	14.9	GALAXY S
MEL 3	11 06 43.	- 77 23			NEBULA
MCG+05-26-067	11 06 45.	+ 29 51	48	15.0	GALAXY
ARC 1177	11 06 46.	+ 21 58		15.7	RICH CLUSTER OF GALAXIES
ZWG 011.009	11 06 48.	+ 00 10		14.5	GALAXY
UGC 06200	11 06 48.	+ 00 10	120	14.5	GALAXY Sa
MCG+00-29-003	11 06 48.	+ 00 10	90	14.5	GALAXY
ZWG 067.013	11 06 48.	+ 10 03		15.7	GALAXY
ZWG 067.014	11 06 48.	+ 11 07		15.0	GALAXY
MCG+02-29-005	11 06 48.	+ 11 07	36	15.0	GALAXY
ZC 1106.8+1513	11 06 48.	+ 15 13	2690		CLUSTER OF GALAXIES
LB 10006	11 06 48.	+ 30 48		17.2	FAINT BLUE STAR
LB 10005	11 06 48.	+ 31 35		17.0	FAINT BLUE STAR
ZWG 241.087	11 06 48.	+ 47 05		14.3	GALAXY
UGC 06201	11 06 48.	+ 47 05	42	14.3	GALAXY
MCG+08-20-093	11 06 48.	+ 47 05	42	15.	GALAXY
ZWG 267.045	11 06 48.	+ 51 13		15.4	GALAXY
UGC 06202	11 06 48.	+ 51 13	60	15.4	GALAXY S
ZWG 011.008	11 06 48.	- 00 35		15.3	GALAXY
KARA.68 075	11 06 48.	- 18 19	34		DWARF GALAXY
MCG+08-20-094	11 06 51.	+ 49 41	42	16.	GALAXY
REIZ 0866	11 06 52.	+ 00 05	24	14.8	GALAXY
IC 0673	11 06 54.	+ 00 11 36.			NONSTELLAR OBJECT
ZWG 096.004	11 06 54.	+ 14 57		15.7	GALAXY
ZWG 125.032	11 06 54.	+ 22 01		15.7	GALAXY
MCG+05-26-068	11 06 54.	+ 28 30 30.	30	15.5	GALAXY
REIZ 0865	11 06 54.	+ 28 56	48	14.4	GALAXY
TON-N 0572	11 06 54.	+ 33 12		15.1	BLUE STAR
MCG+10-16-070	11 06 54.	+ 59 28	36	16.	GALAXY
MCG+11-14-007	11 06 54.	+ 62 34	78	15.	GALAXY
ZWG 314.011	11 06 54.	+ 63 55		14.8	GALAXY
MCG-04-27-001	11 06 54.	- 22 36 30.	60	14.	GALAXY
ACK 285+11.1	11 06 54.	- 47 47			PLANETARY NEBULA
RNGC 3551	11 06 58.	+ 21 59		15.0	GALAXY
RNGC 3544	11 06 58.	- 18 00			NON-EXISTENT OBJECT
ARC 1181	11 06 59.	- 19 31		17.6	RICH CLUSTER OF GALAXIES
ZWG 039.020	11 07 00.	+ 02 41		15.7	GALAXY
ZWG 039.021	11 07 00.	+ 05 38		15.5	GALAXY
ZWG 039.022	11 07 00.	+ 07 31		15.6	GALAXY
ZWG 125.033	11 07 00.	+ 22 02		14.8	GALAXY
UGC 06203	11 07 00.	+ 22 02	84	14.8	GALAXY E?
MCG+08-20-095	11 07 00.	+ 46 21	108	15.	GALAXY
MCG+09-18-095	11 07 00.	+ 52 39	36	15.	GALAXY
FATH 1.378	11 07 00.	+ 59 29	19		NEBULA
MCG+11-14-008	11 07 00.	+ 63 55	18	15.	GALAXY
7ZW 375	11 07 00.	+ 72 30			COMPACT GALAXY
ZC 1107.0-0037	11 07 00.	- 00 37	2820		CLUSTER OF GALAXIES
ZWG 011.010	11 07 00.	- 02 19		15.5	GALAXY
MCG-02-29-006	11 07 00.	- 13 43	54	15.	GALAXY
TON-N 1343	11 07 02.	+ 32 55		17.	BLUE STAR
MCG+04-26-035	11 07 03.	+ 22 00	90	14.8	GALAXY
HOLM 230B	11 07 03.	+ 22 01	18	15.5	PART OF MULTIPLE GALAXY
RNGC 3555	11 07 04.	+ 22 00		15.0	GALAXY
HOLM 230A	11 07 04.	+ 22 02	30	14.5	PART OF MULTIPLE GALAXY
ZWG 039.023	11 07 06.	+ 03 41		15.1	GALAXY
ZWG 125.034	11 07 06.	+ 22 05		15.7	GALAXY
ZC 1107.1+2417	11 07 06.	+ 24 17	2690		CLUSTER OF GALAXIES
ZWG 125.035	11 07 06.	+ 24 32		14.5	GALAXY
UGC 06204	11 07 06.	+ 24 32	60	14.5	GALAXY S
KARA.72 271A	11 07 06.	+ 24 32	60	14.5	PART OF DOUBLE GALAXY
ZC 1107.1+3441	11 07 06.	+ 34 41	1340		CLUSTER OF GALAXIES
MCG+06-25-012	11 07 06.	+ 35 41	30	15.	GALAXY
ZWG 185.012	11 07 06.	+ 37 16		15.5	GALAXY
ZWG 241.088	11 07 06.	+ 46 22		15.6	GALAXY
UGC 06205	11 07 06.	+ 46 22	108	15.6	GALAXY DWRF SP
ZWG 267.046	11 07 06.	+ 52 40		15.5	GALAXY
ZWG 011.011	11 07 06.	- 00 33		15.6	GALAXY
ARC 1178	11 07 07.	+ 34 52		17.8	RICH CLUSTER OF GALAXIES
HOLM 231A	11 07 11.	+ 24 32	36	13.6	PART OF MULTIPLE GALAXY
ZWG 011.013	11 07 12.	+ 01 05		15.6	GALAXY
ZWG 039.024	11 07 12.	+ 05 18		15.6	GALAXY
ZWG 067.015	11 07 12.	+ 09 17		15.5	GALAXY
ZWG 067.016	11 07 12.	+ 11 29		15.7	GALAXY
ZWG 067.017	11 07 12.	+ 13 02		14.2	GALAXY
UGC 06206	11 07 12.	+ 13 02	54	14.2	GALAXY DBL SYS
KARA.72 272B	11 07 12.	+ 13 02	36		PART OF DOUBLE GALAXY
MCG+02-29-006	11 07 12.	+ 13 02	48	14.2	GALAXY
KARA.72 272A	11 07 12.	+ 13 03	30	14.2	PART OF DOUBLE GALAXY
ARC 1179	11 07 12.	+ 24 14		16.6	RICH CLUSTER OF GALAXIES
ZWG 125.036	11 07 12.	+ 24 31		14.6	GALAXY
UGC 06207	11 07 12.	+ 24 31	90	14.6	GALAXY SB?
KARA.72 271B	11 07 12.	+ 24 31	84	14.6	PART OF DOUBLE GALAXY
MCG+04-26-036	11 07 12.	+ 24 33	30	14.5	GALAXY
LB 10009	11 07 12.	+ 28 40		16.7	FAINT BLUE STAR

OBJECT NAME	RIGHT ASCEN.	DECLINATION	DIAM.	MAGN.	TYPE OF OBJECT
LB 10008	11 07 12.	+ 28 42		17.0	FAINT BLUE STAR
LB 10007	11 07 12.	+ 32 16		16.8	FAINT BLUE STAR
ZWG 185.013	11 07 12.	+ 37 13		15.0	GALAXY
MCG+06-25-013	11 07 12.	+ 37 13	42	15.	GALAXY
MCG+06-25-014	11 07 12.	+ 37 16	36	15.	GALAXY
ZWG 185.014	11 07 12.	+ 37 24		15.5	GALAXY
ZWG 241.089	11 07 12.	+ 45 43		15.7	GALAXY
7ZW 376	11 07 12.	+ 60 33			COMPACT GALAXY
ZWG 011.012	11 07 12.	- 03 23		15.4	GALAXY
MCG-02-29-007	11 07 12.	- 13 07	48	14.	GALAXY
MCG-06-25-004	11 07 12.	- 37 05	78	14.	GALAXY
RNGC 3542	11 07 13.	+ 37 13		15.0	GALAXY
TON-N 1344	11 07 14.	+ 20 12		16.8	BLUE STAR
TON-N 1345	11 07 14.	+ 20 24		15.8	BLUE STAR
HOLM 231B	11 07 14.	+ 24 32	72	13.6	PART OF MULTIPLE GALAXY
ARP 301	11 07 14.	+ 24 34			PECULIAR GALAXY
MCG+04-26-037	11 07 15.	+ 24 32 30.	84	14.6	GALAXY
MCG+06-25-015	11 07 15.	+ 37 24	42	15.	GALAXY
MCG-01-29-003	11 07 15.	- 07 18	48	15.	GALAXY
RNGC 3546	11 07 15.	- 13 07		14.0	GALAXY
REIZ 0868	11 07 17.	+ 24 33	42	14.6	GALAXY
REIZ 0867	11 07 17.	+ 24 33	48	14.5	GALAXY
ZWG 039.025	11 07 18.	+ 07 30		15.1	GALAXY
UGC 06208	11 07 18.	+ 07 30	66	15.1	GALAXY Sb
ZWG 039.026	11 07 18.	+ 08 02		15.7	GALAXY
ZWG 067.018	11 07 18.	+ 09 47		15.1	GALAXY
ZWG 067.019	11 07 18.	+ 11 00		12.8	GALAXY
RNGC 3547	11 07 18.	+ 11 00		13.0	GALAXY
UGC 06209	11 07 18.	+ 11 00	126	12.8	GALAXY S
MCG+02-29-007	11 07 18.	+ 11 00	120	12.8	GALAXY
ZWG 067.020	11 07 18.	+ 13 44		15.6	GALAXY
ZWG 067.021	11 07 18.	+ 13 58		15.6	GALAXY
LB 10012	11 07 18.	+ 26 36		15.7	FAINT BLUE STAR
TON-N 0573	11 07 18.	+ 26 38		15.0	BLUE STAR
LB 10011	11 07 18.	+ 30 26		17.4	FAINT BLUE STAR
LB 10010	11 07 18.	+ 32 01		17.2	FAINT BLUE STAR
ZC 1107.3+3513	11 07 18.	+ 35 13	1880		CLUSTER OF GALAXIES
ZWG 242.001	11 07 18.	+ 49 14		15.5	GALAXY
ZWG 241.090	11 07 18.	+ 49 14		15.5	GALAXY
DV.56 N3557B	11 07 18.	- 37 05	60		SA0 GALAXY
RNGC 3557B	11 07 19.	- 37 05			GALAXY
TON-N 1346	11 07 20.	+ 20 23		16.9	BLUE STAR
TON-N 1347	11 07 20.	+ 21 18		14.	BLUE STAR
VV 229B	11 07 21.	+ 24 32 30.	84	14.	INTERACTING GALAXY
VV 229A	11 07 21.	+ 24 32 30.	60	14.	INTERACTING GALAXY
VV 182B	11 07 21.	+ 37 15	18	16.	INTERACTING GALAXY
VV 182A	11 07 21.	+ 37 15	18	16.	INTERACTING GALAXY
VV 182	11 07 21.	+ 37 15	48		INTERACTING GALAXY
ZWG 039.027	11 07 24.	+ 03 45		15.5	GALAXY
ZWG 039.028	11 07 24.	+ 05 35		14.7	GALAXY
UGC 06210	11 07 24.	+ 05 35	108	14.7	GALAXY Sb
MCG+01-29-006	11 07 24.	+ 05 35	72	14.7	GALAXY
ZC 1107.4+1616	11 07 24.	+ 16 16	1550		CLUSTER OF GALAXIES
TON-N 0574	11 07 24.	+ 27 36		16.4	BLUE STAR
LB 10013	11 07 24.	+ 32 06		16.7	FAINT BLUE STAR
ZWG 185.015	11 07 24.	+ 37 14		14.8	GALAXY
KARA.72 273B	11 07 24.	+ 37 14	24		PART OF DOUBLE GALAXY
KARA.72 273A	11 07 24.	+ 37 14	24	14.8	PART OF DOUBLE GALAXY
ZC 1107.4+4540	11 07 24.	+ 45 40	1010		CLUSTER OF GALAXIES
ZWG 242.002	11 07 24.	+ 46 43		15.7	GALAXY
ZWG 241.091	11 07 24.	+ 46 43		15.7	GALAXY
ZWG 314.012	11 07 24.	+ 63 49		15.7	GALAXY
ZC 1107.4+6705	11 07 24.	+ 67 05	1610		CLUSTER OF GALAXIES
ZWG 011.014	11 07 24.	- 02 42		15.6	GALAXY
MCG-04-27-002	11 07 24.	- 23 27 30.	120	12.5	GALAXY
VHA 111	11 07 24.	- 63 35	360		OPEN STAR CLUSTER
REIZ 0869	11 07 25.	+ 30 23	12	14.8	GALAXY
REIZ 0870	11 07 25.	+ 30 30	12	14.7	GALAXY
RNGC 3545B	11 07 25.	+ 37 14			GALAXY
RNGC 3545A	11 07 25.	+ 37 14		15.0	GALAXY
LB 01957	11 07 25.	+ 59 49 42.		15.6	FAINT BLUE STAR
TON-N 1348	11 07 26.	+ 34 53		17.	BLUE STAR
IC 2627	11 07 26.2	- 23 27 18.	448	12.0	GALAXY SA
RNGC 3566	11 07 28.	- 19 46			NON-EXISTENT OBJECT
RNGC 3565	11 07 28.	- 19 46			NON-EXISTENT OBJECT
REIZ 0873	11 07 29.	+ 05 35	60	14.5	GALAXY
KEEL 290	11 07 29.0	+ 55 55 55.		17.	NEBULA
ZCG 1107+01.4	11 07 30.	+ 01 29		15.6	COMPACT GALAXY
ZCG 1107+01.5	11 07 30.	+ 01 32		18.4	COMPACT GALAXY
ZWG 039.029	11 07 30.	+ 03 43		15.6	GALAXY
ZWG 039.030	11 07 30.	+ 07 34		15.4	GALAXY
ZC 1107.5+1221	11 07 30.	+ 12 21	1610		CLUSTER OF GALAXIES
ZC 1107.5+1414	11 07 30.	+ 14 14	810		CLUSTER OF GALAXIES
REIZ 0871	11 07 30.	+ 28 55	24	14.3	GALAXY
ZC 1107.5+2947	11 07 30.	+ 29 47	740		CLUSTER OF GALAXIES
LB 10014	11 07 30.	+ 31 55		16.9	FAINT BLUE STAR
MCG+09-18-096	11 07 30.	+ 55 26	60	16.	GALAXY
MCG+10-16-071	11 07 30.	+ 59 01	27	15.	GALAXY
MCG+11-14-008A	11 07 30.	+ 63 50	36	16.	GALAXY
ZC 1107.5+6738	11 07 30.	+ 67 38	470		CLUSTER OF GALAXIES
REIZ 0872	11 07 31.	+ 30 44	12	15.2	GALAXY
RNGC 3557	11 07 31.	- 37 16		12.0	GALAXY
MCG+06-25-017	11 07 33.	+ 37 14	12	16.	GALAXY
MCG+06-25-016	11 07 33.	+ 37 14	12	16.	GALAXY
LB 01958	11 07 33.	+ 50 55 30.		16.6	FAINT BLUE STAR
ARC 1183	11 07 35.	+ 12 17		17.2	RICH CLUSTER OF GALAXIES
ARC 1182	11 07 35.	+ 32 03		17.5	RICH CLUSTER OF GALAXIES
ZC 1107.6+1041	11 07 36.	+ 10 41	5380		CLUSTER OF GALAXIES
MCG+03-29-003	11 07 36.	+ 15 11	30	15.	GALAXY
LB 10016	11 07 36.	+ 31 04		16.4	FAINT BLUE STAR
LB 10015	11 07 36.	+ 31 20		17.8	FAINT BLUE STAR
ZC 1107.6+3405	11 07 36.	+ 34 05	470		CLUSTER OF GALAXIES
ZWG 213.030	11 07 36.	+ 40 20		15.7	GALAXY
UGC 06211	11 07 36.	+ 55 29	72	16.0	GALAXY Sb/SBc
MCG-06-25-005	11 07 36.	- 37 16	180	12.1	GALAXY
TON-N 1349	11 07 38.	+ 34 04		17.	BLUE STAR
RNGC 3548	11 07 38.	+ 36 18			NON-EXISTENT OBJECT
FATH 1.379	11 07 38.	+ 59 02	22		NEBULA
KEEL 291	11 07 41.4	+ 55 47 54.		15.	SPIRAL NEBULA
ZWG 039.031	11 07 42.	+ 04 58		15.2	GALAXY
REIZ 0874	11 07 42.	+ 29 00	24	14.9	GALAXY
ZC 1107.7+3203	11 07 42.	+ 32 03	2220		CLUSTER OF GALAXIES
ZC 1107.7+3610	11 07 42.	+ 36 10	9340		CLUSTER OF GALAXIES
FATH 1.380	11 07 43.	+ 59 13	8		NEBULA
REIZ 0879	11 07 44.	- 06 18	18	15.2	GALAXY
KEEL 292	11 07 44.3	+ 55 49 25.		17.	NEBULA
HOLM 232B	11 07 46.	+ 05 05	30	14.7	PART OF MULTIPLE GALAXY
REIZ 0880	11 07 47.	+ 04 15	30	14.9	GALAXY

OBJECT NAME	RIGHT ASCEN.	DECLINATION	DIAM.	MAGN.	TYPE OF OBJECT
HOLM 232A	11 07 47.	+ 05 06	42	13.7	PART OF MULTIPLE GALAXY
RNGC 3543	11 07 47.	+ 61 38		15.0	GALAXY
ZWG 039.032	11 07 48.	+ 04 15		15.0	GALAXY
MCG+01-29-007	11 07 48.	+ 04 15	54	15.0	GALAXY
ZWG 039.033	11 07 48.	+ 04 52		15.2	GALAXY
ZWG 039.034	11 07 48.	+ 05 06		14.0	GALAXY
UGC 06212	11 07 48.	+ 05 06	60	14.0	GALAXY Sb
MCG+01-29-008	11 07 48.	+ 05 06	36	14.0	GALAXY
ZWG 039.035	11 07 48.	+ 05 17		15.1	GALAXY
ZWG 039.036	11 07 48.	+ 08 05		15.4	GALAXY
ZWG 067.022	11 07 48.	+ 10 24		15.4	GALAXY
ZC 1107.8+2056	11 07 48.	+ 20 56	2150		CLUSTER OF GALAXIES
LB 10017	11 07 48.	+ 28 38		16.2	FAINT BLUE STAR
REIZ 0876	11 07 48.	+ 28 59	18	14.9	GALAXY
REIZ 0875	11 07 48.	+ 28 59	18	15.0	GALAXY
ZC 1107.8+3230	11 07 48.	+ 32 30	1080		CLUSTER OF GALAXIES
MCG+10-16-072	11 07 48.	+ 57 54	24	16.	GALAXY
7ZW 377	11 07 48.	+ 57 55			COMPACT GALAXY
ZWG 291.033	11 07 48.	+ 57 55		15.2	GALAXY
KARA.73B 0468	11 07 48.	+ 57 55	42	15.2	ISOLATED GALAXY S
ZWG 291.034	11 07 48.	+ 61 38		14.8	GALAXY
UGC 06213	11 07 48.	+ 61 38	84	14.8	GALAXY S
MRSL 290+00/3	11 07 48.	- 59 50	1200		HII REGION
REIZ 0877	11 07 49.	+ 30 14	12	14.8	GALAXY
RNGC 3550	11 07 51.	+ 29 02		14.0	GALAXY
KEEL 293	11 07 53.8	+ 55 43 01.		17.	NEBULA
ZWG 039.037	11 07 54.	+ 03 32		15.2	GALAXY
ZWG 039.038	11 07 54.	+ 04 20		15.3	GALAXY
ZC 1107.9+1009	11 07 54.	+ 10 09	540		CLUSTER OF GALAXIES
ZWG 156.001	11 07 54.	+ 28 35		15.1	GALAXY
ZWG 155.080	11 07 54.	+ 28 35		15.1	GALAXY
ZWG 156.002	11 07 54.	+ 29 00		15.5	GALAXY
ZWG 155.081	11 07 54.	+ 29 00		15.5	GALAXY
REIZ 0881	11 07 54.	+ 29 00	18	15.0	GALAXY
ZWG 156.003	11 07 54.	+ 29 02		14.2	GALAXY
ZWG 155.082	11 07 54.	+ 29 02		14.2	GALAXY
REIZ 0883	11 07 54.	+ 29 02	36	13.8	GALAXY
REIZ 0882	11 07 54.	+ 29 02	12	14.7	GALAXY
UGC 06214	11 07 54.	+ 29 02	72	14.2	GALAXY PECULR
KARA.72 274B	11 07 54.	+ 29 02	36		PART OF DOUBLE GALAXY
KARA.72 274A	11 07 54.	+ 29 02	48	14.2	PART OF DOUBLE GALAXY
ZWG 156.004	11 07 54.	+ 31 56		15.1	GALAXY
ZWG 155.083	11 07 54.	+ 31 56		15.1	GALAXY
ZC 1107.9+4314	11 07 54.	+ 43 14	1080		CLUSTER OF GALAXIES
ZC 1107.9+5828	11 07 54.	+ 58 28	1210		CLUSTER OF GALAXIES
ZWG 314.013	11 07 54.	+ 63 35		15.7	GALAXY
MRSL 290-00/3	11 07 54.	- 60 22	720		HII REGION
REIZ 0884	11 07 55.	+ 30 56	24	14.1	GALAXY
LB 01959	11 07 56.	+ 60 13 06.		14.8	FAINT BLUE STAR
RNGC 3553	11 07 57.	+ 28 58		15.0	GALAXY
RNGC 3552	11 07 57.	+ 28 58		15.0	GALAXY
ZWG 011.015	11 08 00.	+ 01 10		15.7	GALAXY
ZWG 039.039	11 08 00.	+ 04 02		15.2	GALAXY
ZWG 039.040	11 08 00.	+ 04 28		15.3	GALAXY
ZC 1108.0+0914	11 08 00.	+ 09 14	1210		CLUSTER OF GALAXIES
ZWG 067.023	11 08 00.	+ 12 44		15.4	GALAXY
ZWG 096.005	11 08 00.	+ 15 10		15.6	GALAXY
ZC 1108.0+1753	11 08 00.	+ 17 53	940		CLUSTER OF GALAXIES
ZC 1108.0+2555	11 08 00.	+ 25 55	1140		CLUSTER OF GALAXIES
ZWG 156.005	11 08 00.	+ 28 33		15.7	GALAXY
ZWG 155.084	11 08 00.	+ 28 33		15.7	GALAXY
STOCK 24	11 08 00.	+ 28 35			BLUE KNOT NEAR ELLIP GLXY
MCG+05-27-001	11 08 00.	+ 28 35	54	15.1	GALAXY
ZWG 156.006	11 08 00.	+ 28 58		15.1	GALAXY
ZWG 155.085	11 08 00.	+ 28 58		15.1	GALAXY
MCG+05-27-003	11 08 00.	+ 28 58	33	15.1	GALAXY
MCG+05-27-004	11 08 00.	+ 28 59	30	15.1	GALAXY
MCG+05-27-002	11 08 00.	+ 29 02	60	14.2	GALAXY
LB 10019	11 08 00.	+ 30 35		17.1	FAINT BLUE STAR
MCG+05-27-005	11 08 00.	+ 31 57	27	15.1	GALAXY
LB 10018	11 08 00.	+ 32 31		16.0	FAINT BLUE STAR
ZWG 242.003	11 08 00.	+ 47 14		15.4	GALAXY
ZWG 241.092	11 08 00.	+ 47 14		15.4	GALAXY
ZWG 267.047	11 08 00.	+ 53 40		12.8	GALAXY
UGC 06215	11 08 00.	+ 53 40	198	12.8	GALAXY Sc
MCG+10-16-073	11 08 00.	+ 57 24	24	16.	GALAXY
MCG+10-16-074	11 08 00.	+ 58 12	42	15.	GALAXY
MCG+10-16-075	11 08 00.	+ 61 37	72	15.	GALAXY
MCG+11-14-009	11 08 00.	+ 63 35	27	16.	GALAXY
REIZ 0878	11 08 01.	+ 53 39	240	13.0	GALAXY
RNGC 3538	11 08 02.	+ 75 50			NON-EXISTENT OBJECT
REIZ 0887	11 08 02.	- 06 16	24	15.3	GALAXY
KEEL 294	11 08 02.6	+ 55 44 11.			NEBULA
RNGC 3554	11 08 03.	+ 28 56		15.5	GALAXY
RNGC 3549	11 08 03.	+ 53 40		13.0	GALAXY
KLEM 18	11 08 03.	- 37 16	900	10.	BRIGHT GRP OF 3 GALAXIES
AR 34	11 08 03.60	+ 75 50 29.2			NEBULA
ARC 1180	11 08 04.	+ 63 14		17.8	RICH CLUSTER OF GALAXIES
ZWG 039.041	11 08 06.	+ 04 47		15.0	GALAXY
MCG+01-29-009	11 08 06.	+ 04 47	60	15.0	GALAXY
ZWG 039.042	11 08 06.	+ 05 07		14.9	GALAXY
UGC 06216	11 08 06.	+ 05 07	78	14.9	GALAXY S0-a?
MCG+01-29-010	11 08 06.	+ 05 07	72	14.9	GALAXY
ZWG 067.024	11 08 06.	+ 10 23		15.6	GALAXY
RNGC 3560	11 08 06.	+ 11 27			NON-EXISTENT OBJECT
ZWG 067.025	11 08 06.	+ 12 17		13.7	GALAXY
RNGC 3559	11 08 06.	+ 12 17		13.5	GALAXY
UGC 06217	11 08 06.	+ 12 17	84	13.7	GALAXY S
MCG+02-29-008	11 08 06.	+ 12 17	84	13.7	GALAXY
MCG+03-29-004	11 08 06.	+ 14 40	42	15.	GALAXY
ZC 1108.1+1830	11 08 06.	+ 18 30	1140		CLUSTER OF GALAXIES
LB 10020	11 08 06.	+ 27 14		17.0	FAINT BLUE STAR
MCG+05-27-006	11 08 06.	+ 28 32 30.	42	15.7	GALAXY
UGC 06218	11 08 06.	+ 28 33	66	16.0	GALAXY S
ZWG 156.007	11 08 06.	+ 28 56		15.3	GALAXY
ZWG 155.086	11 08 06.	+ 28 56		15.3	GALAXY
MCG+05-27-007	11 08 06.	+ 28 56	24	15.3	GALAXY
REIZ 0885	11 08 06.	+ 28 58	24	14.0	GALAXY
REIZ 0886	11 08 06.	+ 29 00	18	14.7	GALAXY
ZWG 156.008	11 08 06.	+ 30 26		15.7	GALAXY
ZWG 155.087	11 08 06.	+ 30 26		15.7	GALAXY
ZWG 291.035	11 08 06.	+ 58 14		15.5	GALAXY
SC 1105-4550.8	11 08 06.	- 46 07 03.	54		NEBULA
IC 0675	11 08 08.	+ 03 56 58.			NONSTELLAR OBJECT
RNGC 3558	11 08 09.	+ 28 49		15.0	GALAXY
ARC 1185	11 08 09.	+ 28 57		14.3	RICH CLUSTER OF GALAXIES
SEY 090	11 08 09.	+ 58 13 15.		15.3	FAINT GALAXY
ZWG 011.017	11 08 12.	+ 01 30		15.4	GALAXY
SN 1955H	11 08 12.	+ 03 12		18.8	SUPERNOVA
ZWG 039.043	11 08 12.	+ 05 22		15.6	GALAXY
FEIG 037	11 08 12.	+ 07 36		14.3	FAINT BLUE STAR
ZWG 067.026	11 08 12.	+ 10 22		15.6	GALAXY
ZWG 067.027	11 08 12.	+ 11 53		15.4	GALAXY
LB 10022	11 08 12.	+ 27 21		17.1	FAINT BLUE STAR
ZWG 156.009	11 08 12.	+ 27 22		15.5	GALAXY
ZWG 155.088	11 08 12.	+ 27 22		15.5	GALAXY
REIZ 0888	11 08 12.	+ 28 46	42	14.2	GALAXY
ZWG 156.010	11 08 12.	+ 28 49		14.8	GALAXY
ZWG 155.089	11 08 12.	+ 28 49		14.8	GALAXY
MRK 422	11 08 12.	+ 28 49	20	15.5	GALAXY WITH UV CONTINUUM
MCG+05-27-008	11 08 12.	+ 28 49	48	14.8	GALAXY
REIZ 0889	11 08 12.	+ 28 51	24	13.9	GALAXY
REIZ 0890	11 08 12.	+ 28 52	24	14.5	GALAXY
LB 10021	11 08 12.	+ 30 11		16.6	FAINT BLUE STAR
TON-N 0060	11 08 12.	+ 32 33		15.9	BLUE STAR
MCG+09-18-097	11 08 12.	+ 53 39	180	12.8	GALAXY
MCG+10-16-076	11 08 12.	+ 59 45	18	16.	GALAXY
MCG+10-16-077	11 08 12.	+ 62 24	39	15.	GALAXY
ZWG 291.036	11 08 12.	+ 62 25		15.4	GALAXY
ZWG 011.016	11 08 12.	- 02 28		15.5	GALAXY
MCG-06-25-006	11 08 12.	- 37 17	90	13.5	GALAXY
KEEL 295	11 08 12.3	+ 55 32 43.		16.	NEBULA
RNGC 3564	11 08 13.	- 37 16			GALAXY
HN 0352	11 08 14.	- 76 20			NEBULA
CED 112	11 08 14.	- 76 20	120		DIFFUSE GALACTIC NEBULA
IC 2631	11 08 14.	- 76 20			NONSTELLAR OBJECT
HOLM 233A	11 08 15.	+ 19 28	60	14.5	PART OF MULTIPLE GALAXY
MCG+05-27-009	11 08 15.	+ 27 22	36	15.5	GALAXY
MCG-02-29-008	11 08 15.	- 15 27 30.	54	15.	GALAXY
IC 0674	11 08 16.	+ 43 53 46.			NONSTELLAR OBJECT
ARC 1184	11 08 17.	+ 50 34		17.8	RICH CLUSTER OF GALAXIES
SC 1105-2948.3	11 08 17.	- 30 04 34.	18		NEBULA
KEEL 296	11 08 17.2	+ 55 52 31.		17.	NEBULA
ZC 1108.3+0137	11 08 18.	+ 01 37	2550		CLUSTER OF GALAXIES
ZWG 039.044	11 08 18.	+ 05 00		15.5	GALAXY
ZC 1108.3+0727	11 08 18.	+ 07 27	940		CLUSTER OF GALAXIES
ZC 1108.3+1452	11 08 18.	+ 14 52	1080		CLUSTER OF GALAXIES
ZWG 096.006	11 08 18.	+ 19 27		14.9	GALAXY
UGC 06219	11 08 18.	+ 19 27	96	14.9	GALAXY S
KARA.72 275A	11 08 18.	+ 19 27	66	14.9	PART OF DOUBLE GALAXY
ZC 1108.3+2008	11 08 18.	+ 20 08	1080		CLUSTER OF GALAXIES
UGC 06220	11 08 18.	+ 29 37	102	16.5	GALAXY DBL SYS
ZC 1108.3+3132	11 08 18.	+ 31 32	1140		CLUSTER OF GALAXIES
ZWG 213.031	11 08 18.	+ 43 54		14.5	GALAXY
UGC 06221	11 08 18.	+ 43 54	108	14.5	GALAXY Sa-b
MCG+07-23-027	11 08 18.	+ 43 54	90	14.	GALAXY
ZC 1108.3+6315	11 08 18.	+ 63 15	2020		CLUSTER OF GALAXIES
ZC 1108.3+7550	11 08 18.	+ 75 50	2080		CLUSTER OF GALAXIES
MCG-06-25-008	11 08 18.	- 35 04	72	15.	GALAXY
MCG-06-25-007	11 08 18.	- 36 23	30	15.5	GALAXY
ACK 286+10.1	11 08 18.	- 48 50			PLANETARY NEBULA
OCL 0846	11 08 18.	- 59 58	3600	8.8	OPEN STAR CLUSTER
VHA 112	11 08 18.	- 59 58	360		STAR CLSTR IN NEBULOSITY
KEEL 297	11 08 18.3	+ 55 33 05.		15.	SPIRAL NEBULA
TON-N 1350	11 08 20.	+ 19 53		14.9	BLUE STAR
PATH 1.381	11 08 20.	+ 59 46	16		NEBULA
RNGC 3572	11 08 20.	- 59 58		9.0	OPEN CLUSTER
HOLM 233B	11 08 22.	+ 19 29	18	15.4	PART OF MULTIPLE GALAXY
LB 01960	11 08 22.	+ 50 25 24.		16.9	FAINT BLUE STAR
ZWG 011.018	11 08 24.	+ 01 22		15.5	GALAXY
ZWG 039.045	11 08 24.	+ 05 08		15.1	GALAXY
8ZW 1108+05.8	11 08 24.	+ 05 45		17.8	COMPACT GALAXY
ZWG 096.007	11 08 24.	+ 19 28		15.6	GALAXY
KARA.72 275B	11 08 24.	+ 19 28	36	15.6	PART OF DOUBLE GALAXY
MCG+03-29-005	11 08 24.	+ 19 29	90	14.9	GALAXY
ZWG 096.008	11 08 24.	+ 19 39		15.7	GALAXY
ZWG 126.001	11 08 24.	+ 26 12		15.7	GALAXY
ZC 1108.4+3255	11 08 24.	+ 32 55	1140		CLUSTER OF GALAXIES
ZC 1108.4+3347	11 08 24.	+ 33 47	870		CLUSTER OF GALAXIES
UGC 06222	11 08 24.	+ 34 50	90	17.	GALAXY DWARF
ZWG 242.004	11 08 24.	+ 48 28		15.4	GALAXY
ZWG 241.093	11 08 24.	+ 48 28		15.4	GALAXY
ZC 1108.4+5032	11 08 24.	+ 50 32	400		CLUSTER OF GALAXIES
ZWG 314.014	11 08 24.	+ 62 28		15.6	GALAXY
ZWG 291.037	11 08 24.	+ 62 28		15.6	GALAXY
UGC 06223	11 08 24.	+ 62 28	84	15.6	GALAXY Sc?
MCG+10-16-078	11 08 24.	+ 62 28	45	15.	GALAXY
ZC 1108.4-0139	11 08 24.	- 01 39	1080		CLUSTER OF GALAXIES
MCG-06-25-009	11 08 24.	- 37 11	120	13.	GALAXY
SHB 176	11 08 26.	+ 28 57 54.		20.0	QUASI-STELLAR OBJECT
BC QSO1108+285	11 08 26.5	+ 28 57 56.		20.	QUASI-STELLAR OBJECT
RNGC 3561	11 08 27.	+ 28 58		14.5	GALAXY
MCG-04-27-003	11 08 27.	- 21 42	48	15.	GALAXY
BIGO 518	11 08 28.	+ 28 59			NEBULA
ARC 1189	11 08 29.	+ 01 24		17.0	RICH CLUSTER OF GALAXIES
ZWG 039.046	11 08 30.	+ 04 01		15.7	GALAXY
ZWG 039.047	11 08 30.	+ 04 37		15.6	GALAXY
MCG+03-29-006	11 08 30.	+ 19 29 30.	30	15.6	GALAXY
ZC 1108.5+2142	11 08 30.	+ 21 42	2220		CLUSTER OF GALAXIES
ZCG 1108+28	11 08 30.	+ 28 57		18.8	COMPACT GALAXY
STOCK 25	11 08 30.	+ 28 58			BLUE KNOT NEAR ELLIP GLXY
ZWG 156.011	11 08 30.	+ 28 58		14.7	GALAXY
ZWG 155.090	11 08 30.	+ 28 58		14.7	GALAXY
UGC 06224	11 08 30.	+ 28 58	318	14.7	GALAXY DBL SYS
VV 237E	11 08 30.	+ 28 58	6	19.	INTERACTING GALAXY
VV 237D	11 08 30.	+ 28 58	12	19.	INTERACTING GALAXY
VV 237C	11 08 30.	+ 28 58	30	16.	INTERACTING GALAXY
VV 237B	11 08 30.	+ 28 58	30	18.	INTERACTING GALAXY
VV 237A	11 08 30.	+ 28 58	42	15.	INTERACTING GALAXY
VV 237	11 08 30.	+ 28 58	240		INTERACTING GALAXY
MCG+05-27-010	11 08 30.	+ 28 58	36	14.7	GALAXY
SN 1953A	11 08 30.	+ 28 59		16.0	SUPERNOVA
MCG+05-27-011	11 08 30.	+ 28 59	48	14.7	GALAXY
REIZ 0892	11 08 30.	+ 29 00	24	14.2	GALAXY
REIZ 0893	11 08 30.	+ 29 01	30	14.1	GALAXY
VV 237F	11 08 30.	+ 29 02	12	19.	INTERACTING GALAXY
ZWG 185.016	11 08 30.	+ 35 40		15.2	GALAXY
MCG+08-20-096	11 08 30.	+ 47 18 30.	36	14.	GALAXY
MCG+08-20-097	11 08 30.	+ 47 38	36	16.	GALAXY
ZWG 242.005	11 08 30.	+ 47 39		15.3	GALAXY
ZWG 241.094	11 08 30.	+ 47 39		15.3	GALAXY
MCG+08-20-098	11 08 30.	+ 48 28	36	16.	GALAXY
LB 01961	11 08 30.	+ 55 44 24.		16.1	FAINT BLUE STAR
ZWG 268.001	11 08 30.	+ 55 56		10.7	GALAXY
ZWG 267.048	11 08 30.	+ 55 56		10.7	GALAXY
UGC 06225	11 08 30.	+ 55 56	528	10.7	GALAXY Sc

OBJECT NAME	RIGHT ASCEN.	DECLINATION	DIAM.	MAGN.	TYPE OF OBJECT
SN 1969B	11 08 30.	+ 55 56		16.	SUPERNOVA
KARA.73B 0469	11 08 30.	+ 55 56	504	10.7	ISOLATED GALAXY S
MCG+10-16-079	11 08 30.	+ 56 47	72	15.	GALAXY
ZWG 011.019	11 08 30.	- 03 18		15.5	GALAXY
MCG-06-25-010	11 08 30.	- 35 43	60	14.	GALAXY
ARP 105	11 08 31.	+ 28 58			PECULIAR GALAXY
REIZ 0894	11 08 31.	+ 31 52	30	15.0	GALAXY
RNGC 3568	11 08 31.	- 37 11			GALAXY
KEEL 298	11 08 31.5	+ 55 52 02.		16.	NEBULA
ARC 1191	11 08 35.	+ 01 02		17.5	RICH CLUSTER OF GALAXIES
LB 01962	11 08 35.	+ 50 34 48.		17.0	FAINT BLUE STAR
ZWG 011.020	11 08 36.	+ 00 54		15.7	GALAXY
ZC 1108.6+0100	11 08 36.	+ 01 00	1280		CLUSTER OF GALAXIES
ZWG 039.048	11 08 36.	+ 03 21		15.2	GALAXY
ZWG 039.049	11 08 36.	+ 04 13		15.3	GALAXY
UGC 06226	11 08 36.	+ 06 54	60	17.	GALAXY Sc
SN 1955N	11 08 36.	+ 24 27		19.8	SUPERNOVA
ZWG 156.012	11 08 36.	+ 28 32		15.6	GALAXY
MCG+05-27-012	11 08 36.	+ 28 58	30	15.5	GALAXY
REIZ 0895	11 08 36.	+ 28 59	30	14.1	GALAXY
LB 10023	11 08 36.	+ 30 28		16.9	FAINT BLUE STAR
MCG+06-25-018	11 08 36.	+ 35 39	42	15.	GALAXY
ZWG 242.006	11 08 36.	+ 47 19		14.3	GALAXY
ZWG 241.095	11 08 36.	+ 47 19		14.3	GALAXY
UGC 06227	11 08 36.	+ 47 19	54	14.3	GALAXY E-SO
ZWG 291.038	11 08 36.	+ 56 49		15.4	GALAXY
UGC 06228	11 08 36.	+ 56 49	78	15.4	GALAXY SB(c)
MCG+10-16-080	11 08 36.	+ 58 05	15	16.	GALAXY
LB 01963	11 08 36.	+ 61 25 00.		15.5	FAINT BLUE STAR
UGC 06229	11 08 36.	+ 62 37	84	17.	GALAXY DWARF
OCL 0847	11 08 36.	- 60 06	180		OPEN STAR CLUSTER
LB 01964	11 08 37.	+ 56 05 12.		17.0	FAINT BLUE STAR
TON-N 1351	11 08 38.	+ 21 20		14.8	BLUE STAR
RNGC 3556	11 08 38.	+ 55 57		11.0	GALAXY
REIZ 0891	11 08 38.	+ 55 57	420	11.4	GALAXY
KEEL 299	11 08 38.5	+ 55 48 17.		18.	NEBULA
MCG+07-23-028	11 08 39.	+ 44 10	96	14.5	GALAXY
ARC 1188	11 08 40.	+ 21 48		17.5	RICH CLUSTER OF GALAXIES
HOLM 234B	11 08 40.	+ 27 15	12	14.6	PART OF MULTIPLE GALAXY
LB 01965	11 08 40.	+ 53 29 18.		16.3	FAINT BLUE STAR
ZWG 039.050	11 08 42.	+ 03 28		15.4	GALAXY
ZWG 039.051	11 08 42.	+ 06 06		14.4	GALAXY
RNGC 3567	11 08 42.	+ 06 06		14.5	GALAXY
UGC 06230	11 08 42.	+ 06 06	90	14.4	GALAXY
KARA.72 276A	11 08 42.	+ 06 06	42	14.4	PART OF DOUBLE GALAXY
MCG+01-29-011	11 08 42.	+ 06 06	30	14.4	GALAXY
ZWG 067.028	11 08 42.	+ 11 51		15.3	GALAXY
UGC 06231	11 08 42.	+ 11 51	60	15.3	GALAXY
ZC 1108.7+1421	11 08 42.	+ 14 21	1010		CLUSTER OF GALAXIES
HOLM 234A	11 08 42.	+ 27 15	12	14.5	PART OF MULTIPLE GALAXY
ZWG 156.013	11 08 42.	+ 29 00		15.5	GALAXY
ZWG 155.091	11 08 42.	+ 29 00		15.5	GALAXY
ZWG 213.032	11 08 42.	+ 38 40		15.6	GALAXY
ZWG 213.033	11 08 42.	+ 44 10		15.2	GALAXY
UGC 06232	11 08 42.	+ 44 10	108	15.2	GALAXY Sa
MCG+09-18-098	11 08 42.	+ 55 57	480	10.5	GALAXY
MCG+11-14-010	11 08 42.	+ 62 37	39	17.	GALAXY
MCG-02-29-009	11 08 42.	- 09 41	108	14.	GALAXY
TON-N 1352	11 08 44.	+ 19 57		15.9	BLUE STAR
TON-N 1353	11 08 44.	+ 20 43		15.8	BLUE STAR
REIZ 0896	11 08 44.	- 06 05	24	15.8	GALAXY
MCG+05-27-014	11 08 45.	+ 27 13	48	14.6	GALAXY
MCG+05-27-013	11 08 45.	+ 27 13	24	14.6	GALAXY
RNGC 3563B	11 08 45.	+ 27 14		14.5	GALAXY
RNGC 3563A	11 08 45.	+ 27 14		14.5	GALAXY
MCG+05-27-015	11 08 45.	+ 29 00	15	15.5	GALAXY
MCG-01-29-004	11 08 45.	- 04 33	24	15.	GALAXY
KEEL 300	11 08 46.8	+ 56 00 42.		17.	SPIRAL NEBULA
ZWG 039.052	11 08 48.	+ 03 33		15.4	GALAXY
ZWG 039.053	11 08 48.	+ 03 35		15.4	GALAXY
ZWG 039.054	11 08 48.	+ 04 52		15.6	GALAXY
KARA.72 276B	11 08 48.	+ 06 05	30	15.1	PART OF DOUBLE GALAXY
ZWG 039.055	11 08 48.	+ 06 06		15.1	GALAXY
MCG+01-29-012	11 08 48.	+ 06 06	15	15.1	GALAXY
ZWG 039.056	11 08 48.	+ 07 11		14.6	GALAXY
UGC 06233	11 08 48.	+ 07 11	66	14.6	GALAXY Sa
MCG+01-29-013	11 08 48.	+ 07 11	48	14.6	GALAXY
ZWG 039.057	11 08 48.	+ 07 17		15.6	GALAXY
ZC 1108.8+0806	11 08 48.	+ 08 06	2350		CLUSTER OF GALAXIES
ZC 1108.8+1728	11 08 48.	+ 17 28	870		CLUSTER OF GALAXIES
ZWG 126.002	11 08 48.	+ 23 15		15.4	GALAXY
ZWG 156.014	11 08 48.	+ 27 14		14.6	GALAXY
UGC 06234	11 08 48.	+ 27 14	66	14.6	GALAXY SB:0
KARA.72 277B	11 08 48.	+ 27 14	54		PART OF DOUBLE GALAXY
KARA.72 277A	11 08 48.	+ 27 14	36	14.6	PART OF DOUBLE GALAXY
LB 10026	11 08 48.	+ 27 52		17.3	FAINT BLUE STAR
LB 10025	11 08 48.	+ 29 49		16.7	FAINT BLUE STAR
LB 10024	11 08 48.	+ 31 59		17.0	FAINT BLUE STAR
ZWG 213.034	11 08 48.	+ 39 39		15.4	GALAXY
MCG+09-18-100	11 08 48.	+ 52 22	30	16.	GALAXY
MCG+09-18-099	11 08 48.	+ 54 15	30	16.	GALAXY
ZC 1108.8+5908	11 08 48.	+ 59 08	870		CLUSTER OF GALAXIES
ZWG 314.015	11 08 48.	+ 68 01		15.2	GALAXY
KEEL 301	11 08 50.6	+ 55 23 55.			NEBULA
RNGC 3571	11 08 52.	- 18 01		13.0	GALAXY
ARC 1187	11 08 53.	+ 39 51		15.6	RICH CLUSTER OF GALAXIES
KEEL 302	11 08 53.9	+ 55 21 17.			NEBULA
ZWG 039.058	11 08 54.	+ 03 14		15.6	GALAXY
STOCK 26	11 08 54.	+ 03 34			BLUE KNOT NEAR ELLIP GLXY
ZWG 039.059	11 08 54.	+ 06 43		15.5	GALAXY
ZWG 039.060	11 08 54.	+ 07 47		15.5	GALAXY
ZWG 156.015	11 08 54.	+ 29 01		15.7	GALAXY
MCG+05-27-016	11 08 54.	+ 29 01	30	15.7	GALAXY
REIZ 0897	11 08 54.	+ 29 03	18	14.6	GALAXY
MCG+09-18-101	11 08 54.	+ 52 19	30	16.	GALAXY
ZWG 267.049	11 08 54.	+ 52 24		15.5	GALAXY
ZC 1108.9+6629	11 08 54.	+ 66 29	940		CLUSTER OF GALAXIES
UGC 06235	11 08 54.	+ 79 05	72	16.0	GALAXY Sb-c
ZWG 011.021	11 08 54.	- 01 11		15.7	GALAXY
MCG-06-25-011	11 08 54.	- 36 37	72	13.5	GALAXY
VHE 47B	11 08 54.	- 61 05	36		REFLECTION NEBULA
REIZ 0898	11 08 55.	+ 31 50	30	14.9	GALAXY
RNGC 3573	11 08 55.	- 36 35			GALAXY
KEEL 303	11 08 57.7	+ 55 45 23.		16.	SPIRAL NEBULA
ZWG 039.061	11 09 00.	+ 04 56		15.2	GALAXY
ZWG 067.029	11 09 00.	+ 09 59		15.7	GALAXY
ZWG 067.030	11 09 00.	+ 12 24		15.1	GALAXY
ZC 1109.0+2513	11 09 00.	+ 25 13	2020		CLUSTER OF GALAXIES

OBJECT NAME	RIGHT ASCEN.	DECLINATION	DIAM.	MAGN.	TYPE OF OBJECT
ZWG 156.016	11 09 00.	+ 26 35		15.7	GALAXY
REIZ 0899	11 09 00.	+ 28 47	54	13.8	GALAXY
ZWG 185.017	11 09 00.	+ 37 00		15.5	GALAXY
UGC 06236	11 09 00.	+ 37 00	78	15.5	GALAXY Sa
MCG+06-25-019	11 09 00.	+ 37 00	57	15.5	GALAXY
MCG+07-23-029	11 09 00.	+ 39 38	48	15.	GALAXY
MCG+07-23-030	11 09 00.	+ 41 05	18	17.	GALAXY
ARC 1190	11 09 00.	+ 41 07		16.6	RICH CLUSTER OF GALAXIES
MCG+08-21-001	11 09 00.	+ 45 49	78	15.	GALAXY
ZWG 242.007	11 09 00.	+ 45 50		15.2	GALAXY
ZWG 241.096	11 09 00.	+ 45 50		15.2	GALAXY
MCG-03-29-001	11 09 00.	- 18 01	150	13.	GALAXY
IC 2628	11 09 02.	+ 12 23 33.			NONSTELLAR OBJECT
KEEL 304	11 09 02.0	+ 55 37 35.		17.	NEBULA
KEEL 305	11 09 04.8	+ 55 47 07.		17.	NEBULA
ZWG 039.062	11 09 06.	+ 06 56		15.7	GALAXY
ZC 1109.1+1445	11 09 06.	+ 14 45	340		CLUSTER OF GALAXIES
ZCG 1109+27	11 09 06.	+ 27 54		17.6	COMPACT GALAXY
ZWG 156.017	11 09 06.	+ 28 33		15.6	GALAXY
MCG+05-27-017	11 09 06.	+ 28 33	12	15.6	GALAXY
ZC 1109.1+3806	11 09 06.	+ 38 06	3160		CLUSTER OF GALAXIES
MCG+07-23-031	11 09 06.	+ 41 04	42	17.	GALAXY
ZC 1109.1+4239	11 09 06.	+ 42 39	2420		CLUSTER OF GALAXIES
LB 01966	11 09 06.	+ 60 36 06.		14.0	FAINT BLUE STAR
UGC 06237	11 09 06.	+ 65 30	60	16.5	GALAXY DISTRBD
ZWG 334.012	11 09 06.	+ 72 45		15.5	GALAXY
OCL 0848	11 09 06.	- 60 01	1650	4.5	OPEN STAR CLUSTER
MRSL 290+00/1	11 09 06.	- 60 14	43200		HII REGION
MRSL 290-00/4	11 09 06.	- 60 20	780		HII REGION
VHE 47A	11 09 06.	- 61 02	66		REFLECTION NEBULA
REIZ 0900	11 09 07.	+ 30 39	24	14.0	GALAXY
TON-N 1354	11 09 08.	+ 34 10		16.8	BLUE STAR
RNGC 3576	11 09 08.	- 61 06			DIFFUSE NEBULA
KEEL 306	11 09 08.3	+ 55 29 06.			NEBULA
SN 1951D	11 09 12.	+ 03 58		19.6	SUPERNOVA
ZC 1109.2+1659	11 09 12.	+ 16 59	1340		CLUSTER OF GALAXIES
LB 10027	11 09 12.	+ 31 46		17.6	FAINT BLUE STAR
ZC 1109.2+4111	11 09 12.	+ 41 11	1480		CLUSTER OF GALAXIES
MCG+10-16-081	11 09 12.	+ 57 20	27	15.	GALAXY
MCG+11-14-010A	11 09 12.	+ 65 30	39	15.	GALAXY
ZWG 011.022	11 09 12.	- 02 10		14.7	GALAXY
MCG+00-29-004	11 09 12.	- 02 10	24	14.7	GALAXY
LB 01967	11 09 14.	+ 60 50 48.		16.5	FAINT BLUE STAR
ARC 1195	11 09 14.	- 04 40		17.6	RICH CLUSTER OF GALAXIES
KEEL 307	11 09 15.0	+ 55 40 21.		17.	NEBULA
1ZW 025	11 09 18.	+ 28 37			COMPACT GALAXY
MCG+05-27-018	11 09 18.	+ 28 46	45	14.9	GALAXY
ZWG 185.018	11 09 18.	+ 35 43		14.5	GALAXY
UGC 06238	11 09 18.	+ 35 43	72	14.5	GALAXY SO
MCG+09-18-102	11 09 18.	+ 54 17 30.	48	14.	GALAXY
ZWG 291.039	11 09 18.	+ 57 21		15.3	GALAXY
OCL 0850	11 09 18.	- 60 24	2160	8.4	OPEN STAR CLUSTER
VHA 113	11 09 18.	- 60 24	360		OPEN STAR CLUSTER
VHE 47C	11 09 18.	- 61 06			REFLECTION NEBULA
KEEL 308	11 09 19.5	+ 55 40 28.			NEBULA
RNGC 3569	11 09 20.	+ 35 43		14.5	GALAXY
RNGC 3570	11 09 21.	+ 27 51		15.0	GALAXY
KEEL 309	11 09 22.9	+ 55 40 24.		17.	NEBULA
SN 1973V	11 09 23.	+ 54 18		17.0	SUPERNOVA
ZWG 039.063	11 09 24.	+ 03 19		15.4	GALAXY
ZWG 039.064	11 09 24.	+ 03 25		14.6	GALAXY
UGC 06239	11 09 24.	+ 03 25	66	14.6	GALAXY E
MCG+01-29-014	11 09 24.	+ 03 25	36	14.6	GALAXY
MCG+05-27-020	11 09 24.	+ 27 49	18	16.	GALAXY
ZWG 156.018	11 09 24.	+ 27 51		15.0	GALAXY
UGC 06240	11 09 24.	+ 27 51	66	15.0	GALAXY SO
MCG+05-27-019	11 09 24.	+ 27 51	18	15.0	GALAXY
ZWG 156.019	11 09 24.	+ 28 45		14.9	GALAXY
UGC 06241	11 09 24.	+ 28 45	60	14.9	GALAXY S
ZC 1109.4+3358	11 09 24.	+ 33 58	540		CLUSTER OF GALAXIES
MCG+07-23-032	11 09 24.	+ 41 17 30.	36	15.5	GALAXY
ZWG 268.002	11 09 24.	+ 54 18		15.4	GALAXY
ZWG 267.050	11 09 24.	+ 54 18		15.4	GALAXY
ZC 1109.4-0023	11 09 24.	- 00 23	400		CLUSTER OF GALAXIES
OCL 0849	11 09 24.	- 60 06	90		OPEN STAR CLUSTER
KEEL 310	11 09 24.2	+ 55 41 48.		16.	SPIRAL NEBULA
WRAY 19.26	11 09 24.6	- 61 05 27.		10.0	DIFFUSE NEBULA
SN 1973D	11 09 25.	+ 27 51		17.0	SUPERNOVA
SC 1107-6049.2	11 09 25.	- 61 05 29.	36		NEBULA
RNGC 3574	11 09 27.	+ 27 53		15.5	GALAXY
MCG+06-25-020	11 09 27.	+ 35 43	24	14.	GALAXY
ARC 1194	11 09 28.	+ 30 59		17.5	RICH CLUSTER OF GALAXIES
REIZ 0901	11 09 28.	- 00 16	36	14.9	GALAXY
LB 01968	11 09 29.	+ 52 36 24.		16.3	FAINT BLUE STAR
ZWG 039.065	11 09 30.	+ 03 26		15.2	GALAXY
ZWG 067.031	11 09 30.	+ 13 35		15.5	GALAXY
ZC 1109.5+1956	11 09 30.	+ 19 56	1140		CLUSTER OF GALAXIES
ZC 1109.5+2410	11 09 30.	+ 24 10	1010		CLUSTER OF GALAXIES
LB 10028	11 09 30.	+ 27 45		16.9	FAINT BLUE STAR
MCG+05-27-021	11 09 30.	+ 27 50	24	16.	GALAXY
MCG+05-27-022	11 09 30.	+ 27 52 30.	24	15.7	GALAXY
ZWG 156.020	11 09 30.	+ 27 53		15.7	GALAXY
SN 1973A	11 09 30.	+ 27 54		18.0	SUPERNOVA
ZC 1109.5+2906	11 09 30.	+ 29 06	670		CLUSTER OF GALAXIES
ZWG 334.013	11 09 30.	+ 73 09		13.2	GALAXY
UGC 06242	11 09 30.	+ 73 09	108	13.2	GALAXY E
ZWG 351.057	11 09 30.	+ 77 12		15.4	GALAXY
ARC 1193	11 09 31.	+ 42 43		17.2	RICH CLUSTER OF GALAXIES
RNGC 3562	11 09 32.	+ 73 09		13.0	GALAXY
MAI 059	11 09 34.	+ 64 24	53		DWARF SPHEROIDAL GALAXY
LB 10030	11 09 36.	+ 30 39		17.4	FAINT BLUE STAR
LB 10029	11 09 36.	+ 30 46		17.0	FAINT BLUE STAR
ZC 1109.6+3315	11 09 36.	+ 33 15	1010		CLUSTER OF GALAXIES
ZC 1109.6+3414	11 09 36.	+ 34 14	540		CLUSTER OF GALAXIES
ZC 1109.6+4309	11 09 36.	+ 43 09	540		CLUSTER OF GALAXIES
ZWG 242.008	11 09 36.	+ 45 33		15.3	GALAXY
ARC 1192	11 09 36.	+ 59 32		17.6	RICH CLUSTER OF GALAXIES
MCG+12-11-011	11 09 36.	+ 73 10	39	14.	GALAXY
MCG+12-11-012	11 09 36.	+ 73 16	51	16.	GALAXY
ZWG 011.023	11 09 36.	- 02 04		15.0	GALAXY
MCG+00-29-005	11 09 36.	- 02 04	15	15.0	GALAXY
MCG-03-29-002	11 09 36.	- 16 38 30.	30	15.	GALAXY
KEEL 311	11 09 37.9	+ 55 57 16.		16.	NEBULA
KEEL 312	11 09 38.8	+ 55 46 56.		17.	NEBULA
KEEL 313	11 09 40.8	+ 55 48 01.		17.	NEBULA
MCG+03-29-007	11 09 42.	+ 19 57	36	16.	GALAXY
UGC 06243	11 09 42.	+ 31 41	60	16.0	GALAXY Sc
ZC 1109.7+4344	11 09 42.	+ 43 44	670		CLUSTER OF GALAXIES

OBJECT NAME	RIGHT ASCEN.	DECLINATION	DIAM.	MAGN.	TYPE OF OBJECT
MCG+08-21-002	11 09 42.	+ 45 32	30	15.	GALAXY
MRK 164	11 09 42.	+ 51 54	8	17.	GALAXY WITH UV CONTINUUM
ZC 1109.7+5625	11 09 42.	+ 56 25	740		CLUSTER OF GALAXIES
ZWG 011.024	11 09 42.	- 02 07		15.6	GALAXY
ZC 1109.7-0253	11 09 42.	- 02 53	2020		CLUSTER OF GALAXIES
MCG-02-29-010	11 09 42.	- 13 21	24	15.	GALAXY
AR 35	11 09 42.64	+ 77 12 09.9			NEBULA
HOLM 235A	11 09 43.	+ 31 40	36	14.7	PART OF MULTIPLE GALAXY
CED 113B	11 09 43.	- 60 57			DIFFUSE GALACTIC NEBULA
CED 113A	11 09 43.	- 61 06			DIFFUSE GALACTIC NEBULA
RNGC 3579	11 09 44.	- 60 57			DIFFUSE NEBULA
MCG-01-29-005	11 09 45.	- 05 28	60	15.	GALAXY
KEEL 314	11 09 47.0	+ 55 56 07.		17.	SPIRAL NEBULA
ZWG 039.066	11 09 48.	+ 04 02		15.3	GALAXY
SN 19550	11 09 48.	+ 22 03		19.3	SUPERNOVA
ZWG 126.003	11 09 48.	+ 23 33		15.0	GALAXY
MCG+04-27-001	11 09 48.	+ 23 33	42	15.0	GALAXY
LB 10034	11 09 48.	+ 27 32		17.2	FAINT BLUE STAR
ZWG 156.021	11 09 48.	+ 28 18		15.7	GALAXY
UGC 06244	11 09 48.	+ 28 18	84	15.7	GALAXY DBL SYS
MCG+05-27-023A	11 09 48.	+ 28 18		15.7	GALAXY
LB 10033	11 09 48.	+ 30 29		16.7	FAINT BLUE STAR
LB 10032	11 09 48.	+ 30 39		16.2	FAINT BLUE STAR
HOLM 235B	11 09 48.	+ 31 40	30	15.9	PART OF MULTIPLE GALAXY
MCG+05-27-023	11 09 48.	+ 31 40	54	14.7	GALAXY
LB 10031	11 09 48.	+ 31 48		18.0	FAINT BLUE STAR
7ZW 378	11 09 48.	+ 57 33			COMPACT GALAXY
MCG-05-27-001	11 09 48.	- 27 44	36	15.	GALAXY
REIZ 0902	11 09 49.	+ 31 39	24	15.2	GALAXY
REIZ 0903	11 09 49.	+ 32 13	18	14.9	GALAXY
MIL 20	11 09 49.	- 60 22 00.	324		SUPERNOVA REMNANT
CED 113C	11 09 49.	- 61 02			DIFFUSE GALACTIC NEBULA
TON-N 1355	11 09 50.	+ 21 58		16.9	BLUE STAR
ARC 1200	11 09 51.	- 02 54		17.0	RICH CLUSTER OF GALAXIES
KEEL 315	11 09 51.0	+ 55 40 54.		16.	SPIRAL NEBULA
WRAY 19.27	11 09 51.4	- 61 01 38.			DIFFUSE NEBULA
ZWG 039.067	11 09 54.	+ 05 15		15.6	GALAXY
ZC 1109.9+1418	11 09 54.	+ 14 18	1210		CLUSTER OF GALAXIES
ZWG 096.009	11 09 54.	+ 20 05		15.6	GALAXY
MCG+09-18-104	11 09 54.	+ 53 35	36	17.	GALAXY
ZC 1109.9+5405	11 09 54.	+ 54 05	740		CLUSTER OF GALAXIES
MCG+09-18-103	11 09 54.	+ 56 29	30	16.	GALAXY
7ZW 379	11 09 54.	+ 58 37			COMPACT GALAXY
MRSL 290+01/1	11 09 54.	- 58 31	600		HII REGION
MRSL 291-00/1	11 09 54.	- 60 56	1200		HII REGION
ARC 1197	11 09 56.	+ 36 45		17.2	RICH CLUSTER OF GALAXIES
REIZ 0906	11 09 56.	- 06 24	24	14.8	GALAXY
REIZ 0905	11 09 56.	- 06 25	18	14.4	GALAXY
RNGC 3581	11 09 56.	- 61 02			DIFFUSE NEBULA
WRAY 19.28	11 09 56.0	- 61 00 08.			DIFFUSE NEBULA
MCG+03-29-008	11 09 57.	+ 20 07	36	15.6	GALAXY
VDB.66G 092	11 10	+ 53 49	70		DWARF GALAXY
ZWG 011.025	11 10 00.	+ 01 43		15.5	GALAXY
ZC 1110.0+0243	11 10 00.	+ 02 43	2890		CLUSTER OF GALAXIES
ZC 1110.0+1337	11 10 00.	+ 13 37	2490		CLUSTER OF GALAXIES
MCG+03-29-009	11 10 00.	+ 20 09	30	16.	GALAXY
ZC 1110.0+2023	11 10 00.	+ 20 23	1280		CLUSTER OF GALAXIES
ZWG 126.004	11 10 00.	+ 23 35		15.7	GALAXY
MCG+04-27-002	11 10 00.	+ 23 35	30	15.7	GALAXY
TON-N 0061	11 10 00.	+ 24 27		15.1	BLUE STAR
ZWG 126.005	11 10 00.	+ 25 46		15.2	GALAXY
MCG+05-27-024	11 10 00.	+ 28 33 30.	42	16.	GALAXY
ZC 1110.0+3530	11 10 00.	+ 35 30	1280		CLUSTER OF GALAXIES
ZC 1110.0+3646	11 10 00.	+ 36 46	340		CLUSTER OF GALAXIES
ZC 1110.0+3953	11 10 00.	+ 39 53	1810		CLUSTER OF GALAXIES
MCG+08-21-003	11 10 00.	+ 45 20	30	16.	GALAXY
MCG+09-19-001	11 10 00.	+ 55 57 30.	30	16.	GALAXY
MCG+14-05-016	11 10 00.	+ 80 35	24	17.	GALAXY
MCG+14-05-017	11 10 00.	+ 80 36	27	17.	GALAXY
SC 1107-6044.0	11 10 00.	- 61 00 18.	180		NEBULA
IC 2629	11 10 01.	+ 12 22 37.			NONSTELLAR OBJECT
REIZ 0904	11 10 01.	+ 32 15	12	15.7	GALAXY
CED 113D	11 10 01.	- 61 00			DIFFUSE GALACTIC NEBULA
WRAY 19.29	11 10 02.1	- 61 02 18.		9.3	STAR-NEBULA ASSOCIATION
KEEL 316	11 10 02.5	+ 55 57 19.		13.	SPIRAL NEBULA
ARC 1199	11 10 03.	+ 20 21		17.6	RICH CLUSTER OF GALAXIES
ARC 1198	11 10 03.	+ 30 39		17.5	RICH CLUSTER OF GALAXIES
LB 00255	11 10 05.	+ 59 32 24.		15.1	FAINT BLUE STAR
WRAY 19.30	11 10 05.6	- 58 30 41.			DIFFUSE NEBULA
ZWG 067.032	11 10 06.	+ 09 20		13.4	GALAXY
UGC 06245	11 10 06.	+ 09 20	132	13.4	GALAXY SB0
MCG+02-29-009	11 10 06.	+ 09 20	120	13.4	GALAXY
ZWG 067.033	11 10 06.	+ 13 42		15.5	GALAXY
ZWG 156.022	11 10 06.	+ 28 20		15.7	GALAXY
MCG+05-27-025	11 10 06.	+ 28 20	30	15.7	GALAXY
ZC 1110.1+3049	11 10 06.	+ 30 49	3830		CLUSTER OF GALAXIES
KEEL 317	11 10 06.	+ 55 57 43.		16.	NEBULA
MCG+09-19-002	11 10 06.	+ 56 04	30	17.	GALAXY
LB 01969	11 10 06.	+ 59 32 18.		14.9	FAINT BLUE STAR
MCG-06-25-013	11 10 06.	- 36 09	66	13.	GALAXY
MCG-06-25-012	11 10 06.	- 36 33		15.5	GALAXY
IC 0676	11 10 07.	+ 09 20 13.			NONSTELLAR OBJECT
IC 2630	11 10 07.	+ 12 35 37.			NONSTELLAR OBJECT
REIZ 0907	11 10 07.	+ 30 46	18	15.1	GALAXY
CED 113E	11 10 07.	- 60 56			DIFFUSE GALACTIC NEBULA
TON-N 1356	11 10 08.	+ 19 54		15.9	BLUE STAR
ARC 1196	11 10 08.	+ 54 09		17.8	RICH CLUSTER OF GALAXIES
RNGC 3582	11 10 08.	- 61 00			DIFFUSE NEBULA
SC 1108-6040.6	11 10 10.	- 60 56 54.	240		NEBULA
KEEL 318	11 10 11.7	+ 56 04 52.		15.	SPIRAL NEBULA
KARA.68 076	11 10 12.	+ 10 29	60		DWARF GALAXY
ZWG 067.034	11 10 12.	+ 13 41		15.6	GALAXY
ZWG 126.006	11 10 12.	+ 23 32		15.7	GALAXY
UGC 06246	11 10 12.	+ 23 32	90	15.7	GALAXY Sc
MCG+04-27-003	11 10 12.	+ 23 32	84	15.7	GALAXY
ZWG 126.007	11 10 12.	+ 25 53		15.6	GALAXY
MCG+05-27-026	11 10 12.	+ 27 42	96	15.1	GALAXY
ZWG 156.023	11 10 12.	+ 27 43		15.1	GALAXY
UGC 06247	11 10 12.	+ 27 43	90	15.1	GALAXY Sa-b
ZWG 156.024	11 10 12.	+ 28 16		15.5	GALAXY
ZWG 156.025	11 10 12.	+ 28 40		15.7	GALAXY
TON-N 0062	11 10 12.	+ 29 28		14.2	BLUE STAR
LB 10036	11 10 12.	+ 30 46		17.9	FAINT BLUE STAR
LB 10035	11 10 12.	+ 32 18		18.1	FAINT BLUE STAR
ZWG 213.035	11 10 12.	+ 41 54		15.7	GALAXY
MCG+09-19-003	11 10 12.	+ 55 49	30	15.	GALAXY
7ZW 380	11 10 12.	+ 58 37			COMPACT GALAXY
OCL 0851	11 10 12.	- 60 29	180		OPEN STAR CLUSTER
TON-N 1357	11 10 14.	+ 19 26		12.8	BLUE STAR
RNGC 3584	11 10 14.	- 60 56			DIFFUSE NEBULA
UGC 06248	11 10 18.	+ 10 28	96	16.	GALAXY DWRF IR
LB 10038	11 10 18.	+ 29 24		14.6	FAINT BLUE STAR
LB 10037	11 10 18.	+ 29 38		16.8	FAINT BLUE STAR
ZC 1110.3+3404	11 10 18.	+ 34 04	1610		CLUSTER OF GALAXIES
MCG+09-19-004	11 10 18.	+ 52 02 30.	30	16.	GALAXY
MCG+09-19-005	11 10 18.	+ 52 03 30.	18	16.	GALAXY
ZWG 268.003	11 10 18.	+ 55 50		15.7	GALAXY
ZWG 267.051	11 10 18.	+ 55 50		15.7	GALAXY
7ZW 381	11 10 18.	+ 57 16			COMPACT GALAXY
MCG+10-16-082	11 10 18.	+ 59 16	45	16.	GALAXY
MCG+10-16-083	11 10 18.	+ 60 10	78	15.	GALAXY
ZWG 291.040	11 10 18.	+ 60 11		14.6	GALAXY
UGC 06249	11 10 18.	+ 60 11	108	14.6	GALAXY Sc
CED 113F	11 10 19.	- 61 05			DIFFUSE GALACTIC NEBULA
KEEL 319	11 10 19.4	+ 55 49 41.		15.	NEBULA
TON-N 1358	11 10 20.	+ 33 58		17.	BLUE STAR
TON-N 1359	11 10 20.	+ 35 28		16.9	BLUE STAR
SEY 091	11 10 20.	+ 60 11 55.		14.8	FAINT GALAXY
RNGC 3578	11 10 21.	- 15 41			NON-EXISTENT OBJECT
SC 1108-6048.4	11 10 22.	- 61 04 42.	90		NEBULA
ARC 1201	11 10 23.	+ 13 42		17.0	RICH CLUSTER OF GALAXIES
ZWG 011.026	11 10 24.	+ 00 26		15.6	GALAXY
MCG+00-29-006	11 10 24.	+ 00 26	36	15.6	GALAXY
ZWG 039.068	11 10 24.	+ 08 08		15.0	GALAXY
MCG+01-29-015	11 10 24.	+ 08 08	48	15.0	GALAXY
MCG+09-19-006	11 10 24.	+ 53 51	102	15.	GALAXY
MCG+10-16-084	11 10 24.	+ 56 44	27	16.	GALAXY
ZWG 291.041	11 10 24.	+ 56 45		15.3	GALAXY
MCG+10-16-085	11 10 24.	+ 60 11	18	17.	GALAXY
MCG+10-16-086	11 10 24.	+ 60 12	24	16.	GALAXY
ZWG 334.014	11 10 24.	+ 71 20		15.7	GALAXY
ARC 1186	11 10 24.	+ 75 40		16.5	RICH CLUSTER OF GALAXIES
MCG-03-29-003	11 10 24.	- 20 56	42	16.5	GALAXY
LB 01970	11 10 25.	+ 51 13 48.		16.2	FAINT BLUE STAR
WRAY 19.31	11 10 25.7	- 61 08 12.			DIFFUSE NEBULA
RNGC 3586	11 10 26.	- 61 05			DIFFUSE NEBULA
KEEL 320	11 10 27.9	+ 55 50 44.		16.	SPIRAL NEBULA
RNGC 3575	11 10 28.	+ 22 56			NON-EXISTENT OBJECT
IC 2632	11 10 29.	+ 11 56 48.			NONSTELLAR OBJECT
ZWG 011.027	11 10 30.	+ 00 16		15.3	GALAXY
ZWG 039.069	11 10 30.	+ 05 20		14.9	GALAXY
MCG+01-29-016	11 10 30.	+ 05 20	48	14.9	GALAXY
ZWG 039.070	11 10 30.	+ 05 30		15.5	GALAXY
ZWG 039.071	11 10 30.	+ 06 55		15.3	GALAXY
ZWG 039.072	11 10 30.	+ 07 58		15.4	GALAXY
ZC 1110.5+0923	11 10 30.	+ 09 23	670		CLUSTER OF GALAXIES
ZC 1110.5+1448	11 10 30.	+ 14 48	1680		CLUSTER OF GALAXIES
ZWG 156.026	11 10 30.	+ 28 05		14.7	GALAXY
UGC 06250	11 10 30.	+ 28 05	78	14.7	GALAXY
MCG+05-27-027	11 10 30.	+ 28 05	24	14.7	GALAXY
ZC 1110.5+3510	11 10 30.	+ 35 10	340		CLUSTER OF GALAXIES
LB 01971	11 10 30.	+ 52 08 24.		16.1	FAINT BLUE STAR
ZWG 268.004	11 10 30.	+ 53 52		15.7	GALAXY
ZWG 267.052	11 10 30.	+ 53 52		15.7	GALAXY
UGC 06251	11 10 30.	+ 53 52	114	15.7	GALAXY DWRF SP
MCG+10-16-087	11 10 30.	+ 60 10	12	16.	GALAXY
ZC 1110.5+6649	11 10 30.	+ 66 49	1010		CLUSTER OF GALAXIES
LB 01972	11 10 32.	+ 56 37 36.		16.7	FAINT BLUE STAR
IC 2633	11 10 33.	+ 11 52 30.			NONSTELLAR OBJECT
REIZ 0909	11 10 34.	+ 00 17	15	15.5	GALAXY
REIZ 0908	11 10 35.	+ 26 09	42	13.7	GALAXY
LB 01973	11 10 35.	+ 55 50 00.		14.4	FAINT BLUE STAR
ZWG 039.073	11 10 36.	+ 05 28		14.8	GALAXY
MCG+01-29-017	11 10 36.	+ 05 28	48	14.8	GALAXY
ZWG 039.074	11 10 36.	+ 08 06		15.7	GALAXY
ZC 1110.6+1402	11 10 36.	+ 14 02	810		CLUSTER OF GALAXIES
ZWG 126.008	11 10 36.	+ 26 08		15.4	GALAXY
UGC 06252	11 10 36.	+ 26 08	60	15.4	GALAXY Sc
LB 10040	11 10 36.	+ 30 08		17.0	FAINT BLUE STAR
LB 10039	11 10 36.	+ 31 12		18.2	FAINT BLUE STAR
MCG+09-19-007	11 10 36.	+ 55 40	15	17.	GALAXY
ZC 1110.6+5901	11 10 36.	+ 59 01	3970		CLUSTER OF GALAXIES
MCG+10-16-088	11 10 36.	+ 60 23	39	16.	GALAXY
ZC 1110.6+6125	11 10 36.	+ 61 25	3830		CLUSTER OF GALAXIES
ZC 1110.6+6545	11 10 36.	+ 65 45	1080		CLUSTER OF GALAXIES
ZWG 334.015	11 10 36.	+ 73 11		15.6	GALAXY
MRSL 291-00/3	11 10 36.	- 60 28	960		HII REGION
HOLM 236B	11 10 39.	+ 05 28	18	14.3	PART OF MULTIPLE GALAXY
REIZ 0910	11 10 41.	+ 03 55	12	14.0	GALAXY
HOLM 236A	11 10 41.	+ 05 29	66	13.5	PART OF MULTIPLE GALAXY
ZWG 039.075	11 10 42.	+ 03 56		14.7	GALAXY
MCG+01-29-018	11 10 42.	+ 03 56	48	14.7	GALAXY
ZWG 039.076	11 10 42.	+ 05 22		15.5	GALAXY
MCG+04-27-004	11 10 42.	+ 26 09	54	15.4	GALAXY
ZWG 156.027	11 10 42.	+ 30 52		15.1	GALAXY
MCG+05-27-027A	11 10 42.	+ 30 52	36	15.1	GALAXY
ZC 1110.7+4746	11 10 42.	+ 47 46	810		CLUSTER OF GALAXIES
7ZW 382	11 10 42.	+ 61 18			COMPACT GALAXY
RNGC 3580	11 10 43.	+ 03 56		14.5	GALAXY
REIZ 0911	11 10 46.	+ 02 49	60	14.5	GALAXY
ARC 1205	11 10 47.	+ 02 47		16.9	RICH CLUSTER OF GALAXIES
VV 145C	11 10 48.	+ 02 47	12	16.5	INTERACTING GALAXY
VV 145B	11 10 48.	+ 02 47	12	16.5	INTERACTING GALAXY
VV 145A	11 10 48.	+ 02 47	15	16.	INTERACTING GALAXY
ZWG 039.077	11 10 48.	+ 02 49		15.1	GALAXY
MCG+01-29-019	11 10 48.	+ 02 49	48	15.1	GALAXY
ZWG 039.078	11 10 48.	+ 07 42		15.6	GALAXY
REIZ 0912	11 10 48.	+ 10 47	36	14.9	GALAXY
ZWG 096.010	11 10 48.	+ 19 10		15.6	GALAXY
HAW 2	11 10 48.	+ 22 26	15	20.0	GALAXY
KARA.68 077	11 10 48.	+ 22 26	302		DWARF GALAXY
UGC 06253	11 10 48.	+ 22 26	900	12.9	GALAXY DWRF EL
MCG+04-27-005	11 10 48.	+ 22 26	750	12.8	GALAXY
UGC 06254	11 10 48.	+ 23 27	66	16.0	GALAXY S
ZWG 126.009	11 10 48.	+ 23 35		15.7	GALAXY
LB 10043	11 10 48.	+ 26 44		16.8	FAINT BLUE STAR
LB 10042	11 10 48.	+ 27 24		17.0	FAINT BLUE STAR
ZWG 156.028	11 10 48.	+ 28 09		15.7	GALAXY
LB 10041	11 10 48.	+ 30 37		16.6	FAINT BLUE STAR
MCG+06-25-021	11 10 48.	+ 37 49	42	15.5	GALAXY
ZWG 242.009	11 10 48.	+ 47 51		13.6	GALAXY
UGC 06255	11 10 48.	+ 47 51	48	13.6	GALAXY S
MCG+08-21-004	11 10 48.	+ 48 35	36	16.	GALAXY
MCG+09-19-008	11 10 48.	+ 53 20	54	16.	GALAXY
UGC 06256	11 10 48.	+ 65 27	84	16.0	GALAXY Sc
ZWG 334.016	11 10 48.	+ 71 04		15.7	GALAXY

OBJECT NAME	RIGHT ASCEN.	DECLINATION	DIAM.	MAGN.	TYPE OF OBJECT
MCG+13-08-056	11 10 48.	+ 80 01	48	15.	GALAXY
MCG-04-27-004	11 10 48.	- 26 30	90	12.	GALAXY
OCL 0852	11 10 48.	- 60 31	390	8.0	OPEN STAR CLUSTER
VHA 114	11 10 48.	- 60 31	210		OPEN STAR CLUSTER
RNGC 3590	11 10 50.	- 60 31		8.0	OPEN CLUSTER
MCG+06-25-022	11 10 51.	+ 36 09	24	15.5	GALAXY
LB 01974	11 10 52.	+ 52 31 12.		16.1	FAINT BLUE STAR
KEEL 321	11 10 52.9	+ 55 59 40.		16.	NEBULA
IC 2634	11 10 53.	+ 10 45 35.			NONSTELLAR OBJECT
IC 2635	11 10 53.	+ 11 44 17.			NONSTELLAR OBJECT
RNGC 3577	11 10 53.	+ 48 33		14.5	GALAXY
RNGC 3585	11 10 53.	- 26 29		12.0	GALAXY
ZWG 067.035	11 10 54.	+ 10 46		15.0	GALAXY
MCG+02-29-010	11 10 54.	+ 10 46	24	15.0	GALAXY
ZC 1110.9+1251	11 10 54.	+ 12 51	670		CLUSTER OF GALAXIES
MCG+04-27-007	11 10 54.	+ 23 27	48	15.5	GALAXY
MCG+04-27-006	11 10 54.	+ 23 35	54	15.7	GALAXY
REIZ 0913	11 10 54.	+ 28 11	24	15.1	GALAXY
LB 10044	11 10 54.	+ 32 18		17.9	FAINT BLUE STAR
ARC 1202	11 10 54.	+ 47 47		17.0	RICH CLUSTER OF GALAXIES
MCG+08-21-005	11 10 54.	+ 47 51	42	14.	GALAXY
MCG+08-21-006	11 10 54.	+ 48 32	78	14.	GALAXY
ZWG 242.010	11 10 54.	+ 48 33		14.7	GALAXY
UGC 06257	11 10 54.	+ 48 33	96	14.7	GALAXY SBb
MCG+09-19-009	11 10 54.	+ 55 42	36	16.	GALAXY
MCG-02-29-011	11 10 54.	- 15 30	24	15.	GALAXY
OCL 0845	11 10 54.	- 58 39	180	10.	OPEN STAR CLUSTER
KEEL 322	11 10 54.8	+ 55 41 29.		16.	NEBULA
ARC 1204	11 10 55.	+ 17 52		17.8	RICH CLUSTER OF GALAXIES
KEEL 323	11 10 56.7	+ 55 19 59.			NEBULA
LB 01975	11 10 57.	+ 50 55 06.		17.0	FAINT BLUE STAR
IC 2636	11 10 58.	+ 11 43 41.			NONSTELLAR OBJECT
KEEL 324	11 10 58.6	+ 55 44 55.		16.	NEBULA
VDB.66G 093	11 11	+ 22 24	340		DWARF GALAXY
ZWG 039.079	11 11 00.	+ 08 01		15.6	GALAXY
ZC 1111.0+1624	11 11 00.	+ 16 24	1340		CLUSTER OF GALAXIES
ZC 1111.0+1755	11 11 00.	+ 17 55	1810		CLUSTER OF GALAXIES
LB 10045	11 11 00.	+ 28 04		17.9	FAINT BLUE STAR
ZC 1111.0+4032	11 11 00.	+ 40 32	1680		CLUSTER OF GALAXIES
ZC 1111.0+5037	11 11 00.	+ 50 37	1880		CLUSTER OF GALAXIES
ZWG 268.005	11 11 00.	+ 53 22		15.6	GALAXY
ZWG 267.053	11 11 00.	+ 53 22		15.6	GALAXY
ZC 1111.0+5843	11 11 00.	+ 58 43	1410		CLUSTER OF GALAXIES
MCG+11-14-011	11 11 00.	+ 65 26 30.	78	16.	GALAXY
KEEL 325	11 11 00.0	+ 55 35 07.			NEBULA
ARC 1206	11 11 01.	- 05 21		17.6	RICH CLUSTER OF GALAXIES
TON-N 1360	11 11 02.	+ 20 28		13.	BLUE STAR
ZWG 039.080	11 11 06.	+ 06 58		15.2	GALAXY
ZWG 126.010	11 11 06.	+ 21 48		15.6	GALAXY
UGC 06258	11 11 06.	+ 21 48	126	15.6	GALAXY IRR
KARA.73B 0470	11 11 06.	+ 21 48	126	15.6	ISOLATED GALAXY IR
ZWG 185.019	11 11 06.	+ 34 33		15.6	GALAXY
MCG+09-19-010	11 11 06.	+ 55 59	36	16.	GALAXY
MCG+10-16-089	11 11 06.	+ 57 02	39	15.	GALAXY
7ZW 383	11 11 06.	+ 64 13			COMPACT GALAXY
ZWG 314.016	11 11 06.	+ 64 13		15.3	GALAXY
ZWG 011.028	11 11 06.	- 02 20		14.9	GALAXY
MCG+00-29-007	11 11 06.	- 02 20	48	14.9	GALAXY
TON-N 1361	11 11 08.	+ 22 06		16.7	BLUE STAR
TON-N 1362	11 11 08.	+ 22 46		15.8	BLUE STAR
LB 01976	11 11 08.	+ 60 47 54.		16.4	FAINT BLUE STAR
ARC 1203	11 11 10.	+ 40 34		16.6	RICH CLUSTER OF GALAXIES
REIZ 0914	11 11 11.	+ 04 34	24	15.0	GALAXY
ZC 1111.2+0002	11 11 12.	+ 00 02	3290		CLUSTER OF GALAXIES
ZWG 039.081	11 11 12.	+ 07 10		15.6	GALAXY
ZWG 067.036	11 11 12.	+ 09 52		13.9	GALAXY
UGC 06259	11 11 12.	+ 09 52	54	13.9	GALAXY
MCG+02-29-011	11 11 12.	+ 09 52	48	13.9	GALAXY
REIZ 0915	11 11 12.	+ 10 51	24	14.9	GALAXY
MCG+03-29-010	11 11 12.	+ 17 12	36	16.	GALAXY
LB 10049	11 11 12.	+ 29 42		17.7	FAINT BLUE STAR
LB 10048	11 11 12.	+ 30 24		17.1	FAINT BLUE STAR
LB 10047	11 11 12.	+ 30 56		18.0	FAINT BLUE STAR
LB 10046	11 11 12.	+ 32 12		17.5	FAINT BLUE STAR
MCG+06-25-024	11 11 12.	+ 34 19	12	16.	GALAXY
MCG+06-25-023	11 11 12.	+ 34 33	12	16.	GALAXY
ZWG 242.011	11 11 12.	+ 48 05		15.6	GALAXY
MCG+09-19-011	11 11 12.	+ 54 20	12	18.	GALAXY
ZWG 268.006	11 11 12.	+ 55 59		14.9	GALAXY
ZWG 267.054	11 11 12.	+ 55 59		14.9	GALAXY
ZWG 291.042	11 11 12.	+ 57 04		15.3	GALAXY
KEEL 326	11 11 12.7	+ 55 59 00.		12.	SPIRAL NEBULA
IC 2637	11 11 13.	+ 09 51 35.			NONSTELLAR OBJECT
MCG+04-27-008	11 11 15.	+ 21 47	108	15.6	GALAXY
LYNG 05	11 11 15.	- 58 36 12.	1800		OB CONCENTRATION
IC 2638	11 11 16.	+ 10 50 11.			NONSTELLAR OBJECT
ZWG 039.082	11 11 18.	+ 04 33		14.9	GALAXY
UGC 06260	11 11 18.	+ 04 33	78	14.9	GALAXY Sb
MCG+01-29-020	11 11 18.	+ 04 33	72	14.9	GALAXY
ZWG 067.037	11 11 18.	+ 10 50		15.1	GALAXY
UGC 06261	11 11 18.	+ 10 50	66	15.1	GALAXY SBO
MCG+02-29-012	11 11 18.	+ 10 50	36	15.1	GALAXY
ZWG 067.038	11 11 18.	+ 12 35		13.6	GALAXY
UGC 06262	11 11 18.	+ 12 35	96	13.6	GALAXY Sb
MCG+02-29-013	11 11 18.	+ 12 35	90	13.6	GALAXY
TON-N 0575	11 11 18.	+ 27 45		16.8	BLUE STAR
ZC 1111.3+4051	11 11 18.	+ 40 51	6920		CLUSTER OF GALAXIES
ZWG 242.012	11 11 18.	+ 48 36		11.6	GALAXY
UGC 06263	11 11 18.	+ 48 36	162	11.6	GALAXY SBb
MCG+09-19-012	11 11 18.	+ 52 49	24	17.	GALAXY
ZWG 291.043	11 11 18.	+ 62 21		15.2	GALAXY
IC 2639	11 11 19.	+ 09 55 10.			NONSTELLAR OBJECT
IC 0677	11 11 19.	+ 12 34 10.			NONSTELLAR OBJECT
REIZ 0917	11 11 19.	+ 12 35	60	13.7	GALAXY
TON-N 1363	11 11 20.	+ 18 41		16.7	BLUE STAR
TON-N 1364	11 11 20.	+ 35 19		16.	BLUE STAR
SEY 092	11 11 20.	+ 55 59 42.		15.1	FAINT GALAXY
KEEL 327	11 11 20.2	+ 55 56 03.		17.	NEBULA
RNGC 3588	11 11 23.	+ 20 40		15.5	GALAXY
RNGC 3583	11 11 23.	+ 48 35		12.0	GALAXY
ZWG 126.011	11 11 24.	+ 20 40		15.3	GALAXY
UGC 06264	11 11 24.	+ 20 40	66	15.3	GALAXY DBL SYS
MCG+04-27-009	11 11 24.	+ 20 40	30	15.3	GALAXY
ZC 1111.4+2313	11 11 24.	+ 23 13	1480		CLUSTER OF GALAXIES
ZC 1111.4+3313	11 11 24.	+ 33 13	610		CLUSTER OF GALAXIES
ZWG 213.036	11 11 24.	+ 38 41		15.7	GALAXY
MCG+08-21-007	11 11 24.	+ 48 05	60	16.	GALAXY
ZC 1111.4+4934	11 11 24.	+ 49 34	1480		CLUSTER OF GALAXIES

OBJECT NAME	RIGHT ASCEN.	DECLINATION	DIAM.	MAGN.	TYPE OF OBJECT
MCG+10-16-091	11 11 24.	+ 59 21	27	17.	GALAXY
MCG+10-16-090	11 11 24.	+ 62 20	42	15.	GALAXY
REIZ 0916	11 11 26.	+ 48 40	78	12.5	GALAXY
MCG+08-21-008	11 11 27.	+ 48 35	138	12.2	GALAXY
IC 2640	11 11 29.	+ 11 16 22.			NONSTELLAR OBJECT
ZC 1111.5+0438	11 11 30.	+ 04 38	740		CLUSTER OF GALAXIES
ARC 1208	11 11 30.	+ 04 38		17.6	RICH CLUSTER OF GALAXIES
ZWG 039.083	11 11 30.	+ 06 51		15.1	GALAXY
MCG+01-29-021	11 11 30.	+ 06 51	12	15.1	GALAXY
ZWG 156.029	11 11 30.	+ 27 30		15.5	GALAXY
2ZW 048	11 11 30.	+ 30 36			COMPACT GALAXY
ZWG 185.020	11 11 30.	+ 34 26		15.4	GALAXY
MCG+07-23-033	11 11 30.	+ 38 40	36	16.	GALAXY
LB 01977	11 11 30.	+ 59 12 36.		14.5	FAINT BLUE STAR
ZWG 011.029	11 11 30.	- 00 28		15.3	GALAXY
MCG-02-29-012	11 11 30.	- 13 48	60	14.	GALAXY
IC 0678	11 11 31.	+ 06 51 10.			NONSTELLAR OBJECT
TON-N 1365	11 11 32.	+ 34 55		15.5	BLUE STAR
RNGC 3591	11 11 33.	- 13 48		14.0	GALAXY
IC 2641	11 11 35.	+ 09 40 22.			NONSTELLAR OBJECT
REIZ 0918	11 11 35.	+ 28 28	24	14.6	GALAXY
8ZW 1111+04.5	11 11 36.	+ 04 28		17.9	COMPACT GALAXY
ZC 1111.6+1434	11 11 36.	+ 14 34	870		CLUSTER OF GALAXIES
ZWG 156.030	11 11 36.	+ 28 25		15.6	GALAXY
MCG+05-27-028	11 11 36.	+ 28 26	72	15.6	GALAXY
ZWG 156.031	11 11 36.	+ 28 49		15.3	GALAXY
MCG+05-27-030	11 11 36.	+ 28 50	36	15.3	GALAXY
LB 10053	11 11 36.	+ 29 12		18.0	FAINT BLUE STAR
LB 10052	11 11 36.	+ 30 35		18.5	FAINT BLUE STAR
LB 10051	11 11 36.	+ 31 09		17.2	FAINT BLUE STAR
ZWG 156.032	11 11 36.	+ 31 44		15.5	GALAXY
MCG+05-27-029	11 11 36.	+ 31 45	36	15.5	GALAXY
LB 10050	11 11 36.	+ 31 52		17.7	FAINT BLUE STAR
MCG+06-25-025	11 11 36.	+ 34 25	27	15.	GALAXY
ZC 1111.6+3450	11 11 36.	+ 34 50	340		CLUSTER OF GALAXIES
MCG+07-23-034	11 11 36.	+ 38 39	27	17.	GALAXY
ZC 1111.6+4429	11 11 36.	+ 44 29	940		CLUSTER OF GALAXIES
ZWG 242.013	11 11 36.	+ 47 39		15.4	GALAXY
MCG+10-16-092	11 11 36.	+ 58 05	51	15.	GALAXY
FATH 1.382	11 11 36.	+ 59 23	27		NEBULA
ZWG 314.017	11 11 36.	+ 63 38		15.0	GALAXY
UGC 06265	11 11 36.	+ 63 38	66	15.0	GALAXY S
MCG-02-29-013	11 11 36.	- 14 25	36	15.	GALAXY
IC 2642	11 11 39.	+ 10 32 34.			NONSTELLAR OBJECT
KEEL 328	11 11 41.2	+ 55 45 39.		15.	NEBULA
ZWG 039.084	11 11 42.	+ 04 17		15.6	GALAXY
UGC 06266	11 11 42.	+ 43 30	96	17.	GALAXY DWRF SP
ZWG 291.044	11 11 42.	+ 58 06		15.1	GALAXY
MCG+11-14-012	11 11 45.	+ 63 38	39	16.	GALAXY
RNGC 3592	11 11 47.	+ 17 32		15.0	GALAXY
ZWG 039.085	11 11 48.	+ 04 21		15.6	GALAXY
ZWG 039.086	11 11 48.	+ 04 26		15.4	GALAXY
ZWG 096.011	11 11 48.	+ 17 32		14.8	GALAXY
UGC 06267	11 11 48.	+ 17 32	132	14.8	GALAXY Sc
UGC 06268	11 11 48.	+ 22 45	72	16.0	GALAXY Sc
LB 10055	11 11 48.	+ 27 53		18.0	FAINT BLUE STAR
LB 10054	11 11 48.	+ 31 02		18.0	FAINT BLUE STAR
ZC 1111.8+3340	11 11 48.	+ 33 40	400		CLUSTER OF GALAXIES
MCG+07-23-035	11 11 48.	+ 43 30	66	17.	GALAXY
BIGO 519	11 11 48.	+ 55 56			NEBULA
ZWG 291.045	11 11 48.	+ 56 51		15.6	GALAXY
ZWG 334.017	11 11 48.	+ 70 56		15.7	GALAXY
UGC 06269	11 11 48.	+ 70 56	78	15.7	GALAXY S
TON-N 1366	11 11 50.	+ 19 04		16.2	BLUE STAR
TON-N 1367	11 11 50.	+ 21 43		16.5	BLUE STAR
TON-N 1368	11 11 50.	+ 21 59		14.3	BLUE STAR
IC 2643	11 11 51.	+ 10 24 10.			NONSTELLAR OBJECT
RNGC 3587	11 11 51.	+ 55 18		12.0	PLANETARY NEBULA
SEY 093	11 11 52.	+ 58 06 30.		15.0	FAINT GALAXY
ARC 1209	11 11 53.	+ 13 10		17.2	RICH CLUSTER OF GALAXIES
SHB 177	11 11 53.	+ 40 53 42.		18.0	QUASI-STELLAR OBJECT
BC 3CR254	11 11 53.35	+ 40 53 42.0		17.98	QUASI-STELLAR OBJECT
KEEL 329	11 11 53.7	+ 55 21 54.			NEBULA
ZC 1111.9+0743	11 11 54.	+ 07 43	1010		CLUSTER OF GALAXIES
IC 2644	11 11 54.	+ 11 02 33.			NONSTELLAR OBJECT
IC 2645	11 11 54.	+ 12 09 39.			NONSTELLAR OBJECT
ZWG 067.039	11 11 54.	+ 12 10		15.7	GALAXY
MCG+03-29-011	11 11 54.	+ 17 33	96	14.8	GALAXY
UGC 06270	11 11 54.	+ 29 49	66	16.0	GALAXY SB0-a
ZWG 156.033	11 11 54.	+ 30 35		14.6	GALAXY
UGC 06271	11 11 54.	+ 30 35	72	14.6	GALAXY Sa
ZC 1111.9+3300	11 11 54.	+ 33 00	540		CLUSTER OF GALAXIES
MCG+07-23-036	11 11 54.	+ 39 45	30	15.	GALAXY
SN 1972B	11 11 54.	+ 54 46		18.5	SUPERNOVA
MCG+10-16-093	11 11 54.	+ 56 50	36	16.	GALAXY
PK148+57.1	11 11 54.99	+ 55 17 21.3	213	12.0	PLANETARY NEBULA
LB 09849	11 12	- 83 54		13.0	FAINT BLUE STAR
ZWG 039.087	11 12 00.	+ 08 16		15.6	GALAXY
ZWG 067.040	11 12 00.	+ 13 05		11.8	GALAXY
RNGC 3593	11 12 00.	+ 13 05		12.0	GALAXY
UGC 06272	11 12 00.	+ 13 05	312	11.8	GALAXY S0
MCG+02-29-014	11 12 00.	+ 13 05	300	11.8	GALAXY
ZC 1112.0+1310	11 12 00.	+ 13 10	1340		CLUSTER OF GALAXIES
ZC 1112.0+1552	11 12 00.	+ 15 52	940		CLUSTER OF GALAXIES
REIZ 0919	11 12 00.	+ 29 11	24	15.3	GALAXY
MCG+05-27-031	11 12 00.	+ 30 35	66	14.6	GALAXY
ZC 1112.0+3217	11 12 00.	+ 32 17	1080		CLUSTER OF GALAXIES
MCG+09-19-013	11 12 00.	+ 50 37	42	16.	GALAXY
ZC 1112.0+5355	11 12 00.	+ 53 55	1340		CLUSTER OF GALAXIES
MCG+09-19-014	11 12 00.	+ 55 14	24	17.	GALAXY
ZC 1112.0+5805	11 12 00.	+ 58 05	1210		CLUSTER OF GALAXIES
MCG+10-16-094	11 12 00.	+ 61 20	15	17.	GALAXY
MCG+10-16-095	11 12 00.	+ 61 24	12	17.	GALAXY
ZC 1112.0+6752	11 12 00.	+ 67 52	3490		CLUSTER OF GALAXIES
ZC 1112.0+6937	11 12 00.	+ 69 37	1950		CLUSTER OF GALAXIES
MCG+12-11-013	11 12 00.	+ 70 57	66	16.	GALAXY
ZC 1112.0+7121	11 12 00.	+ 71 21	1550		CLUSTER OF GALAXIES
REIZ 0920	11 12 01.	+ 13 06	240	11.7	GALAXY
IC 2647	11 12 02.	+ 12 24 57.			NONSTELLAR OBJECT
IC 2646	11 12 02.	+ 12 48 15.			NONSTELLAR OBJECT
KEEL 330	11 12 05.0	+ 55 13 59.			NEBULA
ZWG 039.088	11 12 06.	+ 04 17		15.1	GALAXY
ZWG 067.041	11 12 06.	+ 10 30		15.7	GALAXY
ZC 1112.1+1706	11 12 06.	+ 17 06	1610		CLUSTER OF GALAXIES
LB 10056	11 12 06.	+ 27 13		17.6	FAINT BLUE STAR
ZC 1112.1+3409	11 12 06.	+ 34 09	740		CLUSTER OF GALAXIES
ZWG 185.021	11 12 06.	+ 35 46		14.4	GALAXY
UGC 06273	11 12 06.	+ 35 46	102	14.4	GALAXY Sa-b

OBJECT NAME	RIGHT ASCEN.	DECLINATION	DIAM.	MAGN.	TYPE OF OBJECT
ZWG 268.007	11 12 06.	+ 50 38		15.7	GALAXY
ZC 1112.1+5939	11 12 06.	+ 59 39	1210		CLUSTER OF GALAXIES
MCG+10-16-096	11 12 06.	+ 60 58	78	12.	GALAXY
MCG+10-16-097	11 12 06.	+ 61 42	27	17.	GALAXY
ZWG 334.018	11 12 06.	+ 69 35		15.5	GALAXY
SN 1954I	11 12 06.	- 20 38		17.3	SUPERNOVA
SC 1109-4413.4	11 12 06.	- 44 29 43.	12		NEBULA
OCL 0843	11 12 06.	- 57 19	1920	9.6	OPEN STAR CLUSTER
TON-N 1369	11 12 08.	+ 18 51		17.	BLUE STAR
TON-N 1370	11 12 08.	+ 19 30		16.6	BLUE STAR
IC 2648	11 12 09.	+ 10 29 57.			NONSTELLAR OBJECT
REIZ 0922	11 12 10.	+ 00 07	18	16.0	GALAXY
RNGC 3597	11 12 10.	- 23 27		13.0	GALAXY
IC 2649	11 12 11.	+ 11 24 09.			NONSTELLAR OBJECT
ZWG 067.042	11 12 12.	+ 11 24		15.0	GALAXY
MCG+02-29-015	11 12 12.	+ 11 24	36	15.0	GALAXY
ZC 1112.2+2000	11 12 12.	+ 20 00	1880		CLUSTER OF GALAXIES
LB 10059	11 12 12.	+ 29 48		18.3	FAINT BLUE STAR
LB 10058	11 12 12.	+ 30 47		18.1	FAINT BLUE STAR
ZWG 156.034	11 12 12.	+ 31 47		15.3	GALAXY
LB 10057	11 12 12.	+ 31 56		16.5	FAINT BLUE STAR
ZWG 185.022	11 12 12.	+ 34 06		14.7	GALAXY
UGC 06274	11 12 12.	+ 34 06	60	14.7	GALAXY Sc
MCG+06-25-027	11 12 12.	+ 35 46	24	16.	GALAXY
MCG+06-25-026	11 12 12.	+ 35 47	90	14.5	GALAXY
ZC 1112.2+3630	11 12 12.	+ 36 30	1210		CLUSTER OF GALAXIES
ZC 1112.2+5339	11 12 12.	+ 53 39	540		CLUSTER OF GALAXIES
VV 160	11 12 12.	+ 53 55 00.	6	19.	INTERACTING GALAXY
MCG+10-16-098	11 12 12.	+ 58 48	30	16.	GALAXY
ZWG 291.046	11 12 12.	+ 60 58		14.5	GALAXY
UGC 06275	11 12 12.	+ 60 58	102	14.5	GALAXY
ZWG 011.030	11 12 12.	- 01 51		15.1	GALAXY
KARA.73B 0471	11 12 12.	- 01 51	48	15.1	ISOLATED GALAXY S
MCG-04-27-005	11 12 12.	- 23 27 30.	42	13.	GALAXY
REIZ 0921	11 12 13.	+ 34 05	24	14.4	GALAXY
RNGC 3589	11 12 13.	+ 60 58		14.5	GALAXY
TON-N 1371	11 12 14.	+ 33 57		14.3	BLUE STAR
TON-N 1372	11 12 14.	+ 34 13		16.8	BLUE STAR
IC 2651	11 12 15.	+ 12 30 57.			NONSTELLAR OBJECT
IC 2650	11 12 15.	+ 14 07 39.			NONSTELLAR OBJECT
KEEL 331	11 12 15.2	+ 55 29 23.			NEBULA
IC 2652	11 12 16.	+ 12 43 15.			NONSTELLAR OBJECT
ARC 1211	11 12 16.	- 11 57		17.5	RICH CLUSTER OF GALAXIES
REIZ 0923	11 12 17.	+ 26 12	54	15.2	GALAXY
IC 2653	11 12 18.	+ 10 49 21.			NONSTELLAR OBJECT
ZWG 096.012	11 12 18.	+ 17 06		15.6	GALAXY
MCG+06-25-028	11 12 18.	+ 34 05	48	14.	GALAXY
ZWG 185.023	11 12 18.	+ 36 49		15.7	GALAXY
ZC 1112.3+5013	11 12 18.	+ 50 13	670		CLUSTER OF GALAXIES
MCG+09-19-015	11 12 18.	+ 54 38	18		GALAXY
ZWG 291.047	11 12 18.	+ 58 49		15.5	GALAXY
ZWG 011.031	11 12 18.	- 02 34		14.9	GALAXY
MCG+00-29-008	11 12 18.	- 02 34	66	14.9	GALAXY
ARC 1210	11 12 19.	+ 17 02		17.3	RICH CLUSTER OF GALAXIES
MCG+03-29-012A	11 12 21.	+ 17 06	42	15.6	GALAXY
MCG+03-29-012B	11 12 21.	+ 17 07	30	15.6	GALAXY
REIZ 0925	11 12 22.	+ 00 27	18	15.4	GALAXY
REIZ 0924	11 12 23.	+ 26 10	18	14.6	GALAXY
ZWG 039.089	11 12 24.	+ 03 53		14.7	GALAXY
MCG+01-29-022	11 12 24.	+ 03 53	21	14.7	GALAXY
8ZW 1112+07.9	11 12 24.	+ 07 56		17.7	COMPACT GALAXY
MCG+04-27-010	11 12 24.	+ 26 10	36	15.5	GALAXY
LB 10062	11 12 24.	+ 28 44		17.9	FAINT BLUE STAR
ZWG 156.035	11 12 24.	+ 29 48		15.5	GALAXY
LB 10061	11 12 24.	+ 30 42		17.9	FAINT BLUE STAR
LB 10060	11 12 24.	+ 31 02		18.0	FAINT BLUE STAR
ZWG 156.036	11 12 24.	+ 31 18		14.4	GALAXY
UGC 06276	11 12 24.	+ 31 18	96	14.4	GALAXY S0
ZC 1112.4+3425	11 12 24.	+ 34 25	470		CLUSTER OF GALAXIES
MCG+09-19-016	11 12 24.	+ 54 42	15	15.	GALAXY
ZWG 268.008	11 12 24.	+ 56 14		15.1	GALAXY
ZWG 267.055	11 12 24.	+ 56 14		15.1	GALAXY
MCG+09-19-017	11 12 24.	+ 56 14	15	16.	GALAXY
SHAH 059	11 12 24.	+ 71 06	138	17.6	GROUP OF COMPACT GALAXIES
MCG-05-27-002	11 12 24.	- 33 40	48	15.	GALAXY
IC 2654	11 12 27.	+ 12 46 39.			NONSTELLAR OBJECT
MCG+06-25-029	11 12 27.	+ 36 48	36	15.5	GALAXY
REIZ 0929	11 12 28.	+ 00 35	24	16.0	GALAXY
IC 2655	11 12 28.	+ 12 26 26.			NONSTELLAR OBJECT
IC 2656	11 12 29.	+ 12 39 14.			NONSTELLAR OBJECT
RNGC 3596	11 12 29.	+ 15 04		12.0	GALAXY
ZWG 039.090	11 12 30.	+ 04 21		15.1	GALAXY
MCG+01-29-023	11 12 30.	+ 04 21	12	15.1	GALAXY
ZWG 096.013	11 12 30.	+ 15 03		11.7	GALAXY
UGC 06277	11 12 30.	+ 15 03	264	11.7	GALAXY Sc
MCG+03-29-013	11 12 30.	+ 15 03	270	11.7	GALAXY
KARA.73B 0472	11 12 30.	+ 15 03	216	11.7	ISOLATED GALAXY S
ZWG 126.012	11 12 30.	+ 26 10		15.5	GALAXY
SN 1962E	11 12 30.	+ 26 10		17.5	SUPERNOVA
REIZ 0926	11 12 30.	+ 29 47	24	14.8	GALAXY
HOLM 237A	11 12 30.	+ 29 49	30	14.6	PART OF MULTIPLE GALAXY
MCG+05-27-033	11 12 30.	+ 29 49	36	15.5	GALAXY
MCG+05-27-032	11 12 30.	+ 31 18	21	14.4	GALAXY
ZC 1112.5+3233	11 12 30.	+ 32 33	740		CLUSTER OF GALAXIES
ZWG 213.037	11 12 30.	+ 38 46		15.6	GALAXY
MCG+09-19-018	11 12 30.	+ 55 43	48	15.	GALAXY
MCG+10-16-099	11 12 30.	+ 58 44	30	17.	GALAXY
IC 2657	11 12 32.	+ 13 58 02.			NONSTELLAR OBJECT
KEEL 332	11 12 32.1	+ 55 07 08.			NEBULA
IC 2658	11 12 33.	+ 13 16 14.			SINGLE STAR
LB 01978	11 12 33.	+ 60 40 36.		16.5	FAINT BLUE STAR
RNGC 3598	11 12 35.	+ 17 32		13.5	GALAXY
ZWG 096.014	11 12 36.	+ 17 32		13.5	GALAXY
UGC 06278	11 12 36.	+ 17 32	96	13.5	GALAXY E-S0
MCG+03-29-014	11 12 36.	+ 17 33	90	13.5	GALAXY
ZWG 156.037	11 12 36.	+ 27 40		15.2	GALAXY
LB 10064	11 12 36.	+ 29 19		15.8	FAINT BLUE STAR
MCG+05-27-033A	11 12 36.	+ 29 49	36	15.9	GALAXY
LB 10063	11 12 36.	+ 31 55		17.1	FAINT BLUE STAR
ZWG 185.024	11 12 36.	+ 35 46		14.8	GALAXY
UGC 06279	11 12 36.	+ 35 46	84	14.8	GALAXY Sa
ZWG 242.014	11 12 36.	+ 47 43		13.0	GALAXY
RNGC 3595	11 12 36.	+ 47 43		13.0	GALAXY
UGC 06280	11 12 36.	+ 47 43	96	13.0	GALAXY
ZWG 268.009	11 12 36.	+ 55 44		15.4	GALAXY
ZWG 267.056	11 12 36.	+ 55 44		15.4	GALAXY
MCG-05-27-003	11 12 36.	- 28 07	120	14.	GALAXY
KEEL 333	11 12 36.2	+ 55 43 43.			NEBULA
HN 0841	11 12 37.	+ 11 50		14.	NEBULA
REIZ 0928	11 12 37.	+ 47 46	30	14.2	GALAXY
ARC 1207	11 12 37.	+ 67 58		17.1	RICH CLUSTER OF GALAXIES
HOLM 237C	11 12 38.	+ 29 49	12	15.8	PART OF MULTIPLE GALAXY
TON-N 1373	11 12 38.	+ 34 28		15.4	BLUE STAR
MAI 060	11 12 38.	+ 58 35	40		DWARF SPHEROIDAL GALAXY
HOLM 237B	11 12 39.	+ 29 49	12	15.3	PART OF MULTIPLE GALAXY
MCG+07-23-037	11 12 39.	+ 38 46	48	16.	GALAXY
MCG+08-21-009	11 12 39.	+ 47 42 30.	60	14.2	GALAXY
SEY 094	11 12 41.	+ 55 44 23.		15.4	FAINT GALAXY
LB 10065	11 12 42.	+ 31 58		18.4	FAINT BLUE STAR
ZC 1112.7+3245	11 12 42.	+ 32 45	1280		CLUSTER OF GALAXIES
MCG+06-25-030	11 12 42.	+ 35 47	66	14.	GALAXY
MCG+09-19-019	11 12 42.	+ 56 11	48	16.	GALAXY
ZC 1112.7+7259	11 12 42.	+ 72 59	7190		CLUSTER OF GALAXIES
ZWG 011.032	11 12 42.	- 03 07		15.7	GALAXY
REIZ 0927	11 12 43.	+ 56 02	30	15.3	GALAXY
RNGC 3599	11 12 47.	+ 18 23		13.0	GALAXY
REIZ 0930	11 12 47.	+ 26 13	18	14.7	GALAXY
REIZ 0931	11 12 47.	+ 26 15	12	15.4	GALAXY
ZWG 011.034	11 12 48.	+ 00 23		15.2	GALAXY
ZC 1112.8+1218	11 12 48.	+ 12 18	2690		CLUSTER OF GALAXIES
ZWG 096.015	11 12 48.	+ 18 23		13.0	GALAXY
UGC 06281	11 12 48.	+ 18 23	180	13.0	GALAXY E/S0
MCG+03-29-015	11 12 48.	+ 18 25	36	13.0	GALAXY
ZWG 126.013	11 12 48.	+ 23 37		15.7	GALAXY
LB 10066	11 12 48.	+ 26 24		17.2	FAINT BLUE STAR
2ZW 049	11 12 48.	+ 32 04			COMPACT GALAXY
ZWG 185.025	11 12 48.	+ 36 47		15.4	GALAXY
ZWG 242.015	11 12 48.	+ 46 25		15.3	GALAXY
ZWG 268.010	11 12 48.	+ 56 11		15.6	GALAXY
ZWG 267.057	11 12 48.	+ 56 11		15.6	GALAXY
ZWG 011.033	11 12 48.	- 03 29		15.2	GALAXY
MCG-02-29-014	11 12 48.	- 13 25 30.	36	14.5	GALAXY
HOLM 238B	11 12 49.	+ 29 49	24	16.1	PART OF MULTIPLE GALAXY
REIZ 0932	11 12 50.	+ 18 23	72	14.5	GALAXY
TON-N 1374	11 12 50.	+ 19 55		15.8	BLUE STAR
KEEL 334	11 12 50.9	+ 55 27 21.			NEBULA
IC 2660	11 12 51.	+ 12 42 38.			NONSTELLAR OBJECT
MCG+08-21-010	11 12 51.	+ 46 23 30.	30	16.	GALAXY
REIZ 0933	11 12 52.	+ 00 02	30	15.1	GALAXY
IC 2659	11 12 52.	+ 13 09 44.			SINGLE STAR
IC 2661	11 12 53.	+ 13 53 02.			NONSTELLAR OBJECT
ZWG 039.091	11 12 54.	+ 05 23		14.1	GALAXY
UGC 06282	11 12 54.	+ 05 23	36	14.1	GALAXY S
MCG+01-29-024	11 12 54.	+ 05 23	30	14.1	GALAXY
IC 2662	11 12 54.	+ 13 02 38.			SINGLE STAR
ZWG 067.043	11 12 54.	+ 13 53		15.4	GALAXY
ZWG 096.016	11 12 54.	+ 17 46		15.7	GALAXY
MCG+03-29-016	11 12 54.	+ 17 47	18	15.7	GALAXY
ZC 1112.9+2600	11 12 54.	+ 26 00	3700		CLUSTER OF GALAXIES
HOLM 238A	11 12 54.	+ 29 48	30	15.9	PART OF MULTIPLE GALAXY
MCG+05-27-034	11 12 54.	+ 29 48	24	15.9	GALAXY
MCG+06-25-031	11 12 54.	+ 36 47	18	15.	GALAXY
8ZW 1112-16.7	11 12 54.	- 16 40		17.4	COMPACT GALAXY
MRSL 291-00/2	11 12 54.	- 60 56	720		HII REGION
OCL 0854	11 12 54.	- 60 59	720	10.3	OPEN STAR CLUSTER
VHA 115	11 12 54.	- 60 59	150		OPEN STAR CLUSTER
RNGC 3601	11 12 55.	+ 05 23		14.0	GALAXY
SEY 095	11 12 55.	+ 56 11 40.		15.2	FAINT GALAXY
RNGC 3603	11 12 55.	- 60 59		10.5	OPEN CLUSTER
IC 2663	11 12 57.	+ 12 52 44.			SINGLE STAR
VV 153	11 12 57.	+ 54 00	30		INTERACTING GALAXY
MCG+09-19-019A	11 12 57.	+ 54 00	30		GALAXY
WRAY 19.32	11 12 58.9	- 60 59 38.		8.0	STAR-NEBULA ASSOCIATION
REIZ 0936	11 12 59.	+ 05 23	30	13.3	GALAXY
ZWG 126.014	11 13 00.	+ 22 21		15.7	GALAXY
ZC 1113.0+2452	11 13 00.	+ 24 52	3360		CLUSTER OF GALAXIES
ZC 1113.0+3335	11 13 00.	+ 33 35	1750		CLUSTER OF GALAXIES
ZC 1113.0+3520	11 13 00.	+ 35 20	610		CLUSTER OF GALAXIES
ZWG 213.038	11 13 00.	+ 41 52		12.6	GALAXY
UGC 06283	11 13 00.	+ 41 52	252	12.6	GALAXY Sa?
SHAH 003	11 13 00.	+ 54 00	24	18.	GROUP OF COMPACT GALAXIES
MCG+09-19-020	11 13 00.	+ 54 04	18	17.	GALAXY
MCG+09-19-021	11 13 00.	+ 55 56	30	16.	GALAXY
MCG+10-16-100	11 13 00.	+ 62 08	18	15.	GALAXY
MCG+13-09-020	11 13 00.	+ 76 31	51	16.	GALAXY
MCG-02-29-015	11 13 00.	- 13 54 30.	60	14.5	GALAXY
REIZ 0934	11 13 01.	+ 33 08	12	15.2	GALAXY
RNGC 3600	11 13 01.	+ 41 52		12.5	GALAXY
IC 2664	11 13 02.	+ 12 50 14.			SINGLE STAR
IC 2668	11 13 02.	- 13 53 47.			NONSTELLAR OBJECT
FATH 1.383	11 13 02.	- 15 31	16		NEBULA
IC 2665	11 13 04.	+ 11 59 49.			NONSTELLAR OBJECT
REIZ 0935	11 13 04.	+ 41 53	108	13.2	GALAXY
ZWG 067.044	11 13 06.	+ 14 03		14.6	GALAXY
MCG+02-29-016	11 13 06.	+ 14 03	42	14.6	GALAXY
2ZW 050	11 13 06.	+ 28 50			COMPACT GALAXY
LB 10067	11 13 06.	+ 30 46		17.8	FAINT BLUE STAR
MCG+07-23-038	11 13 06.	+ 41 51	240	13.	GALAXY
MRSL 291-00/4	11 13 06.	- 60 32	840		HII REGION
IC 2666	11 13 07.	+ 14 03 19.			NONSTELLAR OBJECT
IC 2667	11 13 08.	+ 12 23 25.			NONSTELLAR OBJECT
KEEL 335	11 13 08.2	+ 54 58 37.			NEBULA
REIZ 0938	11 13 10.	+ 01 09	24	15.3	GALAXY
RNGC 3602	11 13 11.	+ 17 41		15.5	GALAXY
REIZ 0937	11 13 11.	+ 26 14	24	14.6	GALAXY
ZWG 011.035	11 13 12.	+ 01 08		15.4	GALAXY
UGC 06284	11 13 12.	+ 01 08	90	15.4	GALAXY DBL SYS
ZWG 096.017	11 13 12.	+ 17 41		15.7	GALAXY
MCG+03-29-017	11 13 12.	+ 17 42	42	15.7	GALAXY
ZWG 126.015	11 13 12.	+ 26 15		15.6	GALAXY
LB 10071	11 13 12.	+ 28 28		16.9	FAINT BLUE STAR
LB 10070	11 13 12.	+ 31 36		16.5	FAINT BLUE STAR
LB 10069	11 13 12.	+ 31 44		17.0	FAINT BLUE STAR
LB 10068	11 13 12.	+ 31 54		18.0	FAINT BLUE STAR
ZC 1113.2+3411	11 13 12.	+ 34 11	470		CLUSTER OF GALAXIES
RNGC 3594	11 13 15.	+ 55 58		15.0	GALAXY
IC 2669	11 13 17.	+ 13 42 13.			NONSTELLAR OBJECT
ZC 1113.3+0445	11 13 18.	+ 04 45	870		CLUSTER OF GALAXIES
8ZW 1113+05.8	11 13 18.	+ 05 50		17.9	COMPACT GALAXY
ZWG 156.038	11 13 18.	+ 29 04		15.6	GALAXY
UGC 06285	11 13 18.	+ 41 40	78	16.0	GALAXY
ZC 1113.3+4509	11 13 18.	+ 45 09	810		CLUSTER OF GALAXIES
ZWG 268.011	11 13 18.	+ 55 58		15.2	GALAXY
ZWG 267.058	11 13 18.	+ 55 58		15.2	GALAXY
UGC 06286	11 13 18.	+ 55 58	84	15.2	GALAXY SB0
MCG+09-19-022	11 13 18.	+ 55 58	60	15.	GALAXY

OBJECT NAME	RIGHT ASCEN.	DECLINATION	DIAM.	MAGN.	TYPE OF OBJECT
LB 01979	11 13 19.	+ 59 59 18.		13.8	FAINT BLUE STAR
TON-N 1375	11 13 20.	+ 21 38		12.8	BLUE STAR
MCG-04-27-006	11 13 21.	- 26 51 30.	48	15.	GALAXY
BIGO 520	11 13 22.	+ 56 00			NEBULA
FATH 1.384	11 13 22.	- 15 52	22		NEBULA
KEEL 336	11 13 22.6	+ 55 03 28.			NEBULA
IC 2670	11 13 23.	+ 12 03 19.			NONSTELLAR OBJECT
ZWG 039.092	11 13 24.	+ 04 56		15.6	GALAXY
REIZ 0939	11 13 24.	+ 10 27	42	15.3	GALAXY
UGC 06287	11 13 24.	+ 24 12	72	17.	GALAXY DWRF IR
LB 10074	11 13 24.	+ 28 58		18.4	FAINT BLUE STAR
ZWG 156.039	11 13 24.	+ 29 43		15.4	GALAXY
LB 10073	11 13 24.	+ 30 02		16.9	FAINT BLUE STAR
LB 10072	11 13 24.	+ 30 34		16.1	FAINT BLUE STAR
MCG+11-14-012A	11 13 24.	+ 64 04	36	16.	GALAXY
ZWG 011.036	11 13 24.	- 00 04		15.5	GALAXY
IC 2671	11 13 27.	+ 13 24 01.			NONSTELLAR OBJECT
IC 2672	11 13 28.	+ 10 25 55.			NONSTELLAR OBJECT
IC 2673	11 13 28.	+ 10 26 13.			NONSTELLAR OBJECT
ZWG 067.045	11 13 30.	+ 10 26		15.3	GALAXY
HOLM 239A	11 13 30.	+ 10 26	66	14.4	PART OF MULTIPLE GALAXY
UGC 06288	11 13 30.	+ 10 26	60	15.3	GALAXY Sc
MCG+02-29-017	11 13 30.	+ 10 26	60	15.3	GALAXY
KARA.73B 0473	11 13 30.	+ 10 26	60	15.3	ISOLATED GALAXY S
ZWG 156.040	11 13 30.	+ 29 30		15.5	GALAXY
SN 19660	11 13 30.	+ 29 38		15.8	SUPERNOVA
ZWG 156.041	11 13 30.	+ 29 40		15.3	GALAXY
MCG+10-16-101	11 13 30.	+ 61 37	36	16.	GALAXY
MCG-06-25-014	11 13 30.	- 33 41	90	14.	GALAXY
HOLM 239B	11 13 33.	+ 10 26	18	15.2	PART OF MULTIPLE GALAXY
IC 2674	11 13 33.	+ 11 19 31.			NONSTELLAR OBJECT
MCG+05-27-035	11 13 33.	+ 29 41	36	15.4	GALAXY
IC 2675	11 13 34.	+ 12 31 25.			NONSTELLAR OBJECT
ZWG 011.038	11 13 36.	+ 00 51		15.5	GALAXY
MCG+00-29-009	11 13 36.	+ 00 51	42	15.5	GALAXY
ZWG 039.093	11 13 36.	+ 03 09		14.5	GALAXY
UGC 06289	11 13 36.	+ 03 09	48	14.5	GALAXY S
MCG+01-29-025	11 13 36.	+ 03 09	54	14.5	GALAXY
ZWG 067.046	11 13 36.	+ 11 19		15.3	GALAXY
UGC 06290	11 13 36.	+ 11 19	84	15.3	GALAXY Sc
KARA.73B 0474	11 13 36.	+ 11 19	66	15.3	ISOLATED GALAXY S
LB 10075	11 13 36.	+ 28 57		17.3	FAINT BLUE STAR
MCG+05-27-036	11 13 36.	+ 29 30 30.	54	15.5	GALAXY
ZWG 156.042	11 13 36.	+ 29 31		15.2	GALAXY
REIZ 0940	11 13 36.	+ 29 31	12	15.7	GALAXY
ZWG 334.019	11 13 36.	+ 69 35		15.7	GALAXY
UGC 06291	11 13 36.	+ 69 35	66	15.7	GALAXY S
ZWG 011.037	11 13 36.	- 03 25		15.6	GALAXY
FATH 1.385	11 13 36.	- 15 09	27		NEBULA
KEEL 337	11 13 37.7	+ 55 33 35.			NEBULA
TON-N 1376	11 13 38.	+ 19 22		17.0	BLUE STAR
ZC 1113.7+0720	11 13 42.	+ 07 20	1410		CLUSTER OF GALAXIES
ZWG 156.043	11 13 42.	+ 29 27		15.2	GALAXY
REIZ 0941	11 13 42.	+ 29 31	30	14.5	GALAXY
MCG+05-27-037	11 13 42.	+ 29 32 30.	54	15.2	GALAXY
ZWG 156.044	11 13 42.	+ 29 36		15.2	GALAXY
UGC 06292	11 13 42.	+ 29 36	108	15.2	GALAXY SBb
ZC 1113.7+3542	11 13 42.	+ 35 42	740		CLUSTER OF GALAXIES
ZWG 213.039	11 13 42.	+ 41 21		15.3	GALAXY
UGC 06293	11 13 42.	+ 41 21	84	15.3	GALAXY Sc+COMP
ZC 1113.7+4545	11 13 42.	+ 45 45	610		CLUSTER OF GALAXIES
MCG+09-19-023	11 13 42.	+ 54 57 30.	42	16.	GALAXY
RNGC 3604	11 13 43.	+ 04 47			NON-EXISTENT OBJECT
IC 2676	11 13 43.	+ 10 05 54.			NONSTELLAR OBJECT
IC 2677	11 13 44.	+ 12 29 18.			NONSTELLAR OBJECT
IC 2678	11 13 45.	+ 12 13 18.			NONSTELLAR OBJECT
REIZ 0946	11 13 46.	+ 03 09	24	13.7	GALAXY
IC 2679	11 13 47.	+ 12 17 30.			NONSTELLAR OBJECT
REIZ 0942	11 13 47.	+ 26 40	72	15.1	GALAXY
REIZ 0943	11 13 47.	+ 29 32	6	15.4	GALAXY
ZWG 011.039	11 13 48.	+ 01 03		15.7	GALAXY
ZC 1113.8+0144	11 13 48.	+ 01 44	2220		CLUSTER OF GALAXIES
ZWG 039.094	11 13 48.	+ 04 36		15.2	GALAXY
ZWG 039.095	11 13 48.	+ 04 59		15.2	GALAXY
ZWG 067.047	11 13 48.	+ 10 05		15.6	GALAXY
ZC 1113.8+1715	11 13 48.	+ 17 15	1080		CLUSTER OF GALAXIES
LB 10079	11 13 48.	+ 26 26		17.0	FAINT BLUE STAR
LB 10078	11 13 48.	+ 26 28		16.6	FAINT BLUE STAR
LB 10077	11 13 48.	+ 26 54		16.7	FAINT BLUE STAR
ZWG 156.045	11 13 48.	+ 29 02		15.2	GALAXY
MCG+05-27-039A	11 13 48.	+ 29 28	24	15.2	GALAXY
MCG+05-27-039	11 13 48.	+ 29 28	15	15.2	GALAXY
REIZ 0944	11 13 48.	+ 29 35	42	14.6	GALAXY
MCG+05-27-038	11 13 48.	+ 29 36 30.	90	15.2	GALAXY
LB 10076	11 13 48.	+ 31 24		17.9	FAINT BLUE STAR
ZWG 185.026	11 13 48.	+ 35 35		15.4	GALAXY
MCG+07-23-039	11 13 48.	+ 41 20 30.	48	15.	GALAXY
MCG+09-19-024	11 13 48.	+ 54 59	30	16.	GALAXY
MCG+12-11-014	11 13 48.	+ 69 35	39	15.	GALAXY
RNGC 3606	11 13 48.	- 33 33			GALAXY
ARC 1213	11 13 49.	+ 29 33		14.5	RICH CLUSTER OF GALAXIES
REIZ 0945	11 13 49.	+ 32 57	24	15.9	GALAXY
IC 2680	11 13 50.	+ 10 04 54.			NONSTELLAR OBJECT
MCG+05-27-041	11 13 51.	+ 29 34	12	15.1	GALAXY
MCG+05-27-040	11 13 51.	+ 29 34	15	15.1	GALAXY
KEEL 338	11 13 51.9	+ 54 58 29.			NEBULA
REIZ 0947	11 13 53.	+ 26 50	18	14.8	GALAXY
REIZ 0948	11 13 53.	+ 29 27	18	14.8	GALAXY
REIZ 0949	11 13 53.	+ 29 29	18	15.1	GALAXY
REIZ 0950	11 13 53.	+ 29 31	6	15.8	GALAXY
REIZ 0951	11 13 53.	+ 29 32	12	15.1	GALAXY
REIZ 0952	11 13 53.	+ 29 33	12	15.2	GALAXY
REIZ 0953	11 13 53.	+ 29 35	18	15.1	GALAXY
MRK 037	11 13 54.	+ 29 03	16	16.	GALAXY WITH UV CONTINUUM
STOCK 27	11 13 54.	+ 29 31			BLUE KNOT NEAR ELLIP GLXY
ZWG 156.046	11 13 54.	+ 29 31		15.1	GALAXY
LB 10080	11 13 54.	+ 32 03		16.2	FAINT BLUE STAR
MCG+06-25-032	11 13 54.	+ 35 34 30.	42	15.	GALAXY
MCG+09-19-025	11 13 54.	+ 55 00	24	17.	GALAXY
ZC 1113.9-0250	11 13 54.	- 02 50	1410		CLUSTER OF GALAXIES
MCG-05-27-004	11 13 54.	- 33 35	36	15.	GALAXY
KEEL 339	11 13 54.4	+ 55 01 26.			NEBULA
LB 01980	11 13 56.	+ 49 31 42.		17.2	FAINT BLUE STAR
IC 2681	11 13 57.	+ 11 28 48.			NONSTELLAR OBJECT
KEEL 340	11 13 57.6	+ 55 00 02.			SPIRAL NEBULA
LB 09850	11 14	- 80 50		12.9	FAINT BLUE STAR
LB 09975	11 14	- 87 09		13.2	FAINT BLUE STAR
ZWG 039.096	11 14 00.	+ 08 14		15.3	GALAXY
ZWG 067.048	11 14 00.	+ 11 28		15.2	GALAXY
ZWG 096.018	11 14 00.	+ 16 00		15.5	GALAXY
HOLM 240C	11 14 00.	+ 18 17	36	13.7	PART OF MULTIPLE GALAXY
MCG+03-29-018	11 14 00.	+ 18 25	48	15.	GALAXY
ZC 1114.0+2102	11 14 00.	+ 21 02	3160		CLUSTER OF GALAXIES
ZC 1114.0+2439	11 14 00.	+ 24 39	810		CLUSTER OF GALAXIES
UGC 06294	11 14 00.	+ 26 38	72	16.0	GALAXY Sb-c
LB 10081	11 14 00.	+ 30 13		16.1	FAINT BLUE STAR
MCG+10-16-102	11 14 00.	+ 57 22	42	16.	GALAXY
7ZW 384	11 14 00.	+ 59 49			COMPACT GALAXY
ZC 1114.0+6453	11 14 00.	+ 64 53	1010		CLUSTER OF GALAXIES
ZWG 011.040	11 14 00.	- 01 00		15.7	GALAXY
IC 2682	11 14 01.	+ 09 41 06.			NONSTELLAR OBJECT
REIZ 0954	11 14 01.	+ 33 42	18	15.1	GALAXY
REIZ 0955	11 14 05.	+ 03 38	30	14.1	GALAXY
RNGC 3605	11 14 05.	+ 18 18		13.5	GALAXY
IC 0679	11 14 05.	- 13 42 06.			NONSTELLAR OBJECT
ZWG 096.019	11 14 06.	+ 18 17		12.7	GALAXY
UGC 06295	11 14 06.	+ 18 17	84	12.7	GALAXY E-S0
ZC 1114.1+1824	11 14 06.	+ 18 24	540		CLUSTER OF GALAXIES
ZWG 156.047	11 14 06.	+ 26 49		15.4	GALAXY
LB 10082	11 14 06.	+ 31 59		17.0	FAINT BLUE STAR
ZC 1114.1+5745	11 14 06.	+ 57 45	1480		CLUSTER OF GALAXIES
ZWG 011.041	11 14 06.	- 00 45		15.7	GALAXY
OCL 0853	11 14 06.	- 60 00	180		OPEN STAR CLUSTER
HOLM 240A	11 14 08.	+ 18 19	108	11.5	PART OF MULTIPLE GALAXY
TON-N 1377	11 14 08.	+ 23 09		16.5	BLUE STAR
HOLM 240B	11 14 11.	+ 18 25	78	12.5	PART OF MULTIPLE GALAXY
FEIG 038	11 14 12.	+ 07 16		12.9	FAINT BLUE STAR
8ZW 1114+07.8	11 14 12.	+ 07 48		17.4	COMPACT GALAXY
ZWG 096.020	11 14 12.	+ 18 04		14.3	GALAXY
UGC 06296	11 14 12.	+ 18 04	102	14.3	GALAXY S
MCG+03-29-019	11 14 12.	+ 18 19	84	12.7	GALAXY
ZWG 156.048	11 14 12.	+ 27 51		15.7	GALAXY
LB 10084	11 14 12.	+ 31 39		17.8	FAINT BLUE STAR
LB 10083	11 14 12.	+ 31 44		16.7	FAINT BLUE STAR
ZWG 185.027	11 14 12.	+ 32 46		15.7	GALAXY
ZWG 185.028	11 14 12.	+ 32 52		15.5	GALAXY
ZWG 185.029	11 14 12.	+ 35 31		15.3	GALAXY
ZC 1114.2+3750	11 14 12.	+ 37 50	340		CLUSTER OF GALAXIES
ZC 1114.2+4702	11 14 12.	+ 47 02	870		CLUSTER OF GALAXIES
ZWG 268.012	11 14 12.	+ 50 53		15.4	GALAXY
TON-N 1378	11 14 14.	+ 21 22		17.	BLUE STAR
MCG+03-29-021	11 14 15.	+ 18 05	72	14.3	GALAXY
MCG+03-29-020	11 14 15.	+ 18 22	108	10.2	GALAXY
MCG+05-27-042	11 14 15.	+ 27 50	48	15.7	GALAXY
MCG+06-25-033	11 14 15.	+ 35 31	18	16.	GALAXY
MCG+06-25-034	11 14 15.	+ 35 31 30.	18	16.	GALAXY
LB 01981	11 14 15.	+ 57 55 00.		14.9	FAINT BLUE STAR
FATH 1.386	11 14 15.	- 15 44	14		NEBULA
RNGC 3620	11 14 16.	- 75 57			UNVERIFIED SOUTHRN OBJECT
RNGC 3607	11 14 17.	+ 18 20		12.0	GALAXY
RNGC 3608	11 14 17.	+ 18 26		12.5	GALAXY
ARC 1212	11 14 17.	+ 57 44		17.2	RICH CLUSTER OF GALAXIES
ZWG 039.097	11 14 18.	+ 05 00		15.6	GALAXY
ZWG 039.098	11 14 18.	+ 05 16		15.7	GALAXY
IC 2683	11 14 18.	+ 12 22 17.			NONSTELLAR OBJECT
ZWG 096.021	11 14 18.	+ 18 19		10.2	GALAXY
UGC 06297	11 14 18.	+ 18 19	300	10.2	GALAXY E
KARA.72 278A	11 14 18.	+ 18 19	288	10.2	PART OF DOUBLE GALAXY
ZC 1114.3+1919	11 14 18.	+ 19 19	1610		CLUSTER OF GALAXIES
LB 10085	11 14 18.	+ 27 46		17.5	FAINT BLUE STAR
VV 198B	11 14 18.	+ 35 33 30.	2	15.5	INTERACTING GALAXY
VV 198A	11 14 18.	+ 35 33 30.	2	15.5	INTERACTING GALAXY
ZWG 185.030	11 14 18.	+ 36 25		15.0	GALAXY
UGC 06298	11 14 18.	+ 36 25	66	15.0	GALAXY S
ZC 1114.3+4335	11 14 18.	+ 43 35	2350		CLUSTER OF GALAXIES
ZWG 242.016	11 14 18.	+ 47 14		15.2	GALAXY
ZC 1114.3+5457	11 14 18.	+ 54 57	7530		CLUSTER OF GALAXIES
REIZ 0957	11 14 20.	- 05 56	18	13.9	GALAXY
MCG+03-29-022	11 14 21.	+ 18 27	84	11.7	GALAXY
MCG+06-25-036	11 14 21.	+ 32 46	30	15.	GALAXY
MCG+06-25-037	11 14 21.	+ 32 51 30.	16	15.	GALAXY
MCG+06-25-035	11 14 21.	+ 36 25	36	15.	GALAXY
LB 01982	11 14 21.	+ 57 37 36.		16.4	FAINT BLUE STAR
WRAY 19.33	11 14 21.4	- 61 13 31.			DIFFUSE NEBULA
IC 2684	11 14 23.	+ 10 22 35.			NONSTELLAR OBJECT
IC 2685	11 14 24.	+ 10 22 11.			NONSTELLAR OBJECT
ZWG 067.049	11 14 24.	+ 13 22		15.4	GALAXY
ZC 1114.4+1624	11 14 24.	+ 16 24	870		CLUSTER OF GALAXIES
ZWG 096.022	11 14 24.	+ 18 25		11.7	GALAXY
UGC 06299	11 14 24.	+ 18 25	192	11.7	GALAXY E
KARA.72 278B	11 14 24.	+ 18 25	180	11.7	PART OF DOUBLE GALAXY
LB 10089	11 14 24.	+ 27 18		18.2	FAINT BLUE STAR
LB 10088	11 14 24.	+ 29 50		17.0	FAINT BLUE STAR
LB 10087	11 14 24.	+ 30 38		16.6	FAINT BLUE STAR
LB 10086	11 14 24.	+ 30 50		17.0	FAINT BLUE STAR
REIZ 0956	11 14 24.	+ 32 46	48	14.8	GALAXY
ZC 1114.4+3733	11 14 24.	+ 37 33	3020		CLUSTER OF GALAXIES
ZWG 213.040	11 14 24.	+ 42 44		15.4	GALAXY
ZC 1114.4+4246	11 14 24.	+ 42 46	670		CLUSTER OF GALAXIES
ZWG 242.017	11 14 24.	+ 47 08		15.4	GALAXY
MCG+08-21-011	11 14 24.	+ 47 08	24	16.	GALAXY
MCG+08-21-012	11 14 24.	+ 47 13	42	15.	GALAXY
SHAH 005	11 14 24.	+ 55 11	42	17.	GROUP OF COMPACT GALAXIES
ZC 1114.4+6916	11 14 24.	+ 69 16	670		CLUSTER OF GALAXIES
ZWG 351.058	11 14 24.	+ 75 20		15.5	GALAXY
MCG-01-29-006	11 14 24.	- 05 56 30.	36	15.	GALAXY
MCG-04-27-007	11 14 24.	- 23 42	24	15.	GALAXY
ARC 1214	11 14 26.	- 05 20		17.5	RICH CLUSTER OF GALAXIES
IC 2686	11 14 27.	+ 13 13 35.			NONSTELLAR OBJECT
LB 01983	11 14 28.	+ 57 35 12.		16.7	FAINT BLUE STAR
KEEL 341	11 14 29.7	+ 55 05 24.			NEBULA
ZWG 126.016	11 14 30.	+ 26 07		15.6	GALAXY
ZC 1114.5+3154	11 14 30.	+ 31 54	2960		CLUSTER OF GALAXIES
ZC 1114.5+3656	11 14 30.	+ 36 56	610		CLUSTER OF GALAXIES
ZWG 011.042	11 14 30.	- 00 58		15.6	GALAXY
LB 01984	11 14 33.	+ 53 34 36.		15.9	FAINT BLUE STAR
ZWG 039.099	11 14 36.	+ 03 18		15.5	GALAXY
IC 2687	11 14 36.	+ 10 25 59.			NONSTELLAR OBJECT
ZWG 096.023	11 14 36.	+ 16 36		15.7	GALAXY
UGC 06300	11 14 36.	+ 16 36	72	15.7	GALAXY
MCG+03-29-023	11 14 36.	+ 16 36	78	15.7	GALAXY
ZC 1114.6+2621	11 14 36.	+ 26 21	740		CLUSTER OF GALAXIES
ZWG 268.013	11 14 36.	+ 51 41		15.6	GALAXY
MCG+09-19-027	11 14 36.	+ 52 14	30	17.	GALAXY
ZWG 268.014	11 14 36.	+ 53 48		15.3	GALAXY
MCG+09-19-026	11 14 36.	+ 53 49	18	16.	GALAXY

OBJECT NAME	RIGHT ASCEN.	DECLINATION	DIAM.	MAGN.	TYPE OF OBJECT
MCG-06-25-015	11 14 36.	- 34 40	78	14.	GALAXY
TON-N 1379	11 14 38.	+ 33 12		17.	BLUE STAR
LB 01985	11 14 41.	+ 53 00 06.		17.1	FAINT BLUE STAR
8ZW 1114+03.1	11 14 42.	+ 03 09		17.8	COMPACT GALAXY
ZWG 039.100	11 14 42.	+ 06 28		15.6	GALAXY
IC 2689	11 14 42.	+ 13 13 59.			NONSTELLAR OBJECT
ZWG 126.017	11 14 42.	+ 22 37		15.7	GALAXY
UGC 06301	11 14 42.	+ 22 37	102	15.7	GALAXY Sc
ZWG 126.018	11 14 42.	+ 25 43		15.6	GALAXY
LB 10091	11 14 42.	+ 27 52		17.9	FAINT BLUE STAR
ZWG 156.049	11 14 42.	+ 27 57		15.7	GALAXY
UGC 06302	11 14 42.	+ 27 57	66	15.7	GALAXY S
LB 10090	11 14 42.	+ 28 56		16.8	FAINT BLUE STAR
REIZ 0958	11 14 42.	+ 32 58	18	15.1	GALAXY
MCG+09-19-028	11 14 42.	+ 51 00	30	16.	GALAXY
ZWG 268.015	11 14 42.	+ 51 33		15.2	GALAXY
8ZW 1114-21.2	11 14 42.	- 21 11		17.5	COMPACT GALAXY
MCG-05-27-005	11 14 42.	- 27 33	108	15.	GALAXY
IC 2688	11 14 43.	+ 13 45 47.			NONSTELLAR OBJECT
IC 2690	11 14 45.	+ 13 15 05.			NONSTELLAR OBJECT
REIZ 0959	11 14 45.	+ 22 37	54	14.7	GALAXY
MCG+04-27-011	11 14 45.	+ 22 37	90	15.7	GALAXY
MCG-02-29-016	11 14 45.	- 14 13	42	15.	GALAXY
ZWG 039.101	11 14 48.	+ 02 43		15.2	GALAXY
ZWG 126.019	11 14 48.	+ 26 02		15.7	GALAXY
ZWG 185.031	11 14 48.	+ 36 20		15.1	GALAXY
UGC 06303	11 14 48.	+ 36 20	78	15.1	GALAXY Sc
ZWG 242.018	11 14 48.	+ 47 22		15.2	GALAXY
ZC 1114.8+5111	11 14 48.	+ 51 11	870		CLUSTER OF GALAXIES
MCG+10-16-103	11 14 48.	+ 58 37	78	15.	GALAXY
UGC 06304	11 14 48.	+ 58 38	84	16.5	GALAXY DWARF
ACK 287+10.1	11 14 48.	- 48 56			PLANETARY NEBULA
IC 2691	11 14 49.	+ 12 18 17.			NONSTELLAR OBJECT
MAI 061	11 14 50.	+ 58 39	47		DWARF SPHEROIDAL GALAXY
REIZ 0961	11 14 52.	+ 02 43	36	15.1	GALAXY
SN 1952D	11 14 53.	- 02 48		19.5	SUPERNOVA
ZWG 039.102	11 14 54.	+ 03 18		15.1	GALAXY
ZWG 039.103	11 14 54.	+ 04 50		12.4	GALAXY
UGC 06305	11 14 54.	+ 04 50	162	12.4	GALAXY Sa
MCG+01-29-026	11 14 54.	+ 04 50	96	12.4	GALAXY
UGC 06306	11 14 54.	+ 04 53	78	17.	GALAXY DWARF
REIZ 0960	11 14 54.	+ 32 53	42	15.4	GALAXY
MCG+06-25-038	11 14 54.	+ 36 20	66	14.5	GALAXY
ZWG 185.032	11 14 54.	+ 38 19		14.8	GALAXY
UGC 06307	11 14 54.	+ 38 19	78	14.8	GALAXY S-IRR
MCG+08-21-013	11 14 54.	+ 47 21	36	15.	GALAXY
MCG+09-19-029	11 14 54.	+ 51 44	72	14.	GALAXY
ZC 1114.9+5215	11 14 54.	+ 52 15	670		CLUSTER OF GALAXIES
MCG+09-19-030	11 14 54.	+ 54 05		16.	GALAXY
ZC 1114.9+5637	11 14 54.	+ 56 37	2550		CLUSTER OF GALAXIES
MCG+10-16-104	11 14 54.	+ 57 24	18	17.	GALAXY
8ZW 1114-21.1	11 14 54.	- 21 09		18.5	COMPACT GALAXY
RNGC 3611	11 14 55.	+ 04 50		12.5	GALAXY
REIN 2.120	11 14 55.59	+ 04 49 44.6			NEBULA
TON-N 1380	11 14 56.	+ 33 16		15.	BLUE STAR
IC 2692	11 14 57.	+ 11 02 34.			NONSTELLAR OBJECT
REIZ 0962	11 14 57.	+ 22 47	24	15.1	GALAXY
REIZ 0964	11 14 59.	+ 04 50	48	12.5	GALAXY
REIZ 0963	11 14 59.	+ 26 55	60	13.6	GALAXY
ZWG 011.043	11 15 00.	+ 00 05		15.6	GALAXY
ZWG 039.104	11 15 00.	+ 07 34		15.1	GALAXY
ZC 1115.0+0927	11 15 00.	+ 09 27	940		CLUSTER OF GALAXIES
ZWG 067.050	11 15 00.	+ 13 39		15.3	GALAXY
MCG+03-29-024	11 15 00.	+ 18 06	102	15.	GALAXY
LB 10092	11 15 00.	+ 27 00		16.9	FAINT BLUE STAR
UGC 06308	11 15 00.	+ 27 22	66	16.0	GALAXY Sb-c
MCG+06-25-039	11 15 00.	+ 38 20	90	14.	GALAXY
ZWG 268.016	11 15 00.	+ 51 46		13.7	GALAXY
UGC 06309	11 15 00.	+ 51 46	96	13.7	GALAXY SB
ZC 1115.0+5157	11 15 00.	+ 51 57	1810		CLUSTER OF GALAXIES
MCG+09-19-031	11 15 00.	+ 54 55		16.	GALAXY
ZC 1115.0+6631	11 15 00.	+ 66 31	740		CLUSTER OF GALAXIES
IC 2693	11 15 01.	+ 13 49 22.			NONSTELLAR OBJECT
IC 2694	11 15 03.	+ 13 38 58.			NONSTELLAR OBJECT
ARC 1217	11 15 04.	- 24 57		17.4	RICH CLUSTER OF GALAXIES
ZWG 039.105	11 15 06.	+ 02 48		15.7	GALAXY
ARC 1215	11 15 06.	+ 03 56		16.7	RICH CLUSTER OF GALAXIES
ZWG 039.106	11 15 06.	+ 07 55		15.7	GALAXY
LB 10094	11 15 06.	+ 27 18		17.8	FAINT BLUE STAR
LB 10093	11 15 06.	+ 30 46		17.0	FAINT BLUE STAR
MCG+09-19-032	11 15 06.	+ 53 43	48	16.	GALAXY
ZWG 011.045	11 15 06.	- 01 02		15.1	GALAXY
ZWG 011.044	11 15 06.	- 02 54		14.8	GALAXY
MCG+00-29-010	11 15 06.	- 02 54	66	14.8	GALAXY
TON-N 1381	11 15 08.	+ 32 03		13.8	BLUE STAR
TON-N 1382	11 15 08.	+ 33 38		16.9	BLUE STAR
RNGC 3609	11 15 10.	+ 26 54		14.0	GALAXY
REIZ 0965	11 15 10.	+ 26 54	36	14.4	GALAXY
HOLM 241A	11 15 10.	+ 26 55	66	13.3	PART OF MULTIPLE GALAXY
HOLM 241C	11 15 10.	+ 26 57	18	15.2	PART OF MULTIPLE GALAXY
HOLM 242B	11 15 10.	+ 32 18	36	14.3	PART OF MULTIPLE GALAXY
ZWG 067.051	11 15 12.	+ 12 10		15.6	GALAXY
IC 2695	11 15 12.	+ 14 00 10.			NONSTELLAR OBJECT
ZC 1115.2+1527	11 15 12.	+ 15 27	1210		CLUSTER OF GALAXIES
ZWG 126.020	11 15 12.	+ 22 46		15.6	GALAXY
LB 10098	11 15 12.	+ 26 40		17.3	FAINT BLUE STAR
LB 10097	11 15 12.	+ 26 52		16.6	FAINT BLUE STAR
MCG+05-27-043	11 15 12.	+ 26 53	72	14.1	GALAXY
ZWG 156.050	11 15 12.	+ 26 54		14.1	GALAXY
UGC 06310	11 15 12.	+ 26 54	72	14.1	GALAXY Sa-b
TON-N 0576	11 15 12.	+ 26 55		15.5	BLUE STAR
MCG+05-27-044	11 15 12.	+ 26 56	21	15.2	GALAXY
LB 10096	11 15 12.	+ 29 32		17.0	FAINT BLUE STAR
ZWG 156.051	11 15 12.	+ 29 35		15.6	GALAXY
ZC 1115.2+3013	11 15 12.	+ 30 13	10420		CLUSTER OF GALAXIES
LB 10095	11 15 12.	+ 30 51		18.1	FAINT BLUE STAR
ZWG 156.052	11 15 12.	+ 32 16		15.7	GALAXY
MCG+06-25-040	11 15 12.	+ 33 55 30.	42	16.	GALAXY
MCG+10-16-105	11 15 12.	+ 58 33	15	16.	GALAXY
MCG-05-27-007	11 15 12.	- 27 37	60	15.5	GALAXY
MCG-05-27-006	11 15 12.	- 30 15	36	15.5	GALAXY
REIZ 0966	11 15 13.	+ 12 10	30	15.6	GALAXY
IC 2696	11 15 13.	+ 13 01 52.			NONSTELLAR OBJECT
ARC 1216	11 15 14.	- 04 12		16.0	RICH CLUSTER OF GALAXIES
IC 2697	11 15 15.	+ 13 40 28.			NONSTELLAR OBJECT
MCG+09-19-033	11 15 15.	+ 55 07 30.	24	16.	GALAXY
IC 2698	11 15 16.	+ 12 09 34.			NONSTELLAR OBJECT
KEEL 342	11 15 16.7	+ 55 06 49.			NEBULA

OBJECT NAME	RIGHT ASCEN.	DECLINATION	DIAM.	MAGN.	TYPE OF OBJECT
IC 2699	11 15 17.	+ 12 11 04.			NONSTELLAR OBJECT
IC 2700	11 15 18.	+ 12 19 46.			NONSTELLAR OBJECT
ZC 1115.3+1717	11 15 18.	+ 17 17	610		CLUSTER OF GALAXIES
MCG+03-29-025	11 15 18.	+ 17 42	36	15.2	GALAXY
ZWG 096.024	11 15 18.	+ 17 43		15.2	GALAXY
MCG+04-27-011A	11 15 18.	+ 22 47	60	15.6	GALAXY
MCG+05-27-045	11 15 18.	+ 28 14 30.	15	16.	GALAXY
LB 10099	11 15 18.	+ 29 05		18.3	FAINT BLUE STAR
REIZ 0967	11 15 18.	+ 32 16	36	14.4	GALAXY
FEIG 039	11 15 18.	+ 44 33		13.4	FAINT BLUE STAR
UGC 00830	11 15 18.	- 02 13	168	15.6	GALAXY Sc
IC 0680	11 15 20.	- 01 40 20.			NONSTELLAR OBJECT
REIZ 0970	11 15 21.	- 01 40	24	14.1	GALAXY
REIZ 0969	11 15 21.	- 01 48	60	13.9	GALAXY
IC 2702	11 15 22.	+ 09 41 10.			NONSTELLAR OBJECT
IC 2701	11 15 22.	+ 11 23 40.			NONSTELLAR OBJECT
RNGC 3615	11 15 22.	+ 23 40		14.0	GALAXY
RNGC 3617	11 15 22.	- 25 51		13.0	GALAXY
ZC 1115.4+0019	11 15 24.	+ 00 19	2890		CLUSTER OF GALAXIES
ZWG 039.107	11 15 24.	+ 07 24		15.5	GALAXY
ZWG 039.108	11 15 24.	+ 08 07		14.9	GALAXY
UGC 06312	11 15 24.	+ 08 07	84	14.9	GALAXY SB?0-a
MCG+01-29-027	11 15 24.	+ 08 07	60	14.9	GALAXY
ZWG 096.025	11 15 24.	+ 17 56		15.7	GALAXY
ZWG 126.021	11 15 24.	+ 23 40		14.0	GALAXY
UGC 06313	11 15 24.	+ 23 40	102	14.0	GALAXY E
MCG+04-27-012	11 15 24.	+ 23 40	27	14.0	GALAXY
LB 10102	11 15 24.	+ 28 25		17.4	FAINT BLUE STAR
LB 10101	11 15 24.	+ 28 38		18.2	FAINT BLUE STAR
LB 10100	11 15 24.	+ 29 04		18.3	FAINT BLUE STAR
ZWG 156.053	11 15 24.	+ 30 41		15.1	GALAXY
UGC 06314	11 15 24.	+ 30 41	60	15.1	GALAXY S
HOLM 242C	11 15 24.	+ 32 16	30	15.4	PART OF MULTIPLE GALAXY
HOLM 242A	11 15 24.	+ 32 17	30	14.4	PART OF MULTIPLE GALAXY
MCG+05-27-046	11 15 24.	+ 32 17 30.	36	15.7	GALAXY
ZWG 268.017	11 15 24.	+ 54 02		15.2	GALAXY
UGC 06315	11 15 24.	+ 54 02	66	15.2	GALAXY DBL SYS
ZWG 314.018	11 15 24.	+ 65 19		14.4	GALAXY
UGC 06316	11 15 24.	+ 65 19	102	14.4	GALAXY DBL SYS
ZWG 011.047	11 15 24.	- 01 40		14.6	GALAXY
MCG+00-29-012	11 15 24.	- 01 40	36	14.6	GALAXY
ZWG 011.046	11 15 24.	- 01 49		14.3	GALAXY
UGC 06311	11 15 24.	- 01 49	90	14.3	GALAXY Sc
MCG+00-29-011	11 15 24.	- 01 49	78	14.3	GALAXY
MCG-04-27-008	11 15 24.	- 25 51	24	13.	GALAXY
RNGC 3614A	11 15 25.	+ 45 59			GALAXY
TON-N 1383	11 15 26.	+ 20 28		13.2	BLUE STAR
MCG+05-27-047	11 15 27.	+ 32 18 30.	36	15.7	GALAXY
MCG+08-21-014	11 15 27.	+ 45 59	36	16.	GALAXY
IC 2703	11 15 27.4	+ 17 55 25.			GALAXY SA(r)
IC 2705	11 15 28.	+ 12 10 40.			NONSTELLAR OBJECT
IC 2704	11 15 28.	+ 12 43 40.			NONSTELLAR OBJECT
REIZ 0974	11 15 28.	+ 23 41	18	14.1	GALAXY
ZWG 039.109	11 15 30.	+ 04 53		15.5	GALAXY
8ZW 1115+04.9	11 15 30.	+ 04 53		15.5	COMPACT GALAXY
UGC 06317	11 15 30.	+ 04 54	66	16.5	GALAXY DWRF IR
ZWG 067.052	11 15 30.	+ 12 44		15.5	GALAXY
KARA.73B 0475	11 15 30.	+ 12 44	36	15.5	ISOLATED GALAXY S
ZC 1115.5+1431	11 15 30.	+ 14 31	740		CLUSTER OF GALAXIES
MCG+03-29-026	11 15 30.	+ 17 56	54	15.7	GALAXY
MCG+03-29-027	11 15 30.	+ 19 05	24	15.	GALAXY
2ZW 051	11 15 30.	+ 30 40			COMPACT GALAXY
ZWG 156.054	11 15 30.	+ 30 40		15.2	GALAXY
MCG+05-27-049	11 15 30.	+ 30 40	42	15.2	GALAXY
HOLM 243A	11 15 30.	+ 30 41	18	15.0	PART OF MULTIPLE GALAXY
MCG+05-27-050	11 15 30.	+ 30 41	15	15.1	GALAXY
ZWG 156.055	11 15 30.	+ 32 15		15.7	GALAXY
REIZ 0971	11 15 30.	+ 32 15	36	14.3	GALAXY
MCG+05-27-048	11 15 30.	+ 32 17	24	15.7	GALAXY
ZC 1115.5+3402	11 15 30.	+ 34 02	540		CLUSTER OF GALAXIES
ZC 1115.5+3511	11 15 30.	+ 35 11	1480		CLUSTER OF GALAXIES
ZWG 242.019	11 15 30.	+ 46 02		12.7	GALAXY
UGC 06318	11 15 30.	+ 46 02	276	12.7	GALAXY Sc
ZC 1115.5+4914	11 15 30.	+ 49 14	130		CLUSTER OF GALAXIES
MRK 038	11 15 30.	+ 54 02	25	16.	GALAXY WITH UV CONTINUUM
MCG+09-19-034	11 15 30.	+ 54 02		15.	GALAXY
MCG+10-16-106	11 15 30.	+ 57 50	18	16.	GALAXY
MCG+10-16-107	11 15 30.	+ 59 02	72	11.9	GALAXY
ZWG 291.048	11 15 30.	+ 59 05		11.4	GALAXY
UGC 06319	11 15 30.	+ 59 05	192	11.4	GALAXY S0
7ZW 385	11 15 30.	+ 61 42			COMPACT GALAXY
ZC 1115.5+6202	11 15 30.	+ 62 02	1210		CLUSTER OF GALAXIES
MCG+11-14-013	11 15 30.	+ 63 22	24	17.	GALAXY
MRK 165	11 15 30.	+ 63 35	13	15.	GALAXY WITH UV CONTINUUM
MCG+11-14-014	11 15 30.	+ 65 18	39	17.	GALAXY
MCG+11-14-015	11 15 30.	+ 65 19	42	15.	GALAXY
REIZ 0975	11 15 31.	+ 12 44	24	15.3	GALAXY
HOLM 243B	11 15 32.	+ 30 40	36	14.7	PART OF MULTIPLE GALAXY
HOLM 243C	11 15 32.	+ 30 41	12	15.4	PART OF MULTIPLE GALAXY
REIZ 0968	11 15 32.	+ 59 04	42	11.9	GALAXY
REIZ 0976	11 15 33.	+ 22 42	30	15.0	GALAXY
MCG+08-21-015	11 15 33.	+ 46 01	300	12.9	GALAXY
MCG+09-19-035	11 15 33.	+ 54 02	30	15.	GALAXY
RNGC 3610	11 15 33.	+ 59 04		12.0	GALAXY
RNGC 3612	11 15 34.	+ 26 54		15.0	GALAXY
SN 1966K	11 15 34.	+ 28 33		17.8	SUPERNOVA
RNGC 3616	11 15 35.	+ 15 00			NON-EXISTENT OBJECT
HOLM 241B	11 15 35.	+ 26 55	48	14.2	PART OF MULTIPLE GALAXY
HOLM 244A	11 15 35.	+ 28 33	54	14.4	PART OF MULTIPLE GALAXY
REIZ 0973	11 15 35.	+ 46 02	72	13.0	GALAXY
ZWG 039.110	11 15 36.	+ 04 05		15.4	GALAXY
ZC 1115.6+0840	11 15 36.	+ 08 40	4500		CLUSTER OF GALAXIES
ZWG 096.026	11 15 36.	+ 19 07		13.6	GALAXY
UGC 06320	11 15 36.	+ 19 07	66	13.6	GALAXY PECULR
ZWG 126.022	11 15 36.	+ 23 42		15.6	GALAXY
MCG+05-27-051	11 15 36.	+ 26 52	60	15.0	GALAXY
ZWG 156.056	11 15 36.	+ 26 54		15.0	GALAXY
UGC 06321	11 15 36.	+ 26 54	60	15.0	GALAXY S-IRR
MCG+05-27-052	11 15 36.	+ 28 29	42	14.8	GALAXY
ZWG 156.057	11 15 36.	+ 28 30		14.8	GALAXY
HOLM 244B	11 15 36.	+ 28 31	54	14.5	PART OF MULTIPLE GALAXY
MCG+05-27-053	11 15 36.	+ 28 31	72	14.7	GALAXY
ZWG 156.058	11 15 36.	+ 28 33		14.7	GALAXY
UGC 06322	11 15 36.	+ 28 33	84	14.7	GALAXY S0
LB 10107	11 15 36.	+ 29 02		18.6	FAINT BLUE STAR
LB 10106	11 15 36.	+ 29 02		18.3	FAINT BLUE STAR
LB 10105	11 15 36.	+ 29 03		17.9	FAINT BLUE STAR
LB 10104	11 15 36.	+ 29 08		17.0	FAINT BLUE STAR

OBJECT NAME	RIGHT ASCEN.	DECLINATION	DIAM.	MAGN.	TYPE OF OBJECT
LB 10103	11 15 36.	+ 29 38		17.0	FAINT BLUE STAR
REIZ 0977	11 15 36.	+ 30 40	18	14.5	GALAXY
REIZ 0979	11 15 36.	+ 30 41	6	15.7	GALAXY
REIZ 0978	11 15 36.	+ 30 41	12	14.7	GALAXY
MRK 039	11 15 36.	+ 54 02	25	17.	GALAXY WITH UV CONTINUUM
ZWG 291.049	11 15 36.	+ 58 17		11.6	GALAXY
UGC 06323	11 15 36.	+ 58 17	222	11.6	GALAXY E-SO
MCG+11-14-016	11 15 36.	+ 63 22	12	16.	GALAXY
ZWG 314.019	11 15 36.	+ 63 33		14.8	GALAXY
ZC 1115.6+6511	11 15 36.	+ 65 11	1010		CLUSTER OF GALAXIES
SN 1971A	11 15 37.	+ 28 33		16.5	SUPERNOVA
RNGC 3614	11 15 37.	+ 46 01		12.5	GALAXY
REIZ 0980	11 15 38.	+ 19 07	42	13.3	GALAXY
LB 01986	11 15 38.	+ 55 09 54.		14.2	FAINT BLUE STAR
REIZ 0981	11 15 41.	+ 28 32	30	14.5	GALAXY
ZWG 039.111	11 15 42.	+ 04 08		15.1	GALAXY
MCG+01-29-028	11 15 42.	+ 04 08	24	15.1	GALAXY
ZC 1115.7+1701	11 15 42.	+ 17 01	1010		CLUSTER OF GALAXIES
ZWG 096.027	11 15 42.	+ 19 00		14.9	GALAXY
UGC 06324	11 15 42.	+ 19 00	78	14.9	GALAXY S0
MCG+03-29-028	11 15 42.	+ 19 08	36	13.6	GALAXY
ZWG 126.023	11 15 42.	+ 22 43		15.5	GALAXY
ZC 1115.7+2538	11 15 42.	+ 25 38	340		CLUSTER OF GALAXIES
REIZ 0982	11 15 42.	+ 33 00	18	15.2	GALAXY
MCG+09-19-036	11 15 42.	+ 54 06	18	16.	GALAXY
MCG-03-29-004	11 15 42.	- 18 27 30.	54	16.	GALAXY
OCL 0855	11 15 42.	- 62 26	1620	8.6	OPEN STAR CLUSTER
VHA 116	11 15 42.	- 62 26	660		OPEN STAR CLUSTER
IC 2714	11 15 42.	- 62 26	960		OPEN CLUSTER
PK295-09.1	11 15 42.	- 70 33	10		PLANETARY NEBULA
REIN 2.121	11 15 42.32	+ 58 16 25.0			NEBULA
REIZ 0972	11 15 43.	+ 58 17	72	12.2	GALAXY
REIZ 0983	11 15 44.	+ 19 01	24	14.5	GALAXY
REIZ 0984	11 15 45.	+ 22 46	18	15.1	GALAXY
RNGC 3613	11 15 45.	+ 58 16		12.0	GALAXY
REIZ 0985	11 15 46.	+ 23 49	12	15.5	GALAXY
HOLM 245A	11 15 46.	+ 25 36	72	14.0	PART OF MULTIPLE GALAXY
SN 1967B	11 15 48.	+ 04 08		14.5	SUPERNOVA
ARC 1219	11 15 48.	+ 16 59		17.5	RICH CLUSTER OF GALAXIES
MCG+03-29-029	11 15 48.	+ 19 02	108	14.9	GALAXY
ZWG 126.024	11 15 48.	+ 25 35		15.5	GALAXY
UGC 06325	11 15 48.	+ 25 35	96	15.5	GALAXY Sc
MCG+04-27-013	11 15 48.	+ 25 36	90	15.5	GALAXY
LB 10108	11 15 48.	+ 27 34		15.7	FAINT BLUE STAR
TON-N 0577	11 15 48.	+ 27 37		11.6	BLUE STAR
UGC 06326	11 15 48.	+ 38 13	66	16.0	GALAXY
MCG+10-16-108	11 15 48.	+ 58 05	24	16.	GALAXY
MCG+10-16-109	11 15 48.	+ 58 15	78	11.8	GALAXY
MCG+10-16-110	11 15 48.	+ 58 51	45	15.	GALAXY
ZWG 291.050	11 15 48.	+ 58 54		15.4	GALAXY
HOLM 245B	11 15 49.	+ 25 37	12	15.4	PART OF MULTIPLE GALAXY
MCG+06-25-041	11 15 51.	+ 38 13 30.	48	15.	GALAXY
RNGC 3618	11 15 52.	+ 23 45		14.5	GALAXY
REIZ 0986	11 15 52.	+ 25 36	48	14.7	GALAXY
IC 2706	11 15 53.	+ 12 49 21.			NONSTELLAR OBJECT
RNGC 3621	11 15 53.	- 32 32		10.0	GALAXY
MCG+04-27-014	11 15 54.	+ 23 44	48	14.4	GALAXY
ZWG 126.025	11 15 54.	+ 23 45		14.4	GALAXY
UGC 06327	11 15 54.	+ 23 45	60	14.4	GALAXY Sb
ZWG 126.026	11 15 54.	+ 25 09		15.7	GALAXY
ZWG 185.033	11 15 54.	+ 33 45		15.6	GALAXY
REIZ 0987	11 15 54.	+ 33 45	18	14.9	GALAXY
ZC 1115.9+3756	11 15 54.	+ 37 56	740		CLUSTER OF GALAXIES
ZC 1115.9+4254	11 15 54.	+ 42 54	1140		CLUSTER OF GALAXIES
MCG+09-19-038	11 15 54.	+ 55 15	42	17.	GALAXY
MCG+09-19-037	11 15 54.	+ 55 27	24	16.	GALAXY
ZWG 351.059	11 15 54.	+ 75 17		15.7	GALAXY
MCG-05-27-008	11 15 54.	- 32 34	540	10.	GALAXY
IC 2707	11 15 55.	+ 09 44 57.			NONSTELLAR OBJECT
TON-N 1384	11 15 56.	+ 20 08		12.3	BLUE STAR
TON-N 1385	11 15 56.	+ 32 39		15.5	BLUE STAR
TON-N 1386	11 15 56.	+ 34 28		17.	BLUE STAR
REIZ 0988	11 15 57.	+ 23 44	42	13.7	GALAXY
KEEL 343	11 15 57.9	+ 12 59 10.			NEBULA
IC 2708	11 15 58.	+ 12 59 03.			NONSTELLAR OBJECT
REIZ 0989	11 15 58.	+ 25 09	18	14.8	GALAXY
KEEL 344	11 15 59.4	+ 13 01 06.			NEBULA
ZWG 011.049	11 16 00.	+ 00 42		15.2	GALAXY
8ZW 1116+05.7	11 16 00.	+ 05 41		18.8	COMPACT GALAXY
ZWG 067.053	11 16 00.	+ 12 59		15.1	GALAXY
ZWG 126.027	11 16 00.	+ 21 51		15.6	GALAXY
MCG+04-27-015	11 16 00.	+ 25 08 30.	30	15.7	GALAXY
ZWG 156.059	11 16 00.	+ 28 48		15.5	GALAXY
LB 10109	11 16 00.	+ 31 48		17.2	FAINT BLUE STAR
STOCK 28	11 16 00.	+ 42 29			BLUE KNOT NEAR ELLIP GLXY
ZC 1116.0+4235	11 16 00.	+ 42 35	470		CLUSTER OF GALAXIES
SHAH 006	11 16 00.	+ 52 01	48	16.	GROUP OF COMPACT GALAXIES
MCG+10-16-111	11 16 00.	+ 58 19	15	16.	GALAXY
ZWG 291.051	11 16 00.	+ 58 20		15.7	GALAXY
ZWG 334.020	11 16 00.	+ 73 05		15.5	GALAXY
ZWG 011.048	11 16 00.	- 01 50		15.4	GALAXY
ZC 1116.0-0410	11 16 00.	- 04 10	19290		CLUSTER OF GALAXIES
MCG-02-29-017	11 16 00.	- 11 50	48	14.5	GALAXY
ARC 1218	11 16 01.	+ 52 00		16.0	RICH CLUSTER OF GALAXIES
REIZ 0990	11 16 03.	+ 22 51	12	14.9	GALAXY
IC 0681	11 16 03.	- 11 52 22.			NONSTELLAR OBJECT
IC 2709	11 16 04.	+ 12 50 15.			NONSTELLAR OBJECT
REIZ 0991	11 16 05.	+ 28 49	24	14.6	GALAXY
ZWG 039.112	11 16 06.	+ 03 58		15.6	GALAXY
ZWG 039.113	11 16 06.	+ 07 46		15.2	GALAXY
ZWG 096.028	11 16 06.	+ 15 27		15.6	GALAXY
ZC 1116.1+1736	11 16 06.	+ 17 36	1080		CLUSTER OF GALAXIES
ZWG 126.028	11 16 06.	+ 23 36		15.5	GALAXY
LB 10112	11 16 06.	+ 28 00		18.3	FAINT BLUE STAR
LB 10111	11 16 06.	+ 31 08		18.4	FAINT BLUE STAR
LB 10110	11 16 06.	+ 31 20		16.6	FAINT BLUE STAR
MCG+10-16-112	11 16 06.	+ 58 18	15	16.	GALAXY
ZC 1116.1+6554	11 16 06.	+ 65 54	670		CLUSTER OF GALAXIES
MCG+12-11-015	11 16 06.	+ 73 06	12	16.	GALAXY
SHB 178	11 16 06.2	- 46 17 50.		17.	QUASI-STELLAR OBJECT
IC 2710	11 16 09.	+ 13 50 32.			NONSTELLAR OBJECT
REIZ 0992	11 16 09.	+ 22 43	18	15.2	GALAXY
REIZ 0993	11 16 09.	+ 23 37	30	14.3	GALAXY
IC 2711	11 16 11.	+ 14 00 50.			NONSTELLAR OBJECT
ZWG 039.114	11 16 12.	+ 07 48		14.7	GALAXY
RNGC 3624	11 16 12.	+ 07 48		14.5	GALAXY
MCG+01-29-029	11 16 12.	+ 07 48	48	14.7	GALAXY
8ZW 1116+07.9	11 16 12.	+ 07 52		17.7	COMPACT GALAXY
MCG+04-27-016	11 16 12.	+ 23 36	36	15.5	GALAXY
ZC 1116.2+2430	11 16 12.	+ 24 30	940		CLUSTER OF GALAXIES
LB 10115	11 16 12.	+ 27 45		17.6	FAINT BLUE STAR
ZWG 156.060	11 16 12.	+ 28 10		15.7	GALAXY
TON-N 0578	11 16 12.	+ 28 57		15.6	BLUE STAR
LB 10114	11 16 12.	+ 29 12		15.3	FAINT BLUE STAR
LB 10113	11 16 12.	+ 30 52		17.0	FAINT BLUE STAR
ZWG 185.034	11 16 12.	+ 36 56		15.5	GALAXY
MCG+06-25-042	11 16 12.	+ 36 57	30	15.	GALAXY
ARC 1220	11 16 12.	+ 37 47		17.3	RICH CLUSTER OF GALAXIES
MCG+10-16-113	11 16 12.	+ 57 28	18	16.	GALAXY
ZWG 291.052	11 16 12.	+ 58 20		15.1	GALAXY
MRK 166	11 16 12.	+ 62 48	8	15.5	GALAXY WITH UV CONTINUUM
ZWG 011.050	11 16 12.	- 00 05		15.4	GALAXY
REIZ 0994	11 16 15.	+ 23 10	24	13.8	GALAXY
HOLM 246B	11 16 16.	+ 13 22	450	11.0	PART OF MULTIPLE GALAXY
ZWG 011.051	11 16 18.	+ 00 54		14.9	GALAXY
MCG+00-29-013	11 16 18.	+ 00 54	24	14.9	GALAXY
ZWG 039.115	11 16 18.	+ 03 52		15.4	GALAXY
IC 2712	11 16 18.	+ 09 54 14.			NONSTELLAR OBJECT
ZWG 067.054	11 16 18.	+ 13 22		9.6	GALAXY
RNGC 3623	11 16 18.	+ 13 22		10.5	GALAXY
UGC 06328	11 16 18.	+ 13 22	570	9.6	GALAXY Sa
MCG+02-29-018	11 16 18.	+ 13 22	540	09.6	GALAXY
ZWG 126.029	11 16 18.	+ 23 10		14.7	GALAXY
MCG+04-27-017	11 16 18.	+ 23 10	36	14.7	GALAXY
MCG+05-27-054	11 16 18.	+ 28 10	30	15.7	GALAXY
LB 10116	11 16 18.	+ 28 54		16.1	FAINT BLUE STAR
MCG+10-16-114	11 16 18.	+ 60 55	24	16.	GALAXY
ZC 1116.3+6256	11 16 18.	+ 62 56	1480		CLUSTER OF GALAXIES
MCG+11-14-017	11 16 18.	+ 63 17 30.	50	17.	GALAXY
REIZ 0997	11 16 19.	+ 13 22	420	10.8	GALAXY
BC PKS1116+12	11 16 20.79	+ 12 51 06.3		19.25	QUASI-STELLAR OBJECT
SHB 179	11 16 20.8	+ 12 51 07.		19.3	QUASI-STELLAR OBJECT
REIZ 0998	11 16 22.	+ 00 27	36	15.1	GALAXY
ZWG 011.052	11 16 24.	+ 00 27		15.4	GALAXY
UGC 06329	11 16 24.	+ 00 27	72	15.4	GALAXY Sc
MCG+00-29-014	11 16 24.	+ 00 27	60	15.4	GALAXY
LB 10119	11 16 24.	+ 29 14		18.5	FAINT BLUE STAR
LB 10118	11 16 24.	+ 29 19		16.4	FAINT BLUE STAR
LB 10117	11 16 24.	+ 30 08		15.1	FAINT BLUE STAR
ZWG 242.020	11 16 24.	+ 47 55		14.7	GALAXY
ZWG 291.053	11 16 24.	+ 60 55		15.6	GALAXY
RNGC 3619	11 16 27.	+ 58 02		12.5	GALAXY
REIN 2.122	11 16 28.39	+ 58 01 54.3			NEBULA
ZWG 039.116	11 16 30.	+ 06 51		15.7	GALAXY
ZC 1116.5+0919	11 16 30.	+ 09 19	670		CLUSTER OF GALAXIES
ZC 1116.5+2136	11 16 30.	+ 21 36	2080		CLUSTER OF GALAXIES
TON-N 0063	11 16 30.	+ 30 12		14.3	BLUE STAR
ZWG 185.035	11 16 30.	+ 34 24		15.5	GALAXY
ZC 1116.5+3743	11 16 30.	+ 37 43	1080		CLUSTER OF GALAXIES
MCG+08-21-016	11 16 30.	+ 47 55	48	14.	GALAXY
REIZ 0995	11 16 30.	+ 57 20	27	14.5	GALAXY
MCG+10-16-115	11 16 30.	+ 58 00	96	12.6	GALAXY
ZWG 291.054	11 16 30.	+ 58 02		12.6	GALAXY
UGC 06330	11 16 30.	+ 58 02	240	12.6	GALAXY S0/Sa
ZC 1116.5+6736	11 16 30.	+ 67 36	610		CLUSTER OF GALAXIES
MCG-05-27-009	11 16 30.	- 29 09	36	15.5	GALAXY
TON-N 1387	11 16 32.	+ 21 22		15.5	BLUE STAR
TON-N 1388	11 16 32.	+ 21 38		14.5	BLUE STAR
REIZ 0999	11 16 33.	+ 23 10	54	14.4	GALAXY
IC 2713	11 16 34.	+ 12 26 20.			NONSTELLAR OBJECT
ZWG 039.117	11 16 36.	+ 03 30		15.1	GALAXY
UGC 06331	11 16 36.	+ 03 30	78	15.1	GALAXY Sc-IRR
ZC 1116.6+0953	11 16 36.	+ 09 53	870		CLUSTER OF GALAXIES
MCG+03-29-030	11 16 36.	+ 18 21	66	16.	GALAXY
MCG+04-27-019	11 16 36.	+ 21 04	72	14.5	GALAXY
ZWG 126.030	11 16 36.	+ 21 05		14.5	GALAXY
UGC 06332	11 16 36.	+ 21 05	78	14.5	GALAXY SBa
ZC 1116.6+2220	11 16 36.	+ 22 20	2350		CLUSTER OF GALAXIES
MCG+04-27-018	11 16 36.	+ 23 09	66	15.4	GALAXY
ZWG 126.031	11 16 36.	+ 23 10		15.4	GALAXY
UGC 06333	11 16 36.	+ 23 10	72	15.4	GALAXY Sc-IRR
LB 10121	11 16 36.	+ 30 02		18.8	FAINT BLUE STAR
LB 10120	11 16 36.	+ 30 02		18.1	FAINT BLUE STAR
ZC 1116.6+3257	11 16 36.	+ 32 57	540		CLUSTER OF GALAXIES
ZC 1116.6+3420	11 16 36.	+ 34 20	940		CLUSTER OF GALAXIES
MCG+06-25-043	11 16 36.	+ 34 24	36	15.	GALAXY
ZWG 185.036	11 16 36.	+ 35 06		15.7	GALAXY
ZC 1116.6+3833	11 16 36.	+ 38 33	740		CLUSTER OF GALAXIES
MCG+09-19-040	11 16 36.	+ 51 04	36	16.	GALAXY
ZC 1116.6+5216	11 16 36.	+ 52 16	1010		CLUSTER OF GALAXIES
MCG+09-19-039	11 16 36.	+ 54 44	42	16.	GALAXY
ZC 1116.6+7110	11 16 36.	+ 71 10	540		CLUSTER OF GALAXIES
ZWG 334.021	11 16 36.	+ 73 09		15.4	GALAXY
REIZ 0996	11 16 37.	+ 58 02	24	13.9	GALAXY
IC 2715	11 16 38.	+ 12 13 32.			NONSTELLAR OBJECT
TON-N 1389	11 16 38.	+ 33 02		17.	BLUE STAR
IC 2716	11 16 41.	+ 11 58 32.			NONSTELLAR OBJECT
ZWG 039.118	11 16 42.	+ 07 41		15.3	GALAXY
LB 10122	11 16 42.	+ 30 12		16.4	FAINT BLUE STAR
REIZ 1000	11 16 42.	+ 32 36	30	14.5	GALAXY
ZC 1116.7+4138	11 16 42.	+ 41 38	1140		CLUSTER OF GALAXIES
MCG+10-16-116	11 16 42.	+ 58 06	27	16.	GALAXY
MCG+10-16-117	11 16 42.	+ 59 32	72	14.	GALAXY
MCG+11-14-018	11 16 42.	+ 63 16	30	17.	GALAXY
MCG-02-29-018	11 16 42.	- 09 34	60	14.5	GALAXY
8ZW 1116-15.5	11 16 42.	- 15 28		12.5	COMPACT GALAXY
IC 2717	11 16 43.	+ 12 19 26.			NONSTELLAR OBJECT
LB 01987	11 16 43.	+ 55 54 06.		16.7	FAINT BLUE STAR
ARP.65 1B	11 16 44.	+ 51 46 05.	1	17.9	COMPACT DWARF GALAXY
ARP.65 1A	11 16 44.	+ 51 46 05.	1	17.8	COMPACT DWARF GALAXY
IC 2718	11 16 45.	+ 12 17 56.			NONSTELLAR OBJECT
FATH 1.387	11 16 45.	- 15 28	16		NEBULA
REIZ 1001	11 16 46.	+ 25 17	12	16.0	GALAXY
ZWG 039.119	11 16 48.	+ 03 28		15.4	GALAXY
ZWG 067.055	11 16 48.	+ 12 18		15.6	GALAXY
ZWG 126.032	11 16 48.	+ 25 15		15.3	GALAXY
MCG+04-27-020	11 16 48.	+ 25 16	12	15.3	GALAXY
LB 10123	11 16 48.	+ 27 35		17.5	FAINT BLUE STAR
MCG+05-27-055	11 16 48.	+ 28 55	156	14.4	GALAXY
ZWG 156.061	11 16 48.	+ 28 56		14.4	GALAXY
UGC 06334	11 16 48.	+ 28 56	174	14.4	GALAXY S0-a
STOCK 29	11 16 48.	+ 29 27			BLUE KNOT NEAR ELLIP GLXY
ZWG 291.055	11 16 48.	+ 59 34		15.1	GALAXY
UGC 06335	11 16 48.	+ 59 34	120	15.1	GALAXY Sc
ZWG 011.053	11 16 48.	- 02 49		15.5	GALAXY
SEY 096	11 16 52.	+ 59 33 49.		14.7	FAINT GALAXY

OBJECT NAME	RIGHT ASCEN.	DECLINATION	DIAM.	MAGN.	TYPE OF OBJECT
ARP 132	11 16 53.	- 02 49			PECULIAR GALAXY
ZWG 011.054	11 16 54.	+ 01 54		15.7	GALAXY
ZWG 039.120	11 16 54.	+ 03 17		15.7	GALAXY
ZWG 156.062	11 16 54.	+ 29 27		15.6	GALAXY
MCG+05-27-056	11 16 54.	+ 29 28	36	15.6	GALAXY
ZC 1116.9+3607	11 16 54.	+ 36 07	1140		CLUSTER OF GALAXIES
MCG+09-19-041	11 16 54.	+ 54 19	36	16.	GALAXY
ARC 1221	11 16 54.	+ 62 58		17.6	RICH CLUSTER OF GALAXIES
IC 2719	11 16 56.	+ 12 20 07.			NONSTELLAR OBJECT
TON-N 1390	11 16 56.	+ 35 00		13.7	BLUE STAR
LB 01989	11 16 57.	+ 52 22 06.		17.1	FAINT BLUE STAR
LB 01988	11 16 57.	+ 53 47 12.		16.2	FAINT BLUE STAR
REIZ 1005	11 16 58.	+ 03 18	12	13.9	GALAXY
REIZ 1002	11 16 59.	+ 30 40	24	14.6	GALAXY
VDB.66G 094	11 17	+ 02 49	130		DWARF GALAXY
LB 09977	11 17	- 87 14		12.7	FAINT BLUE STAR
IC 2720	11 17 00.	+ 12 21 07.			NONSTELLAR OBJECT
ZWG 067.056	11 17 00.	+ 12 22		15.6	GALAXY
ZC 1117.0+1525	11 17 00.	+ 15 25	2890		CLUSTER OF GALAXIES
ZWG 126.033	11 17 00.	+ 25 13		15.6	GALAXY
UGC 06336	11 17 00.	+ 25 13	84	15.6	GALAXY Sa-b
MCG+04-27-021	11 17 00.	+ 25 13	66	15.6	GALAXY
REIZ 1003	11 17 00.	+ 31 32	24	14.3	GALAXY
ZC 1117.0+4315	11 17 00.	+ 43 15	810		CLUSTER OF GALAXIES
ZC 1117.0+4653	11 17 00.	+ 46 53	8600		CLUSTER OF GALAXIES
ZWG 268.018	11 17 00.	+ 54 44		15.4	GALAXY
MCG+09-19-042	11 17 00.	+ 54 45	60	15.	GALAXY
MCG+09-19-043	11 17 00.	+ 56 00	15	17.	GALAXY
MCG+10-16-118	11 17 00.	+ 58 22	57	15.	GALAXY
ZWG 334.022	11 17 00.	+ 73 07		15.7	GALAXY
ZWG 011.055	11 17 00.	- 02 13		15.6	GALAXY
MCG+00-29-015	11 17 00.	- 02 13	54	15.6	GALAXY
TON-N 1391	11 17 02.	+ 22 04		16.1	BLUE STAR
REIZ 1010	11 17 03.	- 05 07	24	14.2	GALAXY
REIZ 1006	11 17 04.	+ 25 12	36	15.0	GALAXY
REIZ 1007	11 17 05.	+ 28 55	42	13.3	GALAXY
REIZ 1008	11 17 05.	+ 29 25	12	15.1	GALAXY
REIZ 1009	11 17 05.	+ 29 27	18	14.8	GALAXY
ZWG 039.121	11 17 06.	+ 07 52		15.5	GALAXY
ZC 1117.1+1641	11 17 06.	+ 16 41	670		CLUSTER OF GALAXIES
LB 10125	11 17 06.	+ 29 24		17.0	FAINT BLUE STAR
LB 10124	11 17 06.	+ 29 24		16.4	FAINT BLUE STAR
ZWG 185.037	11 17 06.	+ 33 21		15.2	GALAXY
UGC 06337	11 17 06.	+ 33 21	60	15.2	GALAXY VY CMPT
IC 2721	11 17 07.	+ 12 25 07.			NONSTELLAR OBJECT
IC 2722	11 17 08.	+ 14 14 31.			NONSTELLAR OBJECT
TON-N 1392	11 17 08.	+ 33 25		17.	BLUE STAR
LB 01990	11 17 09.	+ 53 09 12.		16.2	FAINT BLUE STAR
REIZ 1004	11 17 09.	+ 53 44	48	14.6	GALAXY
RNGC 3622	11 17 10.	+ 67 31		13.5	GALAXY
IC 2724	11 17 12.	+ 10 59 25.			NONSTELLAR OBJECT
ZC 1117.2+1414	11 17 12.	+ 14 14	1880		CLUSTER OF GALAXIES
ZWG 185.038	11 17 12.	+ 36 22		15.0	GALAXY
UGC 06338	11 17 12.	+ 36 22	72	15.0	GALAXY Sc
ZC 1117.2+3935	11 17 12.	+ 39 35	3430		CLUSTER OF GALAXIES
ZC 1117.2+4612	11 17 12.	+ 46 12	1210		CLUSTER OF GALAXIES
ZC 1117.2+5400	11 17 12.	+ 54 00	2020		CLUSTER OF GALAXIES
ZC 1117.2+6531	11 17 12.	+ 65 31	2020		CLUSTER OF GALAXIES
ZWG 314.020	11 17 12.	+ 67 31		13.7	GALAXY
UGC 06339	11 17 12.	+ 67 31	78	13.7	GALAXY S?
ZWG 011.056	11 17 12.	- 03 02		15.7	GALAXY
IC 2723	11 17 13.	+ 12 18 31.			NONSTELLAR OBJECT
LB 01991	11 17 13.	+ 53 29 30.		17.1	FAINT BLUE STAR
MCG+08-21-017A	11 17 15.	+ 47 30	9	17.	GALAXY
MCG+08-21-017	11 17 15.	+ 47 30	9	17.	GALAXY
REIZ 1012	11 17 17.	+ 31 07	18	14.5	GALAXY
LB 10129	11 17 18.	+ 28 42		17.6	FAINT BLUE STAR
LB 10128	11 17 18.	+ 31 03		18.2	FAINT BLUE STAR
LB 10127	11 17 18.	+ 31 14		15.1	FAINT BLUE STAR
TON-N 0064	11 17 18.	+ 31 16		14.3	BLUE STAR
LB 10126	11 17 18.	+ 31 43		16.9	FAINT BLUE STAR
OCL 0856	11 17 18.	- 63 14	720	9.5	OPEN STAR CLUSTER
VHA 117	11 17 18.	- 63 14	240		OPEN STAR CLUSTER
IC 2725	11 17 21.	+ 13 42 07.			NONSTELLAR OBJECT
MCG+06-25-044	11 17 21.	+ 36 22 30.	72	14.5	GALAXY
REIZ 1011	11 17 21.	+ 53 47	30	14.4	GALAXY
REIZ 1013	11 17 21.	- 02 21	48	15.8	GALAXY
A1 001	11 17 21.56	+ 15 29 37.5		17.10	FAINT BLUE OBJECT
IC 2726	11 17 22.	+ 13 41 25.			SINGLE STAR
RNGC 3626	11 17 23.	+ 18 38		12.0	GALAXY
IC 2727	11 17 24.	+ 12 18 25.			NONSTELLAR OBJECT
UGC 06341	11 17 24.	+ 18 32	66	16.0	GALAXY
MCG+03-29-031	11 17 24.	+ 18 33	54	16.	GALAXY
LB 10132	11 17 24.	+ 28 42		16.7	FAINT BLUE STAR
LB 10131	11 17 24.	+ 29 43		17.0	FAINT BLUE STAR
ZWG 156.063	11 17 24.	+ 30 46		15.2	GALAXY
UGC 06342	11 17 24.	+ 30 46	66	15.2	GALAXY E-S0
LB 10130	11 17 24.	+ 31 14		17.8	FAINT BLUE STAR
ZWG 011.059	11 17 24.	- 00 36		14.1	GALAXY
UGC 06340	11 17 24.	- 00 36	114	14.1	GALAXY Sb
MCG+00-29-016	11 17 24.	- 00 36	90	14.1	GALAXY
ZWG 011.058	11 17 24.	- 01 12		15.6	GALAXY
ZWG 011.057	11 17 24.	- 02 21		15.6	GALAXY
REIZ 1015	11 17 26.	+ 18 38	108	11.9	GALAXY
REIZ 1017	11 17 28.	- 00 36	72	13.7	GALAXY
IC 2728	11 17 29.	+ 13 41 54.			SINGLE STAR
ZWG 039.122	11 17 30.	+ 07 36		15.6	GALAXY
ZC 1117.5+1205	11 17 30.	+ 12 05	670		CLUSTER OF GALAXIES
IC 2729	11 17 30.	+ 13 41 00.			NONSTELLAR OBJECT
ZWG 096.029	11 17 30.	+ 18 38		11.2	GALAXY
UGC 06343	11 17 30.	+ 18 38	192	11.2	GALAXY S0/Sa
MCG+03-29-032	11 17 30.	+ 18 39	180	11.2	GALAXY
MCG+05-27-057	11 17 30.	+ 30 46	15	15.2	GALAXY
REIZ 1016	11 17 30.	+ 33 47	18	15.2	GALAXY
UGC 06344	11 17 30.	+ 58 01	72	17.	GALAXY DWRF SP
MCG+10-16-119	11 17 30.	+ 60 07	27	16.	GALAXY
ZWG 291.056	11 17 30.	+ 60 08		15.4	GALAXY
ZWG 011.060	11 17 30.	- 01 25		15.3	GALAXY
FATH 1.388	11 17 30.	- 15 28	11		NEBULA
IC 2730	11 17 32.	+ 12 38 30.			NONSTELLAR OBJECT
SN 1973R	11 17 33.	+ 13 16		14.5	SUPERNOVA
ARP 317	11 17 33.	+ 13 17			PECULIAR GALAXY
ARP 016	11 17 33.	+ 13 18			PECULIAR GALAXY
IC 2731	11 17 34.	+ 13 50 06.			SINGLE STAR
RNGC 3625	11 17 34.	+ 58 03		14.0	GALAXY
ZWG 039.123	11 17 36.	+ 02 48		14.4	GALAXY
8ZW 1117+02.8	11 17 36.	+ 02 48		14.4	COMPACT GALAXY
UGC 06345	11 17 36.	+ 02 48	150	14.4	GALAXY IRR

OBJECT NAME	RIGHT ASCEN.	DECLINATION	DIAM.	MAGN.	TYPE OF OBJECT
MCG+01-29-030	11 17 36.	+ 02 48	132	14.4	GALAXY
IC 2732	11 17 36.	+ 12 40 42.			NONSTELLAR OBJECT
ZWG 067.057	11 17 36.	+ 13 16		8.9	GALAXY
RNGC 3627	11 17 36.	+ 13 16		10.0	GALAXY
UGC 06346	11 17 36.	+ 13 16	540	8.9	GALAXY Sb
MCG+02-29-019	11 17 36.	+ 13 16	456	08.9	GALAXY
VV 308A	11 17 36.	+ 13 17	540	9.5	INTERACTING GALAXY
ZC 1117.6+3352	11 17 36.	+ 33 52	6720		CLUSTER OF GALAXIES
ZWG 185.039	11 17 36.	+ 34 22		15.3	GALAXY
UGC 06347	11 17 36.	+ 34 22	66	15.3	GALAXY Sa-b
ZC 1117.6+5033	11 17 36.	+ 50 33	3230		CLUSTER OF GALAXIES
LB 01992	11 17 36.	+ 52 06 00.		17.0	FAINT BLUE STAR
ZWG 291.057	11 17 36.	+ 58 04		13.9	GALAXY
UGC 06348	11 17 36.	+ 58 04	126	13.9	GALAXY SBb
7ZW 386	11 17 36.	+ 60 25			COMPACT GALAXY
ZWG 011.062	11 17 36.	- 02 46		14.4	GALAXY
MCG+00-29-017	11 17 36.	- 02 46	90	14.4	GALAXY
ZWG 011.061	11 17 36.	- 03 30		15.6	GALAXY
REIZ 1018	11 17 37.	+ 13 16	480	9.8	GALAXY
HOLM 246A	11 17 37.	+ 13 16	510	10.7	PART OF MULTIPLE GALAXY
REIZ 1019	11 17 37.	+ 13 52	120	11.8	GALAXY
HOLM 246C	11 17 37.	+ 13 52	900	11.2	PART OF MULTIPLE GALAXY
MAI 062	11 17 38.	+ 58 03	74		DWARF SPHEROIDAL GALAXY
REIZ 1021	11 17 38.	- 07 49	36	14.6	GALAXY
REIN 2.123	11 17 38.50	+ 58 03 21.1			NEBULA
MCG+06-25-045	11 17 39.	+ 34 22	48	15.	GALAXY
REIZ 1022	11 17 39.	- 02 47	60	13.8	GALAXY
REIZ 1020	11 17 41.	+ 29 16	12	15.3	GALAXY
ZWG 039.124	11 17 42.	+ 03 14		12.8	GALAXY
WEED 3	11 17 42.	+ 03 14		18.5	VERY BLUE STELLAR OBJECT
UGC 06349	11 17 42.	+ 03 14	114	12.8	GALAXY S0-a
MCG+01-29-031	11 17 42.	+ 03 14	96	12.8	GALAXY
ZWG 039.125	11 17 42.	+ 04 36		15.2	GALAXY
ZWG 067.058	11 17 42.	+ 13 52		11.5	GALAXY
RNGC 3628	11 17 42.	+ 13 52		10.5	GALAXY
UGC 06350	11 17 42.	+ 13 52	930	11.5	GALAXY Sb
MCG+02-29-020	11 17 42.	+ 13 52	600	11.5	GALAXY
VV 308B	11 17 42.	+ 13 53	900	11.8	INTERACTING GALAXY
LB 10134	11 17 42.	+ 27 14		17.3	FAINT BLUE STAR
LB 10133	11 17 42.	+ 27 18		17.0	FAINT BLUE STAR
MCG+06-25-046	11 17 42.	+ 35 20	27	16.	GALAXY
ZC 1117.7+4725	11 17 42.	+ 47 25	1140		CLUSTER OF GALAXIES
MCG+10-16-120	11 17 42.	+ 58 02	114	13.3	GALAXY
REIZ 1014	11 17 42.	+ 58 04	72	14.3	GALAXY
7ZW 387	11 17 42.	+ 66 03			COMPACT GALAXY
ZC 1117.7+7655	11 17 42.	+ 76 55	1610		CLUSTER OF GALAXIES
MCG-01-29-007	11 17 42.	- 07 50	96	15.	GALAXY
MCG-03-29-005	11 17 42.	- 21 13	24	15.	GALAXY
RNGC 3630	11 17 43.	+ 03 14		13.0	GALAXY
ARC 1222	11 17 43.	+ 47 27		16.6	RICH CLUSTER OF GALAXIES
RNGC 3638	11 17 44.	- 07 50		15.0	GALAXY
FATH 1.389	11 17 45.	+ 14 48	8		NEBULA
MCG+08-21-018	11 17 45.	+ 47 05	15	16.	GALAXY
REIZ 1024	11 17 46.	+ 02 49	72	13.5	GALAXY
REIZ 1026	11 17 46.	+ 03 15	72	12.7	GALAXY
REIZ 1023	11 17 46.	+ 24 47	18	15.1	GALAXY
IC 2733	11 17 47.	+ 14 08 36.			DOUBLE STAR
IC 2734	11 17 48.	+ 12 43 00.			NONSTELLAR OBJECT
MCG+05-27-058	11 17 48.	+ 27 13	114	12.9	GALAXY
LB 10136	11 17 48.	+ 27 14		16.9	FAINT BLUE STAR
LB 10135	11 17 48.	+ 29 20		17.2	FAINT BLUE STAR
ZWG 242.021	11 17 48.	+ 47 06		15.7	GALAXY
ZWG 268.019	11 17 48.	+ 51 07		15.5	GALAXY
MCG+09-19-044	11 17 48.	+ 52 53	36	15.	GALAXY
ZWG 268.020	11 17 48.	+ 52 54		15.7	GALAXY
ZWG 011.064	11 17 48.	- 02 38		14.9	GALAXY
MCG+00-29-019	11 17 48.	- 02 38	42	14.9	GALAXY
ZWG 011.063	11 17 48.	- 03 26		15.7	GALAXY
MCG+00-29-018	11 17 48.	- 03 26	24	15.7	GALAXY
REIZ 1027	11 17 49.	+ 12 43	42	15.6	GALAXY
REIZ 1030	11 17 50.	- 08 43	18	14.9	GALAXY
REIZ 1029	11 17 50.	- 08 43	9	15.6	GALAXY
RNGC 3629	11 17 52.	+ 27 14		13.0	GALAXY
REIZ 1025	11 17 52.	+ 27 15	96	12.9	GALAXY
HOLM 247A	11 17 52.	+ 27 16	120	12.6	PART OF MULTIPLE GALAXY
A1 002	11 17 52.84	+ 15 09 39.2		17.40	FAINT BLUE OBJECT
RNGC 3632	11 17 53.	+ 18 25			NON-EXISTENT OBJECT
ZWG 039.126	11 17 54.	+ 03 51		14.3	GALAXY
UGC 06351	11 17 54.	+ 03 51	72	14.3	GALAXY Sa
MCG+01-29-032	11 17 54.	+ 03 51	66	14.3	GALAXY
ZWG 039.127	11 17 54.	+ 04 24		15.3	GALAXY
ZWG 039.128	11 17 54.	+ 06 31		15.6	GALAXY
ZWG 156.064	11 17 54.	+ 27 15		12.9	GALAXY
UGC 06352	11 17 54.	+ 27 15	126	12.9	GALAXY Sc
ZC 1117.9+3627	11 17 54.	+ 36 27	670		CLUSTER OF GALAXIES
ZC 1117.9+3704	11 17 54.	+ 37 04	540		CLUSTER OF GALAXIES
MCG+08-21-019	11 17 54.	+ 47 05	18	17.	GALAXY
MCG+10-16-121	11 17 54.	+ 56 58	57	15.	GALAXY
ZWG 291.058	11 17 54.	+ 56 59		15.4	GALAXY
UGC 06353	11 17 54.	+ 56 59	72	15.4	GALAXY Sb
ZC 1117.9+6152	11 17 54.	+ 61 52	610		CLUSTER OF GALAXIES
ZWG 314.021	11 17 54.	+ 63 41		15.3	GALAXY
UGC 06354	11 17 54.	+ 63 41	66	15.3	GALAXY S
RNGC 3633	11 17 55.	+ 03 52		14.5	GALAXY
RNGC 3636	11 17 56.	- 10 01		13.0	GALAXY
MCG-02-29-019	11 17 57.	- 09 59	24	13.	GALAXY
HOLM 247B	11 17 58.	+ 27 19	24	14.8	PART OF MULTIPLE GALAXY
REIZ 1031	11 17 58.	+ 27 34	18	15.4	GALAXY
REIZ 1032	11 17 58.	+ 27 35	18	14.9	GALAXY
HOLM 248A	11 17 59.	+ 31 31	108	14.0	PART OF MULTIPLE GALAXY
LB 10138	11 18 00.	+ 28 16		16.9	FAINT BLUE STAR
ZWG 156.065	11 18 00.	+ 31 30		14.8	GALAXY
UGC 06355	11 18 00.	+ 31 30	126	14.8	GALAXY Sc
MCG+05-27-059	11 18 00.	+ 31 30	120	14.8	GALAXY
LB 10137	11 18 00.	+ 31 44		18.1	FAINT BLUE STAR
ZWG 185.040	11 18 00.	+ 32 34		15.3	GALAXY
UGC 06356	11 18 00.	+ 32 34	66	15.3	GALAXY S
UGC 06357	11 18 00.	+ 33 23	60	16.0	GALAXY Sc
ZC 1118.0+3644	11 18 00.	+ 36 44	610		CLUSTER OF GALAXIES
MCG+08-21-020	11 18 00.	+ 44 51	30	16.	GALAXY
ZC 1118.0+4547	11 18 00.	+ 45 47	2150		CLUSTER OF GALAXIES
ARC 1223	11 18 00.	+ 46 12		17.2	RICH CLUSTER OF GALAXIES
MCG+08-21-021	11 18 00.	+ 47 05	18	17.	GALAXY
MCG+09-19-045	11 18 00.	+ 56 02	30	16.	GALAXY
MCG+10-16-122	11 18 00.	+ 57 11	30	16.	GALAXY
MCG+10-16-123	11 18 00.	+ 57 12	18	15.	GALAXY
ZWG 291.059	11 18 00.	+ 57 14		14.9	GALAXY
ZC 1118.0+6133	11 18 00.	+ 61 33	4500		CLUSTER OF GALAXIES

OBJECT NAME	RIGHT ASCEN.	DECLINATION	DIAM.	MAGN.	TYPE OF OBJECT
LB 01993	11 18 00.	+ 61 40 18.		15.5	FAINT BLUE STAR
MCG+11-14-019	11 18 00.	+ 63 40	54	16.	GALAXY
MCG+12-11-016	11 18 00.	+ 69 02 30.	33	15.	GALAXY
ZC 1118.0+6930	11 18 00.	+ 69 30	1680		CLUSTER OF GALAXIES
ZWG 334.023	11 18 00.	+ 69 44		15.2	GALAXY
ZWG 334.024	11 18 00.	+ 73 05		15.5	GALAXY
UGC 06358	11 18 00.	+ 73 05	78	15.5	GALAXY DBL SYS
ZC 1118.0+7617	11 18 00.	+ 76 17	1810		CLUSTER OF GALAXIES
MCG-01-29-009	11 18 00.	- 08 45	12	16.	GALAXY
MCG-01-29-008	11 18 00.	- 08 45	48	15.5	GALAXY
A1 003	11 18 00.96	+ 15 02 18.5		16.65	FAINT BLUE OBJECT
TON-N 1393	11 18 02.	+ 35 32		15.7	BLUE STAR
RNGC 3635	11 18 02.	- 08 45		16.0	GALAXY
RNGC 3634	11 18 02.	- 08 45		15.5	GALAXY
REIZ 1028	11 18 03.	+ 53 46	36	15.1	GALAXY
REIZ 1036	11 18 05.	+ 03 52	36	13.5	GALAXY
REIZ 1033	11 18 05.	+ 31 31	108	13.4	GALAXY
ZWG 039.129	11 18 06.	+ 07 54		15.3	GALAXY
MCG+03-29-033	11 18 06.	+ 19 38	42	17.	GALAXY
MCG+06-25-047	11 18 06.	+ 32 33	42	15.	GALAXY
REIZ 1034	11 18 06.	+ 32 34	42	13.5	GALAXY
LB 01994	11 18 06.	+ 60 28 42.		16.9	FAINT BLUE STAP
KARA.72 279B	11 18 06.	- 01 12	30		PART OF DOUBLE GALAXY
ZWG 011.066	11 18 06.	- 01 13		14.7	GALAXY
UGC 06359	11 18 06.	- 01 13	66	14.7	GALAXY S0
KARA.72 279A	11 18 06.	- 01 13	48	14.7	PART OF DOUBLE GALAXY
MCG+00-29-020	11 18 06.	- 01 13	15	14.7	GALAXY
ZWG 011.065	11 18 06.	- 03 16		15.5	GALAXY
TON-N 1394	11 18 08.	+ 34 45		16.8	BLUE STAR
RNGC 3637	11 18 08.	- 09 58		13.0	GALAXY
A1 004	11 18 08.04	+ 15 33 08.2		16.90	FAINT BLUE OBJECT
SN 1964A	11 18 09.	+ 53 29		17.0	SUPERNOVA
REIZ 1037	11 18 09.	- 05 23	48	14.7	GALAXY
ARC 1224	11 18 10.	+ 36 43		17.8	RICH CLUSTER OF GALAXIES
SN 1965L	11 18 10.	+ 53 28		16.0	SUPERNOVA
RNGC 3631	11 18 11.	+ 53 27		11.5	GALAXY
ZWG 011.068	11 18 12.	+ 00 51		15.4	GALAXY
ZWG 096.030	11 18 12.	+ 19 38		14.8	GALAXY
MCG+03-29-034	11 18 12.	+ 19 40	48	14.8	GALAXY
LB 10139	11 18 12.	+ 29 12		17.3	FAINT BLUE STAR
ZC 1118.2+3405	11 18 12.	+ 34 05	740		CLUSTER OF GALAXIES
ZC 1118.2+4916	11 18 12.	+ 49 16	940		CLUSTER OF GALAXIES
ZWG 268.021	11 18 12.	+ 53 27		11.0	GALAXY
UGC 06360	11 18 12.	+ 53 27	360	11.0	GALAXY Sc
7ZW 388	11 18 12.	+ 56 57			COMPACT GALAXY
MCG+12-11-017	11 18 12.	+ 73 06	54	16.	GALAXY
ZWG 011.067	11 18 12.	- 03 00		15.5	GALAXY
MCG-01-29-010	11 18 12.	- 05 24	36	15.	GALAXY
MCG-01-29-011	11 18 12.	- 06 44	48	16.	GALAXY
MCG-02-29-020	11 18 12.	- 09 58	36	13.	GALAXY
SN 1971H	11 18 13.	+ 28 33		18.	SUPERNOVA
FATH 1.390	11 18 15.	+ 14 25	11		NEBULA
REIZ 1035	11 18 15.	+ 53 28	240	12.2	GALAXY
MCG+09-19-046	11 18 15.	+ 56 02	30	17.	GALAXY
REIZ 1038	11 18 16.	+ 00 41	12	15.3	GALAXY
REIZ 1039	11 18 16.	+ 00 45	18	14.7	GALAXY
REIZ 1040	11 18 16.	+ 00 51	24	15.4	GALAXY
HOLM 248B	11 18 16.	+ 31 35	24	14.2	PART OF MULTIPLE GALAXY
ARP 027	11 18 16.	+ 53 28			PECULIAR GALAXY
LB 01995	11 18 17.	+ 55 01 48.		15.8	FAINT BLUE STAR
ZWG 011.070	11 18 18.	+ 00 44		14.3	GALAXY
UGC 06361	11 18 18.	+ 00 44	78	14.3	GALAXY TRP SYS
MCG+00-29-022	11 18 18.	+ 00 44	54	14.3	GALAXY
ZWG 096.031	11 18 18.	+ 19 54		14.6	GALAXY
UGC 06362	11 18 18.	+ 19 54	66	14.6	GALAXY S
MCG+03-29-035	11 18 18.	+ 19 56	72	14.6	GALAXY
MCG+04-27-022	11 18 18.	+ 21 35	66	15.2	GALAXY
ZWG 126.034	11 18 18.	+ 21 37		15.2	GALAXY
UGC 06363	11 18 18.	+ 21 37	66	15.2	GALAXY Sc
KARA.72 280A	11 18 18.	+ 21 37	66	15.2	PART OF DOUBLE GALAXY
ZWG 126.035	11 18 18.	+ 24 35		15.4	GALAXY
LB 10140	11 18 18.	+ 30 56		18.3	FAINT BLUE STAR
ZWG 185.041	11 18 18.	+ 34 37		15.4	GALAXY
UGC 06364	11 18 18.	+ 34 37	72	15.4	GALAXY Sa-b
ZC 1118.3+4227	11 18 18.	+ 42 27	870		CLUSTER OF GALAXIES
MCG+09-19-047	11 18 18.	+ 53 28	270	10.9	GALAXY
ZC 1118.3+5528	11 18 18.	+ 55 28	1880		CLUSTER OF GALAXIES
UGC 06365	11 18 18.	+ 63 18	60	16.5	GALAXY
ZWG 334.025	11 18 18.	+ 73 06		15.3	GALAXY
ZWG 011.069	11 18 18.	- 03 22		14.8	GALAXY
MCG+00-29-021	11 18 18.	- 03 22	54	14.8	GALAXY
IC 2736	11 18 19.	+ 12 40 59.			NONSTELLAR OBJECT
TON-N 1395	11 18 20.	+ 33 38		16.8	BLUE STAR
REIZ 1041	11 18 21.	+ 24 35	30	14.5	GALAXY
ARC 1226	11 18 21.	+ 34 00		17.2	RICH CLUSTER OF GALAXIES
REIZ 1042	11 18 22.	+ 00 27	18	15.7	GALAXY
ZWG 011.072	11 18 24.	+ 00 11		15.4	GALAXY
ZWG 011.073	11 18 24.	+ 00 27		15.2	GALAXY
MCG+04-27-023	11 18 24.	+ 21 36	66	15.7	GALAXY
ZWG 126.036	11 18 24.	+ 21 38		15.7	GALAXY
UGC 06366	11 18 24.	+ 21 38	72	15.7	GALAXY Sb-c
KARA.72 280B	11 18 24.	+ 21 38	66	15.7	PART OF DOUBLE GALAXY
MCG+04-27-024	11 18 24.	+ 24 34	60	15.4	GALAXY
ZWG 156.066	11 18 24.	+ 31 31		15.0	GALAXY
UGC 06367	11 18 24.	+ 31 31	72	15.0	GALAXY SBa-b
MCG+06-25-048	11 18 24.	+ 34 36	54	15.	GALAXY
ZC 1118.4+4818	11 18 24.	+ 48 18	4570		CLUSTER OF GALAXIES
ZWG 334.026	11 18 24.	+ 69 34		15.4	GALAXY
ZWG 011.071	11 18 24.	- 01 41		15.6	GALAXY
IC 2735	11 18 25.	+ 34 36 23.			NONSTELLAR OBJECT
REIZ 1044	11 18 28.	+ 00 49	36	14.9	GALAXY
FATH 1.391	11 18 28.	+ 14 09	8		NEBULA
FATH 1.392	11 18 28.	+ 14 11	8		NEBULA
REIZ 1043	11 18 29.	+ 31 32	30	14.0	GALAXY
ZWG 011.075	11 18 30.	+ 00 48		15.1	GALAXY
MCG+01-29-034	11 18 30.	+ 03 28	36	14.	GALAXY
ZWG 039.130	11 18 30.	+ 03 30		11.8	GALAXY
UGC 06368	11 18 30.	+ 03 30	270	11.8	GALAXY E
MCG+01-29-033	11 18 30.	+ 03 30	60	11.8	GALAXY
ZC 1118.5+1056	11 18 30.	+ 10 56	1140		CLUSTER OF GALAXIES
MCG+05-27-060	11 18 30.	+ 31 32	72	15.0	GALAXY
ZWG 213.041	11 18 30.	+ 40 37		15.5	GALAXY
ZWG 242.022	11 18 30.	+ 50 05		15.6	GALAXY
MCG+08-21-022	11 18 30.	+ 50 05	48	16.	GALAXY
MCG+10-16-124	11 18 30.	+ 57 56	57	16.	GALAXY
MCG+10-16-125	11 18 30.	+ 58 00	39	16.	GALAXY
UGC 06369	11 18 30.	+ 58 03	60	17.	GALAXY DWRF SP
MCG+12-11-018	11 18 30.	+ 73 07	21	16.	GALAXY
ZWG 011.074	11 18 30.	- 02 55		15.3	GALAXY
MCG-05-27-010	11 18 30.	- 29 07	36	15.5	GALAXY
RNGC 3640	11 18 31.	+ 03 31		12.0	GALAXY
IC 2737	11 18 32.	+ 14 34 05.			NONSTELLAR OBJECT
ARC 1225	11 18 32.	+ 54 03		15.4	RICH CLUSTER OF GALAXIES
REIZ 1046	11 18 34.	+ 03 29	18	13.4	GALAXY
REIZ 1045	11 18 34.	+ 03 31	60	11.7	GALAXY
IC 0682	11 18 35.	+ 20 29 35.			NONSTELLAR OBJECT
UGC 06370	11 18 36.	+ 03 28	54	14.0	GALAXY E
ZWG 039.131	11 18 36.	+ 04 50		15.6	GALAXY
ZWG 039.132	11 18 36.	+ 05 38		14.9	GALAXY
MCG+01-29-035	11 18 36.	+ 05 38	48	14.9	GALAXY
ZC 1118.6+2550	11 18 36.	+ 25 50	740		CLUSTER OF GALAXIES
ZC 1118.6+2714	11 18 36.	+ 27 14	3090		CLUSTER OF GALAXIES
LB 10141	11 18 36.	+ 29 52		16.6	FAINT BLUE STAR
ZWG 185.042	11 18 36.	+ 34 38		15.3	GALAXY
MCG+07-23-040	11 18 36.	+ 40 36	39	15.	GALAXY
MCG+08-21-023	11 18 36.	+ 46 32 30.	36	16.	GALAXY
ZWG 242.023	11 18 36.	+ 46 35		15.7	GALAXY
7ZW 389	11 18 36.	+ 66 35			COMPACT GALAXY
ZWG 334.027	11 18 36.	+ 70 55		15.4	GALAXY
UGC 06371	11 18 36.	+ 70 55	90	15.4	GALAXY S
RNGC 3641	11 18 37.	+ 03 28		14.5	GALAXY
IC 2739	11 18 37.	+ 12 11 17.			NONSTELLAR OBJECT
MCG+06-25-049	11 18 39.	+ 34 36 30.	30	15.	GALAXY
IC 2740	11 18 41.	+ 09 01 41.			NONSTELLAR OBJECT
ZWG 039.133	11 18 42.	+ 03 57		15.5	GALAXY
IC 2741	11 18 42.	+ 09 25 35.			NONSTELLAR OBJECT
MCG+06-25-050	11 18 42.	+ 33 59 30.	42	15.5	GALAXY
ZWG 185.043	11 18 42.	+ 34 44		15.7	GALAXY
UGC 06372	11 18 42.	+ 39 33	66	16.5	GALAXY Sc
ZC 1118.7+6012	11 18 42.	+ 60 12	3630		CLUSTER OF GALAXIES
MCG+10-16-126	11 18 42.	+ 60 33	24	16.	GALAXY
ZWG 291.060	11 18 42.	+ 60 34		15.5	GALAXY
MCG+11-14-020	11 18 42.	+ 63 17	51	17.	GALAXY
ZWG 011.076	11 18 42.	- 02 42		15.0	GALAXY
MCG+00-29-023	11 18 42.	- 02 42	42	15.0	GALAXY
IC 2742	11 18 43.	+ 10 43 22.			NONSTELLAR OBJECT
IC 2738	11 18 43.	+ 34 37 05.			NONSTELLAR OBJECT
REIZ 1047	11 18 45.	+ 24 42	30	14.8	GALAXY
MCG+08-21-024	11 18 45.	+ 44 51	42	16.	GALAXY
ZWG 039.134	11 18 48.	+ 03 01		15.5	GALAXY
ZWG 039.135	11 18 48.	+ 03 08		15.3	GALAXY
8ZW 1118+03.3	11 18 48.	+ 03 15		17.9	COMPACT GALAXY
ZWG 039.136	11 18 48.	+ 03 17		14.8	GALAXY
MCG+01-29-036	11 18 48.	+ 03 17	36	14.8	GALAXY
ZWG 039.137	11 18 48.	+ 03 43		15.4	GALAXY
ZWG 126.037	11 18 48.	+ 24 41		15.7	GALAXY
LB 10142	11 18 48.	+ 30 40		18.7	FAINT BLUE STAR
ZWG 185.044	11 18 48.	+ 34 14		15.1	GALAXY
MCG+06-25-051	11 18 48.	+ 34 14	30	15.	GALAXY
ZC 1118.8+3812	11 18 48.	+ 38 12	3290		CLUSTER OF GALAXIES
ZC 1118.8+4335	11 18 48.	+ 43 35	1210		CLUSTER OF GALAXIES
MCG+09-19-048	11 18 48.	+ 52 41	18	16.	GALAXY
ARC 1227	11 18 49.	+ 48 19		16.6	RICH CLUSTER OF GALAXIES
IC 2743	11 18 50.	+ 08 58 04.			NONSTELLAR OBJECT
ARC 1228	11 18 51.	+ 34 37		13.8	RICH CLUSTER OF GALAXIES
REIZ 1048	11 18 52.	+ 00 54	60	14.5	GALAXY
REIZ 1049	11 18 52.	+ 03 18	12	14.1	GALAXY
IC 0683	11 18 52.4	+ 03 01 12.			GALAXY Ep
RNGC 3639	11 18 53.	+ 18 44		14.0	GALAXY
ZWG 011.077	11 18 54.	+ 01 48		15.6	GALAXY
ZWG 039.138	11 18 54.	+ 03 01		15.6	GALAXY
ZWG 039.139	11 18 54.	+ 03 05		15.2	GALAXY
UGC 06373	11 18 54.	+ 03 05	96	15.2	GALAXY Sa
ZWG 039.140	11 18 54.	+ 03 09		15.5	GALAXY
FEIG 040	11 18 54.	+ 11 36		11.1	FAINT BLUE STAR
ZWG 067.059	11 18 54.	+ 13 42		15.3	GALAXY
ZC 1118.9+1807	11 18 54.	+ 18 07	1880		CLUSTER OF GALAXIES
ZWG 096.032	11 18 54.	+ 18 44		14.0	GALAXY
UGC 06374	11 18 54.	+ 18 44	60	14.0	GALAXY S?
MCG+03-29-036	11 18 54.	+ 18 44	30	14.0	GALAXY
LB 10143	11 18 54.	+ 29 19		16.9	FAINT BLUE STAR
MCG+08-21-025	11 18 54.	+ 46 28	72	14.	GALAXY
ZWG 242.024	11 18 54.	+ 46 30		15.1	GALAXY
UGC 06375	11 18 54.	+ 46 30	72	15.1	GALAXY Sa-b
ZWG 268.022	11 18 54.	+ 56 01		15.7	GALAXY
7ZW 390	11 18 54.	+ 66 58			COMPACT GALAXY
MCG+12-11-019	11 18 54.	+ 70 55 30.	66	16.	GALAXY
RNGC 3644	11 18 55.	+ 03 05		15.0	GALAXY
IC 2745	11 18 55.	+ 13 42 04.			NONSTELLAR OBJECT
REIZ 1050	11 18 58.	+ 03 03	9	15.5	GALAXY
REIZ 1051	11 18 58.	+ 03 06	24	14.1	GALAXY
MCG+01-29-037	11 19 00.	+ 03 05	90	15.5	GALAXY
IC 0684	11 19 00.	+ 03 06			SAME AS NGC 3644
ZWG 039.141	11 19 00.	+ 03 09		15.4	GALAXY
ZWG 039.142	11 19 00.	+ 03 10		15.6	GALAXY
ZWG 039.143	11 19 00.	+ 03 15		15.5	GALAXY
ZWG 039.144	11 19 00.	+ 04 54		15.5	GALAXY
8ZW 1119+04.9	11 19 00.	+ 04 54		15.5	COMPACT GALAXY
IC 2746	11 19 00.	+ 12 00 46.			NONSTELLAR OBJECT
ZC 1119.0+1534	11 19 00.	+ 15 34	1140		CLUSTER OF GALAXIES
ZWG 096.033	11 19 00.	+ 16 18		15.3	GALAXY
ZWG 096.034	11 19 00.	+ 20 26		11.5	GALAXY
UGC 06376	11 19 00.	+ 20 26	228	11.5	GALAXY Sc
KARA.72 281A	11 19 00.	+ 20 26	210	11.5	PART OF DOUBLE GALAXY
ZWG 156.067	11 19 00.	+ 29 43		15.5	GALAXY
LB 10144	11 19 00.	+ 31 41		17.4	FAINT BLUE STAR
ZWG 185.045	11 19 00.	+ 34 38		15.5	GALAXY
MCG+06-25-052	11 19 00.	+ 34 38	24	15.	GALAXY
UGC 06377	11 19 00.	+ 41 30	60	17.	GALAXY DWRF SP
ZC 1119.0+4621	11 19 00.	+ 46 21	870		CLUSTER OF GALAXIES
ZC 1119.0+4959	11 19 00.	+ 49 59	1080		CLUSTER OF GALAXIES
MCG+10-16-127	11 19 00.	+ 58 10	51	16.	GALAXY
ZWG 334.028	11 19 00.	+ 69 55		14.7	GALAXY
UGC 06378	11 19 00.	+ 69 55	144	14.7	GALAXY Sc
RNGC 3647	11 19 01.	+ 03 10		15.5	GALAXY
RNGC 3645	11 19 01.	+ 03 15		15.5	GALAXY
ARC 1230	11 19 02.	+ 22 37		17.2	RICH CLUSTER OF GALAXIES
IC 2744	11 19 03.	+ 34 37 28.			NONSTELLAR OBJECT
MCG+07-23-041	11 19 03.	+ 38 42 30.	42	14.5	GALAXY
REIZ 1053	11 19 03.	- 05 23	36	15.7	GALAXY
REIZ 1055	11 19 04.	+ 03 11	12	15.4	GALAXY
REIZ 1052	11 19 04.	+ 27 23	42	14.9	GALAXY
IC 2747	11 19 05.	+ 09 04 40.			NONSTELLAR OBJECT
RNGC 3646	11 19 05.	+ 20 27		12.0	GALAXY
ZC 1119.1+0005	11 19 06.	+ 00 05	3360		CLUSTER OF GALAXIES
ZWG 011.078	11 19 06.	+ 01 47		15.6	GALAXY

OBJECT NAME	RIGHT ASCEN.	DECLINATION	DIAM.	MAGN.	TYPE OF OBJECT
UGC 06379	11 19 06.	+ 02 42	60	18.	GALAXY DWARF
ZWG 039.145	11 19 06.	+ 03 51		15.6	GALAXY
ZWG 067.060	11 19 06.	+ 08 51		15.6	GALAXY
ZC 1119.1+1255	11 19 06.	+ 12 55	2350		CLUSTER OF GALAXIES
MCG+03-29-037	11 19 06.	+ 20 28	210	11.5	GALAXY
ZWG 126.038	11 19 06.	+ 22 46		15.6	GALAXY
MCG+09-19-049	11 19 06.	+ 53 14	24	14.	GALAXY
ZWG 268.023	11 19 06.	+ 53 15		14.9	GALAXY
ZC 1119.1+6633	11 19 06.	+ 66 33	3560		CLUSTER OF GALAXIES
RNGC 3643	11 19 07.	+ 03 14		15.0	GALAXY
REIZ 1054	11 19 08.	+ 20 27	240	12.7	GALAXY
IC 2748	11 19 09.	+ 09 04 52.			NONSTELLAR OBJECT
IC 2749	11 19 10.	+ 08 50 52.			NONSTELLAR OBJECT
REIZ 1056	11 19 10.	+ 26 15	18	15.7	GALAXY
ZWG 039.146	11 19 12.	+ 04 24		15.6	GALAXY
FATH 1.393	11 19 12.	+ 14 16	22		NEBULA
ZWG 067.061	11 19 12.	+ 14 17		15.6	GALAXY
ZC 1119.2+1740	11 19 12.	+ 17 40	670		CLUSTER OF GALAXIES
LB 10148	11 19 12.	+ 29 51		17.0	FAINT BLUE STAR
LB 10147	11 19 12.	+ 30 04		18.4	FAINT BLUE STAR
LB 10146	11 19 12.	+ 30 57		18.6	FAINT BLUE STAR
LB 10145	11 19 12.	+ 31 11		17.5	FAINT BLUE STAR
MCG+09-19-050	11 19 12.	+ 50 52	78	14.	GALAXY
ZWG 268.024	11 19 12.	+ 50 53		14.7	GALAXY
UGC 06380	11 19 12.	+ 50 53	96	14.7	GALAXY S0-a
MCG+09-19-051	11 19 12.	+ 51 04	48	15.	GALAXY
ZWG 268.025	11 19 12.	+ 51 06		15.5	GALAXY
MRK 167	11 19 12.	+ 51 25	6	17.5	GALAXY WITH UV CONTINUUM
ZWG 291.061	11 19 12.	+ 58 12		15.7	GALAXY
7ZW 391	11 19 12.	+ 66 02			COMPACT GALAXY
7ZW 392	11 19 12.	+ 67 30			COMPACT GALAXY
UGC 06381	11 19 12.	+ 69 24	102	17.	GALAXY DWRF SP
MCG+12-11-020	11 19 12.	+ 69 55	150	15.	GALAXY
MCG+07-23-042	11 19 15.	+ 43 06	66	14.	GALAXY
REIZ 1058	11 19 15.	- 05 29	18	14.1	GALAXY
ABC 1229	11 19 17.	+ 46 26		17.8	RICH CLUSTER OF GALAXIES
ZWG 039.147	11 19 18.	+ 05 53		15.5	GALAXY
ZWG 185.046	11 19 18.	+ 34 36		15.4	GALAXY
MCG+06-25-053	11 19 18.	+ 35 13	78	15.5	GALAXY
UGC 06382	11 19 18.	+ 35 59	66	17.	GALAXY Sc
ZWG 214.001	11 19 18.	+ 43 05		15.3	GALAXY
ZWG 213.042	11 19 18.	+ 43 05		15.3	GALAXY
UGC 06383	11 19 18.	+ 43 05	72	15.3	GALAXY Sc
KARA.73B 0476	11 19 18.	+ 43 05	90	15.3	ISOLATED GALAXY S
MCG-01-29-012	11 19 18.	- 05 29	30	15.	GALAXY
MCG-02-29-021	11 19 18.	- 12 14	60	15.	GALAXY
LB 03500	11 19 18.	- 75 47		14.2	FAINT BLUE STAR
IC 2750	11 19 19.	+ 18 55 58.			NONSTELLAR OBJECT
TON-N 1396	11 19 20.	+ 35 07		16.9	BLUE STAR
ABC 1233	11 19 20.	- 18 40		17.6	RICH CLUSTER OF GALAXIES
RNGC 3642	11 19 21.	+ 59 21		12.0	GALAXY
REIZ 1059	11 19 22.	+ 26 12	24	15.2	GALAXY
REIZ 1060	11 19 23.	+ 30 48	24	15.0	GALAXY
ZWG 039.148	11 19 24.	+ 08 05		15.3	GALAXY
ZWG 067.062	11 19 24.	+ 08 40		15.1	GALAXY
IC 2753	11 19 24.	+ 10 09 15.			NONSTELLAR OBJECT
IC 2752	11 19 24.	+ 14 24	16		FAINT NEBULA
FATH 2.026	11 19 24.	+ 14 24	16		NEBULA
IC 2754	11 19 24.	+ 14 25	27		FAINT NEBULA
FATH 2.027	11 19 24.	+ 14 25	27		NEBULA
ZWG 096.035	11 19 24.	+ 19 19		15.5	GALAXY
ZWG 126.039	11 19 24.	+ 26 12		15.7	GALAXY
MCG+04-27-025	11 19 24.	+ 26 12	45	15.7	GALAXY
LB 10149	11 19 24.	+ 31 12		18.4	FAINT BLUE STAR
MCG+06-25-054	11 19 24.	+ 34 35	30	15.	GALAXY
ZWG 185.047	11 19 24.	+ 34 39		15.7	GALAXY
UGC 06384	11 19 24.	+ 35 13	78	16.5	GALAXY Sc
ZC 1119.4+4038	11 19 24.	+ 40 38	400		CLUSTER OF GALAXIES
MCG+10-16-128	11 19 24.	+ 59 20	366	11.5	GALAXY
ZWG 291.062	11 19 24.	+ 59 22		11.9	GALAXY
UGC 06385	11 19 24.	+ 59 22	372	11.9	GALAXY Sb/Sc
IC 2756	11 19 26.	+ 10 14 15.			NONSTELLAR OBJECT
IC 2755	11 19 26.	+ 14 04 03.			SINGLE STAR
IC 2764	11 19 26.	- 28 43 57.			NONSTELLAR OBJECT
IC 2757	11 19 27.	+ 08 40 09.			NONSTELLAR OBJECT
REIZ 1061	11 19 27.	- 05 21	42	15.7	GALAXY
IC 2758	11 19 28.	+ 08 05 15.			NONSTELLAR OBJECT
IC 2751	11 19 28.	+ 34 37 46.			NONSTELLAR OBJECT
IC 0685	11 19 29.3	+ 18 01 44.			GALAXY E4
ZWG 039.149	11 19 30.	+ 03 40		15.5	GALAXY
ZWG 039.150	11 19 30.	+ 04 58		15.2	GALAXY
FATH 1.394	11 19 30.	+ 15 00	16		NEBULA
ZC 1119.5+1707	11 19 30.	+ 17 07	2690		CLUSTER OF GALAXIES
ABC 1232	11 19 30.	+ 18 11		17.4	RICH CLUSTER OF GALAXIES
MCG+04-27-026	11 19 30.	+ 24 34	42	15.5	GALAXY
ZWG 126.040	11 19 30.	+ 24 35		15.5	GALAXY
MCG+04-27-027	11 19 30.	+ 24 35	12	15.6	GALAXY
ZWG 126.041	11 19 30.	+ 24 36		15.6	GALAXY
ZC 1119.5+4105	11 19 30.	+ 41 05	2420		CLUSTER OF GALAXIES
ZWG 242.025	11 19 30.	+ 46 59		15.7	GALAXY
ZC 1119.5+5920	11 19 30.	+ 59 20	1080		CLUSTER OF GALAXIES
REIZ 1057	11 19 30.	+ 59 21	42	12.8	GALAXY
ZWG 291.063	11 19 30.	+ 59 40		15.6	GALAXY
ZC 1119.5+6102	11 19 30.	+ 61 02	940		CLUSTER OF GALAXIES
A1 005	11 19 30.38	+ 14 05 52.0		16.45	FAINT BLUE OBJECT
HOLM 249C	11 19 32.	+ 24 35	30	14.7	PART OF MULTIPLE GALAXY
TON-N 1397	11 19 32.	+ 32 44		15.8	BLUE STAR
ABC 1231	11 19 32.	+ 50 00		17.0	RICH CLUSTER OF GALAXIES
REIZ 1062	11 19 33.	+ 24 36	30	14.7	GALAXY
MCG+08-21-026	11 19 33.	+ 46 41	18	16.	GALAXY
MCG+08-21-027	11 19 33.	+ 46 58	42	16.	GALAXY
RNGC 3649	11 19 35.	+ 20 29		14.5	GALAXY
IC 2759	11 19 35.	+ 24 35			NONSTELLAR OBJECT
ZWG 039.151	11 19 36.	+ 05 41		15.4	GALAXY
ZWG 096.036	11 19 36.	+ 20 28		14.7	GALAXY
UGC 06386	11 19 36.	+ 20 28	90	14.7	GALAXY S0
KARA.72 281B	11 19 36.	+ 20 28	78	14.7	PART OF DOUBLE GALAXY
MCG+03-29-038	11 19 36.	+ 20 28	84	14.7	GALAXY
BIGO 521	11 19 36.	+ 38 03			NEBULA
ZWG 242.026	11 19 36.	+ 46 43		15.5	GALAXY
MCG+09-19-052	11 19 36.	+ 53 57 30.	48	15.	GALAXY
ZC 1119.6+6148	11 19 36.	+ 61 48	1080		CLUSTER OF GALAXIES
7ZW 394	11 19 36.	+ 67 27			COMPACT GALAXY
A1 006	11 19 36.60	+ 14 42 48.5		15.75	FAINT BLUE OBJECT
REIZ 1063	11 19 37.	+ 13 20	54	14.9	GALAXY
LB 00187	11 19 37.	+ 14 42 54.		15.2	FAINT BLUE STAR
LB 01996	11 19 37.	+ 52 37 36.		15.7	FAINT BLUE STAR
REIZ 1064	11 19 38.	+ 20 29	12	14.0	GALAXY
IC 2761	11 19 40.	+ 14 27	5		FAINT NEBULA
FATH 2.028	11 19 40.	+ 14 27	5		NEBULA
ZC 1119.7+0305	11 19 42.	+ 03 05	6720		CLUSTER OF GALAXIES
IC 2762	11 19 42.	+ 12 59 51.			NONSTELLAR OBJECT
ZWG 067.063	11 19 42.	+ 13 00		15.5	GALAXY
REIZ 1065	11 19 42.	+ 13 00	36	15.5	GALAXY
IC 2763	11 19 42.	+ 13 20 21.			NONSTELLAR OBJECT
ZWG 067.064	11 19 42.	+ 13 21		14.9	GALAXY
UGC 06387	11 19 42.	+ 13 21	90	14.9	GALAXY Sc?
MCG+02-29-021	11 19 42.	+ 13 21	72	14.9	GALAXY
MCG+04-27-028	11 19 42.	+ 24 34	27	14.6	GALAXY
ZWG 242.027	11 19 42.	+ 47 23		15.7	GALAXY
MCG+09-19-053	11 19 42.	+ 51 48	18	15.	GALAXY
ZWG 268.026	11 19 42.	+ 51 49		15.5	GALAXY
7ZW 395	11 19 42.	+ 57 07			COMPACT GALAXY
MCG-06-25-016	11 19 42.	- 37 48	96	16.	GALAXY
REIZ 1067	11 19 44.	- 05 55	12	14.9	GALAXY
HOLM 249A	11 19 45.	+ 24 35	30	14.3	PART OF MULTIPLE GALAXY
A1 007	11 19 45.34	+ 14 16 50.8		15.65	FAINT BLUE OBJECT
IC 2766	11 19 46.	+ 13 10 39.			NONSTELLAR OBJECT
IC 2765	11 19 46.	+ 14 29	5		FAINT NEBULA
FATH 2.029	11 19 46.	+ 14 29	5		NEBULA
IC 2767	11 19 47.	+ 13 21 09.			NONSTELLAR OBJECT
RNGC 3651	11 19 47.	+ 24 34		14.5	GALAXY
ZC 1119.8+0115	11 19 48.	+ 01 15	3090		CLUSTER OF GALAXIES
IC 2768	11 19 48.	+ 12 48 09.			NONSTELLAR OBJECT
ZWG 067.065	11 19 48.	+ 14 29		15.6	GALAXY
ZC 1119.8+2031	11 19 48.	+ 20 31	1410		CLUSTER OF GALAXIES
ZC 1119.8+2252	11 19 48.	+ 22 52	670		CLUSTER OF GALAXIES
MCG+04-27-029	11 19 48.	+ 24 32 30.	18	15.0	GALAXY
ZWG 126.042	11 19 48.	+ 24 34		14.6	GALAXY
UGC 06388	11 19 48.	+ 24 34	78	14.6	GALAXY E+COMP
MCG+04-27-030	11 19 48.	+ 24 34	15	16.	GALAXY
LB 10154	11 19 48.	+ 27 33		17.0	FAINT BLUE STAR
LB 10153	11 19 48.	+ 28 07		17.8	FAINT BLUE STAR
LB 10152	11 19 48.	+ 29 55		17.9	FAINT BLUE STAR
LB 10151	11 19 48.	+ 32 22		16.5	FAINT BLUE STAR
LB 10150	11 19 48.	+ 32 26		17.7	FAINT BLUE STAR
MCG+07-23-043	11 19 48.	+ 40 08 30.	33	13.5	GALAXY
ZWG 214.002	11 19 48.	+ 40 09		13.5	GALAXY
ZWG 213.043	11 19 48.	+ 40 09		13.5	GALAXY
UGC 06389	11 19 48.	+ 40 09	78	13.5	GALAXY S0
ZC 1119.8+5158	11 19 48.	+ 51 58	940		CLUSTER OF GALAXIES
ZWG 314.022	11 19 48.	+ 64 21		14.7	GALAXY
UGC 06390	11 19 48.	+ 64 21	132	14.7	GALAXY Sc
ZWG 011.079	11 19 48.	- 02 48		15.7	GALAXY
IC 2770	11 19 49.	+ 09 29 51.			NONSTELLAR OBJECT
ABC 1234	11 19 49.	+ 21 41		17.3	RICH CLUSTER OF GALAXIES
HOLM 249B	11 19 49.	+ 24 34	24	14.5	PART OF MULTIPLE GALAXY
IC 2769	11 19 50.	+ 14 28	8		FAINT NEBULA
FATH 2.030	11 19 50.	+ 14 28	8		NEBULA
REIZ 1066	11 19 50.	+ 40 08	30	13.6	GALAXY
RNGC 3648	11 19 50.	+ 40 09		13.5	GALAXY
LB 01997	11 19 50.	+ 61 52 48.		15.1	FAINT BLUE STAR
LB 00256	11 19 50.	+ 61 53 18.		14.3	FAINT BLUE STAR
REIZ 1070	11 19 50.	- 07 22	30	13.9	GALAXY
REIZ 1069	11 19 51.	+ 24 33	24	14.3	GALAXY
REIZ 1068	11 19 51.	+ 24 35	36	14.0	GALAXY
IC 2771	11 19 52.	+ 12 47 39.			NONSTELLAR OBJECT
LB 01998	11 19 52.	+ 50 43 18.		15.7	FAINT BLUE STAR
REIZ 1071	11 19 53.	+ 04 35	18	14.2	GALAXY
RNGC 3653	11 19 53.	+ 24 32		15.0	GALAXY
ZWG 039.152	11 19 54.	+ 04 33		14.8	GALAXY
MCG+01-29-038	11 19 54.	+ 04 33	36	14.8	GALAXY
ZWG 039.153	11 19 54.	+ 07 47		15.5	GALAXY
IC 2772	11 19 54.	+ 13 52 27.			NONSTELLAR OBJECT
ZC 1119.9+1620	11 19 54.	+ 16 20	1210		CLUSTER OF GALAXIES
MCG+04-27-031	11 19 54.	+ 20 57 30.	90	14.6	GALAXY
ZWG 126.043	11 19 54.	+ 20 59		14.6	GALAXY
UGC 06391	11 19 54.	+ 20 59	102	14.6	GALAXY Sb
ZC 1119.9+2214	11 19 54.	+ 22 14	2080		CLUSTER OF GALAXIES
ZWG 126.044	11 19 54.	+ 24 32		15.0	GALAXY
ZC 1119.9+2528	11 19 54.	+ 25 28	610		CLUSTER OF GALAXIES
ZWG 185.048	11 19 54.	+ 32 40		15.4	GALAXY
ZWG 185.049	11 19 54.	+ 38 02		12.6	GALAXY
UGC 06392	11 19 54.	+ 38 02	174	12.6	GALAXY S(Bc)
ZC 1119.9+4656	11 19 54.	+ 46 56	870		CLUSTER OF GALAXIES
7ZW 393	11 19 54.	+ 57 07			COMPACT GALAXY
ZC 1119.9+5708	11 19 54.	+ 57 08	3290		CLUSTER OF GALAXIES
MCG+11-14-021	11 19 54.	+ 64 20	114	15.	GALAXY
MCG-01-29-013	11 19 54.	- 07 23	36	15.	GALAXY
LB 01999	11 19 56.	+ 58 13 06.		15.0	FAINT BLUE STAR
RNGC 3652	11 19 57.	+ 38 02		12.5	GALAXY
MCG+08-21-028	11 19 57.	+ 49 02	36	16.	GALAXY
HELW 434	11 19 58.	- 00 34 02.			NEBULA
RNGC 3650	11 19 59.	+ 20 59		14.5	GALAXY
ZWG 039.154	11 20 00.	+ 03 24		15.5	GALAXY
IC 2773	11 20 00.	+ 13 50 39.			NONSTELLAR OBJECT
ZC 1120.0+2142	11 20 00.	+ 21 42	1140		CLUSTER OF GALAXIES
ZC 1120.0+2508	11 20 00.	+ 25 08	1550		CLUSTER OF GALAXIES
ZC 1120.0+3355	11 20 00.	+ 33 55	870		CLUSTER OF GALAXIES
MCG+06-25-055	11 20 00.	+ 38 02	90	13.	GALAXY
MCG+09-19-054	11 20 00.	+ 51 28	24	16.	GALAXY
ZWG 268.027	11 20 00.	+ 51 30		15.5	GALAXY
SHAH 153	11 20 00.	+ 78 18	66		GROUP OF COMPACT GALAXIES
MCG+14-06-001	11 20 00.	+ 81 11	39	16.	GALAXY
LDN 1311	11 20 00.	+ 85 00	5220		DARK NEBULA
ZWG 011.080	11 20 00.	- 00 34		15.6	GALAXY
MCG-01-29-014	11 20 00.	- 06 47	42	15.	GALAXY
IC 2774	11 20 01.	+ 12 47 21.			NONSTELLAR OBJECT
REIZ 1073	11 20 01.	+ 13 36	42	15.2	GALAXY
HOLM 250A	11 20 01.	+ 13 37	30	14.5	PART OF MULTIPLE GALAXY
REIZ 1072	11 20 01.	+ 38 03	114	13.3	GALAXY
FATH 1.395	11 20 02.	+ 14 34	11		NEBULA
MCG-03-29-006	11 20 03.	- 17 18	24	15.	GALAXY
IC 2777	11 20 04.	+ 12 18 03.			NONSTELLAR OBJECT
IC 2775	11 20 04.	+ 12 47 09.			NONSTELLAR OBJECT
IC 2776	11 20 04.	+ 13 36 51.			NONSTELLAR OBJECT
REIZ 1074	11 20 04.	- 00 33	18	15.7	GALAXY
ZWG 011.081	11 20 06.	+ 01 00		15.7	GALAXY
ZWG 039.155	11 20 06.	+ 07 47		15.5	GALAXY
ZWG 067.066	11 20 06.	+ 12 18		15.4	GALAXY
IC 2778	11 20 06.	+ 12 47 56.			NONSTELLAR OBJECT
ZWG 067.067	11 20 06.	+ 13 36		14.9	GALAXY
MCG+02-29-022	11 20 06.	+ 13 36	48	14.9	GALAXY
HOLM 250B	11 20 06.	+ 13 38	24	15.2	PART OF MULTIPLE GALAXY
ZWG 156.068	11 20 06.	+ 27 52		15.2	GALAXY
LB 10155	11 20 06.	+ 30 52		16.1	FAINT BLUE STAR

OBJECT NAME	RIGHT ASCEN.	DECLINATION	DIAM.	MAGN.	TYPE OF OBJECT
ZC 1120.1+3222	11 20 06.	+ 32 22	1480		CLUSTER OF GALAXIES
ZWG 185.050	11 20 06.	+ 34 37		14.9	GALAXY
UGC 06393	11 20 06.	+ 34 37	60	14.9	GALAXY SBa
ZC 1120.1+4406	11 20 06.	+ 44 06	1880		CLUSTER OF GALAXIES
A1 008	11 20 07.17	+ 14 00 47.5		16.70	FAINT BLUE OBJFCT
REIZ 1078	11 20 08.	- 07 24	60	13.8	GALAXY
IC 2779	11 20 09.	+ 13 37 20.			NONSTELLAR OBJECT
ARC 1236	11 20 10.	+ 00 45		17.2	RICH CLUSTER OF GALAXIES
ZC 1120.2+0045	11 20 12.	+ 00 45	1010		CLUSTER OF GALAXIES
ZC 1120.2+1948	11 20 12.	+ 19 48	5780		CLUSTER OF GALAXIES
MCG+05-27-061	11 20 12.	+ 27 50	15	15.2	GALAXY
LB 10156	11 20 12.	+ 30 21		17.4	FAINT BLUE STAR
REIZ 1075	11 20 12.	+ 34 21	42	13.6	GALAXY
ZWG 185.051	11 20 12.	+ 34 23		14.6	GALAXY
UGC 06394	11 20 12.	+ 34 23	84	14.6	GALAXY E
REIZ 1076	11 20 12.	+ 34 24	18	14.3	GALAXY
MCG+06-25-056	11 20 12.	+ 34 36	54	15.	GALAXY
REIZ 1077	11 20 12.	+ 34 38	18	14.3	GALAXY
ZWG 242.028	11 20 12.	+ 47 20		15.6	GALAXY
MCG+09-19-055	11 20 12.	+ 52 41	42	16.	GALAXY
MCG+09-19-056	11 20 12.	+ 53 59	24	16.	GALAXY
ZC 1120.2+6305	11 20 12.	+ 63 05	1080		CLUSTER OF GALAXIES
ZC 1120.2+6454	11 20 12.	+ 64 54	1410		CLUSTER OF GALAXIES
7ZW 396	11 20 12.	+ 67 23			COMPACT GALAXY
ZC 1120.2+6728	11 20 12.	+ 67 28	2490		CLUSTER OF GALAXIES
MCG-01-29-015	11 20 12.	- 07 24	24	14.	GALAXY
IC 2780	11 20 13.	+ 10 25 26.			NONSTELLAR OBJECT
REIZ 1079	11 20 13.	+ 13 43	24	15.3	GALAXY
REIZ 1081	11 20 13.	- 10 32	36	14.7	GALAXY
REIZ 1083	11 20 14.	- 07 18	12	14.3	GALAXY
REIZ 1082	11 20 14.	- 10 18	24	14.5	GALAXY
A1 009	11 20 14.46	+ 14 53 58.5		16.90	FAINT BLUE OBJECT
IC 2781	11 20 15.	+ 12 37 14.			NONSTELLAR OBJECT
MCG+03-29-039	11 20 15.	+ 16 52	66	11.6	GALAXY
MCG+09-19-057	11 20 15.	+ 51 50	42	15.	GALAXY
A1 010	11 20 16.16	+ 15 33 56.3		16.90	FAINT BLUE OBJFCT
RNGC 3655	11 20 17.	+ 16 52		13.0	GALAXY
SHAH 154	11 20 18.	+ 01 23	120		GROUP OF COMPACT GALAXIES
ZWG 067.068	11 20 18.	+ 13 43		15.2	GALAXY
UGC 06395	11 20 18.	+ 13 43	66	15.2	GALAXY
MCG+02-29-023	11 20 18.	+ 13 43	48	15.2	GALAXY
ZWG 096.037	11 20 18.	+ 16 52		11.6	GALAXY
UGC 06396	11 20 18.	+ 16 52	96	11.6	GALAXY S
KARA.73B 0477	11 20 18.	+ 16 52	96	11.6	ISOLATED GALAXY S
ARC 1235	11 20 18.	+ 19 55		17.0	RICH CLUSTER OF GALAXIES
LB 10157	11 20 18.	+ 26 45		13.6	FAINT BLUE STAR
MCG+06-25-057	11 20 18.	+ 34 22	24	14.5	GALAXY
REIZ 1080	11 20 18.	+ 34 34	18	14.9	GALAXY
ZWG 185.052	11 20 18.	+ 34 47		15.0	GALAXY
UGC 06397	11 20 18.	+ 34 47	108	15.0	GALAXY Sa-b
ZC 1120.3+3725	11 20 18.	+ 37 25	340		CLUSTER OF GALAXIES
ZWG 011.082	11 20 18.	- 02 38		15.3	GALAXY
REIN 2.125	11 20 18.03	+ 16 51 52.2			NEBULA
REIN 2.124	11 20 18.98	+ 38 47 41.0			NEBULA
IC 2783	11 20 19.	+ 09 09 38.			NONSTELLAR OBJECT
IC 2782	11 20 19.	+ 13 42 56.			NONSTELLAR OBJECT
REIZ 1084	11 20 19.	+ 16 52	60	12.8	GALAXY
REIZ 1086	11 20 20.	- 09 28	24	14.0	GALAXY
ARC 1238	11 20 23.	+ 01 23		16.0	RICH CLUSTER OF GALAXIES
ZC 1120.4+1831	11 20 24.	+ 18 31	870		CLUSTER OF GALAXIES
LB 10159	11 20 24.	+ 26 38		18.2	FAINT BLUE STAR
LB 10158	11 20 24.	+ 30 20		17.0	FAINT BLUE STAR
ZWG 156.069	11 20 24.	+ 30 45		15.2	GALAXY
ZWG 185.053	11 20 24.	+ 34 46		15.6	GALAXY
MCG+06-25-058	11 20 24.	+ 34 46	90	14.5	GALAXY
REIZ 1085	11 20 24.	+ 34 47	30	14.3	GALAXY
HOLM 251A	11 20 24.	+ 34 47	30	14.5	PART OF MULTIPLE GALAXY
MCG+07-24-001	11 20 24.	+ 38 48	30	15.	GALAXY
ZWG 242.029	11 20 24.	+ 47 17		15.7	GALAXY
MCG+09-19-058	11 20 24.	+ 51 27	30	16.	GALAXY
MCG+09-19-059	11 20 24.	+ 54 04	36	16.	GALAXY
ZWG 011.083	11 20 24.	- 02 40		15.3	GALAXY
LB 02000	11 20 25.	+ 53 24 24.		14.5	FAINT BLUE STAR
FATH 1.396	11 20 26.	+ 14 50	8		NEBULA
A1 011	11 20 28.00	+ 14 17 55.1		16.30	FAINT BLUE OBJECT
REIZ 1087	11 20 29.	+ 30 47	18	14.2	GALAXY
LB 02001	11 20 29.	+ 53 54 48.		17.0	FAINT BLUE STAR
ZWG 039.156	11 20 30.	+ 03 05		15.5	GALAXY
ZWG 039.157	11 20 30.	+ 05 51		15.6	GALAXY
ZWG 039.158	11 20 30.	+ 05 55		15.3	GALAXY
IC 0686	11 20 30.	+ 05 55 14.			NONSTELLAR OBJECT
ZC 1120.5+1425	11 20 30.	+ 14 25	400		CLUSTER OF GALAXIES
MCG+05-27-062	11 20 30.	+ 29 51	54	15.7	GALAXY
ZWG 156.070	11 20 30.	+ 29 53		15.7	GALAXY
UGC 06398	11 20 30.	+ 29 53	66	15.7	GALAXY Sb-c
MRK 635	11 20 30.	+ 30 13	10	16.	GALAXY WITH UV CONTINUUM
MCG+05-27-063	11 20 30.	+ 30 46	21	15.2	GALAXY
MCG+06-25-059	11 20 30.	+ 34 45 30.	45	15.	GALAXY
REIZ 1088	11 20 30.	+ 34 46	24	14.5	GALAXY
ZC 1120.5+3517	11 20 30.	+ 35 17	740		CLUSTER OF GALAXIES
MCG+06-25-060	11 20 30.	+ 35 57	42	16.	GALAXY
MCG+09-19-060	11 20 30.	+ 51 09	132	14.	GALAXY
ZWG 314.023	11 20 30.	+ 63 39		15.7	GALAXY
HOLM 251B	11 20 33.	+ 34 46	36	14.7	PART OF MULTIPLE GALAXY
A1 012	11 20 33.20	+ 15 02 43.7		17.20	FAINT BLUE OBJECT
LB 02002	11 20 34.	+ 51 16 00.		15.5	FAINT BLUE STAR
IC 2784	11 20 35.	+ 13 23 32.			NONSTELLAR OBJECT
ARC 1242	11 20 35.	+ 17 15		17.2	RICH CLUSTER OF GALAXIES
ZWG 039.159	11 20 36.	+ 02 56		15.6	GALAXY
ZWG 039.160	11 20 36.	+ 03 04		15.5	GALAXY
IC 2760	11 20 36.	+ 12 56 26.			NONSTELLAR OBJECT
ZC 1120.6+2058	11 20 36.	+ 20 58	2220		CLUSTER OF GALAXIES
ZWG 126.045	11 20 36.	+ 24 50		15.7	GALAXY
LB 10163	11 20 36.	+ 28 25		17.1	FAINT BLUE STAR
LB 10162	11 20 36.	+ 29 06		17.2	FAINT BLUE STAR
LB 10161	11 20 36.	+ 29 32		18.0	FAINT BLUE STAR
LB 10160	11 20 36.	+ 29 46		16.8	FAINT BLUE STAR
ZWG 185.054	11 20 36.	+ 34 06		15.5	GALAXY
ZWG 268.028	11 20 36.	+ 51 12		14.9	GALAXY
UGC 06399	11 20 36.	+ 51 12	198	14.9	GALAXY
MCG+09-19-061	11 20 36.	+ 53 45	30	16.	GALAXY
MCG+09-19-062	11 20 36.	+ 53 57 30.	60	15.	GALAXY
UGC 06400	11 20 36.	+ 53 58	66	16.0	GALAXY Sb-c
ZC 1120.6+6023	11 20 36.	+ 60 23	740		CLUSTER OF GALAXIES
MCG+12-11-021	11 20 36.	+ 71 12	36	16.	GALAXY
VHA 118	11 20 36.	- 58 16	150		OPEN STAR CLUSTER
IC 2785	11 20 39.	+ 13 39 56.			NONSTELLAR OBJECT
ARC 1241	11 20 40.	+ 27 30		17.6	RICH CLUSTER OF GALAXIES
ZWG 039.161	11 20 42.	+ 04 14		15.7	GALAXY
MCG+01-29-039	11 20 42.	+ 04 14	18	15.7	GALAXY
ZWG 039.162	11 20 42.	+ 06 08		15.2	GALAXY
ZC 1120.7+1110	11 20 42.	+ 11 10	540		CLUSTER OF GALAXIES
IC 2786	11 20 42.	+ 13 39 56.			NONSTELLAR OBJECT
ZWG 067.069	11 20 42.	+ 13 40		15.2	GALAXY
MCG+02-29-024	11 20 42.	+ 13 40	12	15.2	GALAXY
ZWG 067.070	11 20 42.	+ 13 55		15.5	GALAXY
UGC 06401	11 20 42.	+ 13 55	60	15.5	GALAXY Sc?
LB 10165	11 20 42.	+ 28 56		18.8	FAINT BLUE STAR
LB 10164	11 20 42.	+ 32 26		18.2	FAINT BLUE STAR
ZC 1120.7+3636	11 20 42.	+ 36 36	540		CLUSTER OF GALAXIES
MCG+08-21-029	11 20 42.	+ 45 18	30	16.	GALAXY
IC 2787	11 20 43.	+ 13 54 20.			NONSTELLAR OBJECT
ARC 1237	11 20 44.	+ 43 08		17.3	RICH CLUSTER OF GALAXIES
REIZ 1090	11 20 44.	- 10 13	18	13.8	GALAXY
MCG+06-25-061	11 20 45.	+ 34 06	24	15.	GALAXY
LB 02003	11 20 45.	+ 54 45 18.		16.4	FAINT BLUE STAR
ARP 155	11 20 46.	+ 54 08			PECULIAR GALAXY
HELW 435	11 20 46.	- 00 38 57.			NEBULA
RNGC 3656	11 20 47.	+ 54 07		13.5	GALAXY
ZC 1120.8+2730	11 20 48.	+ 27 30	470		CLUSTER OF GALAXIES
LB 10167	11 20 48.	+ 28 51		17.7	FAINT BLUE STAR
LB 10166	11 20 48.	+ 31 05		16.9	FAINT BLUE STAR
ZC 1120.8+4505	11 20 48.	+ 45 05	670		CLUSTER OF GALAXIES
ZWG 242.030	11 20 48.	+ 45 20		15.6	GALAXY
MCG+09-19-064	11 20 48.	+ 54 06	24	15.	GALAXY
ZWG 268.029	11 20 48.	+ 54 07		13.4	GALAXY
UGC 06403	11 20 48.	+ 54 07	108	13.4	GALAXY PECULR
SN 1973C	11 20 49.	+ 54 07		17.0	SUPERNOVA
SN 1963K	11 20 48.	+ 54 07		15.0	SUPERNOVA
KARA.72 282B	11 20 48.	+ 54 07	60		PART OF DOUBLE GALAXY
KARA.72 282A	11 20 48.	+ 54 07	66	13.4	PART OF DOUBLE GALAXY
MCG+09-19-063	11 20 48.	+ 54 07	48	13.8	GALAXY
7ZW 397	11 20 48.	+ 56 42			COMPACT GALAXY
MCG+10-16-129	11 20 48.	+ 58 03	30	16.	GALAXY
UGC 06404	11 20 48.	+ 77 44	72	16.0	GALAXY S
ZWG 011.084	11 20 48.	- 00 39		14.4	GALAXY
UGC 06402	11 20 48.	- 00 39	54	14.4	GALAXY
MCG+00-29-024	11 20 48.	- 00 39	54	14.4	GALAXY
TON-N 1398	11 20 50.	+ 34 41		16.9	BLUE STAR
ARC 1240	11 20 50.	+ 43 24		17.2	RICH CLUSTER OF GALAXIES
REIZ 1089	11 20 50.	+ 54 07	42	13.8	GALAXY
IC 2788	11 20 51.	+ 12 58 25.			NONSTELLAR OBJECT
REIZ 1091	11 20 52.	- 00 38	60	14.2	GALAXY
8ZW 1120-19.1	11 20 54.	- 19 08		18.0	COMPACT GALAXY
A1 013	11 20 55.76	+ 15 53 10.3		16.25	FAINT BLUE OBJECT
IC 2789	11 20 56.	+ 14 27 49.			DOUBLE STAR
IC 2790	11 20 58.	+ 09 49 43.			NONSTELLAR OBJECT
VDB.66G 095	11 21	+ 03 39	100		DWARF GALAXY
LB 10168	11 21 00.	+ 29 54		17.1	FAINT BLUE STAR
ZWG 185.055	11 21 00.	+ 37 21		15.6	GALAXY
ZC 1121.0+4310	11 21 00.	+ 43 10	1080		CLUSTER OF GALAXIES
ZC 1121.0+4329	11 21 00.	+ 43 29	940		CLUSTER OF GALAXIES
ZC 1121.0+5110	11 21 00.	+ 51 10	200		CLUSTER OF GALAXIES
LB 02004	11 21 00.	+ 56 45 12.		16.2	FAINT BLUE STAR
ZWG 291.064	11 21 00.	+ 58 05		15.5	GALAXY
ZC 1121.0+8044	11 21 00.	+ 80 44	1140		CLUSTER OF GALAXIES
ZWG 011.085	11 21 00.	- 01 17		15.3	GALAXY
HELW 436	11 21 01.	- 01 17 15.			NEBULA
IC 2791	11 21 02.	+ 14 10 13.			NONSTELLAR OBJECT
REIZ 1095	11 21 02.	- 08 22	90	13.3	GALAXY
RNGC 3654	11 21 04.	+ 69 42		13.5	GALAXY
RNGC 3659	11 21 05.	+ 18 06		13.5	GALAXY
ZWG 039.163	11 21 06.	+ 03 17		15.1	GALAXY
IC 2792	11 21 06.	+ 11 40 49.			NONSTELLAR OBJECT
ZWG 096.038	11 21 06.	+ 18 05		12.7	GALAXY
UGC 06405	11 21 06.	+ 18 05	120	12.7	GALAXY Sc-IRR
MCG+03-29-040	11 21 06.	+ 18 06	156	12.7	GALAXY
ARC 1243	11 21 06.	+ 18 29		17.3	RICH CLUSTER OF GALAXIES
TON-N 0065	11 21 06.	+ 23 55		14.3	BLUE STAR
LB 10169	11 21 06.	+ 31 43		17.8	FAINT BLUE STAR
ZWG 268.030	11 21 06.	+ 53 12		13.1	GALAXY
RNGC 3657	11 21 06.	+ 53 12		13.0	GALAXY
UGC 06406	11 21 06.	+ 53 12	174	13.1	GALAXY VY CMPT
MCG+09-19-065	11 21 06.	+ 53 12	60	13.7	GALAXY
ZWG 268.031	11 21 06.	+ 56 14		15.4	GALAXY
MCG+10-16-130	11 21 06.	+ 58 50	42	17.	GALAXY
ZWG 334.029	11 21 06.	+ 69 42		13.4	GALAXY
UGC 06407	11 21 06.	+ 69 42	108	13.4	GALAXY S
UGC 06408	11 21 06.	- 00 50	90	13.8	GALAXY S+COMP
MCG-01-29-016	11 21 06.	- 08 24	138	12.5	GALAXY
MCG-02-29-022	11 21 06.	- 13 32	78	14.	GALAXY
8ZW 1121-19.1	11 21 06.	- 19 05		17.8	COMPACT GALAXY
REIZ 1094	11 21 07.	+ 18 06	48	13.5	GALAXY
REIZ 1092	11 21 07.	+ 53 12	24	13.7	GALAXY
IC 0689	11 21 07.	- 13 33 17.			SAME AS NGC 3661
RNGC 3660	11 21 08.	- 08 24		12.5	GALAXY
ARC 1239	11 21 09.	+ 60 26		17.5	RICH CLUSTER OF GALAXIES
IC 0688	11 21 09.	- 09 31 11.			NONSTELLAR OBJECT
RNGC 3661	11 21 09.	- 13 32		14.0	GALAXY
A1 014	11 21 10.16	+ 14 22 07.1		16.15	FAINT BLUE OBJECT
IC 2793	11 21 12.	+ 09 43 25.			NONSTELLAR OBJECT
LB 10170	11 21 12.	+ 29 48		17.1	FAINT BLUE STAR
MCG+09-19-066	11 21 12.	+ 56 15	48	16.	GALAXY
MCG+09-19-067	11 21 12.	+ 56 17	42	16.	GALAXY
ZWG 351.060	11 21 12.	+ 75 12		15.7	GALAXY
ZWG 011.086	11 21 12.	- 00 50		13.8	GALAXY
MCG+00-29-025	11 21 12.	- 00 50	72	13.8	GALAXY
MCG-04-27-009	11 21 12.	- 21 58	36	15.5	GALAXY
ARC 1246	11 21 13.	+ 21 43		17.6	RICH CLUSTER OF GALAXIES
REIZ 1093	11 21 13.	+ 38 50	18	13.5	GALAXY
RNGC 3662	11 21 13.	- 00 50		14.0	GALAXY
LB 02005	11 21 14.	+ 58 38 54.		16.8	FAINT BLUE STAR
ARC 1248	11 21 14.	- 03 56		17.0	RICH CLUSTER OF GALAXIES
MCG+08-21-030	11 21 15.	+ 46 04	60	14.	GALAXY
REIN 2.126	11 21 15.90	+ 38 50 15.3			NEBULA
REIZ 1096	11 21 16.	- 00 49	60	13.8	GALAXY
LB 00587	11 21 16.	- 12 06 30.		17.5	FAINT BLUE STAR
ZWG 011.087	11 21 18.	+ 02 00		15.3	GALAXY
MCG+00-29-026	11 21 18.	+ 02 00	42	15.3	GALAXY
ZWG 039.164	11 21 18.	+ 03 22		15.3	GALAXY
ZC 1121.3+2014	11 21 18.	+ 20 14	1080		CLUSTER OF GALAXIES
ZC 1121.3+2247	11 21 18.	+ 22 47	1140		CLUSTER OF GALAXIES
ZC 1121.3+3717	11 21 18.	+ 37 17	470		CLUSTER OF GALAXIES
ZC 1121.3+3733	11 21 18.	+ 37 33	810		CLUSTER OF GALAXIES
ZWG 214.003	11 21 18.	+ 38 50		13.3	GALAXY
UGC 06409	11 21 18.	+ 38 50	108	13.3	GALAXY E-S0

OBJECT NAME	RIGHT ASCEN.	DECLINATION	DIAM.	MAGN.	TYPE OF OBJECT
ZWG 242.031	11 21 18.	+ 46 06		14.8	GALAXY
UGC 06410	11 21 18.	+ 46 06	72	14.8	GALAXY Sc
ZWG 242.032	11 21 18.	+ 48 58		15.7	GALAXY
UGC 06411	11 21 18.	+ 48 58	66	15.7	GALAXY Sb-c
UGC 06412	11 21 18.	+ 58 53	66	17.	GALAXY
ZWG 334.030	11 21 18.	+ 71 40		15.7	GALAXY
MCG-03-29-007	11 21 18.	- 20 29	36	15.5	GALAXY
ARC 1245	11 21 19.	+ 32 35		17.2	RICH CLUSTER OF GALAXIES
RNGC 3658	11 21 21.	+ 38 50		13.5	GALAXY
MCG+07-24-002	11 21 21.	+ 38 50	36	13.	GALAXY
FATH 1.397	11 21 22.	+ 14 32	8		NEBULA
ZC 1121.4+0112	11 21 24.	+ 01 12	1210		CLUSTER OF GALAXIES
ZWG 039.165	11 21 24.	+ 02 58		14.8	GALAXY
UGC 06413	11 21 24.	+ 02 58	66	14.8	GALAXY Sb
MCG+01-29-040	11 21 24.	+ 02 58	60	14.8	GALAXY
ZWG 039.166	11 21 24.	+ 07 51		15.7	GALAXY
ZC 1121.4+1537	11 21 24.	+ 15 37	2620		CLUSTER OF GALAXIES
ARC 1247	11 21 24.	+ 20 18		17.2	RICH CLUSTER OF GALAXIES
ZC 1121.4+2147	11 21 24.	+ 21 47	870		CLUSTER OF GALAXIES
MCG+05-27-064	11 21 24.	+ 27 16	120	15.3	GALAXY
LB 10174	11 21 24.	+ 28 27		18.2	FAINT BLUE STAR
LB 10173	11 21 24.	+ 29 08		18.2	FAINT BLUE STAR
LB 10172	11 21 24.	+ 31 07		17.4	FAINT BLUE STAR
LB 10171	11 21 24.	+ 32 21		17.6	FAINT BLUE STAR
ZWG 185.056	11 21 24.	+ 34 19		15.7	GALAXY
ZC 1121.4+3627	11 21 24.	+ 36 27	540		CLUSTER OF GALAXIES
MCG+08-21-031	11 21 24.	+ 48 57	60	15.	GALAXY
ZC 1121.4+5056	11 21 24.	+ 50 56	870		CLUSTER OF GALAXIES
MCG-02-29-023	11 21 24.	- 12 00	96	13.5	GALAXY
RNGC 3663	11 21 26.	- 12 01		13.0	GALAXY
REIZ 1097	11 21 27.	+ 24 23	36	15.0	GALAXY
REIZ 1098	11 21 27.	+ 24 54	66	15.6	GALAXY
ARC 1244	11 21 27.	+ 45 42		17.6	RICH CLUSTER OF GALAXIES
REIZ 1099	11 21 28.	+ 03 00	18	14.9	GALAXY
IC 2795	11 21 28.	+ 12 24 31.			NONSTELLAR OBJECT
IC 2794	11 21 28.	+ 13 04 07.			NONSTELLAR OBJECT
ZWG 039.167	11 21 30.	+ 06 30		15.1	GALAXY
ZC 1121.5+1238	11 21 30.	+ 12 38	470		CLUSTER OF GALAXIES
UGC 06414	11 21 30.	+ 24 54	108	16.0	GALAXY Sb-c
ZWG 156.071	11 21 30.	+ 27 18		15.3	GALAXY
UGC 06415	11 21 30.	+ 27 18	132	15.3	GALAXY Sa-b
ZC 1121.5+2747	11 21 30.	+ 27 47	1550		CLUSTER OF GALAXIES
MCG+06-25-062	11 21 30.	+ 34 18 30.	36	15.	GALAXY
ZWG 185.057	11 21 30.	+ 38 16		15.5	GALAXY
MCG+08-21-032	11 21 30.	+ 48 07	42	15.	GALAXY
ZWG 242.033	11 21 30.	+ 48 08		15.0	GALAXY
LB 02006	11 21 30.	+ 51 29 48.		16.7	FAINT BLUE STAR
MCG+10-16-131	11 21 30.	+ 57 34	30	16.	GALAXY
MCG+12-11-022	11 21 30.	+ 69 41	66	14.	GALAXY
MCG-02-29-024	11 21 30.	- 12 50	36	15.	GALAXY
IC 2796	11 21 33.	+ 09 37 12.			NONSTELLAR OBJECT
MCG+13-08-057	11 21 33.	+ 75 13	12	15.	GALAXY
REIZ 1101	11 21 34.	+ 01 00	60	14.3	GALAXY
REIZ 1100	11 21 34.	+ 27 17	60	14.1	GALAXY
LB 00588	11 21 34.	- 09 12 12.		15.9	FAINT BLUE STAR
HELW 437	11 21 35.	- 00 53 33.			NEBULA
A1 015	11 21 35.23	+ 13 49 48.0		16.40	FAINT BLUE OBJECT
ZWG 011.089	11 21 36.	+ 00 58		14.8	GALAXY
MCG+00-29-028	11 21 36.	+ 00 58	48	14.8	GALAXY
KARA.72 283A	11 21 36.	+ 03 35	66	13.6	PART OF DOUBLE GALAXY
LB 10176	11 21 36.	+ 31 58		17.8	FAINT BLUE STAR
LB 10175	11 21 36.	+ 32 07		17.8	FAINT BLUE STAR
ZC 1121.6+3827	11 21 36.	+ 38 27	400		CLUSTER OF GALAXIES
IC 0687	11 21 36.	+ 48 07 19.			NONSTELLAR OBJECT
MCG+09-19-068	11 21 36.	+ 51 30	48	14.	GALAXY
ZWG 268.032	11 21 36.	+ 51 32		15.1	GALAXY
MCG+12-11-023	11 21 36.	+ 71 40	33	16.	GALAXY
ZWG 011.088	11 21 36.	- 00 53		15.0	GALAXY
MCG+00-29-027	11 21 36.	- 00 53	30	15.0	GALAXY
PK289+07.1	11 21 38.	- 52 35	10		PLANETARY NEBULA
A1 016	11 21 38.97	+ 14 30 13.3		15.65	FAINT BLUE OBJECT
ARP 005	11 21 40.	+ 03 36			PECULIAR GALAXY
REIZ 1103	11 21 40.	- 00 51	12	15.6	GALAXY
REIZ 1102	11 21 40.	- 00 52	12	15.2	GALAXY
VV 251B	11 21 42.	+ 03 35	42	15.	INTERACTING GALAXY
VV 251A	11 21 42.	+ 03 35	60	14.	INTERACTING GALAXY
VV 251	11 21 42.	+ 03 35	90	12.9	INTERACTING GALAXY
KARA.72 283B	11 21 42.	+ 03 36	48		PART OF DOUBLE GALAXY
ZC 1121.7+1748	11 21 42.	+ 17 48	2550		CLUSTER OF GALAXIES
MCG+05-27-065	11 21 42.	+ 27 42	60	15.	GALAXY
ZWG 156.072	11 21 42.	+ 28 35		15.7	GALAXY
MCG+05-27-066	11 21 42.	+ 28 35 30.	21	15.7	GALAXY
UGC 06416	11 21 42.	+ 39 31	78	17.	GALAXY DWARF
ZC 1121.7+4543	11 21 42.	+ 45 43	870		CLUSTER OF GALAXIES
MCG+08-21-034	11 21 42.	+ 49 09	12	17.	GALAXY
MCG+08-21-033	11 21 42.	+ 49 09	12	17.	GALAXY
MCG-02-29-025	11 21 42.	- 13 33	72	13.5	GALAXY
MCG-05-27-011	11 21 42.	- 30 11	18	15.5	GALAXY
HOLM 252A	11 21 43.	- 13 34	24	13.7	PART OF MULTIPLE GALAXY
RNGC 3667A	11 21 44.	- 13 33		13.0	GALAXY
IC 2797	11 21 45.	+ 11 58 54.			NONSTELLAR OBJECT
MCG+05-27-067	11 21 45.	+ 27 53	30	15.2	GALAXY
MCG-02-29-026	11 21 45.	- 13 33	60	15.	GALAXY
REIZ 1106	11 21 46.	+ 00 56	48	14.2	GALAXY
REIZ 1104	11 21 46.	+ 27 44	72	14.7	GALAXY
REIZ 1105	11 21 46.	+ 27 55	36	14.3	GALAXY
IC 2798	11 21 47.	+ 12 41 30.			NONSTELLAR OBJECT
LB 02007	11 21 47.	+ 50 49 42.		14.9	FAINT BLUE STAR
ZWG 011.090	11 21 48.	+ 00 55		14.7	GALAXY
MCG+00-29-029	11 21 48.	+ 00 55	42	14.7	GALAXY
ZWG 039.168	11 21 48.	+ 03 25		15.6	GALAXY
UGC 06417	11 21 48.	+ 03 25	60	15.6	GALAXY Sc
ZWG 039.169	11 21 48.	+ 03 30		15.4	GALAXY
UGC 06418	11 21 48.	+ 03 30	60	15.4	GALAXY IRR
ZWG 039.170	11 21 48.	+ 03 36		13.6	GALAXY
8ZW 1121+03.6	11 21 48.	+ 03 36		13.6	COMPACT GALAXY
UGC 06419	11 21 48.	+ 03 36	114	13.6	GALAXY SB
MCG+01-29-041	11 21 48.	+ 03 36	120	13.6	GALAXY
ZWG 067.071	11 21 48.	+ 11 37		12.5	GALAXY
WEED 4	11 21 48.	+ 11 37		17.5	VERY BLUE STELLAR OBJECT
RNGC 3666	11 21 48.	+ 11 37		12.5	GALAXY
REIZ 1107	11 21 48.	+ 11 37	132	12.6	GALAXY
UGC 06420	11 21 48.	+ 11 37	270	12.5	GALAXY Sc
MCG+02-29-025	11 21 48.	+ 11 37	270	12.5	GALAXY
UGC 06421	11 21 48.	+ 27 44	78	16.0	GALAXY
ZWG 156.073	11 21 48.	+ 27 55		15.2	GALAXY
LB 10177	11 21 48.	+ 31 32		18.1	FAINT BLUE STAR
ZWG 185.058	11 21 48.	+ 34 48		15.5	GALAXY

OBJECT NAME	RIGHT ASCEN.	DECLINATION	DIAM.	MAGN.	TYPE OF OBJECT
ZC 1121.8+3517	11 21 48.	+ 35 17	610		CLUSTER OF GALAXIES
ZC 1121.8+5118	11 21 48.	+ 51 18	670		CLUSTER OF GALAXIES
MCG+09-19-069	11 21 48.	+ 52 48	60	16.	GALAXY
UGC 06422	11 21 48.	+ 54 01	78	16.5	GALAXY
7ZW 398	11 21 48.	+ 60 13			COMPACT GALAXY
HOLM 252B	11 21 48.	- 13 34	18	14.7	PART OF MULTIPLE GALAXY
MCG-06-25-017	11 21 48.	- 34 43	36	14.	GALAXY
VHE 48A	11 21 48.	- 58 38	96		REFLECTION NEBULA
RNGC 3664A	11 21 49.	+ 03 30		15.5	GALAXY
IC 0690	11 21 49.	- 08 04 24.			NONSTELLAR OBJECT
REIN 2.127	11 21 49.47	+ 11 36 43.9			NEBULA
IC 2799	11 21 50.	+ 14 07	22		FAINT NEBULA
FATH 2.031	11 21 50.	+ 14 07	22		NEBULA
LB 02008	11 21 50.	+ 60 37 06.		14.9	FAINT BLUE STAR
RNGC 3667B	11 21 50.	- 13 33		15.0	GALAXY
A1 017	11 21 50.27	+ 15 16 49.3		16.90	FAINT BLUE OBJECT
REIN 2.128	11 21 50.30	+ 11 37 01.4			NEBULA
IC 2800	11 21 51.	+ 12 28 54.			NONSTELLAR OBJECT
REIZ 1108	11 21 51.	+ 23 56	30	15.0	GALAXY
REIZ 1110	11 21 52.	+ 03 37	108	13.6	GALAXY
8ZW 1121+02.8	11 21 54.	+ 02 51		18.0	COMPACT GALAXY
8ZW 1121+06.2	11 21 54.	+ 06 11		18.4	COMPACT GALAXY
IC 2801	11 21 54.	+ 10 27 30.			NONSTELLAR OBJECT
IC 2802	11 21 54.	+ 12 29 00.			NONSTELLAR OBJECT
LB 10178	11 21 54.	+ 30 19		17.5	FAINT BLUE STAR
ZC 1121.9+3648	11 21 54.	+ 36 48	670		CLUSTER OF GALAXIES
SHAH 060	11 21 54.	+ 40 42	96	16.5	GROUP OF COMPACT GALAXIES
ZC 1121.9+4212	11 21 54.	+ 42 12	400		CLUSTER OF GALAXIES
ZWG 291.065	11 21 54.	+ 60 25		15.7	GALAXY
MCG+10-16-132	11 21 54.	+ 60 52	51	16.	GALAXY
UGC 06423	11 21 54.	+ 60 54	60	16.0	GALAXY SB:c
RNGC 3664	11 21 55.	+ 03 36		13.0	GALAXY
ARC 1252	11 21 55.	- 08 31		17.2	RICH CLUSTER OF GALAXIES
FATH 1.398	11 21 56.	+ 14 20	22		NEBULA
IC 2803	11 21 59.	+ 10 07 36.			NONSTELLAR OBJECT
LB 09851	11 22	- 86 49		12.0	FAINT BLUE STAR
ZC 1122.0+1108	11 22 00.	+ 11 08	670		CLUSTER OF GALAXIES
FATH 1.399	11 22 00.	+ 14 41	11		NEBULA
ZC 1122.0+1456	11 22 00.	+ 14 56	1010		CLUSTER OF GALAXIES
ZWG 096.039	11 22 00.	+ 15 13		15.5	GALAXY
UGC 06424	11 22 00.	+ 15 13	78	15.5	GALAXY S-IRR
MCG+03-29-041	11 22 00.	+ 15 13	72	15.5	GALAXY
ZWG 126.046	11 22 00.	+ 23 53		15.4	GALAXY
UGC 06425	11 22 00.	+ 23 53	78	15.4	GALAXY Sb
ZWG 126.047	11 22 00.	+ 24 16		15.7	GALAXY
LB 10179	11 22 00.	+ 32 12		18.5	FAINT BLUE STAR
ZWG 214.004	11 22 00.	+ 39 02		11.6	GALAXY
UGC 06426	11 22 00.	+ 39 02	240	11.6	GALAXY E-S0
MCG+09-19-070	11 22 00.	+ 54 01	30	15.	GALAXY
ZWG 314.024	11 22 00.	+ 64 35		15.6	GALAXY
ZC 1122.0+6811	11 22 00.	+ 68 11	1880		CLUSTER OF GALAXIES
MCG+12-11-024	11 22 00.	+ 71 12	9	16.	GALAXY
ZC 1122.0+8409	11 22 00.	+ 84 09	1210		CLUSTER OF GALAXIES
ZWG 011.091	11 22 00.	- 00 43		15.5	GALAXY
RNGC 3665	11 22 01.	+ 03 02		12.5	GALAXY
REIZ 1111	11 22 01.	+ 39 03	24	12.5	GALAXY
FATH 1.400	11 22 02.	+ 14 45	5		NEBULA
REIZ 1113	11 22 02.	- 06 16	24	14.5	GALAXY
REIZ 1112	11 22 03.	+ 23 53	12	15.1	GALAXY
MCG+07-24-003	11 22 03.	+ 39 02	78	10.	GALAXY
REIZ 1109	11 22 03.	+ 56 31	30	15.0	GALAXY
HELW 438	11 22 04.	- 00 43 04.			NEBULA
ARC 1251	11 22 05.	+ 17 47		17.5	RICH CLUSTER OF GALAXIES
RNGC 3670	11 22 05.	+ 24 13		14.5	GALAXY
MCG+04-27-032	11 22 06.	+ 23 52 30.	72	15.4	GALAXY
ZWG 126.048	11 22 06.	+ 24 13		14.6	GALAXY
UGC 06427	11 22 06.	+ 24 13	78	14.6	GALAXY SB0-a
ZC 1122.1+2442	11 22 06.	+ 24 42	810		CLUSTER OF GALAXIES
MCG+05-27-068	11 22 06.	+ 29 01	36	15.	GALAXY
LB 10180	11 22 06.	+ 31 33		17.6	FAINT BLUE STAR
ZC 1122.1+4156	11 22 06.	+ 41 56	1010		CLUSTER OF GALAXIES
MCG+10-16-133	11 22 06.	+ 56 57	24	16.	GALAXY
A1 018	11 22 07.37	+ 14 20 41.7		16.00	FAINT BLUE OBJECT
A1 019	11 22 07.83	+ 15 28 57.8		15.80	FAINT BLUE OBJECT
LB 02009	11 22 08.	+ 51 43 48.		15.5	FAINT BLUE STAR
REIZ 1114	11 22 09.	+ 23 09	12	15.2	GALAXY
REIZ 1115	11 22 09.	+ 24 14	18	14.1	GALAXY
ZC 1122.2+0814	11 22 12.	+ 08 14	2690		CLUSTER OF GALAXIES
MCG+04-27-033	11 22 12.	+ 24 12	60	14.6	GALAXY
LB 10181	11 22 12.	+ 26 40		15.6	FAINT BLUE STAR
UGC 06428	11 22 12.	+ 38 11	90	17.	GALAXY
ARC 1250	11 22 12.	+ 41 56		17.3	RICH CLUSTER OF GALAXIES
ZWG 242.034	11 22 12.	+ 47 25		15.6	GALAXY
A1 020	11 22 12.46	+ 14 24 42.9		16.80	FAINT BLUE OBJECT
MCG+05-27-069	11 22 15.	+ 27 46	24	16.	GALAXY
A1 021	11 22 17.33	+ 15 51 13.7		17.40	FAINT BLUE OBJECT
ZWG 039.171	11 22 18.	+ 02 57		15.6	GALAXY
ZWG 039.172	11 22 18.	+ 07 54		15.6	GALAXY
ZWG 067.072	11 22 18.	+ 13 29		14.9	GALAXY
MCG+02-29-026	11 22 18.	+ 13 29	48	14.9	GALAXY
REIZ 1116	11 22 18.	+ 13 30	36	15.0	GALAXY
ZWG 185.059	11 22 18.	+ 35 54		15.3	GALAXY
SHAH 061	11 22 18.	+ 41 13			GROUP OF COMPACT GALAXIES
ZC 1122.3+4250	11 22 18.	+ 42 50	1810		CLUSTER OF GALAXIES
ZC 1122.3+5927	11 22 18.	+ 59 27	3090		CLUSTER OF GALAXIES
ZC 1122.3+6317	11 22 18.	+ 63 17	8800		CLUSTER OF GALAXIES
MCG-02-29-028	11 22 18.	- 09 31	228	11.5	GALAXY
MCG-02-29-027	11 22 18.	- 13 17	90	14.	GALAXY
IC 2804	11 22 20.	+ 13 29 53.			NONSTELLAR OBJECT
MCG-01-29-017	11 22 21.	- 05 04	24	15.	GALAXY
MIL 21	11 22 21.	- 58 59 24.	148		SUPERNOVA REMNANT
REIZ 1118	11 22 22.	+ 03 27	18	15.6	GALAXY
LB 02010	11 22 22.	+ 60 08 18.		16.3	FAINT BLUE STAR
ZWG 039.173	11 22 24.	+ 03 27		15.2	GALAXY
IC 2805	11 22 24.	+ 14 17 23.			SINGLE STAR
LB 10183	11 22 24.	+ 29 32		16.3	FAINT BLUE STAR
LB 10182	11 22 24.	+ 31 12		15.9	FAINT BLUE STAR
ZWG 314.025	11 22 24.	+ 64 01		14.1	GALAXY
UGC 06429	11 22 24.	+ 64 01	174	14.1	GALAXY Sc
MCG+11-14-022	11 22 24.	+ 64 01	114	14.	GALAXY
ARC 1249	11 22 25.	+ 68 19		17.2	RICH CLUSTER OF GALAXIES
LB 02011	11 22 28.	+ 55 37 30.		16.2	FAINT BLUE STAR
A1 022	11 22 28.51	+ 15 35 39.1		16.45	FAINT BLUE OBJECT
LB 10184	11 22 30.	+ 30 48		16.8	FAINT BLUE STAR
MCG+06-25-063	11 22 30.	+ 32 31	30	15.5	GALAXY
MCG+09-19-071	11 22 30.	+ 52 21	30	16.	GALAXY
MCG+10-16-134	11 22 30.	+ 57 18	51	15.	GALAXY
7ZW 399	11 22 30.	+ 60 14			COMPACT GALAXY

OBJECT NAME	RIGHT ASCEN.	DECLINATION	DIAM.	MAGN.	TYPE OF OBJECT
ZWG 314.026	11 22 30.	+ 63 43		13.1	GALAXY
UGC 06430	11 22 30.	+ 63 43	126	13.1	GALAXY Sb/Sc
7ZW 400	11 22 30.	+ 64 04			COMPACT GALAXY
ARC 1253	11 22 31.	+ 42 47		17.3	RICH CLUSTER OF GALAXIES
REIZ 1119	11 22 32.	+ 23 05	24	14.6	GALAXY
RNGC 3668	11 22 32.	+ 63 43		13.0	GALAXY
REIZ 1123	11 22 32.	- 09 31	150	11.7	GALAXY
RNGC 3672	11 22 32.	- 09 32		12.0	GALAXY
MCG+06-25-064	11 22 33.	+ 32 36	24	16.	GALAXY
HOLM 253D	11 22 34.	+ 32 31	36	15.0	PART OF MULTIPLE GALAXY
LB 02012	11 22 34.	+ 54 36 36.		16.0	FAINT BLUE STAR
REIZ 1117	11 22 34.	+ 58 00	132	13.4	GALAXY
A1 023	11 22 34.58	+ 14 09 50.4		16.95	FAINT BLUE OBJECT
REIZ 1120	11 22 35.	+ 32 33	30	14.2	GALAXY
REIZ 1121	11 22 35.	+ 32 34	12	14.7	GALAXY
REIZ 1122	11 22 35.	+ 32 38	18	14.5	GALAXY
RNGC 3669	11 22 35.	+ 58 00		13.0	GALAXY
ZWG 039.174	11 22 36.	+ 02 52		15.7	GALAXY
ZWG 096.040	11 22 36.	+ 18 05		15.4	GALAXY
LB 10187	11 22 36.	+ 26 54		18.1	FAINT BLUE STAR
LB 10186	11 22 36.	+ 26 59		14.8	FAINT BLUE STAR
LB 10185	11 22 36.	+ 30 04		18.4	FAINT BLUE STAR
HOLM 253E	11 22 36.	+ 32 36	24	15.4	PART OF MULTIPLE GALAXY
ZWG 291.066	11 22 36.	+ 57 18		15.6	GALAXY
MCG+10-16-135	11 22 36.	+ 57 59	102	13.4	GALAXY
ZWG 291.067	11 22 36.	+ 58 00		12.9	GALAXY
UGC 06431	11 22 36.	+ 58 00	132	12.9	GALAXY IRR
MCG+10-16-136	11 22 36.	+ 59 41	30	16.	GALAXY
MCG+10-16-137	11 22 36.	+ 61 46	36	16.	GALAXY
MCG+12-11-025	11 22 36.	+ 71 12	57	16.	GALAXY
MCG-01-29-018	11 22 36.	- 04 47	30	15.	GALAXY
HOLM 253F	11 22 38.	+ 32 33	30	15.5	PART OF MULTIPLE GALAXY
LB 00589	11 22 38.	- 12 26 42.		17.0	FAINT BLUE STAR
IC 2806	11 22 40.	+ 09 55 47.			NONSTELLAR OBJECT
IC 2807	11 22 41.	+ 11 48 11.			NONSTELLAR OBJECT
ZWG 011.092	11 22 42.	+ 00 37		15.2	GALAXY
UGC 06432	11 22 42.	+ 00 37	60	15.2	GALAXY S
MCG+00-29-030	11 22 42.	+ 00 37	48	15.2	GALAXY
ZWG 039.175	11 22 42.	+ 02 44		15.7	GALAXY
MCG+03-29-042	11 22 42.	+ 18 05	66	15.4	GALAXY
ZC 1122.7+3702	11 22 42.	+ 37 02	610		CLUSTER OF GALAXIES
ZWG 268.033	11 22 42.	+ 51 28		15.7	GALAXY
1ZW 026	11 22 42.	+ 54 40			COMPACT GALAXY
MCG+11-14-023	11 22 42.	+ 63 43	96	13.	GALAXY
MCG+11-14-024	11 22 42.	+ 63 44	18	18.	GALAXY
ZC 1122.7+7122	11 22 42.	+ 71 22	2760		CLUSTER OF GALAXIES
MCG-04-27-010	11 22 42.	- 26 27 30.	210	12.	GALAXY
MCG+09-19-072	11 22 45.	+ 52 49	42	15.	GALAXY
VV 144C	11 22 45.	+ 54 39	12	17.5	INTERACTING GALAXY
VV 144B	11 22 45.	+ 54 39	12	17.5	INTERACTING GALAXY
VV 144A	11 22 45.	+ 54 39	18	17.	INTERACTING GALAXY
VV 144	11 22 45.	+ 54 39	60	16.	INTERACTING GALAXY
REIZ 1126	11 22 46.	+ 00 38	18	14.9	GALAXY
RNGC 3673	11 22 46.	- 26 28		12.5	GALAXY
ARP 151	11 22 47.	+ 54 40			PECULIAR GALAXY
ZWG 039.176	11 22 48.	+ 05 11		15.2	GALAXY
ZC 1122.8+2145	11 22 48.	+ 21 45	1140		CLUSTER OF GALAXIES
ZWG 126.049	11 22 48.	+ 22 59		15.6	GALAXY
LB 10188	11 22 48.	+ 32 22		17.4	FAINT BLUE STAR
MCG+06-25-065	11 22 48.	+ 32 30	24	15.	GALAXY
ZWG 185.060	11 22 48.	+ 35 47		15.4	GALAXY
ZWG 185.061	11 22 48.	+ 38 19		14.7	GALAXY
UGC 06433	11 22 48.	+ 38 19	66	14.7	GALAXY PECULR
ZC 1122.8+4351	11 22 48.	+ 43 51	1140		CLUSTER OF GALAXIES
ZC 1122.8+4900	11 22 48.	+ 49 00	5240		CLUSTER OF GALAXIES
MRK 040	11 22 48.	+ 54 41	10	16.	GALAXY WITH UV CONTINUUM
KW 07	11 22 48.	+ 54 41	16		SEYFERT GALAXY
VVI 45	11 22 48.	+ 54 41	8	16.21	SEYFERT GALAXY
MCG+09-19-073	11 22 48.	+ 54 41	60	15.	GALAXY
REIZ 1125	11 22 49.	+ 38 19	54	14.6	GALAXY
MCG+06-25-066	11 22 51.	+ 38 20	66	14.5	GALAXY
REIZ 1124	11 22 51.	+ 56 30	18	15.2	GALAXY
IC 2808	11 22 52.	+ 09 24 47.			NONSTELLAR OBJECT
ARC 1256	11 22 52.	- 16 02		17.7	RICH CLUSTER OF GALAXIES
REIZ 1127	11 22 53.	+ 32 32	24	14.3	GALAXY
HOLM 253A	11 22 53.	+ 32 32	24	15.0	PART OF MULTIPLE GALAXY
REIZ 1128	11 22 53.	+ 32 34	12	14.4	GALAXY
ZWG 039.177	11 22 54.	+ 02 36		15.6	GALAXY
8ZW 1122+02.6	11 22 54.	+ 02 36		15.6	COMPACT GALAXY
ZWG 039.178	11 22 54.	+ 05 22		15.7	GALAXY
ZWG 126.050	11 22 54.	+ 23 05		14.6	GALAXY
MCG+04-27-034	11 22 54.	+ 23 05 30.	48	14.6	GALAXY
HOLM 253C	11 22 54.	+ 32 30	30	14.7	PART OF MULTIPLE GALAXY
MCG+06-25-067	11 22 54.	+ 32 35	48	15.	GALAXY
ZWG 185.062	11 22 54.	+ 32 36		15.7	GALAXY
UGC 06434	11 22 54.	+ 32 36	60	15.7	GALAXY S
ZC 1122.9+3811	11 22 54.	+ 38 11	1140		CLUSTER OF GALAXIES
VV 087	11 22 54.	+ 38 21	78	14.	INTERACTING GALAXY
MRK 168	11 22 54.	+ 47 17	10	16.	GALAXY WITH UV CONTINUUM
ZC 1122.9+6045	11 22 54.	+ 60 45	940		CLUSTER OF GALAXIES
MCG-01-29-019	11 22 54.	- 09 19	21	15.	GALAXY
REIZ 1130	11 22 56.	+ 23 06	24	14.2	GALAXY
HOLM 254A	11 22 56.	+ 23 06	36	13.4	PART OF MULTIPLE GALAXY
MCG-06-25-018	11 22 57.	- 35 16	30	15.5	GALAXY
RNGC 3671	11 22 58.	+ 60 45		15.5	GALAXY
REIZ 1131	11 22 59.	+ 32 37	54	14.0	GALAXY
LB 02013	11 22 59.	+ 49 41 42.		16.2	FAINT BLUE STAR
HOLM 253B	11 23 00.	+ 32 35	48	14.7	PART OF MULTIPLE GALAXY
MCG+09-19-074	11 23 00.	+ 50 56	42	16.	GALAXY
ZC 1123.0+5938	11 23 00.	+ 59 38	870		CLUSTER OF GALAXIES
ZWG 291.068	11 23 00.	+ 60 45		15.7	GALAXY
ZWG 351.061	11 23 00.	+ 74 28		15.6	GALAXY
ZWG 334.031	11 23 00.	+ 74 28		15.6	GALAXY
ZWG 011.093	11 23 00.	- 00 30		14.2	GALAXY
UGC 06435	11 23 00.	- 00 30	66	14.2	GALAXY S0
MCG+00-29-031	11 23 00.	- 00 30	18	14.2	GALAXY
MCG-02-29-029	11 23 00.	- 10 52	30	15.	GALAXY
HOLM 254B	11 23 01.	+ 23 07	18	15.1	PART OF MULTIPLE GALAXY
REIZ 1132	11 23 02.	+ 23 07	6	15.8	GALAXY
HELW 439	11 23 02.	- 00 29 52.			NEBULA
RNGC 3676	11 23 02.	- 10 21			NON-EXISTENT OBJECT
IC 2809	11 23 03.	+ 08 48 10.			NONSTELLAR OBJECT
MCG+03-29-043	11 23 03.	+ 14 56 30.	72	15.4	GALAXY
REIZ 1133	11 23 03.	+ 24 25	36	14.4	GALAXY
REIZ 1129	11 23 03.	+ 56 36	15	14.8	GALAXY
LB 02014	11 23 05.	+ 59 26 16.		17.0	FAINT BLUE STAR
REIZ 1134	11 23 06.	+ 10 14	24	14.3	GALAXY
MCG+03-29-044	11 23 06.	+ 14 56	24	15.4	GALAXY
ZWG 096.041	11 23 06.	+ 14 57		15.4	GALAXY
UGC 06436	11 23 06.	+ 14 57	72	15.4	GALAXY SB
ZWG 096.042	11 23 06.	+ 19 12		15.3	GALAXY
UGC 06437	11 23 06.	+ 19 12	78	15.3	GALAXY Sb-c
MCG+03-29-045	11 23 06.	+ 19 13	66	15.3	GALAXY
ZWG 126.051	11 23 06.	+ 24 24		15.2	GALAXY
ZC 1123.1+3243	11 23 06.	+ 32 43	1140		CLUSTER OF GALAXIES
ZC 1123.1+4142	11 23 06.	+ 41 42	870		CLUSTER OF GALAXIES
ZC 1123.1+5516	11 23 06.	+ 55 16	2080		CLUSTER OF GALAXIES
MCG-01-29-020	11 23 06.	- 04 42	36	15.	GALAXY
LB 02016	11 23 08.	+ 52 03 54.		15.2	FAINT BLUE STAR
LB 02015	11 23 08.	+ 55 08 30.		15.9	FAINT BLUE STAR
MCG+04-27-035	11 23 09.	+ 24 24	48	15.2	GALAXY
MCG+06-25-068	11 23 09.	+ 33 55	30	15.	GALAXY
IC 2811	11 23 10.	+ 09 26 40.			NONSTELLAR OBJECT
ZWG 039.179	11 23 12.	+ 03 41		15.3	GALAXY
REIZ 1136	11 23 12.	+ 13 56	42	15.3	GALAXY
IC 2810	11 23 12.	+ 14 57	41		FAINT NEBULA
FATH 2.032	11 23 12.	+ 14 57	41		NEBULA
ZWG 096.043	11 23 12.	+ 18 52		15.6	GALAXY
MCG+03-29-046	11 23 12.	+ 18 53	66	15.6	GALAXY
LB 10190	11 23 12.	+ 27 23		16.9	FAINT BLUE STAR
LB 10189	11 23 12.	+ 28 26		16.6	FAINT BLUE STAR
SHAH 062	11 23 12.	+ 38 40	210	17.2	GROUP OF COMPACT GALAXIES
ZC 1123.2+6913	11 23 12.	+ 69 13	1610		CLUSTER OF GALAXIES
REIZ 1137	11 23 16.	+ 03 43	30	15.7	GALAXY
FATH 1.401	11 23 16.	+ 14 56	22		NEBULA
FEIG 041	11 23 18.	+ 06 47		11.0	FAINT BLUE STAR
IC 0692	11 23 18.	+ 10 14			NONSTELLAR OBJECT
ZWG 067.073	11 23 18.	+ 10 16		14.1	GALAXY
UGC 06438	11 23 18.	+ 10 16	48	14.1	GALAXY
MCG+02-29-027	11 23 18.	+ 10 16	36	14.1	GALAXY
LB 10191	11 23 18.	+ 30 56		17.0	FAINT BLUE STAR
MCG+09-19-075	11 23 18.	+ 52 24	30	14.	GALAXY
ZC 1123.3+5757	11 23 18.	+ 57 57	1410		CLUSTER OF GALAXIES
OCL 0823	11 23 18.	- 42 58	840	8.6	OPEN STAR CLUSTER
RNGC 3680	11 23 18.	- 42 58		8.5	OPEN CLUSTER
IC 2812	11 23 20.	+ 11 48 16.			NONSTELLAR OBJECT
LB 00257	11 23 20.	+ 60 35 36.		14.3	FAINT BLUE STAR
REIZ 1135	11 23 21.	+ 56 37	12	14.8	GALAXY
A1 024	11 23 22.38	+ 13 55 09.2		16.25	FAINT BLUE OBJECT
FATH 1.402	11 23 24.	+ 14 16	14		NEBULA
MCG+03-29-047	11 23 24.	+ 15 50	72	17.5	GALAXY
ZC 1123.4+1707	11 23 24.	+ 17 07	610		CLUSTER OF GALAXIES
ZC 1123.4+2530	11 23 24.	+ 25 30	3160		CLUSTER OF GALAXIES
MCG+05-27-070	11 23 24.	+ 27 26	30	15.5	GALAXY
LB 10192	11 23 24.	+ 29 24		18.1	FAINT BLUE STAR
ZC 1123.4+3630	11 23 24.	+ 36 30	810		CLUSTER OF GALAXIES
ZWG 214.005	11 23 24.	+ 43 52		10.4	GALAXY
UGC 06439	11 23 24.	+ 43 52	408	10.4	GALAXY Sb
ZWG 268.034	11 23 24.	+ 52 25		15.0	GALAXY
MCG-06-25-019	11 23 24.	- 35 07	60	13.5	GALAXY
REIN 2.129	11 23 24.59	+ 43 51 38.7			NEBULA
REIZ 1139	11 23 26.	+ 22 07	18	14.7	GALAXY
ARC 1258	11 23 26.	+ 25 43		17.2	RICH CLUSTER OF GALAXIES
ARC 1257	11 23 26.	+ 35 37		15.0	RICH CLUSTER OF GALAXIES
RNGC 3675	11 23 26.	+ 43 52		11.5	GALAXY
REIZ 1138	11 23 26.	+ 43 52	240	12.2	GALAXY
REIZ 1140	11 23 27.	+ 27 29	24	15.2	GALAXY
REIZ 1147	11 23 27.	- 05 34	42	13.6	GALAXY
REIZ 1144	11 23 28.	+ 02 16	42	13.7	GALAXY
REIZ 1141	11 23 28.	+ 31 30	30	14.9	GALAXY
A1 025	11 23 29.80	+ 13 04 56.0		16.35	FAINT BLUE OBJECT
ZWG 011.094	11 23 30.	+ 02 15		14.7	GALAXY
UGC 06440	11 23 30.	+ 02 15	66	14.7	GALAXY Sb
MCG+00-29-032	11 23 30.	+ 02 15	66	14.7	GALAXY
ZC 1123.5+0221	11 23 30.	+ 02 21	1140		CLUSTER OF GALAXIES
ZWG 039.180	11 23 30.	+ 03 46		15.2	GALAXY
ZWG 096.044	11 23 30.	+ 16 40		15.7	GALAXY
ZC 1123.5+2256	11 23 30.	+ 22 56	10010		CLUSTER OF GALAXIES
TON-N 0066	11 23 30.	+ 24 44		15.6	BLUE STAR
ZWG 156.074	11 23 30.	+ 27 02		15.5	GALAXY
ZC 1123.5+4224	11 23 30.	+ 42 24	1010		CLUSTER OF GALAXIES
MCG+07-24-004	11 23 30.	+ 43 52	360	10.	GALAXY
ZWG 214.006	11 23 30.	+ 44 26		15.5	GALAXY
MCG+08-21-035	11 23 30.	+ 47 14	90	13.7	GALAXY
ZWG 242.035	11 23 30.	+ 47 15		13.5	GALAXY
UGC 06441	11 23 30.	+ 47 15	102	13.5	GALAXY Sa
KARA.72 284A	11 23 30.	+ 47 15	102	13.5	PART OF DOUBLE GALAXY
7ZW 401	11 23 30.	+ 62 15			COMPACT GALAXY
SN 1955G	11 23 31.	+ 03 46		19.4	SUPERNOVA
IC 2813	11 23 31.	+ 11 31 52.			NONSTELLAR OBJECT
A1 026	11 23 31.41	+ 13 30 26.2		17.10	FAINT BLUE OBJECT
IC 2814	11 23 33.	+ 09 56 22.			NONSTELLAR OBJECT
REIZ 1145	11 23 33.	+ 27 02	18	14.8	GALAXY
REIZ 1148	11 23 33.	+ 27 28	18	15.0	GALAXY
REIZ 1149	11 23 34.	+ 03 47	18	15.5	GALAXY
RNGC 3678	11 23 34.	+ 28 09		14.0	GALAXY
REIZ 1146	11 23 34.	+ 28 09	36	13.4	GALAXY
REIZ 1143	11 23 34.	+ 47 16	60	13.7	GALAXY
RNGC 3674	11 23 35.	+ 57 19		13.0	GALAXY
ZC 1123.6+0118	11 23 36.	+ 01 18	540		CLUSTER OF GALAXIES
ZC 1123.6+0529	11 23 36.	+ 05 29	1080		CLUSTER OF GALAXIES
ARC 1259	11 23 36.	+ 05 33		18.0	RICH CLUSTER OF GALAXIES
ZWG 039.181	11 23 36.	+ 08 07		15.5	GALAXY
UGC 06442	11 23 36.	+ 08 07	90	15.5	GALAXY Sc
MCG+05-27-071	11 23 36.	+ 28 07	48	14.2	GALAXY
ZWG 156.075	11 23 36.	+ 28 09		14.2	GALAXY
UGC 06443	11 23 36.	+ 28 09	48	14.2	GALAXY Sb-c
LB 10194	11 23 36.	+ 29 01		17.6	FAINT BLUE STAR
LB 10193	11 23 36.	+ 31 18		16.8	FAINT BLUE STAR
ZWG 185.063	11 23 36.	+ 35 37		15.4	GALAXY
ZC 1123.6+3732	11 23 36.	+ 37 32	2020		CLUSTER OF GALAXIES
MCG+07-24-005	11 23 36.	+ 44 26	48	15.	GALAXY
ZWG 268.035	11 23 36.	+ 55 03		15.5	GALAXY
MCG+10-16-138	11 23 36.	+ 57 19	84	13.5	GALAXY
ZWG 291.069	11 23 36.	+ 57 20		13.1	GALAXY
UGC 06444	11 23 36.	+ 57 20	90	13.1	GALAXY S0
ZWG 011.095	11 23 36.	- 01 34		15.2	GALAXY
MCG-01-29-021	11 23 36.	- 05 18	42	15.	GALAXY
RNGC 3685	11 23 37.	+ 04 24			NON-EXISTENT OBJECT
RNGC 3677	11 23 38.	+ 47 15		13.5	GALAXY
LB 02017	11 23 38.	+ 61 43 00.		15.2	FAINT BLUE STAR
RNGC 3679	11 23 38.	- 05 18		15.0	GALAXY
REIZ 1142	11 23 39.	+ 57 20	72	13.5	GALAXY
IC 2815	11 23 40.	+ 13 04 46.			NONSTELLAR OBJECT
LB 02018	11 23 41.	+ 52 29 18.		16.0	FAINT BLUE STAR
SEY 097	11 23 41.	+ 55 03 56.		15.3	FAINT GALAXY

OBJECT NAME	RIGHT ASCEN.	DECLINATION	DIAM.	MAGN.	TYPE OF OBJECT
MCG+04-27-036	11 23 42.	+ 21 20	36	15.5	GALAXY
LB 10195	11 23 42.	+ 27 36		17.2	FAINT BLUE STAR
ZC 1123.7+3205	11 23 42.	+ 32 05	1080		CLUSTER OF GALAXIES
ZC 1123.7+3315	11 23 42.	+ 33 15	1810		CLUSTER OF GALAXIES
ZWG 185.064	11 23 42.	+ 35 36		15.6	GALAXY
MCG+06-25-069	11 23 42.	+ 35 36	15	16.	GALAXY
IC 0691	11 23 42.	+ 59 25 58.			NONSTELLAR OBJECT
IC 2816	11 23 43.	+ 10 54 51.			NONSTELLAR OBJECT
IC 2817	11 23 44.	+ 09 25 33.			NONSTELLAR OBJECT
TON-N 1399	11 23 44.	+ 34 54		17.	BLUE STAR
MCG+09-19-076	11 23 45.	+ 52 27	30	17.	GALAXY
MCG-02-29-030	11 23 45.	- 13 59 30.	15	15.	GALAXY
ARC 1260	11 23 47.	+ 02 21		17.5	RICH CLUSTER OF GALAXIES
ZWG 039.182	11 23 48.	+ 03 12		15.4	GALAXY
KARA.73B 0478	11 23 48.	+ 03 12	42	15.4	ISOLATED GALAXY S
ZWG 096.045	11 23 48.	+ 17 08		12.2	GALAXY
UGC 06445	11 23 48.	+ 17 08	162	12.2	GALAXY Sb/Sc
LB 10198	11 23 48.	+ 26 24		16.4	FAINT BLUE STAR
LB 10197	11 23 48.	+ 28 12		16.8	FAINT BLUE STAR
LB 10196	11 23 48.	+ 32 01		17.4	FAINT BLUE STAR
ZC 1123.8+3224	11 23 48.	+ 32 24	610		CLUSTER OF GALAXIES
MCG+06-25-070	11 23 48.	+ 35 35 30.	30	15.5	GALAXY
REIZ 1150	11 23 48.	+ 35 38	24	14.6	GALAXY
ZWG 242.036	11 23 48.	+ 47 17		15.5	GALAXY
KARA.72 284B	11 23 48.	+ 47 17	42	15.5	PART OF DOUBLE GALAXY
MCG+09-19-078	11 23 48.	+ 51 50	24	17.	GALAXY
ZWG 268.036	11 23 48.	+ 54 01		14.5	GALAXY
UGC 06446	11 23 48.	+ 54 01	246	14.5	GALAXY Sc
MCG+09-19-077	11 23 48.	+ 55 38	54	16.	GALAXY
SEY 098	11 23 50.	+ 54 02 01.		14.1	FAINT GALAXY
IC 2820	11 23 51.	+ 10 30 51.			NONSTELLAR OBJECT
IC 2818	11 23 51.	+ 13 11 51.			NONSTELLAR OBJECT
MCG+03-29-048	11 23 51.	+ 17 08	180	12.2	GALAXY
ARC 1254	11 23 51.	+ 71 22		15.3	RICH CLUSTER OF GALAXIES
IC 2819	11 23 52.	+ 14 07 21.			NONSTELLAR OBJECT
ZC 1123.9+1120	11 23 54.	+ 11 20	740		CLUSTER OF GALAXIES
ZWG 067.074	11 23 54.	+ 14 08		15.7	GALAXY
RNGC 3681	11 23 54.	+ 17 08		13.0	GALAXY
ZC 1123.9+3011	11 23 54.	+ 30 11	670		CLUSTER OF GALAXIES
ZC 1123.9+3541	11 23 54.	+ 35 41	7390		CLUSTER OF GALAXIES
ZWG 185.065	11 23 54.	+ 35 42		15.3	GALAXY
MCG+09-19-079	11 23 54.	+ 54 01	180	13.	GALAXY
ZWG 291.070	11 23 54.	+ 59 26		14.2	GALAXY
UGC 06447	11 23 54.	+ 59 26	48	14.2	GALAXY COMPACT
ZWG 314.027	11 23 54.	+ 64 25		14.9	GALAXY
UGC 06448	11 23 54.	+ 64 25	120	14.9	GALAXY PECULR
ZC 1123.9+7631	11 23 54.	+ 76 31	400		CLUSTER OF GALAXIES
REIZ 1151	11 23 55.	+ 17 10	42	13.4	GALAXY
IC 2822	11 23 59.	+ 11 42 51.			NONSTELLAR OBJECT
IC 2821	11 23 59.	+ 14 14 21.			NONSTELLAR OBJECT
ZWG 067.075	11 24 00.	+ 11 43		15.2	GALAXY
UGC 06449	11 24 00.	+ 11 43	102	15.2	GALAXY SBb-c
LB 10199	11 24 00.	+ 27 56		17.3	FAINT BLUE STAR
MCG+06-25-071	11 24 00.	+ 35 41	45	15.	GALAXY
ZWG 214.007	11 24 00.	+ 39 32		15.0	GALAXY
MRK 169	11 24 00.	+ 59 25	25	14.5	GALAXY WITH UV CONTINUUM
VVI 46	11 24 00.	+ 59 25	27	14.91	SEYFERT GALAXY
MCG+10-16-139	11 24 00.	+ 59 25	27	15.	GALAXY
ZWG 314.028	11 24 00.	+ 63 42		14.7	GALAXY
MCG+11-14-025A	11 24 00.	+ 63 42 30.	24	14.	GALAXY
MRK 170	11 24 00.	+ 64 25	30	16.	GALAXY WITH UV CONTINUUM
MCG+11-14-025	11 24 00.	+ 64 25	57	15.	GALAXY
UGC 06450	11 24 00.	+ 76 54	66	16.5	GALAXY
ZC 1124.0-0003	11 24 00.	- 00 03	740		CLUSTER OF GALAXIES
REIZ 1152	11 24 05.	+ 35 38	18	14.8	GALAXY
REIZ 1153	11 24 05.	+ 35 43	24	14.7	GALAXY
ZWG 067.076	11 24 06.	+ 13 07		15.5	GALAXY
ZWG 185.066	11 24 06.	+ 35 31		14.9	GALAXY
MRK 423	11 24 06.	+ 35 31	20	15.5	GALAXY WITH UV CONTINUUM
MCG+07-24-006	11 24 06.	+ 39 32 30.	48	14.5	GALAXY
MCG+10-16-140	11 24 06.	+ 59 05	39	16.	GALAXY
ZWG 011.096	11 24 06.	- 01 25		15.1	GALAXY
A1 027	11 24 06.34	+ 16 02 24.3		15.70	FAINT BLUE OBJECT
IC 2823	11 24 09.	+ 13 07 27.			NONSTELLAR OBJECT
MCG+06-25-072	11 24 09.	+ 35 30 30.	42	15.	GALAXY
ZWG 126.052	11 24 12.	+ 21 25		15.7	GALAXY
ZWG 126.053	11 24 12.	+ 22 53		15.6	GALAXY
ZWG 185.067	11 24 12.	+ 33 23		15.6	GALAXY
KARA.73B 0479	11 24 12.	+ 33 23	42	15.6	ISOLATED GALAXY S
ZWG 268.037	11 24 12.	+ 50 53		15.6	GALAXY
ZWG 334.032	11 24 12.	+ 71 55		15.7	GALAXY
UGC 06451	11 24 12.	+ 79 54	84	16.5	GALAXY
A1 028	11 24 13.86	+ 13 27 55.4		16.30	FAINT BLUE OBJECT
REIZ 1154	11 24 15.	+ 26 28	24	14.7	GALAXY
MCG+06-25-073	11 24 15.	+ 33 22	42	14.5	GALAXY
IC 0693	11 24 16.	- 04 43 33.			NONSTELLAR OBJECT
A1 029	11 24 16.99	+ 15 30 44.7		16.40	FAINT BLUE OBJECT
REIZ 1155	11 24 17.	+ 33 24	36	14.5	GALAXY
ZWG 096.046	11 24 18.	+ 19 06		15.2	GALAXY
MCG+03-29-049	11 24 18.	+ 19 06	39	15.2	GALAXY
KARA.73B 0480	11 24 18.	+ 19 06	36	15.2	ISOLATED GALAXY S0
LB 10200	11 24 18.	+ 27 10		18.0	FAINT BLUE STAR
ZC 1124.3+3447	11 24 18.	+ 34 47	610		CLUSTER OF GALAXIES
MCG-01-29-022	11 24 18.	- 04 43	36	15.	GALAXY
ACK 283+25.1	11 24 18.	- 34 06			PLANETARY NEBULA
ARC 1263	11 24 19.	- 09 31		17.7	RICH CLUSTER OF GALAXIES
A1 030	11 24 20.28	+ 13 59 59.8		17.50	FAINT BLUE OBJECT
ZC 1124.4+1155	11 24 24.	+ 11 55	1010		CLUSTER OF GALAXIES
LB 10201	11 24 24.	+ 28 14		16.0	FAINT BLUE STAR
ZC 1124.4+3546	11 24 24.	+ 35 46	1140		CLUSTER OF GALAXIES
MCG+10-16-141	11 24 24.	+ 59 25	18	16.	GALAXY
MCG+10-16-142	11 24 24.	+ 59 54	60	14.	GALAXY
ARC 1255	11 24 24.	+ 75 46		16.7	RICH CLUSTER OF GALAXIES
ZWG 011.097	11 24 24.	- 01 03		15.3	GALAXY
MCG+00-29-033	11 24 24.	- 01 03	36	15.3	GALAXY
ARC 1262	11 24 26.	+ 10 55		17.3	RICH CLUSTER OF GALAXIES
REIZ 1157	11 24 27.	- 06 08	54	14.7	GALAXY
IC 2825	11 24 28.	+ 08 43 44.			NONSTELLAR OBJECT
LB 02019	11 24 28.	+ 55 33 24.		16.4	FAINT BLUE STAR
IC 2824	11 24 29.	+ 14 21 44.			NONSTELLAR OBJECT
LB 00258	11 24 29.	+ 62 17 30.		14.0	FAINT BLUE STAR
ZWG 039.183	11 24 30.	+ 07 58		15.6	GALAXY
IC 2826	11 24 30.	+ 13 30 56.			NONSTELLAR OBJECT
ZWG 067.077	11 24 30.	+ 13 31		15.1	GALAXY
MCG+03-29-050	11 24 30.	+ 17 18	240	12.1	GALAXY
ZC 1124.5+1728	11 24 30.	+ 17 28	1340		CLUSTER OF GALAXIES
ZC 1124.5+2401	11 24 30.	+ 24 01	1140		CLUSTER OF GALAXIES
MCG+08-21-036	11 24 30.	+ 50 32 30.	42	16.	GALAXY

OBJECT NAME	RIGHT ASCEN.	DECLINATION	DIAM.	MAGN.	TYPE OF OBJECT
ZC 1124.5+5200	11 24 30.	+ 52 00	1010		CLUSTER OF GALAXIES
ZWG 291.071	11 24 30.	+ 59 55		14.0	GALAXY
UGC 06452	11 24 30.	+ 59 55	66	14.0	GALAXY S
A1 031	11 24 31.20	+ 12 51 07.8		17.00	FAINT BLUE OBJECT
REIZ 1156	11 24 34.	+ 59 55	60	13.8	GALAXY
IC 2827	11 24 35.	+ 11 47 32.			NONSTELLAR OBJECT
ARC 1264	11 24 35.	+ 17 26		17.1	RICH CLUSTER OF GALAXIES
ZWG 067.078	11 24 36.	+ 09 00		14.7	GALAXY
MCG+02-29-028	11 24 36.	+ 09 00	60	14.7	GALAXY
ZWG 067.079	11 24 36.	+ 10 35		15.4	GALAXY
ZC 1124.6+1055	11 24 36.	+ 10 55	2020		CLUSTER OF GALAXIES
ZWG 096.047	11 24 36.	+ 17 18		12.1	GALAXY
RNGC 3684	11 24 36.	+ 17 18		12.5	GALAXY
UGC 06453	11 24 36.	+ 17 18	210	12.1	GALAXY Sc
ZC 1124.6+2246	11 24 36.	+ 22 46	1280		CLUSTER OF GALAXIES
ZC 1124.6+3826	11 24 36.	+ 38 26	610		CLUSTER OF GALAXIES
UGC 06454	11 24 36.	+ 38 56	102	16.0	GALAXY Sc
ZWG 214.008	11 24 36.	+ 40 16		15.6	GALAXY
UGC 06455	11 24 36.	+ 40 16	114	15.6	GALAXY S
VV 265B	11 24 36.	+ 50 32 30.	6	17.	INTERACTING GALAXY
VV 265A	11 24 36.	+ 50 32 30.	6	17.	INTERACTING GALAXY
VV 265	11 24 36.	+ 50 32 30.	54		INTERACTING GALAXY
7ZW 402	11 24 36.	+ 61 44			COMPACT GALAXY
7ZW 403	11 24 36.	+ 79 16			COMPACT GALAXY
ZWG 352.001	11 24 36.	+ 79 16		14.7	GALAXY
ZWG 351.062	11 24 36.	+ 79 16		14.7	GALAXY
UGC 06456	11 24 36.	+ 79 16	96	14.7	GALAXY PECULR
MCG-05-27-012	11 24 36.	- 28 41	60	13.5	GALAXY
IC 2828	11 24 37.	+ 09 00 32.			NONSTELLAR OBJECT
REIZ 1159	11 24 37.	+ 17 19	66	13.4	GALAXY
IC 2829	11 24 39.	+ 10 35 50.			NONSTELLAR OBJECT
REIZ 1160	11 24 39.	+ 27 03	42	14.4	GALAXY
MCG+10-16-143	11 24 39.	+ 57 08	96	13.2	GALAXY
REIZ 1162	11 24 40.	+ 04 02	24	15.5	GALAXY
REIZ 1161	11 24 40.	- 00 43	24	14.7	GALAXY
RNGC 3683	11 24 41.	+ 57 09		13.5	GALAXY
A1 032	11 24 41.67	+ 14 10 45.4		15.80	FAINT BLUE OBJECT
ZWG 011.099	11 24 42.	+ 01 48		15.7	GALAXY
MCG+00-29-035	11 24 42.	+ 01 48	24	15.7	GALAXY
ZWG 039.184	11 24 42.	+ 04 02		15.2	GALAXY
ZC 1124.7+2010	11 24 42.	+ 20 10	1680		CLUSTER OF GALAXIES
MCG+07-24-007	11 24 42.	+ 38 56	90	15.	GALAXY
MCG+07-24-008	11 24 42.	+ 40 17	90	14.5	GALAXY
MCG+09-19-080	11 24 42.	+ 52 10	24	16.	GALAXY
ARP.65 2	11 24 42.	+ 54 11			COMPACT DWARF GALAXY
MCG+09-19-081	11 24 42.	+ 54 12	36	16.	GALAXY
ZWG 291.072	11 24 42.	+ 57 10		12.7	GALAXY
UGC 06458	11 24 42.	+ 57 10	126	12.7	GALAXY
MCG+11-14-026	11 24 42.	+ 63 35	39	16.	GALAXY
ZWG 314.029	11 24 42.	+ 66 52		13.4	GALAXY
UGC 06459	11 24 42.	+ 66 52	138	13.4	GALAXY S0-a
KARA.73B 0481	11 24 42.	+ 66 52	102	13.4	ISOLATED GALAXY S
ZC 1124.7+7557	11 24 42.	+ 75 57	1880		CLUSTER OF GALAXIES
MCG+13-08-058	11 24 42.	+ 79 17	33	14.	GALAXY
ZWG 011.098	11 24 42.	- 00 43		15.0	GALAXY
UGC 06457	11 24 42.	- 00 43	78	15.0	GALAXY IRR
MCG+00-29-034	11 24 42.	- 00 43	48	15.0	GALAXY
SEY 099	11 24 44.	+ 55 09 31.		15.4	FAINT GALAXY
REIZ 1158	11 24 44.	+ 57 10	84	13.0	GALAXY
MCG-02-29-031	11 24 45.	- 10 39 30.	48	15.	GALAXY
IC 2830	11 24 46.	+ 08 05 32.			NONSTELLAR OBJECT
ARC 1261	11 24 46.	+ 48 37		17.2	RICH CLUSTER OF GALAXIES
ZWG 039.185	11 24 48.	+ 08 05		15.2	GALAXY
IC 2831	11 24 48.	+ 09 15 14.			NONSTELLAR OBJECT
ZWG 096.048	11 24 48.	+ 20 07		15.4	GALAXY
LB 10203	11 24 48.	+ 27 12		15.8	FAINT BLUE STAR
LB 10202	11 24 48.	+ 27 42		16.5	FAINT BLUE STAR
ZWG 185.068	11 24 48.	+ 36 19		15.0	GALAXY
KARA.72 285B	11 24 48.	+ 36 19	54	15.0	PART OF DOUBLE GALAXY
ZWG 185.069	11 24 48.	+ 36 20		15.7	GALAXY
KARA.72 285A	11 24 48.	+ 36 20	42	15.7	PART OF DOUBLE GALAXY
REIZ 1164	11 24 48.	+ 36 22	12	14.1	GALAXY
REIZ 1163	11 24 48.	+ 36 22	54	13.9	GALAXY
ZC 1124.8+4325	11 24 48.	+ 43 25	740		CLUSTER OF GALAXIES
MCG+08-21-037	11 24 48.	+ 47 38	36	15.	GALAXY
MCG+10-16-144	11 24 48.	+ 62 04	39	16.	GALAXY
MCG+11-14-027	11 24 48.	+ 66 52	78	12.1	GALAXY
ZC 1124.8-0025	11 24 48.	- 00 25	670		CLUSTER OF GALAXIES
A1 033	11 24 49.82	+ 13 46 01.1		16.05	FAINT BLUE OBJECT
A1 034	11 24 49.88	+ 14 43 42.8		17.20	FAINT BLUE OBJECT
IC 2833	11 24 50.	+ 13 52 44.			NONSTELLAR OBJECT
IC 2832	11 24 50.	+ 14 15 56.			NONSTELLAR OBJECT
RNGC 3682	11 24 50.	+ 66 52		13.5	GALAXY
HOLM 255B	11 24 52.	+ 36 22	24	14.3	PART OF MULTIPLE GALAXY
HOLM 255A	11 24 53.	+ 36 21	60	14.1	PART OF MULTIPLE GALAXY
A1 035	11 24 53.25	+ 16 08 11.9		16.10	FAINT BLUE OBJECT
ZC 1124.9+0026	11 24 54.	+ 00 26	1550		CLUSTER OF GALAXIES
ZWG 156.076	11 24 54.	+ 28 44		15.7	GALAXY
MCG+06-25-075	11 24 54.	+ 36 20	30	16.	GALAXY
MCG+06-25-074	11 24 54.	+ 36 20	48	15.	GALAXY
MCG+07-24-009	11 24 54.	+ 42 49	48	16.5	GALAXY
ZWG 242.037	11 24 54.	+ 46 20		15.6	GALAXY
ZWG 242.038	11 24 54.	+ 47 39		15.3	GALAXY
MCG-05-27-013	11 24 54.	- 28 58	60	15.	GALAXY
IC 2834	11 24 55.	+ 13 50 44.			NONSTELLAR OBJECT
IC 2835	11 24 56.	+ 12 25 14.			NONSTELLAR OBJECT
A1 036	11 24 56.25	+ 15 26 07.0		16.10	FAINT BLUE OBJECT
A1 037	11 24 56.51	+ 13 19 01.8		17.40	FAINT BLUE OBJECT
ZWG 011.100	11 25 00.	+ 00 40		15.3	GALAXY
ZC 1125.0+2150	11 25 00.	+ 21 50	1140		CLUSTER OF GALAXIES
ZWG 126.054	11 25 00.	+ 24 02		15.7	GALAXY
MCG+09-19-082	11 25 00.	+ 56 12 30.	30	16.	GALAXY
MCG+13-08-059	11 25 00.	+ 79 21	21	16.	GALAXY
IC 2836	11 25 02.	+ 09 21 38.			NONSTELLAR OBJECT
A1 038	11 25 02.32	+ 12 56 22.5		16.85	FAINT BLUE OBJECT
PK291+03.1	11 25 04.	- 57 01 14.	10		PLANETARY NEBULA
A1 039	11 25 05.61	+ 16 02 15.5		17.40	FAINT BLUE OBJECT
ZC 1125.1+1619	11 25 06.	+ 16 19	540		CLUSTER OF GALAXIES
MCG+03-29-051	11 25 06.	+ 17 29	192	11.6	GALAXY
ZWG 096.049	11 25 06.	+ 17 30		11.6	GALAXY
RNGC 3686	11 25 06.	+ 17 30		12.0	GALAXY
UGC 06460	11 25 06.	+ 17 30	186	11.6	GALAXY SBb/SBc
UGC 06461	11 25 06.	+ 21 39	72	16.0	GALAXY Sa-b
ZC 1125.1+2413	11 25 06.	+ 24 13	2620		CLUSTER OF GALAXIES
ZWG 156.077	11 25 06.	+ 27 15		15.7	GALAXY
MCG+08-21-038	11 25 06.	+ 47 39 30.	42	15.	GALAXY
ZC 1125.1+6417	11 25 06.	+ 64 17	1810		CLUSTER OF GALAXIES
ZWG 334.033	11 25 06.	+ 73 13		15.3	GALAXY

OBJECT NAME	RIGHT ASCEN.	DECLINATION	DIAM.	MAGN.	TYPE OF OBJECT
MCG+13-08-060	11 25 06.	+ 79 54	39	15.	GALAXY
IC 2837	11 25 07.	+ 10 35 26.			NONSTELLAR OBJECT
REIZ 1165	11 25 09.	+ 27 06	24	15.1	GALAXY
REIZ 1166	11 25 10.	+ 03 12	120	14.5	GALAXY
IC 2839	11 25 10.	+ 11 05 44.			NONSTELLAR OBJECT
IC 2838	11 25 10.	+ 14 17 20.			NONSTELLAR OBJECT
IC 2840	11 25 11.	+ 13 42 02.			NONSTELLAR OBJECT
ZWG 039.186	11 25 12.	+ 06 25		15.2	GALAXY
MCG+01-29-042	11 25 12.	+ 06 25	36	15.2	GALAXY
KARA.73B 0482	11 25 12.	+ 06 25	42	15.2	ISOLATED GALAXY S
ZWG 039.187	11 25 12.	+ 08 16		15.1	GALAXY
UGC 06462	11 25 12.	+ 08 16	102	15.1	GALAXY S
LB 10205	11 25 12.	+ 27 49		16.4	FAINT BLUE STAR
LB 10204	11 25 12.	+ 32 20		17.5	FAINT BLUE STAR
ZWG 242.039	11 25 12.	+ 47 40		14.9	GALAXY
MCG+13-08-061	11 25 12.	+ 79 54	33	15.	GALAXY
IC 2842	11 25 13.	+ 09 55 44.			NONSTELLAR OBJECT
IC 2841	11 25 13.	+ 12 52 32.			NONSTELLAR OBJECT
REIZ 1167	11 25 13.	+ 17 31	114	13.2	GALAXY
RNGC 3688	11 25 14.	- 08 54		15.0	GALAXY
MCG+05-27-072	11 25 15.	+ 27 02	24	16.	GALAXY
REIZ 1168	11 25 15.	+ 27 04	24	15.1	GALAXY
MCG-01-29-023	11 25 15.	- 04 38 30.	78	15.5	GALAXY
MCG-01-29-024	11 25 15.	- 08 54	36	15.	GALAXY
RNGC 3687	11 25 16.	+ 29 47		13.0	GALAXY
ZC 1125.3+1403	11 25 18.	+ 14 03	1010		CLUSTER OF GALAXIES
ZC 1125.3+1830	11 25 18.	+ 18 30	3970		CLUSTER OF GALAXIES
MCG+05-27-074	11 25 18.	+ 27 38	48	15.1	GALAXY
ZWG 156.078	11 25 18.	+ 29 47		13.0	GALAXY
UGC 06463	11 25 18.	+ 29 47	108	13.0	GALAXY Sb
MCG+05-27-073	11 25 18.	+ 29 47	120	13.0	GALAXY
ZC 1125.3+3652	11 25 18.	+ 36 52	270		CLUSTER OF GALAXIES
ZWG 011.102	11 25 18.	- 00 11		15.6	GALAXY
ZWG 011.101	11 25 18.	- 00 57		15.3	GALAXY
KARA.72 286A	11 25 18.	- 00 57	36	15.3	PART OF DOUBLE GALAXY
REIZ 1169	11 25 21.	+ 27 05	18	14.9	GALAXY
ABC 1267	11 25 21.	+ 27 09		15.4	RICH CLUSTER OF GALAXIES
REIZ 1170	11 25 21.	+ 27 40	30	14.9	GALAXY
IC 2843	11 25 22.	+ 13 27 43.			NONSTELLAR OBJECT
REIZ 1171	11 25 22.	+ 29 49	54	12.8	GALAXY
ABC 1265	11 25 22.	+ 41 38		17.8	RICH CLUSTER OF GALAXIES
REIZ 1172	11 25 22.	- 00 57	24	15.1	GALAXY
IC 2844	11 25 23.	+ 11 43 43.			NONSTELLAR OBJECT
LB 02020	11 25 23.	+ 59 36 06.		16.2	FAINT BLUE STAR
ZWG 039.188	11 25 24.	+ 08 15		15.6	GALAXY
ZWG 067.080	11 25 24.	+ 11 26		14.8	GALAXY
MCG+02-29-029	11 25 24.	+ 11 26	48	14.8	GALAXY
ZWG 067.081	11 25 24.	+ 13 28		15.4	GALAXY
ZWG 126.055	11 25 24.	+ 25 33		15.6	GALAXY
ZWG 156.079	11 25 24.	+ 27 40		15.1	GALAXY
LB 10207	11 25 24.	+ 28 29		18.3	FAINT BLUE STAR
LB 10206	11 25 24.	+ 31 30		17.8	FAINT BLUE STAR
ZWG 011.103	11 25 24.	- 00 56		15.4	GALAXY
KARA.72 286B	11 25 24.	- 00 56	36	15.4	PART OF DOUBLE GALAXY
IC 2846	11 25 25.	+ 11 26 07.			NONSTELLAR OBJECT
IC 2845	11 25 25.	+ 12 48 25.			NONSTELLAR OBJECT
ABC 1266	11 25 25.	+ 36 52		17.2	RICH CLUSTER OF GALAXIES
REIZ 1173	11 25 26.	+ 22 16	48	14.0	GALAXY
IC 0695	11 25 26.	- 11 26 29.			NONSTELLAR OBJECT
IC 2847	11 25 27.	+ 14 12 19.			NONSTELLAR OBJECT
RNGC 3689	11 25 29.	+ 25 56		13.0	GALAXY
ZC 1125.5+1700	11 25 30.	+ 17 00	2490		CLUSTER OF GALAXIES
MCG+03-29-053	11 25 30.	+ 17 11	48	13.1	GALAXY
ZWG 096.050	11 25 30.	+ 17 12		13.1	GALAXY
RNGC 3691	11 25 30.	+ 17 12		13.5	GALAXY
UGC 06464	11 25 30.	+ 17 12	72	13.1	GALAXY PECULR
MCG+03-29-052	11 25 30.	+ 17 39	60	16.	GALAXY
ZWG 126.056	11 25 30.	+ 22 16		15.0	GALAXY
UGC 06465	11 25 30.	+ 22 16	78	15.0	GALAXY SB
MCG+05-27-075	11 25 30.	+ 27 00	18	15.6	GALAXY
LB 10208	11 25 30.	+ 27 54		16.9	FAINT BLUE STAR
ZC 1125.5+2759	11 25 30.	+ 27 59	9210		CLUSTER OF GALAXIES
ZC 1125.5+4356	11 25 30.	+ 43 56	1880		CLUSTER OF GALAXIES
LB 02021	11 25 30.	+ 51 20 12.		15.4	FAINT BLUE STAR
REIZ 1175	11 25 31.	+ 17 12	60	13.6	GALAXY
LB 02022	11 25 31.	+ 52 17 18.		15.9	FAINT BLUE STAR
REIZ 1174	11 25 33.	+ 25 57	60	12.7	GALAXY
REIZ 1177	11 25 34.	- 00 57	54	14.8	GALAXY
RNGC 3699	11 25 34.	- 59 40			PLANETARY NEBULA
REIZ 1178	11 25 35.	+ 09 23	36	14.6	GALAXY
ZWG 039.189	11 25 36.	+ 04 36		15.5	GALAXY
UGC 06466	11 25 36.	+ 04 36	78	15.5	GALAXY
KARA.72 287A	11 25 36.	+ 04 36	66	15.5	PART OF DOUBLE GALAXY
MCG+01-29-043	11 25 36.	+ 04 36	48	15.5	GALAXY
ZWG 039.190	11 25 36.	+ 08 17		15.3	GALAXY
ZWG 067.082	11 25 36.	+ 09 20		14.8	GALAXY
MCG+02-29-030	11 25 36.	+ 09 20	36	14.8	GALAXY
MCG+04-27-038	11 25 36.	+ 22 15	54	15.0	GALAXY
ZWG 126.057	11 25 36.	+ 25 56		12.9	GALAXY
UGC 06467	11 25 36.	+ 25 56	96	12.9	GALAXY Sc
MCG+04-27-037	11 25 36.	+ 25 56	78	12.9	GALAXY
ZC 1125.6+2608	11 25 36.	+ 26 08	2550		CLUSTER OF GALAXIES
ZC 1125.6+4136	11 25 36.	+ 41 36	870		CLUSTER OF GALAXIES
ZC 1125.6+4556	11 25 36.	+ 45 56	1140		CLUSTER OF GALAXIES
ZC 1125.6+4749	11 25 36.	+ 47 49	740		CLUSTER OF GALAXIES
MCG+09-19-083	11 25 36.	+ 51 17	42	16.	GALAXY
72W 404	11 25 36.	+ 63 13			COMPACT GALAXY
UGC 06468	11 25 36.	+ 71 07	72	16.5	GALAXY Sc
MCG-02-29-032	11 25 36.	- 12 54	156	13.	GALAXY
MCG-06-25-020	11 25 36.	- 36 15	66	13.5	GALAXY
IC 2849	11 25 37.	+ 09 22 13.			NONSTELLAR OBJECT
IC 2848	11 25 37.	+ 13 18 25.			NONSTELLAR OBJECT
SEY 100	11 25 37.	+ 51 19 24.		15.3	FAINT GALAXY
PK292+01.1	11 25 37.4	- 59 40 46.	72	14.	PLANETARY NEBULA
REIN 7.053	11 25 37.70	+ 09 20 16.2			NEBULA
HOLM 257F	11 25 38.	+ 09 20	30	14.4	PART OF MULTIPLE GALAXY
IC 2850	11 25 38.	+ 09 20 19.			NONSTELLAR OBJECT
REIZ 1179	11 25 38.	+ 22 17	48	14.5	GALAXY
RNGC 3693	11 25 38.	- 12 54		13.0	GALAXY
A1 040	11 25 38.80	+ 17 30 38.6		15.80	FAINT BLUE OBJECT
HOLM 257C	11 25 39.	+ 09 25	66	14.0	PART OF MULTIPLE GALAXY
IC 2852	11 25 39.	+ 10 04 49.			NONSTELLAR OBJECT
REIZ 1180	11 25 39.	+ 27 01	18	14.6	GALAXY
MCG+05-27-076	11 25 39.	+ 27 44	48	15.1	GALAXY
REIN 7.054	11 25 39.59	+ 09 25 22.5			NEBULA
IC 2853	11 25 40.	+ 09 25 25.			NONSTELLAR OBJECT
IC 2851	11 25 40.	+ 11 40 13.			NONSTELLAR OBJECT
CED 114	11 25 40.	- 59 41			DIFFUSE GALACTIC NEBULA

OBJECT NAME	RIGHT ASCEN.	DECLINATION	DIAM.	MAGN.	TYPE OF OBJECT
REIZ 1182	11 25 41.	+ 09 38	108	12.8	GALAXY
RNGC 3690	11 25 41.	+ 58 50		12.5	GALAXY
SC 1123-5924.2	11 25 41.	- 59 40 43.	18		NEBULA
ZWG 011.104	11 25 42.	+ 01 09		15.2	GALAXY
MCG+00-29-036	11 25 42.	+ 01 09	30	15.2	GALAXY
ZWG 039.191	11 25 42.	+ 02 56		14.1	GALAXY
UGC 06469	11 25 42.	+ 02 56	54	14.1	GALAXY S
MCG+01-29-044	11 25 42.	+ 02 56	48	14.1	GALAXY
ZWG 039.192	11 25 42.	+ 04 37		15.3	GALAXY
KARA.72 287B	11 25 42.	+ 04 37	36	15.3	PART OF DOUBLE GALAXY
MCG+01-29-045	11 25 42.	+ 04 37	30	15.3	GALAXY
ZWG 067.083	11 25 42.	+ 09 25		14.6	GALAXY
UGC 06470	11 25 42.	+ 09 25	60	14.6	GALAXY SBa-b
MCG+02-29-031	11 25 42.	+ 09 25	60	14.6	GALAXY
MCG+03-29-054	11 25 42.	+ 17 30	51	15.6	GALAXY
ZWG 096.051	11 25 42.	+ 17 31		15.6	GALAXY
ZWG 156.080	11 25 42.	+ 27 02		15.6	GALAXY
FEIG 042	11 25 42.	+ 39 44		12.7	FAINT BLUE STAR
MCG+07-24-010	11 25 42.	+ 41 43	45	15.	GALAXY
KARA.72 288A	11 25 42.	+ 58 49	78	11.8	PART OF DOUBLE GALAXY
ZWG 291.073	11 25 42.	+ 58 50		11.8	GALAXY
UGC 06472	11 25 42.	+ 58 50	174	11.8	GALAXY DBL SYS
UGC 06471	11 25 42.	+ 58 50	174	11.8	GALAXY DBL SYS
ZWG 334.034	11 25 42.	+ 73 18		15.2	GALAXY
UGC 06473	11 25 42.	+ 73 18	96	15.2	GALAXY Sa-b
MCG-02-29-033	11 25 42.	- 12 36	42	14.5	GALAXY
SC 1123-5924.6	11 25 42.	- 59 41 07.	12		NEBULA
A1 041	11 25 44.77	+ 16 42 21.9		16.40	FAINT BLUE OBJECT
IC 2854	11 25 45.	+ 09 14 43.			NONSTELLAR OBJECT
REIZ 1183	11 25 45.	+ 26 45	18	15.2	GALAXY
REIZ 1176	11 25 45.	+ 58 51	30	13.4	GALAXY
HOLM 256A	11 25 45.	+ 58 51	24	13.1	PART OF MULTIPLE GALAXY
REIZ 1184	11 25 46.	+ 01 09	48	14.5	GALAXY
REIZ 1185	11 25 47.	+ 09 20	84	15.2	GALAXY
ZWG 067.084	11 25 48.	+ 09 41		12.9	GALAXY
RNGC 3692	11 25 48.	+ 09 41		13.0	GALAXY
UGC 06474	11 25 48.	+ 09 41	186	12.9	GALAXY Sb
MCG+02-29-032	11 25 48.	+ 09 41	192	12.9	GALAXY
ZC 1125.8+1511	11 25 48.	+ 15 11	2420		CLUSTER OF GALAXIES
ZWG 096.052	11 25 48.	+ 19 27		14.9	GALAXY
MCG+04-27-039	11 25 48.	+ 21 15	42	15.0	GALAXY
ZWG 126.058	11 25 48.	+ 21 17		15.0	GALAXY
ZC 1125.8+2608	11 25 48.	+ 26 08	810		CLUSTER OF GALAXIES
MCG+05-27-077	11 25 48.	+ 27 09	48	15.3	GALAXY
ZWG 156.081	11 25 48.	+ 27 44		15.1	GALAXY
MCG+10-17-001	11 25 48.	+ 57 56	27	16.	GALAXY
MRK 171	11 25 48.	+ 58 50	60	13.	GALAXY WITH UV CONTINUUM
KARA.72 288B	11 25 48.	+ 58 50	90		PART OF DOUBLE GALAXY
MCG+10-17-003	11 25 48.	+ 58 50	132	12.1	GALAXY
MCG+10-17-002A	11 25 48.	+ 58 50			GALAXY
MCG+10-17-002	11 25 48.	+ 58 50 30.	15	16.	GALAXY
HOLM 256B	11 25 48.	+ 58 51	24	13.3	PART OF MULTIPLE GALAXY
ZWG 011.105	11 25 48.	- 00 59		15.7	GALAXY
REIN 2.130	11 25 48.58	+ 09 40 58.2			NEBULA
ABC 1268	11 25 49.	+ 24 09		17.2	RICH CLUSTER OF GALAXIES
HN 0353	11 25 49.	- 12 37			NEBULA
IC 2856	11 25 49.	- 12 37			NONSTELLAR OBJECT
IC 2855	11 25 50.	+ 09 57 49.			NONSTELLAR OBJECT
REIZ 1186	11 25 51.	+ 27 39	42	15.0	GALAXY
REIZ 1181	11 25 51.	+ 58 51	27	13.6	GALAXY
IC 0694	11 25 53.	+ 58 50			GALAXY S
ZWG 067.085	11 25 54.	+ 09 23		15.3	GALAXY
UGC 06475	11 25 54.	+ 09 23	102	15.3	GALAXY Sc
MCG+02-29-033	11 25 54.	+ 09 23	120	15.3	GALAXY
MCG+03-29-055	11 25 54.	+ 19 27	66	14.9	GALAXY
ZC 1125.9+1934	11 25 54.	+ 19 34	1410		CLUSTER OF GALAXIES
ZWG 126.059	11 25 54.	+ 21 23		15.6	GALAXY
ZWG 126.060	11 25 54.	+ 23 40		14.4	GALAXY
UGC 06476	11 25 54.	+ 23 40	90	14.4	GALAXY Sb
ZWG 156.082	11 25 54.	+ 27 11		15.3	GALAXY
MCG+05-27-078	11 25 54.	+ 27 39	54	15.3	GALAXY
TON-N 0067	11 25 54.	+ 29 35		14.3	BLUE STAR
LB 10209	11 25 54.	+ 30 54		18.0	FAINT BLUE STAR
ZC 1125.9+3427	11 25 54.	+ 34 27	940		CLUSTER OF GALAXIES
ZC 1125.9+4201	11 25 54.	+ 42 01	2490		CLUSTER OF GALAXIES
ZWG 268.038	11 25 54.	+ 51 08		15.1	GALAXY
MCG+09-19-084	11 25 54.	+ 54 27	30	17.	GALAXY
MCG+10-17-004	11 25 54.	+ 58 06	39	15.	GALAXY
ZWG 291.074	11 25 54.	+ 58 07		15.5	GALAXY
VV 118E	11 25 54.	+ 58 49 30.	6	19.	INTERACTING GALAXY
VV 118D	11 25 54.	+ 58 49 30.	12	18.	INTERACTING GALAXY
VV 118C	11 25 54.	+ 58 49 30.	18	17.	INTERACTING GALAXY
VV 118B	11 25 54.	+ 58 49 30.	54	12.	INTERACTING GALAXY
VV 118A	11 25 54.	+ 58 49 30.	78	12.	INTERACTING GALAXY
VV 118	11 25 54.	+ 58 49 30.	150	12.1	INTERACTING GALAXY
REIN 7.055	11 25 55.86	+ 09 22 46.6			NEBULA
HOLM 257E	11 25 56.	+ 09 22	90	14.2	PART OF MULTIPLE GALAXY
IC 2857	11 25 57.	+ 09 22 43.			NONSTELLAR OBJECT
ARP 296	11 25 58.	+ 58 49			PECULIAR GALAXY
REIZ 1187	11 25 59.	+ 09 20	42	14.8	GALAXY
A1 042	11 25 59.61	+ 14 41 29.0		15.80	FAINT BLUE OBJECT
ZC 1126.0+0521	11 26 00.	+ 05 21	2220		CLUSTER OF GALAXIES
ZWG 067.086	11 26 00.	+ 09 22		14.5	GALAXY
UGC 06477	11 26 00.	+ 09 22	60	14.5	GALAXY SBc-IRR
MCG+02-29-034	11 26 00.	+ 09 22	60	14.5	GALAXY
IC 2858	11 26 00.	+ 13 56 19.			NONSTELLAR OBJECT
ZWG 096.053	11 26 00.	+ 20 00		15.5	GALAXY
UGC 06478	11 26 00.	+ 20 00	60	15.5	GALAXY S
MCG+04-27-040	11 26 00.	+ 23 40	84	14.4	GALAXY
ZWG 156.083	11 26 00.	+ 27 41		15.3	GALAXY
ZC 1126.0+3425	11 26 00.	+ 34 25	2220		CLUSTER OF GALAXIES
ZC 1126.0+3635	11 26 00.	+ 36 35	1010		CLUSTER OF GALAXIES
MCG+09-19-085	11 26 00.	+ 52 31	18	17.	GALAXY
ZC 1126.0+5744	11 26 00.	+ 57 44	1140		CLUSTER OF GALAXIES
MCG+10-17-005	11 26 00.	+ 58 52	24	16.	GALAXY
ZC 1126.0+6226	11 26 00.	+ 62 26	940		CLUSTER OF GALAXIES
MCG+12-11-026	11 26 00.	+ 73 19	66	15.	GALAXY
MCG+13-08-062	11 26 00.	+ 77 55	39	15.	GALAXY
ZC 1126.0+8022	11 26 00.	+ 80 22	1480		CLUSTER OF GALAXIES
ZWG 011.106	11 26 00.	- 01 20		15.3	GALAXY
REIZ 1188	11 26 02.	+ 21 17	30	14.9	GALAXY
REIZ 1189	11 26 02.	+ 23 41	24	14.2	GALAXY
REIZ 1190	11 26 02.	+ 25 04	30	14.0	GALAXY
IC 0697	11 26 03.	- 01 21 30.			NONSTELLAR OBJECT
IC 2884	11 26 03.	- 79 28			NONSTELLAR OBJECT
HN 0354	11 26 04.	- 79 28			NEBULA
REIN 7.056	11 26 04.65	+ 09 22 26.1			NEBULA
HOLM 257A	11 26 05.	+ 09 22	36	13.1	PART OF MULTIPLE GALAXY

OBJECT NAME	RIGHT ASCEN.	DECLINATION	DIAM.	MAGN.	TYPE OF OBJECT
IC 0696	11 26 05.	+ 09 22 24.			NONSTELLAR OBJECT
RNGC 3697A	11 26 05.	+ 21 04		14.0	GALAXY
WRAY 19.34	11 26 05.5	- 62 43 41.			DIFFUSE NEBULA
MCG+03-29-056	11 26 06.	+ 20 01	66	15.5	GALAXY
ZWG 126.061	11 26 06.	+ 21 04		14.1	GALAXY
UGC 06479	11 26 06.	+ 21 04	150	14.1	GALAXY Sb/SBb
MCG+04-27-041	11 26 06.	+ 23 11	42	15.5	GALAXY
ZWG 126.062	11 26 06.	+ 23 13		15.5	GALAXY
LB 10210	11 26 06.	+ 31 54		18.4	FAINT BLUE STAR
MCG+08-21-039	11 26 06.	+ 48 47	30	15.	GALAXY
MCG+09-19-086	11 26 06.	+ 52 30	24	17.	GALAXY
LB 02023	11 26 06.	+ 58 00 18.		16.9	FAINT BLUE STAR
REIN 7.057	11 26 06.46	+ 09 23 04.2			NEBULA
IC 2859	11 26 07.	+ 09 23 06.			NONSTELLAR OBJECT
IC 2860	11 26 08.	+ 14 19 00.			NONSTELLAR OBJECT
REIZ 1191	11 26 08.	+ 21 22	42	14.9	GALAXY
HN 0788	11 26 08.	- 62 43	300		NEBULA
CED 115	11 26 08.	- 62 43	300		DIFFUSE GALACTIC NEBULA
IC 2862	11 26 09.	+ 10 24 12.			NONSTELLAR OBJECT
REIZ 1192	11 26 09.	+ 27 38	12	15.6	GALAXY
IC 2872	11 26 09.	- 62 43			NONSTELLAR OBJECT
RNGC 3694	11 26 10.	+ 35 41		13.5	GALAXY
ARP 299	11 26 10.	+ 58 50			PECULIAR GALAXY
A1 043	11 26 10.17	+ 14 33 10.6		16.80	FAINT BLUE OBJECT
REIZ 1193	11 26 11.	+ 35 41	12	13.5	GALAXY
ZWG 126.063	11 26 12.	+ 21 00		15.3	GALAXY
ZWG 126.064	11 26 12.	+ 21 01		15.3	GALAXY
MCG+04-27-042	11 26 12.	+ 21 02	138	14.1	GALAXY
ZWG 126.065	11 26 12.	+ 22 21		15.1	GALAXY
LB 10211	11 26 12.	+ 28 06		17.1	FAINT BLUE STAR
ZWG 185.070	11 26 12.	+ 35 41		13.5	GALAXY
UGC 06480	11 26 12.	+ 35 41	66	13.5	GALAXY
UGC 06481	11 26 12.	+ 79 52	90	16.0	GALAXY S
MCG-02-29-034	11 26 12.	- 13 49	60	14.5	GALAXY
PK290+07.1	11 26 12.	- 52 39	60		PLANETARY NEBULA
MRSL 293-01/3	11 26 12.	- 62 22	1260		HII REGION
MRSL 293-01/1	11 26 12.	- 62 42	120		HII REGION
REIZ 1195	11 26 13.	+ 19 08	18	14.0	GALAXY
HOLM 258A	11 26 14.	+ 21 03	180	13.1	PART OF MULTIPLE GALAXY
REIZ 1198	11 26 14.	+ 21 04	120	14.3	GALAXY
REIZ 1196	11 26 14.	+ 22 21	30	14.5	GALAXY
RNGC 3696	11 26 14.	- 10 59			GALAXY
MCG+06-25-076	11 26 15.	+ 35 40 30.	30	15.	GALAXY
A1 044	11 26 16.34	+ 16 39 46.0		17.20	FAINT BLUE OBJECT
RNGC 3697B	11 26 17.	+ 20 59		15.5	GALAXY
RNGC 3697C	11 26 17.	+ 21 00		15.5	GALAXY
REIZ 1197	11 26 17.	+ 35 45	6	15.3	GALAXY
ZWG 039.193	11 26 18.	+ 03 41		14.9	GALAXY
MCG+01-29-046	11 26 18.	+ 03 41	30	14.9	GALAXY
ZC 1126.3+0913	11 26 18.	+ 09 13	4370		CLUSTER OF GALAXIES
MCG+04-27-044	11 26 18.	+ 20 59 30.	24	15.3	GALAXY
MCG+04-27-045	11 26 18.	+ 21 00	39	15.3	GALAXY
MCG+04-27-046	11 26 18.	+ 22 21	24	15.1	GALAXY
MCG+04-27-043	11 26 18.	+ 23 12 30.	24	16.	GALAXY
LB 10212	11 26 18.	+ 28 51		17.9	FAINT BLUE STAR
ZC 1126.3+3200	11 26 18.	+ 32 00	940		CLUSTER OF GALAXIES
ZWG 214.009	11 26 18.	+ 39 07		15.5	GALAXY
MCG+07-24-011	11 26 18.	+ 39 07 30.	12	14.5	GALAXY
ZWG 242.040	11 26 18.	+ 47 58		15.5	GALAXY
MCG-06-25-021	11 26 18.	- 34 44	36	16.	GALAXY
IC 2863	11 26 19.	+ 09 22 12.			NONSTELLAR OBJECT
IC 2861	11 26 19.	+ 39 07 12.			NONSTELLAR OBJECT
REIZ 1194	11 26 20.	+ 57 23	60	13.4	GALAXY
LB 02024	11 26 21.	+ 57 04 36.		15.2	FAINT BLUE STAR
RNGC 3698	11 26 22.	+ 35 56			NON-EXISTENT OBJECT
REIZ 1199	11 26 23.	+ 09 20	36	14.1	GALAXY
HOLM 258B	11 26 23.	+ 21 00	42	14.4	PART OF MULTIPLE GALAXY
SEY 101	11 26 23.	+ 54 01 42.		15.0	FAINT GALAXY
ZWG 067.087	11 26 24.	+ 09 21		15.6	GALAXY
ZWG 067.088	11 26 24.	+ 09 23		14.4	GALAXY
UGC 06482	11 26 24.	+ 09 23	60	14.4	GALAXY S0?
MCG+02-29-035	11 26 24.	+ 09 23	48	14.4	GALAXY
IC 2864	11 26 24.	+ 12 38 24.			NONSTELLAR OBJECT
MCG+03-29-057	11 26 24.	+ 17 30	144	15.3	GALAXY
ZWG 096.054	11 26 24.	+ 17 31		15.3	GALAXY
UGC 06483	11 26 24.	+ 17 31	126	15.3	GALAXY Sc
ZC 1126.4+2123	11 26 24.	+ 21 23	2220		CLUSTER OF GALAXIES
ZWG 126.066	11 26 24.	+ 21 52		15.4	GALAXY
LB 10214	11 26 24.	+ 29 36		16.4	FAINT BLUE STAR
LB 10213	11 26 24.	+ 31 14		18.0	FAINT BLUE STAR
ZC 1126.4+3412	11 26 24.	+ 34 12	810		CLUSTER OF GALAXIES
ARC 1269	11 26 24.	+ 34 26		17.5	RICH CLUSTER OF GALAXIES
ZC 1126.4+3656	11 26 24.	+ 36 56	3020		CLUSTER OF GALAXIES
MCG+08-21-040	11 26 24.	+ 47 56	48	16.	GALAXY
MCG+08-21-041	11 26 24.	+ 47 57	36	16.	GALAXY
MCG+09-19-087	11 26 24.	+ 51 47	54	16.	GALAXY
ZWG 268.039	11 26 24.	+ 54 00		15.5	GALAXY
ZC 1126.4+5445	11 26 24.	+ 54 45	870		CLUSTER OF GALAXIES
RNGC 3683A	11 26 24.	+ 57 24		12.5	GALAXY
MCG+10-17-006	11 26 24.	+ 57 24	144	13.	GALAXY
ZWG 291.075	11 26 24.	+ 57 25		12.6	GALAXY
UGC 06484	11 26 24.	+ 57 25	144	12.6	GALAXY SBc
MCG-04-27-011	11 26 24.	- 22 11 30.	36	15.	GALAXY
A1 045	11 26 24.06	+ 14 16 47.5		16.20	FAINT BLUE OBJECT
A1 046	11 26 24.75	+ 15 27 49.4		17.30	FAINT BLUE OBJECT
IC 2866	11 26 25.	+ 09 19 06.			SINGLE STAR
IC 2865	11 26 25.	+ 09 23 42.			SINGLE STAR
HOLM 258C	11 26 25.	+ 21 00	30	14.4	PART OF MULTIPLE GALAXY
ARC 1271	11 26 25.	- 09 19		17.7	RICH CLUSTER OF GALAXIES
REIN 7.058	11 26 25.42	+ 09 21 56.1			NEBULA
IC 2867	11 26 26.	+ 09 21 48.			NONSTELLAR OBJECT
HOLM 257G	11 26 26.	+ 09 22	24	15.2	PART OF MULTIPLE GALAXY
REIZ 1200	11 26 27.	+ 29 06	30	15.0	GALAXY
SEY 102	11 26 27.	+ 51 48 35.		15.2	FAINT GALAXY
REIN 7.059	11 26 28.54	+ 09 23 17.1			NEBULA
HOLM 257B	11 26 29.	+ 09 23	24	13.9	PART OF MULTIPLE GALAXY
IC 0698	11 26 29.	+ 09 23 12.			NONSTELLAR OBJECT
REIZ 1201	11 26 29.	+ 35 51	18	14.1	GALAXY
SEY 103	11 26 29.	+ 51 14 23.		15.3	FAINT GALAXY
A1 047	11 26 29.46	+ 15 10 30.9		16.05	FAINT BLUE OBJECT
ZC 1126.5+0050	11 26 30.	+ 00 50	1080		CLUSTER OF GALAXIES
ZWG 067.089	11 26 30.	+ 09 16		14.6	GALAXY
UGC 06485	11 26 30.	+ 09 16	78	14.6	GALAXY Sb
MCG+02-29-036	11 26 30.	+ 09 16	72	14.6	GALAXY
UGC 06486	11 26 30.	+ 12 09	84	18.	GALAXY DWRF IR
ZWG 096.055	11 26 30.	+ 20 02		14.9	GALAXY
ZWG 126.067	11 26 30.	+ 20 51		15.0	GALAXY
UGC 06487	11 26 30.	+ 20 51	84	15.0	GALAXY QUD SYS
TON-N 0579	11 26 30.	+ 25 04		16.2	BLUE STAR
ZWG 156.084	11 26 30.	+ 32 11		15.7	GALAXY
FEIG 043	11 26 30.	+ 38 28		14.1	FAINT BLUE STAR
UGC 06488	11 26 30.	+ 42 03	66	16.0	GALAXY E-S
MCG+07-24-012	11 26 30.	+ 42 03	45	15.	GALAXY
ZWG 214.010	11 26 30.	+ 42 47		15.7	GALAXY
UGC 06489	11 26 30.	+ 42 47	96	15.7	GALAXY S
MCG+07-24-013	11 26 30.	+ 42 48 30.	72	15.	GALAXY
ZC 1126.5+5048	11 26 30.	+ 50 48	940		CLUSTER OF GALAXIES
MCG+09-19-088	11 26 30.	+ 51 12 30.	48	16.	GALAXY
ZWG 268.040	11 26 30.	+ 51 15		15.7	GALAXY
IC 2868	11 26 31.	+ 09 22 12.			SAME AS IC 698
REIZ 1202	11 26 31.	+ 19 01	12	14.7	GALAXY
PK292+01.2	11 26 31.	- 59 50	5		PLANETARY NEBULA
REIN 7.060	11 26 31.28	+ 09 15 52.2			NEBULA
IC 0699	11 26 32.	+ 09 15 48.			NONSTELLAR OBJECT
REIZ 1203	11 26 32.	+ 21 38	12	15.0	GALAXY
REIZ 1204	11 26 32.	+ 21 52	18	15.1	GALAXY
HOLM 257D	11 26 33.	+ 09 16	72	14.2	PART OF MULTIPLE GALAXY
IC 2869	11 26 34.	+ 09 17 36.			NONSTELLAR OBJECT
RNGC 3695	11 26 34.	+ 35 51		15.0	GALAXY
SEY 104	11 26 35.	+ 52 46 23.		15.2	FAINT GALAXY
ZWG 011.108	11 26 36.	+ 02 17		15.6	GALAXY
MCG+00-29-037	11 26 36.	+ 02 17	36	15.6	GALAXY
MCG+03-29-058	11 26 36.	+ 20 04	48	14.9	GALAXY
LB 10215	11 26 36.	+ 29 05		17.1	FAINT BLUE STAR
GCL 017	11 26 36.	+ 29 15	150	14.4	GLOBULAR STAR CLUSTER
ZWG 185.071	11 26 36.	+ 35 51		14.9	GALAXY
UGC 06490	11 26 36.	+ 35 51	66	14.9	GALAXY S
ZC 1126.6+4035	11 26 36.	+ 40 35	940		CLUSTER OF GALAXIES
MCG+09-19-091	11 26 36.	+ 52 30	48	16.	GALAXY
MCG+09-19-090	11 26 36.	+ 52 45	24	16.	GALAXY
MCG+09-19-089	11 26 36.	+ 55 36	36	16.	GALAXY
ZWG 011.107	11 26 36.	- 01 26		15.6	GALAXY
VHE 49	11 26 36.	- 62 49	66		REFLECTION NEBULA
IC 2870	11 26 37.	+ 12 08 30.			NONSTELLAR OBJECT
REIZ 1205	11 26 38.	+ 20 51	36	14.5	GALAXY
WRAY 19.35	11 26 38.3	- 62 22 49.		8.0	STAR-NEBULA ASSOCIATION
SHB 180	11 26 38.6	+ 10 08 30.		18.0	QUASI-STELLAR OBJECT
IC 0700	11 26 39.	+ 20 51 36.			NONSTELLAR OBJECT
CED 116	11 26 39.	- 62 23	930		DIFFUSE GALACTIC NEBULA
REIZ 1206	11 26 41.	+ 35 47	12	14.2	GALAXY
ZWG 067.090	11 26 42.	+ 08 53		15.2	GALAXY
MCG+02-29-037	11 26 42.	+ 08 53	42	15.2	GALAXY
MCG+04-27-047	11 26 42.	+ 20 50	48	15.0	GALAXY
MRK 172	11 26 42.	+ 22 01	8	17.5	GALAXY WITH UV CONTINUUM
ZWG 156.085	11 26 42.	+ 28 49		14.8	GALAXY
ZWG 185.072	11 26 42.	+ 35 08		15.6	GALAXY
UGC 06491	11 26 42.	+ 35 08	120	15.6	GALAXY
MCG+06-25-077	11 26 42.	+ 35 08	78	15.	GALAXY
MCG+06-25-078	11 26 42.	+ 35 50	48	15.	GALAXY
UGC 06492	11 26 42.	+ 52 31	60	16.0	GALAXY S
MCG+09-19-092	11 26 42.	+ 54 27 30.	36	17.	GALAXY
ZWG 011.109	11 26 42.	- 01 26		15.4	GALAXY
REIZ 1207	11 26 44.	+ 22 21	18	14.4	GALAXY
ARC 1270	11 26 44.	+ 54 21		15.4	RICH CLUSTER OF GALAXIES
MCG+05-27-079	11 26 45.	+ 28 48	24	14.8	GALAXY
MCG+07-24-014	11 26 45.	+ 42 08 30.	45	16.	GALAXY
IC 2871	11 26 46.	+ 08 52 42.			NONSTELLAR OBJECT
REIZ 1208	11 26 47.	+ 09 13	60	14.5	GALAXY
RNGC 3701	11 26 47.	+ 24 21		14.0	GALAXY
ZWG 126.068	11 26 48.	+ 24 21		14.1	GALAXY
UGC 06493	11 26 48.	+ 24 21	126	14.1	GALAXY Sb-c
MCG+04-27-048	11 26 48.	+ 24 22	108	14.1	GALAXY
VV 060C	11 26 48.	+ 42 09	6	19.	INTERACTING GALAXY
VV 060B	11 26 48.	+ 42 09	15	16.	INTERACTING GALAXY
VV 060A	11 26 48.	+ 42 09	15	16.	INTERACTING GALAXY
VV 060	11 26 48.	+ 42 09	60		INTERACTING GALAXY
LB 02025	11 26 48.	+ 60 33 12.		15.0	FAINT BLUE STAR
MCG-02-29-035	11 26 48.	- 10 48	48	15.5	GALAXY
REIZ 1209	11 26 50.	+ 24 22	96	13.7	GALAXY
ARC 1273	11 26 50.	- 06 46		17.6	RICH CLUSTER OF GALAXIES
RNGC 3711	11 26 50.	- 10 48		15.0	GALAXY
REIZ 1210	11 26 51.	+ 28 49	18	14.5	GALAXY
IC 2874	11 26 52.	+ 10 54 23.			OPEN CLUSTER
IC 2873	11 26 52.	+ 13 29 41.			NONSTELLAR OBJECT
RNGC 3700	11 26 52.	+ 35 47		15.0	GALAXY
ZWG 067.091	11 26 54.	+ 13 29		14.9	GALAXY
MCG+02-29-038	11 26 54.	+ 13 29	36	14.9	GALAXY
LB 10216	11 26 54.	+ 28 22		16.7	FAINT BLUE STAR
ZWG 185.073	11 26 54.	+ 35 47		15.1	GALAXY
UGC 06494	11 26 54.	+ 35 47	66	15.1	GALAXY SBa-b
ZC 1126.9+3745	11 26 54.	+ 37 45	540		CLUSTER OF GALAXIES
SHAH 063	11 26 54.	+ 42 43	54	16.2	GROUP OF COMPACT GALAXIES
MCG+09-19-093	11 26 54.	+ 51 30	30	16.	GALAXY
ZWG 011.110	11 26 54.	- 01 28		15.5	GALAXY
MCG-01-29-025	11 26 54.	- 07 52	30	15.5	GALAXY
LB 02026	11 26 55.	+ 55 01 24.		16.3	FAINT BLUE STAR
LB 02027	11 26 56.	+ 55 12 06.		16.4	FAINT BLUE STAR
LB 02028	11 26 57.	+ 51 29 00.		16.0	FAINT BLUE STAR
SEY 105	11 26 58.	+ 51 31 59.		15.3	FAINT GALAXY
REIN 7.061A	11 26 58.10	+ 09 17 31.2			NEBULA
REIN 7.061B	11 26 58.10	+ 09 17 33.0			NEBULA
IC 2876	11 26 59.	+ 09 17 41.			NONSTELLAR OBJECT
KEEN 3683A	11 27	+ 57 23		13.11	GALAXY
HOFF L08	11 27	- 60 25	110		DARK HOLE
HOFF L09	11 27	- 60 43	250		DARK HOLE
HOFF L10	11 27	- 60 55	110		DARK HOLE
ZWG 067.092	11 27 00.	+ 13 08		15.7	GALAXY
IC 2875	11 27 00.	+ 13 15 59.			NONSTELLAR OBJECT
ZWG 126.069	11 27 00.	+ 22 24		14.9	GALAXY
UGC 06495	11 27 00.	+ 22 24	60	14.9	GALAXY Sb/SBb
ZC 1127.0+2226	11 27 00.	+ 22 26	740		CLUSTER OF GALAXIES
ZWG 156.086	11 27 00.	+ 28 32		15.4	GALAXY
MCG+05-27-080	11 27 00.	+ 28 32	48	15.4	GALAXY
MCG+06-25-079	11 27 00.	+ 35 46 30.	60	15.	GALAXY
SHAH 064	11 27 00.	+ 42 30	168	17.8	GROUP OF COMPACT GALAXIES
MCG+09-19-094	11 27 00.	+ 51 30	42	16.	GALAXY
ZWG 268.041	11 27 00.	+ 51 32		15.4	GALAXY
ARC 1272	11 27 01.	+ 24 05		17.2	RICH CLUSTER OF GALAXIES
IC 2877	11 27 02.	+ 13 07 47.			NONSTELLAR OBJECT
REIZ 1211	11 27 02.	+ 22 23	30	14.1	GALAXY
RNGC 3703	11 27 02.	- 08 21			NON-EXISTENT OBJECT
IC 2878	11 27 03.	+ 10 14 41.			NONSTELLAR OBJECT
ARC 1274	11 27 05.	+ 20 11		17.2	RICH CLUSTER OF GALAXIES
ZC 1127.1+0309	11 27 06.	+ 03 09	1080		CLUSTER OF GALAXIES
ZC 1127.1+1545	11 27 06.	+ 15 45	870		CLUSTER OF GALAXIES
ZC 1127.1+2512	11 27 06.	+ 25 12	2350		CLUSTER OF GALAXIES

OBJECT NAME	RIGHT ASCEN.	DECLINATION	DIAM.	MAGN.	TYPE OF OBJECT
LB 10217	11 27 06.	+ 30 09		17.3	FAINT BLUE STAR
ZC 1127.1+3850	11 27 06.	+ 38 50	4100		CLUSTER OF GALAXIES
ZC 1127.1+4105	11 27 06.	+ 41 05	940		CLUSTER OF GALAXIES
MCG+08-21-042	11 27 06.	+ 44 56	30	15.	GALAXY
ZC 1127.1+4621	11 27 06.	+ 46 21	670		CLUSTER OF GALAXIES
MCG+09-19-095	11 27 06.	+ 52 27	36	16.	GALAXY
LB 02029	11 27 06.	+ 53 35 24.		16.4	FAINT BLUE STAR
8ZW 1127-18.6	11 27 06.	- 18 39		18.6	COMPACT GALAXY
REIN 7.062	11 27 08.50	+ 09 28 55.2			NEBULA
REIN 7.063A	11 27 09.15	+ 09 17 25.4			NEBULA
REIN 7.063B	11 27 09.16	+ 09 17 27.1			NEBULA
IC 2879	11 27 10.	+ 09 17 29.			NONSTELLAR OBJECT
HOLM 259C	11 27 11.	+ 09 28	24	15.4	PART OF MULTIPLE GALAXY
ZWG 039.194	11 27 12.	+ 04 46		15.4	GALAXY
ZWG 039.195	11 27 12.	+ 06 05		15.6	GALAXY
RNGC 3705B	11 27 12.	+ 09 27		15.5	GALAXY
REIZ 1212	11 27 12.	+ 13 27	42	15.3	GALAXY
ZWG 096.056	11 27 12.	+ 15 36		15.3	GALAXY
MCG+04-27-049	11 27 12.	+ 22 22	54	14.9	GALAXY
ZC 1127.2+2406	11 27 12.	+ 24 06	1680		CLUSTER OF GALAXIES
ZWG 126.070	11 27 12.	+ 25 13		14.5	GALAXY
UGC 06496	11 27 12.	+ 25 13	42	14.5	GALAXY VY CMPT
LB 10221	11 27 12.	+ 27 10		16.0	FAINT BLUE STAR
LB 10220	11 27 12.	+ 28 44		17.4	FAINT BLUE STAR
LB 10219	11 27 12.	+ 30 16		17.6	FAINT BLUE STAR
LB 10218	11 27 12.	+ 31 20		17.7	FAINT BLUE STAR
ZC 1127.2+3318	11 27 12.	+ 33 18	870		CLUSTER OF GALAXIES
ZWG 268.042	11 27 12.	+ 52 28		15.7	GALAXY
MRSL 293-01/2	11 27 12.	- 62 37	900		HII REGION
A1 048	11 27 12.27	+ 17 28 44.4		17.50	FAINT BLUE OBJECT
A1 049	11 27 12.78	+ 17 36 51.4		17.30	FAINT BLUE OBJECT
REIZ 1213	11 27 15.	+ 26 48	24	14.4	GALAXY
RNGC 3706	11 27 17.	- 36 08		12.5	GALAXY
IC 2880	11 27 18.	+ 13 28 29.			NONSTELLAR OBJECT
LB 10223	11 27 18.	+ 27 43		18.1	FAINT BLUE STAR
LB 10222	11 27 18.	+ 27 56		17.9	FAINT BLUE STAR
ZC 1127.3+3456	11 27 18.	+ 34 56	870		CLUSTER OF GALAXIES
ZWG 242.041	11 27 16.	+ 44 58		15.7	GALAXY
KARA.68 078	11 27 18.	+ 52 41	67		DWARF GALAXY
7ZW 405	11 27 18.	+ 58 43			COMPACT GALAXY
MCG-02-29-036	11 27 18.	- 12 06	24	15.	GALAXY
MCG-06-25-022	11 27 18.	- 36 06	120	12.7	GALAXY
ARC 1275	11 27 19.	+ 36 58		15.7	RICH CLUSTER OF GALAXIES
IC 2881	11 27 20.	+ 12 47 11.			NONSTELLAR OBJECT
TON-N 1400	11 27 20.	+ 35 50		17.	BLUE STAR
LB 02030	11 27 21.	+ 57 26 06.		15.2	FAINT BLUE STAR
MCG-05-27-014	11 27 21.	- 31 35	36	15.5	GALAXY
ARC 1276	11 27 23.	+ 33 19		17.6	RICH CLUSTER OF GALAXIES
ZC 1127.4+1519	11 27 24.	+ 15 19	1010		CLUSTER OF GALAXIES
LB 10224	11 27 24.	+ 30 13		17.3	FAINT BLUE STAR
ZC 1127.4+3705	11 27 24.	+ 37 05	1410		CLUSTER OF GALAXIES
ZC 1127.4+3740	11 27 24.	+ 37 40	1750		CLUSTER OF GALAXIES
UGC 06497	11 27 24.	+ 38 54	138	16.0	GALAXY
ZC 1127.4-0045	11 27 24.	- 00 45	3020		CLUSTER OF GALAXIES
MCG-02-29-037	11 27 24.	- 11 14	24	14.5	GALAXY
A1 050	11 27 25.39	+ 13 43 36.0		16.75	FAINT BLUE OBJECT
REIZ 1215	11 27 26.	+ 22 33	12	15.2	GALAXY
RNGC 3704	11 27 26.	- 11 14		14.0	GALAXY
ARC 1277	11 27 27.	+ 13 12		17.8	RICH CLUSTER OF GALAXIES
MCG+07-24-015	11 27 27.	+ 38 53	132	14.5	GALAXY
MCG-03-29-008	11 27 27.	- 20 20	30	15.	GALAXY
REIZ 1214	11 27 28.	+ 62 09	18	14.2	GALAXY
ZWG 067.093	11 27 30.	+ 09 31		11.5	GALAXY
UGC 06498	11 27 30.	+ 09 31	300	11.5	GALAXY Sb
MCG+02-29-039	11 27 30.	+ 09 31	270	11.5	GALAXY
RNGC 3705	11 27 30.	+ 09 33		12.0	GALAXY
ZWG 126.071	11 27 30.	+ 24 26		15.2	GALAXY
ZC 1127.5+4804	11 27 30.	+ 48 04	1010		CLUSTER OF GALAXIES
MCG+10-17-007	11 27 30.	+ 58 24	51	14.	GALAXY
ZWG 292.001	11 27 30.	+ 58 25		15.0	GALAXY
ZWG 291.076	11 27 30.	+ 58 25		15.0	GALAXY
MCG+13-08-063	11 27 30.	+ 79 51	78	15.	GALAXY
REIZ 1216	11 27 32.	+ 24 27	18	14.7	GALAXY
RNGC 3707	11 27 32.	- 11 16			NON-EXISTENT OBJECT
REIN 2.131	11 27 32.07	+ 09 33 14.4			NEBULA
IC 2882	11 27 34.	+ 12 15 59.			NONSTELLAR OBJECT
REIZ 1217	11 27 35.	+ 09 33	180	11.9	GALAXY
HOLM 259A	11 27 35.	+ 09 33	240	12.0	PART OF MULTIPLE GALAXY
ARC 1278	11 27 35.	+ 20 46		17.3	RICH CLUSTER OF GALAXIES
SHB 181	11 27 35.6	- 14 32 57.		16.9	QUASI-STELLAR OBJECT
A1 051	11 27 35.63	+ 15 13 46.6		16.85	FAINT BLUE OBJECT
BC PKS1127-14	11 27 35.70	- 14 32 54.0		16.90	QUASI-STELLAR OBJECT
ZWG 011.111	11 27 36.	+ 02 06		15.6	GALAXY
MCG+00-29-038	11 27 36.	+ 02 06	30	15.6	GALAXY
ZC 1127.6+1314	11 27 36.	+ 13 14	810		CLUSTER OF GALAXIES
LB 10225	11 27 36.	+ 32 28		17.8	FAINT BLUE STAR
UGC 06499	11 27 36.	+ 36 08	102	17.	GALAXY DWARF
ZWG 214.011	11 27 36.	+ 44 26		15.4	GALAXY
MCG+09-19-096	11 27 36.	+ 56 26	36	16.	GALAXY
MCG-06-25-023	11 27 36.	- 35 14	48	14.5	GALAXY
ACK 290+09.1	11 27 36.	- 50 59			PLANETARY NEBULA
RNGC 3702	11 27 38.	- 08 36		15.0	GALAXY
A1 052	11 27 38.62	+ 15 20 15.0		17.00	FAINT BLUE OBJECT
IC 2883	11 27 41.	+ 11 11 10.			NONSTELLAR OBJECT
ZC 1127.7+0600	11 27 42.	+ 06 00	400		CLUSTER OF GALAXIES
ZC 1127.7+1042	11 27 42.	+ 10 42	540		CLUSTER OF GALAXIES
ZC 1127.7+2052	11 27 42.	+ 20 52	1680		CLUSTER OF GALAXIES
MCG+04-27-050	11 27 42.	+ 24 27	42	15.2	GALAXY
ZC 1127.7+3335	11 27 42.	+ 33 35	810		CLUSTER OF GALAXIES
ZC 1127.7+3520	11 27 42.	+ 35 20	1680		CLUSTER OF GALAXIES
ZWG 185.074	11 27 42.	+ 37 00		15.2	GALAXY
MRK 424	11 27 42.	+ 37 00	20	15.5	GALAXY WITH UV CONTINUUM
MCG+09-19-097	11 27 42.	+ 51 42 30.	48	15.	GALAXY
SEY 106	11 27 42.	+ 51 44 41.		14.7	FAINT GALAXY
ZC 1127.7+6104	11 27 42.	+ 61 04	1610		CLUSTER OF GALAXIES
MCG-01-29-026	11 27 42.	- 08 36	66	15.	GALAXY
A1 053	11 27 42.24	+ 13 02 45.8		16.35	FAINT BLUE OBJECT
REIZ 1218	11 27 44.	+ 22 29	24	14.8	GALAXY
MCG+07-24-016	11 27 45.	+ 44 26 30.	36	15.	GALAXY
KN 12.001	11 27 46.4	+ 48 14 45.			NEBULA
REIZ 1219	11 27 47.	+ 09 37	42	15.5	GALAXY
ARC 1281	11 27 47.	+ 33 40		17.6	RICH CLUSTER OF GALAXIES
ARC 1280	11 27 47.	+ 34 58		17.5	RICH CLUSTER OF GALAXIES
ZWG 039.196	11 27 48.	+ 04 43		15.4	GALAXY
ZWG 039.197	11 27 48.	+ 04 58		15.2	GALAXY
ZC 1127.8+0706	11 27 48.	+ 07 06	1480		CLUSTER OF GALAXIES
IC 2885	11 27 48.	+ 10 02 58.			NONSTELLAR OBJECT
ZWG 126.072	11 27 48.	+ 26 05		15.4	GALAXY
ZWG 156.087	11 27 48.	+ 26 30		15.7	GALAXY
LB 10228	11 27 48.	+ 29 51		17.0	FAINT BLUE STAR
LB 10227	11 27 48.	+ 30 28		17.1	FAINT BLUE STAR
LB 10226	11 27 48.	+ 31 56		17.6	FAINT BLUE STAR
ZWG 214.012	11 27 48.	+ 38 31		15.5	GALAXY
ZC 1127.8+4640	11 27 48.	+ 46 40	1950		CLUSTER OF GALAXIES
MRK 173	11 27 48.	+ 48 22	10	16.5	GALAXY WITH UV CONTINUUM
MCG+08-21-043	11 27 48.	+ 49 25	42	16.	GALAXY
ZWG 268.043	11 27 48.	+ 51 45		14.7	GALAXY
MCG+12-11-027	11 27 48.	+ 68 58	12	17.	GALAXY
MCG-02-29-038	11 27 48.	- 12 47 30.	54	14.	GALAXY
KN 12.002	11 27 48.1	+ 49 26 45.			NEBULA
A1 054	11 27 48.74	+ 17 51 02.4		17.00	FAINT BLUE OBJECT
IC 2886	11 27 49.	+ 11 50 16.			NONSTELLAR OBJECT
REIZ 1220	11 27 50.	+ 24 30	36	15.9	GALAXY
A1 055	11 27 52.79	+ 14 22 57.2		17.30	FAINT BLUE OBJECT
ARC 1285	11 27 53.	- 14 17		17.0	RICH CLUSTER OF GALAXIES
ZWG 011.112	11 27 54.	+ 00 46		15.7	GALAXY
RNGC 3705A	11 27 54.	+ 09 38		15.0	GALAXY
ZWG 067.094	11 27 54.	+ 09 40		15.5	GALAXY
ZC 1127.9+3446	11 27 54.	+ 34 46	340		CLUSTER OF GALAXIES
MCG+07-24-017	11 27 54.	+ 38 30	36	15.	GALAXY
MCG+07-24-018	11 27 54.	+ 39 41	30	16.	GALAXY
MCG+08-21-044	11 27 54.	+ 47 16 30.	60	15.	GALAXY
ZWG 242.042	11 27 54.	+ 47 18		14.9	GALAXY
REIN 7.064B	11 27 54.59	+ 09 39 50.3			NEBULA
REIN 7.064A	11 27 54.69	+ 09 39 50.3			NEBULA
IC 2887	11 27 55.	+ 09 39 52.			NONSTELLAR OBJECT
HOLM 259B	11 27 56.	+ 09 40	36	15.2	PART OF MULTIPLE GALAXY
REIZ 1221	11 27 56.	+ 22 38	18	14.9	GALAXY
A1 056	11 27 57.23	+ 14 58 07.6		17.10	FAINT BLUE OBJECT
KEEL 345	11 27 57.6	+ 47 17 35.			NEBULA
A1 057	11 27 57.88	+ 13 45 20.5		16.75	FAINT BLUE OBJECT
ZWG 039.198	11 28 00.	+ 04 01		15.1	GALAXY
ZC 1128.0+0615	11 28 00.	+ 06 15	670		CLUSTER OF GALAXIES
MCG+03-29-059	11 28 00.	+ 18 41	42	15.7	GALAXY
ZWG 096.057	11 28 00.	+ 18 42		15.7	GALAXY
ZC 1128.0+2015	11 28 00.	+ 20 15	2760		CLUSTER OF GALAXIES
LB 10229	11 28 00.	+ 28 02		16.6	FAINT BLUE STAR
ZWG 185.075	11 28 00.	+ 35 47		15.6	GALAXY
ZC 1128.0+4243	11 28 00.	+ 42 43	1810		CLUSTER OF GALAXIES
ZWG 214.013	11 28 00.	+ 44 26		15.3	GALAXY
UGC 06500	11 28 00.	+ 44 26	60	15.3	GALAXY Sa-b
MCG+07-24-019	11 28 00.	+ 44 26 30.	63	15.	GALAXY
MCG+09-19-098	11 28 00.	+ 52 29	36	16.	GALAXY
MCG+09-19-099	11 28 00.	+ 54 40	30	16.	GALAXY
ZWG 268.044	11 28 00.	+ 55 20		15.7	GALAXY
MCG+09-19-100	11 26 00.	+ 55 20	42	16.	GALAXY
SEY 107	11 28 00.	+ 55 20 59.		14.9	FAINT GALAXY
MCG+10-17-008	11 28 00.	+ 60 46	39	15.	GALAXY
UGC 06501	11 28 00.	+ 60 47	72	16.5	GALAXY
UGC 06502	11 28 00.	+ 77 33	84	17.	GALAXY Sc
7ZW 406	11 28 00.	+ 80 35			COMPACT GALAXY
MCG-01-29-027	11 28 00.	- 07 49	30	15.	GALAXY
IC 2888	11 28 01.	+ 10 11 04.			NONSTELLAR OBJECT
LB 02031	11 28 01.	+ 50 31 06.		16.7	FAINT BLUE STAR
SEY 108	11 28 01.	+ 52 29 28.		15.4	FAINT GALAXY
HN 0355	11 28 01.	- 12 49			NEBULA
ARC 1282	11 28 02.	+ 40 17		17.6	RICH CLUSTER OF GALAXIES
IC 2889	11 28 02.	- 12 49			NONSTELLAR OBJECT
A1 058	11 28 05.93	+ 17 19 51.8		17.10	FAINT BLUE OBJECT
LB 10231	11 28 06.	+ 30 05		17.5	FAINT BLUE STAR
LB 10230	11 28 06.	+ 30 48		16.9	FAINT BLUE STAR
ZC 1128.1+5823	11 28 06.	+ 58 23	2690		CLUSTER OF GALAXIES
OCL 0859	11 28 06.	- 63 10	1260	10.	OPEN STAR CLUSTER
VHE 50B	11 28 06.	- 63 33			REFLECTION NEBULA
VHE 50A	11 28 06.	- 63 33			REFLECTION NEBULA
RNGC 3708	11 28 07.	- 02 57			NON-EXISTENT OBJECT
RNGC 3709	11 28 07.	- 02 59			NON-EXISTENT OBJECT
WRAY 19.36	11 28 07.7	- 63 32 39.		7.5	STAR-NEBULA ASSOCIATION
LB 02032	11 28 08.	+ 56 21 36.		16.5	FAINT BLUE STAR
CED 117	11 28 09.	- 63 33	480		DIFFUSE GALACTIC NEBULA
IC 2890	11 28 11.	+ 13 27 40.			NONSTELLAR OBJECT
ARC 1288	11 28 12.	+ 06 16		17.6	RICH CLUSTER OF GALAXIES
IC 2891	11 28 12.	+ 12 57 16.			NONSTELLAR OBJECT
ZWG 126.073	11 28 12.	+ 20 43		15.7	GALAXY
ZWG 126.074	11 28 12.	+ 20 45		14.7	GALAXY
UGC 06503	11 28 12.	+ 20 45	72	14.7	GALAXY SB
ARC 1286	11 28 12.	+ 22 39		17.6	RICH CLUSTER OF GALAXIES
LB 10234	11 28 12.	+ 26 30		16.4	FAINT BLUE STAR
LB 10233	11 28 12.	+ 31 19		18.2	FAINT BLUE STAR
LB 10232	11 28 12.	+ 32 05		16.8	FAINT BLUE STAR
SHAH 065	11 28 12.	+ 35 19	240	17.6	GROUP OF COMPACT GALAXIES
ARC 1284	11 28 12.	+ 35 20		17.5	RICH CLUSTER OF GALAXIES
SHAH 121	11 28 12.	+ 38 08	90	18.0	GROUP OF COMPACT GALAXIES
MCG-02-29-039	11 28 12.	- 15 00	90	13.5	GALAXY
MRSL 294-02/1	11 28 12.	- 63 33	3600		HII REGION
IC 2892	11 28 14.	+ 10 51 52.			NONSTELLAR OBJECT
REIZ 1222	11 28 14.	+ 26 36	42	14.6	GALAXY
REIZ 1223	11 28 17.	+ 35 51	36	14.4	GALAXY
ZWG 039.199	11 28 18.	+ 02 34		15.6	GALAXY
ZWG 067.095	11 28 18.	+ 13 40		15.6	GALAXY
IC 2893	11 28 18.	+ 13 40 10.			NONSTELLAR OBJECT
MCG+03-29-060	11 28 18.	+ 18 40	30	15.5	GALAXY
ZWG 126.075	11 28 18.	+ 20 30		15.4	GALAXY
IC 0701	11 28 18.	+ 20 44 28.			NONSTELLAR OBJECT
ZWG 126.076	11 28 18.	+ 25 29		15.6	GALAXY
MCG+05-27-081	11 28 18.	+ 29 36	42	15.5	GALAXY
MRK 174	11 28 18.	+ 56 23	10	16.5	GALAXY WITH UV CONTINUUM
REIZ 1224	11 28 20.	+ 23 02	12	14.7	GALAXY
ARC 1279	11 28 20.	+ 67 31		16.5	RICH CLUSTER OF GALAXIES
A1 059	11 28 20.95	+ 13 06 28.3		15.90	FAINT BLUE OBJECT
ARP 197	11 28 21.	+ 20 45			PECULIAR GALAXY
IC 2895	11 28 22.	+ 10 15 10.			NONSTELLAR OBJECT
IC 2894	11 28 22.	+ 13 30 40.			NONSTELLAR OBJECT
IC 0702	11 28 22.	- 04 39 02.			NONSTELLAR OBJECT
RNGC 3710	11 28 23.	+ 23 02		14.5	GALAXY
ZC 1128.4+0917	11 28 24.	+ 09 17	470		CLUSTER OF GALAXIES
REIZ 1225	11 28 24.	+ 13 28	42	15.5	GALAXY
ZWG 067.096	11 28 24.	+ 13 30		15.7	GALAXY
ZC 1128.4+1436	11 28 24.	+ 14 36	610		CLUSTER OF GALAXIES
MCG+03-29-061	11 28 24.	+ 20 29	60	15.	GALAXY
MCG+04-27-051	11 28 24.	+ 20 42	36	14.7	GALAXY
VV 003B	11 28 24.	+ 20 43	15	18.	INTERACTING GALAXY
VV 003A	11 28 24.	+ 20 43	36	15.	INTERACTING GALAXY
ZWG 126.077	11 28 24.	+ 21 34		15.5	GALAXY
ZWG 126.078	11 28 24.	+ 23 02		14.5	GALAXY
UGC 06504	11 28 24.	+ 23 02	60	14.5	GALAXY E

OBJECT NAME	RIGHT ASCEN.	DECLINATION	DIAM.	MAGN.	TYPE OF OBJECT
ZWG 126.079	11 28 24.	+ 25 27		15.3	GALAXY
ZWG 156.088	11 28 24.	+ 29 34		15.5	GALAXY
ZWG 185.076	11 28 24.	+ 35 52		15.6	GALAXY
MCG+09-19-101	11 28 24.	+ 51 57	60	15.	GALAXY
UGC 06505	11 28 24.	+ 51 59	66	16.0	GALAXY Sc
ZC 1128.4+5422	11 28 24.	+ 54 22	870		CLUSTER OF GALAXIES
ZC 1128.4+5618	11 28 24.	+ 56 18	3090		CLUSTER OF GALAXIES
REIZ 1226	11 28 26.	+ 23 18	36	15.2	GALAXY
REIZ 1227	11 28 26.	+ 26 19	18	15.3	GALAXY
ARC 1283	11 28 26.	+ 61 03		17.2	RICH CLUSTER OF GALAXIES
RNGC 3712	11 28 29.	+ 28 50		15.5	GALAXY
ZWG 039.200	11 28 30.	+ 03 25		15.1	GALAXY
ZC 1128.5+1930	11 28 30.	+ 19 30	2290		CLUSTER OF GALAXIES
ZWG 156.089	11 28 30.	+ 26 30		15.5	GALAXY
LB 10235	11 28 30.	+ 26 49		15.9	FAINT BLUE STAR
ZWG 156.090	11 28 30.	+ 28 50		15.7	GALAXY
UGC 06506	11 28 30.	+ 28 50	138	15.7	GALAXY B
MCG+05-27-082	11 28 30.	+ 28 50	120	15.7	GALAXY
TON-N 0580	11 28 30.	+ 31 34		16.0	BLUE STAR
ZC 1128.5+3353	11 28 30.	+ 33 53	470		CLUSTER OF GALAXIES
MCG+06-25-080	11 28 30.	+ 34 28	66	15.	GALAXY
ZWG 185.077	11 28 30.	+ 34 29		15.6	GALAXY
MCG+06-25-079A	11 28 30.	+ 35 51	42	15.5	GALAXY
ZC 1128.5+4042	11 28 30.	+ 40 42	1080		CLUSTER OF GALAXIES
ZWG 214.014	11 28 30.	+ 40 55		15.7	GALAXY
MCG+09-19-102	11 28 30.	+ 55 23	54	17.	GALAXY
ZC 1128.5+6007	11 28 30.	+ 60 07	1210		CLUSTER OF GALAXIES
SN 1956E	11 28 30.	- 15 52		19.5	SUPERNOVA
MCG-03-29-009	11 28 30.	- 18 10	36	15.	GALAXY
A1 060	11 28 30.59	+ 15 54 01.9		15.90	FAINT BLUE OBJECT
SN 1960G	11 28 31.	+ 18 40		17.5	SUPERNOVA
REIZ 1229	11 28 31.	+ 21 33	24	14.7	GALAXY
A1 061	11 28 31.38	+ 12 15 32.0		17.10	FAINT BLUE OBJECT
MCG+04-27-052	11 28 33.	+ 23 02	18	14.5	GALAXY
REIZ 1230	11 28 33.	+ 28 50	48	14.7	GALAXY
REIZ 1231	11 28 34.	+ 03 24	30	15.2	GALAXY
ARC 1290	11 28 35.	+ 33 53		17.5	RICH CLUSTER OF GALAXIES
LB 02033	11 28 35.	+ 56 28 54.		16.2	FAINT BLUE STAR
ZWG 039.201	11 28 36.	+ 02 45		15.7	GALAXY
STOCK 30	11 28 36.	+ 06 26			BLUE KNOT NEAR ELLIP GLXY
ZC 1128.6+1604	11 28 36.	+ 16 04	1880		CLUSTER OF GALAXIES
ZC 1128.6+2244	11 28 36.	+ 22 44	1280		CLUSTER OF GALAXIES
MCG+04-27-053	11 28 36.	+ 23 18	36	15.5	GALAXY
ZC 1128.6+2402	11 28 36.	+ 24 02	3090		CLUSTER OF GALAXIES
MCG+05-27-083	11 28 36.	+ 26 32 30.	48	15.5	GALAXY
ZWG 156.091	11 28 36.	+ 26 35		15.5	GALAXY
UGC 06508	11 28 36.	+ 26 35	60	15.5	GALAXY Sb
ZWG 156.092	11 28 36.	+ 28 28		15.2	GALAXY
LB 10236	11 28 36.	+ 31 31		16.0	FAINT BLUE STAR
ZWG 185.078	11 28 36.	+ 32 58		15.4	GALAXY
ZC 1128.6+4546	11 28 36.	+ 45 46	3230		CLUSTER OF GALAXIES
ZC 1128.6+5903	11 28 36.	+ 59 03	1340		CLUSTER OF GALAXIES
REIZ 1232	11 28 38.	+ 26 34	48	14.3	GALAXY
IC 2896	11 28 39.	+ 12 37 39.			NONSTELLAR OBJECT
REIZ 1233	11 28 39.	+ 28 47	12	15.8	GALAXY
A1 062	11 28 39.56	+ 17 01 42.6		15.80	FAINT BLUE OBJECT
ZWG 039.202	11 28 42.	+ 04 56		15.0	GALAXY
MCG+01-29-047	11 28 42.	+ 04 56	30	15.0	GALAXY
MCG+04-27-054	11 28 42.	+ 23 22	138	16.	GALAXY
UGC 06509	11 28 42.	+ 23 23	114	16.0	GALAXY Sc
ZC 1128.7+2540	11 28 42.	+ 25 40	1680		CLUSTER OF GALAXIES
LB 10237	11 28 42.	+ 29 21		16.7	FAINT BLUE STAR
ZC 1128.7+3635	11 28 42.	+ 36 35	670		CLUSTER OF GALAXIES
MCG+09-19-103	11 28 42.	+ 51 40	18	17.	GALAXY
IC 2897	11 28 44.	+ 11 49 33.			NONSTELLAR OBJECT
REIZ 1234	11 28 44.	+ 25 46	30	14.7	GALAXY
IC 2899	11 28 45.	+ 10 54 39.			NONSTELLAR OBJECT
IC 2898	11 28 45.	+ 13 36 45.			NONSTELLAR OBJECT
REIZ 1235	11 28 45.	+ 28 29	24	14.4	GALAXY
LB 00238	11 28 46.	+ 56 50 24.		13.3	FAINT BLUE STAR
MCG+03-29-062	11 28 48.	+ 20 17	54	16.	GALAXY
MCG+04-27-055	11 28 48.	+ 20 45	24	17.	GALAXY
ZC 1128.8+2158	11 28 48.	+ 21 58	2620		CLUSTER OF GALAXIES
LB 10239	11 28 48.	+ 28 09		16.1	FAINT BLUE STAR
LB 10238	11 28 48.	+ 28 54		14.0	FAINT BLUE STAR
MCG-02-29-040	11 28 48.	- 13 31	48	15.	GALAXY
OCL 0858	11 28 48.	- 60 32	120	14.	OPEN STAR CLUSTER
VHA 119	11 28 48.	- 60 32	360		OPEN STAR CLUSTER
ARC 1289	11 28 50.	+ 61 03		17.0	RICH CLUSTER OF GALAXIES
REIZ 1236	11 28 51.	- 02 02	72	14.4	GALAXY
ZWG 039.203	11 28 54.	+ 06 21		14.9	GALAXY
MCG+01-29-048	11 28 54.	+ 06 21	36	14.9	GALAXY
IC 2900	11 28 54.	+ 13 26 45.			NONSTELLAR OBJECT
ZC 1128.9+3124	11 28 54.	+ 31 24	540		CLUSTER OF GALAXIES
ZWG 011.113	11 28 54.	- 01 34		15.5	GALAXY
ARC 1287	11 28 55.	+ 67 04		17.0	RICH CLUSTER OF GALAXIES
REIZ 1237	11 28 56.	+ 26 23	24	14.6	GALAXY
RNGC 3715	11 28 56.	- 13 56		13.0	GALAXY
IC 2901	11 28 57.	+ 12 58 33.			NONSTELLAR OBJECT
IC 2902	11 28 57.	+ 14 29 57.			NONSTELLAR OBJECT
REIZ 1238	11 28 57.	+ 28 25	30	13.7	GALAXY
MCG-02-29-041	11 28 57.	- 13 56	60	13.	GALAXY
KN 12.003	11 28 57.9	+ 47 41 30.			NEBULA
RNGC 3717	11 28 58.	- 29 59		12.5	GALAXY
RNGC 3713	11 28 59.	+ 28 25		14.5	GALAXY
ZWG 126.080	11 29 00.	+ 26 23		15.7	GALAXY
ZWG 156.093	11 29 00.	+ 26 45		15.7	GALAXY
MCG+05-27-084	11 29 00.	+ 28 24	30	14.4	GALAXY
ZWG 156.094	11 29 00.	+ 28 25		14.4	GALAXY
UGC 06511	11 29 00.	+ 28 25	72	14.4	GALAXY E-S0
ZWG 185.079	11 29 00.	+ 34 36		15.3	GALAXY
UGC 06512	11 29 00.	+ 34 36	72	15.3	GALAXY S
ZC 1129.0+3605	11 29 00.	+ 36 05	1280		CLUSTER OF GALAXIES
ZWG 268.045	11 29 00.	+ 54 01		15.7	GALAXY
MCG+10-17-009	11 29 00.	+ 57 26	8	16.	GALAXY
ZC 1129.0+6606	11 29 00.	+ 66 06	1280		CLUSTER OF GALAXIES
ZC 1129.0-0049	11 29 00.	- 00 49	540		CLUSTER OF GALAXIES
ZWG 012.001	11 29 00.	- 02 01		14.0	GALAXY
ZWG 011.114	11 29 00.	- 02 01		14.0	GALAXY
UGC 06510	11 29 00.	- 02 01	126	14.0	GALAXY Sc
MCG+00-30-001	11 29 00.	- 02 01	120	14.0	GALAXY
KN 12.004	11 29 00.1	+ 49 21 49.			NEBULA
REIZ 1240	11 29 02.	+ 26 45	30	14.5	GALAXY
REIZ 1241	11 29 02.	+ 26 57	54	14.8	GALAXY
A1 063	11 29 02.69	+ 15 38 31.3		15.95	FAINT BLUE OBJECT
ARP 203	11 29 03.	+ 28 47			PECULIAR GALAXY
RNGC 3752	11 29 03.	+ 74 54		13.5	GALAXY
IC 2903	11 29 05.	+ 12 55 03.			NONSTELLAR OBJECT
REIZ 1239	11 29 05.	+ 54 07	18	14.9	GALAXY
ZWG 040.001	11 29 06.	+ 03 46		14.5	GALAXY
UGC 06513	11 29 06.	+ 03 46	42	14.5	GALAXY S0?
MCG+01-30-001	11 29 06.	+ 03 46	42	14.5	GALAXY
KARA.73B 0483	11 29 06.	+ 03 46	42	14.5	ISOLATED GALAXY S
ZWG 126.081	11 29 06.	+ 20 40		15.2	GALAXY
LB 10241	11 29 06.	+ 28 21		18.0	FAINT BLUE STAR
LB 10240	11 29 06.	+ 31 29		15.7	FAINT BLUE STAR
ZC 1129.1+3346	11 29 06.	+ 33 46	610		CLUSTER OF GALAXIES
MCG+06-25-081	11 29 06.	+ 34 35 30.	66	15.	GALAXY
ARC 1292	11 29 06.	+ 36 07		17.5	RICH CLUSTER OF GALAXIES
ZC 1129.1+3955	11 29 06.	+ 39 55	1480		CLUSTER OF GALAXIES
MCG+08-21-045	11 29 06.	+ 49 05	42	15.	GALAXY
MCG+09-19-104	11 29 06.	+ 54 01	36	16.	GALAXY
ZWG 334.035	11 29 06.	+ 71 05		15.2	GALAXY
UGC 06514	11 29 06.	+ 71 05	66	15.2	GALAXY CHAIN
ZWG 351.063	11 29 06.	+ 74 54		13.7	GALAXY
UGC 06515	11 29 06.	+ 74 54	108	13.7	GALAXY Sa-b
KARA.73B 0484	11 29 06.	+ 74 54	96	13.7	ISOLATED GALAXY S
ZWG 012.002	11 29 06.	- 03 21		15.7	GALAXY
MCG+00-30-002	11 29 06.	- 03 21	48	15.7	GALAXY
MCG-05-27-015	11 29 06.	- 30 02	330	12.5	GALAXY
ACK 289-10.1	11 29 06.	- 49 42			PLANETARY NEBULA
A1 064	11 29 06.55	+ 13 27 36.8		16.80	FAINT BLUE OBJECT
RNGC 3716	11 29 07.	+ 03 46		14.5	GALAXY
IC 2904	11 29 07.	+ 13 27 45.			NONSTELLAR OBJECT
TON-N 1401	11 29 08.	+ 34 57		15.1	BLUE STAR
LB 02034	11 29 08.	+ 57 31 42.		15.9	FAINT BLUE STAR
ARC 1295	11 29 08.	- 07 15		17.1	RICH CLUSTER OF GALAXIES
KN 12.005	11 29 08.2	+ 49 06 20.			NEBULA
MCG+04-27-056	11 29 09.	+ 20 39	30	15.2	GALAXY
IC 2905	11 29 11.	+ 09 22 57.			NONSTELLAR OBJECT
RNGC 3714	11 29 11.	+ 28 38		14.5	GALAXY
MCG+05-27-085	11 29 12.	+ 28 37	30	14.3	GALAXY
ZWG 156.095	11 29 12.	+ 28 38		14.3	GALAXY
UGC 06516	11 29 12.	+ 28 38	132	14.3	GALAXY PECULR
LB 10242	11 29 12.	+ 29 23		17.1	FAINT BLUE STAR
LB 10243	11 29 12.	+ 31 44		17.3	FAINT BLUE STAR
ZWG 156.096	11 29 12.	+ 32 09		15.4	GALAXY
ZC 1129.2+4805	11 29 12.	+ 48 05	870		CLUSTER OF GALAXIES
MCG+09-19-106	11 29 12.	+ 53 51	24	16.	GALAXY
SN 1973G	11 29 12.	+ 53 52			SUPERNOVA
MCG+09-19-105	11 29 12.	+ 53 59	36	16.	GALAXY
MCG+10-17-010	11 29 12.	+ 57 43	102	14.	GALAXY
VV 172E	11 29 12.	+ 71 05	12	17.5	INTERACTING GALAXY
VV 172D	11 29 12.	+ 71 05	9	17.5	INTERACTING GALAXY
VV 172C	11 29 12.	+ 71 05	15	17.	INTERACTING GALAXY
VV 172B	11 29 12.	+ 71 05	6	17.5	INTERACTING GALAXY
VV 172A	11 29 12.	+ 71 05	6	17.	INTERACTING GALAXY
VV 172	11 29 12.	+ 71 05	66		INTERACTING GALAXY
ARP 329	11 29 12.	+ 71 05			PECULIAR GALAXY
7ZW 407	11 29 12.	+ 71 06			COMPACT GALAXY
ARC 1293	11 29 13.	+ 39 21		17.8	RICH CLUSTER OF GALAXIES
IC 2907	11 29 14.	+ 10 10 33.			NONSTELLAR OBJECT
IC 2906	11 29 14.	+ 13 24 39.			NONSTELLAR OBJECT
REIZ 1243	11 29 14.	+ 26 51	54	14.6	GALAXY
ARC 1296	11 29 14.	- 04 53		17.0	RICH CLUSTER OF GALAXIES
IC 2909	11 29 15.	+ 11 44 51.			NONSTELLAR OBJECT
IC 2908	11 29 15.	+ 13 12 57.			NONSTELLAR OBJECT
REIZ 1244	11 29 15.	+ 28 38	36	13.4	GALAXY
IC 0703	11 29 17.	- 11 19 16.			NONSTELLAR OBJECT
ZWG 156.097	11 29 18.	+ 27 17		15.4	GALAXY
MCG+05-27-086	11 29 18.	+ 32 09	48	15.4	GALAXY
ZWG 185.080	11 29 18.	+ 36 58		14.0	GALAXY
UGC 06517	11 29 18.	+ 36 58	120	14.0	GALAXY Sb
ZWG 292.002	11 29 18.	+ 57 43		15.3	GALAXY
MCG+12-11-028	11 29 18.	+ 71 05	66	14.	GALAXY
MCG-05-27-016	11 29 18.	- 30 08	24	14.	GALAXY
IC 2913	11 29 19.	- 30 09 16.			NONSTELLAR OBJECT
ARC 1291	11 29 20.	+ 56 19		15.4	RICH CLUSTER OF GALAXIES
A1 065	11 29 20.01	+ 12 38 59.2		16.25	FAINT BLUE OBJECT
REIZ 1242	11 29 22.	+ 62 23	12	15.1	GALAXY
IC 0704	11 29 22.	- 11 16 16.			NONSTELLAR OBJECT
REIZ 1245	11 29 23.	+ 36 58	48	13.8	GALAXY
IC 2910	11 29 23.	- 09 28			NONSTELLAR OBJECT
ZC 1129.4+1356	11 29 24.	+ 13 56	270		CLUSTER OF GALAXIES
ZC 1129.4+1501	11 29 24.	+ 15 01	4300		CLUSTER OF GALAXIES
ZC 1129.4+2204	11 29 24.	+ 22 04	270		CLUSTER OF GALAXIES
ZWG 126.082	11 29 24.	+ 25 55		15.3	GALAXY
MCG+04-27-057	11 29 24.	+ 25 55	30	15.3	GALAXY
ZC 1129.4+2858	11 29 24.	+ 28 58	670		CLUSTER OF GALAXIES
MCG+06-25-082	11 29 24.	+ 36 57	90	14.	GALAXY
ZC 1129.4+3925	11 29 24.	+ 39 25	1410		CLUSTER OF GALAXIES
ZC 1129.4+4346	11 29 24.	+ 43 46	870		CLUSTER OF GALAXIES
ZWG 242.043	11 29 24.	+ 49 17		15.4	GALAXY
ZWG 334.036	11 29 24.	+ 70 51		15.7	GALAXY
SHAH 155	11 29 24.	+ 79 01	120		GROUP OF COMPACT GALAXIES
MCG-01-30-001	11 29 24.	- 09 28	15	15.	GALAXY
KEEL 346	11 29 24.3	+ 47 22 36.			NEBULA
KN 12.006	11 29 25.6	+ 49 17 26.			NEBULA
REIZ 1247	11 29 26.	+ 25 55	18	14.3	GALAXY
REIZ 1248	11 29 26.	+ 26 07	48	14.8	GALAXY
REIZ 1228	11 29 26.	+ 26 30	42	14.3	GALAXY
REIZ 1249	11 29 26.	+ 26 55	42	14.4	GALAXY
TON-N 1402	11 29 26.	+ 36 50		16.9	BLUE STAR
MCG+10-17-011	11 29 27.	+ 57 39 30.	45	17.	GALAXY
AR 37	11 29 27.61	+ 74 39 16.6			NEBULA
AR 36	11 29 27.87	+ 74 54 14.5			NEBULA
ARC 1300	11 29 28.	- 19 37		17.6	RICH CLUSTER OF GALAXIES
SN 1955P	11 29 29.	+ 25 15		18.5	SUPERNOVA
REIZ 1246	11 29 29.	+ 54 10	48	14.6	GALAXY
HELW 440	11 29 29.	- 29 57 09.			NEBULA
ZWG 012.004	11 29 30.	+ 00 14		15.6	GALAXY
MCG+00-30-004	11 29 30.	+ 00 14	48	15.6	GALAXY
IC 2911	11 29 30.	+ 13 15 14.			NONSTELLAR OBJECT
MCG+08-21-046	11 29 30.	+ 49 15	18	16.	GALAXY
MCG+08-21-047	11 29 30.	+ 49 15 30.	48	15.	GALAXY
ZWG 268.046	11 29 30.	+ 54 11		14.6	GALAXY
UGC 06518	11 29 30.	+ 54 11	66	14.6	GALAXY S
SEY 109	11 29 30.	+ 54 11 28.		14.2	FAINT GALAXY
ARC 1294	11 29 30.	+ 54 32		18.0	RICH CLUSTER OF GALAXIES
ZC 1129.5+6321	11 29 30.	+ 63 21	2290		CLUSTER OF GALAXIES
MCG+12-11-029	11 29 30.	+ 71 19	21	17.	GALAXY
MCG+13-08-064	11 29 30.	+ 74 55	102	14.	GALAXY
ZWG 012.003	11 29 30.	- 02 38		14.7	GALAXY
MCG+00-30-003	11 29 30.	- 02 38	36	14.7	GALAXY
PK294-04.1	11 29 30.	- 65 42	10		PLANETARY NEBULA
IC 2912	11 29 32.	+ 11 59 08.			NONSTELLAR OBJECT

OBJECT NAME	RIGHT ASCEN.	DECLINATION	DIAM.	MAGN.	TYPE OF OBJECT
A1 066	11 29 32.78	+ 13 51 00.2		17.20	FAINT BLUE OBJECT
LB 02035	11 29 33.	+ 53 03 48.		14.8	FAINT BLUE STAR
A1 067	11 29 33.35	+ 13 13 26.7		16.60	FAINT BLUE OBJECT
A1 068	11 29 33.60	+ 16 29 47.3		16.25	FAINT BLUE OBJECT
ZWG 012.006	11 29 36.	+ 01 30		15.6	GALAXY
UGC 06519	11 29 36.	+ 01 30	78	15.6	GALAXY Sc
ZC 1129.6+0438	11 29 36.	+ 04 38	340		CLUSTER OF GALAXIES
ZWG 068.001	11 29 36.	+ 13 47		15.3	GALAXY
ZC 1129.6+3018	11 29 36.	+ 30 18	2550		CLUSTER OF GALAXIES
LB 10245	11 29 36.	+ 31 54		17.9	FAINT BLUE STAR
LB 10244	11 29 36.	+ 32 19		16.0	FAINT BLUE STAR
ZC 1129.6+3410	11 29 36.	+ 34 10	940		CLUSTER OF GALAXIES
ZC 1129.6+3453	11 29 36.	+ 34 53	340		CLUSTER OF GALAXIES
MCG+09-19-109	11 29 36.	+ 54 11	60	15.	GALAXY
ZC 1129.6+5430	11 29 36.	+ 54 30	540		CLUSTER OF GALAXIES
MCG+09-19-107	11 29 36.	+ 55 16	30	15.	GALAXY
MCG+09-19-108	11 29 36.	+ 56 13	48	16.	GALAXY
MCG+10-17-012	11 29 36.	+ 62 26	30	16.	GALAXY
MRK 175	11 29 36.	+ 62 47	35	14.5	GALAXY WITH UV CONTINUUM
MCG+11-14-028	11 29 36.	+ 62 47	66	15.	GALAXY
ZWG 314.030	11 29 36.	+ 62 48		14.1	GALAXY
UGC 06520	11 29 36.	+ 62 48	90	14.1	GALAXY S
ZWG 012.005	11 29 36.	- 00 39		15.7	GALAXY
A1 069	11 29 36.06	+ 15 36 54.1		17.10	FAINT BLUE OBJECT
HOLM 260B	11 29 37.	+ 01 05	90	13.8	PART OF MULTIPLE GALAXY
IC 2914	11 29 37.	+ 13 46 14.			NONSTELLAR OBJECT
A1 070	11 29 37.89	+ 15 12 59.9		15.35	FAINT BLUE OBJECT
MCG+09-19-110	11 29 39.	+ 56 15	48		GALAXY
REIZ 1253	11 29 40.	+ 01 05	60	13.6	GALAXY
REIZ 1254	11 29 40.	+ 01 29	72	14.7	GALAXY
IC 2915	11 29 40.	+ 14 45 38.			NONSTELLAR OBJECT
ARC 1299	11 29 40.	+ 34 16		17.5	RICH CLUSTER OF GALAXIES
REIZ 1250	11 29 40.	+ 53 20	180	12.9	GALAXY
REIZ 1252	11 29 40.	- 01 02	12	15.2	GALAXY
A1 071	11 29 40.36	+ 14 44 09.2		15.90	FAINT BLUE OBJECT
IC 2916	11 29 41.	+ 11 57 38.			NONSTELLAR OBJECT
ZWG 012.008	11 29 42.	+ 01 06		13.8	GALAXY
UGC 06521	11 29 42.	+ 01 06	126	13.8	GALAXY Sb
KARA.72 289A	11 29 42.	+ 01 06	126	13.8	PART OF DOUBLE GALAXY
MCG+00-30-005	11 29 42.	+ 01 06	84	13.8	GALAXY
MCG+05-27-087	11 29 42.	+ 28 18	24	15.0	GALAXY
ZWG 156.098	11 29 42.	+ 28 19		15.0	GALAXY
UGC 06522	11 29 42.	+ 28 19	90	15.0	GALAXY S0?
ZC 1129.7+3545	11 29 42.	+ 35 45	1140		CLUSTER OF GALAXIES
ZWG 268.047	11 29 42.	+ 55 12		14.9	GALAXY
ZWG 012.007	11 29 42.	- 01 01		15.4	GALAXY
MCG-02-30-001	11 29 42.	- 09 33	48	15.	GALAXY
RNGC 3719	11 29 43.	+ 01 06		14.0	GALAXY
SEY 110	11 29 43.	+ 55 12 28.		14.7	FAINT GALAXY
IC 2917	11 29 44.	+ 11 13 26.			NONSTELLAR OBJECT
MCG+05-27-088	11 29 45.	+ 28 22	48	15.7	GALAXY
MCG+09-19-111	11 29 45.	+ 53 13	60	15.	GALAXY
HOLM 260A	11 29 46.	+ 01 04	18	13.4	PART OF MULTIPLE GALAXY
ARC 1298	11 29 46.	+ 45 06		17.0	RICH CLUSTER OF GALAXIES
KEEL 347	11 29 46.5	+ 47 25 40.			NEBULA
LB 02036	11 29 47.	+ 59 34 48.		17.4	FAINT BLUE STAR
ZWG 012.010	11 29 48.	+ 01 05		13.7	GALAXY
UGC 06523	11 29 48.	+ 01 05	60	13.7	GALAXY
KARA.72 289B	11 29 48.	+ 01 05	78	13.7	PART OF DOUBLE GALAXY
MCG+00-30-006	11 29 48.	+ 01 05	27	13.7	GALAXY
ZWG 040.002	11 29 48.	+ 05 27		15.3	GALAXY
ZWG 156.099	11 29 48.	+ 28 23		15.7	GALAXY
ZC 1129.8+4506	11 29 48.	+ 45 06	1340		CLUSTER OF GALAXIES
MCG+08-21-048	11 29 48.	+ 49 22	18	16.	GALAXY
VV 150C	11 29 48.	+ 53 12	12	16.	INTERACTING GALAXY
VV 150B	11 29 48.	+ 53 12	12	16.	INTERACTING GALAXY
VV 150A	11 29 48.	+ 53 12	24	16.	INTERACTING GALAXY
VV 150	11 29 48.	+ 53 12	120		INTERACTING GALAXY
MRK 176	11 29 48.	+ 53 13	60	15.	GALAXY WITH UV CONTINUUM
KW 18	11 29 48.	+ 53 13	25		SEYFERT GALAXY
VVI 47	11 29 48.	+ 53 13	60	15.50	SEYFERT GALAXY
1ZW 027	11 29 48.	+ 53 14			COMPACT GALAXY
ZWG 268.048	11 29 48.	+ 53 21		11.8	GALAXY
UGC 06524	11 29 48.	+ 53 21	660	11.8	GALAXY S0/SBa
KARA.72 290A	11 29 48.	+ 53 21	306	11.8	PART OF DOUBLE GALAXY
MCG+09-19-112	11 29 48.	+ 55 12	30	16.	GALAXY
MCG+10-17-013	11 29 48.	+ 62 06	60	14.	GALAXY
MCG+11-14-029	11 29 48.	+ 62 42	45	16.	GALAXY
ZWG 314.031	11 29 48.	+ 62 43		15.2	GALAXY
ZC 1129.8+6658	11 29 48.	+ 66 58	3560		CLUSTER OF GALAXIES
7ZW 408	11 29 48.	+ 73 18			COMPACT GALAXY
ZWG 012.009	11 29 48.	- 01 50		15.7	GALAXY
MCG-02-30-002	11 29 48.	- 09 42	36	14.5	GALAXY
RNGC 3720	11 29 49.	+ 01 05		13.5	GALAXY
RNGC 3718	11 29 49.	+ 53 21		11.5	GALAXY
REIZ 1256	11 29 50.	+ 24 12	36	15.5	GALAXY
SEY 111	11 29 50.	+ 53 13 45.		15.1	FAINT GALAXY
RNGC 3723	11 29 50.	- 09 42		14.0	GALAXY
LB 00590	11 29 50.	- 11 18 24.		17.7	FAINT BLUE STAR
KN 12.007	11 29 50.9	+ 49 23 14.			NEBULA
IC 2918	11 29 51.	+ 13 31 32.			NONSTELLAR OBJECT
REIZ 1257	11 29 51.	+ 28 23	24	15.1	GALAXY
REIZ 1259	11 29 52.	+ 01 05	24	13.0	GALAXY
REIZ 1258	11 29 52.	+ 35 36	48	13.7	GALAXY
REIZ 1251	11 29 52.	+ 62 46	42	14.2	GALAXY
KEEL 348	11 29 53.7	+ 46 46 12.			NEBULA
MCG+03-30-002	11 29 54.	+ 18 24	18	15.5	GALAXY
MCG+03-30-001	11 29 54.	+ 18 24	39	15.5	GALAXY
ZWG 097.001	11 29 54.	+ 18 25		15.5	GALAXY
ZC 1129.9+2008	11 29 54.	+ 20 08	1080		CLUSTER OF GALAXIES
ZWG 097.002	11 29 54.	+ 20 25		15.4	GALAXY
ZWG 126.083	11 29 54.	+ 20 43		14.9	GALAXY
UGC 06525	11 29 54.	+ 20 43	84	14.9	GALAXY SBb
MCG+06-25-083	11 29 54.	+ 35 35 30.	84	13.	GALAXY
ZWG 185.081	11 29 54.	+ 35 36		14.1	GALAXY
UGC 06526	11 29 54.	+ 35 36	114	14.1	GALAXY
ZC 1129.9+3643	11 29 54.	+ 36 43	470		CLUSTER OF GALAXIES
ZWG 242.044	11 29 54.	+ 49 23		15.7	GALAXY
MCG+09-19-113	11 29 54.	+ 53 12 30.	60	16.	GALAXY
ZWG 268.049	11 29 54.	+ 53 14		14.7	GALAXY
UGC 06527	11 29 54.	+ 53 14	66	14.7	GALAXY TRP SYS
ARP 214	11 29 54.	+ 53 21			PECULIAR GALAXY
MCG+09-19-114	11 29 54.	+ 53 21	480	11.2	GALAXY
ZWG 292.003	11 29 54.	+ 62 07		14.1	GALAXY
ZWG 291.077	11 29 54.	+ 62 07		14.1	GALAXY
UGC 06528	11 29 54.	+ 62 07	72	14.1	GALAXY Sc
REIZ 1260	11 29 56.	+ 27 06	30	14.9	GALAXY
TON-N 1403	11 29 56.	+ 34 02		16.9	BLUE STAR

OBJECT NAME	RIGHT ASCEN.	DECLINATION	DIAM.	MAGN.	TYPE OF OBJECT
TON-N 1404	11 29 56.	+ 35 59		16.9	BLUE STAR
SEY 112	11 29 57.	+ 54 57 57.		15.2	FAINT GALAXY
REIZ 1255	11 29 57.	+ 62 05	78	13.7	GALAXY
A1 072	11 29 57.93	+ 17 18 32.6		15.60	FAINT BLUE OBJECT
REIZ 1261	11 29 58.	+ 01 04	24	15.0	GALAXY
A1 073	11 29 59.01	+ 17 35 54.5		16.00	FAINT BLUE OBJECT
LB 09852	11 30	- 86 03		14.0	FAINT BLUE STAR
ZWG 012.012	11 30 00.	+ 01 17		15.6	GALAXY
IC 2919	11 30 00.	+ 14 27 56.			NONSTELLAR OBJECT
ZWG 097.003	11 30 00.	+ 18 25		15.5	GALAXY
MCG+04-27-058	11 30 00.	+ 20 41	78	14.9	GALAXY
ZC 1130.0+2453	11 30 00.	+ 24 53	2290		CLUSTER OF GALAXIES
LB 10247	11 30 00.	+ 27 39		16.5	FAINT BLUE STAR
LB 10246	11 30 00.	+ 30 35		16.8	FAINT BLUE STAR
ZWG 214.015	11 30 00.	+ 41 07		15.4	GALAXY
UGC 06529	11 30 00.	+ 41 07	66	15.4	GALAXY Sb-c
UGC 06530	11 30 00.	+ 42 23	72	15.1	GALAXY Sb
ARP 322	11 30 00.	+ 53 14			PECULIAR GALAXY
MCG+09-19-115	11 30 00.	+ 54 57 30.	36	16.	GALAXY
ZWG 292.004	11 30 00.	+ 56 42		15.5	GALAXY
MCG+12-11-030	11 30 00.	+ 72 22	60	15.	GALAXY
ZWG 334.037	11 30 00.	+ 72 23		15.1	GALAXY
MCG+15-01-010	11 30 00.	+ 89 22	66	16.	GALAXY
ZWG 012.011	11 30 00.	- 01 51		15.7	GALAXY
REIZ 1262	11 30 01.	+ 20 43	42	15.1	GALAXY
A1 074	11 30 01.68	+ 13 55 13.2		16.45	FAINT BLUE OBJECT
REIZ 1263	11 30 02.	+ 25 19	30	14.3	GALAXY
ARC 1304	11 30 05.	+ 35 45		17.5	RICH CLUSTER OF GALAXIES
MCG+03-30-003	11 30 06.	+ 20 18	39	15.5	GALAXY
ZWG 126.084	11 30 06.	+ 25 18		15.2	GALAXY
ZC 1130.1+3703	11 30 06.	+ 37 03	740		CLUSTER OF GALAXIES
ARC 1303	11 30 06.	+ 37 05		18.0	RICH CLUSTER OF GALAXIES
ZWG 214.016	11 30 06.	+ 39 21		15.7	GALAXY
UGC 06531	11 30 06.	+ 39 21	120	15.7	GALAXY
MCG+07-24-020	11 30 06.	+ 41 07	54	14.5	GALAXY
ZC 1130.1+4826	11 30 06.	+ 48 26	1480		CLUSTER OF GALAXIES
MCG+10-17-014	11 30 06.	+ 56 42	27	16.	GALAXY
7ZW 409	11 30 06.	+ 57 23			COMPACT GALAXY
MCG-03-30-001	11 30 06.	- 16 28	18	15.	GALAXY
KN 12.008	11 30 06.0	+ 49 22 45.			NEBULA
LB 02037	11 30 09.	+ 58 50 30.		16.6	FAINT BLUE STAR
FEIG 044	11 30 12.	+ 15 34		14.4	FAINT BLUE STAR
ZWG 097.004	11 30 12.	+ 15 57		15.7	GALAXY
ZWG 097.005	11 30 12.	+ 20 18		15.5	GALAXY
MCG+04-27-059	11 30 12.	+ 25 20	30	15.2	GALAXY
LB 10250	11 30 12.	+ 28 52		17.3	FAINT BLUE STAR
LB 10249	11 30 12.	+ 31 34		17.0	FAINT BLUE STAR
LB 10248	11 30 12.	+ 31 46		17.7	FAINT BLUE STAR
ZC 1130.2+3323	11 30 12.	+ 33 23	1340		CLUSTER OF GALAXIES
MCG+07-24-022	11 30 12.	+ 39 21	102	15.	GALAXY
ZWG 214.017	11 30 12.	+ 43 47		14.9	GALAXY
MCG+07-24-021	11 30 12.	+ 43 47 30.	30	15.	GALAXY
MCG+08-21-049	11 30 12.	+ 50 31	18	15.	GALAXY
ZWG 268.050	11 30 12.	+ 50 33		15.1	GALAXY
MCG+12-11-031	11 30 12.	+ 70 33	33	16.	GALAXY
ZWG 334.038	11 30 12.	+ 70 34		15.6	GALAXY
ZWG 334.039	11 30 12.	+ 70 40		15.3	GALAXY
UGC 06532	11 30 12.	+ 70 40	72	15.3	GALAXY SBc
MCG+12-11-032	11 30 12.	+ 70 40	33	15.	GALAXY
MCG+12-11-033	11 30 12.	+ 71 04	21	17.	GALAXY
ZWG 012.015	11 30 12.	- 00 27		15.7	GALAXY
MCG+00-30-007	11 30 12.	- 00 27	48	15.7	GALAXY
ZWG 012.014	11 30 12.	- 01 09		15.3	GALAXY
ZWG 012.013	11 30 12.	- 01 55		15.5	GALAXY
MCG-03-30-002	11 30 12.	- 16 27	30	15.	GALAXY
KEEL 349	11 30 12.1	+ 47 22 21.			SPIRAL NEBULA
KN 12.009	11 30 12.1	+ 50 31 22.			NEBULA
IC 2920	11 30 13.	+ 12 50 08.			NONSTELLAR OBJECT
IC 0705	11 30 13.	+ 50 31 38.			NONSTELLAR OBJECT
IC 2921	11 30 14.	+ 10 34 26.			NONSTELLAR OBJECT
ARC 1307	11 30 15.	+ 14 49		16.8	RICH CLUSTER OF GALAXIES
IC 2924	11 30 17.	+ 09 18 02.			NONSTELLAR OBJECT
IC 2922	11 30 17.	+ 13 12 02.			NONSTELLAR OBJECT
ARC 1305	11 30 17.	+ 35 04		17.2	RICH CLUSTER OF GALAXIES
ZWG 040.003	11 30 18.	+ 06 00		15.2	GALAXY
UGC 06533	11 30 18.	+ 06 00	60	15.2	GALAXY Sb
ZC 1130.3+4654	11 30 18.	+ 46 54	740		CLUSTER OF GALAXIES
MCG+09-19-116	11 30 18.	+ 54 57 30.	30	16.	GALAXY
LB 02038	11 30 18.	+ 55 04 36.		16.4	FAINT BLUE STAR
VHE 51	11 30 18.	- 63 11	96		REFLECTION NEBULA
IC 2923	11 30 19.	+ 13 26 31.			NONSTELLAR OBJECT
ARC 1308	11 30 21.	- 03 42		15.7	RICH CLUSTER OF GALAXIES
ZWG 012.016	11 30 24.	+ 02 06		15.3	GALAXY
MCG+00-30-008	11 30 24.	+ 02 06	30	15.3	GALAXY
ZWG 040.004	11 30 24.	+ 07 00		15.5	GALAXY
KARA.73B 0485	11 30 24.	+ 07 00	24	15.5	ISOLATED GALAXY S
ZWG 097.006	11 30 24.	+ 16 10		15.7	GALAXY
ZWG 126.085	11 30 24.	+ 24 58		15.7	GALAXY
LB 10252	11 30 24.	+ 28 22		17.5	FAINT BLUE STAR
LB 10251	11 30 24.	+ 29 19		16.5	FAINT BLUE STAR
TON-N 0581	11 30 24.	+ 30 00		14.5	BLUE STAR
ZC 1130.4+4052	11 30 24.	+ 40 52	1210		CLUSTER OF GALAXIES
MCG+08-21-050	11 30 24.	+ 50 33 30.	60	15.	GALAXY
MRK 177	11 30 24.	+ 55 20	12	16.	GALAXY WITH UV CONTINUUM
ZWG 314.032	11 30 24.	+ 63 34		13.3	GALAXY
UGC 06534	11 30 24.	+ 63 34	180	13.3	GALAXY Sc
SHB 182	11 30 24.2	+ 10 40 16.		17.5	QUASI-STELLAR OBJECT
BC 4C10.33	11 30 24.9	+ 10 40 18.		17.49	QUASI-STELLAR OBJECT
REIZ 1264	11 30 25.	+ 22 40	12	14.9	GALAXY
KEEL 350	11 30 25.2	+ 47 00 15.			NEBULA
IC 2926	11 30 28.	+ 12 42 49.			NONSTELLAR OBJECT
ARC 1302	11 30 28.	+ 66 42		16.7	RICH CLUSTER OF GALAXIES
A1 075	11 30 28.59	+ 17 37 19.3		17.00	FAINT BLUE OBJECT
ARC 1306	11 30 29.	+ 46 54		17.6	RICH CLUSTER OF GALAXIES
A1 076	11 30 29.27	+ 13 31 58.0		16.90	FAINT BLUE OBJECT
ZWG 012.017	11 30 30.	+ 01 49		15.6	GALAXY
IC 2927	11 30 30.	+ 13 21 49.			NONSTELLAR OBJECT
ZC 1130.5+1740	11 30 30.	+ 17 40	6790		CLUSTER OF GALAXIES
ZWG 097.007	11 30 30.	+ 20 00		15.5	GALAXY
MCG+03-30-004	11 30 30.	+ 20 04	48	15.7	GALAXY
MCG+04-27-060	11 30 30.	+ 24 58	24	15.7	GALAXY
LB 10253	11 30 30.	+ 29 40		17.4	FAINT BLUE STAR
UGC 06535	11 30 30.	+ 50 35	78	15.1	GALAXY Sc
MCG+11-14-030	11 30 30.	+ 63 33	120	14.	GALAXY
MCG-02-30-003	11 30 30.	- 09 58	60	15.	GALAXY
ARC 1309	11 30 30.	- 11 34		17.0	RICH CLUSTER OF GALAXIES
KN 12.010	11 30 30.0	+ 49 22 52.			NEBULA
REIZ 1265	11 30 32.	+ 49 35	18	14.7	GALAXY

OBJECT NAME	RIGHT ASCEN.	DECLINATION	DIAM.	MAGN.	TYPE OF OBJECT
IC 2925	11 30 33.	+ 34 33 01.			NONSTELLAR OBJECT
KN 12.011	11 30 33.6	+ 49 33 45.			NEBULA
SEY 113	11 30 34.	+ 55 21 27.		15.3	FAINT GALAXY
RNGC 3728	11 30 35.	+ 24 43		14.5	GALAXY
HOLM 262B	11 30 35.	+ 32 50	24	15.2	PART OF MULTIPLE GALAXY
A1 077	11 30 35.45	+ 17 38 19.2		16.60	FAINT BLUE OBJECT
ZWG 040.005	11 30 36.	+ 05 05		15.6	GALAXY
ZWG 097.008	11 30 36.	+ 20 04		15.7	GALAXY
ZWG 097.009	11 30 36.	+ 20 05		15.6	GALAXY
MCG+03-30-005	11 30 36.	+ 20 05	39	15.6	GALAXY
ZWG 126.086	11 30 36.	+ 24 25		15.7	GALAXY
ZWG 126.087	11 30 36.	+ 24 43		14.7	GALAXY
UGC 06536	11 30 36.	+ 24 43	150	14.7	GALAXY Sb
ZWG 126.088	11 30 36.	+ 25 30		15.7	GALAXY
LB 10255	11 30 36.	+ 27 24		16.3	FAINT BLUE STAR
LB 10254	11 30 36.	+ 28 28		16.4	FAINT BLUE STAR
MCG+06-25-084	11 30 36.	+ 32 50	30	16.	GALAXY
ZWG 186.001	11 30 36.	+ 34 32		15.6	GALAXY
ZWG 185.082	11 30 36.	+ 34 32		15.6	GALAXY
ZC 1130.6+3514	11 30 36.	+ 35 14	2220		CLUSTER OF GALAXIES
STOCK 31	11 30 36.	+ 38 40			BLUE KNOT NEAR ELLIP GLXY
ZWG 242.045	11 30 36.	+ 47 18		11.2	GALAXY
UGC 06537	11 30 36.	+ 47 18	366	11.2	GALAXY Sc
MCG+08-21-051	11 30 36.	+ 47 18	360	10.8	GALAXY
MCG+08-21-052	11 30 36.	+ 49 32	60	14.8	GALAXY
UGC 06538	11 30 36.	+ 49 34	60	16.0	GALAXY S
ZC 1130.6+6054	11 30 36.	+ 60 54	1480		CLUSTER OF GALAXIES
MCG-02-30-004	11 30 36.	- 13 05	72	15.	GALAXY
REIZ 1267	11 30 38.	+ 23 38	48	14.4	GALAXY
REIZ 1268	11 30 38.	+ 24 55	36	14.9	GALAXY
REIZ 1269	11 30 38.	+ 24 57	24	14.6	GALAXY
REIZ 1266	11 30 38.	+ 49 32	42	14.1	GALAXY
HOLM 263B	11 30 38.	+ 49 33	24	14.8	PART OF MULTIPLE GALAXY
ARC 1297	11 30 38.	+ 76 31		16.1	RICH CLUSTER OF GALAXIES
A1 078	11 30 38.34	+ 13 35 17.8		16.00	FAINT BLUE OBJECT
KN 12.012	11 30 38.5	+ 47 18 21.			NEBULA
MCG+07-24-023	11 30 39.	+ 44 04	18	15.5	GALAXY
RNGC 3726	11 30 39.	+ 47 18		11.0	GALAXY
MCG+04-27-061	11 30 42.	+ 24 43	84	14.7	GALAXY
ZWG 185.083	11 30 42.	+ 32 52		15.4	GALAXY
UGC 06539	11 30 42.	+ 32 52	90	15.4	GALAXY Sb-c
KARA.72 291A	11 30 42.	+ 32 52	84	15.4	PART OF DOUBLE GALAXY
ZWG 186.002	11 30 42.	+ 34 20		15.7	GALAXY
ZWG 185.084	11 30 42.	+ 34 20		15.7	GALAXY
IC 0706	11 30 42.	- 13 04 05.			NONSTELLAR OBJECT
HOLM 261A	11 30 43.	+ 46 08	24	15.0	PART OF MULTIPLE GALAXY
REIZ 1270	11 30 43.	+ 47 19	300	11.6	GALAXY
ARC 1311	11 30 44.	- 23 46		17.4	RICH CLUSTER OF GALAXIES
HOLM 262A	11 30 45.	+ 32 52	96	14.2	PART OF MULTIPLE GALAXY
MCG+06-25-085	11 30 45.	+ 32 52	78	15.	GALAXY
HOLM 264B	11 30 45.	+ 34 34	18	15.3	PART OF MULTIPLE GALAXY
LB 02039	11 30 45.	+ 50 49 42.		16.2	FAINT BLUE STAR
KN 12.013	11 30 45.5	+ 49 31 08.			NEBULA
REIZ 1271	11 30 46.	+ 34 35	36	14.1	GALAXY
RNGC 3725	11 30 47.	+ 62 10		13.5	GALAXY
A1 079	11 30 47.90	+ 13 03 36.9		17.00	FAINT BLUE OBJECT
ZWG 012.018	11 30 48.	+ 01 05		15.5	GALAXY
MCG+04-27-063	11 30 48.	+ 22 42 30.	15	15.3	GALAXY
ZWG 126.089	11 30 48.	+ 22 44		15.3	GALAXY
ZWG 126.090	11 30 48.	+ 25 24		15.3	GALAXY
MCG+04-27-062	11 30 48.	+ 25 25	30	15.3	GALAXY
ZWG 186.003	11 30 48.	+ 33 46		15.4	GALAXY
ZWG 185.085	11 30 48.	+ 33 46		15.4	GALAXY
ZWG 186.004	11 30 48.	+ 34 22		15.7	GALAXY
ZWG 185.086	11 30 48.	+ 34 22		15.7	GALAXY
ZWG 186.005	11 30 48.	+ 34 35		14.7	GALAXY
ZWG 185.087	11 30 48.	+ 34 35		14.7	GALAXY
UGC 06540	11 30 48.	+ 34 35	72	14.7	GALAXY Sb-c
MCG+08-21-053	11 30 48.	+ 49 30	72	13.9	GALAXY
ZWG 242.046	11 30 48.	+ 49 31		13.9	GALAXY
MRK 178	11 30 48.	+ 49 31	30	14.	GALAXY WITH UV CONTINUUM
UGC 06541	11 30 48.	+ 49 31	78	13.9	GALAXY IRR+CMP
ZWG 292.005	11 30 48.	+ 62 10		13.6	GALAXY
ZWG 291.078	11 30 48.	+ 62 10		13.6	GALAXY
UGC 06542	11 30 48.	+ 62 10	96	13.6	GALAXY SBc
ZC 1130.8+6805	11 30 48.	+ 68 05	940		CLUSTER OF GALAXIES
ZWG 352.002	11 30 48.	+ 76 31		15.7	GALAXY
ZWG 351.064	11 30 48.	+ 76 31		15.7	GALAXY
REIZ 1273	11 30 49.	+ 22 43	12	14.8	GALAXY
PATH 1.403	11 30 49.	+ 29 13	14		NEBULA
HOLM 264C	11 30 49.	+ 34 35	18	15.6	PART OF MULTIPLE GALAXY
IC 2928	11 30 49.	+ 34 36 13.			NONSTELLAR OBJECT
ARC 1301	11 30 49.	+ 75 21		16.5	RICH CLUSTER OF GALAXIES
REIZ 1274	11 30 50.	+ 23 42	42	13.9	GALAXY
REIZ 1275	11 30 50.	+ 24 25	24	14.6	GALAXY
REIZ 1276	11 30 50.	+ 24 43	21	14.4	GALAXY
REIZ 1277	11 30 50.	+ 25 25	30	14.4	GALAXY
LB 00591	11 30 50.	+ 27 20 54.		17.2	FAINT BLUE STAR
HOLM 262C	11 30 50.	+ 32 52	18	15.3	PART OF MULTIPLE GALAXY
HOLM 263A	11 30 50.	+ 49 30	48	13.9	PART OF MULTIPLE GALAXY
REIZ 1278	11 30 51.	+ 29 12	18	14.9	GALAXY
MCG+06-25-086	11 30 51.	+ 34 35	60	14.5	GALAXY
HOLM 264A	11 30 52.	+ 34 35	24	14.3	PART OF MULTIPLE GALAXY
HOLM 261B	11 30 52.	+ 46 08	24	15.0	PART OF MULTIPLE GALAXY
ZC 1130.9+3435	11 30 54.	+ 34 35	2820		CLUSTER OF GALAXIES
ZC 1130.9+3810	11 30 54.	+ 38 10	1810		CLUSTER OF GALAXIES
ZC 1130.9+4635	11 30 54.	+ 46 35	670		CLUSTER OF GALAXIES
ZWG 012.019	11 30 54.	- 02 06		15.5	GALAXY
MCG+00-30-009	11 30 54.	- 02 06	48	15.5	GALAXY
IC 2929	11 30 56.	+ 12 25 01.			NONSTELLAR OBJECT
REIZ 1272	11 30 57.	+ 62 08	72	13.5	GALAXY
KN 12.014	11 30 59.9	+ 49 12 43.			NEBULA
ZWG 012.021	11 31 00.	+ 00 40		15.5	GALAXY
ZWG 126.091	11 31 00.	+ 21 39		14.4	GALAXY
UGC 06543	11 31 00.	+ 21 39	36	14.4	GALAXY S
MCG+04-27-064	11 31 00.	+ 21 39	30	14.4	GALAXY
ZWG 126.092	11 31 00.	+ 22 08		15.7	GALAXY
ZWG 126.093	11 31 00.	+ 23 41		15.1	GALAXY
UGC 06544	11 31 00.	+ 23 41	72	15.1	GALAXY Sb-c
ZWG 126.094	11 31 00.	+ 24 57		15.6	GALAXY
LB 10256	11 31 00.	+ 28 36		16.3	FAINT BLUE STAR
ZWG 186.006	11 31 00.	+ 32 54		14.5	GALAXY
ZWG 185.088	11 31 00.	+ 32 54		14.5	GALAXY
UGC 06545	11 31 00.	+ 32 54	66	14.5	GALAXY
KARA.72 291B	11 31 00.	+ 32 54	78	14.5	PART OF DOUBLE GALAXY
ZC 1131.0+4403	11 31 00.	+ 44 03	940		CLUSTER OF GALAXIES
MCG+08-21-054	11 31 00.	+ 49 11	48	16.	GALAXY
MCG+09-19-118	11 31 00.	+ 51 30	36	16.	GALAXY

OBJECT NAME	RIGHT ASCEN.	DECLINATION	DIAM.	MAGN.	TYPE OF OBJECT
MCG+09-19-117	11 31 00.	+ 53 24	150	11.8	GALAXY
LB 00239	11 31 00.	+ 56 24 18.		15.2	FAINT BLUE STAR
MRK 179	11 31 00.	+ 62 10	50	14.	GALAXY WITH UV CONTINUUM
MCG+10-17-015	11 31 00.	+ 62 10	57	14.	GALAXY
ZWG 012.020	11 31 00.	- 02 00		14.9	GALAXY
MCG+00-30-010	11 31 00.	- 02 00	24	14.9	GALAXY
MCG-04-27-012	11 31 00.	- 26 40	60	15.5	GALAXY
MCG-05-27-017	11 31 00.	- 32 53	36	15.5	GALAXY
PK292+04.1	11 31 00.	- 56 49 10.	10		PLANETARY NEBULA
KEEL 351	11 31 01.2	+ 46 42 49.			NEBULA
REIZ 1280	11 31 02.	+ 25 25	30	15.1	GALAXY
HOLM 265B	11 31 02.	+ 32 54	18	14.5	PART OF MULTIPLE GALAXY
TON-N 1405	11 31 02.	+ 33 17		16.8	BLUE STAR
LB 02040	11 31 02.	+ 53 57 24.		16.5	FAINT BLUE STAR
A1 080	11 31 02.74	+ 17 25 38.1		16.90	FAINT BLUE OBJECT
MCG+04-27-065	11 31 03.	+ 23 41	60	15.1	GALAXY
HOLM 265A	11 31 03.	+ 32 54	18	14.3	PART OF MULTIPLE GALAXY
REIZ 1279	11 31 04.	+ 53 24	120	12.8	GALAXY
ZWG 068.002	11 31 06.	+ 10 22		15.7	GALAXY
MCG+03-30-006	11 31 06.	+ 17 39	63	15.5	GALAXY
ZWG 097.010	11 31 06.	+ 17 40		15.5	GALAXY
UGC 06546	11 31 06.	+ 17 40	72	15.5	GALAXY Sb-c
MCG+08-21-055	11 31 06.	+ 45 31	36	16.	GALAXY
ZC 1131.1+4726	11 31 06.	+ 47 26	1480		CLUSTER OF GALAXIES
ZWG 268.051	11 31 06.	+ 53 25		12.2	GALAXY
UGC 06547	11 31 06.	+ 53 25	204	12.2	GALAXY SB
KARA.72 290B	11 31 06.	+ 53 25	192	12.2	PART OF DOUBLE GALAXY
ZC 1131.1+7520	11 31 06.	+ 75 20	2690		CLUSTER OF GALAXIES
SC 1128-5909.1	11 31 06.	- 59 25 40.	18		NEBULA
REIZ 1281	11 31 07.	+ 21 39	24	13.7	GALAXY
ARC 1310	11 31 07.	+ 40 08		17.8	RICH CLUSTER OF GALAXIES
IC 0707	11 31 08.	+ 21 39 25.			NONSTELLAR OBJECT
REIZ 1282	11 31 08.	+ 24 37	12	15.0	GALAXY
REIZ 1283	11 31 08.	+ 24 57	24	14.5	GALAXY
REIZ 1284	11 31 08.	+ 25 25	18	14.7	GALAXY
SEY 114	11 31 08.	+ 51 31 39.		15.0	FAINT GALAXY
RNGC 3729	11 31 08.	+ 53 24		13.0	GALAXY
RNGC 3727	11 31 08.	- 13 36			GALAXY
IC 2930	11 31 10.	+ 10 22 07.			NONSTELLAR OBJECT
KN 12.015	11 31 10.8	+ 49 00 01.			NEBULA
ZWG 126.095	11 31 12.	+ 23 55		15.0	GALAXY
UGC 06548	11 31 12.	+ 23 55	60	15.0	GALAXY S0
MCG+04-27-066	11 31 12.	+ 23 55	48	15.0	GALAXY
MCG+06-26-001	11 31 12.	+ 32 55	54	14.5	GALAXY
ZWG 242.047	11 31 12.	+ 45 33		15.7	GALAXY
ZWG 242.048	11 31 12.	+ 49 20		14.2	GALAXY
UGC 06549	11 31 12.	+ 49 20	102	14.2	GALAXY E
ZC 1131.2+4923	11 31 12.	+ 49 23	5310		CLUSTER OF GALAXIES
MCG+09-19-119	11 31 12.	+ 51 27	30	16.	GALAXY
ZC 1131.2+6009	11 31 12.	+ 60 09	1410		CLUSTER OF GALAXIES
MRK 425	11 31 12.	+ 64 26	7	16.	GALAXY WITH UV CONTINUUM
ZWG 012.022	11 31 12.	- 03 09		15.6	GALAXY
MCG+00-30-011	11 31 12.	- 03 09	66	15.6	GALAXY
KN 12.016	11 31 12.9	+ 49 36 45.			NEBULA
REIZ 1285	11 31 13.	+ 22 35	36	14.6	GALAXY
A1 081	11 31 14.58	+ 12 57 50.5		16.50	FAINT BLUE OBJECT
IC 2931	11 31 15.	+ 12 44 42.			NONSTELLAR OBJECT
KN 12.018	11 31 15.6	+ 45 32 23.			NEBULA
KN 12.017	11 31 16.2	+ 49 20 29.			NEBULA
ZWG 097.011	11 31 18.	+ 20 17		15.6	GALAXY
ZWG 126.096	11 31 18.	+ 22 35		15.5	GALAXY
MCG+08-21-056	11 31 18.	+ 49 19 30.	30	14.	GALAXY
ZWG 268.052	11 31 18.	+ 51 33		15.2	GALAXY
IC 2932	11 31 19.	+ 10 49 18.			NONSTELLAR OBJECT
PATH 1.404	11 31 19.	+ 29 16	14		NEBULA
LB 02041	11 31 19.	+ 55 27 24.		16.2	FAINT BLUE STAR
KN 12.019	11 31 19.1	+ 49 25 54.			NEBULA
REIZ 1286	11 31 20.	+ 25 50	48	15.1	GALAXY
IC 0708	11 31 22.	+ 49 21 12.			NONSTELLAR OBJECT
ZC 1131.4+0528	11 31 24.	+ 05 28	870		CLUSTER OF GALAXIES
MCG+04-27-067	11 31 24.	+ 22 34	30	15.5	GALAXY
LB 10258	11 31 24.	+ 29 20		18.0	FAINT BLUE STAR
LB 10257	11 31 24.	+ 31 38		17.5	FAINT BLUE STAR
REIZ 1287	11 31 26.	+ 25 02	12	15.1	GALAXY
A1 082	11 31 26.57	+ 14 28 12.3		16.50	FAINT BLUE OBJECT
REIZ 1288	11 31 27.	+ 32 48	48	14.1	GALAXY
LB 02042	11 31 27.	+ 54 04 30.		16.6	FAINT BLUE STAR
KN 12.020	11 31 29.4	+ 50 04 19.			NEBULA
ZC 1131.5+2257	11 31 30.	+ 22 57	4700		CLUSTER OF GALAXIES
MCG+04-27-068	11 31 30.	+ 25 48	18	16.5	GALAXY
ZWG 186.007	11 31 30.	+ 34 35		15.2	GALAXY
ZWG 185.089	11 31 30.	+ 34 35		15.2	GALAXY
UGC 06550	11 31 30.	+ 34 35	84	15.2	GALAXY Sa-b
UGC 06551	11 31 30.	+ 36 57	84	16.0	GALAXY Sc
MCG+06-26-002	11 31 30.	+ 36 57 30.	90	15.	GALAXY
ZC 1131.5+3848	11 31 30.	+ 38 48	940		CLUSTER OF GALAXIES
LB 02043	11 31 30.	+ 49 12 18.		16.2	FAINT BLUE STAR
ZWG 242.049	11 31 30.	+ 49 19		15.0	GALAXY
MCG+09-19-120	11 31 30.	+ 54 27 30.	36	16.	GALAXY
ZWG 334.040	11 31 30.	+ 71 49		14.3	GALAXY
UGC 06552	11 31 30.	+ 71 49	72	14.3	GALAXY S
ZWG 012.023	11 31 30.	- 03 19		15.4	GALAXY
MCG+00-30-012	11 31 30.	- 03 19	36	15.4	GALAXY
MCG-02-30-005	11 31 30.	- 09 34 30.	36	13.5	GALAXY
ARC 1312	11 31 31.	+ 50 21		17.6	RICH CLUSTER OF GALAXIES
REIZ 1289	11 31 32.	+ 26 09	24	14.7	GALAXY
TON-N 1406	11 31 32.	+ 32 41		16.9	BLUE STAR
A1 083	11 31 32.68	+ 17 39 59.8			FAINT BLUE OBJECT
ARC 1313	11 31 33.	+ 17 22		17.2	RICH CLUSTER OF GALAXIES
MCG+06-26-003	11 31 33.	+ 33 34 30.	30	15.5	GALAXY
IC 2933	11 31 33.	+ 34 36 00.			NONSTELLAR OBJECT
VV 148	11 31 33.	+ 56 25	60	15.	INTERACTING GALAXY
KEEL 352	11 31 33.2	+ 47 08 12.			NEBULA
REIZ 1290	11 31 34.	+ 34 34	30	14.1	GALAXY
KN 12.021	11 31 34.8	+ 48 49 47.			NEBULA
ZWG 068.003	11 31 36.	+ 12 48		14.3	GALAXY
RNGC 3731	11 31 36.	+ 12 48		14.5	GALAXY
UGC 06553	11 31 36.	+ 12 48	60	14.3	GALAXY E
MCG+02-30-001	11 31 36.	+ 12 48	24	14.3	GALAXY
ZC 1131.6+1252	11 31 36.	+ 12 52	470		CLUSTER OF GALAXIES
MCG+06-26-004	11 31 36.	+ 34 35	66	14.5	GALAXY
ZC 1131.6+4421	11 31 36.	+ 44 21	4910		CLUSTER OF GALAXIES
ZC 1131.6+4753	11 31 36.	+ 47 53	1340		CLUSTER OF GALAXIES
MCG+08-21-057	11 31 36.	+ 49 18 30.	24	15.	GALAXY
IC 0709	11 31 36.	+ 49 20 42.			NONSTELLAR OBJECT
ZWG 012.024	11 31 36.	- 01 18		15.4	GALAXY
A1 084	11 31 39.71	+ 14 12 33.6		16.60	FAINT BLUE OBJECT
LB 02044	11 31 40.	+ 53 49 36.		16.4	FAINT BLUE STAR

OBJECT NAME	RIGHT ASCEN.	DECLINATION	DIAM.	MAGN.	TYPE OF OBJECT
ZWG 068.004	11 31 42.	+ 13 37		14.8	GALAXY
MCG+02-30-002	11 31 42.	+ 13 37	36	14.8	GALAXY
KARA.73B 0486	11 31 42.	+ 13 37	36	14.8	ISOLATED GALAXY S
ZWG 186.008	11 31 42.	+ 38 14		15.7	GALAXY
ZWG 185.090	11 31 42.	+ 38 14		15.7	GALAXY
ZC 1131.7+4102	11 31 42.	+ 41 02	670		CLUSTER OF GALAXIES
MCG+08-21-058	11 31 42.	+ 46 37	18	15.	GALAXY
ZWG 242.050	11 31 42.	+ 46 39		15.7	GALAXY
ZC 1131.7+5028	11 31 42.	+ 50 28	1080		CLUSTER OF GALAXIES
A1 085	11 31 42.36	+ 17 00 16.1		15.80	FAINT BLUE OBJECT
IC 2934	11 31 44.	+ 13 36 00.			NONSTELLAR OBJECT
RNGC 3732	11 31 44.	- 09 34		13.5	GALAXY
REIZ 1293	11 31 44.	- 09 34	24	12.7	GALAXY
A1 086	11 31 44.83	+ 17 19 05.8		16.70	FAINT BLUE OBJECT
REIZ 1291	11 31 45.	+ 32 51	36	14.0	GALAXY
MCG-01-30-002	11 31 45.	- 06 08	72	15.	GALAXY
KN 12.022	11 31 46.5	+ 46 38 08.			NEBULA
ZWG 040.006	11 31 48.	+ 03 32		15.5	GALAXY
REIZ 1292	11 31 48.	+ 12 46	24	14.8	GALAXY
REIZ 1294	11 31 48.	+ 13 35	48	15.2	GALAXY
MCG+03-30-007	11 31 48.	+ 15 54	36	14.6	GALAXY
ZWG 097.012	11 31 48.	+ 15 56		14.6	GALAXY
ZWG 126.097	11 31 48.	+ 20 38		15.5	GALAXY
ZC 1131.8+2555	11 31 48.	+ 25 55	3700		CLUSTER OF GALAXIES
ZWG 126.098	11 31 48.	+ 26 08		15.7	GALAXY
LB 10260	11 31 48.	+ 26 41		16.8	FAINT BLUE STAR
LB 10259	11 31 48.	+ 27 38		16.6	FAINT BLUE STAR
ZWG 186.009	11 31 48.	+ 33 27		15.0	GALAXY
ZC 1131.8+3404	11 31 48.	+ 34 04	1880		CLUSTER OF GALAXIES
ZC 1131.8+3916	11 31 48.	+ 39 16	1680		CLUSTER OF GALAXIES
ZWG 242.051	11 31 48.	+ 46 27		15.6	GALAXY
MCG+08-21-059	11 31 48.	+ 49 12	30	16.	GALAXY
MCG+09-19-121	11 31 48.	+ 55 17	48	17.	GALAXY
MCG+11-14-031	11 31 48.	+ 63 42	27	17.	GALAXY
MCG+12-11-034	11 31 48.	+ 71 49	66	15.	GALAXY
ZWG 012.025	11 31 48.	- 02 15		15.2	GALAXY
MCG+00-30-013	11 31 48.	- 02 15	42	15.2	GALAXY
MCG-01-30-003	11 31 48.	- 09 19	60	15.	GALAXY
MCG-01-30-004	11 31 48.	- 09 19 30.	60	16.	GALAXY
KN 12.023	11 31 48.1	+ 49 24 23.			NEBULA
REIZ 1295	11 31 50.	+ 26 09	30	14.7	GALAXY
RNGC 3721	11 31 50.	- 09 19		15.0	GALAXY
IC 0710	11 31 51.	+ 26 08 30.			NONSTELLAR OBJECT
MCG+06-26-005	11 31 51.	+ 33 27	24	15.	GALAXY
KN 12.024	11 31 51.2	+ 46 26 36.			NEBULA
REIZ 1296	11 31 52.	+ 01 01	12	15.3	GALAXY
ZWG 012.026	11 31 54.	+ 01 02		15.7	GALAXY
MCG+03-30-008	11 31 54.	+ 18 57	54	15.6	GALAXY
ZWG 126.099	11 31 54.	+ 22 47		15.6	GALAXY
ZC 1131.9+2509	11 31 54.	+ 25 09	940		CLUSTER OF GALAXIES
ZC 1131.9+3009	11 31 54.	+ 30 09	340		CLUSTER OF GALAXIES
ZWG 314.033	11 31 54.	+ 63 43		15.7	GALAXY
MCG-01-30-006	11 31 54.	- 09 24 30.	9	17.	GALAXY
MCG-01-30-005	11 31 54.	- 09 25	24	15.	GALAXY
MCG-04-27-013	11 31 54.	- 26 35	30	15.5	GALAXY
REIZ 1297	11 31 55.	+ 22 47	18	15.0	GALAXY
REIZ 1298	11 31 56.	+ 24 48	30	15.6	GALAXY
RNGC 3722	11 31 56.	- 09 23		16.0	GALAXY
RNGC 3724	11 31 56.	- 09 25		15.0	GALAXY
LB 02046	11 31 57.	+ 54 03 42.		16.8	FAINT BLUE STAR
LB 02045	11 31 57.	+ 61 24 48.		15.4	FAINT BLUE STAR
KN 12.025	11 31 58.6	+ 49 13 22.			NEBULA
ZWG 097.013	11 32 00.	+ 18 57		15.6	GALAXY
MCG+03-30-009	11 32 00.	+ 19 18	42	16.	GALAXY
ZC 1132.0+1940	11 32 00.	+ 19 40	1680		CLUSTER OF GALAXIES
LB 10262	11 32 00.	+ 31 34		17.0	FAINT BLUE STAR
LB 10261	11 32 00.	+ 32 05		18.2	FAINT BLUE STAR
ZC 1132.0+3753	11 32 00.	+ 37 53	400		CLUSTER OF GALAXIES
MCG+08-21-061	11 32 00.	+ 47 15	60	15.	GALAXY
ZWG 242.052	11 32 00.	+ 47 16		15.1	GALAXY
MCG+08-21-060	11 32 00.	+ 49 12	24	16.	GALAXY
MCG+08-21-062	11 32 00.	+ 49 12 30.	18	15.	GALAXY
ZWG 242.053	11 32 00.	+ 49 14		15.2	GALAXY
MCG+08-21-063	11 32 00.	+ 49 20	60	14.	GALAXY
ZC 1132.0+6428	11 32 00.	+ 64 28	3090		CLUSTER OF GALAXIES
MCG-01-30-007	11 32 00.	- 09 24	24	15.	GALAXY
REIZ 1299	11 32 01.	+ 20 38	24	14.6	GALAXY
KN 12.026	11 32 02.3	+ 49 32 03.			NEBULA
LB 00592	11 32 03.	+ 29 34 06.		17.3	FAINT BLUE STAR
LB 02047	11 32 03.	+ 50 20 54.		17.2	FAINT BLUE STAR
KN 12.027	11 32 03.6	+ 49 14 05.			NEBULA
ZWG 012.027	11 32 06.	+ 00 25		15.7	GALAXY
ZWG 126.100	11 32 06.	+ 20 45		15.6	GALAXY
LB 10263	11 32 06.	+ 28 43		16.5	FAINT BLUE STAR
ARC 1314	11 32 06.	+ 49 20		13.9	RICH CLUSTER OF GALAXIES
ZWG 242.054	11 32 06.	+ 49 21		14.8	GALAXY
7ZW 410	11 32 06.	+ 79 05			COMPACT GALAXY
ZWG 352.003	11 32 06.	+ 79 05		15.6	GALAXY
ZWG 351.065	11 32 06.	+ 79 05		15.6	GALAXY
MCG-02-30-006	11 32 06.	- 13 50	54	14.5	GALAXY
KN 12.029	11 32 06.2	+ 47 15 59.			NEBULA
KEEL 353	11 32 06.3	+ 47 15 57.			NEBULA
KN 12.028	11 32 06.5	+ 49 21 30.			NEBULA
REIZ 1300	11 32 08.	+ 25 49	42	14.4	GALAXY
RNGC 3730	11 32 08.	- 09 21			GALAXY
RNGC 3734	11 32 08.	- 13 50		14.0	GALAXY
A1 087	11 32 08.73	+ 16 32 05.8		16.65	FAINT BLUE OBJECT
IC 0713	11 32 09.	+ 17 07			NONSTELLAR OBJECT
SEY 115	11 32 09.	+ 51 29 38.		14.2	FAINT GALAXY
MCG-01-30-008	11 32 09.	- 09 23	36	15.	GALAXY
REIZ 1302	11 32 10.	+ 00 23	30	15.4	GALAXY
KN 12.030	11 32 10.3	+ 47 43 14.			NEBULA
IC 0711	11 32 11.	+ 49 14 42.			NONSTELLAR OBJECT
ZWG 012.029	11 32 12.	+ 00 31		15.4	GALAXY
ZWG 040.007	11 32 12.	+ 03 11		15.4	GALAXY
ZWG 068.005	11 32 12.	+ 11 32		15.5	GALAXY
MCG+04-27-069	11 32 12.	+ 20 44	42	15.6	GALAXY
ZWG 126.101	11 32 12.	+ 25 47		15.1	GALAXY
MCG+04-27-070	11 32 12.	+ 25 47	45	15.1	GALAXY
LB 10267	11 32 12.	+ 27 18		18.1	FAINT BLUE STAR
LB 10266	11 32 12.	+ 30 12		17.9	FAINT BLUE STAR
LB 10265	11 32 12.	+ 30 22		16.6	FAINT BLUE STAR
LB 10264	11 32 12.	+ 31 27		15.2	FAINT BLUE STAR
MCG+09-19-122	11 32 12.	+ 51 28	36	15.	GALAXY
ZWG 012.028	11 32 12.	- 03 17		15.7	GALAXY
IC 2935	11 32 14.	+ 10 31 35.			NONSTELLAR OBJECT
SEY 116	11 32 15.	+ 54 56 38.		14.6	FAINT GALAXY
REIZ 1303	11 32 16.	+ 00 30	12	15.2	GALAXY

OBJECT NAME	RIGHT ASCEN.	DECLINATION	DIAM.	MAGN.	TYPE OF OBJECT
IC 0712	11 32 16.	+ 49 22 12.			NONSTELLAR OBJECT
REIZ 1301	11 32 16.	+ 55 07	120	14.7	GALAXY
SHB 183	11 32 16.3	+ 30 22 02.		18.2	QUASI-STELLAR OBJECT
BC 3C261	11 32 16.31	+ 30 22 01.0		18.24	QUASI-STELLAR OBJECT
MCG+03-30-010	11 32 18.	+ 16 21	108	15.6	GALAXY
ZC 1132.3+3739	11 32 18.	+ 37 39	1550		CLUSTER OF GALAXIES
MCG+09-19-124	11 32 18.	+ 51 30	30	16.	GALAXY
ZWG 268.053	11 32 18.	+ 51 31		14.4	GALAXY
UGC 06555	11 32 18.	+ 51 31	84	14.4	GALAXY E-SO
ZWG 268.054	11 32 18.	+ 54 56		14.9	GALAXY
MCG+09-19-125	11 32 18.	+ 54 56	42	15.	GALAXY
MCG+09-19-123	11 32 18.	+ 55 07	300	13.1	GALAXY
ZWG 268.055	11 32 18.	+ 55 08		13.2	GALAXY
UGC 06554	11 32 18.	+ 55 08	288	13.2	GALAXY Sc
ZWG 012.030	11 32 18.	- 03 03		15.6	GALAXY
REIZ 1304	11 32 19.	+ 20 45	36	14.9	GALAXY
REIZ 1305	11 32 19.	+ 22 55	30	14.9	GALAXY
RNGC 3733	11 32 20.	+ 55 08		13.0	GALAXY
LB 02048	11 32 20.	+ 55 58 00.		16.5	FAINT BLUE STAR
IC 2936	11 32 21.	+ 13 17 17.			NONSTELLAR OBJECT
REIZ 1306	11 32 22.	+ 02 59	36	15.2	GALAXY
ZWG 068.006	11 32 24.	+ 08 45		15.4	GALAXY
ZC 1132.4+1312	11 32 24.	+ 13 12	1880		CLUSTER OF GALAXIES
MCG+03-30-011	11 32 24.	+ 15 56	27	15.3	GALAXY
ZWG 097.014	11 32 24.	+ 15 57		15.3	GALAXY
ZWG 097.015	11 32 24.	+ 16 23		15.6	GALAXY
UGC 06556	11 32 24.	+ 16 23	90	15.6	GALAXY Sc
ZWG 126.102	11 32 24.	+ 20 57		15.7	GALAXY
LB 10268	11 32 24.	+ 26 47		16.8	FAINT BLUE STAR
ZC 1132.4+3010	11 32 24.	+ 30 10	2690		CLUSTER OF GALAXIES
UGC 06557	11 32 24.	+ 30 10	60	16.0	GALAXY PECULR
ZC 1132.4+3615	11 32 24.	+ 36 15	2550		CLUSTER OF GALAXIES
ZC 1132.4+4842	11 32 24.	+ 48 42	740		CLUSTER OF GALAXIES
MCG-04-28-001	11 32 24.	- 27 09	24	15.5	GALAXY
KN 12.031	11 32 24.2	+ 47 58 32.			NEBULA
KN 12.032	11 32 25.7	+ 49 34 04.			NEBULA
REIZ 1307	11 32 26.	+ 24 33	12	15.5	GALAXY
LB 02049	11 32 26.	+ 60 29 48.		16.6	FAINT BLUE STAR
REIZ 1308	11 32 28.	+ 02 49	48	15.0	GALAXY
IC 2937	11 32 28.	+ 10 22 53.			NONSTELLAR OBJECT
LB 02050	11 32 29.	+ 56 32 36.		15.1	FAINT BLUE STAR
ZWG 040.008	11 32 30.	+ 02 49		15.4	GALAXY
UGC 06558	11 32 30.	+ 02 49	60	15.4	GALAXY SBc
KARA.73B 0487	11 32 30.	+ 02 49	96	15.4	ISOLATED GALAXY S
ZWG 097.016	11 32 30.	+ 16 14		15.7	GALAXY
UGC 06559	11 32 30.	+ 16 14	102	15.7	GALAXY Sc
MCG+03-30-012	11 32 30.	+ 18 02	63	15.7	GALAXY
ZWG 097.017	11 32 30.	+ 18 03		15.7	GALAXY
FATH 1.405	11 32 30.	+ 30 10			NEBULA
MCG+09-19-127	11 32 30.	+ 51 11	30	16.	GALAXY
ZWG 268.056	11 32 30.	+ 51 27		15.3	GALAXY
MCG+09-19-126	11 32 30.	+ 55 12 30.	30	14.8	GALAXY
ZC 1132.5+5734	11 32 30.	+ 57 34	3830		CLUSTER OF GALAXIES
FATH 1.406	11 32 33.	+ 29 50	14		NEBULA
ZWG 126.103	11 32 36.	+ 20 35		15.7	GALAXY
ZC 1132.6+3019	11 32 36.	+ 30 19	340		CLUSTER OF GALAXIES
ZC 1132.6+4015	11 32 36.	+ 40 15	2690		CLUSTER OF GALAXIES
LB 02051	11 32 36.	+ 49 29 12.		14.6	FAINT BLUE STAR
ZWG 292.006	11 32 36.	+ 57 55		14.8	GALAXY
MCG+10-17-016	11 32 36.	+ 57 56	27	18.	GALAXY
ZWG 314.034	11 32 36.	+ 63 25		15.5	GALAXY
ZWG 334.041	11 32 36.	+ 73 44		15.6	GALAXY
RNGC 3736	11 32 36.	+ 73 44		15.5	GALAXY
UGC 06560	11 32 36.	+ 73 44	96	15.6	GALAXY
KARA.73B 0488	11 32 36.	+ 73 44	60	15.6	ISOLATED GALAXY S
ARC 1317	11 32 36.	- 13 15		16.5	RICH CLUSTER OF GALAXIES
A1 088	11 32 36.61	+ 16 32 25.3		17.40	FAINT BLUE OBJECT
A1 089	11 32 39.03	+ 17 28 09.5		16.80	FAINT BLUE OBJECT
LB 00593	11 32 41.	+ 29 02 30.		17.3	FAINT BLUE STAR
ZWG 012.031	11 32 42.	+ 01 55		15.3	GALAXY
MCG+03-30-013	11 32 42.	+ 16 12	102	15.7	GALAXY
ZC 1132.7+3211	11 32 42.	+ 32 11	940		CLUSTER OF GALAXIES
HOLM 267B	11 32 42.	+ 33 35	18	15.6	PART OF MULTIPLE GALAXY
ZWG 268.057	11 32 42.	+ 55 12		15.4	GALAXY
HOLM 266B	11 32 42.	+ 55 13	18	14.8	PART OF MULTIPLE GALAXY
MCG+10-17-017	11 32 42.	+ 59 10	18	16.	GALAXY
A1 090	11 32 42.11	+ 15 14 52.9		16.40	FAINT BLUE OBJECT
KN 12.033	11 32 43.5	+ 49 21 51.			NEBULA
RNGC 3737A	11 32 44.	+ 55 13			GALAXY
KN 12.035	11 32 45.9	+ 46 57 26.			NEBULA
REIZ 1313	11 32 46.	+ 00 02	78	15.2	GALAXY
REIZ 1309	11 32 46.	+ 55 12	24	15.0	GALAXY
HOLM 266A	11 32 46.	+ 55 14	18	14.0	PART OF MULTIPLE GALAXY
KEEL 354	11 32 46.2	+ 46 57 22.			NEBULA
KN 12.034	11 32 46.6	+ 49 23 08.			NEBULA
ZC 1132.8+0710	11 32 48.	+ 07 10	540		CLUSTER OF GALAXIES
ZC 1132.8+1510	11 32 48.	+ 15 10	1550		CLUSTER OF GALAXIES
ZWG 126.104	11 32 48.	+ 20 47		15.6	GALAXY
TON-N 0582	11 32 48.	+ 27 40		14.2	BLUE STAR
ZWG 186.010	11 32 48.	+ 33 34		15.6	GALAXY
UGC 06561	11 32 48.	+ 33 34	66	15.6	GALAXY Sc-IRR
MCG+06-26-006	11 32 48.	+ 33 35	54	15.	GALAXY
UGC 06562	11 32 48.	+ 43 45	66	16.5	GALAXY Sc
MCG+08-21-064	11 32 48.	+ 49 21 30.	48	16.	GALAXY
ZWG 268.058	11 32 48.	+ 55 13		13.9	GALAXY
UGC 06563	11 32 48.	+ 55 13	96	13.9	GALAXY
ZWG 268.059	11 32 48.	+ 55 48		15.3	GALAXY
LB 02052	11 32 48.	+ 57 41 54.		17.1	FAINT BLUE STAR
MCG+10-17-018	11 32 48.	+ 57 58	24	16.	GALAXY
MCG+12-11-035	11 32 48.	+ 73 42 30.	66	16.	GALAXY
ZWG 352.004	11 32 48.	+ 79 15		15.6	GALAXY
ZWG 351.066	11 32 48.	+ 79 15		15.6	GALAXY
ZWG 012.032	11 32 48.	- 02 29		15.6	GALAXY
REIZ 1311	11 32 49.	+ 49 08	12	14.7	GALAXY
REIZ 1310	11 32 49.	+ 61 26	12	15.1	GALAXY
REIZ 1314	11 32 50.	+ 25 23	48	14.2	GALAXY
HOLM 267A	11 32 51.	+ 33 35	60	14.1	PART OF MULTIPLE GALAXY
MCG+09-19-128	11 32 51.	+ 55 14	36	14.0	GALAXY
REIZ 1315	11 32 51.	- 02 41	30	14.9	GALAXY
REIZ 1312	11 32 52.	+ 55 13	18	14.5	GALAXY
ARC 1316	11 32 53.	+ 37 41		17.6	RICH CLUSTER OF GALAXIES
PK293+01.1	11 32 53.	- 60 00	30		PLANETARY NEBULA
ZC 1132.9+2120	11 32 54.	+ 21 20	1010		CLUSTER OF GALAXIES
ZC 1132.9+2400	11 32 54.	+ 24 00	2290		CLUSTER OF GALAXIES
LB 00594	11 32 54.	+ 25 43 54.		17.1	FAINT BLUE STAR
ZWG 214.018	11 32 54.	+ 42 18		15.7	GALAXY
MCG+09-19-129	11 32 54.	+ 55 49	36	16.	GALAXY
MCG+10-17-019	11 32 54.	+ 57 56	30	15.	GALAXY

OBJECT NAME	RIGHT ASCEN.	DECLINATION	DIAM.	MAGN.	TYPE OF OBJECT
ZWG 292.007	11 32 54.	+ 58 36		15.1	GALAXY
MCG+11-14-032	11 32 54.	+ 63 45	27	17.	GALAXY
TON-N 1407	11 32 56.	+ 32 51		17.	BLUE STAR
RNGC 3737	11 32 56.	+ 55 14		14.0	GALAXY
RNGC 3742	11 32 58.	- 37 40			GALAXY
KN 12.036	11 32 58.5	+ 49 05 02.			NEBULA
RNGC 3739	11 32 59.	+ 25 20		15.5	GALAXY
KN 12.037	11 32 59.5	+ 49 21 26.			NEBULA
ZWG 068.007	11 33 00.	+ 13 58		15.3	GALAXY
MRK 636	11 33 00.	+ 16 15	9	17.	GALAXY WITH UV CONTINUUM
ZWG 126.105	11 33 00.	+ 25 20		15.3	GALAXY
UGC 06564	11 33 00.	+ 25 20	78	15.3	GALAXY Sb-c
TON-N 0583	11 33 00.	+ 27 59		15.5	BLUE STAR
LB 10269	11 33 00.	+ 29 48		15.7	FAINT BLUE STAR
ZC 1133.0+4145	11 33 00.	+ 41 45	1610		CLUSTER OF GALAXIES
MCG+08-21-065	11 33 00.	+ 49 18	66	16.	GALAXY
MCG+08-21-066	11 33 00.	+ 49 25	12	16.	GALAXY
MCG+08-21-067	11 33 00.	+ 50 00	24	15.	GALAXY
ZWG 242.055	11 33 00.	+ 50 01		15.5	GALAXY
ZWG 268.060	11 33 00.	+ 54 47		11.5	GALAXY
UGC 06565	11 33 00.	+ 54 47	210	11.5	GALAXY IRR
UGC 06566	11 33 00.	+ 58 28	78	16.5	GALAXY DWRF SP
ZWG 314.035	11 33 00.	+ 63 46		15.3	GALAXY
ZWG 334.042	11 33 00.	+ 70 48		12.4	GALAXY
UGC 06567	11 33 00.	+ 70 48	270	12.4	GALAXY Sc
ZC 1133.0+8000	11 33 00.	+ 80 00	870		CLUSTER OF GALAXIES
MCG-06-26-001	11 33 00.	- 37 39	120	12.	GALAXY
ASS 62	11 33 00.	- 62 19			OB ASSOCIATION CEN OB2
KN 12.038	11 33 00.2	+ 50 01 08.			NEBULA
A1 091	11 33 00.39	+ 16 15 05.1		16.00	FAINT BLUE OBJECT
KN 12.039	11 33 00.9	+ 49 26 48.			NEBULA
IC 2938	11 33 01.	+ 13 57 29.			NONSTELLAR OBJECT
REIZ 1317	11 33 01.	+ 20 46	36	15.5	GALAXY
KN 12.040	11 33 01.5	+ 49 19 00.			NEBULA
IC 2939	11 33 03.	+ 10 58 29.			NONSTELLAR OBJECT
REIZ 1316	11 33 03.	+ 54 48	60	12.2	GALAXY
A1 092	11 33 03.10	+ 14 00 13.3		17.30	FAINT BLUE OBJECT
KN 12.041	11 33 03.2	+ 49 25 49.			NEBULA
KN 12.042	11 33 03.8	+ 49 07 49.			NEBULA
REIZ 1318	11 33 04.	+ 00 23	36	13.8	GALAXY
LYNG 07	11 33 04.	- 63 13 30.	5040		OB CONCENTRATION
ZWG 012.033	11 33 06.	+ 00 25		14.4	GALAXY
UGC 06568	11 33 06.	+ 00 25	60	14.4	GALAXY S
MCG+00-30-014	11 33 06.	+ 00 25	42	14.4	GALAXY
KARA.73B 0489	11 33 06.	+ 00 25	48	14.4	ISOLATED GALAXY I
RNGC 3746	11 33 06.	+ 22 01			GALAXY
ZC 1133.1+3816	11 33 06.	+ 38 16	1010		CLUSTER OF GALAXIES
ARP 234	11 33 06.	+ 54 48			PECULIAR GALAXY
MCG+09-19-130	11 33 06.	+ 54 48	120	12.0	GALAXY
MCG+10-17-020	11 33 06.	+ 58 28	57	15.	GALAXY
RNGC 3738	11 33 08.	+ 54 48		12.5	GALAXY
RNGC 3735	11 33 08.	+ 70 49		12.5	GALAXY
MCG+04-27-071	11 33 09.	+ 25 21	60	15.3	GALAXY
A1 093	11 33 09.53	+ 14 18 19.3		16.20	FAINT BLUE OBJECT
REIZ 1319	11 33 10.	+ 35 36	18	13.7	GALAXY
ZWG 097.018	11 33 12.	+ 17 43		15.7	GALAXY
UGC 06569	11 33 12.	+ 17 43	60	15.7	GALAXY SB
ZWG 126.106	11 33 12.	+ 22 00		15.6	GALAXY
RNGC 3743	11 33 12.	+ 22 00		15.5	GALAXY
ZWG 126.107	11 33 12.	+ 23 16		15.4	GALAXY
RNGC 3744	11 33 12.	+ 23 16		15.5	GALAXY
ZC 1133.2+2655	11 33 12.	+ 26 55	3430		CLUSTER OF GALAXIES
ZC 1133.2+3116	11 33 12.	+ 31 16	2420		CLUSTER OF GALAXIES
LB 10270	11 33 12.	+ 31 20		16.4	FAINT BLUE STAR
ZWG 186.011	11 33 12.	+ 35 36		14.5	GALAXY
UGC 06570	11 33 12.	+ 35 36	72	14.5	GALAXY S0-a
MCG+06-26-007	11 33 12.	+ 35 37	30	14.5	GALAXY
ZC 1133.2+4259	11 33 12.	+ 42 59	1140		CLUSTER OF GALAXIES
MCG+08-21-068	11 33 12.	+ 45 32	120	13.9	GALAXY
ZWG 268.061	11 33 12.	+ 55 21		15.5	GALAXY
MCG+10-17-021	11 33 12.	+ 57 14	39	16.	GALAXY
ZC 1133.2+6141	11 33 12.	+ 61 41	1140		CLUSTER OF GALAXIES
7ZW 411	11 33 12.	+ 70 57			COMPACT GALAXY
MCG-02-30-007	11 33 12.	- 15 27 30.	54	14.5	GALAXY
ARC 1315	11 33 14.	+ 72 13		16.5	RICH CLUSTER OF GALAXIES
AR 38	11 33 15.14	+ 75 33 50.1			NEBULA
KN 12.043	11 33 16.8	+ 49 32 34.			NEBULA
LB 10271	11 33 18.	+ 29 52		17.7	FAINT BLUE STAR
ZC 1133.3+4104	11 33 18.	+ 41 04	1210		CLUSTER OF GALAXIES
ZWG 242.056	11 33 18.	+ 49 43		15.7	GALAXY
MCG+09-19-131	11 33 18.	+ 55 21	30	16.	GALAXY
MCG+10-17-022	11 33 18.	+ 60 32	39	16.	GALAXY
ZWG 314.036	11 33 18.	+ 62 32		14.6	GALAXY
REIZ 1320	11 33 19.	+ 23 13	12	14.9	GALAXY
ARC 1320	11 33 20.	- 05 32		17.6	RICH CLUSTER OF GALAXIES
RNGC 3741	11 33 21.	+ 45 34		14.0	GALAXY
MCG+11-14-033	11 33 21.	+ 62 31 30.	51	15.	GALAXY
KN 12.044	11 33 21.3	+ 49 43 31.			NEBULA
RNGC 3749	11 33 22.	- 37 43			GALAXY
IC 2940	11 33 23.	+ 22 03			DOUBLE STAR
ZWG 012.034	11 33 24.	+ 00 00		15.7	GALAXY
ZWG 068.008	11 33 24.	+ 08 35		15.3	GALAXY
UGC 06571	11 33 24.	+ 08 35	60	15.3	GALAXY Sa
REIZ 1322	11 33 24.	+ 45 33	30	13.7	GALAXY
ZWG 242.057	11 33 24.	+ 45 34		14.2	GALAXY
UGC 06572	11 33 24.	+ 45 34	102	14.2	GALAXY IRR
MCG+08-21-069	11 33 24.	+ 49 42	24	15.	GALAXY
ZWG 292.008	11 33 24.	+ 60 15		14.9	GALAXY
UGC 06573	11 33 24.	+ 60 15	60	14.9	GALAXY S
MCG+12-11-036	11 33 24.	+ 70 48	234	12.6	GALAXY
MCG-01-30-009	11 33 24.	- 04 20 30.	36	16.	GALAXY
MCG-06-26-002	11 33 24.	- 37 42	90	13.	GALAXY
HN 0789	11 33 24.	- 62 45			NEBULA
KN 12.045	11 33 24.9	+ 45 33 26.			NEBULA
REIZ 1323	11 33 25.	+ 22 03	12	15.3	GALAXY
REIZ 1324	11 33 25.	+ 23 12	6	15.6	GALAXY
REIZ 1325	11 33 25.	+ 23 17	24	14.8	GALAXY
RNGC 3740	11 33 25.	+ 60 15		15.0	GALAXY
REIZ 1326	11 33 27.	+ 33 31	60	14.1	GALAXY
ZC 1133.5+2354	11 33 30.	+ 23 54	670		CLUSTER OF GALAXIES
FEIG 045	11 33 30.	+ 29 18		14.8	FAINT BLUE STAR
MCG+08-21-070	11 33 30.	+ 44 55	30	16.	GALAXY
REIZ 1321	11 33 30.	+ 60 15	42	14.4	GALAXY
MCG+10-17-023	11 33 30.	+ 60 15 30.	42	14.6	GALAXY
MCG+11-14-034	11 33 30.	+ 63 21 30.	39	16.	GALAXY
ZWG 314.037	11 33 30.	+ 63 23		15.1	GALAXY
MRK 180	11 33 30.	+ 70 25	20	15.	GALAXY WITH UV CONTINUUM
VVI 48	11 33 30.	+ 70 25	20	15.06	SEYFERT GALAXY

OBJECT NAME	RIGHT ASCEN.	DECLINATION	DIAM.	MAGN.	TYPE OF OBJECT
ZWG 334.043	11 33 30.	+ 70 27		15.5	GALAXY
MCG-01-30-010	11 33 30.	- 06 25	36	15.5	GALAXY
MCG-02-30-008	11 33 30.	- 12 18 30.	48	14.5	GALAXY
CED 118	11 33 30.	- 62 45	3960		DIFFUSE GALACTIC NEBULA
REIZ 1327	11 33 31.	+ 23 09	18	15.7	GALAXY
RNGC 3777	11 33 32.	- 12 18		14.0	GALAXY
LB 02053	11 33 33.	+ 51 42 48.		15.4	FAINT BLUE STAR
REIZ 1328	11 33 33.	- 01 50	36	14.7	GALAXY
IC 2941	11 33 35.	+ 10 19 58.			NONSTELLAR OBJECT
SN 1966D	11 33 35.	+ 20 48		19.5	SUPERNOVA
ZWG 068.009	11 33 36.	+ 10 20		15.0	GALAXY
MCG+02-30-003	11 33 36.	+ 10 20	48	15.0	GALAXY
ZC 1133.6+1147	11 33 36.	+ 11 47	1210		CLUSTER OF GALAXIES
ZC 1133.6+1346	11 33 36.	+ 13 46	3360		CLUSTER OF GALAXIES
ZWG 126.108	11 33 36.	+ 20 48		15.5	GALAXY
RNGC 3745	11 33 36.	+ 22 02			NON-EXISTENT OBJECT
RNGC 3748	11 33 36.	+ 22 03			NON-EXISTENT OBJECT
ZWG 126.109	11 33 36.	+ 22 42		14.8	GALAXY
MCG+04-27-072	11 33 36.	+ 22 42	42	14.8	GALAXY
ZC 1133.6+2909	11 33 36.	+ 29 09	2150		CLUSTER OF GALAXIES
LB 10272	11 33 36.	+ 29 17		14.8	FAINT BLUE STAR
UGC 06574	11 33 36.	+ 30 03	66	16.5	GALAXY Sb-c
ARC 1319	11 33 36.	+ 40 22		17.8	RICH CLUSTER OF GALAXIES
MCG+09-19-132	11 33 36.	+ 55 16	24	16.	GALAXY
ZWG 292.009	11 33 36.	+ 58 29		14.3	GALAXY
UGC 06575	11 33 36.	+ 58 29	132	14.3	GALAXY Sc
7ZW 412	11 33 36.	+ 70 27			COMPACT GALAXY
ZWG 012.035	11 33 36.	- 02 49		15.0	GALAXY
MCG+00-30-015	11 33 36.	- 02 49	48	15.0	GALAXY
IC 2942	11 33 37.	+ 12 05 46.			NONSTELLAR OBJECT
REIZ 1329	11 33 37.	+ 22 41	24	14.0	GALAXY
REIZ 1330	11 33 37.	+ 23 12	18	15.7	GALAXY
LB 00595	11 33 37.	+ 29 17 54.		15.3	FAINT BLUE STAR
LB 00596	11 33 39.	+ 28 52 54.		17.5	FAINT BLUE STAR
ARC 1318	11 33 39.	+ 55 15		15.0	RICH CLUSTER OF GALAXIES
LB 02054	11 33 40.	+ 51 59 30.		15.3	FAINT BLUE STAR
KN 12.046	11 33 40.2	+ 44 56 32.			NEBULA
MCG+03-30-015	11 33 42.	+ 15 42	42	15.0	GALAXY
ZWG 097.019	11 33 42.	+ 15 45		15.0	GALAXY
MCG+03-30-014	11 33 42.	+ 20 05	15	14.6	GALAXY
ZC 1133.7+2037	11 33 42.	+ 20 37	870		CLUSTER OF GALAXIES
RNGC 3750	11 33 42.	+ 21 59			NON-EXISTENT OBJECT
ZWG 242.058	11 33 42.	+ 44 57		15.7	GALAXY
MCG+09-19-133	11 33 42.	+ 51 15	15	16.	GALAXY
LB 00597	11 33 42.	+ 55 35 54.		17.4	FAINT BLUE STAR
ZWG 012.036	11 33 42.	- 02 35		15.3	GALAXY
REIZ 1332	11 33 43.	+ 20 46	12	15.0	GALAXY
REIZ 1333	11 33 43.	+ 23 09	12	15.4	GALAXY
ARC 1323	11 33 44.	- 07 45		17.2	RICH CLUSTER OF GALAXIES
REIZ 1334	11 33 47.	+ 12 05	36	15.4	GALAXY
REIZ 1331	11 33 47.	+ 58 26	90	14.2	GALAXY
MCG+03-30-016	11 33 48.	+ 15 45	36	15.5	GALAXY
ZWG 097.020	11 33 48.	+ 15 47		15.5	GALAXY
ZWG 097.021	11 33 48.	+ 20 05		14.6	GALAXY
MCG+04-27-073	11 33 48.	+ 21 51	30	14.8	GALAXY
ZWG 126.110	11 33 48.	+ 21 52		14.8	GALAXY
RNGC 3758	11 33 48.	+ 21 52		15.0	GALAXY
RNGC 3751	11 33 48.	+ 21 57			NON-EXISTENT OBJECT
RNGC 3754	11 33 48.	+ 21 59			NON-EXISTENT OBJECT
RNGC 3753	11 33 48.	+ 21 59			NON-EXISTENT OBJECT
ZC 1133.8+2433	11 33 48.	+ 24 33	2220		CLUSTER OF GALAXIES
ZWG 157.001	11 33 48.	+ 27 09		14.9	GALAXY
ZWG 156.100	11 33 48.	+ 27 09		14.9	GALAXY
LB 10273	11 33 48.	+ 27 54		17.7	FAINT BLUE STAR
MCG+08-21-071	11 33 48.	+ 48 04 30.	78	15.	GALAXY
ZWG 242.059	11 33 48.	+ 48 06		15.6	GALAXY
UGC 06576	11 33 48.	+ 48 06	72	15.6	GALAXY SB0
MCG+08-21-072	11 33 48.	+ 49 23	30	15.	GALAXY
ZWG 242.060	11 33 48.	+ 49 24		14.9	GALAXY
LB 02055	11 33 48.	+ 54 55 06.		16.8	FAINT BLUE STAR
MCG+10-17-024	11 33 48.	+ 58 11	39	16.	GALAXY
MCG+10-17-025	11 33 48.	+ 58 29	102	14.	GALAXY
ZWG 012.037	11 33 48.	- 02 34		15.2	GALAXY
MCG-02-30-009	11 33 48.	- 09 34 30.	60	13.5	GALAXY
OCL 0860	11 33 48.	- 61 20	720	4.6	OPEN STAR CLUSTER
VHA 120	11 33 48.	- 61 20	660		OPEN STAR CLUSTER
KN 12.047	11 33 48.3	+ 49 24 44.			NEBULA
KN 12.049	11 33 48.7	+ 47 10 58.			NEBULA
REIZ 1335	11 33 49.	+ 25 00	60	14.9	GALAXY
KN 12.048	11 33 49.0	+ 48 05 43.			NEBULA
LB 02056	11 33 50.	+ 52 11 36.		15.9	FAINT BLUE STAR
LB 00598	11 33 50.	+ 57 27 18.		17.2	FAINT BLUE STAR
RNGC 3763	11 33 50.	- 09 34		13.0	GALAXY
RNGC 3766	11 33 50.	- 61 20		4.5	OPEN CLUSTER
KN 12.050	11 33 51.6	+ 49 08 39.			NEBULA
RNGC 3755	11 33 52.	+ 36 41		14.0	GALAXY
REIZ 1336	11 33 52.	+ 36 41	108	13.9	GALAXY
ARC 1321	11 33 52.	+ 48 22		17.8	RICH CLUSTER OF GALAXIES
A1 094	11 33 52.74	+ 17 00 40.3		16.60	FAINT BLUE OBJECT
LB 02057	11 33 53.	+ 58 32 48.		15.9	FAINT BLUE STAR
ZWG 040.009	11 33 54.	+ 03 01		15.3	GALAXY
MCG+05-27-089	11 33 54.	+ 27 06	48	14.9	GALAXY
ZWG 186.012	11 33 54.	+ 36 40		13.9	GALAXY
UGC 06577	11 33 54.	+ 36 40	216	13.9	GALAXY Sc
MCG+06-26-008	11 33 54.	+ 36 42	192	13.	GALAXY
MCG+08-21-073	11 33 54.	+ 48 05	30	17.	GALAXY
ZC 1133.9+4812	11 33 54.	+ 48 12	1340		CLUSTER OF GALAXIES
MCG+08-21-074	11 33 54.	+ 49 19	30	15.	GALAXY
ZWG 242.061	11 33 54.	+ 49 20		15.5	GALAXY
ZWG 268.062	11 33 54.	+ 55 08		15.0	GALAXY
ZWG 012.038	11 33 54.	- 02 32		15.5	GALAXY
KN 12.051	11 33 54.6	+ 49 20 28.			NEBULA
A1 095	11 33 54.90	+ 13 46 34.1		17.20	FAINT BLUE OBJECT
REIZ 1338	11 33 55.	+ 21 55	18	14.2	GALAXY
REIZ 1339	11 33 55.	+ 23 03	6	15.2	GALAXY
REIZ 1340	11 33 55.	+ 23 04	18	14.8	GALAXY
REIZ 1341	11 33 55.	+ 23 12	12	15.5	GALAXY
REIZ 1342	11 33 55.	+ 23 15	6	16.1	GALAXY
KN 12.052	11 33 55.5	+ 48 06 48.			NEBULA
REIZ 1343	11 33 56.	+ 27 07	24	14.4	GALAXY
A1 096	11 33 56.16	+ 14 30 57.3		16.10	FAINT BLUE OBJECT
REIZ 1344	11 33 57.	+ 32 22	12	15.0	GALAXY
REIZ 1337	11 33 57.	+ 55 07	18	14.7	GALAXY
REIZ 1345	11 33 58.	+ 36 40	84	13.3	GALAXY
IC 2943	11 33 58.	+ 55 07 22.			NONSTELLAR OBJECT
IC 0714	11 33 58.	- 09 34 20.			NONSTELLAR OBJECT
ZWG 012.040	11 34 00.	+ 01 06		15.1	GALAXY
UGC 06578	11 34 00.	+ 01 06	60	15.1	GALAXY DBL SYS

OBJECT NAME	RIGHT ASCEN.	DECLINATION	DIAM.	MAGN.	TYPE OF OBJECT
ZC 1134.0+0726	11 34 00.	+ 07 26	1340		CLUSTER OF GALAXIES
ZC 1134.0+1557	11 34 00.	+ 15 57	470		CLUSTER OF GALAXIES
MCG+03-30-017	11 34 00.	+ 17 54	33	15.2	GALAXY
RNGC 3760	11 34 00.	+ 22 06			NON-EXISTENT OBJECT
ZWG 268.063	11 34 00.	+ 54 34		12.1	GALAXY
UGC 06579	11 34 00.	+ 54 34	300	12.1	GALAXY Sc
MRK 041	11 34 00.	+ 55 08	20	15.	GALAXY WITH UV CONTINUUM
ZC 1134.0+6207	11 34 00.	+ 62 07	1550		CLUSTER OF GALAXIES
ZC 1134.0+6326	11 34 00.	+ 63 26	870		CLUSTER OF GALAXIES
ZWG 334.044	11 34 00.	+ 70 22		15.7	GALAXY
UGC 06580	11 34 00.	+ 70 22	60	15.7	GALAXY Sb-c
ZWG 012.039	11 34 00.	- 02 38		15.6	GALAXY
KARA.73 33	11 34 00.	- 35 50	27		DWARF GALAXY
REIZ 1349	11 34 01.	+ 22 06	24	15.8	GALAXY
CED 119	11 34 01.	- 61 00	18		DIFFUSE GALACTIC NEBULA
A1 097	11 34 02.36	+ 13 18 01.1		16.85	FAINT BLUE OBJECT
A1 098	11 34 02.56	+ 15 05 28.9		15.85	FAINT BLUE OBJECT
REIZ 1346	11 34 03.	+ 54 34	180	12.5	GALAXY
REIZ 1350	11 34 04.	+ 01 05	18	14.0	GALAXY
LB 00599	11 34 04.	+ 56 21 42.		17.4	FAINT BLUE STAR
ZWG 040.010	11 34 06.	+ 06 34		14.8	GALAXY
MCG+01-30-002	11 34 06.	+ 06 34	24	14.8	GALAXY
KARA.73B 0490	11 34 06.	+ 06 34	48	14.8	ISOLATED GALAXY E
ARC 1325	11 34 06.	+ 07 29		17.2	RICH CLUSTER OF GALAXIES
ZWG 097.022	11 34 06.	+ 17 55		15.2	GALAXY
MCG+03-30-018	11 34 06.	+ 20 17	12	15.3	GALAXY
MCG+09-19-134	11 34 06.	+ 54 34	240	12.0	GALAXY
ZWG 268.064	11 34 06.	+ 55 06		14.3	GALAXY
UGC 06581	11 34 06.	+ 55 06	78	14.3	GALAXY S0
MCG+09-19-135	11 34 06.	+ 55 26	60	14.	GALAXY
LB 02058	11 34 06.	+ 56 42 24.		16.8	FAINT BLUE STAR
A1 099	11 34 06.38	+ 15 00 45.6		15.75	FAINT BLUE OBJECT
REIZ 1351	11 34 07.	+ 22 20	12	15.5	GALAXY
REIZ 1352	11 34 07.	+ 23 09	12	15.6	GALAXY
REIZ 1353	11 34 07.	+ 23 13	12	14.9	GALAXY
REIZ 1355	11 34 07.	+ 23 16	18	15.6	GALAXY
RNGC 3756	11 34 08.	+ 54 34		12.5	GALAXY
REIZ 1347	11 34 09.	+ 55 26	60	14.0	GALAXY
REIZ 1356	11 34 10.	+ 01 29	24	15.8	GALAXY
REIZ 1348	11 34 11.	+ 58 40	36	13.7	GALAXY
MCG+03-30-020	11 34 12.	+ 18 09	60	14.9	GALAXY
MCG+03-30-019	11 34 12.	+ 20 15	30	13.9	GALAXY
ZWG 097.023	11 34 12.	+ 20 17		15.3	GALAXY
MCG+03-30-021	11 34 12.	+ 20 17	36	14.6	GALAXY
ZWG 127.001	11 34 12.	+ 23 16		15.0	GALAXY
RNGC 3761	11 34 12.	+ 23 16		15.0	GALAXY
LB 10275	11 34 12.	+ 26 55		16.2	FAINT BLUE STAR
LB 10274	11 34 12.	+ 30 20		16.7	FAINT BLUE STAR
ZWG 242.062	11 34 12.	+ 44 29		15.6	GALAXY
MCG+09-19-136	11 34 12.	+ 55 06	30	13.8	GALAXY
ZWG 268.065	11 34 12.	+ 55 26		14.5	GALAXY
KEEN 3759A	11 34 12.	+ 55 26	78	13.9	E0 GALAXY
UGC 06582	11 34 12.	+ 55 26	84	14.5	GALAXY SBb-c
ARC 1322	11 34 12.	+ 63 31		17.2	RICH CLUSTER OF GALAXIES
REIZ 1357	11 34 13.	+ 23 11	18	16.0	GALAXY
REIZ 1358	11 34 13.	+ 23 12	12	14.9	GALAXY
RNGC 3759	11 34 14.	+ 55 06		14.5	GALAXY
RNGC 3759A	11 34 14.	+ 55 26		14.5	GALAXY
REIZ 1359	11 34 14.	- 09 35	30	14.2	GALAXY
REIZ 1354	11 34 15.	+ 55 05	30	14.4	GALAXY
MCG+09-19-137	11 34 15.	+ 55 52	30	15.	GALAXY
REIZ 1360	11 34 15.	- 08 19	60	14.0	GALAXY
MCG-02-30-010	11 34 15.	- 12 43	60	15.5	GALAXY
KN 12.053	11 34 16.2	+ 50 30 23.			NEBULA
A1 100	11 34 17.13	+ 15 11 23.8		16.35	FAINT BLUE OBJECT
ZWG 040.011	11 34 18.	+ 04 24		15.4	GALAXY
ZWG 068.010	11 34 18.	+ 12 08		15.6	GALAXY
ZWG 097.024	11 34 18.	+ 16 40		15.6	GALAXY
ZC 1134.3+1654	11 34 18.	+ 16 54	1550		CLUSTER OF GALAXIES
2ZW 052	11 34 18.	+ 18 09			COMPACT GALAXY
ZWG 097.025	11 34 18.	+ 18 10		14.9	GALAXY
RNGC 3764	11 34 18.	+ 18 10		15.0	GALAXY
MRK 182	11 34 18.	+ 20 11	6	17.	GALAXY WITH UV CONTINUUM
ZWG 097.026	11 34 18.	+ 20 14		13.9	GALAXY
MRK 181	11 34 18.	+ 20 14	22	14.5	GALAXY WITH UV CONTINUUM
UGC 06583	11 34 18.	+ 20 14	42	13.9	GALAXY
ZWG 097.027	11 34 18.	+ 20 16		14.6	GALAXY
ZC 1134.3+2654	11 34 18.	+ 26 54	1080		CLUSTER OF GALAXIES
HOLM 268B	11 34 18.	+ 39 32	42	14.0	PART OF MULTIPLE GALAXY
ZC 1134.3+5202	11 34 18.	+ 52 02	810		CLUSTER OF GALAXIES
MCG+09-19-138	11 34 18.	+ 55 22 30.	30	16.	GALAXY
ZWG 292.010	11 34 18.	+ 58 42		13.5	GALAXY
UGC 06584	11 34 18.	+ 58 42	66	13.5	GALAXY S0?
ZC 1134.3+5852	11 34 18.	+ 58 52	3160		CLUSTER OF GALAXIES
ZWG 012.041	11 34 18.	- 01 58		15.2	GALAXY
MCG+00-30-016	11 34 18.	- 01 58	48	15.2	GALAXY
MCG-01-30-011	11 34 18.	- 08 20	66	14.5	GALAXY
OCL 0862	11 34 18.	- 62 45	3150	2.9	OPEN STAR CLUSTER
IC 2944	11 34 18.	- 62 45	2400		OPEN CLUSTER
REIZ 1361	11 34 19.	+ 20 15	24	15.4	GALAXY
REIZ 1362	11 34 19.	+ 21 15	12	15.5	GALAXY
RNGC 3757	11 34 19.	+ 58 42		13.5	GALAXY
A1 101	11 34 19.82	+ 16 52 33.8		15.95	FAINT BLUE OBJECT
TON-N 1408	11 34 20.	+ 35 27		16.5	BLUE STAR
MCG+04-28-001	11 34 21.	+ 24 21	42	15.1	GALAXY
MCG-01-30-012	11 34 21.	- 09 18	45	15.	GALAXY
ARC 1326	11 34 23.	+ 40 28		17.8	RICH CLUSTER OF GALAXIES
ARC 1324	11 34 23.	+ 57 23		17.0	RICH CLUSTER OF GALAXIES
IC 0715	11 34 23.	- 08 05 51.			NONSTELLAR OBJECT
ZWG 040.012	11 34 24.	+ 05 46		15.4	GALAXY
ZWG 127.002	11 34 24.	+ 21 17		15.5	GALAXY
ZC 1134.4+2504	11 34 24.	+ 25 04	740		CLUSTER OF GALAXIES
ARC 1327	11 34 24.	+ 26 49		17.5	RICH CLUSTER OF GALAXIES
LB 10277	11 34 24.	+ 29 19		17.8	FAINT BLUE STAR
LB 10276	11 34 24.	+ 30 04		13.3	FAINT BLUE STAR
MCG+10-17-026	11 34 24.	+ 58 42	57	13.7	GALAXY
REIZ 1365	11 34 25.	+ 24 23	18	14.7	GALAXY
LB 00600	11 34 26.	+ 29 03 36.		17.2	FAINT BLUE STAR
A1 102	11 34 27.80	+ 15 07 16.4		15.65	FAINT BLUE OBJECT
REIZ 1363	11 34 28.	+ 37 36	60	14.3	GALAXY
HOLM 268A	11 34 28.	+ 39 32	36	13.8	PART OF MULTIPLE GALAXY
LB 02059	11 34 29.	+ 56 32 30.		16.0	FAINT BLUE STAR
ZWG 040.013	11 34 30.	+ 03 07		15.3	GALAXY
KARA.72 292A	11 34 30.	+ 03 07	36	15.3	PART OF DOUBLE GALAXY
MCG+01-30-003	11 34 30.	+ 03 07	24	15.3	GALAXY
ZWG 068.011	11 34 30.	+ 13 12		15.0	GALAXY
UGC 06585	11 34 30.	+ 13 12	72	15.0	GALAXY DBL SYS
MCG+02-30-004	11 34 30.	+ 13 12	18	15.0	GALAXY
IC 2945	11 34 30.	+ 13 12 15.			NONSTELLAR OBJECT
MCG+03-30-022	11 34 30.	+ 15 49	132	14.5	GALAXY
ZWG 097.028	11 34 30.	+ 15 51		14.5	GALAXY
UGC 06586	11 34 30.	+ 15 51	132	14.5	GALAXY Sb/SBc
ZWG 127.003	11 34 30.	+ 24 23		15.1	GALAXY
RNGC 3765	11 34 30.	+ 24 23		15.0	GALAXY
LB 00601	11 34 30.	+ 30 18 36.		17.6	FAINT BLUE STAR
MCG+08-21-075	11 34 30.	+ 46 00	60	15.	GALAXY
RNGC 3747	11 34 30.	+ 75 15			NON-EXISTENT OBJECT
HOLM 269B	11 34 31.	+ 03 07	36	14.3	PART OF MULTIPLE GALAXY
HOLM 269A	11 34 34.	+ 03 07	48	14.1	PART OF MULTIPLE GALAXY
KN 12.054	11 34 35.0	+ 46 01 59.			NEBULA
ZWG 012.042	11 34 36.	+ 02 25		15.7	GALAXY
ZWG 040.014	11 34 36.	+ 03 06		15.0	GALAXY
UGC 06587	11 34 36.	+ 03 06	60	15.0	GALAXY SBb
KARA.72 292B	11 34 36.	+ 03 06	48	15.0	PART OF DOUBLE GALAXY
MCG+01-30-004	11 34 36.	+ 03 06	48	15.0	GALAXY
ZWG 040.015	11 34 36.	+ 08 22		15.4	GALAXY
ZWG 097.029	11 34 36.	+ 15 42		14.9	GALAXY
UGC 06588	11 34 36.	+ 15 42	84	14.9	GALAXY Sb-c
MCG+03-30-023	11 34 36.	+ 17 08	48	14.5	GALAXY
MCG+03-30-024	11 34 36.	+ 18 06	90	13.7	GALAXY
ZWG 097.030	11 34 36.	+ 18 07		13.7	GALAXY
RNGC 3768	11 34 36.	+ 18 07		13.5	GALAXY
UGC 06589	11 34 36.	+ 18 07	96	13.7	GALAXY S0
ZWG 157.002	11 34 36.	+ 30 26		15.4	GALAXY
ZWG 156.101	11 34 36.	+ 30 26		15.4	GALAXY
ZWG 214.019	11 34 36.	+ 39 32		15.6	GALAXY
KARA.72 293A	11 34 36.	+ 39 32	42	15.6	PART OF DOUBLE GALAXY
MCG+07-24-024	11 34 36.	+ 39 32	36	15.	GALAXY
ZWG 214.020	11 34 36.	+ 39 33		15.6	GALAXY
KARA.72 293B	11 34 36.	+ 39 33	48	15.6	PART OF DOUBLE GALAXY
MCG+07-24-025	11 34 36.	+ 39 33	42	15.	GALAXY
ZC 1134.6+3942	11 34 36.	+ 39 42	2290		CLUSTER OF GALAXIES
ZWG 242.063	11 34 36.	+ 46 03		15.4	GALAXY
REIZ 1364	11 34 36.	+ 62 01	96	13.5	GALAXY
MCG-06-26-003	11 34 36.	- 36 31	66	14.5	GALAXY
FATH 1.407	11 34 38.	+ 30 27	19		NEBULA
LB 02060	11 34 38.	+ 49 41 36.		16.7	FAINT BLUE STAR
MCG+03-30-025	11 34 42.	+ 15 40	72	14.9	GALAXY
ZWG 097.031	11 34 42.	+ 17 09		14.5	GALAXY
RNGC 3767	11 34 42.	+ 17 09		14.5	GALAXY
UGC 06590	11 34 42.	+ 17 09	72	14.5	GALAXY SB0
MCG+04-28-002	11 34 42.	+ 26 00	36	14.5	GALAXY
MCG+05-28-001	11 34 42.	+ 30 26	30	15.4	GALAXY
ZWG 292.011	11 34 42.	+ 62 02		13.3	GALAXY
RNGC 3762	11 34 42.	+ 62 02		13.5	GALAXY
UGC 06591	11 34 42.	+ 62 02	138	13.3	GALAXY Sa
ZWG 334.045	11 34 42.	+ 73 02		15.3	GALAXY
UGC 06592	11 34 42.	+ 73 02	66	15.3	GALAXY S
REIZ 1366	11 34 43.	+ 21 26	30	15.1	GALAXY
TON-N 1409	11 34 44.	+ 21 22		16.2	BLUE STAR
KN 12.055	11 34 44.4	+ 49 13 02.			NEBULA
LB 00602	11 34 45.	+ 28 35 42.		17.3	FAINT BLUE STAR
MCG+03-30-026	11 34 48.	+ 20 27	30	15.	GALAXY
ZWG 127.004	11 34 48.	+ 26 02		14.5	GALAXY
UGC 06593	11 34 48.	+ 26 02	66	14.5	GALAXY PECULR
ZC 1134.8+3310	11 34 48.	+ 33 10	4440		CLUSTER OF GALAXIES
MCG+09-19-139	11 34 48.	+ 54 58	30	16.	GALAXY
LB 02061	11 34 48.	+ 56 08 18.		16.5	FAINT BLUE STAR
MCG+10-17-027	11 34 48.	+ 62 03	78	13.5	GALAXY
7ZW 413	11 34 48.	+ 63 38			COMPACT GALAXY
A1 103	11 34 49.09	+ 16 41 41.8		16.25	FAINT BLUE OBJECT
MCG+04-28-003	11 34 51.	+ 22 40	48	15.4	GALAXY
MCG+05-28-002	11 34 51.	+ 31 39	36	14.6	GALAXY
ARC 1328	11 34 52.	+ 37 40		17.2	RICH CLUSTER OF GALAXIES
KN 12.056	11 34 52.6	+ 45 29 44.			NEBULA
IC 2947	11 34 53.	+ 31 38 21.			NONSTELLAR OBJECT
IC 2946	11 34 53.	+ 32 31 45.			NONSTELLAR OBJECT
FEIG 046	11 34 54.	+ 14 27		13.1	FAINT BLUE STAR
MCG+03-30-027	11 34 54.	+ 16 49	144	14.8	GALAXY
LB 00603	11 34 54.	+ 27 41 00.		17.8	FAINT BLUE STAR
ZC 1134.9+2937	11 34 54.	+ 29 37	740		CLUSTER OF GALAXIES
ZC 1134.9+3133	11 34 54.	+ 31 33	1140		CLUSTER OF GALAXIES
ZWG 157.003	11 34 54.	+ 31 38		14.6	GALAXY
ZWG 156.102	11 34 54.	+ 31 38		14.6	GALAXY
ZWG 186.013	11 34 54.	+ 32 31		15.2	GALAXY
ZWG 012.043	11 34 54.	- 00 34		15.7	GALAXY
REIZ 1367	11 34 55.	+ 22 40	36	14.2	GALAXY
RNGC 3769	11 34 57.	+ 48 10		12.5	GALAXY
REIZ 1368	11 34 58.	- 00 35	12	15.8	GALAXY
KN 12.057	11 34 58.0	+ 47 13 48.			NEBULA
ZWG 068.012	11 35 00.	+ 09 45		15.6	GALAXY
ZWG 097.032	11 35 00.	+ 16 50		14.8	GALAXY
UGC 06594	11 35 00.	+ 16 50	162	14.8	GALAXY Sc
ZWG 097.033	11 35 00.	+ 20 27		15.6	GALAXY
MCG+04-28-005	11 35 00.	+ 22 17	60	15.3	GALAXY
MCG+04-28-004	11 35 00.	+ 22 17 30.	18	15.5	GALAXY
ZWG 127.005	11 35 00.	+ 22 41		15.4	GALAXY
ZC 1135.0+2745	11 35 00.	+ 27 45	3290		CLUSTER OF GALAXIES
LB 10278	11 35 00.	+ 31 51		17.1	FAINT BLUE STAR
ZWG 242.064	11 35 00.	+ 47 14		15.7	GALAXY
MCG+08-21-076	11 35 00.	+ 48 09	150	12.5	GALAXY
ZWG 242.065	11 35 00.	+ 48 10		11.7	GALAXY
UGC 06595	11 35 00.	+ 48 10	198	11.7	GALAXY Sb
KARA.72 294A	11 35 00.	+ 48 10	192	11.7	PART OF DOUBLE GALAXY
HOLM 270A	11 35 00.	+ 48 11	162	12.3	PART OF MULTIPLE GALAXY
ZWG 268.066	11 35 00.	+ 56 25		15.5	GALAXY
UGC 06596	11 35 00.	+ 56 25	72	15.5	GALAXY IRR
MCG+09-19-140	11 35 00.	+ 56 26	60	14.	GALAXY
MRSL 292+04/1	11 35 00.	- 56 40	10800		HII REGION
MRSL 294-01/1	11 35 00.	- 62 54	4800		HII REGION
REIZ 1369	11 35 01.	+ 20 25	24	15.7	GALAXY
REIZ 1370	11 35 01.	+ 22 15	66	14.4	GALAXY
REIZ 1371	11 35 01.	+ 22 19	18	15.6	GALAXY
KN 12.058	11 35 02.6	+ 48 10 20.			NEBULA
MCG+08-21-077	11 35 03.	+ 48 08	48	14.1	GALAXY
RNGC 3769A	11 35 03.	+ 48 10		14.5	GALAXY
KARA.73 34	11 35 03.	- 38 57	34		DWARF GALAXY
MCG+03-30-028	11 35 06.	+ 17 25	39	15.6	GALAXY
ZWG 127.006	11 35 06.	+ 22 17		15.3	GALAXY
UGC 06597	11 35 06.	+ 22 17	78	15.3	GALAXY SBa-b
MCG+04-28-006	11 35 06.	+ 22 58	66	14.4	GALAXY
ZC 1135.1+4518	11 35 06.	+ 45 18	1410		CLUSTER OF GALAXIES
ZWG 242.066	11 35 06.	+ 48 09		14.7	GALAXY
KARA.72 294B	11 35 06.	+ 48 09	66	14.7	PART OF DOUBLE GALAXY
HOLM 270B	11 35 06.	+ 48 10	30	14.1	PART OF MULTIPLE GALAXY
REIZ 1372	11 35 06.	+ 48 11	150	12.8	GALAXY

OBJECT NAME	RIGHT ASCEN.	DECLINATION	DIAM.	MAGN.	TYPE OF OBJECT
ZWG 012.044	11 35 06.	- 00 32		15.6	GALAXY
REIZ 1373	11 35 07.	+ 22 14	108	14.2	GALAXY
REIZ 1375	11 35 07.	+ 22 21	6	15.4	GALAXY
REIZ 1374	11 35 07.	+ 22 21	24	15.1	GALAXY
ARP 280	11 35 07.	+ 48 11			PECULIAR GALAXY
MCG+04-28-007	11 35 09.	+ 22 17 30.	30	15.5	GALAXY
MCG+08-21-078	11 35 09.	+ 47 43	36	16.	GALAXY
KN 12.059	11 35 09.1	+ 48 09 41.			NEBULA
LB 02063	11 35 10.	+ 58 33 00.		15.5	FAINT BLUE STAR
LB 02062	11 35 10.	+ 60 47 48.		16.0	FAINT BLUE STAR
A1 104	11 35 10.01	+ 13 40 48.9		17.20	FAINT BLUE OBJFCT
ZWG 097.034	11 35 12.	+ 17 26		15.6	GALAXY
MCG+04-28-009	11 35 12.	+ 22 12	18	15.5	GALAXY
MCG+04-28-008	11 35 12.	+ 22 15	36	15.2	GALAXY
VV 282C	11 35 12.	+ 22 17	48	15.	INTERACTING GALAXY
ZWG 127.007	11 35 12.	+ 22 18		15.5	GALAXY
ZWG 127.008	11 35 12.	+ 22 58		14.4	GALAXY
RNGC 3772	11 35 12.	+ 22 58		14.5	GALAXY
UGC 06598	11 35 12.	+ 22 58	72	14.4	GALAXY SBa
UGC 06599	11 35 12.	+ 24 25	78	17.	GALAXY DWRF IR
ZC 1135.2+2524	11 35 12.	+ 25 24	940		CLUSTER OF GALAXIES
ZWG 214.021	11 35 12.	+ 42 55		15.7	GALAXY
MCG+08-21-079	11 35 12.	+ 47 43	48	16.	GALAXY
REIZ 1376	11 35 12.	+ 47 45	18	14.8	GALAXY
REIZ 1377	11 35 12.	+ 48 10	42	14.4	GALAXY
LB 02064	11 35 12.	+ 51 54 48.		15.9	FAINT BLUE STAR
MCG+09-19-141	11 35 12.	+ 56 16 30.	48	16.	GALAXY
ZWG 292.012	11 35 12.	+ 59 53		13.5	GALAXY
UGC 06600	11 35 12.	+ 59 53	66	13.5	GALAXY SBa
REIZ 1379	11 35 13.	+ 22 06	12	15.2	GALAXY
REIZ 1380	11 35 13.	+ 22 15	18	15.0	GALAXY
REIZ 1381	11 35 13.	+ 22 17	18	15.0	GALAXY
REIZ 1382	11 35 13.	+ 22 18	12	15.4	GALAXY
REIZ 1383	11 35 13.	+ 22 21	12	15.2	GALAXY
REIZ 1384	11 35 13.	+ 22 57	30	14.4	GALAXY
LB 02065	11 35 13.	+ 55 56 00.		15.2	FAINT BLUE STAR
RNGC 3770	11 35 13.	+ 59 53		13.5	GALAXY
KN 12.060	11 35 13.6	+ 47 44 39.			NEBULA
TON-N 1410	11 35 14.	+ 35 21		16.5	BLUE STAR
LB 02066	11 35 14.	+ 57 56 54.		17.1	FAINT BLUE STAR
KN 12.061	11 35 14.8	+ 46 27 33.			NEBULA
MCG+04-28-010	11 35 15.	+ 22 15	90	14.6	GALAXY
MCG+04-28-011	11 35 15.	+ 22 15 30.	24	14.6	GALAXY
VV 282B	11 35 15.	+ 22 17 30.	24	15.	INTERACTING GALAXY
VV 282A	11 35 15.	+ 22 17 30.	114	15.	INTERACTING GALAXY
ARP 320	11 35 15.	+ 22 18			PECULIAR GALAXY
REIZ 1386	11 35 15.	- 06 23	36	14.3	GALAXY
MCG-01-30-013	11 35 15.	- 07 00	60	14.	GALAXY
KN 12.062	11 35 16.9	+ 47 44 53.			NEBULA
LB 00604	11 35 17.	+ 30 18 00.		17.9	FAINT BLUE STAR
REIZ 1378	11 35 17.	+ 59 53	24	14.1	GALAXY
ZWG 040.016	11 35 18.	+ 05 47		15.5	GALAXY
ZWG 127.009	11 35 18.	+ 22 15		15.2	GALAXY
ZC 1135.3+3742	11 35 18.	+ 37 42	1410		CLUSTER OF GALAXIES
REIZ 1385	11 35 18.	+ 47 11	24	14.8	GALAXY
MCG+08-21-080	11 35 18.	+ 47 44 30.	15	17.	GALAXY
ZWG 242.067	11 35 18.	+ 47 45		15.1	GALAXY
ZC 1135.3+5937	11 35 18.	+ 59 37	1680		CLUSTER OF GALAXIES
7ZW 414	11 35 18.	+ 67 46			COMPACT GALAXY
REIZ 1387	11 35 19.	+ 22 16	12	15.2	GALAXY
REIZ 1388	11 35 19.	+ 22 19	30	14.8	GALAXY
LB 02067	11 35 20.	+ 50 31 00.		16.5	FAINT BLUE STAR
A1 105	11 35 20.14	+ 13 20 30.9		16.15	FAINT BLUE OBJECT
REIZ 1389	11 35 21.	+ 35 28	90	14.0	GALAXY
ZWG 127.010	11 35 24.	+ 21 04		15.7	GALAXY
ZWG 127.011	11 35 24.	+ 22 05		15.3	GALAXY
UGC 06601	11 35 24.	+ 22 05	66	15.3	GALAXY SO?
ZWG 127.012	11 35 24.	+ 22 16		14.6	GALAXY
UGC 06602	11 35 24.	+ 22 16	120	14.6	GALAXY
LB 10279	11 35 24.	+ 32 19		17.4	FAINT BLUE STAR
ZWG 186.014	11 35 24.	+ 35 28		15.0	GALAXY
UGC 06603	11 35 24.	+ 35 28	138	15.0	GALAXY Sc
MCG+06-26-009	11 35 24.	+ 35 28 30.	120	14.	GALAXY
FEIG 047	11 35 24.	+ 45 13		16.6	FAINT BLUE STAR
MCG+09-19-142	11 35 24.	+ 55 45	15	17.	GALAXY
ZWG 292.013	11 35 24.	+ 59 02		14.0	GALAXY
UGC 06604	11 35 24.	+ 59 02	60	14.0	GALAXY
MCG+10-17-028	11 35 24.	+ 59 54	60	14.1	GALAXY
RNGC 3795B	11 35 25.	+ 59 02		14.0	GALAXY
LB 02068	11 35 27.	+ 52 31 12.		16.7	FAINT BLUE STAR
MCG-03-30-003	11 35 27.	- 16 57	108	14.	GALAXY
ZC 1135.5+0138	11 35 30.	+ 01 38	3900		CLUSTER OF GALAXIES
ZC 1135.5+1848	11 35 30.	+ 18 48	1610		CLUSTER OF GALAXIES
ZC 1135.5+2237	11 35 30.	+ 22 37	740		CLUSTER OF GALAXIES
LB 00605	11 35 30.	+ 28 03 30.		18.0	FAINT BLUE STAR
REIZ 1390	11 35 30.	+ 47 43	18	14.3	GALAXY
MCG+08-21-081	11 35 30.	+ 47 45	15	17.	GALAXY
ZC 1135.5+4948	11 35 30.	+ 49 48	540		CLUSTER OF GALAXIES
MCG+10-17-029	11 35 30.	+ 59 02 30.	24	14.	GALAXY
MRK 183	11 35 30.	+ 68 50	13	15.5	GALAXY WITH UV CONTINUUM
ZC 1135.5+7125	11 35 30.	+ 71 25	1210		CLUSTER OF GALAXIES
MCG-01-30-014	11 35 30.	- 05 58	45	15.	GALAXY
REIZ 1391	11 35 31.	+ 22 09	12	15.3	GALAXY
TON-N 1411	11 35 32.	+ 34 14		16.3	BLUE STAR
ARC 1332	11 35 32.	- 09 05		16.0	RICH CLUSTFR OF GALAXIFS
A1 106	11 35 32.56	+ 13 58 27.4		16.90	FAINT BLUE OBJECT
MCG+08-21-082	11 35 33.	+ 47 41	18	16.	GALAXY
KN 12.063	11 35 33.3	+ 47 47 04.			NEBULA
KN 12.064	11 35 34.9	+ 47 42 55.			NEBULA
ZWG 068.013	11 35 36.	+ 11 28		15.3	GALAXY
RNGC 3773	11 35 36.	+ 12 23		13.5	GALAXY
ZWG 068.014	11 35 36.	+ 12 24		13.1	GALAXY
UGC 06605	11 35 36.	+ 12 24	84	13.1	GALAXY SO
MCG+02-30-005	11 35 36.	+ 12 24	21	13.1	GALAXY
ZWG 157.004	11 35 36.	+ 31 56		15.5	GALAXY
ZC 1135.6+4606	11 35 36.	+ 46 06	1550		CLUSTER OF GALAXIES
MCG-05-28-001	11 35 36.	- 32 03	24	15.	GALAXY
MCG-05-28-002	11 35 36.	- 32 17	24	15.	GALAXY
A1 107	11 35 36.78	+ 15 19 41.4		15.75	FAINT BLUE OBJECT
A1 108	11 35 37.61	+ 14 04 59.0		16.20	FAINT BLUE OBJECT
MCG+08-21-083	11 35 39.	+ 47 41	72	15.	GALAXY
ARC 1330	11 35 40.	+ 49 48		17.8	RICH CLUSTER OF GALAXIES
KN 12.065	11 35 40.3	+ 47 42 53.			NEBULA
ZC 1135.7+2055	11 35 42.	+ 20 55	1680		CLUSTER OF GALAXIES
TON-N 0068	11 35 42.	+ 27 02		15.0	BLUE STAR
ZWG 242.068	11 35 42.	+ 47 43		14.9	GALAXY
UGC 06606	11 35 42.	+ 47 43	102	14.9	GALAXY DBL SYS
ZC 1135.7+7658	11 35 42.	+ 76 58	940		CLUSTER OF GALAXIES

OBJECT NAME	RIGHT ASCEN.	DECLINATION	DIAM.	MAGN.	TYPE OF OBJECT
MCG-01-30-015	11 35 42.	- 09 21	36	15.	GALAXY
MCG-02-30-011	11 35 42.	- 12 47	72	15.	GALAXY
KN 12.066	11 35 42.4	+ 45 23 11.			NEBULA
REIZ 1392	11 35 43.	+ 20 48	18	14.7	GALAXY
ARC 1329	11 35 43.	+ 71 24		16.5	RICH CLUSTER OF GALAXIES
RNGC 3776	11 35 43.	- 03 04			GALAXY
RNGC 3775	11 35 44.	- 10 21		15.0	GALAXY
MCG-02-30-012	11 35 45.	- 10 21 30.	60	15.	GALAXY
ZWG 127.013	11 35 48.	+ 20 49		15.5	GALAXY
ZC 1135.8+5645	11 35 48.	+ 56 45	1080		CLUSTER OF GALAXIES
MCG+11-14-035	11 35 48.	+ 66 10	36	16.	GALAXY
ZC 1135.8+7303	11 35 48.	+ 73 03	870		CLUSTER OF GALAXIES
ZWG 012.045	11 35 48.	- 03 04		15.6	GALAXY
MCG-03-30-004	11 35 48.	- 16 10	42	15.	GALAXY
MCG-03-30-005	11 35 48.	- 20 50	36	15.5	GALAXY
MCG-05-28-003	11 35 48.	- 29 25	54	15.	GALAXY
VHA 121	11 35 48.	- 63 04	360		STAR CLSTR IN NEBULOSITY
REIZ 1393	11 35 49.	+ 21 01	36	15.6	GALAXY
RNGC 3771	11 35 50.	- 09 04			GALAXY
KN 12.067	11 35 50.3	+ 46 06 32.			NEBULA
MCG+04-28-013	11 35 51.	+ 20 47	18	14.9	GALAXY
MCG+04-28-012	11 35 51.	+ 21 02	78	15.2	GALAXY
LB 02070	11 35 53.	+ 58 10 30.		15.6	FAINT BLUE STAR
LB 02069	11 35 53.	+ 60 23 06.		17.0	FAINT BLUE STAR
RNGC 3778	11 35 53.	- 50 26			UNVERIFIED SOUTHRN OBJECT
REIZ 1394	11 35 54.	+ 20 48	24	14.9	GALAXY
ZWG 127.014	11 35 54.	+ 21 01		15.2	GALAXY
UGC 06607	11 35 54.	+ 21 01	96	15.2	GALAXY Sc
ZC 1135.9+2723	11 35 54.	+ 27 23	670		CLUSTER OF GALAXIES
ZC 1135.9+3143	11 35 54.	+ 31 43	340		CLUSTER OF GALAXIES
ZWG 186.015	11 35 54.	+ 34 08		15.1	GALAXY
MCG+06-26-010	11 35 54.	+ 34 08 30.	30	15.	GALAXY
MCG+09-19-143	11 35 54.	+ 55 32	24	16.	GALAXY
ZC 1135.9+5706	11 35 54.	+ 57 06	1340		CLUSTER OF GALAXIES
7ZW 415	11 35 54.	+ 58 10			COMPACT GALAXY
ZC 1135.9+6510	11 35 54.	+ 65 10	1550		CLUSTER OF GALAXIES
REIZ 1395	11 35 55.	+ 25 39	42	15.0	GALAXY
TON-N 1412	11 35 56.	+ 21 37		14.3	BLUE STAR
TON-N 1413	11 35 56.	+ 33 16		16.5	BLUE STAR
LB 00606	11 35 57.	+ 26 13 42.		16.7	FAINT BLUE STAR
KN 12.068	11 35 57.2	+ 46 15 20.			NEBULA
ARC 1331	11 35 58.	+ 63 52		17.6	RICH CLUSTER OF GALAXIES
REIZ 1396	11 35 58.	- 00 55	36	13.8	GALAXY
KEEN 3795B	11 36	+ 59 03		13.7	GALAXY
ZWG 040.017	11 36 00.	+ 05 28		15.7	GALAXY
ZWG 040.018	11 36 00.	+ 07 14		15.4	GALAXY
ZWG 097.035	11 36 00.	+ 18 06		15.2	GALAXY
REIZ 1397	11 36 00.	+ 20 48	12	15.5	GALAXY
ZWG 127.015	11 36 00.	+ 20 49		14.9	GALAXY
UGC 06609	11 36 00.	+ 20 49	84	14.9	GALAXY E
MCG+04-28-014	11 36 00.	+ 20 49	15	15.5	GALAXY
ZWG 186.016	11 36 00.	+ 36 02		15.4	GALAXY
SHAH 066	11 36 00.	+ 38 37	282	16.7	GROUP OF COMPACT GALAXIES
MCG+09-19-145	11 36 00.	+ 55 20	30	16.	GALAXY
MCG+09-19-144	11 36 00.	+ 55 32	30	16.	GALAXY
MCG+10-17-030	11 36 00.	+ 57 39	8	17.	GALAXY
ZC 1136.0+8341	11 36 00.	+ 83 41	1410		CLUSTER OF GALAXIES
ZWG 012.046	11 36 00.	- 00 55		14.5	GALAXY
UGC 06608	11 36 00.	- 00 55	66	14.5	GALAXY S
MCG+00-30-017	11 36 00.	- 00 55	66	14.5	GALAXY
KARA.73B 0491	11 36 00.	- 00 55	78	14.5	ISOLATED GALAXY S
MCG-01-30-016	11 36 00.	- 08 43	42	15.	GALAXY
RNGC 3774	11 36 02.	- 08 43		15.0	GALAXY
A1 109	11 36 02.21	+ 14 20 12.1		17.00	FAINT BLUE OBJFCT
REIZ 1399	11 36 04.	+ 04 04	12	16.0	GALAXY
REIZ 1398	11 36 05.	+ 46 12	12	15.6	GALAXY
A1 110	11 36 05.85	+ 16 02 29.9		15.65	FAINT BLUE OBJECT
ZWG 040.019	11 36 06.	+ 04 04		15.4	GALAXY
ZWG 186.017	11 36 06.	+ 34 04		15.3	GALAXY
UGC 06610	11 36 06.	+ 34 04	138	15.3	GALAXY Sc
MCG+06-26-011	11 36 06.	+ 34 05 30.	114	13.5	GALAXY
ZC 1136.1+6353	11 36 06.	+ 63 53	1140		CLUSTER OF GALAXIES
ZWG 334.046	11 36 06.	+ 68 50		15.4	GALAXY
ZWG 352.005	11 36 06.	+ 76 00		15.7	GALAXY
ZWG 351.067	11 36 06.	+ 76 00		15.7	GALAXY
RNGC 3779	11 36 08.	- 10 18		14.0	GALAXY
REIZ 1401	11 36 09.	- 02 16	24	15.2	GALAXY
REIZ 1400	11 36 09.	- 02 21	60	15.5	GALAXY
MCG-02-30-013	11 36 09.	- 10 18 30.	84	14.5	GALAXY
SEY 117	11 36 10.	+ 55 20 24.		15.3	FAINT GALAXY
FATH 1.408	11 36 11.	+ 44 09	5		NEBULA
ZWG 068.015	11 36 12.	+ 11 10		15.7	GALAXY
ZWG 097.036	11 36 12.	+ 19 52		15.7	GALAXY
ZC 1136.2+2248	11 36 12.	+ 22 48	870		CLUSTER OF GALAXIES
ZWG 214.022	11 36 12.	+ 43 26		15.4	GALAXY
UGC 06611	11 36 12.	+ 43 26	102	15.4	GALAXY S
MCG+07-24-026	11 36 12.	+ 43 26	66	15.	GALAXY
MCG+08-21-084	11 36 12.	+ 49 42	24	16.	GALAXY
MCG+08-21-085	11 36 12.	+ 49 55	30	16.	GALAXY
ZWG 242.069	11 36 12.	+ 49 56		15.7	GALAXY
MCG+09-19-146	11 36 12.	+ 52 54	30	16.	GALAXY
MCG+10-17-031	11 36 12.	+ 57 48	18	16.	GALAXY
MCG+10-17-032	11 36 12.	+ 57 49	27	16.	GALAXY
MCG+10-17-033	11 36 12.	+ 62 11	18	17.	GALAXY
KN 12.069	11 36 12.7	+ 49 56 24.			NEBULA
REIZ 1402	11 36 13.	+ 21 14	18	14.5	GALAXY
TON-N 1414	11 36 14.	+ 22 42		16.6	BLUE STAR
FATH 1.409	11 36 15.	+ 44 36	22		NEBULA
REIZ 1403	11 36 16.	+ 05 02	42	13.9	GALAXY
REIZ 1404	11 36 16.	+ 05 38	30	14.6	GALAXY
KN 12.071	11 36 17.1	+ 48 50 29.			NEBULA
KN 12.070	11 36 17.4	+ 50 17 11.			NEBULA
ZWG 040.020	11 36 18.	+ 03 51		14.7	GALAXY
MCG+01-30-005	11 36 18.	+ 03 51	48	14.7	GALAXY
ZWG 068.016	11 36 18.	+ 09 49		15.7	GALAXY
ZC 1136.3+1718	11 36 18.	+ 17 18	1680		CLUSTER OF GALAXIES
REIZ 1405	11 36 18.	+ 19 50	18	15.9	GALAXY
REIZ 1406	11 36 18.	+ 21 12	24	15.3	GALAXY
MRK 637	11 36 18.	+ 21 16	13	16.	GALAXY WITH UV CONTINUUM
ZC 1136.3+4152	11 36 18.	+ 41 52	1210		CLUSTER OF GALAXIES
MCG+08-21-086	11 36 18.	+ 45 51	18	17.	GALAXY
2ZW 053	11 36 18.	+ 45 53			COMPACT GALAXY
ZWG 242.070	11 36 18.	+ 45 54		15.3	GALAXY
LB 02071	11 36 19.	+ 56 08 06.		16.6	FAINT BLUE STAR
LB 02072	11 36 21.	+ 57 17 06.		16.4	FAINT BLUE STAR
ARC 1334	11 36 21.	- 04 03		15.7	RICH CLUSTER OF GALAXIES
A1 111	11 36 21.88	+ 13 42 51.6		17.00	FAINT BLUE OBJECT
MCG+05-28-003	11 36 24.	+ 26 35	36	15.5	GALAXY

OBJECT NAME	RIGHT ASCEN.	DECLINATION	DIAM.	MAGN.	TYPE OF OBJECT
ZWG 157.005	11 36 24.	+ 26 38		14.8	GALAXY
RNGC 3781	11 36 24.	+ 26 38		15.0	GALAXY
ZWG 214.023	11 36 24.	+ 43 31		15.3	GALAXY
ZWG 268.067	11 36 24.	+ 55 56		15.3	GALAXY
OCL 0864	11 36 24.	- 63 15	900		OPEN STAR CLUSTER
VHA 122	11 36 24.	- 63 15	900		STAR CLSTR IN NEBULOSITY
LB 02073	11 36 25.	+ 59 28 18.		16.5	FAINT BLUE STAR
HN 0790	11 36 25.	- 63 15			NEBULA
IC 2948	11 36 25.	- 63 15	3600		DIFFUSE NEBULA
REIZ 1407	11 36 26.	+ 55 54	72	15.0	GALAXY
KN 12.072	11 36 26.7	+ 46 59 28.			NEBULA
MCG+07-24-027	11 36 27.	+ 43 31	36	15.	GALAXY
LB 02074	11 36 27.	+ 55 00 18.		16.7	FAINT BLUE STAR
MCG-03-30-006	11 36 27.	- 17 42	48	15.	GALAXY
REIZ 1408	11 36 28.	+ 00 03	54	14.5	GALAXY
SEY 118	11 36 28.	+ 55 56 24.		14.8	FAINT GALAXY
RNGC 3783	11 36 28.	- 37 28		13.0	GALAXY
A1 112	11 36 28.15	+ 15 57 57.0		16.10	FAINT BLUE OBJECT
KN 12.073	11 36 28.7	+ 46 57 53.			NEBULA
ZWG 012.047	11 36 30.	+ 00 04		14.9	GALAXY
UGC 06612	11 36 30.	+ 00 04	102	14.9	GALAXY Sb
MCG+00-30-018	11 36 30.	+ 00 04	90	14.9	GALAXY
KARA.73B 0492	11 36 30.	+ 00 04	114	14.9	ISOLATED GALAXY S
IC 0716	11 36 30.	+ 00 04 13.			NONSTELLAR OBJECT
ZWG 097.037	11 36 30.	+ 19 38		15.6	GALAXY
ZC 1136.5+2138	11 36 30.	+ 21 38	540		CLUSTER OF GALAXIES
ZWG 214.024	11 36 30.	+ 39 36		15.2	GALAXY
UGC 06613	11 36 30.	+ 39 36	72	15.2	GALAXY SBa
MCG+07-24-028	11 36 30.	+ 39 36	66	15.	GALAXY
MCG+09-19-147	11 36 30.	+ 55 53	30	16.	GALAXY
MCG+09-19-148	11 36 30.	+ 55 56	54	15.	GALAXY
ZWG 314.038	11 36 30.	+ 62 51		15.6	GALAXY
MCG+11-14-036	11 36 30.	+ 66 36	30	16.	GALAXY
VVI 49	11 36 30.	- 37 28	90	13.08	SEYFERT GALAXY
A1 113	11 36 30.18	+ 14 04 06.1		16.90	FAINT BLUE OBJECT
MCG+08-21-087	11 36 33.	+ 46 46	108	12.9	GALAXY
ABC 1333	11 36 33.	+ 50 09		17.4	RICH CLUSTER OF GALAXIES
A1 114	11 36 34.06	+ 16 33 35.8		16.00	FAINT BLUE OBJECT
ZWG 097.038	11 36 36.	+ 15 16		15.6	GALAXY
ZWG 097.039	11 36 36.	+ 17 19		15.7	GALAXY
ZWG 097.040	11 36 36.	+ 17 25		14.8	GALAXY
UGC 06614	11 36 36.	+ 17 25	192	14.8	GALAXY Sa?
MCG+03-30-029	11 36 36.	+ 17 25	78	14.8	GALAXY
REIZ 1410	11 36 36.	+ 19 37	18	14.9	GALAXY
MCG+05-28-004	11 36 36.	+ 26 37	21	14.8	GALAXY
ZC 1136.6+2949	11 36 36.	+ 29 49	2620		CLUSTER OF GALAXIES
ZC 1136.6+3154	11 36 36.	+ 31 54	1480		CLUSTER OF GALAXIES
ZWG 186.018	11 36 36.	+ 34 12		15.4	GALAXY
MCG+06-26-012	11 36 36.	+ 34 13	30	15.	GALAXY
ZWG 292.014	11 36 36.	+ 56 32		12.2	GALAXY
UGC 06615	11 36 36.	+ 56 32	186	12.2	GALAXY Sc
ZWG 292.015	11 36 36.	+ 58 33		14.2	GALAXY
UGC 06616	11 36 36.	+ 58 33	180	14.2	GALAXY Sc
MCG-04-28-002	11 36 36.	- 23 01 30.	36	15.	GALAXY
MCG-06-26-004	11 36 36.	- 37 26	90	12.8	GALAXY
WRAY 19.37	11 36 36.9	- 63 11 50.		10.7	STAR-NEBULA ASSOCIATION
LB 02075	11 36 38.	+ 52 57 18.		15.9	FAINT BLUE STAR
RNGC 3795A	11 36 38.	+ 58 33		14.0	GALAXY
BC PKS1136-13	11 36 38.51	- 13 34 05.9		17.8	QUASI-STELLAR OBJECT
SHB 184	11 36 38.6	- 13 34 09.		17.8	QUASI-STELLAR OBJECT
REIZ 1409	11 36 39.	+ 56 32	210	13.0	GALAXY
MCG-01-30-017	11 36 39.	- 09 04 30.	24	16.	GALAXY
REIZ 1411	11 36 39.	- 09 05	12	14.0	GALAXY
MCG-01-30-018	11 36 39.	- 09 05 30.	24	14.5	GALAXY
RNGC 3782	11 36 40.	+ 46 47		13.5	GALAXY
KN 12.074	11 36 40.0	+ 46 47 31.			NEBULA
LB 02076	11 36 41.	+ 53 28 24.		17.0	FAINT BLUE STAR
ZWG 068.017	11 36 42.	+ 10 14		14.5	GALAXY
UGC 06617	11 36 42.	+ 10 14	54	14.5	GALAXY S0?
MCG+02-30-006	11 36 42.	+ 10 14	30	14.5	GALAXY
MCG+03-30-030	11 36 42.	+ 19 49	48	15.5	GALAXY
REIZ 1412	11 36 42.	+ 20 39	12	15.6	GALAXY
REIZ 1413	11 36 42.	+ 20 42	12	15.1	GALAXY
ZWG 127.016	11 36 42.	+ 20 43		15.6	GALAXY
ZWG 242.071	11 36 42.	+ 46 47		13.1	GALAXY
UGC 06618	11 36 42.	+ 46 47	84	13.1	GALAXY IRR?
ZWG 268.068	11 36 42.	+ 55 26		14.8	GALAXY
ZWG 292.016	11 36 42.	+ 60 49		15.4	GALAXY
UGC 06619	11 36 42.	+ 60 49	84	15.4	GALAXY S
ZC 1136.7+6950	11 36 42.	+ 69 50	3020		CLUSTER OF GALAXIES
MCG-05-28-004	11 36 42.	- 31 42	48	15.5	GALAXY
A1 115	11 36 42.64	+ 14 56 51.3		17.00	FAINT BLUE OBJECT
MAI 063	11 36 43.	+ 59 29	40		DWARF SPHEROIDAL GALAXY
REIZ 1414	11 36 44.	+ 29 08	30	14.0	GALAXY
RNGC 3780	11 36 44.	+ 56 33		12.5	GALAXY
KN 12.075	11 36 44.6	+ 49 14 09.			NEBULA
MCG+09-19-149	11 36 45.	+ 55 27	48	15.	GALAXY
REIZ 1415	11 36 45.	- 07 33	48	14.5	GALAXY
REIZ 1416	11 36 46.	+ 05 39	30	15.0	GALAXY
SEY 119	11 36 46.	+ 55 26 24.		14.4	FAINT GALAXY
RNGC 3793	11 36 47.	+ 32 08		15.5	GALAXY
ZC 1136.8+1026	11 36 48.	+ 10 26	740		CLUSTER OF GALAXIES
ZC 1136.8+1208	11 36 48.	+ 12 08	2350		CLUSTER OF GALAXIES
ZWG 097.041	11 36 48.	+ 19 48		15.5	GALAXY
REIZ 1417	11 36 48.	+ 19 48	24	14.9	GALAXY
REIZ 1418	11 36 49.	+ 21 09	30	15.3	GALAXY
ZWG 157.006	11 36 48.	+ 26 36		15.2	GALAXY
RNGC 3784	11 36 48.	+ 26 36		15.0	GALAXY
ZWG 157.007	11 36 48.	+ 32 09		15.7	GALAXY
MRK 184	11 36 48.	+ 55 15	10	16.	GALAXY WITH UV CONTINUUM
MCG+09-19-150	11 36 48.	+ 56 33 30.	180	12.1	GALAXY
MCG+10-17-034	11 36 48.	+ 58 03	10	17.	GALAXY
MCG+10-17-035	11 36 48.	+ 58 33	114	14.	GALAXY
ZWG 314.039	11 36 48.	+ 62 53		15.5	GALAXY
A1 116	11 36 48.17	+ 14 22 08.2		16.50	FAINT BLUE OBJECT
ABC 1337	11 36 49.	+ 10 26		17.2	RICH CLUSTER OF GALAXIES
HOLM 271A	11 36 49.	+ 26 35	54	14.3	PART OF MULTIPLE GALAXY
REIZ 1419	11 36 49.	+ 28 10	36	15.0	GALAXY
ABC 1336	11 36 49.	+ 32 41		16.0	RICH CLUSTER OF GALAXIES
IC 0717	11 36 49.	- 10 22			MAY NOT EXIST
RNGC 3794	11 36 50.	+ 55 27		15.0	GALAXY
A1 117	11 36 50.15	+ 13 59 48.3		16.25	FAINT BLUE OBJECT
A1 118	11 36 50.45	+ 15 11 11.0		16.00	FAINT BLUE OBJECT
MCG+05-28-005	11 36 51.	+ 32 08 30.	42	15.7	GALAXY
KN 12.076	11 36 52.6	+ 44 59 39.			NEBULA
HOLM 271B	11 36 53.	+ 26 35	66	14.5	PART OF MULTIPLE GALAXY
FATH 2.033	11 36 53.	+ 45 00	16		NEBULA
REIZ 1420	11 36 53.	+ 46 44	60	13.0	GALAXY
LB 02077	11 36 53.	+ 49 03 00.		16.6	FAINT BLUE STAR
PK296-06.1	11 36 53.	- 68 35 25.	5		PLANETARY NEBULA
ZWG 040.021	11 36 54.	+ 03 45		14.7	GALAXY
MCG+01-30-006	11 36 54.	+ 03 45	36	14.7	GALAXY
ZWG 040.022	11 36 54.	+ 08 02		15.0	GALAXY
MCG+01-30-007	11 36 54.	+ 08 02	30	15.0	GALAXY
ZWG 157.008	11 36 54.	+ 26 35		15.3	GALAXY
RNGC 3785	11 36 54.	+ 26 35		15.5	GALAXY
UGC 06620	11 36 54.	+ 26 35	60	15.3	GALAXY S0
ZC 1136.9+5007	11 36 54.	+ 50 07	740		CLUSTER OF GALAXIES
ZWG 012.048	11 36 54.	- 02 25		15.4	GALAXY
MCG+05-28-006	11 36 57.	+ 26 35	48	15.2	GALAXY
LB 02078	11 36 57.	+ 49 49 18.		16.6	FAINT BLUE STAR
REIZ 1421	11 36 58.	+ 00 00	54	15.5	GALAXY
REIZ 1422	11 36 58.	+ 04 44	30	15.4	GALAXY
KEEN 3795A	11 37	+ 58 23			GALAXY
ZWG 012.049	11 37 00.	+ 02 20		15.7	GALAXY
MCG+03-30-031	11 37 00.	+ 20 14	42	14.2	GALAXY
REIZ 1424	11 37 00.	+ 20 44	18	14.4	GALAXY
REIZ 1425	11 37 00.	+ 20 52	18	15.2	GALAXY
ZWG 157.009	11 37 00.	+ 32 11		13.5	GALAXY
UGC 06621	11 37 00.	+ 32 11	132	13.0	GALAXY Sa
KARA.72 295A	11 37 00.	+ 32 11	120	13.5	PART OF DOUBLE GALAXY
ZC 1137.0+3601	11 37 00.	+ 36 01	740		CLUSTER OF GALAXIES
MCG+09-19-151	11 37 00.	+ 52 42	36	16.	GALAXY
ZWG 268.069	11 37 00.	+ 52 43		15.6	GALAXY
MCG+10-17-036	11 37 00.	+ 57 46	24	16.	GALAXY
MCG+10-17-037	11 37 00.	+ 60 50	60	14.	GALAXY
REIZ 1426	11 37 01.	+ 22 47	24	15.5	GALAXY
REIZ 1423	11 37 02.	+ 32 11	54	13.8	GALAXY
MCG+05-28-007	11 37 03.	+ 26 34 30.	42	15.3	GALAXY
ARP 294	11 37 03.	+ 32 12			PECULIAR GALAXY
MCG+05-28-008	11 37 03.	+ 32 12	126	13.5	GALAXY
RNGC 3786	11 37 05.	+ 32 11		13.5	GALAXY
RNGC 3788	11 37 05.	+ 32 13		13.0	GALAXY
ZWG 040.023	11 37 06.	+ 04 45		15.3	GALAXY
UGC 06622	11 37 06.	+ 04 45	66	15.3	GALAXY S
MCG+03-30-032	11 37 06.	+ 17 59	66	14.5	GALAXY
REIZ 1427	11 37 06.	+ 20 09	12	15.5	GALAXY
ZWG 127.017	11 37 06.	+ 20 45		14.7	GALAXY
RNGC 3787	11 37 06.	+ 20 45		14.5	GALAXY
MCG+04-28-015	11 37 06.	+ 20 45	24	14.7	GALAXY
MCG+04-28-016	11 37 06.	+ 22 57 30.	36	15.0	GALAXY
TON-N 0584	11 37 06.	+ 27 39		17.2	BLUE STAR
TON-N 0585	11 37 06.	+ 27 45		17.1	BLUE STAR
HOLM 272B	11 37 06.	+ 32 11	72	13.2	PART OF MULTIPLE GALAXY
VV 228B	11 37 06.	+ 32 11	120	13.2	INTERACTING GALAXY
ZWG 157.010	11 37 06.	+ 32 13		13.2	GALAXY
UGC 06623	11 37 06.	+ 32 13	108	13.2	GALAXY S
VV 228A	11 37 06.	+ 32 13	72	12.6	INTERACTING GALAXY
KARA.72 295B	11 37 06.	+ 32 13	102	13.2	PART OF DOUBLE GALAXY
MCG+05-28-009	11 37 06.	+ 32 13 30.	108	13.2	GALAXY
MCG+06-26-013	11 37 06.	+ 37 17	48	15.5	GALAXY
ZC 1137.1+4255	11 37 06.	+ 42 55	1750		CLUSTER OF GALAXIES
MCG-01-30-019	11 37 06.	- 09 22	15	16.	GALAXY
KN 12.077	11 37 07.8	+ 46 54 07.			NEBULA
REIZ 1428	11 37 08.	+ 32 12	108	13.5	GALAXY
HOLM 272A	11 37 08.	+ 32 12	102	12.6	PART OF MULTIPLE GALAXY
RNGC 3789	11 37 08.	- 09 22		16.0	GALAXY
KN 12.078	11 37 08.0	+ 46 53 36.			NEBULA
REIZ 1429	11 37 09.	- 09 06	12	14.2	GALAXY
BC 3CR263	11 37 09.30	+ 66 04 27.0		16.32	QUASI-STELLAR OBJECT
SHB 185	11 37 09.4	+ 66 04 25.		16.3	QUASI-STELLAR OBJECT
A1 119	11 37 11.23	+ 16 01 27.4		16.75	FAINT BLUE OBJECT
ZWG 097.042	11 37 12.	+ 17 15		15.7	GALAXY
ZWG 097.043	11 37 12.	+ 17 59		14.5	GALAXY
RNGC 3790	11 37 12.	+ 17 59		14.5	GALAXY
UGC 06624	11 37 12.	+ 17 59	66	14.5	GALAXY S0-a
REIZ 1430	11 37 12.	+ 20 10	12	15.4	GALAXY
ZWG 097.044	11 37 12.	+ 20 12		14.2	GALAXY
REIZ 1431	11 37 12.	+ 20 12	18	14.9	GALAXY
UGC 06625	11 37 12.	+ 20 12	48	14.2	GALAXY S
REIZ 1432	11 37 12.	+ 20 40	12	15.5	GALAXY
ZWG 127.018	11 37 12.	+ 22 58		15.0	GALAXY
ZWG 127.019	11 37 12.	+ 23 49		15.6	GALAXY
ZWG 127.020	11 37 12.	+ 25 28		15.6	GALAXY
LB 02079	11 37 12.	+ 55 21 12.		16.2	FAINT BLUE STAR
ZWG 012.050	11 37 12.	- 01 36		15.6	GALAXY
KARA.73B 0493	11 37 12.	- 01 36	36	15.6	ISOLATED GALAXY S
MCG-01-30-020	11 37 12.	- 09 07	22	14.5	GALAXY
RNGC 3792	11 37 13.	+ 05 16			NON-EXISTENT OBJECT
RNGC 3791	11 37 14.	- 09 07		14.5	GALAXY
KN 12.079	11 37 14.3	+ 47 44 37.			NEBULA
ABC 1335	11 37 16.	+ 68 25		17.2	RICH CLUSTER OF GALAXIES
ZWG 068.018	11 37 18.	+ 09 09		14.6	GALAXY
UGC 06626	11 37 18.	+ 09 09	72	14.6	GALAXY IRR
MCG+02-30-007	11 37 18.	+ 09 09	72	14.6	GALAXY
ZC 1137.3+1101	11 37 18.	+ 11 01	1210		CLUSTER OF GALAXIES
ZWG 068.019	11 37 18.	+ 13 45		15.7	GALAXY
UGC 06627	11 37 18.	+ 13 45	66	15.7	GALAXY
MCG+03-30-033	11 37 18.	+ 17 13	18	14.6	GALAXY
ZWG 097.045	11 37 18.	+ 17 14		14.6	GALAXY
REIZ 1434	11 37 18.	+ 17 57	18	13.8	GALAXY
ZWG 097.046	11 37 18.	+ 18 09		15.7	GALAXY
MCG+03-30-034	11 37 18.	+ 18 50	30	17.	GALAXY
REIZ 1435	11 37 18.	+ 20 16	12	15.5	GALAXY
MCG+08-21-088	11 37 18.	+ 50 33	30	15.	GALAXY
ZC 1137.3+6707	11 37 18.	+ 67 07	1010		CLUSTER OF GALAXIES
MCG-01-30-021	11 37 18.	- 07 20	36	15.5	GALAXY
IC 0718	11 37 19.	+ 09 09 01.			NONSTELLAR OBJECT
REIZ 1436	11 37 19.	+ 22 58	18	14.8	GALAXY
REIZ 1433	11 37 21.	+ 58 53	108	14.0	GALAXY
REIZ 1437	11 37 23.	+ 46 50	36	15.0	GALAXY
ZWG 068.020	11 37 24.	+ 11 45		15.7	GALAXY
ZC 1137.4+1448	11 37 24.	+ 14 48	740		CLUSTER OF GALAXIES
REIZ 1438	11 37 24.	+ 20 11	12	15.5	GALAXY
MRK 653	11 37 24.	+ 35 15	11	15.5	GALAXY WITH UV CONTINUUM
MCG+08-21-089	11 37 24.	+ 46 11 30.	150	13.	GALAXY
ZWG 242.072	11 37 24.	+ 46 13		14.6	GALAXY
UGC 06628	11 37 24.	+ 46 13	210	14.6	GALAXY
ZWG 292.017	11 37 24.	+ 58 54		14.1	GALAXY
UGC 06629	11 37 24.	+ 58 54	150	14.1	GALAXY S
ZC 1137.4+6028	11 37 24.	+ 60 28	2550		CLUSTER OF GALAXIES
ZC 1137.4+6842	11 37 24.	+ 68 42	2690		CLUSTER OF GALAXIES
ZWG 012.052	11 37 24.	- 00 36		15.5	GALAXY
ZWG 012.051	11 37 24.	- 03 14		15.7	GALAXY
REIZ 1439	11 37 25.	+ 22 47	24	15.4	GALAXY
KN 12.080	11 37 25.5	+ 46 13 08.			NEBULA

OBJECT NAME	RIGHT ASCEN.	DECLINATION	DIAM.	MAGN.	TYPE OF OBJECT
REIZ 1440	11 37 26.	+ 32 11	18	15.2	GALAXY
RNGC 3795	11 37 26.	+ 58 54		14.0	GALAXY
KN 12.081	11 37 26.6	+ 48 54 15.			NEBULA
KN 12.082	11 37 29.1	+ 50 34 42.			NEBULA
MCG+03-30-035	11 37 30.	+ 17 58	48	16.	GALAXY
REIZ 1441	11 37 30.	+ 20 58	30	15.1	GALAXY
MCG+04-28-017	11 37 30.	+ 25 34	15	15.4	GALAXY
TON-N 0586	11 37 30.	+ 28 50		15.0	BLUE STAR
HOLM 272C	11 37 30.	+ 32 11	12	15.5	PART OF MULTIPLE GALAXY
ZC 1137.5+4356	11 37 30.	+ 43 56	1880		CLUSTER OF GALAXIES
ZC 1137.5+5128	11 37 30.	+ 51 28	1610		CLUSTER OF GALAXIES
ZC 1137.5+5759	11 37 30.	+ 57 59	1280		CLUSTER OF GALAXIES
MCG+10-17-038	11 37 30.	+ 58 54	120	14.0	GALAXY
MCG+04-28-018	11 37 33.	+ 24 57	114	13.9	GALAXY
ARP 083	11 37 34.	+ 15 36			PECULIAR GALAXY
A1 120	11 37 34.26	+ 16 03 35.5		16.20	FAINT BLUE OBJECT
RNGC 3797	11 37 35.	+ 32 11			NON-EXISTENT OBJECT
REIZ 1442	11 37 35.	+ 46 09	120	13.7	GALAXY
ZWG 097.047	11 37 36.	+ 15 36		14.4	GALAXY
RNGC 3799	11 37 36.	+ 15 36		14.5	GALAXY
UGC 06630	11 37 36.	+ 15 36	42	14.4	GALAXY S
VV 350B	11 37 36.	+ 15 36	42	14.	INTERACTING GALAXY
KARA.72 296A	11 37 36.	+ 15 36	42	14.4	PART OF DOUBLE GALAXY
MCG+03-30-037	11 37 36.	+ 15 36	42	14.4	GALAXY
RNGC 3800	11 37 36.	+ 15 37		13.0	GALAXY
ZWG 097.048	11 37 36.	+ 17 35		14.3	GALAXY
UGC 06631	11 37 36.	+ 17 35	48	14.3	GALAXY
MCG+03-30-036	11 37 36.	+ 17 35	54	14.3	GALAXY
MCG+03-30-038	11 37 36.	+ 17 44	48	14.6	GALAXY
HOLM 273A	11 37 36.	+ 18 00	66	13.4	PART OF MULTIPLE GALAXY
ZWG 127.021	11 37 36.	+ 23 48		15.7	GALAXY
ZWG 127.022	11 37 36.	+ 24 59		13.9	GALAXY
RNGC 3798	11 37 36.	+ 24 59		14.0	GALAXY
UGC 06632	11 37 36.	+ 24 59	156	13.9	GALAXY SB0
ZWG 127.023	11 37 36.	+ 25 35		15.4	GALAXY
ZC 1137.6+3922	11 37 36.	+ 39 22	1480		CLUSTER OF GALAXIES
ZWG 012.054	11 37 36.	- 00 34		15.2	GALAXY
ZWG 012.053	11 37 36.	- 00 37		15.1	GALAXY
HN 1355	11 37 36.3	+ 25 35 11.			NEBULA
REIZ 1446	11 37 37.	+ 24 58	18	13.8	GALAXY
HOLM 273B	11 37 38.	+ 18 02	72	13.8	PART OF MULTIPLE GALAXY
REIZ 1445	11 37 38.	+ 32 11	18	15.3	GALAXY
MCG+03-30-039	11 37 39.	+ 15 37	102	13.1	GALAXY
REIZ 1443	11 37 39.	+ 58 53	72	14.0	GALAXY
HARO 27	11 37 41.	+ 28 39			BLUE EMISSION-LINE GALAXY
ZWG 040.024	11 37 42.	+ 03 16		14.8	GALAXY
MCG+01-30-008	11 37 42.	+ 03 16	48	14.8	GALAXY
ZWG 068.021	11 37 42.	+ 09 17		13.6	GALAXY
UGC 06633	11 37 42.	+ 09 17	84	13.6	GALAXY S0?
MCG+02-30-008	11 37 42.	+ 09 17	54	13.6	GALAXY
VV 350A	11 37 42.	+ 15 36 30.	96	13.	INTERACTING GALAXY
ZWG 097.049	11 37 42.	+ 15 37		13.1	GALAXY
UGC 06634	11 37 42.	+ 15 37	114	13.1	GALAXY S
KARA.72 296B	11 37 42.	+ 15 37	102	13.1	PART OF DOUBLE GALAXY
REIZ 1447	11 37 42.	+ 17 33	30	13.6	GALAXY
ZWG 097.050	11 37 42.	+ 17 44		14.6	GALAXY
ZWG 097.051	11 37 42.	+ 18 00		13.3	GALAXY
RNGC 3801	11 37 42.	+ 18 00		13.5	GALAXY
UGC 06635	11 37 42.	+ 18 00	138	13.3	GALAXY S0?
MCG+03-30-040	11 37 42.	+ 18 00	180	13.3	GALAXY
MCG+03-30-041	11 37 42.	+ 18 02	66	14.7	GALAXY
ZWG 097.052	11 37 42.	+ 18 03		14.7	GALAXY
RNGC 3802	11 37 42.	+ 18 03		14.5	GALAXY
UGC 06636	11 37 42.	+ 18 03	72	14.7	GALAXY S
RNGC 3803	11 37 42.	+ 18 05			GALAXY
ZC 1137.7+2505	11 37 42.	+ 25 05	1080		CLUSTER OF GALAXIES
LB 02080	11 37 42.	+ 50 37 54.		17.1	FAINT BLUE STAR
KW 53	11 37 42.	- 37 36	72		SEYFERT GALAXY
A1 121	11 37 42.43	+ 16 00 08.6		16.15	FAINT BLUE OBJECT
IC 0719	11 37 44.	+ 09 17 24.			NONSTELLAR OBJECT
A1 122	11 37 45.26	+ 16 07 52.9		16.20	FAINT BLUE OBJECT
REIZ 1448	11 37 46.	+ 03 16	18	15.3	GALAXY
REIZ 1444	11 37 46.	+ 60 34	24	14.0	GALAXY
KN 12.083	11 37 46.1	+ 45 22 05.			NEBULA
A1 123	11 37 47.20	+ 14 46 41.5		16.90	FAINT BLUE OBJECT
A1 124	11 37 47.99	+ 15 44 46.3		16.15	FAINT BLUE OBJECT
ZWG 068.022	11 37 48.	+ 09 08		15.5	GALAXY
REIZ 1449	11 37 48.	+ 17 58	36	13.5	GALAXY
REIZ 1450	11 37 48.	+ 18 01	72	13.7	GALAXY
ZC 1137.8+1838	11 37 48.	+ 18 38	2350		CLUSTER OF GALAXIES
REIZ 1451	11 37 48.	+ 19 41	12	15.6	GALAXY
ZWG 157.011	11 37 48.	+ 28 40		14.5	GALAXY
UGC 06637	11 37 48.	+ 28 40	54	14.5	GALAXY
ZWG 186.019	11 37 48.	+ 32 47		15.7	GALAXY
3ZW 062	11 37 48.	+ 43 51			COMPACT GALAXY
ZC 1137.8+4425	11 37 48.	+ 44 25	1210		CLUSTER OF GALAXIES
ZC 1137.8+5306	11 37 48.	+ 53 06	810		CLUSTER OF GALAXIES
ZWG 292.018	11 37 48.	+ 60 34		13.4	GALAXY
UGC 06638	11 37 48.	+ 60 34	84	13.4	GALAXY S
MCG+10-17-039	11 37 48.	+ 60 36	72	14.0	GALAXY
ZC 1137.8+6055	11 37 48.	+ 60 55	2550		CLUSTER OF GALAXIES
ZWG 012.055	11 37 48.	- 02 26		15.5	GALAXY
MCG-02-30-014	11 37 48.	- 09 48	120	14.	GALAXY
HN 1356	11 37 48.1	+ 28 39 05.			NEBULA
RNGC 3796	11 37 50.	+ 60 34		13.5	GALAXY
ARC 1338	11 37 51.	+ 18 31		16.9	RICH CLUSTER OF GALAXIES
REIZ 1452	11 37 52.	+ 00 08	42	15.6	GALAXY
LB 02081	11 37 53.	+ 50 02 36.		16.7	FAINT BLUE STAR
ZWG 097.053	11 37 54.	+ 19 41		15.7	GALAXY
MCG+05-28-010	11 37 54.	+ 28 39 30.	30	14.5	GALAXY
ZWG 186.020	11 37 54.	+ 36 24		15.4	GALAXY
UGC 06639	11 37 54.	+ 36 24	60	15.4	GALAXY
MCG+06-26-014	11 37 54.	+ 36 24	36	15.	GALAXY
ZWG 186.021	11 37 54.	+ 36 57		15.5	GALAXY
MCG+06-26-015	11 37 54.	+ 36 58	33	14.5	GALAXY
ZC 1137.9+4505	11 37 54.	+ 45 05	3090		CLUSTER OF GALAXIES
KN 12.084	11 37 55.6	+ 46 24 08.			NEBULA
ZC 1138.0+1040	11 38 00.	+ 10 40	1410		CLUSTER OF GALAXIES
ZWG 157.012	11 38 00.	+ 29 09		15.1	GALAXY
KARA.73B 0494	11 38 00.	+ 29 09	42	15.1	ISOLATED GALAXY S
MCG+08-21-090	11 38 00.	+ 46 48	24	15.	GALAXY
MCG+09-19-152	11 38 00.	+ 55 05	42	16.	GALAXY
ZWG 370.008	11 38 00.	+ 89 22		14.9	GALAXY
ZWG 370.002	11 38 00.	+ 89 22		14.9	GALAXY
ARC 1341	11 38 01.	+ 10 40		17.6	RICH CLUSTER OF GALAXIES
HN 1357	11 38 01.7	+ 29 08 05.	24		NEBULA
MCG+04-28-021	11 38 03.	+ 22 41	90	14.1	GALAXY
REIZ 1453	11 38 04.	+ 05 34	24	14.5	GALAXY
ZWG 040.025	11 38 06.	+ 07 14		15.7	GALAXY
ARC 1342	11 38 06.	+ 10 21		17.2	RICH CLUSTER OF GALAXIES
ZWG 068.023	11 38 06.	+ 12 52		15.0	GALAXY
MCG+02-30-009	11 38 06.	+ 12 52	24	15.0	GALAXY
MCG+03-30-042	11 38 06.	+ 18 05	84	14.6	GALAXY
REIZ 1454	11 38 06.	+ 19 45	12	15.8	GALAXY
MCG+04-28-019	11 38 06.	+ 20 35	30	13.8	GALAXY
REIZ 1456	11 38 06.	+ 20 37	18	13.8	GALAXY
MCG+04-28-020	11 38 06.	+ 22 42	30	15.	GALAXY
ZWG 157.013	11 38 06.	+ 28 11		15.7	GALAXY
ZC 1138.1+3941	11 38 06.	+ 39 41	740		CLUSTER OF GALAXIES
ZC 1138.1+4215	11 38 06.	+ 42 15	1140		CLUSTER OF GALAXIES
ZWG 242.073	11 38 06.	+ 46 49		15.2	GALAXY
ZWG 292.019	11 38 06.	+ 56 28		13.8	GALAXY
ZWG 268.070	11 38 06.	+ 56 28		13.8	GALAXY
UGC 06640	11 38 06.	+ 56 28	150	13.8	GALAXY Sc
KN 12.085	11 38 06.0	+ 45 12 24.			NEBULA
KN 12.086	11 38 07.4	+ 46 49 04.			NEBULA
KN 12.087	11 38 08.6	+ 46 44 41.			NEBULA
MCG+05-28-011	11 38 09.	+ 29 08 30.	36	15.1	GALAXY
REIZ 1458	11 38 10.	+ 02 00	12	15.6	GALAXY
REIZ 1459	11 38 10.	+ 05 09	108	13.9	GALAXY
ZWG 040.026	11 38 12.	+ 05 25		15.5	GALAXY
ZWG 097.054	11 38 12.	+ 18 04		14.6	GALAXY
RNGC 3806	11 38 12.	+ 18 04		14.5	GALAXY
UGC 06641	11 38 12.	+ 18 04	90	14.6	GALAXY Sb/SBb
RNGC 3807	11 38 12.	+ 18 05			NON-EXISTENT OBJECT
REIZ 1460	11 38 12.	+ 20 36	30	13.3	GALAXY
ZWG 127.024	11 38 12.	+ 20 38		13.8	GALAXY
RNGC 3805	11 38 12.	+ 20 38		14.0	GALAXY
UGC 06642	11 38 12.	+ 20 38	96	13.8	GALAXY E-S0
ZWG 127.025	11 38 12.	+ 22 43		14.1	GALAXY
RNGC 3808	11 38 12.	+ 22 43		14.0	GALAXY
UGC 06643	11 38 12.	+ 22 43	150	14.1	GALAXY DBL SYS
KARA.72 297A	11 38 12.	+ 22 43	66	14.1	PART OF DOUBLE GALAXY
VV 300B	11 38 12.	+ 22 44	30	15.	INTERACTING GALAXY
VV 300A	11 38 12.	+ 22 44	78	13.	INTERACTING GALAXY
KARA.72 297B	11 38 12.	+ 22 44	42		PART OF DOUBLE GALAXY
ZC 1138.2+2544	11 38 12.	+ 25 44	1010		CLUSTER OF GALAXIES
MCG+04-28-022	11 38 12.	+ 26 02	54	14.8	GALAXY
ZWG 186.022	11 38 12.	+ 33 57		15.2	GALAXY
ZWG 186.023	11 38 12.	+ 35 28		14.9	GALAXY
MRK 426	11 38 12.	+ 35 28	24	15.	GALAXY WITH UV CONTINUUM
MCG+06-26-016	11 38 12.	+ 35 29	30	15.	GALAXY
ARC 1340	11 38 12.	+ 45 09		17.2	RICH CLUSTER OF GALAXIES
7ZW 416	11 38 12.	+ 69 23			COMPACT GALAXY
VHE 52	11 38 12.	- 64 15	30		REFLECTION NEBULA
REIZ 1457	11 38 13.	+ 55 24	72	15.0	GALAXY
REIZ 1455	11 38 14.	+ 56 29	60	14.0	GALAXY
LB 00607	11 38 15.	+ 27 13 12.		17.6	FAINT BLUE STAR
MCG+05-28-013	11 38 15.	+ 28 10	36	15.7	GALAXY
MCG+05-28-012	11 38 15.	+ 30 51 30.	36	15.6	GALAXY
MCG+06-26-017	11 38 15.	+ 33 58	48	14.5	GALAXY
RNGC 3804	11 38 15.	+ 56 29		14.0	GALAXY
REIZ 1461	11 38 16.	+ 02 00	18	15.8	GALAXY
ZWG 040.027	11 38 18.	+ 05 05		15.1	GALAXY
ZC 1138.3+1024	11 38 18.	+ 10 24	4030		CLUSTER OF GALAXIES
ZWG 068.024	11 38 18.	+ 11 45		11.4	GALAXY
UGC 06644	11 38 18.	+ 11 45	246	11.4	GALAXY Sc
MCG+02-30-010	11 38 18.	+ 11 45	240	11.4	GALAXY
REIZ 1462	11 38 18.	+ 22 43	30	14.1	GALAXY
REIZ 1463	11 38 18.	+ 22 44	12	14.4	GALAXY
ZWG 127.026	11 38 18.	+ 26 04		14.8	GALAXY
UGC 06645	11 38 18.	+ 26 04	102	14.8	GALAXY Sb
TON-N 0587	11 38 18.	+ 29 18		16.8	BLUE STAR
ZWG 157.014	11 38 18.	+ 30 52		15.6	GALAXY
MCG+09-19-153	11 38 18.	+ 56 29	120	12.5	GALAXY
LB 02082	11 38 18.	+ 59 57 12.		14.9	FAINT BLUE STAR
KN 12.088	11 38 19.6	+ 46 26 16.			NEBULA
ARC 1344	11 38 20.	- 10 28		16.6	RICH CLUSTER OF GALAXIES
HN 1358	11 38 20.5	+ 26 03 53.	48		NEBULA
HOLM 274A	11 38 22.	+ 02 02	24	14.9	PART OF MULTIPLE GALAXY
ARP 087	11 38 22.	+ 22 44			PECULIAR GALAXY
PATH 1.410	11 38 22.	+ 44 46	27		NEBULA
REIZ 1464	11 38 23.	+ 11 45	252	11.5	GALAXY
ZWG 040.028	11 38 24.	+ 05 20		15.2	GALAXY
UGC 06646	11 38 24.	+ 05 20	60	15.2	GALAXY S-IRR
MCG+01-30-009	11 38 24.	+ 05 20	60	15.2	GALAXY
MCG+04-28-023	11 38 24.	+ 25 05	24	13.9	GALAXY
ZC 1138.4+3956	11 38 24.	+ 39 56	1210		CLUSTER OF GALAXIES
MCG+09-19-154	11 38 24.	+ 54 36	42	17.	GALAXY
ZC 1138.4+6322	11 38 24.	+ 63 22	1280		CLUSTER OF GALAXIES
IC 2949	11 38 24.	- 46 12			NONSTELLAR OBJECT
HOLM 274B	11 38 25.	+ 02 02	24	15.0	PART OF MULTIPLE GALAXY
RNGC 3810	11 38 25.	+ 11 45		11.5	GALAXY
REIZ 1465	11 38 25.	+ 26 03	42	14.1	GALAXY
REIZ 1467	11 38 28.	+ 01 58	36	15.1	GALAXY
SEY 120	11 38 28.	+ 55 15 35.		15.4	FAINT GALAXY
ARC 1347	11 38 28.	- 25 15		17.0	RICH CLUSTER OF GALAXIES
REIZ 1466	11 38 29.	+ 47 58	60	13.2	GALAXY
ZWG 012.056	11 38 30.	+ 01 46		15.5	GALAXY
KARA.73B 0495	11 38 30.	+ 01 46	24	15.5	ISOLATED GALAXY S
ZWG 012.057	11 38 30.	+ 02 00		15.1	GALAXY
ZWG 040.029	11 38 30.	+ 05 32		15.4	GALAXY
ZWG 068.025	11 38 30.	+ 10 30		15.5	GALAXY
UGC 06647	11 38 30.	+ 10 30	84	15.5	GALAXY Sc
MCG+02-30-011	11 38 30.	+ 10 30	60	15.5	GALAXY
REIZ 1468	11 38 30.	+ 20 39	18	15.2	GALAXY
ZWG 127.027	11 38 30.	+ 25 07		13.9	GALAXY
RNGC 3812	11 38 30.	+ 25 07		14.0	GALAXY
UGC 06648	11 38 30.	+ 25 07	120	13.9	GALAXY E
STOCK 32	11 38 30.	+ 31 23			BLUE KNOT NEAR ELLIP GLXY
ZC 1138.5+4116	11 38 30.	+ 41 16	1010		CLUSTER OF GALAXIES
ZWG 214.025	11 38 30.	+ 43 43		15.5	GALAXY
MCG+07-24-029	11 38 30.	+ 43 43	51	14.5	GALAXY
ZC 1138.5+4501	11 38 30.	+ 45 01	1140		CLUSTER OF GALAXIES
MCG+08-21-091	11 38 30.	+ 47 57	120	13.	GALAXY
LB 02083	11 38 30.	+ 52 45 06.		17.2	FAINT BLUE STAR
ZWG 292.020	11 38 30.	+ 60 10		13.6	GALAXY
UGC 06649	11 38 30.	+ 60 10	84	13.6	GALAXY S0
REIZ 1470	11 38 31.	+ 25 05	36	13.1	GALAXY
RNGC 3809	11 38 32.	+ 60 10		13.5	GALAXY
MCG+05-28-014	11 38 33.	+ 31 24	36	17.	GALAXY
REIZ 1471	11 38 34.	+ 02 49	30	15.8	GALAXY
RNGC 3811	11 38 34.	+ 47 58		13.0	GALAXY
ARC 1343	11 38 34.	+ 60 56		17.2	RICH CLUSTER OF GALAXIES
ARC 1346	11 38 35.	+ 05 58		16.8	RICH CLUSTER OF GALAXIES
SN 1971K	11 38 35.	+ 47 58		16.1	SUPERNOVA

OBJECT NAME	RIGHT ASCEN.	DECLINATION	DIAM.	MAGN.	TYPE OF OBJECT
ZC 1138.6+1102	11 38 36.	+ 11 02	1010		CLUSTER OF GALAXIES
ZWG 068.026	11 38 36.	+ 12 00		15.6	GALAXY
ZC 1138.6+1526	11 38 36.	+ 15 26	810		CLUSTER OF GALAXIES
MCG+03-30-043	11 38 36.	+ 20 16	36	15.3	GALAXY
MCG+08-21-092	11 38 36.	+ 46 39	24	15.	GALAXY
ZWG 242.074	11 38 36.	+ 47 58		13.0	GALAXY
MRK 185	11 38 36.	+ 47 58	55	13.	GALAXY WITH UV CONTINUUM
UGC 06650	11 38 36.	+ 47 58	150	13.0	GALAXY SBc
SN 1969C	11 38 36.	+ 47 58		13.7	SUPERNOVA
ZC 1138.6+5917	11 38 36.	+ 59 17	1210		CLUSTER OF GALAXIES
LB 02084	11 38 36.	+ 59 23 12.		16.3	FAINT BLUE STAR
ZC 1138.6+6305	11 38 36.	+ 63 05	5650		CLUSTER OF GALAXIES
ZWG 012.058	11 38 36.	- 02 32		15.4	GALAXY
KN 12.089	11 38 36.8	+ 47 58 09.			NEBULA
ARC 1345	11 38 37.	+ 10 58		17.2	RICH CLUSTER OF GALAXIES
KN 12.090	11 38 37.9	+ 47 19 55.			NEBULA
TON-N 1415	11 38 38.	+ 21 37		16.2	BLUE STAR
LB 00608	11 38 38.	+ 27 43 18.		17.7	FAINT BLUE STAR
TON-N 1416	11 38 38.	+ 33 41		15.7	BLUE STAR
REIZ 1469	11 38 39.	+ 60 10	48	13.8	GALAXY
MCG+10-17-040	11 38 39.	+ 60 10	36	13.8	GALAXY
REIZ 1472	11 38 40.	+ 44 46	36	14.1	GALAXY
RNGC 3813	11 38 41.	+ 36 49		13.0	GALAXY
ZC 1138.7+0556	11 38 42.	+ 05 56	3020		CLUSTER OF GALAXIES
ZC 1138.7+1457	11 38 42.	+ 14 57	1550		CLUSTER OF GALAXIES
ZWG 097.055	11 38 42.	+ 20 14		15.3	GALAXY
MCG+04-28-024	11 38 42.	+ 25 03	24	15.6	GALAXY
ZC 1138.7+2515	11 38 42.	+ 25 15	400		CLUSTER OF GALAXIES
MCG+06-26-018	11 38 42.	+ 36 00 30.	42	15.	GALAXY
ZWG 186.024	11 38 42.	+ 36 49		12.6	GALAXY
UGC 06651	11 38 42.	+ 36 49	126	12.6	GALAXY S
MCG+06-26-019	11 38 42.	+ 36 50	114	12.	GALAXY
ZWG 242.075	11 38 42.	+ 46 40		15.4	GALAXY
ZC 1138.7+5640	11 38 42.	+ 56 40	1140		CLUSTER OF GALAXIES
ZC 1138.7+5650	11 38 42.	+ 56 50	33060		CLUSTER OF GALAXIES
KN 12.091	11 38 42.2	+ 46 40 13.			NEBULA
ARC 1339	11 38 43.	+ 73 21		17.4	RICH CLUSTER OF GALAXIES
ARC 1348	11 38 43.	- 12 06		17.0	RICH CLUSTER OF GALAXIES
REIZ 1473	11 38 45.	+ 36 49	108	12.8	GALAXY
ZC 1138.8+0055	11 38 48.	+ 00 55	340		CLUSTER OF GALAXIES
ZWG 040.030	11 38 48.	+ 02 59		15.7	GALAXY
ZWG 068.027	11 38 48.	+ 12 07		15.3	GALAXY
HOLM 276B	11 38 48.	+ 25 04	30	15.0	PART OF MULTIPLE GALAXY
ZWG 157.015	11 38 48.	+ 29 08		15.1	GALAXY
UGC 06652	11 38 48.	+ 33 58	90	16.0	GALAXY SB
7ZW 417	11 38 48.	+ 71 50			COMPACT GALAXY
ZWG 334.047	11 38 48.	+ 71 50		15.2	GALAXY
HN 1360	11 38 48.6	+ 25 14 05.			NEBULA
HN 1359	11 38 48.8	+ 28 08 41.			NEBULA
REIZ 1474	11 38 49.	+ 25 13	24	15.0	GALAXY
FATH 1.411	11 38 50.	+ 44 47	5		NEBULA
MCG-01-30-022	11 38 51.	- 06 13	72	13.5	GALAXY
REIZ 1475	11 38 53.	+ 46 36	18	14.7	GALAXY
ZWG 127.028	11 38 54.	+ 25 05		15.6	GALAXY
RNGC 3814	11 38 54.	+ 25 05		15.5	GALAXY
MCG+05-28-015	11 38 54.	+ 28 08	36	15.1	GALAXY
ZWG 186.025	11 38 54.	+ 32 37		15.6	GALAXY
ZWG 012.059	11 38 54.	- 02 13		15.7	GALAXY
REIZ 1476	11 38 55.	+ 25 04	18	14.8	GALAXY
HOLM 275C	11 38 56.	+ 16 15	24	14.7	PART OF MULTIPLE GALAXY
TON-N 1417	11 38 56.	+ 21 22		16.	BLUE STAR
MCG+04-28-025	11 38 57.	+ 25 03	90	14.2	GALAXY
KN 12.092	11 38 57.3	+ 47 21 34.			NEBULA
REIZ 1477	11 38 58.	+ 05 18	30	14.3	GALAXY
A1 125	11 38 58.46	+ 15 43 18.6		16.60	FAINT BLUE OBJECT
LB 09976	11 39	- 88 16		14.6	FAINT BLUE STAR
MCG+03-30-044	11 39 00.	+ 16 14	90	14.6	GALAXY
ZWG 097.056	11 39 00.	+ 16 15		14.6	GALAXY
HOLM 275A	11 39 00.	+ 16 15	60	13.9	PART OF MULTIPLE GALAXY
UGC 06653	11 39 00.	+ 16 15	90	14.6	GALAXY SB?0-a
MCG+03-30-044A	11 39 00.	+ 17 18	48	15.	GALAXY
ZWG 097.057	11 39 00.	+ 17 21		15.2	GALAXY
MCG+03-30-045	11 39 00.	+ 17 21	36	15.2	GALAXY
ZC 1139.0+1809	11 39 00.	+ 18 09	1080		CLUSTER OF GALAXIES
MCG+03-30-046	11 39 00.	+ 20 24	27	13.6	GALAXY
ZWG 127.029	11 39 00.	+ 26 10		15.6	GALAXY
MCG+04-28-026	11 39 00.	+ 26 10	15	15.6	GALAXY
MCG+04-28-027	11 39 00.	+ 26 10 30.	24	15.4	GALAXY
TON-N 0069	11 39 00.	+ 29 02		15.0	BLUE STAR
ZWG 186.026	11 39 00.	+ 38 15		14.8	GALAXY
MRK 638	11 39 00.	+ 38 15	20	15.	GALAXY WITH UV CONTINUUM
IC 2950	11 39 00.	+ 38 16 17.			NONSTELLAR OBJECT
MCG+12-11-037	11 39 00.	+ 71 51	51	16.	GALAXY
HN 1361	11 39 00.6	+ 26 10 23.			NEBULA
REIZ 1478	11 39 01.	+ 25 04	66	13.3	GALAXY
HOLM 276A	11 39 01.	+ 25 04	90	14.0	PART OF MULTIPLE GALAXY
HOLM 277A	11 39 01.	+ 26 09	18	15.3	PART OF MULTIPLE GALAXY
HOLM 277B	11 39 01.	+ 26 10	30	15.4	PART OF MULTIPLE GALAXY
HN 1362	11 39 01.4	+ 26 09 59.			NEBULA
MCG-02-30-015	11 39 03.	- 10 45	30	15.	GALAXY
ZWG 097.058	11 39 06.	+ 17 17		15.3	GALAXY
REIZ 1479	11 39 06.	+ 20 52	12	15.2	GALAXY
ZWG 127.030	11 39 06.	+ 25 05		14.2	GALAXY
RNGC 3815	11 39 06.	+ 25 05		14.0	GALAXY
UGC 06654	11 39 06.	+ 25 05	114	14.2	GALAXY Sa-b
ZWG 012.060	11 39 06.	- 01 58		15.5	GALAXY
TON-N 1418	11 39 08.	+ 34 31		16.	BLUE STAR
A1 126	11 39 09.15	+ 14 50 32.0		16.90	FAINT BLUE OBJECT
ARC 1350	11 39 10.	+ 24 52		17.6	RICH CLUSTER OF GALAXIES
LB 02085	11 39 10.	+ 55 16 48.		16.1	FAINT BLUE STAR
ZWG 097.059	11 39 12.	+ 16 15		14.5	GALAXY
UGC 06655	11 39 12.	+ 16 15	42	14.5	GALAXY
MCG+03-30-047	11 39 12.	+ 16 15	18	14.5	GALAXY
HOLM 275B	11 39 12.	+ 16 16	42	13.6	PART OF MULTIPLE GALAXY
ZWG 097.060	11 39 12.	+ 20 23		13.6	GALAXY
RNGC 3816	11 39 12.	+ 20 23		13.5	GALAXY
REIZ 1480	11 39 12.	+ 20 23	12	14.3	GALAXY
UGC 06656	11 39 12.	+ 20 23	114	13.6	GALAXY S0
MCG+04-28-028	11 39 12.	+ 23 18	36	15.2	GALAXY
ZC 1139.2+2500	11 39 12.	+ 25 00	1080		CLUSTER OF GALAXIES
TON-N 0588	11 39 12.	+ 25 51		16.5	BLUE STAR
MCG+12-11-038	11 39 12.	+ 71 52	24	16.	GALAXY
MCG+04-28-029	11 39 15.	+ 20 34	33	15.	GALAXY
KN 12.093	11 39 15.0	+ 47 13 38.			NEBULA
LB 02086	11 39 16.	+ 53 29 54.		17.3	FAINT BLUE STAR
PK294-00.1	11 39 16.	- 62 12	72		PLANETARY NEBULA
ZWG 040.031	11 39 18.	+ 02 51		15.3	GALAXY
MCG+01-30-010	11 39 18.	+ 02 51	42	15.3	GALAXY
ZWG 068.028	11 39 18.	+ 10 35		14.4	GALAXY
UGC 06657	11 39 18.	+ 10 35	66	14.4	GALAXY SB0-a
MCG+02-30-012	11 39 18.	+ 10 35	60	14.4	GALAXY
ZWG 068.029	11 39 18.	+ 12 52		15.7	GALAXY
ZWG 127.031	11 39 18.	+ 23 20		15.2	GALAXY
REIZ 1481	11 39 18.	+ 23 24	24	15.2	GALAXY
ZC 1139.3+5544	11 39 18.	+ 55 44	2550		CLUSTER OF GALAXIES
RNGC 3817	11 39 19.	+ 10 35		14.5	GALAXY
MCG+05-28-016	11 39 21.	+ 32 17 30.	54	15.0	GALAXY
REIZ 1482	11 39 21.	- 05 54	48	12.8	GALAXY
REIZ 1483	11 39 22.	+ 02 51	24	15.8	GALAXY
ARC 1349	11 39 23.	+ 55 38		16.9	RICH CLUSTER OF GALAXIES
ARC 1352	11 39 23.	- 21 12		17.8	RICH CLUSTER OF GALAXIES
ZWG 157.016	11 39 24.	+ 32 17		15.0	GALAXY
UGC 06658	11 39 24.	+ 32 17	60	15.0	GALAXY S
MCG+06-26-020	11 39 24.	+ 32 49	30	15.5	GALAXY
MCG-01-30-023	11 39 24.	- 05 53	60	13.	GALAXY
HOLM 278B	11 39 25.	+ 32 16	12	15.4	PART OF MULTIPLE GALAXY
RNGC 3818	11 39 25.	- 05 53		13.0	GALAXY
REIZ 1484	11 39 26.	+ 32 17	60	14.2	GALAXY
HOLM 278A	11 39 26.	+ 32 17	48	14.1	PART OF MULTIPLE GALAXY
MCG-02-30-016	11 39 27.	- 10 30	78	15.	GALAXY
ZWG 040.032	11 39 30.	+ 07 56		15.6	GALAXY
ZWG 068.030	11 39 30.	+ 10 38		14.8	GALAXY
MCG+02-30-013	11 39 30.	+ 10 38	18	14.8	GALAXY
ZWG 068.031	11 39 30.	+ 10 40		14.9	GALAXY
MCG+02-30-014	11 39 30.	+ 10 40	36	14.9	GALAXY
ZWG 068.032	11 39 30.	+ 11 49		15.3	GALAXY
MCG+03-30-048	11 39 30.	+ 20 24	30	15.6	GALAXY
MCG+04-28-030	11 39 30.	+ 20 34 30.	84	13.8	GALAXY
ZC 1139.5+2522	11 39 30.	+ 25 22	740		CLUSTER OF GALAXIES
TON-N 0589	11 39 30.	+ 25 58		16.3	BLUE STAR
MCG+04-28-031	11 39 30.	+ 26 16	36	15.0	GALAXY
UGC 06659	11 39 30.	+ 32 49	60	16.0	GALAXY Sc
ZC 1139.5+4948	11 39 30.	+ 49 48	670		CLUSTER OF GALAXIES
MCG+09-19-156	11 39 30.	+ 53 13	18	17.	GALAXY
MCG+09-19-155	11 39 30.	+ 53 14	30	16.	GALAXY
MCG+10-17-041	11 39 30.	+ 58 59	12	16.	GALAXY
UGC 06660	11 39 30.	+ 62 27	78	17.	GALAXY
MCG+10-17-042	11 39 30.	+ 62 27	78	16.	GALAXY
MCG-03-30-007	11 39 30.	- 16 11 30.	48	14.5	GALAXY
RNGC 3819	11 39 31.	+ 10 38		15.0	GALAXY
RNGC 3820	11 39 31.	+ 10 40		15.0	GALAXY
HOLM 279A	11 39 31.	+ 26 15	36	15.0	PART OF MULTIPLE GALAXY
HOLM 279B	11 39 33.	+ 26 15	12	15.4	PART OF MULTIPLE GALAXY
MCG-03-30-008	11 39 33.	- 17 54	90	14.5	GALAXY
KN 12.094	11 39 33.0	+ 50 57 43.			NEBULA
KN 12.095	11 39 33.2	+ 46 13 17.			NEBULA
HN 1363	11 39 34.5	+ 26 15 16.	24		NEBULA
ARP 161	11 39 35.	+ 00 38			PECULIAR GALAXY
REIZ 1485	11 39 35.	+ 11 49	18	15.4	GALAXY
LB 02087	11 39 35.	+ 49 40 36.		15.9	FAINT BLUE STAR
ZC 1139.6+0820	11 39 36.	+ 08 20	2690		CLUSTER OF GALAXIES
ARC 1354	11 39 36.	+ 10 26		17.2	RICH CLUSTER OF GALAXIES
ZWG 068.033	11 39 36.	+ 10 33		13.7	GALAXY
UGC 06661	11 39 36.	+ 10 33	78	13.7	GALAXY S0?
MCG+02-30-015	11 39 36.	+ 10 33	72	13.7	GALAXY
ZC 1139.6+1213	11 39 36.	+ 12 13	1610		CLUSTER OF GALAXIES
ZWG 068.034	11 39 36.	+ 13 13		15.4	GALAXY
UGC 06662	11 39 36.	+ 13 13	60	15.4	GALAXY Sc-IRR
ZC 1139.6+1322	11 39 36.	+ 13 22	1680		CLUSTER OF GALAXIES
ZWG 097.061	11 39 36.	+ 18 41		15.5	GALAXY
MCG+03-30-049	11 39 36.	+ 18 42	12	15.5	GALAXY
ZWG 097.062	11 39 36.	+ 20 15		15.5	GALAXY
ZWG 097.063	11 39 36.	+ 20 19		15.7	GALAXY
ZWG 097.064	11 39 36.	+ 20 22		15.6	GALAXY
REIZ 1486	11 39 36.	+ 20 35	12	14.7	GALAXY
ZWG 127.032	11 39 36.	+ 20 36		13.8	GALAXY
RNGC 3821	11 39 36.	+ 20 36		14.0	GALAXY
UGC 06663	11 39 36.	+ 20 36	90	13.8	GALAXY SBa
MCG+04-28-032	11 39 36.	+ 26 19	24	15.0	GALAXY
TON-N 0590	11 39 36.	+ 28 42		17.0	BLUE STAR
ZC 1139.6+2859	11 39 36.	+ 28 59	2820		CLUSTER OF GALAXIES
ZWG 157.017	11 39 36.	+ 30 31		15.5	GALAXY
UGC 06664	11 39 36.	+ 30 31	96	15.5	GALAXY Sc
MCG+09-19-157	11 39 36.	+ 51 52	198	14.	GALAXY
MCG+09-19-158	11 39 36.	+ 55 02	15	16.	GALAXY
MCG+09-19-159	11 39 36.	+ 55 04	36	16.	GALAXY
ZC 1139.6+7317	11 39 36.	+ 73 17	1880		CLUSTER OF GALAXIES
MCG-02-30-017	11 39 36.	- 13 35 30.	72	14.5	GALAXY
RNGC 3822	11 39 37.	+ 10 33		13.5	GALAXY
HOLM 280A	11 39 37.	+ 26 18	24	15.0	PART OF MULTIPLE GALAXY
A1 127	11 39 37.42	+ 15 36 07.4		16.90	FAINT BLUE OBJECT
SEY 121	11 39 38.	+ 51 52 22.		14.0	FAINT GALAXY
RNGC 3823	11 39 38.	- 13 35		14.0	GALAXY
KN 12.096	11 39 38.0	+ 49 36 56.			NEBULA
MCG+06-26-021	11 39 39.	+ 36 22 30.	42	16.	GALAXY
ARC 1353	11 39 40.	+ 25 14		17.6	RICH CLUSTER OF GALAXIES
HOLM 280B	11 39 41.	+ 26 17	12	15.6	PART OF MULTIPLE GALAXY
ZWG 012.061	11 39 42.	+ 00 37		13.7	GALAXY
UGC 06665	11 39 42.	+ 00 37	30	13.7	GALAXY PECULR
MCG+00-30-019	11 39 42.	+ 00 37	24	13.7	GALAXY
ZWG 068.035	11 39 42.	+ 09 03		14.7	GALAXY
KARA.72 298B	11 39 42.	+ 09 03	30		PART OF DOUBLE GALAXY
KARA.72 298A	11 39 42.	+ 09 03	36	14.7	PART OF DOUBLE GALAXY
MCG+02-30-016	11 39 42.	+ 09 03	60	14.7	GALAXY
ZC 1139.7+1027	11 39 42.	+ 10 27	810		CLUSTER OF GALAXIES
ZC 1139.7+1041	11 39 42.	+ 10 41	1010		CLUSTER OF GALAXIES
ZWG 068.036	11 39 42.	+ 14 21		14.9	GALAXY
MCG+02-30-017	11 39 42.	+ 14 21	15	14.9	GALAXY
MCG+03-30-050	11 39 42.	+ 16 17	66	14.3	GALAXY
ZWG 097.065	11 39 42.	+ 16 18		14.3	GALAXY
UGC 06666	11 39 42.	+ 16 18	66	14.3	GALAXY S
MCG+03-30-051	11 39 42.	+ 20 25	60	14.7	GALAXY
REIZ 1487	11 39 42.	+ 21 48	24	15.3	GALAXY
ZC 1139.7+2205	11 39 42.	+ 22 05	1550		CLUSTER OF GALAXIES
MCG+05-28-017	11 39 42.	+ 30 31	84	15.5	GALAXY
ZWG 268.071	11 39 42.	+ 51 53		14.8	GALAXY
UGC 06667	11 39 42.	+ 51 53	210	14.8	GALAXY Sc
ZWG 268.072	11 39 42.	+ 55 05		15.3	GALAXY
MCG+09-19-160	11 39 42.	+ 55 06	36	16.	GALAXY
MCG+10-17-043	11 39 42.	+ 57 00	30	16.	GALAXY
MCG+10-17-044	11 39 42.	+ 57 34	18	16.	GALAXY
ZC 1139.7+5934	11 39 42.	+ 59 34	540		CLUSTER OF GALAXIES
ZC 1139.7+7423	11 39 42.	+ 74 23	2420		CLUSTER OF GALAXIES
HN 1364	11 39 42.6	+ 26 18 22.	42		NEBULA
REIZ 1488	11 39 43.	+ 20 49	54	14.9	GALAXY
TON-N 1419	11 39 44.	+ 22 20		16.	BLUE STAR

OBJECT NAME	RIGHT ASCEN.	DECLINATION	DIAM.	MAGN.	TYPE OF OBJECT
IC 0720	11 39 48.	+ 09 03 05.			NONSTELLAR OBJECT
ZWG 068.037	11 39 48.	+ 10 33		13.8	GALAXY
UGC 06668	11 39 48.	+ 10 33	84	13.8	GALAXY SBa
MKW 10	11 39 48.	+ 10 33		13.8	POOR GALAXY CLUSTER
MCG+02-30-018	11 39 48.	+ 10 33	72	13.8	GALAXY
UGC 06669	11 39 48.	+ 15 15	96	17.	GALAXY DWRF IR
MCG+03-30-052	11 39 48.	+ 16 22	36	16.	GALAXY
ZWG 097.066	11 39 48.	+ 18 02		15.7	GALAXY
ZWG 097.067	11 39 48.	+ 18 36		14.3	GALAXY
HOLM 281A	11 39 48.	+ 18 36	90	13.2	PART OF MULTIPLE GALAXY
UGC 06670	11 39 48.	+ 18 36	186	14.3	GALAXY IRR
MCG+03-30-053	11 39 48.	+ 18 37	180	14.3	GALAXY
REIZ 1489	11 39 48.	+ 20 23	30	14.7	GALAXY
ZWG 097.068	11 39 48.	+ 20 24		14.7	GALAXY
MCG+04-28-033	11 39 48.	+ 21 50	33	15.5	GALAXY
ZC 1139.8+2246	11 39 48.	+ 22 46	1010		CLUSTER OF GALAXIES
ZWG 157.018	11 39 48.	+ 26 46		14.4	GALAXY
RNGC 3826	11 39 48.	+ 26 46		14.5	GALAXY
UGC 06671	11 39 48.	+ 26 46	60	14.4	GALAXY E
UGC 06672	11 39 48.	+ 33 55	60	16.0	GALAXY Sc
MCG+06-26-022	11 39 48.	+ 36 06	36	15.	GALAXY
MCG+10-17-045	11 39 48.	+ 58 10	45	16.	GALAXY
ZC 1139.8+5845	11 39 48.	+ 58 45	740		CLUSTER OF GALAXIES
ZC 1139.8+5950	11 39 48.	+ 59 50	1340		CLUSTER OF GALAXIES
MCG-01-30-024	11 39 48.	- 08 22	66	15.	GALAXY
RNGC 3825	11 39 49.	+ 10 33		14.0	GALAXY
ARC 1351	11 39 49.	+ 58 49		17.8	RICH CLUSTER OF GALAXIES
TON-N 1420	11 39 50.	+ 21 32		16.9	BLUE STAR
REIZ 1490	11 39 51.	- 07 30	36	14.1	GALAXY
AR 39B	11 39 51.43	+ 77 39 06.5			NEBULA
AR 39A	11 39 51.77	+ 77 39 06.5			NEBULA
HOLM 281B	11 39 53.	+ 18 35	18	15.3	PART OF MULTIPLE GALAXY
ARC 1356	11 39 54.	+ 10 43		17.2	RICH CLUSTER OF GALAXIES
ZWG 097.069	11 39 54.	+ 18 04		15.6	GALAXY
REIZ 1491	11 39 54.	+ 18 35	108	13.5	GALAXY
MCG+03-30-054	11 39 54.	+ 19 08	60	13.6	GALAXY
ZC 1139.9+2542	11 39 54.	+ 25 42	1080		CLUSTER OF GALAXIES
ZC 1139.9+4153	11 39 54.	+ 41 53	2150		CLUSTER OF GALAXIES
MCG-01-30-025	11 39 54.	- 07 30	48	14.5	GALAXY
IC 0721	11 39 56.	- 08 03 20.			NONSTELLAR OBJECT
LB 02088	11 39 57.	+ 55 23 30.		15.6	FAINT BLUE STAR
MCG-01-30-026	11 39 57.	- 08 05	66	15.5	GALAXY
REIZ 1492	11 39 58.	+ 44 52	12	15.3	GALAXY
A1 128	11 39 58.47	+ 15 10 33.5		16.80	FAINT BLUE OBJECT
ZWG 097.070	11 40 00.	+ 19 07		13.6	GALAXY
RNGC 3827	11 40 00.	+ 19 07		13.5	GALAXY
REIZ 1495	11 40 00.	+ 19 07	30	13.3	GALAXY
UGC 06673	11 40 00.	+ 19 07	54	13.6	GALAXY
MCG+03-30-055	11 40 00.	+ 20 20	48	15.0	GALAXY
ZWG 127.033	11 40 00.	+ 25 07		15.2	GALAXY
UGC 06674	11 40 00.	+ 25 07	72	15.2	GALAXY Sc
ZC 1140.0+2715	11 40 00.	+ 27 15	5510		CLUSTER OF GALAXIES
ZC 1140.0+3348	11 40 00.	+ 33 48	1010		CLUSTER OF GALAXIES
ZC 1140.0+3928	11 40 00.	+ 39 28	870		CLUSTER OF GALAXIES
ZC 1140.0+4208	11 40 00.	+ 42 08	810		CLUSTER OF GALAXIES
MCG+10-17-046	11 40 00.	+ 58 54	27	16.	GALAXY
MCG+13-09-001	11 40 00.	+ 77 39 30.	108	14.	GALAXY
ZWG 352.006	11 40 00.	+ 77 40		14.6	GALAXY
ZWG 351.068	11 40 00.	+ 77 40		14.6	GALAXY
RNGC 3901	11 40 00.	+ 77 40		14.5	GALAXY
UGC 06675	11 40 00.	+ 77 40	114	14.6	GALAXY Sc
KARA.73B 0496	11 40 00.	+ 77 40	132	14.6	ISOLATED GALAXY S
MCG+15-01-011	11 40 00.	+ 89 22	60	15.	GALAXY
ZWG 012.062	11 40 00.	- 01 59		15.7	GALAXY
REIZ 1496	11 40 01.	+ 25 05	60	14.1	GALAXY
REIZ 1494	11 40 01.	+ 26 46	24	14.5	GALAXY
TON-N 1421	11 40 02.	+ 21 09		14.7	BLUE STAR
KN 12.097	11 40 02.0	+ 49 49 14.			NEBULA
IC 0722	11 40 03.	+ 09 20 28.			NONSTELLAR OBJECT
ARC 1355	11 40 03.	+ 42 11		17.6	RICH CLUSTER OF GALAXIES
MCG+09-19-161	11 40 03.	+ 53 03	72	13.9	GALAXY
RNGC 3824	11 40 03.	+ 53 04		14.5	GALAXY
HN 1365	11 40 03.3	+ 25 05 52.	48		NEBULA
KN 12.098	11 40 05.1	+ 45 17 12.			NEBULA
ZWG 068.038	11 40 06.	+ 09 09		15.3	GALAXY
ZWG 068.039	11 40 06.	+ 09 15		14.9	GALAXY
MCG+02-30-019	11 40 06.	+ 09 15	60	14.9	GALAXY
MCG+03-30-056	11 40 06.	+ 15 37	30	15.0	GALAXY
ZWG 097.071	11 40 06.	+ 15 38		15.0	GALAXY
ZWG 097.072	11 40 06.	+ 20 18		15.0	GALAXY
REIZ 1497	11 40 06.	+ 20 18	42	14.9	GALAXY
MCG+04-28-034	11 40 06.	+ 25 05	66	15.2	GALAXY
MCG+05-28-018	11 40 06.	+ 26 45	18	14.4	GALAXY
REIZ 1493	11 40 06.	+ 53 03	42	14.4	GALAXY
ZWG 268.073	11 40 06.	+ 53 04		14.6	GALAXY
UGC 06676	11 40 06.	+ 53 04	90	14.6	GALAXY
MCG+10-17-047	11 40 06.	+ 58 31	30	16.	GALAXY
ZC 1140.1+6124	11 40 06.	+ 61 24	1610		CLUSTER OF GALAXIES
MAI 064	11 40 07.	+ 59 24	87		DWARF SPHEROIDAL GALAXY
MCG+04-28-040	11 40 09.	+ 23 00	120	14.0	GALAXY
ARC 1358	11 40 12.	+ 08 30		17.0	RICH CLUSTER OF GALAXIES
ZWG 157.019	11 40 12.	+ 26 50		15.6	GALAXY
UGC 06677	11 40 12.	+ 26 50	78	15.6	GALAXY Sa-b
LB 02089	11 40 12.	+ 59 35 18.		16.6	FAINT BLUE STAR
ZWG 012.063	11 40 12.	- 02 05		15.6	GALAXY
REIZ 1498	11 40 13.	+ 26 49	54	15.5	GALAXY
LB 02090	11 40 14.	+ 48 39 42.		17.0	FAINT BLUE STAR
A1 129	11 40 15.66	+ 14 42 03.6		16.20	FAINT BLUE OBJECT
HN 1366	11 40 16.9	+ 26 48 46.			NEBULA
RNGC 3828	11 40 18.	+ 16 46		15.5	GALAXY
MCG+03-30-057	11 40 18.	+ 16 46	42	15.6	GALAXY
MCG+03-30-058	11 40 18.	+ 19 57	63	15.5	GALAXY
ZWG 097.073	11 40 18.	+ 20 14		15.6	GALAXY
REIZ 1500	11 40 18.	+ 20 14	18	15.6	GALAXY
ZWG 097.074	11 40 18.	+ 20 21		15.4	GALAXY
REIZ 1501	11 40 18.	+ 23 29	18	15.1	GALAXY
MCG+04-28-035	11 40 18.	+ 24 11 30.	54	15.4	GALAXY
ZWG 157.020	11 40 18.	+ 26 33		15.5	GALAXY
UGC 06678	11 40 18.	+ 26 33	60	15.5	GALAXY Sb
ZWG 157.021	11 40 18.	+ 26 34		15.1	GALAXY
ZWG 157.022	11 40 18.	+ 27 41		14.9	GALAXY
UGC 06679	11 40 18.	+ 41 07	78	16.0	GALAXY Sc
MCG-03-30-009	11 40 18.	- 18 48 30.	36	15.	GALAXY
KN 12.099	11 40 19.0	+ 48 15 03.			NEBULA
REIZ 1499	11 40 20.	+ 59 50	60	15.1	GALAXY
MCG+07-24-030	11 40 21.	+ 41 06	72	15.	GALAXY
ARC 1357	11 40 21.	+ 61 34		16.9	RICH CLUSTER OF GALAXIES
REIZ 1502	11 40 23.	+ 49 06	90	14.0	GALAXY

OBJECT NAME	RIGHT ASCEN.	DECLINATION	DIAM.	MAGN.	TYPE OF OBJECT
ZWG 012.064	11 40 24.	+ 00 26		15.5	GALAXY
ZC 1140.4+1118	11 40 24.	+ 11 18	1080		CLUSTER OF GALAXIES
ZWG 068.040	11 40 24.	+ 13 22		15.7	GALAXY
ZWG 097.075	11 40 24.	+ 16 46		15.6	GALAXY
ZWG 097.076	11 40 24.	+ 19 55		15.5	GALAXY
UGC 06680	11 40 24.	+ 19 55	72	15.5	GALAXY Sb
ZWG 127.034	11 40 24.	+ 23 25		15.5	GALAXY
ZWG 127.035	11 40 24.	+ 24 13		15.4	GALAXY
UGC 06681	11 40 24.	+ 24 13	72	15.4	GALAXY S
MCG+04-28-036	11 40 24.	+ 24 16	48	15.5	GALAXY
MCG+05-28-019	11 40 24.	+ 26 48 30.	66	15.6	GALAXY
ZC 1140.4+4839	11 40 24.	+ 48 39	1080		CLUSTER OF GALAXIES
MCG+08-21-093	11 40 24.	+ 48 39	48	15.	GALAXY
ZWG 242.076	11 40 24.	+ 48 40		15.0	GALAXY
REIZ 1503	11 40 24.	+ 53 04	6	15.6	GALAXY
MCG-06-26-005	11 40 24.	- 36 00	24	17.	GALAXY
KN 12.100	11 40 24.9	+ 48 40 38.			NEBULA
IC 0723	11 40 25.	- 08 02 50.			NONSTELLAR OBJECT
HN 1367	11 40 25.0	+ 27 40 34.			NEBULA
HN 1368	11 40 26.7	+ 26 31 46.	36		NEBULA
KN 12.101	11 40 26.7	+ 46 24 12.			NEBULA
VV 353A	11 40 27.	+ 26 32	48	16.	INTERACTING GALAXY
ARP 115	11 40 27.	+ 26 33			PECULIAR GALAXY
MCG-06-26-006	11 40 27.	- 36 00	30	17.5	GALAXY
HN 1369	11 40 27.3	+ 26 32 40.			NEBULA
HN 1370	11 40 27.8	+ 26 32 52.			NEBULA
VV 353B	11 40 28.	+ 26 33	21	16.	INTERACTING GALAXY
ARC 1360	11 40 30.	+ 11 18		17.2	RICH CLUSTER OF GALAXIES
REIZ 1504	11 40 30.	+ 19 55	24	15.4	GALAXY
MCG+03-30-059	11 40 30.	+ 20 02	60	15.2	GALAXY
REIZ 1505	11 40 30.	+ 24 12	36	14.8	GALAXY
MCG+05-28-020	11 40 30.	+ 27 40	18	14.9	GALAXY
ZC 1140.5+5351	11 40 30.	+ 53 51	1080		CLUSTER OF GALAXIES
UGC 06682	11 40 30.	+ 59 23	102	16.5	GALAXY DWRF SP
MCG+10-17-048	11 40 30.	+ 59 50	24	16.	GALAXY
ZC 1140.5+6151	11 40 30.	+ 61 51	1480		CLUSTER OF GALAXIES
MCG-01-30-027	11 40 30.	- 08 05	42	15.	GALAXY
MCG-02-30-019	11 40 30.	- 11 18 30.	30	14.5	GALAXY
MCG-02-30-020	11 40 30.	- 12 21	36	15.	GALAXY
MCG-02-30-018	11 40 30.	- 12 35	72	15.5	GALAXY
REIZ 1506	11 40 31.	+ 26 32	36	14.9	GALAXY
HN 1371	11 40 35.5	+ 26 50 52.			NEBULA
MCG+03-30-060	11 40 36.	+ 18 29	90	15.	GALAXY
ZWG 097.077	11 40 36.	+ 19 39		15.5	GALAXY
ZWG 097.078	11 40 36.	+ 20 01		15.2	GALAXY
UGC 06683	11 40 36.	+ 20 01	60	15.2	GALAXY S0-a
MCG+03-30-061	11 40 36.	+ 20 02	84	15.0	GALAXY
ZWG 097.079	11 40 36.	+ 20 17		15.7	GALAXY
MCG+05-28-022	11 40 36.	+ 26 31 30.	48	15.5	GALAXY
MCG+05-28-021	11 40 36.	+ 26 32 30.	30	15.1	GALAXY
ZWG 157.023	11 40 36.	+ 26 51		14.7	GALAXY
RNGC 3830	11 40 36.	+ 26 51		14.5	GALAXY
SHAH 122	11 40 36.	+ 57 34	180	18.0	GROUP OF COMPACT GALAXIES
MCG+10-17-049	11 40 36.	+ 59 24	90	15.	GALAXY
7ZW 418	11 40 36.	+ 63 28			COMPACT GALAXY
MCG-02-30-022	11 40 36.	- 11 23 30.	48	15.5	GALAXY
MCG-02-30-021	11 40 36.	- 12 29	66	15.	GALAXY
MCG-02-30-023	11 40 36.	- 12 36	150	14.	GALAXY
REIZ 1507	11 40 37.	+ 26 50	18	14.8	GALAXY
RNGC 3831	11 40 38.	- 12 36		14.0	GALAXY
ZC 1140.7+0741	11 40 42.	+ 07 41	1140		CLUSTER OF GALAXIES
ZWG 068.041	11 40 42.	+ 09 15		15.7	GALAXY
KARA.68 079	11 40 42.	+ 14 30	67		DWARF GALAXY
MCG+03-30-062	11 40 42.	+ 16 46	156	15.1	GALAXY
SN 1965C	11 40 42.	+ 18 48		18.0	SUPERNOVA
MCG+03-30-063	11 40 42.	+ 19 55	42	15.2	GALAXY
REIZ 1508	11 40 42.	+ 20 01	24	14.7	GALAXY
REIZ 1509	11 40 42.	+ 20 49	12	15.5	GALAXY
REIZ 1510	11 40 42.	+ 21 52	30	15.1	GALAXY
MRK 639	11 40 42.	+ 24 10	10	16.5	GALAXY WITH UV CONTINUUM
MCG+04-28-037	11 40 42.	+ 25 15 30.	36	15.4	GALAXY
TON-N 0591	11 40 42.	+ 29 12		14.7	BLUE STAR
ZWG 157.024	11 40 42.	+ 31 45		15.5	GALAXY
UGC 06684	11 40 42.	+ 31 45	60	15.5	GALAXY DWARF
MCG+05-28-023	11 40 42.	+ 31 45	60	15.5	GALAXY
ZC 1140.7+4313	11 40 42.	+ 43 13	540		CLUSTER OF GALAXIES
ZC 1140.7+4636	11 40 42.	+ 46 36	1340		CLUSTER OF GALAXIES
MCG+09-19-162	11 40 42.	+ 53 27 30.	36	17.	GALAXY
UGC 06685	11 40 42.	+ 55 47	84	16.0	GALAXY Sc
MCG-02-30-024	11 40 42.	- 12 36 30.	15	15.5	GALAXY
REIZ 1512	11 40 44.	- 12 36	30	14.6	GALAXY
MCG+04-28-038	11 40 45.	+ 21 54 30.	54	15.5	GALAXY
MCG+05-28-025	11 40 45.	+ 26 50	15	14.7	GALAXY
MCG+05-28-024	11 40 45.	+ 26 51	12	14.7	GALAXY
MCG+09-19-163	11 40 45.	+ 55 46	72	15.	GALAXY
RNGC 3829	11 40 46.	+ 52 59		15.0	GALAXY
KN 12.102	11 40 47.3	+ 46 26 54.			NEBULA
KARA.68 080	11 40 48.	+ 08 29	27		DWARF GALAXY
ZWG 068.042	11 40 48.	+ 11 05		15.6	GALAXY
ZC 1140.8+1250	11 40 48.	+ 12 50	1210		CLUSTER OF GALAXIES
ZWG 097.080	11 40 48.	+ 15 43		15.7	GALAXY
ZWG 097.081	11 40 48.	+ 16 46		15.1	GALAXY
UGC 06686	11 40 48.	+ 16 46	174	15.1	GALAXY Sb
UGC 06687	11 40 48.	+ 18 28	96	16.0	GALAXY
MCG+03-30-064	11 40 48.	+ 19 54	8	16.5	GALAXY
ZWG 097.082	11 40 48.	+ 20 01		15.0	GALAXY
REIZ 1513	11 40 48.	+ 20 01	24	14.8	GALAXY
UGC 06688	11 40 48.	+ 20 01	84	15.0	GALAXY Sa
ZWG 127.036	11 40 48.	+ 21 56		15.5	GALAXY
UGC 06689	11 40 48.	+ 21 56	66	15.5	GALAXY S
REIZ 1514	11 40 48.	+ 24 08	6	15.8	GALAXY
REIZ 1515	11 40 48.	+ 25 17	24	14.2	GALAXY
ZWG 127.037	11 40 48.	+ 25 18		15.4	GALAXY
ZWG 186.027	11 40 48.	+ 36 23		15.5	GALAXY
MRK 427	11 40 48.	+ 36 23	20	15.5	GALAXY WITH UV CONTINUUM
ZWG 268.074	11 40 48.	+ 53 00		15.0	GALAXY
REIZ 1511	11 40 48.	+ 53 00	60	14.3	GALAXY
UGC 06690	11 40 48.	+ 53 00	66	15.0	GALAXY SB
MCG+09-19-164	11 40 48.	+ 53 00	54	14.	GALAXY
UGC 06691	11 40 48.	+ 60 58	78	16.5	GALAXY
7ZW 419	11 40 48.	+ 64 16			COMPACT GALAXY
IC 2951	11 40 50.	+ 20 01 40.			NONSTELLAR OBJECT
MCG+04-28-039	11 40 51.	+ 23 17	30	15.3	GALAXY
ARC 1359	11 40 51.	+ 61 56		17.2	RICH CLUSTER OF GALAXIES
MCG-02-30-025	11 40 51.	- 12 08	54	15.	GALAXY
IC 0724	11 40 52.	+ 09 18 52.			NONSTELLAR OBJECT
SEY 122	11 40 52.	+ 55 45 40.		15.3	FAINT GALAXY
ZWG 040.033	11 40 54.	+ 03 40		15.2	GALAXY

OBJECT NAME	RIGHT ASCEN.	DECLINATION	DIAM.	MAGN.	TYPE OF OBJECT
KARA.73B 0497	11 40 54.	+ 03 40	24	15.2	ISOLATED GALAXY IR
ZWG 068.043	11 40 54.	+ 10 26		14.7	GALAXY
UGC 06692	11 40 54.	+ 10 26	90	14.7	GALAXY Sc
MCG+02-30-020	11 40 54.	+ 10 26	84	14.7	GALAXY
ZWG 068.044	11 40 54.	+ 10 37		14.9	GALAXY
MCG+02-30-021	11 40 54.	+ 10 37	36	14.9	GALAXY
MCG+03-30-065	11 40 54.	+ 19 23	78	15.1	GALAXY
ZWG 097.083	11 40 54.	+ 19 54		15.2	GALAXY
REIZ 1516	11 40 54.	+ 20 21	18	15.1	GALAXY
ZWG 127.038	11 40 54.	+ 23 00		14.0	GALAXY
RNGC 3832	11 40 54.	+ 23 00		14.0	GALAXY
UGC 06693	11 40 54.	+ 23 00	132	14.0	GALAXY SBc
REIZ 1517	11 40 54.	+ 23 05	72	13.5	GALAXY
ZWG 127.039	11 40 54.	+ 23 18		15.3	GALAXY
REIZ 1518	11 40 54.	+ 23 22	18	15.2	GALAXY
ZWG 157.025	11 40 54.	+ 31 45		15.3	GALAXY
LB 02091	11 40 54.	+ 60 48 12.		15.8	FAINT BLUE STAR
MCG+10-17-050	11 40 54.	+ 60 58	66	15.	GALAXY
ZWG 012.065	11 40 54.	- 01 23		15.1	GALAXY
MCG-03-30-010	11 40 54.	- 16 32	78	13.5	GALAXY
RNGC 3833	11 40 55.	+ 10 26		14.5	GALAXY
IC 0725	11 40 56.	- 01 23 32.			NONSTELLAR OBJECT
RNGC 3836	11 40 56.	- 16 32		13.0	GALAXY
MIL 22	11 40 57.	- 62 10 24.	840		SUPERNOVA REMNANT
HOLM 282A	11 40 58.	+ 03 40	36	14.3	PART OF MULTIPLE GALAXY
SN 1968F	11 40 59.	+ 19 22		16.2	SUPERNOVA
VDB.66G 096	11 41	+ 59 23	70		DWARF GALAXY
ARC 1362	11 41 00.	+ 07 46		16.0	RICH CLUSTER OF GALAXIES
ZWG 068.045	11 41 00.	+ 09 13		13.8	GALAXY
UGC 06695	11 41 00.	+ 09 13	150	13.8	GALAXY Sa
MCG+02-30-022	11 41 00.	+ 09 13	132	13.8	GALAXY
ZWG 097.084	11 41 00.	+ 19 22		15.1	GALAXY
RNGC 3834	11 41 00.	+ 19 22		15.0	GALAXY
REIZ 1519	11 41 00.	+ 19 22	12	16.0	GALAXY
ZWG 097.085	11 41 00.	+ 19 53		15.7	GALAXY
REIZ 1520	11 41 00.	+ 19 54	12	15.1	GALAXY
MCG+03-30-066	11 41 00.	+ 20 16	96	14.3	GALAXY
ZWG 186.028	11 41 00.	+ 33 41		15.7	GALAXY
UGC 06696	11 41 00.	+ 33 41	66	15.7	GALAXY SBa
SHAH 067	11 41 00.	+ 41 40	78	16.6	GROUP OF COMPACT GALAXIES
ZC 1141.0+5009	11 41 00.	+ 50 09	940		CLUSTER OF GALAXIES
ZWG 268.075	11 41 00.	+ 55 19		15.3	GALAXY
MCG+09-19-165	11 41 00.	+ 55 20	42	15.	GALAXY
UGC 06694	11 41 00.	+ 83 45	84	16.5	GALAXY
KN 12.103	11 41 00.8	+ 46 37 56.			NEBULA
HOLM 282B	11 41 01.	+ 03 42	18	15.0	PART OF MULTIPLE GALAXY
TON-N 1422	11 41 02.	+ 34 59		16.9	BLUE STAR
ARC 1364	11 41 04.	- 01 30		16.0	RICH CLUSTER OF GALAXIES
REIZ 1522	11 41 05.	+ 09 13	108	14.1	GALAXY
ARC 1361	11 41 05.	+ 46 38		17.4	RICH CLUSTER OF GALAXIES
LB 00609	11 41 05.	+ 55 49 06.		17.6	FAINT BLUE STAR
ZWG 068.046	11 41 06.	+ 10 32		14.9	GALAXY
MCG+02-30-023	11 41 06.	+ 10 32	36	14.9	GALAXY
ZC 1141.1+1504	11 41 06.	+ 15 04	870		CLUSTER OF GALAXIES
ZWG 097.086	11 41 06.	+ 20 18		15.7	GALAXY
REIZ 1523	11 41 06.	+ 20 32	12	15.0	GALAXY
ZWG 127.040	11 41 06.	+ 20 34		15.3	GALAXY
REIZ 1524	11 41 06.	+ 20 38	12	15.5	GALAXY
ZC 1141.1+2403	11 41 06.	+ 24 03	3900		CLUSTER OF GALAXIES
ZWG 127.041	11 41 06.	+ 25 44		15.5	GALAXY
SHAH 002	11 41 06.	+ 51 42	60	17.	GROUP OF COMPACT GALAXIES
REIZ 1521	11 41 06.	+ 55 19	30	14.6	GALAXY
VV 320B	11 41 06.	+ 55 19 30.	54	15.	INTERACTING GALAXY
SEY 123	11 41 06.	+ 55 19 52.		14.7	FAINT GALAXY
MCG+10-17-051	11 41 06.	+ 60 31	15	16.	GALAXY
ZWG 012.066	11 41 06.	- 01 21		15.6	GALAXY
OCL 0863	11 41 06.	- 60 52	360	12.	OPEN STAR CLUSTER
HN 1372	11 41 06.6	+ 24 37 22.	72		NEBULA
RNGC 3848	11 41 07.	+ 10 32		15.0	GALAXY
LB 02092	11 41 07.	+ 60 08 00.		15.4	FAINT BLUE STAR
IC 0726	11 41 09.	+ 33 34 52.			NONSTELLAR OBJECT
LB 02093	11 41 39.	+ 51 24 00.		15.7	FAINT BLUE STAR
IC 1049	11 41 09.	+ 62 07 58.			NONSTELLAR OBJECT
LB 02094	11 41 10.	+ 50 26 36.		16.2	FAINT BLUE STAR
HN 1373	11 41 11.4	+ 25 44 28.	18		NEBULA
ZWG 068.047	11 41 12.	+ 12 31		15.7	GALAXY
MCG+03-30-067	11 41 12.	+ 20 04	36	15.2	GALAXY
MCG+03-30-068	11 41 12.	+ 20 11	24	14.2	GALAXY
ZWG 097.087	11 41 12.	+ 20 15		14.3	GALAXY
UGC 06697	11 41 12.	+ 20 15	102	14.3	GALAXY IRR
MCG+03-30-069	11 41 12.	+ 20 20	54	14.9	GALAXY
MCG+03-30-070	11 41 12.	+ 20 22 30.	60	14.7	GALAXY
ZC 1141.2+4409	11 41 12.	+ 44 09	2150		CLUSTER OF GALAXIES
ZC 1141.2+4420	11 41 12.	+ 44 20	610		CLUSTER OF GALAXIES
MCG+10-17-052	11 41 12.	+ 58 51	39	16.	GALAXY
ZWG 334.048	11 41 12.	+ 71 30		15.1	GALAXY
UGC 06698	11 41 12.	+ 71 30	66	15.1	GALAXY S
REIZ 1525	11 41 13.	+ 57 50	30	14.9	GALAXY
MCG+03-30-071	11 41 15.	+ 20 04 30.	48	15.5	GALAXY
ARC 1363	11 41 15.	+ 44 01		17.2	RICH CLUSTER OF GALAXIES
ZWG 040.034	11 41 18.	+ 08 12		14.1	GALAXY
UGC 06699	11 41 18.	+ 08 12	60	14.1	GALAXY S0-a
MCG+01-30-011	11 41 18.	+ 08 12	48	14.1	GALAXY
ZWG 068.048	11 41 18.	+ 11 04		13.6	GALAXY
UGC 06700	11 41 18.	+ 11 04	60	13.6	GALAXY S-IRR
MCG+02-30-024	11 41 18.	+ 11 04	60	13.6	GALAXY
ZWG 097.088	11 41 18.	+ 20 03		15.2	GALAXY
ZWG 097.089	11 41 18.	+ 20 10		14.2	GALAXY
RNGC 3837	11 41 18.	+ 20 10		14.0	GALAXY
REIZ 1527	11 41 18.	+ 20 10	12	14.9	GALAXY
UGC 06701	11 41 18.	+ 20 10	54	14.2	GALAXY E
ZWG 097.090	11 41 18.	+ 20 13		15.3	GALAXY
MCG+03-30-072	11 41 18.	+ 20 15	24	13.3	GALAXY
MCG+03-30-073	11 41 18.	+ 20 16	36	15.0	GALAXY
MCG+03-30-074	11 41 18.	+ 20 17 30.	42	15.1	GALAXY
ZWG 097.091	11 41 18.	+ 20 21		14.7	GALAXY
RNGC 3840	11 41 18.	+ 20 21		14.5	GALAXY
UGC 06702	11 41 18.	+ 20 21	66	14.7	GALAXY Sa
REIZ 1528	11 41 18.	+ 20 27	18	15.7	GALAXY
ZWG 097.092	11 41 18.	+ 20 28		15.5	GALAXY
REIZ 1529	11 41 18.	+ 20 34	12	15.6	GALAXY
HOLM 283B	11 41 18.	+ 26 28	30	15.4	PART OF MULTIPLE GALAXY
MCG+04-28-041	11 41 18.	+ 26 28	30	16.	GALAXY
MCG+04-28-042	11 41 18.	+ 26 29	24	16.	GALAXY
ZC 1141.3+5715	11 41 18.	+ 57 15	1680		CLUSTER OF GALAXIES
ZWG 292.021	11 41 18.	+ 60 23		13.0	GALAXY
UGC 06703	11 41 18.	+ 60 23	138	13.0	GALAXY Sa-b
MCG+10-17-053	11 41 18.	+ 61 02	39	16.	GALAXY

OBJECT NAME	RIGHT ASCEN.	DECLINATION	DIAM.	MAGN.	TYPE OF OBJECT
RNGC 3843	11 41 19.	+ 08 12		14.0	GALAXY
RNGC 3839	11 41 19.	+ 11 04		13.5	GALAXY
HOLM 283A	11 41 19.	+ 26 27	36	15.3	PART OF MULTIPLE GALAXY
HN 1374	11 41 19.2	+ 26 28 46.	18		NEBULA
REIZ 1526	11 41 20.	+ 60 23	90	13.5	GALAXY
LB 02095	11 41 20.	+ 60 47 54.		15.6	FAINT BLUE STAR
HN 1375	11 41 20.1	+ 26 27 34.	12		NEBULA
REIZ 1530	11 41 23.	+ 11 04	42	13.5	GALAXY
ZWG 012.067	11 41 24.	+ 00 42		15.6	GALAXY
ZWG 097.093	11 41 24.	+ 20 03		15.5	GALAXY
ZWG 097.094	11 41 24.	+ 20 05		15.7	GALAXY
ZWG 097.095	11 41 24.	+ 20 13		13.3	GALAXY
RNGC 3842	11 41 24.	+ 20 13		13.5	GALAXY
REIZ 1533	11 41 24.	+ 20 13	12	15.0	GALAXY
UGC 06704	11 41 24.	+ 20 13	90	13.3	GALAXY E
REIZ 1532	11 41 24.	+ 20 14	12	15.6	GALAXY
ZWG 097.096	11 41 24.	+ 20 15		15.0	GALAXY
RNGC 3841	11 41 24.	+ 20 15		15.0	GALAXY
ZWG 097.097	11 41 24.	+ 20 18		14.9	GALAXY
RNGC 3844	11 41 24.	+ 20 18		15.0	GALAXY
UGC 06705	11 41 24.	+ 20 18	84	14.9	GALAXY S0-a
REIZ 1531	11 41 24.	+ 20 21	12	15.6	GALAXY
ZC 1141.4+2445	11 41 24.	+ 24 45	810		CLUSTER OF GALAXIES
ZWG 157.026	11 41 24.	+ 30 20		15.4	GALAXY
MCG+05-28-026	11 41 24.	+ 30 20	54	15.4	GALAXY
ZWG 186.029	11 41 24.	+ 33 48		15.7	GALAXY
MCG+09-19-166	11 41 24.	+ 53 58	18	17.	GALAXY
VV 241B	11 41 24.	+ 58 11 30.		18.	INTERACTING GALAXY
VV 241A	11 41 24.	+ 58 11 30.	60	16.	INTERACTING GALAXY
VV 241	11 41 24.	+ 58 11 30.	84		INTERACTING GALAXY
MCG+10-17-054	11 41 24.	+ 61 55	27	16.	GALAXY
HN 1376	11 41 26.5	+ 27 22 40.			NEBULA
RNGC 3835	11 41 27.	+ 60 24		13.0	GALAXY
REIZ 1536	11 41 28.	- 01 01	60	15.4	GALAXY
ZC 1141.5+1313	11 41 30.	+ 13 13	470		CLUSTER OF GALAXIES
ZWG 097.098	11 41 30.	+ 16 39		15.2	GALAXY
ZWG 097.099	11 41 30.	+ 20 00		15.7	GALAXY
REIZ 1537	11 41 30.	+ 20 15	18	15.3	GALAXY
ZWG 097.100	11 41 30.	+ 20 16		15.1	GALAXY
RNGC 3845	11 41 30.	+ 20 16		15.0	GALAXY
REIZ 1538	11 41 30.	+ 20 16	18	15.7	GALAXY
ZC 1141.5+2045	11 41 30.	+ 20 45	470		CLUSTER OF GALAXIES
ZWG 157.027	11 41 30.	+ 30 10		15.3	GALAXY
ZWG 186.030	11 41 30.	+ 37 27		14.9	GALAXY
MRK 428	11 41 30.	+ 37 27	18	15.5	GALAXY WITH UV CONTINUUM
ZC 1141.5+3915	11 41 30.	+ 39 15	1080		CLUSTER OF GALAXIES
MCG+09-19-167	11 41 30.	+ 51 16	36	15.	GALAXY
MCG+09-19-168	11 41 30.	+ 53 58	24	17.	GALAXY
REIZ 1534	11 41 30.	+ 55 17	60	13.4	GALAXY
ZWG 268.076	11 41 30.	+ 55 18		14.2	GALAXY
UGC 06706	11 41 30.	+ 55 18	138	14.2	GALAXY S-IRR
MCG+09-19-169	11 41 30.	+ 55 19	120	14.	GALAXY
ZC 1141.5+5741	11 41 30.	+ 57 41	1210		CLUSTER OF GALAXIES
ZWG 292.022	11 41 30.	+ 58 14		12.7	GALAXY
UGC 06707	11 41 30.	+ 58 14	90	12.7	GALAXY S0?
MCG+10-17-055	11 41 30.	+ 60 25	72	13.5	GALAXY
7ZW 420	11 41 30.	+ 74 09			COMPACT GALAXY
ZWG 012.068	11 41 30.	- 02 16		15.7	GALAXY
MCG+00-30-020	11 41 30.	- 02 16	54	15.7	GALAXY
LB 00610	11 41 31.	+ 58 07 36.		17.3	FAINT BLUE STAR
LB 02096	11 41 32.	+ 55 38 30.		16.6	FAINT BLUE STAR
RNGC 3846A	11 41 33.	+ 55 19		14.0	GALAXY
RNGC 3838	11 41 33.	+ 58 14		12.5	GALAXY
RNGC 3847	11 41 35.	+ 33 47		14.5	GALAXY
MCG+03-30-075	11 41 36.	+ 16 38	42	15.2	GALAXY
ZWG 097.101	11 41 36.	+ 20 07		15.3	GALAXY
MCG+03-30-076	11 41 36.	+ 20 08	27	15.3	GALAXY
MCG+03-30-077	11 41 36.	+ 20 16	15	15.2	GALAXY
ZWG 127.042	11 41 36.	+ 20 30		15.1	GALAXY
ZWG 097.102	11 41 36.	+ 20 30		15.1	GALAXY
ZC 1141.6+2529	11 41 36.	+ 25 29	1750		CLUSTER OF GALAXIES
ZWG 186.031	11 41 36.	+ 33 37		15.6	GALAXY
MCG+06-26-024	11 41 36.	+ 33 38	15	15.	GALAXY
ZWG 186.032	11 41 36.	+ 33 47		14.6	GALAXY
UGC 06708	11 41 36.	+ 33 47	78	14.6	GALAXY E
MCG+06-26-023	11 41 36.	+ 33 47	18	14.5	GALAXY
ZC 1141.6+3602	11 41 36.	+ 36 02	1210		CLUSTER OF GALAXIES
ZWG 268.077	11 41 36.	+ 51 18		15.6	GALAXY
VV 320A	11 41 36.	+ 55 18 30.	150	13.	INTERACTING GALAXY
OCL 0865	11 41 36.	- 62 13	240	10.	OPEN STAR CLUSTER
HN 1377	11 41 36.6	+ 27 17 10.	18		NEBULA
RNGC 3849	11 41 37.	+ 03 26			NON-EXISTENT OBJECT
REIZ 1535	11 41 37.	+ 58 13	72	13.3	GALAXY
REIZ 1539	11 41 38.	+ 33 48	24	14.0	GALAXY
MCG+03-30-078	11 41 39.	+ 20 07	27	15.4	GALAXY
MCG+10-17-056	11 41 39.	+ 58 14	84	13.3	GALAXY
KN 12.104	11 41 39.9	+ 51 17 57.			NEBULA
HOLM 284B	11 41 41.	+ 20 30	24	14.7	PART OF MULTIPLE GALAXY
IC 2952	11 41 41.	+ 33 36 33.			NONSTELLAR OBJECT
ZWG 097.103	11 41 42.	+ 15 23		15.5	GALAXY
ZWG 097.104	11 41 42.	+ 15 52		15.6	GALAXY
ZC 1141.7+1706	11 41 42.	+ 17 06	610		CLUSTER OF GALAXIES
ZWG 097.105	11 41 42.	+ 20 06		15.4	GALAXY
MCG+03-30-079	11 41 42.	+ 20 07	36	15.5	GALAXY
ZWG 097.106	11 41 42.	+ 20 15		15.2	GALAXY
RNGC 3851	11 41 42.	+ 20 15		15.0	GALAXY
REIZ 1540	11 41 42.	+ 20 29	18	15.5	GALAXY
HOLM 284A	11 41 42.	+ 20 30	36	14.5	PART OF MULTIPLE GALAXY
MCG+04-28-044	11 41 42.	+ 20 30	18	15.1	GALAXY
MCG+04-28-043	11 41 42.	+ 20 30	36	15.1	GALAXY
ZC 1141.7+3113	11 41 42.	+ 31 13	3020		CLUSTER OF GALAXIES
ZWG 186.033	11 41 42.	+ 33 37		15.1	GALAXY
UGC 06709	11 41 42.	+ 33 37	72	15.1	GALAXY SBb
MCG+06-26-025	11 41 42.	+ 33 38	72	14.5	GALAXY
ZC 1141.7+4008	11 41 42.	+ 40 08	1610		CLUSTER OF GALAXIES
MCG+09-19-170	11 41 42.	+ 53 57 30.	21	16.	GALAXY
ZWG 268.078	11 41 42.	+ 55 55		14.7	GALAXY
UGC 06710	11 41 42.	+ 55 55	66	14.7	GALAXY Sb-c
ZWG 334.049	11 41 42.	+ 70 01		13.8	GALAXY
UGC 06711	11 41 42.	+ 70 01	48	13.8	GALAXY
ZC 1141.7-0158	11 41 42.	- 01 58	22310		CLUSTER OF GALAXIES
MCG-03-30-011	11 41 42.	- 17 56	36	15.	GALAXY
REIZ 1541	11 41 43.	+ 33 36	24	15.8	GALAXY
KN 12.105	11 41 44.5	+ 49 07 05.			NEBULA
MCG+08-21-094	11 41 45.	+ 49 06	72	15.	GALAXY
RNGC 3846	11 41 45.	+ 55 56		14.5	GALAXY
HN 1378	11 41 45.9	+ 27 20 21.	18		NEBULA
REIZ 1542	11 41 46.	+ 48 40	30	14.2	GALAXY

OBJECT NAME	RIGHT ASCEN.	DECLINATION	DIAM.	MAGN.	TYPE OF OBJECT
ARC 1365	11 41 47.	+ 31 11		15.7	RICH CLUSTER OF GALAXIES
ZC 1141.8+0601	11 41 48.	+ 06 01	1550		CLUSTER OF GALAXIES
ZWG 040.035	11 41 48.	+ 07 46		15.6	GALAXY
ZWG 040.036	11 41 48.	+ 08 27		14.9	GALAXY
MCG+01-30-012	11 41 48.	+ 08 27	48	14.9	GALAXY
ZWG 068.049	11 41 48.	+ 08 40		15.6	GALAXY
MCG+03-30-080	11 41 48.	+ 15 23	36	15.5	GALAXY
ZWG 097.107	11 41 48.	+ 16 50		13.5	GALAXY
RNGC 3853	11 41 48.	+ 16 50		13.5	GALAXY
UGC 06712	11 41 48.	+ 16 50	96	13.5	GALAXY E
MCG+03-30-081	11 41 48.	+ 16 50	33	13.5	GALAXY
ZWG 097.108	11 41 48.	+ 17 35		15.5	GALAXY
MCG+03-30-082	11 41 48.	+ 17 36	30	15.5	GALAXY
ZWG 097.109	11 41 48.	+ 20 00		15.5	GALAXY
ZWG 097.110	11 41 48.	+ 20 06		15.5	GALAXY
REIZ 1545	11 41 48.	+ 20 14	12	15.6	GALAXY
ZWG 097.111	11 41 48.	+ 20 22		15.5	GALAXY
MCG+03-30-083	11 41 48.	+ 20 22 30.	48	14.9	GALAXY
REIZ 1546	11 41 48.	+ 25 42	24	14.0	GALAXY
ZWG 186.034	11 41 48.	+ 32 56		15.3	GALAXY
MCG+06-26-026	11 41 48.	+ 32 58	24	15.	GALAXY
ZWG 243.001	11 41 48.	+ 49 07		15.5	GALAXY
ZWG 242.077	11 41 48.	+ 49 07		15.5	GALAXY
UGC 06713	11 41 48.	+ 49 07	120	15.5	GALAXY DWRF SP
REIZ 1543	11 41 48.	+ 55 55	48	14.3	GALAXY
MCG+09-19-171	11 41 48.	+ 55 55	60	13.7	GALAXY
ZWG 292.023	11 41 48.	+ 61 49		15.4	GALAXY
ZWG 315.001	11 41 48.	+ 68 13		14.9	GALAXY
ZWG 314.040	11 41 48.	+ 68 13		14.9	GALAXY
UGC 06714	11 41 48.	+ 68 13	60	14.9	GALAXY
KARA.73B 0498	11 41 48.	+ 68 13	36	14.9	ISOLATED GALAXY S
ZWG 012.069	11 41 48.	- 02 54		15.7	GALAXY
RNGC 3852	11 41 49.	+ 10 34			NON-EXISTENT OBJECT
REIZ 1547	11 41 49.	+ 33 36	54	14.7	GALAXY
IC 2953	11 41 49.	+ 33 36 39.			NONSTELLAR OBJECT
LB 02097	11 41 50.	+ 54 04 42.		15.4	FAINT BLUE STAR
MCG+04-28-045	11 41 51.	+ 25 40	42	15.0	GALAXY
REIZ 1548	11 41 53.	+ 11 04	78	15.0	GALAXY
ZWG 068.050	11 41 54.	+ 11 03		15.0	GALAXY
UGC 06715	11 41 54.	+ 11 03	96	15.0	GALAXY Sb
MCG+02-30-025	11 41 54.	+ 11 03	96	15.0	GALAXY
IC 0727	11 41 54.	+ 11 03 45.			NONSTELLAR OBJECT
ZWG 068.051	11 41 54.	+ 11 51		15.3	GALAXY
ZWG 097.112	11 41 54.	+ 20 20		14.9	GALAXY
REIZ 1549	11 41 54.	+ 23 28	18	15.1	GALAXY
ZC 1141.9+3004	11 41 54.	+ 30 04	5240		CLUSTER OF GALAXIES
ZC 1141.9+4543	11 41 54.	+ 45 43	810		CLUSTER OF GALAXIES
ZC 1141.9+5133	11 41 54.	+ 51 33	1950		CLUSTER OF GALAXIES
REIZ 1544	11 41 54.	+ 56 10	108	14.3	GALAXY
MCG+11-14-037	11 41 54.	+ 68 15	36	16.	GALAXY
MCG+12-11-039	11 41 54.	+ 70 00 30.	45	15.	GALAXY
ARC 1367	11 41 56.	+ 20 07		13.5	RICH CLUSTER OF GALAXIES
RNGC 3854	11 41 56.	- 09 06			NON-EXISTENT OBJECT
MCG+10-17-057	11 41 57.	+ 61 49 30.	39	15.	GALAXY
MCG-01-30-027A	11 41 57.	- 03 30	84	14.	GALAXY
HN 1379	11 41 59.1	+ 26 47 45.	12		NEBULA
KEEN 3846A	11 42	+ 55 18		13.5	GALAXY
ZWG 068.052	11 42 00.	+ 11 04		15.6	GALAXY
MCG+02-30-026	11 42 00.	+ 11 04	48	15.6	GALAXY
ZC 1142.0+1130	11 42 00.	+ 11 30	1280		CLUSTER OF GALAXIES
MCG+03-30-084	11 42 00.	+ 19 50	48	15.1	GALAXY
MCG+03-30-085	11 42 00.	+ 19 59	30	15.7	GALAXY
MCG+03-30-086	11 42 00.	+ 20 00	24	17.	GALAXY
RNGC 3860B	11 42 00.	+ 20 04		15.5	GALAXY
MCG+03-30-087	11 42 00.	+ 20 04	27	15.4	GALAXY
RNGC 3860A	11 42 00.	+ 20 05		14.5	GALAXY
MCG+03-30-088	11 42 00.	+ 20 05	42	14.5	GALAXY
MCG+03-30-089	11 42 00.	+ 20 25	72	14.6	GALAXY
REIZ 1550	11 42 00.	+ 23 28	12	15.6	GALAXY
ZWG 127.043	11 42 00.	+ 25 43		15.0	GALAXY
ZWG 186.035	11 42 00.	+ 36 14		15.1	GALAXY
UGC 06716	11 42 00.	+ 36 14	108	15.1	GALAXY Sb
MCG+06-26-027	11 42 00.	+ 36 15	96	14.	GALAXY
ZC 1142.0+4232	11 42 00.	+ 42 32	1550		CLUSTER OF GALAXIES
MCG+09-19-172	11 42 00.	+ 54 44	48	15.	GALAXY
SHAH 123	11 42 00.	+ 57 47	114	16.7	GROUP OF COMPACT GALAXIES
MCG+14-06-002	11 42 00.	+ 83 44	51	16.	GALAXY
ZC 1142.0-0118	11 42 00.	- 01 18	2420		CLUSTER OF GALAXIES
ZWG 012.070	11 42 00.	- 03 17		15.7	GALAXY
HN 1380	11 42 00.4	+ 25 42 09.	30		NEBULA
HN 1381	11 42 01.7	- 02 06 27.			NEBULA
TON-N 1423	11 42 02.	+ 33 40		15.9	BLUE STAR
ZC 1142.1+0410	11 42 06.	+ 04 10	1610		CLUSTER OF GALAXIES
ZC 1142.1+0454	11 42 06.	+ 04 54	1080		CLUSTER OF GALAXIES
ZWG 040.037	11 42 06.	+ 07 33		15.6	GALAXY
ZWG 040.038	11 42 06.	+ 08 14		15.7	GALAXY
ZC 1142.1+1545	11 42 06.	+ 15 45	2960		CLUSTER OF GALAXIES
ZC 1142.1+1612	11 42 06.	+ 16 12	610		CLUSTER OF GALAXIES
ZC 1142.1+1805	11 42 06.	+ 18 05	1680		CLUSTER OF GALAXIES
MCG+03-30-090	11 42 06.	+ 18 10	30	15.5	GALAXY
MCG+03-30-091	11 42 06.	+ 19 46	48	14.9	GALAXY
ZWG 097.113	11 42 06.	+ 20 02		15.6	GALAXY
ZWG 097.114	11 42 06.	+ 20 03		15.4	GALAXY
MCG+03-30-092	11 42 06.	+ 20 04	36	15.6	GALAXY
ZWG 097.115	11 42 06.	+ 20 09		15.5	GALAXY
ZC 1142.1+2126	11 42 06.	+ 21 26	15990		CLUSTER OF GALAXIES
ZWG 186.036	11 42 06.	+ 33 35		15.6	GALAXY
RNGC 3855	11 42 06.	+ 33 35		15.5	GALAXY
MCG+06-26-028	11 42 06.	+ 33 36	30	15.5	GALAXY
RNGC 3856	11 42 06.	+ 33 38			NON-EXISTENT OBJECT
ZWG 292.024	11 42 06.	+ 58 10		15.3	GALAXY
REIZ 1551	11 42 07.	+ 33 35	24	14.9	GALAXY
HOLM 285B	11 42 10.	+ 20 03	78	14.8	PART OF MULTIPLE GALAXY
LB 02098	11 42 10.	+ 53 03 18.		16.4	FAINT BLUE STAR
LB 02099	11 42 11.	+ 49 28 30.		17.5	FAINT BLUE STAR
UGC 06717	11 42 12.	+ 09 28	96	17.	GALAXY DWRF SP
MCG+02-30-027	11 42 12.	+ 09 29	90	16.	GALAXY
ZWG 097.116	11 42 12.	+ 18 09		15.5	GALAXY
ZWG 097.117	11 42 12.	+ 19 49		15.1	GALAXY
RNGC 3857	11 42 12.	+ 19 49		15.0	GALAXY
ZWG 097.118	11 42 12.	+ 19 53		15.7	GALAXY
ZWG 097.119	11 42 12.	+ 19 58		15.7	GALAXY
HOLM 285A	11 42 12.	+ 20 04	72	13.9	PART OF MULTIPLE GALAXY
ZWG 097.120	11 42 12.	+ 20 05		15.5	GALAXY
UGC 06718	11 42 12.	+ 20 05	84	14.5	GALAXY Sa-b
RNGC 3861B	11 42 12.	+ 20 15		15.0	GALAXY
RNGC 3861A	11 42 12.	+ 20 16		14.0	GALAXY
MCG+03-30-093	11 42 12.	+ 20 16	138	14.0	GALAXY
REIZ 1552	11 42 12.	+ 20 23	12	14.9	GALAXY
ZWG 097.121	11 42 12.	+ 20 24		14.6	GALAXY
UGC 06719	11 42 12.	+ 20 24	72	14.6	GALAXY S
ZWG 127.044	11 42 12.	+ 24 13		15.6	GALAXY
ZC 1142.2+3456	11 42 12.	+ 34 56	13040		CLUSTER OF GALAXIES
MCG+10-17-058	11 42 12.	+ 58 09	42	15.	GALAXY
MCG+10-17-059	11 42 12.	+ 58 57	30	16.	GALAXY
ZWG 314.041	11 42 12.	+ 66 00		15.5	GALAXY
MCG+13-09-002	11 42 12.	+ 79 58	39	14.	GALAXY
LB 02100	11 42 13.	+ 50 42 36.		17.5	FAINT BLUE STAR
ARC 1369	11 42 14.	+ 42 38		17.2	RICH CLUSTER OF GALAXIES
ARC 1366	11 42 14.	+ 67 42		16.8	RICH CLUSTER OF GALAXIES
RNGC 3858	11 42 14.	- 09 01			NON-EXISTENT OBJECT
MCG+03-30-094	11 42 15.	+ 20 15	33	14.0	GALAXY
MCG+06-26-029	11 42 15.	+ 34 05 30.	24	16.	GALAXY
LB 00611	11 42 16.	+ 28 53 36.		15.9	FAINT BLUE STAR
REIZ 1553	11 42 16.	- 01 20	30	14.1	GALAXY
ZWG 068.053	11 42 18.	+ 12 02		15.7	GALAXY
ZWG 097.122	11 42 18.	+ 19 44		14.9	GALAXY
RNGC 3859	11 42 18.	+ 19 44		15.0	GALAXY
REIZ 1555	11 42 18.	+ 19 44	36	14.7	GALAXY
UGC 06721	11 42 18.	+ 19 44	60	14.9	GALAXY PECULR
ZWG 097.123	11 42 18.	+ 19 46		15.7	GALAXY
REIZ 1554	11 42 18.	+ 19 48	6	16.0	GALAXY
MCG+03-30-095	11 42 18.	+ 19 54	78	14.0	GALAXY
MCG+03-30-096	11 42 18.	+ 19 55	12	15.2	GALAXY
ZWG 097.124	11 42 18.	+ 20 00		15.3	GALAXY
REIZ 1556	11 42 18.	+ 20 02	24	15.2	GALAXY
ZWG 097.125	11 42 18.	+ 20 03		15.6	GALAXY
REIZ 1557	11 42 18.	+ 21 38	18	15.2	GALAXY
ARC 1368	11 42 18.	+ 51 32		16.9	RICH CLUSTER OF GALAXIES
MCG+09-19-173	11 42 18.	+ 53 55	18	16.	GALAXY
ZWG 012.071	11 42 18.	- 01 19		14.7	GALAXY
UGC 06720	11 42 18.	- 01 19	78	14.7	GALAXY SBb
MCG+00-30-021	11 42 18.	- 01 19	72	14.7	GALAXY
IC 0728	11 42 18.	- 01 19 33.			NONSTELLAR OBJECT
KN 12.106	11 42 18.0	+ 46 30 18.			NEBULA
HOLM 286B	11 42 20.	+ 08 45	18	15.3	PART OF MULTIPLE GALAXY
TON-N 1424	11 42 20.	+ 22 39		16.6	BLUE STAR
HELW 260	11 42 20.	- 08 58 09.			NEBULA
REIZ 1558	11 42 21.	- 08 58	42	13.8	GALAXY
LB 02102	11 42 23.	+ 52 59 00.		17.1	FAINT BLUE STAR
LB 02101	11 42 23.	+ 56 09 36.		15.8	FAINT BLUE STAR
ZWG 097.126	11 42 24.	+ 16 51		15.7	GALAXY
ZWG 157.028	11 42 24.	+ 27 07		15.3	GALAXY
ZWG 157.029	11 42 24.	+ 31 07		15.5	GALAXY
ZWG 012.073	11 42 24.	- 02 07		15.6	GALAXY
ZWG 012.072	11 42 24.	- 02 13		15.6	GALAXY
MCG-01-30-028	11 42 24.	- 08 59	114	13.	GALAXY
HN 1382	11 42 25.3	- 02 13 27.	36		NEBULA
TON-N 1425	11 42 26.	+ 32 36		16.8	BLUE STAR
MCG+04-28-046	11 42 27.	+ 20 41	84	14.5	GALAXY
MCG+08-21-095	11 42 27.	+ 49 59	54	15.	GALAXY
IC 2954	11 42 28.	+ 27 03 39.			NONSTELLAR OBJECT
REIZ 1560	11 42 29.	+ 08 44	90	14.2	GALAXY
HOLM 286A	11 42 29.	+ 08 44	132	14.0	PART OF MULTIPLE GALAXY
ARC 1370	11 42 29.	+ 49 37		17.6	RICH CLUSTER OF GALAXIES
KN 12.107	11 42 29.9	+ 49 59 54.			NEBULA
ZWG 040.039	11 42 30.	+ 07 46		15.7	GALAXY
ZWG 068.054	11 42 30.	+ 08 45		14.0	GALAXY
UGC 06722	11 42 30.	+ 08 45	168	14.0	GALAXY Sb-c
MCG+02-30-028	11 42 30.	+ 08 45	168	14.0	GALAXY
MCG+03-30-097	11 42 30.	+ 19 41	48	15.5	GALAXY
ZWG 097.127	11 42 30.	+ 19 53		14.0	GALAXY
RNGC 3862	11 42 30.	+ 19 53		14.0	GALAXY
UGC 06723	11 42 30.	+ 19 53	78	14.0	GALAXY E
IC 2955	11 42 30.	+ 19 53			NONSTELLAR OBJECT
ZWG 097.128	11 42 30.	+ 19 54		15.2	GALAXY
REIZ 1559	11 42 30.	+ 19 54	12	14.8	GALAXY
MCG+03-30-098	11 42 30.	+ 20 07	18	15.1	GALAXY
ZWG 097.129	11 42 30.	+ 20 15		14.0	GALAXY
HOLM 287A	11 42 30.	+ 20 15	42	14.1	PART OF MULTIPLE GALAXY
UGC 06724	11 42 30.	+ 20 15	144	14.0	GALAXY Sb
KARA.72 299A	11 42 30.	+ 20 15	108	14.0	PART OF DOUBLE GALAXY
REIZ 1561	11 42 30.	+ 20 43	18	14.9	GALAXY
ZWG 127.045	11 42 30.	+ 20 44		14.5	GALAXY
UGC 06725	11 42 30.	+ 20 44	102	14.5	GALAXY S0
ZWG 127.046	11 42 30.	+ 21 42		15.6	GALAXY
ZC 1142.5+3343	11 42 30.	+ 33 43	1750		CLUSTER OF GALAXIES
ZWG 243.002	11 42 30.	+ 50 00		14.8	GALAXY
ZWG 242.078	11 42 30.	+ 50 00		14.8	GALAXY
UGC 06726	11 42 30.	+ 50 00	60	14.8	GALAXY E-S0
ZWG 292.025	11 42 30.	+ 61 59		14.5	GALAXY
UGC 06727	11 42 30.	+ 61 59	54	14.5	GALAXY S
ZC 1142.5+6737	11 42 30.	+ 67 37	3020		CLUSTER OF GALAXIES
ZWG 352.007	11 42 30.	+ 79 58		14.8	GALAXY
ZWG 351.069	11 42 30.	+ 79 58		14.8	GALAXY
UGC 06728	11 42 30.	+ 79 58	78	14.8	GALAXY SB0-a
ZWG 012.074	11 42 30.	- 01 23		15.5	GALAXY
MCG-02-30-026	11 42 30.	- 12 56	36	15.	GALAXY
RNGC 3863	11 42 31.	+ 08 45		14.0	GALAXY
TON-N 1426	11 42 32.	+ 21 29		16.	BLUE STAR
KN 12.108	11 42 32.6	+ 47 44 58.			NEBULA
HOLM 287B	11 42 33.	+ 20 15	42	14.8	PART OF MULTIPLE GALAXY
LB 02103	11 42 33.	+ 48 16 24.		16.4	FAINT BLUE STAR
ZC 1142.6+1741	11 42 36.	+ 17 41	940		CLUSTER OF GALAXIES
ZWG 097.130	11 42 36.	+ 19 40		15.5	GALAXY
RNGC 3864	11 42 36.	+ 19 40		15.5	GALAXY
ZWG 097.131	11 42 36.	+ 20 07		15.1	GALAXY
KARA.72 299B	11 42 36.	+ 20 15	42		PART OF DOUBLE GALAXY
REIZ 1562	11 42 36.	+ 21 05	6	15.9	GALAXY
REIZ 1563	11 42 36.	+ 21 33	18	15.5	GALAXY
ZC 1142.6+2200	11 42 36.	+ 22 00	810		CLUSTER OF GALAXIES
ZWG 157.030	11 42 36.	+ 27 03		14.9	GALAXY
UGC 06729	11 42 36.	+ 27 03	78	14.9	GALAXY Sb/SBc
ZWG 186.037	11 42 36.	+ 33 36		15.5	GALAXY
MCG+08-21-096	11 42 36.	+ 49 49 30.	30	15.	GALAXY
ZWG 243.003	11 42 36.	+ 49 51		15.6	GALAXY
ZWG 242.079	11 42 36.	+ 49 51		15.6	GALAXY
ZC 1142.6+6440	11 42 36.	+ 64 40	1880		CLUSTER OF GALAXIES
ZWG 012.075	11 42 36.	- 01 29		15.4	GALAXY
HN 1383	11 42 37.0	- 00 37 03.	42		NEBULA
KN 12.110	11 42 37.3	+ 45 33 23.			NEBULA
TON-N 1427	11 42 38.	+ 35 48		17.	BLUE STAR
KN 12.109	11 42 38.7	+ 49 51 05.			NEBULA
REIZ 1564	11 42 39.	- 09 02	18	15.2	GALAXY
MCG-02-30-028	11 42 39.	- 09 46	36	15.5	GALAXY
MCG-02-30-027	11 42 39.	- 09 48	156	13.5	GALAXY

OBJECT NAME	RIGHT ASCEN.	DECLINATION	DIAM.	MAGN.	TYPE OF OBJECT
HOLM 288D	11 42 40.	+ 19 39	36	15.2	PART OF MULTIPLE GALAXY
HELW 261	11 42 40.	- 09 02 27.			NEBULA
ZWG 097.132	11 42 42.	+ 14 39		15.7	GALAXY
MCG+03-30-099	11 42 42.	+ 15 46	36	17.	GALAXY
REIZ 1565	11 42 42.	+ 19 40	18	15.7	GALAXY
ZWG 097.133	11 42 42.	+ 20 18		15.7	GALAXY
REIZ 1566	11 42 42.	+ 23 23	30	15.8	GALAXY
IC 2956	11 42 42.	+ 27 02 27.			NONSTELLAR OBJECT
REIZ 1567	11 42 42.	+ 27 03	30	14.2	GALAXY
ZWG 186.038	11 42 42.	+ 32 59		15.4	GALAXY
IC 0729	11 42 42.	+ 33 35 33.			NONSTELLAR OBJECT
ZWG 012.076	11 42 42.	- 01 25		15.3	GALAXY
MCG-01-30-029	11 42 42.	- 09 04	54	14.	GALAXY
RNGC 3865	11 42 44.	- 08 56		13.0	GALAXY
RNGC 3866	11 42 44.	- 09 04		14.0	GALAXY
HN 1384	11 42 44.9	+ 27 08 45.			NEBULA
MCG+08-21-097	11 42 45.	+ 50 09	60	16.	GALAXY
HN 1385	11 42 45.8	+ 28 07 15.	12		NEBULA
REIZ 1568	11 42 46.	+ 49 51	24	14.4	GALAXY
KN 12.111	11 42 46.7	+ 50 09 31.			NEBULA
ZWG 068.055	11 42 48.	+ 09 26		13.4	GALAXY
UGC 06730	11 42 48.	+ 09 26	72	13.4	GALAXY S(a-b)
MCG+02-30-029	11 42 48.	+ 09 26	72	13.4	GALAXY
ZC 1142.8+1147	11 42 48.	+ 11 47	1410		CLUSTER OF GALAXIES
MCG+03-30-101	11 42 48.	+ 14 38	42	15.7	GALAXY
MCG+03-30-100	11 42 48.	+ 15 46	18	17.	GALAXY
MCG+03-30-102	11 42 48.	+ 15 47	24	17.	GALAXY
ZWG 097.134	11 42 48.	+ 19 40		14.6	GALAXY
RNGC 3867	11 42 48.	+ 19 40		14.5	GALAXY
UGC 06731	11 42 48.	+ 19 40	72	14.6	GALAXY S
MCG+03-30-103	11 42 48.	+ 19 41	72	14.6	GALAXY
ZWG 097.135	11 42 48.	+ 19 43		14.8	GALAXY
RNGC 3868	11 42 48.	+ 19 43		15.0	GALAXY
MCG+03-30-104	11 42 48.	+ 19 45	42	14.8	GALAXY
REIZ 1569	11 42 48.	+ 20 36	18	14.5	GALAXY
REIZ 1570	11 42 48.	+ 21 31	24	15.5	GALAXY
MCG+05-28-027	11 42 48.	+ 27 02 30.	66	14.9	GALAXY
7ZW 421	11 42 48.	+ 59 15			COMPACT GALAXY
ZWG 292.026	11 42 48.	+ 59 15		13.5	GALAXY
UGC 06732	11 42 48.	+ 59 15	66	13.5	GALAXY
MCG+10-17-060	11 42 48.	+ 62 00	57	15.	GALAXY
MCG+11-15-001	11 42 48.	+ 64 46	15	16.	GALAXY
ZWG 012.077	11 42 48.	- 01 36		15.5	GALAXY
RNGC 3876	11 42 49.	+ 09 26		13.5	GALAXY
REIZ 1571	11 42 49.	+ 33 26	30	15.2	GALAXY
HN 1386	11 42 49.1	- 01 36 03.	18		NEBULA
TON-N 1428	11 42 50.	+ 32 26		16.8	BLUE STAR
REIZ 1572	11 42 52.	+ 50 09	48	14.4	GALAXY
RNGC 3850	11 42 52.	+ 56 10		14.5	GALAXY
ARC 1373	11 42 52.	- 02 08		17.2	RICH CLUSTER OF GALAXIES
REIZ 1573	11 42 53.	+ 11 50	12	14.9	GALAXY
ZWG 012.078	11 42 54.	+ 00 17		15.3	GALAXY
ZWG 068.056	11 42 54.	+ 10 00		15.3	GALAXY
MCG+02-30-030	11 42 54.	+ 10 00	48	15.3	GALAXY
ZC 1142.9+1005	11 42 54.	+ 10 05	2420		CLUSTER OF GALAXIES
ARC 1372	11 42 54.	+ 11 48		17.2	RICH CLUSTER OF GALAXIES
REIZ 1574	11 42 54.	+ 19 40	48	15.2	GALAXY
HOLM 288A	11 42 54.	+ 19 40	66	14.1	PART OF MULTIPLE GALAXY
REIZ 1575	11 42 54.	+ 19 43	12	15.4	GALAXY
ZWG 127.047	11 42 54.	+ 20 37		14.6	GALAXY
MCG+04-28-047	11 42 54.	+ 20 37	24	14.6	GALAXY
ZWG 127.048	11 42 54.	+ 21 06		15.0	GALAXY
MCG+04-28-048	11 42 54.	+ 21 06	24	15.0	GALAXY
REIZ 1576	11 42 54.	+ 21 39	18	15.6	GALAXY
ZC 1142.9+2603	11 42 54.	+ 26 03	1480		CLUSTER OF GALAXIES
ZWG 186.039	11 42 54.	+ 34 50		15.7	GALAXY
ZC 1142.9+4936	11 42 54.	+ 49 36	1340		CLUSTER OF GALAXIES
ZC 1142.9+5310	11 42 54.	+ 53 10	400		CLUSTER OF GALAXIES
ZWG 268.079	11 42 54.	+ 56 09		14.4	GALAXY
UGC 06733	11 42 54.	+ 56 09	132	14.4	GALAXY SBc
ZWG 314.042	11 42 54.	+ 64 47		15.4	GALAXY
MCG-01-30-030	11 42 54.	- 08 45	36	15.5	GALAXY
KN 12.112	11 42 54.2	+ 49 30 05.			NEBULA
ARC 1371	11 42 55.	+ 15 49		16.6	RICH CLUSTER OF GALAXIES
REIZ 1577	11 42 55.	+ 33 15	30	15.0	GALAXY
MCG+05-28-028	11 42 57.	+ 31 35	18	15.0	GALAXY
REIZ 1579	11 42 57.	- 07 33	42	14.7	GALAXY
REIZ 1578	11 42 57.	- 09 50	72	14.5	GALAXY
RNGC 3889	11 42 58.	+ 56 11			GALAXY
ZC 1143.0+0300	11 43 00.	+ 03 00	810		CLUSTER OF GALAXIES
ZWG 040.040	11 43 00.	+ 03 30		14.7	GALAXY
MCG+01-30-013	11 43 00.	+ 03 30	36	14.7	GALAXY
ZWG 068.057	11 43 00.	+ 09 24		15.4	GALAXY
UGC 06734	11 43 00.	+ 09 24	96	15.4	GALAXY Sb
ZWG 068.058	11 43 00.	+ 12 29		15.7	GALAXY
MCG+02-30-031	11 43 00.	+ 12 29	36	15.7	GALAXY
ZC 1143.0+1444	11 43 00.	+ 14 44	340		CLUSTER OF GALAXIES
MCG+03-30-105	11 43 00.	+ 20 04	60	14.2	GALAXY
MCG+03-30-106	11 43 00.	+ 20 05	78	14.8	GALAXY
ZWG 157.031	11 43 00.	+ 31 35		15.0	GALAXY
MCG+09-19-174	11 43 00.	+ 56 10	120	14.3	GALAXY
MCG+09-19-175	11 43 00.	+ 56 20	30	16.	GALAXY
MCG+10-17-061	11 43 00.	+ 59 16	39	14.	GALAXY
MCG+14-06-003	11 43 00.	+ 82 53	30	16.	GALAXY
ZC 1143.0-0202	11 43 00.	- 02 02	2290		CLUSTER OF GALAXIES
HELW 262	11 43 00.	- 09 49 39.			NEBULA
IC 0730	11 43 01.	+ 03 30 32.			NONSTELLAR OBJECT
IC 2957	11 43 01.	+ 33 34 38.			NONSTELLAR OBJECT
TON-N 1429	11 43 02.	+ 35 38		13.5	BLUE STAR
HN 1387	11 43 02.3	+ 27 12 33.			NEBULA
ZWG 097.136	11 43 06.	+ 18 01		15.4	GALAXY
ZWG 097.137	11 43 06.	+ 20 03		14.2	GALAXY
RNGC 3873	11 43 06.	+ 20 03		14.0	GALAXY
UGC 06735	11 43 06.	+ 20 03	60	14.2	GALAXY E
KARA.72 300A	11 43 06.	+ 20 03	78	14.2	PART OF DOUBLE GALAXY
ZWG 097.138	11 43 06.	+ 20 18		15.5	GALAXY
MCG+06-26-030	11 43 06.	+ 33 26	27	16.	GALAXY
ZC 1143.1+5000	11 43 06.	+ 50 00	1140		CLUSTER OF GALAXIES
ZWG 268.080	11 43 06.	+ 56 20		15.6	GALAXY
MCG+10-17-062	11 43 06.	+ 62 29 30.	18	16.	GALAXY
TON-N 1430	11 43 08.	+ 22 31		14.7	BLUE STAR
IC 2958	11 43 08.	+ 33 25 26.			NONSTELLAR OBJECT
TON-N 1431	11 43 08.	+ 35 13		16.	BLUE STAR
REIZ 1580	11 43 10.	+ 03 18	60	15.2	GALAXY
REIZ 1581	11 43 11.	+ 11 06	90	13.2	GALAXY
ZWG 040.041	11 43 12.	+ 03 18		14.9	GALAXY
UGC 06736	11 43 12.	+ 03 18	90	14.9	GALAXY Sc
MCG+01-30-014	11 43 12.	+ 03 18	84	14.9	GALAXY

OBJECT NAME	RIGHT ASCEN.	DECLINATION	DIAM.	MAGN.	TYPE OF OBJECT
ZWG 068.059	11 43 12.	+ 11 06		13.5	GALAXY
UGC 06737	11 43 12.	+ 11 06	114	13.5	GALAXY Sa
MCG+02-30-032	11 43 12.	+ 11 06	120	13.5	GALAXY
ZWG 068.060	11 43 12.	+ 14 03		12.9	GALAXY
UGC 06738	11 43 12.	+ 14 03	162	12.9	GALAXY E
MCG+02-30-033	11 43 12.	+ 14 03	48	12.9	GALAXY
MCG+03-30-107	11 43 12.	+ 19 50	24	15.4	GALAXY
ZWG 097.139	11 43 12.	+ 20 02		14.8	GALAXY
RNGC 3875	11 43 12.	+ 20 02		15.0	GALAXY
UGC 06739	11 43 12.	+ 20 02	72	14.8	GALAXY S0-a
KARA.72 300B	11 43 12.	+ 20 02	78	14.8	PART OF DOUBLE GALAXY
HOLM 289A	11 43 12.	+ 20 03	18	14.1	PART OF MULTIPLE GALAXY
REIZ 1582	11 43 12.	+ 20 50	18	15.3	GALAXY
ZWG 186.040	11 43 12.	+ 33 15		15.6	GALAXY
ZC 1143.2+4037	11 43 12.	+ 40 37	1010		CLUSTER OF GALAXIES
LB 00240	11 43 12.	+ 55 23 36.		13.4	FAINT BLUE STAR
MCG+09-19-176	11 43 12.	+ 56 22	48	15.	GALAXY
RNGC 3874	11 43 13.	+ 08 49			NON-EXISTENT OBJECT
RNGC 3869	11 43 13.	+ 11 06		13.5	GALAXY
RNGC 3872	11 43 13.	+ 14 03		13.0	GALAXY
HOLM 289B	11 43 15.	+ 20 03	72	14.5	PART OF MULTIPLE GALAXY
RNGC 3870	11 43 16.	+ 50 30		13.0	GALAXY
REIZ 1586	11 43 17.	+ 14 03	30	13.0	GALAXY
ARC 1374	11 43 17.	+ 50 01		17.4	RICH CLUSTER OF GALAXIES
KN 12.113	11 43 17.2	+ 50 28 49.			NEBULA
ZWG 068.061	11 43 18.	+ 10 45		14.8	GALAXY
UGC 06740	11 43 18.	+ 10 45	66	14.8	GALAXY S
MCG+02-30-034	11 43 18.	+ 10 45	60	14.8	GALAXY
ZC 1143.3+1621	11 43 18.	+ 16 21	1080		CLUSTER OF GALAXIES
UGC 06741	11 43 18.	+ 17 28	96	16.0	GALAXY DBL SYS
MCG+03-30-108	11 43 18.	+ 19 44	24	15.6	GALAXY
ZWG 097.140	11 43 18.	+ 19 48		15.4	GALAXY
REIZ 1588	11 43 18.	+ 20 02	48	14.9	GALAXY
REIZ 1587	11 43 18.	+ 20 02	12	14.6	GALAXY
REIZ 1584	11 43 18.	+ 20 42	12	15.2	GALAXY
ZWG 127.049	11 43 18.	+ 20 55		15.5	GALAXY
ZWG 268.081	11 43 18.	+ 50 30		13.2	GALAXY
MRK 186	11 43 18.	+ 50 30	35	13.5	GALAXY WITH UV CONTINUUM
UGC 06742	11 43 18.	+ 50 30	60	13.2	GALAXY S0?
REIZ 1583	11 43 18.	+ 56 21	30	14.9	GALAXY
KN 12.114	11 43 18.0	+ 47 24 03.			NEBULA
SEY 124	11 43 20.	+ 55 41 45.		15.4	FAINT GALAXY
MCG+04-28-049	11 43 21.	+ 21 18	96	14.8	GALAXY
IC 0731	11 43 21.	+ 49 50 32.			NONSTELLAR OBJECT
MCG+08-22-001	11 43 21.	+ 50 29	60	13.1	GALAXY
REIZ 1585	11 43 22.	+ 50 29	36	13.1	GALAXY
ZC 1143.4+1451	11 43 24.	+ 14 51	1880		CLUSTER OF GALAXIES
ZWG 097.141	11 43 24.	+ 19 43		15.6	GALAXY
REIZ 1589	11 43 24.	+ 19 48	6	15.3	GALAXY
HOLM 290B	11 43 24.	+ 20 42	18	15.2	PART OF MULTIPLE GALAXY
REIZ 1591	11 43 24.	+ 20 43	12	15.4	GALAXY
REIZ 1590	11 43 24.	+ 20 43	18	15.2	GALAXY
HOLM 290A	11 43 24.	+ 20 43	24	14.6	PART OF MULTIPLE GALAXY
IC 0732	11 43 24.	+ 20 43			NONSTELLAR OBJECT
ZWG 127.050	11 43 24.	+ 21 19		14.8	GALAXY
REIZ 1592	11 43 24.	+ 21 19	54	13.8	GALAXY
UGC 06743	11 43 24.	+ 21 19	96	14.8	GALAXY Sb/Sc
MCG+10-17-063	11 43 24.	+ 60 41	27	16.	GALAXY
HN 1388	11 43 25.4	+ 26 21 45.			NEBULA
IC 0733	11 43 26.	- 07 52 52.			NONSTELLAR OBJECT
MCG+08-22-002	11 43 27.	+ 47 46	300	12.0	GALAXY
LB 02104	11 43 27.	+ 55 17 36.		15.5	FAINT BLUE STAR
KN 12.115	11 43 28.7	+ 46 15 16.			NEBULA
RNGC 3877	11 43 29.	+ 47 46		12.0	GALAXY
REIN 2.132	11 43 29.11	+ 47 46 19.1			NEBULA
KN 12.116	11 43 29.3	+ 47 46 29.			NEBULA
REIZ 1593	11 43 30.	+ 19 43	6	15.5	GALAXY
ZWG 097.142	11 43 30.	+ 19 55		15.7	GALAXY
ZWG 097.143	11 43 30.	+ 20 04		15.3	GALAXY
MCG+04-28-050	11 43 30.	+ 20 42	30	15.1	GALAXY
REIZ 1594	11 43 30.	+ 20 43	72	14.8	GALAXY
ZWG 127.051	11 43 30.	+ 20 44		15.1	GALAXY
REIZ 1595	11 43 30.	+ 21 14	36	15.4	GALAXY
ZWG 186.041	11 43 30.	+ 33 23		15.4	GALAXY
RNGC 3871	11 43 30.	+ 33 23		15.5	GALAXY
UGC 06744	11 43 30.	+ 33 23	66	15.4	GALAXY S
ZC 1143.5+4634	11 43 30.	+ 46 34	1080		CLUSTER OF GALAXIES
ZWG 243.004	11 43 30.	+ 47 46		11.8	GALAXY
UGC 06745	11 43 30.	+ 47 46	336	11.8	GALAXY Sc
MRK 187	11 43 30.	+ 71 54	7	17.5	GALAXY WITH UV CONTINUUM
IC 0734	11 43 30.	- 07 59 46.			NONSTELLAR OBJECT
TON-N 1432	11 43 32.	+ 32 42		16.9	BLUE STAR
HOLM 291A	11 43 32.	- 18 33	54	14.6	PART OF MULTIPLE GALAXY
HOLM 291B	11 43 32.	- 18 34	30	15.2	PART OF MULTIPLE GALAXY
HN 1389	11 43 32.8	+ 26 20 09.			NEBULA
REIZ 1596	11 43 33.	+ 47 46	288	12.1	GALAXY
ARC 1375	11 43 33.	- 07 59		16.6	RICH CLUSTER OF GALAXIES
REIZ 1597	11 43 34.	+ 03 33	60	14.8	GALAXY
IC 2959	11 43 35.	+ 33 22 38.			NONSTELLAR OBJECT
RNGC 3882	11 43 35.	- 56 05			DIFFUSE NEBULA
HN 1390	11 43 35.8	- 01 07 39.	24		NEBULA
ZWG 040.042	11 43 36.	+ 07 47		15.6	GALAXY
ZC 1143.6+0803	11 43 36.	+ 08 03	3700		CLUSTER OF GALAXIES
ZC 1143.6+0832	11 43 36.	+ 08 32	400		CLUSTER OF GALAXIES
ZWG 068.062	11 43 36.	+ 10 53		14.6	GALAXY
MCG+02-30-035	11 43 36.	+ 10 53	48	14.6	GALAXY
REIZ 1600	11 43 36.	+ 20 40	30	14.7	GALAXY
MCG+04-28-051	11 43 36.	+ 20 40 30.	150	14.0	GALAXY
ZWG 127.052	11 43 36.	+ 20 41		14.0	GALAXY
RNGC 3884	11 43 36.	+ 20 41		14.0	GALAXY
UGC 06746	11 43 36.	+ 20 41	126	14.0	GALAXY Sa
MCG+06-26-031	11 43 36.	+ 33 23 30.	60	15.	GALAXY
ZWG 186.042	11 43 36.	+ 33 28		15.5	GALAXY
RNGC 3878	11 43 36.	+ 33 28		15.5	GALAXY
MCG-01-30-031	11 43 36.	- 08 00 30.	48	16.	GALAXY
REIZ 1598	11 43 37.	+ 33 22	48	14.2	GALAXY
REIZ 1599	11 43 37.	+ 33 29	18	15.0	GALAXY
CED 120	11 43 38.	- 56 06			DIFFUSE GALACTIC NEBULA
MCG+06-26-032	11 43 39.	+ 33 29 30.	24	15.	GALAXY
ARC 1376	11 43 40.	- 00 49		16.6	RICH CLUSTER OF GALAXIES
REIZ 1601	11 43 41.	+ 10 52	24	14.6	GALAXY
HN 1391	11 43 41.8	+ 28 57 51.	24		NEBULA
REIZ 1602	11 43 42.	+ 21 11	18	15.5	GALAXY
ZWG 186.043	11 43 42.	+ 33 26		14.8	GALAXY
RNGC 3880	11 43 42.	+ 33 26		15.0	GALAXY
MCG+06-26-033	11 43 42.	+ 33 26 30.	30	15.	GALAXY
ZWG 186.044	11 43 42.	+ 35 16		15.6	GALAXY
LB 02105	11 43 42.	+ 49 43 18.		16.9	FAINT BLUE STAR

OBJECT NAME	RIGHT ASCEN.	DECLINATION	DIAM.	MAGN.	TYPE OF OBJECT
ZC 1143.7+5952	11 43 42.	+ 59 52	2820		CLUSTER OF GALAXIES
ZWG 012.079	11 43 42.	- 02 54		15.5	GALAXY
KN 12.117	11 43 42.7	+ 50 10 07.			NEBULA
REIZ 1603	11 43 43.	+ 33 27	18	14.3	GALAXY
IC 2960	11 43 44.	+ 35 16 50.			NONSTELLAR OBJECT
LB 00612	11 43 45.	+ 28 21 54.		17.6	FAINT BLUE STAR
MCG+08-22-003	11 43 45.	+ 50 10	24	16.	GALAXY
REIZ 1604	11 43 46.	+ 50 09	12	14.7	GALAXY
LB 02106	11 43 47.	+ 50 20 06.		16.0	FAINT BLUE STAR
ZC 1143.8+1119	11 43 48.	+ 11 19	1610		CLUSTER OF GALAXIES
ZWG 068.063	11 43 48.	+ 14 06		15.7	GALAXY
UGC 06747	11 43 48.	+ 14 06	78	15.7	GALAXY IRR
MCG+03-30-109	11 43 48.	+ 19 24	16	15.3	GALAXY
REIZ 1605	11 43 48.	+ 20 41	60	13.6	GALAXY
ZWG 186.045	11 43 48.	+ 35 07		15.0	GALAXY
MRK 429	11 43 48.	+ 35 07	16	14.5	GALAXY WITH UV CONTINUUM
ZC 1143.8+7142	11 43 48.	+ 71 42	4770		CLUSTER OF GALAXIES
TON-N 1433	11 43 50.	+ 33 36		17.	BLUE STAR
TON-N 1434	11 43 50.	+ 35 33		16.8	BLUE STAR
LB 02107	11 43 52.	+ 49 05 12.		16.1	FAINT BLUE STAR
REIZ 1606	11 43 53.	+ 14 06	36	15.4	GALAXY
ARP 248	11 43 53.	- 03 19			PECULIAR GALAXY
ZWG 186.046	11 43 54.	+ 33 23		15.2	GALAXY
RNGC 3881	11 43 54.	+ 33 23		15.0	GALAXY
MCG+06-26-034	11 43 54.	+ 33 23 30.	30	15.	GALAXY
UGC 06748	11 43 54.	+ 36 00	60	16.0	GALAXY S
VHE 53B	11 43 54.	- 65 17			REFLECTION NEBULA
ZC 1144.0+0442	11 44 00.	+ 04 42	1550		CLUSTER OF GALAXIES
ZC 1144.0+1456	11 44 00.	+ 14 56	340		CLUSTER OF GALAXIES
ZC 1144.0+1702	11 44 00.	+ 17 02	1610		CLUSTER OF GALAXIES
ZWG 097.144	11 44 00.	+ 19 23		15.3	GALAXY
REIZ 1607	11 44 00.	+ 21 29	12	15.1	GALAXY
MCG+04-28-052	11 44 00.	+ 24 12	72	15.4	GALAXY
ZC 1144.0+2547	11 44 00.	+ 25 47	810		CLUSTER OF GALAXIES
ZC 1144.0+4152	11 44 00.	+ 41 52	2820		CLUSTER OF GALAXIES
ZC 1144.0+5555	11 44 00.	+ 55 55	2690		CLUSTER OF GALAXIES
ZWG 292.027	11 44 00.	+ 62 01		15.0	GALAXY
MCG+10-17-064	11 44 00.	+ 62 02	54	14.	GALAXY
UGC 06749	11 44 00.	+ 65 39	60	16.0	GALAXY Sb-c
ZWG 334.050	11 44 00.	+ 70 57		14.8	GALAXY
MCG-01-30-032	11 44 00.	- 03 33 30.	48	15.	GALAXY
VV 035A	11 44 00.	- 03 35	60	14.5	INTERACTING GALAXY
REIZ 1608	11 44 01.	+ 33 23	18	14.3	GALAXY
LB 02108	11 44 01.	+ 57 01 36.		16.6	FAINT BLUE STAR
REIZ 1609	11 44 05.	+ 14 48	60	15.1	GALAXY
REIZ 1610	11 44 06.	+ 20 54	12	15.6	GALAXY
ZWG 127.053	11 44 06.	+ 24 15		15.4	GALAXY
REIZ 1611	11 44 06.	+ 24 15	48	14.7	GALAXY
UGC 06751	11 44 06.	+ 24 15	84	15.4	GALAXY Sb-c
ZWG 186.047	11 44 06.	+ 35 16		15.5	GALAXY
ZC 1144.1+3745	11 44 06.	+ 37 45	3630		CLUSTER OF GALAXIES
MCG+09-19-177	11 44 06.	+ 50 57 30.	84	14.	GALAXY
MCG+12-11-040	11 44 06.	+ 69 39	168	14.	GALAXY
ZWG 334.051	11 44 06.	+ 69 40		13.5	GALAXY
UGC 06752	11 44 06.	+ 69 40	150	13.5	GALAXY Sc-IRR
ZWG 012.080	11 44 06.	- 01 43		15.0	GALAXY
UGC 06750	11 44 06.	- 01 43	72	15.0	GALAXY S
MCG+00-30-022	11 44 06.	- 01 43	66	15.0	GALAXY
VHE 53A	11 44 06.	- 65 18	30		REFLECTION NEBULA
RNGC 3879	11 44 07.	+ 69 40		13.5	GALAXY
MCG+06-26-035	11 44 09.	+ 35 17 30.	30	14.5	GALAXY
ZWG 097.145	11 44 12.	+ 14 45		15.5	GALAXY
MCG+03-30-110	11 44 12.	+ 14 47	66	15.5	GALAXY
ZWG 097.146	11 44 12.	+ 14 49		15.3	GALAXY
UGC 06753	11 44 12.	+ 14 49	66	15.3	GALAXY SB
MCG+04-28-053	11 44 12.	+ 20 56	198	14.2	GALAXY
ZWG 127.054	11 44 12.	+ 20 58		14.2	GALAXY
RNGC 3883	11 44 12.	+ 20 58		14.0	GALAXY
UGC 06754	11 44 12.	+ 20 58	198	14.2	GALAXY Sb
ZWG 127.055	11 44 12.	+ 21 34		15.1	GALAXY
MRK 640	11 44 12.	+ 21 34	15	15.	GALAXY WITH UV CONTINUUM
MCG+04-28-054	11 44 12.	+ 21 34	18	15.1	GALAXY
MCG+06-26-036	11 44 12.	+ 35 13	36	16.	GALAXY
MCG+09-19-178	11 44 12.	+ 55 59	36	16.0	GALAXY
MCG+11-15-002	11 44 12.	+ 65 39	45	16.	GALAXY
VV 035C	11 44 12.	- 03 33	60	15.0	INTERACTING GALAXY
MCG-01-30-033	11 44 12.	- 03 33	90	14.5	GALAXY
VV 035B	11 44 12.	- 03 34	96	14.1	INTERACTING GALAXY
WIL 1	11 44 12.	- 03 34 12.		15.0	3 EMISSION-LINE GALAXIES
MCG-05-28-005	11 44 12.	- 29 48	72	15.	GALAXY
ARC 1378	11 44 14.	+ 23 47		17.5	RICH CLUSTER OF GALAXIES
RNGC 3885	11 44 14.	- 27 39		13.0	GALAXY
LB 02109	11 44 15.	+ 52 50 18.		16.9	FAINT BLUE STAR
MCG-01-30-034	11 44 15.	- 03 31	54	15.	GALAXY
SEY 125	11 44 16.	+ 54 59 15.		14.3	FAINT GALAXY
REIZ 1612	11 44 17.	+ 14 11	24	15.3	GALAXY
ZC 1144.3+1010	11 44 18.	+ 10 10	270		CLUSTER OF GALAXIES
MCG+03-30-111	11 44 18.	+ 20 08	60	14.3	GALAXY
REIZ 1613	11 44 18.	+ 20 58	78	13.6	GALAXY
ZC 1144.3+2731	11 44 18.	+ 27 31	1750		CLUSTER OF GALAXIES
ZWG 268.082	11 44 18.	+ 50 59		14.7	GALAXY
UGC 06755	11 44 18.	+ 50 59	66	14.7	GALAXY S
VV 136C	11 44 18.	+ 58 09	12	17.	INTERACTING GALAXY
VV 136B	11 44 18.	+ 58 09	18	16.5	INTERACTING GALAXY
VV 136A	11 44 18.	+ 58 09	36	16.	INTERACTING GALAXY
VV 136	11 44 18.	+ 58 09	42		INTERACTING GALAXY
KN 12.118	11 44 18.7	+ 50 58 54.			NEBULA
REIZ 1614	11 44 19.	+ 33 17	24	15.4	GALAXY
ARC 1377	11 44 19.	+ 56 01		15.0	RICH CLUSTER OF GALAXIES
TON-N 1435	11 44 20.	+ 20 49		14.8	BLUE STAR
TON-N 1436	11 44 20.	+ 21 18		15.	BLUE STAR
MCG-02-30-029	11 44 21.	- 14 12 30.	72	15.	GALAXY
LB 02110	11 44 22.	+ 58 39 06.		17.4	FAINT BLUE STAR
REIZ 1615	11 44 23.	+ 14 09	42	15.3	GALAXY
LB 02111	11 44 23.	+ 51 43 42.		15.3	FAINT BLUE STAR
ZWG 068.064	11 44 24.	+ 14 09		15.3	GALAXY
MCG+09-19-179	11 44 24.	+ 52 56	30	16.	GALAXY
ZWG 292.028	11 44 24.	+ 57 23		15.3	GALAXY
UGC 06756	11 44 24.	+ 59 23	60	15.2	GALAXY S
UGC 06757	11 44 24.	+ 61 38	66	16.5	GALAXY DWARF
ZC 1144.4+6149	11 44 24.	+ 61 49	1880		CLUSTER OF GALAXIES
MCG+11-15-003	11 44 24.	+ 66 27	45	16.	GALAXY
MCG-05-28-006	11 44 24.	- 27 38	60	13.	GALAXY
TON-N 1437	11 44 26.	+ 36 23		13.9	BLUE STAR
HN 1392	11 44 26.1	- 00 01 40.	24		NEBULA
ARC 1379	11 44 29.	+ 08 14		17.0	RICH CLUSTER OF GALAXIES
REIZ 1616	11 44 29.	+ 13 59	24	14.1	GALAXY
REIZ 1617	11 44 29.	+ 20 07	12	14.5	GALAXY

OBJECT NAME	RIGHT ASCEN.	DECLINATION	DIAM.	MAGN.	TYPE OF OBJECT
ZWG 068.065	11 44 30.	+ 13 59		14.4	GALAXY
UGC 06758	11 44 30.	+ 13 59	120	14.4	GALAXY S
MCG+02-30-036	11 44 30.	+ 13 59	90	14.4	GALAXY
UGC 06759	11 44 30.	+ 16 32	72	16.0	GALAXY S
MCG+03-30-112	11 44 30.	+ 16 32	72	16.	GALAXY
MCG+03-30-113	11 44 30.	+ 19 28	36	15.6	GALAXY
ZWG 097.147	11 44 30.	+ 20 07		14.3	GALAXY
RNGC 3886	11 44 30.	+ 20 07		14.5	GALAXY
UGC 06760	11 44 30.	+ 20 07	78	14.3	GALAXY E-S0
ZWG 157.032	11 44 30.	+ 29 52		15.2	GALAXY
UGC 06761	11 44 30.	+ 29 52	72	15.2	GALAXY S0-a
TON-N 0592	11 44 30.	+ 29 53		13.7	BLUE STAR
TON-N 0593	11 44 30.	+ 32 00		15.4	BLUE STAR
MCG+10-17-065	11 44 30.	+ 57 22	39	15.	GALAXY
MCG+10-17-066	11 44 30.	+ 59 01	18	17.	GALAXY
FEIG 048	11 44 30.	+ 61 33		13.1	FAINT BLUE STAR
MCG+10-17-067	11 44 30.	+ 61 38	39	16.	GALAXY
ZWG 012.081	11 44 30.	- 00 01		15.1	GALAXY
HMS 1.17	11 44 30.	- 03 34			SBb GALAXY
MCG-03-30-012	11 44 30.	- 16 34 30.	180	11.	GALAXY
HN 1393	11 44 30.3	+ 26 41 15.	18		NEBULA
HN 1394	11 44 32.1	- 01 55 46.			NEBULA
MCG+05-28-029	11 44 33.	+ 29 52	48	15.2	GALAXY
RNGC 3835A	11 44 33.	+ 60 34		14.0	GALAXY
REIZ 1618	11 44 34.	+ 03 38	18	15.8	GALAXY
HN 1395	11 44 34.5	- 01 53 04.			NEBULA
LB 02112	11 44 35.	+ 61 32 18.		12.3	FAINT BLUE STAR
ZWG 068.066	11 44 36.	+ 10 48		15.7	GALAXY
ZWG 097.148	11 44 36.	+ 19 12		15.5	GALAXY
ZWG 097.149	11 44 36.	+ 19 27		15.6	GALAXY
REIZ 1619	11 44 36.	+ 30 11	30	14.3	GALAXY
MCG+09-19-180	11 44 36.	+ 52 43	30	16.	GALAXY
ZC 1144.6+5452	11 44 36.	+ 54 52	2550		CLUSTER OF GALAXIES
ZWG 292.029	11 44 36.	+ 60 34		13.9	GALAXY
UGC 06762	11 44 36.	+ 60 34	66	13.9	GALAXY S
RNGC 3887	11 44 38.	- 16 35		12.0	GALAXY
REIZ 1621	11 44 41.	+ 19 30	18	15.8	GALAXY
HMS 1.18	11 44 42.	- 03 34			SB GALAXY
ZWG 097.150	11 44 42.	+ 16 50		15.6	GALAXY
ZC 1144.7+3100	11 44 42.	+ 31 00	870		CLUSTER OF GALAXIES
ZC 1144.7+3348	11 44 42.	+ 33 48	670		CLUSTER OF GALAXIES
ZWG 186.048	11 44 42.	+ 35 18		15.7	GALAXY
ZC 1144.7+3620	11 44 42.	+ 36 20	810		CLUSTER OF GALAXIES
ZWG 214.026	11 44 42.	+ 44 02		15.4	GALAXY
ZWG 268.083	11 44 42.	+ 52 44		15.6	GALAXY
MCG+09-19-181	11 44 42.	+ 55 58	24	16.9	GALAXY
MCG+09-19-182	11 44 42.	+ 56 02	18	16.0	GALAXY
ZC 1144.7+5812	11 44 42.	+ 58 12	1550		CLUSTER OF GALAXIES
REIZ 1620	11 44 42.	+ 60 35	24	14.2	GALAXY
KEEN 3835A	11 44 42.	+ 60 36			GALAXY
TON-N 1438	11 44 44.	+ 21 20		16.6	BLUE STAR
REIZ 1622	11 44 46.	- 01 24	72	15.1	GALAXY
REIZ 1623	11 44 47.	+ 20 12	12	15.1	GALAXY
HMS 1.19	11 44 48.	- 03 35			SC GALAXY
ZWG 097.151	11 44 48.	+ 18 20		15.6	GALAXY
MCG+03-30-114	11 44 48.	+ 20 15	51	15.5	GALAXY
REIZ 1624	11 44 48.	+ 28 24	24	15.5	GALAXY
ZWG 268.084	11 44 48.	+ 50 43		15.6	GALAXY
UGC 06763	11 44 48.	+ 50 43	60	15.6	GALAXY Sb-c
MCG+09-19-183	11 44 48.	+ 56 18	15	16.	GALAXY
UGC 06764	11 44 48.	+ 69 24	108	16.5	GALAXY
MCG-06-26-007	11 44 48.	- 37 16	72	14.5	GALAXY
KN 12.119	11 44 49.8	+ 50 42 35.			NEBULA
LB 02113	11 44 51.	+ 59 55 24.		17.3	FAINT BLUE STAR
RNGC 3888	11 44 52.	+ 56 15		13.5	GALAXY
ZWG 040.043	11 44 54.	+ 03 03		15.3	GALAXY
ZC 1144.9+4442	11 44 54.	+ 44 42	870		CLUSTER OF GALAXIES
ZWG 243.005	11 44 54.	+ 47 58		15.7	GALAXY
MCG+09-19-184	11 44 54.	+ 52 42 30.	24	16.	GALAXY
ZWG 268.085	11 44 54.	+ 56 14		12.6	GALAXY
UGC 06765	11 44 54.	+ 56 14	108	12.6	GALAXY Sc
ZWG 012.082	11 44 54.	- 02 45		15.6	GALAXY
TON-N 1439	11 44 56.	+ 33 30		14.9	BLUE STAR
TON-N 1440	11 44 56.	+ 36 22		15.	BLUE STAR
KN 12.120	11 44 56.4	+ 47 58 36.			NEBULA
ARC 1380	11 44 57.	+ 25 41		16.6	RICH CLUSTER OF GALAXIES
MCG+10-17-068	11 44 57.	+ 60 35	60	14.	GALAXY
REIZ 1625	11 44 58.	+ 50 42	36	14.3	GALAXY
KN 12.121	11 44 58.8	+ 48 56 12.			NEBULA
REIZ 1626	11 44 59.	+ 56 15	48	12.6	GALAXY
REIZ 1627	11 44 59.	+ 57 55	30	14.3	GALAXY
HMS 1145+5559	11 45	+ 55 59			URSA MAJOR GLXY CLSTR 2
VDB.66G 239	11 45	- 28 00	100		DWARF GALAXY
ZC 1145.0+0716	11 45 00.	+ 07 16	1280		CLUSTER OF GALAXIES
ZWG 097.152	11 45 00.	+ 20 13		15.5	GALAXY
ZC 1145.0+2540	11 45 00.	+ 25 40	1550		CLUSTER OF GALAXIES
ZC 1145.0+3142	11 45 00.	+ 31 42	1140		CLUSTER OF GALAXIES
ZWG 186.049	11 45 00.	+ 35 50		15.6	GALAXY
MCG+09-19-187	11 45 00.	+ 51 19	18	16.	GALAXY
MCG+09-19-188	11 45 00.	+ 51 20	24	16.	GALAXY
ZWG 268.086	11 45 00.	+ 54 47		15.6	GALAXY
UGC 06766	11 45 00.	+ 54 47	66	15.6	GALAXY Sb-c
MCG+09-19-186	11 45 00.	+ 54 47	30	16.	GALAXY
MCG+09-19-185	11 45 00.	+ 54 47 30.	84	15.	GALAXY
MRK 188	11 45 00.	+ 56 15	55	13.	GALAXY WITH UV CONTINUUM
MCG+09-19-189	11 45 00.	+ 56 15	102	12.7	GALAXY
ZWG 292.030	11 45 00.	+ 57 55		13.9	GALAXY
UGC 06767	11 45 00.	+ 57 55	48	13.9	GALAXY DBL SYS
KARA.72 301B	11 45 00.	+ 57 55	18		PART OF DOUBLE GALAXY
KARA.72 301A	11 45 00.	+ 57 55	24	13.9	PART OF DOUBLE GALAXY
MCG+10-17-069	11 45 00.	+ 59 45	39	16.	GALAXY
ZWG 292.031	11 45 00.	+ 60 10		15.3	GALAXY
FEIG 049	11 45 00.	- 15 49		11.8	FAINT BLUE STAR
MCG+06-26-037	11 45 06.	+ 32 36	27	16.	GALAXY
ZWG 186.050	11 45 06.	+ 36 43		15.7	GALAXY
SHAH 068	11 45 06.	+ 39 33	120	17.1	GROUP OF COMPACT GALAXIES
ZWG 214.027	11 45 06.	+ 44 01		15.2	GALAXY
UGC 06768	11 45 06.	+ 44 01	60	15.2	GALAXY SB:a-b
ZC 1145.1+4504	11 45 06.	+ 45 04	810		CLUSTER OF GALAXIES
MCG+09-19-190	11 45 06.	+ 56 03	42	16.0	GALAXY
MCG+10-17-070	11 45 06.	+ 57 56	42	14.	GALAXY
VV 270C	11 45 06.	+ 58 22	18	18.	INTERACTING GALAXY
VV 270B	11 45 06.	+ 58 22	24	16.	INTERACTING GALAXY
VV 270A	11 45 06.	+ 58 22	24	16.	INTERACTING GALAXY
VV 270	11 45 06.	+ 58 22	96		INTERACTING GALAXY
MCG+10-17-071	11 45 06.	+ 59 42 30.	24	16.	GALAXY
MCG+12-11-041	11 45 06.	+ 69 22	66	16.	GALAXY
HN 1396	11 45 07.9	+ 02 07 08.	60		NEBULA

OBJECT NAME	RIGHT ASCEN.	DECLINATION	DIAM.	MAGN.	TYPE OF OBJECT
SEY 126	11 45 09.	+ 44 01 26.		14.9	FAINT GALAXY
LB 02114	11 45 09.	+ 55 29 06.		17.0	FAINT BLUE STAR
REIZ 1628	11 45 10.	+ 02 06	54	15.6	GALAXY
ZWG 012.084	11 45 12.	+ 02 06		15.0	GALAXY
UGC 06769	11 45 12.	+ 02 06	66	15.0	GALAXY SB:a-b
MCG+00-30-023	11 45 12.	+ 02 06	66	15.0	GALAXY
KARA.73B 0499	11 45 12.	+ 02 06	66	15.0	ISOLATED GALAXY S
ZWG 097.153	11 45 12.	+ 18 50		15.6	GALAXY
UGC 06770	11 45 12.	+ 18 50	60	15.6	GALAXY DBL SYS
ZWG 157.033	11 45 12.	+ 28 21		15.6	GALAXY
ZWG 157.034	11 45 12.	+ 31 38		15.5	GALAXY
MCG+07-24-031	11 45 12.	+ 44 01 30.	54	15.	GALAXY
MCG+10-17-072	11 45 12.	+ 60 10	18	16.	GALAXY
ZWG 012.083	11 45 12.	- 03 02		15.7	GALAXY
IC 2961	11 45 13.	+ 31 36 55.			NONSTELLAR OBJECT
ZWG 040.044	11 45 18.	+ 04 17		15.2	GALAXY
MCG+03-30-115	11 45 18.	+ 20 19	42	15.0	GALAXY
ZC 1145.3+2239	11 45 18.	+ 22 39	4170		CLUSTER OF GALAXIES
MCG+08-22-004	11 45 18.	+ 50 06	84	14.	GALAXY
MCG+09-19-191	11 45 18.	+ 56 17	30	16.	GALAXY
MRK 189	11 45 18.	+ 67 53	7	16.	GALAXY WITH UV CONTINUUM
MCG-02-30-030	11 45 18.	- 10 41	120	13.	GALAXY
REIZ 1629	11 45 19.	+ 32 37	18	15.1	GALAXY
TON-N 1441	11 45 20.	+ 21 22		14.8	BLUE STAR
REIZ 1630	11 45 21.	+ 50 05	48	14.5	GALAXY
HN 1397	11 45 21.1	+ 28 20 32.	36		NEBULA
KN 12.122	11 45 22.2	+ 50 05 12.			NEBULA
REIZ 1631	11 45 23.	+ 20 16	12	14.8	GALAXY
ZWG 040.045	11 45 24.	+ 04 46		14.4	GALAXY
UGC 06771	11 45 24.	+ 04 46	126	14.4	GALAXY SBa
MCG+01-30-015	11 45 24.	+ 04 46	96	14.4	GALAXY
KARA.73B 0500	11 45 24.	+ 04 46	114	14.4	ISOLATED GALAXY S
ZWG 097.154	11 45 24.	+ 18 55		15.7	GALAXY
ZWG 097.155	11 45 24.	+ 20 17		15.0	GALAXY
MCG+05-28-030	11 45 24.	+ 28 21	42	15.6	GALAXY
ZWG 157.035	11 45 24.	+ 30 39		13.7	GALAXY
RNGC 3891	11 45 24.	+ 30 39		13.5	GALAXY
UGC 06772	11 45 24.	+ 30 39	138	13.7	GALAXY Sb/Sc
ZWG 186.051	11 45 24.	+ 37 43		15.4	GALAXY
ZWG 243.006	11 45 24.	+ 46 28		15.5	GALAXY
ZWG 243.007	11 45 24.	+ 50 05		15.2	GALAXY
UGC 06773	11 45 24.	+ 50 05	108	15.2	GALAXY IRR
MCG+09-19-192	11 45 24.	+ 53 07	36	16.	GALAXY
MCG+09-19-193	11 45 24.	+ 54 55	15	16.	GALAXY
MCG+09-19-194	11 45 24.	+ 55 18	30	16.	GALAXY
UGC 06774	11 45 24.	+ 55 19	126	16.0	GALAXY Sc?
HN 1398	11 45 24.3	- 02 13 46.	18		NEBULA
HOLM 292A	11 45 25.	+ 37 43	12	14.7	PART OF MULTIPLE GALAXY
ARC 1384	11 45 26.	+ 22 50		17.2	RICH CLUSTER OF GALAXIES
KN 12.123	11 45 26.1	+ 46 27 05.			NEBULA
MCG+06-26-038	11 45 27.	+ 37 44 30.	15	15.	GALAXY
REIZ 1634	11 45 27.	- 10 41	24	13.0	GALAXY
HOLM 292B	11 45 28.	+ 37 44	18	14.9	PART OF MULTIPLE GALAXY
LB 02115	11 45 28.	+ 58 27 00.		14.9	FAINT BLUE STAR
ARC 1383	11 45 29.	+ 54 54		15.7	RICH CLUSTER OF GALAXIES
ARC 1385	11 45 30.	+ 11 50		17.2	RICH CLUSTER OF GALAXIES
ZC 1145.5+2157	11 45 30.	+ 21 57	1550		CLUSTER OF GALAXIES
TON-N 0070	11 45 30.	+ 27 34		15.4	BLUE STAR
REIZ 1633	11 45 30.	+ 30 38	36	13.7	GALAXY
MCG+05-28-031	11 45 30.	+ 30 40	108	13.7	GALAXY
MCG+06-26-039	11 45 30.	+ 37 45	30	16.	GALAXY
MCG+08-22-005	11 45 30.	+ 46 27	24	16.	GALAXY
MCG+09-19-195	11 45 30.	+ 55 17	120	15.	GALAXY
MCG+09-19-196	11 45 30.	+ 55 54	30	15.	GALAXY
REIZ 1632	11 45 30.	+ 59 59	18	15.6	GALAXY
ZC 1145.5+7528	11 45 30.	+ 75 28	3160		CLUSTER OF GALAXIES
ZC 1145.5-0218	11 45 30.	- 02 18	400		CLUSTER OF GALAXIES
RNGC 3892	11 45 32.	- 10 41		12.5	GALAXY
REIZ 1635	11 45 35.	+ 13 29	36	15.1	GALAXY
SEY 127	11 45 35.	+ 55 53 56.		15.3	FAINT GALAXY
ZWG 068.067	11 45 36.	+ 13 12		15.3	GALAXY
ZC 1145.6+2438	11 45 36.	+ 24 38	3700		CLUSTER OF GALAXIES
ZWG 157.036	11 45 36.	+ 31 30		15.6	GALAXY
ZWG 214.028	11 45 36.	+ 39 01		15.7	GALAXY
MCG+09-19-197	11 45 36.	+ 54 06	60	18.	GALAXY
MCG+10-17-073	11 45 36.	+ 60 09 30.	30	16.	GALAXY
MCG+10-17-074	11 45 36.	+ 61 10	36	16.	GALAXY
ZC 1145.6-0132	11 45 36.	- 01 32	1550		CLUSTER OF GALAXIES
MCG-02-30-031	11 45 36.	- 11 22	36	14.5	GALAXY
IC 0735	11 45 37.	+ 10 29 25.			NONSTELLAR OBJECT
LB 02116	11 45 37.	+ 55 14 06.		17.0	FAINT BLUE STAR
MCG+07-24-032	11 45 39.	+ 39 01	39	15.	GALAXY
ARC 1382	11 45 39.	+ 71 43		15.9	RICH CLUSTER OF GALAXIES
ARC 1381	11 45 41.	+ 75 30		17.0	RICH CLUSTER OF GALAXIES
ZWG 068.068	11 45 42.	+ 12 59		15.3	GALAXY
MCG+02-30-037	11 45 42.	+ 12 59	24	15.3	GALAXY
ZWG 068.069	11 45 42.	+ 13 29		15.0	GALAXY
UGC 06775	11 45 42.	+ 13 29	84	15.0	GALAXY S
MCG+02-30-038	11 45 42.	+ 13 29	60	15.0	GALAXY
ZC 1145.7+3821	11 45 42.	+ 38 21	1210		CLUSTER OF GALAXIES
ZC 1145.7+5144	11 45 42.	+ 51 44	2350		CLUSTER OF GALAXIES
MCG+09-19-198	11 45 42.	+ 54 45	6	18.	GALAXY
REIZ 1636	11 45 42.	+ 59 59	6	15.5	GALAXY
7ZW 422	11 45 42.	+ 65 28			COMPACT GALAXY
IC 0736	11 45 45.	+ 12 59 37.			NONSTELLAR OBJECT
ARC 1386	11 45 46.	- 01 41		17.2	RICH CLUSTER OF GALAXIES
ZWG 040.046	11 45 48.	+ 06 37		15.7	GALAXY
ZC 1145.8+1030	11 45 48.	+ 10 30	2080		CLUSTER OF GALAXIES
ZWG 068.070	11 45 48.	+ 13 00		14.7	GALAXY
MCG+02-30-039	11 45 48.	+ 13 00	24	14.7	GALAXY
ZC 1145.8+1519	11 45 48.	+ 15 19	810		CLUSTER OF GALAXIES
ZWG 127.056	11 45 48.	+ 21 26		15.7	GALAXY
ZWG 127.057	11 45 48.	+ 26 02		15.3	GALAXY
MCG+10-17-075	11 45 48.	+ 59 42	27	15.	GALAXY
ZWG 012.085	11 45 48.	- 01 21		15.6	GALAXY
HN 1399	11 45 51.1	+ 26 03 44.			NEBULA
HN 1401	11 45 51.6	- 00 15 52.	12		NEBULA
HN 1400	11 45 52.6	+ 26 03 38.			NEBULA
KN 12.124	11 45 52.9	+ 50 18 52.			NEBULA
IC 0737	11 45 53.	+ 13 00 13.			NONSTELLAR OBJECT
REIZ 1637	11 45 53.	+ 21 27	24	15.2	GALAXY
ZC 1145.9+0421	11 45 54.	+ 04 21	2350		CLUSTER OF GALAXIES
ZWG 068.071	11 45 54.	+ 11 22		15.7	GALAXY
ZWG 068.072	11 45 54.	+ 13 00		15.3	GALAXY
MCG+02-30-040	11 45 54.	+ 13 00	36	15.3	GALAXY
ZC 1145.9+4034	11 45 54.	+ 40 34	3090		CLUSTER OF GALAXIES
UGC 06776	11 45 54.	+ 44 00	90	16.5	GALAXY Sc
MCG+08-22-006B	11 45 54.	+ 48 57	6	19.	GALAXY
MCG+08-22-006A	11 45 54.	+ 48 57	12	19.	GALAXY
MCG+08-22-007	11 45 54.	+ 48 59	270	11.0	GALAXY
REIN 2.133	11 45 56.13	+ 48 59 49.9			NEBULA
MCG+06-26-040	11 45 57.	+ 32 55	42	15.	GALAXY
RNGC 3893	11 45 59.	+ 48 59		11.5	GALAXY
VDB.66G 097	11 46	+ 24 08	70		DWARF GALAXY
ZWG 068.073	11 46 00.	+ 12 58		15.3	GALAXY
MCG+02-30-041	11 46 00.	+ 12 58	48	15.3	GALAXY
ZWG 186.052	11 46 00.	+ 32 55		15.5	GALAXY
UGC 06777	11 46 00.	+ 32 55	54	15.5	GALAXY COMPACT
KARA.73B 0501	11 46 00.	+ 32 55	48	15.5	ISOLATED GALAXY E
ZC 1146.0+4227	11 46 00.	+ 42 27	740		CLUSTER OF GALAXIES
MCG+07-24-033	11 46 00.	+ 44 00	84	15.	GALAXY
ZWG 243.008	11 46 00.	+ 48 59		10.6	GALAXY
UGC 06778	11 46 00.	+ 48 59	276	10.6	GALAXY Sc
KARA.72 302A	11 46 00.	+ 48 59	282	10.6	PART OF DOUBLE GALAXY
MCG+09-19-199	11 46 00.	+ 54 48	30	17.	GALAXY
ZC 1146.0+5910	11 46 00.	+ 59 10	1480		CLUSTER OF GALAXIES
ZWG 292.032	11 46 00.	+ 59 20		15.5	GALAXY
MCG+10-17-076	11 46 00.	+ 59 20	27	16.	GALAXY
MCG+10-17-077	11 46 00.	+ 59 30	15	17.	GALAXY
REIZ 1638	11 46 00.	+ 60 24	9	15.7	GALAXY
REIN 2.134	11 46 00.40	+ 48 59 18.3			NEBULA
KN 12.125	11 46 00.5	+ 48 59 20.			NEBULA
REIZ 1639	11 46 03.	+ 48 59	150	11.5	GALAXY
HOLM 293A	11 46 03.	+ 49 00	210	11.1	PART OF MULTIPLE GALAXY
KN 12.126	11 46 03.1	+ 46 52 30.			NEBULA
REIZ 1640	11 46 05.	+ 14 20	48	15.2	GALAXY
HOLM 294A	11 46 05.	+ 59 42	54	13.3	PART OF MULTIPLE GALAXY
HN 1402	11 46 05.4	+ 00 02 08.	30		NEBULA
ZWG 012.086	11 46 06.	+ 00 02		15.7	GALAXY
ZC 1146.1+1135	11 46 06.	+ 11 35	3830		CLUSTER OF GALAXIES
ZWG 186.053	11 46 06.	+ 36 03		15.7	GALAXY
ZC 1146.1+3728	11 46 06.	+ 37 28	940		CLUSTER OF GALAXIES
MCG+09-19-200	11 46 06.	+ 53 21	30	16.	GALAXY
ZWG 292.033	11 46 06.	+ 59 42		12.9	GALAXY
UGC 06779	11 46 06.	+ 59 42	150	12.9	GALAXY S0
KARA.72 303A	11 46 06.	+ 59 42	132	12.9	PART OF DOUBLE GALAXY
ZWG 292.034	11 46 06.	+ 60 28		15.0	GALAXY
VHE 54	11 46 06.	- 61 13	84		REFLECTION NEBULA
LB 02117	11 46 08.	+ 56 29 00.		16.1	FAINT BLUE STAR
REIZ 1642	11 46 10.	+ 04 46	48	14.0	GALAXY
ARC 1387	11 46 10.	+ 51 54		17.0	RICH CLUSTER OF GALAXIES
RNGC 3894	11 46 10.	+ 59 42		13.0	GALAXY
REIZ 1641	11 46 11.	+ 59 41	72	13.4	GALAXY
HN 1403	11 46 11.4	+ 29 57 08.			NEBULA
ZWG 068.074	11 46 12.	+ 14 20		15.4	GALAXY
ZWG 127.058	11 46 12.	+ 22 17		15.7	GALAXY
MCG+10-17-078	11 46 12.	+ 59 41	51	13.3	GALAXY
MCG+10-17-079	11 46 12.	+ 60 29	45	15.	GALAXY
MCG+11-15-004	11 46 12.	+ 62 49	18	16.	GALAXY
ZWG 012.087	11 46 12.	- 00 01		15.6	GALAXY
VHE 55	11 46 12.	- 62 03	48		REFLECTION NEBULA
PK296-03.1	11 46 12.	- 64 51	5		PLANETARY NEBULA
BC M1146+112	11 46 13.7	+ 11 11 42.		18.	QUASI-STELLAR OBJECT
SHB 186	11 46 13.7	+ 11 11 42.		18.	QUASI-STELLAR OBJECT
MCG+05-28-032	11 46 15.	+ 29 57	36	15.5	GALAXY
MCG+08-22-008	11 46 15.	+ 48 57 30.	72	13.9	GALAXY
MCG-04-28-003	11 46 15.	- 27 06	24	14.	GALAXY
REIZ 1644	11 46 17.	+ 22 01	24	15.3	GALAXY
REIZ 1645	11 46 17.	+ 22 17	18	15.3	GALAXY
RNGC 3896	11 46 17.	+ 48 57		14.0	GALAXY
LB 02118	11 46 17.	+ 53 11 00.		16.0	FAINT BLUE STAR
REIZ 1643	11 46 17.	+ 60 39	15	15.8	GALAXY
HN 1404	11 46 17.1	- 01 45 28.	12		NEBULA
ZWG 012.089	11 46 18.	+ 01 20		15.2	GALAXY
ZWG 097.156	11 46 18.	+ 16 10		15.5	GALAXY
MCG+03-30-116	11 46 18.	+ 16 10	48	15.5	GALAXY
ZWG 127.059	11 46 18.	+ 22 02		15.6	GALAXY
TON-N 0594	11 46 18.	+ 27 05		14.8	BLUE STAR
ZC 1146.3+4306	11 46 18.	+ 43 06	1140		CLUSTER OF GALAXIES
ZC 1146.3+4551	11 46 18.	+ 45 51	2890		CLUSTER OF GALAXIES
ZC 1146.3+4800	11 46 18.	+ 48 00	940		CLUSTER OF GALAXIES
ZWG 243.009	11 46 18.	+ 48 56		14.0	GALAXY
UGC 06781	11 46 18.	+ 48 56	108	14.0	GALAXY
KARA.72 302B	11 46 18.	+ 48 56	96	14.0	PART OF DOUBLE GALAXY
ZC 1146.3+5416	11 46 18.	+ 54 16	670		CLUSTER OF GALAXIES
ZC 1146.3+6423	11 46 18.	+ 64 23	1010		CLUSTER OF GALAXIES
ZWG 012.088	11 46 18.	- 01 45		14.7	GALAXY
UGC 06780	11 46 18.	- 01 45	210	14.7	GALAXY Sc
MCG+00-30-024	11 46 18.	- 01 45	192	14.7	GALAXY
KARA.73B 0502	11 46 18.	- 01 45	192	14.7	ISOLATED GALAXY S
HN 1405	11 46 18.5	+ 01 19 44.	18		NEBULA
KN 12.127	11 46 18.5	+ 47 51 36.			NEBULA
HOLM 294B	11 46 19.	+ 59 43	36	14.0	PART OF MULTIPLE GALAXY
KN 12.128	11 46 19.0	+ 48 57 16.			NEBULA
TON-N 1443	11 46 20.	+ 20 59		15.7	BLUE STAR
TON-N 1444	11 46 20.	+ 21 42		15.8	BLUE STAR
TON-N 1442	11 46 20.	+ 22 49		17.6	BLUE STAR
REIZ 1647	11 46 21.	+ 48 58	24	13.7	GALAXY
IC 0738	11 46 21.	- 04 24 00.			NONSTELLAR OBJECT
HN 1406	11 46 21.2	+ 00 47 32.	18		NEBULA
REIZ 1648	11 46 22.	+ 00 48	18	15.4	GALAXY
RNGC 3895	11 46 22.	+ 59 43		14.0	GALAXY
BC PKS1146-037	11 46 22.5	- 03 47 29.			QUASI-STELLAR OBJECT
HOLM 293B	11 46 23.	+ 48 58	30	13.9	PART OF MULTIPLE GALAXY
REIZ 1646	11 46 23.	+ 59 42	48	14.2	GALAXY
ZC 1146.4+2348	11 46 24.	+ 23 48	1480		CLUSTER OF GALAXIES
UGC 06782	11 46 24.	+ 24 07	120	17.	GALAXY DWRF IR
ZWG 157.037	11 46 24.	+ 27 37		15.4	GALAXY
ZC 1146.4+3018	11 46 24.	+ 30 18	470		CLUSTER OF GALAXIES
UGC 06783	11 46 24.	+ 31 35	66	17.	GALAXY DWRF IR
ZWG 186.054	11 46 24.	+ 35 18		14.2	GALAXY
RNGC 3897	11 46 24.	+ 35 18		14.0	GALAXY
UGC 06784	11 46 24.	+ 35 18	126	14.2	GALAXY Sb/Sc
MCG+06-26-041	11 46 24.	+ 35 18	108	13.	GALAXY
MCG+09-19-201	11 46 24.	+ 56 00	36	16.	GALAXY
ZWG 292.035	11 46 24.	+ 59 43		14.0	GALAXY
UGC 06785	11 46 24.	+ 59 43	84	14.0	GALAXY SBa
KARA.72 303B	11 46 24.	+ 59 43	84	14.0	PART OF DOUBLE GALAXY
MCG-05-28-007	11 46 24.	- 27 59	108	14.5	GALAXY
REIZ 1649	11 46 25.	+ 35 17	48	13.4	GALAXY
ARC 1388	11 46 26.	+ 22 40		17.3	RICH CLUSTER OF GALAXIES
HN 1408	11 46 28.6	- 00 49 28.	18		NEBULA
HN 1407	11 46 29.9	+ 27 35 20.			NEBULA
ZC 1146.5+2645	11 46 30.	+ 26 45	2350		CLUSTER OF GALAXIES
ZWG 157.038	11 46 30.	+ 27 19		12.5	GALAXY
UGC 06786	11 46 30.	+ 27 19	192	12.5	GALAXY Sa

OBJECT NAME	RIGHT ASCEN.	DECLINATION	DIAM.	MAGN.	TYPE OF OBJECT
ZC 1146.5+2756	11 46 30.	+ 27 56	1340		CLUSTER OF GALAXIES
MCG+08-22-009	11 46 30.	+ 49 02	24	16.	GALAXY
ZWG 269.001	11 46 30.	+ 55 58		15.7	GALAXY
ZWG 268.087	11 46 30.	+ 55 58		15.7	GALAXY
MCG+09-19-202	11 46 30.	+ 56 00	48	16.	GALAXY
ZWG 269.002	11 46 30.	+ 56 20		11.7	GALAXY
ZWG 268.088	11 46 30.	+ 56 20		11.7	GALAXY
UGC 06787	11 46 30.	+ 56 20	216	11.7	GALAXY Sa
MCG+10-17-080	11 46 30.	+ 59 42 30.	57	14.0	GALAXY
ZWG 012.090	11 46 30.	- 00 50		15.3	GALAXY
MCG-05-28-008	11 46 30.	- 29 53	24	15.	GALAXY
MCG-06-26-008	11 46 30.	- 37 14	60	14.	GALAXY
RNGC 3905	11 46 31.	- 09 29		13.0	GALAXY
SN 1971C	11 46 32.	- 29 03		15.3	SUPERNOVA
MCG+05-28-033	11 46 33.	+ 27 36 30.	12	15.4	GALAXY
IC 2962	11 46 33.	- 12 02 00.			NONSTELLAR OBJECT
RNGC 3903	11 46 33.	- 37 14			GALAXY
YC 1146-37	11 46 33.	- 37 14 30.			UNUSUAL SOUTHERN NEBULA
AR 40	11 46 33.52	+ 74 34 45.3			NEBULA
SEY 128	11 46 34.	+ 55 59 14.		15.4	FAINT GALAXY
REIZ 1650	11 46 34.	+ 56 21	180	11.9	GALAXY
RNGC 3898	11 46 34.	+ 56 22		11.5	GALAXY
MCG+04-28-055	11 46 36.	+ 26 22	90	14.0	GALAXY
RNGC 3899	11 46 36.	+ 26 44			NON-EXISTENT OBJECT
REIZ 1651	11 46 36.	+ 27 17	120	12.2	GALAXY
RNGC 3900	11 46 36.	+ 27 18		12.5	GALAXY
ZWG 243.010	11 46 36.	+ 46 24		15.6	GALAXY
ZC 1146.6+4829	11 46 36.	+ 48 29	1010		CLUSTER OF GALAXIES
ZC 1146.6+4929	11 46 36.	+ 49 29	1210		CLUSTER OF GALAXIES
MCG+09-19-203	11 46 36.	+ 52 33	18	17.	GALAXY
MCG+10-17-081	11 46 36.	+ 57 02	24	18.	GALAXY
ZWG 352.008	11 46 36.	+ 74 36		14.1	GALAXY
RNGC 3890	11 46 36.	+ 74 36		14.0	GALAXY
UGC 06788	11 46 36.	+ 74 36	60	14.1	GALAXY S
ZWG 352.009	11 46 36.	+ 76 17		14.9	GALAXY
UGC 06789	11 46 36.	+ 76 17	66	14.9	GALAXY SBb
MCG-01-30-035	11 46 36.	- 09 27	78	13.	GALAXY
HN 1409	11 46 36.5	+ 30 03 32.			NEBULA
REIN 2.135	11 46 36.65	+ 56 21 45.6			NEBULA
TON-N 1445	11 46 38.	+ 36 33		17.	BLUE STAR
REIZ 1652	11 46 39.	+ 51 42	24	15.2	GALAXY
KN 12.129	11 46 39.5	+ 46 23 57.			NEBULA
MCG+04-28-056	11 46 42.	+ 25 11	60	15.4	GALAXY
ZWG 127.060	11 46 42.	+ 26 24		14.0	GALAXY
RNGC 3902	11 46 42.	+ 26 24		14.0	GALAXY
UGC 06790	11 46 42.	+ 26 24	102	14.0	GALAXY SBb/Sc
ZWG 157.039	11 46 42.	+ 27 01		15.2	GALAXY
UGC 06791	11 46 42.	+ 27 01	114	15.2	GALAXY Sc
REIZ 1653	11 46 42.	+ 27 02	108	14.8	GALAXY
MCG+05-28-034	11 46 42.	+ 27 18	180	12.5	GALAXY
MCG+07-24-034	11 46 42.	+ 40 02 30.	156	14.	GALAXY
ZWG 214.029	11 46 42.	+ 40 03		14.8	GALAXY
UGC 06792	11 46 42.	+ 40 03	156	14.8	GALAXY Sc
MCG+08-22-010	11 46 42.	+ 46 24	36	16.	GALAXY
LB 02119	11 46 42.	+ 60 21 00.		16.1	FAINT BLUE STAR
MCG-05-28-009	11 46 42.	- 28 58	36	12.	GALAXY
TON-N 1446	11 46 44.	+ 34 03		17.	BLUE STAR
RNGC 3904	11 46 44.	- 29 02		12.5	GALAXY
REIZ 1654	11 46 46.	+ 00 49	60	15.1	GALAXY
ARC 1389	11 46 46.	- 01 07		16.6	RICH CLUSTER OF GALAXIES
HN 1410	11 46 47.7	+ 25 13 02.	18		NEBULA
ZC 1146.8+0840	11 46 48.	+ 08 40	1680		CLUSTER OF GALAXIES
ZC 1146.8+1237	11 46 48.	+ 12 37	2550		CLUSTER OF GALAXIES
ZWG 097.157	11 46 48.	+ 15 41		15.1	GALAXY
MCG+03-30-117	11 46 48.	+ 15 41	48	15.1	GALAXY
ZWG 097.158	11 46 48.	+ 16 55		14.7	GALAXY
UGC 06794	11 46 48.	+ 16 55	84	14.7	GALAXY SBc-IRR
MCG+03-30-118	11 46 48.	+ 16 56	84	14.7	GALAXY
REIZ 1656	11 46 48.	+ 25 12	30	15.0	GALAXY
ZWG 127.061	11 46 48.	+ 25 13		15.4	GALAXY
RNGC 3911	11 46 48.	+ 25 13		15.5	GALAXY
UGC 06795	11 46 48.	+ 25 13	72	15.4	GALAXY S
REIZ 1655	11 46 48.	+ 26 24	72	14.3	GALAXY
ZC 1146.8+4332	11 46 48.	+ 43 32	1140		CLUSTER OF GALAXIES
MCG+09-19-205	11 46 48.	+ 50 47 30.	48	16.	GALAXY
LB 00613	11 46 48.	+ 55 28 18.		14.9	FAINT BLUE STAR
MCG+09-19-204	11 46 48.	+ 56 21	210	11.7	GALAXY
ZWG 012.092	11 46 48.	- 00 48		14.8	GALAXY
UGC 06793	11 46 48.	- 00 48	72	14.8	GALAXY Sb
KARA.72 304A	11 46 48.	- 00 48	72	14.6	PART OF DOUBLE GALAXY
MCG+00-30-026	11 46 48.	- 00 48	72	14.8	GALAXY
ZWG 012.091	11 46 48.	- 03 14		15.0	GALAXY
MCG+00-30-025	11 46 48.	- 03 14	18	15.0	GALAXY
MCG-01-30-036	11 46 48.	- 04 50	66	14.5	GALAXY
RNGC 3907B	11 46 49.	- 00 48		15.0	GALAXY
HN 1412	11 46 49.4	- 00 48 04.	48		NEBULA
HN 1411	11 46 49.8	+ 27 00 44.	78		NEBULA
MCG+05-28-035	11 46 51.	+ 27 01	120	15.2	GALAXY
LB 00614	11 46 51.	+ 55 44 00.		17.6	FAINT BLUE STAR
HOLM 295B	11 46 51.	- 00 49	72	14.2	PART OF MULTIPLE GALAXY
IC 2963	11 46 51.	- 04 50			NONSTELLAR OBJECT
REIZ 1657	11 46 52.	+ 00 49	18	14.7	GALAXY
REIZ 1658	11 46 53.	+ 21 20	12	15.4	GALAXY
ZC 1146.9+2055	11 46 54.	+ 20 55	1080		CLUSTER OF GALAXIES
ZWG 127.062	11 46 54.	+ 21 20		15.5	GALAXY
TON-N 0595	11 46 54.	+ 28 12		15.4	BLUE STAR
ZC 1146.9+3704	11 46 54.	+ 37 04	810		CLUSTER OF GALAXIES
MCG+13-09-003	11 46 54.	+ 74 35	51	14.	GALAXY
MCG+13-09-004	11 46 54.	+ 76 17	57	15.	GALAXY
ZWG 012.094	11 46 54.	- 00 48		14.4	GALAXY
UGC 06796	11 46 54.	- 00 48	72	14.4	GALAXY SO
KARA.72 304B	11 46 54.	- 00 48	78	14.4	PART OF DOUBLE GALAXY
MCG+00-30-028	11 46 54.	- 00 48	24	14.4	GALAXY
ZWG 012.093	11 46 54.	- 03 11		14.9	GALAXY
MKW 03	11 46 54.	- 03 11		14.9	POOR GALAXY CLUSTER
MCG+00-30-027	11 46 54.	- 03 11	24	14.9	GALAXY
MCG-06-26-009	11 46 54.	- 38 33	24	15.	GALAXY
RNGC 3907A	11 46 55.	- 00 48		14.5	GALAXY
HOLM 295A	11 46 58.	- 00 49	24	14.1	PART OF MULTIPLE GALAXY
KN 12.130	11 46 58.0	+ 48 06 42.			NEBULA
RNGC 3906	11 46 59.	+ 48 42		14.0	GALAXY
LB 02120	11 46 59.	+ 53 52 00.		17.4	FAINT BLUE STAR
VDB.66G 099	11 47	+ 38 53	230		DWARF GALAXY
ARC 1390	11 47 00.	+ 12 32		16.0	RICH CLUSTER OF GALAXIES
ZC 1147.0+3110	11 47 00.	+ 31 10	2220		CLUSTER OF GALAXIES
MCG+08-22-011	11 47 00.	+ 48 07	24	16.	GALAXY
ZWG 243.011	11 47 00.	+ 48 42		14.1	GALAXY
UGC 06797	11 47 00.	+ 48 42	102	14.1	GALAXY SB

OBJECT NAME	RIGHT ASCEN.	DECLINATION	DIAM.	MAGN.	TYPE OF OBJECT
MCG+08-22-012	11 47 00.	+ 48 42	120	13.4	GALAXY
ZWG 243.012	11 47 00.	+ 50 12		15.6	GALAXY
MCG+08-22-013	11 47 00.	+ 50 13	36	15.	GALAXY
ZWG 292.036	11 47 00.	+ 60 38		15.2	GALAXY
ZC 1147.0+6744	11 47 00.	+ 67 44	1080		CLUSTER OF GALAXIES
ZWG 352.010	11 47 00.	+ 78 07		15.2	GALAXY
ZWG 351.070	11 47 00.	+ 78 07		15.2	GALAXY
UGC 06798	11 47 00.	+ 78 07	78	15.2	GALAXY Sa-b
MCG+13-09-005	11 47 00.	+ 78 07	66	15.	GALAXY
ZWG 012.096	11 47 00.	- 01 10		14.8	GALAXY
MCG+00-30-029	11 47 00.	- 01 10	15	14.8	GALAXY
ZWG 012.095	11 47 00.	- 03 12		15.2	GALAXY
REIZ 1659	11 47 02.	+ 48 06	12	14.4	GALAXY
REIZ 1660	11 47 02.	+ 48 42	66	13.4	GALAXY
HN 1413	11 47 02.8	- 01 10 28.	12		NEBULA
RNGC 3909	11 47 03.	- 47 59			NON-EXISTENT OBJECT
KN 12.131	11 47 03.3	+ 48 42 18.			NEBULA
REIZ 1661	11 47 05.	+ 13 31	24	15.8	GALAXY
KN 12.132	11 47 05.1	+ 50 12 33.			NEBULA
ZWG 068.075	11 47 06.	+ 13 31		15.5	GALAXY
ZWG 268.089	11 47 06.	+ 50 49		15.4	GALAXY
MCG+09-19-206	11 47 06.	+ 56 24	42	16.	GALAXY
ZWG 012.099	11 47 06.	- 03 14		15.5	GALAXY
ZWG 012.098	11 47 06.	- 03 15		15.4	GALAXY
ZWG 012.097	11 47 06.	- 03 17		15.3	GALAXY
KN 12.133	11 47 07.9	+ 50 48 27.			NEBULA
REIZ 1662	11 47 10.	+ 00 53	36	15.8	GALAXY
MCG+04-08-057	11 47 12.	+ 24 05 30.	120	16.	GALAXY
ZWG 157.040	11 47 12.	+ 29 43		15.7	GALAXY
ZWG 243.013	11 47 12.	+ 47 43		15.7	GALAXY
MCG+09-19-207	11 47 12.	+ 50 43	30	17.	GALAXY
MCG+10-17-082	11 47 12.	+ 60 39	45	16.	GALAXY
ZWG 334.052	11 47 12.	+ 71 01		14.9	GALAXY
MCG-06-26-010	11 47 12.	- 38 30	162	12.5	GALAXY
IC 2964	11 47 13.	+ 12 19			NONSTELLAR OBJECT
HN 1414	11 47 13.3	+ 29 29 20.	12		NEBULA
KN 12.134	11 47 13.9	+ 47 43 15.			NEBULA
ARC 1391	11 47 14.	- 12 02		18.0	RICH CLUSTER OF GALAXIES
MCG+05-28-036	11 47 15.	+ 29 42 30.	48	15.7	GALAXY
ZC 1147.3+0549	11 47 18.	+ 05 49	1550		CLUSTER OF GALAXIES
ZWG 040.047	11 47 18.	+ 06 57		15.0	GALAXY
MCG+01-30-016	11 47 18.	+ 06 57	36	15.0	GALAXY
UGC 06799	11 47 18.	+ 29 17	60	16.0	GALAXY Sb-c
MCG+08-22-014	11 47 18.	+ 47 43 30.	24	16.	GALAXY
ZC 1147.3+6537	11 47 18.	+ 65 37	1210		CLUSTER OF GALAXIES
ZWG 012.100	11 47 18.	- 03 14		15.1	GALAXY
KN 12.135	11 47 18.5	+ 50 45 32.			NEBULA
RNGC 3908	11 47 19.	+ 12 27			GALAXY
TON-N 1447	11 47 20.	+ 34 51		17.	BLUE STAR
REIZ 1663	11 47 20.	+ 46 03	12	14.6	GALAXY
HN 1415	11 47 20.2	+ 29 15 44.	24		NEBULA
MCG+04-28-058	11 47 21.	+ 21 35	24	14.4	GALAXY
LB 00615	11 47 21.	+ 56 47 24.		17.3	FAINT BLUE STAR
KN 12.136	11 47 21.5	+ 46 03 29.			NEBULA
HN 1416	11 47 21.8	+ 28 31 26.	12		NEBULA
HN 1417	11 47 22.4	+ 28 13 08.			NEBULA
KN 12.137	11 47 22.7	+ 48 15 13.			NEBULA
REIZ 1664	11 47 23.	+ 21 14	12	15.7	GALAXY
REIZ 1665	11 47 23.	+ 21 37	18	15.1	GALAXY
ZWG 097.159	11 47 24.	+ 15 18		15.4	GALAXY
ZC 1147.4+1825	11 47 24.	+ 18 25	740		CLUSTER OF GALAXIES
ZWG 127.063	11 47 24.	+ 21 38		14.4	GALAXY
RNGC 3910	11 47 24.	+ 21 38		14.5	GALAXY
UGC 06800	11 47 24.	+ 21 38	114	14.4	GALAXY E-SO
MCG+04-28-059	11 47 24.	+ 25 10	30	14.1	GALAXY
ZWG 157.041	11 47 24.	+ 26 45		13.2	GALAXY
UGC 06801	11 47 24.	+ 26 45	108	13.2	GALAXY S-IRR
TON-N 0071	11 47 24.	+ 29 04		15.4	BLUE STAR
ZC 1147.4+4653	11 47 24.	+ 46 53	1410		CLUSTER OF GALAXIES
UGC 06802	11 47 24.	+ 52 09	138	16.0	GALAXY Sc
MCG+09-19-208	11 47 24.	+ 53 42	42	16.	GALAXY
ZC 1147.4+6237	11 47 24.	+ 62 37	1810		CLUSTER OF GALAXIES
MCG+12-11-042	11 47 24.	+ 71 01	39	15.	GALAXY
ZWG 012.101	11 47 24.	- 03 22		15.2	GALAXY
MCG-03-30-013	11 47 24.	- 15 40 30.	12	15.	GALAXY
SEY 129	11 47 25.	+ 52 07 50.		14.8	FAINT GALAXY
REIZ 1666	11 47 27.	+ 52 07	60	15.0	GALAXY
KN 12.138	11 47 27.2	+ 50 29 16.			NEBULA
KN 12.139	11 47 27.8	+ 50 44 50.			NEBULA
KN 12.140	11 47 28.0	+ 49 09 16.			NEBULA
ZC 1147.5+0730	11 47 30.	+ 07 30	540		CLUSTER OF GALAXIES
REIZ 1668	11 47 30.	+ 25 12	36	14.1	GALAXY
ZWG 127.064	11 47 30.	+ 25 13		14.1	GALAXY
RNGC 3920	11 47 30.	+ 25 13		14.0	GALAXY
UGC 06803	11 47 30.	+ 25 13	78	14.1	GALAXY
RNGC 3912	11 47 30.	+ 26 45		13.0	GALAXY
REIZ 1667	11 47 30.	+ 26 46	60	12.9	GALAXY
ZWG 268.090	11 47 30.	+ 50 45		15.7	GALAXY
MCG+09-19-209	11 47 30.	+ 52 07	120	14.	GALAXY
MCG+09-19-210	11 47 30.	+ 53 56	30	16.	GALAXY
MCG+09-19-211	11 47 30.	+ 54 00	18	17.	GALAXY
TON-N 1448	11 47 32.	+ 35 12		17.	BLUE STAR
TON-N 1449	11 47 32.	+ 36 14		15.	BLUE STAR
REIZ 1669	11 47 32.	+ 45 44	18	14.3	GALAXY
MCG+09-19-212	11 47 33.	+ 54 00	30	17.	GALAXY
MCG-02-30-032	11 47 33.	- 12 39	42	15.	GALAXY
ZWG 040.048	11 47 36.	+ 07 16		15.3	GALAXY
UGC 06804	11 47 36.	+ 07 16	108	15.3	GALAXY Sc
MCG+04-28-060	11 47 36.	+ 26 13	114	14.4	GALAXY
MCG+05-28-037	11 47 36.	+ 26 45	72	13.2	GALAXY
ZC 1147.6+2930	11 47 36.	+ 29 30	1140		CLUSTER OF GALAXIES
ZWG 214.030	11 47 36.	+ 42 20		14.2	GALAXY
UGC 06805	11 47 36.	+ 42 20	21	14.2	GALAXY
KARA.73B 0503	11 47 36.	+ 42 20	24	14.2	ISOLATED GALAXY E
ZWG 214.031	11 47 36.	+ 44 00		15.1	GALAXY
ZWG 243.014	11 47 36.	+ 45 45		15.6	GALAXY
MCG+10-17-083	11 47 36.	+ 56 46	18	16.	GALAXY
ZC 1147.6+6807	11 47 36.	+ 68 07	470		CLUSTER OF GALAXIES
MRSL 295+00/1	11 47 36.	- 61 44	3000		HII REGION
KN 12.141	11 47 36.0	+ 45 45 27.			NEBULA
SEY 130	11 47 38.	+ 44 00 25.		15.3	FAINT GALAXY
HOLM 296B	11 47 38.	+ 55 38	24	14.8	PART OF MULTIPLE GALAXY
REIZ 1670	11 47 40.	+ 55 38	18	15.8	GALAXY
REIZ 1671	11 47 42.	+ 26 14	60	14.0	GALAXY
ZWG 127.065	11 47 42.	+ 26 15		14.4	GALAXY
UGC 06806	11 47 42.	+ 26 15	120	14.4	GALAXY S
MCG+04-28-061	11 47 42.	+ 26 15	60	15.	GALAXY
ZC 1147.7+6005	11 47 42.	+ 60 05	1810		CLUSTER OF GALAXIES

OBJECT NAME	RIGHT ASCEN.	DECLINATION	DIAM.	MAGN.	TYPE OF OBJECT
MCG-05-28-010	11 47 42.	- 28 12	36	16.	GALAXY
VHE 56	11 47 42.	- 64 36	156		REFLECTION NEBULA
HN 1418	11 47 42.4	+ 27 24 08.			NEBULA
LB 02121	11 47 43.	+ 52 00 24.		16.7	FAINT BLUE STAR
HN 1419	11 47 43.9	+ 26 43 56.	12		NEBULA
HN 1420	11 47 44.1	+ 26 14 32.	48		NEBULA
HOLM 296A	11 47 45.	+ 55 37	30	14.0	PART OF MULTIPLE GALAXY
HN 1421	11 47 45.1	+ 29 17 20.	18		NEBULA
KN 12.142	11 47 45.1	+ 50 50 11.			NEBULA
HN 1422	11 47 45.5	- 02 32 05.	36		NEBULA
RNGC 3918	11 47 46.	- 56 54		8.5	PLANETARY NEBULA
IC 0739	11 47 47.	+ 24 06 06.			NONSTELLAR OBJECT
HOLM 297B	11 47 47.	+ 50 49	30	15.0	PART OF MULTIPLE GALAXY
ZC 1147.8+1021	11 47 48.	+ 10 21	670		CLUSTER OF GALAXIES
ZWG 127.066	11 47 48.	+ 22 18		15.6	GALAXY
UGC 06807	11 47 48.	+ 26 17	60	16.5	GALAXY DWARF
UGC 06808	11 47 48.	+ 35 32	90	16.5	GALAXY Sc
ZC 1147.8+4843	11 47 48.	+ 48 43	610		CLUSTER OF GALAXIES
FEIG 050	11 47 48.	+ 52 50		11.9	FAINT BLUE STAR
ZC 1147.8+6654	11 47 48.	+ 66 54	1010		CLUSTER OF GALAXIES
ZWG 012.102	11 47 48.	- 02 32		15.0	GALAXY
MCG+00-30-030	11 47 48.	- 02 32	78	15.0	GALAXY
MCG-03-30-014	11 47 48.	- 18 19	48	15.	GALAXY
HOLM 297C	11 47 49.	+ 50 48	18	15.0	PART OF MULTIPLE GALAXY
HOLM 297A	11 47 50.	+ 50 48	30	14.5	PART OF MULTIPLF GALAXY
KN 12.143	11 47 50.6	+ 46 13 00.			NEBULA
PK294+04.1	11 47 52.	- 56 54	25	8.4	PLANETARY NEBULA
SEY 131	11 47 53.	+ 55 20 38.		15.4	FAINT GALAXY
KN 12.144	11 47 53.2	+ 50 49 21.			NEBULA
HN 1423	11 47 53.8	+ 28 52 32.	12		NEBULA
ZWG 040.049	11 47 54.	+ 04 58		15.5	GALAXY
ZC 1147.9+0706	11 47 54.	+ 07 06	470		CLUSTER OF GALAXIES
ZWG 097.160	11 47 54.	+ 18 08		15.6	GALAXY
MCG+03-30-119	11 47 54.	+ 20 20	15	14.5	GALAXY
ZC 1147.9+2114	11 47 54.	+ 21 14	1340		CLUSTER OF GALAXIES
MCG+09-19-213	11 47 54.	+ 55 20 30.	30	16.	GALAXY
MCG+10-17-084	11 47 54.	+ 62 19	39	16.	GALAXY
MCG-02-30-033	11 47 54.	- 12 11	48	15.	GALAXY
MCG-05-28-011	11 47 54.	- 28 14	36	15.5	GALAXY
HN 0791	11 47 56.	- 64 37	30		NEBULA
CED 121	11 47 56.	- 64 37	30		DIFFUSE GALACTIC NEBULA
IC 2966	11 47 56.	- 64 37			NONSTELLAR OBJECT
HN 1424	11 47 56.7	+ 27 55 14.	24		NEBULA
KN 12.145	11 47 56.7	+ 50 48 28.			NEBULA
MCG-01-30-037	11 47 57.	- 04 32 30.	30	15.	GALAXY
IC 2965	11 47 57.	- 19 18 43.			NONSTELLAR OBJECT
KN 12.146	11 47 57.2	+ 46 08 34.			NEBULA
REIZ 1672	11 47 58.	+ 06 51	72	13.8	GALAXY
ARC 1392	11 47 58.	- 00 19		16.6	RICH CLUSTER OF GALAXIES
KN 12.147	11 47 58.8	+ 48 16 17.			NEBULA
LB 02122	11 47 59.	+ 50 36 06.		17.4	FAINT BLUE STAR
RNGC 3921	11 47 59.	+ 55 20		13.5	GALAXY
IC 0741	11 47 59.	- 07 33 25.			NONSTELLAR OBJECT
HN 1425	11 47 59.4	- 02 37 53.	12		NEBULA
VDB.66G 098	11 48	+ 56 43	70		DWARF GALAXY
LB 09853	11 48	- 86 24		13.3	FAINT BLUE STAR
ZWG 040.050	11 48 00.	+ 06 51		13.8	GALAXY
UGC 06809	11 48 00.	+ 06 51	66	13.8	GALAXY SB
MCG+01-30-017	11 48 00.	+ 06 51	72	13.8	GALAXY
ZWG 097.161	11 48 00.	+ 20 17		14.5	GALAXY
UGC 06810	11 48 00.	+ 20 17	54	14.5	GALAXY F
ZWG 127.067	11 48 00.	+ 21 12		15.5	GALAXY
ZC 1148.0+3003	11 48 00.	+ 30 03	1610		CLUSTER OF GALAXIES
REIZ 1673	11 48 00.	+ 32 18	12	15.0	GALAXY
MCG+09-20-002	11 48 00.	+ 50 48	42	15.0	GALAXY
ZWG 269.003	11 48 00.	+ 50 50		14.8	GALAXY
ZWG 268.091	11 48 00.	+ 50 50		14.8	GALAXY
UGC 06811	11 48 00.	+ 50 50	72	14.8	GALAXY S0-a
UGC 06812	11 48 00.	+ 50 53	96	17.	GALAXY
ZWG 269.004	11 48 00.	+ 55 37		14.2	GALAXY
ZWG 268.092	11 48 00.	+ 55 37		14.2	GALAXY
UGC 06813	11 48 00.	+ 55 37	180	14.2	GALAXY Sc
SN 1963J	11 48 00.	+ 55 37		13.7	SUPERNOVA
MCG+09-20-001	11 48 00.	+ 55 39	120	14.0	GALAXY
MCG+10-17-085	11 48 00.	+ 60 24	72	16.	GALAXY
UGC 06814	11 48 00.	+ 64 37	72	17.	GALAXY
ZWG 012.105	11 48 00.	- 00 17		15.4	GALAXY
ZC 1148.0-0020	11 48 00.	- 00 20	1550		CLUSTER OF GALAXIES
ZWG 012.104	11 48 00.	- 01 08		15.5	GALAXY
ZWG 012.103	11 48 00.	- 02 38		14.2	GALAXY
MCG+00-30-031	11 48 00.	- 02 38	60	14.8	GALAXY
MCG-06-26-011	11 48 00.	- 35 13	30	15.5	GALAXY
HN 1426	11 48 00.8	- 00 16 29.	12		NEBULA
RNGC 3914	11 48 01.	+ 06 51		14.0	GALAXY
RNGC 3919	11 48 01.	+ 20 17		14.5	GALAXY
RNGC 3915	11 48 01.	- 04 52			NON-EXISTENT OBJECT
HN 1427	11 48 01.5	- 01 07 53.	18		NEBULA
TON-N 1450	11 48 02.	+ 36 46		16.9	BLUE STAR
HN 1428	11 48 02.8	- 00 18 17.	18		NEBULA
MCG+04-28-062	11 48 03.	+ 21 20 30.	48	15.5	GALAXY
MCG+09-20-003	11 48 03.	+ 50 47	60	14.5	GALAXY
LB 02123	11 48 03.	+ 59 34 30.		16.2	FAINT BLUE STAR
REIZ 1674	11 48 05.	+ 21 12	18	15.3	GALAXY
REIZ 1675	11 48 05.	+ 21 20	12	14.9	GALAXY
REIZ 1676	11 48 05.	+ 22 17	12	15.6	GALAXY
KN 12.148	11 48 05.4	+ 45 46 47.			NEBULA
HN 1429	11 48 05.5	+ 25 48 07.	18		NEBULA
ZWG 068.076	11 48 06.	+ 10 48		15.1	GALAXY
ZWG 186.055	11 48 06.	+ 32 58		15.6	GALAXY
ZWG 269.005	11 48 06.	+ 52 07		12.5	GALAXY
ZWG 268.093	11 48 06.	+ 52 07		12.5	GALAXY
UGC 06815	11 48 06.	+ 52 07	306	12.5	GALAXY Sc
SN 1970F	11 48 06.	+ 53 20		19.0	SUPERNOVA
MCG+09-20-004	11 48 06.	+ 55 07 30.	48	17.	GALAXY
MCG+09-20-005	11 48 06.	+ 55 27	90	13.4	GALAXY
ZWG 292.037	11 48 06.	+ 56 44		15.1	GALAXY
UGC 06816	11 48 06.	+ 56 44	96	15.1	GALAXY IRR
MCG+10-17-086	11 48 06.	+ 56 44	84	15.	GALAXY
MCG-01-30-038	11 48 06.	- 07 20	24	15.	GALAXY
MCG-02-30-034	11 48 06.	- 12 27	36	15.	GALAXY
KN 12.149	11 48 06.1	+ 46 05 03.			NEBULA
KN 12.150	11 48 06.6	+ 46 05 29.			NEBULA
HN 1430	11 48 06.9	- 01 10 47.	12		NEBULA
IC 0740	11 48 07.	+ 55 37 59.			SAME AS NGC 3913
LB 02124	11 48 07.	+ 57 33 12.		16.7	FAINT BLUE STAR
REIZ 1677	11 48 08.	+ 46 04	108	13.7	GALAXY
REIZ 1679	11 48 09.	+ 52 06	270	13.1	GALAXY
REIZ 1678	11 48 09.	+ 55 25	36	14.4	GALAXY

OBJECT NAME	RIGHT ASCEN.	DECLINATION	DIAM.	MAGN.	TYPE OF OBJECT
BC PKS1148-00	11 48 10.17	- 00 07 13.1		17.60	QUASI-STELLAR OBJECT
SHB 187	11 48 10.2	- 00 07 15.		17.6	QUASI-STELLAR OBJECT
KN 12.151	11 48 10.3	+ 46 05 15.			NEBULA
REIZ 1680	11 48 11.	+ 20 18	18	15.2	GALAXY
REIZ 1681	11 48 11.	+ 21 40	24	15.8	GALAXY
RNGC 3916	11 48 11.	+ 55 25		15.0	GALAXY
ZWG 097.162	11 48 12.	+ 20 19		15.7	GALAXY
MCG+04-28-064	11 48 12.	+ 20 45	66	15.4	GALAXY
MCG+04-28-063	11 48 12.	+ 21 25	36	15.3	GALAXY
ZWG 127.068	11 48 12.	+ 21 27		15.3	GALAXY
ZWG 127.069	11 48 12.	+ 21 41		15.7	GALAXY
ZC 1148.2+2940	11 48 12.	+ 29 40	810		CLUSTER OF GALAXIES
ZC 1148.2+3225	11 48 12.	+ 32 25	3290		CLUSTER OF GALAXIES
ZWG 214.032	11 48 12.	+ 39 09		15.1	GALAXY
UGC 06817	11 48 12.	+ 39 09	282	15.1	GALAXY DWRF IR
ZWG 243.015	11 48 12.	+ 46 05		14.6	GALAXY
HOLM 298B	11 48 12.	+ 46 05	18	15.0	PART OF MULTIPLE GALAXY
UGC 06818	11 48 12.	+ 46 05	126	14.6	GALAXY
MCG+08-22-015	11 48 12.	+ 46 05	120	13.9	GALAXY
ZC 1148.2+4717	11 48 12.	+ 47 17	810		CLUSTER OF GALAXIES
MCG+09-20-007	11 48 12.	+ 53 39	36	16.	GALAXY
MCG+09-20-006	11 48 12.	+ 53 40	24	16.	GALAXY
ZWG 269.006	11 48 12.	+ 55 25		14.8	GALAXY
ZWG 268.094	11 48 12.	+ 55 25		14.8	GALAXY
UGC 06819	11 48 12.	+ 55 25	96	14.8	GALAXY Sb
VV 273B	11 48 12.	+ 56 44	24	18.	INTERACTING GALAXY
VV 273A	11 48 12.	+ 56 44	126	16.	INTERACTING GALAXY
VV 273	11 48 12.	+ 56 44	120		INTERACTING GALAXY
ZC 1148.2+7155	11 48 12.	+ 71 55	740		CLUSTER OF GALAXIES
ARC 1393	11 48 13.	+ 47 17		17.8	RICH CLUSTER OF GALAXIES
SN 1974D	11 48 13.	+ 55 26		15.5	SUPERNOVA
HN 1431	11 48 13.6	+ 28 08 07.	12		NEBULA
TON-N 1451	11 48 14.	+ 35 38		16.8	BLUE STAR
TON-N 1452	11 48 14.	+ 35 48		17.	BLUE STAR
HOLM 298A	11 48 14.	+ 46 05	78	13.9	PART OF MULTIPLE GALAXY
MCG+05-28-038	11 48 15.	+ 31 09	18	14.8	GALAXY
MCG+07-24-035	11 48 15.	+ 39 08 30.	222	14.5	GALAXY
REIZ 1682	11 48 15.	+ 53 20	30	15.4	GALAXY
REIZ 1683	11 48 17.	+ 21 26	12	15.1	GALAXY
REIZ 1684	11 48 17.	+ 21 27	18	15.2	GALAXY
REIZ 1685	11 48 17.	+ 22 14	18	16.0	GALAXY
REIZ 1686	11 48 17.	+ 22 16	6	16.3	GALAXY
KN 12.152	11 48 17.0	+ 47 27 38.			NEBULA
ZWG 097.163	11 48 18.	+ 14 53		15.4	GALAXY
ZWG 127.070	11 48 18.	+ 20 47		15.4	GALAXY
UGC 06820	11 48 18.	+ 20 47	78	15.4	GALAXY S0-a
MCG+04-28-066	11 48 18.	+ 21 23 30.	27	15.4	GALAXY
ZWG 127.071	11 48 18.	+ 21 25		15.4	GALAXY
MCG+04-28-065	11 48 18.	+ 22 13 30.	42	15.	GALAXY
ZWG 157.042	11 48 18.	+ 31 08		14.8	GALAXY
ZC 1148.3+4136	11 48 18.	+ 41 36	1140		CLUSTER OF GALAXIES
MCG+09-20-008	11 48 18.	+ 52 06	300	12.8	GALAXY
ZC 1148.3+6408	11 48 18.	+ 64 08	1010		CLUSTER OF GALAXIES
KN 12.154	11 48 19.5	+ 48 48 40.			NEBULA
KN 12.153	11 48 19.9	+ 50 54 02.			NEBULA
IC 2967	11 48 20.	+ 31 07 41.			NONSTELLAR OBJECT
IC 0742	11 48 21.	+ 21 04 53.			NONSTELLAR OBJFCT
ARC 1395	11 48 21.	- 08 13		17.5	RICH CLUSTER OF GALAXIES
RNGC 3960	11 48 22.	- 55 25		9.0	OPEN CLUSTER
REIZ 1687	11 48 23.	+ 21 10	18	15.0	GALAXY
RNGC 3913	11 48 23.	+ 55 40		14.0	GALAXY
ZWG 097.164	11 48 24.	+ 15 01		15.7	GALAXY
MCG+04-28-069	11 48 24.	+ 20 39	84	14.6	GALAXY
ZWG 127.072	11 48 24.	+ 20 41		14.6	GALAXY
UGC 06821	11 48 24.	+ 20 41	84	14.6	GALAXY Sb/SBc
MCG+04-28-068	11 48 24.	+ 21 03	72	15.1	GALAXY
ZWG 127.073	11 48 24.	+ 21 05		15.1	GALAXY
UGC 06822	11 48 24.	+ 21 05	78	15.1	GALAXY SBa-b
MCG+04-28-067	11 48 24.	+ 22 03	30	15.5	GALAXY
ZWG 157.043	11 48 24.	+ 30 51		15.7	GALAXY
ZC 1148.4+4239	11 48 24.	+ 42 39	1010		CLUSTER OF GALAXIES
VV 031	11 48 24.	+ 55 22	72	14.	INTERACTING GALAXY
MCG+09-20-009	11 48 24.	+ 55 22	90	14.1	GALAXY
OCL 0861	11 48 24.	- 55 25	780	9.0	OPEN STAR CLUSTER
VHA 123	11 48 24.	- 55 25	600		OPEN STAR CLUSTER
HN 1433	11 48 25.6	- 00 06 29.	18		NEBULA
HN 1432	11 48 25.7	+ 29 18 31.	12		NEBULA
ARP 224	11 48 27.	+ 55 22			PECULIAR GALAXY
ARC 1398	11 48 27.	- 07 24		17.8	RICH CLUSTER OF GALAXIES
REIZ 1688	11 48 29.	+ 20 40	24	15.4	GALAXY
REIZ 1689	11 48 29.	+ 21 05	36	15.6	GALAXY
ARC 1394	11 48 29.	+ 42 40		17.3	RICH CLUSTER OF GALAXIES
HN 1434	11 48 29.2	+ 28 29 37.	12		NEBULA
STOCK 33	11 48 30.	+ 28 29			BLUE KNOT NEAR ELLIP GLXY
ZC 1148.5+3352	11 48 30.	+ 33 52	1480		CLUSTER OF GALAXIES
3ZW 063	11 48 30.	+ 43 23			COMPACT GALAXY
ZWG 243.016	11 48 30.	+ 45 30		15.7	GALAXY
MCG+09-20-010	11 48 30.	+ 54 30	42	16.	GALAXY
1ZW 028	11 48 30.	+ 55 21			COMPACT GALAXY
ZWG 269.007	11 48 30.	+ 55 21		13.4	GALAXY
ZWG 268.095	11 48 30.	+ 55 21		13.4	GALAXY
MRK 430	11 48 30.	+ 55 21	24	14.5	GALAXY WITH UV CONTINUUM
UGC 06823	11 48 30.	+ 55 21	132	13.4	GALAXY
HN 1435	11 48 30.6	+ 25 06 37.	12		NEBULA
REIZ 1690	11 48 32.	+ 52 16	24	14.5	GALAXY
LB 00616	11 48 32.	+ 54 27 18.		15.9	FAINT BLUE STAR
RNGC 3923	11 48 32.	- 28 33		11.5	GALAXY
ARC 1397	11 48 33.	+ 33 48		17.7	RICH CLUSTER OF GALAXIES
LB 02125	11 48 33.	+ 54 34 30.		16.3	FAINT BLUE STAR
ARC 1399	11 48 34.	- 02 50		16.0	RICH CLUSTER OF GALAXIES
KN 12.155	11 48 34.5	+ 45 30 05.			NEBULA
RNGC 3922	11 48 35.	+ 50 27		14.0	GALAXY
RNGC 3931	11 48 35.	+ 52 17		14.5	GALAXY
RNGC 3917A	11 48 35.	+ 52 17		14.5	GALAXY
KN 12.156	11 48 35.9	+ 50 26 14.			NEBULA
ZC 1148.6+0642	11 48 36.	+ 06 42	2490		CLUSTER OF GALAXIES
ZWG 040.051	11 48 36.	+ 07 02		15.5	GALAXY
ZWG 068.077	11 48 36.	+ 10 55		15.0	GALAXY
MCG+02-30-042	11 48 36.	+ 10 55	39	15.0	GALAXY
ZC 1148.6+1928	11 48 36.	+ 19 28	2490		CLUSTER OF GALAXIES
MCG+04-28-070	11 48 36.	+ 21 15	30	15.0	GALAXY
ZWG 127.074	11 48 36.	+ 21 17		15.0	GALAXY
ZC 1148.6+2555	11 48 36.	+ 25 55	1080		CLUSTER OF GALAXIES
ZC 1148.6+4216	11 48 36.	+ 42 16	810		CLUSTER OF GALAXIES
2ZW 054	11 48 36.	+ 43 22			COMPACT GALAXY
MCG+08-22-016	11 48 36.	+ 48 15	36	16.	GALAXY
MCG+08-22-017	11 48 36.	+ 50 26	96	13.	GALAXY
ZWG 269.008	11 48 36.	+ 50 27		13.8	GALAXY

OBJECT NAME	RIGHT ASCEN.	DECLINATION	DIAM.	MAGN.	TYPE OF OBJECT
ZWG 243.017	11 48 36.	+ 50 27		13.8	GALAXY
UGC 06824	11 48 36.	+ 50 27	126	13.8	GALAXY S0-a
MCG+09-20-011	11 48 36.	+ 52 16	30	14.	GALAXY
ZWG 269.009	11 48 36.	+ 52 17		14.6	GALAXY
ZWG 268.096	11 48 36.	+ 52 17		14.6	GALAXY
UGC 06825	11 48 36.	+ 52 17	66	14.6	GALAXY E-S0
LB 00617	11 48 36.	+ 54 11 30.		16.0	FAINT BLUE STAR
ZC 1148.6+5523	11 48 36.	+ 55 23	3020		CLUSTER OF GALAXIES
ZWG 292.038	11 48 36.	+ 56 40		15.7	GALAXY
ZCG 1148+69.2	11 48 36.	+ 69 36		18.4	COMPACT GALAXY
MCG-05-28-012	11 48 36.	- 28 30	60	11.	GALAXY
TON-N 1453	11 48 38.	+ 33 34		17.	BLUE STAR
REIZ 1691	11 48 38.	+ 50 29	102	13.3	GALAXY
REIZ 1692	11 48 39.	+ 53 43	24	15.3	GALAXY
ARC 1396	11 48 39.	+ 55 08		17.2	RICH CLUSTER OF GALAXIES
KN 12.157	11 48 39.5	+ 48 14 54.			NEBULA
SEY 132	11 48 41.	+ 53 43 43.		15.2	FAINT GALAXY
MCG+04-28-071	11 48 42.	+ 22 08 30.	39	15.3	GALAXY
ZWG 127.075	11 48 42.	+ 22 11		15.3	GALAXY
ZWG 157.044	11 48 42.	+ 27 04		15.4	GALAXY
REIZ 1693	11 48 42.	+ 35 43	42	13.6	GALAXY
ZC 1148.7+3655	11 48 42.	+ 36 55	1080		CLUSTER OF GALAXIES
MCG-01-30-039	11 48 42.	- 05 49	72	15.	GALAXY
OCL 0866	11 48 42.	- 61 48	360	13.	OPEN STAR CLUSTER
RNGC 3925	11 48 43.	+ 22 11		15.5	GALAXY
LB 02126	11 48 43.	+ 54 55 24.		15.4	FAINT BLUE STAR
REIZ 1694	11 48 44.	+ 48 13	42	14.7	GALAXY
MCG+04-28-073	11 48 45.	+ 22 18	15	14.7	GALAXY
HN 1436	11 48 45.7	+ 27 53 43.	12		NEBULA
HN 1437	11 48 46.0	+ 28 01 07.	6		NEBULA
REIZ 1695	11 48 47.	+ 21 10	12	15.1	GALAXY
REIZ 1696	11 48 47.	+ 22 10	18	15.8	GALAXY
REIZ 1697	11 48 47.	+ 24 08	30	15.5	GALAXY
LB 02127	11 48 47.	+ 51 57 18.		14.8	FAINT BLUE STAR
RNGC 3917	11 48 47.	+ 52 06		12.5	GALAXY
MCG+04-28-074	11 48 48.	+ 22 17 30.	15	14.7	GALAXY
VV 218B	11 48 48.	+ 22 18	21	15.	INTERACTING GALAXY
VV 218A	11 48 48.	+ 22 18	21	14.5	INTERACTING GALAXY
VV 218	11 48 48.	+ 22 18	48		INTERACTING GALAXY
MCG+04-28-072	11 48 48.	+ 24 07	66	15.1	GALAXY
MCG+05-28-039	11 48 48.	+ 27 03 30.	36	15.4	GALAXY
UGC 06826	11 48 48.	+ 32 50	78	16.5	GALAXY Sc
ZWG 186.056	11 48 48.	+ 35 29		15.3	GALAXY
MCG+06-26-043	11 48 48.	+ 35 30	18	16.	GALAXY
ZWG 186.057	11 48 48.	+ 35 43		14.3	GALAXY
MRK 431	11 48 48.	+ 35 43	30	14.5	GALAXY WITH UV CONTINUUM
UGC 06827	11 48 48.	+ 35 43	78	14.3	GALAXY S0-a
MCG+06-26-042	11 48 48.	+ 35 44	66	14.5	GALAXY
ZC 1148.8+3952	11 48 48.	+ 39 52	940		CLUSTER OF GALAXIES
ZWG 243.018	11 48 48.	+ 48 11		15.7	GALAXY
MCG+08-22-018	11 48 48.	+ 48 11	36	16.	GALAXY
KEEN 3917A	11 48 48.	+ 52 15		13.9	GALAXY
ZWG 269.010	11 48 48.	+ 53 43		15.5	GALAXY
ZWG 268.097	11 48 48.	+ 53 43		15.5	GALAXY
UGC 06828	11 48 48.	+ 53 43	84	15.5	GALAXY Sb
MCG+09-20-012	11 48 48.	+ 53 43	78	14.	GALAXY
ZWG 292.039	11 48 48.	+ 56 41		15.5	GALAXY
ZC 1148.8+5721	11 48 48.	+ 57 21	2960		CLUSTER OF GALAXIES
MCG-02-30-035	11 48 48.	- 11 08	72	14.5	GALAXY
KN 12.158	11 48 48.0	+ 48 11 35.			NEBULA
HN 1438	11 48 48.4	+ 27 02 55.	24		NEBULA
RNGC 3942	11 48 49.	- 11 08		14.0	GALAXY
HN 1439	11 48 49.5	+ 00 23 01.	12		NEBULA
REIZ 1698	11 48 50.	+ 48 10	30	14.8	GALAXY
LB 00618	11 48 50.	+ 54 28 48.		16.2	FAINT BLUE STAR
ARC 1400	11 48 51.	+ 55 23		17.2	RICH CLUSTER OF GALAXIES
HN 1440	11 48 51.8	+ 26 04 31.	24		NEBULA
REIZ 1699	11 48 53.	+ 22 17	12	15.7	GALAXY
KN 12.159	11 48 53.5	+ 47 15 34.			NEBULA
ZWG 040.052	11 48 54.	+ 07 41		15.7	GALAXY
ZWG 068.078	11 48 54.	+ 09 01		15.6	GALAXY
ZWG 097.165	11 48 54.	+ 16 03		15.3	GALAXY
MCG+03-30-120	11 48 54.	+ 16 04	24	15.3	GALAXY
ZWG 127.076	11 48 54.	+ 22 19		14.7	GALAXY
UGC 06829	11 48 54.	+ 22 19	60	14.7	GALAXY DBL SYS
KARA.72 305B	11 48 54.	+ 22 19	42		PART OF DOUBLE GALAXY
KARA.72 305A	11 48 54.	+ 22 19	36	14.7	PART OF DOUBLE GALAXY
MCG+04-28-075	11 48 54.	+ 22 19	12	15.5	GALAXY
ZC 1148.9+2220	11 48 54.	+ 22 20	810		CLUSTER OF GALAXIES
ZWG 127.077	11 48 54.	+ 24 09		15.1	GALAXY
UGC 06830	11 48 54.	+ 24 09	72	15.1	GALAXY SB:a-b
ZWG 157.045	11 48 54.	+ 27 55		14.8	GALAXY
MCG+09-20-013	11 48 54.	+ 54 30	36	16.	GALAXY
MCG+10-17-087	11 48 54.	+ 56 40	18	16.	GALAXY
ZC 1148.9+6332	11 48 54.	+ 63 32	1140		CLUSTER OF GALAXIES
ZCG 1148+69.1	11 48 54.	+ 69 35		18.1	COMPACT GALAXY
ZC 1148.9-0249	11 48 54.	- 02 49	3490		CLUSTER OF GALAXIES
RNGC 3926B	11 48 55.	+ 22 19		14.5	GALAXY
RNGC 3926A	11 48 55.	+ 22 19		14.5	GALAXY
TON-N 1454	11 48 56.	+ 34 38		16.	BLUE STAR
HN 1442	11 48 56.6	+ 00 12 01.	12		NEBULA
HN 1441	11 48 56.6	+ 27 54 31.			NEBULA
MCG+06-26-044	11 48 57.	+ 36 52 30.	42	15.	GALAXY
LB 02128	11 48 57.	+ 53 13 48.		15.5	FAINT BLUE STAR
HN 1443	11 48 57.1	+ 00 13 37.	12		NEBULA
REIZ 1700	11 48 59.	+ 21 45	18	15.4	GALAXY
ZWG 012.106	11 49 00.	+ 00 11		15.4	GALAXY
ZWG 012.107	11 49 00.	+ 00 13		15.6	GALAXY
ZWG 040.053	11 49 00.	+ 05 22		15.5	GALAXY
ZWG 097.166	11 49 00.	+ 15 45		14.9	GALAXY
UGC 06831	11 49 00.	+ 15 45	60	14.9	GALAXY S0
MCG+03-30-121	11 49 00.	+ 15 46	60	14.9	GALAXY
ZWG 097.167	11 49 00.	+ 16 28		15.2	GALAXY
KARA.73B 0504	11 49 00.	+ 16 28	18	15.2	ISOLATED GALAXY E
ZWG 127.078	11 49 00.	+ 21 45		15.3	GALAXY
ZWG 127.079	11 49 00.	+ 22 20		15.7	GALAXY
MCG+05-28-041	11 49 00.	+ 27 55	30	14.8	GALAXY
MCG+05-28-040	11 49 00.	+ 27 55	24	14.8	GALAXY
RNGC 3927	11 49 00.	+ 28 25			NON-EXISTENT OBJECT
ZWG 186.058	11 49 00.	+ 36 52		15.7	GALAXY
KARA.72 306A	11 49 00.	+ 36 52	48	15.7	PART OF DOUBLE GALAXY
RNGC 3930A	11 49 00.	+ 38 18		16.0	GALAXY
MCG+09-20-015	11 49 00.	+ 55 10	42	17.	GALAXY
MCG+09-20-014	11 49 00.	+ 55 46	24	17.	GALAXY
MCG+10-17-088	11 49 00.	+ 56 41	30	16.	GALAXY
MCG+10-17-089	11 49 00.	+ 58 20	39	16.	GALAXY
MCG+10-17-090	11 49 00.	+ 62 29	18	16.	GALAXY
ZCG 1149+69.4	11 49 00.	+ 69 34		17.9	COMPACT GALAXY
7ZW 423	11 49 00.	+ 69 35			COMPACT GALAXY
KN 12.160	11 49 00.4	+ 50 28 28.			NEBULA
REIZ 1701	11 49 02.	+ 49 03	30	15.1	GALAXY
KN 12.161	11 49 02.5	+ 50 27 55.			NEBULA
KN 12.162	11 49 02.9	+ 50 28 26.			NEBULA
KN 12.163	11 49 05.0	+ 47 02 44.			NEBULA
ZWG 040.054	11 49 06.	+ 06 59		15.2	GALAXY
MCG+04-28-076	11 49 06.	+ 21 15	30	14.5	GALAXY
ZWG 127.080	11 49 06.	+ 21 17		14.5	GALAXY
UGC 06832	11 49 06.	+ 21 17	42	14.5	GALAXY
REIZ 1702	11 49 06.	+ 37 39	36	15.8	GALAXY
ZWG 186.059	11 49 06.	+ 38 16		13.5	GALAXY
UGC 06833	11 49 06.	+ 38 16	270	13.5	GALAXY Sc/SBc
ZC 1149.1+4650	11 49 06.	+ 46 50	1410		CLUSTER OF GALAXIES
ZWG 243.019	11 49 06.	+ 48 57		13.1	GALAXY
MRK 190	11 49 06.	+ 48 57	35	13.	GALAXY WITH UV CONTINUUM
UGC 06834	11 49 06.	+ 48 57	96	13.1	GALAXY
MCG+08-22-019	11 49 06.	+ 48 58	60	13.0	GALAXY
7ZW 424	11 49 06.	+ 60 07			COMPACT GALAXY
RNGC 3929	11 49 07.	+ 21 17		14.5	GALAXY
HOLM 299B	11 49 08.	+ 48 44	12	15.7	PART OF MULTIPLE GALAXY
HOLM 299A	11 49 08.	+ 48 45	24	15.1	PART OF MULTIPLE GALAXY
REIZ 1703	11 49 08.	+ 48 58	36	13.0	GALAXY
HN 1444	11 49 09.0	+ 28 01 43.			NEBULA
KN 12.165	11 49 09.7	+ 48 45 17.			NEBULA
KN 12.164	11 49 10.1	+ 50 41 00.			NEBULA
KN 12.166	11 49 10.2	+ 47 03 50.			NEBULA
KN 12.167	11 49 10.6	+ 48 57 49.			NEBULA
REIZ 1704	11 49 11.	+ 21 16	12	15.0	GALAXY
RNGC 3928	11 49 11.	+ 48 58		13.0	GALAXY
KN 12.168	11 49 11.1	+ 47 09 09.			NEBULA
ZWG 012.108	11 49 12.	+ 02 16		15.5	GALAXY
ZWG 040.055	11 49 12.	+ 03 21		15.5	GALAXY
UGC 06835	11 49 12.	+ 17 08	66	16.0	GALAXY S
ZWG 097.168	11 49 12.	+ 19 38		15.7	GALAXY
ZC 1149.2+2843	11 49 12.	+ 28 43	1280		CLUSTER OF GALAXIES
ZWG 186.060	11 49 12.	+ 36 52		15.6	GALAXY
REIZ 1705	11 49 12.	+ 36 52	12	15.2	GALAXY
UGC 06836	11 49 12.	+ 36 52	60	15.6	GALAXY S
KARA.72 306B	11 49 12.	+ 36 52	42	15.6	PART OF DOUBLE GALAXY
REIZ 1706	11 49 12.	+ 38 17	156	13.0	GALAXY
HOLM 300A	11 49 12.	+ 38 17	180	12.7	PART OF MULTIPLE GALAXY
RNGC 3930	11 49 12.	+ 38 18		13.5	GALAXY
MCG+06-26-045	11 49 12.	+ 38 18	192	12.	GALAXY
ZWG 215.001	11 49 12.	+ 43 42		15.6	GALAXY
ZWG 214.033	11 49 12.	+ 43 42		15.6	GALAXY
MCG+08-22-020	11 49 12.	+ 47 09	24	16.	GALAXY
MCG+10-17-091	11 49 12.	+ 61 48	24	16.	GALAXY
ZWG 292.040	11 49 12.	+ 61 50		15.4	GALAXY
MCG+06-26-047	11 49 15.	+ 35 26	24	16.	GALAXY
MCG+06-26-046	11 49 15.	+ 36 53	48	15.	GALAXY
ZWG 097.169	11 49 18.	+ 18 49		15.7	GALAXY
UGC 06837	11 49 18.	+ 18 49	60	15.7	GALAXY Sc
MCG+04-28-077	11 49 18.	+ 21 21	42	14.7	GALAXY
ZWG 127.081	11 49 18.	+ 23 48		15.7	GALAXY
1ZW 029	11 49 18.	+ 46 02			COMPACT GALAXY
ZWG 243.020	11 49 18.	+ 47 05		15.0	GALAXY
MCG+08-22-021	11 49 18.	+ 47 06	36	16.	GALAXY
MCG+09-20-016	11 49 18.	+ 53 35	24	17.	GALAXY
ZCG 1149+69.3	11 49 18.	+ 69 33		18.4	COMPACT GALAXY
HN 1445	11 49 18.5	+ 27 50 19.	12		NEBULA
REIZ 1707	11 49 19.	+ 47 08	12	14.4	GALAXY
TON-N 1455	11 49 20.	+ 33 21		13.9	BLUE STAR
TON-N 1456	11 49 20.	+ 34 19		16.8	BLUE STAR
KN 12.169	11 49 20.2	+ 50 56 44.			NEBULA
KN 12.170	11 49 20.8	+ 47 05 49.			NEBULA
HN 1446	11 49 21.5	- 02 22 17.	30		NEBULA
KN 12.171	11 49 22.5	+ 47 06 01.			NEBULA
KN 12.172	11 49 23.6	+ 47 09 51.			NEBULA
ZC 1149.4+0821	11 49 24.	+ 08 21	1480		CLUSTER OF GALAXIES
ZWG 097.170	11 49 24.	+ 17 05		14.2	GALAXY
UGC 06839	11 49 24.	+ 17 05	84	14.2	GALAXY S
MCG+03-30-122	11 49 24.	+ 17 06	48	14.2	GALAXY
ZWG 127.082	11 49 24.	+ 21 24		14.7	GALAXY
ZC 1149.4+2734	11 49 24.	+ 27 34	1280		CLUSTER OF GALAXIES
ZWG 157.046	11 49 24.	+ 30 36		15.4	GALAXY
ZC 1149.4+3538	11 49 24.	+ 35 38	940		CLUSTER OF GALAXIES
ZC 1149.4+4623	11 49 24.	+ 46 23	1010		CLUSTER OF GALAXIES
ZWG 243.021	11 49 24.	+ 47 09		15.5	GALAXY
MCG+08-22-022	11 49 24.	+ 47 10	30	16.	GALAXY
ZC 1149.4+4848	11 49 24.	+ 48 48	610		CLUSTER OF GALAXIES
ZWG 269.011	11 49 24.	+ 50 35		15.5	GALAXY
MCG+09-20-018	11 49 24.	+ 52 15	30	16.	GALAXY
MCG+09-20-017	11 49 24.	+ 53 36	24	17.	GALAXY
ZCG 1149+69.1	11 49 24.	+ 69 31		18.2	COMPACT GALAXY
ZWG 012.109	11 49 24.	- 02 22		15.4	GALAXY
UGC 06838	11 49 24.	- 02 22	66	15.4	GALAXY
KARA.73B 0505	11 49 24.	- 02 22	42	15.4	ISOLATED GALAXY E
RNGC 3933	11 49 25.	+ 17 05		14.0	GALAXY
ARC 1401	11 49 28.	+ 37 33		17.0	RICH CLUSTER OF GALAXIES
HN 1447	11 49 28.5	+ 26 01 55.	18		NEBULA
REIZ 1708	11 49 29.	+ 21 16	12	15.1	GALAXY
HOLM 300B	11 49 29.	+ 38 18	12	14.9	PART OF MULTIPLE GALAXY
LB 02129	11 49 29.	+ 56 04 48.		16.6	FAINT BLUE STAR
HN 1448	11 49 29.3	+ 28 23 31.	18		NEBULA
ZWG 040.056	11 49 30.	+ 06 41		14.8	GALAXY
MCG+01-30-018	11 49 30.	+ 06 41	30	14.8	GALAXY
ZWG 068.079	11 49 30.	+ 14 10		15.3	GALAXY
KARA.73B 0506	11 49 30.	+ 14 10	36	15.3	ISOLATED GALAXY S
REIZ 1709	11 49 30.	+ 28 24	12	15.7	GALAXY
MCG+09-20-019	11 49 30.	+ 52 22 30.	72	14.	GALAXY
ZWG 269.012	11 49 30.	+ 52 23		15.6	GALAXY
UGC 06840	11 49 30.	+ 52 23	180	15.6	GALAXY DWRF SP
SEY 133	11 49 30.	+ 52 23 19.		14.7	FAINT GALAXY
ZC 1149.5+6040	11 49 30.	+ 60 40	1480		CLUSTER OF GALAXIES
LB 02130	11 49 30.	+ 61 13 48.		17.0	FAINT BLUE STAR
MCG+10-17-092	11 49 30.	+ 61 51	39	16.	GALAXY
ZCG 1149+69.2	11 49 30.	+ 69 32		18.6	COMPACT GALAXY
SEY 134	11 49 32.	+ 52 16 49.		15.3	FAINT GALAXY
KN 12.173	11 49 34.4	+ 48 53 59.			NEBULA
ZWG 097.171	11 49 36.	+ 17 08		15.0	GALAXY
UGC 06841	11 49 36.	+ 17 08	78	15.0	GALAXY
MCG+03-30-123	11 49 36.	+ 17 08	48	15.0	GALAXY
ZC 1149.6+1746	11 49 36.	+ 17 46	1340		CLUSTER OF GALAXIES
ZWG 097.172	11 49 36.	+ 18 56		15.7	GALAXY
TON-N 0596	11 49 36.	+ 29 22		15.2	BLUE STAR
MCG+06-26-048	11 49 36.	+ 35 05	42	17.	GALAXY
ZC 1149.6+3742	11 49 36.	+ 37 42	3430		CLUSTER OF GALAXIES

OBJECT NAME	RIGHT ASCEN.	DECLINATION	DIAM.	MAGN.	TYPE OF OBJECT
SHAH 069	11 49 36.	+ 40 23	132	17.5	GROUP OF COMPACT GALAXIES
ZC 1149.6+6153	11 49 36.	+ 61 53	1550		CLUSTER OF GALAXIES
RNGC 3934	11 49 37.	+ 17 08		15.0	GALAXY
REIZ 1710	11 49 38.	+ 48 54	24	13.2	GALAXY
ZWG 040.057	11 49 42.	+ 04 23		15.4	GALAXY
ZWG 068.080	11 49 42.	+ 10 53		15.4	GALAXY
MCG+04-28-078	11 49 42.	+ 21 20	30	15.1	GALAXY
UGC 06842	11 49 42.	+ 35 04	72	16.5	GALAXY S
MCG+10-17-093	11 49 42.	+ 62 12	42	16.	GALAXY
ZC 1149.7-0200	11 49 42.	- 02 00	1480		CLUSTER OF GALAXIES
ARC 1403	11 49 44.	+ 28 40		17.6	RICH CLUSTER OF GALAXIES
MCG+06-26-049	11 49 45.	+ 32 40 30.	66	14.5	GALAXY
ARC 1404	11 49 46.	- 02 33		16.6	RICH CLUSTER OF GALAXIES
HN 1449	11 49 46.5	+ 25 22 07.	30		NEBULA
ZWG 040.058	11 49 48.	+ 04 15		15.6	GALAXY
ZWG 127.083	11 49 48.	+ 21 24		15.1	GALAXY
ZWG 127.084	11 49 48.	+ 23 54		15.6	GALAXY
MRK 642	11 49 48.	+ 23 54	9	16.	GALAXY WITH UV CONTINUUM
ZWG 186.061	11 49 48.	+ 32 41		14.0	GALAXY
RNGC 3935	11 49 48.	+ 32 41		14.0	GALAXY
UGC 06843	11 49 48.	+ 32 41	72	14.0	GALAXY S
MRK 641	11 49 48.	+ 35 10	9	16.5	GALAXY WITH UV CONTINUUM
ZC 1149.8+4806	11 49 48.	+ 48 06	2020		CLUSTER OF GALAXIES
MCG+10-17-094	11 49 48.	+ 60 37	27	16.	GALAXY
ZWG 334.053	11 49 48.	+ 70 10		15.5	GALAXY
UGC 06844	11 49 48.	+ 70 10	66	15.5	GALAXY Sa-b
MCG-04-28-004	11 49 48.	- 26 38	210	12.5	GALAXY
IC 2968	11 49 49.	+ 20 54 35.			NONSTELLAR OBJECT
REIZ 1711	11 49 49.	+ 45 57	18	14.7	GALAXY
TON-N 1458	11 49 50.	+ 34 46		15.	BLUE STAR
TON-N 1457	11 49 50.	+ 34 49		16.7	BLUE STAR
MCG-02-30-036	11 49 51.	- 12 22	60	15.	GALAXY
KN 12.174	11 49 51.7	+ 50 32 24.			NEBULA
KN 12.175	11 49 52.7	+ 48 44 20.			NEBULA
REIZ 1712	11 49 53.	+ 20 36	12	15.3	GALAXY
SN 1971B	11 49 53.	+ 49 24		17.5	SUPERNOVA
ARC 1402	11 49 53.	+ 60 42		17.2	RICH CLUSTER OF GALAXIES
ZC 1149.9+0444	11 49 54.	+ 04 44	3160		CLUSTER OF GALAXIES
ZWG 040.059	11 49 54.	+ 08 14		15.5	GALAXY
ZC 1149.9+1356	11 49 54.	+ 13 56	340		CLUSTER OF GALAXIES
ZWG 127.085	11 49 54.	+ 20 55		15.5	GALAXY
MCG+04-28-079	11 49 54.	+ 23 50 30.	48	14.8	GALAXY
MCG+04-28-080	11 49 54.	+ 24 33	66	15.4	GALAXY
REIZ 1713	11 49 54.	+ 32 41	54	13.5	GALAXY
ZC 1149.9+4441	11 49 54.	+ 44 41	470		CLUSTER OF GALAXIES
ZWG 243.022	11 49 54.	+ 48 44		15.1	GALAXY
RNGC 3932	11 49 54.	+ 48 44		15.0	GALAXY
MCG+08-22-023	11 49 54.	+ 48 44 30.	60	15.	GALAXY
MCG+09-20-020	11 49 54.	+ 54 30	24	16.	GALAXY
MCG-01-30-040	11 49 54.	- 03 34	63	14.	GALAXY
KN 12.176	11 49 54.2	+ 45 56 22.			NEBULA
REIZ 1714	11 49 55.	+ 48 45	36	14.0	GALAXY
KN 12.177	11 49 55.4	+ 50 44 16.			NEBULA
ARC 1405	11 49 56.	+ 27 34		17.6	RICH CLUSTER OF GALAXIES
REIZ 1715	11 49 56.	+ 49 26	36	14.5	GALAXY
IC 2969	11 49 56.	- 03 26 56.			NONSTELLAR OBJECT
RNGC 3936	11 49 56.	- 26 37		13.0	GALAXY
MCG+08-22-024	11 49 57.	+ 45 46 30.	42	16.	GALAXY
REIZ 1716	11 49 58.	+ 13 45	30	15.0	GALAXY
LB 02131	11 49 58.	+ 53 19 24.		16.6	FAINT BLUE STAR
KN 12.178	11 49 58.2	+ 49 25 13.			NEBULA
REIZ 1717	11 49 59.	+ 23 47	12	14.9	GALAXY
VDB.66G 100	11 50	+ 52 23	70		DWARF GALAXY
ZC 1150.0+0339	11 50 00.	+ 03 39	1950		CLUSTER OF GALAXIES
ZWG 068.081	11 50 00.	+ 13 28		15.2	GALAXY
ZWG 068.082	11 50 00.	+ 13 45		15.2	GALAXY
UGC 06845	11 50 00.	+ 13 45	66	15.2	GALAXY Sb?
MCG+03-30-124	11 50 00.	+ 20 24	72	15.7	GALAXY
ZWG 097.173	11 50 00.	+ 20 25		15.7	GALAXY
ZC 1150.0+2333	11 50 00.	+ 23 33	940		CLUSTER OF GALAXIES
ZWG 127.086	11 50 00.	+ 23 53		14.8	GALAXY
UGC 06846	11 50 00.	+ 23 53	72	14.8	GALAXY E?
ZWG 127.087	11 50 00.	+ 24 35		15.4	GALAXY
UGC 06847	11 50 00.	+ 24 35	102	15.4	GALAXY Sb-c
KARA.73B 0507	11 50 00.	+ 24 35	78	15.4	ISOLATED GALAXY S
ZC 1150.0+3438	11 50 00.	+ 34 38	940		CLUSTER OF GALAXIES
UGC 06848	11 50 00.	+ 35 39	60	17.	GALAXY
REIZ 1718	11 50 00.	+ 38 04	36	15.4	GALAXY
MCG+06-26-050	11 50 00.	+ 38 05	24	15.	GALAXY
MCG+08-22-025	11 50 00.	+ 49 25 30.	42	16.	GALAXY
ZWG 243.023	11 50 00.	+ 50 18		15.7	GALAXY
RNGC 3924	11 50 00.	+ 50 18		15.5	GALAXY
UGC 06849	11 50 00.	+ 50 18	108	15.7	GALAXY DWRF SP
ZWG 012.110	11 50 00.	- 03 23		15.3	GALAXY
MCG-01-30-041	11 50 00.	- 04 54	36	15.	GALAXY
MCG-01-30-042	11 50 00.	- 08 30	30	15.	GALAXY
MCG-05-28-013	11 50 00.	- 32 34	42	15.5	GALAXY
KN 12.179	11 50 01.8	+ 50 19 08.			NEBULA
LB 02132	11 50 02.	+ 50 56 36.		15.0	FAINT BLUE STAR
MCG+04-28-081	11 50 03.	+ 20 53	24	14.0	GALAXY
MCG+08-22-026	11 50 03.	+ 50 19	102	15.	GALAXY
LB 00619	11 50 03.	+ 54 39 42.		17.6	FAINT BLUE STAR
HN 1451	11 50 03.9	- 02 12 05.	18		NEBULA
HN 1450	11 50 04.1	+ 25 21 07.	12		NEBULA
REIZ 1719	11 50 05.	+ 23 45	12	15.7	GALAXY
REIZ 1720	11 50 05.	+ 24 36	60	15.1	GALAXY
KN 12.180	11 50 05.9	+ 49 15 49.			NEBULA
VV 105C	11 50 06.	+ 02 03	6	17.	INTERACTING GALAXY
VV 105B	11 50 06.	+ 02 03	15	17.	INTFRACTING GALAXY
VV 105A	11 50 06.	+ 02 03	15	16.	INTERACTING GALAXY
ZC 1150.1+1315	11 50 06.	+ 13 15	2490		CLUSTER OF GALAXIES
ZWG 097.174	11 50 06.	+ 18 54		15.7	GALAXY
ZWG 127.088	11 50 06.	+ 20 55		14.0	GALAXY
UGC 06851	11 50 06.	+ 20 55	162	14.0	GALAXY E-S0
MCG+04-28-082	11 50 06.	+ 21 14 30.	84	14.3	GALAXY
ZWG 127.089	11 50 06.	+ 21 17		14.3	GALAXY
UGC 06852	11 50 06.	+ 21 17	102	14.3	GALAXY E
ZWG 157.047	11 50 06.	+ 29 37		14.9	GALAXY
UGC 06853	11 50 06.	+ 29 37	72	14.9	GALAXY S0
ZC 1150.1+3603	11 50 06.	+ 36 03	740		CLUSTER OF GALAXIES
ZC 1150.1+4245	11 50 06.	+ 42 45	940		CLUSTER OF GALAXIES
ZWG 012.111	11 50 06.	- 02 11		14.1	GALAXY
UGC 06850	11 50 06.	- 02 11	42	14.1	GALAXY PECULR
MCG+00-30-032	11 50 06.	- 02 11	24	14.1	GALAXY
RNGC 3937	11 50 07.	+ 20 55		14.0	GALAXY
RNGC 3940	11 50 07.	+ 21 17		14.5	GALAXY
HN 1452	11 50 07.8	+ 02 02 07.	30		NEBULA
MCG+04-28-083	11 50 09.	+ 20 30	30	16.	GALAXY

OBJECT NAME	RIGHT ASCEN.	DECLINATION	DIAM.	MAGN.	TYPE OF OBJECT
REIZ 1721	11 50 10.	+ 02 01	30	15.2	GALAXY
REIZ 1722	11 50 11.	+ 20 25	18	15.2	GALAXY
REIZ 1723	11 50 11.	+ 20 55	12	15.1	GALAXY
REIZ 1724	11 50 11.	+ 29 35	24	14.4	GALAXY
HN 1453	11 50 11.9	+ 29 36 43.			NEBULA
ZWG 012.112	11 50 12.	+ 02 01		14.4	GALAXY
UGC 06854	11 50 12.	+ 02 01	66	14.4	GALAXY SB
KARA.72 307B	11 50 12.	+ 02 01	30		PART OF DOUBLE GALAXY
KARA.72 307A	11 50 12.	+ 02 01	48	14.4	PART OF DOUBLE GALAXY
MCG+00-30-033	11 50 12.	+ 02 01	66	14.4	GALAXY
KARA.73B 0508	11 50 12.	+ 02 01	60	14.4	ISOLATED GALAXY S
UGC 06855	11 50 12.	+ 23 45	72	16.0	GALAXY S0
MCG+05-28-042	11 50 12.	+ 29 37 30.	48	14.9	GALAXY
ZWG 215.002	11 50 12.	+ 44 23		11.0	GALAXY
ZWG 214.034	11 50 12.	+ 44 23		11.0	GALAXY
UGC 06856	11 50 12.	+ 44 23	324	11.0	GALAXY Sc
RNGC 3938	11 50 12.	+ 44 24		11.0	GALAXY
SN 1964L	11 50 12.	+ 44 24		13.3	SUPERNOVA
MCG+07-25-001	11 50 12.	+ 44 24	300	9.	GALAXY
ZC 1150.2+4705	11 50 12.	+ 47 05	1010		CLUSTER OF GALAXIES
7ZW 425	11 50 12.	+ 64 02			COMPACT GALAXY
REIZ 1725	11 50 13.	+ 44 24	270	11.5	GALAXY
REIN 2.136	11 50 13.76	+ 44 23 55.7			NEBULA
KN 12.181	11 50 13.9	+ 44 23 56.			NEBULA
PATH 1.412	11 50 15.	- 00 06	14		NEBULA
SN 1961U	11 50 16.	+ 44 25		13.7	SUPERNOVA
REIZ 1726	11 50 17.	+ 20 31	12	15.4	GALAXY
REIZ 1727	11 50 17.	+ 21 16	18	14.9	GALAXY
ZWG 127.090	11 50 18.	+ 20 46		14.7	GALAXY
MCG+04-28-084	11 50 18.	+ 20 46	60	14.7	GALAXY
TON-N 0597	11 50 18.	+ 33 25		16.0	BLUE STAR
ZWG 186.062	11 50 18.	+ 37 15		11.3	GALAXY
REIZ 1728	11 50 18.	+ 37 15	120	11.3	GALAXY
UGC 06857	11 50 18.	+ 37 15	216	11.3	GALAXY S0
RNGC 3941	11 50 18.	+ 37 16		11.5	GALAXY
MCG+06-26-051	11 50 18.	+ 37 17	102	11.	GALAXY
ZWG 292.041	11 50 18.	+ 61 30		15.1	GALAXY
UGC 06858	11 50 18.	+ 61 30	96	15.1	GALAXY Sc
RNGC 3943	11 50 19.	+ 20 46		14.5	GALAXY
REIZ 1729	11 50 19.	+ 45 44	18	14.4	GALAXY
REIZ 1730	11 50 23.	+ 20 45	12	15.7	GALAXY
ZC 1150.4+0552	11 50 24.	+ 05 52	2690		CLUSTER OF GALAXIES
ZC 1150.4+2015	11 50 24.	+ 20 15	1750		CLUSTER OF GALAXIES
MCG+09-20-021	11 50 24.	+ 53 24	36	16.	GALAXY
MCG-01-30-043	11 50 24.	- 04 08	90	14.	GALAXY
KN 12.182	11 50 25.1	+ 45 44 01.			NEBULA
HN 1454	11 50 26.1	+ 25 06 31.	12		NEBULA
LB 02133	11 50 27.	+ 51 34 48.		16.1	FAINT BLUE STAR
KN 12.183	11 50 28.4	+ 51 17 12.			NEBULA
REIZ 1731	11 50 29.	+ 26 29	30	14.6	GALAXY
MCG+04-28-086	11 50 30.	+ 25 41 30.	48	15.1	GALAXY
MCG+04-28-085	11 50 30.	+ 26 29	30	14.3	GALAXY
ZWG 157.048	11 50 30.	+ 26 30		14.3	GALAXY
ZWG 127.091	11 50 30.	+ 26 30		14.3	GALAXY
UGC 06859	11 50 30.	+ 26 30	84	14.3	GALAXY E-S0
ZWG 186.063	11 50 30.	+ 35 18		14.7	GALAXY
ZC 1150.5+4202	11 50 30.	+ 42 02	610		CLUSTER OF GALAXIES
ZWG 292.042	11 50 30.	+ 60 56		11.6	GALAXY
UGC 06860	11 50 30.	+ 60 56	348	11.6	GALAXY SB0
MCG+10-17-095	11 50 30.	+ 61 30	78	14.	GALAXY
ZWG 334.054	11 50 30.	+ 70 43		15.3	GALAXY
MCG-06-26-012	11 50 30.	- 36 21	150	13.5	GALAXY
RNGC 3944	11 50 31.	+ 26 30		14.5	GALAXY
HN 1455	11 50 31.0	+ 27 35 07.	18		NEBULA
KN 12.184	11 50 31.9	+ 50 45 15.			NEBULA
REIZ 1732	11 50 32.	+ 53 25	36	15.3	GALAXY
RNGC 3939	11 50 32.	+ 75 23			NON-EXISTENT OBJECT
LB 02134	11 50 34.	+ 54 52 36.		16.6	FAINT BLUE STAR
KN 12.185	11 50 34.1	+ 50 32 51.			NEBULA
REIZ 1733	11 50 35.	+ 23 45	18	15.2	GALAXY
RNGC 3945	11 50 35.	+ 60 57		12.0	GALAXY
HN 1456	11 50 35.9	+ 25 43 13.	36		NEBULA
ZWG 040.060	11 50 36.	+ 06 25		15.6	GALAXY
ZWG 127.092	11 50 36.	+ 20 57		15.3	GALAXY
MCG+04-28-087	11 50 36.	+ 23 43	39	15.0	GALAXY
ZWG 127.093	11 50 36.	+ 25 43		15.1	GALAXY
UGC 06861	11 50 36.	+ 25 43	60	15.1	GALAXY S
ZC 1150.6+4920	11 50 36.	+ 49 20	5170		CLUSTER OF GALAXIES
MCG+09-20-022	11 50 36.	+ 51 16	30	16.	GALAXY
MCG+09-20-023	11 50 36.	+ 52 27	24	17.	GALAXY
MCG+09-20-024	11 50 36.	+ 54 29	24	16.	GALAXY
7ZW 426	11 50 36.	+ 70 43			COMPACT GALAXY
MCG-05-28-014	11 50 36.	- 32 17	30	15.	GALAXY
MCG-06-26-013	11 50 36.	- 38 51	54	14.	GALAXY
ACK 293+10.1	11 50 36.	- 50 34			PLANETARY NEBULA
IC 2970	11 50 37.	- 22 50 44.			THREE STARS
HN 1457	11 50 37.2	+ 28 18 37.	12		NEBULA
LB 02135	11 50 38.	+ 50 05 18.		14.7	FAINT BLUE STAR
HN 1458	11 50 38.6	+ 25 51 43.	12		NEBULA
KN 12.186	11 50 38.8	+ 44 59 12.			NEBULA
BC PKS1150+09	11 50 38.9	+ 09 30 53.		17.	QUASI-STELLAR OBJECT
REIZ 1734	11 50 39.	+ 60 57	60	12.4	GALAXY
ARC 1406	11 50 39.	+ 68 10		17.2	RICH CLUSTER OF GALAXIES
KN 12.188	11 50 39.6	+ 45 18 00.			NEBULA
KN 12.187	11 50 39.6	+ 47 36 53.			NEBULA
REIZ 1735	11 50 41.	+ 20 56	12	15.2	GALAXY
HN 1459	11 50 41.2	+ 28 43 43.	18		NEBULA
UGC 06862	11 50 42.	+ 11 56	90	16.0	GALAXY Sc
ZC 1150.7+1821	11 50 42.	+ 18 21	470		CLUSTER OF GALAXIES
MCG+04-28-088	11 50 42.	+ 21 00	78	14.2	GALAXY
ZWG 127.094	11 50 42.	+ 23 45		15.0	GALAXY
MCG+08-22-027	11 50 42.	+ 47 37	30	16.	GALAXY
ZC 1150.7+6807	11 50 42.	+ 68 07	870		CLUSTER OF GALAXIES
MCG-02-30-037	11 50 42.	- 12 58	54	15.	GALAXY
KN 12.189	11 50 42.7	+ 49 35 08.			NEBULA
REIZ 1736	11 50 43.	+ 45 12	18	14.9	GALAXY
HN 1460	11 50 46.6	+ 01 58 43.	12		NEBULA
REIZ 1737	11 50 47.	+ 20 23	12	16.0	GALAXY
REIZ 1739	11 50 47.	+ 21 02	48	15.0	GALAXY
REIZ 1738	11 50 47.	+ 21 18	24	15.5	GALAXY
KN 12.190	11 50 47.0	+ 45 11 48.			NEBULA
ZWG 127.095	11 50 48.	+ 21 02		14.2	GALAXY
UGC 06863	11 50 48.	+ 21 02	84	14.2	GALAXY SBb
MCG+04-28-089	11 50 48.	+ 21 16	30	15.5	GALAXY
ZWG 127.096	11 50 48.	+ 21 19		15.5	GALAXY
ZWG 127.097	11 50 48.	+ 23 21		15.5	GALAXY
BC 4C49.22	11 50 48.	+ 49 47 55.			QUASI-STELLAR OBJECT
LB 02136	11 50 48.	+ 49 48 12.		16.2	FAINT BLUE STAR

OBJECT NAME	RIGHT ASCEN.	DECLINATION	DIAM.	MAGN.	TYPE OF OBJECT
SHB 188	11 50 48.	+ 49 50 00.		16.1	QUASI-STELLAR OBJECT
ZC 1150.8+6057	11 50 48.	+ 60 57	3970		CLUSTER OF GALAXIES
MCG+10-17-096	11 50 48.	+ 60 58	306	11.7	GALAXY
ZC 1150.8-0130	11 50 48.	- 01 30	1410		CLUSTER OF GALAXIES
RNGC 3947	11 50 49.	+ 21 02		14.0	GALAXY
RNGC 3946	11 50 49.	+ 21 19		15.5	GALAXY
SN 1972C	11 50 50.	+ 21 02		16.0	SUPERNOVA
LB 02137	11 50 50.	+ 51 07 06.		16.9	FAINT BLUE STAR
IC 0743	11 50 51.	- 12 58 50.			NONSTELLAR OBJECT
IC 2971	11 50 53.	+ 30 57 52.			NONSTELLAR OBJECT
ZC 1150.9+0820	11 50 54.	+ 08 20	540		CLUSTER OF GALAXIES
TON-N 0598	11 50 54.	+ 29 11		15.4	BLUE STAR
ZWG 157.049	11 50 54.	+ 30 58		15.7	GALAXY
ZC 1150.9+3609	11 50 54.	+ 36 09	3630		CLUSTER OF GALAXIES
ZC 1150.9+6310	11 50 54.	+ 63 10	1080		CLUSTER OF GALAXIES
MCG-05-28-015	11 50 54.	- 28 14	300	13.5	GALAXY
RNGC 3948	11 50 55.	+ 21 14			NON-EXISTENT OBJECT
HELW 263	11 50 56.	- 28 16 41.			NEBULA
KN 12.191	11 50 57.3	+ 47 35 59.			NEBULA
ARC 1407	11 50 58.	- 01 29		17.0	RICH CLUSTER OF GALAXIES
REIZ 1740	11 50 59.	+ 20 52	6	15.8	GALAXY
REIZ 1741	11 50 59.	+ 23 40	42	14.5	GALAXY
ZWG 068.083	11 51 00.	+ 11 09		15.3	GALAXY
UGC 06864	11 51 00.	+ 11 09	72	15.3	GALAXY Sb-c
ZWG 068.084	11 51 00.	+ 12 14		15.6	GALAXY
ZC 1151.0+1510	11 51 00.	+ 15 10	940		CLUSTER OF GALAXIES
ZC 1151.0+1805	11 51 00.	+ 18 05	1340		CLUSTER OF GALAXIES
MCG+04-28-091	11 51 00.	+ 21 08	18	14.4	GALAXY
MCG+04-28-090	11 51 00.	+ 23 38	42	14.5	GALAXY
ZZW 055	11 51 00.	+ 43 44			COMPACT GALAXY
ZWG 215.003	11 51 00.	+ 43 44		14.7	GALAXY
ZWG 214.035	11 51 00.	+ 43 44		14.7	GALAXY
UGC 06865	11 51 00.	+ 43 44	78	14.7	GALAXY DBL SYS
MCG+10-17-097	11 51 00.	+ 59 59 30.	36	15.	GALAXY
HN 1461	11 51 01.2	+ 25 47 31.	6		NEBULA
TON-N 1459	11 51 02.	+ 33 31		17.	BLUE STAR
MCG+07-25-002	11 51 03.	+ 43 44	66	14.5	GALAXY
REIZ 1744	11 51 03.	- 03 44	54	13.0	GALAXY
ARP 062	11 51 04.	+ 43 43			PECULIAR GALAXY
SEY 135	11 51 04.	+ 43 44 18.		14.6	FAINT GALAXY
REIZ 1742	11 51 05.	+ 22 38	30	15.2	GALAXY
KN 12.192	11 51 05.0	+ 48 09 58.			NEBULA
REIN 2.137	11 51 05.29	+ 48 08 13.0			NEBULA
KN 12.193	11 51 05.5	+ 48 08 10.			NEBULA
KN 12.194	11 51 05.9	+ 46 29 28.			NEBULA
ZWG 097.175	11 51 06.	+ 15 27		15.5	GALAXY
ZWG 127.098	11 51 06.	+ 21 10		14.4	GALAXY
UGC 06866	11 51 06.	+ 21 10	84	14.4	GALAXY COMPACT
ZWG 127.099	11 51 06.	+ 23 40		14.5	GALAXY
UGC 06867	11 51 06.	+ 23 40	72	14.5	GALAXY S?
ZC 1151.1+2809	11 51 06.	+ 28 09	3430		CLUSTER OF GALAXIES
UGC 06868	11 51 06.	+ 29 33	60	16.0	GALAXY Sb-c
ZC 1151.1+3106	11 51 06.	+ 31 06	1410		CLUSTER OF GALAXIES
ZWG 186.064	11 51 06.	+ 35 26		15.5	GALAXY
ZC 1151.1+4309	11 51 06.	+ 43 09	740		CLUSTER OF GALAXIES
ZC 1151.1+4613	11 51 06.	+ 46 13	1080		CLUSTER OF GALAXIES
ZWG 243.024	11 51 06.	+ 46 29		15.2	GALAXY
MRK 042	11 51 06.	+ 46 29	12	16.	GALAXY WITH UV CONTINUUM
KW 08	11 51 06.	+ 46 29	44		SEYFERT GALAXY
VVI 50	11 51 06.	+ 46 29	20	16.24	SEYFERT GALAXY
MCG+08-22-028	11 51 06.	+ 46 29 30.	30	16.	GALAXY
ZWG 243.025	11 51 06.	+ 48 08		10.9	GALAXY
RNGC 3949	11 51 06.	+ 48 08		12.0	GALAXY
UGC 06869	11 51 06.	+ 48 08	168	10.9	GALAXY S
MCG+08-22-029	11 51 06.	+ 48 09	180	11.3	GALAXY
HOLM 301B	11 51 06.	+ 48 10	12	15.7	PART OF MULTIPLE GALAXY
RNGC 3950	11 51 06.	+ 48 12		15.5	GALAXY
MCG+08-22-030	11 51 06.	+ 48 12	9	15.7	GALAXY
ZWG 269.013	11 51 06.	+ 52 36		10.8	GALAXY
UGC 06870	11 51 06.	+ 52 36	390	10.8	GALAXY SBb
MRK 191	11 51 06.	+ 70 40	20	15.5	GALAXY WITH UV CONTINUUM
IC 2972	11 51 06.	- 03 41 56.			NONSTELLAR OBJECT
MCG-01-30-044	11 51 06.	- 03 42	54	14.	GALAXY
MCG-02-30-038	11 51 06.	- 15 08	30	15.	GALAXY
HN 1462	11 51 06.1	+ 25 51 55.	12		NEBULA
RNGC 3954	11 51 07.	+ 21 10		14.5	GALAXY
RNGC 3951	11 51 07.	+ 23 40		14.5	GALAXY
REIZ 1745	11 51 07.	+ 46 29	24	14.3	GALAXY
REIZ 1743	11 51 07.	+ 48 08	90	11.3	GALAXY
HOLM 301A	11 51 07.	+ 48 08	132	11.5	PART OF MULTIPLE GALAXY
RNGC 3952	11 51 07.	- 03 43		13.0	GALAXY
VV 286B	11 51 09.	+ 43 45	12	17.	INTERACTING GALAXY
VV 286A	11 51 09.	+ 43 45	48	15.	INTERACTING GALAXY
REIZ 1747	11 51 11.	+ 21 09	12	14.8	GALAXY
ZWG 068.085	11 51 12.	+ 10 41		14.6	GALAXY
UGC 06871	11 51 12.	+ 10 41	60	14.6	GALAXY SBO
MCG+02-30-043	11 51 12.	+ 10 41	60	14.6	GALAXY
ARC 1408	11 51 12.	+ 15 40		16.9	RICH CLUSTER OF GALAXIES
ZC 1151.2+1545	11 51 12.	+ 15 45	2890		CLUSTER OF GALAXIES
ZWG 097.176	11 51 12.	+ 20 01		15.5	GALAXY
ZWG 186.066	11 51 12.	+ 33 38		14.5	GALAXY
UGC 06872	11 51 12.	+ 33 38	90	14.5	GALAXY SBc
ZWG 186.065	11 51 12.	+ 37 26		15.6	GALAXY
RNGC 3953	11 51 12.	+ 52 36		11.0	GALAXY
MCG+09-20-025	11 51 12.	+ 55 38 30.	30	16.	GALAXY
SN 1958D	11 51 12.	- 30 28		19.5	SUPERNOVA
KN 12.195	11 51 12.8	+ 48 09 51.			NEBULA
HN 1463	11 51 12.9	+ 25 25 55.	18		NEBULA
REIZ 1748	11 51 13.	+ 50 30	54	14.3	GALAXY
REIZ 1746	11 51 14.	+ 52 37	420	11.6	GALAXY
BC M1151+102	11 51 14.4	+ 10 12 35.		19.	QUASI-STELLAR OBJECT
SHB 189	11 51 14.4	+ 10 12 35.		19.	QUASI-STELLAR OBJECT
MCG+06-26-052	11 51 15.	+ 33 38	78	14.	GALAXY
IC 2973	11 51 15.	+ 33 38 40.			NONSTELLAR OBJECT
MCG-01-30-045	11 51 15.	- 04 52	126	13.5	GALAXY
REIZ 1749	11 51 15.	- 04 54	120	13.7	GALAXY
REIZ 1750	11 51 17.	+ 25 26	30	15.6	GALAXY
LB 02138	11 51 17.	+ 56 22 36.		15.2	FAINT BLUE STAR
ZWG 068.086	11 51 18.	+ 13 34		15.6	GALAXY
MCG+04-28-093	11 51 18.	+ 20 49 30.	48	14.9	GALAXY
UGC 06873	11 51 18.	+ 21 02	60	16.0	GALAXY S
MCG+04-28-092	11 51 18.	+ 25 57	18	14.9	GALAXY
REIZ 1751	11 51 18.	+ 33 38	66	13.7	GALAXY
ZWG 269.014	11 51 18.	+ 50 29		15.6	GALAXY
ZWG 243.026	11 51 18.	+ 50 29		15.6	GALAXY
UGC 06874	11 51 18.	+ 50 29	66	15.6	GALAXY S
MCG+09-20-026	11 51 18.	+ 52 36	420	10.7	GALAXY
MCG-03-30-015	11 51 18.	- 19 52 30.	90	15.	GALAXY
FATH 1.413	11 51 19.	- 00 14	840		NEBULA
TON-N 1460	11 51 20.	+ 35 55		15.5	BLUE STAR
HN 1464	11 51 20.6	+ 25 44 49.	12		NEBULA
LB 02141	11 51 21.	+ 54 58 12.		14.3	FAINT BLUE STAR
IC 2974	11 51 21.	- 04 51 02.			NONSTELLAR OBJECT
KN 12.196	11 51 21.1	+ 50 27 35.			NEBULA
HN 1465	11 51 23.1	+ 25 58 19.			NEBULA
ZWG 040.061	11 51 24.	+ 06 37		15.3	GALAXY
UGC 06875	11 51 24.	+ 06 37	72	15.3	GALAXY S
ZWG 068.087	11 51 24.	+ 09 54		15.5	GALAXY
ZWG 068.088	11 51 24.	+ 12 29		15.4	GALAXY
ZWG 127.100	11 51 24.	+ 20 52		14.9	GALAXY
UGC 06876	11 51 24.	+ 20 52	66	14.9	GALAXY SB:a-b
ZWG 127.101	11 51 24.	+ 25 58		14.9	GALAXY
ZWG 243.027	11 51 24.	+ 44 55		15.5	GALAXY
MCG-03-30-017	11 51 24.	- 19 18	156	13.	GALAXY
MCG-03-30-016	11 51 24.	- 20 18	180	12.5	GALAXY
MCG-04-28-005	11 51 24.	- 22 53 30.	162	12.	GALAXY
KN 12.197	11 51 26.1	+ 44 55 13.			NEBULA
ARC 1410	11 51 27.	+ 38 00		17.6	RICH CLUSTER OF GALAXIES
MCG+08-22-031	11 51 27.	+ 50 27	60	16.	GALAXY
REIZ 1752	11 51 29.	+ 20 15	24	15.8	GALAXY
REIZ 1753	11 51 29.	+ 20 51	18	15.0	GALAXY
ARC 1409	11 51 29.	+ 49 21		17.2	RICH CLUSTER OF GALAXIES
ZWG 097.177	11 51 30.	+ 14 46		15.6	GALAXY
ZWG 098.001	11 51 30.	+ 20 14		15.6	GALAXY
ZWG 097.178	11 51 30.	+ 20 14		15.6	GALAXY
IC 0744	11 51 30.	+ 23 28 22.			NONSTELLAR OBJECT
ZWG 127.102	11 51 30.	+ 23 29		15.7	GALAXY
MCG+08-22-032	11 51 30.	+ 44 54	60	15.	GALAXY
LB 02139	11 51 31.	+ 56 34 36.		16.0	FAINT BLUE STAR
RNGC 3955	11 51 32.	- 22 54		13.0	GALAXY
HN 1466	11 51 32.5	+ 25 50 43.			NEBULA
IC 0745	11 51 36.	+ 00 24 28.			NONSTELLAR OBJECT
ZC 1151.6+1028	11 51 36.	+ 10 28	400		CLUSTER OF GALAXIES
ZWG 097.179	11 51 36.	+ 15 10		15.7	GALAXY
ZC 1151.6+2934	11 51 36.	+ 29 34	540		CLUSTER OF GALAXIES
MCG+09-20-027	11 51 36.	+ 55 45	42	15.	GALAXY
ZWG 012.113	11 51 36.	- 03 24		15.6	GALAXY
IC 2975	11 51 36.	- 05 30			NONSTELLAR OBJECT
FATH 1.414	11 51 37.	- 00 51	16		NEBULA
TON-N 1461	11 51 38.	+ 36 31		15.4	BLUE STAR
RNGC 3957	11 51 38.	- 19 17		13.0	GALAXY
RNGC 3956	11 51 38.	- 20 18		12.5	GALAXY
MCG+06-26-053	11 51 39.	+ 33 08	48	16.	GALAXY
MCG+08-22-033	11 51 39.	+ 44 38	30	16.	GALAXY
FATH 1.415	11 51 39.	- 00 12	14		NEBULA
MCG-02-30-039	11 51 39.	- 12 11	96	15.	GALAXY
HN 1467	11 51 39.0	+ 25 01 43.	18		NEBULA
LB 02140	11 51 40.	+ 56 07 30.		16.3	FAINT BLUE STAR
SN 1967D	11 51 41.	+ 20 18		19.5	SUPERNOVA
REIZ 1754	11 51 41.	+ 21 02	18	15.8	GALAXY
KN 12.198	11 51 41.0	+ 49 52 00.			NEBULA
ZWG 012.114	11 51 42.	+ 00 25		13.7	GALAXY
UGC 06877	11 51 42.	+ 00 25	48	13.7	GALAXY
MCG+00-30-034	11 51 42.	+ 00 25	30	13.7	GALAXY
FEIG 051	11 51 42.	+ 18 53		10.2	FAINT BLUE STAR
ZWG 098.002	11 51 42.	+ 20 18		15.6	GALAXY
ZWG 097.180	11 51 42.	+ 20 18		15.6	GALAXY
ZC 1151.7+5302	11 51 42.	+ 53 02	3760		CLUSTER OF GALAXIES
ZC 1151.7+6054	11 51 42.	+ 60 54	1010		CLUSTER OF GALAXIES
MCG-03-30-018	11 51 42.	- 18 41	30	15.	GALAXY
HN 1468	11 51 42.6	- 00 10 53.	18		NEBULA
SEY 136	11 51 43.	+ 55 45 31.		15.4	FAINT GALAXY
REIZ 1755	11 51 47.	+ 20 18	36	15.5	GALAXY
REIZ 1756	11 51 47.	+ 27 41	54	15.6	GALAXY
KN 12.199	11 51 47.0	+ 49 49 36.			NEBULA
ZWG 040.062	11 51 48.	+ 05 45		15.6	GALAXY
ZC 1151.8+0944	11 51 48.	+ 09 44	2290		CLUSTER OF GALAXIES
ZWG 069.001	11 51 48.	+ 09 54		15.5	GALAXY
ZWG 068.089	11 51 48.	+ 09 54		15.5	GALAXY
ZWG 098.003	11 51 48.	+ 16 31		14.9	GALAXY
ZWG 097.181	11 51 48.	+ 16 31		14.9	GALAXY
MCG+03-30-125	11 51 48.	+ 16 31	30	14.9	GALAXY
UGC 06878	11 51 48.	+ 33 08	60	16.0	GALAXY Sb-c
BC PKS1151-34	11 51 49.3	- 34 48 48.		17.5	QUASI-STELLAR OBJECT
HN 1469	11 51 51.4	- 02 02 53.	72		NEBULA
LB 00620	11 51 52.	+ 55 38 12.		17.3	FAINT BLUE STAR
LB 02142	11 51 52.	+ 55 43 06.		15.6	FAINT BLUE STAR
REIZ 1757	11 51 52.	- 02 03	108	13.8	GALAXY
KN 12.200	11 51 52.5	+ 48 42 53.			NEBULA
REIZ 1758	11 51 53.	+ 21 04	12	15.7	GALAXY
ZC 1151.9+4933	11 51 54.	+ 49 33	1680		CLUSTER OF GALAXIES
ZWG 292.043	11 51 54.	+ 58 40		13.1	GALAXY
UGC 06880	11 51 54.	+ 58 40	84	13.1	GALAXY Sa
MCG+11-15-005	11 51 54.	+ 65 52	30	18.	GALAXY
MCG+11-15-006	11 51 54.	+ 67 53	27	16.	GALAXY
ZWG 012.115	11 51 54.	- 02 02		14.4	GALAXY
UGC 06879	11 51 54.	- 02 02	114	14.4	GALAXY Sc-IRR
MCG+00-30-035	11 51 54.	- 02 02	102	14.4	GALAXY
KARA.73B 0509	11 51 54.	- 02 02	108	14.4	ISOLATED GALAXY S
KN 12.201	11 51 54.4	+ 48 44 51.			NEBULA
REIZ 1759	11 51 56.	+ 58 38	72	13.3	GALAXY
IC 2976	11 51 56.	- 02 26 44.			NONSTELLAR OBJECT
MCG+10-17-098	11 51 57.	+ 58 39	78	13.3	GALAXY
KN 12.202	11 51 58.0	+ 48 46 52.			NEBULA
REIZ 1760	11 51 59.	+ 32 21	12	15.3	GALAXY
REIZ 1761	11 51 59.	+ 32 22	12	16.0	GALAXY
RNGC 3958	11 51 59.	+ 58 39		13.0	GALAXY
VDB.66G 101	11 52	+ 31 48	70		DWARF GALAXY
ZWG 069.002	11 52 00.	+ 09 50		15.6	GALAXY
ZWG 068.090	11 52 00.	+ 09 50		15.6	GALAXY
ZC 1152.0+1250	11 52 00.	+ 12 50	2220		CLUSTER OF GALAXIES
ZC 1152.0+2142	11 52 00.	+ 21 42	2290		CLUSTER OF GALAXIES
ZC 1152.0+4425	11 52 00.	+ 44 25	610		CLUSTER OF GALAXIES
MCG+08-22-034	11 52 00.	+ 49 35	12	18.	GALAXY
LB 02143	11 52 00.	+ 61 16 48.		16.0	FAINT BLUE STAR
KN 12.203	11 52 00.5	+ 46 12 09.			NEBULA
RNGC 3959	11 52 01.	- 07 28		14.0	GALAXY
HN 1470	11 52 02.6	+ 29 15 19.	18		NEBULA
MCG+08-22-035	11 52 03.	+ 49 34 30.	9	18.	GALAXY
MCG-01-30-046	11 52 03.	- 07 28	42	14.	GALAXY
REIZ 1762	11 52 05.	+ 33 48	54	15.0	GALAXY
HN 1471	11 52 05.0	+ 26 12 55.	18		NEBULA
UGC 06881	11 52 06.	+ 20 22	96	17.	GALAXY DWARF
MCG+10-17-099	11 52 06.	+ 56 52	18	16.	GALAXY
MCG-02-30-040	11 52 06.	- 13 41 30.	48	12.	GALAXY
REIN 2.138	11 52 06.89	- 13 41 49.3			NEBULA

OBJECT NAME	RIGHT ASCEN.	DECLINATION	DIAM.	MAGN.	TYPE OF OBJECT
LB 02144	11 52 07.	+ 54 27 24.		15.5	FAINT BLUE STAR
8ZW 1152+13.8	11 52 12.	+ 13 45		19.6	COMPACT GALAXY
ZWG 157.050	11 52 12.	+ 28 32		15.2	GALAXY
UGC 06882	11 52 12.	+ 33 48	66	16.5	GALAXY S
MCG+13-09-006	11 52 12.	+ 79 42 30.	39	14.	GALAXY
RNGC 3964	11 52 13.	+ 28 32		15.0	GALAXY
RNGC 3962	11 52 13.	- 13 42		12.5	GALAXY
HN 1472	11 52 13.4	+ 28 26 55.	18		NEBULA
MCG-03-30-019	11 52 15.	- 16 35	90	14.5	GALAXY
REIZ 1766	11 52 17.	+ 28 33	30	14.7	GALAXY
ZWG 041.001	11 52 18.	+ 03 14		15.2	GALAXY
ZWG 040.063	11 52 18.	+ 03 14		15.2	GALAXY
MCG+01-30-019	11 52 18.	+ 03 14	36	15.2	GALAXY
KARA.73B 0510	11 52 18.	+ 03 14	48	15.2	ISOLATED GALAXY S
MCG+04-28-094	11 52 18.	+ 26 29	66	15.6	GALAXY
ZWG 157.051	11 52 18.	+ 26 30		15.3	GALAXY
UGC 06883	11 52 18.	+ 26 30	84	15.3	GALAXY SBc
ZC 1152.3+4720	11 52 18.	+ 47 20	2760		CLUSTER OF GALAXIES
KN 12.204	11 52 18.7	+ 50 14 23.			NEBULA
REIZ 1763	11 52 19.	+ 49 03	24	14.5	GALAXY
REIZ 1764	11 52 19.	+ 49 05	24	14.7	GALAXY
FATH 1.416	11 52 19.	- 00 41	19		NEBULA
REIZ 1765	11 52 20.	+ 58 46	132	12.3	GALAXY
MCG+05-28-043	11 52 21.	+ 28 33 30.	48	15.2	GALAXY
REIZ 1767	11 52 22.	+ 03 14	18	15.6	GALAXY
RNGC 3961	11 52 22.	+ 69 36		14.5	GALAXY
KN 12.205	11 52 22.0	+ 49 01 30.			NEBULA
ZWG 041.002	11 52 24.	+ 04 56		15.1	GALAXY
ZWG 040.064	11 52 24.	+ 04 56		15.1	GALAXY
ZWG 041.003	11 52 24.	+ 06 36		15.5	GALAXY
ZWG 040.065	11 52 24.	+ 06 36		15.5	GALAXY
ZWG 127.103	11 52 24.	+ 26 29		15.6	GALAXY
ZWG 157.052	11 52 24.	+ 27 35		15.1	GALAXY
MCG+05-28-044	11 52 24.	+ 27 35	48	15.1	GALAXY
MCG+05-28-045	11 52 24.	+ 32 23 30.	18	15.5	GALAXY
MCG+08-22-036	11 52 24.	+ 49 01	36	16.	GALAXY
ZC 1152.4+5805	11 52 24.	+ 58 05	2550		CLUSTER OF GALAXIES
ZWG 292.044	11 52 24.	+ 58 46		12.2	GALAXY
RNGC 3963	11 52 24.	+ 58 46		12.5	GALAXY
UGC 06884	11 52 24.	+ 58 46	174	12.2	GALAXY SBb
ZWG 334.055	11 52 24.	+ 69 36		14.7	GALAXY
UGC 06885	11 52 24.	+ 69 36	96	14.7	GALAXY SBa
ZWG 352.011	11 52 24.	+ 79 43		15.3	GALAXY
ZWG 351.071	11 52 24.	+ 79 43		15.3	GALAXY
KARA.72 308A	11 52 24.	+ 79 43	36	15.3	PART OF DOUBLE GALAXY
MCG-04-28-006	11 52 24.	- 26 58 30.	72	14.5	GALAXY
MCG-05-28-016	11 52 24.	- 31 17	24	15.5	GALAXY
HN 1473	11 52 24.9	+ 26 28 13.	18		NEBULA
REIZ 1768	11 52 25.	+ 48 59	18	14.4	GALAXY
KN 12.206	11 52 26.0	+ 49 03 59.			NEBULA
KN 12.207	11 52 27.1	+ 48 58 26.			NEBULA
LB 02145	11 52 28.	+ 62 48 36.		15.7	FAINT BLUE STAR
HN 1474	11 52 28.0	+ 27 34 19.	24		NEBULA
KN 12.208	11 52 28.9	+ 46 01 59.			NEBULA
ZC 1152.5+1202	11 52 30.	+ 12 02	1210		CLUSTER OF GALAXIES
ZC 1152.5+1941	11 52 30.	+ 19 41	1080		CLUSTER OF GALAXIES
MCG+04-28-095	11 52 30.	+ 22 58	63	15.5	GALAXY
ZWG 157.053	11 52 30.	+ 32 20		15.5	GALAXY
MCG+08-22-037	11 52 30.	+ 48 06	60	16.	GALAXY
MCG+08-22-038	11 52 30.	+ 49 04	18	17.	GALAXY
MCG+10-17-100	11 52 30.	+ 58 46	156	12.7	GALAXY
ZC 1152.5+6501	11 52 30.	+ 65 01	2080		CLUSTER OF GALAXIES
ZC 1152.5-0110	11 52 30.	- 01 10	1750		CLUSTER OF GALAXIES
ZC 1152.5-0110	11 52 30.	- 01 10	1750		CLUSTER OF GALAXIES
MCG-03-30-020	11 52 30.	- 18 40	54	14.	GALAXY
HN 1475	11 52 30.5	+ 01 58 55.	12		NEBULA
RNGC 3965	11 52 31.	- 10 36			NON-EXISTENT OBJECT
KN 12.209	11 52 31.3	+ 48 06 17.			NEBULA
RNGC 3969	11 52 32.	- 18 40		14.0	GALAXY
KN 12.210	11 52 34.1	+ 50 26 06.			NEBULA
ZWG 013.001	11 52 36.	+ 02 00		14.9	GALAXY
ZWG 012.116	11 52 36.	+ 02 00		14.9	GALAXY
MCG+00-31-001	11 52 36.	+ 02 00	24	14.9	GALAXY
KARA.73B 0511	11 52 36.	+ 02 00	30	14.9	ISOLATED GALAXY E
ZWG 041.004	11 52 36.	+ 06 27		14.5	GALAXY
ZWG 040.066	11 52 36.	+ 06 27		14.5	GALAXY
UGC 06886	11 52 36.	+ 06 27	78	14.5	GALAXY SBb
MCG+01-30-020	11 52 36.	+ 06 27	72	14.5	GALAXY
ZWG 127.104	11 52 36.	+ 22 59		15.5	GALAXY
UGC 06887	11 52 36.	+ 22 59	72	15.5	GALAXY S
KARA.68 081	11 52 36.	+ 44 24	40		DWARF GALAXY
MCG+08-22-039	11 52 36.	+ 50 25	60	16.	GALAXY
ZWG 243.028	11 52 36.	+ 50 26		15.5	GALAXY
UGC 06888	11 52 36.	+ 50 26	60	15.5	GALAXY S
ZC 1152.6+5030	11 52 36.	+ 50 30	1010		CLUSTER OF GALAXIES
ZWG 292.045	11 52 36.	+ 59 35		15.6	GALAXY
ZC 1152.6+5939	11 52 36.	+ 59 39	1480		CLUSTER OF GALAXIES
UGC 06889	11 52 36.	+ 67 41	60	16.5	GALAXY Sc
MCG-01-30-047	11 52 36.	- 07 34	30	15.	GALAXY
RNGC 3967	11 52 37.	- 07 34		15.0	GALAXY
HN 1476	11 52 38.1	+ 25 10 31.			NEBULA
MCG+06-26-054	11 52 39.	+ 33 25	30	15.5	GALAXY
ARC 1411	11 52 40.	- 00 16		17.5	RICH CLUSTER OF GALAXIES
REIZ 1769	11 52 41.	+ 29 33	24	16.2	GALAXY
REIZ 1770	11 52 41.	+ 32 28	12	14.6	GALAXY
HOLM 302B	11 52 41.	+ 33 25	48	15.1	PART OF MULTIPLE GALAXY
REIZ 1771	11 52 41.	+ 33 31	12	16.2	GALAXY
LB 00621	11 52 41.	+ 57 57 12.		16.8	FAINT BLUE STAR
UGC 06890	11 52 42.	+ 00 46	78	17.	GALAXY DWRF IR
ZWG 041.005	11 52 42.	+ 04 52		15.4	GALAXY
ZWG 040.067	11 52 42.	+ 04 52		15.4	GALAXY
ZWG 098.004	11 52 42.	+ 17 46		15.4	GALAXY
ZWG 097.182	11 52 42.	+ 17 46		15.4	GALAXY
UGC 06891	11 52 42.	+ 17 46	102	15.4	GALAXY Sa-b
MCG+03-30-126	11 52 42.	+ 17 47	90	15.4	GALAXY
MCG+06-26-055	11 52 42.	+ 33 24	54	15.	GALAXY
ZWG 186.067	11 52 42.	+ 33 25		15.6	GALAXY
KARA.72 309A	11 52 42.	+ 33 25	54	15.6	PART OF DOUBLE GALAXY
STOCK 34	11 52 42.	+ 37 10			BLUE KNOT NEAR ELLIP GLXY
MCG+09-20-028	11 52 42.	+ 52 35	24	17.	GALAXY
SEY 137	11 52 42.	+ 55 38 31.		15.4	FAINT GALAXY
MCG+10-17-101	11 52 42.	+ 59 36	36	16.	GALAXY
ZWG 352.012	11 52 42.	+ 79 43		15.6	GALAXY
ZWG 351.072	11 52 42.	+ 79 43		15.6	GALAXY
KARA.72 308B	11 52 42.	+ 79 43	36	15.6	PART OF DOUBLE GALAXY
MCG-02-30-041	11 52 42.	- 11 46	60	14.5	GALAXY
KN 12.211	11 52 42.2	+ 48 59 21.			NEBULA
REIZ 1772	11 52 43.	+ 49 00	18	14.8	GALAXY
RNGC 3970	11 52 43.	- 11 46		14.0	GALAXY
IC 2977	11 52 43.	- 37 25 09.			NONSTELLAR OBJECT
KN 12.212	11 52 43.0	+ 51 04 44.			NEBULA
MCG-06-26-014	11 52 45.	- 37 24	78	13.	GALAXY
KN 12.213	11 52 45.9	+ 49 34 53.			NEBULA
LB 02146	11 52 46.	+ 57 47 30.		13.2	FAINT BLUE STAR
REIZ 1773	11 52 47.	+ 28 37	42	15.5	GALAXY
REIZ 1775	11 52 47.	+ 33 24	36	14.1	GALAXY
REIZ 1774	11 52 47.	+ 33 24	42	14.7	GALAXY
HOLM 302A	11 52 47.	+ 33 24	48	14.8	PART OF MULTIPLE GALAXY
KN 12.214	11 52 47.4	+ 48 44 38.			NEBULA
ZC 1152.8+2210	11 52 48.	+ 22 10	1140		CLUSTER OF GALAXIES
ARC 1413	11 52 48.	+ 23 39		17.1	RICH CLUSTER OF GALAXIES
MCG+05-28-046	11 52 48.	+ 28 38	42	16.	GALAXY
ZWG 186.068	11 52 48.	+ 33 24		15.3	GALAXY
UGC 06892	11 52 48.	+ 33 24	66	15.3	GALAXY
KARA.72 309B	11 52 48.	+ 33 24	60	15.3	PART OF DOUBLE GALAXY
ZWG 215.004	11 52 48.	+ 39 30		15.7	GALAXY
UGC 06893	11 52 48.	+ 39 30	72	15.7	GALAXY Sc
MRK 192	11 52 48.	+ 51 25	8	17.	GALAXY WITH UV CONTINUUM
ZWG 269.015	11 52 48.	+ 54 56		15.6	GALAXY
UGC 06894	11 52 48.	+ 54 56	84	15.6	GALAXY Sc
MCG+09-20-029	11 52 48.	+ 54 56	90	14.	GALAXY
SEY 138	11 52 48.	+ 54 56 37.		14.7	FAINT GALAXY
MCG+13-09-007	11 52 48.	+ 79 42 30.	33	14.	GALAXY
ZC 1152.8-0015	11 52 48.	- 00 15	740		CLUSTER OF GALAXIES
MCG-01-30-048	11 52 48.	- 06 12	96	15.	GALAXY
HN 1477	11 52 49.4	+ 28 37 07.	24		NEBULA
MCG+07-25-003	11 52 51.	+ 39 30	54	14.5	GALAXY
KN 12.215	11 52 51.2	+ 48 46 35.			NEBULA
REIZ 1776	11 52 52.	+ 12 15	96	14.3	GALAXY
ZWG 069.003	11 52 54.	+ 12 13		15.7	GALAXY
ZWG 068.091	11 52 54.	+ 12 13		15.7	GALAXY
MCG+02-30-044	11 52 54.	+ 12 13	36	15.7	GALAXY
ZWG 069.004	11 52 54.	+ 12 15		13.3	GALAXY
ZWG 068.092	11 52 54.	+ 12 15		13.3	GALAXY
SCH 17	11 52 54.	+ 12 15		13.3	PECULIAR GALAXY
UGC 06895	11 52 54.	+ 12 15	174	13.3	GALAXY Sb/SBc
MCG+02-30-045	11 52 54.	+ 12 15	156	13.3	GALAXY
ZC 1152.9+1634	11 52 54.	+ 16 34	810		CLUSTER OF GALAXIES
ARC 1414	11 52 54.	+ 16 34		17.2	RICH CLUSTER OF GALAXIES
ZC 1152.9+2336	11 52 54.	+ 23 36	2890		CLUSTER OF GALAXIES
ZWG 127.105	11 52 54.	+ 25 13		15.7	GALAXY
MCG+04-28-096	11 52 54.	+ 26 10	60	14.5	GALAXY
ZWG 186.069	11 52 54.	+ 37 40		15.4	GALAXY
MCG+06-26-056	11 52 54.	+ 37 41 30.	27	15.	GALAXY
MCG+09-20-030	11 52 54.	+ 55 09	30	17.	GALAXY
7ZW 427	11 52 54.	+ 58 53			COMPACT GALAXY
RNGC 3968	11 52 55.	+ 12 15		13.5	GALAXY
FATH 1.417	11 52 55.	- 00 17	11		NEBULA
HN 1479	11 52 55.9	- 00 59 36.	12		NEBULA
HN 1478	11 52 56.6	+ 25 13 24.			NEBULA
SEY 139	11 52 57.	+ 55 11 13.		15.4	FAINT GALAXY
KN 12.216	11 52 57.5	+ 46 30 00.			NEBULA
REIZ 1777	11 52 58.	+ 03 31	18	16.2	GALAXY
REIZ 1778	11 52 59.	+ 30 16	18	13.3	GALAXY
KN 12.218	11 52 59.2	+ 46 30 07.			NEBULA
KN 12.217	11 52 59.5	+ 50 57 18.			NEBULA
HMS 1153+2341	11 53	+ 23 41			CLUSTER OF GALAXIES
ZC 1153.0+0419	11 53 00.	+ 04 19	470		CLUSTER OF GALAXIES
UGC 06897	11 53 00.	+ 10 04	72	16.0	GALAXY SBc
ZC 1153.0+1109	11 53 00.	+ 11 09	1880		CLUSTER OF GALAXIES
ZWG 069.005	11 53 00.	+ 12 17		15.4	GALAXY
ZWG 068.093	11 53 00.	+ 12 17		15.4	GALAXY
MCG+02-31-002	11 53 00.	+ 12 17	6	15.5	GALAXY
MCG+02-31-001	11 53 00.	+ 12 17	24	15.4	GALAXY
ZC 1153.0+2522	11 53 00.	+ 25 22	8400		CLUSTER OF GALAXIES
ZWG 127.106	11 53 00.	+ 26 10		14.5	GALAXY
UGC 06898	11 53 00.	+ 26 10	78	14.5	GALAXY S
ZWG 157.054	11 53 00.	+ 30 16		13.9	GALAXY
UGC 06899	11 53 00.	+ 30 16	84	13.9	GALAXY S0
MCG+05-28-047	11 53 00.	+ 30 18	18	13.9	GALAXY
ZWG 157.055	11 53 00.	+ 31 48		15.7	GALAXY
UGC 06900	11 53 00.	+ 31 48	120	15.7	GALAXY DWARF
ZC 1153.0+3321	11 53 00.	+ 33 21	3430		CLUSTER OF GALAXIES
ZWG 215.005	11 53 00.	+ 43 19		14.4	GALAXY
UGC 06901	11 53 00.	+ 43 19	84	14.4	GALAXY S
ZWG 243.029	11 53 00.	+ 46 30		15.6	GALAXY
HOLM 303A	11 53 00.	+ 46 30	24	14.8	PART OF MULTIPLE GALAXY
UGC 06902	11 53 00.	+ 46 30	60	15.6	GALAXY DBL SYS
REIZ 1779	11 53 00.	+ 49 30	30	14.1	GALAXY
ZC 1153.0+5200	11 53 00.	+ 52 00	1080		CLUSTER OF GALAXIES
MCG+09-20-031	11 53 00.	+ 55 11	30	16.	GALAXY
ZWG 292.046	11 53 00.	+ 56 32		15.6	GALAXY
MCG+11-15-007	11 53 00.	+ 63 43	42	17.	GALAXY
7ZW 428	11 53 00.	+ 64 46			COMPACT GALAXY
SHAH 156	11 53 00.	+ 76 08	138		GROUP OF COMPACT GALAXIES
ZWG 365.002	11 53 00.	+ 80 30		15.2	GALAXY
UGC 06896	11 53 00.	+ 80 30	66	15.2	GALAXY TRP SYS
7ZW 429	11 53 00.	+ 80 31			COMPACT GALAXY
ZC 1153.0+8550	11 53 00.	+ 85 50	2890		CLUSTER OF GALAXIES
ZWG 013.002	11 53 00.	- 00 58		15.3	GALAXY
RNGC 3973	11 53 01.	+ 12 17		15.5	GALAXY
IC 0746	11 53 01.	+ 26 10 03.			NONSTELLAR OBJECT
RNGC 3971	11 53 01.	+ 30 16		14.0	GALAXY
SEY 140	11 53 02.	+ 43 19 18.		14.4	FAINT GALAXY
HOLM 303B	11 53 02.	+ 46 30	24	15.0	PART OF MULTIPLE GALAXY
REIZ 1780	11 53 02.	+ 58 37	48	15.1	GALAXY
HN 1480	11 53 02.9	+ 01 30 18.	18		NEBULA
MCG+05-28-048	11 53 03.	+ 32 29	24	14.7	GALAXY
REIZ 1782	11 53 04.	+ 12 17	12	15.	GALAXY
ARC 1412	11 53 04.	+ 73 45		15.9	RICH CLUSTER OF GALAXIES
KARA.73 35	11 53 04.	- 36 27	27		DWARF GALAXY
HN 1481	11 53 04.5	- 00 29 42.	12		NEBULA
FATH 2.034	11 53 05.	- 00 29	19		NEBULA
ZWG 013.004	11 53 06.	+ 01 32		14.1	GALAXY
UGC 06903	11 53 06.	+ 01 32	162	14.1	GALAXY SBc
MCG+00-31-002	11 53 06.	+ 01 32	156	14.1	GALAXY
KARA.73B 0512	11 53 06.	+ 01 32	168	14.1	ISOLATED GALAXY S
ZWG 069.006	11 53 06.	+ 13 01		15.5	GALAXY
ZWG 068.094	11 53 06.	+ 13 01		15.5	GALAXY
8ZW 1153+13.0	11 53 06.	+ 13 01		15.5	COMPACT GALAXY
MCG+02-31-003	11 53 06.	+ 13 01	6	15.5	GALAXY
8ZW 1153+13.2	11 53 06.	+ 13 11		19.0	COMPACT GALAXY
ZWG 098.005	11 53 06.	+ 18 18		15.5	GALAXY
ZWG 097.183	11 53 06.	+ 18 18		15.5	GALAXY
MCG+05-28-049	11 53 06.	+ 31 50	120	15.7	GALAXY
ZWG 157.056	11 53 06.	+ 32 28		14.7	GALAXY

OBJECT NAME	RIGHT ASCEN.	DECLINATION	DIAM.	MAGN.	TYPE OF OBJECT
MCG+07-25-004	11 53 06.	+ 43 21	60	14.5	GALAXY
ZWG 243.030	11 53 06.	+ 49 34		15.5	GALAXY
ZWG 269.016	11 53 06.	+ 55 35		12.9	GALAXY
UGC 06904	11 53 06.	+ 55 35	246	12.9	GALAXY Sc
MCG+09-20-032	11 53 06.	+ 55 37	210	12.7	GALAXY
MCG+09-20-033	11 53 06.	+ 56 31	18	16.	GALAXY
MRK 193	11 53 06.	+ 57 57	7	16.5	GALAXY WITH UV CONTINUUM
MCG+10-17-102	11 53 06.	+ 58 38	39	16.	GALAXY
ZWG 013.003	11 53 06.	- 00 28		15.7	GALAXY
MCG-02-31-001	11 53 06.	- 11 44	36	14.	GALAXY
RNGC 3966	11 53 07.	+ 32 28		14.5	GALAXY
REIZ 1781	11 53 07.	+ 55 35	210	12.9	GALAXY
HOLM 304A	11 53 07.	+ 55 35	210	12.7	PART OF MULTIPLE GALAXY
RNGC 3974	11 53 07.	- 11 44		14.0	GALAXY
KN 12.219	11 53 07.1	+ 45 21 03.			NEBULA
HN 1482	11 53 07.9	+ 25 11 54.	24		NEBULA
MCG+08-22-040	11 53 09.	+ 49 35	42	16.	GALAXY
KN 12.221	11 53 09.1	+ 46 39 48.			NEBULA
KN 12.220	11 53 09.5	+ 49 34 34.			NEBULA
REIZ 1783	11 53 11.	+ 30 12	54	15.1	GALAXY
ZC 1153.2+0433	11 53 12.	+ 04 33	1410		CLUSTER OF GALAXIES
8ZW 1153+10.0	11 53 12.	+ 09 58		16.6	COMPACT GALAXY
ZWG 098.006	11 53 12.	+ 17 37		15.7	GALAXY
ZWG 097.184	11 53 12.	+ 17 37		15.7	GALAXY
KARA.73B 0513	11 53 12.	+ 17 37	18	15.7	ISOLATED GALAXY E
ZWG 098.007	11 53 12.	+ 18 10		15.5	GALAXY
ZWG 097.185	11 53 12.	+ 18 10		15.5	GALAXY
MCG+03-31-001	11 53 12.	+ 18 11	48	15.5	GALAXY
UGC 06905	11 53 12.	+ 30 13	66	16.0	GALAXY Sb
MCG+05-28-050	11 53 12.	+ 30 15	60	14.5	GALAXY
ZWG 215.006	11 53 12.	+ 39 30		15.6	GALAXY
MCG+07-25-005	11 53 12.	+ 39 30 30.	36	15.	GALAXY
RNGC 3972	11 53 12.	+ 55 36		13.0	GALAXY
ARC 1415	11 53 12.	+ 58 09		17.0	RICH CLUSTER OF GALAXIES
ZC 1153.2+6216	11 53 12.	+ 62 16	1140		CLUSTER OF GALAXIES
ZWG 334.056	11 53 12.	+ 69 12		15.6	GALAXY
LB 00622	11 53 13.	+ 54 22 48.		17.5	FAINT BLUE STAR
KN 12.222	11 53 13.7	+ 46 05 03.			NEBULA
FATH 1.418	11 53 14.	- 00 16	11		NEBULA
HN 1483	11 53 14.7	+ 25 24 30.	18		NEBULA
MCG+08-22-041	11 53 15.	+ 49 31	18	18.	GALAXY
HOLM 306B	11 53 15.	+ 60 48	30	15.2	PART OF MULTIPLE GALAXY
KN 12.223	11 53 15.3	+ 48 31 24.			NEBULA
HN 1484	11 53 16.4	+ 25 02 48.	18		NEBULA
KN 12.224	11 53 16.4	+ 48 30 57.			NEBULA
KN 12.225	11 53 16.8	+ 48 31 55.			NEBULA
ARC 1416	11 53 17.	+ 11 03		17.6	RICH CLUSTER OF GALAXIES
ZWG 041.006	11 53 18.	+ 07 02		12.8	GALAXY
UGC 06906	11 53 18.	+ 07 02	216	12.8	GALAXY Sb
MCG+01-31-001	11 53 18.	+ 07 02	198	12.8	GALAXY
MCG+04-28-097	11 53 18.	+ 23 41	12	18.	GALAXY
ZWG 127.107	11 53 18.	+ 25 25		15.7	GALAXY
KARA.68 082	11 53 18.	+ 31 49	107		DWARF GALAXY
REIZ 1784	11 53 18.	+ 45 09	18	15.0	GALAXY
MCG+08-22-042	11 53 18.	+ 48 31	9	17.	GALAXY
RNGC 3975	11 53 18.	+ 60 48		15.0	GALAXY
ZWG 315.002	11 53 18.	+ 67 30		15.6	GALAXY
REIZ 1785	11 53 20.	+ 60 48	36	15.0	GALAXY
SEY 141	11 53 21.	+ 43 52 12.		15.0	FAINT GALAXY
MCG+08-22-043	11 53 21.	+ 48 32	54	16.	GALAXY
REIZ 1786	11 53 22.	+ 07 01	210	12.2	GALAXY
HOLM 305A	11 53 22.	+ 07 01	240	12.5	PART OF MULTIPLE GALAXY
REIZ 1787	11 53 23.	+ 25 24	18	15.6	GALAXY
ZWG 041.007	11 53 24.	+ 06 58		15.5	GALAXY
MCG+01-31-001A	11 53 24.	+ 06 58	24	15.5	GALAXY
SCH 18	11 53 24.	+ 07 02		12.8	PECULIAR GALAXY
ZC 1153.4+3459	11 53 24.	+ 34 59	870		CLUSTER OF GALAXIES
ZWG 186.070	11 53 24.	+ 37 26		15.7	GALAXY
REIZ 1788	11 53 24.	+ 45 02	24	14.0	GALAXY
MCG+10-17-103	11 53 24.	+ 60 48	39	15.2	GALAXY
ZWG 315.003	11 53 24.	+ 63 55		15.7	GALAXY
MCG-03-31-001A	11 53 24.	- 17 54	84	13.5	GALAXY
HN 1486	11 53 24.9	- 00 43 30.	12		NEBULA
RNGC 3976A	11 53 25.	+ 06 58		15.5	GALAXY
RNGC 3976	11 53 25.	+ 07 02		12.5	GALAXY
REIZ 1789	11 53 25.	+ 54 28	15	15.5	GALAXY
REIZ 1790	11 53 25.	+ 54 30	12	14.8	GALAXY
FATH 1.419	11 53 25.	- 00 18	11		NEBULA
HN 1485	11 53 25.8	+ 25 29 30.	12		NEBULA
LB 02147	11 53 26.	+ 55 00 54.		15.4	FAINT BLUE STAR
ARC 1418	11 53 26.	- 18 23		17.4	RICH CLUSTER OF GALAXIES
FATH 1.420	11 53 27.	+ 00 11	16		NEBULA
TON-N 1462	11 53 27.	+ 34 25		13.5	BLUE STAR
HN 1487	11 53 27.0	+ 27 18 48.			NEBULA
HOLM 305B	11 53 28.	+ 06 57	24	15.3	PART OF MULTIPLE GALAXY
HOLM 304B	11 53 28.	+ 55 40	24	14.0	PART OF MULTIPLE GALAXY
REIZ 1791	11 53 28.	- 02 28	18	14.3	GALAXY
ZWG 041.008	11 53 30.	+ 05 47		15.7	GALAXY
MCG+01-31-002	11 53 30.	+ 05 47	18	15.7	GALAXY
ZC 1153.5+1010	11 53 30.	+ 10 10	810		CLUSTER OF GALAXIES
MCG+03-31-002	11 53 30.	+ 15 25	90	15.2	GALAXY
ZC 1153.5+3449	11 53 30.	+ 34 49	400		CLUSTER OF GALAXIES
UGC 06908	11 53 30.	+ 37 26	66	16.0	GALAXY SB
ZC 1153.5+4041	11 53 30.	+ 40 41	1280		CLUSTER OF GALAXIES
ZC 1153.5+4218	11 53 30.	+ 42 18	870		CLUSTER OF GALAXIES
ZWG 269.017	11 53 30.	+ 55 40		14.7	GALAXY
RNGC 3977	11 53 30.	+ 55 40		14.5	GALAXY
UGC 06909	11 53 30.	+ 55 40	108	14.7	GALAXY Sa
SN 1946A	11 53 30.	+ 55 40		18.0	SUPERNOVA
MCG+09-20-034	11 53 30.	+ 55 40	60	14.0	GALAXY
ZC 1153.5+6004	11 53 30.	+ 60 04	1080		CLUSTER OF GALAXIES
ZWG 292.047	11 53 30.	+ 60 47		13.2	GALAXY
UGC 06910	11 53 30.	+ 60 47	102	13.2	GALAXY Sb
ZWG 013.005	11 53 30.	- 02 26		14.2	GALAXY
UGC 06907	11 53 30.	- 02 26	66	14.2	GALAXY S0?
MCG+00-31-003	11 53 30.	- 02 26	66	14.2	GALAXY
MCG-03-31-001B	11 53 30.	- 18 40	60	15.	GALAXY
ARC 1417	11 53 31.	+ 34 53		17.6	RICH CLUSTER OF GALAXIES
REIZ 1792	11 53 31.	+ 54 22	12	15.3	GALAXY
REIZ 1793	11 53 31.	+ 55 39	24	14.7	GALAXY
RNGC 3979	11 53 31.	- 02 26		14.0	GALAXY
REIZ 1794	11 53 32.	+ 60 48	60	13.4	GALAXY
HOLM 306A	11 53 32.	+ 60 48	36	13.2	PART OF MULTIPLE GALAXY
LB 02149	11 53 33.	+ 54 24 24.		16.8	FAINT BLUE STAR
LB 02148	11 53 33.	+ 57 40 00.		16.2	FAINT BLUE STAR
HN 1488	11 53 33.3	+ 24 40 18.	18		NEBULA
REIZ 1795	11 53 35.	+ 25 28	18	15.7	GALAXY
REIZ 1796	11 53 35.	+ 32 54	12	14.9	GALAXY
ZWG 013.006	11 53 36.	+ 02 24		15.7	GALAXY
ZWG 098.008	11 53 36.	+ 15 24		15.2	GALAXY
UGC 06911	11 53 36.	+ 15 24	84	15.2	GALAXY Sb-c
RNGC 3980	11 53 36.	+ 55 40			NON-EXISTENT OBJECT
ZWG 292.048	11 53 36.	+ 58 29		15.1	GALAXY
UGC 06912	11 53 36.	+ 58 29	144	15.1	GALAXY PECULR
7ZW 430	11 53 36.	+ 58 30			COMPACT GALAXY
RNGC 3978	11 53 36.	+ 60 48		13.0	GALAXY
MCG-03-31-001	11 53 36.	- 19 36	168	12.5	GALAXY
VV 008B	11 53 36.	- 19 37	36	15.	INTERACTING GALAXY
VV 008A	11 53 36.	- 19 37	300	12.7	INTERACTING GALAXY
HN 1489	11 53 36.9	+ 27 48 18.	12		NEBULA
FATH 1.422	11 53 37.	+ 00 04	8		NEBULA
FATH 1.421	11 53 37.	- 00 16	8		NEBULA
ARC 1419	11 53 40.	+ 00 02		17.0	RICH CLUSTER OF GALAXIES
LB 02150	11 53 40.	+ 60 16 00.		16.2	FAINT BLUE STAR
FATH 1.423	11 53 41.	+ 00 05	16		NEBULA
ARP 289	11 53 41.	- 19 37			PECULIAR GALAXY
HN 1490	11 53 41.0	+ 27 23 06.	18		NEBULA
ZWG 098.009	11 53 42.	+ 17 18		15.3	GALAXY
UGC 06913	11 53 42.	+ 17 18	96	15.3	GALAXY Sb
MCG+03-31-003	11 53 42.	+ 17 19	90	15.3	GALAXY
MRK 643	11 53 42.	+ 17 42	9	16.	GALAXY WITH UV CONTINUUM
ZC 1153.7+1756	11 53 42.	+ 17 56	610		CLUSTER OF GALAXIES
ZC 1153.7+1919	11 53 42.	+ 19 19	940		CLUSTER OF GALAXIES
MCG+04-28-098	11 53 42.	+ 24 08	66	14.8	GALAXY
MCG+05-28-051	11 53 42.	+ 32 20	60	15.4	GALAXY
MCG+09-20-035	11 53 42.	+ 52 47	42	16.	GALAXY
MCG+10-17-104	11 53 42.	+ 58 28	114	15.	GALAXY
MCG+10-17-105	11 53 42.	+ 60 49	102	13.2	GALAXY
ZWG 013.007	11 53 42.	- 03 22		15.6	GALAXY
REIZ 1797	11 53 43.	+ 58 28	18	15.4	GALAXY
RNGC 3981	11 53 43.	- 19 37		12.5	GALAXY
KN 12.226	11 53 43.9	+ 47 52 40.			NEBULA
LB 02151	11 53 44.	+ 55 12 24.		15.6	FAINT BLUE STAR
BC 4C31.38	11 53 44.08	+ 31 44 46.9		18.96	QUASI-STELLAR OBJECT
SHB 190	11 53 44.1	+ 31 44 46.		19.0	QUASI-STELLAR OBJECT
REIZ 1798	11 53 46.	+ 15 22	60	14.9	GALAXY
FATH 1.424	11 53 47.	+ 00 05	11		NEBULA
REIZ 1799	11 53 47.	+ 24 09	54	14.7	GALAXY
REIZ 1800	11 53 47.	+ 25 39	30	15.4	GALAXY
REIZ 1801	11 53 47.	+ 32 19	48	14.8	GALAXY
HN 1492	11 53 47.2	- 01 45 18.	12		NEBULA
HN 1491	11 53 47.9	+ 25 38 54.	24		NEBULA
ZWG 041.009	11 53 48.	+ 04 02		15.7	GALAXY
ZWG 127.108	11 53 48.	+ 24 09		14.8	GALAXY
UGC 06914	11 53 48.	+ 24 09	72	14.8	GALAXY S0-a
ZWG 127.109	11 53 48.	+ 25 39		15.7	GALAXY
ZWG 157.057	11 53 48.	+ 32 18		15.4	GALAXY
UGC 06915	11 53 48.	+ 32 18	66	15.4	GALAXY Sc
ZWG 215.007	11 53 48.	+ 40 01		14.7	GALAXY
UGC 06916	11 53 48.	+ 40 01	114	14.7	GALAXY SB:b-c
MCG+08-22-044	11 53 48.	+ 50 09	60	16.	GALAXY
MCG+09-20-036	11 53 48.	+ 55 25	144	11.8	GALAXY
ZWG 292.049	11 53 48.	+ 60 40		15.7	GALAXY
ZC 1153.8+7343	11 53 48.	+ 73 43	2550		CLUSTER OF GALAXIES
KN 12.227	11 53 48.5	+ 50 56 41.			NEBULA
RNGC 3983	11 53 49.	+ 24 09		15.0	GALAXY
IC 2978	11 53 49.	+ 32 18 45.			NONSTELLAR OBJECT
KN 12.228	11 53 49.4	+ 50 52 51.			NEBULA
KN 12.229	11 53 49.8	+ 50 08 42.			NEBULA
MCG+07-25-006	11 53 51.	+ 40 01	90	14.	GALAXY
KN 12.230	11 53 52.0	+ 50 42 21.			NEBULA
REIZ 1802	11 53 53.	+ 28 46	42	15.6	GALAXY
HN 1493	11 53 53.0	+ 24 49 06.	12		NEBULA
KN 12.231	11 53 53.7	+ 50 57 34.			NEBULA
KARA.68 083	11 53 54.	+ 31 36	54		DWARF GALAXY
ZWG 269.018	11 53 54.	+ 50 43		14.0	GALAXY
UGC 06917	11 53 54.	+ 50 43	288	14.0	GALAXY
ZWG 269.019	11 53 54.	+ 55 23		11.6	GALAXY
UGC 06918	11 53 54.	+ 55 23	144	11.6	GALAXY S
RNGC 3982	11 53 54.	+ 55 24		12.5	GALAXY
MCG+10-17-106	11 53 54.	+ 60 41	12	16.	GALAXY
MCG-02-31-002	11 53 54.	- 11 53	54	15.	GALAXY
MCG-03-31-002	11 53 54.	- 19 15	60	15.	GALAXY
HN 1494	11 53 54.4	+ 24 57 48.	18		NEBULA
REIZ 1803	11 53 55.	+ 55 24	60	12.2	GALAXY
HN 1495	11 53 55.8	+ 24 51 54.	12		NEBULA
KN 12.232	11 53 56.3	+ 50 56 56.			NEBULA
HOLM 307A	11 53 57.	- 19 17	60	14.2	PART OF MULTIPLE GALAXY
KN 12.233	11 53 58.0	+ 46 48 32.			NEBULA
VDB.66G 102	11 54	+ 51 08	70		DWARF GALAXY
ARC 1420	11 54 00.	+ 26 22		17.2	RICH CLUSTER OF GALAXIES
KARA.72 310A	11 54 00.	+ 48 36	48	13.0	PART OF DOUBLE GALAXY
MCG+08-22-045	11 54 00.	+ 48 37 30.	60	12.9	GALAXY
MCG+09-20-038	11 54 00.	+ 50 40 30.	168	13.	GALAXY
ZWG 269.020	11 54 00.	+ 55 54		15.4	GALAXY
UGC 06919	11 54 00.	+ 55 54	90	15.4	GALAXY S-IRR
MCG+09-20-037	11 54 00.	+ 55 55	90	14.	GALAXY
MCG+10-17-107	11 54 00.	+ 58 51	18	16.	GALAXY
MCG+10-17-108	11 54 00.	+ 60 46	27	17.	GALAXY
HN 1496	11 54 01.4	+ 28 46 18.	24		NEBULA
HOLM 307B	11 54 03.	- 19 16	30	14.5	PART OF MULTIPLE GALAXY
SVEN 247	11 54 03.	- 19 17	42	14.7	GALAXY
FATH 1.425	11 54 05.	- 00 14	8		NEBULA
HN 1497	11 54 05.0	+ 26 21 42.			NEBULA
ZWG 069.007	11 54 06.	+ 11 41		15.7	GALAXY
MCG+02-31-005	11 54 06.	+ 11 41	9	15.7	GALAXY
MCG+02-31-004	11 54 06.	+ 11 41	30		GALAXY
MCG+05-28-052	11 54 06.	+ 31 49 30.	30	16.	GALAXY
ZWG 157.058	11 54 06.	+ 32 17		14.0	GALAXY
UGC 06920	11 54 06.	+ 32 17	168	14.0	GALAXY Sa
ZWG 186.071	11 54 06.	+ 32 53		15.5	GALAXY
ZWG 243.031	11 54 06.	+ 48 36		13.0	GALAXY
UGC 06921	11 54 06.	+ 48 36	66	13.0	GALAXY S
KARA.72 310B	11 54 06.	+ 48 36	54		PART OF DOUBLE GALAXY
RNGC 3985	11 54 06.	+ 48 37		13.5	GALAXY
REIZ 1804	11 54 06.	+ 48 37	54	12.8	GALAXY
REIZ 1805	11 54 06.	+ 50 25	30	15.0	GALAXY
ZWG 292.050	11 54 06.	+ 62 23		15.7	GALAXY
HN 1498	11 54 06.1	+ 29 07 18.	12		NEBULA
HN 1499	11 54 06.9	+ 24 55 06.	12		NEBULA
LB 02152	11 54 07.	+ 50 25 30.		17.3	FAINT BLUE STAR
REIZ 1806	11 54 07.	+ 55 56	72	15.1	GALAXY
SEY 142	11 54 07.	+ 55 56 13.		14.8	FAINT GALAXY
KN 12.234	11 54 07.1	+ 48 36 50.			NEBULA
HN 1500	11 54 08.5	+ 27 48 54.	12		NEBULA
HN 1501	11 54 08.8	+ 25 58 36.			NEBULA

OBJECT NAME	RIGHT ASCEN.	DECLINATION	DIAM.	MAGN.	TYPE OF OBJECT
HN 1502	11 54 10.5	+ 29 08 42.	12		NEBULA
HN 1504	11 54 10.6	- 01 01 48.	18		NEBULA
HN 1503	11 54 10.8	+ 27 51 54.	12		NEBULA
REIZ 1807	11 54 11.	+ 32 18	120	13.1	GALAXY
ZC 1154.2+2435	11 54 12.	+ 24 35	1210		CLUSTER OF GALAXIES
MCG+05-28-053	11 54 12.	+ 32 20	156	14.0	GALAXY
ZC 1154.2+3625	11 54 12.	+ 36 25	610		CLUSTER OF GALAXIES
ZC 1154.2+3646	11 54 12.	+ 36 46	1210		CLUSTER OF GALAXIES
REIZ 1808	11 54 12.	+ 50 07	36	14.9	GALAXY
REIZ 1809	11 54 12.	+ 50 36	42	15.0	GALAXY
ZWG 269.021	11 54 12.	+ 51 07		15.4	GALAXY
UGC 06922	11 54 12.	+ 51 07	120	15.4	GALAXY S
ZWG 269.022	11 54 12.	+ 53 27		14.1	GALAXY
REIZ 1810	11 54 12.	+ 53 27	60	14.2	GALAXY
UGC 06923	11 54 12.	+ 53 27	132	14.1	GALAXY IRR
ZWG 013.008	11 54 12.	- 01 00		15.4	GALAXY
MCG-01-31-001	11 54 12.	- 03 52	18	15.	GALAXY
MCG-06-26-015	11 54 12.	- 37 55	60	14.	GALAXY
RNGC 3986	11 54 13.	+ 32 18		14.0	GALAXY
LB 02153	11 54 14.	+ 52 23 24.		15.1	FAINT BLUE STAR
SEY 143	11 54 14.	+ 53 27 18.		13.8	FAINT GALAXY
KN 12.235	11 54 14.1	+ 48 52 54.			NEBULA
MCG+05-28-054	11 54 15.	+ 32 28	48	14.5	GALAXY
MCG+09-20-039	11 54 15.	+ 52 46	24	16.	GALAXY
MCG+09-20-040	11 54 15.	+ 53 27	120	13.	GALAXY
KN 12.236	11 54 15.0	+ 46 14 20.			NEBULA
KN 12.237	11 54 16.7	+ 51 05 49.			NEBULA
ZWG 069.008	11 54 18.	+ 13 03		15.3	GALAXY
UGC 06924	11 54 18.	+ 13 03	66	15.3	GALAXY Sc
ZC 1154.3+2645	11 54 18.	+ 26 45	3160		CLUSTER OF GALAXIES
ZWG 157.059	11 54 18.	+ 32 25		14.5	GALAXY
UGC 06925	11 54 18.	+ 32 25	48	14.5	GALAXY SB0
REIZ 1811	11 54 18.	+ 51 05	36	13.9	GALAXY
MCG+09-20-041	11 54 18.	+ 55 47 30.	42	16.	GALAXY
UGC 06926	11 54 18.	+ 57 47	84	16.0	GALAXY
ZWG 292.051	11 54 18.	+ 60 35		15.5	GALAXY
MCG+10-17-109B	11 54 18.	+ 61 43	24	17.	GALAXY
MCG+10-17-109A	11 54 18.	+ 61 43	36	17.	GALAXY
MCG+10-17-110	11 54 18.	+ 62 22	15	16.	GALAXY
ZWG 315.004	11 54 18.	+ 62 27		15.7	GALAXY
ZWG 292.052	11 54 18.	+ 62 27		15.7	GALAXY
HN 1505	11 54 18.6	+ 28 00 30.	12		NEBULA
IC 2979	11 54 19.	+ 33 26 09.			NONSTELLAR OBJECT
SEY 144	11 54 19.	+ 51 05 30.		14.3	FAINT GALAXY
LB 02154	11 54 19.	+ 62 33 42.		17.0	FAINT BLUE STAR
KN 12.238	11 54 19.2	+ 47 43 06.			NEBULA
LB 00241	11 54 21.	+ 55 34 00.		15.6	FAINT BLUE STAR
ARC 1421	11 54 21.	+ 68 15		17.2	RICH CLUSTER OF GALAXIES
PATH 1.426	11 54 22.	- 00 41	19		NEBULA
ZWG 013.009	11 54 24.	+ 01 25		15.5	GALAXY
BZW 1154+13.7	11 54 24.	+ 13 42		17.9	COMPACT GALAXY
MCG+09-20-042	11 54 24.	+ 51 04	72	14.	GALAXY
MCG+10-17-111	11 54 24.	+ 57 48	57	15.	GALAXY
MCG+10-17-112	11 54 24.	+ 60 37	27	16.	GALAXY
KN 12.239	11 54 24.3	+ 51 05 31.			NEBULA
KN 12.240	11 54 26.2	+ 47 49 46.			NEBULA
TON-N 1463	11 54 27.	+ 34 31		15.4	BLUE STAR
SEY 145	11 54 27.	+ 55 48 19.		15.4	FAINT GALAXY
MCG-03-31-003	11 54 27.	- 19 33	48	14.5	GALAXY
KN 12.241	11 54 28.7	+ 46 58 38.			NEBULA
HN 1506	11 54 29.6	+ 28 29 06.			NEBULA
ZWG 013.010	11 54 30.	+ 01 26		15.2	GALAXY
ZC 1154.5+1503	11 54 30.	+ 15 03	2490		CLUSTER OF GALAXIES
ZC 1154.5+2001	11 54 30.	+ 20 01	470		CLUSTER OF GALAXIES
ZWG 157.060	11 54 30.	+ 30 40		14.5	GALAXY
UGC 06927	11 54 30.	+ 30 40	66	14.5	GALAXY S0
MCG+05-28-055	11 54 30.	+ 30 40	60	14.5	GALAXY
REIZ 1812	11 54 30.	+ 49 30	132	12.9	GALAXY
SEY 146	11 54 30.	+ 53 06 24.		15.1	FAINT GALAXY
KARA.68 084	11 54 30.	+ 56 33	27		DWARF GALAXY
ZC 1154.5+6738	11 54 30.	+ 67 38	1210		CLUSTER OF GALAXIES
IC 0747	11 54 31.	- 08 01 03.			NONSTELLAR OBJECT
HN 1507	11 54 31.3	+ 01 24 24.	18		NEBULA
ARC 1422	11 54 33.	- 06 46		17.8	RICH CLUSTER OF GALAXIES
ZC 1154.6+0520	11 54 36.	+ 05 20	2290		CLUSTER OF GALAXIES
ZC 1154.6+1640	11 54 36.	+ 16 40	1550		CLUSTER OF GALAXIES
ZC 1154.6+2140	11 54 36.	+ 21 40	1480		CLUSTER OF GALAXIES
MCG+08-22-046	11 54 36.	+ 49 33	252	13.	GALAXY
REIZ 1813	11 54 36.	+ 54 27	12	15.1	GALAXY
MCG+10-17-113	11 54 36.	+ 61 44	27	16.	GALAXY
MCG+10-17-114	11 54 36.	+ 62 27	15	16.	GALAXY
ZC 1154.6-0008	11 54 36.	- 00 08	4230		CLUSTER OF GALAXIES
HN 1508	11 54 39.4	+ 27 53 00.	12		NEBULA
LB 02155	11 54 40.	+ 61 40 48.		17.1	FAINT BLUE STAR
ZC 1154.7+1726	11 54 42.	+ 17 26	340		CLUSTER OF GALAXIES
ZWG 098.010	11 54 42.	+ 18 35		15.7	GALAXY
MCG+04-28-099	11 54 42.	+ 25 28	135	14.4	GALAXY
ZWG 127.110	11 54 42.	+ 25 29		14.4	GALAXY
UGC 06928	11 54 42.	+ 25 29	138	14.4	GALAXY Sb
MCG+06-26-057	11 54 42.	+ 34 26 30.	36	16.	GALAXY
ZWG 186.072	11 54 42.	+ 36 41		15.3	GALAXY
UGC 06929	11 54 42.	+ 36 41	72	15.3	GALAXY Sc
ZWG 243.032	11 54 42.	+ 49 33		14.2	GALAXY
UGC 06930	11 54 42.	+ 49 33	270	14.2	GALAXY SBc
ZWG 292.053	11 54 42.	+ 58 13		14.5	GALAXY
UGC 06931	11 54 42.	+ 58 13	102	14.5	GALAXY
7ZW 431	11 54 42.	+ 58 50			COMPACT GALAXY
MCG+12-11-043	11 54 42.	+ 72 05	36	16.	GALAXY
ZWG 335.001	11 54 42.	+ 72 06		15.5	GALAXY
ZWG 334.057	11 54 42.	+ 72 06		15.5	GALAXY
MCG-02-31-003	11 54 42.	- 13 58	72	14.5	GALAXY
KN 12.242	11 54 42.4	+ 49 33 54.			NEBULA
RNGC 3987	11 54 43.	+ 25 29		14.5	GALAXY
HN 1509	11 54 43.7	+ 26 47 30.	18		NEBULA
MCG+06-26-058	11 54 45.	+ 36 41	60	14.5	GALAXY
REIZ 1814	11 54 47.	+ 25 28	240	14.6	GALAXY
HOLM 308C	11 54 47.	+ 25 28	150	14.5	PART OF MULTIPLE GALAXY
REIZ 1815	11 54 47.	+ 34 25	18	15.8	GALAXY
REIZ 1816	11 54 47.	+ 36 41	48	14.7	GALAXY
HN 1510	11 54 47.9	+ 25 42 18.	12		NEBULA
ZWG 013.011	11 54 48.	+ 00 47		15.6	GALAXY
ZWG 041.010	11 54 48.	+ 04 49		15.2	GALAXY
ZWG 041.011	11 54 48.	+ 07 45		15.2	GALAXY
MRK 644	11 54 48.	+ 23 39	8	17.	GALAXY WITH UV CONTINUUM
ZWG 127.111	11 54 48.	+ 25 30		15.7	GALAXY
MCG+04-28-100	11 54 48.	+ 25 30	36	15.7	GALAXY
ZWG 157.061	11 54 48.	+ 28 09		14.7	GALAXY
MCG+06-26-060	11 54 48.	+ 32 37	66	14.	GALAXY
ZWG 215.008	11 54 48.	+ 40 02		14.9	GALAXY
LB 02156	11 54 48.	+ 50 25 24.		16.2	FAINT BLUE STAR
OCL 0867	11 54 48.	- 62 22	210	12.	OPEN STAR CLUSTER
RNGC 3989	11 54 49.	+ 25 30		15.5	GALAXY
RNGC 3988	11 54 49.	+ 28 09		14.5	GALAXY
ARC 1423	11 54 49.	+ 33 56		16.5	RICH CLUSTER OF GALAXIES
LB 02157	11 54 49.	+ 51 26 48.		15.1	FAINT BLUE STAR
HN 1512	11 54 49.0	+ 00 45 42.	18		NEBULA
HN 1511	11 54 49.6	+ 26 44 06.	18		NEBULA
MCG+05-28-056	11 54 51.	+ 31 23 30.	54	15.5	GALAXY
MCG+07-25-007	11 54 51.	+ 40 03	42	14.	GALAXY
SVEN 248	11 54 51.	- 18 49	18	14.5	GALAXY
HN 0356	11 54 51.	- 73 25			NEBULA
IC 0748	11 54 52.	+ 07 44 21.			NONSTELLAR OBJECT
IC 2980	11 54 52.	- 73 25			NONSTELLAR OBJECT
HOLM 308D	11 54 53.	+ 25 31	30	15.3	PART OF MULTIPLE GALAXY
REIZ 1817	11 54 53.	+ 25 32	30	15.8	GALAXY
REIZ 1818	11 54 53.	+ 28 09	12	14.9	GALAXY
REIZ 1819	11 54 53.	+ 31 22	42	15.3	GALAXY
HN 1514	11 54 53.0	- 01 58 18.	18		NEBULA
HN 1513	11 54 53.7	+ 25 42 54.	12		NEBULA
MCG+01-31-005	11 54 54.	+ 05 21	15	15.5	GALAXY
MCG+01-31-004	11 54 54.	+ 05 21	9	15.5	GALAXY
MCG+01-31-003	11 54 54.	+ 05 21	9	15.5	GALAXY
MCG+01-31-006	11 54 54.	+ 07 45	15	15.2	GALAXY
BZW 1154+14.1	11 54 54.	+ 14 09		18.3	COMPACT GALAXY
ZC 1154.9+2806	11 54 54.	+ 28 06	4230		CLUSTER OF GALAXIES
MCG+05-28-057	11 54 54.	+ 28 10	15	14.7	GALAXY
ZWG 157.062	11 54 54.	+ 31 21		15.5	GALAXY
UGC 06932	11 54 54.	+ 31 21	66	15.5	GALAXY IRR
MCG+06-26-059	11 54 54.	+ 32 32	42	13.5	GALAXY
ZWG 186.073	11 54 54.	+ 32 36		13.8	GALAXY
UGC 06933	11 54 54.	+ 32 36	84	13.8	GALAXY PECULR
KARA.72 311B	11 54 54.	+ 32 36	36		PART OF DOUBLE GALAXY
KARA.72 311A	11 54 54.	+ 32 36	36	13.8	PART OF DOUBLE GALAXY
ZC 1154.9+3400	11 54 54.	+ 34 00	1810		CLUSTER OF GALAXIES
LB 02158	11 54 54.	+ 50 18 00.		16.5	FAINT BLUE STAR
ZWG 013.012	11 54 54.	- 01 57		15.7	GALAXY
RNGC 3991	11 54 55.	+ 32 37		14.0	GALAXY
HARO 05	11 54 55.	+ 32 37			BLUE EMISSION-LINE GALAXY
HOLM 309C	11 54 56.	+ 32 37	78	14.1	PART OF MULTIPLE GALAXY
LB 02159	11 54 56.	+ 52 06 00.		17.2	FAINT BLUE STAR
HOLM 310B	11 54 57.	+ 55 44	36	13.3	PART OF MULTIPLE GALAXY
HN 1515	11 54 57.4	- 00 59 30.	48		NEBULA
ARC 1424	11 54 58.	+ 05 19		16.6	RICH CLUSTER OF GALAXIES
SVEN 249	11 54 58.	- 00 59 12.	60	15.1	GALAXY
REIZ 1821	11 54 59.	+ 32 37	54	13.8	GALAXY
REIZ 1822	11 54 59.	+ 34 25	24	15.6	GALAXY
VDB.66G 105	11 55	+ 38 18	170		DWARF GALAXY
VDB.66G 103	11 55	- 14 17	70		DWARF GALAXY
VDB.66G 104	11 55	- 14 27	70		DWARF GALAXY
VDB.66G 106	11 55	- 22 12	130		DWARF GALAXY
ZWG 127.112	11 55 00.	+ 25 31		14.8	GALAXY
UGC 06935	11 55 00.	+ 25 31	102	14.8	GALAXY S
MCG+04-28-101	11 55 00.	+ 25 31	90	14.8	GALAXY
ZC 1155.0+3127	11 55 00.	+ 31 27	5780		CLUSTER OF GALAXIES
ZWG 186.074	11 55 00.	+ 32 33		13.7	GALAXY
UGC 06936	11 55 00.	+ 32 33	66	13.7	GALAXY S?
REIZ 1823	11 55 00.	+ 53 38	390	11.3	GALAXY
ZWG 269.023	11 55 00.	+ 53 39		10.7	GALAXY
RNGC 3992	11 55 00.	+ 53 39		11.0	GALAXY
UGC 06937	11 55 00.	+ 53 39	498	10.7	GALAXY SBb
MCG+09-20-044	11 55 00.	+ 53 39 30.	420	10.6	GALAXY
ZWG 269.024	11 55 00.	+ 55 43		13.6	GALAXY
UGC 06938	11 55 00.	+ 55 43	84	13.6	GALAXY S0?
RNGC 3990	11 55 00.	+ 55 44		13.5	GALAXY
REIZ 1820	11 55 00.	+ 55 44	21	13.7	GALAXY
MCG+09-20-043	11 55 00.	+ 55 45	66	13.3	GALAXY
ZWG 292.054	11 55 00.	+ 57 50		14.6	GALAXY
UGC 06939	11 55 00.	+ 57 50	66	14.6	GALAXY
MCG+10-17-115	11 55 00.	+ 58 12	78	14.	GALAXY
ZWG 292.055	11 55 00.	+ 61 28		15.7	GALAXY
ZWG 013.013	11 55 00.	- 00 58		15.2	GALAXY
UGC 06934	11 55 00.	- 00 58	108	15.2	GALAXY Sc
MCG+00-31-004	11 55 00.	- 00 58	114	15.2	GALAXY
KN 12.243	11 55 00.8	+ 46 28 41.			NEBULA
RNGC 3994	11 55 01.	+ 32 33		13.5	GALAXY
REIZ 1824	11 55 01.	+ 58 13	90	14.3	GALAXY
HN 1516	11 55 02.1	+ 01 25 06.	12		NEBULA
HOLM 309B	11 55 03.	+ 32 33	30	13.7	PART OF MULTIPLE GALAXY
SN 1956A	11 55 03.	+ 53 39		12.3	SUPERNOVA
KN 12.244	11 55 03.0	+ 49 11 51.			NEBULA
KN 12.245	11 55 03.9	+ 45 25 33.			NEBULA
REIZ 1825	11 55 04.	+ 20 59	12	15.8	GALAXY
ARP 313	11 55 04.	+ 32 34			PECULIAR GALAXY
REIZ 1826	11 55 05.	+ 25 31	90	14.8	GALAXY
HOLM 308A	11 55 05.	+ 25 31	96	14.7	PART OF MULTIPLE GALAXY
REIZ 1827	11 55 05.	+ 32 34	42	13.0	GALAXY
REIZ 1828	11 55 05.	+ 45 25	54	14.8	GALAXY
VV 249B	11 55 06.	+ 32 34	36	13.7	INTERACTING GALAXY
MCG+06-26-061	11 55 06.	+ 32 34	150	12.5	GALAXY
UGC 06940	11 55 06.	+ 53 32	60	16.0	GALAXY S
MCG+10-17-116	11 55 06.	+ 61 28	24	16.	GALAXY
KN 12.246	11 55 06.2	+ 48 57 55.			NEBULA
RNGC 3993	11 55 07.	+ 25 31		15.0	GALAXY
ARP 194	11 55 10.	+ 36 43			PECULIAR GALAXY
HOLM 309A	11 55 11.	+ 32 34	120	13.3	PART OF MULTIPLE GALAXY
REIZ 1829	11 55 11.	+ 32 35	120	12.6	GALAXY
HN 1517	11 55 11.3	+ 27 10 42.			NEBULA
HN 1518	11 55 11.4	+ 01 37 18.			NEBULA
ZWG 098.011	11 55 12.	+ 14 35		14.4	GALAXY
UGC 06941	11 55 12.	+ 14 35	54	14.4	GALAXY S
MCG+03-31-004	11 55 12.	+ 14 35	60	14.4	GALAXY
ZC 1155.2+1929	11 55 12.	+ 19 29	540		CLUSTER OF GALAXIES
ZWG 127.113	11 55 12.	+ 22 33		15.5	GALAXY
MCG+04-28-102	11 55 12.	+ 25 32	90	14.3	GALAXY
ZWG 127.114	11 55 12.	+ 25 33		14.3	GALAXY
UGC 06942	11 55 12.	+ 25 33	102	14.3	GALAXY SB
ZWG 157.063	11 55 12.	+ 28 08		15.2	GALAXY
ZWG 157.064	11 55 12.	+ 29 19		14.8	GALAXY
UGC 06943	11 55 12.	+ 29 19	72	14.8	GALAXY SBb
ZWG 186.075	11 55 12.	+ 32 34		12.9	GALAXY
UGC 06944	11 55 12.	+ 32 34	168	12.9	GALAXY S
VV 249A	11 55 12.	+ 32 35	48	13.3	INTERACTING GALAXY
ZC 1155.2+4639	11 55 12.	+ 46 39	1810		CLUSTER OF GALAXIES
ZC 1155.2+6817	11 55 12.	+ 68 17	1340		CLUSTER OF GALAXIES
MCG-02-31-004	11 55 12.	- 12 07	36	15.5	GALAXY
VHA 124	11 55 12.	- 62 25	240		OPEN STAR CLUSTER

OBJECT NAME	RIGHT ASCEN.	DECLINATION	DIAM.	MAGN.	TYPE OF OBJECT
RNGC 3996	11 55 13.	+ 14 35		14.5	GALAXY
RNGC 3997	11 55 13.	+ 25 33		14.5	GALAXY
RNGC 3984	11 55 13.	+ 29 19		15.0	GALAXY
RNGC 3995	11 55 13.	+ 32 34		13.0	GALAXY
HOLM 308B	11 55 14.	+ 25 33	96	14.3	PART OF MULTIPLE GALAXY
MCG+05-28-058	11 55 15.	+ 29 20	72	14.8	GALAXY
IC 2981	11 55 15.	+ 33 28 03.			NONSTELLAR OBJECT
MCG+10-17-117	11 55 15.	+ 57 50	24	15.	GALAXY
SVEN 250	11 55 15.	- 19 06	30	15.9	GALAXY
REIZ 1830	11 55 16.	+ 22 30	12	15.4	GALAXY
REIZ 1831	11 55 16.	+ 25 33	72	14.3	GALAXY
IC 2982	11 55 17.	+ 28 06			NONSTELLAR OBJECT
HN 1519	11 55 17.2	+ 25 46 54.	18		NEBULA
ZWG 041.012	11 55 18.	+ 07 24		15.4	GALAXY
MCG+04-28-105	11 55 18.	+ 23 24	66	14.8	GALAXY
MCG+04-28-104	11 55 18.	+ 23 28	36	14.7	GALAXY
MCG+04-28-103	11 55 18.	+ 25 25	60	15.2	GALAXY
MCG+05-28-059	11 55 18.	+ 28 10	24	15.2	GALAXY
ZWG 186.076	11 55 18.	+ 36 40		15.1	GALAXY
UGC 06945	11 55 18.	+ 36 40	72	15.1	GALAXY DBL SYS
ZC 1155.3+3721	11 55 18.	+ 37 21	3430		CLUSTER OF GALAXIES
ZC 1155.3+4055	11 55 18.	+ 40 55	1340		CLUSTER OF GALAXIES
MCG+09-20-045	11 55 18.	+ 53 30	60	15.	GALAXY
ZWG 269.025	11 55 18.	+ 55 43		11.2	GALAXY
UGC 06946	11 55 18.	+ 55 43	180	11.2	GALAXY S0
VVI 51	11 55 18.	+ 55 44	90	11.82	SEYFERT GALAXY
REIZ 1832	11 55 18.	+ 55 44	60	12.1	GALAXY
HOLM 310A	11 55 18.	+ 55 44	60	11.8	PART OF MULTIPLE GALAXY
MCG+09-20-046	11 55 18.	+ 55 45	90	11.8	GALAXY
REIZ 1833	11 55 18.	+ 57 52	24	14.7	GALAXY
RNGC 4004B	11 55 19.	+ 28 10		15.0	GALAXY
HN 1520	11 55 19.0	+ 28 56 12.	12		NEBULA
HOLM 312B	11 55 21.	+ 28 06	18	14.8	PART OF MULTIPLE GALAXY
MCG+06-26-062	11 55 21.	+ 36 40	66	15.	GALAXY
REIZ 1835	11 55 22.	+ 23 25	24	15.3	GALAXY
REIZ 1834	11 55 22.	+ 23 29	18	15.6	GALAXY
REIZ 1836	11 55 22.	+ 25 25	42	15.4	GALAXY
HOLM 311A	11 55 23.	+ 36 41	36	13.9	PART OF MULTIPLE GALAXY
MCG+04-28-106	11 55 24.	+ 21 30	36	15.3	GALAXY
UGC 06947	11 55 24.	+ 22 28	60	16.5	GALAXY Sc
ZWG 127.115	11 55 24.	+ 23 24		14.8	GALAXY
UGC 06948	11 55 24.	+ 23 24	102	14.8	GALAXY SB0
KARA.72 312B	11 55 24.	+ 23 24	78	14.8	PART OF DOUBLE GALAXY
ZWG 127.116	11 55 24.	+ 23 29		14.7	GALAXY
KARA.72 312A	11 55 24.	+ 23 29	72	14.7	PART OF DOUBLE GALAXY
ZWG 127.117	11 55 24.	+ 25 21		15.7	GALAXY
ZWG 127.118	11 55 24.	+ 25 25		15.2	GALAXY
UGC 06949	11 55 24.	+ 25 25	66	15.2	GALAXY S
ZWG 157.065	11 55 24.	+ 28 09		14.0	GALAXY
MRK 432	11 55 24.	+ 28 09	35	14.5	GALAXY WITH UV CONTINUUM
UGC 06950	11 55 24.	+ 28 09	120	14.0	GALAXY PECULR
VV 126D	11 55 24.	+ 36 40	18	13.9	INTERACTING GALAXY
VV 126C	11 55 24.	+ 36 40	36	13.9	INTERACTING GALAXY
VV 126B	11 55 24.	+ 36 40	24	13.9	INTERACTING GALAXY
VV 126A	11 55 24.	+ 36 40	24	14.9	INTERACTING GALAXY
VV 126	11 55 24.	+ 36 40	78	14.	INTERACTING GALAXY
RNGC 3998	11 55 24.	+ 55 44		12.0	GALAXY
MCG+10-17-118	11 55 24.	+ 57 36	24	16.	GALAXY
ZC 1155.4+6636	11 55 24.	+ 66 36	1140		CLUSTER OF GALAXIES
MCG-02-31-005	11 55 24.	- 12 34	60	15.	GALAXY
HN 1521	11 55 24.1	+ 27 17 36.	12		NEBULA
RNGC 4003	11 55 25.	+ 23 24		15.0	GALAXY
RNGC 4002	11 55 25.	+ 23 29		14.5	GALAXY
RNGC 4000	11 55 25.	+ 25 25		15.0	GALAXY
RNGC 4004A	11 55 25.	+ 28 09		14.0	GALAXY
HOLM 311B	11 55 25.	+ 36 41	18	14.9	PART OF MULTIPLE GALAXY
HN 1522	11 55 26.4	+ 00 57 12.	12		NEBULA
REIZ 1837	11 55 28.	+ 14 33	18	12.9	GALAXY
REIZ 1838	11 55 28.	+ 21 30	24	15.5	GALAXY
HOLM 314B	11 55 28.	+ 47 35	18	15.0	PART OF MULTIPLE GALAXY
ZWG 041.013	11 55 30.	+ 06 11		15.2	GALAXY
MCG+01-31-007	11 55 30.	+ 06 11	12	15.2	GALAXY
ZC 1155.5+1953	11 55 30.	+ 19 53	940		CLUSTER OF GALAXIES
ZWG 127.119	11 55 30.	+ 21 32		15.3	GALAXY
MCG+04-28-107	11 55 30.	+ 25 23	48	14.1	GALAXY
VV 230	11 55 30.	+ 28 10	108	14.	INTERACTING GALAXY
HZ 16	11 55 30.	+ 30 14		14.4	BLUE STAR
REIZ 1839	11 55 30.	+ 47 36	18	14.4	GALAXY
MCG+08-22-047	11 55 30.	+ 47 36	36	15.0	GALAXY
ZWG 243.033	11 55 30.	+ 47 37		15.6	GALAXY
MCG+09-20-047	11 55 30.	+ 53 00	24	17.	GALAXY
ZC 1155.5+6528	11 55 30.	+ 65 28	1410		CLUSTER OF GALAXIES
ZWG 013.014	11 55 30.	- 03 27		14.9	GALAXY
MCG+00-31-005	11 55 30.	- 03 27	24	14.9	GALAXY
KARA.73B 0514	11 55 30.	- 03 27	30	14.9	ISOLATED GALAXY S
OCL 0868	11 55 30.	- 64 12	600	9.	OPEN STAR CLUSTER
VHA 125	11 55 30.	- 64 12	420		OPEN STAR CLUSTER
RNGC 4007	11 55 31.	+ 23 24			NON-EXISTENT OBJECT
RNGC 4001	11 55 31.	+ 47 36		15.5	GALAXY
KN 12.247	11 55 32.6	+ 47 36 53.			NEBULA
TON-N 1464	11 55 33.	+ 33 17		16.9	BLUE STAR
MCG-04-28-007	11 55 33.	- 21 31	84	16.	GALAXY
HN 1523	11 55 33.9	+ 26 47 06.			NEBULA
REIZ 1842	11 55 34.	+ 23 24	6	15.4	GALAXY
REIZ 1841	11 55 34.	- 01 52	18	13.8	GALAXY
HN 1524	11 55 34.2	+ 26 42 54.			NEBULA
HOLM 312A	11 55 35.	+ 28 07	120	13.7	PART OF MULTIPLE GALAXY
REIZ 1840	11 55 35.	+ 28 09	60	13.8	GALAXY
HN 1525	11 55 35.9	+ 28 37 54.	24		NEBULA
ZC 1155.6+0834	11 55 36.	+ 08 34	670		CLUSTER OF GALAXIES
ZWG 127.120	11 55 36.	+ 25 24		14.1	GALAXY
UGC 06952	11 55 36.	+ 25 24	66	14.1	GALAXY S
ARC 1425	11 55 36.	+ 26 40		17.2	RICH CLUSTER OF GALAXIES
MCG+05-28-060	11 55 36.	+ 28 10	114	14.0	GALAXY
ZWG 157.066	11 55 36.	+ 28 28		13.1	GALAXY
UGC 06953	11 55 36.	+ 28 28	144	13.1	GALAXY E-S0
ZC 1155.6+4901	11 55 36.	+ 49 01	870		CLUSTER OF GALAXIES
MCG+08-22-048	11 55 36.	+ 49 10	36	16.	GALAXY
ZC 1155.6-0119	11 55 36.	- 01 19	610		CLUSTER OF GALAXIES
ZWG 013.015	11 55 36.	- 01 49		14.2	GALAXY
UGC 06951	11 55 36.	- 01 49	84	14.2	GALAXY E
MCG+00-31-006	11 55 36.	- 01 49	30	14.2	GALAXY
MCG-05-28-017	11 55 36.	- 28 45	9	16.	GALAXY
KN 12.248	11 55 36.7	+ 49 09 47.			NEBULA
RNGC 4005	11 55 37.	+ 25 24		14.0	GALAXY
RNGC 4006	11 55 37.	- 01 49		14.0	GALAXY
ARC 1426	11 55 39.	- 12 44		17.2	RICH CLUSTER OF GALAXIES
REIZ 1843	11 55 41.	+ 28 28	48	13.1	GALAXY
ZC 1155.7+0137	11 55 42.	+ 01 37	1880		CLUSTER OF GALAXIES
ZWG 157.067	11 55 42.	+ 28 38		15.6	GALAXY
ZWG 186.077	11 55 42.	+ 36 54		15.6	GALAXY
ZC 1155.7+4701	11 55 42.	+ 47 01	1010		CLUSTER OF GALAXIES
ZWG 013.016	11 55 42.	- 01 36		15.4	GALAXY
SN 1954A	11 55 42.	- 05 39		17.5	SUPERNOVA
HN 1526	11 55 42.0	- 01 37 06.	12		NEBULA
RNGC 4008	11 55 43.	+ 28 28		13.0	GALAXY
IC 2983	11 55 44.	- 01 48			NONSTELLAR OBJECT
MCG+05-28-061	11 55 45.	+ 28 30	36	13.1	GALAXY
HN 1527	11 55 45.2	- 01 41 06.	12		NEBULA
ARC 1428	11 55 47.	+ 10 08		17.8	RICH CLUSTER OF GALAXIES
ZC 1155.8+0056	11 55 48.	+ 00 56	1880		CLUSTER OF GALAXIES
ZWG 127.121	11 55 48.	+ 25 23		15.7	GALAXY
ZWG 157.068	11 55 48.	+ 27 48		14.6	GALAXY
UGC 06954	11 55 48.	+ 27 48	90	14.6	GALAXY SBc-IRR
MCG+05-28-062	11 55 48.	+ 28 40	36	15.6	GALAXY
ARC 1427	11 55 48.	+ 30 59		17.0	RICH CLUSTER OF GALAXIES
ZC 1155.8+3441	11 55 48.	+ 34 41	2290		CLUSTER OF GALAXIES
ZWG 186.078	11 55 48.	+ 38 20		15.2	GALAXY
UGC 06955	11 55 48.	+ 38 20	360	15.2	GALAXY DWRF IR
ZC 1155.8+3823	11 55 48.	+ 38 23	540		CLUSTER OF GALAXIES
UGC 06956	11 55 48.	+ 51 13	162	17.	GALAXY DWRF SP
REIZ 1844	11 55 48.	+ 54 15	42	14.8	GALAXY
ZWG 292.056	11 55 48.	+ 57 52		14.8	GALAXY
UGC 06957	11 55 48.	+ 57 52	120	14.8	GALAXY SB
ZWG 013.017	11 55 48.	- 01 40		15.5	GALAXY
MCG-06-26-016	11 55 48.	- 39 17	90	15.	GALAXY
ASS 63	11 55 48.	- 62 59			OB ASSOCIATION CEN OB1
RNGC 4011	11 55 49.	+ 25 23		15.5	GALAXY
RNGC 3999	11 55 49.	+ 25 25		15.5	GALAXY
RNGC 4009	11 55 49.	+ 25 29			GALAXY
HN 1528	11 55 49.0	+ 27 54 42.	12		NEBULA
KN 12.249	11 55 49.1	+ 46 20 40.			NEBULA
HN 1529	11 55 50.1	- 01 59 54.	30		NEBULA
SVEN 251	11 55 51.	- 19 15	12	14.4	GALAXY
MCG-04-28-008	11 55 51.	- 22 10 30.	150	15.	GALAXY
KN 12.250	11 55 51.5	+ 51 11 48.			NEBULA
HN 1530	11 55 51.7	+ 25 28 30.	12		NEBULA
REIZ 1846	11 55 52.	+ 10 18	60	14.5	GALAXY
REIZ 1845	11 55 52.	- 02 01	36	14.7	GALAXY
HN 1531	11 55 52.4	+ 24 47 42.			NEBULA
LB 02160	11 55 53.	+ 59 26 00.		16.4	FAINT BLUE STAR
ZWG 013.019	11 55 54.	+ 01 00		15.4	GALAXY
UGC 06959	11 55 54.	+ 01 00	72	15.4	GALAXY CHAIN
MCG+00-31-009	11 55 54.	+ 01 00	30	15.4	GALAXY
MCG+00-31-008	11 55 54.	+ 01 01	18	15.4	GALAXY
ZWG 069.009	11 55 54.	+ 10 18		14.6	GALAXY
UGC 06960	11 55 54.	+ 10 18	126	14.6	GALAXY Sb
MCG+02-31-006	11 55 54.	+ 10 18	108	14.6	GALAXY
ZC 1155.9+3007	11 55 54.	+ 30 07	810		CLUSTER OF GALAXIES
MCG+06-26-063	11 55 54.	+ 38 21	240	14.	GALAXY
ZC 1155.9+4146	11 55 54.	+ 41 46	740		CLUSTER OF GALAXIES
MCG+10-17-119	11 55 54.	+ 57 52	78	15.	GALAXY
ZC 1155.9+6435	11 55 54.	+ 64 35	1550		CLUSTER OF GALAXIES
ZWG 013.018	11 55 54.	- 01 59		14.5	GALAXY
UGC 06958	11 55 54.	- 01 59	78	14.5	GALAXY Sb
MCG+00-31-007	11 55 54.	- 01 59	96	14.5	GALAXY
HN 1532	11 55 54.5	+ 27 47 54.	30		NEBULA
RNGC 4012	11 55 55.	+ 10 18		14.5	GALAXY
RNGC 4016	11 55 55.	+ 27 49		13.5	GALAXY
KN 12.251	11 55 56.2	+ 46 27 19.			NEBULA
MCG+05-28-063	11 55 57.	+ 27 49 30.	90	14.6	GALAXY
REIF 2.139	11 55 57.81	+ 44 13 34.9			NEBULA
ARP 305	11 55 58.	+ 27 51			PECULIAR GALAXY
HN 1533	11 55 58.5	+ 26 24 54.	12		NEBULA
REIZ 1847	11 55 59.	+ 34 37	18	15.7	GALAXY
REIZ 1848	11 55 59.	+ 38 44	42	15.5	GALAXY
HOLM 313A	11 55 59.	+ 43 01	90	12.4	PART OF MULTIPLE GALAXY
REIZ 1849	11 55 59.	+ 44 12	270	12.8	GALAXY
HOLM 314A	11 55 59.	+ 47 31	204	12.5	PART OF MULTIPLE GALAXY
VDB.66G 107	11 56	+ 38 08	100		DWARF GALAXY
FRIG 052	11 56 00.	+ 12 31		12.9	FAINT BLUE STAR
ZWG 098.012	11 56 00.	+ 16 27		13.5	GALAXY
UGC 06961	11 56 00.	+ 16 27	120	13.5	GALAXY S0-a
ZWG 098.013	11 56 00.	+ 19 08		15.1	GALAXY
KARA.73B 0515	11 56 00.	+ 19 08	42	15.1	ISOLATED GALAXY S
MCG+04-28-109	11 56 00.	+ 25 19	24	14.2	GALAXY
MCG+04-28-110	11 56 00.	+ 25 19 30.	48	14.2	GALAXY
MCG+04-28-108	11 56 00.	+ 25 35	90	14.7	GALAXY
ZC 1156.0+3603	11 56 00.	+ 36 03	1210		CLUSTER OF GALAXIES
ZWG 215.009	11 56 00.	+ 43 00		13.4	GALAXY
UGC 06962	11 56 00.	+ 43 00	156	13.4	GALAXY SBc
KARA.72 313A	11 56 00.	+ 43 00	138	13.4	PART OF DOUBLE GALAXY
IC 0749	11 56 00.	+ 43 01	108	13.2	GALAXY Sc
MCG+07-25-008	11 56 00.	+ 43 02	132	12.	GALAXY
ZWG 215.010	11 56 00.	+ 44 13		12.4	GALAXY
UGC 06963	11 56 00.	+ 44 13	318	12.4	GALAXY Sb
MCG+07-25-009	11 56 00.	+ 44 15	270	12.	GALAXY
ZWG 243.034	11 56 00.	+ 47 32		13.1	GALAXY
UGC 06964	11 56 00.	+ 47 32	246	13.1	GALAXY Sc-IRR
MCG+08-22-049	11 56 00.	+ 47 32	270	12.5	GALAXY
ZWG 013.020	11 56 00.	- 01 50		14.7	GALAXY
MCG+00-31-010	11 56 00.	- 01 50	48	14.7	GALAXY
MCG-02-31-006	11 56 00.	- 14 14	60	15.	GALAXY
MCG-02-31-007	11 56 00.	- 14 27	72	17.	GALAXY
KN 12.252	11 56 00.4	+ 46 01 10.			NEBULA
HN 1534	11 56 00.6	- 01 51 54.	12		NEBULA
RNGC 4014	11 56 01.	+ 16 27		13.5	GALAXY
RNGC 4013	11 56 01.	+ 44 14		12.5	GALAXY
RNGC 4010	11 56 01.	+ 47 32		13.0	GALAXY
RNGC 4024	11 56 01.	- 18 05		13.0	GALAXY
MCG-03-31-004	11 56 03.	- 18 04	108	13.	GALAXY
KN 12.253	11 56 03.3	+ 47 32 19.			NEBULA
REIZ 1850	11 56 04.	+ 25 19	18	15.7	GALAXY
REIZ 1851	11 56 04.	+ 25 36	84	14.7	GALAXY
REIZ 1852	11 56 05.	+ 47 31	192	12.5	GALAXY
MCG+03-31-005	11 56 06.	+ 16 27	132	13.5	GALAXY
MCG+03-31-006	11 56 06.	+ 19 10	48	15.1	GALAXY
ZWG 098.014	11 56 06.	+ 19 25		15.7	GALAXY
ZWG 127.122	11 56 06.	+ 25 19		14.2	GALAXY
UGC 06965	11 56 06.	+ 25 19	84	14.2	GALAXY DBL SYS
VV 216B	11 56 06.	+ 25 19	48	15.	INTERACTING GALAXY
VV 216A	11 56 06.	+ 25 19	60	14.	INTERACTING GALAXY
VV 216	11 56 06.	+ 25 19	78		INTERACTING GALAXY
KARA.72 314B	11 56 06.	+ 25 19	54		PART OF DOUBLE GALAXY
KARA.72 314A	11 56 06.	+ 25 19	78	14.2	PART OF DOUBLE GALAXY
ZWG 127.123	11 56 06.	+ 25 36		14.7	GALAXY

OBJECT NAME	RIGHT ASCEN.	DECLINATION	DIAM.	MAGN.	TYPE OF OBJECT
UGC 06966	11 56 06.	+ 25 36	108	14.7	GALAXY Sa-b
ZWG 157.069	11 56 06.	+ 27 43		13.5	GALAXY
UGC 06967	11 56 06.	+ 27 43	108	13.5	GALAXY Sb/SBc
ZWG 157.070	11 56 06.	+ 28 34		15.2	GALAXY
UGC 06968	11 56 06.	+ 28 34	180	15.2	GALAXY S
ZC 1156.1+4909	11 56 06.	+ 49 09	1810		CLUSTER OF GALAXIES
ZWG 269.026	11 56 06.	+ 53 42		15.5	GALAXY
UGC 06969	11 56 06.	+ 53 42	102	15.5	GALAXY IRR
RNGC 4015	11 56 07.	+ 25 19		14.0	GALAXY
RNGC 4018	11 56 07.	+ 25 36		14.5	GALAXY
ARP 138	11 56 10.	+ 25 20			PECULIAR GALAXY
REIZ 1853	11 56 10.	+ 27 50	36	14.5	GALAXY
HN 1535	11 56 10.4	+ 28 34 06.			NEBULA
HN 1536	11 56 10.9	+ 27 10 42.	24		NEBULA
SEY 147	11 56 11.	+ 53 43 18.		14.8	FAINT GALAXY
HN 1537	11 56 11.6	- 02 22 30.	18		NEBULA
BZW 1156+11.1	11 56 12.	+ 11 09		18.6	COMPACT GALAXY
ZWG 098.015	11 56 12.	+ 16 00		15.0	GALAXY
MCG+03-31-007	11 56 12.	+ 16 00	48	15.0	GALAXY
ZC 1156.2+2201	11 56 12.	+ 22 01	5310		CLUSTER OF GALAXIES
ZC 1156.2+2255	11 56 12.	+ 22 55	1480		CLUSTER OF GALAXIES
MCG+05-28-064	11 56 12.	+ 28 36	162	15.2	GALAXY
TON-N 0072	11 56 12.	+ 29 12		15.5	BLUE STAR
REIZ 1854	11 56 12.	+ 57 54	90	14.4	GALAXY
ZC 1156.2+6007	11 56 12.	+ 60 07	1210		CLUSTER OF GALAXIES
MCG+10-17-120	11 56 12.	+ 60 20	45	15.	GALAXY
ZC 1156.2+7202	11 56 12.	+ 72 02	3020		CLUSTER OF GALAXIES
ZWG 013.022	11 56 12.	- 01 10		14.6	GALAXY
UGC 06970	11 56 12.	- 01 10	96	14.6	GALAXY IRR
KARA.72 315B	11 56 12.	- 01 10	12		PART OF DOUBLE GALAXY
KARA.72 315A	11 56 12.	- 01 10	90	14.6	PART OF DOUBLE GALAXY
MCG+00-31-011	11 56 12.	- 01 10	90	14.6	GALAXY
ZWG 013.021	11 56 12.	- 02 21		15.3	GALAXY
HN 1538	11 56 12.1	- 01 11 42.	36		NEBULA
REIN 7.065A	11 56 12.34	- 01 10 59.7			NEBULA
REIN 7.065B	11 56 12.38	- 01 10 59.9			NEBULA
SC 1153-4641.4	11 56 14.	- 46 58 06.	18		NEBULA
KN 12.254	11 56 14.7	+ 45 57 18.			NEBULA
MCG+09-20-048	11 56 15.	+ 53 42	96	14.	GALAXY
SVEN 252	11 56 16.	- 01 12	60	14.9	GALAXY
REIZ 1855	11 56 16.	- 01 12	60	14.6	GALAXY
REIZ 1856	11 56 17.	+ 46 17	42	14.3	GALAXY
MCG+04-28-111	11 56 18.	+ 25 30	54	14.4	GALAXY
MCG+05-28-065	11 56 18.	+ 27 45	108	12.5	GALAXY
ZWG 157.071	11 56 18.	+ 28 40		15.3	GALAXY
ZWG 157.072	11 56 18.	+ 30 41		13.2	GALAXY
UGC 06971	11 56 18.	+ 30 41	120	13.2	GALAXY
MCG+05-28-066	11 56 18.	+ 30 43	114	13.2	GALAXY
ZWG 215.011	11 56 18.	+ 42 50		15.1	GALAXY
UGC 06972	11 56 18.	+ 42 50	84	15.1	GALAXY S
IC 0751	11 56 18.	+ 42 50 57.			NONSTELLAR OBJECT
MCG+07-25-011	11 56 18.	+ 42 52	63	14.	GALAXY
ZWG 215.012	11 56 18.	+ 42 59		12.7	GALAXY
UGC 06973	11 56 18.	+ 42 59	186	12.7	GALAXY S
KARA.72 313B	11 56 18.	+ 42 59	216	12.7	PART OF DOUBLE GALAXY
HOLM 313B	11 56 18.	+ 43 00	48	12.9	PART OF MULTIPLE GALAXY
IC 0750	11 56 18.	+ 43 00	102	13.0	GALAXY
MCG+07-25-010	11 56 18.	+ 43 01	132	13.	GALAXY
KN 12.255	11 56 18.3	+ 46 01 00.			NEBULA
RNGC 4019	11 56 19.	+ 14 29			NON-EXISTENT OBJECT
RNGC 4017	11 56 19.	+ 27 45		13.5	GALAXY
MCG+04-28-112	11 56 21.	+ 25 21 30.	12	15.3	GALAXY
MCG+05-28-067	11 56 21.	+ 28 42	12	15.3	GALAXY
TON-N 1465	11 56 21.	+ 34 51		16.7	BLUE STAR
HN 1539	11 56 21.8	+ 28 40 18.			NEBULA
REIZ 1857	11 56 22.	+ 14 29	18	15.5	GALAXY
HN 1540	11 56 22.3	+ 02 09 06.	12		NEBULA
REIZ 1858	11 56 23.	+ 30 42	84	13.2	GALAXY
KN 12.256	11 56 23.9	+ 51 10 09.			NEBULA
UGC 06974	11 56 24.	+ 21 11	60	16.0	GALAXY S
MCG+04-28-114	11 56 24.	+ 22 04	18	15.0	GALAXY
MCG+04-28-113	11 56 24.	+ 25 16	42	14.6	GALAXY
ZWG 127.124	11 56 24.	+ 25 22		15.3	GALAXY
ZWG 127.125	11 56 24.	+ 25 30		14.4	GALAXY
UGC 06975	11 56 24.	+ 25 30	78	14.4	GALAXY S0
ZWG 215.013	11 56 24.	+ 44 28		15.7	GALAXY
MCG+07-25-012	11 56 24.	+ 44 30	36	15.5	GALAXY
ZC 1156.4+4820	11 56 24.	+ 48 20	1410		CLUSTER OF GALAXIES
ZC 1156.4+5009	11 56 24.	+ 50 09	1550		CLUSTER OF GALAXIES
ZWG 269.027	11 56 24.	+ 54 30		14.7	GALAXY
MRK 433	11 56 24.	+ 54 30	18	15.	GALAXY WITH UV CONTINUUM
REIZ 1859	11 56 24.	+ 54 31	18	14.3	GALAXY
MCG+09-20-049	11 56 24.	+ 54 31	36	14.	GALAXY
MCG-03-31-005	11 56 24.	- 20 03	30	15.	GALAXY
RNGC 4021	11 56 25.	+ 25 22		15.5	GALAXY
RNGC 4022	11 56 25.	+ 25 30		14.5	GALAXY
RNGC 4020	11 56 25.	+ 30 41		13.0	GALAXY
MAI 065	11 56 25.	+ 56 40	33		DWARF SPHEROIDAL GALAXY
KN 12.257	11 56 25.5	+ 49 15 33.			NEBULA
HELW 147	11 56 26.	- 18 44 54.			NEBULA
SVEN 253	11 56 27.	- 18 45	36	13.7	GALAXY
REIZ 1860	11 56 28.	+ 25 16	30	14.4	GALAXY
REIZ 1861	11 56 28.	+ 27 45	72	14.0	GALAXY
SVEN 254	11 56 28.	- 00 48 12.	12	15.9	GALAXY
REIZ 1862	11 56 29.	+ 47 59	18	14.4	GALAXY
HN 1541	11 56 29.0	+ 25 02 36.			NEBULA
ZWG 127.126	11 56 30.	+ 22 06		15.0	GALAXY
UGC 06976	11 56 30.	+ 22 06	96	15.0	GALAXY S0
ZWG 127.127	11 56 30.	+ 25 16		14.6	GALAXY
UGC 06977	11 56 30.	+ 25 16	66	14.6	GALAXY
ZWG 157.073	11 56 30.	+ 30 58		15.6	GALAXY
MCG+05-28-068	11 56 30.	+ 31 00	48	15.6	GALAXY
ARC 1429	11 56 30.	+ 36 02		17.6	RICH CLUSTER OF GALAXIES
ZWG 292.057	11 56 30.	+ 58 37		15.7	GALAXY
ZC 1156.5+6206	11 56 30.	+ 62 06	1080		CLUSTER OF GALAXIES
MCG-03-31-006	11 56 30.	- 18 44	36	14.5	GALAXY
RNGC 4023	11 56 31.	+ 25 16		14.5	GALAXY
KN 12.258	11 56 31.5	+ 46 05 58.			NEBULA
IC 2984	11 56 33.	+ 30 58 32.			NONSTELLAR OBJECT
LB 02162	11 56 33.	+ 49 30 36.		15.5	FAINT BLUE STAR
SN 1964E	11 56 33.	+ 52 58		12.9	SUPERNOVA
LB 02161	11 56 33.	+ 53 23 48.		17.1	FAINT BLUE STAR
REIZ 1863	11 56 34.	+ 22 04	18	15.3	GALAXY
REIZ 1864	11 56 34.	+ 30 59	30	14.9	GALAXY
KN 12.259	11 56 34.5	+ 46 12 37.			NEBULA
KN 12.260	11 56 34.7	+ 46 16 14.			NEBULA
LB 02163	11 56 35.	+ 51 24 18.		14.0	FAINT BLUE STAR
REIZ 1865	11 56 35.	+ 53 00	90	14.0	GALAXY

OBJECT NAME	RIGHT ASCEN.	DECLINATION	DIAM.	MAGN.	TYPE OF OBJECT
SEY 148	11 56 35.	+ 53 00 18.		13.7	FAINT GALAXY
UGC 06980	11 56 36.	+ 24 45	84	17.	GALAXY DWRF IR
ZC 1156.6+2558	11 56 36.	+ 25 58	2220		CLUSTER OF GALAXIES
ZWG 157.074	11 56 36.	+ 31 00		15.2	GALAXY
UGC 06981	11 56 36.	+ 31 00	66	15.2	GALAXY SB
MCG+05-28-069	11 56 36.	+ 31 03	60	15.2	GALAXY
ZWG 186.079	11 56 36.	+ 34 51		15.5	GALAXY
ZWG 186.080	11 56 36.	+ 38 03		14.9	GALAXY
UGC 06982	11 56 36.	+ 38 03	168	14.9	GALAXY SBc
MCG+06-26-064	11 56 36.	+ 38 05	270	14.	GALAXY
ZWG 243.035	11 56 36.	+ 46 16		15.7	GALAXY
ZC 1156.6+5049	11 56 36.	+ 50 49	2290		CLUSTER OF GALAXIES
MCG+09-20-050	11 56 36.	+ 51 04	120	15.	GALAXY
ZWG 269.028	11 56 36.	+ 52 59		14.5	GALAXY
UGC 06983	11 56 36.	+ 52 59	264	14.5	GALAXY SBc
MCG+10-17-121	11 56 36.	+ 58 37	45	15.	GALAXY
MCG+11-15-008	11 56 36.	+ 65 57	24	18.	GALAXY
SHAH 157	11 56 36.	+ 76 04	54		GROUP OF COMPACT GALAXIES
ZWG 013.023	11 56 36.	- 00 13		14.3	GALAXY
UGC 06979	11 56 36.	- 00 13	36	14.3	GALAXY PECULR
MCG+00-31-012	11 56 36.	- 00 13	24	14.3	GALAXY
UGC 06978	11 56 36.	- 02 13	78	17.	GALAXY DWARF
RNGC 4025	11 56 37.	+ 38 04		15.0	GALAXY
HN 1542	11 56 37.8	- 01 25 36.	18		NEBULA
IC 2985	11 56 38.	+ 31 00 32.			NONSTELLAR OBJECT
HN 1543	11 56 38.2	- 00 15 12.	18		NEBULA
REIZ 1866	11 56 40.	+ 31 01	36	14.2	GALAXY
IC 0753	11 56 40.	- 00 14 52.			NONSTELLAR OBJECT
SVEN 255	11 56 40.	- 00 19 12.	12	14.0	GALAXY
REIZ 1867	11 56 41.	+ 34 51	36	14.7	GALAXY
HOLM 315A	11 56 41.	+ 34 51	48	14.8	PART OF MULTIPLE GALAXY
REIZ 1868	11 56 41.	+ 38 05	60	15.3	GALAXY
IC 0752	11 56 41.	+ 42 50 44.			NONSTELLAR OBJECT
SC 1154-4641.7	11 56 41.	- 46 58 24.	18		NEBULA
ZWG 041.014	11 56 42.	+ 06 28		15.3	GALAXY
MCG+06-26-065	11 56 42.	+ 34 52 30.	36	15.	GALAXY
MCG+06-26-066	11 56 42.	+ 34 53	18	15.5	GALAXY
ZWG 215.014	11 56 42.	+ 42 50		15.2	GALAXY
MCG+08-22-050	11 56 42.	+ 46 16	42	16.	GALAXY
MCG+09-20-051	11 56 42.	+ 52 58	180	14.	GALAXY
ZC 1156.7+6050	11 56 42.	+ 60 50	1140		CLUSTER OF GALAXIES
HOLM 315B	11 56 43.	+ 34 52	30	15.1	PART OF MULTIPLE GALAXY
HN 1544	11 56 45.6	+ 26 41 30.	12		NEBULA
KN 12.261	11 56 45.9	+ 47 59 29.			NEBULA
SVEN 256	11 56 46.	- 00 50	12	15.5	GALAXY
SVEN 257	11 56 46.	- 00 53 12.	42	15.1	GALAXY
SVEN 258	11 56 46.	- 01 23	24	14.6	GALAXY
ZWG 013.026	11 56 48.	+ 01 43		15.6	GALAXY
ZC 1156.8+1054	11 56 48.	+ 10 54	4030		CLUSTER OF GALAXIES
ZWG 127.128	11 56 48.	+ 21 18		15.7	GALAXY
ZC 1156.8+3043	11 56 48.	+ 30 43	1210		CLUSTER OF GALAXIES
ZWG 269.029	11 56 48.	+ 51 15		11.5	GALAXY
UGC 06985	11 56 48.	+ 51 15	276	11.5	GALAXY S0
ZC 1156.8+5627	11 56 48.	+ 56 27	3160		CLUSTER OF GALAXIES
ZWG 013.025	11 56 48.	- 01 21		14.5	GALAXY
UGC 06984	11 56 48.	- 01 21	54	14.5	GALAXY P
MCG+00-31-013	11 56 48.	- 01 21	30	14.5	GALAXY
MCG-05-28-018	11 56 48.	- 28 37	30	15.	GALAXY
RNGC 4026	11 56 49.	+ 51 14		12.0	GALAXY
IC 0754	11 56 50.	- 01 22 46.			NONSTELLAR OBJECT
KN 12.262	11 56 50.5	+ 51 14 29.			NEBULA
MCG+06-26-067	11 56 51.	+ 35 10 30.	60	15.	GALAXY
REIN 2.140	11 56 51.05	+ 51 14 25.0			NEBULA
REIZ 1869	11 56 52.	+ 24 46	24	15.7	GALAXY
REIZ 1870	11 56 53.	+ 51 14	228	12.3	GALAXY
HN 1545	11 56 53.6	- 00 52 36.	30		NEBULA
REIN 7.066A	11 56 53.67	- 00 52 36.6			NEBULA
REIN 7.066B	11 56 53.70	- 00 52 36.5			NEBULA
ZWG 041.015	11 56 54.	+ 07 07		15.3	GALAXY
ZWG 098.016	11 56 54.	+ 18 02		15.3	GALAXY
UGC 06986	11 56 54.	+ 18 02	56	15.3	GALAXY S
KARA.73B 0516	11 56 54.	+ 18 02	66	15.3	ISOLATED GALAXY S
ZWG 127.129	11 56 54.	+ 25 45		15.5	GALAXY
ZC 1156.9+2902	11 56 54.	+ 29 02	1340		CLUSTER OF GALAXIES
TON-N 0599	11 56 54.	+ 29 33		15.6	BLUE STAR
ARC 1431	11 56 54.	+ 30 27		17.2	RICH CLUSTER OF GALAXIES
MCG+05-28-070	11 56 54.	+ 30 28	66	15.3	GALAXY
ZWG 186.081	11 56 54.	+ 35 10		15.5	GALAXY
MRK 434	11 56 54.	+ 35 10	18	15.5	GALAXY WITH UV CONTINUUM
MCG+11-15-009	11 56 54.	+ 63 20	39	16.	GALAXY
ZWG 013.027	11 56 54.	- 00 51		15.6	GALAXY
MCG-02-31-008	11 56 54.	- 14 51	48	15.	GALAXY
SC 1154-4636.2	11 56 55.	- 46 52 54.	30		NEBULA
HN 1546	11 56 55.1	+ 25 44 12.			NEBULA
ARC 1430	11 56 56.	+ 50 04		17.6	RICH CLUSTER OF GALAXIES
LB 02164	11 56 56.	+ 58 46 12.		15.9	FAINT BLUE STAR
HELW 148	11 56 56.	- 19 03 06.			NEBULA
HN 1547	11 56 56.8	+ 28 12 12.	24		NEBULA
MCG+03-31-008	11 56 57.	+ 18 03	72	15.3	GALAXY
MCG+09-20-052	11 56 57.	+ 51 12 30.	240	11.7	GALAXY
SVEN 260	11 56 57.	- 19 00	120	11.0	GALAXY
SVEN 259	11 56 57.	- 19 03 12.	48	13.8	GALAXY
MCG-06-26-017	11 56 57.	- 36 26	30	15.	GALAXY
SHB 191	11 56 57.8	+ 29 31 27.		15.6	QUASI-STELLAR OBJECT
REIZ 1871	11 56 58.	+ 30 26	30	14.2	GALAXY
BC 4C29.45	11 56 58.1	+ 29 31 24.		16.	QUASI-STELLAR OBJECT
ARP 022	11 56 59.	- 18 59			PECULIAR GALAXY
ZWG 041.016	11 57 00.	+ 03 36		15.5	GALAXY
ZC 1157.0+0620	11 57 00.	+ 06 20	2080		CLUSTER OF GALAXIES
ZWG 069.010	11 57 00.	+ 14 10		14.9	GALAXY
MCG+02-31-007	11 57 00.	+ 14 10	48	14.9	GALAXY
ZC 1157.0+1512	11 57 00.	+ 15 12	540		CLUSTER OF GALAXIES
MCG+04-28-115	11 57 00.	+ 21 30	30	15.1	GALAXY
ZWG 127.130	11 57 00.	+ 21 32		15.1	GALAXY
ZC 1157.0+2326	11 57 00.	+ 23 26	1480		CLUSTER OF GALAXIES
ZWG 157.075	11 57 00.	+ 26 50		15.7	GALAXY
ZWG 157.076	11 57 00.	+ 30 26		15.3	GALAXY
UGC 06987	11 57 00.	+ 30 26	78	15.3	GALAXY SBb
MCG+14-06-004	11 57 00.	+ 86 29	54	16.	GALAXY
MCG+15-01-012	11 57 00.	+ 88 25	36	15.	GALAXY
ZWG 013.028	11 57 00.	- 03 05		15.6	GALAXY
BZW 1157-19.0	11 57 00.	- 18 59			COMPACT GALAXY
VV 066	11 57 00.	- 18 59	150	11.6	INTERACTING GALAXY
MCG-03-31-008	11 57 00.	- 18 59	180	12.	GALAXY
MCG-03-31-007	11 57 00.	- 19 03	48	15.	GALAXY
RNGC 4027	11 57 01.	- 18 59		12.0	GALAXY
SVEN 261	11 57 03.	- 19 03	6	16.5	GALAXY
SVEN 262	11 57 03.	- 19 04	12	15.5	GALAXY

OBJECT NAME	RIGHT ASCEN.	DECLINATION	DIAM.	MAGN.	TYPE OF OBJECT
REIZ 1872	11 57 04.	+ 26 49	24	15.8	GALAXY
HOLM 316A	11 57 04.	+ 26 50	54	14.9	PART OF MULTIPLE GALAXY
HN 1548	11 57 04.8	+ 28 03 00.	12		NEBULA
ARC 1433	11 57 05.	+ 26 09		17.1	RICH CLUSTER OF GALAXIES
REIZ 1873	11 57 05.	+ 54 26	15	14.7	GALAXY
HN 1549	11 57 05.6	+ 28 27 06.	12		NEBULA
ZC 1157.1+1352	11 57 06.	+ 13 52	400		CLUSTER OF GALAXIES
ZWG 069.011	11 57 06.	+ 14 03		15.6	GALAXY
TON-N 0600	11 57 06.	+ 27 46		16.2	BLUE STAR
ZC 1157.1+3010	11 57 06.	+ 30 10	1140		CLUSTER OF GALAXIES
ZC 1157.1+4045	11 57 06.	+ 40 45	740		CLUSTER OF GALAXIES
ARC 1432	11 57 06.	+ 68 23		17.7	RICH CLUSTER OF GALAXIES
ZWG 013.029	11 57 06.	- 03 23		15.2	GALAXY
HN 1550	11 57 07.0	+ 26 48 42.	18		NEBULA
HOLM 316B	11 57 08.	+ 26 49	30	16.2	PART OF MULTIPLE GALAXY
LB 02165	11 57 08.	+ 50 04 54.		14.4	FAINT BLUE STAR
HN 1551	11 57 09.5	+ 25 53 30.			NEBULA
ZWG 069.012	11 57 12.	+ 08 55		15.5	GALAXY
MCG+04-28-116	11 57 12.	+ 21 32 30.	30	15.	GALAXY
MCG+04-28-117	11 57 12.	+ 21 42	42	15.5	GALAXY
ZWG 127.131	11 57 12.	+ 25 53		15.7	GALAXY
ZWG 157.077	11 57 12.	+ 26 35		15.4	GALAXY
MCG+05-28-071	11 57 12.	+ 26 50	36	15.7	GALAXY
ZWG 157.078	11 57 12.	+ 29 55		15.5	GALAXY
ZWG 157.079	11 57 12.	+ 31 06		15.0	GALAXY
ZC 1157.2+4332	11 57 12.	+ 43 32	10420		CLUSTER OF GALAXIES
MCG+09-20-053	11 57 12.	+ 53 53	30	17.	GALAXY
ZWG 352.013	11 57 12.	+ 76 32		15.3	GALAXY
ZWG 013.030	11 57 12.	- 03 06		15.7	GALAXY
MCG+00-31-014	11 57 12.	- 03 06	60	15.7	GALAXY
MCG-02-31-009	11 57 12.	- 12 07	15	15.5	GALAXY
BC M1157+118	11 57 12.5	+ 11 50 30.		20.	QUASI-STELLAR OBJECT
SMB 192	11 57 12.5	+ 11 50 30.		20.	QUASI-STELLAR OBJECT
HN 1552	11 57 12.7	+ 01 23 48.	12		NEBULA
IC 2986	11 57 15.	+ 31 07 14.			NONSTELLAR OBJECT
MCG+05-28-072	11 57 15.	+ 31 09	36	15.0	GALAXY
MCG+09-20-054	11 57 15.	+ 55 57	78	15.	GALAXY
HN 1553	11 57 15.3	+ 28 07 48.			NEBULA
REIZ 1874	11 57 16.	+ 27 05	24	15.7	GALAXY
FATH 1.427	11 57 16.	+ 29 26	22		NEBULA
SVEN 263	11 57 16.	- 01 07	12	15.5	GALAXY
HN 1555	11 57 16.1	+ 01 26 54.	48		NEBULA
HN 1554	11 57 16.1	+ 27 05 42.	36		NEBULA
HN 1556	11 57 16.8	- 01 06 54.	18		NEBULA
REIN 7.067A	11 57 17.19	- 01 06 13.7			NEBULA
REIN 7.067B	11 57 17.24	- 01 06 13.3			NEBULA
ZC 1157.3+2635	11 57 18.	+ 26 35	1680		CLUSTER OF GALAXIES
MCG+05-28-073	11 57 18.	+ 27 07	54	16.	GALAXY
ZC 1157.3+2758	11 57 18.	+ 27 58	1810		CLUSTER OF GALAXIES
ZWG 186.082	11 57 18.	+ 35 11		15.7	GALAXY
ZWG 215.015	11 57 18.	+ 39 52		14.6	GALAXY
MCG+07-25-013	11 57 18.	+ 39 52 30.	48	14.5	GALAXY
ZC 1157.3+4943	11 57 18.	+ 49 43	1480		CLUSTER OF GALAXIES
UGC 06988	11 57 18.	+ 55 59	78	16.5	GALAXY DWARF
ZWG 013.031	11 57 18.	- 01 05		15.5	GALAXY
HN 1557	11 57 18.3	+ 26 34 06.	30		NEBULA
RNGC 4028	11 57 19.	+ 16 30			NON-EXISTENT OBJECT
REIZ 1875	11 57 22.	+ 32 09	42	16.0	GALAXY
SVEN 264	11 57 22.	- 00 18	30	15.2	GALAXY
HN 1558	11 57 22.0	- 00 16 18.	18		NEBULA
KN 12.263	11 57 22.7	+ 49 50 43.			NEBULA
ZWG 069.013	11 57 24.	+ 09 05		15.5	GALAXY
MCG+02-31-008	11 57 24.	+ 09 05	30	15.5	GALAXY
UGC 06989	11 57 24.	+ 21 55	60	16.0	GALAXY Sc
MCG+04-28-118	11 57 24.	+ 21 55	60	15.5	GALAXY
MCG+10-17-122	11 57 24.	+ 58 52	24	16.	GALAXY
ZWG 013.032	11 57 24.	- 00 15		15.5	GALAXY
MCG+00-31-015	11 57 24.	- 00 15	42	15.5	GALAXY
MCG+08-22-051	11 57 27.	+ 49 50	24	16.	GALAXY
REIZ 1876	11 57 28.	+ 08 28	60	14.3	GALAXY
FATH 1.428	11 57 28.	+ 29 43	14		NEBULA
ZWG 041.017	11 57 30.	+ 08 28		14.5	GALAXY
UGC 06990	11 57 30.	+ 08 28	72	14.5	GALAXY Sb/SBb
MCG+01-31-008	11 57 30.	+ 08 28	72	14.5	GALAXY
8ZW 1157+09.5	11 57 30.	+ 09 28		18.8	COMPACT GALAXY
ZWG 069.014	11 57 30.	+ 09 56		15.6	GALAXY
KARA.68 085	11 57 30.	+ 21 01	27		DWARF GALAXY
RNGC 4029	11 57 31.	+ 08 28		14.5	GALAXY
SEY 149	11 57 31.	+ 53 08 48.		15.4	FAINT GALAXY
LB 02166	11 57 35.	+ 52 23 30.		15.3	FAINT BLUE STAR
ZWG 041.018	11 57 36.	+ 07 22		15.4	GALAXY
8ZW 1157+10.3	11 57 36.	+ 10 18		17.5	COMPACT GALAXY
ZWG 098.017	11 57 36.	+ 19 30		15.7	GALAXY
UGC 06991	11 57 36.	+ 19 30	60	15.7	GALAXY Sb-c
ZC 1157.6+2532	11 57 36.	+ 25 32	940		CLUSTER OF GALAXIES
MCG+10-17-123	11 57 36.	+ 58 52	51	16.	GALAXY
LB 02167	11 57 37.	+ 50 46 36.		16.7	FAINT BLUE STAR
SVEN 265	11 57 39.	- 19 06	12	14.5	GALAXY
SEY 150	11 57 40.	+ 50 55 18.		14.6	FAINT GALAXY
ARC 1434	11 57 40.	- 06 55		17.2	RICH CLUSTER OF GALAXIES
HN 1559	11 57 41.1	+ 26 45 42.	12		NEBULA
ZWG 069.015	11 57 42.	+ 10 59		15.7	GALAXY
MCG+02-31-009	11 57 42.	+ 10 59	9	15.7	GALAXY
MCG+03-31-009	11 57 42.	+ 19 32	60	15.7	GALAXY
ZWG 157.080	11 57 42.	+ 31 29		14.8	GALAXY
MCG+05-28-074	11 57 42.	+ 31 32 30.	42	14.8	GALAXY
ZWG 157.081	11 57 42.	+ 32 05		15.3	GALAXY
ZWG 269.030	11 57 42.	+ 50 57		14.9	GALAXY
UGC 06992	11 57 42.	+ 50 57	90	14.9	GALAXY S-IRR
MCG+09-20-055	11 57 42.	+ 54 32	30	16.	GALAXY
MCG+09-20-056	11 57 42.	+ 56 30	12	16.	GALAXY
MCG+09-20-057B	11 57 42.	+ 56 33	9	16.	GALAXY
ZC 1157.7+6647	11 57 42.	+ 66 47	870		CLUSTER OF GALAXIES
KN 12.264	11 57 43.8	+ 46 28 28.			NEBULA
HN 1560	11 57 44.4	+ 00 34 18.	24		NEBULA
KN 12.265	11 57 44.7	+ 50 55 52.			NEBULA
MCG+09-20-058	11 57 45.	+ 52 56	12	16.	GALAXY
MCG+09-20-057A	11 57 45.	+ 56 33	9	16.	GALAXY
SVEN 266	11 57 45.	- 19 06	30	14.0	GALAXY
HN 1561	11 57 45.2	+ 25 21 06.	12		NEBULA
ARC 1435	11 57 46.	+ 10 58		17.0	RICH CLUSTER OF GALAXIES
REIN 2.141	11 57 46.21	- 00 48 21.1			NEBULA
REIN 2.142	11 57 47.94	- 00 48 35.4			NEBULA
ZWG 013.034	11 57 48.	+ 00 35		15.3	GALAXY
ZC 1157.8+1306	11 57 48.	+ 13 06	1140		CLUSTER OF GALAXIES
ZWG 127.132	11 57 48.	+ 21 51		15.6	GALAXY
ZC 1157.8+4211	11 57 48.	+ 42 11	870		CLUSTER OF GALAXIES
ZC 1157.8+4600	11 57 48.	+ 46 00	3490		CLUSTER OF GALAXIES

OBJECT NAME	RIGHT ASCEN.	DECLINATION	DIAM.	MAGN.	TYPE OF OBJECT
MCG+09-20-059	11 57 48.	+ 50 55	72	14.	GALAXY
7ZW 432	11 57 48.	+ 56 34			COMPACT GALAXY
ZWG 013.033	11 57 48.	- 00 48		12.4	GALAXY
UGC 06993	11 57 48.	- 00 48	252	12.4	GALAXY Sb
MCG+00-31-016	11 57 48.	- 00 48	270	12.4	GALAXY
RNGC 4030	11 57 49.	- 00 49		11.5	GALAXY
KN 12.266	11 57 49.4	+ 48 01 56.			NEBULA
REIN 2.143	11 57 50.08	- 00 49 19.8			NEBULA
HN 1563	11 57 50.3	+ 00 46 18.	24		NEBULA
HN 1562	11 57 50.3	+ 25 45 06.			NEBULA
ARC 1437	11 57 52.	+ 03 37		17.2	RICH CLUSTER OF GALAXIES
LB 02168	11 57 52.	+ 60 15 48.		15.8	FAINT BLUE STAR
SVEN 267	11 57 52.	- 00 49	228	11.0	GALAXY
REIZ 1877	11 57 52.	- 00 50	180	11.3	GALAXY
SVEN 268	11 57 52.	- 00 50 12.	6	16.0	GALAXY
REIZ 1878	11 57 53.	+ 50 58	60	13.9	GALAXY
ZWG 013.035	11 57 54.	+ 00 47		15.3	GALAXY
ZWG 041.019	11 57 54.	+ 04 44		15.7	GALAXY
ZWG 069.016	11 57 54.	+ 09 54		15.7	GALAXY
ZWG 069.017	11 57 54.	+ 12 17		15.2	GALAXY
MCG+02-31-010	11 57 54.	+ 12 17	24	15.2	GALAXY
ZWG 157.082	11 57 54.	+ 32 13		14.7	GALAXY
MCG+05-28-075	11 57 54.	+ 32 16 30.	24	14.7	GALAXY
ZC 1157.9+3620	11 57 54.	+ 36 20	1210		CLUSTER OF GALAXIES
MCG-04-28-009	11 57 54.	- 24 26	90	15.	GALAXY
RNGC 4031	11 57 55.	+ 32 13		14.5	GALAXY
RNGC 4035	11 57 55.	- 15 41		14.0	GALAXY
FATH 1.429	11 57 56.	+ 29 26	14		NEBULA
ARC 1436	11 57 56.	+ 56 32		15.4	RICH CLUSTER OF GALAXIES
TON-N 1466	11 57 57.	+ 34 16		15.2	BLUE STAR
KN 12.267	11 57 57.6	+ 48 47 22.			NEBULA
REIZ 1880	11 57 58.	+ 20 21	48	13.7	GALAXY
REIZ 1879	11 57 58.	+ 32 13	12	14.4	GALAXY
FATH 2.035	11 57 59.	+ 29 17	11		NEBULA
KEEN 4085A	11 58	+ 51 12		11.6	GALAXY
ZWG 069.018	11 58 00.	+ 09 08		15.6	GALAXY
UGC 06994	11 58 00.	+ 09 08	72	15.6	GALAXY Sc
ZWG 098.018	11 58 00.	+ 18 18		15.7	GALAXY
ZWG 098.019	11 58 00.	+ 20 21		12.7	GALAXY
UGC 06995	11 58 00.	+ 20 21	114	12.7	GALAXY IRR
MCG+04-28-119	11 58 00.	+ 25 08 30.	48	15.3	GALAXY
FATH 1.430	11 58 00.	+ 29 16	16		NEBULA
ZC 1158.0+6823	11 58 00.	+ 68 23	1340		CLUSTER OF GALAXIES
UGC 06996	11 58 00.	+ 79 08	90	17.	GALAXY DWARF
ZWG 370.009	11 58 00.	+ 88 24		15.0	GALAXY
ZWG 370.003	11 58 00.	+ 88 24		15.0	GALAXY
MCG-03-31-010	11 58 00.	- 15 40	54	14.	GALAXY
MCG-03-31-009	11 58 00.	- 21 04 30.	36	15.	GALAXY
RNGC 4032	11 58 01.	+ 20 21		12.5	GALAXY
RNGC 4033	11 58 01.	- 17 34		13.0	GALAXY
KN 12.268	11 58 02.6	+ 48 02 45.			NEBULA
MCG-03-31-011	11 58 03.	- 17 33	48	12.5	GALAXY
MCG+03-31-010	11 58 05.	+ 20 23	150	12.7	GALAXY
ARC 1438	11 58 05.	+ 29 58		17.5	RICH CLUSTER OF GALAXIES
8ZW 1158+09.7	11 58 06.	+ 09 43		18.7	COMPACT GALAXY
MCG+02-31-011	11 58 06.	+ 10 53	12	15.2	GALAXY
ZWG 069.019	11 58 06.	+ 10 59		15.2	GALAXY
8ZW 1158+13.8	11 58 06.	+ 13 50		18.6	COMPACT GALAXY
ZC 1158.1+1537	11 58 06.	+ 15 37	2550		CLUSTER OF GALAXIES
ZWG 127.133	11 58 06.	+ 25 08		15.3	GALAXY
ZWG 157.083	11 58 06.	+ 28 36		15.7	GALAXY
MCG+08-22-052	11 58 06.	+ 44 40	30	17.	GALAXY
ZC 1158.1+4845	11 58 06.	+ 48 45	1010		CLUSTER OF GALAXIES
MCG+09-20-060	11 58 06.	+ 54 50	24	18.	GALAXY
MCG-02-31-010	11 58 06.	- 09 31	54	15.	GALAXY
LB 02169	11 58 09.	+ 61 13 18.		14.9	FAINT BLUE STAR
ARC 1440	11 58 09.	- 23 07		17.6	RICH CLUSTER OF GALAXIES
HN 1564	11 58 09.7	+ 25 08 18.	24		NEBULA
REIZ 1881	11 58 10.	+ 25 08	24	14.9	GALAXY
MCG+03-31-011	11 58 11.	+ 18 19	42	15.7	GALAXY
ZC 1158.2+0336	11 58 12.	+ 03 36	1550		CLUSTER OF GALAXIES
ZWG 069.020	11 58 12.	+ 14 26		15.7	GALAXY
ZWG 157.084	11 58 12.	+ 32 09		15.5	GALAXY
UGC 06997	11 58 12.	+ 32 09	66	15.5	GALAXY DWRF SP
MCG+05-28-076	11 58 12.	+ 32 12	42	15.5	GALAXY
MCG+09-20-061	11 58 12.	+ 54 28	15	17.	GALAXY
ZC 1158.2-0230	11 58 12.	- 02 30	1610		CLUSTER OF GALAXIES
FATH 2.036	11 58 13.	+ 29 13	11		NEBULA
ARC 1439	11 58 13.	+ 50 44		17.2	RICH CLUSTER OF GALAXIES
FATH 1.431	11 58 15.	+ 30 10	8		NEBULA
REIZ 1882	11 58 16.	+ 32 09	24	15.8	GALAXY
UGC 06998	11 58 18.	+ 00 16	108	17.	GALAXY DWRF SP
ZWG 069.021	11 58 18.	+ 10 51		15.3	GALAXY
ZWG 069.022	11 58 18.	+ 12 27		15.4	GALAXY
MCG+02-31-012	11 58 18.	+ 12 57	30	15.4	GALAXY
MCG+03-31-012	11 58 18.	+ 15 43	60	15.7	GALAXY
ZWG 098.020	11 58 18.	+ 15 44		15.7	GALAXY
ZC 1158.3+3551	11 58 18.	+ 35 51	2420		CLUSTER OF GALAXIES
ARC 1441	11 58 18.	+ 35 51		17.2	RICH CLUSTER OF GALAXIES
UGC 06999	11 58 18.	+ 50 11	66	18.	GALAXY DWARF
MCG+09-20-062	11 58 18.	+ 54 52 30.	24	17.	GALAXY
ZC 1158.3+5816	11 58 18.	+ 58 16	3020		CLUSTER OF GALAXIES
ZC 1158.3+6235	11 58 18.	+ 62 35	870		CLUSTER OF GALAXIES
KN 12.269	11 58 18.2	+ 45 55 32.			NEBULA
FATH 2.037	11 58 21.	+ 29 09	19		NEBULA
TON-N 1467	11 58 21.	+ 35 24		16.6	BLUE STAR
ZWG 041.020	11 58 24.	+ 07 05		15.3	GALAXY
ZC 1158.4+1227	11 58 24.	+ 12 27	670		CLUSTER OF GALAXIES
ZWG 069.023	11 58 24.	+ 14 01		15.4	GALAXY
MCG+02-31-013	11 58 24.	+ 14 01	48	15.4	GALAXY
ZC 1158.4+2954	11 58 24.	+ 29 54	2080		CLUSTER OF GALAXIES
MCG+09-20-063	11 58 24.	+ 55 19	36	16.	GALAXY
LB 02170	11 58 25.	+ 60 38 36.		15.8	FAINT BLUE STAR
FATH 1.432	11 58 27.	+ 28 52	19		NEBULA
FATH 2.038	11 58 27.	+ 29 19	14		NEBULA
FATH 2.040	11 58 27.	+ 29 58	14		NEBULA
FATH 2.039	11 58 27.	+ 29 59	14		NEBULA
KN 12.270	11 58 27.1	+ 50 11 30.			NEBULA
REIZ 1883	11 58 28.	+ 31 08	36	15.3	GALAXY
ARC 1442	11 58 29.	+ 15 29		17.0	RICH CLUSTER OF GALAXIES
ZWG 013.036	11 58 30.	+ 00 24		15.3	GALAXY
ZWG 041.021	11 58 30.	+ 06 23		15.6	GALAXY
MCG+04-28-120	11 58 30.	+ 22 53	27	15.7	GALAXY
ZWG 157.085	11 58 30.	+ 31 06		15.7	GALAXY
MCG+08-22-053	11 58 30.	+ 50 11	60	16.	GALAXY
MCG+09-20-064	11 58 30.	+ 56 38	12	16.	GALAXY
MCG-06-27-001	11 58 30.	- 34 55	30	16.	GALAXY
MCG-06-27-002	11 58 31.	- 34 56	90	15.5	GALAXY

OBJECT NAME	RIGHT ASCEN.	DECLINATION	DIAM.	MAGN.	TYPE OF OBJECT
HN 1565	11 58 31.3	+ 00 22 42.	12		NEBULA
HOLM 317B	11 58 32.	+ 00 24	30	14.9	PART OF MULTIPLE GALAXY
SFY 151	11 58 32.	+ 55 19 18.		15.0	FAINT GALAXY
FATH 2.041	11 58 33.	+ 29 16	19		NEBULA
SVEN 270	11 58 33.	- 18 49	60	15.8	GALAXY
HOLM 317A	11 58 34.	+ 00 23	30	14.8	PART OF MULTIPLE GALAXY
SVEN 269	11 58 34.	- 01 02	60	14.2	GALAXY
IC 0755	11 58 35.	+ 14 24 14.			NONSTELLAR OBJECT
FATH 2.042	11 58 35.	+ 29 58	8		NEBULA
KN 12.271	11 58 35.1	+ 48 57 28.			NEBULA
KN 12.272	11 58 35.3	+ 47 42 08.			NEBULA
ZC 1158.6+0857	11 58 36.	+ 08 57	670		CLUSTER OF GALAXIES
ZWG 069.024	11 58 36.	+ 14 23		13.9	GALAXY
SCH 19	11 58 36.	+ 14 23		13.9	PECULIAR GALAXY
UGC 07001	11 58 36.	+ 14 23	162	13.9	GALAXY S
MCG+02-31-014	11 58 36.	+ 14 23	132	13.9	GALAXY
ZWG 127.134	11 58 36.	+ 22 54		15.7	GALAXY
FATH 1.433	11 58 36.	+ 29 38	8		NEBULA
ZWG 243.036	11 58 36.	+ 47 42		15.7	GALAXY
ZWG 013.037	11 58 36.	- 01 00		14.4	GALAXY
UGC 07000	11 58 36.	- 01 00	66	14.4	GALAXY IRR
MCG+00-31-017	11 58 36.	- 01 00	66	14.4	GALAXY
MCG-03-31-012	11 58 36.	- 20 12	90	15.	GALAXY
MCG-04-29-001	11 58 36.	- 24 18	132	13.5	GALAXY
HN 1566	11 58 36.6	- 01 01 48.	48		NEBULA
REIN 7.068A	11 58 37.12	- 01 01 06.6			NEBULA
REIN 7.068B	11 58 37.25	- 01 01 06.1			NEBULA
FATH 1.434	11 58 38.	+ 29 34	14		NEBULA
REIZ 1884	11 58 40.	- 01 02	48	13.6	GALAXY
LB 02171	11 58 41.	+ 50 34 12.		17.2	FAINT BLUE STAR
REIZ 1885	11 58 41.	+ 54 31	30	14.8	GALAXY
HN 1567	11 58 41.5	+ 27 49 54.	18		NEBULA
MCG+08-22-054	11 58 42.	+ 47 42	36	15.	GALAXY
ZWG 013.024	11 58 42.	- 03 23		15.1	GALAXY
ZWG 069.025	11 58 42.	+ 10 00		15.5	GALAXY
ZC 1158.7+3435	11 58 42.	+ 34 35	1010		CLUSTER OF GALAXIES
SHAH 070	11 58 42.	+ 41 30	102	18.0	GROUP OF COMPACT GALAXIES
ZC 1158.7+4519	11 58 42.	+ 45 19	810		CLUSTER OF GALAXIES
ZC 1158.7+5153	11 58 42.	+ 51 53	3290		CLUSTER OF GALAXIES
ZWG 292.058	11 58 42.	+ 56 50		15.7	GALAXY
MCG-03-31-013	11 58 42.	- 20 13	36	15.	GALAXY
MCG+09-20-065	11 58 45.	+ 53 45	42	16.	GALAXY
SVEN 271	11 58 45.	- 18 51 12.	18	15.1	GALAXY
REIZ 1886	11 58 47.	+ 62 37	15	15.3	GALAXY
ZWG 069.026	11 58 48.	+ 09 27		15.4	GALAXY
ZWG 069.027	11 58 48.	+ 13 40		13.8	GALAXY
UGC 07002	11 58 48.	+ 13 40	168	13.8	GALAXY SB
MCG+02-31-015	11 58 48.	+ 13 40	132	13.8	GALAXY
UGC 07003	11 58 48.	+ 14 43	108	18.	GALAXY DWARF
ZWG 098.021	11 58 48.	+ 17 49		15.5	GALAXY
ZC 1158.8+3315	11 58 48.	+ 33 15	3630		CLUSTER OF GALAXIES
STOCK 35	11 58 48.	+ 34 33			BLUE KNOT NEAR ELLIP GLXY
3ZW 064	11 58 48.	+ 45 56			COMPACT GALAXY
MCG+09-20-066	11 58 48.	+ 54 32	30	16.	GALAXY
MCG+10-17-124	11 58 48.	+ 56 50	30	16.	GALAXY
RNGC 4037	11 58 49.	+ 13 41		12.5	GALAXY
LB 02172	11 58 51.	+ 59 08 42.		16.5	FAINT BLUE STAR
REIZ 1889	11 58 52.	+ 13 40	108	14.4	GALAXY
REIZ 1887	11 58 52.	+ 14 18	108	13.4	GALAXY
SVEN 272	11 58 52.	- 00 27	30	14.9	GALAXY
ARC 1443	11 58 53.	+ 23 22		17.8	RICH CLUSTER OF GALAXIES
REIZ 1888	11 58 53.	+ 62 10	132	11.4	GALAXY
HN 1568	11 58 53.8	+ 26 41 42.	12		NEBULA
HN 1569	11 58 53.8	- 00 26 42.	18		NEBULA
ZC 1158.9+0006	11 58 54.	+ 00 06	2420		CLUSTER OF GALAXIES
ZWG 069.028	11 58 54.	+ 11 28		15.7	GALAXY
ZWG 069.029	11 58 54.	+ 14 18		15.2	GALAXY
MCG+02-31-016	11 58 54.	+ 14 18	30	15.2	GALAXY
ZC 1158.9+2323	11 58 54.	+ 23 23	870		CLUSTER OF GALAXIES
MCG+06-26-068	11 58 54.	+ 33 37	90	16.	GALAXY
ZWG 292.059	11 58 54.	+ 62 10		11.5	GALAXY
UGC 07005	11 58 54.	+ 62 10	246	11.5	GALAXY S0
MCG+12-11-044	11 58 54.	+ 69 35 30.	102	15.	GALAXY
ZWG 335.002	11 58 54.	+ 69 37		14.5	GALAXY
ZWG 334.058	11 58 54.	+ 69 37		14.5	GALAXY
UGC 07006	11 58 54.	+ 69 37	108	14.5	GALAXY Sc
ZWG 013.038	11 58 54.	- 00 25		15.0	GALAXY
UGC 07004	11 58 54.	- 00 25	78	15.0	GALAXY SBc
MCG+00-31-018	11 58 54.	- 00 25	78	15.0	GALAXY
MCG-04-29-002	11 58 54.	- 23 04	36	15.	GALAXY
RNGC 4036	11 58 55.	+ 62 10		12.0	GALAXY
RNGC 4034	11 58 55.	+ 69 35			GALAXY
TON-N 1468	11 58 57.	+ 36 18		16.5	BLUE STAR
REIZ 1890	11 58 58.	- 00 27	36	14.5	GALAXY
LB 02173	11 58 59.	+ 52 05 42.		16.3	FAINT BLUE STAR
LB 09854	11 59	- 86 47		12.0	FAINT BLUE STAR
MCG+03-31-013	11 59 00.	+ 14 45	30	15.7	GALAXY
MCG+04-28-121	11 59 00.	+ 22 55	48	15.5	GALAXY
FATH 2.043	11 59 00.	+ 29 12	22		NEBULA
UGC 07007	11 59 00.	+ 33 37	120	17.	GALAXY DWRF SP
ZC 1159.0+4424	11 59 00.	+ 44 24	1480		CLUSTER OF GALAXIES
MCG+08-22-055	11 59 00.	+ 45 54	24	16.	GALAXY
ZCG 1159+45	11 59 00.	+ 45 56		17.7	COMPACT GALAXY
ZC 1159.0+4716	11 59 00.	+ 47 16	2690		CLUSTER OF GALAXIES
ZC 1159.0+4800	11 59 00.	+ 48 00	1280		CLUSTER OF GALAXIES
ZWG 013.039	11 59 00.	- 01 44		15.6	GALAXY
MCG+08-22-056	11 59 03.	+ 45 55	12	17.	GALAXY
FATH 1.435	11 59 05.	+ 29 15	14		NEBULA
ARC 1444	11 59 05.	+ 30 18		17.6	RICH CLUSTER OF GALAXIES
ZWG 013.040	11 59 06.	+ 02 17		15.7	GALAXY
ZWG 041.022	11 59 06.	+ 05 02		15.3	GALAXY
UGC 07008	11 59 06.	+ 05 02	78	15.3	GALAXY S
ZWG 098.022	11 59 06.	+ 14 45		15.7	GALAXY
MCG+03-31-014	11 59 06.	+ 19 08	42	16.5	GALAXY
ZWG 127.135	11 59 06.	+ 20 37		15.2	GALAXY
VV 045B	11 59 06.	+ 22 56	9	18.	INTERACTING GALAXY
VV 045A	11 59 06.	+ 22 56	42	16.	INTERACTING GALAXY
ZC 1159.1+2916	11 59 06.	+ 29 16	3430		CLUSTER OF GALAXIES
MCG+10-17-125	11 59 06.	+ 62 10	126	11.4	GALAXY
FATH 1.436	11 59 07.	+ 29 11	19		NEBULA
MCG+04-28-122	11 59 09.	+ 20 35 30.	42	15.2	GALAXY
ARC 1445	11 59 10.	+ 00 07		17.6	RICH CLUSTER OF GALAXIES
REIZ 1891	11 59 10.	+ 33 36	42	16.4	GALAXY
FATH 1.437	11 59 11.	+ 28 57	14		NEBULA
REIZ 1892	11 59 11.	+ 62 36	90	14.0	GALAXY
ZWG 041.023	11 59 12.	+ 06 06		14.9	GALAXY
MCG+01-31-009	11 59 12.	+ 06 06	36	14.9	GALAXY
ZWG 041.024	11 59 12.	+ 06 43		14.7	GALAXY
MCG+01-31-010	11 59 12.	+ 06 43	18	14.7	GALAXY
ZWG 098.023	11 59 12.	+ 18 10		15.1	GALAXY
MCG+03-31-015	11 59 12.	+ 18 12	54	15.1	GALAXY
ZC 1159.2+1850	11 59 12.	+ 18 50	810		CLUSTER OF GALAXIES
ZWG 127.136	11 59 12.	+ 21 22		15.7	GALAXY
ZC 1159.2+3021	11 59 12.	+ 30 21	2150		CLUSTER OF GALAXIES
MCG+11-15-010	11 59 12.	+ 62 36	96	14.	GALAXY
ZWG 315.005	11 59 12.	+ 62 37		14.5	GALAXY
UGC 07009	11 59 12.	+ 62 37	102	14.5	GALAXY IRR
MCG-02-31-011	11 59 12.	- 14 49	42	15.	GALAXY
MCG-05-29-001	11 59 12.	- 31 26	72	15.	GALAXY
FATH 2.044	11 59 13.	+ 29 42	16		NEBULA
FATH 2.045	11 59 13.	+ 29 59	14		NEBULA
FATH 1.438	11 59 14.	+ 29 39	16		NEBULA
VV 185B	11 59 15.	+ 15 43 30.	6	18.	INTERACTING GALAXY
VV 185A	11 59 15.	+ 15 43 30.	9	18.	INTERACTING GALAXY
VV 185	11 59 15.	+ 15 43 30.	42		INTERACTING GALAXY
FATH 1.439	11 59 15.	+ 28 52	14		NEBULA
TON-N 1469	11 59 15.	+ 35 21		14.8	BLUE STAR
ARP 244	11 59 17.	- 18 35			PECULIAR GALAXY
ZWG 098.024	11 59 18.	+ 16 48		15.1	GALAXY
MCG+03-31-016	11 59 18.	+ 16 48	24	15.1	GALAXY
ZWG 098.025	11 59 18.	+ 17 42		15.7	GALAXY
MCG+03-31-017	11 59 18.	+ 17 42	30	15.7	GALAXY
ZWG 127.137	11 59 18.	+ 20 42		15.6	GALAXY
ZWG 127.138	11 59 18.	+ 21 02		15.5	GALAXY
ZWG 127.139	11 59 18.	+ 22 49		15.5	GALAXY
UGC 07010	11 59 18.	+ 22 49	78	15.5	GALAXY Sa
ZC 1159.3+2439	11 59 18.	+ 24 39	1550		CLUSTER OF GALAXIES
ZC 1159.3+3057	11 59 18.	+ 30 57	870		CLUSTER OF GALAXIES
MCG+10-17-126	11 59 18.	+ 58 29	45	16.	GALAXY
MCG+10-17-127	11 59 18.	+ 59 49	45	16.	GALAXY
VV 245B	11 59 18.	- 18 35	144	12.	INTERACTING GALAXY
VV 245A	11 59 18.	- 18 35	138	12.	INTERACTING GALAXY
VV 245	11 59 18.	- 18 35	840	11.	INTERACTING GALAXY
SN 1921A	11 59 18.	- 18 35			SUPERNOVA
OCL 0870	11 59 18.	- 62 55	1320	9.0	OPEN STAR CLUSTER
VHA 126	11 59 18.	- 62 55	360		STAR CLSTR IN NEBULOSITY
ARC 1446	11 59 19.	+ 58 18		17.0	RICH CLUSTER OF GALAXIES
RNGC 4038	11 59 19.	- 18 35		10.5	GALAXY
RNGC 4039	11 59 19.	- 18 36		13.0	GALAXY
RNGC 4052	11 59 19.	- 62 55		9.0	OPEN CLUSTER
REIN 2.144	11 59 19.11	- 18 35 21.4			NEBULA
REIN 2.145	11 59 19.82	- 18 36 23.1			NEBULA
SVEN 274	11 59 21.	- 18 36	150	10.6	GALAXY
HN 1570	11 59 21.0	+ 28 46 30.	12		NEBULA
HN 1571	11 59 21.5	- 00 47 42.	36		NEBULA
SVEN 273	11 59 22.	- 00 48 12.	48	15.1	GALAXY
SN 1974E	11 59 22.	- 18 36		14.0	SUPERNOVA
LB 02174	11 59 23.	+ 52 20 24.		17.1	FAINT BLUE STAR
ZWG 041.025	11 59 24.	+ 06 58		15.1	GALAXY
MCG+01-31-011	11 59 24.	+ 06 58	48	15.1	GALAXY
8ZW 1159+10.8	11 59 24.	+ 10 51		18.1	COMPACT GALAXY
ZC 1159.4+3216	11 59 24.	+ 32 16	470		CLUSTER OF GALAXIES
ZWG 013.041	11 59 24.	- 00 46		15.3	GALAXY
UGC 07011	11 59 24.	- 00 46	66	15.3	GALAXY Sb
MCG+00-31-019	11 59 24.	- 00 46	66	15.3	GALAXY
MCG-03-31-014	11 59 24.	- 18 35	138	11.	GALAXY
MCG-03-31-015	11 59 24.	- 18 36	138	13.	GALAXY
FATH 2.046	11 59 25.	+ 29 48	16		NEBULA
LB 02175	11 59 26.	+ 51 02 06.		16.9	FAINT BLUE STAR
HOLM 318B	11 59 27.	+ 30 06	42	14.5	PART OF MULTIPLE GALAXY
REIZ 1893	11 59 28.	+ 30 07	36	13.9	GALAXY
ARC 1447	11 59 29.	+ 24 06		17.4	RICH CLUSTER OF GALAXIES
ZC 1159.5+0720	11 59 30.	+ 07 20	270		CLUSTER OF GALAXIES
8ZW 1159+13.7	11 59 30.	+ 13 44		18.0	COMPACT GALAXY
ZWG 098.026	11 59 30.	+ 18 11		15.6	GALAXY
ZWG 158.001	11 59 30.	+ 30 07		14.3	GALAXY
ZWG 157.086	11 59 30.	+ 30 07		14.3	GALAXY
UGC 07012	11 59 30.	+ 30 07	126	14.3	GALAXY Sc
KARA.72 316A	11 59 30.	+ 30 07	120	14.3	PART OF DOUBLE GALAXY
MCG+05-28-077	11 59 30.	+ 30 10	120	14.3	GALAXY
ZC 1159.5+6202	11 59 30.	+ 62 02	1080		CLUSTER OF GALAXIES
7ZW 433	11 59 30.	+ 66 40			COMPACT GALAXY
MRK 194	11 59 30.	+ 66 40	10	15.5	GALAXY WITH UV CONTINUUM
HN 1572	11 59 30.2	+ 01 48 30.	12		NEBULA
FATH 1.440	11 59 31.	+ 29 08	8		NEBULA
HN 1573	11 59 31.8	+ 29 07 06.	12		NEBULA
FATH 2.047	11 59 33.	+ 29 22	8		NEBULA
REIZ 1894	11 59 34.	+ 31 15	24	15.4	GALAXY
ZWG 013.042	11 59 36.	+ 01 49		15.7	GALAXY
ZWG 098.027	11 59 36.	+ 17 58		15.7	GALAXY
ZWG 098.028	11 59 36.	+ 18 06		15.0	GALAXY
UGC 07013	11 59 36.	+ 18 06	72	15.0	GALAXY E
MCG+03-31-018	11 59 36.	+ 18 07	102	15.0	GALAXY
ZWG 127.140	11 59 36.	+ 24 35		15.7	GALAXY
MCG+09-20-067	11 59 36.	+ 54 00	12	17.	GALAXY
ZWG 292.060	11 59 36.	+ 56 39		15.2	GALAXY
7ZW 434	11 59 36.	+ 59 27			COMPACT GALAXY
ZWG 292.061	11 59 36.	+ 62 25		11.6	GALAXY
UGC 07014	11 59 36.	+ 62 25	168	11.6	GALAXY Sc
ZC 1159.6+6421	11 59 36.	+ 64 21	1010		CLUSTER OF GALAXIES
ZC 1159.6+6448	11 59 36.	+ 64 48	1080		CLUSTER OF GALAXIES
ZWG 315.006	11 59 36.	+ 66 40		15.2	GALAXY
RNGC 4040	11 59 37.	+ 18 06		15.0	GALAXY
RNGC 4041	11 59 37.	+ 62 25		12.0	GALAXY
FATH 1.441	11 59 38.	+ 29 08	5		NEBULA
HN 1574	11 59 39.3	+ 28 01 42.			NEBULA
REIZ 1895	11 59 40.	+ 41 32	30	14.3	GALAXY
LB 02176	11 59 40.	+ 53 01 12.		17.0	FAINT BLUE STAR
REIZ 1896	11 59 41.	+ 62 25	108	11.8	GALAXY
8ZW 1159+13.3	11 59 42.	+ 13 18		17.5	COMPACT GALAXY
ZC 1159.7+2405	11 59 42.	+ 24 05	2290		CLUSTER OF GALAXIES
ZC 1159.7+5559	11 59 42.	+ 55 59	740		CLUSTER OF GALAXIES
MCG+09-20-068	11 59 42.	+ 56 38	60	15.	GALAXY
MCG+10-17-128	11 59 42.	+ 59 26	8	17.	GALAXY
HN 1575	11 59 42.3	+ 28 46 42.	18		NEBULA
FATH 1.442	11 59 44.	+ 29 47	19		NEBULA
LB 02177	11 59 44.	+ 53 09 48.		16.2	FAINT BLUE STAR
MCG+05-28-078	11 59 45.	+ 30 11	90	14.4	GALAXY
MCG+10-17-129	11 59 45.	+ 62 25	156	12.0	GALAXY
HN 1576	11 59 45.8	+ 27 13 30.	18		NEBULA
HOLM 318A	11 59 46.	+ 30 07	102	14.3	PART OF MULTIPLE GALAXY
REIZ 1897	11 59 46.	+ 30 08	78	13.7	GALAXY
MAI 066	11 59 46.	+ 79 36	40		DWARF SPHEROIDAL GALAXY
ZWG 041.026	11 59 48.	+ 04 37		14.1	GALAXY
UGC 07015	11 59 48.	+ 04 37	42	14.1	GALAXY S
MCG+01-31-012	11 59 48.	+ 04 37	36	14.1	GALAXY

OBJECT NAME	RIGHT ASCEN.	DECLINATION	DIAM.	MAGN.	TYPE OF OBJECT
ZWG 069.030	11 59 48.	+ 12 36		15.3	GALAXY
ZWG 098.029	11 59 48.	+ 15 07		14.7	GALAXY
UGC 07016	11 59 48.	+ 15 07	96	14.7	GALAXY SB:a-b
MCG+03-31-019	11 59 48.	+ 15 07	108	14.7	GALAXY
MCG+04-28-123	11 59 48.	+ 24 06	27	15.0	GALAXY
STOCK 36	11 59 48.	+ 27 48			BLUE KNOT NEAR ELLIP GLXY
ZC 1159.8+2808	11 59 48.	+ 28 08	1280		CLUSTER OF GALAXIES
ZWG 158.002	11 59 48.	+ 30 08		14.4	GALAXY
ZWG 157.087	11 59 48.	+ 30 08		14.4	GALAXY
UGC 07017	11 59 48.	+ 30 08	114	14.4	GALAXY Sb
KARA.72 316B	11 59 48.	+ 30 08	108	14.4	PART OF DOUBLE GALAXY
FEIG 053	11 59 48.	+ 50 12		11.7	FAINT BLUE STAR
SN 19650	11 59 48.	+ 50 12			SUPERNOVA
MCG+10-17-130	11 59 48.	+ 59 25	15	17.	GALAXY
VV 259	11 59 48.	+ 62 41 30.	78	14.	INTERACTING GALAXY
MCG-01-31-002	11 59 48.	- 04 04	78	15.	GALAXY
MCG-05-29-002	11 59 48.	- 29 14	30	15.	GALAXY
RNGC 4043	11 59 49.	+ 04 37		14.0	GALAXY
RNGC 4042	11 59 49.	+ 20 25			NON-EXISTENT OBJECT
FATH 1.443	11 59 51.	+ 29 22	8		NEBULA
HELW 022	11 59 52.	+ 15 08			NEBULA
REIZ 1898	11 59 52.	+ 29 45	36	15.1	GALAXY
REIZ 1899	11 59 52.	+ 36 51	90	12.5	GALAXY
FATH 2.048	11 59 53.	+ 29 39	16		NEBULA
FATH 2.049	11 59 53.	+ 29 45	41		NEBULA
KN 12.273	11 59 53.4	+ 47 45 52.			NEBULA
HN 1577	11 59 53.6	+ 28 46 42.	12		NEBULA
ZWG 013.043	11 59 54.	+ 00 05		14.6	GALAXY
UGC 07018	11 59 54.	+ 00 05	72	14.6	GALAXY E-S0
MCG+00-31-020	11 59 54.	+ 00 05	24	14.6	GALAXY
ZWG 069.031	11 59 54.	+ 09 13		15.5	GALAXY
ZWG 069.032	11 59 54.	+ 11 10		15.5	GALAXY
ZC 1159.9+1449	11 59 54.	+ 14 49	2350		CLUSTER OF GALAXIES
ZC 1159.9+1650	11 59 54.	+ 16 50	1950		CLUSTER OF GALAXIES
ZWG 127.141	11 59 54.	+ 24 06		15.0	GALAXY
ZWG 158.003	11 59 54.	+ 29 45		15.4	GALAXY
ZWG 157.088	11 59 54.	+ 29 45		15.4	GALAXY
MCG+11-15-011	11 59 54.	+ 62 41	78	15.	GALAXY
ZWG 315.007	11 59 54.	+ 62 42		15.6	GALAXY
UGC 07019	11 59 54.	+ 62 42	108	15.6	GALAXY IRR
HN 1578	11 59 54.5	+ 28 48 42.	12		NEBULA
RNGC 4044	11 59 55.	+ 00 05		14.5	GALAXY
FATH 2.050	11 59 57.	+ 29 17	5		NEBULA
FATH 1.444	11 59 57.	+ 29 19	8		NEBULA
HOLM 319B	11 59 58.	+ 28 46	12	15.8	PART OF MULTIPLE GALAXY
HOLM 319A	11 59 58.	+ 28 47	30	15.5	PART OF MULTIPLE GALAXY
SEY 152	11 59 58.	+ 51 09 48.		15.3	FAINT GALAXY
REIZ 1900	11 59 58.	- 04 03	120	14.9	GALAXY
LB 09855	12 00	- 83 15		14.0	FAINT BLUE STAR
MCG+04-29-001	12 00 00.	+ 21 25	39	15.5	GALAXY
ZWG 128.001	12 00 00.	+ 21 26		15.5	GALAXY
FATH 2.051	12 00 00.	+ 29 31	8		NEBULA
ZC 1200.0+3125	12 00 00.	+ 31 25	400		CLUSTER OF GALAXIES
ZWG 215.016	12 00 00.	+ 41 20		14.4	GALAXY
UGC 07020	12 00 00.	+ 41 20	132	14.4	GALAXY Sb
ZC 1200.0+4741	12 00 00.	+ 47 41	670		CLUSTER OF GALAXIES
LB 02178	12 00 00.	+ 50 29 48.		15.5	FAINT BLUE STAR
MRK 195	12 00 00.	+ 64 38	25	14.5	GALAXY WITH UV CONTINUUM
VVI 52	12 00 00.	+ 64 38	60	15.07	SEYFERT GALAXY
MCG+11-15-012	12 00 00.	+ 64 38	57	15.	GALAXY
FATH 1.445	12 00 01.	+ 29 46	19		NEBULA
MCG+07-25-014	12 00 03.	+ 41 19 30.	120	13.	GALAXY
REIZ 1901	12 00 04.	+ 21 27	24	15.5	GALAXY
FATH 1.446	12 00 05.	+ 29 18	8		NEBULA
SEY 153	12 00 05.	+ 41 19 35.		14.0	FAINT GALAXY
MCG+00-31-024	12 00 06.	+ 02 20	21	14.8	GALAXY
ZC 1200.1+0837	12 00 06.	+ 08 37	400		CLUSTER OF GALAXIES
ZC 1200.1+3355	12 00 06.	+ 33 55	400		CLUSTER OF GALAXIES
ZC 1200.1+3410	12 00 06.	+ 34 10	1950		CLUSTER OF GALAXIES
MCG+09-20-069	12 00 06.	+ 56 39	36	16.	GALAXY
MCG+11-15-013	12 00 06.	+ 62 46	45	16.	GALAXY
ZWG 315.008	12 00 06.	+ 64 39		14.3	GALAXY
UGC 07020A	12 00 06.	+ 64 39	72	14.3	GALAXY S0?
MCG-02-31-012	12 00 06.	- 12 36	60	15.	GALAXY
FEIG 054	12 00 06.	- 19 24		15.0	FAINT BLUE STAR
SC 1157-4339.1	12 00 08.	- 43 55 48.	42		NEBULA
MCG+07-25-015	12 00 09.	+ 39 21 30.	36	15.	GALAXY
LB 02179	12 00 09.	+ 54 29 06.		15.8	FAINT BLUE STAR
HN 1579	12 00 09.1	+ 27 25 24.	18		NEBULA
HOLM 320B	12 00 10.	+ 02 12	30	14.8	PART OF MULTIPLE GALAXY
HOLM 320A	12 00 10.	+ 02 14	150	13.1	PART OF MULTIPLE GALAXY
FATH 1.447	12 00 10.	+ 29 23	5		NEBULA
HOLM 321A	12 00 10.	+ 39 22	36	14.3	PART OF MULTIPLE GALAXY
HOLM 321B	12 00 11.	+ 39 23	36	15.1	PART OF MULTIPLE GALAXY
ZWG 013.045	12 00 12.	+ 02 14		15.2	GALAXY
MCG+00-31-021	12 00 12.	+ 02 14	24	15.2	GALAXY
ZWG 013.046	12 00 12.	+ 02 16		13.5	GALAXY
UGC 07021	12 00 12.	+ 02 16	180	13.5	GALAXY Sa
MCG+00-31-022	12 00 12.	+ 02 16	138	13.5	GALAXY
ZC 1200.2+0426	12 00 12.	+ 04 26	670		CLUSTER OF GALAXIES
ZC 1200.2+1757	12 00 12.	+ 17 57	540		CLUSTER OF GALAXIES
FATH 2.052	12 00 12.	+ 29 08	14		NEBULA
UGC 07022	12 00 12.	+ 45 27	60	16.0	GALAXY S
ZWG 013.044	12 00 12.	- 01 52		15.7	GALAXY
RNGC 4046	12 00 13.	+ 02 01			NON-EXISTENT OBJECT
RNGC 4045A	12 00 13.	+ 02 14		15.0	GALAXY
LB 02180	12 00 13.	+ 51 42 54.		17.5	FAINT BLUE STAR
HN 1580	12 00 13.5	+ 00 38 18.	18		NEBULA
REIZ 1902	12 00 16.	+ 19 01	42	13.9	GALAXY
REIZ 1903	12 00 16.	+ 48 55	54	12.7	GALAXY
FATH 1.448	12 00 17.	+ 29 44	8		NEBULA
KN 12.274	12 00 17.3	+ 48 54 59.			NEBULA
REIN 2.146	12 00 17.50	+ 48 54 50.9			NEBULA
ZWG 041.027	12 00 18.	+ 05 53		15.1	GALAXY
MCG+01-31-013	12 00 18.	+ 05 53	36	15.1	GALAXY
ZC 1200.3+1729	12 00 18.	+ 17 29	400		CLUSTER OF GALAXIES
ZWG 098.030	12 00 18.	+ 18 17		14.4	GALAXY
UGC 07023	12 00 18.	+ 18 17	33	14.4	GALAXY
MCG+03-31-020	12 00 18.	+ 18 19	36	14.4	GALAXY
MCG+04-29-002	12 00 18.	+ 21 55 30.	24	15.5	GALAXY
UGC 07024	12 00 18.	+ 22 23	72	16.0	GALAXY Sc
ZWG 158.004	12 00 18.	+ 26 31		15.5	GALAXY
ZWG 157.089	12 00 18.	+ 26 31		15.5	GALAXY
KARA.73B 0517	12 00 18.	+ 26 31	18	15.5	ISOLATED GALAXY E
ZWG 215.017	12 00 18.	+ 39 43		14.7	GALAXY
MCG+08-22-057	12 00 18.	+ 45 28	60	16.	GALAXY
ZWG 243.037	12 00 18.	+ 48 55		12.8	GALAXY
UGC 07025	12 00 18.	+ 48 55	84	12.8	GALAXY S

OBJECT NAME	RIGHT ASCEN.	DECLINATION	DIAM.	MAGN.	TYPE OF OBJECT
MCG+08-22-058	12 00 18.	+ 48 55	72	12.8	GALAXY
LB 00623	12 00 18.	+ 56 45 12.		16.0	FAINT BLUE STAR
ZWG 013.047	12 00 18.	- 01 30		15.4	GALAXY
MCG-05-29-003	12 00 18.	- 28 49	72	14.5	GALAXY
RNGC 4048	12 00 19.	+ 18 17		14.5	GALAXY
HN 1581	12 00 19.2	+ 26 31 24.			NEBULA
FATH 1.449	12 00 21.	+ 28 59	14		NEBULA
FATH 1.450	12 00 21.	+ 29 21	11		NEBULA
ARC 1448	12 00 22.	- 06 34		16.6	RICH CLUSTER OF GALAXIES
LB 02181	12 00 23.	+ 54 54 30.		16.3	FAINT BLUE STAR
ZWG 041.028	12 00 24.	+ 04 31		15.0	GALAXY
MCG+01-31-014	12 00 24.	+ 04 31	12	15.0	GALAXY
ZWG 041.029	12 00 24.	+ 05 07		14.8	GALAXY
UGC 07026	12 00 24.	+ 05 07	126	14.8	GALAXY Sc
MCG+01-31-015	12 00 24.	+ 05 07	84	14.8	GALAXY
IC 0756	12 00 24.	+ 05 08 02.			NONSTELLAR OBJECT
ZC 1200.4+1901	12 00 24.	+ 19 01	740		CLUSTER OF GALAXIES
ZWG 098.031	12 00 24.	+ 19 02		14.2	GALAXY
UGC 07027	12 00 24.	+ 19 02	54	14.2	GALAXY
MCG+03-31-021	12 00 24.	+ 19 03	45	14.2	GALAXY
ZC 1200.4+2342	12 00 24.	+ 23 42	270		CLUSTER OF GALAXIES
ZC 1200.4+3115	12 00 24.	+ 31 15	670		CLUSTER OF GALAXIES
MRK 043	12 00 24.	+ 39 41	20	16.	GALAXY WITH UV CONTINUUM
ZC 1200.4+4020	12 00 24.	+ 40 20	1280		CLUSTER OF GALAXIES
ZC 1200.4+4404	12 00 24.	+ 44 04	940		CLUSTER OF GALAXIES
ZWG 352.014	12 00 24.	+ 78 38		15.7	GALAXY
MCG-03-31-016	12 00 24.	- 16 05	132	12.5	GALAXY
RNGC 4049	12 00 25.	+ 19 02		14.0	GALAXY
FATH 1.451	12 00 25.	+ 28 54	14		NEBULA
RNGC 4050	12 00 25.	- 16 06		12.5	GALAXY
KN 12.275	12 00 26.9	+ 47 56 05.			NEBULA
FATH 1.452	12 00 27.	+ 28 53	11		NEBULA
FATH 1.453	12 00 27.	+ 29 22	5		NEBULA
FATH 1.454	12 00 27.	+ 29 32	8		NEBULA
ARC 1450	12 00 27.	- 23 03		17.6	RICH CLUSTER OF GALAXIES
HN 1582	12 00 27.1	- 00 00 54.	18		NEBULA
LB 02182	12 00 28.	+ 50 05 06.		17.0	FAINT BLUE STAR
ARC 1449	12 00 29.	+ 26 51		17.2	RICH CLUSTER OF GALAXIES
ZWG 041.030	12 00 30.	+ 05 27		15.3	GALAXY
MCG+03-31-022	12 00 30.	+ 16 46	42	17.	GALAXY
ZWG 128.002	12 00 30.	+ 21 05		15.7	GALAXY
MCG+04-29-003	12 00 30.	+ 21 05	24	15.7	GALAXY
ZC 1200.5+2253	12 00 30.	+ 22 53	2020		CLUSTER OF GALAXIES
ZC 1200.5+2809	12 00 30.	+ 28 09	3560		CLUSTER OF GALAXIES
ZC 1200.5+4205	12 00 30.	+ 42 05	940		CLUSTER OF GALAXIES
MCG+09-21-090	12 00 30.	+ 52 02	42	15.	GALAXY
OCL 0872	12 00 30.	- 63 32	240	12.	OPEN STAR CLUSTER
FATH 2.053	12 00 32.	+ 29 44	5		NEBULA
REIZ 1904	12 00 34.	+ 62 29	30	15.0	GALAXY
HN 1583	12 00 34.1	+ 25 46 30.	12		NEBULA
HN 1584	12 00 34.2	+ 25 42 00.	12		NEBULA
UGC 07028	12 00 36.	+ 16 45	72	16.0	GALAXY Sa-b
MCG+03-31-023	12 00 36.	+ 16 45	60	15.	GALAXY
ZWG 098.032	12 00 36.	+ 20 00		14.6	GALAXY
UGC 07029	12 00 36.	+ 20 00	66	14.6	GALAXY S
TON-N 0601	12 00 36.	+ 28 32		15.6	BLUE STAR
ZWG 243.038	12 00 36.	+ 44 48		11.5	GALAXY
VVI 53	12 00 36.	+ 44 48	240	10.81	SEYFERT GALAXY
UGC 07030	12 00 36.	+ 44 48	390	11.5	GALAXY Sb/SBc
MCG+09-20-071	12 00 36.	+ 51 57	15	16.	GALAXY
MCG+09-20-070	12 00 36.	+ 55 08	10	17.	GALAXY
ZWG 292.062	12 00 36.	+ 58 10		15.2	GALAXY
ZWG 292.063	12 00 36.	+ 62 24		15.7	GALAXY
ZC 1200.6+7117	12 00 36.	+ 71 17	610		CLUSTER OF GALAXIES
ZC 1200.6-0145	12 00 36.	- 01 45	940		CLUSTER OF GALAXIES
ZWG 013.048	12 00 36.	- 02 35		15.5	GALAXY
REIN 2.147	12 00 36.63	+ 44 48 32.8			NEBULA
RNGC 4053	12 00 37.	+ 20 00		14.5	GALAXY
RNGC 4054	12 00 37.	+ 58 10		15.0	GALAXY
KN 12.276	12 00 37.1	+ 44 48 35.			NEBULA
HN 1585	12 00 37.9	- 02 36 00.			NEBULA
HN 1586	12 00 38.7	+ 02 14 12.	18		NEBULA
FATH 1.455	12 00 39.	+ 29 33	5		NEBULA
REIZ 1905	12 00 40.	+ 44 48	204	11.3	GALAXY
ARC 1451	12 00 40.	- 21 15		17.3	RICH CLUSTER OF GALAXIES
FATH 1.456	12 00 41.	+ 29 45	8		NEBULA
ZWG 013.049	12 00 42.	+ 02 14		14.7	GALAXY
MCG+00-31-023	12 00 42.	+ 02 14	24	14.7	GALAXY
MCG+03-31-024	12 00 42.	+ 20 02	78	14.6	GALAXY
ZC 1200.7+2846	12 00 42.	+ 28 46	1010		CLUSTER OF GALAXIES
FATH 1.457	12 00 42.	+ 29 33	5		NEBULA
MCG+05-29-001	12 00 42.	+ 29 43	72	15.2	GALAXY
MCG+08-22-059	12 00 42.	+ 44 48	252	10.8	GALAXY
MCG+09-21-091	12 00 42.	+ 51 34	42	14.	GALAXY
SEY 154	12 00 42.	+ 51 40 18.		15.4	FAINT GALAXY
MCG+09-20-072	12 00 42.	+ 52 04	10	16.	GALAXY
FATH 1.458	12 00 43.	+ 29 24	5		NEBULA
LB 02183	12 00 43.	+ 49 23 24.		15.1	FAINT BLUE STAR
REIZ 1906	12 00 46.	+ 29 42	102	14.8	GALAXY
REIZ 1907	12 00 46.	+ 47 40	24	14.3	GALAXY
REIZ 1908	12 00 46.	+ 58 12	60	14.6	GALAXY
SN 1961K	12 00 47.	+ 16 47		16.3	SUPERNOVA
ZC 1200.8+1345	12 00 48.	+ 13 45	740		CLUSTER OF GALAXIES
ZWG 098.033	12 00 48.	+ 16 47		14.6	GALAXY
MCG+03-31-025	12 00 48.	+ 16 47	30	14.6	GALAXY
ZWG 098.034	12 00 48.	+ 20 18		14.8	GALAXY
TON-N 0073	12 00 48.	+ 24 58		16.4	BLUE STAR
ZWG 158.005	12 00 48.	+ 29 42		15.2	GALAXY
UGC 07031	12 00 48.	+ 29 42	102	15.2	GALAXY Sb-c
MRK 044	12 00 48.	+ 39 04	18	16.	GALAXY WITH UV CONTINUUM
ZWG 215.018	12 00 48.	+ 39 05		15.3	GALAXY
MCG+10-17-131	12 00 48.	+ 58 10	27	14.6	GALAXY
FATH 2.054	12 00 49.	+ 29 42	95		NEBULA
FATH 1.459	12 00 50.	+ 29 49	8		NEBULA
MCG+04-29-004	12 00 51.	+ 22 29 30.	30	14.6	GALAXY
LB 02184	12 00 51.	+ 49 52 00.		16.5	FAINT BLUE STAR
MCG-03-31-017	12 00 51.	- 19 14	60	14.5	GALAXY
IC 2987	12 00 52.	+ 39 04 32.			NONSTELLAR OBJECT
KN 12.277	12 00 52.1	+ 47 38 36.			NEBULA
LB 02185	12 00 53.	+ 49 05 18.		14.7	FAINT BLUE STAR
ZWG 069.033	12 00 54.	+ 08 39		15.2	GALAXY
KARA.72 317A	12 00 54.	+ 08 39	48	15.2	PART OF DOUBLE GALAXY
MCG+02-31-017	12 00 54.	+ 08 39	36	15.2	GALAXY
ZWG 098.035	12 00 54.	+ 16 46		14.0	GALAXY
UGC 07032	12 00 54.	+ 16 46	36	14.0	GALAXY
MCG+03-31-027	12 00 54.	+ 16 46	24	14.0	GALAXY
MCG+03-31-026	12 00 54.	+ 16 46		19.	GALAXY
MCG+03-31-028	12 00 54.	+ 17 27	54	14.9	GALAXY

OBJECT NAME	RIGHT ASCEN.	DECLINATION	DIAM.	MAGN.	TYPE OF OBJECT
MCG+03-31-029	12 00 54.	+ 20 20	30	14.8	GALAXY
ZWG 128.003	12 00 54.	+ 22 30		14.6	GALAXY
ZWG 243.039	12 00 54.	+ 47 38		15.6	GALAXY
UGC 07033	12 00 54.	+ 47 38	84	15.6	GALAXY SB
MCG+08-22-060	12 00 54.	+ 47 39	36	15.	GALAXY
ZC 1200.9+5131	12 00 54.	+ 51 31	540		CLUSTER OF GALAXIES
MCG+09-20-074	12 00 54.	+ 51 58	12	16.	GALAXY
IC 0757	12 00 54.	+ 52 56		12.51	DOUBLE STAR
MCG+09-21-093	12 00 54.	+ 54 15	24	17.	GALAXY
MCG+09-20-073	12 00 54.	+ 55 10	30	16.	GALAXY
MCG+09-21-092	12 00 54.	+ 55 59	36	16.	GALAXY
MCG+10-17-132	12 00 54.	+ 62 23	15	16.	GALAXY
LB 02186	12 00 55.	+ 49 35 30.		15.5	FAINT BLUE STAR
SC 1158-4306.1	12 00 57.	- 43 22 48.	36		NEBULA
PATH 1.460	12 00 59.	+ 29 15	8		NEBULA
VDB.66G 108	12 01	- 01 16	70		DWARF GALAXY
VDB.66G 240	12 01	- 27 50	70		DWARF GALAXY
ZWG 013.050	12 01 00.	+ 01 31		15.7	GALAXY
ZWG 069.034	12 01 00.	+ 08 40		15.0	GALAXY
KARA.72 317B	12 01 00.	+ 08 40	48	15.0	PART OF DOUBLE GALAXY
MCG+02-31-018	12 01 00.	+ 08 40	36	15.0	GALAXY
MCG+03-31-030	12 01 00.	+ 16 20	36	14.7	GALAXY
ZWG 098.036	12 01 00.	+ 17 25		14.9	GALAXY
PATH 1.461	12 01 00.	+ 29 03	8		NEBULA
PATH 2.055	12 01 00.	+ 29 59	16		NEBULA
MCG+08-22-061	12 01 00.	+ 48 18	36	16.	GALAXY
MCG+09-20-075	12 01 00.	+ 51 59	15	16.	GALAXY
MCG+09-20-076	12 01 00.	+ 52 03	30	16.	GALAXY
SBY 155	12 01 00.	+ 55 09 36.		15.3	FAINT GALAXY
ZWG 292.064	12 01 00.	+ 60 47		15.6	GALAXY
MCG-04-29-003	12 01 00.	- 25 09	54	15.	GALAXY
HN 1587	12 01 00.8	- 01 40 36.	60		NEBULA
PATH 2.056	12 01 01.	+ 29 56	11		NEBULA
LB 02187	12 01 01.	+ 51 47 06.		17.1	FAINT BLUE STAR
PATH 1.462	12 01 03.	+ 28 58	8		NEBULA
HN 1588	12 01 03.4	+ 02 19 12.	18		NEBULA
ARC 1453	12 01 04.	- 04 23		17.8	RICH CLUSTER OF GALAXIES
ARC 1452	12 01 05.	+ 52 01		15.7	RICH CLUSTER OF GALAXIES
ZC 1201.1+0119	12 01 06.	+ 01 19	470		CLUSTER OF GALAXIES
ZWG 013.051	12 01 06.	+ 02 20		14.8	GALAXY
UGC 07034	12 01 06.	+ 02 20	66	14.8	GALAXY PECULR?
ZWG 041.031	12 01 06.	+ 02 55		14.6	GALAXY
UGC 07035	12 01 06.	+ 02 55	78	14.6	GALAXY S
MCG+01-31-016	12 01 06.	+ 02 55	60	14.6	GALAXY
ZC 1201.1+0357	12 01 06.	+ 03 57	2290		CLUSTER OF GALAXIES
ZWG 098.037	12 01 06.	+ 16 20		14.7	GALAXY
MCG+09-20-077	12 01 06.	+ 51 20	60	15.	GALAXY
ZC 1201.1+6644	12 01 06.	+ 66 44	940		CLUSTER OF GALAXIES
ZC 1201.1+6854	12 01 06.	+ 68 54	1140		CLUSTER OF GALAXIES
RNGC 4045	12 01 07.	+ 02 15		13.0	GALAXY
RNGC 4055	12 01 07.	+ 21 20			NON-EXISTENT OBJECT
PATH 1.463	12 01 07.	+ 29 54	24		NEBULA
IC 2988	12 01 09.	+ 03 41			NONSTELLAR OBJECT
HN 1589	12 01 10.0	+ 02 07 00.	18		NEBULA
HN 1590	12 01 11.7	+ 27 53 36.	18		NEBULA
ZWG 013.052	12 01 12.	+ 01 59		15.0	GALAXY
MCG+00-31-025	12 01 12.	+ 01 59	54	15.0	GALAXY
ZWG 041.032	12 01 12.	+ 03 50		14.0	GALAXY
UGC 07036	12 01 12.	+ 03 50	90	14.0	GALAXY Sa
MCG+01-31-017	12 01 12.	+ 03 50	72	14.0	GALAXY
MRK 645	12 01 12.	+ 23 59	12	15.5	GALAXY WITH UV CONTINUUM
TON-N 0074	12 01 12.	+ 25 49		15.1	BLUE STAR
KARA.73B 0518	12 01 12.	+ 32 10	276	11.9	ISOLATED GALAXY S
MCG+09-21-094	12 01 12.	+ 53 48	48	14.	GALAXY
MRK 045	12 01 12.	+ 60 48	16	16.	GALAXY WITH UV CONTINUUM
MCG+10-17-133	12 01 12.	+ 60 48	39	16.	GALAXY
SN 1955F	12 01 13.	+ 01 59		19.5	SUPERNOVA
RNGC 4058	12 01 13.	+ 03 50		14.0	GALAXY
LB 02188	12 01 14.	+ 52 25 30.		16.7	FAINT BLUE STAR
MCG+03-31-031	12 01 15.	+ 16 50	78	15.6	GALAXY
PATH 2.057	12 01 15.	+ 29 54	11		NEBULA
MCG+05-29-002	12 01 15.	+ 30 02	72	15.	GALAXY
HN 0037	12 01 16.	+ 03 49			NEBULA
ARC 1456	12 01 16.	+ 04 31		17.0	RICH CLUSTER OF GALAXIES
REIZ 1909	12 01 16.	+ 20 33	12	15.9	GALAXY
ARC 1455	12 01 16.	+ 28 16		17.2	RICH CLUSTER OF GALAXIES
REIZ 1910	12 01 16.	+ 29 59	60	15.1	GALAXY
HELW 023	12 01 16.	- 15 32			NEBULA
ARC 1454	12 01 17.	+ 51 18		17.2	RICH CLUSTER OF GALAXIES
HN 1591	12 01 17.1	+ 01 45 24.	12		NEBULA
ZC 1201.3+0151	12 01 18.	+ 01 51	4500		CLUSTER OF GALAXIES
ZWG 041.033	12 01 18.	+ 04 30		14.8	GALAXY
UGC 07037	12 01 18.	+ 04 30	60	14.8	GALAXY S
MCG+01-31-018	12 01 18.	+ 04 30	48	14.8	GALAXY
UGC 07038	12 01 18.	+ 14 49	90	17.	GALAXY DWARF
ZWG 098.038	12 01 18.	+ 16 50		15.6	GALAXY
UGC 07039	12 01 18.	+ 16 50	72	15.6	GALAXY Sc
ZWG 098.039	12 01 18.	+ 20 19		15.7	GALAXY
ZWG 128.004	12 01 18.	+ 25 43		14.9	GALAXY
UGC 07040	12 01 18.	+ 25 43	108	14.9	GALAXY
MCG+04-29-005	12 01 18.	+ 25 43	114	14.9	GALAXY
UGC 07041	12 01 18.	+ 29 59	90	16.0	GALAXY Sc
ZC 1201.3+3530	12 01 18.	+ 35 30	2350		CLUSTER OF GALAXIES
MCG+10-17-134	12 01 18.	+ 61 04	66	16.	GALAXY
MCG-03-31-018	12 01 18.	- 15 32	36	15.	GALAXY
HN 1592	12 01 18.8	+ 28 23 54.	30		NEBULA
RNGC 4059	12 01 19.	+ 21 35			NON-EXISTENT OBJECT
PATH 2.058	12 01 19.	+ 30 00	81		NEBULA
RNGC 4047	12 01 19.	+ 48 55		13.0	GALAXY
HN 1593	12 01 19.8	+ 25 42 12.	24		NEBULA
PATH 1.464	12 01 20.	+ 29 22	3		NEBULA
LB 02189	12 01 20.	+ 50 36 00.		15.5	FAINT BLUE STAR
MCG-04-29-004	12 01 21.	- 27 20	96	15.	GALAXY
HN 1594	12 01 21.6	+ 28 09 24.	30		NEBULA
REIZ 1911	12 01 22.	+ 25 42	60	15.1	GALAXY
REIZ 1912	12 01 22.	+ 28 09	24	15.5	GALAXY
REIZ 1913	12 01 22.	+ 62 48	42	14.3	GALAXY
ZWG 013.053	12 01 24.	+ 02 11		15.4	GALAXY
UGC 07042	12 01 24.	+ 02 11	72	15.4	GALAXY S
ZWG 041.034	12 01 24.	+ 07 10		15.6	GALAXY
MCG+01-31-019	12 01 24.	+ 07 10	48	15.6	GALAXY
ZWG 041.035	12 01 24.	+ 07 34		15.2	GALAXY
MCG+01-31-020	12 01 24.	+ 07 34	36	15.2	GALAXY
ZWG 069.035	12 01 24.	+ 09 29		15.5	GALAXY
ZC 1201.4+1341	12 01 24.	+ 13 41	2350		CLUSTER OF GALAXIES
MCG+04-29-006	12 01 24.	+ 20 30 30.	24	14.4	GALAXY
VV 276	12 01 24.	+ 25 42 30.	120	15.	INTERACTING GALAXY
MCG+05-29-003	12 01 24.	+ 29 16	24	15.7	GALAXY
ZC 1201.4+5115	12 01 24.	+ 51 15	740		CLUSTER OF GALAXIES
MCG+09-20-078	12 01 24.	+ 51 20	24	16.	GALAXY
MCG-05-29-004	12 01 24.	- 29 21	36	15.5	GALAXY
HN 1595	12 01 24.8	+ 01 41 54.	12		NEBULA
RNGC 4056	12 01 25.	+ 20 35			GALAXY
PATH 1.465	12 01 25.	+ 29 12	11		NEBULA
MCG+04-29-007	12 01 27.	+ 20 30	48	14.0	GALAXY
PATH 1.466	12 01 27.	+ 29 27	8		NEBULA
REIZ 1915	12 01 28.	+ 20 30	18	14.8	GALAXY
REIZ 1914	12 01 28.	+ 20 35	12	15.9	GALAXY
REIZ 1916	12 01 28.	+ 32 10	180	11.1	GALAXY
LB 02190	12 01 29.	+ 51 53 12.		15.5	FAINT BLUE STAR
HN 1596	12 01 29.5	+ 28 31 06.	12		NEBULA
ZWG 013.054	12 01 30.	+ 01 41		15.4	GALAXY
ZC 1201.5+0205	12 01 30.	+ 02 05	740		CLUSTER OF GALAXIES
ZWG 013.055	12 01 30.	+ 02 08		15.0	GALAXY
MCG+00-31-026	12 01 30.	+ 02 08	66	15.0	GALAXY
ZC 1201.5+0236	12 01 30.	+ 02 36	1550		CLUSTER OF GALAXIES
UGC 07043	12 01 30.	+ 13 38	96	16.0	GALAXY PECULR
ZC 1201.5+1457	12 01 30.	+ 14 57	870		CLUSTER OF GALAXIES
ZC 1201.5+1525	12 01 30.	+ 15 25	1550		CLUSTER OF GALAXIES
ZWG 128.005	12 01 30.	+ 20 30		14.4	GALAXY
ZWG 098.040	12 01 30.	+ 20 30		14.4	GALAXY
UGC 07044	12 01 30.	+ 20 30	78	14.4	GALAXY (E)
VV 179A	12 01 30.	+ 20 30	36	13.5	INTERACTING GALAXY
MCG+04-29-008	12 01 30.	+ 20 36 30.	60	14.4	GALAXY
ZWG 128.006	12 01 30.	+ 20 37		15.6	GALAXY
ZC 1201.5+2205	12 01 30.	+ 22 05	2290		CLUSTER OF GALAXIES
ZWG 158.006	12 01 30.	+ 28 26		15.3	GALAXY
ZWG 158.007	12 01 30.	+ 29 15		15.7	GALAXY
ZWG 158.008	12 01 30.	+ 32 10		11.9	GALAXY
UGC 07045	12 01 30.	+ 32 10	288	11.9	GALAXY Sc
MCG+05-29-004	12 01 30.	+ 32 12	258	11.9	GALAXY
ZC 1201.5+3916	12 01 30.	+ 39 16	5980		CLUSTER OF GALAXIES
ZWG 243.040	12 01 30.	+ 49 23		15.6	GALAXY
UGC 07046	12 01 30.	+ 49 23	66	15.6	GALAXY Sb-c
MCG+08-22-062	12 01 30.	+ 49 23	72	16.	GALAXY
SN 1970E	12 01 30.	+ 52 12		18.5	SUPERNOVA
ZWG 269.031	12 01 30.	+ 52 52		13.3	GALAXY
UGC 07047	12 01 30.	+ 52 52	180	13.3	GALAXY IRR
MCG+09-20-079	12 01 30.	+ 52 52	210	13.	GALAXY
MCG+09-21-095	12 01 30.	+ 54 02 30.	30	17.	GALAXY
IC 0758	12 01 30.	+ 62 46 14.			NONSTELLAR OBJECT
7ZW 435	12 01 30.	+ 65 22			COMPACT GALAXY
ZC 1201.5+6604	12 01 30.	+ 66 04	1410		CLUSTER OF GALAXIES
MRK 196	12 01 30.	+ 66 51	10	17.	GALAXY WITH UV CONTINUUM
MCG+13-09-008	12 01 30.	+ 79 40	51	16.	GALAXY
RNGC 4063	12 01 31.	+ 02 08		15.0	GALAXY
RNGC 4061	12 01 31.	+ 20 30		14.5	GALAXY
RNGC 4060	12 01 31.	+ 20 37		15.5	GALAXY
RNGC 4062	12 01 31.	+ 32 11		12.0	GALAXY
AMES 0001	12 01 32.	+ 14 20 30.	38	16.4	NEBULA
PATH 2 059	12 01 32.	+ 29 15	19		NEBULA
RNGC 4068	12 01 32.	+ 52 52		13.0	GALAXY
MCG+04-29-009	12 01 33.	+ 20 40	36	14.3	GALAXY
REIZ 1917	12 01 34.	+ 18 43	36	12.9	GALAXY
REIZ 1918	12 01 34.	+ 20 30	18	14.6	GALAXY
REIZ 1919	12 01 34.	+ 20 36	42	15.3	GALAXY
PATH 1.467	12 01 34.	+ 29 14	5		NEBULA
REIZ 1920	12 01 34.	+ 29 15	30	14.9	GALAXY
ARC 1458	12 01 34.	- 04 48		17.3	RICH CLUSTER OF GALAXIES
ARC 1457	12 01 35.	+ 52 42		17.2	RICH CLUSTER OF GALAXIES
ZWG 013.056	12 01 36.	+ 02 07		15.5	GALAXY
ZC 1201.6+0439	12 01 36.	+ 04 39	2220		CLUSTER OF GALAXIES
ZWG 069.036	12 01 36.	+ 11 08		13.2	GALAXY
UGC 07048	12 01 36.	+ 11 08	72	13.2	GALAXY Sb
MCG+02-31-019	12 01 36.	+ 11 08	84	13.2	GALAXY
ZWG 069.037	12 01 36.	+ 14 20		15.5	GALAXY
ZWG 098.041	12 01 36.	+ 20 27		15.7	GALAXY
UGC 07049	12 01 36.	+ 20 27	66	15.7	GALAXY Sc
ZWG 128.007	12 01 36.	+ 20 30		14.0	GALAXY
ZWG 098.042	12 01 36.	+ 20 30		14.0	GALAXY
UGC 07050	12 01 36.	+ 20 30	84	14.0	GALAXY E
VV 179B	12 01 36.	+ 20 30	36	14.	INTERACTING GALAXY
ZWG 128.008	12 01 36.	+ 20 38		14.4	GALAXY
UGC 07051	12 01 36.	+ 20 38	72	14.4	GALAXY E
ZWG 128.009	12 01 36.	+ 20 42		14.3	GALAXY
UGC 07052	12 01 36.	+ 20 42	84	14.3	GALAXY E
MCG+04-29-010	12 01 36.	+ 24 23 30.	24	15.7	GALAXY
MCG+09-21-096	12 01 36.	+ 54 03	18	17.	GALAXY
MCG+10-17-135	12 01 36.	+ 61 05	27	17.	GALAXY
ZC 1201.6+6145	12 01 36.	+ 61 45	1010		CLUSTER OF GALAXIES
RNGC 4071	12 01 36.	- 67 02			PLANETARY NEBULA
RNGC 4067	12 01 37.	+ 11 08		13.0	GALAXY
RNGC 4064	12 01 37.	+ 18 43		12.5	GALAXY
RNGC 4065	12 01 37.	+ 20 30		14.0	GALAXY
RNGC 4069	12 01 37.	+ 20 36			GALAXY
RNGC 4066	12 01 37.	+ 20 38		14.5	GALAXY
RNGC 4070	12 01 37.	+ 20 42		14.5	GALAXY
PATH 2.060	12 01 37.	+ 29 12	19		NEBULA
PATH 2.061	12 01 37.	+ 29 37	8		NEBULA
RNGC 4051	12 01 37.	+ 44 49		11.5	GALAXY
PK298-04.1	12 01 37.	- 67 01 33.	75		PLANETARY NEBULA
REIN 7.069	12 01 37.92	+ 11 07 56.4			NEBULA
VV 046B	12 01 39.	+ 24 22 30.	18	18.	INTERACTING GALAXY
VV 046A	12 01 39.	+ 24 22 30.	36	16.	INTERACTING GALAXY
LB 02192	12 01 39.	+ 48 49 00.		16.5	FAINT BLUE STAR
LB 02191	12 01 39.	+ 50 15 48.		16.7	FAINT BLUE STAR
REIZ 1924	12 01 39.	+ 62 09	42	13.8	GALAXY
ARC 1459	12 01 40.	+ 02 47		17.0	RICH CLUSTER OF GALAXIES
REIZ 1922	12 01 40.	+ 11 08	60	13.6	GALAXY
HELW 024	12 01 40.	+ 14 47			NEBULA
HOLM 322A	12 01 40.	+ 14 47	30	14.8	PART OF MULTIPLE GALAXY
REIZ 1921	12 01 40.	+ 20 38	18	15.0	GALAXY
REIZ 1923	12 01 40.	+ 20 40	18	15.1	GALAXY
HOLM 322B	12 01 41.	+ 14 48	24	15.5	PART OF MULTIPLE GALAXY
PATH 1.468	12 01 41.	+ 29 12	8		NEBULA
ZWG 041.036	12 01 42.	+ 04 09		15.4	GALAXY
ZWG 098.043	12 01 42.	+ 14 46		15.4	GALAXY
MCG+03-31-032	12 01 42.	+ 14 46	24	15.4	GALAXY
ZWG 098.044	12 01 42.	+ 18 43		12.5	GALAXY
UGC 07054	12 01 42.	+ 18 43	258	12.5	GALAXY SBa
MCG+03-31-033	12 01 42.	+ 18 45	270	12.5	GALAXY
ZWG 128.010	12 01 42.	+ 20 29		15.6	GALAXY
ZWG 098.045	12 01 42.	+ 20 29		15.6	GALAXY
ZWG 128.011	12 01 42.	+ 24 25		15.7	GALAXY
UGC 07055	12 01 42.	+ 24 25	60	15.7	GALAXY DBL SYS
ZWG 215.019	12 01 42.	+ 43 57		15.4	GALAXY

OBJECT NAME	RIGHT ASCEN.	DECLINATION	DIAM.	MAGN.	TYPE OF OBJECT
ZC 1201.7+5022	12 01 42.	+ 50 22	1010		CLUSTER OF GALAXIES
MCG+09-20-080	12 01 42.	+ 56 37	84	16.	GALAXY
ZC 1201.7+5912	12 01 42.	+ 59 12	3290		CLUSTER OF GALAXIES
ZWG 315.009	12 01 42.	+ 62 47		14.4	GALAXY
UGC 07056	12 01 42.	+ 62 47	120	14.4	GALAXY SBc
MCG+11-15-014	12 01 42.	+ 62 47	114	14.	GALAXY
MCG+13-09-009	12 01 42.	+ 79 42	66	15.	GALAXY
UGC 07053	12 01 42.	- 01 15	108	17.	GALAXY DWRF IR
HN 1597	12 01 42.0	+ 28 13 42.	30		NEBULA
HN 1598	12 01 42.7	+ 28 28 54.	12		NEBULA
RNGC 4072	12 01 43.	+ 20 29		15.5	GALAXY
HN 1599	12 01 44.3	+ 02 08 36.	12		NEBULA
HN 1600	12 01 44.4	+ 26 25 12.			NEBULA
MCG+07-25-016	12 01 45.	+ 43 58	36	16.	GALAXY
REIZ 1925	12 01 46.	+ 20 28	30	15.4	GALAXY
FATH 1.469	12 01 46.	+ 28 59	14		NEBULA
HN 1601	12 01 46.0	+ 01 50 54.	12		NEBULA
HN 1602	12 01 47.2	+ 01 50 54.	12		NEBULA
ZWG 013.057	12 01 48.	+ 01 51		14.6	GALAXY
UGC 07057	12 01 48.	+ 01 51	120	14.6	GALAXY S
MCG+00-31-027	12 01 48.	+ 01 51	114	14.6	GALAXY
ZWG 013.058	12 01 48.	+ 02 08		15.3	GALAXY
ZWG 041.037	12 01 48.	+ 07 19		15.7	GALAXY
ZWG 069.038	12 01 48.	+ 09 05		14.9	GALAXY
MCG+02-31-020	12 01 48.	+ 09 05	36	14.9	GALAXY
ZWG 069.039	12 01 48.	+ 09 28		15.6	GALAXY
MCG+02-31-021	12 01 48.	+ 09 28	48	15.6	GALAXY
MCG+08-22-063	12 01 48.	+ 50 24 30.	60	16.	GALAXY
7ZW 436	12 01 48.	+ 57 01			COMPACT GALAXY
ZC 1201.8+5705	12 01 48.	+ 57 05	1280		CLUSTER OF GALAXIES
ZWG 352.015	12 01 48.	+ 79 42		15.4	GALAXY
UGC 07059	12 01 48.	+ 79 42	78	15.4	GALAXY Sb/SBb
UGC 07058	12 01 48.	+ 79 42	66	15.4	GALAXY Sb
MCG+00-31-028	12 01 48.	- 01 16	60	17.	GALAXY
MCG-02-31-013	12 01 48.	- 12 31	36	15.	GALAXY
MCG-03-31-020	12 01 48.	- 18 15	24	17.	GALAXY
MCG-03-31-019	12 01 48.	- 18 15	24	16.	GALAXY
LB 02193	12 01 49.	+ 54 15 18.		15.1	FAINT BLUE STAR
HN 1603	12 01 50.6	- 00 21 24.	12		NEBULA
LB 02194	12 01 51.	+ 59 38 48.		15.8	FAINT BLUE STAR
FATH 1.470	12 01 53.	+ 29 32	3		NEBULA
ZWG 013.059	12 01 54.	+ 02 11		13.8	GALAXY
UGC 07060	12 01 54.	+ 02 11	138	13.8	GALAXY P
MKW 04	12 01 54.	+ 02 11		13.8	POOR GALAXY CLUSTER
MCG+00-31-029	12 01 54.	+ 02 11	24	13.8	GALAXY
ZC 1201.9+0303	12 01 54.	+ 03 03	1280		CLUSTER OF GALAXIES
ZWG 069.040	12 01 54.	+ 09 10		15.6	GALAXY
ZWG 069.041	12 01 54.	+ 14 15		15.7	GALAXY
MCG+03-31-034	12 01 54.	+ 20 28	54	14.3	GALAXY
MCG+04-29-011	12 01 54.	+ 20 34 30.	21	15.4	GALAXY
FATH 1.471	12 01 54.	+ 29 36	11		NEBULA
MCG+09-21-097	12 01 54.	+ 54 14	18	15.	GALAXY
VV 269B	12 01 54.	- 18 15	30	16.5	INTERACTING GALAXY
VV 269A	12 01 54.	- 18 15	24	16.	INTERACTING GALAXY
RNGC 4073	12 01 55.	+ 02 11		13.0	GALAXY
AMES 0002	12 01 55.	+ 10 32 48.	28	16.6	NEBULA
FATH 1.472	12 01 55.	+ 29 07	8		NEBULA
AMES 0003	12 01 56.	+ 10 34 42.	31	16.4	NEBULA
FATH 1.473	12 01 56.	+ 29 36	11		NEBULA
FATH 1.474	12 01 57.	+ 29 12	8		NEBULA
REIZ 1926	12 01 58.	+ 20 36	24	15.2	GALAXY
LB 02195	12 01 58.	+ 49 42 18.		17.3	FAINT BLUE STAR
ARC 1460	12 01 58.	+ 53 34		18.0	RICH CLUSTER OF GALAXIES
HN 1604	12 01 58.3	+ 28 25 18.	12		NEBULA
ZWG 013.060	12 02 00.	+ 02 01		15.7	GALAXY
ZWG 013.061	12 02 00.	+ 02 05		14.8	GALAXY
MCG+00-31-030	12 02 00.	+ 02 05	24	14.8	GALAXY
ZC 1202.0+1550	12 02 00.	+ 15 50	870		CLUSTER OF GALAXIES
ZWG 128.012	12 02 00.	+ 20 28		14.3	GALAXY
ZWG 098.046	12 02 00.	+ 20 28		14.3	GALAXY
ZC 1202.0+2028	12 02 00.	+ 20 28	6520		CLUSTER OF GALAXIES
UGC 07061	12 02 00.	+ 20 28	54	14.3	GALAXY S
ZWG 128.013	12 02 00.	+ 20 36		15.4	GALAXY
ZWG 128.014	12 02 00.	+ 22 34		15.7	GALAXY
MCG+09-20-081	12 02 00.	+ 53 14	6	17.	GALAXY
MCG+11-15-015	12 02 00.	+ 64 42	72	14.	GALAXY
ZWG 315.010	12 02 00.	+ 64 43		13.6	GALAXY
UGC 07062	12 02 00.	+ 64 43	108	13.6	GALAXY S
RNGC 4076	12 02 01.	+ 20 28		14.5	GALAXY
RNGC 4074	12 02 01.	+ 20 36		15.5	GALAXY
IC 2989	12 02 02.	+ 02 05			NONSTELLAR OBJECT
RNGC 4081	12 02 02.	+ 64 43		13.5	GALAXY
FATH 1.475	12 02 03.	+ 29 56	5		NEBULA
LB 02196	12 02 03.	+ 60 55 42.		15.8	FAINT BLUE STAR
HN 0792	12 02 04.	+ 11 19		15.	NEBULA
IC 2990	12 02 04.	+ 11 19			NONSTELLAR OBJECT
REIZ 1927	12 02 04.	+ 20 29	24	14.6	GALAXY
ARC 1461	12 02 04.	+ 42 48		16.6	RICH CLUSTER OF GALAXIES
LB 02197	12 02 05.	+ 60 48 24.		13.1	FAINT BLUE STAR
ZC 1202.1+0015	12 02 06.	+ 00 15	3290		CLUSTER OF GALAXIES
ZWG 013.063	12 02 06.	+ 02 04		14.5	GALAXY
UGC 07063	12 02 06.	+ 02 04	66	14.5	GALAXY E-SO
MCG+00-31-031	12 02 06.	+ 02 04	30	14.5	GALAXY
ZWG 013.064	12 02 06.	+ 02 21		14.7	GALAXY
MCG+00-31-032	12 02 06.	+ 02 21	60	14.7	GALAXY
ZWG 069.042	12 02 06.	+ 11 19		15.1	GALAXY
MCG+02-31-022	12 02 06.	+ 11 19	36	15.1	GALAXY
ZWG 158.009	12 02 06.	+ 31 27		14.0	GALAXY
UGC 07064	12 02 06.	+ 31 27	48	14.0	GALAXY S?
ZC 1202.1+3218	12 02 06.	+ 32 18	1010		CLUSTER OF GALAXIES
ZC 1202.1+3735	12 02 06.	+ 37 35	740		CLUSTER OF GALAXIES
MCG+09-21-098	12 02 06.	+ 55 44	42	15.	GALAXY
UGC 07064A	12 02 06.	+ 60 57	120	17.	GALAXY MLT SYS
ZWG 352.016	12 02 06.	+ 76 25		15.7	GALAXY
ZWG 013.062	12 02 06.	- 03 15		15.5	GALAXY
RNGC 4077	12 02 07.	+ 02 04		14.5	GALAXY
RNGC 4075	12 02 07.	+ 02 21		14.5	GALAXY
LB 02198	12 02 07.	+ 54 56 12.		17.0	FAINT BLUE STAR
FATH 1.476	12 02 08.	+ 29 37	8		NEBULA
HOLM 323A	12 02 09.	+ 31 28	42	14.1	PART OF MULTIPLE GALAXY
REIZ 1929	12 02 09.	+ 32 46	30	15.2	GALAXY
ARC 1463	12 02 10.	+ 04 13		17.2	RICH CLUSTER OF GALAXIES
REIZ 1928	12 02 10.	+ 10 49	42	14.9	GALAXY
ARC 1462	12 02 10.	+ 15 20		17.2	RICH CLUSTER OF GALAXIES
HOLM 323C	12 02 11.	+ 31 29	48	14.9	PART OF MULTIPLE GALAXY
ZWG 069.043	12 02 12.	+ 10 52		13.9	GALAXY
UGC 07066	12 02 12.	+ 10 52	78	13.9	GALAXY S0?
MCG+02-31-023	12 02 12.	+ 10 52	60	13.9	GALAXY
MCG+03-31-035	12 02 12.	+ 14 46	72	15.3	GALAXY
ZC 1202.2+1655	12 02 12.	+ 16 55	1080		CLUSTER OF GALAXIES
MCG+05-29-006	12 02 12.	+ 27 16	66	14.0	GALAXY
TON-N 0602	12 02 12.	+ 28 13		15.8	BLUE STAR
ZWG 158.010	12 02 12.	+ 31 26		15.2	GALAXY
ZWG 158.011	12 02 12.	+ 31 28		15.7	GALAXY
MCG+05-29-005	12 02 12.	+ 31 29	42	14.0	GALAXY
ZWG 243.041	12 02 12.	+ 49 28		15.4	GALAXY
MCG+09-20-082	12 02 12.	+ 53 14	24	17.	GALAXY
MCG+09-21-099	12 02 12.	+ 55 53	30	15.	GALAXY
MCG+11-15-016	12 02 12.	+ 63 24	30	15.	GALAXY
KEEN 4125A	12 02 12.	+ 64 42	72	13.8	Sa GALAXY
ZWG 013.065	12 02 12.	- 02 26		14.9	GALAXY
UGC 07065	12 02 12.	- 02 26	102	14.9	GALAXY SBb
MCG+00-31-033	12 02 12.	- 02 26	66	14.9	GALAXY
RNGC 4078	12 02 13.	+ 10 52		14.0	GALAXY
REIN 7.070	12 02 13.12	+ 10 52 26.9			NEBULA
REIN 7.071	12 02 13.92	+ 10 52 24.7			NEBULA
HN 1605	12 02 14.3	- 02 26 24.	12		NEBULA
MCG+04-29-012	12 02 15.	+ 21 30 30.	45	15.	GALAXY
MCG+04-29-013	12 02 15.	+ 21 41	48	15.5	GALAXY
REIZ 1932	12 02 15.	+ 31 24	24	14.2	GALAXY
MCG+08-22-064	12 02 15.	+ 49 28	36	16.	GALAXY
MCG+09-21-100	12 02 15.	+ 55 40	30	15.	GALAXY
REIZ 1930	12 02 16.	+ 10 53	42	13.9	GALAXY
ARC 1464	12 02 16.	+ 27 00		18.0	RICH CLUSTER OF GALAXIES
HOLM 323B	12 02 16.	+ 31 27	36	14.8	PART OF MULTIPLE GALAXY
REIZ 1931	12 02 16.	- 02 07	60	13.5	GALAXY
HELW 441	12 02 17.	- 02 09 36.			NEBULA
ZWG 098.047	12 02 18.	+ 14 46		15.3	GALAXY
ZWG 098.048	12 02 18.	+ 19 31		15.2	GALAXY
MCG+03-31-036	12 02 18.	+ 19 33	60	15.2	GALAXY
MCG+08-22-064A	12 02 18.	+ 48 51		17.	GALAXY
ZWG 292.065	12 02 18.	+ 58 30		15.7	GALAXY
ZWG 013.067	12 02 18.	- 02 05		14.0	GALAXY
UGC 07067	12 02 18.	- 02 05	174	14.0	GALAXY Sb
MCG+00-31-034	12 02 18.	- 02 05	96	14.0	GALAXY
ZWG 013.066	12 02 18.	- 02 09		15.4	GALAXY
MCG-05-29-005	12 02 18.	- 27 50	120	14.5	GALAXY
SC 1159-4310.8	12 02 18.	- 43 27 30.	6		NEBULA
RNGC 4079	12 02 19.	- 02 06		14.0	GALAXY
REIZ 1935	12 02 21.	+ 27 16	66	14.3	GALAXY
REIZ 1934	12 02 22.	+ 21 31	12	15.8	GALAXY
ARC 1466	12 02 22.	+ 22 54		17.2	RICH CLUSTER OF GALAXIES
ARC 1465	12 02 22.	+ 32 24		17.8	RICH CLUSTER OF GALAXIES
LB 02199	12 02 22.	+ 50 23 42.		13.6	FAINT BLUE STAR
REIZ 1933	12 02 22.	- 02 55	18	15.7	GALAXY
AMES 0004	12 02 23.	+ 11 54 18.	18	17.5	NEBULA
ZWG 041.038	12 02 24.	+ 06 35		15.4	GALAXY
ZWG 128.015	12 02 24.	+ 21 31		15.3	GALAXY
ZWG 158.012	12 02 24.	+ 27 16		14.0	GALAXY
UGC 07068	12 02 24.	+ 27 16	78	14.0	GALAXY IRR?
ZWG 187.001	12 02 24.	+ 32 40		15.7	GALAXY
ZWG 215.020	12 02 24.	+ 43 25		15.7	GALAXY
UGC 07069	12 02 24.	+ 43 25	102	15.7	GALAXY SBc
MCG+09-20-083	12 02 24.	+ 54 31	24	17.	GALAXY
ZWG 292.066	12 02 24.	+ 58 23		15.4	GALAXY
UGC 07070	12 02 24.	+ 58 23	120	15.4	GALAXY S+COMP
MCG+10-17-136	12 02 24.	+ 58 29	18	16.	GALAXY
ZWG 315.011	12 02 24.	+ 63 25		15.1	GALAXY
ZWG 352.017	12 02 24.	+ 76 26		15.4	GALAXY
ZWG 013.068	12 02 24.	- 02 55		15.3	GALAXY
SC 1159-4612.2	12 02 24.	- 46 28 54.	18		NEBULA
AMES 0005	12 02 25.	+ 11 46 18.	38	17.3	NEBULA
RNGC 4080	12 02 25.	+ 27 16		14.0	GALAXY
LB 02200	12 02 27.	+ 48 19 12.		15.9	FAINT BLUE STAR
FATH 2.062	12 02 28.	+ 29 08	19		NEBULA
ZC 1202.5+1157	12 02 30.	+ 11 57	540		CLUSTER OF GALAXIES
ZWG 098.049	12 02 30.	+ 15 37		15.6	GALAXY
ZC 1202.5+2256	12 02 30.	+ 22 56	940		CLUSTER OF GALAXIES
ZC 1202.5+2541	12 02 30.	+ 25 41	870		CLUSTER OF GALAXIES
MCG+07-25-017	12 02 30.	+ 43 26	72	14.5	GALAXY
MCG+07-25-018	12 02 30.	+ 43 36 30.	48	16.	GALAXY
ZC 1202.5+4615	12 02 30.	+ 46 15	670		CLUSTER OF GALAXIES
MCG+09-20-084	12 02 30.	+ 54 09 30.	48	17.	GALAXY
MCG+10-17-137	12 02 30.	+ 58 22 30.	60	15.	GALAXY
FEIG 055	12 02 30.	+ 60 51		13.3	FAINT BLUE STAR
AMES 0006	12 02 32.	+ 11 20 06.	25	16.7	NEBULA
LB 02201	12 02 32.	+ 53 26 24.		15.5	FAINT BLUE STAR
REIZ 1936	12 02 33.	+ 29 03	54	15.1	GALAXY
MCG+05-29-007	12 02 33.	+ 29 03	90	15.1	GALAXY
HN 0793	12 02 34.	+ 10 54			NEBULA
IC 2991	12 02 34.	+ 10 54			NONSTELLAR OBJECT
LB 02202	12 02 35.	+ 50 16 24.		16.5	FAINT BLUE STAR
IC 0759	12 02 36.	+ 20 32			NONSTELLAR OBJECT
ZWG 128.016	12 02 36.	+ 22 17		15.2	GALAXY
ZC 1202.6+2700	12 02 36.	+ 27 00	540		CLUSTER OF GALAXIES
ZWG 215.021	12 02 36.	+ 38 31		15.3	GALAXY
UGC 07071	12 02 36.	+ 38 31	60	15.3	GALAXY Sb
2ZW 056	12 02 36.	+ 42 35			COMPACT GALAXY
ZC 1202.6+5127	12 02 36.	+ 51 27	2760		CLUSTER OF GALAXIES
MCG+09-20-085	12 02 36.	+ 54 10	15	17.	GALAXY
MCG-01-31-003	12 02 36.	- 03 39	30	15.	GALAXY
REIN 7.072A	12 02 37.95	+ 10 56 56.4			NEBULA
REIN 7.072B	12 02 38.09	+ 10 56 56.5			NEBULA
REIN 7.073	12 02 38.67	+ 10 55 08.5			NEBULA
HOLM 324B	12 02 39.	+ 10 56	48	14.5	PART OF MULTIPLE GALAXY
AMES 0007	12 02 39.	+ 12 33 00.	23	17.4	NEBULA
FATH 2.063	12 02 39.	+ 29 03	27		NEBULA
FATH 1.477	12 02 39.	+ 29 54	8		NEBULA
MCG+07-25-019	12 02 39.	+ 38 30	54	14.5	GALAXY
REIZ 1939	12 02 40.	+ 10 53	36	15.3	GALAXY
REIZ 1937	12 02 40.	+ 10 54	60	15.8	GALAXY
HOLM 324A	12 02 40.	+ 10 54	66	15.6	PART OF MULTIPLE GALAXY
REIZ 1938	12 02 40.	+ 10 56	48	15.1	GALAXY
IC 2992	12 02 40.	+ 30 37 14.			NONSTELLAR OBJECT
IC 2993	12 02 40.	+ 33 08 26.			NONSTELLAR OBJECT
HZ 17	12 02 40.	+ 40 25		14.4	BLUE STAR
REIN 7.074A	12 02 40.44	+ 10 53 27.6			NEBULA
REIN 7.074B	12 02 40.62	+ 10 53 30.5			NEBULA
HOLM 324C	12 02 42.	+ 10 52	66	14.6	PART OF MULTIPLE GALAXY
ZWG 069.044	12 02 42.	+ 10 53		15.1	GALAXY
MCG+02-31-024	12 02 42.	+ 10 53	24	15.1	GALAXY
ZWG 069.045	12 02 42.	+ 10 55		15.5	GALAXY
MCG+02-31-025	12 02 42.	+ 10 55	36	15.5	GALAXY
ZWG 069.046	12 02 42.	+ 10 57		15.1	GALAXY
MCG+02-31-026	12 02 42.	+ 10 57	48	15.5	GALAXY
MCG+04-29-014	12 02 42.	+ 21 29	24	14.9	GALAXY

OBJECT NAME	RIGHT ASCEN.	DECLINATION	DIAM.	MAGN.	TYPE OF OBJECT
ZWG 128.017	12 02 42.	+ 21 30		14.9	GALAXY
MCG+05-29-009	12 02 42.	+ 28 39	30	15.2	GALAXY
ZWG 158.013	12 02 42.	+ 29 03		15.1	GALAXY
UGC 07072	12 02 42.	+ 29 03	96	15.1	GALAXY
ZWG 158.014	12 02 42.	+ 31 08		14.8	GALAXY
MCG+05-29-008	12 02 42.	+ 31 08	36	14.8	GALAXY
ZC 1202.7+3225	12 02 42.	+ 32 25	2550		CLUSTER OF GALAXIES
ZWG 215.022	12 02 42.	+ 43 26		15.2	GALAXY
MCG+11-15-017	12 02 42.	+ 63 27	15	18.	GALAXY
RNGC 4083	12 02 43.	+ 10 53		15.0	GALAXY
RNGC 4082	12 02 43.	+ 10 57		15.0	GALAXY
RNGC 4057	12 02 43.	+ 20 35			GALAXY
RNGC 4084	12 02 43.	+ 21 30		15.0	GALAXY
REIZ 1940	12 02 45.	+ 21 30	12	14.9	GALAXY
REIZ 1941	12 02 45.	+ 28 38	48	14.6	GALAXY
REIZ 1942	12 02 45.	+ 29 01	48	14.9	GALAXY
TON-N 1470	12 02 45.	+ 34 08		16.8	BLUE STAR
SEY 156	12 02 45.	+ 43 26 47.		14.7	FAINT GALAXY
REIZ 1943	12 02 45.	+ 63 26	54	14.7	GALAXY
AMES 0008	12 02 46.	+ 11 28 00.	31	17.4	NEBULA
HOLM 325A	12 02 46.	+ 18 09	66	13.7	PART OF MULTIPLE GALAXY
HOLM 325B	12 02 46.	+ 18 11	48	14.0	PART OF MULTIPLE GALAXY
HOLM 326B	12 02 46.	+ 50 38	132	12.8	PART OF MULTIPLE GALAXY
ZWG 013.069	12 02 48.	+ 02 22		15.6	GALAXY
ZWG 098.050	12 02 48.	+ 18 09		14.1	GALAXY
SCH 20	12 02 48.	+ 18 09		14.1	PECULIAR GALAXY
UGC 07073	12 02 48.	+ 18 09	72	14.1	GALAXY Sc
KARA.72 318B	12 02 48.	+ 18 09	72	14.1	PART OF DOUBLE GALAXY
MCG+03-31-038	12 02 48.	+ 18 11	66	14.1	GALAXY
ZWG 098.051	12 02 48.	+ 18 12		14.6	GALAXY
UGC 07074	12 02 48.	+ 18 12	72	14.6	GALAXY S
KARA.72 318A	12 02 48.	+ 18 12	66	14.6	PART OF DOUBLE GALAXY
MCG+03-31-037	12 02 48.	+ 18 13	66	14.6	GALAXY
ZC 1202.8+4240	12 02 48.	+ 42 40	400		CLUSTER OF GALAXIES
ZWG 269.032	12 02 48.	+ 50 39		12.8	GALAXY
UGC 07075	12 02 48.	+ 50 39	168	12.8	GALAXY Sc
MCG+09-21-101	12 02 48.	+ 53 56	36	15.	GALAXY
MCG+11-15-019	12 02 48.	+ 63 26	36	15.	GALAXY
MCG+11-15-018	12 02 48.	+ 63 27	18	17.	GALAXY
MCG+13-09-010	12 02 48.	+ 76 24	51	16.	GALAXY
MCG-01-31-004	12 02 48.	- 04 01	48	15.5	GALAXY
MCG-01-31-005	12 02 48.	- 04 02	42	17.	GALAXY
MKSL 297-00/1	12 02 48.	- 62 31	10800		HII REGION
AMES 0009	12 02 49.	+ 10 38 30.	26	17.0	NEBULA
HN 1606	12 02 49.7	+ 28 38 24.	24		NEBULA
RNGC 4085	12 02 50.	+ 50 38		13.0	GALAXY
REIN 7.075A	12 02 50.1	+ 50 37 48.			NEBULA
KN 12.278	12 02 50.1	+ 50 37 57.			NEBULA
REIN 7.075B	12 02 50.66	+ 50 37 53.1			NEBULA
MCG+04-29-015	12 02 51.	+ 20 34 30.	72	15.0	GALAXY
MCG+09-20-086	12 02 51.	+ 50 37 30.	156	12.7	GALAXY
REIZ 1945	12 02 51.	+ 50 38	126	13.1	GALAXY
REIZ 1944	12 02 52.	+ 10 59	30	15.5	GALAXY
AMES 0010	12 02 53.	+ 10 57 36.	20	16.6	NEBULA
ZC 1202.9+1051	12 02 54.	+ 10 51	4700		CLUSTER OF GALAXIES
ZWG 069.047	12 02 54.	+ 12 58		15.6	GALAXY
MCG+04-29-016	12 02 54.	+ 20 30	30	15.1	GALAXY
ZWG 128.018	12 02 54.	+ 20 31		15.1	GALAXY
UGC 07076	12 02 54.	+ 20 31	84	15.1	GALAXY S0
ZWG 128.019	12 02 54.	+ 20 35		15.0	GALAXY
UGC 07077	12 02 54.	+ 20 35	72	15.0	GALAXY Sa-b
ZWG 158.015	12 02 54.	+ 28 39		15.2	GALAXY
TON-N 0075	12 02 54.	+ 30 55		15.3	BLUE STAR
ZC 1202.9+4252	12 02 54.	+ 42 52	2080		CLUSTER OF GALAXIES
KW 54	12 02 54.	+ 44 41	280		SEYFERT GALAXY
ZWG 243.042	12 02 54.	+ 47 03		15.5	GALAXY
UGC 07078	12 02 54.	+ 47 03	78	15.5	GALAXY S
ZWG 315.012	12 02 54.	+ 63 26		14.9	GALAXY
UGC 07079	12 02 54.	+ 63 26	66	14.9	GALAXY S
SN 1973H	12 02 54.	- 04 02		18.5	SUPERNOVA
REIN 7.076A	12 02 54.00	+ 10 58 46.3			NEBULA
REIN 7.076B	12 02 54.03	+ 10 58 48.7			NEBULA
AMES 0011	12 02 55.	+ 10 32 18.	34	17.9	NEBULA
RNGC 4086	12 02 55.	+ 20 31		15.0	GALAXY
RNGC 4090	12 02 55.	+ 20 35		15.0	GALAXY
AMES 0012	12 02 56.	+ 12 51 42.	28	16.4	NEBULA
IC 2994	12 02 57.	+ 12 59			NONSTELLAR OBJECT
REIZ 1947	12 02 57.	+ 20 30	12	14.7	GALAXY
REIZ 1948	12 02 57.	+ 20 34	24	15.3	GALAXY
FATH 1.478	12 02 57.	+ 29 28	11		NEBULA
HOLM 326A	12 02 57.	+ 50 49	300	11.1	PART OF MULTIPLE GALAXY
ARC 1467	12 02 57.	+ 72 53		17.4	RICH CLUSTER OF GALAXIES
HN 0794	12 02 58.	+ 12 59		14.5	NEBULA
FATH 1.479	12 02 58.	+ 29 18	11		NEBULA
REIZ 1946	12 02 58.	- 04 01	36	15.0	GALAXY
ARP 018	12 02 59.	+ 50 49			PECULIAR GALAXY
LB 02203	12 02 59.	+ 56 55 24.		16.2	FAINT BLUE STAR
HN 1607	12 02 59.2	+ 26 51 24.			NEBULA
AMES 0013	12 03 00.	+ 11 38 12.	22	17.8	NEBULA
8ZW 1203+12.8	12 03 00.	+ 12 50		16.5	COMPACT GALAXY
ZWG 098.052	12 03 00.	+ 14 46		15.5	GALAXY
MCG+04-29-017	12 03 00.	+ 20 49 30.	15	14.9	GALAXY
ZWG 128.020	12 03 00.	+ 20 50		14.9	GALAXY
MCG+04-29-018	12 03 00.	+ 25 21 30.	63	15.4	GALAXY
ZWG 128.021	12 03 00.	+ 25 23		15.4	GALAXY
UGC 07080	12 03 00.	+ 25 23	72	15.4	GALAXY Sb-c
MCG+06-27-001	12 03 00.	+ 33 22	72	14.5	GALAXY
MCG+08-22-065	12 03 00.	+ 47 03	60	15.	GALAXY
ZC 1203.0+4841	12 03 00.	+ 48 41	3090		CLUSTER OF GALAXIES
ZWG 269.033	12 03 00.	+ 50 50		11.2	GALAXY
UGC 07081	12 03 00.	+ 50 50	390	11.2	GALAXY Sc
MCG+09-20-088	12 03 00.	+ 51 46	72	15.	GALAXY
ZWG 269.034	12 03 00.	+ 51 48		15.3	GALAXY
UGC 07082	12 03 00.	+ 51 48	102	15.3	GALAXY Sb
MCG+09-20-087	12 03 00.	+ 54 39	12	18.	GALAXY
SHAH 124	12 03 00.	+ 59 40	216	16.5	GROUP OF COMPACT GALAXIES
ZWG 292.067	12 03 00.	+ 62 12		15.5	GALAXY
ZC 1203.0-0246	12 03 00.	- 02 46	670		CLUSTER OF GALAXIES
MCG-04-29-005	12 03 00.	- 26 17 30.	30	13.5	GALAXY
MCG-05-29-006	12 03 00.	- 29 54	36	15.	GALAXY
MCG-05-29-007	12 03 00.	- 31 10	120	15.5	GALAXY
REIN 7.077	12 03 00.03	+ 50 49 28.1			NEBULA
RNGC 4089	12 03 01.	+ 20 50		15.0	GALAXY
RNGC 4087	12 03 01.	- 26 17		13.0	GALAXY
KN 12.279	12 03 01.5	+ 50 49 09.			NEBULA
REIN 7.078A	12 03 01.79	+ 50 49 00.5			NEBULA
AMES 0014	12 03 02.	+ 10 56 54.	35	16.7	NEBULA
RNGC 4088	12 03 02.	+ 50 49		11.5	GALAXY

OBJECT NAME	RIGHT ASCEN.	DECLINATION	DIAM.	MAGN.	TYPE OF OBJECT
REIN 7.078B	12 03 02.05	+ 50 49 03.2			NEBULA
HN 1608	12 03 02.5	+ 25 22 36.	48		NEBULA
MCG+04-29-019	12 03 03.	+ 20 49 30.	48	15.2	GALAXY
REIZ 1950	12 03 03.	+ 20 50	18	14.6	GALAXY
REIZ 1951	12 03 03.	+ 25 22	48	14.5	GALAXY
REIZ 1952	12 03 03.	+ 29 24	18	15.8	GALAXY
MCG+09-20-089	12 03 03.	+ 50 48	330	11.0	GALAXY
REIZ 1949	12 03 03.	+ 50 49	270	11.6	GALAXY
LB 02204	12 03 03.	+ 51 09 30.		16.6	FAINT BLUE STAR
ARC 1468	12 03 03.	+ 51 42		16.0	RICH CLUSTER OF GALAXIES
SEY 157	12 03 03.	+ 51 46 48.		14.5	FAINT GALAXY
HOLM 327B	12 03 04.	+ 09 16	36	15.2	PART OF MULTIPLE GALAXY
HN 1609	12 03 04.6	+ 22 06 30.	12		NEBULA
REIN 7.079	12 03 04.99	+ 50 48 46.7			NEBULA
HN 1610	12 03 05.7	+ 28 08 48.	18		NEBULA
HN 1611	12 03 05.8	+ 28 11 12.			NEBULA
ZWG 013.070	12 03 06.	+ 02 06		15.2	GALAXY
ZWG 069.048	12 03 06.	+ 09 16		15.0	GALAXY
SCH 21	12 03 06.	+ 09 16		15.0	PECULIAR GALAXY
KARA.72 319A	12 03 06.	+ 09 16	42	15.0	PART OF DOUBLE GALAXY
MCG+02-31-027	12 03 06.	+ 09 16	36	15.0	GALAXY
ZC 1203.1+1831	12 03 06.	+ 18 31	1880		CLUSTER OF GALAXIES
ZWG 128.022	12 03 06.	+ 20 50		15.2	GALAXY
UGC 07083	12 03 06.	+ 20 50	66	15.2	GALAXY S
MCG+05-29-011	12 03 06.	+ 31 19	120	15.1	GALAXY
MCG+05-29-012	12 03 06.	+ 31 20	18	15.5	GALAXY
MCG+05-29-010	12 03 06.	+ 31 21	18	15.1	GALAXY
ZWG 187.002	12 03 06.	+ 33 23		14.9	GALAXY
UGC 07084	12 03 06.	+ 33 23	96	14.9	GALAXY S
ZWG 269.035	12 03 06.	+ 53 28		15.7	GALAXY
OCL 0873	12 03 06.	- 62 16	420	12.	OPEN STAR CLUSTER
AMES 0015	12 03 07.	+ 10 38 36.	23	17.8	NEBULA
RNGC 4091	12 03 07.	+ 20 50		15.0	GALAXY
REIN 7.080	12 03 07.68	+ 50 49 04.7			NEBULA
HN 1612	12 03 08.5	+ 01 51 48.	12		NEBULA
HN 1613	12 03 08.6	+ 01 53 00.	12		NEBULA
REIZ 1953	12 03 09.	+ 20 49	36	14.8	GALAXY
REIZ 1954	12 03 09.	+ 31 26	12	14.9	GALAXY
REIZ 1955	12 03 09.	+ 37 01	30	15.6	GALAXY
REIZ 1956	12 03 09.	+ 62 13	18	15.0	GALAXY
HOLM 327A	12 03 10.	+ 09 16	48	14.5	PART OF MULTIPLE GALAXY
ARC 1469	12 03 10.	- 06 51		17.2	RICH CLUSTER OF GALAXIES
HN 0357	12 03 10.	- 29 42			NEBULA
IC 2996	12 03 10.	- 29 42			NONSTELLAR OBJECT
REIN 7.082	12 03 11.31	+ 50 49 58.6			NEBULA
REIN 7.081	12 03 11.83	+ 10 40 29.3			NEBULA
ZWG 013.072	12 03 12.	+ 01 52		15.1	GALAXY
KARA.72 320B	12 03 12.	+ 01 52	60	15.1	PART OF DOUBLE GALAXY
ZWG 013.073	12 03 12.	+ 01 53		15.3	GALAXY
KARA.72 320A	12 03 12.	+ 01 53	36	15.3	PART OF DOUBLE GALAXY
ZWG 069.049	12 03 12.	+ 09 16		14.8	GALAXY
UGC 07085	12 03 12.	+ 09 16	114	14.8	GALAXY S
KARA.72 319B	12 03 12.	+ 09 16	78	14.8	PART OF DOUBLE GALAXY
MCG+02-31-028	12 03 12.	+ 09 16	60	14.8	GALAXY
AMES 0016	12 03 12.	+ 10 39 54.	34	16.4	NEBULA
IC 2997	12 03 12.	+ 20 34			NONSTELLAR OBJECT
MCG+04-29-020	12 03 12.	+ 20 44 30.	60	14.4	GALAXY
MCG+04-29-021	12 03 12.	+ 20 47 30.	15	15.3	GALAXY
ZC 1203.2+2137	12 03 12.	+ 21 37	1010		CLUSTER OF GALAXIES
ZWG 158.016	12 03 12.	+ 31 20		15.1	GALAXY
UGC 07085A	12 03 12.	+ 31 20	138	15.1	GALAXY DBL SYS
MCG+06-27-002	12 03 12.	+ 37 02	27	15.5	GALAXY
LB 02205	12 03 12.	+ 50 47 48.		16.8	FAINT BLUE STAR
MCG+09-20-090	12 03 12.	+ 54 40	24	16.	GALAXY
MCG+10-17-138	12 03 12.	+ 58 19	24	16.	GALAXY
MCG+10-17-139	12 03 12.	+ 61 32	42	16.	GALAXY
MCG+10-17-140	12 03 12.	+ 62 11	18	16.	GALAXY
ZC 1203.2+6332	12 03 12.	+ 63 32	1010		CLUSTER OF GALAXIES
ZWG 352.018	12 03 12.	+ 77 47		15.4	GALAXY
UGC 07086	12 03 12.	+ 77 47	156	15.4	GALAXY Sb
ZWG 013.071	12 03 12.	- 02 47		15.7	GALAXY
MCG-06-27-003	12 03 12.	- 38 34	60	15.	GALAXY
IC 2995	12 03 12.5	- 27 39 42.	204	12.7	GALAXY SB(s)
HN 1614	12 03 13.3	+ 26 18 24.			NEBULA
AMES 0017	12 03 15.	+ 10 47 48.	24	17.6	NEBULA
REIZ 1957	12 03 15.	+ 20 45	30	14.2	GALAXY
MCG+04-29-022	12 03 15.	+ 20 50	15	14.6	GALAXY
HOLM 328B	12 03 15.	+ 31 21	36	14.9	PART OF MULTIPLE GALAXY
VV 013B	12 03 15.	+ 31 21 30.	18	14.9	INTERACTING GALAXY
VV 013A	12 03 15.	+ 31 21 30.	120	14.9	INTERACTING GALAXY
HOLM 328A	12 03 15.	+ 31 22	30	14.9	PART OF MULTIPLE GALAXY
MCG+09-20-091	12 03 15.	+ 53 28	48	15.	GALAXY
HN 1615	12 03 16.6	+ 00 05 24.	18		NEBULA
HN 1616	12 03 17.6	+ 24 38 12.			NEBULA
ZWG 069.050	12 03 18.	+ 10 40		15.5	GALAXY
ZWG 098.053	12 03 18.	+ 16 11		15.4	GALAXY
ZWG 128.023	12 03 18.	+ 20 46		14.4	GALAXY
UGC 07087	12 03 18.	+ 20 46	72	14.4	GALAXY S
ZWG 128.024	12 03 18.	+ 20 49		15.3	GALAXY
ZWG 187.003	12 03 18.	+ 35 27		15.7	GALAXY
MRK 646	12 03 18.	+ 35 27	11	16.	GALAXY WITH UV CONTINUUM
MCG+08-22-066	12 03 18.	+ 48 52	24	16.	GALAXY
ZWG 315.013	12 03 18.	+ 67 31		14.7	GALAXY
UGC 07088	12 03 18.	+ 67 31	90	14.7	GALAXY SBb
KEEN 4108A	12 03 18.	+ 67 32	72		Sb GALAXY
7ZW 437	12 03 18.	+ 79 48			COMPACT GALAXY
ZWG 013.074	12 03 18.	- 02 42		15.7	GALAXY
MCG-05-29-008	12 03 18.	- 27 39	168	13.	GALAXY
MCG-05-29-009	12 03 18.	- 29 43	48	14.5	GALAXY
AMES 0018	12 03 19.	+ 10 40 06.	23	17.8	NEBULA
RNGC 4092	12 03 19.	+ 20 46		14.5	GALAXY
RNGC 4093	12 03 19.	+ 20 49		15.5	GALAXY
FATH 1.480	12 03 19.	+ 28 53	27		NEBULA
RNGC 4094	12 03 19.	- 14 16		13.0	GALAXY
IC 0760	12 03 19.	- 29 00 52.			NONSTELLAR OBJECT
REIZ 1959	12 03 21.	+ 20 48	12	14.5	GALAXY
REIZ 1960	12 03 21.	+ 20 51	12	14.7	GALAXY
FATH 2.064	12 03 21.	+ 29 43	14		NEBULA
IC 0761	12 03 21.	- 12 23 58.			NONSTELLAR OBJECT
REIZ 1958	12 03 22.	+ 10 42	42	15.7	GALAXY
IC 2998	12 03 22.	+ 21 02			NONSTELLAR OBJECT
SHB 193	12 03 22.6	+ 10 59 35.		18.1	QUASI-STELLAR OBJECT
AMES 0019	12 03 23.	+ 11 19 48.	26	17.2	NEBULA
FATH 1.481	12 03 23.	+ 29 47	16		NEBULA
BC 4C10.34	12 03 23.1	+ 10 59 48.		18.08	QUASI-STELLAR OBJECT
ZC 1203.4+0529	12 03 24.	+ 05 29	810		CLUSTER OF GALAXIES
ZC 1203.4+1451	12 03 24.	+ 14 51	1010		CLUSTER OF GALAXIES
ZC 1203.4+1517	12 03 24.	+ 15 17	940		CLUSTER OF GALAXIES

OBJECT NAME	RIGHT ASCEN.	DECLINATION	DIAM.	MAGN.	TYPE OF OBJECT
ZWG 128.025	12 03 24.	+ 20 51		14.6	GALAXY
MCG+05-29-013	12 03 24.	+ 28 31	42	15.6	GALAXY
MCG+05-29-014	12 03 24.	+ 28 32	30	15.6	GALAXY
ZC 1203.4+2925	12 03 24.	+ 29 25	1680		CLUSTER OF GALAXIES
IC 2999	12 03 24.	+ 31 37 08.			NONSTELLAR OBJECT
ZWG 158.017	12 03 24.	+ 31 38		15.6	GALAXY
MCG+06-27-003	12 03 24.	+ 33 06 30.	66	17.	GALAXY
ZWG 215.023	12 03 24.	+ 43 25		14.8	GALAXY
UGC 07089	12 03 24.	+ 43 25	210	14.8	GALAXY
ZWG 243.043	12 03 24.	+ 47 45		11.6	GALAXY
UGC 07090	12 03 24.	+ 47 45	432	11.6	GALAXY Sc
MCG+08-22-067	12 03 24.	+ 47 45	360	10.8	GALAXY
LB 02206	12 03 24.	+ 57 47 48.		16.3	FAINT BLUE STAR
MCG-02-31-015	12 03 24.	- 12 22	24	15.	GALAXY
MCG-02-31-014	12 03 24.	- 12 23	36	15.	GALAXY
MCG-02-31-016	12 03 24.	- 14 15	240	12.	GALAXY
MCG-05-29-010	12 03 24.	- 29 01	72	14.5	GALAXY
RNGC 4095	12 03 25.	+ 20 51		14.5	GALAXY
HN 1617	12 03 25.4	+ 27 59 24.	12		NEBULA
FATH 1.482	12 03 26.	+ 29 28	14		NEBULA
AMES 0020	12 03 27.	+ 10 37 48.	25	17.7	NEBULA
MCG+04-29-023	12 03 27.	+ 20 52	42	14.5	GALAXY
REIZ 1962	12 03 27.	+ 37 09	18	14.2	GALAXY
SEY 158	12 03 27.	+ 43 25 11.		14.1	FAINT GALAXY
REIZ 1961	12 03 27.	+ 47 45	288	11.7	GALAXY
SN 1960H	12 03 27.	+ 47 47		14.5	SUPERNOVA
RNGC 4108A	12 03 27.	+ 67 32		14.5	GALAXY
ARP 097	12 03 29.	+ 31 21			PECULIAR GALAXY
ZWG 041.039	12 03 30.	+ 02 47		15.1	GALAXY
MCG+01-31-021	12 03 30.	+ 02 47	60	15.1	GALAXY
SN 1971J	12 03 30.	+ 14 06		18.	SUPERNOVA
ZWG 128.026	12 03 30.	+ 20 53		14.5	GALAXY
UGC 07091	12 03 30.	+ 20 53	72	14.5	GALAXY S
VV 061B	12 03 30.	+ 20 53	18	16.	INTERACTING GALAXY
VV 061A	12 03 30.	+ 20 53	18	15.	INTERACTING GALAXY
VV 061	12 03 30.	+ 20 53	72		INTERACTING GALAXY
ZC 1203.5+2914	12 03 30.	+ 29 14	540		CLUSTER OF GALAXIES
ZWG 187.004	12 03 30.	+ 37 08		14.6	GALAXY
UGC 07092	12 03 30.	+ 37 08	72	14.6	GALAXY S0
ZWG 215.024	12 03 30.	+ 40 26		15.2	GALAXY
MCG+07-25-020	12 03 30.	+ 43 26	180	13.	GALAXY
MCG+08-22-068	12 03 30.	+ 49 52	300	11.9	GALAXY
MCG+11-15-020	12 03 30.	+ 67 01	30	18.	GALAXY
MCG+11-15-021	12 03 30.	+ 67 31	78	15.	GALAXY
MCG+13-09-011	12 03 30.	+ 77 47	168	13.	GALAXY
RNGC 4098	12 03 31.	+ 20 53		14.5	GALAXY
RNGC 4099	12 03 31.	+ 20 55			NON-EXISTENT OBJECT
HN 1618	12 03 31.7	+ 28 30 24.			NEBULA
RNGC 4097	12 03 32.	+ 37 08		14.5	GALAXY
RNGC 4096	12 03 32.	+ 47 45		12.0	GALAXY
REIZ 1966	12 03 32.	+ 60 24	42	14.0	GALAXY
REIZ 1964	12 03 33.	+ 20 53	36	14.0	GALAXY
HOLM 329A	12 03 33.	+ 28 30	42	14.6	PART OF MULTIPLE GALAXY
HOLM 329B	12 03 33.	+ 28 31	42	14.7	PART OF MULTIPLE GALAXY
MCG+06-27-004	12 03 33.	+ 37 09	48	14.5	GALAXY
REIZ 1965	12 03 33.	+ 49 51	186	11.7	GALAXY
MCG-04-29-006	12 03 33.	- 22 35	120	14.	GALAXY
SEY 159	12 03 34.	+ 43 42 53.		15.3	FAINT GALAXY
REIZ 1963	12 03 34.	- 02 41	24	15.2	GALAXY
HN 0358	12 03 34.	- 29 24			NEBULA
IC 3000	12 03 34.	- 29 24			NONSTELLAR OBJECT
FATH 1.483	12 03 35.	+ 29 40	8		NEBULA
AMES 0021	12 03 36.	+ 12 51 30.	22	17.4	NEBULA
ZWG 128.027	12 03 36.	+ 25 50		14.7	GALAXY
UGC 07093	12 03 36.	+ 25 50	84	14.7	GALAXY S0-a
ZWG 158.018	12 03 36.	+ 28 31		15.6	GALAXY
KARA.72 321A	12 03 36.	+ 28 31	60	15.6	PART OF DOUBLE GALAXY
ZWG 158.019	12 03 36.	+ 28 32		15.6	GALAXY
KARA.72 321B	12 03 36.	+ 28 32	54	15.6	PART OF DOUBLE GALAXY
ZWG 187.005	12 03 36.	+ 33 10		15.5	GALAXY
MCG+06-27-006	12 03 36.	+ 33 11	14	16.5	GALAXY
MCG+06-27-005	12 03 36.	+ 33 11	30	16.	GALAXY
ZC 1203.6+3420	12 03 36.	+ 34 20	1480		CLUSTER OF GALAXIES
MCG+07-25-021	12 03 36.	+ 40 25	42	15.	GALAXY
ZWG 215.025	12 03 36.	+ 43 14		15.6	GALAXY
UGC 07094	12 03 36.	+ 43 14	78	15.6	GALAXY
ZWG 243.044	12 03 36.	+ 49 52		11.7	GALAXY
UGC 07095	12 03 36.	+ 49 52	336	11.7	GALAXY Sb/Sc
MCG+09-20-092	12 03 36.	+ 50 46	24	16.	GALAXY
ZWG 013.075	12 03 36.	- 02 39		15.0	GALAXY
MCG+00-31-035	12 03 36.	- 02 39	48	15.0	GALAXY
HN 1619	12 03 36.1	- 02 40 00.			NEBULA
HOLM 330B	12 03 37.	+ 08 08	30	15.5	PART OF MULTIPLE GALAXY
RNGC 4101	12 03 37.	+ 25 50		14.5	GALAXY
HOLM 331B	12 03 37.	+ 33 10	24	15.4	PART OF MULTIPLE GALAXY
RNGC 4100	12 03 38.	+ 49 52		12.0	GALAXY
MCG+04-29-024	12 03 39.	+ 20 52 30.	36	18.	GALAXY
REIZ 1967	12 03 39.	+ 25 50	24	14.9	GALAXY
MCG+04-29-025	12 03 39.	+ 25 50	60	14.7	GALAXY
HOLM 331A	12 03 39.	+ 33 10	24	15.3	PART OF MULTIPLE GALAXY
SEY 160	12 03 39.	+ 43 13 17.		15.0	FAINT GALAXY
VV 049C	12 03 39.	- 22 34 30.	6	19.	INTERACTING GALAXY
VV 049B	12 03 39.	- 22 34 30.	18	17.	INTERACTING GALAXY
VV 049A	12 03 39.	- 22 34 30.	36	16.	INTERACTING GALAXY
VV 049	12 03 39.	- 22 34 30.	126	16.	INTERACTING GALAXY
HOLM 330A	12 03 40.	+ 08 08	36	15.5	PART OF MULTIPLE GALAXY
REIN 7.083	12 03 40.37	+ 10 49 17.7			NEBULA
LB 02207	12 03 41.	+ 51 24 24.		16.1	FAINT BLUE STAR
VV 062B	12 03 42.	+ 20 32	3	19.	INTERACTING GALAXY
VV 062A	12 03 42.	+ 20 32	18	18.	INTERACTING GALAXY
VV 062	12 03 42.	+ 20 32	30		INTERACTING GALAXY
MRK 647	12 03 42.	+ 21 31	15	16.	GALAXY WITH UV CONTINUUM
ZWG 158.020	12 03 42.	+ 32 15		15.4	GALAXY
MCG+06-27-007	12 03 42.	+ 33 47 30.	42	15.	GALAXY
ZWG 187.006	12 03 42.	+ 33 48		15.7	GALAXY
MCG+07-25-022	12 03 42.	+ 43 14 30.	60	15.	GALAXY
SEY 161	12 03 42.	+ 45 37 18.		15.4	FAINT GALAXY
MCG+09-20-093	12 03 42.	+ 51 07	30	17.	GALAXY
ZWG 292.068	12 03 42.	+ 58 16		15.5	GALAXY
ZWG 292.069	12 03 42.	+ 62 11		15.6	GALAXY
MCG-06-27-004	12 03 42.	- 35 42	30	15.	GALAXY
AMES 0022	12 03 43.	+ 12 23 42.	14	17.1	NEBULA
IC 3001	12 03 44.	+ 33 47 44.			NONSTELLAR OBJECT
REIZ 1969	12 03 44.	+ 62 12	30	15.1	GALAXY
REIZ 1968	12 03 45.	+ 37 18	42	15.5	GALAXY
SC 1201-4353.5	12 03 47.	- 44 10 12.	30		NEBULA
AMES 0023	12 03 48.	+ 10 32 36.	24	17.8	NEBULA
ZC 1203.8+1046	12 03 48.	+ 10 46	340		CLUSTER OF GALAXIES

OBJECT NAME	RIGHT ASCEN.	DECLINATION	DIAM.	MAGN.	TYPE OF OBJECT
ZWG 098.054	12 03 48.	+ 18 29		15.6	GALAXY
ZWG 158.021	12 03 48.	+ 28 14		15.7	GALAXY
MCG+05-29-015	12 03 48.	+ 28 26	30	15.	GALAXY
ZWG 187.007	12 03 48.	+ 37 17		15.6	GALAXY
ZWG 269.036	12 03 48.	+ 53 00		11.8	GALAXY
UGC 07096	12 03 48.	+ 53 00	198	11.8	GALAXY Sb
MCG+10-17-141	12 03 48.	+ 58 16	45	15.	GALAXY
MCG+10-17-142	12 03 48.	+ 59 14	39	16.	GALAXY
MCG+10-17-143	12 03 48.	+ 62 10	39	16.	GALAXY
ZWG 315.014	12 03 48.	+ 63 55		15.6	GALAXY
MCG-05-29-011	12 03 48.	- 27 55	36	15.	GALAXY
AMES 0024	12 03 49.	+ 11 42 42.	29	16.6	NEBULA
LB 02208	12 03 49.	+ 62 00 06.		17.0	FAINT BLUE STAR
HN 1620	12 03 49.2	+ 28 14 24.			NEBULA
REIZ 1972	12 03 50.	+ 67 31	42	14.4	GALAXY
MCG+06-27-008	12 03 51.	+ 37 18	60	15.	GALAXY
REIZ 1971	12 03 51.	+ 52 59	90	12.2	GALAXY
REIZ 1970	12 03 52.	+ 10 51	12	16.2	GALAXY
LB 02209	12 03 52.	+ 51 48 30.		17.1	FAINT BLUE STAR
LB 02210	12 03 53.	+ 50 33 00.		16.5	FAINT BLUE STAR
ZC 1203.9+0409	12 03 54.	+ 04 09	740		CLUSTER OF GALAXIES
AMES 0025	12 03 54.	+ 13 39 36.	34	16.7	NEBULA
ZWG 098.055	12 03 54.	+ 20 17		15.7	GALAXY
MCG+05-29-016	12 03 54.	+ 28 27 30.	42	13.7	GALAXY
MCG+09-20-094	12 03 54.	+ 52 59	180	12.3	GALAXY
ZC 1203.9+5555	12 03 54.	+ 55 55	1080		CLUSTER OF GALAXIES
LB 02211	12 03 54.	+ 57 27 18.		14.9	FAINT BLUE STAR
AMES 0027	12 03 56.	+ 11 43 54.	23	17.8	NEBULA
AMES 0026	12 03 56.	+ 11 44 36.	26	17.6	NEBULA
RNGC 4102	12 03 56.	+ 52 59		12.5	GALAXY
HN 1621	12 03 56.5	+ 28 13 30.			NEBULA
AMES 0028	12 03 57.	+ 10 34 00.	31	17.3	NEBULA
MCG+06-27-009	12 03 57.	+ 33 07	24	16.	GALAXY
HN 1622	12 03 58.2	+ 28 24 54.			NEBULA
LB 02212	12 03 59.	+ 48 36 54.		16.4	FAINT BLUE STAR
VDB.66G 109	12 04	+ 40 03	70		DWARF GALAXY
ZWG 013.076	12 04 00.	+ 01 54		15.4	GALAXY
ZC 1204.0+2223	12 04 00.	+ 22 23	1340		CLUSTER OF GALAXIES
ZWG 128.028	12 04 00.	+ 26 02		15.7	GALAXY
MCG+04-29-026	12 04 00.	+ 26 02	60	15.7	GALAXY
ZWG 158.022	12 04 00.	+ 28 13		15.4	GALAXY
ZWG 158.023	12 04 00.	+ 28 25		15.3	GALAXY
ZC 1204.0+3154	12 04 00.	+ 31 54	670		CLUSTER OF GALAXIES
ZWG 215.026	12 04 00.	+ 39 30		14.5	GALAXY
UGC 07098	12 04 00.	+ 39 30	96	14.5	GALAXY PECULR
MCG+07-25-023	12 04 00.	+ 39 30	54	14.5	GALAXY
ZC 1204.0+4514	12 04 00.	+ 45 14	870		CLUSTER OF GALAXIES
SHAH 125	12 04 00.	+ 53 59	102	17.3	GROUP OF COMPACT GALAXIES
UGC 07097	12 04 00.	+ 81 12	102	16.0	GALAXY Sb-c
MCG-05-29-012	12 04 00.	- 27 55	24	15.5	GALAXY
REIN 7.084A	12 04 01.29	+ 10 50 16.2			NEBULA
REIN 7.084B	12 04 01.37	+ 10 50 15.1			NEBULA
REIZ 1973	12 04 03.	+ 28 27	60	14.0	GALAXY
REIZ 1974	12 04 03.	+ 39 28	30	15.0	GALAXY
HN 1623	12 04 03.2	+ 25 23 18.			NEBULA
BC PKS1204-12	12 04 04.9	- 12 37 27.		17.5	QUASI-STELLAR OBJECT
AMES 0029	12 04 05.	+ 12 18 42.	25	17.3	NEBULA
SC 1201-4647.3	12 04 05.	- 47 04 00.	18		NEBULA
PEIG 056	12 04 06.	+ 11 57		10.9	FAINT BLUE STAR
ZWG 158.024	12 04 06.	+ 28 27		13.7	GALAXY
UGC 07099	12 04 06.	+ 28 27	168	13.7	GALAXY S0
MKW 04S	12 04 06.	+ 28 27		13.7	POOR GALAXY CLUSTER
ZC 1204.1+3043	12 04 06.	+ 30 43	1810		CLUSTER OF GALAXIES
ZWG 158.025	12 04 06.	+ 31 09		15.7	GALAXY
MCG+11-15-022	12 04 06.	+ 65 20	24	17.	GALAXY
OCL 0871	12 04 06.	- 60 58	1140	7.8	OPEN STAR CLUSTER
VHA 127	12 04 06.	- 60 58	360		OPEN STAR CLUSTER
RNGC 4103	12 04 06.	- 60 58		7.5	OPEN CLUSTER
HN 1624	12 04 06.1	+ 26 02 00.			NEBULA
RNGC 4105	12 04 07.	- 29 30		12.5	GALAXY
HN 1625	12 04 07.9	+ 26 02 12.			NEBULA
AMES 0030	12 04 08.	+ 14 05 54.	24	16.7	NEBULA
RNGC 4104	12 04 08.	+ 28 27		13.5	GALAXY
HN 1626	12 04 08.4	+ 25 31 00.			NEBULA
REIZ 1975	12 04 09.	+ 10 52	24	16.1	GALAXY
MCG-03-31-021	12 04 09.	- 19 21	48	15.	GALAXY
MAI 067	12 04 11.	+ 85 56	40		DWARF SPHEROIDAL GALAXY
ZWG 069.051	12 04 12.	+ 14 05		15.7	GALAXY
ZC 1204.2+1411	12 04 12.	+ 14 11	1140		CLUSTER OF GALAXIES
ZWG 098.056	12 04 12.	+ 17 59		14.7	GALAXY
UGC 07100	12 04 12.	+ 17 59	114	14.7	GALAXY S
MCG+03-31-039	12 04 12.	+ 18 00	120	14.7	GALAXY
MCG+04-29-027	12 04 12.	+ 21 05	30	15.6	GALAXY
ZWG 128.029	12 04 12.	+ 25 17		15.7	GALAXY
MCG+04-29-028	12 04 12.	+ 25 17 30.	36	15.7	GALAXY
MCG+08-22-069	12 04 12.	+ 49 01 30.	30	16.	GALAXY
ZWG 292.070	12 04 12.	+ 56 49		15.6	GALAXY
MCG+10-17-144	12 04 12.	+ 56 49	27	16.	GALAXY
ZWG 315.015	12 04 12.	+ 67 26		13.0	GALAXY
UGC 07101	12 04 12.	+ 67 26	108	13.0	GALAXY Sc
MCG-05-29-013	12 04 12.	- 29 29	54	11.5	GALAXY
HN 1627	12 04 12.9	+ 24 53 06.			NEBULA
RNGC 4107	12 04 13.	+ 10 53			NON-EXISTENT OBJECT
RNGC 4106	12 04 13.	- 29 31		12.5	GALAXY
REIZ 1977	12 04 14.	+ 67 26	48	13.1	GALAXY
HN 1628	12 04 14.5	+ 28 11 30.			NEBULA
AMES 0031	12 04 15.	+ 11 40 00.	25	17.8	NEBULA
MCG+04-29-029	12 04 15.	+ 25 17	18	15.7	GALAXY
REIZ 1976	12 04 15.	+ 31 07	36	14.4	GALAXY
MCG+09-20-095	12 04 15.	+ 52 26	30	17.	GALAXY
HOLM 333B	12 04 16.	+ 43 17	30	13.6	PART OF MULTIPLE GALAXY
HN 1629	12 04 16.3	+ 25 17 12.			NEBULA
HN 1630	12 04 17.0	+ 25 17 12.			NEBULA
ZWG 098.057	12 04 18.	+ 17 16		15.7	GALAXY
ZWG 128.030	12 04 18.	+ 21 05		15.6	GALAXY
ZC 1204.3+3802	12 04 18.	+ 38 02	2080		CLUSTER OF GALAXIES
ZWG 215.027	12 04 18.	+ 43 16		15.1	GALAXY
LB 02213	12 04 18.	+ 50 24 18.		16.2	FAINT BLUE STAR
MCG+09-20-096	12 04 18.	+ 55 27 30.	36	15.	GALAXY
ZC 1204.3+7153	12 04 18.	+ 71 53	1010		CLUSTER OF GALAXIES
MCG-02-31-017	12 04 18.	- 10 48	120	14.	GALAXY
MCG-05-29-015	12 04 18.	- 28 03	84	15.	GALAXY
MCG-05-29-014	12 04 18.	- 29 30	42	12.	GALAXY
MCG-05-29-017	12 04 18.	- 31 41	60	16.	GALAXY
MCG-05-29-016	12 04 18.	- 31 41	12	16.	GALAXY
OCL 0869	12 04 18.	- 59 13	720	10.	OPEN STAR CLUSTER
RNGC 4109	12 04 20.	+ 43 16		15.0	GALAXY
LB 02214	12 04 20.	+ 49 42 42.		15.5	FAINT BLUE STAR

OBJECT NAME	RIGHT ASCEN.	DECLINATION	DIAM.	MAGN.	TYPE OF OBJECT
HN 1631	12 04 20.7	+ 26 41 36.			NEBULA
HOLM 332A	12 04 21.	+ 25 16	30	15.1	PART OF MULTIPLE GALAXY
REIZ 1978	12 04 21.	+ 25 59	18	15.0	GALAXY
MCG+07-25-024	12 04 21.	+ 43 16 30.	36	14.5	GALAXY
RNGC 4108	12 04 21.	+ 67 27		13.0	GALAXY
AMES 0032	12 04 23.	+ 12 44 42.	20	17.3	NEBULA
REIN 3.001	12 04 23.86	+ 10 33 44.7			NEBULA
ZWG 069.052	12 04 24.	+ 10 33		15.1	GALAXY
MCG+02-31-029	12 04 24.	+ 10 33	36	15.1	GALAXY
HOLM 332B	12 04 24.	+ 25 16	30	15.3	PART OF MULTIPLE GALAXY
ZWG 128.031	12 04 24.	+ 26 00		15.7	GALAXY
ARC 1470	12 04 24.	+ 71 55		17.4	RICH CLUSTER OF GALAXIES
SEY 162	12 04 26.	+ 55 27 42.		15.2	FAINT GALAXY
HN 1632	12 04 26.7	+ 25 59 24.			NEBULA
REIZ 1979	12 04 27.	+ 26 00	18	14.8	GALAXY
HOLM 333A	12 04 27.	+ 43 21	210	11.4	PART OF MULTIPLE GALAXY
MCG+08-22-070	12 04 27.	+ 50 34	18	17.	GALAXY
HN 1633	12 04 28.7	+ 25 59 30.	18		NEBULA
ZC 1204.5+0441	12 04 30.	+ 04 41	400		CLUSTER OF GALAXIES
ZWG 069.053	12 04 30.	+ 09 23		15.7	GALAXY
AMES 0033	12 04 30.	+ 13 40 42.	36	17.5	NEBULA
ZC 1204.5+1531	12 04 30.	+ 15 31	610		CLUSTER OF GALAXIES
ZWG 098.058	12 04 30.	+ 18 48		14.7	GALAXY
UGC 07102	12 04 30.	+ 18 48	78	14.7	GALAXY SBb
MCG+03-31-040	12 04 30.	+ 18 50	84	14.7	GALAXY
ZWG 187.008	12 04 30.	+ 33 39		15.6	GALAXY
ZWG 215.028	12 04 30.	+ 43 20		11.4	GALAXY
UGC 07103	12 04 30.	+ 43 20	270	11.4	GALAXY S0
MCG+08-22-071	12 04 30.	+ 50 34 30.	24	17.	GALAXY
ZWG 292.071	12 04 30.	+ 59 06		15.3	GALAXY
MCG+11-15-023	12 04 30.	+ 67 26	96	13.1	GALAXY
7ZW 438	12 04 30.	+ 78 19			COMPACT GALAXY
RNGC 4110	12 04 31.	+ 18 48		14.5	GALAXY
IC 3002	12 04 31.	+ 33 38 44.			NONSTELLAR OBJECT
RNGC 4112	12 04 31.	- 39 55			GALAXY
RNGC 4111	12 04 32.	+ 43 21		12.0	GALAXY
HN 1634	12 04 32.3	+ 28 31 24.			NEBULA
MCG+07-25-025	12 04 33.	+ 43 15 30.	45	17.5	GALAXY
MCG-04-29-007	12 04 33.	- 27 28	15	14.5	GALAXY
AMES 0034	12 04 35.	+ 11 24 06.	20	16.4	NEBULA
IC 3003	12 04 35.	+ 33 07 50.			NONSTELLAR OBJECT
LB 02215	12 04 35.	+ 51 22 24.		16.2	FAINT BLUE STAR
AMES 0035	12 04 36.	+ 11 39 06.	29	18.0	NEBULA
ZWG 069.054	12 04 36.	+ 13 31		15.6	GALAXY
MCG+02-31-030	12 04 36.	+ 13 31	54	15.6	GALAXY
UGC 07104	12 04 36.	+ 17 11	72	16.0	GALAXY S
MCG+03-31-043	12 04 36.	+ 17 12	78	17.	GALAXY
MCG+03-31-042	12 04 36.	+ 17 15	42	16.	GALAXY
ZWG 098.059	12 04 36.	+ 17 16		14.9	GALAXY
SN 1960C	12 04 36.	+ 17 16		17.0	SUPERNOVA
MCG+03-31-041	12 04 36.	+ 17 17	78	14.9	GALAXY
ZC 1204.6+2716	12 04 36.	+ 27 16	740		CLUSTER OF GALAXIES
TON-N 0603	12 04 36.	+ 29 18		15.6	BLUE STAR
ZWG 187.009	12 04 36.	+ 33 16		14.8	GALAXY
MCG+06-27-010	12 04 36.	+ 33 39	27	15.	GALAXY
MCG+07-25-026	12 04 36.	+ 43 21	240	11.	GALAXY
MCG+09-20-097	12 04 36.	+ 53 57 30.	15	15.	GALAXY
ZWG 013.077	12 04 36.	- 00 58		15.6	GALAXY
KARA.73B 0519	12 04 36.	- 00 58	36	15.6	ISOLATED GALAXY S
MCG-04-29-008	12 04 36.	- 25 27	15	15.	GALAXY
HN 1635	12 04 36.8	+ 25 55 00.			NEBULA
RNGC 4115	12 04 37.	+ 14 41			NON-EXISTENT OBJECT
SN 1972F	12 04 37.	+ 53 57		16.	SUPERNOVA
AMES 0036	12 04 38.	+ 12 48 30.	19	17.6	NEBULA
RNGC 4122	12 04 38.	+ 33 16		15.0	GALAXY
RNGC 4113	12 04 38.	+ 34 16			NON-EXISTENT OBJECT
AMES 0037	12 04 39.	+ 12 07 00.	22	17.5	NEBULA
HN 0795	12 04 39.	+ 13 32		15.	NEBULA
IC 3004	12 04 39.	+ 13 32			NONSTELLAR OBJECT
REIZ 1980	12 04 39.	+ 14 40	12	14.7	GALAXY
MCG+06-27-011	12 04 39.	+ 33 16	39	14.5	GALAXY
RNGC 4108B	12 04 39.	+ 67 31		14.5	GALAXY
AMES 0038	12 04 40.	+ 10 05 18.	29	16.8	NEBULA
HN 0359	12 04 40.	- 29 45			NEBULA
IC 3005	12 04 40.	- 29 45			NONSTELLAR OBJECT
8ZW 1204+11.4	12 04 42.	+ 11 24		17.8	COMPACT GALAXY
AMES 0039	12 04 42.	+ 13 27 12.	19	17.6	NEBULA
ZC 1204.7+1439	12 04 42.	+ 14 39	1140		CLUSTER OF GALAXIES
ZWG 098.060	12 04 42.	+ 17 15		15.6	GALAXY
ZC 1204.7+2246	12 04 42.	+ 22 46	4500		CLUSTER OF GALAXIES
MCG+05-29-017	12 04 42.	+ 28 07 30.	36	15.5	GALAXY
ZWG 158.026	12 04 42.	+ 32 21		15.2	GALAXY
MCG+05-29-018	12 04 42.	+ 32 22	42	15.2	GALAXY
ZC 1204.7+3319	12 04 42.	+ 33 19	3700		CLUSTER OF GALAXIES
ZWG 187.010	12 04 42.	+ 36 56		15.1	GALAXY
UGC 07105	12 04 42.	+ 36 56	78	15.1	GALAXY SBa
MCG+10-17-145	12 04 42.	+ 59 06	24	16.	GALAXY
MCG+11-15-024	12 04 42.	+ 64 09	24	16.	GALAXY
ZWG 315.016	12 04 42.	+ 67 30		14.5	GALAXY
UGC 07106	12 04 42.	+ 67 30	96	14.5	GALAXY S(Bc)
7ZW 439	12 04 42.	+ 67 31			COMPACT GALAXY
MCG-05-29-018	12 04 42.	- 29 45	132	13.5	GALAXY
RNGC 4114	12 04 43.	- 13 55		13.0	GALAXY
AMES 0040	12 04 44.	+ 13 59 54.	24	17.2	NEBULA
SC 1202-4650.4	12 04 44.	- 47 07 06.	18		NEBULA
AMES 0041	12 04 45.	+ 10 41 42.	32	17.8	NEBULA
MCG-02-31-018	12 04 45.	- 13 54	102	13.5	GALAXY
AMES 0042	12 04 47.	+ 12 51 48.	35	16.8	NEBULA
ZWG 013.078	12 04 48.	+ 00 08		15.5	GALAXY
ZWG 041.040	12 04 48.	+ 06 37		15.6	GALAXY
ZC 1204.8+0739	12 04 48.	+ 07 39	540		CLUSTER OF GALAXIES
ZWG 098.061	12 04 48.	+ 17 07		15.4	GALAXY
MCG+03-31-045	12 04 48.	+ 17 11	90	15.5	GALAXY
ZC 1204.8+1720	12 04 48.	+ 17 20	2290		CLUSTER OF GALAXIES
ZWG 098.062	12 04 48.	+ 17 32		15.5	GALAXY
UGC 07107	12 04 48.	+ 17 32	90	15.5	GALAXY Sb
MCG+03-31-044	12 04 48.	+ 17 33	78	15.5	GALAXY
UGC 07108	12 04 48.	+ 20 52	60	16.5	GALAXY S-IRR
MRK 648	12 04 48.	+ 24 29	12	16.	GALAXY WITH UV CONTINUUM
ZC 1204.8+4707	12 04 48.	+ 47 07	2290		CLUSTER OF GALAXIES
ZC 1204.8+6520	12 04 49.	+ 65 20	17940		CLUSTER OF GALAXIES
KEEN 4108B	12 04 48.	+ 67 30	54		E3 GALAXY
MCG+11-15-025	12 04 48.	+ 67 30	66	15.	GALAXY
ZC 1204.8-0115	12 04 48.	- 01 15	2490		CLUSTER OF GALAXIES
AMES 0043	12 04 49.	+ 11 08 00.	26	17.5	NEBULA
AMES 0044	12 04 50.	+ 10 42 18.	31	17.2	NEBULA
HN 1636	12 04 50.0	+ 28 07 24.	24		NEBULA
HN 0796	12 04 51.	+ 13 16		15.	NEBULA
IC 3006	12 04 51.	+ 13 16			NONSTELLAR OBJECT
REIZ 1981	12 04 51.	+ 28 08	42	15.2	GALAXY
MCG+06-27-012	12 04 51.	+ 36 56	48	14.5	GALAXY
AMES 0045	12 04 54.	+ 11 49 42.	16	17.9	NEBULA
ZWG 098.063	12 04 54.	+ 17 10		15.5	GALAXY
UGC 07109	12 04 54.	+ 17 10	72	15.5	GALAXY S
ZWG 158.027	12 04 54.	+ 31 38		15.5	GALAXY
7ZW 440	12 04 54.	+ 57 02			COMPACT GALAXY
MCG+10-17-146	12 04 54.	+ 57 54	39	16.	GALAXY
MCG+10-17-147	12 04 54.	+ 60 18 30.	30	15.	GALAXY
UGC 07110	12 04 54.	+ 65 41	72	17.	GALAXY DWRF SP
ARC 1471	12 04 56.	+ 52 53		17.8	RICH CLUSTER OF GALAXIES
AMES 0046	12 04 57.	+ 10 31 06.	28	17.7	NEBULA
IC 3007	12 04 57.	+ 31 37 14.			NONSTELLAR OBJECT
HN 1637	12 04 57.1	+ 27 35 36.			NEBULA
AMES 0047	12 04 58.	+ 13 22 06.	37	17.1	NEBULA
LB 02216	12 04 59.	+ 54 18 00.		14.7	FAINT BLUE STAR
ZC 1205.0+2715	12 05 00.	+ 27 15	3090		CLUSTER OF GALAXIES
ZWG 187.011	12 05 00.	+ 33 05		15.0	GALAXY
MCG+06-27-013	12 05 00.	+ 33 05	24	15.	GALAXY
ZC 1205.0+4619	12 05 00.	+ 46 19	2220		CLUSTER OF GALAXIES
ZC 1205.0+5146	12 05 00.	+ 51 46	740		CLUSTER OF GALAXIES
ZC 1205.0+5316	12 05 00.	+ 53 16	1010		CLUSTER OF GALAXIES
ZWG 292.072	12 05 00.	+ 60 42		15.6	GALAXY
AMES 0048	12 05 01.	+ 11 40 54.	28	17.9	NEBULA
LB 02217	12 05 02.	+ 52 54 06.		16.6	FAINT BLUE STAR
REIZ 1984	12 05 03.	+ 13 54	42	14.4	GALAXY
REIZ 1983	12 05 04.	+ 02 57	168	12.2	GALAXY
AMES 0049	12 05 04.	+ 12 53 54.	20	16.6	NEBULA
REIZ 1982	12 05 04.	- 02 06	36	13.3	GALAXY
HELW 025	12 05 04.	- 14 42			NEBULA
ZWG 013.079	12 05 06.	+ 01 51		15.3	GALAXY
ZWG 041.041	12 05 06.	+ 02 58		13.0	GALAXY
UGC 07111	12 05 06.	+ 02 58	228	13.0	GALAXY
KARA.72 322A	12 05 06.	+ 02 58	222	13.0	PART OF DOUBLE GALAXY
MCG+01-31-022	12 05 06.	+ 02 58	240	13.0	GALAXY
MCG+03-34-001	12 05 06.	+ 16 38	36	15.4	GALAXY
ZWG 098.064	12 05 06.	+ 17 35		15.5	GALAXY
ZC 1205.1+4022	12 05 06.	+ 40 22	2490		CLUSTER OF GALAXIES
MCG+10-17-148	12 05 06.	+ 60 53	27	16.	GALAXY
ZC 1205.1+6219	12 05 06.	+ 62 19	1080		CLUSTER OF GALAXIES
RNGC 4116	12 05 07.	+ 02 58		12.5	GALAXY
AMES 0051	12 05 07.	+ 12 19 00.	30	16.6	NEBULA
AMES 0050	12 05 07.	+ 12 53 30.	30	16.6	NEBULA
ARC 1472	12 05 09.	+ 31 05		17.8	RICH CLUSTER OF GALAXIES
HOLM 334A	12 05 09.	+ 43 24	36	13.4	PART OF MULTIPLE GALAXY
REIZ 1985	12 05 10.	- 02 50	18	15.8	GALAXY
AMES 0052	12 05 12.	+ 11 03 06.	16	17.8	NEBULA
ZC 1205.2+2011	12 05 12.	+ 20 11	670		CLUSTER OF GALAXIES
ZC 1205.2+2440	12 05 12.	+ 24 40	670		CLUSTER OF GALAXIES
1ZW 030	12 05 12.	+ 40 27			COMPACT GALAXY
ZWG 215.029	12 05 12.	+ 43 24		14.3	GALAXY
UGC 07112	12 05 12.	+ 43 24	150	14.3	GALAXY S0
LB 02218	12 05 12.	+ 59 49 06.		16.7	FAINT BLUE STAR
MCG-02-31-019	12 05 12.	- 14 42	108	14.	GALAXY
MCG-03-31-022	12 05 12.	- 17 48	48	15.	GALAXY
MCG-05-29-019	12 05 12.	- 31 45	54	16.	GALAXY
AMES 0053	12 05 13.	+ 11 59 30.	18	17.8	NEBULA
RNGC 4117	12 05 14.	+ 43 24		14.5	GALAXY
LB 02219	12 05 14.	+ 50 42 36.		16.5	FAINT BLUE STAR
MCG+07-25-027	12 05 15.	+ 43 24 30.	36	13.5	GALAXY
HOLM 334B	12 05 17.	+ 43 23	24	14.9	PART OF MULTIPLE GALAXY
HN 1638	12 05 17.3	+ 27 32 06.			NEBULA
UGC 07113	12 05 18.	+ 20 02	60	16.0	GALAXY Sc
ZWG 215.030	12 05 18.	+ 43 23		15.7	GALAXY
MCG+10-17-149	12 05 18.	+ 60 42	39	16.	GALAXY
ZWG 315.017	12 05 18.	+ 67 40		14.7	GALAXY
RNGC 4119	12 05 19.	+ 09 49			NON-EXISTENT OBJECT
AMES 0054	12 05 20.	+ 11 27 18.	28	17.8	NEBULA
RNGC 4118	12 05 20.	+ 43 23		15.5	GALAXY
HOLM 335B	12 05 20.	+ 65 23	18	14.1	PART OF MULTIPLE GALAXY
HN 0797	12 05 21.	+ 13 51		13.5	NEBULA
IC 3008	12 05 21.	+ 13 51			NONSTELLAR OBJECT
ARC 1474	12 05 21.	+ 15 14		16.0	RICH CLUSTER OF GALAXIES
REIZ 1986	12 05 21.	+ 25 50	24	15.2	GALAXY
ARC 1473	12 05 21.	+ 30 53		17.8	RICH CLUSTER OF GALAXIES
RNGC 4121	12 05 21.	+ 65 24		14.5	GALAXY
AMES 0055	12 05 22.	+ 11 54 42.	25	17.6	NEBULA
ZWG 013.080	12 05 24.	+ 01 40		15.5	GALAXY
AMES 0056	12 05 24.	+ 11 56 30.	38	17.7	NEBULA
ZWG 069.055	12 05 24.	+ 13 50		15.0	GALAXY
MCG+02-31-031	12 05 24.	+ 13 50	48	15.0	GALAXY
ZC 1205.4+2515	12 05 24.	+ 25 15	6320		CLUSTER OF GALAXIES
MCG+04-29-030	12 05 24.	+ 25 48 30.	36	15.2	GALAXY
ZWG 128.032	12 05 24.	+ 25 50		15.2	GALAXY
ZC 1205.4+3025	12 05 24.	+ 30 25	870		CLUSTER OF GALAXIES
MCG+07-25-028	12 05 24.	+ 43 23	30	16.	GALAXY
MCG-05-29-020	12 05 24.	- 30 04	108	14.	GALAXY
AMES 0058	12 05 26.	+ 10 02 30.	24	17.2	NEBULA
AMES 0057	12 05 26.	+ 11 29 30.	19	16.3	NEBULA
HN 1639	12 05 26.9	+ 25 49 42.			NEBULA
IC 3009	12 05 27.	+ 12 55 32.			MAY NOT EXIST
MCG+03-34-002	12 05 27.	+ 18 42 30.	90	14.4	GALAXY
MCG+04-29-031	12 05 27.	+ 25 30	36	14.4	GALAXY
REIZ 1987	12 05 27.	+ 27 15	24	15.4	GALAXY
AMES 0060	12 05 28.	+ 11 51 06.	14	17.8	NEBULA
AMES 0059	12 05 28.	+ 13 45 54.	24	17.4	NEBULA
LB 02220	12 05 29.	+ 59 49 36.		15.8	FAINT BLUE STAR
IC 3010	12 05 29.	- 30 04 04.			NONSTELLAR OBJECT
ZWG 013.081	12 05 30.	+ 00 59		15.5	GALAXY
ZC 1205.5+0208	12 05 30.	+ 02 08	2080		CLUSTER OF GALAXIES
ZWG 069.056	12 05 30.	+ 10 02		15.6	GALAXY
UGC 07114	12 05 30.	+ 10 02	60	15.6	GALAXY DBL SYS
MCG+02-31-035	12 05 30.	+ 10 02	9	15.6	GALAXY
MCG+02-31-034	12 05 30.	+ 10 02	12	15.6	GALAXY
MCG+02-31-033	12 05 30.	+ 10 02	30	15.6	GALAXY
MCG+02-31-032	12 05 30.	+ 10 02	12	15.6	GALAXY
ZWG 069.057	12 05 30.	+ 11 30		15.3	GALAXY
ZWG 128.033	12 05 30.	+ 25 21		15.6	GALAXY
ZWG 128.034	12 05 30.	+ 25 31		14.4	GALAXY
UGC 07115	12 05 30.	+ 25 31	66	14.4	GALAXY E
MCG+04-29-032	12 05 30.	+ 25 51	15	15.1	GALAXY
ZWG 128.035	12 05 30.	+ 25 53		15.1	GALAXY
ZC 1205.5+6126	12 05 30.	+ 61 26	1810		CLUSTER OF GALAXIES
ZWG 315.018	12 05 30.	+ 65 23		14.6	GALAXY
MRK 197	12 05 30.	+ 67 39	25	14.5	GALAXY WITH UV CONTINUUM
REIZ 1989	12 05 31.	+ 65 23	18	14.4	GALAXY
HOLM 335A	12 05 31.	+ 65 27	108	12.3	PART OF MULTIPLE GALAXY

OBJECT NAME	RIGHT ASCEN.	DECLINATION	DIAM.	MAGN.	TYPE OF OBJECT
HN 1640	12 05 32.3	+ 25 52 06.			NEBULA
REIZ 1988	12 05 33.	+ 14 01	24	15.0	GALAXY
ABC 1475	12 05 33.	+ 24 41		17.8	RICH CLUSTER OF GALAXIES
REIZ 1992	12 05 33.	+ 26 02	24	14.4	GALAXY
REIZ 1990	12 05 33.	+ 33 17	18	15.2	GALAXY
LB 02221	12 05 33.	+ 53 09 30.		16.9	FAINT BLUE STAR
RNGC 4125	12 05 33.	+ 65 27		11.5	GALAXY
HN 1641	12 05 33.4	+ 25 30 36.			NEBULA
REIZ 1991	12 05 34.	+ 03 09	210	12.0	GALAXY
RNGC 4128A	12 05 34.	+ 69 03		15.5	GALAXY
ZWG 041.042	12 05 36.	+ 03 10		13.1	GALAXY
UGC 07116	12 05 36.	+ 03 10	300	13.1	GALAXY SBc
KARA.72 322B	12 05 36.	+ 03 10	276	13.1	PART OF DOUBLE GALAXY
MCG+01-31-023	12 05 36.	+ 03 10	228	13.1	GALAXY
ZWG 069.058	12 05 36.	+ 10 39		12.7	GALAXY
UGC 07117	12 05 36.	+ 10 39	246	12.7	GALAXY S0
MCG+02-31-036	12 05 36.	+ 10 39	48	12.7	GALAXY
IC 3011	12 05 36.	+ 10 39 26.			SAME AS NGC 4124
8ZW 1205+13.9	12 05 36.	+ 13 54		17.8	COMPACT GALAXY
ZWG 128.036	12 05 36.	+ 25 14		15.1	GALAXY
MCG+04-29-033	12 05 36.	+ 25 14	36	15.1	GALAXY
MCG+04-29-034	12 05 36.	+ 26 01 30.	36	14.8	GALAXY
ZWG 128.037	12 05 36.	+ 26 02		14.8	GALAXY
ZC 1205.6+2855	12 05 36.	+ 28 55	1010		CLUSTER OF GALAXIES
MCG+11-15-026	12 05 36.	+ 65 23	27	14.1	GALAXY
ZWG 315.019	12 05 36.	+ 65 27		10.9	GALAXY
UGC 07118	12 05 36.	+ 65 27	360	10.9	GALAXY E
MCG+12-12-002	12 05 36.	+ 69 03	9	15.4	GALAXY
ZC 1205.6-0322	12 05 36.	- 03 22	1010		CLUSTER OF GALAXIES
MCG-03-31-023	12 05 36.	- 15 31 30.	78	14.	GALAXY
REIN 3.002	12 05 36.22	+ 10 39 27.5			NEBULA
RNGC 4123	12 05 37.	+ 03 09		12.0	GALAXY
RNGC 4124	12 05 37.	+ 10 39		12.5	GALAXY
LB 02222	12 05 37.	+ 51 40 42.		17.2	FAINT BLUE STAR
REIZ 1993	12 05 37.	+ 65 27	108	11.6	GALAXY
HN 1642	12 05 37.0	+ 25 53 24.			NEBULA
HN 1643	12 05 37.1	+ 25 21 00.			NEBULA
AMES 0061	12 05 39.	+ 12 13 18.	29	17.9	NEBULA
MCG+04-29-035	12 05 39.	+ 26 05 30.	48	15.3	GALAXY
IC 0762	12 05 40.	+ 26 01 44.			NONSTELLAR OBJECT
HZLW 026	12 05 40.	- 15 27			NEBULA
AMES 0063	12 05 41.	+ 13 13 24.	34	17.9	NEBULA
AMES 0062	12 05 41.	+ 13 53 24.	23	17.0	NEBULA
HN 1644	12 05 41.5	+ 25 13 12.			NEBULA
AMES 0064	12 05 42.	+ 12 41 18.	14	17.8	NEBULA
REA 84	12 05 42.	+ 13 03			DWARF GALAXY
ZC 1205.7+1510	12 05 42.	+ 15 10	2420		CLUSTER OF GALAXIES
ZWG 128.038	12 05 42.	+ 26 06		15.3	GALAXY
MCG+11-15-027	12 05 42.	+ 65 27	156	10.9	GALAXY
ZWG 013.082	12 05 42.	- 02 15		15.5	GALAXY
MCG-05-29-021	12 05 42.	- 31 45	120	15.	GALAXY
AMES 0065	12 05 43.	+ 12 44 30.	19	17.4	NEBULA
AMES 0066	12 05 44.	+ 12 33 30.	22	16.8	NEBULA
IC 0763	12 05 44.	+ 26 04 56.			NONSTELLAR OBJECT
AMES 0067	12 05 45.	+ 12 33 18.	20	16.9	NEBULA
HOLM 336A	12 05 45.	+ 12 34	48	15.2	PART OF MULTIPLE GALAXY
REIZ 1994	12 05 45.	+ 27 13	18	15.5	GALAXY
ABC 1476	12 05 45.	+ 31 15		17.6	RICH CLUSTER OF GALAXIES
MCG+09-20-098	12 05 45.	+ 55 56	42	16.	GALAXY
AMES 0068	12 05 46.	+ 12 43 12.	24	17.4	NEBULA
AMES 0069	12 05 47.	+ 11 42 18.	26	17.7	NEBULA
HOLM 336B	12 05 47.	+ 12 34	30	15.5	PART OF MULTIPLE GALAXY
PK298-01.2	12 05 47.	- 63 55	25		PLANETARY NEBULA
ZWG 013.083	12 05 48.	+ 00 23		15.2	GALAXY
AMES 0071	12 05 48.	+ 11 46 48.	23	17.0	NEBULA
AMES 0070	12 05 48.	+ 11 55 54.	22	17.0	NEBULA
ZC 1205.8+1255	12 05 48.	+ 12 55	2490		CLUSTER OF GALAXIES
ZWG 069.059	12 05 48.	+ 13 57		15.3	GALAXY
ZC 1205.8+1642	12 05 48.	+ 16 42	1340		CLUSTER OF GALAXIES
ZC 1205.8+6005	12 05 48.	+ 60 05	1550		CLUSTER OF GALAXIES
HOLM 337B	12 05 48.	+ 69 04	12	15.4	PART OF MULTIPLE GALAXY
MCG+12-12-001	12 05 48.	+ 69 49	102	15.	GALAXY
MCG-02-31-019A	12 05 48.	- 15 29	60	13.5	GALAXY
AMES 0072	12 05 49.	+ 12 06 18.	30	17.8	NEBULA
AMES 0073	12 05 50.	+ 13 58 06.	37	16.2	NEBULA
HN 0799	12 05 51.	+ 10 17		15.	NEBULA
IC 3013	12 05 51.	+ 10 17			NONSTELLAR OBJECT
HN 0798	12 05 51.	+ 11 27		15.	NEBULA
IC 3012	12 05 51.	+ 11 27			NONSTELLAR OBJECT
REIN 3.003	12 05 52.42	+ 10 17 45.8			NEBULA
ZWG 069.060	12 05 54.	+ 10 18		15.3	GALAXY
MCG+02-31-037	12 05 54.	+ 10 18	48	15.3	GALAXY
ZWG 069.061	12 05 54.	+ 11 27		15.4	GALAXY
8ZW 1205+12.9	12 05 54.	+ 12 54		19.8	COMPACT GALAXY
MCG+11-15-028	12 05 54.	+ 63 12	30	16.	GALAXY
MCG+12-12-002A	12 05 54.	+ 69 02	102	12.9	GALAXY
AMES 0074	12 05 56.	+ 13 48 30.	22	17.0	NEBULA
HN 1645	12 05 56.6	+ 00 25 54.	18		NEBULA
LB 02223	12 05 57.	+ 49 23 18.		15.4	FAINT BLUE STAR
LB 02224	12 05 58.	+ 49 04 18.		16.3	FAINT BLUE STAR
LB 02225	12 05 59.	+ 52 27 36.		17.4	FAINT BLUE STAR
RNGC 4127	12 05 59.	+ 77 05		13.5	GALAXY
MIL 23	12 06	- 52 10	5160		SUPERNOVA REMNANT
ZWG 013.084	12 06 00.	+ 00 25		15.1	GALAXY
ZWG 013.085	12 06 00.	+ 02 11		15.7	GALAXY
MCG+03-31-047	12 06 00.	+ 16 26	54	14.6	GALAXY
MCG+03-31-046	12 06 00.	+ 17 30	36	15.4	GALAXY
ZC 1206.0+3120	12 06 00.	+ 31 20	2150		CLUSTER OF GALAXIES
ZWG 215.031	12 06 00.	+ 39 06		14.4	GALAXY
UGC 07119	12 06 00.	+ 39 06	66	14.4	GALAXY SB
LB 00624	12 06 00.	+ 57 04 12.		16.0	FAINT BLUE STAR
ZC 1206.0+6420	12 06 00.	+ 64 20	1010		CLUSTER OF GALAXIES
ZWG 335.003	12 06 00.	+ 69 03		12.7	GALAXY
UGC 07120	12 06 00.	+ 69 03	150	12.7	GALAXY S0
ZWG 335.004	12 06 00.	+ 69 50		14.1	GALAXY
UGC 07121	12 06 00.	+ 69 50	108	14.1	GALAXY S-IRR
ZWG 352.019	12 06 00.	+ 77 05		13.5	GALAXY
UGC 07122	12 06 00.	+ 77 05	192	13.5	GALAXY Sc
MCG-05-29-022	12 06 00.	- 31 05	36	15.	GALAXY
AMES 0075	12 06 04.	+ 10 55 42.	22	17.3	NEBULA
RNGC 4128	12 06 04.	+ 69 03		13.0	GALAXY
RNGC 4120	12 06 04.	+ 69 49		14.0	GALAXY
REA 09	12 06 06.	+ 15 23			DWARF GALAXY
ZWG 098.065	12 06 06.	+ 16 25		14.6	GALAXY
UGC 07123	12 06 06.	+ 16 25	60	14.6	GALAXY S0
ZWG 098.066	12 06 06.	+ 17 29		15.4	GALAXY
ZWG 098.067	12 06 06.	+ 18 14		15.7	GALAXY
ZWG 128.039	12 06 06.	+ 25 14		15.2	GALAXY
UGC 07124	12 06 06.	+ 25 14	60	15.2	GALAXY S0
MCG+05-29-019	12 06 06.	+ 29 35	72	14.1	GALAXY
ZWG 158.028	12 06 06.	+ 32 26		15.7	GALAXY
ZWG 187.012	12 06 06.	+ 37 05		14.7	GALAXY
UGC 07125	12 06 06.	+ 37 05	282	14.7	GALAXY
MCG+10-17-150	12 06 06.	+ 60 46 30.	12	16.	GALAXY
HOLM 337A	12 06 06.	+ 69 03	120	13.6	PART OF MULTIPLE GALAXY
MCG+13-09-012	12 06 06.	+ 77 05	72	12.	GALAXY
RNGC 4126	12 06 07.	+ 16 25		14.5	GALAXY
AMES 0076	12 06 09.	+ 13 09 30.	18	17.8	NEBULA
MCG+04-29-036	12 06 09.	+ 25 14 30.	36	15.2	GALAXY
REIZ 1995	12 06 09.	+ 37 07	60	14.6	GALAXY
LB 02226	12 06 09.	+ 59 07 36.		16.1	FAINT BLUE STAR
IC 3014	12 06 10.	+ 39 05 38.			NONSTELLAR OBJECT
AMES 0077	12 06 11.	+ 11 54 42.	19	17.7	NEBULA
HOLM 338B	12 06 11.	+ 42 00	12	14.7	PART OF MULTIPLE GALAXY
RNGC 4133	12 06 11.	+ 75 12		13.0	GALAXY
ZWG 158.029	12 06 12.	+ 29 35		14.1	GALAXY
HOLM 339C	12 06 12.	+ 29 35	36	14.3	PART OF MULTIPLE GALAXY
UGC 07126	12 06 12.	+ 29 35	90	14.1	GALAXY S
MCG+07-25-028A	12 06 12.	+ 39 06	45	14.5	GALAXY
ZC 1206.2+6805	12 06 12.	+ 68 05	940		CLUSTER OF GALAXIES
ZWG 352.020	12 06 12.	+ 75 12		13.1	GALAXY
UGC 07127	12 06 12.	+ 75 12	120	13.1	GALAXY Sb/SBb
AMES 0078	12 06 13.	+ 13 50 12.	26	16.6	NEBULA
RNGC 4131	12 06 14.	+ 29 35		14.0	GALAXY
AMES 0079	12 06 15.	+ 11 46 54.	22	17.9	NEBULA
MCG+06-27-014	12 06 15.	+ 37 05 30.	240	14.	GALAXY
MCG+08-22-072	12 06 15.	+ 47 30	30	16.	GALAXY
SN 1954A	12 06 15.	- 08 45		19.9	SUPERNOVA
LB 02227	12 06 17.	+ 60 15 06.		14.2	FAINT BLUE STAR
FEIG 057	12 06 18.	+ 07 44		13.0	FAINT BLUE STAR
ZWG 069.062	12 06 18.	+ 09 24		15.5	GALAXY
ZWG 128.040	12 06 18.	+ 25 28		15.1	GALAXY
MCG+04-29-037	12 06 18.	+ 25 28	24	15.1	GALAXY
MCG+05-29-020	12 06 18.	+ 29 32	60	14.6	GALAXY
ZWG 215.032	12 06 18.	+ 42 31		15.5	GALAXY
SEY 163	12 06 18.	+ 42 32 18.		15.1	FAINT GALAXY
MCG+09-20-099	12 06 18.	+ 53 36	6	17.	GALAXY
UGC 07128	12 06 18.	+ 62 39	72	17.	GALAXY DWARF
OCL 0875	12 06 18.	- 62 36	600	13.	OPEN STAR CLUSTER
REIN 3.004	12 06 18.30	+ 09 24 36.5			NEBULA
RNGC 4130	12 06 19.	- 03 44			NON-EXISTENT OBJECT
RNGC 4129	12 06 19.	- 08 45		13.5	GALAXY
AMES 0080	12 06 20.	+ 12 19 30.	23	17.6	NEBULA
HOLM 338A	12 06 20.	+ 42 01	48	13.4	PART OF MULTIPLE GALAXY
LB 02228	12 06 20.	+ 52 22 54.		16.6	FAINT BLUE STAR
AMES 0081	12 06 21.	+ 13 47 06.	24	17.4	NEBULA
REIZ 1998	12 06 21.	+ 29 34	24	14.3	GALAXY
MCG+07-25-029	12 06 21.	+ 42 01	54	14.	GALAXY
SEY 164	12 06 22.	+ 42 01 30.		14.3	FAINT GALAXY
REIZ 1997	12 06 22.	- 03 45	18	16.2	GALAXY
REIZ 1996	12 06 22.	- 08 46	78	12.5	GALAXY
HN 1646	12 06 22.8	+ 25 28 36.			NEBULA
AMES 0082	12 06 23.	+ 12 23 06.	32	17.2	NEBULA
ABC 1477	12 06 23.	+ 64 21		18.0	RICH CLUSTER OF GALAXIES
ZWG 041.043	12 06 24.	+ 05 28		15.5	GALAXY
ZC 1206.4+1928	12 06 24.	+ 19 28	2220		CLUSTER OF GALAXIES
ZWG 215.033	12 06 24.	+ 42 01		14.3	GALAXY
UGC 07129	12 06 24.	+ 42 01	78	14.3	GALAXY Sa-b
ZC 1206.4+4750	12 06 24.	+ 47 50	1550		CLUSTER OF GALAXIES
ZC 1206.4+5733	12 06 24.	+ 57 33	870		CLUSTER OF GALAXIES
MCG+13-09-013	12 06 24.	+ 75 10	102	13.	GALAXY
ZWG 013.086	12 06 24.	- 02 07		15.5	GALAXY
MCG-01-31-006	12 06 24.	- 08 49	120	13.	GALAXY
MCG-05-29-023	12 06 24.	- 31 15	150	13.5	GALAXY
MCG-06-27-005	12 06 24.	- 36 26	48	15.5	GALAXY
PK298-00.1	12 06 25.	- 62 59	25		PLANETARY NEBULA
HZ 18	12 06 26.	+ 37 19		15.1	DECIDEDLY BLUE STAR
IC 3015	12 06 26.	- 31 14 22.			NONSTELLAR OBJECT
REIZ 1999	12 06 27.	+ 29 31	42	14.3	GALAXY
HOLM 339A	12 06 27.	+ 29 32	48	14.6	PART OF MULTIPLE GALAXY
REIZ 2000	12 06 27.	+ 33 19	12	16.1	GALAXY
AMES 0084	12 06 30.	+ 12 19 41.	34	17.0	NEBULA
AMES 0083	12 06 30.	+ 12 20 35.	16	17.8	NEBULA
MCG+05-29-023	12 06 30.	+ 29 27 30.	114	13.8	GALAXY
ZWG 158.030	12 06 30.	+ 29 31		14.6	GALAXY
MCG+05-29-024	12 06 30.	+ 29 33	48	15.5	GALAXY
MCG+05-29-021	12 06 30.	+ 31 12	72	15.7	GALAXY
MCG+05-29-022	12 06 30.	+ 31 52	18	14.4	GALAXY
ZWG 187.013	12 06 30.	+ 32 56		15.5	GALAXY
MCG+07-25-030	12 06 30.	+ 41 35	30	16.	GALAXY
ZC 1206.5+5640	12 06 30.	+ 56 40	1280		CLUSTER OF GALAXIES
RNGC 4132	12 06 32.	+ 29 31		14.5	GALAXY
LB 02229	12 06 32.	+ 52 37 12.		14.6	FAINT BLUE STAR
AMES 0085	12 06 33.	+ 14 01 11.	30	17.4	NEBULA
LB 02230	12 06 34.	+ 49 59 12.		17.5	FAINT BLUE STAR
PK297+03.1	12 06 34.	- 58 25 51.	5		PLANETARY NEBULA
ZWG 098.068	12 06 36.	+ 20 26		15.7	GALAXY
ZWG 158.031	12 06 36.	+ 29 27		13.8	GALAXY
UGC 07130	12 06 36.	+ 29 27	138	13.8	GALAXY Sb
MCG+05-29-025	12 06 36.	+ 30 12 30.	240	12.1	GALAXY
ZWG 158.032	12 06 36.	+ 31 11		15.7	GALAXY
UGC 07131	12 06 36.	+ 31 11	108	15.7	GALAXY
ZWG 158.033	12 06 36.	+ 31 51		14.4	GALAXY
UGC 07132	12 06 36.	+ 31 51	78	14.4	GALAXY E
MCG+06-27-015	12 06 36.	+ 37 15	45	17.	GALAXY
ZC 1206.6+4053	12 06 36.	+ 40 53	740		CLUSTER OF GALAXIES
ZC 1206.6+4115	12 06 36.	+ 41 15	2960		CLUSTER OF GALAXIES
ZWG 215.034	12 06 36.	+ 44 16		14.8	GALAXY
ZWG 243.045	12 06 36.	+ 47 20		15.2	GALAXY
MRK 198	12 06 36.	+ 47 20	12	15.	GALAXY WITH UV CONTINUUM
KW 19	12 06 36.	+ 47 20	39		SEYFERT GALAXY
VVI 54	12 06 36.	+ 47 20	36	15.28	SEYFERT GALAXY
LB 02231	12 06 36.	+ 53 24 24.		14.8	FAINT BLUE STAR
ZC 1206.6-0203	12 06 36.	- 02 03	1680		CLUSTER OF GALAXIES
AMES 0086	12 06 37.	+ 13 42 41.	16	17.6	NEBULA
HOLM 339B	12 06 37.	+ 29 28	96	13.7	PART OF MULTIPLE GALAXY
RNGC 4134	12 06 38.	+ 29 27		14.0	GALAXY
RNGC 4135	12 06 38.	+ 44 17		15.0	GALAXY
REIZ 2001	12 06 39.	+ 14 20	60	15.2	GALAXY
REIZ 2002	12 06 39.	+ 29 27	96	13.1	GALAXY
AMES 0088	12 06 41.	+ 11 46 29.	20	18.0	NEBULA
AMES 0087	12 06 41.	+ 12 58 53.	17	18.0	NEBULA
SBB 194	12 06 41.7	+ 43 56 05.		18.4	QUASI-STELLAR OBJECT
AMES 0089	12 06 42.	+ 10 51 29.	26	17.8	NEBULA
REA 75	12 06 42.	+ 19 16			DWARF GALAXY
ZWG 128.041	12 06 42.	+ 25 15		15.0	GALAXY

OBJECT NAME	RIGHT ASCEN.	DECLINATION	DIAM.	MAGN.	TYPE OF OBJECT
MCG+04-29-038	12 06 42.	+ 25 15	36	15.0	GALAXY
ZWG 215.035	12 06 42.	+ 41 53		15.2	GALAXY
MCG+07-25-031	12 06 42.	+ 41 53	42	14.5	GALAXY
SEY 165	12 06 42.	+ 41 54 06.		15.2	FAINT GALAXY
BC 3CR268.4	12 06 42.	+ 43 56 03.		18.42	QUASI-STELLAR OBJECT
MCG+07-25-032	12 06 42.	+ 44 17 30.	36	14.5	GALAXY
MCG+08-22-073	12 06 42.	+ 47 21	36	15.	GALAXY
MCG+08-22-074	12 06 42.	+ 47 49	42	16.	GALAXY
LB 02232	12 06 42.	+ 60 22 06.		15.8	FAINT BLUE STAR
ZC 1206.7+6548	12 06 42.	+ 65 48	1550		CLUSTER OF GALAXIES
KARA.68 086	12 06 42.	+ 70 45	47		DWARF GALAXY
AMES 0091	12 06 43.	+ 11 47 05.	14	17.8	NEBULA
AMES 0090	12 06 43.	+ 12 59 41.	16	18.0	NEBULA
LB 02233	12 06 43.	+ 49 32 00.		17.0	FAINT BLUE STAR
AMES 0093	12 06 44.	+ 11 13 17.	17	17.2	NEBULA
AMES 0092	12 06 44.	+ 11 38 17.	22	17.8	NEBULA
HN 0800	12 06 45.	+ 11 42			NEBULA
IC 3016	12 06 45.	+ 11 42			NONSTELLAR OBJECT
AMES 0094	12 06 45.	+ 12 57 29.	31	17.6	NEBULA
REIZ 2003	12 06 45.	+ 26 29	72	14.8	GALAXY
REIZ 2004	12 06 45.	+ 30 12	150	13.2	GALAXY
REIZ 2005	12 06 45.	+ 31 11	60	15.3	GALAXY
REIN 3.005	12 06 45.40	+ 11 42 31.5			NEBULA
AMES 0095	12 06 47.	+ 13 06 23.	22	17.5	NEBULA
LB 02235	12 06 47.	+ 53 58 36.		17.0	FAINT BLUE STAR
LB 02234	12 06 47.	+ 57 13 06.		16.4	FAINT BLUE STAR
HN 1647	12 06 47.5	+ 26 29 24.			NEBULA
ZC 1206.8+0737	12 06 48.	+ 07 37	1080		CLUSTER OF GALAXIES
ZWG 069.063	12 06 48.	+ 11 42		15.6	GALAXY
IC 3019	12 06 48.	+ 14 17			DWARF ELLIP. GALAXY
ZC 1206.8+1434	12 06 48.	+ 14 34	670		CLUSTER OF GALAXIES
ZWG 098.069	12 06 48.	+ 19 16		15.5	GALAXY
UGC 07133	12 06 48.	+ 19 16	96	15.5	GALAXY Sc
MCG+03-31-048	12 06 48.	+ 19 17	120	15.5	GALAXY
MCG+04-29-040	12 06 48.	+ 22 21 30.	18	15.7	GALAXY
MCG+04-29-039	12 06 48.	+ 22 22 30.	54	15.7	GALAXY
MCG+05-29-026	12 06 48.	+ 26 30	90	15.3	GALAXY
ZWG 158.034	12 06 48.	+ 30 12		12.1	GALAXY
UGC 07134	12 06 48.	+ 30 12	252	12.1	GALAXY SBc
ZWG 215.036	12 06 48.	+ 44 21		15.0	GALAXY
UGC 07135	12 06 48.	+ 44 21	66	15.0	GALAXY S+COMP
MCG+07-25-033	12 06 48.	+ 44 22 30.	60	14.	GALAXY
MCG-04-29-009	12 06 48.	- 24 02	24	15.5	GALAXY
MCG-04-29-010	12 06 48.	- 25 58	48	15.	GALAXY
MCG-05-29-024	12 06 48.	- 32 15	84	15.	GALAXY
HOLM 340B	12 06 49.	+ 26 29	30	15.0	PART OF MULTIPLE GALAXY
AMES 0096	12 06 50.	+ 13 34 17.	29	16.5	NEBULA
RNGC 4136	12 06 50.	+ 30 12		12.0	GALAXY
RNGC 4137	12 06 50.	+ 44 22		15.0	GALAXY
HN 0801	12 06 51.	+ 13 53		14.5	NEBULA
IC 3017	12 06 51.	+ 13 53			NONSTELLAR OBJECT
SN 1941C	12 06 51.	+ 30 11		16.8	SUPERNOVA
AMES 0097	12 06 53.	+ 11 57 05.	28	17.6	NEBULA
ZWG 013.087	12 06 54.	+ 00 49		15.4	GALAXY
ZWG 069.064	12 06 54.	+ 13 50		15.3	GALAXY
AMES 0098	12 06 54.	+ 13 51 05.	42	15.7	NEBULA
ZWG 069.065	12 06 54.	+ 14 16		15.0	GALAXY
UGC 07136	12 06 54.	+ 14 16	108	15.0	GALAXY DWRF EL
MCG+02-31-038	12 06 54.	+ 14 16	72	15.0	GALAXY
ZC 1206.9+1746	12 06 54.	+ 17 46	1950		CLUSTER OF GALAXIES
ZC 1206.9+1842	12 06 54.	+ 18 42	670		CLUSTER OF GALAXIES
ZC 1206.9+2200	12 06 54.	+ 22 00	1010		CLUSTER OF GALAXIES
ZWG 128.042	12 06 54.	+ 22 23		15.7	GALAXY
UGC 07137	12 06 54.	+ 22 23	108	15.7	GALAXY DBL SYS
ZWG 158.035	12 06 54.	+ 26 30		15.3	GALAXY
ZWG 128.043	12 06 54.	+ 26 30		15.3	GALAXY
UGC 07138	12 06 54.	+ 26 30	108	15.3	GALAXY Sc
MCG+08-22-075	12 06 54.	+ 49 11	60	15.	GALAXY
LB 02236	12 06 54.	+ 56 30 24.		17.3	FAINT BLUE STAR
7ZW 441	12 06 54.	+ 62 26			COMPACT GALAXY
SEY 166	12 06 55.	+ 41 05 42.		15.2	FAINT GALAXY
AMES 0099	12 06 56.	+ 14 03 41.	20	17.2	NEBULA
ARC 1478	12 06 56.	+ 30 44		17.8	RICH CLUSTER OF GALAXIES
HN 1648	12 06 56.7	+ 26 30 24.			NEBULA
HN 0803	12 06 57.	+ 14 16		14.5	NEBULA
HN 0802	12 06 57.	+ 14 21		13.5	NEBULA
IC 3018	12 06 57.	+ 14 21			MAY NOT EXIST
MCG+03-31-049	12 06 57.	+ 17 19	60	14.8	GALAXY
HOLM 340A	12 06 57.	+ 26 30	108	14.4	PART OF MULTIPLE GALAXY
MCG+07-25-034	12 06 57.	+ 44 21 30.	18	15.	GALAXY
AMES 0100	12 06 58.	+ 11 02 41.	25	17.4	NEBULA
BC PKS1207-399	12 06 59.59	- 39 59 30.6		17.5	QUASI-STELLAR OBJECT
AMES 0101	12 07 00.	+ 11 02 11.	23	17.3	NEBULA
8ZW 1207+11.3	12 07 00.	+ 11 18		18.0	COMPACT GALAXY
2ZW 057	12 07 00.	+ 17 17			COMPACT GALAXY
ZWG 098.070	12 07 00.	+ 17 17		14.8	GALAXY
ZWG 098.071	12 07 00.	+ 18 05		15.5	GALAXY
KARA.72 323A	12 07 00.	+ 18 05	54	15.5	PART OF DOUBLE GALAXY
MCG+03-31-050	12 07 00.	+ 18 06	48	15.5	GALAXY
ZWG 098.072	12 07 00.	+ 19 29		15.7	GALAXY
ZC 1207.0+2439	12 07 00.	+ 24 39	810		CLUSTER OF GALAXIES
ZWG 215.037	12 07 00.	+ 43 57		12.1	GALAXY
UGC 07139	12 07 00.	+ 43 57	180	12.1	GALAXY SO
MCG+07-25-035	12 07 00.	+ 43 58	60	12.	GALAXY
ZWG 243.046	12 07 00.	+ 49 10		15.4	GALAXY
MCG+09-20-100	12 07 00.	+ 51 12	36	16.	GALAXY
LB 02237	12 07 00.	+ 51 46 24.		16.4	FAINT BLUE STAR
MCG+09-20-102	12 07 00.	+ 53 22 30.	132	14.0	GALAXY
ZWG 269.037	12 07 00.	+ 53 24		14.3	GALAXY
UGC 07140	12 07 00.	+ 53 24	132	14.3	GALAXY SBc
MCG+09-20-101	12 07 00.	+ 53 37	42	17.	GALAXY
MCG+10-17-151	12 07 00.	+ 62 26	27	16.	GALAXY
MCG+13-09-014	12 07 00.	+ 75 50	12	16.	GALAXY
ZWG 352.021	12 07 00.	+ 75 52		15.6	GALAXY
ZC 1207.0+7752	12 07 00.	+ 77 52	1810		CLUSTER OF GALAXIES
ZC 1207.0+8229	12 07 00.	+ 82 29	1280		CLUSTER OF GALAXIES
MCG-04-29-011	12 07 00.	- 23 09	72	15.	GALAXY
RNGC 4139	12 07 01.	+ 02 05			NON-EXISTENT OBJECT
RNGC 4138	12 07 02.	+ 43 58		12.5	GALAXY
HN 0804	12 07 03.	+ 14 30		14.	NEBULA
IC 3020	12 07 03.	+ 14 30			NONSTELLAR OBJECT
MCG+04-29-041	12 07 03.	+ 23 33 30.	90	15.7	GALAXY
MCG+07-25-036	12 07 03.	+ 42 49	66	12.	GALAXY
RNGC 4142	12 07 03.	+ 53 23		14.5	GALAXY
REIN 2.148	12 07 04.82	+ 42 48 44.3			NEBULA
ZWG 041.044	12 07 06.	+ 05 32		15.3	GALAXY
ZWG 098.073	12 07 06.	+ 18 08		15.7	GALAXY
KARA.72 323B	12 07 06.	+ 18 08	54	15.7	PART OF DOUBLE GALAXY
RFA 10	12 07 06.	+ 20 19			DWARF GALAXY
ZWG 128.044	12 07 06.	+ 23 34		15.7	GALAXY
UGC 07141	12 07 06.	+ 23 34	102	15.7	GALAXY S
ZWG 215.038	12 07 06.	+ 42 17		15.5	GALAXY
1ZW 031	12 07 06.	+ 42 36			COMPACT GALAXY
ZWG 215.039	12 07 06.	+ 42 48		12.0	GALAXY
UGC 07142	12 07 06.	+ 42 48	174	12.0	GALAXY SO
ZC 1207.1+4751	12 07 06.	+ 47 51	270		CLUSTER OF GALAXIES
MCG+11-15-029	12 07 06.	+ 66 27	39	17.	GALAXY
RNGC 4140	12 07 07.	+ 02 04			NON-EXISTENT OBJECT
SEY 167	12 07 07.	+ 42 17 48.		15.1	FAINT GALAXY
REIZ 2007	12 07 07.	+ 53 24	90	14.0	GALAXY
REIZ 2006	12 07 07.	+ 59 08	60	14.1	GALAXY
HOLM 341A	12 07 08.	+ 36 44	36	14.4	PART OF MULTIPLE GALAXY
RNGC 4143	12 07 08.	+ 42 49		12.5	GALAXY
AMES 0102	12 07 11.	+ 12 24 05.	30	16.4	NEBULA
HN 1649	12 07 11.8	+ 25 18 24.	18		NEBULA
ZC 1207.2+0644	12 07 12.	+ 06 44	2290		CLUSTER OF GALAXIES
ZWG 069.066	12 07 12.	+ 12 24		15.6	GALAXY
MCG+04-29-042	12 07 12.	+ 25 18	54	14.6	GALAXY
ZWG 128.045	12 07 12.	+ 25 19		14.6	GALAXY
UGC 07143	12 07 12.	+ 25 19	60	14.6	GALAXY
ZWG 187.014	12 07 12.	+ 36 45		15.4	GALAXY
MCG+06-27-016	12 07 12.	+ 36 45	36	15.	GALAXY
ZWG 243.047	12 07 12.	+ 47 26		15.4	GALAXY
ZWG 292.073	12 07 12.	+ 56 48		14.3	GALAXY
UGC 07144	12 07 12.	+ 56 48	60	14.3	GALAXY Sb
KARA.73B 0520	12 07 12.	+ 56 48	60	14.3	ISOLATED GALAXY S
MCG+11-15-030	12 07 12.	+ 63 08	24	16.	GALAXY
MCG+07-25-037	12 07 15.	+ 38 30	84	14.5	GALAXY
RNGC 4141	12 07 15.	+ 59 07		14.5	GALAXY
AMES 0103	12 07 16.	+ 11 32 29.	16	17.4	NEBULA
AMES 0104	12 07 17.	+ 13 08 29.	26	17.5	NEBULA
ZWG 128.046	12 07 18.	+ 25 36		15.7	GALAXY
ZWG 215.040	12 07 18.	+ 38 30		15.4	GALAXY
UGC 07145	12 07 18.	+ 38 30	90	15.4	GALAXY Sc
UGC 07146	12 07 18.	+ 43 15	90	16.5	GALAXY DWARF
MCG+07-25-038	12 07 18.	+ 43 31	54	17.	GALAXY
MCG+08-22-076	12 07 18.	+ 47 27	36	15.	GALAXY
MCG+09-20-103	12 07 18.	+ 54 17 30.	42	15.	GALAXY
ZWG 292.074	12 07 18.	+ 59 07		14.5	GALAXY
UGC 07147	12 07 18.	+ 59 07	84	14.5	GALAXY SBc
MCG+11-15-031	12 07 18.	+ 62 32	114	15.	GALAXY
ZWG 315.020	12 07 18.	+ 63 08		15.7	GALAXY
SER 094.06	12 07 18.	- 46 45	200	13.	SO GALAXY
HOLM 341B	12 07 19.	+ 36 42	48	14.7	PART OF MULTIPLE GALAXY
REIZ 2008	12 07 19.	+ 56 47	30	14.5	GALAXY
AMES 0105	12 07 20.	+ 11 06 41.	14	18.0	NEBULA
MCG+06-27-017	12 07 21.	+ 36 44	48	15.	GALAXY
ZWG 013.088	12 07 24.	+ 00 16		15.6	GALAXY
ZWG 013.089	12 07 24.	+ 01 13		15.4	GALAXY
UGC 07148	12 07 24.	+ 01 13	78	15.4	GALAXY SB
ZWG 069.067	12 07 24.	+ 13 18		15.4	GALAXY
UGC 07149	12 07 24.	+ 13 18	108	15.4	GALAXY
MCG+03-31-051	12 07 24.	+ 14 38	84	17.	GALAXY
UGC 07150	12 07 24.	+ 14 39	72	16.0	GALAXY DWARF
ZWG 098.074	12 07 24.	+ 20 11		15.6	GALAXY
ZC 1207.4+2730	12 07 24.	+ 27 30	810		CLUSTER OF GALAXIES
ZWG 187.015	12 07 24.	+ 36 43		15.5	GALAXY
MCG+07-25-039	12 07 24.	+ 39 00	12	14.5	GALAXY
MCG+07-25-040	12 07 24.	+ 40 09	330	12.	GALAXY
MCG+07-25-041	12 07 24.	+ 42 35	42	16.	GALAXY
ZWG 243.048	12 07 24.	+ 46 43		12.3	GALAXY
UGC 07151	12 07 24.	+ 46 43	414	12.3	GALAXY Sc
MCG+09-20-104	12 07 24.	+ 51 14	18	16.	GALAXY
MCG+10-17-153	12 07 24.	+ 56 48	51	14.	GALAXY
MCG+10-17-152	12 07 24.	+ 59 07	72	14.1	GALAXY
ZWG 292.075	12 07 24.	+ 59 43		15.4	GALAXY
UGC 07152	12 07 24.	+ 59 43	72	15.4	GALAXY S
ZWG 315.021	12 07 24.	+ 62 33		15.1	GALAXY
UGC 07153	12 07 24.	+ 62 33	126	15.1	GALAXY Sc
AMES 0106	12 07 26.	+ 10 53 29.	22	16.9	NEBULA
HN 0805	12 07 27.	+ 13 19		14.5	NEBULA
IC 3021	12 07 27.	+ 13 19			NONSTELLAR OBJECT
AMES 0107	12 07 28.	+ 10 59 23.	14	18.0	NEBULA
ARC 1479	12 07 29.	- 16 34		17.7	RICH CLUSTER OF GALAXIES
ZWG 041.045	12 07 30.	+ 05 38		15.3	GALAXY
ZC 1207.5+0542	12 07 30.	+ 05 42	1550		CLUSTER OF GALAXIES
ZC 1207.5+1353	12 07 30.	+ 13 53	1410		CLUSTER OF GALAXIES
ZWG 215.041	12 07 30.	+ 39 00		14.6	GALAXY
IC 3022	12 07 30.	+ 39 01 02.			NONSTELLAR OBJECT
MCG+07-25-042	12 07 30.	+ 39 17	42	15.	GALAXY
ZWG 215.042	12 07 30.	+ 40 10		12.2	GALAXY
UGC 07154	12 07 30.	+ 40 10	390	12.2	GALAXY Sc
KARA.72 324A	12 07 30.	+ 40 10	366	12.2	PART OF DOUBLE GALAXY
ZWG 215.043	12 07 30.	+ 43 43		15.5	GALAXY
MCG+08-22-077	12 07 30.	+ 46 44	330	12.4	GALAXY
ZC 1207.5+4913	12 07 30.	+ 49 13	1010		CLUSTER OF GALAXIES
MCG+08-22-078	12 07 30.	+ 49 32 30.	36	16.	GALAXY
UGC 07155	12 07 30.	+ 49 33	66	16.0	GALAXY Sc
MCG+10-17-154	12 07 30.	+ 59 43 30.	57	15.	GALAXY
UGC 07156	12 07 30.	+ 78 02	78	16.5	GALAXY Sc
ZC 1207.5-0114	12 07 30.	- 01 14	1010		CLUSTER OF GALAXIES
KARA.68 087	12 07 30.	- 19 45	34		DWARF GALAXY
REIN 6.091	12 07 30.39	+ 40 09 41.2			NEBULA
LB 02238	12 07 31.	+ 52 56 36.		15.8	FAINT BLUE STAR
AMES 0109	12 07 32.	+ 11 01 29.	17	17.7	NEBULA
AMES 0108	12 07 32.	+ 11 58 29.	20	18.0	NEBULA
RNGC 4145	12 07 32.	+ 40 10		11.5	GALAXY
REIZ 2011	12 07 32.	+ 40 10	270	12.0	GALAXY
RNGC 4144	12 07 32.	+ 46 44		12.5	GALAXY
REIZ 2010	12 07 32.	+ 46 44	192	12.4	GALAXY
HN 0806	12 07 33.	+ 14 38		15.	NEBULA
IC 3023	12 07 33.	+ 14 38			NONSTELLAR OBJECT
REIZ 2013	12 07 33.	+ 18 49	108	11.0	GALAXY
REIZ 2012	12 07 33.	+ 26 42	24	13.9	GALAXY
REIZ 2009	12 07 34.	- 00 14	48	15.3	GALAXY
ZC 1207.6+1138	12 07 36.	+ 11 38	470		CLUSTER OF GALAXIES
GCL 018	12 07 36.	+ 18 49	204	11.01	GLOBULAR STAR CLUSTER
ZWG 098.075	12 07 36.	+ 20 29		15.7	GALAXY
ZWG 128.047	12 07 36.	+ 25 35		14.5	GALAXY
UGC 07157	12 07 36.	+ 25 35	72	14.5	GALAXY SO
MCG+04-29-043	12 07 36.	+ 25 35 30.	42	14.5	GALAXY
MCG+05-29-028	12 07 36.	+ 26 42	54	13.8	GALAXY
MCG+05-29-027	12 07 36.	+ 32 18	36	16.	GALAXY
ZWG 187.016	12 07 36.	+ 36 09		14.6	GALAXY
UGC 07158	12 07 36.	+ 36 09	108	14.6	GALAXY SO
MCG+06-27-018	12 07 36.	+ 36 10	66	14.	GALAXY

OBJECT NAME	RIGHT ASCEN.	DECLINATION	DIAM.	MAGN.	TYPE OF OBJECT
ZWG 215.044	12 07 36.	+ 39 20		15.6	GALAXY
UGC 07159	12 07 36.	+ 39 20	72	15.6	GALAXY Sc
ZC 1207.6+4145	12 07 36.	+ 41 45	870		CLUSTER OF GALAXIES
MCG+07-25-043	12 07 36.	+ 43 43 30.	42	15.	GALAXY
LB 02239	12 07 36.	+ 50 43 48.		17.2	FAINT BLUE STAR
ZC 1207.6+5735	12 07 36.	+ 57 35	6050		CLUSTER OF GALAXIES
UGC 07160	12 07 36.	+ 70 42	60	17.	GALAXY DWARF
AMES 0110	12 07 38.	+ 14 12 17.	25	17.7	NEBULA
RNGC 4147	12 07 38.	+ 18 49		11.0	GLOBULAR CLUSTER
RNGC 4148	12 07 38.	+ 36 09		14.5	GALAXY
REIZ 2014	12 07 38.	+ 38 39	60	14.8	GALAXY
HOLM 342A	12 07 38.	+ 40 10	300	11.8	PART OF MULTIPLE GALAXY
IC 0764	12 07 38.8	- 29 27 29.	318	12.0	GALAXY SA(s)
REIN 3.006A	12 07 38.94	+ 12 36 14.1			NEBULA
REIN 3.006B	12 07 38.94	+ 12 36 15.7			NEBULA
HN 0807	12 07 39.	+ 12 35		13.5	NEBULA
IC 3024	12 07 39.	+ 12 35			NONSTELLAR OBJECT
LB 02240	12 07 40.	+ 52 12 00.		17.5	FAINT BLUE STAR
ZWG 013.091	12 07 42.	+ 00 55		15.6	GALAXY
ZWG 069.068	12 07 42.	+ 12 35		15.1	GALAXY
UGC 07161	12 07 42.	+ 12 35	72	15.1	GALAXY S?
MCG+02-31-039	12 07 42.	+ 12 35	60	15.1	GALAXY
ZWG 098.076	12 07 42.	+ 19 28		15.5	GALAXY
MCG+03-34-003	12 07 42.	+ 20 01	18	15.1	GALAXY
UGC 07162	12 07 42.	+ 24 10	84	15.7	GALAXY S
MCG+08-22-079	12 07 42.	+ 49 34	60	15.	GALAXY
ZC 1207.7-0004	12 07 42.	- 00 04	2420		CLUSTER OF GALAXIES
ZWG 013.090	12 07 42.	- 01 42		15.6	GALAXY
MCG-05-29-025	12 07 42.	- 29 27	270	13.	GALAXY
AMES 0111	12 07 43.	+ 13 27 29.	25	16.6	NEBULA
MCG+00-31-036	12 07 45.	+ 02 18	60	17.	GALAXY
HOLM 343B	12 07 46.	+ 59 03	24	13.6	PART OF MULTIPLE GALAXY
ZWG 069.069	12 07 48.	+ 10 28		15.3	GALAXY
MCG+02-31-040	12 07 48.	+ 10 28	24	15.3	GALAXY
AMES 0112	12 07 48.	+ 11 37 41.	20	17.9	NEBULA
ZC 1207.8+1611	12 07 48.	+ 16 11	2420		CLUSTER OF GALAXIES
ZWG 158.036	12 07 48.	+ 26 42		13.8	GALAXY
UGC 07163	12 07 48.	+ 26 42	102	13.8	GALAXY SBa
MCG+05-29-029	12 07 48.	+ 30 41	60	12.6	GALAXY
TON-N 0604	12 07 48.	+ 31 12		15.4	BLUE STAR
ZC 1207.8+4534	12 07 48.	+ 45 34	1340		CLUSTER OF GALAXIES
MCG+08-22-080	12 07 48.	+ 47 27	36	15.	GALAXY
MCG+11-15-032	12 07 48.	+ 65 02	24	16.	GALAXY
ZWG 013.092	12 07 48.	- 00 14		15.3	GALAXY
MCG-05-29-026	12 07 48.	- 29 48	72	15.	GALAXY
SN 1963D	12 07 49.	+ 26 42		15.8	SUPERNOVA
REIN 3.007	12 07 49.60	+ 10 28 08.7			NEBULA
RNGC 4146	12 07 50.	+ 26 42		14.0	GALAXY
HOLM 343C	12 07 50.	+ 59 03	18	13.7	PART OF MULTIPLE GALAXY
HN 0808	12 07 51.	+ 10 27			NEBULA
IC 3025	12 07 51.	+ 10 27			NONSTELLAR OBJECT
AMES 0113	12 07 51.	+ 12 58 05.	24	17.4	NEBULA
AMES 0114	12 07 53.	+ 13 26 29.	17	16.5	NEBULA
8ZW 1207+11.1	12 07 54.	+ 11 08		18.7	COMPACT GALAXY
AMES 0116	12 07 54.	+ 11 54 05.	26	18.0	NEBULA
AMES 0115	12 07 54.	+ 11 54 29.	16	18.0	NEBULA
ZC 1207.9+3042	12 07 54.	+ 30 42	810		CLUSTER OF GALAXIES
ZC 1207.9+3843	12 07 54.	+ 38 43	940		CLUSTER OF GALAXIES
REIZ 2015	12 07 54.	+ 58 33	90	13.6	GALAXY
HOLM 343A	12 07 54.	+ 59 04	36	13.5	PART OF MULTIPLE GALAXY
LB 02241	12 07 54.	+ 61 59 06.		16.3	FAINT BLUE STAR
UGC 07164	12 07 54.	+ 70 48	108	17.	GALAXY DWRF SP
MCG-05-29-027	12 07 54.	- 29 49	24	15.	GALAXY
AMES 0117	12 07 55.	+ 11 55 17.	24	17.5	NEBULA
LB 02242	12 07 56.	+ 56 12 18.		11.8	FAINT BLUE STAR
HN 0810	12 07 57.	+ 12 02		15.	NEBULA
IC 3028	12 07 57.	+ 12 02			NONSTELLAR OBJECT
AMES 0118	12 07 57.	+ 13 15 41.	24	17.6	NEBULA
HN 0809	12 07 57.	+ 14 28		14.	NEBULA
IC 3027	12 07 57.	+ 14 28			MAY NOT EXIST
MCG+07-25-044	12 07 57.	+ 39 40	420	11.	GALAXY
RNGC 4149	12 07 57.	+ 58 35		14.0	GALAXY
IC 3026	12 07 58.	- 29 39			SINGLE STAR
AMES 0119	12 07 59.	+ 13 26 29.	12	17.8	NEBULA
IC 0765	12 07 59.	+ 16 25			DOUBLE STAR
HN 0360	12 07 59.	- 29 39			NEBULA
VDB.66G 110	12 08	+ 02 18	70		DWARF GALAXY
VDB.66G 112	12 08	+ 18 18	70		DWARF GALAXY
VDB.66G 111	12 08	+ 50 33	70		DWARF GALAXY
ZWG 013.094	12 08 00.	+ 00 06		15.5	GALAXY
ZWG 013.095	12 08 00.	+ 00 58		15.5	GALAXY
ZWG 128.048	12 08 00.	+ 25 43		15.1	GALAXY
MCG+04-29-045	12 08 00.	+ 25 43	24	15.1	GALAXY
MCG+04-29-044	12 08 00.	+ 26 11 30.	39	15.0	GALAXY
ZWG 128.049	12 08 00.	+ 26 13		15.0	GALAXY
ZWG 158.037	12 08 00.	+ 30 40		12.6	GALAXY
UGC 07165	12 08 00.	+ 30 40	126	12.6	GALAXY S0
MCG+06-27-019	12 08 00.	+ 35 14	48	16.	GALAXY
VV 236B	12 08 00.	+ 35 15	18	16.	INTERACTING GALAXY
VV 236B	12 08 00.	+ 35 15	18	16.	INTERACTING GALAXY
VV 236A	12 08 00.	+ 35 15	18	15.	INTERACTING GALAXY
VV 236	12 08 00.	+ 35 15	54		INTERACTING GALAXY
ZWG 215.045	12 08 00.	+ 39 41		11.2	GALAXY
VVI 55	12 08 00.	+ 39 41	420	11.53	SEYFERT GALAXY
UGC 07166	12 08 00.	+ 39 41	420	11.2	GALAXY Sa/SBb
KARA.72 324B	12 08 00.	+ 39 41	456	11.2	PART OF DOUBLE GALAXY
MCG+10-17-155	12 08 00.	+ 58 34 30.	72	13.6	GALAXY
ZWG 292.076	12 08 00.	+ 58 35		13.9	GALAXY
UGC 07167	12 08 00.	+ 58 35	108	13.9	GALAXY S
MRK 199	12 08 00.	+ 70 37	25	16.	GALAXY WITH UV CONTINUUM
ZWG 335.005	12 08 00.	+ 70 40		15.4	GALAXY
UGC 07168	12 08 00.	+ 70 40	78	15.4	GALAXY PECULR?
ZWG 013.093	12 08 00.	- 02 57		15.6	GALAXY
ZC 1208.0-0327	12 08 00.	- 03 27	340		CLUSTER OF GALAXIES
REIN 2.149	12 08 01.05	+ 39 41 03.3			NEBULA
RNGC 4150	12 08 02.	+ 30 41		12.5	GALAXY
REIZ 2016	12 08 02.	+ 30 41	90	12.2	GALAXY
RNGC 4151	12 08 02.	+ 39 41		11.5	GALAXY
REIZ 2017	12 08 02.	+ 39 41	174	11.4	GALAXY
HOLM 345A	12 08 02.	+ 39 41	180	12.1	PART OF MULTIPLE GALAXY
REIZ 2018	12 08 03.	+ 16 18	54	12.8	GALAXY
MCG+03-31-052	12 08 03.	+ 16 18	156	12.5	GALAXY
HOLM 344A	12 08 03.	+ 18 09	24	14.3	PART OF MULTIPLE GALAXY
REIN 2.150	12 08 04.61	+ 16 18 43.0			NEBULA
AMES 0120	12 08 05.	+ 13 17 41.	34	17.6	NEBULA
ZWG 069.070	12 08 06.	+ 12 04		15.5	GALAXY
ZWG 098.077	12 08 06.	+ 16 18		12.5	GALAXY
UGC 07169	12 08 06.	+ 16 18	132	12.5	GALAXY Sc

OBJECT NAME	RIGHT ASCEN.	DECLINATION	DIAM.	MAGN.	TYPE OF OBJECT
MCG+03-31-053	12 08 06.	+ 16 18	30	17.	GALAXY
SCH 22	12 08 06.	+ 16 19		12.5	PECULIAR GALAXY
ZWG 098.078	12 08 06.	+ 18 09		15.2	GALAXY
MCG+03-31-054	12 08 06.	+ 18 10	36	15.2	GALAXY
MCG+03-34-004	12 08 06.	+ 18 44	48	14.8	GALAXY
ZWG 098.079	12 08 06.	+ 19 06		15.2	GALAXY
UGC 07170	12 08 06.	+ 19 06	180	15.2	GALAXY Sc
MCG+03-31-055	12 08 06.	+ 19 06	180	15.2	GALAXY
ZC 1208.1+3603	12 08 06.	+ 36 03	8200		CLUSTER OF GALAXIES
ZC 1208.1-0315	12 08 06.	- 03 15	810		CLUSTER OF GALAXIES
MCG-06-27-006	12 08 06.	- 33 47	42	15.5	GALAXY
HOLM 344B	12 08 07.	+ 18 10	30	14.3	PART OF MULTIPLE GALAXY
AMES 0121	12 08 08.	+ 11 40 53.	25	17.2	NEBULA
RNGC 4152	12 08 08.	+ 16 19		13.0	GALAXY
LB 02243	12 08 08.	+ 51 08 30.		16.4	FAINT BLUE STAR
HN 0811	12 08 09.	+ 13 36		13.5	NEBULA
REIZ 2019	12 08 09.	+ 13 36	48	14.4	GALAXY
MCG+03-31-057	12 08 09.	+ 16 09	78	15.7	GALAXY
MCG+03-31-056	12 08 09.	+ 18 11	60	15.2	GALAXY
AMES 0122	12 08 10.	+ 13 35 11.	28	16.7	NEBULA
IC 3029	12 08 10.	+ 13 36			NONSTELLAR OBJECT
8ZW 1208+11.7	12 08 12.	+ 11 40		17.0	COMPACT GALAXY
ZWG 069.071	12 08 12.	+ 13 36		15.0	GALAXY
8ZW 1208+13.6	12 08 12.	+ 13 36		15.0	COMPACT GALAXY
UGC 07171	12 08 12.	+ 13 36	90	15.0	GALAXY SB
MCG+02-31-041	12 08 12.	+ 13 36	84	15.0	GALAXY
ZWG 098.080	12 08 12.	+ 16 09		15.7	GALAXY
ZWG 098.081	12 08 12.	+ 18 10		15.2	GALAXY
ZWG 098.082	12 08 12.	+ 19 19		14.7	GALAXY
UGC 07172	12 08 12.	+ 19 19	78	14.7	GALAXY E
MCG+03-31-058	12 08 12.	+ 19 20	24	14.7	GALAXY
ZWG 128.050	12 08 12.	+ 24 00		15.7	GALAXY
ZWG 215.046	12 08 12.	+ 38 37		15.3	GALAXY
LB 02244	12 08 12.	+ 59 54 36.		16.0	FAINT BLUE STAR
ZC 1208.2+6032	12 08 12.	+ 60 32	810		CLUSTER OF GALAXIES
MCG-05-29-028	12 08 12.	- 27 34	48	16.	GALAXY
RNGC 4153	12 08 14.	+ 18 38			NON-EXISTENT OBJECT
RNGC 4155	12 08 14.	+ 19 19		14.5	GALAXY
ARC 1480	12 08 14.	+ 31 08		17.0	RICH CLUSTER OF GALAXIES
MCG+07-25-045	12 08 15.	+ 39 44	72	14.	GALAXY
RNGC 4154	12 08 15.	+ 58 37			NON-EXISTENT OBJECT
SN 1974A	12 08 17.	+ 39 45		20.0	SUPERNOVA
LB 00625	12 08 17.	+ 56 33 12.		16.9	FAINT BLUE STAR
AMES 0123	12 08 18.	+ 13 01 53.	34	17.3	NEBULA
ZWG 187.017	12 08 18.	+ 36 54		14.8	GALAXY
MCG+06-27-020	12 08 18.	+ 36 55 30.	36	14.5	GALAXY
ZWG 215.047	12 08 18.	+ 39 45		14.3	GALAXY
UGC 07173	12 08 18.	+ 39 45	84	14.3	GALAXY SBb
KARA.72 325B	12 08 18.	+ 39 45	24		PART OF DOUBLE GALAXY
KARA.72 325A	12 08 18.	+ 39 45	24	14.3	PART OF DOUBLE GALAXY
MCG+07-25-046	12 08 18.	+ 40 01	102	14.	GALAXY
ZWG 243.049	12 08 18.	+ 44 47		15.4	GALAXY
MCG+08-22-081	12 08 18.	+ 50 34	90	16.	GALAXY
ZWG 352.022	12 08 18.	+ 76 25		14.3	GALAXY
UGC 07174	12 08 18.	+ 76 25	90	14.3	GALAXY S
MCG-02-31-020	12 08 18.	- 12 23	72	14.5	GALAXY
IC 0766	12 08 18.	- 12 23 04.			NONSTELLAR OBJECT
REIN 6.092	12 08 18.39	+ 39 45 02.3			NEBULA
RNGC 4159	12 08 19.	+ 76 25		14.5	GALAXY
RNGC 4156	12 08 20.	+ 39 45		14.5	GALAXY
REIZ 2020	12 08 20.	+ 39 45	48	13.8	GALAXY
HOLM 345B	12 08 20.	+ 39 45	60	14.2	PART OF MULTIPLE GALAXY
RNGC 4145A	12 08 20.	+ 40 01		15.5	GALAXY
REIZ 2021	12 08 20.	+ 40 01	30	15.0	GALAXY
SEY 168	12 08 20.	+ 44 47 18.		15.2	FAINT GALAXY
ARC 1481	12 08 21.	+ 16 09		17.0	RICH CLUSTER OF GALAXIES
TON-N 1471	12 08 21.	+ 33 29		16.9	BLUE STAR
AMES 0126	12 08 22.	+ 09 49 41.	22	17.2	NEBULA
AMES 0125	12 08 22.	+ 13 10 05.	29	17.6	NEBULA
AMES 0124	12 08 22.	+ 13 27 29.	25	17.4	NEBULA
ARC 1482	12 08 22.	- 05 19		17.2	RICH CLUSTER OF GALAXIES
REIN 6.093	12 08 22.83	+ 40 02 07.2			NEBULA
AMES 0127	12 08 23.	+ 12 10 53.	16	17.8	NEBULA
ZWG 069.072	12 08 24.	+ 09 29		15.4	GALAXY
8ZW 1208+09.9	12 08 24.	+ 09 54		18.4	COMPACT GALAXY
ZWG 158.038	12 08 24.	+ 31 56		15.3	GALAXY
MCG+06-27-021	12 08 24.	+ 37 18	24	16.	GALAXY
ZWG 215.048	12 08 24.	+ 40 02		15.3	GALAXY
UGC 07175	12 08 24.	+ 40 02	126	15.3	GALAXY S-IRR
ZC 1208.4+4651	12 08 24.	+ 46 51	1080		CLUSTER OF GALAXIES
UGC 07176	12 08 24.	+ 50 36	96	17.	GALAXY DWARF
AMES 0128	12 08 25.	+ 14 04 47.	20	17.2	NEBULA
ARC 1483	12 08 26.	+ 35 34		17.6	RICH CLUSTER OF GALAXIES
REIZ 2024	12 08 26.	+ 39 57	48	15.5	GALAXY
REIZ 2023	12 08 27.	+ 14 37	18	14.8	GALAXY
REIZ 2022	12 08 28.	+ 01 15	18	14.6	GALAXY
SEY 169	12 08 29.	+ 43 31 00.		15.1	FAINT GALAXY
REIN 3.008A	12 08 29.83	+ 12 22 58.2			NEBULA
REIN 3.008B	12 08 29.84	+ 12 22 56.6			NEBULA
ZWG 013.096	12 08 30.	+ 01 15		15.1	GALAXY
UGC 07177	12 08 30.	+ 01 15	60	15.1	GALAXY VY CMPT
UGC 07178	12 08 30.	+ 02 18	90	17.	GALAXY DWRF IR
ZWG 069.073	12 08 30.	+ 12 22		14.6	GALAXY
MCG+02-31-042	12 08 30.	+ 12 22	36	14.6	GALAXY
IC 0767	12 08 30.	+ 12 22 57.			NONSTELLAR OBJECT
MCG+03-31-059	12 08 30.	+ 14 32	66	15.3	GALAXY
ZC 1208.5+3536	12 08 30.	+ 35 36	1140		CLUSTER OF GALAXIES
ZWG 215.049	12 08 30.	+ 43 31		15.6	GALAXY
MCG+07-25-047	12 08 30.	+ 43 31 30.	36	14.5	GALAXY
MCG+09-20-106	12 08 30.	+ 50 44	360	12.0	GALAXY
MCG+09-20-105	12 08 30.	+ 55 27	36	17.	GALAXY
MCG+11-15-033	12 08 30.	+ 64 11	39	14.	GALAXY
ZWG 315.022	12 08 30.	+ 64 12		14.0	GALAXY
UGC 07179	12 08 30.	+ 64 12	114	14.0	GALAXY S
ZWG 315.023	12 08 30.	+ 67 12		15.0	GALAXY
UGC 07180	12 08 30.	+ 67 12	66	15.0	GALAXY S
AMES 0129	12 08 31.	+ 14 21 29.	34	17.4	NEBULA
HOLM 342B	12 08 31.	+ 40 02	48	14.4	PART OF MULTIPLE GALAXY
REIZ 2025	12 08 32.	+ 31 55	42	14.5	GALAXY
HN 0813	12 08 33.	+ 13 35			NEBULA
HN 0812	12 08 33.	+ 14 25		14.	NEBULA
IC 3031	12 08 34.	+ 13 35			NONSTELLAR OBJECT
IC 3030	12 08 34.	+ 14 25			NONSTELLAR OBJECT
AMES 0130	12 08 35.	+ 11 22 23.	35	18.0	NEBULA
ZWG 041.046	12 08 36.	+ 02 32		15.3	GALAXY
KARA.73B 0521	12 08 36.	+ 02 32	36	15.3	ISOLATED GALAXY S0
ZWG 069.074	12 08 36.	+ 13 52		15.1	GALAXY
UGC 07181	12 08 36.	+ 13 52	72	15.1	GALAXY

OBJECT NAME	RIGHT ASCEN.	DECLINATION	DIAM.	MAGN.	TYPE OF OBJECT
MCG+02-31-043	12 08 36.	+ 13 52	60	15.1	GALAXY
ZWG 098.083	12 08 36.	+ 14 33		15.3	GALAXY
ZWG 098.084	12 08 36.	+ 20 27		13.1	GALAXY
UGC 07182	12 08 36.	+ 20 27	102	13.1	GALAXY S
MCG+03-31-060	12 08 36.	+ 20 29	138	13.1	GALAXY
ZWG 269.038	12 08 36.	+ 50 47		11.9	GALAXY
UGC 07183	12 08 36.	+ 50 47	462	11.9	GALAXY Sb
MCG+09-20-107	12 08 36.	+ 54 59	48	16.	GALAXY
ZC 1208.6+5500	12 08 36.	+ 55 00	670		CLUSTER OF GALAXIES
MCG+10-17-156	12 08 36.	+ 59 22	18	16.	GALAXY
REIZ 2031	12 08 36.	+ 59 38	18	15.3	GALAXY
MCG+11-15-034	12 08 36.	+ 67 12	57	16.	GALAXY
ZC 1208.6+7221	12 08 36.	+ 72 21	2420		CLUSTER OF GALAXIES
ZC 1208.6-0337	12 08 36.	- 03 37	940		CLUSTER OF GALAXIES
MCG-05-29-029	12 08 36.	- 30 52	60	14.	GALAXY
REIZ 2028	12 08 37.	+ 50 46	252	12.5	GALAXY
AMES 0131	12 08 38.	+ 11 43 11.	24	17.2	NEBULA
RNGC 4158	12 08 38.	+ 20 27		13.0	GALAXY
REIZ 2030	12 08 38.	+ 32 52	48	15.0	GALAXY
SN 1937A	12 08 38.	+ 50 48		16.2	SUPERNOVA
HN 0815	12 08 39.	+ 13 51	30	13.5	NEBULA
REIZ 2027	12 08 39.	+ 13 52	60	14.6	GALAXY
HN 0814	12 08 39.	+ 14 33		13.5	NEBULA
REIZ 2029	12 08 39.	+ 20 27	24	13.9	GALAXY
RNGC 4157	12 08 39.	+ 50 46		12.0	GALAXY
LB 02245	12 08 39.	+ 61 38 12.		15.6	FAINT BLUE STAR
AMES 0132	12 08 40.	+ 13 50 35.	32	16.8	NEBULA
IC 3033	12 08 40.	+ 13 51			NONSTELLAR OBJECT
IC 3032	12 08 40.	+ 14 33			NONSTELLAR OBJECT
SN 1955A	12 08 40.	+ 50 48		16.0	SUPERNOVA
REIZ 2026	12 08 40.	- 06 08	24	15.8	GALAXY
AMES 0133	12 08 41.	+ 13 27 53.	24	17.8	NEBULA
ARC 1484	12 08 41.	+ 72 22		17.4	RICH CLUSTER OF GALAXIES
ZWG 098.085	12 08 42.	+ 18 10		14.7	GALAXY
MCG+03-31-061	12 08 42.	+ 18 11	60	14.7	GALAXY
MCG+06-27-022	12 08 42.	+ 32 52	30	15.	GALAXY
ZWG 292.077	12 08 42.	+ 57 45		15.6	GALAXY
MCG+13-09-015	12 08 42.	+ 76 24	66	14.	GALAXY
MCG-01-31-007	12 08 42.	- 06 10	42	16.	GALAXY
MCG+03-31-062	12 08 45.	+ 18 18	132	15.6	GALAXY
MCG+07-25-048	12 08 45.	+ 39 40	60	16.	GALAXY
LB 02246	12 08 45.	+ 50 41 42.		16.3	FAINT BLUE STAR
REIN 6.094	12 08 45.76	+ 39 40 55.0			NEBULA
REIZ 2033	12 08 46.	+ 01 46	48	15.0	GALAXY
AMES 0134	12 08 46.	+ 13 23 53.	43	17.4	NEBULA
LB 02247	12 08 46.	+ 59 39 06.		15.7	FAINT BLUE STAR
REIZ 2032	12 08 46.	- 03 13	24	14.6	GALAXY
AMES 0135	12 08 47.	+ 10 50 35.	23	17.2	NEBULA
ZWG 013.097	12 08 48.	+ 01 20		15.6	GALAXY
ZWG 013.098	12 08 48.	+ 01 46		14.9	GALAXY
UGC 07184	12 08 48.	+ 01 46	102	14.9	GALAXY SBc
MCG+00-31-037	12 08 48.	+ 01 46	90	14.9	GALAXY
ZWG 041.047	12 08 48.	+ 03 12		15.5	GALAXY
UGC 07185	12 08 48.	+ 03 12	78	15.5	GALAXY DWARF
REA 76	12 08 48.	+ 03 12			DWARF GALAXY
MCG+01-31-024	12 08 48.	+ 03 12	72	15.5	GALAXY
ZWG 098.086	12 08 48.	+ 18 18		15.6	GALAXY
UGC 07186	12 08 48.	+ 18 18	114	15.6	GALAXY
ZWG 098.087	12 08 48.	+ 19 22		15.3	GALAXY
MCG+03-31-063	12 08 48.	+ 19 22 30.	60	15.3	GALAXY
ZC 1208.8+1958	12 08 48.	+ 19 58	2080		CLUSTER OF GALAXIES
MCG+05-29-030	12 08 48.	+ 29 22	30	15.1	GALAXY
ZWG 187.018	12 08 48.	+ 36 07		15.0	GALAXY
UGC 07187	12 08 48.	+ 36 07	60	15.0	GALAXY SBa
MCG+06-27-024	12 08 48.	+ 36 08	48	16.	GALAXY
MCG+06-27-023	12 08 48.	+ 36 08	48	14.5	GALAXY
UGC 07188	12 08 48.	+ 39 41	66	16.5	GALAXY
LB 02248	12 08 48.	+ 56 30 06.		16.2	FAINT BLUE STAR
MCG+10-17-157	12 08 48.	+ 57 45	30	16.	GALAXY
ZWG 352.023	12 08 48.	+ 75 06		14.8	GALAXY
UGC 07189	12 08 48.	+ 75 06	120	14.8	GALAXY
AMES 0136	12 08 49.	+ 12 20 41.	19	17.0	NEBULA
AMES 0139	12 08 50.	+ 13 28 53.	28	17.5	NEBULA
AMES 0138	12 08 50.	+ 13 39 05.	20	17.5	NEBULA
AMES 0137	12 08 50.	+ 13 40 53.	29	17.6	NEBULA
AMES 0140	12 08 52.	+ 06 23 29.	28	17.2	NEBULA
ARC 1485	12 08 52.	- 03 37		17.6	RICH CLUSTER OF GALAXIES
ZC 1208.9+0158	12 08 54.	+ 01 58	1340		CLUSTER OF GALAXIES
ZC 1208.9+1332	12 08 54.	+ 13 32	1210		CLUSTER OF GALAXIES
ZC 1208.9+2221	12 08 54.	+ 22 21	870		CLUSTER OF GALAXIES
ZC 1208.9+4721	12 08 54.	+ 47 21	1140		CLUSTER OF GALAXIES
AMES 0141	12 08 56.	+ 13 57 17.	31	17.0	NEBULA
LB 02249	12 08 56.	+ 59 12 06.		16.0	FAINT BLUE STAR
AMES 0142	12 08 57.	+ 13 30 29.	18	17.0	NEBULA
MCG+08-22-082	12 08 57.	+ 50 33	18	16.	GALAXY
RNGC 4161	12 08 57.	+ 58 00		13.5	GALAXY
AMES 0143	12 08 58.	+ 12 26 29.	35	17.6	NEBULA
AMES 0144	12 08 59.	+ 07 56 05.	26	16.9	NEBULA
MCG+03-31-064	12 09 00.	+ 16 45	24	14.8	GALAXY
ZWG 098.088	12 09 00.	+ 20 10		15.7	GALAXY
ZC 1209.0+2722	12 09 00.	+ 27 22	2690		CLUSTER OF GALAXIES
ZWG 158.039	12 09 00.	+ 29 22		15.1	GALAXY
UGC 07190	12 09 00.	+ 29 22	72	15.1	GALAXY S0?
ZC 1209.0+3857	12 09 00.	+ 38 57	870		CLUSTER OF GALAXIES
MCG+09-20-108	12 09 00.	+ 52 07 30.	60	15.	GALAXY
LB 02250	12 09 00.	+ 56 39 24.		16.9	FAINT BLUE STAR
ZWG 292.078	12 09 00.	+ 58 00		13.7	GALAXY
UGC 07191	12 09 00.	+ 58 00	72	13.7	GALAXY S
ZWG 293.001	12 09 00.	+ 59 01		15.4	GALAXY
ZWG 292.079	12 09 00.	+ 59 01		15.4	GALAXY
ZC 1209.0+6027	12 09 00.	+ 60 27	610		CLUSTER OF GALAXIES
ZC 1209.0+6243	12 09 00.	+ 62 43	740		CLUSTER OF GALAXIES
MCG+13-09-016	12 09 00.	+ 75 05	84	13.	GALAXY
LB 02251	12 09 01.	+ 58 08 48.		16.1	FAINT BLUE STAR
RNGC 4160	12 09 03.	+ 44 01			NON-EXISTENT OBJECT
ZC 1209.1+0842	12 09 06.	+ 08 42	740		CLUSTER OF GALAXIES
ZWG 098.089	12 09 06.	+ 16 46		14.8	GALAXY
ZWG 098.090	12 09 06.	+ 20 07		15.7	GALAXY
ZC 1209.1+3016	12 09 06.	+ 30 16	1010		CLUSTER OF GALAXIES
LB 00626	12 09 06.	+ 55 49 36.		17.6	FAINT BLUE STAR
LB 02252	12 09 09.	+ 58 40 48.		16.3	FAINT BLUE STAR
AMES 0145	12 09 10.	+ 11 44 05.	28	18.1	NEBULA
AMES 0146	12 09 12.	+ 12 04 53.	34	17.0	NEBULA
ZC 1209.2+1810	12 09 12.	+ 18 10	740		CLUSTER OF GALAXIES
ZWG 098.091	12 09 12.	+ 19 46		15.6	GALAXY
REIZ 2035	12 09 12.	+ 58 03	66	13.8	GALAXY
MCG+10-18-001	12 09 12.	+ 59 01	36	16.	GALAXY
MCG-06-27-007	12 09 12.	- 38 15	102	14.5	GALAXY
AMES 0147	12 09 14.	+ 10 20 05.	17	16.8	NEBULA
IC 0768	12 09 14.	+ 12 25 21.			NONSTELLAR OBJECT
REIN 3.009A	12 09 14.79	+ 12 25 19.9			NEBULA
REIN 3.009B	12 09 14.85	+ 12 25 19.9			NEBULA
AMES 0148	12 09 15.	+ 10 44 53.	19	17.2	NEBULA
REIZ 2034	12 09 15.	+ 12 25	60	14.4	GALAXY
MCG+04-29-046	12 09 15.	+ 24 23	138	12.6	GALAXY
AMES 0149	12 09 16.	+ 13 40 05.	28	16.6	NEBULA
AMES 0150	12 09 17.	+ 13 57 17.	32	17.4	NEBULA
ZWG 069.075	12 09 18.	+ 12 25		14.8	GALAXY
UGC 07192	12 09 18.	+ 12 25	96	14.8	GALAXY Sc
MCG+02-31-044	12 09 18.	+ 12 25	90	14.8	GALAXY
ZC 1209.3+1357	12 09 18.	+ 13 57	1340		CLUSTER OF GALAXIES
MCG+03-31-066	12 09 18.	+ 15 40	90	15.2	GALAXY
MCG+03-31-065	12 09 18.	+ 16 30	96	15.2	GALAXY
ZWG 128.051	12 09 18.	+ 24 24		12.6	GALAXY
UGC 07193	12 09 18.	+ 24 24	150	12.6	GALAXY Sc
MCG+05-29-031	12 09 18.	+ 27 55	36	15.3	GALAXY
KW 55	12 09 18.	+ 39 33	450		SEYFERT GALAXY
ZC 1209.3+4039	12 09 18.	+ 40 39	940		CLUSTER OF GALAXIES
LB 02253	12 09 18.	+ 55 29 06.		16.8	FAINT BLUE STAR
MCG+10-18-002	12 09 18.	+ 58 01	60	13.8	GALAXY
MCG+10-18-003	12 09 18.	+ 58 32	24	16.	GALAXY
MCG+10-18-004	12 09 18.	+ 60 20	39	16.	GALAXY
ZC 1209.3+6427	12 09 18.	+ 64 27	540		CLUSTER OF GALAXIES
ZC 1209.3+6654	12 09 18.	+ 66 54	1410		CLUSTER OF GALAXIES
ZWG 013.099	12 09 18.	- 02 11		15.5	GALAXY
AMES 0151	12 09 19.	+ 09 52 17.	35	18.0	NEBULA
SN 1965G	12 09 19.	+ 24 24		14.0	SUPERNOVA
AMES 0152	12 09 20.	+ 10 43 47.	48	17.8	NEBULA
RNGC 4162	12 09 20.	+ 24 24		13.0	GALAXY
ARC 1487	12 09 20.	+ 30 16		17.8	RICH CLUSTER OF GALAXIES
ARC 1486	12 09 20.	+ 30 51		17.2	RICH CLUSTER OF GALAXIES
AMES 0153	12 09 21.	+ 06 57 05.	25	17.2	NEBULA
HN 0816	12 09 21.	+ 14 28			NEBULA
AMES 0154	12 09 22.	+ 13 42 41.	22	17.3	NEBULA
IC 3034	12 09 22.	+ 14 28			NONSTELLAR OBJECT
AMES 0155	12 09 23.	+ 13 20 35.	35	16.9	NEBULA
ZWG 013.100	12 09 24.	+ 01 38		15.4	GALAXY
MCG+03-31-067	12 09 24.	+ 15 11	48	15.7	GALAXY
ZWG 098.092	12 09 24.	+ 16 30		15.2	GALAXY
UGC 07194	12 09 24.	+ 16 30	66	15.2	GALAXY Sc
ZWG 098.093	12 09 24.	+ 20 18		15.3	GALAXY
ZWG 158.040	12 09 24.	+ 27 55		15.3	GALAXY
ZC 1209.4+3813	12 09 24.	+ 38 13	670		CLUSTER OF GALAXIES
ZWG 215.050	12 09 24.	+ 40 56		15.0	GALAXY
MRK 435	12 09 24.	+ 40 56	14	15.	GALAXY WITH UV CONTINUUM
ZC 1209.4+4556	12 09 24.	+ 45 56	3360		CLUSTER OF GALAXIES
MCG+08-22-083	12 09 24.	+ 47 16	54	17.	GALAXY
MRK 200	12 09 24.	+ 48 48	8	17.	GALAXY WITH UV CONTINUUM
ZWG 315.024	12 09 24.	+ 68 12		15.4	GALAXY
UGC 07195	12 09 24.	+ 68 12	60	15.4	GALAXY SB
MCG+13-09-017	12 09 24.	+ 75 15	21	16.	GALAXY
AMES 0156	12 09 25.	+ 06 45 53.	29	17.3	NEBULA
REIZ 2039	12 09 26.	+ 33 01	36	15.2	GALAXY
LB 02254	12 09 26.	+ 51 09 48.		16.0	FAINT BLUE STAR
REIZ 2037	12 09 27.	+ 15 40	54	14.1	GALAXY
REIZ 2038	12 09 27.	+ 24 25	120	12.9	GALAXY
REIZ 2036	12 09 28.	- 04 36	60	15.0	GALAXY
ZWG 069.076	12 09 30.	+ 13 28		15.7	GALAXY
ZWG 098.094	12 09 30.	+ 15 11		15.7	GALAXY
ZWG 098.095	12 09 30.	+ 15 41		15.2	GALAXY
UGC 07196	12 09 30.	+ 15 41	90	15.2	GALAXY S
ZWG 187.019	12 09 30.	+ 33 00		15.6	GALAXY
MCG+06-27-025	12 09 30.	+ 33 00	36	15.	GALAXY
UGC 07197	12 09 30.	+ 61 37	60	16.0	GALAXY S
ZWG 013.101	12 09 30.	- 00 19		15.6	GALAXY
MCG-01-31-008	12 09 30.	- 04 38	60	15.	GALAXY
AMES 0157	12 09 31.	+ 06 48 17.	28	17.6	NEBULA
AMES 0158	12 09 32.	+ 11 51 17.	30	18.1	NEBULA
RNGC 4164	12 09 32.	+ 13 28		15.5	GALAXY
AMES 0159	12 09 32.	+ 13 50 35.	20	17.3	NEBULA
REIN 3.010	12 09 32.75	+ 13 29 01.3			NEBULA
REIZ 2040	12 09 33.	+ 13 30	60	15.1	GALAXY
REIN 3.011A	12 09 33.90	+ 13 28 41.6			NEBULA
REIN 3.011B	12 09 33.96	+ 13 28 41.9			NEBULA
AMES 0160	12 09 35.	+ 13 20 53.	18	17.6	NEBULA
ZWG 098.096	12 09 36.	+ 18 02		14.3	GALAXY
UGC 07198	12 09 36.	+ 18 02	66	14.3	GALAXY SB0
MCG+03-31-068	12 09 36.	+ 18 03	66	14.3	GALAXY
MCG+05-29-032	12 09 36.	+ 29 28	42	12.9	GALAXY
ZC 1209.6+3049	12 09 36.	+ 30 49	2080		CLUSTER OF GALAXIES
ZWG 187.020	12 09 36.	+ 36 27		13.7	GALAXY
UGC 07199	12 09 36.	+ 36 27	132	13.7	GALAXY IRR
MCG+06-27-026	12 09 36.	+ 36 28	102	13.	GALAXY
MCG+10-18-005	12 09 36.	+ 61 36	54	16.	GALAXY
MCG+11-15-035	12 09 36.	+ 68 12	54	16.	GALAXY
AMES 0161	12 09 37.	+ 13 20 17.	22	16.9	NEBULA
LB 02255	12 09 37.	+ 51 54 12.		17.4	FAINT BLUE STAR
AMES 0163	12 09 38.	+ 07 09 41.	16	17.8	NEBULA
AMES 0162	12 09 38.	+ 13 09 53.	37	17.6	NEBULA
RNGC 4165	12 09 38.	+ 13 31		14.5	GALAXY
RNGC 4166	12 09 38.	+ 18 02		14.5	GALAXY
HOLM 346A	12 09 38.	+ 29 30	210	13.5	PART OF MULTIPLE GALAXY
RNGC 4163	12 09 38.	+ 36 27		13.5	GALAXY
RNGC 4167	12 09 38.	+ 36 47			NON-EXISTENT OBJECT
LB 02256	12 09 38.	+ 50 25 30.		16.7	FAINT BLUE STAR
AMES 0164	12 09 39.	+ 13 46 53.	19	17.6	NEBULA
REIZ 2042	12 09 39.	+ 18 01	36	12.7	GALAXY
REIN 3.012A	12 09 39.00	+ 13 31 26.7			NEBULA
REIN 3.012B	12 09 39.12	+ 13 31 28.1			NEBULA
AMES 0166	12 09 40.	+ 05 49 35.	28	17.6	NEBULA
AMES 0165	12 09 40.	+ 13 24 05.	16	17.7	NEBULA
IC 3035	12 09 40.	+ 13 31 33.			SAME AS NGC 4165
REIZ 2041	12 09 40.	- 06 07	60	15.3	GALAXY
AMES 0168	12 09 41.	+ 12 52 41.	17	16.9	NEBULA
AMES 0167	12 09 41.	+ 13 43 17.	25	17.2	NEBULA
ZWG 069.077	12 09 42.	+ 12 45		15.3	GALAXY
UGC 07200	12 09 42.	+ 12 45	108	15.3	GALAXY
SN 1971G	12 09 42.	+ 13 30		13.6	SUPERNOVA
ZWG 069.078	12 09 42.	+ 13 31		14.7	GALAXY
UGC 07201	12 09 42.	+ 13 31	78	14.8	GALAXY Sb
MCG+02-31-045	12 09 42.	+ 13 31	72	14.8	GALAXY
AMES 0169	12 09 42.	+ 13 46 29.	23	17.5	NEBULA
VV 147E	12 09 42.	+ 18 23	6	18.5	INTERACTING GALAXY
VV 147D	12 09 42.	+ 18 23	6	18.5	INTERACTING GALAXY
VV 147C	12 09 42.	+ 18 23	6	18.5	INTERACTING GALAXY
VV 147B	12 09 42.	+ 18 23	12	17.	INTERACTING GALAXY

OBJECT NAME	RIGHT ASCEN.	DECLINATION	DIAM.	MAGN.	TYPE OF OBJECT
VV 147A	12 09 42.	+ 18 23	24	17.	INTERACTING GALAXY
VV 147	12 09 42.	+ 18 23	48		INTERACTING GALAXY
REA 04	12 09 42.	+ 18 23			DWARF GALAXY
MCG+03-31-069	12 09 42.	+ 18 23	48	17.	GALAXY
MCG+05-29-035	12 09 42.	+ 29 06	30	14.8	GALAXY
ZWG 158.041	12 09 42.	+ 29 27		12.9	GALAXY
UGC 07202	12 09 42.	+ 29 27	120	12.9	GALAXY S0
MCG+05-29-034	12 09 42.	+ 29 27	36	14.3	GALAXY
MCG+05-29-033	12 09 42.	+ 29 30 30.	270	13.7	GALAXY
MCG+06-27-027	12 09 42.	+ 37 19	120	16.	GALAXY
ZC 1209.7+4429	12 09 42.	+ 44 29	3830		CLUSTER OF GALAXIES
ZC 1209.7+4758	12 09 42.	+ 47 58	1410		CLUSTER OF GALAXIES
REIZ 2044	12 09 42.	+ 56 27	72	13.4	GALAXY
MCG+09-20-109	12 09 42.	+ 56 27 30.	72	13.4	GALAXY
MCG-01-31-009	12 09 42.	- 06 09	60	15.5	GALAXY
MCG-04-29-012	12 09 42.	- 26 52	36	15.5	GALAXY
REIN 3.013	12 09 42.00	+ 12 46 03.1			NEBULA
AMES 0170	12 09 44.	+ 13 23 53.	14	17.5	NEBULA
RNGC 4168	12 09 44.	+ 13 29		13.0	GALAXY
RNGC 4169	12 09 44.	+ 29 27		13.0	GALAXY
REIN 3.014A	12 09 44.47	+ 13 28 59.2			NEBULA
REIN 3.014B	12 09 44.51	+ 13 29 00.1			NEBULA
HN 0817	12 09 45.	+ 12 45			NEBULA
REIZ 2043	12 09 45.	+ 13 29	48	13.1	GALAXY
RNGC 4172	12 09 45.	+ 56 27		14.5	GALAXY
IC 3036	12 09 46.	+ 12 45			NONSTELLAR OBJECT
LB 02257	12 09 46.	+ 60 51 42.		15.4	FAINT BLUE STAR
REIN 4.073	12 09 47.17	+ 10 15 52.2			NEBULA
REA 77	12 09 48.	+ 03 05			DWARF GALAXY
ZC 1209.8+0649	12 09 48.	+ 06 49	1680		CLUSTER OF GALAXIES
ZWG 069.079	12 09 48.	+ 09 03		15.6	GALAXY
ZWG 069.080	12 09 48.	+ 10 16		15.5	GALAXY
ZWG 069.081	12 09 48.	+ 13 29		12.7	GALAXY
UGC 07203	12 09 48.	+ 13 29	180	12.7	GALAXY E
MCG+02-31-046	12 09 48.	+ 13 29	42	12.7	GALAXY
MCG+03-31-071	12 09 48.	+ 15 45	60	17.	GALAXY
MCG+03-31-070	12 09 48.	+ 17 15	24	16.5	GALAXY
ZWG 158.042	12 09 48.	+ 29 05		14.8	GALAXY
MCG+05-29-036	12 09 48.	+ 29 27 30.	90	14.2	GALAXY
ZWG 158.043	12 09 48.	+ 29 29		13.7	GALAXY
UGC 07204	12 09 48.	+ 29 29	306	13.7	GALAXY
LB 02258	12 09 48.	+ 50 25 48.		17.0	FAINT BLUE STAR
ZWG 292.080	12 09 48.	+ 56 27		14.4	GALAXY
ZWG 269.039	12 09 48.	+ 56 27		14.4	GALAXY
UGC 07205	12 09 48.	+ 56 27	84	14.4	GALAXY S
LB 02259	12 09 48.	+ 57 04 54.		16.2	FAINT BLUE STAR
ZC 1209.8+5920	12 09 48.	+ 59 20	3900		CLUSTER OF GALAXIES
AMES 0171	12 09 49.	+ 13 32 17.	25	17.7	NEBULA
HOLM 346B	12 09 49.	+ 29 28	78	14.0	PART OF MULTIPLE GALAXY
LB 02260	12 09 49.	+ 57 04 42.		16.8	FAINT BLUE STAR
AMES 0173	12 09 50.	+ 08 54 23.	14	17.0	NEBULA
AMES 0172	12 09 50.	+ 09 03 11.	38	16.8	NEBULA
RNGC 4173	12 09 50.	+ 29 27			NON-EXISTENT OBJECT
RNGC 4171	12 09 50.	+ 29 28			NON-EXISTENT OBJECT
RNGC 4170	12 09 50.	+ 29 29		13.5	GALAXY
AMES 0175	12 09 51.	+ 07 00 47.	30	17.7	NEBULA
HN 0818	12 09 51.	+ 10 16		16.	NEBULA
AMES 0174	12 09 51.	+ 12 00 05.	25	17.9	NEBULA
LB 02261	12 09 51.	+ 57 04 36.		17.0	FAINT BLUE STAR
AMES 0177	12 09 52.	+ 06 55 47.	24	17.7	NEBULA
IC 3037	12 09 52.	+ 10 16			NONSTELLAR OBJECT
AMES 0176	12 09 52.	+ 12 14 47.	28	17.9	NEBULA
LB 02262	12 09 52.	+ 62 17 30.		16.7	FAINT BLUE STAR
REIZ 2045	12 09 52.	- 06 12	36	16.3	GALAXY
ARC 1488	12 09 53.	- 11 33		17.8	RICH CLUSTER OF GALAXIES
MCG+03-31-072	12 09 54.	+ 15 32	48	15.2	GALAXY
ZWG 098.097	12 09 54.	+ 15 40		15.4	GALAXY
ZWG 158.044	12 09 54.	+ 29 25		14.3	GALAXY
UGC 07206	12 09 54.	+ 29 25	48	14.3	GALAXY
UGC 07207	12 09 54.	+ 37 17	150	17.	GALAXY DWRF IR
MCG+07-25-049	12 09 54.	+ 39 22	72	14.	GALAXY
IZW 032	12 09 54.	+ 39 23			COMPACT GALAXY
ZWG 215.051	12 09 54.	+ 39 23		14.6	GALAXY
UGC 07208	12 09 54.	+ 39 23	108	14.6	GALAXY
ZWG 243.050	12 09 54.	+ 45 58		15.5	GALAXY
MCG+09-20-110	12 09 54.	+ 53 51	60	17.	GALAXY
REIZ 2049	12 09 54.	+ 57 59	36	14.3	GALAXY
ZC 1209.9+7440	12 09 54.	+ 74 40	1950		CLUSTER OF GALAXIES
AMES 0178	12 09 55.	+ 08 46 29.	29	17.2	NEBULA
LB 02263	12 09 55.	+ 60 46 54.		15.3	FAINT BLUE STAR
RNGC 4174	12 09 56.	+ 29 25		14.5	GALAXY
REIZ 2047	12 09 56.	+ 29 27	180	13.1	GALAXY
REIZ 2048	12 09 56.	+ 39 22	12	14.1	GALAXY
REIN 6.095	12 09 56.97	+ 39 23 20.5			NEBULA
HN 0819	12 09 57.	+ 11 38		16.	NEBULA
REIZ 2046	12 09 57.	+ 12 23	108	14.2	GALAXY
TON-N 1472	12 09 57.	+ 33 41		16.9	BLUE STAR
TON-N 1473	12 09 57.	+ 35 27		15.2	BLUE STAR
AMES 0179	12 09 58.	+ 06 21 11.	36	17.6	NEBULA
IC 3038	12 09 58.	+ 11 38			NONSTELLAR OBJECT
LB 02264	12 09 58.	+ 52 41 00.		16.4	FAINT BLUE STAR
AMES 0180	12 09 59.	+ 13 38 17.	23	17.6	NEBULA
REIN 3.015A	12 09 59.31	+ 12 24 07.3			NEBULA
REIN 3.015B	12 09 59.31	+ 12 24 08.6			NEBULA
REIN 3.016	12 09 59.66	+ 12 35 20.4			NEBULA
AMES 0181	12 10 00.	+ 11 26 29.	24	17.8	NEBULA
ZWG 069.082	12 10 00.	+ 11 37		15.7	GALAXY
ZWG 069.083	12 10 00.	+ 12 24		14.1	GALAXY
UGC 07209	12 10 00.	+ 12 24	150	14.1	GALAXY Sb/Sc
MCG+02-31-047	12 10 00.	+ 12 24	168	14.1	GALAXY
ZWG 069.084	12 10 00.	+ 12 35		15.2	GALAXY
MCG+02-31-048	12 10 00.	+ 12 35	60	15.2	GALAXY
ZWG 098.098	12 10 00.	+ 15 33		15.2	GALAXY
UGC 07210	12 10 00.	+ 15 33	60	15.2	GALAXY S
ZC 1210.0+1550	12 10 00.	+ 15 50	1480		CLUSTER OF GALAXIES
ZWG 098.099	12 10 00.	+ 16 42		15.7	GALAXY
MCG+03-31-073	12 10 00.	+ 16 42	30	15.7	GALAXY
ZWG 128.052	12 10 00.	+ 23 04		15.6	GALAXY
ZWG 158.045	12 10 00.	+ 29 26		14.2	GALAXY
UGC 07211	12 10 00.	+ 29 26	126	14.2	GALAXY S
MCG+06-27-029	12 10 00.	+ 34 58	66	15.	GALAXY
MCG+06-27-028	12 10 00.	+ 34 59	30	15.	GALAXY
ZC 1210.0+4116	12 10 00.	+ 41 16	870		CLUSTER OF GALAXIES
ZWG 013.102	12 10 00.	- 02 27		15.5	GALAXY
MCG-05-29-030	12 10 00.	- 32 20	60	15.5	GALAXY
REIZ 2053	12 10 01.	+ 40 09	12	15.7	GALAXY
ARC 1490	12 10 02.	+ 18 54		17.5	RICH CLUSTER OF GALAXIES
ARC 1489	12 10 02.	+ 27 46		17.8	RICH CLUSTER OF GALAXIES

OBJECT NAME	RIGHT ASCEN.	DECLINATION	DIAM.	MAGN.	TYPE OF OBJECT
RNGC 4175	12 10 02.	+ 29 26		14.0	GALAXY
REIZ 2052	12 10 02.	+ 29 26	72	13.8	GALAXY
HN 0821	12 10 03.	+ 11 21		16.	NEBULA
HN 0820	12 10 03.	+ 12 35	18		NEBULA
REIZ 2051	12 10 03.	+ 12 35	30	15.0	GALAXY
AMES 0184	12 10 04.	+ 07 08 17.	24	17.6	NEBULA
AMES 0183	12 10 04.	+ 07 57 41.	24	17.6	NEBULA
AMES 0182	12 10 04.	+ 10 18 35.	19	16.8	NEBULA
IC 3040	12 10 04.	+ 11 21			NONSTELLAR OBJECT
IC 3039	12 10 04.	+ 12 35			NONSTELLAR OBJECT
REIZ 2050	12 10 04.	- 06 14	18	16.0	GALAXY
ZWG 069.085	12 10 06.	+ 10 18		15.6	GALAXY
ZWG 069.086	12 10 06.	+ 11 21		15.6	GALAXY
MCG+02-31-049	12 10 06.	+ 11 21	36	15.6	GALAXY
IC 0769	12 10 06.	+ 12 25			DWARF SPIRAL GALAXY
LB 00001	12 10 06.	+ 26 54		13.3	FAINT BLUE STAR
ZWG 187.021	12 10 06.	+ 34 59		15.4	GALAXY
UGC 07212	12 10 06.	+ 34 59	78	15.4	GALAXY Sc
ZWG 215.052	12 10 06.	+ 40 50		15.5	GALAXY
UGC 07213	12 10 06.	+ 40 50	84	15.5	GALAXY SBc
MCG+07-25-050	12 10 06.	+ 40 50	63	14.	GALAXY
ZC 1210.1+4945	12 10 06.	+ 49 45	940		CLUSTER OF GALAXIES
ZWG 293.002	12 10 06.	+ 58 43		15.7	GALAXY
ZWG 292.081	12 10 06.	+ 58 43		15.7	GALAXY
AMES 0185	12 10 07.	+ 07 57 59.	16	17.8	NEBULA
RNGC 4176	12 10 07.	- 08 52			GALAXY
AMES 0186	12 10 09.	+ 11 50 17.	23	16.8	NEBULA
HN 0822	12 10 09.	+ 13 01	18		NEBULA
REIZ 2054	12 10 09.	+ 14 15	96	14.3	GALAXY
LB 02265	12 10 09.	+ 58 58 24.		15.8	FAINT BLUE STAR
REIN 3.017	12 10 09.42	+ 11 07 13.8			NEBULA
REIN 3.018	12 10 09.44	+ 11 07 54.1			NEBULA
IC 3041	12 10 10.	+ 13 01			NONSTELLAR OBJECT
LB 02266	12 10 11.	+ 55 51 42.		17.0	FAINT BLUE STAR
ZC 1210.2+0832	12 10 12.	+ 08 32	1010		CLUSTER OF GALAXIES
ZWG 069.087	12 10 12.	+ 11 50		15.6	GALAXY
ZC 1210.2+2055	12 10 12.	+ 20 55	2890		CLUSTER OF GALAXIES
VV 345B	12 10 12.	+ 35 59	78	15.	INTERACTING GALAXY
VV 345A	12 10 12.	+ 36 00	36	15.	INTERACTING GALAXY
HZ 20	12 10 12.	+ 42 56		14.9	DECIDEDLY BLUE STAR
MCG+10-18-006	12 10 12.	+ 58 43	39	15.	GALAXY
ZWG 293.003	12 10 12.	+ 62 14		15.7	GALAXY
ZWG 292.082	12 10 12.	+ 62 14		15.7	GALAXY
ZWG 013.103	12 10 12.	- 02 27		15.6	GALAXY
MCG-02-31-021	12 10 12.	- 13 44	96	13.	GALAXY
AMES 0187	12 10 13.	+ 07 25 11.	25	17.8	NEBULA
IC 3042	12 10 13.	+ 11 08 33.			SAME AS NGC 4178
RNGC 4178	12 10 13.	+ 11 09		12.0	GALAXY
RNGC 4177	12 10 13.	- 13 44		13.0	GALAXY
REIN 3.019	12 10 13.33	+ 11 08 38.8			NEBULA
AMES 0188	12 10 14.	+ 13 37 05.	26	17.8	NEBULA
HN 0823	12 10 15.	+ 10 17	30	15.	NEBULA
REIZ 2055	12 10 15.	+ 11 09	300	11.5	GALAXY
MCG+09-20-111	12 10 15.	+ 53 10	18	15.2	GALAXY
IC 3044	12 10 15.5	+ 14 15 16.			GALAXY
AMES 0189	12 10 16.	+ 08 29 29.	38	17.6	NEBULA
IC 3043	12 10 16.	+ 10 17			NONSTELLAR OBJECT
SN 1963I	12 10 17.	+ 11 08		14.3	SUPERNOVA
AMES 0191	12 10 17.	+ 12 39 29.	29	18.0	NEBULA
AMES 0190	12 10 17.	+ 13 31 05.	22	17.0	NEBULA
ZWG 013.104	12 10 18.	+ 01 35		12.8	GALAXY
UGC 07214	12 10 18.	+ 01 35	240	12.8	GALAXY S0
MCG+00-31-038	12 10 18.	+ 01 35	240	12.8	GALAXY
ZC 1210.3+0810	12 10 18.	+ 08 10	1080		CLUSTER OF GALAXIES
ZWG 069.088	12 10 18.	+ 11 08		12.9	GALAXY
SCH 23	12 10 18.	+ 11 08		12.9	PECULIAR GALAXY
UGC 07215	12 10 18.	+ 11 08	342	12.9	GALAXY SBc
MCG+02-31-050	12 10 18.	+ 11 08	276	12.9	GALAXY
ZWG 069.089	12 10 18.	+ 14 15		14.7	GALAXY
UGC 07216	12 10 18.	+ 14 15	132	14.7	GALAXY S
MCG+02-31-051	12 10 18.	+ 14 15	108	14.7	GALAXY
ZWG 243.051	12 10 18.	+ 47 02		14.8	GALAXY
VV 183B	12 10 18.	+ 60 11	8	17.	INTERACTING GALAXY
VV 183A	12 10 18.	+ 60 11	12	16.	INTERACTING GALAXY
VV 183	12 10 18.	+ 60 11	72		INTERACTING GALAXY
REIN 2.151	12 10 18.63	+ 01 34 37.8			NEBULA
RNGC 4179	12 10 19.	+ 01 35		12.0	GALAXY
HZ 19	12 10 19.	+ 34 10		12.8	BLUE STAR
REIZ 2057	12 10 19.	+ 40 04	18	15.6	GALAXY
ARC 1491	12 10 21.	+ 08 10		17.2	RICH CLUSTER OF GALAXIES
MCG-03-31-024	12 10 21.	- 20 09 30.	102	15.	GALAXY
REIZ 2056	12 10 22.	+ 01 34	108	12.3	GALAXY
AMES 0192	12 10 23.	+ 12 18 41.	26	17.4	NEBULA
REA 23	12 10 24.	+ 05 39	42		DWARF GALAXY
ZC 1210.4+0619	12 10 24.	+ 06 19	1550		CLUSTER OF GALAXIES
AMES 0193	12 10 24.	+ 07 34 47.	46	17.2	NEBULA
ZWG 128.053	12 10 24.	+ 25 34		15.6	GALAXY
UGC 07217	12 10 24.	+ 25 34	102	15.6	GALAXY S
LB 00036	12 10 24.	+ 25 58		17.8	FAINT BLUE STAR
ZWG 269.040	12 10 24.	+ 52 33		15.0	GALAXY
UGC 07218	12 10 24.	+ 52 33	84	15.0	GALAXY IRR
ZWG 269.041	12 10 24.	+ 53 11		15.0	GALAXY
ZWG 315.025	12 10 24.	+ 66 05		15.3	GALAXY
OCL 0877	12 10 24.	- 62 21	300	13.	OPEN STAR CLUSTER
VBA 128	12 10 24.	- 62 21	300		OPEN STAR CLUSTER
REIZ 2058	12 10 27.	+ 07 00	24	15.1	GALAXY
AMES 0194	12 10 27.	+ 08 10 47.	24	17.6	NEBULA
AMES 0195	12 10 27.	+ 11 39 05.	11	18.0	NEBULA
IC 3045	12 10 27.	+ 13 03 27.			MAY NOT EXIST
IC 0770	12 10 29.	- 04 16 45.			NONSTELLAR OBJECT
ZWG 041.048	12 10 30.	+ 07 19		13.2	GALAXY
UGC 07219	12 10 30.	+ 07 19	102	13.2	GALAXY S
MCG+01-31-025	12 10 30.	+ 07 19	84	13.2	GALAXY
ZC 1210.5+1220	12 10 30.	+ 12 20	1210		CLUSTER OF GALAXIES
MCG+03-31-074	12 10 30.	+ 16 21	30	15.5	GALAXY
MCG+03-31-075	12 10 30.	+ 16 34	30	14.8	GALAXY
ZWG 098.100	12 10 30.	+ 17 22		15.4	GALAXY
MCG+05-29-037	12 10 30.	+ 29 07 30.	42	15.0	GALAXY
ZC 1210.5+3420	12 10 30.	+ 34 20	1480		CLUSTER OF GALAXIES
MCG+09-20-112	12 10 30.	+ 52 05	30	17.	GALAXY
MCG+09-20-113	12 10 30.	+ 52 32	72	15.	GALAXY
MCG+11-15-036	12 10 30.	+ 66 05	18	15.	GALAXY
AMES 0197	12 10 31.	+ 11 06 29.	37	17.4	NEBULA
AMES 0196	12 10 31.	+ 11 43 35.	20	17.7	NEBULA
ARC 1492	12 10 31.	+ 34 27		17.2	RICH CLUSTER OF GALAXIES
RNGC 1480	12 10 31.	- 07 19			NON-EXISTENT OBJECT
AMES 0198	12 10 32.	+ 05 58 17.	31	17.4	NEBULA
REIZ 2059	12 10 33.	+ 07 19	60	12.9	GALAXY

OBJECT NAME	RIGHT ASCEN.	DECLINATION	DIAM.	MAGN.	TYPE OF OBJECT
AMES 0199	12 10 33.	+ 08 11 17.	23	17.8	NEBULA
TON-N 1474	12 10 33.	+ 35 01		15.3	BLUE STAR
IC 3046	12 10 35.	+ 13 11 51.			NONSTELLAR OBJECT
AMES 0200	12 10 35.	+ 13 51 41.	50	17.7	NEBULA
REIN 3.020	12 10 35.08	+ 13 11 43.3			NEBULA
ZWG 069.090	12 10 36.	+ 13 12		15.1	GALAXY
UGC 07220	12 10 36.	+ 13 12	72	15.1	GALAXY S
MCG+02-31-052	12 10 36.	+ 13 12	84	15.1	GALAXY
ZWG 098.101	12 10 36.	+ 16 22		15.5	GALAXY
ZWG 098.102	12 10 36.	+ 16 34		14.8	GALAXY
ZWG 187.022	12 10 36.	+ 32 52		15.7	GALAXY
LB 02267	12 10 36.	+ 49 35 48.		17.0	FAINT BLUE STAR
REIZ 2060	12 10 36.	+ 53 10	12	15.2	GALAXY
MCG+10-18-007	12 10 36.	+ 57 54	24	16.	GALAXY
MCG+10-18-008	12 10 36.	+ 62 14	30	16.	GALAXY
RNGC 4180	12 10 37.	+ 07 19		13.0	GALAXY
AMES 0201	12 10 37.	+ 12 17 41.	16	17.8	NEBULA
ARC 1493	12 10 39.	+ 06 19		17.2	RICH CLUSTER OF GALAXIES
AMES 0202	12 10 39.	+ 12 13 53.	20	17.6	NEBULA
SN 1968U	12 10 41.	+ 44 00		14.5	SUPERNOVA
8ZW 1210+13.4	12 10 42.	+ 13 26		17.3	COMPACT GALAXY
MCG+04-29-047	12 10 42.	+ 21 53 30.	21	14.6	GALAXY
MCG+05-29-038	12 10 42.	+ 28 47 30.	144	13.5	GALAXY
ZWG 158.046	12 10 42.	+ 29 07		15.0	GALAXY
UGC 07221	12 10 42.	+ 29 07	90	15.0	GALAXY S
ZWG 215.053	12 10 42.	+ 43 58		13.5	GALAXY
UGC 07222	12 10 42.	+ 43 58	330	13.5	GALAXY Sc
AMES 0203	12 10 43.	+ 07 30 17.	29	17.4	NEBULA
ARC 1494	12 10 44.	+ 24 13		17.6	RICH CLUSTER OF GALAXIES
AMES 0204	12 10 45.	+ 11 25 29.	29	17.9	NEBULA
HN 0824	12 10 45.	+ 13 15		14.5	NEBULA
IC 3047	12 10 45.	+ 13 15			NONSTELLAR OBJECT
MCG+05-29-039	12 10 45.	+ 30 13	60	15.7	GALAXY
MCG+09-20-114	12 10 45.	+ 51 02	18	16.	GALAXY
AMES 0205	12 10 46.	+ 08 20 17.	25	17.6	NEBULA
AMES 0207	12 10 47.	+ 07 16 35.	22	17.4	NEBULA
APP.65 3	12 10 47.	+ 13 50			COMPACT DWARF GALAXY
AMES 0206	12 10 47.	+ 14 08 59.	29	17.4	NEBULA
2ZW 058	12 10 48.	+ 13 27			COMPACT GALAXY
MCG+03-31-076	12 10 48.	+ 15 02	78	17.5	GALAXY
UGC 07223	12 10 48.	+ 15 03	84	16.0	GALAXY
ZWG 098.103	12 10 48.	+ 17 14		15.7	GALAXY
ZWG 098.104	12 10 48.	+ 20 12		15.6	GALAXY
ZWG 128.054	12 10 48.	+ 21 55		14.6	GALAXY
UGC 07224	12 10 48.	+ 21 55	72	14.6	GALAXY E
ZWG 158.047	12 10 48.	+ 28 47		13.5	GALAXY
UGC 07225	12 10 48.	+ 28 47	174	13.5	GALAXY Sb
ZC 1210.8+4040	12 10 48.	+ 40 40	870		CLUSTER OF GALAXIES
LB 02269	12 10 48.	+ 52 25 06.		16.1	FAINT BLUE STAR
LB 02268	12 10 48.	+ 52 29 54.		16.5	FAINT BLUE STAR
RNGC 4182	12 10 49.	+ 04 19			NON-EXISTENT OBJECT
IC 3048	12 10 49.	+ 13 20 51.			TWO STARS
RNGC 4192A	12 10 50.	+ 15 02		17.0	GALAXY
RNGC 4185	12 10 50.	+ 28 47		13.5	GALAXY
REIZ 2061	12 10 50.	+ 28 47	108	13.4	GALAXY
MCG+07-25-051	12 10 51.	+ 43 58	276	12.	GALAXY
RNGC 4183	12 10 51.	+ 43 59		12.5	GALAXY
LB 02270	12 10 52.	+ 62 13 54.		14.3	FAINT BLUE STAR
AMES 0209	12 10 53.	+ 11 06 35.	26	17.7	NEBULA
AMES 0208	12 10 53.	+ 13 50 29.	28	17.6	NEBULA
HOLM 348C	12 10 53.	+ 15 02	24	15.1	PART OF MULTIPLE GALAXY
HOLM 347B	12 10 53.	+ 51 01	12	15.3	PART OF MULTIPLE GALAXY
ZC 1210.9+1831	12 10 54.	+ 18 31	340		CLUSTER OF GALAXIES
ZWG 158.048	12 10 54.	+ 30 12		15.7	GALAXY
ZWG 187.023	12 10 54.	+ 37 58		15.7	GALAXY
HOLM 347A	12 10 54.	+ 51 00	18	14.4	PART OF MULTIPLE GALAXY
MCG+09-20-115	12 10 54.	+ 51 15	42	16.	GALAXY
MCG+10-18-009	12 10 54.	+ 56 51	15	17.	GALAXY
7ZW 442	12 10 54.	+ 57 03			COMPACT GALAXY
ZWG 352.024	12 10 54.	+ 75 20		15.4	GALAXY
UGC 07226	12 10 54.	+ 75 20	114	15.4	GALAXY S IV-V
ZC 1210.9-0329	12 10 54.	- 03 29	2290		CLUSTER OF GALAXIES
AMES 0210	12 10 55.	+ 11 54 29.	18	16.6	NEBULA
ARC 1495	12 10 55.	+ 29 31		17.0	RICH CLUSTER OF GALAXIES
ARC 1496	12 10 56.	+ 59 33		16.0	RICH CLUSTER OF GALAXIES
AMES 0211	12 10 57.	+ 11 30 53.	20	17.9	NEBULA
RNGC 4187A	12 10 57.	+ 51 00		14.5	GALAXY
RNGC 4187B	12 10 57.	+ 51 01		13.5	GALAXY
RNGC 4181	12 10 57.	+ 51 02		14.5	GALAXY
LB 02272	12 10 57.	+ 53 20 30.		13.8	FAINT BLUE STAR
LB 02271	12 10 57.	+ 58 05 00.		15.6	FAINT BLUE STAR
AMES 0212	12 10 58.	+ 10 40 53.	29	18.0	NEBULA
REIZ 2062	12 10 58.	- 06 06	24	15.5	GALAXY
RNGC 4184	12 10 58.	- 62 26			NON-EXISTENT OBJECT
AMES 0215	12 10 59.	+ 05 12 41.	30	17.6	NEBULA
AMES 0214	12 10 59.	+ 11 38 47.	16	18.0	NEBULA
AMES 0213	12 10 59.	+ 13 04 41.	22	17.7	NEBULA
SHB 195	12 10 59.2	+ 13 24 01.		18.1	QUASI-STELLAR OBJECT
BC 4C13.46	12 10 59.5	+ 13 23 58.		18.09	QUASI-STELLAR OBJECT
VDB.66G 114	12 11	+ 13 03	70		DWARF GALAXY
VDB.66G 113	12 11	+ 36 28	70		DWARF GALAXY
SER 090.02	12 11	- 43 34	20	17.6	LOW SURFACE BRIGHT. GALXY
AMES 0217	12 11 00.	+ 06 34 47.	25	17.5	NEBULA
AMES 0216	12 11 00.	+ 12 50 53.	12	17.6	NEBULA
ZWG 098.105	12 11 00.	+ 14 46		15.7	GALAXY
UGC 07227	12 11 00.	+ 14 46	66	15.7	GALAXY DBL SYS
VV 128B	12 11 00.	+ 16 24	36	15.	INTERACTING GALAXY
VV 128A	12 11 00.	+ 16 24	78	14.	INTERACTING GALAXY
VV 128	12 11 00.	+ 16 24	90		INTERACTING GALAXY
MCG+03-31-077	12 11 00.	+ 16 24	90	14.8	GALAXY
ZC 1211.0+1830	12 11 00.	+ 18 30	5910		CLUSTER OF GALAXIES
ZC 1211.0+2415	12 11 00.	+ 24 15	2150		CLUSTER OF GALAXIES
ZWG 243.052	12 11 00.	+ 46 45		15.1	GALAXY
UGC 07228	12 11 00.	+ 46 45	78	15.1	GALAXY
MCG+09-20-118	12 11 00.	+ 50 47	24	16.	GALAXY
REIZ 2063	12 11 00.	+ 51 00	18	14.6	GALAXY
MCG+09-20-117	12 11 00.	+ 51 00	60	14.4	GALAXY
MCG+09-20-116	12 11 00.	+ 51 01	12	13.3	GALAXY
ZWG 269.042	12 11 00.	+ 51 02		14.5	GALAXY
UGC 07229	12 11 00.	+ 51 02	90	14.5	GALAXY E
ZC 1211.0+5556	12 11 00.	+ 55 56	2080		CLUSTER OF GALAXIES
MCG+13-09-018	12 11 00.	+ 75 18	66	15.	GALAXY
MCG-02-31-022	12 11 00.	- 12 55	30	15.	GALAXY
MCG-06-27-008	12 11 00.	- 34 12	24	15.	GALAXY
KARA.73 36	12 11 01.	- 37 56	54		DWARF GALAXY
AMES 0219	12 11 02.	+ 06 21 41.	28	17.4	NEBULA
AMES 0218	12 11 02.	+ 13 21 53.	25	17.8	NEBULA
AMES 0220	12 11 03.	+ 12 09 23.	23	17.7	NEBULA

OBJECT NAME	RIGHT ASCEN.	DECLINATION	DIAM.	MAGN.	TYPE OF OBJECT
HN 0825	12 11 03.	+ 14 45		16.	NEBULA
IC 3049	12 11 03.	+ 14 45			NONSTELLAR OBJECT
MCG+03-31-078	12 11 03.	+ 15 43	66	15.5	GALAXY
MCG+03-34-005	12 11 03.	+ 16 15	90	14.9	GALAXY
ARP 260	12 11 05.	+ 16 25			PECULIAR GALAXY
ZWG 013.105	12 11 06.	+ 02 28		15.6	GALAXY
MCG+00-31-039	12 11 06.	+ 02 28	30	15.6	GALAXY
ZC 1211.1+1501	12 11 06.	+ 15 01	2550		CLUSTER OF GALAXIES
ZWG 098.106	12 11 06.	+ 15 44		15.5	GALAXY
ZWG 098.107	12 11 06.	+ 16 24		14.8	GALAXY
SCH 24	12 11 06.	+ 16 24		14.8	PECULIAR GALAXY
UGC 07230	12 11 06.	+ 16 24	96	14.8	GALAXY MLT SYS
KARA.72 326A	12 11 06.	+ 16 24	54	14.8	PART OF DOUBLE GALAXY
MCG+08-22-085	12 11 06.	+ 46 47 30.	60	15.	GALAXY
SN 1973K	12 11 06.	- 05 42		19.1	SUPERNOVA
REIZ 2064	12 11 07.	+ 46 46	36	15.0	GALAXY
TON-N 1475	12 11 09.	+ 35 56		15.4	BLUE STAR
MCG+06-27-030	12 11 09.	+ 36 56	90	12.	GALAXY
AMES 0222	12 11 10.	+ 06 00 29.	32	17.8	NEBULA
AMES 0221	12 11 10.	+ 13 21 23.	26	17.6	NEBULA
HOLM 349A	12 11 11.	+ 57 25	78	13.2	PART OF MULTIPLE GALAXY
SCH 25	12 11 12.	+ 13 42		12.7	PECULIAR GALAXY
ZWG 098.108	12 11 12.	+ 15 10		11.0	GALAXY
SCH 26	12 11 12.	+ 15 10		11.0	PECULIAR GALAXY
UGC 07231	12 11 12.	+ 15 10	612	11.0	GALAXY Sb
MCG+03-31-079	12 11 12.	+ 15 11	630	11.0	GALAXY
KARA.72 326B	12 11 12.	+ 16 24	36		PART OF DOUBLE GALAXY
ZC 1211.2+2930	12 11 12.	+ 29 30	2620		CLUSTER OF GALAXIES
VV 104	12 11 12.	+ 36 54	78	13.2	INTERACTING GALAXY
ZWG 187.024	12 11 12.	+ 36 55		13.5	GALAXY
UGC 07232	12 11 12.	+ 36 55	108	13.5	GALAXY IRR
LB 02273	12 11 13.	+ 62 45 00.		16.3	FAINT BLUE STAR
RNGC 4189	12 11 14.	+ 13 42		13.0	GALAXY
RNGC 4190	12 11 14.	+ 36 55		13.5	GALAXY
REIN 3.021A	12 11 14.71	+ 13 42 10.0			NEBULA
REIN 3.021B	12 11 14.84	+ 13 42 09.8			NEBULA
AMES 0223	12 11 15.	+ 07 43 53.	28	17.4	NEBULA
REIZ 2065	12 11 15.	+ 13 42	108	13.3	GALAXY
IC 3050	12 11 15.	+ 13 42 15.			SAME AS NGC 4189
REIZ 2066	12 11 15.	+ 15 10	390	11.6	GALAXY
HOLM 348A	12 11 15.	+ 15 10	420	10.9	PART OF MULTIPLE GALAXY
AMES 0224	12 11 16.	+ 06 20 47.	30	17.8	NEBULA
AMES 0225	12 11 17.	+ 12 37 53.	16	17.0	NEBULA
SN 1966E	12 11 17.	+ 13 43		14.8	SUPERNOVA
REIN 3.022A	12 11 17.43	+ 13 42 53.0			NEBULA
REIN 3.022B	12 11 17.44	+ 13 42 53.6			NEBULA
ZC 1211.3+0410	12 11 18.	+ 04 10	270		CLUSTER OF GALAXIES
REA 05	12 11 18.	+ 05 37			DWARF GALAXY
AMES 0226	12 11 18.	+ 05 39 05.	43	17.3	NEBULA
ZWG 041.049	12 11 18.	+ 07 28		13.9	GALAXY
UGC 07233	12 11 18.	+ 07 28	78	13.9	GALAXY S0
MCG+01-31-026	12 11 18.	+ 07 28	48	13.9	GALAXY
ZWG 069.091	12 11 18.	+ 13 27		13.4	GALAXY
UGC 07234	12 11 18.	+ 13 27	138	13.4	GALAXY Sb
MCG+02-31-053	12 11 18.	+ 13 27	144	13.4	GALAXY
ZWG 069.092	12 11 18.	+ 13 42		12.7	GALAXY
UGC 07235	12 11 18.	+ 13 42	162	12.7	GALAXY Sc
MCG+02-31-054	12 11 18.	+ 13 42	120	12.7	GALAXY
ZWG 098.109	12 11 18.	+ 14 30		15.7	GALAXY
ZWG 128.055	12 11 18.	+ 22 02		15.7	GALAXY
UGC 07236	12 11 18.	+ 24 33	72	16.5	GALAXY DWARF
ZC 1211.3+2452	12 11 18.	+ 24 52	740		CLUSTER OF GALAXIES
ZC 1211.3+4214	12 11 18.	+ 42 14	870		CLUSTER OF GALAXIES
ZC 1211.3+4501	12 11 18.	+ 45 01	940		CLUSTER OF GALAXIES
ZC 1211.3+6428	12 11 18.	+ 64 28	940		CLUSTER OF GALAXIES
RNGC 4191	12 11 19.	+ 07 28		14.0	GALAXY
AMES 0227	12 11 19.	+ 07 43 53.	23	17.9	NEBULA
AMES 4193	12 11 19.	+ 13 26 59.	149	13.4	NEBULA
REIZ 2068	12 11 19.	+ 36 55	60	13.7	GALAXY
RNGC 4192	12 11 20.	+ 15 11		11.0	GALAXY
REIN 2.152	12 11 20.70	+ 13 27 01.9			NEBULA
REIN 3.023	12 11 20.82	+ 13 27 01.8			NEBULA
HN 0826	12 11 21.	+ 12 57		16.5	NEBULA
IC 3052	12 11 21.	+ 12 57			NONSTELLAR OBJECT
REIZ 2067	12 11 21.	+ 13 27	84	13.6	GALAXY
IC 3051	12 11 21.	+ 13 27 03.			SAME AS NGC 4193
AMES 0228	12 11 21.	+ 13 57 17.	17	17.6	NEBULA
HOLM 349B	12 11 23.	+ 57 24	30	13.9	PART OF MULTIPLE GALAXY
FEIG 058	12 11 24.	+ 02 17		11.8	FAINT BLUE STAR
MCG+03-34-006	12 11 24.	+ 17 21	48	14.7	GALAXY
ZC 1211.4+3228	12 11 24.	+ 32 28	810		CLUSTER OF GALAXIES
ZC 1211.4+5529	12 11 24.	+ 55 29	740		CLUSTER OF GALAXIES
ZC 1211.4+6013	12 11 24.	+ 60 13	6920		CLUSTER OF GALAXIES
HZ 21	12 11 25.	+ 33 12		14.2	VERY BLUE STAR
AMES 0230	12 11 26.	+ 08 22 47.	20	17.9	NEBULA
AMES 0229	12 11 26.	+ 14 28 05.	24	16.4	NEBULA
AMES 0231	12 11 27.	+ 10 56 53.	25	17.8	NEBULA
HN 0827	12 11 27.	+ 14 29		16.	NEBULA
IC 3053	12 11 27.	+ 14 29			NONSTELLAR OBJECT
REIZ 2069	12 11 27.	+ 17 49	42	15.4	GALAXY
AMES 0232	12 11 28.	+ 14 05 29.	20	17.7	NEBULA
ZWG 013.106	12 11 30.	+ 00 12		15.6	GALAXY
ZWG 041.050	12 11 30.	+ 07 00		15.2	GALAXY
MCG+01-31-027	12 11 30.	+ 07 00	54	15.2	GALAXY
MCG+03-31-081	12 11 30.	+ 15 00	66	14.9	GALAXY
ZWG 098.110	12 11 30.	+ 17 49		15.4	GALAXY
UGC 07237	12 11 30.	+ 18 11	60	16.0	GALAXY
MCG+03-31-080	12 11 30.	+ 18 11	72	17.	GALAXY
ZWG 128.056	12 11 30.	+ 24 20		15.5	GALAXY
ZC 1211.5+2657	12 11 30.	+ 26 57	1080		CLUSTER OF GALAXIES
MCG+06-27-031	12 11 30.	+ 34 48 30.	42	15.	GALAXY
MCG+06-27-032	12 11 30.	+ 35 19	42	15.5	GALAXY
ZC 1211.5+4305	12 11 30.	+ 43 05	740		CLUSTER OF GALAXIES
ZC 1211.5+4927	12 11 30.	+ 49 27	1340		CLUSTER OF GALAXIES
MCG+13-09-019	12 11 30.	+ 74 46	84	15.	GALAXY
UGC 07238	12 11 30.	+ 74 47	84	16.0	GALAXY Sc
AMES 0234	12 11 31.	+ 07 00 29.	58	16.4	NEBULA
AMES 0233	12 11 31.	+ 14 29 29.	29	16.3	NEBULA
AMES 0236	12 11 32.	+ 05 38 41.	40	17.4	NEBULA
AMES 0235	12 11 32.	+ 14 26 05.	34	16.2	NEBULA
RNGC 4192B	12 11 32.	+ 15 00		15.0	GALAXY
HOLM 350A	12 11 32.	+ 35 18	60	14.9	PART OF MULTIPLE GALAXY
LB 02274	12 11 32.	+ 49 32 30.		16.7	FAINT BLUE STAR
BC B21211+33	12 11 32.6	+ 33 26 18.0		17.	QUASI-STELLAR OBJECT
REIZ 2070	12 11 33.	+ 07 00	24	15.3	GALAXY
REIZ 2071	12 11 33.	+ 15 00	18	14.9	GALAXY
TON-N 1476	12 11 33.	+ 36 42		17.	BLUE STAR
ARC 1500	12 11 33.	+ 74 40		15.6	RICH CLUSTER OF GALAXIES

OBJECT NAME	RIGHT ASCEN.	DECLINATION	DIAM.	MAGN.	TYPE OF OBJECT
AMES 0237	12 11 35.	+ 14 20 52.	31	16.8	NEBULA
HOLM 348B	12 11 35.	+ 15 00	30	14.3	PART OF MULTIPLE GALAXY
REIN 3.024	12 11 35.71	+ 08 03 22.2			NEBULA
REA 78	12 11 36.	+ 08 02			DWARF GALAXY
ZWG 041.051	12 11 36.	+ 08 03		15.0	GALAXY
UGC 07239	12 11 36.	+ 08 03	150	15.0	GALAXY DWARF
MCG+01-31-028	12 11 36.	+ 08 03	90	15.0	GALAXY
AMES 0238	12 11 36.	+ 10 45 40.	24	17.0	NEBULA
ZWG 069.093	12 11 36.	+ 13 52		15.6	GALAXY
MCG+02-31-055	12 11 36.	+ 13 52	15	15.6	GALAXY
ZC 1211.6+1355	12 11 36.	+ 13 55	1550		CLUSTER OF GALAXIES
ZWG 069.094	12 11 36.	+ 14 26		15.3	GALAXY
ZWG 098.111	12 11 36.	+ 15 00		14.9	GALAXY
UGC 07240	12 11 36.	+ 15 00	60	14.9	GALAXY Sa
KARA.68 088	12 11 36.	+ 16 14	54		DWARF GALAXY
ZWG 098.112	12 11 36.	+ 19 36		15.6	GALAXY
ZWG 158.049	12 11 36.	+ 30 34		15.7	GALAXY
ZC 1211.6+3155	12 11 36.	+ 31 55	3290		CLUSTER OF GALAXIES
MCG+06-27-033	12 11 36.	+ 32 41	30	15.	GALAXY
ZWG 187.025	12 11 36.	+ 32 43		15.1	GALAXY
MRK 436	12 11 36.	+ 37 02	7	16.	GALAXY WITH UV CONTINUUM
1ZW 033	12 11 36.	+ 54 48			COMPACT GALAXY
ZWG 269.043	12 11 36.	+ 54 48		13.0	GALAXY
MRK 201	12 11 36.	+ 54 48	30	13.	GALAXY WITH UV CONTINUUM
UGC 07241	12 11 36.	+ 54 48	138	13.0	GALAXY PECULR
VV 261	12 11 36.	+ 54 49	150	14.	INTERACTING GALAXY
MCG+09-20-119	12 11 36.	+ 54 49	120	13.3	GALAXY
MCG-02-31-023	12 11 36.	- 12 17	36	14.	GALAXY
ARC 1497	12 11 37.	+ 26 56		17.8	RICH CLUSTER OF GALAXIES
ARP 160	12 11 37.	+ 54 49			PECULIAR GALAXY
RNGC 4188	12 11 37.	- 12 17		14.0	GALAXY
RNGC 4186	12 11 38.	+ 15 00		15.5	GALAXY
REIZ 2072	12 11 39.	+ 08 03	90	15.3	GALAXY
AMES 0240	12 11 39.	+ 13 03 28.	23	17.4	NEBULA
AMES 0239	12 11 39.	+ 13 56 04.	18	17.8	NEBULA
RNGC 4194	12 11 40.	+ 54 48		13.0	GALAXY
HOLM 350B	12 11 41.	+ 35 19	12	15.4	PART OF MULTIPLE GALAXY
REIZ 2073	12 11 41.	+ 54 49	48	13.3	GALAXY
ARC 1501	12 11 41.	+ 63 30		17.2	RICH CLUSTER OF GALAXIES
AMES 0241	12 11 42.	+ 08 23 52.	28	17.1	NEBULA
REA 24	12 11 42.	+ 16 15	66		DWARF GALAXY
MCG+06-27-035	12 11 42.	+ 32 43	18	16.	GALAXY
ZWG 187.026	12 11 42.	+ 34 54		15.5	GALAXY
MCG+06-27-034	12 11 42.	+ 34 54	30	15.	GALAXY
REIZ 2075	12 11 42.	+ 52 31	42	14.7	GALAXY
ZWG 315.026	12 11 42.	+ 66 22		15.0	GALAXY
UGC 07242	12 11 42.	+ 66 22	102	15.0	GALAXY Sc
ZWG 013.107	12 11 42.	- 00 33		15.5	GALAXY
AMES 0242	12 11 43.	+ 10 02 04.	32	17.0	NEBULA
ARC 1498	12 11 43.	+ 31 56		17.6	RICH CLUSTER OF GALAXIES
LB 02275	12 11 43.	+ 59 55 42.		15.7	FAINT BLUE STAR
HN 0828	12 11 45.	+ 13 48		16.5	NEBULA
IC 3054	12 11 45.	+ 13 48			NONSTELLAR OBJECT
ARC 1499	12 11 45.	+ 15 02		16.6	RICH CLUSTER OF GALAXIES
REIZ 2074	12 11 45.	+ 15 56	60	14.6	GALAXY
MCG+06-27-036	12 11 45.	+ 32 40 30.	42	15.	GALAXY
TON-N 1477	12 11 45.	+ 34 34		16.7	BLUE STAR
MCG+09-20-120	12 11 45.	+ 53 13	30	16.	GALAXY
RNGC 4195	12 11 46.	+ 59 53		15.5	GALAXY
AMES 0244	12 11 48.	+ 10 26 46.	28	18.0	NEBULA
AMES 0243	12 11 48.	+ 12 24 46.	16	17.8	NEBULA
ZWG 128.057	12 11 48.	+ 22 28		15.6	GALAXY
MCG+04-29-048	12 11 48.	+ 24 26 30.	36	15.7	GALAXY
MCG+05-29-040	12 11 48.	+ 28 42	48	13.7	GALAXY
ZWG 187.027	12 11 48.	+ 32 42		14.6	GALAXY
UGC 07243	12 11 48.	+ 32 42	60	14.6	GALAXY S0
ZC 1211.8+4411	12 11 48.	+ 44 11	810		CLUSTER OF GALAXIES
ZWG 269.044	12 11 48.	+ 51 36		15.1	GALAXY
MCG+09-20-121	12 11 48.	+ 53 12 30.	30	16.	GALAXY
ZWG 293.004	12 11 48.	+ 59 53		15.5	GALAXY
ZWG 292.083	12 11 48.	+ 59 53		15.5	GALAXY
UGC 07244	12 11 48.	+ 59 53	102	15.5	GALAXY SBc
MCG+10-18-010	12 11 48.	+ 59 53	90	14.	GALAXY
MCG+11-15-037	12 11 48.	+ 66 22	114	13.	GALAXY
AMES 0245	12 11 49.	+ 13 34 52.	24	17.8	NEBULA
AMES 0246	12 11 51.	+ 10 43 04.	29	17.4	NEBULA
HN 0829	12 11 51.	+ 12 21		15.	NEBULA
IC 3055	12 11 51.	+ 12 21			NONSTELLAR OBJECT
AMES 0247	12 11 52.	+ 07 02 46.	37	17.2	NEBULA
RNGC 4198	12 11 52.	+ 56 17		14.5	GALAXY
AMES 0248	12 11 53.	+ 13 27 04.	29	16.8	NEBULA
AMES 0249	12 11 54.	+ 11 54 46.	17	17.0	NEBULA
ZC 1211.9+1426	12 11 54.	+ 14 26	1280		CLUSTER OF GALAXIES
ZC 1211.9+2144	12 11 54.	+ 21 44	540		CLUSTER OF GALAXIES
ZWG 128.058	12 11 54.	+ 24 28		15.7	GALAXY
ZWG 158.050	12 11 54.	+ 28 42		13.7	GALAXY
UGC 07245	12 11 54.	+ 28 42	120	13.7	GALAXY S0?
MCG+09-20-122	12 11 54.	+ 51 34	36	16.	GALAXY
ZWG 269.045	12 11 54.	+ 56 17		14.6	GALAXY
UGC 07246	12 11 54.	+ 56 17	66	14.6	GALAXY S0-a
MCG+09-20-123	12 11 54.	+ 56 17 30.	60	14.5	GALAXY
KARA.68 089	12 11 54.	- 12 03	54		DWARF GALAXY
AMES 0251	12 11 55.	+ 06 11 28.	31	17.2	NEBULA
AMES 0250	12 11 55.	+ 08 48 04.	26	17.5	NEBULA
AMES 0253	12 11 56.	+ 08 39 40.	30	17.0	NEBULA
AMES 0252	12 11 56.	+ 14 03 28.	24	16.6	NEBULA
RNGC 4196	12 11 56.	+ 28 42		13.5	GALAXY
TON-N 1478	12 11 57.	+ 33 02		16.9	BLUE STAR
LB 02276	12 11 57.	+ 62 26 18.		15.4	FAINT BLUE STAR
LB 02277	12 11 58.	+ 57 51 06.		14.3	FAINT BLUE STAR
AMES 0254	12 11 59.	+ 08 39 40.	28	17.4	NEBULA
LB 02278	12 11 59.	+ 62 29 48.		15.9	FAINT BLUE STAR
VDB.66G 115	12 12	+ 13 48	70		DWARF GALAXY
ZWG 013.108	12 12 00.	+ 01 06		15.0	GALAXY
MCG+00-31-040	12 12 00.	+ 01 06	42	15.0	GALAXY
ZWG 041.052	12 12 00.	+ 06 05		13.8	GALAXY
UGC 07247	12 12 00.	+ 06 05	228	13.8	GALAXY Sc
MCG+01-31-029	12 12 00.	+ 06 05	192	13.8	GALAXY
ZWG 098.113	12 12 00.	+ 18 31		15.3	GALAXY
ZC 1212.0+2409	12 12 00.	+ 24 09	6380		CLUSTER OF GALAXIES
MCG+04-29-049	12 12 00.	+ 24 33	54	15.6	GALAXY
ZWG 128.059	12 12 00.	+ 24 35		15.6	GALAXY
UGC 07248	12 12 00.	+ 24 35	60	15.6	GALAXY SBb
ZC 1212.0+2601	12 12 00.	+ 26 01	2890		CLUSTER OF GALAXIES
ZWG 335.006	12 12 00.	+ 68 40		15.4	GALAXY
RNGC 4197	12 12 01.	+ 06 05		14.0	GALAXY
AMES 0255	12 12 01.	+ 06 41 16.	17	17.5	NEBULA
AMES 0258	12 12 02.	+ 09 28 28.	55	16.9	NEBULA

OBJECT NAME	RIGHT ASCEN.	DECLINATION	DIAM.	MAGN.	TYPE OF OBJECT
AMES 0257	12 12 02.	+ 09 58 28.	40	17.3	NEBULA
AMES 0256	12 12 02.	+ 12 08 16.	37	17.6	NEBULA
REIZ 2076	12 12 03.	+ 06 05	126	13.8	GALAXY
HN 0830	12 12 03.	+ 14 04	60		NEBULA
IC 3056	12 12 03.	+ 14 04			MAY NOT EXIST
MCG+04-29-050	12 12 03.	+ 24 07	36	15.5	GALAXY
TON-N 1479	12 12 03.	+ 35 57		17.	BLUE STAR
LB 00627	12 12 03.	+ 58 29 06.		16.1	FAINT BLUE STAR
AMES 0260	12 12 04.	+ 10 05 52.	25	17.8	NEBULA
AMES 0259	12 12 04.	+ 13 05 22.	89	16.2	NEBULA
REIZ 2078	12 12 04.	+ 60 14	12	14.8	GALAXY
AMES 0261	12 12 05.	+ 07 03 22.	18	16.6	NEBULA
REIZ 2077	12 12 05.	+ 56 17	36	14.5	GALAXY
ZC 1212.1+0520	12 12 06.	+ 05 20	1340		CLUSTER OF GALAXIES
ZWG 069.095	12 12 06.	+ 13 05		15.3	GALAXY
UGC 07249	12 12 06.	+ 13 05	84	15.3	GALAXY IRR
MCG+02-31-056	12 12 06.	+ 13 05	90	15.3	GALAXY
AMES 0262	12 12 06.	+ 13 21 16.	18	17.8	NEBULA
ZWG 215.054	12 12 06.	+ 43 48		15.3	GALAXY
ZC 1212.1+6329	12 12 06.	+ 63 29	1340		CLUSTER OF GALAXIES
ZWG 013.109	12 12 06.	- 02 10		14.6	GALAXY
MCG+00-31-041	12 12 06.	- 02 10	42	14.6	GALAXY
MCG-02-31-024	12 12 06.	- 11 18	48	14.	GALAXY
AMES 0263	12 12 07.	+ 05 45 58.	29	17.0	NEBULA
RNGC 4201	12 12 07.	- 11 18		14.0	GALAXY
AMES 0264	12 12 08.	+ 10 18 22.	16	17.7	NEBULA
REIZ 2079	12 12 09.	+ 12 27	18	14.3	GALAXY
AMES 0265	12 12 09.	+ 13 15 58.	30	16.6	NEBULA
HOLM 351A	12 12 09.	+ 13 36	36	14.7	PART OF MULTIPLE GALAXY
FATH 1.484	12 12 09.	+ 15 12	19		NEBULA
AMES 0269	12 12 10.	+ 05 46 10.	30	17.5	NEBULA
AMES 0268	12 12 10.	+ 10 35 52.	23	17.8	NEBULA
AMES 0267	12 12 10.	+ 13 14 16.	23	16.8	NEBULA
AMES 0266	12 12 10.	+ 13 15 10.	26	17.2	NEBULA
SEY 170	12 12 10.	+ 43 48 13.		15.1	FAINT GALAXY
AMES 0270	12 12 11.	+ 08 44 52.	24	17.8	NEBULA
REIN 3.025	12 12 11.53	+ 12 27 31.3			NEBULA
ZWG 013.110	12 12 12.	+ 01 01		15.3	GALAXY
UGC 07250	12 12 12.	+ 01 01	60	15.3	GALAXY S
ZWG 069.096	12 12 12.	+ 12 27		14.1	GALAXY
UGC 07251	12 12 12.	+ 12 27	96	14.1	GALAXY S0
MCG+02-31-057	12 12 12.	+ 12 27	22	14.1	GALAXY
FATH 1.485	12 12 12.	+ 14 16	14		NEBULA
UGC 07252	12 12 12.	+ 35 48	72	16.5	GALAXY Sc
MCG+09-20-124	12 12 12.	+ 53 11	42	17.	GALAXY
MCG+10-18-011	12 12 12.	+ 60 10	27	14.8	GALAXY
ZWG 335.007	12 12 12.	+ 72 37		15.7	GALAXY
MCG-06-27-009	12 12 12.	- 35 13	156	12.	GALAXY
AMES 0272	12 12 13.	+ 10 42 58.	17	18.0	NEBULA
AMES 0271	12 12 13.	+ 12 14 10.	17	17.6	NEBULA
FATH 1.486	12 12 13.	+ 14 23	33		NEBULA
REIN 3.026A	12 12 13.14	+ 13 36 12.3			NEBULA
REIN 3.026B	12 12 13.23	+ 13 36 14.8			NEBULA
RNGC 4200	12 12 14.	+ 12 27		14.0	GALAXY
HOLM 351B	12 12 14.	+ 13 36	24	15.4	PART OF MULTIPLE GALAXY
AMES 0273	12 12 15.	+ 11 11 40.	14	16.7	NEBULA
MCG+06-27-037	12 12 15.	+ 33 17	27	16.	GALAXY
AMES 0275	12 12 16.	+ 10 42 52.	28	17.2	NEBULA
AMES 0274	12 12 16.	+ 13 35 40.	28	16.8	NEBULA
AMES 0277	12 12 17.	+ 11 28 52.	12	17.6	NEBULA
AMES 0276	12 12 17.	+ 11 48 46.	22	18.2	NEBULA
AMES 0278	12 12 18.	+ 10 42 22.	22	17.0	NEBULA
MCG+02-31-060	12 12 18.	+ 13 14	12	15.6	GALAXY
ZWG 069.097	12 12 18.	+ 13 15		15.6	GALAXY
MCG+02-31-059	12 12 18.	+ 13 15	12	15.6	GALAXY
MCG+02-31-058	12 12 18.	+ 13 16	12	15.6	GALAXY
ZWG 069.098	12 12 18.	+ 13 35		15.6	GALAXY
ZWG 187.028	12 12 18.	+ 33 18		15.7	GALAXY
HZ 22	12 12 18.	+ 36 56		12.7	DECIDEDLY BLUE STAR
MCG+09-20-125	12 12 18.	+ 56 07	36	17.	GALAXY
ZC 1212.3+5711	12 12 18.	+ 57 11	2760		CLUSTER OF GALAXIES
7ZW 443	12 12 18.	+ 64 44			COMPACT GALAXY
MCG+12-12-003	12 12 18.	+ 70 04	33	16.	GALAXY
ZWG 013.111	12 12 18.	- 03 12		15.4	GALAXY
AMES 0279	12 12 19.	+ 09 57 10.	26	17.8	NEBULA
RNGC 4202	12 12 19.	- 02 11		14.5	GALAXY
AMES 0280	12 12 20.	+ 06 19 34.	17	17.8	NEBULA
HN 0831	12 12 21.	+ 14 21		16.	NEBULA
IC 3058	12 12 21.	+ 14 21			NONSTELLAR OBJECT
LB 02279	12 12 22.	+ 50 39 00.		16.2	FAINT BLUE STAR
RNGC 4199	12 12 22.	+ 60 10		15.5	GALAXY
AMES 0281	12 12 23.	+ 08 28 22.	22	17.8	NEBULA
LB 02280	12 12 23.	+ 58 39 12.		15.5	FAINT BLUE STAR
REA 11	12 12 24.	+ 09 51			DWARF GALAXY
ZC 1212.4+1635	12 12 24.	+ 16 35	1140		CLUSTER OF GALAXIES
MCG+03-34-007	12 12 24.	+ 17 30	102	14.7	GALAXY
ZC 1212.4+2654	12 12 24.	+ 26 54	2020		CLUSTER OF GALAXIES
KARA.68 090	12 12 24.	+ 35 29	114		DWARF GALAXY
LB 02281	12 12 24.	+ 49 42 54.		13.5	FAINT BLUE STAR
7ZW 444	12 12 24.	+ 59 36			COMPACT GALAXY
ZWG 293.005	12 12 24.	+ 60 10		15.5	GALAXY
ZWG 292.084	12 12 24.	+ 60 10		15.5	GALAXY
UGC 07253	12 12 24.	+ 60 10	66	15.5	GALAXY DBL SYS
ZC 1212.4+6749	12 12 24.	+ 67 49	940		CLUSTER OF GALAXIES
AMES 0282	12 12 25.	+ 10 46 52.	34	17.5	NEBULA
HN 0361	12 12 25.	- 44 12			NEBULA
IC 3057	12 12 25.	- 44 12			NONSTELLAR OBJECT
AMES 0283	12 12 26.	+ 05 20 28.	26	17.6	NEBULA
HN 0832	12 12 27.	+ 13 43			NEBULA
IC 3059	12 12 27.	+ 13 43			MAY NOT EXIST
MCG+06-27-040	12 12 27.	+ 33 28	48	11.	GALAXY
MCG+06-27-039	12 12 27.	+ 36 15	66	14.	GALAXY
MCG+06-27-038	12 12 27.	+ 36 15	9	16.	GALAXY
AMES 0285	12 12 28.	+ 07 24 52.	32	17.8	NEBULA
AMES 0284	12 12 28.	+ 13 33 16.	29	17.1	NEBULA
IC 3060	12 12 29.	+ 12 49 28.			NONSTELLAR OBJECT
REIZ 2084	12 12 29.	+ 52 49	24	15.1	GALAXY
RNGC 4205	12 12 29.	+ 64 04		14.0	GALAXY
REIN 3.027	12 12 29.40	+ 12 49 29.3			NEBULA
ZWG 069.099	12 12 30.	+ 12 48		15.1	GALAXY
MCG+02-31-061	12 12 30.	+ 12 48	15	15.1	GALAXY
ZWG 069.100	12 12 30.	+ 13 44		15.3	GALAXY
UGC 07254	12 12 30.	+ 13 44	120	15.3	GALAXY DWARF
MCG+02-31-062	12 12 30.	+ 13 44	84	15.3	GALAXY
ZWG 069.101	12 12 30.	+ 14 19		14.9	GALAXY
UGC 07255	12 12 30.	+ 14 19	144	14.9	GALAXY Sc
MCG+02-31-063	12 12 30.	+ 14 19	108	14.9	GALAXY
FATH 2.065	12 12 30.	+ 14 19	122		NEBULA

OBJECT NAME	RIGHT ASCEN.	DECLINATION	DIAM.	MAGN.	TYPE OF OBJECT
MCG+03-34-009	12 12 30.	+ 16 21	42	15.7	GALAXY
MCG+03-34-008	12 12 30.	+ 17 30	18	15.2	GALAXY
ZC 1212.5+1954	12 12 30.	+ 19 54	3090		CLUSTER OF GALAXIES
ZWG 187.029	12 12 30.	+ 33 28		11.8	GALAXY
UGC 07256	12 12 30.	+ 33 28	222	11.8	GALAXY S0
ZWG 187.030	12 12 30.	+ 36 14		14.2	GALAXY
UGC 07257	12 12 30.	+ 36 14	96	14.2	GALAXY S-IRR
ZWG 269.046	12 12 30.	+ 53 14		15.7	GALAXY
ZC 1212.5+6110	12 12 30.	+ 61 10	1410		CLUSTER OF GALAXIES
MCG+11-15-038	12 12 30.	+ 64 03	78	13.6	GALAXY
ZWG 315.027	12 12 30.	+ 64 04		13.8	GALAXY
UGC 07258	12 12 30.	+ 64 04	126	13.8	GALAXY S
IC 3061	12 12 31.4	+ 14 18 30.	122		GALAXY
AMES 0286	12 12 33.	+ 10 01 40.	30	17.1	NEBULA
REIZ 2082	12 12 33.	+ 12 14	36	14.9	GALAXY
REIZ 2083	12 12 33.	+ 12 17	24	14.6	GALAXY
IC 3062	12 12 33.	+ 13 52 22.			NONSTELLAR OBJECT
REIZ 2081	12 12 33.	+ 13 53	36	14.4	GALAXY
REIZ 2080	12 12 33.	+ 14 18	90	14.1	GALAXY
PK298-01.1	12 12 34.	- 63 22	5		PLANETARY NEBULA
REIN 3.028	12 12 34.14	+ 12 17 41.1			NEBULA
AMES 0288	12 12 35.	+ 05 35 16.	24	17.6	NEBULA
AMES 0287	12 12 35.	+ 07 12 46.	19	17.7	NEBULA
ARC 1502	12 12 35.	- 07 58		17.2	RICH CLUSTER OF GALAXIES
ZWG 041.053	12 12 36.	+ 04 51		15.3	GALAXY
AMES 0289	12 12 36.	+ 08 42 52.	20	17.2	NEBULA
KARA.68 091	12 12 36.	+ 09 51	54		DWARF GALAXY
ZWG 069.102	12 12 36.	+ 12 17		14.9	GALAXY
UGC 07259	12 12 36.	+ 12 17	60	14.9	GALAXY Sa
MCG+02-31-064	12 12 36.	+ 12 17	60	14.9	GALAXY
ZWG 069.103	12 12 36.	+ 13 52		14.9	GALAXY
MCG+02-31-065	12 12 36.	+ 13 52	60	14.9	GALAXY
MCG+03-31-082	12 12 36.	+ 14 42	60	14.7	GALAXY
ZWG 098.114	12 12 36.	+ 14 43		14.7	GALAXY
MCG+04-29-051	12 12 36.	+ 20 55	240	14.3	GALAXY
ZWG 187.031	12 12 36.	+ 34 53		15.3	GALAXY
MCG+06-27-041	12 12 36.	+ 34 53 30.	24	15.	GALAXY
ZC 1212.6+4110	12 12 36.	+ 41 10	740		CLUSTER OF GALAXIES
SEY 171	12 12 36.	+ 41 50 13.		15.4	FAINT GALAXY
ZWG 013.112	12 12 36.	- 03 09		14.8	GALAXY
MCG+00-31-042	12 12 36.	- 03 09	66	14.8	GALAXY
MCG-02-31-025	12 12 36.	- 12 56	72	15.	GALAXY
MCG-05-29-031	12 12 36.	- 30 31	96	13.5	GALAXY
AMES 0290	12 12 37.	+ 08 41 40.	23	17.1	NEBULA
AMES 0291	12 12 38.	+ 08 33 28.	19	17.5	NEBULA
RNGC 4203	12 12 38.	+ 33 29		12.0	GALAXY
REIZ 2086	12 12 39.	+ 04 50	18	15.0	GALAXY
HN 0833	12 12 39.	+ 12 17		14.5	NEBULA
IC 3063	12 12 39.	+ 12 17			NONSTELLAR OBJECT
HOLM 353D	12 12 39.	+ 13 28	36	14.4	PART OF MULTIPLE GALAXY
IC 0771	12 12 39.	+ 13 28			NONSTELLAR OBJECT
REIZ 2087	12 12 39.	+ 14 42	18	14.7	GALAXY
TON-N 1480	12 12 39.	+ 33 26		17.	BLUE STAR
REIZ 2089	12 12 39.	+ 64 04	60	13.6	GALAXY
FATH 1.487	12 12 40.	+ 14 36	22		NEBULA
REIZ 2085	12 12 40.	- 03 11	60	14.3	GALAXY
REIN 3.029A	12 12 40.70	+ 13 27 43.5			NEBULA
REIN 3.029B	12 12 40.76	+ 13 27 43.9			NEBULA
AMES 0293	12 12 41.	+ 04 50 16.	38	16.4	NEBULA
AMES 0292	12 12 41.	+ 14 02 52.	17	17.2	NEBULA
IC 3065	12 12 41.	+ 14 43	27		BRIGHT NEBULA
FATH 2.066	12 12 41.	+ 14 43	27		NEBULA
HOLM 352B	12 12 41.	+ 51 38	30	14.8	PART OF MULTIPLE GALAXY
MCG+01-31-030	12 12 42.	+ 06 03	24	16.	GALAXY
KARA.68 092	12 12 42.	+ 09 25	27		DWARF GALAXY
AMES 0294	12 12 42.	+ 10 00 04.	17	16.9	NEBULA
ZWG 069.104	12 12 42.	+ 13 18		13.8	GALAXY
UGC 07260	12 12 42.	+ 13 18	348	13.8	GALAXY Sc
MCG+02-31-066	12 12 42.	+ 13 18	318	13.8	GALAXY
ZWG 069.105	12 12 42.	+ 13 28		14.9	GALAXY
MCG+02-31-067	12 12 42.	+ 13 28	48	14.9	GALAXY
ZWG 128.060	12 12 42.	+ 20 56		14.3	GALAXY
SCH 27	12 12 42.	+ 20 56		14.3	PECULIAR GALAXY
UGC 07261	12 12 42.	+ 20 56	282	14.3	GALAXY
ZWG 128.061	12 12 42.	+ 24 14		15.5	GALAXY
ZC 1212.7+2750	12 12 42.	+ 27 50	1680		CLUSTER OF GALAXIES
ZC 1212.7+4747	12 12 42.	+ 47 47	470		CLUSTER OF GALAXIES
MCG+09-20-126	12 12 42.	+ 50 57	24	16.	GALAXY
MCG+09-20-127	12 12 42.	+ 54 55	36	16.	GALAXY
MCG-04-29-013	12 12 42.	- 27 24 30.	30	15.	GALAXY
AMES 0295	12 12 43.	+ 13 08 28.	22	16.6	NEBULA
AMES 0296	12 12 44.	+ 08 33 40.	28	17.4	NEBULA
RNGC 4206	12 12 44.	+ 13 18		14.0	GALAXY
IC 3064	12 12 44.	+ 13 18 04.			SAME AS NGC 4206
HOLM 353B	12 12 44.	+ 13 27	270	12.5	PART OF MULTIPLE GALAXY
ARC 1503	12 12 44.	+ 19 43		17.2	RICH CLUSTER OF GALAXIES
RNGC 4204	12 12 44.	+ 20 56		14.5	GALAXY
REIZ 2088	12 12 44.	+ 20 56	60	14.6	GALAXY
IC 3067	12 12 44.	+ 24 12 40.			NONSTELLAR OBJECT
LB 02282	12 12 44.	+ 50 19 06.		16.8	FAINT BLUE STAR
REIN 3.030A	12 12 44.16	+ 13 18 05.7			NEBULA
REIN 3.030B	12 12 44.25	+ 13 18 04.9			NEBULA
HN 0835	12 12 45.	+ 13 44	60		NEBULA
IC 3066	12 12 45.	+ 13 44			NONSTELLAR OBJECT
HN 0834	12 12 45.	+ 14 41		14.	NEBULA
IC 0772	12 12 45.	+ 24 16			NONSTELLAR OBJECT
TON-N 1481	12 12 45.	+ 33 46		16.	BLUE STAR
TON-N 1482	12 12 45.	+ 34 09		17.	BLUE STAR
TON-N 1483	12 12 45.	+ 35 37		14.4	BLUE STAR
TON-N 1484	12 12 45.	+ 36 52		16.9	BLUE STAR
AMES 0297	12 12 48.	+ 08 59 16.	32	17.0	NEBULA
ZWG 069.106	12 12 48.	+ 13 45		15.2	GALAXY
UGC 07262	12 12 48.	+ 13 45	60	15.2	GALAXY S
MCG+02-31-068	12 12 48.	+ 13 45	60	15.2	GALAXY
ZC 1212.8+1404	12 12 48.	+ 14 04	1010		CLUSTER OF GALAXIES
ZWG 098.115	12 12 48.	+ 17 45		15.7	GALAXY
ZWG 098.116	12 12 48.	+ 19 34		14.9	GALAXY
UGC 07263	12 12 48.	+ 19 34	66	14.9	GALAXY SB?b-c
ZC 1212.8+2130	12 12 48.	+ 21 30	2350		CLUSTER OF GALAXIES
ZC 1212.8+3214	12 12 48.	+ 32 14	1140		CLUSTER OF GALAXIES
MCG+10-18-012	12 12 48.	+ 58 08	24	17.	GALAXY
ZWG 315.028	12 12 48.	+ 66 15		13.4	GALAXY
UGC 07264	12 12 48.	+ 66 15	126	13.4	GALAXY SBb
UGC 07265	12 12 48.	+ 76 32	72	16.5	GALAXY
ZC 1212.8-0110	12 12 48.	- 01 10	1480		CLUSTER OF GALAXIES
AMES 0298	12 12 49.	+ 07 37 52.	18	17.6	NEBULA
ARC 1504	12 12 49.	+ 27 48		17.6	RICH CLUSTER OF GALAXIES
AMES 0301	12 12 50.	+ 07 01 10.	18	17.4	NEBULA
AMES 0300	12 12 50.	+ 08 11 04.	22	17.8	NEBULA
AMES 0299	12 12 50.	+ 12 25 22.	22	17.8	NEBULA
REIZ 2092	12 12 50.	+ 66 15	60	13.5	GALAXY
HARO 06	12 12 51.	+ 06 03			BLUE EMISSION-LINE GALAXY
HN 0837	12 12 51.	+ 10 27		15.5	NEBULA
IC 3069	12 12 51.	+ 10 27			NONSTELLAR OBJECT
HN 0836	12 12 51.	+ 11 47		16.	NEBULA
IC 3068	12 12 51.	+ 11 47			NONSTELLAR OBJECT
REIZ 2090	12 12 51.	+ 13 19	240	13.5	GALAXY
MCG+03-31-083	12 12 51.	+ 19 34	78	14.9	GALAXY
MCG+04-29-052	12 12 51.	+ 24 20 30.	60	15.6	GALAXY
IC 3070	12 12 52.	+ 13 19 04.			SINGLE STAR
FATH 2.067	12 12 52.	+ 14 12	11		NEBULA
AMES 0302	12 12 53.	+ 08 03 22.	25	17.0	NEBULA
SCH 28	12 12 54.	+ 06 05			PECULIAR GALAXY
AMES 0303	12 12 54.	+ 12 32 46.	30	17.4	NEBULA
ZWG 128.062	12 12 54.	+ 24 22		15.6	GALAXY
UGC 07266	12 12 54.	+ 24 22	78	15.6	GALAXY S0
ZWG 158.051	12 12 54.	+ 27 10		15.3	GALAXY
MCG+05-29-041	12 12 54.	+ 27 18	18	15.4	GALAXY
MCG+05-29-043	12 12 54.	+ 28 27	12	15.5	GALAXY
MCG+05-29-042	12 12 54.	+ 28 27 30.	54	14.4	GALAXY
REIZ 2091	12 12 54.	+ 50 23	72	13.8	GALAXY
MCG+09-20-128	12 12 54.	+ 51 37	132	13.6	GALAXY
ZWG 269.047	12 12 54.	+ 51 39		14.6	GALAXY
UGC 07267	12 12 54.	+ 51 39	120	14.6	GALAXY
ZC 1212.9+6150	12 12 54.	+ 61 50	1480		CLUSTER OF GALAXIES
RNGC 4210	12 12 54.	+ 66 16		13.5	GALAXY
ZWG 352.025	12 12 54.	+ 75 55		15.3	GALAXY
MCG-06-27-010	12 12 54.	- 37 52	102	15.	GALAXY
AMES 0304	12 12 55.	+ 10 17 22.	19	17.2	NEBULA
FATH 1.488	12 12 55.	+ 14 40	14		NEBULA
AMES 0307	12 12 56.	+ 05 53 40.	32	17.9	NEBULA
AMES 0306	12 12 56.	+ 08 06 52.	20	17.8	NEBULA
AMES 0305	12 12 56.	+ 10 47 04.	22	17.8	NEBULA
RNGC 4208	12 12 56.	+ 14 11			NON-EXISTENT OBJECT
RNGC 4209	12 12 56.	+ 28 47			NON-EXISTENT OBJECT
AMES 0308	12 12 57.	+ 07 03 04.	20	17.6	NEBULA
FATH 1.489	12 12 57.	+ 14 58	14		NEBULA
MCG+04-29-053	12 12 57.	+ 22 05	90	15.3	GALAXY
MCG-06-27-011	12 12 57.	- 35 21	156	12.	GALAXY
REIN 3.031	12 12 57.45	+ 09 51 50.8			NEBULA
AMES 0309	12 12 58.	+ 11 08 04.	17	17.6	NEBULA
IC 3071	12 12 59.	+ 09 49 28.			MAY NOT EXIST
ARP 106	12 12 59.	+ 28 28			PECULIAR GALAXY
HOLM 352A	12 12 59.	+ 51 37	72	13.6	PART OF MULTIPLE GALAXY
VDB.66G 116	12 13	- 11 12	100		DWARF GALAXY
ZWG 069.107	12 13 00.	+ 09 52		13.7	GALAXY
UGC 07268	12 13 00.	+ 09 52	96	13.7	GALAXY
MCG+02-31-069	12 13 00.	+ 09 52	96	13.7	GALAXY
AMES 0310	12 13 00.	+ 13 45 52.	28	17.8	NEBULA
FATH 1.490	12 13 00.	+ 14 13	16		NEBULA
UGC 07269	12 13 00.	+ 15 17	60	17.	GALAXY DWARF
ZC 1213.0+1535	12 13 00.	+ 15 35	1340		CLUSTER OF GALAXIES
MCG+03-31-084	12 13 00.	+ 17 06	54	15.5	GALAXY
ZWG 128.063	12 13 00.	+ 22 06		15.3	GALAXY
UGC 07270	12 13 00.	+ 22 06	102	15.3	GALAXY S
MCG+04-29-055	12 13 00.	+ 22 10	36	15.1	GALAXY
MCG+04-29-054	12 13 00.	+ 24 14 30.	18	14.3	GALAXY
LB 00037	12 13 00.	+ 26 19		16.4	FAINT BLUE STAR
ZC 1213.0+4213	12 13 00.	+ 42 13	3490		CLUSTER OF GALAXIES
ZWG 215.055	12 13 00.	+ 43 42		15.4	GALAXY
UGC 07271	12 13 00.	+ 43 42	144	15.4	GALAXY
UGC 07272	12 13 00.	+ 52 13	60	16.0	GALAXY Sc
MCG+11-15-039	12 13 00.	+ 66 15	114	13.5	GALAXY
SHB 196	12 13 01.5	+ 53 52 35.		17.	QUASI-STELLAR OBJECT
AMES 0311	12 13 02.	+ 06 09 16.	48	17.3	NEBULA
RNGC 4207	12 13 02.	+ 09 52		13.5	GALAXY
LB 02283	12 13 02.	+ 51 15 18.		16.7	FAINT BLUE STAR
AMES 0314	12 13 03.	+ 07 58 40.	32	16.8	NEBULA
AMES 0313	12 13 03.	+ 08 23 04.	26	17.8	NEBULA
AMES 0312	12 13 03.	+ 12 29 52.	28	17.0	NEBULA
SVEN 275	12 13 03.	+ 14 11	108	12.3	GALAXY
TON-N 1485	12 13 03.	+ 32 50		16.6	BLUE STAR
IC 3072	12 13 04.	+ 09 50 04.			MAY NOT EXIST
AMES 0316	12 13 05.	+ 12 29 46.	26	17.2	NEBULA
AMES 0315	12 13 05.	+ 13 04 34.	23	17.4	NEBULA
REIN 3.032	12 13 05.98	+ 08 33 47.1			NEBULA
AMES 0319	12 13 06.	+ 07 58 16.	26	17.2	NEBULA
ZWG 041.054	12 13 06.	+ 08 25		15.3	GALAXY
UGC 07273	12 13 06.	+ 08 25	60	15.3	GALAXY S
ZWG 069.108	12 13 06.	+ 08 34		15.7	GALAXY
AMES 0318	12 13 06.	+ 11 35 10.	28	17.8	NEBULA
AMES 0317	12 13 06.	+ 13 02 04.	19	17.2	NEBULA
ZWG 069.109	12 13 06.	+ 13 53		15.6	GALAXY
UGC 07274	12 13 06.	+ 13 53	60	15.6	GALAXY DWARF
ZWG 069.110	12 13 06.	+ 14 11		11.9	GALAXY
UGC 07275	12 13 06.	+ 14 11	180	11.9	GALAXY Sc
MCG+02-31-070	12 13 06.	+ 14 11	168	11.9	GALAXY
REA 25	12 13 06.	+ 15 17	48		DWARF GALAXY
ZWG 098.117	12 13 06.	+ 17 06		15.5	GALAXY
ZWG 128.064	12 13 06.	+ 22 11		15.1	GALAXY
ZWG 128.065	12 13 06.	+ 24 15		14.3	GALAXY
UGC 07276	12 13 06.	+ 24 15	102	14.3	GALAXY E
ZWG 158.052	12 13 06.	+ 27 17		15.4	GALAXY
KARA.72 327B	12 13 06.	+ 28 26	36		PART OF DOUBLE GALAXY
ZWG 158.053	12 13 06.	+ 28 27		14.4	GALAXY
UGC 07277	12 13 06.	+ 28 27	120	14.4	GALAXY DBL SYS
KARA.72 327A	12 13 06.	+ 28 27	60	14.4	PART OF DOUBLE GALAXY
MCG+06-27-042	12 13 06.	+ 36 36 30.	450	9.	GALAXY
ZWG 187.032	12 13 06.	+ 36 37		10.3	GALAXY
UGC 07278	12 13 06.	+ 36 37	660	10.3	GALAXY IRR
SEY 172	12 13 06.	+ 43 41 37.		15.0	FAINT GALAXY
MCG+07-25-052	12 13 06.	+ 43 43	108	14.	GALAXY
MCG+08-22-086	12 13 06.	+ 47 01	42	16.	GALAXY
MCG+09-20-129	12 13 06.	+ 52 11	72	16.	GALAXY
MCG-01-31-010	12 13 06.	- 07 32	60	16.	GALAXY
MCG-06-27-012	12 13 06.	- 34 37	42	14.	GALAXY
REIN 2.153	12 13 06.69	+ 14 10 46.2			NEBULA
AMES 0320	12 13 07.	+ 10 15 16.	24	17.9	NEBULA
FATH 2.068	12 13 07.	+ 14 12	122		NEBULA
KEEL 355	12 13 07.6	+ 38 13 55.			NEBULA
AMES 0322	12 13 08.	+ 05 50 04.	30	17.6	NEBULA
AMES 0321	12 13 08.	+ 05 51 52.	36	17.6	NEBULA
RNGC 4212	12 13 08.	+ 14 11		12.5	GALAXY
ARC 1505	12 13 08.	+ 18 58		17.5	RICH CLUSTER OF GALAXIES
RNGC 4213	12 13 08.	+ 24 15		14.5	GALAXY
RNGC 4211	12 13 08.	+ 28 27		14.5	GALAXY

OBJECT NAME	RIGHT ASCEN.	DECLINATION	DIAM.	MAGN.	TYPE OF OBJECT
REIN 3.033	12 13 08.39	+ 08 24 45.0			NEBULA
REIN 6.096A	12 13 08.68	+ 36 36 16.6			NEBULA
REIN 6.096B	12 13 08.83	+ 36 36 17.6			NEBULA
REIZ 2094	12 13 09.	+ 04 35	12	16.3	GALAXY
AMES 0325	12 13 09.	+ 05 51 04.	29	17.4	NEBULA
AMES 0324	12 13 09.	+ 06 16 40.	26	17.5	NEBULA
AMES 0323	12 13 09.	+ 12 30 40.	22	16.8	NEBULA
HN 0838	12 13 09.	+ 13 53		16.	NEBULA
IC 3073	12 13 09.	+ 13 53			NONSTELLAR OBJECT
REIZ 2095	12 13 09.	+ 14 10	120	12.3	GALAXY
RNGC 4214	12 13 09.	+ 36 36		10.5	GALAXY
REIN 6.097	12 13 09.97	+ 36 36 13.9			NEBULA
AMES 0326	12 13 10.	+ 05 27 58.	52	17.4	NEBULA
FATH 1.491	12 13 10.	+ 14 42	22		NEBULA
SN 1954A	12 13 10.	+ 36 33		9.8	SUPERNOVA
REIZ 2093	12 13 10.	- 03 09	36	15.6	GALAXY
REIN 6.098A	12 13 10.03	+ 36 35 52.4			NEBULA
REIN 6.098B	12 13 10.26	+ 36 35 46.9			NEBULA
ZWG 069.111	12 13 12.	+ 10 58		15.1	GALAXY
UGC 07279	12 13 12.	+ 10 58	132	15.1	GALAXY
MCG+02-31-071	12 13 12.	+ 10 58	132	15.1	GALAXY
MCG+03-34-010	12 13 12.	+ 15 45	24	15.7	GALAXY
VV 199B	12 13 12.	+ 28 27	30	16.	INTERACTING GALAXY
VV 199A	12 13 12.	+ 28 27	60	15.5	INTERACTING GALAXY
LB 02284	12 13 12.	+ 51 32 36.		15.8	FAINT BLUE STAR
ZWG 269.048	12 13 12.	+ 55 08		15.5	GALAXY
ZWG 293.006	12 13 12.	+ 62 10		14.9	GALAXY
ZWG 292.085	12 13 12.	+ 62 10		14.9	GALAXY
KARA.73B 0522	12 13 12.	+ 62 10	24	14.9	ISOLATED GALAXY E
OCL 0876	12 13 12.	- 58 08	162	13.	OPEN STAR CLUSTER
AMES 0327	12 13 13.	+ 08 44 10.	31	17.7	NEBULA
IC 3074	12 13 13.	+ 10 58 40.			NONSTELLAR OBJECT
REIZ 2096	12 13 13.	+ 36 36	90	10.7	GALAXY
REIN 3.034	12 13 13.25	+ 10 58 41.3			NEBULA
RNGC 4218	12 13 15.	+ 48 25		13.0	GALAXY
MCG+09-20-131	12 13 15.	+ 52 40	72	15.	GALAXY
MCG+09-20-130	12 13 15.	+ 55 08	60	14.	GALAXY
LB 02285	12 13 15.	+ 58 09 36.		16.0	FAINT BLUE STAR
KEEL 356	12 13 16.2	+ 37 48 03.			NEBULA
AMES 0328	12 13 17.	+ 07 10 04.	20	17.6	NEBULA
ZWG 013.113	12 13 18.	+ 00 40		14.8	GALAXY
UGC 07280	12 13 18.	+ 00 40	78	14.8	GALAXY S
MCG+00-31-043	12 13 18.	+ 00 40	66	14.8	GALAXY
AMES 0329	12 13 18.	+ 05 34 04.	26	17.6	NEBULA
ZC 1213.3+0606	12 13 18.	+ 06 06	340		CLUSTER OF GALAXIES
ZWG 041.055	12 13 18.	+ 06 40		13.0	GALAXY
UGC 07281	12 13 18.	+ 06 40	102	13.0	GALAXY S0-a
MCG+01-31-031	12 13 18.	+ 06 40	72	13.0	GALAXY
MCG+03-31-085	12 13 18.	+ 14 42	96	15.4	GALAXY
FATH 1.492	12 13 18.	+ 15 08	19		NEBULA
ZC 1213.3+1900	12 13 18.	+ 19 00	1140		CLUSTER OF GALAXIES
MCG+04-29-056	12 13 18.	+ 23 52	42	15.1	GALAXY
MCG+05-29-044	12 13 18.	+ 27 43	66	15.3	GALAXY
ARC 1506	12 13 18.	+ 32 03		18.0	RICH CLUSTER OF GALAXIES
MRK 437	12 13 18.	+ 41 09	6	16.	GALAXY WITH UV CONTINUUM
ZWG 243.053	12 13 18.	+ 47 21		12.4	GALAXY
UGC 07282	12 13 18.	+ 47 21	330	12.4	GALAXY Sb
HOLM 354A	12 13 18.	+ 47 22	270	12.0	PART OF MULTIPLE GALAXY
ZWG 243.054	12 13 18.	+ 48 23		13.2	GALAXY
UGC 07283	12 13 18.	+ 48 23	60	13.2	GALAXY
ZWG 269.049	12 13 18.	+ 52 40		15.3	GALAXY
RNGC 4216	12 13 20.	+ 13 25		11.0	GALAXY
FATH 1.493	12 13 20.	+ 14 19	35		NEBULA
REIN 3.035	12 13 20.70	+ 13 29 34.5			NEBULA
REIZ 2097	12 13 21.	+ 06 40	78	13.2	GALAXY
REIZ 2098	12 13 21.	+ 13 25	390	10.8	GALAXY
HOLM 353A	12 13 21.	+ 13 26	450	10.9	PART OF MULTIPLE GALAXY
RNGC 4217	12 13 21.	+ 47 22		12.0	GALAXY
MCG+08-22-087	12 13 21.	+ 47 23	300	11.9	GALAXY
HARO 28	12 13 21.	+ 48 30			BLUE EMISSION-LINE GALAXY
REIN 3.036A	12 13 21.71	+ 13 25 39.6			NEBULA
REIN 3.036B	12 13 21.77	+ 13 25 38.2			NEBULA
KEEL 357	12 13 22.6	+ 38 17 22.			NEBULA
FATH 1.495	12 13 23.	+ 14 17	19		NEBULA
FATH 1.494	12 13 23.	+ 14 29	14		NEBULA
FATH 1.496	12 13 23.	+ 14 42	16		NEBULA
FATH 1.497	12 13 23.	+ 14 44	54		NEBULA
IC 3075	12 13 23.	+ 23 51 58.			NONSTELLAR OBJECT
ARC 1507	12 13 23.	+ 60 15		15.8	RICH CLUSTER OF GALAXIES
ZWG 041.056	12 13 24.	+ 03 35		15.2	GALAXY
MCG+01-31-032	12 13 24.	+ 03 35	24	15.2	GALAXY
ZWG 041.057	12 13 24.	+ 04 56		14.8	GALAXY
MCG+01-31-033	12 13 24.	+ 04 56	36	14.8	GALAXY
ZC 1213.4+0545	12 13 24.	+ 05 45	2020		CLUSTER OF GALAXIES
ZWG 069.112	12 13 24.	+ 13 26		11.2	GALAXY
SCH 29	12 13 24.	+ 13 26		11.2	PECULIAR GALAXY
UGC 07284	12 13 24.	+ 13 26	510	11.2	GALAXY Sb
MCG+02-31-072	12 13 24.	+ 13 26	438	11.2	GALAXY
ZWG 069.113	12 13 24.	+ 13 29		15.4	GALAXY
FATH 1.498	12 13 24.	+ 14 32	11		NEBULA
ZWG 098.118	12 13 24.	+ 14 42		15.4	GALAXY
UGC 07285	12 13 24.	+ 14 42	84	15.4	GALAXY Sc
ZC 1213.4+1600	12 13 24.	+ 16 00	1680		CLUSTER OF GALAXIES
ZWG 128.066	12 13 24.	+ 23 53		15.1	GALAXY
ZWG 158.054	12 13 24.	+ 26 56		14.6	GALAXY
MCG+05-29-045	12 13 24.	+ 26 56	36	14.6	GALAXY
ZWG 158.055	12 13 24.	+ 27 43		15.3	GALAXY
UGC 07286	12 13 24.	+ 27 43	90	15.3	GALAXY Sa-b
MRK 046	12 13 24.	+ 41 02	11	17.	GALAXY WITH UV CONTINUUM
REIZ 2100	12 13 24.	+ 47 22	234	12.2	GALAXY
MCG+08-22-088	12 13 24.	+ 48 25	48	13.6	GALAXY
MCG+09-20-132	12 13 24.	+ 52 23	60	16.	GALAXY
MCG+09-20-133	12 13 24.	+ 55 00	36	16.	GALAXY
MCG-05-29-032	12 13 24.	- 28 47	60	15.	GALAXY
MCG-06-27-013	12 13 24.	- 37 49 30.	66	16.	GALAXY
REIN 3.037A	12 13 24.19	+ 13 25 38.2			NEBULA
REIN 3.037B	12 13 24.25	+ 13 25 36.2			NEBULA
RNGC 4215	12 13 25.	+ 06 41		13.0	GALAXY
AMES 0330	12 13 25.	+ 10 32 52.	31	18.0	NEBULA
AMES 0331	12 13 26.	+ 05 32 10.	32	17.3	NEBULA
AMES 0333	12 13 27.	+ 04 55 10.	26	16.4	NEBULA
AMES 0332	12 13 27.	+ 10 32 52.	28	17.8	NEBULA
FATH 1.500	12 13 27.	+ 14 18	24		NEBULA
FATH 1.499	12 13 27.	+ 14 31	14		NEBULA
REIZ 2099	12 13 27.	+ 14 42	30	15.2	GALAXY
TON-N 1486	12 13 27.	+ 35 26		17.	BLUE STAR
LB 02286	12 13 27.	+ 56 02 36.		16.7	FAINT BLUE STAR
REIN 3.038	12 13 27.41	+ 08 28 50.0			NEBULA

OBJECT NAME	RIGHT ASCEN.	DECLINATION	DIAM.	MAGN.	TYPE OF OBJECT
REIN 3.039	12 13 27.51	+ 12 57 53.5			NEBULA
AMES 0335	12 13 29.	+ 12 46 40.	28	16.8	NEBULA
AMES 0334	12 13 29.	+ 13 05 16.	14	18.0	NEBULA
FATH 1.501	12 13 29.	+ 14 28	16		NEBULA
AMES 0336	12 13 30.	+ 05 58 28.	24	17.6	NEBULA
ZWG 041.058	12 13 30.	+ 08 29		15.4	GALAXY
ZWG 069.114	12 13 30.	+ 12 58		15.2	GALAXY
MCG+02-31-073	12 13 30.	+ 12 58	30	15.2	GALAXY
ZWG 069.115	12 13 30.	+ 14 28		15.4	GALAXY
ZWG 098.119	12 13 30.	+ 17 48		15.7	GALAXY
MCG+03-31-086	12 13 30.	+ 17 48	48	15.7	GALAXY
MCG+05-29-046	12 13 30.	+ 28 25	36	15.5	GALAXY
MCG+05-29-047	12 13 30.	+ 29 03	72	15.5	GALAXY
MRK 047	12 13 30.	+ 40 51	9	16.	GALAXY WITH UV CONTINUUM
REIZ 2102	12 13 30.	+ 48 25	36	13.6	GALAXY
MCG+09-20-134	12 13 30.	+ 55 00	36	16.	GALAXY
ZWG 293.007	12 13 30.	+ 59 47		15.6	GALAXY
AMES 0337	12 13 31.	+ 08 30 40.	25	17.4	NEBULA
AMES 0338	12 13 32.	+ 05 54 40.	31	17.8	NEBULA
IC 3076	12 13 32.	+ 09 21 22.			MAY NOT EXIST
IC 3079	12 13 33.	+ 11 49			NONSTELLAR OBJECT
HN 0840	12 13 33.	+ 12 56		14.	NEBULA
IC 3078	12 13 33.	+ 12 56			NONSTELLAR OBJECT
HN 0839	12 13 33.	+ 14 41		15.	NEBULA
KARA.68 093	12 13 33.	+ 15 49	34		DWARF GALAXY
REIZ 2101	12 13 34.	- 03 18	42	14.7	GALAXY
AMES 0339	12 13 35.	+ 08 28 52.	30	17.8	NEBULA
IC 3077	12 13 35.	+ 14 41	14		FAINT NEBULA
FATH 2.069	12 13 35.	+ 14 41	14		NEBULA
ZWG 013.116	12 13 36.	+ 01 27		15.1	GALAXY
REA 26	12 13 36.	+ 08 40	36		DWARF GALAXY
ZWG 069.116	12 13 36.	+ 11 48		15.5	GALAXY
ZWG 069.117	12 13 36.	+ 12 58		15.6	GALAXY
MCG+02-31-074	12 13 36.	+ 12 58	30	15.6	GALAXY
ZWG 098.120	12 13 36.	+ 19 43		15.7	GALAXY
ZC 1213.6+2149	12 13 36.	+ 21 49	740		CLUSTER OF GALAXIES
ZC 1213.6+2429	12 13 36.	+ 24 29	670		CLUSTER OF GALAXIES
ZWG 158.056	12 13 36.	+ 28 24		15.5	GALAXY
UGC 07287	12 13 36.	+ 28 24	66	15.5	GALAXY SBc
LB 02288	12 13 36.	+ 50 46 36.		16.2	FAINT BLUE STAR
MCG+09-20-135	12 13 36.	+ 55 45	30	16.	GALAXY
LB 00628	12 13 36.	+ 57 26 18.		17.7	FAINT BLUE STAR
LB 02287	12 13 36.	+ 62 43 48.		15.0	FAINT BLUE STAR
ZWG 315.029	12 13 36.	+ 66 30		13.6	GALAXY
UGC 07288	12 13 36.	+ 66 30	114	13.6	GALAXY SB0
RNGC 4221	12 13 36.	+ 66 31		13.5	GALAXY
ZWG 013.115	12 13 36.	- 03 01		15.7	GALAXY
ZWG 013.114	12 13 36.	- 03 17		14.8	GALAXY
MCG+00-31-044	12 13 36.	- 03 17	66	14.8	GALAXY
REIN 3.040	12 13 36.36	+ 12 58 09.0			NEBULA
AMES 0340	12 13 37.	+ 08 34 52.	23	17.8	NEBULA
FATH 1.502	12 13 37.	+ 14 17	41		NEBULA
AMES 0341	12 13 38.	+ 08 01 40.	14	18.0	NEBULA
ARC 1508	12 13 38.	+ 17 46		17.2	RICH CLUSTER OF GALAXIES
HN 0843	12 13 39.	+ 12 57		15.	NEBULA
IC 3081	12 13 39.	+ 12 57			NONSTELLAR OBJECT
AMES 0342	12 13 39.	+ 13 42 34.	61	16.5	NEBULA
HN 0842	12 13 39.	+ 14 26		14.	NEBULA
RNGC 4220	12 13 39.	+ 48 10		12.5	GALAXY
REIN 3.041	12 13 39.95	+ 08 12 23.5			NEBULA
AMES 0346	12 13 41.	+ 05 43 28.	17	17.8	NEBULA
AMES 0345	12 13 41.	+ 06 59 52.	17	17.6	NEBULA
AMES 0344	12 13 41.	+ 13 04 58.	16	17.3	NEBULA
AMES 0343	12 13 41.	+ 13 51 22.	24	17.2	NEBULA
IC 3080	12 13 41.	+ 14 30	16		FAINT NEBULA
FATH 2.070	12 13 41.	+ 14 30	16		NEBULA
IC 3082	12 13 41.	+ 24 07 10.			NONSTELLAR OBJECT
AMES 0347	12 13 42.	+ 05 54 40.	26	18.0	NEBULA
ZWG 041.059	12 13 42.	+ 08 12		15.7	GALAXY
FATH 1.503	12 13 42.	+ 14 32	14		NEBULA
REA 27	12 13 42.	+ 15 51	42		DWARF GALAXY
ZWG 098.121	12 13 42.	+ 18 05		15.4	GALAXY
ZWG 098.122	12 13 42.	+ 18 41		15.7	GALAXY
ZWG 158.057	12 13 42.	+ 29 02		15.5	GALAXY
UGC 07289	12 13 42.	+ 29 02	90	15.5	GALAXY S
ZC 1213.7+3604	12 13 42.	+ 36 04	2420		CLUSTER OF GALAXIES
ZWG 243.055	12 13 42.	+ 48 09		12.4	GALAXY
UGC 07290	12 13 42.	+ 48 09	234	12.4	GALAXY Sa
MCG+09-20-136	12 13 42.	+ 51 05	48	17.	GALAXY
LB 02289	12 13 42.	+ 56 19 06.		17.1	FAINT BLUE STAR
MCG+10-18-014	12 13 42.	+ 59 47	24	16.	GALAXY
MCG+10-18-013	12 13 42.	+ 59 47	27	16.	GALAXY
MCG+10-18-015	12 13 42.	+ 60 15	24	16.	GALAXY
MCG-04-29-014	12 13 42.	- 26 23 30.	15	15.5	GALAXY
AMES 0348	12 13 43.	+ 12 54 40.	26	17.0	NEBULA
REIZ 2106	12 13 43.	+ 66 30	12	13.9	GALAXY
FATH 1.505	12 13 45.	+ 14 26	19		NEBULA
FATH 1.504	12 13 45.	+ 14 30	11		NEBULA
MCG+03-31-087	12 13 45.	+ 15 31	60	15.3	GALAXY
MCG+08-22-089	12 13 45.	+ 48 10	180	12.2	GALAXY
AMES 0352	12 13 46.	+ 05 25 10.	18	17.8	NEBULA
AMES 0351	12 13 46.	+ 05 47 04.	28	17.4	NEBULA
AMES 0350	12 13 46.	+ 05 47 34.	24	17.7	NEBULA
AMES 0349	12 13 46.	+ 05 56 52.	23	17.8	NEBULA
AMES 0353	12 13 47.	+ 05 35 04.	13	17.8	NEBULA
ARC 1509	12 13 47.	+ 35 54		17.0	RICH CLUSTER OF GALAXIES
RNGC 4219	12 13 47.	- 43 03		12.5	GALAXY
KEEL 358	12 13 47.7	+ 47 34 55.		15.	NEBULA
ZWG 069.118	12 13 48.	+ 11 05		15.5	GALAXY
ZWG 098.123	12 13 48.	+ 15 32		15.3	GALAXY
MCG+04-29-057	12 13 48.	+ 24 10 30.	24	15.7	GALAXY
ZWG 128.067	12 13 48.	+ 24 12		15.7	GALAXY
REIZ 2104	12 13 48.	+ 46 36	54	15.5	GALAXY
REIZ 2105	12 13 48.	+ 48 09	102	12.8	GALAXY
MCG+10-18-016	12 13 48.	+ 60 15 30.	24	16.	GALAXY
7ZW 445	12 13 48.	+ 60 46			COMPACT GALAXY
MCG+10-18-017	12 13 48.	+ 60 46	45	16.	GALAXY
ZCG 199.01	12 13 48.	+ 63 43		16.0	COMPACT GALAXY
MCG+11-15-040	12 13 48.	+ 66 30	102	13.9	GALAXY
ZC 1213.8-0005	12 13 48.	- 00 05	1680		CLUSTER OF GALAXIES
AGU 34	12 13 48.	- 43 03 00.	102	12.5	PECULIAR GALAXY
REIN 3.042	12 13 48.52	+ 11 04 59.2			NEBULA
AMES 0358	12 13 49.	+ 05 44 22.	20	17.8	NEBULA
AMES 0357	12 13 49.	+ 05 50 04.	31	17.7	NEBULA
AMES 0356	12 13 49.	+ 06 57 46.	28	17.8	NEBULA
AMES 0355	12 13 49.	+ 12 31 22.	18	17.6	NEBULA
AMES 0354	12 13 49.	+ 12 48 16.	26	17.5	NEBULA
FATH 1.506	12 13 49.	+ 14 22	8		NEBULA

OBJECT NAME	RIGHT ASCEN.	DECLINATION	DIAM.	MAGN.	TYPE OF OBJECT
AMES 0360	12 13 50.	+ 11 23 40.	16	17.8	NEBULA
AMES 0359	12 13 50.	+ 12 26 28.	32	17.6	NEBULA
RNGC 4222	12 13 50.	+ 13 35		14.5	GALAXY
HOLM 353C	12 13 50.	+ 13 36	150	13.3	PART OF MULTIPLE GALAXY
REIN 3.043	12 13 50.62	+ 13 35 11.2			NEBULA
HN 0844	12 13 51.	+ 12 52		15.	NEBULA
IC 3083	12 13 51.	+ 12 52			MAY NOT EXIST
REIZ 2107	12 13 51.	+ 13 37	180	14.2	GALAXY
REIZ 2103	12 13 51.	+ 15 32	42	14.7	GALAXY
MCG+04-29-058	12 13 51.	+ 24 06	24	15.7	GALAXY
RNGC 4226	12 13 51.	+ 47 18		14.5	GALAXY
AMES 0361	12 13 52.	+ 06 20 16.	22	17.6	NEBULA
IC 3084	12 13 52.	+ 24 11 40.			NONSTELLAR OBJECT
IC 3085	12 13 53.	+ 09 44 52.			SINGLE STAR
AMES 0362	12 13 53.	+ 12 51 16.	25	17.9	NEBULA
FATH 1.507	12 13 53.	+ 14 47	8		NEBULA
AMES 0365	12 13 54.	+ 05 33 04.	34	17.7	NEBULA
AMES 0364	12 13 54.	+ 11 10 16.	20	16.4	NEBULA
AMES 0363	12 13 54.	+ 13 16 16.	23	17.4	NEBULA
ZWG 069.119	12 13 54.	+ 13 35		14.6	GALAXY
UGC 07291	12 13 54.	+ 13 35	186	14.6	GALAXY Sc
MCG+02-31-075	12 13 54.	+ 13 35	192	14.6	GALAXY
ZWG 098.124	12 13 54.	+ 14 46		15.1	GALAXY
MCG+03-31-088	12 13 54.	+ 14 46	60	15.1	GALAXY
ZWG 128.068	12 13 54.	+ 24 07		15.7	GALAXY
ZC 1213.9+2612	12 13 54.	+ 26 12	540		CLUSTER OF GALAXIES
MCG+08-22-090	12 13 54.	+ 47 18 30.	60	13.6	GALAXY
ZWG 243.056	12 13 54.	+ 50 17		15.7	GALAXY
MCG+08-22-091	12 13 54.	+ 50 19	18	16.	GALAXY
ZC 1213.9+7307	12 13 54.	+ 73 07	1610		CLUSTER OF GALAXIES
MCG-02-31-026	12 13 54.	- 11 14	72	16.	GALAXY
REIN 3.044	12 13 54.40	+ 13 34 10.1			NEBULA
RNGC 4223	12 13 55.	+ 06 58			NON-EXISTENT OBJECT
IC 3086	12 13 55.	+ 09 17 10.			MAY NOT EXIST
AMES 0366	12 13 55.	+ 10 40 40.	29	17.4	NEBULA
IC 3087	12 13 55.	+ 13 33 58.			TWO STARS
HOLM 354B	12 13 55.	+ 47 18	48	13.6	PART OF MULTIPLE GALAXY
AMES 0367	12 13 56.	+ 05 38 04.	25	17.7	NEBULA
IC 3088	12 13 56.	+ 09 44 22.			SINGLE STAR
KEEL 359	12 13 56.9	+ 47 32 21.		15.	SPIRAL NEBULA
AMES 0369	12 13 57.	+ 05 41 04.	22	17.4	NEBULA
AMES 0368	12 13 57.	+ 12 21 40.	23	17.7	NEBULA
FATH 1.508	12 13 57.	+ 14 19	19		NEBULA
MCG+03-34-011	12 13 57.	+ 19 20	36	15.	GALAXY
MCG+06-27-043	12 13 57.	+ 33 48	90	14.	GALAXY
IC 3089	12 13 58.	+ 24 06 22.			NONSTELLAR OBJECT
IC 3090	12 13 59.	+ 09 43 04.			TWO STARS
VDB.66G 117	12 14	+ 29 03	70		DWARF GALAXY
VDB.66G 118	12 14	- 11 22	70		DWARF GALAXY
ZC 1214.0+0057	12 14 00.	+ 00 57	340		CLUSTER OF GALAXIES
AMES 0372	12 14 00.	+ 05 38 04.	29	17.8	NEBULA
KARA.68 094	12 14 00.	+ 07 07	27		DWARF GALAXY
ZWG 041.060	12 14 00.	+ 07 44		13.3	GALAXY
UGC 07292	12 14 00.	+ 07 44	138	13.3	GALAXY Sa
MCG+01-31-034	12 14 00.	+ 07 44	120	13.3	GALAXY
AMES 0543	12 14 00.	+ 08 09 49.	19	17.6	NEBULA
AMES 0371	12 14 00.	+ 13 01 04.	16	17.4	NEBULA
ZWG 069.120	12 14 00.	+ 13 19		15.2	GALAXY
MCG+02-31-076	12 14 00.	+ 13 19	18	15.2	GALAXY
AMES 0370	12 14 00.	+ 13 45 16.	26	17.3	NEBULA
ZWG 069.121	12 14 00.	+ 14 17		14.7	GALAXY
MCG+02-31-077	12 14 00.	+ 14 17	36	14.7	GALAXY
ZWG 098.125	12 14 00.	+ 17 45		15.4	GALAXY
UGC 07293	12 14 00.	+ 17 45	66	15.4	GALAXY S
ZC 1214.0+2729	12 14 00.	+ 27 29	740		CLUSTER OF GALAXIES
UGC 07294	12 14 00.	+ 30 07	66	16.0	GALAXY Sc
UGC 07295	12 14 00.	+ 33 42	60	16.5	GALAXY
ZWG 187.033	12 14 00.	+ 33 48		13.8	GALAXY
UGC 07296	12 14 00.	+ 33 48	96	13.8	GALAXY S0-a
MCG+06-27-044	12 14 00.	+ 33 50	60	14.5	GALAXY
ZWG 243.057	12 14 00.	+ 47 17		14.4	GALAXY
REIZ 2110	12 14 00.	+ 47 17	36	14.3	GALAXY
UGC 07297	12 14 00.	+ 47 17	72	14.4	GALAXY S
MCG+09-20-137	12 14 00.	+ 52 30	72	15.	GALAXY
UGC 07298	12 14 00.	+ 52 32	72	17.	GALAXY DWARF
ZWG 293.008	12 14 00.	+ 58 10		15.6	GALAXY
MCG+10-18-018	12 14 00.	+ 60 38	30	16.	GALAXY
ZCG 199.02	12 14 00.	+ 63 44		19.5	COMPACT GALAXY
ZWG 013.117	12 14 00.	- 00 40		15.5	GALAXY
MCG-01-31-011	12 14 00.	- 04 51	18	15.5	GALAXY
REIN 3.045A	12 14 00.86	+ 07 44 24.5			NEBULA
REIN 3.045B	12 14 00.91	+ 07 44 21.3			NEBULA
RNGC 4224	12 14 01.	+ 07 44		13.0	GALAXY
ABC 1510	12 14 01.	+ 27 28		18.0	RICH CLUSTER OF GALAXIES
HOLM 355A	12 14 01.	+ 33 48	48	14.1	PART OF MULTIPLE GALAXY
REIN 3.046A	12 14 01.17	+ 13 18 34.2			NEBULA
REIN 3.046B	12 14 01.18	+ 13 18 32.6			NEBULA
AMES 0374	12 14 02.	+ 05 37 04.	18	17.8	NEBULA
AMES 0373	12 14 02.	+ 12 51 46.	14	17.8	NEBULA
REIZ 2109	12 14 03.	+ 07 44	144	13.2	GALAXY
HN 0846	12 14 03.	+ 10 20		15.5	NEBULA
IC 3092	12 14 03.	+ 10 20			NONSTELLAR OBJECT
FATH 1.509	12 14 03.	+ 14 12	8		NEBULA
HN 0845	12 14 03.	+ 14 16		14.	NEBULA
IC 3091	12 14 03.	+ 14 16			TWO STARS
MCG+03-31-089	12 14 03.	+ 17 46	60	15.4	GALAXY
MCG+05-29-048	12 14 03.	+ 29 01	78	16.	GALAXY
RNGC 4227	12 14 03.	+ 33 48		14.0	GALAXY
RNGC 4228	12 14 03.	+ 36 36			NON-EXISTENT OBJECT
LB 02290	12 14 03.	+ 60 02 36.		15.7	FAINT BLUE STAR
AMES 0376	12 14 04.	+ 07 40 46.	20	17.8	NEBULA
AMES 0375	12 14 04.	+ 08 49 04.	18	17.9	NEBULA
LB 00629	12 14 04.	+ 59 34 06.		15.5	FAINT BLUE STAR
REIZ 2108	12 14 04.	- 04 50	18	14.4	GALAXY
FATH 1.510	12 14 04.	- 15 43	22		NEBULA
AMES 0377	12 14 06.	+ 05 37 40.	16	17.8	NEBULA
ZWG 098.126	12 14 06.	+ 14 33		15.4	GALAXY
ZWG 098.127	12 14 06.	+ 18 39		15.7	GALAXY
ZC 1214.1+2525	12 14 06.	+ 25 25	610		CLUSTER OF GALAXIES
LB 00002	12 14 06.	+ 26 45		16.2	FAINT BLUE STAR
ZWG 187.034	12 14 06.	+ 33 50		14.3	GALAXY
HOLM 355B	12 14 06.	+ 33 50	30	14.7	PART OF MULTIPLE GALAXY
UGC 07299	12 14 06.	+ 33 50	84	14.3	GALAXY S
ZC 1214.1+4058	12 14 06.	+ 40 58	740		CLUSTER OF GALAXIES
ZC 1214.1-0258	12 14 06.	- 02 58	1810		CLUSTER OF GALAXIES
MCG-02-31-027	12 14 06.	- 12 02	36	14.5	GALAXY
FATH 1.511	12 14 06.	- 15 38	19		NEBULA
IC 0783B	12 14 06.8	+ 16 01 24.			GALAXY S0
AMES 0380	12 14 07.	+ 05 36 58.	16	18.0	NEBULA
AMES 0379	12 14 07.	+ 10 29 16.	17	17.6	NEBULA
AMES 0378	12 14 07.	+ 10 55 40.	23	17.6	NEBULA
FATH 1.513	12 14 07.	+ 14 20	11		NEBULA
FATH 1.512	12 14 07.	+ 14 49	16		NEBULA
RNGC 4225	12 14 07.	- 12 02		14.0	GALAXY
AMES 0381	12 14 08.	+ 08 24 28.	20	17.5	NEBULA
LB 02291	12 14 08.	+ 61 34 18.		15.0	FAINT BLUE STAR
AMES 0383	12 14 09.	+ 08 17 04.	20	17.8	NEBULA
AMES 0382	12 14 09.	+ 12 15 22.	31	17.4	NEBULA
SVEN 276	12 14 09.	+ 14 34	18	14.0	GALAXY
FATH 1.514	12 14 09.	+ 14 37	14		NEBULA
TON-N 1487	12 14 09.	+ 33 37		16.6	BLUE STAR
RNGC 4229	12 14 09.	+ 33 50		14.5	GALAXY
IC 3093	12 14 10.	+ 14 34			BRIGHT NEBULA
FATH 2.071	12 14 10.	+ 14 34	19		NEBULA
LB 02292	12 14 10.	+ 56 52 18.		16.3	FAINT BLUE STAR
REIZ 2111	12 14 10.	- 00 42	24	15.1	GALAXY
AMES 0385	12 14 12.	+ 08 41 16.	16	17.7	NEBULA
AMES 0384	12 14 12.	+ 09 04 58.	44	17.8	NEBULA
MCG+03-34-012	12 14 12.	+ 14 41	48	15.5	GALAXY
MCG+03-31-090	12 14 12.	+ 14 47	72	15.4	GALAXY
UGC 07300	12 14 12.	+ 29 00	102	17.	GALAXY DWRF IR
REIZ 2112	12 14 12.	+ 46 20	84	14.4	GALAXY
ZWG 243.058	12 14 12.	+ 46 21		15.7	GALAXY
UGC 07301	12 14 12.	+ 46 21	126	15.7	GALAXY Sc
MCG+08-22-092	12 14 12.	+ 46 22	90	15.	GALAXY
ZCG 199.03B	12 14 12.	+ 63 44		18.3	COMPACT GALAXY
ZCG 199.03A	12 14 12.	+ 63 44		19.0	COMPACT GALAXY
ZCG 199.04	12 14 12.	+ 63 47		17.9	COMPACT GALAXY
MCG+12-12-004	12 14 12.	+ 69 45	1350	10.0	GALAXY
ZWG 013.118	12 14 12.	- 03 09		15.7	GALAXY
ABC 1511	12 14 12.	- 18 59		17.4	RICH CLUSTER OF GALAXIES
AMES 0387	12 14 13.	+ 09 00 16.	16	17.6	NEBULA
AMES 0386	12 14 13.	+ 09 10 58.	30	17.0	NEBULA
AMES 0388	12 14 14.	+ 08 19 16.	19	16.9	NEBULA
HN 0847	12 14 15.	+ 14 32		14.	NEBULA
RNGC 4232	12 14 16.	+ 47 43		14.5	GALAXY
RNGC 4231	12 14 16.	+ 47 44		14.5	GALAXY
KEEL 360	12 14 16.1	+ 47 49 21.		16.	NEBULA
AMES 0389	12 14 17.	+ 01 49 38.	20	17.6	NEBULA
REIZ 2115	12 14 17.	+ 47 44	30	14.0	GALAXY
REIZ 2114	12 14 17.	+ 47 45	18	13.8	GALAXY
AMES 0390	12 14 18.	+ 08 38 52.	19	17.8	NEBULA
AMES 0389	12 14 18.	+ 13 19 16.	20	17.6	NEBULA
MCG+03-34-013	12 14 18.	+ 14 40	30	16.	GALAXY
ZWG 098.128	12 14 18.	+ 14 47		15.4	GALAXY
ZWG 128.069	12 14 18.	+ 24 14		15.6	GALAXY
UGC 07302	12 14 18.	+ 30 33	60	17.	GALAXY DWARF
ZWG 215.056	12 14 18.	+ 39 28		15.7	GALAXY
ZWG 243.059	12 14 18.	+ 47 42		14.6	GALAXY
UGC 07303	12 14 18.	+ 47 42	90	14.6	GALAXY SBa-b
ZWG 243.060	12 14 18.	+ 47 43		14.5	GALAXY
UGC 07304	12 14 18.	+ 47 43	78	14.5	GALAXY
MCG+08-22-093	12 14 18.	+ 47 43	66	14.0	GALAXY
MCG+08-22-094	12 14 18.	+ 47 44	42	13.8	GALAXY
LB 02293	12 14 18.	+ 55 43 42.		17.1	FAINT BLUE STAR
MRK 048	12 14 18.	+ 58 09	10	16.	GALAXY WITH UV CONTINUUM
KEEL 361	12 14 18.0	+ 47 38 19.		15.	SPIRAL NEBULA
AMES 0391	12 14 19.	+ 06 27 28.	17	17.8	NEBULA
RNGC 4236	12 14 19.	+ 69 45		10.5	GALAXY
IC 3096	12 14 20.	+ 14 48	54		BRIGHT NEBULA
FATH 2.072	12 14 20.	+ 14 48	54		NEBULA
KEEL 362	12 14 20.7	+ 38 10 53.			NEBULA
AMES 0393	12 14 21.	+ 05 32 58.	22	17.6	NEBULA
AMES 0392	12 14 21.	+ 11 45 28.	29	17.4	NEBULA
REIZ 2113	12 14 21.	+ 14 47	42	14.7	GALAXY
MCG+04-29-059	12 14 21.	+ 24 13	27	15.6	GALAXY
ABC 1512	12 14 21.	+ 45 30		17.8	RICH CLUSTER OF GALAXIES
AMES 0394	12 14 22.	+ 08 37 52.	35	15.7	NEBULA
HOLM 356B	12 14 23.	+ 47 45	42	14.0	PART OF MULTIPLE GALAXY
HOLM 356A	12 14 23.	+ 47 46	24	13.8	PART OF MULTIPLE GALAXY
HOLM 357A	12 14 23.	+ 69 45	900	10.9	PART OF MULTIPLE GALAXY
ZWG 069.122	12 14 24.	+ 08 38		15.4	GALAXY
RMB 161	12 14 24.	+ 13 01			FAINT BLUE OBJECT
ZWG 069.123	12 14 24.	+ 13 54		14.1	GALAXY
UGC 07305	12 14 24.	+ 13 54	36	14.1	GALAXY
MCG+02-31-078	12 14 24.	+ 13 54	36	14.1	GALAXY
IC 3094	12 14 24.	+ 13 54 16.			NONSTELLAR OBJECT
FEIG 059	12 14 24.	+ 15 51		11.6	FAINT BLUE STAR
IC 3095	12 14 24.	+ 24 14 10.			NONSTELLAR OBJECT
TON-N 0605	12 14 24.	+ 30 23		15.1	BLUE STAR
ZC 1214.4+4955	12 14 24.	+ 49 55	810		CLUSTER OF GALAXIES
ZC 1214.4+6122	12 14 24.	+ 61 22	1340		CLUSTER OF GALAXIES
MCG+11-15-041	12 14 24.	+ 63 41	108	14.	GALAXY
ZC 1214.4+6540	12 14 24.	+ 65 40	940		CLUSTER OF GALAXIES
ZWG 315.030	12 14 24.	+ 66 06		15.5	GALAXY
ZWG 335.008	12 14 24.	+ 69 45		10.7	GALAXY
UGC 07306	12 14 24.	+ 69 45	1380	10.7	GALAXY
KARA.73B 0523	12 14 24.	+ 69 45	1476	10.7	ISOLATED GALAXY S
MCG-04-29-015	12 14 24.	- 25 56 30.	24	16.	GALAXY
AMES 0395	12 14 25.	+ 13 58 28.	24	17.2	NEBULA
AMES 0396	12 14 26.	+ 10 42 04.	26	18.0	NEBULA
AMES 0397	12 14 27.	+ 13 19 04.	17	17.5	NEBULA
HN 0848	12 14 27.	+ 14 46		14.5	NEBULA
TON-N 1488	12 14 27.	+ 32 44		15.8	BLUE STAR
TON-N 1489	12 14 27.	+ 33 05		16.9	BLUE STAR
VV 333A	12 14 27.	- 25 55	36	16.	INTERACTING GALAXY
IC 3097	12 14 28.	+ 09 41 04.			NONSTELLAR OBJECT
HOLM 360C	12 14 28.	+ 12 34	30	14.8	PART OF MULTIPLE GALAXY
REIN 3.047	12 14 28.28	+ 09 41 05.7			NEBULA
BC M1214+106	12 14 28.7	+ 10 36 36.		20.	QUASI-STELLAR OBJECT
SHB 197	12 14 28.7	+ 10 36 36.		20.	QUASI-STELLAR OBJECT
LB 00630	12 14 29.	+ 55 26 06.		16.3	FAINT BLUE STAR
ZWG 069.124	12 14 30.	+ 09 41		15.3	GALAXY
8ZW 1214+10.2	12 14 30.	+ 10 13		18.0	COMPACT GALAXY
UGC 07307	12 14 30.	+ 10 17	120	16.5	GALAXY DWARF
REA 12	12 14 30.	+ 10 17			DWARF GALAXY
RMB 164	12 14 30.	+ 11 53			FAINT BLUE OBJECT
ZWG 315.031	12 14 30.	+ 63 42		14.2	GALAXY
ZCG 199.05	12 14 30.	+ 63 42		14.2	COMPACT GALAXY
RNGC 4238	12 14 30.	+ 63 42		14.0	GALAXY
UGC 07308	12 14 30.	+ 63 42	108	14.2	GALAXY Sc
MCG+11-15-042	12 14 30.	+ 66 06	36	16.	GALAXY
ZC 1214.5+7633	12 14 30.	+ 76 33	1210		CLUSTER OF GALAXIES
VV 333B	12 14 30.	- 25 55	42	16.	INTERACTING GALAXY
MCG-06-27-014	12 14 30.	- 3 4	36	14.5	GALAXY
AMES 0398	12 14 31.	+ 08 43 04.	28	17.0	NEBULA

OBJECT NAME	RIGHT ASCEN.	DECLINATION	DIAM.	MAGN.	TYPE OF OBJECT
HOLM 360B	12 14 31.	+ 12 44	120	14.2	PART OF MULTIPLE GALAXY
FATH 1.515	12 14 31.	+ 14 27	14		NEBULA
HOLM 358B	12 14 32.	+ 03 58	18	14.8	PART OF MULTIPLE GALAXY
REIZ 2120	12 14 32.	+ 63 41	72	13.7	GALAXY
REIN 3.048	12 14 32.70	+ 12 34 01.1			NEBULA
REIZ 2116	12 14 33.	+ 07 54	120	13.0	GALAXY
MCG+03-31-091	12 14 33.	+ 15 35	138	12.3	GALAXY
TON-N 1490	12 14 33.	+ 36 50		17.0	BLUE STAR
MCG-04-29-016	12 14 33.	- 25 56 30.	24	16.	GALAXY
RNGC 4230	12 14 34.	- 54 51		9.5	OPEN CLUSTER
REIN 3.049A	12 14 34.63	+ 07 54 10.5			NEBULA
REIN 3.049B	12 14 34.78	+ 07 54 05.4			NEBULA
AMES 0401	12 14 35.	+ 08 07 40.	17	17.5	NEBULA
AMES 0400	12 14 35.	+ 12 16 16.	22	17.2	NEBULA
AMES 0399	12 14 35.	+ 12 49 16.	24	17.6	NEBULA
ZWG 041.061	12 14 36.	+ 03 58		13.4	GALAXY
UGC 07309	12 14 36.	+ 03 58	72	13.4	GALAXY IRR
MCG+01-31-035	12 14 36.	+ 03 58	60	13.4	GALAXY
ZWG 041.062	12 14 36.	+ 07 28		13.2	GALAXY
UGC 07310	12 14 36.	+ 07 28	234	13.2	GALAXY Sa
MCG+01-31-036	12 14 36.	+ 07 28	180	13.2	GALAXY
ZWG 041.063	12 14 36.	+ 07 54		13.2	GALAXY
UGC 07311	12 14 36.	+ 07 54	120	13.2	GALAXY S0
MCG+01-31-037	12 14 36.	+ 07 54	96	13.2	GALAXY
ZWG 069.125	12 14 36.	+ 12 34		15.3	GALAXY
UGC 07312	12 14 36.	+ 12 34	102	15.3	GALAXY
ZWG 069.126	12 14 36.	+ 12 44		15.1	GALAXY
UGC 07313	12 14 36.	+ 12 44	126	15.1	GALAXY Sc
MCG+02-31-079	12 14 36.	+ 12 44	120	15.1	GALAXY
ZC 1214.6+2459	12 14 36.	+ 24 59	810		CLUSTER OF GALAXIES
ZWG 158.058	12 14 36.	+ 30 55		15.7	GALAXY
ZWG 215.057	12 14 36.	+ 43 13		15.4	GALAXY
ZC 1214.6+4531	12 14 36.	+ 45 31	740		CLUSTER OF GALAXIES
LB 02294	12 14 36.	+ 62 28 24.		15.4	FAINT BLUE STAR
OCL 0874	12 14 36.	- 54 51	540	9.4	OPEN STAR CLUSTER
KEEL 363	12 14 36.3	+ 47 49 18.		15.	NEBULA
REIN 3.051	12 14 36.65	+ 12 43 52.9			NEBULA
REIN 3.050A	12 14 36.92	+ 07 28 11.2			NEBULA
RNGC 4234	12 14 37.	+ 03 58		13.0	GALAXY
RNGC 4235	12 14 37.	+ 07 28		12.5	GALAXY
IC 3098	12 14 37.	+ 07 28 11.			SAME AS NGC 4235
RNGC 4233	12 14 37.	+ 07 54		13.0	GALAXY
IC 3099	12 14 37.	+ 12 43 53.			NONSTELLAR OBJECT
REIN 3.050B	12 14 37.05	+ 07 28 09.0			NEBULA
RNGC 4237	12 14 38.	+ 15 36		13.0	GALAXY
REIZ 2117	12 14 39.	+ 03 57	60	12.7	GALAXY
HOLM 358A	12 14 39.	+ 03 58	60	12.4	PART OF MULTIPLE GALAXY
REIZ 2118	12 14 39.	+ 07 28	210	12.8	GALAXY
HOLM 359A	12 14 39.	+ 07 28	240	12.2	PART OF MULTIPLE GALAXY
HN 0849	12 14 39.	+ 12 32		14.5	NEBULA
IC 3100	12 14 39.	+ 12 32			NONSTELLAR OBJECT
REIZ 2119	12 14 39.	+ 15 36	84	13.1	GALAXY
MCG+03-31-093	12 14 39.	+ 16 38	42	15.2	GALAXY
MCG+03-31-092	12 14 39.	+ 16 47 30.	96	13.5	GALAXY
LB 00631	12 14 39.	+ 72 27 24.		17.0	FAINT BLUE STAR
AMES 0403	12 14 40.	+ 05 55 28.	20	17.8	NEBULA
AMES 0402	12 14 40.	+ 11 20 22.	23	18.0	NEBULA
AMES 0406	12 14 41.	+ 08 33 46.	13	17.4	NEBULA
AMES 0405	12 14 41.	+ 09 12 58.	76	16.3	NEBULA
AMES 0404	12 14 41.	+ 10 07 52.	23	17.8	NEBULA
FATH 1.516	12 14 41.	+ 14 51	16		NEBULA
ARC 1513	12 14 41.	+ 73 06		17.1	RICH CLUSTER OF GALAXIES
ZWG 069.127	12 14 42.	+ 09 14		15.5	GALAXY
UGC 07314	12 14 42.	+ 09 14	66	15.5	GALAXY
ZC 1214.7+1102	12 14 42.	+ 11 02	400		CLUSTER OF GALAXIES
ZWG 098.130	12 14 42.	+ 15 36		12.3	GALAXY
UGC 07315	12 14 42.	+ 15 36	132	12.3	GALAXY Sb
ZWG 098.129	12 14 42.	+ 16 48		13.5	GALAXY
UGC 07316	12 14 42.	+ 16 48	96	13.5	GALAXY E
ZC 1214.7+4052	12 14 42.	+ 40 52	3560		CLUSTER OF GALAXIES
ZWG 215.058	12 14 42.	+ 41 35		15.7	GALAXY
UGC 07317	12 14 42.	+ 41 35	72	15.7	GALAXY
MCG+07-25-053	12 14 42.	+ 41 35 30.	30	15.	GALAXY
UGC 07318	12 14 42.	+ 49 46	60	17.	GALAXY Sc
MCG-02-31-028	12 14 42.	- 11 23	60	17.	GALAXY
AMES 0407	12 14 43.	+ 12 30 28.	25	17.8	NEBULA
AMES 0408	12 14 44.	+ 13 04 04.	26	16.2	NEBULA
REIZ 2121	12 14 44.	+ 16 38	18	14.6	GALAXY
REIZ 2122	12 14 44.	+ 16 47	24	13.2	GALAXY
RNGC 4239	12 14 44.	+ 16 48		13.5	GALAXY
KEEL 364	12 14 44.8	+ 14 28 42.		17.	NEBULA
HARO 07	12 14 45.	+ 03 57			BLUE EMISSION-LINE GALAXY
AMES 0409	12 14 45.	+ 07 57 40.	31	17.8	NEBULA
HN 0850	12 14 45.	+ 12 12		15.	NEBULA
IC 3101	12 14 45.	+ 12 12			NONSTELLAR OBJECT
FATH 2.073	12 14 45.	+ 14 29	8		NEBULA
AMES 0410	12 14 46.	+ 06 02 28.	28	17.2	NEBULA
AMES 0412	12 14 47.	+ 07 18 40.	17	16.9	NEBULA
AMES 0411	12 14 47.	+ 08 59 52.	34	17.6	NEBULA
REIN 3.052	12 14 47.16	+ 12 13 16.3			NEBULA
ZWG 041.064	12 14 48.	+ 02 52		15.7	GALAXY
ZWG 069.128	12 14 48.	+ 12 13		15.5	GALAXY
ZWG 069.129	12 14 48.	+ 13 04		15.4	GALAXY
FATH 1.517	12 14 48.	+ 14 42	8		NEBULA
ZWG 098.131	12 14 48.	+ 16 39		15.2	GALAXY
MCG+06-27-045	12 14 48.	+ 38 06	960	9.	GALAXY
ZC 1214.8+4456	12 14 48.	+ 44 56	2690		CLUSTER OF GALAXIES
7ZW 446	12 14 48.	+ 69 41			COMPACT GALAXY
MCG-02-31-029	12 14 48.	- 09 40	54	14.	GALAXY
RNGC 4240	12 14 49.	- 09 40		14.0	GALAXY
KEEL 365	12 14 49.0	+ 14 53 46.		18.	NEBULA
AMES 0413	12 14 50.	+ 09 08 46.	31	17.4	NEBULA
FATH 2.074	12 14 50.	+ 14 54	5		NEBULA
FATH 1.518	12 14 50.	+ 14 57	5		NEBULA
AMES 0414	12 14 51.	+ 12 19 52.	24	17.8	NEBULA
TON-N 1491	12 14 51.	+ 21 20		17.0	BLUE STAR
LB 02295	12 14 52.	+ 60 39 12.		16.5	FAINT BLUE STAR
KEEL 366	12 14 52.3	+ 14 37 26.		18.	NEBULA
AMES 0417	12 14 53.	+ 05 36 04.	23	17.9	NEBULA
IC 3102	12 14 53.	+ 06 57 59.			SAME AS NGC 4241
AMES 0416	12 14 53.	+ 07 35 46.	34	17.5	NEBULA
AMES 0415	12 14 53.	+ 12 55 40.	23	17.9	NEBULA
FATH 2.075	12 14 53.	+ 14 38	16		NEBULA
HZ 23	12 14 53.	+ 32 26		14.2	BLUE STAR
ZWG 041.065	12 14 54.	+ 06 58		13.6	GALAXY
UGC 07319	12 14 54.	+ 06 58	150	13.6	GALAXY S0
MCG+01-31-038	12 14 54.	+ 06 58	96	13.6	GALAXY
ZWG 098.132	12 14 54.	+ 17 55		14.7	GALAXY
MCG+03-31-094	12 14 54.	+ 17 55	42	14.7	GALAXY
MCG+04-29-060	12 14 54.	+ 22 48 30.	330	14.5	GALAXY
MCG+05-29-049	12 14 54.	+ 29 54	180	12.4	GALAXY
UGC 07320	12 14 54.	+ 45 05	66	16.5	GALAXY
REIZ 2125	12 14 54.	+ 45 54	228	12.0	GALAXY
MCG+10-18-019	12 14 54.	+ 60 23	27	16.	GALAXY
RNGC 4241	12 14 55.	+ 06 58		13.5	GALAXY
KEEL 367	12 14 55.8	+ 14 54 54.		18.	NEBULA
AMES 0419	12 14 56.	+ 08 48 16.	34	16.8	NEBULA
IC 3103	12 14 56.	+ 09 38 17.			SINGLE STAR
AMES 0418	12 14 56.	+ 13 12 04.	24	17.2	NEBULA
FATH 2.076	12 14 56.	+ 14 56	8		NEBULA
LB 02296	12 14 56.	+ 57 18 06.		16.8	FAINT BLUE STAR
KEEL 368	12 14 56.1	+ 14 23 05.		18.	NEBULA
REIZ 2123	12 14 57.	+ 04 26	30	15.5	GALAXY
REIZ 2124	12 14 57.	+ 06 57	90	13.3	GALAXY
AMES 0422	12 14 57.	+ 08 13 04.	19	17.8	NEBULA
AMES 0421	12 14 57.	+ 08 20 52.	22	17.6	NEBULA
HOLM 360A	12 14 57.	+ 12 40	150	14.3	PART OF MULTIPLE GALAXY
AMES 0420	12 14 57.	+ 12 54 40.	20	17.0	NEBULA
FATH 2.077	12 14 57.	+ 14 23	19		NEBULA
FATH 1.519	12 14 58.	+ 15 10	19		NEBULA
RNGC 4242	12 14 58.	+ 45 54		11.5	GALAXY
KEEL 369	12 14 58.3	+ 14 45 00.		18.	NEBULA
AMES 0423	12 14 59.	+ 05 08 52.	35	17.4	NEBULA
FATH 2.078	12 14 59.	+ 14 46	5		NEBULA
REIZ 2130	12 14 59.	+ 47 05	18	15.3	GALAXY
KEEN 4250A	12 15	+ 71 09		13.5	GALAXY
ZC 1215.0+0427	12 15 00.	+ 04 27	340		CLUSTER OF GALAXIES
RMB 168	12 15 00.	+ 12 18			FAINT BLUE OBJECT
REA 85	12 15 00.	+ 14 37			DWARF GALAXY
MCG+03-31-095	12 15 00.	+ 16 59	120	15.7	GALAXY
ZC 1215.0+2240	12 15 00.	+ 22 40	4770		CLUSTER OF GALAXIES
ZWG 128.070	12 15 00.	+ 22 50		14.5	GALAXY
UGC 07321	12 15 00.	+ 22 50	330	14.0	GALAXY Sc
KARA.73B 0524	12 15 00.	+ 22 50	348	14.5	ISOLATED GALAXY S
ZWG 187.035	12 15 00.	+ 38 05		10.8	GALAXY
REIZ 2129	12 15 00.	+ 38 05	120	11.1	GALAXY
UGC 07322	12 15 00.	+ 38 05	1110	10.8	GALAXY Sc
MCG+08-22-095	12 15 00.	+ 45 05	42	16.	GALAXY
ZWG 243.061	12 15 00.	+ 45 53		11.9	GALAXY
UGC 07323	12 15 00.	+ 45 53	342	11.9	GALAXY
MCG+08-22-096	12 15 00.	+ 46 41	60	16.	GALAXY
ZWG 243.062	12 15 00.	+ 46 50		15.4	GALAXY
UGC 07324	12 15 00.	+ 46 50	66	15.4	GALAXY DISRPTD
ZWG 243.063	12 15 00.	+ 47 05		15.3	GALAXY
UGC 07325	12 15 00.	+ 47 05	96	15.3	GALAXY Sa-b
MCG+08-22-097	12 15 00.	+ 47 07	48	15.	GALAXY
LB 02297	12 15 00.	+ 52 35 48.		16.3	FAINT BLUE STAR
ZC 1215.0+5257	12 15 00.	+ 52 57	1080		CLUSTER OF GALAXIES
MCG+09-20-138	12 15 00.	+ 53 50	36	15.	GALAXY
ZCG 199.06	12 15 00.	+ 63 47		18.4	COMPACT GALAXY
MCG+12-12-005	12 15 00.	+ 71 05	102	13.	GALAXY
AMES 0425	12 15 01.	+ 05 38 52.	25	17.7	NEBULA
AMES 0424	12 15 01.	+ 14 05 22.	29	17.4	NEBULA
FATH 1.520	12 15 01.	+ 14 12	14		NEBULA
RNGC 4243	12 15 01.	- 11 03			NON-EXISTENT OBJECT
REIN 3.053	12 15 01.35	+ 12 39 57.7			NEBULA
AMES 0426	12 15 02.	+ 08 12 52.	24	17.8	NEBULA
IC 3105	12 15 02.	+ 12 40 05.			NONSTELLAR OBJECT
REIZ 2127	12 15 02.	+ 17 00	48	15.6	GALAXY
REIZ 2128	12 15 02.	+ 17 55	24	14.4	GALAXY
REIZ 2126	12 15 03.	+ 03 56	36	16.0	GALAXY
AMES 0428	12 15 03.	+ 06 51 16.	23	17.8	NEBULA
AMES 0427	12 15 03.	+ 13 13 40.	34	17.4	NEBULA
FATH 1.521	12 15 03.	+ 14 14	16		NEBULA
TON-N 1492	12 15 03.	+ 34 42		16.8	BLUE STAR
RNGC 4244	12 15 03.	+ 38 05		10.5	GALAXY
MCG+08-22-098	12 15 03.	+ 45 53 30.	270	11.3	GALAXY
AMES 0431	12 15 04.	+ 09 35 28.	34	17.4	NEBULA
AMES 0430	12 15 04.	+ 12 32 22.	23	17.5	NEBULA
AMES 0429	12 15 04.	+ 12 35 10.	16	17.8	NEBULA
LB 02298	12 15 04.	+ 59 44 06.		15.4	FAINT BLUE STAR
AMES 0432	12 15 05.	+ 10 39 52.	16	17.8	NEBULA
REIN 3.054	12 15 05.02	+ 13 09 41.5			NEBULA
KEEL 370	12 15 05.5	+ 47 48 45.		16.	NEBULA
ZC 1215.1+0400	12 15 06.	+ 04 00	2820		CLUSTER OF GALAXIES
AMES 0433	12 15 06.	+ 06 43 34.	17	18.0	NEBULA
RMB 228	12 15 06.	+ 09 53			FAINT BLUE OBJECT
RMB 165	12 15 06.	+ 12 04			FAINT BLUE OBJECT
ZWG 069.130	12 15 06.	+ 12 40		15.0	GALAXY
UGC 07326	12 15 06.	+ 12 40	108	15.0	GALAXY IRR
MCG+02-31-080	12 15 06.	+ 12 40	84	15.0	GALAXY
ZWG 098.133	12 15 06.	+ 17 00		15.7	GALAXY
UGC 07327	12 15 06.	+ 17 00	114	15.7	GALAXY Sc
MCG+05-29-049A	12 15 06.	+ 28 29	60	16.	GALAXY
ZWG 158.059	12 15 06.	+ 29 53		12.4	GALAXY
UGC 07328	12 15 06.	+ 29 53	210	12.4	GALAXY SB0/SBa
SN 1960E	12 15 06.	+ 48 10		16.5	SUPERNOVA
MCG+08-22-098A	12 15 06.	+ 48 10		17.8	GALAXY
ZC 1215.1+5033	12 15 06.	+ 50 33	1410		CLUSTER OF GALAXIES
7ZW 447	12 15 06.	+ 71 05			COMPACT GALAXY
ZWG 335.009	12 15 06.	+ 71 05		13.0	GALAXY
UGC 07329	12 15 06.	+ 71 05	150	13.0	GALAXY S0/SBa
AMES 0435	12 15 08.	+ 05 54 04.	18	17.2	NEBULA
AMES 0434	12 15 08.	+ 12 32 16.	13	18.0	NEBULA
RNGC 4245	12 15 08.	+ 29 53		12.5	GALAXY
RNGC 4250	12 15 08.	+ 71 05		13.0	GALAXY
REIZ 2131	12 15 09.	+ 03 55	30	16.1	GALAXY
HOLM 361B	12 15 09.	+ 17 45	30	15.7	PART OF MULTIPLE GALAXY
KARA.68 095	12 15 09.	+ 28 45	27		DWARF GALAXY
AMES 0436	12 15 10.	+ 06 22 03.	29	17.6	NEBULA
FATH 1.522	12 15 10.	+ 15 01	16		NEBULA
AMES 0437	12 15 11.	+ 06 25 27.	26	17.8	NEBULA
FATH 1.523	12 15 11.	+ 14 15	14		NEBULA
HOLM 357B	12 15 11.	+ 69 53	24	15.1	PART OF MULTIPLE GALAXY
REIN 3.055	12 15 11.69	+ 13 26 54.0			NEBULA
ZWG 041.066	12 15 12.	+ 04 46		15.7	GALAXY
AMES 0438	12 15 12.	+ 09 04 51.	20	17.6	NEBULA
RMB 166	12 15 12.	+ 12 06			FAINT BLUE OBJECT
RMB 167	12 15 12.	+ 12 11			FAINT BLUE OBJECT
ZWG 069.131	12 15 12.	+ 13 27		15.1	GALAXY
MCG+02-31-081	12 15 12.	+ 13 27	42	15.1	GALAXY
ZC 1215.2+2056	12 15 12.	+ 20 56	1080		CLUSTER OF GALAXIES
MCG+04-29-061	12 15 12.	+ 26 18	48	15.3	GALAXY
MCG+11-15-043	12 15 12.	+ 62 54	30	16.	GALAXY
ZCG 199.07	12 15 12.	+ 63 47		18.5	COMPACT GALAXY
MCG+11-15-044	12 15 12.	+ 67 50	36	16.	GALAXY

OBJECT NAME	RIGHT ASCEN.	DECLINATION	DIAM.	MAGN.	TYPE OF OBJECT
MCG+13-09-021	12 15 12.	+ 77 33	33	14.	GALAXY
ZWG 352.026	12 15 12.	+ 77 34		15.0	GALAXY
IC 3106	12 15 13.	+ 09 53 29.			NONSTELLAR OBJECT
AMES 0439	12 15 13.	+ 13 35 15.	13	17.6	NEBULA
AMES 0440	12 15 14.	+ 05 44 39.	19	17.8	NEBULA
IC 3107	12 15 14.	+ 11 07 23.			NONSTELLAR OBJECT
HOLM 361A	12 15 14.	+ 17 43	72	14.0	PART OF MULTIPLE GALAXY
IC 3110	12 15 14.	+ 37 39 29.			NONSTELLAR OBJECT
REIN 3.056	12 15 14.12	+ 11 07 22.7			NEBULA
KEEL 371	12 15 14.6	+ 14 29 11.		18.	NEBULA
KEEL 374	12 15 14.8	+ 47 21 37.		16.	NEBULA
HN 0852	12 15 15.	+ 13 25		14.	NEBULA
IC 3109	12 15 15.	+ 13 25			NONSTELLAR OBJECT
HN 0851	12 15 15.	+ 13 38		14.	NEBULA
IC 3108	12 15 15.	+ 13 38			NONSTELLAR OBJECT
TON-N 1493	12 15 15.	+ 35 32		15.9	BLUE STAR
KEEL 373	12 15 15.5	+ 37 40 29.			NEBULA
AMES 0441	12 15 16.	+ 04 44 33.	44	16.6	NEBULA
PATH 2.079	12 15 16.	+ 14 29	8		NEBULA
AMES 0443	12 15 17.	+ 07 03 33.	22	18.0	NEBULA
AMES 0442	12 15 17.	+ 08 23 39.	19	17.8	NEBULA
PATH 2.080	12 15 17.	+ 14 22	22		NEBULA
IC 3112	12 15 17.	+ 26 18 29.			NONSTELLAR OBJECT
REIZ 2135	12 15 17.	+ 47 42	72	13.5	GALAXY
RNGC 4219A	12 15 17.	- 43 14			GALAXY
KEEL 372	12 15 17.0	+ 14 21 28.		18.	NEBULA
PRA 3	12 15 17.8	+ 30 49 06.		14.5	NONSTELLAR BLUE OBJECT
REA 28	12 15 18.	+ 05 18	30		DWARF GALAXY
KARA.68 096	12 15 18.	+ 05 19	34		DWARF GALAXY
ZWG 041.067	12 15 18.	+ 06 09		15.5	GALAXY
IC 3111	12 15 18.	+ 08 42 23.			NONSTELLAR OBJECT
ZWG 069.132	12 15 18.	+ 08 43		15.5	GALAXY
ZWG 069.133	12 15 18.	+ 11 07		14.5	GALAXY
UGC 07330	12 15 18.	+ 11 07	90	14.5	GALAXY S
MCG+02-31-082	12 15 18.	+ 11 07	84	14.5	GALAXY
ZWG 069.134	12 15 18.	+ 13 29		15.3	GALAXY
ZWG 099.001	12 15 18.	+ 17 42		14.7	GALAXY
ZWG 098.134	12 15 18.	+ 17 42		14.7	GALAXY
SCH 30	12 15 18.	+ 17 42		14.7	PECULIAR GALAXY
UGC 07331	12 15 18.	+ 17 42	96	14.7	GALAXY SBb
MCG+03-31-096	12 15 18.	+ 17 43	90	14.7	GALAXY
ZWG 128.071	12 15 18.	+ 26 18		15.3	GALAXY
SN 1963G	12 15 18.	+ 26 18		15.8	SUPERNOVA
ZC 1215.3+4745	12 15 18.	+ 47 45	740		CLUSTER OF GALAXIES
ZC 1215.3+4858	12 15 18.	+ 48 58	3560		CLUSTER OF GALAXIES
ZWG 013.119	12 15 18.	- 00 22		15.2	GALAXY
DV.56 N4219A	12 15 18.	- 43 14	72		S GALAXY
AMES 0444	12 15 19.	+ 12 38 03.	37	17.8	NEBULA
HOLM 363B	12 15 19.	+ 47 41	84	13.2	PART OF MULTIPLE GALAXY
AMES 0447	12 15 20.	+ 04 47 39.	24	17.5	NEBULA
AMES 0446	12 15 20.	+ 11 23 03.	17	18.0	NEBULA
AMES 0445	12 15 20.	+ 13 26 51.	30	17.1	NEBULA
REIZ 2134	12 15 20.	+ 17 43	60	14.7	GALAXY
AMES 0448	12 15 21.	+ 06 23 51.	24	17.3	NEBULA
REIZ 2132	12 15 21.	+ 07 28	102	13.8	GALAXY
HN 0035	12 15 21.	+ 07 34			NEBULA
REIZ 2133	12 15 21.	+ 07 34	24	14.0	GALAXY
TON-N 1494	12 15 21.	+ 33 52		16.9	BLUE STAR
TON-N 1495	12 15 21.	+ 34 05		16.9	BLUE STAR
MCG+08-22-099	12 15 21.	+ 47 41 30.	120	13.0	GALAXY
SC 1212-4259.1	12 15 21.	- 43 15 46.	66		NEBULA
BC B21215+30	12 15 21.2	+ 30 23 39.		14.	QUASI-STELLAR OBJECT
AMES 0449	12 15 22.	+ 06 09 27.	37	16.6	NEBULA
RNGC 4248	12 15 22.	+ 47 41		14.0	GALAXY
AMES 0450	12 15 23.	+ 08 25 57.	20	17.4	NEBULA
KEEL 376	12 15 23.5	+ 47 28 51.		17.	NEBULA
ZWG 013.120	12 15 24.	+ 00 43		15.4	GALAXY
UGC 07332	12 15 24.	+ 00 43	138	15.4	GALAXY DWRF IP
MCG+00-31-045	12 15 24.	+ 00 43	150	15.4	GALAXY
ZWG 041.068	12 15 24.	+ 05 52		14.8	GALAXY
MCG+01-31-039	12 15 24.	+ 05 52	24	14.8	GALAXY
ZWG 041.069	12 15 24.	+ 06 56		14.4	GALAXY
UGC 07333	12 15 24.	+ 06 56	102	14.4	GALAXY SBc
MCG+01-31-040	12 15 24.	+ 06 56	72	14.4	GALAXY
ZWG 041.070	12 15 24.	+ 07 28		14.0	GALAXY
UGC 07334	12 15 24.	+ 07 28	162	14.0	GALAXY Sc
MCG+01-31-041	12 15 24.	+ 07 28	120	13.3	GALAXY
ZWG 041.071	12 15 24.	+ 07 33		14.7	GALAXY
MCG+01-31-042	12 15 24.	+ 07 33	15	14.7	GALAXY
IC 3114	12 15 24.	+ 09 24 41.			MAY NOT EXIST
FBB 160	12 15 24.	+ 13 12			FAINT BLUE OBJECT
ZWG 099.002	12 15 24.	+ 18 40		15.5	GALAXY
ZWG 098.135	12 15 24.	+ 18 40		15.5	GALAXY
TON-N 0606	12 15 24.	+ 30 50		16.2	BLUE STAR
MCG+07-25-054	12 15 24.	+ 41 34	39	16.	GALAXY
ZWG 244.001	12 15 24.	+ 47 41		13.9	GALAXY
ZWG 243.064	12 15 24.	+ 47 41		13.9	GALAXY
UGC 07335	12 15 24.	+ 47 41	192	13.9	GALAXY S
LB 02299	12 15 24.	+ 50 33 06.		17.5	FAINT BLUE STAR
MCG+09-20-139	12 15 24.	+ 54 43	36	16.	GALAXY
ZCG 199.08	12 15 24.	+ 64 32		18.5	COMPACT GALAXY
MCG-05-29-033	12 15 24.	- 29 21	48	15.	GALAXY
KEEL 375	12 15 24.1	+ 14 28 02.		18.	NEBULA
REIN 3.057A	12 15 24.98	+ 07 33 08.7			NEBULA
RNGC 4249	12 15 25.	+ 05 52		15.0	GALAXY
IC 3113	12 15 25.	+ 07 27 53.			SAME AS NGC 4246
RNGC 4246	12 15 25.	+ 07 28		14.0	GALAXY
RNGC 4247	12 15 25.	+ 07 33		14.5	GALAXY
AMES 0452	12 15 25.	+ 08 22 21.	17	17.8	NEBULA
AMES 0451	12 15 25.	+ 08 27 57.	16	17.3	NEBULA
PATH 2.081	12 15 25.	+ 14 28	14		NEBULA
ABC 1514	12 15 25.	+ 20 56		17.6	RICH CLUSTER OF GALAXIES
REIN 3.057B	12 15 25.13	+ 07 33 07.2			NEBULA
REIN 3.058A	12 15 25.31	+ 07 27 47.9			NEBULA
REIN 3.058B	12 15 25.37	+ 07 27 51.3			NEBULA
AMES 0453	12 15 26.	+ 12 39 21.	19	17.8	NEBULA
IC 3116	12 15 26.	+ 25 21 17.			NONSTELLAR OBJECT
IC 3115	12 15 26.7	+ 06 55 56.			GALAXY
AMES 0455	12 15 27.	+ 05 34 15.	19	17.4	NEBULA
REIZ 2136	12 15 27.	+ 05 52	12	14.8	GALAXY
REIZ 2137	12 15 27.	+ 06 54	108	14.9	GALAXY
HOLM 359C	12 15 27.	+ 07 33	60	14.0	PART OF MULTIPLE GALAXY
AMES 0454	12 15 27.	+ 12 36 21.	23	17.8	NEBULA
PATH 2.082	12 15 27.	+ 14 38	5		NEBULA
MCG-03-31-025	12 15 27.	- 20 36	24	15.	GALAXY
KEEL 377	12 15 27.2	+ 14 37 20.		18.	NEBULA
HOLM 359B	12 15 28.	+ 07 28	108	12.7	PART OF MULTIPLE GALAXY
AMES 0456	12 15 28.	+ 08 26 27.	30	18.0	NEBULA
KEEL 378	12 15 28.8	+ 14 43 22.		17.	NEBULA
AMES 0457	12 15 29.	+ 05 41 27.	28	17.4	NEBULA
ZWG 041.072	12 15 30.	+ 05 58		15.3	GALAXY
MCG+01-31-043	12 15 30.	+ 05 58	12	15.3	GALAXY
ZWG 099.003	12 15 30.	+ 16 14		15.4	GALAXY
ZWG 098.136	12 15 30.	+ 16 14		15.4	GALAXY
UGC 07336	12 15 30.	+ 16 14	60	15.4	GALAXY S
MCG+03-31-097	12 15 30.	+ 16 14	54	15.4	GALAXY
MCG+05-29-050	12 15 30.	+ 28 27	240	11.5	GALAXY
TON-N 0076	12 15 30.	+ 30 11		15.0	BLUE STAR
ZWG 269.050	12 15 30.	+ 50 45		15.7	GALAXY
MCG+09-20-140	12 15 30.	+ 51 58	30	17.	GALAXY
ZWG 293.009	12 15 30.	+ 58 55		15.6	GALAXY
MRK 202	12 15 30.	+ 58 58	7	16.5	GALAXY WITH UV CONTINUUM
ZC 1215.5+6449	12 15 30.	+ 64 49	1340		CLUSTER OF GALAXIES
AMES 0459	12 15 31.	+ 06 21 15.	19	17.6	NEBULA
AMES 0458	12 15 31.	+ 07 36 03.	36	18.0	NEBULA
IC 3117	12 15 31.	+ 09 20 59.			MAY NOT EXIST
PATH 1.524	12 15 31.	+ 14 23	5		NEBULA
PATH 2.083	12 15 31.	+ 14 44	5		NEBULA
AMES 0461	12 15 32.	+ 08 10 15.	18	17.8	NEBULA
AMES 0460	12 15 32.	+ 11 00 33.	30	16.9	NEBULA
REIZ 2139	12 15 32.	+ 16 15	42	14.6	GALAXY
HOLM 362B	12 15 32.	+ 44 27	18	14.4	PART OF MULTIPLE GALAXY
AMES 0463	12 15 33.	+ 05 18 51.	20	17.7	NEBULA
REIZ 2138	12 15 33.	+ 05 57	18	15.7	GALAXY
AMES 0462	12 15 33.	+ 11 09 39.	23	16.2	NEBULA
PATH 1.525	12 15 33.	+ 15 02	14		NEBULA
TON-N 1496	12 15 33.	+ 36 26		16.8	BLUE STAR
AMES 0465	12 15 34.	+ 05 57 45.	35	16.3	NEBULA
AMES 0464	12 15 34.	+ 06 13 03.	34	17.6	NEBULA
KEEL 379	12 15 34.5	+ 14 45 48.		17.	NEBULA
IC 0773	12 15 34.8	+ 06 25 04.			GALAXY
PATH 2.084	12 15 35.	+ 14 47	8		NEBULA
LB 02300	12 15 35.	+ 60 09 42.		16.8	FAINT BLUE STAR
ZWG 013.122	12 15 36.	+ 02 10		15.7	GALAXY
AMES 0468	12 15 36.	+ 05 38 15.	23	17.6	NEBULA
ZWG 041.073	12 15 36.	+ 06 25		15.2	GALAXY
MCG+01-31-044	12 15 36.	+ 06 25	42	15.2	GALAXY
ZWG 041.074	12 15 36.	+ 07 56		15.6	GALAXY
AMES 0467	12 15 36.	+ 08 35 27.	28	16.8	NEBULA
ZC 1215.6+1058	12 15 36.	+ 10 58	1140		CLUSTER OF GALAXIES
ZC 1215.6+2437	12 15 36.	+ 24 37	870		CLUSTER OF GALAXIES
ZWG 128.072	12 15 36.	+ 24 58		15.4	GALAXY
LB 00038	12 15 36.	+ 26 20		16.9	FAINT BLUE STAR
ZWG 158.060	12 15 36.	+ 28 28		11.5	GALAXY
UGC 07338	12 15 36.	+ 28 28	216	11.5	GALAXY S0
ZC 1215.6+3646	12 15 36.	+ 36 46	1080		CLUSTER OF GALAXIES
ZWG 243.065	12 15 36.	+ 44 40		15.3	GALAXY
ZCG 199.10	12 15 36.	+ 64 08		18.3	COMPACT GALAXY
ZCG 199.09	12 15 36.	+ 64 32		18.1	COMPACT GALAXY
ZWG 013.121	12 15 36.	- 00 47		14.7	GALAXY
UGC 07337	12 15 36.	- 00 47	66	14.7	GALAXY Sb-c
MCG+00-31-046	12 15 36.	- 00 47	66	14.7	GALAXY
AMES 0466	12 15 37.	+ 13 52 51.	14	17.8	NEBULA
IC 3119	12 15 37.	+ 24 57 53.			NONSTELLAR OBJECT
REIZ 2142	12 15 37.	+ 28 27	168	12.6	GALAXY
HN 0362	12 15 37.	- 79 26			NEBULA
IC 3104	12 15 37.	- 79 26			OPEN CLUSTER
AMES 0471	12 15 38.	+ 07 55 15.	55	16.6	NEBULA
IC 3118	12 15 38.	+ 09 46 41.			NONSTELLAR OBJECT
AMES 0470	12 15 38.	+ 13 18 45.	35	16.8	NEBULA
RNGC 4251	12 15 38.	+ 28 27		12.0	GALAXY
KEEL 380	12 15 38.1	+ 14 58 24.		18.	NEBULA
REIN 3.059	12 15 38.31	+ 09 46 39.2			NEBULA
A3 001	12 15 38.6	+ 14 55 59.		17.2	FAINT BLUE OBJECT
REIZ 2141	12 15 39.	+ 06 53	18	15.4	GALAXY
AMES 0469	12 15 39.	+ 13 51 15.	26	17.2	NEBULA
PATH 1.526	12 15 39.	+ 14 14	8		NEBULA
TON-N 1497	12 15 39.	+ 33 43		14.9	BLUE STAR
TON-N 1498	12 15 39.	+ 34 40		16.8	BLUE STAR
AMES 0472	12 15 40.	+ 04 54 51.	35	17.4	NEBULA
PATH 2.085	12 15 40.	+ 14 58	19		NEBULA
REIZ 2140	12 15 40.	- 00 48	42	14.7	GALAXY
KEEL 381	12 15 40.4	+ 14 43 28.		18.	NEBULA
AMES 0473	12 15 41.	+ 06 53 03.	36	16.4	NEBULA
PATH 2.086	12 15 41.	+ 14 44	8		NEBULA
ZWG 041.075	12 15 42.	+ 06 53		15.5	GALAXY
ZWG 069.135	12 15 42.	+ 09 47		15.2	GALAXY
UGC 07339	12 15 42.	+ 09 47	96	15.2	GALAXY DWARF
MCG+02-31-083	12 15 42.	+ 09 47	84	15.2	GALAXY
AMES 0475	12 15 42.	+ 10 38 21.	42	17.7	NEBULA
ZWG 069.136	12 15 42.	+ 11 11		15.6	GALAXY
AMES 0474	12 15 42.	+ 12 11 21.	29	17.4	NEBULA
BZW 1215+13.5	12 15 42.	+ 13 33		18.1	COMPACT GALAXY
KARA.68 097	12 15 42.	+ 13 38	27		DWARF GALAXY
RBB 058	12 15 42.	+ 14 01			FAINT BLUE OBJECT
ZWG 099.004	12 15 42.	+ 20 17		15.6	GALAXY
ZWG 098.137	12 15 42.	+ 20 17		15.6	GALAXY
LB 00003	12 15 42.	+ 28 23		17.7	FAINT BLUE STAR
MCG+05-29-051	12 15 42.	+ 30 07	48	13.7	GALAXY
ZWG 243.066	12 15 42.	+ 44 27		14.2	GALAXY
ZWG 215.059	12 15 42.	+ 44 27		14.2	GALAXY
MRK 203	12 15 42.	+ 44 27	13	15.5	GALAXY WITH UV CONTINUUM
HOLM 362A	12 15 42.	+ 44 27	30	13.4	PART OF MULTIPLE GALAXY
UGC 07340	12 15 42.	+ 44 27	90	14.2	GALAXY DBL SYS
MCG+08-22-100	12 15 42.	+ 44 40	18	16.	GALAXY
MCG+13-09-022	12 15 42.	+ 77 32	21	16.	GALAXY
KEEL 382	12 15 42.0	+ 14 56 03.		17.	NEBULA
AMES 0476	12 15 43.	+ 11 29 51.	36	18.0	NEBULA
PATH 2.087	12 15 43.	+ 14 57	14		NEBULA
SEY 173	12 15 43.	+ 44 40 19.		15.0	FAINT GALAXY
KEEL 383	12 15 43.5	+ 37 50 06.			NEBULA
AMES 0477	12 15 44.	+ 12 56 15.	19	17.2	NEBULA
AMES 0479	12 15 45.	+ 08 15 21.	13	18.0	NEBULA
PK275+72.1	12 15 45.	+ 11 19	774		PLANETARY NEBULA
AMES 0478	12 15 45.	+ 11 38 27.	23	16.8	NEBULA
MCG+04-29-062	12 15 45.	+ 25 28 30.	84	14.7	GALAXY
SEY 174	12 15 45.	+ 35 53 26.		15.4	FAINT GALAXY
MCG+07-25-055	12 15 45.	+ 44 27	39	15.	GALAXY
AMES 0480	12 15 46.	+ 05 30 39.	22	17.8	NEBULA
AMES 0484	12 15 47.	+ 05 11 57.	22	17.6	NEBULA
AMES 0483	12 15 47.	+ 05 24 15.	16	17.9	NEBULA
AMES 0482	12 15 47.	+ 08 16 03.	17	17.9	NEBULA
AMES 0481	12 15 47.	+ 08 23 03.	12	17.6	NEBULA
ZWG 069.137	12 15 48.	+ 14 02		15.7	GALAXY
ZC 1215.8+1907	12 15 48.	+ 19 07	940		CLUSTER OF GALAXIES
ZWG 128.073	12 15 48.	+ 25 30		14.7	GALAXY

OBJECT NAME	RIGHT ASCEN.	DECLINATION	DIAM.	MAGN.	TYPE OF OBJECT
UGC 07341	12 15 48.	+ 25 30	90	14.7	GALAXY Sb
UGC 07342	12 15 48.	+ 29 33	66	16.0	GALAXY
ZCG 199.11	12 15 48.	+ 63 37		18.3	COMPACT GALAXY
RNGC 4897	12 15 49.	- 13 15			GALAXY
AMES 0485	12 15 50.	+ 05 08 03.	23	17.4	NEBULA
IC 3122	12 15 50.	+ 25 29 41.			NONSTELLAR OBJECT
REIZ 2143	12 15 50.	+ 25 31	54	15.3	GALAXY
LB 02301	12 15 50.	+ 53 13 12.		16.7	FAINT BLUE STAR
AMES 0487	12 15 51.	+ 07 36 39.	19	17.5	NEBULA
AMES 0486	12 15 51.	+ 12 14 03.	31	17.8	NEBULA
HN 0854	12 15 51.	+ 13 31		15.	NEBULA
IC 3121	12 15 51.	+ 13 31			NONSTELLAR OBJECT
HN 0853	12 15 51.	+ 13 59		14.5	NEBULA
IC 3120	12 15 51.	+ 13 59			NONSTELLAR OBJECT
KARA.68 098	12 15 51.	+ 28 56	34		DWARF GALAXY
AMES 0488	12 15 53.	+ 08 47 03.	30	17.8	NEBULA
BC M1215+114	12 15 53.5	+ 11 21 44.		17.	QUASI-STELLAR OBJECT
SHB 198	12 15 53.5	+ 11 21 44.		17.	QUASI-STELLAR OBJECT
AMES 0490	12 15 54.	+ 05 14 39.	40	17.2	NEBULA
ZC 1215.9+0534	12 15 54.	+ 05 34	1610		CLUSTER OF GALAXIES
ZWG 041.076	12 15 54.	+ 05 50		15.0	GALAXY
UGC 07343	12 15 54.	+ 05 50	72	15.0	GALAXY S
MCG+01-31-045	12 15 54.	+ 05 50	72	15.0	GALAXY
AMES 0489	12 15 54.	+ 06 41 15.	26	17.2	NEBULA
IC 3123	12 15 54.	+ 08 20 35.			SINGLE STAR
RMB 158	12 15 54.	+ 12 51			FAINT BLUE OBJECT
ZWG 099.005	12 15 54.	+ 17 32		15.7	GALAXY
ZWG 098.138	12 15 54.	+ 17 32		15.7	GALAXY
ZWG 099.006	12 15 54.	+ 20 24		15.5	GALAXY
ZWG 098.139	12 15 54.	+ 20 24		15.5	GALAXY
IC 3125	12 15 54.	+ 24 38 35.			NONSTELLAR OBJECT
ZWG 158.061	12 15 54.	+ 30 05		13.7	GALAXY
UGC 07344	12 15 54.	+ 30 05	54	13.7	GALAXY SBa
MCG+10-18-020	12 15 54.	+ 60 00	30	16.	GALAXY
ZCG 199.13	12 15 54.	+ 63 45		18.7	COMPACT GALAXY
ZCG 199.12	12 15 54.	+ 64 33		18.3	COMPACT GALAXY
ZC 1215.9-0036	12 15 54.	- 00 36	3760		CLUSTER OF GALAXIES
AMES 0493	12 15 55.	+ 05 53 27.	24	17.2	NEBULA
AMES 0492	12 15 55.	+ 08 13 27.	22	18.0	NEBULA
IC 3124	12 15 55.	+ 09 51 53.			SINGLE STAR
AMES 0491	12 15 55.	+ 12 58 27.	49	17.0	NEBULA
REIZ 2144	12 15 57.	+ 05 51	66	14.8	GALAXY
RNGC 4253	12 15 57.	+ 30 05		13.5	GALAXY
TON-N 1499	12 15 57.	+ 33 38		16.8	BLUE STAR
AMES 0495	12 15 58.	+ 05 30 03.	20	17.8	NEBULA
AMES 0494	12 15 58.	+ 06 08 09.	22	17.8	NEBULA
A3 002	12 15 58.5	+ 14 49 20.		16.5	FAINT BLUE OBJECT
LB 02302	12 15 59.	+ 56 25 36.		15.6	FAINT BLUE STAR
LB 09856	12 16	- 86 10		13.6	FAINT BLUE STAR
AMES 0496	12 16 00.	+ 06 26 39.	17	17.8	NEBULA
ZWG 042.001	12 16 00.	+ 07 00		15.5	GALAXY
ZWG 070.001	12 16 00.	+ 12 08		15.7	GALAXY
ZWG 069.138	12 16 00.	+ 12 08		15.7	GALAXY
MCG+02-31-084	12 16 00.	+ 12 08	36	15.7	GALAXY
RMB 057	12 16 00.	+ 14 05			FAINT BLUE OBJECT
ZWG 099.007	12 16 00.	+ 16 32		15.7	GALAXY
ZWG 098.140	12 16 00.	+ 16 32		15.7	GALAXY
ZC 1216.0+3347	12 16 00.	+ 33 47	1950		CLUSTER OF GALAXIES
ZWG 187.036	12 16 00.	+ 35 54		15.5	GALAXY
ZC 1216.0+4030	12 16 00.	+ 40 30	400		CLUSTER OF GALAXIES
MCG+09-20-142	12 16 00.	+ 51 52	30	16.	GALAXY
MCG+09-20-141	12 16 00.	+ 55 36	24	17.	GALAXY
ZCG 199.14	12 16 00.	+ 63 46		19.5	COMPACT GALAXY
MCG-04-29-017	12 16 00.	- 27 12	24	16.	GALAXY
KEEL 384	12 16 00.4	+ 37 48 03.			NEBULA
RNGC 4252	12 16 01.	+ 05 50		15.0	GALAXY
AMES 0498	12 16 01.	+ 06 01 39.	17	17.4	NEBULA
AMES 0497	12 16 01.	+ 13 08 03.	20	17.1	NEBULA
PATH 1.527	12 16 01.	+ 15 06	11		NEBULA
AMES 0499	12 16 02.	+ 09 29 51.	24	17.9	NEBULA
REIZ 2145	12 16 02.	+ 16 31	18	15.1	GALAXY
REIN 3.060	12 16 02.68	+ 12 08 53.2			NEBULA
AMES 0501	12 16 03.	+ 10 28 09.	20	17.2	NEBULA
AMES 0500	12 16 03.	+ 11 13 33.	28	18.1	NEBULA
HN 0856	12 16 03.	+ 12 08		16.	NEBULA
IC 3127	12 16 03.	+ 12 08			NONSTELLAR OBJECT
HN 0855	12 16 03.	+ 14 05		15.	NEBULA
IC 3126	12 16 03.	+ 14 05			NONSTELLAR OBJECT
AMES 0504	12 16 04.	+ 04 51 03.	25	17.4	NEBULA
AMES 0503	12 16 04.	+ 05 53 39.	28	17.5	NEBULA
AMES 0502	12 16 04.	+ 08 39 27.	23	17.4	NEBULA
KEEL 385	12 16 04.1	+ 37 40 49.			NEBULA
AMES 0505	12 16 05.	+ 06 06 09.	28	17.6	NEBULA
AMES 0506	12 16 06.	+ 06 58 15.	56	16.6	NEBULA
8ZW 1215+11.01	12 16 06.	+ 10 55		19.2	COMPACT GALAXY
ZWG 070.002	12 16 06.	+ 12 01		15.2	GALAXY
ZWG 069.139	12 16 06.	+ 12 01		15.2	GALAXY
MCG+02-31-086	12 16 06.	+ 12 01	24	16.	GALAXY
MCG+02-31-085	12 16 06.	+ 12 01	36	15.2	GALAXY
8ZW 1216+13.5	12 16 06.	+ 13 32		18.5	COMPACT GALAXY
ZWG 099.008	12 16 06.	+ 16 34		15.6	GALAXY
ZWG 098.141	12 16 06.	+ 16 34		15.6	GALAXY
MCG+03-31-098	12 16 06.	+ 17 59	120	15.5	GALAXY
REA 01	12 16 06.	+ 18 00			DWARF GALAXY
ZWG 099.009	12 16 06.	+ 18 28		15.6	GALAXY
ZWG 098.142	12 16 06.	+ 18 28		15.6	GALAXY
MCG+07-25-056	12 16 06.	+ 42 17	42	16.	GALAXY
MCG+08-22-101	12 16 06.	+ 49 09	24	17.	GALAXY
MCG+08-22-102	12 16 06.	+ 49 10	30	16.	GALAXY
7ZW 448	12 16 06.	+ 57 31			COMPACT GALAXY
AMES 0508	12 16 07.	+ 05 16 03.	26	17.6	NEBULA
AMES 0507	12 16 07.	+ 05 27 39.	18	17.6	NEBULA
AMES 0509	12 16 08.	+ 07 38 15.	31	17.8	NEBULA
REIZ 2146	12 16 08.	+ 16 33	24	15.2	GALAXY
AMES 0512	12 16 09.	+ 06 04 33.	18	17.2	NEBULA
AMES 0511	12 16 09.	+ 08 26 27.	24	17.6	NEBULA
HN 0857	12 16 09.	+ 12 00		14.5	NEBULA
IC 3128	12 16 09.	+ 12 00			NONSTELLAR OBJECT
AMES 0510	12 16 09.	+ 12 31 27.	16	17.5	NEBULA
TON-N 1500	12 16 09.	+ 32 38		16.8	BLUE STAR
KEEL 386	12 16 09.1	+ 14 48 01.		18.	NEBULA
REIN 3.061	12 16 09.27	+ 12 00 35.9			NEBULA
KEEL 387	12 16 09.8	+ 14 30 52.		18.	NEBULA
AMES 0513	12 16 10.	+ 11 51 27.	17	17.4	NEBULA
PATH 2.088	12 16 10.	+ 14 48	8		NEBULA
AMES 0514	12 16 11.	+ 07 53 57.	16	17.9	NEBULA
PATH 2.089	12 16 11.	+ 14 31	5		NEBULA
IC 3129	12 16 12.	+ 09 52 23.			SINGLE STAR
RMB 159	12 16 12.	+ 12 32			FAINT BLUE OBJECT
AMES 0515	12 16 12.	+ 12 38 03.	22	17.8	NEBULA
ZWG 069.140	12 16 12.	+ 13 13		15.7	GALAXY
RMB 102	12 16 12.	+ 13 27			FAINT BLUE OBJECT
ZC 1216.2+1337	12 16 12.	+ 13 37	540		CLUSTER OF GALAXIES
MCG+03-31-099	12 16 12.	+ 14 41 30.	360	10.2	GALAXY
UGC 07345	12 16 12.	+ 14 42	300	10.2	GALAXY Sc
ZWG 099.010	12 16 12.	+ 17 59		15.5	GALAXY
ZWG 098.143	12 16 12.	+ 17 59		15.5	GALAXY
UGC 07346	12 16 12.	+ 17 59	72	15.5	GALAXY
LB 00004	12 16 12.	+ 27 04		18.8	FAINT BLUE STAR
ABC 1515	12 16 12.	+ 28 15		17.6	RICH CLUSTER OF GALAXIES
LB 02303	12 16 12.	+ 50 31 48.		17.3	FAINT BLUE STAR
MCG+09-20-143	12 16 12.	+ 51 41	36	16.	GALAXY
ZC 1216.2+5855	12 16 12.	+ 58 55	1410		CLUSTER OF GALAXIES
AMES 0518	12 16 13.	+ 10 11 03.	34	17.3	NEBULA
AMES 0517	12 16 13.	+ 10 45 27.	30	17.8	NEBULA
AMES 0516	12 16 13.	+ 12 09 27.	26	16.9	NEBULA
PATH 1.528	12 16 13.	+ 15 08	5		NEBULA
AMES 0519	12 16 14.	+ 05 22 51.	17	17.8	NEBULA
REIZ 2147	12 16 14.	+ 14 41	300	11.2	GALAXY
A3 003	12 16 14.3	+ 14 40 14.		17.3	UV CONCENTRATION
A3 004	12 16 14.5	+ 14 40 30.		16.9	UV CONCENTRATION
AMES 0520	12 16 15.	+ 10 47 21.	25	17.8	NEBULA
SVEN 277	12 16 15.	+ 14 42	300	10.3	GALAXY
IC 3130	12 16 16.	+ 08 30 29.			MAY NOT EXIST
IC 0774	12 16 16.	- 06 29 07.			NONSTELLAR OBJECT
KEEL 388	12 16 16.3	+ 14 30 48.		18.	NEBULA
AMES 0521	12 16 17.	+ 07 03 15.	18	17.8	NEBULA
PATH 2.090	12 16 17.	+ 14 31	5		NEBULA
REIN 3.063	12 16 17.85	+ 13 10 28.7			NEBULA
ZWG 042.002	12 16 18.	+ 03 17		15.4	GALAXY
AMES 0523	12 16 18.	+ 05 45 39.	19	17.5	NEBULA
ZWG 042.003	12 16 18.	+ 08 08		14.8	GALAXY
MCG+01-31-046	12 16 18.	+ 08 08	36	14.8	GALAXY
IC 3132	12 16 18.	+ 08 08 23.			SAME AS IC 3131
RMB 221	12 16 18.	+ 10 32			FAINT BLUE OBJECT
8ZW 1215+11.02	12 16 18.	+ 11 06		17.3	COMPACT GALAXY
UGC 07347	12 16 18.	+ 12 45	72	16.5	GALAXY Sc
AMES 0522	12 16 18.	+ 12 52 33.	38	17.6	NEBULA
ZWG 099.011	12 16 18.	+ 14 42		10.2	GALAXY
ZWG 098.144	12 16 18.	+ 14 42		10.2	GALAXY
SCH 31	12 16 18.	+ 14 42		10.2	PECULIAR GALAXY
SEY 175	12 16 18.	+ 45 03 20.		14.8	FAINT GALAXY
ZWG 244.002	12 16 18.	+ 45 04		15.4	GALAXY
MCG+08-22-103	12 16 18.	+ 45 04	36	16.	GALAXY
ZCG 199.15	12 16 18.	+ 63 47		17.4	COMPACT GALAXY
REIN 3.062	12 16 18.01	+ 08 08 24.3			NEBULA
AMES 0526	12 16 19.	+ 05 31 15.	17	16.8	NEBULA
AMES 0525	12 16 19.	+ 10 56 39.	12	18.0	NEBULA
AMES 0524	12 16 19.	+ 11 02 15.	20	17.7	NEBULA
PATH 2.091	12 16 19.	+ 14 42	257		NEBULA
SN 1972Q	12 16 19.	+ 14 44		15.8	SUPERNOVA
RNGC 4256	12 16 19.	+ 66 11		12.5	GALAXY
RNGC 4254	12 16 20.	+ 14 42		10.5	GALAXY
REIZ 2148	12 16 21.	+ 05 04	60	13.5	GALAXY
ABC 1516	12 16 21.	+ 05 31		16.6	RICH CLUSTER OF GALAXIES
IC 0775	12 16 21.0	+ 13 11 31.			GALAXY
REIN 3.064	12 16 21.27	+ 13 11 22.9			NEBULA
AMES 0530	12 16 22.	+ 06 16 57.	18	17.4	NEBULA
IC 3133	12 16 22.	+ 07 55 11.			MAY NOT EXIST
AMES 0529	12 16 22.	+ 08 44 45.	19	17.6	NEBULA
AMES 0528	12 16 22.	+ 10 17 03.	25	16.5	NEBULA
IC 3135	12 16 22.	+ 27 46 17.			NONSTELLAR OBJECT
AMES 0532	12 16 23.	+ 08 26 51.	22	17.8	NEBULA
IC 3134	12 16 23.	+ 09 14 17.			NONSTELLAR OBJECT
AMES 0527	12 16 23.	+ 13 44 39.	38	17.7	NEBULA
SN 1967H	12 16 23.	+ 14 42		14.6	SUPERNOVA
REIN 3.065	12 16 23.24	+ 09 14 20.9			NEBULA
ZWG 042.004	12 16 24.	+ 05 04		13.5	GALAXY
UGC 07348	12 16 24.	+ 05 04	72	13.5	GALAXY S0
MCG+01-31-047	12 16 24.	+ 05 04	60	13.5	GALAXY
ZWG 042.005	12 16 24.	+ 06 28		14.7	GALAXY
UGC 07349	12 16 24.	+ 06 28	60	14.7	GALAXY S
MCG+01-31-048	12 16 24.	+ 06 28	48	14.7	GALAXY
ZWG 070.003	12 16 24.	+ 09 15		15.4	GALAXY
RMB 162	12 16 24.	+ 11 58			FAINT BLUE OBJECT
8ZW 1216+12.0	12 16 24.	+ 12 02		18.8	COMPACT GALAXY
8ZW 1216+12.6	12 16 24.	+ 12 38		17.7	COMPACT GALAXY
ZWG 070.004	12 16 24.	+ 13 11		14.8	GALAXY
UGC 07350	12 16 24.	+ 13 11	84	14.8	GALAXY E-S0
MCG+02-31-087	12 16 24.	+ 13 11	48	14.8	GALAXY
AMES 0531	12 16 24.	+ 13 43 03.	20	17.4	NEBULA
RMB 055	12 16 24.	+ 14 42			FAINT BLUE OBJECT
ZWG 128.074	12 16 24.	+ 24 28		15.7	GALAXY
ZC 1216.4+2717	12 16 24.	+ 27 17	1810		CLUSTER OF GALAXIES
ZWG 215.060	12 16 24.	+ 43 53		15.5	GALAXY
LB 02305	12 16 24.	+ 60 32 12.		15.4	FAINT BLUE STAR
LB 02304	12 16 24.	+ 60 44 12.		15.2	FAINT BLUE STAR
ZCG 199.14A	12 16 24.	+ 64 37		16.8	COMPACT GALAXY
ZWG 315.032	12 16 24.	+ 66 10		12.7	GALAXY
REIZ 2152	12 16 24.	+ 66 10	240	13.0	GALAXY
UGC 07351	12 16 24.	+ 66 10	270	12.7	GALAXY Sb
MCG+11-15-045	12 16 24.	+ 66 10	228	13.0	GALAXY
MCG-03-31-026	12 16 24.	- 19 49	36	14.5	GALAXY
REIN 3.066	12 16 24.31	+ 09 14 49.1			NEBULA
A3 007	12 16 24.6	+ 14 41 58.		15.5	UV CONCENTRATION
RNGC 4255	12 16 25.	+ 05 04		13.5	GALAXY
AMES 0535	12 16 25.	+ 05 22 27.	14	17.6	NEBULA
AMES 0534	12 16 25.	+ 05 31 39.	30	17.4	NEBULA
AMES 0533	12 16 25.	+ 06 22 51.	34	16.6	NEBULA
RNGC 3172	12 16 25.	+ 89 22		15.0	GALAXY
IC 3136	12 16 25.0	+ 06 27 45.			GALAXY
KEEL 389	12 16 25.6	+ 14 50 08.		18.	NEBULA
AMES 0538	12 16 26.	+ 08 25 39.	20	17.6	NEBULA
AMES 0537	12 16 26.	+ 12 17 51.	22	17.2	NEBULA
KEEL 390	12 16 26.1	+ 14 47 20.		17.	NEBULA
SVEN 278	12 16 27.	+ 05 03	42	13.9	GALAXY
REIZ 2150	12 16 27.	+ 06 28	60	14.0	GALAXY
REIZ 2149	12 16 27.	+ 08 08	42	14.3	GALAXY
IC 0776	12 16 27.	+ 09 06 29.			NONSTELLAR OBJECT
HN 0859	12 16 27.	+ 12 42		15.5	NEBULA
IC 3138	12 16 27.	+ 12 42			NONSTELLAR OBJECT
HN 0858	12 16 27.	+ 12 44	60		NEBULA
IC 3137	12 16 27.	+ 12 44			NONSTELLAR OBJECT
AMES 0536	12 16 27.	+ 14 14 39.	30	17.2	NEBULA
PATH 2.092	12 16 27.	+ 14 48	5		NEBULA
PATH 2.093	12 16 27.	+ 14 50	11		NEBULA

OBJECT NAME	RIGHT ASCEN.	DECLINATION	DIAM.	MAGN.	TYPE OF OBJECT
IC 3141	12 16 27.	+ 24 27 53.			NONSTELLAR OBJECT
IC 3140	12 16 27.	+ 27 24 29.			NONSTELLAR OBJECT
IC 3139	12 16 28.	+ 09 24 23.			SINGLE STAR
RNGC 4258	12 16 28.	+ 47 35		9.5	GALAXY
AMES 0541	12 16 29.	+ 12 27 15.	26	17.4	NEBULA
AMES 0539	12 16 29.	+ 13 20 27.	16	17.8	NEBULA
HOLM 363A	12 16 29.	+ 47 35	600	09.8	PART OF MULTIPLE GALAXY
KEEL 391	12 16 29.9	+ 14 46 54.		18.	NEBULA
ZWG 042.006	12 16 30.	+ 06 00		15.0	GALAXY
MCG+01-31-049	12 16 30.	+ 06 00	48	15.0	GALAXY
ZWG 042.007	12 16 30.	+ 06 22		15.3	GALAXY
ZWG 070.005	12 16 30.	+ 09 08		14.9	GALAXY
UGC 07352	12 16 30.	+ 09 08	126	14.9	GALAXY
MCG+02-31-088	12 16 30.	+ 09 08	108	14.9	GALAXY
8ZW 1215+11.03	12 16 30.	+ 10 57		19.0	COMPACT GALAXY
FMB 101	12 16 30.	+ 13 35			FAINT BLUE OBJECT
AMES 0540	12 16 30.	+ 13 36 15.	26	18.0	NEBULA
FATH 1.529	12 16 30.	+ 14 16	16		NEBULA
MCG+03-34-014	12 16 30.	+ 15 02	30	15.3	GALAXY
FEIG 060	12 16 30.	+ 15 24		13.1	FAINT BLUE STAR
ZWG 099.012	12 16 30.	+ 17 15		15.3	GALAXY
ZWG 098.145	12 16 30.	+ 17 15		15.3	GALAXY
TON-N 0607	12 16 30.	+ 26 03		15.3	BLUE STAR
ZC 1216.5+3049	12 16 30.	+ 30 49	2290		CLUSTER OF GALAXIES
ZC 1216.5+3740	12 16 30.	+ 37 40	740		CLUSTER OF GALAXIES
MCG+07-25-057	12 16 30.	+ 43 54	30	15.5	GALAXY
ZWG 244.003	12 16 30.	+ 47 35		9.6	GALAXY
ZWG 243.067	12 16 30.	+ 47 35		9.6	GALAXY
UGC 07353	12 16 30.	+ 47 35	1320	9.6	GALAXY Sb
MCG+08-22-104	12 16 30.	+ 47 35	720	8.9	GALAXY
MCG+09-20-144	12 16 30.	+ 51 32 30.	24	17.	GALAXY
MCG+10-18-021	12 16 30.	+ 60 27	39	15.	GALAXY
ZC 1216.5+6148	12 16 30.	+ 61 48	6850		CLUSTER OF GALAXIES
ZCG 199.16	12 16 30.	+ 63 46		18.4	COMPACT GALAXY
MRSL 297+10/1	12 16 30.	- 52 00	15000		HII REGION
REIN 3.067	12 16 30.33	+ 09 08 01.0			NEBULA
FATH 2.094	12 16 31.	+ 14 47	5		NEBULA
SEY 176	12 16 31.	+ 43 53 14.		15.1	FAINT GALAXY
KEEL 392	12 16 31.1	+ 14 24 04.		19.	NEBULA
AMES 0545	12 16 33.	+ 05 06 21.	23	17.6	NEBULA
AMES 0544	12 16 33.	+ 05 53 09.	19	17.6	NEBULA
REIZ 2151	12 16 33.	+ 06 22	12	15.6	GALAXY
AMES 0543	12 16 33.	+ 07 53 09.	19	17.6	NEBULA
8ZW 1215+11.04	12 16 33.	+ 10 52		18.9	COMPACT GALAXY
AMES 0542	12 16 33.	+ 13 22 21.	19	17.8	NEBULA
HN 0860	12 16 33.	+ 14 15		14.5	NEBULA
MCG+02-31-089	12 16 33.	+ 14 15	24	18.	GALAXY
IC 3142	12 16 33.	+ 14 16			FAINT NEBULA
FATH 2.095	12 16 33.	+ 14 16	22		NEBULA
MCG+04-29-063	12 16 33.	+ 22 41 30.	90	15.5	GALAXY
FATH 1.530	12 16 33.	- 15 52	16		NEBULA
AMES 0546	12 16 34.	+ 11 34 21.	19	17.8	NEBULA
REIZ 2156	12 16 34.	+ 49 37	108	13.6	GALAXY
A3 009	12 16 34.0	+ 15 17 04.		17.1	FAINT BLUE OBJECT
IC 3143	12 16 35.	+ 27 34 35.			NONSTELLAR OBJECT
REIZ 2155	12 16 35.	+ 47 35	540	10.1	GALAXY
ARC 1517	12 16 35.	- 04 45		16.6	RICH CLUSTER OF GALAXIES
ZWG 042.008	12 16 36.	+ 04 08		14.5	GALAXY
MRK 049	12 16 36.	+ 04 08	13	15.	GALAXY WITH UV CONTINUUM
ZCG 1216+04	12 16 36.	+ 04 08		14.5	COMPACT GALAXY
UGC 07354	12 16 36.	+ 04 08	42	14.5	GALAXY
MCG+01-31-050	12 16 36.	+ 04 08	24	14.5	GALAXY
ZWG 042.009	12 16 36.	+ 06 40		15.2	GALAXY
RMB 157	12 16 36.	+ 12 44			FAINT BLUE OBJECT
AMES 0547	12 16 36.	+ 13 36 45.	29	17.9	NEBULA
ZWG 070.006	12 16 36.	+ 14 15		15.4	GALAXY
UGC 07355	12 16 36.	+ 14 15	96	15.4	GALAXY DBL SYS
MCG+02-31-090	12 16 36.	+ 14 15	30	16.	GALAXY
ZWG 099.013	12 16 36.	+ 19 32		15.7	GALAXY
ZWG 098.146	12 16 36.	+ 19 32		15.7	GALAXY
MCG+03-31-100	12 16 36.	+ 19 32	36	15.7	GALAXY
ZWG 158.062	12 16 36.	+ 27 35		15.7	GALAXY
ZC 1216.6+3548	12 16 36.	+ 35 48	1140		CLUSTER OF GALAXIES
UGC 07356	12 16 36.	+ 47 22	66	17.	GALAXY DWARF?
ARC 1518	12 16 36.	+ 63 47		17.0	RICH CLUSTER OF GALAXIES
RNGC 4257	12 16 37.	+ 06 00		15.0	GALAXY
AMES 0550	12 16 37.	+ 06 39 57.	12	16.6	NEBULA
AMES 0549	12 16 37.	+ 11 23 21.	20	17.8	NEBULA
REIN 3.068	12 16 37.41	+ 11 59 15.3			NEBULA
KEEL 393	12 16 37.7	+ 14 24 30.		18.	NEBULA
AMES 0553	12 16 38.	+ 08 43 51.	23	17.8	NEBULA
AMES 0552	12 16 38.	+ 12 15 33.	17	17.0	NEBULA
AMES 0548	12 16 38.	+ 13 09 45.	48	17.0	NEBULA
SVEN 279	12 16 38.	+ 14 49	6	14.8	GALAXY
LB 02306	12 16 38.	+ 54 09 30.		16.0	FAINT BLUE STAR
KEEL 394	12 16 38.7	+ 14 48 51.		19.	NEBULA
HARO 08	12 16 39.	+ 04 08			BLUE EMISSION-LINE GALAXY
REIZ 2153	12 16 39.	+ 04 08	24	14.1	GALAXY
REIZ 2154	12 16 39.	+ 06 00	42	14.6	GALAXY
AMES 0554	12 16 39.	+ 10 05 39.	19	17.8	NEBULA
AMES 0551	12 16 39.	+ 13 07 45.	30	17.4	NEBULA
FATH 2.096	12 16 39.	+ 14 25	14		NEBULA
IC 3144	12 16 39.	+ 25 34 29.			NONSTELLAR OBJECT
AMES 0557	12 16 40.	+ 06 34 45.	20	16.8	NEBULA
AMES 0556	12 16 40.	+ 06 35 33.	19	16.7	NEBULA
IC 3145	12 16 40.	+ 24 34 17.			NONSTELLAR OBJECT
AMES 0555	12 16 41.	+ 14 08 51.	19	16.2	NEBULA
FATH 1.531	12 16 41.	+ 14 10	11		NEBULA
FATH 2.097	12 16 41.	+ 14 49	16		NEBULA
KEEL 395	12 16 41.4	+ 47 22 05.		17.	SPIRAL NEBULA
AMES 0558	12 16 42.	+ 06 09 57.	19	17.2	NEBULA
KARA.68 099	12 16 42.	+ 06 16	27		DWARF GALAXY
ZWG 042.010	12 16 42.	+ 06 34		15.3	GALAXY
KARA.72 328B	12 16 42.	+ 06 34	30	15.3	PART OF DOUBLE GALAXY
ZWG 042.011	12 16 42.	+ 06 35		15.3	GALAXY
KARA.72 328A	12 16 42.	+ 06 35	30	15.3	PART OF DOUBLE GALAXY
8ZW 1215+11.05	12 16 42.	+ 10 53		18.8	COMPACT GALAXY
ZWG 070.007	12 16 42.	+ 12 20		15.5	GALAXY
MCG+02-31-092	12 16 42.	+ 12 20	15	15.5	GALAXY
MCG+02-31-091	12 16 42.	+ 12 20	30	15.5	GALAXY
RMB 056	12 16 42.	+ 14 09			FAINT BLUE OBJECT
ZWG 128.075	12 16 42.	+ 22 43		15.5	GALAXY
UGC 07357	12 16 42.	+ 22 43	102	15.5	GALAXY Sc
IC 3146	12 16 42.	+ 25 59 29.			NONSTELLAR OBJECT
MCG+05-29-052	12 16 42.	+ 28 36	66	14.5	GALAXY
ZC 1216.7+3008	12 16 42.	+ 30 08	2150		CLUSTER OF GALAXIES
ZWG 244.004	12 16 42.	+ 49 38		14.3	GALAXY
ZWG 243.068	12 16 42.	+ 49 38		14.3	GALAXY
UGC 07358	12 16 42.	+ 49 38	102	14.3	GALAXY Sc
ZC 1216.7+5542	12 16 42.	+ 55 42	6250		CLUSTER OF GALAXIES
ZCG 199.17	12 16 42.	+ 63 51		18.6	COMPACT GALAXY
MRSL 299-00/1	12 16 42.	- 62 41	180		HII REGION
REIN 3.069	12 16 44.83	+ 12 17 38.2			NEBULA
REIZ 2157	12 16 45.	+ 06 01	24	16.5	GALAXY
REIZ 2158	12 16 45.	+ 12 15	54	14.7	GALAXY
MCG+08-22-106	12 16 45.	+ 49 38	90	14.	GALAXY
AMES 0559	12 16 46.	+ 12 13 15.	19	17.4	NEBULA
IC 3147	12 16 46.	+ 12 17 42.			NONSTELLAR OBJECT
REIN 3.070	12 16 46.29	+ 12 17 45.7			NEBULA
AMES 0561	12 16 47.	+ 05 44 45.	24	17.6	NEBULA
AMES 0560	12 16 47.	+ 06 11 57.	26	15.6	NEBULA
IC 3148	12 16 47.	+ 08 08 48.			NONSTELLAR OBJECT
AMES 0565	12 16 48.	+ 05 24 45.	16	17.7	NEBULA
AMES 0564	12 16 48.	+ 05 29 33.	12	17.8	NEBULA
ZWG 042.012	12 16 48.	+ 05 39		14.5	GALAXY
UGC 07359	12 16 48.	+ 05 39	60	14.5	GALAXY S0
MCG+01-31-051	12 16 48.	+ 05 39	36	14.5	GALAXY
ZWG 042.013	12 16 48.	+ 06 06		12.0	GALAXY
UGC 07360	12 16 48.	+ 06 06	210	12.0	GALAXY E
MCG+01-31-052	12 16 48.	+ 06 06	42	12.0	GALAXY
ZWG 042.014	12 16 48.	+ 06 11		14.8	GALAXY
MCG+01-31-053	12 16 48.	+ 06 11	30	14.8	GALAXY
ZWG 042.015	12 16 48.	+ 06 22		13.1	GALAXY
UGC 07361	12 16 48.	+ 06 22	156	13.1	GALAXY SBa
MCG+01-31-054	12 16 48.	+ 06 22	120	13.1	GALAXY
AMES 0563	12 16 48.	+ 06 31 03.	20	17.6	NEBULA
ZWG 042.016	12 16 48.	+ 08 09		15.3	GALAXY
MCG+01-31-055	12 16 48.	+ 08 09	48	15.3	GALAXY
ZWG 070.008	12 16 48.	+ 12 21		15.6	GALAXY
ZWG 070.009	12 16 48.	+ 12 35		15.6	GALAXY
MCG+02-31-093	12 16 48.	+ 12 35	30	15.6	GALAXY
KARA.68 100	12 16 48.	+ 14 18	47		DWARF GALAXY
ZC 1216.8+2056	12 16 48.	+ 20 56	540		CLUSTER OF GALAXIES
UGC 07362	12 16 48.	+ 26 13	66	16.5	GALAXY Sb-c
ARC 1519	12 16 48.	+ 27 12		17.5	RICH CLUSTER OF GALAXIES
ZWG 158.063	12 16 48.	+ 30 37		15.6	GALAXY
KARA.68 101	12 16 48.	+ 47 21	67		DWARF GALAXY
MCG+10-18-022	12 16 48.	+ 61 12	27	16.	GALAXY
MCG+11-15-046	12 16 48.	+ 67 25	27	17.	GALAXY
ZWG 014.001	12 16 48.	- 00 20		15.7	GALAXY
REIN 3.071	12 16 48.89	+ 08 08 54.3			NEBULA
RNGC 4259	12 16 49.	+ 05 39		14.5	GALAXY
RNGC 4261	12 16 49.	+ 06 06		11.5	GALAXY
RNGC 4260	12 16 49.	+ 06 23		12.5	GALAXY
AMES 0568	12 16 49.	+ 06 57 45.	26	17.8	NEBULA
AMES 0567	12 16 49.	+ 08 29 33.	17	17.7	NEBULA
AMES 0566	12 16 49.	+ 08 43 09.	20	17.0	NEBULA
AMES 0562	12 16 49.	+ 13 45 33.	24	17.6	NEBULA
REIN 3.072	12 16 50.64	+ 12 19 08.5			NEBULA
REIZ 2159	12 16 51.	+ 05 39	42	14.6	GALAXY
REIZ 2161	12 16 51.	+ 06 06	60	12.3	GALAXY
REIZ 2160	12 16 51.	+ 06 22	120	12.9	GALAXY
AMES 0569	12 16 51.	+ 07 26 21.	16	17.6	NEBULA
REIZ 2162	12 16 51.	+ 08 09	42	15.6	GALAXY
HOLM 364B	12 16 51.	+ 17 32	24	16.0	PART OF MULTIPLE GALAXY
TON-N 1501	12 16 51.	+ 21 31		17.0	BLUE STAR
LB 02307	12 16 51.	+ 48 39 12.		15.6	FAINT BLUE STAR
REIN 3.073	12 16 51.84	+ 12 34 38.6			NEBULA
HOLM 368F	12 16 52.	+ 05 39	18	13.8	PART OF MULTIPLE GALAXY
AMES 0570	12 16 52.	+ 12 20 03.	18	17.7	NEBULA
IC 3149	12 16 52.	+ 12 34 54.			NONSTELLAR OBJECT
AMES 0572	12 16 53.	+ 05 19 09.	37	15.8	NEBULA
AMES 0571	12 16 53.	+ 06 33 27.	23	17.8	NEBULA
FATH 1.532	12 16 53.	+ 14 46	5		NEBULA
AMES 0575	12 16 54.	+ 04 51 45.	29	17.2	NEBULA
ZWG 042.018	12 16 54.	+ 05 19		15.2	GALAXY
AMES 0574	12 16 54.	+ 08 04 33.	28	16.4	NEBULA
ZWG 042.017	12 16 54.	+ 08 05		15.5	GALAXY
AMES 0573	12 16 54.	+ 12 02 33.	35	17.7	NEBULA
MCG+03-31-101	12 16 54.	+ 15 09	42	12.3	GALAXY
ZWG 158.064	12 16 54.	+ 28 35		14.5	GALAXY
UGC 07363	12 16 54.	+ 28 35	78	14.5	GALAXY SB
ZC 1216.9+3711	12 16 54.	+ 37 11	870		CLUSTER OF GALAXIES
AMES 0576	12 16 55.	+ 05 27 51.	14	17.8	NEBULA
REIZ 2165	12 16 55.	+ 28 28	60	15.4	GALAXY
KEEL 396	12 16 55.2	+ 14 51 51.		18.	NEBULA
IC 3150	12 16 56.	+ 08 04 30.			MAY NOT EXIST
AMES 0578	12 16 56.	+ 10 37 03.	34	16.7	NEBULA
AMES 0577	12 16 56.	+ 11 48 27.	24	17.8	NEBULA
REIZ 2164	12 16 56.	+ 15 09	30	12.3	GALAXY
SVEN 280	12 16 56.	+ 15 10	18	13.3	GALAXY
HOLM 364A	12 16 56.	+ 17 31	60	14.7	PART OF MULTIPLE GALAXY
REIZ 2163	12 16 57.	+ 05 19	42	15.7	GALAXY
HN 0861	12 16 57.	+ 09 41		13.5	NEBULA
IC 3151	12 16 57.	+ 09 41			NONSTELLAR OBJECT
FATH 1.533	12 16 57.	+ 14 16	27		NEBULA
IC 0778	12 16 57.	+ 56 16 42.			NONSTELLAR OBJECT
KEEL 397	12 16 57.7	+ 14 41 08.		17.	NEBULA
AMES 0580	12 16 58.	+ 05 21 15.	20	17.1	NEBULA
AMES 0579	12 16 58.	+ 08 28 57.	20	17.6	NEBULA
AMES 0581	12 16 59.	+ 11 16 45.	18	17.7	NEBULA
FATH 2.098	12 16 59.	+ 14 41	11		NEBULA
VDB.66G 119	12 17	+ 46 38	100		DWARF GALAXY
ZC 1217.0+0417	12 17 00.	+ 04 17	340		CLUSTER OF GALAXIES
ZWG 042.019	12 17 00.	+ 05 40		15.2	GALAXY
ZWG 042.020	12 17 00.	+ 06 07		13.9	GALAXY
UGC 07364	12 17 00.	+ 06 07	48	13.9	GALAXY SB0
MCG+01-32-001	12 17 00.	+ 06 07	36	13.9	GALAXY
ZWG 070.010	12 17 00.	+ 09 38		15.1	GALAXY
MCG+02-32-002	12 17 00.	+ 09 38	48	15.1	GALAXY
KARA.68 102	12 17 00.	+ 14 16	34		DWARF GALAXY
REA 29	12 17 00.	+ 14 17	54		DWARF GALAXY
MCG+02-32-001	12 17 00.	+ 14 17	48	17.	GALAXY
ZWG 099.014	12 17 00.	+ 15 09		12.3	GALAXY
UGC 07365	12 17 00.	+ 15 09	114	12.3	GALAXY SB0
FATH 2.099	12 17 00.	+ 15 10	33		NEBULA
ZC 1217.0+1646	12 17 00.	+ 16 46	4300		CLUSTER OF GALAXIES
ZWG 099.015	12 17 00.	+ 17 30		15.5	GALAXY
UGC 07366	12 17 00.	+ 17 30	66	15.5	GALAXY
MCG+03-32-001	12 17 00.	+ 17 30	72	15.5	GALAXY
TON-N 0608	12 17 00.	+ 24 21		16.5	BLUE STAR
MCG+05-29-056	12 17 00.	+ 29 00	12	15.5	GALAXY
MCG+05-29-055	12 17 00.	+ 29 04	12	15.7	GALAXY
MCG+05-29-054	12 17 00.	+ 29 07 30.	24	15.4	GALAXY
MCG+05-29-053	12 17 00.	+ 30 10	54	15.4	GALAXY
ZWG 244.005	12 17 00.	+ 50 05		13.7	GALAXY

OBJECT NAME	RIGHT ASCEN.	DECLINATION	DIAM.	MAGN.	TYPE OF OBJECT
ZWG 243.069	12 17 00.	+ 50 05		13.7	GALAXY
UGC 07367	12 17 00.	+ 50 05	96	13.7	GALAXY S0-a
MCG+08-23-001	12 17 00.	+ 50 06 30.	72	14.	GALAXY
MCG+10-18-023	12 17 00.	+ 59 58	30	16.	GALAXY
ZCG 199.19	12 17 00.	+ 63 48		18.0	COMPACT GALAXY
ZCG 199.18	12 17 00.	+ 63 50		18.1	COMPACT GALAXY
ZWG 014.002	12 17 00.	- 02 32		15.5	GALAXY
ARC 1520	12 17 00.	- 13 00		16.8	RICH CLUSTER OF GALAXIES
ARC 1521	12 17 00.	- 13 26		16.8	RICH CLUSTER OF GALAXIES
MCG-04-29-018	12 17 00.	- 25 52 30.	30	13.	GALAXY
REIN 3.074	12 17 00.15	+ 09 41 30.2			NEBULA
RNGC 4264	12 17 01.	+ 06 07		14.0	GALAXY
AMES 0582	12 17 02.	+ 13 22 03.	12	17.4	NEBULA
RNGC 4262	12 17 02.	+ 15 09		12.5	GALAXY
IC 0777	12 17 02.	+ 28 34 36.			NONSTELLAR OBJECT
KARA.73 37	12 17 02.	- 34 21	27		DWARF GALAXY
REIZ 2166	12 17 03.	+ 06 07	36	14.0	GALAXY
AMES 0584	12 17 03.	+ 11 16 15.	30	17.4	NEBULA
TON-N 1502	12 17 03.	+ 21 07		13.8	BLUE STAR
IC 3154	12 17 03.	+ 25 51 48.			NONSTELLAR OBJECT
MCG+05-29-057	12 17 03.	+ 29 09	60	15.3	GALAXY
KEEL 398	12 17 03.5	+ 14 42 10.		18.	NEBULA
IC 3153	12 17 04.	+ 05 40 30.			NONSTELLAR OBJECT
AMES 0585	12 17 04.	+ 05 51 45.	23	17.8	NEBULA
AMES 0583	12 17 04.	+ 13 44 21.	23	17.0	NEBULA
IC 3152	12 17 04.	- 25 54 00.			NONSTELLAR OBJECT
AMES 0587	12 17 05.	+ 12 15 57.	25	17.3	NEBULA
FATH 2.100	12 17 05.	+ 14 42	11		NEBULA
ZWG 014.005	12 17 06.	+ 02 19		15.6	GALAXY
ZWG 042.021	12 17 06.	+ 05 49		15.0	GALAXY
UGC 07368	12 17 06.	+ 05 49	126	15.0	GALAXY S
MCG+01-32-002	12 17 06.	+ 05 49	96	15.0	GALAXY
AMES 0590	12 17 06.	+ 05 54 27.	14	17.5	NEBULA
ZWG 042.022	12 17 06.	+ 06 17		15.0	GALAXY
MCG+01-32-003	12 17 06.	+ 06 17	36	15.0	GALAXY
AMES 0589	12 17 06.	+ 11 19 21.	34	17.0	NEBULA
AMES 0586	12 17 06.	+ 12 39 45.	18	17.4	NEBULA
RMB 156	12 17 06.	+ 12 49			FAINT BLUE OBJECT
RMB 155	12 17 06.	+ 13 01			FAINT BLUE OBJECT
MCG+03-34-015	12 17 06.	+ 17 15	66	15.2	GALAXY
MCG+05-29-058	12 17 06.	+ 27 54	48	13.4	GALAXY
ZWG 158.065	12 17 06.	+ 28 46		15.7	GALAXY
ZWG 158.066	12 17 06.	+ 30 10		15.4	GALAXY
UGC 07369	12 17 06.	+ 30 10	66	15.4	GALAXY
ZC 1217.1+4929	12 17 06.	+ 49 29	1210		CLUSTER OF GALAXIES
ZCG 199.20	12 17 06.	+ 63 48		18.1	COMPACT GALAXY
ZWG 335.010	12 17 06.	+ 69 00		15.7	GALAXY
ZWG 014.004	12 17 06.	- 00 27		15.7	GALAXY
ZWG 014.003	12 17 06.	- 02 30		15.7	GALAXY
ZC 1217.1-0301	12 17 06.	- 03 01	470		CLUSTER OF GALAXIES
MCG-02-32-001	12 17 06.	- 11 58	60	14.	GALAXY
ACK 299-00.2	12 17 06.	- 62 38			PLANETARY NEBULA
HOLM 368G	12 17 07.	+ 05 40	18	14.5	PART OF MULTIPLE GALAXY
AMES 0588	12 17 07.	+ 12 40 33.	20	17.3	NEBULA
HOLM 367C	12 17 07.	+ 35 48	24	15.1	PART OF MULTIPLE GALAXY
HOLM 367B	12 17 07.	+ 35 48	24	15.1	PART OF MULTIPLE GALAXY
RNGC 4263	12 17 07.	- 11 58		14.0	GALAXY
RNGC 4265	12 17 07.	- 11 59			NON-EXISTENT OBJECT
AMES 0591	12 17 08.	+ 06 49 09.	16	17.8	NEBULA
HOLM 366A	12 17 08.	+ 57 00	36	13.4	PART OF MULTIPLE GALAXY
REIZ 2175	12 17 08.	+ 57 01	18	14.7	GALAXY
REIZ 2174	12 17 08.	+ 57 01	42	14.0	GALAXY
HOLM 366B	12 17 08.	+ 57 01	24	13.5	PART OF MULTIPLE GALAXY
AMES 0593	12 17 09.	+ 05 49 03.	103	15.0	NEBULA
REIZ 2167	12 17 09.	+ 05 50	72	14.8	GALAXY
TON-N 1503	12 17 09.	+ 22 09		17.0	BLUE STAR
MCG+05-29-059	12 17 09.	+ 30 38	24	14.2	GALAXY
FAI 068	12 17 09.	+ 75 25	40		DWARF SPHEROIDAL GALAXY
AMES 0592	12 17 10.	+ 13 42 57.	28	16.8	NEBULA
SC 1214-4254.5	12 17 10.	- 43 11 09.	30		NEBULA
HOLM 365B	12 17 11.	+ 06 17	24	14.2	PART OF MULTIPLE GALAXY
AMES 0594	12 17 11.	+ 08 41 09.	23	17.8	NEBULA
IC 3156	12 17 11.	+ 09 25 36.			NONSTELLAR OBJECT
RNGC 4271	12 17 11.	+ 57 01		13.5	GALAXY
REIN 3.075	12 17 11.38	+ 09 25 32.5			NEBULA
ZWG 014.007	12 17 12.	+ 02 22		14.6	GALAXY
UGC 07370	12 17 12.	+ 02 22	72	14.6	GALAXY Sb-c
MCG+00-32-001	12 17 12.	+ 02 22	66	14.6	GALAXY
ZWG 042.023	12 17 12.	+ 05 34		13.9	GALAXY
UGC 07371	12 17 12.	+ 05 34	78	13.9	GALAXY S0-a
MCG+01-32-004	12 17 12.	+ 05 34	72	13.9	GALAXY
ZWG 042.024	12 17 12.	+ 06 18		13.9	GALAXY
UGC 07372	12 17 12.	+ 06 18	84	13.9	GALAXY S0-a
MCG+01-32-005	12 17 12.	+ 06 18	24	13.9	GALAXY
ZWG 042.025	12 17 12.	+ 07 16		15.0	GALAXY
MCG+01-32-006	12 17 12.	+ 07 16	30	15.0	GALAXY
ZWG 070.011	12 17 12.	+ 09 26		15.0	GALAXY
MCG+02-32-003	12 17 12.	+ 09 26	36	15.0	GALAXY
ZWG 070.012	12 17 12.	+ 12 42		15.5	GALAXY
ZWG 070.013	12 17 12.	+ 13 05		12.4	GALAXY
UGC 07373	12 17 12.	+ 13 05	216	12.4	GALAXY S0
MCG+02-32-004	12 17 12.	+ 13 05	150	12.4	GALAXY
RMB 154	12 17 12.	+ 13 08			FAINT BLUE OBJECT
RMB 053	12 17 12.	+ 15 09			FAINT BLUE OBJECT
ZC 1217.2+1906	12 17 12.	+ 19 06	5110		CLUSTER OF GALAXIES
ZWG 158.067	12 17 12.	+ 29 00		15.5	GALAXY
ZWG 158.068	12 17 12.	+ 29 04		15.7	GALAXY
ZWG 158.069	12 17 12.	+ 29 07		15.4	GALAXY
ZWG 158.070	12 17 12.	+ 29 09		15.3	GALAXY
UGC 07374	12 17 12.	+ 29 09	66	15.3	GALAXY S
MCG+09-20-145	12 17 12.	+ 53 20	60	15.	GALAXY
ZWG 293.010	12 17 12.	+ 57 01		13.7	GALAXY
UGC 07375	12 17 12.	+ 57 01	90	13.7	GALAXY E-S0
ZWG 014.006	12 17 12.	- 00 28		15.3	GALAXY
IC 3155	12 17 12.3	+ 06 17 01.			GALAXY
RNGC 4268	12 17 13.	+ 05 34		14.0	GALAXY
RNGC 4266	12 17 13.	+ 05 49		15.0	GALAXY
AMES 0596	12 17 13.	+ 12 08 21.	26	17.9	NEBULA
AMES 0595	12 17 13.	+ 13 16 21.	19	17.1	NEBULA
REIN 3.076	12 17 13.00	+ 13 04 31.2			NEBULA
AMES 0598	12 17 14.	+ 05 27 09.	25	17.4	NEBULA
AMES 0597	12 17 14.	+ 08 21 27.	48	18.0	NEBULA
RNGC 4267	12 17 14.	+ 13 05		12.5	GALAXY
A3 012	12 17 14.2	+ 15 09 37.		17.6	FAINT BLUE OBJECT
REIZ 2171	12 17 15.	+ 05 33	60	13.5	GALAXY
REIZ 2173	12 17 15.	+ 05 44	72	13.0	GALAXY
REIZ 2169	12 17 15.	+ 06 17	36	15.1	GALAXY
REIZ 2172	12 17 15.	+ 06 18	24	14.2	GALAXY

OBJECT NAME	RIGHT ASCEN.	DECLINATION	DIAM.	MAGN.	TYPE OF OBJECT
HOLM 365A	12 17 15.	+ 06 18	12	13.6	PART OF MULTIPLE GALAXY
HN 0862	12 17 15.	+ 12 41		14.	NEBULA
IC 3157	12 17 15.	+ 12 41			NONSTELLAR OBJECT
REIZ 2170	12 17 15.	+ 13 03	24	12.4	GALAXY
FATH 2.101	12 17 15.	+ 14 34	11		NEBULA
MCG-01-32-001	12 17 15.	- 06 35 30.	66	15.	GALAXY
REIN 3.077	12 17 15.61	+ 12 41 56.1			NEBULA
AMES 0600	12 17 16.	+ 05 26 27.	19	17.6	NEBULA
AMES 0599	12 17 16.	+ 06 02 15.	17	17.6	NEBULA
IC 3158	12 17 16.	+ 09 34 12.			SINGLE STAR
REIZ 2168	12 17 16.	- 06 35	48	15.6	GALAXY
KEEL 399	12 17 16.4	+ 14 34 10.		18.	NEBULA
AMES 0601	12 17 17.	+ 05 31 21.	20	17.6	NEBULA
HOLM 368D	12 17 18.	+ 05 33	60	13.7	PART OF MULTIPLE GALAXY
ZWG 042.026	12 17 18.	+ 05 45		13.3	GALAXY
UGC 07376	12 17 18.	+ 05 45	102	13.3	GALAXY S0
MCG+01-32-007	12 17 18.	+ 05 45	84	13.3	GALAXY
SCH 32	12 17 18.	+ 06 18		13.9	PECULIAR GALAXY
ZWG 042.027	12 17 18.	+ 08 00		15.6	GALAXY
RMB 100	12 17 18.	+ 13 25			FAINT BLUE OBJECT
ZWG 128.076	12 17 18.	+ 21 08		15.7	GALAXY
ZWG 158.071	12 17 18.	+ 29 54		11.1	GALAXY
WEED 5	12 17 18.	+ 29 54		17.5	VERY BLUE STELLAR OBJECT
UGC 07377	12 17 18.	+ 29 54	438	11.1	GALAXY SBa
MCG+05-29-060	12 17 18.	+ 29 54	450	11.1	GALAXY
IC 0779	12 17 18.	+ 30 09 54.			NONSTELLAR OBJECT
UGC 07378	12 17 18.	+ 30 27	66	14.2	GALAXY E
ZWG 158.072	12 17 18.	+ 30 37		14.2	GALAXY
HOLM 367A	12 17 18.	+ 35 48	84	14.8	PART OF MULTIPLE GALAXY
UGC 07379	12 17 18.	+ 53 23	66	16.0	GALAXY Sc
MCG+10-18-024	12 17 18.	+ 56 41	24	16.	GALAXY
MCG+10-18-025	12 17 18.	+ 57 00	30	13.4	GALAXY
ZC 1217.3+6408	12 17 18.	+ 64 08	870		CLUSTER OF GALAXIES
ZC 1217.3+6733	12 17 18.	+ 67 33	1080		CLUSTER OF GALAXIES
ZWG 014.008	12 17 18.	- 00 35		15.4	GALAXY
MCG-06-27-015	12 17 18.	- 36 10	90	14.5	GALAXY
A3 013	12 17 18.4	+ 14 51 48.		16.4	FAINT BLUE OBJECT
RNGC 4270	12 17 19.	+ 05 45		13.0	GALAXY
RNGC 4269	12 17 19.	+ 06 18		14.0	GALAXY
AMES 0604	12 17 19.	+ 08 48 57.	23	17.8	NEBULA
AMES 0603	12 17 19.	+ 11 44 57.	26	17.8	NEBULA
HOLM 368C	12 17 20.	+ 05 44	72	13.2	PART OF MULTIPLE GALAXY
AMES 0607	12 17 20.	+ 07 15 09.	26	16.2	NEBULA
AMES 0606	12 17 20.	+ 11 17 45.	28	17.7	NEBULA
AMES 0602	12 17 20.	+ 12 51 15.	28	16.9	NEBULA
SVEN 281	12 17 20.	+ 14 09	12	14.2	GALAXY
REIZ 2176	12 17 21.	+ 05 37	90	12.4	GALAXY
AMES 0609	12 17 21.	+ 08 00 45.	18	17.0	NEBULA
AMES 0608	12 17 21.	+ 11 44 45.	28	17.7	NEBULA
HN 0863	12 17 21.	+ 11 56		15.	NEBULA
IC 3159	12 17 21.	+ 11 56			NONSTELLAR OBJECT
AMES 0605	12 17 21.	+ 13 50 57.	19	17.6	NEBULA
MCG+04-29-064	12 17 21.	+ 26 02	84	14.5	GALAXY
RNGC 4274	12 17 21.	+ 29 53		11.5	GALAXY
RNGC 4272	12 17 21.	+ 30 37		14.0	GALAXY
TON-N 1504	12 17 21.	+ 32 50		16.9	BLUE STAR
TON-N 1505	12 17 21.	+ 34 01		15.9	BLUE STAR
MCG-03-32-001	12 17 21.	- 21 17	24	16.	GALAXY
MCG-03-32-002	12 17 21.	- 21 18	24	15.	GALAXY
KEEL 400	12 17 21.6	+ 14 54 58.		18.	NEBULA
ZWG 014.009	12 17 24.	+ 02 04		15.2	GALAXY
ZWG 042.028	12 17 24.	+ 05 37		12.3	GALAXY
UGC 07380	12 17 24.	+ 05 37	150	12.3	GALAXY Sc
SN 1936A	12 17 24.	+ 05 37		14.4	SUPERNOVA
MCG+01-32-008	12 17 24.	+ 05 37	102	12.3	GALAXY
MCG+03-34-016	12 17 24.	+ 16 06	96	14.8	GALAXY
ZWG 099.016	12 17 24.	+ 17 53		15.2	GALAXY
ZWG 128.077	12 17 24.	+ 26 03		14.5	GALAXY
UGC 07381	12 17 24.	+ 26 03	78	14.5	GALAXY E-S0
ZWG 158.073	12 17 24.	+ 27 54		13.4	GALAXY
UGC 07382	12 17 24.	+ 27 54	48	13.4	GALAXY S?
MCG+05-29-061	12 17 24.	+ 28 15 30.	114	14.9	GALAXY
ZWG 158.074	12 17 24.	+ 28 40		15.6	GALAXY
ZWG 158.075	12 17 24.	+ 28 42		15.6	GALAXY
MCG+05-29-062	12 17 24.	+ 29 34	75	11.2	GALAXY
MCG+06-27-046	12 17 24.	+ 35 49	42	15.	GALAXY
ARC 1522	12 17 24.	+ 49 28		17.8	RICH CLUSTER OF GALAXIES
MCG-06-27-016	12 17 24.	- 35 41	48	16.	GALAXY
RNGC 4273	12 17 25.	+ 05 37		12.5	GALAXY
REIZ 2177	12 17 25.	+ 26 02	48	15.5	GALAXY
REIZ 2178	12 17 25.	+ 28 15	120	15.5	GALAXY
HOLM 368A	12 17 27.	+ 05 37	102	12.2	PART OF MULTIPLE GALAXY
IC 3160	12 17 27.	+ 09 22 42.			SINGLE STAR
IC 0780	12 17 27.	+ 26 02 18.			NONSTELLAR OBJECT
RNGC 4275	12 17 27.	+ 27 54		13.5	GALAXY
LB 00259	12 17 27.	+ 77 26 42.		15.8	FAINT BLUE STAR
MCG-06-27-017	12 17 27.	- 35 40	24	17.	GALAXY
IC 3161	12 17 28.	+ 09 16 30.			TWO STARS
LB 00632	12 17 28.	+ 59 07 54.		16.4	FAINT BLUE STAR
REIN 3.078	12 17 23.56	+ 08 53 08.2			NEBULA
ZWG 042.029	12 17 30.	+ 05 37		15.0	GALAXY
MCG+01-32-009	12 17 30.	+ 05 37	42	15.0	GALAXY
AMES 0611	12 17 30.	+ 05 41 45.	23	17.6	NEBULA
ZWG 042.030	12 17 30.	+ 08 06		15.5	GALAXY
ZWG 070.014	12 17 30.	+ 08 53		14.8	GALAXY
UGC 07383	12 17 30.	+ 08 53	72	14.8	GALAXY Sa-b
MCG+02-32-005	12 17 30.	+ 08 53	36	14.8	GALAXY
IC 3162	12 17 30.	+ 09 16 30.			TWO STARS
IC 3163	12 17 30.	+ 09 32 00.			TWO STARS
AMES 0610	12 17 30.	+ 13 25 21.	24	17.0	NEBULA
MCG+03-32-002	12 17 30.	+ 15 13	60	14.8	GALAXY
ZC 1217.5+2538	12 17 30.	+ 25 38	610		CLUSTER OF GALAXIES
ZWG 158.076	12 17 30.	+ 28 15		14.9	GALAXY
UGC 07384	12 17 30.	+ 28 15	120	14.9	GALAXY SBb
ZC 1217.5+2915	12 17 30.	+ 29 15	20770		CLUSTER OF GALAXIES
ZC 1217.5+3419	12 17 30.	+ 34 19	1010		CLUSTER OF GALAXIES
ZC 1217.5+3450	12 17 30.	+ 34 50	1010		CLUSTER OF GALAXIES
MCG+08-23-002	12 17 30.	+ 49 10	42	16.	GALAXY
ZC 1217.5+5743	12 17 30.	+ 57 43	540		CLUSTER OF GALAXIES
IC 3166	12 17 30.	+ 60 58 18.			NONSTELLAR OBJECT
MCG-03-32-003	12 17 30.	- 17 06	21	15.	GALAXY
RNGC 4277	12 17 31.	+ 05 37		15.0	GALAXY
AMES 0614	12 17 32.	+ 05 51 57.	24	17.4	NEBULA
AMES 0613	12 17 32.	+ 10 20 51.	23	16.4	NEBULA
LB 02308	12 17 32.	+ 50 02 00.		15.4	FAINT BLUE STAR
REIZ 2179	12 17 33.	+ 07 57	60	13.5	GALAXY
AMES 0612	12 17 33.	+ 12 38 57.	20	17.6	NEBULA
IC 0781	12 17 33.	+ 15 15	33		BRIGHT NEBULA

OBJECT NAME	RIGHT ASCEN.	DECLINATION	DIAM.	MAGN.	TYPE OF OBJECT
FATH 2.102	12 17 33.	+ 15 15	33		NEBULA
AMES 0615	12 17 34.	+ 08 05 51.	28	16.2	NEBULA
AMES 0615	12 17 34.	+ 08 05 51.	26	16.2	NEBULA
IC 3164	12 17 34.	+ 25 14 00.			NONSTELLAR OBJECT
IC 3165	12 17 34.	+ 28 15 12.			NONSTELLAR OBJECT
REIN 3.079	12 17 34.79	+ 07 58 12.9			NEBULA
HOLM 368F	12 17 35.	+ 05 37	24	13.9	PART OF MULTIPLE GALAXY
AMES 0618	12 17 35.	+ 08 09 21.	19	17.8	NEBULA
AMES 0617	12 17 35.	+ 08 41 21.	13	17.5	NEBULA
AMES 0616	12 17 35.	+ 08 50 45.	17	17.6	NEBULA
A3 014	12 17 35.5	+ 15 21 01.		17.4	FAINT BLUE OBJECT
ZWG 042.031	12 17 36.	+ 06 55		15.4	GALAXY
ZWG 042.032	12 17 36.	+ 07 58		14.1	GALAXY
UGC 07385	12 17 36.	+ 07 58	102	14.1	GALAXY S?
MCG+01-32-010	12 17 36.	+ 07 58	96	14.1	GALAXY
AMES 0620	12 17 36.	+ 08 08 45.	28	17.7	NEBULA
AMES 0619	12 17 36.	+ 08 17 45.	17	17.8	NEBULA
ZWG 070.015	12 17 36.	+ 08 55		15.1	GALAXY
MCG+02-32-006	12 17 36.	+ 08 55	36	15.1	GALAXY
ZWG 099.017	12 17 36.	+ 15 14		14.8	GALAXY
RMB 052	12 17 36.	+ 15 52			FAINT BLUE OBJECT
MCG+03-34-017	12 17 36.	+ 16 09	48	15.0	GALAXY
ZC 1217.6+1726	12 17 36.	+ 17 26	670		CLUSTER OF GALAXIES
MCG+05-29-064	12 17 36.	+ 28 12	42	15.7	GALAXY
ZWG 158.077	12 17 36.	+ 29 33		11.2	GALAXY
UGC 07386	12 17 36.	+ 29 33	216	11.2	GALAXY E
MCG+05-29-063	12 17 36.	+ 29 36	36	13.1	GALAXY
1ZW 034	12 17 36.	+ 33 32			COMPACT GALAXY
MCG+09-20-148	12 17 36.	+ 56 07	36	16.	GALAXY
MCG+09-20-147	12 17 36.	+ 56 23 30.	24	16.	GALAXY
MCG+09-20-146	12 17 36.	+ 56 27	42	16.	GALAXY
ZC 1217.6-0101	12 17 36.	- 01 01	1610		CLUSTER OF GALAXIES
MCG-04-29-019	12 17 36.	- 25 48	138	14.	GALAXY
AMES 0621	12 17 37.	+ 05 10 21.	28	17.4	NEBULA
FATH 2.103	12 17 37.	+ 14 25	16		NEBULA
LB 00633	12 17 37.	+ 73 17 36.		16.4	FAINT BLUE STAR
REIN 3.080	12 17 37.07	+ 08 55 20.6			NEBULA
KEEL 401	12 17 37.3	+ 14 36 24.		18.	NEBULA
KEEL 402	12 17 37.4	+ 14 24 50.		18.	NEBULA
RNGC 4276	12 17 38.	+ 07 58		14.0	GALAXY
BC PKS1217+02	12 17 38.35	+ 02 20 20.9		16.53	QUASI-STELLAR OBJECT
SHB 199	12 17 38.4	+ 02 20 20.		16.5	QUASI-STELLAR OBJECT
REIZ 2180	12 17 39.	+ 05 37	12	15.4	GALAXY
REIZ 2181	12 17 39.	+ 06 54	18	15.5	GALAXY
AMES 0622	12 17 39.	+ 07 19 57.	17	18.0	NEBULA
RNGC 4278	12 17 39.	+ 29 34		11.5	GALAXY
MCG-02-32-002	12 17 39.	- 12 59 30.	30	15.	GALAXY
FATH 1.534	12 17 39.	- 15 15	8		NEBULA
AMES 0623	12 17 41.	+ 06 53 57.	44	16.6	NEBULA
ZWG 042.033	12 17 42.	+ 04 29		15.2	GALAXY
UGC 07387	12 17 42.	+ 04 29	126	15.2	GALAXY Sc
MCG+01-32-011	12 17 42.	+ 04 29	108	15.2	GALAXY
ZWG 070.016	12 17 42.	+ 08 49		15.6	GALAXY
ZWG 070.017	12 17 42.	+ 09 49		15.1	GALAXY
RMB 172	12 17 42.	+ 12 25			FAINT BLUE OBJECT
RMB 171	12 17 42.	+ 12 25			FAINT BLUE OBJECT
ZWG 158.078	12 17 42.	+ 28 40		15.7	GALAXY
UGC 07388	12 17 42.	+ 33 55	60	16.0	GALAXY SB
MCG+06-27-047	12 17 42.	+ 33 55	66	15.	GALAXY
MCG-03-32-004	12 17 42.	- 16 49	84	15.	GALAXY
AMES 0624	12 17 43.	+ 13 41 21.	17	16.7	NEBULA
FATH 2.104	12 17 43.	+ 14 59	5		NEBULA
REIZ 2185	12 17 43.	+ 28 15	60	15.8	GALAXY
HOLM 369A	12 17 43.	+ 29 33	120	11.7	PART OF MULTIPLE GALAXY
KEEL 404	12 17 43.5	+ 14 56 57.		17.	NEBULA
AMES 0626	12 17 44.	+ 09 30 33.	19	17.2	NEBULA
LB 02309	12 17 44.	+ 48 31 48.		15.9	FAINT BLUE STAR
KEEL 403	12 17 44.3	+ 04 28 42.			NEBULA
SVEN 282	12 17 45.	+ 04 29	96	14.4	GALAXY
REIZ 2183	12 17 45.	+ 04 29	90	15.0	GALAXY
REIZ 2184	12 17 45.	+ 06 53	18	15.8	GALAXY
AMES 0625	12 17 45.	+ 14 06 45.	37	17.3	NEBULA
MCG+08-23-003	12 17 45.	+ 48 24	60	16.	GALAXY
A3 015	12 17 45.7	+ 15 22 41.		15.0	FAINT BLUE OBJECT
REIN 3.081	12 17 45.93	+ 08 48 41.1			NEBULA
IC 3167	12 17 46.	+ 09 49 24.			NONSTELLAR OBJECT
LB 02310	12 17 46.	+ 59 44 12.		16.6	FAINT BLUE STAR
REIZ 2182	12 17 46.	- 06 35	60	15.8	GALAXY
REIN 3.082	12 17 46.03	+ 09 49 19.5			NEBULA
AMES 0629	12 17 47.	+ 09 12 03.	19	16.6	NEBULA
AMES 0627	12 17 47.	+ 13 45 57.	22	17.1	NEBULA
ZWG 014.010	12 17 48.	+ 00 40		15.5	GALAXY
ZWG 042.034	12 17 48.	+ 05 40		12.5	GALAXY
UGC 07389	12 17 48.	+ 05 40	162	12.5	GALAXY S0
MCG+01-32-012	12 17 48.	+ 05 40	48	12.5	GALAXY
ZWG 042.035	12 17 48.	+ 05 51		14.7	GALAXY
MCG+01-32-013	12 17 48.	+ 05 51	36	14.7	GALAXY
ZWG 042.036	12 17 48.	+ 07 11		15.4	GALAXY
AMES 0628	12 17 48.	+ 12 27 21.	29	17.6	NEBULA
RMB 169	12 17 48.	+ 12 28			FAINT BLUE OBJECT
REA 30	12 17 48.	+ 14 59	42		DWARF GALAXY
ZWG 099.018	12 17 48.	+ 17 38		14.7	GALAXY
MCG+03-32-003	12 17 48.	+ 17 38	30	14.7	GALAXY
MCG+04-29-065	12 17 48.	+ 25 49	18	14.8	GALAXY
IC 3168	12 17 48.	+ 28 11 54.			NONSTELLAR OBJECT
ZWG 158.079	12 17 48.	+ 28 12		15.7	GALAXY
ZWG 158.080	12 17 48.	+ 29 35		13.1	GALAXY
UGC 07390	12 17 48.	+ 29 35	78	13.1	GALAXY E
ZWG 215.061	12 17 48.	+ 39 30		15.6	GALAXY
UGC 07391	12 17 48.	+ 46 12	66	16.5	GALAXY DBL SYS
UGC 07392	12 17 48.	+ 48 25	78	16.5	GALAXY SB?c
MCG+09-20-149	12 17 48.	+ 56 03	24	16.	GALAXY
ZWG 293.011	12 17 48.	+ 58 22		14.7	GALAXY
RNGC 4284	12 17 48.	+ 58 22		14.5	GALAXY
UGC 07393	12 17 48.	+ 58 22	180	14.7	GALAXY Sb-c
KARA.72 329A	12 17 48.	+ 58 22	162	14.7	PART OF DOUBLE GALAXY
ZWG 352.027	12 17 48.	+ 75 47		15.2	GALAXY
MCG-02-32-003	12 17 48.	- 11 24 30.	54	14.5	GALAXY
RNGC 4281	12 17 49.	+ 05 40		12.5	GALAXY
RNGC 4282	12 17 49.	+ 05 51		14.5	GALAXY
AMES 0630	12 17 49.	+ 09 08 33.	22	16.8	NEBULA
RNGC 4279	12 17 49.	- 11 24		14.0	GALAXY
RNGC 4280	12 17 49.	- 11 25			NON-EXISTENT OBJECT
AMES 0632	12 17 50.	+ 09 04 15.	17	17.2	NEBULA
IC 3169	12 17 50.	+ 26 52 30.			NONSTELLAR OBJECT
HOLM 373B	12 17 50.	+ 58 22	48	14.0	PART OF MULTIPLE GALAXY
REIZ 2189	12 17 50.	+ 58 24	48	14.3	GALAXY
FATH 1.535	12 17 50.	- 15 13	8		NEBULA

OBJECT NAME	RIGHT ASCEN.	DECLINATION	DIAM.	MAGN.	TYPE OF OBJECT
A3 016	12 17 50.8	+ 14 56 06.		16.5	FAINT BLUE OBJECT
REIZ 2186	12 17 51.	+ 01 45	60	14.5	GALAXY
REIZ 2187	12 17 51.	+ 05 39	60	12.5	GALAXY
REIZ 2188	12 17 51.	+ 05 53	18	15.2	GALAXY
AMES 0631	12 17 51.	+ 14 09 09.	31	17.7	NEBULA
MCG+03-34-018	12 17 51.	+ 14 46	78	15.4	GALAXY
RNGC 4283	12 17 51.	+ 29 35		13.0	GALAXY
MCG+08-22-004	12 17 51.	+ 46 10 30.	36	16.	GALAXY
AMES 0634	12 17 52.	+ 08 15 33.	22	17.4	NEBULA
AMES 0633	12 17 52.	+ 12 51 21.	23	16.8	NEBULA
HOLM 368B	12 17 53.	+ 05 40	84	12.5	PART OF MULTIPLE GALAXY
REIN 3.083	12 17 53.95	+ 09 42 04.8			NEBULA
ZWG 014.011	12 17 54.	+ 01 46		15.2	GALAXY
UGC 07394	12 17 54.	+ 01 46	108	15.2	GALAXY Sc
AMES 0636	12 17 54.	+ 07 11 09.	52	16.5	NEBULA
ZWG 070.018	12 17 54.	+ 09 42		15.0	GALAXY
MCG+02-32-007	12 17 54.	+ 09 42	24	15.0	GALAXY
IC 3170	12 17 54.	+ 09 42 00.			NONSTELLAR OBJECT
ZWG 070.019	12 17 54.	+ 10 07		15.5	GALAXY
ZWG 070.020	12 17 54.	+ 10 31		15.6	GALAXY
AMES 0635	12 17 54.	+ 12 49 33.	22	17.8	NEBULA
IC 3171	12 17 54.	+ 25 50 18.			NONSTELLAR OBJECT
ZWG 128.078	12 17 54.	+ 25 51		14.8	GALAXY
IC 3172	12 17 54.	+ 28 05 48.			NONSTELLAR OBJECT
ZC 1217.9+2918	12 17 54.	+ 29 18	2020		CLUSTER OF GALAXIES
ZWG 158.081	12 17 54.	+ 31 27		14.5	GALAXY
UGC 07395	12 17 54.	+ 31 27	33	14.5	GALAXY
MCG+05-29-064A	12 17 54.	+ 31 27	30	14.5	GALAXY
MCG+09-20-150	12 17 54.	+ 56 02	36	16.	GALAXY
MCG+10-18-026	12 17 54.	+ 58 22	114	14.0	GALAXY
MCG+13-09-023	12 17 54.	+ 75 47	39	15.	GALAXY
AMES 0638	12 17 55.	+ 08 53 33.	19	17.2	NEBULA
FATH 2.105	12 17 55.	+ 14 50	11		NEBULA
SEY 177	12 17 55.	+ 31 26 33.		14.5	FAINT GALAXY
KEEL 405	12 17 55.1	+ 14 50 05.		17.	NEBULA
AMES 0637	12 17 56.	+ 13 25 21.	19	18.0	NEBULA
HOLM 370B	12 17 56.	- 06 40	30	15.3	PART OF MULTIPLE GALAXY
AMES 0639	12 17 57.	+ 08 06 09.	25	17.4	NEBULA
HN 0865	12 17 57.	+ 10 07		13.	NEBULA
IC 3175	12 17 57.	+ 10 07			NONSTELLAR OBJECT
IC 3174	12 17 57.	+ 10 31 12.			NONSTELLAR OBJECT
HN 0864	12 17 57.	+ 11 36		13.	NEBULA
IC 3173	12 17 57.	+ 11 36			NONSTELLAR OBJECT
HOLM 369B	12 17 57.	+ 29 35	54	12.8	PART OF MULTIPLE GALAXY
TON-N 1506	12 17 57.	+ 34 00		17.0	BLUE STAR
AMES 0644	12 17 58.	+ 06 39 57.	17	17.9	NEBULA
AMES 0643	12 17 58.	+ 07 45 57.	34	17.4	NEBULA
AMES 0642	12 17 58.	+ 08 24 57.	24	17.9	NEBULA
AMES 0641	12 17 58.	+ 10 22 27.	19	16.8	NEBULA
REIZ 2191	12 17 58.	+ 46 34	60	13.6	GALAXY
AMES 0645	12 17 59.	+ 09 10 03.	22	17.9	NEBULA
AMES 0640	12 17 59.	+ 12 39 33.	23	17.1	NEBULA
LB 02311	12 17 59.	+ 50 32 18.		15.8	FAINT BLUE STAR
VDB.66G 120	12 18	+ 46 08	70		DWARF GALAXY
ZWG 014.012	12 18 00.	+ 01 05		15.0	GALAXY
UGC 07396	12 18 00.	+ 01 05	102	15.0	GALAXY Sc-IRR
MCG+00-32-002	12 18 00.	+ 01 05	90	15.0	GALAXY
AMES 0646	12 18 00.	+ 06 40 09.	20	17.9	NEBULA
ZWG 070.021	12 18 00.	+ 11 37		15.6	GALAXY
IC 3176	12 18 00.	+ 25 47 30.			NONSTELLAR OBJECT
MCG+05-29-065	12 18 00.	+ 29 39	96	14.7	GALAXY
MCG+06-27-048	12 18 00.	+ 34 42	27	15.	GALAXY
MCG+07-25-058	12 18 00.	+ 39 34	39	15.	GALAXY
ZWG 215.062	12 18 00.	+ 39 55		15.5	GALAXY
MCG+12-12-005	12 18 00.	+ 74 20	18	16.	GALAXY
MCG+13-09-024	12 18 00.	+ 75 39	45	12.4	GALAXY
ZWG 352.028	12 18 00.	+ 75 40		12.3	GALAXY
UGC 07397	12 18 00.	+ 75 40	120	12.3	GALAXY E
MCG-01-32-003	12 18 00.	- 06 40	60	15.5	GALAXY
MCG-01-32-002	12 18 00.	- 06 41	24	16.	GALAXY
REIN 3.084	12 18 00.66	+ 10 07 49.7			NEBULA
KARA.73 38	12 18 01.	- 34 28	81		DWARF GALAXY
AMES 0648	12 18 02.	+ 06 05 21.	20	17.6	NEBULA
HN 0866	12 18 02.	+ 14 24	150		NEBULA
IC 3177	12 18 02.	+ 14 24			NONSTELLAR OBJECT
AMES 0650	12 18 03.	+ 08 20 09.	24	17.8	NEBULA
AMES 0647	12 18 03.	+ 13 39 57.	29	16.4	NEBULA
MCG+08-23-005	12 18 03.	+ 48 59	48	16.	GALAXY
A3 017	12 18 03.0	+ 14 47 06.		16.6	FAINT BLUE OBJECT
KEEL 406	12 18 03.6	+ 14 30 55.		18.	NEBULA
SHE 200	12 18 03.9	+ 33 59 50.		18.6	QUASI-STELLAR OBJECT
REIZ 2190	12 18 04.	+ 01 03	60	14.1	GALAXY
AMES 0649	12 18 04.	+ 12 14 21.	36	17.6	NEBULA
IC 3178	12 18 04.	+ 26 26 54.			NONSTELLAR OBJECT
SEY 178	12 18 04.	+ 31 03 45.		15.1	FAINT GALAXY
BC 3CR270.1	12 18 04.00	+ 33 59 50.0		18.61	QUASI-STELLAR OBJECT
AMES 0651	12 18 05.	+ 05 22 15.	22	17.4	NEBULA
FATH 2.106	12 18 05.	+ 14 30	8		NEBULA
ZWG 158.082	12 18 06.	+ 31 04		15.0	GALAXY
MCG+05-29-066	12 18 06.	+ 31 04	36	15.0	GALAXY
ZWG 269.051	12 18 06.	+ 52 32		15.6	GALAXY
MCG+09-20-151	12 18 06.	+ 52 32	60	15.	GALAXY
MCG+10-18-027	12 18 06.	+ 58 08	24	15.	GALAXY
MCG+11-15-047	12 18 06.	+ 63 22	36	17.	GALAXY
ZCG 199.23	12 18 06.	+ 63 51		18.7	COMPACT GALAXY
ZCG 199.21	12 18 06.	+ 63 59		21.3	COMPACT GALAXY
ZCG 199.22	12 18 06.	+ 64 07		18.3	COMPACT GALAXY
RNGC 4291	12 18 06.	+ 75 39		13.0	GALAXY
MCG-02-32-004	12 18 06.	- 11 23	48	14.5	GALAXY
AMES 0656	12 18 07.	+ 05 26 32.	20	17.4	NEBULA
AMES 0655	12 18 07.	+ 06 14 56.	22	17.5	NEBULA
AMES 0654	12 18 07.	+ 08 06 20.	38	17.8	NEBULA
IC 3179	12 18 07.	+ 26 26 36.			NONSTELLAR OBJECT
RNGC 4285	12 18 07.	- 11 23		14.0	GALAXY
A3 018	12 18 07.6	+ 14 01 55.		17.5	FAINT BLUE OBJECT
AMES 0659	12 18 08.	+ 07 41 08.	16	17.8	NEBULA
AMES 0658	12 18 08.	+ 11 40 08.	24	17.3	NEBULA
AMES 0653	12 18 08.	+ 12 31 20.	19	17.4	NEBULA
AMES 0652	12 18 08.	+ 13 55 50.	16	17.6	NEBULA
A3 019	12 18 08.7	+ 14 50 47.		16.7	FAINT BLUE OBJECT
AMES 0660	12 18 09.	+ 06 15 20.	32	17.7	NEBULA
AMES 0657	12 18 09.	+ 14 09 32.	26	17.0	NEBULA
LB 02312	12 18 09.	+ 50 25 24.		16.0	FAINT BLUE STAR
MCG-02-32-005	12 18 09.	- 13 33	90	14.5	GALAXY
MCG-03-32-005	12 18 09.	- 18 23 30.	84	14.5	GALAXY
AMES 0662	12 18 10.	+ 06 16 02.	25	17.4	NEBULA
AMES 0661	12 18 10.	+ 08 52 44.	24	17.5	NEBULA
RNGC 4288A	12 18 10.	+ 46 31		15.0	GALAXY

OBJECT NAME	RIGHT ASCEN.	DECLINATION	DIAM.	MAGN.	TYPE OF OBJECT
RNGC 4288	12 18 10.	+ 46 34		13.5	GALAXY
HOLM 371A	12 18 10.	+ 46 34	60	13.3	PART OF MULTIPLE GALAXY
HOLM 370A	12 18 10.	- 06 42	90	15.0	PART OF MULTIPLE GALAXY
AMES 0663	12 18 11.	+ 05 40 50.	22	17.6	NEBULA
KEEL 407	12 18 11.6	+ 14 54 31.		18.	NEBULA
ZWG 042.037	12 18 12.	+ 05 55		15.2	GALAXY
MCG+01-32-014	12 18 12.	+ 05 55	60	15.2	GALAXY
ZWG 099.019	12 18 12.	+ 19 23		15.7	GALAXY
ZWG 128.079	12 18 12.	+ 25 12		15.6	GALAXY
IC 3181	12 18 12.	+ 29 37 24.			SAME AS NGC 4286
ZWG 158.083	12 18 12.	+ 29 38		14.7	GALAXY
UGC 07398	12 18 12.	+ 29 38	102	14.7	GALAXY S0?
HOLM 371B	12 18 12.	+ 46 32	18	15.0	PART OF MULTIPLE GALAXY
MCG+08-23-006	12 18 12.	+ 46 34	120	13.3	GALAXY
ZWG 244.006	12 18 12.	+ 46 35		13.6	GALAXY
UGC 07399	12 18 12.	+ 46 35	180	13.6	GALAXY SBc
ZC 1218.2+4824	12 18 12.	+ 48 24	610		CLUSTER OF GALAXIES
RNGC 4287	12 18 13.	+ 05 55		15.0	GALAXY
AMES 0664	12 18 13.	+ 09 06 44.	25	17.0	NEBULA
A3 020	12 18 13.7	+ 14 17 27.		16.6	FAINT BLUE OBJECT
REIZ 2193	12 18 14.	+ 13 16	54	15.0	GALAXY
LB 02313	12 18 14.	+ 56 29 18.		15.9	FAINT BLUE STAR
AMES 0665	12 18 15.	+ 12 09 32.	37	16.6	NEBULA
RNGC 4286	12 18 15.	+ 29 37		14.5	GALAXY
TON-N 1507	12 18 15.	+ 33 52		16.0	BLUE STAR
TON-N 1508	12 18 15.	+ 33 55		16.9	BLUE STAR
MCG+08-23-007	12 18 15.	+ 46 31	15	15.0	GALAXY
MCG-01-32-004	12 18 15.	- 06 42	78	15.	GALAXY
IC 3183	12 18 16.	+ 06 57 54.			SINGLE STAR
AMES 0666	12 18 16.	+ 11 36 44.	26	17.9	NEBULA
IC 3182	12 18 16.	+ 13 00 18.			THREE STARS
REIZ 2192	12 18 16.	- 06 42	54	15.4	GALAXY
AMES 0669	12 18 17.	+ 07 00 44.	29	17.4	NEBULA
AMES 0668	12 18 17.	+ 08 59 20.	19	17.3	NEBULA
AMES 0667	12 18 17.	+ 08 59 20.	26	16.9	NEBULA
IC 3184	12 18 17.	+ 25 11 30.			NONSTELLAR OBJECT
A3 021	12 18 17.8	+ 14 52 50.		16.9	FAINT BLUE OBJECT
KARA.68 103	12 18 18.	+ 05 30	27		DWARF GALAXY
AMES 0670	12 18 18.	+ 07 19 44.	20	17.3	NEBULA
RMB 170	12 18 18.	+ 12 21			FAINT BLUE OBJECT
REA 31	12 18 18.	+ 15 17	54		DWARF GALAXY
ZWG 099.020	12 18 18.	+ 17 18		15.1	GALAXY
MCG+03-32-005	12 18 18.	+ 17 18	24	15.1	GALAXY
ZWG 099.021	12 18 18.	+ 17 46		15.1	GALAXY
UGC 07399A	12 18 18.	+ 17 46	84	15.1	GALAXY DWRF EL
MCG+03-32-004	12 18 18.	+ 17 46	96	15.1	GALAXY
MCG+09-20-152	12 18 18.	+ 52 53	30	16.	GALAXY
UGC 07400	12 18 18.	+ 63 22	60	17.	GALAXY
ZCG 199.24	12 18 18.	+ 63 51		18.4	COMPACT GALAXY
AMES 0674	12 18 19.	+ 07 25 56.	16	18.0	NEBULA
AMES 0673	12 18 19.	+ 08 19 44.	23	17.9	NEBULA
AMES 0672	12 18 19.	+ 08 26 44.	25	17.9	NEBULA
AMES 0671	12 18 19.	+ 08 36 20.	28	16.8	NEBULA
REIZ 2194	12 18 20.	+ 17 45	48	14.7	GALAXY
LB 02314	12 18 20.	+ 49 54 24.		16.4	FAINT BLUE STAR
IC 3185	12 18 21.	+ 25 42 24.			NONSTELLAR OBJECT
AMES 0676	12 18 22.	+ 05 47 08.	26	17.4	NEBULA
AMES 0675	12 18 22.	+ 06 38 08.	35	16.8	NEBULA
REIN 3.085	12 18 22.50	+ 11 17 12.0			NEBULA
AMES 0677	12 18 23.	+ 08 18 32.	19	17.8	NEBULA
ZWG 070.022	12 18 24.	+ 11 18		15.1	GALAXY
MCG+02-32-008	12 18 24.	+ 11 18	24	15.1	GALAXY
ZWG 070.023	12 18 24.	+ 11 27		15.6	GALAXY
RMB 051	12 18 24.	+ 15 51			FAINT BLUE OBJECT
ZWG 128.080	12 18 24.	+ 24 57		15.0	GALAXY
MCG+04-29-066	12 18 24.	+ 24 57	24	15.0	GALAXY
ZWG 128.081	12 18 24.	+ 25 43		15.7	GALAXY
MCG+05-29-067	12 18 24.	+ 28 11	48	15.7	GALAXY
ZWG 215.063	12 18 24.	+ 38 42		15.4	GALAXY
MCG+08-23-008	12 18 24.	+ 48 05 30.	60	16.	GALAXY
UGC 07401	12 18 24.	+ 48 07	66	16.5	GALAXY DWRF IR
LB 02315	12 18 24.	+ 50 15 36.		17.0	FAINT BLUE STAR
ZWG 293.012	12 18 24.	+ 58 22		12.8	GALAXY
RNGC 4290	12 18 24.	+ 58 22		12.5	GALAXY
UGC 07402	12 18 24.	+ 58 22	150	12.8	GALAXY SBb
KARA.72 329B	12 18 24.	+ 58 22	150	12.8	PART OF DOUBLE GALAXY
A3 023	12 18 24.8	+ 15 50 34.		16.6	FAINT BLUE OBJECT
AMES 0678	12 18 25.	+ 07 56 14.	16	17.7	NEBULA
IC 3186	12 18 25.	+ 24 56 42.			NONSTELLAR OBJECT
IC 3189	12 18 25.	+ 25 42 12.			NONSTELLAR OBJECT
HOLM 372A	12 18 25.	+ 28 15	42	14.7	PART OF MULTIPLE GALAXY
REIZ 2197	12 18 25.	+ 58 22	108	13.0	GALAXY
HOLM 373A	12 18 25.	+ 58 22	72	13.3	PART OF MULTIPLE GALAXY
AMES 0679	12 18 26.	+ 08 01 32.	31	17.8	NEBULA
HOLM 372B	12 18 26.	+ 28 14	36	14.7	PART OF MULTIPLE GALAXY
LB 02316	12 18 26.	+ 55 09 30.		17.2	FAINT BLUE STAR
A3 024	12 18 26.1	+ 14 11 59.		16.8	FAINT BLUE OBJECT
REIZ 2195	12 18 27.	+ 03 59	90	14.8	GALAXY
AMES 0681	12 18 27.	+ 08 22 44.	18	17.8	NEBULA
HW 0868	12 18 27.	+ 11 17		13.5	NEBULA
IC 3188	12 18 27.	+ 11 17			NONSTELLAR OBJECT
HW 0867	12 18 27.	+ 11 26		14.5	NEBULA
IC 3187	12 18 27.	+ 11 26			NONSTELLAR OBJECT
MCG+07-25-059	12 18 27.	+ 40 07 30.	30	15.	GALAXY
MCG+07-25-060	12 18 27.	+ 40 09	33	15.	GALAXY
A3 025	12 18 27.6	+ 15 41 40.		16.6	FAINT BLUE OBJECT
AMES 0683	12 18 28.	+ 09 24 44.	48	17.2	NEBULA
AMES 0680	12 18 28.	+ 13 44 32.	38	17.1	NEBULA
AMES 0682	12 18 29.	+ 12 59 38.	24	17.1	NEBULA
ZWG 042.038	12 18 30.	+ 04 00		15.0	GALAXY
UGC 07403	12 18 30.	+ 04 00	234	15.0	GALAXY Sc
MCG+01-32-015	12 18 30.	+ 04 00	240	15.0	GALAXY
AMES 0687	12 18 30.	+ 06 20 32.	16	17.3	NEBULA
ZWG 042.039	12 18 30.	+ 07 21		15.7	GALAXY
AMES 0686	12 18 30.	+ 08 11 50.	41	17.8	NEBULA
AMES 0685	12 18 30.	+ 08 37 44.	12	17.7	NEBULA
IC 3190	12 18 30.	+ 09 50 48.			SINGLE STAR
ZWG 158.084	12 18 30.	+ 28 10		15.7	GALAXY
ZC 1218.5+3217	12 18 30.	+ 32 17	1410		CLUSTER OF GALAXIES
ZC 1218.5+3518	12 18 30.	+ 35 18	670		CLUSTER OF GALAXIES
ZWG 215.064	12 18 30.	+ 40 08		15.1	GALAXY
MCG+07-25-061	12 18 30.	+ 40 09	45	15.	GALAXY
ZWG 215.065	12 18 30.	+ 40 11		15.4	GALAXY
MCG+08-23-010	12 18 30.	+ 46 25	9	16.	GALAXY
MCG+08-23-009	12 18 30.	+ 46 25 30.	9	16.	GALAXY
MRK 204	12 18 30.	+ 62 17	10	16.	GALAXY WITH UV CONTINUUM
RNGC 4289	12 18 31.	+ 04 00		15.0	GALAXY
AMES 0690	12 18 31.	+ 07 20 56.	34	16.9	NEBULA
AMES 0689	12 18 31.	+ 08 53 02.	42	17.0	NEBULA
AMES 0684	12 18 31.	+ 12 47 08.	23	17.6	NEBULA
IC 3193	12 18 31.	+ 28 10 36.			NONSTELLAR OBJECT
REIZ 2196	12 18 31.	+ 28 15	72	15.5	GALAXY
IC 3191	12 18 32.	+ 07 58 54.			MAY NOT EXIST
AMES 0688	12 18 32.	+ 11 53 44.	35	16.5	NEBULA
LB 02317	12 18 32.	+ 50 40 06.		16.2	FAINT BLUE STAR
AMES 0691	12 18 33.	+ 11 51 38.	54	16.9	NEBULA
HW 0869	12 18 33.	+ 12 01		15.	NEBULA
IC 3192	12 18 33.	+ 12 01			NONSTELLAR OBJECT
TON-N 1509	12 18 33.	+ 20 52		17.0	BLUE STAR
AMES 0692	12 18 34.	+ 08 16 02.	17	17.7	NEBULA
HOLM 374C	12 18 34.	+ 40 11	24	14.2	PART OF MULTIPLE GALAXY
REIZ 2198	12 18 34.	+ 46 04	60	14.4	GALAXY
AMES 0694	12 18 35.	+ 08 45 02.	22	17.8	NEBULA
HOLM 374A	12 18 35.	+ 40 09	24	14.1	PART OF MULTIPLE GALAXY
ZWG 014.014	12 18 36.	+ 00 51		15.5	GALAXY
RMB 184	12 18 36.	+ 11 31			FAINT BLUE OBJECT
RMB 185	12 18 36.	+ 11 41			FAINT BLUE OBJECT
AMES 0693	12 18 36.	+ 12 20 08.	20	17.6	NEBULA
ZWG 215.066	12 18 36.	+ 40 10		15.2	GALAXY
MCG+08-23-011	12 18 36.	+ 46 23	18	16.	GALAXY
MCG+10-18-028	12 18 36.	+ 58 04	24	15.	GALAXY
MCG+10-18-029	12 18 36.	+ 58 22	114	12.7	GALAXY
MCG+10-18-030	12 18 36.	+ 61 11	27	16.	GALAXY
ZWG 014.013	12 18 36.	- 02 45		15.4	GALAXY
LB 02318	12 18 37.	+ 49 44 24.		16.0	FAINT BLUE STAR
IC 3194	12 18 38.	+ 25 24 37.			NONSTELLAR OBJECT
HOLM 374B	12 18 38.	+ 40 11	24	14.2	PART OF MULTIPLE GALAXY
TON-N 1510	12 18 39.	+ 33 19		17.0	BLUE STAR
AMES 0698	12 18 40.	+ 05 36 32.	19	17.8	NEBULA
AMES 0697	12 18 40.	+ 08 19 56.	24	17.7	NEBULA
AMES 0696	12 18 40.	+ 09 00 32.	35	16.7	NEBULA
AMES 0695	12 18 40.	+ 09 14 56.	29	17.8	NEBULA
AMES 0700	12 18 41.	+ 08 22 32.	26	17.2	NEBULA
ZWG 042.040	12 18 42.	+ 04 52		14.1	GALAXY
UGC 07404	12 18 42.	+ 04 52	114	14.1	GALAXY S0
MCG+01-32-016	12 18 42.	+ 04 52	72	14.1	GALAXY
AMES 0702	12 18 42.	+ 05 14 08.	20	17.5	NEBULA
AMES 0701	12 18 42.	+ 07 11 08.	17	17.8	NEBULA
8ZW 1218+11.9	12 18 42.	+ 11 54		19.0	COMPACT GALAXY
AMES 0699	12 18 42.	+ 13 37 08.	38	17.6	NEBULA
ZWG 099.022	12 18 42.	+ 17 55		15.2	GALAXY
ZWG 099.023	12 18 42.	+ 18 40		11.6	GALAXY
SCH 33	12 18 42.	+ 18 40		11.6	PECULIAR GALAXY
UGC 07405	12 18 42.	+ 18 40	348	11.6	GALAXY SB0/SBa
MCG+03-32-006	12 18 42.	+ 18 40	336	11.6	GALAXY
ZWG 158.085	12 18 42.	+ 28 26		15.0	GALAXY
MCG+05-29-068	12 18 42.	+ 28 26	18	15.0	GALAXY
MCG+09-20-153	12 18 42.	+ 52 10	36	15.	GALAXY
ZWG 293.013	12 18 42.	+ 61 22		14.5	GALAXY
UGC 07406	12 18 42.	+ 61 22	66	14.5	GALAXY SBc
MCG+10-18-031	12 18 42.	+ 61 22	51	14.	GALAXY
ZWG 014.015	12 18 42.	- 00 34		15.6	GALAXY
RNGC 4292	12 18 43.	+ 04 52			GALAXY
RNGC 4292A	12 18 43.	+ 04 54		15.0	GALAXY
SEY 179	12 18 43.	+ 32 59 09.		15.3	FAINT GALAXY
KEEL 408	12 18 43.7	+ 04 54 36.			NEBULA
AMES 0703	12 18 44.	+ 04 54 44.	24	16.6	NEBULA
REIZ 2201	12 18 44.	+ 18 39	168	12.3	GALAXY
RNGC 4293	12 18 44.	+ 18 40		11.5	GALAXY
SVEN 283	12 18 45.	+ 04 51	60	14.2	GALAXY
REIZ 2199	12 18 45.	+ 04 52	48	14.6	GALAXY
HOLM 375A	12 18 45.	+ 04 52	60	13.7	PART OF MULTIPLE GALAXY
SVEN 284	12 18 45.	+ 04 54	18	15.3	GALAXY
REIZ 2200	12 18 45.	+ 04 54	24	16.0	GALAXY
HOLM 375B	12 18 45.	+ 04 54	24	14.9	PART OF MULTIPLE GALAXY
ABC 1523	12 18 45.	+ 06 25		17.2	RICH CLUSTER OF GALAXIES
AMES 0704	12 18 45.	+ 06 26 32.	17	17.8	NEBULA
REIZ 2202	12 18 45.	+ 11 47	120	12.8	GALAXY
RNGC 4295	12 18 45.	+ 28 26		15.0	GALAXY
TON-N 1511	12 18 45.	+ 34 22		16.0	BLUE STAR
MCG-03-32-006	12 18 45.	- 17 40	60	15.	GALAXY
REIN 3.086	12 18 45.35	+ 11 47 13.9			NEBULA
KEEL 409	12 18 45.7	+ 04 21 13.			NEBULA
A3 026	12 18 46.0	+ 14 28 40.		16.6	FAINT BLUE OBJECT
IC 3195	12 18 47.	+ 26 05 07.			NONSTELLAR OBJECT
A3 027	12 18 47.7	+ 16 00 08.		17.0	FAINT BLUE OBJECT
ZC 1218.8+0033	12 18 48.	+ 00 33	1410		CLUSTER OF GALAXIES
ZWG 070.024	12 18 48.	+ 11 47		12.6	GALAXY
UGC 07407	12 18 48.	+ 11 47	180	12.6	GALAXY Sc
KARA.72 330B	12 18 48.	+ 11 47	132		PART OF DOUBLE GALAXY
MCG+02-32-009	12 18 48.	+ 11 47	180	12.6	GALAXY
KARA.72 330A	12 18 48.	+ 11 48	48	12.6	PART OF DOUBLE GALAXY
RMB 059	12 18 48.	+ 14 29			FAINT BLUE OBJECT
ZC 1218.8+1736	12 18 48.	+ 17 36	1080		CLUSTER OF GALAXIES
LB 00039	12 18 48.	+ 26 43		18.6	FAINT BLUE STAR
ZWG 187.037	12 18 48.	+ 32 59		15.2	GALAXY
ZWG 187.038	12 18 48.	+ 36 37		15.6	GALAXY
ZWG 244.007	12 18 48.	+ 46 06		14.8	GALAXY
UGC 07408	12 18 48.	+ 46 06	162	14.8	GALAXY DWRF IR
MCG+09-20-154	12 18 48.	+ 54 57	36	17.	GALAXY
ZCG 199.25	12 18 48.	+ 63 34		17.5	COMPACT GALAXY
AMES 0705	12 18 49.	+ 12 03 56.	17	17.8	NEBULA
RNGC 4294	12 18 50.	+ 11 47		13.0	GALAXY
HOLM 376A	12 18 51.	+ 11 47	180	11.8	PART OF MULTIPLE GALAXY
MCG+08-23-012	12 18 51.	+ 50 30 30.	48	16.	GALAXY
AMES 0706	12 18 52.	+ 11 00 44.	19	17.6	NEBULA
AMES 0707	12 18 53.	+ 08 13 38.	17	17.6	NEBULA
A3 030	12 18 53.3	+ 15 50 53.		17.2	FAINT BLUE OBJECT
KARA.72 331A	12 18 54.	+ 06 55	36	14.0	PART OF DOUBLE GALAXY
ZWG 042.041	12 18 54.	+ 06 56		14.0	GALAXY
UGC 07409	12 18 54.	+ 06 56	102	14.0	GALAXY S0
KARA.72 331B	12 18 54.	+ 06 56	84		PART OF DOUBLE GALAXY
MCG+01-32-018	12 18 54.	+ 06 56	24	16.	GALAXY
MCG+01-32-017	12 18 54.	+ 06 56	48	14.0	GALAXY
KARA.68 104	12 18 54.	+ 15 16	47		DWARF GALAXY
REA 86	12 18 54.	+ 15 19			DWARF GALAXY
REA 32	12 18 54.	+ 15 54	30		DWARF GALAXY
ZC 1218.9+1915	12 18 54.	+ 19 15	810		CLUSTER OF GALAXIES
ZC 1218.9+2743	12 18 54.	+ 27 43	2420		CLUSTER OF GALAXIES
MCG+08-23-013	12 18 54.	+ 46 05	150	14.	GALAXY
ZC 1218.9+5350	12 18 54.	+ 53 50	1210		CLUSTER OF GALAXIES
AGU 35	12 18 54.	- 39 30 54.	90	14.5	PECULIAR GALAXY
AMES 0710	12 18 55.	+ 05 24 32.	23	17.5	NEBULA
AMES 0709	12 18 55.	+ 06 57 38.	30	16.5	NEBULA
AMES 0708	12 18 55.	+ 08 26 20.	19	17.8	NEBULA
IC 3197	12 18 55.	+ 25 43 13.			NONSTELLAR OBJECT

OBJECT NAME	RIGHT ASCEN.	DECLINATION	DIAM.	MAGN.	TYPE OF OBJECT
RNGC 4297	12 18 56.	+ 06 56		14.0	GALAXY
RNGC 4296	12 18 56.	+ 06 56		14.0	GALAXY
SVEN 285	12 18 56.	+ 14 53	120	12.4	GALAXY
HOLM 377A	12 18 56.	+ 14 53	192	11.6	PART OF MULTIPLE GALAXY
SVEN 286	12 18 57.	+ 05 03	48	14.4	GALAXY
REIZ 2203	12 18 57.	+ 06 55	24	13.6	GALAXY
AMES 0711	12 18 57.	+ 07 49 20.	24	17.4	NEBULA
HN 0870	12 18 57.	+ 12 01		14.	NEBULA
IC 3196	12 18 57.	+ 12 01			NONSTELLAR OBJECT
AMES 0713	12 18 58.	+ 08 29 38.	14	17.8	NEBULA
AMES 0712	12 18 59.	+ 12 47 44.	26	17.1	NEBULA
VDB.66G 121	12 19	+ 00 43	70		DWARF GALAXY
ZC 1219.0+0158	12 19 00.	+ 01 58	610		CLUSTER OF GALAXIES
ZWG 042.042	12 19 00.	+ 05 03		14.9	GALAXY
UGC 07411	12 19 00.	+ 05 03	78	14.9	GALAXY S0-a
MCG+01-32-019	12 19 00.	+ 05 03	48	14.9	GALAXY
ZWG 042.043	12 19 00.	+ 06 02		15.1	GALAXY
MCG+01-32-020	12 19 00.	+ 06 02	36	15.1	GALAXY
IC 0782	12 19 00.	+ 06 02 37.			NONSTELLAR OBJECT
ZC 1219.0+0624	12 19 00.	+ 06 24	1550		CLUSTER OF GALAXIES
AMES 0714	12 19 00.	+ 08 38 56.	25	17.8	NEBULA
ZC 1219.0+1405	12 19 00.	+ 14 05	1410		CLUSTER OF GALAXIES
MCG+03-32-007	12 19 00.	+ 14 52	180	12.2	GALAXY
ZWG 099.024	12 19 00.	+ 14 53		12.2	GALAXY
SCH 34	12 19 00.	+ 14 53		12.2	PECULIAR GALAXY
UGC 07412	12 19 00.	+ 14 53	192	12.2	GALAXY Sc
KARA.72 332A	12 19 00.	+ 14 53	168	12.2	PART OF DOUBLE GALAXY
LB 00040	12 19 00.	+ 25 27		15.5	FAINT BLUE STAR
IC 3198	12 19 00.	+ 26 38 37.			NONSTELLAR OBJECT
MCG+10-18-032	12 19 00.	+ 58 21	30	16.	GALAXY
ZCG 199.26	12 19 00.	+ 64 17		18.6	COMPACT GALAXY
UGC 07410	12 19 00.	+ 82 58	96	16.0	GALAXY S0?
ZWG 014.016	12 19 00.	- 00 33		15.6	GALAXY
MCG-01-32-005	12 19 00.	- 05 22	48	16.	GALAXY
MCG-06-27-018	12 19 00.	- 39 29	90	14.	GALAXY
BC 0N231	12 19 01.06	+ 28 30 35.9		14.	QUASI-STELLAR OBJECT
AMES 0716	12 19 02.	+ 08 44 56.	16	17.6	NEBULA
AMES 0715	12 19 02.	+ 14 02 08.	26	17.3	NEBULA
RNGC 4298	12 19 02.	+ 14 53		12.5	GALAXY
REIZ 2206	12 19 02.	+ 14 53	168	13.0	GALAXY
MAI 069	12 19 02.	+ 57 06	40		DWARF SPHEROIDAL GALAXY
HOLM 378A	12 19 03.	+ 05 03	18	14.2	PART OF MULTIPLE GALAXY
REIZ 2204	12 19 03.	+ 06 01	30	15.3	GALAXY
REIZ 2205	12 19 03.	+ 06 58	54	15.6	GALAXY
KEEL 410	12 19 03.3	+ 16 15 37.		16.	SPIRAL NEBULA
A3 033	12 19 03.5	+ 13 47 20.		16.6	FAINT BLUE OBJECT
HOLM 378B	12 19 04.	+ 05 03	24	14.3	PART OF MULTIPLE GALAXY
AMES 0717	12 19 04.	+ 06 37 20.	22	17.9	NEBULA
AMES 0718	12 19 05.	+ 06 47 44.	12	17.6	NEBULA
LB 02320	12 19 05.	+ 53 29 06.		16.7	FAINT BLUE STAR
LB 02319	12 19 05.	+ 53 29 06.		16.4	FAINT BLUE STAR
ZWG 042.044	12 19 06.	+ 05 39		13.9	GALAXY
UGC 07413	12 19 06.	+ 05 39	102	13.9	GALAXY Sa
MCG+01-32-021	12 19 06.	+ 05 39	72	13.9	GALAXY
AMES 0719	12 19 06.	+ 07 31 44.	18	17.9	NEBULA
RMB 183	12 19 06.	+ 11 46			FAINT BLUE OBJECT
ZWG 070.025	12 19 06.	+ 11 47		12.8	GALAXY
UGC 07414	12 19 06.	+ 11 47	102	12.8	GALAXY IRR
MCG+02-32-010	12 19 06.	+ 11 47	90	12.8	GALAXY
HOLM 377B	12 19 06.	+ 14 53	300	12.0	PART OF MULTIPLE GALAXY
RMB 050	12 19 06.	+ 15 37			FAINT BLUE OBJECT
ZWG 099.025	12 19 06.	+ 16 00		15.0	GALAXY
UGC 07415	12 19 06.	+ 16 00	72	15.0	GALAXY
MCG+03-32-008	12 19 06.	+ 16 01	84	15.0	GALAXY
ZWG 099.026	12 19 06.	+ 16 15		15.7	GALAXY
ZC 1219.1+2322	12 19 06.	+ 23 22	1340		CLUSTER OF GALAXIES
ZC 1219.1+2713	12 19 06.	+ 27 13	870		CLUSTER OF GALAXIES
KARA.68 105	12 19 06.	+ 38 20	40		DWARF GALAXY
ZWG 216.001	12 19 06.	+ 41 08		14.2	GALAXY
ZWG 215.067	12 19 06.	+ 41 08		14.2	GALAXY
UGC 07416	12 19 06.	+ 41 08	90	14.2	GALAXY SBb
KARA.73B 0525	12 19 06.	+ 41 08	96	14.2	ISOLATED GALAXY S
ZC 1219.1+4910	12 19 06.	+ 49 10	470		CLUSTER OF GALAXIES
ZCG 199.27	12 19 06.	+ 64 01		18.6	COMPACT GALAXY
MCG-02-32-006	12 19 06.	- 12 13	48	15.	GALAXY
SER 090.03	12 19 06.	- 43 04	80		PEC ONE-ARMED SPIRAL GLXY
KEEL 411	12 19 06.8	+ 16 01 21.		16.	SPIRAL NEBULA
RNGC 4300	12 19 07.	+ 05 39		14.0	GALAXY
IC 3200	12 19 07.	+ 27 02 19.			NONSTELLAR OBJECT
AMES 0721	12 19 08.	+ 06 16 26.	14	17.6	NEBULA
AMES 0720	12 19 08.	+ 08 03 14.	23	17.8	NEBULA
RNGC 4299	12 19 08.	+ 11 47		13.5	GALAXY
REIZ 2209	12 19 08.	+ 14 52	240	13.3	GALAXY
REIZ 2210	12 19 08.	+ 16 01	24	14.7	GALAXY
HOLM 387C	12 19 08.	+ 16 02	60	14.0	PART OF MULTIPLE GALAXY
REIN 3.087	12 19 08.17	+ 11 46 48.7			NEBULA
KEEL 412	12 19 08.8	+ 15 56 51.		18.	NEBULA
REIZ 2208	12 19 09.	+ 05 39	90	13.2	GALAXY
HN 0871	12 19 09.	+ 10 52		14.	NEBULA
IC 3199	12 19 09.	+ 10 52			NONSTELLAR OBJECT
REIZ 2207	12 19 09.	+ 11 47	60	13.0	GALAXY
MCG+04-29-067	12 19 09.	+ 26 09 30.	90	15.7	GALAXY
TON-N 1512	12 19 09.	+ 34 52		17.0	BLUE STAR
AMES 0727	12 19 10.	+ 08 07 44.	20	16.8	NEBULA
AMES 0726	12 19 10.	+ 09 33 44.	20	17.6	NEBULA
AMES 0722	12 19 10.	+ 12 53 32.	28	17.2	NEBULA
IC 3201	12 19 10.	+ 26 00 07.			NONSTELLAR OBJECT
AMES 0729	12 19 11.	+ 06 42 20.	26	17.8	NEBULA
AMES 0725	12 19 11.	+ 12 31 56.	26	17.1	NEBULA
AMES 0724	12 19 11.	+ 13 05 32.	17	16.7	NEBULA
AMES 0723	12 19 11.	+ 14 12 56.	30	17.6	NEBULA
AMES 0731	12 19 12.	+ 06 34 44.	24	17.8	NEBULA
ZC 1219.2+0806	12 19 12.	+ 08 06	1410		CLUSTER OF GALAXIES
KARA.68 106	12 19 12.	+ 08 49	27		DWARF GALAXY
AMES 0730	12 19 12.	+ 09 30 50.	25	17.6	NEBULA
ZWG 070.026	12 19 12.	+ 10 53		15.4	GALAXY
UGC 07417	12 19 12.	+ 10 53	66	15.4	GALAXY S
RMB 182	12 19 12.	+ 11 41			FAINT BLUE OBJECT
SCH 35	12 19 12.	+ 11 47		12.8	PECULIAR GALAXY
AMES 0728	12 19 12.	+ 14 00 44.	19	17.3	NEBULA
MCG+03-32-009	12 19 12.	+ 14 51	300	13.4	GALAXY
RMB 054	12 19 12.	+ 14 52			FAINT BLUE OBJECT
ZWG 099.027	12 19 12.	+ 14 53		13.4	GALAXY
UGC 07418	12 19 12.	+ 14 53	330	13.4	GALAXY Sc
KARA.72 332B	12 19 12.	+ 14 53	294	13.4	PART OF DOUBLE GALAXY
KARA.68 107	12 19 12.	+ 15 20	34		DWARF GALAXY
REA 33	12 19 12.	+ 15 22	42		DWARF GALAXY
ZWG 128.082	12 19 12.	+ 26 10		15.7	GALAXY

OBJECT NAME	RIGHT ASCEN.	DECLINATION	DIAM.	MAGN.	TYPE OF OBJECT
UGC 07419	12 19 12.	+ 26 10	90	15.7	GALAXY Sb
MCG+07-26-001	12 19 12.	+ 41 07	96	13.5	GALAXY
MCG+09-20-155	12 19 12.	+ 56 05	60	15.	GALAXY
7ZW 449	12 19 12.	+ 63 15			COMPACT GALAXY
ZCG 199.28	12 19 12.	+ 63 15		17.3	COMPACT GALAXY
MCG-04-29-021	12 19 12.	- 23 54	108	13.	GALAXY
MCG-04-29-020	12 19 12.	- 24 48 30.	48	15.	GALAXY
AMES 0733	12 19 13.	+ 06 43 44.	29	17.7	NEBULA
AMES 0732	12 19 13.	+ 07 35 08.	23	17.7	NEBULA
RNGC 4302	12 19 14.	+ 14 53		12.5	GALAXY
IC 3202	12 19 14.	+ 27 20 01.			NONSTELLAR OBJECT
ARC 1524	12 19 15.	+ 08 07		17.2	RICH CLUSTER OF GALAXIES
HOLM 376B	12 19 15.	+ 11 47	78	12.0	PART OF MULTIPLE GALAXY
IC 3203	12 19 15.	+ 26 09 43.			NONSTELLAR OBJECT
TON-N 1513	12 19 15.	+ 33 39		17.0	BLUE STAR
SN 1964F	12 19 16.	+ 04 45		14.0	SUPERNOVA
AMES 0735	12 19 16.	+ 08 20 44.	35	16.6	NEBULA
SN 1926A	12 19 17.	+ 04 46		14.0	SUPERNOVA
AMES 0734	12 19 17.	+ 13 50 08.	22	17.0	NEBULA
ZWG 042.045	12 19 18.	+ 04 45		10.9	GALAXY
UGC 07420	12 19 18.	+ 04 45	396	10.9	GALAXY Sb/SBc
MCG+01-32-022	12 19 18.	+ 04 45	360	10.9	GALAXY
AMES 0736	12 19 18.	+ 10 00 56.	36	17.4	NEBULA
UGC 07421	12 19 18.	+ 12 14	66	16.5	GALAXY DWARF
RMB 153	12 19 18.	+ 13 24			FAINT BLUE OBJECT
MCG+03-34-019	12 19 18.	+ 14 34	48	15.6	GALAXY
REA 87	12 19 18.	+ 18 43			DWARF GALAXY
ZWG 158.086	12 19 18.	+ 26 38		15.7	GALAXY
ZWG 158.087	12 19 18.	+ 26 39		15.7	GALAXY
MCG+05-29-069	12 19 18.	+ 30 21	24	14.3	GALAXY
ZC 1219.3+3550	12 19 18.	+ 35 50	1010		CLUSTER OF GALAXIES
FEIG 061	12 19 18.	+ 48 04		11.9	FAINT BLUE STAR
ZWG 293.014	12 19 18.	+ 58 46		15.5	GALAXY
LB 02321	12 19 18.	+ 59 12 00.		15.8	FAINT BLUE STAR
ZC 1219.3+6520	12 19 18.	+ 65 20	1210		CLUSTER OF GALAXIES
MCG-06-27-019	12 19 18.	- 38 31	30	15.5	GALAXY
IC 1219-43	12 19 18.	- 43 03 30.			UNUSUAL SOUTHERN GALAXY
AGB 36	12 19 18.	- 43 05 00.	54	13.5	PECULIAR GALAXY
AMES 0738	12 19 19.	+ 08 21 44.	25	17.6	NEBULA
AMES 0737	12 19 19.	+ 08 35 32.	18	17.2	NEBULA
IC 3204	12 19 19.	+ 24 31 31.			NONSTELLAR OBJECT
REIZ 2212	12 19 19.	+ 26 09	42	14.6	GALAXY
SC 1216-4425.0	12 19 19.	- 44 41 39.	12		NEBULA
REIN 3.088	12 19 19.35	+ 13 05 20.2			NEBULA
IC 3205	12 19 20.	+ 26 37 07.			NONSTELLAR OBJECT
IC 3206	12 19 20.	+ 26 38 19.			NONSTELLAR OBJECT
SVEN 287	12 19 21.	+ 04 44 51.	300	10.7	GALAXY
REIZ 2211	12 19 21.	+ 04 45	210	11.2	GALAXY
HOLM 379A	12 19 21.	+ 04 45	210	10.4	PART OF MULTIPLE GALAXY
AMES 0743	12 19 21.	+ 05 08 32.	25	17.5	NEBULA
AMES 0742	12 19 21.	+ 08 26 20.	23	17.4	NEBULA
AMES 0740	12 19 21.	+ 12 05 02.	28	17.6	NEBULA
AMES 0739	12 19 21.	+ 13 32 44.	16	17.6	NEBULA
IC 3207	12 19 21.	+ 24 38 01.			NONSTELLAR OBJECT
MCG+05-29-070	12 19 21.	+ 28 43 30.	18	15.5	GALAXY
MCG+06-27-049	12 19 21.	+ 35 20	48	15.	GALAXY
A3 034	12 19 21.0	+ 15 54 51.		17.0	FAINT BLUE OBJECT
KEEL 413	12 19 21.7	+ 15 55 21.		17.	NEBULA
AMES 0744	12 19 22.	+ 08 22 32.	24	17.4	NEBULA
AMES 0741	12 19 22.	+ 13 55 06.	32	17.2	NEBULA
LB 02322	12 19 22.	+ 57 02 24.		16.0	FAINT BLUE STAR
A3 035	12 19 22.4	+ 15 42 22.		17.3	FAINT BLUE OBJECT
SN 1961I	12 19 23.	+ 04 45		13.0	SUPERNOVA
AMES 0746	12 19 23.	+ 05 41 44.	26	17.4	NEBULA
AMES 0745	12 19 23.	+ 06 44 20.	88	16.8	NEBULA
SEY 180	12 19 23.	+ 35 20 21.		15.4	FAINT GALAXY
SC 1216-4246.6	12 19 23.	- 43 03 15.	24		NEBULA
REIN 3.089	12 19 23.34	+ 08 57 41.1			NEBULA
ZWG 042.046	12 19 24.	+ 02 37		15.5	GALAXY
ZWG 042.047	12 19 24.	+ 05 23		14.9	GALAXY
UGC 07422	12 19 24.	+ 05 23	60	14.9	GALAXY S
MCG+01-32-023	12 19 24.	+ 05 23	54	14.9	GALAXY
ZWG 042.048	12 19 24.	+ 06 43		15.3	GALAXY
UGC 07423	12 19 24.	+ 06 43	78	15.3	GALAXY
MCG+01-32-024	12 19 24.	+ 06 43	48	15.3	GALAXY
ZWG 070.027	12 19 24.	+ 08 57		15.6	GALAXY
UGC 07424	12 19 24.	+ 08 57	72	15.6	GALAXY DWARF
MCG+02-32-011	12 19 24.	+ 08 57	60	15.6	GALAXY
RMB 220	12 19 24.	+ 10 38			FAINT BLUE OBJECT
UGC 07425	12 19 24.	+ 15 55	72	16.5	GALAXY
MCG+03-32-010	12 19 24.	+ 15 55	72	14.9	GALAXY
KARA.68 108	12 19 24.	+ 18 09	47		DWARF GALAXY
MCG+05-29-071	12 19 24.	+ 28 28	27	15.7	GALAXY
ZWG 158.088	12 19 24.	+ 30 20		14.3	GALAXY
UGC 07426	12 19 24.	+ 30 20	48	14.3	GALAXY E
UGC 07427	12 19 24.	+ 35 18	78	16.5	GALAXY DWRF IR
MCG+08-23-014	12 19 24.	+ 49 59	24	17.	GALAXY
ZWG 269.052	12 19 24.	+ 51 20		15.7	GALAXY
ZC 1219.4+6241	12 19 24.	+ 62 41	1610		CLUSTER OF GALAXIES
ZC 1219.4-0053	12 19 24.	- 00 53	470		CLUSTER OF GALAXIES
MCG-06-27-020	12 19 24.	- 35 30	192	12.5	GALAXY
RNGC 4303	12 19 25.	+ 04 45		10.5	GALAXY
HOLM 387G	12 19 25.	+ 15 56	36	14.9	PART OF MULTIPLE GALAXY
REIN 3.090	12 19 25.05	+ 08 57 01.6			NEBULA
AMES 0747	12 19 26.	+ 11 52 26.	38	17.6	NEBULA
HN 0872	12 19 26.	+ 12 14	90		NEBULA
REIZ 2214	12 19 26.	+ 13 01	48	14.6	GALAXY
REIZ 2215	12 19 26.	+ 13 03	24	14.7	GALAXY
REIZ 2216	12 19 26.	+ 15 54	12	15.4	GALAXY
LB 02324	12 19 26.	+ 58 44 18.		16.0	FAINT BLUE STAR
LB 02323	12 19 26.	+ 61 08 42.		15.8	FAINT BLUE STAR
SC 1216-4539.6	12 19 26.	- 45 56 15.	12		NEBULA
REIZ 2213	12 19 27.	+ 06 43	18	16.3	GALAXY
AMES 0748	12 19 27.	+ 06 55 20.	22	17.8	NEBULA
IC 3208	12 19 27.	+ 12 14			NONSTELLAR OBJECT
RNGC 4308	12 19 27.	+ 30 20		14.5	GALAXY
MCG+05-29-072	12 19 27.	+ 32 23 30.	72	14.3	GALAXY
HOLM 383B	12 19 27.	+ 51 19	12	15.2	PART OF MULTIPLE GALAXY
LB 02325	12 19 27.	+ 53 33 48.		15.2	FAINT BLUE STAR
AMES 0750	12 19 28.	+ 05 22 02.	60	15.4	NEBULA
ARC 1525	12 19 28.	- 00 53		18.0	RICH CLUSTER OF GALAXIES
AMES 0753	12 19 29.	+ 06 23 14.	26	16.8	NEBULA
AMES 0752	12 19 29.	+ 10 53 44.	30	16.6	NEBULA
AMES 0749	12 19 29.	+ 13 27 56.	30	17.6	NEBULA
SEY 181	12 19 29.	+ 32 21 39.		14.5	FAINT GALAXY
LB 04365	12 19 30.	+ 13 06		18.9	FAINT BLUE STAR
AMES 0751	12 19 30.	+ 13 49 20.	23	17.8	NEBULA
LB 04364	12 19 30.	+ 13 58		17.9	FAINT BLUE STAR

OBJECT NAME	RIGHT ASCEN.	DECLINATION	DIAM.	MAGN.	TYPE OF OBJECT
LB 04363	12 19 30.	+ 14 01		17.0	FAINT BLUE STAR
ZWG 158.089	12 19 30.	+ 28 43		15.5	GALAXY
ZWG 158.090	12 19 30.	+ 32 22		14.3	GALAXY
UGC 07428	12 19 30.	+ 32 22	84	14.3	GALAXY IRR
MCG+07-26-002	12 19 30.	+ 42 35 30.	30	14.5	GALAXY
LB 02326	12 19 30.	+ 48 42 00.		15.9	FAINT BLUE STAR
MCG+09-20-156	12 19 30.	+ 51 19	42	14.0	GALAXY
ZC 1219.5+5720	12 19 30.	+ 57 20	2760		CLUSTER OF GALAXIES
MRK 205	12 19 30.	+ 75 35	6	14.5	GALAXY WITH UV CONTINUUM
VVI 56	12 19 30.	+ 75 35	6	15.83	SEYFERT GALAXY
MCG+13-09-025	12 19 30.	+ 75 36	156	12.	GALAXY
ZWG 352.029	12 19 30.	+ 75 37		13.0	GALAXY
UGC 07429	12 19 30.	+ 75 37	204	13.0	GALAXY SBb
ZWG 014.017	12 19 30.	- 02 19		15.5	GALAXY
VHE 57A	12 19 30.	- 63 01	30		REFLECTION NEBULA
AMES 0754	12 19 31.	+ 06 53 38.	13	17.8	NEBULA
LB 02327	12 19 31.	+ 48 54 06.		14.6	FAINT BLUE STAR
REIN 3.091	12 19 31.20	+ 13 01 03.9			NEBULA
SHB 201	12 19 31.8	+ 75 35 10.		14.5	QUASI-STELLAR OBJECT
REIN 3.092	12 19 31.92	+ 13 03 54.9			NEBULA
RNGC 4307	12 19 32.	+ 09 19		13.0	GALAXY
RNGC 4305	12 19 32.	+ 13 01		14.0	GALAXY
HOLM 381A	12 19 32.	+ 13 01	102	13.5	PART OF MULTIPLE GALAXY
RNGC 4306	12 19 32.	+ 13 04		14.5	GALAXY
HOLM 381B	12 19 32.	+ 13 04	48	14.0	PART OF MULTIPLE GALAXY
IC 3210	12 19 32.	+ 28 42 31.			NONSTELLAR OBJECT
REIN 3.093	12 19 32.94	+ 09 19 16.4			NEBULA
REIZ 2217	12 19 33.	+ 05 22	48	14.9	GALAXY
AMES 0760	12 19 33.	+ 06 55 32.	17	17.8	NEBULA
AMES 0759	12 19 33.	+ 08 07 08.	13	17.0	NEBULA
REIZ 2218	12 19 33.	+ 09 19	204	12.8	GALAXY
HOLM 380A	12 19 33.	+ 09 19	180	12.2	PART OF MULTIPLE GALAXY
HN 0873	12 19 33.	+ 12 01	60		NEBULA
IC 3209	12 19 33.	+ 12 01			NONSTELLAR OBJECT
AMES 0755	12 19 33.	+ 13 59 08.	29	17.7	NEBULA
HOLM 383A	12 19 33.	+ 51 20	42	14.0	PART OF MULTIPLE GALAXY
REIN 3.094	12 19 33.71	+ 12 01 56.5			NEBULA
AMES 0761	12 19 34.	+ 07 04 14.	29	17.6	NEBULA
AMES 0758	12 19 34.	+ 12 10 56.	26	16.4	NEBULA
AMES 0757	12 19 34.	+ 14 00 32.	28	17.1	NEBULA
AMES 0756	12 19 34.	+ 14 15 38.	40	17.4	NEBULA
IC 3212	12 19 34.	+ 28 27 49.			NONSTELLAR OBJECT
REIN 3.095	12 19 34.63	+ 09 16 02.6			NEBULA
A3 037	12 19 34.7	+ 15 05 35.		16.8	FAINT BLUE OBJECT
AMES 0763	12 19 35.	+ 05 13 08.	36	16.5	NEBULA
HOLM 380B	12 19 35.	+ 09 16	42	14.6	PART OF MULTIPLE GALAXY
IC 3211	12 19 35.	+ 09 16 01.			NONSTELLAR OBJECT
RNGC 4304	12 19 35.	- 33 12		12.5	GALAXY
KEEL 414	12 19 35.6	+ 16 04 35.		18.	NEBULA
ZWG 042.049	12 19 36.	+ 05 14		15.6	GALAXY
ZWG 042.050	12 19 36.	+ 06 23		15.6	GALAXY
AMES 0764	12 19 36.	+ 07 46 14.	14	17.2	NEBULA
ZWG 070.028	12 19 36.	+ 09 16		15.5	GALAXY
UGC 07430	12 19 36.	+ 09 16	66	15.5	GALAXY Sc
MCG+02-32-012	12 19 36.	+ 09 16	30	15.5	GALAXY
ZWG 070.029	12 19 36.	+ 09 19		13.4	GALAXY
UGC 07431	12 19 36.	+ 09 19	210	13.4	GALAXY Sb
MCG+02-32-012A	12 19 36.	+ 09 19	210	13.4	GALAXY
ZWG 070.030	12 19 36.	+ 12 02		15.3	GALAXY
AMES 0762	12 19 36.	+ 12 38 56.	22	18.0	NEBULA
LB 04370	12 19 36.	+ 12 44		16.8	FAINT BLUE STAR
ZWG 070.031	12 19 36.	+ 13 01		13.8	GALAXY
UGC 07432	12 19 36.	+ 13 01	126	13.8	GALAXY Sa
KARA.72 333A	12 19 36.	+ 13 01	90	13.8	PART OF DOUBLE GALAXY
MCG+02-32-013	12 19 36.	+ 13 01	108	13.8	GALAXY
ZWG 070.032	12 19 36.	+ 13 04		14.4	GALAXY
UGC 07433	12 19 36.	+ 13 04	84	14.4	GALAXY S0
KARA.72 333B	12 19 36.	+ 13 04	72	14.4	PART OF DOUBLE GALAXY
MCG+02-32-014	12 19 36.	+ 13 04	18	14.4	GALAXY
LB 04369	12 19 36.	+ 13 29		18.7	FAINT BLUE STAR
LB 04368	12 19 36.	+ 13 42		18.2	FAINT BLUE STAR
LB 04367	12 19 36.	+ 13 54		17.9	FAINT BLUE STAR
LB 04366	12 19 36.	+ 14 14		18.5	FAINT BLUE STAR
MCG+03-32-011	12 19 36.	+ 16 04	60	19.	GALAXY
REA 34	12 19 36.	+ 16 05	36		DWARF GALAXY
ZC 1219.6+1859	12 19 36.	+ 18 59	470		CLUSTER OF GALAXIES
ZWG 128.083	12 19 36.	+ 24 09		15.7	GALAXY
ZC 1219.6+2444	12 19 36.	+ 24 44	940		CLUSTER OF GALAXIES
ZWG 128.084	12 19 36.	+ 26 20		15.7	GALAXY
UGC 07434	12 19 36.	+ 26 20	120	15.7	GALAXY
MCG+04-29-068	12 19 36.	+ 26 20	96	15.7	GALAXY
ZWG 158.091	12 19 36.	+ 28 28		15.7	GALAXY
MCG+05-29-073	12 19 36.	+ 29 07 30.	48	15.5	GALAXY
ZC 1219.6+4536	12 19 36.	+ 45 36	870		CLUSTER OF GALAXIES
ZCG 199.29	12 19 36.	+ 63 43		17.7	COMPACT GALAXY
MCG-05-29-034	12 19 36.	- 33 15	120	13.	GALAXY
A3 038	12 19 36.3	+ 15 39 20.		16.5	FAINT BLUE OBJECT
AMES 0766	12 19 37.	+ 08 41 32.	36	17.6	NEBULA
KEEL 415	12 19 37.	+ 15 39 51.		17.	SPIRAL NEBULA
IC 3213	12 19 37.	+ 24 08 07.			NONSTELLAR OBJECT
RNGC 4319	12 19 37.	+ 75 36		13.0	GALAXY
KEEL 416	12 19 37.3	+ 16 06 41.		18.	NEBULA
RNGC 4309A	12 19 38.	+ 07 26			GALAXY
AMES 0767	12 19 38.	+ 08 29 20.	12	17.8	NEBULA
RNGC 4307A	12 19 38.	+ 09 16		15.5	GALAXY
ARC 1526	12 19 38.	+ 14 01		16.6	RICH CLUSTER OF GALAXIES
AMES 0765	12 19 38.	+ 14 02 26.	22	17.3	NEBULA
LB 02328	12 19 38.	+ 53 36 00.		16.9	FAINT BLUE STAR
A3 039	12 19 38.2	+ 14 12 44.		17.1	FAINT BLUE OBJECT
HOLM 382A	12 19 39.	+ 07 25	60	13.6	PART OF MULTIPLE GALAXY
IC 3214	12 19 39.	+ 27 30 49.			NONSTELLAR OBJECT
LB 02329	12 19 39.	+ 61 22 48.		16.1	FAINT BLUE STAR
AMES 0768	12 19 40.	+ 13 40 02.	25	16.9	NEBULA
IC 3215	12 19 40.	+ 26 19 49.			NONSTELLAR OBJECT
LB 02330	12 19 40.	+ 59 33 24.		15.5	FAINT BLUE STAR
PK299-00.1	12 19 40.	- 63 00 47.	10		PLANETARY NEBULA
HOLM 382B	12 19 41.	+ 07 27	18	15.3	PART OF MULTIPLE GALAXY
IC 3216	12 19 41.	+ 25 33 49.			NONSTELLAR OBJECT
MCG+01-32-026	12 19 42.	+ 07 12	36	15.6	GALAXY
ZWG 042.051	12 19 42.	+ 07 25		14.3	GALAXY
UGC 07435	12 19 42.	+ 07 25	108	14.3	GALAXY S0-a
MCG+01-32-025	12 19 42.	+ 07 25	72	14.3	GALAXY
RMB 174	12 19 42.	+ 12 30			FAINT BLUE OBJECT
ZC 1219.7+1237	12 19 42.	+ 12 37	1280		CLUSTER OF GALAXIES
LB 04371	12 19 42.	+ 12 49		18.8	FAINT BLUE STAR
LB 04372	12 19 42.	+ 13 32		17.8	FAINT BLUE STAR
AMES 0769	12 19 42.	+ 14 19 56.	26	17.3	NEBULA
RMB 049	12 19 42.	+ 15 53			FAINT BLUE OBJECT
TON-N 0077	12 19 42.	+ 27 37		16.0	BLUE STAR
ZC 1219.7+3833	12 19 42.	+ 38 33	1080		CLUSTER OF GALAXIES
ZC 1219.7+4042	12 19 42.	+ 40 42	1550		CLUSTER OF GALAXIES
MCG+08-22-105	12 19 42.	+ 47 22	48	16.	GALAXY
MCG+09-20-157	12 19 42.	+ 56 07	48	15.	GALAXY
REIZ 2224	12 19 42.	+ 59 35	24	15.0	GALAXY
ZC 1219.7+6821	12 19 42.	+ 68 21	1010		CLUSTER OF GALAXIES
A3 041	12 19 42.8	+ 14 16 15.		16.8	FAINT BLUE OBJECT
REIZ 2220	12 19 43.	+ 26 20	90	14.8	GALAXY
IC 3217	12 19 43.	+ 26 39 49.			NONSTELLAR OBJECT
RNGC 4309	12 19 44.	+ 07 25		14.5	GALAXY
AMES 0771	12 19 44.	+ 12 40 38.	25	16.2	NEBULA
AMES 0770	12 19 44.	+ 13 57 26.	22	17.2	NEBULA
LB 02331	12 19 44.	+ 61 50 06.		16.6	FAINT BLUE STAR
SVEN 288	12 19 45.	+ 04 56	12	13.5	GALAXY
REIZ 2219	12 19 45.	+ 07 24	84	13.7	GALAXY
AMES 0775	12 19 45.	+ 07 54 14.	18	17.5	NEBULA
AMES 0774	12 19 45.	+ 10 18 32.	24	16.7	NEBULA
AMES 0773	12 19 45.	+ 12 12 44.	16	17.0	NEBULA
AMES 0772	12 19 45.	+ 14 02 44.	23	17.2	NEBULA
IC 3219	12 19 45.	+ 26 13 49.			NONSTELLAR OBJECT
TON-N 1514	12 19 45.	+ 34 28		16.3	BLUE STAR
KEEL 417	12 19 45.3	+ 04 33 44.			NEBULA
AMES 0778	12 19 46.	+ 06 43 14.	25	17.7	NEBULA
A3 042	12 19 46.0	+ 15 49 39.		17.0	FAINT BLUE OBJECT
KEEL 418	12 19 46.4	+ 16 04 37.		17.	NEBULA
IC 3218	12 19 47.	+ 07 12 31.			NONSTELLAR OBJECT
AMES 0777	12 19 47.	+ 12 49 50.	17	17.8	NEBULA
AMES 0776	12 19 47.	+ 13 34 50.	29	17.5	NEBULA
KEEL 419	12 19 47.7	+ 16 00 39.		16.	SPIRAL NEBULA
IC 0783A	12 19 47.9	+ 16 00 36.			GALAXY
AMES 0779	12 19 48.	+ 05 57 56.	20	17.6	NEBULA
ZWG 042.052	12 19 48.	+ 07 12		15.6	GALAXY
LB 04374	12 19 48.	+ 12 46		18.7	FAINT BLUE STAR
LB 04373	12 19 48.	+ 14 08		16.5	FAINT BLUE STAR
MCG+03-32-012	12 19 48.	+ 15 01	96	15.2	GALAXY
ZWG 099.028	12 19 48.	+ 15 02		15.2	GALAXY
UGC 07436	12 19 48.	+ 15 02	78	15.2	GALAXY S
MCG+03-32-013	12 19 48.	+ 16 00	30	14.7	GALAXY
IC 3222	12 19 48.	+ 29 06 31.			NONSTELLAR OBJECT
UGC 07437	12 19 48.	+ 29 07	66	16.0	GALAXY SBc
MCG+05-29-074	12 19 48.	+ 29 30	120	13.5	GALAXY
MCG+05-29-075	12 19 48.	+ 30 11	270	11.5	GALAXY
UGC 07438	12 19 48.	+ 30 21	90	16.0	GALAXY
ZWG 187.039	12 19 48.	+ 35 25		15.6	GALAXY
SEY 182	12 19 48.	+ 35 25 52.		15.2	FAINT GALAXY
MCG+06-27-050	12 19 48.	+ 35 26	54	15.	GALAXY
MCG-04-29-022	12 19 48.	- 22 05	90	15.	GALAXY
BC PKS1219+04	12 19 48.04	+ 04 29 59.		17.5	QUASI-STELLAR OBJECT
AMES 0781	12 19 49.	+ 05 47 56.	20	16.6	NEBULA
AMES 0780	12 19 49.	+ 08 25 56.	14	17.6	NEBULA
IC 3221	12 19 49.	+ 25 33 37.			NONSTELLAR OBJECT
REIN 3.096	12 19 49.26	+ 10 52 42.2			NEBULA
REIZ 2223	12 19 50.	+ 16 00	12	15.1	GALAXY
REIZ 2221	12 19 51.	+ 04 48	36	15.0	GALAXY
SVEN 289	12 19 51.	+ 04 50	54	13.8	GALAXY
REIZ 2222	12 19 51.	+ 04 50	54	14.2	GALAXY
AMES 0784	12 19 51.	+ 06 54 08.	17	17.9	NEBULA
AMES 0783	12 19 51.	+ 08 09 32.	20	17.2	NEBULA
HN 0874	12 19 51.	+ 10 52		15.5	NEBULA
IC 3220	12 19 51.	+ 10 52			NONSTELLAR OBJECT
AMES 0782	12 19 51.	+ 13 44 02.	23	16.5	NEBULA
HOLM 387E	12 19 51.	+ 16 01	30	14.7	PART OF MULTIPLE GALAXY
TON-N 1515	12 19 51.	+ 34 22		16.9	BLUE STAR
AMES 0787	12 19 53.	+ 04 50 08.	102	14.0	NEBULA
AMES 0786	12 19 53.	+ 06 47 20.	16	17.2	NEBULA
ZWG 042.053	12 19 54.	+ 04 50		14.9	GALAXY
UGC 07439	12 19 54.	+ 04 50	96	14.9	GALAXY SB
MCG+01-32-027	12 19 54.	+ 04 50	96	14.9	GALAXY
HOLM 379B	12 19 54.	+ 04 51	66	13.5	PART OF MULTIPLE GALAXY
AMES 0788	12 19 54.	+ 08 37 38.	23	17.8	NEBULA
LB 04377	12 19 54.	+ 12 34		20.3	FAINT BLUE STAR
RMB 173	12 19 54.	+ 12 43			FAINT BLUE OBJECT
LB 04376	12 19 54.	+ 13 51		16.6	FAINT BLUE STAR
AMES 0785	12 19 54.	+ 13 57 20.	23	17.0	NEBULA
LB 04375	12 19 54.	+ 14 26		19.2	FAINT BLUE STAR
KARA.68 109	12 19 54.	+ 14 27	27		DWARF GALAXY
KARA.68 110	12 19 54.	+ 15 44	27		DWARF GALAXY
REA 35	12 19 54.	+ 15 45	48		DWARF GALAXY
KARA.68 111	12 19 54.	+ 17 19	47		DWARF GALAXY
ZWG 158.092	12 19 54.	+ 29 29		13.5	GALAXY
UGC 07440	12 19 54.	+ 29 29	138	13.5	GALAXY S
MCG+06-27-051	12 19 54.	+ 36 54	30	16.	GALAXY
ZC 1219.9+4206	12 19 54.	+ 42 06	5850		CLUSTER OF GALAXIES
7ZW 450	12 19 54.	+ 57 09			COMPACT GALAXY
KEEL 420	12 19 54.2	+ 04 50 35.			SPIRAL NEBULA
A3 045	12 19 54.3	+ 13 50 56.		17.3	FAINT BLUE OBJECT
RNGC 4303A	12 19 55.	+ 04 50		15.0	GALAXY
RNGC 4301	12 19 55.	+ 04 51		15.0	GALAXY
AMES 0789	12 19 55.	+ 06 47 20.	20	17.6	NEBULA
LB 02332	12 19 55.	+ 51 19 54.		16.1	FAINT BLUE STAR
AMES 0791	12 19 56.	+ 07 24 20.	23	17.0	NEBULA
AMES 0790	12 19 57.	+ 13 56 20.	29	16.7	NEBULA
RNGC 4311	12 19 57.	+ 29 29			NON-EXISTENT OBJECT
RNGC 4310	12 19 57.	+ 29 29		13.5	GALAXY
LB 02333	12 19 57.	+ 61 35 06.		15.1	FAINT BLUE STAR
MCG-01-32-006	12 19 57.	- 04 22	90	14.	GALAXY
MCG-03-32-007	12 19 57.	- 20 55	48	15.	GALAXY
IC 3223	12 19 58.	+ 09 45 49.			SINGLE STAR
HOLM 384A	12 19 58.	- 04 23	126	13.4	PART OF MULTIPLE GALAXY
AMES 0794	12 19 59.	+ 05 47 44.	29	17.6	NEBULA
LB 09857	12 20	- 84 19		13.3	FAINT BLUE STAR
ZWG 042.054	12 20 00.	+ 06 57		14.9	GALAXY
ZC 1220.0+0657	12 20 00.	+ 06 57	1810		CLUSTER OF GALAXIES
UGC 07441	12 20 00.	+ 06 57	108	14.9	GALAXY
MCG+01-32-028	12 20 00.	+ 06 57	96	14.9	GALAXY
AMES 0796	12 20 00.	+ 09 05 08.	18	17.2	NEBULA
LB 04380	12 20 00.	+ 12 38		19.3	FAINT BLUE STAR
LB 04379	12 20 00.	+ 13 46		17.3	FAINT BLUE STAR
AMES 0793	12 20 00.	+ 13 55 44.	20	17.4	NEBULA
AMES 0792	12 20 00.	+ 14 02 32.	20	16.6	NEBULA
LB 04378	12 20 00.	+ 14 15		20.1	FAINT BLUE STAR
MCG+03-32-014	12 20 00.	+ 15 48	300	12.9	GALAXY
ZWG 099.029	12 20 00.	+ 15 49		12.9	GALAXY
UGC 07442	12 20 00.	+ 15 49	252	12.9	GALAXY Sa
REA 36	12 20 00.	+ 17 19	54		DWARF GALAXY
LB 00005	12 20 00.	+ 24 48		17.6	FAINT BLUE STAR
ZWG 158.093	12 20 00.	+ 30 10		11.5	GALAXY

OBJECT NAME	RIGHT ASCEN.	DECLINATION	DIAM.	MAGN.	TYPE OF OBJECT
UGC 07443	12 20 00.	+ 30 10	276	11.5	GALAXY SBa
ZWG 244.008	12 20 00.	+ 45 20		14.7	GALAXY
UGC 07444	12 20 00.	+ 45 20	84	14.7	GALAXY Sa
MCG+08-23-015	12 20 00.	+ 45 20	72	14.	GALAXY
ZCG 199.30	12 20 00.	+ 64 18		18.2	COMPACT GALAXY
ZWG 352.030	12 20 00.	+ 75 07		15.1	GALAXY
IC 0784	12 20 00.	- 04 22 35.			NONSTELLAR OBJECT
AMES 0795	12 20 01.	+ 12 38 08.	32	17.8	NEBULA
LB 02334	12 20 01.	+ 57 53 30.		16.6	FAINT BLUE STAR
HOLM 384B	12 20 01.	- 04 23	18	14.9	PART OF MULTIPLE GALAXY
AMES 0797	12 20 02.	+ 11 16 08.	41	17.7	NEBULA
REIZ 2226	12 20 02.	+ 12 04	180	12.9	GALAXY
HN 0875	12 20 02.	+ 12 26		14.5	NEBULA
IC 3224	12 20 02.	+ 12 26			NONSTELLAR OBJECT
RNGC 4312	12 20 02.	+ 15 49		13.0	GALAXY
REIZ 2225	12 20 02.	+ 15 49	150	13.4	GALAXY
SC 1217-4639.5	12 20 02.	- 46 56 09.	60		NEBULA
KEEL 421	12 20 02.7	+ 04 25 14.			NEBULA
AMES 0800	12 20 03.	+ 06 18 08.	36	17.6	NEBULA
HOLM 387B	12 20 03.	+ 15 50	222	12.1	PART OF MULTIPLE GALAXY
RNGC 4314	12 20 03.	+ 30 10		11.5	GALAXY
AMES 0799	12 20 04.	+ 13 43 20.	17	16.5	NEBULA
AMES 0798	12 20 04.	+ 13 59 02.	22	16.8	NEBULA
FATH 1.536	12 20 04.	+ 75 07	8		NEBULA
A3 047	12 20 04.2	+ 14 13 45.		16.9	FAINT BLUE OBJECT
A3 048	12 20 04.4	+ 13 43 11.		16.9	FAINT BLUE OBJECT
A3 049	12 20 04.9	+ 13 43 28.		16.3	FAINT BLUE OBJECT
AMES 0804	12 20 05.	+ 04 52 56.	32	17.8	NEBULA
AMES 0803	12 20 05.	+ 06 48 32.	14	17.6	NEBULA
IC 3225	12 20 05.	+ 06 57 13.			NONSTELLAR OBJECT
AMES 0802	12 20 05.	+ 08 35 32.	58	16.9	NEBULA
AMES 0801	12 20 05.	+ 09 08 44.	31	17.5	NEBULA
IC 3227	12 20 05.	+ 24 21 43.			NONSTELLAR OBJECT
IC 3226	12 20 05.	+ 26 20 37.			NONSTELLAR OBJECT
A3 050	12 20 05.1	+ 14 43 36.		16.9	FAINT BLUE OBJECT
ZWG 042.055	12 20 06.	+ 03 25		15.3	GALAXY
ZWG 042.056	12 20 06.	+ 05 55		15.6	GALAXY
ZWG 042.057	12 20 06.	+ 06 30		15.5	GALAXY
AMES 0805	12 20 06.	+ 06 30 50.	64	16.3	NEBULA
HOLM 386B	12 20 06.	+ 06 57	108	14.5	PART OF MULTIPLE GALAXY
REA 13	12 20 06.	+ 08 12			DWARF GALAXY
ZWG 042.058	12 20 06.	+ 08 14		15.3	GALAXY
ZWG 070.033	12 20 06.	+ 08 29		14.8	GALAXY
MCG+02-32-015	12 20 06.	+ 08 30	24	14.8	GALAXY
ZWG 070.034	12 20 06.	+ 12 05		13.2	GALAXY
UGC 07445	12 20 06.	+ 12 05	210	13.2	GALAXY Sa-b
MCG+02-32-016	12 20 06.	+ 12 05	210	13.2	GALAXY
RMB 175	12 20 06.	+ 12 26			FAINT BLUE OBJECT
LB 04384	12 20 06.	+ 13 28		20.6	FAINT BLUE STAR
LB 04383	12 20 06.	+ 13 40		19.4	FAINT BLUE STAR
LB 04382	12 20 06.	+ 14 00		17.2	FAINT BLUE STAR
LB 04381	12 20 06.	+ 14 19		17.4	FAINT BLUE STAR
ZWG 158.094	12 20 06.	+ 29 43		15.3	GALAXY
ZWG 269.053	12 20 06.	+ 51 27		15.7	GALAXY
FEIG 062	12 20 06.	+ 57 09		13.8	FAINT BLUE STAR
A3 051	12 20 06.1	+ 16 19 24.		16.5	FAINT BLUE OBJECT
REIN 3.097	12 20 06.27	+ 12 04 42.4			NEBULA
HELW 056	12 20 07.	- 18 36 15.			NEBULA
AMES 0806	12 20 08.	+ 08 13 32.	48	16.3	NEBULA
RNGC 4315	12 20 08.	+ 09 35			NON-EXISTENT OBJECT
RNGC 4313	12 20 08.	+ 12 05		13.0	GALAXY
HOLM 385A	12 20 08.	+ 12 05	192	12.0	PART OF MULTIPLE GALAXY
IC 3228	12 20 08.	+ 24 36 25.			NONSTELLAR OBJECT
A3 052	12 20 08.7	+ 15 19 00.		16.7	FAINT BLUE OBJECT
REIZ 2227	12 20 09.	+ 06 57	96	14.9	GALAXY
REIZ 2228	12 20 09.	+ 09 36	90	13.5	GALAXY
IC 3230	12 20 09.	+ 28 01 19.			NONSTELLAR OBJECT
RNGC 4317	12 20 09.	+ 31 19			NON-EXISTENT OBJECT
REIN 3.098A	12 20 09.74	+ 09 36 34.5			NEBULA
REIN 3.098B	12 20 09.75	+ 09 36 32.7			NEBULA
AMES 0808	12 20 10.	+ 05 55 08.	24	16.6	NEBULA
KEEL 422	12 20 10.2	+ 04 24 34.			NEBULA
AMES 0809	12 20 11.	+ 06 07 32.	20	17.4	NEBULA
IC 3229	12 20 11.	+ 06 57 19.			NONSTELLAR OBJECT
HOLM 385B	12 20 11.	+ 12 09	30	15.6	PART OF MULTIPLE GALAXY
AMES 0807	12 20 11.	+ 14 13 44.	47	16.9	NEBULA
ZWG 014.018	12 20 12.	+ 01 34		15.5	GALAXY
ZC 1220.2+0225	12 20 12.	+ 02 25	1480		CLUSTER OF GALAXIES
ZWG 042.059	12 20 12.	+ 08 29		14.1	GALAXY
UGC 07446	12 20 12.	+ 08 29	48	14.1	GALAXY E?
ZWG 070.035	12 20 12.	+ 09 37		14.0	GALAXY
UGC 07447	12 20 12.	+ 09 37	162	14.0	GALAXY Sc?
MCG+02-32-017	12 20 12.	+ 09 37	150	14.0	GALAXY
AMES 0811	12 20 12.	+ 10 16 20.	28	17.0	NEBULA
RMB 181	12 20 12.	+ 11 56			FAINT BLUE OBJECT
8ZW 1220+12.4	12 20 12.	+ 12 25		17.0	COMPACT GALAXY
LB 04388	12 20 12.	+ 12 53		16.8	FAINT BLUE STAR
LB 04387	12 20 12.	+ 13 20		17.3	FAINT BLUE STAR
LB 04386	12 20 12.	+ 13 40		20.8	FAINT BLUE STAR
LB 04385	12 20 12.	+ 14 01		18.4	FAINT BLUE STAR
MCG+09-20-158	12 20 12.	+ 51 26	36	16.	GALAXY
MCG+09-20-159	12 20 12.	+ 55 22 30.	30	16.	GALAXY
ZCG 199.31	12 20 12.	+ 64 11		17.5	COMPACT GALAXY
MCG+13-09-026	12 20 12.	+ 76 27	132	14.	GALAXY
A3 053	12 20 12.9	+ 15 16 34.		17.4	FAINT BLUE OBJECT
AMES 0813	12 20 13.	+ 06 25 20.	14	17.4	NEBULA
AMES 0812	12 20 13.	+ 07 12 38.	14	18.0	NEBULA
AMES 0810	12 20 13.	+ 12 34 20.	30	17.2	NEBULA
IC 3231	12 20 13.	+ 25 05 49.			NONSTELLAR OBJECT
REIZ 2231	12 20 13.	+ 27 49	90	15.0	GALAXY
RNGC 4318	12 20 14.	+ 08 29		14.0	GALAXY
RNGC 4316	12 20 14.	+ 09 37		14.0	GALAXY
PK299-01.1	12 20 14.	- 63 44 47.	25		PLANETARY NEBULA
REIZ 2229	12 20 15.	+ 03 23	18	14.7	GALAXY
AMES 0816	12 20 15.	+ 05 13 56.	22	17.6	NEBULA
AMES 0815	12 20 15.	+ 05 53 14.	17	16.5	NEBULA
REIZ 2230	12 20 15.	+ 08 28	30	14.3	GALAXY
MCG+03-34-020	12 20 15.	+ 17 55	66	15.6	GALAXY
AMES 0814	12 20 16.	+ 12 24 44.	28	17.7	NEBULA
AMES 0817	12 20 17.	+ 12 31 44.	19	17.8	NEBULA
SN 1901B	12 20 17.	+ 16 06		15.6	SUPERNOVA
IC 3232	12 20 17.	+ 24 42 13.			NONSTELLAR OBJECT
A3 055	12 20 17.1	+ 15 47 22.		17.4	FAINT BLUE OBJECT
ZWG 042.060	12 20 18.	+ 03 24		14.9	GALAXY
MCG+01-32-029	12 20 18.	+ 03 24	36	14.9	GALAXY
REA 37	12 20 18.	+ 05 33	30		DWARF GALAXY
ZWG 042.061	12 20 18.	+ 06 57		15.0	GALAXY
UGC 07448	12 20 18.	+ 06 57	66	15.0	GALAXY Sb-c

OBJECT NAME	RIGHT ASCEN.	DECLINATION	DIAM.	MAGN.	TYPE OF OBJECT
MCG+01-32-030	12 20 18.	+ 06 57	60	15.0	GALAXY
AMES 0820	12 20 18.	+ 08 51 08.	30	17.8	NEBULA
LB 04395	12 20 18.	+ 12 27		19.1	FAINT BLUE STAR
LB 04394	12 20 18.	+ 12 47		18.9	FAINT BLUE STAR
AMES 0819	12 20 18.	+ 12 50 50.	18	17.0	NEBULA
LB 04393	12 20 18.	+ 13 13		20.3	FAINT BLUE STAR
RMB 152	12 20 18.	+ 13 15			FAINT BLUE OBJECT
RMB 099	12 20 18.	+ 13 44			FAINT BLUE OBJECT
LB 04392	12 20 18.	+ 13 46		20.6	FAINT BLUE STAR
LB 04391	12 20 18.	+ 13 55		17.5	FAINT BLUE STAR
AMES 0818	12 20 18.	+ 14 07 44.	31	17.1	NEBULA
LB 04390	12 20 18.	+ 14 31		17.5	FAINT BLUE STAR
LB 04389	12 20 18.	+ 14 34		17.8	FAINT BLUE STAR
KARA.68 112	12 20 18.	+ 15 32	27		DWARF GALAXY
KARA.68 113	12 20 18.	+ 15 46	67		DWARF GALAXY
REIZ 2234	12 20 18.	+ 30 04	120	15.0	GALAXY
MCG+09-20-160	12 20 18.	+ 51 24	30	16.	GALAXY
ZWG 269.054	12 20 18.	+ 51 26		15.6	GALAXY
ZCG 199.32	12 20 18.	+ 63 23		19.0	COMPACT GALAXY
7ZW 451	12 20 18.	+ 76 28			COMPACT GALAXY
ZWG 352.031	12 20 18.	+ 76 28		14.8	GALAXY
UGC 07449	12 20 18.	+ 76 28	144	14.8	GALAXY IRR
ZWG 014.019	12 20 18.	- 02 23		15.6	GALAXY
A3 056	12 20 19.8	+ 14 17 36.		16.7	FAINT BLUE OBJECT
A3 057	12 20 19.9	+ 13 44 22.		16.1	FAINT BLUE OBJECT
AMES 0823	12 20 20.	+ 06 06 32.	30	16.8	NEBULA
ABC 1527	12 20 20.	+ 12 30		17.2	RICH CLUSTER OF GALAXIES
AMES 0822	12 20 20.	+ 12 31 08.	11	17.8	NEBULA
HN 0876	12 20 20.	+ 12 51		15.	NEBULA
IC 3233	12 20 20.	+ 12 51			NONSTELLAR OBJECT
AMES 0821	12 20 20.	+ 13 52 08.	35	16.9	NEBULA
RNGC 4323	12 20 20.	+ 16 10			NON-EXISTENT OBJECT
RNGC 4331	12 20 20.	+ 76 28		15.0	GALAXY
REIZ 2233	12 20 21.	+ 06 57	54	15.2	GALAXY
HOLM 386A	12 20 21.	+ 06 57	60	14.1	PART OF MULTIPLE GALAXY
LB 02335	12 20 21.	+ 57 07 30.		13.7	FAINT BLUE STAR
IC 3234	12 20 22.	+ 28 23 20.			NONSTELLAR OBJECT
AMES 0824	12 20 23.	+ 12 34 20.	22	17.4	NEBULA
REIZ 2232	12 20 23.	- 09 02	54	15.4	GALAXY
ZWG 042.062	12 20 24.	+ 03 02		15.3	GALAXY
MCG+01-32-031	12 20 24.	+ 03 02	42	15.3	GALAXY
REA 38	12 20 24.	+ 08 37	48		DWARF GALAXY
AMES 0828	12 20 24.	+ 10 24 44.	29	17.8	NEBULA
AMES 0826	12 20 24.	+ 11 13 14.	28	17.4	NEBULA
AMES 0825	12 20 24.	+ 12 30 20.	14	17.5	NEBULA
LB 04400	12 20 24.	+ 12 32		19.7	FAINT BLUE STAR
LB 04399	12 20 24.	+ 12 48		20.1	FAINT BLUE STAR
LB 04398	12 20 24.	+ 13 57		20.7	FAINT BLUE STAR
LB 04397	12 20 24.	+ 14 08		18.0	FAINT BLUE STAR
LB 04396	12 20 24.	+ 14 18		18.3	FAINT BLUE STAR
REA 39	12 20 24.	+ 15 47	54		DWARF GALAXY
RMB 223	12 20 24.	+ 16 05			FAINT BLUE OBJECT
MCG+03-32-015	12 20 24.	+ 16 05	528	10.6	GALAXY
ZWG 099.030	12 20 24.	+ 16 06		10.6	GALAXY
UGC 07450	12 20 24.	+ 16 06	408	10.6	GALAXY Sc
MRK 438	12 20 24.	+ 22 42	12	16.5	GALAXY WITH UV CONTINUUM
TON-N 0609	12 20 24.	+ 32 41		15.1	BLUE STAR
ZCG 199.33	12 20 24.	+ 64 13		18.5	COMPACT GALAXY
MCG-02-32-007	12 20 24.	- 12 57 30.	48	15.	GALAXY
AMES 0827	12 20 25.	+ 10 44 56.	34	16.8	NEBULA
REIN 3.099	12 20 25.38	+ 10 49 35.2			NEBULA
AMES 0830	12 20 26.	+ 08 18 08.	14	17.6	NEBULA
HN 0877	12 20 26.	+ 13 49		15.	NEBULA
IC 3235	12 20 26.	+ 13 49			NONSTELLAR OBJECT
SN 1914A	12 20 26.	+ 16 04		15.7	SUPERNOVA
RNGC 4321	12 20 26.	+ 16 06		10.5	GALAXY
REIZ 2235	12 20 26.	+ 16 06	330	10.8	GALAXY
HOLM 387A	12 20 26.	+ 16 07	360	10.5	PART OF MULTIPLE GALAXY
LB 02336	12 20 26.	+ 51 28 06.		16.9	FAINT BLUE STAR
RNGC 4332	12 20 26.	+ 66 07		13.0	GALAXY
AMES 0832	12 20 27.	+ 08 01 26.	14	18.0	NEBULA
HN 0878	12 20 27.	+ 10 22		14.	NEBULA
IC 3236	12 20 27.	+ 10 22			NONSTELLAR OBJECT
AMES 0829	12 20 27.	+ 13 04 32.	25	17.6	NEBULA
REIZ 2242	12 20 27.	+ 66 07	60	13.7	GALAXY
AMES 0837	12 20 28.	+ 05 37 44.	19	17.3	NEBULA
AMES 0836	12 20 28.	+ 06 34 20.	18	17.9	NEBULA
AMES 0835	12 20 28.	+ 08 37 50.	22	18.0	NEBULA
AMES 0831	12 20 28.	+ 10 49 32.	17	18.0	NEBULA
SN 1959E	12 20 28.	+ 16 06		17.5	SUPERNOVA
IC 3237	12 20 28.	+ 28 46 14.			NONSTELLAR OBJECT
IC 0785	12 20 28.	- 12 56 59.			NONSTELLAR OBJECT
AMES 0838	12 20 29.	+ 06 04 14.	26	17.7	NEBULA
AMES 0834	12 20 29.	+ 14 03 32.	23	17.2	NEBULA
AMES 0833	12 20 29.	+ 16 48 02.	34	17.2	NEBULA
ARC 1528	12 20 29.	+ 59 11		17.8	RICH CLUSTER OF GALAXIES
KEEL 423	12 20 29.9	+ 16 10 57.		17.	SPIRAL NEBULA
ZWG 042.063	12 20 30.	+ 05 32		12.5	GALAXY
UGC 07451	12 20 30.	+ 05 32	168	12.5	GALAXY S0
MCG+01-32-032	12 20 30.	+ 05 32	96	12.5	GALAXY
ZWG 070.036	12 20 30.	+ 10 49		15.3	GALAXY
8ZW 1220+10.8	12 20 30.	+ 10 49		15.3	COMPACT GALAXY
UGC 07452	12 20 30.	+ 10 49	72	15.3	GALAXY PECULR
MCG+02-32-018	12 20 30.	+ 10 49	42	15.3	GALAXY
LB 04404	12 20 30.	+ 12 47		20.0	FAINT BLUE STAR
LB 04403	12 20 30.	+ 13 42		19.8	FAINT BLUE STAR
LB 04402	12 20 30.	+ 13 50		19.0	FAINT BLUE STAR
LB 04401	12 20 30.	+ 14 06		19.7	FAINT BLUE STAR
MCG+03-32-016	12 20 30.	+ 16 10	54	15.7	GALAXY
ZWG 099.031	12 20 30.	+ 16 11		15.7	GALAXY
MCG+05-29-076	12 20 30.	+ 29 39	54	16.	GALAXY
ZC 1220.5+3610	12 20 30.	+ 36 10	1080		CLUSTER OF GALAXIES
ZCG 199.34	12 20 30.	+ 63 57		18.2	COMPACT GALAXY
MCG+11-15-048	12 20 30.	+ 66 06	102	13.7	GALAXY
ZWG 315.033	12 20 30.	+ 66 07		13.2	GALAXY
UGC 07453	12 20 30.	+ 66 07	144	13.2	GALAXY SBa
ZWG 315.034	12 20 30.	+ 67 08		15.4	GALAXY
HOLM 388B	12 20 31.	+ 05 31	12	14.9	PART OF MULTIPLE GALAXY
RNGC 4345	12 20 31.	+ 75 36			NON-EXISTENT OBJECT
RNGC 4320	12 20 32.	+ 10 49		15.5	GALAXY
RNGC 4322	12 20 32.	+ 16 11		15.5	GALAXY
REIZ 2238	12 20 32.	+ 16 11	18	14.6	GALAXY
A3 060	12 20 32.0	+ 14 20 07.		17.0	FAINT BLUE OBJECT
REIZ 2236	12 20 33.	+ 05 31	90	12.5	GALAXY
HOLM 388A	12 20 33.	+ 05 32	54	12.5	PART OF MULTIPLE GALAXY
AMES 0840	12 20 33.	+ 08 37 32.	24	17.8	NEBULA
REIZ 2237	12 20 33.	+ 10 54	12	14.6	GALAXY
HOLM 387F	12 20 33.	+ 16 12	30	14.7	PART OF MULTIPLE GALAXY

OBJECT NAME	RIGHT ASCEN.	DECLINATION	DIAM.	MAGN.	TYPE OF OBJECT
RNGC 4338	12 20 33.	+ 29 11		15.5	GALAXY
MCG+05-29-077	12 20 33.	+ 29 11	120	15.6	GALAXY
AMES 0839	12 20 34.	+ 12 35 32.	31	17.0	NEBULA
REIN 3.100	12 20 34.14	+ 10 53 56.2			NEBULA
AMES 0841	12 20 35.	+ 09 12 20.	24	17.3	NEBULA
SN 1955E	12 20 35.	+ 58 44		16.3	SUPERNOVA
ZWG 042.064	12 20 36.	+ 06 21		15.1	GALAXY
UGC 07454	12 20 36.	+ 06 21	102	13.1	GALAXY S
MCG+01-32-033	12 20 36.	+ 06 21	72	15.1	GALAXY
KARA.68 114	12 20 36.	+ 08 37	34		DWARF GALAXY
RMB 219	12 20 36.	+ 10 06			FAINT BLUE OBJECT
ZWG 070.037	12 20 36.	+ 10 54		15.0	GALAXY
MCG+02-32-019	12 20 36.	+ 10 54	36	15.0	GALAXY
ZWG 070.038	12 20 36.	+ 12 00		15.6	GALAXY
LB 04411	12 20 36.	+ 12 29		17.5	FAINT BLUE STAR
LB 04410	12 20 36.	+ 12 59		18.5	FAINT BLUE STAR
LB 04409	12 20 36.	+ 13 15		19.6	FAINT BLUE STAR
LB 04408	12 20 36.	+ 13 58		17.2	FAINT BLUE STAR
LB 04407	12 20 36.	+ 14 07		20.5	FAINT BLUE STAR
LB 04406	12 20 36.	+ 14 10		17.7	FAINT BLUE STAR
LB 04405	12 20 36.	+ 14 32		15.8	FAINT BLUE STAR
MCG+03-32-017	12 20 36.	+ 14 42	33	15.2	GALAXY
ZWG 099.032	12 20 36.	+ 14 44		15.2	GALAXY
ZC 1220.6+2115	12 20 36.	+ 21 15	3160		CLUSTER OF GALAXIES
MCG+09-20-161	12 20 36.	+ 51 30	30	16.	GALAXY
MCG+10-18-033	12 20 36.	+ 57 07	27	16.	GALAXY
ZWG 293.015	12 20 36.	+ 58 44		13.7	GALAXY
RNGC 4335	12 20 36.	+ 58 44		13.5	GALAXY
UGC 07455	12 20 36.	+ 58 44	114	13.7	GALAXY E
ZC 1220.6+5909	12 20 36.	+ 59 09	2550		CLUSTER OF GALAXIES
ZCG 199.35	12 20 36.	+ 63 30		15.8	COMPACT GALAXY
ZCG 199.36	12 20 36.	+ 64 13		18.3	COMPACT GALAXY
MCG+11-15-049	12 20 36.	+ 67 07	15	16.	GALAXY
ZWG 315.035	12 20 36.	+ 68 01		15.7	GALAXY
KARA.73B 0526	12 20 36.	+ 68 01	36	15.7	ISOLATED GALAXY S
MCG-02-32-008	12 20 36.	- 12 56 30.	60	15.	GALAXY
RNGC 4324	12 20 37.	+ 05 32		12.5	GALAXY
LB 02337	12 20 37.	+ 52 03 54.		16.1	FAINT BLUE STAR
IC 0786	12 20 37.	- 12 55 52.			NONSTELLAR OBJECT
REIN 3.101	12 20 37.27	+ 12 00 06.9			NEBULA
RNGC 4326	12 20 38.	+ 06 21		15.0	GALAXY
RNGC 4325	12 20 38.	+ 10 54		15.0	GALAXY
HN 0880	12 20 38.	+ 11 59		15.	NEBULA
IC 3239	12 20 38.	+ 11 59			NONSTELLAR OBJECT
AMES 0842	12 20 38.	+ 13 18 26.	16	17.1	NEBULA
HN 0879	12 20 38.	+ 14 43		14.	NEBULA
IC 3238	12 20 38.	+ 14 43			NONSTELLAR OBJECT
RNGC 4327	12 20 38.	+ 16 03			NON-EXISTENT OBJECT
REIZ 2241	12 20 38.	+ 16 05	18	14.4	GALAXY
HOLM 389A	12 20 38.	+ 19 43	60	12.4	PART OF MULTIPLE GALAXY
KEEL 424	12 20 38.3	+ 04 38 14.			NEBULA
REIZ 2239	12 20 39.	+ 03 23	24	15.8	GALAXY
REIZ 2240	12 20 39.	+ 06 21	12	15.1	GALAXY
AMES 0843	12 20 39.	+ 07 05 07.	12	17.6	NEBULA
HN 0881	12 20 39.	+ 10 37		15.	NEBULA
IC 3240	12 20 39.	+ 10 37			NONSTELLAR OBJECT
IC 3241	12 20 39.	+ 27 10 56.			NONSTELLAR OBJECT
A3 063	12 20 39.9	+ 15 09 10.		16.3	FAINT BLUE OBJECT
AMES 0845	12 20 40.	+ 05 48 43.	28	17.2	NEBULA
IC 3242	12 20 40.	+ 26 31 38.			NONSTELLAR OBJECT
A3 064	12 20 40.1	+ 14 39 58.		15.0	FAINT BLUE OBJECT
AMES 0844	12 20 41.	+ 13 05 55.	17	18.0	NEBULA
HOLM 389B	12 20 41.	+ 19 45	36	14.8	PART OF MULTIPLE GALAXY
IC 3243	12 20 41.	+ 28 02 38.			NONSTELLAR OBJECT
A3 066	12 20 41.5	+ 14 55 16.		16.4	FAINT BLUE OBJECT
MRK 050	12 20 42.	+ 02 57	12	15.5	GALAXY WITH UV CONTINUUM
KW 09	12 20 42.	+ 02 57	16		SEYFERT GALAXY
VVI 57	12 20 42.	+ 02 57	12	15.5	SEYFERT GALAXY
ZWG 070.039	12 20 42.	+ 11 39		14.0	GALAXY
UGC 07456	12 20 42.	+ 11 39	270	14.0	GALAXY Sc?
MCG+02-32-020	12 20 42.	+ 11 39	240	14.0	GALAXY
RMB 151	12 20 42.	+ 12 59			FAINT BLUE OBJECT
LB 04413	12 20 42.	+ 13 48		15.6	FAINT BLUE STAR
LB 04412	12 20 42.	+ 14 27		19.2	FAINT BLUE STAR
MCG+03-32-018	12 20 42.	+ 14 38	36	15.3	GALAXY
ZWG 099.033	12 20 42.	+ 14 40		15.3	GALAXY
AMES 0846	12 20 42.	+ 16 23 43.	34	17.2	NEBULA
ZC 1220.7+1709	12 20 42.	+ 17 09	1410		CLUSTER OF GALAXIES
TON-N 0610	12 20 42.	+ 23 27		14.6	BLUE STAR
ZWG 158.095	12 20 42.	+ 28 02		15.7	GALAXY
UGC 07457	12 20 42.	+ 29 38	72	16.0	GALAXY S
REIZ 2248	12 20 42.	+ 58 43	24	13.7	GALAXY
ZCG 199.37	12 20 42.	+ 63 23		19.2	COMPACT GALAXY
ZCG 199.38	12 20 42.	+ 64 14		18.7	COMPACT GALAXY
ZCG 199.39	12 20 42.	+ 64 34		17.4	COMPACT GALAXY
AMES 0847	12 20 43.	+ 10 47 01.	25	17.0	NEBULA
REIN 3.102	12 20 43.97	+ 11 38 38.0			NEBULA
REIZ 2244	12 20 44.	+ 11 38	180	13.3	GALAXY
RNGC 4330	12 20 44.	+ 11 39		14.0	GALAXY
HN 0882	12 20 44.	+ 14 39		14.	NEBULA
IC 3244	12 20 44.	+ 14 39			NONSTELLAR OBJECT
IC 3247	12 20 44.	+ 29 10 14.			NONSTELLAR OBJECT
AMES 0849	12 20 45.	+ 05 53 25.	31	17.0	NEBULA
HN 0883	12 20 45.	+ 09 24	90	13.	NEBULA
IC 3245	12 20 45.	+ 09 24			MAY NOT EXIST
IC 3246	12 20 45.	+ 13 19 44.			NONSTELLAR OBJECT
LB 02338	12 20 45.	+ 54 10 24.		16.3	FAINT BLUE STAR
AMES 0853	12 20 46.	+ 08 39 37.	26	17.8	NEBULA
AMES 0852	12 20 46.	+ 08 40 07.	22	17.4	NEBULA
AMES 0851	12 20 46.	+ 08 58 43.	29	17.6	NEBULA
AMES 0848	12 20 46.	+ 12 11 37.	19	16.8	NEBULA
IC 3248	12 20 46.	+ 25 49 44.			NONSTELLAR OBJECT
AMES 0855	12 20 47.	+ 05 57 37.	20	17.8	NEBULA
AMES 0850	12 20 47.	+ 12 13 43.	23	17.2	NEBULA
REIZ 2243	12 20 47.	- 12 16	42	14.2	GALAXY
ZWG 014.020	12 20 48.	+ 01 46		15.5	GALAXY
ZWG 042.065	12 20 48.	+ 06 19		14.8	GALAXY
MCG+01-32-034	12 20 48.	+ 06 19	48	14.8	GALAXY
ZWG 042.066	12 20 48.	+ 07 45		14.9	GALAXY
UGC 07458	12 20 48.	+ 07 45	138	14.9	GALAXY SB0
MCG+01-32-035	12 20 48.	+ 07 45	108	14.9	GALAXY
AMES 0854	12 20 48.	+ 12 16 19.	17	17.9	NEBULA
LB 04420	12 20 48.	+ 12 27		18.0	FAINT BLUE STAR
LB 04419	12 20 48.	+ 12 28		20.4	FAINT BLUE STAR
LB 04418	12 20 48.	+ 13 02		17.4	FAINT BLUE STAR
LB 04417	12 20 48.	+ 13 04		17.6	FAINT BLUE STAR
LB 04416	12 20 48.	+ 13 07		20.0	FAINT BLUE STAR
LB 04415	12 20 48.	+ 13 21		18.5	FAINT BLUE STAR

OBJECT NAME	RIGHT ASCEN.	DECLINATION	DIAM.	MAGN.	TYPE OF OBJECT
LB 04414	12 20 48.	+ 13 56		18.6	FAINT BLUE STAR
ZWG 099.034	12 20 48.	+ 16 05		15.0	GALAXY
IC 3249	12 20 48.	+ 25 43 26.			NONSTELLAR OBJECT
IC 3250	12 20 48.	+ 25 54 20.			NONSTELLAR OBJECT
ZWG 158.096	12 20 48.	+ 29 10		15.6	GALAXY
REIZ 2247	12 20 48.	+ 29 10	120	14.7	GALAXY
UGC 07459	12 20 48.	+ 29 10	144	15.6	GALAXY Sc
MCG+10-18-034	12 20 48.	+ 57 45	45	16.	GALAXY
MCG+10-18-035	12 20 48.	+ 58 43	30	13.7	GALAXY
ZCG 199.40	12 20 48.	+ 63 47		15.8	COMPACT GALAXY
RNGC 4329	12 20 48.	- 12 15		14.0	GALAXY
MCG-02-32-009	12 20 48.	- 12 18	36	14.5	GALAXY
MRSL 299+01/1	12 20 48.	- 61 07	1200		HII REGION
KEEL 425	12 20 48.2	+ 16 05 51.		15.	SPIRAL NEBULA
AMES 0858	12 20 49.	+ 06 53 19.	14	17.8	NEBULA
IC 3251	12 20 49.	+ 25 55 50.			NONSTELLAR OBJECT
KEEL 426	12 20 49.4	+ 16 08 43.		18.	SPIRAL NEBULA
RNGC 4333	12 20 50.	+ 06 19		15.0	GALAXY
AMES 0857	12 20 50.	+ 11 51 13.	30	16.9	NEBULA
AMES 0856	12 20 50.	+ 15 26 07.	19	16.3	NEBULA
RNGC 4328	12 20 50.	+ 16 05		15.0	GALAXY
REIZ 2245	12 20 51.	+ 06 19	18	14.4	GALAXY
REIZ 2246	12 20 51.	+ 07 45	36	14.9	GALAXY
AMES 0860	12 20 51.	+ 08 49 19.	22	17.8	NEBULA
AMES 0861	12 20 52.	+ 09 35 19.	30	17.8	NEBULA
AMES 0859	12 20 52.	+ 13 13 43.	20	16.4	NEBULA
HOLM 387D	12 20 52.	+ 16 07	30	14.2	PART OF MULTIPLE GALAXY
KEEL 427	12 20 52.5	+ 04 44 01.			NEBULA
A3 068	12 20 53.5	+ 14 42 49.		16.5	FAINT BLUE OBJECT
ZWG 042.067	12 20 54.	+ 06 05		15.7	GALAXY
AMES 0862	12 20 54.	+ 06 05 49.	35	16.6	NEBULA
LB 04421	12 20 54.	+ 13 59		18.5	FAINT BLUE STAR
MCG+03-32-019	12 20 54.	+ 16 05	78	15.0	GALAXY
ZC 1220.9+2028	12 20 54.	+ 20 28	2550		CLUSTER OF GALAXIES
ZC 1220.9+2038	12 20 54.	+ 20 38	2550		CLUSTER OF GALAXIES
ZWG 293.016	12 20 54.	+ 57 45		15.7	GALAXY
UGC 07460	12 20 54.	+ 57 45	72	15.7	GALAXY S
ZCG 199.41	12 20 54.	+ 63 24		18.2	COMPACT GALAXY
IC 3252	12 20 55.	+ 28 53 44.			OPEN CLUSTER
A3 069	12 20 55.0	+ 15 51 13.		16.5	FAINT BLUE OBJECT
RNGC 4334	12 20 56.	+ 07 45		15.0	GALAXY
AMES 0863	12 20 56.	+ 11 02 13.	29	17.6	NEBULA
REIZ 2249	12 20 56.	+ 19 42	24	13.8	GALAXY
TON-N 1516	12 20 57.	+ 34 47		16.6	BLUE STAR
ABC 1529	12 20 57.	+ 61 30		18.0	RICH CLUSTER OF GALAXIES
HOLM 390B	12 20 58.	+ 17 49	30	15.0	PART OF MULTIPLE GALAXY
LB 02339	12 20 58.	+ 53 10 48.		16.8	FAINT BLUE STAR
RNGC 4346	12 20 59.	+ 47 16		12.5	GALAXY
LB 09978	12 21	- 88 26		13.5	FAINT BLUE STAR
ZWG 042.068	12 21 00.	+ 06 21		13.1	GALAXY
UGC 07461	12 21 00.	+ 06 21	138	13.1	GALAXY E
MCG+01-32-036	12 21 00.	+ 06 21	90	13.1	GALAXY
AMES 0864	12 21 00.	+ 08 39 07.	17	17.4	NEBULA
LB 04426	12 21 00.	+ 12 34		20.6	FAINT BLUE STAR
LB 04425	12 21 00.	+ 13 12		16.0	FAINT BLUE STAR
LB 04424	12 21 00.	+ 13 38		19.8	FAINT BLUE STAR
LB 04423	12 21 00.	+ 13 58		15.5	FAINT BLUE STAR
LB 04422	12 21 00.	+ 14 28		20.8	FAINT BLUE STAR
ZWG 099.035	12 21 00.	+ 19 43		13.6	GALAXY
UGC 07462	12 21 00.	+ 19 43	96	13.6	GALAXY SB0/SBa
MCG+03-32-020	12 21 00.	+ 19 43	138	13.6	GALAXY
ZC 1221.0+2217	12 21 00.	+ 22 17	3830		CLUSTER OF GALAXIES
ZWG 244.009	12 21 00.	+ 47 16		12.3	GALAXY
UGC 07463	12 21 00.	+ 47 16	192	12.3	GALAXY S0
MCG+08-23-016	12 21 00.	+ 47 17	102	12.4	GALAXY
ZC 1221.0+4915	12 21 00.	+ 49 15	1550		CLUSTER OF GALAXIES
MCG+09-20-162	12 21 00.	+ 55 11	30	16.	GALAXY
MCG+09-20-163	12 21 00.	+ 55 57 30.	36	16.	GALAXY
ZC 1221.0+5759	12 21 00.	+ 57 59	1480		CLUSTER OF GALAXIES
ZC 1221.0+6130	12 21 00.	+ 61 30	610		CLUSTER OF GALAXIES
MCG+14-06-005	12 21 00.	+ 82 57	51	15.	GALAXY
ZWG 014.021	12 21 00.	- 00 11		15.3	GALAXY
IC 3254	12 21 01.	+ 19 44			SAME AS NGC 4336
A3 070	12 21 01.5	+ 13 57 46.		16.5	FAINT BLUE OBJECT
RNGC 4339	12 21 02.	+ 06 22		12.5	GALAXY
AMES 0867	12 21 02.	+ 10 59 31.	29	17.5	NEBULA
AMES 0866	12 21 02.	+ 13 15 25.	26	17.6	NEBULA
AMES 0865	12 21 02.	+ 13 17 13.	19	17.7	NEBULA
REIZ 2251	12 21 02.	+ 17 00	132	12.0	GALAXY
REIZ 2254	12 21 02.	+ 19 37	48	13.2	GALAXY
RNGC 4336	12 21 02.	+ 19 43		13.5	GALAXY
HN 0884	12 21 02.	+ 19 44	24		NEBULA
REIN 3.103A	12 21 02.14	+ 09 55 34.2			NEBULA
REIN 3.103B	12 21 02.25	+ 09 55 32.5			NEBULA
REIZ 2250	12 21 03.	+ 06 21	60	13.0	GALAXY
REIZ 2252	12 21 03.	+ 07 15	120	13.1	GALAXY
REIZ 2253	12 21 03.	+ 07 21	42	12.1	GALAXY
AMES 0870	12 21 03.	+ 07 52 31.	20	18.0	NEBULA
HN 0885	12 21 03.	+ 09 55		13.	NEBULA
IC 3255	12 21 03.	+ 09 55			NONSTELLAR OBJECT
AMES 0869	12 21 03.	+ 13 10 43.	40	17.3	NEBULA
AMES 0868	12 21 03.	+ 16 48 31.	35	17.6	NEBULA
HOLM 391B	12 21 03.	+ 17 02	150	12.8	PART OF MULTIPLE GALAXY
REIZ 2259	12 21 03.	+ 47 16	126	12.8	GALAXY
A3 071	12 21 03.0	+ 15 21 57.		16.4	FAINT BLUE OBJECT
LB 02340	12 21 04.	+ 49 18 42.		15.1	FAINT BLUE STAR
AMES 0872	12 21 05.	+ 07 48 01.	19	17.8	NEBULA
PK299+02.1	12 21 05.	- 59 57	30		PLANETARY NEBULA
ZWG 014.022	12 21 06.	+ 01 37		15.5	GALAXY
ZWG 042.069	12 21 06.	+ 03 14		15.2	GALAXY
UGC 07464	12 21 06.	+ 03 14	96	15.2	GALAXY Sb
MCG+01-32-037	12 21 06.	+ 03 14	78	15.2	GALAXY
ZWG 042.070	12 21 06.	+ 07 14		13.5	GALAXY
UGC 07465	12 21 06.	+ 07 14	156	13.5	GALAXY Sb
MCG+01-32-038	12 21 06.	+ 07 14	156	13.5	GALAXY
ZWG 042.071	12 21 06.	+ 07 20		13.0	GALAXY
UGC 07466	12 21 06.	+ 07 20	66	13.0	GALAXY S?
MCG+01-32-039	12 21 06.	+ 07 20	60	13.0	GALAXY
IC 3256	12 21 06.	+ 07 20		13.0	GALAXY IN VIRGO CLSTR
ZWG 070.040	12 21 06.	+ 09 54		15.3	GALAXY
8ZW 1221+09.9	12 21 06.	+ 09 54		15.3	COMPACT GALAXY
RMB 176	12 21 06.	+ 12 22			FAINT BLUE OBJECT
LB 04431	12 21 06.	+ 13 00		20.1	FAINT BLUE STAR
LB 04430	12 21 06.	+ 13 37		18.5	FAINT BLUE STAR
LB 04429	12 21 06.	+ 14 06		18.4	FAINT BLUE STAR
LB 04428	12 21 06.	+ 14 14		16.7	FAINT BLUE STAR
LB 04427	12 21 06.	+ 14 16		17.3	FAINT BLUE STAR
AMES 0871	12 21 06.	+ 16 38 43.	26	17.6	NEBULA

OBJECT NAME	RIGHT ASCEN.	DECLINATION	DIAM.	MAGN.	TYPE OF OBJECT
ZWG 099.036	12 21 06.	+ 17 00		12.4	GALAXY
SCH 36	12 21 06.	+ 17 00		12.4	PECULIAR GALAXY
UGC 07467	12 21 06.	+ 17 00	216	12.4	GALAXY SB0
MCG+03-32-021	12 21 06.	+ 17 00	168	12.4	GALAXY
ZWG 099.037	12 21 06.	+ 17 49		13.7	GALAXY
UGC 07468	12 21 06.	+ 17 49	84	13.7	GALAXY S0?
ZC 1221.1+2359	12 21 06.	+ 23 59	1010		CLUSTER OF GALAXIES
MCG+11-15-050	12 21 06.	+ 66 51	27	17.	GALAXY
IC 3253	12 21 06.	- 34 21	150	12.3	GALAXY S
KEEL 428	12 21 06.3	+ 18 38 07.		18.	NEBULA
IC 3257	12 21 07.	+ 07 33			MAY NOT EXIST
AMES 0875	12 21 07.	+ 08 48 55.	19	17.5	NEBULA
AMES 0874	12 21 07.	+ 10 14 07.	28	17.2	NEBULA
A3 073	12 21 07.5	+ 14 20 41.		16.5	FAINT BLUE OBJECT
RNGC 4343	12 21 08.	+ 07 14		13.5	GALAXY
RNGC 4342	12 21 08.	+ 07 20		13.5	GALAXY
AMES 0877	12 21 08.	+ 09 10 31.	26	17.5	NEBULA
AMES 0873	12 21 08.	+ 10 57 55.	20	17.6	NEBULA
REIZ 2257	12 21 08.	+ 12 45	48	14.8	GALAXY
RNGC 4340	12 21 08.	+ 17 00		12.5	GALAXY
RNGC 4344	12 21 08.	+ 17 49		13.5	GALAXY
REIZ 2258	12 21 08.	+ 17 49	24	13.3	GALAXY
HOLM 390A	12 21 08.	+ 17 50	72	13.5	PART OF MULTIPLE GALAXY
HN 0363	12 21 08.	- 34 21			NEBULA
REIZ 2255	12 21 09.	+ 03 14	72	14.2	GALAXY
REIZ 2256	12 21 09.	+ 03 55	48	14.9	GALAXY
AMES 0879	12 21 09.	+ 07 03 31.	19	17.6	NEBULA
AMES 0878	12 21 09.	+ 07 08 55.	13	17.6	NEBULA
AMES 0876	12 21 09.	+ 12 11 01.	26	17.7	NEBULA
MCG+03-32-022	12 21 09.	+ 17 49	90	13.7	GALAXY
TON-N 1517	12 21 09.	+ 34 39		15.2	BLUE STAR
REIZ 2264	12 21 09.	+ 64 40	15	14.3	GALAXY
MCG-06-27-021	12 21 09.	- 34 19 30.	150	12.	GALAXY
LB 02341	12 21 10.	+ 56 06 06.		14.7	FAINT BLUE STAR
HELF 149	12 21 10.	- 18 27 45.			NEBULA
REIN 3.105	12 21 11.87	+ 12 45 15.6			NEBULA
REIN 3.104A	12 21 11.99	+ 09 36 53.5			NEBULA
AMES 0880	12 21 12.	+ 04 54 55.	28	17.5	NEBULA
ZWG 042.072	12 21 12.	+ 07 28		14.7	GALAXY
UGC 07469	12 21 12.	+ 07 28	108	14.7	GALAXY
MCG+01-32-040	12 21 12.	+ 07 28	96	14.7	GALAXY
ZWG 070.041	12 21 12.	+ 09 36		15.2	GALAXY
MCG+02-32-022	12 21 12.	+ 09 36	48	15.2	GALAXY
RMB 180	12 21 12.	+ 12 02			FAINT BLUE OBJECT
LB 04434	12 21 12.	+ 12 32		17.8	FAINT BLUE STAR
LB 04433	12 21 12.	+ 12 33		18.1	FAINT BLUE STAR
ZWG 070.042	12 21 12.	+ 12 45		14.3	GALAXY
SCH 37	12 21 12.	+ 12 45		14.3	PECULIAR GALAXY
UGC 07470	12 21 12.	+ 12 45	114	14.3	GALAXY IRR
MCG+02-32-021	12 21 12.	+ 12 45	78	14.3	GALAXY
LB 04432	12 21 12.	+ 13 14		20.4	FAINT BLUE STAR
ZC 1221.2+2317	12 21 12.	+ 23 17	870		CLUSTER OF GALAXIES
UGC 07471	12 21 12.	+ 32 02	72	16.5	GALAXY S
ZWG 187.040	12 21 12.	+ 36 25		15.5	GALAXY
MCG+06-27-052	12 21 12.	+ 36 26	36	15.	GALAXY
MCG+10-18-036	12 21 12.	+ 58 43	30	16.	GALAXY
ZWG 352.032	12 21 12.	+ 75 14		14.6	GALAXY
OCL 0878	12 21 12.	- 57 51	540	10.0	OPEN STAR CLUSTER
VHA 129	12 21 12.	- 57 51	360		OPEN STAR CLUSTER
MRSL 299+00/1	12 21 12.	- 61 37	1080		HII REGION
KEEL 429	12 21 12.1	+ 18 33 03.		18.	NEBULA
REIN 3.104B	12 21 12.10	+ 09 36 53.9			NEBULA
IC 3258	12 21 12.2	+ 12 45 19.		14.0	GALAXY Sc
AMES 0881	12 21 13.	+ 07 00 55.	12	18.0	NEBULA
BC 4C18.34	12 21 13.8	+ 18 37 30.		18.74	QUASI-STELLAR OBJECT
AMES 0882	12 21 14.	+ 07 01 07.	29	17.9	NEBULA
RNGC 4337	12 21 14.	- 57 50		10.0	OPEN CLUSTER
SBB 202	12 21 14.7	+ 18 37 43.		18.8	QUASI-STELLAR OBJECT
AFC 1530	12 21 15.	+ 02 22		17.8	RICH CLUSTER OF GALAXIES
SVEF 290	12 21 15.	+ 04 22	30	15.4	GALAXY
REIZ 2260	12 21 15.	+ 07 28	48	14.5	GALAXY
TON-N 1518	12 21 15.	+ 22 27		17.0	BLUE STAR
AMES 0885	12 21 16.	+ 06 20 31.	24	17.3	NEBULA
AMES 0884	12 21 16.	+ 09 01 13.	31	16.5	NEBULA
AMES 0883	12 21 16.	+ 13 00 25.	19	17.4	NEBULA
IC 3259	12 21 16.3	+ 07 27 58.			GALAXY
AMES 0886	12 21 17.	+ 07 43 43.	12	17.5	NEBULA
ZWG 014.024	12 21 18.	+ 01 40		15.6	GALAXY
ZWG 042.073	12 21 18.	+ 03 21		15.4	GALAXY
ZWG 042.074	12 21 18.	+ 04 22		15.3	GALAXY
RMB 218	12 21 18.	+ 10 45			FAINT BLUE OBJECT
LB 04442	12 21 18.	+ 12 44		18.8	FAINT BLUE STAR
LB 04441	12 21 18.	+ 12 57		20.5	FAINT BLUE STAR
LB 04440	12 21 18.	+ 13 28		18.8	FAINT BLUE STAR
LB 04439	12 21 18.	+ 13 36		20.0	FAINT BLUE STAR
LB 04438	12 21 18.	+ 13 54		20.8	FAINT BLUE STAR
LB 04437	12 21 18.	+ 14 29		19.0	FAINT BLUE STAR
LB 04436	12 21 18.	+ 14 30		17.3	FAINT BLUE STAR
LB 04435	12 21 18.	+ 14 32		16.4	FAINT BLUE STAR
PEA 40	12 21 18.	+ 15 24	36		DWARF GALAXY
KARA.68 115	12 21 18.	+ 17 05	40		DWARF GALAXY
ZWG 158.097	12 21 18.	+ 27 40		15.3	GALAXY
IC 3262	12 21 18.	+ 27 40 08.			OPEN CLUSTER
ZWG 158.098	12 21 18.	+ 28 29		15.1	GALAXY
MCG+05-29-078	12 21 18.	+ 28 29	36	15.1	GALAXY
ZC 1221.3+7950	12 21 18.	+ 79 50	1010		CLUSTER OF GALAXIES
ZWG 014.023	12 21 18.	- 03 09		13.6	GALAXY
MCG+00-32-003	12 21 18.	- 03 09	180	13.6	GALAXY
KARA.73B 0527	12 21 18.	- 03 09	210	13.6	ISOLATED GALAXY S
AMES 0888	12 21 19.	+ 05 41 13.	17	17.2	NEBULA
AMES 0887	12 21 19.	+ 08 00 19.	16	17.2	NEBULA
RNGC 4363	12 21 19.	+ 75 14		14.5	GALAXY
RNGC 4347	12 21 19.	- 02 58			NON-EXISTENT OBJECT
RNGC 4348	12 21 19.	- 03 10		13.0	GALAXY
HN 0886	12 21 20.	+ 11 45	60		NEBULA
IC 3261	12 21 20.	+ 11 45			NONSTELLAR OBJECT
IC 3263	12 21 20.	+ 28 28 38.			NONSTELLAR OBJECT
REIZ 2263	12 21 21.	+ 04 21	18	15.8	GALAXY
REIZ 2261	12 21 21.	+ 07 23	36	14.3	GALAXY
AMES 0889	12 21 21.	+ 11 45 31.	24	17.2	NEBULA
A3 076	12 21 21.4	+ 15 04 03.		16.5	FAINT BLUE OBJECT
IC 3264	12 21 22.	+ 25 50 08.			NONSTELLAR OBJECT
LB 02342	12 21 22.	+ 57 56 36.		13.2	FAINT BLUE STAR
REIZ 2262	12 21 22.	- 03 11	108	12.9	GALAXY
PATH 2.107	12 21 23.	+ 75 14	27		NEBULA
ZWG 042.075	12 21 24.	+ 05 27		15.3	GALAXY
MCG+01-32-041	12 21 24.	+ 05 27	48	15.3	GALAXY
ZWG 042.076	12 21 24.	+ 07 23		14.5	GALAXY
UGC 07472	12 21 24.	+ 07 23	96	14.5	GALAXY S0
MCG+01-32-042	12 21 24.	+ 07 23	60	14.5	GALAXY
IC 3260	12 21 24.	+ 07 23		14.5	GALAXY IN VIRGO CLSTR
ZWG 042.077	12 21 24.	+ 08 04		14.6	GALAXY
MCG+01-32-043	12 21 24.	+ 08 04	48	14.6	GALAXY
ZWG 070.043	12 21 24.	+ 09 31		15.5	GALAXY
RMB 177	12 21 24.	+ 12 30			FAINT BLUE OBJECT
LB 04448	12 21 24.	+ 12 42		17.9	FAINT BLUE STAR
LB 04447	12 21 24.	+ 13 30		18.0	FAINT BLUE STAR
LB 04446	12 21 24.	+ 13 38		19.2	FAINT BLUE STAR
LB 04445	12 21 24.	+ 14 04		17.0	FAINT BLUE STAR
LB 04444	12 21 24.	+ 14 23		18.5	FAINT BLUE STAR
LB 04443	12 21 24.	+ 14 31		18.8	FAINT BLUE STAR
RMB 048	12 21 24.	+ 15 04			FAINT BLUE OBJECT
ZWG 099.038	12 21 24.	+ 16 58		11.5	GALAXY
UGC 07473	12 21 24.	+ 16 58	162	11.5	GALAXY S0
ZC 1221.4+2411	12 21 24.	+ 24 11	5780		CLUSTER OF GALAXIES
FEIG 063	12 21 24.	+ 57 57		13.5	FAINT BLUE STAR
AMES 0891	12 21 25.	+ 06 35 37.	17	17.8	NEBULA
RNGC 4349	12 21 25.	- 61 37		8.0	OPEN CLUSTER
AMES 0895	12 21 26.	+ 05 35 19.	41	16.6	NEBULA
RNGC 4341	12 21 26.	+ 07 23		14.5	GALAXY
AMES 0894	12 21 26.	+ 07 43 43.	23	17.8	NEBULA
IC 3265	12 21 26.	+ 08 04 50.			MAY NOT EXIST
AMES 0893	12 21 26.	+ 08 14 07.	20	17.8	NEBULA
AMES 0892	12 21 26.	+ 08 26 07.	28	17.0	NEBULA
AMES 0890	12 21 26.	+ 13 10 13.	24	17.4	NEBULA
RNGC 4350	12 21 26.	+ 16 58		12.0	GALAXY
HOLM 391A	12 21 26.	+ 17 00	138	12.0	PART OF MULTIPLE GALAXY
REIZ 2265	12 21 27.	+ 05 35	24	16.0	GALAXY
IC 3266	12 21 27.	+ 08 03 44.			SAME AS NGC 4353
MCG+03-32-023	12 21 27.	+ 16 58	138	11.5	GALAXY
REIZ 2272	12 21 27.	+ 49 03	78	13.4	GALAXY
REIN 3.106	12 21 27.62	+ 08 03 42.0			NEBULA
AMES 0898	12 21 28.	+ 05 27 31.	48	16.3	NEBULA
AMES 0897	12 21 28.	+ 09 12 31.	29	17.5	NEBULA
AMES 0899	12 21 29.	+ 05 48 31.	28	17.4	NEBULA
AMES 0896	12 21 29.	+ 12 18 07.	22	18.0	NEBULA
REIN 3.107	12 21 29.28	+ 12 28 57.2			NEBULA
ZWG 042.078	12 21 30.	+ 05 35		15.5	GALAXY
AMES 0901	12 21 30.	+ 07 10 25.	16	17.8	NEBULA
ZWG 042.079	12 21 30.	+ 07 19		14.6	GALAXY
UGC 07474	12 21 30.	+ 07 19	66	14.6	GALAXY S
MCG+01-32-044	12 21 30.	+ 07 19	48	14.6	GALAXY
KARA.68 116	12 21 30.	+ 10 20	27		DWARF GALAXY
ZWG 070.044	12 21 30.	+ 11 30		14.0	GALAXY
UGC 07475	12 21 30.	+ 11 30	96	14.0	GALAXY S0
MCG+02-32-023	12 21 30.	+ 11 30	72	14.0	GALAXY
RMB 179	12 21 30.	+ 12 04			FAINT BLUE OBJECT
ZWG 070.045	12 21 30.	+ 12 29		13.5	GALAXY
SCH 38	12 21 30.	+ 12 29		13.5	PECULIAR GALAXY
UGC 07476	12 21 30.	+ 12 29	114	13.5	GALAXY S
MCG+02-32-024	12 21 30.	+ 12 29	102	13.5	GALAXY
LB 04450	12 21 30.	+ 12 52		18.0	FAINT BLUE STAR
RMB 150	12 21 30.	+ 13 05			FAINT BLUE OBJECT
LB 04449	12 21 30.	+ 14 25		20.0	FAINT BLUE STAR
ZC 1221.5+1920	12 21 30.	+ 19 20	1340		CLUSTER OF GALAXIES
MCG+05-29-079	12 21 30.	+ 31 49	198	13.9	GALAXY
MCG+08-23-017	12 21 30.	+ 49 04	210	13.4	GALAXY
REIZ 2275	12 21 30.	+ 58 39	12	14.7	GALAXY
AMES 0903	12 21 31.	+ 08 42 01.	23	17.8	NEBULA
AMES 0900	12 21 31.	+ 10 22 19.	60	17.1	NEBULA
RNGC 4353	12 21 32.	+ 08 04		14.5	GALAXY
AMES 0905	12 21 32.	+ 08 35 01.	22	17.2	NEBULA
REIZ 2270	12 21 32.	+ 11 29	60	13.4	GALAXY
RNGC 4352	12 21 32.	+ 11 30		14.5	GALAXY
RNGC 4354	12 21 32.	+ 12 28			NON-EXISTENT OBJECT
RNGC 4351	12 21 32.	+ 12 29		13.5	GALAXY
REIZ 2269	12 21 32.	+ 12 29	54	13.3	GALAXY
AMES 0902	12 21 32.	+ 15 13 43.	24	17.2	NEBULA
REIZ 2268	12 21 32.	+ 16 58	120	11.5	GALAXY
LB 02343	12 21 32.	+ 49 06 54.		15.7	FAINT BLUE STAR
A3 080	12 21 32.5	+ 15 09 44.		17.0	FAINT BLUE OBJECT
REIN 3.108	12 21 32.71	+ 11 29 43.5			NEBULA
IC 3267	12 21 32.8	+ 07 19 10.			GALAXY
A3 081	12 21 32.8	+ 15 14 10.		17.5	FAINT BLUE OBJECT
REIZ 2266	12 21 33.	+ 05 27	24	15.8	GALAXY
AMES 0907	12 21 33.	+ 06 20 43.	16	18.0	NEBULA
REIZ 2267	12 21 33.	+ 07 19	42	14.7	GALAXY
REIZ 2271	12 21 33.	+ 08 05	60	14.0	GALAXY
AMES 0904	12 21 33.	+ 16 36 19.	23	16.6	NEBULA
MCG+07-26-003	12 21 33.	+ 42 35	27	17.	GALAXY
REIN 7.085	12 21 33.40	+ 46 13 16.1			NEBULA
AMES 0906	12 21 34.	+ 12 00 37.	18	17.1	NEBULA
IC 3269	12 21 34.	+ 27 42 26.			NONSTELLAR OBJECT
AMES 0910	12 21 35.	+ 05 38 07.	28	17.6	NEBULA
AMES 0909	12 21 35.	+ 10 39 07.	37	17.7	NEBULA
AMES 0908	12 21 35.	+ 13 34 55.	31	17.8	NEBULA
RNGC 4357	12 21 35.	+ 49 03		13.5	GALAXY
REIN 3.109A	12 21 35.16	+ 09 39 38.8			NEBULA
REIN 3.109B	12 21 35.23	+ 09 39 37.1			NEBULA
A3 084	12 21 35.8	+ 15 14 21.		18.0	FAINT BLUE OBJECT
MCG+00-32-004	12 21 36.	+ 00 51	60	16.	GALAXY
ZWG 014.042	12 21 36.	+ 01 29		15.3	GALAXY
2ZW 059	12 21 36.	+ 04 49			COMPACT GALAXY
ZWG 042.080	12 21 36.	+ 06 53		14.2	GALAXY
SCH 39	12 21 36.	+ 06 53		14.2	PECULIAR GALAXY
UGC 07477	12 21 36.	+ 06 53	42	14.2	GALAXY PECULR
MCG+01-32-045	12 21 36.	+ 06 53	42	14.2	GALAXY
IC 3268	12 21 36.	+ 06 53		14.2	GALAXY IN VIRGO CLSTR
ZWG 070.046	12 21 36.	+ 09 39		15.2	GALAXY
MCG+02-32-025	12 21 36.	+ 09 39	24	15.2	GALAXY
ZWG 070.047	12 21 36.	+ 09 45		15.6	GALAXY
LB 04453	12 21 36.	+ 12 27		17.4	FAINT BLUE STAR
RMB 149	12 21 36.	+ 13 11			FAINT BLUE OBJECT
LB 04452	12 21 36.	+ 13 11		17.8	FAINT BLUE STAR
LB 04451	12 21 36.	+ 14 00		17.4	FAINT BLUE STAR
ZC 1221.6+1620	12 21 36.	+ 16 20	1480		CLUSTER OF GALAXIES
MCG+04-29-069	12 21 36.	+ 20 32 30.	36	15.	GALAXY
IC 3270	12 21 36.	+ 27 51 20.			NONSTELLAR OBJECT
ZWG 244.010	12 21 36.	+ 49 03		13.5	GALAXY
UGC 07478	12 21 36.	+ 49 03	234	13.5	GALAXY Sb
KARA.73B 0528	12 21 36.	+ 49 03	222	13.5	ISOLATED GALAXY S
REIZ 2277	12 21 36.	+ 58 38	210	14.8	GALAXY
ZWG 293.017	12 21 36.	+ 58 40		14.3	GALAXY
RNGC 4364	12 21 36.	+ 58 40		14.5	GALAXY
RNGC 4358	12 21 36.	+ 58 40		14.5	GALAXY
UGC 07479	12 21 36.	+ 58 40	78	14.3	GALAXY S0

OBJECT NAME	RIGHT ASCEN.	DECLINATION	DIAM.	MAGN.	TYPE OF OBJECT
UGC 07480	12 21 36.	+ 71 34	78	17.	GALAXY Sc
AMES 0911	12 21 37.	+ 12 23 07.	23	17.9	NEBULA
RNGC 4355	12 21 37.	- 00 44			NON-EXISTENT OBJECT
HOLM 393B	12 21 38.	+ 09 31	24	14.1	PART OF MULTIPLE GALAXY
A3 085	12 21 38.0	+ 14 27 11.		17.3	FAINT BLUE OBJECT
REIZ 2273	12 21 39.	+ 06 53	36	13.6	GALAXY
HOLM 392A	12 21 39.	+ 06 53	30	13.4	PART OF MULTIPLE GALAXY
HOLM 392B	12 21 39.	+ 06 54	12	15.0	PART OF MULTIPLE GALAXY
REIZ 2274	12 21 39.	+ 08 48	180	13.9	GALAXY
AMES 0912	12 21 39.	+ 09 31 01.	28	16.9	NEBULA
IC 3272	12 21 39.	+ 23 34 08.			NONSTELLAR OBJECT
RNGC 4359	12 21 39.	+ 31 48		14.0	GALAXY
A3 086	12 21 40.8	+ 16 11 19.		17.1	FAINT BLUE OBJECT
IC 3271	12 21 41.	+ 08 13 50.			NONSTELLAR OBJECT
ARC 1531	12 21 41.	+ 57 59		17.8	RICH CLUSTER OF GALAXIES
A3 087	12 21 41.2	+ 15 27 49.		16.5	FAINT BLUE OBJECT
REIN 3.110	12 21 41.23	+ 08 13 51.9			NEBULA
ZWG 014.044	12 21 42.	+ 01 30		15.4	GALAXY
MCG+00-32-005	12 21 42.	+ 01 30	48	15.4	GALAXY
ZWG 014.025	12 21 42.	+ 01 41		15.6	GALAXY
ZWG 042.081	12 21 42.	+ 04 30		15.2	GALAXY
MRK 051	12 21 42.	+ 04 30	20	15.	GALAXY WITH UV CONTINUUM
MCG+01-32-046	12 21 42.	+ 04 30	48	15.2	GALAXY
AMES 0913	12 21 42.	+ 05 01 19.	24	17.0	NEBULA
ZWG 042.082	12 21 42.	+ 08 14		15.0	GALAXY
UGC 07481	12 21 42.	+ 08 14	66	15.0	GALAXY Sc
MCG+01-32-047	12 21 42.	+ 08 14	60	15.0	GALAXY
ZWG 070.048	12 21 42.	+ 08 49		14.3	GALAXY
UGC 07482	12 21 42.	+ 08 49	156	14.3	GALAXY Sc
MCG+02-32-026	12 21 42.	+ 08 49	180	14.3	GALAXY
ZWG 070.049	12 21 42.	+ 09 29		15.7	GALAXY
ZWG 070.050	12 21 42.	+ 09 33		15.2	GALAXY
MCG+02-32-027	12 21 42.	+ 09 33	18	15.2	GALAXY
ZWG 070.051	12 21 42.	+ 09 37		15.3	GALAXY
LB 04458	12 21 42.	+ 12 57		19.7	FAINT BLUE STAR
LB 04457	12 21 42.	+ 13 28		18.7	FAINT BLUE STAR
LB 04456	12 21 42.	+ 13 42		18.9	FAINT BLUE STAR
LB 04455	12 21 42.	+ 13 48		17.2	FAINT BLUE STAR
LB 04454	12 21 42.	+ 14 18		19.8	FAINT BLUE STAR
ZWG 158.099	12 21 42.	+ 31 47		13.9	GALAXY
UGC 07483	12 21 42.	+ 31 47	216	13.9	GALAXY S
MCG+10-18-038	12 21 42.	+ 58 40	45	14.	GALAXY
MCG+10-18-037	12 21 42.	+ 58 40	15	14.7	GALAXY
ZCG 199.42	12 21 42.	+ 63 30		18.0	COMPACT GALAXY
ZCG 199.43	12 21 42.	+ 63 33		17.8	COMPACT GALAXY
OCL 0882	12 21 42.	- 61 37	1740	8.1	OPEN STAR CLUSTER
VHA 130	12 21 42.	- 61 37	720		OPEN STAR CLUSTER
REIN 3.111A	12 21 42.12	+ 09 32 38.4			NEBULA
REIN 3.111B	12 21 42.16	+ 09 32 37.9			NEBULA
REIN 3.112	12 21 42.35	+ 08 48 50.9			NEBULA
IC 3273	12 21 43.	+ 08 48 56.			SAME AS NGC 4356
IC 3274	12 21 43.	+ 09 32 38.			NONSTELLAR OBJECT
IC 3276	12 21 43.	+ 26 05 44.			NONSTELLAR OBJECT
RNGC 4356	12 21 44.	+ 08 49		14.5	GALAXY
RNGC 4360B	12 21 44.	+ 09 33		15.0	GALAXY
AMES 0914	12 21 44.	+ 11 47 19.	19	17.9	NEBULA
LB 02344	12 21 44.	+ 50 17 06.		17.2	FAINT BLUE STAR
ARC 1534	12 21 44.	+ 61 47		17.0	RICH CLUSTER OF GALAXIES
REIZ 2276	12 21 45.	+ 08 13	54	14.7	GALAXY
HOLM 393A	12 21 45.	+ 09 33	36	13.0	PART OF MULTIPLE GALAXY
AMES 0916	12 21 45.	+ 09 46 13.	16	16.6	NEBULA
HN 0887	12 21 45.	+ 10 42		14.5	NEBULA
IC 3275	12 21 45.	+ 10 42			NONSTELLAR OBJECT
AMES 0915	12 21 45.	+ 11 16 49.	22	17.7	NEBULA
IC 3278	12 21 45.	+ 27 42 02.			NONSTELLAR OBJECT
MCG+09-20-164	12 21 45.	+ 51 17	36	16.	GALAXY
REIN 3.113A	12 21 45.05	+ 09 36 40.4			NEBULA
REIN 3.113B	12 21 45.17	+ 09 36 40.0			NEBULA
IC 3277	12 21 46.	+ 25 50 26.			NONSTELLAR OBJECT
HELW 057	12 21 47.	- 18 25 08.			NEBULA
BC M1221+114	12 21 47.8	+ 11 24 01.		18.	QUASI-STELLAR OBJECT
SHB 203	12 21 47.8	+ 11 24 01.		18.	QUASI-STELLAR OBJECT
2ZW 060	12 21 48.	+ 05 21			COMPACT GALAXY
AMES 0918	12 21 48.	+ 06 34 43.	24	17.7	NEBULA
ZWG 070.052	12 21 48.	+ 09 34		13.9	GALAXY
UGC 07484	12 21 48.	+ 09 34	84	13.9	GALAXY E
MCG+02-32-028	12 21 48.	+ 09 34	18	13.9	GALAXY
AMES 0917	12 21 48.	+ 11 52 07.	25	17.6	NEBULA
ZWG 070.053	12 21 48.	+ 13 19		15.6	GALAXY
MCG+02-32-029	12 21 48.	+ 13 19	15	15.6	GALAXY
LB 04462	12 21 48.	+ 13 47		19.1	FAINT BLUE STAR
LB 04461	12 21 48.	+ 14 12		18.7	FAINT BLUE STAR
LB 04460	12 21 48.	+ 14 19		17.6	FAINT BLUE STAR
LB 04459	12 21 48.	+ 14 29		20.8	FAINT BLUE STAR
ZWG 128.085	12 21 48.	+ 21 26		15.6	GALAXY
UGC 07485	12 21 48.	+ 21 26	60	15.6	GALAXY
REIZ 2279	12 21 48.	+ 31 48	120	13.1	GALAXY
ZWG 187.041	12 21 48.	+ 33 28		15.4	GALAXY
UGC 07486	12 21 48.	+ 45 13	60	16.0	GALAXY Sc
MCG+08-23-018	12 21 48.	+ 46 34 30.	60	16.	GALAXY
ZWG 293.018	12 21 48.	+ 58 38		15.2	GALAXY
ZCG 199.44	12 21 48.	+ 63 34		18.6	COMPACT GALAXY
ZCG 199.45	12 21 48.	+ 63 56		17.9	COMPACT GALAXY
MCG+12-12-007	12 21 48.	+ 74 12	21	15.	GALAXY
ZWG 014.026	12 21 48.	- 00 24		15.7	GALAXY
RNGC 4362	12 21 49.	+ 58 38		15.0	GALAXY
REIN 3.114A	12 21 49.13	+ 09 34 11.3			NEBULA
REIN 3.114B	12 21 49.14	+ 09 34 11.6			NEBULA
KEEL 430	12 21 49.3	+ 16 20 59.		17.	SPIRAL NEBULA
RNGC 4360A	12 21 50.	+ 09 34		14.0	GALAXY
HN 0888	12 21 50.	+ 13 29	12		NEBULA
LB 02345	12 21 50.	+ 55 07 42.		17.2	FAINT BLUE STAR
REIZ 2278	12 21 52.	+ 00 24	42	14.4	GALAXY
AMES 0919	12 21 52.	+ 09 20 01.	22	17.4	NEBULA
IC 3279	12 21 52.	+ 13 07 44.			SINGLE STAR
IC 3280	12 21 53.	+ 13 29 51.			NONSTELLAR OBJECT
ZWG 014.027	12 21 54.	+ 00 27		14.9	GALAXY
UGC 07487	12 21 54.	+ 00 27	60	14.9	GALAXY SB
MCG+00-32-006	12 21 54.	+ 00 27	36	14.9	GALAXY
KARA.68 117	12 21 54.	+ 07 24	27		DWARF GALAXY
ZWG 042.083	12 21 54.	+ 07 35		11.5	GALAXY
UGC 07488	12 21 54.	+ 07 35	330	11.5	GALAXY E
MCG+01-32-048	12 21 54.	+ 07 35	84	11.5	GALAXY
AMES 0920	12 21 54.	+ 11 03 37.	23	17.6	NEBULA
LB 04463	12 21 54.	+ 13 10		19.8	FAINT BLUE STAR
ZWG 070.054	12 21 54.	+ 13 31		15.4	GALAXY
MCG+02-32-030	12 21 54.	+ 13 31	42	15.4	GALAXY
RMB 047	12 21 54.	+ 15 26			FAINT BLUE OBJECT

OBJECT NAME	RIGHT ASCEN.	DECLINATION	DIAM.	MAGN.	TYPE OF OBJECT
ARC 1532	12 21 54.	+ 21 04		17.3	RICH CLUSTER OF GALAXIES
ZC 1221.9+4347	12 21 54.	+ 43 47	3760		CLUSTER OF GALAXIES
MCG+08-23-019	12 21 54.	+ 45 12	42	15.	GALAXY
MCG+10-18-039	12 21 54.	+ 58 38	39	15.0	GALAXY
ZWG 315.036	12 21 54.	+ 67 43		15.5	GALAXY
ZWG 335.011	12 21 54.	+ 74 10		15.7	GALAXY
ZC 1221.9-0012	12 21 54.	- 00 12	1340		CLUSTER OF GALAXIES
MCG-06-27-022	12 21 54.	- 35 07	78	15.5	GALAXY
REIN 2.154	12 21 54.84	- 18 30 28.0			NEBULA
IC 3281	12 21 55.	+ 08 05 57.			MAY NOT EXIST
PK294+43.1	12 21 55.11	- 18 30 28.2	115	10.8	PLANETARY NEBULA
RNGC 4365	12 21 56.	+ 07 36		11.0	GALAXY
AMES 0921	12 21 56.	+ 13 39 25.	35	17.0	NEBULA
A3 091	12 21 56.2	+ 13 39 42.		17.0	FAINT BLUE OBJECT
REIZ 2280	12 21 57.	+ 07 35	180	11.5	GALAXY
IC 3282	12 21 57.	+ 25 56 51.			NONSTELLAR OBJECT
TON-N 1519	12 21 57.	+ 34 22		15.1	BLUE STAR
REIZ 2283	12 21 57.	+ 46 18	30	14.1	GALAXY
ARC 1533	12 21 58.	+ 01 11		18.0	RICH CLUSTER OF GALAXIES
AMES 0922	12 21 58.	+ 08 33 37.	20	17.5	NEBULA
IC 3283	12 21 58.	+ 27 29 21.			NONSTELLAR OBJECT
AMES 0923	12 21 59.	+ 06 45 55.	24	17.8	NEBULA
KARA.68 118	12 22 00.	+ 00 22	34		DWARF GALAXY
ZC 1222.0+0114	12 22 00.	+ 01 14	870		CLUSTER OF GALAXIES
ZWG 042.084	12 22 00.	+ 07 00		15.4	GALAXY
AMES 0925	12 22 00.	+ 08 27 31.	18	17.6	NEBULA
LB 04467	12 22 00.	+ 12 39		18.2	FAINT BLUE STAR
RMB 148	12 22 00.	+ 12 49			FAINT BLUE OBJECT
LB 04466	12 22 00.	+ 13 07		19.2	FAINT BLUE STAR
LB 04465	12 22 00.	+ 13 48		17.7	FAINT BLUE STAR
RMB 061	12 22 00.	+ 13 54			FAINT BLUE OBJECT
LB 04464	12 22 00.	+ 14 34		19.0	FAINT BLUE STAR
ZWG 244.011	12 22 00.	+ 46 18		15.0	GALAXY
ZC 1222.0+5632	12 22 00.	+ 56 32	1080		CLUSTER OF GALAXIES
ZWG 293.019	12 22 00.	+ 60 23		15.7	GALAXY
MCG+10-18-040	12 22 00.	+ 60 23	18	16.	GALAXY
ZC 1222.0+6312	12 22 00.	+ 63 12	3090		CLUSTER OF GALAXIES
ZCG 199.46	12 22 00.	+ 64 18		17.2	COMPACT GALAXY
MRK 206	12 22 00.	+ 67 41	12	15.5	GALAXY WITH UV CONTINUUM
MCG+12-12-008	12 22 00.	+ 70 37	150	14.	GALAXY
RNGC 4361	12 22 00.	- 18 30		11.0	PLANETARY NEBULA
OCL 0879	12 22 00.	- 60 10	216	13.	OPEN STAR CLUSTER
AMES 0928	12 22 01.	+ 06 59 19.	31	16.3	NEBULA
AMES 0927	12 22 01.	+ 08 46 55.	24	17.8	NEBULA
AMES 0924	12 22 01.	+ 12 59 31.	23	16.3	NEBULA
IC 3284	12 22 02.	+ 11 05 51.			NONSTELLAR OBJECT
HN 0889	12 22 02.	+ 11 06		14.	NEBULA
RNGC 4367	12 22 02.	+ 12 27			NON-EXISTENT OBJECT
AMES 0926	12 22 02.	+ 17 25 31.	24	16.8	NEBULA
MAI 070	12 22 02.	+ 61 34	53		DWARF SPHEROIDAL GALAXY
REIN 7.086A	12 22 02.07	+ 46 17 44.9			NEBULA
REIN 7.086B	12 22 02.18	+ 46 17 44.9			NEBULA
REIZ 2281	12 22 03.	+ 04 16	42	15.5	GALAXY
AMES 0932	12 22 03.	+ 06 53 31.	20	17.4	NEBULA
REIZ 2282	12 22 03.	+ 07 00	18	15.5	GALAXY
AMES 0931	12 22 03.	+ 07 49 13.	20	17.6	NEBULA
AMES 0930	12 22 03.	+ 08 47 31.	34	17.0	NEBULA
AMES 0929	12 22 03.	+ 11 09 07.	30	17.0	NEBULA
TON-N 1520	12 22 03.	+ 34 09		17.0	BLUE STAR
AMES 0933	12 22 04.	+ 08 25 19.	19	18.0	NEBULA
IC 3286	12 22 04.	+ 24 01 39.			NONSTELLAR OBJECT
IC 3285	12 22 04.	+ 25 08 15.			NONSTELLAR OBJECT
RNGC 4369	12 22 04.	+ 39 40		13.0	GALAXY
A3 092	12 22 04.4	+ 13 58 40.		16.8	FAINT BLUE OBJECT
A3 093	12 22 04.7	+ 14 59 42.		16.8	FAINT BLUE OBJECT
REIZ 2285	12 22 05.	+ 39 38	30	12.9	GALAXY
LB 02346	12 22 05.	+ 49 52 48.		16.9	FAINT BLUE STAR
ZWG 014.028	12 22 06.	+ 01 50		15.7	GALAXY
ZWG 014.029	12 22 06.	+ 02 11		15.6	GALAXY
ZWG 042.085	12 22 06.	+ 03 35		14.9	GALAXY
MCG+01-32-049	12 22 06.	+ 03 35	72	14.9	GALAXY
ZWG 042.086	12 22 06.	+ 04 16		15.3	GALAXY
2ZW 061	12 22 06.	+ 04 32			COMPACT GALAXY
AMES 0935	12 22 06.	+ 08 10 07.	14	17.2	NEBULA
ZWG 070.055	12 22 06.	+ 11 08		15.6	GALAXY
LB 04472	12 22 06.	+ 12 42		18.0	FAINT BLUE STAR
LB 04471	12 22 06.	+ 12 48		20.5	FAINT BLUE STAR
LB 04470	12 22 06.	+ 13 09		20.4	FAINT BLUE STAR
LB 04469	12 22 06.	+ 13 15		18.4	FAINT BLUE STAR
8ZW 1222+13.7	12 22 06.	+ 13 42		19.2	COMPACT GALAXY
RMB 060	12 22 06.	+ 13 59			FAINT BLUE OBJECT
LB 04468	12 22 06.	+ 14 00		18.4	FAINT BLUE STAR
ZC 1222.1+1510	12 22 06.	+ 15 10	610		CLUSTER OF GALAXIES
ZC 1222.1+3534	12 22 06.	+ 35 34	200		CLUSTER OF GALAXIES
MCG+07-26-004	12 22 06.	+ 39 39	144	12.5	GALAXY
ZWG 216.002	12 22 06.	+ 39 40		12.3	GALAXY
MRK 439	12 22 06.	+ 39 40	36	12.5	GALAXY WITH UV CONTINUUM
UGC 07489	12 22 06.	+ 39 40	150	12.3	GALAXY S0/Sa
MCG+08-23-020	12 22 06.	+ 48 18	36	15.	GALAXY
MCG+09-20-165	12 22 06.	+ 55 35	48	16.	GALAXY
ZWG 335.012	12 22 06.	+ 70 37		14.6	GALAXY
UGC 07490	12 22 06.	+ 70 37	240	14.6	GALAXY DWRF SP
ZWG 352.033	12 22 06.	+ 75 49		12.6	GALAXY
UGC 07491	12 22 06.	+ 75 49	168	12.6	GALAXY S0
AMES 0938	12 22 07.	+ 09 03 55.	23	17.8	NEBULA
AMES 0937	12 22 07.	+ 09 26 31.	28	17.6	NEBULA
AMES 0934	12 22 07.	+ 12 37 25.	16	17.5	NEBULA
IC 3287	12 22 07.	+ 24 52 15.			NONSTELLAR OBJECT
AMES 0936	12 22 08.	+ 10 51 49.	29	17.0	NEBULA
RNGC 4368	12 22 08.	+ 10 52			NON-EXISTENT OBJECT
REIZ 2284	12 22 09.	+ 03 38	24	14.6	GALAXY
AMES 0939	12 22 09.	+ 05 42 07.	23	17.4	NEBULA
TON-N 1521	12 22 09.	+ 20 07		16.0	BLUE STAR
TON-N 1522	12 22 09.	+ 21 36		16.9	BLUE STAR
TON-N 1523	12 22 09.	+ 21 44		17.0	BLUE STAR
IC 3288	12 22 09.	+ 25 13 33.			NONSTELLAR OBJECT
TON-N 1524	12 22 09.	+ 33 42		16.9	BLUE STAR
AMES 0940	12 22 11.	+ 08 02 55.	13	17.6	NEBULA
AMES 0941	12 22 12.	+ 04 46 13.	50	17.1	NEBULA
ZWG 042.087	12 22 12.	+ 07 38		15.2	GALAXY
MCG+01-32-050	12 22 12.	+ 07 38	18	15.2	GALAXY
LB 04474	12 22 12.	+ 12 35		19.0	FAINT BLUE STAR
LB 04473	12 22 12.	+ 13 00		18.5	FAINT BLUE STAR
TON-N 0611	12 22 12.	+ 24 54		14.0	BLUE STAR
ZC 1222.2+3644	12 22 12.	+ 36 44	1080		CLUSTER OF GALAXIES
MCG+09-20-166	12 22 12.	+ 52 12	36	16.	GALAXY
MCG+10-18-041	12 22 12.	+ 61 46	12	16.	GALAXY
ZCG 199.47	12 22 12.	+ 63 47		18.5	COMPACT GALAXY

OBJECT NAME	RIGHT ASCEN.	DECLINATION	DIAM.	MAGN.	TYPE OF OBJECT
MCG+11-15-051B	12 22 12.	+ 66 49	24	17.	GALAXY
MCG+11-15-051A	12 22 12.	+ 66 49	18	17.	GALAXY
MCG+13-09-027	12 22 12.	+ 75 48	51	12.8	GALAXY
MCG-06-27-023	12 22 12.	- 38 23	36	15.	GALAXY
HZ 24	12 22 13.	+ 39 25		11.4	BLUE STAR
RNGC 4366	12 22 14.	+ 07 38		15.0	GALAXY
AMES 0943	12 22 14.	+ 08 42 55.	14	17.6	NEBULA
AMES 0942	12 22 14.	+ 16 21 37.	25	17.4	NEBULA
LB 02347	12 22 14.	+ 59 48 42.		16.0	FAINT BLUE STAR
HOLM 394A	12 22 15.	+ 05 14	18	14.8	PART OF MULTIPLE GALAXY
REIZ 2286	12 22 15.	+ 07 02	30	15.3	GALAXY
TON-N 1525	12 22 15.	+ 19 41		15.0	BLUE STAR
MCG+07-26-005	12 22 15.	+ 38 38 30.	36	15.	GALAXY
A3 096	12 22 15.2	+ 13 06 39.		16.9	FAINT BLUE OBJECT
A3 097	12 22 15.8	+ 14 37 08.		16.3	FAINT BLUE OBJECT
A3 098	12 22 16.3	+ 14 32 14.		16.5	FAINT BLUE OBJECT
REIN 3.115	12 22 16.57	+ 09 04 38.4			NEBULA
AMES 0946	12 22 17.	+ 08 53 55.	14	17.5	NEBULA
AMES 0944	12 22 17.	+ 12 46 13.	29	17.4	NEBULA
A3 099	12 22 17.1	+ 13 40 33.		17.0	FAINT BLUE OBJECT
KEEL 431	12 22 17.1	+ 18 28 23.		15.	NEBULA
HOLM 394B	12 22 18.	+ 05 14	12	14.8	PART OF MULTIPLE GALAXY
AMES 0948	12 22 18.	+ 05 18 49.	19	17.4	NEBULA
ZWG 042.088	12 22 18.	+ 07 02		15.5	GALAXY
AMES 0947	12 22 18.	+ 07 02 07.	42	16.6	NEBULA
ZWG 070.056	12 22 18.	+ 09 04		15.3	GALAXY
KARA.68 119	12 22 18.	+ 09 17	27		DWARF GALAXY
KARA.68 120	12 22 18.	+ 09 46	34		DWARF GALAXY
REA 41	12 22 18.	+ 09 46	36		DWARF GALAXY
MCG+02-32-031	12 22 18.	+ 09 46	30	18.	GALAXY
ZC 1222.3+1121	12 22 18.	+ 11 21	340		CLUSTER OF GALAXIES
AMES 0945	12 22 18.	+ 12 26 07.	30	16.7	NEBULA
LB 04478	12 22 18.	+ 12 52		20.6	FAINT BLUE STAR
LB 04477	12 22 18.	+ 12 54		18.8	FAINT BLUE STAR
LB 04476	12 22 18.	+ 13 11		17.9	FAINT BLUE STAR
KARA.68 121	12 22 18.	+ 13 24	34		DWARF GALAXY
LB 04475	12 22 18.	+ 14 32		17.0	FAINT BLUE STAR
ZWG 099.039	12 22 18.	+ 18 28		15.5	GALAXY
TON-N 0612	12 22 18.	+ 25 58		14.0	BLUE STAR
MCG+10-18-042	12 22 18.	+ 58 26	24	16.	GALAXY
MCG+11-15-052	12 22 18.	+ 64 27	36	17.	GALAXY
MCG+13-09-028	12 22 18.	+ 77 42 30.	33	16.	GALAXY
IC 3294	12 22 19.	+ 25 52 33.			NONSTELLAR OBJECT
IC 3295	12 22 19.	+ 28 59 03.			NONSTELLAR OBJECT
AMES 0951	12 22 20.	+ 07 50 07.	19	17.8	NEBULA
HN 0892	12 22 20.	+ 12 17		14.	NEBULA
IC 3291	12 22 20.	+ 12 17			NONSTELLAR OBJECT
AMES 0950	12 22 20.	+ 12 32 19.	14	17.4	NEBULA
AMES 0949	12 22 20.	+ 17 20 07.	26	17.0	NEBULA
HN 0891	12 22 20.	+ 17 43	6		NEBULA
IC 3293	12 22 20.	+ 17 43			NONSTELLAR OBJECT
HN 0890	12 22 20.	+ 18 29	12		NEBULA
IC 3292	12 22 20.	+ 18 29			NONSTELLAR OBJECT
LB 02349	12 22 20.	+ 50 41 48.		15.8	FAINT BLUE STAR
LB 02348	12 22 20.	+ 57 29 30.		16.9	FAINT BLUE STAR
MCG+05-29-080	12 22 21.	+ 28 50 30.	84	13.9	GALAXY
REIZ 2289	12 22 21.	+ 45 31	24	14.9	GALAXY
LB 02350	12 22 21.	+ 50 41 54.		16.7	FAINT BLUE STAR
IC 3289	12 22 21.	- 25 45 15.			NONSTELLAR OBJECT
REIN 3.116	12 22 22.31	+ 07 43 16.4			NEBULA
AMES 0952	12 22 23.	+ 12 23 19.	28	16.8	NEBULA
HOLM 403B	12 22 23.	+ 13 07	150	11.1	PART OF MULTIPLE GALAXY
BC 4C21.35	12 22 23.	+ 21 39 27.		17.50	QUASI-STELLAR OBJECT
REIN 3.117	12 22 23.46	+ 11 58 52.2			NEBULA
REIN 2.155	12 22 23.66	+ 11 59 52.3			NEBULA
AMES 0955	12 22 24.	+ 07 00 07.	20	17.5	NEBULA
AMES 0954	12 22 24.	+ 07 17 43.	24	17.6	NEBULA
ZWG 042.089	12 22 24.	+ 07 43		14.1	GALAXY
SCH 40	12 22 24.	+ 07 43		14.1	PECULIAR GALAXY
UGC 07492	12 22 24.	+ 07 43	90	14.1	GALAXY Sa
MCG+01-32-051	12 22 24.	+ 07 43	72	14.1	GALAXY
ZC 1222.4+1033	12 22 24.	+ 10 33	400		CLUSTER OF GALAXIES
AMES 0953	12 22 24.	+ 11 25 07.	19	17.9	NEBULA
ZWG 070.057	12 22 24.	+ 11 59		12.1	GALAXY
SCH 41	12 22 24.	+ 11 59		12.1	PECULIAR GALAXY
UGC 07493	12 22 24.	+ 11 59	270	12.1	GALAXY SB0
MCG+02-32-033	12 22 24.	+ 11 59	150	12.1	GALAXY
KARA.68 122	12 22 24.	+ 12 07	27		DWARF GALAXY
LB 04485	12 22 24.	+ 12 53		19.2	FAINT BLUE STAR
LB 04484	12 22 24.	+ 12 56		19.0	FAINT BLUE STAR
LB 04483	12 22 24.	+ 13 08		17.6	FAINT BLUE STAR
LB 04482	12 22 24.	+ 13 11		18.0	FAINT BLUE STAR
REA 42	12 22 24.	+ 13 24	42		DWARF GALAXY
MCG+02-32-032	12 22 24.	+ 13 24	42		GALAXY
LB 04481	12 22 24.	+ 13 29		19.3	FAINT BLUE STAR
LB 04480	12 22 24.	+ 13 44		17.8	FAINT BLUE STAR
LB 04479	12 22 24.	+ 14 21		18.9	FAINT BLUE STAR
TON-N 0613	12 22 24.	+ 28 13		15.0	BLUE STAR
7ZW 452	12 22 24.	+ 61 11			COMPACT GALAXY
ZCG 199.48	12 22 24.	+ 63 40		17.8	COMPACT GALAXY
MCG-04-29-023	12 22 24.	- 25 45	48	14.5	GALAXY
SHB 204	12 22 24.6	+ 21 39 23.		17.5	QUASI-STELLAR OBJECT
AMES 0956	12 22 25.	+ 09 27 43.	24	17.8	NEBULA
RNGC 4370	12 22 26.	+ 07 43		14.0	GALAXY
RNGC 4371	12 22 26.	+ 11 59		12.0	GALAXY
REIZ 2288	12 22 26.	+ 11 59	132	12.2	GALAXY
RNGC 4386	12 22 26.	+ 75 48		13.0	GALAXY
REIZ 2287	12 22 27.	+ 07 43	48	13.5	GALAXY
AMES 0958	12 22 27.	+ 08 57 19.	28	17.5	NEBULA
AMES 0957	12 22 27.	+ 16 56 25.	20	17.0	NEBULA
RNGC 4375	12 22 27.	+ 28 50		14.0	GALAXY
A3 103	12 22 27.9	+ 14 17 18.		16.2	FAINT BLUE OBJECT
IC 3296	12 22 28.	+ 24 39 39.			NONSTELLAR OBJECT
IC 3297	12 22 28.	+ 27 02 45.			NONSTELLAR OBJECT
AMES 0961	12 22 29.	+ 05 30 49.	25	17.4	NEBULA
AMES 0960	12 22 29.	+ 06 24 07.	20	17.3	NEBULA
AMES 0959	12 22 29.	+ 07 47 01.	22	16.9	NEBULA
SN 1957B	12 22 29.	+ 13 11		12.5	SUPERNOVA
ZWG 042.090	12 22 30.	+ 05 37		15.4	GALAXY
LB 04495	12 22 30.	+ 12 36		19.1	FAINT BLUE STAR
LB 04494	12 22 30.	+ 12 48		19.4	FAINT BLUE STAR
LB 04493	12 22 30.	+ 12 56		19.0	FAINT BLUE STAR
LB 04492	12 22 30.	+ 13 06		18.2	FAINT BLUE STAR
ZWG 070.058	12 22 30.	+ 13 10		10.8	GALAXY
UGC 07494	12 22 30.	+ 13 10	360	10.8	GALAXY S0
MCG+02-32-034	12 22 30.	+ 13 10	78	10.8	GALAXY
LB 04491	12 22 30.	+ 13 17		19.4	FAINT BLUE STAR
LB 04490	12 22 30.	+ 13 18		19.2	FAINT BLUE STAR
LB 04489	12 22 30.	+ 13 46		15.4	FAINT BLUE STAR
LB 04488	12 22 30.	+ 14 07		18.5	FAINT BLUE STAR
LB 04487	12 22 30.	+ 14 08		18.8	FAINT BLUE STAR
LB 04486	12 22 30.	+ 14 20		19.1	FAINT BLUE STAR
ZWG 128.086	12 22 30.	+ 24 40		15.6	GALAXY
KARA.73B 0529	12 22 30.	+ 24 40	24	15.6	ISOLATED GALAXY E
ZWG 128.087	12 22 30.	+ 26 14		15.3	GALAXY
UGC 07495	12 22 30.	+ 26 14	66	15.3	GALAXY Sc
MCG+04-29-070	12 22 30.	+ 26 15	60	15.3	GALAXY
ZWG 158.100	12 22 30.	+ 28 50		13.9	GALAXY
REIZ 2291	12 22 30.	+ 28 50	48	13.4	GALAXY
UGC 07496	12 22 30.	+ 28 50	84	13.9	GALAXY Sa-b
MCG+08-23-021	12 22 30.	+ 45 31	36	16.	GALAXY
ZC 1222.5+5433	12 22 30.	+ 54 33	740		CLUSTER OF GALAXIES
MCG+10-18-043	12 22 30.	+ 61 16	30	16.	GALAXY
ZC 1222.5+7723	12 22 30.	+ 77 23	1750		CLUSTER OF GALAXIES
IC 3290	12 22 30.	- 39 29 51.			NONSTELLAR OBJECT
A3 104	12 22 30.8	+ 15 15 00.		17.1	FAINT BLUE OBJECT
A3 105	12 22 30.9	+ 13 43 23.		16.7	FAINT BLUE OBJECT
AMES 0964	12 22 31.	+ 05 36 37.	30	15.8	NEBULA
AMES 0963	12 22 31.	+ 13 30 55.	19	17.0	NEBULA
AMES 0962	12 22 31.	+ 15 06 07.	29	17.2	NEBULA
IC 3298	12 22 31.	+ 17 19			NONSTELLAR OBJECT
REIN 3.118	12 22 31.75	+ 13 09 48.2			NEBULA
RNGC 4374	12 22 32.	+ 13 10		11.0	GALAXY
REIZ 2290	12 22 32.	+ 13 10	150	11.3	GALAXY
HN 0893	12 22 32.	+ 17 19	30		NEBULA
SN 1960J	12 22 32.	+ 28 50		18.5	SUPERNOVA
LB 02351	12 22 32.	+ 57 31 12.		17.6	FAINT BLUE STAR
KEEL 432	12 22 32.6	+ 15 59 19.		18.	NEBULA
TON-N 1526	12 22 33.	+ 33 57		16.8	BLUE STAR
AMES 0966	12 22 34.	+ 04 44 43.	49	16.7	NEBULA
AMES 0965	12 22 34.	+ 07 40 55.	25	17.6	NEBULA
IC 3300	12 22 34.	+ 26 14 09.			NONSTELLAR OBJECT
IC 3299	12 22 34.	+ 27 39 09.			NONSTELLAR OBJECT
A3 106	12 22 34.1	+ 14 34 23.		16.9	FAINT BLUE OBJECT
A3 108	12 22 34.7	+ 15 51 03.		16.2	FAINT BLUE OBJECT
ZWG 042.091	12 22 36.	+ 04 45		15.5	GALAXY
REA 14	12 22 36.	+ 07 00			DWARF GALAXY
LB 04500	12 22 36.	+ 13 10		18.2	FAINT BLUE STAR
LB 04499	12 22 36.	+ 13 12		19.8	FAINT BLUE STAR
LB 04498	12 22 36.	+ 13 28		17.4	FAINT BLUE STAR
AMES 0967	12 22 36.	+ 13 43 13.	38	16.6	NEBULA
LB 04497	12 22 36.	+ 14 04		17.4	FAINT BLUE STAR
LB 04496	12 22 36.	+ 14 06		17.7	FAINT BLUE STAR
HOLM 395C	12 22 36.	+ 16 22	24	15.5	PART OF MULTIPLE GALAXY
ZWG 099.040	12 22 36.	+ 17 17		15.3	GALAXY
MCG+03-32-024	12 22 36.	+ 17 17	66	15.3	GALAXY
OCL 0558	12 22 36.	+ 26 23	18000	2.7	OPEN STAR CLUSTER
MCG+05-29-081	12 22 36.	+ 27 00	72	15.4	GALAXY
ZC 1222.6+4548	12 22 36.	+ 45 48	7190		CLUSTER OF GALAXIES
MCG+10-18-044	12 22 36.	+ 61 21	39	15.	GALAXY
ZCG 199.49	12 22 36.	+ 63 41		18.3	COMPACT GALAXY
ZCG 199.50	12 22 36.	+ 64 29		18.2	COMPACT GALAXY
MCG-06-27-024	12 22 36.	- 39 30	60	14.	GALAXY
REIZ 2294	12 22 37.	+ 26 14	66	14.4	GALAXY
HN 0894	12 22 38.	+ 14 26	12		NEBULA
IC 3301	12 22 38.	+ 14 26			NONSTELLAR OBJECT
HOLM 395A	12 22 38.	+ 16 24	24	15.1	PART OF MULTIPLE GALAXY
A3 112	12 22 38.2	+ 16 51 04.		16.4	FAINT BLUE OBJECT
REIZ 2292	12 22 39.	+ 04 45	60	15.9	GALAXY
REIZ 2293	12 22 39.	+ 06 00	60	14.1	GALAXY
AMES 0969	12 22 39.	+ 11 24 19.	32	17.8	NEBULA
AMES 0968	12 22 39.	+ 14 18 49.	40	16.8	NEBULA
TON-N 1527	12 22 39.	+ 22 31		17.0	BLUE STAR
A3 113	12 22 39.2	+ 14 33 46.		17.0	FAINT BLUE OBJECT
AMES 0971	12 22 40.	+ 07 13 37.	14	17.8	NEBULA
AMES 0970	12 22 40.	+ 09 18 25.	28	17.4	NEBULA
IC 3302	12 22 40.	+ 26 09 21.			NONSTELLAR OBJECT
RNGC 4373	12 22 40.	- 39 28		12.5	GALAXY
KEEL 433	12 22 40.9	+ 16 22 13.		17.	SPIRAL NEBULA
IC 3304	12 22 41.	+ 25 42 03.			NONSTELLAR OBJECT
RNGC 4381	12 22 41.	+ 49 06			NON-EXISTENT OBJECT
A3 115	12 22 41.0	+ 16 25 22.		17.6	FAINT BLUE OBJECT
ZWG 042.092	12 22 42.	+ 05 12		13.2	GALAXY
UGC 07497	12 22 42.	+ 05 12	198	13.2	GALAXY Sa
MCG+01-32-052	12 22 42.	+ 05 12	138	13.2	GALAXY
ZWG 042.093	12 22 42.	+ 06 01		13.9	GALAXY
UGC 07498	12 22 42.	+ 06 01	96	13.9	GALAXY IRR
MCG+01-32-053	12 22 42.	+ 06 01	72	13.9	GALAXY
ZC 1222.7+0946	12 22 42.	+ 09 46	1140		CLUSTER OF GALAXIES
ZWG 070.059	12 22 42.	+ 12 07		15.4	GALAXY
UGC 07499	12 22 42.	+ 12 07	66	15.4	GALAXY S
MCG+02-32-036	12 22 42.	+ 12 07	54	15.4	GALAXY
RMB 145	12 22 42.	+ 12 35			FAINT BLUE OBJECT
LB 04505	12 22 42.	+ 12 35		16.7	FAINT BLUE STAR
LB 04504	12 22 42.	+ 12 52		17.8	FAINT BLUE STAR
ZWG 070.060	12 22 42.	+ 13 00		15.1	GALAXY
UGC 07500	12 22 42.	+ 13 00	66	15.1	GALAXY S0?
MCG+02-32-035	12 22 42.	+ 13 00	18	15.1	GALAXY
LB 04503	12 22 42.	+ 13 01		19.5	FAINT BLUE STAR
RMB 147	12 22 42.	+ 13 09			FAINT BLUE OBJECT
LB 04502	12 22 42.	+ 13 09		17.5	FAINT BLUE STAR
LB 04501	12 22 42.	+ 13 12		19.5	FAINT BLUE STAR
MCG+03-32-025	12 22 42.	+ 15 01	48	12.5	GALAXY
3ZW 065	12 22 42.	+ 15 02			COMPACT GALAXY
ZWG 099.041	12 22 42.	+ 15 02		12.5	GALAXY
UGC 07501	12 22 42.	+ 15 02	90	12.5	GALAXY S0
MCG+03-32-026	12 22 42.	+ 15 52	120	12.6	GALAXY
ZWG 099.042	12 22 42.	+ 15 54		12.6	GALAXY
SCH 42	12 22 42.	+ 15 54		12.6	PECULIAR GALAXY
UGC 07502	12 22 42.	+ 15 54	96	12.6	GALAXY S0
MCG+05-29-082	12 22 42.	+ 28 40	78	15.7	GALAXY
ZC 1222.7+4026	12 22 42.	+ 40 26	1410		CLUSTER OF GALAXIES
MCG+09-20-167	12 22 42.	+ 52 43	60	16.	GALAXY
RNGC 4384	12 22 42.	+ 54 47		13.5	GALAXY
ZC 1222.7+5607	12 22 42.	+ 56 07	870		CLUSTER OF GALAXIES
MCG-06-27-025	12 22 42.	- 39 29	78	12.	GALAXY
REIN 3.119	12 22 42.55	+ 12 07 31.1			NEBULA
KEEL 434	12 22 42.7	+ 16 23 49.		17.	SPIRAL NEBULA
IC 3303	12 22 42.8	+ 12 59 29.			GALAXY
RNGC 4378	12 22 43.	+ 05 12		12.5	GALAXY
RNGC 4376	12 22 43.	+ 05 12		14.0	GALAXY
IC 3306	12 22 43.	+ 27 40 45.			NONSTELLAR OBJECT
REIN 3.120	12 22 43.32	+ 12 59 27.6			NEBULA
IC 3305	12 22 44.	+ 12 06 51.			NONSTELLAR OBJECT
HN 0895	12 22 44.	+ 12 07		15.	NEBULA
RNGC 4377	12 22 44.	+ 15 02		13.0	GALAXY

OBJECT NAME	RIGHT ASCEN.	DECLINATION	DIAM.	MAGN.	TYPE OF OBJECT
RNGC 4379	12 22 44.	+ 15 53		13.0	GALAXY
A3 116	12 22 44.3	+ 16 56 54.		17.8	FAINT BLUE OBJECT
REIN 7.087	12 22 44.76	+ 46 15 11.3			NEBULA
REIZ 2295	12 22 45.	+ 05 12	180	12.9	GALAXY
AMES 0973	12 22 45.	+ 11 52 55.	38	17.0	NEBULA
AMES 0972	12 22 45.	+ 15 02 25.	31	16.7	NEBULA
TON-N 1528	12 22 45.	+ 20 42		17.0	BLUE STAR
TON-N 1529	12 22 45.	+ 22 06		16.0	BLUE STAR
REIZ 2299	12 22 45.	+ 46 16	30	15.6	GALAXY
MCG+09-20-168	12 22 45.	+ 54 48	72	13.	GALAXY
AMES 0975	12 22 46.	+ 06 37 43.	18	17.3	NEBULA
AMES 0974	12 22 46.	+ 09 25 43.	37	17.2	NEBULA
RNGC 4373A	12 22 46.	- 39 02			GALAXY
RNGC 4392	12 22 47.	+ 46 08		14.5	GALAXY
ABC 1536	12 22 47.	+ 77 24		16.9	RICH CLUSTER OF GALAXIES
A3 118	12 22 47.6	+ 16 54 47.		17.0	FAINT BLUE OBJECT
RMB 217	12 22 48.	+ 10 16			FAINT BLUE OBJECT
ZWG 070.061	12 22 48.	+ 10 17		13.4	GALAXY
WEEL 6	12 22 48.	+ 10 17		18.	VERY BLUE STELLAR OBJECT
UGC 07503	12 22 48.	+ 10 17	210	13.4	GALAXY Sa
MCG+02-32-037	12 22 48.	+ 10 17	180	13.4	GALAXY
LB 04511	12 22 48.	+ 12 30		19.2	FAINT BLUE STAR
LB 04510	12 22 48.	+ 12 30		18.8	FAINT BLUE STAR
LB 04509	12 22 48.	+ 12 42		19.5	FAINT BLUE STAR
LB 04508	12 22 48.	+ 13 04		18.8	FAINT BLUE STAR
AMES 0976	12 22 48.	+ 13 20 25.	26	17.5	NEBULA
LB 04507	12 22 48.	+ 13 30		19.7	FAINT BLUE STAR
LB 04506	12 22 48.	+ 13 49		18.5	FAINT BLUE STAR
UGC 07504	12 22 48.	+ 16 43	60	16.0	GALAXY
SCH 43	12 22 48.	+ 18 28		10.2	PECULIAR GALAXY
ZWG 158.101	12 22 48.	+ 26 59		15.4	GALAXY
UGC 07505	12 22 48.	+ 26 59	90	15.4	GALAXY S-IRR
IC 3308	12 22 48.	+ 26 59 33.			NONSTELLAR OBJECT
72W 035	12 22 48.	+ 46 08			COMPACT GALAXY
ZWG 244.012	12 22 48.	+ 46 08		14.6	GALAXY
MCG+08-23-022	12 22 48.	+ 46 15	24	16.	GALAXY
ZWG 269.055	12 22 48.	+ 54 47		13.5	GALAXY
MRK 207	12 22 48.	+ 54 47	30	13.5	GALAXY WITH UV CONTINUUM
UGC 07506	12 22 48.	+ 54 47	84	13.5	GALAXY Sa
ZCG 199.51	12 22 48.	+ 64 17		18.6	COMPACT GALAXY
DV.56 N4373A	12 22 48.	- 39 02	108		S GALAXY
REIN 7.088	12 22 48.40	+ 46 13 10.8			NEBULA
AMES 0980	12 22 49.	+ 05 49 31.	26	17.1	NEBULA
AMES 0979	12 22 49.	+ 08 44 31.	18	17.6	NEBULA
AMES 0978	12 22 49.	+ 09 22 13.	19	17.1	NEBULA
ABC 1535	12 22 49.	- 15 24		16.6	RICH CLUSTER OF GALAXIES
REIN 3.121	12 22 49.82	+ 10 17 37.9			NEBULA
REIZ 2297	12 22 50.	+ 10 17	108	13.7	GALAXY
RNGC 4380	12 22 50.	+ 10 18		12.5	GALAXY
HN 0896	12 22 50.	+ 14 26		15.	NEBULA
IC 3307	12 22 50.	+ 14 26			SINGLE STAR
HOLM 395B	12 22 50.	+ 16 24	48	14.7	PART OF MULTIPLE GALAXY
AMES 0977	12 22 50.	+ 16 42 43.	36	16.7	NEBULA
REIZ 2298	12 22 50.	+ 18 28	240	10.8	GALAXY
REIZ 2296	12 22 51.	+ 02 26	60	16.2	GALAXY
MCG+03-32-027	12 22 51.	+ 16 42	66	17.	GALAXY
IC 3309	12 22 51.	+ 28 39 27.			NONSTELLAR OBJECT
REIZ 2303	12 22 51.	+ 45 30	30	15.1	GALAXY
REIZ 2304	12 22 51.	+ 46 08	24	13.8	GALAXY
REIZ 2305	12 22 51.	+ 46 14	12	15.7	GALAXY
MCG-01-32-007	12 22 51.	- 06 57	42	15.	GALAXY
HOLM 396A	12 22 52.	- 06 58	42	14.5	PART OF MULTIPLE GALAXY
REIN 7.089	12 22 52.51	+ 46 07 28.2			NEBULA
KEEL 435	12 22 52.8	+ 18 25 02.		18.	NEBULA
KEEL 436	12 22 53.6	+ 16 24 02.		18.	NEBULA
A3 119	12 22 53.8	+ 16 44 49.		13.5	UV CONCENTRATION
AMES 0983	12 22 54.	+ 08 37 01.	19	17.0	NEBULA
AMES 0982	12 22 54.	+ 08 39 01.	16	17.6	NEBULA
ZWG 070.062	12 22 54.	+ 09 09		15.6	GALAXY
LB 04518	12 22 54.	+ 12 47		18.3	FAINT BLUE STAR
RMB 146	12 22 54.	+ 12 58			FAINT BLUE OBJECT
LB 04517	12 22 54.	+ 12 59		16.8	FAINT BLUE STAR
LB 04516	12 22 54.	+ 13 47		18.0	FAINT BLUE STAR
LB 04515	12 22 54.	+ 13 52		17.2	FAINT BLUE STAR
LB 04514	12 22 54.	+ 14 04		15.8	FAINT BLUE STAR
LB 04513	12 22 54.	+ 14 05		20.6	FAINT BLUE STAR
LB 04512	12 22 54.	+ 14 14		19.0	FAINT BLUE STAR
ZWG 099.043	12 22 54.	+ 16 24		15.4	GALAXY
ZWG 099.044	12 22 54.	+ 16 45		12.3	GALAXY
UGC 07507	12 22 54.	+ 16 45	114	12.3	GALAXY PECULR
MCG+03-32-028	12 22 54.	+ 18 26	18	18.5	GALAXY
ZWG 099.045	12 22 54.	+ 18 28		10.2	GALAXY
UGC 07508	12 22 54.	+ 18 28	480	10.2	GALAXY S0
KARA.72 334A	12 22 54.	+ 18 28	390	10.2	PART OF DOUBLE GALAXY
TON-N 0614	12 22 54.	+ 24 54		15.4	BLUE STAR
ZC 1222.9+2551	12 22 54.	+ 25 51	4170		CLUSTER OF GALAXIES
ZWG 158.102	12 22 54.	+ 28 40		15.7	GALAXY
UGC 07509	12 22 54.	+ 28 40	84	15.7	GALAXY S
REIZ 2302	12 22 54.	+ 30 10	42	14.9	GALAXY
ZWG 158.103	12 22 54.	+ 32 05		15.7	GALAXY
MCG+08-23-023	12 22 54.	+ 46 07 30.	36	14.	GALAXY
ZC 1222.9+4908	12 22 54.	+ 49 08	810		CLUSTER OF GALAXIES
72W 453	12 22 54.	+ 61 57			COMPACT GALAXY
ZCG 199.52	12 22 54.	+ 63 47		18.7	COMPACT GALAXY
MCG+11-15-053	12 22 54.	+ 65 11	45	14.4	GALAXY
KEEL 437	12 22 54.3	+ 18 45 54.		18.	NEBULA
AMES 0985	12 22 55.	+ 09 21 42.	17	17.6	NEBULA
AMES 0981	12 22 55.	+ 11 56 48.	22	17.4	NEBULA
SN 1960R	12 22 55.	+ 18 26		12.0	SUPERNOVA
HZ 25	12 22 55.	+ 36 15		10.0	VERY BLUE STAR
REIZ 2300	12 22 56.	+ 12 32	78	14.3	GALAXY
AMES 0984	12 22 56.	+ 13 46 06.	34	17.4	NEBULA
REIZ 2301	12 22 56.	+ 16 44	60	11.8	GALAXY
RNGC 4383	12 22 56.	+ 16 45		13.0	GALAXY
RNGC 4382	12 22 56.	+ 18 28		10.5	GALAXY
HOLM 397A	12 22 56.	+ 18 28	270	11.5	PART OF MULTIPLE GALAXY
HOLM 396B	12 22 56.	- 06 58	18	15.7	PART OF MULTIPLE GALAXY
A3 121	12 22 56.4	+ 13 34 02.		17.6	FAINT BLUE OBJECT
AMES 0988	12 22 57.	+ 05 29 42.	41	17.2	NEBULA
AMES 0987	12 22 57.	+ 07 54 42.	22	15.4	NEBULA
AMES 0986	12 22 57.	+ 13 33 42.	20	17.8	NEBULA
MCG+03-32-031	12 22 57.	+ 16 23	60	15.4	GALAXY
IC 0787	12 22 57.	+ 16 24 03.			NONSTELLAR OBJECT
MCG+03-32-030	12 22 57.	+ 16 45	138	12.3	GALAXY
MCG+03-32-029	12 22 57.	+ 18 29	450	10.2	GALAXY
TON-N 1530	12 22 57.	+ 22 53		16.8	BLUE STAR
REIZ 2307	12 22 57.	+ 47 33	24	13.9	GALAXY
A3 124	12 22 57.6	+ 14 37 20.		17.9	FAINT BLUE OBJECT

OBJECT NAME	RIGHT ASCEN.	DECLINATION	DIAM.	MAGN.	TYPE OF OBJECT
A3 125	12 22 57.7	+ 14 34 27.		16.0	FAINT BLUE OBJECT
IC 3310	12 22 58.	+ 15 57			SINGLE STAR
REIZ 2311	12 22 58.	+ 62 12	36	14.2	GALAXY
AMES 0989	12 22 59.	+ 15 03 06.	34	16.7	NEBULA
VDB.66G 123	12 23	+ 58 33	130		DWARF GALAXY
VDB.66G 122	12 23	+ 70 38	100		DWARF GALAXY
LB 09858	12 23	- 81 08		12.5	FAINT BLUE STAR
ZWG 042.094	12 23 00.	+ 03 43		15.3	GALAXY
AMES 0992	12 23 00.	+ 08 19 54.	24	17.0	NEBULA
ZWG 070.063	12 23 00.	+ 12 32		15.0	GALAXY
UGC 07510	12 23 00.	+ 12 32	126	15.0	GALAXY Sc-IRR
MCG+02-32-038	12 23 00.	+ 12 32	108	15.0	GALAXY
LB 04523	12 23 00.	+ 12 36		19.9	FAINT BLUE STAR
AMES 0990	12 23 00.	+ 12 42 18.	48	17.6	NEBULA
LB 04522	12 23 00.	+ 13 02		19.9	FAINT BLUE STAR
LB 04521	12 23 00.	+ 13 10		17.4	FAINT BLUE STAR
LB 04520	12 23 00.	+ 13 13		19.0	FAINT BLUE STAR
RMB 098	12 23 00.	+ 13 34			FAINT BLUE OBJECT
LB 04519	12 23 00.	+ 13 34		17.0	FAINT BLUE STAR
KARA.68 123	12 23 00.	+ 14 25	34		DWARF GALAXY
HOLM 400B	12 23 00.	+ 15 57	48	15.1	PART OF MULTIPLE GALAXY
IC 3312	12 23 00.	+ 23 51 33.			NONSTELLAR OBJECT
LB 00006	12 23 00.	+ 27 28		18.8	FAINT BLUE STAR
ZC 1223.0+3752	12 23 00.	+ 37 52	670		CLUSTER OF GALAXIES
MCG+08-23-024	12 23 00.	+ 47 32	48	15.	GALAXY
ZWG 244.013	12 23 00.	+ 47 33		15.1	GALAXY
MCG+09-20-169	12 23 00.	+ 53 15	48	17.	GALAXY
ZCG 199.53	12 23 00.	+ 63 18		18.1	COMPACT GALAXY
72W 454	12 23 00.	+ 65 13			COMPACT GALAXY
ZWG 315.037	12 23 00.	+ 65 13		13.8	GALAXY
UGC 07511	12 23 00.	+ 65 13	72	13.8	GALAXY (S0)
ZWG 014.031	12 23 00.	- 02 33		15.3	GALAXY
ZWG 014.030	12 23 00.	- 02 37		15.2	GALAXY
HELW 150	12 23 00.	- 18 10 02.			NEBULA
MCG-06-27-026	12 23 00.	- 39 03	90	13.	GALAXY
MESL 300+00/2	12 23 00.	- 62 00	2700		HII REGION
GCL 019	12 23 00.	- 72 24	1188	9.1	GLOBULAR STAR CLUSTER
A3 126	12 23 00.1	+ 14 47 05.		16.4	FAINT BLUE OBJECT
REIN 3.122	12 23 00.86	+ 12 32 13.6			NEBULA
AMES 0993	12 23 01.	+ 05 40 00.	26	17.7	NEBULA
IC 3311	12 23 01.	+ 12 32 09.			NONSTELLAR OBJECT
AMES 0991	12 23 01.	+ 13 29 54.	24	17.7	NEBULA
IC 3313	12 23 01.	+ 16 07			NONSTELLAR OBJECT
IC 3314	12 23 01.	+ 23 52 03.			NONSTELLAR OBJECT
HELW 151	12 23 01.	- 18 10 14.			NEBULA
HN 0897	12 23 02.	+ 16 07	12		NEBULA
REIZ 2306	12 23 02.	+ 16 23	18	14.5	GALAXY
REIZ 2318	12 23 02.	+ 65 13	18	14.4	GALAXY
RNGC 4372	12 23 02.	- 72 24		9.0	GLOBULAR CLUSTER
AMES 0994	12 23 03.	+ 07 59 42.	12	17.8	NEBULA
RNGC 4391	12 23 03.	+ 65 13		14.0	GALAXY
PPA 4	12 23 03.0	+ 24 52 06.		15.6	NONSTELLAR BLUE OBJECT
AMES 0996	12 23 04.	+ 07 02 24.	22	17.4	NEBULA
A3 127	12 23 04.3	+ 16 28 16.		16.8	FAINT BLUE OBJECT
MCG+03-32-032	12 23 05.	+ 16 06 30.	60	15.5	GALAXY
AMES 0995	12 23 05.	+ 16 08 18.	58	17.5	NEBULA
RNGC 4389	12 23 05.	+ 45 58		13.0	GALAXY
A3 128	12 23 05.2	+ 15 41 59.		16.7	FAINT BLUE OBJECT
A3 129	12 23 05.3	+ 16 16 39.		16.6	FAINT BLUE OBJECT
ZWG 014.033	12 23 06.	+ 02 27		15.7	GALAXY
UGC 07512	12 23 06.	+ 02 27	72	15.7	GALAXY DWRF IR
MCG+00-32-008	12 23 06.	+ 02 27	66	15.7	GALAXY
ZWG 042.095	12 23 06.	+ 07 30		14.4	GALAXY
UGC 07513	12 23 06.	+ 07 30	210	14.4	GALAXY Sc
MCG+01-32-054	12 23 06.	+ 07 30	216	14.4	GALAXY
ZWG 070.064	12 23 06.	+ 09 18		15.6	GALAXY
KARA.68 124	12 23 06.	+ 10 52	34		DWARF GALAXY
RMB 215	12 23 06.	+ 11 12			FAINT BLUE OBJECT
LB 04528	12 23 06.	+ 12 40		17.8	FAINT BLUE STAR
LB 04527	12 23 06.	+ 12 48		18.3	FAINT BLUE STAR
LB 04526	12 23 06.	+ 12 51		19.2	FAINT BLUE STAR
LB 04525	12 23 06.	+ 13 30		19.7	FAINT BLUE STAR
LB 04524	12 23 06.	+ 14 04		15.7	FAINT BLUE STAR
RMB 224	12 23 06.	+ 15 45			FAINT BLUE OBJECT
MCG+03-32-033	12 23 06.	+ 16 05 30.	60	15.6	GALAXY
ZWG 099.046	12 23 06.	+ 16 06		15.6	GALAXY
TON-N 0615	12 23 06.	+ 24 32		15.6	BLUE STAR
IC 3316	12 23 06.	+ 26 26 21.			NONSTELLAR OBJECT
ZWG 244.014	12 23 06.	+ 45 58		12.8	GALAXY
UGC 07514	12 23 06.	+ 45 58	162	12.8	GALAXY SB
MCG+08-23-025	12 23 06.	+ 46 20	30	15.0	GALAXY
ZWG 269.056	12 23 06.	+ 50 37		15.4	GALAXY
ZWG 014.032	12 23 06.	- 02 40		14.9	GALAXY
MCG+00-32-007	12 23 06.	- 02 40	84	14.9	GALAXY
MESL 299+02/1	12 23 06.	- 60 20	1200		HII REGION
KEEL 438	12 23 06.7	+ 18 10 14.		18.	NEBULA
RNGC 4385	12 23 07.	+ 00 51		13.0	GALAXY
AMES 0999	12 23 07.	+ 07 28 12.	28	17.3	NEBULA
AMES 0998	12 23 07.	+ 10 50 54.	48	17.3	NEBULA
HOLM 403C	12 23 07.	+ 12 54	240	11.7	PART OF MULTIPLE GALAXY
AMES 0997	12 23 07.	+ 13 59 36.	32	17.6	NEBULA
REIN 3.123	12 23 07.04	+ 09 11 19.4			NEBULA
REIN 7.090	12 23 07.34	+ 46 20 39.4			NEBULA
HN 0898	12 23 08.	+ 12 35		15.	NEBULA
IC 3315	12 23 08.	+ 12 35			NONSTELLAR OBJECT
REIZ 2309	12 23 08.	+ 13 05	36	13.7	GALAXY
REIZ 2310	12 23 08.	+ 16 06	24	15.5	GALAXY
HOLM 398A	12 23 08.	+ 16 06	36	14.8	PART OF MULTIPLE GALAXY
HOLM 398B	12 23 08.	+ 16 07	42	15.5	PART OF MULTIPLE GALAXY
IC 3317	12 23 08.	+ 25 37 09.			NONSTELLAR OBJECT
REIZ 2317	12 23 08.	+ 50 23	24	13.8	GALAXY
REIZ 2324	12 23 08.	+ 64 43	60	13.6	GALAXY
REIN 7.091A	12 23 08.31	+ 45 57 42.6			NEBULA
REIN 7.091B	12 23 08.31	+ 45 57 43.7			NEBULA
REIZ 2316	12 23 09.	+ 45 58	108	13.0	GALAXY
MCG+08-23-026	12 23 09.	+ 46 20	30	14.8	GALAXY
IC 3322A	12 23 09.7	+ 07 29 31.			GALAXY
REIN 3.124	12 23 09.90	+ 13 05 11.9			NEBULA
REIZ 2308	12 23 10.	+ 00 50	108	13.3	GALAXY
AMES 1002	12 23 10.	+ 06 49 54.	13	17.9	NEBULA
AMES 1001	12 23 10.	+ 07 26 42.	23	16.9	NEBULA
A3 131	12 23 10.1	+ 17 05 42.		15.4	FAINT BLUE OBJECT
REIN 7.092	12 23 10.47	+ 46 20 28.7			NEBULA
A3 132	12 23 10.6	+ 14 10 28.		16.8	FAINT BLUE OBJECT
REIN 3.125	12 23 10.62	+ 09 18 10.7			NEBULA
AMES 1003	12 23 11.	+ 07 10 06.	25	17.6	NEBULA
AMES 1000	12 23 11.	+ 17 38 54.	28	16.7	NEBULA
HOLM 399B	12 23 11.	+ 46 20	12	15.0	PART OF MULTIPLE GALAXY

OBJECT NAME	RIGHT ASCEN.	DECLINATION	DIAM.	MAGN.	TYPE OF OBJECT
ZWG 014.034	12 23 12.	+ 00 52		13.4	GALAXY
MRK 052	12 23 12.	+ 00 52	10	14.5	GALAXY WITH UV CONTINUUM
UGC 07515	12 23 12.	+ 00 52	120	12.4	GALAXY SB0
MCG+00-32-009	12 23 12.	+ 00 52	102	12.4	GALAXY
ZWG 042.096	12 23 12.	+ 04 47		15.0	GALAXY
UGC 07516	12 23 12.	+ 04 47	78	14.4	GALAXY Sc
MCG+01-32-055	12 23 12.	+ 04 47	60	15.0	GALAXY
VVI 58	12 23 12.	+ 05 57	100	14.71	SEYFERT GALAXY
AMES 1004	12 23 12.	+ 07 29 30.	222	13.9	NEBULA
KARA.68 125	12 23 12.	+ 10 20	27		DWARF GALAXY
ZWG 070.065	12 23 12.	+ 13 05		13.2	GALAXY
UGC 07517	12 23 12.	+ 13 05	90	13.2	GALAXY E
MCG+02-32-039	12 23 12.	+ 13 05	84	13.2	GALAXY
LB 04533	12 23 12.	+ 13 32		18.2	FAINT BLUE STAR
RMB 097	12 23 12.	+ 13 35			FAINT BLUE OBJECT
LB 04532	12 23 12.	+ 13 35		18.8	FAINT BLUE STAR
LB 04531	12 23 12.	+ 14 18		17.8	FAINT BLUE STAR
LB 04530	12 23 12.	+ 14 26		18.6	FAINT BLUE STAR
LB 04529	12 23 12.	+ 14 27		19.2	FAINT BLUE STAR
BC TON1530	12 23 12.	+ 22 51			QUASI-STELLAR OBJECT
SHB 205	12 23 12.	+ 22 51		17.0	QUASI-STELLAR OBJECT
BC 4C25.40	12 23 12.	+ 25 14 30.			QUASI-STELLAR OBJECT
SHB 206	12 23 12.	+ 25 14 30.		16.0	QUASI-STELLAR OBJECT
TON-N 0616	12 23 12.	+ 25 16		15.7	BLUE STAR
MCG+05-29-083	12 23 12.	+ 27 51	180	13.8	GALAXY
MCG+08-23-027	12 23 12.	+ 45 42	66	16.	GALAXY
MCG+08-23-028	12 23 12.	+ 45 58	150	12.8	GALAXY
MCG+09-20-170	12 23 12.	+ 51 05	36	17.	GALAXY
MCG+09-20-171	12 23 12.	+ 56 01	36	16.	GALAXY
LB 02352	12 23 12.	+ 61 12 42.		15.8	FAINT BLUE STAR
ZC 1223.2+6147	12 23 12.	+ 61 47	2350		CLUSTER OF GALAXIES
ZC 1223.2+6715	12 23 12.	+ 67 15	810		CLUSTER OF GALAXIES
KEEL 439	12 23 12.3	+ 18 47 38.		18.	NEBULA
AMES 1007	12 23 13.	+ 04 47 18.	78	15.5	NEBULA
AMES 1006	12 23 13.	+ 09 16 48.	26	17.4	NEBULA
A3 134	12 23 13.2	+ 17 08 01.		16.4	FAINT BLUE OBJECT
REIN 7.093	12 23 13.66	+ 46 05 52.6			NEBULA
RNGC 4388	12 23 14.	+ 12 56		12.0	GALAXY
REIZ 2314	12 23 14.	+ 12 56	300	12.3	GALAXY
RNGC 4387	12 23 14.	+ 13 05		13.0	GALAXY
REIZ 2315	12 23 14.	+ 13 05	24	15.8	GALAXY
REIN 7.094	12 23 14.23	+ 46 02 55.1			NEBULA
REIZ 2312	12 23 15.	+ 04 47	42	15.2	GALAXY
REIZ 2313	12 23 15.	+ 07 29	192	14.1	GALAXY
AMES 1009	12 23 15.	+ 16 54 54.	42	17.4	NEBULA
AMES 1005	12 23 15.	+ 17 08 30.	26	16.6	NEBULA
RNGC 4393	12 23 15.	+ 27 50		14.0	GALAXY
RNGC 4395	12 23 15.	+ 33 49		10.5	GALAXY
HOLM 399A	12 23 15.	+ 46 20	18	14.8	PART OF MULTIPLE GALAXY
REIZ 2323	12 23 15.	+ 46 21	18	14.7	GALAXY
REIZ 2322	12 23 15.	+ 46 21	12	14.6	GALAXY
MCG+08-23-029	12 23 15.	+ 50 22	54	15.	GALAXY
MCG+09-20-172	12 23 15.	+ 51 56	60	15.	GALAXY
REIN 3.126	12 23 15.01	+ 12 56 17.3			NEBULA
A3 135	12 23 15.7	+ 15 13 47.		16.9	FAINT BLUE OBJECT
AMES 1013	12 23 16.	+ 05 43 54.	24	17.7	NEBULA
AMES 1012	12 23 16.	+ 06 35 30.	38	17.8	NEBULA
AMES 1010	12 23 16.	+ 15 13 42.	34	17.0	NEBULA
AMES 1008	12 23 16.	+ 17 44 18.	28	17.1	NEBULA
IC 3321	12 23 16.	+ 26 21 33.			NONSTELLAR OBJECT
AMES 1011	12 23 17.	+ 16 05 30.	23	17.8	NEBULA
AMES 1015	12 23 18.	+ 05 30 00.	32	17.7	NEBULA
ZWG 042.097	12 23 18.	+ 06 05		15.2	GALAXY
MCG+01-32-056	12 23 18.	+ 06 05	36	15.2	GALAXY
ZWG 042.098	12 23 18.	+ 07 50		14.7	GALAXY
MCG+01-32-057	12 23 18.	+ 07 50	144	14.7	GALAXY
ZWG 070.066	12 23 18.	+ 09 13		15.6	GALAXY
IC 3318	12 23 18.	+ 10 02 27.			SINGLE STAR
AMES 1014	12 23 18.	+ 10 26 18.	34	17.4	NEBULA
IC 3319	12 23 18.	+ 10 40 03.			SAME AS NGC 4390
ZWG 070.067	12 23 18.	+ 10 44		13.7	GALAXY
SCH 44	12 23 18.	+ 10 44		13.7	PECULIAR GALAXY
UGC 07519	12 23 18.	+ 10 44	108	13.7	GALAXY Sc
MCG+02-32-040	12 23 18.	+ 10 44	78	13.7	GALAXY
IC 3320	12 23 18.	+ 10 44 03.			SAME AS NGC 4390
RMB 216	12 23 18.	+ 11 30			FAINT BLUE OBJECT
LB 04538	12 23 18.	+ 12 38		18.0	FAINT BLUE STAR
ZWG 070.068	12 23 18.	+ 12 56		12.2	GALAXY
UGC 07520	12 23 18.	+ 12 56	372	12.2	GALAXY Sb
MCG+02-32-041	12 23 18.	+ 12 56	300	12.2	GALAXY
LB 04537	12 23 18.	+ 13 28		18.6	FAINT BLUE STAR
RMB 096	12 23 18.	+ 13 30			FAINT BLUE OBJECT
LB 04536	12 23 18.	+ 13 32		19.6	FAINT BLUE STAR
LB 04535	12 23 18.	+ 13 43		18.9	FAINT BLUE STAR
LB 04534	12 23 18.	+ 13 59		18.6	FAINT BLUE STAR
RMB 046	12 23 18.	+ 15 13			FAINT BLUE OBJECT
ZWG 158.104	12 23 18.	+ 27 50		13.8	GALAXY
REIZ 2321	12 23 18.	+ 27 50	180	12.9	GALAXY
UGC 07521	12 23 18.	+ 27 50	210	13.8	GALAXY Sc
ZWG 244.015	12 23 18.	+ 45 34		15.4	GALAXY
MRK 208	12 23 18.	+ 47 36	9	16.5	GALAXY WITH UV CONTINUUM
ZWG 269.057	12 23 18.	+ 51 57		14.9	GALAXY
MCG-05-29-035	12 23 18.	- 30 26	36	15.5	GALAXY
REIN 3.127	12 23 18.07	+ 10 44 09.2			NEBULA
REIN 7.095A	12 23 18.10	+ 46 00 30.9			NEBULA
REIN 7.095B	12 23 18.12	+ 46 00 30.9			NEBULA
REIN 3.128	12 23 18.53	+ 09 11 50.4			NEBULA
AMES 1016	12 23 19.	+ 06 35 30.	24	17.6	NEBULA
IC 3323	12 23 19.	+ 27 49 10.			NONSTELLAR OBJECT
AMES 1018	12 23 20.	+ 06 05 06.	38	15.8	NEBULA
AMES 1017	12 23 20.	+ 07 01 18.	19	17.2	NEBULA
RNGC 4390	12 23 20.	+ 10 44		13.5	GALAXY
REIZ 2320	12 23 20.	+ 10 44	60	13.8	GALAXY
IC 3324	12 23 20.	+ 27 00 58.			NONSTELLAR OBJECT
REIZ 2319	12 23 21.	+ 06 05	54	15.1	GALAXY
AMES 1020	12 23 21.	+ 07 10 30.	29	17.8	NEBULA
TON-N 1531	12 23 21.	+ 21 11		17.0	BLUE STAR
IC 3325	12 23 21.	+ 24 10 22.			NONSTELLAR OBJECT
MCG+08-23-030	12 23 21.	+ 45 21	12	16.	GALAXY
REIZ 2331	12 23 21.	+ 45 33	30	14.8	GALAXY
HOLM 401A	12 23 21.	+ 46 10	36	14.3	PART OF MULTIPLE GALAXY
REIZ 2332	12 23 21.	+ 46 12	36	14.9	GALAXY
MCG-03-32-008	12 23 21.	- 21 28	42	15.	GALAXY
REIN 7.096	12 23 21.28	+ 46 11 13.3			NEBULA
IC 3322	12 23 21.3	+ 07 49 55.			GALAXY
REIN 7.097	12 23 21.39	+ 46 05 08.8			NEBULA
REIN 3.129	12 23 21.48	+ 07 49 53.2			NEBULA
AMES 1021	12 23 22.	+ 07 29 36.	24	17.0	NEBULA

OBJECT NAME	RIGHT ASCEN.	DECLINATION	DIAM.	MAGN.	TYPE OF OBJECT
AMES 1019	12 23 22.	+ 11 14 06.	32	17.2	NEBULA
IC 3326	12 23 23.	+ 24 02 40.			NONSTELLAR OBJECT
REIZ 2328	12 23 23.	+ 33 49	480	12.5	GALAXY
ZWG 042.099	12 23 24.	+ 03 42		15.0	GALAXY
UGC 07522	12 23 24.	+ 03 42	168	15.0	GALAXY Sc
MCG+01-32-058	12 23 24.	+ 03 42	180	15.0	GALAXY
ZC 1223.4+0815	12 23 24.	+ 08 15	610		CLUSTER OF GALAXIES
REA 43	12 23 24.	+ 09 15	60		DWARF GALAXY
ZWG 070.069	12 23 24.	+ 10 20		14.9	GALAXY
MCG+02-32-042	12 23 24.	+ 10 20	36	14.9	GALAXY
LB 04545	12 23 24.	+ 12 28		17.7	FAINT BLUE STAR
LB 04544	12 23 24.	+ 12 37		19.0	FAINT BLUE STAR
LB 04543	12 23 24.	+ 12 38		19.8	FAINT BLUE STAR
LB 04542	12 23 24.	+ 12 50		20.7	FAINT BLUE STAR
LB 04541	12 23 24.	+ 12 54		18.9	FAINT BLUE STAR
LB 04540	12 23 24.	+ 13 23		17.9	FAINT BLUE STAR
LB 04539	12 23 24.	+ 13 26		18.4	FAINT BLUE STAR
ZWG 099.047	12 23 24.	+ 18 30		11.9	GALAXY
UGC 07523	12 23 24.	+ 18 30	234	11.9	GALAXY SBb
KARA.72 234B	12 23 24.	+ 18 30	198	11.9	PART OF DOUBLE GALAXY
ZC 1223.4+1941	12 23 24.	+ 19 41	2350		CLUSTER OF GALAXIES
REIZ 2330	12 23 24.	+ 27 50	168	15.4	GALAXY
MCG+05-29-084	12 23 24.	+ 31 08 30.	60	15.1	GALAXY
MCG+06-27-053	12 23 24.	+ 33 47	720	12.	GALAXY
ZWG 187.042	12 23 24.	+ 33 49		11.7	GALAXY
UGC 07524	12 23 24.	+ 33 49	960	11.7	GALAXY S IV-V
ZWG 244.016	12 23 24.	+ 45 33		15.5	GALAXY
UGC 07525	12 23 24.	+ 45 33	72	15.5	GALAXY SBc
MCG+08-23-031	12 23 24.	+ 45 33	42	15.	GALAXY
MCG+08-23-032	12 23 24.	+ 46 11	42	14.3	GALAXY
ZWG 244.017	12 23 24.	+ 50 22		14.9	GALAXY
MCG+09-20-173	12 23 24.	+ 55 26	48	18.	GALAXY
MCG+11-15-054	12 23 24.	+ 62 41	54	15.	GALAXY
ZWG 014.035	12 23 24.	- 03 11		15.0	GALAXY
MCG+00-32-010	12 23 24.	- 03 11	30	15.0	GALAXY
A3 140	12 23 24.4	+ 14 37 51.		17.5	FAINT BLUE OBJECT
REIN 7.098	12 23 24.77	+ 46 12 00.4			NEBULA
AMES 1024	12 23 25.	+ 07 30 48.	34	17.8	NEBULA
IC 3327	12 23 25.	+ 15 10			NONSTELLAR OBJECT
REIZ 2327	12 23 25.	+ 18 29	180	11.9	GALAXY
REIN 3.130	12 23 25.40	+ 10 19 49.8			NEBULA
IC 3328	12 23 26.	+ 10 19 52.			NONSTELLAR OBJECT
AMES 1023	12 23 26.	+ 12 14 00.	23	17.0	NEBULA
HOLM 403D	12 23 26.	+ 13 21	210	12.1	PART OF MULTIPLE GALAXY
HN 0899	12 23 26.	+ 15 10		15.	NEBULA
REIZ 2329	12 23 26.	+ 15 56	120	13.4	GALAXY
HOLM 400A	12 23 26.	+ 15 57	180	12.4	PART OF MULTIPLE GALAXY
AMES 1022	12 23 26.	+ 16 43 42.	23	17.8	NEBULA
RNGC 4394	12 23 26.	+ 18 29		12.0	GALAXY
HOLM 397B	12 23 26.	+ 18 29	114	13.3	PART OF MULTIPLE GALAXY
RNGC 4397	12 23 26.	+ 18 35			NON-EXISTENT OBJECT
A3 141	12 23 26.2	+ 13 47 40.		16.5	FAINT BLUE OBJECT
A3 142	12 23 26.5	+ 16 15 10.		17.1	FAINT BLUE OBJECT
AMES 1027	12 23 27.	+ 07 02 36.	18	17.4	NEBULA
REIZ 2325	12 23 27.	+ 07 42	72	14.1	GALAXY
REIZ 2326	12 23 27.	+ 07 50	120	14.6	GALAXY
MCG+03-32-034	12 23 27.	+ 15 56	210	13.7	GALAXY
MCG+03-32-035	12 23 27.	+ 18 31	210	11.9	GALAXY
IC 3329	12 23 27.	+ 27 50 22.			NONSTELLAR OBJECT
REIZ 2333	12 23 27.	+ 45 33	42	14.5	GALAXY
A3 143	12 23 27.0	+ 14 54 08.		16.4	FAINT BLUE OBJECT
AMES 1026	12 23 28.	+ 12 35 36.	22	16.6	NEBULA
IC 3330	12 23 28.	+ 31 07 34.			NONSTELLAR OBJECT
RNGC 4401	12 23 28.	+ 33 47			NON-EXISTENT OBJECT
RNGC 4400	12 23 28.	+ 33 50			NON-EXISTENT OBJECT
RNGC 4399	12 23 28.	+ 33 50			NON-EXISTENT OBJECT
LB 00634	12 23 28.	+ 56 53 24.		17.5	FAINT BLUE STAR
A3 144	12 23 28.8	+ 13 30 09.		16.7	FAINT BLUE OBJECT
AMES 1025	12 23 29.	+ 18 02 24.	26	16.9	NEBULA
LB 02353	12 23 29.	+ 59 29 42.		16.9	FAINT BLUE STAR
A3 145	12 23 29.0	+ 17 19 17.		16.9	FAINT BLUE OBJECT
ZWG 042.100	12 23 30.	+ 04 45		15.1	GALAXY
MCG+01-32-059	12 23 30.	+ 04 45	12	15.1	GALAXY
AMES 1029	12 23 30.	+ 08 32 00.	20	17.8	NEBULA
AMES 1028	12 23 30.	+ 09 21 30.	31	17.8	NEBULA
ZWG 070.070	12 23 30.	+ 12 05		15.2	GALAXY
MCG+02-32-043	12 23 30.	+ 12 05	12	15.2	GALAXY
LB 04551	12 23 30.	+ 12 43		19.6	FAINT BLUE STAR
LB 04550	12 23 30.	+ 12 55		18.7	FAINT BLUE STAR
LB 04549	12 23 30.	+ 13 18		17.8	FAINT BLUE STAR
LB 04548	12 23 30.	+ 13 19		17.8	FAINT BLUE STAR
LB 04547	12 23 30.	+ 13 22		17.3	FAINT BLUE STAR
LB 04546	12 23 30.	+ 14 19		19.4	FAINT BLUE STAR
ZWG 099.048	12 23 30.	+ 15 09		15.5	GALAXY
ZWG 099.049	12 23 30.	+ 15 57		13.7	GALAXY
UGC 07526	12 23 30.	+ 15 57	210	13.7	GALAXY
ZC 1223.5+1627	12 23 30.	+ 16 27	870		CLUSTER OF GALAXIES
ZWG 158.105	12 23 30.	+ 31 06		15.1	GALAXY
UGC 07527	12 23 30.	+ 31 06	72	15.1	GALAXY SB
HOLM 404A	12 23 30.	+ 31 07	36	13.7	PART OF MULTIPLE GALAXY
MCG+08-23-033	12 23 30.	+ 45 33	60	15.	GALAXY
ZC 1223.5+5130	12 23 30.	+ 51 30	870		CLUSTER OF GALAXIES
7ZW 455	12 23 30.	+ 66 31			COMPACT GALAXY
MCG-06-27-027	12 23 30.	- 38 51	42	15.	GALAXY
A3 146	12 23 31.3	+ 15 55 58.		14.0	UV CONCENTRATION
AMES 1030	12 23 32.	+ 08 43 42.	20	16.8	NEBULA
RNGC 4398	12 23 32.	+ 10 58			NON-EXISTENT OBJECT
HOLM 403A	12 23 32.	+ 13 11	210	11.1	PART OF MULTIPLE GALAXY
RNGC 4396	12 23 32.	+ 15 57		13.5	GALAXY
LB 02354	12 23 32.	+ 52 55 36.		16.6	FAINT BLUE STAR
A3 148	12 23 32.1	+ 13 17 25.		16.9	FAINT BLUE OBJECT
AMES 1032	12 23 33.	+ 06 34 30.	26	17.4	NEBULA
LB 02356	12 23 33.	+ 51 51 18.		16.8	FAINT BLUE STAR
LB 02355	12 23 33.	+ 58 38 12.		16.5	FAINT BLUE STAR
REIN 3.131	12 23 33.42	+ 12 05 20.0			NEBULA
IC 3331	12 23 34.	+ 12 05 22.			NONSTELLAR OBJECT
AMES 1031	12 23 34.	+ 12 50 18.	38	17.2	NEBULA
REIZ 2340	12 23 34.	+ 59 42	24	15.2	GALAXY
HOLM 402A	12 23 34.	- 07 25	54	13.9	PART OF MULTIPLE GALAXY
IC 0788	12 23 35.	+ 16 28 28.			SAME AS NGC 4405
IC 3332	12 23 35.	+ 25 33 22.			NONSTELLAR OBJECT
REIN 3.132	12 23 35.99	+ 13 23 17.1			NEBULA
AMES 1037	12 23 36.	+ 06 17 24.	17	17.9	NEBULA
UGC 07518	12 23 36.	+ 07 50	144	14.7	GALAXY Sc-IRR
2ZW 062	12 23 36.	+ 07 56			COMPACT GALAXY
AMES 1036	12 23 36.	+ 08 30 48.	12	17.8	NEBULA
KARA.68 126	12 23 36.	+ 09 15	67		DWARF GALAXY
AMES 1034	12 23 36.	+ 11 08 30.	24	17.6	NEBULA

OBJECT NAME	RIGHT ASCEN.	DECLINATION	DIAM.	MAGN.	TYPE OF OBJECT
REA 15	12 23 36.	+ 12 51			DWARF GALAXY
MCG+02-32-045	12 23 36.	+ 12 51	18	17.	GALAXY
LB 04554	12 23 36.	+ 12 52		18.0	FAINT BLUE STAR
ZWG 070.071	12 23 36.	+ 13 23		13.6	GALAXY
UGC 07528	12 23 36.	+ 13 23	228	13.6	GALAXY S
MCG+02-32-044	12 23 36.	+ 13 24	198	13.6	GALAXY
LB 04553	12 23 36.	+ 13 29		17.3	FAINT BLUE STAR
LB 04552	12 23 36.	+ 14 27		18.2	FAINT BLUE STAR
KARA.68 127	12 23 36.	+ 15 11	34		DWARF GALAXY
MCG+03-32-036	12 23 36.	+ 16 27	96	12.9	GALAXY
ZWG 099.050	12 23 36.	+ 16 28		12.9	GALAXY
UGC 07529	12 23 36.	+ 16 28	102	12.9	GALAXY S0-a
ZWG 128.088	12 23 36.	+ 20 37		15.7	GALAXY
HOLM 404B	12 23 36.	+ 31 07	30	14.8	PART OF MULTIPLE GALAXY
HOLM 401B	12 23 36.	+ 46 09	48	14.4	PART OF MULTIPLE GALAXY
UGC 07530	12 23 36.	+ 46 11	66	16.0	GALAXY Sb
ZWG 315.038	12 23 36.	+ 62 43		15.0	GALAXY
ZCG 199.54	12 23 36.	+ 63 45		18.3	COMPACT GALAXY
ZWG 014.036	12 23 36.	- 02 04		15.6	GALAXY
REIN 7.099	12 23 36.03	+ 46 10 06.8			NEBULA
AMES 1033	12 23 37.	+ 18 01 18.	26	17.1	NEBULA
REIZ 2338	12 23 38.	+ 13 13	132	11.2	GALAXY
RNGC 4402	12 23 38.	+ 13 23		13.5	GALAXY
REIZ 2334	12 23 38.	+ 13 24	180	13.0	GALAXY
IC 3333	12 23 38.	+ 13 24 40.			SINGLE STAR
RNGC 4405	12 23 38.	+ 16 27		13.0	GALAXY
REIZ 2337	12 23 38.	+ 16 27	48	13.3	GALAXY
AMES 1035	12 23 38.	+ 17 01 42.	30	16.9	NEBULA
MAI 071	12 23 38.	+ 58 36	94		DWARF SPHEROIDAL GALAXY
HOLM 402B	12 23 38.	- 07 25	30	14.1	PART OF MULTIPLE GALAXY
A3 149	12 23 38.0	+ 13 56 37.		17.1	FAINT BLUE OBJECT
AMES 1038	12 23 39.	+ 06 19 18.	18	17.8	NEBULA
REIZ 2339	12 23 39.	+ 46 11	60	14.4	GALAXY
REIZ 2342	12 23 39.	+ 62 12	18	14.8	GALAXY
A3 151	12 23 39.4	+ 13 45 22.		16.1	FAINT BLUE OBJECT
REIN 3.133	12 23 39.85	+ 13 13 20.7			NEBULA
AMES 1040	12 23 40.	+ 07 02 06.	16	17.6	NEBULA
AMES 1039	12 23 40.	+ 09 07 54.	32	17.6	NEBULA
IC 3334	12 23 40.	+ 28 44 22.			NONSTELLAR OBJECT
REIZ 2336	12 23 40.	- 07 25	18	14.1	GALAXY
REIZ 2335	12 23 40.	- 07 25	60	13.8	GALAXY
A3 152	12 23 40.1	+ 16 42 24.		16.8	FAINT BLUE OBJECT
ZWG 014.038	12 23 42.	+ 00 33		15.5	GALAXY
ZWG 042.101	12 23 42.	+ 05 45		15.2	GALAXY
MCG+01-32-060	12 23 42.	+ 05 45	24	15.2	GALAXY
AMES 1041	12 23 42.	+ 08 20 06.	23	17.5	NEBULA
LB 04561	12 23 42.	+ 12 24		18.3	FAINT BLUE STAR
LB 04560	12 23 42.	+ 12 26		17.8	FAINT BLUE STAR
LB 04559	12 23 42.	+ 12 29		19.8	FAINT BLUE STAR
RMB 144	12 23 42.	+ 12 35			FAINT BLUE OBJECT
ZWG 070.072	12 23 42.	+ 13 14		10.9	GALAXY
UGC 07532	12 23 42.	+ 13 14	720	10.9	GALAXY E
MCG+02-32-046	12 23 42.	+ 13 14	90	10.9	GALAXY
LB 04558	12 23 42.	+ 13 22		18.0	FAINT BLUE STAR
LB 04557	12 23 42.	+ 14 02		19.5	FAINT BLUE STAR
LB 04556	12 23 42.	+ 14 10		17.1	FAINT BLUE STAR
LB 04555	12 23 42.	+ 14 20		16.2	FAINT BLUE STAR
REA 44	12 23 42.	+ 15 12	48		DWARF GALAXY
MCG+05-29-085	12 23 42.	+ 31 31	180	10.9	GALAXY
MCG+08-23-034	12 23 42.	+ 45 10	60	14.4	GALAXY
MCG+09-20-174	12 23 42.	+ 55 48	36	15.	GALAXY
7ZW 456	12 23 42.	+ 61 56			COMPACT GALAXY
ZWG 014.037	12 23 42.	- 01 01		14.9	GALAXY
UGC 07531	12 23 42.	- 01 01	66	14.9	GALAXY MLT SYS
MCG+00-32-011	12 23 42.	- 01 01	60	14.9	GALAXY
ZC 1223.7-0338	12 23 42.	- 03 38	1810		CLUSTER OF GALAXIES
MCG-01-32-008	12 23 42.	- 07 25	90	14.	GALAXY
MCG-02-32-010	12 23 42.	- 13 25	36	15.	GALAXY
A3 153	12 23 42.1	+ 16 00 16.		17.0	FAINT BLUE OBJECT
HOLM 403E	12 23 43.	+ 13 21	24	15.5	PART OF MULTIPLE GALAXY
RNGC 4403	12 23 43.	- 07 25		14.0	GALAXY
RNGC 4407	12 23 44.	+ 12 55			NON-EXISTENT OBJECT
RNGC 4406	12 23 44.	+ 13 13		11.0	GALAXY
HARO 29	12 23 44.	+ 48 44			BLUE EMISSION-LINE GALAXY
RNGC 4408	12 23 45.	+ 28 09		15.0	GALAXY
MCG-01-32-009	12 23 45.	- 07 25	30	14.	GALAXY
REIN 3.134	12 23 45.99	+ 08 37 39.0			NEBULA
AMES 1043	12 23 46.	+ 06 57 06.	18	17.0	NEBULA
AMES 1042	12 23 46.	+ 17 41 42.	26	17.4	NEBULA
REIZ 2345	12 23 46.	+ 59 30	42	15.9	GALAXY
LB 02357	12 23 47.	+ 55 47 36.		17.0	FAINT BLUE STAR
REIN 3.135	12 23 47.91	+ 07 44 12.3			NEBULA
ZWG 042.102	12 23 48.	+ 07 44		15.2	GALAXY
UGC 07533	12 23 48.	+ 07 44	66	15.2	GALAXY S0?
MCG+01-32-061	12 23 48.	+ 07 44	18	15.2	GALAXY
IC 0789	12 23 48.	+ 07 44		15.2	GALAXY IN VIRGO CLSTR
AMES 1047	12 23 48.	+ 08 57 06.	28	17.6	NEBULA
RMB 209	12 23 48.	+ 10 02			FAINT BLUE OBJECT
REA 67	12 23 48.	+ 10 13	24		DWARF GALAXY
AMES 1046	12 23 48.	+ 12 03 00.	25	17.1	NEBULA
LB 04566	12 23 48.	+ 12 45		19.0	FAINT BLUE STAR
LB 04565	12 23 48.	+ 13 03		19.2	FAINT BLUE STAR
LB 04564	12 23 48.	+ 13 35		20.4	FAINT BLUE STAR
LB 04563	12 23 48.	+ 13 39		20.0	FAINT BLUE STAR
LB 04562	12 23 48.	+ 14 08		18.9	FAINT BLUE STAR
ZC 1223.8+2035	12 23 48.	+ 20 35	540		CLUSTER OF GALAXIES
ZWG 158.106	12 23 48.	+ 27 06		15.5	GALAXY
SN 1970O	12 23 48.	+ 28 02		13.7	SUPERNOVA
REIZ 2341	12 23 48.	+ 28 08	18	15.2	GALAXY
ZWG 158.107	12 23 48.	+ 28 09		15.2	GALAXY
1ZW 036	12 23 48.	+ 48 46			COMPACT GALAXY
MCG+08-23-035	12 23 48.	+ 48 47	48	15.	GALAXY
ZWG 293.020	12 23 48.	+ 58 35		15.3	GALAXY
UGC 07534	12 23 48.	+ 58 35	210	15.3	GALAXY IRR
MCG+10-18-045	12 23 48.	+ 58 36	132	14.	GALAXY
MRSL 300-00/1	12 23 48.	- 62 33	720		HII REGION
VHA 131	12 23 48.	- 63 09	270		OPEN STAR CLUSTER
AMES 1048	12 23 49.	+ 08 28 12.	16	17.5	NEBULA
AMES 1045	12 23 49.	+ 18 00 18.	23	16.5	NEBULA
AMES 1044	12 23 49.	+ 18 03 36.	22	17.2	NEBULA
RNGC 4404	12 23 49.	- 07 25		14.0	GALAXY
HELW 058	12 23 49.	- 18 05 31.			NEBULA
A3 156	12 23 49.6	+ 15 50 39.		17.0	FAINT BLUE OBJECT
IC 3335	12 23 50.	+ 26 24 22.			NONSTELLAR OBJECT
REIZ 2344	12 23 50.	+ 48 46	36	14.9	GALAXY
IC 3337	12 23 51.	+ 25 35 22.			NONSTELLAR OBJECT
IC 3336	12 23 51.	+ 27 06 58.			NONSTELLAR OBJECT
ARC 1538	12 23 51.	+ 57 10		18.0	RICH CLUSTER OF GALAXIES

OBJECT NAME	RIGHT ASCEN.	DECLINATION	DIAM.	MAGN.	TYPE OF OBJECT
AMES 1050	12 23 52.	+ 08 42 30.	16	17.2	NEBULA
IC 3338	12 23 52.	+ 26 09 52.			NONSTELLAR OBJECT
MAI 073	12 23 52.	+ 70 37	121		DWARF SPHEROIDAL GALAXY
AMES 1052	12 23 53.	+ 06 59 24.	34	17.6	NEBULA
A3 157	12 23 53.0	+ 13 27 53.		17.3	FAINT BLUE OBJECT
A3 158	12 23 53.3	+ 14 35 40.		16.6	FAINT BLUE OBJECT
KEEL 440	12 23 53.3	+ 18 43 21.		18.	NEBULA
A3 159	12 23 53.9	+ 13 51 12.		16.6	FAINT BLUE OBJECT
ZWG 042.103	12 23 54.	+ 08 20		15.4	GALAXY
ZWG 070.073	12 23 54.	+ 09 17		13.6	GALAXY
SCH 45	12 23 54.	+ 09 17		13.6	PECULIAR GALAXY
UGC 07535	12 23 54.	+ 09 17	60	13.6	GALAXY DBL SYS
KARA.72 335B	12 23 54.	+ 09 17	36		PART OF DOUBLE GALAXY
KARA.72 335A	12 23 54.	+ 09 17	42	13.6	PART OF DOUBLE GALAXY
MCG+02-32-047	12 23 54.	+ 09 17	60	13.6	GALAXY
RMB 208	12 23 54.	+ 10 01			FAINT BLUE OBJECT
LB 04571	12 23 54.	+ 12 54		19.8	FAINT BLUE STAR
AMES 1051	12 23 54.	+ 13 03 18.	19	17.9	NEBULA
RMB 095	12 23 54.	+ 13 14			FAINT BLUE OBJECT
LB 04570	12 23 54.	+ 13 17		18.0	FAINT BLUE STAR
LB 04569	12 23 54.	+ 13 28		17.0	FAINT BLUE STAR
LB 04568	12 23 54.	+ 13 49		17.6	FAINT BLUE STAR
LB 04567	12 23 54.	+ 14 05		16.8	FAINT BLUE STAR
RMB 045	12 23 54.	+ 15 50			FAINT BLUE OBJECT
AMES 1049	12 23 54.	+ 17 55 30.	26	16.7	NEBULA
IC 3341	12 23 54.	+ 28 01 28.			NONSTELLAR OBJECT
ZWG 244.018	12 23 54.	+ 48 46		15.3	GALAXY
MRK 209	12 23 54.	+ 48 46	11	15.5	GALAXY WITH UV CONTINUUM
RNGC 4409	12 23 55.	+ 02 46			NON-EXISTENT OBJECT
AMES 1054	12 23 55.	+ 06 45 18.	18	17.4	NEBULA
SN 1965A	12 23 55.	+ 09 17		16.0	SUPERNOVA
REIN 3.136	12 23 55.91	+ 09 17 46.6			NEBULA
RNGC 4411A	12 23 56.	+ 09 09		14.5	GALAXY
RNGC 4410A	12 23 56.	+ 09 18		13.5	GALAXY
AMES 1053	12 23 56.	+ 12 27 12.	26	17.0	NEBULA
HOLM 403F	12 23 56.	+ 12 51	90	12.3	PART OF MULTIPLE GALAXY
HN 0900	12 23 56.	+ 17 08	12		NEBULA
IC 3340	12 23 56.	+ 17 08			NONSTELLAR OBJECT
LB 02358	12 23 56.	+ 59 59 48.		16.2	FAINT BLUE STAR
A3 160	12 23 56.3	+ 13 00 14.		16.9	FAINT BLUE OBJECT
REIN 3.137	12 23 56.69	+ 09 08 52.4			NEBULA
AMES 1056	12 23 57.	+ 07 20 18.	26	15.9	NEBULA
IC 3339	12 23 57.	+ 09 08 52.			SAME AS NGC 4411
REIZ 2343	12 23 57.	+ 09 18	48	14.5	GALAXY
AMES 1055	12 23 57.	+ 16 29 30.	22	17.6	NEBULA
RNGC 4414	12 23 57.	+ 31 30		11.5	GALAXY
REIN 3.138	12 23 57.21	+ 09 17 43.8			NEBULA
AMES 1057	12 23 58.	+ 07 17 54.	17	15.7	NEBULA
IC 3342	12 23 58.	+ 27 24 58.			NONSTELLAR OBJECT
MAI 072	12 23 58.	+ 62 43	47		DWARF SPHEROIDAL GALAXY
ARC 1539	12 23 58.	+ 62 50		17.2	RICH CLUSTER OF GALAXIES
ARC 1537	12 23 58.	- 25 34		17.6	RICH CLUSTER OF GALAXIES
RNGC 4373B	12 23 58.	- 38 51			GALAXY
VDB.66G 124	12 24	+ 13 23	70		DWARF GALAXY
VDB.66G 126	12 24	+ 37 28	130		DWARF GALAXY
VDB.66G 125	12 24	+ 43 43	200		DWARF GALAXY
ZWG 042.104	12 24 00.	+ 04 14		13.2	GALAXY
UGC 07536	12 24 00.	+ 04 14	102	13.2	GALAXY SBc?
MCG+01-32-062	12 24 00.	+ 04 14	54	13.2	GALAXY
ZWG 070.074	12 24 00.	+ 09 09		14.4	GALAXY
UGC 07537	12 24 00.	+ 09 09	138	14.4	GALAXY SBc
KARA.72 336A	12 24 00.	+ 09 09	120	14.4	PART OF DOUBLE GALAXY
MCG+02-32-048	12 24 00.	+ 09 09	120	14.4	GALAXY
ZWG 070.075	12 24 00.	+ 09 18		15.2	GALAXY
MCG+02-32-051	12 24 00.	+ 09 18	30	15.2	GALAXY
LB 04577	12 24 00.	+ 12 28		19.4	FAINT BLUE STAR
ZWG 070.076	12 24 00.	+ 12 53		13.6	GALAXY
UGC 07538	12 24 00.	+ 12 53	144	13.6	GALAXY SBa
MCG+02-32-049	12 24 00.	+ 12 53	120	13.6	GALAXY
LB 04576	12 24 00.	+ 13 43		17.3	FAINT BLUE STAR
LB 04575	12 24 00.	+ 13 44		18.5	FAINT BLUE STAR
LB 04574	12 24 00.	+ 13 50		14.7	FAINT BLUE STAR
ZWG 070.077	12 24 00.	+ 13 51		15.2	GALAXY
MCG+02-32-050	12 24 00.	+ 13 51	30	15.2	GALAXY
LB 04573	12 24 00.	+ 13 58		17.7	FAINT BLUE STAR
LB 04572	12 24 00.	+ 14 04		15.0	FAINT BLUE STAR
ZWG 099.051	12 24 00.	+ 16 55		15.7	GALAXY
ZWG 099.052	12 24 00.	+ 17 07		15.5	GALAXY
MCG+03-32-037	12 24 00.	+ 17 07	42	15.5	GALAXY
ZC 1224.0+1805	12 24 00.	+ 18 05	1480		CLUSTER OF GALAXIES
ZWG 158.108	12 24 00.	+ 31 30		10.9	GALAXY
REIZ 2348	12 24 00.	+ 31 30	150	12.0	GALAXY
UGC 07539	12 24 00.	+ 31 30	288	10.9	GALAXY Sc
LB 06384	12 24 00.	+ 31 54		18.9	FAINT BLUE STAR
ZC 1224.0+5415	12 24 00.	+ 54 15	1280		CLUSTER OF GALAXIES
MCG+09-20-176	12 24 00.	+ 54 28	60	15.	GALAXY
MCG+09-20-175	12 24 00.	+ 56 08	30	17.	GALAXY
ZCG 199.55	12 24 00.	+ 64 03		18.2	COMPACT GALAXY
DV.56 N4373B	12 24 00.	- 38 51	63		Sd GALAXY
MCG-06-27-027A	12 24 00.	- 38 51	48	17.	GALAXY
A3 162	12 24 00.0	+ 12 53 16.		14.0	UV CONCENTRATION
A3 163	12 24 00.1	+ 13 21 03.		16.7	FAINT BLUE OBJECT
REIN 3.139	12 24 00.42	+ 12 53 13.1			NEBULA
REIN 2.156	12 24 00.43	+ 12 53 13.6			NEBULA
RNGC 4412	12 24 01.	+ 04 14		13.0	GALAXY
AMES 1058	12 24 01.	+ 16 54 42.	24	16.3	NEBULA
REIN 3.140	12 24 01.67	+ 13 01 06.2			NEBULA
RNGC 4410B	12 24 02.	+ 09 18		13.5	GALAXY
RNGC 4413	12 24 02.	+ 12 53		13.5	GALAXY
REIZ 2347	12 24 02.	+ 12 53	54	13.5	GALAXY
HN 0901	12 24 02.	+ 13 50		14.5	NEBULA
AMES 1059	12 24 02.	+ 16 48 54.	28	16.4	NEBULA
IC 3345	12 24 02.	+ 24 38 46.			NONSTELLAR OBJECT
SN 1974G	12 24 02.	+ 31 29		13.0	SUPERNOVA
REIZ 2346	12 24 03.	+ 04 14	60	13.5	GALAXY
IC 3343	12 24 03.	+ 09 09 10.			MAY NOT EXIST
AMES 1060	12 24 03.	+ 18 33 00.	38	17.4	NEBULA
TON-N 1532	12 24 03.	+ 21 50		17.0	BLUE STAR
REIN 3.141	12 24 03.09	+ 09 18 42.7			NEBULA
A3 164	12 24 03.2	+ 12 49 58.		16.6	FAINT BLUE OBJECT
AMES 1063	12 24 04.	+ 06 43 42.	20	17.8	NEBULA
IC 0790	12 24 04.	+ 09 18			NONSTELLAR OBJECT
AMES 1061	12 24 04.	+ 11 07 18.	30	16.8	NEBULA
A3 166	12 24 04.0	+ 13 49 18.		17.3	FAINT BLUE OBJECT
AMES 1062	12 24 05.	+ 12 45 48.	46	16.6	NEBULA
IC 3344	12 24 05.	+ 13 50 52.			NONSTELLAR OBJECT
A3 168	12 24 05.5	+ 14 07 33.		16.3	FAINT BLUE OBJECT
AMES 1066	12 24 06.	+ 06 17 42.	23	17.6	NEBULA

OBJECT NAME	RIGHT ASCEN.	DECLINATION	DIAM.	MAGN.	TYPE OF OBJECT
ZWG 070.078	12 24 06.	+ 08 42		14.2	GALAXY
UGC 07540	12 24 06.	+ 08 42	84	14.2	GALAXY S0-a
MCG+02-32-052	12 24 06.	+ 08 42	60	14.2	GALAXY
ZC 1224.1+0914	12 24 06.	+ 09 14	1610		CLUSTER OF GALAXIES
LB 04584	12 24 06.	+ 12 26		19.5	FAINT BLUE STAR
LB 04583	12 24 06.	+ 12 29		17.3	FAINT BLUE STAR
RMB 143	12 24 06.	+ 12 47			FAINT BLUE OBJECT
LB 04582	12 24 06.	+ 12 47		16.6	FAINT BLUE STAR
LB 04581	12 24 06.	+ 12 50		18.7	FAINT BLUE STAR
LB 04580	12 24 06.	+ 13 20		18.6	FAINT BLUE STAR
AMES 1064	12 24 06.	+ 13 30 54.	17	17.6	NEBULA
LB 04579	12 24 06.	+ 13 49		14.0	FAINT BLUE STAR
LB 04578	12 24 06.	+ 14 05		15.9	FAINT BLUE STAR
TON-N 0617	12 24 06.	+ 30 58		15.2	BLUE STAR
LB 02359	12 24 06.	+ 52 50 06.		16.9	FAINT BLUE STAR
MCG+09-20-177	12 24 06.	+ 53 17	60	15.	GALAXY
ZC 1224.1+5703	12 24 06.	+ 57 03	1880		CLUSTER OF GALAXIES
MCG+11-15-055	12 24 06.	+ 62 38	39	17.	GALAXY
A3 169	12 24 06.1	+ 16 41 35.		16.2	FAINT BLUE OBJECT
AMES 1065	12 24 07.	+ 15 53 54.	14	17.6	NEBULA
RNGC 4415	12 24 08.	+ 08 42		14.0	GALAXY
AMES 1070	12 24 08.	+ 09 00 54.	25	16.8	NEBULA
HN 0903	12 24 08.	+ 11 10		15.5	NEBULA
IC 3347	12 24 08.	+ 11 10			NONSTELLAR OBJECT
HN 0902	12 24 08.	+ 11 38	6		NEBULA
IC 3346	12 24 08.	+ 11 38			NONSTELLAR OBJECT
AMES 1069	12 24 08.	+ 12 57 30.	29	17.0	NEBULA
AMES 1068	12 24 08.	+ 13 30 06.	19	17.7	NEBULA
IC 3348	12 24 08.	+ 25 54 04.			NONSTELLAR OBJECT
REIZ 2351	12 24 08.	+ 49 28	30	14.3	GALAXY
A3 170	12 24 08.3	+ 14 05 18.		17.4	FAINT BLUE OBJECT
REIN 3.142	12 24 08.35	+ 08 42 47.7			NEBULA
REIZ 2349	12 24 09.	+ 08 42	30	14.2	GALAXY
AMES 1067	12 24 09.	+ 17 00 00.	22	17.6	NEBULA
TON-N 1533	12 24 09.	+ 19 26		17.1	BLUE STAR
LB 00242	12 24 09.	+ 58 16 24.		17.1	FAINT BLUE STAR
HOLM 403G	12 24 10.	+ 13 25	66	14.3	PART OF MULTIPLE GALAXY
REIZ 2354	12 24 10.	+ 59 40	36	15.3	GALAXY
KEEL 441	12 24 10.6	+ 18 24 52.		18.	NEBULA
A3 171	12 24 10.8	+ 16 05 26.		16.2	FAINT BLUE OBJECT
AMES 1072	12 24 11.	+ 07 56 36.	12	16.7	NEBULA
AMES 1071	12 24 11.	+ 11 02 54.	39	17.0	NEBULA
IC 3351	12 24 11.	+ 27 52 58.			NONSTELLAR OBJECT
REIN 3.143	12 24 11.87	+ 09 19 29.8			NEBULA
ZWG 042.105	12 24 12.	+ 08 12		13.5	GALAXY
UGC 07541	12 24 12.	+ 08 12	108	13.5	GALAXY SBc
MCG+01-32-063	12 24 12.	+ 08 12	96	13.5	GALAXY
AMES 1075	12 24 12.	+ 09 05 54.	24	17.8	NEBULA
ZWG 070.079	12 24 12.	+ 09 19		15.3	GALAXY
MCG+02-32-054	12 24 12.	+ 09 19	36	15.3	GALAXY
ZC 1224.2+0931	12 24 12.	+ 09 31	1010		CLUSTER OF GALAXIES
ZWG 070.080	12 24 12.	+ 09 51		12.2	GALAXY
UGC 07542	12 24 12.	+ 09 51	192	12.2	GALAXY S0
MCG+02-32-053	12 24 12.	+ 09 51	180	12.2	GALAXY
KARA.68 128	12 24 12.	+ 11 11	369		DWARF GALAXY
KARA.68 129	12 24 12.	+ 11 16	27		DWARF GALAXY
ZWG 070.081	12 24 12.	+ 12 44		15.3	GALAXY
LB 04592	12 24 12.	+ 12 48		19.8	FAINT BLUE STAR
LB 04591	12 24 12.	+ 12 50		19.4	FAINT BLUE STAR
LB 04590	12 24 12.	+ 13 28		17.7	FAINT BLUE STAR
LB 04589	12 24 12.	+ 13 40		18.1	FAINT BLUE STAR
LB 04588	12 24 12.	+ 13 45		19.3	FAINT BLUE STAR
LB 04587	12 24 12.	+ 13 52		18.4	FAINT BLUE STAR
LB 04586	12 24 12.	+ 14 05		20.0	FAINT BLUE STAR
LB 04585	12 24 12.	+ 14 28		17.3	FAINT BLUE STAR
ZC 1224.2+2256	12 24 12.	+ 22 56	2690		CLUSTER OF GALAXIES
IC 3353	12 24 12.	+ 28 11 22.			NONSTELLAR OBJECT
LB 06386	12 24 12.	+ 30 27		18.1	FAINT BLUE STAR
LB 06385	12 24 12.	+ 31 18		18.5	FAINT BLUE STAR
UGC 07543	12 24 12.	+ 53 18	60	16.0	GALAXY Sb
ZC 1224.2+6030	12 24 12.	+ 60 30	1880		CLUSTER OF GALAXIES
UGC 07544	12 24 12.	+ 62 29	66	17.	GALAXY DWARF
ZCG 199.56	12 24 12.	+ 64 04		18.4	COMPACT GALAXY
VHA 132	12 24 12.	- 63 48	180		OPEN STAR CLUSTER
A3 172	12 24 12.0	+ 14 39 44.		16.7	FAINT BLUE OBJECT
REIN 3.144	12 24 12.20	+ 11 11 43.1			NEBULA
AMES 1074	12 24 13.	+ 12 31 30.	20	17.1	NEBULA
RNGC 4416	12 24 14.	+ 08 12		13.5	GALAXY
RNGC 4411B	12 24 14.	+ 09 10		14.5	GALAXY
HN 0904	12 24 14.	+ 12 43	12		NEBULA
IC 3349	12 24 14.	+ 12 43			NONSTELLAR OBJECT
AMES 1073	12 24 14.	+ 17 32 30.	14	17.5	NEBULA
REIN 3.145	12 24 14.44	+ 08 11 47.0			NEBULA
REIN 3.146	12 24 14.81	+ 09 09 38.6			NEBULA
REIZ 2350	12 24 15.	+ 08 12	60	13.1	GALAXY
AMES 1080	12 24 15.	+ 08 51 18.	14	18.0	NEBULA
IC 3350	12 24 15.	+ 09 43 16.			SINGLE STAR
AMES 1076	12 24 15.	+ 11 11 18.	22	17.8	NEBULA
REIN 3.147	12 24 15.50	+ 12 43 48.1			NEBULA
IC 3352	12 24 16.	+ 09 02 04.			MAY NOT EXIST
AMES 1079	12 24 16.	+ 09 14 06.	22	17.6	NEBULA
AMES 1078	12 24 16.	+ 15 09 42.	31	17.6	NEBULA
AMES 1081	12 24 17.	+ 06 31 54.	26	17.4	NEBULA
AMES 1077	12 24 17.	+ 17 04 54.	28	17.3	NEBULA
AMES 1083	12 24 18.	+ 06 04 54.	28	17.6	NEBULA
ZWG 070.082	12 24 18.	+ 09 09		14.4	GALAXY
UGC 07546	12 24 18.	+ 09 09	192	14.4	GALAXY Sc
KARA.72 336B	12 24 18.	+ 09 09	156	14.4	PART OF DOUBLE GALAXY
MCG+02-32-055	12 24 18.	+ 09 09	120	14.4	GALAXY
ZWG 070.083	12 24 18.	+ 10 02		15.4	GALAXY
REA 68	12 24 18.	+ 10 10	36		DWARF GALAXY
ZWG 070.084	12 24 18.	+ 11 51		15.5	GALAXY
UGC 07547	12 24 18.	+ 11 51	96	15.5	GALAXY DWRF IR
REA 45	12 24 18.	+ 11 51	36		DWARF GALAXY
LB 04602	12 24 18.	+ 12 28		19.8	FAINT BLUE STAR
LB 04601	12 24 18.	+ 12 34		18.9	FAINT BLUE STAR
LB 04600	12 24 18.	+ 12 44		18.8	FAINT BLUE STAR
LB 04599	12 24 18.	+ 13 12		19.5	FAINT BLUE STAR
LB 04598	12 24 18.	+ 13 25		17.3	FAINT BLUE STAR
ZWG 070.085	12 24 18.	+ 13 27		15.2	GALAXY
UGC 07548	12 24 18.	+ 13 27	78	15.2	GALAXY IRR
LB 04597	12 24 18.	+ 13 48		20.4	FAINT BLUE STAR
MCG+02-32-056	12 24 18.	+ 13 57	60	15.2	GALAXY
LB 04596	12 24 18.	+ 14 05		19.5	FAINT BLUE STAR
LB 04595	12 24 18.	+ 14 18		18.6	FAINT BLUE STAR
LB 04594	12 24 18.	+ 14 20		18.8	FAINT BLUE STAR
LB 04593	12 24 18.	+ 14 29		18.4	FAINT BLUE STAR
ZZW 063	12 24 18.	+ 16 33			COMPACT GALAXY

OBJECT NAME	RIGHT ASCEN.	DECLINATION	DIAM.	MAGN.	TYPE OF OBJECT
ZWG 099.053	12 24 18.	+ 16 33		15.6	GALAXY
LB 06387	12 24 18.	+ 29 40		19.0	FAINT BLUE STAR
MCG+09-20-178	12 24 18.	+ 53 27 30.	42	16.	GALAXY
MCG+09-20-179	12 24 18.	+ 53 40	30	16.	GALAXY
MCG+09-20-180	12 24 18.	+ 56 22 30.	60	16.	GALAXY
MCG+10-18-046	12 24 18.	+ 58 07	39	16.	GALAXY
ZWG 014.039	12 24 18.	- 00 35		14.2	GALAXY
UGC 07545	12 24 18.	- 00 35	84	14.2	GALAXY Sa
KARA.72 337A	12 24 18.	- 00 35	96	14.2	PART OF DOUBLE GALAXY
MCG+00-32-012	12 24 18.	- 00 35	78	14.2	GALAXY
REIN 3.148	12 24 18.51	+ 09 51 42.7			NEBULA
A3 173	12 24 18.9	+ 13 27 15.		16.5	FAINT BLUE OBJECT
RNGC 4418	12 24 19.	- 00 35		14.0	GALAXY
AMES 1086	12 24 20.	+ 05 45 06.	32	17.2	NEBULA
AMES 1085	12 24 20.	+ 06 25 54.	17	17.6	NEBULA
REIZ 2352	12 24 20.	+ 09 51	96	12.4	GALAXY
RNGC 4417	12 24 20.	+ 09 52		12.5	GALAXY
HN 0906	12 24 20.	+ 10 02		15.	NEBULA
IC 3357	12 24 20.	+ 10 02			NONSTELLAR OBJECT
HN 0905	12 24 20.	+ 11 49		16.	NEBULA
IC 3356	12 24 20.	+ 11 49			SINGLE STAR
IC 3354	12 24 20.	+ 12 22 22.			SINGLE STAR
IC 3355	12 24 20.	+ 13 27 10.			NONSTELLAR OBJECT
AMES 1082	12 24 20.	+ 17 20 36.	24	17.8	NEBULA
LB 02360	12 24 20.	+ 54 09 54.		16.5	FAINT BLUE STAR
AMES 1084	12 24 21.	+ 09 17 30.	19	17.5	NEBULA
IC 3359	12 24 21.	+ 23 46 28.			NONSTELLAR OBJECT
IC 3360	12 24 21.	+ 26 19 28.			NONSTELLAR OBJECT
IC 3358	12 24 22.	+ 11 56 22.			NONSTELLAR OBJECT
AMES 1087	12 24 22.	+ 13 50 18.	30	16.8	NEBULA
REIZ 2353	12 24 22.	- 00 37	36	13.8	GALAXY
REIN 7.100A	12 24 22.33	+ 46 09 15.5			NEBULA
REIN 7.100B	12 24 22.39	+ 46 09 15.5			NEBULA
REIN 3.149	12 24 22.58	+ 11 56 25.2			NEBULA
ZWG 042.106	12 24 24.	+ 02 46		12.7	GALAXY
UGC 07549	12 24 24.	+ 02 46	132	12.7	GALAXY Sc
MCG+01-32-064	12 24 24.	+ 02 46	120	12.7	GALAXY
KARA.68 130	12 24 24.	+ 11 50	87		DWARF GALAXY
ZWG 070.086	12 24 24.	+ 11 56		15.0	GALAXY
UGC 07550	12 24 24.	+ 11 56	72	15.0	GALAXY
MCG+02-32-057	12 24 24.	+ 11 56	48	15.0	GALAXY
LB 04617	12 24 24.	+ 12 44		18.1	FAINT BLUE STAR
LB 04616	12 24 24.	+ 12 51		18.4	FAINT BLUE STAR
LB 04615	12 24 24.	+ 12 52		20.4	FAINT BLUE STAR
LB 04614	12 24 24.	+ 12 57		20.3	FAINT BLUE STAR
LB 04613	12 24 24.	+ 12 59		17.9	FAINT BLUE STAR
LB 04612	12 24 24.	+ 13 02		19.3	FAINT BLUE STAR
LB 04611	12 24 24.	+ 13 15		20.5	FAINT BLUE STAR
LB 04610	12 24 24.	+ 13 28		19.2	FAINT BLUE STAR
LB 04609	12 24 24.	+ 13 42		19.9	FAINT BLUE STAR
LB 04608	12 24 24.	+ 13 59		19.9	FAINT BLUE STAR
LB 04607	12 24 24.	+ 14 07		18.4	FAINT BLUE STAR
LB 04606	12 24 24.	+ 14 26		18.5	FAINT BLUE STAR
LB 04605	12 24 24.	+ 14 28		18.8	FAINT BLUE STAR
LB 04604	12 24 24.	+ 14 28		17.6	FAINT BLUE STAR
LB 04603	12 24 24.	+ 14 34		19.7	FAINT BLUE STAR
MCG+03-32-038	12 24 24.	+ 15 18	180	11.6	GALAXY
ZWG 099.054	12 24 24.	+ 15 19		11.6	GALAXY
UGC 07551	12 24 24.	+ 15 19	192	11.6	GALAXY Sa
ZC 1224.4+1539	12 24 24.	+ 15 39	270		CLUSTER OF GALAXIES
RMB 044	12 24 24.	+ 15 50			FAINT BLUE OBJECT
ZC 1224.4+2005	12 24 24.	+ 20 05	940		CLUSTER OF GALAXIES
LB 11210	12 24 24.	+ 30 47		17.6	FAINT BLUE STAR
ZWG 187.043	12 24 24.	+ 38 10		14.7	GALAXY
MCG+09-20-181	12 24 24.	+ 52 21	30	15.	GALAXY
MCG+09-20-182	12 24 24.	+ 53 23	36	16.	GALAXY
UGC 07552	12 24 24.	+ 53 29	60	16.0	GALAXY
MCG+10-18-047	12 24 24.	+ 59 48	39	16.	GALAXY
ZCG 199.57	12 24 24.	+ 63 41		19.0	COMPACT GALAXY
ZCG 199.58	12 24 24.	+ 63 42		18.9	COMPACT GALAXY
ZC 1224.4+6517	12 24 24.	+ 65 17	1410		CLUSTER OF GALAXIES
LB 00635	12 24 24.	+ 73 43 54.		17.2	FAINT BLUE STAR
MCG-02-32-011	12 24 24.	- 11 15	36	15.	GALAXY
VHA 133	12 24 24.	- 60 30	360		OPEN STAR CLUSTER
REIN 2.157	12 24 24.59	+ 15 19 25.9			NEBULA
RNGC 4420	12 24 25.	+ 02 46		13.0	GALAXY
AMES 1088	12 24 25.	+ 05 19 06.	29	17.6	NEBULA
IC 3362	12 24 25.	+ 26 58 04.			NONSTELLAR OBJECT
HN 0907	12 24 26.	+ 10 55		15.5	NEBULA
IC 3361	12 24 26.	+ 10 55			NONSTELLAR OBJECT
RNGC 4419	12 24 26.	+ 15 19		12.5	GALAXY
REIZ 2355	12 24 26.	+ 15 19	126	12.5	GALAXY
A3 175	12 24 26.5	+ 13 11 18.		16.2	FAINT BLUE OBJECT
KEEL 442	12 24 26.7	+ 18 19 00.		17.	NEBULA
REIZ 2356	12 24 27.	+ 02 46	72	12.6	GALAXY
MCG+08-23-036	12 24 27.	+ 46 09	30	17.	GALAXY
KEEL 443	12 24 27.8	+ 18 47 37.		18.	NEBULA
AMES 1089	12 24 29.	+ 16 37 12.	28	17.4	NEBULA
REA 79	12 24 30.	+ 07 32			DWARF GALAXY
ZWG 070.087	12 24 30.	+ 09 13		15.6	GALAXY
ZWG 070.088	12 24 30.	+ 12 50		15.5	GALAXY
LB 04624	12 24 30.	+ 12 57		19.6	FAINT BLUE STAR
LB 04623	12 24 30.	+ 13 11		16.6	FAINT BLUE STAR
LB 04622	12 24 30.	+ 13 17		20.4	FAINT BLUE STAR
LB 04621	12 24 30.	+ 13 44		18.5	FAINT BLUE STAR
LB 04620	12 24 30.	+ 13 49		18.9	FAINT BLUE STAR
LB 04619	12 24 30.	+ 14 12		18.3	FAINT BLUE STAR
LB 04618	12 24 30.	+ 14 19		17.3	FAINT BLUE STAR
MCG+03-32-039	12 24 30.	+ 15 43	138	12.9	GALAXY
ZWG 099.055	12 24 30.	+ 15 45		12.9	GALAXY
UGC 07554	12 24 30.	+ 15 45	168	12.9	GALAXY SB0/SBa
AMES 1090	12 24 30.	+ 16 37 00.	18	17.8	NEBULA
ZWG 128.089	12 24 30.	+ 22 55		14.2	GALAXY
UGC 07555	12 24 30.	+ 22 55	72	14.2	GALAXY SBa
LB 00007	12 24 30.	+ 27 28		18.0	FAINT BLUE STAR
MCG+06-27-054	12 24 30.	+ 38 13	48	14.	GALAXY
MCG+08-23-037	12 24 30.	+ 48 33	48	15.	GALAXY
ZWG 270.001	12 24 30.	+ 52 21		15.4	GALAXY
ZWG 269.058	12 24 30.	+ 52 21		15.4	GALAXY
ZWG 270.002	12 24 30.	+ 53 58		15.7	GALAXY
ZWG 269.059	12 24 30.	+ 53 58		15.7	GALAXY
ZCG 199.59	12 24 30.	+ 63 41		19.0	COMPACT GALAXY
ZWG 014.040	12 24 30.	- 00 37		15.0	GALAXY
KARA.72 337B	12 24 30.	- 00 37	48	15.0	PART OF DOUBLE GALAXY
MCG+00-32-013	12 24 30.	- 00 37	54	15.0	GALAXY
UGC 07553	12 24 30.	- 01 15	66	17.	GALAXY
REIZ 2.158	12 24 30.80	+ 15 44 15.8			NEBULA
IC 0791	12 24 31.	+ 22 54 34.			NONSTELLAR OBJECT

OBJECT NAME	RIGHT ASCEN.	DECLINATION	DIAM.	MAGN.	TYPE OF OBJECT
KEEL 444	12 24 31.2	+ 18 25 23.		18.	NEBULA
REIN 3.150	12 24 31.49	+ 12 50 11.5			NEBULA
HN 0908	12 24 32.	+ 12 49		15.	NEBULA
IC 3363	12 24 32.	+ 12 49			NONSTELLAR OBJECT
RNGC 4421	12 24 32.	+ 15 44		13.0	GALAXY
REIZ 2359	12 24 32.	+ 15 44	48	13.2	GALAXY
REIZ 2360	12 24 32.	+ 16 10	108	15.1	GALAXY
AMES 1091	12 24 32.	+ 16 31 36.	17	17.1	NEBULA
AMES 1094	12 24 33.	+ 07 11 30.	16	17.2	NEBULA
LB 02361	12 24 33.	+ 61 34 36.		16.3	FAINT BLUE STAR
A3 176	12 24 33.8	+ 15 25 19.		16.7	FAINT BLUE OBJECT
AMES 1098	12 24 34.	+ 06 42 00.	35	16.9	NEBULA
AMES 1097	12 24 34.	+ 07 54 48.	36	17.8	NEBULA
AMES 1093	12 24 34.	+ 16 03 00.	17	17.8	NEBULA
REIZ 2358	12 24 34.	- 00 39	24	15.1	GALAXY
AMES 1100	12 24 35.	+ 08 40 00.	26	17.2	NEBULA
AMES 1096	12 24 35.	+ 09 29 36.	44	17.6	NEBULA
HOLM 403E	12 24 35.	+ 12 59	180	12.3	PART OF MULTIPLE GALAXY
AMES 1092	12 24 35.	+ 18 55 24.	29	17.8	NEBULA
IC 3364	12 24 35.	+ 25 50 22.			NONSTELLAR OBJECT
REIZ 2357	12 24 35.	- 07 26	18	14.4	GALAXY
ZWG 042.107	12 24 36.	+ 06 09		14.4	GALAXY
UGC 07556	12 24 36.	+ 06 09	132	14.4	GALAXY Sc-IRR
MCG+01-32-065	12 24 36.	+ 06 09	120	14.4	GALAXY
ZWG 042.108	12 24 36.	+ 07 32		15.3	GALAXY
UGC 07557	12 24 36.	+ 07 32	192	15.3	GALAXY DWRF SP
MCG+01-32-066	12 24 36.	+ 07 32	180	15.3	GALAXY
REA 69	12 24 36.	+ 08 30	18		DWARF GALAXY
MCG+02-32-058	12 24 36.	+ 09 42	168	13.1	GALAXY
ZWG 070.089	12 24 36.	+ 09 58		15.5	GALAXY
LB 04628	12 24 36.	+ 12 28		18.4	FAINT BLUE STAR
LB 04627	12 24 36.	+ 13 56		18.7	FAINT BLUE STAR
LB 04626	12 24 36.	+ 14 06		19.8	FAINT BLUE STAR
RMB 062	12 24 36.	+ 14 12			FAINT BLUE OBJECT
LB 04625	12 24 36.	+ 14 23		17.2	FAINT BLUE STAR
MCG+03-32-040	12 24 36.	+ 16 36	96	15.1	GALAXY
ZWG 099.056	12 24 36.	+ 16 37		15.1	GALAXY
UGC 07558	12 24 36.	+ 16 37	108	15.1	GALAXY Sb
AMES 1095	12 24 36.	+ 16 52 00.	24	16.8	NEBULA
ZC 1224.6+2157	12 24 36.	+ 21 57	740		CLUSTER OF GALAXIES
LB 06390	12 24 36.	+ 29 58		18.0	FAINT BLUE STAR
ZC 1224.6+3131	12 24 36.	+ 31 31	8000		CLUSTER OF GALAXIES
ZC 1224.6+3334	12 24 36.	+ 33 34	2550		CLUSTER OF GALAXIES
ZWG 187.044	12 24 36.	+ 37 25		15.6	GALAXY
UGC 07559	12 24 36.	+ 37 25	270	15.6	GALAXY DWRF IR
MCG+06-27-055	12 24 36.	+ 37 26	150	15.5	GALAXY
LB 06389	12 24 36.	+ 39 37		18.7	FAINT BLUE STAR
ZWG 244.019	12 24 36.	+ 48 33		14.7	GALAXY
MRK 210	12 24 36.	+ 48 33	13	14.5	GALAXY WITH UV CONTINUUM
UGC 07560	12 24 36.	+ 48 33	60	14.7	GALAXY SB
SN 1960I	12 24 36.	+ 48 33		18.5	SUPERNOVA
ZC 1224.6+5011	12 24 36.	+ 50 11	1210		CLUSTER OF GALAXIES
MCG+09-20-183	12 24 36.	+ 53 58	42	15.	GALAXY
ZWG 014.041	12 24 36.	- 01 04		15.5	GALAXY
HOLM 405E	12 24 37.	+ 16 11	150	13.9	PART OF MULTIPLE GALAXY
IC 3365	12 24 37.	+ 16 11			NONSTELLAR OBJECT
AMES 1099	12 24 37.	+ 17 18 00.	22	17.5	NEBULA
RNGC 4423	12 24 38.	+ 06 09		14.5	GALAXY
AMES 1103	12 24 38.	+ 07 32 48.	200	16.2	NEBULA
REIZ 2362	12 24 38.	+ 09 42	120	13.3	GALAXY
HN 0909	12 24 38.	+ 16 11	60		NEBULA
HOLM 405A	12 24 38.	+ 16 16	48	15.2	PART OF MULTIPLE GALAXY
REIZ 2363	12 24 38.	+ 16 35	54	14.4	GALAXY
IC 0792	12 24 38.	+ 16 36 22.			NONSTELLAR OBJECT
AMES 1101	12 24 38.	+ 18 12 00.	25	17.8	NEBULA
LB 02362	12 24 38.	+ 60 40 06.		15.9	FAINT BLUE STAR
ABC 1540	12 24 39.	+ 04 30		17.8	RICH CLUSTER OF GALAXIES
REIZ 2361	12 24 39.	+ 06 08	126	14.1	GALAXY
AMES 1105	12 24 39.	+ 07 56 48.	36	17.6	NEBULA
AMES 1102	12 24 39.	+ 11 37 12.	23	17.6	NEBULA
TON-N 1534	12 24 39.	+ 22 08		15.8	BLUE STAR
MCG+04-29-071	12 24 39.	+ 22 55	72	14.2	GALAXY
LB 02364	12 24 39.	+ 53 23 30.		16.1	FAINT BLUE STAR
LB 02363	12 24 39.	+ 56 26 30.		16.8	FAINT BLUE STAR
REIN 3.151	12 24 39.55	+ 09 41 53.1			NEBULA
AMES 1104	12 24 40.	+ 10 44 48.	17	17.2	NEBULA
IC 3367	12 24 40.	+ 27 14 05.			NONSTELLAR OBJECT
REIN 3.152	12 24 41.52	+ 13 00 38.7			NEBULA
ZC 1224.7+0431	12 24 42.	+ 04 31	1080		CLUSTER OF GALAXIES
KARA.68 131	12 24 42.	+ 08 29	27		DWARF GALAXY
ZWG 070.090	12 24 42.	+ 09 42		13.1	GALAXY
UGC 07561	12 24 42.	+ 09 42	216	13.1	GALAXY S
RMB 213	12 24 42.	+ 11 12			FAINT BLUE OBJECT
LB 04635	12 24 42.	+ 12 25		18.6	FAINT BLUE STAR
LB 04634	12 24 42.	+ 12 39		18.2	FAINT BLUE STAR
LB 04633	12 24 42.	+ 12 41		18.7	FAINT BLUE STAR
LB 04632	12 24 42.	+ 12 45		18.2	FAINT BLUE STAR
ZWG 070.091	12 24 42.	+ 13 01		13.3	GALAXY
UGC 07562	12 24 42.	+ 13 01	174	13.3	GALAXY S0-a
MCG+02-32-059	12 24 42.	+ 13 01	156	13.3	GALAXY
LB 04631	12 24 42.	+ 13 22		20.3	FAINT BLUE STAR
LB 04630	12 24 42.	+ 13 33		18.9	FAINT BLUE STAR
LB 04629	12 24 42.	+ 13 43		18.7	FAINT BLUE STAR
3ZW 066	12 24 42.	+ 14 24			COMPACT GALAXY
MCG+03-32-041	12 24 42.	+ 16 10	150	15.0	GALAXY
ZWG 099.057	12 24 42.	+ 16 12		15.0	GALAXY
UGC 07563	12 24 42.	+ 16 12	126	15.0	GALAXY IRR
MCG+03-32-042	12 24 42.	+ 16 14	27	15.2	GALAXY
AMES 1106	12 24 42.	+ 16 16 00.	38	16.7	NEBULA
HOLM 405C	12 24 42.	+ 16 19	48	14.3	PART OF MULTIPLE GALAXY
LB 06388	12 24 42.	+ 26 49		19.5	FAINT BLUE STAR
ZWG 158.109	12 24 42.	+ 27 14		15.7	GALAXY
ZWG 014.043	12 24 42.	- 01 45		15.2	GALAXY
MCG-01-32-010	12 24 42.	- 05 33	30	14.5	GALAXY
MCG-01-32-011	12 24 42.	- 07 25	30	14.5	GALAXY
MCG-05-30-001	12 24 42.	- 33 17	6	16.	GALAXY
IC 3366	12 24 43.	+ 09 41 23.			MAY NOT EXIST
AMES 1108	12 24 43.	+ 17 17 30.	12	17.2	NEBULA
AMES 1107	12 24 43.	+ 18 06 24.	37	17.4	NEBULA
RNGC 4422	12 24 43.	- 05 33		14.5	GALAXY
AMES 1109	12 24 44.	+ 09 14 42.	23	17.2	NEBULA
RNGC 4424	12 24 44.	+ 09 42		12.5	GALAXY
RNGC 4425	12 24 44.	+ 13 01		13.0	GALAXY
REIZ 2364	12 24 44.	+ 13 01	120	12.9	GALAXY
HN 0911	12 24 44.	+ 16 19		14.	NEBULA
IC 3369	12 24 44.	+ 16 19			NONSTELLAR OBJECT
HN 0910	12 24 44.	+ 16 42	12		NEBULA
IC 3368	12 24 44.	+ 16 42			NONSTELLAR OBJECT
REIZ 2366	12 24 44.	+ 46 07	18	15.1	GALAXY
A3 181	12 24 44.6	+ 16 17 35.		17.4	FAINT BLUE OBJECT
KEEL 445	12 24 44.6	+ 18 32 30.		17.	NEBULA
A3 182	12 24 44.9	+ 16 46 31.		17.3	FAINT BLUE OBJECT
AMES 1113	12 24 45.	+ 08 53 30.	30	17.0	NEBULA
TON-N 1535	12 24 45.	+ 22 25		16.8	BLUE STAR
RNGC 4427	12 24 45.	+ 28 07			NON-EXISTENT OBJECT
RNGC 4426	12 24 45.	+ 28 07			NON-EXISTENT OBJECT
AMES 1112	12 24 46.	+ 10 00 36.	28	17.2	NEBULA
AMES 1110	12 24 46.	+ 16 46 24.	26	17.0	NEBULA
A3 184	12 24 46.5	+ 14 12 54.		17.1	FAINT BLUE OBJECT
SN 1895A	12 24 47.	+ 09 42		12.5	SUPERNOVA
AMES 1115	12 24 47.	+ 12 03 12.	22	17.9	NEBULA
AMES 1111	12 24 47.	+ 17 18 36.	32	16.7	NEBULA
A3 186	12 24 47.4	+ 14 09 49.		16.7	FAINT BLUE OBJECT
ZC 1224.8+0251	12 24 48.	+ 02 51	1010		CLUSTER OF GALAXIES
ZWG 042.109	12 24 48.	+ 03 34		15.4	GALAXY
UGC 07564	12 24 48.	+ 03 34	66	15.4	GALAXY SBb
AMES 1118	12 24 48.	+ 07 10 36.	22	17.4	NEBULA
ZC 1224.8+0957	12 24 48.	+ 09 57	870		CLUSTER OF GALAXIES
ZWG 070.092	12 24 48.	+ 11 09		15.0	GALAXY
UGC 07565	12 24 48.	+ 11 09	114	15.0	GALAXY Sc
MCG+02-32-060	12 24 48.	+ 11 09	96	15.0	GALAXY
AMES 1116	12 24 48.	+ 12 13 48.	17	17.8	NEBULA
LB 04642	12 24 48.	+ 12 36		18.1	FAINT BLUE STAR
LB 04641	12 24 48.	+ 12 48		18.9	FAINT BLUE STAR
LB 04640	12 24 48.	+ 12 57		18.8	FAINT BLUE STAR
LB 04639	12 24 48.	+ 13 10		18.9	FAINT BLUE STAR
LB 04638	12 24 48.	+ 13 12		18.6	FAINT BLUE STAR
LB 04637	12 24 48.	+ 13 14		18.7	FAINT BLUE STAR
LB 04636	12 24 48.	+ 14 13		15.8	FAINT BLUE STAR
MCG+03-32-043	12 24 48.	+ 16 16	18	15.1	GALAXY
ZWG 099.058	12 24 48.	+ 16 17		15.1	GALAXY
AMES 1114	12 24 48.	+ 18 46 24.	22	17.4	NEBULA
ZC 1224.8+2402	12 24 48.	+ 24 02	740		CLUSTER OF GALAXIES
LB 06392	12 24 48.	+ 27 47		18.5	FAINT BLUE STAR
LB 06391	12 24 48.	+ 27 53		19.1	FAINT BLUE STAR
ZC 1224.8+2806	12 24 48.	+ 28 06	1280		CLUSTER OF GALAXIES
ZWG 158.110	12 24 48.	+ 30 54		15.7	GALAXY
MCG+10-18-048	12 24 48.	+ 58 03	30	16.	GALAXY
AMES 1119	12 24 49.	+ 08 55 24.	20	17.2	NEBULA
AMES 1121	12 24 50.	+ 07 02 24.	23	17.4	NEBULA
IC 3371	12 24 50.	+ 11 07 53.			NONSTELLAR OBJECT
HN 0912	12 24 50.	+ 11 08	60		NEBULA
AMES 1117	12 24 50.	+ 18 59 36.	34	17.1	NEBULA
REIN 3.153	12 24 50.20	+ 11 08 37.1			NEBULA
HOLM 406A	12 24 51.	+ 06 32	132	12.2	PART OF MULTIPLE GALAXY
TON-N 1536	12 24 51.	+ 22 25		16.5	BLUE STAR
MCG-06-27-028	12 24 51.	- 38 40	30	15.5	GALAXY
AMES 1123	12 24 52.	+ 09 07 06.	17	17.7	NEBULA
AMES 1122	12 24 52.	+ 13 59 18.	31	17.5	NEBULA
AMES 1120	12 24 52.	+ 17 17 54.	14	16.8	NEBULA
A3 187	12 24 52.6	+ 15 38 15.		17.3	FAINT BLUE OBJECT
REIZ 2365	12 24 53.	- 07 55	60	12.9	GALAXY
ZWG 042.110	12 24 54.	+ 03 32		15.4	GALAXY
AMES 1128	12 24 54.	+ 05 31 06.	23	17.6	NEBULA
ZWG 042.111	12 24 54.	+ 06 32		13.4	GALAXY
SCH 46	12 24 54.	+ 06 32		13.4	PECULIAR GALAXY
UGC 07566	12 24 54.	+ 06 32	180	13.4	GALAXY SBb
KARA.72 338A	12 24 54.	+ 06 32	150	13.4	PART OF DOUBLE GALAXY
MCG+01-32-067	12 24 54.	+ 06 32	120	13.4	GALAXY
ZWG 042.112	12 24 54.	+ 07 55		15.3	GALAXY
UGC 07567	12 24 54.	+ 07 55	66	15.3	GALAXY DWARF
ZWG 042.113	12 24 54.	+ 08 15		15.6	GALAXY
AMES 1127	12 24 54.	+ 09 14 24.	25	17.3	NEBULA
REA 46	12 24 54.	+ 09 53	60		DWARF GALAXY
MCG+02-32-063	12 24 54.	+ 09 53	120	18.	GALAXY
RMB 210	12 24 54.	+ 10 54			FAINT BLUE OBJECT
ZWG 070.093	12 24 54.	+ 11 23		11.4	GALAXY
UGC 07568	12 24 54.	+ 11 23	360	11.4	GALAXY S0
MCG+02-32-061	12 24 54.	+ 11 23	258	11.4	GALAXY
RMB 141	12 24 54.	+ 11 55			FAINT BLUE OBJECT
HOLM 408C	12 24 54.	+ 12 31	66	14.1	PART OF MULTIPLE GALAXY
ZWG 070.094	12 24 54.	+ 12 34		14.5	GALAXY
LB 04651	12 24 54.	+ 12 34		19.2	FAINT BLUE STAR
UGC 07569	12 24 54.	+ 12 34	102	14.5	GALAXY S0
MCG+02-32-062	12 24 54.	+ 12 34	36	14.5	GALAXY
LB 04650	12 24 54.	+ 12 52		18.7	FAINT BLUE STAR
LB 04649	12 24 54.	+ 13 08		19.2	FAINT BLUE STAR
LB 04648	12 24 54.	+ 13 09		19.4	FAINT BLUE STAR
LB 04647	12 24 54.	+ 13 10		18.8	FAINT BLUE STAR
LB 04646	12 24 54.	+ 13 23		18.7	FAINT BLUE STAR
LB 04645	12 24 54.	+ 13 24		18.0	FAINT BLUE STAR
LB 04644	12 24 54.	+ 13 57		18.6	FAINT BLUE STAR
AMES 1126	12 24 54.	+ 14 01 24.	36	17.4	NEBULA
LB 04643	12 24 54.	+ 14 18		20.7	FAINT BLUE STAR
AMES 1124	12 24 54.	+ 17 00 48.	28	17.6	NEBULA
LB 11211	12 24 54.	+ 31 30		13.4	FAINT BLUE STAR
ZCG 199.60	12 24 54.	+ 63 44		18.7	COMPACT GALAXY
HOLM 407B	12 24 54.	- 07 54	78	12.1	PART OF MULTIPLE GALAXY
REIN 3.154	12 24 54.44	+ 11 23 04.5			NEBULA
AMES 1130	12 24 55.	+ 06 50 12.	36	17.6	NEBULA
AMES 1125	12 24 55.	+ 16 54 12.	32	18.0	NEBULA
IC 3372	12 24 55.	+ 25 33 47.			NONSTELLAR OBJECT
RNGC 4428	12 24 55.	- 07 54		13.5	GALAXY
REIN 3.155	12 24 55.62	+ 12 33 59.4			NEBULA
A3 189	12 24 55.7	+ 14 44 01.		17.0	FAINT BLUE OBJECT
RNGC 4430	12 24 56.	+ 06 32		13.5	GALAXY
ABC 1541	12 24 56.	+ 09 07		16.0	RICH CLUSTER OF GALAXIES
RNGC 4429	12 24 56.	+ 11 23		11.5	GALAXY
REIZ 2369	12 24 56.	+ 11 23	210	11.9	GALAXY
RNGC 4431	12 24 56.	+ 12 34		14.5	GALAXY
REIZ 2371	12 24 56.	+ 12 34	36	14.4	GALAXY
A3 183	12 24 56.6	+ 13 12 00.		17.8	FAINT BLUE OBJECT
AMES 1135	12 24 57.	+ 06 16 12.	30	17.3	NEBULA
REIZ 2370	12 24 57.	+ 06 32	120	12.9	GALAXY
REIZ 2367	12 24 57.	+ 07 55	60	15.4	GALAXY
REIZ 2368	12 24 57.	+ 08 30	60	15.4	GALAXY
AMES 1134	12 24 57.	+ 09 19 00.	26	16.5	NEBULA
AMES 1133	12 24 57.	+ 09 37 06.	29	17.4	NEBULA
AMES 1132	12 24 57.	+ 10 27 06.	20	17.0	NEBULA
AMES 1129	12 24 57.	+ 17 14 42.	23	17.2	NEBULA
MCG-01-32-012	12 24 57.	- 07 54	96	14.5	GALAXY
REIN 3.156	12 24 57.16	+ 07 55 14.3			NEBULA
REIN 3.157	12 24 57.92	+ 09 06 03.0			NEBULA
AMES 1138	12 24 58.	+ 08 14 36.	28	16.6	NEBULA
AMES 1131	12 24 58.	+ 17 18 54.	26	17.6	NEBULA
IC 3373	12 24 58.	+ 25 43 47.			NONSTELLAR OBJECT

OBJECT NAME	RIGHT ASCEN.	DECLINATION	DIAM.	MAGN.	TYPE OF OBJECT
HOLM 406B	12 24 59.	+ 06 30	36	14.3	PART OF MULTIPLE GALAXY
AMES 1139	12 24 59.	+ 06 49 48.	23	17.6	NEBULA
AMES 1137	12 24 59.	+ 09 07 24.	22	16.9	NEBULA
KEEL 523	12 24 59.9	+ 26 20 59.		18.	NEBULA
VDB.66G 127	12 25	+ 37 28	70		DWARF GALAXY
LB 09859	12 25	- 84 57		15.2	FAINT BLUE STAR
ZWG 042.114	12 25 00.	+ 06 30		15.0	GALAXY
UGC 07570	12 25 00.	+ 06 30	60	15.0	GALAXY Sb
KARA.72 338B	12 25 00.	+ 06 30	60	15.0	PART OF DOUBLE GALAXY
MCG+01-32-068	12 25 00.	+ 06 30	60	15.0	GALAXY
ZWG 042.115	12 25 00.	+ 08 26		13.2	GALAXY
UGC 07571	12 25 00.	+ 08 26	84	13.2	GALAXY E
MCG+01-32-069	12 25 00.	+ 08 26	24	13.2	GALAXY
KARA.68 132	12 25 00.	+ 09 53	67		DWARF GALAXY
ZWG 070.095	12 25 00.	+ 10 16		15.6	GALAXY
LB 04656	12 25 00.	+ 12 57		17.6	FAINT BLUE STAR
LB 04655	12 25 00.	+ 13 17		19.5	FAINT BLUE STAR
LB 04654	12 25 00.	+ 13 30		19.1	FAINT BLUE STAR
LB 04653	12 25 00.	+ 14 13		19.9	FAINT BLUE STAR
LB 04652	12 25 00.	+ 14 22		18.7	FAINT BLUE STAR
AMES 1136	12 25 00.	+ 17 09 24.	26	17.8	NEBULA
LB 06395	12 25 00.	+ 27 56		19.2	FAINT BLUE STAR
MCG+05-29-086	12 25 00.	+ 28 58	96	16.	GALAXY
LB 06394	12 25 00.	+ 29 34		19.1	FAINT BLUE STAR
LB 06393	12 25 00.	+ 31 12		18.5	FAINT BLUE STAR
MRK 440	12 25 00.	+ 36 58	8	16.5	GALAXY WITH UV CONTINUUM
ZC 1225.0+4943	12 25 00.	+ 49 43	1480		CLUSTER OF GALAXIES
ZCG 199.61	12 25 00.	+ 63 43		18.5	COMPACT GALAXY
ZC 1225.0+6445	12 25 00.	+ 64 45	1810		CLUSTER OF GALAXIES
MCG+11-15-056	12 25 00.	+ 65 03	78	14.4	GALAXY
ZWG 315.039	12 25 00.	+ 65 05		13.5	GALAXY
UGC 07572	12 25 00.	+ 65 05	270	13.5	GALAXY PECULR
IC 3370	12 25 00.	- 39 04	102	12.4	GALAXY E2
MCG-06-27-029	12 25 00.	- 39 05	90	12.	GALAXY
RNGC 4433	12 25 01.	- 08 01		13.0	GALAXY
RNGC 4432	12 25 02.	+ 06 30		15.0	GALAXY
RNGC 4434	12 25 02.	+ 08 26		13.0	GALAXY
HN 0513	12 25 02.	+ 10 15		15.5	NEBULA
IC 3374	12 25 02.	+ 10 15			NONSTELLAR OBJECT
HOLM 409B	12 25 02.	+ 13 19	54	11.9	PART OF MULTIPLE GALAXY
A3 190	12 25 02.0	+ 13 41 18.		16.9	FAINT BLUE OBJECT
REIZ 2372	12 25 03.	+ 06 30	30	15.4	GALAXY
AMES 1146	12 25 03.	+ 08 17 29.	22	17.4	NEBULA
REIZ 2374	12 25 03.	+ 08 26	18	13.6	GALAXY
REIZ 2376	12 25 03.	+ 43 46	180	13.7	GALAXY
RNGC 4441	12 25 03.	+ 65 05		13.5	GALAXY
AMES 1145	12 25 04.	+ 09 09 23.	31	18.0	NEBULA
AMES 1144	12 25 04.	+ 10 14 59.	25	16.8	NEBULA
AMES 1143	12 25 04.	+ 12 32 47.	24	17.9	NEBULA
AMES 1141	12 25 04.	+ 17 16 59.	24	17.4	NEBULA
AMES 1140	12 25 04.	+ 18 11 35.	19	17.2	NEBULA
A3 191	12 25 04.2	+ 12 59 35.		16.4	FAINT BLUE OBJECT
REIN 3.158	12 25 04.54	+ 08 25 55.0			NEBULA
AMES 1142	12 25 05.	+ 16 38 47.	25	17.0	NEBULA
REIZ 2377	12 25 05.	+ 31 11	24	14.5	GALAXY
REIZ 2373	12 25 05.	- 08 01	78	12.9	GALAXY
HOLM 407A	12 25 05.	- 08 01	72	12.0	PART OF MULTIPLE GALAXY
AMES 1148	12 25 06.	+ 10 14 11.	29	16.6	NEBULA
AMES 1147	12 25 06.	+ 12 03 59.	16	17.7	NEBULA
LB 04660	12 25 06.	+ 12 42		17.9	FAINT BLUE STAR
LB 04659	12 25 06.	+ 12 53		17.6	FAINT BLUE STAR
LB 04658	12 25 06.	+ 13 08		19.6	FAINT BLUE STAR
MCG+02-32-064	12 25 06.	+ 13 21	66	11.9	GALAXY
LB 04657	12 25 06.	+ 13 54		20.2	FAINT BLUE STAR
ZWG 099.059	12 25 06.	+ 18 43		15.7	GALAXY
MCG+03-32-044	12 25 06.	+ 18 44	66	15.7	GALAXY
MCG+03-32-045	12 25 06.	+ 18 50	24	17.	GALAXY
ZWG 159.001	12 25 06.	+ 31 13		15.3	GALAXY
ZWG 158.111	12 25 06.	+ 31 13		15.3	GALAXY
MCG+10-18-049	12 25 06.	+ 56 51	39	16.	GALAXY
MCG+10-18-050	12 25 06.	+ 57 11	18	16.	GALAXY
ZWG 293.021	12 25 06.	+ 57 12		15.4	GALAXY
ZCG 199.62	12 25 06.	+ 63 21		16.6	COMPACT GALAXY
ZC 1225.1+6334	12 25 06.	+ 63 34	2020		CLUSTER OF GALAXIES
72W 457	12 25 06.	+ 78 12			COMPACT GALAXY
MCG-01-32-013	12 25 06.	- 08 01	102	14.	GALAXY
MCG-06-27-030	12 25 06.	- 34 07	36	15.5	GALAXY
REIZ 2381	12 25 07.	+ 65 04	24	14.4	GALAXY
AMES 1155	12 25 08.	+ 08 45 59.	25	17.6	NEBULA
AMES 1152	12 25 08.	+ 09 16 23.	28	17.7	NEBULA
HOLM 408A	12 25 08.	+ 12 33	60	14.2	PART OF MULTIPLE GALAXY
REIZ 2376	12 25 08.	+ 12 35	36	14.4	GALAXY
HOLM 409A	12 25 08.	+ 13 15	540	11.8	PART OF MULTIPLE GALAXY
RNGC 4435	12 25 08.	+ 13 21		12.0	GALAXY
REIZ 2375	12 25 08.	+ 13 21	60	11.6	GALAXY
AMES 1149	12 25 08.	+ 16 07 53.	19	17.5	NEBULA
REIN 3.159	12 25 08.80	+ 13 21 17.7			NEBULA
AMES 1157	12 25 09.	+ 08 49 11.	48	17.6	NEBULA
AMES 1154	12 25 09.	+ 12 22 35.	18	17.0	NEBULA
AMES 1151	12 25 09.	+ 16 38 23.	19	17.8	NEBULA
AMES 1150	12 25 09.	+ 18 42 47.	40	16.6	NEBULA
REIN 3.160	12 25 09.49	+ 12 35 31.7			NEBULA
AMES 1156	12 25 10.	+ 10 53 59.	36	17.8	NEBULA
AMES 1153	12 25 10.	+ 17 09 35.	38	17.3	NEBULA
AMES 1158	12 25 11.	+ 08 38 35.	19	17.8	NEBULA
IC 3375	12 25 11.	+ 27 38 35.			NONSTELLAR OBJECT
SEY 183	12 25 11.	+ 31 11 48.		15.4	FAINT GALAXY
RNGC 4456	12 25 11.	- 29 51		15.0	GALAXY
A3 192	12 25 11.5	+ 14 11 37.		16.5	FAINT BLUE OBJECT
ZWG 070.096	12 25 12.	+ 12 35		14.8	GALAXY
UGC 07573	12 25 12.	+ 12 35	96	14.8	GALAXY S0
MCG+02-32-066	12 25 12.	+ 12 35	72	14.8	GALAXY
RMB 142	12 25 12.	+ 12 37			FAINT BLUE OBJECT
LB 04664	12 25 12.	+ 12 44		19.8	FAINT BLUE STAR
ZWG 070.097	12 25 12.	+ 13 17		12.0	GALAXY
UGC 07574	12 25 12.	+ 13 17	582	12.0	GALAXY S
MCG+02-32-065	12 25 12.	+ 13 17	540	12.0	GALAXY
ZWG 070.098	12 25 12.	+ 13 21		11.9	GALAXY
SCH 47	12 25 12.	+ 13 21		11.9	PECULIAR GALAXY
UGC 07575	12 25 12.	+ 13 21	192	11.9	GALAXY SB0
LB 04663	12 25 12.	+ 13 27		19.4	FAINT BLUE STAR
LB 04662	12 25 12.	+ 13 56		17.5	FAINT BLUE STAR
ZC 1225.2+1417	12 25 12.	+ 14 17	670		CLUSTER OF GALAXIES
LB 04661	12 25 12.	+ 14 32		19.2	FAINT BLUE STAR
LB 00041	12 25 12.	+ 24 40		17.8	FAINT BLUE STAR
LB 06397	12 25 12.	+ 26 48		18.4	FAINT BLUE STAR
LB 06396	12 25 12.	+ 27 01		18.8	FAINT BLUE STAR
MCG+05-29-087	12 25 12.	+ 27 16 30.	96	14.4	GALAXY

OBJECT NAME	RIGHT ASCEN.	DECLINATION	DIAM.	MAGN.	TYPE OF OBJECT
MCG+05-29-088	12 25 12.	+ 28 54 30.	60	16.	GALAXY
UGC 07576	12 25 12.	+ 28 58	84	16.0	GALAXY DBL SYS
ZC 1225.2+3045	12 25 12.	+ 30 45	2080		CLUSTER OF GALAXIES
ZWG 216.003	12 25 12.	+ 40 26		15.0	GALAXY
ZWG 216.004	12 25 12.	+ 43 46		14.7	GALAXY
UGC 07577	12 25 12.	+ 43 46	270	14.7	GALAXY DWRF IR
MCG+09-20-184	12 25 12.	+ 53 47	24	16.	GALAXY
ZCG 199.63	12 25 12.	+ 63 37		19.1	COMPACT GALAXY
MCG-05-30-002	12 25 12.	- 29 51	60	15.	GALAXY
RNGC 4437	12 25 13.	+ 00 24			NON-EXISTENT OBJECT
AMES 1159	12 25 13.	+ 09 14 59.	20	17.2	NEBULA
ABC 1542	12 25 13.	+ 49 43		17.2	RICH CLUSTER OF GALAXIES
REIN 3.161	12 25 13.93	+ 13 17 04.6			NEBULA
AMES 1160	12 25 14.	+ 11 42 41.	30	18.0	NEBULA
RNGC 4436	12 25 14.	+ 12 36		15.0	GALAXY
RNGC 4438	12 25 14.	+ 13 17		11.0	GALAXY
REIZ 2379	12 25 14.	+ 13 17	192	12.1	GALAXY
A3 195	12 25 14.7	+ 12 59 11.		16.6	FAINT BLUE OBJECT
A3 194	12 25 14.7	+ 14 49 15.		16.6	FAINT BLUE OBJECT
TON-N 1537	12 25 15.	+ 19 49		15.9	BLUE STAR
MCG+07-26-007	12 25 15.	+ 40 26	36	15.	GALAXY
MCG+07-26-006	12 25 15.	+ 43 47	240	13.	GALAXY
MCG+09-20-185	12 25 15.	+ 50 40	15	17.	GALAXY
A3 196	12 25 15.1	+ 13 20 40.		17.1	FAINT BLUE OBJECT
A3 197	12 25 15.8	+ 13 21 25.		16.5	FAINT BLUE OBJECT
AMES 1161	12 25 16.	+ 18 48 59.	44	16.4	NEBULA
AMES 1162	12 25 17.	+ 10 57 29.	23	17.5	NEBULA
ARP 120	12 25 17.	+ 13 17			PECULIAR GALAXY
RMB 094	12 25 18.	+ 12 59			FAINT BLUE OBJECT
LB 04669	12 25 18.	+ 13 03		20.4	FAINT BLUE STAR
LB 04668	12 25 18.	+ 13 05		20.0	FAINT BLUE STAR
LB 04667	12 25 18.	+ 13 15		17.8	FAINT BLUE STAR
VV 188	12 25 18.	+ 13 17	120	11.2	INTERACTING GALAXY
LB 04666	12 25 18.	+ 13 36		19.4	FAINT BLUE STAR
LB 04665	12 25 18.	+ 14 08		17.8	FAINT BLUE STAR
ZC 1225.3+2700	12 25 18.	+ 27 00	1680		CLUSTER OF GALAXIES
ZWG 158.112	12 25 18.	+ 27 16		14.4	GALAXY
UGC 07578	12 25 18.	+ 27 16	108	14.4	GALAXY SBa
LB 06399	12 25 18.	+ 28 11		18.2	FAINT BLUE STAR
LB 06398	12 25 18.	+ 29 13		18.7	FAINT BLUE STAR
ZC 1225.3+3503	12 25 18.	+ 35 03	1340		CLUSTER OF GALAXIES
ZC 1225.3+4442	12 25 18.	+ 44 42	940		CLUSTER OF GALAXIES
MCG+09-20-187	12 25 18.	+ 50 39	60	15.	GALAXY
MCG+09-20-186	12 25 18.	+ 50 40	18	17.	GALAXY
ZWG 270.003	12 25 18.	+ 53 47		15.7	GALAXY
ZWG 269.060	12 25 18.	+ 53 47		15.7	GALAXY
A3 200	12 25 18.4	+ 13 28 15.		17.2	FAINT BLUE OBJECT
AMES 1164	12 25 19.	+ 14 50 11.	25	17.2	NEBULA
A3 201	12 25 19.2	+ 13 19 54.		17.1	FAINT BLUE OBJECT
AMES 1167	12 25 20.	+ 09 19 35.	25	16.3	NEBULA
AMES 1166	12 25 20.	+ 11 53 35.	23	17.6	NEBULA
REIZ 2380	12 25 20.	+ 12 34	42	13.3	GALAXY
AMES 1165	12 25 20.	+ 15 32 23.	31	17.6	NEBULA
AMES 1163	12 25 20.	+ 17 50 35.	22	17.3	NEBULA
ABC 1543	12 25 20.	+ 30 39		17.2	RICH CLUSTER OF GALAXIES
AMES 1169	12 25 21.	+ 06 56 35.	28	17.7	NEBULA
AMES 1168	12 25 21.	+ 07 42 47.	43	17.9	NEBULA
HOLM 408B	12 25 21.	+ 12 32	60	12.9	PART OF MULTIPLE GALAXY
IC 3376	12 25 21.	+ 27 16 17.			NONSTELLAR OBJECT
SEY 184	12 25 21.	+ 35 34 24.		15.4	FAINT GALAXY
REIN 3.162	12 25 21.79	+ 12 34 09.6			NEBULA
AMES 1171	12 25 22.	+ 05 59 59.	92	15.6	NEBULA
IC 3377	12 25 22.	+ 25 13 05.			NONSTELLAR OBJECT
REIN 3.163	12 25 22.84	+ 08 22 06.6			NEBULA
AMES 1170	12 25 23.	+ 13 42 23.	23	17.3	NEBULA
ZWG 042.116	12 25 24.	+ 04 34		15.7	GALAXY
HOLM 410B	12 25 24.	+ 05 59	60	14.8	PART OF MULTIPLE GALAXY
ZWG 042.117	12 25 24.	+ 06 00		15.0	GALAXY
UGC 07579	12 25 24.	+ 06 00	78	15.0	GALAXY S-IRR
MCG+01-32-070	12 25 24.	+ 06 00	60	15.0	GALAXY
ZWG 042.118	12 25 24.	+ 08 22		15.2	GALAXY
UGC 07580	12 25 24.	+ 08 22	78	15.2	GALAXY
MCG+01-32-071	12 25 24.	+ 08 22	12	15.2	GALAXY
ZWG 070.099	12 25 24.	+ 12 34		13.0	GALAXY
UGC 07581	12 25 24.	+ 12 34	108	13.0	GALAXY SBa
MCG+02-32-067	12 25 24.	+ 12 34	96	13.0	GALAXY
LB 04676	12 25 24.	+ 12 58		18.5	FAINT BLUE STAR
LB 04675	12 25 24.	+ 13 19		18.5	FAINT BLUE STAR
LB 04674	12 25 24.	+ 13 23		19.3	FAINT BLUE STAR
LB 04673	12 25 24.	+ 13 34		18.8	FAINT BLUE STAR
LB 04672	12 25 24.	+ 13 36		18.6	FAINT BLUE STAR
LB 04671	12 25 24.	+ 13 41		19.0	FAINT BLUE STAR
LB 04670	12 25 24.	+ 14 09		17.0	FAINT BLUE STAR
RMB 043	12 25 24.	+ 15 03			FAINT BLUE OBJECT
LB 06400	12 25 24.	+ 30 00		18.6	FAINT BLUE STAR
ZWG 270.004	12 25 24.	+ 50 41		15.6	GALAXY
ZWG 269.061	12 25 24.	+ 50 41		15.6	GALAXY
UGC 07582	12 25 24.	+ 50 41	72	15.6	GALAXY SBb
MCG-02-32-012	12 25 24.	- 13 15 30.	36	15.5	GALAXY
A3 202	12 25 24.8	+ 15 39 50.		17.5	FAINT BLUE OBJECT
RNGC 4439	12 25 25.	- 59 49		8.5	OPEN CLUSTER
A3 203	12 25 25.0	+ 15 02 52.		17.1	FAINT BLUE OBJECT
AMES 1173	12 25 26.	+ 08 27 47.	26	16.9	NEBULA
RNGC 4440	12 25 26.	+ 12 34		13.0	GALAXY
ABC 1544	12 25 26.	+ 63 42		17.2	RICH CLUSTER OF GALAXIES
REIZ 2382	12 25 27.	+ 05 59	54	14.7	GALAXY
HOLM 410A	12 25 27.	+ 05 59	90	13.9	PART OF MULTIPLE GALAXY
AMES 1172	12 25 27.	+ 11 09 17.	26	17.4	NEBULA
IC 0793	12 25 28.	+ 09 42 23.			SAME AS NGC 4445
ZWG 070.100	12 25 30.	+ 10 05		11.2	GALAXY
UGC 07583	12 25 30.	+ 10 05	270	11.2	GALAXY SB0
MCG+02-32-068	12 25 30.	+ 10 05	210	11.2	GALAXY
LB 04680	12 25 30.	+ 12 42		18.8	FAINT BLUE STAR
LB 04679	12 25 30.	+ 13 15		18.8	FAINT BLUE STAR
LB 04678	12 25 30.	+ 13 17		18.2	FAINT BLUE STAR
LB 04677	12 25 30.	+ 14 20		18.6	FAINT BLUE STAR
ZWG 099.060	12 25 30.	+ 17 34		15.3	GALAXY
MCG+03-32-046	12 25 30.	+ 20 26	42	15.5	GALAXY
UGC 07584	12 25 30.	+ 22 52	60	16.0	GALAXY Sc-IRR
LB 06403	12 25 30.	+ 26 58		19.5	FAINT BLUE STAR
LB 11212	12 25 30.	+ 27 14		12.8	FAINT BLUE STAR
LB 06402	12 25 30.	+ 28 25		20.2	FAINT BLUE STAR
LB 06401	12 25 30.	+ 28 33		19.5	FAINT BLUE STAR
VV 279A	12 25 30.	+ 28 54 30.	48	15.	INTERACTING GALAXY
VV 279B	12 25 30.	+ 28 55 30.	36	16.	INTERACTING GALAXY
MCG+09-20-188	12 25 30.	+ 54 01	30	16.	GALAXY
ZC 1225.5+5751	12 25 30.	+ 57 51	2620		CLUSTER OF GALAXIES
ZCG 199.64	12 25 30.	+ 63 46		18.6	COMPACT GALAXY

OBJECT NAME	RIGHT ASCEN.	DECLINATION	DIAM.	MAGN.	TYPE OF OBJECT
REIZ 2386	12 25 30.	+ 65 17	90	15.0	GALAXY
ZWG 014.045	12 25 30.	- 01 22		15.4	GALAXY
AMES 1176	12 25 31.	+ 06 55 59.	24	17.8	NEBULA
LB 02365	12 25 31.	+ 54 22 54.		16.1	FAINT BLUE STAR
REIN 3.164	12 25 31.88	+ 10 04 52.2			NEBULA
RNGC 4442	12 25 32.	+ 10 05		11.5	GALAXY
REIZ 2383	12 25 32.	+ 10 05	132	11.9	GALAXY
AMES 1175	12 25 32.	+ 12 22 05.	25	17.0	NEBULA
RNGC 4443	12 25 32.	+ 13 24			NON-EXISTENT OBJECT
AMES 1174	12 25 32.	+ 13 51 05.	32	17.4	NEBULA
AMES 1181	12 25 33.	+ 07 14 35.	16	17.7	NEBULA
AMES 1177	12 25 33.	+ 09 06 47.	16	17.8	NEBULA
MCG+05-29-089	12 25 33.	+ 28 54	192	11.9	GALAXY
AMES 1180	12 25 34.	+ 12 15 11.	29	17.6	NEBULA
AMES 1179	12 25 34.	+ 13 10 53.	30	17.2	NEBULA
A3 204	12 25 34.0	+ 14 24 23.		18.0	FAINT BLUE OBJECT
AMES 1178	12 25 35.	+ 18 53 11.	47	16.9	NEBULA
ZWG 070.101	12 25 36.	+ 10 03		15.6	GALAXY
KARA.68 134	12 25 36.	+ 10 39	34		DWARF GALAXY
REA 16	12 25 36.	+ 10 39			DWARF GALAXY
MCG+02-32-071	12 25 36.	+ 10 39	48	18.	GALAXY
KARA.68 133	12 25 36.	+ 10 49	27		DWARF GALAXY
ZWG 070.102	12 25 36.	+ 12 22		15.1	GALAXY
UGC 07585	12 25 36.	+ 12 22	90	15.1	GALAXY E
MCG+02-32-070	12 25 36.	+ 12 22	18	15.1	GALAXY
LB 04688	12 25 36.	+ 12 25		19.0	FAINT BLUE STAR
LB 04687	12 25 36.	+ 12 38		17.9	FAINT BLUE STAR
LB 04686	12 25 36.	+ 13 06		18.8	FAINT BLUE STAR
LB 04685	12 25 36.	+ 13 27		19.1	FAINT BLUE STAR
LB 04684	12 25 36.	+ 13 33		17.7	FAINT BLUE STAR
LB 04683	12 25 36.	+ 13 52		17.3	FAINT BLUE STAR
LB 04682	12 25 36.	+ 14 00		17.3	FAINT BLUE STAR
ZWG 070.103	12 25 36.	+ 14 11		15.0	GALAXY
UGC 07586	12 25 36.	+ 14 11	66	15.0	GALAXY Sc
KARA.72 339A	12 25 36.	+ 14 11	60	15.0	PART OF DOUBLE GALAXY
MCG+02-32-069	12 25 36.	+ 14 11	72	15.0	GALAXY
LB 04681	12 25 36.	+ 14 13		18.0	FAINT BLUE STAR
ZC 1225.6+1905	12 25 36.	+ 19 05	2350		CLUSTER OF GALAXIES
ZWG 126.090	12 25 36.	+ 20 27		15.5	GALAXY
ZWG 099.061	12 25 36.	+ 20 27		15.5	GALAXY
LB 06407	12 25 36.	+ 26 28		19.3	FAINT BLUE STAR
LB 06406	12 25 36.	+ 26 44		19.5	FAINT BLUE STAR
IC 3380	12 25 36.	+ 26 56 59.			NONSTELLAR OBJECT
LB 06404	12 25 36.	+ 27 57		19.7	FAINT BLUE STAR
LB 06405	12 25 36.	+ 29 41		19.7	FAINT BLUE STAR
MRK 211	12 25 36.	+ 44 45	8	17.	GALAXY WITH UV CONTINUUM
MCG+09-20-189	12 25 36.	+ 52 38	24	17.	GALAXY
ZWG 270.005	12 25 36.	+ 54 28		15.5	GALAXY
ZWG 269.062	12 25 36.	+ 54 28		15.5	GALAXY
MCG+09-20-190	12 25 36.	+ 54 29	36	16.	GALAXY
ZCG 199.65	12 25 36.	+ 63 38		18.7	COMPACT GALAXY
ZCG 199.66	12 25 36.	+ 63 43		18.6	COMPACT GALAXY
ZCG 199.67	12 25 36.	+ 63 56		19.5	COMPACT GALAXY
MCG-01-32-014	12 25 36.	- 04 40	30	16.	GALAXY
OCL 0884	12 25 36.	- 59 49	480	9.2	OPEN STAR CLUSTER
VHA 134	12 25 36.	- 59 49	180		OPEN STAR CLUSTER
IC 0794	12 25 36.6	+ 12 22 11.			GALAXY
REIN 3.165	12 25 36.91	+ 12 22 10.4			NEBULA
HN 0914	12 25 37.	+ 17 35	60	15.5	NEBULA
A3 205	12 25 37.7	+ 15 06 47.		17.5	FAINT BLUE OBJECT
AMES 1183	12 25 38.	+ 13 46 41.	33	17.3	NEBULA
RNGC 4446	12 25 38.	+ 14 11		15.0	GALAXY
AMES 1182	12 25 38.	+ 15 00 05.	20	17.0	NEBULA
IC 3379	12 25 38.	+ 17 35			NONSTELLAR OBJECT
IC 3378	12 25 38.	+ 17 35			NONSTELLAR OBJECT
AMES 1184	12 25 39.	+ 06 33 47.	38	17.5	NEBULA
REIZ 2384	12 25 39.	+ 07 52	24	15.2	GALAXY
MAI 074	12 25 39.	+ 75 54	33		DWARF SPHEROIDAL GALAXY
REIN 3.166	12 25 39.88	+ 10 34 26.8			NEBULA
AMES 1186	12 25 40.	+ 07 05 29.	30	17.0	NEBULA
AMES 1185	12 25 40.	+ 07 53 23.	25	15.6	NEBULA
BC PKS1225+20	12 25 41.1	+ 20 40 03.		18.	QUASI-STELLAR OBJECT
ZWG 042.119	12 25 42.	+ 07 06		15.6	GALAXY
ZWG 042.120	12 25 42.	+ 07 53		15.2	GALAXY
MCG+01-32-072	12 25 42.	+ 07 53	15	15.2	GALAXY
MCG+02-32-072	12 25 42.	+ 09 42	144	13.7	GALAXY
ZWG 070.104	12 25 42.	+ 09 43		13.7	GALAXY
UGC 07587	12 25 42.	+ 09 43	156	13.7	GALAXY S
ZWG 070.105	12 25 42.	+ 10 34		15.5	GALAXY
MCG+02-32-075	12 25 42.	+ 10 34	42	15.5	GALAXY
ZWG 070.106	12 25 42.	+ 12 04		15.1	GALAXY
UGC 07589	12 25 42.	+ 12 04	78	15.1	GALAXY E
MCG+02-32-074	12 25 42.	+ 12 04	42	15.1	GALAXY
LB 04696	12 25 42.	+ 13 07		19.7	FAINT BLUE STAR
LB 04695	12 25 42.	+ 13 12		18.8	FAINT BLUE STAR
LB 04694	12 25 42.	+ 13 30		18.8	FAINT BLUE STAR
LB 04693	12 25 42.	+ 13 38		20.2	FAINT BLUE STAR
LB 04692	12 25 42.	+ 13 43		19.9	FAINT BLUE STAR
LB 04691	12 25 42.	+ 13 43		19.6	FAINT BLUE STAR
UGC 07588	12 25 42.	+ 13 50	60	16.0	GALAXY Sc
ZWG 070.107	12 25 42.	+ 14 10		15.2	GALAXY
KARA.72 339B	12 25 42.	+ 14 10	42	15.2	PART OF DOUBLE GALAXY
MCG+02-32-073	12 25 42.	+ 14 10	15	15.2	GALAXY
LB 04690	12 25 42.	+ 14 20		20.1	FAINT BLUE STAR
LB 04689	12 25 42.	+ 14 22		20.0	FAINT BLUE STAR
RMB 042	12 25 42.	+ 15 16			FAINT BLUE OBJECT
RMB 041	12 25 42.	+ 15 36			FAINT BLUE OBJECT
LB 06408	12 25 42.	+ 27 24		18.3	FAINT BLUE STAR
MCG+07-26-008	12 25 42.	+ 42 09	27	16.	GALAXY
ZCG 199.68	12 25 42.	+ 63 42		19.5	COMPACT GALAXY
7ZW 458	12 25 42.	+ 63 48			COMPACT GALAXY
ARC 1546	12 25 42.	+ 64 53		18.0	RICH CLUSTER OF GALAXIES
ZC 1225.7+6820	12 25 42.	+ 68 20	1410		CLUSTER OF GALAXIES
MCG-01-32-015	12 25 42.	- 04 41	18	15.	GALAXY
MCG-06-27-031	12 25 42.	- 35 41	36	14.	GALAXY
IC 3382	12 25 43.	+ 13 50			NONSTELLAR OBJECT
AMES 1188	12 25 43.	+ 17 09 59.	20	17.6	NEBULA
AMES 1187	12 25 43.	+ 17 53 35.	29	17.0	NEBULA
IC 3384	12 25 43.	+ 25 21 59.			NONSTELLAR OBJECT
IC 3381	12 25 43.0	+ 12 03 58.			GALAXY
REIN 3.167	12 25 43.07	+ 12 03 58.3			NEBULA
REIZ 2385	12 25 44.	+ 09 42	78	13.9	GALAXY
RNGC 4445	12 25 44.	+ 09 43		13.5	GALAXY
HN 0916	12 25 44.	+ 10 33		15.	NEBULA
IC 3383	12 25 44.	+ 10 33			NONSTELLAR OBJECT
HN 0915	12 25 44.	+ 13 50	48		NEBULA
RNGC 4447	12 25 44.	+ 14 10		15.0	GALAXY
AMES 1189	12 25 44.	+ 15 02 47.	37	17.4	NEBULA

OBJECT NAME	RIGHT ASCEN.	DECLINATION	DIAM.	MAGN.	TYPE OF OBJECT
REIN 3.168	12 25 44.07	+ 09 42 49.0			NEBULA
A3 210	12 25 44.2	+ 15 10 46.		16.9	FAINT BLUE OBJECT
IC 3385	12 25 45.	+ 25 42 29.			NONSTELLAR OBJECT
RNGC 4448	12 25 45.	+ 28 54		12.0	GALAXY
A3 211	12 25 45.9	+ 14 09 23.		17.1	FAINT BLUE OBJECT
A3 212	12 25 46.1	+ 15 53 41.		15.6	FAINT BLUE OBJECT
REIN 3.169	12 25 46.40	+ 09 00 21.4			NEBULA
A3 214	12 25 46.7	+ 16 01 01.		17.2	FAINT BLUE OBJECT
AMES 1191	12 25 47.	+ 05 43 11.	25	17.6	NEBULA
RNGC 4449	12 25 47.	+ 44 22		10.5	GALAXY
AMES 1192	12 25 48.	+ 08 08 11.	23	17.6	NEBULA
ZWG 070.108	12 25 48.	+ 09 00		15.0	GALAXY
UGC 07590	12 25 48.	+ 09 00	90	15.0	GALAXY S
MCG+02-32-076	12 25 48.	+ 09 00	78	15.0	GALAXY
RMB 204	12 25 48.	+ 10 17			FAINT BLUE OBJECT
REA 70	12 25 48.	+ 10 45	24		DWARF GALAXY
AMES 1190	12 25 48.	+ 12 12 53.	17	17.6	NEBULA
LB 04702	12 25 48.	+ 13 12		18.7	FAINT BLUE STAR
LB 04701	12 25 48.	+ 13 28		19.1	FAINT BLUE STAR
LB 04700	12 25 48.	+ 13 35		19.0	FAINT BLUE STAR
LB 04699	12 25 48.	+ 13 40		19.1	FAINT BLUE STAR
LB 04698	12 25 48.	+ 13 48		20.6	FAINT BLUE STAR
LB 04697	12 25 48.	+ 13 49		20.4	FAINT BLUE STAR
ZC 1225.8+2149	12 25 48.	+ 21 49	540		CLUSTER OF GALAXIES
LB 00008	12 25 43.	+ 26 52		18.5	FAINT BLUE STAR
LB 06410	12 25 48.	+ 28 43		19.5	FAINT BLUE STAR
ZWG 159.002	12 25 48.	+ 28 54		11.9	GALAXY
ZWG 158.113	12 25 48.	+ 28 54		11.9	GALAXY
REIZ 2387	12 25 48.	+ 28 54	180	13.2	GALAXY
UGC 07591	12 25 48.	+ 28 54	240	11.9	GALAXY Sa
LB 06409	12 25 48.	+ 30 22		18.9	FAINT BLUE STAR
ZWG 216.005	12 25 48.	+ 44 22		10.0	GALAXY
UGC 07592	12 25 48.	+ 44 22	360	10.0	GALAXY IRR
MCG+07-26-009	12 25 48.	+ 44 23	330	10.	GALAXY
ZWG 244.020	12 25 48.	+ 44 43		14.8	GALAXY
MRK 212	12 25 48.	+ 44 43	13	15.	GALAXY WITH UV CONTINUUM
UGC 07593	12 25 48.	+ 44 43	72	14.8	GALAXY DBL SYS
KARA.72 340B	12 25 48.	+ 44 43	18		PART OF DOUBLE GALAXY
KARA.72 340A	12 25 48.	+ 44 43	18	14.8	PART OF DOUBLE GALAXY
1ZW 037	12 25 48.	+ 44 44			COMPACT GALAXY
ZC 1225.8+4743	12 25 48.	+ 47 43	740		CLUSTER OF GALAXIES
ZC 1225.8+5006	12 25 48.	+ 50 06	870		CLUSTER OF GALAXIES
ZCG 199.69	12 25 48.	+ 63 48		18.5	COMPACT GALAXY
ZCG 199.70	12 25 48.	+ 63 49		18.2	COMPACT GALAXY
ZC 1225.8+7050	12 25 48.	+ 70 50	1950		CLUSTER OF GALAXIES
OCL 0883	12 25 48.	- 59 32	180		OPEN STAR CLUSTER
A3 216	12 25 48.8	+ 15 20 06.		16.6	FAINT BLUE OBJECT
AMES 1193	12 25 49.	+ 08 16 59.	24	17.9	NEBULA
BB 1.09	12 25 49.85	+ 02 25 48.5			GALAXY NEAR QSO 3C273
HN 0917	12 25 50.	+ 13 28	42		NEBULA
IC 3386	12 25 50.	+ 13 28			NONSTELLAR OBJECT
IC 3387	12 25 50.	+ 28 16 23.			NONSTELLAR OBJECT
ARC 1545	12 25 50.	+ 47 41		17.8	RICH CLUSTER OF GALAXIES
A3 218	12 25 50.7	+ 15 12 29.		17.7	FAINT BLUE OBJECT
A3 219	12 25 50.9	+ 14 57 31.		16.3	FAINT BLUE OBJECT
AMES 1194	12 25 51.	+ 06 10 29.	29	17.6	NEBULA
REIZ 2389	12 25 51.	+ 44 22	252	10.7	GALAXY
RNGC 4444	12 25 52.	- 42 59			GALAXY
ZWG 014.046	12 25 54.	+ 00 25		15.6	GALAXY
KARA.68 135	12 25 54.	+ 11 51	27		DWARF GALAXY
RMB 140	12 25 54.	+ 12 15			FAINT BLUE OBJECT
LB 04706	12 25 54.	+ 12 52		18.5	FAINT BLUE STAR
ZWG 070.109	12 25 54.	+ 13 06		15.4	GALAXY
LB 04705	12 25 54.	+ 13 36		19.2	FAINT BLUE STAR
SN 1952F	12 25 54.	+ 13 37		16.9	SUPERNOVA
LB 04704	12 25 54.	+ 13 55		18.4	FAINT BLUE STAR
LB 04703	12 25 54.	+ 14 19		19.4	FAINT BLUE STAR
ZWG 099.062	12 25 54.	+ 17 21		11.2	GALAXY
SCH 48	12 25 54.	+ 17 21		11.2	PECULIAR GALAXY
UGC 07594	12 25 54.	+ 17 21	390	11.2	GALAXY Sb
AMES 1195	12 25 54.	+ 17 24 23.	20	17.3	NEBULA
ZWG 099.063	12 25 54.	+ 18 41		13.9	GALAXY
SCH 49	12 25 54.	+ 18 41		13.9	PECULIAR GALAXY
UGC 07595	12 25 54.	+ 18 41	66	13.9	GALAXY Sc
MCG+03-32-047	12 25 54.	+ 18 42 30.	66	13.9	GALAXY
MCG+08-23-039	12 25 54.	+ 44 42 30.	48	16.	GALAXY
MCG+08-23-038	12 25 54.	+ 44 42 30.	30	16.	GALAXY
MCG+10-18-051	12 25 54.	+ 59 47	42	16.	GALAXY
ZCG 199.71	12 25 54.	+ 63 36		18.6	COMPACT GALAXY
MCG-04-30-001	12 25 54.	- 22 14	48	15.	GALAXY
OCL 0881	12 25 54.	- 56 11	1320	7.	OPEN STAR CLUSTER
A3 222	12 25 54.7	+ 14 59 16.		17.1	FAINT BLUE OBJECT
AMES 1197	12 25 55.	+ 07 13 23.	23	17.8	NEBULA
REIZ 2390	12 25 55.	+ 17 21	192	11.5	GALAXY
REIZ 2388	12 25 55.	+ 18 40	60	13.5	GALAXY
IC 3389	12 25 55.	+ 28 07 18.			NONSTELLAR OBJECT
YC 1225-43	12 25 55.	- 43 59 06.			UNUSUAL SOUTHERN GALAXY
AMES 1199	12 25 56.	+ 07 36 29.	36	17.4	NEBULA
HN 0918	12 25 56.	+ 13 05	12		NEBULA
IC 3388	12 25 56.	+ 13 05			NONSTELLAR OBJECT
PK300+00.1	12 25 56.	- 61 49	25		PLANETARY NEBULA
REIN 3.170	12 25 56.43	+ 13 05 58.0			NEBULA
MCG+03-32-048	12 25 57.	+ 17 22	240	11.2	GALAXY
AMES 1196	12 25 57.	+ 17 37 11.	23	17.8	NEBULA
BB 1.08	12 25 57.0	+ 02 23 05.			GALAXY NEAR QSO 3C273
A3 225	12 25 57.9	+ 12 47 07.		16.1	FAINT BLUE OBJECT
AMES 1198	12 25 58.	+ 16 39 11.	30	17.8	NEBULA
BB 1.15	12 25 58.2	+ 02 15 41.			GALAXY NEAR QSO 3C273
A3 226	12 25 58.8	+ 13 13 01.		17.0	FAINT BLUE OBJECT
IC 3390	12 25 59.	+ 25 05 12.			NONSTELLAR OBJECT
REIZ 2391	12 25 59.	+ 32 50	30	15.2	GALAXY
VDB.66G 128	12 26	+ 03 03	100		DWARF GALAXY
VDB.66G 130	12 26	+ 11 43	70		DWARF GALAXY
VDB.66G 129	12 26	+ 43 28	130		DWARF GALAXY
2ZW 064	12 26 00.	+ 05 51			COMPACT GALAXY
ZWG 070.110	12 26 00.	+ 08 54		15.3	GALAXY
UGC 07596	12 26 00.	+ 08 54	114	15.3	GALAXY DWARF
MCG+02-32-078	12 26 00.	+ 08 54	90	15.3	GALAXY
ZC 1226.0+0859	12 26 00.	+ 08 59	870		CLUSTER OF GALAXIES
REA 47	12 26 00.	+ 09 21	48		DWARF GALAXY
MCG+02-32-077	12 26 00.	+ 09 21	36	18.	GALAXY
RMB 212	12 26 00.	+ 11 20			FAINT BLUE OBJECT
LB 04720	12 26 00.	+ 12 26		17.8	FAINT BLUE STAR
LB 04719	12 26 00.	+ 12 30		17.6	FAINT BLUE STAR
LB 04718	12 26 00.	+ 12 51		19.7	FAINT BLUE STAR
LB 04717	12 26 00.	+ 12 53		18.0	FAINT BLUE STAR
LB 04716	12 26 00.	+ 12 55		18.3	FAINT BLUE STAR
LB 04715	12 26 00.	+ 12 59		19.6	FAINT BLUE STAR

OBJECT NAME	RIGHT ASCEN.	DECLINATION	DIAM.	MAGN.	TYPE OF OBJECT
RMB 093	12 26 00.	+ 13 13			FAINT BLUE OBJECT
LB 04714	12 26 00.	+ 13 17		19.0	FAINT BLUE STAR
LB 04713	12 26 00.	+ 13 29		18.0	FAINT BLUE STAR
LB 04712	12 26 00.	+ 13 32		17.2	FAINT BLUE STAR
LB 04711	12 26 00.	+ 13 34		17.8	FAINT BLUE STAR
LB 04710	12 26 00.	+ 13 46		19.4	FAINT BLUE STAR
LB 04709	12 26 00.	+ 14 03		17.7	FAINT BLUE STAR
LB 04708	12 26 00.	+ 14 22		20.2	FAINT BLUE STAR
KARA.68 136	12 26 00.	+ 14 25	34		DWARF GALAXY
LB 04707	12 26 00.	+ 14 26		20.6	FAINT BLUE STAR
ZWG 129.001	12 26 00.	+ 23 35		15.3	GALAXY
ZWG 128.091	12 26 00.	+ 23 35		15.3	GALAXY
LB 06413	12 26 00.	+ 27 08		17.8	FAINT BLUE STAR
LB 06412	12 26 00.	+ 28 12		19.2	FAINT BLUE STAR
TON-N 0078	12 26 00.	+ 28 31		16.2	BLUE STAR
UGC 07597	12 26 00.	+ 28 57	78	16.0	GALAXY S
MCG+05-30-001	12 26 00.	+ 28 57	72	15.5	GALAXY
LB 06411	12 26 00.	+ 29 07		18.3	FAINT BLUE STAR
TON-N 0618	12 26 00.	+ 31 46		15.4	BLUE STAR
ZWG 159.003	12 26 00.	+ 32 05		15.7	GALAXY
ZWG 158.114	12 26 00.	+ 32 05		15.7	GALAXY
MCG+06-27-056	12 26 00.	+ 32 49	36	15.	GALAXY
ZWG 187.045	12 26 00.	+ 32 50		15.6	GALAXY
UGC 07598	12 26 00.	+ 32 50	90	15.6	GALAXY SBc
ZWG 188.001	12 26 00.	+ 37 30		15.6	GALAXY
ZWG 187.046	12 26 00.	+ 37 30		15.6	GALAXY
UGC 07599	12 26 00.	+ 37 30	120	15.6	GALAXY DWRF SP
MCG+06-27-057	12 26 00.	+ 37 31 30.	120	15.	GALAXY
ZC 1226.0+4334	12 26 00.	+ 43 34	2620		CLUSTER OF GALAXIES
ZWG 244.021	12 26 00.	+ 44 54		15.1	GALAXY
MCG+09-21-001	12 26 00.	+ 53 52 30.	24	15.	GALAXY
MCG+10-18-052	12 26 00.	+ 57 00	51	16.	GALAXY
ZCG 199.72	12 26 00.	+ 63 37		18.6	COMPACT GALAXY
ZC 1226.0+6604	12 26 00.	+ 66 04	1550		CLUSTER OF GALAXIES
A3 227	12 26 00.8	+ 13 07 42.		17.1	FAINT BLUE OBJECT
AMES 1201	12 26 01.	+ 06 53 11.	23	17.9	NEBULA
AMES 1200	12 26 01.	+ 07 43 59.	26	17.4	NEBULA
HN 0919	12 26 01.	+ 18 41	42		NEBULA
IC 3391	12 26 01.	+ 18 41			NONSTELLAR OBJECT
PK300-00.1	12 26 01.	- 63 27 31.			PLANETARY NEBULA
A3 228	12 26 01.1	+ 15 14 56.		17.6	FAINT BLUE OBJECT
REIN 3.171	12 26 01.59	+ 08 54 53.5			NEBULA
RNGC 4450	12 26 02.	+ 17 22		11.5	GALAXY
IC 0795	12 26 02.	+ 23 35 12.			NONSTELLAR OBJECT
A3 230	12 26 02.4	+ 15 01 27.		16.1	FAINT BLUE OBJECT
A3 231	12 26 02.7	+ 13 26 46.		16.8	FAINT BLUE OBJECT
AMES 1205	12 26 03.	+ 07 06 35.	25	17.8	NEBULA
AMES 1204	12 26 03.	+ 07 46 05.	29	17.4	NEBULA
AMES 1203	12 26 03.	+ 12 16 41.	34	17.1	NEBULA
ABC 1547	12 26 03.	+ 27 02		17.2	RICH CLUSTER OF GALAXIES
AMES 1202	12 26 04.	+ 18 51 23.	18	17.0	NEBULA
AMES 1207	12 26 05.	+ 07 19 59.	18	18.0	NEBULA
REIZ 2396	12 26 05.	+ 31 44	78	15.8	GALAXY
AMES 1208	12 26 06.	+ 06 25 47.	24	17.4	NEBULA
ZWG 070.111	12 26 06.	+ 09 32		13.4	GALAXY
UGC 07600	12 26 06.	+ 09 32	90	13.4	GALAXY
MCG+02-32-079	12 26 06.	+ 09 32	36	13.4	GALAXY
KARA.68 137	12 26 06.	+ 10 00	34		DWARF GALAXY
REA 71	12 26 06.	+ 10 01	48		DWARF GALAXY
AMES 1206	12 26 06.	+ 11 53 05.	23	18.0	NEBULA
LB 04732	12 26 06.	+ 12 36		18.3	FAINT BLUE STAR
LB 04731	12 26 06.	+ 12 43		17.4	FAINT BLUE STAR
LB 04730	12 26 06.	+ 12 58		19.8	FAINT BLUE STAR
LB 04729	12 26 06.	+ 13 15		19.0	FAINT BLUE STAR
LB 04728	12 26 06.	+ 13 21		18.7	FAINT BLUE STAR
LB 04727	12 26 06.	+ 13 24		19.6	FAINT BLUE STAR
LB 04726	12 26 06.	+ 13 47		17.6	FAINT BLUE STAR
LB 04725	12 26 06.	+ 13 52		18.5	FAINT BLUE STAR
LB 04724	12 26 06.	+ 14 06		15.8	FAINT BLUE STAR
LB 04723	12 26 06.	+ 14 10		20.7	FAINT BLUE STAR
LB 04722	12 26 06.	+ 14 12		19.8	FAINT BLUE STAR
LB 04721	12 26 06.	+ 14 14		18.1	FAINT BLUE STAR
MCG+03-32-049	12 26 06.	+ 15 16	132	13.3	GALAXY
ZWG 099.064	12 26 06.	+ 19 44		15.2	GALAXY
REIZ 2394	12 26 06.	+ 26 39	60	15.8	GALAXY
REIZ 2395	12 26 06.	+ 26 41	48	15.9	GALAXY
MCG+05-30-002	12 26 06.	+ 32 07	48	15.7	GALAXY
MCG+08-23-040	12 26 06.	+ 44 54	36	14.	GALAXY
ZWG 270.006	12 26 06.	+ 53 53		15.2	GALAXY
ZWG 269.063	12 26 06.	+ 53 53		15.2	GALAXY
7ZW 459	12 26 06.	+ 62 15			COMPACT GALAXY
ZCG 199.73	12 26 06.	+ 63 45		18.8	COMPACT GALAXY
SER 090.01	12 26 06.	- 41 45	15	16.	IRR. MAGELLANIC GALAXY
SEY 185	12 26 07.	+ 35 59 24.		14.8	FAINT GALAXY
AMES 1211	12 26 08.	+ 08 05 11.	32	17.8	NEBULA
RNGC 4451	12 26 08.	+ 09 32		13.5	GALAXY
REIZ 2392	12 26 08.	+ 09 32	36	13.6	GALAXY
HN 0921	12 26 08.	+ 13 11	30		NEBULA
REIZ 2393	12 26 08.	+ 15 16	84	13.3	GALAXY
HN 0920	12 26 08.	+ 15 17	90		NEBULA
REIN 3.172	12 26 08.71	+ 09 32 14.8			NEBULA
AMES 1210	12 26 09.	+ 10 40 47.	23	17.0	NEBULA
AMES 1209	12 26 09.	+ 12 18 35.	22	17.8	NEBULA
IC 3393	12 26 09.8	+ 13 11 32.			GALAXY
REIN 3.173	12 26 09.88	+ 13 11 31.0			NEBULA
A3 235	12 26 10.3	+ 13 57 39.		16.6	FAINT BLUE OBJECT
AMES 1212	12 26 11.	+ 10 57 47.	26	17.5	NEBULA
REIN 3.174	12 26 11.41	+ 12 01 52.5			NEBULA
IC 3392	12 26 11.5	+ 15 16 30.			GALAXY
A3 237	12 26 11.6	+ 16 55 55.		16.3	FAINT BLUE OBJECT
ZWG 042.121	12 26 12.	+ 06 47		15.4	GALAXY
SN 1966F	12 26 12.	+ 06 47		17.5	SUPERNOVA
MCG+01-32-073	12 26 12.	+ 06 47	30	15.4	GALAXY
KARA.68 138	12 26 12.	+ 09 18	27		DWARF GALAXY
ZWG 070.112	12 26 12.	+ 12 02		13.1	GALAXY
UGC 07601	12 26 12.	+ 12 02	162	13.1	GALAXY S0?
MCG+02-32-080	12 26 12.	+ 12 02	90	13.1	GALAXY
LB 04735	12 26 12.	+ 12 52		19.1	FAINT BLUE STAR
LB 04734	12 26 12.	+ 13 08		19.2	FAINT BLUE STAR
ZWG 070.113	12 26 12.	+ 13 11		15.1	GALAXY
MCG+02-32-081	12 26 12.	+ 13 11	48	15.1	GALAXY
LB 04733	12 26 12.	+ 14 06		18.9	FAINT BLUE STAR
ZWG 099.065	12 26 12.	+ 15 16		13.3	GALAXY
UGC 07602	12 26 12.	+ 15 16	126	13.3	GALAXY Sa-b
ZWG 129.002	12 26 12.	+ 23 06		13.0	GALAXY
UGC 07603	12 26 12.	+ 23 06	162	13.0	GALAXY Sc-IRR
MCG+04-30-001	12 26 12.	+ 23 07	150	13.0	GALAXY
REIZ 2404	12 26 12.	+ 26 40	24	16.0	GALAXY

OBJECT NAME	RIGHT ASCEN.	DECLINATION	DIAM.	MAGN.	TYPE OF OBJECT
IC 3394	12 26 12.	+ 27 04 30.			NONSTELLAR OBJECT
LB 06416	12 26 12.	+ 28 15		18.5	FAINT BLUE STAR
LB 11213	12 26 12.	+ 28 37		13.7	FAINT BLUE STAR
LB 06415	12 26 12.	+ 29 30		19.5	FAINT BLUE STAR
UGC 07604	12 26 12.	+ 31 46	72	16.5	GALAXY Sc
LB 06414	12 26 12.	+ 31 54		17.6	FAINT BLUE STAR
MCG+06-27-058	12 26 12.	+ 35 59 30.	60	14.	GALAXY
ZWG 188.002	12 26 12.	+ 36 00		15.3	GALAXY
ZWG 187.047	12 26 12.	+ 36 00		15.3	GALAXY
UGC 07605	12 26 12.	+ 36 00	96	15.3	GALAXY DWRF IR
KARA.73B 0530	12 26 12.	+ 36 00	66	15.3	ISOLATED GALAXY IR
ZC 1226.2+5447	12 26 12.	+ 54 47	270		CLUSTER OF GALAXIES
ZCG 199.74	12 26 12.	+ 63 44		18.6	COMPACT GALAXY
OCL 0880	12 26 12.	- 60 29	420	9.1	OPEN STAR CLUSTER
REIZ 2403	12 26 13.	+ 23 05	126	12.7	GALAXY
AMES 1218	12 26 14.	+ 06 42 59.	23	17.9	NEBULA
RNGC 4453	12 26 14.	+ 06 47		15.5	GALAXY
AMES 1217	12 26 14.	+ 07 44 47.	20	17.7	NEBULA
RNGC 4452	12 26 14.	+ 12 02		13.0	GALAXY
REIZ 2399	12 26 14.	+ 12 02	90	12.6	GALAXY
REIZ 2402	12 26 14.	+ 15 16	102	14.1	GALAXY
REIZ 2400	12 26 15.	+ 06 47	24	15.3	GALAXY
AMES 1215	12 26 15.	+ 16 29 35.	34	17.3	NEBULA
AMES 1214	12 26 15.	+ 17 06 47.	32	17.1	NEBULA
AMES 1213	12 26 15.	+ 19 45 05.	37	16.1	NEBULA
RNGC 4455	12 26 15.	+ 23 06		13.0	GALAXY
IC 3395	12 26 15.	+ 25 18 42.			NONSTELLAR OBJECT
IC 3396	12 26 15.	+ 25 19 30.			NONSTELLAR OBJECT
HZ 26	12 26 15.	+ 28 37		13.7	BLUE STAR
AMES 1216	12 26 16.	+ 18 23 35.	22	17.8	NEBULA
IC 3397	12 26 16.	+ 26 00 30.			NONSTELLAR OBJECT
REIZ 2405	12 26 16.	+ 36 01	78	14.7	GALAXY
REIZ 2401	12 26 16.	- 01 41	60	13.3	GALAXY
REIZ 2398	12 26 16.	- 01 48	30	14.9	GALAXY
AMES 1220	12 26 17.	+ 06 25 47.	29	16.8	NEBULA
RNGC 4460	12 26 17.	+ 45 08		12.5	GALAXY
REIZ 2397	12 26 17.	- 11 22	24	13.9	GALAXY
A3 240	12 26 17.8	+ 15 56 43.		16.5	FAINT BLUE OBJECT
ZWG 042.122	12 26 18.	+ 04 34		15.5	GALAXY
UGC 07607	12 26 18.	+ 04 34	126	15.5	GALAXY Sc
MCG+01-32-074	12 26 18.	+ 04 34	150	15.5	GALAXY
ZWG 042.123	12 26 18.	+ 06 25		15.6	GALAXY
RMB 205	12 26 18.	+ 10 33			FAINT BLUE OBJECT
LB 04738	12 26 18.	+ 13 28		19.9	FAINT BLUE STAR
LB 04737	12 26 18.	+ 13 40		17.9	FAINT BLUE STAR
RMB 063	12 26 18.	+ 13 55			FAINT BLUE OBJECT
LB 04736	12 26 18.	+ 14 13		19.0	FAINT BLUE STAR
RMB 040	12 26 18.	+ 15 38			FAINT BLUE OBJECT
RMB 038	12 26 18.	+ 15 52			FAINT BLUE OBJECT
AMES 1219	12 26 18.	+ 16 38 05.	18	16.9	NEBULA
ZWG 099.066	12 26 18.	+ 19 45		15.7	GALAXY
LB 06420	12 26 18.	+ 26 31		19.4	FAINT BLUE STAR
LB 06419	12 26 18.	+ 27 04		18.5	FAINT BLUE STAR
MCG+05-30-003	12 26 18.	+ 27 04	24	16.	GALAXY
LB 06418	12 26 18.	+ 30 35		17.7	FAINT BLUE STAR
LB 06417	12 26 18.	+ 30 58		19.3	FAINT BLUE STAR
UGC 07608	12 26 18.	+ 43 31	240	16.0	GALAXY DWRF IR
ZWG 216.006	12 26 18.	+ 43 55		15.5	GALAXY
MCG+09-21-002	12 26 18.	+ 53 40	24	16.	GALAXY
ZCG 199.75	12 26 18.	+ 64 13		18.1	COMPACT GALAXY
ZC 1226.3+6648	12 26 18.	+ 66 48	2150		CLUSTER OF GALAXIES
ZWG 014.048	12 26 18.	- 01 39		13.5	GALAXY
UGC 07606	12 26 18.	- 01 39	162	13.5	GALAXY SB0/SBa
MCG+00-32-014	12 26 18.	- 01 39	78	13.5	GALAXY
ZWG 014.047	12 26 18.	- 01 46		15.1	GALAXY
RNGC 4484	12 26 18.	- 11 23		14.0	GALAXY
MCG-02-32-013	12 26 18.	- 11 23	90	14.	GALAXY
A3 241	12 26 18.1	+ 15 52 11.		16.3	FAINT BLUE OBJECT
BB 1.01	12 26 18.5	+ 02 22 36.5			GALAXY NEAR QSO 3C273
A3 242	12 26 18.5	+ 16 28 38.		16.8	FAINT BLUE OBJECT
RNGC 4454	12 26 19.	- 01 40		13.0	GALAXY
A3 244	12 26 19.8	+ 15 51 02.		17.5	FAINT BLUE OBJECT
HOLM 411B	12 26 20.	+ 13 29	48	12.5	PART OF MULTIPLE GALAXY
AMES 1221	12 26 20.	+ 19 45 53.	29	16.5	NEBULA
REIZ 2409	12 26 20.	+ 45 08	162	12.8	GALAXY
BB 1.14	12 26 20.85	+ 02 14 08.			GALAXY NEAR QSO 3C273
AMES 1222	12 26 21.	+ 17 39 05.	24	17.0	NEBULA
MCG+07-26-010	12 26 21.	+ 43 31	180	14.	GALAXY
A3 245	12 26 21.1	+ 16 47 26.		17.3	FAINT BLUE OBJECT
AMES 1224	12 26 22.	+ 15 51 23.	24	17.1	NEBULA
AMES 1223	12 26 22.	+ 16 40 35.	24	17.4	NEBULA
A3 247	12 26 23.7	+ 16 43 42.		17.9	FAINT BLUE OBJECT
ZWG 042.124	12 26 24.	+ 03 51		11.9	GALAXY
UGC 07609	12 26 24.	+ 03 51	204	11.9	GALAXY S0/SBa
MCG+01-32-075	12 26 24.	+ 03 51	132	11.9	GALAXY
ZWG 042.125	12 26 24.	+ 08 08		15.2	GALAXY
MCG+01-32-076	12 26 24.	+ 08 08	48	15.2	GALAXY
AMES 1225	12 26 24.	+ 09 42 05.	31	16.6	NEBULA
RMB 211	12 26 24.	+ 11 15			FAINT BLUE OBJECT
LB 04747	12 26 24.	+ 12 39		18.7	FAINT BLUE STAR
LB 04746	12 26 24.	+ 12 40		20.3	FAINT BLUE STAR
LB 04745	12 26 24.	+ 12 42		18.4	FAINT BLUE STAR
LB 04744	12 26 24.	+ 13 07		18.7	FAINT BLUE STAR
ZWG 070.114	12 26 24.	+ 13 31		13.3	GALAXY
UGC 07610	12 26 24.	+ 13 31	90	13.3	GALAXY E
MCG+02-32-082	12 26 24.	+ 13 31	30	13.3	GALAXY
LB 04743	12 26 24.	+ 13 32		18.8	FAINT BLUE STAR
LB 04742	12 26 24.	+ 13 38		18.6	FAINT BLUE STAR
LB 04741	12 26 24.	+ 13 51		16.3	FAINT BLUE STAR
LB 04740	12 26 24.	+ 14 16		18.3	FAINT BLUE STAR
LB 04739	12 26 24.	+ 14 35		18.5	FAINT BLUE STAR
RMB 024	12 26 24.	+ 15 24			FAINT BLUE OBJECT
RMB 037	12 26 24.	+ 15 53			FAINT BLUE OBJECT
LB 00042	12 26 24.	+ 25 49		18.7	FAINT BLUE STAR
LB 06425	12 26 24.	+ 26 57		19.5	FAINT BLUE STAR
LB 06424	12 26 24.	+ 27 28		18.4	FAINT BLUE STAR
LB 06423	12 26 24.	+ 28 20		18.8	FAINT BLUE STAR
LB 11214	12 26 24.	+ 29 24		13.8	FAINT BLUE STAR
LB 06422	12 26 24.	+ 29 59		19.3	FAINT BLUE STAR
LB 06421	12 26 24.	+ 31 09		18.6	FAINT BLUE STAR
MCG+08-23-041	12 26 24.	+ 45 07	120	12.5	GALAXY
ZWG 244.022	12 26 24.	+ 45 09		12.5	GALAXY
UGC 07611	12 26 24.	+ 45 09	252	12.5	GALAXY S0
LB 02366	12 26 24.	+ 57 09 12.		15.7	FAINT BLUE STAR
ZCG 199.76	12 26 24.	+ 64 06		18.0	COMPACT GALAXY
ZWG 335.013	12 26 24.	+ 69 10		15.6	GALAXY
BB 1.11	12 26 24.4	+ 02 26 42.5			GALAXY NEAR QSO 3C273
A3 248	12 26 24.9	+ 13 41 09.		16.5	FAINT BLUE OBJECT

OBJECT NAME	RIGHT ASCEN.	DECLINATION	DIAM.	MAGN.	TYPE OF OBJECT
RNGC 4457	12 26 25.	+ 03 51		12.0	GALAXY
REIN 3.175	12 26 25.96	+ 13 31 02.3			NEBULA
AMES 1228	12 26 26.	+ 07 48 23.	18	18.0	NEBULA
AMES 1227	12 26 26.	+ 08 56 35.	23	17.6	NEBULA
AMES 1226	12 26 26.	+ 10 47 59.	32	17.2	NEBULA
HOLM 411A	12 26 26.	+ 13 26	180	11.9	PART OF MULTIPLE GALAXY
RNGC 4458	12 26 26.	+ 13 31		13.5	GALAXY
REIZ 2408	12 26 26.	+ 13 31	30	13.5	GALAXY
IC 3399	12 26 26.	+ 25 58 24.			NONSTELLAR OBJECT
A3 250	12 26 26.7	+ 14 32 26.		16.6	FAINT BLUE OBJECT
A3 251	12 26 26.9	+ 14 54 25.		17.0	FAINT BLUE OBJECT
REIZ 2406	12 26 27.	+ 03 51	78	11.8	GALAXY
REIZ 2407	12 26 27.	+ 08 06	36	15.4	GALAXY
IC 3398	12 26 27.	+ 13 50 30.			SINGLE STAR
REIN 3.176	12 26 27.31	+ 08 07 38.3			NEBULA
A3 252	12 26 27.6	+ 16 18 45.		17.3	FAINT BLUE OBJECT
AMES 1232	12 26 28.	+ 07 13 59.	22	17.8	NEBULA
A3 253	12 26 28.7	+ 16 49 48.		17.4	FAINT BLUE OBJECT
AMES 1231	12 26 29.	+ 08 34 59.	25	17.0	NEBULA
AMES 1230	12 26 29.	+ 12 56 17.	16	17.0	NEBULA
ARC 1548	12 26 29.	+ 19 42		17.5	RICH CLUSTER OF GALAXIES
IC 3401	12 26 29.	+ 26 44 12.			NONSTELLAR OBJECT
SVEN 291	12 26 29.	- 08 02 07.	30	14.9	GALAXY
ZWG 042.126	12 26 30.	+ 03 00		15.2	GALAXY
UGC 07612	12 26 30.	+ 03 00	132	15.2	GALAXY DWRF SP
MCG+01-32-077	12 26 30.	+ 03 00	120	15.2	GALAXY
RFA 80	12 26 30.	+ 03 01			DWARF GALAXY
AMES 1234	12 26 30.	+ 08 48 11.	28	17.6	NEBULA
AMES 1233	12 26 30.	+ 08 55 17.	20	17.0	NEBULA
IC 3400	12 26 30.	+ 09 41 00.			MAY NOT EXIST
KARA.68 139	12 26 30.	+ 09 42	40		DWARF GALAXY
MCG+02-32-085	12 26 30.	+ 09 43	48	17.	GALAXY
RMB 222	12 26 30.	+ 10 57			FAINT BLUE OBJECT
LB 04754	12 26 30.	+ 12 29		19.9	FAINT BLUE STAR
LB 04753	12 26 30.	+ 12 39		18.9	FAINT BLUE STAR
LB 04752	12 26 30.	+ 12 59		20.3	FAINT BLUE STAR
LB 04751	12 26 30.	+ 13 24		18.6	FAINT BLUE STAR
ZWG 070.115	12 26 30.	+ 13 28		12.2	GALAXY
UGC 07613	12 26 30.	+ 13 28	216	12.2	GALAXY S0
MCG+02-32-084	12 26 30.	+ 13 28	132	12.2	GALAXY
LB 04750	12 26 30.	+ 13 31		19.4	FAINT BLUE STAR
LB 04749	12 26 30.	+ 13 42		17.4	FAINT BLUE STAR
LB 04748	12 26 30.	+ 14 12		18.2	FAINT BLUE STAR
ZWG 070.116	12 26 30.	+ 14 15		11.6	GALAXY
UGC 07614	12 26 30.	+ 14 15	210	11.6	GALAXY S0
MCG+02-32-083	12 26 30.	+ 14 15	66	11.6	GALAXY
ZC 1226.5+1641	12 26 30.	+ 16 41	1080		CLUSTER OF GALAXIES
AMES 1229	12 26 30.	+ 18 41 35.	29	17.4	NEBULA
ZC 1226.5+1940	12 26 30.	+ 19 40	1140		CLUSTER OF GALAXIES
ZWG 129.003	12 26 30.	+ 24 55		15.7	GALAXY
LB 00043	12 26 30.	+ 25 24		14.2	FAINT BLUE STAR
ZWG 159.004	12 26 30.	+ 27 55		15.7	GALAXY
LB 06428	12 26 30.	+ 28 01		19.4	FAINT BLUE STAR
ZWG 159.005	12 26 30.	+ 28 04		14.7	GALAXY
UGC 07615	12 26 30.	+ 28 04	72	14.7	GALAXY S
LB 06427	12 26 30.	+ 28 13		18.9	FAINT BLUE STAR
ZWG 159.006	12 26 30.	+ 29 08		15.7	GALAXY
ZWG 158.115	12 26 30.	+ 29 08		15.7	GALAXY
UGC 07616	12 26 30.	+ 29 08	60	15.7	GALAXY Sb-c
IC 3402	12 26 30.	+ 29 08 36.			NONSTELLAR OBJECT
LB 06426	12 26 30.	+ 30 20		18.5	FAINT BLUE STAR
IC 3405	12 26 30.	+ 38 00 24.			NONSTELLAR OBJECT
MCG+07-26-011	12 26 30.	+ 42 28 30.	42	15.	GALAXY
A3 254	12 26 30.7	+ 15 52 34.		17.6	FAINT BLUE OBJECT
REIN 3.177	12 26 31.42	+ 13 27 33.5			NEBULA
AMES 1239	12 26 32.	+ 08 57 05.	23	17.0	NEBULA
AMES 1238	12 26 32.	+ 08 58 41.	19	17.7	NEBULA
AMES 1237	12 26 32.	+ 11 04 59.	34	17.7	NEBULA
AMES 1236	12 26 32.	+ 11 47 29.	25	17.6	NEBULA
AMES 1235	12 26 32.	+ 12 40 35.	26	17.2	NEBULA
REIZ 2411	12 26 32.	+ 13 27	120	12.3	GALAXY
RNGC 4461	12 26 32.	+ 13 28		12.5	GALAXY
RNGC 4459	12 26 32.	+ 14 15		12.0	GALAXY
IC 3403	12 26 32.	+ 24 54 30.			NONSTELLAR OBJECT
ARC 1549	12 26 32.	+ 29 13		17.3	RICH CLUSTER OF GALAXIES
BB 1.02	12 26 32.3	+ 02 17 11.			GALAXY NEAR QSO 3C273
AMES 1241	12 26 33.	+ 07 33 11.	25	17.6	NEBULA
BC 3C273	12 26 33.3	+ 02 19 43.7		12.80	QUASI-STELLAR OBJECT
BB 1.03	12 26 33.3	+ 02 20 46.5			GALAXY NEAR QSO 3C273
SHB 207	12 26 33.4	+ 02 19 42.		12.8	QUASI-STELLAR OBJECT
BB 1.13	12 26 33.5	+ 02 30 22.			GALAXY NEAR QSO 3C273
A3 257	12 26 33.8	+ 15 02 48.		16.7	FAINT BLUE OBJECT
IC 3406	12 26 34.	+ 27 55 06.			NONSTELLAR OBJECT
AMES 1240	12 26 35.	+ 17 45 23.	14	17.6	NEBULA
IC 3407	12 26 35.	+ 28 03 18.			NONSTELLAR OBJECT
REIZ 2410	12 26 35.	- 08 02	24	15.9	GALAXY
ZWG 014.049	12 26 36.	+ 00 52		15.5	GALAXY
AMES 1243	12 26 36.	+ 07 31 29.	26	17.8	NEBULA
RMB 178	12 26 36.	+ 12 00			FAINT BLUE OBJECT
LB 04758	12 26 36.	+ 12 27		18.6	FAINT BLUE STAR
LB 04757	12 26 36.	+ 12 54		17.9	FAINT BLUE STAR
LB 04756	12 26 36.	+ 13 08		18.9	FAINT BLUE STAR
LB 04755	12 26 36.	+ 13 44		20.1	FAINT BLUE STAR
RMB 039	12 26 36.	+ 15 39			FAINT BLUE OBJECT
ZWG 099.067	12 26 36.	+ 19 17		15.7	GALAXY
LB 06430	12 26 36.	+ 27 14		19.4	FAINT BLUE STAR
REIZ 2414	12 26 36.	+ 27 55	42	14.5	GALAXY
MCG+05-30-004	12 26 36.	+ 27 55	30	15.7	GALAXY
REIZ 2415	12 26 36.	+ 28 03	48	14.2	GALAXY
MCG+05-30-005	12 26 36.	+ 28 04	48	14.7	GALAXY
LB 06429	12 26 36.	+ 29 48		16.0	FAINT BLUE STAR
LB 11216	12 26 36.	+ 30 03		17.6	FAINT BLUE STAR
LB 11215	12 26 36.	+ 32 33		12.6	FAINT BLUE STAR
ZC 1226.6+5136	12 26 36.	+ 51 36	3290		CLUSTER OF GALAXIES
ZC 1226.6-0254	12 26 36.	- 02 54	1610		CLUSTER OF GALAXIES
A3 258	12 26 36.1	+ 14 34 59.		16.8	FAINT BLUE OBJECT
BB 1.04	12 26 36.5	+ 02 23 24.			GALAXY NEAR QSO 3C273
BB 1.10	12 26 36.95	+ 02 27 07.			GALAXY NEAR QSO 3C273
AMES 1242	12 26 37.	+ 08 57 59.	25	17.0	NEBULA
REIZ 2413	12 26 37.	+ 20 43	48	15.6	GALAXY
IC 3404	12 26 38.	+ 07 25 48.			MAY NOT EXIST
A3 260	12 26 38.4	+ 17 07 40.		15.9	FAINT BLUE OBJECT
REIZ 2412	12 26 39.	+ 08 10	12	16.0	GALAXY
AMES 1244	12 26 39.	+ 14 48 11.	24	17.5	NEBULA
KARA.68 140	12 26 39.	+ 16 57	27		DWARF GALAXY
MCG-04-30-002	12 26 39.	- 22 54	192	13.	GALAXY
AMES 1245	12 26 41.	+ 08 09 35.	35	17.4	NEBULA
RNGC 4462	12 26 41.	- 22 54		13.0	GALAXY
A3 262	12 26 41.1	+ 12 40 37.		16.9	FAINT BLUE OBJECT
AMES 1246	12 26 42.	+ 06 38 35.	23	17.4	NEBULA
MCG+02-32-087	12 26 42.	+ 09 08 30.	2	17.	GALAXY
ZWG 070.117	12 26 42.	+ 09 09		15.6	GALAXY
MCG+02-32-086	12 26 42.	+ 09 09	24	15.6	GALAXY
ZWG 070.118	12 26 42.	+ 10 24		15.4	GALAXY
RMB 138	12 26 42.	+ 11 58			FAINT BLUE OBJECT
LB 04764	12 26 42.	+ 13 29		18.2	FAINT BLUE STAR
LB 04763	12 26 42.	+ 13 44		18.2	FAINT BLUE STAR
LB 04762	12 26 42.	+ 13 50		20.6	FAINT BLUE STAR
LB 04761	12 26 42.	+ 14 19		18.5	FAINT BLUE STAR
LB 04760	12 26 42.	+ 14 22		18.8	FAINT BLUE STAR
LB 04759	12 26 42.	+ 14 29		18.8	FAINT BLUE STAR
LB 06434	12 26 42.	+ 26 46		20.0	FAINT BLUE STAR
LB 06433	12 26 42.	+ 26 49		19.1	FAINT BLUE STAR
LB 11217	12 26 42.	+ 28 03		16.9	FAINT BLUE STAR
LB 06432	12 26 42.	+ 28 28		20.1	FAINT BLUE STAR
LB 06431	12 26 42.	+ 30 17		19.1	FAINT BLUE STAR
ZWG 244.023	12 26 42.	+ 44 55		15.6	GALAXY
UGC 07617	12 26 42.	+ 44 55	150	15.6	GALAXY Sc
ZC 1226.7+4519	12 26 42.	+ 45 19	810		CLUSTER OF GALAXIES
UGC 07618	12 26 42.	+ 58 12	84	16.0	GALAXY Sc
ZC 1226.7+5846	12 26 42.	+ 58 46	2150		CLUSTER OF GALAXIES
LB 00636	12 26 42.	+ 72 28 06.		15.8	FAINT BLUE STAR
ZWG 014.050	12 26 42.	- 01 04		15.7	GALAXY
IC 3409	12 26 43.	+ 15 04			NONSTELLAR OBJECT
HN 0922	12 26 43.	+ 19 16		15.5	NEBULA
IC 3410	12 26 43.	+ 19 16			NONSTELLAR OBJECT
IC 3411	12 26 43.	+ 24 51 36.			NONSTELLAR OBJECT
AMES 1249	12 26 44.	+ 06 40 47.	17	17.5	NEBULA
IC 3408	12 26 44.	+ 12 09 06.			SINGLE STAR
HN 0923	12 26 44.	+ 15 04		15.	NEBULA
AMES 1250	12 26 45.	+ 05 53 59.	47	17.4	NEBULA
AMES 1248	12 26 45.	+ 09 05 29.	38	17.2	NEBULA
AMES 1247	12 26 45.	+ 13 15 05.	28	16.6	NEBULA
KARA.68 141	12 26 45.	+ 17 05	34		DWARF GALAXY
MCG+08-23-042	12 26 45.	+ 44 55	144	14.	GALAXY
BB 1.12	12 26 45.5	+ 02 29 23.			GALAXY NEAR QSO 3C273
BB 1.05	12 26 46.1	+ 02 21 44.5			GALAXY NEAR QSO 3C273
AMES 1252	12 26 47.	+ 06 41 41.	24	17.7	NEBULA
ZWG 014.051	12 26 48.	+ 01 20		15.0	GALAXY
MCG+00-32-015	12 26 48.	+ 01 20	48	15.0	GALAXY
ZWG 042.127	12 26 48.	+ 08 18		15.4	GALAXY
ZWG 042.128	12 26 48.	+ 08 26		13.5	GALAXY
UGC 07619	12 26 48.	+ 08 26	60	13.5	GALAXY S?
MCG+01-32-078	12 26 48.	+ 08 26	48	13.5	GALAXY
8ZW 1226+10.2	12 26 48.	+ 10 12		15.2	COMPACT GALAXY
ZWG 070.119	12 26 48.	+ 10 16		15.4	GALAXY
AMES 1251	12 26 48.	+ 10 25 23.	25	16.6	NEBULA
RMB 214	12 26 48.	+ 11 14			FAINT BLUE OBJECT
ZWG 070.120	12 26 48.	+ 11 43		15.2	GALAXY
UGC 07620	12 26 48.	+ 11 43	72	15.2	GALAXY E
MCG+02-32-088	12 26 48.	+ 11 43	15	15.2	GALAXY
LB 04772	12 26 48.	+ 12 37		17.6	FAINT BLUE STAR
LB 04771	12 26 48.	+ 12 47		20.7	FAINT BLUE STAR
LB 04770	12 26 48.	+ 12 50		18.2	FAINT BLUE STAR
LB 04769	12 26 48.	+ 12 54		18.4	FAINT BLUE STAR
LB 04768	12 26 48.	+ 13 18		18.1	FAINT BLUE STAR
LB 04767	12 26 48.	+ 13 25		17.6	FAINT BLUE STAR
LB 04766	12 26 48.	+ 13 36		18.7	FAINT BLUE STAR
LB 04765	12 26 48.	+ 13 46		19.3	FAINT BLUE STAR
ZC 1226.8+2113	12 26 48.	+ 21 13	2080		CLUSTER OF GALAXIES
LB 06437	12 26 48.	+ 28 44		19.6	FAINT BLUE STAR
LB 06436	12 26 48.	+ 28 51		18.4	FAINT BLUE STAR
LB 11218	12 26 48.	+ 30 47		13.3	FAINT BLUE STAR
LB 06435	12 26 48.	+ 31 32		20.1	FAINT BLUE STAR
ZC 1226.8+3525	12 26 48.	+ 35 25	740		CLUSTER OF GALAXIES
ZC 1226.8+5611	12 26 48.	+ 56 11	740		CLUSTER OF GALAXIES
MCG+10-18-053	12 26 48.	+ 58 01	84	15.	GALAXY
LB 00637	12 26 48.	+ 61 15 24.		17.3	FAINT BLUE STAR
REIN 3.178	12 26 48.83	+ 08 25 58.8			NEBULA
AMES 1254	12 26 49.	+ 06 46 17.	35	17.4	NEBULA
REIZ 2418	12 26 49.	+ 16 41	42	14.2	GALAXY
ARC 1550	12 26 49.	+ 47 59		17.8	RICH CLUSTER OF GALAXIES
A3 265	12 26 49.3	+ 16 06 49.		15.9	FAINT BLUE OBJECT
RNGC 4465	12 26 50.	+ 08 18		15.5	GALAXY
RNGC 4464	12 26 50.	+ 08 26		13.5	GALAXY
IC 3412	12 26 50.	+ 10 15 42.			NONSTELLAR OBJECT
IC 3413	12 26 50.	+ 11 42 36.			NONSTELLAR OBJECT
AMES 1253	12 26 50.	+ 17 05 35.	41	17.4	NEBULA
LB 02367	12 26 50.	+ 52 07 00.		17.2	FAINT BLUE STAR
IC 3418	12 26 50.5	+ 11 40 35.			DWARF GALAXY
REIN 3.180	12 26 50.65	+ 11 42 36.4			NEBULA
BB 1.06	12 26 50.7	+ 02 24 45.5			GALAXY NEAR QSO 3C273
AMES 1258	12 26 51.	+ 05 37 35.	29	17.1	NEBULA
REIZ 2417	12 26 51.	+ 08 18	12	15.7	GALAXY
REIZ 2416	12 26 51.	+ 08 26	24	13.9	GALAXY
MCG+03-32-050	12 26 51.	+ 15 03	36	17.	GALAXY
A3 267	12 26 51.0	+ 16 02 22.		17.2	FAINT BLUE OBJECT
REIN 3.179	12 26 51.01	+ 08 18 09.4			NEBULA
A3 268	12 26 51.1	+ 14 00 04.		16.4	FAINT BLUE OBJECT
HOLM 413D	12 26 52.	+ 08 19	18	15.1	PART OF MULTIPLE GALAXY
AMES 1257	12 26 52.	+ 10 10 23.	40	16.8	NEBULA
AMES 1255	12 26 52.	+ 17 22 47.	14	17.7	NEBULA
AMES 1256	12 26 53.	+ 18 08 35.	22	17.8	NEBULA
IC 3415	12 26 53.	+ 27 02 36.			NONSTELLAR OBJECT
A3 270	12 26 53.5	+ 16 36 04.		13.8	FAINT BLUE OBJECT
A3 271	12 26 53.8	+ 13 40 29.		16.3	FAINT BLUE OBJECT
BB 1.07	12 26 53.95	+ 02 20 05.			GALAXY NEAR QSO 3C273
AMES 1262	12 26 54.	+ 06 38 23.	34	17.2	NEBULA
ZWG 042.129	12 26 54.	+ 07 03		14.2	GALAXY
SCH 50	12 26 54.	+ 07 03		14.2	PECULIAR GALAXY
UGC 07621	12 26 54.	+ 07 03	114	14.2	GALAXY SBc/IRR
MCG+01-32-079	12 26 54.	+ 07 03	84	14.2	GALAXY
IC 3414	12 26 54.	+ 07 03		14.2	GALAXY IN VIRGO CLSTR
ZWG 042.130	12 26 54.	+ 08 16		15.2	GALAXY
MCG+01-32-080	12 26 54.	+ 08 16	18	15.2	GALAXY
ZWG 070.121	12 26 54.	+ 09 01		12.6	GALAXY
UGC 07622	12 26 54.	+ 09 01	192	12.6	GALAXY S0-a
MCG+02-32-089	12 26 54.	+ 09 01	198	12.6	GALAXY
ZC 1226.9+1205	12 26 54.	+ 12 05	2490		CLUSTER OF GALAXIES
LB 04780	12 26 54.	+ 12 33		19.7	FAINT BLUE STAR
LB 04779	12 26 54.	+ 12 40		18.8	FAINT BLUE STAR
LB 04778	12 26 54.	+ 12 42		17.5	FAINT BLUE STAR
AMES 1259	12 26 54.	+ 12 43 23.	25	17.0	NEBULA
LB 04777	12 26 54.	+ 13 24		17.8	FAINT BLUE STAR
LB 04776	12 26 54.	+ 14 03		18.3	FAINT BLUE STAR
LB 04775	12 26 54.	+ 14 08		19.0	FAINT BLUE STAR

OBJECT NAME	RIGHT ASCEN.	DECLINATION	DIAM.	MAGN.	TYPE OF OBJECT
LB 04774	12 26 54.	+ 14 14		18.9	FAINT BLUE STAR
LB 04773	12 26 54.	+ 14 25		18.4	FAINT BLUE STAR
MCG+03-32-051	12 26 54.	+ 16 40	66	13.9	GALAXY
ZWG 099.068	12 26 54.	+ 16 41		13.9	GALAXY
UGC 07623	12 26 54.	+ 16 41	84	13.9	GALAXY S0-a
ZC 1226.9+1735	12 26 54.	+ 17 35	2220		CLUSTER OF GALAXIES
ZC 1226.9+2710	12 26 54.	+ 27 10	740		CLUSTER OF GALAXIES
TON-N 0079	12 26 54.	+ 45 30		16.1	BLUE STAR
ZC 1226.9+5925	12 26 54.	+ 59 25	1140		CLUSTER OF GALAXIES
ZCG 199.77	12 26 54.	+ 63 38		19.3	COMPACT GALAXY
A3 272	12 26 54.8	+ 14 31 10.		16.3	FAINT BLUE OBJECT
A3 273	12 26 54.9	+ 16 40 53.		14.2	UV CONCENTRATION
AMES 1264	12 26 55.	+ 06 31 10.	28	17.4	NEBULA
AMES 1261	12 26 55.	+ 11 50 23.	20	17.9	NEBULA
AMES 1260	12 26 55.	+ 12 04 23.	25	18.0	NEBULA
LB 02368	12 26 55.	+ 56 46 18.		17.1	FAINT BLUE STAR
REIN 3.181A	12 26 55.75	+ 09 01 31.2			NEBULA
REIN 3.181B	12 26 55.81	+ 09 01 31.1			NEBULA
RNGC 4469	12 26 56.	+ 09 02		12.5	GALAXY
AMES 1263	12 26 56.	+ 12 57 52.	25	17.6	NEBULA
IC 0796	12 26 56.	+ 16 41 18.			NONSTELLAR OBJECT
REIZ 2426	12 26 56.	+ 47 09	24	15.3	GALAXY
REIZ 2419	12 26 57.	+ 07 02	60	14.0	GALAXY
REIZ 2420	12 26 57.	+ 07 58	54	14.7	GALAXY
HOLM 412A	12 26 57.	+ 07 58	60	13.7	PART OF MULTIPLE GALAXY
REIZ 2421	12 26 57.	+ 08 16	18	15.1	GALAXY
REIN 3.182	12 26 57.76	+ 08 16 09.7			NEBULA
A3 276	12 26 57.8	+ 12 39 15.		17.0	FAINT BLUE OBJECT
REIN 3.183	12 26 57.97	+ 07 58 23.5			NEBULA
HOLM 413C	12 26 58.	+ 08 16	24	14.9	PART OF MULTIPLE GALAXY
AMES 1266	12 26 58.	+ 11 11 58.	23	17.6	NEBULA
SEY 186	12 26 58.	+ 31 43 30.		15.4	FAINT GALAXY
A3 278	12 26 58.7	+ 16 11 54.		17.1	FAINT BLUE OBJECT
HOLM 412B	12 26 59.	+ 08 00	24	15.1	PART OF MULTIPLE GALAXY
AMES 1265	12 26 59.	+ 17 38 58.	24	17.5	NEBULA
REIZ 2425	12 26 59.	+ 31 41	18	15.0	GALAXY
SVEN 292	12 26 59.	- 08 12	48	14.1	GALAXY
UGC 07625	12 27 00.	+ 01 07	60	17.	GALAXY
ZWG 042.131	12 27 00.	+ 07 58		14.7	GALAXY
UGC 07626	12 27 00.	+ 07 58	72	14.7	GALAXY S
MCG+01-32-081	12 27 00.	+ 07 58	60	14.7	GALAXY
ZWG 042.132	12 27 00.	+ 08 06		12.9	GALAXY
UGC 07627	12 27 00.	+ 08 06	90	12.9	GALAXY S
MCG+01-32-082	12 27 00.	+ 08 06	66	12.9	GALAXY
RMB 136	12 27 00.	+ 11 41			FAINT BLUE OBJECT
LB 04785	12 27 00.	+ 12 28		19.7	FAINT BLUE STAR
AMES 1269	12 27 00.	+ 12 31 40	18	17.6	NEBULA
LB 04784	12 27 00.	+ 13 06		19.0	FAINT BLUE STAR
LB 04783	12 27 00.	+ 13 19		19.0	FAINT BLUE STAR
LB 04732	12 27 00.	+ 13 31		18.7	FAINT BLUE STAR
LB 04781	12 27 00.	+ 13 40		19.9	FAINT BLUE STAR
ZWG 070.122	12 27 00.	+ 14 20		14.2	GALAXY
UGC 07628	12 27 00.	+ 14 20	90	14.2	GALAXY E?
MCG+02-32-090	12 27 00.	+ 14 20	24	14.2	GALAXY
AMES 1267	12 27 00.	+ 18 28 22.	13	17.4	NEBULA
LB 06438	12 27 00.	+ 27 54		19.0	FAINT BLUE STAR
ZWG 159.007	12 27 00.	+ 31 44		15.1	GALAXY
ZWG 158.116	12 27 00.	+ 31 44		15.1	GALAXY
MCG+05-30-006	12 27 00.	+ 31 44	15	15.1	GALAXY
MCG+06-28-001	12 27 00.	+ 36 02	24	16.	GALAXY
ZC 1227.0+4555	12 27 00.	+ 45 55	1480		CLUSTER OF GALAXIES
ZC 1227.0+6411	12 27 00.	+ 64 11	2150		CLUSTER OF GALAXIES
ZC 1227.0+7655	12 27 00.	+ 76 55	1680		CLUSTER OF GALAXIES
UGC 07624	12 27 00.	+ 82 10	72	17.	GALAXY Sc
A3 279	12 27 00.2	+ 12 38 05.		16.6	FAINT BLUE OBJECT
AMES 1270	12 27 01.	+ 11 29 04.	20	15.2	NEBULA
AMES 1268	12 27 01.	+ 17 49 46.	16	18.0	NEBULA
RNGC 4466	12 27 02.	+ 07 58		14.5	GALAXY
RNGC 4467	12 27 02.	+ 08 16		15.0	GALAXY
REIZ 2423	12 27 02.	+ 09 02	180	12.5	GALAXY
HN 0924	12 27 02.	+ 11 03	36		NEBULA
IC 3416	12 27 02.	+ 11 03			NONSTELLAR OBJECT
RNGC 4468	12 27 02.	+ 14 20		14.0	GALAXY
REIZ 2424	12 27 03.	+ 08 06	30	12.9	GALAXY
AMES 1273	12 27 03.	+ 08 12 58.	25	16.8	NEBULA
AMES 1271	12 27 03.	+ 16 42 28.	22	17.6	NEBULA
TON-N 1538	12 27 03.	+ 21 04		16.6	BLUE STAR
SEY 187	12 27 03.	+ 36 02 37.		15.4	FAINT GALAXY
LB 00638	12 27 03.	+ 56 56 06.		15.9	FAINT BLUE STAR
REIZ 2431	12 27 03.	+ 59 49	42	14.4	GALAXY
REIN 3.184A	12 27 03.17	+ 11 04 10.7			NEBULA
REIN 3.184B	12 27 03.23	+ 11 04 11.1			NEBULA
A3 280	12 27 03.3	+ 15 31 27.		16.8	FAINT BLUE OBJECT
REIZ 2427	12 27 04.	+ 36 02	30	15.0	GALAXY
A3 281	12 27 04.5	+ 15 07 33.		17.4	FAINT BLUE OBJECT
REIN 3.185	12 27 04.73	+ 08 12 32.9			NEBULA
AMES 1277	12 27 05.	+ 06 56 46.	22	17.8	NEBULA
AMES 1276	12 27 05.	+ 06 59 10.	26	17.7	NEBULA
AMES 1275	12 27 05.	+ 07 23 04.	35	17.6	NEBULA
AMES 1272	12 27 05.	+ 15 04 16.	14	17.2	NEBULA
SVEN 293	12 27 05.	- 07 55	18	15.8	GALAXY
REIZ 2422	12 27 05.	- 08 11	42	13.9	GALAXY
RNGC 4463	12 27 05.	- 64 30		8.0	OPEN CLUSTER
REIN 3.186	12 27 05.42	+ 08 05 59.4			NEBULA
REIN 4.074	12 27 05.42	+ 09 46 19.8			NEBULA
ZWG 042.133	12 27 06.	+ 03 54		15.6	GALAXY
ZWG 070.123	12 27 06.	+ 09 46		15.4	GALAXY
ZWG 070.124	12 27 06.	+ 11 04		15.3	GALAXY
MCG+02-32-091	12 27 06.	+ 11 04	6	15.3	GALAXY
RMB 135	12 27 06.	+ 12 44			FAINT BLUE OBJECT
LB 04791	12 27 06.	+ 12 44		18.1	FAINT BLUE STAR
LB 04790	12 27 06.	+ 13 12		18.8	FAINT BLUE STAR
LB 04789	12 27 06.	+ 13 49		18.2	FAINT BLUE STAR
LB 04787	12 27 06.	+ 14 13		18.9	FAINT BLUE STAR
LB 04786	12 27 06.	+ 14 15		18.2	FAINT BLUE STAR
RMB 034	12 27 06.	+ 15 07			FAINT BLUE OBJECT
AMES 1274	12 27 06.	+ 18 23 52.	13	17.3	NEBULA
LB 06439	12 27 06.	+ 32 20		19.2	FAINT BLUE STAR
ZWG 188.003	12 27 06.	+ 33 11		15.7	GALAXY
ZWG 187.048	12 27 06.	+ 33 11		15.7	GALAXY
ZWG 014.052	12 27 06.	- 03 28		15.5	GALAXY
MCG-01-32-016	12 27 06.	- 08 11	51	14.5	GALAXY
MCG-03-32-009	12 27 06.	- 18 07	54	14.5	GALAXY
MCG-05-30-003	12 27 06.	- 32 55	48	17.	GALAXY
OCL 0885	12 27 06.	- 64 31	540	8.5	OPEN STAR CLUSTER
VHA 135	12 27 06.	- 64 31	180		OPEN STAR CLUSTER
A3 282	12 27 06.2	+ 12 44 03.		17.1	FAINT BLUE OBJECT
IC 3417	12 27 07.	+ 08 08 18.			SINGLE STAR

OBJECT NAME	RIGHT ASCEN.	DECLINATION	DIAM.	MAGN.	TYPE OF OBJECT
AMES 1279	12 27 07.	+ 08 56 34.	28	17.2	NEBULA
AMES 1278	12 27 07.	+ 09 46 40.	48	15.8	NEBULA
LB 02370	12 27 07.	+ 56 20 48.		16.0	FAINT BLUE STAR
LB 02369	12 27 07.	+ 62 11 36.		16.8	FAINT BLUE STAR
RNGC 4470	12 27 08.	+ 08 06		13.0	GALAXY
RNGC 4471	12 27 08.	+ 08 11			NON-EXISTENT OBJECT
HN 0926	12 27 08.	+ 11 40			NEBULA
HN 0925	12 27 08.	+ 15 18		16.	NEBULA
IC 3419	12 27 08.	+ 15 18			NONSTELLAR OBJECT
HOLM 413B	12 27 09.	+ 08 12	12	14.8	PART OF MULTIPLE GALAXY
AMES 1280	12 27 09.	+ 12 07 22.	18	17.6	NEBULA
IC 3421	12 27 09.	+ 26 30 25.			OPEN CLUSTER
A3 284	12 27 09.6	+ 13 11 11.		16.7	FAINT BLUE OBJECT
AMES 1281	12 27 10.	+ 11 55 52.	24	17.7	NEBULA
REIN 3.187	12 27 10.75	+ 11 40 38.5			NEBULA
REIN 3.188A	12 27 10.95	+ 11 01 57.6			NEBULA
IC 3420	12 27 11.	+ 13 43 13.			SINGLE STAR
REIZ 2430	12 27 11.	+ 33 12	30	15.1	GALAXY
REIN 3.188B	12 27 11.04	+ 11 01 57.8			NEBULA
AMES 1284	12 27 12.	+ 06 12 22.	30	17.6	NEBULA
ZWG 042.134	12 27 12.	+ 08 16		10.2	GALAXY
UGC 07629	12 27 12.	+ 08 16	480	10.2	GALAXY E
MCG+01-32-083	12 27 12.	+ 08 16	132	10.2	GALAXY
8ZW 1227+11.5	12 27 12.	+ 11 28		14.3	COMPACT GALAXY
UGC 07630	12 27 12.	+ 11 40	90	16.5	GALAXY DWRF IR
MCG+02-32-092	12 27 12.	+ 11 41	72	17.	GALAXY
RMB 139	12 27 12.	+ 12 11			FAINT BLUE OBJECT
LB 04798	12 27 12.	+ 12 34		20.4	FAINT BLUE STAR
LB 04797	12 27 12.	+ 12 41		14.8	FAINT BLUE STAR
LB 04796	12 27 12.	+ 12 52		17.6	FAINT BLUE STAR
LB 04795	12 27 12.	+ 12 58		18.6	FAINT BLUE STAR
RMB 092	12 27 12.	+ 13 11			FAINT BLUE OBJECT
LB 04794	12 27 12.	+ 13 15		19.8	FAINT BLUE STAR
LB 04793	12 27 12.	+ 13 37		18.5	FAINT BLUE STAR
LB 04792	12 27 12.	+ 14 10		17.2	FAINT BLUE STAR
RMB 186	12 27 12.	+ 14 35			FAINT BLUE OBJECT
ZC 1227.2+1440	12 27 12.	+ 14 40	1140		CLUSTER OF GALAXIES
ZC 1227.2+2248	12 27 12.	+ 22 48	2150		CLUSTER OF GALAXIES
MCG+05-30-007	12 27 12.	+ 26 30	36	15.5	GALAXY
LB 11220	12 27 12.	+ 26 51		12.0	FAINT BLUE STAR
LB 11219	12 27 12.	+ 28 54		13.2	FAINT BLUE STAR
LB 06444	12 27 12.	+ 29 03		19.2	FAINT BLUE STAR
LB 06443	12 27 12.	+ 29 44		18.1	FAINT BLUE STAR
LB 06442	12 27 12.	+ 30 35		19.0	FAINT BLUE STAR
LB 06441	12 27 12.	+ 31 20		19.3	FAINT BLUE STAR
LB 06440	12 27 12.	+ 32 14		19.3	FAINT BLUE STAR
ARC 1551	12 27 12.	+ 36 56		17.2	RICH CLUSTER OF GALAXIES
IC 3422	12 27 13.	+ 14 58			NONSTELLAR OBJECT
REIZ 2429	12 27 13.	+ 20 41	30	15.5	GALAXY
AMES 1286	12 27 14.	+ 07 14 46.	24	17.8	NEBULA
RNGC 4472	12 27 14.	+ 08 17		10.0	GALAXY
HN 0927	12 27 14.	+ 14 58		15.	NEBULA
AMES 1283	12 27 14.	+ 18 20 52.	17	17.2	NEBULA
AMES 1282	12 27 14.	+ 18 31 28.	22	17.2	NEBULA
IC 3424	12 27 14.	+ 24 41 01.			NONSTELLAR OBJECT
A3 289	12 27 14.0	+ 12 49 05.		16.9	FAINT BLUE OBJECT
REIN 3.189	12 27 14.35	+ 08 16 36.4			NEBULA
AMES 1288	12 27 15.	+ 05 57 10.	16	16.9	NEBULA
AMES 1287	12 27 15.	+ 06 38 46.	60	17.5	NEBULA
REIZ 2428	12 27 15.	+ 08 16	300	10.7	GALAXY
HOLM 413A	12 27 15.	+ 08 17	240	10.9	PART OF MULTIPLE GALAXY
IC 3423	12 27 15.	+ 13 56 07.			MAY NOT EXIST
RNGC 4475	12 27 15.	+ 27 32		14.5	GALAXY
AMES 1285	12 27 16.	+ 18 31 04.	18	17.6	NEBULA
ARP 134	12 27 17.	+ 08 17			PECULIAR GALAXY
REIN 3.190	12 27 17.76	+ 08 16 36.7			NEBULA
REA 06	12 27 18.	+ 08 12			DWARF GALAXY
AMES 1290	12 27 18.	+ 08 36 34.	29	17.0	NEBULA
KARA.68 142	12 27 18.	+ 10 37	27		DWARF GALAXY
AMES 1289	12 27 18.	+ 12 19 22.	18	17.2	NEBULA
LB 04804	12 27 18.	+ 12 32		19.5	FAINT BLUE STAR
LB 04803	12 27 18.	+ 13 28		20.5	FAINT BLUE STAR
LB 04802	12 27 18.	+ 13 37		19.1	FAINT BLUE STAR
ZWG 070.125	12 27 18.	+ 13 42		11.2	GALAXY
WEED 7	12 27 18.	+ 13 42		18.5	VERY BLUE STELLAR OBJECT
UGC 07631	12 27 18.	+ 13 42	216	11.2	GALAXY E
MCG+02-32-093	12 27 18.	+ 13 42	78	11.2	GALAXY
LB 04801	12 27 18.	+ 13 48		17.7	FAINT BLUE STAR
LB 04800	12 27 18.	+ 13 52		19.2	FAINT BLUE STAR
LB 04799	12 27 18.	+ 14 24		17.8	FAINT BLUE STAR
ZC 1227.3+1827	12 27 18.	+ 18 27	2490		CLUSTER OF GALAXIES
ZWG 129.004	12 27 18.	+ 22 39		15.2	GALAXY
LB 06446	12 27 18.	+ 26 48		19.2	FAINT BLUE STAR
ZWG 159.008	12 27 18.	+ 27 32		14.6	GALAXY
REIZ 2433	12 27 18.	+ 27 32	72	14.0	GALAXY
UGC 07632	12 27 18.	+ 27 32	120	14.6	GALAXY Sb
LB 06445	12 27 18.	+ 30 28		19.3	FAINT BLUE STAR
ZCG 199.78	12 27 18.	+ 63 34		19.0	COMPACT GALAXY
7ZW 460	12 27 18.	+ 65 49			COMPACT GALAXY
AMES 1293	12 27 19.	+ 10 18 22.	23	17.2	NEBULA
AMES 1292	12 27 19.	+ 11 35 04.	23	17.6	NEBULA
ARC 1552	12 27 19.	+ 12 01		16.6	RICH CLUSTER OF GALAXIES
AMES 1291	12 27 19.	+ 14 08 16.	30	16.4	NEBULA
AMES 1295	12 27 20.	+ 06 08 10.	24	17.6	NEBULA
AMES 1294	12 27 20.	+ 11 47 34.	28	17.2	NEBULA
RNGC 4473	12 27 20.	+ 13 42		12.0	GALAXY
REIZ 2432	12 27 20.	+ 13 42	60	11.2	GALAXY
REIN 3.191	12 27 20.47	+ 11 57 11.5			NEBULA
TON-N 1539	12 27 21.	+ 20 25		16.7	BLUE STAR
A3 294	12 27 21.8	+ 12 43 09.		16.9	FAINT BLUE OBJECT
AMES 1298	12 27 22.	+ 06 23 34.	22	17.7	NEBULA
REIZ 2436	12 27 22.	+ 57 05	36	14.0	GALAXY
REIN 3.192	12 27 22.10	+ 10 36 52.9			NEBULA
A3 295	12 27 22.2	+ 12 20 16.		16.7	FAINT BLUE OBJECT
AMES 1297	12 27 23.	+ 08 42 46.	28	16.6	NEBULA
AMES 1296	12 27 23.	+ 11 43 34.	30	17.6	NEBULA
ZWG 070.126	12 27 24.	+ 10 53		15.3	GALAXY
UGC 07633	12 27 24.	+ 10 53	114	15.3	GALAXY S0?
MCG+02-32-095	12 27 24.	+ 10 53	48	15.3	GALAXY
RMB 137	12 27 24.	+ 11 55			FAINT BLUE OBJECT
LB 04815	12 27 24.	+ 12 39		20.7	FAINT BLUE STAR
LB 04814	12 27 24.	+ 12 47		18.7	FAINT BLUE STAR
LB 04813	12 27 24.	+ 12 49		17.8	FAINT BLUE STAR
LB 04812	12 27 24.	+ 12 50		19.6	FAINT BLUE STAR
LB 04811	12 27 24.	+ 13 13		19.2	FAINT BLUE STAR
LB 04810	12 27 24.	+ 13 37		17.7	FAINT BLUE STAR
LB 04809	12 27 24.	+ 13 46		20.8	FAINT BLUE STAR
LB 04808	12 27 24.	+ 13 47		17.8	FAINT BLUE STAR

OBJECT NAME	RIGHT ASCEN.	DECLINATION	DIAM.	MAGN.	TYPE OF OBJECT
LB 04788	12 27 24.	+ 13 57		16.7	FAINT BLUE STAR
LB 04807	12 27 24.	+ 14 08		19.5	FAINT BLUE STAR
ZWG 070.127	12 27 24.	+ 14 21		12.6	GALAXY
UGC 07634	12 27 24.	+ 14 21	126	12.6	GALAXY SO
MCG+02-32-094	12 27 24.	+ 14 21	108	12.6	GALAXY
LB 04806	12 27 24.	+ 14 28		18.1	FAINT BLUE STAR
LB 04805	12 27 24.	+ 14 31		17.6	FAINT BLUE STAR
MCG+05-30-008	12 27 24.	+ 27 31	108	14.6	GALAXY
LB 11222	12 27 24.	+ 28 17		12.4	FAINT BLUE STAR
LB 11221	12 27 24.	+ 30 14		13.0	FAINT BLUE STAR
ZWG 293.022	12 27 24.	+ 57 06		14.6	GALAXY
UGC 07635	12 27 24.	+ 57 06	72	14.6	GALAXY SO
MCG+10-18-054	12 27 24.	+ 58 18	24	17.	GALAXY
ZCG 199.79	12 27 24.	+ 63 33		18.0	COMPACT GALAXY
ZCG 199.80	12 27 24.	+ 64 08		18.9	COMPACT GALAXY
MCG-01-32-017	12 27 24.	- 05 42	48	14.5	GALAXY
REIN 3.193A	12 27 24.28	+ 10 53 30.5			NEBULA
A3 296	12 27 24.3	+ 13 34 45.		16.4	FAINT BLUE OBJECT
REIN 3.193B	12 27 24.57	+ 10 53 29.7			NEBULA
AMES 1301	12 27 25.	+ 05 19 40.	44	17.0	NEBULA
AMES 1300	12 27 25.	+ 11 58 28.	14	18.0	NEBULA
REIN 3.194	12 27 25.19	+ 11 11 26.8			NEBULA
HN 0928	12 27 26.	+ 10 53		14.	NEBULA
IC 3425	12 27 26.	+ 10 53			NONSTELLAR OBJECT
RNGC 4474	12 27 26.	+ 14 21		13.0	GALAXY
REIZ 2434	12 27 26.	+ 15 24	60	14.8	GALAXY
AMES 1299	12 27 26.	+ 16 22 10.	23	17.2	NEBULA
PK300-01.1	12 27 26.	- 63 36 24.	10		PLANETARY NEBULA
REIN 4.075	12 27 26.97	+ 08 42 32.6			NEBULA
AMES 1302	12 27 27.	+ 08 12 46.	16	17.0	NEBULA
REIN 3.195	12 27 27.38	+ 12 37 27.0			NEBULA
AMES 1303	12 27 29.	+ 11 55 46.	16	18.0	NEBULA
ZWG 042.135	12 27 30.	+ 08 12		15.4	GALAXY
UGC 07636	12 27 30.	+ 08 12	72	15.4	GALAXY DWRF IR
MCG+01-32-084	12 27 30.	+ 08 12	42	15.4	GALAXY
8ZW 1227+08.3	12 27 30.	+ 08 21		17.5	COMPACT GALAXY
ZWG 070.128	12 27 30.	+ 12 37		13.3	GALAXY
UGC 07637	12 27 30.	+ 12 37	102	13.3	GALAXY SO
MCG+02-32-096	12 27 30.	+ 12 37	36	13.3	GALAXY
LB 04820	12 27 30.	+ 12 43		18.6	FAINT BLUE STAR
LB 04819	12 27 30.	+ 13 22		19.7	FAINT BLUE STAR
LB 04818	12 27 30.	+ 13 42		15.6	FAINT BLUE STAR
IC 3426	12 27 30.	+ 13 52 43.			SINGLE STAR
ZWG 070.129	12 27 30.	+ 13 55		11.9	GALAXY
UGC 07638	12 27 30.	+ 13 55	228	11.9	GALAXY SBO
MCG+02-32-097	12 27 30.	+ 13 55	174	11.9	GALAXY
LB 04817	12 27 30.	+ 14 09		20.7	FAINT BLUE STAR
LB 04816	12 27 30.	+ 14 18		18.9	FAINT BLUE STAR
8ZW 1227+26.0	12 27 30.	+ 26 00		16.6	COMPACT GALAXY
LB 06448	12 27 30.	+ 26 53		19.3	FAINT BLUE STAR
LB 11224	12 27 30.	+ 27 10		14.9	FAINT BLUE STAR
LB 06447	12 27 30.	+ 28 04		19.5	FAINT BLUE STAR
LB 11223	12 27 30.	+ 29 46		13.4	FAINT BLUE STAR
ZWG 244.024	12 27 30.	+ 47 48		14.9	GALAXY
UGC 07639	12 27 30.	+ 47 48	216	14.9	GALAXY IRR
MCG+08-23-043	12 27 30.	+ 47 48	90	14.	GALAXY
MCG+10-18-055	12 27 30.	+ 57 06	45	15.	GALAXY
ZCG 199.81	12 27 30.	+ 64 07		18.9	COMPACT GALAXY
MCG+11-15-057	12 27 30.	+ 64 17	39	14.3	GALAXY
7ZW 461	12 27 30.	+ 71 14			COMPACT GALAXY
AMES 1307	12 27 31.	+ 08 21 16.	29	17.0	NEBULA
AMES 1306	12 27 31.	+ 11 35 10.	16	17.9	NEBULA
AMES 1304	12 27 31.	+ 15 09 28.	24	16.6	NEBULA
REIZ 2439	12 27 31.	+ 47 48	54	14.5	GALAXY
RNGC 4476	12 27 32.	+ 12 37		13.5	GALAXY
REIZ 2435	12 27 32.	+ 12 37	36	13.7	GALAXY
RNGC 4477	12 27 32.	+ 13 55		11.5	GALAXY
AMES 1305	12 27 32.	+ 14 58 40.	22	17.5	NEBULA
LB 02371	12 27 32.	+ 52 34 48.		17.1	FAINT BLUE STAR
A3 298	12 27 32.1	+ 13 41 51.		16.6	FAINT BLUE OBJECT
AMES 1308	12 27 33.	+ 16 38 34.	28	16.2	NEBULA
AMES 1309	12 27 34.	+ 15 32 40.	17	17.4	NEBULA
RNGC 4481	12 27 34.	+ 64 19		15.0	GALAXY
A3 300	12 27 34.0	+ 13 03 20.		17.0	FAINT BLUE OBJECT
AMES 1313	12 27 35.	+ 06 02 04.	41	17.3	NEBULA
AMES 1311	12 27 35.	+ 11 30 34.	24	17.3	NEBULA
SVEN 294	12 27 35.	- 08 10	90	13.9	GALAXY
ZWG 042.136	12 27 36.	+ 03 51		15.5	GALAXY
REA 48	12 27 36.	+ 10 27	54		DWARF GALAXY
KARA.68 143	12 27 36.	+ 10 28	34		DWARF GALAXY
ZWG 070.130	12 27 36.	+ 11 03		14.2	GALAXY
UGC 07640	12 27 36.	+ 11 03	96	14.2	GALAXY E
MCG+02-32-098	12 27 36.	+ 11 03	84	14.2	GALAXY
IC 3427	12 27 36.	+ 11 03		14.2	GALAXY IN VIRGO CLSTR
RMB 134	12 27 36.	+ 12 31			FAINT BLUE OBJECT
KARA.68 144	12 27 36.	+ 12 39	27		DWARF GALAXY
AMES 1312	12 27 36.	+ 12 45 16.	23	17.8	NEBULA
LB 04823	12 27 36.	+ 12 58		18.2	FAINT BLUE STAR
LB 04822	12 27 36.	+ 14 21		19.6	FAINT BLUE STAR
LB 04821	12 27 36.	+ 14 26		20.8	FAINT BLUE STAR
ZWG 099.069	12 27 36.	+ 16 39		15.6	GALAXY
AMES 1310	12 27 36.	+ 17 40 34.	48	17.0	NEBULA
ZC 1227.6+2151	12 27 36.	+ 21 51	2420		CLUSTER OF GALAXIES
ZC 1227.6+2520	12 27 36.	+ 25 20	2960		CLUSTER OF GALAXIES
LB 06451	12 27 36.	+ 26 40		19.4	FAINT BLUE STAR
LB 06450	12 27 36.	+ 28 54		19.2	FAINT BLUE STAR
LB 11225	12 27 36.	+ 29 20		15.6	FAINT BLUE STAR
LB 06449	12 27 36.	+ 31 44		19.7	FAINT BLUE STAR
UGC 07641	12 27 36.	+ 37 48	84	17.	GALAXY
MCG+06-28-002	12 27 36.	+ 37 51	72	17.	GALAXY
MCG+10-18-056	12 27 36.	+ 58 00	15	16.	GALAXY
ZCG 199.82	12 27 36.	+ 64 15		18.5	COMPACT GALAXY
REIZ 2443	12 27 36.	+ 64 18	30	14.3	GALAXY
ZWG 315.040	12 27 36.	+ 64 19		14.9	GALAXY
AMES 1314	12 27 37.	+ 11 09 46.	22	17.8	NEBULA
REIZ 2438	12 27 38.	+ 11 01	30	14.3	GALAXY
RNGC 4482	12 27 38.	+ 11 03		14.0	GALAXY
AMES 1315	12 27 38.	+ 11 49 16.	26	18.0	NEBULA
IC 3429	12 27 38.	+ 23 49 13.			NONSTELLAR OBJECT
IC 3428	12 27 38.	+ 23 57 01.			NONSTELLAR OBJECT
REIN 3.196A	12 27 38.25	+ 11 03 20.5			NEBULA
REIN 3.196B	12 27 38.46	+ 11 03 22.0			NEBULA
AMES 1317	12 27 39.	+ 12 05 22.	20	18.0	NEBULA
PK300-02.1	12 27 39.	- 64 35	5		PLANETARY NEBULA
A3 302	12 27 39.7	+ 17 02 40.		16.6	FAINT BLUE OBJECT
A3 303	12 27 39.8	+ 14 44 46.		16.9	FAINT BLUE OBJECT
AMES 1318	12 27 40.	+ 11 10 16.	16	17.4	NEBULA
AMES 1316	12 27 40.	+ 17 02 58.	25	17.6	NEBULA

OBJECT NAME	RIGHT ASCEN.	DECLINATION	DIAM.	MAGN.	TYPE OF OBJECT
REIN 3.197	12 27 40.25	+ 12 01 13.0			NEBULA
REIZ 2437	12 27 41.	- 08 08	90	14.0	GALAXY
A3 305	12 27 41.4	+ 13 04 08.		16.7	FAINT BLUE OBJECT
ZWG 042.137	12 27 42.	+ 02 54		15.3	GALAXY
UGC 07642	12 27 42.	+ 02 54	66	15.3	GALAXY DWARF
MCG+01-32-085	12 27 42.	+ 02 54	60	15.3	GALAXY
AMES 1320	12 27 42.	+ 06 57 58.	22	17.4	NEBULA
ZWG 070.131	12 27 42.	+ 08 47		15.6	GALAXY
ZWG 070.132	12 27 42.	+ 09 21		15.4	GALAXY
UGC 07643	12 27 42.	+ 09 21	72	15.4	GALAXY
AMES 1319	12 27 42.	+ 11 19 28.	23	17.6	NEBULA
LB 04829	12 27 42.	+ 12 43		17.9	FAINT BLUE STAR
LB 04828	12 27 42.	+ 12 45		19.0	FAINT BLUE STAR
LB 04827	12 27 42.	+ 12 55		16.3	FAINT BLUE STAR
LB 04826	12 27 42.	+ 13 03		19.9	FAINT BLUE STAR
LB 04825	12 27 42.	+ 13 31		20.0	FAINT BLUE STAR
LB 04824	12 27 42.	+ 14 19		18.6	FAINT BLUE STAR
ZC 1227.7+1637	12 27 42.	+ 16 37	1080		CLUSTER OF GALAXIES
TON-N 0619	12 27 42.	+ 25 37		15.8	BLUE STAR
LB 11227	12 27 42.	+ 26 49		12.3	FAINT BLUE STAR
LB 06459	12 27 42.	+ 26 50		18.7	FAINT BLUE STAR
LB 06458	12 27 42.	+ 26 58		19.4	FAINT BLUE STAR
LB 06457	12 27 42.	+ 27 25		19.2	FAINT BLUE STAR
ZC 1227.7+2811	12 27 42.	+ 28 11	2150		CLUSTER OF GALAXIES
LB 06456	12 27 42.	+ 29 16		20.4	FAINT BLUE STAR
LB 06455	12 27 42.	+ 30 12		18.7	FAINT BLUE STAR
LB 06454	12 27 42.	+ 31 10		20.2	FAINT BLUE STAR
LB 06453	12 27 42.	+ 31 27		18.8	FAINT BLUE STAR
LB 11226	12 27 42.	+ 32 14		13.5	FAINT BLUE STAR
LB 06452	12 27 42.	+ 32 18		18.3	FAINT BLUE STAR
ZCG 199.83	12 27 42.	+ 64 16		18.5	COMPACT GALAXY
MCG-01-32-018	12 27 42.	- 05 41	48	15.	GALAXY
A3 307	12 27 42.3	+ 17 03 31.		16.8	FAINT BLUE OBJECT
AMES 1322	12 27 44.	+ 12 03 10.	26	17.5	NEBULA
REIZ 2441	12 27 44.	+ 13 51	18	14.3	GALAXY
REIN 3.198A	12 27 44.44	+ 09 21 41.8			NEBULA
REIN 3.198B	12 27 44.85	+ 09 21 39.3			NEBULA
REIZ 2440	12 27 45.	+ 04 01	90	14.6	GALAXY
AMES 1324	12 27 45.	+ 07 44 58.	22	17.2	NEBULA
IC 3430	12 27 45.	+ 09 21 43.			NONSTELLAR OBJECT
AMES 1321	12 27 45.	+ 18 45 10.	19	17.0	NEBULA
MCG-01-32-019	12 27 45.	- 08 08	132	13.	GALAXY
A3 308	12 27 45.6	+ 12 36 17.		16.0	UV CONCENTRATION
REIN 3.199	12 27 45.66	+ 12 36 16.3			NEBULA
AMES 1323	12 27 46.	+ 11 35 34.	19	17.6	NEBULA
AMES 1327	12 27 47.	+ 07 45 34.	24	17.3	NEBULA
AMES 1325	12 27 47.	+ 11 54 34.	24	18.0	NEBULA
A3 310	12 27 47.0	+ 13 32 43.		16.6	FAINT BLUE OBJECT
A3 312	12 27 47.8	+ 13 17 19.		16.6	FAINT BLUE OBJECT
ZWG 042.138	12 27 48.	+ 04 01		15.1	GALAXY
UGC 07644	12 27 48.	+ 04 01	132	15.1	GALAXY Sc
MCG+01-32-086	12 27 48.	+ 04 01	150	15.1	GALAXY
AMES 1329	12 27 48.	+ 06 18 10.	37	17.2	NEBULA
AMES 1328	12 27 49.	+ 07 17 28.	34	17.3	NEBULA
LB 04833	12 27 48.	+ 12 32		18.4	FAINT BLUE STAR
LB 04832	12 27 48.	+ 12 33		19.8	FAINT BLUE STAR
ZWG 070.133	12 27 48.	+ 12 36		12.2	GALAXY
UGC 07645	12 27 48.	+ 12 36	102	12.2	GALAXY E
MCG+02-32-099	12 27 48.	+ 12 36	48	12.2	GALAXY
LB 04831	12 27 48.	+ 13 33		19.6	FAINT BLUE STAR
ZWG 070.134	12 27 48.	+ 13 51		13.9	GALAXY
UGC 07646	12 27 48.	+ 13 51	90	13.9	GALAXY SBO
MCG+02-32-100	12 27 48.	+ 13 51	72	13.9	GALAXY
LB 04830	12 27 48.	+ 14 08		19.0	FAINT BLUE STAR
RMB 031	12 27 48.	+ 14 11			FAINT BLUE OBJECT
RMB 035	12 27 48.	+ 15 45			FAINT BLUE OBJECT
ZC 1227.8+2557	12 27 48.	+ 25 57	740		CLUSTER OF GALAXIES
LB 06461	12 27 48.	+ 27 56		19.3	FAINT BLUE STAR
LB 06460	12 27 48.	+ 27 58		19.1	FAINT BLUE STAR
MCG+05-30-009	12 27 48.	+ 30 40	48	15.5	GALAXY
LB 11228	12 27 48.	+ 31 18		13.7	FAINT BLUE STAR
FEIG 064	12 27 48.	+ 46 53		12.3	FAINT BLUE STAR
ZWG 244.025	12 27 48.	+ 47 17		15.6	GALAXY
MCG+08-23-044	12 27 48.	+ 47 17 30.	36	16.	GALAXY
ZC 1227.8+6232	12 27 48.	+ 62 32	2690		CLUSTER OF GALAXIES
MCG-01-32-020	12 27 48.	- 05 44	36	15.	GALAXY
A3 313	12 27 48.8	+ 13 13 55.		16.8	FAINT BLUE OBJECT
AMES 1326	12 27 49.	+ 18 18 46.	24	17.6	NEBULA
REIZ 2445	12 27 49.	+ 47 16	12	15.0	GALAXY
HN 0930	12 27 50.	+ 11 52	12		NEBULA
IC 3431	12 27 50.	+ 11 52			SINGLE STAR
IC 3131	12 27 50.	+ 11 52			NONSTELLAR OBJECT
RNGC 4478	12 27 50.	+ 12 36		12.5	GALAXY
RNGC 4479	12 27 50.	+ 13 51		14.0	GALAXY
HN 0929	12 27 50.	+ 14 25	18		NEBULA
IC 3432	12 27 50.	+ 14 25			NONSTELLAR OBJECT
REIZ 2442	12 27 51.	+ 04 31	90	13.7	GALAXY
AMES 1331	12 27 51.	+ 10 42 16.	14	17.1	NEBULA
TON-N 1540	12 27 51.	+ 21 07		16.9	BLUE STAR
A3 314	12 27 51.1	+ 13 49 02.		16.2	FAINT BLUE OBJECT
A3 315	12 27 51.4	+ 16 23 07.		16.8	FAINT BLUE OBJECT
AMES 1333	12 27 52.	+ 11 17 16.	17	17.1	NEBULA
AMES 1330	12 27 52.	+ 18 01 46.	23	17.2	NEBULA
AMES 1334	12 27 53.	+ 11 27 58.	22	17.9	NEBULA
AMES 1332	12 27 53.	+ 18 33 46.	35	16.7	NEBULA
LB 02373	12 27 53.	+ 51 59 06.		16.2	FAINT BLUE STAR
LB 02372	12 27 53.	+ 53 04 18.		16.6	FAINT BLUE STAR
A3 316	12 27 53.7	+ 12 53 38.		15.9	FAINT BLUE OBJECT
ZWG 042.139	12 27 54.	+ 04 31		13.4	GALAXY
UGC 07647	12 27 54.	+ 04 31	156	13.4	GALAXY Sc
MCG+01-32-087	12 27 54.	+ 04 31	144	13.4	GALAXY
ZC 1227.9+1059	12 27 54.	+ 10 59	2350		CLUSTER OF GALAXIES
MCG+02-32-101	12 27 54.	+ 12 47	18	15.	GALAXY
RMB 030	12 27 54.	+ 14 06			FAINT BLUE OBJECT
LB 04834	12 27 54.	+ 14 12		17.6	FAINT BLUE STAR
ZWG 070.135	12 27 54.	+ 14 27		15.1	GALAXY
MCG+02-32-102	12 27 54.	+ 14 27	36	15.1	GALAXY
KARA.68 145	12 27 54.	+ 15 57	34		DWARF GALAXY
REA 49	12 27 54.	+ 15 59	48		DWARF GALAXY
ZC 1227.9+3006	12 27 54.	+ 30 06	2620		CLUSTER OF GALAXIES
LB 11231	12 27 54.	+ 30 44		16.4	FAINT BLUE STAR
LB 06462	12 27 54.	+ 31 47		17.1	FAINT BLUE STAR
LB 11230	12 27 54.	+ 31 50		12.8	FAINT BLUE STAR
LB 11229	12 27 54.	+ 31 52		14.7	FAINT BLUE STAR
MCG+10-18-057	12 27 54.	+ 59 12	18	16.	GALAXY
ZCG 199.84	12 27 54.	+ 64 12		18.0	COMPACT GALAXY
RNGC 4480	12 27 55.	+ 04 31		13.5	GALAXY
AMES 1338	12 27 55.	+ 07 27 46.	35	17.2	NEBULA

OBJECT NAME	RIGHT ASCEN.	DECLINATION	DIAM.	MAGN.	TYPE OF OBJECT
AMES 1336	12 27 55.	+ 10 48 10.	19	16.8	NEBULA
HN 0931	12 27 55.	+ 17 34		14.5	NEBULA
IC 3433	12 27 55.	+ 17 34			NONSTELLAR OBJECT
AMES 1335	12 27 55.	+ 18 35 10.	16	17.6	NEBULA
REIZ 2444	12 27 55.	+ 19 56	30	13.7	GALAXY
REIZ 2447	12 27 55.	+ 47 29	24	15.2	GALAXY
A3 318	12 27 55.1	+ 16 44 02.		16.6	FAINT BLUE OBJECT
A3 319	12 27 55.5	+ 16 56 18.		16.6	FAINT BLUE OBJECT
AMES 1337	12 27 56.	+ 10 21 46.	22	17.2	NEBULA
A3 320	12 27 56.1	+ 14 25 53.		15.0	UV CONCENTRATION
KEEL 446	12 27 56.3	+ 14 26 11.		14.	NEBULA
AMES 1340	12 27 58.	+ 07 51 10.	22	18.0	NEBULA
AMES 1339	12 27 59.	+ 16 45 22.	25	17.4	NEBULA
HMS 1228+1050	12 28	+ 10 50			CLUSTER OF GALAXIES
VDB.66G 131	12 28	+ 29 58	70		DWARF GALAXY
LB 04842	12 28 00.	+ 12 25		17.9	FAINT BLUE STAR
1ZW 038	12 28 00.	+ 12 45			COMPACT GALAXY
LB 04841	12 28 00.	+ 13 08		17.9	FAINT BLUE STAR
LB 04840	12 28 00.	+ 13 22		20.1	FAINT BLUE STAR
LB 04839	12 28 00.	+ 13 24		20.5	FAINT BLUE STAR
LB 04838	12 28 00.	+ 13 32		20.4	FAINT BLUE STAR
LB 04837	12 28 00.	+ 13 51		19.5	FAINT BLUE STAR
LB 04836	12 28 00.	+ 13 57		18.7	FAINT BLUE STAR
LB 04835	12 28 00.	+ 14 18		18.5	FAINT BLUE STAR
ZWG 099.070	12 28 00.	+ 19 05		15.4	GALAXY
ZWG 099.071	12 28 00.	+ 19 56		15.0	GALAXY
MCG+03-32-052	12 28 00.	+ 19 56	21	15.0	GALAXY
KARA.73B 0531	12 28 00.	+ 19 56	36	15.0	ISOLATED GALAXY S
LB 11232	12 28 00.	+ 29 43		14.0	FAINT BLUE STAR
ZWG 216.007	12 28 00.	+ 41 59		12.4	GALAXY
UGC 07648	12 28 00.	+ 41 59	180	12.4	GALAXY IRR
KARA.72 341A	12 28 00.	+ 41 59	162	12.4	PART OF DOUBLE GALAXY
MCG+07-26-012	12 28 00.	+ 43 11	42	15.	GALAXY
MCG+10-18-058	12 28 00.	+ 56 58	18	16.	GALAXY
ZWG 293.023	12 28 00.	+ 57 04		15.6	GALAXY
7ZW 462	12 28 00.	+ 64 12			COMPACT GALAXY
ZC 1228.0+8245	12 28 00.	+ 82 45	1410		CLUSTER OF GALAXIES
A3 321	12 28 00.7	+ 16 54 22.		17.0	FAINT BLUE OBJECT
HN 0933	12 28 01.	+ 15 24	24		NEBULA
HN 0932	12 28 01.	+ 19 04		15.	NEBULA
IC 3434	12 28 01.	+ 19 04			NONSTELLAR OBJECT
REIZ 2446	12 28 01.	+ 19 06	18	15.1	GALAXY
A3 322	12 28 01.9	+ 15 24 25.		16.5	FAINT BLUE OBJECT
AMES 1341	12 28 02.	+ 11 47 34.	23	17.8	NEBULA
RNGC 4486B	12 28 02.	+ 12 46			GALAXY
IC 3435	12 28 02.	+ 15 24			NONSTELLAR OBJECT
HOLM 414B	12 28 03.	+ 41 57	66	11.9	PART OF MULTIPLE GALAXY
A3 323	12 28 03.9	+ 16 48 16.		14.2	FAINT BLUE OBJECT
AMES 1342	12 28 04.	+ 11 27 58.	14	18.0	NEBULA
RNGC 4485	12 28 05.	+ 41 59		13.0	GALAXY
SVEN 296	12 28 05.	- 07 18 06.	36	15.1	GALAXY
SVEN 295	12 28 05.	- 07 41 06.	24	15.2	GALAXY
ZWG 070.136	12 28 06.	+ 09 17		13.4	GALAXY
UGC 07649	12 28 06.	+ 09 17	102	13.4	GALAXY SB0
MCG+02-32-103	12 28 06.	+ 09 17	90	13.4	GALAXY
RMB 202	12 28 06.	+ 10 02			FAINT BLUE OBJECT
RMB 203	12 28 06.	+ 10 03			FAINT BLUE OBJECT
RMB 206	12 28 06.	+ 10 53			FAINT BLUE OBJECT
RMB 225	12 28 06.	+ 11 00			FAINT BLUE OBJECT
LB 04847	12 28 06.	+ 12 49		18.5	FAINT BLUE STAR
LB 04846	12 28 06.	+ 13 03		19.8	FAINT BLUE STAR
LB 04845	12 28 06.	+ 13 09		19.7	FAINT BLUE STAR
LB 04844	12 28 06.	+ 13 10		19.9	FAINT BLUE STAR
LB 04843	12 28 06.	+ 13 46		20.2	FAINT BLUE STAR
MCG+03-32-053	12 28 06.	+ 15 23	78	15.7	GALAXY
ZWG 099.072	12 28 06.	+ 15 24		15.7	GALAXY
UGC 07650	12 28 06.	+ 15 24	72	15.7	GALAXY S
RMB 036	12 28 06.	+ 15 59			FAINT BLUE OBJECT
KARA.68 146	12 28 06.	+ 30 20	34		DWARF GALAXY
LB 06464	12 28 06.	+ 31 08		20.3	FAINT BLUE STAR
LB 06463	12 28 06.	+ 31 56		19.5	FAINT BLUE STAR
ZWG 216.008	12 28 06.	+ 41 55		10.1	GALAXY
UGC 07651	12 28 06.	+ 41 55	420	10.1	GALAXY
KARA.72 341B	12 28 06.	+ 41 55	372	10.1	PART OF DOUBLE GALAXY
MCG+07-26-013	12 28 06.	+ 41 59	96	13.	GALAXY
MCG+09-21-003	12 28 06.	+ 51 23	30	16.	GALAXY
MCG+10-18-059	12 28 06.	+ 57 03	27	16.	GALAXY
HN 0934	12 28 07.	+ 19 56		14.	NEBULA
IC 3436	12 28 07.	+ 19 56			NONSTELLAR OBJECT
KEEL 447	12 28 07.2	+ 14 51 58.		18.	NEBULA
REIZ 2448	12 28 08.	+ 09 17	30	13.0	GALAXY
RNGC 4483	12 28 08.	+ 09 18		13.5	GALAXY
AMES 1343	12 28 08.	+ 18 13 22.	26	16.7	NEBULA
REIN 3.200A	12 28 08.35	+ 09 17 30.1			NEBULA
REIN 3.200B	12 28 08.43	+ 09 17 29.7			NEBULA
AMES 1344	12 28 09.	+ 10 10 40.	23	16.6	NEBULA
TON-N 1541	12 28 09.	+ 21 27		17.0	BLUE STAR
HOLM 414A	12 28 09.	+ 41 54	270	9.8	PART OF MULTIPLE GALAXY
REIZ 2449	12 28 09.	+ 41 58	48	12.7	GALAXY
A3 325	12 28 09.5	+ 15 07 25.		16.5	FAINT BLUE OBJECT
AMES 1345	12 28 10.	+ 08 13 22.	17	17.7	NEBULA
YC 1228-43	12 28 10.	- 43 57 42.			UNUSUAL SOUTHERN NEBULA
A3 326	12 28 10.4	+ 16 05 43.		16.8	FAINT BLUE OBJECT
AMES 1346	12 28 11.	+ 10 09 34.	31	17.6	NEBULA
RNGC 4490	12 28 11.	+ 41 55		11.0	GALAXY
A3 327	12 28 11.2	+ 16 41 49.		17.0	FAINT BLUE OBJECT
AMES 1348	12 28 12.	+ 07 42 10.	40	17.8	NEBULA
RMB 133	12 28 12.	+ 12 24			FAINT BLUE OBJECT
LB 04854	12 28 12.	+ 12 26		18.6	FAINT BLUE STAR
UGC 07652	12 28 12.	+ 12 39	66		GALAXY DBL SYS
LB 04853	12 28 12.	+ 12 48		18.2	FAINT BLUE STAR
LB 04852	12 28 12.	+ 12 49		17.2	FAINT BLUE STAR
LB 04851	12 28 12.	+ 12 52		20.5	FAINT BLUE STAR
LB 04850	12 28 12.	+ 13 05		17.3	FAINT BLUE STAR
KARA.68 147	12 28 12.	+ 13 30	34		DWARF GALAXY
LB 04849	12 28 12.	+ 13 44		19.1	FAINT BLUE STAR
LB 04848	12 28 12.	+ 14 24		19.3	FAINT BLUE STAR
LB 06468	12 28 12.	+ 27 48		18.8	FAINT BLUE STAR
LB 06467	12 28 12.	+ 28 21		19.4	FAINT BLUE STAR
LB 06466	12 28 12.	+ 29 50		17.5	FAINT BLUE STAR
LB 06465	12 28 12.	+ 30 20		18.9	FAINT BLUE STAR
MCG+07-26-014	12 28 12.	+ 41 55 30.	360	10.	GALAXY
VV 030B	12 28 12.	+ 41 58	108	11.9	INTERACTING GALAXY
7ZW 463	12 28 12.	+ 76 56			COMPACT GALAXY
ACK 299+18.1	12 28 12.	- 43 58			PLANETARY NEBULA
A3 328	12 28 12.6	+ 12 58 11.		16.6	FAINT BLUE OBJECT
AMES 1351	12 28 13.	+ 06 06 40.	52	17.6	NEBULA
AMES 1350	12 28 13.	+ 07 52 34.	40	17.4	NEBULA

OBJECT NAME	RIGHT ASCEN.	DECLINATION	DIAM.	MAGN.	TYPE OF OBJECT
AMES 1347	12 28 13.	+ 17 30 46.	41	16.5	NEBULA
HN 0935	12 28 14.	+ 11 36		15.	NEBULA
IC 3437	12 28 14.	+ 11 36			NONSTELLAR OBJECT
AMES 1349	12 28 14.	+ 12 23 10.	22	17.8	NEBULA
A3 330	12 28 14.4	+ 13 21 22.		17.1	FAINT BLUE OBJECT
AMES 1353	12 28 15.	+ 08 10 34.	30	17.8	NEBULA
REIZ 2454	12 28 15.	+ 41 55	276	10.6	GALAXY
MCG-03-32-010	12 28 15.	- 19 38 30.	36	16.	GALAXY
AMES 1355	12 28 16.	+ 06 20 46.	22	17.4	NEBULA
AMES 1354	12 28 16.	+ 10 20 46.	26	17.3	NEBULA
AMES 1352	12 28 16.	+ 16 32 58.	23	17.0	NEBULA
A3 331	12 28 16.9	+ 13 14 32.		14.5	FAINT BLUE OBJECT
ARP 152	12 28 17.	+ 12 41			PECULIAR GALAXY
SN 1919A	12 28 17.	+ 12 42		12.3	SUPERNOVA
REIN 3.201	12 28 17.72	+ 12 40 01.2			NEBULA
ZWG 070.137	12 28 18.	+ 08 38		13.8	GALAXY
UGC 07653	12 28 18.	+ 08 38	276	13.8	GALAXY SB
MCG+02-32-104	12 28 18.	+ 08 38	150	13.8	GALAXY
ZWG 070.138	12 28 18.	+ 11 37		15.5	GALAXY
RMB 132	12 28 18.	+ 12 19			FAINT BLUE OBJECT
ZWG 070.139	12 28 18.	+ 12 40		10.4	GALAXY
UGC 07654	12 28 18.	+ 12 40	420	10.4	GALAXY E
MCG+02-32-105	12 28 18.	+ 12 40	360	10.4	GALAXY
LB 04855	12 28 18.	+ 13 20		20.4	FAINT BLUE STAR
RMB 029	12 28 18.	+ 14 11			FAINT BLUE OBJECT
RMB 163	12 28 18.	+ 14 53			FAINT BLUE OBJECT
RMB 033	12 28 18.	+ 14 53			FAINT BLUE OBJECT
ZWG 099.073	12 28 18.	+ 17 02		13.2	GALAXY
UGC 07655	12 28 18.	+ 17 02	114	13.2	GALAXY E
MCG+03-32-054	12 28 18.	+ 17 02	84	13.2	GALAXY
8ZW 1228+24.5	12 28 18.	+ 24 33		17.8	COMPACT GALAXY
LB 06470	12 28 18.	+ 26 48		19.3	FAINT BLUE STAR
LB 06469	12 28 18.	+ 27 54		20.0	FAINT BLUE STAR
TON-N 0620	12 28 18.	+ 30 48		15.2	BLUE STAR
VV 030A	12 28 18.	+ 41 55	360	10.0	INTERACTING GALAXY
ZC 1228.3+4734	12 28 18.	+ 47 34	470		CLUSTER OF GALAXIES
ZWG 293.024	12 28 18.	+ 57 35		15.3	GALAXY
KARA.72 342A	12 28 18.	+ 57 35	36	15.3	PART OF DOUBLE GALAXY
KEEL 448	12 28 18.6	+ 14 54 38.		15.	NEBULA
REIN 4.076	12 28 18.94	+ 08 38 08.6			NEBULA
ARC 1553	12 28 19.	+ 10 51		17.8	RICH CLUSTER OF GALAXIES
REIZ 2453	12 28 19.	+ 17 02	48	13.7	GALAXY
RNGC 4487	12 28 19.	- 07 48		12.0	GALAXY
RNGC 4488	12 28 20.	+ 08 38		14.0	GALAXY
REIZ 2452	12 28 20.	+ 08 38	144	13.1	GALAXY
RNGC 4486	12 28 20.	+ 12 40		11.0	GALAXY
REIZ 2450	12 28 20.	+ 12 40	138	10.8	GALAXY
RNGC 4489	12 28 20.	+ 17 02		13.0	GALAXY
ARP 269	12 28 20.	+ 41 55			PECULIAR GALAXY
AMES 1357	12 28 21.	+ 06 23 28.	26	17.0	NEBULA
KEEL 449	12 28 21.4	+ 14 49 42.		16.	NEBULA
AMES 1359	12 28 22.	+ 05 41 10.	25	17.1	NEBULA
AMES 1358	12 28 23.	+ 11 54 10.	22	17.2	NEBULA
AMES 1356	12 28 23.	+ 18 24 22.	38	17.4	NEBULA
SVEN 297	12 28 23.	- 07 49	240	11.6	GALAXY
REIZ 2451	12 28 23.	- 07 49	210	12.0	GALAXY
ZWG 042.140	12 28 24.	+ 03 46		15.2	GALAXY
MCG+01-32-088	12 28 24.	+ 03 46	15	15.2	GALAXY
ZWG 042.141	12 28 24.	+ 08 21		14.1	GALAXY
UGC 07656	12 28 24.	+ 08 21	150	14.1	GALAXY S
MCG+01-32-089	12 28 24.	+ 08 21	90	14.1	GALAXY
MCG+02-32-109	12 28 24.	+ 10 50	6	20.	GALAXY
MCG+02-32-108	12 28 24.	+ 10 50	6	20.	GALAXY
RMB 207	12 28 24.	+ 11 17			FAINT BLUE OBJECT
RMB 129	12 28 24.	+ 11 46			FAINT BLUE OBJECT
ZWG 070.140	12 28 24.	+ 11 46		13.7	GALAXY
UGC 07657	12 28 24.	+ 11 46	108	13.7	GALAXY
MCG+02-32-107	12 28 24.	+ 11 46	84	13.7	GALAXY
ZWG 070.141	12 28 24.	+ 12 33		11.2	GALAXY
UGC 07658	12 28 24.	+ 12 33	60	11.2	GALAXY (E)
MCG+02-32-110	12 28 24.	+ 12 33	30	11.2	GALAXY
LB 04859	12 28 24.	+ 12 56		18.4	FAINT BLUE STAR
REA 50	12 28 24.	+ 13 30	36		DWARF GALAXY
MCG+02-32-106	12 28 24.	+ 13 30	36	18.	GALAXY
LB 04858	12 28 24.	+ 13 54		20.5	FAINT BLUE STAR
LB 04857	12 28 24.	+ 13 59		18.6	FAINT BLUE STAR
LB 04856	12 28 24.	+ 14 02		18.5	FAINT BLUE STAR
LB 06474	12 28 24.	+ 27 16		19.4	FAINT BLUE STAR
LB 06473	12 28 24.	+ 28 15		19.5	FAINT BLUE STAR
LB 06472	12 28 24.	+ 30 56		19.8	FAINT BLUE STAR
LB 11233	12 28 24.	+ 31 40		15.3	FAINT BLUE STAR
ZC 1228.4+3228	12 28 24.	+ 32 28	740		CLUSTER OF GALAXIES
LB 06471	12 28 24.	+ 32 29		17.8	FAINT BLUE STAR
ZC 1228.4+4105	12 28 24.	+ 41 05	3360		CLUSTER OF GALAXIES
MCG+09-21-004	12 28 24.	+ 53 12	36	16.	GALAXY
ZWG 293.025	12 28 24.	+ 57 34		14.9	GALAXY
UGC 07659	12 28 24.	+ 57 34	66	14.9	GALAXY S
KARA.72 342B	12 28 24.	+ 57 34	60	14.9	PART OF DOUBLE GALAXY
MCG+10-18-060	12 28 24.	+ 57 35	27	15.	GALAXY
ZCG 199.85	12 28 24.	+ 63 53		17.6	COMPACT GALAXY
KEEL 450	12 28 24.5	+ 15 03 01.		16.	NEBULA
REIZ 2457	12 28 25.	+ 20 23	60	15.6	GALAXY
REIN 3.202	12 28 25.43	+ 11 45 33.8			NEBULA
AMES 1362	12 28 26.	+ 07 19 34.	20	17.7	NEBULA
AMES 1361	12 28 26.	+ 07 59 34.	30	18.0	NEBULA
REIZ 2456	12 28 26.	+ 08 21	48	13.7	GALAXY
REIZ 2455	12 28 26.	+ 11 45	48	13.6	GALAXY
RNGC 4491	12 28 26.	+ 11 46		13.5	GALAXY
RNGC 4486A	12 28 26.	+ 12 33			GALAXY
AMES 1360	12 28 26.	+ 16 01 09.	60	16.6	NEBULA
AMES 1360	12 28 26.	+ 16 01 10.	60	16.6	NEBULA
IC 3438	12 28 27.	+ 08 21 20.			SAME AS NGC 4492
LB 02374	12 28 27.	+ 58 28 36.		16.1	FAINT BLUE STAR
REIN 3.203	12 28 27.27	+ 08 21 14.0			NEBULA
A3 334	12 28 28.8	+ 12 57 41.		16.9	FAINT BLUE OBJECT
ZWG 014.054	12 28 30.	+ 01 57		15.6	GALAXY
MCG+00-32-016	12 28 30.	+ 01 57	42	15.6	GALAXY
8ZW 1228+10.8	12 28 30.	+ 10 45		19.0	COMPACT GALAXY
LB 04863	12 28 30.	+ 12 28		16.3	FAINT BLUE STAR
LB 04862	12 28 30.	+ 12 51		18.8	FAINT BLUE STAR
LB 04861	12 28 30.	+ 13 12		17.3	FAINT BLUE STAR
LB 04860	12 28 30.	+ 14 18		18.4	FAINT BLUE STAR
RMB 032	12 28 30.	+ 14 32			FAINT BLUE OBJECT
MCG+03-32-055	12 28 30.	+ 16 00	72	17.5	GALAXY
ARC 1554	12 28 30.	+ 16 14		17.6	RICH CLUSTER OF GALAXIES
ZC 1228.5+1615	12 28 30.	+ 16 15	1080		CLUSTER OF GALAXIES
KARA.68 148	12 28 30.	+ 18 48	34		DWARF GALAXY
IC 3439	12 28 30.	+ 25 50 14.			NONSTELLAR OBJECT

OBJECT NAME	RIGHT ASCEN.	DECLINATION	DIAM.	MAGN.	TYPE OF OBJECT
LB 06478	12 28 30.	+ 27 30		19.3	FAINT BLUE STAR
LB 06477	12 28 30.	+ 27 44		19.1	FAINT BLUE STAR
LB 11235	12 28 30.	+ 28 10		12.5	FAINT BLUE STAR
LB 06476	12 28 30.	+ 28 31		18.7	FAINT BLUE STAR
LB 06475	12 28 30.	+ 31 11		19.2	FAINT BLUE STAR
LB 11234	12 28 30.	+ 31 22		14.4	FAINT BLUE STAR
UGC 06507	12 28 30.	+ 34 29	90	15.6	GALAXY SBc
MCG+09-21-006	12 28 30.	+ 51 23	60	16.	GALAXY
MCG+09-21-005	12 28 30.	+ 53 10	24	17.	GALAXY
ZWG 014.053	12 28 30.	- 02 04		15.5	GALAXY
MCG-01-32-021	12 28 30.	- 07 48	210	11.5	GALAXY
AMES 1363	12 28 31.	+ 18 17 28.	32	17.0	NEBULA
REIZ 2459	12 28 31.	+ 20 09	18	14.3	GALAXY
AMES 1364	12 28 32.	+ 06 31 40.	24	17.8	NEBULA
RNGC 4492	12 28 32.	+ 08 21		14.0	GALAXY
A3 336	12 28 32.1	+ 12 27 48.		16.8	FAINT BLUE OBJECT
KEEL 452	12 28 32.4	+ 15 03 25.		17.	NEBULA
KEEL 451	12 28 32.5	+ 14 37 54.		16.	NEBULA
AMES 1365	12 28 33.	+ 04 52 28.	36	16.5	NEBULA
IC 3440	12 28 33.	+ 12 18 08.			NONSTELLAR OBJECT
REIZ 2458	12 28 34.	+ 00 53	18	14.6	GALAXY
A3 337	12 28 34.0	+ 12 46 10.		16.5	FAINT BLUE OBJECT
AMES 1370	12 28 35.	+ 05 27 58.	58	17.2	NEBULA
AMES 1366	12 28 35.	+ 12 15 34.	12	17.3	NEBULA
A3 338	12 28 35.8	+ 15 44 11.		16.6	FAINT BLUE OBJECT
ZWG 014.055	12 28 36.	+ 00 44		15.7	GALAXY
ZWG 014.056	12 28 36.	+ 00 53		14.9	GALAXY
MCG+00-32-017	12 28 36.	+ 00 53	48	14.9	GALAXY
ZWG 042.142	12 28 36.	+ 04 52		15.4	GALAXY
AMES 1369	12 28 36.	+ 08 02 22.	20	17.6	NEBULA
AMES 1368	12 28 36.	+ 10 23 58.	30	17.0	NEBULA
LB 04866	12 28 36.	+ 12 46		19.0	FAINT BLUE STAR
RMB 085	12 28 36.	+ 13 30			FAINT BLUE OBJECT
LB 04865	12 28 36.	+ 13 44		19.0	FAINT BLUE STAR
LB 04864	12 28 36.	+ 13 55		20.2	FAINT BLUE STAR
RMB 022	12 28 36.	+ 15 41			FAINT BLUE OBJECT
LB 06481	12 28 36.	+ 26 57		19.4	FAINT BLUE STAR
LB 06480	12 28 36.	+ 27 12		19.7	FAINT BLUE STAR
IC 3441	12 28 36.	+ 29 07 50.			NONSTELLAR OBJECT
LB 06479	12 28 36.	+ 30 48		19.4	FAINT BLUE STAR
MCG+07-26-015	12 28 36.	+ 42 48 30.	42	15.	GALAXY
MCG+08-23-046	12 28 36.	+ 47 09	24	16.	GALAXY
MCG+08-23-045	12 28 36.	+ 47 09	15	16.	GALAXY
MCG+09-21-007	12 28 36.	+ 51 35	18	16.	GALAXY
MCG+10-18-061	12 28 36.	+ 57 34	60	15.	GALAXY
MCG-04-30-003	12 28 36.	- 26 02	42	15.	GALAXY
KEEL 453	12 28 36.3	+ 14 32 18.		18.	NEBULA
A3 339	12 28 36.5	+ 13 14 38.		14.5	FAINT BLUE OBJECT
A3 340	12 28 36.7	+ 14 27 49.		16.6	FAINT BLUE OBJECT
RNGC 4493	12 28 37.	+ 00 53		15.0	GALAXY
AMES 1367	12 28 37.	+ 15 58 46.	20	17.3	NEBULA
A3 341	12 28 39.2	+ 12 43 35.		17.4	FAINT BLUE OBJECT
KEEL 454	12 28 39.6	+ 14 38 39.		16.	NEBULA
AMES 1371	12 28 40.	+ 10 55 22.	14	16.8	NEBULA
A3 342	12 28 41.9	+ 13 09 07.		16.5	FAINT BLUE OBJECT
ZWG 042.143	12 28 42.	+ 05 21		15.4	GALAXY
LB 04869	12 28 42.	+ 13 03		20.1	FAINT BLUE STAR
RMB 091	12 28 42.	+ 13 09			FAINT BLUE OBJECT
LB 04868	12 28 42.	+ 13 28		20.5	FAINT BLUE STAR
KARA.68 149	12 28 42.	+ 14 06	40		DWARF GALAXY
LB 04867	12 28 42.	+ 14 29		16.9	FAINT BLUE STAR
LB 06484	12 28 42.	+ 27 56		18.9	FAINT BLUE STAR
MCG+05-30-010	12 28 42.	+ 29 08	24	15.5	GALAXY
LB 06483	12 28 42.	+ 29 14		18.8	FAINT BLUE STAR
LB 11236	12 28 42.	+ 31 08		13.8	FAINT BLUE STAR
LB 06482	12 28 42.	+ 31 18		19.8	FAINT BLUE STAR
ZWG 188.004	12 28 42.	+ 36 53		15.3	GALAXY
UGC 07660	12 28 42.	+ 36 53	66	15.3	GALAXY Sa-b
MCG+09-21-008	12 28 42.	+ 51 35	30	15.	GALAXY
MCG+09-21-009	12 28 42.	+ 51 51	42	15.	GALAXY
ZWG 270.007	12 28 42.	+ 52 42		15.7	GALAXY
UGC 07661	12 28 42.	+ 52 42	78	15.7	GALAXY SBb
A3 343	12 28 42.6	+ 12 46 02.		15.4	FAINT BLUE OBJECT
AMES 1372	12 28 43.	+ 13 11 22.	16	16.7	NEBULA
KEEL 455	12 28 43.8	+ 14 28 24.		12.	NEBULA
REIN 3.204	12 28 43.99	+ 12 36 26.3			NEBULA
AMES 1373	12 28 44.	+ 05 21 51.	44	16.4	NEBULA
HN 0937	12 28 44.	+ 12 36		15.5	NEBULA
IC 3443	12 28 44.	+ 12 36			NONSTELLAR OBJECT
HN 0936	12 28 44.	+ 14 23	12		NEBULA
IC 3444	12 28 45.	+ 27 49 26.			NONSTELLAR OBJECT
A3 344	12 28 45.8	+ 14 06 18.		16.5	FAINT BLUE OBJECT
A3 345	12 28 46.6	+ 12 35 19.		16.5	FAINT BLUE OBJECT
AMES 1377	12 28 48.	+ 07 42 21.	23	17.8	NEBULA
ZWG 070.142	12 28 48.	+ 11 46		15.3	GALAXY
RMB 131	12 28 48.	+ 12 17			FAINT BLUE OBJECT
ZWG 070.143	12 28 48.	+ 12 36		15.6	GALAXY
MCG+02-32-112	12 28 48.	+ 12 36	12	15.6	GALAXY
LB 04876	12 28 48.	+ 12 52		17.0	FAINT BLUE STAR
LB 04875	12 28 48.	+ 13 00		18.1	FAINT BLUE STAR
LB 04874	12 28 48.	+ 13 11		17.8	FAINT BLUE STAR
LB 04873	12 28 48.	+ 13 24		17.0	FAINT BLUE STAR
LB 04872	12 28 48.	+ 13 30		19.0	FAINT BLUE STAR
LB 04871	12 28 48.	+ 13 48		17.2	FAINT BLUE STAR
LB 04870	12 28 48.	+ 14 11		18.3	FAINT BLUE STAR
ZWG 070.144	12 28 48.	+ 14 24		15.4	GALAXY
MCG+02-32-111	12 28 48.	+ 14 24	48	15.4	GALAXY
RMB 021	12 28 48.	+ 15 43			FAINT BLUE OBJECT
AMES 1374	12 28 48.	+ 16 38 09.	22	17.4	NEBULA
KARA.68 150	12 28 48.	+ 18 37	34		DWARF GALAXY
TON-N 0080	12 28 48.	+ 21 20		15.7	BLUE STAR
LB 06489	12 28 48.	+ 27 13		16.0	FAINT BLUE STAR
LB 06488	12 28 48.	+ 27 24		18.8	FAINT BLUE STAR
LB 06487	12 28 48.	+ 30 00		18.0	FAINT BLUE STAR
LB 06486	12 28 48.	+ 30 52		19.8	FAINT BLUE STAR
LB 06485	12 28 48.	+ 32 30		17.7	FAINT BLUE STAR
MCG+06-28-003	12 28 48.	+ 36 55	60	15.	GALAXY
TON-N 0081	12 28 48.	+ 39 15		15.2	BLUE STAR
MCG+08-23-047	12 28 48.	+ 47 09	36	16.	GALAXY
MCG+09-21-010	12 28 48.	+ 52 40	72	15.	GALAXY
ZCG 199.86	12 28 48.	+ 64 01		19.0	COMPACT GALAXY
ZWG 014.057	12 28 48.	- 03 23		15.4	GALAXY
IC 3442	12 28 48.7	+ 14 23 28.			GALAXY
KEEL 456	12 28 48.8	+ 14 28 20.		16.	NEBULA
KEEL 457	12 28 48.9	+ 14 23 31.		16.	SPIRAL NEBULA
AMES 1376	12 28 49.	+ 12 08 45.	23	18.0	NEBULA
IC 3445	12 28 49.	+ 13 00			NONSTELLAR OBJECT
AMES 1375	12 28 49.	+ 19 19 57.	31	16.7	NEBULA
REIZ 2462	12 28 49.	+ 47 17	24	15.2	GALAXY
HN 0940	12 28 50.	+ 10 57	6		NEBULA
IC 3447	12 28 50.	+ 10 57			SINGLE STAR
HN 0939	12 28 50.	+ 11 45		14.	NEBULA
IC 3446	12 28 50.	+ 11 45			NONSTELLAR OBJECT
HN 0938	12 28 50.	+ 13 00	6		NEBULA
AMES 1379	12 28 50.	+ 13 26 27.	12	18.0	NEBULA
AMES 1380	12 28 51.	+ 12 07 45.	25	17.8	NEBULA
AMES 1378	12 28 51.	+ 17 18 45.	24	16.6	NEBULA
RNGC 4494	12 28 51.	+ 26 03		11.0	GALAXY
MCG+04-30-002	12 28 51.	+ 26 03	72	10.7	GALAXY
LB 02375	12 28 51.	+ 57 55 12.		16.4	FAINT BLUE STAR
REIN 3.205	12 28 51.40	+ 11 46 06.8			NEBULA
RNGC 4495	12 28 52.	+ 29 25		14.0	GALAXY
A3 348	12 28 52.1	+ 17 09 45.		16.6	FAINT BLUE OBJECT
REIZ 2461	12 28 53.	+ 29 25	60	13.5	GALAXY
RMB 128	12 28 54.	+ 11 46			FAINT BLUE OBJECT
LB 04878	12 28 54.	+ 12 33		19.3	FAINT BLUE STAR
REA 51	12 28 54.	+ 14 06	54		DWARF GALAXY
LB 04877	12 28 54.	+ 14 12		17.6	FAINT BLUE STAR
RMB 020	12 28 54.	+ 15 47			FAINT BLUE OBJECT
AMES 1382	12 28 54.	+ 17 09 33.	32	17.0	NEBULA
AMES 1381	12 28 54.	+ 17 12 45.	22	17.6	NEBULA
ZWG 129.005	12 28 54.	+ 26 03		10.7	GALAXY
REIZ 2460	12 28 54.	+ 26 03	60	11.2	GALAXY
UGC 07662	12 28 54.	+ 26 03	270	10.7	GALAXY E
IC 3449	12 28 54.	+ 26 11 20.			TWO STARS
UGC 07663	12 28 54.	+ 28 25	96	14.1	GALAXY Sa-b
MCG+05-30-011	12 28 54.	+ 29 08	30	15.5	GALAXY
ZWG 159.009	12 28 54.	+ 29 25		14.1	GALAXY
MCG+05-30-012	12 28 54.	+ 29 25	72	14.1	GALAXY
LB 06492	12 28 54.	+ 30 14		19.4	FAINT BLUE STAR
ZC 1228.9+3101	12 28 54.	+ 31 01	1080		CLUSTER OF GALAXIES
LB 06491	12 28 54.	+ 31 19		20.1	FAINT BLUE STAR
LB 11237	12 28 54.	+ 31 20		14.0	FAINT BLUE STAR
LB 06490	12 28 54.	+ 32 10		18.0	FAINT BLUE STAR
ZC 1228.9+4034	12 28 54.	+ 40 34	740		CLUSTER OF GALAXIES
ZC 1228.9+5758	12 28 54.	+ 57 58	1010		CLUSTER OF GALAXIES
ZC 1228.9+6307	12 28 54.	+ 63 07	2020		CLUSTER OF GALAXIES
MRSL 300+01/1	12 28 54.	- 61 24	900		HII REGION
AMES 1384	12 28 55.	+ 10 51 45.	22	17.5	NEBULA
HN 0941	12 28 55.	+ 17 28			NEBULA
IC 3448	12 28 55.	+ 17 28			NONSTELLAR OBJECT
AMES 1383	12 28 55.	+ 17 48 21.	24	17.6	NEBULA
LB 02376	12 28 55.	+ 58 27 12.		16.1	FAINT BLUE STAR
IC 3450	12 28 55.9	+ 27 04 19.			GALAXY S0
AMES 1388	12 28 56.	+ 06 52 57.	41	17.1	NEBULA
IC 3451	12 28 56.	+ 29 07 56.			NONSTELLAR OBJECT
AMES 1390	12 28 57.	+ 07 34 33.	29	17.8	NEBULA
AMES 1387	12 28 57.	+ 08 23 03.	24	17.8	NEBULA
A3 352	12 28 57.5	+ 16 02 37.		17.1	FAINT BLUE OBJECT
AMES 1389	12 28 58.	+ 13 07 03.	13	17.0	NEBULA
AMES 1386	12 28 58.	+ 17 19 09.	24	17.5	NEBULA
AMES 1385	12 28 58.	+ 19 21 57.	24	16.6	NEBULA
VDB.66G 132	12 29	+ 13 03	70		DWARF GALAXY
ZWG 070.145	12 29 00.	+ 11 54		13.8	GALAXY
UGC 07665	12 29 00.	+ 11 54	120	13.8	GALAXY S0
MCG+02-32-113	12 29 00.	+ 11 54	108	13.8	GALAXY
LB 04884	12 29 00.	+ 12 28		17.6	FAINT BLUE STAR
LB 04883	12 29 00.	+ 12 34		17.9	FAINT BLUE STAR
LB 04882	12 29 00.	+ 13 10		17.4	FAINT BLUE STAR
LB 04881	12 29 00.	+ 13 42		20.5	FAINT BLUE STAR
LB 04880	12 29 00.	+ 13 43		18.9	FAINT BLUE STAR
REA 88	12 29 00.	+ 14 20			DWARF GALAXY
LB 04879	12 29 00.	+ 14 33		18.1	FAINT BLUE STAR
ZWG 099.074	12 29 00.	+ 15 08		15.2	GALAXY
UGC 07666	12 29 00.	+ 15 08	78	15.2	GALAXY IRR
LB 00009	12 29 00.	+ 22 55		14.7	FAINT BLUE STAR
LB 06497	12 29 00.	+ 26 31		19.1	FAINT BLUE STAR
LB 06496	12 29 00.	+ 26 35		19.0	FAINT BLUE STAR
LB 06495	12 29 00.	+ 26 58		18.0	FAINT BLUE STAR
LB 06494	12 29 00.	+ 27 18		20.1	FAINT BLUE STAR
IC 3454	12 29 00.	+ 27 46 20.			NONSTELLAR OBJECT
LB 06499	12 29 00.	+ 30 43		16.5	FAINT BLUE STAR
LB 06493	12 29 00.	+ 31 58		19.6	FAINT BLUE STAR
ZWG 293.026	12 29 00.	+ 58 15		13.2	GALAXY
UGC 07667	12 29 00.	+ 58 15	102	13.2	GALAXY SBa
ZC 1229.0+6456	12 29 00.	+ 64 56	1480		CLUSTER OF GALAXIES
MCG+14-06-006	12 29 00.	+ 85 03	57	16.	GALAXY
UGC 07664	12 29 00.	+ 85 04	66	16.5	GALAXY S
REIN 3.206	12 29 00.97	+ 11 54 01.1			NEBULA
AMES 1391	12 29 01.	+ 11 30 09.	19	17.6	NEBULA
IC 3452	12 29 01.	+ 11 54 02.			SAME AS NGC 4497
HN 0942	12 29 01.	+ 15 08	36		NEBULA
A3 355	12 29 01.9	+ 17 07 32.		16.8	FAINT BLUE OBJECT
REIZ 2464	12 29 02.	+ 11 53	66	14.2	GALAXY
RNGC 4497	12 29 02.	+ 11 54		14.0	GALAXY
IC 3453	12 29 02.	+ 15 08			NONSTELLAR OBJECT
RNGC 4500	12 29 02.	+ 58 14		13.0	GALAXY
A3 356	12 29 02.4	+ 14 38 42.		16.5	FAINT BLUE OBJECT
A3 357	12 29 02.9	+ 15 21 50.		16.4	FAINT BLUE OBJECT
REIZ 2463	12 29 03.	+ 04 12	270	12.8	GALAXY
A3 358	12 29 03.5	+ 16 19 38.		15.8	FAINT BLUE OBJECT
AMES 1392	12 29 04.	+ 17 52 45.	24	17.2	NEBULA
AMES 1393	12 29 05.	+ 07 05 45.	38	17.3	NEBULA
VV 076B	12 29 06.	+ 04 12	60	12.5	INTERACTING GALAXY
VV 076A	12 29 06.	+ 04 12	240	12.	INTERACTING GALAXY
VV 076	12 29 06.	+ 04 12	240	12.0	INTERACTING GALAXY
ZWG 042.144	12 29 06.	+ 04 13		13.3	GALAXY
UGC 07668	12 29 06.	+ 04 13	240	13.3	GALAXY SBc+CMP
KARA.72 343A	12 29 06.	+ 04 13	240	13.3	PART OF DOUBLE GALAXY
MCG+01-32-090	12 29 06.	+ 04 13	210	13.3	GALAXY
KARA.68 151	12 29 06.	+ 11 07	67		DWARF GALAXY
REA 52	12 29 06.	+ 11 08	54		DWARF GALAXY
ZC 1229.1+1112	12 29 06.	+ 11 12	670		CLUSTER OF GALAXIES
LB 04886	12 29 06.	+ 13 40		20.5	FAINT BLUE STAR
LB 04885	12 29 06.	+ 13 42		16.8	FAINT BLUE STAR
RMB 025	12 29 06.	+ 15 08			FAINT BLUE OBJECT
ZWG 099.075	12 29 06.	+ 17 08		12.8	GALAXY
SCH 51	12 29 06.	+ 17 08		12.8	PECULIAR GALAXY
UGC 07669	12 29 06.	+ 17 08	210	12.8	GALAXY SBc
ZWG 159.010	12 29 06.	+ 27 47		15.7	GALAXY
UGC 07670	12 29 06.	+ 27 47	78	15.7	GALAXY Sb
LB 06502	12 29 06.	+ 28 24		19.7	FAINT BLUE STAR
LB 06501	12 29 06.	+ 28 28		19.5	FAINT BLUE STAR
LB 06500	12 29 06.	+ 30 11		18.6	FAINT BLUE STAR
LB 06498	12 29 06.	+ 31 21		19.9	FAINT BLUE STAR
ZWG 188.005	12 29 06.	+ 38 15		15.2	GALAXY

OBJECT NAME	RIGHT ASCEN.	DECLINATION	DIAM.	MAGN.	TYPE OF OBJECT
KARA.73B 0532	12 29 06.	+ 38 15	36	15.2	ISOLATED GALAXY S
MRK 213	12 29 06.	+ 58 14	50	13.	GALAXY WITH UV CONTINUUM
MCG+10-18-062	12 29 06.	+ 58 14	96	12.8	GALAXY
ZCG 199.87	12 29 06.	+ 63 32		19.1	COMPACT GALAXY
7ZW 464	12 29 06.	+ 66 00			COMPACT GALAXY
MRSL 300+00/1	12 29 06.	- 62 25	2520		HII REGION
KEEL 458	12 29 06.1	+ 15 08 14.		14.	NEBULA
A3 360	12 29 06.8	+ 15 07 55.		17.8	UV CONCENTRATION
RNGC 4496B	12 29 07.	+ 04 12		13.5	GALAXY
RNGC 4496A	12 29 07.	+ 04 13		12.0	GALAXY
REIZ 2466	12 29 07.	+ 17 07	84	13.6	GALAXY
A3 361	12 29 08.5	+ 17 07 43.		14.9	UV CONCENTRATION
REIZ 2465	12 29 09.	+ 04 13	60	13.0	GALAXY
HOLM 415A	12 29 09.	+ 04 13	240	12.4	PART OF MULTIPLE GALAXY
SN 1960F	12 29 09.	+ 04 13		11.6	SUPERNOVA
AMES 1394	12 29 09.	+ 09 59 33.	28	16.8	NEBULA
MCG+03-32-057	12 29 09.	+ 15 07	45	15.2	GALAXY
RNGC 4498	12 29 09.	+ 17 08		13.0	GALAXY
MCG+03-32-056	12 29 09.	+ 17 08	180	12.8	GALAXY
MCG+04-30-003	12 29 09.	+ 20 45	42	15.6	GALAXY
HOLM 415B	12 29 11.	+ 04 12	48	13.4	PART OF MULTIPLE GALAXY
REIZ 2468	12 29 11.	+ 27 45	30	14.4	GALAXY
SVEN 298	12 29 11.	- 07 37 05.	24	14.9	GALAXY
KARA.72 343B	12 29 12.	+ 04 12	48		PART OF DOUBLE GALAXY
AMES 1396	12 29 12.	+ 10 43 15.	17	17.8	NEBULA
AMES 1395	12 29 12.	+ 11 02 21.	19	17.6	NEBULA
REA 53	12 29 12.	+ 11 18	30		DWARF GALAXY
8ZW 1229+14.0	12 29 12.	+ 14 01		18.4	COMPACT GALAXY
ZWG 129.006	12 29 12.	+ 20 46		15.6	GALAXY
LB 06504	12 29 12.	+ 26 33		18.8	FAINT BLUE STAR
MCG+05-30-013	12 29 12.	+ 27 46	66	15.7	GALAXY
LB 06503	12 29 12.	+ 32 00		17.9	FAINT BLUE STAR
ZWG 188.006	12 29 12.	+ 33 52		15.6	GALAXY
UGC 07671	12 29 12.	+ 33 52	66	15.6	GALAXY S
MCG+06-28-004	12 29 12.	+ 33 52	42	15.	GALAXY
ZWG 014.059	12 29 12.	- 00 44		15.6	GALAXY
KARA.73B 0533	12 29 12.	- 00 44	42	15.6	ISOLATED GALAXY S
ZWG 014.058	12 29 12.	- 02 42		14.9	GALAXY
MCG+00-32-018	12 29 12.	- 02 42	15	14.9	GALAXY
REIZ 2467	12 29 13.	+ 15 38	72	14.9	GALAXY
AMES 1398	12 29 15.	+ 11 00 51.	16	17.2	NEBULA
AMES 1397	12 29 15.	+ 19 03 45.	23	17.2	NEBULA
RNGC 4529	12 29 15.	+ 20 46		15.0	GALAXY
IC 3455	12 29 15.	+ 26 03 38.			MAY NOT EXIST
IC 3456	12 29 15.	+ 28 38 02.			NONSTELLAR OBJECT
MCG+06-28-005	12 29 15.	+ 38 17	27	15.5	GALAXY
A3 367	12 29 15.7	+ 15 47 28.		16.6	FAINT BLUE OBJECT
IC 3458	12 29 16.	+ 28 25 20.			NONSTELLAR OBJECT
REIZ 2470	12 29 16.	+ 33 52	42	15.8	GALAXY
HZ 27	12 29 16.	+ 39 16		12.5	BLUE STAR
REIZ 2469	12 29 17.	+ 28 23	72	15.8	GALAXY
RMB 130	12 29 18.	+ 12 00			FAINT BLUE OBJECT
ZWG 070.146	12 29 18.	+ 12 56		15.4	GALAXY
UGC 07672	12 29 18.	+ 12 56	66	15.4	GALAXY DWARF
MCG+02-32-114	12 29 18.	+ 12 56	72	15.4	GALAXY
ZC 1229.3+1717	12 29 18.	+ 17 17	1010		CLUSTER OF GALAXIES
LB 11240	12 29 18.	+ 26 23		13.4	FAINT BLUE STAR
UGC 07673	12 29 18.	+ 30 00	120	17.	GALAXY DWRF IR
LB 11239	12 29 18.	+ 30 03		13.6	FAINT BLUE STAR
LB 06507	12 29 18.	+ 30 46		19.0	FAINT BLUE STAR
LB 11238	12 29 18.	+ 31 08		14.3	FAINT BLUE STAR
LB 06506	12 29 18.	+ 32 20		19.5	FAINT BLUE STAR
LB 06505	12 29 18.	+ 32 20		19.4	FAINT BLUE STAR
MCG+07-26-016	12 29 18.	+ 40 52	18	17.	GALAXY
ZC 1229.3+4225	12 29 18.	+ 42 25	870		CLUSTER OF GALAXIES
REIZ 2472	12 29 18.	+ 48 13	24	15.1	GALAXY
ZC 1229.3+5946	12 29 18.	+ 59 46	2290		CLUSTER OF GALAXIES
ZCG 199.88	12 29 18.	+ 63 50		17.9	COMPACT GALAXY
ZC 1229.3+6733	12 29 18.	+ 67 33	4570		CLUSTER OF GALAXIES
AMES 1399	12 29 19.	+ 12 45 39.	29	16.8	NEBULA
IC 3457	12 29 19.5	+ 12 56 00.			GALAXY
REIN 3.207	12 29 19.77	+ 12 56 00.2			NEBULA
AMES 1402	12 29 20.	+ 08 52 27.	24	17.2	NEBULA
HN 0943	12 29 20.	+ 12 55	18		NEBULA
AMES 1401	12 29 20.	+ 13 32 21.	54	17.2	NEBULA
REIZ 2473	12 29 20.	+ 58 14	60	13.3	GALAXY
A3 371	12 29 20.7	+ 15 07 39.		16.9	FAINT BLUE OBJECT
AMES 1400	12 29 21.	+ 17 18 09.	19	17.4	NEBULA
IC 3460	12 29 21.	+ 27 39 51.			NONSTELLAR OBJECT
KN 16.001	12 29 21.9	+ 27 39 43.			NEBULA
AMES 1405	12 29 22.	+ 11 38 51.	22	18.0	NEBULA
IC 0797	12 29 22.	+ 15 24 14.			NONSTELLAR OBJECT
AMES 1404	12 29 22.	+ 19 22 45.	31	16.8	NEBULA
AMES 1403	12 29 22.	+ 19 39 33.	26	17.0	NEBULA
KEEL 459	12 29 22.6	+ 14 34 59.		15.	SPIRAL NEBULA
AMES 1407	12 29 23.	+ 05 19 45.	41	17.2	NEBULA
AMES 1406	12 29 23.	+ 09 04 21.	34	17.3	NEBULA
SVEN 299	12 29 23.	- 07 33	24	15.0	GALAXY
A3 373	12 29 23.2	+ 13 02 54.		17.7	FAINT BLUE OBJECT
A3 374	12 29 23.7	+ 15 23 57.		18.0	UV CONCENTRATION
REA 72	12 29 24.	+ 09 12	24		DWARF GALAXY
ZWG 070.147	12 29 24.	+ 12 27		15.5	GALAXY
UGC 07674	12 29 24.	+ 12 27	72	15.5	GALAXY DWARF
MCG+02-32-115	12 29 24.	+ 12 27	60	15.5	GALAXY
IC 3459	12 29 24.	+ 12 27 03.			NONSTELLAR OBJECT
MCG+03-32-059	12 29 24.	+ 14 41	360	10.6	GALAXY
ZWG 099.076	12 29 24.	+ 14 42		10.6	GALAXY
SCH 52	12 29 24.	+ 14 42		10.6	PECULIAR GALAXY
UGC 07675	12 29 24.	+ 14 42	402	10.6	GALAXY Sb/Sc
RMB 026	12 29 24.	+ 14 43			FAINT BLUE OBJECT
MCG+03-32-058	12 29 24.	+ 15 23	90	13.9	GALAXY
ZWG 099.077	12 29 24.	+ 15 24		13.9	GALAXY
UGC 07676	12 29 24.	+ 15 24	72	13.9	GALAXY SBc
MCG+07-26-017	12 29 24.	+ 40 51 30.	18	17.	GALAXY
MCG+08-23-048	12 29 24.	+ 48 12	42	16.	GALAXY
ZCG 199.89	12 29 24.	+ 63 28		17.8	COMPACT GALAXY
MRK 214	12 29 24.	+ 66 02	10	16.	GALAXY WITH UV CONTINUUM
ZWG 014.060	12 29 24.	- 02 36		15.6	GALAXY
REIN 3.208	12 29 24.66	+ 12 26 56.4			NEBULA
REIZ 2471	12 29 25.	+ 20 46	12	14.3	GALAXY
ABC 1555	12 29 25.	- 13 08		17.2	RICH CLUSTER OF GALAXIES
A3 375	12 29 25.8	+ 14 42 46.		17.5	UV CONCENTRATION
BC PKS1229-02	12 29 25.88	- 02 07 31.9		16.75	QUASI-STELLAR OBJECT
SHB 208	12 29 25.9	- 02 07 31.		16.8	QUASI-STELLAR OBJECT
AMES 1408	12 29 26.	+ 18 29 45.	31	16.6	NEBULA
A3 376	12 29 26.9	+ 12 59 14.		16.9	FAINT BLUE OBJECT
A3 377	12 29 27.0	+ 14 32 13.		16.6	FAINT BLUE OBJECT
AMES 1410	12 29 28.	+ 13 26 51.	23	17.8	NEBULA
LB 00639	12 29 28.	+ 59 45 18.		17.3	FAINT BLUE STAR
RNGC 4510	12 29 28.	+ 64 31		14.0	GALAXY
RNGC 4499	12 29 28.	- 39 42			GALAXY
AMES 1409	12 29 29.	+ 16 14 57.	19	17.9	NEBULA
SVEN 300	12 29 29.	- 07 53	18	15.9	GALAXY
KARA.68 152	12 29 30.	+ 08 57	27		DWARF GALAXY
ZWG 070.148	12 29 30.	+ 12 10		15.4	GALAXY
MCG+02-32-116	12 29 30.	+ 12 10	36	15.4	GALAXY
IC 3461	12 29 30.	+ 12 10 03.			NONSTELLAR OBJECT
ZWG 099.078	12 29 30.	+ 16 58		14.8	GALAXY
UGC 07677	12 29 30.	+ 16 58	78	14.8	GALAXY Sc
MCG+03-32-060	12 29 30.	+ 16 58	66	14.8	GALAXY
ZWG 099.079	12 29 30.	+ 20 25		15.1	GALAXY
MCG+03-32-061	12 29 30.	+ 20 25	12	15.1	GALAXY
LB 06509	12 29 30.	+ 27 58		19.7	FAINT BLUE STAR
LB 06508	12 29 30.	+ 31 55		19.4	FAINT BLUE STAR
ZWG 216.009	12 29 30.	+ 40 07		13.9	GALAXY
UGC 07678	12 29 30.	+ 40 07	96	13.9	GALAXY SB?
MCG+07-26-018	12 29 30.	+ 40 51	15	18.	GALAXY
ZC 1229.5+4153	12 29 30.	+ 41 53	2350		CLUSTER OF GALAXIES
ZWG 270.008	12 29 30.	+ 52 41		15.4	GALAXY
MCG+11-15-058	12 29 30.	+ 64 29	30	14.2	GALAXY
ZWG 315.041	12 29 30.	+ 64 31		14.2	GALAXY
UGC 07679	12 29 30.	+ 64 31	90	14.2	GALAXY F
REIZ 2474	12 29 31.	+ 14 42	300	10.7	GALAXY
HN 0944	12 29 31.	+ 15 34	6		NEBULA
IC 3462	12 29 31.	+ 15 34			NONSTELLAR OBJECT
REIZ 2475	12 29 31.	+ 16 58	42	15.0	GALAXY
IC 3464	12 29 31.	+ 26 16 51.			SINGLE STAR
LB 02377	12 29 31.	+ 58 38 54.		17.1	FAINT BLUE STAR
REIN 3.209	12 29 31.38	+ 12 09 56.9			NEBULA
RNGC 4501	12 29 32.	+ 14 42		11.0	GALAXY
IC 3463	12 29 33.	+ 12 35 51.			TWO STARS
RNGC 4502	12 29 33.	+ 16 58		15.0	GALAXY
TON-N 1542	12 29 33.	+ 20 27		16.6	BLUE STAR
TON-N 1543	12 29 33.	+ 22 58		16.9	BLUE STAR
MCG+08-23-049	12 29 33.	+ 49 55	72	15.	GALAXY
MCG-03-32-011	12 29 33.	- 20 18 30.	24	15.5	GALAXY
AMES 1413	12 29 34.	+ 10 42 57.	12	17.3	NEBULA
AMES 1412	12 29 34.	+ 11 02 33.	25	17.4	NEBULA
AMES 1411	12 29 34.	+ 12 13 21.	25	17.5	NEBULA
REIZ 2484	12 29 34.	+ 64 30	18	14.2	GALAXY
REIN 3.210	12 29 34.27	+ 12 05 35.8			NEBULA
REIN 3.211	12 29 34.46	+ 11 27 08.2			NEBULA
A3 379	12 29 34.6	+ 12 58 21.		17.3	FAINT BLUE OBJECT
KN 16.002	12 29 34.8	+ 28 00 20.			NEBULA
AMES 1414	12 29 35.	+ 13 06 39.	23	17.8	NEBULA
LB 00640	12 29 35.	+ 61 11 30.		17.1	FAINT BLUE STAR
AMES 1415	12 29 36.	+ 05 24 45.	38	17.4	NEBULA
ZWG 070.149	12 29 36.	+ 11 27		12.4	GALAXY
UGC 07680	12 29 36.	+ 11 27	216	12.4	GALAXY SO
MCG+02-32-118	12 29 36.	+ 11 27	144	12.4	GALAXY
ZWG 070.150	12 29 36.	+ 12 06		15.3	GALAXY
MCG+02-32-117	12 29 36.	+ 12 06	30	15.3	GALAXY
LB 06515	12 29 36.	+ 26 54		19.6	FAINT BLUE STAR
LB 06514	12 29 36.	+ 27 55		18.7	FAINT BLUE STAR
LB 06513	12 29 36.	+ 28 18		20.0	FAINT BLUE STAR
LB 06512	12 29 36.	+ 29 24		18.7	FAINT BLUE STAR
MCG+05-30-014	12 29 36.	+ 30 00	66	15.	GALAXY
LB 06511	12 29 36.	+ 31 19		19.5	FAINT BLUE STAR
LB 06510	12 29 36.	+ 32 23		19.5	FAINT BLUE STAR
MCG+07-26-019	12 29 36.	+ 40 06	72	14.	GALAXY
ZC 1229.6+5029	12 29 36.	+ 50 29	810		CLUSTER OF GALAXIES
MCG+09-21-011	12 29 36.	+ 52 39	48	14.	GALAXY
IC 3466	12 29 37.	+ 12 04			NONSTELLAR OBJECT
IC 3465	12 29 37.	+ 12 20			NONSTELLAR OBJECT
A3 380	12 29 37.2	+ 12 54 35.		15.7	FAINT BLUE OBJECT
RNGC 4503	12 29 38.	+ 11 27		12.5	GALAXY
REIZ 2476	12 29 38.	+ 11 27	90	13.0	GALAXY
AMES 1416	12 29 38.	+ 11 49 45.	22	17.5	NEBULA
HN 0946	12 29 38.	+ 12 04		15.	NEBULA
HN 0945	12 29 38.	+ 12 20		16.	NEBULA
RNGC 4506	12 29 38.	+ 13 42		14.0	GALAXY
REIZ 2480	12 29 38.	+ 42 55	54	13.2	GALAXY
REIZ 2479	12 29 39.	+ 40 04	42	14.1	GALAXY
MCG+08-23-050	12 29 39.	+ 46 37	60	16.	GALAXY
REIZ 2485	12 29 39.	+ 56 44	60	14.0	GALAXY
AMES 1417	12 29 40.	+ 12 53 21.	20	18.0	NEBULA
REIZ 2478	12 29 40.	+ 35 10	24	14.9	GALAXY
A3 382	12 29 40.9	+ 15 17 53.		17.2	FAINT BLUE OBJECT
AMES 1418	12 29 41.	+ 06 29 03.	40	17.8	NEBULA
IC 3469	12 29 41.	+ 26 04 45.			NONSTELLAR OBJECT
RNGC 4513	12 29 41.	+ 66 36		14.0	GALAXY
SVEN 302	12 29 41.	- 07 11	18	15.5	GALAXY
SVEN 301	12 29 41.	- 07 17	210	11.8	GALAXY
REIZ 2477	12 29 41.	- 07 17	168	12.6	GALAXY
ZWG 070.151	12 29 42.	+ 10 31		14.6	GALAXY
UGC 07681	12 29 42.	+ 10 31	84	14.6	GALAXY E
MCG+02-32-119	12 29 42.	+ 10 31	48	14.6	GALAXY
RMB 126	12 29 42.	+ 12 46			FAINT BLUE OBJECT
ZWG 070.152	12 29 42.	+ 13 42		14.2	GALAXY
UGC 07682	12 29 42.	+ 13 42	102	14.2	GALAXY SB?
MCG+02-32-120	12 29 42.	+ 13 42	60	14.2	GALAXY
LB 06518	12 29 42.	+ 26 37		19.4	FAINT BLUE STAR
LB 11241	12 29 42.	+ 27 15		16.6	FAINT BLUE STAR
LB 06517	12 29 42.	+ 28 36		19.4	FAINT BLUE STAR
LB 06516	12 29 42.	+ 29 33		19.6	FAINT BLUE STAR
ZC 1229.7+3325	12 29 42.	+ 33 25	1610		CLUSTER OF GALAXIES
ZWG 315.042	12 29 42.	+ 66 36		14.1	GALAXY
UGC 07683	12 29 42.	+ 66 36	144	14.1	GALAXY SO
REIN 4.077	12 29 42.18	+ 10 31 37.3			NEBULA
RNGC 4505	12 29 43.	+ 04 15			NON-EXISTENT OBJECT
AMES 1419	12 29 43.	+ 10 37 21.	24	17.8	NEBULA
IC 3467	12 29 43.	+ 12 03			NONSTELLAR OBJECT
RNGC 4504	12 29 43.	- 07 17		12.0	GALAXY
HN 0948	12 29 44.	+ 10 31		13.5	NEBULA
IC 3468	12 29 44.	+ 10 31			GALAXY
HN 0947	12 29 44.	+ 12 03	48		NEBULA
REIZ 2491	12 29 44.	+ 66 37	9	14.5	GALAXY
AMES 1420	12 29 45.	+ 16 54 27.	22	17.1	NEBULA
MCG-01-32-022	12 29 45.	- 07 18	180	12.	GALAXY
REIN 4.078	12 29 45.16	+ 00 33 48.9			NEBULA
A3 383	12 29 45.5	+ 12 57 32.		16.5	FAINT BLUE OBJECT
A3 384	12 29 45.8	+ 12 50 32.		17.6	FAINT BLUE OBJECT
AMES 1424	12 29 46.	+ 12 30 51.	17	17.2	NEBULA
AMES 1423	12 29 46.	+ 12 36 33.	22	17.0	NEBULA
AMES 1421	12 29 46.	+ 16 20 33.	29	17.8	NEBULA
AMES 1422	12 29 47.	+ 17 48 57.	26	17.4	NEBULA

OBJECT NAME	RIGHT ASCEN.	DECLINATION	DIAM.	MAGN.	TYPE OF OBJECT
SVEN 303	12 29 47.	- 08 15	24	15.7	GALAXY
ZWG 014.061	12 29 48.	+ 00 17		15.7	GALAXY
MCG+03-32-063	12 29 48.	+ 16 17 30.	42	15.7	GALAXY
ZWG 099.080	12 29 48.	+ 16 18		15.7	GALAXY
2ZW 065	12 29 48.	+ 16 35			COMPACT GALAXY
ZWG 099.081	12 29 48.	+ 18 31		15.4	GALAXY
UGC 07684	12 29 48.	+ 18 31	66	15.4	GALAXY
MCG+03-32-062	12 29 48.	+ 18 32	72	15.4	GALAXY
BC TON1542	12 29 48.	+ 20 25 00.			QUASI-STELLAR OBJECT
SHB 209	12 29 48.	+ 20 25 00.		15.3	QUASI-STELLAR OBJECT
REIZ 2482	12 29 48.	+ 24 58	24	15.0	GALAXY
REIZ 2483	12 29 48.	+ 26 03	66	14.1	GALAXY
LB 06521	12 29 48.	+ 26 26		16.5	FAINT BLUE STAR
LB 06520	12 29 48.	+ 27 08		18.0	FAINT BLUE STAR
LB 11242	12 29 48.	+ 28 18		12.4	FAINT BLUE STAR
LB 06519	12 29 48.	+ 29 37		18.8	FAINT BLUE STAR
MCG+06-28-006	12 29 48.	+ 35 10	30	15.	GALAXY
MCG+08-23-051	12 29 48.	+ 47 37	18	15.	GALAXY
ZWG 293.027	12 29 48.	+ 56 45		14.6	GALAXY
ZC 1229.8+5659	12 29 48.	+ 56 59	2220		CLUSTER OF GALAXIES
LB 02378	12 29 48.	+ 58 01 30.		17.0	FAINT BLUE STAR
ZCG 199.90	12 29 48.	+ 64 23		17.8	COMPACT GALAXY
7ZW 465	12 29 48.	+ 65 30			COMPACT GALAXY
MCG+11-15-059	12 29 48.	+ 66 35	27	15.	GALAXY
7ZW 466	12 29 48.	+ 66 40			COMPACT GALAXY
ZWG 315.043	12 29 48.	+ 66 40		15.6	GALAXY
MCG-06-28-001	12 29 48.	- 36 05	48	15.	GALAXY
A3 385	12 29 48.1	+ 15 03 20.		16.0	FAINT BLUE OBJECT
KN 16.003	12 29 48.9	+ 27 34 02.			NEBULA
AMES 1427	12 29 49.	+ 10 23 45.	20	17.4	NEBULA
HW 0949	12 29 49.	+ 16 18	12		NEBULA
REIZ 2481	12 29 49.	+ 16 18	12	15.7	GALAXY
IC 3471	12 29 49.	+ 16 18			NONSTELLAR OBJECT
RNGC 4508	12 29 50.	+ 06 06			NON-EXISTENT OBJECT
HW 0950	12 29 50.	+ 11 32		13.5	NEBULA
IC 3470	12 29 50.	+ 11 32			GALAXY
AMES 1426	12 29 50.	+ 17 34 03.	19	17.6	NEBULA
AMES 1425	12 29 50.	+ 19 15 09.	20	17.0	NEBULA
IC 3472	12 29 50.	+ 25 00 15.			NONSTELLAR OBJECT
RNGC 4511	12 29 50.	+ 56 45		14.5	GALAXY
LB 02379	12 29 50.	+ 57 21 12.		16.4	FAINT BLUE STAR
AMES 1429	12 29 51.	+ 11 51 45.	20	18.0	NEBULA
REIZ 2488	12 29 51.	+ 40 06	12	15.2	GALAXY
REIZ 2490	12 29 51.	+ 56 55	42	14.2	GALAXY
KN 16.004	12 29 51.4	+ 27 48 26.			NEBULA
REIN 3.212	12 29 51.64	+ 11 32 19.5			NEBULA
AMES 1428	12 29 52.	+ 16 35 09.	14	16.3	NEBULA
AMES 1430	12 29 53.	+ 18 12 09.	17	17.4	NEBULA
SVEN 304	12 29 53.	- 07 37	42	14.9	GALAXY
A3 387	12 29 53.0	+ 13 44 14.		16.5	FAINT BLUE OBJECT
REIN 3.213	12 29 53.08	+ 12 03 46.4			NEBULA
KN 16.005	12 29 53.8	+ 27 45 08.			NEBULA
SC 1227+0055.2	12 29 54.	+ 00 38 38.	6		NEBULA
ZWG 014.062	12 29 54.	+ 00 40		14.1	GALAXY
UGC 07685	12 29 54.	+ 00 40	282	14.1	GALAXY
KARA.72 344A	12 29 54.	+ 00 40	288	14.1	PART OF DOUBLE GALAXY
MCG+00-32-019	12 29 54.	+ 00 40	252	14.1	GALAXY
ZWG 070.153	12 29 54.	+ 11 32		15.0	GALAXY
MCG+02-32-122	12 29 54.	+ 11 32	36	15.0	GALAXY
ZWG 070.154	12 29 54.	+ 12 04		15.4	GALAXY
UGC 07686	12 29 54.	+ 12 04	78	15.4	GALAXY Sc
MCG+02-32-121	12 29 54.	+ 12 04	60	15.4	GALAXY
KARA.68 153	12 29 54.	+ 12 09	40		DWARF GALAXY
RMB 127	12 29 54.	+ 12 51			FAINT BLUE OBJECT
ZWG 099.082	12 29 54.	+ 16 35		15.7	GALAXY
LB 06526	12 29 54.	+ 27 56		20.2	FAINT BLUE STAR
LB 06525	12 29 54.	+ 29 24		19.7	FAINT BLUE STAR
LB 06524	12 29 54.	+ 29 31		19.0	FAINT BLUE STAR
LB 06523	12 29 54.	+ 30 32		19.8	FAINT BLUE STAR
LB 06522	12 29 54.	+ 31 28		19.0	FAINT BLUE STAR
ZWG 244.026	12 29 54.	+ 47 37		15.5	GALAXY
ZC 1229.9+6153	12 29 54.	+ 61 53	2550		CLUSTER OF GALAXIES
REIN 4.079	12 29 54.67	+ 00 39 03.0			NEBULA
REIN 4.080	12 29 54.83	+ 00 39 54.7			NEBULA
RNGC 4517A	12 29 55.	+ 00 40		13.5	GALAXY
AMES 1432	12 29 55.	+ 10 00 39.	34	17.2	NEBULA
HW 0951	12 29 55.	+ 18 30	18		NEBULA
REIZ 2487	12 29 55.	+ 18 30	18	15.6	GALAXY
IC 3473	12 29 55.	+ 18 30			NONSTELLAR OBJECT
REIZ 2489	12 29 55.	+ 46 38	24	14.0	GALAXY
AMES 1431	12 29 56.	+ 18 22 09.	35	17.1	NEBULA
A3 390	12 29 56.3	+ 16 16 21.		17.1	FAINT BLUE OBJECT
AMES 1434	12 29 57.	+ 11 44 21.	16	17.0	NEBULA
AMES 1433	12 29 57.	+ 12 45 21.	18	17.6	NEBULA
LB 02380	12 29 57.	+ 51 08 48.		14.8	FAINT BLUE STAR
LB 00641	12 29 57.	+ 57 11 18.		17.5	FAINT BLUE STAR
REIZ 2486	12 29 58.	+ 00 38	60	13.0	GALAXY
AMES 1435	12 29 58.	+ 11 38 33.	26	17.1	NEBULA
AMES 1436	12 29 59.	+ 10 22 33.	26	17.2	NEBULA
KEEL 460	12 29 59.7	+ 02 52 02.		17.	SPIRAL NEBULA
VDB.66G 133	12 30	+ 31 54	200		DWARF GALAXY
VDB.66G 134	12 30	- 02 21	70		DWARF GALAXY
ZWG 042.145	12 30 00.	+ 02 56		15.0	GALAXY
UGC 07687	12 30 00.	+ 02 56	132	15.0	GALAXY Sc
MCG+01-32-091	12 30 00.	+ 02 56	120	15.0	GALAXY
ZWG 042.146	12 30 00.	+ 08 19		15.3	GALAXY
UGC 07688	12 30 00.	+ 08 19	60	15.3	GALAXY DWARF
MCG+01-32-092	12 30 00.	+ 08 19	48	15.3	GALAXY
ZWG 070.155	12 30 00.	+ 09 26		15.4	GALAXY
AMES 1437	12 30 00.	+ 11 56 57.	17	17.9	NEBULA
REA 54	12 30 00.	+ 12 11	54		DWARF GALAXY
ZWG 099.083	12 30 00.	+ 15 41		15.3	GALAXY
LB 11243	12 30 00.	+ 27 46		13.6	FAINT BLUE STAR
LB 06530	12 30 00.	+ 28 04		18.7	FAINT BLUE STAR
LB 06529	12 30 00.	+ 28 38		18.5	FAINT BLUE STAR
LB 06528	12 30 00.	+ 28 45		19.4	FAINT BLUE STAR
LB 06527	12 30 00.	+ 32 29		19.8	FAINT BLUE STAR
UGC 07689	12 30 00.	+ 39 52	108	16.0	GALAXY Sc
ZWG 216.010	12 30 00.	+ 42 59		13.7	GALAXY
UGC 07690	12 30 00.	+ 42 59	138	13.7	GALAXY IRR
ZC 1230.0+4314	12 30 00.	+ 43 14	740		CLUSTER OF GALAXIES
ZC 1230.0+4530	12 30 00.	+ 45 30	740		CLUSTER OF GALAXIES
MCG+09-21-013	12 30 00.	+ 51 18	42	16.	GALAXY
MCG+09-21-012	12 30 00.	+ 53 21	30	17.	GALAXY
MCG+10-18-063	12 30 00.	+ 56 45	42	14.0	GALAXY
ZWG 293.028	12 30 00.	+ 56 56		14.3	GALAXY
UGC 07691	12 30 00.	+ 56 56	48	14.3	GALAXY
ZCG 199.91	12 30 00.	+ 63 28		19.1	COMPACT GALAXY
ZCG 199.92	12 30 00.	+ 64 14		17.4	COMPACT GALAXY
7ZW 467	12 30 00.	+ 66 40			COMPACT GALAXY
ZWG 315.044	12 30 00.	+ 66 40		15.7	GALAXY
REIN 3.214	12 30 00.14	+ 08 19 16.6			NEBULA
A3 391	12 30 00.4	+ 14 00 22.		16.8	FAINT BLUE OBJECT
AMES 1438	12 30 01.	+ 10 35 33.	22	17.4	NEBULA
IC 0798	12 30 01.	+ 15 42 09.			NONSTELLAR OBJECT
HZ 28	12 30 02.	+ 41 45		15.2	DECIDEDLY BLUE STAR
REIN 3.215	12 30 02.69	+ 09 40 47.7			NEBULA
AMES 1439	12 30 03.	+ 17 08 09.	12	17.9	NEBULA
MCG+07-26-020	12 30 03.	+ 39 51 30.	90	15.	GALAXY
MCG+07-26-021	12 30 03.	+ 42 59	84	13.5	GALAXY
KEEL 462	12 30 03.1	+ 14 51 22.		16.	NEBULA
REIN 3.216	12 30 03.33	+ 08 22 05.9			NEBULA
A3 392	12 30 03.4	+ 13 43 42.		17.1	FAINT BLUE OBJECT
IC 3474	12 30 03.5	+ 02 56 16.			GALAXY
KEEL 461	12 30 03.6	+ 02 56 21.		15.	NEBULA
AMES 1440	12 30 04.	+ 18 18 27.	41	16.1	NEBULA
RNGC 4514	12 30 04.	+ 30 00		14.0	GALAXY
KEEL 463	12 30 04.9	+ 03 17 39.		17.	NEBULA
AMES 1441	12 30 05.	+ 10 09 33.	41	17.2	NEBULA
A3 393	12 30 05.8	+ 14 27 28.		17.8	FAINT BLUE OBJECT
MCG+02-32-124	12 30 06.	+ 09 27	15	15.4	GALAXY
ZWG 070.156	12 30 06.	+ 13 03		15.4	GALAXY
UGC 07692	12 30 06.	+ 13 03	150	15.4	GALAXY DWRF EL
MCG+02-32-123	12 30 06.	+ 13 03	90	15.4	GALAXY
RMB 027	12 30 06.	+ 14 28			FAINT BLUE OBJECT
ZWG 099.084	12 30 06.	+ 18 17		15.3	GALAXY
ZC 1230.1+2345	12 30 06.	+ 23 45	1280		CLUSTER OF GALAXIES
LB 06532	12 30 06.	+ 27 30		19.7	FAINT BLUE STAR
ZWG 159.011	12 30 06.	+ 30 00		14.2	GALAXY
UGC 07693	12 30 06.	+ 30 00	72	14.2	GALAXY Sb-c
LB 06531	12 30 06.	+ 31 44		17.5	FAINT BLUE STAR
MCG+06-28-007	12 30 06.	+ 33 30	27	16.	GALAXY
MCG+10-18-064	12 30 06.	+ 56 56	39	15.	GALAXY
ZC 1230.1+5815	12 30 06.	+ 58 15	1810		CLUSTER OF GALAXIES
MCG+11-15-060	12 30 06.	+ 64 08	114	14.	GALAXY
A3 394	12 30 06.7	+ 12 55 39.		16.5	FAINT BLUE OBJECT
REIN 3.217	12 30 07.82	+ 13 02 55.8			NEBULA
AMES 1442	12 30 08.	+ 10 21 45.	40	17.4	NEBULA
HW 0952	12 30 08.	+ 13 02	60		NEBULA
IC 3477	12 30 09.	+ 26 19 03.			SINGLE STAR
IC 3475	12 30 09.5	+ 13 02 54.		15.	GALAXY
A3 395	12 30 09.9	+ 14 31 22.		15.6	FAINT BLUE OBJECT
REIZ 2492	12 30 10.	+ 00 21	504	11.2	GALAXY
A3 396	12 30 10.6	+ 14 19 32.		17.8	UV CONCENTRATION
IC 3476	12 30 10.7	+ 14 19 32.	90	13.4	GALAXY Sc
KEEL 464	12 30 10.7	+ 14 19 42.		11.	NEBULA
SN 1970A	12 30 11.	+ 14 20		14.0	SUPERNOVA
AMES 1443	12 30 11.	+ 16 11 27.	32	16.8	NEBULA
IC 3479	12 30 11.	+ 25 40 51.			NONSTELLAR OBJECT
A3 397	12 30 11.8	+ 12 57 44.		17.1	FAINT BLUE OBJECT
ZWG 014.063	12 30 12.	+ 00 24		12.4	GALAXY
UGC 07694	12 30 12.	+ 00 24	648	12.4	GALAXY Sc
KARA.72 344B	12 30 12.	+ 00 24	756	12.4	PART OF DOUBLE GALAXY
MCG+00-32-020	12 30 12.	+ 00 24	630	12.4	GALAXY
RMB 201	12 30 12.	+ 10 42			FAINT BLUE OBJECT
SCH 53	12 30 12.	+ 14 19		13.5	PECULIAR GALAXY
ZWG 070.157	12 30 12.	+ 14 20		13.5	GALAXY
UGC 07695	12 30 12.	+ 14 20	162	13.5	GALAXY IRR
MCG+02-32-125	12 30 12.	+ 14 20	90	13.5	GALAXY
ZWG 070.158	12 30 12.	+ 14 28		15.0	GALAXY
UGC 07696	12 30 12.	+ 14 28	60	15.0	GALAXY S0?
MCG+02-32-126	12 30 12.	+ 14 28	15	15.0	GALAXY
AMES 1444	12 30 12.	+ 19 01 15.	14	17.8	NEBULA
REIZ 2494	12 30 12.	+ 25 41	12	16.0	GALAXY
LB 06543	12 30 12.	+ 26 27		19.1	FAINT BLUE STAR
LB 06542	12 30 12.	+ 26 43		19.5	FAINT BLUE STAR
LB 06541	12 30 12.	+ 26 53		19.1	FAINT BLUE STAR
LB 06540	12 30 12.	+ 27 12		19.2	FAINT BLUE STAR
LB 06539	12 30 12.	+ 27 48		19.3	FAINT BLUE STAR
LB 06538	12 30 12.	+ 28 25		19.6	FAINT BLUE STAR
LB 06537	12 30 12.	+ 28 25		19.1	FAINT BLUE STAR
ZWG 159.012	12 30 12.	+ 29 40		15.7	GALAXY
MCG+05-30-015	12 30 12.	+ 30 00	78	14.2	GALAXY
LB 06536	12 30 12.	+ 30 09		19.6	FAINT BLUE STAR
LB 06535	12 30 12.	+ 31 25		18.7	FAINT BLUE STAR
LB 06534	12 30 12.	+ 31 28		20.0	FAINT BLUE STAR
LB 11244	12 30 12.	+ 31 46		14.7	FAINT BLUE STAR
LB 06533	12 30 12.	+ 31 58		19.1	FAINT BLUE STAR
ZC 1230.2+4350	12 30 12.	+ 43 50	2620		CLUSTER OF GALAXIES
MCG+08-23-052	12 30 12.	+ 46 02	18	16.	GALAXY
ZWG 244.027	12 30 12.	+ 46 03		14.6	GALAXY
MRK 215	12 30 12.	+ 46 03	13	14.5	GALAXY WITH UV CONTINUUM
VVI 59	12 30 12.	+ 46 03	18	15.03	SEYFERT GALAXY
REIN 4.081	12 30 12.27	+ 00 23 15.3			NEBULA
RNGC 4517	12 30 13.	+ 00 23		11.5	GALAXY
REIZ 2493	12 30 13.	+ 18 18	30	15.3	GALAXY
AMES 1445	12 30 13.	+ 18 59 09.	24	17.7	NEBULA
IC 3490	12 30 13.	+ 27 06 15.			TWO STARS
IC 3480	12 30 13.	+ 27 06 15.			NONSTELLAR OBJECT
IC 3478	12 30 13.0	+ 14 28 20.			GALAXY
KEEL 465	12 30 13.1	+ 14 28 21.		11.	NEBULA
A3 398	12 30 13.3	+ 16 32 17.		17.3	FAINT BLUE OBJECT
A3 399	12 30 13.8	+ 14 21 58.		16.0	FAINT BLUE OBJECT
KN 16.006	12 30 15.0	+ 28 44 38.			NEBULA
REIN 4.082	12 30 15.29	+ 00 24 02.3			NEBULA
KN 16.007	12 30 15.9	+ 27 19 32.			NEBULA
AMES 1447	12 30 16.	+ 08 04 21.	36	17.8	NEBULA
AMES 1446	12 30 16.	+ 18 54 27.	22	17.8	NEBULA
AMES 1450	12 30 17.	+ 06 04 21.	48	16.9	NEBULA
REIZ 2497	12 30 17.	+ 29 59	48	14.1	GALAXY
A3 401	12 30 17.3	+ 14 30 57.		17.0	FAINT BLUE OBJECT
KEEL 466	12 30 17.4	+ 03 04 26.		17.	NEBULA
KN 16.008	12 30 17.7	+ 27 14 38.			NEBULA
ZWG 042.147	12 30 18.	+ 03 34		15.3	GALAXY
MCG+01-32-093	12 30 18.	+ 03 34	48	15.3	GALAXY
REA 02	12 30 18.	+ 03 39			DWARF GALAXY
ZWG 042.148	12 30 18.	+ 06 04		15.4	GALAXY
AMES 1449	12 30 18.	+ 11 37 45.	24	18.0	NEBULA
KARA.68 154	12 30 18.	+ 13 03	94		DWARF GALAXY
RMB 023	12 30 18.	+ 15 18			FAINT BLUE OBJECT
AMES 1448	12 30 18.	+ 16 15 51.	24	16.8	NEBULA
LB 06547	12 30 18.	+ 26 36		18.7	FAINT BLUE STAR
LB 06546	12 30 18.	+ 26 58		18.9	FAINT BLUE STAR
LB 06545	12 30 18.	+ 28 08		18.6	FAINT BLUE STAR
LB 06544	12 30 18.	+ 29 53		19.6	FAINT BLUE STAR
MCG+07-26-022	12 30 18.	+ 39 47	30	16.	GALAXY

OBJECT NAME	RIGHT ASCEN.	DECLINATION	DIAM.	MAGN.	TYPE OF OBJECT
ZC 1230.3+7450	12 30 18.	+ 74 50	11830		CLUSTER OF GALAXIES
A3 403	12 30 18.0	+ 13 23 14.		15.9	FAINT BLUE OBJECT
KEEL 467	12 30 18.2	+ 02 54 21.		17.	NEBULA
A3 404	12 30 18.3	+ 12 45 38.		16.5	FAINT BLUE OBJECT
REIN 3.218	12 30 19.31	+ 09 00 56.2			NEBULA
REIZ 2496	12 30 20.	+ 09 01	18	16.0	GALAXY
AMES 1453	12 30 20.	+ 11 17 27.	16	16.8	NEBULA
HN 0953	12 30 20.	+ 11 40		13.	NEBULA
AMES 1451	12 30 20.	+ 17 06 21.	17	17.3	NEBULA
REIN 3.219	12 30 20.61	+ 11 40 47.2			NEBULA
A3 405	12 30 20.7	+ 13 40 11.		17.3	FAINT BLUE OBJECT
IC 3481	12 30 20.9	+ 11 40 58.			GALAXY
REIZ 2495	12 30 21.	+ 06 03	18	16.4	GALAXY
HOLM 416B	12 30 21.	+ 06 03	24	16.4	PART OF MULTIPLE GALAXY
HOLM 416A	12 30 21.	+ 06 04	42	15.0	PART OF MULTIPLE GALAXY
AMES 1454	12 30 21.	+ 10 14 57.	28	17.0	NEBULA
AMES 1452	12 30 21.	+ 16 58 33.	24	16.8	NEBULA
MCG+03-32-064	12 30 21.	+ 20 29	120	15.3	GALAXY
ARC 1557	12 30 21.	+ 63 08		18.0	RICH CLUSTER OF GALAXIES
RNGC 4512	12 30 22.	+ 64 10		14.5	GALAXY
ARP 175	12 30 23.	+ 11 41			PECULIAR GALAXY
REIZ 2498	12 30 23.	+ 29 18	48	15.2	GALAXY
SVEN 305	12 30 23.	- 07 10	36	15.1	GALAXY
MCG+02-32-128	12 30 24.	+ 11 40	9	17.	GALAXY
ZWG 070.159	12 30 24.	+ 11 41		14.8	GALAXY
MCG+02-32-127	12 30 24.	+ 11 41	30	14.8	GALAXY
ZWG 099.085	12 30 24.	+ 20 27		15.3	GALAXY
UGC 07697	12 30 24.	+ 20 27	126	15.3	GALAXY Sc
LB 11246	12 30 24.	+ 26 21		14.2	FAINT BLUE STAR
LB 06551	12 30 24.	+ 26 40		18.8	FAINT BLUE STAR
LB 06550	12 30 24.	+ 27 32		18.6	FAINT BLUE STAR
LB 11245	12 30 24.	+ 30 32		17.3	FAINT BLUE STAR
LB 06549	12 30 24.	+ 30 34		20.3	FAINT BLUE STAR
LB 06548	12 30 24.	+ 31 43		20.0	FAINT BLUE STAR
ZWG 159.013	12 30 24.	+ 31 49		15.6	GALAXY
UGC 07698	12 30 24.	+ 31 49	420	15.6	GALAXY DWRF IR
ZWG 188.007	12 30 24.	+ 37 54		13.4	GALAXY
UGC 07699	12 30 24.	+ 37 54	240	13.4	GALAXY SBc
KARA.73B 0534	12 30 24.	+ 37 54	234	13.4	ISOLATED GALAXY S
ZC 1230.4+6326	12 30 24.	+ 63 26	1610		CLUSTER OF GALAXIES
ZWG 315.045	12 30 24.	+ 64 10		14.7	GALAXY
UGC 07700	12 30 24.	+ 64 10	132	14.7	GALAXY
KARA.72 345A	12 30 24.	+ 64 10	120	14.7	PART OF DOUBLE GALAXY
MCG+11-15-061	12 30 24.	+ 64 11	144	13.6	GALAXY
7ZW 468	12 30 24.	+ 66 41			COMPACT GALAXY
REIN 3.220	12 30 25.09	+ 11 39 55.0			NEBULA
IC 3481A	12 30 25.4	+ 11 40 06.			GALAXY
A3 406	12 30 25.4	+ 14 33 54.		16.8	FAINT BLUE OBJECT
AMES 1455	12 30 26.	+ 12 44 03.	24	17.0	NEBULA
KN 16.009	12 30 26.3	+ 27 54 08.			NEBULA
MCG+06-28-008	12 30 27.	+ 37 55	210	12.5	GALAXY
REIZ 2502	12 30 27.	+ 37 56	138	13.1	GALAXY
LB 02381	12 30 27.	+ 61 22 12.		14.5	FAINT BLUE STAR
AMES 1456	12 30 28.	+ 16 53 14.	23	17.5	NEBULA
RNGC 4509	12 30 28.	+ 32 25		14.0	GALAXY
KEEL 468	12 30 28.3	+ 02 36 39.		18.	NEBULA
AMES 1457	12 30 29.	+ 11 25 20.	14	16.6	NEBULA
SVEN 306	12 30 29.	- 07 04	42	14.8	GALAXY
HMS 1.20	12 30 30.	+ 11 40			S0 GALAXY
ZWG 099.086	12 30 30.	+ 16 32		13.3	GALAXY
UGC 07701	12 30 30.	+ 16 32	84	13.3	GALAXY E-S0
ZC 1230.5+2609	12 30 30.	+ 26 09	1550		CLUSTER OF GALAXIES
LB 06552	12 30 30.	+ 26 38		19.5	FAINT BLUE STAR
LB 11248	12 30 30.	+ 28 25		13.0	FAINT BLUE STAR
LB 11247	12 30 30.	+ 31 50		16.8	FAINT BLUE STAR
MCG+05-30-016	12 30 30.	+ 31 50	360	15.6	GALAXY
MCG+05-30-017	12 30 30.	+ 32 26	42	15.	GALAXY
ZC 1230.5+4639	12 30 30.	+ 46 39	1480		CLUSTER OF GALAXIES
MCG+11-15-062	12 30 30.	+ 64 34	36	17.	GALAXY
UGC 07702	12 30 30.	+ 65 33	60	16.5	GALAXY Sb-c
REIZ 2501	12 30 31.	+ 16 33	18	13.7	GALAXY
REIZ 2499	12 30 32.	+ 09 40	18	15.8	GALAXY
REIZ 2500	12 30 32.	+ 09 41	12	15.4	GALAXY
AMES 1458	12 30 32.	+ 16 23 50.	29	16.3	NEBULA
KN 16.010	12 30 32.6	+ 28 06 14.			NEBULA
AMES 1459	12 30 33.	+ 10 57 38.	18	17.8	NEBULA
MCG+03-32-067	12 30 33.	+ 14 50	102	13.9	GALAXY
RNGC 4515	12 30 33.	+ 16 32		13.5	GALAXY
MCG+03-32-065	12 30 33.	+ 16 32	30	13.3	GALAXY
MCG+03-32-066	12 30 33.	+ 17 41	36	15.4	GALAXY
IC 3482	12 30 33.	+ 28 06 28.			NONSTELLAR OBJECT
REIZ 2506	12 30 34.	+ 32 25	36	15.8	GALAXY
RNGC 4521	12 30 35.	+ 64 13		13.0	GALAXY
A3 408	12 30 35.2	+ 12 54 09.		16.8	FAINT BLUE OBJECT
A3 409	12 30 35.9	+ 12 52 06.		16.3	FAINT BLUE OBJECT
ZWG 042.149	12 30 36.	+ 08 06		15.2	GALAXY
MCG+01-32-094	12 30 36.	+ 08 06	60	15.2	GALAXY
ZWG 042.150	12 30 36.	+ 08 07		15.0	GALAXY
MCG+01-32-095	12 30 36.	+ 08 07	60	15.0	GALAXY
RMB 199	12 30 36.	+ 10 11			FAINT BLUE OBJECT
ZWG 070.160	12 30 36.	+ 11 37		15.4	GALAXY
MCG+02-32-129	12 30 36.	+ 11 37	24	15.4	GALAXY
AMES 1461	12 30 36.	+ 13 48 20.	22	17.4	NEBULA
ZWG 099.087	12 30 36.	+ 14 51		13.9	GALAXY
UGC 07703	12 30 36.	+ 14 51	120	13.9	GALAXY S
ZWG 099.088	12 30 36.	+ 17 40		15.4	GALAXY
ZWG 159.014	12 30 36.	+ 26 38		15.6	GALAXY
LB 06556	12 30 36.	+ 27 10		17.6	FAINT BLUE STAR
LB 06555	12 30 36.	+ 28 18		19.7	FAINT BLUE STAR
LB 06554	12 30 36.	+ 30 07		19.1	FAINT BLUE STAR
LB 06553	12 30 36.	+ 31 11		19.9	FAINT BLUE STAR
ZWG 159.015	12 30 36.	+ 32 22		14.1	GALAXY
UGC 07704	12 30 36.	+ 32 22	60	14.1	GALAXY
MCG+05-30-018	12 30 36.	+ 32 23	30	14.1	GALAXY
ZC 1230.6+4020	12 30 36.	+ 40 20	1140		CLUSTER OF GALAXIES
REIZ 2511	12 30 36.	+ 48 21	42	15.1	GALAXY
ZC 1230.6+5342	12 30 36.	+ 53 42	1080		CLUSTER OF GALAXIES
UGC 07705	12 30 36.	+ 57 09	60	16.0	GALAXY Sc
ZWG 315.046	12 30 36.	+ 64 13		13.0	GALAXY
UGC 07706	12 30 36.	+ 64 13	162	13.0	GALAXY S0-a
KARA.72 345B	12 30 36.	+ 64 13	144	13.0	PART OF DOUBLE GALAXY
ZCG 199.93	12 30 36.	+ 64 24		18.8	COMPACT GALAXY
ZCG 199.94	12 30 36.	+ 64 25		16.8	COMPACT GALAXY
A3 410	12 30 36.1	+ 16 27 02.		16.7	FAINT BLUE OBJECT
HOLM 417B	12 30 37.	+ 08 06	18	15.4	PART OF MULTIPLE GALAXY
AMES 1462	12 30 37.	+ 12 27 56.	19	17.3	NEBULA
HN 0954	12 30 37.	+ 17 40	18		NEBULA
REIZ 2505	12 30 37.	+ 17 40	24	15.2	GALAXY
IC 3484	12 30 37.	+ 17 40			NONSTELLAR OBJECT
AMES 1460	12 30 37.	+ 19 39 32.	26	16.8	NEBULA
RNGC 4518B	12 30 38.	+ 08 06		15.0	GALAXY
REIZ 2503	12 30 38.	+ 08 06	12	15.6	GALAXY
RNGC 4518A	12 30 38.	+ 08 07		15.0	GALAXY
REIZ 2504	12 30 38.	+ 08 07	24	15.2	GALAXY
HOLM 417A	12 30 38.	+ 08 07	30	14.7	PART OF MULTIPLE GALAXY
HN 0955	12 30 38.	+ 11 37		14.	NEBULA
RNGC 4516	12 30 38.	+ 14 51		14.0	GALAXY
REIN 3.221	12 30 38.16	+ 08 06 36.2			NEBULA
REIN 3.222	12 30 38.38	+ 11 37 23.4			NEBULA
IC 3483	12 30 38.6	+ 11 37 31.			GALAXY Sc
IC 3485	12 30 39.	+ 09 29 40.			SINGLE STAR
REIN 3.223	12 30 39.43	+ 08 07 38.5			NEBULA
IC 3488	12 30 39.7	+ 26 37 32.			GALAXY SA0
REIN 3.224	12 30 39.80	+ 08 05 47.6			NEBULA
KN 16.011	12 30 39.9	+ 26 37 38.			NEBULA
IC 3486	12 30 40.	+ 13 08 10.			NONSTELLAR OBJECT
AMES 1463	12 30 40.	+ 13 49 26.	17	17.6	NEBULA
REIZ 2510	12 30 40.	+ 32 22	36	14.6	GALAXY
ARC 1556	12 30 40.	- 21 44		17.6	RICH CLUSTER OF GALAXIES
IC 3491	12 30 40.3	+ 27 22 13.			GALAXY Sb
KN 16.012	12 30 40.5	+ 27 22 15.			NEBULA
IC 3487	12 30 41.	+ 09 40 22.			NONSTELLAR OBJECT
AMES 1464	12 30 41.	+ 16 23 44.	18	18.0	NEBULA
REIZ 2508	12 30 41.	+ 26 37	18	14.6	GALAXY
REIZ 2509	12 30 41.	+ 27 20	36	14.9	GALAXY
REIZ 2518	12 30 41.	+ 62 53	12	14.3	GALAXY
REIN 3.225	12 30 41.59	+ 09 40 22.1			NEBULA
ZWG 070.161	12 30 42.	+ 09 40		15.1	GALAXY
MCG+02-32-133	12 30 42.	+ 09 40	18	15.1	GALAXY
ZWG 070.162	12 30 42.	+ 11 12		15.6	GALAXY
ZWG 070.163	12 30 42.	+ 12 31		15.2	GALAXY
MCG+02-32-130	12 30 42.	+ 12 31	36	15.2	GALAXY
ZWG 070.164	12 30 42.	+ 13 08		15.3	GALAXY
MCG+02-32-132	12 30 42.	+ 13 08	18	15.	GALAXY
MCG+05-30-019	12 30 42.	+ 26 37	42	15.6	GALAXY
LB 06559	12 30 42.	+ 26 44		18.1	FAINT BLUE STAR
LB 06558	12 30 42.	+ 30 44		19.3	FAINT BLUE STAR
LB 06557	12 30 42.	+ 32 23		18.6	FAINT BLUE STAR
7ZW 469	12 30 42.	+ 73 43			COMPACT GALAXY
KARA.68 155	12 30 42.	- 00 16	74		DWARF GALAXY
MCG-05-30-004	12 30 42.	- 31 06	6	16.	GALAXY
REIN 3.226	12 30 42.57	+ 13 07 57.7			NEBULA
IC 3489	12 30 43.	+ 12 31			NONSTELLAR OBJECT
AMES 1465	12 30 43.	+ 16 25 20.	14	17.7	NEBULA
KEEL 469	12 30 43.4	+ 02 25 29.		17.	NEBULA
KN 16.013	12 30 43.5	+ 28 51 45.			NEBULA
REIZ 2507	12 30 44.	+ 09 41	24	15.3	GALAXY
HN 0957	12 30 44.	+ 11 12	36		NEBULA
HN 0956	12 30 44.	+ 12 31		13.	NEBULA
AMES 1466	12 30 44.	+ 16 21 50.	12	16.7	NEBULA
MCG+02-32-131	12 30 45.	+ 13 08	18	15.3	GALAXY
LB 02382	12 30 45.	+ 58 54 54.		16.3	FAINT BLUE STAR
IC 3494	12 30 45.3	+ 27 51 34.			GALAXY E1
IC 3492	12 30 46.	+ 13 07 52.			NONSTELLAR OBJECT
REIN 3.227	12 30 46.20	+ 09 01 32.4			NEBULA
A3 414	12 30 46.5	+ 13 03 07.		15.4	FAINT BLUE OBJECT
REIZ 2514	12 30 47.	+ 27 01	18	15.4	GALAXY
REIZ 2515	12 30 47.	+ 27 04	12	15.4	GALAXY
REIZ 2516	12 30 47.	+ 27 50	12	15.0	GALAXY
ZWG 014.064	12 30 48.	+ 01 47		15.3	GALAXY
RMB 200	12 30 48.	+ 10 52			FAINT BLUE OBJECT
AMES 1471	12 30 48.	+ 12 39 08.	18	17.0	NEBULA
AMES 1470	12 30 48.	+ 13 00 08.	25	17.8	NEBULA
AMES 1467	12 30 48.	+ 13 29 44.	26	17.1	NEBULA
RMB 084	12 30 48.	+ 13 34			FAINT BLUE OBJECT
UGC 07707	12 30 48.	+ 17 45	60	17.	GALAXY DWARF
ZC 1230.8+1946	12 30 48.	+ 19 46	470		CLUSTER OF GALAXIES
LB 06562	12 30 48.	+ 26 52		19.4	FAINT BLUE STAR
LB 06561	12 30 48.	+ 28 20		19.5	FAINT BLUE STAR
LB 06560	12 30 48.	+ 30 10		19.4	FAINT BLUE STAR
MCG+08-23-053	12 30 48.	+ 48 20	60	16.	GALAXY
UGC 07708	12 30 48.	+ 48 21	60	16.0	GALAXY Sb
MCG+09-21-014	12 30 48.	+ 50 48	30	16.	GALAXY
MCG+09-21-015	12 30 48.	+ 52 58	36	16.	GALAXY
MCG+10-18-065	12 30 48.	+ 57 08	51	15.	GALAXY
MCG-04-30-004	12 30 48.	- 27 24 30.	30	16.	GALAXY
REIN 3.229	12 30 48.37	+ 13 07 43.4			NEBULA
REIN 3.228	12 30 48.44	+ 08 07 46.0			NEBULA
IC 3493	12 30 49.	+ 09 39 58.			SINGLE STAR
AMES 1469	12 30 49.	+ 16 35 14.	24	17.7	NEBULA
AMES 1468	12 30 49.	+ 18 57 20.	34	17.8	NEBULA
A3 416	12 30 49.3	+ 13 30 05.		17.3	FAINT BLUE OBJECT
REIZ 2512	12 30 50.	+ 08 07	12	16.5	GALAXY
RNGC 4519A	12 30 50.	+ 08 56		15.5	GALAXY
REIZ 2513	12 30 50.	+ 08 57	30	15.9	GALAXY
HOLM 418B	12 30 50.	+ 08 58	24	15.1	PART OF MULTIPLE GALAXY
AMES 1472	12 30 50.	+ 17 44 08.	46	17.2	NEBULA
MCG+02-32-134	12 30 51.	+ 08 56	30	15.3	GALAXY
LB 02383	12 30 51.	+ 57 51 36.		13.9	FAINT BLUE STAR
KEEL 470	12 30 51.5	+ 02 25 50.		16.	NEBULA
LB 02384	12 30 52.	+ 59 32 42.		16.1	FAINT BLUE STAR
REIZ 2524	12 30 52.	+ 64 13	72	13.6	GALAXY
REIN 3.230	12 30 52.56	+ 08 57 57.8			NEBULA
AMES 1476	12 30 53.	+ 06 53 56.	49	17.0	NEBULA
AMES 1474	12 30 53.	+ 13 30 08.	20	15.6	NEBULA
AMES 1473	12 30 53.	+ 16 21 20.	24	16.9	NEBULA
REIZ 2517	12 30 53.	+ 27 00	42	14.8	GALAXY
A3 418	12 30 53.9	+ 13 30 31.		15.5	UV CONCENTRATION
ZWG 042.151	12 30 54.	+ 04 04		15.7	GALAXY
MCG+01-32-096	12 30 54.	+ 04 04	48	15.7	GALAXY
MCG+02-32-135	12 30 54.	+ 08 56	162	12.8	GALAXY
ZWG 070.165	12 30 54.	+ 08 57		15.3	GALAXY
REA 73	12 30 54.	+ 09 44	24		DWARF GALAXY
ZWG 070.166	12 30 54.	+ 13 32		15.4	GALAXY
MCG+02-32-136	12 30 54.	+ 13 32	15	15.4	GALAXY
ZC 1230.9+1444	12 30 54.	+ 14 44	670		CLUSTER OF GALAXIES
ZC 1230.9+2221	12 30 54.	+ 22 21	1010		CLUSTER OF GALAXIES
LB 06564	12 30 54.	+ 27 25		19.6	FAINT BLUE STAR
LB 06563	12 30 54.	+ 31 22		19.4	FAINT BLUE STAR
MCG+06-28-009	12 30 54.	+ 33 38		16.	GALAXY
TON-N 0082	12 30 54.	+ 46 31		15.2	BLUE STAR
MRK 216	12 30 54.	+ 52 04	9	16.5	GALAXY WITH UV CONTINUUM
A3 419	12 30 54.7	+ 16 41 35.		16.9	FAINT BLUE OBJECT
AMES 1475	12 30 55.	+ 18 57 56.	19	17.6	NEBULA
LB 02385	12 30 55.	+ 50 28 06.		17.1	FAINT BLUE STAR
ARC 1559	12 30 55.	+ 67 24		17.2	RICH CLUSTER OF GALAXIES

OBJECT NAME	RIGHT ASCEN.	DECLINATION	DIAM.	MAGN.	TYPE OF OBJECT
HOLM 418A	12 30 56.	+ 08 56	150	12.1	PART OF MULTIPLE GALAXY
MCG+08-23-054	12 20 57.	+ 46 54	36	16.	GALAXY
AMES 1477	12 30 58.	+ 13 38 20.	16	17.8	NEBULA
REIZ 2522	12 30 58.	+ 33 37	42	16.1	GALAXY
REIN 3.231	12 30 58.28	+ 08 55 49.0			NEBULA
AMES 1478	12 30 59.	+ 11 22 38.	19	17.8	NEBULA
IC 3497	12 30 59.	+ 25 45 52.			NONSTELLAR OBJECT
A3 422	12 30 59.5	+ 14 08 52.		15.9	FAINT BLUE OBJECT
A3 423	12 30 59.6	+ 15 56 23.		17.1	FAINT BLUE OBJECT
KEEL 471	12 30 59.7	+ 02 34 13.		17.	NEBULA
VDB.66G 135	12 31	+ 15 29	100		DWARF GALAXY
LB 09860	12 31	- 83 56		14.7	FAINT BLUE STAR
MCG+01-32-097	12 31 00.	+ 08 08	54	15.4	GALAXY
ZWG 070.167	12 31 00.	+ 08 55		12.8	GALAXY
UGC 07709	12 31 00.	+ 08 55	228	12.8	GALAXY Sc/SBc
ZC 1231.0+1740	12 31 00.	+ 17 40	1210		CLUSTER OF GALAXIES
LB 06571	12 31 00.	+ 26 58		19.8	FAINT BLUE STAR
IC 3498	12 31 00.	+ 27 00 46.			SINGLE STAR
LB 06570	12 31 00.	+ 27 38		19.4	FAINT BLUE STAR
LB 06569	12 31 00.	+ 28 42		19.5	FAINT BLUE STAR
LB 06568	12 31 00.	+ 28 57		18.6	FAINT BLUE STAR
LB 06567	12 31 00.	+ 29 07		18.5	FAINT BLUE STAR
LB 06566	12 31 00.	+ 30 32		19.0	FAINT BLUE STAR
LB 06565	12 31 00.	+ 32 03		19.0	FAINT BLUE STAR
ZWG 244.028	12 31 00.	+ 46 54		15.7	GALAXY
REIZ 2526	12 31 00.	+ 46 54	24	14.0	GALAXY
MCG+10-18-066	12 31 00.	+ 58 12	36	16.	GALAXY
ZWG 293.029	12 31 00.	+ 58 14		15.7	GALAXY
ZC 1231.0+6342	12 31 00.	+ 63 42	1210		CLUSTER OF GALAXIES
7ZW 470	12 31 00.	+ 67 24			COMPACT GALAXY
IC 3496	12 31 00.1	+ 27 00 48.			GALAXY Sb
KN 16.014	12 31 00.3	+ 27 00 57.			NEBULA
IC 3495	12 31 00.8	+ 27 03 18.			GALAXY S0/a
REIZ 2519	12 31 02.	+ 08 08	54	15.9	GALAXY
RNGC 4519	12 31 02.	+ 08 56		12.5	GALAXY
REIZ 2520	12 31 02.	+ 08 56	138	12.8	GALAXY
ARC 1561	12 31 02.	+ 69 39		17.4	RICH CLUSTER OF GALAXIES
KN 16.015	12 31 02.4	+ 27 49 21.			NEBULA
A3 424	12 31 02.7	+ 13 01 23.		16.6	FAINT BLUE OBJECT
AMES 1480	12 31 03.	+ 11 31 50.	23	16.8	NEBULA
TON-N 1544	12 31 03.	+ 21 44		15.7	BLUE STAR
TON-N 1545	12 31 03.	+ 22 44		16.5	BLUE STAR
MCG-01-32-023	12 31 03.	- 04 36	114	15.	GALAXY
A3 425	12 31 03.4	+ 16 45 32.		16.4	FAINT BLUE OBJECT
AMES 1479	12 31 04.	+ 18 36 50.	19	16.6	NEBULA
AMES 1482	12 31 05.	+ 11 51 20.	28	17.2	NEBULA
REIZ 2523	12 31 05.	+ 25 46	12	15.8	GALAXY
REIZ 2521	12 31 05.	- 07 07	24	14.0	GALAXY
REIN 3.232	12 31 05.31	+ 08 08 49.0			NEBULA
ZWG 042.152	12 31 06.	+ 08 08		15.4	GALAXY
ZWG 070.168	12 31 06.	+ 09 26		13.6	GALAXY
UGC 07711	12 31 06.	+ 09 26	252	13.6	GALAXY Sc
RMB 124	12 31 06.	+ 12 23			FAINT BLUE OBJECT
RMB 028	12 31 06.	+ 14 16			FAINT BLUE OBJECT
AMES 1481	12 31 06.	+ 18 37 44.	16	17.2	NEBULA
LB 06575	12 31 06.	+ 30 01		19.6	FAINT BLUE STAR
LB 06574	12 31 06.	+ 30 05		18.9	FAINT BLUE STAR
LB 11249	12 31 06.	+ 30 12		13.7	FAINT BLUE STAR
TON-N 0083	12 31 06.	+ 31 18		15.5	BLUE STAR
LB 06573	12 31 06.	+ 31 44		19.4	FAINT BLUE STAR
LB 06572	12 31 06.	+ 32 16		19.5	FAINT BLUE STAR
MCG+08-23-055	12 31 06.	+ 49 24 30.	36	16.	GALAXY
UGC 07710	12 31 06.	- 02 23	84	16.5	GALAXY DWRF IR
MCG+00-32-021	12 31 06.	- 02 23		16.	GALAXY
AMES 1484	12 31 07.	+ 11 23 32.	25	17.4	NEBULA
AMES 1483	12 31 07.	+ 12 18 08.	16	17.2	NEBULA
RNGC 4520	12 31 07.	- 07 07			GALAXY
IC 0799	12 31 07.	- 07 07			SAME AS NGC 4520
REIN 3.233	12 31 07.72	+ 09 26 55.8			NEBULA
RNGC 4522	12 31 08.	+ 09 27		12.5	GALAXY
REIZ 2525	12 31 08.	+ 09 27	180	12.8	GALAXY
A3 428	12 31 08.0	+ 14 16 16.		17.0	FAINT BLUE OBJECT
KEEL 473	12 31 08.7	+ 14 53 58.		16.	SPIRAL NEBULA
KEEL 472	12 31 09.2	+ 02 24 46.		17.	NEBULA
KN 16.016	12 31 09.9	+ 27 22 27.			NEBULA
AMES 1485	12 31 10.	+ 13 39 56.	14	17.8	NEBULA
KN 16.017	12 31 10.1	+ 28 23 45.			NEBULA
REIZ 2528	12 31 11.	+ 26 52	12	15.7	GALAXY
SVEN 307	12 31 11.	- 07 05 04.	18	14.9	GALAXY
SN 1969G	12 31 12.	+ 06 10		18.0	SUPERNOVA
KARA.68 156	12 31 12.	+ 08 03	34		DWARF GALAXY
SCH 54	12 31 12.	+ 08 56		12.8	PECULIAR GALAXY
MCG+02-32-137	12 31 12.	+ 09 27	180	13.6	GALAXY
ZWG 070.169	12 31 12.	+ 11 16		14.5	GALAXY
UGC 07712	12 31 12.	+ 11 16	96	14.5	GALAXY S0-a
MCG+02-32-138	12 31 12.	+ 11 16	60	14.5	GALAXY
IC 3499	12 31 12.	+ 11 16		14.5	GALAXY IN VIRGO CLSTR
LB 06576	12 31 12.	+ 27 15		18.4	FAINT BLUE STAR
MCG+10-18-067	12 31 12.	+ 59 21	18	16.	GALAXY
ZWG 352.034	12 31 12.	+ 78 15		15.6	GALAXY
KARA.73B 0535	12 31 12.	+ 78 15	36	15.6	ISOLATED GALAXY S
MCG-02-32-015	12 31 12.	- 10 24	120	15.	GALAXY
RNGC 4524	12 31 12.	- 11 45		14.0	GALAXY
MCG-02-32-014	12 31 12.	- 11 45	72	14.5	GALAXY
MCG-03-32-012	12 31 12.	- 21 23	72	15.	GALAXY
MCG-06-28-003	12 31 12.	- 34 50	66	15.	GALAXY
MCG-06-28-002	12 31 12.	- 37 29	36	15.5	GALAXY
OCL 0886	12 31 12.	- 61 17	780	12.	OPEN STAR CLUSTER
HN 0958	12 31 13.	+ 14 13	30		NEBULA
REIN 3.234	12 31 13.18	+ 11 16 15.3			NEBULA
A3 432	12 31 13.2	+ 13 29 52.		16.3	FAINT BLUE OBJECT
IC 3502	12 31 13.6	+ 26 53 29.			GALAXY Sd
HN 0959	12 31 14.	+ 11 16	18		NEBULA
IC 3500	12 31 14.	+ 14 13			NONSTELLAR OBJECT
A3 433	12 31 14.7	+ 16 51 40.		16.6	FAINT BLUE OBJECT
MCG+03-32-068	12 31 15.	+ 15 26	138	15.1	GALAXY
LB 02386	12 31 15.	+ 50 36 00.		16.7	FAINT BLUE STAR
A3 434	12 31 15.8	+ 13 10 16.		16.9	FAINT BLUE OBJECT
AMES 1487	12 31 16.	+ 12 53 44.	22	17.3	NEBULA
AMES 1486	12 31 16.	+ 16 46 02.	20	17.8	NEBULA
RNGC 4525	12 31 16.	+ 30 34		13.0	GALAXY
AMES 1488	12 31 17.	+ 12 53 20.	18	16.6	NEBULA
IC 3501	12 31 17.	+ 13 36 10.			NONSTELLAR OBJECT
REIZ 2527	12 31 17.	- 11 45	48	13.8	GALAXY
KEEL 474	12 31 17.1	+ 02 09 19.		16.	NEBULA
KEEL 475	12 31 17.2	+ 02 33 18.		17.	NEBULA
REA 03	12 31 18.	+ 03 50			DWARF GALAXY
ZWG 042.153	12 31 18.	+ 08 18		15.2	GALAXY
MCG+01-32-098	12 31 18.	+ 08 18	21	15.2	GALAXY
REA 74	12 31 18.	+ 10 50	24		DWARF GALAXY
ZWG 070.170	12 31 18.	+ 13 36		15.0	GALAXY
MCG+02-32-139	12 31 18.	+ 13 36	36	15.0	GALAXY
ZWG 070.171	12 31 18.	+ 14 15		15.3	GALAXY
ZC 1231.3+1437	12 31 18.	+ 14 37	470		CLUSTER OF GALAXIES
ZWG 099.089	12 31 18.	+ 15 26		15.1	GALAXY
UGC 07713	12 31 18.	+ 15 26	150	15.1	GALAXY IRR
ZC 1231.3+2601	12 31 18.	+ 26 01	400		CLUSTER OF GALAXIES
LB 06579	12 31 18.	+ 28 02		19.0	FAINT BLUE STAR
LB 06578	12 31 18.	+ 29 28		19.8	FAINT BLUE STAR
ZWG 159.016	12 31 18.	+ 30 34		13.0	GALAXY
UGC 07714	12 31 18.	+ 30 34	192	13.0	GALAXY Sc
LB 06577	12 31 18.	+ 31 04		20.0	FAINT BLUE STAR
LB 11250	12 31 18.	+ 31 51		18.0	FAINT BLUE STAR
ZC 1231.3+5848	12 31 18.	+ 58 48	1340		CLUSTER OF GALAXIES
ZC 1231.3+6246	12 31 18.	+ 62 46	610		CLUSTER OF GALAXIES
MCG-01-32-024	12 31 18.	- 08 27	42	16.	GALAXY
A3 435	12 31 19.0	+ 16 14 25.		16.8	FAINT BLUE OBJECT
REIN 3.235	12 31 19.05	+ 08 18 00.6			NEBULA
REIZ 2529	12 31 20.	+ 08 17	18	15.3	GALAXY
RNGC 4523	12 31 20.	+ 15 27		15.0	GALAXY
AMES 1490	12 31 20.	+ 16 41 44.	26	17.6	NEBULA
LB 00642	12 31 20.	+ 72 00 24.		17.3	FAINT BLUE STAR
AMES 1491	12 31 21.	+ 10 20 38.	28	17.4	NEBULA
AMES 1489	12 31 21.	+ 18 48 50.	13	17.4	NEBULA
IC 3503	12 31 21.	+ 38 03 16.			NONSTELLAR OBJECT
A3 437	12 31 21.6	+ 12 55 35.		17.5	FAINT BLUE OBJECT
A3 436	12 31 21.6	+ 15 11 38.		17.1	FAINT BLUE OBJECT
AMES 1493	12 31 22.	+ 11 44 20.	22	17.3	NEBULA
REIZ 2530	12 31 22.	+ 30 32	90	13.5	GALAXY
A3 438	12 31 22.2	+ 13 27 54.		15.9	FAINT BLUE OBJECT
KN 16.018	12 31 22.2	+ 28 46 27.			NEBULA
AMES 1494	12 31 23.	+ 10 22 02.	24	17.0	NEBULA
AMES 1492	12 31 23.	+ 17 10 56.	16	18.0	NEBULA
RNGC 4530	12 31 23.	+ 41 37			NON-EXISTENT OBJECT
ZWG 042.154	12 31 24.	+ 03 49		15.3	GALAXY
UGC 07715	12 31 24.	+ 03 49	60	15.3	GALAXY DWARF
MCG+01-32-099	12 31 24.	+ 03 49	60	15.3	GALAXY
ZC 1231.4+1007	12 31 24.	+ 10 07	1340		CLUSTER OF GALAXIES
AMES 1495	12 31 24.	+ 10 17 44.	25	17.6	NEBULA
MCG+03-32-069	12 31 24.	+ 15 37	90	14.3	GALAXY
RMB 019	12 31 24.	+ 15 38			FAINT BLUE OBJECT
ZWG 099.090	12 31 24.	+ 15 38		14.3	GALAXY
UGC 07716	12 31 24.	+ 15 38	108	14.3	GALAXY SBc
LB 06581	12 31 24.	+ 27 09		19.7	FAINT BLUE STAR
LB 06580	12 31 24.	+ 27 21		19.5	FAINT BLUE STAR
LB 11251	12 31 24.	+ 27 23		14.0	FAINT BLUE STAR
TON-N 0621	12 31 24.	+ 29 26		16.0	BLUE STAR
MCG+05-30-020	12 31 24.	+ 30 34	150	13.0	GALAXY
MCG+07-26-023	12 31 24.	+ 39 47	48	15.	GALAXY
ZC 1231.4+3954	12 31 24.	+ 39 54	4030		CLUSTER OF GALAXIES
ZC 1231.4+5044	12 31 24.	+ 50 44	5240		CLUSTER OF GALAXIES
MCG+09-21-016	12 31 24.	+ 50 53	24	16.	GALAXY
ZWG 270.009	12 31 24.	+ 52 32		14.8	GALAXY
UGC 07717	12 31 24.	+ 52 32	72	14.8	GALAXY Sa?
IC 0801	12 31 24.	+ 52 32 35.			NONSTELLAR OBJECT
LB 02387	12 31 24.	+ 57 29 42.		15.4	FAINT BLUE STAR
ZC 1231.4+6432	12 31 24.	+ 64 32	810		CLUSTER OF GALAXIES
MCG-02-32-016	12 31 24.	- 09 31	24	15.5	GALAXY
MCG-06-28-004	12 31 24.	- 34 51	60	15.5	GALAXY
KEEL 476	12 31 24.0	+ 02 07 45.		18.	NEBULA
A3 439	12 31 24.3	+ 14 17 06.		16.7	FAINT BLUE OBJECT
A3 440	12 31 24.7	+ 13 21 08.		15.8	FAINT BLUE OBJECT
AMES 1497	12 31 25.	+ 06 28 44.	32	17.4	NEBULA
A3 441	12 31 25.1	+ 16 12 55.		16.8	FAINT BLUE OBJECT
KEEL 477	12 31 25.6	+ 14 55 16.		18.	NEBULA
IC 0800	12 31 25.7	+ 15 37 49.	72		GALAXY SB(s)
A3 442	12 31 25.8	+ 15 37 43.		14.5	UV CONCENTRATION
SC 1228+0100.9	12 31 26.	+ 00 44 21.	18		NEBULA
AMES 1496	12 31 26.	+ 10 22 56.	20	17.2	NEBULA
ARC 1558	12 31 26.	- 13 18		17.2	RICH CLUSTER OF GALAXIES
MCG+07-26-024	12 31 27.	+ 39 53 30.	45	15.	GALAXY
A3 443	12 31 27.4	+ 13 53 24.		18.0	FAINT BLUE OBJECT
ARC 1571	12 31 28.	+ 83 38		17.6	RICH CLUSTER OF GALAXIES
A3 445	12 31 28.5	+ 13 15 19.		15.9	FAINT BLUE OBJECT
KEEL 478	12 31 28.8	+ 02 02 29.		18.	NEBULA
SN 1969E	12 31 29.	+ 07 58		16.0	SUPERNOVA
AMES 1499	12 31 29.	+ 13 16 56.	20	16.7	NEBULA
AMES 1498	12 31 29.	+ 16 32 38.	30	17.6	NEBULA
ZWG 042.155	12 31 30.	+ 07 58		10.6	GALAXY
UGC 07718	12 31 30.	+ 07 58	450	10.6	GALAXY S0
MCG+01-32-100	12 31 30.	+ 07 58	300	10.6	GALAXY
RMB 082	12 31 30.	+ 13 53			FAINT BLUE OBJECT
AMES 1500	12 31 30.	+ 16 37 20.	16	16.9	NEBULA
TON-N 0622	12 31 30.	+ 23 27		14.5	BLUE STAR
ZWG 216.011	12 31 30.	+ 39 18		15.4	GALAXY
UGC 07719	12 31 30.	+ 39 18	114	15.4	GALAXY
MCG+09-21-017	12 31 30.	+ 52 30	72	14.	GALAXY
MCG-04-30-005	12 31 30.	- 26 02	72	14.5	GALAXY
MRSL 300+01/2	12 31 30.	- 61 18	660		HII REGION
REIN 3.236	12 31 30.64	+ 07 58 30.7			NEBULA
A3 446	12 31 30.7	+ 14 47 20.		16.8	FAINT BLUE OBJECT
AMES 1502	12 31 31.	+ 11 16 56.	26	17.2	NEBULA
AMES 1501	12 31 31.	+ 16 57 08.	19	17.7	NEBULA
REIN 4.083	12 31 31.38	- 00 04 09.1			NEBULA
KEEL 479	12 31 31.8	+ 02 25 21.		18.	NEBULA
REIZ 2531	12 31 32.	+ 07 58	240	10.8	GALAXY
RNGC 4526	12 31 32.	+ 07 59		11.0	GALAXY
REIZ 2532	12 31 33.	+ 02 56	360	12.0	GALAXY
AMES 1503	12 31 33.	+ 16 38 14.	24	18.0	NEBULA
AMES 1504	12 31 34.	+ 16 33 20.	14	17.6	NEBULA
LB 02388	12 31 34.	+ 61 12 54.		16.3	FAINT BLUE STAR
REIN 3.237	12 31 34.36	+ 11 35 48.3			NEBULA
AMES 1505	12 31 35.	+ 13 33 56.	37	16.6	NEBULA
ARC 1560	12 31 35.	+ 15 28		18.1	RICH CLUSTER OF GALAXIES
REIZ 2534	12 31 35.	+ 25 38	12	15.7	GALAXY
IC 3507	12 31 35.	+ 25 38 17.			NONSTELLAR OBJECT
REIZ 2535	12 31 35.	+ 26 56	18	14.4	GALAXY
RNGC 4534	12 31 35.	+ 35 48		13.0	GALAXY
LB 02389	12 31 35.	+ 51 38 48.		16.0	FAINT BLUE STAR
SC 1229+0011.8	12 31 35.	- 00 04 46.	18		NEBULA
SC 1229-0223.2	12 31 35.	- 02 39 46.	18		NEBULA
KEEL 480	12 31 35.0	+ 02 07 09.		17.	NEBULA
REIN 4.084	12 31 35.17	- 00 04 49.1			NEBULA
REIN 6.099	12 31 35.40	+ 08 26 39.5			NEBULA
ZWG 014.066	12 31 36.	+ 00 45		15.5	GALAXY
ZWG 042.156	12 31 36.	+ 02 56		12.4	GALAXY

OBJECT NAME	RIGHT ASCEN.	DECLINATION	DIAM.	MAGN.	TYPE OF OBJECT
UGC 07721	12 31 36.	+ 02 56	402	12.4	GALAXY Sb
MCG+01-32-101	12 31 36.	+ 02 56	360	12.4	GALAXY
IC 3504	12 31 36.	+ 07 09 41.			MAY NOT EXIST
ZWG 070.172	12 31 36.	+ 11 36		12.9	GALAXY
UGC 07722	12 31 36.	+ 11 36	96	12.9	GALAXY S0
MCG+02-32-140	12 31 36.	+ 11 36	36	12.9	GALAXY
RMB 125	12 31 36.	+ 12 50			FAINT BLUE OBJECT
ZWG 099.091	12 31 36.	+ 16 14		15.3	GALAXY
MCG+03-32-070	12 31 36.	+ 16 14	48	15.3	GALAXY
TON-N 0084	12 31 36.	+ 17 13		13.7	BLUE STAR
8ZW 1231+21.8	12 31 36.	+ 21 49		18.3	COMPACT GALAXY
ZC 1231.6+2440	12 31 36.	+ 24 40	870		CLUSTER OF GALAXIES
MCG+05-30-021	12 31 36.	+ 26 56	42	15.4	GALAXY
ZWG 159.017	12 31 36.	+ 26 57		15.4	GALAXY
LB 06584	12 31 36.	+ 28 26		19.6	FAINT BLUE STAR
LB 06583	12 31 36.	+ 31 02		19.8	FAINT BLUE STAR
LB 06582	12 31 36.	+ 31 33		18.0	FAINT BLUE STAR
ZWG 188.008	12 31 36.	+ 35 48		13.2	GALAXY
UGC 07723	12 31 36.	+ 35 48	270	13.2	GALAXY
KARA.73B 0536	12 31 36.	+ 35 48	204	13.2	ISOLATED GALAXY S
MCG+07-26-025	12 31 36.	+ 39 16 30.	114	14.5	GALAXY
ZWG 216.012	12 31 36.	+ 43 42		15.2	GALAXY
MCG+08-23-056	12 31 36.	+ 47 16	48	16.	GALAXY
ZWG 014.065	12 31 36.	- 00 05		14.8	GALAXY
UGC 07720	12 31 36.	- 00 05	66	14.8	GALAXY SBa-b
MCG+00-32-022	12 31 36.	- 00 05	36	14.8	GALAXY
MCG-03-32-014	12 31 36.	- 20 46 30.	30	16.	GALAXY
MCG-03-32-013	12 31 36.	- 20 47	36	15.	GALAXY
RNGC 4527	12 31 37.	+ 02 56		11.5	GALAXY
AMES 1509	12 31 37.	+ 11 47 56.	24	17.5	NEBULA
HN 0960	12 31 37.	+ 16 14	30		NEBULA
IC 3505	12 31 37.	+ 16 14			NONSTELLAR OBJECT
AMES 1506	12 31 37.	+ 19 44 44.	30	16.7	NEBULA
REIN 3.238	12 31 37.65	+ 08 32 42.8			NEBULA
A3 447	12 31 37.8	+ 13 51 26.		16.8	FAINT BLUE OBJECT
RNGC 4528	12 31 38.	+ 11 36		13.0	GALAXY
HN 0961	12 31 38.	+ 13 00			NEBULA
IC 3506	12 31 38.	+ 13 00			NONSTELLAR OBJECT
AMES 1508	12 31 38.	+ 16 33 02.	22	17.7	NEBULA
AMES 1507	12 31 38.	+ 18 57 20.	19	17.6	NEBULA
KN 16.019	12 31 38.4	+ 26 56 45.			NEBULA
IC 3508	12 31 38.5	+ 26 56 46.			GALAXY SB0
REIZ 2533	12 31 39.	+ 01 48	60	14.8	GALAXY
SN 1915A	12 31 39.	+ 02 56		15.5	SUPERNOVA
AMES 1510	12 31 39.	+ 15 25 56.	29	17.5	NEBULA
REIZ 2542	12 31 39.	+ 35 47	60	13.6	GALAXY
HOLM 419A	12 31 39.	+ 35 47	54	12.8	PART OF MULTIPLE GALAXY
MCG+07-26-026	12 31 39.	+ 38 56	30	16.	GALAXY
A3 448	12 31 39.6	+ 13 25 44.		14.5	FAINT BLUE OBJECT
AMES 1511	12 31 40.	+ 16 49 32.	26	17.4	NEBULA
LB 02390	12 31 40.	+ 59 58 42.		16.7	FAINT BLUE STAR
KEEL 481	12 31 40.2	+ 14 49 50.		16.	SPIRAL NEBULA
A3 449	12 31 40.5	+ 15 00 06.		16.0	FAINT BLUE OBJECT
REIN 3.239	12 31 40.60	+ 08 30 29.6			NEBULA
KN 16.020	12 31 40.8	+ 28 24 15.			NEBULA
AMES 1513	12 31 41.	+ 10 19 56.	18	17.8	NEBULA
AMES 1512	12 31 41.	+ 12 04 44.	14	16.6	NEBULA
REIZ 2541	12 31 41.	+ 27 37	12	15.7	GALAXY
IC 3511	12 31 41.	+ 27 37 29.			MAY NOT EXIST
IC 3512	12 31 41.	+ 27 38 17.			MAY NOT EXIST
KN 16.022	12 31 41.3	+ 27 00 57.			NEBULA
KN 16.021	12 31 41.3	+ 27 38 21.			NEBULA
REIN 6.100	12 31 41.74	+ 08 32 55.6			NEBULA
REIN 3.240	12 31 41.76	+ 08 32 57.3			NEBULA
AMES 1514	12 31 42.	+ 11 39 56.	24	17.5	NEBULA
RMB 090	12 31 42.	+ 13 05			FAINT BLUE OBJECT
LB 06587	12 31 42.	+ 26 28		19.6	FAINT BLUE STAR
LB 06586	12 31 42.	+ 27 34		19.3	FAINT BLUE STAR
UGC 07724	12 31 42.	+ 27 43	78	16.0	GALAXY Sb-c
LB 06585	12 31 42.	+ 32 16		18.8	FAINT BLUE STAR
HOLM 419B	12 31 42.	+ 35 47	18	15.0	PART OF MULTIPLE GALAXY
ZC 1231.7+4250	12 31 42.	+ 42 50	2960		CLUSTER OF GALAXIES
MCG+07-26-027	12 31 42.	+ 43 42	48	14.5	GALAXY
ZC 1231.7+4407	12 31 42.	+ 44 07	540		CLUSTER OF GALAXIES
KN 16.023	12 31 42.1	+ 26 39 33.			NEBULA
A3 450	12 31 42.6	+ 13 05 05.		17.0	FAINT BLUE OBJECT
REIN 3.241	12 31 42.66	+ 08 26 45.1			NEBULA
IC 3510	12 31 43.	+ 11 21			NONSTELLAR OBJECT
IC 3509	12 31 43.	+ 12 19			NONSTELLAR OBJECT
IC 3513	12 31 43.	+ 27 36 17.			MAY NOT EXIST
REIN 3.242	12 31 43.09	+ 11 20 48.9			NEBULA
KN 16.024	12 31 43.2	+ 27 56 45.			NEBULA
A3 451	12 31 43.3	+ 16 18 50.		16.0	FAINT BLUE OBJECT
KN 16.025	12 31 43.8	+ 27 41 21.			NEBULA
RNGC 4535A	12 31 44.	+ 08 32		14.5	GALAXY
REIZ 2538	12 31 44.	+ 08 33	30	15.3	GALAXY
HOLM 420B	12 31 44.	+ 08 33	36	14.7	PART OF MULTIPLE GALAXY
HN 0963	12 31 44.	+ 11 21		15.	NEBULA
HN 0962	12 31 44.	+ 12 19		14.	NEBULA
RNGC 4531	12 31 44.	+ 13 21		13.5	GALAXY
REIZ 2539	12 31 44.	+ 13 21	42	13.7	GALAXY
AMES 1515	12 31 44.	+ 16 32 08.	24	17.4	NEBULA
REIZ 2548	12 31 44.	+ 39 20	78	14.6	GALAXY
REIN 3.243	12 31 44.19	+ 11 20 57.4			NEBULA
REIN 3.245	12 31 44.58	+ 13 21 00.9			NEBULA
REIZ 2536	12 31 45.	+ 01 49	18	15.5	GALAXY
REIZ 2537	12 31 45.	+ 02 36	60	14.5	GALAXY
REIZ 2540	12 31 45.	+ 06 44	120	12.3	GALAXY
MCG+06-28-010	12 31 45.	+ 35 49	150	12.5	GALAXY
REIN 3.244	12 31 45.15	+ 08 28 08.4			NEBULA
A3 452	12 31 45.3	+ 14 23 50.		15.9	FAINT BLUE OBJECT
AMES 1516	12 31 46.	+ 15 33 20.	25	17.7	NEBULA
IC 3568	12 31 46.6	+ 82 50 21.9	18	10.8	PLANETARY NEBULA
PK123+34.1	12 31 46.64	+ 82 50 21.9	18	11.6	PLANETARY NEBULA
KN 16.026	12 31 46.9	+ 28 42 39.			NEBULA
AMES 1517	12 31 47.	+ 16 02 38.	22	17.8	NEBULA
REIZ 2545	12 31 47.	+ 26 58	12	15.4	GALAXY
REIZ 2546	12 31 47.	+ 27 42	54	15.1	GALAXY
REIZ 2547	12 31 47.	+ 28 07	18	15.1	GALAXY
IC 3515	12 31 47.	+ 28 08 17.			NONSTELLAR OBJECT
ARC 1563	12 31 47.	+ 54 21		17.6	RICH CLUSTER OF GALAXIES
KN 16.025A	12 31 47.2	+ 26 58 33.			NEBULA
REIN 3.246	12 31 47.34	+ 08 27 15.7			NEBULA
KN 16.027	12 31 47.5	+ 28 16 27.			NEBULA
KN 16.028	12 31 47.8	+ 28 08 21.			NEBULA
KN 16.029	12 31 47.9	+ 28 27 21.			NEBULA
ZWG 042.157	12 31 48.	+ 02 36		14.7	GALAXY
UGC 07725	12 31 48.	+ 02 36	150	14.7	GALAXY S-IRR

OBJECT NAME	RIGHT ASCEN.	DECLINATION	DIAM.	MAGN.	TYPE OF OBJECT
MCG+01-32-102	12 31 48.	+ 02 36	84	14.7	GALAXY
ZWG 042.158	12 31 48.	+ 06 45		12.3	GALAXY
SCH 55	12 31 48.	+ 06 45		12.3	PECULIAR GALAXY
UGC 07726	12 31 48.	+ 06 45	210	12.3	GALAXY IRR
MCG+01-32-103	12 31 48.	+ 06 45	168	12.3	GALAXY
ZWG 042.159	12 31 48.	+ 08 28		11.1	GALAXY
UGC 07727	12 31 48.	+ 08 28	468	11.1	GALAXY Sc/SBc
MCG+01-32-104	12 31 48.	+ 08 28	360	11.1	GALAXY
ZWG 070.173	12 31 48.	+ 11 21		15.2	GALAXY
UGC 07728	12 31 48.	+ 11 21	66	15.2	GALAXY S
MCG+02-32-142	12 31 48.	+ 11 21	18	15.2	GALAXY
ZWG 070.174	12 31 48.	+ 12 21		15.3	GALAXY
ZWG 070.175	12 31 48.	+ 13 21		13.3	GALAXY
UGC 07729	12 31 48.	+ 13 21	216	13.3	GALAXY SB:0
MCG+02-32-141	12 31 48.	+ 13 21	120	13.3	GALAXY
REA 55	12 31 48.	+ 16 20	54		DWARF GALAXY
TON-N 0085	12 31 48.	+ 25 07		15.6	BLUE STAR
IC 3514	12 31 48.	+ 26 58 29.			TWO STARS
MCG+05-30-022	12 31 48.	+ 27 43	90	15.5	GALAXY
LB 06588	12 31 48.	+ 28 47		19.7	FAINT BLUE STAR
ZWG 244.029	12 31 48.	+ 48 00		15.7	GALAXY
MCG+08-23-057	12 31 48.	+ 48 01	48	16.	GALAXY
ZC 1231.8+4810	12 31 48.	+ 48 10	6050		CLUSTER OF GALAXIES
UGC 07730	12 31 48.	+ 64 49	96	16.0	GALAXY DWRF SP
MCG+11-15-063	12 31 48.	+ 64 49	51	16.	GALAXY
REIN 3.247	12 31 48.05	+ 08 28 26.1			NEBULA
REIN 6.101	12 31 48.07	+ 08 28 23.9			NEBULA
REIN 3.248	12 31 48.47	+ 08 26 16.0			NEBULA
IC 3516	12 31 48.6	+ 27 43 37.			GALAXY Sbc
KN 16.030	12 31 48.7	+ 27 43 39.			NEBULA
REIN 3.249	12 31 48.79	+ 08 29 23.2			NEBULA
KEEL 483	12 31 48.9	+ 03 22 42.		18.	SPIRAL NEBULA
KEEL 482	12 31 48.9	+ 03 23 06.		17.	NEBULA
RNGC 4533	12 31 49.	+ 02 36		14.5	GALAXY
AMES 1518	12 31 49.	+ 16 44 02.	12	17.5	NEBULA
ARC 1562	12 31 49.	+ 41 28		17.5	RICH CLUSTER OF GALAXIES
REIN 3.250	12 31 49.20	+ 07 48 35.3			NEBULA
REIN 3.251	12 31 49.30	+ 08 30 58.9			NEBULA
RNGC 4532	12 31 50.	+ 06 45		12.5	GALAXY
REIZ 2543	12 31 50.	+ 08 28	450	11.4	GALAXY
HOLM 420A	12 31 50.	+ 08 28	360	11.6	PART OF MULTIPLE GALAXY
RNGC 4535	12 31 50.	+ 08 29		11.0	GALAXY
AMES 1519	12 31 50.	+ 16 46 20.	16	17.8	NEBULA
A3 454	12 31 50.2	+ 14 23 25.		17.3	FAINT BLUE OBJECT
REIN 3.252	12 31 50.73	+ 08 29 04.8			NEBULA
REIZ 2544	12 31 51.	+ 02 28	420	12.0	GALAXY
AMES 1521	12 31 51.	+ 11 19 02.	18	17.5	NEBULA
AMES 1520	12 31 51.	+ 17 12 44.	20	17.6	NEBULA
KN 16.031	12 31 51.2	+ 27 38 39.			NEBULA
AMES 1522	12 31 52.	+ 12 19 56.	18	17.4	NEBULA
KN 16.032	12 31 52.3	+ 28 16 46.			NEBULA
KN 16.033	12 31 52.5	+ 28 26 52.			NEBULA
RMB 198	12 31 54.	+ 10 28			FAINT BLUE OBJECT
KARA.68 157	12 31 54.	+ 12 00	27		DWARF GALAXY
KARA.68 158	12 31 54.	+ 13 10	47		DWARF GALAXY
RMB 083	12 31 54.	+ 13 33			FAINT BLUE OBJECT
AMES 1523	12 31 54.	+ 13 52 56.	20	17.8	NEBULA
ZC 1231.9+1530	12 31 54.	+ 15 30	1880		CLUSTER OF GALAXIES
LB 06590	12 31 54.	+ 31 03		19.7	FAINT BLUE STAR
LB 06589	12 31 54.	+ 32 08		19.4	FAINT BLUE STAR
ZC 1231.9+3732	12 31 54.	+ 37 32	4970		CLUSTER OF GALAXIES
TON-N 0086	12 31 54.	+ 37 55		14.7	BLUE STAR
ZC 1231.9+4722	12 31 54.	+ 47 22	1010		CLUSTER OF GALAXIES
ZC 1231.9+6435	12 31 54.	+ 64 35	1680		CLUSTER OF GALAXIES
ZC 1231.9+6814	12 31 54.	+ 68 14	940		CLUSTER OF GALAXIES
7ZW 471	12 31 54.	+ 77 59			COMPACT GALAXY
ZWG 352.035	12 31 54.	+ 77 59		15.5	GALAXY
ZWG 014.067	12 31 54.	- 03 00		15.7	GALAXY
A3 455	12 31 54.8	+ 15 33 20.		12.4	FAINT BLUE OBJECT
RNGC 4536	12 31 55.	+ 02 28		11.0	GALAXY
SHB 210	12 31 55.1	+ 24 44 42.		17.5	QUASI-STELLAR OBJECT
REIN 3.253	12 31 55.20	+ 08 27 30.7			NEBULA
AMES 1524	12 31 56.	+ 16 33 56.	36	17.7	NEBULA
BC 4C24.27	12 31 56.13	+ 24 48 16.9		17.5	QUASI-STELLAR OBJECT
KN 16.034	12 31 57.5	+ 27 16 28.			NEBULA
A3 458	12 31 57.7	+ 13 33 32.		17.7	FAINT BLUE OBJECT
REIN 3.254	12 31 57.94	+ 08 27 36.2			NEBULA
IC 3517	12 31 59.	+ 09 25 47.			NONSTELLAR OBJECT
REIN 3.255	12 31 59.00	+ 09 25 48.0			NEBULA
KN 16.035	12 31 59.4	+ 28 59 40.			NEBULA
REIN 3.256	12 31 59.84	+ 08 27 45.0			NEBULA
VDB.66G 137	12 32	+ 06 34	70		DWARF GALAXY
VDB.66G 136	12 32	+ 15 29	70		DWARF GALAXY
ZWG 014.068	12 32 00.	+ 02 28		12.3	GALAXY
UGC 07732	12 32 00.	+ 02 28	462	12.3	GALAXY Sb/Sc
MCG+00-32-023	12 32 00.	+ 02 28	420	12.3	GALAXY
ZWG 042.160	12 32 00.	+ 06 17		15.4	GALAXY
ZWG 070.176	12 32 00.	+ 09 26		15.3	GALAXY
UGC 07733	12 32 00.	+ 09 26	90	15.3	GALAXY
MCG+02-32-143	12 32 00.	+ 09 26	60	15.3	GALAXY
ZWG 070.177	12 32 00.	+ 09 54		15.3	GALAXY
UGC 07734	12 32 00.	+ 09 54	78	15.3	GALAXY
ZWG 070.178	12 32 00.	+ 13 47		15.4	GALAXY
ZWG 099.092	12 32 00.	+ 18 28		13.5	GALAXY
UGC 07735	12 32 00.	+ 18 28	192	13.5	GALAXY SB?0-a
MCG+03-32-071	12 32 00.	+ 18 29	228	13.5	GALAXY
LB 06591	12 32 00.	+ 26 45		17.4	FAINT BLUE STAR
LB 06593	12 32 00.	+ 28 30		18.0	FAINT BLUE STAR
LB 06592	12 32 00.	+ 31 34		19.9	FAINT BLUE STAR
ZWG 270.010	12 32 00.	+ 50 58		15.7	GALAXY
MCG+14-06-007	12 32 00.	+ 81 45	15	16.	GALAXY
ZC 1232.0+8222	12 32 00.	+ 82 22	3020		CLUSTER OF GALAXIES
UGC 07731	12 32 00.	+ 82 52	27	14.	GALAXY CMPT?
ACK 301+01.1	12 32 00.	- 61 22			PLANETARY NEBULA
KEEL 484	12 32 00.2	+ 02 41 31.		18.	NEBULA
HN 0965	12 32 01.	+ 13 46			NEBULA
HN 0964	12 32 01.	+ 15 51	12		NEBULA
IC 3519	12 32 01.	+ 15 51			MAY NOT EXIST
AMES 1526	12 32 01.	+ 16 18 44.	26	16.3	NEBULA
AMES 1525	12 32 01.	+ 17 04 20.	28	18.0	NEBULA
KN 16.036	12 32 01.1	+ 28 22 52.			NEBULA
KEEL 485	12 32 01.9	+ 02 50 41.		17.	SPIRAL NEBULA
HN 0966	12 32 02.	+ 09 53	48		NEBULA
IC 3518	12 32 02.	+ 09 53			NONSTELLAR OBJECT
IC 3520	12 32 02.	+ 13 46			NONSTELLAR OBJECT
AMES 1529	12 32 03.	+ 10 49 14.	31	17.4	NEBULA
AMES 1528	12 32 03.	+ 16 34 56.	17	17.8	NEBULA
AMES 1527	12 32 03.	+ 16 47 44.	19	18.0	NEBULA

OBJECT NAME	RIGHT ASCEN.	DECLINATION	DIAM.	MAGN.	TYPE OF OBJECT
KEEL 486	12 32 03.8	+ 02 11 16.		17.	NEBULA
AMES 1531	12 32 05.	+ 16 36 43.	29	17.6	NEBULA
AMES 1530	12 32 05.	+ 17 16 07.	13	18.1	NEBULA
REIZ 2551	12 32 05.	+ 26 54	6	15.6	GALAXY
ZWG 042.161	12 32 06.	+ 03 36		14.8	GALAXY
MCG+01-32-105	12 32 06.	+ 03 36	30	14.8	GALAXY
REA 81	12 32 06.	+ 06 35			DWARF GALAXY
ZWG 042.162	12 32 06.	+ 07 26		14.2	GALAXY
UGC 07736	12 32 06.	+ 07 26	78	14.2	GALAXY IRR
MCG+01-32-106	12 32 06.	+ 07 26	60	14.2	GALAXY
IC 3521	12 32 06.	+ 07 26			GALAXY IN VIRGO CLSTR
REA 56	12 32 06.	+ 12 01	42		DWARF GALAXY
UGC 07737	12 32 06.	+ 15 29	72	17.	GALAXY DWRF IR
MCG+03-32-072	12 32 06.	+ 15 29	90	17.	GALAXY
AMES 1532	12 32 06.	+ 16 48 25.	22	16.6	NEBULA
ZC 1232.1+2319	12 32 06.	+ 23 19	940		CLUSTER OF GALAXIES
LB 06596	12 32 06.	+ 29 30		17.9	FAINT BLUE STAR
LB 06595	12 32 06.	+ 29 42		18.5	FAINT BLUE STAR
LB 06594	12 32 06.	+ 31 00		19.5	FAINT BLUE STAR
UGC 07738	12 32 06.	+ 42 44	78	16.0	GALAXY Sc
MCG+11-15-064	12 32 06.	+ 63 47	132	13.5	GALAXY
A3 459	12 32 06.5	+ 15 24 25.		16.7	FAINT BLUE OBJECT
RNGC 4538	12 32 07.	+ 03 36		15.0	GALAXY
AMES 1535	12 32 07.	+ 11 42 07.	29	17.7	NEBULA
HW 0968	12 32 07.	+ 14 17			NEBULA
IC 3523	12 32 07.	+ 14 17			NONSTELLAR OBJECT
HW 0967	12 32 07.	+ 15 28	36		NEBULA
REIZ 2550	12 32 07.	+ 18 29	120	13.7	GALAXY
KN 16.037	12 32 07.9	+ 27 41 58.			NEBULA
REIZ 2549	12 32 08.	+ 07 26	42	14.1	GALAXY
AMES 1534	12 32 08.	+ 12 48 55.	32	17.5	NEBULA
AMES 1533	12 32 08.	+ 16 55 19.	22	17.8	NEBULA
MAI 075	12 32 08.	+ 64 57	40		DWARF SPHEROIDAL GALAXY
RNGC 4539	12 32 09.	+ 18 29		13.5	GALAXY
AMES 1539	12 32 10.	+ 06 34 43.	78	16.2	NEBULA
AMES 1536	12 32 10.	+ 18 42 49.	25	17.8	NEBULA
SC 1229-0033.8	12 32 10.	- 00 50 21.	12		NEBULA
IC 3524	12 32 11.	+ 14 30 17.			TWO STARS
KN 16.037A	12 32 11.7	+ 25 57 40.			NEBULA
ZWG 042.163	12 32 12.	+ 06 34		15.3	GALAXY
UGC 07739	12 32 12.	+ 06 34	84	15.3	GALAXY DWRF IR
MCG+01-32-107	12 32 12.	+ 06 34	60	15.3	GALAXY
ZWG 070.179	12 32 12.	+ 09 17		15.4	GALAXY
SN 1960B	12 32 12.	+ 09 17		16.0	SUPERNOVA
MCG+02-32-144	12 32 12.	+ 09 17	36	15.4	GALAXY
AMES 1538	12 32 12.	+ 13 57 31.	26	18.0	NEBULA
KARA.68 159	12 32 12.	+ 14 28	34		DWARF GALAXY
ZZW 066	12 32 12.	+ 14 50			COMPACT GALAXY
AMES 1537	12 32 12.	+ 16 49 07.	29	17.9	NEBULA
ZC 1232.2+2157	12 32 12.	+ 21 57	870		CLUSTER OF GALAXIES
ZC 1232.2+2556	12 32 12.	+ 25 56	1080		CLUSTER OF GALAXIES
IC 3526	12 32 12.	+ 25 57 35.			NONSTELLAR OBJECT
LB 11252	12 32 12.	+ 27 15		12.3	FAINT BLUE STAR
LB 06598	12 32 12.	+ 27 40		19.6	FAINT BLUE STAR
LB 06597	12 32 12.	+ 31 37		19.5	FAINT BLUE STAR
ZC 1232.2+4220	12 32 12.	+ 42 20	1210		CLUSTER OF GALAXIES
UGC 07740	12 32 12.	+ 49 38	60	16.0	GALAXY S
MCG+09-21-018	12 32 12.	+ 50 50	36	16.	GALAXY
ZC 1232.2+5237	12 32 12.	+ 52 37	940		CLUSTER OF GALAXIES
UGC 07741	12 32 12.	+ 70 08	66	16.0	GALAXY Sb
MRSL 301+01/1	12 32 12.	- 61 32	9000		HII REGION
KN 16.038	12 32 12.6	+ 28 27 52.			NEBULA
AMES 1540	12 32 13.	+ 14 49 13.	12	17.3	NEBULA
KN 16.038A	12 32 13.7	+ 26 25 46.			NEBULA
KEEL 487	12 32 13.9	+ 02 17 08.		17.	SPIRAL NEBULA
HW 0969	12 32 14.	+ 10 26			NEBULA
IC 3525	12 32 14.	+ 10 26			NONSTELLAR OBJECT
AMES 1542	12 32 14.	+ 16 58 42.	17	18.0	NEBULA
AMES 1541	12 32 14.	+ 17 03 43.	22	17.6	NEBULA
IC 3527	12 32 14.	+ 26 25 59.			TWO STARS
KN 16.039	12 32 14.5	+ 27 55 10.			NEBULA
IC 3522	12 32 14.7	+ 15 29 46.			GALAXY IN VIRGO CLSTR
MCG+03-32-073	12 32 15.	+ 18 06	78	15.2	GALAXY
KEEL 488	12 32 15.9	+ 02 39 01.		17.	NEBULA
AMES 1543	12 32 16.	+ 13 31 25.	18	16.5	NEBULA
A3 460	12 32 16.1	+ 15 27 46.		16.4	FAINT BLUE OBJECT
AMES 1545	12 32 17.	+ 16 54 49.	17	17.6	NEBULA
AMES 1544	12 32 17.	+ 17 03 55.	23	18.0	NEBULA
RNGC 4545	12 32 17.	+ 63 48		13.0	GALAXY
A3 461	12 32 17.6	+ 14 35 15.		17.0	FAINT BLUE OBJECT
RMB 123	12 32 18.	+ 12 15			FAINT BLUE OBJECT
MCG+03-32-074	12 32 18.	+ 15 49	126	12.5	GALAXY
ZWG 099.093	12 32 18.	+ 15 50		12.5	GALAXY
SCH 56	12 32 18.	+ 15 50		12.5	PECULIAR GALAXY
UGC 07742	12 32 18.	+ 15 50	138	12.5	GALAXY
AMES 1546	12 32 18.	+ 17 29 07.	24	17.2	NEBULA
ZWG 099.094	12 32 18.	+ 18 05		15.2	GALAXY
LB 06600	12 32 18.	+ 27 00		19.6	FAINT BLUE STAR
UGC 07743	12 32 18.	+ 27 55	60	17.	GALAXY Sc
LB 06599	12 32 18.	+ 32 14		19.7	FAINT BLUE STAR
ZWG 244.030	12 32 18.	+ 44 40		15.2	GALAXY
MCG+08-23-059	12 32 18.	+ 47 14 30.	60	16.	GALAXY
UGC 07744	12 32 18.	+ 47 15	60	16.5	GALAXY Sc
MCG+08-23-060	12 32 18.	+ 49 37	60	15.	GALAXY
MCG+09-21-019	12 32 18.	+ 50 55	24	16.	GALAXY
MCG+10-18-068	12 32 18.	+ 59 13	39	15.	GALAXY
UGC 07745	12 32 18.	+ 72 58	66	17.	GALAXY DWARF
ZC 1232.3-0033	12 32 18.	- 00 33	1280		CLUSTER OF GALAXIES
MCG-06-28-005	12 32 18.	- 36 34	24	15.5	GALAXY
KEEL 489	12 32 18.7	+ 02 09 31.		17.	NEBULA
AMES 1548	12 32 19.	+ 15 15 31.	29	17.0	NEBULA
HW 0970	12 32 19.	+ 15 49		14.	NEBULA
REIZ 2552	12 32 19.	+ 15 50	60	13.0	GALAXY
HOLM 421A	12 32 19.	+ 15 50	72	12.4	PART OF MULTIPLE GALAXY
AMES 1547	12 32 19.	+ 16 19 31.	19	17.0	NEBULA
KN 16.040A	12 32 19.9	+ 28 45 52.			NEBULA
KN 16.040	12 32 19.9	+ 28 46 28.			NEBULA
KN 16.040B	12 32 20.9	+ 25 58 28.			NEBULA
ABC 1564	12 32 21.	+ 02 08		16.6	RICH CLUSTER OF GALAXIES
RNGC 4540	12 32 21.	+ 15 50		13.0	GALAXY
AMES 1549	12 32 21.	+ 17 14 43.	12	17.9	NEBULA
IC 3529	12 32 21.	+ 25 58 29.			NONSTELLAR OBJECT
REIZ 2560	12 32 21.	+ 63 47	90	13.5	GALAXY
AMES 1550	12 32 22.	+ 15 26 31.	16	17.4	NEBULA
REIZ 2557	12 32 22.	+ 51 05	18	14.7	GALAXY
KN 16.041	12 32 22.9	+ 27 52 40.			NEBULA
SC 1229+0023.9	12 32 23.	+ 00 07 21.	12		NEBULA
AMES 1551	12 32 23.	+ 12 54 37.	16	17.8	NEBULA
REIZ 2554	12 32 23.	+ 25 57	9	15.0	GALAXY
REIZ 2555	12 32 23.	+ 26 34	9	15.0	GALAXY
RMB 018	12 32 24.	+ 15 50			FAINT BLUE OBJECT
ZWG 099.095	12 32 24.	+ 15 51		15.2	GALAXY
HOLM 421B	12 32 24.	+ 15 51	24	14.4	PART OF MULTIPLE GALAXY
MCG+03-32-074A	12 32 24.	+ 15 51	24	15.2	GALAXY
IC 3530	12 32 24.	+ 18 04			NONSTELLAR OBJECT
LB 06601	12 32 24.	+ 26 28		19.8	FAINT BLUE STAR
LB 11253	12 32 24.	+ 32 37		18.4	FAINT BLUE STAR
MCG+07-26-028	12 32 24.	+ 39 51	27	15.	GALAXY
ZWG 244.031	12 32 24.	+ 48 01		15.3	GALAXY
MCG+08-23-061	12 32 24.	+ 48 02 30.	12	15.	GALAXY
ZWG 270.011	12 32 24.	+ 51 06		15.5	GALAXY
UGC 07746	12 32 24.	+ 51 06	72	15.5	GALAXY S
ZWG 315.047	12 32 24.	+ 63 48		13.1	GALAXY
UGC 07747	12 32 24.	+ 63 48	174	13.1	GALAXY SBc
SN 1940D	12 32 24.	+ 63 48		15.0	SUPERNOVA
MCG+12-12-009	12 32 24.	+ 70 06	66	16.	GALAXY
ZWG 014.069	12 32 24.	- 03 24		15.7	GALAXY
MCG-01-32-025	12 32 24.	- 05 12 30.	48	15.	GALAXY
KEEL 490	12 32 24.5	+ 03 07 56.		17.	NEBULA
A3 462	12 32 24.9	+ 15 50 26.		14.0	UV CONCENTRATION
IC 3528	12 32 24.9	+ 15 50 27.			GALAXY
AMES 1555	12 32 25.	+ 11 47 37.	17	17.5	NEBULA
AMES 1553	12 32 25.	+ 16 04 43.	24	17.9	NEBULA
AMES 1552	12 32 25.	+ 16 53 19.	12	18.0	NEBULA
HW 0971	12 32 25.	+ 18 04	18		NEBULA
REIZ 2553	12 32 25.	+ 18 05	24	15.0	GALAXY
LB 02391	12 32 25.	+ 50 11 24.		14.5	FAINT BLUE STAR
RNGC 4542	12 32 25.	+ 51 06		15.5	GALAXY
KN 16.042	12 32 25.8	+ 27 13 28.			NEBULA
AMES 1554	12 32 26.	+ 16 05 19.	28	17.7	NEBULA
HZ 29	12 32 27.	+ 37 55		13.6	DECIDEDLY BLUE STAR
A3 463	12 32 27.7	+ 13 35 35.		16.5	FAINT BLUE OBJECT
KN 16.043	12 32 27.8	+ 26 54 10.			NEBULA
IC 3531	12 32 27.9	+ 26 54 05.			GALAXY E1
KEEL 491	12 32 28.6	+ 02 04 27.		17.	NEBULA
SC 1229+0035.0	12 32 29.	+ 00 18 27.	18		NEBULA
IC 3532	12 32 29.	+ 26 09 24.			NONSTELLAR OBJECT
KEEL 492	12 32 29.8	+ 02 43 49.		18.	NEBULA
ZWG 042.164	12 32 30.	+ 05 42		15.7	GALAXY
ZWG 042.165	12 32 30.	+ 06 09		15.4	GALAXY
ZWG 129.007	12 32 30.	+ 26 03		15.7	GALAXY
KARA.73B 0537	12 32 30.	+ 26 03	18	15.7	ISOLATED GALAXY E
LB 06604	12 32 30.	+ 27 16		18.1	FAINT BLUE STAR
LB 06603	12 32 30.	+ 28 54		18.4	FAINT BLUE STAR
LB 06602	12 32 30.	+ 29 24		19.9	FAINT BLUE STAR
1ZW 039	12 32 30.	+ 48 01			COMPACT GALAXY
LB 02392	12 32 30.	+ 50 03 24.		14.2	FAINT BLUE STAR
MCG+12-12-010	12 32 30.	+ 68 36	45	16.	GALAXY
ZWG 335.014	12 32 30.	+ 68 38		15.6	GALAXY
UGC 07748	12 32 30.	+ 68 38	84	15.6	GALAXY Sc-IRR
ZWG 014.070	12 32 30.	- 03 19		15.6	GALAXY
KEEL 493	12 32 30.4	+ 02 03 40.		17.	NEBULA
KN 16.044	12 32 30.6	+ 26 11 40.			NEBULA
AMES 1559	12 32 31.	+ 06 09 07.	55	16.8	NEBULA
AMES 1556	12 32 31.	+ 16 03 37.	23	16.6	NEBULA
KN 16.045	12 32 31.1	+ 27 09 40.			NEBULA
KEEL 494	12 32 31.8	+ 02 02 41.		18.	NEBULA
AMES 1558	12 32 32.	+ 12 13 01.	32	17.5	NEBULA
IC 3533	12 32 32.	+ 26 03 18.			NONSTELLAR OBJECT
A3 464	12 32 32.0	+ 16 27 01.		16.6	FAINT BLUE OBJECT
KEEL 495	12 32 32.5	+ 26 03 21.		15.	NEBULA
KN 16.047	12 32 32.5	+ 28 10 16.			NEBULA
KN 16.046	12 32 32.6	+ 26 03 16.			NEBULA
AMES 1561	12 32 33.	+ 11 53 13.	36	17.0	NEBULA
AMES 1557	12 32 33.	+ 16 50 37.	26	17.4	NEBULA
RNGC 4549	12 32 33.	+ 59 12		15.5	GALAXY
RNGC 4547	12 32 33.	+ 59 12		15.5	GALAXY
REIZ 2556	12 32 34.	+ 00 02	48	13.8	GALAXY
AMES 1560	12 32 34.	+ 13 03 25.	18	17.2	NEBULA
A3 466	12 32 34.7	+ 14 27 20.		16.4	FAINT BLUE OBJECT
KN 16.048	12 32 34.7	+ 27 21 16.			NEBULA
REIZ 2558	12 32 35.	+ 26 03	24	14.6	GALAXY
REIZ 2559	12 32 35.	+ 26 08	12	15.0	GALAXY
KN 16.049	12 32 35.5	+ 25 52 04.			NEBULA
ZWG 014.071	12 32 36.	+ 00 03		14.0	GALAXY
UGC 07749	12 32 36.	+ 00 03	96	14.0	GALAXY Sb
MCG+00-32-024	12 32 36.	+ 00 03	102	14.0	GALAXY
AMES 1563	12 32 36.	+ 17 33 01.	13	17.8	NEBULA
LB 06610	12 32 36.	+ 27 10		19.8	FAINT BLUE STAR
LB 06609	12 32 36.	+ 27 58		19.8	FAINT BLUE STAR
ZWG 159.018	12 32 36.	+ 30 02		15.1	GALAXY
UGC 07750	12 32 36.	+ 30 02	84	15.1	GALAXY S
LB 06608	12 32 36.	+ 30 30		19.2	FAINT BLUE STAR
LB 06607	12 32 36.	+ 31 21		19.7	FAINT BLUE STAR
LB 06606	12 32 36.	+ 32 09		17.0	FAINT BLUE STAR
LB 06605	12 32 36.	+ 32 26		19.9	FAINT BLUE STAR
ZC 1232.6+4354	12 32 36.	+ 43 54	870		CLUSTER OF GALAXIES
MCG+09-21-020	12 32 36.	+ 53 47 30.	30	16.	GALAXY
ZWG 293.030	12 32 36.	+ 59 12		15.3	GALAXY
AMES 1566	12 32 37.	+ 16 48 43.	23	17.5	NEBULA
AMES 1565	12 32 37.	+ 16 50 37.	20	18.0	NEBULA
AMES 1564	12 32 37.	+ 17 33 01.	16	17.7	NEBULA
AMES 1562	12 32 37.	+ 19 25 13.	32	16.7	NEBULA
RNGC 4541	12 32 37.	- 00 03		14.0	GALAXY
REIF 4.085	12 32 37.11	+ 00 03 12.9			NEBULA
KEEL 496	12 32 37.5	+ 02 03 34.		18.	NEBULA
KN 16.050	12 32 38.7	+ 27 11 04.			NEBULA
AMES 1569	12 32 39.	+ 16 26 31.	25	18.0	NEBULA
AMES 1568	12 32 39.	+ 16 51 07.	19	17.8	NEBULA
AMES 1567	12 32 39.	+ 17 42 43.	17	17.1	NEBULA
REIZ 2562	12 32 39.	+ 34 23	36	15.1	GALAXY
KEEL 497	12 32 39.7	+ 03 16 33.		17.	NEBULA
REIZ 2561	12 32 40.	+ 30 01	60	14.7	GALAXY
KN 16.051	12 32 40.6	+ 28 19 58.			NEBULA
KN 16.052	12 32 41.2	+ 29 00 46.			NEBULA
KN 16.051A	12 32 41.7	+ 26 00 28.			NEBULA
KEEL 500	12 32 41.9	+ 26 05 51.		17.	NEBULA
ZWG 042.166	12 32 42.	+ 06 49		15.2	GALAXY
MCG+01-32-108	12 32 42.	+ 06 49	48	15.2	GALAXY
RMB 197	12 32 42.	+ 10 46			FAINT BLUE OBJECT
AMES 1572	12 32 42.	+ 13 06 37.	19	17.7	NEBULA
KARA.68 160	12 32 42.	+ 14 13	34		DWARF GALAXY
REA 17	12 32 42.	+ 14 16			DWARF GALAXY
MCG+02-32-145	12 32 42.	+ 14 16	48	18.	GALAXY
AMES 1571	12 32 42.	+ 16 39 13.	22	18.0	NEBULA
AMES 1570	12 32 42.	+ 16 40 25.	24	18.0	NEBULA

OBJECT NAME	RIGHT ASCEN.	DECLINATION	DIAM.	MAGN.	TYPE OF OBJECT
ZC 1232.7+1653	12 32 42.	+ 16 53	3090		CLUSTER OF GALAXIES
IC 3535	12 32 42.	+ 26 00 24.			NONSTELLAR OBJECT
MCG+05-30-024	12 32 42.	+ 26 48	30	15.5	GALAXY
LB 06613	12 32 42.	+ 29 33		19.6	FAINT BLUE STAR
LB 11254	12 32 42.	+ 29 53		17.3	FAINT BLUE STAR
MCG+05-30-023	12 32 42.	+ 30 02	84	15.1	GALAXY
LB 06612	12 32 42.	+ 30 09		19.1	FAINT BLUE STAR
LB 06611	12 32 42.	+ 31 45		19.8	FAINT BLUE STAR
ZWG 188.009	12 32 42.	+ 34 24		15.2	GALAXY
UGC 07751	12 32 42.	+ 41 20	66	16.5	GALAXY IRR
MCG+09-21-021	12 32 42.	+ 51 03	60	14.7	GALAXY
ZC 1232.7+7005	12 32 42.	+ 70 05	1480		CLUSTER OF GALAXIES
KN 16.053	12 32 42.5	+ 26 46 10.			NEBULA
HN 0972	12 32 43.	+ 15 16	42		NEBULA
IC 3534	12 32 43.	+ 15 16			NONSTELLAR OBJECT
KN 16.054	12 32 43.8	+ 26 48 40.			NEBULA
IC 3536	12 32 43.9	+ 26 48 31.			GALAXY SBb
AMES 1573	12 32 44.	+ 17 01 49.	22	17.4	NEBULA
KEEL 498	12 32 44.3	+ 03 14 03.		17.	NEBULA
AMES 1574	12 32 45.	+ 17 03 25.	12	17.6	NEBULA
MCG-01-32-026	12 32 45.	- 04 36	60	15.5	GALAXY
KEEL 499	12 32 45.7	+ 02 08 34.		18.	NEBULA
KN 16.055	12 32 45.7	+ 25 54 40.			NEBULA
A3 468	12 32 45.9	+ 14 25 36.		16.9	FAINT BLUE OBJECT
AMES 1576	12 32 46.	+ 06 49 19.	38	16.0	NEBULA
IC 3538	12 32 46.	+ 26 30 42.			SINGLE STAR
ARC 1566	12 32 46.	+ 64 40		16.9	RICH CLUSTER OF GALAXIES
SC 1230+0022.0	12 32 47.	+ 00 05 27.	18		NEBULA
REIZ 2563	12 32 47.	+ 26 00	12	14.8	GALAXY
REIZ 2564	12 32 47.	+ 26 20	30	14.2	GALAXY
REIZ 2565	12 32 47.	+ 26 48	36	15.0	GALAXY
SN 1954G	12 32 47.	- 18 58		16.6	SUPERNOVA
ZC 1232.8+0206	12 32 48.	+ 02 06	2020		CLUSTER OF GALAXIES
ZWG 042.167	12 32 48.	+ 06 23		14.6	GALAXY
MCG+01-32-109	12 32 48.	+ 06 23	24	14.6	GALAXY
AMES 1575	12 32 48.	+ 17 15 43.	14	17.7	NEBULA
ZC 1232.8+1931	12 32 48.	+ 19 31	870		CLUSTER OF GALAXIES
LB 06618	12 32 48.	+ 26 28		17.6	FAINT BLUE STAR
LB 06617	12 32 48.	+ 27 22		19.2	FAINT BLUE STAR
LB 06616	12 32 48.	+ 29 40		17.0	FAINT BLUE STAR
LB 06615	12 32 48.	+ 30 52		19.5	FAINT BLUE STAR
LB 06614	12 32 48.	+ 32 23		18.8	FAINT BLUE STAR
MCG+07-26-029	12 32 48.	+ 41 19 30.	48	15.	GALAXY
ZC 1232.8+4130	12 32 48.	+ 41 30	2020		CLUSTER OF GALAXIES
ZC 1232.8+4414	12 32 48.	+ 44 14	940		CLUSTER OF GALAXIES
MCG+09-21-022	12 32 48.	+ 51 06	54	16.	GALAXY
MCG+10-18-070	12 32 48.	+ 59 11	18	16.	GALAXY
MCG+10-18-069	12 32 48.	+ 59 11	18	15.	GALAXY
MRK 217	12 32 48.	+ 66 37	8	16.	GALAXY WITH UV CONTINUUM
7ZW 472	12 32 48.	+ 66 39			COMPACT GALAXY
MRSL 301+01/2	12 32 48.	- 61 35	60		HII REGION
KEEL 501	12 32 48.1	+ 02 09 56.		18.	NEBULA
AMES 1578	12 32 49.	+ 10 58 13.	17	17.5	NEBULA
ARC 1565	12 32 49.	+ 41 39		17.6	RICH CLUSTER OF GALAXIES
RNGC 4537	12 32 49.	+ 51 06		16.0	GALAXY
KN 16.056	12 32 49.1	+ 29 05 34.			NEBULA
RNGC 4543	12 32 50.	+ 06 23		14.5	GALAXY
IC 3537	12 32 50.	+ 07 55 42.			TWO STARS
AMES 1577	12 32 50.	+ 17 38 01.	24	17.0	NEBULA
IC 3539	12 32 50.	+ 24 15 30.			NONSTELLAR OBJECT
MCG+06-28-011	12 32 51.	+ 34 25	36	14.5	GALAXY
LB 02393	12 32 51.	+ 62 20 36.		16.6	FAINT BLUE STAR
MCG-01-32-027	12 32 51.	- 03 31	108	11.	GALAXY
RNGC 4507	12 32 51.	- 39 38		13.0	GALAXY
KN 16.057	12 32 51.0	+ 26 18 10.			NEBULA
AMES 1579	12 32 52.	+ 17 02 19.	24	17.9	NEBULA
LB 02394	12 32 52.	+ 59 25 54.		16.2	FAINT BLUE STAR
KN 16.058	12 32 52.7	+ 27 50 22.			NEBULA
AMES 1580	12 32 53.	+ 16 35 49.	20	17.0	NEBULA
IC 3541	12 32 53.	+ 24 15 00.			NONSTELLAR OBJECT
AMES 1584	12 32 54.	+ 06 36 37.	52	16.6	NEBULA
SCH 57	12 32 54.	+ 12 30		12.5	PECULIAR GALAXY
ZWG 070.180	12 32 54.	+ 13 02		14.8	GALAXY
MCG+02-32-146	12 32 54.	+ 13 02	24	14.8	GALAXY
8ZW 1232+13.4	12 32 54.	+ 13 27		17.7	COMPACT GALAXY
MCG+03-32-075	12 32 54.	+ 14 45	300	11.5	GALAXY
ZWG 099.096	12 32 54.	+ 14 46		11.5	GALAXY
UGC 07753	12 32 54.	+ 14 46	390	11.5	GALAXY SBb
AMES 1582	12 32 54.	+ 15 25 43.	20	17.8	NEBULA
AMES 1581	12 32 54.	+ 16 52 07.	17	17.2	NEBULA
ZWG 159.019	12 32 54.	+ 29 47		14.9	GALAXY
UGC 07754	12 32 54.	+ 29 47	66	14.9	GALAXY S
MCG+05-30-025	12 32 54.	+ 29 47	66	14.9	GALAXY
LB 06619	12 32 54.	+ 30 15		19.3	FAINT BLUE STAR
7ZW 473	12 32 54.	+ 59 19			COMPACT GALAXY
ZWG 315.048	12 32 54.	+ 64 14		15.7	GALAXY
7ZW 474	12 32 54.	+ 72 14			COMPACT GALAXY
ZWG 014.072	12 32 54.	- 02 03		15.4	GALAXY
UGC 07752	12 32 54.	- 02 03	102	15.4	GALAXY S
KARA.68 161	12 32 54.	- 12 49	40		DWARF GALAXY
SA 2	12 32 54.	- 39 31	72	12.9	GALAXY
IC 3540	12 32 55.	+ 13 01 42.			NONSTELLAR OBJECT
REIZ 2567	12 32 55.	+ 14 46	270	12.4	GALAXY
AMES 1583	12 32 55.	+ 16 08 01.	20	17.8	NEBULA
RNGC 4546	12 32 55.	- 03 31		12.0	GALAXY
REIN 4.086	12 32 55.31	- 03 31 03.8			NEBULA
RNGC 4548	12 32 56.	+ 14 46		11.5	GALAXY
SC 1230-0147.3	12 32 56.	- 02 03 51.	12		NEBULA
KN 16.059	12 32 56.2	+ 29 45 52.			NEBULA
A3 470	12 32 56.9	+ 16 38 19.		17.8	FAINT BLUE OBJECT
AMES 1587	12 32 58.	+ 11 46 13.	18	17.7	NEBULA
AMES 1585	12 32 58.	+ 16 24 25.	24	18.1	NEBULA
REIZ 2566	12 32 58.	- 03 32	78	11.5	GALAXY
AMES 1586	12 32 59.	+ 16 55 37.	18	17.9	NEBULA
REIZ 2569	12 32 59.	+ 24 12	18	15.4	GALAXY
REIZ 2570	12 32 59.	+ 24 13	18	14.9	GALAXY
REIZ 2571	12 32 59.	+ 26 25	12	15.3	GALAXY
REIN 3.257	12 32 59.19	+ 12 29 42.9			NEBULA
BC PKS1233-24	12 32 59.4	- 24 55 46.		17.25	QUASI-STELLAR OBJECT
STB 211	12 32 59.4	- 24 55 46.		17.2	QUASI-STELLAR OBJECT
VDB.66G 138	12 33	+ 06 54	70		DWARF GALAXY
UGC 07755	12 33 00.	+ 00 04	60	16.5	GALAXY S
ZWG 042.168	12 33 00.	+ 03 19		14.4	GALAXY
UGC 07756	12 33 00.	+ 03 19	144	14.4	GALAXY S
MCG+01-32-110	12 33 00.	+ 03 19	72	14.4	GALAXY
ZWG 042.169	12 33 00.	+ 06 36		15.4	GALAXY
ZWG 070.181	12 33 00.	+ 11 58		15.6	GALAXY
RMB 109	12 33 00.	+ 12 30			FAINT BLUE OBJECT
ZWG 070.182	12 33 00.	+ 12 30		12.5	GALAXY
WEED 8	12 33 00.	+ 12 30		18.5	VERY BLUE STELLAR OBJECT
UGC 07757	12 33 00.	+ 12 30	198	12.5	GALAXY S0
MCG+02-32-147	12 33 00.	+ 12 30	120	12.5	GALAXY
ZC 1233.0+1334	12 33 00.	+ 13 34	1410		CLUSTER OF GALAXIES
RMB 081	12 33 00.	+ 13 49			FAINT BLUE OBJECT
AMES 1588	12 33 00.	+ 16 05 01.	29	17.4	NEBULA
ZWG 129.008	12 33 00.	+ 26 08		14.6	GALAXY
UGC 07758	12 33 00.	+ 26 08	144	14.6	GALAXY Sc
LB 06625	12 33 00.	+ 26 34		19.5	FAINT BLUE STAR
LB 06624	12 33 00.	+ 26 48		19.8	FAINT BLUE STAR
LB 06623	12 33 00.	+ 27 16		18.0	FAINT BLUE STAR
LB 06622	12 33 00.	+ 27 22		17.9	FAINT BLUE STAR
LB 06621	12 33 00.	+ 29 53		19.4	FAINT BLUE STAR
LB 11255	12 33 00.	+ 31 46		13.4	FAINT BLUE STAR
LB 06620	12 33 00.	+ 32 07		19.1	FAINT BLUE STAR
KARA.68 162	12 33 00.	+ 58 40	34		DWARF GALAXY
KEEL 502	12 33 00.6	+ 02 09 43.		18.	NEBULA
RNGC 4544	12 33 01.	+ 03 19		14.5	GALAXY
KEEL 503	12 33 01.7	+ 02 10 08.		17.	NEBULA
RNGC 4550	12 33 02.	+ 12 30		12.5	GALAXY
REIZ 2568	12 33 02.	+ 12 30	54	12.3	GALAXY
HOLM 422A	12 33 02.	+ 12 30	168	12.7	PART OF MULTIPLE GALAXY
AMES 1590	12 33 02.	+ 17 03 43.	34	17.2	NEBULA
AMES 1589	12 33 02.	+ 17 38 19.	17	17.4	NEBULA
A3 472	12 33 02.2	+ 13 56 20.		17.1	FAINT BLUE OBJECT
KEEL 504	12 33 02.3	+ 02 03 03.		16.	NEBULA
AMES 1592	12 33 03.	+ 11 24 49.	18	17.4	NEBULA
AMES 1591	12 33 03.	+ 17 44 01.	14	17.8	NEBULA
MCG-01-32-028	12 33 03.	- 07 37	240	14.	GALAXY
KEEL 505	12 33 03.7	+ 03 18 33.		15.	SPIRAL NEBULA
RNGC 4565A	12 33 04.	+ 26 08		14.5	GALAXY
RNGC 4562	12 33 04.	+ 26 08		14.5	GALAXY
AMES 1595	12 33 05.	+ 11 22 31.	16	17.6	NEBULA
AMES 1593	12 33 05.	+ 12 45 13.	29	17.3	NEBULA
REIZ 2573	12 33 05.	+ 26 07	78	13.8	GALAXY
KN 16.060	12 33 05.2	+ 27 49 35.			NEBULA
KEEL 506	12 33 05.5	+ 02 05 44.		18.	NEBULA
KN 16.062	12 33 05.6	+ 28 57 47.			NEBULA
KEEL 508	12 33 05.9	+ 26 07 32.		14.	NEBULA
REIN 3.258	12 33 05.90	+ 12 39 23.9			NEBULA
ZWG 014.073	12 33 06.	+ 00 03		15.1	GALAXY
SC 1230+0020.2	12 33 06.	+ 00 03 39.	30		NEBULA
ZWG 070.183	12 33 06.	+ 12 32		13.1	GALAXY
UGC 07759	12 33 06.	+ 12 32	108	13.1	GALAXY E
MCG+02-32-148	12 33 06.	+ 12 32	36	13.1	GALAXY
ZWG 070.184	12 33 06.	+ 12 50		11.1	GALAXY
UGC 07760	12 33 06.	+ 12 50	216	11.1	GALAXY E
MCG+02-32-149	12 33 06.	+ 12 50	66	11.1	GALAXY
RMB 089	12 33 06.	+ 13 17			FAINT BLUE OBJECT
AMES 1594	12 33 06.	+ 16 57 25.	16	17.7	NEBULA
ZC 1233.1+1929	12 33 06.	+ 19 29	270		CLUSTER OF GALAXIES
MCG+04-30-004	12 33 06.	+ 26 08	150	14.6	GALAXY
MCG+04-30-005	12 33 06.	+ 26 29	42	15.3	GALAXY
LB 06627	12 33 06.	+ 31 17		19.8	FAINT BLUE STAR
LB 06626	12 33 06.	+ 31 57		20.0	FAINT BLUE STAR
ZC 1233.1+4200	12 33 06.	+ 42 00	810		CLUSTER OF GALAXIES
MCG+08-23-062	12 33 06.	+ 48 10	48	16.	GALAXY
MCG+10-18-071	12 33 06.	+ 59 19	30	16.	GALAXY
ZWG 335.015	12 33 06.	+ 72 30		14.3	GALAXY
UGC 07761	12 33 06.	+ 72 30	42	14.3	GALAXY SB
KARA.73B 0538	12 33 06.	+ 72 30	48	14.3	ISOLATED GALAXY S
MCG-05-30-005	12 33 06.	- 32 07	24	16.	GALAXY
KN 16.061	12 33 06.3	+ 26 07 41.			NEBULA
REIN 3.259	12 33 06.58	+ 12 32 19.1			NEBULA
KN 16.063	12 33 06.8	+ 27 41 47.			NEBULA
LB 02395	12 33 07.	+ 60 06 00.		16.5	FAINT BLUE STAR
KN 16.064	12 33 07.3	+ 27 07 11.			NEBULA
KN 16.065	12 33 07.9	+ 26 30 35.			NEBULA
RNGC 4554	12 33 08.	+ 11 27			NON-EXISTENT OBJECT
RNGC 4551	12 33 08.	+ 12 32		13.0	GALAXY
REIZ 2572	12 33 08.	+ 12 32	30	12.7	GALAXY
RNGC 4552	12 33 08.	+ 12 50		11.5	GALAXY
HOLM 426B	12 33 08.	+ 26 07	120	13.9	PART OF MULTIPLE GALAXY
REIN 3.260	12 33 08.44	+ 12 49 52.2			NEBULA
REIN 4.087	12 33 08.74	+ 00 03 36.5			NEBULA
SC 1230+0019.9	12 33 09.	+ 00 03 21.	12		NEBULA
AMES 1596	12 33 09.	+ 16 43 25.	32	17.6	NEBULA
HOLM 426C	12 33 09.	+ 26 29	36	14.3	PART OF MULTIPLE GALAXY
KEEL 507	12 33 09.0	+ 02 21 07.		18.	NEBULA
KN 16.067	12 33 09.6	+ 27 06 47.			NEBULA
KN 16.066	12 33 09.6	+ 27 10 53.			NEBULA
AMES 1597	12 33 10.	+ 14 40 55.	40	16.0	NEBULA
RNGC 4565B	12 33 10.	+ 26 29		15.5	GALAXY
RNGC 4565C	12 33 10.	+ 26 33		15.5	GALAXY
RNGC 4555	12 33 10.	+ 26 48		13.5	GALAXY
MAI 076	12 33 10.	+ 73 58	33		DWARF SPHEROIDAL GALAXY
HOLM 422B	12 33 11.	+ 12 34	42	12.8	PART OF MULTIPLE GALAXY
HOLM 426D	12 33 11.	+ 26 33	36	15.3	PART OF MULTIPLE GALAXY
REIZ 2574	12 33 11.	+ 26 45	12	15.2	GALAXY
REIZ 2575	12 33 11.	+ 26 48	24	13.1	GALAXY
REIZ 2576	12 33 11.	+ 27 08	60	15.7	GALAXY
REIZ 2577	12 33 11.	+ 27 12	54	15.1	GALAXY
REIZ 2578	12 33 11.	+ 27 13	42	16.1	GALAXY
KN 16.068	12 33 11.9	+ 27 12 17.			NEBULA
ZWG 042.170	12 33 12.	+ 03 29		15.7	GALAXY
REA 57	12 33 12.	+ 09 54	42		DWARF GALAXY
ZWG 099.097	12 33 12.	+ 14 41		15.7	GALAXY
ZC 1233.2+2111	12 33 12.	+ 21 11	2420		CLUSTER OF GALAXIES
ZWG 159.020	12 33 12.	+ 26 30		15.3	GALAXY
ZWG 129.009	12 33 12.	+ 26 30		15.3	GALAXY
LB 06631	12 33 12.	+ 26 42		19.6	FAINT BLUE STAR
IC 3545	12 33 12.	+ 26 47 54.			SAME AS NGC 4555
ZWG 159.021	12 33 12.	+ 26 48		13.5	GALAXY
MCG+05-30-026	12 33 12.	+ 26 48	30	13.5	GALAXY
LB 06630	12 33 12.	+ 27 07		19.2	FAINT BLUE STAR
UGC 07762	12 33 12.	+ 29 48	96	13.5	GALAXY E
LB 06629	12 33 12.	+ 30 23		19.7	FAINT BLUE STAR
LB 06628	12 33 12.	+ 30 46		18.8	FAINT BLUE STAR
ZWG 014.075	12 33 12.	- 02 38		14.9	GALAXY
MCG+00-32-025	12 33 12.	- 02 38	60	14.9	GALAXY
KARA.68 163	12 33 12.	- 03 19	27		DWARF GALAXY
ZWG 014.074	12 33 12.	- 03 29		15.7	GALAXY
MCG-02-32-017	12 33 12.	- 12 49	84	14.5	GALAXY
KN 16.069	12 33 12.1	+ 26 55 11.			NEBULA
IC 3543	12 33 12.4	+ 26 33 37.			GALAXY SBd
KN 16.072	12 33 12.5	+ 27 28 59.			NEBULA
KN 16.071	12 33 12.7	+ 26 29 59.			NEBULA

OBJECT NAME	RIGHT ASCEN.	DECLINATION	DIAM.	MAGN.	TYPE OF OBJECT
KEEL 512	12 33 12.7	+ 26 33 45.		17.	NEBULA
KN 16.070	12 33 12.7	+ 26 33 47.			NEBULA
KN 16.075	12 33 12.7	+ 27 38 11.			NEBULA
KEEL 509	12 33 12.8	+ 02 17 53.		18.	NEBULA
KN 16.073	12 33 12.8	+ 26 45 23.			NEBULA
KN 16.073A	12 33 12.8	+ 26 47 53.			NEBULA
KN 16.076	12 33 12.9	+ 27 55 59.			NEBULA
AMES 1598	12 33 13.	+ 17 01 19.	13	17.8	NEBULA
KEEL 510	12 33 13.0	+ 02 13 23.		18.	NEBULA
KN 16.074	12 33 13.1	+ 26 09 29.			NEBULA
KEEL 513	12 33 13.2	+ 26 29 53.		15.	NEBULA
IC 3546	12 33 13.3	+ 26 29 51.			GALAXY S0
KN 16.077	12 33 13.6	+ 27 10 59.			NEBULA
HN 0973	12 33 14.	+ 11 56		14.5	NEBULA
IC 3542	12 33 14.	+ 11 56			NONSTELLAR OBJECT
AMES 1599	12 33 14.	+ 11 56 49.	17	16.8	NEBULA
KN 16.078	12 33 14.1	+ 28 58 11.			NEBULA
MCG+08-23-063	12 33 15.	+ 48 02	36	16.	GALAXY
KN 16.079	12 33 15.3	+ 27 44 23.			NEBULA
KEEL 511	12 33 15.9	+ 02 08 42.		18.	NEBULA
IC 3544	12 33 16.	+ 14 34 30.			TWO STARS
AMES 1601	12 33 16.	+ 16 44 55.	23	18.0	NEBULA
AMES 1600	12 33 16.	+ 17 18 01.	17	17.3	NEBULA
RNGC 4556	12 33 16.	+ 27 12		14.5	GALAXY
REIN 4.088	12 33 16.75	- 03 29 28.2			NEBULA
AMES 1602	12 33 17.	+ 17 10 13.	24	18.0	NEBULA
REIZ 2580	12 33 17.	+ 26 36	9	16.0	GALAXY
REIZ 2581	12 33 17.	+ 26 40	12	15.3	GALAXY
REIZ 2582	12 33 17.	+ 27 11	24	13.5	GALAXY
KN 16.080	12 33 17.3	+ 27 11 05.			NEBULA
KN 16.081	12 33 17.7	+ 27 25 29.			NEBULA
ZWG 042.171	12 33 18.	+ 06 02		15.5	GALAXY
RMB 017	12 33 18.	+ 14 12			FAINT BLUE OBJECT
AMES 1603	12 33 18.	+ 16 13 01.	17	17.0	NEBULA
ZC 1233.3+2632	12 33 18.	+ 26 32	670		CLUSTER OF GALAXIES
UGC 07764	12 33 18.	+ 26 33	60	16.5	GALAXY Sc
MCG+05-30-027	12 33 18.	+ 27 10	30	14.4	GALAXY
ZWG 159.022	12 33 18.	+ 27 12		14.4	GALAXY
UGC 07765	12 33 18.	+ 27 12	84	14.4	GALAXY E-S0
HOLM 423D	12 33 18.	+ 28 12	30	15.5	PART OF MULTIPLE GALAXY
LB 06634	12 33 18.	+ 30 00		19.8	FAINT BLUE STAR
LB 06633	12 33 18.	+ 31 37		17.2	FAINT BLUE STAR
LB 06632	12 33 18.	+ 32 20		19.7	FAINT BLUE STAR
MCG+09-21-023	12 33 18.	+ 51 13	30	16.	GALAXY
MCG+10-18-072	12 33 18.	+ 59 13	27	16.	GALAXY
ZWG 352.036	12 33 18.	+ 76 12		14.7	GALAXY
ZWG 014.077	12 33 18.	- 01 35		15.6	GALAXY
UGC 07763	12 33 18.	- 01 35	60	15.6	GALAXY S
ZWG 014.076	12 33 18.	- 02 07		15.1	GALAXY
AMES 1604	12 33 19.	+ 15 38 43.	14	17.2	NEBULA
SC 1230-0118.1	12 33 19.	- 01 34 39.	30		NEBULA
A3 476	12 33 19.0	+ 15 59 51.		16.9	FAINT BLUE OBJECT
IC 3547	12 33 20.	+ 26 36 18.			TWO STARS
SC 1230-0150.9	12 33 20.	- 02 07 27.	12		NEBULA
KN 16.082	12 33 20.4	+ 27 12 23.			NEBULA
A3 477	12 33 20.5	+ 14 39 56.		16.8	FAINT BLUE OBJECT
REIZ 2579	12 33 21.	+ 01 31	48	14.6	GALAXY
AMES 1605	12 33 21.	+ 12 38 25.	22	16.9	NEBULA
LB 02396	12 33 21.	+ 52 22 00.		17.2	FAINT BLUE STAR
SC 1230-0221.1	12 33 21.	- 02 37 39.	24		NEBULA
RNGC 4553	12 33 21.	- 39 10			GALAXY
KN 16.083	12 33 21.2	+ 29 28 17.			NEBULA
REIN 4.089	12 33 21.28	- 02 37 32.7			NEBULA
KN 16.084	12 33 21.9	+ 29 27 53.			NEBULA
IC 3549	12 33 22.	+ 26 40 19.			SINGLE STAR
RNGC 4558	12 33 22.	+ 27 17		15.0	GALAXY
SN 1941A	12 33 22.	+ 28 14		13.2	SUPERNOVA
AMES 1608	12 33 23.	+ 11 35 37.	32	17.4	NEBULA
REIZ 2584	12 33 23.	+ 27 15	24	14.0	GALAXY
REIZ 2585	12 33 23.	+ 28 11	18	15.7	GALAXY
HOLM 423A	12 33 23.	+ 28 14	600	10.8	PART OF MULTIPLE GALAXY
KN 16.085	12 33 23.0	+ 27 46 05.			NEBULA
REA 58	12 33 24.	+ 07 53	30		DWARF GALAXY
AMES 1611	12 33 24.	+ 11 32 43.	16	17.2	NEBULA
RMB 121	12 33 24.	+ 11 49			FAINT BLUE OBJECT
AMES 1607	12 33 24.	+ 16 59 13.	24	18.0	NEBULA
MCG+05-30-028	12 33 24.	+ 27 15	18	15.1	GALAXY
ZWG 159.023	12 33 24.	+ 27 17		15.1	GALAXY
IC 3550	12 33 24.	+ 28 12 31.			NONSTELLAR OBJECT
ZWG 159.024	12 33 24.	+ 28 14		10.7	GALAXY
UGC 07766	12 33 24.	+ 28 14	780	10.7	GALAXY Sc
LB 11256	12 33 24.	+ 28 22		13.0	FAINT BLUE STAR
LB 06639	12 33 24.	+ 28 35		18.8	FAINT BLUE STAR
LB 06638	12 33 24.	+ 28 44		19.3	FAINT BLUE STAR
LB 06637	12 33 24.	+ 29 09		18.9	FAINT BLUE STAR
LB 06636	12 33 24.	+ 29 22		17.4	FAINT BLUE STAR
LB 06635	12 33 24.	+ 31 17		19.6	FAINT BLUE STAR
ZC 1233.4+3448	12 33 24.	+ 34 48	3360		CLUSTER OF GALAXIES
MCG+07-26-030	12 33 24.	+ 38 37 30.	48	14.5	GALAXY
ZWG 216.013	12 33 24.	+ 38 40		15.1	GALAXY
FEIG 065	12 33 24.	+ 42 39		11.9	FAINT BLUE STAR
ZC 1233.4+4458	12 33 24.	+ 44 58	1480		CLUSTER OF GALAXIES
ZC 1233.4+5706	12 33 24.	+ 57 06	1550		CLUSTER OF GALAXIES
ZC 1233.4+6038	12 33 24.	+ 60 38	2290		CLUSTER OF GALAXIES
ZWG 315.049	12 33 24.	+ 64 15		15.1	GALAXY
ZWG 335.016	12 33 24.	+ 73 57		13.5	GALAXY
UGC 07767	12 33 24.	+ 73 57	54	13.5	GALAXY E
KN 16.087	12 33 24.0	+ 28 12 35.			NEBULA
KN 16.086	12 33 24.2	+ 27 15 59.			NEBULA
KN 16.088	12 33 24.3	+ 27 56 05.			NEBULA
SC 1230+0120.5	12 33 25.	+ 01 03 58.	12		NEBULA
IC 3548	12 33 25.	+ 11 13			NONSTELLAR OBJECT
AMES 1612	12 33 25.	+ 11 32 31.	13	16.9	NEBULA
AMES 1606	12 33 25.	+ 18 13 13.	23	17.0	NEBULA
KN 16.089	12 33 25.0	+ 26 29 41.			NEBULA
KEEL 514	12 33 25.3	+ 02 06 24.		17.	NEBULA
KEEL 515	12 33 25.7	+ 03 03 27.		17.	SPIRAL NEBULA
HN 0974	12 33 26.	+ 11 13	6		NEBULA
AMES 1615	12 33 26.	+ 11 51 01.	16	17.5	NEBULA
AMES 1610	12 33 26.	+ 17 53 49.	23	18.0	NEBULA
AMES 1609	12 33 26.	+ 18 13 25.	16	18.0	NEBULA
IC 3551	12 33 26.	+ 28 14 25.			NONSTELLAR OBJECT
IC 3552	12 33 26.	+ 28 16 19.			NONSTELLAR OBJECT
AMES 1614	12 33 27.	+ 13 08 01.	24	17.9	NEBULA
AMES 1613	12 33 27.	+ 16 38 55.	23	17.8	NEBULA
IC 3553	12 33 27.	+ 26 28 13.			MAY NOT EXIST
MCG-06-28-006	12 33 27.	- 39 10	78	12.5	GALAXY
KN 16.089A	12 33 27.6	+ 28 12 11.			NEBULA

OBJECT NAME	RIGHT ASCEN.	DECLINATION	DIAM.	MAGN.	TYPE OF OBJECT
AMES 1616	12 33 28.	+ 13 12 13.	17	17.2	NEBULA
RNGC 4563	12 33 28.	+ 27 15		15.5	GALAXY
IC 3554	12 33 28.	+ 28 12 13.			NONSTELLAR OBJECT
RNGC 4559	12 33 28.	+ 28 14		10.5	GALAXY
ARC 1568	12 33 28.	+ 53 42		17.8	RICH CLUSTER OF GALAXIES
REIZ 2583	12 33 28.	- 03 00	18	15.4	GALAXY
AMES 1617	12 33 29.	+ 16 52 43.	22	17.6	NEBULA
REIZ 2586	12 33 29.	+ 27 13	30	14.3	GALAXY
REIZ 2587	12 33 29.	+ 27 16	24	15.7	GALAXY
REIZ 2588	12 33 29.	+ 28 11	12	15.6	GALAXY
REIZ 2589	12 33 29.	+ 28 14	480	10.6	GALAXY
IC 3555	12 33 29.	+ 28 16 01.			NONSTELLAR OBJECT
KN 16.090	12 33 29.6	+ 28 14 05.			NEBULA
AMES 1618	12 33 30.	+ 12 49 37.	26	17.1	NEBULA
RMB 016	12 33 30.	+ 14 08			FAINT BLUE OBJECT
MCG+05-30-029	12 33 30.	+ 27 14	48	15.6	GALAXY
ZWG 159.025	12 33 30.	+ 27 15		15.6	GALAXY
LB 06640	12 33 30.	+ 27 52		19.3	FAINT BLUE STAR
SCH 58	12 33 30.	+ 28 14		10.7	PECULIAR GALAXY
MCG+05-30-030	12 33 30.	+ 28 14	720	10.7	GALAXY
LB 11257	12 33 30.	+ 29 57		17.5	FAINT BLUE STAR
ZC 1233.5+5344	12 33 30.	+ 53 44	1080		CLUSTER OF GALAXIES
ZWG 315.050	12 33 30.	+ 64 16		15.3	GALAXY
MCG+12-12-011	12 33 30.	+ 73 57 30.	27	15.	GALAXY
KN 16.091	12 33 30.0	+ 27 14 29.			NEBULA
IC 3556	12 33 30.3	+ 27 14 23.			GALAXY SA(s)
REIN 3.261	12 33 31.75	+ 11 32 01.1			NEBULA
RNGC 4560	12 33 32.	+ 07 57			NON-EXISTENT OBJECT
AMES 1619	12 33 32.	+ 16 50 25.	19	17.8	NEBULA
REIZ 2597	12 33 32.	+ 64 16	30	14.8	GALAXY
LB 00643	12 33 32.	+ 72 48 36.		17.5	FAINT BLUE STAR
KN 16.092	12 33 33.6	+ 27 03 41.			NEBULA
AMES 1621	12 33 34.	+ 16 11 13.	17	17.8	NEBULA
AMES 1620	12 33 34.	+ 16 49 31.	20	17.6	NEBULA
IC 3559	12 33 34.9	+ 27 15 37.			GALAXY S0
KN 16.093	12 33 34.9	+ 27 15 41.			NEBULA
REIZ 2591	12 33 35.	+ 27 09	18	14.4	GALAXY
REIZ 2592	12 33 35.	+ 27 14	24	14.9	GALAXY
REIZ 2593	12 33 35.	+ 27 20	12	14.9	GALAXY
IC 3560	12 33 35.5	+ 27 21 09.			GALAXY SB0
KN 16.094	12 33 35.7	+ 27 21 11.			NEBULA
ZWG 042.172	12 33 36.	+ 05 21		15.6	GALAXY
AMES 1622	12 33 36.	+ 11 18 24.	20	17.5	NEBULA
KARA.68 164	12 33 36.	+ 13 50	40		DWARF GALAXY
ZC 1233.6+1546	12 33 36.	+ 15 46	1010		CLUSTER OF GALAXIES
IC 0557	12 33 36.	+ 16 53			NONSTELLAR OBJECT
MCG+03-32-077	12 33 36.	+ 16 55	42	15.5	GALAXY
REIZ 2590	12 33 36.	+ 19 35	72	12.8	GALAXY
ZWG 099.098	12 33 36.	+ 19 36		12.7	GALAXY
SCH 59	12 33 36.	+ 19 36		12.7	PECULIAR GALAXY
UGC 07768	12 33 36.	+ 19 36	84	12.7	GALAXY SBc
KARA.72 346A	12 33 36.	+ 19 36	60	12.7	PART OF DOUBLE GALAXY
MCG+03-32-076	12 33 36.	+ 19 36	78	12.7	GALAXY
ZC 1233.6+2204	12 33 36.	+ 22 04	1340		CLUSTER OF GALAXIES
LB 11260	12 33 36.	+ 26 28		17.4	FAINT BLUE STAR
LB 06643	12 33 36.	+ 26 37		20.1	FAINT BLUE STAR
LB 06642	12 33 36.	+ 26 52		19.9	FAINT BLUE STAR
LB 11259	12 33 36.	+ 26 58		16.0	FAINT BLUE STAR
MCG+05-30-032	12 33 36.	+ 27 10	36	15.7	GALAXY
ZWG 159.026	12 33 36.	+ 27 11		15.7	GALAXY
MCG+05-30-031	12 33 36.	+ 27 15 30.	36	15.6	GALAXY
LB 06641	12 33 36.	+ 27 42		20.1	FAINT BLUE STAR
LB 11258	12 33 36.	+ 28 27		14.7	FAINT BLUE STAR
ZC 1233.6+3825	12 33 36.	+ 38 25	940		CLUSTER OF GALAXIES
TON-N 0087	12 33 36.	+ 41 05		16.8	BLUE STAR
ZC 1233.6+5313	12 33 36.	+ 53 13	1480		CLUSTER OF GALAXIES
ZWG 270.012	12 33 36.	+ 54 30		13.9	GALAXY
UGC 07769	12 33 36.	+ 54 30	72	13.9	GALAXY S
KARA.73B 0539	12 33 36.	+ 54 30	66	13.9	ISOLATED GALAXY S
A3 480	12 33 36.0	+ 13 41 59.		15.7	FAINT BLUE OBJECT
KN 16.095	12 33 36.4	+ 27 10 41.			NEBULA
IC 3561	12 33 36.5	+ 27 10 26.			GALAXY S0
HN 0975	12 33 37.	+ 16 53		15.5	NEBULA
ARC 1567	12 33 37.	+ 27 02		17.2	RICH CLUSTER OF GALAXIES
HOLM 424A	12 33 38.	+ 07 48	42	15.2	PART OF MULTIPLE GALAXY
HN 0976	12 33 38.	+ 12 07	12		NEBULA
IC 3558	12 33 38.	+ 12 07			NONSTELLAR OBJECT
AMES 1624	12 33 38.	+ 16 56 00.	19	17.5	NEBULA
RNGC 4566	12 33 38.	+ 54 30		14.0	GALAXY
AMES 1623	12 33 39.	+ 17 40 00.	26	17.8	NEBULA
RNGC 4561	12 33 39.	+ 19 36		13.5	GALAXY
KN 16.096	12 33 39.7	+ 28 01 23.			NEBULA
AMES 1625	12 33 40.	+ 12 42 36.	24	18.0	NEBULA
RNGC 4557	12 33 40.	+ 27 13		15.5	GALAXY
AMES 1626	12 33 41.	+ 17 01 54.	24	17.6	NEBULA
IC 3563	12 33 41.	+ 28 12 07.			NONSTELLAR OBJECT
HOLM 424B	12 33 42.	+ 07 48	18	15.3	PART OF MULTIPLE GALAXY
ZWG 070.185	12 33 42.	+ 10 12		15.6	GALAXY
RMB 118	12 33 42.	+ 11 35			FAINT BLUE OBJECT
RMB 120	12 33 42.	+ 11 46			FAINT BLUE OBJECT
AMES 1628	12 33 42.	+ 14 18 00.	19	17.4	NEBULA
KARA.68 165	12 33 42.	+ 14 56	47		DWARF GALAXY
KARA.72 346B	12 33 42.	+ 19 36	42		PART OF DOUBLE GALAXY
LB 06652	12 33 42.	+ 27 22		18.6	FAINT BLUE STAR
LB 06651	12 33 42.	+ 28 00		19.0	FAINT BLUE STAR
IC 3564	12 33 42.	+ 28 12 07.			NONSTELLAR OBJECT
LB 06650	12 33 42.	+ 28 13		19.6	FAINT BLUE STAR
LB 06649	12 33 42.	+ 28 56		18.7	FAINT BLUE STAR
LB 06648	12 33 42.	+ 29 03		18.1	FAINT BLUE STAR
LB 06647	12 33 42.	+ 30 20		19.8	FAINT BLUE STAR
LB 06646	12 33 42.	+ 30 27		19.5	FAINT BLUE STAR
LB 11261	12 33 42.	+ 30 58		16.4	FAINT BLUE STAR
LB 06645	12 33 42.	+ 31 04		18.7	FAINT BLUE STAR
LB 06644	12 33 42.	+ 32 23		18.6	FAINT BLUE STAR
MCG+08-23-064	12 33 42.	+ 48 11	36	17.	GALAXY
MCG+09-21-024	12 33 42.	+ 54 29	90	13.	GALAXY
MCG+12-12-012	12 33 42.	+ 74 31	78	15.	GALAXY
7ZW 475	12 33 42.	+ 81 53			COMPACT GALAXY
KN 16.097	12 33 42.4	+ 26 58 17.			NEBULA
HOLM 425A	12 33 43.	+ 16 54	48	15.5	PART OF MULTIPLE GALAXY
HOLM 425B	12 33 43.	+ 16 55	36	15.7	PART OF MULTIPLE GALAXY
AMES 1627	12 33 43.	+ 17 44 24.	20	17.6	NEBULA
KN 16.098	12 33 43.9	+ 27 01 47.			NEBULA
HN 0977	12 33 44.	+ 10 11	24		NEBULA
IC 3562	12 33 44.	+ 10 11			NONSTELLAR OBJECT
AMES 1630	12 33 44.	+ 16 34 48.	22	17.8	NEBULA
AMES 1629	12 33 44.	+ 16 56 12.	19	17.5	NEBULA
IC 3565	12 33 44.1	+ 27 01 51.			GALAXY S0

OBJECT NAME	RIGHT ASCEN.	DECLINATION	DIAM.	MAGN.	TYPE OF OBJECT
KN 16.099	12 33 44.3	+ 27 01 41.			NEBULA
KN 16.100	12 33 44.4	+ 27 13 05.			NEBULA
MCG-01-32-029	12 33 45.	- 03 37 30.	60	15.	GALAXY
AMES 1631	12 33 46.	+ 16 41 36.	17	17.5	NEBULA
AMES 1632	12 33 47.	+ 16 14 00.	22	17.2	NEBULA
ARC 1569	12 33 47.	+ 16 53		17.2	RICH CLUSTER OF GALAXIES
REIZ 2594	12 33 47.	+ 24 20	9	16.0	GALAXY
REIZ 2595	12 33 47.	+ 27 00	36	15.3	GALAXY
REIZ 2596	12 33 47.	+ 27 13	30	13.7	GALAXY
REIN 4.090	12 33 47.75	- 03 38 46.9			NEBULA
RMB 122	12 33 48.	+ 11 57			FAINT BLUE OBJECT
RMB 088	12 33 48.	+ 13 19			FAINT BLUE OBJECT
RMB 015	12 33 48.	+ 14 05			FAINT BLUE OBJECT
AMES 1633	12 33 48.	+ 16 21 12.	26	18.0	NEBULA
HN 0978	12 33 48.	+ 19 34	42		NEBULA
UGC 07770	12 33 48.	+ 21 15	66	16.5	GALAXY Sc
MCG+04-30-007	12 33 48.	+ 21 54	156	14.	GALAXY
ZC 1233.8+2206	12 33 48.	+ 22 06	270		CLUSTER OF GALAXIES
UGC 07771	12 33 48.	+ 23 27	60	16.0	GALAXY S
ZC 1233.8+2423	12 33 48.	+ 24 23	4100		CLUSTER OF GALAXIES
LB 00010	12 33 48.	+ 25 33		17.8	FAINT BLUE STAR
ZWG 129.010	12 33 48.	+ 26 15		10.3	GALAXY
UGC 07772	12 33 48.	+ 26 15	930	10.3	GALAXY Sb
MCG+04-30-006	12 33 48.	+ 26 15	720	10.3	GALAXY
LB 06654	12 33 48.	+ 26 38		19.1	FAINT BLUE STAR
MCG+05-30-033	12 33 48.	+ 27 12	12	15.7	GALAXY
LB 06653	12 33 48.	+ 29 17		19.5	FAINT BLUE STAR
LB 02397	12 33 48.	+ 56 49 00.		17.5	FAINT BLUE STAR
LB 00644	12 33 48.	+ 62 04 48.		16.9	FAINT BLUE STAR
IC 0802	12 33 48.	+ 74 32	60		FAINT NEBULA
FATH 2.108	12 33 48.	+ 74 32	60		NEBULA
MCG-03-32-015	12 33 48.	- 19 07	30	14.5	GALAXY
IC 3566	12 33 49.	+ 11 26			MAY NOT EXIST
HN 0979	12 33 49.	+ 13 52	12		NEBULA
IC 3567	12 33 49.	+ 13 52			NONSTELLAR OBJECT
IC 3569	12 33 49.	+ 19 34			SAME AS NGC 4561
IC 3570	12 33 49.	+ 24 21 13.			NONSTELLAR OBJECT
A3 481	12 33 49.7	+ 14 31 10.		16.9	FAINT BLUE OBJECT
HN 0980	12 33 50.	+ 11 26	60		NEBULA
AMES 1635	12 33 50.	+ 12 43 36.	16	17.8	NEBULA
AMES 1634	12 33 50.	+ 14 06 00.	36	17.1	NEBULA
REIZ 2603	12 33 50.	+ 54 29	36	13.9	GALAXY
LB 02398	12 33 50.	+ 59 14 24.		12.8	FAINT BLUE STAR
KN 16.101	12 33 50.7	+ 28 01 35.			NEBULA
AMES 1637	12 33 51.	+ 13 11 36.	22	17.4	NEBULA
AMES 1636	12 33 51.	+ 16 55 36.	19	18.0	NEBULA
ARC 1570	12 33 51.	+ 20 06		17.8	RICH CLUSTER OF GALAXIES
KEEL 516	12 33 51.4	+ 26 21 35.		18.	NEBULA
KN 16.102	12 33 51.4	+ 26 21 35.			NEBULA
IC 3571	12 33 51.6	+ 26 21 29.			GALAXY IRR
A3 482	12 33 51.7	+ 15 02 57.		16.8	FAINT BLUE OBJECT
KN 16.103	12 33 51.9	+ 26 15 41.			NEBULA
KN 16.104	12 33 51.9	+ 29 19 41.			NEBULA
RNGC 4565	12 33 52.	+ 26 16		10.5	GALAXY
HELW 264	12 33 52.	- 03 38 32.			NEBULA
AMES 1638	12 33 53.	+ 10 11 24.	18	16.8	NEBULA
REIZ 2599	12 33 53.	+ 26 15	180	10.3	GALAXY
HOLM 426A	12 33 53.	+ 26 15	900	10.3	PART OF MULTIPLE GALAXY
REIZ 2601	12 33 53.	+ 26 21	24	15.3	GALAXY
KN 16.105	12 33 53.8	+ 25 42 53.			NEBULA
ZWG 070.186	12 33 54.	+ 11 43		12.2	GALAXY
UGC 07773	12 33 54.	+ 11 43	180	12.2	GALAXY E
SN 1961H	12 33 54.	+ 11 43		11.2	SUPERNOVA
MCG+02-32-150	12 33 54.	+ 11 43	60	12.2	GALAXY
RMB 087	12 33 54.	+ 13 18			FAINT BLUE OBJECT
ZWG 070.187	12 33 54.	+ 13 53		15.3	GALAXY
ZC 1233.9+2002	12 33 54.	+ 20 02	1010		CLUSTER OF GALAXIES
ZC 1233.9+2536	12 33 54.	+ 25 36	340		CLUSTER OF GALAXIES
LB 06656	12 33 54.	+ 26 39		19.4	FAINT BLUE STAR
LB 11262	12 33 54.	+ 27 33		13.0	FAINT BLUE STAR
LB 06655	12 33 54.	+ 28 28		18.4	FAINT BLUE STAR
LB 06657	12 33 54.	+ 32 13		18.0	FAINT BLUE STAR
ZWG 216.014	12 33 54.	+ 40 17		15.2	GALAXY
UGC 07774	12 33 54.	+ 40 17	222	15.2	GALAXY Sc
ZWG 244.032	12 33 54.	+ 48 10		15.6	GALAXY
MRK 218	12 33 54.	+ 48 10	12	15.5	GALAXY WITH UV CONTINUUM
MCG+08-23-065	12 33 54.	+ 48 10	30	15.	GALAXY
MCG+10-18-073	12 33 54.	+ 58 21	39	16.	GALAXY
ZWG 352.037	12 33 54.	+ 74 31		14.9	GALAXY
UGC 07775	12 33 54.	+ 74 31	108	14.9	GALAXY S
AMES 1639	12 33 55.	+ 17 06 00.	20	17.1	NEBULA
RNGC 4572	12 33 55.	+ 74 31		15.0	GALAXY
REIN 3.262	12 33 55.39	+ 11 42 49.5			NEBULA
REIZ 2600	12 33 56.	+ 11 32	96	13.0	GALAXY
RNGC 4564	12 23 56.	+ 11 43		12.5	GALAXY
REIZ 2598	12 33 56.	+ 11 43	108	12.4	GALAXY
IC 3572	12 33 56.	+ 11 53 37.			SINGLE STAR
HN 0981	12 33 56.	+ 12 01	12		NEBULA
IC 3573	12 33 56.	+ 12 01			NONSTELLAR OBJECT
REIZ 2602	12 33 56.	+ 40 17	72	14.2	GALAXY
IC 3574	12 33 57.	+ 12 40 43.			NONSTELLAR OBJECT
AMES 1640	12 33 57.	+ 16 49 24.	32	17.0	NEBULA
AMES 1641	12 33 58.	+ 13 15 48.	24	17.8	NEBULA
KN 16.106	12 33 59.6	+ 26 30 47.			NEBULA
KN 16.107	12 33 59.9	+ 28 01 53.			NEBULA
VDB.66G 139	12 34	+ 07 24	70		DWARF GALAXY
ZWG 042.173	12 34 00.	+ 06 27		15.3	GALAXY
ZWG 042.174	12 34 00.	+ 08 20		15.0	GALAXY
MCG+01-32-111	12 34 00.	+ 08 20	36	15.0	GALAXY
RMB 195	12 24 00.	+ 10 02			FAINT BLUE OBJECT
KARA.68 166	12 34 00.	+ 11 26	27		DWARF GALAXY
ZWG 070.188	12 34 00.	+ 11 31		12.5	GALAXY
UGC 07776	12 34 00.	+ 11 31	306	12.5	GALAXY Sc
KARA.72 347B	12 34 00.	+ 11 31	252	12.5	PART OF DOUBLE GALAXY
MCG+02-32-152	12 34 00.	+ 11 31	210	12.5	GALAXY
ZWG 070.189	12 34 00.	+ 11 32		12.5	GALAXY
SCH 60	12 34 00.	+ 11 32		12.5	PECULIAR GALAXY
HOLM 427B	12 34 00.	+ 11 32	120	11.5	PART OF MULTIPLE GALAXY
UGC 07777	12 34 00.	+ 11 32	180	12.5	GALAXY Sc
VV 219B	12 34 00.	+ 11 32	144	11.5	INTERACTING GALAXY
KARA.72 347A	12 34 00.	+ 11 32	156	12.5	PART OF DOUBLE GALAXY
MCG+02-32-151	12 34 00.	+ 11 32	168	12.5	GALAXY
KARA.68 167	12 34 00.	+ 12 09	34		DWARF GALAXY
REA 59	12 34 00.	+ 13 53	48		DWARF GALAXY
AMES 1644	12 34 00.	+ 13 55 12.	60	16.8	NEBULA
AMES 1643	12 34 00.	+ 16 21 30.	23	17.4	NEBULA
AMES 1642	12 34 00.	+ 16 48 18.	31	17.8	NEBULA
8ZW 1234+23.7	12 34 00.	+ 23 42		17.5	COMPACT GALAXY

OBJECT NAME	RIGHT ASCEN.	DECLINATION	DIAM.	MAGN.	TYPE OF OBJECT
LB 00044	12 34 00.	+ 25 32		17.7	FAINT BLUE STAR
LB 00011	12 34 00.	+ 26 07		16.4	FAINT BLUE STAR
ZWG 129.011	12 34 00.	+ 26 28		14.3	GALAXY
MRK 649	12 34 00.	+ 26 28	20	14.5	GALAXY WITH UV CONTINUUM
UGC 07778	12 34 00.	+ 26 28	24	14.3	GALAXY VY CMPT
MCG+04-30-008	12 34 00.	+ 26 28	24	14.3	GALAXY
LB 06661	12 34 00.	+ 26 56		19.6	FAINT BLUE STAR
LB 11263	12 34 00.	+ 27 35		11.5	FAINT BLUE STAR
ZC 1234.0+2916	12 34 00.	+ 29 16	1340		CLUSTER OF GALAXIES
LB 06660	12 34 00.	+ 30 02		19.6	FAINT BLUE STAR
LB 06659	12 34 00.	+ 31 32		18.5	FAINT BLUE STAR
LB 06658	12 34 00.	+ 31 47		19.9	FAINT BLUE STAR
ZWG 159.027	12 34 00.	+ 32 21		15.4	GALAXY
UGC 07779	12 34 00.	+ 32 21	66	15.4	GALAXY S
MCG+07-26-031	12 34 00.	+ 40 16	204	14.	GALAXY
MCG+08-23-066	12 34 00.	+ 48 47	15	16.	GALAXY
TON-N 0088	12 34 00.	+ 53 08		16.1	BLUE STAR
KN 16.108	12 34 00.7	+ 27 35 59.			NEBULA
AMES 1648	12 34 01.	+ 10 44 30.	34	17.0	NEBULA
HN 0982	12 34 01.	+ 14 01	12		NEBULA
IC 3575	12 34 01.	+ 14 01			NONSTELLAR OBJECT
AMES 1645	12 34 01.	+ 14 59 00.	26	17.0	NEBULA
SEY 188	12 34 01.	+ 34 50 35.		15.4	FAINT GALAXY
REIN 3.263	12 34 01.09	+ 11 31 57.6			NEBULA
KEEL 517	12 34 01.6	+ 26 31 26.		18.	NEBULA
IC 3582	12 34 01.7	+ 26 28 27.			GALAXY Sab
AMES 1649	12 34 02.	+ 10 34 30.	22	17.5	NEBULA
RNGC 4568	12 34 02.	+ 11 31		12.0	GALAXY
HOLM 427A	12 34 02.	+ 11 31	210	11.3	PART OF MULTIPLE GALAXY
RNGC 4567	12 34 02.	+ 11 32		12.5	GALAXY
AMES 1647	12 34 02.	+ 16 56 24.	24	18.0	NEBULA
KN 16.109	12 34 02.3	+ 26 29 53.			NEBULA
BC B21234+26	12 34 02.4	+ 26 51 46.4		20.4	QUASI-STELLAR OBJECT
REIN 3.264	12 34 02.53	+ 11 30 48.1			NEBULA
AMES 1646	12 34 03.	+ 17 59 18.	28	18.0	NEBULA
MCG+06-28-012	12 34 03.	+ 34 51	27	15.5	GALAXY
LB 02399	12 34 03.	+ 51 54 18.		16.9	FAINT BLUE STAR
IC 3579	12 34 04.	+ 26 22 43.			SINGLE STAR
REIZ 2606	12 34 04.	+ 32 20	30	15.7	GALAXY
KN 16.110	12 34 04.2	+ 28 18 23.			NEBULA
IC 3577	12 34 05.	+ 12 10 19.			TWO STARS
REIZ 2604	12 34 05.	+ 24 41	36	13.6	GALAXY
REIZ 2605	12 34 05.	+ 26 22	18	15.2	GALAXY
IC 3576	12 34 05.4	+ 06 53 47.	60	15.2	GALAXY S
ZWG 042.175	12 34 06.	+ 03 23		15.7	GALAXY
UGC 07780	12 34 06.	+ 03 23	102	16.5	GALAXY DWARF
ZWG 042.176	12 34 06.	+ 06 54		15.2	GALAXY
UGC 07781	12 34 06.	+ 06 54	180	15.2	GALAXY DWRF SP
MCG+01-32-112	12 34 06.	+ 06 54	102	15.2	GALAXY
RMB 196	12 34 06.	+ 10 13			FAINT BLUE OBJECT
ZWG 070.190	12 34 06.	+ 11 23		15.1	GALAXY
UGC 07782	12 34 06.	+ 11 23	60	15.1	GALAXY Sc
MCG+02-32-153	12 34 06.	+ 11 23	42	15.1	GALAXY
VV 219A	12 34 06.	+ 11 31	240	11.3	INTERACTING GALAXY
KARA.68 168	12 34 06.	+ 13 52	34		DWARF GALAXY
HN 0983	12 34 06.	+ 18 33	12		NEBULA
TON-N 0089	12 34 06.	+ 22 28		15.2	BLUE STAR
TON-N 0623	12 34 06.	+ 23 22		14.6	BLUE STAR
MCG+04-30-009	12 34 06.	+ 24 41	24	14.9	GALAXY
ZWG 129.012	12 34 06.	+ 24 42		14.9	GALAXY
ZWG 159.028	12 34 06.	+ 27 07		15.0	GALAXY
UGC 07783	12 34 06.	+ 27 07	66	15.0	GALAXY S0
LB 06662	12 34 06.	+ 30 18		19.1	FAINT BLUE STAR
MCG+05-30-034	12 34 06.	+ 32 21	54	15.4	GALAXY
ZC 1234.1+4032	12 34 06.	+ 40 32	870		CLUSTER OF GALAXIES
ZC 1234.1+4717	12 34 06.	+ 47 17	670		CLUSTER OF GALAXIES
TON-N 0090	12 34 06.	+ 50 36		14.5	BLUE STAR
ZC 1234.1+7805	12 34 06.	+ 78 05	1210		CLUSTER OF GALAXIES
IC 3578	12 34 07.	+ 11 22			NONSTELLAR OBJECT
AMES 1653	12 34 07.	+ 16 45 48.	24	17.0	NEBULA
AMES 1652	12 34 07.	+ 16 50 48.	24	17.2	NEBULA
AMES 1651	12 34 07.	+ 16 53 24.	25	17.8	NEBULA
AMES 1650	12 34 07.	+ 16 56 00.	14	17.4	NEBULA
IC 3580	12 34 07.	+ 18 33			NONSTELLAR OBJECT
REIN 3.265	12 34 07.69	+ 11 22 36.6			NEBULA
HN 0984	12 34 08.	+ 11 22	24		NEBULA
AMES 1654	12 34 08.	+ 16 47 36.	19	16.5	NEBULA
ARC 1572	12 34 08.	- 14 25		17.3	RICH CLUSTER OF GALAXIES
ARC 1573	12 34 08.	- 14 40		17.3	RICH CLUSTER OF GALAXIES
KEEL 518	12 34 08.5	+ 26 30 37.		16.	NEBULA
KN 16.111	12 34 08.5	+ 26 30 42.			NEBULA
IC 3581	12 34 09.	+ 24 42 13.			NONSTELLAR OBJECT
A3 483	12 34 09.4	+ 14 29 23.		16.8	FAINT BLUE OBJECT
AMES 1655	12 34 10.	+ 14 56 12.	28	17.0	NEBULA
REIZ 2607	12 34 11.	+ 26 30	12	15.4	GALAXY
REIZ 2608	12 34 11.	+ 27 05	24	14.0	GALAXY
IC 3585	12 34 11.6	+ 27 06 12.			GALAXY SA(s)
KN 16.112	12 34 11.6	+ 27 06 18.			NEBULA
ZWG 042.177	12 34 12.	+ 04 23		15.4	GALAXY
MCG+01-32-113	12 34 12.	+ 04 23	30	15.4	GALAXY
AMES 1657	12 34 12.	+ 13 12 18.	28	17.8	NEBULA
RMB 086	12 34 12.	+ 13 29			FAINT BLUE OBJECT
ZWG 070.191	12 34 12.	+ 13 32		15.0	GALAXY
UGC 07784	12 34 12.	+ 13 32	150	15.0	GALAXY IRR
MCG+02-32-154	12 34 12.	+ 13 32	90	15.0	GALAXY
AMES 1656	12 34 12.	+ 15 01 54.	40	17.3	NEBULA
MCG+05-30-035	12 34 12.	+ 27 06	54	15.0	GALAXY
LB 06664	12 34 12.	+ 28 17		18.9	FAINT BLUE STAR
LB 06663	12 34 12.	+ 29 53		18.9	FAINT BLUE STAR
LB 11266	12 34 12.	+ 30 21		15.4	FAINT BLUE STAR
LB 11265	12 34 12.	+ 31 56		16.9	FAINT BLUE STAR
LB 11264	12 34 12.	+ 32 24		15.6	FAINT BLUE STAR
LB 02401	12 34 12.	+ 49 23 54.		14.8	FAINT BLUE STAR
LB 02400	12 34 12.	+ 58 06 12.		14.9	FAINT BLUE STAR
IC 3583	12 34 12.5	+ 13 32 01.	156	14.2	GALAXY IRR
AMES 1659	12 34 13.	+ 11 41 42.	24	17.8	NEBULA
IC 3584	12 34 14.	+ 12 30 25.			SINGLE STAR
AMES 1658	12 34 14.	+ 16 29 36.	13	17.8	NEBULA
KN 16.113	12 34 14.4	+ 27 41 00.			NEBULA
KN 16.115	12 34 14.9	+ 28 09 18.			NEBULA
AMES 1664	12 34 15.	+ 11 05 24.	11	17.3	NEBULA
AMES 1662	12 34 15.	+ 15 28 18.	22	17.4	NEBULA
LB 00260	12 34 15.	+ 74 40 30.		16.0	FAINT BLUE STAR
KEEL 519	12 34 15.0	+ 26 23 19.		17.	NEBULA
KN 16.114	12 34 15.2	+ 27 21 06.			NEBULA
KN 16.116	12 34 15.8	+ 28 02 12.			NEBULA
AMES 1663	12 34 16.	+ 13 11 24.	17	18.0	NEBULA
AMES 1661	12 34 16.	+ 18 42 12.	22	17.6	NEBULA

OBJECT NAME	RIGHT ASCEN.	DECLINATION	DIAM.	MAGN.	TYPE OF OBJECT
AMES 1660	12 34 16.	+ 19 18 48.	25	17.1	NEBULA
KN 16.117	12 34 17.3	+ 29 07 42.			NEBULA
KN 16.118	12 34 17.8	+ 27 50 36.			NEBULA
ZWG 042.178	12 34 18.	+ 07 31		11.8	GALAXY
UGC 07785	12 34 18.	+ 07 31	252	11.8	GALAXY S0
MCG+01-32-114	12 34 18.	+ 07 31	180	11.8	GALAXY
RMB 119	12 34 18.	+ 11 47			FAINT BLUE OBJECT
ZWG 070.192	12 34 18.	+ 13 26		11.8	GALAXY
SCH 61	12 34 18.	+ 13 26		11.8	PECULIAR GALAXY
UGC 07786	12 34 18.	+ 13 26	684	11.8	GALAXY Sb
ARP 076	12 34 18.	+ 13 26			PECULIAR GALAXY
MCG+02-32-155	12 34 18.	+ 13 26	660	11.8	GALAXY
MCG+02-32-156	12 34 18.	+ 14 28	210	13.6	GALAXY
AMES 1665	12 34 18.	+ 15 31 12.	20	17.5	NEBULA
LB 06669	12 34 18.	+ 26 44		19.2	FAINT BLUE STAR
LB 06668	12 34 18.	+ 27 00		19.7	FAINT BLUE STAR
LB 06667	12 34 18.	+ 27 18		19.4	FAINT BLUE STAR
UGC 07787	12 34 18.	+ 27 49	90	16.0	GALAXY Sc
LB 06666	12 34 18.	+ 29 15		19.4	FAINT BLUE STAR
LB 06665	12 34 18.	+ 29 42		18.6	FAINT BLUE STAR
ZWG 159.029	12 34 18.	+ 32 23		15.4	GALAXY
TON-N 0091	12 34 18.	+ 47 07		15.6	BLUE STAR
LB 02402	12 34 18.	+ 57 35 24.		16.2	FAINT BLUE STAR
A3 484	12 34 18.5	+ 15 31 28.		18.0	FAINT BLUE OBJECT
AMES 1666	12 34 19.	+ 12 41 36.	19	17.8	NEBULA
REIZ 2609	12 34 19.	+ 13 26	300	11.4	GALAXY
ARC 1574	12 34 19.	- 10 26		16.8	RICH CLUSTER OF GALAXIES
RNGC 4570	12 34 20.	+ 07 31		12.0	GALAXY
REIZ 2610	12 34 20.	+ 07 31	126	12.1	GALAXY
AMES 1667	12 34 20.	+ 11 40 24.	29	16.8	NEBULA
RNGC 4569	12 34 20.	+ 13 26		11.0	GALAXY
KN 16.119	12 34 20.0	+ 27 49 30.			NEBULA
IC 3587	12 34 21.	+ 27 49 26.			NONSTELLAR OBJECT
KARA.68 169	12 34 21.	+ 32 19	47		DWARF GALAXY
AMES 1668	12 34 22.	+ 16 23 36.	18	17.9	NEBULA
REIZ 2614	12 34 22.	+ 28 01	18	14.8	GALAXY
HOLM 423C	12 34 22.	+ 28 02	30	14.9	PART OF MULTIPLE GALAXY
REIZ 2615	12 34 22.	+ 28 07	24	14.5	GALAXY
REIZ 2616	12 34 22.	+ 32 22	36	15.6	GALAXY
LB 02403	12 34 22.	+ 61 20 48.		16.6	FAINT BLUE STAR
A3 486	12 34 22.2	+ 14 14 36.		16.9	FAINT BLUE OBJECT
KN 16.120	12 34 22.5	+ 27 33 12.			NEBULA
AMES 1669	12 34 23.	+ 12 22 00.	29	17.8	NEBULA
REIZ 2612	12 34 23.	+ 27 32	60	14.8	GALAXY
IC 3590	12 34 23.	+ 27 33 20.			NONSTELLAR OBJECT
REIZ 2613	12 34 23.	+ 27 48	60	15.1	GALAXY
HOLM 423B	12 34 23.	+ 28 08	42	14.7	PART OF MULTIPLE GALAXY
IC 3586	12 34 24.	+ 12 47 38.			NONSTELLAR OBJECT
ZWG 070.193	12 34 24.	+ 12 48		15.3	GALAXY
MCG+02-32-157	12 34 24.	+ 12 48	36	15.3	GALAXY
ZWG 070.194	12 34 24.	+ 14 29		13.6	GALAXY
UGC 07788	12 34 24.	+ 14 29	270	13.6	GALAXY Sc
TON-N 0092	12 34 24.	+ 15 32		14.2	BLUE STAR
MCG+05-30-037	12 34 24.	+ 27 49	12	17.	GALAXY
MCG+05-30-036	12 34 24.	+ 27 49	84	15.	GALAXY
ZWG 159.030	12 34 24.	+ 28 02		15.4	GALAXY
ZWG 159.031	12 34 24.	+ 28 09		15.3	GALAXY
UGC 07789	12 34 24.	+ 28 09	78	15.3	GALAXY Sa?
LB 06674	12 34 24.	+ 28 44		19.2	FAINT BLUE STAR
LB 06673	12 34 24.	+ 28 54		19.0	FAINT BLUE STAR
TON-N 0624	12 34 24.	+ 28 56		15.0	BLUE STAR
LB 06672	12 34 24.	+ 30 01		18.7	FAINT BLUE STAR
LB 06671	12 34 24.	+ 31 04		19.4	FAINT BLUE STAR
LB 06670	12 34 24.	+ 31 16		19.4	FAINT BLUE STAR
ZC 1234.4+6327	12 34 24.	+ 63 27	1010		CLUSTER OF GALAXIES
MCG-05-30-006	12 34 24.	- 28 13	108	14.5	GALAXY
AMES 1671	12 34 25.	+ 10 42 12.	14	16.5	NEBULA
IC 3588	12 34 25.	+ 14 29 38.			MAY NOT EXIST
AMES 1670	12 34 25.	+ 16 24 48.	20	17.4	NEBULA
IC 3593	12 34 25.	+ 28 01 26.			NONSTELLAR OBJECT
IC 3592	12 34 25.	+ 28 08 14.			NONSTELLAR OBJECT
REIZ 2619	12 34 25.	+ 40 37	12	14.2	GALAXY
KN 16.121	12 34 25.5	+ 28 08 18.			NEBULA
KN 16.122	12 34 25.9	+ 28 01 24.			NEBULA
REIZ 2611	12 34 26.	+ 07 12	12	15.2	GALAXY
RNGC 4571	12 34 26.	+ 14 30		12.5	GALAXY
AMES 1672	12 34 27.	+ 16 12 00.	22	17.5	NEBULA
IC 3589	12 34 28.	+ 07 12 44.			NONSTELLAR OBJECT
IC 3594	12 34 28.	+ 26 23 20.			SINGLE STAR
RNGC 4559B	12 34 28.	+ 28 02		15.5	GALAXY
RNGC 4559A	12 34 28.	+ 28 08		15.5	GALAXY
RNGC 4559C	12 34 28.	+ 28 12		15.5	GALAXY
KN 16.123	12 34 28.2	+ 25 58 12.			NEBULA
AMES 1673	12 34 29.	+ 16 46 48.	17	17.0	NEBULA
REIZ 2618	12 34 29.	+ 26 23	18	15.4	GALAXY
KN 16.124	12 34 29.2	+ 26 41 24.			NEBULA
ZWG 014.078	12 34 30.	+ 02 05		15.5	GALAXY
ZWG 042.179	12 34 30.	+ 07 12		14.6	GALAXY
UGC 07790	12 34 30.	+ 07 12	60	14.6	GALAXY IRR
MCG+01-32-115	12 34 30.	+ 07 12	48	14.6	GALAXY
IC 3591	12 34 30.	+ 07 12 02.			NONSTELLAR OBJECT
HOLM 428B	12 34 30.	+ 07 13	24	14.5	PART OF MULTIPLE GALAXY
KARA.68 170	12 34 30.	+ 16 34	47		DWARF GALAXY
LB 11267	12 34 30.	+ 26 26		16.8	FAINT BLUE STAR
MCG+05-30-039	12 34 30.	+ 28 02	24	15.4	GALAXY
MCG+05-30-038	12 34 30.	+ 28 08	30	15.3	GALAXY
LB 06675	12 34 30.	+ 28 30		19.5	FAINT BLUE STAR
ZWG 216.015	12 34 30.	+ 40 38		15.3	GALAXY
MCG+08-23-067	12 34 30.	+ 45 55	36	15.	GALAXY
ZWG 244.033	12 34 30.	+ 45 56		15.2	GALAXY
HELW 152	12 34 30.	- 05 07 56.			NEBULA
AMES 1677	12 34 31.	+ 10 41 06.	14	17.0	NEBULA
AMES 1674	12 34 31.	+ 17 16 12.	14	18.0	NEBULA
HOLM 428A	12 34 32.	+ 07 12	66	14.3	PART OF MULTIPLE GALAXY
REIZ 2617	12 34 32.	+ 07 13	54	15.0	GALAXY
AMES 1676	12 34 32.	+ 13 12 30.	22	17.6	NEBULA
AMES 1675	12 34 32.	+ 18 28 54.	14	17.0	NEBULA
LB 02404	12 34 33.	+ 58 10 48.		15.7	FAINT BLUE STAR
KN 16.125	12 34 33.7	+ 27 58 12.			NEBULA
KN 16.126	12 34 33.8	+ 27 33 30.			NEBULA
ARC 1576	12 34 34.	+ 63 29		18.0	RICH CLUSTER OF GALAXIES
KN 16.127	12 34 34.6	+ 27 55 54.			NEBULA
AMES 1678	12 34 35.	+ 18 22 54.	14	17.8	NEBULA
LB 02405	12 34 35.	+ 61 49 54.		16.0	FAINT BLUE STAR
KEEL 520	12 34 35.7	+ 26 21 56.		17.	NEBULA
ZWG 042.180	12 34 36.	+ 05 42		15.7	GALAXY
ZC 1234.6+1919	12 34 36.	+ 19 19	1550		CLUSTER OF GALAXIES
LB 00045	12 34 36.	+ 26 35		13.0	FAINT BLUE STAR

OBJECT NAME	RIGHT ASCEN.	DECLINATION	DIAM.	MAGN.	TYPE OF OBJECT
ZC 1234.6+2705	12 34 36.	+ 27 05	1280		CLUSTER OF GALAXIES
LB 06679	12 34 36.	+ 28 10		18.8	FAINT BLUE STAR
LB 06678	12 34 36.	+ 29 17		18.7	FAINT BLUE STAR
LB 06677	12 34 36.	+ 29 58		18.9	FAINT BLUE STAR
LB 06676	12 34 36.	+ 30 54		19.6	FAINT BLUE STAR
TON-N 0625	12 34 36.	+ 33 26		16.4	BLUE STAR
ZC 1234.6+4148	12 34 36.	+ 41 48	740		CLUSTER OF GALAXIES
MCG+08-23-068	12 34 36.	+ 48 22	30	16.	GALAXY
AMES 1681	12 34 37.	+ 15 24 24.	19	17.4	NEBULA
AMES 1679	12 34 37.	+ 17 37 48.	19	17.7	NEBULA
IC 3595	12 34 37.	+ 24 03 50.			NONSTELLAR OBJECT
AMES 1684	12 34 38.	+ 10 32 48.	42	17.0	NEBULA
AMES 1683	12 34 38.	+ 15 42 12.	25	17.0	NEBULA
AMES 1680	12 34 38.	+ 18 14 12.	22	18.0	NEBULA
A3 490	12 34 38.2	+ 15 02 26.		17.7	FAINT BLUE OBJECT
A3 491	12 34 38.8	+ 14 59 35.		17.1	FAINT BLUE OBJECT
REIZ 2621	12 34 39.	+ 04 38	54	14.0	GALAXY
AMES 1682	12 34 39.	+ 17 38 00.	17	18.0	NEBULA
MCG-03-32-016	12 34 39.	- 19 46 30.	36	14.5	GALAXY
AMES 1685	12 34 41.	+ 17 40 24.	31	17.9	NEBULA
REIZ 2622	12 34 41.	+ 24 03	48	15.3	GALAXY
LB 02406	12 34 41.	+ 60 02 00.		15.2	FAINT BLUE STAR
REIZ 2620	12 34 41.	- 05 12	36	12.9	GALAXY
AMES 1686	12 34 42.	+ 16 51 24.	24	17.5	NEBULA
ZWG 099.099	12 34 42.	+ 20 25		15.3	GALAXY
MCG+03-32-078	12 34 42.	+ 20 27	54	15.3	GALAXY
LB 06682	12 34 42.	+ 26 42		19.4	FAINT BLUE STAR
LB 06681	12 34 42.	+ 26 55		18.9	FAINT BLUE STAR
ZWG 159.032	12 34 42.	+ 29 55		15.7	GALAXY
ZC 1234.7+3139	12 34 42.	+ 31 39	2020		CLUSTER OF GALAXIES
LB 06680	12 34 42.	+ 31 57		17.0	FAINT BLUE STAR
ZC 1234.7+6701	12 34 42.	+ 67 01	670		CLUSTER OF GALAXIES
MCG-02-32-018	12 34 42.	- 09 28	60	15.5	GALAXY
KN 16.128	12 34 43.0	+ 27 27 12.			NEBULA
KN 16.129	12 34 43.4	+ 27 54 36.			NEBULA
A3 492	12 34 44.0	+ 14 55 59.		17.6	FAINT BLUE OBJECT
AMES 1687	12 34 46.	+ 17 17 36.	24	18.1	NEBULA
REIZ 2623	12 34 46.	+ 28 28	30	14.5	GALAXY
KN 16.130	12 34 46.5	+ 29 54 18.			NEBULA
KN 16.131	12 34 47.7	+ 27 25 54.			NEBULA
AMES 1688	12 34 48.	+ 10 47 24.	25	17.8	NEBULA
RMB 117	12 34 48.	+ 11 31			FAINT BLUE OBJECT
RMB 014	12 34 48.	+ 14 11			FAINT BLUE OBJECT
ZC 1234.8+2638	12 34 48.	+ 26 38	1010		CLUSTER OF GALAXIES
LB 06688	12 34 48.	+ 27 12		18.0	FAINT BLUE STAR
ZWG 159.033	12 34 48.	+ 28 30		15.0	GALAXY
UGC 07791	12 34 48.	+ 28 30	96	15.0	GALAXY Sa-b
LB 06687	12 34 48.	+ 28 44		19.0	FAINT BLUE STAR
LB 06686	12 34 48.	+ 29 03		19.0	FAINT BLUE STAR
LB 06685	12 34 48.	+ 29 32		19.5	FAINT BLUE STAR
LB 06684	12 34 48.	+ 30 57		19.1	FAINT BLUE STAR
LB 06683	12 34 48.	+ 32 12		19.5	FAINT BLUE STAR
LB 11268	12 34 48.	+ 32 21		17.0	FAINT BLUE STAR
VV 042B	12 34 48.	+ 39 00	18	18.	INTERACTING GALAXY
VV 042A	12 34 48.	+ 39 00	15	17.5	INTERACTING GALAXY
MCG+08-23-069	12 34 48.	+ 49 44	48	15.	GALAXY
ZC 1234.8+6137	12 34 48.	+ 61 37	1280		CLUSTER OF GALAXIES
HOLM 429B	12 34 49.	+ 09 49	24	15.4	PART OF MULTIPLE GALAXY
ARC 1575	12 34 49.	+ 27 06		17.8	RICH CLUSTER OF GALAXIES
HZ 30	12 34 49.	+ 38 52		13.4	DECIDEDLY BLUE STAR
KEEL 521	12 34 49.9	+ 26 15 25.		18.	NEBULA
AMES 1692	12 34 50.	+ 11 14 36.	19	16.6	NEBULA
AMES 1690	12 34 50.	+ 12 34 00.	46	17.6	NEBULA
AMES 1689	12 34 50.	+ 13 20 42.	22	17.9	NEBULA
ARP 211	12 34 50.	+ 39 02			PECULIAR GALAXY
KN 16.132	12 34 50.2	+ 29 09 00.			NEBULA
AMES 1691	12 34 51.	+ 16 29 12.	14	17.9	NEBULA
IC 3596	12 34 51.	+ 26 47 44.			TWO STARS
MCG+07-26-032	12 34 51.	+ 39 44	42	15.	GALAXY
KN 16.133	12 34 51.1	+ 26 26 06.			NEBULA
KN 16.134	12 34 51.4	+ 26 49 42.			NEBULA
KEEL 522	12 34 51.5	+ 26 26 05.		18.	NEBULA
AMES 1694	12 34 53.	+ 16 37 48.	30	17.4	NEBULA
REIZ 2624	12 34 53.	+ 24 08	60	15.2	GALAXY
IC 3598	12 34 53.	+ 28 29 02.			NONSTELLAR OBJECT
SVEN 308	12 34 53.	- 05 08 02.	12	15.2	GALAXY
KN 16.135	12 34 53.2	+ 28 28 54.			NEBULA
UGC 08612	12 34 54.	+ 04 21	60	14.8	GALAXY SBb
ZWG 042.181	12 34 54.	+ 05 01		15.7	GALAXY
REA 18	12 34 54.	+ 12 39			DWARF GALAXY
MCG+02-32-158	12 34 54.	+ 12 39	60	18.	GALAXY
RMB 080	12 34 54.	+ 13 38			FAINT BLUE OBJECT
AMES 1693	12 34 54.	+ 17 18 18.	14	17.2	NEBULA
FEIG 066	12 34 54.	+ 25 21		10.0	FAINT BLUE STAR
MCG+05-30-040	12 34 54.	+ 28 28	84	15.0	GALAXY
LB 11269	12 34 54.	+ 30 17		14.7	FAINT BLUE STAR
MCG+07-26-034	12 34 54.	+ 39 00	12	16.	GALAXY
MCG+07-26-033	12 34 54.	+ 39 00	18	17.	GALAXY
ZWG 244.034	12 34 54.	+ 49 43		15.0	GALAXY
OCL 0889	12 34 54.	- 68 12	420	10.8	OPEN STAR CLUSTER
VHA 136	12 34 54.	- 68 12	480		OPEN STAR CLUSTER
A3 495	12 34 54.5	+ 14 21 52.		16.0	FAINT BLUE OBJECT
IC 3597	12 34 55.	+ 24 08 20.			NONSTELLAR OBJECT
RNGC 4577	12 34 56.	+ 06 20			NON-EXISTENT OBJECT
AMES 1695	12 34 56.	+ 13 53 00.	25	17.8	NEBULA
REIZ 2628	12 34 56.	+ 63 13	18	14.6	GALAXY
RNGC 4573	12 34 56.	- 43 21			GALAXY
MCG+07-26-035	12 34 57.	+ 43 08	39	16.	GALAXY
KN 16.136	12 34 57.3	+ 28 37 36.			NEBULA
REIN 4.091	12 34 58.73	+ 09 49 48.0			NEBULA
AMES 1696	12 34 59.	+ 16 49 30.	24	16.5	NEBULA
ZWG 042.182	12 35 00.	+ 04 38		14.7	GALAXY
UGC 07792	12 35 00.	+ 04 38	78	14.7	GALAXY Sb
MCG+01-32-116	12 35 00.	+ 04 38	72	14.7	GALAXY
ZWG 070.195	12 35 00.	+ 09 50		12.9	GALAXY
UGC 07793	12 35 00.	+ 09 50	192	12.9	GALAXY S0
MCG+02-32-159	12 35 00.	+ 09 50	132	12.9	GALAXY
ZC 1235.0+2605	12 35 00.	+ 26 05	2080		CLUSTER OF GALAXIES
LB 11270	12 35 00.	+ 26 22		13.3	FAINT BLUE STAR
LB 06690	12 35 00.	+ 26 59		19.5	FAINT BLUE STAR
LB 06689	12 35 00.	+ 29 52		19.8	FAINT BLUE STAR
ZC 1235.0+3057	12 35 00.	+ 30 57	1080		CLUSTER OF GALAXIES
ZC 1235.0+3852	12 35 00.	+ 38 52	1080		CLUSTER OF GALAXIES
ZWG 014.079	12 35 00.	- 03 18		15.4	GALAXY
KN 16.137	12 35 00.5	+ 28 36 54.			NEBULA
A3 496	12 35 01.8	+ 14 47 44.		16.9	FAINT BLUE OBJECT
RNGC 4576	12 35 02.	+ 04 39		14.5	GALAXY
RNGC 4578	12 35 02.	+ 09 50		12.0	GALAXY

OBJECT NAME	RIGHT ASCEN.	DECLINATION	DIAM.	MAGN.	TYPE OF OBJECT
HOLM 429A	12 35 02.	+ 09 50	120	13.4	PART OF MULTIPLE GALAXY
AMES 1697	12 35 03.	+ 16 37 48.	34	17.0	NEBULA
RNGC 4575	12 35 03.	- 40 16			GALAXY
KEEL 524	12 35 03.0	+ 26 11 53.		16.	NEBULA
RNGC 4574	12 35 04.	- 35 14			GALAXY
ZWG 070.196	12 35 06.	+ 08 50		15.1	GALAXY
RMB 079	12 35 06.	+ 13 35			FAINT BLUE OBJECT
ZWG 159.034	12 35 06.	+ 26 59		15.7	GALAXY
ZWG 159.035	12 35 06.	+ 27 24		15.5	GALAXY
HARO 30	12 35 06.	+ 27 24			BLUE EMISSION-LINE GALAXY
MRK 650	12 35 06.	+ 27 24	10	15.5	GALAXY WITH UV CONTINUUM
LB 06695	12 35 06.	+ 27 24		18.4	FAINT BLUE STAR
LB 06694	12 35 06.	+ 27 57		19.1	FAINT BLUE STAR
LB 06693	12 35 06.	+ 29 04		19.4	FAINT BLUE STAR
LB 06692	12 35 06.	+ 32 13		20.2	FAINT BLUE STAR
LB 11271	12 35 06.	+ 32 14		17.0	FAINT BLUE STAR
LB 06691	12 35 06.	+ 32 33		18.6	FAINT BLUE STAR
MCG-06-28-007	12 35 06.	- 35 13	108	12.5	GALAXY
AGU 37	12 35 06.	- 40 16 00.	54	12.5	PECULIAR GALAXY
REIN 4.094	12 35 06.26	+ 09 54 43.1			NEBULA
REIZ 2625	12 35 08.	+ 08 50	60	14.7	GALAXY
YC 1235-40	12 35 08.	- 40 15 42.			UNUSUAL SOUTHERN GALAXY
AMES 1699	12 35 09.	+ 11 41 11.	26	17.1	NEBULA
REIZ 2627	12 35 09.	+ 32 21	42	15.2	GALAXY
MCG+08-23-058	12 35 09.	+ 49 53	12	16.	GALAXY
REIN 3.266	12 35 09.19	+ 08 50 04.4			NEBULA
AMES 1698	12 35 10.	+ 17 30 35.	22	17.7	NEBULA
KN 16.138	12 35 10.9	+ 25 57 49.			NEBULA
REIZ 2626	12 35 11.	+ 27 24	24	14.5	GALAXY
ZWG 042.183	12 35 12.	+ 05 38		13.1	GALAXY
SCH 62	12 35 12.	+ 05 38		13.1	PECULIAR GALAXY
UGC 07794	12 35 12.	+ 05 38	156	13.1	GALAXY Sa-b
MCG+01-32-117	12 35 12.	+ 05 38	96	13.1	GALAXY
FEA 82	12 35 12.	+ 07 22			DWARF GALAXY
ZWG 042.184	12 35 12.	+ 07 23		15.4	GALAXY
UGC 07795	12 35 12.	+ 07 23	78	15.4	GALAXY DWRF IR
MCG+01-32-118	12 35 12.	+ 07 23	72	15.4	GALAXY
MCG+02-32-161	12 35 12.	+ 08 50	60	15.1	GALAXY
ZWG 070.197	12 35 12.	+ 12 05		11.5	GALAXY
UGC 07796	12 35 12.	+ 12 05	378	11.5	GALAXY Sb
MCG+02-32-160	12 35 12.	+ 12 05	300	11.5	GALAXY
RMB 013	12 35 12.	+ 14 01			FAINT BLUE OBJECT
ZC 1235.2+2144	12 35 12.	+ 21 44	870		CLUSTER OF GALAXIES
LB 00013	12 35 12.	+ 24 38		14.7	FAINT BLUE STAR
LB 00012	12 35 12.	+ 25 43		16.5	FAINT BLUE STAR
LB 06699	12 35 12.	+ 28 24		18.8	FAINT BLUE STAR
TON-N 0626	12 35 12.	+ 29 27		16.7	BLUE STAR
LB 06698	12 35 12.	+ 29 33		19.7	FAINT BLUE STAR
LB 11272	12 35 12.	+ 29 38		18.4	FAINT BLUE STAR
LB 06697	12 35 12.	+ 30 55		19.7	FAINT BLUE STAR
LB 06696	12 35 12.	+ 31 25		19.7	FAINT BLUE STAR
ZWG 159.036	12 35 12.	+ 32 22		15.4	GALAXY
ZWG 014.080	12 35 12.	- 01 05		15.3	GALAXY
ZC 1235.2-0114	12 35 12.	- 01 14	1680		CLUSTER OF GALAXIES
REIN 3.267	12 35 12.10	+ 12 05 33.1			NEBULA
KN 16.141	12 35 12.6	+ 29 29 19.			NEBULA
AMES 1700	12 35 13.	+ 18 32 35.	19	17.8	NEBULA
IC 2600	12 35 13.	+ 27 24 15.			NONSTELLAR OBJECT
LB 02407	12 35 13.	+ 55 47 18.		15.5	FAINT BLUE STAR
KN 16.139	12 35 13.0	+ 27 24 13.			NEBULA
IC 3599	12 35 13.1	+ 26 58 55.			GALAXY S
KN 16.140	12 35 13.2	+ 26 59 07.			NEBULA
RNGC 4579	12 35 14.	+ 12 06		11.0	GALAXY
HOLM 432B	12 35 14.	+ 70 18	18	15.5	PART OF MULTIPLE GALAXY
KN 16.142	12 35 14.9	+ 28 11 19.			NEBULA
ARC 1577	12 35 16.	- 00 00		17.8	RICH CLUSTER OF GALAXIES
AMES 1702	12 35 17.	+ 11 01 41.	19	17.8	NEBULA
REIZ 2630	12 35 17.	+ 26 58	18	14.4	GALAXY
KN 16.144	12 35 17.7	+ 29 48 31.			NEBULA
KN 16.143	12 35 17.9	+ 28 12 19.			NEBULA
RMB 077	12 35 18.	+ 13 08			FAINT BLUE OBJECT
ZWG 099.100	12 35 18.	+ 14 33		15.2	GALAXY
AMES 1701	12 35 18.	+ 18 43 11.	19	17.2	NEBULA
TON-N 0627	12 35 18.	+ 26 35		14.7	BLUE STAR
LB 06705	12 35 18.	+ 26 36		19.5	FAINT BLUE STAR
LB 00014	12 35 18.	+ 26 53		17.5	FAINT BLUE STAR
LB 06704	12 35 18.	+ 27 02		19.8	FAINT BLUE STAR
MCG+05-30-041	12 35 18.	+ 27 24	30	15.5	GALAXY
LB 06703	12 35 18.	+ 28 58		19.9	FAINT BLUE STAR
LB 06702	12 35 18.	+ 30 46		19.0	FAINT BLUE STAR
LB 06701	12 35 18.	+ 32 04		18.8	FAINT BLUE STAR
LB 06700	12 35 18.	+ 32 11		18.9	FAINT BLUE STAR
MCG+05-30-041A	12 35 18.	+ 32 22	48	15.4	GALAXY
ZC 1235.3+5749	12 35 18.	+ 57 49	2490		CLUSTER OF GALAXIES
HOLM 432A	12 35 18.	+ 70 19	18	15.5	PART OF MULTIPLE GALAXY
KEEL 525	12 35 18.4	+ 26 34 28.		18.	NEBULA
HN 0985	12 35 19.	+ 15 28		15.	NEBULA
IC 3601	12 35 19.	+ 15 28			PLATE DEFECT
RNGC 4580	12 35 20.	+ 05 39		13.0	GALAXY
REIZ 2629	12 35 21.	+ 05 38	54	13.0	GALAXY
ARC 1580	12 35 21.	+ 78 13		17.5	RICH CLUSTER OF GALAXIES
AMES 1703	12 35 22.	+ 14 34 05.	23	15.4	NEBULA
KN 16.145	12 35 23.1	+ 28 16 31.			NEBULA
A3 498	12 35 23.6	+ 14 44 40.		16.5	FAINT BLUE OBJECT
ZC 1235.4+0000	12 35 24.	+ 00 00	1080		CLUSTER OF GALAXIES
AMES 1705	12 35 24.	+ 16 36 23.	16	17.1	NEBULA
LB 00015	12 35 24.	+ 25 27		16.2	FAINT BLUE STAR
LB 06709	12 35 24.	+ 27 07		19.7	FAINT BLUE STAR
LB 06708	12 35 24.	+ 27 07		19.6	FAINT BLUE STAR
LB 06707	12 35 24.	+ 27 42		18.7	FAINT BLUE STAR
LB 06706	12 35 24.	+ 28 06		18.8	FAINT BLUE STAR
LB 11273	12 35 24.	+ 28 39		16.0	FAINT BLUE STAR
MCG+06-28-013	12 35 24.	+ 33 43 30.	24	16.5	GALAXY
ZC 1235.4+4322	12 35 24.	+ 43 22	740		CLUSTER OF GALAXIES
ZC 1235.4+4510	12 35 24.	+ 45 10	610		CLUSTER OF GALAXIES
ZWG 352.038	12 35 24.	+ 74 28		12.0	GALAXY
ZWG 335.017	12 35 24.	+ 74 28		12.0	GALAXY
UGC 07797	12 35 24.	+ 74 28	198	12.0	GALAXY E
KN 16.146	12 35 24.8	+ 28 13 31.			NEBULA
AMES 1708	12 35 25.	+ 10 37 41.	22	16.8	NEBULA
AMES 1704	12 35 25.	+ 17 31 29.	24	17.5	NEBULA
AMES 1709	12 35 27.	+ 12 43 23.	17	17.6	NEBULA
AMES 1707	12 35 27.	+ 17 18 53.	19	17.1	NEBULA
AMES 1706	12 35 27.	+ 19 08 29.	22	17.2	NEBULA
KN 16.147	12 35 27.5	+ 27 54 01.			NEBULA
AMES 1711	12 35 29.	+ 15 38 59.	24	17.8	NEBULA
REIZ 2633	12 35 29.	+ 26 37	60	12.4	GALAXY
HOLM 430C	12 35 29.	+ 34 18	18	15.7	PART OF MULTIPLE GALAXY
REIZ 2631	12 35 29.	- 08 04	42	15.7	GALAXY
REIN 4.097	12 35 29.38	- 01 59 22.1			NEBULA
RMB 009	12 35 30.	+ 15 25			FAINT BLUE OBJECT
AMES 1710	12 35 30.	+ 18 05 47.	19	18.0	NEBULA
ZC 1235.5+1836	12 35 30.	+ 18 36	1410		CLUSTER OF GALAXIES
LB 06713	12 35 30.	+ 31 01		20.0	FAINT BLUE STAR
LB 06712	12 35 30.	+ 31 25		20.0	FAINT BLUE STAR
LB 06711	12 35 30.	+ 31 42		20.2	FAINT BLUE STAR
LB 06710	12 35 30.	+ 31 56		20.2	FAINT BLUE STAR
ZWG 188.010	12 35 30.	+ 34 17		15.5	GALAXY
UGC 07799	12 35 30.	+ 34 17	96	15.5	GALAXY S
SBY 189	12 35 30.	+ 34 17 29.		15.1	FAINT GALAXY
TON-N 0093	12 35 30.	+ 43 49		15.6	BLUE STAR
ZC 1235.5+6239	12 35 30.	+ 62 39	1010		CLUSTER OF GALAXIES
ZWG 014.081	12 35 30.	- 02 00		14.1	GALAXY
UGC 07798	12 35 30.	- 02 00	54	14.1	GALAXY IRR
MCG+00-32-026	12 35 30.	- 02 00	48	14.1	GALAXY
KARA.73B 0540	12 35 30.	- 02 00	60	14.1	ISOLATED GALAXY IR
MCG-01-32-030	12 35 30.	- 08 04	96	15.	GALAXY
MCG-03-32-017	12 35 30.	- 19 51 30.	12	15.	GALAXY
FATH 2.109	12 35 31.	+ 74 28	33		NEBULA
SC 1232-0142.4	12 35 31.	- 01 58 55.	30		NEBULA
REIN 4.099	12 35 31.77	+ 01 45 07.6			NEBULA
REIN 4.098	12 35 31.79	+ 00 14 51.3			NEBULA
RNGC 4589	12 35 32.	+ 74 28		12.5	GALAXY
SC 1232+0031.4	12 35 33.	+ 00 14 53.	60		NEBULA
REIZ 2632	12 35 33.	+ 01 44	24	13.3	GALAXY
AMES 1712	12 35 33.	+ 16 54 17.	13	17.8	NEBULA
REIZ 2636	12 35 33.	+ 33 43	24	14.9	GALAXY
HOLM 430D	12 35 33.	+ 34 16	60	16.0	PART OF MULTIPLE GALAXY
REIZ 2637	12 35 33.	+ 34 17	120	15.1	GALAXY
HOLM 430A	12 35 33.	+ 34 18	108	14.2	PART OF MULTIPLE GALAXY
LB 00645	12 25 33.	+ 56 39 42.		17.3	FAINT BLUE STAR
REIZ 2635	12 35 34.	+ 29 13	48	13.8	GALAXY
KN 16.148	12 35 34.3	+ 27 53 19.			NEBULA
AMES 1713	12 35 35.	+ 17 24 05.	19	17.2	NEBULA
RNGC 4583	12 35 35.	+ 33 44		14.5	GALAXY
ZWG 014.082	12 35 36.	+ 00 14		14.9	GALAXY
UGC 07800	12 35 36.	+ 00 14	66	14.9	GALAXY S
MCG+00-32-027	12 35 36.	+ 00 14	60	14.9	GALAXY
ZWG 014.083	12 35 36.	+ 01 44		13.4	GALAXY
UGC 07801	12 35 36.	+ 01 44	108	13.4	GALAXY E
MCG+00-32-028	12 35 36.	+ 01 44	48	13.4	GALAXY
MCG+01-32-119	12 35 36.	+ 08 05	42	15.5	GALAXY
ZWG 070.198	12 35 36.	+ 10 21		15.5	GALAXY
ZWG 129.013	12 35 36.	+ 22 58		15.7	GALAXY
KARA.73B 0541	12 35 36.	+ 22 58	54	15.7	ISOLATED GALAXY S
ZC 1235.6+2332	12 35 36.	+ 23 32	3160		CLUSTER OF GALAXIES
LB 06722	12 35 36.	+ 26 32		17.4	FAINT BLUE STAR
LB 06721	12 35 36.	+ 27 00		19.4	FAINT BLUE STAR
LB 06720	12 35 36.	+ 27 03		19.5	FAINT BLUE STAR
LB 11275	12 35 36.	+ 28 13		15.5	FAINT BLUE STAR
LB 06719	12 35 36.	+ 28 50		19.5	FAINT BLUE STAR
LB 11274	12 35 36.	+ 28 56		13.2	FAINT BLUE STAR
LB 06718	12 35 36.	+ 30 14		19.6	FAINT BLUE STAR
LB 06717	12 35 36.	+ 30 37		19.0	FAINT BLUE STAR
LB 06716	12 35 36.	+ 30 39		18.7	FAINT BLUE STAR
LB 06715	12 35 36.	+ 30 51		20.2	FAINT BLUE STAR
LB 06714	12 35 36.	+ 30 57		17.5	FAINT BLUE STAR
ZWG 188.011	12 35 36.	+ 33 44		14.7	GALAXY
MCG+06-28-014	12 35 36.	+ 34 18	9	18.	GALAXY
MCG+06-28-015	12 35 36.	+ 34 18 30.	78	14.5	GALAXY
MCG+12-12-013	12 35 36.	+ 74 27	78	12.0	GALAXY
RNGC 4582	12 35 37.	+ 00 27			NON-EXISTENT OBJECT
RNGC 4581	12 35 37.	+ 01 44		13.5	GALAXY
IC 3602	12 35 37.	+ 10 21			NONSTELLAR OBJECT
AMES 1714	12 35 37.	+ 18 00 35.	17	17.4	NEBULA
A3 500	12 35 37.5	+ 14 41 45.		16.6	FAINT BLUE OBJECT
REIZ 2634	12 35 38.	+ 08 04	12	16.3	GALAXY
HN 0986	12 35 38.	+ 10 21	6		NEBULA
AMES 1717	12 35 38.	+ 13 16 59.	26	17.9	NEBULA
AMES 1716	12 35 38.	+ 13 49 11.	18	17.8	NEBULA
REIN 3.268	12 35 38.40	+ 08 05 10.5			NEBULA
AMES 1715	12 35 39.	+ 17 18 41.	26	17.2	NEBULA
MCG+04-30-010	12 35 39.	+ 22 58 30.	42	15.7	GALAXY
HOLM 430B	12 35 39.	+ 34 18	18	15.3	PART OF MULTIPLE GALAXY
MCG+07-26-036	12 35 39.	+ 41 34	39	17.	GALAXY
REIN 3.269	12 35 39.85	+ 08 05 14.0			NEBULA
HN 0020	12 35 40.	+ 00 27			NEBULA
HOLM 431B	12 35 40.	+ 08 04	24	15.2	PART OF MULTIPLE GALAXY
RNGC 4585	12 35 40.	+ 29 13		14.5	GALAXY
ARC 1578	12 35 40.	+ 43 21		17.2	RICH CLUSTER OF GALAXIES
ARC 1579	12 35 40.	+ 66 08		17.6	RICH CLUSTER OF GALAXIES
LB 02408	12 35 41.	+ 54 46 24.		16.0	FAINT BLUE STAR
A3 501	12 35 41.1	+ 16 03 48.		16.9	FAINT BLUE OBJECT
ZWG 042.185	12 35 42.	+ 08 05		15.5	GALAXY
MCG+01-32-120	12 35 42.	+ 08 05	24	15.5	GALAXY
ZC 1235.7+0950	12 35 42.	+ 09 50	870		CLUSTER OF GALAXIES
RMB 008	12 35 42.	+ 15 30			FAINT BLUE OBJECT
IC 3603	12 35 42.	+ 15 48			NONSTELLAR OBJECT
AMES 1719	12 35 42.	+ 17 08 17.	24	17.8	NEBULA
AMES 1718	12 35 42.	+ 17 19 41.	19	17.6	NEBULA
ZWG 159.037	12 35 42.	+ 29 13		14.6	GALAXY
MCG+05-30-042	12 35 42.	+ 29 13	36	14.6	GALAXY
LB 11276	12 35 42.	+ 31 54		15.5	FAINT BLUE STAR
LB 06723	12 35 42.	+ 32 25		19.2	FAINT BLUE STAR
VV 203	12 35 42.	+ 33 26 00.	2	18.	INTERACTING GALAXY
MCG+06-28-017	12 35 42.	+ 33 45	60	14.	GALAXY
MCG+06-28-016	12 35 42.	+ 34 18 30.	12	17.	GALAXY
MCG-04-30-006	12 35 42.	- 23 23	36	16.	GALAXY
MCG-05-30-007	12 35 42.	- 28 39	36	16.	GALAXY
MCG-06-28-008	12 35 42.	- 35 20	60	16.	GALAXY
AMES 1721	12 35 43.	+ 11 35 59.	16	17.1	NEBULA
HN 0987	12 35 43.	+ 15 48		15.	NEBULA
REIN 3.270	12 35 43.68	+ 08 05 44.6			NEBULA
REIZ 2638	12 35 44.	+ 08 05	18	16.0	GALAXY
HOLM 431A	12 35 44.	+ 08 05	30	15.0	PART OF MULTIPLE GALAXY
AMES 1720	12 35 44.	+ 17 27 35.	26	17.2	NEBULA
A3 502	12 35 45.3	+ 14 43 26.		17.1	FAINT BLUE OBJECT
KN 16.149	12 35 45.7	+ 29 12 43.			NEBULA
AMES 1722	12 35 47.	+ 15 45 29.	25	18.0	NEBULA
ZWG 042.186	12 35 48.	+ 08 10		15.3	GALAXY
UGC 07802	12 35 48.	+ 08 10	108	15.3	GALAXY Sc
MCG+01-32-121	12 35 48.	+ 08 10	36	15.3	GALAXY
RMB 108	12 35 48.	+ 12 35			FAINT BLUE OBJECT
ZWG 070.199	12 35 48.	+ 13 23		14.2	GALAXY
UGC 07803	12 35 48.	+ 13 23	96	14.2	GALAXY SB
MCG+02-32-162	12 35 48.	+ 13 23	60	14.2	GALAXY

OBJECT NAME	RIGHT ASCEN.	DECLINATION	DIAM.	MAGN.	TYPE OF OBJECT
HN 0988	12 35 48.	+ 19 47	12		NEBULA
IC 3605	12 35 48.	+ 19 47			NONSTELLAR OBJECT
8ZW 1235+24.0	12 35 48.	+ 24 02		16.5	COMPACT GALAXY
LB 06730	12 35 48.	+ 27 02		18.7	FAINT BLUE STAR
ZC 1235.8+2731	12 35 48.	+ 27 31	870		CLUSTER OF GALAXIES
LB 06729	12 35 48.	+ 28 15		18.5	FAINT BLUE STAR
LB 06728	12 35 48.	+ 28 38		19.6	FAINT BLUE STAR
LB 06727	12 35 48.	+ 28 54		19.1	FAINT BLUE STAR
LB 06726	12 35 48.	+ 29 10		19.3	FAINT BLUE STAR
LB 06725	12 35 48.	+ 29 23		18.1	FAINT BLUE STAR
LB 06724	12 35 48.	+ 29 30		19.4	FAINT BLUE STAR
LB 11277	12 35 48.	+ 30 40		16.6	FAINT BLUE STAR
ZC 1235.8+3234	12 35 48.	+ 32 34	1080		CLUSTER OF GALAXIES
ZC 1235.8+5639	12 35 48.	+ 56 39	2020		CLUSTER OF GALAXIES
ZC 1235.8+5836	12 35 48.	+ 58 36	2080		CLUSTER OF GALAXIES
ZWG 014.084	12 35 48.	- 00 42		15.7	GALAXY
MCG-05-30-008	12 35 48.	- 28 40	36	16.	GALAXY
REIF 3.271	12 35 48.65	+ 08 09 58.2			NEBULA
A3 503	12 35 48.8	+ 15 24 43.		17.3	FAINT BLUE OBJECT
HN 0989	12 35 49.	+ 11 59	12		NEBULA
SC 1233-0024.8	12 35 49.	- 00 41 19.	12		NEBULA
REIN 3.272	12 35 49.21	+ 12 00 19.5			NEBULA
REIZ 2639	12 35 50.	+ 08 10	72	15.0	GALAXY
AMES 1729	12 35 50.	+ 10 09 23.	31	17.6	NEBULA
IC 3604	12 35 50.	+ 11 59			NONSTELLAR OBJECT
RNGC 4584	12 35 50.	+ 13 23		14.0	GALAXY
AMES 1725	12 35 50.	+ 19 17 47.	28	16.9	NEBULA
AMES 1724	12 35 50.	+ 19 36 11.	30	16.9	NEBULA
AMES 1723	12 35 50.	+ 19 48 11.	55	16.5	NEBULA
REIF 4.100	12 35 50.67	+ 09 48 11.6			NEBULA
AMES 1726	12 35 51.	+ 17 26 47.	17	17.8	NEBULA
AMES 1728	12 35 52.	+ 17 14 59.	26	17.4	NEBULA
AMES 1727	12 35 52.	+ 17 26 35.	17	17.9	NEBULA
REIZ 2641	12 35 52.	+ 47 48	12	14.8	GALAXY
AMES 1733	12 35 53.	+ 12 29 23.	31	17.6	NEBULA
AMES 1732	12 35 53.	+ 16 19 47.	17	17.3	NEBULA
ZWG 042.187	12 35 54.	+ 04 35		13.5	GALAXY
UGC 07804	12 35 54.	+ 04 35	234	13.5	GALAXY Sa
MCG+01-32-122	12 35 54.	+ 04 35	240	13.5	GALAXY
ZWG 070.200	12 35 54.	+ 09 48		15.5	GALAXY
AMES 1735	12 35 54.	+ 10 54 23.	29	17.4	NEBULA
RMB 078	12 35 54.	+ 13 23			FAINT BLUE OBJECT
AMES 1734	12 35 54.	+ 14 21 59.	26	17.7	NEBULA
AMES 1731	12 35 54.	+ 17 16 59.	26	17.5	NEBULA
AMES 1730	12 35 54.	+ 17 59 29.	26	18.1	NEBULA
TON-N 0094	12 35 54.	+ 20 17		15.5	BLUE STAR
LB 06737	12 35 54.	+ 27 32		19.3	FAINT BLUE STAR
LB 06736	12 35 54.	+ 27 54		19.4	FAINT BLUE STAR
LB 06735	12 35 54.	+ 28 39		20.0	FAINT BLUE STAR
LB 06734	12 35 54.	+ 29 01		19.2	FAINT BLUE STAR
LB 06733	12 35 54.	+ 31 59		19.1	FAINT BLUE STAR
LB 06732	12 35 54.	+ 32 04		17.3	FAINT BLUE STAR
LB 06731	12 35 54.	+ 32 10		18.5	FAINT BLUE STAR
ZWG 216.016	12 35 54.	+ 42 29		15.6	GALAXY
ZC 1235.9+6436	12 35 54.	+ 64 36	1480		CLUSTER OF GALAXIES
SC 1233+0218.9	12 35 55.	+ 02 02 23.	18		NEBULA
HN 0990	12 35 55.	+ 12 52			NEBULA
IC 3606	12 35 55.	+ 12 52			NONSTELLAR OBJECT
REIN 4.101	12 35 55.41	+ 02 01 46.4			NEBULA
RNGC 4586	12 35 56.	+ 04 36		12.5	GALAXY
KN 16.150	12 35 56.1	+ 28 07 55.			NEBULA
A3 504	12 35 56.8	+ 15 15 17.		17.4	FAINT BLUE OBJECT
REIZ 2640	12 35 57.	+ 04 35	120	13.0	GALAXY
AMES 1737	12 35 57.	+ 17 31 29.	29	17.3	NEBULA
AMES 1736	12 35 57.	+ 17 32 35.	25	17.4	NEBULA
KN 16.151	12 35 57.2	+ 28 46 31.			NEBULA
KN 16.152	12 35 57.6	+ 28 15 01.			NEBULA
AMES 1739	12 35 58.	+ 17 30 35.	14	17.8	NEBULA
AMES 1738	12 35 58.	+ 17 42 05.	16	17.6	NEBULA
AMES 1741	12 35 59.	+ 16 49 23.	24	17.8	NEBULA
AMES 1740	12 35 59.	+ 17 26 59.	20	18.0	NEBULA
KN 16.153	12 35 59.5	+ 28 45 13.			NEBULA
VDB.66G 140	12 36	+ 08 14	70		DWARF GALAXY
ZWG 014.085	12 36 00.	+ 02 01		14.7	GALAXY
MCG+00-32-029	12 36 00.	+ 02 01	15	14.7	GALAXY
ZWG 042.188	12 36 00.	+ 02 56		14.4	GALAXY
UGC 07805	12 36 00.	+ 02 56	72	14.4	GALAXY S0
MCG+01-32-123	12 36 00.	+ 02 56	48	14.4	GALAXY
RMB 116	12 36 00.	+ 11 32			FAINT BLUE OBJECT
ZWG 099.101	12 36 00.	+ 14 38		15.5	GALAXY
ZC 1236.0+1536	12 36 00.	+ 15 36	2420		CLUSTER OF GALAXIES
AMES 1742	12 36 00.	+ 17 22 23.	22	17.8	NEBULA
LB 06740	12 36 00.	+ 27 40		19.0	FAINT BLUE STAR
LB 06739	12 36 00.	+ 29 31		19.4	FAINT BLUE STAR
LB 06738	12 36 00.	+ 30 57		19.3	FAINT BLUE STAR
LB 11278	12 36 00.	+ 31 00		15.8	FAINT BLUE STAR
LB 06741	12 36 00.	+ 31 38		16.3	FAINT BLUE STAR
ZC 1236.0+5054	12 36 00.	+ 50 54	1010		CLUSTER OF GALAXIES
MCG+09-21-025	12 36 00.	+ 53 19	24	16.	GALAXY
MCG+12-12-014	12 36 00.	+ 71 41	45	15.	GALAXY
HELW 153	12 36 00.	- 04 39 19.			NEBULA
OCL 0887	12 36 00.	- 50 53	180	15.	OPEN STAR CLUSTER
KN 16.154	12 36 00.2	+ 27 39 13.			NEBULA
RNGC 4587	12 36 01.	+ 02 56		14.5	GALAXY
IC 3607	12 36 01.	+ 10 38			MAY NOT EXIST
AMES 1745	12 36 01.	+ 12 28 05.	22	17.6	NEBULA
AMES 1744	12 36 01.	+ 17 25 47.	22	17.4	NEBULA
AMES 1743	12 36 01.	+ 19 07 23.	34	17.8	NEBULA
SC 1233+0157.8	12 36 02.	+ 01 41 17.	60		NEBULA
AMES 1746	12 36 02.	+ 10 21 35.	20	16.8	NEBULA
HN 0991	12 36 02.	+ 10 38	6		NEBULA
LB 02409	12 36 02.	+ 57 03 06.		15.9	FAINT BLUE STAR
KN 16.155	12 36 02.7	+ 29 19 43.			NEBULA
REIN 4.102	12 36 02.70	+ 01 40 29.8			NEBULA
A3 505	12 36 03.7	+ 14 37 36.		14.8	UV CONCENTRATION
SVEN 309	12 36 04.	- 04 40	60	14.8	GALAXY
SC 1233+0202.0	12 36 05.	+ 01 45 29.	'2		NEBULA
REIN 4.103	12 36 05.24	+ 01 44 53.6			NEBULA
A3 506	12 36 05.7	+ 14 30 16.		17.0	FAINT BLUE OBJECT
ZWG 014.086	12 36 06.	+ 01 39		15.2	GALAXY
UGC 07806	12 36 06.	+ 01 39	114	15.2	GALAXY Sc
MCG+00-32-030	12 36 06.	+ 01 39	114	15.2	GALAXY
ZWG 014.087	12 36 06.	+ 01 44		15.3	GALAXY
UGC 07807	12 36 06.	+ 01 44	60	15.3	GALAXY S
ZWG 070.201	12 36 06.	+ 10 45		15.4	GALAXY
UGC 07808	12 36 06.	+ 10 45	210	15.4	GALAXY Sb
RMB 115	12 36 06.	+ 11 58			FAINT BLUE OBJECT
8ZW 1236+14.3	12 36 06.	+ 14 18		19.3	COMPACT GALAXY

OBJECT NAME	RIGHT ASCEN.	DECLINATION	DIAM.	MAGN.	TYPE OF OBJECT
TON-N 0095	12 36 06.	+ 14 22		15.6	BLUE STAR
AMES 1747	12 36 06.	+ 17 09 29.	19	18.0	NEBULA
LB 06742	12 36 06.	+ 28 24		19.2	FAINT BLUE STAR
LB 11279	12 36 06.	+ 28 31		14.5	FAINT BLUE STAR
MRK 219	12 36 06.	+ 56 11	20	16.	GALAXY WITH UV CONTINUUM
ZWG 335.018	12 36 06.	+ 71 42		15.5	GALAXY
UGC 07809	12 36 06.	+ 71 42	66	15.5	GALAXY SBc
MCG-07-26-019	12 36 06.	- 41 57	72	16.	GALAXY
KN 16.156	12 36 06.4	+ 28 07 25.			NEBULA
KN 16.157	12 36 06.9	+ 26 44 37.			NEBULA
IC 3608	12 36 07.	+ 10 44			NONSTELLAR OBJECT
HN 0992	12 36 07.	+ 14 37	12		NEBULA
IC 3609	12 36 07.	+ 14 37			NONSTELLAR OBJECT
HN 0993	12 36 08.	+ 10 44	60		NEBULA
HOLM 433C	12 36 10.	+ 44 56	18	15.3	PART OF MULTIPLE GALAXY
AMES 1749	12 36 11.	+ 14 00 23.	23	16.6	NEBULA
ZWG 014.088	12 36 12.	+ 00 35		15.7	GALAXY
SC 1233+0052.5	12 36 12.	+ 00 35 59.	12		NEBULA
ZWG 042.189	12 36 12.	+ 07 02		15.1	GALAXY
UGC 07810	12 36 12.	+ 07 02	84	15.1	GALAXY Sc
MCG+01-32-124	12 36 12.	+ 07 02	66	15.1	GALAXY
AMES 1748	12 36 12.	+ 16 45 35.	20	17.6	NEBULA
LB 06747	12 36 12.	+ 27 12		19.5	FAINT BLUE STAR
LB 06746	12 36 12.	+ 29 01		19.0	FAINT BLUE STAR
LB 06745	12 36 12.	+ 29 42		19.5	FAINT BLUE STAR
LB 06744	12 36 12.	+ 31 56		19.7	FAINT BLUE STAR
LB 06743	12 36 12.	+ 32 17		17.8	FAINT BLUE STAR
OCL 0379	12 36 12.	+ 36 35	900		OPEN STAR CLUSTER
LB 02410	12 36 12.	+ 52 50 00.		15.8	FAINT BLUE STAR
ZC 1236.2+6806	12 36 12.	+ 68 06	1550		CLUSTER OF GALAXIES
HOLM 434B	12 36 12.	+ 71 40	18	15.4	PART OF MULTIPLE GALAXY
MCG-04-30-007	12 36 12.	- 27 03	48	15.	GALAXY
REIN 4.104	12 36 12.97	+ 00 35 47.5			NEBULA
RNGC 4588	12 36 14.	+ 07 02		15.0	GALAXY
AMES 1751	12 36 14.	+ 10 42 29.	29	17.2	NEBULA
HOLM 433B	12 36 14.	+ 44 57	18	15.3	PART OF MULTIPLE GALAXY
SC 1233+0049.5	12 36 15.	+ 00 32 59.	12		NEBULA
AMES 1750	12 36 15.	+ 11 59 05.	16	17.8	NEBULA
AMES 1753	12 36 16.	+ 16 19 35.	26	17.0	NEBULA
REIZ 2644	12 36 16.	+ 27 08	42	15.9	GALAXY
REIZ 2646	12 36 16.	+ 47 49	12	14.8	GALAXY
KN 16.158	12 36 16.9	+ 27 59 50.			NEBULA
AMES 1755	12 36 17.	+ 12 52 11.	30	17.8	NEBULA
AMES 1752	12 36 17.	+ 17 31 59.	14	16.9	NEBULA
HOLM 433A	12 36 17.	+ 44 56	18	15.2	PART OF MULTIPLE GALAXY
SVEN 310	12 36 17.	- 04 47 01.	24	15.1	GALAXY
SVEN 311	12 36 17.	- 05 26	24	15.1	GALAXY
ZWG 042.190	12 36 18.	+ 07 24		15.5	GALAXY
REA 19	12 36 18.	+ 10 06			DWARF GALAXY
MCG+02-32-163	12 36 18.	+ 10 06	18	18.	GALAXY
RMB 114	12 36 18.	+ 11 59			FAINT BLUE OBJECT
AMES 1754	12 36 18.	+ 18 02 11.	14	17.4	NEBULA
8ZW 1236+25.1	12 36 18.	+ 25 07		17.4	COMPACT GALAXY
8ZW 1236+25.4	12 36 18.	+ 25 25		17.6	COMPACT GALAXY
LB 06749	12 36 18.	+ 27 11		18.6	FAINT BLUE STAR
LB 06748	12 36 18.	+ 29 42		18.9	FAINT BLUE STAR
LB 11280	12 36 18.	+ 30 06		17.0	FAINT BLUE STAR
MCG+09-21-026	12 36 18.	+ 53 43	18	18.	GALAXY
ZC 1236.3+5553	12 36 18.	+ 55 53	1340		CLUSTER OF GALAXIES
MCG-04-30-008	12 36 18.	- 25 53	15	16.	GALAXY
AMES 1756	12 36 19.	+ 17 31 53.	19	17.1	NEBULA
IC 3610	12 36 19.	+ 27 08 52.			NONSTELLAR OBJECT
HELW 154	12 36 19.	- 05 25 31.			NEBULA
KN 16.159	12 36 19.2	+ 27 08 56.			NEBULA
REIZ 2643	12 36 20.	+ 07 04	48	15.4	GALAXY
AMES 1759	12 36 20.	+ 13 49 59.	18	17.2	NEBULA
AMES 1758	12 36 20.	+ 15 47 59.	24	17.9	NEBULA
SEY 190	12 36 20.	+ 32 21 48.		14.0	FAINT GALAXY
KN 16.160	12 36 20.3	+ 29 06 56.			NEBULA
AMES 1757	12 36 21.	+ 17 15 23.	24	17.6	NEBULA
REIZ 2645	12 36 21.	+ 32 21	48	14.7	GALAXY
KN 16.161	12 36 21.7	+ 25 25 14.			NEBULA
AMES 1762	12 36 22.	+ 12 11 47.	11	17.2	NEBULA
AMES 1761	12 36 22.	+ 15 16 23.	25	17.1	NEBULA
REIZ 2642	12 36 22.	- 02 44	18	15.8	GALAXY
KN 16.162	12 36 22.3	+ 25 07 08.			NEBULA
AMES 1760	12 36 23.	+ 16 48 35.	20	17.6	NEBULA
LB 02411	12 36 23.	+ 57 38 36.		16.0	FAINT BLUE STAR
RMB 107	12 36 24.	+ 12 25			FAINT BLUE OBJECT
AMES 1765	12 36 24.	+ 16 41 23.	22	17.6	NEBULA
AMES 1764	12 36 24.	+ 17 45 35.	20	17.2	NEBULA
AMES 1763	12 36 24.	+ 18 06 59.	19	17.4	NEBULA
ZC 1236.4+2515	12 36 24.	+ 25 15	1340		CLUSTER OF GALAXIES
LB 06756	12 36 24.	+ 27 02		18.8	FAINT BLUE STAR
LB 06755	12 36 24.	+ 27 14		18.8	FAINT BLUE STAR
LB 06754	12 36 24.	+ 27 56		19.9	FAINT BLUE STAR
LB 06753	12 36 24.	+ 27 58		19.8	FAINT BLUE STAR
LB 11283	12 36 24.	+ 30 01		17.0	FAINT BLUE STAR
LB 06752	12 36 24.	+ 30 44		18.7	FAINT BLUE STAR
LB 06751	12 36 24.	+ 30 53		19.6	FAINT BLUE STAR
LB 06750	12 36 24.	+ 31 11		18.9	FAINT BLUE STAR
LB 11282	12 36 24.	+ 32 13		12.0	FAINT BLUE STAR
ZWG 159.038	12 36 24.	+ 32 16		14.6	GALAXY
UGC 07811	12 36 24.	+ 32 16	96	14.6	GALAXY S?
MCG+05-30-043	12 36 24.	+ 32 16	72	14.6	GALAXY
ZWG 159.039	12 36 24.	+ 32 23		14.0	GALAXY
LB 11281	12 36 24.	+ 32 23		15.0	FAINT BLUE STAR
UGC 07812	12 36 24.	+ 32 23	60	14.0	GALAXY S
ZC 1236.4+4049	12 36 24.	+ 40 49	1410		CLUSTER OF GALAXIES
KARA.68 171	12 36 24.	- 00 23	40		DWARF GALAXY
KN 16.163	12 36 24.4	+ 29 31 50.			NEBULA
AMES 1769	12 36 26.	+ 11 23 59.	25	16.8	NEBULA
AMES 1768	12 36 27.	+ 17 02 59.	14	17.0	NEBULA
AMES 1767	12 36 27.	+ 17 09 59.	17	17.0	NEBULA
AMES 1766	12 36 27.	+ 17 31 47.	23	17.2	NEBULA
MCG+05-30-044	12 36 27.	+ 32 23 30.	60	14.0	GALAXY
REIF 4.105	12 36 27.51	+ 00 38 23.8			NEBULA
SC 1233+0037.0	12 36 28.	+ 00 20 30.	6		NEBULA
AMES 1770	12 36 28.	+ 17 53 47.	22	17.8	NEBULA
AMES 1772	12 36 29.	+ 12 17 35.	25	17.8	NEBULA
AMES 1771	12 36 29.	+ 13 15 05.	25	17.3	NEBULA
KN 16.164	12 36 29.8	+ 26 54 50.			NEBULA
ZWG 014.090	12 36 30.	+ 00 38		14.4	GALAXY
UGC 07813	12 36 30.	+ 00 38	54	14.4	GALAXY COMPACT
MCG+00-32-031	12 36 30.	+ 00 38	15	14.4	GALAXY
RMB 192	12 36 30.	+ 10 57			FAINT BLUE OBJECT
RMB 076	12 36 30.	+ 13 08			FAINT BLUE OBJECT
RMB 075	12 36 30.	+ 13 13			FAINT BLUE OBJECT

OBJECT NAME	RIGHT ASCEN.	DECLINATION	DIAM.	MAGN.	TYPE OF OBJECT
REA 89	12 36 30.	+ 13 13			DWARF GALAXY
AMES 1775	12 36 30.	+ 13 19 41.	19	17.3	NEBULA
ZWG 099.102	12 36 30.	+ 15 00		15.4	GALAXY
UGC 07814	12 36 30.	+ 15 00	66	15.4	GALAXY S
ZWG 099.103	12 36 30.	+ 18 01		15.7	GALAXY
ZWG 099.104	12 36 30.	+ 18 27		15.4	GALAXY
REIZ 2647	12 36 30.	+ 18 27	54	15.6	GALAXY
UGC 07815	12 36 30.	+ 18 27	78	15.4	GALAXY Sc
MCG+03-32-079	12 36 30.	+ 18 28	66	15.4	GALAXY
8ZW 1236+25.6	12 36 30.	+ 25 34		15.4	COMPACT GALAXY
LB 00046	12 36 30.	+ 27 34		16.2	FAINT BLUE STAR
MRK 651	12 36 30.	+ 28 36	10	16.5	GALAXY WITH UV CONTINUUM
LB 06761	12 36 30.	+ 29 15		18.8	FAINT BLUE STAR
LB 06760	12 36 30.	+ 30 15		20.0	FAINT BLUE STAR
LB 06759	12 36 30.	+ 30 22		18.6	FAINT BLUE STAR
LB 06758	12 36 30.	+ 30 56		19.4	FAINT BLUE STAR
LB 11284	12 36 30.	+ 30 59		15.8	FAINT BLUE STAR
LB 06757	12 36 30.	+ 31 13		19.4	FAINT BLUE STAR
ZC 1236.5+3453	12 36 30.	+ 34 53	870		CLUSTER OF GALAXIES
ZWG 188.012	12 36 30.	+ 38 22		15.4	GALAXY
UGC 07816	12 36 30.	+ 38 22	108	15.4	GALAXY PECULR
ZWG 014.089	12 36 30.	- 02 15		15.3	GALAXY
MCG-02-32-019	12 36 30.	- 09 55	66	15.	GALAXY
MCG-06-28-009	12 36 30.	- 34 30	36	15.	GALAXY
AMES 1774	12 36 31.	+ 16 35 05.	31	16.6	NEBULA
AMES 1773	12 36 31.	+ 19 05 22.	18	16.5	NEBULA
HOLM 434A	12 36 32.	+ 71 41	36	14.6	PART OF MULTIPLE GALAXY
REIN 4.106	12 36 32.25	- 02 14 02.7			NEBULA
SC 1233+0100.2	12 36 33.	+ 00 43 42.	12		NEBULA
IC 3611	12 36 33.	+ 13 38 10.			GALAXY
AMES 1776	12 36 33.	+ 16 31 04.	23	17.6	NEBULA
IC 3614	12 36 33.	+ 26 34 34.			NONSTELLAR OBJECT
KN 16.165	12 36 33.2	+ 26 34 38.			NEBULA
KN 16.166	12 36 33.2	+ 27 52 50.			NEBULA
KN 16.167	12 36 33.7	+ 26 34 20.			NEBULA
AMES 1777	12 36 34.	+ 18 02 10.	22	17.0	NEBULA
REIZ 2651	12 36 34.	+ 26 33	54	15.3	GALAXY
SVEN 313	12 36 34.	- 04 33 01.	18	15.0	GALAXY
HELW 155	12 36 34.	- 04 33 12.			NEBULA
AMES 1780	12 36 35.	+ 16 30 10.	25	18.0	NEBULA
AMES 1779	12 36 35.	+ 18 01 46.	17	16.5	NEBULA
AMES 1778	12 36 35.	+ 19 08 28.	17	16.7	NEBULA
REIZ 2650	12 36 35.	+ 23 25	72	14.8	GALAXY
SVEN 312	12 36 35.	- 04 49	18	15.2	GALAXY
KN 16.168	12 36 35.0	+ 26 33 50.			NEBULA
AMES 1782	12 36 36.	+ 10 30 34.	22	16.8	NEBULA
ZWG 070.202	12 36 36.	+ 13 38		14.7	GALAXY
UGC 07817	12 36 36.	+ 13 38	114	14.7	GALAXY S
MCG+02-32-164	12 36 36.	+ 13 38	42	14.7	GALAXY
KARA.68 172	12 36 36.	+ 15 53	40		DWARF GALAXY
HN 0994	12 36 36.	+ 18 27	30		NEBULA
IC 3615	12 36 36.	+ 18 27			NONSTELLAR OBJECT
TON-N 0628	12 36 36.	+ 25 30		14.5	BLUE STAR
LB 06763	12 36 36.	+ 26 28		17.9	FAINT BLUE STAR
LB 11286	12 36 36.	+ 26 48		14.4	FAINT BLUE STAR
LB 06762	12 36 36.	+ 26 59		20.0	FAINT BLUE STAR
LB 11285	12 36 36.	+ 30 08		14.7	FAINT BLUE STAR
ZWG 159.040	12 36 36.	+ 30 41		15.2	GALAXY
UGC 07818	12 36 36.	+ 30 41	66	15.2	GALAXY Sa?
TON-N 0096	12 36 36.	+ 47 56		15.1	BLUE STAR
MCG-04-30-009	12 36 36.	- 26 38 30.	60	15.5	GALAXY
SC 1234+0116.5	12 36 37.	+ 01 00 00.	48		NEBULA
HN 0996	12 36 37.	+ 14 00	6		NEBULA
IC 3613	12 36 37.	+ 14 00			NONSTELLAR OBJECT
HN 0995	12 36 37.	+ 14 59	12		NEBULA
IC 3612	12 36 37.	+ 14 59			NONSTELLAR OBJECT
REIN 4.107	12 36 37.70	+ 00 59 30.2			NEBULA
REIZ 2649	12 36 38.	+ 06 17	36	13.9	GALAXY
AMES 1783	12 36 38.	+ 11 24 28.	20	17.7	NEBULA
AMES 1781	12 36 38.	+ 19 29 16.	36	17.4	NEBULA
REIZ 2648	12 36 39.	+ 00 59	30	14.8	GALAXY
AMES 1790	12 36 39.	+ 11 07 58.	32	17.6	NEBULA
KN 16.169	12 36 39.8	+ 27 46 08.			NEBULA
SC 1234+0200.0	12 36 40.	+ 01 43 30.	18		NEBULA
AMES 1789	12 36 40.	+ 13 44 40.	20	17.1	NEBULA
AMES 1786	12 36 40.	+ 16 35 04.	29	17.2	NEBULA
AMES 1785	12 36 40.	+ 16 52 10.	24	17.4	NEBULA
AMES 1784	12 36 40.	+ 18 02 40.	26	17.2	NEBULA
AMES 1788	12 36 41.	+ 17 48 46.	25	17.6	NEBULA
AMES 1787	12 36 41.	+ 18 00 16.	22	17.6	NEBULA
REIZ 2654	12 36 41.	+ 23 21	78	15.0	GALAXY
ZWG 014.092	12 36 42.	+ 00 59		14.6	GALAXY
UGC 07820	12 36 42.	+ 00 59	108	14.6	GALAXY Sc
MCG+00-32-033	12 36 42.	+ 00 59	114	14.6	GALAXY
ZWG 014.093	12 36 42.	+ 01 42		15.4	GALAXY
ZWG 042.191	12 36 42.	+ 06 17		14.1	GALAXY
UGC 07821	12 36 42.	+ 06 17	108	14.1	GALAXY S
MCG+01-32-125	12 36 42.	+ 06 17	72	14.1	GALAXY
KARA.68 173	12 36 42.	+ 08 21	27		DWARF GALAXY
8ZW 1236+12.5	12 36 42.	+ 12 28		17.9	COMPACT GALAXY
ZC 1236.7+1603	12 36 42.	+ 16 03	810		CLUSTER OF GALAXIES
ZC 1236.7+1707	12 36 42.	+ 17 07	3160		CLUSTER OF GALAXIES
AMES 1791	12 36 42.	+ 18 02 04.	24	17.6	NEBULA
ZWG 159.041	12 36 42.	+ 26 57		15.5	GALAXY
LB 06769	12 36 42.	+ 27 23		19.5	FAINT BLUE STAR
ZWG 159.042	12 36 42.	+ 28 01		15.4	GALAXY
MCG+05-30-046	12 36 42.	+ 28 02	15	15.4	GALAXY
LB 06768	12 36 42.	+ 28 11		19.3	FAINT BLUE STAR
MRK 652	12 36 42.	+ 28 17	9	16.	GALAXY WITH UV CONTINUUM
LB 06767	12 36 42.	+ 29 12		19.6	FAINT BLUE STAR
LB 06766	12 36 42.	+ 29 27		19.2	FAINT BLUE STAR
MCG+05-30-045	12 36 42.	+ 30 41	42	15.2	GALAXY
LB 06765	12 36 42.	+ 31 26		19.3	FAINT BLUE STAR
LB 06764	12 36 42.	+ 31 53		19.8	FAINT BLUE STAR
7ZW 476	12 36 42.	+ 61 03			COMPACT GALAXY
ZWG 014.091	12 36 42.	- 00 15		12.6	GALAXY
UGC 07819	12 36 42.	- 00 15	306	12.6	GALAXY Sc-IRR
MCG+00-32-032	12 36 42.	- 00 15	288	12.6	GALAXY
HN 0997	12 36 43.	+ 15 00		14.	NEBULA
IC 3616	12 36 43.	+ 15 00			MAY NOT EXIST
AMES 1792	12 36 43.	+ 18 00 04.	22	18.0	NEBULA
RNGC 4592	12 36 43.	- 00 15		12.5	GALAXY
RNGC 4591	12 36 44.	+ 06 17		14.0	GALAXY
KN 16.170	12 36 44.4	+ 28 00 56.			NEBULA
REIN 4.108	12 36 44.68	- 00 15 26.4			NEBULA
REIZ 2653	12 36 45.	+ 05 12	30	15.7	GALAXY
AMES 1796	12 36 45.	+ 14 10 52.	29	17.2	NEBULA
AMES 1794	12 36 45.	+ 18 38 22.	38	16.5	NEBULA

OBJECT NAME	RIGHT ASCEN.	DECLINATION	DIAM.	MAGN.	TYPE OF OBJECT
AMES 1793	12 36 45.	+ 19 08 40.	22	17.6	NEBULA
LB 02412	12 36 45.	+ 57 28 42.		16.7	FAINT BLUE STAR
MCG-01-32-031	12 36 45.	- 03 31	60	17.	GALAXY
KN 16.171	12 36 45.7	+ 28 05 14.			NEBULA
SC 1234+0159.2	12 36 46.	+ 01 42 42.	18		NEBULA
AMES 1797	12 36 46.	+ 14 21 40.	17	16.8	NEBULA
AMES 1795	12 36 46.	+ 18 17 40.	35	16.9	NEBULA
REIZ 2656	12 36 46.	+ 26 56	18	14.5	GALAXY
REIZ 2652	12 36 46.	- 00 16	180	12.6	GALAXY
RNGC 4590	12 36 46.	- 26 29		9.0	GLOBULAR CLUSTER
REIN 4.109	12 36 46.25	+ 01 42 20.5			NEBULA
LB 02413	12 36 47.	+ 61 28 00.		16.5	FAINT BLUE STAR
ZWG 014.094	12 36 48.	+ 00 43		15.6	GALAXY
ZWG 014.095	12 36 48.	+ 01 41		15.6	GALAXY
ZC 1236.8+0305	12 36 48.	+ 03 05	2290		CLUSTER OF GALAXIES
ZWG 042.192	12 36 48.	+ 05 13		15.4	GALAXY
MCG+01-32-126	12 36 48.	+ 05 13	60	15.4	GALAXY
AMES 1800	12 36 48.	+ 11 02 10.	29	17.5	NEBULA
RMB 191	12 36 48.	+ 11 03			FAINT BLUE OBJECT
AMES 1799	12 36 48.	+ 16 33 28.	24	17.4	NEBULA
AMES 1798	12 36 48.	+ 16 40 04.	14	17.8	NEBULA
ZC 1236.8+1901	12 36 48.	+ 19 01	740		CLUSTER OF GALAXIES
LB 06772	12 36 48.	+ 26 56		19.5	FAINT BLUE STAR
TON-N 0629	12 36 48.	+ 27 55		15.5	BLUE STAR
LB 06771	12 36 48.	+ 27 56		18.8	FAINT BLUE STAR
VV 287B	12 36 48.	+ 27 59	9	18.	INTERACTING GALAXY
VV 287A	12 36 48.	+ 27 59	18	18.	INTERACTING GALAXY
MCG+05-30-049	12 36 48.	+ 27 59	36	18.	GALAXY
ZWG 159.043	12 36 48.	+ 28 03		15.3	GALAXY
MCG+05-30-048	12 36 48.	+ 28 03	36	15.3	GALAXY
ZWG 159.044	12 36 48.	+ 28 11		15.6	GALAXY
ZWG 159.045	12 36 48.	+ 29 11		15.6	GALAXY
LB 06770	12 36 48.	+ 31 48		19.9	FAINT BLUE STAR
LB 11287	12 36 48.	+ 32 11		15.7	FAINT BLUE STAR
MCG+05-30-047	12 36 48.	+ 32 17	36	15.6	GALAXY
ZC 1236.8+4149	12 36 48.	+ 41 49	2960		CLUSTER OF GALAXIES
GCL 020	12 36 48.	- 26 28	588	9.12	GLOBULAR STAR CLUSTER
OCL 0888	12 36 48.	- 60 20	1440	10.3	OPEN STAR CLUSTER
VHA 137	12 36 48.	- 60 20	420		OPEN STAR CLUSTER
KN 16.172	12 36 48.6	+ 27 59 32.			NEBULA
REIN 4.110	12 36 48.87	+ 00 44 03.4			NEBULA
SC 1234+0101.0	12 36 49.	+ 00 44 30.	18		NEBULA
IC 3619	12 36 49.	+ 26 57 10.			NONSTELLAR OBJECT
IC 3618	12 36 49.	+ 26 57 10.			NONSTELLAR OBJECT
KN 16.173	12 36 49.3	+ 26 57 14.			NEBULA
REIZ 2655	12 36 50.	+ 08 14	72	15.1	GALAXY
IC 3620	12 36 50.	+ 28 11 04.			NONSTELLAR OBJECT
KN 16.174	12 36 50.5	+ 28 10 56.			NEBULA
AMES 1802	12 36 51.	+ 18 36 34.	14	17.8	NEBULA
AMES 1801	12 36 51.	+ 18 38 46.	31	16.7	NEBULA
KN 16.176	12 36 51.0	+ 29 10 50.			NEBULA
KN 16.175	12 36 51.4	+ 28 02 56.			NEBULA
AMES 1804	12 36 52.	+ 16 19 04.	19	17.2	NEBULA
AMES 1803	12 36 52.	+ 17 28 52.	14	16.7	NEBULA
REIZ 2659	12 36 52.	+ 27 22	48	14.4	GALAXY
REIZ 2660	12 36 52.	+ 28 11	42	14.6	GALAXY
REIN 3.273	12 36 52.55	+ 08 14 24.5			NEBULA
KN 16.178	12 36 52.6	+ 27 52 44.			NEBULA
KN 16.177	12 36 52.9	+ 25 31 44.			NEBULA
IC 3617	12 36 53.	+ 08 14 28.			NONSTELLAR OBJECT
REIZ 2658	12 36 53.	+ 24 24	18	14.8	GALAXY
KARA.68 174	12 36 54.	+ 01 49	27		DWARF GALAXY
ZWG 042.193	12 36 54.	+ 04 33		15.7	GALAXY
ZWG 042.194	12 36 54.	+ 08 14		14.8	GALAXY
UGC 07822	12 36 54.	+ 08 14	72	14.8	GALAXY IRR
MCG+01-32-127	12 36 54.	+ 08 14	72	14.8	GALAXY
AMES 1806	12 36 54.	+ 13 39 52.	13	17.7	NEBULA
ZWG 159.046	12 36 54.	+ 27 23		15.2	GALAXY
LB 06773	12 36 54.	+ 30 04		19.3	FAINT BLUE STAR
LB 11288	12 36 54.	+ 30 46		14.8	FAINT BLUE STAR
ZWG 159.047	12 36 54.	+ 32 16		15.6	GALAXY
DV.56 N4603A	12 36 54.	- 40 26	132		S GALAXY
MCG-07-26-020	12 36 54.	- 40 28	60	15.	GALAXY
AMES 1805	12 36 55.	+ 17 56 52.	28	17.2	NEBULA
AMES 1808	12 36 56.	+ 16 42 28.	31	16.7	NEBULA
AMES 1807	12 36 56.	+ 17 30 04.	25	17.6	NEBULA
AMES 1809	12 36 57.	+ 18 39 16.	28	17.9	NEBULA
HELW 156	12 36 57.	- 05 36 00.			NEBULA
RNGC 4603A	12 36 57.	- 40 28			GALAXY
AMES 1811	12 36 58.	+ 12 56 52.	36	17.8	NEBULA
REIZ 2657	12 36 58.	- 00 45	30	15.7	GALAXY
AMES 1810	12 36 59.	+ 18 46 46.	25	17.4	NEBULA
SVEN 314	12 36 59.	- 05 02 00.	12	15.4	GALAXY
HELW 157	12 36 59.	- 05 02 06.			NEBULA
LB 09861	12 37	- 84 18		13.6	FAINT BLUE STAR
TON-N 0097	12 37 00.	+ 13 18		14.6	BLUE STAR
RMB 012	12 37 00.	+ 14 25			FAINT BLUE OBJECT
IC 3622	12 37 00.	+ 15 41			NONSTELLAR OBJECT
IC 3621	12 37 00.	+ 15 45			NONSTELLAR OBJECT
ARP 149	12 37 00.	+ 16 53			PECULIAR GALAXY
AMES 1812	12 37 00.	+ 17 49 22.	20	17.6	NEBULA
ZC 1237.0+2024	12 37 00.	+ 20 24	400		CLUSTER OF GALAXIES
ZWG 129.014	12 37 00.	+ 25 57		15.7	GALAXY
LB 06777	12 37 00.	+ 26 43		19.6	FAINT BLUE STAR
IC 3623	12 37 00.	+ 27 22 41.			NONSTELLAR OBJECT
MCG+05-30-050	12 37 00.	+ 27 23	21	15.2	GALAXY
LB 06776	12 37 00.	+ 28 08		19.5	FAINT BLUE STAR
LB 06775	12 37 00.	+ 28 34		19.3	FAINT BLUE STAR
LB 06774	12 37 00.	+ 28 58		18.9	FAINT BLUE STAR
LB 11291	12 37 00.	+ 30 30		14.8	FAINT BLUE STAR
LB 11290	12 37 00.	+ 30 32		13.3	FAINT BLUE STAR
LB 11289	12 37 00.	+ 31 39		14.4	FAINT BLUE STAR
ZC 1237.0+4429	12 37 00.	+ 44 29	740		CLUSTER OF GALAXIES
7ZW 477	12 37 00.	+ 59 42			COMPACT GALAXY
LB 00647	12 37 00.	+ 62 18 00.		16.4	FAINT BLUE STAR
KN 16.179	12 37 00.0	+ 27 22 44.			NEBULA
HN 0999	12 37 01.	+ 15 41	18		NEBULA
HN 0998	12 37 01.	+ 15 45		14.	NEBULA
RNGC 4593	12 37 01.	- 05 04		12.0	GALAXY
KN 16.180	12 37 01.2	+ 26 09 26.			NEBULA
KN 16.181	12 37 01.2	+ 26 54 44.			NEBULA
SC 1234+1018.7	12 37 02.	+ 10 02 13.	18		NEBULA
AMES 1813	12 37 02.	+ 12 43 16.	28	16.4	NEBULA
KN 16.182	12 37 03.8	+ 25 57 02.			NEBULA
HOLM 435C	12 37 04.	+ 16 52	12	15.4	PART OF MULTIPLE GALAXY
IC 3626	12 37 04.	+ 25 57 11.			NONSTELLAR OBJECT
IC 3627	12 37 04.	+ 27 46 17.			NONSTELLAR OBJECT
REIZ 2662	12 37 04.	+ 28 41	12	15.0	GALAXY

OBJECT NAME	RIGHT ASCEN.	DECLINATION	DIAM.	MAGN.	TYPE OF OBJECT
KN 16.183	12 37 04.4	+ 27 46 20.			NEBULA
AMES 1814	12 37 05.	+ 16 19 04.	20	16.7	NEBULA
REIZ 2661	12 37 05.	- 05 04	120	12.5	GALAXY
SVEN 315	12 37 05.	- 05 05	210	11.5	GALAXY
ZWG 014.096	12 37 06.	+ 01 01		15.4	GALAXY
KARA.68 175	12 37 06.	+ 07 17	40		DWARF GALAXY
ZWG 042.195	12 37 06.	+ 07 27		15.4	GALAXY
AMES 1817	12 37 06.	+ 10 51 04.	16	17.2	NEBULA
ZWG 070.203	12 37 06.	+ 11 15		15.5	GALAXY
MCG+02-32-166	12 37 06.	+ 11 15	3	15.5	GALAXY
MCG+02-32-165	12 37 06.	+ 11 15	12	15.5	GALAXY
RMB 113	12 37 06.	+ 11 57			FAINT BLUE OBJECT
RMB 106	12 37 06.	+ 12 23			FAINT BLUE OBJECT
RMB 074	12 37 06.	+ 13 06			FAINT BLUE OBJECT
ZWG 099.105	12 37 06.	+ 16 51		15.3	GALAXY
IC 0803	12 37 06.	+ 16 51 05.			NONSTELLAR OBJECT
HOLM 435A	12 37 06.	+ 16 52	24	15.6	PART OF MULTIPLE GALAXY
MCG+03-32-080	12 37 06.	+ 16 52	42	15.3	GALAXY
ZC 1237.1+1801	12 37 06.	+ 18 01	2350		CLUSTER OF GALAXIES
LB 06781	12 37 06.	+ 27 58		19.6	FAINT BLUE STAR
LB 11292	12 37 06.	+ 28 45		14.7	FAINT BLUE STAR
LB 06780	12 37 06.	+ 31 19		19.5	FAINT BLUE STAR
LB 06779	12 37 06.	+ 31 49		20.1	FAINT BLUE STAR
LB 06778	12 37 06.	+ 32 20		20.0	FAINT BLUE STAR
ZC 1237.1+6008	12 37 06.	+ 60 08	870		CLUSTER OF GALAXIES
MCG-01-32-032	12 37 06.	- 05 03	180	11.	GALAXY
MCG-07-26-021	12 37 06.	- 40 50	42	14.5	GALAXY
IC 3624	12 37 07.	+ 12 15			NONSTELLAR OBJECT
HN 1000	12 37 07.	+ 12 16	24		NEBULA
HOLM 435B	12 37 07.	+ 16 52	18	15.4	PART OF MULTIPLE GALAXY
IC 3625	12 37 08.	+ 11 14			NONSTELLAR OBJECT
AMES 1818	12 37 08.	+ 11 14 52.	35	17.2	NEBULA
HN 1001	12 37 08.	+ 11 15			NEBULA
AMES 1816	12 37 08.	+ 16 40 52.	23	17.8	NEBULA
AMES 1815	12 37 08.	+ 18 07 16.	37	17.0	NEBULA
ABC 1582	12 37 09.	+ 52 23		17.8	RICH CLUSTER OF GALAXIES
REIN 4.111	12 37 09.21	- 00 45 56.9			NEBULA
REIZ 2665	12 37 10.	+ 26 30	12	15.4	GALAXY
KN 16.183A	12 37 10.8	+ 26 30 56.			NEBULA
REIZ 2664	12 37 11.	+ 25 57	60	14.7	GALAXY
IC 3628	12 37 11.	+ 26 30 59.			NONSTELLAR OBJECT
SC 1234+0127.5	12 37 12.	+ 01 11 00.	12		NEBULA
AMES 1823	12 37 12.	+ 10 34 52.	19	16.8	NEBULA
RMB 190	12 37 12.	+ 11 04			FAINT BLUE OBJECT
AMES 1820	12 37 12.	+ 13 52 34.	24	17.6	NEBULA
RMB 007	12 37 12.	+ 15 08			FAINT BLUE OBJECT
8ZW 1237+25.3	12 37 12.	+ 25 18		17.5	COMPACT GALAXY
LB 06784	12 37 12.	+ 27 16		19.7	FAINT BLUE STAR
LB 06783	12 37 12.	+ 27 32		20.0	FAINT BLUE STAR
LB 06782	12 37 12.	+ 27 58		18.9	FAINT BLUE STAR
LB 11293	12 37 12.	+ 31 57		14.6	FAINT BLUE STAR
ZWG 244.035	12 37 12.	+ 47 54		14.9	GALAXY
UGC 07823	12 37 12.	+ 47 54	66	14.9	GALAXY Sc
7ZW 478	12 37 12.	+ 59 44			COMPACT GALAXY
KN 16.184	12 37 12.2	+ 27 16 32.			NEBULA
AMES 1822	12 37 13.	+ 13 44 52.	17	17.5	NEBULA
AMES 1821	12 37 13.	+ 15 08 16.	22	17.1	NEBULA
AMES 1819	12 37 13.	+ 17 19 52.	17	17.6	NEBULA
AMES 1824	12 37 14.	+ 13 18 52.	23	17.8	NEBULA
ABC 1581	12 37 15.	+ 03 02		17.2	RICH CLUSTER OF GALAXIES
MCG+08-23-070	12 37 15.	+ 47 55	60	14.	GALAXY
REIN 4.112	12 37 15.11	- 00 42 25.5			NEBULA
AMES 1826	12 37 16.	+ 12 43 52.	26	17.8	NEBULA
AMES 1825	12 37 16.	+ 16 19 40.	38	17.8	NEBULA
REIZ 2670	12 37 16.	+ 47 53	48	13.4	GALAXY
REIZ 2663	12 37 16.	- 00 47	18	15.3	GALAXY
KN 16.185	12 37 16.1	+ 28 28 27.			NEBULA
A3 509	12 37 16.2	+ 15 35 12.		16.8	UV CONCENTRATION
REIZ 2667	12 37 17.	+ 25 41	9	15.3	GALAXY
SVEN 316	12 37 17.	- 05 05	24	14.6	GALAXY
KN 16.186	12 37 17.1	+ 28 06 09.			NEBULA
REIN 4.113	12 37 17.34	+ 01 56 48.5			NEBULA
ZWG 014.097	12 37 18.	+ 01 56		15.5	GALAXY
UGC 07824	12 37 18.	+ 01 56	96	15.5	GALAXY
ZWG 042.196	12 37 18.	+ 04 05		15.6	GALAXY
KARA.73B 0542	12 37 18.	+ 04 05	24	15.6	ISOLATED GALAXY S
RMB 111	12 37 18.	+ 11 57			FAINT BLUE OBJECT
ZWG 070.204	12 37 18.	+ 13 15		14.5	GALAXY
UGC 07825	12 37 18.	+ 13 15	60	14.5	GALAXY
MCG+02-32-169	12 37 18.	+ 13 15	42	14.5	GALAXY
ZWG 070.205	12 37 18.	+ 13 48		15.2	GALAXY
MCG+02-32-167	12 37 18.	+ 13 48	42	15.2	GALAXY
MCG+02-32-168	12 37 18.	+ 13 49	12	15.2	GALAXY
IC 3631	12 37 18.	+ 13 15		14.5	GALAXY IN VIRGO CLSTR
MCG+03-32-081	12 37 18.	+ 15 33	84	12.8	GALAXY
ZWG 099.106	12 37 18.	+ 15 34		12.8	GALAXY
SCH 63	12 37 18.	+ 15 34		12.8	PECULIAR GALAXY
UGC 07826	12 37 18.	+ 15 34	102	12.8	GALAXY S
LB 06788	12 37 18.	+ 27 06		19.4	FAINT BLUE STAR
LB 06787	12 37 18.	+ 28 11		19.7	FAINT BLUE STAR
LB 06786	12 37 18.	+ 29 22		20.1	FAINT BLUE STAR
LB 06785	12 37 18.	+ 30 49		18.8	FAINT BLUE STAR
LB 11295	12 37 18.	+ 31 05		18.1	FAINT BLUE STAR
LB 11294	12 37 18.	+ 31 07		17.9	FAINT BLUE STAR
ZC 1237.3+3758	12 37 18.	+ 37 58	1810		CLUSTER OF GALAXIES
MCG+08-23-071	12 37 18.	+ 45 06	78	15.	GALAXY
UGC 07827	12 37 18.	+ 45 07	72	17.	GALAXY IRR
KARA.68 176	12 37 18.	- 00 12	34		DWARF GALAXY
MCG-01-32-033	12 37 18.	- 05 04	18	15.5	GALAXY
RNGC 4594	12 37 18.	- 11 21		9.5	GALAXY
MCG-02-32-020	12 37 18.	- 11 21	480	9.	GALAXY
MRSL 301+01/3	12 37 18.	- 61 36	900		HII REGION
IC 3629	12 37 19.	+ 13 47			NONSTELLAR OBJECT
HN 1002	12 37 19.	+ 13 48	30		NEBULA
AMES 1830	12 37 19.	+ 13 49 04.	23	16.7	NEBULA
AMES 1828	12 37 19.	+ 17 23 46.	11	17.2	NEBULA
AMES 1827	12 37 19.	+ 18 40 40.	23	18.0	NEBULA
IC 3630	12 37 19.	+ 25 42 23.			NONSTELLAR OBJECT
AMES 1833	12 37 20.	+ 13 11 16.	18	17.5	NEBULA
AMES 1829	12 37 20.	+ 16 39 16.	16	17.8	NEBULA
HELW 158	12 37 20.	- 05 04 42.			NEBULA
RNGC 4595	12 37 21.	+ 15 34		13.0	GALAXY
AMES 1832	12 37 21.	+ 16 41 58.	34	17.5	NEBULA
AMES 1831	12 37 21.	+ 18 02 22.	28	17.4	NEBULA
LB 02414	12 37 21.	+ 54 49 42.		16.0	FAINT BLUE STAR
KN 16.187	12 37 21.9	+ 26 32 09.			NEBULA
SC 1234-0030.6	12 37 22.	- 00 47 06.	24		NEBULA
KN 16.188	12 37 22.1	+ 26 32 33.			NEBULA

OBJECT NAME	RIGHT ASCEN.	DECLINATION	DIAM.	MAGN.	TYPE OF OBJECT
AMES 1835	12 37 23.	+ 12 02 10.	20	17.0	NEBULA
REIZ 2673	12 37 23.	+ 45 07	60	14.6	GALAXY
SC 1234+0041.0	12 37 24.	+ 00 24 30.	12		NEBULA
ZWG 070.206	12 37 24.	+ 10 27		12.4	GALAXY
UGC 07828	12 37 24.	+ 10 27	270	12.4	GALAXY SB0
MCG+02-32-170	12 37 24.	+ 10 27	138	12.4	GALAXY
RMB 193	12 37 24.	+ 10 43			FAINT BLUE OBJECT
TON-N 0098	12 37 24.	+ 11 52		15.4	BLUE STAR
RMB 227	12 37 24.	+ 12 56			FAINT BLUE OBJECT
RMB 073	12 37 24.	+ 13 06			FAINT BLUE OBJECT
ZC 1237.4+1504	12 37 24.	+ 15 04	810		CLUSTER OF GALAXIES
RMB 003	12 37 24.	+ 15 46			FAINT BLUE OBJECT
AMES 1834	12 37 24.	+ 16 26 04.	34	17.8	NEBULA
LB 06791	12 37 24.	+ 26 28		18.7	FAINT BLUE STAR
LB 06790	12 37 24.	+ 27 30		19.6	FAINT BLUE STAR
LB 06789	12 37 24.	+ 32 28		17.7	FAINT BLUE STAR
ZWG 188.013	12 37 24.	+ 35 15		15.5	GALAXY
ZC 1237.4+7120	12 37 24.	+ 71 20	400		CLUSTER OF GALAXIES
ZWG 014.098	12 37 24.	- 03 20		15.6	GALAXY
REIZ 2666	12 37 24.	- 11 20	360	8.7	GALAXY
SVEN 317	12 37 24.	- 11 21 01.	300	8.3	GALAXY
MCG-06-28-010	12 37 24.	- 33 56	30	15.5	GALAXY
MCG-07-26-022	12 37 24.	- 41 47	72	14.5	GALAXY
REIN 3.274	12 37 24.46	+ 10 26 59.9			NEBULA
HN 1003	12 37 25.	+ 13 15		13.	NEBULA
REIZ 2668	12 37 25.	+ 15 35	66	12.9	GALAXY
AMES 1836	12 37 25.	+ 16 55 22.	23	17.1	NEBULA
RNGC 4596	12 37 26.	+ 10 27		12.0	GALAXY
REIZ 2669	12 37 26.	+ 10 27	108	12.1	GALAXY
AMES 1838	12 37 26.	+ 12 28 40.	22	17.2	NEBULA
AMES 1837	12 37 26.	+ 16 54 10.	22	17.4	NEBULA
KN 16.189	12 37 27.5	+ 28 06 09.			NEBULA
REIZ 2672	12 37 28.	+ 26 57	48	15.3	GALAXY
RMB 112	12 37 30.	+ 11 45			FAINT BLUE OBJECT
RMB 226	12 37 30.	+ 12 51			FAINT BLUE OBJECT
RMB 072	12 37 30.	+ 13 09			FAINT BLUE OBJECT
KARA.68 177	12 37 30.	+ 14 04	34		DWARF GALAXY
ZC 1237.5+2720	12 37 30.	+ 27 20	1410		CLUSTER OF GALAXIES
MCG+05-30-051	12 37 30.	+ 28 06	18	17.	GALAXY
ZC 1237.5+2836	12 37 30.	+ 28 36	870		CLUSTER OF GALAXIES
LB 06797	12 37 30.	+ 28 48		18.8	FAINT BLUE STAR
LB 11296	12 37 30.	+ 29 11		14.3	FAINT BLUE STAR
LB 06796	12 37 30.	+ 29 30		19.6	FAINT BLUE STAR
LB 06795	12 37 30.	+ 29 31		19.3	FAINT BLUE STAR
LB 06794	12 37 30.	+ 29 53		18.8	FAINT BLUE STAR
LB 06793	12 37 30.	+ 30 52		18.5	FAINT BLUE STAR
LB 06792	12 37 30.	+ 32 00		18.8	FAINT BLUE STAR
TON-N 0099	12 37 30.	+ 43 20		16.8	BLUE STAR
RNGC 4597	12 37 31.	- 05 32		12.5	GALAXY
AMES 1839	12 37 32.	+ 11 20 22.	22	16.6	NEBULA
IC 3632	12 37 32.	+ 26 57 29.			NONSTELLAR OBJECT
KN 16.190	12 37 32.1	+ 26 57 33.			NEBULA
AMES 1841	12 37 34.	+ 12 24 58.	26	17.2	NEBULA
AMES 1840	12 37 34.	+ 18 51 40.	24	18.0	NEBULA
REIZ 2671	12 37 35.	- 05 31	240	12.5	GALAXY
SVEN 318	12 37 35.	- 05 32	150	12.0	GALAXY
REIN 4.114	12 37 35.10	- 00 41 05.9			NEBULA
REIN 4.115	12 37 35.23	- 00 04 37.0			NEBULA
ZWG 042.197	12 37 36.	+ 05 39		15.7	GALAXY
RMB 006	12 37 36.	+ 15 08			FAINT BLUE OBJECT
AMES 1843	12 37 36.	+ 16 56 04.	24	17.6	NEBULA
AMES 1842	12 37 36.	+ 18 58 52.	24	17.0	NEBULA
ZC 1237.6+3109	12 37 36.	+ 31 09	1080		CLUSTER OF GALAXIES
KARA.68 178	12 37 36.	+ 32 58	47		DWARF GALAXY
ZC 1237.6+4330	12 37 36.	+ 43 30	610		CLUSTER OF GALAXIES
MCG-04-30-010	12 37 36.	- 25 03 30.	90	14.	GALAXY
MCG-07-26-023	12 37 36.	- 40 30	36	15.	GALAXY
MCG-07-26-024	12 37 36.	- 42 57	42	14.5	GALAXY
SC 1235+0115.2	12 37 38.	+ 00 58 42.	12		NEBULA
REIZ 2676	12 37 38.	+ 08 40	30	14.6	GALAXY
AMES 1844	12 37 38.	+ 18 20 46.	40	16.5	NEBULA
REIN 4.116	12 37 38.58	- 00 46 00.8			NEBULA
KN 16.191	12 37 38.6	+ 28 53 33.			NEBULA
AMES 1845	12 37 39.	+ 12 24 40.	35	17.5	NEBULA
MCG-01-32-034	12 37 39.	- 05 31	180	12.5	GALAXY
REIN 3.276	12 37 39.76	+ 10 10 12.1			NEBULA
REIZ 2675	12 37 40.	- 00 47	24	15.9	GALAXY
HELW 159	12 37 40.	- 05 06 48.			NEBULA
REIN 3.275	12 37 40.01	+ 08 39 29.3			NEBULA
AMES 1847	12 37 41.	+ 13 44 28.	22	17.7	NEBULA
AMES 1846	12 37 41.	+ 18 54 40.	17	17.0	NEBULA
SVEN 321	12 37 41.	- 05 03	18	15.4	GALAXY
HELW 160	12 37 41.	- 05 04 54.			NEBULA
SVEN 320	12 37 41.	- 05 05 00.	18	14.9	GALAXY
SVEN 319	12 37 41.	- 05 07 00.	12	15.2	GALAXY
REIZ 2674	12 37 41.	- 09 01	60	13.3	GALAXY
ZWG 070.207	12 37 42.	+ 08 39		14.1	GALAXY
UGC 07829	12 37 42.	+ 08 39	102	14.1	GALAXY SB0
MCG+02-32-171	12 37 42.	+ 08 39	72	14.1	GALAXY
KARA.68 179	12 37 42.	+ 10 07	27		DWARF GALAXY
ZWG 070.208	12 37 42.	+ 10 09		15.4	GALAXY
TON-N 0100	12 37 42.	+ 10 19		15.7	BLUE STAR
RMB 071	12 37 42.	+ 13 06			FAINT BLUE OBJECT
ZWG 070.209	12 37 42.	+ 13 09		15.5	GALAXY
UGC 07830	12 37 42.	+ 13 09	72	15.5	GALAXY Sc?
RMB 064	12 37 42.	+ 13 26			FAINT BLUE OBJECT
TON-N 0630	12 37 42.	+ 24 43		15.4	BLUE STAR
LB 06802	12 37 42.	+ 27 06		19.5	FAINT BLUE STAR
LB 06801	12 37 42.	+ 27 09		18.8	FAINT BLUE STAR
LB 06800	12 37 42.	+ 27 34		19.1	FAINT BLUE STAR
LB 06799	12 37 42.	+ 28 23		19.3	FAINT BLUE STAR
LB 06798	12 37 42.	+ 28 35		19.5	FAINT BLUE STAR
ZWG 159.048	12 37 42.	+ 31 27		15.5	GALAXY
SEY 191	12 37 42.	+ 31 27 07.		15.4	FAINT GALAXY
LB 11297	12 37 42.	+ 31 53		15.0	FAINT BLUE STAR
ZWG 293.031	12 37 42.	+ 61 53		10.8	GALAXY
UGC 07831	12 37 42.	+ 61 53	420	10.8	GALAXY S
KARA.73B 0543	12 37 42.	+ 61 53	348	10.8	ISOLATED GALAXY S
MCG-01-32-035	12 37 42.	- 09 02	60	14.	GALAXY
DV.56 N4603B	12 37 42.	- 40 28	108		S GALAXY
IC 3634	12 37 43.	+ 10 06			NONSTELLAR OBJECT
IC 3633	12 37 43.	+ 10 09			NONSTELLAR OBJECT
IC 3635	12 37 43.	+ 13 08			NONSTELLAR OBJECT
HN 1004	12 37 43.	+ 13 09	12		NEBULA
AMES 1849	12 37 43.	+ 13 38 04.	16	17.6	NEBULA
AMES 1848	12 37 43.	+ 13 45 04.	19	17.2	NEBULA
REIZ 2677	12 37 43.	+ 16 13	12	15.0	GALAXY
HELW 161	12 37 43.	- 05 03 18.			NEBULA

OBJECT NAME	RIGHT ASCEN.	DECLINATION	DIAM.	MAGN.	TYPE OF OBJECT
RNGC 4598	12 37 44.	+ 08 39		14.0	GALAXY
HN 1006	12 37 44.	+ 10 07			NEBULA
HN 1005	12 37 44.	+ 10 10	12		NEBULA
RNGC 4603B	12 37 44.	- 40 29			GALAXY
KN 16.192	12 37 44.9	+ 28 56 15.			NEBULA
SC 1235+0111.2	12 37 45.	+ 00 54 43.	18		NEBULA
MCG+03-32-082	12 37 45.	+ 16 12	24	14.9	GALAXY
AMES 1850	12 37 45.	+ 18 06 10.	18	16.8	NEBULA
REIN 3.277	12 37 45.04	+ 10 47 30.1			NEBULA
AMES 1853	12 37 46.	+ 10 40 40.	31	16.7	NEBULA
AMES 1852	12 37 46.	+ 12 15 46.	24	17.3	NEBULA
IC 3636	12 37 46.	+ 22 20 54.			NONSTELLAR OBJECT
REIN 3.278	12 37 46.12	+ 10 12 50.4			NEBULA
AMES 1851	12 37 47.	+ 18 32 58.	23	18.2	NEBULA
ZWG 042.198	12 37 48.	+ 03 24		13.7	GALAXY
UGC 07832	12 37 48.	+ 03 24	84	13.7	GALAXY S0
MCG+01-32-128	12 37 48.	+ 03 24	36	13.7	GALAXY
ZWG 070.210	12 37 48.	+ 10 47		14.6	GALAXY
MCG+02-32-172	12 37 48.	+ 10 47	36	14.6	GALAXY
AMES 1854	12 37 48.	+ 12 47 52.	24	18.0	NEBULA
IC 3637	12 37 48.	+ 14 58			NONSTELLAR OBJECT
ZWG 099.107	12 37 48.	+ 14 59		15.6	GALAXY
ZWG 099.108	12 37 48.	+ 16 12		14.9	GALAXY
ZWG 099.109	12 37 48.	+ 17 02		15.6	GALAXY
8ZW 1237+26.2	12 37 48.	+ 26 10		19.1	COMPACT GALAXY
LB 06804	12 37 48.	+ 29 17		19.0	FAINT BLUE STAR
LB 11298	12 37 48.	+ 30 34		14.9	FAINT BLUE STAR
LB 06803	12 37 48.	+ 32 15		19.0	FAINT BLUE STAR
RNGC 4605	12 37 48.	+ 61 53		11.5	GALAXY
ZC 1237.8+6313	12 37 48.	+ 63 13	2350		CLUSTER OF GALAXIES
ZC 1237.8+6457	12 37 48.	+ 64 57	1080		CLUSTER OF GALAXIES
REIN 3.279	12 37 48.48	+ 10 11 16.5			NEBULA
KN 16.193	12 37 48.6	+ 25 50 51.			NEBULA
RNGC 4600	12 37 49.	+ 03 24		13.5	GALAXY
HN 1007	12 37 49.	+ 14 59	12		NEBULA
KN 16.194	12 37 49.0	+ 25 57 09.			NEBULA
KN 16.195	12 37 49.2	+ 26 44 45.			NEBULA
IC 3638	12 37 50.	+ 10 46			NONSTELLAR OBJECT
HN 1008	12 37 50.	+ 10 47	18		NEBULA
AMES 1855	12 37 50.	+ 16 12 34.	35	15.4	NEBULA
REIZ 2684	12 37 50.	+ 61 54	210	11.4	GALAXY
KN 16.196	12 37 50.1	+ 25 50 09.			NEBULA
AMES 1856	12 37 51.	+ 13 52 16.	20	17.4	NEBULA
LB 02415	12 37 51.	+ 61 27 54.		15.2	FAINT BLUE STAR
ARC 1583	12 37 51.	- 15 41		17.8	RICH CLUSTER OF GALAXIES
KN 16.197	12 37 51.0	+ 25 52 33.			NEBULA
AMES 1858	12 37 52.	+ 13 39 22.	17	17.2	NEBULA
AMES 1857	12 37 53.	+ 18 31 51.	24	18.0	NEBULA
REIZ 2678	12 37 53.	+ 22 22	24	14.9	GALAXY
SVEN 322	12 37 53.	- 04 41	60	14.4	GALAXY
REIN 4.117	12 37 53.62	+ 01 28 14.4			NEBULA
ZWG 014.099	12 37 54.	+ 01 27		13.7	GALAXY
UGC 07833	12 37 54.	+ 01 27	120	13.7	GALAXY Sa
MCG+00-32-034	12 37 54.	+ 01 27	102	13.7	GALAXY
ZWG 042.199	12 37 54.	+ 05 19		15.7	GALAXY
ZWG 042.200	12 37 54.	+ 08 27		15.4	GALAXY
MCG+01-32-131	12 37 54.	+ 08 27	7	15.4	GALAXY
MCG+01-32-130	12 37 54.	+ 08 27	6	15.4	GALAXY
MCG+01-32-129	12 37 54.	+ 08 27	15	15.4	GALAXY
RMB 194	12 37 54.	+ 10 16			FAINT BLUE OBJECT
RMB 011	12 37 54.	+ 14 34			FAINT BLUE OBJECT
ZC 1237.9+2053	12 37 54.	+ 20 53	940		CLUSTER OF GALAXIES
8ZW 1237+25.2	12 37 54.	+ 25 10		16.9	COMPACT GALAXY
LB 06807	12 37 54.	+ 26 38		18.9	FAINT BLUE STAR
LB 06806	12 37 54.	+ 26 58		18.7	FAINT BLUE STAR
LB 06805	12 37 54.	+ 30 00		18.8	FAINT BLUE STAR
ZC 1237.9+4248	12 37 54.	+ 42 48	3700		CLUSTER OF GALAXIES
MCG+10-18-074	12 37 54.	+ 61 53	318	10.9	GALAXY
RNGC 4599	12 37 55.	+ 01 27		13.5	GALAXY
KN 16.198	12 37 55.2	+ 25 49 33.			NEBULA
KN 16.199	12 37 55.9	+ 27 44 03.			NEBULA
AMES 1859	12 37 56.	+ 18 10 27.	20	17.7	NEBULA
AMES 1862	12 37 57.	+ 17 38 09.	28	17.6	NEBULA
AMES 1861	12 37 57.	+ 18 09 39.	26	17.4	NEBULA
AMES 1860	12 37 57.	+ 18 55 03.	17	17.2	NEBULA
IC 3641	12 37 57.	+ 26 48 00.			NONSTELLAR OBJECT
IC 3640	12 37 57.	+ 26 48 00.			NONSTELLAR OBJECT
KN 16.200	12 37 57.7	+ 26 47 45.			NEBULA
REIZ 2681	12 37 58.	+ 26 48	30	15.0	GALAXY
REIZ 2682	12 37 58.	+ 27 00	48	15.1	GALAXY
IC 3642	12 37 58.	+ 27 00 18.			NONSTELLAR OBJECT
REIZ 2690	12 37 58.	+ 58 15	24	15.0	GALAXY
KN 16.201	12 37 58.3	+ 27 00 27.			NEBULA
ZWG 042.201	12 38 00.	+ 02 45		15.5	GALAXY
MCG+01-32-132	12 38 00.	+ 02 45	60	15.5	GALAXY
ZWG 042.202	12 38 00.	+ 05 38		15.7	GALAXY
ZC 1238.0+1020	12 38 00.	+ 10 20	1340		CLUSTER OF GALAXIES
AMES 1866	12 38 00.	+ 11 11 51.	16	17.0	NEBULA
RMB 105	12 38 00.	+ 12 25			FAINT BLUE OBJECT
TON-N 0101	12 38 00.	+ 23 23		14.3	BLUE STAR
LB 06813	12 38 00.	+ 26 30		18.9	FAINT BLUE STAR
ZWG 159.049	12 38 00.	+ 26 48		15.4	GALAXY
LB 06812	12 38 00.	+ 27 01		18.7	FAINT BLUE STAR
LB 06811	12 38 00.	+ 27 32		19.5	FAINT BLUE STAR
LB 06810	12 38 00.	+ 27 41		18.5	FAINT BLUE STAR
LB 06809	12 38 00.	+ 29 00		18.2	FAINT BLUE STAR
TON-N 0631	12 38 00.	+ 29 12		14.8	BLUE STAR
TON-N 0632	12 38 00.	+ 30 19		15.1	BLUE STAR
LB 06808	12 38 00.	+ 31 34		18.3	FAINT BLUE STAR
ZC 1238.0+4305	12 38 00.	+ 43 05	1210		CLUSTER OF GALAXIES
ZC 1238.0+5043	12 38 00.	+ 50 43	2080		CLUSTER OF GALAXIES
TON-N 0102	12 38 00.	+ 51 36		13.7	BLUE STAR
LB 00648	12 38 00.	+ 55 44 30.		17.6	FAINT BLUE STAR
ZC 1238.0+6359	12 38 00.	+ 63 59	2350		CLUSTER OF GALAXIES
MCG-02-32-021	12 38 00.	- 10 33	36	15.	GALAXY
MCG-03-32-018	12 38 00.	- 20 17	15	15.	GALAXY
DV.56 N4603C	12 38 00.	- 40 29			S0 GALAXY
MCG-07-26-025	12 38 00.	- 40 30	78	14.	GALAXY
AMES 1867	12 38 01.	+ 13 03 03.	26	17.6	NEBULA
AMES 1865	12 38 01.	+ 17 31 51.	19	17.2	NEBULA
AMES 1864	12 38 01.	+ 18 09 33.	19	17.9	NEBULA
AMES 1863	12 38 01.	+ 18 23 03.	23	18.0	NEBULA
RNGC 4602	12 38 01.	- 04 52		12.5	GALAXY
KN 16.202	12 38 02.5	+ 29 15 57.			NEBULA
AMES 1868	12 38 03.	+ 19 30 15.	29	17.0	NEBULA
MCG-01-32-036	12 38 03.	- 04 51	156	11.5	GALAXY
KN 16.204	12 38 03.4	+ 29 28 15.			NEBULA
AMES 1871	12 38 04.	+ 12 02 33.	25	17.4	NEBULA
REIZ 2680	12 38 04.	- 00 43	18	15.1	GALAXY
KLEM 19	12 38 04.	- 40 37	1800	13.	GROUP OF 9 GALAXIES
KN 16.203	12 38 04.5	+ 25 31 57.			NEBULA
AMES 1870	12 38 05.	+ 17 18 51.	28	17.6	NEBULA
AMES 1869	12 38 05.	+ 19 43 03.	14	17.4	NEBULA
SVEN 323	12 38 05.	- 04 51	168	11.7	GALAXY
REIZ 2679	12 38 05.	- 04 51	120	12.5	GALAXY
TON-N 0103	12 38 06.	+ 14 11		16.1	BLUE STAR
RMB 002	12 38 06.	+ 15 46			FAINT BLUE OBJECT
RMB 001	12 38 06.	+ 16 04			FAINT BLUE OBJECT
AMES 1872	12 38 06.	+ 17 32 09.	22	17.5	NEBULA
ZC 1238.1+1834	12 38 06.	+ 18 34	340		CLUSTER OF GALAXIES
LB 06816	12 38 06.	+ 28 22		19.9	FAINT BLUE STAR
LB 06815	12 38 06.	+ 29 17		19.5	FAINT BLUE STAR
LB 06814	12 38 06.	+ 32 00		19.0	FAINT BLUE STAR
MCG-07-26-026	12 38 06.	- 40 38	60	15.	GALAXY
RNGC 4604	12 38 07.	- 04 53			NON-EXISTENT OBJECT
KEEL 526	12 38 07.7	+ 32 51 19.		16.	NEBULA
AMES 1874	12 38 08.	+ 13 20 39.	23	18.0	NEBULA
AMES 1873	12 38 08.	+ 17 15 15.	26	17.6	NEBULA
RNGC 4603C	12 38 08.	- 40 29			GALAXY
RNGC 4601	12 38 08.	- 40 39			GALAXY
KN 16.205	12 38 08.5	+ 26 46 45.			NEBULA
IC 3644	12 38 09.	+ 26 46 54.			NONSTELLAR OBJECT
KN 16.206	12 38 09.2	+ 27 21 15.			NEBULA
REIZ 2687	12 38 10.	+ 26 47	48	14.9	GALAXY
REIZ 2688	12 38 10.	+ 26 48	42	14.5	GALAXY
IC 3645	12 38 10.	+ 26 48 54.			NONSTELLAR OBJECT
REIZ 2689	12 38 10.	+ 26 49	12	15.2	GALAXY
ARC 1584	12 38 10.	- 18 18		16.9	RICH CLUSTER OF GALAXIES
KN 16.207	12 38 10.8	+ 27 10 03.			NEBULA
KN 16.208	12 38 10.9	+ 26 48 09.			NEBULA
IC 3646	12 38 11.	+ 26 48 00.			NONSTELLAR OBJECT
SVEN 324	12 38 11.	- 04 49 00.	18	15.4	GALAXY
SVEN 325	12 38 11.	- 05 01	48	13.7	GALAXY
HELW 162	12 38 11.	- 05 01 35.			NEBULA
IC 3639	12 38 11.	- 36 28 54.			NONSTELLAR OBJECT
KN 16.209	12 38 11.1	+ 27 14 39.			NEBULA
KN 16.210	12 38 11.3	+ 27 06 39.			NEBULA
ZC 1238.2+1556	12 38 12.	+ 15 56	870		CLUSTER OF GALAXIES
MCG+05-30-052	12 38 12.	+ 26 46	48	15.5	GALAXY
MCG+05-30-053	12 38 12.	+ 26 47	36	15.4	GALAXY
LB 06820	12 38 12.	+ 27 55		19.5	FAINT BLUE STAR
LB 06819	12 38 12.	+ 29 37		19.2	FAINT BLUE STAR
LB 06818	12 38 12.	+ 29 56		19.0	FAINT BLUE STAR
LB 06817	12 38 12.	+ 31 52		19.6	FAINT BLUE STAR
TON-N 0104	12 38 12.	+ 34 12		14.7	BLUE STAR
ZC 1238.2+6244	12 38 12.	+ 62 44	870		CLUSTER OF GALAXIES
MCG-01-32-037	12 38 12.	- 05 01	54	15.	GALAXY
REIZ 2683	12 38 12.	- 09 58	18	14.3	GALAXY
MCG-02-32-022	12 38 12.	- 14 16 30.	72	15.	GALAXY
MCG-06-28-011	12 38 12.	- 36 28	66	12.5	GALAXY
MCG-07-26-027	12 38 12.	- 41 21	84	14.	GALAXY
IC 3643	12 38 13.	+ 12 40			NONSTELLAR OBJECT
HN 1009	12 38 13.	+ 12 41			NEBULA
ARC 1590	12 38 13.	+ 73 26		16.8	RICH CLUSTER OF GALAXIES
REIZ 2686	12 38 14.	+ 07 45	24	14.3	GALAXY
AMES 1875	12 38 14.	+ 18 42 27.	12	17.4	NEBULA
RNGC 4603	12 38 14.	- 40 42		12.0	GALAXY
MCG+08-23-072	12 38 15.	+ 46 54	36	16.	GALAXY
MCG-07-26-028	12 38 15.	- 40 43	150	12.5	GALAXY
REIZ 2685	12 38 16.	- 00 49	36	15.0	GALAXY
KN 16.211	12 38 16.6	+ 27 36 58.			NEBULA
AMES 1877	12 38 17.	+ 16 25 57.	26	17.6	NEBULA
AMES 1876	12 38 17.	+ 19 08 15.	26	17.9	NEBULA
SVEN 326	12 38 17.	- 05 07	24	16.4	GALAXY
ZWG 042.203	12 38 18.	+ 04 48		15.7	GALAXY
KARA.68 180	12 38 18.	+ 10 45	67		DWARF GALAXY
ZWG 070.211	12 38 18.	+ 10 46		15.3	GALAXY
UGC 07834	12 38 18.	+ 10 46	108	15.3	GALAXY DWARF
MCG+02-32-173	12 38 18.	+ 10 46	48	15.3	GALAXY
8ZW 1238+12.1	12 38 18.	+ 12 06		17.5	COMPACT GALAXY
AMES 1878	12 38 18.	+ 12 25 21.	30	17.8	NEBULA
RMB 070	12 38 18.	+ 13 10			FAINT BLUE OBJECT
ZWG 159.050	12 38 18.	+ 27 00		14.4	GALAXY
UGC 07835	12 38 18.	+ 27 00	60	14.4	GALAXY S0
LB 11300	12 38 18.	+ 28 14		16.2	FAINT BLUE STAR
LB 11299	12 38 18.	+ 31 37		13.1	FAINT BLUE STAR
ZC 1238.3+4725	12 38 18.	+ 47 25	3160		CLUSTER OF GALAXIES
MCG-06-28-012	12 38 18.	- 36 26	72	14.	GALAXY
KN 16.212	12 38 18.0	+ 28 24 10.			NEBULA
REIZ 2691	12 38 19.	+ 12 11	48	13.4	GALAXY
AMES 1879	12 38 19.	+ 13 00 09.	12	18.0	NEBULA
LB 02416	12 38 19.	+ 57 07 06.		16.0	FAINT BLUE STAR
KN 16.213	12 38 19.0	+ 27 50 16.			NEBULA
KN 16.214	12 38 19.7	+ 26 29 58.			NEBULA
SC 1235+0134.0	12 38 20.	+ 01 17 31.	12		NEBULA
IC 3647	12 38 20.	+ 10 44			MAY NOT EXIST
HN 1010	12 38 20.	+ 10 45	48		NEBULA
IC 3649	12 38 20.	+ 21 22 48.			NONSTELLAR OBJECT
IC 3650	12 38 20.	+ 26 44 54.			NONSTELLAR OBJECT
AMES 1883	12 38 21.	+ 13 00 45.	23	18.1	NEBULA
AMES 1882	12 38 21.	+ 18 49 15.	19	18.0	NEBULA
AMES 1881	12 38 21.	+ 18 52 51.	10	18.0	NEBULA
AMES 1880	12 38 21.	+ 18 54 03.	14	16.9	NEBULA
MCG+11-16-001	12 38 21.	+ 63 37	57	16.	GALAXY
KN 16.216	12 38 21.1	+ 28 02 58.			NEBULA
KN 16.215	12 38 21.6	+ 26 01 22.			NEBULA
IC 3648	12 38 22.	+ 13 15 36.			MAY NOT EXIST
AMES 1884	12 38 22.	+ 13 44 21.	20	18.0	NEBULA
REIZ 2692	12 38 22.	+ 26 45	9	15.4	GALAXY
KN 16.217	12 38 22.1	+ 27 50 28.			NEBULA
ZWG 014.100	12 38 24.	+ 01 16		15.7	GALAXY
SCH 64	12 38 24.	+ 12 10		12.7	PECULIAR GALAXY
AMES 1887	12 38 24.	+ 13 01 03.	25	18.0	NEBULA
AMES 1886	12 38 24.	+ 18 23 51.	14	18.1	NEBULA
AMES 1885	12 38 24.	+ 18 59 03.	19	17.0	NEBULA
MCG+05-30-054	12 38 24.	+ 26 59	18	14.4	GALAXY
LB 06827	12 38 24.	+ 27 17		18.0	FAINT BLUE STAR
LB 06826	12 38 24.	+ 27 22		18.5	FAINT BLUE STAR
TON-N 0105	12 38 24.	+ 28 03		14.8	BLUE STAR
ZWG 159.051	12 38 24.	+ 28 15		15.4	GALAXY
LB 06825	12 38 24.	+ 28 35		18.7	FAINT BLUE STAR
ZWG 159.052	12 38 24.	+ 29 45		14.9	GALAXY
UGC 07836	12 38 24.	+ 29 45	84	14.9	GALAXY Sc
LB 06824	12 38 24.	+ 29 52		18.6	FAINT BLUE STAR
LB 06823	12 38 24.	+ 30 07		19.0	FAINT BLUE STAR
LB 06822	12 38 24.	+ 30 32		19.5	FAINT BLUE STAR

OBJECT NAME	RIGHT ASCEN.	DECLINATION	DIAM.	MAGN.	TYPE OF OBJECT
LB 06821	12 38 24.	+ 31 48		18.8	FAINT BLUE STAR
ZWG 316.001	12 38 24.	+ 63 37		15.4	GALAXY
ZWG 315.051	12 38 24.	+ 63 37		15.4	GALAXY
UGC 07837	12 38 24.	+ 63 37	72	15.4	GALAXY Sc
MCG-04-30-011	12 38 24.	- 25 41 30.	60	15.	GALAXY
KN 16.218	12 38 24.9	+ 25 41 46.			NEBULA
HOLM 436A	12 38 25.	+ 12 12	78	12.4	PART OF MULTIPLE GALAXY
AMES 1891	12 38 25.	+ 15 32 39.	34	17.8	NEBULA
IC 3651	12 38 25.	+ 27 00 06.			NONSTELLAR OBJECT
KN 16.219	12 38 25.4	+ 27 00 22.			NEBULA
REIN 3.280	12 38 25.61	+ 12 10 38.6			NEBULA
RNGC 4606	12 38 26.	+ 12 11		12.5	GALAXY
AMES 1890	12 38 26.	+ 17 50 03.	20	17.6	NEBULA
AMES 1889	12 38 26.	+ 18 51 57.	25	16.7	NEBULA
AMES 1888	12 38 26.	+ 18 54 45.	14	17.7	NEBULA
REIN 3.281	12 38 26.37	+ 12 11 06.1			NEBULA
AMES 1894	12 38 27.	+ 15 37 51.	24	18.0	NEBULA
AMES 1893	12 38 27.	+ 18 48 45.	17	17.7	NEBULA
AMES 1892	12 38 27.	+ 19 10 09.	18	17.6	NEBULA
LB 02417	12 38 27.	+ 48 32 54.		13.4	FAINT BLUE STAR
SC 1235-0050.1	12 38 27.	- 01 06 35.	12		NEBULA
REIN 3.282	12 38 27.36	+ 11 27 29.9			NEBULA
REIN 3.283	12 38 27.64	+ 10 28 07.8			NEBULA
KN 16.220	12 38 27.9	+ 28 48 04.			NEBULA
AMES 1895	12 38 28.	+ 18 49 39.	17	17.0	NEBULA
KN 16.221	12 38 28.9	+ 29 44 16.			NEBULA
HARO 31	12 38 29.	+ 28 14			BLUE EMISSION-LINE GALAXY
HELW 163	12 38 29.	- 04 56 47.			NEBULA
SVEN 327	12 38 29.	- 04 56 59.	12	15.5	GALAXY
ZWG 070.212	12 38 30.	+ 11 27		15.1	GALAXY
UGC 07838	12 38 30.	+ 11 27	60	15.1	GALAXY E
MCG+02-32-175	12 38 30.	+ 11 27	15	15.1	GALAXY
ZWG 070.213	12 38 30.	+ 12 11		12.7	GALAXY
UGC 07839	12 38 30.	+ 12 11	168	12.7	GALAXY S
MCG+02-32-174	12 38 30.	+ 12 11	120	12.7	GALAXY
RMB 104	12 38 30.	+ 12 37			FAINT BLUE OBJECT
AMES 1897	12 38 30.	+ 18 40 21.	19	17.3	NEBULA
AMES 1896	12 38 30.	+ 18 50 03.	20	16.6	NEBULA
ZC 1238.5+2545	12 38 30.	+ 25 45	4100		CLUSTER OF GALAXIES
MCG+05-30-055	12 38 30.	+ 28 15	30	15.4	GALAXY
LB 06829	12 38 30.	+ 28 48		18.0	FAINT BLUE STAR
LB 06828	12 38 30.	+ 29 34		19.7	FAINT BLUE STAR
MCG+05-30-056	12 38 30.	+ 29 45	84	14.9	GALAXY
MCG+09-21-027	12 38 30.	+ 53 22 30.	24	17.	GALAXY
KN 16.222	12 38 30.7	+ 28 15 04.			NEBULA
HN 1011	12 38 31.	+ 11 27	18		NEBULA
AMES 1898	12 38 31.	+ 19 04 33.	24	17.4	NEBULA
KN 16.223	12 38 31.5	+ 28 14 16.			NEBULA
IC 3652	12 38 32.	+ 11 26			NONSTELLAR OBJECT
SC 1235-0046.8	12 38 32.	- 01 03 17.	18		NEBULA
REIN 3.284	12 38 32.72	+ 12 13 13.6			NEBULA
AMES 1900	12 38 33.	+ 16 33 03.	28	16.4	NEBULA
AMES 1899	12 38 33.	+ 18 37 51.	22	18.0	NEBULA
REIN 3.285	12 38 33.43	+ 10 30 39.5			NEBULA
AMES 1902	12 38 34.	+ 18 54 33.	13	18.1	NEBULA
AMES 1901	12 38 34.	+ 18 59 45.	12	17.4	NEBULA
REIZ 2693	12 38 34.	- 02 41	30	14.7	GALAXY
KN 16.224	12 38 35.3	+ 26 45 34.			NEBULA
REIN 3.286	12 38 35.58	+ 10 26 52.1			NEBULA
ARC 1586	12 38 36.	+ 10 14		18.0	RICH CLUSTER OF GALAXIES
RMB 189	12 38 36.	+ 11 24			FAINT BLUE OBJECT
ZC 1238.6+1637	12 38 36.	+ 16 37	1210		CLUSTER OF GALAXIES
AMES 1903	12 38 36.	+ 18 47 09.	17	17.6	NEBULA
LB 06838	12 38 36.	+ 26 54		19.3	FAINT BLUE STAR
LB 06837	12 38 36.	+ 26 56		19.9	FAINT BLUE STAR
LB 06836	12 38 36.	+ 27 03		19.3	FAINT BLUE STAR
LB 06835	12 38 36.	+ 27 44		19.3	FAINT BLUE STAR
LB 06834	12 38 36.	+ 28 09		19.3	FAINT BLUE STAR
TON-N 0633	12 38 36.	+ 28 31		14.7	BLUE STAR
TON-N 0634	12 38 36.	+ 28 35		14.7	BLUE STAR
LB 06833	12 38 36.	+ 29 28		19.4	FAINT BLUE STAR
ZWG 159.053	12 38 36.	+ 29 49		15.7	GALAXY
LB 06832	12 38 36.	+ 30 08		18.8	FAINT BLUE STAR
LB 06831	12 38 36.	+ 30 29		18.9	FAINT BLUE STAR
LB 06830	12 38 36.	+ 31 02		19.2	FAINT BLUE STAR
SC 1236+0003.3	12 38 36.	- 00 13 11.	12		NEBULA
ZWG 014.101	12 38 36.	- 01 20		15.7	GALAXY
UGC 07840	12 38 36.	- 01 20	60	15.7	GALAXY Sc
REIZ 2694	12 38 37.	+ 12 09	180	13.8	GALAXY
AMES 1910	12 38 37.	+ 12 31 33.	20	17.3	NEBULA
AMES 1904	12 38 37.	+ 18 31 51.	20	17.0	NEBULA
LB 02418	12 38 37.	+ 59 55 06.		16.5	FAINT BLUE STAR
SC 1236+0158.4	12 38 38.	+ 01 41 55.	18		NEBULA
AMES 1909	12 38 38.	+ 17 15 51.	17	17.4	NEBULA
AMES 1908	12 38 38.	+ 18 07 09.	19	17.8	NEBULA
AMES 1907	12 38 38.	+ 18 37 09.	18	17.8	NEBULA
AMES 1906	12 38 38.	+ 19 18 39.	22	17.6	NEBULA
AMES 1905	12 38 38.	+ 19 29 39.	24	17.8	NEBULA
LB 02421	12 38 38.	+ 55 46 30.		16.0	FAINT BLUE STAR
LB 02420	12 38 38.	+ 57 30 48.		15.9	FAINT BLUE STAR
LB 02419	12 38 38.	+ 61 51 18.		16.6	FAINT BLUE STAR
FATH 1.537	12 38 38.	- 00 28	8		NEBULA
REIN 4.118	12 38 38.20	+ 01 41 03.6			NEBULA
KN 16.225	12 38 38.9	+ 27 05 34.			NEBULA
AMES 1912	12 38 39.	+ 18 29 51.	14	18.0	NEBULA
AMES 1911	12 38 39.	+ 18 47 03.	22	16.3	NEBULA
HOLM 436B	12 38 40.	+ 12 11	168	12.7	PART OF MULTIPLE GALAXY
AMES 1915	12 38 40.	+ 16 50 27.	24	17.6	NEBULA
AMES 1914	12 38 40.	+ 18 54 45.	12	17.5	NEBULA
AMES 1913	12 38 40.	+ 19 16 51.	19	17.0	NEBULA
ARC 1585	12 38 40.	- 16 13		16.9	RICH CLUSTER OF GALAXIES
KN 16.226	12 38 40.0	+ 27 44 16.			NEBULA
AMES 1916	12 38 41.	+ 17 23 21.	22	17.6	NEBULA
ARC 1587	12 38 41.	+ 27 50		17.6	RICH CLUSTER OF GALAXIES
SVEN 328	12 38 41.	- 04 43 59.	36	13.8	GALAXY
REIN 3.287	12 38 41.19	+ 12 09 34.0			NEBULA
IC 0804	12 38 41.4	- 04 44 06.			GALAXY SA(r)
REIN 3.288	12 38 41.89	+ 10 25 45.9			NEBULA
ZWG 014.103	12 38 42.	+ 01 40		14.3	GALAXY
UGC 07841	12 38 42.	+ 01 40	60	14.3	GALAXY
MCG+00-32-035	12 38 42.	+ 01 40	24	14.3	GALAXY
KARA.68 181	12 38 42.	+ 08 30	27		DWARF GALAXY
ZWG 070.214	12 38 42.	+ 10 25		12.6	GALAXY
UGC 07842	12 38 42.	+ 10 25	210	12.6	GALAXY SB0
MCG+02-32-177	12 38 42.	+ 10 26	120	12.6	GALAXY
ZWG 070.215	12 38 42.	+ 11 40		14.7	GALAXY
MCG+02-32-178	12 38 42.	+ 11 40	30	14.7	GALAXY
IC 3653	12 38 42.	+ 11 40		14.7	GALAXY IN VIRGO CLSTR
ZWG 070.216	12 38 42.	+ 12 09		14.7	GALAXY
UGC 07843	12 38 42.	+ 12 09	192	14.7	GALAXY S
MCG+02-32-176	12 38 42.	+ 12 09	132	14.7	GALAXY
RMB 103	12 38 42.	+ 12 43			FAINT BLUE OBJECT
AMES 1917	12 38 42.	+ 18 46 09.	22	17.6	NEBULA
MCG+03-32-083	12 38 42.	+ 18 51	4	18.	GALAXY
MCG+05-30-057	12 38 42.	+ 27 00	48	15.5	GALAXY
ZWG 159.054	12 38 42.	+ 27 01		15.5	GALAXY
LB 11301	12 38 42.	+ 28 49		16.1	FAINT BLUE STAR
SHAH 126	12 38 42.	+ 53 07	180	17.5	GROUP OF COMPACT GALAXIES
MCG+10-18-075	12 38 42.	+ 60 10	24	16.	GALAXY
ZC 1238.7+6610	12 38 42.	+ 66 10	4910		CLUSTER OF GALAXIES
UGC 07844	12 38 42.	+ 74 00	72	16.5	GALAXY Sc
ZWG 014.102	12 38 42.	- 00 30		15.7	GALAXY
HOLM 437A	12 38 42.	- 12 20	84	13.5	PART OF MULTIPLE GALAXY
MCG-02-32-023	12 38 42.	- 12 20 30.	90	14.	GALAXY
KN 16.228	12 38 42.2	+ 29 48 52.			NEBULA
REIZ 2695	12 38 43.	+ 11 26	36	15.4	GALAXY
HN 1012	12 38 43.	+ 11 40		13.	NEBULA
AMES 1919	12 38 43.	+ 18 48 03.	20	17.1	NEBULA
AMES 1918	12 38 43.	+ 18 53 21.	25	16.8	NEBULA
IC 3654	12 38 43.	+ 22 51 49.			NONSTELLAR OBJECT
KN 16.227	12 38 43.3	+ 26 29 10.			NEBULA
RNGC 4608	12 38 44.	+ 10 26		12.5	GALAXY
RNGC 4607	12 38 44.	+ 12 10		14.5	GALAXY
AMES 1927	12 38 44.	+ 13 15 15.	30	16.8	NEBULA
AMES 1924	12 38 44.	+ 16 58 21.	18	17.7	NEBULA
AMES 1923	12 38 44.	+ 18 28 21.	28	17.0	NEBULA
AMES 1922	12 38 44.	+ 18 33 51.	14	17.9	NEBULA
AMES 1921	12 38 44.	+ 18 34 57.	24	17.4	NEBULA
AMES 1920	12 38 44.	+ 19 00 03.	14	17.4	NEBULA
REIN 3.289	12 38 44.52	+ 11 39 40.0			NEBULA
AMES 1928	12 38 45.	+ 11 34 21.	29	16.8	NEBULA
AMES 1926	12 38 45.	+ 18 49 51.	17	18.0	NEBULA
AMES 1925	12 38 45.	+ 19 11 09.	13	18.0	NEBULA
IC 3655	12 38 45.	+ 20 56 25.			NONSTELLAR OBJECT
IC 3656	12 38 45.	+ 22 52 07.			NONSTELLAR OBJECT
MCG+08-23-073	12 38 45.	+ 48 37 30.	36	15.	GALAXY
IC 0805	12 38 46.	+ 14 00 43.			SAME AS NGC 4611
REIZ 2697	12 38 46.	+ 27 01	36	15.0	GALAXY
HOLM 437B	12 38 46.	- 12 23	36	14.4	PART OF MULTIPLE GALAXY
KN 16.229	12 38 46.5	+ 27 00 28.			NEBULA
AMES 1929	12 38 47.	+ 17 51 27.	20	17.3	NEBULA
SC 1236-0200.6	12 38 47.	- 02 17 05.	42		NEBULA
ZWG 042.204	12 38 48.	+ 06 58		15.2	GALAXY
MCG+01-32-133	12 38 48.	+ 06 58	36	15.2	GALAXY
8ZW 1238+09.1	12 38 48.	+ 09 08		18.2	COMPACT GALAXY
ZC 1238.8+1116	12 38 48.	+ 11 16	1340		CLUSTER OF GALAXIES
ZWG 099.110	12 38 48.	+ 14 58		15.5	GALAXY
ZC 1238.8+1817	12 38 48.	+ 18 17	670		CLUSTER OF GALAXIES
AMES 1931	12 38 48.	+ 18 51 33.	28	16.1	NEBULA
MCG+03-32-084	12 38 48.	+ 18 52	18	17.	GALAXY
AMES 1930	12 38 48.	+ 19 05 15.	19	17.5	NEBULA
LB 06841	12 38 48.	+ 27 01		18.8	FAINT BLUE STAR
LB 06840	12 38 48.	+ 28 06		18.7	FAINT BLUE STAR
ZWG 159.055	12 38 48.	+ 28 08		15.6	GALAXY
UGC 07845	12 38 48.	+ 28 08	78	15.6	GALAXY Sb-c
MCG+05-30-058	12 38 48.	+ 28 08	72	15.6	GALAXY
LB 11302	12 38 48.	+ 29 25		13.8	FAINT BLUE STAR
ZWG 159.056	12 38 48.	+ 30 24		15.4	GALAXY
KARA.73B 0544	12 38 48.	+ 30 24	48	15.4	ISOLATED GALAXY S
LB 06839	12 38 48.	+ 32 14		19.2	FAINT BLUE STAR
ZWG 244.036	12 38 48.	+ 48 21		15.6	GALAXY
UGC 07846	12 38 48.	+ 48 38	84	15.6	GALAXY S
ZWG 270.013	12 38 48.	+ 50 42		14.2	GALAXY
UGC 07847	12 38 48.	+ 50 42	186	14.2	GALAXY Sb
KARA.73B 0545	12 38 48.	+ 50 42	186	14.2	ISOLATED GALAXY S
ZC 1238.8+5259	12 38 48.	+ 52 59	740		CLUSTER OF GALAXIES
ZWG 293.032	12 38 48.	+ 60 10		15.6	GALAXY
ZWG 316.002	12 38 48.	+ 63 47		14.8	GALAXY
ZWG 315.052	12 38 48.	+ 63 47		14.8	GALAXY
UGC 07848	12 38 48.	+ 63 47	156	14.8	GALAXY Sc
ZWG 014.104	12 38 48.	- 02 47		14.9	GALAXY
MCG+00-32-036	12 38 48.	- 02 47	66	14.9	GALAXY
REIN 4.119	12 38 48.61	- 02 47 00.9			NEBULA
KN 16.230	12 38 49.1	+ 28 07 46.			NEBULA
REIN 3.290	12 38 49.72	+ 10 25 24.1			NEBULA
RNGC 4610	12 38 50.	+ 07 59			NON-EXISTENT OBJECT
AMES 1933	12 38 50.	+ 18 50 03.	26	18.1	NEBULA
AMES 1932	12 38 50.	+ 18 56 27.	26	16.6	NEBULA
IC 3657	12 38 50.	+ 21 57 01.			NONSTELLAR OBJECT
RNGC 4617	12 38 50.	+ 50 42		14.0	GALAXY
AMES 1935	12 38 51.	+ 12 34 15.	23	17.8	NEBULA
MCG+11-16-002	12 38 51.	+ 63 47	114	14.	GALAXY
AMES 1934	12 38 52.	+ 18 35 15.	22	17.8	NEBULA
REIZ 2696	12 38 52.	- 02 46	72	15.3	GALAXY
AMES 1937	12 38 53.	+ 18 25 39.	17	18.1	NEBULA
AMES 1936	12 38 53.	+ 18 52 21.	22	17.1	NEBULA
ZL 092	12 38 53.	+ 27 54 06.		20.4	ULTRAFAINT BLUE STAR
KN 16.231	12 38 53.3	+ 30 23 46.			NEBULA
REIN 3.291	12 38 53.49	+ 12 26 57.9			NEBULA
KARA.68 182	12 38 54.	+ 08 38	27		DWARF GALAXY
ZWG 070.217	12 38 54.	+ 12 31		15.4	GALAXY
ZWG 070.218	12 38 54.	+ 14 00		15.1	GALAXY
UGC 07849	12 38 54.	+ 14 00	78	15.1	GALAXY S
MCG+02-32-179	12 38 54.	+ 14 00	66	15.1	GALAXY
IC 3658	12 38 54.	+ 14 58 01.			NONSTELLAR OBJECT
AMES 1939	12 38 54.	+ 18 55 15.	14	17.8	NEBULA
AMES 1938	12 38 54.	+ 19 29 51.	34	16.7	NEBULA
LB 06846	12 38 54.	+ 26 44		19.7	FAINT BLUE STAR
LB 06845	12 38 54.	+ 26 53		19.6	FAINT BLUE STAR
LB 06844	12 38 54.	+ 27 24		19.6	FAINT BLUE STAR
LB 06843	12 38 54.	+ 27 51		19.3	FAINT BLUE STAR
MCG+05-30-059	12 38 54.	+ 30 25	48	15.4	GALAXY
LB 06842	12 38 54.	+ 31 02		18.9	FAINT BLUE STAR
LB 11303	12 38 54.	+ 32 28		16.6	FAINT BLUE STAR
MCG+09-21-028	12 38 54.	+ 50 38	162	13.	GALAXY
ZWG 335.019	12 38 54.	+ 72 03		15.4	GALAXY
HELW 164	12 38 54.	- 11 05 59.			NEBULA
HN 1013	12 38 55.	+ 14 59	24		NEBULA
AMES 1943	12 38 55.	+ 16 37 39.	23	17.8	NEBULA
AMES 1942	12 38 55.	+ 18 35 39.	24	16.9	NEBULA
AMES 1941	12 38 55.	+ 18 43 09.	13	17.7	NEBULA
AMES 1940	12 38 55.	+ 18 55 27.	17	17.8	NEBULA
REIN 3.292	12 38 55.79	+ 10 13 31.9			NEBULA
RNGC 4611	12 38 57.	+ 14 00		15.0	GALAXY
AMES 1945	12 38 57.	+ 18 53 51.	23	17.0	NEBULA
AMES 1944	12 38 57.	+ 18 56 03.	17	17.0	NEBULA

OBJECT NAME	RIGHT ASCEN.	DECLINATION	DIAM.	MAGN.	TYPE OF OBJECT
AMES 1947	12 38 58.	+ 12 38 15.	22	17.0	NEBULA
AMES 1946	12 38 58.	+ 16 42 15.	19	18.0	NEBULA
RNGC 4614	12 38 58.	+ 26 18		14.0	GALAXY
RNGC 4613	12 38 58.	+ 26 21		15.5	GALAXY
REIZ 2698	12 38 58.	+ 27 04	60	16.3	GALAXY
KN 16.232	12 38 58.4	+ 26 39 46.			NEBULA
REIN 3.293	12 38 58.51	+ 10 24 16.1			NEBULA
IC 3659	12 38 59.	+ 23 12 19.			NONSTELLAR OBJECT
ARC 1588	12 38 59.	- 04 31		17.2	RICH CLUSTER OF GALAXIES
HMS 1239+1852	12 39	+ 18 52			CLUSTER OF GALAXIES
VDB.66G 141	12 39	+ 38 44	130		DWARF GALAXY
LB 09862	12 39	- 82 14		13.7	FAINT BLUE STAR
ZWG 042.205	12 39 00.	+ 07 35		12.9	GALAXY
UGC 07850	12 39 00.	+ 07 35	114	12.9	GALAXY S0
MCG+01-32-134	12 39 00.	+ 07 35	36	12.9	GALAXY
AMES 1953	12 39 00.	+ 11 01 21.	28	17.2	NEBULA
KARA.68 183	12 39 00.	+ 11 24	27		DWARF GALAXY
RMB 069	12 39 00.	+ 13 11			FAINT BLUE OBJECT
ZC 1239.0+1744	12 39 00.	+ 17 44	1410		CLUSTER OF GALAXIES
ZWG 129.015	12 39 00.	+ 26 18		14.2	GALAXY
UGC 07851	12 39 00.	+ 26 18	66	14.2	GALAXY SB0-a
KARA.72 348A	12 39 00.	+ 26 18	60	14.2	PART OF DOUBLE GALAXY
ZWG 129.016	12 39 00.	+ 26 21		15.5	GALAXY
MCG+04-30-011	12 39 00.	+ 26 21	24	15.5	GALAXY
LB 11305	12 39 00.	+ 26 56		12.6	FAINT BLUE STAR
LB 06851	12 39 00.	+ 26 58		19.6	FAINT BLUE STAR
LB 06850	12 39 00.	+ 27 50		18.9	FAINT BLUE STAR
TON-N 0635	12 39 00.	+ 29 10		16.8	BLUE STAR
LB 06849	12 39 00.	+ 29 22		19.7	FAINT BLUE STAR
LB 11304	12 39 00.	+ 30 12		14.6	FAINT BLUE STAR
LB 06848	12 39 00.	+ 31 09		20.2	FAINT BLUE STAR
LB 06847	12 39 00.	+ 31 36		18.8	FAINT BLUE STAR
HOLM 438A	12 39 00.	+ 41 29	168	11.8	PART OF MULTIPLE GALAXY
MCG+10-18-076	12 39 00.	+ 61 59	36	16.	GALAXY
ZWG 293.033	12 39 00.	+ 62 00		15.5	GALAXY
KARA.73B 0546	12 39 00.	+ 62 00	54	15.5	ISOLATED GALAXY S
MCG+12-12-015	12 39 00.	+ 72 02	39	14.3	GALAXY
MCG+12-12-016	12 39 00.	+ 72 04	51	14.3	GALAXY
ZWG 335.020	12 39 00.	+ 72 05		15.5	GALAXY
HELW 165	12 39 00.	- 11 07 23.			NEBULA
KN 16.234	12 39 00.6	+ 29 03 58.			NEBULA
REIN 3.294	12 39 00.62	+ 07 35 19.7			NEBULA
AMES 1952	12 39 01.	+ 16 54 03.	22	17.7	NEBULA
AMES 1951	12 39 01.	+ 18 30 51.	36	17.6	NEBULA
AMES 1950	12 39 01.	+ 18 58 03.	26	17.7	NEBULA
AMES 1949	12 39 01.	+ 18 59 15.	12	17.6	NEBULA
AMES 1948	12 39 01.	+ 19 28 51.	29	16.7	NEBULA
ARP 034	12 39 01.	+ 26 21			PECULIAR GALAXY
HOLM 439C	12 39 01.	+ 26 22	30	15.2	PART OF MULTIPLE GALAXY
KEEL 527	12 39 01.0	+ 32 39 48.		17.	NEBULA
KN 16.233	12 39 01.3	+ 26 21 46.			NEBULA
FATH 1.538	12 39 02.	+ 00 09			NEBULA
RNGC 4612	12 39 02.	+ 07 35		12.5	GALAXY
AMES 1955	12 39 02.	+ 16 58 27.	19	17.8	NEBULA
AMES 1954	12 39 02.	+ 17 40 03.	20	16.5	NEBULA
REIN 4.120	12 39 02.34	+ 00 09 19.0			NEBULA
SC 1236+0026.2	12 39 03.	+ 00 09 43.	6		NEBULA
AMES 1957	12 39 03.	+ 12 40 21.	18	17.8	NEBULA
MCG+04-30-012	12 39 03.	+ 26 18 30.	48	14.2	GALAXY
HOLM 439B	12 39 03.	+ 26 19	30	14.6	PART OF MULTIPLE GALAXY
ARC 1591	12 39 03.	+ 50 12		17.6	RICH CLUSTER OF GALAXIES
KN 16.236	12 39 03.3	+ 29 40 10.			NEBULA
KN 16.235	12 39 03.9	+ 26 19 04.			NEBULA
AMES 1956	12 39 04.	+ 18 30 03.	19	17.8	NEBULA
REIZ 2701	12 39 04.	+ 26 19	36	14.3	GALAXY
REIZ 2699	12 39 04.	+ 26 19	60	14.5	GALAXY
RNGC 4615	12 39 04.	+ 26 20		14.0	GALAXY
REIZ 2700	12 39 04.	+ 26 22	18	14.9	GALAXY
ARC 1597	12 39 04.	+ 72 31		17.5	RICH CLUSTER OF GALAXIES
AMES 1958	12 39 05.	+ 16 37 27.	24	16.6	NEBULA
AMES 1963	12 39 06.	+ 11 35 51.	31	16.8	NEBULA
RMB 005	12 39 06.	+ 15 05			FAINT BLUE OBJECT
KARA.68 184	12 39 06.	+ 16 01	54		DWARF GALAXY
REA 60	12 39 06.	+ 16 05	48		DWARF GALAXY
AMES 1962	12 39 06.	+ 18 48 57.	20	17.6	NEBULA
AMES 1961	12 39 06.	+ 18 59 39.	19	17.6	NEBULA
AMES 1960	12 39 06.	+ 19 06 51.	26	17.8	NEBULA
AMES 1959	12 39 06.	+ 19 15 15.	17	17.9	NEBULA
MCG+04-30-014	12 39 06.	+ 23 41	36	15.2	GALAXY
ZWG 129.017	12 39 06.	+ 23 42		15.2	GALAXY
ZWG 129.018	12 39 06.	+ 26 20		13.8	GALAXY
UGC 07852	12 39 06.	+ 26 20	96	13.8	GALAXY Sc
KARA.72 348B	12 39 06.	+ 26 20	90	13.8	PART OF DOUBLE GALAXY
LB 06855	12 39 06.	+ 27 18		19.1	FAINT BLUE STAR
LB 06854	12 39 06.	+ 27 31		17.6	FAINT BLUE STAR
ZC 1239.1+2757	12 39 06.	+ 27 57	1610		CLUSTER OF GALAXIES
LB 06853	12 39 06.	+ 29 06		19.2	FAINT BLUE STAR
ZC 1239.1+3004	12 39 06.	+ 30 04	1010		CLUSTER OF GALAXIES
LB 06852	12 39 06.	+ 31 40		18.8	FAINT BLUE STAR
ZWG 216.017	12 39 06.	+ 41 26		11.5	GALAXY
RNGC 4618	12 39 06.	+ 41 26		11.5	GALAXY
REIZ 2704	12 39 06.	+ 41 26	180	12.0	GALAXY
UGC 07853	12 39 06.	+ 41 26	270	11.5	GALAXY
KARA.72 349A	12 39 06.	+ 41 26	282	11.5	PART OF DOUBLE GALAXY
ZC 1239.1+5015	12 39 06.	+ 50 15	1080		CLUSTER OF GALAXIES
ZC 1239.1+5854	12 39 06.	+ 58 54	670		CLUSTER OF GALAXIES
KN 16.237	12 39 06.8	+ 26 20 10.			NEBULA
IC 3661	12 39 07.	+ 22 46 07.			NONSTELLAR OBJECT
AMES 1965	12 39 08.	+ 16 36 27.	17	17.8	NEBULA
AMES 1964	12 39 08.	+ 18 07 39.	22	17.2	NEBULA
IC 3660	12 39 08.	+ 21 22 01.			NONSTELLAR OBJECT
ZL 093	12 39 08.	+ 27 51 12.		20.9	ULTRAFAINT BLUE STAR
KN 16.238	12 39 08.1	+ 29 07 04.			NEBULA
REIN 3.295	12 39 08.35	+ 12 31 15.3			NEBULA
AMES 1967	12 39 09.	+ 17 20 27.	25	18.0	NEBULA
AMES 1966	12 39 09.	+ 18 02 57.	23	17.4	NEBULA
ARC 1589	12 39 09.	+ 18 53		16.6	RICH CLUSTER OF GALAXIES
IC 3662	12 39 09.	+ 23 42 07.			NONSTELLAR OBJECT
MCG+07-26-037	12 39 09.	+ 41 24	240	12.	GALAXY
IC 3668	12 39 09.	+ 41 24 01.			NONSTELLAR OBJECT
IC 3667	12 39 09.	+ 41 25 25.			NONSTELLAR OBJECT
ARP 023	12 39 09.	+ 41 26			PECULIAR GALAXY
MCG-02-32-024	12 39 09.	- 14 15	48	15.	GALAXY
KN 16.239	12 39 09.6	+ 26 20 59.			NEBULA
AMES 1968	12 39 10.	+ 18 26 39.	14	17.2	NEBULA
REIZ 2703	12 39 10.	+ 26 21	120	13.4	GALAXY
HOLM 439A	12 39 10.	+ 26 21	108	13.8	PART OF MULTIPLE GALAXY
KN 16.240	12 39 10.6	+ 26 18 17.			NEBULA

OBJECT NAME	RIGHT ASCEN.	DECLINATION	DIAM.	MAGN.	TYPE OF OBJECT
KN 16.243	12 39 10.7	+ 28 50 35.			NEBULA
AMES 1969	12 39 11.	+ 18 34 51.	24	17.9	NEBULA
REIZ 2702	12 39 11.	+ 23 43	24	13.9	GALAXY
KN 16.241	12 39 11.2	+ 26 19 47.			NEBULA
KN 16.242	12 39 11.6	+ 26 20 23.			NEBULA
KEEL 528	12 39 11.8	+ 32 40 20.		18.	NEBULA
REA 61	12 39 12.	+ 11 32	48		DWARF GALAXY
RMB 068	12 39 12.	+ 13 07			FAINT BLUE OBJECT
ZC 1239.2+1900	12 39 12.	+ 19 00	3490		CLUSTER OF GALAXIES
AMES 1970	12 39 12.	+ 19 10 03.	19	18.0	NEBULA
IC 3664	12 39 12.	+ 20 13 01.			NONSTELLAR OBJECT
MCG+04-30-013	12 39 12.	+ 26 20	90	13.8	GALAXY
LB 06859	12 39 12.	+ 26 59		19.4	FAINT BLUE STAR
LB 06858	12 39 12.	+ 27 28		19.3	FAINT BLUE STAR
LB 06857	12 39 12.	+ 27 37		19.4	FAINT BLUE STAR
LB 11306	12 39 12.	+ 27 51		14.5	FAINT BLUE STAR
LB 06856	12 39 12.	+ 31 22		18.8	FAINT BLUE STAR
VV 073B	12 39 12.	+ 41 25	102		INTERACTING GALAXY
VV 073A	12 39 12.	+ 41 25	300		INTERACTING GALAXY
VV 073	12 39 12.	+ 41 25	240	11.5	INTERACTING GALAXY
IC 3669	12 39 12.	+ 41 25 01.			NONSTELLAR OBJECT
KN 16.244	12 39 12.7	+ 26 36 05.			NEBULA
IC 3663	12 39 13.	+ 12 30			NONSTELLAR OBJECT
HN 1014	12 39 13.	+ 12 31	18		NEBULA
REIN 3.296	12 39 13.25	+ 10 21 34.1			NEBULA
AMES 1972	12 39 14.	+ 18 29 38.	19	17.8	NEBULA
AMES 1971	12 39 14.	+ 19 09 26.	24	17.8	NEBULA
REIZ 2705	12 39 14.	+ 33 41	12	16.0	GALAXY
FATH 1.539	12 39 14.	- 00 35	14		NEBULA
AMES 1974	12 39 15.	+ 16 06 08.	18	17.6	NEBULA
AMES 1973	12 39 15.	+ 18 53 26.	19	17.2	NEBULA
MCG+08-23-074	12 39 15.	+ 44 46	36	16.	GALAXY
KN 16.245	12 39 15.8	+ 28 08 05.			NEBULA
AMES 1977	12 39 16.	+ 18 28 44.	22	18.0	NEBULA
AMES 1976	12 39 16.	+ 18 54 26.	22	16.9	NEBULA
AMES 1975	12 39 16.	+ 18 57 08.	14	17.6	NEBULA
HOLM 443B	12 39 16.	+ 72 04	36	14.3	PART OF MULTIPLE GALAXY
AMES 1979	12 39 17.	+ 11 05 14.	19	17.0	NEBULA
RNGC 4619	12 39 17.	+ 35 20		13.5	GALAXY
HOLM 438B	12 39 17.	+ 41 36	72	12.7	PART OF MULTIPLE GALAXY
REIZ 2709	12 39 17.	+ 44 46	18	14.8	GALAXY
HOLM 443A	12 39 17.	+ 72 06	54	14.3	PART OF MULTIPLE GALAXY
ZWG 070.219	12 39 18.	+ 09 41		15.3	GALAXY
UGC 07854	12 39 18.	+ 09 41	60	15.3	GALAXY
AMES 1981	12 39 18.	+ 11 28 08.	30	17.1	NEBULA
KARA.68 185	12 39 18.	+ 11 30	40		DWARF GALAXY
ZWG 070.220	12 39 18.	+ 11 46		15.3	GALAXY
UGC 07855	12 39 18.	+ 11 46	60	15.3	GALAXY DWARF
MCG+02-32-180	12 39 18.	+ 11 46	42	15.3	GALAXY
AMES 1978	12 39 18.	+ 19 14 56.	18	16.5	NEBULA
ZWG 129.019	12 39 18.	+ 23 47		15.5	GALAXY
LB 06864	12 39 18.	+ 27 46		20.0	FAINT BLUE STAR
LB 06863	12 39 18.	+ 27 49		18.8	FAINT BLUE STAR
LB 06862	12 39 18.	+ 29 37		18.0	FAINT BLUE STAR
LB 06861	12 39 18.	+ 30 04		19.0	FAINT BLUE STAR
LB 06860	12 39 18.	+ 30 46		19.2	FAINT BLUE STAR
MCG+05-30-060	12 39 18.	+ 32 17	30	16.	GALAXY
ZWG 188.014	12 39 18.	+ 35 20		13.5	GALAXY
UGC 07856	12 39 18.	+ 35 20	96	13.5	GALAXY SBb
ZC 1239.3+7231	12 39 18.	+ 72 31	740		CLUSTER OF GALAXIES
IC 3665	12 39 19.	+ 11 45			NONSTELLAR OBJECT
HN 1015	12 39 19.	+ 11 46	30		NEBULA
AMES 1980	12 39 19.	+ 18 49 44.	14	18.0	NEBULA
ZL 094	12 39 19.	+ 27 46 18.		20.7	ULTRAFAINT BLUE STAR
AMES 1982	12 39 20.	+ 17 32 50.	19	17.8	NEBULA
ZL 096	12 39 20.	+ 27 44 18.		21.5	ULTRAFAINT BLUE STAR
ZL 095	12 39 20.	+ 27 51 36.		21.3	ULTRAFAINT BLUE STAR
REIZ 2706	12 39 20.	+ 35 21	42	13.3	GALAXY
AMES 1984	12 39 21.	+ 18 08 02.	19	18.0	NEBULA
AMES 1983	12 39 21.	+ 19 14 08.	13	16.7	NEBULA
MCG+04-30-015	12 39 21.	+ 23 47	30	15.5	GALAXY
REIZ 2708	12 39 21.	+ 32 16	42	16.0	GALAXY
KEEL 529	12 39 21.3	+ 33 07 55.		17.	NEBULA
IC 3666	12 39 22.	+ 08 07 07.			NONSTELLAR OBJECT
KN 16.246	12 39 22.6	+ 29 49 17.			NEBULA
REIN 3.297	12 39 23.85	+ 12 02 52.1			NEBULA
ZWG 070.221	12 39 24.	+ 09 51		15.3	GALAXY
RMB 067	12 39 24.	+ 13 00			FAINT BLUE OBJECT
AMES 1986	12 39 24.	+ 13 05 02.	16	17.5	NEBULA
REA 62	12 39 24.	+ 13 21	66		DWARF GALAXY
AMES 1985	12 39 24.	+ 14 02 26	60	16.2	NEBULA
ZWG 070.222	12 39 24.	+ 14 03		15.3	GALAXY
UGC 07857	12 39 24.	+ 14 03	78	15.3	GALAXY Sc
MCG+02-32-181	12 39 24.	+ 14 03	36	15.3	GALAXY
FEIG 067	12 39 24.	+ 17 49		11.5	FAINT BLUE STAR
ZC 1239.4+2006	12 39 24.	+ 20 06	470		CLUSTER OF GALAXIES
TON-N 0106	12 39 24.	+ 22 36		15.3	BLUE STAR
IC 3671	12 39 24.	+ 23 47 02.			NONSTELLAR OBJECT
LB 06871	12 39 24.	+ 27 06		19.3	FAINT BLUE STAR
LB 06870	12 39 24.	+ 27 17		18.8	FAINT BLUE STAR
LB 06869	12 39 24.	+ 27 40		19.7	FAINT BLUE STAR
LB 06868	12 39 24.	+ 27 56		18.6	FAINT BLUE STAR
LB 06867	12 39 24.	+ 28 04		19.4	FAINT BLUE STAR
LB 06866	12 39 24.	+ 29 22		20.0	FAINT BLUE STAR
LB 11307	12 39 24.	+ 29 37		14.5	FAINT BLUE STAR
LB 06865	12 39 24.	+ 32 10		19.2	FAINT BLUE STAR
MCG+06-28-018	12 39 24.	+ 35 21	66	13.	GALAXY
ZC 1239.4+5416	12 39 24.	+ 54 16	1210		CLUSTER OF GALAXIES
ZC 1239.4+6014	12 39 24.	+ 60 14	1010		CLUSTER OF GALAXIES
ZWG 014.106	12 39 24.	- 00 55		15.6	GALAXY
ZWG 014.105	12 39 24.	- 03 19		15.3	GALAXY
OCL 0890	12 39 24.	- 62 42	660	8.9	OPEN STAR CLUSTER
VHA 138	12 39 24.	- 62 42	180		OPEN STAR CLUSTER
REIZ 2707	12 39 25.	+ 11 55	60	11.9	GALAXY
IC 3670	12 39 25.	+ 12 01			NONSTELLAR OBJECT
HN 1016	12 39 25.	+ 12 02	12		NEBULA
AMES 1987	12 39 26.	+ 18 46 20.	32	16.6	NEBULA
FATH 1.540	12 39 26.	- 00 54			NEBULA
RNGC 4616	12 39 26.	- 40 22			GALAXY
RNGC 4603D	12 39 26.	- 40 33			GALAXY
RNGC 4609	12 39 26.	- 62 42		9.0	OPEN CLUSTER
AMES 1988	12 39 27.	+ 18 32 50.	24	17.9	NEBULA
IC 0808	12 39 27.	+ 20 13			OPEN CLUSTER
REIZ 2716	12 39 27.	+ 59 19	54	15.3	GALAXY
AMES 1989	12 39 28.	+ 18 38 08.	22	17.7	NEBULA
ZL 097	12 39 28.	+ 27 53 30.		19.8	ULTRAFAINT BLUE STAR
KN 16.247	12 39 28.8	+ 27 14 35.			NEBULA
REIZ 2710	12 39 29.	+ 23 48	30	13.9	GALAXY

OBJECT NAME	RIGHT ASCEN.	DECLINATION	DIAM.	MAGN.	TYPE OF OBJECT
IC 3675	12 39 29.	+ 41 32 56.			SAME AS NGC 4625
KN 16.248	12 39 29.5	+ 27 36 17.			NEBULA
ZC 1239.5+0949	12 39 30.	+ 09 49	1340		CLUSTER OF GALAXIES
AMES 1991	12 39 30.	+ 11 14 38.	23	17.7	NEBULA
SN 1939B	12 39 30.	+ 11 54		11.9	SUPERNOVA
ZWG 070.223	12 39 30.	+ 11 55		11.0	GALAXY
UGC 07858	12 39 30.	+ 11 55	270	11.0	GALAXY E
MCG+02-32-183	12 39 30.	+ 11 55	84	11.0	GALAXY
ZWG 070.224	12 39 30.	+ 13 13		14.0	GALAXY
UGC 07859	12 39 30.	+ 13 13	108	14.0	GALAXY S0
MCG+02-32-182	12 39 30.	+ 13 13	78	14.0	GALAXY
KARA.68 186	12 39 30.	+ 13 21	34		DWARF GALAXY
ZC 1239.5+2058	12 39 30.	+ 20 58	810		CLUSTER OF GALAXIES
LB 06875	12 39 30.	+ 26 46		17.0	FAINT BLUE STAR
LB 06874	12 39 30.	+ 28 30		19.4	FAINT BLUE STAR
LB 06873	12 39 30.	+ 29 16		19.1	FAINT BLUE STAR
LB 11308	12 39 30.	+ 30 12		16.6	FAINT BLUE STAR
LB 06872	12 39 30.	+ 32 24		18.0	FAINT BLUE STAR
ZWG 188.015	12 39 30.	+ 32 51		13.3	GALAXY
UGC 07860	12 39 30.	+ 32 51	126	13.3	GALAXY E?
MCG+06-28-019	12 39 30.	+ 32 52	42	13.	GALAXY
MCG+07-26-038	12 39 30.	+ 41 32	198	13.	GALAXY
RNGC 4625	12 39 30.	+ 41 33		13.0	GALAXY
ZWG 216.018	12 39 30.	+ 41 34		13.0	GALAXY
REIZ 2712	12 39 30.	+ 41 34	36	13.2	GALAXY
UGC 07861	12 39 30.	+ 41 34	90	13.0	GALAXY
KARA.72 349B	12 39 30.	+ 41 34	216	13.0	PART OF DOUBLE GALAXY
ZC 1239.5+5804	12 39 30.	+ 58 04	1340		CLUSTER OF GALAXIES
SEA G-1	12 39 30.	- 40 22 00.	30	13.91	PART OF CHAIN OF GALAXIES
DV.56 N4603D	12 39 30.	- 40 33	90		Sc GALAXY
SEA G-2	12 39 30.	- 40 33 00.	102	13.95	PART OF CHAIN OF GALAXIES
MCG-07-26-029	12 39 30.	- 40 34	72	14.5	GALAXY
AMES 1990	12 39 31.	+ 17 42 38.	23	18.0	NEBULA
REIN 3.298	12 39 31.04	+ 11 55 14.5			NEBULA
KN 16.249	12 39 31.7	+ 28 04 47.			NEBULA
RNGC 4621	12 39 32.	+ 11 55		11.5	GALAXY
AMES 1994	12 39 32.	+ 12 43 56.	13	16.8	NEBULA
IC 0806	12 39 32.	- 17 04 41.			NONSTELLAR OBJECT
RNGC 4620	12 39 33.	+ 13 13		14.0	GALAXY
AMES 1993	12 39 33.	+ 15 54 14.	29	16.3	NEBULA
AMES 1992	12 39 33.	+ 19 15 50.	38	16.5	NEBULA
IC 0809	12 39 34.	+ 12 00 56.			NONSTELLAR OBJECT
ARC 1592	12 39 34.	+ 30 05		18.0	RICH CLUSTER OF GALAXIES
AMES 1997	12 39 35.	+ 18 28 08.	20	18.0	NEBULA
AMES 1996	12 39 35.	+ 19 07 14.	26	17.6	NEBULA
AMES 1995	12 39 35.	+ 19 32 20.	26	17.6	NEBULA
RNGC 4627	12 39 35.	+ 32 51		13.5	GALAXY
SVEN 329	12 39 35.	- 04 51	18	15.4	GALAXY
KARA.68 187	12 39 36.	+ 03 45	40		DWARF GALAXY
TON-N 0108	12 39 36.	+ 04 23		15.0	BLUE STAR
ZWG 042.206	12 39 36.	+ 05 48		15.6	GALAXY
ZWG 042.207	12 39 36.	+ 07 57		13.6	GALAXY
UGC 07862	12 39 36.	+ 07 57	132	13.6	GALAXY S0-a
MCG+01-32-135	12 39 36.	+ 07 57	120	13.6	GALAXY
REA 07	12 39 36.	+ 11 44			DWARF GALAXY
ZWG 070.225	12 39 36.	+ 12 01		15.1	GALAXY
UGC 07863	12 39 36.	+ 12 01	66	15.1	GALAXY E
MCG+02-32-184	12 39 36.	+ 12 01	18	15.1	GALAXY
ZWG 070.226	12 39 36.	+ 12 52		14.7	GALAXY
UGC 07864	12 39 36.	+ 12 52	102	14.7	GALAXY
MCG+02-32-185	12 39 36.	+ 12 52	72	14.7	GALAXY
IC 0810	12 39 36.	+ 12 52 14.			NONSTELLAR OBJECT
KARA.68 188	12 39 36.	+ 14 30	27		DWARF GALAXY
RMB 010	12 39 36.	+ 14 36			FAINT BLUE OBJECT
ZC 1239.6+1616	12 39 36.	+ 16 16	870		CLUSTER OF GALAXIES
AMES 1999	12 39 36.	+ 17 32 14.	24	17.5	NEBULA
AMES 1998	12 39 36.	+ 18 57 26.	24	17.8	NEBULA
IC 3673	12 39 36.	+ 21 24 44.			NONSTELLAR OBJECT
IC 3674	12 39 36.	+ 22 47 08.			NONSTELLAR OBJECT
ZC 1239.6+2456	12 39 36.	+ 24 56	940		CLUSTER OF GALAXIES
LB 06882	12 39 36.	+ 28 47		19.7	FAINT BLUE STAR
LB 06881	12 39 36.	+ 29 03		20.0	FAINT BLUE STAR
LB 06880	12 39 36.	+ 29 54		19.4	FAINT BLUE STAR
LB 06879	12 39 36.	+ 30 03		19.0	FAINT BLUE STAR
LB 06878	12 39 36.	+ 30 13		19.5	FAINT BLUE STAR
LB 06877	12 39 36.	+ 31 18		19.9	FAINT BLUE STAR
LB 11309	12 39 36.	+ 31 24		14.8	FAINT BLUE STAR
LB 06876	12 39 36.	+ 31 49		19.7	FAINT BLUE STAR
MCG+06-28-020	12 39 36.	+ 32 50	840	9.	GALAXY
ZC 1239.6+3832	12 39 36.	+ 38 32	1950		CLUSTER OF GALAXIES
IC 3680	12 39 36.	+ 39 22 38.			NONSTELLAR OBJECT
TON-N 0107	12 39 36.	+ 43 58		15.8	BLUE STAR
LB 02422	12 39 36.	+ 60 17 18.		15.1	FAINT BLUE STAR
ZC 1239.6+6102	12 39 36.	+ 61 02	1480		CLUSTER OF GALAXIES
MCG-03-32-019	12 39 36.	- 17 03 30.	36	15.	GALAXY
IC 0807	12 39 36.	- 17 07 58.			NONSTELLAR OBJECT
MCG-07-26-030	12 39 36.	- 40 23	42	14.	GALAXY
IC 3672	12 39 37.	+ 12 01 38.			GALAXY S0
HOLM 442B	12 39 37.	+ 32 52	90	13.3	PART OF MULTIPLE GALAXY
IC 3681	12 39 37.	+ 39 21 26.			NONSTELLAR OBJECT
KN 16.250	12 39 37.1	+ 27 40 17.			NEBULA
REIN 3.299	12 39 37.46	+ 12 01 39.4			NEBULA
RNGC 4623	12 39 38.	+ 07 57		13.0	GALAXY
REIZ 2711	12 39 38.	+ 07 57	72	13.2	GALAXY
AMES 2003	12 39 38.	+ 12 06 26.	26	17.2	NEBULA
AMES 2002	12 39 38.	+ 18 05 56.	19	17.6	NEBULA
AMES 2001	12 39 38.	+ 18 53 38.	24	17.7	NEBULA
REIZ 2714	12 39 38.	+ 32 52	42	13.7	GALAXY
ARC 1593	12 39 38.	+ 33 36		17.2	RICH CLUSTER OF GALAXIES
REIN 3.301	12 39 38.09	+ 12 52 14.6			NEBULA
REIN 3.300	12 39 38.64	+ 07 57 02.8			NEBULA
AMES 2000	12 39 39.	+ 19 43 26.	24	17.4	NEBULA
REIN 2.159	12 39 39.30	+ 32 49 27.6			NEBULA
AMES 2008	12 39 40.	+ 17 20 32.	14	18.0	NEBULA
AMES 2007	12 39 40.	+ 18 24 50.	35	17.5	NEBULA
AMES 2006	12 39 40.	+ 18 53 50.	22	18.0	NEBULA
AMES 2005	12 39 40.	+ 19 04 26.	12	17.8	NEBULA
AMES 2004	12 39 40.	+ 19 19 50.	24	17.0	NEBULA
KN 16.251	12 39 40.2	+ 28 02 53.			NEBULA
KEEL 530	12 39 40.7	+ 33 02 07.		15.	SPIRAL NEBULA
AMES 2009	12 39 41.	+ 16 52 44.	24	18.0	NEBULA
ARC 1594	12 39 41.	+ 27 47		17.8	RICH CLUSTER OF GALAXIES
RNGC 4631	12 39 41.	+ 32 49		10.0	GALAXY
HOLM 440A	12 39 41.	- 05 30	30	14.3	PART OF MULTIPLE GALAXY
SVEN 331	12 39 41.	- 05 30 59.	12	15.4	GALAXY
SVEN 330	12 39 41.	- 05 30 59.	18	14.7	GALAXY
ZWG 014.107	12 39 42.	+ 00 09		15.6	GALAXY
RMB 188	12 39 42.	+ 11 19			FAINT BLUE OBJECT
KARA.68 189	12 39 42.	+ 12 49	27		DWARF GALAXY
IC 3676	12 39 42.	+ 13 49 56.			SINGLE STAR
RMB 004	12 39 42.	+ 15 05			FAINT BLUE OBJECT
AMES 2010	12 39 42.	+ 18 37 20.	16	17.4	NEBULA
LB 06887	12 39 42.	+ 26 47		19.8	FAINT BLUE STAR
LB 06886	12 39 42.	+ 27 42		19.2	FAINT BLUE STAR
LB 11312	12 39 42.	+ 28 32		16.4	FAINT BLUE STAR
LB 06885	12 39 42.	+ 28 37		19.5	FAINT BLUE STAR
LB 06884	12 39 42.	+ 29 52		20.0	FAINT BLUE STAR
LB 11311	12 39 42.	+ 31 12		13.0	FAINT BLUE STAR
LB 06883	12 39 42.	+ 32 05		18.8	FAINT BLUE STAR
LB 11310	12 39 42.	+ 32 13		12.5	FAINT BLUE STAR
ZWG 188.016	12 39 42.	+ 32 49		9.8	GALAXY
UGC 07865	12 39 42.	+ 32 49	1020	9.8	GALAXY Sc
KARA.72 350A	12 39 42.	+ 32 49	858	9.8	PART OF DOUBLE GALAXY
ZCG 1239+34	12 39 42.	+ 34 02		18.3	COMPACT GALAXY
ZWG 216.019	12 39 42.	+ 38 46		15.5	GALAXY
UGC 07866	12 39 42.	+ 38 46	240	15.5	GALAXY DWRF IR
MCG+10-18-077	12 39 42.	+ 60 08	27	16.	GALAXY
ZC 1239.7+6027	12 39 42.	+ 60 27	470		CLUSTER OF GALAXIES
MCG+13-09-029	12 39 42.	+ 74 40	96	12.3	GALAXY
MCG-01-32-038	12 39 42.	- 05 30 30.	18	15.	GALAXY
HELW 166	12 39 42.	- 05 30 52.			NEBULA
MCG-02-32-025	12 39 42.	- 12 23 30.	60	15.	GALAXY
MCG-03-32-020	12 39 42.	- 17 06	18	15.	GALAXY
IC 3678	12 39 43.	+ 21 09 14.			NONSTELLAR OBJECT
IC 3677	12 39 43.	+ 21 09 32.			NONSTELLAR OBJECT
IC 3679	12 39 43.	+ 23 05 26.			NONSTELLAR OBJECT
HOLM 440B	12 39 43.	- 05 29	36	14.3	PART OF MULTIPLE GALAXY
RNGC 4624	12 39 44.	+ 03 20			NON-EXISTENT OBJECT
AMES 2012	12 39 44.	+ 17 48 32.	42	16.3	NEBULA
AMES 2011	12 39 44.	+ 19 20 14.	38	17.8	NEBULA
ZL 098	12 39 44.	+ 27 51 48.		21.5	ULTRAFAINT BLUE STAR
REIZ 2718	12 39 44.	+ 32 50	90	9.5	GALAXY
HOLM 442A	12 39 44.	+ 32 50	870	8.9	PART OF MULTIPLE GALAXY
KN 16.252	12 39 44.3	+ 27 04 05.			NEBULA
KN 16.253	12 39 44.9	+ 26 12 23.			NEBULA
SC 1237+0111.2	12 39 45.	+ 00 54 44.	18		NEBULA
AMES 2015	12 39 45.	+ 17 59 20.	14	17.8	NEBULA
AMES 2014	12 39 45.	+ 18 31 02.	22	18.0	NEBULA
AMES 2013	12 39 45.	+ 19 12 50.	23	17.2	NEBULA
MCG-01-32-039	12 39 45.	- 05 29	36	15.	GALAXY
REIN 2.160	12 39 45.44	+ 32 49 05.6			NEBULA
AMES 2016	12 39 46.	+ 18 55 02.	19	16.7	NEBULA
AMES 2017	12 39 47.	+ 19 18 56.	24	18.0	NEBULA
SVEN 332	12 39 47.	- 05 29 58.	30	14.5	GALAXY
REIZ 2715	12 39 47.	- 06 41	90	13.5	GALAXY
REIZ 2713	12 39 47.	- 06 45	60	13.7	GALAXY
TON-N 0109	12 39 48.	+ 12 05		15.2	BLUE STAR
FEIG 068	12 39 48.	+ 16 41		12.2	FAINT BLUE STAR
REIZ 2717	12 39 48.	+ 17 47	12	15.9	GALAXY
AMES 2018	12 39 48.	+ 18 51 26.	17	17.6	NEBULA
LB 06891	12 39 48.	+ 28 18		20.2	FAINT BLUE STAR
LB 06890	12 39 48.	+ 28 31		19.1	FAINT BLUE STAR
LB 06889	12 39 48.	+ 30 57		18.5	FAINT BLUE STAR
LB 06888	12 39 48.	+ 31 56		18.7	FAINT BLUE STAR
ZCG 1239+33	12 39 48.	+ 33 59		18.8	COMPACT GALAXY
MCG+07-26-039	12 39 48.	+ 38 45	180	14.	GALAXY
ZC 1239.8+4429	12 39 48.	+ 44 29	1080		CLUSTER OF GALAXIES
MCG+08-23-075	12 39 48.	+ 50 19	42	16.	GALAXY
ZWG 293.034	12 39 48.	+ 60 08		15.7	GALAXY
MCG+13-09-030	12 39 48.	+ 75 34	84	16.	GALAXY
HELW 167	12 39 48.	- 05 29 52.			NEBULA
RNGC 4628	12 39 48.	- 06 41		14.0	GALAXY
RNGC 4626	12 39 48.	- 06 46		14.0	GALAXY
AMES 2021	12 39 49.	+ 15 57 26.	22	17.6	NEBULA
AMES 2020	12 39 49.	+ 17 47 02.	36	16.7	NEBULA
AMES 2019	12 39 49.	+ 19 10 50.	19	17.6	NEBULA
ARP 281	12 39 50.	+ 32 49			PECULIAR GALAXY
KN 16.254	12 39 50.6	+ 26 08 47.			NEBULA
IC 3682	12 39 51.	+ 21 08 20.			NONSTELLAR OBJECT
MCG-01-32-041	12 39 51.	- 06 41	78	14.	GALAXY
MCG-01-32-040	12 39 51.	- 06 46	60	14.	GALAXY
AMES 2025	12 39 52.	+ 11 19 14.	25	17.0	NEBULA
AMES 2023	12 39 52.	+ 16 18 32.	19	17.6	NEBULA
AMES 2022	12 39 52.	+ 19 19 02.	29	17.5	NEBULA
IC 3683	12 39 52.	+ 21 08 44.			NONSTELLAR OBJECT
IC 3687	12 39 52.	+ 38 46 32.			NONSTELLAR OBJECT
RNGC 4648	12 39 52.	+ 74 42		12.5	GALAXY
PATH 2.111	12 39 52.	+ 74 42	22		NEBULA
HOLM 441A	12 39 53.	- 06 41	66	13.4	PART OF MULTIPLE GALAXY
ZWG 014.108	12 39 54.	+ 02 19		15.7	GALAXY
ZC 1239.9+0258	12 39 54.	+ 02 58	1480		CLUSTER OF GALAXIES
RMB 066	12 39 54.	+ 13 06			FAINT BLUE OBJECT
TON-N 0110	12 39 54.	+ 18 27		14.0	BLUE STAR
AMES 2026	12 39 54.	+ 18 47 50.	22	18.0	NEBULA
AMES 2024	12 39 54.	+ 19 47 26.	26	16.9	NEBULA
ZC 1239.9+2001	12 39 54.	+ 20 01	540		CLUSTER OF GALAXIES
LB 11314	12 39 54.	+ 27 44		14.5	FAINT BLUE STAR
TON-N 0636	12 39 54.	+ 28 12		14.7	BLUE STAR
LB 06893	12 39 54.	+ 30 23		19.4	FAINT BLUE STAR
LB 06892	12 39 54.	+ 30 36		19.5	FAINT BLUE STAR
LB 11313	12 39 54.	+ 31 42		14.7	FAINT BLUE STAR
ZWG 216.020	12 39 54.	+ 42 48		15.6	GALAXY
UGC 07867	12 39 54.	+ 42 48	60	15.6	GALAXY S
MCG+08-23-077	12 39 54.	+ 45 44	24	17.	GALAXY
MCG+08-23-076	12 39 54.	+ 45 44	18	17.	GALAXY
ZWG 352.039	12 39 54.	+ 74 43		12.6	GALAXY
UGC 07868	12 39 54.	+ 74 43	102	12.6	GALAXY E
HOLM 441B	12 39 54.	- 06 45	66	13.4	PART OF MULTIPLE GALAXY
MCG-04-30-012	12 39 54.	- 23 30	30	15.5	GALAXY
SEA G0	12 39 54.	- 40 28 00.	108	13.47	PART OF CHAIN OF GALAXIES
MCG-07-26-031	12 39 54.	- 40 29	84	13.5	GALAXY
IC 3684	12 39 55.	+ 11 59			NONSTELLAR OBJECT
HN 1017	12 39 55.	+ 12 00	12		NEBULA
RNGC 4629	12 39 55.	- 01 32			NON-EXISTENT OBJECT
AMES 2029	12 39 56.	+ 12 24 08.	28	17.3	NEBULA
AMES 2028	12 39 56.	+ 15 47 26.	24	17.3	NEBULA
AMES 2027	12 39 56.	+ 16 57 50.	18	17.9	NEBULA
ARC 1596	12 39 56.	+ 20 02		17.8	RICH CLUSTER OF GALAXIES
RNGC 4622	12 39 56.	- 40 28			GALAXY
KN 16.255	12 39 56.9	+ 29 48 35.			NEBULA
REIZ 2720	12 39 57.	+ 01 40	30	15.6	GALAXY
REIZ 2721	12 39 57.	+ 04 14	96	13.2	GALAXY
AMES 2030	12 39 57.	+ 11 10 26.	23	16.8	NEBULA
MCG+07-26-040	12 39 57.	+ 42 48	60	15.	GALAXY
PATH 2.110	12 39 58.	+ 00 12	136		NEBULA
AMES 2031	12 39 58.	+ 10 47 50.	26	17.4	NEBULA

OBJECT NAME	RIGHT ASCEN.	DECLINATION	DIAM.	MAGN.	TYPE OF OBJECT
ARC 1598	12 39 58.	+ 29 24		18.0	RICH CLUSTER OF GALAXIES
REIZ 2719	12 39 58.	- 01 01	24	15.0	GALAXY
ARC 1595	12 39 58.	- 16 09		17.7	RICH CLUSTER OF GALAXIES
REIN 4.121	12 39 58.47	+ 00 11 26.6			NEBULA
REIN 4.122	12 39 58.86	- 01 04 39.4			NEBULA
ZWG 014.110	12 40 00.	+ 00 10		12.6	GALAXY
UGC 07870	12 40 00.	+ 00 10	192	12.6	GALAXY Sc
MCG+00-32-038	12 40 00.	+ 00 10	180	12.6	GALAXY
ZWG 043.001	12 40 00.	+ 04 14		13.4	GALAXY
ZWG 042.208	12 40 00.	+ 04 14		13.4	GALAXY
UGC 07871	12 40 00.	+ 04 14	108	13.4	GALAXY IRR
MCG+01-32-136	12 40 00.	+ 04 14	72	13.4	GALAXY
IC 3685	12 40 00.	+ 07 08 38.			NONSTELLAR OBJECT
3ZW 067	12 40 00.	+ 10 10			COMPACT GALAXY
RMB 110	12 40 00.	+ 12 00			FAINT BLUE OBJECT
AMES 2033	12 40 00.	+ 12 28 08.	17	16.8	NEBULA
MCG+03-32-085	12 40 00.	+ 14 37	96	14.7	GALAXY
SCH 65	12 40 00.	+ 14 38		14.7	PECULIAR GALAXY
ZC 1240.0+2427	12 40 00.	+ 24 27	870		CLUSTER OF GALAXIES
LB 00047	12 40 00.	+ 26 08		17.9	FAINT BLUE STAR
LB 11315	12 40 00.	+ 27 55		12.8	FAINT BLUE STAR
LB 06894	12 40 00.	+ 28 52		20.2	FAINT BLUE STAR
TON-N 0111	12 40 00.	+ 41 52		14.7	BLUE STAR
MCG+08-23-078	12 40 00.	+ 48 38	36	17.	GALAXY
ZC 1240.0+6625	12 40 00.	+ 66 25	810		CLUSTER OF GALAXIES
UGC 07872	12 40 00.	+ 75 35	120	16.0	GALAXY DWRF IR
ZC 1240.0+7621	12 40 00.	+ 76 21	2350		CLUSTER OF GALAXIES
ZWG 014.109	12 40 00.	- 01 05		14.3	GALAXY
UGC 07869	12 40 00.	- 01 05	90	14.3	GALAXY IRR
MCG+00-32-037	12 40 00.	- 01 05	66	14.3	GALAXY
SN 1946B	12 40 01.	+ 00 10		15.7	SUPERNOVA
RNGC 4632	12 40 01.	+ 00 11		12.5	GALAXY
HN 1018	12 40 01.	+ 10 50	24		NEBULA
HOLM 444B	12 40 01.	+ 10 52	18	15.6	PART OF MULTIPLE GALAXY
HOLM 444A	12 40 01.	+ 10 52	24	15.1	PART OF MULTIPLE GALAXY
AMES 2032	12 40 01.	+ 18 37 02.	20	17.6	NEBULA
SC 1237+0157.0	12 40 02.	+ 01 40 32.	12		NEBULA
RNGC 4630	12 40 02.	+ 04 14		13.5	GALAXY
IC 3686	12 40 02.	+ 10 49			NONSTELLAR OBJECT
ZL 099	12 40 02.	+ 27 57 30.		21.1	ULTRAFAINT BLUE STAR
SC 1237-0048.5	12 40 02.	- 01 04 58.	48		NEBULA
REIN 4.123	12 40 03.24	+ 01 39 47.5			NEBULA
REIZ 2722	12 40 04.	+ 00 12	108	12.2	GALAXY
REIZ 2725	12 40 04.	+ 24 45	42	15.3	GALAXY
ZL 100	12 40 04.	+ 27 57 54.		22.0	ULTRAFAINT BLUE STAR
REIN 3.302	12 40 04.46	+ 10 50 22.1			NEBULA
REIN 4.124	12 40 04.69	- 01 16 44.3			NEBULA
AMES 2034	12 40 05.	+ 17 27 50.	24	17.8	NEBULA
SC 1237+0154.4	12 40 06.	+ 01 37 56.	60		NEBULA
ZWG 015.001	12 40 06.	+ 01 38		15.3	GALAXY
UGC 07873	12 40 06.	+ 01 38	108	15.3	GALAXY Sc
MCG+00-33-001	12 40 06.	+ 01 38	66	15.3	GALAXY
ZWG 071.001	12 40 06.	+ 10 50		15.6	GALAXY
ZWG 070.227	12 40 06.	+ 10 50		15.6	GALAXY
MCG+02-32-186	12 40 06.	+ 10 50	90	15.6	GALAXY
RMB 187	12 40 06.	+ 11 13			FAINT BLUE OBJECT
MCG+03-32-086	12 40 06.	+ 14 33	102	13.6	GALAXY
ZWG 100.001	12 40 06.	+ 14 38		14.7	GALAXY
ZWG 099.111	12 40 06.	+ 14 38		14.7	GALAXY
UGC 07874	12 40 06.	+ 14 38	120	14.7	GALAXY
KARA.72 351A	12 40 06.	+ 14 38	114	14.7	PART OF DOUBLE GALAXY
IC 3688	12 40 06.	+ 14 38 02.			SAME AS NGC 4633
ZC 1240.1+1510	12 40 06.	+ 15 10	670		CLUSTER OF GALAXIES
MCG+03-32-087	12 40 06.	+ 20 14	114	13.7	GALAXY
LB 06895	12 40 06.	+ 26 28		19.2	FAINT BLUE STAR
LB 11316	12 40 06.	+ 32 35		17.2	FAINT BLUE STAR
1ZW 040	12 40 06.	+ 34 00			COMPACT GALAXY
ZC 1240.1+4944	12 40 06.	+ 49 44	1010		CLUSTER OF GALAXIES
MCG-02-32-026	12 40 06.	- 12 07	72	15.	GALAXY
MCG-05-30-009	12 40 06.	- 30 08	60	15.	GALAXY
REIN 4.125	12 40 07.46	+ 01 37 06.2			NEBULA
AMES 2037	12 40 08.	+ 13 31 38.	32	16.7	NEBULA
HOLM 445B	12 40 08.	+ 14 39	126	13.5	PART OF MULTIPLE GALAXY
AMES 2036	12 40 08.	+ 16 33 50.	24	18.0	NEBULA
AMES 2035	12 40 08.	+ 18 50 26.	11	17.4	NEBULA
IC 3689	12 40 08.	+ 21 07 27.			NONSTELLAR OBJECT
REIN 4.126	12 40 08.71	- 01 11 31.1			NEBULA
REIZ 2723	12 40 09.	+ 01 37	48	15.3	GALAXY
REIZ 2724	12 40 09.	+ 04 22	24	13.5	GALAXY
RNGC 4633	12 40 09.	+ 14 38		14.5	GALAXY
RNGC 4635	12 40 09.	+ 20 13		13.5	GALAXY
AMES 2038	12 40 10.	+ 12 26 26.	24	18.0	NEBULA
REIZ 2726	12 40 11.	+ 20 14	96	12.9	GALAXY
ZL 101	12 40 11.	+ 27 52 48.		19.6	ULTRAFAINT BLUE STAR
RBA 63	12 40 12.	+ 07 37	48		DWARF GALAXY
ZWG 071.002	12 40 12.	+ 09 47		15.6	GALAXY
ZWG 070.228	12 40 12.	+ 09 47		15.6	GALAXY
AMES 2040	12 40 12.	+ 11 13 02.	12	17.4	NEBULA
RMB 065	12 40 12.	+ 13 07			FAINT BLUE OBJECT
ZWG 071.003	12 40 12.	+ 13 32		15.5	GALAXY
ZWG 100.002	12 40 12.	+ 14 34		13.6	GALAXY
ZWG 099.112	12 40 12.	+ 14 34		13.6	GALAXY
UGC 07875	12 40 12.	+ 14 34	168	13.6	GALAXY Sc
KARA.72 351B	12 40 12.	+ 14 34	162	13.6	PART OF DOUBLE GALAXY
ZWG 100.003	12 40 12.	+ 20 12		13.7	GALAXY
ZWG 099.113	12 40 12.	+ 20 12		13.7	GALAXY
SCH 66	12 40 12.	+ 20 12		13.7	PECULIAR GALAXY
UGC 07876	12 40 12.	+ 20 12	120	13.7	GALAXY Sc
KARA.73B 0547	12 40 12.	+ 20 12	126	13.7	ISOLATED GALAXY S
ZC 1240.2+2223	12 40 12.	+ 22 23	2290		CLUSTER OF GALAXIES
LB 06898	12 40 12.	+ 26 50		19.8	FAINT BLUE STAR
ZWG 159.057	12 40 12.	+ 27 24		15.6	GALAXY
UGC 07877	12 40 12.	+ 27 32	72	16.5	GALAXY S
ZL 102	12 40 12.	+ 27 50 36.		22.5	ULTRAFAINT BLUE STAR
ZC 1240.2+2921	12 40 12.	+ 29 21	2550		CLUSTER OF GALAXIES
LB 06897	12 40 12.	+ 29 25		19.1	FAINT BLUE STAR
LB 06896	12 40 12.	+ 31 54		19.4	FAINT BLUE STAR
TON-N 0112	12 40 12.	+ 33 41		16.0	BLUE STAR
ZC 1240.2+3549	12 40 12.	+ 35 49	3090		CLUSTER OF GALAXIES
MCG+08-23-079	12 40 12.	+ 48 50	24	16.	GALAXY
MCG+09-21-029	12 40 12.	+ 53 27 30.	42	15.	GALAXY
7ZW 479	12 40 12.	+ 63 31			COMPACT GALAXY
ZC 1240.2-0058	12 40 12.	- 00 58	1880		CLUSTER OF GALAXIES
AMES 2041	12 40 13.	+ 14 00 02.	22	16.8	NEBULA
HOLM 445A	12 40 13.	+ 14 35	138	12.8	PART OF MULTIPLE GALAXY
AMES 2039	12 40 14.	+ 19 20 14.	38	17.6	NEBULA
ZL 103	12 40 14.	+ 28 00 00.		16.2	ULTRAFAINT BLUE STAR
REIZ 2734	12 40 14.	+ 65 40	18	14.8	GALAXY

OBJECT NAME	RIGHT ASCEN.	DECLINATION	DIAM.	MAGN.	TYPE OF OBJECT
REIZ 2727	12 40 15.	+ 02 58	180	12.6	GALAXY
ARC 1599	12 40 15.	+ 03 06		17.2	RICH CLUSTER OF GALAXIES
RNGC 4634	12 40 15.	+ 14 34		13.5	GALAXY
REIN 4.127	12 40 15.82	+ 00 17 49.2			NEBULA
SN 1939A	12 40 16.	+ 02 58		12.2	SUPERNOVA
AMES 2042	12 40 16.	+ 13 15 14.	20	17.8	NEBULA
REIZ 2729	12 40 16.	+ 24 43	54	15.1	GALAXY
LB 00261	12 40 16.	+ 75 26 00.		14.9	FAINT BLUE STAR
REIN 3.303	12 40 16.23	+ 11 42 57.3			NEBULA
AMES 2043	12 40 17.	+ 11 15 38.	14	17.3	NEBULA
KN 16.256	12 40 17.7	+ 25 41 30.			NEBULA
KN 16.257	12 40 17.8	+ 27 23 36.			NEBULA
ZWG 043.002	12 40 18.	+ 02 58		11.8	GALAXY
UGC 07878	12 40 18.	+ 02 58	420	11.8	GALAXY E
MCG+01-32-137	12 40 18.	+ 02 58	300	11.8	GALAXY
ZWG 071.004	12 40 18.	+ 10 38		15.3	GALAXY
UGC 07879	12 40 18.	+ 10 38	78	15.3	GALAXY S
ZWG 071.005	12 40 18.	+ 10 57		15.4	GALAXY
ZWG 071.006	12 40 18.	+ 11 43		12.2	GALAXY
ZWG 070.229	12 40 18.	+ 11 43		12.2	GALAXY
UGC 07881	12 40 18.	+ 11 43	84	16.0	GALAXY
UGC 07880	12 40 18.	+ 11 43	174	12.2	GALAXY S0
MCG+02-32-188	12 40 18.	+ 11 43	54	16.	GALAXY
MCG+02-32-187	12 40 18.	+ 11 43	132	12.2	GALAXY
SCH 67	12 40 18.	+ 13 31		12.4	PECULIAR GALAXY
ZWG 159.058	12 40 18.	+ 26 55		15.5	GALAXY
MCG+05-30-061	12 40 18.	+ 27 23	30	15.6	GALAXY
ZWG 188.017	12 40 18.	+ 33 33		15.7	GALAXY
UGC 07882	12 40 18.	+ 33 33	90	15.7	GALAXY Sc-IRR
MRK 654	12 40 18.	+ 34 22	9	16.5	GALAXY WITH UV CONTINUUM
TON-N 0113	12 40 18.	+ 37 42		15.5	BLUE STAR
DV.56 N4645A	12 40 18.	- 41 05	198		S0 GALAXY
ACK 302-05.1	12 40 18.	- 67 56			PLANETARY NEBULA
RNGC 4636	12 40 19.	+ 02 58		11.0	GALAXY
HN 1019	12 40 19.	+ 10 37	12		NEBULA
AMES 2045	12 40 19.	+ 12 33 14.	19	16.9	NEBULA
AMES 2044	12 40 19.	+ 14 06 32.	31	17.8	NEBULA
IC 3690	12 40 20.	+ 10 36			NONSTELLAR OBJECT
RNGC 4638	12 40 20.	+ 11 43		12.5	GALAXY
RNGC 4645A	12 40 20.	- 41 05			GALAXY
KN 16.258	12 40 20.1	+ 27 32 48.			NEBULA
REIN 4.128	12 40 20.83	- 01 07 02.5			NEBULA
AMES 2047	12 40 21.	+ 11 41 44.	41	16.6	NEBULA
IC 3691	12 40 21.	+ 23 03 45.			NONSTELLAR OBJECT
LB 02423	12 40 21.	+ 57 03 06.		16.8	FAINT BLUE STAR
KN 16.259	12 40 21.5	+ 26 54 42.			NEBULA
AMES 2046	12 40 22.	+ 18 27 50.	24	17.9	NEBULA
REIZ 2731	12 40 22.	+ 24 42	48	15.3	GALAXY
RNGC 4644	12 40 22.	+ 55 25		15.0	GALAXY
REIZ 2728	12 40 22.	- 00 53	60	15.1	GALAXY
PATH 1.541	12 40 22.	- 00 57	108		NEBULA
ARC 1600	12 40 22.	- 17 00		17.7	RICH CLUSTER OF GALAXIES
REIZ 2730	12 40 23.	+ 23 04	36	14.6	GALAXY
HOLM 447A	12 40 23.	+ 55 27	60	13.6	PART OF MULTIPLE GALAXY
REIN 4.129	12 40 23.38	- 00 57 19.1			NEBULA
KN 16.260	12 40 23.6	+ 30 05 36.			NEBULA
ZWG 043.003	12 40 24.	+ 03 57		15.1	GALAXY
ZWG 071.007	12 40 24.	+ 11 42		15.6	GALAXY
KARA.68 190	12 40 24.	+ 12 35	27		DWARF GALAXY
ZWG 071.008	12 40 24.	+ 13 32		12.4	GALAXY
ZWG 070.230	12 40 24.	+ 13 32		12.4	GALAXY
UGC 07884	12 40 24.	+ 13 32	192	12.4	GALAXY Sb
MCG+02-32-189	12 40 24.	+ 13 32	180	12.4	GALAXY
AMES 2049	12 40 24.	+ 14 04 56.	25	17.2	NEBULA
AMES 2048	12 40 24.	+ 17 53 14.	30	17.7	NEBULA
MCG+04-30-016	12 40 24.	+ 21 15	54	14.8	GALAXY
ZWG 129.020	12 40 24.	+ 21 16		14.8	GALAXY
UGC 07885	12 40 24.	+ 21 16	60	14.8	GALAXY SBa
KARA.73B 0548	12 40 24.	+ 21 16	54	14.8	ISOLATED GALAXY S
LB 06906	12 40 24.	+ 26 36		19.1	FAINT BLUE STAR
LB 06905	12 40 24.	+ 27 57		18.7	FAINT BLUE STAR
LB 11319	12 40 24.	+ 28 36		13.5	FAINT BLUE STAR
LB 06904	12 40 24.	+ 28 50		18.7	FAINT BLUE STAR
LB 06903	12 40 24.	+ 29 15		19.4	FAINT BLUE STAR
LB 06902	12 40 24.	+ 29 49		18.4	FAINT BLUE STAR
LB 11318	12 40 24.	+ 30 26		12.6	FAINT BLUE STAR
LB 11317	12 40 24.	+ 30 54		14.2	FAINT BLUE STAR
LB 06901	12 40 24.	+ 31 11		20.4	FAINT BLUE STAR
LB 06900	12 40 24.	+ 31 27		19.9	FAINT BLUE STAR
UGC 07886	12 40 24.	+ 32 10	60	16.5	GALAXY Sc
LB 06899	12 40 24.	+ 32 22		19.9	FAINT BLUE STAR
MCG+06-28-021	12 40 24.	+ 33 35	66	14.5	GALAXY
ZC 1240.4+3956	12 40 24.	+ 39 56	4500		CLUSTER OF GALAXIES
MCG+08-22-084	12 40 24.	+ 47 04	36	14.	GALAXY
ZWG 270.014	12 40 24.	+ 55 25		14.8	GALAXY
UGC 07887	12 40 24.	+ 55 25	102	14.8	GALAXY SB:b
KARA.72 352A	12 40 24.	+ 55 25	72	14.8	PART OF DOUBLE GALAXY
MCG+09-21-030	12 40 24.	+ 55 25	90	13.6	GALAXY
ZWG 015.002	12 40 24.	- 00 57		14.3	GALAXY
UGC 07883	12 40 24.	- 00 57	192	14.3	GALAXY Sc
MCG+00-33-002	12 40 24.	- 00 57	132	14.3	GALAXY
MCG-07-26-032	12 40 24.	- 41 06	150	13.	GALAXY
MCG-07-26-033	12 40 24.	- 42 39	72	15.	GALAXY
HOLM 446A	12 40 25.	+ 12 34	60	14.4	PART OF MULTIPLE GALAXY
IC 3692	12 40 25.	+ 21 15 51.			NONSTELLAR OBJECT
KN 16.262	12 40 25.6	+ 28 57 48.			NEBULA
RNGC 4637	12 40 26.	+ 11 43		12.0	GALAXY
RNGC 4640B	12 40 26.	+ 12 34		16.5	GALAXY
AMES 2050	12 40 26.	+ 12 36 26.	22	17.6	NEBULA
REIZ 2732	12 40 26.	+ 33 34	66	15.2	GALAXY
KN 16.261	12 40 26.6	+ 25 56 12.			NEBULA
REIN 3.305	12 40 26.81	+ 12 33 39.6			NEBULA
REIN 3.304	12 40 26.83	+ 10 57 22.0			NEBULA
RNGC 4639	12 40 27.	+ 13 32		12.5	GALAXY
SC 1237-0041.2	12 40 27.	- 00 57 40.	120		NEBULA
SC 1237-0044.3	12 40 28.	- 01 00 46.	30		NEBULA
HOLM 446B	12 40 29.	+ 12 34	36	15.6	PART OF MULTIPLE GALAXY
REIN 4.130	12 40 29.16	- 01 00 33.6			NEBULA
KARA.68 191	12 40 30.	+ 11 58	34		DWARF GALAXY
ZWG 071.009	12 40 30.	+ 12 34		15.2	GALAXY
UGC 07888	12 40 30.	+ 12 34	96	15.2	GALAXY S
MCG+02-32-190	12 40 30.	+ 12 34	60	15.2	GALAXY
AMES 2051	12 40 30.	+ 13 26 55.	18	17.6	NEBULA
TON-N 0114	12 40 30.	+ 21 15		15.5	BLUE STAR
LB 06910	12 40 30.	+ 27 53		18.9	FAINT BLUE STAR
LB 06911	12 40 30.	+ 29 40		16.7	FAINT BLUE STAR
LB 06909	12 40 30.	+ 31 12		20.2	FAINT BLUE STAR
LB 06908	12 40 30.	+ 31 35		20.2	FAINT BLUE STAR

OBJECT NAME	RIGHT ASCEN.	DECLINATION	DIAM.	MAGN.	TYPE OF OBJECT
LB 06907	12 40 30.	+ 32 12		18.8	FAINT BLUE STAR
LB 11320	12 40 30.	+ 32 20		12.8	FAINT BLUE STAR
ZC 1240.5+3745	12 40 30.	+ 37 45	1550		CLUSTER OF GALAXIES
TON-N 0115	12 40 30.	+ 50 47		16.6	BLUE STAR
MCG+09-21-031	12 40 30.	+ 55 07	36	13.8	GALAXY
KARA.72 352B	12 40 30.	+ 55 25	60		PART OF DOUBLE GALAXY
MCG+09-21-032	12 40 30.	+ 55 25	60	14.3	GALAXY
ZWG 316.003	12 40 30.	+ 64 52		15.5	GALAXY
ZWG 315.053	12 40 30.	+ 64 52		15.5	GALAXY
ZWG 015.004	12 40 30.	- 00 09		15.7	GALAXY
ZWG 015.003	12 40 30.	- 01 00		15.5	GALAXY
MCG+00-33-003	12 40 30.	- 01 00	24	15.5	GALAXY
HN 1020	12 40 31.	+ 10 57	12		NEBULA
KN 16.263	12 40 31.6	+ 29 40 30.			NEBULA
IC 3693	12 40 32.	+ 10 56			NONSTELLAR OBJECT
RNGC 4640A	12 40 32.	+ 12 34		15.0	GALAXY
HZ 31	12 40 32.	+ 32 21		12.8	BLUE STAR
SC 1237+0006.2	12 40 32.	- 00 10 16.	12		NEBULA
REIN 4.131	12 40 32.69	- 00 09 52.6			NEBULA
KN 16.265	12 40 33.3	+ 30 23 54.			NEBULA
KN 16.264	12 40 33.6	+ 29 20 00.			NEBULA
REIZ 2733	12 40 34.	+ 28 00	120	15.9	GALAXY
IC 3697	12 40 34.	+ 40 07 15.			NONSTELLAR OBJECT
RNGC 4646	12 40 34.	+ 55 07		14.0	GALAXY
HOLM 447B	12 40 34.	+ 55 27	30	14.3	PART OF MULTIPLE GALAXY
REIZ 2742	12 40 35.	+ 55 09	30	13.8	GALAXY
LB 02424	12 40 35.	+ 59 29 16.		13.1	FAINT BLUE STAR
ZWG 043.004	12 40 36.	+ 03 51		15.4	GALAXY
SN 1956C	12 40 36.	+ 04 05		17.6	SUPERNOVA
ZWG 043.005	12 40 36.	+ 07 44		15.5	GALAXY
ZWG 071.010	12 40 36.	+ 11 29		15.3	GALAXY
MCG+02-32-192	12 40 36.	+ 11 29	42	15.3	GALAXY
ZWG 071.011	12 40 36.	+ 12 19		14.9	GALAXY
UGC 07889	12 40 36.	+ 12 19	78	14.9	GALAXY S0
MCG+02-32-191	12 40 36.	+ 12 19	48	14.9	GALAXY
MCG+04-30-017	12 40 36.	+ 24 42 30.	48	15.5	GALAXY
LB 06913	12 40 36.	+ 27 28		19.5	FAINT BLUE STAR
ZC 1240.6+2731	12 40 36.	+ 27 31	1280		CLUSTER OF GALAXIES
ZWG 159.059	12 40 36.	+ 28 00		14.5	GALAXY
LB 11322	12 40 36.	+ 28 00		15.0	FAINT BLUE STAR
UGC 07890	12 40 36.	+ 28 00	36	14.5	GALAXY
MCG+05-30-062	12 40 36.	+ 28 00	36	14.5	GALAXY
LB 06912	12 40 36.	+ 29 08		19.0	FAINT BLUE STAR
ZWG 159.060	12 40 36.	+ 30 40		15.5	GALAXY
SCH 68	12 40 36.	+ 30 40		15.5	PECULIAR GALAXY
UGC 07891	12 40 36.	+ 30 40	66	15.5	GALAXY DBL SYS
LB 11321	12 40 36.	+ 32 05		14.5	FAINT BLUE STAR
MCG+07-26-041	12 40 36.	+ 40 06	42	17.	GALAXY
MCG+08-23-080	12 40 36.	+ 48 52	30	15.	GALAXY
ZWG 270.015	12 40 36.	+ 55 07		13.8	GALAXY
UGC 07892	12 40 36.	+ 55 07	42	13.8	GALAXY
ZC 1240.6+5720	12 40 36.	+ 57 20	3090		CLUSTER OF GALAXIES
ZWG 015.006	12 40 36.	- 01 09		15.6	GALAXY
ZWG 015.005	12 40 36.	- 02 14		15.6	GALAXY
MCG-02-32-028	12 40 36.	- 10 37 30.	12	18.	GALAXY
MCG-02-32-027	12 40 36.	- 10 37 30.	13	16.	GALAXY
OCL 0891	12 40 36.	- 62 50	120		OPEN STAR CLUSTER
VHA 139	12 40 36.	- 62 50	90		OPEN STAR CLUSTER
REIN 4.132	12 40 36.04	- 01 10 04.1			NEBULA
KEEL 531	12 40 36.2	+ 32 36 11.		18.	NEBULA
REIN 3.306	12 40 36.22	+ 11 29 07.1			NEBULA
REIN 3.307	12 40 36.46	+ 12 19 28.1			NEBULA
AMES 2053	12 40 37.	+ 10 57 19.	35	17.0	NEBULA
IC 3694	12 40 37.	+ 11 28			NONSTELLAR OBJECT
HN 1021	12 40 37.	+ 11 29		13.5	NEBULA
RNGC 4641	12 40 38.	+ 11 19		15.0	GALAXY
AMES 2052	12 40 38.	+ 18 25 31.	19	18.0	NEBULA
SC 1238-0156.7	12 40 38.	- 02 13 10.	18		NEBULA
KN 16.266	12 40 38.0	+ 27 59 24.			NEBULA
AMES 2054	12 40 39.	+ 13 09 07.	18	16.8	NEBULA
IC 3695	12 40 39.	+ 23 00 57.			NONSTELLAR OBJECT
REIZ 2738	12 40 39.	+ 30 39	60	15.3	GALAXY
REIZ 2746	12 40 39.	+ 64 54	24	15.0	GALAXY
REIZ 2737	12 40 40.	+ 24 31	60	15.2	GALAXY
KEEL 532	12 40 40.0	+ 32 49 45.		18.	NEBULA
IC 3696	12 40 41.	+ 20 12 03.			NONSTELLAR OBJECT
REIZ 2744	12 40 41.	+ 55 26	42	14.7	GALAXY
KEEL 533	12 40 41.2	+ 32 45 54.		14.	NEBULA
ZWG 043.006	12 40 42.	+ 04 21		15.6	GALAXY
ZC 1240.7+0918	12 40 42.	+ 09 18	940		CLUSTER OF GALAXIES
ZC 1240.7+1722	12 40 42.	+ 17 22	1010		CLUSTER OF GALAXIES
ZC 1240.7+2134	12 40 42.	+ 21 34	940		CLUSTER OF GALAXIES
VV 151D	12 40 42.	+ 30 39	18	17.	INTERACTING GALAXY
VV 151C	12 40 42.	+ 30 39	18	16.	INTERACTING GALAXY
VV 151B	12 40 42.	+ 30 39	42	15.	INTERACTING GALAXY
VV 151A	12 40 42.	+ 30 39	36	15.	INTERACTING GALAXY
VV 151	12 40 42.	+ 30 39	114	14.	INTERACTING GALAXY
MCG+05-30-063	12 40 42.	+ 30 40	72	15.5	GALAXY
LB 06914	12 40 42.	+ 32 23		19.7	FAINT BLUE STAR
MCG+06-28-022	12 40 42.	+ 32 47	15	15.	GALAXY
TON-N 0116	12 40 42.	+ 36 44		15.7	BLUE STAR
ZC 1240.7+6728	12 40 42.	+ 67 28	2080		CLUSTER OF GALAXIES
UGC 07894	12 40 42.	+ 72 11	60	16.5	GALAXY SB?c
ZWG 015.007	12 40 42.	- 00 22		13.8	GALAXY
UGC 07893	12 40 42.	- 00 22	120	13.8	GALAXY S
MCG+00-33-004	12 40 42.	- 00 22	90	13.8	GALAXY
AMES 2056	12 40 43.	+ 13 10 07.	14	16.7	NEBULA
AMES 2055	12 40 43.	+ 16 15 07.	22	17.4	NEBULA
RNGC 4642	12 40 43.	- 00 22		14.0	GALAXY
PATE 2.112	12 40 43.	- 00 22	81		NEBULA
REIN 4.133	12 40 43.99	- 00 22 18.3			NEBULA
REIZ 2736	12 40 45.	+ 02 16	90	12.2	GALAXY
REIN 4.134	12 40 45.00	- 00 22 09.9			NEBULA
AMES 2058	12 40 46.	+ 12 05 31.	28	16.8	NEBULA
AMES 2057	12 40 46.	+ 18 33 31.	19	17.4	NEBULA
REIZ 2741	12 40 46.	+ 24 30	24	15.0	GALAXY
REIZ 2735	12 40 46.	- 00 22	54	13.3	GALAXY
REIN 3.308	12 40 46.20	+ 11 29 03.6			NEBULA
ZWG 015.008	12 40 48.	+ 02 16		11.9	GALAXY
UGC 07895	12 40 48.	+ 02 16	240	11.9	GALAXY SB0
MCG+00-33-005	12 40 48.	+ 02 16	78	11.9	GALAXY
ZWG 043.007	12 40 48.	+ 03 49		15.4	GALAXY
ZWG 043.008	12 40 48.	+ 06 57		15.6	GALAXY
ZWG 071.012	12 40 48.	+ 11 29		15.4	GALAXY
MCG+02-32-193	12 40 48.	+ 11 29	30	15.4	GALAXY
REIZ 2740	12 40 48.	+ 19 16	24	14.9	GALAXY
IC 3699	12 40 48.	+ 19 16 33.			NONSTELLAR OBJECT
ZC 1240.8+2351	12 40 48.	+ 23 51	2020		CLUSTER OF GALAXIES
LB 06920	12 40 48.	+ 27 08		19.3	FAINT BLUE STAR
LB 06919	12 40 48.	+ 27 08		18.5	FAINT BLUE STAR
LB 06918	12 40 48.	+ 27 40		19.3	FAINT BLUE STAR
LB 06917	12 40 48.	+ 27 50		17.9	FAINT BLUE STAR
LB 06916	12 40 48.	+ 28 22		18.9	FAINT BLUE STAR
LB 06915	12 40 48.	+ 28 46		18.8	FAINT BLUE STAR
SEY 192	12 40 48.	+ 31 21 33.		15.3	FAINT GALAXY
ZWG 159.061	12 40 48.	+ 31 22		14.8	GALAXY
MCG+05-30-064	12 40 48.	+ 31 23	48	14.8	GALAXY
7ZW 480	12 40 48.	+ 63 33			COMPACT GALAXY
DV.56 N4645B	12 40 48.	- 41 05	84		S0 GALAXY
MCG-07-26-034	12 40 48.	- 41 06	120	13.5	GALAXY
KN 16.267	12 40 48.2	+ 27 21 36.			NEBULA
RNGC 4643	12 40 49.	+ 02 15		12.0	GALAXY
IC 3698	12 40 49.	+ 11 28			NONSTELLAR OBJECT
HN 1022	12 40 49.	+ 11 29	18		NEBULA
IC 3700	12 40 50.	+ 19 32 21.			NONSTELLAR OBJECT
REIZ 2748	12 40 50.	+ 59 15	66	14.2	GALAXY
RNGC 4645B	12 40 50.	- 41 05			GALAXY
REIN 3.309	12 40 50.59	+ 11 28 37.1			NEBULA
AMES 2059	12 40 51.	+ 17 28 07.	12	16.6	NEBULA
LB 02425	12 40 51.	+ 57 39 54.		17.0	FAINT BLUE STAR
AMES 2060	12 40 52.	+ 13 46 43.	28	17.6	NEBULA
LB 02426	12 40 52.	+ 58 33 18.		16.6	FAINT BLUE STAR
REIZ 2739	12 40 52.	- 00 52	48	15.0	GALAXY
KN 16.268	12 40 52.3	+ 26 15 30.			NEBULA
ZC 1240.9+1429	12 40 54.	+ 14 29	610		CLUSTER OF GALAXIES
REIZ 2743	12 40 54.	+ 19 33	6	16.0	GALAXY
LB 06924	12 40 54.	+ 27 33		19.3	FAINT BLUE STAR
LB 06923	12 40 54.	+ 28 47		18.8	FAINT BLUE STAR
LB 06922	12 40 54.	+ 30 10		17.8	FAINT BLUE STAR
LB 06921	12 40 54.	+ 32 24		18.6	FAINT BLUE STAR
ZC 1240.9+5000	12 40 54.	+ 50 00	870		CLUSTER OF GALAXIES
MCG+08-23-081	12 40 54.	+ 50 10	15	17.	GALAXY
MCG-06-28-013	12 40 54.	- 35 27	36	15.	GALAXY
KN 16.269	12 40 54.5	+ 26 21 54.			NEBULA
LB 02427	12 40 55.	+ 58 29 42.		16.8	FAINT BLUE STAR
IC 3703	12 40 57.	+ 38 14 58.			NONSTELLAR OBJECT
REIN 3.310	12 40 57.24	+ 11 08 49.9			NEBULA
KN 16.270	12 40 57.6	+ 26 10 48.			NEBULA
AMES 2063	12 40 58.	+ 10 21 07.	40	16.4	NEBULA
HOLM 448B	12 40 59.	+ 11 52	96	11.4	PART OF MULTIPLE GALAXY
AMES 2064	12 40 59.	+ 13 08 19.	26	17.9	NEBULA
AMES 2062	12 40 59.	+ 17 16 55.	26	16.5	NEBULA
AMES 2061	12 40 59.	+ 17 54 13.	14	17.7	NEBULA
ABC 1602	12 40 59.	+ 27 35		17.8	RICH CLUSTER OF GALAXIES
KEEL 534	12 40 59.4	+ 33 09 36.		15.	NEBULA
VDB.66G 144	12 41	+ 00 44	70		DWARF GALAXY
VDB.66G 143	12 41	+ 34 39	130		DWARF GALAXY
VDB.66G 142	12 41	- 05 26	130		DWARF GALAXY
ZWG 043.009	12 41 00.	+ 07 43		15.3	GALAXY
MCG+01-33-001	12 41 00.	+ 07 43	18	15.3	GALAXY
ABC 1601	12 41 00.	+ 09 16		17.2	RICH CLUSTER OF GALAXIES
ZWG 071.013	12 41 00.	+ 10 22		15.6	GALAXY
ZWG 071.014	12 41 00.	+ 11 09		15.5	GALAXY
SCH 69	12 41 00.	+ 11 51		12.5	PECULIAR GALAXY
VV 206B	12 41 00.	+ 11 51	180	12.1	INTERACTING GALAXY
ZWG 071.015	12 41 00.	+ 11 52		12.5	GALAXY
UGC 07896	12 41 00.	+ 11 52	180	12.5	GALAXY Sc
KARA.72 353A	12 41 00.	+ 11 52	168	12.5	PART OF DOUBLE GALAXY
ARP 116	12 41 00.	+ 11 52			PECULIAR GALAXY
MCG+02-33-001	12 41 00.	+ 11 52	180	12.5	GALAXY
LB 06928	12 41 00.	+ 26 58		19.3	FAINT BLUE STAR
ZWG 159.062	12 41 00.	+ 29 45		15.7	GALAXY
UGC 07897	12 41 00.	+ 29 45	60	15.7	GALAXY S
MCG+05-30-065	12 41 00.	+ 29 45	60	15.7	GALAXY
LB 06927	12 41 00.	+ 30 18		19.4	FAINT BLUE STAR
LB 06926	12 41 00.	+ 31 56		19.0	FAINT BLUE STAR
LB 06925	12 41 00.	+ 32 12		19.3	FAINT BLUE STAR
ZC 1241.0+4705	12 41 00.	+ 47 05	870		CLUSTER OF GALAXIES
MCG-03-33-001	12 41 00.	- 20 35	60	15.	GALAXY
IC 3702	12 41 01.	+ 11 07			NONSTELLAR OBJECT
HN 1024	12 41 01.	+ 11 08	12		NEBULA
IC 3701	12 41 01.	+ 11 18			NONSTELLAR OBJECT
HN 1023	12 41 01.	+ 11 19	12		NEBULA
AMES 2065	12 41 01.	+ 15 35 19.	22	17.4	NEBULA
REIN 3.311	12 41 01.37	+ 11 51 20.4			NEBULA
KEEL 535	12 41 01.8	+ 32 44 25.		16.	NEBULA
AMES 2067	12 41 02.	+ 11 32 49.	31	16.8	NEBULA
RNGC 4647	12 41 02.	+ 11 51		12.0	GALAXY
AMES 2066	12 41 02.	+ 17 35 55.	19	17.0	NEBULA
ABC 1607	12 41 02.	+ 76 26		16.7	RICH CLUSTER OF GALAXIES
KN 16.271	12 41 02.1	+ 29 44 31.			NEBULA
MCG+08-23-082	12 41 03.	+ 50 10	24	16.	GALAXY
LB 02428	12 41 03.	+ 51 12 24.		16.1	FAINT BLUE STAR
KN 16.272	12 41 03.7	+ 30 05 25.			NEBULA
IC 3707	12 41 04.	+ 38 15 22.			NONSTELLAR OBJECT
REIZ 2745	12 41 04.	- 01 27	18	14.8	GALAXY
AMES 2068	12 41 05.	+ 10 19 55.	22	17.0	NEBULA
RNGC 4652	12 41 05.	+ 59 14		15.5	GALAXY
SC 1238+0105.9	12 41 06.	+ 00 49 27.	12		NEBULA
AMES 2071	12 41 06.	+ 11 35 55.	18	17.6	NEBULA
VV 206A	12 41 06.	+ 11 49	300	9.9	INTERACTING GALAXY
ZWG 071.016	12 41 06.	+ 11 50		10.3	GALAXY
UGC 07898	12 41 06.	+ 11 50	540	10.3	GALAXY E
KARA.72 353B	12 41 06.	+ 11 50	402	10.3	PART OF DOUBLE GALAXY
MCG+02-33-002	12 41 06.	+ 11 50	180	10.3	GALAXY
AMES 2070	12 41 06.	+ 12 50 19.	13	18.0	NEBULA
ZC 1241.1+1348	12 41 06.	+ 13 48	610		CLUSTER OF GALAXIES
LB 00048	12 41 06.	+ 26 10		18.0	FAINT BLUE STAR
TON-N 0117	12 41 06.	+ 29 49		14.0	BLUE STAR
LB 06929	12 41 06.	+ 30 29		18.4	FAINT BLUE STAR
LB 11324	12 41 06.	+ 32 17		17.2	FAINT BLUE STAR
LB 11323	12 41 06.	+ 46 53		14.3	FAINT BLUE STAR
ZWG 293.035	12 41 06.	+ 59 14		15.5	GALAXY
ZC 1241.1+6425	12 41 06.	+ 64 25	1880		CLUSTER OF GALAXIES
SEA G1	12 41 06.	- 40 26 18.	18	14.19	PART OF CHAIN OF GALAXIES
SEA G2	12 41 06.	- 40 26 24.	84	14.98	PART OF CHAIN OF GALAXIES
KN 16.273	12 41 06.2	+ 29 44 37.			NEBULA
AMES 2072	12 41 07.	+ 10 45 25.	22	17.0	NEBULA
HOLM 448A	12 41 07.	+ 11 50	162	10.3	PART OF MULTIPLE GALAXY
AMES 2069	12 41 07.	+ 18 12 43.	17	17.2	NEBULA
KEEL 536	12 41 07.9	+ 32 40 20.		18.	NEBULA
RNGC 4649	12 41 08.	+ 11 50		10.5	GALAXY
REIN 3.312	12 41 08.86	+ 11 49 33.0			NEBULA
AMES 2074	12 41 09.	+ 12 30 43.	22	17.0	NEBULA
AMES 2073	12 41 09.	+ 17 19 07.	13	17.7	NEBULA
MCG-07-26-036	12 41 09.	- 40 27	48	14.5	GALAXY

OBJECT NAME	RIGHT ASCEN.	DECLINATION	DIAM.	MAGN.	TYPE OF OBJECT
MCG-07-26-035	12 41 09.	- 40 27	48	15.5	GALAXY
KN 16.274	12 41 09.6	+ 29 47 19.			NEBULA
REIZ 2757	12 41 10.	+ 56 24	24	14.0	GALAXY
REIZ 2747	12 41 10.	- 01 24	12	15.0	GALAXY
AMES 2076	12 41 11.	+ 14 11 43.	35	17.8	NEBULA
SC 1238+0010.9	12 41 11.	- 00 05 33.	12		NEBULA
REIN 4.135	12 41 11.34	- 00 05 03.5			NEBULA
KEEL 537	12 41 11.5	+ 32 46 26.		18.	NEBULA
ZWG 071.017	12 41 12.	+ 11 03		14.7	GALAXY
UGC 07899	12 41 12.	+ 11 03	78	14.7	GALAXY Sb-c
MCG+02-33-003	12 41 12.	+ 11 03	78	14.7	GALAXY
AMES 2077	12 41 12.	+ 11 57 19.	32	16.9	NEBULA
MCG+03-33-001	12 41 12.	+ 16 37	210	11.3	GALAXY
SCH 70	12 41 12.	+ 16 40		11.3	PECULIAR GALAXY
REIZ 2750	12 41 12.	+ 16 40	180	11.2	GALAXY
ARP 189	12 41 12.	+ 16 40			PECULIAR GALAXY
AMES 2075	12 41 12.	+ 17 25 07.	22	17.6	NEBULA
IC 3705	12 41 12.	+ 19 35 58.			NONSTELLAR OBJECT
LB 06936	12 41 12.	+ 26 48		19.5	FAINT BLUE STAR
LB 06935	12 41 12.	+ 28 27		19.6	FAINT BLUE STAR
LB 06934	12 41 12.	+ 30 23		18.7	FAINT BLUE STAR
LB 06933	12 41 12.	+ 30 32		20.3	FAINT BLUE STAR
LB 06932	12 41 12.	+ 30 45		19.5	FAINT BLUE STAR
LB 06931	12 41 12.	+ 30 46		16.5	FAINT BLUE STAR
LB 06930	12 41 12.	+ 31 55		17.2	FAINT BLUE STAR
MCG+07-26-042	12 41 12.	+ 41 17	54	14.5	GALAXY
ZWG 216.021	12 41 12.	+ 41 18		15.3	GALAXY
RNGC 4655	12 41 12.	+ 41 18		15.5	GALAXY
ZWG 244.037	12 41 12.	+ 50 08		15.6	GALAXY
7ZW 481	12 41 12.	+ 56 56			COMPACT GALAXY
MCG+10-18-078	12 41 12.	+ 59 13	57	14.2	GALAXY
REIN 3.313	12 41 12.36	+ 11 50 40.8			NEBULA
KN 16.275	12 41 13.8	+ 25 45 13.			NEBULA
IC 3704	12 41 14.	+ 11 02 46.			NONSTELLAR OBJECT
REIN 3.314	12 41 14.49	+ 11 02 36.2			NEBULA
KN 16.276	12 41 14.5	+ 25 45 07.			NEBULA
RNGC 4651	12 41 15.	+ 16 40		12.0	GALAXY
KN 16.278	12 41 15.0	+ 27 38 55.			NEBULA
KN 16.277	12 41 15.6	+ 25 44 43.			NEBULA
IC 3706	12 41 16.	+ 09 29 58.			SINGLE STAR
AMES 2078	12 41 16.	+ 17 41 43.	12	17.3	NEBULA
REIZ 2752	12 41 16.	+ 24 03	18	15.8	GALAXY
REIZ 2753	12 41 16.	+ 24 04	24	15.2	GALAXY
LB 02429	12 41 16.	+ 49 52 06.		16.2	FAINT BLUE STAR
REIZ 2749	12 41 16.	- 01 21	18	14.7	GALAXY
ARC 1603	12 41 16.	- 15 17		17.2	RICH CLUSTER OF GALAXIES
PATH 2.113	12 41 17.	- 00 17	108		NEBULA
REIN 4.136	12 41 17.12	- 00 17 17.2			NEBULA
ZC 1241.3+0125	12 41 18.	+ 01 25	5510		CLUSTER OF GALAXIES
ZWG 071.018	12 41 18.	+ 11 44		15.7	GALAXY
REA 20	12 41 18.	+ 12 06			DWARF GALAXY
ZWG 100.004	12 41 18.	+ 16 40		11.3	GALAXY
UGC 07901	12 41 18.	+ 16 40	240	11.3	GALAXY Sc
KARA.73B 0549	12 41 18.	+ 16 40	282	11.3	ISOLATED GALAXY S
VV 056	12 41 18.	+ 16 42	210	11.8	INTERACTING GALAXY
LB 00049	12 41 18.	+ 25 22		15.3	FAINT BLUE STAR
ZWG 159.063	12 41 18.	+ 27 08		15.7	GALAXY
LB 06941	12 41 18.	+ 27 49		17.6	FAINT BLUE STAR
ZC 1241.3+2753	12 41 18.	+ 27 53	1080		CLUSTER OF GALAXIES
LB 06940	12 41 18.	+ 28 35		18.8	FAINT BLUE STAR
LB 06939	12 41 18.	+ 29 31		19.6	FAINT BLUE STAR
LB 06938	12 41 18.	+ 29 33		18.4	FAINT BLUE STAR
LB 06937	12 41 18.	+ 29 50		19.6	FAINT BLUE STAR
ZC 1241.3+3244	12 41 18.	+ 32 44	2220		CLUSTER OF GALAXIES
ZC 1241.3+5030	12 41 18.	+ 50 30	740		CLUSTER OF GALAXIES
ZWG 015.009	12 41 18.	- 00 16		13.7	GALAXY
UGC 07900	12 41 18.	- 00 16	150	13.7	GALAXY Sc
MCG+00-33-006	12 41 18.	- 00 16	150	13.7	GALAXY
MCG-03-33-002	12 41 18.	- 19 53 30.	54	15.5	GALAXY
MCG-06-28-014	12 41 18.	- 33 56	48	13.	GALAXY
MRSL 302.04/1	12 41 18.	- 58 35	14400		HII REGION
RNGC 4653	12 41 19.	- 00 17		13.0	GALAXY
ARC 1604	12 41 19.	- 22 50		17.6	RICH CLUSTER OF GALAXIES
RNGC 4645	12 41 20.	- 41 29		13.0	GALAXY
REIN 3.315	12 41 20.02	+ 11 44 24.9			NEBULA
AMES 2079	12 41 21.	+ 18 19 49.	24	17.2	NEBULA
HOLM 449B	12 41 21.	+ 44 23	18	15.3	PART OF MULTIPLE GALAXY
IC 3708	12 41 22.	+ 13 24 40.			SAME AS NGC 4654
REIZ 2755	12 41 22.	+ 24 01	42	15.8	GALAXY
REIZ 2756	12 41 22.	+ 24 03	18	15.8	GALAXY
HOLM 449A	12 41 22.	+ 44 23	18	15.3	PART OF MULTIPLE GALAXY
REIZ 2751	12 41 22.	- 01 20	18	14.6	GALAXY
AMES 2080	12 41 23.	+ 11 01 43.	24	17.1	NEBULA
REIZ 2766	12 41 23.	+ 62 56	42	14.0	GALAXY
ZWG 043.010	12 41 24.	+ 03 53		15.5	GALAXY
SCH 71	12 41 24.	+ 13 23		11.8	PECULIAR GALAXY
ZWG 071.019	12 41 24.	+ 13 25		11.8	GALAXY
UGC 07902	12 41 24.	+ 13 25	330	11.8	GALAXY Sc
MCG+02-33-004	12 41 24.	+ 13 25	300	11.8	GALAXY
FEIG 069	12 41 24.	+ 16 12		12.7	FAINT BLUE STAR
LB 06948	12 41 24.	+ 27 20		19.6	FAINT BLUE STAR
LB 06947	12 41 24.	+ 27 53		20.3	FAINT BLUE STAR
LB 06946	12 41 24.	+ 27 58		19.4	FAINT BLUE STAR
LB 06945	12 41 24.	+ 28 54		19.6	FAINT BLUE STAR
LB 06944	12 41 24.	+ 30 57		19.2	FAINT BLUE STAR
LB 06943	12 41 24.	+ 31 53		18.0	FAINT BLUE STAR
LB 06942	12 41 24.	+ 32 12		16.7	FAINT BLUE STAR
MCG+06-28-023	12 41 24.	+ 34 05	36	16.	GALAXY
MCG+07-26-043	12 41 24.	+ 44 23	18	15.	GALAXY
MCG+07-26-044	12 41 24.	+ 44 23 30.	24	15.	GALAXY
UGC 07903	12 41 24.	+ 54 16	60	17.	GALAXY DWARF
MCG-01-33-001	12 41 24.	- 05 24	180	14.5	GALAXY
ARC 1605	12 41 24.	- 20 39		17.0	RICH CLUSTER OF GALAXIES
SC 1238+0008.6	12 41 25.	- 00 07 51.	30		NEBULA
KN 16.280	12 41 25.5	+ 30 03 13.			NEBULA
KN 16.279	12 41 25.8	+ 27 08 25.			NEBULA
AMES 2081	12 41 26.	+ 16 48 55.	30	17.8	NEBULA
REIZ 2761	12 41 26.	+ 34 04	30	16.0	GALAXY
HARO 32	12 41 26.	+ 55 11			BLUE EMISSION-LINE GALAXY
REIZ 2770	12 41 26.	+ 64 57	30	14.8	GALAXY
RNGC 4654	12 41 27.	+ 13 24		11.5	GALAXY
SHB 212	12 41 27.6	+ 16 39 17.		19.0	QUASI-STELLAR OBJECT
RC 3CR275.1	12 41 27.68	+ 16 39 18.7		19.00	QUASI-STELLAR OBJECT
REIZ 2760	12 41 28.	+ 24 01	36	15.9	GALAXY
REIZ 2754	12 41 28.	- 00 17	60	13.0	GALAXY
KN 16.281	12 41 28.2	+ 28 35 37.			NEBULA
KN 16.282	12 41 28.4	+ 29 54 07.			NEBULA
RNGC 4656	12 41 29.	+ 32 27		11.0	GALAXY

OBJECT NAME	RIGHT ASCEN.	DECLINATION	DIAM.	MAGN.	TYPE OF OBJECT
LB 02430	12 41 29.	+ 52 38 18.		15.1	FAINT BLUE STAR
ZWG 071.020	12 41 30.	+ 09 20		15.3	GALAXY
LB 06952	12 41 30.	+ 26 52		19.3	FAINT BLUE STAR
LB 06951	12 41 30.	+ 27 41		19.2	FAINT BLUE STAR
LB 06950	12 41 30.	+ 28 00		18.7	FAINT BLUE STAR
TON-N 0118	12 41 30.	+ 29 00		14.7	BLUE STAR
LB 06949	12 41 30.	+ 29 25		17.6	FAINT BLUE STAR
LB 11325	12 41 30.	+ 32 26		15.0	FAINT BLUE STAR
MCG+05-30-066	12 41 30.	+ 32 26	1200	10.6	GALAXY
ZWG 216.022	12 41 30.	+ 41 01		15.7	GALAXY
UGC 07904	12 41 30.	+ 41 01	78	15.7	GALAXY SB?c
ZC 1241.5+5205	12 41 30.	+ 52 05	1080		CLUSTER OF GALAXIES
1ZW 041	12 41 30.	+ 55 10			COMPACT GALAXY
ZWG 270.016	12 41 30.	+ 55 10		14.1	GALAXY
MRK 220	12 41 30.	+ 55 10	15	14.5	GALAXY WITH UV CONTINUUM
UGC 07905	12 41 30.	+ 55 10	114	14.1	GALAXY DBL SYS
KARA.72 354A	12 41 30.	+ 55 10	48	14.1	PART OF DOUBLE GALAXY
MCG+09-21-034	12 41 30.	+ 55 10	24	13.9	GALAXY
MRK 221	12 41 30.	+ 55 11	10	16.	GALAXY WITH UV CONTINUUM
KARA.72 354B	12 41 30.	+ 55 11	48		PART OF DOUBLE GALAXY
MCG+09-21-033	12 41 30.	+ 55 11	54	14.7	GALAXY
MCG-07-26-037	12 41 30.	- 41 29	120	13.1	GALAXY
MRSL 302+00/1	12 41 30.	- 62 18	300		HII REGION
IC 3709	12 41 32.	+ 09 20 10.			NONSTELLAR OBJECT
REIZ 2763	12 41 32.	+ 32 27	480	11.0	GALAXY
LB 02431	12 41 32.	+ 59 39 18.		15.7	FAINT BLUE STAR
REIZ 2769	12 41 34.	+ 56 35	30	14.1	GALAXY
REIZ 2759	12 41 34.	- 00 50	18	15.4	GALAXY
REIZ 2758	12 41 34.	- 00 50	30	15.3	GALAXY
REIZ 2767	12 41 35.	+ 55 12	24	14.0	GALAXY
HOLM 452A	12 41 35.	+ 55 12	24	13.9	PART OF MULTIPLE GALAXY
REIZ 2768	12 41 35.	+ 55 13	24	15.0	GALAXY
IC 3710	12 41 36.	+ 12 24			GALAXY IN VIRGO CLSTR
UGC 07906	12 41 36.	+ 12 25	84	16.5	GALAXY DWRF IP
TON-N 0119	12 41 36.	+ 13 08		14.6	BLUE STAR
LB 00050	12 41 36.	+ 26 57		18.0	FAINT BLUE STAR
LB 06957	12 41 36.	+ 27 06		19.5	FAINT BLUE STAR
LB 06956	12 41 36.	+ 27 26		20.3	FAINT BLUE STAR
ZWG 159.064	12 41 36.	+ 29 10		15.6	GALAXY
MRK 655	12 41 36.	+ 29 10	26	16.	GALAXY WITH UV CONTINUUM
MCG+05-30-067	12 41 36.	+ 29 11	36	15.6	GALAXY
LB 06955	12 41 36.	+ 29 17		18.9	FAINT BLUE STAR
LB 06954	12 41 36.	+ 31 30		17.9	FAINT BLUE STAR
ZWG 159.065	12 41 36.	+ 32 27		10.6	GALAXY
UGC 07907	12 41 36.	+ 32 27	1320	10.6	GALAXY
KARA.72 350B	12 41 36.	+ 32 27	534	10.6	PART OF DOUBLE GALAXY
LB 06953	12 41 36.	+ 32 30		19.4	FAINT BLUE STAR
ZC 1241.6+3345	12 41 36.	+ 33 45	4170		CLUSTER OF GALAXIES
MCG+07-26-045	12 41 36.	+ 41 00	78	15.	GALAXY
ZWG 216.023	12 41 36.	+ 41 27		15.5	GALAXY
KARA.68 192	12 41 36.	+ 54 14	47		DWARF GALAXY
MCG+09-21-035	12 41 36.	+ 54 14	30	16.	GALAXY
HOLM 452E	12 41 36.	+ 55 13	24	14.7	PART OF MULTIPLE GALAXY
SEA G3	12 41 36.	- 40 27 18.	126	12.81	PART OF CHAIN OF GALAXIES
MCG-07-26-038	12 41 36.	- 40 28	66	13.5	GALAXY
KN 16.284	12 41 36.7	+ 30 11 19.			NEBULA
KN 16.283	12 41 36.9	+ 29 10 43.			NEBULA
AMES 2082	12 41 38.	+ 12 57 31.	17	17.7	NEBULA
RNGC 4650	12 41 38.	- 40 27			GALAXY
REIZ 2762	12 41 39.	+ 04 43	18	16.0	GALAXY
AMES 2083	12 41 40.	+ 13 12 19.	19	16.8	NEBULA
LB 02432	12 41 40.	+ 54 36 24.		16.0	FAINT BLUE STAR
HOLM 450A	12 41 41.	+ 22 50	36	14.8	PART OF MULTIPLE GALAXY
RNGC 4657	12 41 41.	+ 32 29			GALAXY
IC 3713	12 41 41.	+ 41 26 29.			NONSTELLAR OBJECT
LB 00243	12 41 41.	+ 54 36 30.		16.0	FAINT BLUE STAR
REA 64	12 41 42.	+ 10 00	42		DWARF GALAXY
ZWG 071.021	12 41 42.	+ 13 13		15.6	GALAXY
AMES 2084	12 41 42.	+ 16 36 07.	24	16.9	NEBULA
LB 06960	12 41 42.	+ 26 44		19.0	FAINT BLUE STAR
LB 06959	12 41 42.	+ 27 16		19.4	FAINT BLUE STAR
LB 06958	12 41 42.	+ 28 36		19.7	FAINT BLUE STAR
TON-N 0637	12 41 42.	+ 30 20		15.3	BLUE STAR
MCG+07-26-046	12 41 42.	+ 41 26	36	15.	GALAXY
MCG+08-23-083	12 41 42.	+ 48 56	48	16.	GALAXY
MCG+11-16-003	12 41 42.	+ 62 36	45	16.	GALAXY
MCG+12-12-017	12 41 42.	+ 73 52	84	16.	GALAXY
ZWG 335.021	12 41 42.	+ 73 54		15.3	GALAXY
UGC 07908	12 41 42.	+ 73 54	96	15.3	GALAXY Sc
SEA G4	12 41 42.	- 40 23 06.	48	15.44	PART OF CHAIN OF GALAXIES
IC 3711	12 41 43.	+ 11 26			NONSTELLAR OBJECT
HN 1026	12 41 43.	+ 11 27	12		NEBULA
HN 1025	12 41 43.	+ 12 23	36		NEBULA
KN 16.285	12 41 44.2	+ 28 38 19.			NEBULA
KN 16.286	12 41 44.5	+ 29 16 55.			NEBULA
REIZ 2764	12 41 45.	+ 00 45	48	15.9	GALAXY
REIZ 2765	12 41 45.	+ 04 43	18	15.6	GALAXY
HOLM 450B	12 41 45.	+ 22 50	36	15.4	PART OF MULTIPLE GALAXY
IC 3712	12 41 46.	+ 10 38 53.			MAY NOT EXIST
AMES 2086	12 41 46.	+ 10 52 42.	17	17.3	NEBULA
AMES 2085	12 41 46.	+ 17 09 54.	16	17.3	NEBULA
KN 16.287	12 41 47.2	+ 28 40 43.			NEBULA
ZWG 043.011	12 41 48.	+ 04 42		15.3	GALAXY
UGC 07909	12 41 48.	+ 04 42	66	15.3	GALAXY DBL SYS
MCG+01-33-003	12 41 48.	+ 04 42	18	15.3	GALAXY
MCG+01-33-002	12 41 48.	+ 04 42	30	15.3	GALAXY
VV 064B	12 41 48.	+ 04 42 30.	24	14.6	INTERACTING GALAXY
AMES 2087	12 41 48.	+ 12 07 06.	22	17.8	NEBULA
LB 00016	12 41 48.	+ 23 31		15.2	FAINT BLUE STAR
LB 06964	12 41 48.	+ 26 58		18.6	FAINT BLUE STAR
LB 06963	12 41 48.	+ 28 05		20.2	FAINT BLUE STAR
LB 06962	12 41 48.	+ 29 17		17.1	FAINT BLUE STAR
LB 06961	12 41 48.	+ 31 11		19.7	FAINT BLUE STAR
TON-N 0638	12 41 48.	+ 31 14		14.9	BLUE STAR
ZWG 244.038	12 41 48.	+ 45 10		15.7	GALAXY
MCG+08-23-084	12 41 48.	+ 45 10	42	16.	GALAXY
ZWG 244.039	12 41 48.	+ 45 17		15.4	GALAXY
UGC 07910	12 41 48.	+ 45 17	66	15.4	GALAXY MLT SYS
MCG+08-23-085	12 41 48.	+ 45 17	60	14.	GALAXY
ZC 1241.8+4311	12 41 48.	+ 48 11	740		CLUSTER OF GALAXIES
MCG+08-23-086	12 41 48.	+ 48 53	30	16.	GALAXY
TON-N 0120	12 41 48.	- 01 00		14.1	BLUE STAR
HOLM 451B	12 41 49.	+ 04 41	24	14.6	PART OF MULTIPLE GALAXY
AMES 2089	12 41 50.	+ 12 33 18.	25	17.8	NEBULA
AMES 2088	12 41 50.	+ 18 36 18.	18	17.8	NEBULA
HOLM 451A	12 41 51.	+ 04 41	24	14.5	PART OF MULTIPLE GALAXY
VV 064A	12 41 51.	+ 04 43	24	14.5	INTERACTING GALAXY
MCG-04-30-013	12 41 51.	- 24 00	48	15.	GALAXY

OBJECT NAME	RIGHT ASCEN.	DECLINATION	DIAM.	MAGN.	TYPE OF OBJECT
REIN 3.316	12 41 51.79	+ 10 27 43.5			NEBULA
AMES 2090	12 41 52.	+ 13 16 18.	24	17.3	NEBULA
IC 3715	12 41 53.	+ 20 07 53.			NONSTELLAR OBJECT
LB 02433	12 41 53.	+ 57 18 36.		17.2	FAINT BLUE STAR
SC 1239-0158.0	12 41 53.	- 02 14 27.	18		NEBULA
REIN 4.137	12 41 53.80	- 02 43 53.4			NEBULA
KN 16.288	12 41 53.9	+ 27 21 08.			NEBULA
ZWG 015.012	12 41 54.	+ 00 45		14.9	GALAXY
UGC 07911	12 41 54.	+ 00 45	174	14.9	GALAXY DWRF SP
MCG+00-33-007	12 41 54.	+ 00 45	120	14.9	GALAXY
ZWG 043.012	12 41 54.	+ 03 48		15.3	GALAXY
UGC 07912	12 41 54.	+ 03 48	90	15.3	GALAXY DBL SYS
MCG+01-33-004	12 41 54.	+ 03 48	24	15.3	GALAXY
ZWG 071.022	12 41 54.	+ 10 27		15.6	GALAXY
MCG+02-33-005	12 41 54.	+ 10 27	48	15.6	GALAXY
TON-N 0121	12 41 54.	+ 26 02		15.4	BLUE STAR
LB 06968	12 41 54.	+ 28 13		19.5	FAINT BLUE STAR
LB 06967	12 41 54.	+ 29 34		18.9	FAINT BLUE STAR
TON-N 0122	12 41 54.	+ 29 39		15.5	BLUE STAR
LB 06966	12 41 54.	+ 30 20		18.0	FAINT BLUE STAR
LB 06965	12 41 54.	+ 31 20		19.1	FAINT BLUE STAR
1ZW 042	12 41 54.	+ 34 40			COMPACT GALAXY
VV 127F	12 41 54.	+ 34 40	18	19.5	INTERACTING GALAXY
VV 127E	12 41 54.	+ 34 40	6	19.5	INTERACTING GALAXY
VV 127D	12 41 54.	+ 34 40	18	19.5	INTERACTING GALAXY
VV 127C	12 41 54.	+ 34 40	18	19.5	INTERACTING GALAXY
VV 127B	12 41 54.	+ 34 40	6	19.	INTERACTING GALAXY
VV 127A	12 41 54.	+ 34 40	6	19.	INTERACTING GALAXY
VV 127	12 41 54.	+ 34 40	150		INTERACTING GALAXY
LB 00649	12 41 54.	+ 58 39 30.		16.7	FAINT BLUE STAR
ZWG 015.011	12 41 54.	- 00 10		15.7	GALAXY
ZWG 015.010	12 41 54.	- 02 44		15.1	GALAXY
MCG-06-28-015	12 41 54.	- 33 36	72	15.5	GALAXY
AGU 38	12 41 54.	- 40 25 30.	60	13.5	PECULIAR GALAXY
HN 1027	12 41 55.	+ 10 27	18		NEBULA
AMES 2091	12 41 55.	+ 18 06 12.	24	17.6	NEBULA
REIN 4.138	12 41 55.09	+ 00 44 23.8			NEBULA
IC 3714	12 41 56.	+ 10 26			NONSTELLAR OBJECT
FATH 1.542	12 41 56.	- 00 11			NEBULA
MCG+08-23-087	12 41 57.	+ 50 31	36	16.	GALAXY
HARO 33	12 41 58.	+ 28 45			BLUE EMISSION-LINE GALAXY
KN 16.289	12 41 59.2	+ 27 27 56.			NEBULA
VDB.66G 145	12 42	+ 12 19	70		DWARF GALAXY
VDB.66G 146	12 42	- 05 51	170		DWARF GALAXY
KARA.68 193	12 42 00.	+ 02 01	34		DWARF GALAXY
ZWG 071.023	12 42 00.	+ 11 28		12.1	GALAXY
UGC 07914	12 42 00.	+ 11 28	144	12.1	GALAXY E
MCG+02-33-006	12 42 00.	+ 11 28	60	12.1	GALAXY
AMES 2092	12 42 00.	+ 12 54 42.	24	17.8	NEBULA
ZWG 071.024	12 42 00.	+ 13 47		13.3	GALAXY
UGC 07915	12 42 00.	+ 13 47	138	13.3	GALAXY S0-a
MCG+02-33-007	12 42 00.	+ 13 47	36	13.3	GALAXY
LB 06971	12 42 00.	+ 26 29		18.0	FAINT BLUE STAR
LB 06970	12 42 00.	+ 28 25		19.4	FAINT BLUE STAR
ZC 1242.0+2843	12 42 00.	+ 28 43	1410		CLUSTER OF GALAXIES
LB 06969	12 42 00.	+ 28 49		19.3	FAINT BLUE STAR
UGC 07916	12 42 00.	+ 34 40	156	17.	GALAXY DWRF IR
MCG+06-28-024	12 42 00.	+ 34 41	138	15.	GALAXY
MCG+07-26-047	12 42 00.	+ 39 46 30.	36	16.	GALAXY
IC 3717	12 42 00.	+ 39 47 59.			NONSTELLAR OBJECT
ZC 1242.0+5935	12 42 00.	+ 59 35	670		CLUSTER OF GALAXIES
MCG+11-16-004	12 42 00.	+ 64 05	72	16.	GALAXY
ZWG 015.013	12 42 00.	- 02 02		15.7	GALAXY
UGC 07913	12 42 00.	- 02 02	72	15.7	GALAXY DWARF
MCG-02-33-001	12 42 00.	- 09 48	108	12.5	GALAXY
SEA G5	12 42 00.	- 40 26 12.	78	14.33	PART OF CHAIN OF GALAXIES
REIN 3.317	12 42 00.94	+ 11 27 50.2			NEBULA
REIZ 2773	12 42 01.	+ 37 25	60	14.7	GALAXY
RNGC 4660	12 42 02.	+ 11 28		12.5	GALAXY
ABC 1606	12 42 02.	- 11 43		17.0	RICH CLUSTER OF GALAXIES
RNGC 4659	12 42 03.	+ 13 47		13.5	GALAXY
SC 1239+0049.4	12 42 04.	+ 00 32 58.	12		NEBULA
AMES 2093	12 42 04.	+ 14 02 30.	28	17.6	NEBULA
SEY 193	12 42 04.	+ 31 27 46.		15.3	FAINT GALAXY
KN 16.291	12 42 04.2	+ 29 18 32.			NEBULA
KN 16.290	12 42 04.2	+ 29 30 26.			NEBULA
LB 02434	12 42 05.	+ 57 17 18.		16.1	FAINT BLUE STAR
ZWG 043.013	12 42 06.	+ 03 07		15.6	GALAXY
ZWG 043.014	12 42 06.	+ 06 58		15.6	GALAXY
KARA.68 194	12 42 06.	+ 17 43	34		DWARF GALAXY
ZC 1242.1+2043	12 42 06.	+ 20 43	870		CLUSTER OF GALAXIES
ZWG 159.066	12 42 06.	+ 26 42		15.3	GALAXY
LB 06973	12 42 06.	+ 27 55		19.5	FAINT BLUE STAR
ZWG 159.067	12 42 06.	+ 28 45		14.7	GALAXY
BFGS 2	12 42 06.	+ 28 45			GALAXY WITH UV KNOTS
LB 06972	12 42 06.	+ 32 22		19.3	FAINT BLUE STAR
ZWG 188.018	12 42 06.	+ 37 23		14.1	GALAXY
RNGC 4662	12 42 06.	+ 37 23		14.0	GALAXY
UGC 07917	12 42 06.	+ 37 23	150	14.1	GALAXY SBb
KARA.73B 0550	12 42 06.	+ 37 23	102	14.1	ISOLATED GALAXY S
MCG+06-28-025	12 42 06.	+ 37 25	120	13.	GALAXY
MCG+07-26-048	12 42 06.	+ 41 00	18	16.	GALAXY
ZWG 216.024	12 42 06.	+ 41 01		15.7	GALAXY
MRK 441	12 42 06.	+ 41 01	18	15.5	GALAXY WITH UV CONTINUUM
MCG+09-21-036	12 42 06.	+ 56 17	24	17.	GALAXY
ZC 1242.1+6133	12 42 06.	+ 61 33	1610		CLUSTER OF GALAXIES
MCG+11-16-005	12 42 06.	+ 62 32	27	16.	GALAXY
UGC 07918	12 42 06.	+ 64 04	90	16.5	GALAXY DWRF IR
REIZ 2771	12 42 06.	- 09 48	90	12.6	GALAXY
RNGC 4658	12 42 06.	- 09 49		13.0	GALAXY
SEA G6	12 42 06.	- 40 30 00.	48	16.56	PART OF CHAIN OF GALAXIES
KN 16.292	12 42 06.0	+ 29 17 02.			NEBULA
IC 3723	12 42 08.	+ 41 00 47.			NONSTELLAR OBJECT
KN 16.293	12 42 09.9	+ 29 02 26.			NEBULA
AMES 2094	12 42 10.	+ 16 39 06.	24	17.3	NEBULA
IC 0811	12 42 11.	- 09 56			SAME AS NGC 4663
KN 16.294	12 42 11.6	+ 28 44 44.			NEBULA
KN 16.295	12 42 11.9	+ 28 11 50.			NEBULA
ZWG 043.015	12 42 12.	+ 04 20		15.6	GALAXY
ZWG 071.025	12 42 12.	+ 10 02		15.6	GALAXY
AMES 2095	12 42 12.	+ 16 32 48.	19	17.2	NEBULA
MCG+05-30-069	12 42 12.	+ 26 41	42	15.3	GALAXY
LB 06980	12 42 12.	+ 27 12		17.9	FAINT BLUE STAR
ZWG 159.068	12 42 12.	+ 28 10		15.7	GALAXY
MCG+05-30-068	12 42 12.	+ 28 12	36	16.5	GALAXY
MCG+05-30-070	12 42 12.	+ 28 45	39	14.7	GALAXY
LB 06979	12 42 12.	+ 29 07		18.6	FAINT BLUE STAR
LB 06978	12 42 12.	+ 29 24		19.5	FAINT BLUE STAR

OBJECT NAME	RIGHT ASCEN.	DECLINATION	DIAM.	MAGN.	TYPE OF OBJECT
LB 06977	12 42 12.	+ 30 09		19.4	FAINT BLUE STAR
LB 06976	12 42 12.	+ 30 56		19.5	FAINT BLUE STAR
LB 06975	12 42 12.	+ 31 27		20.3	FAINT BLUE STAR
LB 06974	12 42 12.	+ 32 14		20.0	FAINT BLUE STAR
ZC 1242.2+3810	12 42 12.	+ 38 10	1340		CLUSTER OF GALAXIES
REIZ 2772	12 42 12.	- 09 55	18	13.3	GALAXY
IC 3716	12 42 13.	+ 08 22 29.			NONSTELLAR OBJECT
FIT.N	12 42 13.	- 78 33	330	21.	BRIGHT RIMMED DARK NEBULA
KN 16.296	12 42 13.9	+ 26 41 32.			NEBULA
REIN 3.318	12 42 14.38	+ 12 37 27.2			NEBULA
SC 1239+0049.5	12 42 15.	+ 00 33 04.	12		NEBULA
IC 3718	12 42 15.	+ 12 37 29.			NONSTELLAR OBJECT
AMES 2096	12 42 15.	+ 18 14 42.	14	17.7	NEBULA
MCG-03-33-003	12 42 15.	- 20 10	54	14.	GALAXY
MCG-03-33-004	12 42 15.	- 20 31 30.	78	15.	GALAXY
KN 16.297	12 42 15.5	+ 27 17 20.			NEBULA
REIN 3.319	12 42 15.62	+ 08 22 50.7			NEBULA
IC 3719	12 42 16.	+ 08 22 53.			NONSTELLAR OBJECT
IC 3720	12 42 16.5	+ 12 20 13.			GALAXY
IC 3721	12 42 17.	+ 19 01			NONSTELLAR OBJECT
IC 0812	12 42 17.	- 04 09 31.			NONSTELLAR OBJECT
KN 16.298	12 42 17.2	+ 28 10 14.			NEBULA
SC 1239+0214.5	12 42 18.	+ 01 58 04.	12		NEBULA
ZWG 043.016	12 42 18.	+ 08 22		15.5	GALAXY
UGC 07919	12 42 18.	+ 12 21	150	16.0	GALAXY DWRF EL
ZWG 071.026	12 42 18.	+ 12 37		14.7	GALAXY
UGC 07920	12 42 18.	+ 12 37	162	14.7	GALAXY S
MCG+02-33-008	12 42 18.	+ 12 37	180	14.7	GALAXY
HN 1028	12 42 18.	+ 19 02		13.5	NEBULA
LB 06985	12 42 18.	+ 27 08		20.0	FAINT BLUE STAR
LB 06984	12 42 18.	+ 27 48		18.8	FAINT BLUE STAR
MCG+05-30-071	12 42 18.	+ 28 10	48	15.7	GALAXY
LB 06983	12 42 18.	+ 29 23		19.1	FAINT BLUE STAR
LB 06982	12 42 18.	+ 29 59		18.2	FAINT BLUE STAR
LB 06981	12 42 18.	+ 31 10		19.9	FAINT BLUE STAR
MCG+07-26-049	12 42 18.	+ 40 56 30.	72	14.5	GALAXY
ZWG 216.025	12 42 18.	+ 40 58		15.6	GALAXY
UGC 07921	12 42 18.	+ 40 58	102	15.6	GALAXY Sc
ZWG 270.017	12 42 18.	+ 56 25		15.3	GALAXY
UGC 07922	12 42 18.	+ 56 25	126	15.3	GALAXY SB?b-c
MCG+09-21-037	12 42 18.	+ 56 27	90	13.9	GALAXY
KARA.68 195	12 42 18.	+ 71 03	47		DWARF GALAXY
KN 16.299	12 42 18.6	+ 27 46 26.			NEBULA
KN 16.300	12 42 18.7	+ 29 25 44.			NEBULA
IC 3722	12 42 20.	+ 12 03 05.			SINGLE STAR
IC 3726	12 42 20.	+ 40 57 06.			NONSTELLAR OBJECT
HOLM 454B	12 42 20.	+ 56 28	18	15.1	PART OF MULTIPLE GALAXY
MAI 077	12 42 20.	+ 64 17	33		DWARF SPHEROIDAL GALAXY
RNGC 4661	12 42 20.	- 40 45			GALAXY
MCG+03-33-002	12 42 21.	+ 19 00	54	14.4	GALAXY
HOLM 454A	12 42 22.	+ 56 27	30	13.9	PART OF MULTIPLE GALAXY
REIN 3.320	12 42 22.56	+ 10 33 20.3			NEBULA
SC 1239+0029.8	12 42 23.	+ 00 13 22.	6		NEBULA
AMES 2098	12 42 23.	+ 14 30 06.	48	17.6	NEBULA
AMES 2097	12 42 23.	+ 16 41 42.	31	17.2	NEBULA
IC 3725	12 42 23.	+ 19 01 42.			SAME AS IC 3721
ZWG 015.014	12 42 24.	+ 00 14		15.6	GALAXY
FATH 1.543	12 42 24.	+ 00 14			NEBULA
ZWG 043.017	12 42 24.	+ 02 58		15.7	GALAXY
ZWG 071.027	12 42 24.	+ 10 33		15.6	GALAXY
ZWG 100.005	12 42 24.	+ 19 01		14.4	GALAXY
UGC 07923	12 42 24.	+ 19 01	60	14.4	GALAXY S
TON-N 0639	12 42 24.	+ 23 48		16.0	BLUE STAR
LB 11329	12 42 24.	+ 27 02		11.8	FAINT BLUE STAR
LB 11328	12 42 24.	+ 28 12		12.2	FAINT BLUE STAR
LB 06992	12 42 24.	+ 28 16		20.0	FAINT BLUE STAR
LB 06991	12 42 24.	+ 29 02		19.3	FAINT BLUE STAR
LB 06990	12 42 24.	+ 29 25		17.6	FAINT BLUE STAR
LB 06989	12 42 24.	+ 29 33		19.2	FAINT BLUE STAR
LB 06988	12 42 24.	+ 29 44		18.7	FAINT BLUE STAR
LB 11327	12 42 24.	+ 29 59		17.1	FAINT BLUE STAR
LB 06987	12 42 24.	+ 30 44		18.8	FAINT BLUE STAR
LB 11326	12 42 24.	+ 31 06		17.0	FAINT BLUE STAR
LB 06986	12 42 24.	+ 31 25		20.0	FAINT BLUE STAR
TON-N 0123	12 42 24.	+ 44 11		15.6	BLUE STAR
MCG+09-21-038	12 42 24.	+ 55 08	72	14.2	GALAXY
RNGC 4663	12 42 24.	- 09 54		14.0	GALAXY
MCG-02-33-002	12 42 24.	- 09 54	48	14.5	GALAXY
MCG-07-26-039	12 42 24.	- 40 45	9	16.	GALAXY
KN 16.301	12 42 24.5	+ 26 52 08.			NEBULA
KN 16.302	12 42 24.6	+ 28 46 38.			NEBULA
HN 1029	12 42 25.	+ 10 33	18		NEBULA
RNGC 4664	12 42 26.	+ 03 29			NON-EXISTENT OBJECT
IC 3724	12 42 26.	+ 10 32			NONSTELLAR OBJECT
KN 16.303	12 42 26.9	+ 29 00 38.			NEBULA
LB 02435	12 42 28.	+ 50 33 48.		15.4	FAINT BLUE STAR
RNGC 4669	12 42 28.	+ 55 09		15.0	GALAXY
REIZ 2780	12 42 28.	+ 55 09	90	14.2	GALAXY
AMES 2099	12 42 29.	+ 17 32 24.	17	17.5	NEBULA
REIZ 2777	12 42 29.	+ 19 01	48	15.1	GALAXY
REIZ 2778	12 42 29.	+ 19 06	24	15.4	GALAXY
ZWG 043.018	12 42 30.	+ 03 20		12.4	GALAXY
UGC 07924	12 42 30.	+ 03 20	288	12.4	GALAXY SB0
MCG+01-33-005	12 42 30.	+ 03 20	180	12.4	GALAXY
REA 65	12 42 30.	+ 12 38	48		DWARF GALAXY
LB 06995	12 42 30.	+ 26 59		19.2	FAINT BLUE STAR
LB 06994	12 42 30.	+ 29 05		17.8	FAINT BLUE STAR
LB 06993	12 42 30.	+ 30 32		18.9	FAINT BLUE STAR
ZC 1242.5+3155	12 42 30.	+ 31 55	1280		CLUSTER OF GALAXIES
LB 11330	12 42 30.	+ 32 00		13.7	FAINT BLUE STAR
MCG+07-26-050	12 42 30.	+ 39 36	12	16.	GALAXY
IC 3729	12 42 30.	+ 39 37 42.			NONSTELLAR OBJECT
ZWG 270.018	12 42 30.	+ 55 09		15.1	GALAXY
UGC 07925	12 42 30.	+ 55 09	120	15.1	GALAXY S
ZC 1242.5+5840	12 42 30.	+ 58 40	2420		CLUSTER OF GALAXIES
ZC 1242.5+5919	12 42 30.	+ 59 19	1210		CLUSTER OF GALAXIES
MCG-03-33-005	12 42 30.	- 15 32 30.	24	15.	GALAXY
SEA G7	12 42 30.	- 40 33 00.	48	14.83	PART OF CHAIN OF GALAXIES
MCG-07-26-040	12 42 30.	- 40 34	30	15.5	GALAXY
KN 16.304	12 42 30.3	+ 25 02 44.			NEBULA
AMES 2100	12 42 31.	+ 16 44 54.	22	16.9	NEBULA
REIZ 2775	12 42 33.	+ 03 20	120	12.8	GALAXY
HOLM 453A	12 42 34.	- 00 10	240	11.8	PART OF MULTIPLE GALAXY
REIZ 2776	12 42 34.	- 00 11	210	11.1	GALAXY
HOLM 453B	12 42 34.	- 00 15	60	12.6	PART OF MULTIPLE GALAXY
REIZ 2774	12 42 34.	- 01 42	30	15.3	GALAXY
REIN 3.322	12 42 34.52	+ 12 43 08.4			NEBULA
REIN 3.321	12 42 34.59	+ 11 10 26.1			NEBULA

OBJECT NAME	RIGHT ASCEN.	DECLINATION	DIAM.	MAGN.	TYPE OF OBJECT
REIN 4.139	12 42 34.73	- 00 11 18.6			NEBULA
IC 3728	12 42 35.	+ 21 14 48.			NONSTELLAR OBJECT
SN 1965H	12 42 35.	- 00 10		14.0	SUPERNOVA
FATH 2.114	12 42 35.	- 00 11	272		NEBULA
A2 001	12 42 35.3	+ 28 53 44.		18.4	FAINT BLUE OBJECT
ZC 1242.6+0523	12 42 36.	+ 05 23	740		CLUSTER OF GALAXIES
ZWG 071.028	12 42 36.	+ 11 10		15.6	GALAXY
UGC 07927	12 42 36.	+ 11 10	96	15.6	GALAXY Sc
MCG+02-33-009	12 42 36.	+ 11 10	72	15.6	GALAXY
AMES 2101	12 42 36.	+ 16 36 18.	24	17.7	NEBULA
MCG+04-30-018	12 42 36.	+ 21 25	18	15.3	GALAXY
ZWG 129.021	12 42 36.	+ 21 27		15.3	GALAXY
LB 00051	12 42 36.	+ 23 53		11.0	FAINT BLUE STAR
LB 11333	12 42 36.	+ 27 25		12.8	FAINT BLUE STAR
LB 11332	12 42 36.	+ 27 53		13.8	FAINT BLUE STAR
LB 06998	12 42 36.	+ 28 46		19.0	FAINT BLUE STAR
LB 06997	12 42 36.	+ 29 23		19.1	FAINT BLUE STAR
LB 11331	12 42 36.	+ 30 18		11.8	FAINT BLUE STAR
LB 06996	12 42 36.	+ 32 17		20.4	FAINT BLUE STAR
LB 00650	12 42 36.	+ 60 07 48.		18.6	FAINT BLUE STAR
ZWG 015.015	12 42 36.	- 00 10		12.0	GALAXY
UGC 07926	12 42 36.	- 00 10	282	12.0	GALAXY Sc
MCG+00-33-008	12 42 36.	- 00 10	252	12.0	GALAXY
MCG-06-28-016	12 42 36.	- 34 24	36	15.5	GALAXY
IC 3727	12 42 37.	+ 11 09			NONSTELLAR OBJECT
HN 1030	12 42 37.	+ 11 10	30		NEBULA
HARO 34	12 42 37.	+ 21 26			BLUE EMISSION-LINE GALAXY
RNGC 4666	12 42 37.	- 00 11		11.5	GALAXY
RNGC 4665	12 42 38.	+ 03 20		11.5	GALAXY
AMES 2102	12 42 38.	+ 16 35 42.	29	17.3	NEBULA
IC 3730	12 42 39.	+ 21 27 00.			NONSTELLAR OBJECT
AMES 2105	12 42 40.	+ 16 18 00.	22	17.2	NEBULA
AMES 2104	12 42 40.	+ 16 22 54	24	17.2	NEBULA
AMES 2103	12 42 40.	+ 16 34 42.	24	17.2	NEBULA
KN 16.306	12 42 40.7	+ 30 11 14.			NEBULA
REIN 3.323	12 42 40.77	+ 10 35 51.4			NEBULA
AMES 2107	12 42 41.	+ 11 49 30.	24	17.4	NEBULA
AMES 2106	12 42 41.	+ 17 37 06.	18	17.4	NEBULA
REIZ 2779	12 42 41.	+ 21 27	12	14.4	GALAXY
IC 3734	12 42 41.	+ 23 18 30.			NONSTELLAR OBJECT
KN 16.305	12 42 41.3	+ 27 51 02.			NEBULA
A2 002	12 42 41.4	+ 29 09 15.		17.1	FAINT BLUE OBJECT
KN 16.307	12 42 41.4	+ 29 50 44.			NEBULA
ZWG 071.029	12 42 42.	+ 10 28		15.5	GALAXY
MCG+04-30-019	12 42 42.	+ 23 18	48	14.4	GALAXY
ZWG 129.022	12 42 42.	+ 23 19		14.4	GALAXY
UGC 07928	12 42 42.	+ 23 19	66	14.4	GALAXY S
LB 07002	12 42 42.	+ 27 16		19.2	FAINT BLUE STAR
MCG+05-30-072	12 42 42.	+ 27 24	96	12.6	GALAXY
LB 07001	12 42 42.	+ 27 32		19.9	FAINT BLUE STAR
LB 11334	12 42 42.	+ 27 44		12.3	FAINT BLUE STAR
LB 07000	12 42 42.	+ 29 14		18.3	FAINT BLUE STAR
LB 06999	12 42 42.	+ 31 04		19.7	FAINT BLUE STAR
LB 02436	12 42 42.	+ 52 53 36.		15.6	FAINT BLUE STAR
HOLM 456B	12 42 42.	+ 66 23	54	14.9	PART OF MULTIPLE GALAXY
KARA.68 196	12 42 42.	- 07 57	27		DWARF GALAXY
IC 3732	12 42 43.	+ 10 34			NONSTELLAR OBJECT
HN 1032	12 42 43.	+ 10 35	12		NEBULA
IC 3731	12 42 43.	+ 12 41			NONSTELLAR OBJECT
HN 1031	12 42 43.	+ 12 42		14.5	NEBULA
RNGC 4667	12 42 44.	+ 11 42			NON-EXISTENT OBJECT
IC 0813	12 42 44.	+ 23 18 30.			NONSTELLAR OBJECT
KN 16.308	12 42 44.2	+ 30 14 45.			NEBULA
REIN 3.324	12 42 44.51	+ 10 27 58.7			NEBULA
IC 3733	12 42 45.	+ 07 13 54.			NONSTELLAR OBJECT
AMES 2108	12 42 45.	+ 17 52 42.	22	18.0	NEBULA
LB 02437	12 42 45.	+ 53 42 30.		14.7	FAINT BLUE STAR
AMES 2109	12 42 46.	+ 13 29 06.	14	17.5	NEBULA
REIZ 2783	12 42 46.	+ 23 19	18	14.2	GALAXY
RNGC 4670	12 42 46.	+ 27 24		13.5	GALAXY
HARO 09	12 42 46.	+ 27 24			BLUE EMISSION-LINE GALAXY
MAI 078	12 42 46.	+ 71 05	47		DWARF SPHEROIDAL GALAXY
REIZ 2782	12 42 47.	+ 19 01	30	15.7	GALAXY
A2 003	12 42 47.0	+ 29 11 21.		17.3	FAINT BLUE OBJECT
ZWG 071.030	12 42 48.	+ 09 22		15.7	GALAXY
AMES 2110	12 42 48.	+ 12 07 00.	23	17.1	NEBULA
ZWG 071.031	12 42 48.	+ 13 58		15.1	GALAXY
MCG+02-33-010	12 42 48.	+ 13 58	24	15.1	GALAXY
UGC 07929	12 42 48.	+ 21 43	60	17.	GALAXY DWARF
MCG+04-30-020	12 42 48.	+ 21 47 30.	27	15.7	GALAXY
ZWG 129.023	12 42 48.	+ 21 49		15.7	GALAXY
TON-N 0640	12 42 48.	+ 24 16		14.5	BLUE STAR
LB 11335	12 42 48.	+ 27 11		17.6	FAINT BLUE STAR
LB 07009	12 42 48.	+ 27 22		18.9	FAINT BLUE STAR
VVI 60	12 42 48.	+ 27 23	100	13.26	SEYFERT GALAXY
ZWG 159.069	12 42 48.	+ 27 24		12.6	GALAXY
UGC 07930	12 42 48.	+ 27 24	102	12.6	GALAXY PECULR
LB 07008	12 42 48.	+ 29 54		19.9	FAINT BLUE STAR
LB 07007	12 42 48.	+ 30 02		19.4	FAINT BLUE STAR
LB 07006	12 42 48.	+ 30 34		20.2	FAINT BLUE STAR
LB 07005	12 42 48.	+ 30 52		20.4	FAINT BLUE STAR
LB 07004	12 42 48.	+ 30 58		18.9	FAINT BLUE STAR
LB 07003	12 42 48.	+ 31 26		19.3	FAINT BLUE STAR
MCG+10-18-079	12 42 48.	+ 56 48	24	16.	GALAXY
KN 16.309	12 42 48.1	+ 26 21 27.			NEBULA
ARP 163	12 42 49.	+ 27 24			PECULIAR GALAXY
KN 16.310	12 42 49.9	+ 27 24 03.			NEBULA
IC 3735	12 42 50.	+ 13 57 48.			NONSTELLAR OBJECT
IC 3736	12 42 51.	+ 21 48 30.			NONSTELLAR OBJECT
PK302-00.1	12 42 51.	- 62 44	10		PLANETARY NEBULA
IC 3737	12 42 52.	+ 22 13 54.			NONSTELLAR OBJECT
REIZ 2781	12 42 52.	- 01 42	24	15.4	GALAXY
KN 16.311	12 42 52.0	+ 29 43 51.			NEBULA
AMES 2111	12 42 53.	+ 12 29 18.	18	17.5	NEBULA
LB 02438	12 42 53.	+ 59 14 12.		15.2	FAINT BLUE STAR
HOLM 456A	12 42 54.	+ 66 23	60	14.6	PART OF MULTIPLE GALAXY
MCG-04-30-014	12 42 54.	- 25 58	132	14.	GALAXY
KN 16.312	12 42 54.8	+ 28 51 45.			NEBULA
AMES 2113	12 42 55.	+ 11 15 18.	34	17.2	NEBULA
LB 02439	12 42 55.	+ 51 02 36.		14.3	FAINT BLUE STAR
AMES 2112	12 42 56.	+ 16 15 54.	16	17.8	NEBULA
IC 3738	12 42 56.	+ 19 30 06.			NONSTELLAR OBJECT
LB 00651	12 42 56.	+ 60 27 12.		16.0	FAINT BLUE STAR
A2 004	12 42 57.2	+ 28 13 59.		17.2	FAINT BLUE OBJECT
FATH 2.115	12 42 58.	- 00 16	68		NEBULA
IC 0814	12 42 58.	- 07 49 18.			NONSTELLAR OBJECT
REIN 4.140	12 42 58.20	- 00 15 45.2			NEBULA
REIZ 2785	12 42 59.	+ 19 30	24	15.1	GALAXY

OBJECT NAME	RIGHT ASCEN.	DECLINATION	DIAM.	MAGN.	TYPE OF OBJECT
ZC 1243.0+0508	12 43 00.	+ 05 08	540		CLUSTER OF GALAXIES
ZWG 071.032	12 43 00.	+ 13 36		14.6	GALAXY
UGC 07932	12 43 00.	+ 13 36	102	14.6	GALAXY SB
MCG+02-33-011	12 43 00.	+ 13 36	108	14.6	GALAXY
ZWG 129.024	12 43 00.	+ 21 05		15.6	GALAXY
LB 11336	12 43 00.	+ 26 48		14.2	FAINT BLUE STAR
LB 07015	12 43 00.	+ 26 48		19.5	FAINT BLUE STAR
LB 07014	12 43 00.	+ 26 54		20.2	FAINT BLUE STAR
LB 07013	12 43 00.	+ 27 25		18.8	FAINT BLUE STAR
TON-N 0641	12 43 00.	+ 29 43		14.8	BLUE STAR
LB 07012	12 43 00.	+ 30 08		18.9	FAINT BLUE STAR
LB 07011	12 43 00.	+ 30 46		17.9	FAINT BLUE STAR
LB 07010	12 43 00.	+ 32 13		18.7	FAINT BLUE STAR
MCG+10-18-080	12 43 00.	+ 61 59	24	16.	GALAXY
ZWG 015.016	12 43 00.	- 00 15		13.5	GALAXY
UGC 07931	12 43 00.	- 00 15	72	13.5	GALAXY IRR
MCG+00-33-009	12 43 00.	- 00 15	90	13.5	GALAXY
MCG-04-30-015	12 43 00.	- 26 21	90	14.5	GALAXY
IC 3739	12 43 01.	+ 13 16 18.			MAY NOT EXIST
RNGC 4668	12 43 01.	- 00 16		13.5	GALAXY
KN 16.314	12 43 01.4	+ 27 06 57.			NEBULA
KN 16.313	12 43 01.6	+ 26 24 45.			NEBULA
IC 3740	12 43 02.	+ 21 05 24.			NONSTELLAR OBJECT
MCG-01-33-002	12 43 03.	- 07 28	54	15.	GALAXY
IC 3741	12 43 04.	+ 19 28 42.			NONSTELLAR OBJECT
RNGC 4673	12 43 04.	+ 27 20		13.5	GALAXY
REIZ 2784	12 43 04.	- 00 16	54	13.0	GALAXY
KN 16.315	12 43 04.3	+ 26 11 39.			NEBULA
REIZ 2789	12 43 05.	+ 19 28	9	16.0	GALAXY
IC 3742	12 43 06.	+ 13 35			NONSTELLAR OBJECT
AMES 2116	12 43 06.	+ 13 35 41.	30	17.2	NEBULA
AMES 2114	12 43 06.	+ 14 04 41.	20	16.9	NEBULA
REIZ 2788	12 43 06.	+ 18 06	42	13.8	GALAXY
MCG+05-30-073	12 43 06.	+ 27 19	27	13.7	GALAXY
ZWG 159.070	12 43 06.	+ 27 20		13.7	GALAXY
MRK 656	12 43 06.	+ 27 20	20	14.5	GALAXY WITH UV CONTINUUM
UGC 07933	12 43 06.	+ 27 20	60	13.7	GALAXY S0
LB 11337	12 43 06.	+ 30 47		13.0	FAINT BLUE STAR
MCG+05-30-074	12 43 06.	+ 30 55	24	15.5	GALAXY
HOLM 455B	12 43 06.	+ 35 25	24	15.0	PART OF MULTIPLE GALAXY
ZWG 293.036	12 43 06.	+ 61 59		15.6	GALAXY
MCG-01-33-003	12 43 06.	- 05 48	180	14.5	GALAXY
MCG-02-33-003	12 43 06.	- 10 25 30.	30	15.	GALAXY
KN 16.317	12 43 06.8	+ 26 14 21.			NEBULA
KN 16.316	12 43 06.8	+ 26 16 51.			NEBULA
HN 1033	12 43 07.	+ 13 36	78		NEBULA
KN 16.318	12 43 07.8	+ 27 20 09.			NEBULA
AMES 2115	12 43 08.	+ 18 22 17.	22	17.2	NEBULA
KN 16.319	12 43 08.3	+ 28 40 03.			NEBULA
KN 16.320	12 43 08.8	+ 29 02 09.			NEBULA
MCG+03-33-003	12 43 09.	+ 19 26	18	15.0	GALAXY
IC 3746	12 43 09.	+ 38 05 49.			NONSTELLAR OBJECT
KN 16.321	12 43 09.7	+ 28 23 51.			NEBULA
IC 3743	12 43 10.	+ 11 22 31.			SINGLE STAR
REIZ 2786	12 43 11.	- 06 47	48	13.8	GALAXY
KN 16.323	12 43 11.2	+ 30 08 51.			NEBULA
ZWG 043.019	12 43 12.	+ 06 36		15.7	GALAXY
AMES 2117	12 43 12.	+ 17 46 35.	24	17.8	NEBULA
ZWG 100.006	12 43 12.	+ 19 27		15.0	GALAXY
8ZW 1243+25.2	12 43 12.	+ 25 13		17.5	COMPACT GALAXY
LB 07020	12 43 12.	+ 27 26		19.6	FAINT BLUE STAR
LB 07019	12 43 12.	+ 27 43		19.8	FAINT BLUE STAR
LB 07018	12 43 12.	+ 27 58		18.8	FAINT BLUE STAR
LB 07017	12 43 12.	+ 28 35		19.5	FAINT BLUE STAR
LB 07016	12 43 12.	+ 29 42		19.0	FAINT BLUE STAR
ZWG 159.071	12 43 12.	+ 29 43		15.5	GALAXY
LB 11339	12 43 12.	+ 31 30		16.6	FAINT BLUE STAR
LB 11338	12 43 12.	+ 32 00		13.5	FAINT BLUE STAR
MCG+06-28-026	12 43 12.	+ 33 46	48	14.	GALAXY
UGC 07934	12 43 12.	+ 35 23	72	16.0	GALAXY S
ZWG 188.019	12 43 12.	+ 38 05		15.6	GALAXY
IC 3747	12 43 12.	+ 38 14 31.			NONSTELLAR OBJECT
ZC 1243.2+4143	12 43 12.	+ 41 43	7120		CLUSTER OF GALAXIES
TON-N 0124	12 43 12.	+ 42 54		16.1	BLUE STAR
MCG+09-21-039	12 43 12.	+ 55 00	66	14.2	GALAXY
RNGC 4671	12 43 12.	- 06 48		14.5	GALAXY
MCG-01-33-004	12 43 12.	- 06 48	60	14.5	GALAXY
MCG-02-33-004	12 43 12.	- 10 33	60	15.5	GALAXY
SER 094.03	12 43 12.	- 45 11			PEC. GALAXY
BC PKS1243-412	12 43 12.39	- 41 12 22.8		18.	QUASI-STELLAR OBJECT
AMES 2118	12 43 13.	+ 11 41 11.	24	16.8	NEBULA
IC 3744	12 43 13.	+ 19 46 19.			NONSTELLAR OBJECT
ARC 1608	12 43 13.	+ 33 41		17.2	RICH CLUSTER OF GALAXIES
HOLM 455A	12 43 13.	+ 35 22	60	14.3	PART OF MULTIPLE GALAXY
KN 16.322	12 43 13.0	+ 25 01 15.			NEBULA
KN 16.324	12 43 13.6	+ 25 13 45.			NEBULA
SC 1240-0216.1	12 43 14.	- 02 32 32.	18		NEBULA
AMES 2120	12 43 15.	+ 15 59 05.	24	17.1	NEBULA
AMES 2119	12 43 15.	+ 17 43 29.	29	16.7	NEBULA
SC 1240+0006.1	12 43 15.	- 00 10 20.	12		NEBULA
IC 3745	12 43 16.	+ 19 27 01.			NONSTELLAR OBJECT
RNGC 4675	12 43 16.	+ 55 00		15.5	GALAXY
REIZ 2793	12 43 16.	+ 55 02	60	14.2	GALAXY
KN 16.325	12 43 16.4	+ 25 34 09.			NEBULA
KN 16.326	12 43 16.5	+ 29 36 15.			NEBULA
REIZ 2790	12 43 17.	+ 19 27	24	14.4	GALAXY
KN 16.327	12 43 17.2	+ 29 42 27.			NEBULA
ZWG 043.020	12 43 18.	+ 06 28		15.5	GALAXY
AMES 2121	12 43 18.	+ 15 59 59.	24	17.9	NEBULA
LB 07023	12 43 18.	+ 27 23		20.0	FAINT BLUE STAR
LB 11341	12 43 18.	+ 28 28		13.7	FAINT BLUE STAR
MCG+05-30-075	12 43 18.	+ 29 43	48	15.5	GALAXY
LB 11340	12 43 18.	+ 30 06		13.8	FAINT BLUE STAR
LB 07022	12 43 18.	+ 31 07		20.0	FAINT BLUE STAR
LB 07021	12 43 18.	+ 31 08		20.1	FAINT BLUE STAR
MCG+06-28-027	12 43 18.	+ 35 22	66	15.	GALAXY
MCG+09-21-040	12 43 18.	+ 54 08	42	17.	GALAXY
ZWG 270.019	12 43 18.	+ 55 00		15.4	GALAXY
UGC 07935	12 43 18.	+ 55 00	96	15.4	GALAXY SBb
7ZW 482	12 43 18.	+ 60 27			COMPACT GALAXY
ZWG 015.017	12 43 18.	- 01 24		15.5	GALAXY
KARA.68 197	12 43 18.	- 13 17	54		DWARF GALAXY
MCG-06-28-017	12 43 18.	- 33 33	90	16.	GALAXY
AMES 2122	12 43 20.	+ 18 40 05.	24	17.4	NEBULA
KN 16.329	12 43 21.3	+ 28 41 39.			NEBULA
KN 16.328	12 43 21.8	+ 26 47 09.			NEBULA
AMES 2123	12 43 22.	+ 11 42 11.	25	17.2	NEBULA
IC 3748	12 43 22.	+ 19 42 07.			NONSTELLAR OBJECT

OBJECT NAME	RIGHT ASCEN.	DECLINATION	DIAM.	MAGN.	TYPE OF OBJECT
IC 3751	12 43 22.	+ 38 05 49.			NONSTELLAR OBJECT
IC 3749	12 43 23.	+ 19 48 37.			NONSTELLAR OBJECT
SN 1907A	12 43 23.	- 08 23		13.5	SUPERNOVA
ZWG 043.021	12 43 24.	+ 07 37		15.7	GALAXY
ZC 1243.4+2025	12 43 24.	+ 20 25	340		CLUSTER OF GALAXIES
LB 07027	12 43 24.	+ 26 30		18.8	FAINT BLUE STAR
ZC 1243.4+2658	12 43 24.	+ 26 58	3290		CLUSTER OF GALAXIES
LB 07026	12 43 24.	+ 27 00		18.0	FAINT BLUE STAR
LB 07025	12 43 24.	+ 27 08		18.8	FAINT BLUE STAR
LB 07024	12 43 24.	+ 27 14		19.3	FAINT BLUE STAR
MRK 222	12 43 24.	+ 47 21	8	16.5	GALAXY WITH UV CONTINUUM
RNGC 4674	12 43 24.	- 08 23		15.0	GALAXY
MCG-03-33-006	12 43 24.	- 20 56	30	15.	GALAXY
AMES 2124	12 43 26.	+ 11 09 41.	29	16.6	NEBULA
REIZ 2797	12 43 26.	+ 47 22	30	14.8	GALAXY
MCG+08-23-088	12 43 27.	+ 47 49	42	16.	GALAXY
MCG-01-33-005	12 43 27.	- 08 23	72	15.	GALAXY
REIZ 2792	12 43 29.	+ 19 22	12	15.4	GALAXY
IC 3750	12 43 29.	+ 19 22 37.			NONSTELLAR OBJECT
REIZ 2787	12 43 29.	- 08 22	78	13.8	GALAXY
KN 16.330	12 43 29.7	+ 27 15 27.			NEBULA
ZWG 071.033	12 43 30.	+ 08 45		15.2	GALAXY
MCG+02-33-012	12 43 30.	+ 08 45	18	15.2	GALAXY
ZC 1243.5+1748	12 43 30.	+ 17 48	940		CLUSTER OF GALAXIES
LB 07030	12 43 30.	+ 27 08		19.8	FAINT BLUE STAR
LB 07029	12 43 30.	+ 27 48		19.4	FAINT BLUE STAR
LB 07028	12 43 30.	+ 28 09		17.0	FAINT BLUE STAR
LB 11342	12 43 30.	+ 30 04		17.4	FAINT BLUE STAR
ZWG 216.026	12 43 30.	+ 41 03		15.5	GALAXY
MCG+09-21-041	12 43 30.	+ 55 07 30.	60	15.	GALAXY
MRK 223	12 43 30.	+ 71 36	12	15.	GALAXY WITH UV CONTINUUM
MCG+13-09-031	12 43 30.	+ 74 44	15	15.	GALAXY
AMES 2126	12 43 31.	+ 12 03 41.	31	16.9	NEBULA
AMES 2125	12 43 31.	+ 16 22 53.	16	17.2	NEBULA
KN 16.332	12 43 31.5	+ 30 19 33.			NEBULA
REIZ 2798	12 43 32.	+ 31 00	120	15.2	GALAXY
RNGC 4672	12 43 32.	- 41 27			GALAXY
HELW 265	12 43 33.	- 41 00 19.			NEBULA
KN 16.331	12 43 33.0	+ 26 12 15.			NEBULA
REIZ 2791	12 43 34.	- 03 34	24	13.9	GALAXY
KN 16.333	12 43 34.9	+ 29 05 10.			NEBULA
REIZ 2794	12 43 35.	+ 19 17	6	15.6	GALAXY
IC 3752	12 43 35.	+ 19 17 01.			NONSTELLAR OBJECT
REIZ 2795	12 43 35.	+ 19 23	6	15.0	GALAXY
REIZ 2796	12 43 35.	+ 19 26	18	15.1	GALAXY
HOLM 457B	12 43 35.	- 10 43	18	15.3	PART OF MULTIPLE GALAXY
LB 07034	12 43 36.	+ 27 24		18.8	FAINT BLUE STAR
LB 07033	12 43 36.	+ 28 18		18.4	FAINT BLUE STAR
LB 07032	12 43 36.	+ 29 32		18.7	FAINT BLUE STAR
ZC 1243.6+2939	12 43 36.	+ 29 39	1750		CLUSTER OF GALAXIES
LB 07031	12 43 36.	+ 30 20		17.8	FAINT BLUE STAR
LB 11343	12 43 36.	+ 31 55		13.2	FAINT BLUE STAR
ZWG 244.040	12 43 36.	+ 45 28		15.2	GALAXY
UGC 07936	12 43 36.	+ 45 28	78	15.2	GALAXY MLT SYS
HOLM 457A	12 43 36.	- 10 43	24	14.9	PART OF MULTIPLE GALAXY
MCG-02-33-005	12 43 36.	- 10 43	30	15.5	GALAXY
MCG-07-26-041	12 43 36.	- 41 27	72	14.5	GALAXY
IC 3753	12 43 37.	+ 19 23 43.			NONSTELLAR OBJECT
IC 3757	12 43 37.	+ 38 47 13.			NONSTELLAR OBJECT
A2 008	12 43 37.1	+ 28 29 05.		17.1	FAINT BLUE OBJECT
REIZ 2799	12 43 38.	+ 30 59	30	15.0	GALAXY
IC 3758	12 43 38.	+ 41 02 55.			NONSTELLAR OBJECT
AMES 2128	12 43 39.	+ 12 02 17.	18	17.6	NEBULA
HOLM 458A	12 43 39.	+ 45 29	66	14.6	PART OF MULTIPLE GALAXY
HELW 266	12 43 39.	- 40 28 43.			NEBULA
IC 3755	12 43 40.	+ 19 25 49.			NONSTELLAR OBJECT
A2 009	12 43 40.3	+ 28 17 21.		18.5	FAINT BLUE OBJECT
AMES 2127	12 43 41.	+ 18 33 41.	29	17.4	NEBULA
RNGC 4676A	12 43 41.	+ 31 00		14.0	GALAXY
ZWG 071.034	12 43 42.	+ 08 37		14.9	GALAXY
UGC 07937	12 43 42.	+ 08 37	84	14.9	GALAXY Sa-b
MCG+02-33-013	12 43 42.	+ 08 37	72	14.9	GALAXY
LB 00052	12 43 42.	+ 25 12		16.0	FAINT BLUE STAR
8ZW 1243+25.4	12 43 42.	+ 25 26		19.3	COMPACT GALAXY
LB 07037	12 43 42.	+ 27 39		18.4	FAINT BLUE STAR
LB 07036	12 43 42.	+ 28 05		19.2	FAINT BLUE STAR
LB 07035	12 43 42.	+ 28 13		19.5	FAINT BLUE STAR
LB 11344	12 43 42.	+ 28 57		11.0	FAINT BLUE STAR
ZWG 159.072	12 43 42.	+ 31 00		14.1	GALAXY
SCH 72	12 43 42.	+ 31 00		14.1	PECULIAR GALAXY
UGC 07939	12 43 42.	+ 31 00	114	14.1	GALAXY DBL SYS
UGC 07938	12 43 42.	+ 31 00	132	14.1	GALAXY DBL SYS
VV 224B	12 43 42.	+ 31 00	30	14.7	INTERACTING GALAXY
VV 224A	12 43 42.	+ 31 00	36	14.4	INTERACTING GALAXY
VV 224	12 43 42.	+ 31 00	270	15.0	INTERACTING GALAXY
KARA.72 355B	12 43 42.	+ 31 00	48		PART OF DOUBLE GALAXY
KARA.72 355A	12 43 42.	+ 31 01	48	14.1	PART OF DOUBLE GALAXY
MCG+08-23-089	12 43 42.	+ 45 29	60	14.6	GALAXY
HOLM 458B	12 43 42.	+ 45 30	12	15.3	PART OF MULTIPLE GALAXY
7ZW 463	12 43 42.	+ 71 35			COMPACT GALAXY
ZWG 335.022	12 43 42.	+ 71 36		15.3	GALAXY
ZC 1243.7-0139	12 43 42.	- 01 39	5850		CLUSTER OF GALAXIES
MCG-02-33-006	12 43 42.	- 13 05	48	15.	GALAXY
A2 010	12 43 42.7	+ 29 13 53.		16.7	FAINT BLUE GALAXY
KN 16.336	12 43 42.9	+ 29 13 52.			NEBULA
IC 3754	12 43 43.	+ 08 36 31.			NONSTELLAR OBJECT
IC 3756	12 43 43.	+ 12 10			NONSTELLAR OBJECT
HN 1034	12 43 43.	+ 12 11		15.	NEBULA
KN 16.334	12 43 43.1	+ 26 32 46.			NEBULA
KN 16.335	12 43 43.7	+ 26 31 16.			NEBULA
REIF 3.325	12 43 43.77	+ 08 37 17.6			NEBULA
ARP 242	12 43 44.	+ 31 01			PECULIAR GALAXY
KN 16.337	12 43 44.0	+ 28 12 40.			NEBULA
AMES 2129	12 43 45.	+ 11 43 41.	28	16.9	NEBULA
RNGC 4676B	12 43 47.	+ 31 00		14.0	GALAXY
MCG+02-33-014	12 43 48.	+ 12 09	15	15.5	GALAXY
TON-N 0642	12 43 48.	+ 27 34		13.7	BLUE STAR
LB 07042	12 43 48.	+ 27 58		17.6	FAINT BLUE STAR
LB 07041	12 43 48.	+ 28 27		18.8	FAINT BLUE STAR
LB 07040	12 43 48.	+ 28 37		19.6	FAINT BLUE STAR
LB 07039	12 43 48.	+ 29 52		19.6	FAINT BLUE STAR
LB 07038	12 43 48.	+ 30 44		19.7	FAINT BLUE STAR
HOLM 459B	12 43 48.	+ 30 59	48	14.7	PART OF MULTIPLE GALAXY
MCG+05-30-077	12 43 48.	+ 31 01	138	14.1	GALAXY
MCG+05-30-076	12 43 48.	+ 31 01	120	14.1	GALAXY
STOCK 37	12 43 48.	+ 36 20			BLUE KNOT NEAR ELLIP GLXY
ZWG 244.041	12 43 48.	+ 46 59		15.5	GALAXY
UGC 07940	12 43 48.	+ 46 59	78	15.5	GALAXY S-IRR
MCG+08-23-090	12 43 48.	+ 48 58	36	16.	GALAXY
LB 02440	12 43 48.	+ 53 46 42.		17.1	FAINT BLUE STAR
MCG-07-26-042	12 43 48.	- 40 29	36	14.5	GALAXY
KN 16.338	12 43 48.2	+ 26 42 28.			NEBULA
IC 3759	12 43 49.	+ 21 03 26.			NONSTELLAR OBJECT
HOLM 459A	12 43 50.	+ 30 58	48	14.4	PART OF MULTIPLE GALAXY
AMES 2130	12 43 52.	+ 12 39 35.	31	16.6	NEBULA
LB 02441	12 43 52.	+ 52 47 06.		14.7	FAINT BLUE STAR
KN 16.340	12 43 52.0	+ 29 28 46.			NEBULA
KN 16.341	12 43 52.8	+ 27 48 40.			NEBULA
IC 0815	12 43 53.	+ 12 08 56.			NONSTELLAR OBJECT
KN 16.339	12 43 53.1	+ 26 42 34.			NEBULA
ZC 1243.9+0837	12 43 54.	+ 08 37	870		CLUSTER OF GALAXIES
ZWG 071.035	12 43 54.	+ 12 09		15.5	GALAXY
MCG+02-33-015	12 43 54.	+ 12 09	24	15.5	GALAXY
LB 11345	12 43 54.	+ 30 43		16.8	FAINT BLUE STAR
LB 07044	12 43 54.	+ 31 07		19.9	FAINT BLUE STAR
LB 07043	12 43 54.	+ 32 24		18.9	FAINT BLUE STAR
MCG+08-23-091	12 43 54.	+ 47 00	60	15.	GALAXY
ZC 1243.9+4846	12 43 54.	+ 48 46	1210		CLUSTER OF GALAXIES
ZWG 244.042	12 43 54.	+ 48 56		15.7	GALAXY
MCG+10-18-081	12 43 54.	+ 59 03	18	16.	GALAXY
ZWG 352.040	12 43 54.	+ 74 46		15.4	GALAXY
IC 3760	12 43 55.	+ 12 08			SAME AS IC 815
HN 1035	12 43 55.	+ 12 09		14.	NEBULA
LB 02442	12 43 55.	+ 54 55 24.		17.0	FAINT BLUE STAR
KN 16.342	12 43 55.1	+ 26 43 34.			NEBULA
HOLM 460B	12 43 56.	+ 71 28	24	15.4	PART OF MULTIPLE GALAXY
KN 16.343	12 43 56.0	+ 26 42 28.			NEBULA
KN 16.344	12 43 57.6	+ 26 47 58.			NEBULA
ABC 1609	12 43 58.	+ 26 43		16.8	RICH CLUSTER OF GALAXIES
SC 1241-0053.2	12 43 58.	- 01 09 37.	12		NEBULA
KN 16.345	12 43 58.4	+ 26 59 04.			NEBULA
IC 3761	12 43 59.	+ 20 33 38.			NONSTELLAR OBJECT
BFG 001	12 43 59.05	+ 33 15 20.7		19.40	ULTRAVIOLET-EXCESS OBJECT
VDB.66G 147	12 44	+ 36 44	70		DWARF GALAXY
AMES 2131	12 44 00.	+ 12 10 29.	34	17.0	NEBULA
ZC 1244.0+1550	12 44 00.	+ 15 50	870		CLUSTER OF GALAXIES
ZC 1244.0+2119	12 44 00.	+ 21 19	4230		CLUSTER OF GALAXIES
LB 00053	12 44 00.	+ 24 01		16.3	FAINT BLUE STAR
LB 07047	12 44 00.	+ 28 24		18.3	FAINT BLUE STAR
LB 07046	12 44 00.	+ 30 03		18.7	FAINT BLUE STAR
LB 07045	12 44 00.	+ 31 51		19.3	FAINT BLUE STAR
MCG+08-23-092	12 44 00.	+ 48 31	48	16.	GALAXY
MCG+10-18-082	12 44 00.	+ 59 53	8	16.	GALAXY
REIZ 2804	12 44 00.	+ 60 04	24	14.4	GALAXY
ZWG 316.004	12 44 00.	+ 64 50		15.0	GALAXY
UGC 07941	12 44 00.	+ 64 50	282	15.0	GALAXY
KARA.73B 0551	12 44 00.	+ 64 50	186	15.0	ISOLATED GALAXY S
MCG+11-16-006	12 44 00.	+ 64 51	96	14.	GALAXY
ZC 1244.0+7116	12 44 00.	+ 71 16	740		CLUSTER OF GALAXIES
ZC 1244.0+7729	12 44 00.	+ 77 29	870		CLUSTER OF GALAXIES
MCG+13-09-032	12 44 00.	+ 77 50	45	17.	GALAXY
AMES 2132	12 44 01.	+ 12 10 41.	23	17.4	NEBULA
KN 16.346	12 44 01.9	+ 26 59 28.			NEBULA
AMES 2134	12 44 02.	+ 12 09 11.	22	17.6	NEBULA
KARA.73 39	12 44 02.	- 37 43	54		DWARF GALAXY
A2 012	12 44 02.9	+ 28 28 26.		17.1	FAINT BLUE OBJECT
AMES 2133	12 44 03.	+ 17 46 47.	26	17.4	NEBULA
AMES 2135	12 44 04.	+ 12 09 53.	32	16.4	NEBULA
ZWG 071.036	12 44 06.	+ 09 35		15.6	GALAXY
UGC 07942	12 44 06.	+ 09 35	66	15.6	GALAXY
MCG+02-33-016	12 44 06.	+ 09 35	48	15.6	GALAXY
ZWG 071.037	12 44 06.	+ 12 10		15.7	GALAXY
MCG+02-33-018	12 44 06.	+ 12 10	24	15.7	GALAXY
MCG+02-33-017	12 44 06.	+ 12 10	48	15.7	GALAXY
LB 07051	12 44 06.	+ 27 09		16.0	FAINT BLUE STAR
LB 07050	12 44 06.	+ 29 02		19.5	FAINT BLUE STAR
TON-N 0643	12 44 06.	+ 29 23		15.4	BLUE STAR
LB 07049	12 44 06.	+ 31 08		18.0	FAINT BLUE STAR
LB 07048	12 44 06.	+ 31 41		20.0	FAINT BLUE STAR
ZWG 216.027	12 44 06.	+ 41 06		15.7	GALAXY
MRK 224	12 44 06.	+ 48 31	13	16.	GALAXY WITH UV CONTINUUM
ZWG 270.020	12 44 06.	+ 51 05		15.5	GALAXY
REIZ 2806	12 44 06.	+ 60 03	18	15.1	GALAXY
ZC 1244.1+6958	12 44 06.	+ 69 58	1340		CLUSTER OF GALAXIES
KARA.68 198	12 44 06.	- 03 46	54		DWARF GALAXY
KLEM 20	12 44 06.	- 40 49	10800	11.	RICH CLSTR OF 100 GLXIES
KN 16.347	12 44 06.1	+ 27 58 28.			NEBULA
RNGC 4678	12 44 07.	- 04 19			NON-EXISTENT OBJECT
AMES 2136	12 44 09.	+ 17 35 04.	31	17.1	NEBULA
IC 3762	12 44 09.	+ 22 31 02.			NONSTELLAR OBJECT
SC 1241-0049.1	12 44 09.	- 01 05 31.	30		NEBULA
KN 16.348	12 44 09.3	+ 26 59 34.			NEBULA
AMES 2137	12 44 10.	+ 17 39 40.	29	17.7	NEBULA
REIZ 2800	12 44 11.	- 08 30	30	13.7	GALAXY
ZWG 043.022	12 44 12.	+ 03 16		15.5	GALAXY
MCG+01-33-006	12 44 12.	+ 03 16	36	15.5	GALAXY
ZWG 043.023	12 44 12.	+ 06 14		14.7	GALAXY
UGC 07943	12 44 12.	+ 06 14	150	14.7	GALAXY Sc
REA 83	12 44 12.	+ 06 14			DWARF GALAXY
MCG+01-33-007	12 44 12.	+ 06 14	120	14.7	GALAXY
ZC 1244.2+0842	12 44 12.	+ 08 42	2420		CLUSTER OF GALAXIES
ZWG 071.038	12 44 12.	+ 10 07		15.1	GALAXY
UGC 07944	12 44 12.	+ 10 07	60	15.1	GALAXY SB
MCG+02-33-019	12 44 12.	+ 10 07	60	15.1	GALAXY
IC 0816	12 44 12.	+ 10 07 32.			NONSTELLAR OBJECT
KARA.68 199	12 44 12.	+ 22 04	40		DWARF GALAXY
LB 07054	12 44 12.	+ 27 00		17.5	FAINT BLUE STAR
LB 07053	12 44 12.	+ 27 38		18.0	FAINT BLUE STAR
LB 07052	12 44 12.	+ 27 48		20.2	FAINT BLUE STAR
ZWG 159.073	12 44 12.	+ 30 00		15.1	GALAXY
IC 0818	12 44 12.	+ 30 00			GALAXY S0
KN 16.349	12 44 12.0	+ 30 07 22.			NEBULA
KN 16.350	12 44 12.3	+ 29 28 10.			NEBULA
BFG 002	12 44 12.66	+ 36 58 32.8		18.69	ULTRAVIOLET-EXCESS OBJECT
AMES 2139	12 44 13.	+ 10 57 22.	28	16.2	NEBULA
AMES 2138	12 44 13.	+ 17 38 16.	23	17.2	NEBULA
IC 3765	12 44 13.	+ 38 50 50.			NONSTELLAR OBJECT
LB 00652	12 44 13.	+ 75 37 18.		16.8	FAINT BLUE STAR
HELW 267	12 44 13.	- 41 13 19.			NEBULA
RNGC 4696A	12 44 13.	- 41 14			GALAXY
RNGC 4677	12 44 13.	- 41 19			GALAXY
KN 16.351	12 44 14.1	+ 28 14 04.			NEBULA
REIN 3.326	12 44 15.03	+ 10 07 24.4			NEBULA
KN 16.352	12 44 15.2	+ 27 14 16.			NEBULA
AMES 2140	12 44 16.	+ 15 46 04.	36	17.0	NEBULA
LB 02443	12 44 16.	+ 50 08 54.		15.5	FAINT BLUE STAR

OBJECT NAME	RIGHT ASCEN.	DECLINATION	DIAM.	MAGN.	TYPE OF OBJECT
KN 16.353	12 44 16.7	+ 26 40 40.			NEBULA
REIZ 2803	12 44 17.	+ 19 18	12	15.3	GALAXY
ARC 1614	12 44 17.	+ 69 58		17.5	RICH CLUSTER OF GALAXIES
IC 3763	12 44 18.	+ 22 15 20.			NONSTELLAR OBJECT
ZC 1244.3+2617	12 44 18.	+ 26 17	610		CLUSTER OF GALAXIES
LB 07055	12 44 18.	+ 28 48		19.7	FAINT BLUE STAR
LB 11347	12 44 18.	+ 31 50		12.8	FAINT BLUE STAR
LB 11346	12 44 18.	+ 32 15		15.2	FAINT BLUE STAR
MCG+09-21-043	12 44 18.	+ 51 03	30	15.	GALAXY
MCG+09-21-042	12 44 18.	+ 51 10	30	15.	GALAXY
MCG+09-21-044	12 44 18.	+ 54 48	120	13.6	GALAXY
RNGC 4680	12 44 18.	- 11 21		13.0	GALAXY
REIZ 2801	12 44 18.	- 11 22	48	13.2	GALAXY
MCG-03-33-007	12 44 18.	- 21 08	60	15.	GALAXY
MCG-07-26-043	12 44 18.	- 41 14	78	14.	GALAXY
MCG-07-26-044	12 44 18.	- 41 19	60	15.	GALAXY
KN 16.354	12 44 18.2	+ 30 00 34.			NEBULA
SC 1241+0050.8	12 44 19.	+ 00 34 23.	6		NEBULA
IC 3768	12 44 19.	+ 40 52 08.			NONSTELLAR OBJECT
KN 16.355	12 44 20.2	+ 29 39 40.			NEBULA
KN 16.356	12 44 20.9	+ 28 13 16.			NEBULA
MCG-02-33-007	12 44 21.	- 11 21 30.	90	13.	GALAXY
RNGC 4686	12 44 22.	+ 54 49		13.5	GALAXY
REIZ 2805	12 44 23.	+ 21 05	30	15.3	GALAXY
ZWG 071.039	12 44 24.	+ 10 08		15.4	GALAXY
MCG+02-33-020	12 44 24.	+ 10 08	18	15.4	GALAXY
AMES 2141	12 44 24.	+ 17 24 28.	31	17.6	NEBULA
LB 11350	12 44 24.	+ 26 25		13.7	FAINT BLUE STAR
MCG+05-30-079	12 44 24.	+ 26 49	18	15.0	GALAXY
ZWG 159.074	12 44 24.	+ 26 50		15.0	GALAXY
LB 11349	12 44 24.	+ 27 44		12.9	FAINT BLUE STAR
LB 07060	12 44 24.	+ 27 56		19.2	FAINT BLUE STAR
LB 07059	12 44 24.	+ 28 29		19.5	FAINT BLUE STAR
LB 11348	12 44 24.	+ 29 03		13.8	FAINT BLUE STAR
LB 07058	12 44 24.	+ 29 23		17.7	FAINT BLUE STAR
MCG+05-30-078	12 44 24.	+ 30 01	48	15.1	GALAXY
LB 07057	12 44 24.	+ 31 57		20.0	FAINT BLUE STAR
LB 07056	12 44 24.	+ 32 06		19.5	FAINT BLUE STAR
TON-N 0125	12 44 24.	+ 49 53		16.2	BLUE STAR
ZWG 270.021	12 44 24.	+ 54 49		13.7	GALAXY
UGC 07946	12 44 24.	+ 54 49	132	13.7	GALAXY Sa
ZWG 015.018	12 44 24.	- 01 17		15.6	GALAXY
UGC 07945	12 44 24.	- 01 17	90	15.6	GALAXY DWRF SP
MCG+00-33-010	12 44 24.	- 01 17	72	15.6	GALAXY
REIZ 2802	12 44 24.	- 11 20	18	15.6	GALAXY
KN 16.357	12 44 24.1	+ 26 38 46.			NEBULA
AMES 2142	12 44 25.	+ 17 29 16.	26	17.4	NEBULA
IC 3766	12 44 25.	+ 19 22 56.			NONSTELLAR OBJECT
REIN 3.327	12 44 25.57	+ 10 07 48.0			NEBULA
KN 16.358	12 44 25.9	+ 26 59 11.			NEBULA
IC 3764	12 44 26.	+ 10 07 50.			SAME AS IC 816
IC 0817	12 44 26.	+ 10 08 02.			NONSTELLAR OBJECT
IC 3769	12 44 26.	+ 40 44 33.			NONSTELLAR OBJECT
A2 014	12 44 26.0	+ 28 03 46.		17.3	FAINT BLUE GALAXY
REIN 4.141	12 44 26.88	- 01 18 21.8			NEBULA
A2 015	12 44 26.9	+ 28 58 28.		16.2	FAINT BLUE OBJECT
REIZ 2810	12 44 28.	+ 54 49	90	13.6	GALAXY
BPG 003	12 44 28.04	+ 35 45 20.3		17.87	ULTRAVIOLET-EXCESS OBJECT
KN 16.359	12 44 28.6	+ 26 50 17.			NEBULA
BPG 004	12 44 28.74	+ 38 06 21.8		18.14	ULTRAVIOLET-EXCESS OBJECT
KN 16.360	12 44 29.2	+ 25 53 41.			NEBULA
KN 16.361	12 44 29.6	+ 30 07 23.			NEBULA
LB 11351	12 44 30.	+ 27 13		16.0	FAINT BLUE STAR
LB 07063	12 44 30.	+ 27 51		18.7	FAINT BLUE STAR
LB 07062	12 44 30.	+ 28 20		18.9	FAINT BLUE STAR
LB 07061	12 44 30.	+ 29 38		19.1	FAINT BLUE STAR
ZC 1244.5+3303	12 44 30.	+ 33 03	2550		CLUSTER OF GALAXIES
UGC 07947	12 44 30.	+ 36 33	66	16.0	GALAXY S
IC 3771	12 44 30.	+ 39 26 45.			NONSTELLAR OBJECT
ZC 1244.5+4100	12 44 30.	+ 41 00	740		CLUSTER OF GALAXIES
MCG+09-21-045	12 44 30.	+ 51 17	36	16.	GALAXY
TON-N 0126	12 44 30.	- 00 20		15.5	BLUE STAR
A2 016	12 44 30.2	+ 29 01 06.		18.2	FAINT BLUE OBJECT
IC 3767	12 44 31.	+ 10 26			NONSTELLAR OBJECT
HN 1036	12 44 31.	+ 10 27		16.	NEBULA
KN 16.362	12 44 31.4	+ 28 44 59.			NEBULA
REIN 3.328	12 44 31.87	+ 10 19 21.7			NEBULA
LB 00653	12 44 32.	+ 61 00 42.		17.0	FAINT BLUE STAR
HOLE 460A	12 44 32.	+ 71 27	126	12.8	PART OF MULTIPLE GALAXY
AMES 2143	12 44 33.	+ 17 45 58.	32	17.4	NEBULA
IC 3772	12 44 33.	+ 36 48 21.			NONSTELLAR OBJECT
KN 16.363	12 44 34.6	+ 28 13 47.			NEBULA
UGC 07948	12 44 36.	+ 09 17	60	16.0	GALAXY Sc
AMES 2144	12 44 36.	+ 17 40 46.	29	17.1	NEBULA
MCG+03-33-004	12 44 36.	+ 19 44	78	13.8	GALAXY
LB 07071	12 44 36.	+ 26 30		19.2	FAINT BLUE STAR
LB 07070	12 44 36.	+ 26 44		18.9	FAINT BLUE STAR
LB 07069	12 44 36.	+ 26 57		20.0	FAINT BLUE STAR
LB 07068	12 44 36.	+ 27 54		19.2	FAINT BLUE STAR
LB 07067	12 44 36.	+ 28 05		19.0	FAINT BLUE STAR
LB 07066	12 44 36.	+ 28 52		18.5	FAINT BLUE STAR
LB 07065	12 44 36.	+ 29 52		18.7	FAINT BLUE STAR
LB 07064	12 44 36.	+ 30 04		19.5	FAINT BLUE STAR
UGC 07949	12 44 36.	+ 36 45	126	17.	GALAXY DWARF
ZWG 188.020	12 44 36.	+ 36 47		15.7	GALAXY
ZWG 216.028	12 44 36.	+ 40 53		15.6	GALAXY
MCG+07-26-051	12 44 36.	+ 42 32	30	16.	GALAXY
ZC 1244.6+4630	12 44 36.	+ 46 30	1550		CLUSTER OF GALAXIES
MCG+09-21-046	12 44 36.	+ 50 59	42	15.	GALAXY
ZWG 270.022	12 44 36.	+ 51 55		14.1	GALAXY
UGC 07950	12 44 36.	+ 51 55	108	14.1	GALAXY IRR
REIZ 2807	12 44 36.	- 09 47	120	13.0	GALAXY
MCG-07-26-045	12 44 36.	- 40 58	36	14.	GALAXY
KN 16.364	12 44 36.2	+ 26 58 59.			NEBULA
A2 018	12 44 36.3	+ 29 00 40.		18.4	FAINT BLUE GALAXY
IC 3774	12 44 37.	+ 36 33 51.			NONSTELLAR OBJECT
HELW 268	12 44 37.	- 40 57 36.			NEBULA
KN 16.365	12 44 38.7	+ 26 38 59.			NEBULA
KN 16.366	12 44 38.9	+ 26 59 05.			NEBULA
KN 16.367	12 44 38.9	+ 30 18 35.			NEBULA
LB 00654	12 44 39.	+ 62 21 18.		17.1	FAINT BLUE STAR
A2 019	12 44 39.1	+ 29 44 55.		17.7	FAINT BLUE GALAXY
RNGC 4685	12 44 40.	+ 19 44		14.0	GALAXY
IC 3778	12 44 40.	+ 40 52 09.			NONSTELLAR OBJECT
HARO 36	12 44 40.	+ 51 52			BLUE EMISSION-LINE GALAXY
REIZ 2808	12 44 40.	- 02 27	42	12.2	GALAXY
KN 16.368	12 44 40.6	+ 28 12 53.			NEBULA
REIZ 2809	12 44 41.	+ 19 44	24	13.8	GALAXY

OBJECT NAME	RIGHT ASCEN.	DECLINATION	DIAM.	MAGN.	TYPE OF OBJECT
ZWG 071.040	12 44 42.	+ 10 29		14.3	GALAXY
UGC 07952	12 44 42.	+ 10 29	126	14.3	GALAXY E
MCG+02-33-021	12 44 42.	+ 10 29	42	14.3	GALAXY
UGC 07953	12 44 42.	+ 12 03	60	16.0	GALAXY Sc
ZC 1244.7+1913	12 44 42.	+ 19 13	1010		CLUSTER OF GALAXIES
ZWG 100.007	12 44 42.	+ 19 44		13.8	GALAXY
UGC 07954	12 44 42.	+ 19 44	96	13.8	GALAXY E-S0
LB 11353	12 44 42.	+ 26 32		12.2	FAINT BLUE STAR
UGC 07955	12 44 42.	+ 26 58	84	16.0	GALAXY Sc
MCG+05-30-080	12 44 42.	+ 26 58	84	15.	GALAXY
MCG+05-30-081	12 44 42.	+ 28 04	30	16.	GALAXY
ZC 1244.7+2915	12 44 42.	+ 29 15	1210		CLUSTER OF GALAXIES
LB 07073	12 44 42.	+ 30 56		19.7	FAINT BLUE STAR
LB 07072	12 44 42.	+ 30 57		19.9	FAINT BLUE STAR
ZC 1244.7+3153	12 44 42.	+ 31 53	870		CLUSTER OF GALAXIES
LB 11352	12 44 42.	+ 32 09		13.7	FAINT BLUE STAR
MCG+06-28-028	12 44 42.	+ 36 50	36	16.	GALAXY
MCG+09-21-047	12 44 42.	+ 51 51	60	13.	GALAXY
ZC 1244.7+5805	12 44 42.	+ 58 05	1410		CLUSTER OF GALAXIES
ZC 1244.7+6159	12 44 42.	+ 61 59	1280		CLUSTER OF GALAXIES
ZWG 015.019	12 44 42.	- 02 27		12.4	GALAXY
UGC 07951	12 44 42.	- 02 27	168	12.4	GALAXY S0
MCG+00-33-011	12 44 42.	- 02 27	114	12.4	GALAXY
MCG-02-33-008	12 44 42.	- 09 47	150	13.	GALAXY
RNGC 4682	12 44 42.	- 09 48			GALAXY
MCG-02-33-009	12 44 42.	- 13 46	36	15.5	GALAXY
KN 16.369	12 44 42.8	+ 28 03 59.			NEBULA
REIN 4.142	12 44 42.82	- 02 27 29.1			NEBULA
RNGC 4684	12 44 43.	- 02 28		12.5	GALAXY
RNGC 4696B	12 44 43.	- 40 58			GALAXY
RNGC 4681	12 44 43.	- 43 04			GALAXY
REIN 4.143	12 44 43.21	- 02 27 15.9			NEBULA
KN 16.371	12 44 43.6	+ 30 25 05.			NEBULA
IC 3770	12 44 44.	+ 09 28 33.			NONSTELLAR OBJECT
ARC 1610	12 44 44.	+ 30 19		17.2	RICH CLUSTER OF GALAXIES
IC 0819	12 44 44.	+ 31 00 27.			NONSTELLAR OBJECT
RNGC 4679	12 44 44.	- 39 18		13.0	GALAXY
A2 020	12 44 44.0	+ 28 05 17.		17.5	FAINT BLUE OBJECT
REIN 3.329	12 44 44.07	+ 10 28 35.6			NEBULA
REIN 4.144	12 44 44.15	- 02 26 43.7			NEBULA
REIN 4.145	12 44 44.44	- 02 27 51.0			NEBULA
KN 16.372	12 44 44.8	+ 29 43 23.			NEBULA
KN 16.370	12 44 44.9	+ 26 59 11.			NEBULA
IC 3773	12 44 45.	+ 10 29 27.			NONSTELLAR OBJECT
IC 3776	12 44 45.	+ 22 45 27.			NONSTELLAR OBJECT
HARO 35	12 44 45.	+ 28 04			BLUE EMISSION-LINE GALAXY
IC 0820	12 44 45.	+ 30 59 39.			SAME AS NGC 4676
MCG+06-28-029	12 44 45.	+ 36 35	48	16.	GALAXY
KARA.68 200	12 44 45.	+ 36 47	101		DWARF GALAXY
MCG+06-28-030	12 44 45.	+ 36 47	108	16.	GALAXY
MCG-01-33-006	12 44 45.	- 09 14	36	15.	GALAXY
KN 16.373	12 44 45.3	+ 29 27 41.			NEBULA
AMES 2145	12 44 46.	+ 11 45 22.	32	17.1	NEBULA
IC 3780	12 44 46.	+ 40 30 27.			NONSTELLAR OBJECT
KN 16.374	12 44 46.3	+ 29 43 53.			NEBULA
SC 1242+0057.2	12 44 47.	+ 00 40 48.	12		NEBULA
REIZ 2813	12 44 47.	+ 20 55	36	14.8	GALAXY
ZWG 015.020	12 44 48.	+ 00 41		15.6	GALAXY
ZWG 071.041	12 44 48.	+ 12 26		15.5	GALAXY
ZC 1244.8+1506	12 44 48.	+ 15 06	540		CLUSTER OF GALAXIES
ZC 1244.8+1741	12 44 48.	+ 17 41	2550		CLUSTER OF GALAXIES
ZC 1244.8+1847	12 44 48.	+ 18 47	1010		CLUSTER OF GALAXIES
ZC 1244.8+2122	12 44 48.	+ 21 22	810		CLUSTER OF GALAXIES
LB 07083	12 44 48.	+ 26 50		19.5	FAINT BLUE STAR
LB 07082	12 44 48.	+ 26 59		17.7	FAINT BLUE STAR
LB 07081	12 44 48.	+ 27 01		20.0	FAINT BLUE STAR
LB 07080	12 44 48.	+ 27 07		19.2	FAINT BLUE STAR
LB 07079	12 44 48.	+ 27 33		17.1	FAINT BLUE STAR
LB 07078	12 44 48.	+ 29 02		19.4	FAINT BLUE STAR
LB 07077	12 44 48.	+ 29 33		19.4	FAINT BLUE STAR
LB 11355	12 44 48.	+ 29 38		13.0	FAINT BLUE STAR
LB 11354	12 44 48.	+ 29 55		16.8	FAINT BLUE STAR
LB 07076	12 44 48.	+ 30 14		17.6	FAINT BLUE STAR
LB 07075	12 44 48.	+ 30 41		18.1	FAINT BLUE STAR
LB 07074	12 44 48.	+ 31 34		18.9	FAINT BLUE STAR
ZC 1244.8+5630	12 44 48.	+ 56 30	1340		CLUSTER OF GALAXIES
MCG-06-28-018	12 44 48.	- 39 18	120	13.	GALAXY
MCG-07-26-046	12 44 48.	- 43 04	60	14.	GALAXY
BPG 005	12 44 48.53	+ 37 23 25.5		19.41	ULTRAVIOLET-EXCESS OBJECT
KN 16.375	12 44 48.9	+ 28 10 23.			NEBULA
IC 3775	12 44 49.	+ 12 00			NONSTELLAR OBJECT
HN 1037	12 44 49.	+ 12 01		16.	NEBULA
AMES 2146	12 44 49.	+ 15 58 58.	29	16.9	NEBULA
AMES 2147	12 44 50.	+ 13 34 34.	23	17.0	NEBULA
ARC 1611	12 44 50.	+ 19 12		17.8	RICH CLUSTER OF GALAXIES
KN 16.376	12 44 50.5	+ 26 50 53.			NEBULA
KN 16.377	12 44 51.8	+ 28 00 11.			NEBULA
A2 022	12 44 51.9	+ 29 42 25.		16.7	FAINT BLUE OBJECT
REIZ 2812	12 44 52.	- 03 41	54	13.3	GALAXY
REIZ 2816	12 44 53.	+ 20 24	60	14.5	GALAXY
A2 023	12 44 53.7	+ 28 55 24.		16.5	FAINT BLUE OBJECT
ZWG 043.024	12 44 54.	+ 03 18		15.0	GALAXY
MCG+01-33-008	12 44 54.	+ 03 18	18	15.0	GALAXY
ZWG 043.025	12 44 54.	+ 03 52		15.5	GALAXY
MCG+01-33-009	12 44 54.	+ 03 52	12	15.5	GALAXY
ZWG 043.026	12 44 54.	+ 05 02		15.6	GALAXY
MCG+01-33-011	12 44 54.	+ 05 02	12	15.6	GALAXY
MCG+01-33-010	12 44 54.	+ 05 02	30	15.6	GALAXY
IC 3777	12 44 54.	+ 09 24 57.			NONSTELLAR OBJECT
ZC 1244.9+2355	12 44 54.	+ 23 55	3020		CLUSTER OF GALAXIES
ZWG 159.075	12 44 54.	+ 27 44		15.2	GALAXY
LB 11356	12 44 54.	+ 27 54		13.2	FAINT BLUE STAR
LB 07086	12 44 54.	+ 29 50		19.4	FAINT BLUE STAR
LB 07085	12 44 54.	+ 31 12		19.1	FAINT BLUE STAR
LB 07084	12 44 54.	+ 31 48		18.7	FAINT BLUE STAR
IC 3782	12 44 54.	+ 40 39 03.			NONSTELLAR OBJECT
MRK 225	12 44 54.	+ 47 24	11	16.5	GALAXY WITH UV CONTINUUM
ZC 1244.9+5514	12 44 54.	+ 55 14	4440		CLUSTER OF GALAXIES
7ZW 484	12 44 54.	+ 58 03			COMPACT GALAXY
REIZ 2811	12 44 54.	- 10 07	30	15.8	GALAXY
IC 3779	12 44 55.	+ 12 25			NONSTELLAR OBJECT
HN 1038	12 44 55.	+ 12 26		15.	NEBULA
RNGC 4683	12 44 55.	- 41 14			GALAXY
KN 16.378	12 44 55.2	+ 28 16 41.			NEBULA
BPG 006	12 44 55.40	+ 32 25 22.8		18.69	ULTRAVIOLET-EXCESS OBJECT
KN 16.381	12 44 55.8	+ 30 04 17.			NEBULA
BC B21244+32B	12 44 55.8	+ 32 25 21.5		18.	QUASI-STELLAR OBJECT
KN 16.379	12 44 56.3	+ 25 32 59.			NEBULA

OBJECT NAME	RIGHT ASCEN.	DECLINATION	DIAM.	MAGN.	TYPE OF OBJECT
KN 16.380	12 44 56.9	+ 26 49 35.			NEBULA
REIZ 2815	12 44 57.	+ 03 19	24	14.1	GALAXY
AMES 2148	12 44 57.	+ 13 36 40.	28	17.2	NEBULA
IC 3781	12 44 57.	+ 22 50 33.			NONSTELLAR OBJECT
MCG-04-30-016	12 44 57.	- 25 56	54	15.	GALAXY
ARC 1612	12 44 58.	- 01 25		17.6	RICH CLUSTER OF GALAXIES
REIZ 2814	12 44 58.	- 03 39	12	15.6	GALAXY
ARC 1613	12 44 59.	+ 35 51		17.6	RICH CLUSTER OF GALAXIES
LB 00655	12 44 59.	+ 60 35 00.		17.3	FAINT BLUE STAR
KN 16.382	12 44 59.2	+ 30 11 29.			NEBULA
VDB.66G 148	12 45	- 05 01	100		DWARF GALAXY
ZC 1245.0+1215	12 45 00.	+ 12 15	540		CLUSTER OF GALAXIES
MCG+05-30-082	12 45 00.	+ 27 44	30	15.2	GALAXY
ZWG 159.076	12 45 00.	+ 30 04		14.5	GALAXY
UGC 07957	12 45 00.	+ 30 04	72	14.5	GALAXY SBb/Sc
MCG+05-30-083	12 45 00.	+ 30 04	72	14.5	GALAXY
IC 0821	12 45 00.	+ 30 04			GALAXY SB
LB 07088	12 45 00.	+ 30 58		19.3	FAINT BLUE STAR
LB 07087	12 45 00.	+ 31 35		19.0	FAINT BLUE STAR
ZWG 188.021	12 45 00.	+ 35 37		14.3	GALAXY
MRK 442	12 45 00.	+ 35 37	18	15.	GALAXY WITH UV CONTINUUM
UGC 07958	12 45 00.	+ 35 37	60	14.3	GALAXY
RNGC 4687	12 45 00.	+ 35 38		14.5	GALAXY
ZC 1245.0+3546	12 45 00.	+ 35 46	2960		CLUSTER OF GALAXIES
ZWG 216.029	12 45 00.	+ 40 51		15.6	GALAXY
ZC 1245.0+5942	12 45 00.	+ 59 42	2220		CLUSTER OF GALAXIES
ZWG 365.003	12 45 00.	+ 83 42		15.5	GALAXY
UGC 07956	12 45 00.	+ 83 42	114	15.5	GALAXY S
UGC 07956A	12 45 00.	+ 88 05	78	16.0	GALAXY Sc
ZC 1245.0-0002	12 45 00.	- 00 02	4700		CLUSTER OF GALAXIES
DV.56 N4696A	12 45 00.	- 41 12	108		S GALAXY
MCG-07-26-047	12 45 00.	- 41 16	72	14.5	GALAXY
KN 16.383	12 45 00.2	+ 30 03 35.			NEBULA
REIZ 2818	12 45 01.	+ 35 38	30	14.2	GALAXY
LB 02444	12 45 01.	+ 55 03 36.		16.1	FAINT BLUE STAR
KN 16.384	12 45 01.8	+ 27 43 53.			NEBULA
HELW 269	12 45 02.	- 41 15 30.			NEBULA
BC B19	12 45 02.7	+ 34 31 29.1		17.94	QUASI-STELLAR OBJECT
MCG+06-28-031	12 45 03.	+ 35 38	36	14.	GALAXY
MCG+07-26-052	12 45 03.	+ 40 50	36	14.5	GALAXY
MCG-01-33-007	12 45 03.	- 05 36	24	15.	GALAXY
VV 290A	12 45 03.	- 25 55	66	15.	INTERACTING GALAXY
KN 16.385	12 45 03.2	+ 26 48 17.			NEBULA
SHB 213	12 45 03.2	+ 34 31 31.		17.9	QUASI-STELLAR OBJECT
BFG 007	12 45 03.20	+ 34 31 31.5		17.94	ULTRAVIOLET-EXCESS OBJECT
ZWG 159.077	12 45 06.	+ 27 16		14.7	GALAXY
UGC 07959	12 45 06.	+ 27 16	72	14.7	GALAXY S0
LB 07092	12 45 06.	+ 27 23		19.3	FAINT BLUE STAR
LB 07091	12 45 06.	+ 28 46		18.6	FAINT BLUE STAR
LB 07090	12 45 06.	+ 30 06		19.3	FAINT BLUE STAR
ZC 1245.1+3030	12 45 06.	+ 30 30	4770		CLUSTER OF GALAXIES
TON-N 0644	12 45 06.	+ 30 46		15.4	BLUE STAR
LB 07089	12 45 06.	+ 32 22		19.0	FAINT BLUE STAR
IC 3783	12 45 06.	+ 40 50 16.			NONSTELLAR OBJECT
ZC 1245.1+4133	12 45 06.	+ 41 33	1210		CLUSTER OF GALAXIES
ZC 1245.1-0230	12 45 06.	- 02 30	740		CLUSTER OF GALAXIES
MCG-06-28-019	12 45 06.	- 35 47	48	15.5	GALAXY
A2 025	12 45 08.4	+ 29 11 18.		17.2	FAINT BLUE OBJECT
MCG-04-30-018	12 45 09.	- 21 38 30.	54	15.	GALAXY
MCG-04-30-017	12 45 09.	- 25 56	48	16.	GALAXY
KN 16.387	12 45 09.4	+ 27 53 05.			NEBULA
A2 026	12 45 09.6	+ 27 43 55.		17.1	FAINT BLUE OBJECT
A2 027	12 45 09.6	+ 28 58 18.		17.8	FAINT BLUE OBJECT
SC 1242+0025.5	12 45 10.	+ 00 09 06.	6		NEBULA
SN 1966B	12 45 10.	+ 04 35		15.0	SUPERNOVA
KN 16.386	12 45 10.0	+ 25 13 17.			NEBULA
REIZ 2817	12 45 11.	- 04 37	18	15.8	GALAXY
UGC 07960	12 45 12.	+ 04 05	66	15.2	GALAXY Sa-b
ZWG 043.027	12 45 12.	+ 04 09		15.2	GALAXY
MCG+01-33-012	12 45 12.	+ 04 09	60	15.2	GALAXY
ZWG 043.028	12 45 12.	+ 04 36		14.5	GALAXY
UGC 07961	12 45 12.	+ 04 36	270	14.5	GALAXY SBc
MCG+01-33-013	12 45 12.	+ 04 36	180	14.5	GALAXY
SCH 73	12 45 12.	+ 14 01		12.8	PECULIAR GALAXY
LB 11357	12 45 12.	+ 26 37		13.0	FAINT BLUE STAR
LB 07095	12 45 12.	+ 27 12		17.9	FAINT BLUE STAR
MCG+05-30-084	12 45 12.	+ 27 15	78	14.7	GALAXY
LB 07094	12 45 12.	+ 27 28		18.8	FAINT BLUE STAR
TON-N 0645	12 45 12.	+ 28 26		15.5	BLUE STAR
LB 07093	12 45 12.	+ 32 02		19.5	FAINT BLUE STAR
TON-N 0646	12 45 12.	+ 33 24		14.2	BLUE STAR
MCG+09-21-048	12 45 12.	+ 54 38	60	14.0	GALAXY
ZC 1245.2+6346	12 45 12.	+ 63 46	4230		CLUSTER OF GALAXIES
MCG+12-12-018	12 45 12.	+ 71 27 30.	150	15.	GALAXY
ZWG 335.023	12 45 12.	+ 71 28		14.0	GALAXY
UGC 07962	12 45 12.	+ 71 28	144	14.0	GALAXY Sc
MCG-04-30-019	12 45 12.	- 22 00	90	13.5	GALAXY
VV 290B	12 45 12.	- 25 55	54	16.	INTERACTING GALAXY
KN 16.388	12 45 12.0	+ 26 49 29.			NEBULA
KN 16.389	12 45 13.5	+ 25 49 11.			NEBULA
A2 028	12 45 13.8	+ 29 05 16.		16.6	FAINT BLUE OBJECT
RNGC 4688	12 45 14.	+ 04 37		12.5	GALAXY
IC 3786	12 45 14.	+ 39 19 04.			NONSTELLAR OBJECT
REIZ 2819	12 45 15.	+ 04 09	60	14.9	GALAXY
REIZ 2821	12 45 15.	+ 04 37	60	14.3	GALAXY
REIZ 2820	12 45 15.	+ 04 37	24	14.9	GALAXY
HOLM 461A	12 45 15.	+ 04 37	120	13.0	PART OF MULTIPLE GALAXY
RNGC 4693	12 45 15.	+ 71 28		14.0	GALAXY
KN 16.390	12 45 15.3	+ 27 15 12.			NEBULA
REIZ 2831	12 45 16.	+ 54 39	48	14.0	GALAXY
LB 00656	12 45 16.	+ 56 46 24.		17.1	FAINT BLUE STAR
HELW 270	12 45 16.	- 40 32 30.			NEBULA
RNGC 4695	12 45 17.	+ 54 39		14.5	GALAXY
IC 3791	12 45 17.	+ 54 43 22.			NONSTELLAR OBJECT
SCH 74	12 45 18.	+ 04 36		14.5	PECULIAR GALAXY
ZWG 071.042	12 45 18.	+ 10 09		15.7	GALAXY
ZWG 071.043	12 45 18.	+ 14 02		12.8	GALAXY
UGC 07965	12 45 18.	+ 14 02	240	12.8	GALAXY Sc
MCG+02-33-022	12 45 18.	+ 14 02	300	12.8	GALAXY
LB 07098	12 45 18.	+ 26 38		19.6	FAINT BLUE STAR
LB 07097	12 45 18.	+ 28 09		18.7	FAINT BLUE STAR
LB 07096	12 45 18.	+ 30 53		19.1	FAINT BLUE STAR
LB 11358	12 45 18.	+ 32 20		18.2	FAINT BLUE STAR
ZC 1245.3+4741	12 45 18.	+ 47 41	1080		CLUSTER OF GALAXIES
LB 02445	12 45 18.	+ 49 21 30.		15.8	FAINT BLUE STAR
ZWG 270.023	12 45 18.	+ 54 39		14.5	GALAXY
UGC 07966	12 45 18.	+ 54 39	72	14.5	GALAXY S
ZC 1245.3+6422	12 45 18.	+ 64 22	470		CLUSTER OF GALAXIES
MRK 226	12 45 18.	+ 72 11	7	17.5	GALAXY WITH UV CONTINUUM
ZWG 015.022	12 45 18.	- 00 55		15.7	GALAXY
UGC 07963	12 45 18.	- 00 55	90	15.7	GALAXY Sc
ZWG 015.021	12 45 18.	- 01 22		14.0	GALAXY
UGC 07964	12 45 18.	- 01 22	66	14.0	GALAXY E-S0
MCG+00-33-012	12 45 18.	- 01 22	18	14.0	GALAXY
MCG-04-30-020	12 45 18.	- 27 18	84	14.5	GALAXY
MCG-06-28-020	12 45 18.	- 39 21	48	15.	GALAXY
KN 16.391	12 45 18.2	+ 26 49 24.			NEBULA
IC 0822	12 45 19.	+ 30 20			NONSTELLAR OBJECT
IC 3787	12 45 19.	+ 42 53 40.			NONSTELLAR OBJECT
RNGC 4690	12 45 19.	- 01 22		14.0	GALAXY
RNGC 4696C	12 45 19.	- 40 32			GALAXY
KN 16.393	12 45 19.6	+ 30 21 00.			NEBULA
KN 16.392	12 45 20.5	+ 27 51 00.			NEBULA
REIZ 2823	12 45 21.	+ 04 37	108	13.2	GALAXY
RNGC 4689	12 45 21.	+ 14 02		12.0	GALAXY
MCG-07-26-048	12 45 21.	- 40 33	48	15.	GALAXY
KN 16.394	12 45 21.0	+ 28 13 12.			NEBULA
REIF 4.146	12 45 21.42	- 01 23 02.4			NEBULA
IC 3784	12 45 22.	+ 19 39 22.			NONSTELLAR OBJECT
IC 3785	12 45 23.	+ 19 32 52.			NONSTELLAR OBJECT
REIZ 2827	12 45 23.	+ 19 40	18	15.2	GALAXY
IC 0823	12 45 23.	+ 27 28			MAY NOT EXIST
ARC 1616	12 45 23.	+ 55 20		16.0	RICH CLUSTER OF GALAXIES
ARC 1617	12 45 23.	+ 59 29		17.6	RICH CLUSTER OF GALAXIES
ZWG 043.029	12 45 24.	+ 04 42		15.5	GALAXY
MCG+01-33-014	12 45 24.	+ 04 42	48	15.5	GALAXY
ZWG 159.078	12 45 24.	+ 27 30		14.0	GALAXY
UGC 07967	12 45 24.	+ 27 30	72	14.0	GALAXY E-S0
MCG+05-30-086	12 45 24.	+ 27 30	33	14.0	GALAXY
MCG+05-30-085	12 45 24.	+ 30 21	24	16.	GALAXY
ZC 1245.4+3129	12 45 24.	+ 31 29	1750		CLUSTER OF GALAXIES
ZC 1245.4+4031	12 45 24.	+ 40 31	1480		CLUSTER OF GALAXIES
TON-N 0127	12 45 24.	+ 46 06		15.6	BLUE STAR
ARC 1615	12 45 24.	+ 49 10		18.0	RICH CLUSTER OF GALAXIES
LB 02446	12 45 25.	+ 52 35 36.		16.9	FAINT BLUE STAR
REIZ 2822	12 45 25.	- 12 51	72	15.4	GALAXY
HOLM 462B	12 45 26.	- 09 54	42	14.9	PART OF MULTIPLE GALAXY
HOLM 461B	12 45 28.	+ 04 43	42	14.4	PART OF MULTIPLE GALAXY
REIZ 2826	12 45 28.	- 01 23	18	14.3	GALAXY
KN 16.395	12 45 28.6	+ 27 29 42.			NEBULA
RNGC 4692	12 45 29.	+ 27 30		14.0	GALAXY
REIZ 2825	12 45 29.	- 04 37	48	15.6	GALAXY
ZWG 043.030	12 45 30.	+ 04 58		15.7	GALAXY
KARA.68 201	12 45 30.	+ 08 35	67		DWARF GALAXY
REA 90	12 45 30.	+ 13 58			DWARF GALAXY
ZC 1245.5+1535	12 45 30.	+ 15 35	540		CLUSTER OF GALAXIES
LB 07105	12 45 30.	+ 28 29		19.0	FAINT BLUE STAR
LB 07104	12 45 30.	+ 29 14		19.5	FAINT BLUE STAR
LB 07103	12 45 30.	+ 30 00		19.8	FAINT BLUE STAR
LB 07102	12 45 30.	+ 30 41		18.4	FAINT BLUE STAR
LB 07101	12 45 30.	+ 30 58		20.2	FAINT BLUE STAR
LB 07100	12 45 30.	+ 31 09		19.0	FAINT BLUE STAR
LB 07099	12 45 30.	+ 32 30		18.5	FAINT BLUE STAR
ZC 1245.5+4909	12 45 30.	+ 49 09	1010		CLUSTER OF GALAXIES
REIZ 2824	12 45 30.	- 09 53	18	15.9	GALAXY
HOLM 462A	12 45 30.	- 09 54	54	14.3	PART OF MULTIPLE GALAXY
DV.56 N4696B	12 45 30.	- 40 58	84		S0 GALAXY
ACK 302-00.1	12 45 30.	- 63 34			PLANETARY NEBULA
AMES 2149	12 45 31.	+ 13 08 21.	25	17.4	NEBULA
REIZ 2841	12 45 32.	+ 56 00	42	14.4	GALAXY
REIZ 2830	12 45 33.	+ 04 43	48	14.2	GALAXY
AMES 2150	12 45 35.	+ 13 17 03.	29	17.6	NEBULA
REIZ 2835	12 45 35.	+ 19 08	24	14.9	GALAXY
REIZ 2836	12 45 35.	+ 20 53	9	15.4	GALAXY
REIZ 2829	12 45 35.	- 04 05	42	15.4	GALAXY
ZWG 043.031	12 45 36.	+ 07 15		15.7	GALAXY
KARA.68 202	12 45 36.	+ 08 32	27		DWARF GALAXY
8ZW 1245+18.3	12 45 36.	+ 18 18		18.3	COMPACT GALAXY
MCG+03-33-005	12 45 36.	+ 19 08	27	17.	GALAXY
ZC 1245.6+2300	12 45 36.	+ 23 00	2150		CLUSTER OF GALAXIES
LB 07112	12 45 36.	+ 26 33		19.5	FAINT BLUE STAR
LB 07111	12 45 36.	+ 26 50		18.8	FAINT BLUE STAR
LB 07110	12 45 36.	+ 27 15		19.8	FAINT BLUE STAR
LB 07109	12 45 36.	+ 27 42		20.1	FAINT BLUE STAR
TON-N 0647	12 45 36.	+ 28 04		16.5	BLUE STAR
LB 07108	12 45 36.	+ 28 50		18.7	FAINT BLUE STAR
LB 11359	12 45 36.	+ 29 09		15.5	FAINT BLUE STAR
LB 07107	12 45 36.	+ 29 23		18.9	FAINT BLUE STAR
LB 07106	12 45 36.	+ 29 41		18.8	FAINT BLUE STAR
MRK 443	12 45 36.	+ 33 36	10	16.5	GALAXY WITH UV CONTINUUM
ZWG 216.030	12 45 36.	+ 41 00		15.7	GALAXY
7ZW 485	12 45 36.	+ 67 39			COMPACT GALAXY
ZWG 015.023	12 45 36.	- 03 03		12.0	GALAXY
MCG+00-33-013	12 45 36.	- 03 03	132	12.0	GALAXY
MCG-01-33-008	12 45 36.	- 04 04	60	15.5	GALAXY
REIZ 2828	12 45 36.	- 09 54	36	15.5	GALAXY
MCG-02-33-010	12 45 36.	- 09 54	72	15.5	GALAXY
MCG-04-30-021	12 45 36.	- 26 12	78	14.	GALAXY
KN 16.396	12 45 36.0	+ 27 32 54.			NEBULA
RNGC 4691	12 45 37.	- 03 04		12.0	GALAXY
RNGC 4696D	12 45 37.	- 41 26			GALAXY
KN 16.397	12 45 37.2	+ 28 26 30.			NEBULA
A2 031	12 45 37.9	+ 28 32 32.		16.4	FAINT BLUE OBJECT
A2 032	12 45 38.0	+ 28 37 30.		17.9	FAINT BLUE OBJECT
KN 16.398	12 45 38.2	+ 26 55 00.			NEBULA
REIN 4.147	12 45 38.96	- 03 03 36.8			NEBULA
IC 3788	12 45 39.	+ 19 08 28.			NONSTELLAR OBJECT
IC 3789	12 45 39.	+ 20 27 58.			NONSTELLAR OBJECT
KN 16.399	12 45 39.8	+ 29 43 00.			NEBULA
REIZ 2834	12 45 40.	- 01 31	24	15.7	GALAXY
REIZ 2833	12 45 40.	- 03 03	60	12.1	GALAXY
HELW 271	12 45 40.	- 40 39 36.			NEBULA
REIZ 2839	12 45 41.	+ 19 26	24	15.0	GALAXY
REIZ 2832	12 45 41.	- 04 28	24	14.0	GALAXY
TON-N 0129	12 45 42.	+ 04 07		16.3	BLUE STAR
TON-N 0128	12 45 42.	+ 14 48		14.6	BLUE STAR
8ZW 1245+13.0	12 45 42.	+ 18 03		16.5	COMPACT GALAXY
LB 07118	12 45 42.	+ 26 44		19.5	FAINT BLUE STAR
LB 07117	12 45 42.	+ 27 33		19.4	FAINT BLUE STAR
LB 11360	12 45 42.	+ 28 08		13.3	FAINT BLUE STAR
LB 07116	12 45 42.	+ 28 10		19.0	FAINT BLUE STAR
LB 07115	12 45 42.	+ 28 15		18.9	FAINT BLUE STAR
MCG+05-30-087	12 45 42.	+ 29 44	42	15.5	GALAXY
LB 07114	12 45 42.	+ 30 00		18.9	FAINT BLUE STAR
LB 07113	12 45 42.	+ 31 51		19.6	FAINT BLUE STAR
MCG+07-26-053	12 45 42.	+ 40 59	27	16.	GALAXY

OBJECT NAME	RIGHT ASCEN.	DECLINATION	DIAM.	MAGN.	TYPE OF OBJECT
UGC 07968	12 45 42.	+ 60 34	72	16.5	GALAXY
MCG-07-26-049	12 45 42.	- 41 27 30.	66	13.5	GALAXY
KN 16.400	12 45 42.7	+ 27 12 42.			NEBULA
REIZ 2838	12 45 43.	+ 11 16	54	13.9	GALAXY
IC 3793	12 45 43.	+ 19 25 22.			NONSTELLAR OBJECT
HELW 442	12 45 43.	- 11 20 30.			NEBULA
RNGC 4696E	12 45 43.	- 40 39			GALAXY
KN 16.401	12 45 43.9	+ 27 55 42.			NEBULA
RNGC 4694	12 45 44.	+ 11 15		12.5	GALAXY
IC 3790	12 45 44.	+ 11 22 52.			SINGLE STAR
IC 3795	12 45 44.	+ 40 59 28.			NONSTELLAR OBJECT
KN 16.402	12 45 44.3	+ 30 14 00.			NEBULA
REIN 3.330	12 45 44.33	+ 11 15 20.7			NEBULA
IC 3792	12 45 45.	+ 11 21 16.			MAY NOT EXIST
MCG-07-26-050	12 45 45.	- 40 41	42	15.	GALAXY
AMES 2151	12 45 46.	+ 10 50 15.	47	16.4	NEBULA
REIZ 2837	12 45 46.	- 03 34	30	15.7	GALAXY
KN 16.403	12 45 46.7	+ 27 11 30.			NEBULA
KARA.68 203	12 45 48.	+ 10 01	47		DWARF GALAXY
ZWG 071.044	12 45 48.	+ 11 15		12.4	GALAXY
UGC 07969	12 45 48.	+ 11 15	210	12.4	GALAXY SB?0
MCG+02-33-023	12 45 48.	+ 11 15	90	12.4	GALAXY
ZC 1245.8+1355	12 45 48.	+ 13 55	470		CLUSTER OF GALAXIES
8ZW 1245+22.5	12 45 48.	+ 22 30		17.7	COMPACT GALAXY
LB 07124	12 45 48.	+ 26 59		18.8	FAINT BLUE STAR
LB 07123	12 45 48.	+ 27 08		18.7	FAINT BLUE STAR
TON-N 0648	12 45 48.	+ 28 03		17.0	BLUE STAR
LB 07122	12 45 48.	+ 28 07		19.8	FAINT BLUE STAR
LB 07121	12 45 48.	+ 29 10		19.5	FAINT BLUE STAR
LB 07120	12 45 48.	+ 31 11		18.8	FAINT BLUE STAR
LB 07119	12 45 48.	+ 32 24		18.7	FAINT BLUE STAR
MCG+14-06-008	12 45 48.	+ 83 40	39	15.	GALAXY
MCG-01-33-009	12 45 48.	- 04 57	24	17.	GALAXY
MCG-03-33-008	12 45 48.	- 20 02	72	15.	GALAXY
A2 035	12 45 49.0	+ 27 47 05.		17.5	FAINT BLUE OBJECT
KN 16.404	12 45 49.4	+ 29 30 36.			NEBULA
LB 02447	12 45 50.	+ 59 00 54.		17.2	FAINT BLUE STAR
REIN 3.331	12 45 51.27	+ 08 45 36.2			NEBULA
REIN 2.161	12 45 51.29	+ 08 45 37.0			NEBULA
IC 3794	12 45 52.	+ 19 26 35.			NONSTELLAR OBJECT
REIZ 2840	12 45 52.	- 03 11	30	15.7	GALAXY
KN 16.405	12 45 52.1	+ 29 13 30.			NEBULA
REIZ 2843	12 45 53.	+ 19 27	12	16.0	GALAXY
ZWG 043.032	12 45 54.	+ 03 42		15.6	GALAXY
ZWG 071.045	12 45 54.	+ 08 45		12.1	GALAXY
UGC 07970	12 45 54.	+ 08 45	258	12.1	GALAXY Sa
MCG+02-33-024	12 45 54.	+ 08 45	240	12.1	GALAXY
KARA.68 204	12 45 54.	+ 08 49	34		DWARF GALAXY
ZC 1245.9+2511	12 45 54.	+ 25 11	4370		CLUSTER OF GALAXIES
MCG+05-30-088	12 45 54.	+ 27 07	54	15.2	GALAXY
ZWG 159.079	12 45 54.	+ 27 08		15.2	GALAXY
LB 07129	12 45 54.	+ 28 08		19.9	FAINT BLUE STAR
LB 07128	12 45 54.	+ 28 28		20.1	FAINT BLUE STAR
LB 07127	12 45 54.	+ 29 01		18.5	FAINT BLUE STAR
LB 07126	12 45 54.	+ 29 44		18.8	FAINT BLUE STAR
LB 07125	12 45 54.	+ 31 02		19.2	FAINT BLUE STAR
KN 16.406	12 45 54.7	+ 29 11 48.			NEBULA
RNGC 4698	12 45 56.	+ 08 46		12.0	GALAXY
MCG-01-33-010	12 45 57.	- 05 31	132	11.	GALAXY
KN 16.407	12 45 57.3	+ 27 08 00.			NEBULA
REIN 4.148	12 45 57.32	- 03 11 47.7			NEBULA
A2 038	12 45 57.8	+ 29 26 38.		18.4	FAINT BLUE OBJECT
AMES 2152	12 45 58.	+ 13 34 51.	36	17.4	NEBULA
KN 16.408	12 45 58.6	+ 28 34 42.			NEBULA
IC 3796	12 45 59.	+ 20 18 35.			NONSTELLAR OBJECT
REIZ 2842	12 45 59.	- 05 31	210	11.1	GALAXY
VDB.66G 150	12 46	+ 51 24	100		DWARF GALAXY
VDB.66G 149	12 46	- 03 46	70		DWARF GALAXY
LB 09863	12 46	- 81 40		14.8	FAINT BLUE STAR
ZWG 071.046	12 46 00.	+ 10 50		15.7	GALAXY
ZC 1246.0+1723	12 46 00.	+ 17 23	870		CLUSTER OF GALAXIES
LB 07133	12 46 00.	+ 29 06		18.4	FAINT BLUE STAR
LB 07132	12 46 00.	+ 29 32		19.3	FAINT BLUE STAR
LB 07131	12 46 00.	+ 30 00		19.6	FAINT BLUE STAR
LB 07130	12 46 00.	+ 31 08		19.6	FAINT BLUE STAR
ZC 1246.0+3924	12 46 00.	+ 39 24	3020		CLUSTER OF GALAXIES
ZC 1246.0+4427	12 46 00.	+ 44 27	740		CLUSTER OF GALAXIES
MRK 227	12 46 00.	+ 51 16	8	16.	GALAXY WITH UV CONTINUUM
ZWG 270.024	12 46 00.	+ 54 18		14.9	GALAXY
MCG+10-18-083	12 46 00.	+ 60 32	54	15.	GALAXY
ZWG 015.024	12 46 00.	- 03 11		15.6	GALAXY
KN 16.410	12 46 00.2	+ 30 31 12.			NEBULA
KN 16.409	12 46 00.6	+ 27 51 12.			NEBULA
RNGC 4697	12 46 01.	- 05 32		10.5	GALAXY
KN 16.411	12 46 01.8	+ 29 30 43.			NEBULA
REIZ 2844	12 46 02.	+ 08 46	180	12.2	GALAXY
REIZ 2845	12 46 03.	+ 26 23	12	15.8	GALAXY
IC 3797	12 46 04.	+ 11 52 11.			MAY NOT EXIST
IC 3800	12 46 04.	+ 36 50 59.			NONSTELLAR OBJECT
RNGC 4707	12 46 04.	+ 51 27		15.0	GALAXY
KN 16.412	12 46 04.5	+ 29 31 55.			NEBULA
REIN 3.332	12 46 04.73	+ 11 08 51.5			NEBULA
AMES 2153	12 46 05.	+ 11 09 09.	42	16.4	NEBULA
ZWG 015.025	12 46 06.	+ 01 06		15.3	GALAXY
ZWG 071.047	12 46 06.	+ 09 24		15.7	GALAXY
MCG+02-33-026	12 46 06.	+ 09 24	15	15.7	GALAXY
MCG+02-33-025	12 46 06.	+ 09 24	36	15.7	GALAXY
ZWG 071.048	12 46 06.	+ 11 11		15.6	GALAXY
LB 00017	12 46 06.	+ 23 02		14.0	FAINT BLUE STAR
LB 11361	12 46 06.	+ 27 10		16.6	FAINT BLUE STAR
ZC 1246.1+4344	12 46 06.	+ 43 44	1410		CLUSTER OF GALAXIES
1ZW 043	12 46 06.	+ 51 27			COMPACT GALAXY
ZWG 270.025	12 46 06.	+ 51 27		15.2	GALAXY
UGC 07971	12 46 06.	+ 51 27	150	15.2	GALAXY DWRF SP
LB 02448	12 46 06.	+ 59 37 12.		16.4	FAINT BLUE STAR
KARA.68 205	12 46 06.	- 04 57	81		DWARF GALAXY
MCG-07-26-051	12 46 06.	- 41 03	180	12.2	GALAXY
KN 16.414	12 46 06.1	+ 30 29 55.			NEBULA
KN 16.413	12 46 06.2	+ 26 23 13.			NEBULA
REIN 3.333	12 46 06.20	+ 08 45 56.0			NEBULA
LB 02449	12 46 07.	+ 58 36 54.		15.4	FAINT BLUE STAR
RNGC 4696	12 46 07.	- 41 02		12.0	GALAXY
KN 16.415	12 46 08.9	+ 26 24 25.			NEBULA
KN 16.416	12 46 08.9	+ 29 06 55.			NEBULA
A2 042	12 46 10.5	+ 29 58 57.		17.5	FAINT BLUE OBJECT
IC 3798	12 46 12.	+ 09 30 47.			NONSTELLAR OBJECT
MCG+05-30-089	12 46 12.	+ 26 41	48	15.7	GALAXY
ZWG 159.080	12 46 12.	+ 26 42		15.7	GALAXY
LB 07139	12 46 12.	+ 27 28		19.8	FAINT BLUE STAR
LB 07138	12 46 12.	+ 27 30		19.6	FAINT BLUE STAR
LB 07137	12 46 12.	+ 27 38		19.8	FAINT BLUE STAR
TON-N 0649	12 46 12.	+ 27 39		17.0	BLUE STAR
LB 07136	12 46 12.	+ 27 53		18.6	FAINT BLUE STAR
LB 07135	12 46 12.	+ 30 15		20.2	FAINT BLUE STAR
LB 07134	12 46 12.	+ 32 29		18.2	FAINT BLUE STAR
MCG+09-21-050	12 46 12.	+ 51 25	120	14.	GALAXY
MCG+09-21-049	12 46 12.	+ 54 29	42	16.	GALAXY
ZC 1246.2+6855	12 46 12.	+ 68 55	1280		CLUSTER OF GALAXIES
MCG-01-33-011	12 46 12.	- 04 58	90	15.	GALAXY
DV.56 N4696C	12 46 12.	- 40 32	108		S GALAXY
KN 16.417	12 46 12.2	+ 29 27 49.			NEBULA
KN 16.417A	12 46 14.3	+ 26 03 25.			NEBULA
REIZ 2846	12 46 15.	+ 26 41	36	15.1	GALAXY
KN 16.418	12 46 15.4	+ 26 41 19.			NEBULA
KN 16.419	12 46 16.7	+ 29 30 25.			NEBULA
HARO 37	12 46 17.	+ 34 45			BLUE EMISSION-LINE GALAXY
TON-N 0650	12 46 18.	+ 27 40		16.7	BLUE STAR
LB 07142	12 46 18.	+ 28 00		19.3	FAINT BLUE STAR
LB 11362	12 46 18.	+ 31 46		12.7	FAINT BLUE STAR
LB 07141	12 46 18.	+ 31 51		19.8	FAINT BLUE STAR
LB 07140	12 46 18.	+ 32 22		17.6	FAINT BLUE STAR
MRK 444	12 46 18.	+ 34 46	10	15.5	GALAXY WITH UV CONTINUUM
REIZ 2850	12 46 18.	+ 35 37	120	13.5	GALAXY
ZWG 216.031	12 46 18.	+ 42 12		14.8	GALAXY
MRK 228	12 46 18.	+ 42 12	6	16.5	GALAXY WITH UV CONTINUUM
UGC 07972	12 46 18.	+ 42 12	60	14.8	GALAXY SBb
ZC 1246.3+5900	12 46 18.	+ 59 00	870		CLUSTER OF GALAXIES
HELW 443	12 46 18.	- 11 24 11.			NEBULA
IC 3802	12 46 20.	+ 38 31 05.			NONSTELLAR OBJECT
KN 16.420	12 46 20.0	+ 27 01 13.			NEBULA
REIN 3.334	12 46 20.59	+ 08 24 57.6			NEBULA
MCG+06-28-032	12 46 21.	+ 34 46	12	16.	GALAXY
IC 3805	12 46 21.	+ 38 31 29.			NONSTELLAR OBJECT
MCG-02-33-011	12 46 21.	- 14 07	150	14.	GALAXY
KN 16.421	12 46 21.2	+ 25 50 31.			NEBULA
KN 16.422	12 46 21.8	+ 27 18 25.			NEBULA
AMES 2154	12 46 22.	+ 16 38 57.	28	16.7	NEBULA
IC 3799	12 46 22.	- 14 07 31.			NONSTELLAR OBJECT
KN 16.423	12 46 22.2	+ 26 53 43.			NEBULA
IC 3804	12 46 23.	+ 35 36 17.			SAME AS NGC 4711
KN 16.424	12 46 23.1	+ 27 18 37.			NEBULA
ZWG 071.049	12 46 24.	+ 10 38		15.6	GALAXY
ZC 1246.4+1109	12 46 24.	+ 11 09	1210		CLUSTER OF GALAXIES
8ZW 1246+17.2	12 46 24.	+ 17 13		18.5	COMPACT GALAXY
LB 07148	12 46 24.	+ 26 27		19.3	FAINT BLUE STAR
LB 07147	12 46 24.	+ 27 56		19.0	FAINT BLUE STAR
LB 07146	12 46 24.	+ 28 39		20.0	FAINT BLUE STAR
LB 07145	12 46 24.	+ 29 20		19.0	FAINT BLUE STAR
LB 07144	12 46 24.	+ 30 32		20.1	FAINT BLUE STAR
LB 07143	12 46 24.	+ 30 40		17.4	FAINT BLUE STAR
ZWG 188.022	12 46 24.	+ 35 36		14.4	GALAXY
UGC 07973	12 46 24.	+ 35 36	84	14.4	GALAXY S
TON-N 0130	12 46 24.	+ 42 11		15.7	BLUE STAR
MCG+07-26-054	12 46 24.	+ 42 11	60	14.	GALAXY
ZC 1246.4+6614	12 46 24.	+ 66 14	1410		CLUSTER OF GALAXIES
ZC 1246.4+7859	12 46 24.	+ 78 59	1210		CLUSTER OF GALAXIES
MCG-02-33-012	12 46 24.	- 12 36 30.	48	15.	GALAXY
MCG-07-26-052	12 46 24.	- 41 35	36	15.5	GALAXY
KN 16.425	12 46 24.2	+ 26 54 37.			NEBULA
RNGC 4704	12 46 25.	+ 42 12		15.0	GALAXY
KN 16.427	12 46 25.1	+ 29 14 25.			NEBULA
IC 3806	12 46 25.2	+ 15 10 48.			GALAXY
KN 16.426	12 46 25.6	+ 26 53 55.			NEBULA
BFG 008	12 46 26.24	+ 35 41 44.7		15.85	ULTRAVIOLET-EXCESS OBJECT
KN 16.428	12 46 26.5	+ 26 56 07.			NEBULA
MAI 079	12 46 27.	+ 75 43	33		DWARF SPHEROIDAL GALAXY
MCG-01-33-012	12 46 27.	- 09 12	42	15.5	GALAXY
REIZ 2855	12 46 28.	+ 42 12	60	14.9	GALAXY
BC BS01	12 46 28.7	+ 37 47 01.		16.98	QUASI-STELLAR OBJECT
SHB 214	12 46 28.7	+ 37 47 01.		17.9	QUASI-STELLAR OBJECT
BFG 009	12 46 28.74	+ 37 46 49.7		17.06	ULTRAVIOLET-EXCESS OBJECT
BFG 010	12 46 28.92	+ 33 34 12.3		19.03	ULTRAVIOLET-EXCESS OBJECT
ARC 1618	12 46 29.	+ 11 09		17.5	RICH CLUSTER OF GALAXIES
BSO 01	12 46 29.	+ 37 46 25.		16.98	BLUE STELLAR OBJECT
BC B46	12 46 29.2	+ 34 40 49.		17.83	QUASI-STELLAR OBJECT
SHB 215	12 46 29.6	+ 34 40 49.		17.8	QUASI-STELLAR OBJECT
BFG 011	12 46 29.64	+ 34 40 49.3		17.75	ULTRAVIOLET-EXCESS OBJECT
ZC 1246.5+0623	12 46 30.	+ 06 23	610		CLUSTER OF GALAXIES
ZWG 043.033	12 46 30.	+ 07 31		15.7	GALAXY
IC 3801	12 46 30.	+ 11 13 41.			MAY NOT EXIST
MCG+03-33-006	12 46 30.	+ 15 09	84	14.6	GALAXY
ZWG 100.008	12 46 30.	+ 15 10		14.6	GALAXY
UGC 07974	12 46 30.	+ 15 10	84	14.6	GALAXY S
ZC 1246.5+1940	12 46 30.	+ 19 40	810		CLUSTER OF GALAXIES
ZC 1246.5+2020	12 46 30.	+ 20 20	610		CLUSTER OF GALAXIES
MRK 657	12 46 30.	+ 27 28	12	15.5	GALAXY WITH UV CONTINUUM
MCG+05-30-090	12 46 30.	+ 27 38	48	16.	GALAXY
LB 07149	12 46 30.	+ 29 13		18.7	FAINT BLUE STAR
RNGC 4711	12 46 30.	+ 35 37		14.5	GALAXY
MCG+06-28-033	12 46 30.	+ 35 37	72	13.	GALAXY
ZWG 216.032	12 46 30.	+ 40 52		15.7	GALAXY
MRK 445	12 46 30.	+ 40 52	24	15.5	GALAXY WITH UV CONTINUUM
MRK 229	12 46 30.	+ 47 58	8	17.	GALAXY WITH UV CONTINUUM
REIZ 2847	12 46 30.	- 08 23	216	10.6	GALAXY
SN 1948A	12 46 30.	- 08 23		17.0	SUPERNOVA
RNGC 4699	12 46 30.	- 08 24		11.0	GALAXY
MCG-01-33-013	12 46 30.	- 08 24	198	11.	GALAXY
REIZ 2849	12 46 30.	- 09 12	24	15.2	GALAXY
RNGC 4700	12 46 30.	- 11 08		12.5	GALAXY
REIZ 2848	12 46 30.	- 11 08	150	12.0	GALAXY
MCG-02-33-013	12 46 30.	- 11 08	168	12.	GALAXY
DV.56 N4696D	12 46 30.	- 41 26	84		SO GALAXY
KEEL 538	12 46 31.3	+ 41 22 22.		18.	NEBULA
KN 16.429	12 46 31.4	+ 29 24 43.			NEBULA
REIZ 2854	12 46 32.	+ 31 12	42	14.9	GALAXY
KN 16.430	12 46 32.2	+ 28 06 37.			NEBULA
KN 16.431	12 46 32.9	+ 27 38 49.			NEBULA
MCG+08-23-094	12 46 33.	+ 47 58	48	16.	GALAXY
MCG+08-23-093	12 46 33.	+ 49 35 30.	36	16.	GALAXY
ARC 1621	12 46 33.	+ 62 58		16.5	RICH CLUSTER OF GALAXIES
IC 3803	12 46 34.	+ 10 54 12.			NONSTELLAR OBJECT
REIZ 2851	12 46 34.	- 01 29	18	15.8	GALAXY
RNGC 4702	12 46 35.	+ 27 27		15.0	GALAXY
KN 16.432	12 46 35.1	+ 27 27 07.			NEBULA
KN 16.433	12 46 35.9	+ 27 09 37.			NEBULA
HN 1039	12 46 36.	+ 15 11	12		NEBULA

OBJECT NAME	RIGHT ASCEN.	DECLINATION	DIAM.	MAGN.	TYPE OF OBJECT
MCG+03-33-007	12 46 36.	+ 17 16	18	15.5	GALAXY
ZWG 100.009	12 46 36.	+ 17 17		15.5	GALAXY
LB 00018	12 46 36.	+ 27 04		13.0	FAINT BLUE STAR
MCG+05-30-091	12 46 36.	+ 27 27	30	15.5	GALAXY
LB 07152	12 46 36.	+ 29 24		19.4	FAINT BLUE STAR
LB 07151	12 46 36.	+ 30 22		19.4	FAINT BLUE STAR
ZWG 159.081	12 46 36.	+ 31 11		15.5	GALAXY
MCG+05-30-092	12 46 36.	+ 31 13	42	15.5	GALAXY
LB 07150	12 46 36.	+ 31 21		18.8	FAINT BLUE STAR
ZWG 188.023	12 46 36.	+ 35 08		15.7	GALAXY
MCG+07-26-055	12 46 36.	+ 40 52	36	15.	GALAXY
MCG+09-21-051	12 46 36.	+ 51 54	36	16.	GALAXY
ZC 1246.6+6126	12 46 36.	+ 61 26	1810		CLUSTER OF GALAXIES
DV.56 N4696E	12 46 36.	- 40 39	108		SO GALAXY
AMES 2155	12 46 37.	+ 17 17 08.	30	16.4	NEBULA
IC 3808	12 46 37.	+ 40 52 00.			NONSTELLAR OBJECT
HELW 272	12 46 37.	- 41 14 23.			NEBULA
RNGC 4701	12 46 38.	+ 03 40		13.0	GALAXY
REIZ 2853	12 46 39.	+ 03 40	42	12.6	GALAXY
KN 16.434	12 46 39.0	+ 27 54 43.			NEBULA
BPG 012	12 46 39.55	+ 35 29 22.3		18.42	ULTRAVIOLET-EXCESS OBJECT
HELW 273	12 46 41.	- 41 12 29.			NEBULA
KN 16.435	12 46 41.3	+ 28 04 07.			NEBULA
ZWG 043.034	12 46 42.	+ 03 40		13.1	GALAXY
UGC 07975	12 46 42.	+ 03 40	216	13.1	GALAXY Sc
MCG+01-33-015	12 46 42.	+ 03 40	150	13.1	GALAXY
ZWG 043.035	12 46 42.	+ 04 56		15.5	GALAXY
UGC 07976	12 46 42.	+ 04 56	60	15.5	GALAXY
MCG+01-33-016	12 46 42.	+ 04 56	60	15.5	GALAXY
MCG+03-33-008	12 46 42.	+ 17 57	15	15.7	GALAXY
ZWG 100.010	12 46 42.	+ 17 58		15.7	GALAXY
ZC 1246.7+2429	12 46 42.	+ 24 29	940		CLUSTER OF GALAXIES
LB 07157	12 46 42.	+ 26 55		19.5	FAINT BLUE STAR
LB 07156	12 46 42.	+ 27 02		19.0	FAINT BLUE STAR
LB 07155	12 46 42.	+ 28 16		19.5	FAINT BLUE STAR
LB 07153	12 46 42.	+ 29 15		19.5	FAINT BLUE STAR
IC 3809	12 46 42.	+ 36 45 42.			NONSTELLAR OBJECT
IC 3810	12 46 42.	+ 40 55 06.			NONSTELLAR OBJECT
REIZ 2852	12 46 42.	- 09 52	180	12.7	GALAXY
MCG-04-30-022	12 46 42.	- 22 44 30.	24	15.5	GALAXY
HELW 274	12 46 42.	- 41 10 41.			NEBULA
KN 16.436	12 46 43.0	+ 27 22 19.			NEBULA
REIZ 2857	12 46 44.	+ 04 56	42	15.0	GALAXY
MCG-01-33-014	12 46 45.	- 03 43	84	15.5	GALAXY
MCG-01-33-015	12 46 45.	- 08 51	138	14.5	GALAXY
MCG-02-33-014	12 46 45.	- 13 04	54	15.	GALAXY
KN 16.437	12 46 45.1	+ 27 39 31.			NEBULA
SVEN 333	12 46 47.	- 06 10	24	15.3	GALAXY
A2 050	12 46 47.3	+ 29 10 27.		17.7	FAINT BLUE OBJECT
ZC 1246.8+2039	12 46 48.	+ 20 39	810		CLUSTER OF GALAXIES
LB 00054	12 46 48.	+ 26 07		15.7	FAINT BLUE STAR
LB 11363	12 46 48.	+ 26 58		15.0	FAINT BLUE STAR
LB 07168	12 46 48.	+ 27 06		18.7	FAINT BLUE STAR
LB 07167	12 46 48.	+ 27 40		18.7	FAINT BLUE STAR
LB 07166	12 46 48.	+ 28 48		18.5	FAINT BLUE STAR
LB 07165	12 46 48.	+ 29 06		18.8	FAINT BLUE STAR
TON-N 0651	12 46 48.	+ 29 07		15.1	BLUE STAR
LB 07164	12 46 48.	+ 29 20		20.2	FAINT BLUE STAR
LB 07163	12 46 48.	+ 29 30		19.5	FAINT BLUE STAR
LB 07162	12 46 48.	+ 30 08		20.3	FAINT BLUE STAR
LB 07161	12 46 48.	+ 30 16		20.1	FAINT BLUE STAR
LB 07160	12 46 48.	+ 32 06		19.4	FAINT BLUE STAR
LB 07159	12 46 48.	+ 32 15		19.5	FAINT BLUE STAR
LB 07158	12 46 48.	+ 32 21		18.9	FAINT BLUE STAR
ZC 1246.8+4121	12 46 48.	+ 41 21	2490		CLUSTER OF GALAXIES
ZC 1246.8+6448	12 46 48.	+ 64 48	1210		CLUSTER OF GALAXIES
MCG-01-33-016	12 46 48.	- 04 54	150	14.	GALAXY
RNGC 4703	12 46 48.	- 08 51		14.5	GALAXY
REIZ 2856	12 46 48.	- 08 51	72	14.2	GALAXY
MCG-02-33-015	12 46 48.	- 09 50	270	13.	GALAXY
SA 3	12 46 48.	- 09 51	210	12.5	GALAXY
HELW 275	12 46 48.	- 40 46 17.			NEBULA
BPG 013	12 46 48.41	+ 34 23 22.6		17.37	ULTRAVIOLET-EXCESS OBJECT
KN 16.438	12 46 49.1	+ 29 19 08.			NEBULA
KN 16.438A	12 46 49.7	+ 28 46 08.			NEBULA
ARC 1619	12 46 50.	+ 28 48		17.8	RICH CLUSTER OF GALAXIES
MCG-07-26-053	12 46 51.	- 41 38	36	15.5	GALAXY
A2 051	12 46 51.0	+ 30 07 45.		17.0	FAINT BLUE OBJECT
LB 02450	12 46 52.	+ 51 35 48.		16.3	FAINT BLUE STAR
KEEL 539	12 46 52.4	+ 41 21 54.		15.	SPIRAL NEBULA
A2 052	12 46 52.6	+ 28 34 00.		18.0	FAINT BLUE OBJECT
BPG 014	12 46 52.98	+ 32 20 07.9		18.64	ULTRAVIOLET-EXCESS OBJECT
REIZ 2859	12 46 53.	- 04 08	12	15.1	GALAXY
REIZ 2858	12 46 53.	- 04 55	120	13.6	GALAXY
HELW 276	12 46 53.	- 41 06 35.			NEBULA
KEEL 540	12 46 53.7	+ 41 25 23.		18.	NEBULA
ZWG 071.050	12 46 54.	+ 10 05		15.7	GALAXY
KARA.73B 0552	12 46 54.	+ 10 05	54	15.7	ISOLATED GALAXY S
ZC 1246.9+1635	12 46 54.	+ 16 35	610		CLUSTER OF GALAXIES
LB 00055	12 46 54.	+ 25 40		17.5	FAINT BLUE STAR
LB 07171	12 46 54.	+ 26 58		17.8	FAINT BLUE STAR
LB 07170	12 46 54.	+ 28 03		19.0	FAINT BLUE STAR
ZC 1246.9+2850	12 46 54.	+ 28 50	1210		CLUSTER OF GALAXIES
LB 11364	12 46 54.	+ 31 22		17.3	FAINT BLUE STAR
TON-N 0652	12 46 54.	+ 31 35		15.3	BLUE STAR
LB 07169	12 46 54.	+ 32 20		16.8	FAINT BLUE STAR
IC 3807	12 46 54.	- 04 07 48.			NONSTELLAR OBJECT
MCG-04-30-023	12 46 54.	- 24 53 30.	30	15.5	GALAXY
MCG-07-26-054	12 46 54.	- 41 08	66	15.	GALAXY
A2 054	12 46 54.7	+ 29 21 14.		18.4	FAINT BLUE OBJECT
LB 02451	12 46 55.	+ 56 08 06.		15.2	FAINT BLUE STAR
RNGC 4705	12 46 55.	- 04 55		14.0	GALAXY
AMES 2156	12 46 56.	+ 16 11 14.	34	17.4	NEBULA
IC 3811	12 46 57.	+ 21 44 06.			NONSTELLAR OBJECT
KN 16.439	12 46 57.7	+ 29 35 32.			NEBULA
REIZ 2862	12 46 58.	- 03 33	24	13.5	GALAXY
KN 16.441	12 46 58.1	+ 29 35 14.			NEBULA
KEEL 541	12 46 58.2	+ 41 23 09.		18.	NEBULA
KN 16.440	12 46 58.9	+ 27 03 38.			NEBULA
REIZ 2866	12 46 59.	+ 20 59	30	15.8	GALAXY
REIZ 2861	12 46 59.	- 03 50	18	13.7	GALAXY
KN 16.442	12 46 59.1	+ 29 12 32.			NEBULA
A2 056	12 46 59.4	+ 27 42 54.		18.4	FAINT BLUE OBJECT
VDB.66G 151	12 47	- 10 36	230		DWARF GALAXY
LB 09864	12 47	- 84 16		14.0	FAINT BLUE STAR
ZWG 043.036	12 47 00.	+ 03 52		15.4	GALAXY
ZC 1247.0+1806	12 47 00.	+ 18 06	1410		CLUSTER OF GALAXIES
LB 07176	12 47 00.	+ 27 34		19.0	FAINT BLUE STAR

OBJECT NAME	RIGHT ASCEN.	DECLINATION	DIAM.	MAGN.	TYPE OF OBJECT
LB 07175	12 47 00.	+ 29 43		19.3	FAINT BLUE STAR
LB 11366	12 47 00.	+ 30 31		12.4	FAINT BLUE STAR
LB 07174	12 47 00.	+ 30 31		17.0	FAINT BLUE STAR
LB 07173	12 47 00.	+ 30 45		18.7	FAINT BLUE STAR
LB 07172	12 47 00.	+ 30 58		19.7	FAINT BLUE STAR
LB 11365	12 47 00.	+ 31 54		11.6	FAINT BLUE STAR
MCG-01-33-017	12 47 00.	- 09 28	72	15.	GALAXY
REIZ 2860	12 47 00.	- 10 49	12	15.2	GALAXY
KEEL 542	12 47 00.9	+ 41 25 29.		18.	NEBULA
KN 16.443	12 47 01.3	+ 27 32 26.			NEBULA
SC 1244+0018.0	12 47 03.	+ 00 01 38.	6		NEBULA
RNGC 4710	12 47 03.	+ 15 26		12.0	GALAXY
REIZ 2869	12 47 03.	+ 26 16	12	15.9	GALAXY
A2 058	12 47 03.3	+ 27 52 52.		16.5	FAINT BLUE OBJECT
KN 16.443A	12 47 03.8	+ 29 29 08.			NEBULA
IC 3814	12 47 04.	+ 20 19 18.			NONSTELLAR OBJECT
RNGC 4712	12 47 04.	+ 25 45		13.0	GALAXY
LB 00244	12 47 04.	+ 57 34 24.		16.2	FAINT BLUE STAR
HOLM 463B	12 47 04.	- 10 49	12	14.5	PART OF MULTIPLE GALAXY
REIZ 2868	12 47 05.	+ 19 33	18	15.0	GALAXY
IC 3816	12 47 05.	+ 37 30 30.			NONSTELLAR OBJECT
REIZ 2865	12 47 05.	- 04 18	48	14.4	GALAXY
KN 16.444	12 47 05.5	+ 26 15 50.			NEBULA
ZWG 043.037	12 47 06.	+ 03 01		15.7	GALAXY
SCH 75	12 47 06.	+ 15 26		11.6	PECULIAR GALAXY
ZWG 129.025	12 47 06.	+ 25 44		13.5	GALAXY
UGC 07977	12 47 06.	+ 25 44	156	13.5	GALAXY Sc
MCG+04-30-021	12 47 06.	+ 25 44	138	13.5	GALAXY
LB 07178	12 47 06.	+ 29 50		17.0	FAINT BLUE STAR
LB 11367	12 47 06.	+ 29 52		14.5	FAINT BLUE STAR
ZWG 159.082	12 47 06.	+ 31 07		14.8	GALAXY
UGC 07978	12 47 06.	+ 31 07	78	14.8	GALAXY Sc
LB 07177	12 47 06.	+ 31 58		19.7	FAINT BLUE STAR
ZWG 216.033	12 47 06.	+ 41 58		15.3	GALAXY
MCG-01-33-018	12 47 06.	- 04 18	30	15.	GALAXY
REIZ 2864	12 47 06.	- 09 29	60	13.7	GALAXY
RNGC 4708	12 47 06.	- 10 49		14.0	GALAXY
REIZ 2863	12 47 06.	- 10 49	30	14.5	GALAXY
HOLM 463A	12 47 06.	- 10 49	30	13.9	PART OF MULTIPLE GALAXY
MCG-02-33-016	12 47 06.	- 10 49	72	14.	GALAXY
HELW 277	12 47 06.	- 40 56 34.			NEBULA
A2 060	12 47 06.4	+ 29 38 49.		17.2	FAINT BLUE OBJECT
IC 0824	12 47 07.	- 04 18 06.			NONSTELLAR OBJECT
KN 16.445	12 47 07.4	+ 25 44 38.			NEBULA
REIZ 2874	12 47 08.	+ 31 06	36	13.6	GALAXY
LB 00657	12 47 08.	+ 55 45 24.		17.1	FAINT BLUE STAR
KEEL 543	12 47 08.4	+ 41 33 04.		17.	NEBULA
KN 16.446	12 47 08.8	+ 27 28 26.			NEBULA
REIZ 2873	12 47 09.	+ 25 45	168	13.4	GALAXY
LB 00245	12 47 09.	+ 55 32 00.		16.3	FAINT BLUE STAR
KEEL 545	12 47 09.5	+ 41 07 29.		18.	NEBULA
IC 3815	12 47 10.	+ 19 32 42.			NONSTELLAR OBJECT
HOLM 468B	12 47 10.	+ 25 45	120	12.2	PART OF MULTIPLE GALAXY
LB 02453	12 47 10.	+ 56 30 24.		16.9	FAINT BLUE STAR
ARC 1620	12 47 10.	- 01 19		17.2	RICH CLUSTER OF GALAXIES
HOLM 464B	12 47 11.	+ 10 30	60	14.7	PART OF MULTIPLE GALAXY
HOLM 465B	12 47 11.	+ 32 20	36	15.9	PART OF MULTIPLE GALAXY
HN 0364	12 47 11.	- 06 27			NEBULA
IC 3812	12 47 11.	- 06 27			NONSTELLAR OBJECT
ZC 1247.2+0200	12 47 12.	+ 02 00	1010		CLUSTER OF GALAXIES
ZWG 043.038	12 47 12.	+ 04 07		15.7	GALAXY
ZWG 043.039	12 47 12.	+ 05 02		15.6	GALAXY
ZWG 071.051	12 47 12.	+ 12 54		15.3	GALAXY
UGC 07979	12 47 12.	+ 12 54	96	15.3	GALAXY
MCG+02-33-027	12 47 12.	+ 12 54	90	15.3	GALAXY
MCG+03-33-009	12 47 12.	+ 15 24	300	11.6	GALAXY
ZWG 100.011	12 47 12.	+ 15 26		11.6	GALAXY
UGC 07980	12 47 12.	+ 15 26	270	11.6	GALAXY S0-a
LB 07183	12 47 12.	+ 26 59		19.8	FAINT BLUE STAR
ZWG 159.083	12 47 12.	+ 27 10		14.9	GALAXY
LB 07182	12 47 12.	+ 27 33		19.2	FAINT BLUE STAR
LB 07181	12 47 12.	+ 28 02		19.4	FAINT BLUE STAR
UGC 07981	12 47 12.	+ 31 01	66	16.0	GALAXY Sc
MCG+05-30-093	12 47 12.	+ 31 08	72	14.8	GALAXY
LB 07180	12 47 12.	+ 31 50		17.1	FAINT BLUE STAR
LB 07179	12 47 12.	+ 32 27		16.8	FAINT BLUE STAR
MCG+06-28-034	12 47 12.	+ 37 31	30	15.	GALAXY
ZC 1247.2+4758	12 47 12.	+ 47 58	670		CLUSTER OF GALAXIES
ZC 1247.2+5008	12 47 12.	+ 50 08	1410		CLUSTER OF GALAXIES
TON-N 0131	12 47 12.	+ 52 50		15.6	BLUE STAR
ZC 1247.2+5400	12 47 12.	+ 54 00	1340		CLUSTER OF GALAXIES
REIZ 2867	12 47 12.	- 11 06	60	13.6	GALAXY
HELW 444	12 47 12.	- 11 06 46.			NEBULA
MCG-07-26-055	12 47 12.	- 41 01	60	15.	GALAXY
REIZ 2872	12 47 13.	+ 12 55	120	14.8	GALAXY
HOLM 465A	12 47 13.	+ 32 21	48	14.7	PART OF MULTIPLE GALAXY
RNGC 4706	12 47 13.	- 40 59			GALAXY
KN 16.448	12 47 13.0	+ 29 36 02.			NEBULA
A2 062	12 47 13.1	+ 28 43 43.		18.0	FAINT BLUE OBJECT
REIZ 2877	12 47 14.	+ 31 01	48	15.6	GALAXY
KEEL 544	12 47 14.0	+ 26 02 42.		16.	NEBULA
KN 16.447	12 47 14.3	+ 26 02 38.			NEBULA
KEEL 546	12 47 14.3	+ 41 14 21.		17.	NEBULA
REIZ 2871	12 47 15.	+ 03 07	210	13.8	GALAXY
REIZ 2876	12 47 15.	+ 26 04	42	15.2	GALAXY
KN 16.449	12 47 15.8	+ 27 09 56.			NEBULA
WRAY 19.38	12 47 15.9	- 61 18 26.		9.3	STAR-NEBULA ASSOCIATION
IC 3817	12 47 16.	+ 23 06 13.			NONSTELLAR OBJECT
IC 3820	12 47 16.	+ 37 23 31.			NONSTELLAR OBJECT
ARC 1622	12 47 16.	+ 50 07		18.0	RICH CLUSTER OF GALAXIES
SVEN 334	12 47 17.	- 06 25	54	14.0	GALAXY
REIZ 2870	12 47 17.	- 06 27	48	14.3	GALAXY
KEEL 547	12 47 17.2	+ 41 30 08.		18.	NEBULA
ZWG 043.040	12 47 18.	+ 03 08		14.6	GALAXY
UGC 07982	12 47 18.	+ 03 08	204	14.6	GALAXY S
MCG+01-33-017	12 47 18.	+ 03 08	210	14.6	GALAXY
UGC 07983	12 47 18.	+ 04 07	78	16.5	GALAXY DWRF IR
LB 00056	12 47 18.	+ 26 54		18.0	FAINT BLUE STAR
MCG+05-30-094	12 47 18.	+ 27 09	24	14.9	GALAXY
LB 07188	12 47 18.	+ 28 50		19.4	FAINT BLUE STAR
LB 07187	12 47 18.	+ 29 09		19.8	FAINT BLUE STAR
LB 07186	12 47 18.	+ 30 07		16.8	FAINT BLUE STAR
LB 07185	12 47 18.	+ 30 10		18.9	FAINT BLUE STAR
LB 07184	12 47 18.	+ 32 04		20.0	FAINT BLUE STAR
ZWG 159.084	12 47 18.	+ 32 20		15.2	GALAXY
LB 02454	12 47 18.	+ 51 07 06.		14.9	FAINT BLUE STAR
UGC 07984	12 47 18.	+ 69 54	66	16.5	GALAXY
MCG-01-33-019	12 47 18.	- 06 26	48	15.	GALAXY

OBJECT NAME	RIGHT ASCEN.	DECLINATION	DIAM.	MAGN.	TYPE OF OBJECT
A2 064	12 47 18.2	+ 29 04 30.		16.8	FAINT BLUE OBJECT
BPG 015	12 47 18.30	+ 35 05 48.7		19.06	ULTRAVIOLET-EXCESS OBJECT
KEEL 548	12 47 18.4	+ 41 09 31.		18.	NEBULA
KN 16.450	12 47 18.9	+ 28 52 08.			NEBULA
HOLM 464A	12 47 19.	+ 10 32	60	14.7	PART OF MULTIPLE GALAXY
IC 3818	12 47 19.	+ 22 01 25.			NONSTELLAR OBJECT
RNGC 4709	12 47 19.	- 41 06			GALAXY
A2 065	12 47 20.5	+ 29 18 32.		17.8	FAINT BLUE OBJECT
MCG+08-23-095	12 47 21.	+ 49 44	30	16.	GALAXY
MCG-04-30-024	12 47 21.	- 25 39	72	14.5	GALAXY
IC 3813	12 47 22.	- 25 39 00.			NONSTELLAR OBJECT
KN 16.451	12 47 22.0	+ 25 06 50.			NEBULA
RNGC 4715	12 47 23.	+ 28 05		15.5	GALAXY
IC 3823	12 47 23.	+ 41 09 25.			NONSTELLAR OBJECT
REIZ 2875	12 47 23.	- 07 09	36	15.0	GALAXY
HELW 278	12 47 23.	- 41 06 46.			NEBULA
SPA 3C	12 47 23.8	- 14 02 09.	18	14.5	SPHEROIDAL GALAXY
ZWG 043.041	12 47 24.	+ 05 35		12.3	GALAXY
UGC 07985	12 47 24.	+ 05 35	192	12.3	GALAXY Sc
MCG+01-33-018	12 47 24.	+ 05 35	150	12.3	GALAXY
LB 00057	12 47 24.	+ 26 33		15.8	FAINT BLUE STAR
LB 07192	12 47 24.	+ 26 51		19.3	FAINT BLUE STAR
LB 07191	12 47 24.	+ 27 20		18.6	FAINT BLUE STAR
LB 07190	12 47 24.	+ 27 50		19.5	FAINT BLUE STAR
ZWG 159.085	12 47 24.	+ 28 05		15.4	GALAXY
UGC 07986	12 47 24.	+ 28 05	108	15.4	GALAXY S0
LB 07189	12 47 24.	+ 28 17		19.9	FAINT BLUE STAR
LB 11369	12 47 24.	+ 29 50		14.4	FAINT BLUE STAR
LB 11368	12 47 24.	+ 31 05		16.5	FAINT BLUE STAR
MCG+05-30-095	12 47 24.	+ 32 22	42	15.2	GALAXY
ABC 1623	12 47 24.	+ 48 01		17.5	RICH CLUSTER OF GALAXIES
ZWG 245.001	12 47 24.	+ 49 43		15.3	GALAXY
ZWG 244.043	12 47 24.	+ 49 43		15.3	GALAXY
MCG-02-33-017	12 47 24.	- 14 28	108	13.5	GALAXY
HELW 279	12 47 24.	- 40 56 22.			NEBULA
MCG-07-26-056	12 47 24.	- 41 07	66	13.	GALAXY
MRSL 302+01/1	12 47 24.	- 61 18	180		HII REGION
KN 16.452	12 47 25.4	+ 27 10 44.			NEBULA
RNGC 4713	12 47 26.	+ 05 35		12.5	GALAXY
REIZ 2878	12 47 26.	+ 05 35	96	12.3	GALAXY
MCG+03-33-010	12 47 27.	+ 18 30	30	17.	GALAXY
HELW 280	12 47 27.	- 41 13 52.			NEBULA
REIZ 2880	12 47 28.	+ 21 15	24	15.6	GALAXY
IC 3821	12 47 29.	+ 21 14 25.			NONSTELLAR OBJECT
LB 02455	12 47 29.	+ 51 30 24.		15.1	FAINT BLUE STAR
KN 16.453	12 47 29.3	+ 25 55 44.			NEBULA
KARA.68 206	12 47 30.	+ 02 31	27		DWARF GALAXY
SCH 76	12 47 30.	+ 05 35		12.3	PECULIAR GALAXY
ZWG 071.052	12 47 30.	+ 11 32		15.7	GALAXY
LB 11371	12 47 30.	+ 27 40		17.4	FAINT BLUE STAR
MCG+05-30-096	12 47 30.	+ 28 05	90	15.4	GALAXY
LB 07194	12 47 30.	+ 28 12		19.9	FAINT BLUE STAR
LB 11370	12 47 30.	+ 29 36		13.6	FAINT BLUE STAR
LB 07193	12 47 30.	+ 31 38		18.9	FAINT BLUE STAR
MCG-07-26-057	12 47 30.	- 41 08	18	16.	GALAXY
MCG-07-26-058	12 47 30.	- 41 15	42	15.	GALAXY
KEEL 549	12 47 30.1	+ 41 38 18.		18.	NEBULA
LB 02456	12 47 31.	+ 61 03 06.		17.6	FAINT BLUE STAR
A2 066	12 47 31.6	+ 28 57 22.		16.7	FAINT BLUE OBJECT
KN 16.454	12 47 31.7	+ 28 05 38.			NEBULA
BPG 016	12 47 31.85	+ 34 56 22.8		18.95	ULTRAVIOLET-EXCESS OBJECT
REIZ 2882	12 47 32.	+ 28 06	12	14.3	GALAXY
KN 16.455	12 47 34.6	+ 27 25 26.			NEBULA
REIZ 2879	12 47 35.	- 03 50	12	15.4	GALAXY
KN 16.456	12 47 35.1	+ 29 05 21.			NEBULA
ZC 1247.6+2328	12 47 36.	+ 23 28	940		CLUSTER OF GALAXIES
ZWG 129.026	12 47 36.	+ 25 17		15.4	GALAXY
LB 11373	12 47 36.	+ 26 47		14.2	FAINT BLUE STAR
LB 07201	12 47 36.	+ 26 54		18.7	FAINT BLUE STAR
LB 07200	12 47 36.	+ 26 57		18.0	FAINT BLUE STAR
LB 07199	12 47 36.	+ 27 01		18.9	FAINT BLUE STAR
LB 07198	12 47 36.	+ 27 15		19.0	FAINT BLUE STAR
LB 07197	12 47 36.	+ 28 17		19.6	FAINT BLUE STAR
LB 07196	12 47 36.	+ 28 48		19.2	FAINT BLUE STAR
TON-N 0653	12 47 36.	+ 29 33		15.3	BLUE STAR
LB 07195	12 47 36.	+ 30 11		19.1	FAINT BLUE STAR
LB 11372	12 47 36.	+ 31 02		14.2	FAINT BLUE STAR
MCG+09-21-052	12 47 36.	+ 51 47 30.	30	16.	GALAXY
ZWG 335.024	12 47 36.	+ 73 16		15.6	GALAXY
IC 3819	12 47 38.	- 14 06 29.			NONSTELLAR OBJECT
KEEL 550	12 47 38.4	+ 41 07 04.		16.	SPIRAL NEBULA
KN 16.457	12 47 39.3	+ 25 17 51.			NEBULA
LB 02457	12 47 41.	+ 57 25 54.		17.1	FAINT BLUE STAR
IC 0825	12 47 41.	- 05 06 23.			NONSTELLAR OBJECT
ZC 1247.7+1104	12 47 42.	+ 11 04	1140		CLUSTER OF GALAXIES
ZC 1247.7+1318	12 47 42.	+ 13 18	1340		CLUSTER OF GALAXIES
LB 00019	12 47 42.	+ 26 47		15.6	FAINT BLUE STAR
TON-N 0654	12 47 42.	+ 27 03		15.5	BLUE STAR
LB 07204	12 47 42.	+ 29 08		19.5	FAINT BLUE STAR
LB 07203	12 47 42.	+ 31 00		19.2	FAINT BLUE STAR
LB 07202	12 47 42.	+ 31 03		19.3	FAINT BLUE STAR
ZWG 188.024	12 47 42.	+ 33 25		14.2	GALAXY
MRK 446	12 47 42.	+ 33 25	18	15.	GALAXY WITH UV CONTINUUM
RNGC 4719	12 47 42.	+ 33 25		14.0	GALAXY
UGC 07987	12 47 42.	+ 33 25	102	14.2	GALAXY SBb
KARA.73B 0553	12 47 42.	+ 33 25	90	14.2	ISOLATED GALAXY S
RNGC 4714	12 47 42.	- 13 02		14.0	GALAXY
MCG-02-33-018	12 47 42.	- 13 02	24	14.	GALAXY
REIZ 2885	12 47 43.	+ 33 26	36	13.6	GALAXY
REIZ 2881	12 47 43.	- 13 03	12	15.5	GALAXY
KEEL 552	12 47 43.3	+ 41 09 46.		16.	NEBULA
KEEL 551	12 47 44.1	+ 25 33 38.		17.	NEBULA
MCG+06-28-035	12 47 45.	+ 33 26	90	14.	GALAXY
MCG-02-33-019	12 47 45.	- 14 02 30.	66	15.	GALAXY
IC 3822	12 47 45.	- 14 02 59.			NONSTELLAR OBJECT
MCG-04-30-025	12 47 45.	- 26 35	72	14.	GALAXY
KN 16.458	12 47 45.9	+ 26 39 39.			NEBULA
RNGC 4721	12 47 47.	+ 27 36		15.0	GALAXY
HELW 281	12 47 47.	- 41 16 10.			NEBULA
KN 16.459	12 47 47.0	+ 27 02 57.			NEBULA
ABC 1624	12 47 48.	+ 08 56		17.8	RICH CLUSTER OF GALAXIES
ZC 1247.8+0857	12 47 48.	+ 08 57	470		CLUSTER OF GALAXIES
ZWG 071.053	12 47 48.	+ 10 49		15.6	GALAXY
BZW 1247+16.3	12 47 48.	+ 16 15		19.6	COMPACT GALAXY
LB 07210	12 47 48.	+ 27 00		17.6	FAINT BLUE STAR
LB 07209	12 47 48.	+ 27 08		19.3	FAINT BLUE STAR
LB 07208	12 47 48.	+ 27 33		19.9	FAINT BLUE STAR
ZWG 159.086	12 47 48.	+ 27 36		15.2	GALAXY
MCG+05-30-097	12 47 48.	+ 27 36	36	15.2	GALAXY
LB 07207	12 47 48.	+ 27 41		18.5	FAINT BLUE STAR
LB 07206	12 47 48.	+ 27 50		18.0	FAINT BLUE STAR
LB 07205	12 47 48.	+ 29 19		18.8	FAINT BLUE STAR
LB 11374	12 47 48.	+ 29 48		13.0	FAINT BLUE STAR
MCG+09-21-053	12 47 48.	+ 53 07	30	14.0	GALAXY
KEEL 553	12 47 48.8	+ 41 12 59.		18.	SPIRAL NEBULA
RNGC 4729	12 47 49.	- 41 11			GALAXY
HELW 282	12 47 49.	- 41 11 22.			NEBULA
KN 16.460	12 47 51.4	+ 26 35 57.			NEBULA
SPA 3B	12 47 51.4	- 14 00 51.	12	15.	SPHEROIDAL GALAXY
RNGC 4732	12 47 52.	+ 53 08		15.0	GALAXY
IC 3824	12 47 52.	- 14 09 17.			NONSTELLAR OBJECT
REIZ 2884	12 47 53.	- 03 47	18	14.9	GALAXY
HELW 445	12 47 53.	- 10 34 58.			NEBULA
A2 070	12 47 53.8	+ 30 13 16.		16.7	FAINT BLUE OBJECT
KN 16.461	12 47 53.9	+ 27 01 27.			NEBULA
KN 16.462	12 47 53.9	+ 27 35 45.			NEBULA
ZWG 043.042	12 47 54.	+ 07 05		15.7	GALAXY
TON-N 0132	12 47 54.	+ 17 42		15.2	BLUE STAR
ZC 1247.9+2356	12 47 54.	+ 23 56	870		CLUSTER OF GALAXIES
LB 07213	12 47 54.	+ 28 24		18.8	FAINT BLUE STAR
LB 07212	12 47 54.	+ 28 32		18.6	FAINT BLUE STAR
LB 11375	12 47 54.	+ 30 33		16.9	FAINT BLUE STAR
TON-N 0655	12 47 54.	+ 30 57		16.0	BLUE STAR
LB 07211	12 47 54.	+ 31 56		17.5	FAINT BLUE STAR
ZWG 270.026	12 47 54.	+ 53 08		15.2	GALAXY
UGC 07988	12 47 54.	+ 53 08	72	15.2	GALAXY E
ZC 1247.9+5619	12 47 54.	+ 56 19	1280		CLUSTER OF GALAXIES
MCG-01-33-020	12 47 54.	- 05 00	96	15.	GALAXY
RNGC 4716A	12 47 54.	- 09 10		15.0	GALAXY
MCG-01-33-021	12 47 54.	- 09 10	24	15.	GALAXY
RNGC 4716B	12 47 54.	- 09 13		15.0	GALAXY
MCG-01-33-022	12 47 54.	- 09 13	60	15.	GALAXY
REIZ 2883	12 47 54.	- 10 34	120	13.5	GALAXY
MCG-02-33-020	12 47 54.	- 10 34	210	13.	GALAXY
SVEN 335	12 47 54.	- 10 34 52.	102	14.3	GALAXY
HOLM 467B	12 47 55.	+ 01 47	36	15.4	PART OF MULTIPLE GALAXY
REIZ 2893	12 47 55.	+ 31 06	24	15.4	GALAXY
RNGC 4730	12 47 55.	- 41 09			GALAXY
KEEL 554	12 47 55.3	+ 25 34 17.		18.	NEBULA
HELW 283	12 47 56.	- 41 08 46.			NEBULA
KEEL 555	12 47 56.3	+ 25 57 07.		17.	NEBULA
KN 16.463	12 47 56.5	+ 27 27 09.			NEBULA
REIZ 2898	12 47 57.	+ 53 10	24	14.0	GALAXY
MAI 080	12 47 57.	+ 78 44	67		DWARF SPHEROIDAL GALAXY
KEEL 556	12 47 57.5	+ 25 57 49.		17.	NEBULA
KN 16.464	12 47 58.3	+ 29 15 57.			NEBULA
RNGC 4725	12 47 59.	+ 25 49		10.0	GALAXY
RNGC 4728	12 47 59.	+ 27 42		15.5	GALAXY
IC 3828	12 47 59.	+ 38 13 14.			TWO STARS
IC 3825	12 47 59.	- 14 13 17.			NONSTELLAR OBJECT
KN 16.465	12 47 59.9	+ 25 46 15.			NEBULA
LB 09865	12 48	- 86 29		11.5	FAINT BLUE STAR
MCG+03-33-011	12 48 00.	+ 17 44	54	18.	GALAXY
ZWG 129.027	12 48 00.	+ 25 46		10.2	GALAXY
UGC 07989	12 48 00.	+ 25 46	720	10.2	GALAXY Sb/SBb
MCG+04-30-022	12 48 00.	+ 25 47	660	10.2	GALAXY
ZC 1248.0+2628	12 48 00.	+ 26 28	870		CLUSTER OF GALAXIES
MCG+05-30-098	12 48 00.	+ 27 42	36	15.6	GALAXY
ZWG 159.087	12 48 00.	+ 27 43		15.6	GALAXY
UGC 07990	12 48 00.	+ 28 37	66	17.	GALAXY DWRF IR
LB 07215	12 48 00.	+ 28 50		19.0	FAINT BLUE STAR
ZWG 159.088	12 48 00.	+ 31 07		15.4	GALAXY
LB 07214	12 48 00.	+ 31 11		18.8	FAINT BLUE STAR
ZC 1248.0+3136	12 48 00.	+ 31 36	870		CLUSTER OF GALAXIES
MCG+08-23-096	12 48 00.	+ 48 22	18	16.	GALAXY
LB 02458	12 48 00.	+ 61 45 18.		15.9	FAINT BLUE STAR
ZC 1248.0+6250	12 48 00.	+ 62 50	3490		CLUSTER OF GALAXIES
REIZ 2887	12 48 00.	- 08 43	18	15.6	GALAXY
RNGC 4717	12 48 00.	- 09 10		15.0	GALAXY
MCG-01-33-023	12 48 00.	- 09 10	78	15.	GALAXY
REIZ 2886	12 48 00.	- 09 11	18	14.0	GALAXY
HOLM 466A	12 48 00.	- 09 11	66	13.8	PART OF MULTIPLE GALAXY
MCG-07-26-059	12 48 00.	- 42 17	36	14.5	GALAXY
KN 16.467	12 48 00.1	+ 27 00 27.			NEBULA
KN 16.466	12 48 00.2	+ 26 59 03.			NEBULA
BC B21248+30	12 48 00.4	+ 30 32 48.0		18.	QUASI-STELLAR OBJECT
KN 16.467A	12 48 00.6	+ 25 55 09.			NEBULA
SN 1969H	12 48 01.	+ 25 46		15.0	SUPERNOVA
SPA 3	12 48 01.	+ 48	1470		COMPACT CLSTR OF GALAXIES
RNGC 4718	12 48 01.	- 05 00		15.0	GALAXY
HOLM 466B	12 48 01.	- 09 15	90	14.5	PART OF MULTIPLE GALAXY
KN 16.468	12 48 01.7	+ 26 13 03.			NEBULA
KN 16.469	12 48 01.9	+ 27 42 33.			NEBULA
HOLM 469A	12 48 02.	+ 27 43	18	14.7	PART OF MULTIPLE GALAXY
REIZ 2892	12 48 03.	+ 01 44	72	15.5	GALAXY
REIZ 2897	12 48 03.	+ 25 47	480	11.0	GALAXY
HOLM 468A	12 48 03.	+ 25 47	600	10.8	PART OF MULTIPLE GALAXY
LB 02459	12 48 03.	+ 57 47 12.		15.7	FAINT BLUE STAR
IC 3826	12 48 04.	- 08 45			NONSTELLAR OBJECT
KEEL 557	12 48 04.8	+ 41 27 50.		17.	NEBULA
SC 1245+0201.4	12 48 05.	+ 01 45 03.	72		NEBULA
REIZ 2891	12 48 05.	- 03 50	12	15.0	GALAXY
REIZ 2890	12 48 05.	- 03 52	30	13.3	GALAXY
KN 16.470	12 48 05.4	+ 27 35 09.			NEBULA
ZWG 015.027	12 48 06.	+ 01 44		14.8	GALAXY
UGC 07991	12 48 06.	+ 01 44	114	14.8	GALAXY Sc
MCG+00-33-014	12 48 06.	+ 01 44	78	14.8	GALAXY
SCH 77	12 48 06.	+ 25 46		10.2	PECULIAR GALAXY
SN 1940B	12 48 06.	+ 25 48		12.8	SUPERNOVA
LB 11376	12 48 06.	+ 27 26		12.0	FAINT BLUE STAR
LB 07216	12 48 06.	+ 27 41		18.8	FAINT BLUE STAR
UGC 07992	12 48 06.	+ 27 43	60	16.0	GALAXY S
ABC 1626	12 48 06.	+ 31 37		17.8	RICH CLUSTER OF GALAXIES
ZWG 245.002	12 48 06.	+ 48 20		15.7	GALAXY
ZWG 244.044	12 48 06.	+ 48 20		15.7	GALAXY
ZC 1248.1+7157	12 48 06.	+ 71 57	1210		CLUSTER OF GALAXIES
ZWG 015.026	12 48 06.	- 02 33		15.7	GALAXY
MCG-01-33-024	12 48 06.	- 03 51	39	15.	GALAXY
MCG-01-33-025	12 48 06.	- 08 44	36	15.	GALAXY
REIZ 2889	12 48 06.	- 08 46	24	13.9	GALAXY
REIZ 2888	12 48 06.	- 09 12	48	14.3	GALAXY
KN 16.471	12 48 06.4	+ 29 01 21.			NEBULA
RNGC 4720	12 48 07.	- 03 51		15.0	GALAXY
KN 16.472	12 48 07.0	+ 29 01 51.			NEBULA
SPA 3A	12 48 07.8	- 14 00 37.	48	16.	SPINDLE GALAXY
HOLM 467A	12 48 09.	+ 01 45	84	14.3	PART OF MULTIPLE GALAXY

OBJECT NAME	RIGHT ASCEN.	DECLINATION	DIAM.	MAGN.	TYPE OF OBJECT
LB 02460	12 48 09.	+ 51 33 06.		16.3	FAINT BLUE STAR
MCG-03-33-009	12 48 09.	- 20 04	48	15.	GALAXY
RNGC 4728A	12 48 11.	+ 27 41		14.5	GALAXY
RNGC 4728B	12 48 11.	+ 27 45		16.0	GALAXY
HOLM 469C	12 48 11.	+ 27 46	36	15.1	PART OF MULTIPLE GALAXY
REIZ 2896	12 48 11.	- 03 49	30	16.0	GALAXY
REIZ 2895	12 48 11.	- 03 53	24	16.1	GALAXY
KN 16.473	12 48 11.1	+ 27 45 03.			NEBULA
KN 16.474	12 48 11.4	+ 27 41 39.			NEBULA
ZC 1248.2+1602	12 48 12.	+ 16 02	810		CLUSTER OF GALAXIES
8ZW 1248+25.2	12 48 12.	+ 25 14		18.0	COMPACT GALAXY
LB 07220	12 48 12.	+ 26 27		19.5	FAINT BLUE STAR
LB 07219	12 48 12.	+ 26 32		16.4	FAINT BLUE STAR
LB 07218	12 48 12.	+ 27 34		19.5	FAINT BLUE STAR
MCG+05-30-100	12 48 12.	+ 27 41	72	14.6	GALAXY
HOLM 469B	12 48 12.	+ 27 42	42	14.6	PART OF MULTIPLE GALAXY
LB 07217	12 48 12.	+ 28 54		19.5	FAINT BLUE STAR
MCG+05-30-099	12 48 12.	+ 31 08	42	15.4	GALAXY
ZWG 270.027	12 48 12.	+ 52 24		15.3	GALAXY
UGC 07993	12 48 12.	+ 52 24	138	15.3	GALAXY Sc
ZWG 335.025	12 48 12.	+ 73 09		11.8	GALAXY
UGC 07994	12 48 12.	+ 73 09	144	11.8	GALAXY Sb
MCG+13-09-033	12 48 12.	+ 78 15	33	17.	GALAXY
REIZ 2894	12 48 12.	- 08 30	18	14.6	GALAXY
MCG-02-33-022	12 48 12.	- 14 03	18	15.	GALAXY
MCG-02-33-021	12 48 12.	- 14 13	48	14.	GALAXY
HELW 284	12 48 12.	- 41 06 45.			NEBULA
SBP 094.07	12 48 12.	- 47 22	70	15.	HIGH SURFACE BRIGHT. GLXY
KN 16.475	12 48 13.9	+ 26 14 39.			NEBULA
HOLM 470B	12 48 15.	- 14 03	30	13.7	PART OF MULTIPLE GALAXY
RNGC 4727	12 48 17.	- 14 04		13.0	GALAXY
RNGC 4724	12 48 17.	- 14 04		15.0	GALAXY
KN 16.476	12 48 17.4	+ 28 50 09.			NEBULA
KN 16.477	12 48 17.5	+ 29 15 15.			NEBULA
BFG 017	12 48 17.68	+ 33 47 11.3		18.80	ULTRAVIOLET-EXCESS OBJECT
BC BS02	12 48 17.7	+ 33 47 04.		18.64	QUASI-STELLAR OBJECT
SHB 216	12 48 17.7	+ 33 47 04.		18.6	QUASI-STELLAR OBJECT
LB 07224	12 48 18.	+ 26 51		20.0	FAINT BLUE STAR
LB 07223	12 48 18.	+ 28 03		19.4	FAINT BLUE STAR
ZWG 159.089	12 48 18.	+ 28 06		14.8	GALAXY
LB 07222	12 48 18.	+ 28 17		19.3	FAINT BLUE STAR
LB 07221	12 48 18.	+ 29 19		19.5	FAINT BLUE STAR
TON-N 0656	12 48 18.	+ 32 29		16.1	BLUE STAR
MCG+09-21-054	12 48 18.	+ 52 22	120	14.	GALAXY
MCG+12-12-019	12 48 18.	+ 73 08	114	12.2	GALAXY
RNGC 4750	12 48 18.	+ 73 09		12.5	GALAXY
MCG-02-33-023	12 48 18.	- 14 03	72	13.5	GALAXY
SN 1965B	12 48 18.	- 14 05		16.0	SUPERNOVA
MCG-04-30-026	12 48 18.	- 26 24	54	15.5	GALAXY
HOLM 470A	12 48 19.	- 14 03	78	12.7	PART OF MULTIPLE GALAXY
IC 3827	12 48 19.	- 14 13 28.			NONSTELLAR OBJECT
ABC 1625	12 48 19.	- 20 31		17.0	RICH CLUSTER OF GALAXIES
BFG 018	12 48 19.75	+ 36 48 14.7		19.33	ULTRAVIOLET-EXCESS OBJECT
SN 1961D	12 48 20.	+ 28 06		16.5	SUPERNOVA
LB 00658	12 48 20.	+ 62 21 54.		17.0	FAINT BLUE STAR
KN 16.477A	12 48 20.6	+ 28 57 15.			NEBULA
BSO 02	12 48 21.	+ 33 47 06.		18.64	BLUE STELLAR OBJECT
MCG-02-33-025	12 48 21.	- 09 34 30.	60	14.5	GALAXY
MCG-02-33-024	12 48 21.	- 13 11	60	15.5	GALAXY
KN 16.479	12 48 21.8	+ 28 47 28.			NEBULA
KN 16.478	12 48 22.7	+ 26 13 40.			NEBULA
BFG 019	12 48 22.96	+ 35 56 07.4		19.45	ULTRAVIOLET-EXCESS OBJECT
LB 02461	12 48 23.	+ 49 57 30.		16.5	FAINT BLUE STAR
SVEN 336	12 48 23.	- 06 06 52.	360	11.6	GALAXY
REIZ 2899	12 48 23.	- 06 07	300	12.2	GALAXY
HOLM 472A	12 48 23.	- 06 07	420	11.2	PART OF MULTIPLE GALAXY
KEEL 558	12 48 23.6	+ 25 32 53.		16.	NEBULA
A2 076	12 48 23.6	+ 29 21 56.		18.5	FAINT BLUE OBJECT
ZWG 043.043	12 48 24.	+ 03 28		15.4	GALAXY
ZC 1248.4+0945	12 48 24.	+ 09 45	810		CLUSTER OF GALAXIES
IC 3830	12 48 24.	+ 20 06 32.			NONSTELLAR OBJECT
LB 07228	12 48 24.	+ 26 28		19.5	FAINT BLUE STAR
LB 07227	12 48 24.	+ 27 00		18.7	FAINT BLUE STAR
LB 11378	12 48 24.	+ 27 51		12.8	FAINT BLUE STAR
LB 07226	12 48 24.	+ 28 45		18.6	FAINT BLUE STAR
LB 07225	12 48 24.	+ 30 22		18.7	FAINT BLUE STAR
LB 11377	12 48 24.	+ 31 42		16.0	FAINT BLUE STAR
MCG+08-23-097	12 48 24.	+ 48 13	36	16.	GALAXY
MCG+12-12-020	12 48 24.	+ 71 54	102	15.	GALAXY
LB 00659	12 48 24.	+ 72 57 36.		17.7	FAINT BLUE STAR
UGC 07995	12 48 24.	+ 78 40	120	16.5	GALAXY DWRF IR
MCG-01-33-026	12 48 24.	- 06 06	360	12.	GALAXY
RNGC 4731	12 48 24.	- 06 08		11.5	GALAXY
MCG-02-33-026	12 48 24.	- 12 57	54	15.	GALAXY
MCG-03-33-010	12 48 24.	- 20 06	24	15.	GALAXY
MCG-04-30-027	12 48 24.	- 22 17	30	15.5	GALAXY
LB 02462	12 48 25.	+ 54 16 24.		15.1	FAINT BLUE STAR
HOLM 471A	12 48 25.	- 12 58	60	13.6	PART OF MULTIPLE GALAXY
KEEL 559	12 48 25.4	+ 25 31 52.		16.	NEBULA
HOLM 471B	12 48 26.	- 13 00	90	14.1	PART OF MULTIPLE GALAXY
KN 16.480	12 48 27.9	+ 28 06 52.			NEBULA
IC 3832	12 48 28.	+ 40 04 51.			NONSTELLAR OBJECT
HELW 285	12 48 28.	- 40 32 09.			NEBULA
KN 16.481	12 48 29.1	+ 28 56 22.			NEBULA
ZC 1248.5+1340	12 48 30.	+ 13 40	1210		CLUSTER OF GALAXIES
ZC 1248.5+1404	12 48 30.	+ 14 04	1080		CLUSTER OF GALAXIES
ZC 1248.5+1424	12 48 30.	+ 14 24	810		CLUSTER OF GALAXIES
ZC 1248.5+1659	12 48 30.	+ 16 59	940		CLUSTER OF GALAXIES
LB 07231	12 48 30.	+ 26 42		20.0	FAINT BLUE STAR
TON-N 0657	12 48 30.	+ 27 34		15.1	BLUE STAR
ZWG 159.090	12 48 30.	+ 27 39		15.5	GALAXY
MCG+05-30-101	12 48 30.	+ 28 06	18	14.8	GALAXY
LB 07230	12 48 30.	+ 28 18		19.6	FAINT BLUE STAR
LB 07229	12 48 30.	+ 29 53		19.4	FAINT BLUE STAR
TON-N 0133	12 48 30.	+ 30 45		16.2	BLUE STAR
ZWG 188.025	12 48 30.	+ 34 25		15.3	GALAXY
RNGC 4737	12 48 30.	+ 34 25		15.5	GALAXY
ZWG 217.001	12 48 30.	+ 41 23		8.7	GALAXY
ZWG 216.034	12 48 30.	+ 41 23		8.7	GALAXY
UGC 07996	12 48 30.	+ 41 23	900	8.7	GALAXY Sb
ZC 1248.5+5329	12 48 30.	+ 53 29	870		CLUSTER OF GALAXIES
RNGC 4723	12 48 30.	- 12 59		15.0	GALAXY
REIZ 2903	12 48 31.	+ 11 11	18	14.8	GALAXY
RNGC 4736	12 48 31.	+ 41 24		9.5	GALAXY
KEEL 561	12 48 31.9	+ 41 09 36.		18.	NEBULA
A2 079	12 48 32.1	+ 29 32 33.		17.4	FAINT BLUE OBJECT
REIN 4.149	12 48 32.47	+ 11 11 01.5			NEBULA
SC 1245+0047.0	12 48 33.	+ 00 30 39.	12		NEBULA
REIZ 2902	12 48 33.	+ 03 21	90	12.4	GALAXY
HOLM 473B	12 48 33.	+ 11 12	18	14.7	PART OF MULTIPLE GALAXY
MCG+06-28-036	12 48 33.	+ 34 26	30	14.5	GALAXY
KN 16.482	12 48 33.1	+ 26 42 46.			NEBULA
KN 16.483	12 48 33.5	+ 27 04 10.			NEBULA
ABC 1627	12 48 34.	+ 13 37		18.0	RICH CLUSTER OF GALAXIES
REIZ 2910	12 48 34.	+ 41 24	372	9.1	GALAXY
RNGC 4735	12 48 35.	+ 29 12		15.0	GALAXY
IC 3835	12 48 35.	+ 40 28 39.			NONSTELLAR OBJECT
REIZ 2901	12 48 35.	- 03 56	24	14.9	GALAXY
SVEN 337	12 48 35.	- 06 17	48	14.0	GALAXY
REIZ 2900	12 48 35.	- 06 17	30	14.0	GALAXY
REIN 4.150	12 48 35.92	+ 11 11 02.7			NEBULA
ZWG 043.044	12 48 36.	+ 04 51		15.6	GALAXY
ZWG 071.054	12 48 36.	+ 11 11		13.2	GALAXY
UGC 07997	12 48 36.	+ 11 11	132	13.2	GALAXY E
MCG+02-33-028	12 48 36.	+ 11 11	54	13.2	GALAXY
MCG+03-33-012	12 48 36.	+ 17 39	36	17.	GALAXY
LB 07238	12 48 36.	+ 27 00		19.1	FAINT BLUE STAR
MCG+05-30-102	12 48 36.	+ 27 38	42	15.5	GALAXY
LB 07237	12 48 36.	+ 28 13		19.5	FAINT BLUE STAR
MCG+05-30-103	12 48 36.	+ 29 03	120	14.9	GALAXY
ZWG 159.091	12 48 36.	+ 29 12		15.1	GALAXY
SCH 78	12 48 36.	+ 29 12		15.1	PECULIAR GALAXY
MCG+05-30-104	12 48 36.	+ 29 12	36	15.1	GALAXY
LB 07236	12 48 36.	+ 29 18		19.5	FAINT BLUE STAR
LB 07235	12 48 36.	+ 30 10		18.5	FAINT BLUE STAR
LB 07234	12 48 36.	+ 31 07		19.3	FAINT BLUE STAR
LB 07233	12 48 36.	+ 31 18		20.2	FAINT BLUE STAR
LB 07232	12 48 36.	+ 31 56		18.1	FAINT BLUE STAR
ZC 1248.6+5919	12 48 36.	+ 59 19	2080		CLUSTER OF GALAXIES
RNGC 4731A	12 48 36.	- 06 17		14.5	GALAXY
HOLM 472B	12 48 36.	- 06 17	54	12.9	PART OF MULTIPLE GALAXY
MCG-01-33-027	12 48 36.	- 06 17	36	14.5	GALAXY
HELW 446	12 48 36.	- 06 17 09.			NEBULA
MCG-05-30-010	12 48 36.	- 27 31	30	15.5	GALAXY
KN 16.484	12 48 36.0	+ 29 11 58.			NEBULA
REIZ 2905	12 48 37.	+ 11 11	60	13.1	GALAXY
HOLM 473A	12 48 37.	+ 11 12	36	13.0	PART OF MULTIPLE GALAXY
LB 02463	12 48 37.	+ 55 46 24.		16.9	FAINT BLUE STAR
KN 16.485	12 48 37.2	+ 26 43 04.			NEBULA
KN 16.486	12 48 37.6	+ 27 38 34.			NEBULA
KN 16.487	12 48 37.8	+ 27 41 34.			NEBULA
REIZ 2906	12 48 38.	+ 05 08	42	13.4	GALAXY
ABC 1628	12 48 38.	+ 28 54		17.4	RICH CLUSTER OF GALAXIES
REIZ 2909	12 48 38.	+ 29 13	30	14.3	GALAXY
HZ 32	12 48 38.	+ 37 27		15.2	BLUE STAR
BFG 021	12 48 38.73	+ 37 27 07.8		15.89	ULTRAVIOLET-EXCESS OBJECT
BFG 020	12 48 38.91	+ 35 35 31.6		19.60	ULTRAVIOLET-EXCESS OBJECT
RNGC 4733	12 48 39.	+ 11 11		13.0	GALAXY
RNGC 4741	12 48 39.	+ 47 56		14.5	GALAXY
MCG-02-33-027	12 48 39.	- 14 18	84	14.	GALAXY
A2 068	12 48 39.0	+ 29 49 55.		18.0	FAINT BLUE GALAXY
KN 16.488	12 48 39.5	+ 28 04 40.			NEBULA
A2 080	12 48 40.1	+ 29 23 54.		18.0	FAINT BLUE OBJECT
RNGC 4738	12 48 41.	+ 29 04		15.0	GALAXY
HZ 33	12 48 41.	+ 33 54		14.6	BLUE STAR
LB 02464	12 48 41.	+ 49 38 48.		15.1	FAINT BLUE STAR
REIZ 2904	12 48 41.	- 07 30	72	13.6	GALAXY
IC 3831	12 48 41.	- 14 18 09.			NONSTELLAR OBJECT
IC 3829	12 48 41.	- 27 33 58.			NONSTELLAR OBJECT
KN 16.489	12 48 41.0	+ 28 54 46.			NEBULA
ZWG 043.045	12 48 42.	+ 05 08		14.3	GALAXY
UGC 07998	12 48 42.	+ 05 08	66	14.3	GALAXY Sc
MCG+01-33-019	12 48 42.	+ 05 08	54	14.3	GALAXY
KARA.68 207	12 48 42.	+ 11 31	40		DWARF GALAXY
ZC 1248.7+1722	12 48 42.	+ 17 22	1080		CLUSTER OF GALAXIES
ZWG 159.092	12 48 42.	+ 29 04		14.9	GALAXY
UGC 07999	12 48 42.	+ 29 04	132	14.9	GALAXY Sc?
LB 07240	12 48 42.	+ 29 34		17.2	FAINT BLUE STAR
LB 07239	12 48 42.	+ 30 26		19.0	FAINT BLUE STAR
LB 11379	12 48 42.	+ 32 28		10.6	FAINT BLUE STAR
REIZ 2911	12 48 42.	+ 34 26	30	14.5	GALAXY
ZWG 245.003	12 48 42.	+ 47 56		14.5	GALAXY
ZWG 244.045	12 48 42.	+ 47 56		14.5	GALAXY
UGC 08000	12 48 42.	+ 47 56	90	14.5	GALAXY Sc
REIZ 2916	12 48 42.	+ 47 57	48	13.2	GALAXY
MCG+08-23-098	12 48 42.	+ 47 58	72	13.2	GALAXY
MCG-01-33-028	12 48 42.	- 07 28	48	14.5	GALAXY
SVEN 338	12 48 42.	- 10 27 52.	54	15.5	GALAXY
MCG-04-30-028	12 48 42.	- 26 33	60	15.	GALAXY
REIZ 2908	12 48 43.	+ 11 31	30	14.3	GALAXY
IC 3836	12 48 43.	+ 40 27 27.			NONSTELLAR OBJECT
HELW 168	12 48 43.	- 10 28 03.			NEBULA
KN 16.491	12 48 43.0	+ 29 03 22.			NEBULA
KN 16.490	12 48 43.4	+ 27 40 46.			NEBULA
KEEL 562	12 48 43.6	+ 41 39 46.		15.	SPIRAL NEBULA
RNGC 4734	12 48 44.	+ 05 08		14.5	GALAXY
REIZ 2912	12 48 44.	+ 29 04	72	13.7	GALAXY
LB 00660	12 48 44.	+ 54 50 12.		17.3	FAINT BLUE STAR
SC 1246-0049.8	12 48 44.	- 01 06 09.	12		NEBULA
KEEL 563	12 48 44.4	+ 41 05 56.		18.	NEBULA
LB 02465	12 48 45.	+ 58 51 18.		14.4	FAINT BLUE STAR
LB 02466	12 48 46.	+ 51 01 06.		17.2	FAINT BLUE STAR
KN 16.492	12 48 46.3	+ 27 04 04.			NEBULA
SC 1246+0104.2	12 48 47.	+ 00 47 51.	12		NEBULA
RNGC 4745B	12 48 47.	+ 27 42		16.0	GALAXY
HELW 286	12 48 47.	- 40 56 33.			NEBULA
KEEL 564	12 48 47.8	+ 41 25 22.		17.	NEBULA
ZWG 071.055	12 48 48.	+ 10 13		15.6	GALAXY
ZC 1248.8+2012	12 48 48.	+ 20 12	270		CLUSTER OF GALAXIES
MCG+05-30-104A	12 48 48.	+ 27 22	33	15.3	GALAXY
ZWG 159.093	12 48 48.	+ 27 23		15.3	GALAXY
MCG+05-30-105	12 48 48.	+ 27 42	48	15.4	GALAXY
LB 07243	12 48 48.	+ 29 22		19.0	FAINT BLUE STAR
LB 07242	12 48 48.	+ 29 40		19.3	FAINT BLUE STAR
LB 07241	12 48 48.	+ 29 44		19.5	FAINT BLUE STAR
LB 11381	12 48 48.	+ 31 15		12.0	FAINT BLUE STAR
LB 11380	12 48 48.	+ 32 12		9.6	FAINT BLUE STAR
ZC 1248.8+4039	12 48 48.	+ 40 39	3700		CLUSTER OF GALAXIES
TON-N 0134	12 48 48.	+ 41 15		15.2	BLUE STAR
ZC 1248.8+4230	12 48 48.	+ 42 30	1210		CLUSTER OF GALAXIES
REIZ 2907	12 48 48.	- 09 07	30	15.1	GALAXY
MCG-02-33-028	12 48 48.	- 14 56	24	15.5	GALAXY
MCG-04-30-029	12 48 48.	- 26 11	36	13.	GALAXY
A2 082	12 48 48.0	+ 28 02 34.		17.5	FAINT BLUE OBJECT
KEEL 565	12 48 48.1	+ 41 19 18.		18.	NEBULA
REIZ 2915	12 48 49.	+ 31 18	18	13.6	GALAXY

OBJECT NAME	RIGHT ASCEN.	DECLINATION	DIAM.	MAGN.	TYPE OF OBJECT
BFG 022	12 48 49.15	+ 36 11 16.3		18.81	ULTRAVIOLET-EXCESS OBJECT
HELW 447	12 48 50.	- 10 29 45.			NEBULA
KN 16.493	12 48 50.3	+ 28 18 16.			NEBULA
KN 16.494	12 48 50.6	+ 29 02 22.			NEBULA
MCG-02-33-029	12 48 51.	- 11 30	36	16.	GALAXY
HELW 287	12 48 51.	- 41 00 57.			NEBULA
SPA 3D	12 48 51.5	- 14 04 32.	18	14.5	SPHEROIDAL GALAXY
KN 16.495	12 48 51.9	+ 26 15 34.			NEBULA
KN 16.496	12 48 51.9	+ 27 22 40.			NEBULA
RNGC 4745A	12 48 53.	+ 27 42		15.5	GALAXY
RNGC 4726	12 48 53.	- 13 57		14.0	GALAXY
ZC 1248.9+0413	12 48 54.	+ 04 13	1210		CLUSTER OF GALAXIES
ZWG 071.056	12 48 54.	+ 10 11		15.6	GALAXY
UGC 08001	12 48 54.	+ 10 52	60	16.5	GALAXY
ZWG 071.057	12 48 54.	+ 14 21		15.6	GALAXY
TON-N 0658	12 48 54.	+ 27 32		14.2	BLUE STAR
LB 07249	12 48 54.	+ 27 39		19.4	FAINT BLUE STAR
MCG+05-30-105	12 48 54.	+ 27 41	36	15.4	GALAXY
ZWG 159.094	12 48 54.	+ 27 42		15.4	GALAXY
LB 07248	12 48 54.	+ 28 14		20.1	FAINT BLUE STAR
LB 07247	12 48 54.	+ 30 00		19.3	FAINT BLUE STAR
LB 07246	12 48 54.	+ 30 32		18.7	FAINT BLUE STAR
LB 07245	12 48 54.	+ 30 37		19.3	FAINT BLUE STAR
LB 11382	12 48 54.	+ 30 46		17.3	FAINT BLUE STAR
ZWG 159.095	12 48 54.	+ 31 19		14.9	GALAXY
MCG+05-30-106	12 48 54.	+ 31 20	36	14.9	GALAXY
LB 07244	12 48 54.	+ 32 20		19.0	FAINT BLUE STAR
TON-N 0135	12 48 54.	+ 38 03		14.6	BLUE STAR
ZC 1248.9+3908	12 48 54.	+ 39 08	870		CLUSTER OF GALAXIES
SVEN 339	12 48 54.	- 09 57 51.	36	15.9	GALAXY
RNGC 4722	12 48 54.	- 13 03		13.0	GALAXY
MCG-02-33-031	12 48 54.	- 13 03	108	13.5	GALAXY
MCG-02-33-030	12 48 54.	- 13 57	42	14.5	GALAXY
MCG-04-30-030	12 48 54.	- 25 51	96	14.5	GALAXY
IC 0826	12 48 55.	+ 31 19 45.			NONSTELLAR OBJECT
KN 16.497	12 48 55.9	+ 27 43 04.			NEBULA
HELW 448	12 48 56.	- 09 58 03.			NEBULA
IC 3833	12 48 56.	- 12 33			NONSTELLAR OBJECT
IC 3834	12 48 56.	- 13 57			NONSTELLAR OBJECT
HOLM 474B	12 48 57.	+ 27 43	18	15.4	PART OF MULTIPLE GALAXY
LB 02467	12 48 58.	+ 60 53 30.		15.0	FAINT BLUE STAR
KEEL 566	12 48 59.2	+ 41 19 45.		18.	NEBULA
KN 16.498	12 48 59.8	+ 28 15 16.			NEBULA
VDB.66G 152	12 49	- 06 01	70		DWARF GALAXY
ZC 1249.0+0310	12 49 00.	+ 03 10	610		CLUSTER OF GALAXIES
ZC 1249.0+0447	12 49 00.	+ 04 47	2080		CLUSTER OF GALAXIES
ZWG 043.046	12 49 00.	+ 08 18		15.5	GALAXY.
ZWG 071.058	12 49 00.	+ 14 03		15.4	GALAXY
ZWG 100.012	12 49 00.	+ 18 20		15.3	GALAXY
MCG+03-33-013	12 49 00.	+ 18 20	18	15.3	GALAXY
ZC 1249.0+2444	12 49 00.	+ 24 44	940		CLUSTER OF GALAXIES
TON-N 0659	12 49 00.	+ 27 13		14.3	BLUE STAR
TON-N 0660	12 49 00.	+ 27 20		16.5	BLUE STAR
LB 07253	12 49 00.	+ 28 35		18.4	FAINT BLUE STAR
LB 07252	12 49 00.	+ 29 12		19.1	FAINT BLUE STAR
LB 07251	12 49 00.	+ 30 58		19.7	FAINT BLUE STAR
LB 07250	12 49 00.	+ 32 20		19.5	FAINT BLUE STAR
ZC 1249.0+4540	12 49 00.	+ 45 40	5170		CLUSTER OF GALAXIES
UGC 08003	12 49 00.	+ 53 38	66	14.3	GALAXY
MCG+09-21-055	12 49 00.	+ 53 57 30.	48	14.	GALAXY
ZWG 270.028	12 49 00.	+ 53 58		14.3	GALAXY
ZC 1249.0+5720	12 49 00.	+ 57 20	2690		CLUSTER OF GALAXIES
ZWG 352.041	12 49 00.	+ 77 49		15.4	GALAXY
MCG+13-09-034	12 49 00.	+ 77 49	54	16.	GALAXY
ZWG 365.004	12 49 00.	+ 80 39		15.7	GALAXY
MCG+14-06-009	12 49 00.	+ 80 52	66	15.	GALAXY
ZWG 365.005	12 49 00.	+ 80 53		15.6	GALAXY
UGC 08002	12 49 00.	+ 80 53	90	15.6	GALAXY Sb
RNGC 4739	12 49 00.	- 08 08		14.0	GALAXY
REIZ 2914	12 49 00.	- 08 08	24	13.4	GALAXY
MCG-01-33-029	12 49 00.	- 08 08	60	14.	GALAXY
MCG-06-28-021	12 49 00.	- 38 32	84	14.	GALAXY
HELW 288	12 49 00.	- 40 50 51.			NEBULA
MCG-07-27-001	12 49 00.	- 42 20	48	15.	GALAXY
KN 16.499	12 49 00.2	+ 27 41 34.			NEBULA
REIZ 2913	12 49 01.	- 12 28	30	14.7	GALAXY
SPA 3E	12 49 01.1	- 14 04 08.	24	15.	ELLIPSOIDAL GALAXY
REIZ 2919	12 49 02.	+ 27 42	12	15.9	GALAXY
HOLM 474A	12 49 02.	+ 27 42	24	14.7	PART OF MULTIPLE GALAXY
LB 02468	12 49 03.	+ 55 50 00.		15.1	FAINT BLUE STAR
IC 3837	12 49 05.	+ 19 59 39.			NONSTELLAR OBJECT
REIZ 2917	12 49 05.	- 04 24	30	15.0	GALAXY
ZWG 100.013	12 49 06.	+ 18 55		15.7	GALAXY
LB 07258	12 49 06.	+ 26 53		19.0	FAINT BLUE STAR
LB 07257	12 49 06.	+ 27 06		18.9	FAINT BLUE STAR
LB 11383	12 49 06.	+ 27 09		12.7	FAINT BLUE STAR
LB 07256	12 49 06.	+ 27 20		19.1	FAINT BLUE STAR
TON-N 0661	12 49 06.	+ 29 34		15.5	BLUE STAR
TON-N 0662	12 49 06.	+ 29 40		17.0	BLUE STAR
LB 07255	12 49 06.	+ 31 34		19.4	FAINT BLUE STAR
LB 07254	12 49 06.	+ 31 52		18.9	FAINT BLUE STAR
MCG+09-21-056	12 49 06.	+ 52 45	30	16.	GALAXY
ZC 1249.1+5820	12 49 06.	+ 58 20	2690		CLUSTER OF GALAXIES
ZC 1249.1+6726	12 49 06.	+ 67 26	4370		CLUSTER OF GALAXIES
MCG-03-33-012	12 49 06.	- 16 00 30.	30	15.	GALAXY
MCG-03-33-011	12 49 06.	- 19 37	10	15.	GALAXY
MCG-07-27-002	12 49 06.	- 40 52	42	14.	GALAXY
REIZ 2921	12 49 07.	+ 31 36	54	14.3	GALAXY
KN 16.500	12 49 07.3	+ 27 02 29.			NEBULA
ARC 1629	12 49 08.	+ 04 13		17.5	RICH CLUSTER OF GALAXIES
KN 16.501	12 49 09.4	+ 26 54 05.			NEBULA
A2 087	12 49 09.8	+ 28 02 23.		17.4	FAINT BLUE OBJECT
IC 0830	12 49 11.	+ 53 57 10.			NONSTELLAR OBJECT
SVEN 341	12 49 11.	- 06 19	24	15.4	GALAXY
REIN 2.162	12 49 11.78	- 10 11 00.4			NEBULA
ZWG 071.059	12 49 12.	+ 13 22		15.7	GALAXY
ZC 1249.2+2404	12 49 12.	+ 24 04	1280		CLUSTER OF GALAXIES
LB 07266	12 49 12.	+ 26 24		19.2	FAINT BLUE STAR
LB 07265	12 49 12.	+ 26 32		18.0	FAINT BLUE STAR
LB 07264	12 49 12.	+ 26 56		18.7	FAINT BLUE STAR
LB 07263	12 49 12.	+ 27 38		18.9	FAINT BLUE STAR
LB 11384	12 49 12.	+ 29 06		16.0	FAINT BLUE STAR
LB 07262	12 49 12.	+ 29 08		19.0	FAINT BLUE STAR
LB 07261	12 49 12.	+ 30 03		19.4	FAINT BLUE STAR
ZWG 159.096	12 49 12.	+ 31 37		15.1	GALAXY
UGC 08004	12 49 12.	+ 31 37	114	15.1	GALAXY Sc
LB 07260	12 49 12.	+ 32 13		20.2	FAINT BLUE STAR
LB 07259	12 49 12.	+ 32 16		18.6	FAINT BLUE STAR
ZWG 352.042	12 49 12.	+ 77 35		15.7	GALAXY
SC 1246-0023.2	12 49 12.	- 00 39 32.	12		NEBULA
HELW 449	12 49 12.	- 06 19 14.			NEBULA
SVEN 340	12 49 12.	- 10 10 51.	60	12.0	GALAXY
REIZ 2918	12 49 12.	- 10 11	60	12.2	GALAXY
RNGC 4742	12 49 12.	- 10 12		12.5	GALAXY
MCG-04-30-031	12 49 12.	- 25 49	78	14.	GALAXY
KN 16.502	12 49 12.0	+ 27 34 59.			NEBULA
KEEL 567	12 49 13.5	+ 25 53 16.		16.	NEBULA
ARC 1630	12 49 14.	+ 04 51		16.7	RICH CLUSTER OF GALAXIES
HELW 289	12 49 14.	- 40 51 39.			NEBULA
A2 092	12 49 14.4	+ 29 06 15.		17.7	FAINT BLUE OBJECT
IC 3842	12 49 15.	+ 40 38 34.			NONSTELLAR OBJECT
MCG-02-33-032	12 49 15.	- 10 10	48	12.5	GALAXY
A2 094	12 49 15.5	+ 28 03 36.		17.5	FAINT BLUE OBJECT
BFG 023	12 49 15.70	+ 36 46 40.4		18.49	ULTRAVIOLET-EXCESS OBJECT
HOLM 475B	12 49 16.	+ 12 46	18	14.8	PART OF MULTIPLE GALAXY
RNGC 4747	12 49 17.	+ 26 03		12.5	GALAXY
IC 3843	12 49 17.	+ 39 16 22.			NONSTELLAR OBJECT
SVEN 342	12 49 17.	- 06 24	24	15.4	GALAXY
REIZ 2920	12 49 17.	- 07 28	36	15.8	GALAXY
HELW 290	12 49 17.	- 41 10 32.			NEBULA
KN 16.503	12 49 17.7	+ 26 22 59.			NEBULA
KN 16.504	12 49 17.9	+ 28 14 11.			NEBULA
HOLM 475A	12 49 18.	+ 12 45	24	14.7	PART OF MULTIPLE GALAXY
IC 3839	12 49 18.	+ 20 41 28.			NONSTELLAR OBJECT
IC 3840	12 49 18.	+ 22 00 28.			NONSTELLAR OBJECT
ZWG 129.028	12 49 18.	+ 26 02		13.2	GALAXY
UGC 08005	12 49 18.	+ 26 02	210	13.2	GALAXY IRR
MCG+05-30-108	12 49 18.	+ 28 14	18	15.5	GALAXY
LB 07268	12 49 18.	+ 28 36		19.5	FAINT BLUE STAR
LB 07267	12 49 18.	+ 28 54		19.1	FAINT BLUE STAR
LB 11385	12 49 18.	+ 30 36		14.4	FAINT BLUE STAR
MCG+05-30-107	12 49 18.	+ 31 38	108	15.1	GALAXY
ZC 1249.3+3534	12 49 18.	+ 35 34	1210		CLUSTER OF GALAXIES
ZWG 335.026	12 49 18.	+ 71 55		14.2	GALAXY
UGC 08006	12 49 18.	+ 71 55	114	14.2	GALAXY S
ZWG 015.028	12 49 18.	- 01 48		14.8	GALAXY
MCG+00-33-015	12 49 18.	- 01 48	30	14.8	GALAXY
MCG-01-33-030	12 49 18.	- 04 16	42	15.	GALAXY
MCG-04-30-032	12 49 18.	- 26 22	72	14.5	GALAXY
MCG-07-27-003	12 49 18.	- 40 53	42	14.	GALAXY
KN 16.506	12 49 18.7	+ 27 17 35.			NEBULA
ARP 159	12 49 19.	+ 26 04			PECULIAR GALAXY
KN 16.505	12 49 19.1	+ 26 02 47.			NEBULA
BFG 024	12 49 19.44	+ 32 57 51.3		18.28	ULTRAVIOLET-EXCESS OBJECT
HELW 450	12 49 20.	- 06 24 14.			NEBULA
BFG 025	12 49 20.67	+ 35 39 48.5		19.26	ULTRAVIOLET-EXCESS OBJECT
KN 16.507	12 49 20.9	+ 25 48 35.			NEBULA
REIZ 2927	12 49 21.	+ 26 02	180	12.9	GALAXY
HOLM 468C	12 49 21.	+ 26 03	210	12.3	PART OF MULTIPLE GALAXY
MCG+04-30-023	12 49 21.	+ 26 03	210	13.2	GALAXY
IC 3838	12 49 21.	- 14 11			NONSTELLAR OBJECT
REIN 4.151	12 49 21.83	- 01 48 23.1			NEBULA
IC 0827	12 49 22.	+ 16 32 40.			NONSTELLAR OBJECT
BSO 03	12 49 22.	+ 36 08 06.		17.30	BLUE STELLAR OBJECT
REIZ 2922	12 49 22.	- 01 48	24	14.8	GALAXY
SC 1246-0131.8	12 49 22.	- 01 48 08.	18		NEBULA
IC 3841	12 49 23.	+ 22 36 58.			NONSTELLAR OBJECT
RNGC 4749	12 49 23.	+ 71 54			GALAXY
BFG 026	12 49 23.91	+ 35 08 27.2		18.90	ULTRAVIOLET-EXCESS OBJECT
ZWG 043.047	12 49 24.	+ 04 50		15.6	GALAXY
ZWG 071.060	12 49 24.	+ 12 21		13.3	GALAXY
UGC 08007	12 49 24.	+ 12 21	132	13.3	GALAXY S
MCG+02-33-029	12 49 24.	+ 12 21	200	13.3	GALAXY
LB 11389	12 49 24.	+ 26 21		12.5	FAINT BLUE STAR
LB 00020	12 49 24.	+ 26 41		18.3	FAINT BLUE STAR
LB 07270	12 49 24.	+ 27 13		19.6	FAINT BLUE STAR
LB 11388	12 49 24.	+ 27 20		12.8	FAINT BLUE STAR
TON-N 0663	12 49 24.	+ 27 28		16.8	BLUE STAR
LB 07269	12 49 24.	+ 28 49		19.7	FAINT BLUE STAR
LB 11387	12 49 24.	+ 30 13		11.7	FAINT BLUE STAR
LB 11386	12 49 24.	+ 31 34		14.2	FAINT BLUE STAR
ZC 1249.4+4636	12 49 24.	+ 46 36	1080		CLUSTER OF GALAXIES
ZC 1249.4+4654	12 49 24.	+ 46 54	1340		CLUSTER OF GALAXIES
ZC 1249.4+4812	12 49 24.	+ 48 12	1950		CLUSTER OF GALAXIES
MCG-02-33-033	12 49 24.	- 15 11	18	15.	GALAXY
MCG-04-30-033	12 49 24.	- 21 34 30.	30	14.	GALAXY
MCG-05-30-011	12 49 24.	- 29 34	102	14.	GALAXY
MCG-07-27-004	12 49 24.	- 41 12	42	15.	GALAXY
KN 16.508	12 49 24.3	+ 27 38 11.			NEBULA
A2 096	12 49 24.5	+ 26 55 13.		17.9	FAINT BLUE OBJECT
BFG 027	12 49 24.53	+ 38 13 40.8		18.87	ULTRAVIOLET-EXCESS OBJECT
KN 16.509	12 49 24.9	+ 26 55 11.			NEBULA
REIZ 2926	12 49 25.	+ 12 21	120	13.3	GALAXY
REIN 4.152	12 49 25.02	+ 12 21 13.2			NEBULA
KN 16.510	12 49 26.4	+ 28 20 53.			NEBULA
KEEL 568	12 49 26.6	+ 41 06 11.		18.	NEBULA
RNGC 4746	12 49 27.	+ 12 21		13.5	GALAXY
MCG+03-33-014	12 49 27.	+ 16 32	54	15.0	GALAXY
KN 16.511	12 49 27.4	+ 26 20 17.			NEBULA
ZWG 043.048	12 49 30.	+ 06 49		15.4	GALAXY
ZWG 071.061	12 49 30.	+ 09 48		15.6	GALAXY
ZWG 100.014	12 49 30.	+ 16 33		15.0	GALAXY
UGC 08008	12 49 30.	+ 16 33	60	15.0	GALAXY S
LB 07273	12 49 30.	+ 27 38		20.2	FAINT BLUE STAR
LB 07272	12 49 30.	+ 28 23		18.5	FAINT BLUE STAR
TON-N 0664	12 49 30.	+ 29 41		15.6	BLUE STAR
LB 07271	12 49 30.	+ 31 17		19.0	FAINT BLUE STAR
LB 11390	12 49 30.	+ 31 35		14.5	FAINT BLUE STAR
ZC 1249.5+3232	12 49 30.	+ 32 32	1480		CLUSTER OF GALAXIES
ZC 1249.5+4712	12 49 30.	+ 47 12	870		CLUSTER OF GALAXIES
MCG+10-18-084	12 49 30.	+ 61 25	10	16.	GALAXY
MCG+14-06-010	12 49 30.	+ 80 37 30.	30	16.	GALAXY
REIZ 2925	12 49 30.	- 08 03	24	15.8	GALAXY
REIZ 2924	12 49 30.	- 09 03	30	15.5	GALAXY
REIZ 2923	12 49 30.	- 09 12	30	15.8	GALAXY
A2 098	12 49 30.1	+ 30 12 36.		18.2	FAINT BLUE OBJECT
RNGC 4743	12 49 31.	- 41 07			GALAXY
A2 100	12 49 32.5	+ 30 10 58.		18.2	FAINT BLUE GALAXY
MCG-04-30-034	12 49 33.	- 26 02	24	14.	GALAXY
A2 101	12 49 34.0	+ 26 53 48.		18.6	FAINT BLUE OBJECT
KN 16.512	12 49 34.4	+ 26 25 47.			NEBULA
KEEL 569	12 49 34.4	+ 41 38 26.		17.	NEBULA
FATH 1.544	12 49 35.	+ 44 22	27		NEBULA
RNGC 4748	12 49 35.	- 13 08			GALAXY
ZWG 159.097	12 49 36.	+ 27 18		15.4	GALAXY
MCG+05-30-109	12 49 36.	+ 27 44	12	15.5	GALAXY

OBJECT NAME	RIGHT ASCEN.	DECLINATION	DIAM.	MAGN.	TYPE OF OBJECT
STOCK 38	12 49 36.	+ 27 45			BLUE KNOT NEAR ELLIP GLXY
ZWG 159.098	12 49 36.	+ 27 51		15.5	GALAXY
MCG+05-30-110	12 49 36.	+ 27 51	18	15.5	GALAXY
LB 07276	12 49 36.	+ 28 32		18.7	FAINT BLUE STAR
LB 07275	12 49 36.	+ 29 15		19.0	FAINT BLUE STAR
LB 07274	12 49 36.	+ 30 57		18.8	FAINT BLUE STAR
MCG+09-21-057	12 49 36.	+ 52 07	42	14.	GALAXY
LB 00661	12 49 36.	+ 74 30 06.		16.0	FAINT BLUE STAR
REIZ 2928	12 49 36.	- 09 11	24	15.0	GALAXY
MCG-01-33-031	12 49 36.	- 09 13	42	15.	GALAXY
MCG-02-33-034	12 49 36.	- 13 08	42	14.5	GALAXY
MCG-07-27-006	12 49 36.	- 40 48	90	13.5	GALAXY
MCG-07-27-005	12 49 36.	- 41 07	66	14.	GALAXY
KN 16.513	12 49 36.0	+ 29 06 05.			NEBULA
KEEL 570	12 49 36.0	+ 41 08 32.		17.	NEBULA
KN 16.514	12 49 36.8	+ 27 26 11.			NEBULA
BFG 028	12 49 36.86	+ 36 56 52.2		17.97	ULTRAVIOLET-EXCESS OBJECT
RNGC 4744	12 49 37.	- 40 46			GALAXY
KN 16.515	12 49 38.9	+ 26 53 41.			NEBULA
A2 103	12 49 39.5	+ 26 57 40.		18.4	FAINT BLUE GALAXY
KN 16.516	12 49 39.7	+ 27 52 11.			NEBULA
KN 16.518	12 49 40.6	+ 27 17 59.			NEBULA
BC B86	12 49 40.6	+ 33 54 42.		17.58	QUASI-STELLAR OBJECT
SHB 217	12 49 40.6	+ 33 54 46.		17.6	QUASI-STELLAR OBJECT
BFG 029	12 49 40.64	+ 33 54 46.5		17.58	ULTRAVIOLET-EXCESS OBJECT
SN 1965I	12 49 41.	- 00 55		13.5	SUPERNOVA
HELW 451	12 49 41.	- 06 14 56.			NEBULA
KN 16.517	12 49 41.0	+ 25 53 47.			NEBULA
8ZW 1249+21.9	12 49 42.	+ 21 53		15.5	COMPACT GALAXY
LB 11391	12 49 42.	+ 26 26		14.3	FAINT BLUE STAR
LB 07279	12 49 42.	+ 26 47		18.0	FAINT BLUE STAR
MCG+05-30-111	12 49 42.	+ 27 48	36	15.5	GALAXY
LB 07278	12 49 42.	+ 28 07		19.1	FAINT BLUE STAR
LB 07277	12 49 42.	+ 32 29		17.6	FAINT BLUE STAR
LB 02469	12 49 42.	+ 57 01 06.		15.1	FAINT BLUE STAR
7ZW 486	12 49 42.	+ 66 57			COMPACT GALAXY
MCG-02-33-035	12 49 42.	- 15 13 30.	6	17.	GALAXY
MCG-02-33-036	12 49 42.	- 15 14 30.	30	15.	GALAXY
MCG-07-27-007	12 49 42.	- 40 27		14.5	GALAXY
KN 16.519	12 49 43.1	+ 27 44 53.			NEBULA
IC 3844	12 49 44.	+ 40 05 17.			NONSTELLAR OBJECT
KN 16.520	12 49 44.0	+ 27 50 53.			NEBULA
KN 16.521	12 49 44.9	+ 27 44 17.			NEBULA
HOLM 478B	12 49 45.	+ 11 36	180	12.4	PART OF MULTIPLE GALAXY
RNGC 4752	12 49 45.	+ 13 45			NON-EXISTENT OBJECT
MCG-02-33-037	12 49 45.	- 15 14 30.	15	15.	GALAXY
KEEL 571	12 49 45.5	+ 41 30 10.		18.	NEBULA
KN 16.522	12 49 45.8	+ 27 09 05.			NEBULA
HELW 169	12 49 46.	- 12 50 44.			NEBULA
KN 16.524	12 49 46.8	+ 28 39 29.			NEBULA
REIN 4.154	12 49 46.88	+ 11 35 07.8			NEBULA
IC 3845	12 49 47.	+ 38 53 23.			NONSTELLAR OBJECT
KN 16.523	12 49 47.2	+ 26 38 17.			NEBULA
KARA.68 208	12 49 48.	+ 02 34	27		DWARF GALAXY
ZWG 043.049	12 49 48.	+ 07 22		15.7	GALAXY
KARA.68 209	12 49 48.	+ 11 28	27		DWARF GALAXY
ZWG 071.062	12 49 48.	+ 11 35		11.6	GALAXY
UGC 08010	12 49 48.	+ 11 35	300	11.6	GALAXY SB0
KARA.72 356A	12 49 48.	+ 11 35	288	11.6	PART OF DOUBLE GALAXY
MCG+02-33-030	12 49 48.	+ 11 35	270	11.6	GALAXY
ZC 1249.8+1840	12 49 48.	+ 18 40	1080		CLUSTER OF GALAXIES
UGC 08011	12 49 48.	+ 21 55	90	18.	GALAXY DWRF IR
LB 07285	12 49 48.	+ 26 42		19.4	FAINT BLUE STAR
LB 07284	12 49 48.	+ 26 52		19.6	FAINT BLUE STAR
LB 07283	12 49 48.	+ 27 48		19.2	FAINT BLUE STAR
LB 07282	12 49 48.	+ 28 20		18.2	FAINT BLUE STAR
LB 07281	12 49 48.	+ 30 47		19.5	FAINT BLUE STAR
LB 07280	12 49 48.	+ 31 36		18.8	FAINT BLUE STAR
ZWG 015.029	12 49 48.	- 00 56		11.7	GALAXY
UGC 08009	12 49 48.	- 00 56	360	11.7	GALAXY
MCG+00-33-016	12 49 48.	- 00 56	240	11.7	GALAXY
REIZ 2929	12 49 48.	- 09 23	108	13.4	GALAXY
MCG-01-33-032	12 49 48.	- 09 28	120	14.5	GALAXY
MCG-02-33-038	12 49 48.	- 15 14 30.	24	15.5	GALAXY
MCG-04-30-035	12 49 48.	- 21 38	24	15.5	GALAXY
REIN 4.153	12 49 48.02	- 00 55 42.0			NEBULA
REIZ 2931	12 49 49.	+ 11 35	240	12.0	GALAXY
RNGC 4753	12 49 49.	- 00 56		11.0	GALAXY
KN 16.525	12 49 49.1	+ 26 49 53.			NEBULA
A2 108	12 49 49.1	+ 29 34 37.		17.8	FAINT BLUE OBJECT
KN 16.527	12 49 49.4	+ 29 41 59.			NEBULA
IC 0829	12 49 50.	- 15 15			NONSTELLAR OBJECT
KN 16.526	12 49 50.3	+ 27 48 24.			NEBULA
KN 16.528	12 49 50.3	+ 28 39 48.			NEBULA
KEEL 572	12 49 50.9	+ 41 39 41.		18.	NEBULA
RNGC 4754	12 49 51.	+ 11 35		12.0	GALAXY
MCG-04-30-036	12 49 51.	- 21 54 30.	36	15.5	GALAXY
KN 16.528A	12 49 51.2	+ 27 36 48.			NEBULA
KN 16.529	12 49 51.7	+ 27 21 30.			NEBULA
REIZ 2930	12 49 52.	- 00 55	120	11.0	GALAXY
KN 16.530	12 49 52.0	+ 28 40 30.			NEBULA
A2 110	12 49 53.5	+ 26 57 09.		17.5	FAINT BLUE OBJECT
KN 16.531	12 49 53.8	+ 28 38 48.			NEBULA
ZC 1249.9+0722	12 49 54.	+ 07 22	1550		CLUSTER OF GALAXIES
KARA.68 210	12 49 54.	+ 10 43	40		DWARF GALAXY
LB 11395	12 49 54.	+ 27 44		13.2	FAINT BLUE STAR
TON-N 0665	12 49 54.	+ 28 08		16.5	BLUE STAR
TON-N 0666	12 49 54.	+ 28 38		13.9	BLUE STAR
TON-N 0667	12 49 54.	+ 28 40		17.0	BLUE STAR
LB 07287	12 49 54.	+ 28 42		18.9	FAINT BLUE STAR
LB 11394	12 49 54.	+ 29 29		16.2	FAINT BLUE STAR
LB 07286	12 49 54.	+ 30 39		18.8	FAINT BLUE STAR
LB 11392	12 49 54.	+ 31 36		17.5	FAINT BLUE STAR
ZWG 270.029	12 49 54.	+ 51 58		14.3	GALAXY
UGC 08012	12 49 54.	+ 51 58	48	14.3	GALAXY S
ZWG 293.037	12 49 54.	+ 60 48		15.6	GALAXY
MCG+10-18-085	12 49 54.	+ 60 49	27	16.	GALAXY
7ZW 487	12 49 54.	+ 67 00			COMPACT GALAXY
HOLM 476A	12 49 54.	- 09 29	120	13.8	PART OF MULTIPLE GALAXY
MCG-06-28-022	12 49 54.	- 38 44	72	13.5	GALAXY
KN 16.532	12 49 55.5	+ 28 38 06.			NEBULA
KN 16.534	12 49 56.6	+ 28 38 00.			NEBULA
KN 16.533	12 49 57.1	+ 26 55 00.			NEBULA
REIZ 2932	12 49 58.	- 01 45	48	14.6	GALAXY
KEEL 573	12 49 58.5	+ 25 28 44.		16.	NEBULA
KN 16.535	12 49 59.0	+ 26 45 18.			NEBULA
VDB.66G 153	12 50	- 11 51	100		DWARF GALAXY
LB 09866	12 50	- 86 08		13.3	FAINT BLUE STAR
ZWG 043.050	12 50 00.	+ 03 31		15.7	GALAXY
ZC 1250.0+0415	12 50 00.	+ 04 15	870		CLUSTER OF GALAXIES
ZC 1250.0+1959	12 50 00.	+ 19 59	2220		CLUSTER OF GALAXIES
LB 07289	12 50 00.	+ 27 05		18.0	FAINT BLUE STAR
LB 11396	12 50 00.	+ 30 55		13.5	FAINT BLUE STAR
LB 07288	12 50 00.	+ 32 12		20.4	FAINT BLUE STAR
MCG+08-23-099	12 50 00.	+ 48 28	36	16.	GALAXY
LB 02470	12 50 00.	+ 53 24 18.		14.2	FAINT BLUE STAR
ZWG 353.001	12 50 00.	+ 78 59		15.6	GALAXY
ZWG 352.043	12 50 00.	+ 78 59		15.6	GALAXY
MCG+15-01-013	12 50 00.	+ 88 04	42	16.	GALAXY
KARA.68 211	12 50 00.	- 06 00	67		DWARF GALAXY
MCG-01-33-033	12 50 00.	- 06 00	102	16.	GALAXY
MCG-01-33-034	12 50 00.	- 09 30	60	15.	GALAXY
MRSL 303-00/1	12 50 00.	- 63 00	5400		HII REGION
VHA 140	12 50 00.	- 66 55	270		OPEN STAR CLUSTER
A2 112	12 50 00.0	+ 29 29 33.		17.2	FAINT BLUE OBJECT
KEEL 574	12 50 00.0	+ 41 31 00.		18.	NEBULA
KN 16.536	12 50 00.5	+ 26 46 06.			NEBULA
KN 16.537	12 50 01.7	+ 29 16 12.			NEBULA
LB 02471	12 50 02.	+ 57 10 00.		13.8	FAINT BLUE STAR
HOLM 476B	12 50 03.	- 09 30	18	13.9	PART OF MULTIPLE GALAXY
A2 114	12 50 03.5	+ 29 37 00.		17.4	FAINT BLUE OBJECT
REIN 4.155	12 50 04.07	+ 12 31 43.1			NEBULA
KEEL 575	12 50 04.8	+ 41 26 29.		18.	NEBULA
BFG 030	12 50 04.89	+ 36 28 53.2		19.02	ULTRAVIOLET-EXCESS OBJECT
REIZ 2933	12 50 05.	- 07 10	54	14.3	GALAXY
ZC 1250.1+0804	12 50 06.	+ 08 04	540		CLUSTER OF GALAXIES
ZWG 071.063	12 50 06.	+ 12 31		15.6	GALAXY
MCG+02-33-031	12 50 06.	+ 12 31	36	15.6	GALAXY
REIZ 2936	12 50 06.	+ 12 32	42	15.1	GALAXY
ZC 1250.1+1613	12 50 06.	+ 16 13	1010		CLUSTER OF GALAXIES
8ZW 1250+22.1	12 50 06.	+ 22 05		16.7	COMPACT GALAXY
IC 3847	12 50 06.	+ 22 20 05.			NONSTELLAR OBJECT
ZC 1250.1+2509	12 50 06.	+ 25 09	810		CLUSTER OF GALAXIES
LB 11401	12 50 06.	+ 26 20		14.4	FAINT BLUE STAR
LB 11400	12 50 06.	+ 26 51		11.3	FAINT BLUE STAR
MCG+05-30-112	12 50 06.	+ 27 00	90	15.7	GALAXY
ZWG 159.099	12 50 06.	+ 27 01		15.7	GALAXY
UGC 08013	12 50 06.	+ 27 01	84	15.7	GALAXY S
LB 11399	12 50 06.	+ 29 03		12.5	FAINT BLUE STAR
ZC 1250.1+2905	12 50 06.	+ 29 05	2620		CLUSTER OF GALAXIES
LB 07293	12 50 06.	+ 29 14		19.3	FAINT BLUE STAR
LB 11398	12 50 06.	+ 30 06		9.8	FAINT BLUE STAR
LB 11397	12 50 06.	+ 30 48		13.2	FAINT BLUE STAR
LB 07292	12 50 06.	+ 31 34		19.3	FAINT BLUE STAR
LB 07291	12 50 06.	+ 31 59		16.7	FAINT BLUE STAR
LB 07290	12 50 06.	+ 32 26		17.5	FAINT BLUE STAR
REIZ 2942	12 50 06.	+ 56 54	90	15.2	GALAXY
MCG+13-09-035	12 50 06.	+ 79 01	12	16.	GALAXY
MCG-07-27-008	12 50 06.	- 40 11	90	13.5	GALAXY
RNGC 4751	12 50 06.	- 42 24			GALAXY
KN 16.538	12 50 06.8	+ 26 54 30.			NEBULA
KN 16.538A	12 50 07.9	+ 27 35 12.			NEBULA
FATH 1.545	12 50 08.	+ 44 08	5		NEBULA
IC 3846	12 50 09.	+ 13 55 11.			NONSTELLAR OBJECT
RNGC 4758	12 50 09.	+ 16 07		14.0	GALAXY
KN 16.539	12 50 10.2	+ 27 01 18.			NEBULA
KN 16.540	12 50 10.2	+ 27 54 12.			NEBULA
RNGC 4740	12 50 11.	- 14 12			GALAXY
ARC 1631	12 50 11.	- 15 10		15.4	RICH CLUSTER OF GALAXIES
KN 16.540A	12 50 11.5	+ 26 31 18.			NEBULA
KN 16.541	12 50 11.9	+ 26 54 18.			NEBULA
ZC 1250.2+0506	12 50 12.	+ 05 06	1010		CLUSTER OF GALAXIES
ZWG 100.015	12 50 12.	+ 16 07		14.1	GALAXY
UGC 08014	12 50 12.	+ 16 07	180	14.1	GALAXY IRR
IC 3848	12 50 12.	+ 21 41 05.			NONSTELLAR OBJECT
LB 11402	12 50 12.	+ 26 33		10.5	FAINT BLUE STAR
ZWG 159.100	12 50 12.	+ 26 45		15.5	GALAXY
LB 07295	12 50 12.	+ 28 17		18.4	FAINT BLUE STAR
LB 07294	12 50 12.	+ 31 19		20.0	FAINT BLUE STAR
MCG-01-33-035	12 50 12.	- 07 07	72	13.	GALAXY
REIZ 2935	12 50 12.	- 09 13	24	15.6	GALAXY
REIZ 2934	12 50 12.	- 10 02	18	14.3	GALAXY
MCG-02-33-040	12 50 12.	- 10 02	66	15.	GALAXY
MCG-02-33-039	12 50 12.	- 15 08 30.	90	13.	GALAXY
MCG-07-27-010	12 50 12.	- 41 02	90	16.	GALAXY
MCG-07-27-009	12 50 12.	- 41 03	90	14.5	GALAXY
MCG-07-27-011	12 50 12.	- 42 23	120	12.	GALAXY
A2 117	12 50 12.8	+ 30 18 26.		18.2	FAINT BLUE OBJECT
KN 16.542	12 50 13.1	+ 27 09 18.			NEBULA
KEEL 576	12 50 14.5	+ 41 31 51.		18.	NEBULA
KN 16.543	12 50 15.0	+ 26 38 18.			NEBULA
SHB 218	12 50 15.1	+ 56 50 37.		17.9	QUASI-STELLAR OBJECT
BC 3CR277.1	12 50 15.28	+ 56 50 36.7		17.93	QUASI-STELLAR OBJECT
KN 16.544	12 50 15.8	+ 29 14 42.			NEBULA
IC 3849	12 50 16.	+ 41 02 42.			NONSTELLAR OBJECT
KN 16.546	12 50 16.2	+ 29 13 48.			NEBULA
A2 120	12 50 16.8	+ 28 29 08.		18.2	FAINT BLUE OBJECT
RNGC 4756	12 50 17.	- 15 08		13.5	GALAXY
KN 16.545	12 50 17.0	+ 26 54 12.			NEBULA
ZWG 043.051	12 50 18.	+ 07 24		15.6	GALAXY
MCG+03-33-015	12 50 18.	+ 16 06 30.	180	14.1	GALAXY
ZC 1250.3+2131	12 50 18.	+ 21 31	470		CLUSTER OF GALAXIES
MCG+05-30-113	12 50 18.	+ 26 43	30	15.5	GALAXY
IC 0831	12 50 18.	+ 26 45		15.5	GALAXY IN COMA CLSTR
ZWG 159.101	12 50 18.	+ 27 40		15.3	GALAXY
LB 07299	12 50 18.	+ 28 08		19.5	FAINT BLUE STAR
LB 07298	12 50 18.	+ 28 35		19.4	FAINT BLUE STAR
LB 07297	12 50 18.	+ 28 41		19.0	FAINT BLUE STAR
LB 07296	12 50 18.	+ 32 26		19.4	FAINT BLUE STAR
MCG+07-27-001	12 50 18.	+ 40 22	54	15.	GALAXY
ZWG 293.038	12 50 18.	+ 60 20		15.7	GALAXY
MCG+10-18-086	12 50 18.	+ 60 21	27	16.	GALAXY
SVEN 343	12 50 18.	- 10 02	66	14.8	GALAXY
RNGC 4757	12 50 18.	- 10 02		15.0	GALAXY
DV.56 N4767A	12 50 18.	- 39 35	60		S GALAXY
MCG-07-27-011A	12 50 18.	- 39 35	36	16.	GALAXY
KN 16.547	12 50 18.1	+ 26 44 30.			NEBULA
KN 16.547A	12 50 18.4	+ 29 13 30.			NEBULA
IC 3850	12 50 19.	+ 40 22 24.			NONSTELLAR OBJECT
RNGC 4767A	12 50 19.	- 39 35			GALAXY
KN 16.548	12 50 20.4	+ 28 10 48.			NEBULA
A2 121	12 50 20.7	+ 28 22 55.		17.7	FAINT BLUE OBJECT
BFG 031	12 50 21.72	+ 36 02 19.8		18.78	ULTRAVIOLET-EXCESS OBJECT
HELW 452	12 50 22.	- 10 10 43.			NEBULA
REIN 4.156	12 50 22.39	+ 10 15 40.1			NEBULA
KN 16.549	12 50 22.7	+ 28 03 36.			NEBULA

OBJECT NAME	RIGHT ASCEN.	DECLINATION	DIAM.	MAGN.	TYPE OF OBJECT
KEEL 577	12 50 22.7	+ 41 18 04.		16.	NEBULA
HELW 453	12 50 23.	- 10 01 43.			NEBULA
KN 16.550	12 50 23.1	+ 27 40 24.			NEBULA
ZWG 043.052	12 50 24.	+ 07 46		15.7	GALAXY
ZWG 071.064	12 50 24.	+ 10 16		15.1	GALAXY
UGC 08015	12 50 24.	+ 10 16	108	15.1	GALAXY Sa-b
MCG+02-33-032	12 50 24.	+ 10 16	120	15.1	GALAXY
ZWG 071.065	12 50 24.	+ 11 30		11.1	GALAXY
UGC 08016	12 50 24.	+ 11 30	570	11.1	GALAXY SB?0
KARA.72 356B	12 50 24.	+ 11 30	324	11.1	PART OF DOUBLE GALAXY
MCG+02-33-033	12 50 24.	+ 11 30	540	11.1	GALAXY
REA 21	12 50 24.	+ 14 40			DWARF GALAXY
LB 07301	12 50 24.	+ 27 23		19.3	FAINT BLUE STAR
MCG+05-30-114	12 50 24.	+ 28 38	45	14.5	GALAXY
ZWG 159.102	12 50 24.	+ 28 39		14.5	GALAXY
UGC 08017	12 50 24.	+ 28 39	54	14.5	GALAXY
LB 11404	12 50 24.	+ 30 19		14.7	FAINT BLUE STAR
LB 11403	12 50 24.	+ 30 47		14.0	FAINT BLUE STAR
LB 07300	12 50 24.	+ 31 18		19.8	FAINT BLUE STAR
MCG+13-09-036	12 50 24.	+ 79 02	15	17.	GALAXY
SVEN 345	12 50 24.	- 10 01 50.	24	15.9	GALAXY
REIZ 2937	12 50 24.	- 10 10	18	15.8	GALAXY
SVEN 344	12 50 24.	- 10 10 50.	6	16.0	GALAXY
AGU 40	12 50 24.	- 41 20 24.	126	12.5	PECULIAR GALAXY
REIZ 2940	12 50 25.	+ 10 16	60	14.1	GALAXY
REIZ 2941	12 50 25.	+ 11 31	270	11.7	GALAXY
HOLM 478A	12 50 25.	+ 11 31	270	11.8	PART OF MULTIPLE GALAXY
LB 02472	12 50 25.	+ 54 23 00.		17.0	FAINT BLUE STAR
REIN 4.157	12 50 25.41	+ 11 30 07.3			NEBULA
KN 16.551	12 50 26.4	+ 27 27 18.			NEBULA
RNGC 4762	12 50 27.	+ 11 30		11.5	GALAXY
KN 16.552	12 50 27.9	+ 28 38 30.			NEBULA
HELW 454	12 50 28.	- 10 05 37.			NEBULA
A2 123	12 50 29.0	+ 26 49 28.		16.8	FAINT BLUE OBJECT
ZWG 043.053	12 50 30.	+ 07 57		15.5	GALAXY
ZC 1250.5+1351	12 50 30.	+ 13 51	340		CLUSTER OF GALAXIES
ZC 1250.5+1923	12 50 30.	+ 19 23	810		CLUSTER OF GALAXIES
LB 07305	12 50 30.	+ 26 33		19.0	FAINT BLUE STAR
LB 07304	12 50 30.	+ 26 34		19.0	FAINT BLUE STAR
LB 07303	12 50 30.	+ 26 55		19.8	FAINT BLUE STAR
LB 07302	12 50 30.	+ 30 12		20.3	FAINT BLUE STAR
HOLM 479A	12 50 30.	+ 32 42	42	15.3	PART OF MULTIPLE GALAXY
ZC 1250.5+4001	12 50 30.	+ 40 01	540		CLUSTER OF GALAXIES
ZWG 245.004	12 50 30.	+ 47 57		15.7	GALAXY
ZWG 244.046	12 50 30.	+ 47 57		15.7	GALAXY
LB 02473	12 50 30.	+ 56 45 18.		16.3	FAINT BLUE STAR
RNGC 4761	12 50 30.	- 08 55		14.5	GALAXY
RNGC 4759	12 50 30.	- 08 55		15.0	GALAXY
HOLM 477A	12 50 30.	- 08 55	24	13.6	PART OF MULTIPLE GALAXY
MCG-01-33-037	12 50 30.	- 08 55	24	14.5	GALAXY
MCG-01-33-036	12 50 30.	- 08 55	24	15.	GALAXY
REIZ 2939	12 50 30.	- 08 56	24	14.4	GALAXY
SVEN 347	12 50 30.	- 09 51 50.	30	15.6	GALAXY
HELW 455	12 50 30.	- 09 51 55.			NEBULA
MCG-02-33-042	12 50 30.	- 10 05	60	15.	GALAXY
SVEN 346	12 50 30.	- 10 05 50.	6	15.8	GALAXY
SVEN 349	12 50 30.	- 10 06	42	14.9	GALAXY
RNGC 4766	12 50 30.	- 10 06		15.0	GALAXY
MCG-02-33-041	12 50 30.	- 10 12	24	13.	GALAXY
SVEN 348	12 50 30.	- 10 13	36	12.6	GALAXY
REIZ 2938	12 50 30.	- 10 13	24	13.4	GALAXY
RNGC 4760	12 50 30.	- 10 14		13.0	GALAXY
MCG-04-31-001	12 50 30.	- 27 13	48	15.	GALAXY
REIN 2.163	12 50 30.86	- 10 13 23.1			NEBULA
HOLM 479B	12 50 31.	+ 32 43	18	16.1	PART OF MULTIPLE GALAXY
SC 1247+0000.2	12 50 31.	- 00 16 07.	12		NEBULA
A2 124	12 50 31.2	+ 27 47 02.		16.8	FAINT BLUE GALAXY
HELW 170	12 50 32.	- 09 51 37.			NEBULA
MCG+08-23-100	12 50 33.	+ 47 58 30.	12	16.	GALAXY
KN 16.553	12 50 33.3	+ 26 59 36.			NEBULA
BFG 032	12 50 33.54	+ 33 32 13.2		18.92	ULTRAVIOLET-EXCESS OBJECT
HOLM 477B	12 50 34.	- 08 55	18	14.2	PART OF MULTIPLE GALAXY
A2 126	12 50 34.4	+ 26 43 30.		17.6	FAINT BLUE GALAXY
KN 16.554	12 50 34.4	+ 26 43 37.			NEBULA
REIZ 2948	12 50 35.	+ 36 04	42	14.8	GALAXY
HELW 456	12 50 35.	- 06 08 37.			NEBULA
RNGC 4763	12 50 35.	- 16 43		14.0	GALAXY
KN 16.555	12 50 35.4	+ 29 20 55.			NEBULA
REIN 4.158	12 50 35.51	+ 10 11 55.8			NEBULA
ZWG 071.066	12 50 36.	+ 10 12		15.6	GALAXY
ZC 1250.6+2401	12 50 36.	+ 24 01	740		CLUSTER OF GALAXIES
LB 07313	12 50 36.	+ 26 27		19.3	FAINT BLUE STAR
LB 07312	12 50 36.	+ 27 12		18.4	FAINT BLUE STAR
LB 11406	12 50 36.	+ 28 17		13.3	FAINT BLUE STAR
LB 11405	12 50 36.	+ 28 27		12.8	FAINT BLUE STAR
LB 07311	12 50 36.	+ 28 42		18.7	FAINT BLUE STAR
LB 07310	12 50 36.	+ 29 07		19.5	FAINT BLUE STAR
LB 07309	12 50 36.	+ 30 15		20.1	FAINT BLUE STAR
LB 07308	12 50 36.	+ 30 22		19.5	FAINT BLUE STAR
LB 07307	12 50 36.	+ 31 46		19.7	FAINT BLUE STAR
LB 07306	12 50 36.	+ 32 10		19.8	FAINT BLUE STAR
ZWG 159.103	12 50 36.	+ 32 22		14.8	GALAXY
MCG+05-30-115	12 50 36.	+ 32 22 30.	36	14.8	GALAXY
1ZW 044	12 50 36.	+ 47 01			COMPACT GALAXY
ZWG 293.039	12 50 36.	+ 59 32		15.7	GALAXY
KARA.73B 0554	12 50 36.	+ 59 32	42	15.7	ISOLATED GALAXY S
MCG-01-33-038	12 50 36.	- 06 37	42	16.	GALAXY
RNGC 4764	12 50 36.	- 08 55		16.0	GALAXY
REIZ 2945	12 50 36.	- 08 55	30	15.2	GALAXY
MCG-01-33-039	12 50 36.	- 08 55	30	16.	GALAXY
REIZ 2944	12 50 36.	- 10 03	12	14.4	GALAXY
REIZ 2943	12 50 36.	- 10 03	9	14.7	GALAXY
OCL 0892	12 50 36.	- 60 04	1260	5.2	OPEN STAR CLUSTER
VHA 141	12 50 36.	- 60 04	420		OPEN STAR CLUSTER
RNGC 4755	12 50 36.	- 60 05		5.0	OPEN CLUSTER
A2 127	12 50 36.4	+ 28 12 11.		16.4	FAINT BLUE OBJECT
REIZ 2946	12 50 37.	+ 10 13	18	14.6	GALAXY
IC 3851	12 50 37.	+ 22 10 54.			NONSTELLAR OBJECT
ARC 1632	12 50 37.	+ 29 06		17.2	RICH CLUSTER OF GALAXIES
HELW 457	12 50 37.	- 06 07 01.			NEBULA
HELW 171	12 50 38.	- 10 45 49.			NEBULA
BFG 033	12 50 38.36	+ 35 17 37.9		19.14	ULTRAVIOLET-EXCESS OBJECT
KEEL 578	12 50 38.6	+ 41 11 15.		18.	NEBULA
LB 02474	12 50 39.	+ 51 42 42.		15.4	FAINT BLUE STAR
IC 0828	12 50 40.	- 07 51 24.			NONSTELLAR OBJECT
IC 3852	12 50 41.	+ 36 02 42.			NONSTELLAR OBJECT
LB 02475	12 50 41.	+ 60 05 12.		15.5	FAINT BLUE STAR
RNGC 4785	12 50 41.	- 48 30			GALAXY

OBJECT NAME	RIGHT ASCEN.	DECLINATION	DIAM.	MAGN.	TYPE OF OBJECT
KN 16.556	12 50 41.4	+ 27 18 43.			NEBULA
REIN 4.159	12 50 41.71	+ 10 12 48.2			NEBULA
KN 16.559	12 50 41.8	+ 28 30 49.			NEBULA
KN 16.557	12 50 41.9	+ 26 44 55.			NEBULA
ZWG 043.054	12 50 42.	+ 04 44		13.0	GALAXY
UGC 08018	12 50 42.	+ 04 44	96	13.0	GALAXY
MCG+01-33-020	12 50 42.	+ 04 44	48	13.0	GALAXY
ZC 1250.7+0539	12 50 42.	+ 05 39	1280		CLUSTER OF GALAXIES
ZWG 129.029	12 50 42.	+ 25 32		15.5	GALAXY
KARA.73B 0555	12 50 42.	+ 25 32	18	15.5	ISOLATED GALAXY E
ZWG 159.104	12 50 42.	+ 27 21		15.0	GALAXY
LB 07316	12 50 42.	+ 27 51		18.8	FAINT BLUE STAR
LB 07315	12 50 42.	+ 28 12		19.1	FAINT BLUE STAR
LB 07314	12 50 42.	+ 30 57		19.6	FAINT BLUE STAR
ZC 1250.7+3335	12 50 42.	+ 33 35	3290		CLUSTER OF GALAXIES
UGC 08019	12 50 42.	+ 36 03	84	16.0	GALAXY SBc
1ZW 045	12 50 42.	+ 37 05			COMPACT GALAXY
ZC 1250.7+4136	12 50 42.	+ 41 36	1210		CLUSTER OF GALAXIES
MCG+10-18-087	12 50 42.	+ 59 32	39	15.	GALAXY
RNGC 4769	12 50 42.	- 09 16			NON-EXISTENT OBJECT
RNGC 4768	12 50 42.	- 09 16			NON-EXISTENT OBJECT
KARA.68 212	12 50 42.	- 21 29	67		DWARF GALAXY
MCG-04-31-002	12 50 42.	- 22 26	30	15.5	GALAXY
MCG-07-27-012	12 50 42.	- 40 56	66	13.	GALAXY
MCG-07-27-013	12 50 42.	- 41 22	30	14.	GALAXY
SER 093.02	12 50 42.	- 48 40	40	15.	LOW SURFACE BRIGHT. GALXY
KN 16.558	12 50 42.0	+ 27 11 13.			NEBULA
A2 128	12 50 43.2	+ 28 11 22.		18.6	FAINT BLUE OBJECT
RNGC 4765	12 50 44.	+ 04 44		13.0	GALAXY
REIZ 2947	12 50 44.	+ 04 44	30	13.0	GALAXY
A2 129	12 50 44.0	+ 26 31 39.		16.5	FAINT BLUE OBJECT
HELW 172	12 50 45.	- 12 00 55.			NEBULA
LB 02476	12 50 46.	+ 52 06 00.		15.8	FAINT BLUE STAR
KEEL 579	12 50 46.3	+ 41 32 21.		18.	NEBULA
KN 16.560	12 50 46.7	+ 27 35 37.			NEBULA
REIZ 2954	12 50 47.	+ 37 06	30	14.2	GALAXY
A2 130	12 50 47.4	+ 29 50 33.		18.2	FAINT BLUE OBJECT
ZWG 015.031	12 50 48.	+ 01 32		13.3	GALAXY
UGC 08020	12 50 48.	+ 01 32	240	13.3	GALAXY Sc
MCG+00-33-017	12 50 48.	+ 01 32	240	13.3	GALAXY
LB 07322	12 50 48.	+ 26 25		19.9	FAINT BLUE STAR
LB 11410	12 50 48.	+ 26 44		13.7	FAINT BLUE STAR
MCG+05-30-116	12 50 48.	+ 27 21	48	15.0	GALAXY
LB 07321	12 50 48.	+ 27 38		18.1	FAINT BLUE STAR
LB 07320	12 50 48.	+ 28 37		20.4	FAINT BLUE STAR
LB 11409	12 50 48.	+ 30 23		14.5	FAINT BLUE STAR
LB 07319	12 50 48.	+ 30 47		19.2	FAINT BLUE STAR
LB 11408	12 50 48.	+ 31 22		15.8	FAINT BLUE STAR
LB 11407	12 50 48.	+ 31 40		13.0	FAINT BLUE STAR
LB 07318	12 50 48.	+ 31 56		19.2	FAINT BLUE STAR
LB 07317	12 50 48.	+ 32 08		20.2	FAINT BLUE STAR
MCG+06-28-038	12 50 48.	+ 36 03	66	15.	GALAXY
ZWG 188.026	12 50 48.	+ 37 05		14.6	GALAXY
MCG+06-28-037	12 50 48.	+ 37 06	30	14.	GALAXY
ZC 1250.8+6306	12 50 48.	+ 63 06	610		CLUSTER OF GALAXIES
ZWG 015.030	12 50 48.	- 00 09		15.6	GALAXY
KARA.68 213	12 50 48.	- 04 42	40		DWARF GALAXY
MCG-03-33-013	12 50 48.	- 16 44	72	14.	GALAXY
RNGC 4771	12 50 49.	+ 01 32		13.0	GALAXY
RNGC 4774	12 50 49.	+ 37 05		14.5	GALAXY
IC 3853	12 50 49.	+ 39 05 54.			NONSTELLAR OBJECT
KN 16.561	12 50 50.2	+ 27 21 55.			NEBULA
REIZ 2951	12 50 51.	+ 01 33	240	12.7	GALAXY
HOLM 481A	12 50 51.	+ 24 48	48	12.4	PART OF MULTIPLE GALAXY
REIZ 2957	12 50 51.	+ 40 41	36	14.2	GALAXY
KN 16.561A	12 50 51.3	+ 28 44 31.			NEBULA
KEEL 580	12 50 51.9	+ 41 07 12.		17.	NEBULA
HOLM 481B	12 50 52.	+ 24 50	24	15.4	PART OF MULTIPLE GALAXY
LB 02477	12 50 52.	+ 50 38 48.		17.0	FAINT BLUE STAR
KEEL 581	12 50 52.1	+ 41 36 58.		18.	NEBULA
KN 16.562	12 50 53.2	+ 28 28 55.			NEBULA
KEEL 582	12 50 53.9	+ 41 07 11.		16.	NEBULA
ZWG 015.032	12 50 54.	+ 02 26		12.9	GALAXY
UGC 08021	12 50 54.	+ 02 26	270	12.9	GALAXY Sa
MCG+00-33-018	12 50 54.	+ 02 26	156	12.9	GALAXY
TON-N 0668	12 50 54.	+ 26 46		15.8	BLUE STAR
LB 07325	12 50 54.	+ 26 54		19.1	FAINT BLUE STAR
LB 07324	12 50 54.	+ 27 06		18.9	FAINT BLUE STAR
HZ 35	12 50 54.	+ 30 23		15.0	DECIDEDLY BLUE STAR
LB 07323	12 50 54.	+ 31 19		19.6	FAINT BLUE STAR
IC 3854	12 50 54.	+ 41 07 01.			NONSTELLAR OBJECT
LB 00662	12 50 54.	+ 61 59 06.		17.4	FAINT BLUE STAR
MCG-01-33-042	12 50 54.	- 08 21	15	16.	GALAXY
MCG-01-33-041	12 50 54.	- 08 21	30	15.	GALAXY
RNGC 4770	12 50 54.	- 09 15		14.0	GALAXY
REIZ 2950	12 50 54.	- 09 15	30	13.8	GALAXY
MCG-01-33-040	12 50 54.	- 09 15	36	14.	GALAXY
SVEN 350	12 50 54.	- 10 10	24	16.3	GALAXY
HELW 458	12 50 54.	- 10 10 13.			NEBULA
PK303+40.1	12 50 54.	- 22 36	938	12.0	PLANETARY NEBULA
MCG-04-31-003	12 50 54.	- 26 03 30.	48	15.	GALAXY
RNGC 4772	12 50 55.	+ 02 26		12.5	GALAXY
REIZ 2949	12 50 55.	- 12 31	30	14.9	GALAXY
A2 135	12 50 56.9	+ 26 37 36.		17.1	FAINT BLUE OBJECT
REIZ 2953	12 50 57.	+ 02 24	180	12.9	GALAXY
MCG-02-33-043	12 50 57.	- 11 44	48	15.5	GALAXY
MCG-04-31-004	12 50 57.	- 25 09	36	15.	GALAXY
KN 16.563	12 50 57.1	+ 27 18 19.			NEBULA
A2 136	12 50 58.8	+ 28 37 39.		18.1	FAINT BLUE OBJECT
VDB.66G 154	12 51	+ 27 24	200		DWARF GALAXY
ZWG 071.067	12 51 00.	+ 09 49		15.7	GALAXY
8ZW 1251+21.7	12 51 00.	+ 21 43		16.5	COMPACT GALAXY
LB 00021	12 51 00.	+ 26 19		17.4	FAINT BLUE STAR
LB 07328	12 51 00.	+ 27 13		17.8	FAINT BLUE STAR
LB 07327	12 51 00.	+ 27 20		19.6	FAINT BLUE STAR
LB 11411	12 51 00.	+ 29 56		13.0	FAINT BLUE STAR
LB 07326	12 51 00.	+ 31 33		20.2	FAINT BLUE STAR
RNGC 4773	12 51 00.	- 08 23		15.0	GALAXY
REIZ 2952	12 51 00.	- 08 24	30	13.8	GALAXY
HELW 459	12 51 00.	- 10 01 43.			NEBULA
SVEN 351	12 51 00.	- 10 01 49.	6	15.9	GALAXY
KN 16.564	12 51 00.4	+ 28 33 01.			NEBULA
REIZ 2955	12 51 01.	+ 09 47	9	15.4	GALAXY
REIZ 2956	12 51 01.	+ 10 10	18	15.0	GALAXY
IC 3855	12 51 01.	+ 37 03 43.			NONSTELLAR OBJECT
HOLM 480A	12 51 01.	- 11 44	48	14.1	PART OF MULTIPLE GALAXY
BFG 034	12 51 02.32	+ 34 12 36.8		19.09	ULTRAVIOLET-EXCESS OBJECT
HOLM 480B	12 51 04.	- 11 46	48	15.0	PART OF MULTIPLE GALAXY

OBJECT NAME	RIGHT ASCEN.	DECLINATION	DIAM.	MAGN.	TYPE OF OBJECT
HELW 173	12 51 04.	- 11 46 43.			NEBULA
REIN 4.160	12 51 04.58	+ 09 48 36.6			NEBULA
KN 16.565	12 51 04.6	+ 27 56 43.			NEBULA
SVEN 352	12 51 05.	- 06 20	102	11.9	GALAXY
A2 137	12 51 05.1	+ 28 38 49.		17.8	FAINT BLUE OBJECT
ZWG 043.055	12 51 06.	+ 07 04		15.5	GALAXY
LB 11416	12 51 06.	+ 27 11		14.5	FAINT BLUE STAR
LB 11415	12 51 06.	+ 27 53		12.2	FAINT BLUE STAR
LB 11414	12 51 06.	+ 27 54		13.0	FAINT BLUE STAR
LB 11413	12 51 06.	+ 27 58		14.5	FAINT BLUE STAR
LB 07332	12 51 06.	+ 28 28		18.8	FAINT BLUE STAR
LB 07331	12 51 06.	+ 29 22		18.8	FAINT BLUE STAR
LB 07330	12 51 06.	+ 30 50		19.3	FAINT BLUE STAR
LB 11412	12 51 06.	+ 31 12		12.5	FAINT BLUE STAR
LB 07329	12 51 06.	+ 32 01		18.6	FAINT BLUE STAR
ZC 1251.1+4910	12 51 06.	+ 49 10	1010		CLUSTER OF GALAXIES
MCG+09-21-058	12 51 06.	+ 50 40	18	16.	GALAXY
RNGC 4775	12 51 06.	- 06 21		12.0	GALAXY
SVEN 353	12 51 06.	- 10 19	12	16.0	GALAXY
MCG-06-28-023	12 51 06.	- 39 26 30.	120	13.	GALAXY
REIZ 2960	12 51 07.	+ 09 49	24	15.1	GALAXY
SC 1248+0102.4	12 51 08.	+ 00 46 05.	18		NEBULA
MCG-02-33-044	12 51 09.	- 11 01 30.	60	15.	GALAXY
KN 16.566	12 51 09.4	+ 28 01 49.			NEBULA
HELW 174	12 51 10.	- 12 35 07.			NEBULA
REIZ 2959	12 51 11.	- 06 21	108	12.0	GALAXY
HELW 175	12 51 11.	- 12 35 13.			NEBULA
KN 16.567	12 51 11.8	+ 27 53 07.			NEBULA
8ZW 1251+24.8	12 51 12.	+ 24 49		17.5	COMPACT GALAXY
TON-N 0136	12 51 12.	+ 26 42		14.8	BLUE STAR
LB 07337	12 51 12.	+ 26 44		19.0	FAINT BLUE STAR
LB 07336	12 51 12.	+ 27 52		18.6	FAINT BLUE STAR
LB 07335	12 51 12.	+ 28 42		19.2	FAINT BLUE STAR
LB 07334	12 51 12.	+ 29 09		19.6	FAINT BLUE STAR
LB 11418	12 51 12.	+ 29 17		12.3	FAINT BLUE STAR
LB 07333	12 51 12.	+ 30 10		19.0	FAINT BLUE STAR
LB 11417	12 51 12.	+ 31 29		12.0	FAINT BLUE STAR
LB 02478	12 51 12.	+ 50 30 36.		14.6	FAINT BLUE STAR
MCG-01-33-043	12 51 12.	- 06 20	120	11.	GALAXY
MCG-02-33-045	12 51 12.	- 12 34	30	15.	GALAXY
KN 16.568	12 51 12.2	+ 28 03 25.			NEBULA
A2 138	12 51 12.2	+ 28 21 14.		17.0	FAINT BLUE OBJECT
REIZ 2958	12 51 13.	- 12 37	30	14.6	GALAXY
RNGC 4767	12 51 13.	- 39 27		13.0	GALAXY
BFG 035	12 51 14.01	+ 32 36 21.4		19.01	ULTRAVIOLET-EXCESS OBJECT
MCG-02-33-046	12 51 15.	- 13 17	24	15.	GALAXY
BFG 036	12 51 16.15	+ 36 22 04.6		16.79	ULTRAVIOLET-EXCESS OBJECT
HELW 177	12 51 17.	- 11 57 25.			NEBULA
HELW 176	12 51 17.	- 12 01 43.			NEBULA
ARC 1633	12 51 17.	- 26 07		17.6	RICH CLUSTER OF GALAXIES
KN 16.569	12 51 17.9	+ 26 44 38.			NEBULA
ZWG 043.056	12 51 18.	+ 03 43		15.7	GALAXY
ZWG 071.068	12 51 18.	+ 10 00		13.5	GALAXY
SCH 79	12 51 18.	+ 10 00		13.5	PECULIAR GALAXY
UGC 08022	12 51 18.	+ 10 00	126	13.5	GALAXY SBc
MCG+02-33-034	12 51 18.	+ 10 00	132	13.5	GALAXY
IC 3856	12 51 18.	+ 20 21 49.			NONSTELLAR OBJECT
LB 07341	12 51 18.	+ 27 54		19.2	FAINT BLUE STAR
LB 07340	12 51 18.	+ 28 54		18.8	FAINT BLUE STAR
LB 07339	12 51 18.	+ 29 33		18.6	FAINT BLUE STAR
LB 07338	12 51 18.	+ 29 56		20.2	FAINT BLUE STAR
LB 02479	12 51 18.	+ 52 36 18.		17.0	FAINT BLUE STAR
MCG-02-33-047	12 51 18.	- 11 49 30.	120	15.	GALAXY
KN 16.570	12 51 18.2	+ 26 41 56.			NEBULA
KN 16.571	12 51 18.2	+ 28 03 20.			NEBULA
REIZ 2961	12 51 19.	+ 09 59	72	13.3	GALAXY
REIZ 2965	12 51 19.	+ 29 52	18	14.8	GALAXY
A2 139	12 51 19.0	+ 26 54 13.		16.4	FAINT BLUE OBJECT
KN 16.572	12 51 19.6	+ 27 31 08.			NEBULA
REIN 4.161	12 51 19.73	+ 09 58 54.3			NEBULA
RNGC 4779	12 51 20.	+ 10 00		13.5	GALAXY
KN 16.573	12 51 20.6	+ 27 39 26.			NEBULA
BFG 037	12 51 20.85	+ 37 51 40.8		19.33	ULTRAVIOLET-EXCESS OBJECT
HOLM 484B	12 51 22.	+ 29 52	24	14.2	PART OF MULTIPLE GALAXY
ARC 1636	12 51 22.	+ 63 06		17.8	RICH CLUSTER OF GALAXIES
KN 16.574	12 51 23.4	+ 29 51 32.			NEBULA
A2 140	12 51 23.7	+ 27 46 00.		18.2	FAINT BLUE GALAXY
ZWG 043.057	12 51 24.	+ 05 18		15.7	GALAXY
ZWG 159.105	12 51 24.	+ 26 43		15.0	GALAXY
IC 0832	12 51 24.	+ 26 43		15.0	GALAXY E
LB 07345	12 51 24.	+ 28 07		18.6	FAINT BLUE STAR
LB 07344	12 51 24.	+ 29 02		19.7	FAINT BLUE STAR
ZWG 160.001	12 51 24.	+ 29 15		15.6	GALAXY
ZWG 159.106	12 51 24.	+ 29 15		15.6	GALAXY
ZWG 160.002	12 51 24.	+ 29 51		15.3	GALAXY
ZWG 159.107	12 51 24.	+ 29 51		15.3	GALAXY
SCH 80	12 51 24.	+ 29 51		15.3	PECULIAR GALAXY
KARA.72 357A	12 51 24.	+ 29 51	42	15.3	PART OF DOUBLE GALAXY
MCG+05-30-117	12 51 24.	+ 29 52	36	15.3	GALAXY
ZC 1251.4+3103	12 51 24.	+ 31 03	1810		CLUSTER OF GALAXIES
LB 07343	12 51 24.	+ 31 14		18.0	FAINT BLUE STAR
ZWG 160.003	12 51 24.	+ 31 22		15.3	GALAXY
ZWG 159.108	12 51 24.	+ 31 22		15.3	GALAXY
LB 07342	12 51 24.	+ 31 50		19.0	FAINT BLUE STAR
LB 11420	12 51 24.	+ 32 31		13.2	FAINT BLUE STAR
LB 11419	12 51 24.	+ 32 36		15.0	FAINT BLUE STAR
UGC 08023	12 51 24.	+ 36 22	60	16.0	GALAXY DBL SYS
MCG-01-33-044	12 51 24.	- 08 30	102	14.5	GALAXY
RNGC 4777	12 51 24.	- 08 32		14.5	GALAXY
HELW 178	12 51 24.	- 12 03 30.			NEBULA
REIZ 2964	12 51 25.	+ 09 47	18	15.0	GALAXY
ARC 1634	12 51 25.	- 06 25		17.6	RICH CLUSTER OF GALAXIES
LB 02480	12 51 26.	+ 58 21 18.		16.5	FAINT BLUE STAR
KN 16.575	12 51 26.0	+ 29 15 02.			NEBULA
HELW 180	12 51 27.	- 11 50 00.			NEBULA
MCG-02-33-048	12 51 27.	- 12 42	42	16.	GALAXY
HELW 181	12 51 27.	- 12 42 42.			NEBULA
HELW 179	12 51 27.	- 12 43 00.			NEBULA
A2 142	12 51 27.6	+ 29 56 22.		18.0	FAINT BLUE GALAXY
REIN 4.162	12 51 27.83	+ 09 46 15.1			NEBULA
IC 3857	12 51 28.	+ 19 52 43.			NONSTELLAR OBJECT
SC 1248-0058.5	12 51 28.	- 01 14 48.	18		NEBULA
HOLM 482B	12 51 28.	- 08 23	24	14.9	PART OF MULTIPLE GALAXY
KN 16.576	12 51 28.4	+ 27 14 14.			NEBULA
KN 16.578	12 51 28.7	+ 28 27 32.			NEBULA
A2 144	12 51 28.9	+ 28 52 22.		16.6	FAINT BLUE OBJECT
IC 3858	12 51 29.	+ 21 03 25.			NONSTELLAR OBJECT
KN 16.577	12 51 29.2	+ 27 22 32.			NEBULA
ZWG 071.069	12 51 30.	+ 09 47		15.7	GALAXY
MCG+05-30-119	12 51 30.	+ 26 41	18	15.0	GALAXY
LB 07347	12 51 30.	+ 27 07		19.8	FAINT BLUE STAR
LB 11421	12 51 30.	+ 29 54		13.2	FAINT BLUE STAR
LB 07346	12 51 30.	+ 31 18		18.7	FAINT BLUE STAR
MCG+05-30-118	12 51 30.	+ 31 24	42	15.3	GALAXY
VV 266B	12 51 30.	+ 36 21	18	18.	INTERACTING GALAXY
VV 266A	12 51 30.	+ 36 21	18	18.	INTERACTING GALAXY
VV 266	12 51 30.	+ 36 21	54		INTERACTING GALAXY
ZC 1251.5+3636	12 51 30.	+ 36 36	870		CLUSTER OF GALAXIES
IC 3861	12 51 30.	+ 38 33 08.			NONSTELLAR OBJECT
ZC 1251.5+5204	12 51 30.	+ 52 04	740		CLUSTER OF GALAXIES
ZCG 1251+73	12 51 30.	+ 73 14		16.0	COMPACT GALAXY
ZWG 015.033	12 51 30.	- 01 14		15.5	GALAXY
KARA.73B 0556	12 51 30.	- 01 14	18	15.5	ISOLATED GALAXY E
MCG-01-33-045	12 51 30.	- 08 20	108	14.	GALAXY
RNGC 4780	12 51 30.	- 08 21		14.0	GALAXY
HOLM 482B	12 51 30.	- 08 21	66	13.2	PART OF MULTIPLE GALAXY
RNGC 4780A	12 51 30.	- 08 22			GALAXY
REIZ 2963	12 51 30.	- 08 23	60	14.0	GALAXY
REIZ 2962	12 51 30.	- 08 31	18	14.5	GALAXY
RNGC 4778	12 51 30.	- 08 56			NON-EXISTENT OBJECT
RNGC 4776	12 51 30.	- 08 57			NON-EXISTENT OBJECT
MCG-03-33-014	12 51 30.	- 19 55	60	14.	GALAXY
REIZ 2967	12 51 31.	+ 29 53	78	13.7	GALAXY
IC 3862	12 51 31.	+ 36 21 26.			NONSTELLAR OBJECT
HELW 460	12 51 32.	- 06 23 42.			NEBULA
KN 16.578A	12 51 32.0	+ 26 51 56.			NEBULA
KN 16.579	12 51 32.7	+ 27 27 44.			NEBULA
A2 145	12 51 32.7	+ 28 14 03.		16.7	FAINT BLUE OBJECT
ARP 265	12 51 33.	+ 36 22			PECULIAR GALAXY
IC 3863	12 51 33.	+ 38 45 08.			NONSTELLAR OBJECT
KN 16.580	12 51 33.2	+ 26 42 50.			NEBULA
REIZ 2966	12 51 34.	+ 19 34	18	15.2	GALAXY
MCG+05-30-121	12 51 36.	+ 27 20	72	15.5	GALAXY
ZWG 160.004	12 51 36.	+ 27 25		14.9	GALAXY
ZWG 159.109	12 51 36.	+ 27 25		14.9	GALAXY
UGC 08024	12 51 36.	+ 27 25	150	14.9	GALAXY DWRF IR
MCG+05-30-120	12 51 36.	+ 27 25	180	14.9	GALAXY
LB 11422	12 51 36.	+ 29 08		12.0	FAINT BLUE STAR
ZWG 160.005	12 51 36.	+ 29 52		14.8	GALAXY
ZWG 159.110	12 51 36.	+ 29 52		14.8	GALAXY
UGC 08025	12 51 36.	+ 29 52	126	14.8	GALAXY Sb
KARA.72 357B	12 51 36.	+ 29 52	102	14.8	PART OF DOUBLE GALAXY
MCG+05-30-122	12 51 36.	+ 29 53	114	14.8	GALAXY
LB 07351	12 51 36.	+ 30 49		19.3	FAINT BLUE STAR
LB 07350	12 51 36.	+ 31 42		19.6	FAINT BLUE STAR
LB 07349	12 51 36.	+ 32 01		19.3	FAINT BLUE STAR
LB 07348	12 51 36.	+ 32 24		17.0	FAINT BLUE STAR
MCG+06-28-039	12 51 36.	+ 36 22	60	17.	GALAXY
MCG+09-21-059	12 51 36.	+ 54 07	30	16.	GALAXY
SC 1249-0043.3	12 51 36.	- 00 59 36.	18		NEBULA
SVEN 354	12 51 36.	- 10 09	18	16.4	GALAXY
MCG-07-27-014	12 51 36.	- 41 33	60	14.5	GALAXY
HOLM 484A	12 51 37.	+ 29 53	90	14.1	PART OF MULTIPLE GALAXY
KN 16.582	12 51 37.2	+ 29 52 26.			NEBULA
BFG 038	12 51 37.25	+ 37 34 28.8		18.82	ULTRAVIOLET-EXCESS OBJECT
KN 16.581	12 51 37.8	+ 28 03 26.			NEBULA
LB 02481	12 51 38.	+ 52 18 30.		17.3	FAINT BLUE STAR
IC 3860	12 51 39.	+ 19 34 14.			NONSTELLAR OBJECT
KN 16.583	12 51 39.2	+ 27 24 38.			NEBULA
KN 16.584	12 51 39.7	+ 27 20 26.			NEBULA
REIZ 2971	12 51 41.	+ 19 14	18	15.5	GALAXY
RNGC 4787	12 51 41.	+ 27 20		15.5	GALAXY
RNGC 4789A	12 51 41.	+ 27 25		15.0	GALAXY
BFG 039	12 51 41.48	+ 35 23 09.1		19.02	ULTRAVIOLET-EXCESS OBJECT
A2 148	12 51 41.7	+ 27 13 54.		17.3	FAINT BLUE OBJECT
KN 16.586	12 51 41.9	+ 28 24 08.			NEBULA
ZWG 015.035	12 51 42.	+ 01 04		15.7	GALAXY
ZWG 043.058	12 51 42.	+ 05 37		15.5	GALAXY
ZWG 071.070	12 51 42.	+ 09 43		15.7	GALAXY
ZWG 160.006	12 51 42.	+ 27 20		15.5	GALAXY
ZWG 159.111	12 51 42.	+ 27 20		15.5	GALAXY
UGC 08026	12 51 42.	+ 27 20	60	15.5	GALAXY S0-a
MCG+05-30-123	12 51 42.	+ 27 34	24	15.4	GALAXY
LB 11423	12 51 42.	+ 29 18		13.8	FAINT BLUE STAR
ARC 1637	12 51 42.	+ 51 06		17.0	RICH CLUSTER OF GALAXIES
ZWG 015.034	12 51 42.	- 03 27		15.0	GALAXY
MCG+00-33-019	12 51 42.	- 03 27	66	15.0	GALAXY
MCG-02-33-049	12 51 42.	- 10 15	210	11.5	GALAXY
HELW 182	12 51 42.	- 12 46 06.			NEBULA
KN 16.585	12 51 42.0	+ 26 41 56.			NEBULA
BFG 040	12 51 42.19	+ 38 08 58.8		19.04	ULTRAVIOLET-EXCESS OBJECT
KN 16.587	12 51 42.4	+ 27 36 14.			NEBULA
REIZ 2970	12 51 43.	+ 09 43	36	14.9	GALAXY
REIN 2.164	12 51 43.41	- 10 15 51.4			NEBULA
REIN 4.163	12 51 43.77	+ 09 42 47.7			NEBULA
IC 3864	12 51 44.	+ 19 13 14.			NONSTELLAR OBJECT
REIZ 2974	12 51 44.	+ 27 21	24	15.7	GALAXY
REIZ 2975	12 51 44.	+ 27 25	78	15.6	GALAXY
ARC 1635	12 51 44.	- 08 40		17.2	RICH CLUSTER OF GALAXIES
HELW 183	12 51 44.	- 12 13 48.			NEBULA
KLEM 21	12 51 44.	- 28 42	720	16.	CLUSTER OF 15 GALAXIES
A2 150	12 51 44.2	+ 27 56 04.		17.9	FAINT BLUE OBJECT
KN 16.588	12 51 44.3	+ 28 21 50.			NEBULA
IC 3859	12 51 45.	- 08 50			OPEN CLUSTER
IC 3865	12 51 46.	+ 19 08 20.			NONSTELLAR OBJECT
KN 16.589	12 51 46.3	+ 27 53 56.			NEBULA
REIZ 2973	12 51 47.	+ 19 09	18	15.2	GALAXY
RNGC 4788	12 51 47.	+ 27 25		15.5	GALAXY
REIN 2.165	12 51 47.49	- 10 15 57.0			NEBULA
ZC 1251.8+0129	12 51 48.	+ 01 29	400		CLUSTER OF GALAXIES
ZWG 043.059	12 51 48.	+ 05 32		15.6	GALAXY
LB 07357	12 51 48.	+ 26 32		16.8	FAINT BLUE STAR
MCG+05-30-124	12 51 48.	+ 27 20	42	13.3	GALAXY
ZWG 160.007	12 51 48.	+ 27 35		15.4	GALAXY
ZWG 159.112	12 51 48.	+ 27 35		15.4	GALAXY
LB 07356	12 51 48.	+ 27 42		19.4	FAINT BLUE STAR
LB 07355	12 51 48.	+ 31 01		20.0	FAINT BLUE STAR
LB 07354	12 51 48.	+ 31 25		20.2	FAINT BLUE STAR
LB 07353	12 51 48.	+ 31 40		19.7	FAINT BLUE STAR
LB 07352	12 51 48.	+ 31 48		19.7	FAINT BLUE STAR
LB 11424	12 51 48.	+ 32 05		13.2	FAINT BLUE STAR
UGC 08027	12 51 48.	+ 42 58	60	16.5	GALAXY Sb-c
MCG+08-24-001	12 51 48.	+ 49 03	36	17.	GALAXY
ZC 1251.8+6515	12 51 48.	+ 65 15	1950		CLUSTER OF GALAXIES
REIZ 2969	12 51 48.	- 08 51	18	14.3	GALAXY
SVEN 356	12 51 48.	- 09 56	48	16.0	GALAXY

OBJECT NAME	RIGHT ASCEN.	DECLINATION	DIAM.	MAGN.	TYPE OF OBJECT
HELW 461	12 51 48.	- 09 56 30.			NEBULA
REIZ 2968	12 51 48.	- 10 15	150	11.7	GALAXY
HOLM 483A	12 51 48.	- 10 15	192	11.4	PART OF MULTIPLE GALAXY
SVEN 355	12 51 48.	- 10 15 48.	186	11.6	GALAXY
RNGC 4781	12 51 48.	- 10 16		12.0	GALAXY
MCG-03-33-015	12 51 48.	- 16 05	18	15.	GALAXY
MCG-05-31-001	12 51 48.	- 28 56	30	16.	GALAXY
IC 3866	12 51 49.	+ 22 37 50.			OPEN CLUSTER
HELW 184	12 51 49.	- 12 17 42.			NEBULA
KN 16.590	12 51 50.3	+ 27 34 26.			NEBULA
IC 3867	12 51 51.	+ 19 12 44.			NONSTELLAR OBJECT
FATH 1.546	12 51 51.	+ 44 18	14		NEBULA
KN 16.591	12 51 51.1	+ 28 20 50.			NEBULA
REIZ 2980	12 51 52.	+ 19 15	24	15.2	GALAXY
IC 3868	12 51 52.	+ 19 15 56.			NONSTELLAR OBJECT
REIZ 2981	12 51 52.	+ 19 16	12	15.5	GALAXY
REIZ 2979	12 51 53.	+ 19 13	30	15.0	GALAXY
IC 3869	12 51 53.	+ 19 14 32.			NONSTELLAR OBJECT
RNGC 4789	12 51 53.	+ 27 20		13.5	GALAXY
FATH 1.547	12 51 53.	+ 44 12	8		NEBULA
HELW 462	12 51 53.	- 09 53 18.			NEBULA
KN 16.592	12 51 53.3	+ 27 20 32.			NEBULA
KN 16.593	12 51 53.8	+ 28 21 38.			NEBULA
ZC 1251.9+2231	12 51 54.	+ 22 31	3490		CLUSTER OF GALAXIES
LB 11429	12 51 54.	+ 26 36		16.0	FAINT BLUE STAR
LB 07365	12 51 54.	+ 27 16		19.5	FAINT BLUE STAR
ZWG 160.008	12 51 54.	+ 27 20		13.3	GALAXY
ZWG 159.113	12 51 54.	+ 27 20		13.3	GALAXY
UGC 08028	12 51 54.	+ 27 20	84	13.3	GALAXY E-SO
LB 07364	12 51 54.	+ 29 00		19.6	FAINT BLUE STAR
TON-N 0669	12 51 54.	+ 29 01		15.0	BLUE STAR
LB 11428	12 51 54.	+ 29 33		14.0	FAINT BLUE STAR
LB 07363	12 51 54.	+ 29 43		20.2	FAINT BLUE STAR
LB 07362	12 51 54.	+ 29 48		18.0	FAINT BLUE STAR
LB 07361	12 51 54.	+ 30 15		19.3	FAINT BLUE STAR
LB 07360	12 51 54.	+ 30 33		19.4	FAINT BLUE STAR
LB 11427	12 51 54.	+ 31 00		14.0	FAINT BLUE STAR
LB 11426	12 51 54.	+ 31 42		17.5	FAINT BLUE STAR
LB 07359	12 51 54.	+ 31 48		19.8	FAINT BLUE STAR
LB 11425	12 51 54.	+ 31 49		14.3	FAINT BLUE STAR
LB 07358	12 51 54.	+ 32 09		19.0	FAINT BLUE STAR
MCG+08-24-002	12 51 54.	+ 48 32	36	16.	GALAXY
ZC 1251.9-0205	12 51 54.	- 02 05	2890		CLUSTER OF GALAXIES
MCG-01-33-046	12 51 54.	- 06 34	42	14.	GALAXY
SVEN 357	12 51 54.	- 09 53	30	16.3	GALAXY
VV 201A	12 51 54.	- 12 11	30	12.8	INTERACTING GALAXY
MCG-02-33-051	12 51 54.	- 12 17	30	13.	GALAXY
MCG-02-33-050	12 51 54.	- 12 17	30	13.	GALAXY
SC 1249+0100.9	12 51 55.	+ 00 44 36.	48		NEBULA
IC 3870	12 51 55.	+ 22 39 08.			NONSTELLAR OBJECT
LB 02482	12 51 55.	+ 60 30 12.		14.8	FAINT BLUE STAR
REIZ 2972	12 51 55.	- 12 18	24	13.0	GALAXY
A2 152	12 51 56.2	+ 27 54 54.		18.0	FAINT BLUE OBJECT
KN 16.594	12 51 56.4	+ 27 21 08.			NEBULA
A2 153	12 51 56.8	+ 26 36 20.		17.8	FAINT BLUE OBJECT
IC 3871	12 51 57.	+ 19 12 08.			NONSTELLAR OBJECT
SN 1956B	12 51 57.	- 12 16		18.6	SUPERNOVA
REIZ 2982	12 51 58.	+ 19 13	18	15.4	GALAXY
REIZ 2983	12 51 58.	+ 19 15	24	15.0	GALAXY
LB 02483	12 51 58.	+ 52 55 36.		16.0	FAINT BLUE STAR
A2 154	12 51 58.1	+ 29 22 42.		17.4	FAINT BLUE OBJECT
BFG 041	12 51 58.56	+ 37 37 55.9		18.42	ULTRAVIOLET-EXCESS OBJECT
FATH 1.548	12 51 59.	+ 45 17	27		NEBULA
REIZ 2978	12 51 59.	- 06 35	36	12.6	GALAXY
LB 09979	12 52	- 87 07		12.0	FAINT BLUE STAR
ZWG 100.016	12 52 00.	+ 15 25		15.6	GALAXY
UGC 08030	12 52 00.	+ 26 34	60	18.	GALAXY DWRF IR
LB 07368	12 52 00.	+ 27 54		16.0	FAINT BLUE STAR
LB 07367	12 52 00.	+ 29 05		19.0	FAINT BLUE STAR
LB 07366	12 52 00.	+ 31 06		17.7	FAINT BLUE STAR
LB 11430	12 52 00.	+ 32 23		12.3	FAINT BLUE STAR
UGC 08031	12 52 00.	+ 35 34	66	16.0	GALAXY S
ZC 1252.0+3617	12 52 00.	+ 36 17	1010		CLUSTER OF GALAXIES
MCG+08-24-003	12 52 00.	+ 45 17	48	16.	GALAXY
ZC 1252.0+5112	12 52 00.	+ 51 12	1880		CLUSTER OF GALAXIES
7ZW 488	12 52 00.	+ 74 17			COMPACT GALAXY
MCG+13-09-037	12 52 00.	+ 74 55	12	15.	GALAXY
ZWG 365.006	12 52 00.	+ 80 36		15.6	GALAXY
UGC 08029	12 52 00.	+ 82 40	78	16.0	GALAXY Sb
RNGC 4786	12 52 00.	- 06 35		13.0	GALAXY
SVEN 358	12 52 00.	- 10 20	96	14.6	GALAXY
MCG-02-33-053	12 52 00.	- 10 20	54	15.	GALAXY
RNGC 4784	12 52 00.	- 10 21		15.0	GALAXY
REIZ 2977	12 52 00.	- 10 21	18	14.6	GALAXY
MCG-02-33-052	12 52 00.	- 11 20	108	16.	GALAXY
VV 201B	12 52 00.	- 12 12	24	13.2	INTERACTING GALAXY
RNGC 4783	12 52 00.	- 12 18		13.0	GALAXY
RNGC 4782	12 52 00.	- 12 19		13.0	GALAXY
MCG-03-33-016	12 52 00.	- 18 02 30.	60	15.	GALAXY
DV.56 N4767B	12 52 00.	- 39 35	72		SB(s) GALAXY
MCG-07-27-015	12 52 00.	- 39 35	66	14.5	GALAXY
HOLM 485B	12 52 01.	- 12 17	24	13.2	PART OF MULTIPLE GALAXY
REIZ 2976	12 52 01.	- 12 18	30	12.8	GALAXY
HOLM 485A	12 52 01.	- 12 18	30	12.8	PART OF MULTIPLE GALAXY
RNGC 4767B	12 52 01.	- 39 35			GALAXY
IC 3872	12 52 02.	+ 19 13 56.			NONSTELLAR OBJECT
HOLM 483B	12 52 02.	- 10 20	72	14.2	PART OF MULTIPLE GALAXY
A2 156	12 52 02.7	+ 26 36 55.		17.5	FAINT BLUE OBJECT
IC 3873	12 52 03.	+ 19 09 14.			NONSTELLAR OBJECT
KN 16.595	12 52 03.2	+ 26 34 27.			NEBULA
REIZ 2986	12 52 04.	+ 19 14	12	14.9	GALAXY
REIZ 2985	12 52 05.	+ 19 10	30	15.2	GALAXY
HELW 185	12 52 05.	- 12 20 18.			NEBULA
KARA.73 40	12 52 05.	- 44 58	54		DWARF GALAXY
KN 16.596	12 52 05.5	+ 26 36 51.			NEBULA
ZWG 015.036	12 52 06.	+ 02 22		14.8	GALAXY
MCG+00-33-020	12 52 06.	+ 02 22	15	14.8	GALAXY
IC 3874	12 52 06.	+ 19 13 38.			NONSTELLAR OBJECT
ZC 1252.1+2609	12 52 06.	+ 26 09	2150		CLUSTER OF GALAXIES
LB 07371	12 52 06.	+ 27 18		18.6	FAINT BLUE STAR
ZWG 160.009	12 52 06.	+ 28 39		15.5	GALAXY
ZWG 159.114	12 52 06.	+ 28 39		15.5	GALAXY
LB 07370	12 52 06.	+ 29 52		18.8	FAINT BLUE STAR
LB 07369	12 52 06.	+ 30 46		17.0	FAINT BLUE STAR
MCG+07-27-002	12 52 06.	+ 38 54	36	16.	GALAXY
FATH 1.549	12 52 06.	+ 44 12	11		NEBULA
ZWG 352.044	12 52 06.	+ 74 55		15.5	GALAXY
MCG-02-33-054	12 52 06.	- 12 29	18	15.	GALAXY
DG 125	12 52 06.	- 22 41	780		REFLECTION NEBULA
HELW 186	12 52 07.	- 12 29 42.			NEBULA
KN 16.598	12 52 07.4	+ 28 38 51.			NEBULA
KN 16.597	12 52 07.6	+ 27 54 09.			NEBULA
SHB 219	12 52 07.7	+ 11 57 21.		16.6	QUASI-STELLAR OBJECT
BC PKS1252+11	12 52 07.86	+ 11 57 20.8		16.64	QUASI-STELLAR OBJECT
HELW 187	12 52 08.	- 12 19 42.			NEBULA
IC 3878	12 52 09.	+ 40 20 33.			NONSTELLAR OBJECT
HELW 188	12 52 09.	- 12 31 36.			NEBULA
MCG-02-33-055	12 52 09.	- 12 49	30	15.	GALAXY
REIZ 2990	12 52 10.	+ 22 18	9	15.3	GALAXY
IC 3875	12 52 10.	+ 22 18 21.			NONSTELLAR OBJECT
IC 3879	12 52 10.	+ 38 53 57.			NONSTELLAR OBJECT
HELW 189	12 52 10.	- 12 49 42.			NEBULA
RNGC 4793	12 52 11.	+ 29 13		12.5	GALAXY
KN 16.599	12 52 11.0	+ 27 12 15.			NEBULA
A2 160	12 52 11.8	+ 27 59 10.		17.6	FAINT BLUE OBJECT
ZWG 043.060	12 52 12.	+ 08 19		15.1	GALAXY
MCG+01-33-021	12 52 12.	+ 08 19	72	15.1	GALAXY
ZC 1252.2+1020	12 52 12.	+ 10 20	1480		CLUSTER OF GALAXIES
ZWG 071.071	12 52 12.	+ 13 30		14.8	GALAXY
UGC 08032	12 52 12.	+ 13 30	174	14.8	GALAXY S
MCG+02-33-035	12 52 12.	+ 13 30	168	14.8	GALAXY
REIZ 2989	12 52 12.	+ 13 31	108	13.7	GALAXY
ZC 1252.2+1914	12 52 12.	+ 19 14	1610		CLUSTER OF GALAXIES
ARC 1638	12 52 12.	+ 19 16		16.0	RICH CLUSTER OF GALAXIES
MCG+03-33-016	12 52 12.	+ 19 27	210	14.1	GALAXY
8ZW 1252+24.4	12 52 12.	+ 24 27		17.4	COMPACT GALAXY
ZC 1252.2+2541	12 52 12.	+ 25 41	1550		CLUSTER OF GALAXIES
LB 11432	12 52 12.	+ 26 42		13.4	FAINT BLUE STAR
ZWG 160.010	12 52 12.	+ 27 12		15.6	GALAXY
ZWG 159.115	12 52 12.	+ 27 12		15.6	GALAXY
TON-N 0670	12 52 12.	+ 28 38		15.0	BLUE STAR
MCG+05-31-001	12 52 12.	+ 28 39	42	15.5	GALAXY
LB 07375	12 52 12.	+ 28 51		17.8	FAINT BLUE STAR
ZWG 160.011	12 52 12.	+ 29 12		12.3	GALAXY
ZWG 159.116	12 52 12.	+ 29 12		12.3	GALAXY
UGC 08033	12 52 12.	+ 29 12	204	12.3	GALAXY Sc
LB 11431	12 52 12.	+ 30 36		14.0	FAINT BLUE STAR
LB 07374	12 52 12.	+ 31 09		17.9	FAINT BLUE STAR
LB 07373	12 52 12.	+ 31 14		19.8	FAINT BLUE STAR
LB 07372	12 52 12.	+ 31 47		18.1	FAINT BLUE STAR
ZWG 188.027	12 52 12.	+ 35 33		15.6	GALAXY
ZWG 293.040	12 52 12.	+ 59 10		15.0	GALAXY
MCG+10-18-088	12 52 12.	+ 59 10	36	15.	GALAXY
RNGC 4790	12 52 12.	- 09 58		13.0	GALAXY
REIZ 2984	12 52 12.	- 09 58	72	12.3	GALAXY
MCG-02-33-056	12 52 12.	- 09 58	90	12.5	GALAXY
BFG 042	12 52 12.73	+ 37 26 16.8		16.61	ULTRAVIOLET-EXCESS OBJECT
REIN 4.164	12 52 12.74	+ 08 19 27.6			NEBULA
REIZ 2988	12 52 13.	+ 08 20	30	13.9	GALAXY
REIZ 2996	12 52 13.	+ 29 14	66	12.5	GALAXY
A2 161	12 52 13.6	+ 26 15 10.		17.1	FAINT BLUE OBJECT
RNGC 4791	12 52 14.	+ 08 19		15.0	GALAXY
REIZ 2995	12 52 14.	+ 27 12	18	15.8	GALAXY
REIN 4.165	12 52 14.23	+ 13 30 21.8			NEBULA
MCG+06-28-040	12 52 15.	+ 35 34	42	15.	GALAXY
RNGC 4800	12 52 15.	+ 46 48		13.0	GALAXY
MCG-02-33-058	12 52 15.	- 11 37	84	14.5	GALAXY
HELW 190	12 52 15.	- 12 29 48.			NEBULA
MCG-02-33-057	12 52 15.	- 13 50	18	15.	GALAXY
KN 16.600	12 52 15.6	+ 28 22 33.			NEBULA
KN 16.601	12 52 15.6	+ 29 12 33.			NEBULA
REIN 2.166	12 52 16.03	+ 46 48 19.4			NEBULA
KN 16.602	12 52 17.4	+ 27 43 27.			NEBULA
ZWG 043.061	12 52 18.	+ 02 55		14.8	GALAXY
KARA.72 358A	12 52 18.	+ 02 55	48	14.8	PART OF DOUBLE GALAXY
MCG+01-33-022	12 52 18.	+ 02 55	96	14.8	GALAXY
ZWG 043.062	12 52 18.	+ 02 56		14.9	GALAXY
UGC 08034	12 52 18.	+ 02 56	108	14.9	GALAXY IRR
KARA.72 358B	12 52 18.	+ 02 56	96	14.9	PART OF DOUBLE GALAXY
MCG+01-33-023	12 52 18.	+ 02 56	36	14.9	GALAXY
ZWG 043.063	12 52 18.	+ 04 08		15.5	GALAXY
LB 07378	12 52 18.	+ 27 02		19.5	FAINT BLUE STAR
MCG+05-31-002	12 52 18.	+ 27 11	30	15.6	GALAXY
LB 11433	12 52 18.	+ 27 34		13.9	FAINT BLUE STAR
MCG+05-31-003	12 52 18.	+ 29 12 30.	156	12.3	GALAXY
SCH 81	12 52 18.	+ 29 13		12.3	PECULIAR GALAXY
LB 07377	12 52 18.	+ 29 29		19.0	FAINT BLUE STAR
LB 07376	12 52 18.	+ 30 08		19.3	FAINT BLUE STAR
ZWG 160.012	12 52 18.	+ 30 48		15.7	GALAXY
ZWG 159.117	12 52 18.	+ 30 48		15.7	GALAXY
TON-N 0137	12 52 18.	+ 41 10		16.2	BLUE STAR
ZWG 245.005	12 52 18.	+ 46 48		12.0	GALAXY
UGC 08035	12 52 18.	+ 46 48	102	12.0	GALAXY S
FEIG 070	12 52 18.	+ 52 01		11.8	FAINT BLUE STAR
ZC 1252.3+5855	12 52 18.	+ 58 55	1340		CLUSTER OF GALAXIES
REIZ 2987	12 52 18.	- 09 47	24	14.5	GALAXY
SVEN 360	12 52 18.	- 09 48	12	15.4	GALAXY
SVEN 359	12 52 18.	- 09 58	96	12.8	GALAXY
DV.56 N4767C	12 52 18.	- 39 04	48		GALAXY
MCG-07-27-016	12 52 18.	- 40 42	66	15.	GALAXY
A2 164	12 52 18.9	+ 27 56 24.		17.2	FAINT BLUE OBJECT
REIZ 2994	12 52 19.	+ 08 21	24	14.7	GALAXY
FATH 1.550	12 52 19.	+ 44 08			NEBULA
HELW 463	12 52 19.	- 09 48 06.			NEBULA
RNGC 4810	12 52 20.	+ 02 55		15.0	GALAXY
RNGC 4809	12 52 20.	+ 02 55		15.0	GALAXY
IC 3876	12 52 20.	+ 19 17 09.			NONSTELLAR OBJECT
IC 3877	12 52 20.	+ 19 33 51.			NONSTELLAR OBJECT
IC 3880	12 52 20.	+ 22 46 15.			NONSTELLAR OBJECT
REIN 4.166	12 52 20.11	+ 08 20 38.4			NEBULA
REIN 2.167	12 52 20.71	+ 46 48 07.6			NEBULA
BFG 043	12 52 20.74	+ 35 32 46.4		18.72	ULTRAVIOLET-EXCESS OBJECT
REIZ 2993	12 52 21.	+ 02 52	36	14.1	GALAXY
REIZ 2992	12 52 21.	+ 02 53	72	13.6	GALAXY
HOLM 486B	12 52 21.	+ 02 55	36	13.2	PART OF MULTIPLE GALAXY
HOLM 486A	12 52 21.	+ 02 56	120	13.1	PART OF MULTIPLE GALAXY
IC 3885	12 52 21.	+ 37 25 33.			NONSTELLAR OBJECT
LB 02484	12 52 21.	+ 52 02 18.		12.0	FAINT BLUE STAR
ARC 1640	12 52 21.	+ 62 50		17.8	RICH CLUSTER OF GALAXIES
MCG-02-33-059	12 52 21.	- 09 47	12	15.	GALAXY
REIZ 2997	12 52 22.	+ 19 18	18	15.4	GALAXY
REIZ 2998	12 52 22.	+ 19 27	72	13.5	GALAXY
KN 16.603	12 52 22.0	+ 27 36 39.			NEBULA
IC 3887	12 52 23.	+ 40 34 39.			NONSTELLAR OBJECT
RNGC 4801	12 52 23.	+ 53 22		15.5	GALAXY
RNGC 4792	12 52 23.	- 12 16			GALAXY

OBJECT NAME	RIGHT ASCEN.	DECLINATION	DIAM.	MAGN.	TYPE OF OBJECT
HELW 191	12 52 23.	- 12 31 24.			NEBULA
KARA.68 214	12 52 24.	+ 02 14	34		DWARF GALAXY
ZC 1252.4+1325	12 52 24.	+ 13 25	540		CLUSTER OF GALAXIES
ZWG 100.017	12 52 24.	+ 19 27		14.1	GALAXY
UGC 08036	12 52 24.	+ 19 27	240	14.1	GALAXY SBc
LB 07383	12 52 24.	+ 28 04		17.8	FAINT BLUE STAR
LB 07382	12 52 24.	+ 28 39		18.9	FAINT BLUE STAR
LB 07381	12 52 24.	+ 30 37		19.2	FAINT BLUE STAR
LB 07380	12 52 24.	+ 31 32		20.3	FAINT BLUE STAR
LB 11434	12 52 24.	+ 31 33		14.0	FAINT BLUE STAR
LB 07379	12 52 24.	+ 32 20		17.6	FAINT BLUE STAR
MCG+07-27-003	12 52 24.	+ 39 50	54	16.	GALAXY
ZC 1252.4+4326	12 52 24.	+ 43 26	1410		CLUSTER OF GALAXIES
ZWG 270.030	12 52 24.	+ 53 22		15.7	GALAXY
ZC 1252.4+5914	12 52 24.	+ 59 14	1080		CLUSTER OF GALAXIES
MCG+10-18-089	12 52 24.	+ 62 29 30.	15	15.	GALAXY
ZWG 353.002	12 52 24.	+ 77 41		15.5	GALAXY
ZWG 352.045	12 52 24.	+ 77 41		15.5	GALAXY
MCG-01-33-047	12 52 24.	- 07 27	24	17.	GALAXY
SVEN 361	12 52 24.	- 10 18	18	16.7	GALAXY
KN 16.604	12 52 24.2	+ 26 36 09.			NEBULA
BFG 044	12 52 24.56	+ 36 58 52.6		19.15	ULTRAVIOLET-EXCESS OBJECT
LB 00663	12 52 25.	+ 72 07 42.		17.1	FAINT BLUE STAR
HELW 464	12 52 25.	- 10 18 17.			NEBULA
REIZ 2991	12 52 25.	- 12 16	18	14.2	GALAXY
IC 3881	12 52 26.	+ 19 25 15.			NONSTELLAR OBJECT
IC 3882	12 52 26.	+ 22 50 39.			NONSTELLAR OBJECT
IC 3888	12 52 26.	+ 39 50 33.			NONSTELLAR OBJECT
MCG+08-24-004	12 52 27.	+ 46 48	90	12.2	GALAXY
MCG+08-24-005	12 52 27.	+ 49 39	60	16.	GALAXY
IC 3890	12 52 28.	+ 37 27 39.			NONSTELLAR OBJECT
LB 02485	12 52 28.	+ 56 01 00.		17.3	FAINT BLUE STAR
LB 00664	12 52 28.	+ 57 49 42.		17.0	FAINT BLUE STAR
KN 16.605	12 52 28.1	+ 28 41 21.			NEBULA
ARC 1639	12 52 29.	+ 10 26		17.5	RICH CLUSTER OF GALAXIES
RNGC 4797	12 52 29.	+ 27 37			NON-EXISTENT OBJECT
RNGC 4798	12 52 29.	+ 27 41		14.5	GALAXY
IC 3889	12 52 29.	+ 36 17 15.			NONSTELLAR OBJECT
HELW 465	12 52 29.	- 06 06 05.			NEBULA
KN 16.606	12 52 29.6	+ 27 41 03.			NEBULA
ZWG 043.064	12 52 30.	+ 08 20		13.5	GALAXY
UGC 08037	12 52 30.	+ 08 20	144	13.5	GALAXY SB0/SBa
KARA.72 359A	12 52 30.	+ 08 20	108	13.5	PART OF DOUBLE GALAXY
MCG+01-33-024	12 52 30.	+ 08 20	36	13.5	GALAXY
ZWG 071.072	12 52 30.	+ 10 58		15.7	GALAXY
IC 3884	12 52 30.	+ 19 57 09.			NONSTELLAR OBJECT
LB 11436	12 52 30.	+ 27 03		14.7	FAINT BLUE STAR
ZWG 160.013	12 52 30.	+ 27 41		14.3	GALAXY
ZWG 159.118	12 52 30.	+ 27 41		14.3	GALAXY
UGC 08038	12 52 30.	+ 27 41	60	14.3	GALAXY
ZWG 160.014	12 52 30.	+ 28 41		15.7	GALAXY
ZWG 159.119	12 52 30.	+ 28 41		15.7	GALAXY
LB 11435	12 52 30.	+ 29 52		12.2	FAINT BLUE STAR
ZC 1252.5+3533	12 52 30.	+ 35 33	2420		CLUSTER OF GALAXIES
MCG+06-28-041	12 52 30.	+ 35 40	15	17.	GALAXY
UGC 08039	12 52 30.	+ 44 25	66	16.0	GALAXY S
MCG+09-21-060	12 52 30.	+ 53 21	42	14.6	GALAXY
ZWG 293.041	12 52 30.	+ 59 02		14.4	GALAXY
UGC 08040	12 52 30.	+ 59 02	102	14.4	GALAXY S
ZCG 1252+74.3	12 52 30.	+ 74 17		18.2	COMPACT GALAXY
ZCG 1252+74.4	12 52 30.	+ 74 20		13.0	COMPACT GALAXY
MCG+14-06-011	12 52 30.	+ 82 38	54	16.	GALAXY
KN 16.607	12 52 30.1	+ 29 17 09.			NEBULA
PATH 1.551	12 52 31.	+ 44 26	41		NEBULA
REIN 4.167	12 52 31.55	+ 08 20 11.0			NEBULA
KN 16.608	12 52 31.7	+ 28 23 51.			NEBULA
RNGC 4796	12 52 32.	+ 08 20		13.5	GALAXY
RNGC 4795	12 52 32.	+ 08 20		13.5	GALAXY
IC 3886	12 52 32.	+ 19 16 57.			NONSTELLAR OBJECT
KN 16.609	12 52 32.4	+ 28 34 51.			NEBULA
BFG 045	12 52 32.77	+ 37 36 19.2		18.53	ULTRAVIOLET-EXCESS OBJECT
HELW 291	12 52 33.	- 08 35 23.			NEBULA
REIN 4.168	12 52 33.37	+ 08 20 14.2			NEBULA
KN 16.610	12 52 34.6	+ 28 03 21.			NEBULA
KN 16.611	12 52 34.9	+ 27 33 51.			NEBULA
RNGC 4794	12 52 35.	- 12 21		14.0	GALAXY
KN 16.612	12 52 35.3	+ 27 45 09.			NEBULA
UGC 08041	12 52 36.	+ 00 20	222	13.6	GALAXY Sc
ZWG 015.037	12 52 36.	+ 00 23		13.6	GALAXY
SA 4	12 52 36.	+ 00 23	198	12.9	GALAXY
MCG+00-33-021	12 52 36.	+ 00 23	156	13.6	GALAXY
ZWG 043.065	12 52 36.	+ 08 11		15.5	GALAXY
UGC 08042	12 52 36.	+ 08 11	84	15.5	GALAXY S
KARA.72 359B	12 52 36.	+ 08 20	30		PART OF DOUBLE GALAXY
MCG+05-31-004	12 52 36.	+ 27 40	42	14.3	GALAXY
ZC 1252.6+2837	12 52 36.	+ 28 37	540		CLUSTER OF GALAXIES
MCG+05-31-005	12 52 36.	+ 28 41	30	15.7	GALAXY
LB 07388	12 52 36.	+ 29 15		18.7	FAINT BLUE STAR
LB 07387	12 52 36.	+ 30 04		19.1	FAINT BLUE STAR
LB 07386	12 52 36.	+ 30 22		19.4	FAINT BLUE STAR
LB 07385	12 52 36.	+ 30 26		19.3	FAINT BLUE STAR
LB 11437	12 52 36.	+ 31 47		13.7	FAINT BLUE STAR
LB 07384	12 52 36.	+ 32 12		19.2	FAINT BLUE STAR
ZWG 188.028	12 52 36.	+ 35 39		14.8	GALAXY
MCG+06-28-042	12 52 36.	+ 35 40	24	14.5	GALAXY
IC 3891	12 52 36.	+ 36 19 27.			NONSTELLAR OBJECT
MCG+07-27-004	12 52 36.	+ 44 26	60	15.5	GALAXY
ZC 1252.6+4803	12 52 36.	+ 48 03	1210		CLUSTER OF GALAXIES
MCG+10-18-090	12 52 36.	+ 59 02	78	14.	GALAXY
ZWG 294.001	12 52 36.	+ 59 09		15.1	GALAXY
ZWG 293.042	12 52 36.	+ 59 09		15.1	GALAXY
ZWG 316.005	12 52 36.	+ 62 30		15.1	GALAXY
MCG-01-33-049	12 52 36.	- 07 50	48	15.	GALAXY
REIZ 3000	12 52 36.	- 07 51	48	13.2	GALAXY
MCG-01-33-048	12 52 36.	- 08 34	24	15.	GALAXY
REIZ 3001	12 52 36.	- 08 52	84	13.5	GALAXY
MCG-02-33-060	12 52 36.	- 12 20	102	14.	GALAXY
SER 094.01	12 52 36.	- 44 33	55	14.	2 HIGH SURF BRGHTNSS GLXY
SER 094.02	12 52 36.	- 45 01	132		IRR. MAGELLANIC GALAXY
SER 093.01	12 52 36.	- 49 47	170		LOW SURFACE BRIGHT. GALXY
DV.56 I3896A	12 52 36.	- 49 49	66		SB:d GALAXY
REIZ 2999	12 52 37.	- 12 21	24	14.4	GALAXY
A2 170	12 52 37.3	+ 26 47 05.		16.6	FAINT BLUE OBJECT
HELW 292	12 52 38.	- 08 30 17.			NEBULA
HELW 466	12 52 39.	- 06 16 53.			NEBULA
REIZ 3002	12 52 39.	- 20 01	12	15.3	GALAXY
REIN 4.169	12 52 39.01	+ 08 11 22.7			NEBULA
KN 16.613	12 52 39.2	+ 26 59 45.			NEBULA

OBJECT NAME	RIGHT ASCEN.	DECLINATION	DIAM.	MAGN.	TYPE OF OBJECT
KN 16.614	12 52 39.9	+ 28 00 39.			NEBULA
SC 1250+0039.6	12 52 40.	+ 00 23 19.	120		NEBULA
REIZ 3007	12 52 40.	- 00 56	180	13.0	GALAXY
REIZ 3003	12 52 40.	- 23 23	24	15.2	GALAXY
A2 173	12 52 40.9	+ 26 54 36.		16.9	FAINT BLUE OBJECT
HOLM 487B	12 52 41.	+ 39 30	36	15.2	PART OF MULTIPLE GALAXY
REIZ 3006	12 52 41.	- 06 16	54	15.3	GALAXY
ZWG 015.038	12 52 42.	+ 00 30		15.6	GALAXY
KARA.73B 0557	12 52 42.	+ 00 30	18	15.6	ISOLATED GALAXY E
ZWG 043.066	12 52 42.	+ 03 10		14.4	GALAXY
UGC 08043	12 52 42.	+ 03 10	96	14.4	GALAXY S
MCG+01-33-025	12 52 42.	+ 03 10	66	14.4	GALAXY
ZC 1252.7+2001	12 52 42.	+ 20 01	740		CLUSTER OF GALAXIES
LB 07392	12 52 42.	+ 26 42		17.7	FAINT BLUE STAR
LB 07391	12 52 42.	+ 26 58		17.9	FAINT BLUE STAR
LB 07390	12 52 42.	+ 28 07		18.7	FAINT BLUE STAR
LB 07389	12 52 42.	+ 29 58		18.3	FAINT BLUE STAR
TON-N 0138	12 52 42.	+ 35 32		16.2	BLUE STAR
MCG+07-27-005	12 52 42.	+ 39 30	72	15.	GALAXY
ZC 1252.7+6249	12 52 42.	+ 62 49	540		CLUSTER OF GALAXIES
HN 0365	12 52 42.	- 07 52			NEBULA
IC 3883	12 52 42.	- 07 52			NONSTELLAR OBJECT
REIZ 3005	12 52 42.	- 08 36	21	13.9	GALAXY
REIZ 3008	12 52 42.	- 08 52	12	14.2	GALAXY
MCG-05-31-002	12 52 42.	- 28 11	30	16.	GALAXY
RNGC 4799	12 52 44.	+ 03 10		14.5	GALAXY
HOLM 487A	12 52 45.	+ 39 29	54	14.9	PART OF MULTIPLE GALAXY
IC 3892	12 52 45.	+ 39 29 40.			NONSTELLAR OBJECT
LB 02486	12 52 45.	+ 52 33 42.		16.2	FAINT BLUE STAR
REIZ 3009	12 52 45.	- 19 49	12	15.2	GALAXY
REIZ 3010	12 52 45.	- 20 29	36	14.7	GALAXY
IC 3893	12 52 46.	+ 38 50 10.			NONSTELLAR OBJECT
A2 177	12 52 46.5	+ 26 23 17.		16.3	FAINT BLUE OBJECT
KN 16.615	12 52 46.8	+ 28 20 52.			NEBULA
LB 07397	12 52 48.	+ 27 06		19.5	FAINT BLUE STAR
LB 07396	12 52 48.	+ 27 42		16.8	FAINT BLUE STAR
LB 07395	12 52 48.	+ 28 35		19.0	FAINT BLUE STAR
LB 07394	12 52 48.	+ 28 51		19.0	FAINT BLUE STAR
LB 07393	12 52 48.	+ 30 51		19.4	FAINT BLUE STAR
MCG+07-27-006	12 52 48.	+ 39 28 30.	42	14.5	GALAXY
ZWG 217.002	12 52 48.	+ 39 30		15.7	GALAXY
UGC 08044	12 52 48.	+ 39 30	96	15.7	GALAXY S?
KARA.72 360A	12 52 48.	+ 39 30	78	15.7	PART OF DOUBLE GALAXY
MCG+10-19-001	12 52 48.	+ 59 08	36	15.	GALAXY
ZWG 015.039	12 52 48.	- 02 45		15.5	GALAXY
MCG-01-33-050	12 52 48.	- 07 28	36	16.5	GALAXY
MCG-01-33-051	12 52 48.	- 07 30	48	15.	GALAXY
REIZ 3011	12 52 48.	- 08 43	30	14.3	GALAXY
KN 16.616	12 52 48.8	+ 27 58 52.			NEBULA
IC 3895	12 52 49.	+ 39 28 28.			NONSTELLAR OBJECT
A2 178	12 52 49.2	+ 27 40 40.		16.9	FAINT BLUE OBJECT
LB 02487	12 52 50.	+ 52 18 36.		16.7	FAINT BLUE STAR
REIZ 3004	12 52 50.	- 47 20	54	12.7	GALAXY
MCG-04-31-005	12 52 51.	- 26 35 30.	60	14.5	GALAXY
A2 180	12 52 51.3	+ 27 28 42.		18.0	FAINT BLUE OBJECT
REIN 4.170	12 52 52.20	+ 08 10 52.3			NEBULA
A2 182	12 52 52.6	+ 28 58 47.		17.7	FAINT BLUE OBJECT
BFG 046	12 52 52.73	+ 35 28 27.3		17.53	ULTRAVIOLET-EXCESS OBJECT
BSO 04	12 52 53.	+ 35 28 51.			BLUE STELLAR OBJECT
LB 02488	12 52 53.	+ 55 24 54.		15.6	FAINT BLUE STAR
ZWG 043.067	12 52 54.	+ 02 30		15.7	GALAXY
MCG+01-33-026	12 52 54.	+ 02 30	36	15.7	GALAXY
ZWG 043.068	12 52 54.	+ 08 11		15.3	GALAXY
UGC 08045	12 52 54.	+ 08 11	60	15.3	GALAXY IRR
LB 07399	12 52 54.	+ 27 18		19.6	FAINT BLUE STAR
LB 11438	12 52 54.	+ 27 57		12.6	FAINT BLUE STAR
LB 07398	12 52 54.	+ 28 47		19.1	FAINT BLUE STAR
HZ 34	12 52 54.	+ 37 49		14.7	DECIDEDLY BLUE STAR
ZWG 217.003	12 52 54.	+ 39 28		15.7	GALAXY
KARA.72 360B	12 52 54.	+ 39 28	48	15.7	PART OF DOUBLE GALAXY
ZWG 294.002	12 52 54.	+ 59 03		15.5	GALAXY
ZWG 293.043	12 52 54.	+ 59 03		15.5	GALAXY
UGC 08046	12 52 54.	+ 59 03	66	15.5	GALAXY
REIZ 3014	12 52 54.	- 07 31	48	14.4	GALAXY
MCG-01-33-052	12 52 54.	- 07 46	48	15.	GALAXY
REIZ 3013	12 52 54.	- 07 47	60	13.9	GALAXY
MCG-06-28-024	12 52 54.	- 39 16 30.	48	15.	GALAXY
MCG-07-27-017	12 52 54.	- 44 32	48	14.5	GALAXY
KN 16.618	12 52 54.4	+ 28 08 16.			NEBULA
KN 16.617	12 52 54.6	+ 27 04 10.			NEBULA
A2 183	12 52 54.7	+ 27 41 41.		18.2	FAINT BLUE OBJECT
KN 16.619	12 52 55.3	+ 27 39 04.			NEBULA
HOLM 488A	12 52 56.	+ 27 48	66	14.3	PART OF MULTIPLE GALAXY
HOLM 488B	12 52 57.	+ 27 50	18	15.0	PART OF MULTIPLE GALAXY
KN 16.620	12 52 57.4	+ 28 44 22.			NEBULA
BC B114	12 52 57.7	+ 35 55 26.		17.92	QUASI-STELLAR OBJECT
SHB 220	12 52 57.9	+ 35 55 24.		17.9	QUASI-STELLAR OBJECT
BFG 047	12 52 57.94	+ 35 55 24.2		18.07	ULTRAVIOLET-EXCESS OBJECT
IC 3897	12 52 58.	+ 39 56 22.			NONSTELLAR OBJECT
IC 3894	12 52 59.	+ 19 20 28.			NONSTELLAR OBJECT
RNGC 4807B	12 52 59.	+ 27 49			GALAXY
RNGC 4805	12 52 59.	+ 28 13			GALAXY
LB 02489	12 52 59.	+ 52 36 54.		16.2	FAINT BLUE STAR
BFG 048	12 52 59.06	+ 35 39 57.1		18.46	ULTRAVIOLET-EXCESS OBJECT
KN 16.621	12 52 59.4	+ 27 55 04.			NEBULA
KN 16.622	12 52 59.5	+ 28 04 10.			NEBULA
A2 186	12 52 59.8	+ 27 54 51.		18.0	FAINT BLUE OBJECT
EMS 1253+4422	12 53	+ 44 22			CLUSTER OF GALAXIES
LB 09867	12 53	- 84 14		14.7	FAINT BLUE STAR
MCG+05-31-006	12 53 00.	+ 27 47	60	14.4	GALAXY
ZWG 160.015	12 53 00.	+ 28 05		15.5	GALAXY
LB 07401	12 53 00.	+ 28 30		17.7	FAINT BLUE STAR
ZWG 160.016	12 53 00.	+ 28 44		15.7	GALAXY
LB 11440	12 53 00.	+ 30 17		13.7	FAINT BLUE STAR
LB 11439	12 53 00.	+ 30 49		14.3	FAINT BLUE STAR
MCG+08-24-006	12 53 00.	+ 46 19	24	16.	GALAXY
ZWG 270.031	12 53 00.	+ 55 25		15.2	GALAXY
UGC 08047	12 53 00.	+ 55 25	84	15.2	GALAXY Sb-c
MCG+10-19-002	12 53 00.	+ 59 02	57	15.	GALAXY
MRK 230	12 53 00.	+ 64 31	12	16.	GALAXY WITH UV CONTINUUM
MCG+12-12-021	12 53 00.	+ 73 27	102	15.	GALAXY
ZWG 015.040	12 53 00.	- 00 37		15.7	GALAXY
REIZ 3016	12 53 00.	- 07 40	54	15.0	GALAXY
REIZ 3017	12 53 00.	- 08 43	42	13.9	GALAXY
MCG-04-31-006	12 53 00.	- 26 34	30	15.	GALAXY
HELW 293	12 53 01.	- 07 46 05.			NEBULA
A2 187	12 53 01.5	+ 27 24 50.		17.4	FAINT BLUE OBJECT
IC 3898	12 53 02.	+ 37 51 10.			NONSTELLAR OBJECT

OBJECT NAME	RIGHT ASCEN.	DECLINATION	DIAM.	MAGN.	TYPE OF OBJECT
LB 00665	12 53 02.	+ 61 07 42.		16.4	FAINT BLUE STAR
SC 1250-0021.2	12 53 02.	- 00 37 29.	24		NEBULA
KN 16.623	12 53 02.1	+ 27 55 34.			NEBULA
REIN 4.171	12 53 02.45	+ 08 30 41.2			NEBULA
HELW 294	12 53 03.	- 08 01 59.			NEBULA
HELW 192	12 53 03.	- 11 42 41.			NEBULA
KN 16.624	12 53 03.6	+ 27 47 34.			NEBULA
REIZ 3015	12 53 04.	- 40 01	36	14.8	GALAXY
A2 188	12 53 04.2	+ 26 16 33.		18.7	FAINT BLUE OBJECT
KN 16.624A	12 53 04.2	+ 27 56 10.			NEBULA
KN 16.625	12 53 04.3	+ 29 16 28.			NEBULA
RNGC 4807A	12 53 05.	+ 27 47		14.5	GALAXY
SC 1250-0046.9	12 53 05.	- 01 03 11.	12		NEBULA
KN 16.626	12 53 05.0	+ 27 49 04.			NEBULA
KN 16.625A	12 53 05.0	+ 27 50 22.			NEBULA
BPG 049	12 53 05.71	+ 37 08 11.8		18.75	ULTRAVIOLET-EXCESS OBJECT
UGC 08048	12 53 06.	+ 00 00	120	17.	GALAXY DWARF
ZWG 071.073	12 53 06.	+ 08 30		15.0	GALAXY
ZWG 043.069	12 53 06.	+ 08 30		15.0	GALAXY
MCG+02-33-036	12 53 06.	+ 08 30	30	15.0	GALAXY
KARA.68 215	12 53 06.	+ 19 29	60		DWARF GALAXY
ZWG 160.017	12 53 06.	+ 27 47		14.4	GALAXY
UGC 08049	12 53 06.	+ 27 47	48	14.4	GALAXY COMPACT
MCG+05-31-008	12 53 06.	+ 27 54	36	15.3	GALAXY
ZWG 160.018	12 53 06.	+ 27 56		15.3	GALAXY
MCG+05-31-007	12 53 06.	+ 28 03	48	15.5	GALAXY
LB 07402	12 53 06.	+ 28 33		19.2	FAINT BLUE STAR
MCG+09-21-061	12 53 06.	+ 51 30	30	16.	GALAXY
REIZ 3020	12 53 06.	- 07 36	12	15.9	GALAXY
REIZ 3021	12 53 06.	- 09 03	18	13.7	GALAXY
REIZ 3012	12 53 06.	- 53 54	42	14.6	GALAXY
A2 189	12 53 06.0	+ 30 17 56.		18.6	FAINT BLUE GALAXY
REIZ 3019	12 53 07.	- 12 14	30	14.8	GALAXY
RNGC 4803	12 53 08.	+ 08 30		15.0	GALAXY
PATH 1.552	12 53 08.	+ 44 12	33		NEBULA
LB 02490	12 53 08.	+ 59 01 48.		16.4	FAINT BLUE STAR
HELW 193	12 53 08.	- 12 11 59.			NEBULA
KN 16.627	12 53 08.1	+ 28 12 46.			NEBULA
KN 16.628	12 53 08.5	+ 28 06 46.			NEBULA
REIZ 3022	12 53 09.	- 19 53	12	15.7	GALAXY
KN 16.631	12 53 09.6	+ 29 16 16.			NEBULA
KN 16.630	12 53 09.9	+ 28 02 22.			NEBULA
REIZ 3018	12 53 10.	- 40 00	48	14.6	GALAXY
KN 16.629	12 53 10.0	+ 27 24 04.			NEBULA
KN 16.632	12 53 10.4	+ 27 49 16.			NEBULA
HELW 295	12 53 11.	- 07 50 59.			NEBULA
ZC 1253.2+0959	12 53 12.	+ 09 59	740		CLUSTER OF GALAXIES
ZC 1253.2+1724	12 53 12.	+ 17 24	870		CLUSTER OF GALAXIES
LB 00058	12 53 12.	+ 23 26		18.5	FAINT BLUE STAR
LB 07407	12 53 12.	+ 26 53		19.1	FAINT BLUE STAR
LB 07406	12 53 12.	+ 27 03		18.1	FAINT BLUE STAR
LB 07405	12 53 12.	+ 30 29		19.7	FAINT BLUE STAR
LB 07404	12 53 12.	+ 31 41		19.5	FAINT BLUE STAR
MCG+09-21-062	12 53 12.	+ 52 31	66	14.	GALAXY
ZWG 270.032	12 53 12.	+ 52 32		15.1	GALAXY
UGC 08050	12 53 12.	+ 52 32	84	15.1	GALAXY SBb
MCG+09-21-063	12 53 12.	+ 55 29	60	14.	GALAXY
ZWG 294.003	12 53 12.	+ 58 36		12.4	GALAXY
ZWG 293.044	12 53 12.	+ 58 36		12.4	GALAXY
UGC 08051	12 53 12.	+ 58 36	210	12.4	GALAXY Sb
ZWG 335.027	12 53 12.	+ 73 28		14.4	GALAXY
UGC 08052	12 53 12.	+ 73 28	120	14.4	GALAXY S
ZWG 015.041	12 53 12.	- 00 43		15.5	GALAXY
MCG-01-33-053	12 53 12.	- 06 10	36	17.	GALAXY
REIZ 3025	12 53 12.	- 07 51	18	15.2	GALAXY
REIZ 3024	12 53 12.	- 08 49	18	14.7	GALAXY
MCG-02-33-062	12 53 12.	- 11 17	54	15.	GALAXY
RNGC 4802	12 53 12.	- 11 47		12.0	GALAXY
MCG-02-33-061	12 53 12.	- 11 47	48	12.	GALAXY
MCG-02-33-063	12 53 12.	- 13 02 30.	36	15.	GALAXY
IC 3899	12 53 13.	+ 20 54 22.			NONSTELLAR OBJECT
REIZ 3023	12 53 13.	- 11 47	30	13.8	GALAXY
A2 190	12 53 13.8	+ 28 06 20.		16.8	FAINT BLUE OBJECT
PATH 1.553	12 53 14.	+ 44 34	22		NEBULA
RNGC 4814	12 53 14.	+ 58 37		12.5	GALAXY
REIN 2.168	12 53 14.39	+ 58 36 53.3			NEBULA
KN 16.633	12 53 15.9	+ 27 31 10.			NEBULA
IC 3902	12 53 16.	+ 36 15 53.			NONSTELLAR OBJECT
LB 02491	12 53 16.	+ 51 26 36.		15.8	FAINT BLUE STAR
ARP 277	12 53 17.	+ 02 48			PECULIAR GALAXY
HOLM 489B	12 53 17.	+ 33 57	18	15.8	PART OF MULTIPLE GALAXY
HOLM 489A	12 53 17.	+ 33 57	24	15.8	PART OF MULTIPLE GALAXY
REIZ 3027	12 53 17.	- 06 10	30	15.2	GALAXY
RNGC 4804	12 53 17.	- 12 46			NON-EXISTENT OBJECT
VV 313B	12 53 18.	+ 02 49	42	13.2	INTERACTING GALAXY
VV 313A	12 53 18.	+ 02 49	66	13.1	INTERACTING GALAXY
ZWG 043.070	12 53 18.	+ 04 17		15.0	GALAXY
UGC 08053	12 53 18.	+ 04 17	108	15.0	GALAXY
MCG+01-33-027	12 53 18.	+ 04 17	96	15.0	GALAXY
ZWG 043.071	12 53 18.	+ 04 34		12.5	GALAXY
UGC 08054	12 53 18.	+ 04 34	156	12.5	GALAXY Sc
MCG+01-33-028	12 53 18.	+ 04 34	138	12.5	GALAXY
ZC 1253.3+1417	12 53 18.	+ 14 17	940		CLUSTER OF GALAXIES
ZC 1253.3+1842	12 53 18.	+ 18 42	540		CLUSTER OF GALAXIES
LB 11442	12 53 18.	+ 27 28		11.8	FAINT BLUE STAR
MCG+05-31-009	12 53 18.	+ 27 30	30	14.8	GALAXY
ZWG 160.019	12 53 18.	+ 27 31		14.8	GALAXY
IC 3900	12 53 19.	+ 27 31		14.8	GALAXY S0
LB 07410	12 53 18.	+ 27 50		18.1	FAINT BLUE STAR
TON-N 0671	12 53 18.	+ 28 20		15.3	BLUE STAR
LB 11441	12 53 18.	+ 29 19		12.8	FAINT BLUE STAR
LB 07409	12 53 18.	+ 31 06		19.1	FAINT BLUE STAR
LB 07408	12 53 18.	+ 32 07		17.6	FAINT BLUE STAR
IC 3903	12 53 18.	+ 40 40 11.			NONSTELLAR OBJECT
MCG+08-24-007	12 53 18.	+ 48 45	36	16.	GALAXY
MCG+09-21-064	12 53 18.	+ 51 31	60	16.	GALAXY
MCG+09-21-065	12 53 18.	+ 52 15	42	17.	GALAXY
MCG+10-19-003	12 53 18.	+ 58 36	192	12.7	GALAXY
REIZ 3026	12 53 19.	- 09 00	24	16.2	GALAXY
KEEL 583	12 53 19.2	+ 22 09 39.			NEBULA
BPG 050	12 53 18.83	+ 37 13 25.0		18.92	ULTRAVIOLET-EXCESS OBJECT
RNGC 4808	12 53 20.	+ 04 34		12.5	GALAXY
PATH 1.554	12 53 21.	+ 44 31	8		NEBULA
LB 02492	12 53 22.	+ 56 24 36.		16.7	FAINT BLUE STAR
HELW 194	12 53 22.	- 12 32 11.			NEBULA
KN 16.634	12 53 22.0	+ 28 31 29.			NEBULA
BPG 051	12 53 22.91	+ 35 12 43.0		18.93	ULTRAVIOLET-EXCESS OBJECT
HOLM 489C	12 53 23.	+ 33 58	18	15.8	PART OF MULTIPLE GALAXY
REIZ 3031	12 53 23.	- 04 49	72	15.0	GALAXY
REIZ 3032	12 53 23.	- 05 07	120	12.3	GALAXY
HELW 296	12 53 23.	- 08 17 53.			NEBULA
A2 191	12 53 23.0	+ 27 19 32.		17.9	FAINT BLUE OBJECT
A2 192	12 53 23.2	+ 28 11 13.		17.9	FAINT BLUE OBJECT
KN 16.635	12 53 23.7	+ 28 10 41.			NEBULA
IC 3901	12 53 24.	+ 22 12 29.			SINGLE STAR
ABC 1641	12 53 24.	+ 28 43		17.6	RICH CLUSTER OF GALAXIES
LB 07411	12 53 24.	+ 31 56		17.0	FAINT BLUE STAR
LB 11443	12 53 24.	+ 32 15		13.3	FAINT BLUE STAR
ZWG 189.001	12 53 24.	+ 36 33		15.7	GALAXY
ZWG 188.029	12 53 24.	+ 36 33		15.7	GALAXY
KARA.73B 0558	12 53 24.	+ 36 33	36	15.7	ISOLATED GALAXY S
IC 3904	12 53 24.	+ 36 33 53.			NONSTELLAR OBJECT
ZWG 015.042	12 53 24.	- 01 05		15.3	GALAXY
HELW 297	12 53 24.	- 07 47 59.			NEBULA
REIZ 3030	12 53 24.	- 07 48	12	15.1	GALAXY
REIZ 3029	12 53 24.	- 07 51	18	15.0	GALAXY
MCG-03-33-017	12 53 24.	- 19 00	150	14.5	GALAXY
MCG-04-31-007	12 53 24.	- 24 46	15	15.5	GALAXY
MCG-05-31-003	12 53 24.	- 29 14	60	14.5	GALAXY
KN 16.636	12 53 24.2	+ 27 53 59.			NEBULA
BPG 052	12 53 25.20	+ 37 14 52.0		17.20	ULTRAVIOLET-EXCESS OBJECT
SC 1250-0046.9	12 53 26.	- 01 03 11.	18		NEBULA
HELW 298	12 53 26.	- 07 51 04.			NEBULA
A2 194	12 53 26.2	+ 27 32 59.		17.7	FAINT BLUE OBJECT
KN 16.637	12 53 26.3	+ 28 42 59.			NEBULA
A2 195	12 53 26.3	+ 29 26 04.		16.6	FAINT BLUE OBJECT
A2 196	12 53 26.7	+ 30 21 42.		18.4	FAINT BLUE OBJECT
REIN 4.172	12 53 26.72	+ 08 52 49.4			NEBULA
MCG-02-33-064	12 53 27.	- 11 19	36	15.	GALAXY
REIZ 3033	12 53 27.	- 20 40	24	15.0	GALAXY
RNGC 4806	12 53 27.	- 29 14		14.0	GALAXY
A2 197	12 53 27.6	+ 27 40 21.		18.2	FAINT BLUE OBJECT
PATH 1.555	12 53 29.	+ 44 16	19		NEBULA
UGC 08055	12 53 30.	+ 04 05	60	17.	GALAXY DWRF IR
ZC 1253.5+1805	12 53 30.	+ 18 05	740		CLUSTER OF GALAXIES
LB 07412	12 53 30.	+ 29 53		19.8	FAINT BLUE STAR
MCG+06-28-043	12 53 30.	+ 36 34	30	15.5	GALAXY
ZC 1253.5+3635	12 53 30.	+ 36 35	1950		CLUSTER OF GALAXIES
IC 3906	12 53 30.	+ 40 44 11.			NONSTELLAR OBJECT
ZC 1253.5+4421	12 53 30.	+ 44 21	870		CLUSTER OF GALAXIES
MCG+08-24-008	12 53 30.	+ 47 02 30.	54	16.	GALAXY
ZC 1253.5+4704	12 53 30.	+ 47 04	470		CLUSTER OF GALAXIES
MCG+08-24-009	12 53 30.	+ 48 33	36	16.	GALAXY
HELW 195	12 53 30.	- 12 26 40.			NEBULA
BPG 053	12 53 30.31	+ 37 22 12.5		18.31	ULTRAVIOLET-EXCESS OBJECT
KN 16.638	12 53 30.7	+ 27 12 47.			NEBULA
LB 02493	12 53 31.	+ 58 00 54.		16.6	FAINT BLUE STAR
A2 198	12 53 31.2	+ 28 11 13.		18.0	FAINT BLUE OBJECT
A2 199	12 53 31.8	+ 30 07 58.		18.3	FAINT BLUE OBJECT
KN 16.639	12 53 31.9	+ 28 10 35.			NEBULA
PATH 1.556	12 53 32.	+ 44 16	22		NEBULA
HELW 196	12 53 32.	- 12 07 22.			NEBULA
REIZ 3028	12 53 32.	- 46 52	18	14.9	GALAXY
MCG-02-33-065	12 53 33.	- 14 08 30.	36	15.	GALAXY
A2 201	12 53 33.0	+ 29 51 25.		18.2	FAINT BLUE OBJECT
REIN 4.173	12 53 33.15	+ 10 26 23.9			NEBULA
LB 02494	12 53 34.	+ 61 14 18.		16.0	FAINT BLUE STAR
HELW 299	12 53 34.	- 07 52 34.			NEBULA
BC 3C279	12 53 35.82	- 05 31 07.6		17.75	QUASI-STELLAR OBJECT
SHB 221	12 53 35.9	- 05 31 08.		17.8	QUASI-STELLAR OBJECT
ZWG 071.074	12 53 36.	+ 12 46		15.4	GALAXY
TON-N 0672	12 53 36.	+ 26 33		16.7	BLUE STAR
TON-N 0139	12 53 36.	+ 28 23		13.1	BLUE STAR
ZC 1253.6+2846	12 53 36.	+ 28 46	670		CLUSTER OF GALAXIES
LB 07413	12 53 36.	+ 29 59		19.1	FAINT BLUE STAR
LB 11444	12 53 36.	+ 31 19		13.4	FAINT BLUE STAR
ABC 1643	12 53 36.	+ 44 21		17.7	RICH CLUSTER OF GALAXIES
ZC 1253.6+5946	12 53 36.	+ 59 46	1810		CLUSTER OF GALAXIES
REIZ 3035	12 53 36.	- 07 52	48	14.1	GALAXY
MCG-01-33-054	12 53 36.	- 07 52	48	14.	GALAXY
KN 16.640	12 53 36.5	+ 28 18 41.			NEBULA
BC P1253+104	12 53 36.6	+ 10 25 08.		18.	QUASI-STELLAR OBJECT
SHB 222	12 53 36.6	+ 10 25 08.		18.	QUASI-STELLAR OBJECT
BPG 054	12 53 38.56	+ 35 48 42.5		19.02	ULTRAVIOLET-EXCESS OBJECT
A2 202	12 53 38.9	+ 29 00 09.		17.0	FAINT BLUE OBJECT
MCG-04-31-008	12 53 39.	- 25 32	48	15.5	GALAXY
REIN 4.174	12 53 39.19	+ 12 45 24.4			NEBULA
IC 3905	12 53 40.	+ 20 07 23.			NONSTELLAR OBJECT
SC 1251-0136.1	12 53 40.	- 01 52 22.	30		NEBULA
KN 16.641	12 53 40.5	+ 27 56 53.			NEBULA
KN 16.642	12 53 40.9	+ 27 54 59.			NEBULA
A2 203	12 53 41.5	+ 26 55 35.		16.8	FAINT BLUE OBJECT
ZWG 160.020	12 53 42.	+ 27 57		15.5	GALAXY
MRK 053	12 53 42.	+ 27 57	13	15.5	GALAXY WITH UV CONTINUUM
TON-N 0673	12 53 42.	+ 28 23		13.2	BLUE STAR
IC 3909	12 53 42.	+ 40 39 23.			NONSTELLAR OBJECT
ZC 1253.7+6559	12 53 42.	+ 65 59	2620		CLUSTER OF GALAXIES
ZC 1253.7+7006	12 53 42.	+ 70 06	810		CLUSTER OF GALAXIES
ZWG 335.028	12 53 42.	+ 71 06		14.9	GALAXY
MCG-06-28-025	12 53 42.	- 36 04	36	14.5	GALAXY
KN 16.643	12 53 42.8	+ 27 29 29.			NEBULA
ABC 1642	12 53 43.	+ 06 40		17.3	RICH CLUSTER OF GALAXIES
KEEL 584	12 53 43.2	+ 22 04 44.			NEBULA
REIZ 3034	12 53 44.	- 46 43	24	14.6	GALAXY
KN 16.644	12 53 44.1	+ 28 34 47.			NEBULA
A2 205	12 53 44.2	+ 29 52 26.		17.3	FAINT BLUE OBJECT
KN 16.645	12 53 44.4	+ 28 06 53.			NEBULA
IC 3910	12 53 45.	+ 39 59 41.			NONSTELLAR OBJECT
MCG+08-24-010	12 53 45.	+ 48 43	30	16.	GALAXY
ABC 1646	12 53 45.	+ 62 26		16.9	RICH CLUSTER OF GALAXIES
HN 1040	12 53 46.	- 50 03		14.	NEBULA
IC 3896	12 53 46.	- 50 03	78	13.0	GALAXY E1
A2 206	12 53 46.1	+ 28 48 47.		18.0	FAINT BLUE GALAXY
KN 16.646	12 53 46.8	+ 28 00 59.			NEBULA
RNGC 4816	12 53 47.	+ 28 01		15.0	GALAXY
RNGC 4817	12 53 47.	+ 28 16			NON-EXISTENT OBJECT
ZWG 043.072	12 53 48.	+ 05 36		15.7	GALAXY
ZWG 043.073	12 53 48.	+ 05 45		15.2	GALAXY
MCG+01-33-029	12 53 48.	+ 05 45	30	15.2	GALAXY
ZWG 043.074	12 53 48.	+ 07 43		15.5	GALAXY
MCG+01-33-030	12 53 48.	+ 07 43	18	15.5	GALAXY
ZWG 071.075	12 53 48.	+ 10 28		15.3	GALAXY
UGC 08056	12 53 48.	+ 10 28	78	15.3	GALAXY SBc
MCG+05-31-010	12 53 48.	+ 28 00	21	14.8	GALAXY
ZWG 160.021	12 53 48.	+ 28 01		14.8	GALAXY
UGC 08057	12 53 48.	+ 28 01	78	14.8	GALAXY E-S0

OBJECT NAME	RIGHT ASCEN.	DECLINATION	DIAM.	MAGN.	TYPE OF OBJECT
TON-N 0674	12 53 48.	+ 29 44		16.7	BLUE STAR
LB 07414	12 53 48.	+ 31 52		19.6	FAINT BLUE STAR
IC 3911	12 53 48.	+ 35 54 24.			NONSTELLAR OBJECT
IC 3912	12 53 48.	+ 40 10 48.			NONSTELLAR OBJECT
ZC 1253.8+5744	12 53 48.	+ 57 44	1810		CLUSTER OF GALAXIES
7ZW 489	12 53 48.	+ 71 06			COMPACT GALAXY
ZWG 353.003	12 53 48.	+ 79 15		15.7	GALAXY
ZWG 352.046	12 53 48.	+ 79 15		15.7	GALAXY
REIN 4.175	12 53 49.16	+ 10 27 34.6			NEBULA
IC 3907	12 53 50.	+ 19 03 17.			NONSTELLAR OBJECT
BPG 055	12 53 50.82	+ 36 38 49.2		17.33	ULTRAVIOLET-EXCESS OBJECT
ZL 104	12 53 51.	+ 26 56 42.		19.7	ULTRAFAINT BLUE STAR
A2 207	12 53 51.3	+ 27 51 04.		16.8	FAINT BLUE OBJECT
KN 16.646A	12 53 51.6	+ 27 42 59.			NEBULA
ZL 105	12 53 52.	+ 26 48 12.		19.9	ULTRAFAINT BLUE STAR
KN 16.647	12 53 52.7	+ 26 37 47.			NEBULA
ZC 1253.9+0455	12 53 54.	+ 04 55	1280		CLUSTER OF GALAXIES
ZWG 043.075	12 53 54.	+ 06 20		15.6	GALAXY
KARA.68 216	12 53 54.	+ 15 20	74		DWARF GALAXY
REA 66	12 53 54.	+ 15 20	66		DWARF GALAXY
TON-N 0140	12 53 54.	+ 28 00		14.8	BLUE STAR
LB 07415	12 53 54.	+ 31 30		19.6	FAINT BLUE STAR
MCG+07-27-007	12 53 54.	+ 44 20		20.	GALAXY
LB 02495	12 53 54.	+ 58 03 06.		16.4	FAINT BLUE STAR
ZC 1253.9+6217	12 53 54.	+ 62 17	1550		CLUSTER OF GALAXIES
MCG+11-16-007	12 53 54.	+ 63 53	39	15.	GALAXY
MCG+13-09-038	12 53 54.	+ 79 16	10	17.	GALAXY
MCG-02-33-066	12 53 54.	- 13 07	84	15.	GALAXY
MCG-06-28-026	12 53 54.	- 34 25	54	15.	GALAXY
KN 16.649	12 53 54.2	+ 28 01 11.			NEBULA
KN 16.648	12 53 54.6	+ 26 58 35.			NEBULA
LB 02496	12 53 55.	+ 52 01 06.		16.9	FAINT BLUE STAR
REIZ 3037	12 53 55.	- 11 00	90	14.3	GALAXY
BPG 056	12 53 55.13	+ 33 24 07.2		18.85	ULTRAVIOLET-EXCESS OBJECT
A2 209	12 53 55.6	+ 29 53 17.		18.4	FAINT BLUE OBJECT
HZ 36	12 53 57.	+ 32 42		14.3	BLUE STAR
REIZ 3038	12 53 57.	- 20 25	12	15.2	GALAXY
A2 210	12 53 57.5	+ 28 52 29.		17.8	FAINT BLUE OBJECT
RNGC 4819	12 53 59.	+ 27 15		14.0	GALAXY
REIZ 3040	12 53 59.	- 06 32	36	13.2	GALAXY
VDB.66G 156	12 54	+ 02 59	70		DWARF GALAXY
MCG+05-31-011	12 54 00.	+ 26 37	30	15.2	GALAXY
ZWG 160.022	12 54 00.	+ 26 38		15.2	GALAXY
ZWG 160.023	12 54 00.	+ 28 01		15.5	GALAXY
ZC 1254.0+3039	12 54 00.	+ 30 39	610		CLUSTER OF GALAXIES
LB 07400	12 54 00.	+ 31 11		19.5	FAINT BLUE STAR
ZC 1254.0+3600	12 54 00.	+ 36 00	1140		CLUSTER OF GALAXIES
ZC 1254.0+4728	12 54 00.	+ 47 28	810		CLUSTER OF GALAXIES
ZWG 294.004	12 54 00.	+ 57 09		14.1	GALAXY
ZWG 293.045	12 54 00.	+ 57 09		14.1	GALAXY
UGC 08058	12 54 00.	+ 57 09	102	14.1	GALAXY S
ZC 1254.0+5829	12 54 00.	+ 58 29	1480		CLUSTER OF GALAXIES
ZWG 316.006	12 54 00.	+ 63 53		14.6	GALAXY
UGC 08059	12 54 00.	+ 63 53	84	14.6	GALAXY S
ZWG 353.004	12 54 00.	+ 79 43		15.5	GALAXY
ZWG 352.047	12 54 00.	+ 79 43		15.5	GALAXY
RNGC 4813	12 54 00.	- 06 32		15.0	GALAXY
MCG-01-33-055	12 54 00.	- 06 32	60	15.	GALAXY
IC 3914	12 54 01.	+ 36 38 06.			NONSTELLAR OBJECT
REIZ 3039	12 54 01.	- 12 14	30	14.5	GALAXY
KN 16.651	12 54 01.5	+ 28 06 05.			NEBULA
KN 16.650	12 54 01.8	+ 27 02 11.			NEBULA
ZL 106	12 54 02.	+ 26 53 54.		21.3	ULTRAFAINT BLUE STAR
HOLM 490A	12 54 02.	+ 27 16	60	14.0	PART OF MULTIPLE GALAXY
IC 0836	12 54 02.	+ 63 53 18.			NONSTELLAR OBJECT
HELW 467	12 54 02.	- 06 33 40.			NEBULA
KN 16.652	12 54 02.2	+ 27 15 29.			NEBULA
A2 211	12 54 02.4	+ 27 58 46.		15.4	FAINT BLUE OBJECT
HOLM 490B	12 54 03.	+ 27 14	12	14.5	PART OF MULTIPLE GALAXY
KN 16.654	12 54 03.4	+ 27 33 42.			NEBULA
KN 16.653	12 54 03.5	+ 27 13 42.			NEBULA
KN 16.655	12 54 04.5	+ 28 12 36.			NEBULA
ZL 107	12 54 05.	+ 26 51 18.		20.9	ULTRAFAINT BLUE STAR
RNGC 4821	12 54 05.	+ 27 14		15.0	GALAXY
HELW 197	12 54 05.	- 12 12 04.			NEBULA
REIZ 3036	12 54 05.	- 59 09	60	12.2	GALAXY
A2 213	12 54 05.8	+ 29 32 33.		17.2	FAINT BLUE OBJECT
ZC 1254.1+0639	12 54 06.	+ 06 39	1610		CLUSTER OF GALAXIES
IC 0834	12 54 06.	+ 26 36 12.			NONSTELLAR OBJECT
IC 3913	12 54 06.	+ 27			GALAXY IN COMA CLSTR
ZWG 160.024	12 54 06.	+ 27 13		15.0	GALAXY
ZWG 160.025	12 54 06.	+ 27 15		14.0	GALAXY
UGC 08060	12 54 06.	+ 27 15	66	14.1	GALAXY SBa
ZWG 160.026	12 54 06.	+ 27 33		15.5	GALAXY
MCG+05-31-012	12 54 06.	+ 27 33	36	15.5	GALAXY
ZWG 160.027	12 54 06.	+ 28 06		15.6	GALAXY
MCG+05-31-013	12 54 06.	+ 28 10	36	16.	GALAXY
LB 00059	12 54 06.	+ 30 07		15.9	FAINT BLUE STAR
MCG+07-27-008	12 54 06.	+ 38 53	54	14.5	GALAXY
MCG+09-21-066	12 54 06.	+ 51 22	24	17.	GALAXY
7ZW 490	12 54 06.	+ 57 09			COMPACT GALAXY
IC 0833	12 54 06.	- 06 27 24.			NONSTELLAR OBJECT
MCG-01-33-056	12 54 06.	- 07 16	96	14.	GALAXY
HN C366	12 54 06.	- 07 18			NEBULA
IC 3908	12 54 06.	- 07 18			NONSTELLAR OBJECT
RNGC 4811	12 54 06.	- 41 32			GALAXY
RNGC 4812	12 54 06.	- 41 33			GALAXY
BPG 057	12 54 07.47	+ 33 43 35.3		18.56	ULTRAVIOLET-EXCESS OBJECT
KN 16.655A	12 54 08.7	+ 27 48 36.			NEBULA
KN 16.656	12 54 08.9	+ 27 29 48.			NEBULA
MCG-03-33-018	12 54 09.	- 20 46	21	15.	GALAXY
KN 16.657	12 54 09.8	+ 28 32 48.			NEBULA
IC 3916	12 54 10.	+ 38 53 00.			NONSTELLAR OBJECT
IC 3915	12 54 11.	+ 20 23 30.			NONSTELLAR OBJECT
RNGC 4824	12 54 11.	+ 27 47			GALAXY
RNGC 4834	12 54 11.	+ 52 34		15.5	GALAXY
RNGC 4835A	12 54 11.	- 46 07			GALAXY
A2 216	12 54 11.0	+ 27 17 49.		17.0	FAINT BLUE OBJECT
A2 217	12 54 11.3	+ 26 52 56.		18.2	FAINT BLUE OBJECT
ZWG 043.076	12 54 12.	+ 08 04		15.5	GALAXY
UGC 08061	12 54 12.	+ 12 12	60	16.5	GALAXY DWARF
ZWG 130.001	12 54 12.	+ 21 57		8.9	GALAXY
UGC 08062	12 54 12.	+ 21 57	660	8.9	GALAXY Sb
KARA.73B 0559	12 54 12.	+ 21 57	780	8.9	ISOLATED GALAXY S
MCG+05-31-015	12 54 12.	+ 27 12	27	15.0	GALAXY
MCG+05-31-014	12 54 12.	+ 27 14	48	14.0	GALAXY
LB 07403	12 54 12.	+ 31 44		20.0	FAINT BLUE STAR
ZWG 217.004	12 54 12.	+ 38 53		15.6	GALAXY

OBJECT NAME	RIGHT ASCEN.	DECLINATION	DIAM.	MAGN.	TYPE OF OBJECT
UGC 08063	12 54 12.	+ 38 53	66	15.6	GALAXY S
ZWG 270.033	12 54 12.	+ 51 22		15.6	GALAXY
MCG+09-21-067	12 54 12.	+ 52 33	48	15.	GALAXY
ZWG 270.034	12 54 12.	+ 52 34		15.5	GALAXY
MRK 231	12 54 12.	+ 57 08	12	14.	GALAXY WITH UV CONTINUUM
VVI 61	12 54 12.	+ 57 08	65	14.63	SEYFERT GALAXY
MCG+10-19-004	12 54 12.	+ 57 08	27	14.	GALAXY
ZC 1254.2+6042	12 54 12.	+ 60 42	1280		CLUSTER OF GALAXIES
REIZ 3041	12 54 12.	- 08 15	180	12.1	GALAXY
MCG-07-27-019	12 54 12.	- 41 31	60	14.	GALAXY
MCG-07-27-018	12 54 12.	- 41 32	72	14.5	GALAXY
DV.56 N4835A	12 54 12.	- 46 07	162		S GALAXY
LB 00246	12 54 13.	+ 55 13 48.		15.4	FAINT BLUE STAR
KN 16.658	12 54 13.4	+ 28 21 00.			NEBULA
ZL 108	12 54 14.	+ 26 52 30.		15.8	ULTRAFAINT BLUE STAR
FATH 1.557	12 54 14.	+ 44 57	27		NEBULA
ZL 109	12 54 15.	+ 27 01 00.		20.7	ULTRAFAINT BLUE STAR
RNGC 4826	12 54 16.	+ 21 57		9.0	GALAXY
REIZ 3042	12 54 16.	- 23 11	24	14.7	GALAXY
RNGC 4827	12 54 17.	+ 27 27		14.0	GALAXY
RNGC 4828	12 54 17.	+ 28 17		15.5	GALAXY
KN 16.659	12 54 17.4	+ 28 17 24.			NEBULA
KN 16.660	12 54 17.9	+ 27 26 54.			NEBULA
ZWG 071.076	12 54 18.	+ 11 21		15.7	GALAXY
UGC 08064	12 54 18.	+ 11 21	60	15.7	GALAXY Sb
ZWG 071.077	12 54 18.	+ 11 26		15.7	GALAXY
MCG+04-31-001	12 54 18.	+ 21 57	600	8.9	GALAXY
8ZW 1254+25.3	12 54 18.	+ 25 17		17.2	COMPACT GALAXY
ZWG 160.028	12 54 18.	+ 27 26		14.1	GALAXY
UGC 08065	12 54 18.	+ 27 26	84	14.1	GALAXY S0
MCG+05-31-016	12 54 18.	+ 27 26	24	14.1	GALAXY
ZWG 160.029	12 54 18.	+ 28 17		15.4	GALAXY
MCG+05-31-017	12 54 18.	+ 28 17	30	15.4	GALAXY
ZWG 160.030	12 54 18.	+ 30 59		15.1	GALAXY
MCG+09-21-068	12 54 18.	+ 51 20	42	15.	GALAXY
MCG-01-33-057	12 54 18.	- 08 14	180	11.	GALAXY
RNGC 4818	12 54 18.	- 08 15		12.0	GALAXY
MCG-03-33-019	12 54 18.	- 20 13 30.	10	15.	GALAXY
MCG-07-27-020	12 54 18.	- 42 51	72	14.	GALAXY
A2 218	12 54 18.8	+ 30 00 37.		18.4	FAINT BLUE OBJECT
A2 219	12 54 20.7	+ 28 08 03.		17.4	FAINT BLUE OBJECT
A2 220	12 54 21.1	+ 29 26 55.		16.4	FAINT BLUE OBJECT
RNGC 4820	12 54 23.	- 13 26		15.0	GALAXY
SV 280	12 54 23.44	+ 47 40 13.4		19.48	BLUE STELLAR OBJECT
ZWG 043.077	12 54 24.	+ 04 20		15.3	GALAXY
ZWG 043.078	12 54 24.	+ 08 25		15.4	GALAXY
ZWG 130.002	12 54 24.	+ 22 39		15.6	GALAXY
KARA.73B 0560	12 54 24.	+ 22 39	24	15.6	ISOLATED GALAXY S
ZWG 160.031	12 54 24.	+ 27 22		15.7	GALAXY
7ZW 491	12 54 24.	+ 62 23			COMPACT GALAXY
ZWG 015.043	12 54 24.	- 02 55		15.7	GALAXY
RNGC 4822	12 54 24.	- 10 29		14.0	GALAXY
REIZ 3044	12 54 24.	- 10 29	24	14.2	GALAXY
MCG-02-33-067	12 54 24.	- 13 26	60	15.	GALAXY
MCG-03-33-023	12 54 24.	- 17 02 30.	36	15.	GALAXY
MCG-03-33-022	12 54 24.	- 19 15	72	16.	GALAXY
MCG-03-33-021	12 54 24.	- 20 14	12	15.	GALAXY
MCG-03-33-020	12 54 24.	- 20 15	12	15.	GALAXY
MCG-04-31-009	12 54 24.	- 22 34	42	15.5	GALAXY
MCG-05-31-004	12 54 24.	- 32 16	30	15.5	GALAXY
KN 16.661	12 54 24.2	+ 27 21 48.			NEBULA
REIZ 3043	12 54 25.	- 12 02	30	14.9	GALAXY
KN 16.662	12 54 25.7	+ 27 10 06.			NEBULA
IC 3917	12 54 26.3	+ 22 38 41.			GALAXY
KN 16.663	12 54 26.8	+ 26 45 18.			NEBULA
IC 3918	12 54 27.	+ 22 38 37.			SAME AS IC 3917
ZL 110	12 54 27.	+ 26 59 36.		20.8	ULTRAFAINT BLUE STAR
MCG-02-33-069	12 54 27.	- 10 29	18	14.5	GALAXY
MCG-02-33-068	12 54 27.	- 12 00	84	14.5	GALAXY
MCG-05-31-005	12 54 27.	- 32 16	9	15.5	GALAXY
REIN 4.176	12 54 27.02	+ 12 38 00.7			NEBULA
IC 3919	12 54 28.	+ 38 48 01.			NONSTELLAR OBJECT
REIZ 3045	12 54 28.	- 22 30	420	8.6	GALAXY
IC 2920	12 54 29.	+ 40 13 43.			NONSTELLAR OBJECT
REIZ 3049	12 54 29.	- 04 53	48	14.3	GALAXY
REIZ 3048	12 54 29.	- 06 25	48	15.8	GALAXY
RNGC 4825	12 54 29.	- 13 24		13.0	GALAXY
RNGC 4823	12 54 29.	- 13 24			NON-EXISTENT OBJECT
REIZ 3046	12 54 29.	- 25 05	18	14.9	GALAXY
A2 221	12 54 29.0	+ 30 29 31.		18.3	FAINT BLUE OBJECT
BPG 058	12 54 29.13	+ 35 55 23.0		19.09	ULTRAVIOLET-EXCESS OBJECT
UGC 08066	12 54 30.	+ 01 18	108	17.	GALAXY DWRF SP
ZWG 043.079	12 54 30.	+ 04 52		15.7	GALAXY
ZC 1254.5+1459	12 54 30.	+ 14 59	670		CLUSTER OF GALAXIES
ZWG 160.032	12 54 30.	+ 26 45		14.9	GALAXY
IC 0835	12 54 30.	+ 26 45		14.9	GALAXY IN COMA CLSTR
TON-N 0675	12 54 30.	+ 27 08		15.5	BLUE STAR
ZWG 160.033	12 54 30.	+ 27 10		15.1	GALAXY
MCG+05-31-018	12 54 30.	+ 27 10	18	15.1	GALAXY
MCG+05-31-019	12 54 30.	+ 27 21	48	15.7	GALAXY
ZWG 160.034	12 54 30.	+ 29 12		15.2	GALAXY
MCG+05-31-020	12 54 30.	+ 29 12	42	15.2	GALAXY
HMS 1.21	12 54 30.	+ 32 42			PEC GALAXY
MCG+06-28-044	12 54 30.	+ 32 42	42	15.5	GALAXY
MCG+09-21-069	12 54 30.	+ 53 55	36	16.	GALAXY
SN 1973P	12 54 30.	- 04 45		16.2	SUPERNOVA
MCG-01-33-058	12 54 30.	- 06 31	60	17.	GALAXY
REIZ 3047	12 54 31.	- 10 46	18	14.8	GALAXY
LB 00247	12 54 32.	+ 58 25 48.		14.0	FAINT BLUE STAR
HELW 198	12 54 32.	- 12 00 03.			NEBULA
ZL 111	12 54 33.	+ 26 45 12.		18.7	ULTRAFAINT BLUE STAR
HZ 46	12 54 33.	+ 32 42		14.7	BLUE STAR
MCG-02-33-070	12 54 33.	- 13 23	30	12.5	GALAXY
RNGC 4837	12 54 34.	+ 48 34		14.5	GALAXY
KLEM 22	12 54 34.	- 30 09	1200	15.	RICH CLSTR OF 50 GLXIES
BPG 059	12 54 34.20	+ 36 49 13.8		19.06	ULTRAVIOLET-EXCESS OBJECT
ZC 1254.6+1705	12 54 36.	+ 17 05	470		CLUSTER OF GALAXIES
MCG+05-31-021	12 54 36.	+ 26 45	30	14.9	GALAXY
LB 00022	12 54 36.	+ 28 23		15.8	FAINT BLUE STAR
ZWG 160.035	12 54 36.	+ 29 19		15.4	GALAXY
ZWG 160.036	12 54 36.	+ 30 59		14.7	GALAXY
MCG+05-31-022	12 54 36.	+ 30 59	18	14.7	GALAXY
TON-N 0676	12 54 36.	+ 31 37		15.7	BLUE STAR
MRK 054	12 54 36.	+ 32 42	13	15.	GALAXY WITH UV CONTINUUM
ZWG 189.002	12 54 36.	+ 32 43		15.1	GALAXY
ZWG 188.030	12 54 36.	+ 32 43		15.1	GALAXY
ZCG 1254+32	12 54 36.	+ 32 43		15.1	COMPACT GALAXY
ZC 1254.6+3651	12 54 36.	+ 36 51	810		CLUSTER OF GALAXIES

OBJECT NAME	RIGHT ASCEN.	DECLINATION	DIAM.	MAGN.	TYPE OF OBJECT
IC 3921	12 54 36.	+ 38 54 31.			NONSTELLAR OBJECT
1ZW 046	12 54 36.	+ 48 34			COMPACT GALAXY
ZWG 245.006	12 54 36.	+ 48 34		14.4	GALAXY
UGC 08068	12 54 36.	+ 48 34	78	14.4	GALAXY PECULR
MCG+08-24-012	12 54 36.	+ 48 35	18	16.	GALAXY
MCG+08-24-011	12 54 36.	+ 48 35	60	14.	GALAXY
LB 02497	12 54 36.	+ 57 50 54.		16.4	FAINT BLUE STAR
ZWG 015.045	12 54 36.	- 00 23		15.6	GALAXY
ZWG 015.044	12 54 36.	- 01 26		14.3	GALAXY
UGC 08067	12 54 36.	- 01 26	126	14.3	GALAXY S
MCG+00-33-022	12 54 36.	- 01 26	114	14.3	GALAXY
REIZ 3050	12 54 36.	- 10 20	12	14.7	GALAXY
HELW 199	12 54 36.	- 12 34 15.			NEBULA
MCG-02-33-071	12 54 36.	- 14 46	60	14.5	GALAXY
MCG-05-31-006	12 54 36.	- 27 56	42	15.	GALAXY
REIN 4.177	12 54 36.44	+ 09 08 44.1			NEBULA
A2 222	12 54 36.5	+ 28 42 29.		17.0	FAINT BLUE OBJECT
IC 3922	12 54 37.	+ 38 44 55.			NONSTELLAR OBJECT
ARC 1644	12 54 37.	- 17 06		15.7	RICH CLUSTER OF GALAXIES
REIN 4.178	12 54 37.11	+ 09 47 48.7			NEBULA
A2 223	12 54 37.4	+ 29 32 41.		18.0	FAINT BLUE GALAXY
A2 224	12 54 39.5	+ 28 55 13.		18.3	FAINT BLUE OBJECT
IC 3923	12 54 40.	+ 38 13 31.			NONSTELLAR OBJECT
RNGC 4829	12 54 41.	- 13 28			GALAXY
ARC 1645	12 54 41.	- 14 40		17.8	RICH CLUSTER OF GALAXIES
ZWG 071.078	12 54 42.	+ 09 43		15.7	GALAXY
ZC 1254.7+6427	12 54 42.	+ 64 27	3560		CLUSTER OF GALAXIES
MCG-01-33-059	12 54 42.	- 05 03	78	14.5	GALAXY
REIZ 3054	12 54 43.	+ 09 43	18	15.1	GALAXY
BFG 060	12 54 43.10	+ 37 30 21.7		19.14	ULTRAVIOLET-EXCESS OBJECT
A2 225	12 54 43.3	+ 29 12 07.		18.1	FAINT BLUE OBJECT
REIZ 3056	12 54 44.	+ 24 47	12	15.7	GALAXY
KN 16.664	12 54 45.3	+ 27 40 25.			NEBULA
REIN 4.180	12 54 45.36	+ 10 35 02.2			NEBULA
REIN 4.179	12 54 45.50	+ 09 42 59.9			NEBULA
RNGC 4830	12 54 46.	- 19 26		13.0	GALAXY
REIZ 3053	12 54 47.	- 05 03	42	13.6	GALAXY
BFG 061	12 54 47.38	+ 35 38 36.0		18.53	ULTRAVIOLET-EXCESS OBJECT
ZWG 071.079	12 54 48.	+ 14 30		15.6	GALAXY
ZC 1254.8+2024	12 54 48.	+ 20 24	1080		CLUSTER OF GALAXIES
ZWG 160.037	12 54 48.	+ 27 44		15.0	GALAXY
MCG+05-31-023	12 54 48.	+ 27 44	24	15.0	GALAXY
ZWG 160.038	12 54 48.	+ 29 18		14.8	GALAXY
UGC 08069	12 54 48.	+ 29 18	90	14.8	GALAXY SB?
ZWG 015.047	12 54 48.	- 00 21		15.6	GALAXY
ZWG 015.046	12 54 48.	- 03 13		14.3	GALAXY
MCG+00-33-023	12 54 48.	- 03 13	15	14.3	GALAXY
ZCG 1254-19	12 54 48.	- 19 24		18.5	COMPACT GALAXY
MCG-03-33-024	12 54 48.	- 19 26	72	13.5	GALAXY
A2 226	12 54 48.3	+ 27 10 36.		17.1	FAINT BLUE OBJECT
REIZ 3052	12 54 49.	- 11 50	24	15.8	GALAXY
REIZ 3051	12 54 49.	- 11 51	30	15.7	GALAXY
A2 227	12 54 50.1	+ 28 46 29.		17.8	FAINT BLUE GALAXY
FATH 1.558	12 54 51.	+ 44 47	14		NEBULA
RNGC 4831	12 54 51.	- 27 02		14.0	GALAXY
ZL 112	12 54 52.	+ 26 49 30.		20.5	ULTRAFAINT BLUE STAR
RNGC 4836	12 54 53.	- 12 29		13.0	GALAXY
MCG+05-31-024	12 54 54.	+ 29 19	36	14.8	GALAXY
IC 3925	12 54 54.	+ 36 41 38.			NONSTELLAR OBJECT
MRK 232	12 54 54.	+ 59 17	11	16.5	GALAXY WITH UV CONTINUUM
KW 20	12 54 54.	+ 59 17	27		SEYFERT GALAXY
MCG-04-31-010	12 54 54.	- 27 04	24	14.	GALAXY
OCL 0893	12 54 54.	- 64 41	600	10.6	OPEN STAR CLUSTER
VHA 142	12 54 54.	- 64 41	360		OPEN STAR CLUSTER
BFG 062	12 54 54.99	+ 37 03 27.4		17.84	ULTRAVIOLET-EXCESS OBJECT
REIZ 3058	12 54 55.	+ 09 46	36	15.2	GALAXY
SHB 223	12 54 55.	+ 37 03 27.		17.8	QUASI-STELLAR OBJECT
REIZ 3055	12 54 55.	- 12 28	30	14.1	GALAXY
BC B142	12 54 55.0	+ 37 03 29.1		17.84	QUASI-STELLAR OBJECT
ZL 113	12 54 56.	+ 26 55 00.		20.6	ULTRAFAINT BLUE STAR
RNGC 4815	12 54 56.	- 64 41		10.5	OPEN CLUSTER
IC 3924	12 54 57.	+ 19 03 01.			NONSTELLAR OBJECT
A2 228	12 54 57.1	+ 27 17 53.		18.4	FAINT BLUE OBJECT
KN 16.665	12 54 57.5	+ 27 45 43.			NEBULA
FATH 1.559	12 54 58.	+ 44 52	14		NEBULA
KN 16.666	12 54 58.1	+ 28 02 13.			NEBULA
ARC 1647	12 54 59.	+ 20 22		17.8	RICH CLUSTER OF GALAXIES
RNGC 4839	12 54 59.	+ 27 46		13.5	GALAXY
RNGC 4842	12 54 59.	+ 27 49		15.0	GALAXY
KN 16.667	12 54 59.1	+ 27 45 55.			NEBULA
VDB.66G 158	12 55	+ 03 04	70		DWARF GALAXY
VDB.66G 155	12 55	+ 14 29	70		DWARF GALAXY
VDB.66G 157	12 55	+ 15 09	70		DWARF GALAXY
MRK 055	12 55 00.	+ 27 40	12	16.	GALAXY WITH UV CONTINUUM
ZWG 160.039	12 55 00.	+ 27 46		13.6	GALAXY
UGC 08070	12 55 00.	+ 27 46	240	13.6	GALAXY E
MCG+05-31-025	12 55 00.	+ 27 46	36	13.6	GALAXY
ZWG 160.040	12 55 00.	+ 27 49		15.3	GALAXY
TON-N 0677	12 55 00.	+ 28 44		14.9	BLUE STAR
MCG+13-09-039	12 55 00.	+ 79 18	12	16.	GALAXY
MCG+14-06-012	12 55 00.	+ 80 32	27	16.	GALAXY
REIZ 3057	12 55 00.	- 10 27	48	14.0	GALAXY
MCG-02-33-073	12 55 00.	- 10 27	60	14.	GALAXY
MCG-02-33-072	12 55 00.	- 12 28	72	13.5	GALAXY
MCG-05-31-007	12 55 00.	- 29 30	30	15.5	GALAXY
MCG-06-29-001	12 55 00.	- 39 29	36	14.	GALAXY
KN 16.668	12 55 00.0	+ 27 40 31.			NEBULA
REIZ 3060	12 55 01.	+ 27 45	84	13.8	GALAXY
REIZ 3061	12 55 01.	+ 28 27	18	15.6	GALAXY
IC 3928	12 55 01.	+ 40 41 50.			NONSTELLAR OBJECT
BFG 063	12 55 01.18	+ 33 53 11.0		19.26	ULTRAVIOLET-EXCESS OBJECT
BC B154	12 55 02.1	+ 33 21 21.		18.56	QUASI-STELLAR OBJECT
SHB 224	12 55 02.1	+ 35 21 20.		18.6	QUASI-STELLAR OBJECT
BFG 064	12 55 02.14	+ 35 21 20.6		18.80	ULTRAVIOLET-EXCESS OBJECT
HOLM 491A	12 55 03.	+ 23 07	12	14.1	PART OF MULTIPLE GALAXY
IC 3930	12 55 03.	+ 39 01 50.			NONSTELLAR OBJECT
IC 3926	12 55 04.	+ 23 04 56.			NONSTELLAR OBJECT
KN 16.669	12 55 04.8	+ 28 27 31.			NEBULA
HOLM 491B	12 55 05.	+ 23 07	12	14.5	PART OF MULTIPLE GALAXY
RNGC 4840	12 55 05.	+ 27 53		15.0	GALAXY
REIN 4.181	12 55 05.29	+ 09 10 00.9			NEBULA
ZC 1255.1+0135	12 55 06.	+ 01 35	270		CLUSTER OF GALAXIES
ZWG 160.041	12 55 06.	+ 26 46		15.4	GALAXY
IC 0837	12 55 06.	+ 26 46		15.4	GALAXY IN COMA CLSTR
ZWG 160.042	12 55 06.	+ 27 53		14.8	GALAXY
ZWG 160.043	12 55 06.	+ 28 28		15.4	GALAXY
UGC 08071	12 55 06.	+ 28 28	66	15.4	GALAXY S
ZWG 160.044	12 55 06.	+ 28 45		13.5	GALAXY

OBJECT NAME	RIGHT ASCEN.	DECLINATION	DIAM.	MAGN.	TYPE OF OBJECT
RNGC 4841A	12 55 06.	+ 28 45		14.5	GALAXY
UGC 08073	12 55 06.	+ 28 45	96	13.5	GALAXY E+E
UGC 08072	12 55 06.	+ 28 45	96	13.5	GALAXY E+E
KARA.72 361A	12 55 06.	+ 28 45	90	13.5	PART OF DOUBLE GALAXY
ZWG 353.005	12 55 06.	+ 79 18		15.5	GALAXY
ZWG 352.048	12 55 06.	+ 79 18		15.5	GALAXY
REIZ 3059	12 55 06.	- 09 22	120	13.6	GALAXY
MCG-04-31-011	12 55 06.	- 23 56	48	15.	GALAXY
RNGC 4832	12 55 06.	- 40 28			NON-EXISTENT OBJECT
HOLM 492A	12 55 07.	+ 28 45	30	14.5	PART OF MULTIPLE GALAXY
KN 16.670	12 55 07.6	+ 27 52 55.			NEBULA
A2 232	12 55 08.0	+ 27 49 08.		17.9	FAINT BLUE OBJECT
HOLM 492B	12 55 10.	+ 28 45	24	14.7	PART OF MULTIPLE GALAXY
REIN 4.182	12 55 10.12	+ 08 56 07.1			NEBULA
KN 16.670A	12 55 10.4	+ 27 45 49.			NEBULA
KN 16.671	12 55 10.9	+ 27 17 55.			NEBULA
ZL 114	12 55 11.	+ 26 51 36.		17.7	ULTRAFAINT BLUE STAR
ZWG 043.080	12 55 12.	+ 02 58		15.4	GALAXY
UGC 08074	12 55 12.	+ 02 58	96	15.4	GALAXY DWRF SP
MCG+01-33-031	12 55 12.	+ 02 58	60	15.4	GALAXY
ZWG 071.080	12 55 12.	+ 08 56		15.3	GALAXY
MCG+02-33-037	12 55 12.	+ 08 56	60	15.3	GALAXY
MCG+05-31-028	12 55 12.	+ 26 46	48	15.4	GALAXY
ZWG 160.045	12 55 12.	+ 27 07		15.5	GALAXY
MCG+05-31-031	12 55 12.	+ 27 44 30.	12	14.9	GALAXY
ZWG 160.046	12 55 12.	+ 27 45		14.9	GALAXY
MCG+05-31-030	12 55 12.	+ 27 45	18	14.9	GALAXY
MCG+05-31-029	12 55 12.	+ 27 53	24	14.8	GALAXY
RNGC 4841B	12 55 12.	+ 28 45		14.5	GALAXY
SCH 82	12 55 12.	+ 28 45		13.5	PECULIAR GALAXY
MCG+05-31-027	12 55 12.	+ 28 45	18	13.5	GALAXY
MCG+05-31-026	12 55 12.	+ 28 45	24	13.5	GALAXY
KARA.72 361B	12 55 12.	+ 28 46	66		PART OF DOUBLE GALAXY
MCG+06-29-001	12 55 12.	+ 35 49 30.	48	15.5	GALAXY
UGC 08075	12 55 12.	+ 41 44	66	16.5	GALAXY Sc
ZC 1255.2+5654	12 55 12.	+ 56 54	1480		CLUSTER OF GALAXIES
MCG-01-33-060	12 55 12.	- 09 22	180	13.5	GALAXY
SEE 094.05	12 55 12.	- 46 02	1800	13.	LOOSE GROUP OF 8 GALAXIES
LB 02499	12 55 13.	+ 51 08 24.		14.4	FAINT BLUE STAR
LB 02498	12 55 13.	+ 59 05 54.		16.9	FAINT BLUE STAR
IC 3929	12 55 14.	+ 20 40 02.			NONSTELLAR OBJECT
MCG+08-24-013	12 55 15.	+ 49 25	36	16.	GALAXY
A2 233	12 55 15.4	+ 27 16 14.		16.1	FAINT BLUE OBJECT
RNGC 4838	12 55 17.	- 12 47		13.0	GALAXY
RNGC 4835	12 55 17.	- 45 59		12.5	GALAXY
A2 234	12 55 17.0	+ 27 59 56.		17.0	FAINT BLUE OBJECT
REIZ 3064	12 55 18.	+ 12 31	12	14.9	GALAXY
ZC 1255.3+1251	12 55 18.	+ 12 51	3360		CLUSTER OF GALAXIES
ZWG 160.047	12 55 18.	+ 28 06		15.7	GALAXY
ZWG 160.048	12 55 18.	+ 28 09		15.3	GALAXY
SN 1963C	12 55 18.	+ 28 09		15.7	SUPERNOVA
1ZW 047	12 55 18.	+ 31 52			COMPACT GALAXY
ZWG 189.003	12 55 18.	+ 32 49		15.6	GALAXY
ZWG 188.031	12 55 18.	+ 32 49		15.6	GALAXY
HOLM 493A	12 55 18.	+ 32 49	66	15.2	PART OF MULTIPLE GALAXY
ZWG 245.007	12 55 18.	+ 48 45		15.6	GALAXY
MCG+12-12-022	12 55 18.	+ 70 27 30.	78	15.	GALAXY
REIZ 3062	12 55 19.	- 12 47	48	13.7	GALAXY
KN 16.672	12 55 20.9	+ 28 01 31.			NEBULA
MCG-02-33-074	12 55 21.	- 12 47 30.	84	13.5	GALAXY
HOLM 493B	12 55 22.	+ 32 49	18	15.8	PART OF MULTIPLE GALAXY
REIZ 3068	12 55 22.	+ 37 39	60	14.3	GALAXY
LB 02500	12 55 22.	+ 54 45 06.		17.3	FAINT BLUE STAR
REIN 4.183	12 55 22.38	+ 12 31 58.1			NEBULA
REIZ 3063	12 55 23.	- 03 20	90	13.2	GALAXY
SN 1958C	12 55 23.	- 30 18		19.6	SUPERNOVA
KN 16.673	12 55 23.4	+ 28 26 50.			NEBULA
REIN 4.184	12 55 23.72	+ 10 21 15.9			NEBULA
ZWG 043.081	12 55 24.	+ 02 37		15.7	GALAXY
ZWG 043.082	12 55 24.	+ 04 00		15.7	GALAXY
ZWG 071.081	12 55 24.	+ 09 31		15.4	GALAXY
LB 00060	12 55 24.	+ 23 47		17.7	FAINT BLUE STAR
MRK 447	12 55 24.	+ 24 39	10	16.5	GALAXY WITH UV CONTINUUM
TON-N 0678	12 55 24.	+ 27 00		14.9	BLUE STAR
HOLM 494B	12 55 24.	+ 27 46	24	14.9	PART OF MULTIPLE GALAXY
MCG+05-31-032	12 55 24.	+ 28 10	24	15.5	GALAXY
ZWG 160.049	12 55 24.	+ 28 27		15.5	GALAXY
ZWG 160.050	12 55 24.	+ 29 55		15.2	GALAXY
UGC 08076	12 55 24.	+ 29 55	78	15.2	GALAXY Sc
ZC 1255.4+4501	12 55 24.	+ 45 01	1880		CLUSTER OF GALAXIES
MCG+08-24-014	12 55 24.	+ 48 46	42	15.	GALAXY
ZC 1255.4+6946	12 55 24.	+ 69 46	610		CLUSTER OF GALAXIES
ZWG 335.029	12 55 24.	+ 70 28		14.7	GALAXY
RNGC 4857	12 55 24.	+ 70 28		14.5	GALAXY
UGC 08077	12 55 24.	+ 70 28	78	14.7	GALAXY Sb/SBb
ZWG 015.048	12 55 24.	- 03 21		14.1	GALAXY
MCG+00-33-024	12 55 24.	- 03 21	120	14.1	GALAXY
REIZ 3066	12 55 25.	+ 09 31	18	14.5	GALAXY
RNGC 4843	12 55 25.	- 03 21		14.0	GALAXY
HELW 200	12 55 25.	- 14 24 08.			NEBULA
BFG 065	12 55 25.47	+ 33 18 52.0		19.00	ULTRAVIOLET-EXCESS OBJECT
A2 235	12 55 25.6	+ 26 52 41.		18.3	FAINT BLUE OBJECT
REIN 4.185	12 55 26.97	+ 09 30 36.7			NEBULA
REIZ 3065	12 55 27.	+ 01 51	174	12.5	GALAXY
MCG+05-31-033	12 55 27.	+ 31 27	42	15.5	GALAXY
MCG+06-29-002	12 55 27.	+ 36 39	66	14.5	GALAXY
REIZ 3067	12 55 28.	+ 19 22	30	15.0	GALAXY
RNGC 4844	12 55 29.	- 12 48			NON-EXISTENT OBJECT
KN 16.674	12 55 29.2	+ 27 45 26.			NEBULA
KN 16.675	12 55 29.5	+ 27 52 56.			NEBULA
ZWG 015.049	12 55 30.	+ 01 50		12.9	GALAXY
UGC 08078	12 55 30.	+ 01 50	318	12.9	GALAXY Sb
MCG+00-33-025	12 55 30.	+ 01 50	252	12.9	GALAXY
ZC 1255.5+1950	12 55 30.	+ 19 50	1010		CLUSTER OF GALAXIES
SN 1963M	12 55 30.	+ 28 20		15.9	SUPERNOVA
MCG+05-31-035	12 55 30.	+ 28 20	24	16.	GALAXY
MCG+05-31-034	12 55 30.	+ 29 56	42	15.2	GALAXY
ZWG 189.004	12 55 30.	+ 36 38		14.6	GALAXY
ZWG 188.032	12 55 30.	+ 36 38		14.6	GALAXY
UGC 08079	12 55 30.	+ 36 38	90	14.6	GALAXY S
KARA.73B 0561	12 55 30.	+ 36 38	66	14.6	ISOLATED GALAXY S
MCG+09-21-070	12 55 30.	+ 53 54	30	15.	GALAXY
MCG+15-01-014	12 55 30.	+ 87 35	18	17.	GALAXY
IC 3927	12 55 30.	- 22 36 22.			NONSTELLAR OBJECT
RNGC 4845	12 55 31.	+ 01 51		12.5	GALAXY
IC 3931	12 55 31.	+ 19 53 14.			NONSTELLAR OBJECT
HOLM 494A	12 55 31.	+ 27 46	24	14.5	PART OF MULTIPLE GALAXY
RNGC 4846	12 55 31.	+ 36 38		14.5	GALAXY

OBJECT NAME	RIGHT ASCEN.	DECLINATION	DIAM.	MAGN.	TYPE OF OBJECT
LB 02501	12 55 31.	+ 54 59 54.		14.9	FAINT BLUE STAR
A2 236	12 55 31.0	+ 27 51 04.		17.5	FAINT BLUE GALAXY
BFG 066	12 55 33.92	+ 32 45 38.3		18.61	ULTRAVIOLET-EXCESS OBJECT
REIZ 3070	12 55 34.	+ 19 53	30	14.6	GALAXY
A2 238	12 55 34.4	+ 30 01 50.		18.3	FAINT BLUE OBJECT
A2 239	12 55 34.8	+ 28 58 56.		16.8	FAINT BLUE OBJECT
IC 3933	12 55 35.	+ 36 54 51.			NONSTELLAR OBJECT
A2 240	12 55 35.5	+ 30 14 41.		18.3	FAINT BLUE OBJECT
3ZW 068	12 55 36.	+ 27 07			COMPACT GALAXY
ZWG 160.051	12 55 36.	+ 27 08		15.2	GALAXY
UGC 08080	12 55 36.	+ 27 08	66	15.2	GALAXY S0-a
ZWG 160.052	12 55 36.	+ 27 11		15.5	GALAXY
MCG+05-31-036	12 55 36.	+ 27 45	24	14.9	GALAXY
ZWG 160.053	12 55 36.	+ 28 05		15.7	GALAXY
RNGC 4850	12 55 36.	+ 28 14		15.5	GALAXY
ZWG 160.054	12 55 36.	+ 31 26		15.5	GALAXY
ZC 1255.6+7858	12 55 36.	+ 78 58	8060		CLUSTER OF GALAXIES
MCG-04-31-012	12 55 36.	- 22 37	24	14.	GALAXY
MCG-04-31-013	12 55 36.	- 26 06	18	15.	GALAXY
KN 16.676	12 55 36.4	+ 27 45 26.			NEBULA
A2 241	12 55 36.8	+ 28 38 43.		16.0	FAINT BLUE OBJECT
REIZ 3073	12 55 37.	+ 28 31	54	14.0	GALAXY
A2 242	12 55 37.4	+ 28 06 10.		17.7	FAINT BLUE OBJECT
IC 3932	12 55 38.	+ 19 51 15.			NONSTELLAR OBJECT
REIZ 3072	12 55 40.	+ 19 51	18	15.2	GALAXY
BC B185	12 55 40.2	+ 37 15 19.1		18.12	QUASI-STELLAR OBJECT
BFG 067	12 55 40.26	+ 37 15 17.4		18.12	ULTRAVIOLET-EXCESS OBJECT
SHB 225	12 55 40.3	+ 37 15 17.		18.1	QUASI-STELLAR OBJECT
KN 16.677	12 55 40.5	+ 28 30 44.			NEBULA
ZWG 043.083	12 55 42.	+ 03 48		15.6	GALAXY
ZC 1255.7+1436	12 55 42.	+ 14 36	400		CLUSTER OF GALAXIES
UGC 08081	12 55 42.	+ 15 07	60	17.	GALAXY DWARF
REA 22	12 55 42.	+ 15 07			DWARF GALAXY
MCG+03-33-017	12 55 42.	+ 15 07 30.	120	18.	GALAXY
MCG+05-31-037	12 55 42.	+ 27 45	24	14.5	GALAXY
ZWG 160.055	12 55 42.	+ 28 31		14.2	GALAXY
RNGC 4848	12 55 42.	+ 28 31		14.0	GALAXY
UGC 08082	12 55 42.	+ 28 31	102	14.2	GALAXY SB:a-b
ZC 1255.7+4047	12 55 42.	+ 40 47	3490		CLUSTER OF GALAXIES
UGC 08083	12 55 42.	+ 75 10	66	16.0	GALAXY S
MCG-04-31-014	12 55 42.	- 24 56 30.	54	14.5	GALAXY
VHE 58	12 55 42.	- 66 02	30		REFLECTION NEBULA
REIZ 3069	12 55 43.	- 10 54	12	14.1	GALAXY
MCG+05-31-038	12 55 45.	+ 27 06	60	14.5	GALAXY
MCG+05-31-039	12 55 45.	+ 28 30	84	14.2	GALAXY
IC 3935	12 55 46.	+ 26 38 45.			SAME AS NGC 4849
A2 246	12 55 46.6	+ 29 08 43.		18.2	FAINT BLUE OBJECT
RNGC 4849	12 55 47.	+ 26 40		14.5	GALAXY
ZWG 043.084	12 55 48.	+ 03 04		15.0	GALAXY
UGC 08084	12 55 48.	+ 03 04	102	15.0	GALAXY
MCG+01-33-032	12 55 48.	+ 03 04	84	15.0	GALAXY
ZWG 043.085	12 55 48.	+ 05 09		15.3	GALAXY
REIZ 3075	12 55 48.	+ 12 39	18	14.9	GALAXY
ZWG 071.082	12 55 49.	+ 13 39		14.9	GALAXY
MCG+02-33-038	12 55 48.	+ 13 39	48	14.9	GALAXY
ZWG 100.018	12 55 48.	+ 14 50		14.9	GALAXY
UGC 08085	12 55 48.	+ 14 50	150	14.9	GALAXY SB?c
ZWG 160.056	12 55 48.	+ 26 40		14.5	GALAXY
UGC 08086	12 55 48.	+ 26 40	102	14.5	GALAXY S0
IC 0838	12 55 48.	+ 26 40		14.5	GALAXY IN COMA CLSTR
MCG+05-31-040	12 55 48.	+ 28 14	60	15.3	GALAXY
ZWG 160.057	12 55 48.	+ 28 24		15.5	GALAXY
ZWG 160.058	12 55 48.	+ 28 59		15.5	GALAXY
MCG+05-31-041	12 55 48.	+ 28 59	45	15.5	GALAXY
ZWG 160.059	12 55 48.	+ 29 13		15.2	GALAXY
ZWG 160.060	12 55 48.	+ 32 17		15.0	GALAXY
MCG+05-31-042	12 55 48.	+ 32 17	30	15.0	GALAXY
ZWG 353.006	12 55 48.	+ 78 53		15.6	GALAXY
ZWG 352.049	12 55 48.	+ 78 53		15.6	GALAXY
UGC 08087	12 55 49.	+ 78 53	84	15.6	GALAXY S
REIZ 3071	12 55 48.	- 10 18	48	14.0	GALAXY
MCG-02-33-075	12 55 48.	- 10 18	42	14.	GALAXY
MCG-06-29-002	12 55 48.	- 37 18	60	15.	GALAXY
REIN 4.186	12 55 48.11	+ 12 39 35.8			NEBULA
A2 249	12 55 48.9	+ 26 31 21.		17.6	FAINT BLUE OBJECT
REIN 4.187	12 55 49.14	+ 13 39 40.5			NEBULA
IC 3934	12 55 50.	+ 19 05 45.			NONSTELLAR OBJECT
REIZ 3078	12 55 50.	+ 24 30	24	14.8	GALAXY
HOLM 495A	12 55 50.	+ 26 40	18	14.4	PART OF MULTIPLE GALAXY
REIZ 3074	12 55 51.	+ 03 03	48	14.9	GALAXY
HOLM 495B	12 55 51.	+ 26 42	42	14.9	PART OF MULTIPLE GALAXY
IC 3936	12 55 52.	+ 19 19 09.			NONSTELLAR OBJECT
REIZ 3077	12 55 52.	+ 20 48	60	15.5	GALAXY
LB 02502	12 55 52.	+ 50 29 30.		16.5	FAINT BLUE STAR
RNGC 4849A	12 55 53.	+ 26 41		16.0	GALAXY
SN 1968H	12 55 53.	+ 27 24		16.6	SUPERNOVA
LB 02503	12 55 53.	+ 62 04 06.		16.9	FAINT BLUE STAR
RNGC 4847	12 55 53.	- 12 52			GALAXY
ZWG 071.083	12 55 54.	+ 09 05		15.7	GALAXY
ZWG 071.084	12 55 54.	+ 09 54		15.6	GALAXY
ZWG 071.085	12 55 54.	+ 14 09		15.2	GALAXY
UGC 08088	12 55 54.	+ 14 09	60	15.2	GALAXY S
MCG+02-33-039	12 55 54.	+ 14 09	66	15.2	GALAXY
MCG+03-33-018	12 55 54.	+ 14 48	150	14.9	GALAXY
MCG+05-31-044	12 55 54.	+ 26 39	96	14.5	GALAXY
MCG+05-31-043	12 55 54.	+ 26 41	30	14.9	GALAXY
RB 182	12 55 54.	+ 28 00	18		GALAXY IN COMA CLUSTER
RB 183	12 55 54.	+ 28 02	14		GALAXY IN COMA CLUSTER
RB 181	12 55 54.	+ 28 04	8		GALAXY IN COMA CLUSTER
RB 240	12 55 54.	+ 28 09	3		GALAXY IN COMA CLUSTER
RB 239	12 55 54.	+ 28 10	15		GALAXY IN COMA CLUSTER
RB 238	12 55 54.	+ 28 19	6		GALAXY IN COMA CLUSTER
ZWG 160.061	12 55 54.	+ 28 25		15.2	GALAXY
RNGC 4851	12 55 54.	+ 28 25		15.0	GALAXY
ZWG 160.062	12 55 54.	+ 29 24		15.0	GALAXY
SCH 83	12 55 54.	+ 29 24		15.0	PECULIAR GALAXY
MCG+05-31-045	12 55 54.	+ 29 24	42	15.0	GALAXY
ZC 1255.9+3729	12 55 54.	+ 37 29	2550		CLUSTER OF GALAXIES
MCG-02-33-076	12 55 54.	- 15 16	54	14.5	GALAXY
A2 251	12 55 54.6	+ 30 21 20.		18.4	FAINT BLUE OBJECT
REIN 4.188	12 55 54.97	+ 09 53 57.9			NEBULA
REIZ 3076	12 55 55.	+ 09 54	24	15.1	GALAXY
IC 3940	12 55 55.	+ 36 06 21.			NONSTELLAR OBJECT
IC 3941	12 55 55.	+ 40 02 27.			NONSTELLAR OBJECT
IC 0839	12 55 56.4	+ 28 25 04.			GALAXY SB0
BFG 068	12 55 56.72	+ 36 44 22.8		18.56	ULTRAVIOLET-EXCESS OBJECT
IC 3937	12 55 57.	+ 19 05 21.			NONSTELLAR OBJECT
DUN 01	12 55 57.	+ 28 31	40		SMALL NEBULA
REIN 1.014	12 55 57.02	+ 28 14 17.0			NEBULA
IC 3938	12 55 58.	+ 19 01 15.			NONSTELLAR OBJECT
IC 3942	12 55 58.	+ 36 22 39.			NONSTELLAR OBJECT
REIN 4.189	12 55 59.24	+ 09 48 51.0			NEBULA
ZWG 015.050	12 56 00.	+ 01 58		15.7	GALAXY
UGC 08089	12 56 00.	+ 09 48	60	16.0	GALAXY
RB 184	12 56 00.	+ 27 59	4		GALAXY IN COMA CLUSTER
ZWG 160.063	12 56 00.	+ 28 14		15.3	GALAXY
ZC 1256.0+3135	12 56 00.	+ 31 35	3560		CLUSTER OF GALAXIES
GCL 021	12 56 00.	- 70 36	762	8.5	GLOBULAR STAR CLUSTER
REIN 4.190	12 56 00.64	+ 08 29 13.3			NEBULA
REIZ 3079	12 56 01.	+ 09 49	48	15.0	GALAXY
IC 3939	12 56 01.	+ 19 01 15.			NONSTELLAR OBJECT
RNGC 4833	12 56 02.	- 70 36		8.5	GLOBULAR CLUSTER
KARA.73 41	12 56 03.	- 37 30	34		DWARF GALAXY
ARC 1649	12 56 05.	+ 10 02		17.2	RICH CLUSTER OF GALAXIES
REIN 1.017	12 56 05.39	+ 28 17 05.4			NEBULA
ZWG 043.086	12 56 06.	+ 08 29		15.2	GALAXY
MCG+01-33-033	12 56 06.	+ 08 29	48	15.2	GALAXY
8ZW 1256+14.2	12 56 06.	+ 14 14		15.5	COMPACT GALAXY
REA 08	12 56 06.	+ 14 29			DWARF GALAXY
MCG+03-33-019	12 56 06.	+ 16 39	42	17.5	GALAXY
ZWG 160.064	12 56 06.	+ 27 31		15.4	GALAXY
MRK 056	12 56 06.	+ 27 31	10	16.5	GALAXY WITH UV CONTINUUM
RB 188	12 56 06.	+ 27 57	7		GALAXY IN COMA CLUSTER
RB 187	12 56 06.	+ 28 04	2		GALAXY IN COMA CLUSTER
RB 185	12 56 06.	+ 28 04	2		GALAXY IN COMA CLUSTER
RB 186	12 56 06.	+ 28 05	9		GALAXY IN COMA CLUSTER
RB 190	12 56 06.	+ 28 07	3		GALAXY IN COMA CLUSTER
RB 189	12 56 06.	+ 28 07	3		GALAXY IN COMA CLUSTER
RB 242	12 56 06.	+ 28 10	3		GALAXY IN COMA CLUSTER
ZWG 160.065	12 56 06.	+ 28 17		15.0	GALAXY
RB 241	12 56 06.	+ 28 17	12		GALAXY IN COMA CLUSTER
MCG+05-31-046	12 56 06.	+ 28 17	48	15.0	GALAXY
RB 243	12 56 06.	+ 28 19	7		GALAXY IN COMA CLUSTER
ZC 1256.1+6537	12 56 06.	+ 65 37	1080		CLUSTER OF GALAXIES
MCG+13-09-040	12 56 06.	+ 78 52 30.	72	16.	GALAXY
REIN 1.019	12 56 07.01	+ 27 56 39.6			NEBULA
REIN 1.015	12 56 07.26	+ 14 30 32.0			NEBULA
SHB 226	12 56 07.8	+ 35 44 53.		18.0	QUASI-STELLAR OBJECT
BC B194	12 56 07.8	+ 35 44 56.		17.96	QUASI-STELLAR OBJECT
BFG 069	12 56 07.84	+ 35 44 53.7		18.30	ULTRAVIOLET-EXCESS OBJECT
LB 02504	12 56 09.	+ 53 21 00.		16.1	FAINT BLUE STAR
REIN 1.016	12 56 09.56	+ 14 29 57.0			NEBULA
IC 3945	12 56 10.	+ 40 12 22.			NONSTELLAR OBJECT
A2 255	12 56 10.4	+ 31 23 37.		18.0	FAINT BLUE OBJECT
A2 256	12 56 10.6	+ 30 35 24.		17.1	FAINT BLUE OBJECT
REIN 1.018	12 56 10.99	+ 14 29 14.1			NEBULA
RNGC 4853	12 56 11.	+ 27 52		14.0	GALAXY
REA.66 1	12 56 11.	+ 28 12 36.			DWARF GALAXY
ARC 1650	12 56 11.	- 01 30		17.0	RICH CLUSTER OF GALAXIES
IC 3943	12 56 11.5	+ 28 22 54.			GALAXY S0
REIN 4.191	12 56 11.58	+ 10 53 08.9			NEBULA
ZWG 071.086	12 56 12.	+ 10 53		15.1	GALAXY
UGC 08090	12 56 12.	+ 10 53	60	15.1	GALAXY SB:a
MCG+02-33-040	12 56 12.	+ 10 53	78	15.1	GALAXY
IC 0840	12 56 12.	+ 10 54 16.			NONSTELLAR OBJECT
ZWG 071.087	12 56 12.	+ 14 29		15.3	GALAXY
UGC 08091	12 56 12.	+ 14 29	66	15.3	GALAXY IRR
MCG+02-33-041	12 56 12.	+ 14 29	60	15.3	GALAXY
ZWG 160.066	12 56 12.	+ 27 22		15.3	GALAXY
ZWG 160.067	12 56 12.	+ 27 26		15.4	GALAXY
MRK 057	12 56 12.	+ 27 26	15	15.5	GALAXY WITH UV CONTINUUM
2ZW 067	12 56 12.	+ 27 51			COMPACT GALAXY
ZWG 160.068	12 56 12.	+ 27 51		14.2	GALAXY
UGC 08092	12 56 12.	+ 27 51	48	14.2	GALAXY VY CMPT
RB 193	12 56 12.	+ 27 54	3		GALAXY IN COMA CLUSTER
RB 191	12 56 12.	+ 27 57	3		GALAXY IN COMA CLUSTER
RB 196	12 56 12.	+ 28 06	11		GALAXY IN COMA CLUSTER
RB 195	12 56 12.	+ 28 07	6		GALAXY IN COMA CLUSTER
RB 192	12 56 12.	+ 28 07	15		GALAXY IN COMA CLUSTER
REA.66 2	12 56 12.	+ 28 08 48.			DWARF GALAXY
RB 194	12 56 12.	+ 28 09	4		GALAXY IN COMA CLUSTER
RB 245	12 56 12.	+ 28 13	12		GALAXY IN COMA CLUSTER
RB 244	12 56 12.	+ 28 13	9		GALAXY IN COMA CLUSTER
ZWG 160.069	12 56 12.	+ 28 23		15.6	GALAXY
LB 00023	12 56 12.	+ 30 36		17.1	FAINT BLUE STAR
MCG+08-24-015	12 56 12.	+ 45 32	30	16.	GALAXY
ZWG 270.035	12 56 12.	+ 52 22		15.7	GALAXY
MCG+09-21-071	12 56 12.	+ 52 48	24	16.	GALAXY
MCG+10-19-005	12 56 12.	+ 61 06	42	16.	GALAXY
REIZ 3083	12 56 13.	+ 10 53	36	14.6	GALAXY
LB 02505	12 56 14.	+ 49 56 30.		15.7	FAINT BLUE STAR
MCG+05-31-047	12 56 15.	+ 27 22	30	15.3	GALAXY
MCG+05-31-048	12 56 15.	+ 27 51	54	14.2	GALAXY
ARC 1655	12 56 15.	+ 65 39		18.0	RICH CLUSTER OF GALAXIES
A2 257	12 56 15.7	+ 26 51 41.		17.1	FAINT BLUE OBJECT
REIZ 3082	12 56 17.	- 05 51	108	14.1	GALAXY
REIZ 3081	12 56 17.	- 05 54	120	13.7	GALAXY
ZWG 071.088	12 56 18.	+ 09 55		15.0	GALAXY
SCH 84	12 56 18.	+ 09 55		15.0	PECULIAR GALAXY
UGC 08093	12 56 18.	+ 09 55	60	15.0	GALAXY S
MCG+02-33-042	12 56 18.	+ 09 55	60	15.0	GALAXY
IC 3944	12 56 18.	+ 24 02 58.			NONSTELLAR OBJECT
RB 197	12 56 18.	+ 27 56	2		GALAXY IN COMA CLUSTER
RB 202	12 56 18.	+ 28 02	4		GALAXY IN COMA CLUSTER
RB 199	12 56 18.	+ 28 02	6		GALAXY IN COMA CLUSTER
RB 200	12 56 18.	+ 28 03	12		GALAXY IN COMA CLUSTER
RB 198	12 56 18.	+ 28 06	11		GALAXY IN COMA CLUSTER
RB 201	12 56 18.	+ 28 09	4		GALAXY IN COMA CLUSTER
RB 246	12 56 18.	+ 28 11	6		GALAXY IN COMA CLUSTER
RB 247	12 56 18.	+ 28 12	3		GALAXY IN COMA CLUSTER
RB 248	12 56 18.	+ 28 19	5		GALAXY IN COMA CLUSTER
1ZW 048	12 56 18.	+ 36 24			COMPACT GALAXY
MCG+08-24-016	12 56 18.	+ 49 22	36	16.	GALAXY
MCG+10-19-006	12 56 18.	+ 59 24	30	16.	GALAXY
ZC 1256.3-0129	12 56 18.	- 01 29	2960		CLUSTER OF GALAXIES
MCG-01-33-061	12 56 18.	- 05 49	96	14.5	GALAXY
ARC 1648	12 56 18.	- 26 22		16.9	RICH CLUSTER OF GALAXIES
MCG-07-27-021	12 56 18.	- 43 35	36	15.	GALAXY
A2 258	12 56 18.4	+ 29 51 41.		17.6	FAINT BLUE OBJECT
BFG 070	12 56 18.40	+ 37 23 43.6		18.90	ULTRAVIOLET-EXCESS OBJECT
REIZ 3084	12 56 19.	+ 09 56	48	13.8	GALAXY
REIZ 3080	12 56 19.	- 10 27	30	16.0	GALAXY
REIN 4.192	12 56 20.26	+ 09 55 25.2			NEBULA
A2 259	12 56 21.6	+ 28 48 45.		18.3	FAINT BLUE GALAXY
REIN 1.020	12 56 22.75	+ 27 56 41.2			NEBULA
REIN 4.194	12 56 23.19	+ 13 25 16.5			NEBULA

OBJECT NAME	RIGHT ASCEN.	DECLINATION	DIAM.	MAGN.	TYPE OF OBJECT
A2 260	12 56 23.3	+ 26 46 38.		18.3	FAINT BLUE OBJECT
IC 3946	12 56 23.4	+ 28 04 42.		15.3	GALAXY E6
BFG 071	12 56 23.75	+ 33 38 44.2		19.13	ULTRAVIOLET-EXCESS OBJECT
REIN 4.193	12 56 23.81	+ 09 34 14.2			NEBULA
REIN 1.021	12 56 23.99	+ 28 04 50.3			NEBULA
ZWG 071.089	12 56 24.	+ 09 35		15.6	GALAXY
ZC 1256.4+1001	12 56 24.	+ 10 01	1280		CLUSTER OF GALAXIES
ZWG 071.090	12 56 24.	+ 13 25		15.3	GALAXY
MCG+02-33-043	12 56 24.	+ 13 25	54	15.3	GALAXY
ZWG 160.070	12 56 24.	+ 27 57		15.2	GALAXY
RNGC 4854	12 56 24.	+ 27 57		15.0	GALAXY
RB 206	12 56 24.	+ 28 05	2		GALAXY IN COMA CLUSTER
RB 205	12 56 24.	+ 28 07	7		GALAXY IN COMA CLUSTER
RB 203	12 56 24.	+ 28 08	3		GALAXY IN COMA CLUSTER
RB 204	12 56 24.	+ 28 10	8		GALAXY IN COMA CLUSTER
RB 250	12 56 24.	+ 28 11	4		GALAXY IN COMA CLUSTER
RB 207	12 56 24.	+ 28 11	14		GALAXY IN COMA CLUSTER
RB 253	12 56 24.	+ 28 16	4		GALAXY IN COMA CLUSTER
RB 254	12 56 24.	+ 28 17	3		GALAXY IN COMA CLUSTER
RB 249	12 56 24.	+ 28 17	4		GALAXY IN COMA CLUSTER
RB 251	12 56 24.	+ 28 19	5		GALAXY IN COMA CLUSTER
RB 255	12 56 24.	+ 28 21	5		GALAXY IN COMA CLUSTER
RB 252	12 56 24.	+ 28 21	26		GALAXY IN COMA CLUSTER
LB 00061	12 56 24.	+ 29 29		17.3	FAINT BLUE STAR
ZC 1256.4+4701	12 56 24.	+ 47 01	4700		CLUSTER OF GALAXIES
UGC 08094	12 56 24.	+ 50 15	60	16.0	GALAXY S
MCG+09-21-072	12 56 24.	+ 53 40	30	15.	GALAXY
MRK 233	12 56 24.	+ 59 22	7	17.	GALAXY WITH UV CONTINUUM
ZWG 335.030	12 56 24.	+ 71 42		15.2	GALAXY
UGC 08095	12 56 24.	+ 71 42	78	15.2	GALAXY Sa-b
SHAF 158	12 56 24.	+ 78 32	162		GROUP OF COMPACT GALAXIES
REIN 1.022	12 56 26.02	+ 28 21 13.2			NEBULA
REIN 1.023	12 56 26.13	+ 28 02 51.4			NEBULA
MCG+08-24-017	12 56 27.	+ 50 16 30.	60	15.	GALAXY
IC 3947	12 56 27.2	+ 28 03 12.		15.6	GALAXY E4
REIN 1.024	12 56 27.34	+ 28 03 19.6			NEBULA
LB 02506	12 56 28.	+ 57 36 06.		15.7	FAINT BLUE STAR
A2 262	12 56 28.2	+ 30 48 34.		17.7	FAINT BLUE OBJECT
REIN 1.025	12 56 28.29	+ 28 05 00.3			NEBULA
REIN 1.026	12 56 28.64	+ 28 23 44.4			NEBULA
A2 263	12 56 28.9	+ 26 15 11.		17.2	FAINT BLUE OBJECT
BFG 072	12 56 28.95	+ 33 37 33.4		18.91	ULTRAVIOLET-EXCESS OBJECT
BFG 073	12 56 29.62	+ 36 39 35.3		18.92	ULTRAVIOLET-EXCESS OBJECT
A2 266	12 56 29.8	+ 26 45 21.		18.7	FAINT BLUE OBJECT
MCG+05-31-049	12 56 30.	+ 27 56	15	15.2	GALAXY
RB 214	12 56 30.	+ 28 04	23		GALAXY IN COMA CLUSTER
RB 210	12 56 30.	+ 28 04	8		GALAXY IN COMA CLUSTER
RB 212	12 56 30.	+ 28 05	3		GALAXY IN COMA CLUSTER
RB 209	12 56 30.	+ 28 05	18		GALAXY IN COMA CLUSTER
MCG+05-31-050	12 56 30.	+ 28 05	30	15.3	GALAXY
TON-N 0679	12 56 30.	+ 28 06		15.5	BLUE STAR
RB 213	12 56 30.	+ 28 06	9		GALAXY IN COMA CLUSTER
RB 211	12 56 30.	+ 28 06	6		GALAXY IN COMA CLUSTER
UGC 08096	12 56 30.	+ 28 06	60	14.9	GALAXY S
RB 208	12 56 30.	+ 28 10	3		GALAXY IN COMA CLUSTER
RB 258	12 56 30.	+ 28 11	3		GALAXY IN COMA CLUSTER
RB 257	12 56 30.	+ 28 14	10		GALAXY IN COMA CLUSTER
RB 256	12 56 30.	+ 28 20	19		GALAXY IN COMA CLUSTER
ZWG 270.036	12 56 30.	+ 53 15		15.6	GALAXY
MCG+12-12-023	12 56 30.	+ 71 42	66	15.	GALAXY
ZWG 015.051	12 56 30.	- 02 53		15.6	GALAXY
REIZ 3085	12 56 31.	- 10 38	24	14.7	GALAXY
A2 268	12 56 31.0	+ 26 51 04.		17.6	FAINT BLUE OBJECT
IC 3949	12 56 31.1	+ 28 06 05.		14.9	GALAXY S0
REIN 1.027	12 56 31.28	+ 28 06 12.6			NEBULA
IC 3948	12 56 32.	+ 24 19 52.			NONSTELLAR OBJECT
IC 3952	12 56 32.	+ 39 08 17.			NONSTELLAR OBJECT
LB 02507	12 56 32.	+ 50 38 54.		16.5	FAINT BLUE STAR
REIN 1.028	12 56 32.81	+ 28 15 42.9			NEBULA
ARP 266	12 56 33.	+ 35 08			PECULIAR GALAXY
A2 270	12 56 33.4	+ 28 31 38.		16.2	FAINT BLUE OBJECT
BFG 074	12 56 34.63	+ 32 22 36.5		18.96	ULTRAVIOLET-EXCESS OBJECT
RNGC 4859	12 56 35.	+ 27 05		15.0	GALAXY
IC 3956	12 56 35.	+ 37 39 59.			NONSTELLAR OBJECT
REIN 1.029	12 56 35.33	+ 28 14 13.3			NEBULA
TON-N 0680	12 56 36.	+ 26 52		15.5	BLUE STAR
ZWG 160.071	12 56 36.	+ 27 05		14.8	GALAXY
UGC 08097	12 56 36.	+ 27 05	96	14.8	GALAXY S0-a
RB 216	12 56 36.	+ 28 02	9		GALAXY IN COMA CLUSTER
MCG+05-31-052	12 56 36.	+ 28 06	42	14.9	GALAXY
RB 215	12 56 36.	+ 28 10	3		GALAXY IN COMA CLUSTER
RB 259	12 56 36.	+ 28 12	5		GALAXY IN COMA CLUSTER
RB 260	12 56 36.	+ 28 14	7		GALAXY IN COMA CLUSTER
MCG+05-31-051	12 56 36.	+ 28 22 30.	18	15.5	GALAXY
RNGC 4858	12 56 36.	+ 28 23		15.5	GALAXY
LB 00062	12 56 36.	+ 28 29		16.5	FAINT BLUE STAR
ZWG 160.072	12 56 36.	+ 31 31		15.7	GALAXY
MCG+06-29-003	12 56 36.	+ 35 08	210	13.	GALAXY
REIZ 3087	12 56 37.	+ 28 11	36	14.5	GALAXY
REIN 1.030	12 56 37.21	+ 28 29 44.2			NEBULA
REIN 1.031	12 56 37.26	+ 28 23 05.9			NEBULA
A2 273	12 56 37.5	+ 28 03 58.		16.6	FAINT BLUE OBJECT
A2 275	12 56 37.8	+ 30 48 46.		18.7	FAINT BLUE OBJECT
REIN 1.032	12 56 38.73	+ 28 13 33.2			NEBULA
IC 3950	12 56 39.	+ 19 00 17.			NONSTELLAR OBJECT
MCG+05-31-053	12 56 39.	+ 27 04 30.	48	14.8	GALAXY
DUN 02	12 56 39.	+ 28 08	55		SMALL NEBULA
MCG+05-31-054	12 56 39.	+ 28 23	42	14.7	GALAXY
REIN 1.033	12 56 39.12	+ 28 23 35.6			NEBULA
REIN 1.034	12 56 39.35	+ 28 13 43.8			NEBULA
REIN 1.035	12 56 39.99	+ 28 19 12.0			NEBULA
REIZ 3088	12 56 40.	+ 35 08	72	13.4	GALAXY
IC 3961	12 56 40.	+ 35 08 05.			NONSTELLAR OBJECT
REIN 4.195	12 56 40.41	+ 10 46 12.3			NEBULA
A2 276	12 56 40.6	+ 30 06 43.		16.4	FAINT BLUE OBJECT
REIN 1.036	12 56 40.78	+ 28 07 08.9			NEBULA
RNGC 4855	12 56 41.	- 12 57		14.0	GALAXY
RNGC 4856	12 56 41.	- 14 46		11.5	GALAXY
REIN 1.037	12 56 41.27	+ 28 15 58.8			NEBULA
IC 3955	12 56 41.3	+ 28 15 55.		15.6	GALAXY SB0
ZWG 071.091	12 56 42.	+ 09 15		15.7	GALAXY
MCG+02-33-044	12 56 42.	+ 09 15	120	15.7	GALAXY
TON-N 0681	12 56 42.	+ 26 53		14.9	BLUE STAR
MCG+05-31-057	12 56 42.	+ 27 54 30.	30	15.1	GALAXY
ZWG 160.073	12 56 42.	+ 27 55		15.1	GALAXY
MRK 058	12 56 42.	+ 27 55	15	15.5	GALAXY WITH UV CONTINUUM
RB 219	12 56 42.	+ 27 55	25		GALAXY IN COMA CLUSTER
RB 221	12 56 42.	+ 28 03	5		GALAXY IN COMA CLUSTER

OBJECT NAME	RIGHT ASCEN.	DECLINATION	DIAM.	MAGN.	TYPE OF OBJECT
RB 217	12 56 42.	+ 28 04	6		GALAXY IN COMA CLUSTER
RB 222	12 56 42.	+ 28 05	5		GALAXY IN COMA CLUSTER
RB 220	12 56 42.	+ 28 07	5		GALAXY IN COMA CLUSTER
RB 218	12 56 42.	+ 28 07	2		GALAXY IN COMA CLUSTER
MCG+05-31-055	12 56 42.	+ 28 07	18	15.5	GALAXY
MCG+05-31-056	12 56 42.	+ 28 08	4	17.	GALAXY
RB 265	12 56 42.	+ 28 10	4		GALAXY IN COMA CLUSTER
RB 262	12 56 42.	+ 28 11	18		GALAXY IN COMA CLUSTER
RB 261	12 56 42.	+ 28 14	7		GALAXY IN COMA CLUSTER
RB 266	12 56 42.	+ 28 19	11		GALAXY IN COMA CLUSTER
RB 263	12 56 42.	+ 28 19	7		GALAXY IN COMA CLUSTER
RB 264	12 56 42.	+ 28 20	5		GALAXY IN COMA CLUSTER
RNGC 4860	12 56 42.	+ 28 24		14.5	GALAXY
LB 07154	12 56 42.	+ 29 10		19.2	FAINT BLUE STAR
1ZW 049	12 56 42.	+ 35 08			COMPACT GALAXY
ZWG 189.005	12 56 42.	+ 35 08		12.8	GALAXY
MRK 059	12 56 42.	+ 35 08	20	14.	GALAXY WITH UV CONTINUUM
UGC 08098	12 56 42.	+ 35 08	270	12.8	GALAXY DBL SYS
KARA.72 362A	12 56 42.	+ 35 08	66	13.2	PART OF DOUBLE GALAXY
KARA.72 362B	12 56 42.	+ 35 09	168	14.1	PART OF DOUBLE GALAXY
ZC 1256.7+4147	12 56 42.	+ 41 47	610		CLUSTER OF GALAXIES
MCG+09-21-073	12 56 42.	+ 50 48	24	17.	GALAXY
ZWG 294.005	12 56 42.	+ 59 32		15.6	GALAXY
MCG+10-19-007	12 56 42.	+ 59 32	27	16.	GALAXY
MCG-02-33-077	12 56 42.	- 12 57	24	14.	GALAXY
MCG-02-33-078	12 56 42.	- 14 47	240	11.	GALAXY
MCG-04-31-015	12 56 42.	- 27 12	72	14.5	GALAXY
IC 3957	12 56 42.2	+ 28 02 13.		15.6	GALAXY E0
IC 3960	12 56 42.8	+ 28 07 26.		15.5	GALAXY E0
IC 3959	12 56 42.9	+ 28 03 12.		15.2	GALAXY E2
REIZ 3086	12 56 43.	+ 08 46	30	15.7	GALAXY
IC 3951	12 56 43.	+ 19 02 05.			NONSTELLAR OBJECT
IC 3953	12 56 43.	+ 23 21 17.			NONSTELLAR OBJECT
RNGC 4861	12 56 43.	+ 35 08		12.5	GALAXY
REIN 1.038	12 56 43.24	+ 28 07 28.7			NEBULA
REIN 2.169	12 56 43.29	- 14 46 20.1			NEBULA
REIN 1.039	12 56 44.71	+ 28 18 37.6			NEBULA
REIN 2.170	12 56 44.87	- 14 46 30.3			NEBULA
REIN 1.040	12 56 44.88	+ 28 08 14.1			NEBULA
IC 3960A	12 56 44.9	+ 28 08 12.			GALAXY
IC 3954	12 56 45.	+ 19 32 29.			NONSTELLAR OBJECT
IC 3958	12 56 45.	+ 24 17 29.			NONSTELLAR OBJECT
ZL 115	12 56 45.	+ 28 14 54.		17.5	ULTRAFAINT BLUE STAR
MCG+05-31-058	12 56 45.	+ 28 15	27	14.8	GALAXY
A2 277	12 56 45.8	+ 29 11 04.		18.3	FAINT BLUE OBJECT
ZL 116	12 56 46.	+ 28 12 06.		20.0	ULTRAFAINT BLUE STAR
REIN 2.171	12 56 46.56	- 14 45 51.6			NEBULA
REIN 1.041	12 56 46.83	+ 28 16 44.9			NEBULA
RNGC 4862	12 56 47.	- 13 51		15.0	GALAXY
TON-N 0682	12 56 48.	+ 26 55		14.7	BLUE STAR
ZWG 160.074	12 56 48.	+ 27 40		15.4	GALAXY
RB 224	12 56 48.	+ 27 54	9		GALAXY IN COMA CLUSTER
MCG+05-31-060	12 56 48.	+ 28 02	18	15.6	GALAXY
MCG+05-31-059	12 56 48.	+ 28 03	24	15.2	GALAXY
RB 223	12 56 48.	+ 28 08	4		GALAXY IN COMA CLUSTER
RB 269	12 56 48.	+ 28 10	4		GALAXY IN COMA CLUSTER
TON-N 0683	12 56 48.	+ 28 12		16.5	BLUE STAR
RB 001	12 56 48.	+ 28 12	3		GALAXY IN COMA CLUSTER
RNGC 4864	12 56 48.	+ 28 15		15.0	GALAXY
TON-N 0684	12 56 48.	+ 28 16		15.5	BLUE STAR
RB 267	12 56 48.	+ 28 17	6		GALAXY IN COMA CLUSTER
ZL 117	12 56 48.	+ 28 18 18.		17.2	ULTRAFAINT BLUE STAR
RB 268	12 56 48.	+ 28 21	13		GALAXY IN COMA CLUSTER
ZWG 189.006	12 56 48.	+ 36 11		15.4	GALAXY
ZWG 189.007	12 56 48.	+ 36 23		15.7	GALAXY
ZWG 189.008	12 56 48.	+ 37 34		12.9	GALAXY
UGC 08099	12 56 48.	+ 37 34	96	12.9	GALAXY Sa
MCG+09-21-074	12 56 48.	+ 53 20	30	16.	GALAXY
MCG+09-21-075	12 56 48.	+ 53 21	30	16.	GALAXY
7ZW 492	12 56 48.	+ 59 32			COMPACT GALAXY
ABC 1651	12 56 48.	- 03 56		16.0	RICH CLUSTER OF GALAXIES
MCG-05-31-008	12 56 48.	- 28 59	9	15.5	GALAXY
IC 3963	12 56 48.5	+ 28 02 42.		15.7	GALAXY SB0
REIN 1.042	12 56 48.53	+ 28 14 45.9			NEBULA
IC 3962	12 56 49.	+ 23 56 11.			NONSTELLAR OBJECT
IC 3964	12 56 49.	+ 28 07 23.			SINGLE STAR
REIN 1.043	12 56 49.41	+ 28 20 46.4			NEBULA
REA.66 3	12 56 50.	+ 28 10 24.			DWARF GALAXYY
ZL 118	12 56 50.	+ 28 10 54.		18.8	ULTRAFAINT BLUE STAR
RNGC 4868	12 56 50.	+ 37 35		13.0	GALAXY
REIN 1.045	12 56 50.55	+ 28 14 24.7			NEBULA
ZL 119	12 56 51.	+ 28 09 48.		18.2	ULTRAFAINT BLUE STAR
MCG+06-29-005	12 56 51.	+ 36 23	24	15.	GALAXY
IC 3967	12 56 51.	+ 36 23 59.			NONSTELLAR OBJECT
REIZ 3093	12 56 51.	+ 37 35	66	13.2	GALAXY
MCG+06-29-004	12 56 51.	+ 37 35	72	12.5	GALAXY
MCG+07-27-009	12 56 51.	+ 39 09	30	15.5	GALAXY
MCG-02-33-079	12 56 51.	- 13 51 30.	48	15.	GALAXY
SHB 227	12 56 51.1	+ 36 48 07.		19.2	QUASI-STELLAR OBJECT
BC B189	12 56 51.1	+ 36 48 10.		19.22	QUASI-STELLAR OBJECT
BFG 075	12 56 51.11	+ 36 48 07.6		19.26	ULTRAVIOLET-EXCESS OBJECT
IC 3966	12 56 52.	+ 36 07 17.			NONSTELLAR OBJECT
IC 3970	12 56 53.	+ 40 40 23.			NONSTELLAR OBJECT
A2 280	12 56 53.6	+ 28 49 03.		17.3	FAINT BLUE OBJECT
RB 227	12 56 54.	+ 28 00	4		GALAXY IN COMA CLUSTER
MCG+05-31-061	12 56 54.	+ 28 02 30.	30	15.7	GALAXY
RB 225	12 56 54.	+ 28 03	2		GALAXY IN COMA CLUSTER
RB 228	12 56 54.	+ 28 05	5		GALAXY IN COMA CLUSTER
RB 006	12 56 54.	+ 28 09	16		GALAXY IN COMA CLUSTER
RB 003	12 56 54.	+ 28 10	3		GALAXY IN COMA CLUSTER
RB 004	12 56 54.	+ 28 14	6		GALAXY IN COMA CLUSTER
MCG+05-31-062	12 56 54.	+ 28 14	30	15.5	GALAXY
RB 007	12 56 54.	+ 28 15	12		GALAXY IN COMA CLUSTER
RB 005	12 56 54.	+ 28 15	6		GALAXY IN COMA CLUSTER
RB 002	12 56 54.	+ 28 15	6		GALAXY IN COMA CLUSTER
RNGC 4867	12 56 54.	+ 28 16		15.5	GALAXY
MCG+05-31-063	12 56 54.	+ 28 20	18	15.6	GALAXY
RNGC 4865	12 56 54.	+ 28 21		14.5	GALAXY
RB 271	12 56 54.	+ 28 21	10		GALAXY IN COMA CLUSTER
UGC 08100	12 56 54.	+ 28 21	78	14.6	GALAXY E?
ZL 120	12 56 54.	+ 28 21 30.		22.4	ULTRAFAINT BLUE STAR
RB 270	12 56 54.	+ 28 22	2		GALAXY IN COMA CLUSTER
RB 135	12 56 54.	+ 28 27	3		GALAXY IN COMA CLUSTER
RB 136	12 56 54.	+ 28 28	12		GALAXY IN COMA CLUSTER
RB 226	12 56 54.	+ 28 59	6		GALAXY IN COMA CLUSTER
TON-N 0685	12 56 54.	+ 29 10		14.6	BLUE STAR
MCG+06-29-007	12 56 54.	+ 36 08	24	15.	GALAXY
MCG+06-29-006	12 56 54.	+ 36 11	27	15.	GALAXY

OBJECT NAME	RIGHT ASCEN.	DECLINATION	DIAM.	MAGN.	TYPE OF OBJECT
MCG-07-27-022	12 56 54.	- 44 10	48	15.	GALAXY
REIN 1.044	12 56 54.77	+ 14 26 39.7			NEBULA
ZL 121	12 56 55.	+ 28 22 12.		18.0	ULTRAFAINT BLUE STAR
RNGC 4870	12 56 55.	+ 37 16			GALAXY
REIN 1.047	12 56 55.35	+ 28 21 15.6			NEBULA
REIN 1.048	12 56 55.63	+ 28 09 21.9			NEBULA
REIN 1.049	12 56 55.67	+ 28 20 39.7			NEBULA
REIN 1.050	12 56 55.71	+ 28 28 04.8			NEBULA
IC 3965	12 56 56.	+ 19 06 41.			NONSTELLAR OBJECT
IC 3972	12 56 56.	+ 37 32 59.			NONSTELLAR OBJECT
IC 3975	12 56 56.	+ 39 08 59.			NONSTELLAR OBJECT
RNGC 4866	12 56 57.	+ 14 26		12.0	GALAXY
REIZ 3092	12 56 57.	+ 22 06	18	16.0	GALAXY
A2 282	12 56 57.2	+ 29 51 23.		18.0	FAINT BLUE OBJECT
REIN 1.046	12 56 57.71	+ 14 26 27.6			NEBULA
ABC 1654	12 56 58.	+ 30 17		16.9	RICH CLUSTER OF GALAXIES
REIN 1.051	12 56 58.04	+ 28 11 07.2			NEBULA
REIN 1.052	12 56 58.08	+ 28 10 00.8			NEBULA
A2 284	12 56 58.6	+ 30 02 51.		18.0	FAINT BLUE OBJECT
REIN 1.053	12 56 58.85	+ 28 10 53.7			NEBULA
REIZ 3091	12 56 59.	+ 14 27	210	12.3	GALAXY
IC 3977	12 56 59.	+ 37 03 59.			NONSTELLAR OBJECT
IC 3980	12 56 59.	+ 39 25 05.			NONSTELLAR OBJECT
IC 3982	12 56 59.	+ 40 20 59.			NONSTELLAR OBJECT
RNGC 4863	12 56 59.	- 13 45		15.0	GALAXY
ZWG 071.092	12 57 00.	+ 14 27		11.9	GALAXY
UGC 08102	12 57 00.	+ 14 27	378	11.9	GALAXY Sa
MCG+02-33-045	12 57 00.	+ 14 27	360	11.9	GALAXY
ZC 1257.0+2425	12 57 00.	+ 24 25	4440		CLUSTER OF GALAXIES
HZ 47	12 57 00.	+ 27 37		14.7	DECIDEDLY BLUE STAR
HZ 38	12 57 00.	+ 27 49		13.9	VERY BLUE STAR
RB 229	12 57 00.	+ 27 59	6		GALAXY IN COMA CLUSTER
RB 230	12 57 00.	+ 28 01	24		GALAXY IN COMA CLUSTER
RB 231	12 57 00.	+ 28 04	12		GALAXY IN COMA CLUSTER
RB 017	12 57 00.	+ 28 08	2		GALAXY IN COMA CLUSTER
RB 010	12 57 00.	+ 28 10	12		GALAXY IN COMA CLUSTER
RB 008	12 57 00.	+ 28 10	12		GALAXY IN COMA CLUSTER
MCG+05-31-065	12 57 00.	+ 28 10	30	14.9	GALAXY
RNGC 4869	12 57 00.	+ 28 11		15.0	GALAXY
RB 009	12 57 00.	+ 28 11	3		GALAXY IN COMA CLUSTER
RB 011	12 57 00.	+ 28 12	4		GALAXY IN COMA CLUSTER
RB 013	12 57 00.	+ 28 14	17		GALAXY IN COMA CLUSTER
RB 019	12 57 00.	+ 28 15	2		GALAXY IN COMA CLUSTER
RB 016	12 57 00.	+ 28 15	3		GALAXY IN COMA CLUSTER
RB 014	12 57 00.	+ 28 15	15		GALAXY IN COMA CLUSTER
RB 018	12 57 00.	+ 28 16	12		GALAXY IN COMA CLUSTER
RB 012	12 57 00.	+ 28 16	4		GALAXY IN COMA CLUSTER
RB 015	12 57 00.	+ 28 21	4		GALAXY IN COMA CLUSTER
MCG+05-31-064	12 57 00.	+ 28 21	36	14.6	GALAXY
RB 137	12 57 00.	+ 28 24	6		GALAXY IN COMA CLUSTER
RB 140	12 57 00.	+ 28 27	5		GALAXY IN COMA CLUSTER
RB 138	12 57 00.	+ 28 32	3		GALAXY IN COMA CLUSTER
RB 139	12 57 00.	+ 28 33	12		GALAXY IN COMA CLUSTER
IC 3979	12 57 00.	+ 36 35 41.			NONSTELLAR OBJECT
IC 3981	12 57 00.	+ 37 29 53.			NONSTELLAR OBJECT
MCG+09-21-076	12 57 00.	+ 56 12	30	15.	GALAXY
MCG+14-06-013	12 57 00.	+ 84 21	57	14.	GALAXY
ZWG 366.001	12 57 00.	+ 84 23		15.4	GALAXY
ZWG 365.007	12 57 00.	+ 84 23		15.4	GALAXY
UGC 08101	12 57 00.	+ 84 23	78	15.4	GALAXY S?
KARA.73B 0562	12 57 00.	+ 84 23	66	15.4	ISOLATED GALAXY S
REIZ 3089	12 57 00.	- 08 43	24	15.8	GALAXY
MCG-02-33-080	12 57 00.	- 14 42	90	14.	GALAXY
MCG-03-33-025	12 57 00.	- 20 50	42	15.	GALAXY
MCG-05-31-009	12 57 00.	- 29 20	84	15.5	GALAXY
IC 3968	12 57 00.8	+ 28 14 30.			GALAXY E2
REIN 1.054	12 57 00.87	+ 28 14 16.9			NEBULA
IC 3983	12 57 01.	+ 39 30 59.			NONSTELLAR OBJECT
REIN 1.055	12 57 01.02	+ 28 14 35.3			NEBULA
A2 285	12 57 01.6	+ 30 46 00.		18.6	FAINT BLUE OBJECT
HELW 201	12 57 02.	- 14 41 49.			NEBULA
BPG 076	12 57 02.52	+ 36 35 34.9		17.98	ULTRAVIOLET-EXCESS OBJECT
REIZ 3095	12 57 03.	+ 22 34	60	14.8	GALAXY
MCG-02-33-081	12 57 03.	- 13 45	108	15.	GALAXY
PK049+88.1	12 57 03.0	+ 27 54 22.		16.	PLANETARY NEBULA
ABC 1653	12 57 04.	+ 11 37		17.7	RICH CLUSTER OF GALAXIES
IC 3976	12 57 04.	+ 28 07 04.		15.5	GALAXY E5
BSO 05	12 57 04.	+ 36 35 10.		17.48	BLUE STELLAR OBJECT
LB 02508	12 57 04.	+ 54 28 24.		16.9	FAINT BLUE STAR
KARA.73 42	12 57 04.	- 45 06	81		DWARF GALAXY
A2 286	12 57 04.0	+ 29 01 59.		17.8	FAINT BLUE OBJECT
REIN 1.056	12 57 04.83	+ 28 07 11.4			NEBULA
IC 3969	12 57 05.	+ 19 55 23.			NONSTELLAR OBJECT
IC 3971	12 57 05.	+ 23 06 47.			NONSTELLAR OBJECT
IC 3987	12 57 05.	+ 39 00 06.			NONSTELLAR OBJECT
REIZ 3090	12 57 05.	- 05 37	90	14.0	GALAXY
ABC 1652	12 57 05.	- 13 14		17.4	RICH CLUSTER OF GALAXIES
RNGC 4852	12 57 05.	- 59 20		9.0	OPEN CLUSTER
REIN 1.057	12 57 05.51	+ 28 13 34.8			NEBULA
ZWG 043.097	12 57 06.	+ 03 30		15.7	GALAXY
KARA.73B 0563	12 57 06.	+ 03 30	42	15.7	ISOLATED GALAXY S
ZWG 043.088	12 57 06.	+ 07 19		15.7	GALAXY
MCG+01-33-034	12 57 06.	+ 07 19	36	15.7	GALAXY
ZC 1257.1+1221	12 57 06.	+ 12 21	400		CLUSTER OF GALAXIES
MCG+03-33-020	12 57 06.	+ 15 19	51	15.7	GALAXY
ZWG 100.019	12 57 06.	+ 15 20		15.7	GALAXY
ZC 1257.1+1824	12 57 06.	+ 18 24	1550		CLUSTER OF GALAXIES
RB 236	12 57 06.	+ 27 59	3		GALAXY IN COMA CLUSTER
RB 235	12 57 06.	+ 27 59	6		GALAXY IN COMA CLUSTER
RB 237	12 57 06.	+ 28 00	6		GALAXY IN COMA CLUSTER
RB 232	12 57 06.	+ 28 00	6		GALAXY IN COMA CLUSTER
RB 234	12 57 06.	+ 28 04	5		GALAXY IN COMA CLUSTER
RB 233	12 57 06.	+ 28 05	2		GALAXY IN COMA CLUSTER
ZC 1257.1+2806	12 57 06.	+ 28 06	19150		CLUSTER OF GALAXIES
RB 029	12 57 06.	+ 28 08	3		GALAXY IN COMA CLUSTER
RB 030	12 57 06.	+ 28 11	3		GALAXY IN COMA CLUSTER
TON-N 0686	12 57 06.	+ 28 13		17.0	BLUE STAR
RB 025	12 57 06.	+ 28 13	6		GALAXY IN COMA CLUSTER
RB 023	12 57 06.	+ 28 13	3		GALAXY IN COMA CLUSTER
RNGC 4871	12 57 06.	+ 28 14		15.0	GALAXY
MCG+05-31-066	12 57 06.	+ 28 14	24	15.1	GALAXY
RNGC 4873	12 57 06.	+ 28 15		15.5	GALAXY
RB 024	12 57 06.	+ 28 17	3		GALAXY IN COMA CLUSTER
RB 020	12 57 06.	+ 28 17	3		GALAXY IN COMA CLUSTER
RB 027	12 57 06.	+ 28 19	3		GALAXY IN COMA CLUSTER
RB 026	12 57 06.	+ 28 19	18		GALAXY IN COMA CLUSTER
RB 022	12 57 06.	+ 28 19	15		GALAXY IN COMA CLUSTER
RB 021	12 57 06.	+ 28 21	10		GALAXY IN COMA CLUSTER

OBJECT NAME	RIGHT ASCEN.	DECLINATION	DIAM.	MAGN.	TYPE OF OBJECT
RB 143	12 57 06.	+ 28 22	1		GALAXY IN COMA CLUSTER
RB 142	12 57 06.	+ 28 22	1		GALAXY IN COMA CLUSTER
RB 028	12 57 06.	+ 28 22	5		GALAXY IN COMA CLUSTER
RB 141	12 57 06.	+ 28 30	3		GALAXY IN COMA CLUSTER
ZWG 160.075	12 57 06.	+ 31 36		15.6	GALAXY
IC 3988	12 57 06.	+ 37 31 00.			NONSTELLAR OBJECT
ZWG 245.008	12 57 06.	+ 48 37		15.7	GALAXY
MCG+08-24-018	12 57 06.	+ 48 37	24	16.	GALAXY
LB 02509	12 57 06.	+ 58 05 00.		15.8	FAINT BLUE STAR
OCL 0894	12 57 06.	- 59 20	1020	9.0	OPEN STAR CLUSTER
VHA 143	12 57 06.	- 59 20	420		OPEN STAR CLUSTER
IC 3973	12 57 06.0	+ 28 09 09.		15.2	GALAXY SB0
REIN 1.058	12 57 06.28	+ 28 09 14.0			NEBULA
HELW 202	12 57 07.	- 14 35 55.			NEBULA
IC 3989	12 57 08.	+ 37 03 24.			NONSTELLAR OBJECT
REIN 1.059	12 57 08.33	+ 28 15 12.9			NEBULA
DUN 03	12 57 09.	+ 28 15	30		SMALL NEBULA
MCG-04-31-016	12 57 09.	- 21 53 30.	48	14.5	GALAXY
REIN 1.060	12 57 09.60	+ 28 12 58.9			NEBULA
IC 3978	12 57 10.	+ 19 53 42.			NONSTELLAR OBJECT
BPG 077	12 57 10.59	+ 33 27 40.3		18.81	ULTRAVIOLET-EXCESS OBJECT
SN 1968B	12 57 11.	+ 28 14		17.4	SUPERNOVA
REIN 1.061	12 57 11.22	+ 28 13 44.9			NEBULA
ZWG 043.089	12 57 12.	+ 07 46		15.6	GALAXY
LB 00024	12 57 12.	+ 27 40		16.0	FAINT BLUE STAR
RB 036	12 57 12.	+ 28 08	2		GALAXY IN COMA CLUSTER
RB 031	12 57 12.	+ 28 08	6		GALAXY IN COMA CLUSTER
MCG+05-31-069	12 57 12.	+ 28 09	36	15.2	GALAXY
RB 033	12 57 12.	+ 28 10	3		GALAXY IN COMA CLUSTER
RNGC 4876	12 57 12.	+ 28 11		15.0	GALAXY
RNGC 4875	12 57 12.	+ 28 11		15.5	GALAXY
RB 034	12 57 12.	+ 28 11	2		GALAXY IN COMA CLUSTER
RB 032	12 57 12.	+ 28 12	6		GALAXY IN COMA CLUSTER
RNGC 4872	12 57 12.	+ 28 13		15.5	GALAXY
MCG+05-31-068	12 57 12.	+ 28 13	15	15.3	GALAXY
ZCG 1257+28	12 57 12.	+ 28 14		13.7	COMPACT GALAXY
RNGC 4874	12 57 12.	+ 28 14		13.5	GALAXY
UGC 08103	12 57 12.	+ 28 14	144	13.7	GALAXY S0
RB 038	12 57 12.	+ 28 15	17		GALAXY IN COMA CLUSTER
RB 037	12 57 12.	+ 28 16	10		GALAXY IN COMA CLUSTER
RB 035	12 57 12.	+ 28 17	3		GALAXY IN COMA CLUSTER
RB 144	12 57 12.	+ 28 26	6		GALAXY IN COMA CLUSTER
RB 145	12 57 12.	+ 28 30	6		GALAXY IN COMA CLUSTER
ZWG 160.076	12 57 12.	+ 28 54		15.6	GALAXY
ZWG 160.077	12 57 12.	+ 29 10		15.0	GALAXY
IC 3990	12 57 12.	+ 29 10		15.0	GALAXY S0
ZWG 160.078	12 57 12.	+ 29 12		15.5	GALAXY
IC 3991	12 57 12.	+ 29 12		15.5	GALAXY IN COMA CLSTR
MCG+05-31-067	12 57 12.	+ 32 19	48	14.7	GALAXY
IC 3992	12 57 12.	+ 37 02 30.			NONSTELLAR OBJECT
MCG+07-27-010	12 57 12.	+ 39 05	42	15.5	GALAXY
IC 3993	12 57 12.	+ 40 52 12.			NONSTELLAR OBJECT
REIZ 3094	12 57 12.	- 08 28	36	14.2	GALAXY
A2 289	12 57 12.2	+ 26 44 58.		17.0	FAINT BLUE OBJECT
REIZ 3096	12 57 13.	+ 28 14	24	13.9	GALAXY
IC 3995	12 57 13.	+ 39 09 24.			NONSTELLAR OBJECT
IC 3996	12 57 13.	+ 40 44 12.			NONSTELLAR OBJECT
LB 00666	12 57 13.	+ 56 35 30.		17.8	FAINT BLUE STAR
REIN 1.062	12 57 13.07	+ 28 26 09.1			NEBULA
REIN 1.063	12 57 13.38	+ 28 10 37.1			NEBULA
A2 290	12 57 14.3	+ 30 06 04.		18.4	FAINT BLUE OBJECT
IC 3984	12 57 15.	+ 19 53 42.			NONSTELLAR OBJECT
DUN 04	12 57 15.	+ 28 00	45		SMALL NEBULA
ZL 122	12 57 15.	+ 28 08 24.		19.3	ULTRAFAINT BLUE STAR
MCG+05-31-070	12 57 15.	+ 28 14	27	13.7	GALAXY
MCG+05-31-071	12 57 15.	+ 29 10	48	15.0	GALAXY
MCG+05-31-072	12 57 15.	+ 29 12	36	15.5	GALAXY
MCG+07-27-011	12 57 15.	+ 39 05	42	16.	GALAXY
IC 3985	12 57 16.	+ 19 51 36.			NONSTELLAR OBJECT
IC 3997	12 57 16.	+ 36 58 06.			NONSTELLAR OBJECT
IC 4000	12 57 16.	+ 39 51 18.			NONSTELLAR OBJECT
HELW 300	12 57 16.	- 08 28 31.			NEBULA
A2 293	12 57 16.7	+ 27 34 33.		18.2	FAINT BLUE OBJECT
ZWG 015.052	12 57 18.	+ 01 54		15.4	GALAXY
ZWG 043.090	12 57 18.	+ 04 35		15.6	GALAXY
ZC 1257.3+1133	12 57 18.	+ 11 33	1340		CLUSTER OF GALAXIES
ZWG 130.003	12 57 18.	+ 22 05		15.4	GALAXY
MCG+04-31-002	12 57 18.	+ 22 05	48	15.4	GALAXY
TON-N 0687	12 57 18.	+ 26 43		14.8	BLUE STAR
ZWG 160.079	12 57 18.	+ 27 58		15.1	GALAXY
RB 041	12 57 18.	+ 28 07	10		GALAXY IN COMA CLUSTER
RB 039	12 57 18.	+ 28 07	10		GALAXY IN COMA CLUSTER
RB 047	12 57 18.	+ 28 08	8		GALAXY IN COMA CLUSTER
RB 044	12 57 18.	+ 28 08	15		GALAXY IN COMA CLUSTER
RB 048	12 57 18.	+ 28 10	4		GALAXY IN COMA CLUSTER
MCG+05-31-073	12 57 18.	+ 28 11	18	15.1	GALAXY
RB 043	12 57 18.	+ 28 12	7		GALAXY IN COMA CLUSTER
RB 040	12 57 18.	+ 28 13	16		GALAXY IN COMA CLUSTER
RB 046	12 57 18.	+ 28 14	15		GALAXY IN COMA CLUSTER
RB 042	12 57 18.	+ 28 14	15		GALAXY IN COMA CLUSTER
RB 045	12 57 18.	+ 28 16	10		GALAXY IN COMA CLUSTER
RB 147	12 57 18.	+ 28 25	5		GALAXY IN COMA CLUSTER
RB 149	12 57 18.	+ 28 27	4		GALAXY IN COMA CLUSTER
RB 148	12 57 18.	+ 28 27	6		GALAXY IN COMA CLUSTER
RB 146	12 57 18.	+ 28 27	7		GALAXY IN COMA CLUSTER
RB 150	12 57 18.	+ 28 32	5		GALAXY IN COMA CLUSTER
IC 4001	12 57 18.	+ 39 08 24.			NONSTELLAR OBJECT
ZC 1257.3+3925	12 57 18.	+ 39 25	4570		CLUSTER OF GALAXIES
UGC 08104	12 57 18.	+ 43 02	66	17.	GALAXY Sc
ZWG 294.006	12 57 18.	+ 59 17		15.4	GALAXY
MCG+10-19-008	12 57 18.	+ 59 18	24	16.	GALAXY
ZWG 353.007	12 57 18.	+ 78 31		15.7	GALAXY
ZWG 352.050	12 57 18.	+ 78 31		15.7	GALAXY
ZZW 493	12 57 18.	+ 79 02			COMPACT GALAXY
MCG-01-33-062	12 57 18.	- 08 28	42	15.	GALAXY
A2 294	12 57 18.2	+ 29 58 13.		18.4	FAINT BLUE OBJECT
A2 295	12 57 18.9	+ 29 07 05.		16.8	FAINT BLUE OBJECT
IC 4003	12 57 19.	+ 39 05 06.			NONSTELLAR OBJECT
REIN 1.064	12 57 19.74	+ 28 13 41.9			NEBULA
REIN 1.065	12 57 19.88	+ 28 10 54.9			NEBULA
BEA.66 4	12 57 20.	+ 28 31 30.			DWARF GALAXY
IC 4002	12 57 20.	+ 37 02 30.			NONSTELLAR OBJECT
HOLM 496B	12 57 20.	+ 39 05	12	15.5	PART OF MULTIPLE GALAXY
HOLM 496A	12 57 20.	+ 39 05	18	15.2	PART OF MULTIPLE GALAXY
IC 0841	12 57 21.	+ 22 04 54.			NONSTELLAR OBJECT
ZL 124	12 57 21.	+ 28 08 12.		19.5	ULTRAFAINT BLUE STAR
ZL 123	12 57 21.	+ 28 13 06.		16.4	ULTRAFAINT BLUE STAR
HELW 203	12 57 21.	- 14 15 18.			NEBULA

OBJECT NAME	RIGHT ASCEN.	DECLINATION	DIAM.	MAGN.	TYPE OF OBJECT
REIN 1.066	12 57 21.68	+ 27 59 05.0			NEBULA
IC 3998	12 57 21.9	+ 28 14 33.		15.6	GALAXY SB0
REIN 1.067	12 57 22.30	+ 28 14 38.1			NEBULA
REIN 1.068	12 57 22.56	+ 27 58 47.3			NEBULA
ZL 125	12 57 23.	+ 28 08 30.		19.0	ULTRAFAINT BLUE STAR
IC 4004	12 57 23.	+ 39 04 48.			NONSTELLAR OBJECT
A2 296	12 57 23.8	+ 26 55 31.		18.6	FAINT BLUE OBJECT
ZWG 015.053	12 57 24.	+ 02 18		15.5	GALAXY
UGC 08105	12 57 24.	+ 02 18	78	15.5	GALAXY DBL SYS
ZWG 043.091	12 57 24.	+ 05 19		15.5	GALAXY
ZWG 043.092	12 57 24.	+ 07 20		15.6	GALAXY
IC 3994	12 57 24.	+ 22 59 06.			NONSTELLAR OBJECT
MCG+05-31-074	12 57 24.	+ 27 58	42	15.1	GALAXY
RB 060	12 57 24.	+ 28 06	12		GALAXY IN COMA CLUSTER
RB 053	12 57 24.	+ 28 07	4		GALAXY IN COMA CLUSTER
RB 049	12 57 24.	+ 28 08	25		GALAXY IN COMA CLUSTER
RB 057	12 57 24.	+ 28 11	5		GALAXY IN COMA CLUSTER
RB 056	12 57 24.	+ 28 11	3		GALAXY IN COMA CLUSTER
RB 055	12 57 24.	+ 28 12	14		GALAXY IN COMA CLUSTER
ARC 1656	12 57 24.	+ 28 15		13.5	RICH CLUSTER OF GALAXIES
RB 054	12 57 24.	+ 28 15	4		GALAXY IN COMA CLUSTER
RB 051	12 57 24.	+ 28 16	4		GALAXY IN COMA CLUSTER
RB 050	12 57 24.	+ 28 16	3		GALAXY IN COMA CLUSTER
ZL 126	12 57 24.	+ 28 20 30.		22.0	ULTRAFAINT BLUE STAR
RB 059	12 57 24.	+ 28 21	9		GALAXY IN COMA CLUSTER
RB 058	12 57 24.	+ 28 22	5		GALAXY IN COMA CLUSTER
RB 052	12 57 24.	+ 28 22	5		GALAXY IN COMA CLUSTER
RB 153	12 57 24.	+ 28 25	12		GALAXY IN COMA CLUSTER
RB 152	12 57 24.	+ 28 32	3		GALAXY IN COMA CLUSTER
RB 151	12 57 24.	+ 28 33	3		GALAXY IN COMA CLUSTER
RB 154	12 57 24.	+ 28 34	3		GALAXY IN COMA CLUSTER
ZWG 160.080	12 57 24.	+ 32 18		14.7	GALAXY
ZWG 217.005	12 57 24.	+ 39 05		15.4	GALAXY
LB 02510	12 57 24.	+ 53 22 36.		15.9	FAINT BLUE STAR
ZWG 294.007	12 57 24.	+ 60 45		15.2	GALAXY
MCG+10-19-009	12 57 24.	+ 60 46	30	16.	GALAXY
MCG+10-19-010	12 57 24.	+ 61 11	36	16.	GALAXY
ZWG 294.008	12 57 24.	+ 61 12		15.7	GALAXY
MRK 234	12 57 24.	+ 64 43	13	16.5	GALAXY WITH UV CONTINUUM
HZ 48	12 57 25.	+ 41 17		13.6	BLUE STAR
LB 02511	12 57 25.	+ 52 02 36.		15.9	FAINT BLUE STAR
REIN 1.069	12 57 25.77	+ 28 24 49.8			NEBULA
BC P201	12 57 26.6	+ 34 39 34.		16.79	QUASI-STELLAR OBJECT
BFG 078	12 57 26.68	+ 34 39 31.4		17.09	ULTRAVIOLET-EXCESS OBJECT
SHB 228	12 57 26.7	+ 34 39 31.		17.0	QUASI-STELLAR OBJECT
REIZ 3098	12 57 27.	+ 22 33	24	15.7	GALAXY
MCG-02-33-082	12 57 27.	- 15 06	60	14.5	GALAXY
IC 4006	12 57 28.	+ 37 16 42.			NONSTELLAR OBJECT
ARC 1657	12 57 29.	+ 19 52		17.5	RICH CLUSTER OF GALAXIES
ZC 1257.5+1641	12 57 30.	+ 16 41	3290		CLUSTER OF GALAXIES
ZC 1257.5+1856	12 57 30.	+ 18 56	1140		CLUSTER OF GALAXIES
LB 00025	12 57 30.	+ 22 14		17.3	FAINT BLUE STAR
RB 062	12 57 30.	+ 28 07	3		GALAXY IN COMA CLUSTER
RB 063	12 57 30.	+ 28 11	4		GALAXY IN COMA CLUSTER
RB 065	12 57 30.	+ 28 12	6		GALAXY IN COMA CLUSTER
RB 064	12 57 30.	+ 28 12	17		GALAXY IN COMA CLUSTER
RB 061	12 57 30.	+ 28 14	5		GALAXY IN COMA CLUSTER
RNGC 4882	12 57 30.	+ 28 15		15.5	GALAXY
RNGC 4883	12 57 30.	+ 28 13		15.0	GALAXY
RB 066	12 57 30.	+ 28 20	11		GALAXY IN COMA CLUSTER
RB 155	12 57 30.	+ 28 24	12		GALAXY IN COMA CLUSTER
RNGC 4881	12 57 30.	+ 28 31		14.5	GALAXY
UGC 08106	12 57 30.	+ 28 31	66	14.7	GALAXY E
RB 156	12 57 30.	+ 28 32	3		GALAXY IN COMA CLUSTER
ZWG 270.037	12 57 30.	+ 53 37		15.0	GALAXY
UGC 08107	12 57 30.	+ 53 37	162	15.0	GALAXY IRR
MCG+09-21-077	12 57 30.	+ 53 37	120	13.	GALAXY
ZC 1257.5+6015	12 57 30.	+ 60 15	1480		CLUSTER OF GALAXIES
ZC 1257.5+6046	12 57 30.	+ 60 46	1950		CLUSTER OF GALAXIES
7ZW 494	12 57 30.	+ 61 25			COMPACT GALAXY
MCG-02-33-083	12 57 30.	- 12 44	36	15.5	GALAXY
REIN 1.070	12 57 30.74	+ 28 23 53.0			NEBULA
REIN 1.071	12 57 31.61	+ 28 18 14.1			NEBULA
IC 4009	12 57 32.	+ 36 55 48.			NONSTELLAR OBJECT
MCG+05-31-075	12 57 33.	+ 28 31	15	14.7	GALAXY
REIN 1.072	12 57 33.36	+ 28 30 58.9			NEBULA
IC 4010	12 57 34.	+ 38 08 54.			NONSTELLAR OBJECT
REIN 1.073	12 57 34.29	+ 28 29 12.7			NEBULA
RNGC 4892	12 57 35.	+ 27 10		14.5	GALAXY
LB 02512	12 57 35.	+ 55 16 24.		15.8	FAINT BLUE STAR
ZWG 071.093	12 57 36.	+ 08 58		15.2	GALAXY
MCG+02-33-046	12 57 36.	+ 08 58	36	15.2	GALAXY
IC 4005	12 57 36.	+ 22 54 24.			NONSTELLAR OBJECT
ZWG 160.081	12 57 36.	+ 27 10		14.7	GALAXY
UGC 08108	12 57 36.	+ 27 10	96	14.7	GALAXY S
RB 071	12 57 36.	+ 28 13	3		GALAXY IN COMA CLUSTER
RB 068	12 57 36.	+ 28 13	3		GALAXY IN COMA CLUSTER
RNGC 4884	12 57 36.	+ 28 15			NON-EXISTENT OBJECT
RB 070	12 57 36.	+ 28 16	3		GALAXY IN COMA CLUSTER
RB 072	12 57 36.	+ 28 18	3		GALAXY IN COMA CLUSTER
RB 069	12 57 36.	+ 28 21	5		GALAXY IN COMA CLUSTER
RB 067	12 57 36.	+ 28 22	3		GALAXY IN COMA CLUSTER
RB 158	12 57 36.	+ 28 27	2		GALAXY IN COMA CLUSTER
RB 159	12 57 36.	+ 28 29	3		GALAXY IN COMA CLUSTER
RB 160	12 57 36.	+ 28 31	10		GALAXY IN COMA CLUSTER
RB 157	12 57 36.	+ 28 35	6		GALAXY IN COMA CLUSTER
HOLM 498C	12 57 36.	+ 37 28	18	15.9	PART OF MULTIPLE GALAXY
ZC 1257.6+6605	12 57 36.	+ 66 05	670		CLUSTER OF GALAXIES
SN 1952C	12 57 36.	- 03 05		18.7	SUPERNOVA
MCG-02-33-084	12 57 36.	- 12 44	60	15.	GALAXY
MCG-05-31-010	12 57 36.	- 31 11	30	16.	GALAXY
A2 297	12 57 36.4	+ 29 35 49.		16.3	FAINT BLUE OBJECT
REIN 4.196	12 57 36.44	+ 08 58 03.8			NEBULA
REIN 1.074	12 57 36.80	+ 28 30 22.9			NEBULA
REIN 4.197	12 57 36.89	+ 08 58 33.8			NEBULA
IC 4013	12 57 37.	+ 37 28 13.			NONSTELLAR OBJECT
REIZ 3097	12 57 37.	- 12 05	72	13.8	GALAXY
IC 4018	12 57 38.	+ 40 45 13.			NONSTELLAR OBJECT
LB 02513	12 57 38.	+ 60 55 18.		16.7	FAINT BLUE STAR
IC 3974	12 57 38.	- 35 05 06.			NONSTELLAR OBJECT
REIN 4.198	12 57 38.00	+ 08 57 37.0			NEBULA
REIN 1.075	12 57 38.64	+ 28 30 34.5			NEBULA
REIN 1.076	12 57 38.75	+ 28 29 16.6			NEBULA
RNGC 4880	12 57 39.	+ 12 45		12.5	GALAXY
HOLM 497B	12 57 39.	+ 12 47	24	15.5	PART OF MULTIPLE GALAXY
IC 4008	12 57 39.	+ 22 37 12.			NONSTELLAR OBJECT
MCG+05-31-076	12 57 39.	+ 28 15	36	15.1	GALAXY
DUN 05	12 57 39.	+ 28 16	30		SMALL NEBULA
IC 4016	12 57 39.	+ 37 27 43.			SAME AS NGC 4893
IC 4015	12 57 39.	+ 37 27 55.			SAME AS NGC 4893
REIZ 3105	12 57 39.	+ 37 28	18	15.8	GALAXY
REIZ 3104	12 57 39.	+ 37 28	18	15.3	GALAXY
REIZ 3103	12 57 39.	+ 37 28	12	15.8	GALAXY
HOLM 498B	12 57 39.	+ 37 28	18	15.5	PART OF MULTIPLE GALAXY
HOLM 498A	12 57 39.	+ 37 28	18	15.2	PART OF MULTIPLE GALAXY
MCG-02-33-085	12 57 39.	- 12 04	96	14.5	GALAXY
A2 299	12 57 39.3	+ 29 36 45.		18.2	FAINT BLUE OBJECT
REIN 1.077	12 57 39.83	+ 28 30 58.2			NEBULA
IC 4007	12 57 40.	+ 20 13 54.			NONSTELLAR OBJECT
RNGC 4901	12 57 40.	+ 47 28		15.5	GALAXY
REIZ 3110	12 57 40.	+ 47 29	18	14.3	GALAXY
REIN 1.078	12 57 40.03	+ 28 15 24.8			NEBULA
A2 300	12 57 40.1	+ 30 02 33.		17.5	FAINT BLUE OBJECT
REIN 4.199	12 57 40.68	+ 12 45 08.0			NEBULA
BFG 079	12 57 40.89	+ 35 06 51.7		18.96	ULTRAVIOLET-EXCESS OBJECT
REIN 1.080	12 57 40.93	+ 28 17 36.4			NEBULA
REIN 1.079	12 57 40.98	+ 28 04 36.7			NEBULA
BFG 080	12 57 41.02	+ 33 39 44.3		18.68	ULTRAVIOLET-EXCESS OBJECT
BFG 081	12 57 41.55	+ 33 36 09.2		18.95	ULTRAVIOLET-EXCESS OBJECT
REIN 1.081	12 57 41.70	+ 28 14 50.4			NEBULA
REIN 1.082	12 57 41.78	+ 28 02 42.4			NEBULA
IC 4011	12 57 41.9	+ 28 16 20.		15.6	GALAXY E0
REIN 1.083	12 57 41.94	+ 28 16 24.2			NEBULA
ZWG 071.094	12 57 42.	+ 12 45		13.3	GALAXY
REIZ 3101	12 57 42.	+ 12 45	120	13.2	GALAXY
UGC 08109	12 57 42.	+ 12 45	210	13.3	GALAXY S0
MCG+02-33-047	12 57 42.	+ 12 45	126	13.3	GALAXY
HOLM 497A	12 57 42.	+ 12 45	60	12.7	PART OF MULTIPLE GALAXY
ZC 1257.7+2000	12 57 42.	+ 20 00	2350		CLUSTER OF GALAXIES
MCG+05-31-078	12 57 42.	+ 27 10	66	14.7	GALAXY
MRK 060	12 57 42.	+ 28 08	12	15.5	GALAXY WITH UV CONTINUUM
RB 082	12 57 42.	+ 28 08	11		GALAXY IN COMA CLUSTER
RB 081	12 57 42.	+ 28 08	5		GALAXY IN COMA CLUSTER
RB 078	12 57 42.	+ 28 12	3		GALAXY IN COMA CLUSTER
RB 076	12 57 42.	+ 28 12	8		GALAXY IN COMA CLUSTER
RB 080	12 57 42.	+ 28 14	3		GALAXY IN COMA CLUSTER
RB 079	12 57 42.	+ 28 14	3		GALAXY IN COMA CLUSTER
RB 073	12 57 42.	+ 28 14	8		GALAXY IN COMA CLUSTER
MCG+05-31-077	12 57 42.	+ 28 14	48	13.0	GALAXY
RNGC 4889	12 57 42.	+ 28 15		12.5	GALAXY
RNGC 4886	12 57 42.	+ 28 15		15.0	GALAXY
RB 077	12 57 42.	+ 28 15	21		GALAXY IN COMA CLUSTER
RB 075	12 57 42.	+ 28 15	4		GALAXY IN COMA CLUSTER
UGC 08110	12 57 42.	+ 28 15	168	13.0	GALAXY E
RB 074	12 57 42.	+ 28 18	18		GALAXY IN COMA CLUSTER
RB 168	12 57 42.	+ 28 25	3		GALAXY IN COMA CLUSTER
RB 164	12 57 42.	+ 28 25	3		GALAXY IN COMA CLUSTER
RB 163	12 57 42.	+ 28 25	6		GALAXY IN COMA CLUSTER
RNGC 4895A	12 57 42.	+ 28 26		15.5	GALAXY
RB 167	12 57 42.	+ 28 26	22		GALAXY IN COMA CLUSTER
RB 166	12 57 42.	+ 28 26	16		GALAXY IN COMA CLUSTER
RB 162	12 57 42.	+ 28 26	5		GALAXY IN COMA CLUSTER
RB 165	12 57 42.	+ 28 31	8		GALAXY IN COMA CLUSTER
RB 161	12 57 42.	+ 28 32	7		GALAXY IN COMA CLUSTER
MRK 235	12 57 42.	+ 33 41	20	15.	GALAXY WITH UV CONTINUUM
MCG+06-29-010	12 57 42.	+ 33 41	24	15.	GALAXY
ZWG 189.009	12 57 42.	+ 33 42		15.2	GALAXY
ZWG 189.010	12 57 42.	+ 37 27		15.6	GALAXY
UGC 08111	12 57 42.	+ 37 27	60	15.6	GALAXY DBL SYS
VV 222B	12 57 42.	+ 37 27	24	16.5	INTERACTING GALAXY
VV 222A	12 57 42.	+ 37 27	24	16.	INTERACTING GALAXY
VV 222	12 57 42.	+ 37 27	42		INTERACTING GALAXY
MCG+06-29-009	12 57 42.	+ 37 28	18	16.	GALAXY
MCG+06-29-008	12 57 42.	+ 37 28	24	15.5	GALAXY
IC 4020	12 57 42.	+ 38 52 37.			NONSTELLAR OBJECT
ZWG 245.009	12 57 42.	+ 47 28		15.5	GALAXY
UGC 08112	12 57 42.	+ 47 28	72	15.5	GALAXY
MCG+08-24-019	12 57 42.	+ 47 29	48	14.3	GALAXY
MCG+09-21-078	12 57 42.	+ 51 13	24	15.	GALAXY
MCG+09-21-079	12 57 42.	+ 51 42 30.	60	16.	GALAXY
RNGC 4878	12 57 42.	- 05 49		14.5	GALAXY
MCG-01-33-064	12 57 42.	- 05 49	60	14.5	GALAXY
MCG-01-33-063	12 57 42.	- 07 49	66	15.	GALAXY
MCG-04-31-017	12 57 42.	- 22 26 30.	18	15.	GALAXY
REIZ 3102	12 57 43.	+ 28 16	42	13.8	GALAXY
IC 4012	12 57 43.4	+ 28 20 49.		15.7	GALAXY E1
REIN 1.084	12 57 43.64	+ 28 20 54.0			NEBULA
REIN 1.085	12 57 43.69	+ 28 14 46.4			NEBULA
RNGC 4893A	12 57 44.	+ 37 27			GALAXY
RNGC 4893	12 57 44.	+ 37 27		15.5	GALAXY
IC 3986	12 57 44.	- 32 00 00.			NONSTELLAR OBJECT
SHB 229	12 57 44.0	- 23 02 08.		18.5	QUASI-STELLAR OBJECT
A2 302	12 57 44.4	+ 26 54 24.		18.2	FAINT BLUE OBJECT
REIN 1.086	12 57 44.68	+ 28 08 09.9			NEBULA
DUN 06	12 57 45.	+ 28 29	60		SMALL NEBULA
IC 4022	12 57 45.	+ 38 44 55.			NONSTELLAR OBJECT
IC 4024	12 57 45.	+ 40 46 43.			NONSTELLAR OBJECT
A2 305	12 57 45.5	+ 28 42 01.		16.5	FAINT BLUE OBJECT
REIN 1.087	12 57 45.83	+ 28 08 01.5			NEBULA
IC 4014	12 57 47.	+ 22 46 01.			NONSTELLAR OBJECT
REIZ 3100	12 57 47.	- 05 49	30	14.6	GALAXY
REIZ 3099	12 57 47.	- 05 50	6	16.1	GALAXY
RNGC 4877	12 57 47.	- 15 01		13.0	GALAXY
ZC 1257.8+2030	12 57 48.	+ 20 30	400		CLUSTER OF GALAXIES
LB 00063	12 57 48.	+ 24 47		16.9	FAINT BLUE STAR
ZWG 160.082	12 57 48.	+ 27 39		15.6	GALAXY
RB 090	12 57 48.	+ 28 08	4		GALAXY IN COMA CLUSTER
RB 083	12 57 48.	+ 28 08	8		GALAXY IN COMA CLUSTER
RB 084	12 57 48.	+ 28 14	10		GALAXY IN COMA CLUSTER
RB 086	12 57 48.	+ 28 16	3		GALAXY IN COMA CLUSTER
RB 089	12 57 48.	+ 28 19	11		GALAXY IN COMA CLUSTER
RB 085	12 57 48.	+ 28 20	5		GALAXY IN COMA CLUSTER
RB 087	12 57 48.	+ 28 21	17		GALAXY IN COMA CLUSTER
UGC 08113	12 57 48.	+ 28 21	108	14.3	GALAXY S0
RB 088	12 57 48.	+ 28 22	5		GALAXY IN COMA CLUSTER
RB 169	12 57 48.	+ 28 30	5		GALAXY IN COMA CLUSTER
ZWG 160.083	12 57 48.	+ 29 06		15.4	GALAXY
MCG+05-31-079	12 57 48.	+ 29 06	42	15.4	GALAXY
MCG+09-21-080	12 57 48.	+ 52 51	42	17.	GALAXY
ZC 1257.8+6327	12 57 48.	+ 63 27	340		CLUSTER OF GALAXIES
MCG-01-33-064A	12 57 48.	- 05 47	30	17.	GALAXY
MCG-02-33-086	12 57 48.	- 15 01 30.	132	13.	GALAXY
MCG-05-31-011	12 57 48.	- 32 34	42	15.	GALAXY
MCG-07-27-023	12 57 48.	- 43 58	36	16.	GALAXY
REIN 1.088	12 57 48.20	+ 28 03 05.5			NEBULA
REIN 1.089	12 57 48.50	+ 28 20 42.8			NEBULA

OBJECT NAME	RIGHT ASCEN.	DECLINATION	DIAM.	MAGN.	TYPE OF OBJECT
IC 4017	12 57 49.	+ 22 49 31.			NONSTELLAR OBJECT
ZL 127	12 57 49.	+ 28 17 06.		20.4	ULTRAFAINT BLUE STAR
REIZ 3109	12 57 49.	+ 28 29	30	14.0	GALAXY
IC 4021	12 57 50.0	+ 28 18 28.		15.6	GALAXY E0
REIN 1.090	12 57 50.36	+ 28 18 39.2			NEBULA
IC 4019	12 57 51.	+ 23 59 13.			NONSTELLAR OBJECT
MCG+05-31-080	12 57 51.	+ 28 18 30.	12	15.6	GALAXY
LB 02514	12 57 51.	+ 58 13 54.		15.4	FAINT BLUE STAR
MCG-04-31-018	12 57 51.	- 21 50	30	15.5	GALAXY
HELW 204	12 57 52.	- 14 02 18.			NEBULA
REIN 1.091	12 57 52.07	+ 28 14 12.0			NEBULA
IC 3999	12 57 52.2	- 14 02 03.			GALAXY Sb
REIN 1.092	12 57 52.65	+ 28 20 00.8			NEBULA
IC 4027	12 57 53.	+ 37 24 37.			NONSTELLAR OBJECT
REIN 1.093	12 57 53.39	+ 28 13 29.5			NEBULA
ZWG 071.095	12 57 54.	+ 13 57		15.4	GALAXY
UGC 08114	12 57 54.	+ 13 57	96	15.4	GALAXY PECULR
MCG+02-33-048	12 57 54.	+ 13 57	90	15.4	GALAXY
TON-N 0688	12 57 54.	+ 27 36		15.5	BLUE STAR
RNGC 4898	12 57 54.	+ 28 13		14.5	GALAXY
RB 094	12 57 54.	+ 28 13	12		GALAXY IN COMA CLUSTER
RNGC 4894	12 57 54.	+ 28 14		15.5	GALAXY
RB 101	12 57 54.	+ 28 15	3		GALAXY IN COMA CLUSTER
RB 102	12 57 54.	+ 28 16	3		GALAXY IN COMA CLUSTER
RB 093	12 57 54.	+ 28 20	4		GALAXY IN COMA CLUSTER
RB 091	12 57 54.	+ 28 20	19		GALAXY IN COMA CLUSTER
RB 092	12 57 54.	+ 28 21	3		GALAXY IN COMA CLUSTER
RB 171	12 57 54.	+ 28 22	6		GALAXY IN COMA CLUSTER
RB 172	12 57 54.	+ 28 24	3		GALAXY IN COMA CLUSTER
RB 170	12 57 54.	+ 28 26	4		GALAXY IN COMA CLUSTER
RNGC 4895	12 57 54.	+ 28 28		14.5	GALAXY
MCG+05-31-081	12 57 54.	+ 28 28	120	14.3	GALAXY
RB 173	12 57 54.	+ 28 32	4		GALAXY IN COMA CLUSTER
TON-N 0689	12 57 54.	+ 29 20		14.4	BLUE STAR
MCG+06-29-011	12 57 54.	+ 35 13	36	15.	GALAXY
ZWG 189.011	12 57 54.	+ 36 31		15.5	GALAXY
UGC 08115	12 57 54.	+ 36 31	72	15.5	GALAXY SBc
MCG+06-29-012	12 57 54.	+ 36 31	48	15.	GALAXY
RNGC 4879	12 57 54.	- 05 47		15.0	GALAXY
MCG-01-33-066	12 57 54.	- 05 47	45	15.	GALAXY
RNGC 4885	12 57 54.	- 06 34		15.0	GALAXY
MCG-01-33-065	12 57 54.	- 06 34	24	15.	GALAXY
REIN 1.094	12 57 54.10	+ 28 05 06.9			NEBULA
A2 311	12 57 54.2	+ 29 34 52.		16.8	FAINT BLUE OBJECT
REIN 1.095	12 57 54.36	+ 28 12 21.8			NEBULA
REIZ 3115	12 57 55.	+ 28 28	114	13.9	GALAXY
IC 4028	12 57 55.	+ 36 31 25.			NONSTELLAR OBJECT
IC 4029	12 57 55.	+ 39 01 43.			NONSTELLAR OBJECT
IC 4031	12 57 55.	+ 39 24 49.			NONSTELLAR OBJECT
REIN 4.200	12 57 55.51	+ 13 56 22.5			NEBULA
ZL 128	12 57 56.	+ 28 14 48.		20.6	ULTRAFAINT BLUE STAR
LB 02515	12 57 56.	+ 51 53 54.		16.9	FAINT BLUE STAR
ZL 129	12 57 57.	+ 28 20 30.		22.2	ULTRAFAINT BLUE STAR
DUN 07	12 57 57.	+ 28 21	20		SMALL NEBULA
IC 4035	12 57 57.	+ 40 34 01.			NONSTELLAR OBJECT
IC 4026	12 57 57.3	+ 28 18 54.		15.5	GALAXY E1
REIN 1.096	12 57 57.82	+ 28 19 00.8			NEBULA
A2 313	12 57 57.9	+ 30 22 56.		17.0	FAINT BLUE OBJECT
BON 2	12 57 58.	+ 28 40 12.	15	17.	VARIABLE GALAXY
IC 4034	12 57 58.	+ 37 18 55.			NONSTELLAR OBJECT
IC 4023	12 57 59.	+ 19 21 55.			NONSTELLAR OBJECT
REIZ 3108	12 57 59.	- 05 48	42	14.2	GALAXY
SN 1970C	12 57 59.	- 06 12		16.	SUPERNOVA
REIZ 3107	12 57 59.	- 06 13	30	13.8	GALAXY
RNGC 4887	12 57 59.	- 14 24		14.0	GALAXY
VDB.66G 160	12 58	- 04 26	70		DWARF GALAXY
VDB.66G 159	12 58	- 15 26	100		DWARF GALAXY
LB 09868	12 58	- 84 05		13.2	FAINT BLUE STAR
ZWG 071.096	12 58 00.	+ 10 24		15.1	GALAXY
MCG+02-33-049	12 58 00.	+ 10 24	36	15.1	GALAXY
REIZ 3114	12 58 00.	+ 13 46	138	13.8	GALAXY
ZWG 160.084	12 58 00.	+ 26 56		15.2	GALAXY
RB 095	12 58 00.	+ 28 10	11		GALAXY IN COMA CLUSTER
RB 096	12 58 00.	+ 28 12	14		GALAXY IN COMA CLUSTER
RB 097	12 58 00.	+ 28 13	5		GALAXY IN COMA CLUSTER
MCG+05-31-082	12 58 00.	+ 28 13	24	14.7	GALAXY
RB 098	12 58 00.	+ 28 17	4		GALAXY IN COMA CLUSTER
RB 103	12 58 00.	+ 28 20	2		GALAXY IN COMA CLUSTER
RB 177	12 58 00.	+ 28 27	3		GALAXY IN COMA CLUSTER
RB 175	12 58 00.	+ 28 29	2		GALAXY IN COMA CLUSTER
RB 176	12 58 00.	+ 28 30	7		GALAXY IN COMA CLUSTER
RB 174	12 58 00.	+ 28 31	13		GALAXY IN COMA CLUSTER
VVI 61A	12 58 00.	+ 28 40		14.17	SEYFERT GAL1XY
ZWG 160.085	12 58 00.	+ 29 08		15.4	GALAXY
IC 4032	12 58 00.	+ 29 08		15.4	GALAXY IN COMA CLSTE
MCG+05-31-083	12 58 00.	+ 29 09	24	15.4	GALAXY
IC 4036	12 58 00.	+ 37 10 49.			NONSTELLAR OBJECT
IC 4037	12 58 00.	+ 39 16 19.			NONSTELLAR OBJECT
MCG-01-33-067	12 58 00.	- 04 18	42	14.	GALAXY
MCG-01-33-068	12 58 00.	- 06 11	60	15.	GALAXY
REIZ 3106	12 58 00.	- 06 35	18	14.6	GALAXY
MCG-02-33-087	12 58 00.	- 14 24	60	14.	GALAXY
MCG-02-33-088	12 58 00.	- 15 26	90	16.	GALAXY
KARA.68 217	12 58 00.	- 15 28	101		DWARF GALAXY
MCG-04-31-019	12 58 00.	- 21 55	36	15.5	GALAXY
IC 4025	12 58 01.	+ 19 22 31.			NONSTELLAR OBJECT
REIZ 3119	12 58 01.	+ 28 26	18	14.5	GALAXY
IC 4038	12 58 01.	+ 37 18 37.			NONSTELLAR OBJECT
RNGC 4890	12 58 01.	- 04 18		14.0	GALAXY
REIN 4.201	12 58 02.83	+ 10 23 56.4			NEBULA
BFG 082	12 58 02.89	+ 37 12 40.0		18.68	ULTRAVIOLET-EXCESS OBJECT
HELW 205	12 58 03.	- 14 00 12.			NEBULA
IC 4030	12 58 03.2	+ 28 13 25.			GALAXY E2
REIN 1.097	12 58 03.55	+ 28 13 30.7			NEBULA
IC 4033	12 58 04.0	+ 28 14 21.			GALAXY E6
SHB 230	12 58 04.0	+ 28 46 21.		18.	QUASI-STELLAR OBJECT
REIN 1.098	12 58 04.04	+ 28 14 29.1			NEBULA
BC 5C4.105	12 58 04.29	+ 28 46 15.8		18.1	QUASI-STELLAR OBJECT
A2 316	12 58 04.3	+ 28 46 16.		18.0	FAINT BLUE OBJECT
REIZ 3113	12 58 05.	- 04 06	18	15.1	GALAXY
REIZ 3112	12 58 05.	- 04 19	42	13.1	GALAXY
RNGC 4891	12 58 05.	- 13 09		13.0	GALAXY
ZWG 043.093	12 58 06.	+ 02 46		12.8	GALAXY
UGC 08116	12 58 06.	+ 02 46	162	12.8	GALAXY Sc
MCG+01-33-035	12 58 06.	+ 02 46	120	12.8	GALAXY
ZWG 071.097	12 58 06.	+ 08 58		15.6	GALAXY
TON-N 0690	12 58 06.	+ 27 35		17.0	BLUE STAR
ZWG 160.086	12 58 06.	+ 27 55		15.4	GALAXY
RB 099	12 58 06.	+ 28 14	19		GALAXY IN COMA CLUSTER
RB 100	12 58 06.	+ 28 15	28		GALAXY IN COMA CLUSTER
RB 180	12 58 06.	+ 28 26	2		GALAXY IN COMA CLUSTER
RB 179	12 58 06.	+ 28 26	4		GALAXY IN COMA CLUSTER
RB 178	12 58 06.	+ 28 26	4		GALAXY IN COMA CLUSTER
ZWG 160.087	12 58 06.	+ 28 37		15.1	GALAXY
RNGC 4896	12 58 06.	+ 28 37		15.0	GALAXY
UGC 08117	12 58 06.	+ 28 37	66	15.1	GALAXY E-S0
MCG+05-31-084	12 58 06.	+ 28 37	30	15.1	GALAXY
REIZ 3111	12 58 06.	- 07 33	18	15.8	GALAXY
REIN 2.172	12 58 06.39	+ 02 46 15.1			NEBULA
REIN 4.202	12 58 06.92	+ 08 57 59.5			NEBULA
REIZ 3118	12 58 07.	+ 10 24	24	15.1	GALAXY
LB 02516	12 58 07.	+ 54 42 30.		14.6	FAINT BLUE STAR
SN 1964D	12 58 07.	- 14 24		16.5	SUPERNOVA
RNGC 4900	12 58 08.	+ 02 46		12.0	GALAXY
LB 02517	12 58 08.	+ 51 46 30.		17.2	FAINT BLUE STAR
REIN 2.173	12 58 08.36	+ 02 45 39.6			NEBULA
REIZ 3117	12 58 09.	+ 02 46	90	11.6	GALAXY
DUN 08	12 58 09.	+ 28 24	40		SMALL NEBULA
A2 320	12 58 09.7	+ 28 22 38.		16.3	FAINT BLUE OBJECT
A2 322	12 58 10.8	+ 31 07 34.		17.8	FAINT BLUE OBJECT
REIN 1.099	12 58 11.29	+ 28 24 56.0			NEBULA
A2 323	12 58 11.8	+ 29 25 28.		16.3	FAINT BLUE OBJECT
ZC 1258.2+0531	12 58 12.	+ 05 31	1410		CLUSTER OF GALAXIES
ZC 1258.2+0833	12 58 12.	+ 08 33	540		CLUSTER OF GALAXIES
RB 105	12 58 12.	+ 28 06	6		GALAXY IN COMA CLUSTER
RB 106	12 58 12.	+ 28 12	8		GALAXY IN COMA CLUSTER
RB 104	12 58 12.	+ 28 12	3		GALAXY IN COMA CLUSTER
RB 108	12 58 12.	+ 28 15	4		GALAXY IN COMA CLUSTER
RB 110	12 58 12.	+ 28 17	22		GALAXY IN COMA CLUSTER
RB 109	12 58 12.	+ 28 17	7		GALAXY IN COMA CLUSTER
RB 107	12 58 12.	+ 28 17	2		GALAXY IN COMA CLUSTER
MCG+05-31-085	12 58 12.	+ 28 20	48	15.1	GALAXY
ZWG 160.088	12 58 12.	+ 29 17		14.6	GALAXY
UGC 08118	12 58 12.	+ 29 17	78	14.6	GALAXY S
IC 0842	12 58 12.	+ 29 17		14.6	GALAXY IN COMA CLSTR
ZWG 245.010	12 58 12.	+ 48 34		15.7	GALAXY
UGC 08119	12 58 12.	+ 48 34	102	15.7	GALAXY S
ZC 1258.2+7125	12 58 12.	+ 71 25	1080		CLUSTER OF GALAXIES
MCG+12-12-024	12 58 12.	+ 73 59	84	15.	GALAXY
HELW 206	12 58 12.	- 14 20 06.			NEBULA
MCG-06-29-003	12 58 12.	- 35 40	24	15.	GALAXY
REIN 1.100	12 58 12.38	+ 28 24 38.2			NEBULA
IC 4040	12 58 12.8	+ 28 19 27.		15.1	GALAXY Sa
IC 4039	12 58 13.	+ 21 57 38.			NONSTELLAR OBJECT
REIZ 3121	12 58 13.	+ 28 20	24	14.6	GALAXY
REIZ 3122	12 58 13.	+ 28 26	18	15.8	GALAXY
REIZ 3116	12 58 13.	- 13 11	48	13.5	GALAXY
HELW 207	12 58 13.	- 14 43 42.			NEBULA
REIN 1.101	12 58 13.59	+ 28 19 37.4			NEBULA
IC 4043	12 58 14.	+ 37 20 32.			NONSTELLAR OBJECT
BC 3CR280.1	12 58 14.15	+ 40 25 15.4		19.44	QUASI-STELLAR OBJECT
SHB 231	12 58 14.2	+ 40 25 15.		19.4	QUASI-STELLAR OBJECT
REIN 1.102	12 58 14.40	+ 28 17 02.5			NEBULA
REIN 1.103	12 58 14.56	+ 28 17 12.9			NEBULA
DUN 09	12 58 15.	+ 28 06	35		SMALL NEBULA
MCG-02-33-089	12 58 15.	- 13 11	150	13.	GALAXY
MCG-06-29-004	12 58 15.	- 35 21	24	15.	GALAXY
REIN 1.104	12 58 15.35	+ 28 11 34.7			NEBULA
IC 4041	12 58 16.4	+ 28 15 56.		15.7	GALAXY E4
REIN 1.105	12 58 16.50	+ 28 15 58.0			NEBULA
REIN 1.106	12 58 16.81	+ 28 09 51.2			NEBULA
REIZ 3120	12 58 17.	- 06 12	24	13.9	GALAXY
RNGC 4899	12 58 17.	- 13 41		12.5	GALAXY
RNGC 4902	12 58 17.	- 14 15		12.0	GALAXY
ZWG 100.020	12 58 18.	+ 15 59		15.5	GALAXY
MCG+03-33-021	12 58 18.	+ 19 59	24	15.7	GALAXY
RB 112	12 58 18.	+ 28 10	12		GALAXY IN COMA CLUSTER
RNGC 4906	12 58 18.	+ 28 12		15.0	GALAXY
RB 117	12 58 18.	+ 28 12	2		GALAXY IN COMA CLUSTER
RB 115	12 58 18.	+ 28 14	5		GALAXY IN COMA CLUSTER
RB 113	12 58 18.	+ 28 14	17		GALAXY IN COMA CLUSTER
MCG+05-31-086	12 58 18.	+ 28 14	30	15.7	GALAXY
RB 114	12 58 18.	+ 28 19	7		GALAXY IN COMA CLUSTER
RB 111	12 58 18.	+ 28 19	4		GALAXY IN COMA CLUSTER
RB 116	12 58 18.	+ 28 22	19		GALAXY IN COMA CLUSTER
ZWG 160.089	12 58 18.	+ 28 36		15.6	GALAXY
WEI 61972	12 58 18.	+ 28 39		17.75	FAINT BLUE OBJECT
TON-N 0691	12 58 18.	+ 29 10		15.4	BLUE STAR
MCG+05-31-087	12 58 18.	+ 29 17	60	14.6	GALAXY
REIZ 3127	12 58 18.	+ 29 18	60	14.0	GALAXY
MCG+07-27-012	12 58 18.	+ 40 01	54	15.	GALAXY
ZC 1258.3+4730	12 58 18.	+ 47 30	1080		CLUSTER OF GALAXIES
MCG+08-24-020	12 58 18.	+ 48 35	102	15.	GALAXY
MCG+08-24-021	12 58 18.	+ 50 16	42	16.	GALAXY
MCG+10-19-011	12 58 18.	+ 61 55	30	16.	GALAXY
MCG+10-19-012	12 58 18.	+ 62 31	39	16.	GALAXY
ZWG 336.001	12 58 18.	+ 73 58		14.7	GALAXY
ZWG 335.031	12 58 18.	+ 73 58		14.7	GALAXY
UGC 08120	12 58 18.	+ 73 58	126	14.7	GALAXY Sc
KARA.73B 0564	12 58 18.	+ 73 58	114	14.7	ISOLATED GALAXY S
ZWG 015.054	12 58 18.	- 01 05		15.2	GALAXY
MCG-02-33-090	12 58 18.	- 13 41	138	12.5	GALAXY
MCG-05-31-012	12 58 18.	- 32 12	18	15.	GALAXY
MCG-07-27-024	12 58 18.	- 40 08	36	15.	GALAXY
IC 4042	12 58 18.0	+ 28 14 20.		15.5	GALAXY SA0
REIN 1.107	12 58 18.41	+ 28 14 25.0			NEBULA
REIN 1.108	12 58 18.42	+ 28 13 55.6			NEBULA
REIN 2.174	12 58 18.79	- 13 40 28.8			NEBULA
REIZ 3125	12 58 19.	+ 28 17	12	15.3	GALAXY
REIZ 3126	12 58 19.	+ 28 26	18	14.7	GALAXY
IC 4046	12 58 19.	+ 36 57 14.			NONSTELLAR OBJECT
IC 4048	12 58 19.	+ 40 05 50.			NONSTELLAR OBJECT
RNGC 4916	12 58 20.	+ 37 38			GALAXY
LB 02518	12 58 20.	+ 53 25 00.		15.3	FAINT BLUE STAR
REIN 1.109	12 58 20.29	+ 28 22 11.6			NEBULA
ZL 130	12 58 21.	+ 28 15 24.		21.7	ULTRAFAINT BLUE STAR
LB 02519	12 58 21.	+ 56 28 00.		14.5	FAINT BLUE STAR
REIN 2.175	12 58 21.77	- 14 14 39.1			NEBULA
IC 4049	12 58 22.	+ 36 36 50.			NONSTELLAR OBJECT
IC 4052	12 58 22.	+ 39 56 08.			NONSTELLAR OBJECT
IC 4044	12 58 22.9	+ 28 11 25.			GALAXY E4
HOLM 499B	12 58 23.	+ 28 05	18	14.9	PART OF MULTIPLE GALAXY
IC 4050	12 58 23.	+ 37 00 26.			NONSTELLAR OBJECT
REIN 1.110	12 58 23.00	+ 28 11 27.9			NEBULA
A2 326	12 58 23.6	+ 29 35 22.		16.0	FAINT BLUE OBJECT
A2 327	12 58 23.8	+ 28 39 28.		17.6	FAINT BLUE OBJECT

OBJECT NAME	RIGHT ASCEN.	DECLINATION	DIAM.	MAGN.	TYPE OF OBJECT
ZWG 015.055	12 58 24.	+ 00 14		13.2	GALAXY
UGC 08121	12 58 24.	+ 00 14	150	13.2	GALAXY SBc
MCG+00-33-026	12 58 24.	+ 00 14	114	13.2	GALAXY
ZC 1258.4+0118	12 58 24.	+ 01 18	1880		CLUSTER OF GALAXIES
ZWG 043.094	12 58 24.	+ 07 19		15.6	GALAXY
ZWG 160.090	12 58 24.	+ 27 40		15.5	GALAXY
UGC 08122	12 58 24.	+ 27 40	78	15.5	GALAXY S0-a
RB 118	12 58 24.	+ 28 06	7		GALAXY IN COMA CLUSTER
RB 119	12 58 24.	+ 28 11	17		GALAXY IN COMA CLUSTER
RNGC 4908	12 58 24.	+ 28 19		15.0	GALAXY
RB 120	12 58 24.	+ 28 19	5		GALAXY IN COMA CLUSTER
MCG+05-31-088	12 58 24.	+ 28 21	30	15.1	GALAXY
RNGC 4907	12 58 24.	+ 28 26		14.5	GALAXY
ZWG 189.012	12 58 24.	+ 36 37		15.3	GALAXY
UGC 08124	12 58 24.	+ 36 37	72	15.3	GALAXY S0?
MCG+06-29-013	12 58 24.	+ 36 37	24	15.	GALAXY
UGC 08123	12 58 24.	+ 37 22	60	16.5	GALAXY Sc-IRR
ZWG 189.013	12 58 24.	+ 37 35		12.7	GALAXY
UGC 08125	12 58 24.	+ 37 35	228	12.7	GALAXY E
MCG+06-29-014	12 58 24.	+ 37 36	54	12.5	GALAXY
UGC 08126	12 58 24.	+ 40 01	66	16.0	GALAXY SBb-c
ZWG 294.009	12 58 24.	+ 59 15		15.7	GALAXY
MCG-02-33-091	12 58 24.	- 14 03 30.	72	15.	GALAXY
MCG-02-33-092	12 58 24.	- 14 14	180	11.5	GALAXY
IC 4045	12 58 24.2	+ 28 21 30.		15.1	GALAXY E3
REIN 1.111	12 58 24.40	+ 28 21 36.3			NEBULA
REIN 1.112	12 58 24.50	+ 28 25 39.4			NEBULA
A2 328	12 58 24.7	+ 30 11 03.		18.0	FAINT BLUE OBJECT
RNGC 4904	12 58 25.	+ 00 14		13.0	GALAXY
HOLM 499A	12 58 25.	+ 28 05	36	13.2	PART OF MULTIPLE GALAXY
REIZ 3129	12 58 25.	+ 28 22	12	14.7	GALAXY
REIZ 3130	12 58 25.	+ 28 27	42	14.2	GALAXY
RNGC 4914	12 58 26.	+ 37 35		13.0	GALAXY
RNGC 4913	12 58 26.	+ 37 37			NON-EXISTENT OBJECT
RNGC 4912	12 58 26.	+ 37 39			NON-EXISTENT OBJECT
IC 4056	12 58 26.	+ 40 01 14.			NONSTELLAR OBJECT
REIZ 3133	12 58 27.	+ 37 35	48	14.0	GALAXY
HELW 208	12 58 27.	- 14 01 29.			NEBULA
REIN 1.113	12 58 27.28	+ 28 18 44.0			NEBULA
REIZ 3124	12 58 28.	+ 00 15	120	12.4	GALAXY
REIZ 3137	12 58 28.	+ 47 29	54	13.9	GALAXY
HELW 209	12 58 28.	- 14 03 29.			NEBULA
REIZ 3123	12 58 29.	- 03 53	24	13.4	GALAXY
A2 329	12 58 29.5	+ 30 57 37.		16.8	FAINT BLUE OBJECT
REIN 1.114	12 58 29.76	+ 28 03 10.6			NEBULA
BPG 083	12 58 29.96	+ 36 51 32.5		19.49	ULTRAVIOLET-EXCESS OBJECT
ZC 1258.5+0816	12 58 30.	+ 08 16	1140		CLUSTER OF GALAXIES
ZWG 100.021	12 58 30.	+ 19 57		15.7	GALAXY
TON-N 0692	12 58 30.	+ 27 20		15.4	BLUE STAR
MCG+05-31-091	12 58 30.	+ 27 40	24	15.5	GALAXY
RNGC 4911	12 58 30.	+ 28 03		13.5	GALAXY
MCG+05-31-093	12 58 30.	+ 28 03 30.	48	13.7	GALAXY
UGC 08128	12 58 30.	+ 28 04	78	13.7	GALAXY S
RB 122	12 58 30.	+ 28 06	9		GALAXY IN COMA CLUSTER
RB 123	12 58 30.	+ 28 10	8		GALAXY IN COMA CLUSTER
RB 121	12 58 30.	+ 28 14	5		GALAXY IN COMA CLUSTER
UGC 08129	12 58 30.	+ 28 17	66	14.8	GALAXY E
SN 1950A	12 58 30.	+ 28 17		17.7	SUPERNOVA
MCG+05-31-092	12 58 30.	+ 28 17	18	14.9	GALAXY
MCG+05-31-090	12 58 30.	+ 28 18 30.	42	14.8	GALAXY
MCG+05-31-089	12 58 30.	+ 28 26	60	14.6	GALAXY
WEI 62385	12 58 30.	+ 28 36		17.79	FAINT BLUE OBJECT
ZWG 160.091	12 58 30.	+ 28 38		14.9	GALAXY
MCG+07-27-013	12 58 30.	+ 40 51 30.	39	15.	GALAXY
MCG+08-24-022	12 58 30.	+ 48 50	36	16.	GALAXY
MCG+10-19-013	12 58 30.	+ 59 13	30	16.	GALAXY
MRK 236	12 58 30.	+ 62 02	11	17.	GALAXY WITH UV CONTINUUM
ZWG 015.056	12 58 30.	- 01 42		15.3	GALAXY
UGC 08127	12 58 30.	- 01 42	102	15.3	GALAXY DWARF
MCG+00-33-027	12 58 30.	- 01 42	84	15.3	GALAXY
RNGC 4888	12 58 30.	- 05 48			NON-EXISTENT OBJECT
MCG-03-33-026	12 58 30.	- 17 55	66	14.5	GALAXY
IC 4051	12 58 30.0	+ 28 16 34.		14.8	GALAXY SA0
REIN 1.115	12 58 30.18	+ 28 16 37.0			NEBULA
IC 4047	12 58 31.	+ 19 57 20.			NONSTELLAR OBJECT
REIZ 3131	12 58 31.	+ 28 05	30	13.4	GALAXY
REIZ 3132	12 58 31.	+ 23 40	48	14.1	GALAXY
LB 00667	12 58 31.	+ 59 52 42.		17.0	FAINT BLUE STAR
BC B246	12 58 31.08	+ 34 04 48.3		18.18	QUASI-STELLAR OBJECT
BPG 084	12 58 31.36	+ 34 04 43.1		18.18	ULTRAVIOLET-EXCESS OBJECT
SHB 232	12 58 31.4	+ 34 04 43.		18.2	QUASI-STELLAR OBJECT
REIN 1.116	12 58 31.52	+ 28 03 36.4			NEBULA
LB 02520	12 58 32.	+ 59 20 24.		15.7	FAINT BLUE STAR
IC 4060	12 58 34.	+ 40 51 20.			NONSTELLAR OBJECT
REIN 1.117	12 58 34.89	+ 28 10 09.4			NEBULA
IC 4054	12 58 35.	+ 23 10 20.			NONSTELLAR OBJECT
IC 4053	12 58 35.	+ 23 11 44.			NONSTELLAR OBJECT
REIZ 3128	12 58 35.	- 05 16	18	13.4	GALAXY
ZWG 071.098	12 58 36.	+ 10 02		15.6	GALAXY
MCG+02-33-050	12 58 36.	+ 10 02	60	15.6	GALAXY
ZC 1258.6+2235	12 58 36.	+ 22 35	1810		CLUSTER OF GALAXIES
IC 4055	12 58 36.	+ 23 10 38.			NONSTELLAR OBJECT
RNGC 4911A	12 58 36.	+ 28 04		15.0	GALAXY
MCG+05-31-094	12 58 36.	+ 28 04	24	14.9	GALAXY
RB 125	12 58 36.	+ 28 10	6		GALAXY IN COMA CLUSTER
RB 124	12 58 36.	+ 28 10	36		GALAXY IN COMA CLUSTER
MCG+05-31-095	12 58 36.	+ 28 38	42	14.9	GALAXY
LB 00064	12 58 36.	+ 30 57		16.7	FAINT BLUE STAR
MCG+07-27-014	12 58 36.	+ 40 07 30.	21	16.5	GALAXY
ZC 1258.6+4531	12 58 36.	+ 45 31	810		CLUSTER OF GALAXIES
LB 02521	12 58 36.	+ 49 18 54.		14.7	FAINT BLUE STAR
ARC 1658	12 58 36.	- 03 11		17.2	RICH CLUSTER OF GALAXIES
MCG-05-31-013	12 58 36.	- 30 40	72	15.	GALAXY
MCG-05-31-014	12 58 36.	- 32 09	48	16.	GALAXY
REIZ 3136	12 58 37.	+ 28 17	18	14.9	GALAXY
HELW 210	12 58 37.	- 05 17 11.			NEBULA
RNGC 4903	12 58 38.	- 30 40		15.0	GALAXY
IC 4057	12 58 39.	+ 23 25 38.			NONSTELLAR OBJECT
DUN 10	12 58 39.	+ 28 11	60		SMALL NEBULA
IC 4061	12 58 39.	+ 39 51 15.			NONSTELLAR OBJECT
REIZ 3134	12 58 40.	+ 18 42	18	15.2	GALAXY
REIZ 3135	12 58 40.	+ 19 46	24	14.8	GALAXY
IC 4062	12 58 40.	+ 40 07 39.			NONSTELLAR OBJECT
RNGC 4917	12 58 40.	+ 47 30		15.0	GALAXY
SHB 233	12 58 41.7	+ 35 38 44.		18.3	QUASI-STELLAR OBJECT
BPG 086	12 58 41.74	+ 35 38 44.0		18.27	ULTRAVIOLET-EXCESS OBJECT
BPG 085	12 58 41.85	+ 34 27 45.3		18.76	ULTRAVIOLET-EXCESS OBJECT
IC 4058	12 58 42.	+ 19 45 44.			NONSTELLAR OBJECT
8ZW 1258+24.1	12 58 42.	+ 24 04		17.8	COMPACT GALAXY
ZWG 160.092	12 58 42.	+ 28 05		15.7	GALAXY
RB 126	12 58 42.	+ 28 10	6		GALAXY IN COMA CLUSTER
RB 128	12 58 42.	+ 28 18	3		GALAXY IN COMA CLUSTER
RB 127	12 58 42.	+ 28 20	7		GALAXY IN COMA CLUSTER
ZWG 245.011	12 58 42.	+ 47 30		15.0	GALAXY
UGC 08130	12 58 42.	+ 47 30	102	15.0	GALAXY SBb
MCG+08-24-023	12 58 42.	+ 47 30	72	13.9	GALAXY
ZWG 316.007	12 58 42.	+ 64 44		15.6	GALAXY
MCG-05-31-015	12 58 42.	- 30 36	18	15.5	GALAXY
MCG-05-31-016	12 58 42.	- 32 03	15	15.	GALAXY
BC B196	12 58 42.0	+ 35 38 48.2		18.27	QUASI-STELLAR OBJECT
RNGC 4905	12 58 44.	- 30 36		15.0	GALAXY
A2 333	12 58 44.2	+ 29 42 18.		17.9	FAINT BLUE OBJECT
MCG+05-31-096	12 58 45.	+ 28 05	12	15.7	GALAXY
MCG+07-27-015	12 58 45.	+ 40 07	78	14.5	GALAXY
REIZ 3141	12 58 46.	+ 19 33	18	15.3	GALAXY
IC 4063	12 58 47.	+ 39 30 51.			NONSTELLAR OBJECT
ZWG 043.095	12 58 48.	+ 07 35		15.5	GALAXY
REIZ 3140	12 58 48.	+ 12 05	42	14.9	GALAXY
ZC 1258.8+1505	12 58 48.	+ 15 05	1610		CLUSTER OF GALAXIES
IC 4059	12 58 48.	+ 19 32 27.			NONSTELLAR OBJECT
8ZW 1258+21.5	12 58 48.	+ 21 31		18.1	COMPACT GALAXY
ZWG 160.093	12 58 48.	+ 28 04		15.7	GALAXY
RB 129	12 58 48.	+ 28 11	6		GALAXY IN COMA CLUSTER
RB 130	12 58 48.	+ 28 21	4		GALAXY IN COMA CLUSTER
ZWG 217.006	12 58 48.	+ 40 07		15.0	GALAXY
UGC 08131	12 58 48.	+ 40 07	102	15.0	GALAXY S0
UGC 08132	12 58 48.	+ 55 07	60	16.0	GALAXY S
ZWG 335.032	12 58 48.	+ 70 28		15.3	GALAXY
MCG+12-12-025	12 58 48.	+ 70 28	57	16.	GALAXY
MCG-01-33-069	12 58 48.	- 04 14	60	14.	GALAXY
MCG-01-33-070	12 58 48.	- 04 28	72	15.5	GALAXY
MCG-05-31-017	12 58 48.	- 29 47	24	15.	GALAXY
RNGC 4910	12 58 49.	+ 01 56			NON-EXISTENT OBJECT
IC 4064	12 58 49.	+ 40 06 21.			NONSTELLAR OBJECT
RNGC 4915	12 58 49.	- 04 16		13.0	GALAXY
SHB 234	12 58 49.5	+ 34 22 38.		17.7	QUASI-STELLAR OBJECT
BPG 087	12 58 49.50	+ 34 22 38.1		17.66	ULTRAVIOLET-EXCESS OBJECT
BC B471	12 58 49.8	+ 34 22 42.6		17.66	QUASI-STELLAR OBJECT
MCG+07-27-016	12 58 51.	+ 40 00 30.	42	15.5	GALAXY
IC 4065	12 58 52.	+ 40 00 51.			NONSTELLAR OBJECT
REIZ 3139	12 58 53.	- 04 16	18	13.0	GALAXY
ZC 1258.9+0359	12 58 54.	+ 03 59	1480		CLUSTER OF GALAXIES
ZC 1258.9+1736	12 58 54.	+ 17 36	670		CLUSTER OF GALAXIES
ZWG 160.094	12 58 54.	+ 28 04		14.9	GALAXY
RNGC 4919	12 58 54.	+ 28 04		15.0	GALAXY
UGC 08133	12 58 54.	+ 28 04	66	14.9	GALAXY S0
RB 131	12 58 54.	+ 28 08	5		GALAXY IN COMA CLUSTER
RB 132	12 58 54.	+ 28 14	4		GALAXY IN COMA CLUSTER
MCG+09-21-081	12 58 54.	+ 55 06	24	16.	GALAXY
MCG+12-12-026	12 58 54.	+ 70 31 30.	11	17.	GALAXY
REIZ 3138	12 58 54.	- 09 06	36	16.0	GALAXY
MCG-05-31-018	12 58 54.	- 30 39	24	15.	GALAXY
BC 5C4.127	12 58 55.57	+ 28 37 48.7		19.	QUASI-STELLAR OBJECT
SHB 235	12 58 55.6	+ 28 37 48.		19.	QUASI-STELLAR OBJECT
A2 338	12 58 57.3	+ 28 11 02.		16.9	FAINT BLUE OBJECT
REIZ 3143	12 58 59.	- 04 30	48	15.2	GALAXY
ZC 1259.0+1920	12 59 00.	+ 19 20	1610		CLUSTER OF GALAXIES
MCG+05-31-097	12 59 00.	+ 28 05	48	14.9	GALAXY
ZWG 160.095	12 59 00.	+ 28 08		13.7	GALAXY
UGC 08134	12 59 00.	+ 28 08	150	13.7	GALAXY Sa
RNGC 4921	12 59 00.	+ 28 09		13.5	GALAXY
RB 133	12 59 00.	+ 28 16	7		GALAXY IN COMA CLUSTER
ZWG 160.096	12 59 00.	+ 29 35		14.2	GALAXY
RNGC 4922	12 59 00.	+ 29 35		14.0	GALAXY
SCH 85	12 59 00.	+ 29 35		14.2	PECULIAR GALAXY
UGC 08135	12 59 00.	+ 29 35	138	14.2	GALAXY DBL SYS
KARA.72 363B	12 59 00.	+ 29 35	36		PART OF DOUBLE GALAXY
KARA.72 363A	12 59 00.	+ 29 35	66	14.2	PART OF DOUBLE GALAXY
MCG+07-27-017	12 59 00.	+ 40 10	30	16.	GALAXY
IC 4067	12 59 00.	+ 40 12 27.			NONSTELLAR OBJECT
ZWG 245.012	12 59 00.	+ 48 19		15.1	GALAXY
MRK 237	12 59 00.	+ 48 19	13	15.	GALAXY WITH UV CONTINUUM
KW 21	12 59 00.	+ 48 19	30		SEYFERT GALAXY
VVI 62	12 59 00.	+ 48 19		17.0	SEYFERT GALAXY
MCG+08-24-024	12 59 00.	+ 48 20	42	17.	GALAXY
ZC 1259.0+4830	12 59 00.	+ 48 30	740		CLUSTER OF GALAXIES
MCG+09-21-083	12 59 00.	+ 54 05	30	15.	GALAXY
MCG+09-21-084	12 59 00.	+ 54 08	42	15.	GALAXY
MCG+09-21-082	12 59 00.	+ 55 05	60	15.	GALAXY
UGC 08136	12 59 00.	+ 77 51	72	16.0	GALAXY Sa
MCG+13-09-041	12 59 00.	+ 80 23	66	15.	GALAXY
MCG-04-31-020	12 59 00.	- 23 44	12	15.5	GALAXY
A2 339	12 59 00.1	+ 28 23 30.		17.3	FAINT BLUE GALAXY
BE 2.15	12 59 00.7	+ 36 01 03.			GALAXY NEAR QSO B234
SN 1959B	12 59 01.	+ 28 07		18.5	SUPERNOVA
REIZ 3144	12 59 01.	+ 28 10	96	14.1	GALAXY
IC 4068	12 59 01.	+ 40 10 03.			NONSTELLAR OBJECT
REIZ 3142	12 59 01.	- 11 29	90	15.3	GALAXY
REIN 1.118	12 59 01.77	+ 28 09 18.0			NEBULA
ARC 1659	12 59 02.	+ 03 57		17.5	RICH CLUSTER OF GALAXIES
ARC 1660	12 59 02.	+ 50 36		17.8	RICH CLUSTER OF GALAXIES
A2 340	12 59 02.6	+ 28 02 49.		17.6	FAINT BLUE OBJECT
REIZ 3151	12 59 03.	+ 37 09	42	15.4	GALAXY
REIZ 3157	12 59 04.	+ 54 08	54	14.4	GALAXY
A2 342	12 59 04.6	+ 26 34 04.		17.5	FAINT BLUE OBJECT
A2 343	12 59 04.8	+ 26 36 29.		17.5	FAINT BLUE OBJECT
IC 4069	12 59 05.	+ 36 21 09.			NONSTELLAR OBJECT
BE 2.14	12 59 05.45	+ 36 14 14.			GALAXY NEAR QSO B234
ZWG 160.097	12 59 06.	+ 28 06		14.7	GALAXY
RNGC 4923	12 59 06.	+ 28 06		14.5	GALAXY
MCG+05-31-098	12 59 06.	+ 28 10	138	13.7	GALAXY
RB 134	12 59 06.	+ 28 12	6		GALAXY IN COMA CLUSTER
ZWG 160.098	12 59 06.	+ 28 57		15.3	GALAXY
REIZ 3150	12 59 06.	+ 29 35	54	13.6	GALAXY
MCG+05-31-099	12 59 06.	+ 29 35	78	14.2	GALAXY
MRK 448	12 59 06.	+ 30 19	7	17.	GALAXY WITH UV CONTINUUM
MCG+06-29-015	12 59 06.	+ 35 02	30	15.	GALAXY
IC 4072	12 59 06.	+ 37 37 33.			NONSTELLAR OBJECT
IC 4073	12 59 06.	+ 40 10 51.			NONSTELLAR OBJECT
MCG+09-21-085	12 59 06.	+ 54 04	36	17.	GALAXY
ZWG 270.038	12 59 06.	+ 54 08		15.7	GALAXY
MCG+10-19-014	12 59 06.	+ 58 21	24	15.	GALAXY
MCG-07-27-025	12 59 06.	- 41 07	36	15.5	GALAXY
MCG-07-27-026	12 59 06.	- 41 35	90	15.5	GALAXY
LB 02522	12 59 08.	+ 59 18 18.		16.2	FAINT BLUE STAR
LB 00668	12 59 09.	+ 57 58 54.		17.1	FAINT BLUE STAR

OBJECT NAME	RIGHT ASCEN.	DECLINATION	DIAM.	MAGN.	TYPE OF OBJECT
MCG-04-31-021	12 59 09.	- 26 52 30.	72	15.	GALAXY
A2 344	12 59 10.9	+ 28 32 36.		17.3	FAINT BLUE GALAXY
LB 02523	12 59 11.	+ 50 52 30.		12.8	FAINT BLUE STAR
RNGC 4909	12 59 11.	- 42 30			GALAXY
REIN 4.203	12 59 11.14	+ 10 17 06.9			NEBULA
ZWG 043.096	12 59 12.	+ 03 33		15.4	GALAXY
ZWG 043.097	12 59 12.	+ 04 57		15.4	GALAXY
ZWG 043.098	12 59 12.	+ 05 18		15.6	GALAXY
ZWG 071.099	12 59 12.	+ 10 17		15.6	GALAXY
ZC 1259.2+1340	12 59 12.	+ 13 40	470		CLUSTER OF GALAXIES
ZC 1259.2+1837	12 59 12.	+ 18 37	1550		CLUSTER OF GALAXIES
UGC 08137	12 59 12.	+ 26 24	60	14.8	GALAXY S0
MCG+05-31-101	12 59 12.	+ 28 08	48	14.7	GALAXY
ZWG 160.099	12 59 12.	+ 29 24		14.8	GALAXY
MCG+05-31-100	12 59 12.	+ 29 24	60	14.8	GALAXY
IC 0843	12 59 12.	+ 29 24		14.8	GALAXY IN COMA CLSTR
REIZ 3153	12 59 12.	+ 29 25	48	13.9	GALAXY
MCG+06-29-016	12 59 12.	+ 36 38	54	15.	GALAXY
MCG-01-33-071	12 59 12.	- 08 04	150	14.	GALAXY
MCG-07-27-027	12 59 12.	- 40 47	66	15.	GALAXY
REIZ 3149	12 59 13.	+ 10 17	30	15.3	GALAXY
IC 4066	12 59 13.	+ 19 32 27.			NONSTELLAR OBJECT
RNGC 4918	12 59 13.	- 04 14			GALAXY
BFG 088	12 59 13.16	+ 37 12 32.9		18.67	ULTRAVIOLET-EXCESS OBJECT
IC 4077	12 59 14.	+ 37 39 28.			NONSTELLAR OBJECT
IC 4078	12 59 15.	+ 36 52 46.			NONSTELLAR OBJECT
IC 4070	12 59 16.	+ 19 34 10.			NONSTELLAR OBJECT
REIZ 3152	12 59 16.	+ 19 35	24	14.8	GALAXY
REIZ 3148	12 59 17.	- 04 13	36	14.9	GALAXY
RNGC 4920	12 59 17.	- 11 06		14.0	GALAXY
ZWG 043.099	12 59 18.	+ 04 36		15.5	GALAXY
MCG+01-33-036	12 59 18.	+ 04 36	36	15.5	GALAXY
VV 283B	12 59 18.	+ 04 37 30.	18	17.	INTERACTING GALAXY
VV 283A	12 59 18.	+ 04 37 30.	18	16.	INTERACTING GALAXY
VV 283	12 59 18.	+ 04 37 30.	36		INTERACTING GALAXY
ZWG 043.100	12 59 18.	+ 05 16		15.4	GALAXY
UGC 08138	12 59 18.	+ 05 16	60	15.4	GALAXY Sa-b
ZC 1259.3+1829	12 59 18.	+ 18 29	340		CLUSTER OF GALAXIES
TON-N 0693	12 59 18.	+ 27 44		15.6	BLUE STAR
HOLM 500A	12 59 18.	+ 29 19	84	15.1	PART OF MULTIPLE GALAXY
MCG+05-31-102	12 59 18.	+ 29 19	78	14.8	GALAXY
MRK 061	12 59 18.	+ 29 40	10	17.	GALAXY WITH UV CONTINUUM
UGC 08139	12 59 18.	+ 36 38	66	16.5	GALAXY S
IC 4082	12 59 18.	+ 37 36 28.			NONSTELLAR OBJECT
IC 4085	12 59 18.	+ 39 58 16.			NONSTELLAR OBJECT
ZC 1259.3+4201	12 59 18.	+ 42 01	870		CLUSTER OF GALAXIES
MRK 238	12 59 18.	+ 65 18	8	16.5	GALAXY WITH UV CONTINUUM
MCG+11-16-008	12 59 18.	+ 65 18	45	16.	GALAXY
MCG-01-33-072	12 59 18.	- 06 38	78	15.	GALAXY
REIZ 3147	12 59 18.	- 06 39	60	14.3	GALAXY
REIZ 3146	12 59 18.	- 08 05	144	14.0	GALAXY
MCG-05-31-019	12 59 18.	- 32 32	72	15.5	GALAXY
HOLM 500B	12 59 19.	+ 29 16	30	15.8	PART OF MULTIPLE GALAXY
IC 4083	12 59 19.	+ 38 24 40.			NONSTELLAR OBJECT
LB 02524	12 59 19.	+ 56 35 42.		15.3	FAINT BLUE STAR
REIZ 3145	12 59 19.	- 11 13	36	13.4	GALAXY
A2 347	12 59 20.7	+ 28 27 41.		18.4	FAINT BLUE OBJECT
BFG 089	12 59 20.88	+ 36 46 01.8		17.83	ULTRAVIOLET-EXCESS OBJECT
SHB 236	12 59 20.9	+ 36 46 01.		17.8	QUASI-STELLAR OBJECT
IC 4075	12 59 21.	+ 20 13 58.			NONSTELLAR OBJECT
IC 4076	12 59 21.	+ 23 39 34.			NONSTELLAR OBJECT
IC 4084	12 59 21.	+ 37 14 10.			NONSTELLAR OBJECT
BC B228	12 59 21.4	+ 36 46 22.6		17.83	QUASI-STELLAR OBJECT
IC 4074	12 59 22.	+ 19 16 34.			NONSTELLAR OBJECT
ARC 1661	12 59 22.	+ 29 21		17.6	RICH CLUSTER OF GALAXIES
IC 4086	12 59 22.	+ 36 54 58.			NONSTELLAR OBJECT
ZWG 043.101	12 59 24.	+ 03 22		15.6	GALAXY
MCG+01-33-037	12 59 24.	+ 03 22	42	15.6	GALAXY
ZWG 043.102	12 59 24.	+ 07 52		15.6	GALAXY
ZWG 160.100	12 59 24.	+ 28 09		15.5	GALAXY
ZWG 160.101	12 59 24.	+ 28 22		15.2	GALAXY
ZWG 160.102	12 59 24.	+ 29 19		14.8	GALAXY
REIZ 3158	12 59 24.	+ 29 19	84	13.9	GALAXY
UGC 08140	12 59 24.	+ 29 19	102	14.8	GALAXY Sa-b
IC 4088	12 59 24.	+ 29 19		14.8	GALAXY IN COMA CLSTR
ZC 1259.4+2920	12 59 24.	+ 29 20	540		CLUSTER OF GALAXIES
ZWG 189.014	12 59 24.	+ 36 54		15.7	GALAXY
UGC 08141	12 59 24.	+ 36 54	60	15.7	GALAXY Sc
MCG+06-29-017	12 59 24.	+ 36 56	42	14.5	GALAXY
ZC 1259.4+4113	12 59 24.	+ 41 13	1080		CLUSTER OF GALAXIES
MCG+09-21-086	12 59 24.	+ 54 40	45	14.	GALAXY
ZWG 316.008	12 59 24.	+ 65 17		15.2	GALAXY
MCG-01-33-073	12 59 24.	- 07 19	30	15.	GALAXY
MCG-02-33-095	12 59 24.	- 10 09	78	14.	GALAXY
MCG-02-33-094	12 59 24.	- 11 06	60	14.	GALAXY
MCG-02-33-093	12 59 24.	- 14 37	48	14.	GALAXY
MCG-05-31-020	12 59 24.	- 32 52	72	15.	GALAXY
MCG-07-27-028	12 59 24.	- 42 29	156	15.	GALAXY
HELW 211	12 59 25.	- 14 36 46.			NEBULA
IC 4090	12 59 26.	+ 37 06 22.			NONSTELLAR OBJECT
A2 348	12 59 26.2	+ 29 09 59.		17.6	FAINT BLUE OBJECT
A2 349	12 59 26.5	+ 29 18 04.		16.4	FAINT BLUE OBJECT
IC 4079	12 59 29.	+ 19 31 10.			NONSTELLAR OBJECT
IC 4081	12 59 29.	+ 23 02 22.			NONSTELLAR OBJECT
BSO 06	12 59 29.	+ 34 27 00.		17.87	BLUE STELLAR OBJECT
ZWG 043.103	12 59 30.	+ 05 20		15.5	GALAXY
ZWG 043.104	12 59 30.	+ 06 33		15.5	GALAXY
IC 4080	12 59 30.	+ 19 31 16.			NONSTELLAR OBJECT
LB 00026	12 59 30.	+ 25 10		17.9	FAINT BLUE STAR
ZWG 160.103	12 59 30.	+ 27 53		14.1	GALAXY
UGC 08142	12 59 30.	+ 27 53	66	14.1	GALAXY E-S0
RNGC 4926	12 59 30.	+ 27 54		14.0	GALAXY
FEIG 071	12 59 30.	+ 29 51		10.4	FAINT BLUE STAR
ZC 1259.5+3341	12 59 30.	+ 33 41	1140		CLUSTER OF GALAXIES
SHAH 071	12 59 30.	+ 38 20	210	17.2	GROUP OF COMPACT GALAXIES
ZC 1259.5+4646	12 59 30.	+ 46 46	1210		CLUSTER OF GALAXIES
ZWG 270.039	12 59 30.	+ 54 39		15.4	GALAXY
UGC 08143	12 59 30.	+ 54 39	60	15.4	GALAXY Sb-c
7ZW 495	12 59 30.	+ 71 57			COMPACT GALAXY
MCG+13-09-042	12 59 30.	+ 80 21	51	16.	GALAXY
HN 0367	12 59 30.	- 07 20			NEBULA
REIZ 3156	12 59 30.	- 07 20	24	14.1	GALAXY
IC 4071	12 59 30.	- 07 20			NONSTELLAR OBJECT
RNGC 4925	12 59 30.	- 07 26		14.0	GALAXY
MCG-01-33-074	12 59 30.	- 07 26	36	14.	GALAXY
REIZ 3155	12 59 30.	- 07 27	30	13.6	GALAXY
MCG-05-31-021	12 59 30.	- 32 30	48	16.	GALAXY
BC BS06	12 59 30.5	+ 34 27 15.		17.87	QUASI-STELLAR OBJECT
SHB 237	12 59 30.5	+ 34 27 15.		17.9	QUASI-STELLAR OBJECT
BFG 090	12 59 30.92	+ 34 27 08.8		18.26	ULTRAVIOLET-EXCESS OBJECT
REIZ 3154	12 59 31.	- 11 05	36	14.8	GALAXY
BFG 091	12 59 32.05	+ 32 21 54.7		16.81	ULTRAVIOLET-EXCESS OBJECT
BC B264	12 59 32.05	+ 32 21 55.		16.89	QUASI-STELLAR OBJECT
SHB 238	12 59 32.1	+ 32 21 54.		17.0	QUASI-STELLAR OBJECT
IC 4087	12 59 33.	+ 20 15 46.			NONSTELLAR OBJECT
MCG-03-33-027	12 59 33.	- 16 57	60	16.	GALAXY
BB 2.12	12 59 33.3	+ 36 29 50.			GALAXY NEAR QSO B234
BFG 092	12 59 33.81	+ 33 11 58.3			ULTRAVIOLET-EXCESS OBJECT
REIZ 3160	12 59 34.	+ 19 47	18	15.8	GALAXY
RNGC 4924	12 59 35.	- 14 42		14.0	GALAXY
ZWG 043.105	12 59 36.	+ 03 22		15.3	GALAXY
ZWG 043.106	12 59 36.	+ 03 42		15.7	GALAXY
MCG+05-31-103	12 59 36.	+ 27 53	24	14.1	GALAXY
ZWG 160.104	12 59 36.	+ 28 03		15.4	GALAXY
ZWG 160.105	12 59 36.	+ 28 16		14.8	GALAXY
RNGC 4927	12 59 36.	+ 28 16		15.0	GALAXY
MCG+05-31-104	12 59 36.	+ 28 16	24	14.8	GALAXY
ZC 1259.6+3241	12 59 36.	+ 32 41	610		CLUSTER OF GALAXIES
ZC 1259.6+4226	12 59 36.	+ 42 26	540		CLUSTER OF GALAXIES
MCG-02-33-096	12 59 36.	- 14 42 30.	30	14.	GALAXY
A2 330	12 59 36.6	+ 28 35 52.		17.4	FAINT BLUE OBJECT
A2 352	12 59 36.9	+ 27 33 46.		18.0	FAINT BLUE OBJECT
IC 4089	12 59 37.	+ 19 46 16.			NONSTELLAR OBJECT
A2 353	12 59 38.9	+ 27 14 32.		16.5	FAINT BLUE OBJECT
IC 4094	12 59 39.	+ 38 03 46.			NONSTELLAR OBJECT
MCG-02-33-097	12 59 39.	- 10 58	36	15.	GALAXY
REIZ 3161	12 59 40.	+ 18 56	18	14.5	GALAXY
ZWG 043.107	12 59 42.	+ 04 30		15.7	GALAXY
ZWG 043.108	12 59 42.	+ 06 47		15.6	GALAXY
ZWG 160.106	12 59 42.	+ 27 55		15.1	GALAXY
RNGC 4926A	12 59 42.	+ 27 55		15.0	GALAXY
ZWG 160.107	12 59 42.	+ 29 31		14.9	GALAXY
REIZ 3162	12 59 42.	+ 29 31	42	13.9	GALAXY
ZWG 217.007	12 59 42.	+ 40 41		15.1	GALAXY
UGC 08144	12 59 42.	+ 40 41	90	15.1	GALAXY Sc
MCG-02-33-098	12 59 42.	- 15 30 30.	90	14.5	GALAXY
MCG-02-33-099	12 59 42.	- 15 31	24	18.	GALAXY
IC 4093	12 59 43.	+ 29 16			NONSTELLAR OBJECT
REIZ 3159	12 59 43.	- 12 05	42	14.0	GALAXY
IC 4097	12 59 45.	+ 36 52 23.			NONSTELLAR OBJECT
IC 4098	12 59 45.	+ 38 14 59.			NONSTELLAR OBJECT
LB 02525	12 59 45.	+ 58 38 42.		15.7	FAINT BLUE STAR
MCG-02-33-100	12 59 45.	- 12 04	48	14.5	GALAXY
MCG-03-33-028	12 59 45.	- 17 24	156	15.	GALAXY
IC 4091	12 59 46.	+ 20 09 41.			NONSTELLAR OBJECT
IC 4100	12 59 46.	+ 40 40 35.			NONSTELLAR OBJECT
IC 4092	12 59 47.	+ 19 27 11.			NONSTELLAR OBJECT
ZC 1259.8+1351	12 59 48.	+ 13 51	270		CLUSTER OF GALAXIES
MCG+05-31-107	12 59 48.	+ 27 55	30	15.1	GALAXY
ZWG 160.108	12 59 48.	+ 28 29		15.5	GALAXY
REIZ 3164	12 59 48.	+ 28 38	18	15.0	GALAXY
ZWG 160.109	12 59 48.	+ 28 40		15.5	GALAXY
MCG+05-31-105	12 59 48.	+ 28 40	30	15.5	GALAXY
MCG+05-31-106	12 59 48.	+ 29 31	48	14.0	GALAXY
ZC 1259.8+5913	12 59 48.	+ 59 13	340		CLUSTER OF GALAXIES
BFG 093	12 59 48.76	+ 37 00 17.4		18.97	ULTRAVIOLET-EXCESS OBJECT
A2 354	12 59 49.2	+ 30 57 28.		18.4	FAINT BLUE OBJECT
IC 4096	12 59 51.	+ 24 16 41.			NONSTELLAR OBJECT
BFG 094	12 59 51.68	+ 35 07 40.8		18.48	ULTRAVIOLET-EXCESS OBJECT
REIZ 3163	12 59 52.	+ 19 23	30	14.9	GALAXY
IC 4101	12 59 52.	+ 40 12 29.			NONSTELLAR OBJECT
HELW 212	12 59 52.	- 05 38 46.			NEBULA
IC 4095	12 59 53.	+ 19 22 05.			NONSTELLAR OBJECT
ZC 1259.9+0835	12 59 54.	+ 08 35	870		CLUSTER OF GALAXIES
MCG+05-31-108	12 59 54.	+ 28 27	30	15.5	GALAXY
MCG+06-29-018	12 59 54.	+ 33 10	72	14.5	GALAXY
LB 02526	12 59 54.	+ 54 20 30.		16.0	FAINT BLUE STAR
A2 356	12 59 54.8	+ 28 18 09.		18.0	FAINT BLUE OBJECT
IC 4099	12 59 57.	+ 24 21 17.			NONSTELLAR OBJECT
MCG+08-24-025	12 59 57.	+ 50 34	36	16.	GALAXY
IC 4102	12 59 58.	+ 36 25 11.			NONSTELLAR OBJECT
ARC 1662	12 59 59.	+ 08 35		17.2	RICH CLUSTER OF GALAXIES
REIZ 3165	12 59 59.	+ 33 09	60	14.3	GALAXY
HOLM 501A	12 59 59.	+ 33 09	72	15.0	PART OF MULTIPLE GALAXY
IC 4103	12 59 59.	+ 38 17 11.			NONSTELLAR OBJECT
IC 4105	12 59 59.	+ 38 32 29.			NONSTELLAR OBJECT
IC 4104	12 59 59.	+ 38 51 35.			NONSTELLAR OBJECT
REIZ 3170	12 59 59.	+ 58 59	108	14.2	GALAXY
LB 00669	12 59 59.	+ 59 10 48.		17.4	FAINT BLUE STAR
VDB.66G 161	13 00	- 17 11	440		DWARF GALAXY
LB 09980	13 00	- 87 49		12.2	FAINT BLUE STAR
ZWG 015.057	13 00 00.	+ 00 36		15.7	GALAXY
ZC 1300.0+1445	13 00 00.	+ 14 45	670		CLUSTER OF GALAXIES
ZWG 160.110	13 00 00.	+ 28 30		15.2	GALAXY
ZWG 160.111	13 00 00.	+ 28 32		15.7	GALAXY
ZWG 189.015	13 00 00.	+ 33 09		15.5	GALAXY
UGC 08145	13 00 00.	+ 33 09	96	15.5	GALAXY Sa-b
MCG+06-29-018A	13 00 00.	+ 33 11	24	16.	GALAXY
ZWG 294.010	13 00 00.	+ 58 58		14.6	GALAXY
UGC 08146	13 00 00.	+ 58 58	228	14.6	GALAXY Sc
MCG+12-12-027	13 00 00.	+ 70 04	45	17.	GALAXY
ZWG 352.051	13 00 00.	+ 80 19		15.4	GALAXY
UGC 08147	13 00 00.	+ 80 19	66	15.4	GALAXY S
MCG-05-31-022	13 00 00.	- 32 26	24	15.5	GALAXY
BB 2.11	13 00 00.2	+ 36 33 25.5			GALAXY NEAR QSO B234
MCG+06-29-019	13 00 03.	+ 38 18	42	15.	GALAXY
BFG 095	13 00 03.23	+ 33 45 38.5		18.39	ULTRAVIOLET-EXCESS OBJECT
HOLM 501B	13 00 04.	+ 33 11	36	15.8	PART OF MULTIPLE GALAXY
LB 02527	13 00 05.	+ 60 08 48.		15.2	FAINT BLUE STAR
MCG+10-19-015	13 00 06.	+ 58 57 30.	192	13.	GALAXY
KLEM 23	13 00 06.	- 32 30	1200	13.	LOOSE GRP OF 8 GALAXIES
A2 359	13 00 06.6	+ 27 10 47.		18.5	FAINT BLUE OBJECT
LB 02528	13 00 07.	+ 50 27 18.		15.6	FAINT BLUE STAR
A2 361	13 00 07.5	+ 28 53 42.		16.7	FAINT BLUE OBJECT
MCG-04-31-022	13 00 09.	- 21 48	42	15.	GALAXY
ARC 1663	13 00 11.	- 02 16		17.0	RICH CLUSTER OF GALAXIES
ZWG 015.058	13 00 12.	+ 00 55		15.6	GALAXY
ZWG 043.109	13 00 12.	+ 03 38		15.5	GALAXY
8ZW 1300+25.6	13 00 12.	+ 25 38		16.4	COMPACT GALAXY
WEI 64217	13 00 12.	+ 27 56		14.39	FAINT BLUE OBJECT
ZWG 160.112	13 00 12.	+ 28 22		15.5	GALAXY
IC 4108	13 00 12.	+ 38 44 48.			NONSTELLAR OBJECT
ZC 1300.2+4534	13 00 12.	+ 45 34	1410		CLUSTER OF GALAXIES
MCG+08-24-026	13 00 12.	+ 50 14 30.	30	15.	GALAXY
MCG+09-21-087	13 00 12.	+ 52 28	42	15.	GALAXY
ZWG 335.033	13 00 12.	+ 70 04		15.6	GALAXY

OBJECT NAME	RIGHT ASCEN.	DECLINATION	DIAM.	MAGN.	TYPE OF OBJECT
ZC 1300.2+7554	13 00 12.	+ 75 54	1080		CLUSTER OF GALAXIES
UGC 08148	13 00 12.	+ 80 23	72	15.	GALAXY Sb
A2 363	13 00 12.6	+ 30 27 38.		16.7	FAINT BLUE GALAXY
IC 4106	13 00 13.9	+ 28 22 55.		15.5	GALAXY SB(rs)
LB 00670	13 00 14.	+ 57 16 42.		17.6	FAINT BLUE STAR
MCG-04-31-023	13 00 15.	- 23 40	72	15.	GALAXY
IC 4107	13 00 16.	+ 22 16 12.			NONSTELLAR OBJECT
LB 00671	13 00 16.	+ 57 39 12.		17.4	FAINT BLUE STAR
BFG 096	13 00 16.27	+ 33 02 03.7		19.29	ULTRAVIOLET-EXCESS OBJECT
BB 2.05	13 00 16.3	+ 36 10 45.5			GALAXY NEAR QSO B234
A2 366	13 00 16.9	+ 27 20 30.		18.5	FAINT BLUE OBJECT
ZC 1300.3+0546	13 00 18.	+ 05 46	2150		CLUSTER OF GALAXIES
ZWG 043.110	13 00 18.	+ 07 03		15.6	GALAXY
UGC 08149	13 00 18.	+ 07 03	66	15.6	GALAXY Sc
ZWG 071.100	13 00 18.	+ 10 22		15.2	GALAXY
MCG+02-33-051	13 00 18.	+ 10 22	24	15.2	GALAXY
ZWG 100.022	13 00 18.	+ 15 47		14.6	GALAXY
MCG+03-33-022	13 00 18.	+ 15 47	39	14.6	GALAXY
ZC 1300.3+2703	13 00 18.	+ 27 03	2290		CLUSTER OF GALAXIES
LB 00027	13 00 18.	+ 27 57		13.0	FAINT BLUE STAR
ZWG 160.113	13 00 18.	+ 28 18		14.9	GALAXY
RNGC 4929	13 00 18.	+ 28 18		15.0	GALAXY
MCG+05-31-109	13 00 18.	+ 28 23	48	15.5	GALAXY
ZWG 160.114	13 00 18.	+ 28 39		15.6	GALAXY
REIZ 3169	13 00 18.	+ 28 39	30	14.4	GALAXY
MCG+09-21-088	13 00 18.	+ 52 02	42	16.	GALAXY
MCG+11-16-009	13 00 18.	+ 67 22	45	16.	GALAXY
ZWG 352.052	13 00 18.	+ 75 05		15.7	GALAXY
RNGC 4928	13 00 18.	- 07 49		13.5	GALAXY
MCG-05-31-023	13 00 18.	- 30 32	36	15.	GALAXY
BFG 097	13 00 18.05	+ 35 18 12.3		18.74	ULTRAVIOLET-EXCESS OBJECT
BB 2.13	13 00 18.9	+ 36 19 02.5			GALAXY NEAR QSO B234
HZ 37	13 00 20.	+ 38 43		12.0	DECIDEDLY BLUE STAR
MCG+05-31-110	13 00 21.	+ 28 39	48	15.6	GALAXY
REIN 4.204	13 00 21.54	+ 10 21 22.4			NEBULA
REIN 4.205	13 00 22.08	+ 10 21 46.3			NEBULA
BFG 098	13 00 22.97	+ 38 02 55.4		18.82	ULTRAVIOLET-EXCESS OBJECT
ZWG 043.111	13 00 24.	+ 05 47		15.3	GALAXY
MCG+01-33-038	13 00 24.	+ 05 47		15.3	GALAXY
ZWG 043.112	13 00 24.	+ 07 57		15.6	GALAXY
REIZ 3168	13 00 24.	+ 10 22	18	15.2	GALAXY
ZWG 160.115	13 00 24.	+ 28 07		15.4	GALAXY
REIZ 3171	13 00 24.	+ 28 19	24	14.6	GALAXY
MCG+05-31-111	13 00 24.	+ 28 20	48	14.9	GALAXY
ZC 1300.4+3213	13 00 24.	+ 32 13	1610		CLUSTER OF GALAXIES
IC 4112	13 00 24.	+ 37 28 42.			NONSTELLAR OBJECT
IC 4114	13 00 24.	+ 40 22 18.			NONSTELLAR OBJECT
MCG+09-21-089	13 00 24.	+ 50 42	72	13.	GALAXY
ZWG 270.040	13 00 24.	+ 50 43		14.7	GALAXY
RNGC 4932	13 00 24.	+ 50 43		14.5	GALAXY
UGC 08150	13 00 24.	+ 50 43	108	14.7	GALAXY Sc
LB 02529	13 00 24.	+ 52 23 06.		16.0	FAINT BLUE STAR
ZC 1300.4+5805	13 00 24.	+ 58 05	2150		CLUSTER OF GALAXIES
REIZ 3166	13 00 24.	- 07 49	60	12.5	GALAXY
MCG-01-33-075	13 00 24.	- 07 49	54	13.5	GALAXY
A2 368	13 00 24.0	+ 30 29 32.		16.7	FAINT BLUE OBJECT
A2 369	13 00 25.8	+ 27 15 29.		18.4	FAINT BLUE OBJECT
IC 4115	13 00 28.	+ 37 29 30.			NONSTELLAR OBJECT
REIZ 3167	13 00 29.	- 05 43	12	15.5	GALAXY
BFG 099	13 00 29.41	+ 38 08 02.9		16.94	ULTRAVIOLET-EXCESS OBJECT
ZWG 043.113	13 00 30.	+ 08 07		15.7	GALAXY
ZWG 071.101	13 00 30.	+ 10 46		15.6	GALAXY
SN 1962I	13 00 30.	+ 27 47		17.2	SUPERNOVA
MCG+05-31-112	13 00 30.	+ 27 47	18	18.	GALAXY
ZWG 160.116	13 00 30.	+ 28 20		15.7	GALAXY
REIZ 3173	13 00 30.	+ 28 21	36	14.9	GALAXY
MCG+06-29-020	13 00 30.	+ 37 30 30.	24	14.5	GALAXY
IC 4117	13 00 30.	+ 40 47 36.			NONSTELLAR OBJECT
ZWG 270.041	13 00 30.	+ 52 03		15.6	GALAXY
UGC 08151	13 00 30.	+ 52 03	66	15.6	GALAXY Sc
UGC 08152	13 00 30.	+ 67 22	72	16.0	GALAXY Sc
7ZW 496	13 00 30.	+ 80 17			COMPACT GALAXY
MCG+13-09-043	13 00 30.	+ 80 19	24	15.	GALAXY
MCG-03-33-029	13 00 30.	- 17 05	36	17.	GALAXY
MCG-05-31-024	13 00 30.	- 30 15	66	15.	GALAXY
A2 370	13 00 30.8	+ 30 05 22.		18.5	FAINT BLUE OBJECT
IC 4109	13 00 31.	+ 19 16 12.			NONSTELLAR OBJECT
IC 4110	13 00 31.	+ 19 29 42.			NONSTELLAR OBJECT
HELW 213	13 00 31.	- 05 42 39.			NEBULA
IC 4111	13 00 32.5	+ 28 20 15.		15.7	GALAXY E5
REIZ 3172	13 00 34.	+ 19 17	60	14.9	GALAXY
IC 4118	13 00 34.	+ 38 33 42.			NONSTELLAR OBJECT
ARC 1666	13 00 34.	+ 52 10		16.8	RICH CLUSTER OF GALAXIES
IC 0844	13 00 34.0	- 30 15 10.	102		GALAXY S0
A2 371	13 00 35.7	+ 27 40 04.		18.3	FAINT BLUE OBJECT
ZWG 043.114	13 00 36.	+ 04 16		15.2	GALAXY
UGC 08153	13 00 36.	+ 04 16	114	15.2	GALAXY Sc
MCG+01-33-039	13 00 36.	+ 04 16	96	15.2	GALAXY
ZWG 043.115	13 00 36.	+ 07 49		15.5	GALAXY
8ZW 1300+15.7	13 00 36.	+ 15 41		18.2	COMPACT GALAXY
8ZW 1300+25.7	13 00 36.	+ 25 44		16.2	COMPACT GALAXY
ZWG 160.117	13 00 36.	+ 26 47		15.7	GALAXY
ZWG 160.118	13 00 36.	+ 28 17		14.4	GALAXY
UGC 08154	13 00 36.	+ 28 17	96	14.4	GALAXY S0
MCG+05-31-113	13 00 36.	+ 28 20	24	15.7	GALAXY
ZC 1300.6+5355	13 00 36.	+ 53 55	1080		CLUSTER OF GALAXIES
MCG-03-33-030	13 00 36.	- 17 08	240	14.	GALAXY
MCG-05-31-025	13 00 36.	- 29 33	150	15.	GALAXY
MCG-05-31-026	13 00 36.	- 31 58	72	15.	GALAXY
IC 4113	13 00 37.	+ 20 44 30.			NONSTELLAR OBJECT
REIZ 3175	13 00 39.	+ 20 45	36	15.3	GALAXY
REIZ 3174	13 00 40.	+ 19 22	30	15.0	GALAXY
BB 2.06	13 00 40.85	+ 36 07 18.			GALAXY NEAR QSO B234
IC 4120	13 00 41.	+ 37 21 07.			NONSTELLAR OBJECT
ZWG 015.059	13 00 42.	+ 01 43		15.5	GALAXY
ZWG 043.116	13 00 42.	+ 08 04		14.9	GALAXY
UGC 08155	13 00 42.	+ 08 04	192	14.9	GALAXY S
MCG+01-33-040	13 00 42.	+ 08 04	180	14.9	GALAXY
REIZ 3176	13 00 42.	+ 28 18	30	13.8	GALAXY
MCG+05-31-114	13 00 42.	+ 28 18	78	14.4	GALAXY
WEI 62579	13 00 42.	+ 28 26		17.51	FAINT BLUE OBJECT
MCG+08-24-027	13 00 42.	+ 47 45	60	16.	GALAXY
MCG+13-09-044	13 00 42.	+ 75 40	45	15.	GALAXY
MCG-05-31-027	13 00 42.	- 32 37	48	16.	GALAXY
SHB 239	13 00 42.5	+ 36 07 33.		17.5	QUASI-STELLAR OBJECT
BFG 100	13 00 42.51	+ 36 07 33.5		17.70	ULTRAVIOLET-EXCESS OBJECT
IC 4116	13 00 43.	+ 19 21 00.			NONSTELLAR OBJECT
BC E234	13 00 43.2	+ 36 07 48.		17.52	QUASI-STELLAR OBJECT

OBJECT NAME	RIGHT ASCEN.	DECLINATION	DIAM.	MAGN.	TYPE OF OBJECT
A2 374	13 00 44.9	+ 27 10 53.		17.6	FAINT BLUE OBJECT
IC 4123	13 00 46.	+ 38 35 01.			NONSTELLAR OBJECT
ARC 1665	13 00 47.	+ 26 57		17.7	RICH CLUSTER OF GALAXIES
RNGC 4954	13 00 47.	+ 75 40		14.0	GALAXY
BB 2.01	13 00 47.0	+ 36 16 29.5			GALAXY NEAR QSO B234
ZWG 043.117	13 00 48.	+ 04 52		15.4	GALAXY
UGC 08156	13 00 48.	+ 04 52	72	15.4	GALAXY Sb-c
MCG+01-33-042	13 00 48.	+ 04 52	30	15.4	GALAXY
MCG+01-33-041	13 00 48.	+ 04 52	36	15.4	GALAXY
ZWG 043.118	13 00 48.	+ 07 17		15.7	GALAXY
TON-N 0694	13 00 48.	+ 28 27		16.7	BLUE STAR
ZWG 160.119	13 00 48.	+ 28 50		15.2	GALAXY
TON-N 0141	13 00 48.	+ 28 58		14.9	BLUE STAR
ZWG 270.042	13 00 48.	+ 51 36		15.3	GALAXY
RNGC 4938	13 00 48.	+ 51 36		15.5	GALAXY
ZWG 353.008	13 00 48.	+ 75 40		14.2	GALAXY
ZWG 352.053	13 00 48.	+ 75 40		14.2	GALAXY
UGC 08157	13 00 48.	+ 75 40	66	14.2	GALAXY S0?
UGC 08158	13 00 48.	+ 80 13	84	16.0	GALAXY
IC 4119	13 00 49.	+ 19 30 13.			NONSTELLAR OBJECT
LB 00672	13 00 50.	+ 72 14 30.		17.6	FAINT BLUE STAR
MAI 081	13 00 50.	+ 89 05	53		DWARF SPHEROIDAL GALAXY
BFG 101	13 00 50.49	+ 32 35 16.3		19.07	ULTRAVIOLET-EXCESS OBJECT
RNGC 4935	13 00 51.	+ 14 39		14.0	GALAXY
MCG+06-29-021	13 00 51.	+ 32 58 30.	48	15.	GALAXY
BFG 102	13 00 51.62	+ 35 09 56.6		18.93	ULTRAVIOLET-EXCESS OBJECT
REIZ 3182	13 00 52.	+ 32 59	12	15.7	GALAXY
TON-N 1546	13 00 52.	+ 34 22		17.0	BLUE STAR
REIZ 3177	13 00 53.	+ 14 40	48	13.8	GALAXY
ZWG 100.023	13 00 54.	+ 14 39		13.9	GALAXY
UGC 08159	13 00 54.	+ 14 39	72	13.9	GALAXY SBb
MCG+03-33-023	13 00 54.	+ 14 39	72	13.9	GALAXY
ZWG 100.024	13 00 54.	+ 20 27		15.5	GALAXY
TON-N 0142	13 00 54.	+ 25 26		15.4	BLUE STAR
ZWG 160.120	13 00 54.	+ 28 17		15.0	GALAXY
RNGC 4934	13 00 54.	+ 28 17		15.5	GALAXY
UGC 08160	13 00 54.	+ 28 17	60	15.0	GALAXY S
REIZ 3181	13 00 54.	+ 28 18	42	13.9	GALAXY
ZC 1300.9+4216	13 00 54.	+ 42 16	870		CLUSTER OF GALAXIES
ZC 1300.9+5929	13 00 54.	+ 59 29	1010		CLUSTER OF GALAXIES
ZC 1300.9+6050	13 00 54.	+ 60 50	400		CLUSTER OF GALAXIES
ZC 1300.9-0244	13 00 54.	- 02 44	3760		CLUSTER OF GALAXIES
BB 2.08	13 00 54.5	+ 35 57 54.			GALAXY NEAR QSO B234
IC 4121	13 00 55.	+ 19 32 55.			NONSTELLAR OBJECT
IC 4122	13 00 55.	+ 20 27 55.			NONSTELLAR OBJECT
ARC 1667	13 00 56.	+ 32 05		17.6	RICH CLUSTER OF GALAXIES
BFG 103	13 00 56.16	+ 37 20 12.7		18.53	ULTRAVIOLET-EXCESS OBJECT
IC 4127	13 00 58.	+ 38 18 55.			NONSTELLAR OBJECT
A2 378	13 00 59.5	+ 27 24 01.		17.7	FAINT BLUE OBJECT
ZWG 043.119	13 01 00.	+ 08 14		15.6	GALAXY
ZC 1301.0+2312	13 01 00.	+ 23 12	2490		CLUSTER OF GALAXIES
ZWG 160.121	13 01 00.	+ 26 49		15.5	GALAXY
UGC 08161	13 01 00.	+ 26 49	66	15.5	GALAXY S
MCG+05-31-115	13 01 00.	+ 28 18	60	15.0	GALAXY
REIZ 3187	13 01 00.	+ 28 35	24	15.1	GALAXY
MCG+06-29-022	13 01 00.	+ 38 20	24	15.	GALAXY
ZWG 245.013	13 01 00.	+ 47 40		15.2	GALAXY
MCG+08-24-028	13 01 00.	+ 47 40	36	15.	GALAXY
MCG+08-24-029	13 01 00.	+ 49 04	30	15.	GALAXY
MCG+10-19-016	13 01 00.	+ 60 31	24	16.	GALAXY
ZWG 353.009	13 01 00.	+ 80 17		15.3	GALAXY
ZWG 352.054	13 01 00.	+ 80 17		15.3	GALAXY
ARC 1664	13 01 00.	- 23 58		17.3	RICH CLUSTER OF GALAXIES
BB 2.04	13 01 01.45	+ 36 12 40.5			GALAXY NEAR QSO B234
REIZ 3186	13 01 02.	+ 23 05	30	15.1	GALAXY
LS 02530	13 01 02.	+ 53 23 12.		14.7	FAINT BLUE STAR
REIZ 3185	13 01 03.	+ 20 28	12	15.0	GALAXY
LB 02531	13 01 03.	+ 55 40 06.		15.8	FAINT BLUE STAR
IC 4124	13 01 05.	+ 23 06 55.			NONSTELLAR OBJECT
BB 2.03	13 01 05.9	+ 36 23 54.5			GALAXY NEAR QSO B234
ZWG 043.120	13 01 06.	+ 02 50		15.4	GALAXY
ZWG 043.121	13 01 06.	+ 03 27		15.5	GALAXY
KARA.73B 0565	13 01 06.	+ 03 27	30	15.5	ISOLATED GALAXY S
ZWG 043.122	13 01 06.	+ 08 10		15.4	GALAXY
MCG+05-31-116	13 01 06.	+ 26 49	66	15.5	GALAXY
IC 4131	13 01 06.	+ 39 13 01.			NONSTELLAR OBJECT
ZC 1301.1+4105	13 01 06.	+ 41 05	740		CLUSTER OF GALAXIES
MCG+08-24-030	13 01 06.	+ 48 35 30.	48	16.	GALAXY
ZWG 270.043	13 01 06.	+ 50 54		15.6	GALAXY
UGC 08162	13 01 06.	+ 50 54	72	15.6	GALAXY Sc
REIZ 3180	13 01 07.	- 11 14	36	13.6	GALAXY
REIZ 3179	13 01 07.	- 11 15	24	14.3	GALAXY
REIZ 3178	13 01 07.	- 11 21	36	15.8	GALAXY
A2 380	13 01 07.5	+ 27 31 48.		18.0	FAINT BLUE OBJECT
IC 4125	13 01 08.	+ 19 04 13.			NONSTELLAR OBJECT
IC 4126	13 01 08.	+ 19 35 19.			NONSTELLAR OBJECT
BFG 104	13 01 08.25	+ 32 23 33.3		18.80	ULTRAVIOLET-EXCESS OBJECT
REIZ 3184	13 01 09.	+ 02 49	18	14.4	GALAXY
MCG+03-33-024	13 01 09.	+ 19 33	48	17.5	GALAXY
ARP 176	13 01 09.	- 11 14			PECULIAR GALAXY
A2 381	13 01 09.4	+ 28 46 47.		17.5	FAINT BLUE OBJECT
BB 2.02	13 01 09.75	+ 36 17 40.			GALAXY NEAR QSO B234
LB 02533	13 01 11.	+ 52 36 18.		17.0	FAINT BLUE STAR
LB 02532	13 01 11.	+ 53 12 24.		17.3	FAINT BLUE STAR
HELW 059	13 01 11.	- 04 52			NEBULA
REIZ 3183	13 01 11.	- 04 53	48	13.6	GALAXY
RNGC 4933B	13 01 11.	- 11 14		15.0	GALAXY
RNGC 4933A	13 01 11.	- 11 14		13.0	GALAXY
REIN 4.206	13 01 11.29	- 04 51 55.2			NEBULA
ZWG 043.123	13 01 12.	+ 07 47		15.4	GALAXY
ZWG 100.025	13 01 12.	+ 15 11		15.6	GALAXY
3ZW 069	13 01 12.	+ 27 27			COMPACT GALAXY
ZC 1301.2+5954	13 01 12.	+ 59 54	1750		CLUSTER OF GALAXIES
MCG+13-09-045	13 01 12.	+ 78 48 30.	72	15.	GALAXY
MCG-01-33-076	13 01 12.	- 04 51	90	15.	GALAXY
MCG-06-29-005	13 01 12.	- 37 56	42	15.	GALAXY
A2 384	13 01 12.6	+ 29 34 11.		16.6	FAINT BLUE OBJECT
BB 2.07	13 01 13.35	+ 36 02 26.5			GALAXY NEAR QSO B234
IC 4128	13 01 15.	+ 20 29 01.			NONSTELLAR OBJECT
REIZ 3189	13 01 15.	+ 20 30	18	14.9	GALAXY
IC 4132	13 01 15.	+ 38 38 50.			NONSTELLAR OBJECT
IC 4129	13 01 17.	+ 19 08 43.			NONSTELLAR OBJECT
REIZ 3188	13 01 17.	- 03 46	30	15.9	GALAXY
HOLM 502B	13 01 17.	- 11 15	24	14.1	PART OF MULTIPLE GALAXY
RNGC 4930	13 01 17.	- 41 09			GALAXY
ZWG 043.124	13 01 18.	+ 08 28		15.7	GALAXY
ZC 1301.3+2516	13 01 18.	+ 25 16	870		CLUSTER OF GALAXIES
ZWG 130.004	13 01 18.	+ 26 21		15.7	GALAXY

OBJECT NAME	RIGHT ASCEN.	DECLINATION	DIAM.	MAGN.	TYPE OF OBJECT
MCG+04-31-003	13 01 18.	+ 26 22	42	15.7	GALAXY
IC 4135	13 01 18.	+ 40 30 56.			NONSTELLAR OBJECT
UGC 08163	13 01 18.	+ 40 31	60	16.5	GALAXY S
ZWG 270.044	13 01 18.	+ 53 48		15.6	GALAXY
LB C0673	13 01 18.	+ 59 50 00.		16.9	FAINT BLUE STAR
ZWG 353.010	13 01 18.	+ 78 48		15.2	GALAXY
ZWG 352.055	13 01 18.	+ 78 48		15.2	GALAXY
UGC 08164	13 01 18.	+ 78 48	120	15.2	GALAXY Sb
MCG-02-33-102	13 01 18.	- 11 14	132	13.	GALAXY
MCG-02-33-101	13 01 18.	- 11 14 30.	12	15.	GALAXY
A2 387	13 01 18.9	+ 27 46 35.		18.3	FAINT BLUE OBJECT
IC 4130	13 01 19.	+ 19 32 20.			NONSTELLAR OBJECT
HOLM 503A	13 01 19.	+ 26 22	42	14.3	PART OF MULTIPLE GALAXY
HOLM 502A	13 01 19.	- 11 14	42	13.4	PART OF MULTIPLE GALAXY
A2 389	13 01 19.5	+ 29 53 43.		17.6	FAINT BLUE OBJECT
HELW 027	13 01 20.	- 30 06			NEBULA
MCG-02-33-103	13 01 21.	- 11 14	24	17.	GALAXY
MCG-07-27-029	13 01 21.	- 41 08	180	13.	GALAXY
A2 390	13 01 21.7	+ 26 49 21.		17.9	FAINT BLUE OBJECT
ABC 1668	13 01 22.	+ 19 32		16.6	RICH CLUSTER OF GALAXIES
REIZ 3200	13 01 22.	+ 53 48	108	14.2	GALAXY
RNGC 4933C	13 01 23.	- 11 14		17.0	GALAXY
A2 391	13 01 23.0	+ 27 02 44.		15.3	FAINT BLUE OBJECT
ZWG 043.125	13 01 24.	+ 05 58		15.5	GALAXY
UGC 08165	13 01 24.	+ 05 58	66	15.5	GALAXY S
ZWG 071.102	13 01 24.	+ 11 15		15.7	GALAXY
UGC 08166	13 01 24.	+ 11 15	90	15.7	GALAXY Sc
KARA.73B 0566	13 01 24.	+ 11 15	96	15.7	ISOLATED GALAXY S
MCG+03-33-025	13 01 24.	+ 16 27	36	15.4	GALAXY
HOLM 503B	13 01 24.	+ 26 24	66	15.1	PART OF MULTIPLE GALAXY
TON-N 0143	13 01 24.	+ 27 03		15.1	BLUE STAR
ZWG 160.122	13 01 24.	+ 28 21		15.6	GALAXY
RNGC 4943	13 01 24.	+ 28 21		15.5	GALAXY
RNGC 4944	13 01 24.	+ 28 27		13.5	GALAXY
IC 4133	13 01 26.8	+ 28 15 21.		15.4	GALAXY E0
TON-N 1547	13 01 27.	+ 21 37		17.0	BLUE STAR
IC 4143	13 01 27.	+ 40 28 32.			NONSTELLAR OBJECT
LB 00248	13 01 27.	+ 54 29 06.		15.3	FAINT BLUE STAR
LB 02534	13 01 27.	+ 58 52 48.		16.8	FAINT BLUE STAR
REIZ 3192	13 01 28.	+ 16 27	24	15.5	GALAXY
IC 4142	13 01 28.	+ 38 27 44.			NONSTELLAR OBJECT
LB 00674	13 01 28.	+ 57 00 00.		16.3	FAINT BLUE STAR
REIZ 3190	13 01 29.	- 04 38	18	13.6	GALAXY
ZWG 043.126	13 01 30.	+ 05 08		15.7	GALAXY
ZC 1301.5+1306	13 01 30.	+ 13 06	400		CLUSTER OF GALAXIES
ZWG 100.026	13 01 30.	+ 15 10		15.3	GALAXY
MCG+03-33-026	13 01 30.	+ 15 10	24	15.3	GALAXY
ZWG 100.027	13 01 30.	+ 16 27		15.4	GALAXY
KARA.73B 0567	13 01 30.	+ 16 27	42	15.4	ISOLATED GALAXY S
MCG+03-33-027	13 01 30.	+ 19 30	36	17.5	GALAXY
ZWG 160.123	13 01 30.	+ 28 15		15.4	GALAXY
MCG+05-31-117	13 01 30.	+ 28 15	15	15.4	GALAXY
MCG+05-31-118	13 01 30.	+ 28 27	84	13.3	GALAXY
ZWG 160.124	13 01 30.	+ 28 28		13.3	GALAXY
REIZ 3197	13 01 30.	+ 28 28	54	13.4	GALAXY
UGC 08167	13 01 30.	+ 28 28	96	13.3	GALAXY Sa-b
WEI 34497	13 01 30.	+ 29 53		17.97	FAINT BLUE OBJECT
IC 4144	13 01 30.	+ 37 12 44.			NONSTELLAR OBJECT
IC 4145	13 01 30.	+ 38 33 20.			NONSTELLAR OBJECT
ZC 1301.5+5135	13 01 30.	+ 51 35	2220		CLUSTER OF GALAXIES
1ZW 050	13 01 30.	+ 51 46			COMPACT GALAXY
UGC 08168	13 01 30.	+ 51 48	66	16.0	GALAXY PECULR
MCG-05-31-028	13 01 30.	- 30 16	30	12.	GALAXY
DV.56 N4947A	13 01 30.	- 34 58	72		SBd GALAXY
MCG-06-29-005A	13 01 30.	- 34 58	60	16.	GALAXY
RNGC 4947A	13 01 31.	- 34 58			GALAXY
REIZ 3196	13 01 32.	+ 22 59	12	15.0	GALAXY
SN 1973F	13 01 32.	+ 28 28		17.0	SUPERNOVA
REIZ 3199	13 01 32.	+ 38 36	42	15.4	GALAXY
RNGC 4936	13 01 32.	- 30 15		13.0	GALAXY
MCG+06-29-023	13 01 33.	+ 37 14	66	15.	GALAXY
MCG-02-33-104	13 01 33.	- 10 04	318	11.5	GALAXY
BPG 105	13 01 33.09	+ 34 40 18.0		18.77	ULTRAVIOLET-EXCESS OBJECT
IC 4137	13 01 34.	+ 23 00 26.			NONSTELLAR OBJECT
SHB 240	13 01 34.6	+ 37 30 06.		17.3	QUASI-STELLAR OBJECT
BPG 106	13 01 34.60	+ 37 30 06.9		17.25	ULTRAVIOLET-EXCESS OBJECT
BC B272	13 01 35.1	+ 37 30 48.3		17.25	QUASI-STELLAR OBJECT
ZWG 043.127	13 01 36.	+ 08 11		15.2	GALAXY
MCG+01-33-043	13 01 36.	+ 08 11	36	15.2	GALAXY
ZC 1301.6+1918	13 01 36.	+ 19 18	810		CLUSTER OF GALAXIES
IC 4138	13 01 36.	+ 20 55 56.			NONSTELLAR OBJECT
MCG+04-31-004	13 01 36.	+ 22 32	24	15.5	GALAXY
RNGC 4931	13 01 36.	+ 28 18		14.5	GALAXY
WEI 33211	13 01 36.	+ 30 43		17.56	FAINT BLUE OBJECT
UGC 08169	13 01 36.	+ 37 13	66	16.0	GALAXY Sb
ZWG 245.014	13 01 36.	+ 47 39		15.3	GALAXY
MCG-01-33-077	13 01 36.	- 05 15	150	12.	GALAXY
RNGC 4941	13 01 36.	- 05 17		12.5	GALAXY
MCG-02-33-105	13 01 36.	- 11 14 30.	30	16.	GALAXY
REIN 4.208	13 01 36.91	- 04 37 23.0			NEBULA
REIZ 3195	13 01 37.	+ 09 15	24	14.7	GALAXY
IC 4139	13 01 37.	+ 19 33 44.			NONSTELLAR OBJECT
REIZ 3191	13 01 37.	- 10 05	210	12.4	GALAXY
BPG 107	13 01 37.00	+ 36 46 46.7		18.85	ULTRAVIOLET-EXCESS OBJECT
BPG 108	13 01 37.00	+ 36 52 12.6		18.94	ULTRAVIOLET-EXCESS OBJECT
REIN 4.209	13 01 37.84	- 05 17 00.6			NEBULA
LB 02535	13 01 38.	+ 51 30 18.		17.2	FAINT BLUE STAR
BIGO 522	13 01 38.	+ 56 28			NEBULA
HELW 028	13 01 38.	- 29 54			NEBULA
A2 399	13 01 38.1	+ 28 49 13.		16.1	FAINT BLUE OBJECT
A2 400	13 01 38.5	+ 28 47 45.		17.0	FAINT BLUE OBJECT
IC 4140	13 01 39.	+ 20 21 50.			NONSTELLAR OBJECT
IC 4151	13 01 39.	+ 37 07 32.			NONSTELLAR OBJECT
IC 4141	13 01 40.	+ 19 28 38.			NONSTELLAR OBJECT
IC 4152	13 01 40.	+ 38 28 02.			NONSTELLAR OBJECT
SN 1968X	13 01 40.	- 10 05		16.0	SUPERNOVA
REIZ 3193	13 01 41.	- 05 17	120	12.7	GALAXY
BPG 109	13 01 41.98	+ 35 49 21.5		18.65	ULTRAVIOLET-EXCESS OBJECT
ZWG 043.128	13 01 42.	+ 08 10		15.5	GALAXY
ZC 1301.7+2037	13 01 42.	+ 20 37	1080		CLUSTER OF GALAXIES
ZWG 130.005	13 01 42.	+ 22 34		15.5	GALAXY
8ZW 1301+22.6	13 01 42.	+ 22 34		15.5	COMPACT GALAXY
REIZ 3201	13 01 42.	+ 28 35	24	15.4	GALAXY
SHB 241	13 01 42.	+ 35 49 22.		18.7	QUASI-STELLAR OBJECT
ZC 1301.7+4605	13 01 42.	+ 46 05	1480		CLUSTER OF GALAXIES
MCG+08-24-031	13 01 42.	+ 47 40	42	15.	GALAXY
7ZW 497	13 01 42.	+ 65 17			COMPACT GALAXY
RNGC 4942	13 01 42.	- 07 24		14.0	GALAXY
REIZ 3194	13 01 42.	- 07 24	60	13.4	GALAXY
RNGC 4939	13 01 42.	- 10 05		11.5	GALAXY
VVI 63	13 01 42.	- 10 05	320	12.2	SEYFERT GALAXY
MCG-05-31-029	13 01 42.	- 29 54	24	15.	GALAXY
ASS 64	13 01 42.	- 61 48	21600		OB ASSOCIATION CEN OB1
REIZ 3198	13 01 43.	+ 09 30	48	13.8	GALAXY
IC 4146	13 01 43.	+ 19 32 44.			NONSTELLAR OBJECT
IC 4147	13 01 43.	+ 20 31 14.			NONSTELLAR OBJECT
ABC 1674	13 01 43.	+ 67 46		17.2	RICH CLUSTER OF GALAXIES
SN 1973J	13 01 43.	- 10 05		16.	SUPERNOVA
SVEN 362	13 01 43.	- 28 07 38.	42	15.8	GALAXY
BC B286	13 01 43.0	+ 35 49 32.		18.65	QUASI-STELLAR OBJECT
IC 4148	13 01 44.	+ 19 31 32.			NONSTELLAR OBJECT
REIN 4.210	13 01 44.26	+ 09 29 28.4			NEBULA
IC 4149	13 01 45.	+ 22 33 32.			NONSTELLAR OBJECT
MCG-01-33-078	13 01 45.	- 07 22	78	14.	GALAXY
ABC 1669	13 01 46.	+ 19 21		17.8	RICH CLUSTER OF GALAXIES
IC 4134	13 01 46.	- 11 11			NONSTELLAR OBJECT
IC 4150	13 01 47.	+ 22 15 15.			NONSTELLAR OBJECT
LB 02536	13 01 47.	+ 51 21 06.		17.2	FAINT BLUE STAR
HN 0368	13 01 47.	- 05 42			NEBULA
IC 4136	13 01 47.	- 05 42			MAY NOT EXIST
ZWG 071.103	13 01 48.	+ 09 16		15.7	GALAXY
ZWG 071.104	13 01 48.	+ 09 29		14.8	GALAXY
UGC 08170	13 01 48.	+ 09 29	72	14.8	GALAXY Sc
MCG+02-33-052	13 01 48.	+ 09 29	66	14.8	GALAXY
KARA.73B 0568	13 01 48.	+ 09 29	60	14.8	ISOLATED GALAXY S
ZWG 160.125	13 01 48.	+ 28 31		15.4	GALAXY
RNGC 4949	13 01 48.	+ 29 18			GALAXY
MCG+08-24-032	13 01 48.	+ 49 44	60	16.	GALAXY
ZC 1301.8+6053	13 01 48.	+ 60 53	2020		CLUSTER OF GALAXIES
HOLM 504A	13 01 49.	+ 26 24	18	15.5	PART OF MULTIPLE GALAXY
REIZ 3202	13 01 50.	+ 22 32	30	14.2	GALAXY
REIZ 3203	13 01 50.	+ 22 35	18	15.7	GALAXY
REIZ 3204	13 01 50.	+ 22 36	36	15.2	GALAXY
HELW 029	13 01 50.	- 29 54			NEBULA
IC 4155	13 01 51.	+ 40 16 57.			NONSTELLAR OBJECT
BPG 110	13 01 51.54	+ 37 49 50.5		19.19	ULTRAVIOLET-EXCESS OBJECT
ABC 1670	13 01 52.	+ 20 36		17.6	RICH CLUSTER OF GALAXIES
HOLM 504B	13 01 52.	+ 26 24	30	15.7	PART OF MULTIPLE GALAXY
RNGC 4937	13 01 52.	- 46 57			NON-EXISTENT OBJECT
A2 404	13 01 52.0	+ 27 15 03.		18.3	FAINT BLUE OBJECT
BPG 111	13 01 53.01	+ 37 34 50.8		18.98	ULTRAVIOLET-EXCESS OBJECT
A2 405	13 01 53.3	+ 27 19 14.		17.0	FAINT BLUE OBJECT
ZWG 043.129	13 01 54.	+ 08 11		15.6	GALAXY
MCG+01-33-044	13 01 54.	+ 08 11	9	15.6	GALAXY
ZC 1301.9+1931	13 01 54.	+ 19 31	3230		CLUSTER OF GALAXIES
MCG+06-29-024	13 01 54.	+ 35 40 30.	30	16.	GALAXY
ZWG 189.016	13 01 54.	+ 35 41		15.7	GALAXY
ZC 1301.9+5001	13 01 54.	+ 50 01	7800		CLUSTER OF GALAXIES
ZC 1301.9+6745	13 01 54.	+ 67 45	1950		CLUSTER OF GALAXIES
ZWG 015.060	13 01 54.	- 03 18		14.3	GALAXY
MCG+00-33-028	13 01 54.	- 03 18	198	14.3	GALAXY
KARA.73B 0569	13 01 54.	- 03 18	204	14.3	ISOLATED GALAXY S
LB 00675	13 01 55.	+ 54 19 30.		17.2	FAINT BLUE STAR
IC 4157	13 01 59.	+ 38 55 57.			NONSTELLAR OBJECT
VDB.66G 163	13 02	- 07 41	70		DWARF GALAXY
VDB.66G 162	13 02	- 07 56	70		DWARF GALAXY
ZWG 043.130	13 02 00.	+ 08 04		15.7	GALAXY
ZWG 043.131	13 02 00.	+ 08 15		15.6	GALAXY
ZC 1302.0+0852	13 02 00.	+ 08 52	1140		CLUSTER OF GALAXIES
ZC 1302.0+1644	13 02 00.	+ 16 44	1010		CLUSTER OF GALAXIES
IC 4153	13 02 00.	+ 19 19 03.			NONSTELLAR OBJECT
ZWG 160.126	13 02 00.	+ 26 56		15.3	GALAXY
ZC 1302.0+2731	13 02 00.	+ 27 31	470		CLUSTER OF GALAXIES
ZWG 160.127	13 02 00.	+ 27 34		15.5	GALAXY
ZWG 160.128	13 02 00.	+ 29 05		15.3	GALAXY
SHAH 159	13 02 00.	+ 86 38	54		GROUP OF COMPACT GALAXIES
MCG-05-31-030	13 02 00.	- 29 58	30	15.	GALAXY
BPG 113	13 02 00.85	+ 36 13 17.7		16.55	ULTRAVIOLET-EXCESS OBJECT
BPG 112	13 02 01.07	+ 37 56 11.7		19.00	ULTRAVIOLET-EXCESS OBJECT
BB 2.10	13 02 01.8	+ 36 31 47.5			GALAXY NEAR QSO B234
IC 4154	13 02 03.	+ 23 50 33.			NONSTELLAR OBJECT
MCG+05-31-119	13 02 03.	+ 27 34	60	15.5	GALAXY
RNGC 4940	13 02 03.	- 47 00			GALAXY
A2 407	13 02 03.7	+ 28 58 22.		16.6	FAINT BLUE OBJECT
IC 4158	13 02 05.	+ 36 44 57.			NONSTELLAR OBJECT
ZWG 043.132	13 02 06.	+ 03 14		15.4	GALAXY
KARA.73B 0570	13 02 06.	+ 03 14	18	15.4	ISOLATED GALAXY S0
ZWG 043.133	13 02 06.	+ 03 59		15.5	GALAXY
MCG+01-33-046	13 02 06.	+ 03 59	24	15.5	GALAXY
MCG+01-33-045	13 02 06.	+ 03 59	42	15.5	GALAXY
ZWG 043.134	13 02 06.	+ 04 02		15.5	GALAXY
MCG+01-33-047	13 02 06.	+ 04 02	12	15.5	GALAXY
ZWG 071.105	13 02 06.	+ 08 48		15.6	GALAXY
MCG+03-33-028	13 02 06.	+ 18 43	48	15.7	GALAXY
MCG+05-31-120	13 02 06.	+ 26 56	48	15.3	GALAXY
TON-N 0695	13 02 06.	+ 27 16		15.7	BLUE STAR
ZC 1302.1+3350	13 02 06.	+ 33 50	1410		CLUSTER OF GALAXIES
ZC 1302.1+4211	13 02 06.	+ 42 11	610		CLUSTER OF GALAXIES
ZWG 217.008	13 02 06.	+ 44 05		15.5	GALAXY
MCG+08-24-033	13 02 06.	+ 45 19	36	15.	GALAXY
ZWG 245.015	13 02 06.	+ 45 20		15.6	GALAXY
ZWG 245.016	13 02 06.	+ 47 52		15.4	GALAXY
REIZ 3206	13 02 08.	+ 23 50	24	14.7	GALAXY
HELW 030	13 02 08.	- 29 58			NEBULA
A2 409	13 02 09.2	+ 28 25 26.		17.3	FAINT BLUE OBJECT
BPG 114	13 02 11.01	+ 32 22 39.7		18.51	ULTRAVIOLET-EXCESS OBJECT
ZWG 043.135	13 02 12.	+ 04 10		15.6	GALAXY
MCG+01-33-048	13 02 12.	+ 04 10	42	15.6	GALAXY
ZWG 043.136	13 02 12.	+ 08 14		15.6	GALAXY
ZWG 100.028	13 02 12.	+ 19 42		15.7	GALAXY
UGC 08171	13 02 12.	+ 18 42	60	15.7	GALAXY Sc
MCG+08-24-034	13 02 12.	+ 47 52 30.	36	15.	GALAXY
MCG+10-19-017	13 02 12.	+ 61 16	27	16.	GALAXY
UGC 08172	13 02 12.	+ 61 28	60	16.0	GALAXY SBb
ZC 1302.2+6243	13 02 12.	+ 62 43	9410		CLUSTER OF GALAXIES
A2 410	13 02 12.0	+ 28 53 41.		15.5	FAINT BLUE GALAXY
A2 411	13 02 12.1	+ 28 19 03.		17.2	FAINT BLUE OBJECT
ABC 1673	13 02 16.	+ 51 47		16.9	RICH CLUSTER OF GALAXIES
IC 4161	13 02 17.	+ 40 14 46.			NONSTELLAR OBJECT
BPG 115	13 02 17.07	+ 35 45 11.4		18.39	ULTRAVIOLET-EXCESS OBJECT
SHB 242	13 02 17.1	+ 35 45 11.		18.4	QUASI-STELLAR OBJECT
LB 00065	13 02 18.	+ 25 56		17.8	FAINT BLUE STAR
LB 00028	13 02 18.	+ 31 45		16.2	FAINT BLUE STAR
ZC 1302.3+5201	13 02 18.	+ 52 01	400		CLUSTER OF GALAXIES
ZWG 271.001	13 02 18.	+ 55 37		15.4	GALAXY
ZWG 270.045	13 02 18.	+ 55 37		15.4	GALAXY

OBJECT NAME	RIGHT ASCEN.	DECLINATION	DIAM.	MAGN.	TYPE OF OBJECT
LB 02537	13 02 18.	+ 59 20 00.		15.4	FAINT BLUE STAR
MCG-01-33-079	13 02 18.	- 07 40	78	14.	GALAXY
RNGC 4948	13 02 18.	- 07 41		14.0	GALAXY
REIZ 3205	13 02 18.	- 07 41	72	13.6	GALAXY
HOLM 505A	13 02 18.	- 07 41	78	12.9	PART OF MULTIPLE GALAXY
MCG-04-31-024	13 02 18.	- 24 35	30	15.	GALAXY
BC R288	13 02 18.0	+ 35 45 21.5		18.39	QUASI-STELLAR OBJECT
LB 02538	13 02 19.	+ 50 14 48.		15.8	FAINT BLUE STAR
REIZ 3216	13 02 19.	+ 55 54	42	14.3	GALAXY
IC 4159	13 02 20.	+ 22 30 34.			NONSTELLAR OBJECT
REIZ 3209	13 02 20.	+ 23 46	90	15.3	GALAXY
BIGO 523	13 02 20.	+ 55 54			NEBULA
HOLM 505B	13 02 20.	- 07 42	30	14.6	PART OF MULTIPLE GALAXY
A2 414	13 02 20.5	+ 28 18 46.		18.1	FAINT BLUE OBJECT
RNGC 4945	13 02 21.	- 49 13		9.5	GALAXY
IC 4160	13 02 22.	+ 23 09 34.			NONSTELLAR OBJECT
HZ 39	13 02 22.	+ 28 23		14.8	BLUE STAR
BFG 116	13 02 22.86	+ 35 18 59.0		18.83	ULTRAVIOLET-EXCESS OBJECT
HN 0369	13 02 23.	- 06 00			NEBULA
REIN 4.212	13 02 23.84	+ 13 30 17.1			NEBULA
ZWG 043.137	13 02 24.	+ 04 05		15.3	GALAXY
UGC 08173	13 02 24.	+ 04 05	60	15.3	GALAXY S
MCG+01-33-049	13 02 24.	+ 04 05	60	15.3	GALAXY
ZWG 043.138	13 02 24.	+ 05 48		15.4	GALAXY
REIZ 3208	13 02 24.	+ 12 20	18	14.8	GALAXY
ZWG 071.106	13 02 24.	+ 13 30		15.3	GALAXY
MCG+03-33-029	13 02 24.	+ 17 51	30	15.3	GALAXY
ARC 1672	13 02 24.	+ 33 50		17.2	RICH CLUSTER OF GALAXIES
ZWG 271.002	13 02 24.	+ 55 52		15.6	GALAXY
ZWG 270.046	13 02 24.	+ 55 52		15.6	GALAXY
UGC 08174	13 02 24.	+ 55 52	72	15.6	GALAXY E-SO
ZC 1302.4+6235	13 02 24.	+ 62 35	1340		CLUSTER OF GALAXIES
ZC 1302.4+7134	13 02 24.	+ 71 34	1140		CLUSTER OF GALAXIES
IC 4156	13 02 24.	- 06 00			MAY NOT EXIST
8ZW 1302-29.6	13 02 24.	- 29 36		18.0	COMPACT GALAXY
SVEN 363	13 02 25.	- 28 07 37.	66	14.7	GALAXY
LB 02539	13 02 27.	+ 59 42 54.		14.8	FAINT BLUE STAR
MCG-04-31-025	13 02 27.	- 22 08	48	15.	GALAXY
REIN 4.215	13 02 27.83	+ 12 20 46.4			NEBULA
REIN 4.211	13 02 27.88	- 06 13 18.7			NEBULA
HELW 301	13 02 28.	- 28 07 37.			NEBULA
REIZ 3214	13 02 29.	+ 31 08	12	16.0	GALAXY
A2 417	13 02 29.3	+ 27 53 53.		16.9	FAINT BLUE OBJECT
ZWG 071.107	13 02 30.	+ 12 21		15.5	GALAXY
MCG+02-33-053	13 02 30.	+ 12 21	18	15.5	GALAXY
ZWG 100.029	13 02 30.	+ 17 50		15.3	GALAXY
8ZW 1302+24.4	13 02 30.	+ 24 23		15.9	COMPACT GALAXY
ZC 1302.5+7214	13 02 30.	+ 72 14	940		CLUSTER OF GALAXIES
RNGC 4951	13 02 30.	- 06 14		12.5	GALAXY
MCG-01-33-081	13 02 30.	- 06 14	180	13.	GALAXY
RNGC 4948A	13 02 30.	- 07 53		15.0	GALAXY
HOLM 506A	13 02 30.	- 07 53	72	13.2	PART OF MULTIPLE GALAXY
MCG-01-33-080	13 02 30.	- 07 53	66	15.	GALAXY
REIZ 3207	13 02 30.	- 07 54	36	13.7	GALAXY
MCG-06-29-006	13 02 30.	- 35 05	108	12.6	GALAXY
LB 02540	13 02 32.	+ 53 59 18.		16.3	FAINT BLUE STAR
REIN 4.214	13 02 32.18	- 06 13 35.8			NEBULA
HOLM 506B	13 02 34.	- 07 56	102	15.0	PART OF MULTIPLE GALAXY
REIZ 3212	13 02 35.	- 04 35	36	14.9	GALAXY
REIZ 3211	13 02 35.	- 05 46	18	15.0	GALAXY
RNGC 4946	13 02 35.	- 43 20			GALAXY
A2 418	13 02 35.7	+ 27 24 43.		17.3	FAINT BLUE OBJECT
IC 4162	13 02 36.	+ 20 49 16.			NONSTELLAR OBJECT
ZWG 160.129	13 02 36.	+ 29 23		13.6	GALAXY
UGC 08175	13 02 36.	+ 29 23	84	13.6	GALAXY E
RNGC 4952	13 02 36.	+ 29 24		13.5	GALAXY
MCG+05-31-121	13 02 36.	+ 29 24	27	13.6	GALAXY
REIZ 3210	13 02 36.	- 06 14	120	12.8	GALAXY
MCG-01-33-082	13 02 36.	- 07 37	78	16.	GALAXY
MCG-03-33-031	13 02 36.	- 16 36	84	16.	GALAXY
MCG-05-31-031	13 02 36.	- 29 50	24	15.5	GALAXY
REIZ 3213	13 02 37.	+ 08 50	30	14.6	GALAXY
RNGC 4947	13 02 37.	- 35 04		12.5	GALAXY
A2 419	13 02 37.8	+ 29 27 20.		18.3	FAINT BLUE OBJECT
REIZ 3221	13 02 39.	+ 35 27	48	13.3	GALAXY
IC 4165	13 02 39.	+ 40 11 29.			NONSTELLAR OBJECT
MIL 24	13 02 39.	- 62 26 42.	366		SUPERNOVA REMNANT
IC 4163	13 02 41.	+ 21 02 22.			NONSTELLAR OBJECT
REIZ 3220	13 02 41.	+ 30 47	24	15.7	GALAXY
RNGC 4950	13 02 41.	- 43 15			GALAXY
ZWG 100.030	13 02 42.	+ 15 42		15.7	GALAXY
MCG+03-33-030	13 02 42.	+ 17 45	30	17.	GALAXY
ZWG 189.017	13 02 42.	+ 35 27		13.5	GALAXY
UGC 08177	13 02 42.	+ 35 27	102	13.5	GALAXY SO
MCG+06-29-025	13 02 42.	+ 35 27	36	14.	GALAXY
ZWG 217.009	13 02 42.	+ 43 49		15.7	GALAXY
ZWG 245.017	13 02 42.	+ 47 46		15.0	GALAXY
UGC 08176	13 02 42.	- 00 05	60	18.	GALAXY
MCG-05-31-032	13 02 42.	- 28 11	72	16.	GALAXY
BFG 117	13 02 42.64	+ 33 44 45.7		19.20	ULTRAVIOLET-EXCESS OBJECT
REIZ 3215	13 02 43.	+ 08 51	18	15.0	GALAXY
REIZ 3218	13 02 44.	+ 23 44	24	15.7	GALAXY
REIZ 3219	13 02 44.	+ 23 46	30	15.2	GALAXY
RNGC 4956	13 02 44.	+ 35 27		13.5	GALAXY
PK304+05.1	13 02 44.	- 57 23 11.	10		PLANETARY NEBULA
A2 421	13 02 44.2	+ 26 57 20.		17.6	FAINT BLUE OBJECT
A2 422	13 02 44.3	+ 29 27 58.		16.6	FAINT BLUE OBJECT
MCG+08-24-035	13 02 45.	+ 47 47	54	15.	GALAXY
ARC 1676	13 02 45.	+ 48 03		17.5	RICH CLUSTER OF GALAXIES
LB 02541	13 02 45.	+ 53 38 06.		16.7	FAINT BLUE STAR
BFG 118	13 02 46.42	+ 34 22 50.8		16.53	ULTRAVIOLET-EXCESS OBJECT
LB 02542	13 02 47.	+ 57 52 30.		16.1	FAINT BLUE STAR
ARC 1671	13 02 47.	- 22 24		17.4	RICH CLUSTER OF GALAXIES
ZWG 043.139	13 02 48.	+ 04 55		15.6	GALAXY
8ZW 1302+18.1	13 02 48.	+ 18 08		15.7	COMPACT GALAXY
IC 4164	13 02 48.	+ 20 48 47.			OPEN CLUSTER
ZWG 130.006	13 02 48.	+ 26 13		15.0	GALAXY
8ZW 1302+26.2	13 02 48.	+ 26 13		15.0	COMPACT GALAXY
MCG+04-31-005	13 02 48.	+ 26 13	30	15.0	GALAXY
ZWG 160.130	13 02 48.	+ 27 50		14.2	GALAXY
RNGC 4957	13 02 48.	+ 27 50		14.0	GALAXY
UGC 08178	13 02 48.	+ 27 50	78	14.2	GALAXY E
RNGC 4972	13 02 48.	+ 75 34			NON-EXISTENT OBJECT
MCG-02-33-106	13 02 48.	- 11 00	30	16.	GALAXY
MCG-06-29-007	13 02 48.	- 37 25	42	15.5	GALAXY
MCG-07-27-030	13 02 48.	- 43 18	72	13.5	GALAXY
REIZ 3217	13 02 49.	+ 09 28	18	15.0	GALAXY
IC 0845	13 02 49.	+ 12 23			NONSTELLAR OBJECT
REIZ 3223	13 02 50.	+ 23 22	18	14.4	GALAXY
LB 02543	13 02 51.	+ 52 16 18.		16.0	FAINT BLUE STAR
IC 4168	13 02 52.	+ 40 29 11.			NONSTELLAR OBJECT
ARC 1675	13 02 53.	+ 34 49		17.2	RICH CLUSTER OF GALAXIES
ZWG 043.140	13 02 54.	+ 07 54		15.6	GALAXY
MCG+01-33-050	13 02 54.	+ 07 54	18	15.6	GALAXY
ZWG 130.007	13 02 54.	+ 23 22		15.5	GALAXY
MCG+05-31-124	13 02 54.	+ 27 50	24	14.2	GALAXY
MRK 062	13 02 54.	+ 30 33	10	16.5	GALAXY WITH UV CONTINUUM
IC 4166	13 02 54.	+ 31 42 35.			NONSTELLAR OBJECT
MCG+05-31-122	13 02 54.	+ 31 43 30.	54	15.2	GALAXY
ZWG 160.131	13 02 54.	+ 32 16		15.1	GALAXY
SCH 86	13 02 54.	+ 32 16		15.1	PECULIAR GALAXY
UGC 08179	13 02 54.	+ 32 16	90	15.1	GALAXY Sb
MCG+05-31-123	13 02 54.	+ 32 18	96	15.1	GALAXY
IC 4169	13 02 54.	+ 39 02 29.			NONSTELLAR OBJECT
ZWG 271.003	13 02 54.	+ 53 55		15.4	GALAXY
ZWG 270.047	13 02 54.	+ 53 55		15.4	GALAXY
MCG-01-33-083	13 02 54.	- 09 16	42	15.	GALAXY
MCG-05-31-033	13 02 54.	- 28 08	48	15.5	GALAXY
HELW 302	13 02 54.	- 28 11 18.			NEBULA
MCG-07-27-031	13 02 54.	- 43 13	30	15.5	GALAXY
SVEN 364	13 02 55.	- 28 11	72	14.4	GALAXY
IC 0846	13 02 56.	+ 23 21 47.			NONSTELLAR OBJECT
LB 02544	13 02 56.	+ 53 10 00.		16.2	FAINT BLUE STAR
IC 4171	13 02 58.	+ 36 22 17.			NONSTELLAR OBJECT
ARC 1678	13 02 59.	+ 62 31		17.4	RICH CLUSTER OF GALAXIES
FEIG 072	13 03 00.	+ 06 15		11.0	FAINT BLUE STAR
ZC 1303.0+0823	13 03 00.	+ 08 23	1410		CLUSTER OF GALAXIES
ZC 1303.0+1819	13 03 00.	+ 18 19	3160		CLUSTER OF GALAXIES
TON-N 0696	13 03 00.	+ 26 55		15.5	BLUE STAR
ZWG 160.132	13 03 00.	+ 29 34		15.5	GALAXY
MCG+05-31-125	13 03 00.	+ 29 34	36	15.5	GALAXY
ZWG 160.133	13 03 00.	+ 31 43		15.2	GALAXY
UGC 08180	13 03 00.	+ 31 43	60	15.2	GALAXY SB?a-b
MCG+06-29-027	13 03 00.	+ 33 09	66	14.5	GALAXY
MCG+06-29-026	13 03 00.	+ 36 22	48	14.5	GALAXY
MCG+09-22-001	13 03 00.	+ 53 57	24	17.	GALAXY
MCG+13-10-001	13 03 00.	+ 78 39 30.	84	15.	GALAXY
REIZ 3222	13 03 00.	- 07 15	48	14.2	GALAXY
EB 2.09	13 03 01.1	+ 36 08 21.			GALAXY NEAR QSO B234
BFG 119	13 03 02.70	+ 37 43 21.3		19.14	ULTRAVIOLET-EXCESS OBJECT
HOLM 507B	13 03 03.	+ 33 12	36	16.0	PART OF MULTIPLE GALAXY
HOLM 507A	13 03 04.	+ 33 10	60	15.0	PART OF MULTIPLE GALAXY
IC 4167	13 03 05.	+ 22 10 35.			NONSTELLAR OBJECT
HELW 303	13 03 05.	- 28 09 06.			NEBULA
ZC 1303.1+1116	13 03 06.	+ 11 16	610		CLUSTER OF GALAXIES
8ZW 1303+18.6	13 03 06.	+ 18 34		17.5	COMPACT GALAXY
8ZW 1303+21.4	13 03 06.	+ 21 24		18.6	COMPACT GALAXY
LB 00029	13 03 06.	+ 21 32		17.3	FAINT BLUE STAR
UGC 08181	13 03 06.	+ 33 10	96	16.0	GALAXY Sc-IRR
MCG+06-29-028	13 03 06.	+ 34 07	42	15.	GALAXY
UGC 08182	13 03 06.	+ 36 23	60	16.0	GALAXY Sc
MCG+09-22-002	13 03 06.	+ 53 52	24	15.	GALAXY
ZWG 353.011	13 03 06.	+ 78 38		15.5	GALAXY
ZWG 352.056	13 03 06.	+ 78 38		15.5	GALAXY
UGC 08183	13 03 06.	+ 78 38	96	15.5	GALAXY Sb
KARA.68 218	13 03 06.	- 07 28	60		DWARF GALAXY
RNGC 4958	13 03 06.	- 07 45		12.0	GALAXY
A2 425	13 03 06.4	+ 28 59 08.		18.7	FAINT BLUE GALAXY
MAI 082	13 03 07.	+ 67 59	202		DWARF SPHEROIDAL GALAXY
SVEN 365	13 03 07.	- 28 09	120	13.6	GALAXY
IC 4172	13 03 08.	+ 23 07 05.			NONSTELLAR OBJECT
IC 4170	13 03 09.	+ 21 24 05.			NONSTELLAR OBJECT
IC 4174	13 03 09.	+ 36 40 00.			NONSTELLAR OBJECT
HELW 214	13 03 09.	- 05 18 30.			NEBULA
A2 426	13 03 09.2	+ 27 32 07.		17.7	FAINT BLUE OBJECT
A2 427	13 03 11.1	+ 27 51 30.		16.6	FAINT BLUE OBJECT
ZWG 043.141	13 03 12.	+ 07 49		15.5	GALAXY
ZC 1303.2+3107	13 03 12.	+ 31 07	1680		CLUSTER OF GALAXIES
ZWG 294.011	13 03 12.	+ 56 36		14.0	GALAXY
UGC 08184	13 03 12.	+ 56 36	78	14.0	GALAXY S
KARA.73B 0571	13 03 12.	+ 56 36	60	14.0	ISOLATED GALAXY S
MCG-01-33-084	13 03 12.	- 07 44	90	12.	GALAXY
REIZ 3224	13 03 12.	- 07 45	138	11.4	GALAXY
MCG-04-31-026	13 03 12.	- 25 20	54	15.5	GALAXY
BFG 120	13 03 13.17	+ 37 37 22.4		19.34	ULTRAVIOLET-EXCESS OBJECT
A2 428	13 03 13.7	+ 29 16 08.		18.2	FAINT BLUE GALAXY
LB 02545	13 03 15.	+ 51 32 00.		17.1	FAINT BLUE STAR
REIN 4.216	13 03 15.57	- 06 16 12.7			NEBULA
REIZ 3226	13 03 16.	+ 33 27	12	14.7	GALAXY
ZWG 160.134	13 03 18.	+ 28 00		13.5	GALAXY
UGC 08185	13 03 18.	+ 28 00	96	13.5	GALAXY Sc
MCG+05-31-126	13 03 18.	+ 28 00	96	13.5	GALAXY
WEI 22722	13 03 18.	+ 30 50		17.85	FAINT BLUE OBJECT
ZC 1303.3+3206	13 03 18.	+ 32 06	2550		CLUSTER OF GALAXIES
ZWG 189.018	13 03 18.	+ 33 27		15.4	GALAXY
MCG+09-22-005	13 03 18.	+ 53 50	18	14.2	GALAXY
ZWG 271.004	13 03 18.	+ 53 52		15.3	GALAXY
ZWG 270.048	13 03 18.	+ 53 52		15.3	GALAXY
MRK 239	13 03 18.	+ 53 52	11	15.5	GALAXY WITH UV CONTINUUM
MCG+09-22-004	13 03 18.	+ 53 53	24	17.	GALAXY
MCG+09-22-003	13 03 18.	+ 53 53	18	16.	GALAXY
MCG+09-22-006	13 03 18.	+ 53 57	36	14.	GALAXY
REIZ 3233	13 03 18.	+ 56 35	42	13.5	GALAXY
ZC 1303.3+5832	13 03 18.	+ 58 32	2150		CLUSTER OF GALAXIES
MCG+10-19-018	13 03 18.	+ 59 10	39	16.	GALAXY
MCG+10-19-019	13 03 18.	+ 61 24	57	16.	GALAXY
MCG-05-31-034	13 03 18.	- 29 29	18	15.	GALAXY
MCG-06-29-008	13 03 18.	- 34 03 30.	24	15.5	GALAXY
RNGC 4953	13 03 18.	- 37 18			GALAXY
REIN 4.217	13 03 18.92	+ 10 29 24.0			NEBULA
RNGC 4959	13 03 19.	+ 33 27		15.5	GALAXY
IC 4175	13 03 20.	+ 20 38 24.			NONSTELLAR OBJECT
LB 02546	13 03 20.	+ 50 35 54.		16.0	FAINT BLUE STAR
RNGC 4955	13 03 20.	- 29 29		15.0	GALAXY
MCG+06-29-029	13 03 21.	+ 33 26	42	14.5	GALAXY
IC 4178	13 03 21.	+ 36 17 12.			NONSTELLAR OBJECT
RNGC 4964	13 03 21.	+ 56 34		13.5	GALAXY
REIN 4.218	13 03 21.32	+ 10 30 02.5			NEBULA
BFG 121	13 03 21.49	+ 32 37 42.5		19.11	ULTRAVIOLET-EXCESS OBJECT
BFG 122	13 03 22.41	+ 33 51 34.3		18.07	ULTRAVIOLET-EXCESS OBJECT
ARC 1681	13 03 23.	+ 72 08		17.1	RICH CLUSTER OF GALAXIES
ZWG 015.061	13 03 24.	+ 02 02		15.1	GALAXY
ZWG 043.142	13 03 24.	+ 04 13		15.4	GALAXY
UGC 08186	13 03 24.	+ 04 13	108	15.4	GALAXY Sb-c
MCG+01-33-051	13 03 24.	+ 04 13	90	15.4	GALAXY
REIZ 3225	13 03 24.	+ 10 30	18	15.7	GALAXY

OBJECT NAME	RIGHT ASCEN.	DECLINATION	DIAM.	MAGN.	TYPE OF OBJECT
RNGC 4960	13 03 24.	+ 27 49			GALAXY
RNGC 4961	13 03 24.	+ 28 00		13.5	GALAXY
REIZ 3227	13 03 24.	+ 28 00	66	13.0	GALAXY
1ZW 051	13 03 24.	+ 28 22			COMPACT GALAXY
ZWG 189.019	13 03 24.	+ 33 06		15.5	GALAXY
UGC 08187	13 03 24.	+ 36 16	72	16.0	GALAXY IRR
MCG+06-29-030	13 03 24.	+ 36 17	66	15.	GALAXY
MRK 240	13 03 24.	+ 53 54	10	17.	GALAXY WITH UV CONTINUUM
ZWG 271.005	13 03 24.	+ 53 57		15.1	GALAXY
ZWG 270.049	13 03 24.	+ 53 57		15.1	GALAXY
MCG+09-22-008	13 03 24.	+ 55 51	24	18.	GALAXY
MCG+09-22-007	13 03 24.	+ 56 34	48	13.5	GALAXY
REIZ 3239	13 03 24.	+ 56 36	30	14.0	GALAXY
MCG+10-19-020	13 03 24.	+ 61 24	27	16.	GALAXY
ZWG 294.012	13 03 24.	+ 61 25		15.4	GALAXY
7ZW 498	13 03 24.	+ 70 31			COMPACT GALAXY
MCG-01-33-085	13 03 24.	- 07 13	60	15.5	GALAXY
MCG-06-29-009	13 03 24.	- 37 19	36	15.	GALAXY
LB 02547	13 03 25.	+ 56 38 12.		17.3	FAINT BLUE STAR
LB 02548	13 03 26.	+ 56 56 48.		15.7	FAINT BLUE STAR
IC 4179	13 03 27.	+ 37 27 48.			NONSTELLAR OBJECT
IC 4182	13 03 27.	+ 37 52	246	13.5	GALAXY Sc
REIZ 3237	13 03 27.	+ 53 50	18	14.2	GALAXY
REIZ 3238	13 03 27.	+ 53 57	24	14.3	GALAXY
RNGC 4962	13 03 30.	+ 29 20			NON-EXISTENT OBJECT
MCG+05-31-127	13 03 30.	+ 29 32 30.	30	15.0	GALAXY
MCG+05-31-128	13 03 30.	+ 31 09	6	18.	GALAXY
ZWG 189.020	13 03 30.	+ 37 52		14.0	GALAXY
UGC 08188	13 03 30.	+ 37 52	420	14.0	GALAXY DWRF SP
ZWG 245.018	13 03 30.	+ 46 44		15.7	GALAXY
UGC 08189	13 03 30.	+ 46 44	102	15.7	GALAXY Sb-c
ZWG 245.019	13 03 30.	+ 49 41		15.5	GALAXY
MCG+08-24-036	13 03 30.	+ 49 42	30	15.	GALAXY
ZWG 271.006	13 03 30.	+ 53 50		15.2	GALAXY
ZWG 270.050	13 03 30.	+ 53 50		15.2	GALAXY
MCG+09-22-009	13 03 30.	+ 53 56	36	14.	GALAXY
MCG+10-19-021	13 03 30.	+ 60 27	36		GALAXY
ZC 1303.5+6910	13 03 30.	+ 69 10	1950		CLUSTER OF GALAXIES
ARC 1677	13 03 31.	+ 31 10		17.7	RICH CLUSTER OF GALAXIES
LB 02549	13 03 31.	+ 59 07 06.		16.6	FAINT BLUE STAR
SN 1937C	13 03 32.	+ 37 53		8.4	SUPERNOVA
RNGC 4967	13 03 32.	+ 53 50		15.0	GALAXY
IC 4184	13 03 33.	+ 39 06 18.			NONSTELLAR OBJECT
RNGC 4963	13 03 33.	+ 41 59		14.0	GALAXY
HELW 215	13 03 33.	- 05 06 54.			NEBULA
REIZ 3232	13 03 34.	+ 33 07	18	15.4	GALAXY
REIZ 3231	13 03 34.	+ 33 10	48	15.0	GALAXY
FATH 2.116	13 03 35.	+ 29 17	35		NEBULA
REIZ 3236	13 03 35.	+ 41 59	18	14.3	GALAXY
ZC 1303.6+1354	13 03 36.	+ 13 54	340		CLUSTER OF GALAXIES
8ZW 1303+22.8	13 03 36.	+ 22 50		17.0	COMPACT GALAXY
ZC 1303.6+2647	13 03 36.	+ 26 47	810		CLUSTER OF GALAXIES
ZWG 160.135	13 03 36.	+ 28 51		15.5	GALAXY
ZWG 160.136	13 03 36.	+ 29 33		15.0	GALAXY
MCG+06-29-031	13 03 36.	+ 37 53	300	13.	GALAXY
ZWG 217.010	13 03 36.	+ 41 59		14.2	GALAXY
UGC 08190	13 03 36.	+ 41 59	54	14.2	GALAXY
MCG+08-24-037	13 03 36.	+ 46 44	90	15.	GALAXY
MCG+10-19-022	13 03 36.	+ 60 27	30	16.	GALAXY
MCG-07-27-032	13 03 36.	- 40 07	36	14.	GALAXY
BFG 123	13 03 36.05	+ 33 18 53.8		19.12	ULTRAVIOLET-EXCESS OBJECT
BFG 124	13 03 36.94	+ 34 34 00.0		18.94	ULTRAVIOLET-EXCESS OBJECT
FATH 2.117	13 03 37.	+ 29 33	14		NEBULA
IC 4173	13 03 37.	- 11 18			NONSTELLAR OBJECT
IC 4186	13 03 38.	+ 37 15 07.			NONSTELLAR OBJECT
RNGC 4945A	13 03 38.	- 49 24			GALAXY
A2 432	13 03 38.4	+ 27 28 05.		18.0	FAINT BLUE OBJECT
RNGC 5463B	13 03 39.	+ 09 37		16.0	GALAXY
HOLM 508B	13 03 40.	+ 30 29	12	16.0	PART OF MULTIPLE GALAXY
TON-N 1548	13 03 40.	+ 34 34		17.0	BLUE STAR
IC 4187	13 03 40.	+ 36 34 07.			NONSTELLAR OBJECT
IC 4176	13 03 40.	- 11 17			NONSTELLAR OBJECT
IC 4181	13 03 41.	+ 21 45 36.			NONSTELLAR OBJECT
HOLM 508A	13 03 41.	+ 30 28	54	15.4	PART OF MULTIPLE GALAXY
LB 02550	13 03 41.	+ 50 31 36.		13.6	FAINT BLUE STAR
TON-N 0697	13 03 42.	+ 27 23		15.5	BLUE STAR
REIZ 3235	13 03 42.	+ 27 50	36	15.2	GALAXY
TON-N 0144	13 03 42.	+ 27 52		15.5	BLUE STAR
MCG+05-31-129	13 03 42.	+ 30 28	60	15.4	GALAXY
MRK 241	13 03 42.	+ 33 17	13	16.	GALAXY WITH UV CONTINUUM
ZWG 189.021	13 03 42.	+ 36 14		14.5	GALAXY
UGC 08191	13 03 42.	+ 36 14	96	14.5	GALAXY Sc
MCG+06-29-034	13 03 42.	+ 36 14	72	12.5	GALAXY
MCG+06-29-032	13 03 42.	+ 36 34 30.	12	15.	GALAXY
MCG+06-29-033	13 03 42.	+ 36 36	48	15.5	GALAXY
REIZ 3230	13 03 42.	- 07 21	12	15.3	GALAXY
REIZ 3229	13 03 42.	- 07 29	12	15.5	GALAXY
REIZ 3228	13 03 42.	- 07 30	18	15.5	GALAXY
MCG-02-33-107	13 03 42.	- 13 19	48	15.	GALAXY
MCG-04-31-027	13 03 42.	- 23 54 30.	36	15.	GALAXY
MCG-07-27-033	13 03 42.	- 40 01 30.	24	16.	GALAXY
DV.56 N4945A	13 03 42.	- 49 24	132		S GALAXY
IC 4188	13 03 43.	+ 36 35 55.			NONSTELLAR OBJECT
IC 4189	13 03 44.	+ 36 14 55.			NONSTELLAR OBJECT
IC 4183	13 03 46.	+ 21 46 19.			NONSTELLAR OBJECT
IC 4185	13 03 46.	+ 22 02 37.			NONSTELLAR OBJECT
IC 4190	13 03 46.	+ 37 42 49.			MAY NOT EXIST
REIZ 3241	13 03 47.	+ 30 29	36	15.8	GALAXY
ZC 1303.8+0530	13 03 48.	+ 05 30	400		CLUSTER OF GALAXIES
ZWG 072.001	13 03 48.	+ 10 38		15.0	GALAXY
ZWG 071.108	13 03 48.	+ 10 38		15.0	GALAXY
UGC 08192	13 03 48.	+ 10 38	96	15.0	GALAXY Sa-b
KARA.72 364A	13 03 48.	+ 10 38	54	15.0	PART OF DOUBLE GALAXY
MCG+02-33-054	13 03 48.	+ 10 38	108	15.0	GALAXY
REIZ 3234	13 03 48.	+ 10 39	18	14.3	GALAXY
ZWG 130.008	13 03 48.	+ 25 43		14.9	GALAXY
MCG+04-31-006	13 03 48.	+ 25 43	24	14.9	GALAXY
UGC 08193	13 03 48.	+ 30 29	60	16.0	GALAXY Sc
MRK 063	13 03 48.	+ 30 51	8	17.	GALAXY WITH UV CONTINUUM
ZC 1303.8+3656	13 03 48.	+ 36 56	810		CLUSTER OF GALAXIES
IC 4193	13 03 48.	+ 39 41 25.			NONSTELLAR OBJECT
ZWG 217.011	13 03 48.	+ 39 42		15.6	GALAXY
MCG+08-24-038	13 03 48.	+ 49 38	42	16.	GALAXY
ZWG 271.007	13 03 48.	+ 53 56		14.9	GALAXY
ZWG 270.051	13 03 48.	+ 53 56		14.9	GALAXY
MCG+09-22-010	13 03 48.	+ 55 56	42	13.	GALAXY
ZWG 015.062	13 03 48.	- 01 25		14.8	GALAXY
MCG+00-33-029	13 03 48.	- 01 25	36	14.8	GALAXY
MCG-03-33-032	13 03 48.	- 17 16	90	18.	GALAXY
MCG-06-29-010	13 03 48.	- 38 01		15.	GALAXY
REIN 4.219	13 03 48.19	+ 10 38 40.4			NEBULA
IC 4192	13 03 49.	+ 37 52 25.			MAY NOT EXIST
IC 4194	13 03 49.	+ 39 08 31.			NONSTELLAR OBJECT
LB 02551	13 03 49.	+ 51 43 36.		17.4	FAINT BLUE STAR
IC 0847	13 03 49.	+ 53 57 01.			SAME AS NGC 4974
RNGC 4973	13 03 50.	+ 53 56		15.0	GALAXY
HN 0370	13 03 50.	- 13 18			NEBULA
IC 4177	13 03 50.	- 13 18			NONSTELLAR OBJECT
REIZ 3243	13 03 51.	+ 53 56	24	14.3	GALAXY
FATH 1.560	13 03 52.	+ 29 20	16		NEBULA
A2 433	13 03 52.3	+ 29 10 33.		18.2	FAINT BLUE OBJECT
REIN 4.220	13 03 53.41	+ 10 42 00.8			NEBULA
ZC 1303.9+0027	13 03 54.	+ 00 27	1550		CLUSTER OF GALAXIES
ZWG 072.002	13 03 54.	+ 10 42		15.7	GALAXY
ZWG 071.109	13 03 54.	+ 10 42		15.7	GALAXY
KARA.72 364B	13 03 54.	+ 10 42	42	15.7	PART OF DOUBLE GALAXY
REIZ 3240	13 03 54.	+ 10 43	84	15.0	GALAXY
ZWG 160.137	13 03 54.	+ 29 20		13.9	GALAXY
RNGC 4966	13 03 54.	+ 29 20		14.0	GALAXY
UGC 08194	13 03 54.	+ 29 20	60	13.9	GALAXY S
LB 00030	13 03 54.	+ 31 22		16.7	FAINT BLUE STAR
WEI 21541	13 03 54.	+ 31 23		17.72	FAINT BLUE OBJECT
ZC 1303.9+5156	13 03 54.	+ 51 56	1340		CLUSTER OF GALAXIES
MCG+09-22-011	13 03 54.	+ 53 53	36	14.3	GALAXY
ZC 1303.9+5507	13 03 54.	+ 55 07	740		CLUSTER OF GALAXIES
MCG+10-19-023	13 03 54.	+ 57 49	39	16.	GALAXY
ZC 1303.9+5900	13 03 54.	+ 59 00	1610		CLUSTER OF GALAXIES
REIN 4.221	13 03 54.55	+ 10 12 37.1			NEBULA
FATH 2.118	13 03 55.	+ 29 20	46		NEBULA
IC 4195	13 03 56.	+ 37 18 31.			NONSTELLAR OBJECT
REIZ 3246	13 03 57.	+ 53 53	42	14.9	GALAXY
A2 434	13 03 57.9	+ 27 36 59.		18.1	FAINT BLUE OBJECT
A2 435	13 03 58.8	+ 29 24 06.		18.2	FAINT BLUE OBJECT
FATH 2.119	13 03 59.	+ 29 26	11		NEBULA
FATH 2.120	13 03 59.	+ 29 55	68		NEBULA
BB 3.09	13 03 59.5	+ 34 39 04.5			GALAXY NEAR QSO B340
HMS 1304+3110	13 04	+ 31 10			CLUSTER OF GALAXIES
VDB.66G 165	13 04	+ 67 59	170		DWARF GALAXY
VDB.66G 164	13 04	- 17 16	70		DWARF GALAXY
ZC 1304.0+1150	13 04 00.	+ 11 50	940		CLUSTER OF GALAXIES
MCG+05-31-131	13 04 00.	+ 29 20	60	13.9	GALAXY
UGC 08195	13 04 00.	+ 29 57	84	16.0	GALAXY
MCG+05-31-130	13 04 00.	+ 29 57	66	15.	GALAXY
ZC 1304.0+4639	13 04 00.	+ 46 39	670		CLUSTER OF GALAXIES
MRK 242	13 04 00.	+ 53 46	11	16.	GALAXY WITH UV CONTINUUM
ZWG 271.008	13 04 00.	+ 53 53		15.7	GALAXY
ZWG 270.052	13 04 00.	+ 53 53		15.7	GALAXY
ZWG 271.009	13 04 00.	+ 55 55		14.5	GALAXY
ZWG 270.053	13 04 00.	+ 55 55		14.5	GALAXY
UGC 08196	13 04 00.	+ 55 55	120	14.5	GALAXY Sb
MCG+10-19-024	13 04 00.	+ 59 03	51	16.	GALAXY
UGC 08197	13 04 00.	+ 59 04	66	16.5	GALAXY S
ZWG 294.013	13 04 00.	+ 59 29		15.5	GALAXY
ZC 1304.0+7245	13 04 00.	+ 72 45	1210		CLUSTER OF GALAXIES
RNGC 4974	13 04 02.	+ 53 53		15.5	GALAXY
RNGC 4977	13 04 03.	+ 55 55		14.5	GALAXY
BFG 125	13 04 03.95	+ 34 17 37.5		17.97	ULTRAVIOLET-EXCESS OBJECT
BFG 126	13 04 04.59	+ 33 35 23.1		13.61	ULTRAVIOLET-EXCESS OBJECT
ZWG 160.138	13 04 06.	+ 27 26		15.7	GALAXY
ZC 1304.1+4005	13 04 06.	+ 40 05	1810		CLUSTER OF GALAXIES
UGC 08198	13 04 06.	+ 49 14	60	17.	GALAXY Sc-IRR
MCG+08-24-039	13 04 06.	+ 49 14	60	16.	GALAXY
MCG+08-24-040	13 04 06.	+ 49 35	48	15.	GALAXY
MCG+10-19-025	13 04 06.	+ 59 28	24	16.	GALAXY
IC 4180	13 04 06.	- 23 38 11.			NONSTELLAR OBJECT
MCG-05-31-035	13 04 06.	- 28 16	84	14.5	GALAXY
BFG 127	13 04 06.02	+ 34 20 42.4		18.77	ULTRAVIOLET-EXCESS OBJECT
BFG 128	13 04 07.28	+ 36 45 06.5		18.71	ULTRAVIOLET-EXCESS OBJECT
REIZ 3242	13 04 08.	+ 23 07	30	14.7	GALAXY
FATH 1.561	13 04 09.	+ 29 10	11		NEBULA
MCG-04-31-028	13 04 09.	- 22 35	60	15.5	GALAXY
BFG 129	13 04 09.02	+ 36 53 32.1		19.07	ULTRAVIOLET-EXCESS OBJECT
FATH 2.121	13 04 11.	+ 29 27	19		NEBULA
HELW 304	13 04 11.	- 28 17 17.			NEBULA
ZWG 044.001	13 04 12.	+ 02 47		15.6	GALAXY
MCG+01-33-052	13 04 12.	+ 02 47	24	15.6	GALAXY
8ZW 1304+20.5	13 04 12.	+ 20 28		17.5	COMPACT GALAXY
ZC 1304.2+2741	13 04 12.	+ 27 41	1410		CLUSTER OF GALAXIES
REIZ 3245	13 04 12.	+ 28 07	36	15.2	GALAXY
ZWG 160.139	13 04 12.	+ 29 06		15.0	GALAXY
SCH 87	13 04 12.	+ 29 06		15.0	PECULIAR GALAXY
FATH 2.122	13 04 12.	+ 29 38	16		NEBULA
ARC 1679	13 04 12.	+ 32 04		17.5	RICH CLUSTER OF GALAXIES
ZC 1304.2+5706	13 04 12.	+ 57 06	1550		CLUSTER OF GALAXIES
MCG+10-19-026	13 04 12.	+ 60 29	27	16.	GALAXY
SVEN 366	13 04 13.	- 28 17	180	13.3	GALAXY
BFG 130	13 04 14.00	+ 32 27 37.1		18.70	ULTRAVIOLET-EXCESS OBJECT
MCG+05-31-132	13 04 15.	+ 28 09	60	15.	GALAXY
FATH 2.123	13 04 15.	+ 29 07	41		NEBULA
SVEN 367	13 04 15.	- 15 14	60	14.4	GALAXY
MCG-04-31-029	13 04 15.	- 23 40	48	14.5	GALAXY
SN 1962A	13 04 17.	+ 28 08		15.6	SUPERNOVA
REIZ 3247	13 04 17.	+ 29 06	60	15.2	GALAXY
ZWG 044.002	13 04 18.	+ 07 43		15.5	GALAXY
8ZW 1304+09.9	13 04 18.	+ 09 56		17.5	COMPACT GALAXY
LB 00262	13 04 18.	+ 28 49 24.		15.0	FAINT BLUE STAR
MCG+05-31-133	13 04 18.	+ 29 08	48	15.0	GALAXY
ZC 1304.3+4123	13 04 18.	+ 41 23	1080		CLUSTER OF GALAXIES
MCG+08-24-041	13 04 18.	+ 45 33	48	15.	GALAXY
MCG+10-19-027	13 04 18.	+ 62 17	57	15.	GALAXY
SHAH 127	13 04 18.	+ 71 50	66	17.8	GROUP OF COMPACT GALAXIES
MCG-06-29-011	13 04 18.	- 33 36	30	15.5	GALAXY
A2 438	13 04 18.8	+ 27 59 10.		18.1	FAINT BLUE OBJECT
BB 3.01	13 04 19.25	+ 34 38 19.			GALAXY NEAR QSO B340
BB 3.02	13 04 19.35	+ 34 37 00.			GALAXY NEAR QSO B340
FATH 2.124	13 04 20.	+ 29 38	5		NEBULA
BFG 131	13 04 20.44	+ 37 20 14.7		18.84	ULTRAVIOLET-EXCESS OBJECT
MCG+06-29-035	13 04 21.	+ 35 22	36	14.	GALAXY
RNGC 4968	13 04 21.	- 23 25		14.0	GALAXY
MCG-06-29-012	13 04 21.	- 33 36 30.	12	15.5	GALAXY
REIN 4.222	13 04 21.34	+ 09 54 26.7			NEBULA
ARC 1680	13 04 23.	+ 40 04		17.0	RICH CLUSTER OF GALAXIES
A2 440	13 04 23.7	+ 29 15 45.		17.2	FAINT BLUE OBJECT
ZWG 044.003	13 04 24.	+ 05 00		15.6	GALAXY
ZC 1304.4+0605	13 04 24.	+ 06 05	1480		CLUSTER OF GALAXIES
ZC 1304.4+3430	13 04 24.	+ 34 30	1550		CLUSTER OF GALAXIES

OBJECT NAME	RIGHT ASCEN.	DECLINATION	DIAM.	MAGN.	TYPE OF OBJECT
ZWG 189.022	13 04 24.	+ 35 22		14.7	GALAXY
UGC 08199	13 04 24.	+ 35 22	84	14.7	GALAXY
VV 292A	13 04 24.	+ 35 23	48	14.	INTERACTING GALAXY
ZWG 294.014	13 04 24.	+ 57 20		15.5	GALAXY
MCG+11-16-010	13 04 24.	+ 67 59	192	14.	GALAXY
REIZ 3244	13 04 24.	- 07 26	18	15.3	GALAXY
MCG-05-31-036	13 04 24.	- 27 56	132	14.	GALAXY
MCG-07-27-034	13 04 24.	- 40 03 30.	42	15.	GALAXY
FATH 2.125	13 04 25.	+ 29 24	14		NEBULA
MCG-04-31-030	13 04 27.	- 23 25	72	14.	GALAXY
A2 441	13 04 27.6	+ 29 15 47.		18.2	FAINT BLUE OBJECT
A2 442	13 04 28.2	+ 27 59 20.		17.7	FAINT BLUE OBJECT
BIGO 524	13 04 29.	+ 13 54			NEBULA
BB 3.03	13 04 29.4	+ 34 36 39.5			GALAXY NEAR QSO B340
8ZW 1304+00.3	13 04 30.	+ 00 17		18.9	COMPACT GALAXY
8ZW 1304+09.9	13 04 30.	+ 09 54		17.6	COMPACT GALAXY
ZWG 101.001	13 04 30.	+ 15 04		15.7	GALAXY
MCG+03-33-031	13 04 30.	+ 16 18	45	15.4	GALAXY
ZWG 160.140	13 04 30.	+ 28 48		15.0	GALAXY
RNGC 4971	13 04 30.	+ 28 48		15.0	GALAXY
MCG+05-31-134	13 04 30.	+ 28 48	48	15.0	GALAXY
ZC 1304.5+4555	13 04 30.	+ 45 55	1340		CLUSTER OF GALAXIES
FEIG 073	13 04 30.	+ 49 06		13.8	FAINT BLUE STAR
REIZ 3250	13 04 30.	+ 55 57	24	14.5	GALAXY
MCG+10-19-028	13 04 30.	+ 56 54	18	16.	GALAXY
MCG+10-19-029	13 04 30.	+ 57 19	42	15.	GALAXY
ZWG 294.015	13 04 30.	+ 59 01		15.3	GALAXY
7ZW 499	13 04 30.	+ 67 59			COMPACT GALAXY
ZWG 016.001	13 04 30.	- 00 54		15.0	GALAXY
MCG+00-34-001	13 04 30.	- 00 54	15	15.0	GALAXY
MCG-04-31-031	13 04 30.	- 23 52	96	16.	GALAXY
LB 02552	13 04 31.	+ 56 57 48.		15.8	FAINT BLUE STAR
SVEN 368	13 04 31.	- 27 57	180	12.8	GALAXY
BB 3.10	13 04 31.6	+ 34 39 51.			GALAXY NEAR QSO B340
RNGC 4965	13 04 32.	- 27 57		14.0	GALAXY
REIN 4.223	13 04 32.79	+ 09 52 55.9			NEBULA
RNGC 4969	13 04 33.	+ 13 55		15.0	GALAXY
MCG+06-29-036	13 04 33.	+ 35 24	48	14.5	GALAXY
VV 292B	13 04 33.	+ 35 25	66	15.	INTERACTING GALAXY
ARC 1682	13 04 33.	+ 46 49		17.5	RICH CLUSTER OF GALAXIES
ARC 1683	13 04 33.	+ 72 08		17.1	RICH CLUSTER OF GALAXIES
REIN 4.224	13 04 33.66	+ 13 20 29.9			NEBULA
REIZ 3248	13 04 34.	+ 16 16	18	15.4	GALAXY
IC 0848	13 04 34.	+ 16 16 02.			NONSTELLAR OBJECT
BB 3.11	13 04 34.15	+ 34 41 52.5			GALAXY NEAR QSO B340
A2 443	13 04 35.7	+ 27 45 37.		18.5	FAINT BLUE GALAXY
ZWG 072.003	13 04 36.	+ 13 21		15.3	GALAXY
ZWG 072.004	13 04 36.	+ 13 55		14.9	GALAXY
KARA.72 365B	13 04 36.	+ 13 55	30		PART OF DOUBLE GALAXY
KARA.72 365A	13 04 36.	+ 13 55	36	14.9	PART OF DOUBLE GALAXY
MCG+02-33-055	13 04 36.	+ 13 55	36	14.9	GALAXY
ZWG 101.002	13 04 36.	+ 16 16		15.4	GALAXY
UGC 08200	13 04 36.	+ 35 24	90	16.0	GALAXY
ZWG 294.016	13 04 36.	+ 61 18		15.6	GALAXY
MCG+10-19-030	13 04 36.	+ 61 44	27	16.	GALAXY
ZC 1304.6+6411	13 04 36.	+ 64 11	1140		CLUSTER OF GALAXIES
ZWG 316.009	13 04 36.	+ 67 58		14.1	GALAXY
UGC 08201	13 04 36.	+ 67 58	222	14.1	GALAXY DWRF IR
8ZW 1304-01.3	13 04 36.	- 01 19		18.1	COMPACT GALAXY
FATH 2.126	13 04 39.	+ 30 10	11		NEBULA
A2 445	13 04 39.9	+ 27 52 45.		16.6	FAINT BLUE OBJECT
LB 00067	13 04 42.	+ 25 49		16.7	FAINT BLUE STAR
MRK 064	13 04 42.	+ 34 40	7	17.	GALAXY WITH UV CONTINUUM
KW 10	13 04 42.	+ 34 40	8		SEYFERT GALAXY
MCG+08-24-042	13 04 42.	+ 45 32	48	16.	GALAXY
MCG+10-19-031	13 04 42.	+ 59 01	30	16.	GALAXY
IC 4196	13 04 42.	- 23 45 10.			NONSTELLAR OBJECT
BB 3.08	13 04 42.7	+ 34 33 01.5			GALAXY NEAR QSO B340
A2 446	13 04 43.3	+ 27 25 31.		16.2	FAINT BLUE OBJECT
REIN 4.225	13 04 43.76	+ 10 02 47.3			NEBULA
REIZ 3249	13 04 44.	+ 23 09	12	15.3	GALAXY
A2 447	13 04 44.3	+ 28 47 38.		16.5	FAINT BLUE OBJECT
MCG+08-24-043	13 04 45.	+ 45 32	18	17.	GALAXY
SVEN 369	13 04 45.	- 15 14	30	15.4	GALAXY
RNGC 4970	13 04 45.	- 23 46		14.0	GALAXY
TON-N 1549	13 04 46.	+ 34 34		16.6	BLUE STAR
BFG 132	13 04 46.51	+ 33 59 15.6		18.82	ULTRAVIOLET-EXCESS OBJECT
BC B340	13 04 47.1	+ 34 40 39.		16.97	QUASI-STELLAR OBJECT
ZWG 016.002	13 04 48.	+ 02 17		15.2	GALAXY
ZWG 072.005	13 04 48.	+ 13 55		15.6	GALAXY
MCG+02-33-056	13 04 48.	+ 13 55	30	15.6	GALAXY
ZC 1304.8+1845	13 04 48.	+ 18 45	470		CLUSTER OF GALAXIES
LB 00031	13 04 48.	+ 27 19		15.7	FAINT BLUE STAR
ZWG 160.141	13 04 48.	+ 28 18		15.5	GALAXY
MCG+06-29-037	13 04 48.	+ 34 33	30	15.	GALAXY
ZWG 189.023	13 04 48.	+ 34 34		15.7	GALAXY
SHB 243	13 04 48.	+ 34 40 24.		17.0	QUASI-STELLAR OBJECT
MCG-04-31-032	13 04 48.	- 27 09	36	15.	GALAXY
BFG 133	13 04 48.01	+ 34 40 24.2		16.48	ULTRAVIOLET-EXCESS OBJECT
MCG-04-31-033	13 04 51.	- 23 46	72	14.	GALAXY
REIN 4.226	13 04 51.49	+ 10 08 10.1			NEBULA
TON-N 1550	13 04 52.	+ 35 01		16.5	BLUE STAR
BFG 134	13 04 52.08	+ 37 28 38.3		19.30	ULTRAVIOLET-EXCESS OBJECT
SHB 244	13 04 52.1	+ 37 28 38.		19.1	QUASI-STELLAR OBJECT
BC B312	13 04 53.1	+ 37 29 33.		19.08	QUASI-STELLAR OBJECT
8ZW 1304+02.0	13 04 54.	+ 02 03		18.3	COMPACT GALAXY
MCG+03-33-032	13 04 54.	+ 18 05	48	16.	GALAXY
REIZ 3251	13 04 54.	+ 28 18	18	15.0	GALAXY
MCG+08-24-044	13 04 54.	+ 45 47	42	16.	GALAXY
MCG+09-22-012	13 04 54.	+ 52 32	18	17.	GALAXY
MCG+10-19-032	13 04 54.	+ 60 27	12	16.	GALAXY
MCG-04-31-034	13 04 54.	- 23 19	48	15.	GALAXY
MCG-07-27-035	13 04 54.	- 44 43	72	15.	GALAXY
LB 02553	13 04 56.	+ 53 23 48.		16.4	FAINT BLUE STAR
MCG+06-29-038	13 04 57.	+ 33 07	90	14.5	GALAXY
BB 3.12	13 04 58.1	+ 34 48 33.5			GALAXY NEAR QSO B340
LB 09869	13 05	- 80 20		13.1	FAINT BLUE STAR
LB 09670	13 05	- 83 07		13.6	FAINT BLUE STAR
FEIG 074	13 05 00.	+ 19 53		11.6	FAINT BLUE STAR
ZWG 160.142	13 05 00.	+ 26 59		15.5	GALAXY
UGC 08203	13 05 00.	+ 33 07	108	16.0	GALAXY Sc
MCG+06-29-039	13 05 00.	+ 36 03	30	15.	GALAXY
ZC 1305.0+3630	13 05 00.	+ 36 30	870		CLUSTER OF GALAXIES
ZC 1305.0+4652	13 05 00.	+ 46 52	1410		CLUSTER OF GALAXIES
ZC 1305.0+5336	13 05 00.	+ 53 36	540		CLUSTER OF GALAXIES
MCG+10-19-033	13 05 00.	+ 56 48	30	16.	GALAXY
ZWG 016.003	13 05 00.	- 00 40		14.0	GALAXY
UGC 08202	13 05 00.	- 00 40	72	14.0	GALAXY Sc
MCG+00-34-002	13 05 00.	- 00 40	48	14.0	GALAXY
MCG-02-34-001	13 05 00.	- 12 18	36	15.	GALAXY
A2 449	13 05 00.1	+ 29 40 04.		16.8	FAINT BLUE GALAXY
A2 450	13 05 01.2	+ 28 07 51.		16.6	FAINT BLUE OBJECT
ARP 139	13 05 02.	+ 26 59			PECULIAR GALAXY
TON-N 1551	13 05 03.	+ 21 01		16.8	BLUE STAR
MCG-04-31-035	13 05 03.	- 22 36	180	14.	GALAXY
BFG 135	13 05 03.82	+ 37 13 38.6		18.38	ULTRAVIOLET-EXCESS OBJECT
REIZ 3254	13 05 04.	+ 33 08	78	15.8	GALAXY
REIZ 3258	13 05 04.	+ 58 26	30	14.3	GALAXY
IC 0849	13 05 04.	- 00 39 33.			NONSTELLAR OBJECT
BB 3.07	13 05 04.6	+ 34 30 56.			GALAXY NEAR QSO B340
LB 02554	13 05 05.	+ 54 25 54.		16.0	FAINT BLUE STAR
BFG 136	13 05 05.43	+ 37 27 57.9		18.52	ULTRAVIOLET-EXCESS OBJECT
BFG 137	13 05 05.94	+ 37 26 43.8		19.00	ULTRAVIOLET-EXCESS OBJECT
ZWG 044.004	13 05 06.	+ 06 36		15.3	GALAXY
UGC 08204	13 05 06.	+ 06 36	120	15.3	GALAXY S0
MCG+01-34-001	13 05 06.	+ 06 36	36	15.3	GALAXY
ZWG 101.003	13 05 06.	+ 16 37		15.4	GALAXY
REIZ 3253	13 05 06.	+ 26 59	48	15.6	GALAXY
MCG+05-31-135	13 05 06.	+ 26 59	54	15.5	GALAXY
MCG+09-22-013	13 05 06.	+ 53 52	18	14.	GALAXY
ZWG 294.017	13 05 06.	+ 58 24		14.8	GALAXY
UGC 08205	13 05 06.	+ 58 24	66	14.8	GALAXY Sb
MCG-01-34-002	13 05 06.	- 04 44	24	15.5	GALAXY
MCG-01-34-001	13 05 06.	- 04 44	36	16.	GALAXY
A2 453	13 05 06.3	+ 29 10 55.		16.8	FAINT BLUE OBJECT
A2 454	13 05 07.2	+ 28 55 36.		17.0	FAINT BLUE OBJECT
BB 3.04	13 05 07.65	+ 34 39 23.			GALAXY NEAR QSO B340
TON-N 1552	13 05 09.	+ 20 49		15.9	BLUE STAR
BFG 138	13 05 10.69	+ 36 48 49.0		19.33	ULTRAVIOLET-EXCESS OBJECT
ZC 1305.2+2013	13 05 12.	+ 20 13	870		CLUSTER OF GALAXIES
8ZW 1305+20.4	13 05 12.	+ 20 23		16.4	COMPACT GALAXY
ZWG 130.009	13 05 12.	+ 25 05		15.3	GALAXY
RNGC 4979	13 05 12.	+ 25 05		15.5	GALAXY
ZWG 160.143	13 05 12.	+ 27 45		15.7	GALAXY
UGC 08206	13 05 12.	+ 27 45	60	15.7	GALAXY S0-a
IC 4199	13 05 12.	+ 36 07 28.			NONSTELLAR OBJECT
MCG+06-29-040	13 05 12.	+ 36 38 30.	24	15.5	GALAXY
UGC 08207	13 05 12.	+ 36 40	66	16.5	GALAXY SBc
ZWG 245.020	13 05 12.	+ 45 29		15.6	GALAXY
ZWG 271.010	13 05 12.	+ 53 51		15.6	GALAXY
MCG+10-19-034	13 05 12.	+ 58 23	42	14.	GALAXY
ZWG 016.004	13 05 12.	- 00 35		14.8	GALAXY
MCG+00-34-003	13 05 12.	- 00 35	42	14.8	GALAXY
MCG-03-34-001	13 05 12.	- 19 06 30.	90	14.5	GALAXY
A2 456	13 05 12.2	+ 28 24 05.		18.0	FAINT BLUE OBJECT
REIZ 3255	13 05 13.	+ 25 04	48	14.3	GALAXY
LB 02555	13 05 13.	+ 60 47 42.		15.8	FAINT BLUE STAR
SN 1956D	13 05 13.	- 00 35		19.2	SUPERNOVA
A2 458	13 05 13.1	+ 28 00 29.		16.3	FAINT BLUE OBJECT
REIZ 3259	13 05 14.	+ 53 51	24	14.7	GALAXY
MCG+06-29-041	13 05 15.	+ 33 01	54	14.5	GALAXY
REIZ 3257	13 05 16.	+ 32 58	42	15.1	GALAXY
IC 0850	13 05 16.	- 00 35 15.			NONSTELLAR OBJECT
IC 4198	13 05 17.	+ 25 04 28.			NONSTELLAR OBJECT
FATH 1.562	13 05 17.	+ 29 06	14		NEBULA
REIZ 3252	13 05 17.	- 04 46	18	14.2	GALAXY
ZWG 044.005	13 05 18.	+ 03 28		15.6	GALAXY
UGC 08208	13 05 18.	+ 03 28	60	15.6	GALAXY SB
MCG+04-31-007	13 05 18.	+ 25 03	60	15.3	GALAXY
UGC 08209	13 05 18.	+ 25 05	66	15.3	GALAXY SB
UGC 08210	13 05 18.	+ 33 02	78	16.0	GALAXY Sc
ZWG 245.021	13 05 18.	+ 45 25		15.7	GALAXY
MCG+08-24-045	13 05 18.	+ 45 25	42	15.	GALAXY
MCG+08-24-046	13 05 18.	+ 45 29	30	16.	GALAXY
ZWG 271.011	13 05 18.	+ 52 40		15.7	GALAXY
UGC 08211	13 05 18.	+ 52 40	78	15.7	GALAXY Sc
LB 02556	13 05 19.	+ 54 32 24.		15.5	FAINT BLUE STAR
BB 3.06	13 05 19.2	+ 34 35 34.			GALAXY NEAR QSO B340
BFG 139	13 05 20.58	+ 35 17 55.3		19.15	ULTRAVIOLET-EXCESS OBJECT
REIZ 3256	13 05 21.	+ 18 41	18	14.2	GALAXY
RNGC 4978	13 05 22.	+ 18 41		14.5	GALAXY
A2 461	13 05 22.0	+ 27 53 13.		17.3	FAINT BLUE OBJECT
SHB 245	13 05 22.5	+ 06 58 13.		17.0	QUASI-STELLAR OBJECT
BC 3C281	13 05 22.52	+ 06 58 16.4		17.02	QUASI-STELLAR OBJECT
IC 4197	13 05 23.	- 23 31 45.			NONSTELLAR OBJECT
BFG 140	13 05 23.34	+ 37 26 30.3		18.90	ULTRAVIOLET-EXCESS OBJECT
A2 464	13 05 23.6	+ 29 19 32.		17.3	FAINT BLUE OBJECT
ZWG 101.004	13 05 24.	+ 18 41		14.4	GALAXY
UGC 08212	13 05 24.	+ 18 41	108	14.4	GALAXY S0-a
ZC 1305.4+2941	13 05 24.	+ 29 41	740		CLUSTER OF GALAXIES
MCG+09-22-014	13 05 24.	+ 52 40	36	14.	GALAXY
8ZW 1305-03.4	13 05 24.	- 03 27		18.5	COMPACT GALAXY
RNGC 4975	13 05 24.	- 04 45		15.5	GALAXY
MCG-03-34-002	13 05 24.	- 16 24	24	15.	GALAXY
KARA.68 219	13 05 24.	- 20 16	47		DWARF GALAXY
MCG-04-31-036	13 05 24.	- 23 33	27	14.	GALAXY
SMI 01	13 05 24.	- 62 00	360		FAINT NEBULOSITY
BB 3.05	13 05 24.25	+ 34 39 43.			GALAXY NEAR QSO B340
BFG 141	13 05 24.65	+ 36 25 21.2		18.00	ULTRAVIOLET-EXCESS OBJECT
SHB 246	13 05 24.7	+ 36 25 21.		18.0	QUASI-STELLAR OBJECT
SN 1968A	13 05 25.	- 04 44		15.	SUPERNOVA
BC B330	13 05 26.0	+ 36 25 54.		18.01	QUASI-STELLAR OBJECT
SVEN 370	13 05 27.	- 15 31 33.	12	15.9	GALAXY
HN 0107	13 05 29.	- 67 22			NEBULA
PK304-04.1	13 05 29.	- 67 22	5	12.0	PLANETARY NEBULA
IC 4191	13 05 29.	- 67 22	5	12.0	PLANETARY NEBULA
SHB 247	13 05 29.3	+ 35 17 41.		17.6	QUASI-STELLAR OBJECT
BFG 142	13 05 29.33	+ 35 17 41.1		17.62	ULTRAVIOLET-EXCESS OBJECT
ZWG 016.005	13 05 30.	+ 01 14		15.6	GALAXY
ZWG 044.006	13 05 30.	+ 03 52		15.6	GALAXY
ZWG 044.007	13 05 30.	+ 06 45		15.2	GALAXY
8ZW 1305+20.9	13 05 30.	+ 20 52		17.2	COMPACT GALAXY
MCG+06-29-043	13 05 30.	+ 34 20 30.	36	15.	GALAXY
ZWG 189.024	13 05 30.	+ 34 21		15.5	GALAXY
ZWG 189.025	13 05 30.	+ 35 11		15.3	GALAXY
MCG+06-29-042	13 05 30.	+ 36 06	24	15.	GALAXY
ZC 1305.5+4322	13 05 30.	+ 43 22	1080		CLUSTER OF GALAXIES
MCG+10-19-035	13 05 30.	+ 60 26	18	15.	GALAXY
ZC 1305.5+6119	13 05 30.	+ 61 19	1950		CLUSTER OF GALAXIES
BC B337	13 05 30.2	+ 35 17 50.1		17.62	QUASI-STELLAR OBJECT
IC 4201	13 05 31.	+ 36 06 05.			NONSTELLAR OBJECT
REIN 4.227	13 05 32.16	+ 10 13 23.4			NEBULA
SVEN 372	13 05 33.	- 15 12 33.	18	16.5	GALAXY
SVEN 371	13 05 33.	- 15 44 33.	48	14.4	GALAXY
FATH 2.127	13 05 35.	+ 28 58	14		NEBULA
ZC 1305.6+1451	13 05 36.	+ 14 51	810		CLUSTER OF GALAXIES

OBJECT NAME	RIGHT ASCEN.	DECLINATION	DIAM.	MAGN.	TYPE OF OBJECT
ZC 1305.6+1926	13 05 36.	+ 19 26	2220		CLUSTER OF GALAXIES
ZWG 160.144	13 05 36.	+ 27 01		15.6	GALAXY
MCG+05-31-136	13 05 36.	+ 27 02	42	15.6	GALAXY
ZWG 160.145	13 05 36.	+ 28 58		15.4	GALAXY
ZWG 271.012	13 05 36.	+ 53 33		15.1	GALAXY
ZWG 294.018	13 05 36.	+ 60 25		14.8	GALAXY
UGC 08213	13 05 36.	+ 60 25	66	14.8	GALAXY E-S0
MCG+10-19-036	13 05 36.	+ 62 30	42	14.	GALAXY
VHE 59	13 05 36.	- 62 02	48		REFLECTION NEBULA
LB 02557	13 05 38.	+ 52 31 42.		16.8	FAINT BLUE STAR
SVEN 373	13 05 39.	- 15 01 33.	12	15.9	GALAXY
PATH 1.563	13 05 42.	+ 29 16	22		NEBULA
ZWG 189.026	13 05 42.	+ 34 16		15.6	GALAXY
ZWG 294.019	13 05 42.	+ 62 29		14.3	GALAXY
UGC 08214	13 05 42.	+ 62 29	48	14.3	GALAXY SB
MCG-02-34-002	13 05 42.	- 14 17	36	15.	GALAXY
LB 02558	13 05 47.	+ 58 19 48.		16.1	FAINT BLUE STAR
ZWG 160.146	13 05 48.	+ 27 46		15.4	GALAXY
UGC 08215	13 05 48.	+ 47 06	60	17.	GALAXY DWRF IR
ZWG 271.013	13 05 48.	+ 52 12		14.6	GALAXY
UGC 08216	13 05 48.	+ 52 12	72	14.6	GALAXY E
MCG+09-22-015	13 05 48.	+ 52 12	30	14.1	GALAXY
MCG-06-29-013	13 05 48.	- 34 19	24	15.	GALAXY
RNGC 4987	13 05 49.	+ 52 12		14.5	GALAXY
PATH 2.128	13 05 50.	+ 29 07	27		NEBULA
IC 0852	13 05 50.	+ 60 25 30.			NONSTELLAR OBJECT
MCG+08-24-047	13 05 51.	+ 50 28 30.	42	16.	GALAXY
RNGC 4985	13 05 52.	+ 41 57		15.0	GALAXY
REIZ 3265	13 05 52.	+ 52 12	36	14.1	GALAXY
REIZ 3261	13 05 53.	+ 41 55	12	14.5	GALAXY
LB 02559	13 05 53.	+ 54 01 12.		15.8	FAINT BLUE STAR
ZWG 044.008	13 05 54.	+ 04 38		15.5	GALAXY
UGC 08217	13 05 54.	+ 04 38	66	15.5	GALAXY DBL SYS
ZWG 130.010	13 05 54.	+ 22 03		15.7	GALAXY
MCG+05-31-137	13 05 54.	+ 27 47	24	15.4	GALAXY
WEI 51148	13 05 54.	+ 28 38		17.80	FAINT BLUE OBJECT
ZWG 217.012	13 05 54.	+ 41 57		15.1	GALAXY
UGC 08218	13 05 54.	+ 41 57	78	15.1	GALAXY S0
MCG+08-24-048	13 05 54.	+ 47 06	60	15.	GALAXY
ZWG 271.014	13 05 54.	+ 54 12		15.3	GALAXY
ZC 1305.9+5505	13 05 54.	+ 55 05	940		CLUSTER OF GALAXIES
ZWG 316.010	13 05 54.	+ 62 37		15.7	GALAXY
MCG-07-27-036	13 05 54.	- 41 12	72	15.	GALAXY
PATH 1.564	13 05 55.	+ 29 17	14		NEBULA
A2 487	13 05 55.1	+ 28 03 51.		18.1	FAINT BLUE OBJECT
RNGC 4976	13 05 56.	- 49 14		11.5	GALAXY
SVEN 374	13 05 57.	- 14 56	24	16.3	GALAXY
A2 468	13 05 57.1	+ 28 38 43.		17.7	FAINT BLUE OBJECT
LB 02560	13 05 58.	+ 51 31 42.		16.6	FAINT BLUE STAR
8ZW 1306+19.3	13 06 00.	+ 19 20		17.6	COMPACT GALAXY
ZWG 130.011	13 06 00.	+ 21 19		14.9	GALAXY
UGC 08219	13 06 00.	+ 21 19	66	14.9	GALAXY S
ZWG 130.012	13 06 00.	+ 24 58		15.2	GALAXY
UGC 08220	13 06 00.	+ 24 58	108	15.2	GALAXY Sb-c
ZC 1306.0+2716	13 06 00.	+ 27 16	940		CLUSTER OF GALAXIES
ZWG 160.147	13 06 00.	+ 28 35		14.9	GALAXY
RNGC 4983	13 06 00.	+ 28 35		15.0	GALAXY
ZWG 189.027	13 06 00.	+ 35 28		14.2	GALAXY
UGC 08221	13 06 00.	+ 35 28	102	14.2	GALAXY SBb
MCG+06-29-044	13 06 00.	+ 35 28	102	14.	GALAXY
MCG+09-22-017	13 06 00.	+ 50 55	48	13.	GALAXY
ZWG 271.015	13 06 00.	+ 50 56		15.2	GALAXY
UGC 08222	13 06 00.	+ 52 17	60	17.	GALAXY
MCG+09-22-016	13 06 00.	+ 54 13	36	14.	GALAXY
ZC 1306.0+5941	13 06 00.	+ 59 41	2620		CLUSTER OF GALAXIES
MCG+10-19-037	13 06 00.	+ 62 37	27	16.	GALAXY
MCG+14-06-014	13 06 00.	+ 80 37	9	16.	GALAXY
MCG+14-06-015	13 06 00.	+ 83 08	12	15.	GALAXY
MCG-02-34-003	13 06 00.	- 09 50	36	15.5	GALAXY
KARA.73 43	13 06 00.	- 42 40	47		DWARF GALAXY
REIN 4.228	13 06 00.80	+ 10 14 35.5			NEBULA
IC 4203	13 06 01.	+ 40 41 36.			NONSTELLAR OBJECT
RNGC 4998	13 06 01.	+ 50 56		15.0	GALAXY
BFG 143	13 06 01.52	+ 34 32 45.5		18.22	ULTRAVIOLET-EXCESS OBJECT
RNGC 4986	13 06 02.	+ 35 28		14.0	GALAXY
SVEN 375	13 06 03.	- 15 06	18	15.6	GALAXY
IC 4204	13 06 04.	+ 39 43 36.			NONSTELLAR OBJECT
MCG+04-31-009	13 06 06.	+ 21 18	60	14.9	GALAXY
MCG+04-31-008	13 06 06.	+ 24 57 30.	96	15.2	GALAXY
MCG+05-31-138	13 06 06.	+ 28 35	60	14.9	GALAXY
WEI 23694	13 06 06.	+ 30 22		17.17	FAINT BLUE OBJECT
UGC 08224	13 06 06.	+ 39 43	90	16.0	GALAXY SB?c
ZWG 245.022	13 06 06.	+ 45 38		14.6	GALAXY
UGC 08225	13 06 06.	+ 45 38	72	14.6	GALAXY S0
MCG+09-22-018	13 06 06.	+ 52 17	48	15.	GALAXY
MCG+10-19-038	13 06 06.	+ 61 51	45	15.	GALAXY
ZWG 016.006	13 06 06.	- 01 52		14.7	GALAXY
UGC 08223	13 06 06.	- 01 52	60	14.7	GALAXY SBc
MCG+00-34-004	13 06 06.	- 01 52	54	14.7	GALAXY
RNGC 4981	13 06 06.	- 06 31		12.5	GALAXY
SN 1968I	13 06 06.	- 06 31		13.5	SUPERNOVA
REIZ 3264	13 06 07.	+ 24 58	108	13.9	GALAXY
IC 4202	13 06 07.	+ 24 58 06.			NONSTELLAR OBJECT
REIZ 3263	13 06 08.	+ 21 19	60	14.5	GALAXY
REIZ 3267	13 06 08.	+ 35 28	42	13.8	GALAXY
IC 0851	13 06 09.	+ 21 19 00.			NONSTELLAR OBJECT
TON-N 1553	13 06 09.	+ 21 56		17.0	BLUE STAR
SVEN 376	13 06 09.	- 14 59	12	16.4	GALAXY
REIZ 3266	13 06 11.	+ 28 35	18	14.9	GALAXY
LB 02561	13 06 11.	+ 52 04 36.		13.8	FAINT BLUE STAR
RNGC 4982	13 06 11.	- 10 19			NON-EXISTENT OBJECT
ZWG 044.009	13 06 12.	+ 04 25		15.4	GALAXY
MCG+01-34-002	13 06 12.	+ 04 25	36	15.4	GALAXY
ZWG 044.010	13 06 12.	+ 04 52		15.4	GALAXY
MCG+06-29-045	13 06 12.	+ 34 13	60	15.	GALAXY
MCG+08-24-049	13 06 12.	+ 45 38	66	14.	GALAXY
MCG+08-24-050	13 06 12.	+ 48 25	42	16.	GALAXY
ZWG 294.020	13 06 12.	+ 61 50		15.4	GALAXY
UGC 08226	13 06 12.	+ 61 50	72	15.4	GALAXY SB0-a
KARA.73B 0572	13 06 12.	+ 61 50	18	15.4	ISOLATED GALAXY E
ZWG 016.007	13 06 12.	- 00 32		14.6	GALAXY
MCG+00-34-005	13 06 12.	- 00 32	15	14.6	GALAXY
REIZ 3260	13 06 12.	- 06 31	90	12.5	GALAXY
REIN 2.176	13 06 12.79	- 06 30 39.9			NEBULA
LB 02562	13 06 13.	+ 52 49 30.		16.6	FAINT BLUE STAR
REIN 2.177	13 06 14.25	- 06 31 41.1			NEBULA
REIZ 3269	13 06 15.	+ 34 14	54	14.1	GALAXY
MCG-02-34-004	13 06 15.	- 15 16	120	12.	GALAXY
REIN 4.229	13 06 16.70	+ 10 11 03.4			NEBULA
REIZ 3262	13 06 17.	- 03 28	42	14.5	GALAXY
8ZW 1306+01.5	13 06 18.	+ 01 32		17.0	COMPACT GALAXY
ZC 1306.3+1610	13 06 18.	+ 16 10	340		CLUSTER OF GALAXIES
ZWG 189.028	13 06 18.	+ 34 15		15.7	GALAXY
UGC 08227	13 06 18.	+ 34 15	78	15.7	GALAXY S
ZC 1306.3+4232	13 06 18.	+ 42 32	1140		CLUSTER OF GALAXIES
MCG+08-24-051	13 06 18.	+ 46 50	30	16.	GALAXY
ZWG 016.008	13 06 18.	- 00 31		15.1	GALAXY
MCG+00-34-006	13 06 18.	- 00 31	36	15.1	GALAXY
MCG-05-31-037	13 06 18.	- 28 22	96	15.	GALAXY
BFG 144	13 06 18.87	+ 37 56 29.0		18.95	ULTRAVIOLET-EXCESS OBJECT
PATH 1.565	13 06 19.	+ 29 09	11		NEBULA
SVEN 378	13 06 21.	- 15 15	240	12.2	GALAXY
BFG 146	13 06 21.60	+ 35 23 01.4		19.05	ULTRAVIOLET-EXCESS OBJECT
BFG 145	13 06 22.98	+ 33 45 26.0		18.94	ULTRAVIOLET-EXCESS OBJECT
PATH 1.566	13 06 23.	+ 29 08	19		NEBULA
RNGC 4984	13 06 23.	- 15 15		12.0	GALAXY
LB 00066	13 06 24.	+ 23 08		17.5	FAINT BLUE STAR
ZC 1306.4+2629	13 06 24.	+ 26 29	670		CLUSTER OF GALAXIES
MCG+05-31-139	13 06 24.	+ 29 20	48	15.5	GALAXY
ZC 1306.4+3503	13 06 24.	+ 35 03	1340		CLUSTER OF GALAXIES
ZWG 217.013	13 06 24.	+ 44 00		15.3	GALAXY
UGC 08228	13 06 24.	+ 44 00	84	15.3	GALAXY Sa
ZWG 271.016	13 06 24.	+ 55 05		15.2	GALAXY
ZWG 016.009	13 06 24.	- 03 25		15.3	GALAXY
MCG+00-34-007	13 06 24.	- 03 25	36	15.3	GALAXY
KARA.73B 0573	13 06 24.	- 03 25	54	15.3	ISOLATED GALAXY S
MCG-01-34-003	13 06 24.	- 06 31	150	12.	GALAXY
SVEN 377	13 06 26.	- 28 23	60	14.0	GALAXY
RNGC 4980	13 06 26.	- 28 23		15.0	GALAXY
REIZ 3275	13 06 27.	+ 53 04	120	14.0	GALAXY
SVEN 379	13 06 27.	- 15 29 32.	30	16.3	GALAXY
PATH 1.567	13 06 28.	+ 29 09	22		NEBULA
REIN 4.230	13 06 28.53	- 05 09 07.4			NEBULA
SHB 248	13 06 28.6	+ 12 15 48.		18.5	QUASI-STELLAR OBJECT
REIZ 3268	13 06 29.	- 03 55	24	14.8	GALAXY
ZWG 044.011	13 06 30.	+ 07 14		15.5	GALAXY
ZWG 160.148	13 06 30.	+ 28 27		14.3	GALAXY
UGC 08229	13 06 30.	+ 28 27	96	14.3	GALAXY SBb
MCG+05-31-142	13 06 30.	+ 28 27	78	14.3	GALAXY
MCG+05-31-141	13 06 30.	+ 28 36	36	15.5	GALAXY
MCG+05-31-140	13 06 30.	+ 29 01	42	15.5	GALAXY
ZWG 271.017	13 06 30.	+ 52 17		14.7	GALAXY
MCG+09-22-020	13 06 30.	+ 52 17	48	14.	GALAXY
ZWG 271.018	13 06 30.	+ 53 02		15.0	GALAXY
UGC 08230	13 06 30.	+ 53 02	66	15.0	GALAXY Sb/SBb
MCG+09-22-019	13 06 30.	+ 53 03	60	13.	GALAXY
ZWG 271.019	13 06 30.	+ 54 20		14.2	GALAXY
UGC 08231	13 06 30.	+ 54 20	102	14.2	GALAXY SB
MCG+09-22-021	13 06 30.	+ 54 21	96	13.	GALAXY
ZWG 016.010	13 06 30.	- 00 29		14.8	GALAXY
MCG+00-34-008	13 06 30.	- 00 29	30	14.8	GALAXY
PK305+01.1	13 06 30.	- 61 03 35.	10		PLANETARY NEBULA
REIZ 3277	13 06 31.	+ 54 22	90	13.9	GALAXY
PATH 2.129	13 06 32.	+ 29 18	27		NEBULA
PATH 2.130	13 06 32.	+ 29 49	14		NEBULA
LB 02563	13 06 32.	+ 59 18 24.		16.5	FAINT BLUE STAR
RNGC 4992	13 06 33.	+ 11 54		14.5	GALAXY
ARC 1685	13 06 33.	+ 35 02		17.2	RICH CLUSTER OF GALAXIES
SVEN 380	13 06 33.	- 15 27	12	15.4	GALAXY
HN 1041	13 06 33.	- 51 42		14.5	NEBULA
IC 4200	13 06 33.	- 51 42			NONSTELLAR OBJECT
REIZ 3273	13 06 35.	+ 28 26	54	13.8	GALAXY
REIZ 3274	13 06 35.	+ 29 00	36	15.7	GALAXY
PATH 2.131	13 06 35.	+ 29 00	41		NEBULA
IC 4205	13 06 35.	+ 53 07 49.			NONSTELLAR OBJECT
REIN 4.232	13 06 35.99	+ 11 53 59.5			NEBULA
ZWG 044.012	13 06 36.	+ 06 07		15.6	GALAXY
ZWG 072.006	13 06 36.	+ 11 54		14.6	GALAXY
REIZ 3272	13 06 36.	+ 11 54	30	13.8	GALAXY
UGC 08232	13 06 36.	+ 11 54	72	14.6	GALAXY Sa
MCG+02-34-001	13 06 36.	+ 11 54	72	14.6	GALAXY
ZC 1306.6+2018	13 06 36.	+ 20 18	4230		CLUSTER OF GALAXIES
ZWG 160.149	13 06 36.	+ 28 18		15.6	GALAXY
ZWG 160.150	13 06 36.	+ 29 18		15.7	GALAXY
MCG+10-19-039	13 06 36.	+ 59 22	39	16.	GALAXY
REIN 4.231	13 06 36.60	- 04 58 39.2			NEBULA
ARC 1684	13 06 39.	+ 10 42		17.2	RICH CLUSTER OF GALAXIES
LB 02564	13 06 39.	+ 50 17 24.		15.8	FAINT BLUE STAR
SVEN 382	13 06 39.	- 15 10 32.	72	14.4	GALAXY
SVEN 381	13 06 39.	- 15 29	24	16.1	GALAXY
MCG-04-31-037	13 06 39.	- 24 08 30.	72	15.	GALAXY
IC 0853	13 06 40.	+ 53 01 31.			NONSTELLAR OBJECT
REIN 2.178	13 06 40.57	- 05 07 47.8			NEBULA
REIZ 3271	13 06 41.	- 05 00	12	13.8	GALAXY
REIZ 3270	13 06 41.	- 05 08	18	13.1	GALAXY
REIN 4.233	13 06 41.90	- 05 00 22.9			NEBULA
ZWG 044.013	13 06 42.	+ 02 37		15.5	GALAXY
ZWG 072.007	13 06 42.	+ 08 48		15.2	GALAXY
ZC 1306.7+1046	13 06 42.	+ 10 46	1880		CLUSTER OF GALAXIES
ZWG 130.013	13 06 42.	+ 22 44		15.6	GALAXY
8ZW 1306+22.7	13 06 42.	+ 22 44		15.6	COMPACT GALAXY
REIN 2.179	13 06 42.38	- 06 36 28.6			NEBULA
RNGC 4991	13 06 44.	+ 02 37		15.5	GALAXY
REIN 4.234	13 06 44.23	+ 08 47 52.1			NEBULA
SVEN 383	13 06 45.	- 15 04 32.	36	15.5	GALAXY
PATH 2.132	13 06 46.	+ 29 09	11		NEBULA
LB 02565	13 06 46.	+ 55 48 48.		17.0	FAINT BLUE STAR
REIZ 3276	13 06 47.	+ 28 56	18	15.8	GALAXY
ZWG 044.014	13 06 48.	+ 07 16		15.1	GALAXY
UGC 08233	13 06 48.	+ 07 16	102	15.1	GALAXY Sc
MCG+01-34-003	13 06 48.	+ 07 16	102	15.1	GALAXY
ZWG 044.015	13 06 48.	+ 07 46		15.7	GALAXY
MCG+04-31-010	13 06 48.	+ 22 42	30	15.6	GALAXY
ZC 1306.8+3113	13 06 48.	+ 31 13	610		CLUSTER OF GALAXIES
ZC 1306.8+4845	13 06 48.	+ 48 45	870		CLUSTER OF GALAXIES
MCG+08-24-052	13 06 48.	+ 50 02	12	16.	GALAXY
LB 02566	13 06 48.	+ 61 22 06.		15.0	FAINT BLUE STAR
ZWG 316.011	13 06 48.	+ 62 33		13.8	GALAXY
UGC 08234	13 06 48.	+ 62 33	90	13.8	GALAXY S0-a
MCG+10-19-040	13 06 48.	+ 62 33	60	12.	GALAXY
MCG+10-19-041	13 06 48.	+ 62 35	51	14.	GALAXY
RNGC 4990	13 06 48.	- 05 00		15.0	GALAXY
MCG-01-34-004	13 06 48.	- 05 00	18	15.	GALAXY
RNGC 4989	13 06 48.	- 05 09		14.0	GALAXY
MCG-01-34-005	13 06 48.	- 05 09	78	14.	GALAXY
MCG-01-34-006	13 06 48.	- 05 10	30	16.	GALAXY

OBJECT NAME	RIGHT ASCEN.	DECLINATION	DIAM.	MAGN.	TYPE OF OBJECT
LB 02567	13 06 50.	+ 55 14 36.		14.7	FAINT BLUE STAR
BIGO 525	13 06 51.	+ 62 32			NEBULA
SVEN 385	13 06 51.	- 15 17	12	16.1	GALAXY
SVEN 384	13 06 51.	- 15 17 32.	24	16.1	GALAXY
MCG-04-31-038	13 06 51.	- 26 54	36	15.5	GALAXY
FATH 2.133	13 06 52.	+ 29 38	14		NEBULA
TON-N 1554	13 06 52.	+ 31 56		17.0	BLUE STAR
TON-N 1555	13 06 52.	+ 35 00		16.7	BLUE STAR
PK305-00.1	13 06 52.	- 62 55	5		PLANETARY NEBULA
NAB 1306+35	13 06 53.	+ 35 01 32.		17.5	QUASI-STELLAR OBJECT
BC B382	13 06 53.0	+ 35 01 32.		17.55	QUASI-STELLAR OBJECT
ZWG 160.151	13 06 54.	+ 29 38		15.1	GALAXY
MCG+05-31-143	13 06 54.	+ 29 38 30.	24	15.1	GALAXY
ZC 1306.9+3125	13 06 54.	+ 31 25	2960		CLUSTER OF GALAXIES
ZWG 336.002	13 06 54.	+ 73 50		15.5	GALAXY
ZWG 335.034	13 06 54.	+ 73 50		15.5	GALAXY
SHB 249	13 06 54.2	+ 35 01 08.		17.6	QUASI-STELLAR OBJECT
BFG 147	13 06 54.22	+ 35 01 08.5		17.55	ULTRAVIOLET-EXCESS OBJECT
BFG 148	13 06 55.27	+ 34 32 51.8		18.95	ULTRAVIOLET-EXCESS OBJECT
LB 00676	13 06 56.	+ 56 11 48.		17.6	FAINT BLUE STAR
TON-N 1556	13 06 57.	+ 20 18		17.0	BLUE STAR
SVEN 386	13 06 57.	- 15 08 31.	12	16.2	GALAXY
MCG-03-34-003	13 06 57.	- 18 42	30	14.5	GALAXY
RNGC 4993	13 06 57.	- 23 08		14.0	GALAXY
RNGC 4988	13 06 58.	- 42 50			GALAXY
BIGO 526	13 06 59.	+ 62 33			NEBULA
ZWG 016.011	13 07 00.	+ 01 08		14.4	GALAXY
UGC 08235	13 07 00.	+ 01 08	120	14.4	GALAXY SB0/SBa
MCG+00-34-009	13 07 00.	+ 01 08	72	14.4	GALAXY
ZWG 016.012	13 07 00.	+ 01 57		13.5	GALAXY
UGC 08236	13 07 00.	+ 01 57	144	13.5	GALAXY SBb
MCG+00-34-010	13 07 00.	+ 01 57	138	13.5	GALAXY
ZWG 044.016	13 07 00.	+ 05 13		15.3	GALAXY
LB 00068	13 07 00.	+ 23 12		16.4	FAINT BLUE STAR
ZC 1307.0+3944	13 07 00.	+ 39 44	5780		CLUSTER OF GALAXIES
ZWG 217.014	13 07 00.	+ 43 36		15.7	GALAXY
ZC 1307.0+4555	13 07 00.	+ 45 55	1550		CLUSTER OF GALAXIES
MCG+08-24-053	13 07 00.	+ 46 30	18	17.	GALAXY
LB 02568	13 07 00.	+ 53 52 18.		15.0	FAINT BLUE STAR
1ZW 052	13 07 00.	+ 54 12			COMPACT GALAXY
ZWG 316.012	13 07 00.	+ 62 35		13.7	GALAXY
UGC 08237	13 07 00.	+ 62 35	60	13.7	GALAXY SBb
ZWG 365.008	13 07 00.	+ 83 08		15.7	GALAXY
KARA.73B 0574	13 07 00.	+ 83 08	36	15.7	ISOLATED GALAXY E
RNGC 4995	13 07 00.	- 07 34		12.0	GALAXY
MCG-02-34-005	13 07 00.	- 12 04	36	15.5	GALAXY
RNGC 4996	13 07 01.	+ 01 09		14.5	GALAXY
RNGC 4999	13 07 01.	+ 01 56		12.5	GALAXY
BSO 07	13 07 03.	+ 35 55 11.		18.34	BLUE STELLAR OBJECT
BC BSO7	13 07 03.	+ 35 55 11.		18.34	QUASI-STELLAR OBJECT
SVEN 387	13 07 03.	- 15 34	60	14.9	GALAXY
MCG-04-31-039	13 07 03.	- 23 08 30.	24	14.	GALAXY
BFG 149	13 07 04.18	+ 35 55 32.5		18.62	ULTRAVIOLET-EXCESS OBJECT
REIN 2.180	13 07 04.38	- 07 34 02.2			NEBULA
A2 479	13 07 04.5	+ 29 02 30.		17.7	FAINT BLUE OBJECT
REIZ 3279	13 07 05.	+ 29 15	18	16.0	GALAXY
IC 4206	13 07 05.	+ 39 17 20.			NONSTELLAR OBJECT
BC 4C12.46	13 07 05.5	+ 12 10 21.		18.49	QUASI-STELLAR OBJECT
SHB 250	13 07 05.6	+ 35 55 58.		18.3	QUASI-STELLAR OBJECT
UGC 08239	13 07 06.	+ 46 30	72	16.5	GALAXY Sc
MCG+08-24-054	13 07 06.	+ 46 30	72	16.	GALAXY
ZWG 016.013	13 07 06.	- 00 46		14.9	GALAXY
UGC 08238	13 07 06.	- 00 46	66	14.9	GALAXY Sc
MCG+00-34-011	13 07 06.	- 00 46	54	14.9	GALAXY
REIZ 3278	13 07 06.	- 07 34	90	11.8	GALAXY
MCG-02-34-006	13 07 06.	- 10 03	150	14.	GALAXY
FATH 2.134	13 07 08.	+ 29 15	14		NEBULA
IC 4207	13 07 09.	+ 38 05 20.			NONSTELLAR OBJECT
SVEN 388	13 07 09.	- 15 02 31.	12	16.3	GALAXY
RNGC 4994	13 07 09.	- 22 17			NON-EXISTENT OBJECT
REIN 4.235	13 07 09.53	- 07 34 12.4			NEBULA
A2 480	13 07 10.5	+ 28 52 42.		17.1	FAINT BLUE OBJECT
REIN 4.236	13 07 10.87	- 07 37 26.8			NEBULA
REIZ 3282	13 07 11.	+ 28 41	18	15.6	GALAXY
REIZ 3283	13 07 11.	+ 29 15	18	15.4	GALAXY
LB 00677	13 07 11.	+ 56 27 24.		17.3	FAINT BLUE STAR
ZWG 044.017	13 07 12.	+ 02 44		15.5	GALAXY
MCG+01-34-004	13 07 12.	+ 02 44	24	15.5	GALAXY
ZWG 044.018	13 07 12.	+ 06 29		15.6	GALAXY
ZWG 044.019	13 07 12.	+ 06 45		15.7	GALAXY
ZC 1307.2+0816	13 07 12.	+ 08 16	540		CLUSTER OF GALAXIES
ZC 1307.2+1324	13 07 12.	+ 13 24	740		CLUSTER OF GALAXIES
8ZW 1307+24.9	13 07 12.	+ 24 55		18.5	COMPACT GALAXY
MCG+10-19-042	13 07 12.	+ 62 27 30.	39	14.	GALAXY
MCG-03-34-004	13 07 12.	- 16 19	84	14.	GALAXY
MCG-04-31-040	13 07 12.	- 23 59	102	14.	GALAXY
MCG-07-27-037	13 07 12.	- 42 48 30.	66	14.	GALAXY
SN 1952E	13 07 13.	- 03 13		19.8	SUPERNOVA
MCG-02-34-007	13 07 15.	- 10 27 30.	42	15.	GALAXY
ZWG 044.020	13 07 18.	+ 06 43		15.4	GALAXY
REIZ 3281	13 07 18.	+ 11 55	18	14.7	GALAXY
ZWG 072.008	13 07 18.	+ 11 56		15.3	GALAXY
ZC 1307.3+1701	13 07 18.	+ 17 01	670		CLUSTER OF GALAXIES
ZWG 189.029	13 07 18.	+ 35 08		15.7	GALAXY
MCG+06-29-046	13 07 18.	+ 35 08	42	15.	GALAXY
MCG+08-24-055	13 07 18.	+ 47 48	30	16.	GALAXY
ZC 1307.3+5756	13 07 18.	+ 57 56	940		CLUSTER OF GALAXIES
MCG+10-19-043	13 07 18.	+ 58 07	39	16.	GALAXY
ZWG 294.021	13 07 18.	+ 62 26		14.2	GALAXY
UGC 08240	13 07 18.	+ 62 26	54	14.2	GALAXY E-S0
MCG-01-34-007	13 07 18.	- 07 34	132	12.	GALAXY
SER 102.04	13 07 18.	- 43 00	246		INTERACTING GALAXIES
REIN 4.237	13 07 18.97	+ 11 54 59.5			NEBULA
IC 4208	13 07 19.	+ 37 31 21.			NONSTELLAR OBJECT
RNGC 5007	13 07 20.	+ 62 26		14.0	GALAXY
TON-N 1557	13 07 21.	+ 22 50		16.5	BLUE STAR
SVEN 389	13 07 21.	- 15 29 31.	120	13.8	GALAXY
A2 482	13 07 21.9	+ 28 43 32.		18.5	FAINT BLUE OBJECT
RNGC 4997	13 07 22.	- 16 14		13.0	GALAXY
REIZ 3286	13 07 23.	+ 29 11	54	13.7	GALAXY
ZWG 130.014	13 07 24.	+ 24 51		15.1	GALAXY
MCG+04-31-011	13 07 24.	+ 24 51	36	15.1	GALAXY
HOLM 509A	13 07 24.	+ 25 50	36	14.2	PART OF MULTIPLE GALAXY
ZWG 160.152	13 07 24.	+ 29 10		14.0	GALAXY
UGC 08241	13 07 24.	+ 29 10	102	14.0	GALAXY SBb
MCG+05-31-144	13 07 24.	+ 29 10	102	14.0	GALAXY
FATH 2.135	13 07 24.	+ 29 10	108		NEBULA
ZWG 189.030	13 07 24.	+ 35 15		15.6	GALAXY

OBJECT NAME	RIGHT ASCEN.	DECLINATION	DIAM.	MAGN.	TYPE OF OBJECT
UGC 08242	13 07 24.	+ 35 15	66	15.6	GALAXY S
MCG+06-29-047	13 07 24.	+ 35 15	54	15.	GALAXY
ZWG 245.023	13 07 24.	+ 45 32		15.6	GALAXY
ZWG 271.020	13 07 24.	+ 53 45		14.6	GALAXY
UGC 08243	13 07 24.	+ 53 45	78	14.6	GALAXY SB
MCG+10-19-045	13 07 24.	+ 61 30	27	16.	GALAXY
MCG+10-19-044	13 07 24.	+ 62 26	30	16.	GALAXY
REIZ 3280	13 07 24.	- 07 22	24	15.7	GALAXY
MCG-03-34-005	13 07 24.	- 16 14	33	13.5	GALAXY
REIN 4.239	13 07 24.67	+ 09 25 56.1			NEBULA
IC 0854	13 07 25.	+ 24 50 39.			NONSTELLAR OBJECT
REIZ 3285	13 07 25.	+ 24 51	42	14.1	GALAXY
RNGC 5000	13 07 25.	+ 29 10		14.0	GALAXY
RNGC 5001	13 07 26.	+ 53 45		14.5	GALAXY
HOLM 509B	13 07 27.	+ 25 49	24	15.1	PART OF MULTIPLE GALAXY
MCG+05-31-145	13 07 27.	+ 30 20	54	15.4	GALAXY
TON-N 1558	13 07 28.	+ 32 08		17.	BLUE STAR
REIZ 3289	13 07 29.	+ 29 10	48	14.9	GALAXY
HOLM 510A	13 07 29.	+ 29 11	90	13.5	PART OF MULTIPLE GALAXY
ZWG 072.009	13 07 30.	+ 11 43		15.5	GALAXY
ZC 1307.5+1415	13 07 30.	+ 14 15	870		CLUSTER OF GALAXIES
MCG+05-31-146	13 07 30.	+ 28 40	96	15.	GALAXY
ZC 1307.5+3810	13 07 30.	+ 38 10	940		CLUSTER OF GALAXIES
MCG+08-24-056	13 07 30.	+ 45 32 30.	48	15.	GALAXY
ZWG 245.024	13 07 30.	+ 46 11		15.3	GALAXY
MCG+09-22-022	13 07 30.	+ 53 46	60	13.9	GALAXY
SER 094.04	13 07 30.	- 46 13	210	13.	GALAXY
REIN 4.238	13 07 30.11	- 07 20 00.7			NEBULA
A2 484	13 07 30.5	+ 28 45 24.		17.8	FAINT BLUE OBJECT
LB 02569	13 07 31.	+ 51 55 12.		16.1	FAINT BLUE STAR
REIZ 3293	13 07 31.	+ 53 47	42	13.9	GALAXY
BFG 150	13 07 31.35	+ 37 31 29.6		17.77	ULTRAVIOLET-EXCESS OBJECT
REIN 4.240	13 07 32.41	+ 09 20 52.3			NEBULA
REIN 4.241	13 07 32.99	+ 11 43 01.7			NEBULA
HOLM 510B	13 07 33.	+ 29 11	42	15.6	PART OF MULTIPLE GALAXY
LB 02570	13 07 33.	+ 59 49 00.		15.0	FAINT BLUE STAR
REIZ 3291	13 07 35.	+ 28 39	30	15.8	GALAXY
REIZ 3288	13 07 36.	+ 11 43	18	15.4	GALAXY
ZWG 101.005	13 07 36.	+ 19 59		15.1	GALAXY
UGC 08244	13 07 36.	+ 28 39	78	16.0	GALAXY Sc
FATH 1.568	13 07 36.	+ 29 05	14		NEBULA
ZWG 160.153	13 07 36.	+ 30 18		15.4	GALAXY
ZC 1307.6+3245	13 07 36.	+ 32 45	1140		CLUSTER OF GALAXIES
MCG+08-24-057	13 07 36.	+ 46 17 30.	30	16.	GALAXY
ZWG 294.022	13 07 36.	+ 59 33		15.4	GALAXY
ZWG 294.023	13 07 36.	+ 62 25		15.6	GALAXY
ZWG 353.012	13 07 36.	+ 79 12		14.7	GALAXY
ZWG 352.057	13 07 36.	+ 79 12		14.7	GALAXY
UGC 08245	13 07 36.	+ 79 12	108	14.7	GALAXY IRR
8ZW 1307-00.4	13 07 36.	- 00 22		19.4	COMPACT GALAXY
REIZ 3284	13 07 36.	- 07 11	18	13.6	GALAXY
FATH 2.136	13 07 38.	+ 30 18	24		NEBULA
BIGO 527	13 07 39.	+ 62 22			NEBULA
SVEN 390	13 07 39.	- 14 52 31.	48	14.2	GALAXY
TON-N 1559	13 07 40.	+ 35 24		15.0	BLUE STAR
REIN 4.242	13 07 40.91	- 07 11 19.6			NEBULA
MCG+03-36-067	13 07 42.	+ 17 47 30.	18	15.7	GALAXY
MCG+03-36-068	13 07 42.	+ 17 59	300	14.3	GALAXY
8ZW 1307+24.4	13 07 42.	+ 24 25		17.6	COMPACT GALAXY
MCG+06-29-048	13 07 42.	+ 34 26	180	13.	GALAXY
ZWG 189.031	13 07 42.	+ 34 27		15.1	GALAXY
UGC 08246	13 07 42.	+ 34 27	198	15.1	GALAXY SBc
MCG+08-24-058	13 07 42.	+ 46 10	36	16.	GALAXY
ZWG 271.021	13 07 42.	+ 55 45		15.7	GALAXY
UGC 08247	13 07 42.	+ 55 45	66	15.7	GALAXY S0
ZC 1307.7+5829	13 07 42.	+ 58 29	1550		CLUSTER OF GALAXIES
ZWG 294.024	13 07 42.	+ 59 17		15.3	GALAXY
MCG+10-19-046	13 07 42.	+ 59 33	27	16.	GALAXY
REIZ 3287	13 07 42.	- 06 54	48	13.9	GALAXY
MCG-03-34-006	13 07 42.	- 21 24 30.	48	15.	GALAXY
MCG-06-29-014	13 07 42.	- 39 20	60	14.	GALAXY
REIZ 3292	13 07 44.	+ 34 27	180	13.0	GALAXY
BFG 151	13 07 44.39	+ 33 02 20.5		18.95	ULTRAVIOLET-EXCESS OBJECT
MCG+05-31-147	13 07 45.	+ 31 44	30	15.0	GALAXY
SVEN 391	13 07 45.	- 14 48 30.	30	15.0	GALAXY
SVEN 392	13 07 45.	- 15 17 30.	36	15.1	GALAXY
ABC 1687	13 07 46.	+ 58 41		17.5	RICH CLUSTER OF GALAXIES
IC 4209	13 07 46.0	- 06 54 19.			GALAXY
ZWG 016.015	13 07 48.	+ 02 18		15.4	GALAXY
MCG+00-34-013	13 07 48.	+ 02 18	36	15.4	GALAXY
ZC 1307.8+0835	13 07 48.	+ 08 35	1210		CLUSTER OF GALAXIES
MCG+03-36-069	13 07 48.	+ 17 51	60	15.2	GALAXY
MCG+03-36-070	13 07 48.	+ 18 36	120	14.5	GALAXY
TON-N 0145	13 07 48.	+ 32 16		15.6	BLUE STAR
MCG+08-24-059	13 07 48.	+ 48 15	24	16.	GALAXY
MCG+08-24-060	13 07 48.	+ 49 56	24	16.	GALAXY
ZC 1307.8+5902	13 07 48.	+ 59 02	2150		CLUSTER OF GALAXIES
MCG+13-10-002	13 07 48.	+ 79 12	102	14.	GALAXY
ZWG 016.014	13 07 48.	- 00 44		14.6	GALAXY
MCG+00-34-012	13 07 48.	- 00 44	27	14.6	GALAXY
MCG-01-34-008	13 07 48.	- 07 11	42	15.	GALAXY
REIZ 3290	13 07 48.	- 07 23	24	13.8	GALAXY
VV 047B	13 07 48.	- 21 19	9	18.	INTERACTING GALAXY
VV 047A	13 07 48.	- 21 19	24	17.	INTERACTING GALAXY
MCG-03-34-007	13 07 48.	- 21 19	30	16.	GALAXY
MRSL 305-00/2	13 07 48.	- 62 33	900		HII REGION
BFG 152	13 07 48.62	+ 37 30 01.5		18.62	ULTRAVIOLET-EXCESS OBJECT
REIZ 3296	13 07 51.	+ 32 43	42	15.3	GALAXY
MCG+06-29-049	13 07 51.	+ 32 43 30.	72	15.	GALAXY
MCG-03-34-008	13 07 51.	- 16 38 30.	72	14.5	GALAXY
REIN 4.243	13 07 52.17	- 07 23 06.5			NEBULA
ZWG 044.021	13 07 54.	+ 06 36		15.7	GALAXY
ZC 1307.9+1114	13 07 54.	+ 11 14	810		CLUSTER OF GALAXIES
ZC 1307.9+1258	13 07 54.	+ 12 58	1280		CLUSTER OF GALAXIES
ZC 1307.9+2115	13 07 54.	+ 21 15	2550		CLUSTER OF GALAXIES
MCG+10-19-047	13 07 54.	+ 59 17	30	16.	GALAXY
MCG-01-34-009	13 07 54.	- 06 55	72	14.5	GALAXY
MCG-03-34-009	13 07 54.	- 21 28	96	14.5	GALAXY
MCG-07-27-038	13 07 54.	- 41 18	60	14.	GALAXY
SN 1954H	13 07 56.	- 07 23		17.4	SUPERNOVA
REIZ 3295	13 07 57.	+ 18 42	42	14.6	GALAXY
LB 02571	13 07 57.	+ 55 32 06.		16.5	FAINT BLUE STAR
REIN 4.244	13 07 59.95	+ 09 17 21.3			NEBULA
8ZW 1308+08.1	13 08 00.	+ 08 05		16.0	COMPACT GALAXY
ZWG 072.010	13 08 00.	+ 11 18		15.7	GALAXY
MCG+03-36-071	13 08 00.	+ 16 34	72	15.6	GALAXY
ZWG 101.006	13 08 00.	+ 18 43		14.8	GALAXY
UGC 08248	13 08 00.	+ 18 43	66	14.8	GALAXY S

OBJECT NAME	RIGHT ASCEN.	DECLINATION	DIAM.	MAGN.	TYPE OF OBJECT
8ZW 1308+23.7	13 08 00.	+ 23 44		17.3	COMPACT GALAXY
UGC 08249	13 08 00.	+ 25 12	60	17.	GALAXY DWRF IR
ZC 1308.0+2705	13 08 00.	+ 27 05	1810		CLUSTER OF GALAXIES
ZWG 160.154	13 08 00.	+ 31 43		15.0	GALAXY
ZWG 189.032	13 08 00.	+ 32 45		15.6	GALAXY
UGC 08250	13 08 00.	+ 32 45	96	15.6	GALAXY Sc
MCG+09-22-024	13 08 00.	+ 50 46	42	15.	GALAXY
MCG+09-22-023	13 08 00.	+ 52 47 30.	36	14.	GALAXY
LB 02572	13 08 00.	+ 54 21 48.		16.2	FAINT BLUE STAR
ZWG 294.025	13 08 00.	+ 60 29		15.6	GALAXY
ZWG 353.013	13 08 00.	+ 80 02		15.5	GALAXY
ZWG 352.058	13 08 00.	+ 80 02		15.5	GALAXY
MCG+14-06-016	13 08 00.	+ 84 55	30	15.	GALAXY
HN 0371	13 08 00.	- 06 54			NEBULA
MCG-01-34-010	13 08 00.	- 07 23	42	15.5	GALAXY
IC 0855	13 08 02.	- 04 13 09.			NONSTELLAR OBJECT
MCG-04-31-041	13 08 03.	- 23 37	72	14.	GALAXY
REIN 4.245	13 08 04.52	+ 09 21 25.0			NEBULA
REIZ 3294	13 08 05.	- 03 56	24	14.1	GALAXY
LB 00032	13 08 06.	+ 28 24		16.8	FAINT BLUE STAR
ZC 1308.1+3031	13 08 06.	+ 30 31	270		CLUSTER OF GALAXIES
MCG+10-19-048	13 08 06.	+ 60 29	30	16.	GALAXY
8ZW 1308-01.1	13 08 06.	- 01 05		19.7	COMPACT GALAXY
ZWG 016.016	13 08 12.	+ 00 18		15.4	GALAXY
8ZW 1308+12.0	13 08 12.	+ 11 58		16.6	COMPACT GALAXY
ZC 1308.2+1700	13 08 12.	+ 17 00	5780		CLUSTER OF GALAXIES
ZWG 130.015	13 08 12.	+ 20 49		15.2	GALAXY
MCG+06-29-050	13 08 12.	+ 34 53	27	15.5	GALAXY
ZWG 189.033	13 08 12.	+ 34 54		15.7	GALAXY
UGC 08251	13 08 12.	+ 34 54	72	15.7	GALAXY S0
ZC 1308.2+3510	13 08 12.	+ 35 10	940		CLUSTER OF GALAXIES
ZC 1308.2+3531	13 08 12.	+ 35 31	9610		CLUSTER OF GALAXIES
UGC 08252	13 08 12.	+ 44 15	60	16.5	GALAXY Sc
MCG+11-16-011	13 08 12.	+ 63 22	45	16.	GALAXY
ZC 1308.2+6656	13 08 12.	+ 66 56	1140		CLUSTER OF GALAXIES
SN 1956J	13 08 13.	+ 00 18		19.8	SUPERNOVA
LB 02573	13 08 14.	+ 52 07 06.		16.4	FAINT BLUE STAR
REIN 4.246	13 08 14.56	+ 11 58 20.4			NEBULA
SVEN 393	13 08 15.	- 15 24 30.	24	15.7	GALAXY
IC 0856	13 08 16.	+ 20 48 16.			NONSTELLAR OBJECT
LB 02574	13 08 16.	+ 51 41 54.		17.0	FAINT BLUE STAR
SHB 251	13 08 16.6	+ 38 12 42.		17.6	QUASI-STELLAR OBJECT
BFG 154	13 08 16.63	+ 38 12 42.1		17.56	ULTRAVIOLET-EXCESS OBJECT
BC B360	13 08 17.6	+ 38 14 17.4		17.56	QUASI-STELLAR OBJECT
UGC 08253	13 08 18.	+ 11 58	96	17.	GALAXY DWRF SP
ZWG 072.011	13 08 18.	+ 12 08		15.6	GALAXY
ZWG 189.034	13 08 18.	+ 36 53		14.7	GALAXY
UGC 08254	13 08 18.	+ 36 53	102	14.7	GALAXY
MCG+06-29-051	13 08 18.	+ 36 54	96	13.5	GALAXY
ZC 1308.3+4456	13 08 18.	+ 44 56	13640		CLUSTER OF GALAXIES
MCG-05-31-038	13 08 18.	- 27 44	48	16.	GALAXY
BFG 153	13 08 18.88	+ 33 39 10.5		19.06	ULTRAVIOLET-EXCESS OBJECT
ARC 1686	13 08 19.	+ 22 09		18.0	RICH CLUSTER OF GALAXIES
REIZ 3299	13 08 19.	+ 36 55	54	14.8	GALAXY
LB 02575	13 08 20.	+ 56 07 36.		17.1	FAINT BLUE STAR
RNGC 5002	13 08 21.	+ 36 53		14.5	GALAXY
RNGC 5003	13 08 22.	+ 42 33			GALAXY
REIZ 3297	13 08 23.	+ 12 09	30	14.6	GALAXY
BFG 155	13 08 23.40	+ 33 57 43.7		18.36	ULTRAVIOLET-EXCESS OBJECT
BFG 156	13 08 23.58	+ 34 39 50.0		18.45	ULTRAVIOLET-EXCESS OBJECT
ZWG 016.017	13 08 24.	+ 01 25		15.4	GALAXY
ZWG 072.012	13 08 24.	+ 11 44		14.4	GALAXY
UGC 08255	13 08 24.	+ 11 44	102	14.4	GALAXY Sc
MCG+02-34-002	13 08 24.	+ 11 44	84	14.4	GALAXY
FATH 1.569	13 08 24.	+ 29 15	19		NEBULA
ZWG 160.155	13 08 24.	+ 29 59		15.3	GALAXY
MCG+05-31-148	13 08 24.	+ 30 00	42	15.3	GALAXY
ZC 1308.4+4849	13 08 24.	+ 48 49	3560		CLUSTER OF GALAXIES
MCG-07-27-039	13 08 24.	- 44 17	126	14.	GALAXY
MRSL 305+03/1	13 08 24.	- 59 00	7200		HII REGION
BIGO 528	13 08 25.	+ 29 58			NEBULA
HOLM 511C	13 08 25.	+ 29 59	36	14.2	PART OF MULTIPLE GALAXY
RNGC 5004B	13 08 25.	+ 30 00		15.5	GALAXY
REIN 4.247	13 08 25.31	+ 11 59 43.6			NEBULA
IC 4210	13 08 26.	+ 29 58	41		FAINT NEBULA
FATH 2.137	13 08 26.	+ 29 58	41		NEBULA
REIN 4.248	13 08 26.88	+ 11 44 34.4			NEBULA
TON-N 1560	13 08 27.	+ 22 23		17.0	BLUE STAR
SC 1306+4618.8	13 08 27.	+ 46 02 48.	48		NEBULA
REIZ 3300	13 08 28.	+ 29 59	36	14.3	GALAXY
BFG 157	13 08 28.76	+ 34 42 56.0		19.31	ULTRAVIOLET-EXCESS OBJECT
BC 4C18.36	13 08 29.47	+ 18 15 33.8		17.5	QUASI-STELLAR OBJECT
SHB 252	13 08 29.5	+ 18 15 33.		17.5	QUASI-STELLAR OBJECT
8ZW 1308+03.9	13 08 30.	+ 03 54		18.2	COMPACT GALAXY
REIZ 3298	13 08 30.	+ 11 44	72	13.8	GALAXY
ZC 1308.5+2215	13 08 30.	+ 22 15	270		CLUSTER OF GALAXIES
ZWG 217.015	13 08 30.	+ 44 17		15.3	GALAXY
MRSL 305-01/1	13 08 30.	- 64 25	21600		HII REGION
LB 02576	13 08 31.	+ 56 05 36.		17.3	FAINT BLUE STAR
TON-N 1561	13 08 33.	+ 19 33		15.8	BLUE STAR
REIN 4.249	13 08 34.28	+ 09 04 25.3			NEBULA
ZWG 072.013	13 08 36.	+ 09 06		15.1	GALAXY
LB 00033	13 08 36.	+ 24 11		18.2	FAINT BLUE STAR
LB 00069	13 08 36.	+ 27 15		15.1	FAINT BLUE STAR
MCG+05-31-150	13 08 36.	+ 29 51	78	15.3	GALAXY
MCG+05-31-149	13 08 36.	+ 29 55	42	14.3	GALAXY
ZWG 189.035	13 08 36.	+ 37 19		10.6	GALAXY
UGC 08256	13 08 36.	+ 37 19	378	10.6	GALAXY Sb
MCG+06-29-052	13 08 36.	+ 37 20	300	10.	GALAXY
ZC 1308.6+4621	13 08 36.	+ 46 21	1950		CLUSTER OF GALAXIES
UGC 08257	13 08 36.	+ 50 10	72	16.0	GALAXY Sc
ZWG 245.025	13 08 36.	+ 50 21		15.6	GALAXY
UGC 08258	13 08 36.	+ 50 21	78	15.6	GALAXY SBb
7ZW 500	13 08 36.	+ 59 31			COMPACT GALAXY
MRK 243	13 08 36.	+ 60 49	12	16.	GALAXY WITH UV CONTINUUM
REIN 1.119	13 08 36.32	+ 37 19 24.2			NEBULA
RNGC 5004C	13 08 37.	+ 29 51		15.5	GALAXY
RNGC 5009	13 08 37.	+ 50 21		15.5	GALAXY
REIN 2.181	13 08 37.85	+ 37 19 29.0			NEBULA
IC 4211	13 08 38.	+ 37 26 29.			SINGLE STAR
REIN 1.120	13 08 38.02	+ 37 19 30.5			NEBULA
FATH 2.139	13 08 39.	+ 29 50	38		NEBULA
FATH 2.138	13 08 39.	+ 29 54	16		NEBULA
RNGC 5005	13 08 39.	+ 37 19		11.5	GALAXY
MCG+08-24-061	13 08 39.	+ 50 22 30.	60	15.	GALAXY
REIN 1.121	13 08 39.70	+ 37 20 35.1			NEBULA
HOLM 511B	13 08 40.	+ 29 51	54	14.5	PART OF MULTIPLE GALAXY
REIZ 3301	13 08 40.	+ 29 55	18	14.0	GALAXY
HOLM 511A	13 08 40.	+ 29 55	60	14.2	PART OF MULTIPLE GALAXY
TON-N 1562	13 08 40.	+ 31 40		16.9	BLUE STAR
REIN 1.122	13 08 40.97	+ 37 19 14.8			NEBULA
LB 02577	13 08 41.	+ 57 09 36.		14.6	FAINT BLUE STAR
BFG 158	13 08 41.80	+ 32 58 27.0			ULTRAVIOLET-EXCESS OBJECT
8ZW 1308+06.6	13 08 42.	+ 06 34		17.2	COMPACT GALAXY
ZWG 160.156	13 08 42.	+ 29 50		15.3	GALAXY
UGC 08259	13 08 42.	+ 29 50	84	15.3	GALAXY SBa-b
ZWG 160.157	13 08 42.	+ 29 54		14.3	GALAXY
UGC 08260	13 08 42.	+ 29 54	96	14.3	GALAXY S0
UGC 08261	13 08 42.	+ 35 46	60	16.0	GALAXY IRR
MCG+06-29-053	13 08 42.	+ 35 46	54	15.	GALAXY
ZC 1308.7+3843	13 08 42.	+ 38 43	400		CLUSTER OF GALAXIES
MCG+08-24-062	13 08 42.	+ 50 08	18	16.	GALAXY
MCG+08-24-063	13 08 42.	+ 50 10	60	16.	GALAXY
REIN 1.123	13 08 42.83	+ 37 20 06.7			NEBULA
RNGC 5004A	13 08 43.	+ 29 54		14.5	GALAXY
REIZ 3303	13 08 43.	+ 37 20	300	11.6	GALAXY
BFG 159	13 08 43.76	+ 32 54 18.7		16.45	ULTRAVIOLET-EXCESS OBJECT
LB 02578	13 08 45.	+ 55 35 36.		16.9	FAINT BLUE STAR
REIZ 3302	13 08 46.	+ 29 51	54	14.7	GALAXY
BFG 160	13 08 47.03	+ 33 12 49.1		19.29	ULTRAVIOLET-EXCESS OBJECT
ZWG 016.018	13 08 48.	+ 00 01		14.3	GALAXY
UGC 08262	13 08 48.	+ 00 01	78	14.3	GALAXY S
MCG+00-34-014	13 08 48.	+ 00 01	36	14.3	GALAXY
ZWG 044.022	13 08 48.	+ 03 40		15.4	GALAXY
UGC 08263	13 08 48.	+ 03 40	72	15.4	GALAXY SBc
SN 1959C	13 08 48.	+ 03 40		13.6	SUPERNOVA
MCG+01-34-005	13 08 48.	+ 03 40	90	15.4	GALAXY
ZWG 072.014	13 08 48.	+ 10 21		15.7	GALAXY
8ZW 1308+11.3	13 08 48.	+ 11 18		18.0	COMPACT GALAXY
TON-N 0146	13 08 48.	+ 28 53		14.7	BLUE STAR
LB 00263	13 08 48.	+ 30 31 18.		15.9	FAINT BLUE STAR
ZC 1308.8-0103	13 08 48.	- 01 03	810		CLUSTER OF GALAXIES
REIN 4.250	13 08 48.41	+ 10 20 49.0			NEBULA
REIN 4.251	13 08 48.49	+ 10 20 29.4			NEBULA
REIN 4.252	13 08 50.31	+ 09 05 51.9			NEBULA
ARC 1690	13 08 51.	+ 19 40		17.2	RICH CLUSTER OF GALAXIES
8ZW 1308+03.9	13 08 54.	+ 03 57		16.2	COMPACT GALAXY
RNGC 5008	13 08 54.	+ 25 39			NON-EXISTENT OBJECT
MCG+10-19-049	13 08 54.	+ 56 52	57	15.	GALAXY
8ZW 1308-00.3	13 08 54.	- 00 18		17.6	COMPACT GALAXY
ARC 1688	13 08 55.	- 04 25		17.6	RICH CLUSTER OF GALAXIES
LYNG 08	13 08 55.	- 60 55 36.	3600		OB CONCENTRATION
BFG 161	13 08 55.58	+ 33 27 36.0		18.91	ULTRAVIOLET-EXCESS OBJECT
SVEN 394	13 08 57.	- 15 06	48	14.4	GALAXY
ARC 1689	13 08 58.	- 01 06		17.6	RICH CLUSTER OF GALAXIES
HMS 1309-0105	13 09	- 01 05			CLUSTER OF GALAXIES
ZWG 016.019	13 09 00.	+ 00 56		14.4	GALAXY
UGC 08265	13 09 00.	+ 00 56	48	14.4	GALAXY SB
MCG+00-34-015	13 09 00.	+ 00 56	36	14.4	GALAXY
ZWG 072.015	13 09 00.	+ 09 00		15.1	GALAXY
ZC 1309.0+1936	13 09 00.	+ 19 36	1080		CLUSTER OF GALAXIES
MCG+06-29-054	13 09 00.	+ 34 40	48	15.	GALAXY
ZWG 189.036	13 09 00.	+ 34 41		15.5	GALAXY
UGC 08266	13 09 00.	+ 34 41	84	15.5	GALAXY E-S0
STOCK 39	13 09 00.	+ 39 35			BLUE KNOT NEAR ELLIP GLXY
ZWG 217.016	13 09 00.	+ 44 00		15.7	GALAXY
UGC 08267	13 09 00.	+ 44 00	78	15.7	GALAXY S
ZWG 245.026	13 09 00.	+ 45 56		15.4	GALAXY
UGC 08268	13 09 00.	+ 45 56	66	15.4	GALAXY SBb
MCG+08-24-064	13 09 00.	+ 46 54	48	16.	GALAXY
UGC 08269	13 09 00.	+ 46 58	60	16.0	GALAXY SB
7ZW 501	13 09 00.	+ 84 53			COMPACT GALAXY
ZWG 366.002	13 09 00.	+ 84 53		14.5	GALAXY
ZWG 365.009	13 09 00.	+ 84 53		14.5	GALAXY
UGC 08264	13 09 00.	+ 84 53	120	14.5	GALAXY PECULR
REIN 4.253	13 09 01.00	+ 09 00 32.3			NEBULA
LB 02579	13 09 02.	+ 54 48 36.		16.7	FAINT BLUE STAR
LB 02580	13 09 03.	+ 51 19 18.		16.7	FAINT BLUE STAR
TON-N 1563	13 09 04.	+ 32 03		14.8	BLUE STAR
ARC 1691	13 09 04.	+ 39 29		15.4	RICH CLUSTER OF GALAXIES
ARC 1693	13 09 04.	+ 48 46		17.2	RICH CLUSTER OF GALAXIES
LB 02581	13 09 04.	+ 52 39 54.		16.1	FAINT BLUE STAR
8ZW 1309+06.6	13 09 06.	+ 06 34		18.6	COMPACT GALAXY
ZWG 072.016	13 09 06.	+ 10 23		15.3	GALAXY
REIZ 3304	13 09 06.	+ 10 23	18	14.5	GALAXY
ZC 1309.1+1934	13 09 06.	+ 19 34	3560		CLUSTER OF GALAXIES
ZC 1309.1+2216	13 09 06.	+ 22 16	870		CLUSTER OF GALAXIES
ZWG 130.016	13 09 06.	+ 23 12		13.6	GALAXY
UGC 08270	13 09 06.	+ 23 12	180	13.6	GALAXY SBc
TON-N 0698	13 09 06.	+ 27 18		16.8	BLUE STAR
MCG+08-24-065	13 09 06.	+ 46 59	60	16.	GALAXY
MCG+08-24-066	13 09 06.	+ 47 01	42	16.	GALAXY
ZC 1309.1+6920	13 09 06.	+ 69 20	870		CLUSTER OF GALAXIES
MCG-03-34-010	13 09 06.	- 16 07	48	15.5	GALAXY
REIN 4.254	13 09 06.51	+ 10 23 40.5			NEBULA
MAI 083	13 09 08.	+ 85 50	40		DWARF SPHEROIDAL GALAXY
SVEN 395	13 09 09.	- 15 11	30	14.3	GALAXY
RNGC 5006	13 09 10.	- 18 58		14.0	GALAXY
RNGC 5012	13 09 11.	+ 23 11		12.5	GALAXY
REIN 2.182	13 09 11.76	+ 23 10 54.3			NEBULA
ZWG 072.017	13 09 12.	+ 09 22		15.4	GALAXY
REIZ 3305	13 09 12.	+ 10 51	48	14.4	GALAXY
MCG+04-31-012	13 09 12.	+ 23 12	168	13.6	GALAXY
STOCK 40	13 09 12.	+ 31 47			BLUE KNOT NEAR ELLIP GLXY
MCG+05-31-151	13 09 12.	+ 31 48	21	14.6	GALAXY
ZWG 189.037	13 09 12.	+ 36 32		13.5	GALAXY
MRK 449	13 09 12.	+ 36 32	60	14.	GALAXY WITH UV CONTINUUM
UGC 08271	13 09 12.	+ 36 32	102	13.5	GALAXY S
MCG+06-29-055	13 09 12.	+ 36 32	60	13.	GALAXY
MCG+08-24-067	13 09 12.	+ 45 56	60	15.	GALAXY
UGC 08272	13 09 12.	+ 47 49	102	16.0	GALAXY SBc
MCG+08-24-068	13 09 12.	+ 47 49	24	15.	GALAXY
ZWG 271.022	13 09 12.	+ 53 28		15.5	GALAXY
ZC 1309.2+7244	13 09 12.	+ 72 44	610		CLUSTER OF GALAXIES
ZC 1309.2+7529	13 09 12.	+ 75 29	1210		CLUSTER OF GALAXIES
ZC 1309.2-0032	13 09 12.	- 00 32	2350		CLUSTER OF GALAXIES
MCG-03-34-011	13 09 12.	- 18 58 30.	90	14.	GALAXY
REIZ 3306	13 09 13.	+ 36 33	60	14.6	GALAXY
BFG 163	13 09 13.95	+ 37 50 03.3		17.65	ULTRAVIOLET-EXCESS OBJECT
REIZ 3308	13 09 14.	+ 57 49	60	14.8	GALAXY
TON-N 1564	13 09 15.	+ 22 23		17.0	BLUE STAR
BSO 08	13 09 15.	+ 34 02 19.		17.43	BLUE STELLAR OBJECT
BC BSO8	13 09 15.	+ 34 02 19.		17.43	QUASI-STELLAR OBJECT
RNGC 5014	13 09 15.	+ 36 33		13.5	GALAXY
SHB 253	13 09 15.2	+ 34 03 08.		17.4	QUASI-STELLAR OBJECT
LB 02582	13 09 16.	+ 52 24 18.		15.2	FAINT BLUE STAR

OBJECT NAME	RIGHT ASCEN.	DECLINATION	DIAM.	MAGN.	TYPE OF OBJECT
BFG 162	13 09 16.86	+ 34 02 44.5		17.84	ULTRAVIOLET-EXCESS OBJECT
BFG 164	13 09 17.00	+ 34 50 53.9		18.59	ULTRAVIOLET-EXCESS OBJECT
ZWG 072.018	13 09 18.	+ 10 52		15.2	GALAXY
ZWG 101.007	13 09 18.	+ 14 50		15.7	GALAXY
ZC 1309.3+2255	13 09 18.	+ 22 55	10750		CLUSTER OF GALAXIES
ZC 1309.3+3145	13 09 18.	+ 31 45	740		CLUSTER OF GALAXIES
ZWG 160.158	13 09 18.	+ 31 46		14.6	GALAXY
ZWG 245.027	13 09 18.	+ 46 37		15.4	GALAXY
MCG+08-24-069	13 09 18.	+ 47 49	30	15.	GALAXY
MCG+00-34-016	13 09 18.	- 01 04	6	20.	GALAXY
MCG-02-34-008	13 09 18.	- 11 48	48	15.	GALAXY
REIN 4.255	13 09 18.36	+ 10 08 42.6			NEBULA
REIN 4.256	13 09 18.86	+ 10 52 02.3			NEBULA
ARC 1694	13 09 21.	+ 34 17		17.2	RICH CLUSTER OF GALAXIES
RNGC 5011A	13 09 22.	- 43 03			GALAXY
HN 0372	13 09 23.	- 04 04			NEBULA
TON-N 0699	13 09 24.	+ 29 29		16.6	BLUE STAR
MCG+08-24-070	13 09 24.	+ 46 36	42	15.	GALAXY
MCG-07-27-040	13 09 24.	- 39 38	42	14.5	GALAXY
MCG-07-27-041	13 09 24.	- 43 02	66	15.	GALAXY
DV.56 N5011A	13 09 24.	- 43 03	96		Sc GALAXY
AGU 44	13 09 24.	- 43 03 00.	66	13.5	PECULIAR GALAXY
LB 02583	13 09 25.	+ 53 06 24.		13.8	FAINT BLUE STAR
REIZ 3311	13 09 26.	+ 57 48	60	14.6	GALAXY
REIN 4.257	13 09 26.10	+ 11 36 37.0			NEBULA
ZWG 016.020	13 09 30.	+ 01 55		15.5	GALAXY
ZWG 072.019	13 09 30.	+ 12 37		15.7	GALAXY
ZWG 101.008	13 09 30.	+ 14 50		15.7	GALAXY
ZWG 130.017	13 09 30.	+ 21 04		15.5	GALAXY
UGC 08273	13 09 30.	+ 21 04	60	15.5	GALAXY S0
MCG+06-29-056	13 09 30.	+ 34 37	48	14.5	GALAXY
MCG+08-24-071	13 09 30.	+ 45 11	18	18.	GALAXY
MCG+08-24-072	13 09 30.	+ 50 31	60	16.	GALAXY
UGC 08274	13 09 30.	+ 50 32	72	17.	GALAXY
ZC 1309.5+6403	13 09 30.	+ 64 03	2490		CLUSTER OF GALAXIES
MCG-01-34-011	13 09 30.	- 06 44	120	14.5	GALAXY
MCG-03-34-012	13 09 30.	- 15 57	54	14.	GALAXY
MCG+08-24-073	13 09 33.	+ 45 11	15	18.	GALAXY
ZWG 044.023	13 09 36.	+ 03 15		15.6	GALAXY
UGC 08275	13 09 36.	+ 03 15	72	15.6	GALAXY
ZWG 044.024	13 09 36.	+ 03 28		15.0	GALAXY
MCG+01-34-007	13 09 36.	+ 03 28	42	15.0	GALAXY
MCG+01-34-006	13 09 36.	+ 03 40	84	15.6	GALAXY
ZWG 044.025	13 09 36.	+ 04 08		15.3	GALAXY
UGC 08276	13 09 36.	+ 05 44	60	18.	GALAXY DWARF
ZWG 072.020	13 09 36.	+ 11 29		15.7	GALAXY
ZWG 189.038	13 09 36.	+ 34 38		14.8	GALAXY
UGC 08277	13 09 36.	+ 41 09	66	16.5	GALAXY Sc
MCG+08-24-074	13 09 36.	+ 45 00	30	16.	GALAXY
MCG+08-24-075	13 09 36.	+ 45 21	30	17.	GALAXY
MCG-02-34-009	13 09 36.	- 12 37 30.	24	15.	GALAXY
RNGC 5013	13 09 38.	+ 03 28		15.0	GALAXY
LB 00264	13 09 39.	+ 29 03 42.		15.2	FAINT BLUE STAR
ARC 1692	13 09 40.	- 00 40		17.2	RICH CLUSTER OF GALAXIES
REIN 4.258	13 09 40.03	+ 11 30 03.4			NEBULA
ZC 1309.7+0601	13 09 42.	+ 06 01	610		CLUSTER OF GALAXIES
UGC 08278	13 09 42.	+ 21 40	78	17.	GALAXY DWARF
ZWG 130.018	13 09 42.	+ 22 42		15.7	GALAXY
8ZW 1309+23.9	13 09 42.	+ 23 53		17.1	COMPACT GALAXY
MCG+04-31-013	13 09 42.	+ 24 20	90	14.3	GALAXY
ZWG 130.019	13 09 42.	+ 24 21		14.3	GALAXY
UGC 08279	13 09 42.	+ 24 21	120	14.3	GALAXY Sb-c
KARA.73B 0575	13 09 42.	+ 24 21	114	14.3	ISOLATED GALAXY S
RNGC 5016	13 09 42.	+ 24 22		13.0	GALAXY
ZWG 160.159	13 09 42.	+ 27 35		15.0	GALAXY
MCG+05-31-152	13 09 42.	+ 27 35	15	15.0	GALAXY
LB 00070	13 09 42.	+ 27 45		17.0	FAINT BLUE STAR
MCG+08-24-076	13 09 42.	+ 47 12	24	16.	GALAXY
ZC 1309.7+6158	13 09 42.	+ 61 58	2020		CLUSTER OF GALAXIES
MCG-03-34-013	13 09 42.	- 19 10	36	15.	GALAXY
REIN 2.183	13 09 42.00	+ 24 21 36.2			NEBULA
BIGO 529	13 09 43.	+ 23 06			NEBULA
REIZ 3310	13 09 43.	+ 24 21	60	12.6	GALAXY
REIZ 3313	13 09 43.	+ 46 27	90	13.5	GALAXY
MCG+08-24-077	13 09 45.	+ 45 20	6	16.	GALAXY
LB 02584	13 09 46.	+ 55 24 36.		15.4	FAINT BLUE STAR
LB 00678	13 09 46.	+ 74 11 00.		16.2	FAINT BLUE STAR
REIZ 3307	13 09 47.	- 04 05	60	13.5	GALAXY
REIZ 3309	13 09 48.	+ 10 20	48	14.7	GALAXY
ZC 1309.8+2444	13 09 48.	+ 24 44	810		CLUSTER OF GALAXIES
ZWG 189.039	13 09 48.	+ 35 56		14.1	GALAXY
UGC 08280	13 09 48.	+ 35 56	168	14.1	GALAXY Sc
UGC 08281	13 09 48.	+ 45 05	72	16.0	GALAXY Sc
MCG+08-24-078	13 09 48.	+ 45 42	42	16.	GALAXY
ZWG 294.026	13 09 48.	+ 60 30		15.0	GALAXY
UGC 08282	13 09 48.	+ 60 30	120	15.0	GALAXY Sc
MCG+10-19-050	13 09 48.	+ 60 30	78	15.	GALAXY
MCG+11-16-012	13 09 48.	+ 62 55	39	17.	GALAXY
ZWG 016.022	13 09 48.	- 02 11		15.5	GALAXY
ZWG 016.021	13 09 48.	- 02 31		15.7	GALAXY
MCG+00-34-017	13 09 48.	- 02 31	42	15.7	GALAXY
RNGC 5015	13 09 48.	- 04 03		13.0	GALAXY
MCG-01-34-012	13 09 48.	- 04 03	78	13.	GALAXY
MCG+08-24-079	13 09 51.	+ 45 04 30.	24	16.	GALAXY
LB 02585	13 09 51.	+ 51 37 48.		16.2	FAINT BLUE STAR
RNGC 5010	13 09 52.	- 15 31		14.0	GALAXY
REIN 4.259	13 09 52.40	+ 12 26 17.3			NEBULA
ARC 1695	13 09 53.	+ 61 57		17.8	RICH CLUSTER OF GALAXIES
ZWG 072.021	13 09 54.	+ 08 55		15.4	GALAXY
ZWG 072.022	13 09 54.	+ 12 26		15.4	GALAXY
ZWG 160.160	13 09 54.	+ 26 57		14.6	GALAXY
MCG+05-31-153	13 09 54.	+ 26 57	42	14.6	GALAXY
MCG+06-29-057	13 09 54.	+ 35 56	144	13.5	GALAXY
ZWG 245.028	13 09 54.	+ 44 57		15.5	GALAXY
UGC 08283	13 09 54.	+ 44 57	60	15.5	GALAXY Sb
MCG+08-24-080	13 09 54.	+ 45 05	66	16.	GALAXY
ZWG 245.029	13 09 54.	+ 45 42		15.6	GALAXY
ZWG 245.030	13 09 54.	+ 46 28		14.3	GALAXY
RNGC 5021	13 09 54.	+ 46 28		14.5	GALAXY
UGC 08284	13 09 54.	+ 46 28	96	14.3	GALAXY SBb
7ZW 502	13 09 54.	+ 76 09			COMPACT GALAXY
HN 0373	13 09 54.	- 06 30			NEBULA
IC 4212	13 09 54.	- 06 30			NONSTELLAR OBJECT
MCG-03-34-015	13 09 54.	- 15 31	60	14.	GALAXY
MCG-03-34-014	13 09 54.	- 17 15 30.	150	12.5	GALAXY
LB 02586	13 09 58.	+ 51 09 36.		16.5	FAINT BLUE STAR
RNGC 5011	13 09 58.	- 42 50		13.0	GALAXY
RNGC 5023	13 09 59.	+ 44 18		13.0	GALAXY
VDB.66G 166	13 10	+ 36 35	100		DWARF GALAXY
ZWG 044.026	13 10 00.	+ 07 27		15.2	GALAXY
UGC 08285	13 10 00.	+ 07 27	120	15.2	GALAXY S-IRR
MCG+01-34-008	13 10 00.	+ 07 27	84	15.2	GALAXY
ZWG 072.023	13 10 00.	+ 11 59		15.2	GALAXY
REIZ 3315	13 10 00.	+ 26 57	42	14.7	GALAXY
ZWG 160.161	13 10 00.	+ 28 47		15.7	GALAXY
TON-N 0147	13 10 00.	+ 30 25		15.0	BLUE STAR
ZWG 217.017	13 10 00.	+ 44 18		13.2	GALAXY
UGC 08286	13 10 00.	+ 44 18	450	13.2	GALAXY Sc
MCG+08-24-081	13 10 00.	+ 44 57	60	15.	GALAXY
ZWG 245.031	13 10 00.	+ 45 07		15.5	GALAXY
MCG+08-24-082	13 10 00.	+ 45 37	42	15.	GALAXY
MCG+08-24-083	13 10 00.	+ 45 42	18	16.	GALAXY
MCG+08-24-084	13 10 00.	+ 46 28	90	13.5	GALAXY
ZWG 353.014	13 10 00.	+ 78 40		15.0	GALAXY
ZWG 352.059	13 10 00.	+ 78 40		15.0	GALAXY
UGC 08287	13 10 00.	+ 78 40	126	15.0	GALAXY SBa
MCG+13-10-003	13 10 00.	+ 78 40	66	15.	GALAXY
REIZ 3312	13 10 01.	+ 07 27	60	14.7	GALAXY
IC 4213	13 10 02.	+ 35 56 03.			NONSTELLAR OBJECT
REIZ 3317	13 10 02.	+ 44 18	240	13.3	GALAXY
MCG+08-24-085	13 10 03.	+ 46 40	24	16.	GALAXY
TON-N 1565	13 10 04.	+ 35 32		15.9	BLUE STAR
REIZ 3314	13 10 05.	+ 12 57	90	14.8	GALAXY
LB 00034	13 10 06.	+ 24 32		17.4	FAINT BLUE STAR
ZWG 189.040	13 10 06.	+ 34 20		15.7	GALAXY
SHAH 072	13 10 06.	+ 39 15	198	17.3	GROUP OF COMPACT GALAXIES
MRK 244	13 10 06.	+ 50 42	11	17.	GALAXY WITH UV CONTINUUM
MCG-07-27-042	13 10 06.	- 42 47 30.	90	12.9	GALAXY
RNGC 5020	13 10 09.	+ 12 52		13.5	GALAXY
MCG+06-29-058	13 10 09.	+ 34 18	27	15.5	GALAXY
REIN 4.260	13 10 10.74	+ 12 51 51.1			NEBULA
REIZ 3316	13 10 11.	+ 12 52	132	12.8	GALAXY
ZWG 044.027	13 10 12.	+ 05 00		14.5	GALAXY
UGC 08288	13 10 12.	+ 05 00	54	14.5	GALAXY S
MCG+01-34-009	13 10 12.	+ 05 00	48	14.5	GALAXY
ZWG 072.024	13 10 12.	+ 12 52		13.4	GALAXY
UGC 08289	13 10 12.	+ 12 52	198	13.4	GALAXY Sb/SBc
MCG+02-34-003	13 10 12.	+ 12 52	180	13.4	GALAXY
ZWG 130.020	13 10 12.	+ 23 06		14.8	GALAXY
8ZW 1310+23.1	13 10 12.	+ 23 06		14.8	COMPACT GALAXY
UGC 08290	13 10 12.	+ 23 06	96	14.8	GALAXY PECULR
MCG+08-24-086	13 10 12.	+ 45 06	42	15.	GALAXY
MCG-05-31-039	13 10 12.	- 32 25	144	15.	GALAXY
LB 02587	13 10 13.	+ 60 27 48.		16.0	FAINT BLUE STAR
RNGC 5019	13 10 14.	+ 05 00		14.5	GALAXY
LB 02588	13 10 14.	+ 58 40 30.		16.6	FAINT BLUE STAR
RNGC 5017	13 10 16.	- 16 30		13.5	GALAXY
RNGC 5018	13 10 16.	- 19 15		12.0	GALAXY
BFG 165	13 10 17.53	+ 33 23 18.7		19.32	ULTRAVIOLET-EXCESS OBJECT
ZC 1310.3+1443	13 10 18.	+ 14 43	1610		CLUSTER OF GALAXIES
MCG+04-31-014	13 10 18.	+ 23 06 30.	90	14.8	GALAXY
ZC 1310.3+5407	13 10 18.	+ 54 07	1280		CLUSTER OF GALAXIES
MRK 245	13 10 18.	+ 67 47	10	16.	GALAXY WITH UV CONTINUUM
MCG-06-29-015	13 10 18.	- 36 28	60	13.	GALAXY
REIN 2.184	13 10 18.56	+ 23 05 58.6			NEBULA
REIZ 3318	13 10 21.	+ 32 05	108	13.1	GALAXY
GCL 022	13 10 24.	+ 18 26	864	8.68	GLOBULAR STAR CLUSTER
UGC 08291	13 10 24.	+ 22 04	60	17.	GALAXY DWARF
MCG+05-31-154	13 10 24.	+ 27 25	48	15.5	GALAXY
ZWG 160.162	13 10 24.	+ 32 04		14.6	GALAXY
UGC 08292	13 10 24.	+ 32 04	132	14.6	GALAXY Sb
MCG+05-31-155	13 10 24.	+ 32 06	114	14.6	GALAXY
MCG+06-29-059	13 10 24.	+ 34 19	48	15.5	GALAXY
ZWG 245.032	13 10 24.	+ 47 20		14.5	GALAXY
RNGC 5029	13 10 24.	+ 47 20		14.5	GALAXY
UGC 08293	13 10 24.	+ 47 20	102	14.5	GALAXY E
REIZ 3323	13 10 24.	+ 47 21	18	14.7	GALAXY
ZC 1310.4+4943	13 10 24.	+ 49 43	1550		CLUSTER OF GALAXIES
MCG+09-22-025	13 10 24.	+ 51 02	42	16.	GALAXY
MCG+09-22-026	13 10 24.	+ 53 12	36	15.	GALAXY
MCG+10-19-051	13 10 24.	+ 58 48	27	16.	GALAXY
MCG-03-34-016	13 10 24.	- 16 30	60	14.	GALAXY
DV.56 N5011B	13 10 24.	- 42 57			E GALAXY
DV.56 N5011C	13 10 24.	- 43 01	36		GALAXY
ARC 1696	13 10 25.	+ 49 50		17.7	RICH CLUSTER OF GALAXIES
RNGC 5025	13 10 26.	+ 32 04		14.5	GALAXY
MCG-03-34-017	13 10 27.	- 19 14	54	12.	GALAXY
MCG-03-34-018	13 10 27.	- 19 43	168	13.5	GALAXY
RNGC 5024	13 10 29.	+ 18 26		8.5	GLOBULAR CLUSTER
TON-N 0700	13 10 30.	+ 26 42		16.2	BLUE STAR
MCG+08-24-087	13 10 30.	+ 47 20	48	14.7	GALAXY
MCG-03-34-019	13 10 30.	- 16 12	66	15.	GALAXY
MCG-07-27-043	13 10 30.	- 41 59	60	14.	GALAXY
MCG-07-27-044	13 10 30.	- 42 51 30.	36	15.	GALAXY
LB 02589	13 10 31.	+ 55 37 54.		17.0	FAINT BLUE STAR
REIZ 3322	13 10 33.	+ 31 30	42	14.2	GALAXY
8ZW 1310+04.1	13 10 36.	+ 04 08		18.7	COMPACT GALAXY
ZC 1310.6+1733	13 10 36.	+ 17 33	1610		CLUSTER OF GALAXIES
ZWG 160.163	13 10 36.	+ 27 25		15.7	GALAXY
ZWG 160.164	13 10 36.	+ 31 31		15.2	GALAXY
UGC 08294	13 10 36.	+ 31 31	60	15.2	GALAXY Sc
MCG+05-31-156	13 10 36.	+ 31 32 30.	66	15.2	GALAXY
TON-N 0148	13 10 36.	+ 31 59		14.6	BLUE STAR
ZC 1310.6+3536	13 10 36.	+ 35 36	940		CLUSTER OF GALAXIES
ZWG 217.018	13 10 36.	+ 40 48		15.0	GALAXY
MCG+08-24-088	13 10 36.	+ 47 12	30	16.	GALAXY
ZWG 336.003	13 10 36.	+ 70 55		14.1	GALAXY
RNGC 5034	13 10 36.	+ 70 55		14.0	GALAXY
UGC 08295	13 10 36.	+ 70 55	60	14.1	GALAXY S
REIZ 3319	13 10 37.	+ 06 32	30	14.8	GALAXY
TON-N 1566	13 10 39.	+ 21 48		17.0	BLUE STAR
MCG-03-34-020	13 10 39.	- 15 51	36	15.	GALAXY
RNGC 5022	13 10 40.	- 19 15		13.0	GALAXY
ZWG 072.025	13 10 42.	+ 09 57		15.0	GALAXY
REIZ 3321	13 10 42.	+ 09 57	30	14.0	GALAXY
UGC 08296	13 10 42.	+ 09 57	66	15.0	GALAXY S0-a
MCG+02-34-004	13 10 42.	+ 09 57	60	15.0	GALAXY
ZC 1310.7+3510	13 10 42.	+ 35 10	400		CLUSTER OF GALAXIES
ZC 1310.7+3719	13 10 42.	+ 37 19	470		CLUSTER OF GALAXIES
ARC 1697	13 10 42.	+ 46 32		17.5	RICH CLUSTER OF GALAXIES
MCG+09-22-027	13 10 42.	+ 51 57 30.	30	16.	GALAXY
MCG+12-13-001	13 10 42.	+ 70 56	54	15.	GALAXY
MCG-07-27-045	13 10 42.	- 43 06	66	15.	GALAXY
MCG-07-27-046	13 10 42.	- 43 06 30.	30	16.	GALAXY
SER 102.03	13 10 42.	- 43 12			PEC. GALAXY
REIZ 3320	13 10 43.	+ 06 18	30	14.5	GALAXY

OBJECT NAME	RIGHT ASCEN.	DECLINATION	DIAM.	MAGN.	TYPE OF OBJECT
LB 00249	13 10 44.	+ 54 48 42.		15.5	FAINT BLUE STAR
ZWG 044.028	13 10 48.	+ 06 20		14.8	GALAXY
UGC 08297	13 10 48.	+ 06 20	90	14.8	GALAXY SB:b
MCG+01-34-010	13 10 49.	+ 06 20	48	14.8	GALAXY
ZWG 044.029	13 10 48.	+ 07 12		15.4	GALAXY
ZWG 072.026	13 10 48.	+ 10 27		15.2	GALAXY
REIZ 3324	13 10 48.	+ 10 27	36	14.3	GALAXY
UGC 08298	13 10 48.	+ 10 27	66	15.2	GALAXY IRR
ZWG 072.027	13 10 48.	+ 10 35		15.7	GALAXY
REIZ 3325	13 10 48.	+ 10 35	30	14.4	GALAXY
ZC 1310.8+3230	13 10 48.	+ 32 30	3290		CLUSTER OF GALAXIES
MCG+06-29-060	13 10 48.	+ 34 14	51	14.5	GALAXY
ZWG 189.041	13 10 48.	+ 34 15		15.0	GALAXY
UGC 08299	13 10 48.	+ 34 15	60	15.0	GALAXY SBb
ZWG 217.019	13 10 48.	+ 43 27		15.6	GALAXY
ZC 1310.8+4631	13 10 48.	+ 46 31	1210		CLUSTER OF GALAXIES
ZWG 245.033	13 10 48.	+ 47 44		15.5	GALAXY
MCG+08-24-089	13 10 48.	+ 47 44	48	15.	GALAXY
MCG+10-19-052	13 10 48.	+ 61 14	30	16.	GALAXY
MCG-02-34-010	13 10 48.	- 15 10	96	14.5	GALAXY
MCG-07-27-047	13 10 48.	- 43 15	48	14.5	GALAXY
REIN 4.261	13 10 49.51	+ 10 27 30.3			NEBULA
RNGC 5027	13 10 50.	+ 06 20		15.0	GALAXY
LB 02591	13 10 50.	+ 55 23 30.		17.1	FAINT BLUE STAR
REIN 4.262	13 10 50.52	+ 10 35 11.4			NEBULA
HZ 42	13 10 53.	+ 31 38		14.2	BLUE STAR
ZWG 044.030	13 10 54.	+ 08 18		15.7	GALAXY
ZWG 072.028	13 10 54.	+ 09 14		15.7	GALAXY
7ZW 503	13 10 54.	+ 57 13			COMPACT GALAXY
ZC 1310.9+5726	13 10 54.	+ 57 26	1550		CLUSTER OF GALAXIES
MCG-03-34-021	13 10 54.	- 19 16	132	13.5	GALAXY
MCG-03-34-022	13 10 57.	- 16 47	48	14.5	GALAXY
TON-N 1567	13 10 58.	+ 32 33		16.6	BLUE STAR
VDB.66G 167	13 11	+ 46 35	70		DWARF GALAXY
ZC 1311.0+0706	13 11 00.	+ 07 06	1750		CLUSTER OF GALAXIES
ZWG 044.031	13 11 00.	+ 07 09		15.7	GALAXY
ZWG 160.165	13 11 00.	+ 28 01		15.4	GALAXY
KARA.72 366A	13 11 00.	+ 28 01	66	15.4	PART OF DOUBLE GALAXY
ZWG 160.166	13 11 00.	+ 28 04		13.6	GALAXY
UGC 08300	13 11 00.	+ 28 04	132	13.6	GALAXY SBb
KARA.72 366B	13 11 00.	+ 28 04	120	13.6	PART OF DOUBLE GALAXY
MCG+05-31-157	13 11 00.	+ 30 36	42	16.	GALAXY
UGC 08301	13 11 00.	+ 30 37	60	16.5	GALAXY
MCG+05-31-158	13 11 00.	+ 30 50	48	15.	GALAXY
UGC 08302	13 11 00.	+ 33 33	66	16.5	GALAXY
ZWG 189.042	13 11 00.	+ 36 28		14.7	GALAXY
UGC 08303	13 11 00.	+ 36 28	168	14.7	GALAXY DWRF IR
MCG+06-29-061	13 11 00.	+ 36 28	132	13.	GALAXY
ZC 1311.0+5040	13 11 00.	+ 50 40	940		CLUSTER OF GALAXIES
MCG+09-22-028	13 11 00.	+ 50 50	120	14.	GALAXY
ZWG 271.023	13 11 00.	+ 50 52		15.0	GALAXY
UGC 08304	13 11 00.	+ 50 52	138	15.0	GALAXY Sb
MRK 246	13 11 00.	+ 56 19	9	16.	GALAXY WITH UV CONTINUUM
ZWG 336.004	13 11 00.	+ 71 28		15.4	GALAXY
UGC 08305	13 11 00.	+ 71 28	72	15.4	GALAXY Sb-c
MCG+12-13-002	13 11 00.	+ 71 29	54	16.	GALAXY
MCG-02-34-011	13 11 00.	- 12 47 30.	30	14.	GALAXY
SMI 02	13 11 00.	- 60 53	360		FAINT NEBULOSITY
RNGC 5032B	13 11 01.	+ 28 02		15.5	GALAXY
RNGC 5032A	13 11 01.	+ 28 04		13.5	GALAXY
LB 02592	13 11 01.	+ 55 12 48.		17.0	FAINT BLUE STAR
MCG+05-31-159	13 11 03.	+ 28 02	36	15.4	GALAXY
HOLM 513B	13 11 04.	+ 28 01	24	14.9	PART OF MULTIPLE GALAXY
REIZ 3329	13 11 04.	+ 30 34	30	15.5	GALAXY
REIZ 3330	13 11 04.	+ 30 49	42	14.8	GALAXY
HOLM 513A	13 11 05.	+ 28 03	60	13.7	PART OF MULTIPLE GALAXY
REIZ 3328	13 11 05.	+ 28 04	66	13.6	GALAXY
RNGC 5028	13 11 05.	- 12 47		14.0	GALAXY
ZWG 072.029	13 11 06.	+ 13 19		15.7	GALAXY
ZWG 101.009	13 11 06.	+ 16 15		14.9	GALAXY
UGC 08306	13 11 06.	+ 16 15	96	14.9	GALAXY Sb
KARA.73B 0576	13 11 06.	+ 16 15	102	14.9	ISOLATED GALAXY S
ZC 1311.1+1802	13 11 06.	+ 18 02	470		CLUSTER OF GALAXIES
MCG+05-31-160	13 11 06.	+ 28 05	120	13.6	GALAXY
LB 02593	13 11 06.	+ 52 30 06.		15.7	FAINT BLUE STAR
UGC 08537	13 11 06.	- 01 38	66	15.3	GALAXY S
REIN 2.185	13 11 06.36	+ 36 52 43.4			NEBULA
RNGC 5033	13 11 09.	+ 36 52		10.5	GALAXY
REIN 2.186	13 11 09.11	+ 36 51 36.2			NEBULA
REIZ 3327	13 11 10.	+ 16 14	42	15.0	GALAXY
HOLM 512A	13 11 10.	+ 16 16	78	14.2	PART OF MULTIPLE GALAXY
RNGC 5030	13 11 10.	- 16 14		14.0	GALAXY
HZ 40	13 11 11.	+ 37 14		13.8	DECIDEDLY BLUE STAR
SN 1956K	13 11 11.	- 03 20		19.6	SUPERNOVA
ZWG 044.032	13 11 12.	+ 07 07		15.5	GALAXY
ZWG 044.033	13 11 12.	+ 07 15		15.1	GALAXY
WEI 13323	13 11 12.	+ 30 15		17.99	FAINT BLUE OBJECT
ZWG 189.043	13 11 12.	+ 36 51		10.9	GALAXY
UGC 08307	13 11 12.	+ 36 51	690	10.9	GALAXY Sc
MCG+06-29-062	13 11 12.	+ 36 52	600	11.	GALAXY
UGC 08308	13 11 12.	+ 46 35	78	17.	GALAXY DWRF IR
UGC 08309	13 11 12.	+ 51 45	66	16.5	GALAXY Sa-b
MCG+09-22-029	13 11 12.	+ 53 16	48	16.	GALAXY
ZWG 336.005	13 11 12.	+ 69 38		15.3	GALAXY
MCG+12-13-003	13 11 12.	+ 69 38	51	16.	GALAXY
ZWG 016.023	13 11 12.	- 03 20		15.1	GALAXY
REIZ 3326	13 11 13.	+ 07 15	48	14.7	GALAXY
SHB 254	13 11 13.5	+ 34 45 21.		19.1	QUASI-STELLAR OBJECT
LB 02594	13 11 14.	+ 51 31 42.		17.4	FAINT BLUE STAR
BC BS09	13 11 14C	ε 3 46 48		18H41	-U1 -003--A- OBJ&60
MCG+08-24-090	13 11 15.	+ 46 36	60	16.	GALAXY
MCG+09-22-030	13 11 15.	+ 51 42 30.	48	15.	GALAXY
BFG 166	13 11 15.61	+ 34 45 52.9		19.05	ULTRAVIOLET-EXCESS OBJECT
BSO 09	13 11 16.	+ 34 46 04.		19.10	BLUE STELLAR OBJECT
BSO 11	13 11 16.	+ 36 16 30.		18.41	BLUE STELLAR OBJECT
RNGC 5026	13 11 16.	- 42 42			GALAXY
ZWG 016.024	13 11 18.	+ 02 24		15.2	GALAXY
ZWG 044.034	13 11 18.	+ 02 49		15.5	GALAXY
ZWG 044.035	13 11 18.	+ 07 13		15.4	GALAXY
ZWG 072.030	13 11 18.	+ 12 57		15.6	GALAXY
HOLM 512B	13 11 18.	+ 16 14	18	15.1	PART OF MULTIPLE GALAXY
ZWG 130.021	13 11 18.	+ 25 15		15.4	GALAXY
MCG+09-22-031	13 11 18.	+ 51 31	54	14.3	GALAXY
LB 02595	13 11 18.	+ 52 18 54.		17.0	FAINT BLUE STAR
BFG 167	13 11 19.04	+ 33 34 39.4		19.21	ULTRAVIOLET-EXCESS OBJECT
BFG 168	13 11 19.51	+ 36 15 40.7		18.62	ULTRAVIOLET-EXCESS OBJECT
TON-N 1568	13 11 21.	+ 20 51		16.8	BLUE STAR
MCG-03-34-023	13 11 21.	- 16 13	72	14.	GALAXY
BC BS011	13 11 22.1	+ 36 16 30.		18.41	QUASI-STELLAR OBJECT
SHB 255	13 11 22.1	+ 36 16 30.		18.4	QUASI-STELLAR OBJECT
IC 0857	13 11 23.	+ 17 20 35.			NONSTELLAR OBJECT
ZWG 044.036	13 11 24.	+ 07 13		15.1	GALAXY
ZWG 101.010	13 11 24.	+ 17 20		14.7	GALAXY
UGC 08310	13 11 24.	+ 17 20	78	14.7	GALAXY SBb
ZWG 130.022	13 11 24.	+ 23 31		15.7	GALAXY
UGC 08311	13 11 24.	+ 23 31	60	15.7	GALAXY S?
MCG+06-29-063	13 11 24.	+ 34 14	33	15.	GALAXY
SN 1950C	13 11 24.	+ 36 50		18.2	SUPERNOVA
ZWG 271.024	13 11 24.	+ 51 32		15.1	GALAXY
ZC 1311.4+6109	13 11 24.	+ 61 09	2550		CLUSTER OF GALAXIES
REIZ 3331	13 11 25.	+ 07 13	30	14.6	GALAXY
RNGC 5040	13 11 26.	+ 51 32		15.0	GALAXY
BSO 10	13 11 27.	+ 33 02 28.			BLUE STELLAR OBJECT
BFG 169	13 11 27.83	+ 33 02 16.9		18.66	ULTRAVIOLET-EXCESS OBJECT
RNGC 5031	13 11 28.	- 15 50		14.0	GALAXY
ZWG 072.031	13 11 30.	+ 12 52		15.7	GALAXY
ZC 1311.5+2055	13 11 30.	+ 20 55	1140		CLUSTER OF GALAXIES
ZC 1311.5+2442	13 11 30.	+ 24 42	610		CLUSTER OF GALAXIES
LB 02596	13 11 30.	+ 51 08 18.		11.6	FAINT BLUE STAR
UGC 08312	13 11 30.	+ 72 52	78	16.0	GALAXY
MCG-03-34-024	13 11 30.	- 15 50 30.	90	14.	GALAXY
MCG-07-27-048	13 11 30.	- 42 40	120	13.	GALAXY
ARC 1698	13 11 32.	- 06 45		17.6	RICH CLUSTER OF GALAXIES
KEEL 585	13 11 32.6	+ 42 18 01.		18.	NEBULA
LB 02597	13 11 34.	+ 57 54 12.		17.1	FAINT BLUE STAR
REIZ 3332	13 11 35.	- 05 20	66	14.3	GALAXY
LB 00035	13 11 36.	+ 28 03		17.2	FAINT BLUE STAR
1ZW 053	13 11 36.	+ 35 35			COMPACT GALAXY
ZWG 217.020	13 11 36.	+ 42 28		14.9	GALAXY
UGC 08313	13 11 36.	+ 42 28	120	14.9	GALAXY S
MCG+09-22-032	13 11 36.	+ 55 06	60	16.	GALAXY
ARC 1701	13 11 36.	+ 61 17		17.0	RICH CLUSTER OF GALAXIES
ZWG 016.025	13 11 36.	- 00 10		15.3	GALAXY
MCG+00-34-018	13 11 36.	- 00 10	36	15.3	GALAXY
MCG-01-34-013	13 11 36.	- 05 20	78	15.	GALAXY
MCG-04-31-042	13 11 36.	- 26 21	15	15.	GALAXY
AGU 45	13 11 36.	- 42 24 30.	60	12.5	PECULIAR GALAXY
BFG 170	13 11 36.41	+ 37 22 55.4		18.68	ULTRAVIOLET-EXCESS OBJECT
REIZ 3337	13 11 37.	+ 52 58	9	15.8	GALAXY
REIZ 3333	13 11 39.	+ 30 58	18	15.9	GALAXY
KEEL 586	13 11 39.6	+ 42 28 26.		14.	NEBULA
KEEL 587	13 11 40.3	+ 42 14 06.		18.	NEBULA
LB 02598	13 11 41.	+ 56 39 06.		16.3	FAINT BLUE STAR
MCG+10-19-053	13 11 42.	+ 60 44	45	16.	GALAXY
MRSL 305+02/1	13 11 42.	- 60 29	4680		HII REGION
REIZ 3340	13 11 44.	+ 51 33	36	14.3	GALAXY
MCG+09-22-033	13 11 45.	+ 52 13	30	16.	GALAXY
BFG 171	13 11 45.46	+ 37 10 31.1		18.59	ULTRAVIOLET-EXCESS OBJECT
KEEL 588	13 11 45.8	+ 42 02 10.		17.	NEBULA
REIZ 3334	13 11 46.	+ 30 49	18	14.7	GALAXY
8ZW 1311+25.1	13 11 48.	+ 25 04		17.3	COMPACT GALAXY
ZC 1311.8+2647	13 11 48.	+ 26 47	670		CLUSTER OF GALAXIES
UGC 08314	13 11 48.	+ 36 35	66	17.	GALAXY DWRF IR
ZWG 217.021	13 11 48.	+ 38 33		15.1	GALAXY
ZWG 245.034	13 11 48.	+ 45 59		15.6	GALAXY
MCG+08-24-091	13 11 48.	+ 49 59	18	16.	GALAXY
ZWG 016.026	13 11 49.	- 01 52		15.7	GALAXY
VHA 144	13 11 48.	- 65 40	90		OPEN STAR CLUSTER
KEEL 589	13 11 50.3	+ 42 04 52.		17.	NEBULA
TON-N 1569	13 11 51.	+ 22 28		17.0	BLUE STAR
KEEL 590	13 11 51.4	+ 42 01 33.		17.	NEBULA
BFG 172	13 11 51.99	+ 32 40 19.3			ULTRAVIOLET-EXCESS OBJECT
KEEL 591	13 11 53.8	+ 42 03 10.		18.	NEBULA
ZWG 072.032	13 11 54.	+ 13 46		15.7	GALAXY
ZWG 217.022	13 11 54.	+ 39 24		15.6	GALAXY
UGC 08315	13 11 54.	+ 39 24	78	15.6	GALAXY DBL SYS
MCG+08-24-092	13 11 54.	+ 47 39	36	16.	GALAXY
ZWG 245.035	13 11 54.	+ 48 25		15.7	GALAXY
UGC 08316	13 11 54.	+ 48 25	60	15.7	GALAXY S
MCG-03-34-025	13 11 54.	- 17 16	30	15.	GALAXY
MCG-06-29-016	13 11 54.	- 36 54	24	15.5	GALAXY
MCG-07-27-049	13 11 54.	- 42 22	60	14.	GALAXY
REIZ 3336	13 11 56.	+ 19 58	24	13.6	GALAXY
TON-N 1570	13 11 57.	+ 22 07		17.0	BLUE STAR
KEEL 592	13 11 58.7	+ 42 00 01.		18.	NEBULA
VDB.66G 168	13 12	+ 46 10	200		DWARF GALAXY
ZWG 044.037	13 12 00.	+ 03 18		15.5	GALAXY
ZC 1312.0+1248	13 12 00.	+ 12 48	610		CLUSTER OF GALAXIES
ZC 1312.0+2744	13 12 00.	+ 27 44	1010		CLUSTER OF GALAXIES
ZC 1312.0+2857	13 12 00.	+ 28 57	1610		CLUSTER OF GALAXIES
ZWG 160.167	13 12 00.	+ 30 45		15.0	GALAXY
UGC 08317	13 12 00.	+ 30 45	66	15.0	GALAXY S
MCG+05-31-161	13 12 00.	+ 30 46	60	15.0	GALAXY
MCG+09-22-034	13 12 00.	+ 52 15	30	16.	GALAXY
MCG+11-16-013	13 12 00.	+ 62 52 30.	57	17.	GALAXY
ZWG 016.028	13 12 00.	- 03 02		15.7	GALAXY
ZWG 016.027	13 12 00.	- 03 22		15.7	GALAXY
FATH 1.570	13 12 01.	+ 14 26	33		NEBULA
LB 02599	13 12 01.	+ 51 59 00.		15.2	FAINT BLUE STAR
KEEL 593	13 12 01.9	+ 42 12 18.		17.	NEBULA
ARP 196	13 12 03.	+ 26 24			PECULIAR GALAXY
REIZ 3341	13 12 03.	+ 30 45	48	13.8	GALAXY
KEEL 594	13 12 03.0	+ 42 02 06.		16.	NEBULA
TON-N 1571	13 12 04.	+ 34 47		16.9	BLUE STAR
TON-N 1572	13 12 04.	+ 35 15		17.0	BLUE STAR
RNGC 5035	13 12 04.	- 16 15		14.0	GALAXY
REIZ 3335	13 12 05.	- 03 56	12	14.7	GALAXY
8ZW 1312+24.6	13 12 06.	+ 24 35		18.1	COMPACT GALAXY
MCG+05-31-162	13 12 06.	+ 31 00	84	14.2	GALAXY
ZC 1312.1+3349	13 12 06.	+ 33 49	1140		CLUSTER OF GALAXIES
ZWG 189.044	13 12 06.	+ 35 38		15.4	GALAXY
UGC 08318	13 12 06.	+ 35 38	132	15.4	GALAXY SBc
ZC 1312.1+7311	13 12 06.	+ 73 11	1210		CLUSTER OF GALAXIES
RNGC 5036	13 12 06.	- 03 55			GALAXY
MCG-03-34-026	13 12 06.	- 18 31 30.	30	15.	GALAXY
BFG 173	13 12 06.86	+ 34 26 19.5		19.11	ULTRAVIOLET-EXCESS OBJECT
KEEL 595	13 12 08.4	+ 42 00 37.		18.	NEBULA
KEEL 596	13 12 08.6	+ 41 56 25.		18.	NEBULA
REIZ 3343	13 12 09.	+ 30 58	48	14.0	GALAXY
MCG+06-29-064	13 12 09.	+ 35 38	66	14.5	GALAXY
REIZ 3342	13 12 10.	+ 30 30	12	15.6	GALAXY
LB 00679	13 12 11.	+ 53 43 24.		16.8	FAINT BLUE STAR
REIZ 3339	13 12 11.	- 03 54	18	14.8	GALAXY
REIZ 3338	13 12 11.	- 04 02	21	15.0	GALAXY
KEEL 597	13 12 11.3	+ 42 12 46.		16.	NEBULA
ZWG 044.038	13 12 12.	+ 06 35		15.5	GALAXY

OBJECT NAME	RIGHT ASCEN.	DECLINATION	DIAM.	MAGN.	TYPE OF OBJECT
ZCG 1312+26	13 12 12.	+ 26 23		16.0	COMPACT GALAXY
STOCK 41	13 12 12.	+ 26 24			BLUE KNOT NEAR ELLIP GLXY
ZWG 160.168	13 12 12.	+ 30 58		14.2	GALAXY
UGC 08319	13 12 12.	+ 30 58	102	14.2	GALAXY Sc
ZC 1312.2+4529	13 12 12.	+ 45 29	1610		CLUSTER OF GALAXIES
ZWG 245.036	13 12 12.	+ 46 11		14.0	GALAXY
UGC 08320	13 12 12.	+ 46 11	258	14.0	GALAXY IRR
LB 02600	13 12 12.	+ 50 51 00.		16.1	FAINT BLUE STAR
MCG-03-34-027	13 12 12.	- 20 08	30	15.	GALAXY
MCG-06-29-017	13 12 12.	- 36 53	36	14.	GALAXY
KEEL 598	13 12 12.3	+ 42 20 40.		18.	NEBULA
RNGC 5041	13 12 13.	+ 30 58		14.0	GALAXY
REIZ 3345	13 12 13.	+ 57 27	24	15.1	GALAXY
ARC 1699	13 12 13.	- 21 47		17.4	RICH CLUSTER OF GALAXIES
ARP 060	13 12 15.	+ 26 22			PECULIAR GALAXY
KEEL 599	13 12 16.5	+ 41 54 19.		14.	NEBULA
ZWG 160.169	13 12 18.	+ 30 15		15.6	GALAXY
ZC 1312.3+4115	13 12 18.	+ 41 15	6180		CLUSTER OF GALAXIES
MCG+08-24-093	13 12 18.	+ 46 12	180	13.	GALAXY
MCG+09-22-035	13 12 18.	+ 51 13 30.	36	14.	GALAXY
ZWG 271.025	13 12 18.	+ 51 15		15.7	GALAXY
MRK 247	13 12 18.	+ 55 05	10	15.	GALAXY WITH UV CONTINUUM
VVI 64	13 12 18.	+ 55 05	10	15.53	SEYFERT GALAXY
ZWG 294.027	13 12 18.	+ 62 28		15.7	GALAXY
ARC 1705	13 12 18.	+ 73 09		17.7	RICH CLUSTER OF GALAXIES
RNGC 5039	13 12 18.	- 03 54			GALAXY
MCG-03-34-028	13 12 18.	- 16 13 30.	60	14.	GALAXY
ARC 1700	13 12 19.	+ 29 00		17.2	RICH CLUSTER OF GALAXIES
ARC 1702	13 12 19.	+ 45 24		17.8	RICH CLUSTER OF GALAXIES
KEEL 600	13 12 19.5	+ 42 24 11.		18.	NEBULA
KEEL 601	13 12 20.2	+ 42 31 18.		16.	NEBULA
KEEL 604	13 12 20.7	+ 42 22 02.		18.	NEBULA
KEEL 602	13 12 20.8	+ 42 06 16.		18.	NEBULA
KEEL 603	13 12 20.9	+ 42 02 53.		17.	NEBULA
TON-N 1573	13 12 21.	+ 19 52		17.0	BLUE STAR
RNGC 5037	13 12 22.	- 16 20		13.0	GALAXY
KEEL 605	13 12 22.0	+ 42 15 22.		17.	NEBULA
ZWG 101.011	13 12 24.	+ 17 29		14.7	GALAXY
UGC 08321	13 12 24.	+ 17 29	102	14.7	GALAXY S0?
KARA.72 367A	13 12 24.	+ 17 29	66	14.7	PART OF DOUBLE GALAXY
ZC 1312.4+2005	13 12 24.	+ 20 05	1550		CLUSTER OF GALAXIES
MRK 658	13 12 24.	+ 24 15	12	16.	GALAXY WITH UV CONTINUUM
TON-N 0149	13 12 24.	+ 26 35		15.1	BLUE STAR
MCG+08-24-094	13 12 24.	+ 44 46	60	16.	GALAXY
MCG+10-19-054	13 12 24.	+ 57 26	24	16.	GALAXY
MCG+10-19-055	13 12 24.	+ 58 49	45	16.	GALAXY
MCG+11-16-014	13 12 24.	+ 62 55	30	17.	GALAXY
ZC 1312.4+6451	13 12 24.	+ 64 51	2550		CLUSTER OF GALAXIES
MCG-03-34-029	13 12 24.	- 16 19	108	13.	GALAXY
MCG-03-34-030	13 12 24.	- 17 00	30	15.	GALAXY
MCG-07-27-050	13 12 24.	- 42 00	42	15.	GALAXY
IC 0858	13 12 25.	+ 17 29 26.			NONSTELLAR OBJECT
KEEL 606	13 12 26.4	+ 42 19 51.		18.	NEBULA
KEEL 607	13 12 26.7	+ 42 14 50.		18.	NEBULA
MCG+06-29-065	13 12 27.	+ 35 08	48	14.	GALAXY
KEEL 608	13 12 27.4	+ 42 14 13.		18.	NEBULA
BFG 174	13 12 27.60	+ 37 05 17.8		18.65	ULTRAVIOLET-EXCESS OBJECT
KEEL 609	13 12 27.7	+ 42 23 50.		18.	NEBULA
RNGC 5038	13 12 28.	- 15 40		14.0	GALAXY
KEEL 610	13 12 29.1	+ 42 15 14.		18.	NEBULA
ZWG 044.039	13 12 30.	+ 03 17		15.3	GALAXY
ZWG 072.033	13 12 30.	+ 10 17		15.4	GALAXY
ZC 1312.5+1113	13 12 30.	+ 11 13	2020		CLUSTER OF GALAXIES
ZWG 072.034	13 12 30.	+ 12 58		14.7	GALAXY
UGC 08322	13 12 30.	+ 12 58	78	14.7	GALAXY Sa?
MCG+02-34-005	13 12 30.	+ 12 58	72	14.7	GALAXY
ZWG 101.012	13 12 30.	+ 17 29		15.2	GALAXY
KARA.72 367B	13 12 30.	+ 17 29	48	15.2	PART OF DOUBLE GALAXY
IC 0859	13 12 30.	+ 17 29 20.			NONSTELLAR OBJECT
MCG+05-31-163	13 12 30.	+ 27 17	48	15.5	GALAXY
MCG+06-29-066	13 12 30.	+ 34 56 30.	15	15.5	GALAXY
ZWG 189.045	13 12 30.	+ 35 08		14.9	GALAXY
MRK 450	13 12 30.	+ 35 08	6	17.	GALAXY WITH UV CONTINUUM
UGC 08323	13 12 30.	+ 35 08	66	14.9	GALAXY (IRR)
ZC 1312.5+4139	13 12 30.	+ 41 39	1680		CLUSTER OF GALAXIES
LB 02602	13 12 30.	+ 51 50 00.		16.2	FAINT BLUE STAR
LB 02601	13 12 30.	+ 57 27 36.		16.0	FAINT BLUE STAR
ZWG 353.015	13 12 30.	+ 78 30		15.7	GALAXY
ZWG 352.060	13 12 30.	+ 78 30		15.7	GALAXY
MCG-03-34-031	13 12 30.	- 15 40	72	14.	GALAXY
MCG-06-29-018	13 12 30.	- 34 40	42	14.5	GALAXY
KEEL 611	13 12 31.2	+ 42 14 16.		18.	NEBULA
REIZ 3344	13 12 33.	+ 31 10	36	14.8	GALAXY
ARC 1704	13 12 35.	+ 64 52		17.8	RICH CLUSTER OF GALAXIES
ZWG 044.040	13 12 36.	+ 03 18		15.3	GALAXY
UGC 08324	13 12 36.	+ 03 18	78	15.3	GALAXY Sc
MCG+01-34-011	13 12 36.	+ 03 18	90	15.3	GALAXY
ZWG 044.041	13 12 36.	+ 06 48		15.6	GALAXY
ZWG 072.035	13 12 36.	+ 09 37		15.5	GALAXY
ZWG 101.013	13 12 36.	+ 16 21		15.7	GALAXY
8ZW 1312+22.9	13 12 36.	+ 22 52		17.7	COMPACT GALAXY
ZWG 130.023	13 12 36.	+ 24 53		14.8	GALAXY
UGC 08325	13 12 36.	+ 27 17	84	16.0	GALAXY Sa-b
TON-N 0150	13 12 36.	+ 32 30		15.2	BLUE STAR
ZWG 271.026	13 12 36.	+ 54 05		15.0	GALAXY
ZC 1312.6+5802	13 12 36.	+ 58 02	1550		CLUSTER OF GALAXIES
MCG+13-10-004	13 12 36.	+ 78 31	42	16.	GALAXY
MCG-03-34-032	13 12 36.	- 17 42	48	15.5	GALAXY
MCG-05-31-040	13 12 36.	- 28 00	24	16.	GALAXY
KEEL 612	13 12 36.1	+ 41 58 39.		14.	NEBULA
KEEL 613	13 12 37.8	+ 41 54 58.		18.	NEBULA
IC 0860	13 12 39.	+ 24 52 27.			NONSTELLAR OBJECT
MCG+06-29-067	13 12 39.	+ 38 18	27	14.5	GALAXY
ARP 238	13 12 39.	+ 62 24			PECULIAR GALAXY
LB 00265	13 12 39.	- 17 21 36.		15.8	FAINT BLUE STAR
MCG+04-31-015	13 12 42.	+ 24 52	30	14.8	GALAXY
STOCK 42	13 12 42.	+ 26 51			BLUE KNOT NEAR ELLIP GLXY
MCG+09-22-036	13 12 42.	+ 54 05 30.	54	14.	GALAXY
MCG-03-34-033	13 12 42.	- 16 12 30.	36	15.	GALAXY
MCG+06-29-068	13 12 45.	+ 34 34 30.	30	15.	GALAXY
RNGC 5042	13 12 45.	- 23 43		13.0	GALAXY
RNGC 5044	13 12 46.	- 16 08		12.0	GALAXY
KEEL 614	13 12 46.2	+ 42 16 34.		17.	NEBULA
KEEL 615	13 12 47.2	+ 42 29 57.		18.	NEBULA
ZWG 016.030	13 12 48.	+ 01 35		15.3	GALAXY
8ZW 1312+01.6	13 12 48.	+ 01 35		15.3	COMPACT GALAXY
ZWG 016.031	13 12 48.	+ 02 12		15.0	GALAXY
ZWG 160.170	13 12 48.	+ 30 39		15.4	GALAXY
ZC 1312.8+3046	13 12 48.	+ 30 46	870		CLUSTER OF GALAXIES
ZWG 189.046	13 12 48.	+ 34 35		15.6	GALAXY
UGC 08326	13 12 48.	+ 34 35	66	15.6	GALAXY S0-a
IC 0861	13 12 48.	+ 34 35 33.			NONSTELLAR OBJECT
7ZW 504	13 12 48.	+ 77 04			COMPACT GALAXY
ZWG 016.029	13 12 48.	- 02 25		15.6	GALAXY
MCG-03-34-034	13 12 48.	- 16 07	48	12.	GALAXY
MCG-04-31-043	13 12 48.	- 23 45	240	13.	GALAXY
REIZ 3346	13 12 52.	+ 29 54	30	15.1	GALAXY
KEEL 616	13 12 53.2	+ 42 06 25.		18.	NEBULA
KEEL 617	13 12 53.9	+ 41 58 29.		18.	NEBULA
ZWG 016.032	13 12 54.	+ 02 20		15.3	GALAXY
8ZW 1312+11.1	13 12 54.	+ 11 07		17.8	COMPACT GALAXY
ZWG 072.036	13 12 54.	+ 13 10		15.3	GALAXY
ZWG 072.037	13 12 54.	+ 13 21		15.5	GALAXY
ZC 1312.9+4738	13 12 54.	+ 47 38	1480		CLUSTER OF GALAXIES
ZC 1312.9+5206	13 12 54.	+ 52 06	940		CLUSTER OF GALAXIES
MPSL 305-00/1	13 12 54.	- 63 05	4500		HII REGION
ARC 1703	13 12 56.	+ 52 06		16.0	RICH CLUSTER OF GALAXIES
KEEL 618	13 12 56.2	+ 42 01 57.		17.	NEBULA
TON-N 1574	13 12 57.	+ 20 43		17.0	BLUE STAR
RNGC 5046	13 12 58.	- 16 04		15.0	GALAXY
KEEL 619	13 12 58.4	+ 42 08 40.		18.	NEBULA
VDB.66G 170	13 13	+ 25 40	70		DWARF GALAXY
VDB.66G 169	13 13	+ 47 45	200		DWARF GALAXY
ZWG 016.033	13 13 00.	+ 00 45		15.3	GALAXY
MCG+00-34-019	13 13 00.	+ 00 45	36	15.3	GALAXY
ZWG 044.042	13 13 00.	+ 03 46		15.4	GALAXY
ZC 1313.0+0849	13 13 00.	+ 08 49	3020		CLUSTER OF GALAXIES
ZC 1313.0+1256	13 13 00.	+ 12 56	1680		CLUSTER OF GALAXIES
ZWG 101.014	13 13 00.	+ 15 41		15.6	GALAXY
ZWG 245.037	13 13 00.	+ 44 40		14.9	GALAXY
MRK 248	13 13 00.	+ 44 40	12	16.5	GALAXY WITH UV CONTINUUM
UGC 08327	13 13 00.	+ 44 40	96	14.9	GALAXY DBL SYS
KARA.72 368A	13 13 00.	+ 44 40	24	14.9	PART OF DOUBLE GALAXY
ZC 1313.0+5410	13 13 00.	+ 54 10	18350		CLUSTER OF GALAXIES
KEEL 620	13 13 02.8	+ 41 55 46.		18.	NEBULA
BFG 175	13 13 03.87	+ 38 04 53.7		19.21	ULTRAVIOLET-EXCESS OBJECT
TON-N 1575	13 13 04.	+ 34 10		17.0	BLUE STAR
LB 02603	13 13 04.	+ 55 37 36.		17.2	FAINT BLUE STAR
RNGC 5047	13 13 04.	- 16 14		13.0	GALAXY
ZC 1313.1+1008	13 13 06.	+ 10 08	2220		CLUSTER OF GALAXIES
FATH 1.571	13 13 06.	+ 14 49	14		NEBULA
UGC 08328	13 13 06.	+ 27 34	66	16.5	GALAXY Sc
KARA.72 368B	13 13 06.	+ 44 40	36		PART OF DOUBLE GALAXY
MCG+08-24-095	13 13 06.	+ 44 40	18	16.	GALAXY
ZC 1313.1+7823	13 13 06.	+ 78 23	1010		CLUSTER OF GALAXIES
MCG-04-31-044	13 13 06.	- 24 00 30.	36	15.5	GALAXY
KEEL 621	13 13 06.0	+ 42 35 20.		16.	NEBULA
LB 00266	13 13 07.	- 14 48 48.		16.0	FAINT BLUE STAR
KEEL 622	13 13 08.4	+ 42 36 10.		17.	NEBULA
MCG+08-24-096	13 13 09.	+ 44 40	42	16.	GALAXY
ZWG 044.043	13 13 12.	+ 03 08		14.7	GALAXY
UGC 08329	13 13 12.	+ 03 08	84	14.7	GALAXY S0-a
MCG+01-34-012	13 13 12.	+ 03 08	60	14.7	GALAXY
TON-N 0701	13 13 12.	+ 26 34		15.9	BLUE STAR
ZC 1313.2+2805	13 13 12.	+ 28 05	870		CLUSTER OF GALAXIES
ZWG 160.171	13 13 12.	+ 29 55		14.6	GALAXY
UGC 08330	13 13 12.	+ 29 55	84	14.6	GALAXY S0-a
MCG+05-31-165	13 13 12.	+ 29 58	72	14.6	GALAXY
MCG+05-31-164	13 13 12.	+ 32 09 30.	54	15.1	GALAXY
7ZW 505	13 13 12.	+ 61 44			COMPACT GALAXY
MCG+11-16-015	13 13 12.	+ 62 47	30	16.	GALAXY
ZWG 016.034	13 13 12.	- 00 11		15.6	GALAXY
MCG+00-34-020	13 13 12.	- 00 11	36	15.6	GALAXY
MCG-06-29-019	13 13 12.	- 37 04	24	15.	GALAXY
KEEL 623	13 13 12.6	+ 41 55 49.		18.	NEBULA
RNGC 5052	13 13 13.	+ 29 55		14.5	GALAXY
MCG-06-29-020	13 13 13.	- 37 03	18	15.5	GALAXY
KEEL 624	13 13 13.0	+ 42 06 57.		17.	NEBULA
RNGC 5050	13 13 14.	+ 03 08		14.5	GALAXY
MCG-03-34-035	13 13 15.	- 16 03	36	15.	GALAXY
REIZ 3347	13 13 16.	+ 29 57	24	14.2	GALAXY
RNGC 5049	13 13 16.	- 16 08		14.0	GALAXY
SVEN 396	13 13 17.	- 20 24	36	15.5	GALAXY
ZWG 044.044	13 13 18.	+ 08 18		15.2	GALAXY
ZWG 072.038	13 13 18.	+ 10 16		15.6	GALAXY
ZWG 101.015	13 13 18.	+ 15 46		15.7	GALAXY
ZWG 245.038	13 13 18.	+ 47 45		15.6	GALAXY
UGC 08331	13 13 18.	+ 47 45	180	15.6	GALAXY IRR
UGC 08332	13 13 18.	+ 52 16	60	17.	GALAXY Sb-c
ZWG 316.013	13 13 18.	+ 62 48		15.4	GALAXY
ZWG 016.035	13 13 18.	- 03 28		15.2	GALAXY
MCG-03-34-036	13 13 18.	- 16 14	132	13.5	GALAXY
MCG-05-31-041	13 13 18.	- 28 07	15	15.5	GALAXY
KEEL 625	13 13 18.1	+ 42 04 53.		18.	NEBULA
RNGC 5048	13 13 19.	- 28 07		15.0	GALAXY
KEEL 626	13 13 20.8	+ 42 02 05.		18.	NEBULA
REIZ 3348	13 13 22.	+ 29 53	30	15.5	GALAXY
KEEL 627	13 13 23.2	+ 41 56 29.		17.	NEBULA
ZWG 044.045	13 13 24.	+ 07 35		15.5	GALAXY
ZWG 072.039	13 13 24.	+ 13 36		15.7	GALAXY
ZC 1313.4+2835	13 13 24.	+ 28 35	1010		CLUSTER OF GALAXIES
ZWG 160.172	13 13 24.	+ 32 06		15.1	GALAXY
MCG+08-24-097	13 13 24.	+ 47 47	120	15.	GALAXY
VV 250B	13 13 24.	+ 62 22	15	15.	INTERACTING GALAXY
VV 250A	13 13 24.	+ 62 22	24	15.	INTERACTING GALAXY
VV 250	13 13 24.	+ 62 22	120		INTERACTING GALAXY
ZWG 016.036	13 13 24.	- 03 23		15.4	GALAXY
MCG-05-31-042	13 13 24.	- 28 00	36	15.	GALAXY
KEEL 628	13 13 24.4	+ 42 04 51.		18.	NEBULA
RNGC 5051	13 13 26.	- 28 00		15.0	GALAXY
MCG-03-34-037	13 13 27.	- 16 07	102	14.	GALAXY
KEEL 629	13 13 29.9	+ 42 03 16.		18.	NEBULA
FEIG 075	13 13 30.	+ 13 14		14.2	FAINT BLUE STAR
ZWG 101.016	13 13 30.	+ 17 25		15.7	GALAXY
OCL 0970	13 13 30.	+ 17 57	540	11.4	OPEN STAR CLUSTER
ZWG 101.017	13 13 30.	+ 18 01		15.7	GALAXY
KARA.73B 0577	13 13 30.	+ 18 01	18	15.7	ISOLATED GALAXY S0
REIZ 3349	13 13 30.	+ 25 42	60	15.0	GALAXY
UGC 08333	13 13 30.	+ 25 42	78	17.	GALAXY DWARF
8ZW 1313+25.8	13 13 30.	+ 25 48		18.2	COMPACT GALAXY
TON-N 0151	13 13 30.	+ 29 40		15.6	BLUE STAR
SN 1971I	13 13 30.	+ 42 15		11.5	SUPERNOVA
ZWG 217.023	13 13 30.	+ 42 17		9.7	GALAXY
UGC 08334	13 13 30.	+ 42 17	960	9.7	GALAXY Sb
KARA.72 369A	13 13 30.	+ 62 23	30	14.4	PART OF DOUBLE GALAXY
MCG+10-19-057	13 13 30.	+ 62 23	51	15.	GALAXY

OBJECT NAME	RIGHT ASCEN.	DECLINATION	DIAM.	MAGN.	TYPE OF OBJECT
MCG+10-19-056	13 13 30.	+ 62 23	51	15.	GALAXY
RNGC 5043	13 13 31.	- 59 48			NON-EXISTENT OBJECT
REIZ 3350	13 13 33.	+ 42 18	180	11.8	GALAXY
KEEL 630	13 13 34.6	+ 41 54 07.		18.	NEBULA
RNGC 5055	13 13 35.	+ 42 18		10.0	GALAXY
SVEN 397	13 13 35.	- 20 26	36	15.4	GALAXY
ZC 1313.6+0935	13 13 36.	+ 09 35	1610		CLUSTER OF GALAXIES
ZWG 072.040	13 13 36.	+ 13 40		15.7	GALAXY
ZC 1313.6+2328	13 13 36.	+ 23 28	400		CLUSTER OF GALAXIES
MCG+04-31-016	13 13 36.	+ 25 41	72	15.5	GALAXY
ZWG 271.027	13 13 36.	+ 52 28		15.6	GALAXY
ZWG 294.028	13 13 36.	+ 62 23		14.4	GALAXY
UGC 08335	13 13 36.	+ 62 23	102	14.4	GALAXY DBL SYS
KARA.72 369B	13 13 36.	+ 62 23	30		PART OF DOUBLE GALAXY
7ZW 506	13 13 36.	+ 62 24			COMPACT GALAXY
SMI 03	13 13 36.	- 60 52	1080		FAINT NEBULOSITY
KEEL 631	13 13 36.0	+ 42 11 21.		17.	NEBULA
FATH 1.572	13 13 37.	+ 14 23	11		NEBULA
LB 02604	13 13 38.	+ 51 32 18.		16.3	FAINT BLUE STAR
LB 02605	13 13 40.	+ 53 33 30.		16.1	FAINT BLUE STAR
ARC 1707	13 13 41.	+ 58 30		17.6	RICH CLUSTER OF GALAXIES
ZC 1313.7+0721	13 13 42.	+ 07 21	2890		CLUSTER OF GALAXIES
8ZW 1313+24.1	13 13 42.	+ 24 05		18.0	COMPACT GALAXY
MCG+06-29-069	13 13 42.	+ 35 17	66	14.	GALAXY
MCG+09-22-037	13 13 42.	+ 52 37 30.	42	15.	GALAXY
KEEL 632	13 13 42.2	+ 42 11 15.		17.	NEBULA
8ZW 1313+01.11	13 13 44.	+ 01 07		18.4	COMPACT GALAXY
KEEL 633	13 13 44.5	+ 42 06 26.		18.	NEBULA
KEEL 634	13 13 45.4	+ 42 07 35.		17.	NEBULA
LB 02606	13 13 46.	+ 57 29 24.		16.8	FAINT BLUE STAR
RNGC 5045	13 13 46.	- 63 09			NON-EXISTENT OBJECT
REIZ 3352	13 13 47.	+ 25 56	60	15.3	GALAXY
8ZW 1313+01.12	13 13 48.	+ 01 07		18.3	COMPACT GALAXY
ZWG 044.046	13 13 48.	+ 07 19		15.5	GALAXY
8ZW 1313+07.3	13 13 48.	+ 07 19		17.8	COMPACT GALAXY
MCG+01-34-013	13 13 48.	+ 07 19	36	15.5	GALAXY
ZC 1313.8+1927	13 13 48.	+ 19 27	670		CLUSTER OF GALAXIES
ZWG 101.018	13 13 48.	+ 20 18		15.2	GALAXY
KARA.73B 0578	13 13 48.	+ 20 18	18	15.2	ISOLATED GALAXY E
ZWG 130.024	13 13 48.	+ 25 40		15.0	GALAXY
UGC 08336	13 13 48.	+ 25 40	108	15.0	GALAXY Sa-b
ZWG 160.173	13 13 48.	+ 31 12		13.6	GALAXY
UGC 08337	13 13 48.	+ 31 12	114	13.6	GALAXY Sc
MCG+05-31-166	13 13 48.	+ 31 14	96	13.6	GALAXY
MCG+06-29-070	13 13 48.	+ 34 18	30	15.5	GALAXY
ZWG 189.047	13 13 48.	+ 35 18		14.6	GALAXY
UGC 08338	13 13 48.	+ 35 18	72	14.6	GALAXY Sb-c
ZWG 217.024	13 13 48.	+ 41 45		15.2	GALAXY
ZC 1313.8+5453	13 13 48.	+ 54 53	1080		CLUSTER OF GALAXIES
ZWG 294.029	13 13 48.	+ 57 05		15.4	GALAXY
UGC 08339	13 13 48.	+ 57 05	102	15.4	GALAXY Sc
7ZW 507	13 13 48.	+ 61 45			COMPACT GALAXY
MCG-03-34-038	13 13 48.	- 16 24	48	15.	GALAXY
IC 0862	13 13 49.	+ 20 18 47.			NONSTELLAR OBJECT
REIZ 3358	13 13 49.	+ 57 05	60	14.3	GALAXY
REIZ 3351	13 13 50.	+ 20 19	24	14.9	GALAXY
RNGC 5056	13 13 50.	+ 31 12		13.5	GALAXY
LB 02607	13 13 50.	+ 60 10 54.		16.7	FAINT BLUE STAR
REIZ 3354	13 13 51.	+ 31 13	90	13.2	GALAXY
LB 02608	13 13 51.	+ 56 38 42.		15.1	FAINT BLUE STAR
KEEL 635	13 13 51.6	+ 42 02 06.		17.	NEBULA
KEEL 636	13 13 52.8	+ 42 24 29.		18.	NEBULA
RNGC 5053	13 13 53.	+ 17 56		11.0	GLOBULAR CLUSTER
IC 4215	13 13 53.	+ 25 40 00.			NONSTELLAR OBJECT
KEEL 637	13 13 53.3	+ 42 06 52.		18.	NEBULA
ZWG 044.047	13 13 54.	+ 07 15		15.6	GALAXY
ZWG 072.041	13 13 54.	+ 13 46		15.2	GALAXY
FATH 1.573	13 13 54.	+ 14 23	11		NEBULA
GCL 023	13 13 54.	+ 17 57			GLOBULAR STAR CLUSTER
REIZ 3353	13 13 54.	+ 25 40	42	13.8	GALAXY
ZWG 160.174	13 13 54.	+ 30 31		14.9	GALAXY
MCG+05-31-167	13 13 54.	+ 30 31	24	14.9	GALAXY
MCG+06-29-071	13 13 54.	+ 36 20	48	14.5	GALAXY
MCG+10-19-058	13 13 54.	+ 57 03	96	14.	GALAXY
KEEL 638	13 13 54.7	+ 42 28 06.		18.	NEBULA
IC 4214	13 13 56.	- 31 49 43.			NONSTELLAR OBJECT
REIZ 3356	13 13 57.	+ 30 57	36	14.4	GALAXY
KEEL 639	13 13 57.6	+ 42 11 13.		17.	NEBULA
8ZW 1313+01.13	13 13 58.	+ 01 08		18.0	COMPACT GALAXY
KEEL 640	13 13 58.8	+ 42 05 19.		18.	NEBULA
LB 09871	13 14	- 83 31		13.3	FAINT BLUE STAR
8ZW 1314+01.1	13 14 00.	+ 01 08		18.4	COMPACT GALAXY
ZWG 044.048	13 14 00.	+ 03 20		15.4	GALAXY
8ZW 1314+13.2	13 14 00.	+ 13 12		14.7	COMPACT GALAXY
MCG+04-31-017	13 14 00.	+ 25 40	90	15.0	GALAXY
HZ 43	13 14 00.	+ 29 22		12.5	BLUE STAR
ZWG 160.175	13 14 00.	+ 30 55		15.1	GALAXY
MCG+05-31-168	13 14 00.	+ 30 58	30	15.1	GALAXY
STOCK 44	13 14 00.	+ 31 38			BLUE KNOT NEAR ELLIP GLXY
MCG+06-29-072	13 14 00.	+ 36 00	27	14.5	GALAXY
UGC 08341	13 14 00.	+ 36 20	66	16.0	GALAXY Sc
STOCK 43	13 14 00.	+ 37 15			BLUE KNOT NEAR ELLIP GLXY
ZC 1314.0+3745	13 14 00.	+ 37 45	2420		CLUSTER OF GALAXIES
ZWG 217.025	13 14 00.	+ 41 45		15.4	GALAXY
MCG+09-22-038	13 14 00.	+ 53 10	24	17.	GALAXY
1ZW 054	13 14 00.	+ 54 55			COMPACT GALAXY
MCG+14-06-017	13 14 00.	+ 84 58	42	16.	GALAXY
8ZW 1314-00.7	13 14 00.	- 00 40		16.9	COMPACT GALAXY
ZWG 016.037	13 14 00.	- 01 50		14.5	GALAXY
UGC 08340	13 14 00.	- 01 50	72	14.5	GALAXY Sc?
MCG+00-34-021	13 14 00.	- 01 50	54	14.5	GALAXY
8ZW 1313+01.14	13 14 02.	+ 01 08		17.5	COMPACT GALAXY
MCG+05-31-169	13 14 03.	+ 31 20	72	14.6	GALAXY
KEEL 641	13 14 04.5	+ 42 20 55.		17.	NEBULA
ARC 1706	13 14 05.	+ 41 40		17.2	RICH CLUSTER OF GALAXIES
SVEN 398	13 14 05.	- 20 42 22.	12	16.0	GALAXY
ZC 1314.1+0114	13 14 06.	+ 01 14	2220		CLUSTER OF GALAXIES
8ZW 1314+07.3	13 14 06.	+ 07 21		18.0	COMPACT GALAXY
8ZW 1314+08.1	13 14 06.	+ 08 06		17.4	COMPACT GALAXY
ZC 1314.1+2925	13 14 06.	+ 29 25	2290		CLUSTER OF GALAXIES
ZWG 160.176	13 14 06.	+ 31 17		14.6	GALAXY
UGC 08342	13 14 06.	+ 31 17	78	14.6	GALAXY S0
KEEL 642	13 14 06.3	+ 42 04 25.		17.	NEBULA
RNGC 5057	13 14 08.	+ 31 17		14.5	GALAXY
KEEL 643	13 14 08.6	+ 42 05 32.		18.	NEBULA
REIZ 3357	13 14 09.	+ 31 18	42	14.1	GALAXY
LB 02609	13 14 09.	+ 54 19 48.		15.5	FAINT BLUE STAR
KEEL 644	13 14 09.6	+ 42 04 49.		18.	NEBULA

OBJECT NAME	RIGHT ASCEN.	DECLINATION	DIAM.	MAGN.	TYPE OF OBJECT
TON-N 1576	13 14 10.	+ 23 03		17.0	BLUE STAR
REIZ 3355	13 14 11.	- 02 48	30	15.1	GALAXY
KEEL 645	13 14 11.7	+ 42 04 15.		17.	NEBULA
ZWG 044.049	13 14 12.	+ 07 57		15.6	GALAXY
8ZW 1314+08.0	13 14 12.	+ 07 59		17.8	COMPACT GALAXY
VV 298B	13 14 12.	+ 14 42	4	19.	INTERACTING GALAXY
VV 298A	13 14 12.	+ 14 42	42	15.	INTERACTING GALAXY
FATH 1.574	13 14 12.	+ 14 51	11		NEBULA
MCG+04-31-018	13 14 12.	+ 22 12	60	15.5	GALAXY
ZWG 130.025	13 14 12.	+ 22 15		15.5	GALAXY
UGC 08343	13 14 12.	+ 22 15	78	15.5	GALAXY S
KARA.73B 0579	13 14 12.	+ 22 15	66	15.5	ISOLATED GALAXY S
ZC 1314.2+2440	13 14 12.	+ 24 40	2290		CLUSTER OF GALAXIES
MCG-02-34-012	13 14 12.	- 13 20	36	15.	GALAXY
ARP 057	13 14 13.	+ 14 41			PECULIAR GALAXY
KEEL 646	13 14 13.2	+ 42 10 02.		18.	NEBULA
LB 02610	13 14 14.	+ 55 07 18.		16.8	FAINT BLUE STAR
RNGC 5054	13 14 16.	- 16 23		11.5	GALAXY
ZWG 101.019	13 14 18.	+ 14 41		15.5	GALAXY
8ZW 1314+14.7	13 14 18.	+ 14 41		18.7	COMPACT GALAXY
ZC 1314.3+2613	13 14 18.	+ 26 13	740		CLUSTER OF GALAXIES
ZC 1314.3+5233	13 14 18.	+ 52 33	1010		CLUSTER OF GALAXIES
ZWG 016.038	13 14 18.	- 02 48		15.5	GALAXY
MCG-04-31-045	13 14 18.	- 25 06	60	15.	GALAXY
FATH 1.575	13 14 19.	+ 14 41	41		NEBULA
RNGC 5058	13 14 22.	+ 12 49		14.5	GALAXY
FATH 1.576	13 14 23.	+ 14 40	14		NEBULA
REIZ 3359	13 14 23.	+ 53 12	18	15.2	GALAXY
ZWG 044.050	13 14 24.	+ 08 06		15.5	GALAXY
UGC 08344	13 14 24.	+ 08 06	60	15.5	GALAXY S
ZWG 072.042	13 14 24.	+ 12 49		14.6	GALAXY
UGC 08345	13 14 24.	+ 12 49	60	14.6	GALAXY S?
KARA.72 370B	13 14 24.	+ 12 49	24		PART OF DOUBLE GALAXY
KARA.72 370A	13 14 24.	+ 12 49	12	14.6	PART OF DOUBLE GALAXY
MCG+02-34-006	13 14 24.	+ 12 49	36	14.6	GALAXY
ZC 1314.4+3030	13 14 24.	+ 30 30	810		CLUSTER OF GALAXIES
UGC 08346	13 14 24.	+ 31 39	66	16.0	GALAXY SB
ZC 1314.4+4432	13 14 24.	+ 44 32	1410		CLUSTER OF GALAXIES
ZWG 271.028	13 14 24.	+ 51 43		15.6	GALAXY
ZWG 271.029	13 14 24.	+ 53 12		15.5	GALAXY
UGC 08347	13 14 24.	+ 53 12	72	15.5	GALAXY (S0)
MCG+09-22-039	13 14 24.	+ 53 12 30.	60	15.	GALAXY
MCG-02-34-013	13 14 24.	- 10 30	108	13.	GALAXY
MCG-03-34-040	13 14 24.	- 16 20	48	15.	GALAXY
MCG-03-34-039	13 14 24.	- 16 22	300	12.5	GALAXY
MCG-06-29-021	13 14 24.	- 34 30	18	15.5	GALAXY
HN 0374	13 14 25.	- 10 31			NEBULA
IC 4216	13 14 25.	- 10 31			NONSTELLAR OBJECT
RNGC 5059	13 14 27.	+ 08 06		15.5	GALAXY
MCG+06-29-073	13 14 27.	+ 34 13	12	15.5	GALAXY
HN 0375	13 14 28.	- 02 00			NEBULA
IC 4218	13 14 29.	- 02 00			NONSTELLAR OBJECT
KEEL 647	13 14 29.8	+ 42 24 45.		18.	NEBULA
ZWG 016.040	13 14 30.	+ 02 05		15.2	GALAXY
ZWG 044.051	13 14 30.	+ 05 55		15.5	GALAXY
ZWG 044.052	13 14 30.	+ 06 37		15.0	GALAXY
UGC 08349	13 14 30.	+ 06 37	84	15.0	GALAXY SBb
MCG+01-34-014	13 14 30.	+ 06 37	48	15.0	GALAXY
ZWG 189.048	13 14 30.	+ 34 15		15.7	GALAXY
MCG+09-22-041	13 14 30.	+ 51 41	18	15.	GALAXY
MCG+09-22-040	13 14 30.	+ 53 11	24	16.	GALAXY
LB 02611	13 14 30.	+ 61 01 48.		16.0	FAINT BLUE STAR
ZWG 016.039	13 14 30.	- 02 00		14.4	GALAXY
UGC 08348	13 14 30.	- 02 00	84	14.4	GALAXY S
MCG+00-34-022	13 14 30.	- 02 00	66	14.4	GALAXY
HF 0376	13 14 32.	- 12 53			NEBULA
IC 0863	13 14 32.	- 16 59 30.			NONSTELLAR OBJECT
IC 4217	13 14 33.	- 12 53			NONSTELLAR OBJECT
MCG-03-34-041	13 14 33.	- 15 58	120	15.	GALAXY
SN 1956G	13 14 35.	+ 02 36		19.3	SUPERNOVA
ZWG 016.042	13 14 36.	+ 00 47		15.7	GALAXY
MCG+00-34-023	13 14 36.	+ 00 47	30	15.7	GALAXY
ZWG 016.043	13 14 36.	+ 01 10		15.5	GALAXY
UGC 08350	13 14 36.	+ 08 40	60	16.0	GALAXY Sc
ZWG 160.177	13 14 36.	+ 31 50		15.4	GALAXY
ZWG 160.178	13 14 36.	+ 31 52		15.6	GALAXY
MCG+06-29-074	13 14 36.	+ 34 20	48	15.	GALAXY
ZWG 016.041	13 14 36.	- 01 56		15.4	GALAXY
8ZW 1314-01.9	13 14 36.	- 01 56		15.4	COMPACT GALAXY
MCG-02-34-014	13 14 36.	- 12 53	36	15.	GALAXY
MCG-03-34-042	13 14 36.	- 15 57	24	15.5	GALAXY
MCG-03-34-043	13 14 36.	- 16 58	36	14.5	GALAXY
LB 02612	13 14 37.	+ 61 38 36.		17.1	FAINT BLUE STAR
KEEL 648	13 14 38.5	+ 42 19 31.		17.	NEBULA
ARC 1708	13 14 40.	+ 46 46		17.8	RICH CLUSTER OF GALAXIES
ZWG 044.053	13 14 42.	+ 06 18		14.2	GALAXY
UGC 08351	13 14 42.	+ 06 18	84	14.2	GALAXY SBb
MCG+01-34-015	13 14 42.	+ 06 18	48	14.2	GALAXY
ZWG 044.054	13 14 42.	+ 06 40		15.5	GALAXY
ZWG 044.055	13 14 42.	+ 07 56		15.7	GALAXY
IC 0864	13 14 42.	+ 20 57 13.			NONSTELLAR OBJECT
ZWG 189.049	13 14 42.	+ 34 22		15.4	GALAXY
UGC 08352	13 14 42.	+ 34 22	72	15.4	GALAXY S
KEEL 649	13 14 43.5	+ 42 38 24.		17.	NEBULA
RNGC 5060	13 14 44.	+ 06 18		14.0	GALAXY
LB 02613	13 14 44.	+ 51 31 24.		16.7	FAINT BLUE STAR
MCG+04-31-019	13 14 45.	+ 20 55	66	15.6	GALAXY
MCG-06-29-022	13 14 45.	- 37 01	120	15.5	GALAXY
LYNG 09	13 14 47.	- 62 07 12.	1440		OB CONCENTRATION
ZWG 044.056	13 14 48.	+ 07 39		15.7	GALAXY
FEIG 076	13 14 48.	+ 13 03		15.0	FAINT BLUE STAR
MCG+04-31-020	13 14 48.	+ 20 52	84	15.5	GALAXY
ZWG 130.026	13 14 48.	+ 20 55		15.5	GALAXY
UGC 08353	13 14 48.	+ 20 55	90	15.5	GALAXY Sc/SBc
ZWG 130.027	13 14 48.	+ 20 58		15.6	GALAXY
UGC 08354	13 14 48.	+ 20 58	66	15.6	GALAXY SB
ZC 1314.8+2544	13 14 48.	+ 25 44	470		CLUSTER OF GALAXIES
ZC 1314.8+2631	13 14 48.	+ 26 31	1210		CLUSTER OF GALAXIES
ZWG 189.050	13 14 48.	+ 38 12		15.7	GALAXY
MCG+06-29-075	13 14 48.	+ 38 12	36	15.	GALAXY
ZWG 316.014	13 14 48.	+ 62 55		15.5	GALAXY
MCG-05-31-043	13 14 48.	- 31 50	66	13.	GALAXY
MCG-06-29-023	13 14 48.	- 34 05	102	15.5	GALAXY
MCG-06-29-024	13 14 48.	- 37 23	18	15.5	GALAXY
MRSL 306+01/1	13 14 48.	- 61 15	7200		HII REGION
HZ 41	13 14 49.	+ 39 19		13.3	BLUE STAR
LB 02614	13 14 49.	+ 58 55 48.		16.3	FAINT BLUE STAR
IC 0866	13 14 50.	+ 20 57 02.			NONSTELLAR OBJECT

OBJECT NAME	RIGHT ASCEN.	DECLINATION	DIAM.	MAGN.	TYPE OF OBJECT
LB 02615	13 14 51.	+ 50 42 00.		15.8	FAINT BLUE STAR
IC 4219	13 14 51.	- 31 23 41.			NONSTELLAR OBJECT
KEEL 650	13 14 51.6	+ 42 26 47.		16.	NEBULA
IC 0867	13 14 53.	+ 20 54 08.			NONSTELLAR OBJECT
KEEL 651	13 14 53.2	+ 42 19 56.		18.	NEBULA
KEEL 652	13 14 53.8	+ 42 33 54.		17.	NEBULA
ZWG 072.043	13 14 54.	+ 09 35		15.6	GALAXY
ZWG 130.028	13 14 54.	+ 20 53		15.4	GALAXY
ZWG 160.179	13 14 54.	+ 30 47		15.3	GALAXY
ZC 1314.9+4644	13 14 54.	+ 46 44	1340		CLUSTER OF GALAXIES
ZWG 016.044	13 14 54.	- 02 55		14.9	GALAXY
MCG+00-34-024	13 14 54.	- 02 55	36	14.9	GALAXY
REIZ 3363	13 14 55.	+ 57 07	48	14.6	GALAXY
MCG+04-31-021	13 14 57.	+ 20 50 30.	30	15.4	GALAXY
REIZ 3362	13 14 59.	+ 52 56	90	15.6	GALAXY
KEEL 653	13 14 59.0	+ 42 21 32.		18.	NEBULA
VDB.66G 171	13 15	- 08 10	70		DWARF GALAXY
MCG+04-31-022	13 15 00.	+ 20 49 30.	36	15.4	GALAXY
ZWG 130.029	13 15 00.	+ 20 52		15.4	GALAXY
ZWG 160.180	13 15 00.	+ 31 18		15.4	GALAXY
MCG+08-24-098	13 15 00.	+ 46 18	36	16.	GALAXY
ZC 1315.0+8206	13 15 00.	+ 82 06	2220		CLUSTER OF GALAXIES
ZC 1315.0+8540	13 15 00.	+ 85 40	2020		CLUSTER OF GALAXIES
IC 0865	13 15 00.	- 05 33 52.			NONSTELLAR OBJECT
MCG-02-34-016	13 15 00.	- 15 15	12	15.	GALAXY
MCG-02-34-015	13 15 00.	- 15 16	66	15.	GALAXY
MCG-04-31-046	13 15 00.	- 24 04	12	15.	GALAXY
IC 0868	13 15 01.	+ 20 52 26.			NONSTELLAR OBJECT
MCG+08-24-099	13 15 03.	+ 47 37 30.	30	15.	GALAXY
IC 0869	13 15 04.	+ 20 59 32.			NONSTELLAR OBJECT
TON-N 1577	13 15 04.	+ 34 52		16.6	BLUE STAR
IC 0870	13 15 05.	+ 20 51 44.			NONSTELLAR OBJECT
FATH 1.577	13 15 05.	- 15 16	8		NEBULA
ZWG 044.057	13 15 06.	+ 04 45		15.6	GALAXY
MCG+05-31-170	13 15 06.	+ 31 22 30.	60	14.3	GALAXY
ZWG 294.030	13 15 06.	+ 57 48		13.9	GALAXY
UGC 08355	13 15 06.	+ 57 48	96	13.9	GALAXY S0
IC 0875	13 15 06.	+ 57 48 27.			NONSTELLAR OBJECT
ZWG 016.045	13 15 06.	- 00 44		15.5	GALAXY
FATH 1.578	13 15 06.	- 15 14	8		NEBULA
MCG-06-29-025	13 15 06.	- 36 41	60	14.	GALAXY
REIZ 3360	13 15 09.	+ 16 46	12	15.5	GALAXY
FATH 1.579	13 15 10.	+ 14 31	8		NEBULA
TON-N 1578	13 15 10.	+ 22 23		16.8	BLUE STAR
KEEL 654	13 15 10.8	+ 42 22 44.		18.	NEBULA
8ZW 1315+25.8	13 15 12.	+ 25 47		16.4	COMPACT GALAXY
ZWG 160.181	13 15 12.	+ 31 20		14.3	GALAXY
UGC 08356	13 15 12.	+ 31 20	90	14.3	GALAXY Sc
ZC 1315.2+4216	13 15 12.	+ 42 16	940		CLUSTER OF GALAXIES
MRK 249	13 15 12.	+ 57 48	30	14.	GALAXY WITH UV CONTINUUM
MCG+10-19-059	13 15 12.	+ 57 48	39	14.	GALAXY
SN 1956I	13 15 12.	- 01 18		20.2	SUPERNOVA
SVEN 399	13 15 12.	- 20 55	12	16.3	GALAXY
FATH 1.580	13 15 13.	+ 14 11	16		NEBULA
SVEN 400	13 15 13.	- 24 03	18	15.0	GALAXY
RNGC 5065	13 15 14.	+ 31 20		14.5	GALAXY
IC 4220	13 15 14.	- 13 21			NONSTELLAR OBJECT
REIZ 3361	13 15 15.	+ 31 21	54	13.4	GALAXY
HN 0377	13 15 15.	- 13 21			NEBULA
8ZW 1315+02.5	13 15 18.	+ 02 29		17.8	COMPACT GALAXY
ZWG 245.039	13 15 18.	+ 48 03		15.7	GALAXY
MCG-02-34-017	13 15 18.	- 13 20	42	14.5	GALAXY
MCG-04-31-047	13 15 18.	- 21 37	48	17.	GALAXY
VHE 60C	13 15 18.	- 62 26			REFLECTION NEBULA
KEEL 655	13 15 18.3	+ 42 24 56.		18.	NEBULA
RNGC 5061	13 15 20.	- 26 36		12.0	GALAXY
REIN 2.187	13 15 20.48	- 26 34 27.3			NEBULA
FATH 1.581	13 15 21.	- 15 03	27		NEBULA
MCG-04-31-048	13 15 21.	- 26 36	60	12.	GALAXY
FATH 1.582	13 15 23.	- 15 03	30		NEBULA
REIN 2.188	13 15 23.34	- 26 34 08.7			NEBULA
ZWG 044.058	13 15 24.	+ 04 40		14.8	GALAXY
UGC 08358	13 15 24.	+ 04 40	126	14.8	GALAXY Sb
MCG+01-34-016	13 15 24.	+ 04 40	72	14.8	GALAXY
ZWG 160.182	13 15 24.	+ 27 49		15.0	GALAXY
UGC 08359	13 15 24.	+ 27 49	96	15.0	GALAXY Sa-b
MCG+05-31-171	13 15 24.	+ 27 50	54	15.0	GALAXY
ZWG 217.026	13 15 24.	+ 44 04		15.5	GALAXY
MRK 250	13 15 24.	+ 44 04	13	16.	GALAXY WITH UV CONTINUUM
MCG+08-24-100	13 15 24.	+ 47 57 30.	12	16.	GALAXY
MCG+08-24-101	13 15 24.	+ 48 04	36	15.	GALAXY
ZC 1315.4+5648	13 15 24.	+ 56 48	1210		CLUSTER OF GALAXIES
ZWG 016.046	13 15 24.	- 00 03		14.5	GALAXY
UGC 08357	13 15 24.	- 00 03	102	14.5	GALAXY TRP SYS
MCG+00-34-025	13 15 24.	- 00 03	54	14.5	GALAXY
VHA 145	13 15 24.	- 66 48	300		OPEN STAR CLUSTER
OCL 0895	13 15 24.	- 66 56	480		OPEN STAR CLUSTER
KEEL 657	13 15 25.9	+ 42 27 06.		18.	NEBULA
KEEL 656	13 15 26.1	+ 41 57 39.		17.	NEBULA
TON-N 1579	13 15 27.	+ 20 54		17.0	BLUE STAR
IC 0871	13 15 28.	+ 04 39 39.			NONSTELLAR OBJECT
ZC 1315.5+1244	13 15 30.	+ 12 44	610		CLUSTER OF GALAXIES
ZC 1315.5+1851	13 15 30.	+ 18 51	1750		CLUSTER OF GALAXIES
8ZW 1315+23.0	13 15 30.	+ 23 02		17.9	COMPACT GALAXY
ZC 1315.5+4354	13 15 30.	+ 43 54	1410		CLUSTER OF GALAXIES
MCG+09-22-042	13 15 30.	+ 56 32	36	16.	GALAXY
MCG-02-34-018	13 15 30.	- 15 26	54	14.5	GALAXY
MCG-06-29-027	13 15 30.	- 35 06	90	12.5	GALAXY
MCG-06-29-026	13 15 30.	- 35 12	90	13.	GALAXY
REIZ 3364	13 15 32.	+ 04 39	18	14.3	GALAXY
FATH 1.584	13 15 32.	+ 14 57	19		NEBULA
KEEL 658	13 15 33.8	+ 42 11 22.		18.	NEBULA
FATH 1.585	13 15 34.	+ 14 42	8		NEBULA
TON-N 1580	13 15 34.	+ 22 03		17.0	BLUE STAR
TON-N 1581	13 15 34.	+ 35 08		16.9	BLUE STAR
FATH 1.583	13 15 34.	- 15 26	41		NEBULA
RNGC 5063	13 15 35.	- 35 04			GALAXY
RNGC 5062	13 15 35.	- 35 10			GALAXY
KEEL 659	13 15 35.4	+ 42 32 46.		18.	NEBULA
ZWG 016.048	13 15 36.	+ 00 29		15.6	GALAXY
ZWG 016.047	13 15 36.	- 00 58		14.3	GALAXY
UGC 08360	13 15 36.	- 00 58	90	14.3	GALAXY Sc
MCG+00-34-026	13 15 36.	- 00 58	78	14.3	GALAXY
MCG-05-31-044	13 15 36.	- 31 22	54	14.	GALAXY
KEEL 660	13 15 36.2	+ 42 09 54.		18.	NEBULA
KEEL 661	13 15 36.8	+ 42 28 44.		17.	NEBULA
ARC 1710	13 15 37.	+ 43 55		17.2	RICH CLUSTER OF GALAXIES
LB 02616	13 15 39.	+ 52 27 06.		16.8	FAINT BLUE STAR
IC 0872	13 15 41.	+ 06 37 10.			NONSTELLAR OBJECT
ZC 1315.7+1411	13 15 42.	+ 14 11	2020		CLUSTER OF GALAXIES
MCG+11-16-016	13 15 42.	+ 64 37	39	16.	GALAXY
SN 1973I	13 15 42.	- 06 40		20.5	SUPERNOVA
MCG-02-34-019	13 15 42.	- 15 30	36	15.	GALAXY
VHE 60A	13 15 42.	- 62 18			REFLECTION NEBULA
VHE 60B	13 15 42.	- 62 19			REFLECTION NEBULA
KEEL 662	13 15 42.2	+ 42 31 59.		18.	NEBULA
FATH 1.586	13 15 44.	- 15 31	41		NEBULA
HN 0378	13 15 45.	- 14 22			NEBULA
IC 4221	13 15 45.	- 14 22			NONSTELLAR OBJECT
LB 00267	13 15 45.	- 16 08 30.		15.9	FAINT BLUE STAR
IC 0873	13 15 46.	+ 04 43 22.			NONSTELLAR OBJECT
FATH 1.587	13 15 46.	+ 14 32	5		NEBULA
RNGC 5067	13 15 47.	- 09 52			NON-EXISTENT OBJECT
RNGC 5066	13 15 47.	- 09 58		14.0	GALAXY
ZWG 044.059	13 15 48.	+ 04 44		15.3	GALAXY
ZWG 044.060	13 15 48.	+ 06 35		15.1	GALAXY
UGC 08361	13 15 48.	+ 06 35	60	15.1	GALAXY Sa-b
ZC 1315.8+1451	13 15 48.	+ 14 51	610		CLUSTER OF GALAXIES
LB 02617	13 15 48.	+ 50 30 06.		16.4	FAINT BLUE STAR
MCG-02-34-020	13 15 48.	- 09 58	30	14.	GALAXY
MCG-02-34-021	13 15 48.	- 14 20	66	13.5	GALAXY
MCG-03-34-044	13 15 48.	- 17 21 30.	36	14.5	GALAXY
SVEN 401	13 15 48.	- 21 02 20.	60	14.8	GALAXY
OCL 0898	13 15 48.	- 62 18	180	10.	OPEN STAR CLUSTER
REIZ 3365	13 15 49.	- 09 59	18	14.0	GALAXY
KEEL 663	13 15 50.3	+ 42 26 20.		17.	NEBULA
8ZW 1315+22.1	13 15 54.	+ 22 04		17.2	COMPACT GALAXY
1ZW 055	13 15 54.	+ 35 39			COMPACT GALAXY
ZC 1315.9+3623	13 15 54.	+ 36 23	870		CLUSTER OF GALAXIES
ZC 1315.9+5249	13 15 54.	+ 52 49	1010		CLUSTER OF GALAXIES
REIZ 3366	13 15 55.	- 09 57	9	15.7	GALAXY
FATH 1.589	13 15 58.	+ 14 19	14		NEBULA
FATH 1.590	13 15 59.	+ 14 08	14		NEBULA
LB 02618	13 15 59.	+ 53 02 42.		16.5	FAINT BLUE STAR
VDB.66G 172	13 16	+ 42 15	100		DWARF GALAXY
ZWG 044.061	13 16 00.	+ 04 45		14.8	GALAXY
MCG+01-34-017	13 16 00.	+ 04 45	36	14.8	GALAXY
ZC 1316.0+1535	13 16 00.	+ 15 35	740		CLUSTER OF GALAXIES
TON-N 0702	13 16 00.	+ 27 15		14.7	BLUE STAR
MCG+05-31-172	13 16 00.	+ 31 46	36	14.7	GALAXY
ZC 1316.0+3153	13 16 00.	+ 31 53	610		CLUSTER OF GALAXIES
ZC 1316.0+3351	13 16 00.	+ 33 51	470		CLUSTER OF GALAXIES
MCG+09-22-043	13 16 00.	+ 54 31	36	16.	GALAXY
MCG+14-06-018	13 16 00.	+ 86 06	27	16.	GALAXY
KARA.68 220	13 16 00.	- 17 18	27		DWARF GALAXY
MCG-03-34-045	13 16 00.	- 18 50	30	15.	GALAXY
SVEN 402	13 16 00.	- 20 43	24	15.5	GALAXY
FATH 1.588	13 16 01.	- 14 48	41		NEBULA
ARC 1709	13 16 01.	- 21 12		16.4	RICH CLUSTER OF GALAXIES
RNGC 5064	13 16 01.	- 47 39		13.0	GALAXY
REIZ 3367	13 16 02.	+ 04 44	24	14.5	GALAXY
IC 0876	13 16 05.	+ 04 44 35.			NONSTELLAR OBJECT
IC 3160	13 16 05.	+ 60 59 48.			NONSTELLAR OBJECT
RNGC 5069	13 16 05.	- 09 56			NON-EXISTENT OBJECT
KARA.73 44	13 16 05.	- 37 15	27		DWARF GALAXY
ZWG 016.049	13 16 06.	+ 00 05		15.6	GALAXY
8ZW 1316+00.1	13 16 06.	+ 00 05		15.6	COMPACT GALAXY
8ZW 1316+01.5	13 16 06.	+ 01 30		17.8	COMPACT GALAXY
ZWG 044.062	13 16 06.	+ 08 12		15.5	GALAXY
ZWG 044.063	13 16 06.	+ 08 16		15.7	GALAXY
ZC 1316.1+1115	13 16 06.	+ 11 15	2020		CLUSTER OF GALAXIES
ZWG 161.001	13 16 06.	+ 31 44		14.7	GALAXY
ZWG 160.183	13 16 06.	+ 31 44		14.7	GALAXY
ZWG 161.002	13 16 06.	+ 32 02		15.6	GALAXY
ZWG 160.184	13 16 06.	+ 32 02		15.6	GALAXY
UGC 08362	13 16 06.	+ 33 07	72	16.0	GALAXY SBb
ARC 1712	13 16 06.	+ 34 33		18.0	RICH CLUSTER OF GALAXIES
ZC 1316.1+3438	13 16 06.	+ 34 38	610		CLUSTER OF GALAXIES
8ZW 1316-00.1	13 16 06.	- 00 06		17.5	COMPACT GALAXY
SVEN 403	13 16 06.	- 20 47	432	10.9	GALAXY
ACK 305-03.1	13 16 06.	- 65 53			PLANETARY NEBULA
FATH 1.591	13 16 08.	+ 14 18	14		NEBULA
RNGC 5074	13 16 08.	+ 31 44		14.5	GALAXY
REIZ 3370	13 16 08.	+ 31 44	42	13.6	GALAXY
RNGC 5071	13 16 09.	+ 08 12		15.5	GALAXY
RNGC 5068	13 16 09.	- 20 47		11.0	GALAXY
KEEL 664	13 16 10.8	+ 42 31 15.		18.	NEBULA
ZWG 072.044	13 16 12.	+ 08 57		15.7	GALAXY
ZWG 101.020	13 16 12.	+ 19 40		15.6	GALAXY
ZWG 161.003	13 16 12.	+ 31 36		15.2	GALAXY
ZWG 160.185	13 16 12.	+ 31 36		15.2	GALAXY
MCG+08-24-102	13 16 12.	+ 47 30	12	17.	GALAXY
MCG+08-24-103	13 16 12.	+ 47 31	15	17.	GALAXY
ZC 1316.2+5146	13 16 12.	+ 51 46	1480		CLUSTER OF GALAXIES
ZWG 271.030	13 16 12.	+ 54 14		15.6	GALAXY
MCG+09-22-044	13 16 12.	+ 54 15	30	15.	GALAXY
7ZW 508	13 16 12.	+ 58 14			COMPACT GALAXY
HELW 468	13 16 12.	- 24 26 07.			NEBULA
MCG-04-31-049	13 16 12.	- 24 26 30.	48	15.	GALAXY
LB 02619	13 16 13.	+ 51 38 12.		16.9	FAINT BLUE STAR
SVEN 404	13 16 13.	- 24 26 19.	54	14.9	GALAXY
ARC 1711	13 16 14.	+ 11 15		17.2	RICH CLUSTER OF GALAXIES
LB 02620	13 16 16.	+ 52 06 30.		15.9	FAINT BLUE STAR
IC 0874	13 16 16.	- 27 22 01.			NONSTELLAR OBJECT
KEEL 665	13 16 16.3	+ 42 29 05.		17.	NEBULA
ZWG 072.045	13 16 18.	+ 12 40		15.7	GALAXY
ZC 1316.3+1613	13 16 18.	+ 16 13	1080		CLUSTER OF GALAXIES
ZWG 161.004	13 16 18.	+ 31 07		15.5	GALAXY
ZWG 160.186	13 16 18.	+ 31 07		15.5	GALAXY
ZWG 161.005	13 16 18.	+ 31 49		15.6	GALAXY
ZWG 160.187	13 16 18.	+ 31 49		15.6	GALAXY
ZC 1316.3+5818	13 16 18.	+ 58 18	2690		CLUSTER OF GALAXIES
MCG-01-34-014	13 16 18.	- 08 11	72	14.5	GALAXY
KARA.68 221	13 16 18.	- 17 21	81		DWARF GALAXY
MCG-03-34-046	13 16 18.	- 20 47	420	10.	GALAXY
MCG-04-31-050	13 16 18.	- 27 23 30.	24	14.5	GALAXY
REIZ 3369	13 16 19.	+ 08 02	24	15.2	GALAXY
LB 02621	13 16 19.	+ 56 43 18.		17.2	FAINT BLUE STAR
MCG+08-24-104	13 16 21.	+ 47 28	9	17.	GALAXY
BIGO 530	13 16 21.	- 12 28			NEBULA
RNGC 5090A	13 16 21.	- 43 24			GALAXY
REIZ 3372	13 16 22.	+ 28 00	36	15.8	GALAXY
HELW 469	13 16 23.	- 24 09 19.			NEBULA
ZWG 016.050	13 16 24.	+ 00 04		15.7	GALAXY
ZWG 044.064	13 16 24.	+ 08 03		15.4	GALAXY
ZC 1316.4+2126	13 16 24.	+ 21 26	2620		CLUSTER OF GALAXIES

OBJECT NAME	RIGHT ASCEN.	DECLINATION	DIAM.	MAGN.	TYPE OF OBJECT
ZWG 161.006	13 16 24.	+ 28 00		15.6	GALAXY
ZWG 160.188	13 16 24.	+ 28 00		15.6	GALAXY
UGC 08363	13 16 24.	+ 28 00	78	15.6	GALAXY IRR
MCG+05-31-173	13 16 24.	+ 28 00	48	15.6	GALAXY
FEIG 077	13 16 24.	+ 29 02		15.0	FAINT BLUE STAR
ZC 1316.4+3328	13 16 24.	+ 33 28	610		CLUSTER OF GALAXIES
UGC 08364	13 16 24.	+ 47 24	84	17.	GALAXY Sc
MCG+08-24-105	13 16 24.	+ 47 25 30.	9	17.	GALAXY
ZWG 271.031	13 16 24.	+ 54 20		15.5	GALAXY
ZC 1316.4-0044	13 16 24.	- 00 44	2220		CLUSTER OF GALAXIES
MCG-03-34-047	13 16 24.	- 18 20	54	15.	GALAXY
DV.56 N5090A	13 16 24.	- 43 24	78		S GALAXY
REIZ 3371	13 16 25.	+ 08 04	48	14.5	GALAXY
SVEN 405	13 16 25.	- 24 09 19.	60	15.3	GALAXY
IC 0877	13 16 26.	+ 06 20 42.			NONSTELLAR OBJECT
IC 4223	13 16 26.	+ 08 04			NONSTELLAR OBJECT
REIZ 3368	13 16 26.	- 12 17	18	14.9	GALAXY
MCG+08-24-106	13 16 27.	+ 47 25	60	17.	GALAXY
IC 0878	13 16 29.	+ 06 23 00.			NONSTELLAR OBJECT
RNGC 5072	13 16 29.	- 12 16		14.0	GALAXY
8ZW 1316+07.8	13 16 30.	+ 07 46		14.3	COMPACT GALAXY
ZC 1316.5+2939	13 16 30.	+ 29 39	1010		CLUSTER OF GALAXIES
ZWG 217.027	13 16 30.	+ 42 12		15.5	GALAXY
UGC 08365	13 16 30.	+ 42 12	150	15.5	GALAXY SBc
MCG+09-22-046	13 16 30.	+ 54 21	42	16.	GALAXY
MCG+09-22-045	13 16 30.	+ 55 46	24	17.	GALAXY
ZWG 016.051	13 16 30.	- 02 15		14.6	GALAXY
MCG+00-34-027	13 16 30.	- 02 15	48	14.6	GALAXY
MCG-02-34-022	13 16 30.	- 12 16	21	14.5	GALAXY
FATH 1.592	13 16 30.	- 15 46	27		NEBULA
MRSL 306+00/1	13 16 30.	- 62 15	1080		HII REGION
LB 00268	13 16 31.	- 14 31 54.		15.3	FAINT BLUE STAR
HN 0379	13 16 34.	- 02 15			NEBULA
IC 4224	13 16 34.	- 02 15			NONSTELLAR OBJECT
RNGC 5070	13 16 35.	- 12 13		15.0	GALAXY
IC 0880	13 16 36.	+ 06 22 30.			NONSTELLAR OBJECT
ZWG 044.065	13 16 36.	+ 08 05		15.1	GALAXY
ZWG 101.021	13 16 36.	+ 15 03		15.3	GALAXY
KARA.73B 0580	13 16 36.	+ 15 03	36	15.3	ISOLATED GALAXY S
8ZW 1316-00.6	13 16 36.	- 00 38		17.4	COMPACT GALAXY
MCG-02-34-023	13 16 36.	- 12 13	36	15.	GALAXY
MCG-02-34-024	13 16 36.	- 14 54	48	14.5	GALAXY
FATH 1.593	13 16 36.	- 15 40	16		NEBULA
MCG-07-27-051	13 16 36.	- 43 20	60	13.5	GALAXY
FATH 1.595	13 16 37.	- 14 53	41		NEBULA
REIZ 3378	13 16 38.	+ 31 25	18	14.9	GALAXY
FATH 1.594	13 16 38.	- 15 34	8		NEBULA
RNGC 5075	13 16 39.	+ 08 05		15.0	GALAXY
LB 02622	13 16 40.	+ 51 14 06.		16.1	FAINT BLUE STAR
RNGC 5073	13 16 40.	- 14 36		13.0	GALAXY
ZWG 044.066	13 16 42.	+ 04 11		15.6	GALAXY
ZWG 044.067	13 16 42.	+ 07 42		15.2	GALAXY
8ZW 1316+21.9	13 16 42.	+ 21 53		17.6	COMPACT GALAXY
ZWG 161.007	13 16 42.	+ 28 09		15.7	GALAXY
ZWG 160.189	13 16 42.	+ 28 09		15.7	GALAXY
ZWG 161.008	13 16 42.	+ 31 02		15.3	GALAXY
ZWG 160.190	13 16 42.	+ 31 02		15.3	GALAXY
ZWG 161.009	13 16 42.	+ 31 04		15.7	GALAXY
ZWG 160.191	13 16 42.	+ 31 04		15.7	GALAXY
MCG-02-34-025	13 16 42.	- 14 36	210	13.	GALAXY
VHE 61	13 16 42.	- 62 45	48		REFLECTION NEBULA
FATH 1.596	13 16 44.	- 14 53	11		NEBULA
MCG+05-31-174	13 16 45.	+ 28 47	108	14.3	GALAXY
MCG-04-31-051	13 16 45.	- 26 46	72	15.	GALAXY
IC 4222	13 16 45.	- 28 09 54.			NONSTELLAR OBJECT
REIZ 3379	13 16 46.	+ 28 46	90	13.8	GALAXY
RNGC 5083	13 16 46.	+ 39 52		15.5	GALAXY
RNGC 5076	13 16 47.	- 12 29		13.0	GALAXY
REIZ 3377	13 16 48.	+ 08 37	30	14.2	GALAXY
ZWG 072.046	13 16 48.	+ 08 41		14.6	GALAXY
MCG+02-34-007	13 16 48.	+ 08 41	18	14.6	GALAXY
ZWG 072.047	13 16 48.	+ 09 01		15.5	GALAXY
ZWG 161.010	13 16 48.	+ 28 46		14.3	GALAXY
ZWG 160.192	13 16 48.	+ 28 46		14.3	GALAXY
UGC 08366	13 16 48.	+ 28 46	132	14.3	GALAXY SBb
KARA.73B 0581	13 16 48.	+ 28 46	162	14.3	ISOLATED GALAXY S
ZWG 161.011	13 16 48.	+ 31 24		15.6	GALAXY
ZWG 160.193	13 16 48.	+ 31 24		15.6	GALAXY
ZC 1316.8+3129	13 16 48.	+ 31 29	1680		CLUSTER OF GALAXIES
ZWG 217.028	13 16 48.	+ 39 52		15.4	GALAXY
UGC 08367	13 16 48.	+ 39 52	84	15.4	GALAXY SBc
UGC 08368	13 16 48.	+ 43 29	60	16.0	GALAXY S
MCG-02-34-027	13 16 48.	- 12 23	48	12.	GALAXY
MCG-02-34-026	13 16 48.	- 12 29	60	13.5	GALAXY
MCG-02-34-028	13 16 48.	- 14 31	12	15.5	GALAXY
HELW 470	13 16 48.	- 24 13 18.			NEBULA
8ZW 1316-28.5	13 16 48.	- 28 29		17.8	COMPACT GALAXY
RNGC 5081	13 16 49.	+ 28 46		14.5	GALAXY
FATH 1.597	13 16 49.	- 15 36	8		NEBULA
SVEN 406	13 16 49.	- 24 13 18.	12	14.8	GALAXY
SVEN 407	13 16 49.	- 24 14	12	15.4	GALAXY
REIZ 3380	13 16 50.	+ 31 19	66	14.9	GALAXY
REIZ 3375	13 16 50.	- 12 24	18	14.6	GALAXY
REIZ 3374	13 16 50.	- 12 24	24	12.8	GALAXY
REIZ 3373	13 16 50.	- 12 29	18	14.1	GALAXY
HELW 471	13 16 50.	- 24 13 42.			NEBULA
RNGC 5080	13 16 51.	+ 08 41		14.5	GALAXY
TON-N 1582	13 16 52.	+ 33 15		17.0	BLUE STAR
RNGC 5077	13 16 53.	- 12 24		12.5	GALAXY
ZC 1316.9+0017	13 16 54.	+ 00 17	1210		CLUSTER OF GALAXIES
8ZW 1316+10.3	13 16 54.	+ 10 17		18.0	COMPACT GALAXY
HOLM 514C	13 16 54.	- 12 29	24	14.0	PART OF MULTIPLE GALAXY
MCG-04-31-052	13 16 54.	- 27 12	48	14.5	GALAXY
DVDV 2	13 16 54.	- 35 42	210		GALAXY
HOLM 514D	13 16 55.	- 12 24	18	14.3	PART OF MULTIPLE GALAXY
HOLM 514B	13 16 55.	- 12 24	30	13.0	PART OF MULTIPLE GALAXY
IC 0879	13 16 55.	- 27 10 00.			NONSTELLAR OBJECT
REIZ 3376	13 16 56.	- 12 27	60	13.2	GALAXY
RNGC 5079	13 16 59.	- 12 27		12.0	GALAXY
ZC 1317.0+0627	13 17 00.	+ 06 27	1480		CLUSTER OF GALAXIES
ZWG 044.068	13 17 00.	+ 07 45		15.5	GALAXY
ZWG 101.022	13 17 00.	+ 14 50		15.6	GALAXY
ZC 1317.0+1643	13 17 00.	+ 16 43	2620		CLUSTER OF GALAXIES
TON-N 0152	13 17 00.	+ 28 48		16.5	BLUE STAR
ZC 1317.0+3208	13 17 00.	+ 32 08	470		CLUSTER OF GALAXIES
ZC 1317.0+3347	13 17 00.	+ 33 47	740		CLUSTER OF GALAXIES
UGC 08369	13 17 00.	+ 81 39	66	16.5	GALAXY Sc
MCG+14-06-019	13 17 00.	+ 84 56	66	16.	GALAXY

OBJECT NAME	RIGHT ASCEN.	DECLINATION	DIAM.	MAGN.	TYPE OF OBJECT
ZWG 016.052	13 17 00.	- 02 39		14.9	GALAXY
MCG+00-34-028	13 17 00.	- 02 39	15	14.9	GALAXY
MCG-02-34-029	13 17 00.	- 11 12	24	14.5	GALAXY
MCG-02-34-030	13 17 00.	- 12 26	84	12.	GALAXY
FATH 1.598	13 17 02.	+ 14 49	16		NEBULA
HOLM 514A	13 17 02.	- 12 26	90	12.7	PART OF MULTIPLE GALAXY
KARA.73 45	13 17 04.	- 36 46	101		DWARF GALAXY
SHB 256	13 17 04.5	- 00 34 12.		17.3	QUASI-STELLAR OBJECT
BC PKS1317-00	13 17 04.5	- 00 34 18.		17.32	QUASI-STELLAR OBJECT
REIN 2.189	13 17 05.14	- 27 08 49.6			NEBULA
ZWG 044.069	13 17 06.	+ 03 17		15.1	GALAXY
ZWG 101.023	13 17 06.	+ 17 15		15.2	GALAXY
UGC 08370	13 17 06.	+ 17 15	60	15.2	GALAXY Sb-c
MCG+08-24-107	13 17 06.	+ 48 11	15	16.	GALAXY
ZC 1317.1+5222	13 17 06.	+ 52 22	1010		CLUSTER OF GALAXIES
MCG-04-32-001	13 17 06.	- 27 08 30.	138	12.5	GALAXY
RNGC 5078	13 17 07.	- 27 09		12.0	GALAXY
SC 1314+4820.4	13 17 08.	+ 48 04 36.	36		NEBULA
MCG-04-32-002	13 17 09.	- 27 37	30	14.	GALAXY
LB 02623	13 17 11.	+ 50 07 06.		17.2	FAINT BLUE STAR
ZWG 072.048	13 17 12.	+ 12 04		15.6	GALAXY
ZC 1317.2+4005	13 17 12.	+ 40 05	1340		CLUSTER OF GALAXIES
ZWG 271.032	13 17 12.	+ 51 24		15.7	GALAXY
MCG-04-32-003	13 17 12.	- 22 00	96	13.5	GALAXY
OCL 0896	13 17 12.	- 64 41	354	12.	OPEN STAR CLUSTER
VHA 146	13 17 12.	- 64 41	150		OPEN STAR CLUSTER
LB 02624	13 17 14.	+ 60 53 12.		16.2	FAINT BLUE STAR
ARC 1713	13 17 15.	+ 58 21		17.2	RICH CLUSTER OF GALAXIES
RNGC 5090B	13 17 15.	- 43 36			GALAXY
LB 02625	13 17 16.	+ 52 14 18.		15.7	FAINT BLUE STAR
LB 02626	13 17 17.	+ 55 39 54.			FAINT BLUE STAR
ZC 1317.3+1327	13 17 18.	+ 13 27	540		CLUSTER OF GALAXIES
ZWG 161.012	13 17 18.	+ 30 31		14.4	GALAXY
ZWG 160.194	13 17 18.	+ 30 31		14.4	GALAXY
UGC 08371	13 17 18.	+ 30 31	126	14.4	GALAXY S
MCG+05-31-175	13 17 18.	+ 30 32 30.	90	14.4	GALAXY
ZC 1317.3+3605	13 17 18.	+ 36 05	1010		CLUSTER OF GALAXIES
LB 02627	13 17 18.	+ 52 15 18.		17.3	FAINT BLUE STAR
ZC 1317.3+7018	13 17 18.	+ 70 18	1880		CLUSTER OF GALAXIES
MCG-02-34-031	13 17 18.	- 12 13	36	14.	GALAXY
MCG-06-29-028	13 17 18.	- 35 47	120	16.	GALAXY
DV.56 N5090B	13 17 18.	- 43 36	108		S GALAXY
RNGC 5089	13 17 20.	+ 30 31		14.5	GALAXY
REIZ 3381	13 17 20.	- 12 14	15	14.4	GALAXY
FATH 1.599	13 17 20.	- 14 58	8		NEBULA
REIZ 3383	13 17 21.	+ 30 30	42	14.1	GALAXY
REIZ 3386	13 17 21.	+ 40 40	36	14.5	GALAXY
RNGC 5093	13 17 23.	+ 40 39		15.0	GALAXY
ZWG 044.070	13 17 24.	+ 06 04		15.3	GALAXY
UGC 08372	13 17 24.	+ 06 04	60	15.3	GALAXY PECULR
ZWG 101.024	13 17 24.	+ 14 39		15.7	GALAXY
MCG+04-31-023	13 17 24.	+ 23 14	15	14.7	GALAXY
TON-N 0153	13 17 24.	+ 27 45		15.3	BLUE STAR
BC TON153	13 17 24.	+ 27 45		15.3	QUASI-STELLAR OBJECT
SHB 257	13 17 24.	+ 27 45 00.		15.3	QUASI-STELLAR OBJECT
FEIG 078	13 17 24.	+ 29 40		13.8	FAINT BLUE STAR
ZWG 217.029	13 17 24.	+ 40 39		14.8	GALAXY
UGC 08373	13 17 24.	+ 40 39	96	14.8	GALAXY Sa
FEIG 079	13 17 24.	+ 56 55		16.0	FAINT BLUE STAR
ZWG 336.006	13 17 24.	+ 73 02		14.8	GALAXY
UGC 08374	13 17 24.	+ 73 02	84	14.8	GALAXY SBb/Sc
MCG-02-34-032	13 17 24.	- 14 30	48	15.	GALAXY
SN 1969F	13 17 24.	- 16 52		16.0	SUPERNOVA
FATH 1.600	13 17 25.	+ 14 38	16		NEBULA
LB 02628	13 17 25.	+ 55 27 00.		16.4	FAINT BLUE STAR
IC 0881	13 17 27.	+ 16 06 56.			NONSTELLAR OBJECT
RNGC 5084	13 17 27.	- 21 34		12.5	GALAXY
ZWG 101.025	13 17 30.	+ 16 07		14.8	GALAXY
UGC 08375	13 17 30.	+ 16 07	102	14.8	GALAXY Sa
ZC 1317.5+2207	13 17 30.	+ 22 07	740		CLUSTER OF GALAXIES
REIZ 3385	13 17 30.	+ 23 15	24	14.1	GALAXY
ZWG 131.001	13 17 30.	+ 23 16		14.7	GALAXY
ZWG 130.030	13 17 30.	+ 23 16		14.7	GALAXY
RNGC 5092	13 17 30.	+ 23 16		14.5	GALAXY
UGC 08376	13 17 30.	+ 23 16	66	14.7	GALAXY E
TON-N 0154	13 17 30.	+ 24 53		14.6	BLUE STAR
ZWG 161.013	13 17 30.	+ 31 05		15.7	GALAXY
ZWG 160.195	13 17 30.	+ 31 05		15.7	GALAXY
ZC 1317.5+3105	13 17 30.	+ 31 05	940		CLUSTER OF GALAXIES
ZC 1317.5+4310	13 17 30.	+ 43 10	1410		CLUSTER OF GALAXIES
ZC 1317.5+5111	13 17 30.	+ 51 11	940		CLUSTER OF GALAXIES
MCG+12-13-004	13 17 30.	+ 73 02	72	15.	GALAXY
MCG-03-34-049	13 17 30.	- 16 52	36	15.	GALAXY
MCG-03-34-048	13 17 30.	- 16 52	24	15.	GALAXY
MCG-07-27-052	13 17 30.	- 43 33	60	14.	GALAXY
HOLM 515B	13 17 31.	- 12 21	18	14.5	PART OF MULTIPLE GALAXY
SVEN 409	13 17 31.	- 23 48 17.	36	15.5	GALAXY
SVEN 408	13 17 31.	- 24 11	162	12.0	GALAXY
REIZ 3382	13 17 32.	- 12 20	12	14.8	GALAXY
REIN 2.190	13 17 34.33	- 21 33 48.4			NEBULA
FEIG 080	13 17 36.	+ 12 20		11.3	FAINT BLUE STAR
ZWG 161.014	13 17 36.	+ 29 41		15.4	GALAXY
ZWG 160.196	13 17 36.	+ 29 41		15.4	GALAXY
ZC 1317.6+2943	13 17 36.	+ 29 43	340		CLUSTER OF GALAXIES
ZWG 161.015	13 17 36.	+ 30 23		15.2	GALAXY
ZWG 160.197	13 17 36.	+ 30 23		15.2	GALAXY
UGC 08377	13 17 36.	+ 30 23	60	15.2	GALAXY
MCG+05-31-176	13 17 36.	+ 30 24	15	15.2	GALAXY
MCG+05-31-177	13 17 36.	+ 32 17	60	15.1	GALAXY
ZWG 245.040	13 17 36.	+ 47 59		15.7	GALAXY
MCG+08-24-108	13 17 36.	+ 48 29 30.	36	15.	GALAXY
ZC 1317.6+4905	13 17 36.	+ 49 05	1080		CLUSTER OF GALAXIES
MCG-02-34-034	13 17 36.	- 12 18	132	12.5	GALAXY
MCG-02-34-033	13 17 36.	- 14 33	18	15.	GALAXY
MCG-04-32-004	13 17 36.	- 21 33	192	12.	GALAXY
MCG-04-32-005	13 17 36.	- 24 09	168	12.	GALAXY
IC 0882	13 17 38.	+ 16 09 51.			NONSTELLAR OBJECT
RNGC 5085	13 17 38.	- 24 09		12.0	GALAXY
RNGC 5087	13 17 39.	- 20 21		13.0	GALAXY
IC 4225	13 17 41.	+ 32 14 39.			NONSTELLAR OBJECT
RNGC 5088	13 17 41.	- 12 19		13.0	GALAXY
BC 4C52.27	13 17 41.13	+ 52 03 51.6		17.	QUASI-STELLAR OBJECT
ZWG 101.026	13 17 42.	+ 14 49		15.6	GALAXY
ZWG 101.027	13 17 42.	+ 16 10		15.0	GALAXY
ZWG 161.016	13 17 42.	+ 32 14		15.1	GALAXY
ZWG 160.198	13 17 42.	+ 32 14		15.1	GALAXY
UGC 08378	13 17 42.	+ 32 14	66	15.1	GALAXY S0-a
MCG+08-24-109	13 17 42.	+ 48 00	48	15.	GALAXY

OBJECT NAME	RIGHT ASCEN.	DECLINATION	DIAM.	MAGN.	TYPE OF OBJECT
ZWG 271.033	13 17 42.	+ 52 20		14.8	GALAXY
MCG+09-22-047	13 17 42.	+ 52 20	36	15.	GALAXY
ZC 1317.7+6708	13 17 42.	+ 67 08	1010		CLUSTER OF GALAXIES
ZWG 016.053	13 17 42.	- 03 23		15.4	GALAXY
MCG-03-34-051	13 17 42.	- 16 17	30	15.	GALAXY
SVEN 412	13 17 42.	- 20 20 17.	84	12.4	GALAXY
MCG-03-34-050	13 17 42.	- 20 22	150	13.	GALAXY
SVEN 411	13 17 42.	- 21 15	12	15.3	GALAXY
SVEN 410	13 17 42.	- 21 15	42	15.5	GALAXY
SHB 258	13 17 42.1	+ 52 01 42.		17.	QUASI-STELLAR OBJECT
REIZ 3384	13 17 44.	- 12 19	90	13.2	GALAXY
HOLM 515A	13 17 44.	- 12 19	96	13.3	PART OF MULTIPLE GALAXY
MCG+06-29-076	13 17 45.	+ 33 20	78	15.	GALAXY
MCG+09-22-048	13 17 45.	+ 52 20	30	16.	GALAXY
MCG-04-32-006	13 17 45.	- 25 49	30	15.	GALAXY
ZC 1317.8+0525	13 17 48.	+ 05 25	470		CLUSTER OF GALAXIES
ZWG 044.071	13 17 48.	+ 06 36		15.3	GALAXY
UGC 08379	13 17 48.	+ 06 36	78	15.3	GALAXY Sb-c
ZC 1317.8+2824	13 17 48.	+ 28 24	940		CLUSTER OF GALAXIES
ZWG 161.017	13 17 48.	+ 30 43		15.7	GALAXY
ZWG 160.199	13 17 48.	+ 30 43		15.7	GALAXY
ZWG 189.051	13 17 48.	+ 33 21		15.1	GALAXY
MCG+06-29-077	13 17 48.	+ 33 22	42	15.	GALAXY
SHAH 128	13 17 48.	+ 56 01	48	17.2	GROUP OF COMPACT GALAXIES
ABC 1718	13 17 49.	+ 67 07		17.8	RICH CLUSTER OF GALAXIES
RNGC 5096	13 17 51.	+ 33 21		15.0	GALAXY
HDB E18	13 17 51.	- 41 48			DIFFUSE NEBULA
TON-N 1583	13 17 52.	+ 34 13		16.9	BLUE STAR
ZC 1317.9+2537	13 17 54.	+ 25 37	3160		CLUSTER OF GALAXIES
ZWG 161.018	13 17 54.	+ 31 10		15.5	GALAXY
ZWG 160.200	13 17 54.	+ 31 10		15.5	GALAXY
ZWG 161.019	13 17 54.	+ 31 15		15.1	GALAXY
ZWG 160.201	13 17 54.	+ 31 15		15.1	GALAXY
MCG+06-29-078	13 17 54.	+ 33 22	42	15.	GALAXY
ZC 1317.9+3350	13 17 54.	+ 33 50	5650		CLUSTER OF GALAXIES
MRK 251	13 17 54.	+ 52 19	8	16.	GALAXY WITH UV CONTINUUM
MCG+09-22-049	13 17 54.	+ 52 19	15	16.	GALAXY
MCG+09-22-050	13 17 54.	+ 53 14	36	15.	GALAXY
ZC 1317.9+6025	13 17 54.	+ 60 25	3090		CLUSTER OF GALAXIES
MCG-03-34-052	13 17 54.	- 16 40	36	17.	GALAXY
MCG-06-29-029	13 17 54.	- 35 42	30	15.5	GALAXY
MCG-07-27-053	13 17 54.	- 43 23	72	15.0	GALAXY
RNGC 5082	13 17 57.	- 43 26			GALAXY
VDB.66G 173	13 18	+ 10 00	100		DWARF GALAXY
ZC 1318.0+0209	13 18 00.	+ 02 09	740		CLUSTER OF GALAXIES
ZC 1318.0+0248	13 18 00.	+ 02 48	1750		CLUSTER OF GALAXIES
ZWG 044.072	13 18 00.	+ 05 40		15.3	GALAXY
UGC 08382	13 18 00.	+ 05 40	66	15.3	GALAXY IRR
MCG+01-34-018	13 18 00.	+ 05 40	60	15.3	GALAXY
ZWG 044.073	13 18 00.	+ 07 48		15.6	GALAXY
ZWG 101.028	13 18 00.	+ 14 48		15.4	GALAXY
UGC 08383	13 18 00.	+ 14 48	66	15.4	GALAXY Sb
ZC 1318.0+2850	13 18 00.	+ 28 50	1340		CLUSTER OF GALAXIES
ZWG 161.020	13 18 00.	+ 31 47		14.9	GALAXY
ZWG 160.202	13 18 00.	+ 31 47		14.9	GALAXY
MCG+05-32-001	13 18 00.	+ 31 48	48	14.9	GALAXY
ZWG 189.052	13 18 00.	+ 33 24		15.0	GALAXY
ZC 1318.0+4146	13 18 00.	+ 41 46	3830		CLUSTER OF GALAXIES
UGC 08380	13 18 00.	+ 84 57	72	16.5	GALAXY Sc
ZWG 016.054	13 18 00.	- 02 01		14.8	GALAXY
UGC 08381	13 18 00.	- 02 01	84	14.8	GALAXY S
MCG+00-34-029	13 18 00.	- 02 01	54	14.8	GALAXY
MCG-02-34-036	13 18 00.	- 13 50	30	15.	GALAXY
MCG-02-34-035	13 18 00.	- 13 50	30	16.	GALAXY
MCG-05-32-001	13 18 00.	- 29 12	30	15.	GALAXY
RNGC 5095	13 18 01.	- 02 01		15.0	GALAXY
RNGC 5098	13 18 03.	+ 33 24		15.0	GALAXY
REIZ 3387	13 18 03.	- 13 50	24	14.0	GALAXY
MCG-03-34-053	13 18 03.	- 16 38 30.	54	15.	GALAXY
TON-N 1584	13 18 04.	+ 35 00		14.8	BLUE STAR
RNGC 5094	13 18 04.	- 13 49		14.0	GALAXY
ZC 1318.1+1740	13 18 06.	+ 17 40	1750		CLUSTER OF GALAXIES
ZWG 161.021	13 18 06.	+ 30 51		15.4	GALAXY
ZWG 160.203	13 18 06.	+ 30 51		15.4	GALAXY
ZWG 161.022	13 18 06.	+ 31 37		15.6	GALAXY
ZWG 160.204	13 18 06.	+ 31 37		15.6	GALAXY
ZWG 161.023	13 18 06.	+ 32 15		15.2	GALAXY
ZWG 160.205	13 18 06.	+ 32 15		15.2	GALAXY
ZWG 217.030	13 18 06.	+ 41 44		15.7	GALAXY
UGC 08384	13 18 06.	+ 41 44	60	15.7	GALAXY S
MCG+09-22-051	13 18 06.	+ 52 01	24	16.	GALAXY
ZWG 016.055	13 18 06.	- 01 11		15.6	GALAXY
MCG-02-34-037	13 18 06.	- 13 49	48	14.	GALAXY
MCG-03-34-054	13 18 06.	- 16 37 30.	54	15.	GALAXY
REIZ 3389	13 18 07.	+ 43 21	48	13.3	GALAXY
MCG-04-32-007	13 18 09.	- 21 45	108	14.	GALAXY
IC 4226	13 18 11.	+ 32 16 04.			NONSTELLAR OBJECT
ABC 1722	13 18 11.	+ 70 22		17.7	RICH CLUSTER OF GALAXIES
SN 1956E	13 18 12.	+ 02 32		19.3	SUPERNOVA
ZWG 044.074	13 18 12.	+ 05 44		15.4	GALAXY
8ZW 1318+07.0	13 18 12.	+ 06 59		18.0	COMPACT GALAXY
ZWG 072.049	13 18 12.	+ 10 03		14.6	GALAXY
UGC 08385	13 18 12.	+ 10 03	156	14.6	GALAXY S(B) IV
MCG+02-34-008	13 18 12.	+ 10 03	120	14.6	GALAXY
ZC 1318.2+1149	13 18 12.	+ 11 49	1340		CLUSTER OF GALAXIES
MCG+05-32-002	13 18 12.	+ 32 14	24	15.2	GALAXY
MCG+06-29-079	13 18 12.	+ 33 31	42	14.5	GALAXY
ZC 1318.2+3801	13 18 12.	+ 38 01	870		CLUSTER OF GALAXIES
MCG+09-22-052	13 18 12.	+ 50 58	72	15.	GALAXY
UGC 08386	13 18 12.	+ 51 00	78	16.0	GALAXY Sc
MRK 252	13 18 12.	+ 55 39	6	17.	GALAXY WITH UV CONTINUUM
RNGC 5090	13 18 14.	- 43 28		13.0	GALAXY
RNGC 5086	13 18 14.	- 43 30			GALAXY
LB 02629	13 18 15.	+ 59 26 06.		16.0	FAINT BLUE STAR
IC 0883	13 18 16.	+ 34 24 11.			NONSTELLAR OBJECT
ARP 193	13 18 17.	+ 34 23			PECULIAR GALAXY
RNGC 5097	13 18 17.	- 12 13			GALAXY
ZWG 189.053	13 18 18.	+ 33 33		15.4	GALAXY
ABC 1714	13 18 18.	+ 33 44		18.0	RICH CLUSTER OF GALAXIES
1ZW 056	13 18 18.	+ 34 24			COMPACT GALAXY
ZWG 189.054	13 18 18.	+ 34 25		14.8	GALAXY
UGC 08387	13 18 18.	+ 34 25	102	14.8	GALAXY PECULR
ZWG 218.001	13 18 18.	+ 43 20		13.6	GALAXY
ZWG 217.031	13 18 18.	+ 43 20		13.6	GALAXY
RNGC 5103	13 18 18.	+ 43 20		13.5	GALAXY
UGC 08388	13 18 18.	+ 43 20	90	13.6	GALAXY PECULR
ZWG 294.031	13 18 18.	+ 56 43		15.7	GALAXY
ZC 1318.3+6646	13 18 18.	+ 66 46	4500		CLUSTER OF GALAXIES

OBJECT NAME	RIGHT ASCEN.	DECLINATION	DIAM.	MAGN.	TYPE OF OBJECT
8ZW 1318-01.0	13 18 18.	- 01 01		20.3	COMPACT GALAXY
SER 102.08	13 18 18.	- 43 27	102	15.	INTERACTING GALAXIES
SER 102.07	13 18 18.	- 43 27	1980	14.	LOOSE GROUP OF 10 GALXIES
LB 02630	13 18 20.	+ 51 38 18.		17.0	FAINT BLUE STAR
RNGC 5091	13 18 20.	- 43 24			GALAXY
ZWG 016.056	13 18 24.	+ 01 45		14.9	GALAXY
MCG+00-34-030	13 18 24.	+ 01 45	24	14.9	GALAXY
ZC 1318.4+2926	13 18 24.	+ 29 26	1810		CLUSTER OF GALAXIES
ZC 1318.4+5436	13 18 24.	+ 54 36	3230		CLUSTER OF GALAXIES
ZC 1318.4+7517	13 18 24.	+ 75 17	2220		CLUSTER OF GALAXIES
MCG-07-27-054	13 18 24.	- 43 23	102	12.9	GALAXY
LB 02631	13 18 26.	+ 51 06 24.		16.9	FAINT BLUE STAR
REIZ 3388	13 18 26.	- 12 16	24	13.1	GALAXY
ZWG 072.050	13 18 30.	+ 09 14		15.1	GALAXY
8ZW 1318+09.2	13 18 30.	+ 09 14		17.9	COMPACT GALAXY
UGC 08389	13 18 30.	+ 09 14	78	15.1	GALAXY DBL SYS
MCG+02-34-009	13 18 30.	+ 09 14	48	15.1	GALAXY
ZWG 072.051	13 18 30.	+ 14 25		15.5	GALAXY
ZWG 161.024	13 18 30.	+ 31 38		15.6	GALAXY
ZWG 160.206	13 18 30.	+ 31 38		15.6	GALAXY
ZC 1318.5+3316	13 18 30.	+ 33 16	1550		CLUSTER OF GALAXIES
MCG+10-19-060	13 18 30.	+ 56 41	24	16.	GALAXY
MCG-07-27-055	13 18 30.	- 43 24	84	13.5	GALAXY
ABC 1715	13 18 31.	+ 38 00		17.0	RICH CLUSTER OF GALAXIES
FATH 1.601	13 18 31.	- 15 34	11		NEBULA
RNGC 5100	13 18 33.	+ 09 14		15.0	GALAXY
IC 4227	13 18 34.	+ 32 26 29.			NONSTELLAR OBJECT
ABC 1716	13 18 35.	+ 34 10		17.8	RICH CLUSTER OF GALAXIES
FATH 1.602	13 18 35.	- 15 34	16		NEBULA
ZC 1318.6+1914	13 18 36.	+ 19 14	1550		CLUSTER OF GALAXIES
ZWG 161.025	13 18 36.	+ 32 27		15.6	GALAXY
ZC 1318.6+3410	13 18 36.	+ 34 10	670		CLUSTER OF GALAXIES
MCG+06-29-080	13 18 36.	+ 36 51	30	15.	GALAXY
ZWG 189.055	13 18 36.	+ 36 52		15.6	GALAXY
FATH 1.603	13 18 36.	- 15 39	14		NEBULA
MCG-06-29-030	13 18 36.	- 35 32	24	15.5	GALAXY
ABC 1717	13 18 39.	+ 41 42		17.2	RICH CLUSTER OF GALAXIES
LB 02632	13 18 39.	+ 51 18 48.		16.8	FAINT BLUE STAR
REIZ 3390	13 18 39.	- 12 50	24	14.5	GALAXY
MCG-03-34-055	13 18 39.	- 15 37 30.	36	15.	GALAXY
RNGC 5099	13 18 41.	- 12 48			GALAXY
ZWG 072.052	13 18 42.	+ 10 13		15.6	GALAXY
ZC 1318.7+1600	13 18 42.	+ 16 00	1480		CLUSTER OF GALAXIES
MCG+06-29-081	13 18 42.	+ 33 34	15	15.	GALAXY
ZWG 189.056	13 18 42.	+ 33 36		15.5	GALAXY
UGC 08390	13 18 42.	+ 36 02	66	16.5	GALAXY Sc-IRR
MRK 253	13 18 42.	+ 56 43	18	16.	GALAXY WITH UV CONTINUUM
ZC 1318.7+5709	13 18 42.	+ 57 09	3090		CLUSTER OF GALAXIES
ZC 1318.7-0041	13 18 42.	- 00 41	2550		CLUSTER OF GALAXIES
REIZ 3393	13 18 46.	+ 57 54	120	13.5	GALAXY
FATH 1.604	13 18 46.	- 15 51	16		NEBULA
LB 02633	13 18 47.	+ 52 52 48.		17.0	FAINT BLUE STAR
ZWG 016.057	13 18 48.	+ 00 36		14.5	GALAXY
UGC 08391	13 18 48.	+ 00 36	78	14.5	GALAXY Sa
MCG+00-34-031	13 18 48.	+ 00 36	66	14.5	GALAXY
ZWG 072.053	13 18 48.	+ 12 29		15.6	GALAXY
TON-N 0155	13 18 48.	+ 29 07		16.6	BLUE STAR
ZWG 218.002	13 18 48.	+ 42 43		15.5	GALAXY
ZWG 217.032	13 18 48.	+ 42 43		15.5	GALAXY
ZC 1318.8+4543	13 18 48.	+ 45 43	3900		CLUSTER OF GALAXIES
ZWG 336.007	13 18 48.	+ 71 00		15.5	GALAXY
MCG-03-34-056	13 18 48.	- 16 02	48	15.	GALAXY
RNGC 5104	13 18 49.	+ 00 36		14.5	GALAXY
SHB 259	13 18 49.8	+ 11 22 33.		19.1	QUASI-STELLAR OBJECT
BC 4C11.45	13 18 50.0	+ 11 22 35.		19.13	QUASI-STELLAR OBJECT
ZC 1318.9+0323	13 18 54.	+ 03 23	1340		CLUSTER OF GALAXIES
TON-N 0156	13 18 54.	+ 29 06		16.0	BLUE STAR
BC TON156	13 18 54.	+ 29 06		16.0	QUASI-STELLAR OBJECT
SHB 260	13 18 54.	+ 29 06 00.		16.4	QUASI-STELLAR OBJECT
SHB 261	13 18 54.	+ 29 06 35.		16.9	QUASI-STELLAR OBJECT
ZWG 161.026	13 18 54.	+ 31 29		15.3	GALAXY
ZWG 160.207	13 18 54.	+ 31 29		15.3	GALAXY
UGC 08392	13 18 54.	+ 31 29	96	15.3	GALAXY Sc
ZC 1318.9+3645	13 18 54.	+ 36 45	1210		CLUSTER OF GALAXIES
ZC 1318.9+4715	13 18 54.	+ 47 15	470		CLUSTER OF GALAXIES
ZC 1318.9+5515	13 18 54.	+ 55 15	1140		CLUSTER OF GALAXIES
ZWG 294.032	13 18 54.	+ 57 55		13.6	GALAXY
UGC 08393	13 18 54.	+ 57 55	108	13.6	GALAXY S
ZWG 016.058	13 18 54.	- 01 54		15.7	GALAXY
VV 184B	13 18 54.	- 16 00	15	17.	INTERACTING GALAXY
VV 184A	13 18 54.	- 16 00	24	16.	INTERACTING GALAXY
VV 184	13 18 54.	- 16 00	78		INTERACTING GALAXY
LB 02634	13 18 55.	+ 51 59 18.		16.1	FAINT BLUE STAR
RNGC 5109	13 18 55.	+ 57 55		13.5	GALAXY
FATH 1.605	13 18 56.	- 15 14	41		NEBULA
KLEM 24	13 18 56.	- 35 34	600	15.	GROUP OF 10 GALAXIES
FATH 1.606	13 18 59.	- 15 42	16		NEBULA
8ZW 1319+03.0	13 19 00.	+ 03 00		17.9	COMPACT GALAXY
ZC 1319.0+0346	13 19 00.	+ 03 46	1550		CLUSTER OF GALAXIES
ZWG 072.054	13 19 00.	+ 12 36		15.6	GALAXY
ZWG 101.029	13 19 00.	+ 15 00		15.5	GALAXY
MCG+04-32-001	13 19 00.	+ 25 46	36	15.4	GALAXY
MCG+05-32-003	13 19 00.	+ 31 30	96	15.3	GALAXY
ZWG 161.027	13 19 00.	+ 31 39		15.0	GALAXY
ZWG 160.208	13 19 00.	+ 31 39		15.0	GALAXY
ZWG 161.028	13 19 00.	+ 31 48		15.4	GALAXY
ZWG 160.209	13 19 00.	+ 31 48		15.4	GALAXY
ZC 1319.0+3736	13 19 00.	+ 37 36	3560		CLUSTER OF GALAXIES
LB 02635	13 19 00.	+ 53 36 42.		16.0	FAINT BLUE STAR
MCG+10-19-061	13 19 00.	+ 57 54	102	13.5	GALAXY
ZC 1319.0+7756	13 19 00.	+ 77 56	870		CLUSTER OF GALAXIES
MCG+14-06-020	13 19 00.	+ 84 45	18	16.	GALAXY
7ZW 509	13 19 00.	+ 84 46			COMPACT GALAXY
ZWG 366.003	13 19 00.	+ 84 46		15.7	GALAXY
ZWG 365.010	13 19 00.	+ 84 46		15.7	GALAXY
UGC 08394	13 19 00.	+ 84 46	60	15.7	GALAXY DBL SYS
MCG-02-34-038	13 19 00.	- 13 50	24	15.	GALAXY
MCG-03-34-058	13 19 00.	- 16 00	15	15.	GALAXY
MCG-03-34-057	13 19 00.	- 16 00 30.	12	16.	GALAXY
MCG-04-32-008	13 19 00.	- 27 11	360	12.	GALAXY
REIN 2.191	13 19 00.77	- 27 10 07.6			NEBULA
RNGC 5101	13 19 01.	- 27 11		12.0	GALAXY
ABC 1719	13 19 03.	+ 36 39		18.0	RICH CLUSTER OF GALAXIES
RNGC 5105	13 19 05.	- 12 57		13.0	GALAXY
RNGC 5102	13 19 05.	- 36 23		10.5	GALAXY
ZWG 072.055	13 19 06.	+ 12 27		15.7	GALAXY
UGC 08395	13 19 06.	+ 12 27	60	15.7	GALAXY Sa-b
MCG+05-32-004	13 19 06.	+ 31 40	39	15.0	GALAXY

OBJECT NAME	RIGHT ASCEN.	DECLINATION	DIAM.	MAGN.	TYPE OF OBJECT
MCG+07-28-001	13 19 06.	+ 38 47 30.	96	14.	GALAXY
ZWG 218.003	13 19 06.	+ 38 48		13.7	GALAXY
ZWG 217.033	13 19 06.	+ 38 48		13.7	GALAXY
UGC 08396	13 19 06.	+ 38 48	108	13.7	GALAXY
ZWG 294.033	13 19 06.	+ 58 48		15.4	GALAXY
MCG+10-19-062	13 19 06.	+ 58 48	39	16.	GALAXY
MRK 065	13 19 06.	+ 59 21	13	15.5	GALAXY WITH UV CONTINUUM
MCG-02-34-039	13 19 06.	- 12 57	120	13.	GALAXY
MCG-03-34-059	13 19 06.	- 16 10	9	15.5	GALAXY
MCG-06-29-031	13 19 06.	- 36 21	360	10.8	GALAXY
SER 102.06	13 19 06.	- 45 42	120		INTERACTING GALAXIES
RNGC 5106	13 19 09.	+ 08 45			NON-EXISTENT OBJECT
IC 4228	13 19 09.	+ 25 46 31.			NONSTELLAR OBJECT
LB 02636	13 19 09.	+ 51 29 18.		16.9	FAINT BLUE STAR
REIZ 3391	13 19 09.	- 12 58	60	13.7	GALAXY
MCG-03-34-060	13 19 09.	- 16 10	48	15.	GALAXY
RNGC 5107	13 19 10.	+ 38 48		13.5	GALAXY
REIZ 3394	13 19 10.	+ 38 49	60	14.0	GALAXY
FATH 1.607	13 19 11.	+ 58 48	11		NEBULA
ZWG 016.059	13 19 12.	+ 00 28		15.2	GALAXY
ZWG 101.030	13 19 12.	+ 15 35		15.5	GALAXY
ZWG 131.002	13 19 12.	+ 25 47		15.4	GALAXY
MCG-06-29-032	13 19 12.	- 37 08	36	15.	GALAXY
OCL 0897	13 19 12.	- 65 51	900	9.2	OPEN STAR CLUSTER
PK306-00.1	13 19 14.	- 63 04 40.	25		PLANETARY NEBULA
ZWG 016.060	13 19 18.	+ 00 24		15.2	GALAXY
MCG+04-32-002	13 19 18.	+ 22 42	42	14.9	GALAXY
ZWG 161.029	13 19 18.	+ 26 34		15.7	GALAXY
ZWG 161.030	13 19 18.	+ 31 37		14.8	GALAXY
UGC 08397	13 19 18.	+ 31 37	84	14.8	GALAXY Sb-c
UGC 08398	13 19 18.	+ 42 59	60	17.	GALAXY DWARF
MCG+10-19-063	13 19 18.	+ 58 09	192	16.	GALAXY
MCG+11-16-017	13 19 18.	+ 63 00	45	17.	GALAXY
MCG-03-34-061	13 19 18.	- 17 04	36	15.	GALAXY
TON-N 0703	13 19 24.	+ 29 09		15.0	BLUE STAR
ZWG 161.031	13 19 24.	+ 31 30		14.9	GALAXY
UGC 08399	13 19 24.	+ 31 30	66	14.9	GALAXY SBb
ZWG 218.004	13 19 24.	+ 42 32		14.5	GALAXY
ZWG 217.034	13 19 24.	+ 42 32		14.5	GALAXY
UGC 08400	13 19 24.	+ 42 32	48	14.5	GALAXY (SB)?
MCG+07-28-002	13 19 24.	+ 42 33	45	14.5	GALAXY
ZC 1319.4+4940	13 19 24.	+ 49 40	1010		CLUSTER OF GALAXIES
MCG+10-19-064	13 19 24.	+ 57 57	45	15.	GALAXY
ZWG 294.034	13 19 24.	+ 57 58		15.2	GALAXY
8ZW 1319-01.1	13 19 24.	- 01 04		19.4	COMPACT GALAXY
RNGC 5113	13 19 25.	+ 57 58		15.0	GALAXY
REIZ 3392	13 19 27.	- 12 56	42	14.9	GALAXY
ZWG 101.031	13 19 30.	+ 14 36		15.6	GALAXY
ZC 1319.5+1945	13 19 30.	+ 19 45	1210		CLUSTER OF GALAXIES
ZWG 131.003	13 19 30.	+ 22 42		14.9	GALAXY
UGC 08401	13 19 30.	+ 28 57	60	17.	GALAXY DWF PEC
MCG+05-32-005	13 19 30.	+ 31 38	72	14.8	GALAXY
MCG+14-06-021	13 19 30.	+ 84 46	36	16.	GALAXY
MCG-05-32-002	13 19 30.	- 29 16	36	15.5	GALAXY
ARC 1720	13 19 31.	+ 03 21		17.6	RICH CLUSTER OF GALAXIES
MCG+05-32-006	13 19 33.	+ 31 30	48	14.9	GALAXY
MCG+07-28-003	13 19 33.	+ 38 59	228	13.	GALAXY
ZWG 044.075	13 19 36.	+ 06 22		15.4	GALAXY
ZWG 072.056	13 19 36.	+ 10 58		15.6	GALAXY
MCG+05-32-007	13 19 36.	+ 29 07	24	15.4	GALAXY
ZC 1319.6+3135	13 19 36.	+ 31 35	10890		CLUSTER OF GALAXIES
MCG+06-29-082	13 19 36.	+ 35 35 30.	27	14.5	GALAXY
ZWG 189.057	13 19 36.	+ 35 37		15.5	GALAXY
LB 02637	13 19 36.	+ 49 55 48.		15.5	FAINT BLUE STAR
8ZW 1319-02.6	13 19 36.	- 02 39		17.7	COMPACT GALAXY
ZWG 016.061	13 19 36.	- 02 50		15.5	GALAXY
MCG-05-32-003	13 19 36.	- 30 48	30	15.	GALAXY
IC 4230	13 19 38.	+ 26 59 38.			NONSTELLAR OBJECT
MCG+05-32-008	13 19 39.	+ 27 00	54	15.2	GALAXY
REIZ 3395	13 19 39.	+ 39 00	192	13.0	GALAXY
MCG-03-34-062	13 19 39.	- 16 38	24	15.	GALAXY
TON-N 1585	13 19 40.	+ 21 10		15.8	BLUE STAR
TON-N 1586	13 19 40.	+ 21 16		14.5	BLUE STAR
RNGC 5112	13 19 40.	+ 39 00		12.0	GALAXY
ZWG 161.032	13 19 42.	+ 27 00		15.2	GALAXY
UGC 08402	13 19 42.	+ 27 00	66	15.2	GALAXY Sa?
ZWG 161.033	13 19 42.	+ 29 06		15.4	GALAXY
ZWG 161.034	13 19 42.	+ 31 32		15.7	GALAXY
ZWG 189.058	13 19 42.	+ 35 46		15.4	GALAXY
ZWG 218.005	13 19 42.	+ 39 00		12.5	GALAXY
UGC 08403	13 19 42.	+ 39 00	240	12.5	GALAXY SBc
ZC 1319.7+6450	13 19 42.	+ 64 50	2150		CLUSTER OF GALAXIES
8ZW 1319-01.5	13 19 42.	- 01 32		19.7	COMPACT GALAXY
8ZW 1319-03.4	13 19 42.	- 03 22		19.6	COMPACT GALAXY
MCG+06-29-083	13 19 45.	+ 35 44	42	14.5	GALAXY
MCG-02-34-040	13 19 45.	- 10 29	42	15.5	GALAXY
MCG-03-34-063	13 19 45.	- 16 27	102	14.5	GALAXY
TON-N 1587	13 19 46.	+ 21 28		15.9	BLUE STAR
MCG+04-32-003	13 19 48.	+ 22 02	12	16.5	GALAXY
MCG+07-28-004	13 19 48.	+ 44 32 30.	45	15.	GALAXY
ZWG 016.062	13 19 48.	- 02 09		14.2	GALAXY
UGC 08404	13 19 48.	- 02 09	60	14.2	GALAXY SB
MCG+00-34-032	13 19 48.	- 02 09	54	14.2	GALAXY
MCG-03-34-064	13 19 48.	- 16 28	60	14.5	GALAXY
ARC 1721	13 19 49.	+ 19 39		17.8	RICH CLUSTER OF GALAXIES
ARC 1723	13 19 49.	+ 37 28		17.0	RICH CLUSTER OF GALAXIES
MCG+04-32-004	13 19 51.	+ 22 03	9	16.	GALAXY
RNGC 5110	13 19 53.	- 12 49			GALAXY
MCG+04-32-005	13 19 54.	+ 22 05	12	16.	GALAXY
MCG+04-32-006	13 19 57.	+ 22 04 30.	24	16.	GALAXY
LB 02638	13 19 57.	+ 53 48 36.		16.4	FAINT BLUE STAR
TON-N 1588	13 19 58.	+ 21 44		12.6	BLUE STAR
HN 0380	13 19 58.	- 02 09			NEBULA
IC 4229	13 19 59.	- 02 08			NONSTELLAR OBJECT
LB 09872	13 20	- 83 51		14.3	FAINT BLUE STAR
UGC 08405	13 20 00.	+ 20 34	60	17.	GALAXY
ZWG 131.004	13 20 00.	+ 21 41		15.1	GALAXY
MRK 659	13 20 00.	+ 21 41	19	15.	GALAXY WITH UV CONTINUUM
ZC 1320.0+2320	13 20 00.	+ 23 20	2290		CLUSTER OF GALAXIES
ZC 1320.0+6142	13 20 00.	+ 61 42	2220		CLUSTER OF GALAXIES
ZC 1320.0+6215	13 20 00.	+ 62 15	1210		CLUSTER OF GALAXIES
ZC 1320.0+7314	13 20 00.	+ 73 14	1280		CLUSTER OF GALAXIES
MCG+14-06-022	13 20 00.	+ 84 39	96	17.	GALAXY
8ZW 1320-03.4	13 20 00.	- 03 26		18.3	COMPACT GALAXY
LB 02639	13 20 04.	+ 57 39 06.		17.0	FAINT BLUE STAR
8ZW 1320+12.9	13 20 06.	+ 12 57		16.8	COMPACT GALAXY
ZWG 131.005	13 20 06.	+ 21 35		14.9	GALAXY
MCG+04-32-007	13 20 06.	+ 21 41	36	15.1	GALAXY
ZC 1320.1+5858	13 20 06.	+ 58 58	4030		CLUSTER OF GALAXIES
ZC 1320.1+6318	13 20 06.	+ 63 18	1340		CLUSTER OF GALAXIES
MCG-06-29-033	13 20 06.	- 35 30	60	15.	GALAXY
IC 0885	13 20 08.	+ 21 35 33.			NONSTELLAR OBJECT
TON-N 1589	13 20 10.	+ 21 15		16.9	BLUE STAR
ARC 1724	13 20 10.	+ 23 16		17.5	RICH CLUSTER OF GALAXIES
RNGC 5111	13 20 11.	- 12 42		13.0	GALAXY
ZC 1320.2+1523	13 20 12.	+ 15 23	1610		CLUSTER OF GALAXIES
MCG+04-32-008	13 20 12.	+ 21 35	48	14.9	GALAXY
TON-N 0704	13 20 12.	+ 32 25		16.4	BLUE STAR
MCG-02-34-041	13 20 12.	- 12 42	24	13.	GALAXY
LB 02640	13 20 14.	+ 58 20 18.		15.8	FAINT BLUE STAR
LB 02641	13 20 15.	+ 55 36 00.		16.1	FAINT BLUE STAR
REIZ 3396	13 20 15.	- 12 45	18	13.9	GALAXY
IC 0884	13 20 16.	- 12 28 03.			NONSTELLAR OBJECT
8ZW 1320+08.0	13 20 18.	+ 07 58		17.9	COMPACT GALAXY
ZWG 101.032	13 20 18.	+ 17 56		15.6	GALAXY
UGC 08406	13 20 18.	+ 17 56	66	15.6	GALAXY S
MCG+06-29-084	13 20 18.	+ 33 07	27	15.	GALAXY
ZC 1320.3+3649	13 20 18.	+ 36 49	1010		CLUSTER OF GALAXIES
MCG-03-34-065	13 20 18.	- 16 46	36	15.	GALAXY
MCG-05-32-004	13 20 18.	- 31 58	30	15.	GALAXY
SC 1317-3143.3	13 20 18.	- 31 59 02.	12		NEBULA
LB 02642	13 20 20.	+ 50 03 42.		15.8	FAINT BLUE STAR
ZWG 044.076	13 20 24.	+ 08 25		15.4	GALAXY
ZC 1320.4+1121	13 20 24.	+ 11 21	3090		CLUSTER OF GALAXIES
ZWG 101.033	13 20 24.	+ 19 57		15.7	GALAXY
UGC 08407	13 20 24.	+ 19 57	72	15.7	GALAXY Sc
ZC 1320.4+3155	13 20 24.	+ 31 55	740		CLUSTER OF GALAXIES
ZWG 189.059	13 20 24.	+ 33 08		15.7	GALAXY
ZC 1320.4+4700	13 20 24.	+ 47 00	540		CLUSTER OF GALAXIES
MCG+09-22-053	13 20 24.	+ 55 05	36	14.	GALAXY
RNGC 5108	13 20 24.	- 32 04		15.0	GALAXY
MCG-05-32-005	13 20 24.	- 32 04	66	15.5	GALAXY
SER 102.02	13 20 24.	- 42 27	30	16.5	PEC. COMET-LIKE GALAXY
RNGC 5115	13 20 28.	+ 14 13		15.0	GALAXY
LB 02643	13 20 28.	+ 52 11 12.		17.0	FAINT BLUE STAR
SN 1956F	13 20 30.	+ 07 04		19.6	SUPERNOVA
8ZW 1320+07.1	13 20 30.	+ 07 06		17.9	COMPACT GALAXY
ZWG 044.077	13 20 30.	+ 08 21		15.6	GALAXY
ZWG 072.057	13 20 30.	+ 14 13		14.8	GALAXY
UGC 08408	13 20 30.	+ 14 13	96	14.8	GALAXY SBc
MCG+02-34-010	13 20 30.	+ 14 13	72	14.8	GALAXY
MCG+03-34-021	13 20 30.	+ 19 57	66	15.7	GALAXY
ZC 1320.5+2945	13 20 30.	+ 29 45	610		CLUSTER OF GALAXIES
ZWG 161.035	13 20 30.	+ 32 05		15.4	GALAXY
ZWG 189.060	13 20 30.	+ 32 53		15.6	GALAXY
ZWG 271.034	13 20 30.	+ 55 05		15.4	GALAXY
MCG-04-32-009	13 20 30.	- 26 02 30.	96	14.	GALAXY
LB 02644	13 20 31.	+ 49 26 12.		16.8	FAINT BLUE STAR
SC 1317-3056.3	13 20 31.	- 31 12 01.	18		NEBULA
MCG+05-32-009	13 20 33.	+ 27 15	108	13.7	GALAXY
REIZ 3398	13 20 33.	+ 28 36	108	13.2	GALAXY
IC 4231	13 20 33.	- 26 02			NONSTELLAR OBJECT
HN 1042	13 20 33.	- 26 03	90		NEBULA
REIZ 3397	13 20 34.	+ 27 18	72	12.8	GALAXY
ZWG 131.006	13 20 36.	+ 23 34		15.4	GALAXY
UGC 08409	13 20 36.	+ 23 34	156	15.4	GALAXY
MCG+04-32-009	13 20 36.	+ 23 34	180	15.4	GALAXY
ZWG 161.036	13 20 36.	+ 27 15		13.7	GALAXY
UGC 08410	13 20 36.	+ 27 15	138	13.7	GALAXY Sc
ZWG 161.037	13 20 36.	+ 28 35		14.5	GALAXY
UGC 08411	13 20 36.	+ 28 35	132	14.5	GALAXY SBc?
MCG+05-32-010	13 20 36.	+ 28 35	132	14.5	GALAXY
MCG+06-29-085	13 20 36.	+ 35 23	60	15.	GALAXY
UGC 08412	13 20 36.	+ 35 24	66	16.0	GALAXY Sc
ZC 1320.6+5825	13 20 36.	+ 58 25	1080		CLUSTER OF GALAXIES
MCG-03-34-066	13 20 36.	- 19 22	48	15.	GALAXY
RNGC 5116	13 20 37.	+ 27 15		13.0	GALAXY
RNGC 5117	13 20 37.	+ 28 35		14.5	GALAXY
IC 4234	13 20 38.	+ 27 21 29.			NONSTELLAR OBJECT
ZC 1320.7+2611	13 20 42.	+ 26 11	1080		CLUSTER OF GALAXIES
ZWG 161.038	13 20 42.	+ 27 23		14.9	GALAXY
MCG+05-32-011	13 20 42.	+ 27 23	36	14.9	GALAXY
ZWG 016.063	13 20 42.	- 01 54		15.7	GALAXY
KARA.73B 0582	13 20 42.	- 01 54	18	15.7	ISOLATED GALAXY E
MCG-04-32-010	13 20 42.	- 25 50	60	14.5	GALAXY
MCG-06-29-034	13 20 42.	- 36 49	30	15.	GALAXY
LB 00680	13 20 44.	+ 53 55 48.		17.2	FAINT BLUE STAR
IC 4232	13 20 45.	- 25 50			NONSTELLAR OBJECT
HN 1043	13 20 45.	- 25 51		14.	NEBULA
ARC 1731	13 20 47.	+ 58 26		17.2	RICH CLUSTER OF GALAXIES
8ZW 1320+00.7	13 20 48.	+ 00 43		19.5	COMPACT GALAXY
TON-N 0705	13 20 48.	+ 31 25		15.6	BLUE STAR
TON-N 0706	13 20 48.	+ 32 10		14.8	BLUE STAR
ZWG 271.035	13 20 48.	+ 52 00		15.4	GALAXY
MRK 254	13 20 48.	+ 52 00	20	15.5	GALAXY WITH UV CONTINUUM
ZWG 353.016	13 20 48.	+ 79 24		15.6	GALAXY
ZWG 352.061	13 20 48.	+ 79 24		15.6	GALAXY
ARC 1726	13 20 50.	+ 17 21		17.2	RICH CLUSTER OF GALAXIES
LB 02645	13 20 50.	+ 50 35 36.		17.1	FAINT BLUE STAR
REIZ 3403	13 20 51.	+ 53 14	24	15.2	GALAXY
LB 02647	13 20 52.	+ 51 11 24.		15.8	FAINT BLUE STAR
LB 02646	13 20 52.	+ 53 54 06.		17.2	FAINT BLUE STAR
ZWG 044.078	13 20 54.	+ 06 39		14.4	GALAXY
UGC 08413	13 20 54.	+ 06 39	54	14.4	GALAXY Sc
MCG+01-34-019	13 20 54.	+ 06 39	48	14.4	GALAXY
ZC 1320.9+1730	13 20 54.	+ 17 30	5710		CLUSTER OF GALAXIES
ZC 1320.9+1750	13 20 54.	+ 17 50	400		CLUSTER OF GALAXIES
MCG+09-22-054	13 20 54.	+ 53 14	30	15.	GALAXY
REIZ 3400	13 20 55.	+ 06 38	30	14.5	GALAXY
REIZ 3399	13 20 55.	- 09 24	30	14.8	GALAXY
REIN 5.003	13 20 56.55	+ 12 01 50.5			NEBULA
REIN 5.005	13 20 56.91	+ 11 56 13.7			NEBULA
RNGC 5118	13 20 57.	+ 06 39		14.5	GALAXY
REIN 5.006	13 20 57.06	+ 11 43 37.4			NEBULA
REIN 5.007	13 20 58.55	+ 11 42 22.9			NEBULA
VDB.66G 241	13 21	- 24 24	70		DWARF GALAXY
ZWG 072.058	13 21 00.	+ 14 02		15.5	GALAXY
ZC 1321.0+1724	13 21 00.	+ 17 24	540		CLUSTER OF GALAXIES
TON-N 0158	13 21 00.	+ 26 34		15.1	BLUE STAR
MCG+05-32-012	13 21 00.	+ 26 48 30.	42	15.6	GALAXY
ZC 1321.0+2904	13 21 00.	+ 29 04	1880		CLUSTER OF GALAXIES
TON-N 0157	13 21 00.	+ 29 27		16.0	BLUE STAR
BC TON157	13 21 00.	+ 29 27		16.0	QUASI-STELLAR OBJECT
SHB 262	13 21 00.	+ 29 27 00.		16.0	QUASI-STELLAR OBJECT
ZWG 161.039	13 21 00.	+ 32 19		15.6	GALAXY
ZWG 218.006	13 21 00.	+ 43 20		13.5	GALAXY

OBJECT NAME	RIGHT ASCEN.	DECLINATION	DIAM.	MAGN.	TYPE OF OBJECT
RNGC 5123	13 21 00.	+ 43 20		13.5	GALAXY
UGC 08415	13 21 00.	+ 43 20	84	13.5	GALAXY Sc
REIZ 3402	13 21 00.	+ 43 21	24	13.8	GALAXY
MCG+07-28-005	13 21 00.	+ 43 21	66	13.	GALAXY
MRK 255	13 21 00.	+ 53 12	10	16.	GALAXY WITH UV CONTINUUM
ZWG 271.036	13 21 00.	+ 53 15		15.7	GALAXY
ZWG 365.011	13 21 00.	+ 80 38		15.3	GALAXY
UGC 08414	13 21 00.	+ 84 42	90	17.	GALAXY
REIN 5.009	13 21 01.84	+ 11 49 10.5			NEBULA
LB 02648	13 21 03.	+ 56 26 24.		16.3	FAINT BLUE STAR
IC 4236	13 21 04.	+ 06 29 35.			NONSTELLAR OBJECT
ZWG 016.064	13 21 06.	+ 01 37		15.6	GALAXY
ZWG 044.079	13 21 06.	+ 06 40		15.5	GALAXY
ZC 1321.1+1958	13 21 06.	+ 19 58	1210		CLUSTER OF GALAXIES
ZWG 161.040	13 21 06.	+ 26 48		15.6	GALAXY
REIZ 3406	13 21 09.	+ 52 55	12	15.2	GALAXY
ARC 1725	13 21 10.	- 16 33		17.7	RICH CLUSTER OF GALAXIES
ARC 1730	13 21 11.	+ 21 43		17.6	RICH CLUSTER OF GALAXIES
ZWG 072.059	13 21 12.	+ 13 23		15.7	GALAXY
ZWG 072.060	13 21 12.	+ 14 14		15.3	GALAXY
1ZW 057	13 21 12.	+ 52 55			COMPACT GALAXY
ZWG 271.037	13 21 12.	+ 52 55		15.6	GALAXY
UGC 08416	13 21 12.	+ 52 55	90	15.6	GALAXY DBL SYS
VV 235B	13 21 12.	+ 52 55	30	15.	INTERACTING GALAXY
VV 235A	13 21 12.	+ 52 55	24	15.	INTERACTING GALAXY
VV 235	13 21 12.	+ 52 55	84		INTERACTING GALAXY
MCG+09-22-056	13 21 12.	+ 52 55	66	15.	GALAXY
MCG+09-22-055	13 21 12.	+ 52 55	30	15.	GALAXY
7ZW 510	13 21 12.	+ 61 17			COMPACT GALAXY
SHAH 160	13 21 12.	+ 77 39	132		GROUP OF COMPACT GALAXIES
IC 4233	13 21 12.	- 30 03 19.			NONSTELLAR OBJECT
RNGC 5114	13 21 12.	- 32 05		14.0	GALAXY
MCG-05-32-006	13 21 12.	- 32 05	24	14.5	GALAXY
ACK 307+05.1	13 21 12.	- 57 16			PLANETARY NEBULA
ARC 1728	13 21 13.	+ 11 32		17.6	RICH CLUSTER OF GALAXIES
IC 4235	13 21 14.	- 12 29			NONSTELLAR OBJECT
HN 0381	13 21 15.	- 12 30			NEBULA
RNGC 5119	13 21 17.	- 12 01		14.0	GALAXY
ZWG 161.041	13 21 18.	+ 31 54		15.5	GALAXY
ZC 1321.3+4843	13 21 18.	+ 48 43	870		CLUSTER OF GALAXIES
MCG+08-24-110	13 21 18.	+ 49 18	42	16.	GALAXY
MCG-02-34-042	13 21 18.	- 12 01	72	14.	GALAXY
MCG-05-32-007	13 21 18.	- 31 24	24	15.5	GALAXY
HZ 44	13 21 20.	+ 36 23		11.0	VERY BLUE STAR
LB 02649	13 21 20.	+ 52 32 30.		16.0	FAINT BLUE STAR
REIZ 3401	13 21 20.	- 12 02	48	13.4	GALAXY
CED 122	13 21 20.	- 63 46	9000		DIFFUSE GALACTIC NEBULA
MCG+03-34-022	13 21 21.	+ 15 39	42	15.7	GALAXY
IC 0886	13 21 21.	- 04 07 54.			NONSTELLAR OBJECT
REIN 5.013	13 21 22.03	+ 12 02 05.3			NEBULA
REIN 5.014	13 21 22.14	+ 12 04 23.4			NEBULA
ZWG 072.061	13 21 24.	+ 13 58		15.6	GALAXY
ZC 1321.4+1358	13 21 24.	+ 13 58	5380		CLUSTER OF GALAXIES
UGC 08417	13 21 24.	+ 13 58	66	15.6	GALAXY Sb-c
ZC 1321.4+2140	13 21 24.	+ 21 40	1950		CLUSTER OF GALAXIES
UGC 08418	13 21 24.	+ 30 49	66	16.5	GALAXY Sc
ZWG 161.042	13 21 24.	+ 31 50		13.9	GALAXY
UGC 08419	13 21 24.	+ 31 50	150	13.9	GALAXY E
ZWG 190.001	13 21 24.	+ 37 24		15.6	GALAXY
ZWG 189.061	13 21 24.	+ 37 24		15.6	GALAXY
ZWG 336.008	13 21 24.	+ 70 47		13.2	GALAXY
MRK 256	13 21 24.	+ 70 47	50	13.	GALAXY WITH UV CONTINUUM
UGC 08420	13 21 24.	+ 70 47	78	13.2	GALAXY S+COMP
MCG+12-13-005	13 21 24.	+ 70 47	72	12.8	GALAXY
ARC 1729	13 21 24.	- 03 07		17.2	RICH CLUSTER OF GALAXIES
REIZ 3405	13 21 25.	+ 31 49	60	13.5	GALAXY
RNGC 5127	13 21 26.	+ 31 50		14.0	GALAXY
RNGC 5125	13 21 27.	+ 09 58		13.5	GALAXY
ARC 1727	13 21 28.	- 22 48		17.4	RICH CLUSTER OF GALAXIES
RNGC 5144	13 21 29.	+ 70 47		13.0	GALAXY
RNGC 5122	13 21 29.	- 10 23		14.0	GALAXY
ZC 1321.5+0115	13 21 30.	+ 01 15	2220		CLUSTER OF GALAXIES
ZWG 072.062	13 21 30.	+ 09 58		13.5	GALAXY
UGC 08421	13 21 30.	+ 09 58	120	13.5	GALAXY Sb
MCG+02-34-011	13 21 30.	+ 09 58	84	13.5	GALAXY
ZWG 072.063	13 21 30.	+ 12 19		15.6	GALAXY
ZWG 101.034	13 21 30.	+ 14 38		15.6	GALAXY
ZWG 101.035	13 21 30.	+ 15 40		15.7	GALAXY
TON-N 0159	13 21 30.	+ 26 32		16.0	BLUE STAR
7ZW 511	13 21 30.	+ 70 47			COMPACT GALAXY
MCG-02-34-043	13 21 30.	- 10 23	48	14.5	GALAXY
REIN 5.015	13 21 31.04	+ 12 00 05.2			NEBULA
REIN 5.016	13 21 32.58	+ 12 06 24.7			NEBULA
MCG+05-32-013	13 21 33.	+ 31 51	30	13.9	GALAXY
IC 0887	13 21 33.	- 12 12 00.			NONSTELLAR OBJECT
LB 00681	13 21 35.	+ 57 27 42.		16.2	FAINT BLUE STAR
ZWG 044.080	13 21 36.	+ 05 31		15.4	GALAXY
ZWG 044.081	13 21 36.	+ 07 10		15.5	GALAXY
ZWG 161.043	13 21 36.	+ 31 15		14.4	GALAXY
UGC 08422	13 21 36.	+ 31 15	132	14.4	GALAXY Sa
ZC 1321.6+3206	13 21 36.	+ 32 06	740		CLUSTER OF GALAXIES
ZWG 218.007	13 21 36.	+ 43 33		15.7	GALAXY
MAI 084	13 21 36.	+ 58 04	128		DWARF SPHEROIDAL GALAXY
8ZW 1321-01.8	13 21 36.	- 01 51		19.0	COMPACT GALAXY
MCG-03-34-067	13 21 36.	- 16 26	48	15.	GALAXY
REIZ 3404	13 21 37.	+ 05 32	60	15.3	GALAXY
REIN 5.017	13 21 37.64	+ 11 36 32.7			NEBULA
RNGC 5131	13 21 38.	+ 31 15		14.5	GALAXY
REIZ 3407	13 21 38.	+ 31 15	96	13.8	GALAXY
RNGC 5129	13 21 40.	+ 14 15		13.5	GALAXY
IC 4238	13 21 41.	+ 31 11 26.			NONSTELLAR OBJECT
RNGC 5130	13 21 41.	- 09 57		14.0	GALAXY
REIN 5.018	13 21 41.43	+ 11 58 45.3			NEBULA
ZWG 044.082	13 21 42.	+ 07 31		15.6	GALAXY
ZWG 072.064	13 21 42.	+ 09 55		15.4	GALAXY
ZWG 072.065	13 21 42.	+ 14 15		13.3	GALAXY
UGC 08423	13 21 42.	+ 14 15	102	13.3	GALAXY E
MCG+02-34-012	13 21 42.	+ 14 15	23	13.3	GALAXY
MCG+03-34-024	13 21 42.	+ 15 48	42	15.7	GALAXY
MCG+03-34-023	13 21 42.	+ 15 49	84	15.6	GALAXY
ZWG 131.007	13 21 42.	+ 22 58		15.7	GALAXY
UGC 08424	13 21 42.	+ 22 58	60	15.7	GALAXY S
MCG+04-32-010	13 21 42.	+ 22 59	45	15.7	GALAXY
MCG+05-32-014	13 21 42.	+ 31 16	126	14.4	GALAXY
ZC 1321.7-0318	13 21 42.	- 03 18	1950		CLUSTER OF GALAXIES
MCG-02-34-044	13 21 42.	- 09 57	48	14.5	GALAXY
KARA.68 222	13 21 42.	- 12 55	40		DWARF GALAXY
MCG-05-32-008	13 21 42.	- 30 10	48	15.	GALAXY
REIN 5.019	13 21 42.69	+ 11 43 38.5			NEBULA
REIN 5.020	13 21 43.87	+ 11 37 16.3			NEBULA
MCG+07-28-006	13 21 45.	+ 43 35	42	15.	GALAXY
SN 1962H	13 21 47.	- 20 52		13.0	SUPERNOVA
ZWG 072.066	13 21 48.	+ 14 12		15.6	GALAXY
MCG+02-34-013	13 21 48.	+ 14 12	48	15.6	GALAXY
ZWG 101.036	13 21 48.	+ 15 50		15.7	GALAXY
KARA.72 371B	13 21 48.	+ 15 50	42	15.7	PART OF DOUBLE GALAXY
ZWG 101.037	13 21 48.	+ 15 52		15.6	GALAXY
UGC 08425	13 21 48.	+ 15 52	84	15.6	GALAXY S
KARA.72 371A	13 21 48.	+ 15 52	66	15.6	PART OF DOUBLE GALAXY
ZC 1321.8+1650	13 21 48.	+ 16 50	870		CLUSTER OF GALAXIES
SHAH 009	13 21 48.	+ 19 16	150	16.	GROUP OF COMPACT GALAXIES
ZC 1321.8+2612	13 21 48.	+ 26 12	740		CLUSTER OF GALAXIES
TON-N 0707	13 21 48.	+ 30 00		15.0	BLUE STAR
7ZW 512	13 21 48.	+ 57 45			COMPACT GALAXY
ZWG 016.065	13 21 48.	- 02 50		15.7	GALAXY
8ZW 1321-03.1	13 21 48.	- 03 08		19.3	COMPACT GALAXY
MCG-03-34-068	13 21 48.	- 20 53 30.	90	13.	GALAXY
MCG-06-29-035	13 21 48.	- 37 24	90	12.5	GALAXY
IC 4237	13 21 49.9	- 20 52 38.			GALAXY
LB 02650	13 21 50.	+ 51 21 36.		17.2	FAINT BLUE STAR
SC 1319-3148.4	13 21 52.	- 32 04 05.	6		NEBULA
RNGC 5121	13 21 52.	- 37 25		12.5	GALAXY
REIN 5.021	13 21 53.26	+ 11 43 18.5			NEBULA
ZWG 072.067	13 21 54.	+ 14 12		15.4	GALAXY
ZWG 161.044	13 21 54.	+ 31 36		15.3	GALAXY
UGC 08426	13 21 54.	+ 31 36	90	15.3	GALAXY SB?c
ARC 1734	13 21 54.	+ 55 28		17.3	RICH CLUSTER OF GALAXIES
MCG+13-10-005	13 21 54.	+ 77 09	33	16.	GALAXY
MCG-02-34-045	13 21 54.	- 10 14	60	15.5	GALAXY
RNGC 5132	13 21 58.	+ 14 21		14.5	GALAXY
LB 02651	13 21 59.	+ 53 23 12.		15.0	FAINT BLUE STAR
ZWG 016.066	13 22 00.	+ 01 50		15.5	GALAXY
ZWG 044.083	13 22 00.	+ 06 47		14.7	GALAXY
UGC 08427	13 22 00.	+ 06 47	90	14.7	GALAXY S
MCG+01-34-020	13 22 00.	+ 06 47	72	14.7	GALAXY
ZWG 072.068	13 22 00.	+ 14 21		14.3	GALAXY
UGC 08428	13 22 00.	+ 14 21	90	14.3	GALAXY SB0
MCG+02-34-014	13 22 00.	+ 14 21	72	14.3	GALAXY
ZWG 101.038	13 22 00.	+ 17 21		15.6	GALAXY
UGC 08429	13 22 00.	+ 17 21	60	15.6	GALAXY Sc
ZC 1322.0+2433	13 22 00.	+ 24 33	4370		CLUSTER OF GALAXIES
MCG+06-30-001	13 22 00.	+ 33 05	18	15.	GALAXY
ZC 1322.0+3750	13 22 00.	+ 37 50	810		CLUSTER OF GALAXIES
ZC 1322.0+4145	13 22 00.	+ 41 45	1010		CLUSTER OF GALAXIES
UGC 08430	13 22 00.	+ 42 04	78	16.0	GALAXY Sc
MCG+14-06-023	13 22 00.	+ 80 37 30.	30	15.	GALAXY
MCG-03-34-069	13 22 00.	- 19 27	150	14.5	GALAXY
RNGC 5124	13 22 00.	- 30 03		13.0	GALAXY
REIZ 3408	13 22 01.	+ 06 47	60	13.5	GALAXY
FATH 1.608	13 22 01.	+ 59 32	8		NEBULA
MCG+07-28-007	13 22 03.	+ 42 06	60	15.	GALAXY
FATH 1.609	13 22 03.	+ 58 40	5		NEBULA
SC 1319-2946.7	13 22 03.	- 30 02 23.	60		NEBULA
MCG-05-32-009	13 22 03.	- 30 03	90	13.5	GALAXY
LB 02652	13 22 04.	+ 53 02 18.		17.1	FAINT BLUE STAR
KARA.73 46	13 22 04.	- 37 06	27		DWARF GALAXY
ZWG 101.039	13 22 06.	+ 18 30		15.7	GALAXY
ZC 1322.1+2845	13 22 06.	+ 28 45	670		CLUSTER OF GALAXIES
ZWG 161.045	13 22 06.	+ 31 13		15.3	GALAXY
IC 4239	13 22 06.	+ 31 13 03.			NONSTELLAR OBJECT
TON-N 0160	13 22 06.	+ 32 20		16.0	BLUE STAR
ZWG 190.002	13 22 06.	+ 33 06		15.6	GALAXY
ZWG 189.062	13 22 06.	+ 33 06		15.6	GALAXY
UGC 08431	13 22 06.	+ 33 06	60	15.6	GALAXY S0-a
ZWG 190.003	13 22 06.	+ 36 10		15.3	GALAXY
ZWG 189.063	13 22 06.	+ 36 10		15.3	GALAXY
UGC 08432	13 22 06.	+ 36 10	66	15.3	GALAXY S
MCG+06-30-002	13 22 06.	+ 36 10	48	14.5	GALAXY
ZWG 190.004	13 22 06.	+ 36 51		14.6	GALAXY
ZWG 189.064	13 22 06.	+ 36 51		14.6	GALAXY
MRK 451	13 22 06.	+ 36 51	12	15.	GALAXY WITH UV CONTINUUM
MCG+10-19-065	13 22 06.	+ 57 42	30	16.	GALAXY
SC 1319-2948.2	13 22 06.	- 30 03 53.	54		NEBULA
RNGC 5126	13 22 06.	- 30 04		14.0	GALAXY
MCG-05-32-010	13 22 06.	- 30 04	72	14.5	GALAXY
IC 4240	13 22 08.	+ 31 14 09.			NONSTELLAR OBJECT
FATH 1.610	13 22 09.	+ 59 02	16		NEBULA
ZC 1322.2+0600	13 22 12.	+ 06 00	1340		CLUSTER OF GALAXIES
TON-N 0708	13 22 12.	+ 28 35		14.3	BLUE STAR
ZC 1322.2+5028	13 22 12.	+ 50 28	810		CLUSTER OF GALAXIES
REIN 5.026	13 22 12.67	+ 12 01 43.4			NEBULA
MCG+05-32-015	13 22 15.	+ 31 14	42	15.3	GALAXY
LB 02653	13 22 16.	+ 52 37 42.		17.3	FAINT BLUE STAR
MCG+03-34-025	13 22 17.	+ 16 23	15	15.5	GALAXY
SHB 270	13 22 17.	+ 55 15 48.		16.	QUASI-STELLAR OBJECT
KARA.72 372B	13 22 18.	+ 16 23	12	15.5	PART OF DOUBLE GALAXY
KARA.72 372A	13 22 18.	+ 16 23	18	15.5	PART OF DOUBLE GALAXY
MCG+03-34-026	13 22 18.	+ 16 23	12	15.5	GALAXY
ZWG 218.008	13 22 18.	+ 43 01		15.6	GALAXY
LB 02654	13 22 18.	+ 54 06 00.		16.0	FAINT BLUE STAR
RNGC 5133	13 22 18.	- 03 46		15.0	GALAXY
MCG-01-34-015	13 22 18.	- 03 46	30	15.	GALAXY
MCG-05-32-011	13 22 18.	- 33 25	96	15.	GALAXY
ARC 1732	13 22 19.	- 19 59		17.5	RICH CLUSTER OF GALAXIES
RNGC 5128	13 22 20.	- 42 45		7.5	GALAXY
RNGC 5136	13 22 22.	+ 14 00		14.5	GALAXY
RNGC 5137	13 22 22.	+ 14 20		15.5	GALAXY
IC 4242	13 22 22.	+ 31 17 03.			NONSTELLAR OBJECT
RNGC 5121A	13 22 22.	- 37 06			GALAXY
LB 02655	13 22 23.	+ 53 36 00.		15.8	FAINT BLUE STAR
ZWG 072.069	13 22 24.	+ 13 51		15.6	GALAXY
ZWG 072.070	13 22 24.	+ 14 00		14.7	GALAXY
MCG+02-34-015	13 22 24.	+ 14 00	18	14.7	GALAXY
ZWG 072.071	13 22 24.	+ 14 20		15.6	GALAXY
ZWG 161.046	13 22 24.	+ 27 00		15.5	GALAXY
MCG+05-32-016	13 22 24.	+ 27 00	27	15.5	GALAXY
MCG+06-30-003	13 22 24.	+ 32 47	39	14.5	GALAXY
ZC 1322.4+4737	13 22 24.	+ 47 37	1140		CLUSTER OF GALAXIES
MCG+10-19-066	13 22 24.	+ 57 36	18	16.	GALAXY
MCG-05-32-012	13 22 24.	- 30 16	36	15.	GALAXY
DV.56 N5121A	13 22 24.	- 37 06	84		S GALAXY
MCG-06-30-001	13 22 24.	- 37 06	42	15.5	GALAXY
IC 4241	13 22 25.	+ 26 59 46.			NONSTELLAR OBJECT
SC 1319-3001.3	13 22 26.	- 30 16 59.	18		NEBULA
ARP 153	13 22 26.	- 42 45			PECULIAR GALAXY
REIN 5.028	13 22 28.81	+ 11 54 44.4			NEBULA

OBJECT NAME	RIGHT ASCEN.	DECLINATION	DIAM.	MAGN.	TYPE OF OBJECT
REIN 5.029	13 22 28.59	+ 11 56 09.8			NEBULA
FATH 1.611	13 22 29.	+ 59 17	5		NEBULA
8ZW 1322+00.2	13 22 30.	+ 00 10		17.8	COMPACT GALAXY
TON-N 0709	13 22 30.	+ 24 23		14.6	BLUE STAR
ZWG 190.005	13 22 30.	+ 32 48		15.3	GALAXY
MCG+07-28-008	13 22 30.	+ 43 02	48	15.	GALAXY
ZC 1322.5+4951	13 22 30.	+ 49 51	2760		CLUSTER OF GALAXIES
MCG+09-22-057	13 22 30.	+ 53 13	78	14.	GALAXY
ZWG 016.067	13 22 30.	- 00 39		15.1	GALAXY
MCG-03-34-071	13 22 30.	- 19 31	72	15.	GALAXY
MCG-03-34-070	13 22 30.	- 19 31	72	14.	GALAXY
MCG-07-28-001	13 22 30.	- 42 45	840	7.8	GALAXY
ARC 1733	13 22 32.	+ 02 27		18.0	RICH CLUSTER OF GALAXIES
RNGC 5120	13 22 32.	- 63 09			NON-EXISTENT OBJECT
MCG+03-34-027	13 22 33.	+ 16 05	36	15.6	GALAXY
MCG-03-34-072	13 22 33.	- 20 25	30	15.5	GALAXY
RNGC 5134	13 22 33.	- 20 51		12.5	GALAXY
RNGC 5141	13 22 34.	+ 36 38		14.0	GALAXY
ZWG 161.047	13 22 36.	+ 26 43		15.6	GALAXY
ZWG 190.006	13 22 36.	+ 36 38		13.9	GALAXY
ZWG 189.065	13 22 36.	+ 36 38		13.9	GALAXY
UGC 08433	13 22 36.	+ 36 38	90	13.9	GALAXY S0
KARA.72 373A	13 22 36.	+ 36 38	102	13.9	PART OF DOUBLE GALAXY
ZWG 218.009	13 22 36.	+ 43 56		15.5	GALAXY
LB 02656	13 22 36.	+ 49 57 42.		16.8	FAINT BLUE STAR
ZWG 271.038	13 22 36.	+ 53 14		15.5	GALAXY
ZC 1322.6+6044	13 22 36.	+ 60 44	1210		CLUSTER OF GALAXIES
MCG-03-34-073	13 22 36.	- 20 53	150	12.	GALAXY
MCG-04-32-011	13 22 36.	- 25 05	60	15.5	GALAXY
MCG-06-30-001A	13 22 36.	- 33 31	30	15.	GALAXY
IC 4244	13 22 37.	+ 26 43 04.			NONSTELLAR OBJECT
LB 02657	13 22 37.	+ 49 13 12.		16.6	FAINT BLUE STAR
MCG+06-30-004	13 22 39.	+ 36 38	54	14.	GALAXY
8ZW 1322+06.8	13 22 42.	+ 06 48		14.7	COMPACT GALAXY
ZC 1322.7+1325	13 22 42.	+ 13 25	1210		CLUSTER OF GALAXIES
ZWG 101.040	13 22 42.	+ 16 06		15.6	GALAXY
MCG+03-34-028	13 22 42.	+ 17 18	36	15.	GALAXY
ZWG 294.035	13 22 42.	+ 59 35		15.7	GALAXY
MCG+10-19-067	13 22 42.	+ 59 35	39	16.	GALAXY
MCG+11-16-018	13 22 42.	+ 63 22	54	15.	GALAXY
UGC 08434	13 22 42.	+ 70 48	60	16.5	GALAXY Sc
ARC 1741	13 22 42.	+ 71 44		17.1	RICH CLUSTER OF GALAXIES
OCL 0899	13 22 42.	- 63 09	168	13.	OPEN STAR CLUSTER
MCG+09-22-058	13 22 45.	+ 53 12 30.	30	16.	GALAXY
FATH 1.612	13 22 45.	+ 59 35	19		NEBULA
RNGC 5142	13 22 46.	+ 36 40		14.0	GALAXY
RNGC 5143	13 22 46.	+ 36 42		15.5	GALAXY
ZWG 101.041	13 22 48.	+ 14 45		15.7	GALAXY
ZWG 101.042	13 22 48.	+ 17 19		15.2	GALAXY
TON-N 0710	13 22 48.	+ 30 04		14.1	BLUE STAR
MCG+06-30-006	13 22 48.	+ 36 39	42	14.	GALAXY
ZWG 190.007	13 22 48.	+ 36 40		14.0	GALAXY
ZWG 189.066	13 22 48.	+ 36 40		14.0	GALAXY
MRK 452	13 22 48.	+ 36 40	24	15.5	GALAXY WITH UV CONTINUUM
UGC 08435	13 22 48.	+ 36 40	60	14.0	GALAXY S0
KARA.72 373B	13 22 48.	+ 36 40	90	14.0	PART OF DOUBLE GALAXY
MCG+06-30-005	13 22 48.	+ 36 41	24	15.	GALAXY
ZWG 190.008	13 22 48.	+ 36 42		15.5	GALAXY
ZWG 189.067	13 22 48.	+ 36 42		15.5	GALAXY
ZWG 316.015	13 22 48.	+ 63 21		15.1	GALAXY
UGC 08436	13 22 48.	+ 63 21	66	15.1	GALAXY Sc
MCG+12-13-006	13 22 48.	+ 70 48	66	17.	GALAXY
MCG-02-34-046	13 22 48.	- 14 06	36	15.	GALAXY
LB 02658	13 22 49.	+ 53 33 18.		17.1	FAINT BLUE STAR
FATH 1.613	13 22 50.	+ 58 45	11		NEBULA
REIN 5.030	13 22 51.51	+ 11 54 40.2			NEBULA
FATH 1.614	13 22 52.	+ 58 44	19		NEBULA
REIN 5.032	13 22 53.23	+ 12 02 26.3			NEBULA
REIN 5.033	13 22 53.99	+ 12 02 38.3			NEBULA
ZC 1322.9+0435	13 22 54.	+ 04 35	200		CLUSTER OF GALAXIES
8ZW 1322+07.9	13 22 54.	+ 07 52		17.5	COMPACT GALAXY
MCG+03-34-029	13 22 54.	+ 18 42	66	15.0	GALAXY
ZWG 101.043	13 22 54.	+ 18 43		15.0	GALAXY
UGC 08437	13 22 54.	+ 18 43	60	15.0	GALAXY Sa
ZC 1322.9+5615	13 22 54.	+ 56 15	1140		CLUSTER OF GALAXIES
MCG-05-32-014	13 22 54.	- 29 16	36	15.	GALAXY
RNGC 5135	13 22 54.	- 29 34		13.0	GALAXY
MCG-05-32-013	13 22 54.	- 29 34	96	13.	GALAXY
SMI 04	13 22 54.	- 63 18	360		FAINT NEBULOSITY
MCG+03-34-030	13 22 57.	+ 16 25	54	15.7	GALAXY
REIZ 3409	13 22 59.	+ 43 31	24	13.5	GALAXY
LB 02659	13 22 59.	+ 51 46 12.		17.2	FAINT BLUE STAR
MIL 25	13 22 59.	- 61 07 24.	78		SUPERNOVA REMNANT
VDB.66G 175	13 23	+ 58 05	100		DWARF GALAXY
VV 039B	13 23	+ 84 46	12	17.	INTERACTING GALAXY
VV 039A	13 23	+ 84 46	24	16.	INTERACTING GALAXY
ZWG 072.072	13 23 00.	+ 12 05		15.7	GALAXY
ZWG 101.044	13 23 00.	+ 16 27		15.7	GALAXY
UGC 08438	13 23 00.	+ 16 27	60	15.7	GALAXY S0
ZWG 218.010	13 23 00.	+ 43 31		13.6	GALAXY
RNGC 5145	13 23 00.	+ 43 31		13.5	GALAXY
UGC 08439	13 23 00.	+ 43 31	138	13.6	GALAXY S
LB 02660	13 23 01.	+ 50 15 06.		15.5	FAINT BLUE STAR
REIN 5.034	13 23 02.48	+ 11 59 16.8			NEBULA
REIN 5.035	13 23 02.84	+ 12 05 06.1			NEBULA
ZC 1323.1+5756	13 23 06.	+ 57 56	2690		CLUSTER OF GALAXIES
MCG+07-28-009	13 23 09.	+ 43 33	108	14.	GALAXY
MCG-03-34-074	13 23 09.	- 20 20	36	15.5	GALAXY
IC 4243	13 23 10.	- 27 22			NONSTELLAR OBJECT
HN 1044	13 23 10.	- 27 23		14.	NEBULA
REIN 5.036	13 23 10.44	+ 11 53 22.7			NEBULA
ARC 1738	13 23 11.	+ 57 52		16.6	RICH CLUSTER OF GALAXIES
REIN 5.037	13 23 11.67	+ 11 49 11.8			NEBULA
ZC 1323.2+5256	13 23 12.	+ 52 56	1010		CLUSTER OF GALAXIES
ZC 1323.2+6248	13 23 12.	+ 62 48	2620		CLUSTER OF GALAXIES
ZWG 016.068	13 23 12.	- 03 15		15.6	GALAXY
REIN 5.038	13 23 12.39	+ 11 59 14.9			NEBULA
REIZ 3411	13 23 13.	+ 54 06	36	15.0	GALAXY
FATH 1.615	13 23 14.	+ 58 44	8		NEBULA
ARC 1737	13 23 17.	+ 49 49		17.3	RICH CLUSTER OF GALAXIES
ZC 1323.3+1907	13 23 18.	+ 19 07	740		CLUSTER OF GALAXIES
MCG+06-30-007	13 23 18.	+ 33 55	48	15.	GALAXY
MCG+10-19-068	13 23 18.	+ 57 50	12	16.	GALAXY
MCG+10-19-069	13 23 18.	+ 57 54	27	16.	GALAXY
LB 02661	13 23 18.	+ 59 46 24.		13.6	FAINT BLUE STAR
MCG-02-34-047	13 23 18.	- 11 53	48	15.	GALAXY
MCG-05-32-015	13 23 18.	- 31 53	48	15.5	GALAXY
MCG+06-30-008	13 23 21.	+ 36 37	30	15.	GALAXY
MCG-04-32-012	13 23 21.	- 26 24	12	15.5	GALAXY
IC 4246	13 23 21.	- 26 24			NONSTELLAR OBJECT
IC 4245	13 23 21.	- 26 24			NONSTELLAR OBJECT
HN 1045	13 23 21.	- 26 25	18		NEBULA
REIN 5.040	13 23 21.77	+ 11 56 28.6			NEBULA
IC 0886	13 23 23.	+ 14 00 18.			NONSTELLAR OBJECT
RNGC 5140	13 23 23.	- 33 37		13.0	GALAXY
REIN 5.041	13 23 23.50	+ 11 52 43.9			NEBULA
ZWG 072.073	13 23 24.	+ 13 34		15.6	GALAXY
ZC 1323.4+1936	13 23 24.	+ 19 36	1880		CLUSTER OF GALAXIES
ZC 1323.4+2847	13 23 24.	+ 28 47	1140		CLUSTER OF GALAXIES
ZWG 190.009	13 23 24.	+ 33 56		15.3	GALAXY
UGC 08440	13 23 24.	+ 36 39	60	16.5	GALAXY SBc
ZC 1323.4+5704	13 23 24.	+ 57 04	1750		CLUSTER OF GALAXIES
HELW 472	13 23 24.	- 20 46 27.			NEBULA
MCG-03-34-075	13 23 24.	- 20 47	72	15.	GALAXY
MCG-04-32-013	13 23 24.	- 26 24	30	15.5	GALAXY
MCG-05-32-016	13 23 27.	- 33 37	30	13.5	GALAXY
ZC 1323.5+1239	13 23 30.	+ 12 39	1080		CLUSTER OF GALAXIES
MCG+09-22-059	13 23 30.	+ 55 40	30	17.	GALAXY
MCG+10-19-070	13 23 30.	+ 58 04	132	15.	GALAXY
ZWG 294.036	13 23 30.	+ 58 05		15.6	GALAXY
UGC 08441	13 23 30.	+ 58 05	210	15.6	GALAXY DWRF IR
MCG+10-19-071	13 23 30.	+ 59 52	45	15.	GALAXY
ZC 1323.5-0005	13 23 30.	- 00 05	1410		CLUSTER OF GALAXIES
MCG-03-34-076	13 23 30.	- 16 15	72	14.	GALAXY
HZ 45	13 23 32.	+ 40 23		12.4	DECIDEDLY BLUE STAR
MCG+06-30-009	13 23 33.	+ 34 45	24	15.	GALAXY
ZC 1323.6+0949	13 23 36.	+ 09 49	470		CLUSTER OF GALAXIES
ARC 1735	13 23 36.	+ 12 39		17.8	RICH CLUSTER OF GALAXIES
REIZ 3410	13 23 36.	+ 21 19	54	13.6	GALAXY
ZC 1323.6+2815	13 23 36.	+ 28 15	940		CLUSTER OF GALAXIES
ZC 1323.6+2942	13 23 36.	+ 29 42	1480		CLUSTER OF GALAXIES
ZWG 161.048	13 23 36.	+ 31 53		15.1	GALAXY
MCG+05-32-017	13 23 36.	+ 31 53	48	15.1	GALAXY
MCG-03-34-077	13 23 36.	- 19 24	72	15.	GALAXY
MCG-04-32-014	13 23 36.	- 22 23	72	15.	GALAXY
MCG-02-34-048	13 23 39.	- 12 22	90	15.	GALAXY
SC 1320-3013.4	13 23 39.	- 30 29 03.	18		NEBULA
ZC 1323.7+1008	13 23 42.	+ 10 08	400		CLUSTER OF GALAXIES
UGC 08442	13 23 42.	+ 22 11	60	16.5	GALAXY Sc
MCG+08-24-111	13 23 42.	+ 46 26	36	16.	GALAXY
REIZ 3416	13 23 42.	+ 54 32	30	15.2	GALAXY
REIZ 3414	13 23 46.	+ 36 11	60	13.9	GALAXY
ZWG 016.069	13 23 48.	+ 02 21		12.7	GALAXY
UGC 08443	13 23 48.	+ 02 21	108	12.7	GALAXY
MCG+00-34-033	13 23 48.	+ 02 21	96	12.7	GALAXY
MRK 453	13 23 48.	+ 33 16	24	16.5	GALAXY WITH UV CONTINUUM
ZC 1323.8+4844	13 23 48.	+ 48 44	940		CLUSTER OF GALAXIES
MCG-03-34-078	13 23 48.	- 19 25	30	15.5	GALAXY
MCG-04-32-015	13 23 48.	- 26 28 30.	9	15.5	GALAXY
RNGC 5139	13 23 48.	- 47 03		4.0	GLOBULAR CLUSTER
GCL 024	13 23 48.	- 47 13	3924	4.25	GLOBULAR STAR CLUSTER
RNGC 5147	13 23 50.	+ 02 22		12.5	GALAXY
IC 4250	13 23 50.	+ 26 43 55.			NONSTELLAR OBJECT
REIZ 3413	13 23 51.	+ 02 21	60	11.9	GALAXY
MCG+06-30-010	13 23 51.	+ 36 10	90	13.5	GALAXY
REIN 5.042	13 23 51.57	+ 12 03 54.4			NEBULA
RNGC 5149	13 23 52.	+ 36 12		14.0	GALAXY
RNGC 5146	13 23 53.	- 12 03		14.0	GALAXY
HN 1046	13 23 53.	- 30 06	30	14.	NEBULA
ZWG 044.084	13 23 54.	+ 04 44		15.5	GALAXY
KARA.72 374A	13 23 54.	+ 04 44	36	15.5	PART OF DOUBLE GALAXY
ZWG 101.045	13 23 54.	+ 18 01		15.5	GALAXY
ZWG 161.049	13 23 54.	+ 26 45		15.2	GALAXY
ZWG 190.010	13 23 54.	+ 36 12		13.8	GALAXY
UGC 08444	13 23 54.	+ 36 12	96	13.8	GALAXY SBb
KARA.72 375A	13 23 54.	+ 36 12	96	13.8	PART OF DOUBLE GALAXY
ZWG 218.011	13 23 54.	+ 40 20		15.5	GALAXY
ARC 1740	13 23 54.	+ 41 55		17.6	RICH CLUSTER OF GALAXIES
ZC 1323.9+4157	13 23 54.	+ 41 57	1210		CLUSTER OF GALAXIES
MCG+09-22-060	13 23 54.	+ 51 32 30.	30	15.	GALAXY
MCG-02-34-050	13 23 54.	- 10 12	54	15.5	GALAXY
MCG-02-34-049	13 23 54.	- 12 03	36	14.	GALAXY
MCG-04-32-016	13 23 54.	- 27 11	18	15.	GALAXY
IC 4247	13 23 54.	- 30 05			NONSTELLAR OBJECT
MCG-05-32-017	13 23 54.	- 30 06	72	15.	GALAXY
SHB 263	13 23 54.1	+ 03 42 08.		18.	QUASI-STELLAR OBJECT
LB 02662	13 23 55.	+ 51 58 36.		17.2	FAINT BLUE STAR
ARC 1739	13 23 56.	+ 29 42		17.6	RICH CLUSTER OF GALAXIES
MCG+03-34-031	13 23 57.	+ 17 13	30	15.6	GALAXY
REIZ 3412	13 23 57.	- 12 04	18	13.8	GALAXY
MCG-04-32-017	13 23 57.	- 21 58	90	15.5	GALAXY
ARC 1744	13 23 59.	+ 59 34		17.2	RICH CLUSTER OF GALAXIES
IC 4248	13 23 59.	- 29 38			NONSTELLAR OBJECT
HN 1047	13 23 59.	- 29 39		13.	NEBULA
VV 039C	13 24	+ 84 46	42	15.	INTERACTING GALAXY
VDB.66G 174	13 24	- 22 00	70		DWARF GALAXY
MCG+01-34-021	13 24 00.	+ 02 34	42	15.4	GALAXY
ZC 1324.0+0409	13 24 00.	+ 04 09	540		CLUSTER OF GALAXIES
ZWG 044.085	13 24 00.	+ 04 43		15.3	GALAXY
UGC 08445	13 24 00.	+ 04 43	60	15.3	GALAXY S
KARA.72 374B	13 24 00.	+ 04 43	66	15.3	PART OF DOUBLE GALAXY
ZWG 101.046	13 24 00.	+ 17 14		15.6	GALAXY
ZWG 161.050	13 24 00.	+ 31 58		15.6	GALAXY
ZWG 271.039	13 24 00.	+ 51 35		15.7	GALAXY
UGC 08446	13 24 00.	+ 51 35	60	15.7	GALAXY Sc
KARA.73B 0583	13 24 00.	+ 51 35	42	15.7	ISOLATED GALAXY S
ZC 1324.0+6714	13 24 00.	+ 67 14	1340		CLUSTER OF GALAXIES
ZWG 016.070	13 24 00.	- 03 00		15.7	GALAXY
MCG+00-34-034	13 24 00.	- 03 00	78	15.7	GALAXY
MCG-02-34-051	13 24 00.	- 10 57 30.	84	14.5	GALAXY
MCG-05-32-018	13 24 00.	- 29 36	54	14.	GALAXY
MCG-06-30-002	13 24 00.	- 36 39	48	15.	GALAXY
HOLM 516A	13 24 03.	- 12 04	42	13.7	PART OF MULTIPLE GALAXY
HOLM 516B	13 24 04.	- 12 05	18	14.5	PART OF MULTIPLE GALAXY
FATH 1.616	13 24 05.	+ 59 21	3		NEBULA
RNGC 5138	13 24 05.	- 58 45		10.0	OPEN CLUSTER
ZWG 044.086	13 24 06.	+ 02 35		15.4	GALAXY
ZWG 072.074	13 24 06.	+ 12 08		15.5	GALAXY
ZWG 101.047	13 24 06.	+ 14 41		15.3	GALAXY
ZC 1324.1+3520	13 24 06.	+ 35 20	1210		CLUSTER OF GALAXIES
DVDV 3	13 24 06.	- 41 12	39		GALAXY
OCL 0902	13 24 06.	- 58 45	840	9.8	OPEN STAR CLUSTER
RNGC 5148	13 24 08.	+ 02 34		15.5	GALAXY
ARC 1736	13 24 08.	- 26 52		14.8	RICH CLUSTER OF GALAXIES
REIZ 3415	13 24 09.	+ 02 33	48	15.1	GALAXY
IC 0889	13 24 09.	+ 12 08 08.			NONSTELLAR OBJECT

OBJECT NAME	RIGHT ASCEN.	DECLINATION	DIAM.	MAGN.	TYPE OF OBJECT
LB 02663	13 24 09.	+ 52 48 42.		17.1	FAINT BLUE STAR
REIN 5.043	13 24 09.09	+ 12 07 42.9			NEBULA
RNGC 5154	13 24 10.	+ 36 16		15.0	GALAXY
REIZ 3418	13 24 10.	+ 36 16	48	14.4	GALAXY
RNGC 5151	13 24 11.	+ 17 08		15.0	GALAXY
MCG+03-34-032	13 24 12.	+ 17 06	42	14.9	GALAXY
ZWG 101.048	13 24 12.	+ 17 08		14.9	GALAXY
ZWG 190.011	13 24 12.	+ 36 16		14.9	GALAXY
UGC 08447	13 24 12.	+ 36 16	84	14.9	GALAXY Sc
KARA.72 375B	13 24 12.	+ 36 16	90	14.9	PART OF DOUBLE GALAXY
MCG+06-30-011	13 24 12.	+ 36 16	72	14.	GALAXY
MRK 066	13 24 12.	+ 57 32	15	15.	GALAXY WITH UV CONTINUUM
MCG+10-19-072	13 24 12.	+ 57 32	24	16.	GALAXY
LB 02664	13 24 13.	+ 60 27 42.		17.2	FAINT BLUE STAR
REIZ 3417	13 24 14.	+ 17 08	12	15.0	GALAXY
SC 1321-2937.4	13 24 14.	- 29 53 02.	30		NEBULA
REIN 5.044	13 24 16.88	+ 12 44 48.6			NEBULA
REIN 5.045	13 24 17.88	+ 12 47 15.8			NEBULA
ZC 1324.3+0349	13 24 18.	+ 03 49	1410		CLUSTER OF GALAXIES
SHAH 073	13 24 18.	+ 16 26	48	18.5	GROUP OF COMPACT GALAXIES
MCG+03-34-033	13 24 18.	+ 20 12	72	14.9	GALAXY
ZWG 101.049	13 24 18.	+ 20 13		14.9	GALAXY
UGC 08448	13 24 18.	+ 20 13	66	14.9	GALAXY Sb/Sc
ZC 1324.3+2136	13 24 18.	+ 21 36	3020		CLUSTER OF GALAXIES
BIGO 531	13 24 18.	+ 36 10			NEBULA
LB 02665	13 24 18.	+ 53 20 42.		16.2	FAINT BLUE STAR
MCG+10-19-073	13 24 18.	+ 57 36	36	16.	GALAXY
MCG-05-32-019	13 24 18.	- 28 55	36	15.	GALAXY
LB 02668	13 24 19.	+ 50 26 06.		17.0	FAINT BLUE STAR
LB 02667	13 24 19.	+ 50 37 42.		16.3	FAINT BLUE STAR
LB 02666	13 24 19.	+ 60 19 06.		16.5	FAINT BLUE STAR
REIN 5.046	13 24 22.61	+ 12 46 34.6			NEBULA
ZWG 072.075	13 24 24.	+ 11 50		15.4	GALAXY
MCG+03-34-034	13 24 24.	+ 20 14	42	16.	GALAXY
ZC 1324.4+2421	13 24 24.	+ 24 21	940		CLUSTER OF GALAXIES
ZWG 161.051	13 24 24.	+ 30 46		15.6	GALAXY
MCG+05-32-018	13 24 24.	+ 30 46	18	15.6	GALAXY
UGC 08449	13 24 24.	+ 43 01	84	16.0	GALAXY
7ZW 513	13 24 24.	+ 57 43			COMPACT GALAXY
ZC 1324.4+5931	13 24 24.	+ 59 31	2020		CLUSTER OF GALAXIES
MCG-02-34-052	13 24 24.	- 11 32	18	15.	GALAXY
MCG-05-32-020	13 24 24.	- 27 40	24	14.5	GALAXY
SER 102.01	13 24 24.	- 41 17	20		LOW SURF. BRGHT. IRR GLXY
REIN 5.048	13 24 25.36	+ 12 50 44.5			NEBULA
LB 02669	13 24 26.	+ 53 57 06.		16.8	FAINT BLUE STAR
REIN 5.049	13 24 26.56	+ 12 51 05.4			NEBULA
IC 4249	13 24 28.	- 27 41			NONSTELLAR OBJECT
HN 1048	13 24 28.	- 27 42		13.	NEBULA
UGC 08450	13 24 30.	+ 10 14	66	17.	GALAXY DWRF IR
ZWG 072.076	13 24 30.	+ 10 19		15.6	GALAXY
KARA.73B 0584	13 24 30.	+ 10 19	60	15.6	ISOLATED GALAXY IR
ZC 1324.5+3225	13 24 30.	+ 32 25	810		CLUSTER OF GALAXIES
MCG+07-28-010	13 24 30.	+ 43 03	78	15.	GALAXY
MCG+09-22-061	13 24 30.	+ 52 02	48	16.	GALAXY
MCG+10-19-074	13 24 30.	+ 59 59	30	16.	GALAXY
MCG+14-06-024	13 24 30.	+ 84 46		16.	GALAXY
REIN 5.051	13 24 30.41	+ 12 50 36.6			NEBULA
REIN 5.052	13 24 34.69	+ 12 38 15.6			NEBULA
IC 4251	13 24 35.	- 29 12			NONSTELLAR OBJECT
HN 1049	13 24 35.	- 29 13		14.	NEBULA
ZC 1324.6+0229	13 24 36.	+ 02 29	610		CLUSTER OF GALAXIES
MCG+03-34-035	13 24 36.	+ 15 19	36	14.1	GALAXY
TON-N 0711	13 24 36.	+ 23 18		14.6	BLUE STAR
ZC 1324.6+2602	13 24 36.	+ 26 02	4230		CLUSTER OF GALAXIES
ZWG 161.052	13 24 36.	+ 26 51		15.1	GALAXY
MRK 454	13 24 36.	+ 26 51	10	16.	GALAXY WITH UV CONTINUUM
MCG+05-32-020	13 24 36.	+ 26 51	48	15.1	GALAXY
REIZ 3420	13 24 36.	+ 32 26	60	13.5	GALAXY
ZWG 161.053	13 24 36.	+ 32 28		14.6	GALAXY
UGC 08451	13 24 36.	+ 32 28	108	14.6	GALAXY Sc
MCG+05-32-019	13 24 36.	+ 32 28 30.	108	14.6	GALAXY
MCG+07-28-011	13 24 36.	+ 41 06	36	16.5	GALAXY
ZC 1324.6+4912	13 24 36.	+ 49 12	1410		CLUSTER OF GALAXIES
ZC 1324.6+5314	13 24 36.	+ 53 14	1480		CLUSTER OF GALAXIES
ZC 1324.6-0222	13 24 36.	- 02 22	1410		CLUSTER OF GALAXIES
MCG-05-32-021	13 24 36.	- 29 10	24	15.5	GALAXY
MCG-06-30-003	13 24 36.	- 37 55	150	14.5	GALAXY
LB 00269	13 24 37.	+ 57 18 12.		14.6	FAINT BLUE STAR
HOLM 517A	13 24 41.	- 02 19	48	14.5	PART OF MULTIPLE GALAXY
ZC 1324.7+1351	13 24 42.	+ 13 51	1080		CLUSTER OF GALAXIES
ZWG 101.050	13 24 42.	+ 15 21		14.1	GALAXY
UGC 08452	13 24 42.	+ 15 21	48	14.1	GALAXY S0
ZC 1324.7+1907	13 24 42.	+ 19 07	870		CLUSTER OF GALAXIES
REIZ 3419	13 24 42.	+ 21 11	12	14.3	GALAXY
TON-N 0712	13 24 42.	+ 29 41		16.8	BLUE STAR
TON-N 0161	13 24 42.	+ 31 14		15.6	BLUE STAR
ZWG 294.037	13 24 42.	+ 59 59		15.6	GALAXY
ZC 1324.7+6208	13 24 42.	+ 62 08	1480		CLUSTER OF GALAXIES
ARP 204	13 24 42.	+ 84 45			PECULIAR GALAXY
MCG-02-34-053	13 24 42.	- 11 33	24	15.	GALAXY
MCG-05-32-022	13 24 42.	- 27 35	60	15.5	GALAXY
MCG-07-28-001A	13 24 42.	- 40 49	78	15.5	GALAXY
MRSL 307-01/1	13 24 42.	- 63 19	2700		HII REGION
IC 4256	13 24 43.	+ 31 18 04.			NONSTELLAR OBJECT
REIN 5.053	13 24 43.65	+ 12 38 34.9			NEBULA
HOLM 517B	13 24 45.	- 02 22	24	14.8	PART OF MULTIPLE GALAXY
KLEM 25	13 24 45.	- 40 48	60	16.	CMPT GROUP OF 5 GALAXIES
ARC 1742	13 24 47.	+ 13 50		17.5	RICH CLUSTER OF GALAXIES
ZWG 044.087	13 24 48.	+ 05 54		15.7	GALAXY
KARA.73B 0585	13 24 48.	+ 05 54	36	15.7	ISOLATED GALAXY S
ZC 1324.8+1423	13 24 48.	+ 14 23	1340		CLUSTER OF GALAXIES
ZWG 101.051	13 24 48.	+ 20 04		15.7	GALAXY
ZWG 161.054	13 24 48.	+ 31 14		15.5	GALAXY
ZWG 246.001	13 24 48.	+ 45 05		15.6	GALAXY
MCG+09-22-062	13 24 48.	+ 53 00	30	14.7	GALAXY
FATH 1.617	13 24 48.	+ 59 25	11		NEBULA
MCG-05-32-023	13 24 48.	- 29 17	60	14.	GALAXY
SER 102.05	13 24 48.	- 43 13	100		PEC. GALAXY
ARC 1745	13 24 49.	+ 54 05		18.4	RICH CLUSTER OF GALAXIES
MCG-04-32-018	13 24 51.	- 27 04	30	15.5	GALAXY
IC 4252	13 24 52.	- 27 04			NONSTELLAR OBJECT
HN 1050	13 24 52.	- 27 05		13.5	NEBULA
IC 4253	13 24 52.	- 27 36			NONSTELLAR OBJECT
HN 1051	13 24 52.	- 27 37			NEBULA
RNGC 5163	13 24 53.	+ 53 01		15.0	GALAXY
LB 02670	13 24 53.	+ 55 51 30.		16.5	FAINT BLUE STAR
ZWG 072.077	13 24 54.	+ 12 37		15.7	GALAXY
ZWG 218.012	13 24 54.	+ 42 13		15.5	GALAXY
MCG+07-28-012	13 24 54.	+ 43 04	33	16.	GALAXY
MCG+08-25-001	13 24 54.	+ 45 04	12	17.	GALAXY
ZWG 271.040	13 24 54.	+ 53 01		14.9	GALAXY
UGC 08453	13 24 54.	+ 53 01	66	14.9	GALAXY E
MCG+10-19-075	13 24 54.	+ 58 21	39	15.	GALAXY
ZC 1324.9+7146	13 24 54.	+ 71 46	3230		CLUSTER OF GALAXIES
RNGC 5150	13 24 54.	- 29 18		14.0	GALAXY
REIN 5.055	13 24 54.29	+ 12 36 48.8			NEBULA
MCG+07-28-013	13 24 57.	+ 42 14	42	15.	GALAXY
LB 02671	13 24 58.	+ 51 19 24.		16.2	FAINT BLUE STAR
LB 09873	13 25	- 83 13		14.2	FAINT BLUE STAR
ZC 1325.0+1944	13 25 00.	+ 19 44	670		CLUSTER OF GALAXIES
ZC 1325.0+2346	13 25 00.	+ 23 46	3290		CLUSTER OF GALAXIES
ZWG 161.055	13 25 00.	+ 32 03		15.7	GALAXY
REIZ 3422	13 25 00.	+ 32 16	60	13.9	GALAXY
ZWG 161.056	13 25 00.	+ 32 17		14.4	GALAXY
UGC 08455	13 25 00.	+ 32 17	96	14.4	GALAXY SBa
MCG+05-32-021	13 25 00.	+ 32 18	66	14.4	GALAXY
7ZW 514	13 25 00.	+ 84 47			COMPACT GALAXY
ZWG 366.004	13 25 00.	+ 84 47		15.6	GALAXY
ZWG 365.012	13 25 00.	+ 84 47		15.6	GALAXY
UGC 08454	13 25 00.	+ 84 47	150	15.6	GALAXY DBL SYS
ARC 1743	13 25 01.	+ 03 52		17.8	RICH CLUSTER OF GALAXIES
LB 02672	13 25 01.	+ 51 39 54.		15.7	FAINT BLUE STAR
REIZ 3421	13 25 03.	+ 00 42	24	15.8	GALAXY
RNGC 5157	13 25 03.	+ 32 17		14.5	GALAXY
IC 4254	13 25 04.	- 26 58			NONSTELLAR OBJECT
HN 1052	13 25 04.	- 26 59		14.5	NEBULA
HN 1650	13 25 04.3	+ 00 43 54.	30		NEBULA
ZWG 016.071	13 25 06.	+ 00 42		15.3	GALAXY
ZWG 101.052	13 25 06.	+ 15 26		15.7	GALAXY
UGC 08456	13 25 06.	+ 15 26	90	15.7	GALAXY Sc
MCG+03-34-036	13 25 06.	+ 19 35 30.	21	15.2	GALAXY
ZWG 101.053	13 25 06.	+ 19 36		15.2	GALAXY
ZWG 131.008	13 25 06.	+ 21 09		15.6	GALAXY
UGC 08457	13 25 06.	+ 21 09	78	15.6	GALAXY Sb-c
TON-N 0713	13 25 06.	+ 29 39		16.7	BLUE STAR
RNGC 5152	13 25 06.	- 29 21		14.0	GALAXY
MCG-05-32-024	13 25 06.	- 29 21	96	14.	GALAXY
MCG+03-34-037	13 25 07.	+ 19 35	30	15.2	GALAXY
SC 1322-3059.1	13 25 08.	- 31 14 42.	18		NEBULA
IC 4257	13 25 10.	+ 47 07 42.			NONSTELLAR OBJECT
ARC 1747	13 25 10.	+ 52 53		18.0	RICH CLUSTER OF GALAXIES
PK307-01.1	13 25 11.	- 63 33	10	14.8	PLANETARY NEBULA
ZC 1325.2+2725	13 25 12.	+ 27 25	870		CLUSTER OF GALAXIES
ZC 1325.2+2916	13 25 12.	+ 29 16	810		CLUSTER OF GALAXIES
ZWG 271.041	13 25 12.	+ 55 45		14.6	GALAXY
MRK 257	13 25 12.	+ 55 45	15	14.5	GALAXY WITH UV CONTINUUM
UGC 08458	13 25 12.	+ 55 45	60	14.6	GALAXY SBb
KARA.72 376B	13 25 12.	+ 55 45	42		PART OF DOUBLE GALAXY
KARA.72 376A	13 25 12.	+ 55 45	48	14.6	PART OF DOUBLE GALAXY
7ZW 515	13 25 12.	+ 61 24			COMPACT GALAXY
MCG-02-34-054	13 25 12.	- 13 10	102	13.	GALAXY
MCG-04-32-019	13 25 12.	- 25 34 30.	120	14.5	GALAXY
RNGC 5153	13 25 12.	- 29 21		14.0	GALAXY
MCG-05-32-025	13 25 12.	- 29 21	24	14.	GALAXY
MCG-05-32-026	13 25 12.	- 31 14	48	15.	GALAXY
KEEL 666	13 25 12.9	+ 47 07 33.			NEBULA
RNGC 5164	13 25 13.	+ 55 45		14.5	GALAXY
REIN 5.058	13 25 13.79	+ 12 51 31.1			NEBULA
REIZ 3426	13 25 14.	+ 53 03	12	14.7	GALAXY
REIN 5.059	13 25 14.33	+ 12 50 35.8			NEBULA
MCG+09-22-063	13 25 15.	+ 55 45	60	14.0	GALAXY
MCG-04-32-020	13 25 15.	- 27 06	18	15.	GALAXY
LB 02673	13 25 16.	+ 52 19 24.		16.4	FAINT BLUE STAR
REIZ 3427	13 25 16.	+ 55 44	48	14.0	GALAXY
IC 4255	13 25 16.	- 27 06			NONSTELLAR OBJECT
HN 1053	13 25 16.	- 27 07		13.5	NEBULA
RNGC 5158	13 25 17.	+ 18 02		14.0	GALAXY
MCG+03-34-038	13 25 18.	+ 18 01	72	13.8	GALAXY
ZWG 101.054	13 25 18.	+ 18 02		13.8	GALAXY
UGC 08459	13 25 18.	+ 18 02	84	13.8	GALAXY SB?a-b
ZC 1325.3+2028	13 25 18.	+ 20 28	1080		CLUSTER OF GALAXIES
REIZ 3424	13 25 20.	+ 18 02	18	14.3	GALAXY
REIN 5.062	13 25 21.78	+ 12 42 43.1			NEBULA
HN 1651	13 25 23.3	- 01 39 30.	18		NEBULA
TON-N 0162	13 25 24.	+ 28 01		15.0	BLUE STAR
MCG-03-34-079	13 25 24.	- 16 05	15	15.	GALAXY
MCG-03-34-080	13 25 24.	- 20 44	24	15.	GALAXY
MCG+08-25-002	13 25 27.	+ 45 41	66	16.	GALAXY
REIZ 3423	13 25 27.	- 13 10	60	14.1	GALAXY
REIN 5.065	13 25 28.39	+ 12 57 29.0			NEBULA
8ZW 1325+03.7	13 25 30.	+ 03 42		17.5	COMPACT GALAXY
MCG+14-06-025	13 25 30.	+ 84 46 30.	42	16.	GALAXY
MCG-04-32-021	13 25 30.	- 24 41 30.	60	15.5	GALAXY
MCG-05-32-027	13 25 30.	- 27 42	12	15.5	GALAXY
MCG-07-28-002	13 25 30.	- 43 55	60	15.	GALAXY
ARC 1746	13 25 31.	+ 35 43		17.2	RICH CLUSTER OF GALAXIES
REIN 5.067	13 25 31.48	+ 12 44 01.6			NEBULA
IC 4258	13 25 34.	+ 28 39 00.			NONSTELLAR OBJECT
FATH 1.618	13 25 34.	+ 59 41	19		NEBULA
SC 1322-3120.1	13 25 34.	- 31 35 42.	60		NEBULA
ZC 1325.6+1331	13 25 36.	+ 13 31	400		CLUSTER OF GALAXIES
ZWG 161.057	13 25 36.	+ 28 46		15.2	GALAXY
MCG+05-32-022	13 25 36.	+ 28 46	30	15.2	GALAXY
ZWG 161.058	13 25 36.	+ 30 38		15.4	GALAXY
LB 02674	13 25 36.	+ 52 19 36.		15.7	FAINT BLUE STAR
REIN 5.068	13 25 36.34	+ 11 57 06.3			NEBULA
MCG-02-34-055	13 25 39.	- 11 31 30.	120	15.	GALAXY
RNGC 5156	13 25 41.	- 48 39		13.0	GALAXY
ZWG 044.088	13 25 42.	+ 03 15		15.2	GALAXY
UGC 08460	13 25 42.	+ 03 15	90	15.2	GALAXY Sc
MCG+01-34-022	13 25 42.	+ 03 15	72	15.2	GALAXY
MCG+05-32-023	13 25 42.	+ 30 38	39	15.4	GALAXY
ZWG 161.059	13 25 42.	+ 32 07		15.7	GALAXY
ZWG 161.060	13 25 42.	+ 32 13		15.7	GALAXY
MCG+09-22-065	13 25 42.	+ 51 22 30.	36	15.	GALAXY
MCG+09-22-064	13 25 42.	+ 55 03	72	14.	GALAXY
ZC 1325.7+6643	13 25 42.	+ 66 43	1080		CLUSTER OF GALAXIES
MCG-03-34-081	13 25 42.	- 20 42 30.	30	15.	GALAXY
MCG-05-32-029	13 25 42.	- 27 54	54	15.5	GALAXY
MCG-05-32-028	13 25 42.	- 31 35	48	15.5	GALAXY
REIN 5.069	13 25 42.08	+ 11 09 06.8			NEBULA
REIZ 3425	13 25 44.	+ 03 13	84	15.0	GALAXY
RNGC 5159	13 25 44.	+ 03 15		15.0	GALAXY
IC 0890	13 25 44.	- 15 49 54.			NONSTELLAR OBJECT
RNGC 5160	13 25 45.	+ 06 15			NON-EXISTENT OBJECT
MCG+06-30-012	13 25 45.	+ 34 36	18	15.	GALAXY

OBJECT NAME	RIGHT ASCEN.	DECLINATION	DIAM.	MAGN.	TYPE OF OBJECT
REIN 5.070	13 25 45.34	+ 12 47 28.7			NEBULA
REIZ 3428	13 25 47.	+ 34 35	24	15.0	GALAXY
REIZ 3429	13 25 47.	+ 34 36	18	14.9	GALAXY
LB 02676	13 25 47.	+ 49 47 42.		16.1	FAINT BLUE STAR
LB 02675	13 25 47.	+ 51 51 24.		16.8	FAINT BLUE STAR
8ZW 1325+00.5	13 25 48.	+ 00 33		18.5	COMPACT GALAXY
MCG+06-30-013	13 25 48.	+ 34 34	48	15.	GALAXY
UGC 08461	13 25 48.	+ 34 35	126	16.0	GALAXY PECULR
ZWG 271.042	13 25 48.	+ 55 04		15.7	GALAXY
UGC 08462	13 25 48.	+ 55 04	72	15.7	GALAXY Sa-b
MCG+11-16-019	13 25 48.	+ 63 02	57	16.	GALAXY
7ZW 516	13 25 48.	+ 65 48			COMPACT GALAXY
MCG-05-32-030	13 25 48.	- 27 54	30	15.5	GALAXY
MCG-06-30-004	13 25 48.	- 37 19	60	15.	GALAXY
REIN 5.071	13 25 49.60	+ 12 51 51.7			NEBULA
REIN 5.072	13 25 50.95	+ 11 10 32.7			NEBULA
ZWG 016.072	13 25 54.	+ 00 08		15.6	GALAXY
ZWG 044.089	13 25 54.	+ 07 01		15.5	GALAXY
KARA.73B 0586	13 25 54.	+ 07 01	42	15.5	ISOLATED GALAXY S
ZWG 161.061	13 25 54.	+ 28 56		15.6	GALAXY
ZWG 161.062	13 25 54.	+ 32 17		14.3	GALAXY
REIZ 3431	13 25 54.	+ 32 17	108	13.6	GALAXY
HOLM 519A	13 25 54.	+ 32 17	120	13.7	PART OF MULTIPLE GALAXY
UGC 08463	13 25 54.	+ 32 17	144	14.3	GALAXY Sb
ZWG 218.013	13 25 54.	+ 40 36		15.7	GALAXY
MCG+08-25-003	13 25 54.	+ 46 00	9	17.	GALAXY
REIN 5.074	13 25 54.29	+ 11 42 20.4			NEBULA
HN 1652	13 25 55.4	+ 00 08 55.	18		NEBULA
REIN 5.075	13 25 55.66	+ 11 16 13.4			NEBULA
HOLM 518A	13 25 56.	+ 28 56	48	14.5	PART OF MULTIPLE GALAXY
REIZ 3435	13 25 56.	+ 46 53	60	14.2	GALAXY
LB 00682	13 25 56.	+ 54 46 42.		17.0	FAINT BLUE STAR
REIN 5.076	13 25 56.62	+ 12 02 53.6			NEBULA
HOLM 518B	13 25 57.	+ 28 57	18	15.1	PART OF MULTIPLE GALAXY
RNGC 5166A	13 25 57.	+ 32 17		14.5	GALAXY
PATH 1.619	13 25 57.	+ 58 42	8		NEBULA
REIZ 3432	13 25 59.	+ 34 26	42	15.0	GALAXY
8ZW 1326+01.4	13 26 00.	+ 01 24		18.4	COMPACT GALAXY
ZC 1326.0+1300	13 26 00.	+ 13 00	740		CLUSTER OF GALAXIES
ZWG 101.055	13 26 00.	+ 16 06		15.7	GALAXY
SHAH 019	13 26 00.	+ 16 06	21	16.5	GROUP OF COMPACT GALAXIES
MCG+03-34-039	13 26 00.	+ 16 06	21	15.7	GALAXY
ZWG 101.056	13 26 00.	+ 16 43		14.9	GALAXY
MCG+03-34-040	13 26 00.	+ 16 43	21	14.9	GALAXY
ZC 1326.0+1741	13 26 00.	+ 17 41	1610		CLUSTER OF GALAXIES
MCG+05-32-024	13 26 00.	+ 28 56	21	15.6	GALAXY
MCG+05-32-025	13 26 00.	+ 28 57	30	15.1	GALAXY
UGC 08464	13 26 00.	+ 30 17	72	17.	GALAXY Sc
MCG+05-32-026	13 26 00.	+ 32 18	132	14.3	GALAXY
ZWG 246.002	13 26 00.	+ 46 56		14.7	GALAXY
UGC 08465	13 26 00.	+ 46 56	126	14.7	GALAXY Sc
MCG+09-22-066	13 26 00.	+ 54 42 30.	42	16.	GALAXY
ZWG 271.043	13 26 00.	+ 56 16		15.5	GALAXY
KARA.72 377A	13 26 00.	+ 56 16	42	15.5	PART OF DOUBLE GALAXY
PATH 1.620	13 26 00.	+ 58 42	8		NEBULA
ZWG 316.016	13 26 00.	+ 63 01		14.9	GALAXY
ZWG 336.009	13 26 00.	+ 72 12		15.7	GALAXY
OCL 0901	13 26 00.	- 60 57	240	13.	OPEN STAR CLUSTER
REIN 5.077	13 26 03.78	+ 12 42 08.0			NEBULA
REIN 5.078	13 26 05.76	+ 12 32 38.6			NEBULA
ZC 1326.1+1236	13 26 06.	+ 12 36	3020		CLUSTER OF GALAXIES
ZC 1326.1+2243	13 26 06.	+ 22 43	1810		CLUSTER OF GALAXIES
ZWG 161.063	13 26 06.	+ 31 05		15.5	GALAXY
UGC 08466	13 26 06.	+ 31 05	78	15.5	GALAXY Sb-c
MCG+08-25-004	13 26 06.	+ 46 56	132	14.2	GALAXY
ZWG 271.044	13 26 06.	+ 56 17		15.4	GALAXY
KARA.72 377B	13 26 06.	+ 56 17	42	15.4	PART OF DOUBLE GALAXY
MCG+09-22-067	13 26 06.	+ 56 17	36	15.	GALAXY
UGC 08467	13 26 06.	+ 63 01	60	14.9	GALAXY SBb
ZC 1326.1+6533	13 26 06.	+ 65 33	400		CLUSTER OF GALAXIES
ZWG 016.073	13 26 06.	- 01 47		14.6	GALAXY
MCG+00-34-035	13 26 06.	- 01 47	30	14.6	GALAXY
REIZ 3434	13 26 07.	+ 31 05	30	14.2	GALAXY
RNGC 5169	13 26 08.	+ 46 56		14.5	GALAXY
RRS 1.001	13 26 09.	+ 28 45		18.78	BLUE OBJECT
MCG-03-34-082	13 26 09.	- 17 15	48	15.	GALAXY
RNGC 5165	13 26 10.	+ 11 39		14.5	GALAXY
RNGC 5167	13 26 10.	+ 12 58		14.5	GALAXY
REIZ 3430	13 26 10.	- 01 48	24	14.9	GALAXY
REIN 5.079	13 26 10.71	+ 11 38 41.9			NEBULA
HN 1653	13 26 11.1	- 01 46 52.	30		NEBULA
8ZW 1326+00.8	13 26 12.	+ 00 49		19.6	COMPACT GALAXY
ZWG 072.078	13 26 12.	+ 11 39		14.6	GALAXY
MCG+02-34-016	13 26 12.	+ 11 39	36	14.6	GALAXY
ZWG 072.079	13 26 12.	+ 11 52		15.6	GALAXY
ZWG 072.080	13 26 12.	+ 12 58		14.7	GALAXY
MCG+02-34-017	13 26 12.	+ 12 58	48	14.7	GALAXY
REIZ 3433	13 26 12.	+ 20 35	30	13.7	GALAXY
REIZ 3436	13 26 12.	+ 32 19	24	14.2	GALAXY
ZWG 161.064	13 26 12.	+ 32 20		15.6	GALAXY
MCG+11-17-001	13 26 12.	+ 66 07	39	17.	GALAXY
MCG-07-28-003	13 26 12.	- 41 44 30.	72	14.	GALAXY
REIN 5.080	13 26 12.14	+ 12 58 12.3			NEBULA
HOLM 519B	13 26 13.	+ 32 19	36	14.5	PART OF MULTIPLE GALAXY
RNGC 5155	13 26 13.	- 63 09			NON-EXISTENT OBJECT
REIN 5.081	13 26 13.25	+ 11 51 59.3			NEBULA
REIN 5.083	13 26 14.96	+ 11 20 53.1			NEBULA
HOLM 520B	13 26 15.	+ 11 51	24	14.9	PART OF MULTIPLE GALAXY
MCG+05-32-028	13 26 15.	+ 31 05	48	15.5	GALAXY
RNGC 5166B	13 26 15.	+ 32 19		15.5	GALAXY
MCG+05-32-027	13 26 15.	+ 32 19	30	15.6	GALAXY
HOLM 520A	13 26 16.	+ 11 52	24	14.8	PART OF MULTIPLE GALAXY
SN 1974B	13 26 17.	- 32 53		14.5	SUPERNOVA
RNGC 5161	13 26 17.	- 32 54		12.5	GALAXY
REIN 5.085	13 26 17.15	+ 12 54 16.0			NEBULA
8ZW 1326+10.8	13 26 18.	+ 10 45		17.2	COMPACT GALAXY
ZWG 246.003	13 26 18.	+ 46 51		13.5	GALAXY
UGC 08468	13 26 18.	+ 46 51	72	13.5	GALAXY E?
OCL 0900	13 26 18.	- 63 56	420	10.4	OPEN STAR CLUSTER
HN 1654	13 26 19.2	- 00 04 04.	18		NEBULA
REIN 5.087	13 26 19.79	+ 11 17 26.7			NEBULA
REIZ 3437	13 26 20.	+ 46 50	42	13.2	GALAXY
RNGC 5173	13 26 20.	+ 46 51		13.5	GALAXY
REIN 5.088	13 26 20.25	+ 12 01 50.8			NEBULA
MCG-05-32-031	13 26 21.	- 32 56	318	12.	GALAXY
REIN 5.089	13 26 21.92	+ 10 20 19.7			NEBULA
REIN 5.090	13 26 22.15	+ 11 26 15.2			NEBULA
REIN 5.091	13 26 22.63	+ 11 21 48.6			NEBULA

OBJECT NAME	RIGHT ASCEN.	DECLINATION	DIAM.	MAGN.	TYPE OF OBJECT
REIN 5.093	13 26 23.58	+ 11 21 16.0			NEBULA
ZC 1326.4+1612	13 26 24.	+ 16 12	3230		CLUSTER OF GALAXIES
ZC 1326.4+2710	13 26 24.	+ 27 10	870		CLUSTER OF GALAXIES
1ZW 058	13 26 24.	+ 38 00			COMPACT GALAXY
MCG+07-28-014	13 26 24.	+ 39 14	30	16.	GALAXY
MCG+08-25-005	13 26 24.	+ 46 50	36	13.8	GALAXY
IC 4263	13 26 24.	+ 47 10 57.			NONSTELLAR OBJECT
MCG+09-22-068	13 26 24.	+ 56 18	36	15.	GALAXY
REIN 5.094	13 26 24.02	+ 12 00 47.5			NEBULA
MAI 085	13 26 25.	+ 68 00	53		DWARF SPHEROIDAL GALAXY
REIN 5.095	13 26 25.67	+ 10 17 43.6			NEBULA
REIN 5.096	13 26 26.22	+ 10 21 48.0			NEBULA
KEEL 667	13 26 26.4	+ 47 11 09.			NEBULA
REIN 5.097	13 26 27.07	+ 11 08 10.9			NEBULA
REIN 5.098	13 26 27.07	+ 11 47 20.2			NEBULA
REIN 5.099	13 26 27.99	+ 10 17 27.2			NEBULA
REIN 5.100	13 26 29.21	+ 12 27 32.8			NEBULA
ZWG 016.074	13 26 30.	+ 00 12		15.7	GALAXY
ZWG 072.081	13 26 30.	+ 10 18		15.6	GALAXY
UGC 08469	13 26 30.	+ 10 18	60	15.6	GALAXY Sc-IRR
ZWG 072.082	13 26 30.	+ 12 28		15.6	GALAXY
ZC 1326.5+2731	13 26 30.	+ 27 31	670		CLUSTER OF GALAXIES
ZC 1326.5+2812	13 26 30.	+ 28 12	1210		CLUSTER OF GALAXIES
MCG+08-25-006	13 26 30.	+ 46 45	60	16.	GALAXY
MCG+08-25-007	13 26 30.	+ 47 11	120	15.	GALAXY
ZWG 246.004	13 26 30.	+ 47 12		15.4	GALAXY
UGC 08470	13 26 30.	+ 47 12	114	15.4	GALAXY SBc
ZC 1326.5+5454	13 26 30.	+ 54 54	2760		CLUSTER OF GALAXIES
HN 1655	13 26 30.4	+ 00 12 32.	18		NEBULA
LB 02677	13 26 31.	+ 53 43 42.		16.5	FAINT BLUE STAR
LB 02678	13 26 32.	+ 56 50 42.		16.5	FAINT BLUE STAR
MCG+07-28-015	13 26 33.	+ 39 00	39	15.	GALAXY
REIN 5.104	13 26 34.43	+ 11 45 26.7			NEBULA
REIN 5.103	13 26 34.66	+ 10 51 34.1			NEBULA
REIN 5.105	13 26 35.96	+ 11 29 26.2			NEBULA
MCG+00-34-036	13 26 36.	+ 00 58	42	15.7	GALAXY
ZWG 072.083	13 26 36.	+ 12 30		15.7	GALAXY
MRK 660	13 26 36.	+ 22 34	9	16.5	GALAXY WITH UV CONTINUUM
ZC 1326.6+3750	13 26 36.	+ 37 50	2420		CLUSTER OF GALAXIES
ZWG 218.014	13 26 36.	+ 38 50		15.0	GALAXY
UGC 08471	13 26 36.	+ 38 50	78	15.0	GALAXY S
MCG+07-28-016	13 26 36.	+ 38 50	66	14.5	GALAXY
MRK 258	13 26 36.	+ 53 42	9	16.	GALAXY WITH UV CONTINUUM
MCG+10-19-076	13 26 36.	+ 59 21	30	16.	GALAXY
PATH 1.621	13 26 36.	+ 59 21	14		NEBULA
MCG-02-34-056	13 26 36.	- 10 17	42	15.	GALAXY
MCG-06-30-005	13 26 36.	- 34 00	120	14.5	GALAXY
REIN 5.106	13 26 36.41	+ 12 29 48.6			NEBULA
RES 1.002	13 26 37.	+ 30 50		17.35	BLUE OBJECT
REIN 5.107	13 26 38.83	+ 12 37 45.2			NEBULA
RRS 1.003	13 26 39.	+ 31 24		17.53	BLUE OBJECT
MCG+07-28-017	13 26 39.	+ 38 50	24	16.	GALAXY
RNGC 5162	13 26 40.	+ 11 32		15.5	GALAXY
IC 4260	13 26 40.	- 28 01			NONSTELLAR OBJECT
HN 1054	13 26 41.	- 28 02		14.5	NEBULA
IC 4259	13 26 41.	- 29 52			NONSTELLAR OBJECT
ZWG 016.075	13 26 42.	+ 00 58		15.7	GALAXY
ZC 1326.7+1000	13 26 42.	+ 10 00	2350		CLUSTER OF GALAXIES
ZWG 072.084	13 26 42.	+ 11 32		15.3	GALAXY
UGC 08472	13 26 42.	+ 11 32	60	15.3	GALAXY S0
ZWG 072.085	13 26 42.	+ 12 12		15.7	GALAXY
ZC 1326.7+1938	13 26 42.	+ 19 38	2020		CLUSTER OF GALAXIES
ZWG 246.005	13 26 42.	+ 46 31		15.1	GALAXY
ZC 1326.7+5752	13 26 42.	+ 57 52	1080		CLUSTER OF GALAXIES
8ZW 1326-02.9	13 26 42.	- 02 53		19.5	COMPACT GALAXY
MCG+00-34-037	13 26 42.	- 03 28	24	14.5	GALAXY
HN 1055	13 26 42.	- 29 53		14.	NEBULA
REIN 5.109	13 26 43.49	+ 11 39 17.2			NEBULA
REIN 5.110	13 26 44.13	+ 10 21 55.1			NEBULA
MCG-02-34-057	13 26 45.	- 12 14	54	15.	GALAXY
SC 1323-3002.3	13 26 45.	- 30 17 52.	18		NEBULA
REIN 5.111	13 26 45.16	+ 11 32 00.0			NEBULA
REIN 5.112	13 26 45.17	+ 12 11 40.8			NEBULA
REIN 5.113	13 26 46.90	+ 11 27 38.6			NEBULA
ZWG 072.086	13 26 48.	+ 11 28		15.4	GALAXY
MCG+03-34-041	13 26 48.	+ 17 17	210	12.7	GALAXY
ZWG 161.065	13 26 48.	+ 28 18		15.7	GALAXY
MCG+08-25-008	13 26 48.	+ 45 59	72	16.	GALAXY
MCG+08-25-009	13 26 48.	+ 46 30 30.	30	16.	GALAXY
ZWG 016.076	13 26 48.	- 03 25		14.5	GALAXY
REIZ 3438	13 26 50.	+ 17 18	66	14.2	GALAXY
IC 4266	13 26 51.	+ 37 52 46.			NONSTELLAR OBJECT
LB 02679	13 26 51.	+ 49 58 06.		15.4	FAINT BLUE STAR
HN 1656	13 26 51.6	- 00 08 15.	36		NEBULA
RNGC 5175	13 26 52.	+ 11 16		13.5	GALAXY
RNGC 5174	13 26 52.	+ 11 16		13.5	GALAXY
RNGC 5171	13 26 52.	+ 12 00		14.5	GALAXY
RNGC 5176	13 26 52.	+ 12 03		15.5	GALAXY
RNGC 5177	13 26 52.	+ 12 04		15.5	GALAXY
REIN 5.114	13 26 52.11	+ 12 10 25.9			NEBULA
REIN 5.115	13 26 52.34	+ 11 59 52.9			NEBULA
REIN 5.116	13 26 52.67	+ 11 35 38.8			NEBULA
RNGC 5172	13 26 53.	+ 17 19		12.5	GALAXY
REIN 5.117	13 26 53.24	+ 11 59 34.7			NEBULA
REIN 5.118	13 26 53.73	+ 12 07 29.0			NEBULA
KARA.68 223	13 26 54.	+ 01 09	60		DWARF GALAXY
UGC 08474	13 26 54.	+ 01 10	60	18.	GALAXY DWARF
ZWG 072.087	13 26 54.	+ 11 16		13.7	GALAXY
UGC 08475	13 26 54.	+ 11 16	222	13.7	GALAXY Sc
MCG+02-34-018	13 26 54.	+ 11 16	168	13.7	GALAXY
ZWG 072.088	13 26 54.	+ 11 55		15.7	GALAXY
ZWG 072.089	13 26 54.	+ 12 00		14.7	GALAXY
UGC 08476	13 26 54.	+ 12 00	66	14.7	GALAXY E-S0
MKW 11	13 26 54.	+ 12 00		14.7	POOR GALAXY CLUSTER
MCG+02-34-020	13 26 54.	+ 12 00	15	14.7	GALAXY
ZWG 072.090	13 26 54.	+ 12 03		15.4	GALAXY
MCG+02-34-021	13 26 54.	+ 12 03	12	15.4	GALAXY
ZWG 072.091	13 26 54.	+ 12 04		15.4	GALAXY
MCG+02-34-019	13 26 54.	+ 12 04	36	15.4	GALAXY
ZWG 072.092	13 26 54.	+ 14 09		15.6	GALAXY
MCG+03-34-042	13 26 54.	+ 17 03	90	14.3	GALAXY
ZWG 101.057	13 26 54.	+ 17 19		12.7	GALAXY
UGC 08477	13 26 54.	+ 17 19	198	12.7	GALAXY Sc
TON-N 0714	13 26 54.	+ 23 05		14.6	BLUE STAR
MRK 259	13 26 54.	+ 44 09	9	16.5	GALAXY WITH UV CONTINUUM
UGC 08473	13 26 54.	- 00 08	66	17.	GALAXY DWARF
REIN 5.119	13 26 54.57	+ 11 49 31.8			NEBULA
REIN 5.120	13 26 55.42	+ 10 15 51.7			NEBULA

OBJECT NAME	RIGHT ASCEN.	DECLINATION	DIAM.	MAGN.	TYPE OF OBJECT
REIN 5.121	13 26 55.94	+ 10 54 15.1			NEBULA
REIN 5.123	13 26 55.95	+ 12 03 17.2			NEBULA
REIZ 3440	13 26 56.	+ 17 04	48	14.4	GALAXY
RRS 1.004	13 26 56.	+ 28 39		18.37	BLUE OBJECT
REIN 5.122	13 26 56.21	+ 10 51 49.9			NEBULA
REIN 5.124	13 26 56.56	+ 12 02 21.7			NEBULA
HOLM 521B	13 26 57.	+ 12 03	42	14.5	PART OF MULTIPLE GALAXY
MCG+07-28-018	13 26 57.	+ 41 34	30	15.5	GALAXY
REIN 5.125	13 26 57.11	+ 11 15 55.7			NEBULA
REIN 5.126	13 26 57.49	+ 11 15 09.7			NEBULA
REIN 5.127	13 26 57.55	+ 12 10 05.0			NEBULA
RNGC 5178	13 26 58.	+ 11 53		15.0	GALAXY
RNGC 5179	13 26 58.	+ 12 01		15.0	GALAXY
HOLM 521A	13 26 58.	+ 12 02	36	14.5	PART OF MULTIPLE GALAXY
IC 4268	13 26 58.	+ 37 55 40.			NONSTELLAR OBJECT
RNGC 5180	13 26 59.	+ 17 05		14.5	GALAXY
REIN 5.128	13 26 59.95	+ 12 01 52.6			NEBULA
ZWG 072.093	13 27 00.	+ 11 53		15.0	GALAXY
UGC 08478	13 27 00.	+ 11 53	78	15.0	GALAXY S0-a
MCG+02-34-022	13 27 00.	+ 11 53	36	15.0	GALAXY
ZWG 072.094	13 27 00.	+ 12 01		14.9	GALAXY
MCG+02-34-023	13 27 00.	+ 12 01	36	14.9	GALAXY
ZWG 072.095	13 27 00.	+ 12 10		15.6	GALAXY
ZWG 101.058	13 27 00.	+ 17 05		14.3	GALAXY
UGC 08479	13 27 00.	+ 17 05	102	14.3	GALAXY S0?
MCG+07-28-019	13 27 00.	+ 41 33	36	15.5	GALAXY
ZWG 218.015	13 27 00.	+ 42 00		15.7	GALAXY
MCG+07-28-020	13 27 00.	+ 42 00	24	15.	GALAXY
ZC 1327.0+6958	13 27 00.	+ 69 58	1480		CLUSTER OF GALAXIES
MCG-03-34-083	13 27 00.	- 20 56 30.	36	15.	GALAXY
SC 1324-3047.0	13 27 00.	- 31 02 33.	18		NEBULA
REIN 5.132	13 27 00.89	+ 11 52 56.3			NEBULA
REIN 5.129	13 27 00.98	+ 10 38 10.9			NEBULA
REIN 5.130	13 27 01.27	+ 10 48 20.7			NEBULA
REIN 5.133	13 27 01.33	+ 11 57 31.0			NEBULA
REIN 5.131	13 27 01.36	+ 10 38 56.0			NEBULA
REIN 5.134	13 27 02.50	+ 12 00 12.6			NEBULA
BC B21327+31	13 27 02.8	+ 31 20 32.3		19.6	QUASI-STELLAR OBJECT
RNGC 5170	13 27 03.	- 17 42		12.0	GALAXY
HOLM 522A	13 27 04.	+ 11 52	42	14.5	PART OF MULTIPLE GALAXY
SC 1324+4600.4	13 27 04.	+ 45 44 51.	72		NEBULA
HN 1056	13 27 04.	- 27 46		14.	NEBULA
IC 4261	13 27 04.	- 27 46			NONSTELLAR OBJECT
REIN 5.136	13 27 04.03	+ 12 26 50.7			NEBULA
REIN 5.137	13 27 04.22	+ 12 48 26.2			NEBULA
REIN 5.135	13 27 04.71	+ 10 38 26.1			NEBULA
SC 1324-3049.1	13 27 05.	- 31 04 39.	18		NEBULA
REIN 5.139	13 27 05.83	+ 11 53 38.3			NEBULA
ZWG 072.096	13 27 06.	+ 11 58		15.7	GALAXY
ZC 1327.1+2259	13 27 06.	+ 22 59	670		CLUSTER OF GALAXIES
ZC 1327.1+3134	13 27 06.	+ 31 34	470		CLUSTER OF GALAXIES
ZWG 190.012	13 27 06.	+ 37 40		15.6	GALAXY
IC 4269	13 27 06.	+ 37 53 23.			NONSTELLAR OBJECT
ZWG 246.006	13 27 06.	+ 46 46		15.7	GALAXY
MCG+10-19-077	13 27 06.	+ 62 10	54	15.	GALAXY
MCG+00-34-038	13 27 06.	- 01 12	30	14.6	GALAXY
MCG-02-34-058	13 27 06.	- 10 38	72	15.	GALAXY
MCG-05-32-032	13 27 06.	- 27 43	12	15.5	GALAXY
REIN 5.140	13 27 07.15	+ 11 57 19.4			NEBULA
REIZ 3439	13 27 08.	- 09 51	30	14.8	GALAXY
REIN 5.141	13 27 08.05	+ 12 04 19.0			NEBULA
HOLM 522B	13 27 09.	+ 11 53	30	15.2	PART OF MULTIPLE GALAXY
FATH 1.622	13 27 09.	+ 59 17	5		NEBULA
FATH 1.623	13 27 09.	+ 59 33	14		NEBULA
MCG-03-34-084	13 27 09.	- 17 43	540	12.	GALAXY
REIN 5.142	13 27 09.10	+ 10 49 48.8			NEBULA
RNGC 5181	13 27 10.	+ 13 34		14.5	GALAXY
IC 4271	13 27 10.	+ 37 40 05.			NONSTELLAR OBJECT
ZWG 072.097	13 27 12.	+ 10 11		15.7	GALAXY
ZWG 072.098	13 27 12.	+ 13 34		14.7	GALAXY
MCG+02-34-024	13 27 12.	+ 13 34	15	14.7	GALAXY
REIZ 3443	13 27 12.	+ 32 38	24	14.0	GALAXY
VV 355B	13 27 12.	+ 37 40	18	16.	INTERACTING GALAXY
VV 355A	13 27 12.	+ 37 40	36	15.	INTERACTING GALAXY
ARP 040	13 27 12.	+ 37 40			PECULIAR GALAXY
MCG+06-30-015	13 27 12.	+ 37 41	42	15.	GALAXY
MCG+06-30-014	13 27 12.	+ 37 54	18	15.	GALAXY
MCG+08-25-010	13 27 12.	+ 46 45	36	16.	GALAXY
LB 02680	13 27 12.	+ 52 42 42.		15.1	FAINT BLUE STAR
ZWG 271.045	13 27 12.	+ 53 20		14.3	GALAXY
RNGC 5201	13 27 12.	+ 53 20		14.5	GALAXY
UGC 08480	13 27 12.	+ 53 20	108	14.3	GALAXY S
MCG+09-22-069	13 27 12.	+ 53 20	90	14.5	GALAXY
ZWG 016.077	13 27 12.	- 00 02		15.6	GALAXY
MCG-02-34-059	13 27 12.	- 10 31 30.	60	15.	GALAXY
MCG-03-34-085	13 27 12.	- 20 27	36	15.	GALAXY
REIN 5.144	13 27 12.30	+ 12 46 36.1			NEBULA
REIN 5.145	13 27 12.80	+ 11 38 42.5			NEBULA
REIN 5.147	13 27 13.18	+ 11 25 34.6			NEBULA
REIN 5.149	13 27 13.27	+ 12 33 43.6			NEBULA
REIN 5.150	13 27 13.32	+ 12 34 18.8			NEBULA
REIN 5.152	13 27 14.58	+ 12 26 53.2			NEBULA
MCG+06-30-016	13 27 15.	+ 32 38	66	14.5	GALAXY
SC 1324-3053.9	13 27 15.	- 31 09 27.	60		NEBULA
REIN 5.151	13 27 15.31	+ 10 11 27.5			NEBULA
REIN 5.154	13 27 15.68	+ 10 16 31.6			NEBULA
REIN 5.155	13 27 15.89	+ 11 00 00.4			NEBULA
HN 1657	13 27 15.9	- 01 10 03.	18		NEBULA
REIZ 3446	13 27 16.	+ 34 52	18	14.8	GALAXY
ABC 1749	13 27 16.	+ 37 53		16.0	RICH CLUSTER OF GALAXIES
REIZ 3442	13 27 16.	- 01 11	24	14.4	GALAXY
HN 1658	13 27 16.0	- 00 02 15.	48		NEBULA
REIN 5.156	13 27 16.30	+ 12 58 22.9			NEBULA
ZWG 072.099	13 27 18.	+ 10 17		15.6	GALAXY
ZC 1327.3+1145	13 27 18.	+ 11 45	7060		CLUSTER OF GALAXIES
ZWG 072.100	13 27 18.	+ 12 27		15.7	GALAXY
ZWG 072.101	13 27 18.	+ 12 37		15.6	GALAXY
ZWG 101.059	13 27 18.	+ 17 16		15.7	GALAXY
ZWG 161.066	13 27 18.	+ 26 40		15.7	GALAXY
UGC 08482	13 27 18.	+ 26 40	60	15.7	GALAXY S
TON-N 0163	13 27 18.	+ 26 51		15.1	BLUE STAR
ZWG 161.067	13 27 18.	+ 30 02		15.5	GALAXY
UGC 08483	13 27 18.	+ 30 02	60	15.5	GALAXY S
ZWG 190.013	13 27 18.	+ 32 40		15.1	GALAXY
UGC 08484	13 27 18.	+ 32 40	66	15.1	GALAXY SBb
VV 325A	13 27 18.	+ 33 36	24	16.	INTERACTING GALAXY
MCG+09-22-070	13 27 18.	+ 53 30	36	16.	GALAXY
ZWG 294.038	13 27 18.	+ 62 10		15.5	GALAXY

OBJECT NAME	RIGHT ASCEN.	DECLINATION	DIAM.	MAGN.	TYPE OF OBJECT
UGC 08481	13 27 18.	- 00 02	60	16.5	GALAXY Sc
ZWG 016.078	13 27 18.	- 01 10		14.6	GALAXY
MCG-03-34-086	13 27 18.	- 20 19	60	15.	GALAXY
OCL 0904	13 27 18.	- 61 01	690	9.6	OPEN STAR CLUSTER
REIN 5.157	13 27 18.64	+ 10 56 27.2			NEBULA
REIZ 3450	13 27 19.	+ 53 19	48	14.5	GALAXY
REIN 5.159	13 27 19.46	+ 11 58 00.9			NEBULA
REIN 5.158	13 27 19.80	+ 10 11 43.2			NEBULA
REIZ 3441	13 27 20.	- 09 51	9	15.4	GALAXY
MCG+06-30-017	13 27 21.	+ 34 51 30.	30	15.	GALAXY
REIN 5.161	13 27 22.37	+ 11 33 37.5			NEBULA
REIN 5.162	13 27 22.54	+ 12 36 13.9			NEBULA
LB 00683	13 27 23.	+ 54 05 24.		16.3	FAINT BLUE STAR
SHB 264	13 27 23.2	- 21 26 34.		16.7	QUASI-STELLAR OBJECT
BC PKS1327-21	13 27 23.36	- 21 26 33.8		16.74	QUASI-STELLAR OBJECT
REIN 5.163	13 27 23.45	+ 11 31 55.5			NEBULA
MCG+00-34-040	13 27 24.	+ 00 33	15	14.7	GALAXY
ABC 1748	13 27 24.	+ 18 40		17.4	RICH CLUSTER OF GALAXIES
REIZ 3448	13 27 24.	+ 32 13	18	14.6	GALAXY
ZWG 161.068	13 27 24.	+ 32 15		15.7	GALAXY
VV 325B	13 27 24.	+ 33 37	54	16.	INTERACTING GALAXY
ZWG 190.014	13 27 24.	+ 34 52		15.5	GALAXY
MRK 260	13 27 24.	+ 45 16	7	17.	GALAXY WITH UV CONTINUUM
MCG+00-34-039	13 27 24.	- 01 30	108	13.6	GALAXY
REIN 5.164	13 27 25.93	+ 10 22 20.4			NEBULA
FATH 1.624	13 27 26.	+ 59 18	8		NEBULA
FATH 1.625	13 27 26.	+ 59 19	14		NEBULA
RS 01	13 27 26.1	+ 26 50 09.		15.59	BLUE STELLAR OBJECT
HOLM 524B	13 27 27.	+ 33 35	30	15.0	PART OF MULTIPLE GALAXY
MCG+06-30-019	13 27 27.	+ 33 35	12	16.	GALAXY
MCG+06-30-018	13 27 27.	+ 37 10	24	15.	GALAXY
REIN 5.165	13 27 27.91	+ 11 19 44.1			NEBULA
IC 0891	13 27 28.	+ 00 32 53.			NONSTELLAR OBJECT
REIN 5.166	13 27 28.05	+ 12 34 45.0			NEBULA
HOLM 524A	13 27 29.	+ 33 36	36	15.0	PART OF MULTIPLE GALAXY
HN 1057	13 27 29.	- 28 02	6		NEBULA
IC 4262	13 27 29.	- 28 02			NONSTELLAR OBJECT
REIN 5.167	13 27 29.62	+ 12 35 06.0			NEBULA
ZWG 016.080	13 27 30.	+ 00 33		14.7	GALAXY
ZWG 072.102	13 27 30.	+ 11 20		15.6	GALAXY
UGC 08486	13 27 30.	+ 11 20	72	15.6	GALAXY Sc
ZWG 161.069	13 27 30.	+ 31 23		14.6	GALAXY
ZWG 190.015	13 27 30.	+ 33 36		15.4	GALAXY
MCG+06-30-020	13 27 30.	+ 33 36	42	16.	GALAXY
MCG+09-22-071	13 27 30.	+ 55 59	18	17.	GALAXY
MCG+00-34-041	13 27 30.	- 01 25	108	13.7	GALAXY
ZWG 016.079	13 27 30.	- 01 28		13.6	GALAXY
UGC 08485	13 27 30.	- 01 28	132	13.6	GALAXY S
KARA.72 378A	13 27 30.	- 01 28	108	13.6	PART OF DOUBLE GALAXY
MCG-04-32-022	13 27 30.	- 23 53	42	15.	GALAXY
REIN 5.168	13 27 30.81	+ 12 31 02.8			NEBULA
REIZ 3449	13 27 31.	+ 31 23	18	13.5	GALAXY
RNGC 5183	13 27 31.	- 01 28		13.5	GALAXY
LB 02681	13 27 32.	+ 56 52 30.		16.8	FAINT BLUE STAR
REIN 5.169	13 27 32.12	+ 11 08 29.1			NEBULA
RNGC 5187	13 27 33.	+ 31 23		14.5	GALAXY
RNGC 5186	13 27 34.	+ 12 26		15.5	GALAXY
RNGC 5185	13 27 34.	+ 13 41		14.5	GALAXY
FATH 1.626	13 27 34.	+ 59 17	5		NEBULA
REIZ 3445	13 27 34.	- 01 29	90	13.7	GALAXY
HOLM 523B	13 27 34.	- 01 29	78	13.6	PART OF MULTIPLE GALAXY
HN 1058	13 27 34.	- 27 41		14.5	NEBULA
IC 4264	13 27 34.	- 27 41			NONSTELLAR OBJECT
HN 1659	13 27 34.8	- 01 50 50.	18		NEBULA
REIN 5.172	13 27 35.55	+ 12 25 58.0			NEBULA
ZWG 016.082	13 27 36.	+ 00 46		15.5	GALAXY
ZWG 072.103	13 27 36.	+ 12 26		15.6	GALAXY
ZWG 072.104	13 27 36.	+ 13 41		14.7	GALAXY
MCG+02-34-025	13 27 36.	+ 13 41	108	14.7	GALAXY
MCG+05-32-029	13 27 36.	+ 31 23 30.	90	14.6	GALAXY
UGC 08488	13 27 36.	+ 31 41	114	14.7	GALAXY Sb
ZWG 246.007	13 27 36.	+ 45 39		15.2	GALAXY
UGC 08489	13 27 36.	+ 45 39	138	15.2	GALAXY
MCG+08-25-011	13 27 36.	+ 45 39	120	15.	GALAXY
LB 02682	13 27 36.	+ 52 14 42.		17.2	FAINT BLUE STAR
ZWG 016.081	13 27 36.	- 01 25		13.7	GALAXY
UGC 08487	13 27 36.	- 01 25	132	13.7	GALAXY Sb
KARA.72 378B	13 27 36.	- 01 25	126	13.7	PART OF DOUBLE GALAXY
ZC 1327.6-0134	13 27 36.	- 01 34	3230		CLUSTER OF GALAXIES
BZW 1327-01.9	13 27 36.	- 01 52		19.2	COMPACT GALAXY
MCG-01-34-016	13 27 36.	- 05 07	42	15.	GALAXY
MCG-02-34-060	13 27 36.	- 09 48	72	14.	GALAXY
KARA.68 224	13 27 36.	- 17 40	34		DWARF GALAXY
MCG-05-32-033	13 27 36.	- 27 38	60	15.	GALAXY
REIN 5.171	13 27 36.32	+ 10 11 31.3			NEBULA
RNGC 5184	13 27 37.	- 01 25		13.5	GALAXY
REIZ 3444	13 27 38.	- 09 48	60	13.9	GALAXY
MCG-04-32-023	13 27 39.	- 21 31	36	15.5	GALAXY
HN 1059	13 27 39.	- 25 31		14.	NEBULA
IC 4265	13 27 39.	- 25 31			NONSTELLAR OBJECT
REIN 5.173	13 27 39.93	+ 11 13 42.2			NEBULA
RNGC 5204	13 27 40.	+ 58 41		12.0	GALAXY
REIZ 3447	13 27 40.	- 01 25	78	13.5	GALAXY
HOLM 523A	13 27 40.	- 01 25	108	13.3	PART OF MULTIPLE GALAXY
REIN 5.174	13 27 40.92	+ 11 29 13.4			NEBULA
REIN 5.175	13 27 41.37	+ 12 34 35.5			NEBULA
ZC 1327.7+1355	13 27 42.	+ 13 55	400		CLUSTER OF GALAXIES
ZC 1327.7+2544	13 27 42.	+ 25 44	1140		CLUSTER OF GALAXIES
MCG+07-28-021	13 27 42.	+ 38 39	42	14.5	GALAXY
ZWG 218.016	13 27 42.	+ 38 40		15.7	GALAXY
ZWG 294.039	13 27 42.	+ 58 40		11.7	GALAXY
UGC 08490	13 27 42.	+ 58 40	318	11.7	GALAXY
MCG+10-19-078	13 27 42.	+ 58 40	270	11.6	GALAXY
FATH 1.627	13 27 42.	+ 58 41	95		NEBULA
UGC 08491	13 27 42.	+ 63 10	66	16.5	GALAXY IRR
REIZ 3451	13 27 43.	+ 31 16	18	14.2	GALAXY
REIZ 3452	13 27 43.	+ 47 26	450	10.5	GALAXY
HOLM 526A	13 27 43.	+ 47 27	480	10.4	PART OF MULTIPLE GALAXY
SC 1324-3103.9	13 27 43.	- 31 19 26.	54		NEBULA
REIN 5.176	13 27 43.08	+ 11 04 24.6			NEBULA
REIN 5.177	13 27 43.30	+ 10 57 33.7			NEBULA
REIN 5.178	13 27 44.15	+ 10 48 50.2			NEBULA
REIN 2.192A	13 27 44.36	+ 58 40 35.3			NEBULA
REIN 2.192B	13 27 44.37	+ 58 40 40.1			NEBULA
RRS 1.005	13 27 45.	+ 30 49		18.72	BLUE OBJECT
ARP 085	13 27 45.	+ 47 27			PECULIAR GALAXY
REIZ 3460	13 27 45.	+ 59 33	48	14.7	GALAXY
REIN 5.181	13 27 47.28	+ 10 26 28.8			NEBULA

OPJPCT NAME	RIGHT ASCEN.	DECLINATION	DIAM.	MAGN.	TYPE OF OBJECT
ZWG 044.090	13 27 48.	+ 07 45		15.3	GALAXY
MCG+01-34-023	13 27 48.	+ 07 45	60	15.3	GALAXY
ZWG 073.001	13 27 48.	+ 11 47		15.5	GALAXY
ZWG 072.105	13 27 48.	+ 11 47		15.5	GALAXY
ZC 1327.8+1839	13 27 48.	+ 18 39	1210		CLUSTER OF GALAXIES
ZC 1327.8+2620	13 27 48.	+ 26 20	1880		CLUSTER OF GALAXIFS
ZWG 161.070	13 27 48.	+ 31 39		15.0	GALAXY
UGC 08492	13 27 48.	+ 31 39	60	15.0	GALAXY COMPACT
MCG+05-32-030	13 27 48.	+ 31 40	36	15.0	GALAXY
ZWG 218.017	13 27 48.	+ 41 53		15.6	GALAXY
MCG+07-28-022	13 27 48.	+ 41 54	48	15.	GALAXY
ZWG 246.008	13 27 48.	+ 47 27		8.8	GALAXY
UGC 08493	13 27 48.	+ 47 27	540	8.8	GALAXY Sc
VV 001A	13 27 48.	+ 47 27	600	8.6	INTERACTING GALAXY
KARA.72 379A	13 27 48.	+ 47 27	546	8.8	PART OF DOUBLE GALAXY
MCG+08-25-012	13 27 48.	+ 47 27	540	8.8	GALAXY
MCG+08-25-013	13 27 48.	+ 49 06	60	17.	GALAXY
MCG+09-22-073	13 27 48.	+ 51 07	54	15.	GALAXY
ZWG 271.046	13 27 48.	+ 55 52		15.2	GALAXY
MCG+09-22-074	13 27 48.	+ 55 52	54	15.	GALAXY
MCG+09-22-072	13 27 48.	+ 55 56	18	17.	GALAXY
MCG+10-19-079	13 27 48.	+ 58 00	15	15.	GALAXY
ZWG 016.083	13 27 48.	- 02 39		15.3	GALAXY
REIN 5.182	13 27 48.21	+ 11 34 03.9			NEBULA
HN 1660	13 27 48.8	- 02 39 02.	18		NEBULA
REIZ 3456	13 27 49.	+ 47 31	90	12.0	GALAXY
HOLM 526B	13 27 49.	+ 47 31	126	12.2	PART OF MULTIPLE GALAXY
RRS 1.006	13 27 50.	+ 29 21		16.04	BLUE OBJECT
REIN 5.183	13 27 50.07	+ 10 36 13.9			NEBULA
REIN 5.185	13 27 50.60	+ 11 46 09.2			NEBULA
RNGC 5194	13 27 51.	+ 47 27		10.0	GALAXY
MCG+11-17-002	13 27 51.	+ 63 06	90	16.	GALAXY
HOLM 525B	13 27 51.	- 08 21	60	15.0	PART OF MULTIPLE GALAXY
MCG-04-32-024	13 27 51.	- 26 01	72	15.	GALAXY
HN 1060	13 27 52.	- 26 01	42		NEBULA
IC 4267	13 27 52.	- 26 01			NONSTELLAR OBJFCT
REIN 5.186	13 27 53.08	+ 11 46 17.6			NEBULA
REIN 5.187	13 27 53.58	+ 11 45 41.8			NEBULA
ZC 1327.9+1856	13 27 54.	+ 18 56	670		CLUSTER OF GALAXIES
ZC 1327.9+3059	13 27 54.	+ 30 59	740		CLUSTER OF GALAXIES
MCG+05-32-031	13 27 54.	+ 31 37	72	14.9	GALAXY
ZWG 246.009	13 27 54.	+ 47 31		10.6	GALAXY
UGC 08494	13 27 54.	+ 47 31	480	10.6	GALAXY
VV 001B	13 27 54.	+ 47 31	300	10.7	INTERACTING GALAXY
SN 1945A	13 27 54.	+ 47 31		14.0	SUPERNOVA
KARA.72 379B	13 27 54.	+ 47 31	282	10.6	PART OF DOUBLF GALAXY
MCG+08-25-014	13 27 54.	+ 47 32	270	10.4	GALAXY
ZWG 271.047	13 27 54.	+ 51 10		15.7	GALAXY
UGC 08495	13 27 54.	+ 51 10	60	15.7	GALAXY
KARA.73B 0587	13 27 54.	+ 51 10	48	15.7	ISOLATED GALAXY S
LB 02683	13 27 54.	+ 52 51 48.		16.8	FAINT BLUE STAR
ZC 1327.9+6615	13 27 54.	+ 66 15	2080		CLUSTER OF GALAXIES
ZWG 016.084	13 27 54.	- 00 21		15.6	GALAXY
8ZW 1327-00.3	13 27 54.	- 00 21		15.6	COMPACT GALAXY
MCG-04-32-025	13 27 54.	- 22 14	36	15.	GALAXY
RNGC 5182	13 27 54.	- 27 52		13.0	GALAXY
MCG-05-32-034	13 27 54.	- 27 52	84	13.5	GALAXY
OCL 0905	13 27 54.	- 60 41	600	11.5	OPEN STAR CLUSTER
VHA 147	13 27 54.	- 60 41	240		OPEN STAR CLUSTER
HOLM 525A	13 27 55.	- 08 20	66	14.5	PART OF MULTIPLE GALAXY
HN 1661	13 27 56.1	- 00 21 13.	12		NEBULA
REIN 5.188	13 27 56.64	+ 10 45 59.1			NEBULA
RNGC 5195	13 27 57.	+ 47 32		11.5	GALAXY
REIZ 3469	13 27 57.	+ 59 36	18	14.9	GALAXY
RNGC 5168	13 27 57.	- 60 41		11.5	OPEN CLUSTER
REIN 5.189	13 27 57.29	+ 10 08 47.6			NEBULA
REIN 5.190	13 27 58.02	+ 11 56 25.6			NEBULA
SC 1325-3110.1	13 27 59.	- 31 25 38.	36		NEBULA
REIN 5.191	13 27 59.74	+ 10 31 30.1			NEBULA
VDB.66G 176	13 28	+ 45 35	100		DWARF GALAXY
ZC 1328.0+3026	13 28 00.	+ 30 26	540		CLUSTER OF GALAXIES
ZWG 161.071	13 28 00.	+ 31 35		14.9	GALAXY
REIZ 3455	13 28 00.	+ 31 35	30	14.0	GALAXY
UGC 08496	13 28 00.	+ 31 35	78	14.9	GALAXY MLT SYS
ZC 1328.0+6031	13 28 00.	+ 60 31	2760		CLUSTER OF GALAXIES
ZC 1328.0+6156	13 28 00.	+ 61 56	1010		CLUSTER OF GALAXIES
MCG+13-10-006	13 28 00.	+ 78 45 30.	9	17.	GALAXY
MCG-01-34-017	13 28 00.	- 05 49	24	15.	GALAXY
MCG-04-32-026	13 28 00.	- 22 08 30.	72	14.	GALAXY
REIZ 3454	13 28 01.	+ 30 16	42	14.8	GALAXY
REIN 5.192	13 28 01.69	+ 11 01 45.8			NEBULA
SC 1325-3138.3	13 28 02.	- 31 53 50.	12		NEBULA
VV 069	13 28 03.	+ 31 35 30.	42	15.	INTERACTING GALAXY
HN 1061	13 28 03.	- 25 05		14.	NEBULA
MCG-04-32-027	13 28 03.	- 25 05	48	14.	GALAXY
IC 4270	13 28 03.	- 25 05			NONSTELLAR OBJFCT
REIN 5.193	13 28 03.17	+ 10 48 29.9			NEBULA
REIN 5.194	13 28 03.17	+ 12 46 36.7			NEBULA
RS 02	13 28 04.4	+ 25 54 51.		16.43	BLUE STELLAR OBJECT
REIN 5.195	13 28 04.40	+ 10 56 19.4			NEBULA
RRS 1.007	13 28 05.	+ 30 24		17.81	BLUE OBJECT
REIN 2.193	13 28 05.53	+ 46 55 41.7			NEBULA
ZWG 045.001	13 28 06.	+ 07 48		15.4	GALAXY
ZC 1328.1+1148	13 28 06.	+ 11 48	540		CLUSTER OF GALAXIES
ZWG 161.072	13 28 06.	+ 30 17		15.1	GALAXY
UGC 08497	13 28 06.	+ 30 17	66	15.1	GALAXY S
ZWG 161.073	13 28 06.	+ 31 53		14.2	GALAXY
UGC 08498	13 28 06.	+ 31 53	168	14.2	GALAXY Sb
REIZ 3458	13 28 06.	+ 32 45	18	15.3	GALAXY
1ZW 059	13 28 06.	+ 46 55			COMPACT GALAXY
ZWG 246.010	13 28 06.	+ 46 56		13.2	GALAXY
UGC 08499	13 28 06.	+ 46 56	132	13.2	GALAXY E
MCG+08-25-015	13 28 06.	+ 46 56	42	13.0	GALAXY
REIZ 3463	13 28 07.	+ 46 55	24	12.9	GALAXY
REIN 5.196	13 28 07.11	+ 10 51 39.1			NEBULA
LB 02684	13 28 08.	+ 55 36 48.		15.0	FAINT BLUE STAR
RNGC 5198	13 28 09.	+ 46 56		13.0	GALAXY
HN 1662	13 28 09.2	- 00 49 31.	18		NEBULA
KEEL 668	13 28 10.5	+ 47 34 11.			NEBULA
RNGC 5190	13 28 11.	+ 18 24		13.5	GALAXY
RRS 1.008	13 28 11.	+ 29 44		18.54	BLUE OBJECT
ARP 334	13 28 11.	+ 31 53			PECULIAR GALAXY
REIN 5.197	13 28 11.40	+ 11 51 07.3			NEBULA
ZWG 073.002	13 28 12.	+ 11 52		15.6	GALAXY
ZWG 102.001	13 28 12.	+ 18 24		13.7	GALAXY
ZWG 101.060	13 28 12.	+ 18 24		13.7	GALAXY
UGC 08500	13 28 12.	+ 18 24	66	13.7	GALAXY SB:b
MCG+03-34-043	13 28 12.	+ 18 24	60	13.7	GALAXY

OBJECT NAME	RIGHT ASCEN.	DECLIFATION	DIAM.	MAGN.	TYPE OF OBJECT
TON-N 0715	13 28 12.	+ 29 00		15.5	BLUE STAR
MCG+05-32-032	13 28 12.	+ 30 18	60	15.1	GALAXY
REIZ 3459	13 28 12.	+ 31 52	54	13.3	GALAXY
ZWG 218.018	13 28 12.	+ 40 15		15.7	GALAXY
MCG+07-28-023	13 28 12.	+ 40 15 30.	36	15.	GALAXY
ZC 1328.2+4215	13 28 12.	+ 42 15	12500		CLUSTER OF GALAXIES
IC 4277	13 28 12.	+ 47 34 14.			NONSTELLAR OBJFCT
ZWG 316.017	13 28 12.	+ 62 46		13.5	GALAXY
UGC 08501	13 28 12.	+ 62 46	210	13.5	GALAXY S
REIE 5.198	13 28 12.42	+ 10 44 12.3			NEBULA
REIZ 3457	13 28 13.	+ 18 23	24	14.2	GALAXY
RNGC 5205	13 28 14.	+ 62 46		13.5	GALAXY
RRS 1.009	13 28 15.	+ 30 45		19.03	BLUE OBJECT
REIZ 3474	13 28 15.	+ 62 46	42	13.3	GALAXY
REIN 5.199	13 28 15.68	+ 10 20 24.8			NEBULA
BC 3CR287	13 28 15.92	+ 25 24 37.7		17.67	QUASI-STELLAR OBJECT
RNGC 5191	13 28 16.	+ 11 28		15.0	GALAXY
REIZ 3453	13 28 16.	- 01 32	18	15.0	GALAXY
SHB 265	13 28 16.1	+ 25 24 37.		17.7	QUASI-STELLAR OBJECT
HN 1663	13 28 16.5	- 01 31 13.	18		NEBULA
REIN 5.200	13 28 16.60	+ 10 20 42.1			NEBULA
REIZ 3473	13 28 17.	+ 58 40	180	12.2	GALAXY
ARC 1750	13 28 17.	- 01 36		15.9	RICH CLUSTER OF GALAYIFS
REIN 5.201	13 28 17.01	+ 11 20 57.5			NEBULA
ZWG 073.003	13 28 18.	+ 11 28		14.9	GALAXY
MCG+02-34-026	13 28 18.	+ 11 28	36	14.9	GALAXY
ZWG 161.074	13 28 18.	+ 31 32		14.6	GALAXY
MRK 455	13 28 18.	+ 31 32	25	15.5	GALAXY WITH UV CONTINUUM
REIZ 3462	13 28 18.	+ 31 32	36	13.9	GALAXY
UGC 08502	13 28 18.	+ 31 32	60	14.6	GALAXY DBL SYS
MCG+05-32-034	13 28 18.	+ 31 33 30.	24	14.6	GALAXY
MCG+05-32-033	13 28 18.	+ 31 53 30.	162	14.2	GALAXY
ZC 1328.3+3905	13 28 18.	+ 39 05	740		CLUSTER OF GALAXIES
LB 02685	13 28 18.	+ 53 25 00.		16.2	FAINT BLUE STAP
ZC 1328.3+6509	13 28 18.	+ 65 09	1280		CLUSTER OF GALAXIES
7ZW 517	13 28 18.	+ 73 25			COMPACT GALAXY
ZWG 017.001	13 28 18.	- 01 31		15.2	GALAXY
MCG-04-32-028	13 28 18.	- 21 36	60	15.	GALAXY
REIN 5.202	13 28 18.69	+ 11 27 28.6			NEBULA
RNGC 5192	13 28 19.	- 01 31		15.0	GALAXY
REIN 5.203	13 28 19.52	+ 10 26 53.0			NEBULA
REIN 5.204	13 28 20.26	+ 10 11 05.7			NEBULA
MCG+05-32-035	13 28 21.	+ 31 33 30.	36	14.6	GALAXY
MCG+06-30-021	13 28 21.	+ 35 10	36	15.	GALAXY
IC 4278	13 28 21.	+ 47 30 03.			NONSTELLAR OBJECT
REIN 5.205	13 28 21.53	+ 10 23 56.6			NEBULA
KEEL 669	13 28 21.6	+ 47 30 17.			NEBULA
RNGC 5199	13 28 22.	+ 35 05		15.0	GALAXY
REIZ 3468	13 28 22.	+ 35 10	48	15.0	GALAXY
RRS 1.010	13 28 23.	+ 28 29		18.02	BLUE OBJECT
IC 4272	13 28 23.	- 29 43			NONSTELLAR OBJFCT
REIN 5.206	13 28 23.99	+ 12 21 14.6			NEBULA
ZWG 073.004	13 28 24.	+ 11 14		15.6	GALAXY
ZWG 161.075	13 28 24.	+ 26 47		15.7	GALAXY
REIZ 3467	13 28 24.	+ 32 32	18	14.7	GALAXY
UGC 08503	13 28 24.	+ 33 00	66	17.	GALAXY IRR
MCG+06-30-022	13 28 24.	+ 33 12	24	16.	GALAXY
UGC 08505	13 28 24.	+ 34 09	78	16.5	GALAXY S
ZWG 190.016	13 28 24.	+ 35 05		15.1	GALAXY
UGC 08504	13 28 24.	+ 35 05	72	15.1	GALAXY COMPACT
MCG-02-34-061	13 28 24.	- 14 52	84	15.	GALAXY
HN 1062	13 28 24.	- 29 43		14.	NEBULA
REIN 5.207	13 28 24.75	+ 10 57 34.5			NEBULA
RRS 1.011	13 28 25.	+ 28 31		17.15	BLUF OBJECT
RS 03	13 28 25.3	+ 28 30 08.		17.89	BLUE STELLAR OBJECT
REIN 5.208	13 28 26.69	+ 10 27 06.8			NEBULA
REIN 5.209	13 28 26.77	+ 11 58 00.7			NEBULA
MCG+06-30-023	13 28 27.	+ 34 10	66	17.	GALAXY
MCG+06-30-024	13 28 27.	+ 35 05 30.	24	14.5	GALAXY
MCG+07-28-024	13 28 27.	+ 41 26	48	16.	GALAXY
MCG+11-17-003	13 28 27.	+ 62 46	174	12.	GALAXY
MCG-03-34-087	13 28 27.	- 15 52	42	15.5	GALAXY
REIZ 3471	13 28 28.	+ 35 05	18	14.4	GALAXY
LB 02686	13 28 28.	+ 53 01 30.		17.1	FAINT BLUE STAR
REIN 5.211	13 28 29.43	+ 12 37 54.7			NEBULA
REIN 5.210	13 28 29.89	+ 10 54 50.1			NEBULA
FAI 1328+09	13 28 30.	+ 09 21			COMPACT GALAXY
ZWG 073.005	13 28 30.	+ 11 29		15.7	GALAXY
ZWG 073.006	13 28 30.	+ 11 38		15.5	GALAXY
VV 326B	13 28 30.	+ 31 32 30.	30	16.	INTERACTING GALAXY
VV 326A	13 28 30.	+ 31 32 30.	36	15.	INTERACTING GALAXY
ZC 1328.5+3205	13 28 30.	+ 32 05	3360		CLUSTER OF GALAXIES
UGC 08506	13 28 30.	+ 49 24	66	17.	GALAXY Sc
KARA.68 225	13 28 30.	- 13 41	60		DWARF GALAXY
MCG-04-32-029	13 28 30.	- 25 07 30.	36	15.	GALAXY
MCG-06-30-006	13 28 30.	- 34 49	72	15.	GALAXY
REIN 5.212	13 28 30.05	+ 11 28 41.5			NEBULA
REIN 5.213	13 28 30.76	+ 10 55 58.0			NEBULA
REIN 5.214	13 28 31.24	+ 10 30 44.0			NEBULA
REIN 5.216	13 28 31.83	+ 11 40 17.4			NEBULA
RRS 1.014	13 28 32.	+ 29 24		16.03	BLUE OBJECT
RRS 1.013	13 28 32.	+ 30 53		17.28	BLUE OBJECT
RRS 1.012	13 28 32.	+ 30 57		18.48	BLUF OBJECT
REIN 5.215	13 28 32.09	+ 11 00 29.4			NEBULA
MCG+06-30-025	13 28 33.	+ 34 54	30	16.	GALAXY
REIN 5.217	13 28 33.94	+ 11 37 25.9			NEBULA
ARC 1752	13 28 34.	+ 32 01		17.2	RICH CLUSTER OF GALAYIES
REIZ 3472	13 28 34.	+ 34 54	18	14.9	GALAXY
REIZ 3461	13 28 34.	- 01 30	30	15.4	GALAXY
RNGC 5188	13 28 34.	- 34 31			GALAXY
RS 04	13 28 34.1	+ 27 50 20.		18.21	BLUE STELLAR OBJECT
HN 1063	13 28 35.	- 28 39		13.	NEBULA
IC 4273	13 28 35.	- 28 39			NONSTELLAR OBJFCT
REIN 5.218	13 28 35.04	+ 11 17 01.4			NEBULA
ZWG 045.002	13 28 36.	+ 07 52		15.6	GALAXY
8ZW 1328+12.8	13 28 36.	+ 12 51		17.1	COMPACT GALAXY
ZC 1328.6+1725	13 28 36.	+ 17 25	2620		CLUSTER OF GALAXIES
ZWG 102.002	13 28 36.	+ 19 42		14.0	GALAXY
ZWG 101.061	13 28 36.	+ 19 42		14.0	GALAXY
8ZW 1328+19.7	13 28 36.	+ 19 42		14.0	COMPACT GALAXY
UGC 08507	13 28 36.	+ 19 42	96	14.0	GALAXY (IRR)
MCG+03-34-044	13 28 36.	+ 19 42	96	14.0	GALAXY
KARA.73B 0588	13 28 36.	+ 19 42	66	14.0	ISOLATED GALAXY IR
MCG+08-25-016	13 28 36.	+ 49 24	90	16.	GALAXY
ZC 1328.6+5715	13 28 36.	+ 57 15	3490		CLUSTER OF GALAXIES
ZWG 294.040	13 28 36.	+ 58 35		15.7	GALAXY
KARA.73B 0589	13 28 36.	+ 58 35	18	15.7	ISOLATED GALAXY E
ZC 1328.6+6224	13 28 36.	+ 62 24	1280		CLUSTER OF GALAXIES

OBJECT NAME	RIGHT ASCEN.	DECLINATION	DIAM.	MAGN.	TYPE OF OBJECT
8ZW 1328-01.5	13 28 36.	- 01 30		19.5	COMPACT GALAXY
IC 4274	13 28 37.	- 25 44			PLANETARY NEBULA
REIN 5.219	13 28 37.11	+ 10 26 06.5			NEBULA
REIN 5.220	13 28 37.97	+ 10 24 32.7			NEBULA
FATH 1.628	13 28 38.	+ 58 36	5		NEBULA
REIN 5.222	13 28 38.27	+ 11 14 32.4			NEBULA
REIN 5.221	13 28 38.46	+ 10 26 34.1			NEBULA
REIZ 3466	13 28 40.	- 01 23	12	14.9	GALAXY
REIZ 3465	13 28 41.	- 02 11	48	15.9	GALAXY
REIZ 3464	13 28 41.	- 02 18	60	15.1	GALAXY
REIN 5.223	13 28 41.46	+ 11 41 27.3			NEBULA
ZWG 073.007	13 28 42.	+ 11 42		15.5	GALAXY
ZWG 073.008	13 28 42.	+ 12 49		15.6	GALAXY
VV 088	13 28 42.	+ 19 41	72	15.	INTERACTING GALAXY
ZC 1328.7+4633	13 28 42.	+ 46 33	1140		CLUSTER OF GALAXIES
LB 02687	13 28 42.	+ 51 12 54.		17.0	FAINT BLUE STAR
MCG+10-19-080	13 28 42.	+ 58 35	18	16.	GALAXY
MCG-05-32-035	13 28 42.	- 28 36	24	15.	GALAXY
MCG-06-30-007	13 28 42.	- 34 31	180	12.7	GALAXY
REIN 5.225	13 28 42.84	+ 11 01 55.4			NEBULA
LB 00684	13 28 43.	+ 55 01 00.		17.0	FAINT BLUE STAR
ARC 1751	13 28 44.	- 05 30		17.6	RICH CLUSTER OF GALAXIES
REIN 5.226	13 28 44.41	+ 11 02 12.0			NEBULA
REIN 5.227	13 28 44.94	+ 10 28 18.4			NEBULA
ARC 1756	13 28 45.	+ 62 29		17.6	RICH CLUSTER OF GALAXIES
REIZ 3470	13 28 46.	- 01 28	9	15.3	GALAXY
ZC 1328.8+0510	13 28 48.	+ 05 10	1140		CLUSTER OF GALAXIES
ZWG 073.010	13 28 48.	+ 10 05		15.3	GALAXY
ZWG 073.009	13 28 48.	+ 10 14		15.1	GALAXY
ZWG 073.011	13 28 48.	+ 10 59		15.5	GALAXY
ZC 1328.8+2743	13 28 48.	+ 27 43	940		CLUSTER OF GALAXIES
RRS 1.015	13 28 48.	+ 30 33		17.29	BLUE OBJECT
ZWG 190.017	13 28 48.	+ 34 56		15.7	GALAXY
1ZW 060	13 28 48.	+ 55 10			COMPACT GALAXY
ZWG 271.048	13 28 48.	+ 55 10		14.8	GALAXY
UGC 08508	13 28 48.	+ 55 10	102	14.8	GALAXY IRR
MCG+09-22-075	13 28 48.	+ 55 10	102	13.	GALAXY
UGC 08509	13 28 48.	+ 67 55	66	17.	GALAXY DWRF IR
ZWG 336.010	13 28 48.	+ 71 12		15.7	GALAXY
ZWG 017.002	13 28 48.	- 01 21		15.5	GALAXY
MCG-05-32-036	13 28 48.	- 32 59 30.	36	15.	GALAXY
OCL 0903	13 28 48.	- 62 32	780	9.6	OPEN STAR CLUSTER
VHA 148	13 28 48.	- 62 32	240		OPEN STAR CLUSTER
RRS 1.017	13 28 49.	+ 29 36		16.60	BLUE OBJECT
RRS 1.016	13 28 49.	+ 30 35		17.07	BLUE OBJECT
RNGC 5196	13 28 49.	- 01 21		15.5	GALAXY
REIN 5.229	13 28 49.63	+ 09 49 13.6			NEBULA
EC 3CR286	13 28 49.67	+ 30 45 58.8		17.25	QUASI-STELLAR OBJECT
SHB 266	13 28 49.7	+ 30 45 59.		17.3	QUASI-STELLAR OBJECT
REIN 5.230	13 28 49.71	+ 09 48 34.0			NEBULA
RRS 1.018	13 28 50.	+ 29 40		16.54	BLUE OBJECT
REIN 5.231	13 28 50.82	+ 09 52 28.1			NEBULA
REIN 5.232	13 28 51.31	+ 11 29 42.1			NEBULA
REIZ 3475	13 28 52.	+ 34 58	18	14.7	GALAXY
REIN 5.234	13 28 52.57	+ 11 16 39.4			NEBULA
REIN 5.233	13 28 52.60	+ 10 13 37.7			NEBULA
IC 4275	13 28 53.	- 29 29			NONSTELLAR OBJECT
REIN 5.235	13 28 53.19	+ 10 58 58.8			NEBULA
ZC 1328.9+2727	13 28 54.	+ 27 27	540		CLUSTER OF GALAXIES
ZWG 161.076	13 28 54.	+ 29 38		15.6	GALAXY
UGC 08510	13 28 54.	+ 29 38	72	15.6	GALAXY Sb-c
RRS 1.019	13 28 54.	+ 30 04		18.37	BLUE OBJECT
MCG+06-30-026	13 28 54.	+ 34 56	30	15.	GALAXY
ZC 1328.9+6518	13 28 54.	+ 65 18	3230		CLUSTER OF GALAXIES
ZWG 017.003	13 28 54.	- 01 26		15.4	GALAXY
HN 1064	13 28 54.	- 29 29		12.5	NEBULA
MCG-05-32-037	13 28 54.	- 32 59	30	13.5	GALAXY
OCL 0907	13 28 54.	- 58 14	750	9.	OPEN STAR CLUSTER
REIN 5.236	13 28 54.47	+ 10 04 52.3			NEBULA
RNGC 5197	13 28 55.	- 01 26		15.5	GALAXY
REIN 5.237	13 28 55.68	+ 09 54 01.1			NEBULA
REIN 5.238	13 28 55.78	+ 10 12 54.6			NEBULA
HN 1664	13 28 56.7	- 01 45 36.	30		NEBULA
REIN 5.239	13 28 57.08	+ 11 17 42.9			NEBULA
RS 05	13 28 57.8	+ 26 24 56.		18.45	BLUE STELLAR OBJECT
REIZ 3478	13 28 58.	+ 55 09	72	14.2	GALAXY
LB 09874	13 29	- 82 12		14.0	FAINT BLUE STAR
LB 09875	13 29	- 85 32		12.4	FAINT BLUE STAR
ZC 1329.0+1852	13 29 00.	+ 18 52	740		CLUSTER OF GALAXIES
MCG+04-32-011	13 29 00.	+ 25 51 30.	36	15.3	GALAXY
ZWG 131.009	13 29 00.	+ 25 52		15.3	GALAXY
MCG+05-32-036	13 29 00.	+ 29 37	60	15.6	GALAXY
MCG+10-19-081	13 29 00.	+ 62 29	54	15.	GALAXY
7ZW 518	13 29 00.	+ 75 50			COMPACT GALAXY
ZWG 353.017	13 29 00.	+ 75 50		15.3	GALAXY
ZWG 017.004	13 29 00.	- 01 46		15.6	GALAXY
MPSL 307+00/1	13 29 00.	- 61 50	1500		HII REGION
REIN 5.240	13 29 00.90	+ 09 55 38.0			NEBULA
REIN 5.241	13 29 01.09	+ 10 21 15.7			NEBULA
REIN 5.243	13 29 02.32	+ 11 57 39.9			NEBULA
FATH 1.629	13 29 03.	+ 59 08	5		NEBULA
REIN 5.242	13 29 03.14	+ 10 08 05.6			NEBULA
REIN 5.244	13 29 04.81	+ 11 07 55.4			NEBULA
SC 1326-3115.9	13 29 05.	- 31 31 24.	18		NEBULA
RNGC 5193	13 29 05.	- 32 58		13.0	GALAXY
ARC 1753	13 29 06.	+ 05 07		17.3	RICH CLUSTER OF GALAXIES
RRS 1.020	13 29 06.	+ 29 25		15.65	BLUE OBJECT
ZC 1329.1+3015	13 29 06.	+ 30 15	340		CLUSTER OF GALAXIES
ZWG 317.001	13 29 06.	+ 62 29		15.6	GALAXY
ZWG 316.018	13 29 06.	+ 62 29		15.6	GALAXY
UGC 08511	13 29 06.	+ 62 29	60	15.6	GALAXY S
MCG-05-32-038	13 29 06.	- 29 27	48	15.	GALAXY
RNGC 5200	13 29 07.	+ 00 14			NON-EXISTENT OBJECT
REIN 5.245	13 29 07.65	+ 11 28 24.1			NEBULA
FATH 1.630	13 29 08.	+ 59 11	5		NEBULA
MIL 26	13 29 08.	- 62 32	222		SUPERNOVA REMNANT
REIN 5.246	13 29 08.98	+ 09 38 49.1			NEBULA
HN 0018	13 29 09.	+ 00 14			NEBULA
SC 1327+4646.7	13 29 10.	+ 46 31 13.	36		NEBULA
IC 0892	13 29 11.	- 02 28 14.			NONSTELLAR OBJECT
REIN 5.247	13 29 11.02	+ 10 12 19.6			NEBULA
REIN 5.248	13 29 11.70	+ 10 23 10.1			NEBULA
ZC 1329.2+1310	13 29 12.	+ 13 10	1010		CLUSTER OF GALAXIES
ZC 1329.2+1621	13 29 12.	+ 16 21	1680		CLUSTER OF GALAXIES
ZWG 102.003	13 29 12.	+ 19 10		15.7	GALAXY
ZC 1329.2+2400	13 29 12.	+ 24 00	940		CLUSTER OF GALAXIES
MCG+06-30-027	13 29 12.	+ 33 47	24	15.	GALAXY
MCG+10-19-082	13 29 12.	+ 62 17	60	15.	GALAXY
ZWG 336.011	13 29 12.	+ 74 02		15.6	GALAXY
MRK 261	13 29 12.	+ 75 53	15	15.5	GALAXY WITH UV CONTINUUM
ZWG 017.006	13 29 12.	- 02 21		14.9	GALAXY
UGC 08513	13 29 12.	- 02 21	72	14.9	GALAXY Sa-b
MCG+00-35-002	13 29 12.	- 02 21	66	14.9	GALAXY
IC 0893	13 29 12.	- 02 22 08.			NONSTELLAR OBJECT
ZWG 017.005	13 29 12.	- 02 28		14.5	GALAXY
UGC 08512	13 29 12.	- 02 28	78	14.5	GALAXY S0
MCG+00-35-001	13 29 12.	- 02 28	30	14.5	GALAXY
MCG-03-35-001	13 29 12.	- 19 56	36	15.	GALAXY
REIN 5.249	13 29 12.43	+ 11 01 56.3			NEBULA
HN 1665	13 29 12.6	- 02 15 05.	24		NEBULA
IC 4282	13 29 13.	+ 47 26 05.			NONSTELLAR OBJECT
HN 1666	13 29 13.7	- 01 37 53.			NEBULA
KEEL 670	13 29 14.0	+ 47 26 25.			NEBULA
REIN 5.250	13 29 15.81	+ 09 50 24.6			NEBULA
REIN 5.251	13 29 17.30	+ 10 21 08.2			NEBULA
ZWG 073.012	13 29 18.	+ 14 23		15.7	GALAXY
KARA.72 380A	13 29 18.	+ 14 23	36	15.7	PART OF DOUBLE GALAXY
ZC 1329.3+3119	13 29 18.	+ 31 19	610		CLUSTER OF GALAXIES
1ZW 061	13 29 18.	+ 54 21			COMPACT GALAXY
ZWG 294.041	13 29 18.	+ 62 17		14.9	GALAXY
UGC 08514	13 29 18.	+ 62 17	78	14.9	GALAXY Sa-b
ZWG 017.008	13 29 18.	- 01 38		15.4	GALAXY
ZWG 017.007	13 29 18.	- 02 14		15.6	GALAXY
ARC 1755	13 29 20.	+ 16 14		17.5	RICH CLUSTER OF GALAXIES
HN 1667	13 29 20.2	- 01 44 29.	18		NEBULA
RRS 1.021	13 29 21.	+ 29 45		16.86	BLUE OBJECT
REIZ 3476	13 29 22.	- 01 27	24	16.1	GALAXY
REIN 5.253	13 29 22.37	+ 11 30 25.7			NEBULA
HN 1668	13 29 22.5	- 02 40 53.	24		NEBULA
RRS 1.022	13 29 23.	+ 28 27		18.56	BLUE OBJECT
SC 1327+3106.5	13 29 23.	+ 30 51 01.	12		NEBULA
HN 1065	13 29 23.	- 27 54	12		NEBULA
IC 4276	13 29 23.	- 27 55			NONSTELLAR OBJECT
ZWG 073.013	13 29 24.	+ 11 32		15.1	GALAXY
UGC 08515	13 29 24.	+ 11 32	66	15.1	GALAXY Sa-b
ZWG 073.014	13 29 24.	+ 14 24		15.3	GALAXY
KARA.72 380B	13 29 24.	+ 14 24	36	15.3	PART OF DOUBLE GALAXY
FEIG 081	13 29 24.	+ 15 56		13.3	FAINT BLUE STAR
ZC 1329.4+4834	13 29 24.	+ 48 34	1010		CLUSTER OF GALAXIES
ZWG 017.009	13 29 24.	- 01 45		15.3	GALAXY
8ZW 1329-32.6	13 29 24.	- 32 35		18.0	COMPACT GALAXY
ARC 1754	13 29 25.	- 11 24		17.0	RICH CLUSTER OF GALAXIES
RS 06	13 29 25.8	+ 28 27 32.		18.49	BLUE STELLAR OBJECT
SC 1327+3127.1	13 29 26.	+ 31 11 38.	30		NEBULA
IC 4284	13 29 26.	+ 47 02 48.			NONSTELLAR OBJECT
LB 02688	13 29 26.	+ 51 19 30.		15.4	FAINT BLUE STAR
KEEL 671	13 29 26.0	+ 47 03 11.			NEBULA
MCG-04-32-031	13 29 27.	- 22 40 30.	72	15.	GALAXY
MCG-04-32-030	13 29 27.	- 24 34 30.	72	15.	GALAXY
ZC 1329.5+1931	13 29 30.	+ 19 31	400		CLUSTER OF GALAXIES
ZWG 102.004	13 29 30.	+ 20 15		13.8	GALAXY
UGC 08516	13 29 30.	+ 20 15	72	13.8	GALAXY Sc
KARA.73B 0590	13 29 30.	+ 20 15	60	13.8	ISOLATED GALAXY S
REIZ 3479	13 29 30.	+ 20 16	42	13.9	GALAXY
ZC 1329.5+2456	13 29 30.	+ 24 56	1010		CLUSTER OF GALAXIES
ZWG 246.011	13 29 30.	+ 46 25		15.0	GALAXY
ZC 1329.5+5825	13 29 30.	+ 58 25	1410		CLUSTER OF GALAXIES
MRK 262	13 29 30.	+ 75 53	13	16.	GALAXY WITH UV CONTINUUM
ZWG 017.010	13 29 30.	- 01 26		15.6	GALAXY
MCG-06-30-008	13 29 30.	- 37 55	60	14.	GALAXY
REIN 5.254	13 29 30.61	+ 09 52 33.3			NEBULA
RNGC 5202	13 29 31.	- 01 26		15.5	GALAXY
REIN 5.255	13 29 31.68	+ 09 47 41.4			NEBULA
HN 1670	13 29 34.0	+ 01 50 43.	18		NEBULA
HN 1669	13 29 34.3	- 02 30 11.	18		NEBULA
RNGC 5203	13 29 35.	- 08 32		14.0	GALAXY
ZWG 017.012	13 29 36.	+ 01 52		14.8	GALAXY
MCG+00-35-004	13 29 36.	+ 01 52	36	14.8	GALAXY
ZWG 073.015	13 29 36.	+ 10 43		15.6	GALAXY
ZWG 073.016	13 29 36.	+ 13 26		15.3	GALAXY
MCG+03-35-001	13 29 36.	+ 20 17	66	13.8	GALAXY
ZC 1329.6+2755	13 29 36.	+ 27 55	470		CLUSTER OF GALAXIES
FEIG 082	13 29 36.	+ 32 02		10.4	FAINT BLUE STAR
MCG+08-25-017	13 29 36.	+ 46 24	42	15.	GALAXY
ZC 1329.6+6102	13 29 36.	+ 61 02	1610		CLUSTER OF GALAXIES
ZWG 017.011	13 29 36.	- 02 30		15.6	GALAXY
MCG+00-35-003	13 29 36.	- 02 30	36	15.6	GALAXY
MCG-01-35-001	13 29 36.	- 08 32	108	14.	GALAXY
REIZ 3477	13 29 37.	- 08 31	30	13.0	GALAXY
REIN 5.258	13 29 38.56	+ 09 52 50.5			NEBULA
IC 0894	13 29 39.	+ 17 18 42.			NONSTELLAR OBJECT
SC 1327+2836.6	13 29 39.	+ 28 21 08.	12		NEBULA
KEEL 672	13 29 39.6	+ 47 04 46.			NEBULA
IC 4285	13 29 40.	+ 47 04 19.			NONSTELLAR OBJECT
RRS 1.023	13 29 41.	+ 29 55		16.57	BLUE OBJECT
ZWG 045.003	13 29 42.	+ 06 25		15.5	GALAXY
ZWG 073.017	13 29 42.	+ 10 40		15.3	GALAXY
ZWG 102.005	13 29 42.	+ 14 41		15.7	GALAXY
ZWG 102.006	13 29 42.	+ 17 18		14.9	GALAXY
MCG+03-35-002	13 29 42.	+ 17 18	60	14.9	GALAXY
KARA.73B 0591	13 29 42.	+ 17 18	66	14.9	ISOLATED GALAXY S
UGC 08517	13 29 42.	+ 31 17	60	16.0	GALAXY S
MCG+05-32-037	13 29 42.	+ 31 18	51	15.5	GALAXY
ZWG 218.019	13 29 42.	+ 44 24		15.6	GALAXY
LB 02689	13 29 42.	+ 52 41 30.		15.5	FAINT BLUE STAR
MCG+09-22-076	13 29 42.	+ 54 41	36	16.	GALAXY
MCG-03-35-002	13 29 42.	- 20 57 30.	60	15.	GALAXY
MCG-04-32-032	13 29 42.	- 26 52	30	15.5	GALAXY
REIZ 3482	13 29 44.	+ 28 40	36	13.8	GALAXY
IC 4280	13 29 44.	- 23 57 25.			NONSTELLAR OBJECT
RS 07	13 29 45.8	+ 27 56 26.		17.28	BLUE STELLAR OBJECT
RNGC 5207	13 29 46.	+ 14 10		14.5	GALAXY
HN 1066	13 29 46.	- 26 52	18		NEBULA
IC 4279	13 29 46.	- 26 53			NONSTELLAR OBJECT
REIN 5.261	13 29 46.33	+ 11 21 09.7			NEBULA
REIN 5.262	13 29 47.54	+ 09 48 40.8			NEBULA
RS 08	13 29 47.7	+ 27 52 14.		19.71	BLUE STELLAR OBJECT
ZWG 017.014	13 29 48.	+ 01 50		14.9	GALAXY
MCG+00-35-005	13 29 48.	+ 01 50	60	14.9	GALAXY
ZC 1329.8+0306	13 29 48.	+ 03 06	340		CLUSTER OF GALAXIES
ZWG 045.004	13 29 48.	+ 03 18		15.6	GALAXY
KARA.73B 0592	13 29 48.	+ 03 18	24	15.6	ISOLATED GALAXY S
8ZW 1329+11.4	13 29 48.	+ 11 23		14.5	COMPACT GALAXY
ZWG 073.018	13 29 48.	+ 14 10		14.7	GALAXY
UGC 08518	13 29 48.	+ 14 10	120	14.7	GALAXY Sb
MCG+02-35-001	13 29 48.	+ 14 10	84	14.7	GALAXY

OBJECT NAME	RIGHT ASCEN.	DECLINATION	DIAM.	MAGN.	TYPE OF OBJECT
ZC 1329.8+1725	13 29 48.	+ 17 25	540		CLUSTER OF GALAXIES
MCG+07-28-025	13 29 48.	+ 44 25	42	16.	GALAXY
ZC 1329.8+6250	13 29 48.	+ 62 50	1410		CLUSTER OF GALAXIES
ZWG 017.013	13 29 48.	- 02 14		15.7	GALAXY
ZC 1329.8-0251	13 29 48.	- 02 51	3160		CLUSTER OF GALAXIES
HN 1671	13 29 48.6	- 02 13 58.	18		NEBULA
HN 1672	13 29 48.7	+ 01 48 56.	18		NEBULA
IC 4283	13 29 49.	+ 28 38 42.			NONSTELLAR OBJECT
REIN 5.263	13 29 49.14	+ 09 49 33.3			NEBULA
REIN 5.266	13 29 50.46	+ 11 25 00.0			NEBULA
REIN 5.265	13 29 50.66	+ 10 12 26.5			NEBULA
REIZ 3480	13 29 51.	+ 01 50	18	14.7	GALAXY
MCG-04-32-033	13 29 51.	- 26 54	36	15.	GALAXY
HN 1067	13 29 52.	- 26 54		13.5	NEBULA
IC 4281	13 29 52.	- 26 55			NONSTELLAR OBJECT
CED 123	13 29 53.	- 65 42	180		DIFFUSE GALACTIC NEBULA
HN 0093	13 29 53.	- 65 43			NEBULA
REIN 5.268	13 29 53.59	+ 10 09 06.2			NEBULA
ZWG 045.005	13 29 54.	+ 03 51		15.7	GALAXY
ZWG 045.006	13 29 54.	+ 05 03		15.6	GALAXY
ZWG 045.007	13 29 54.	+ 07 35		14.4	GALAXY
UGC 08519	13 29 54.	+ 07 35	102	14.4	GALAXY S0
MCG+01-35-001	13 29 54.	+ 07 35	84	14.4	GALAXY
ZWG 161.077	13 29 54.	+ 27 13		15.7	GALAXY
MRK 661	13 29 54.	+ 27 13	20	16.5	GALAXY WITH UV CONTINUUM
ZWG 161.078	13 29 54.	+ 28 39		15.4	GALAXY
ZC 1329.9+4119	13 29 54.	+ 41 19	810		CLUSTER OF GALAXIES
MCG+07-28-026	13 29 54.	+ 44 24	24	16.	GALAXY
ZC 1329.9+5234	13 29 54.	+ 52 34	1480		CLUSTER OF GALAXIES
ZC 1329.9+5539	13 29 54.	+ 55 39	740		CLUSTER OF GALAXIES
ZC 1329.9+5848	13 29 54.	+ 58 48	1480		CLUSTER OF GALAXIES
ZWG 294.042	13 29 54.	+ 59 26		15.1	GALAXY
KLEM 26	13 29 54.	- 23 18	900	13.	GROUP OF 8 GALAXIES
PK307-03.1	13 29 54.1	- 65 43 06.	185	10.3	PLANETARY NEBULA
REIN 5.269	13 29 54.88	+ 10 07 10.6			NEBULA
REIN 5.270	13 29 55.44	+ 11 21 43.5			NEBULA
REIN 5.272	13 29 56.70	+ 11 24 39.7			NEBULA
REIZ 3481	13 29 57.	+ 02 07	36	13.9	GALAXY
RNGC 5208	13 29 57.	+ 07 35		14.5	GALAXY
HN 1673	13 29 58.3	+ 02 06 56.	30		NEBULA
RRS 1.025	13 29 59.	+ 28 28		18.13	BLUE OBJECT
RRS 1.024	13 29 59.	+ 29 16		18.87	BLUE OBJECT
SC 1327-3118.9	13 29 59.	- 31 34 23.	18		NEBULA
ZWG 017.015	13 30 00.	+ 02 06		14.5	GALAXY
UGC 08521	13 30 00.	+ 02 06	66	14.5	GALAXY SBa-b
MCG+00-35-006	13 30 00.	+ 02 06	60	14.5	GALAXY
ZC 1330.0+0605	13 30 00.	+ 06 05	540		CLUSTER OF GALAXIES
ZWG 045.008	13 30 00.	+ 07 34		15.5	GALAXY
REIZ 3483	13 30 00.	+ 07 35	48	14.4	GALAXY
ZWG 073.019	13 30 00.	+ 13 05		15.4	GALAXY
ZWG 102.007	13 30 00.	+ 18 24		15.3	GALAXY
ZC 1330.0+2027	13 30 00.	+ 20 27	1280		CLUSTER OF GALAXIES
MCG+07-28-027	13 30 00.	+ 40 46	36	16.	GALAXY
MCG+10-19-083	13 30 00.	+ 59 26	36	15.	GALAXY
ZWG 336.012	13 30 00.	+ 72 56		15.3	GALAXY
ZWG 365.013	13 30 00.	+ 80 45		15.5	GALAXY
UGC 08520	13 30 00.	+ 80 45	78	15.5	GALAXY Sc
MCG+14-06-026	13 30 00.	+ 80 46	54	15.	GALAXY
MCG-02-35-001	13 30 00.	- 10 13	36	15.	GALAXY
MCG-04-32-034	13 30 00.	- 23 26	36	15.	GALAXY
RNGC 5212	13 30 03.	+ 07 34		15.5	GALAXY
IC 0895	13 30 03.	+ 35 54 55.			NONSTELLAR OBJECT
REIN 5.273	13 30 03.93	+ 09 46 47.0			NEBULA
RRS 1.026	13 30 05.	+ 30 23		18.05	BLUE OBJECT
REIN 5.274	13 30 05.25	+ 09 47 01.3			NEBULA
REIN 5.275	13 30 05.62	+ 09 47 28.2			NEBULA
ZWG 017.016	13 30 06.	+ 01 07		15.4	GALAXY
ZWG 045.009	13 30 06.	+ 07 36		14.7	GALAXY
UGC 08522	13 30 06.	+ 07 36	78	14.7	GALAXY E
MCG+01-35-002	13 30 06.	+ 07 36	21	14.7	GALAXY
ZWG 073.020	13 30 06.	+ 11 52		15.6	GALAXY
ZWG 073.021	13 30 06.	+ 14 12		15.6	GALAXY
ZWG 102.008	13 30 06.	+ 18 18		15.7	GALAXY
MCG+03-35-003	13 30 06.	+ 18 25	12	16.	GALAXY
ZWG 102.009	13 30 06.	+ 18 28		15.7	GALAXY
MCG+03-35-004	13 30 06.	+ 18 28	42	15.7	GALAXY
TON-N 0716	13 30 06.	+ 27 38		16.1	BLUE STAR
MCG+10-19-084	13 30 06.	+ 60 50	18	16.	GALAXY
HN 1674	13 30 06.2	+ 01 06 56.	18		NEBULA
REIN 5.277	13 30 06.58	+ 10 06 33.9			NEBULA
RRS 1.027	13 30 07.	+ 29 31		19.24	BLUE OBJECT
REIN 5.278	13 30 07.58	+ 10 17 45.7			NEBULA
RNGC 5209	13 30 09.	+ 07 36		14.5	GALAXY
MCG-04-32-035	13 30 09.	- 24 54	36	14.5	GALAXY
REIN 5.279	13 30 09.11	+ 09 34 37.7			NEBULA
REIN 5.280	13 30 09.74	+ 09 47 43.3			NEBULA
REIN 5.281	13 30 10.33	+ 09 48 32.5			NEBULA
REIN 5.282	13 30 10.33	+ 10 09 14.7			NEBULA
REIZ 3484	13 30 12.	+ 07 36	48	14.7	GALAXY
ZC 1330.2+0949	13 30 12.	+ 09 49	2150		CLUSTER OF GALAXIES
ZWG 190.018	13 30 12.	+ 33 46		15.7	GALAXY
ZWG 218.020	13 30 12.	+ 38 37		15.6	GALAXY
FEIG 083	13 30 12.	+ 53 38		13.4	FAINT BLUE STAR
MCG+13-10-007	13 30 12.	+ 78 55	36	16.	GALAXY
MCG-04-32-036	13 30 12.	- 23 58	48	13.5	GALAXY
KARA.68 226	13 30 12.	- 24 25	107		DWARF GALAXY
REIN 5.283	13 30 12.77	+ 09 50 25.0			NEBULA
IC 4287	13 30 14.	+ 25 39 56.			NONSTELLAR OBJECT
REIZ 3488	13 30 14.	+ 37 07	60	14.9	GALAXY
RNGC 5189	13 30 14.	- 65 44			PLANETARY NEBULA
MCG+04-32-012	13 30 15.	+ 22 49	30	15.5	GALAXY
HN 1675	13 30 15.4	- 02 49 34.	48		NEBULA
HN 1676	13 30 17.6	- 01 00 40.	18		NEBULA
ZWG 045.010	13 30 18.	+ 07 26		14.4	GALAXY
UGC 08523	13 30 18.	+ 07 26	108	14.4	GALAXY Sa
MCG+01-35-003	13 30 18.	+ 07 26	60	14.4	GALAXY
REIZ 3486	13 30 18.	+ 07 31	36	14.8	GALAXY
ZWG 045.011	13 30 18.	+ 07 35		15.6	GALAXY
ZWG 045.012	13 30 18.	+ 07 39		15.4	GALAXY
8ZW 1330+10.3	13 30 18.	+ 10 16		16.0	COMPACT GALAXY
VV 043A	13 30 18.	+ 11 41	96	15.0	INTERACTING GALAXY
ZWG 102.010	13 30 18.	+ 18 26		15.6	GALAXY
ZWG 131.010	13 30 18.	+ 25 41		15.3	GALAXY
ZWG 190.019	13 30 18.	+ 37 06		15.7	GALAXY
UGC 08524	13 30 18.	+ 37 06	72	15.7	GALAXY SB:b
MCG+07-28-028	13 30 18.	+ 40 04	30	16.	GALAXY
ZWG 336.013	13 30 18.	+ 71 44		14.4	GALAXY
UGC 08525	13 30 18.	+ 71 44	66	14.4	GALAXY E-S0?
MCG+12-13-007	13 30 18.	+ 71 45	21	16.	GALAXY
ZWG 017.017	13 30 18.	- 02 50		15.2	GALAXY
MCG+00-35-007	13 30 18.	- 02 50	84	15.2	GALAXY
REIN 5.285	13 30 19.28	+ 09 41 44.5			NEBULA
REIZ 3493	13 30 20.	+ 62 58	60	13.5	GALAXY
REIN 5.286	13 30 20.23	+ 09 40 05.8			NEBULA
REIN 5.287	13 30 20.57	+ 09 45 21.6			NEBULA
HN 1677	13 30 20.6	- 00 54 28.	36		NEBULA
KW 69	13 30 21.	+ 02 19 18.	8		SEYFERT GALAXY
RNGC 5210	13 30 21.	+ 07 26		14.5	GALAXY
RRS 1.028	13 30 21.	+ 30 51		19.29	BLUE OBJECT
MCG+06-30-028	13 30 21.	+ 37 07	66	14.5	GALAXY
SC 1327-3109.4	13 30 21.	- 31 24 52.	12		NEBULA
HN 1678	13 30 21.5	- 02 46 16.	18		NEBULA
REIZ 3485	13 30 22.	- 00 54	30	14.2	GALAXY
REIN 5.288	13 30 22.98	+ 10 14 44.3			NEBULA
ZWG 045.013	13 30 24.	+ 05 12		15.7	GALAXY
ZWG 045.014	13 30 24.	+ 07 33		15.4	GALAXY
ZWG 073.022	13 30 24.	+ 09 47		15.0	GALAXY
UGC 08527	13 30 24.	+ 09 47	66	15.0	GALAXY Sc-IRR
MCG+02-35-002	13 30 24.	+ 09 47	72	15.0	GALAXY
VV 043B	13 30 24.	+ 11 40	48	16.0	INTERACTING GALAXY
MCG+03-35-005	13 30 24.	+ 18 27	72	15.6	GALAXY
ZC 1330.4+2345	13 30 24.	+ 23 45	1010		CLUSTER OF GALAXIES
1ZW 062	13 30 24.	+ 50 47			COMPACT GALAXY
ZC 1330.4+5047	13 30 24.	+ 50 47	1410		CLUSTER OF GALAXIES
ZWG 317.002	13 30 24.	+ 62 57		14.0	GALAXY
ZWG 316.019	13 30 24.	+ 62 57		14.0	GALAXY
UGC 08528	13 30 24.	+ 62 57	180	14.0	GALAXY
VV 033A	13 30 24.	+ 62 58	102	13.3	INTERACTING GALAXY
ZWG 017.020	13 30 24.	- 00 54		14.5	GALAXY
UGC 08526	13 30 24.	- 00 54	54	14.5	GALAXY
MCG+00-35-008	13 30 24.	- 00 54	36	14.5	GALAXY
ZWG 017.019	13 30 24.	- 01 00		15.4	GALAXY
ZWG 017.018	13 30 24.	- 02 46		15.1	GALAXY
MCG-04-32-037	13 30 24.	- 22 20	24	15.	GALAXY
REIN 5.289	13 30 24.51	+ 09 37 53.5			NEBULA
ARP 104	13 30 25.	+ 62 58			PECULIAR GALAXY
REIZ 3496	13 30 25.	+ 63 02	102	13.0	GALAXY
REIN 5.291	13 30 25.17	+ 09 46 36.9			NEBULA
RNGC 5216	13 30 26.	+ 62 57		14.0	GALAXY
SC 1327-3057.7	13 30 27.	- 31 13 10.	12		NEBULA
REIN 2.194	13 30 27.00	+ 63 01 25.4			NEBULA
ABC 1758	13 30 29.	+ 50 47		18.0	RICH CLUSTER OF GALAXIES
ZWG 045.015	13 30 30.	+ 05 15		15.6	GALAXY
ZWG 102.011	13 30 30.	+ 16 54		14.7	GALAXY
MCG+03-35-006	13 30 30.	+ 16 54	66	14.7	GALAXY
ZC 1330.5+2046	13 30 30.	+ 20 46	610		CLUSTER OF GALAXIES
ZC 1330.5+3915	13 30 30.	+ 39 15	940		CLUSTER OF GALAXIES
ZC 1330.5+4725	13 30 30.	+ 47 25	1280		CLUSTER OF GALAXIES
ZWG 294.043	13 30 30.	+ 60 39		15.5	GALAXY
MCG+10-19-085	13 30 30.	+ 60 39	45	15.	GALAXY
MCG+11-17-004	13 30 30.	+ 62 57	114	13.0	GALAXY
ZWG 317.003	13 30 30.	+ 63 01		13.1	GALAXY
ZWG 316.020	13 30 30.	+ 63 01		13.1	GALAXY
UGC 08529	13 30 30.	+ 63 01	120	13.1	GALAXY S
VV 033B	13 30 30.	+ 63 01	72	13.5	INTERACTING GALAXY
MCG+11-17-005	13 30 30.	+ 63 01 30.	102	13.5	GALAXY
RS 09	13 30 31.1	+ 28 52 01.		16.62	BLUE STELLAR OBJECT
RNGC 5218	13 30 32.	+ 63 01		13.0	GALAXY
HN 1679	13 30 32.1	- 00 09 27.	12		NEBULA
REIN 5.294	13 30 32.34	+ 09 45 03.9			NEBULA
RRS 1.029	13 30 33.	+ 31 08		18.70	BLUE OBJECT
REIZ 3487	13 30 34.	- 00 47	60	13.9	GALAXY
REIN 5.295	13 30 34.86	+ 09 45 31.7			NEBULA
VV 043B	13 30 36.	+ 11 37	96	15.6	INTERACTING GALAXY
ZWG 073.023	13 30 36.	+ 14 27		15.3	GALAXY
ZWG 102.012	13 30 36.	+ 17 51		15.4	GALAXY
TON-N 0718	13 30 36.	+ 28 49		15.9	BLUE STAR
TON-N 0717	13 30 36.	+ 28 54		16.7	BLUE STAR
ZC 1330.6+3011	13 30 36.	+ 30 11	610		CLUSTER OF GALAXIES
MCG+06-30-029	13 30 36.	+ 32 51	24	15.	GALAXY
KARA.72 381A	13 30 36.	+ 42 07	30	14.4	PART OF DOUBLE GALAXY
ZWG 218.021	13 30 36.	+ 42 08		14.4	GALAXY
UGC 08531	13 30 36.	+ 42 08	66	14.4	GALAXY Sc
KARA.72 381B	13 30 36.	+ 42 08	66		PART OF DOUBLE GALAXY
ZC 1330.6+5440	13 30 36.	+ 54 40	2290		CLUSTER OF GALAXIES
ZWG 336.014	13 30 36.	+ 72 51		15.4	GALAXY
ZWG 017.021	13 30 36.	- 00 46		13.9	GALAXY
UGC 08530	13 30 36.	- 00 46	144	13.9	GALAXY Sb/SBb
MCG+00-35-009	13 30 36.	- 00 46	120	13.9	GALAXY
MCG-05-32-039	13 30 36.	- 32 28	48	15.	GALAXY
REIN 5.296	13 30 36.49	+ 10 09 21.4			NEBULA
REIN 5.297	13 30 36.98	+ 09 44 23.1			NEBULA
MCG+07-28-029	13 30 37.	+ 42 07 30.	24	16.	GALAXY
RNGC 5214	13 30 37.	+ 42 08		14.5	GALAXY
LB C2690	13 30 37.	+ 50 58 06.		15.8	FAINT BLUE STAR
RNGC 5211	13 30 37.	- 00 46		14.0	GALAXY
REIN 5.299	13 30 38.97	+ 09 45 57.6			NEBULA
MCG+07-28-030	13 30 39.	+ 42 08	66	14.	GALAXY
MCG-03-35-003	13 30 39.	- 15 52	30	15.	GALAXY
SC 1327-3212.9	13 30 40.	- 32 28 22.	60		NEBULA
REIZ 3492	13 30 41.	+ 42 08	42	14.1	GALAXY
RNGC 5206	13 30 41.	- 47 53			GALAXY
ZWG 045.016	13 30 42.	+ 06 01		15.3	GALAXY
UGC 08532	13 30 42.	+ 06 01	66	15.3	GALAXY S
MCG+01-35-004	13 30 42.	+ 06 01	36	15.3	GALAXY
ZC 1330.7+1347	13 30 42.	+ 13 47	5710		CLUSTER OF GALAXIES
MCG+06-30-030	13 30 42.	+ 32 50	18	15.	GALAXY
ZWG 190.020	13 30 42.	+ 32 52		15.5	GALAXY
MCG+10-19-086	13 30 42.	+ 56 59	51	15.	GALAXY
ZWG 294.044	13 30 42.	+ 57 00		15.6	GALAXY
MCG-03-35-004	13 30 42.	- 15 51	120	14.	GALAXY
SC 1327-2941.4	13 30 42.	- 29 56 51.	30		NEBULA
MRSL 308+01/1	13 30 42.	- 61 12	9720		HII REGION
REIN 5.300	13 30 42.83	+ 09 44 50.2			NEBULA
RRS 1.030	13 30 43.	+ 29 03		17.72	BLUE OBJECT
SC 1327-3109.6	13 30 44.	- 31 25 03.	24		NEBULA
RRS 1.031	13 30 45.	+ 30 52		18.61	BLUE OBJECT
HOLM 528B	13 30 45.	+ 34 49	54	15.3	PART OF MULTIPLE GALAXY
ZL 131	13 30 46.	+ 20 20 18.		21.6	ULTRAFAINT BLUE STAR
RRS 1.032	13 30 46.	+ 30 12		18.24	BLUE OBJECT
REIZ 3491	13 30 46.	+ 34 49	18	15.7	GALAXY
REIZ 3490	13 30 47.	+ 33 24	36	14.5	GALAXY
ZC 1330.8+0138	13 30 48.	+ 01 38	540		CLUSTER OF GALAXIES
ZWG 045.017	13 30 48.	+ 04 10		15.6	GALAXY
ZWG 045.018	13 30 48.	+ 06 35		15.2	GALAXY
ZWG 073.024	13 30 43.	+ 09 47		15.7	GALAXY

OBJECT NAME	RIGHT ASCEN.	DECLINATION	DIAM.	MAGN.	TYPE OF OBJECT
ZWG 073.025	13 30 48.	+ 11 22		15.6	GALAXY
ZWG 102.013	13 30 48.	+ 17 23		15.7	GALAXY
ZC 1330.8+1931	13 30 48.	+ 19 31	3160		CLUSTER OF GALAXIES
ZC 1330.8+2147	13 30 48.	+ 21 47	870		CLUSTER OF GALAXIES
MCG+06-30-031	13 30 48.	+ 32 56	48	15.	GALAXY
ZWG 190.021	13 30 48.	+ 37 27		15.6	GALAXY
MRK 456	13 30 48.	+ 37 27	12	15.5	GALAXY WITH UV CONTINUUM
MRK 457	13 30 48.	+ 39 17	8	16.	GALAXY WITH UV CONTINUUM
ZC 1330.8+6203	13 30 48.	+ 62 03	1080		CLUSTER OF GALAXIES
7ZW 519	13 30 48.	+ 78 06			COMPACT GALAXY
ZWG 353.018	13 30 48.	+ 78 06		15.2	GALAXY
ARC 1757	13 30 48.	- 23 01		17.0	RICH CLUSTER OF GALAXIES
REIN 5.301	13 30 50.72	+ 09 43 37.5			NEBULA
MCG+06-30-034	13 30 51.	+ 33 24	42	14.5	GALAXY
MCG+06-30-033	13 30 51.	+ 37 28	9	16.	GALAXY
MCG+06-30-032	13 30 51.	+ 37 28 30.	48	15.	GALAXY
RRS 1.033	13 30 52.	+ 30 08		18.50	BLUE OBJECT
ARC 1761	13 30 52.	+ 57 55		17.8	RICH CLUSTER OF GALAXIES
HOLM 527A	13 30 53.	+ 33 25	30	14.3	PART OF MULTIPLE GALAXY
HN 1068	13 30 53.	- 27 22		14.	NEBULA
IC 4286	13 30 53.	- 27 23			NONSTELLAR OBJECT
REIN 5.302	13 30 53.16	+ 09 47 01.7			NEBULA
REIN 5.303	13 30 53.29	+ 09 59 30.4			NEBULA
ZWG 045.019	13 30 54.	+ 05 44		15.4	GALAXY
UGC 08523	13 30 54.	+ 05 44	66	15.4	GALAXY S
KARA.72 382A	13 30 54.	+ 05 44	60	15.4	PART OF DOUBLE GALAXY
ZWG 190.022	13 30 54.	+ 33 25		15.5	GALAXY
7ZW 520	13 30 54.	+ 60 23			COMPACT GALAXY
MCG-06-30-009	13 30 54.	- 37 29	48	15.	GALAXY
MCG+06-30-035	13 30 57.	+ 33 21	48	14.5	GALAXY
REIZ 3495	13 30 58.	+ 34 48	30	14.9	GALAXY
HOLM 528A	13 30 58.	+ 34 48	48	15.2	PART OF MULTIPLE GALAXY
RRS 1.034	13 30 59.	+ 31 04		18.13	BLUE OBJECT
REIZ 3494	13 30 59.	+ 33 22	36	14.3	GALAXY
ZWG 017.022	13 31 00.	+ 01 34		15.2	GALAXY
UGC 08534	13 31 00.	+ 01 34	78	15.2	GALAXY SBc
ZWG 045.020	13 31 00.	+ 05 45		15.3	GALAXY
KARA.72 382B	13 31 00.	+ 05 45	42	15.3	PART OF DOUBLE GALAXY
ZWG 045.021	13 31 00.	+ 07 24		15.5	GALAXY
ZWG 102.014	13 31 00.	+ 17 44		15.1	GALAXY
UGC 08535	13 31 00.	+ 17 44	84	15.1	GALAXY S0-a
ZC 1331.0+2009	13 31 00.	+ 20 09	1210		CLUSTER OF GALAXIES
ZL 132	13 31 00.	+ 20 28 00.		22.0	ULTRAFAINT BLUE STAR
ZWG 190.023	13 31 00.	+ 33 22		15.6	GALAXY
HOLM 527B	13 31 00.	+ 33 22	36	14.7	PART OF MULTIPLE GALAXY
UGC 08536	13 31 00.	+ 34 48	66	16.0	GALAXY
MCG+06-30-036	13 31 00.	+ 34 48	54	15.5	GALAXY
STOCK 45	13 31 00.	+ 37 20			BLUE KNOT NEAR ELLIP GLXY
MCG-04-32-038	13 31 00.	- 24 29 30.	54	14.5	GALAXY
SC 1328-3057.8	13 31 01.	- 31 13 15.	12		NEBULA
HN 1680	13 31 01.0	+ 01 33 16.	48		NEBULA
REIZ 3469	13 31 03.	+ 01 35	42	14.8	GALAXY
HN 1681	13 31 04.5	- 01 38 50.	42		NEBULA
ZL 133	13 31 05.	+ 20 27 24.		21.7	ULTRAFAINT BLUE STAR
HN 1682	13 31 05.5	- 02 29 20.	18		NEBULA
ZWG 073.026	13 31 06.	+ 13 30		15.5	GALAXY
ZC 1331.1+1818	13 31 06.	+ 18 18	1080		CLUSTER OF GALAXIES
ZC 1331.1+1842	13 31 06.	+ 18 42	610		CLUSTER OF GALAXIES
ZC 1331.1+2711	13 31 06.	+ 27 11	1080		CLUSTER OF GALAXIES
UGC 08538	13 31 06.	+ 46 07	72	17.	GALAXY Sc
ZWG 271.049	13 31 06.	+ 55 05		15.7	GALAXY
ZWG 017.024	13 31 06.	- 01 38		15.3	GALAXY
ZWG 017.023	13 31 06.	- 02 29		15.5	GALAXY
RS 10	13 31 07.6	+ 28 20 43.		18.58	BLUE STELLAR OBJECT
REIN 5.304	13 31 08.29	+ 09 45 15.3			NEBULA
ZL 134	13 31 10.	+ 20 23 48.		21.5	ULTRAFAINT BLUE STAR
BC M1331+170	13 31 10.1	+ 17 04 24.		16.	QUASI-STELLAR OBJECT
SHB 267	13 31 10.1	+ 17 04 24.		16.	QUASI-STELLAR OBJECT
RS 11	13 31 10.3	+ 26 05 49.		18.32	BLUE STELLAR OBJECT
REIN 5.305	13 31 10.53	+ 09 40 09.7			NEBULA
SHB 268	13 31 11.	+ 28 08 24.		18.6	QUASI-STELLAR OBJECT
REIZ 3497	13 31 11.	+ 33 17	48	14.0	GALAXY
RS 12	13 31 11.1	+ 28 08 25.		18.15	BLUE STELLAR OBJECT
ZWG 073.027	13 31 12.	+ 11 43		15.5	GALAXY
MCG+03-35-007	13 31 12.	+ 17 44	84	15.1	GALAXY
ZWG 190.024	13 31 12.	+ 33 18		14.4	GALAXY
UGC 08539	13 31 12.	+ 33 18	84	14.4	GALAXY S
MCG+06-30-037	13 31 12.	+ 33 18	66	14.	GALAXY
MCG+09-22-078	13 31 12.	+ 51 44	36	13.9	GALAXY
MCG+09-22-077	13 31 12.	+ 55 04	30	16.	GALAXY
ZC 1331.2+7206	13 31 12.	+ 72 06	740		CLUSTER OF GALAXIES
MCG-02-35-002	13 31 12.	- 11 52	42	15.	GALAXY
MRSL 307-00/1	13 31 12.	- 62 12	1200		HII REGION
SC 1328-3050.9	13 31 13.	- 31 06 21.	36		NEBULA
ZL 135	13 31 16.	+ 20 33 12.		21.0	ULTRAFAINT BLUE STAR
ZWG 017.025	13 31 18.	+ 02 18		15.5	GALAXY
ZWG 045.022	13 31 18.	+ 03 34		15.4	GALAXY
FAI 1331+08	13 31 18.	+ 08 32			COMPACT GALAXY
ZWG 102.015	13 31 18.	+ 17 13		15.5	GALAXY
MCG+07-28-031	13 31 18.	+ 38 51	42	15.	GALAXY
ZWG 218.022	13 31 18.	+ 40 47		15.6	GALAXY
MCG+07-28-032	13 31 18.	+ 40 47	36	15.	GALAXY
ZWG 246.012	13 31 18.	+ 49 21		14.9	GALAXY
MCG+08-25-018	13 31 18.	+ 49 22	60	15.	GALAXY
ZWG 271.050	13 31 18.	+ 51 46		14.4	GALAXY
RNGC 5225	13 31 18.	+ 51 46		14.5	GALAXY
UGC 08540	13 31 18.	+ 51 46	42	14.4	GALAXY
7ZW 521	13 31 18.	+ 60 23			COMPACT GALAXY
ZWG 353.019	13 31 18.	+ 77 06		15.5	GALAXY
UGC 08541	13 31 18.	+ 77 06	60	15.5	GALAXY S
OCL 0908	13 31 18.	- 59 52	420	10.2	OPEN STAR CLUSTER
REIN 5.306	13 31 18.68	+ 09 55 20.1			NEBULA
REIN 5.307	13 31 18.93	+ 09 54 08.6			NEBULA
RRS 1.036	13 31 19.	+ 29 03		18.79	BLUE OBJECT
RRS 1.035	13 31 19.	+ 29 42		17.69	BLUE OBJECT
REIN 5.308	13 31 19.58	+ 09 56 33.4			NEBULA
REIN 5.309	13 31 21.06	+ 09 38 15.2			NEBULA
MAI 086	13 31 23.	+ 82 03	40		DWARF SPHEROIDAL GALAXY
ZWG 045.023	13 31 24.	+ 03 32		14.6	GALAXY
MCG+01-35-005	13 31 24.	+ 03 32	48	14.6	GALAXY
ZC 1331.4+1156	13 31 24.	+ 11 56	1340		CLUSTER OF GALAXIES
ZWG 073.028	13 31 24.	+ 13 32		15.6	GALAXY
ZWG 131.011	13 31 24.	+ 21 29		15.6	GALAXY
ZWG 218.023	13 31 24.	+ 38 52		15.7	GALAXY
UGC 08542	13 31 24.	+ 38 52	72	15.7	GALAXY S-IRR
MCG+09-22-079	13 31 24.	+ 55 12 30.	42	14.	GALAXY
ZWG 271.051	13 31 24.	+ 55 13		15.1	GALAXY
HN 1683	13 31 27.3	- 02 30 56.	12		NEBULA
ZWG 045.024	13 31 30.	+ 03 36		15.7	GALAXY
ZWG 045.025	13 31 30.	+ 05 00		15.3	GALAXY
UGC 08543	13 31 30.	+ 05 00	144	15.3	GALAXY Sb
MCG+01-35-006	13 31 30.	+ 05 00	60	15.3	GALAXY
ZWG 045.026	13 31 30.	+ 05 21		15.4	GALAXY
ZWG 102.016	13 31 30.	+ 17 58		15.4	GALAXY
UGC 08544	13 31 30.	+ 17 58	60	15.4	GALAXY
ZWG 102.017	13 31 30.	+ 18 10		15.7	GALAXY
ZWG 102.018	13 31 30.	+ 18 34		15.6	GALAXY
SHB 269	13 31 30.	+ 27 45 36.		17.9	QUASI-STELLAR OBJECT
RRS 1.037	13 31 30.	+ 31 05		16.61	BLUE OBJECT
ZC 1331.5+3432	13 31 30.	+ 34 32	9680		CLUSTER OF GALAXIES
MCG+06-30-038	13 31 30.	+ 36 35	36	15.	GALAXY
RS 13	13 31 30.0	+ 27 45 37.		17.87	BLUE STELLAR OBJECT
REIN 5.310	13 31 30.16	+ 10 01 36.1			NEBULA
REIN 5.311	13 31 30.42	+ 09 31 36.8			NEBULA
RS 14	13 31 30.9	+ 28 02 25.		18.53	BLUE STELLAR OBJECT
ZL 136	13 31 31.	+ 20 23 30.		21.2	ULTRAFAINT BLUE STAR
SC 1328-3114.1	13 31 31.	- 31 29 32.	6		NEBULA
RS 15	13 31 31.4	+ 26 27 19.		17.96	BLUE STELLAR OBJECT
REIZ 3498	13 31 33.	+ 02 01	24	15.0	GALAXY
SC 1328-3113.6	13 31 33.	- 31 29 02.	18		NEBULA
HN 1684	13 31 33.9	+ 02 00 17.	18		NEBULA
IC 0897	13 31 34.	+ 17 57			NONSTELLAR OBJECT
ARC 1759	13 31 34.	+ 20 31		17.6	RICH CLUSTER OF GALAXIES
LB 02691	13 31 35.	+ 52 00 12.		16.8	FAINT BLUE STAR
REIN 5.312	13 31 35.96	+ 09 54 35.1			NEBULA
ZWG 017.026	13 31 36.	+ 02 01		15.2	GALAXY
ZWG 045.027	13 31 36.	+ 05 07		15.0	GALAXY
UGC 08545	13 31 36.	+ 05 07	60	15.0	GALAXY E
MCG+01-35-007	13 31 36.	+ 05 07	60	15.0	GALAXY
ZC 1331.6+1759	13 31 36.	+ 17 59	810		CLUSTER OF GALAXIES
MCG+03-35-008	13 31 36.	+ 17 59	54	15.4	GALAXY
ZL 137	13 31 36.	+ 20 32 24.		21.9	ULTRAFAINT BLUE STAR
MCG+10-19-087	13 31 36.	+ 58 33	36	16.	GALAXY
MRK 263	13 31 36.	+ 69 10	10	16.5	GALAXY WITH UV CONTINUUM
REIZ 3499	13 31 37.	+ 18 07	36	14.3	GALAXY
IC 0896	13 31 40.	+ 05 06 41.			NONSTELLAR OBJECT
ARC 1760	13 31 40.	+ 20 29		17.2	RICH CLUSTER OF GALAXIES
RNGC 5217	13 31 41.	+ 18 07		14.0	GALAXY
ZL 138	13 31 41.	+ 20 37 48.		21.3	ULTRAFAINT BLUE STAR
ZWG 073.029	13 31 42.	+ 09 54		15.2	GALAXY
ZWG 073.030	13 31 42.	+ 10 55		15.6	GALAXY
ZWG 073.031	13 31 42.	+ 13 26		15.6	GALAXY
IC 0898	13 31 42.	+ 13 31 24.			NONSTELLAR OBJECT
ZWG 073.032	13 31 42.	+ 13 32		15.4	GALAXY
ZWG 102.019	13 31 42.	+ 18 07		14.0	GALAXY
UGC 08546	13 31 42.	+ 18 07	66	14.0	GALAXY E
ZC 1331.7+3805	13 31 42.	+ 38 05	8330		CLUSTER OF GALAXIES
REIN 5.313	13 31 44.72	+ 09 53 28.3			NEBULA
MCG+03-35-009	13 31 45.	+ 18 08	78	14.0	GALAXY
RRS 1.038	13 31 45.	+ 28 29		17.63	BLUE OBJECT
ZL 139	13 31 46.	+ 20 35 36.		21.0	ULTRAFAINT BLUE STAR
REIZ 3509	13 31 47.	+ 57 45	42	13.8	GALAXY
HN 1069	13 31 47.	- 27 03		14.	NEBULA
IC 4288	13 31 47.	- 27 04			NONSTELLAR OBJECT
REIN 5.314	13 31 47.83	+ 10 00 34.7			NEBULA
ZC 1331.8+0114	13 31 48.	+ 01 14	400		CLUSTER OF GALAXIES
ZWG 073.033	13 31 48.	+ 10 49		15.7	GALAXY
ZC 1331.8+2030	13 31 48.	+ 20 30	1410		CLUSTER OF GALAXIES
ZC 1331.8+2050	13 31 48.	+ 20 50	740		CLUSTER OF GALAXIES
TON-N 0164	13 31 48.	+ 24 19		14.8	BLUE STAR
MCG+10-19-088	13 31 48.	+ 58 32	18	16.	GALAXY
MCG-04-32-039	13 31 48.	- 27 04	36	15.	GALAXY
MCG-07-28-004	13 31 48.	- 45 18	180	14.	GALAXY
REIZ 3505	13 31 49.	+ 51 46	42	13.9	GALAXY
RRS 1.039	13 31 52.	+ 29 35		18.61	BLUE OBJECT
SC 1329-3117.3	13 31 53.	- 31 32 44.	12		NEBULA
SC 1329-3151.7	13 31 53.	- 32 07 08.	18		NEBULA
BC BSII12	13 31 53.0	+ 28 03 48.		18.61	QUASI-STELLAR OBJECT
ZWG 102.020	13 31 54.	+ 18 06		15.6	GALAXY
MCG+05-32-038	13 31 54.	+ 31 41 30.	78	15.5	GALAXY
ZC 1331.9+3234	13 31 54.	+ 32 34	540		CLUSTER OF GALAXIES
UGC 08547	13 31 54.	+ 35 00	60	16.0	GALAXY Sb
MCG+10-19-089	13 31 54.	+ 60 38	39	16.	GALAXY
MCG-04-32-040	13 31 54.	- 23 25	36	15.	GALAXY
REIN 5.315	13 31 54.37	+ 09 40 03.2			NEBULA
RRS 1.040	13 31 55.	+ 31 18		17.78	BLUE OBJECT
LB 02692	13 31 56.	+ 51 17 06.		17.1	FAINT BLUE STAR
REIZ 3501	13 31 57.	+ 35 03	42	15.1	GALAXY
RNGC 5229	13 31 58.	+ 48 10		14.5	GALAXY
ARP 036	13 31 59.	+ 31 41			PECULIAR GALAXY
LB 02693	13 31 59.	+ 51 53 12.		16.8	FAINT BLUE STAR
LB 00270	13 31 59.	+ 59 05 24.		14.5	FAINT BLUE STAR
SC 1329-3114.1	13 31 59.	- 31 29 31.	12		NEBULA
ZWG 073.034	13 32 00.	+ 08 58		15.6	GALAXY
ZWG 073.035	13 32 00.	+ 09 00		15.6	GALAXY
ZWG 073.036	13 32 00.	+ 09 04		15.3	GALAXY
MCG+02-35-003	13 32 00.	+ 09 04	48	15.3	GALAXY
ZC 1332.0+2605	13 32 00.	+ 26 05	5170		CLUSTER OF GALAXIES
ZWG 161.079	13 32 00.	+ 27 32		15.6	GALAXY
VV 004C	13 32 00.	+ 31 40 30.	12	18.	INTERACTING GALAXY
VV 004B	13 32 00.	+ 31 40 30.	9	16.5	INTERACTING GALAXY
VV 004A	13 32 00.	+ 31 40 30.	42	16.	INTERACTING GALAXY
ZWG 161.080	13 32 00.	+ 31 41		15.5	GALAXY
UGC 08548	13 32 00.	+ 31 41	78	15.5	GALAXY SB?
UGC 08549	13 32 00.	+ 32 28	72	16.5	GALAXY Sb
ZWG 246.013	13 32 00.	+ 48 10		14.6	GALAXY
UGC 08550	13 32 00.	+ 48 10	222	14.6	GALAXY SBc
MCG+08-25-019	13 32 00.	+ 48 10	180	14.	GALAXY
MCG+09-22-080	13 32 00.	+ 52 57	90	14.	GALAXY
UGC 08551	13 32 00.	+ 52 59	96	16.0	GALAXY Sb
ZWG 017.027	13 32 00.	- 00 05		15.4	GALAXY
KARA.68 227	13 32 00.	- 12 03	87		DWARF GALAXY
MCG-06-30-010	13 32 00.	- 34 02 30.	72	14.	GALAXY
MCG-06-30-011	13 32 00.	- 36 15	18	16.	GALAXY
ZL 140	13 32 02.	+ 20 27 24.		21.8	ULTRAFAINT BLUE STAR
REIZ 3504	13 32 02.	+ 37 27	54	15.0	GALAXY
HN 1685	13 32 02.3	- 00 06 01.	24		NEBULA
ZL 141	13 32 03.	+ 20 25 24.		22.2	ULTRAFAINT BLUE STAR
REIZ 3502	13 32 03.	+ 35 08	12	15.2	GALAXY
REIZ 3503	13 32 03.	+ 35 12	24	14.6	GALAXY
REIN 5.316	13 32 03.13	+ 09 39 52.8			NEBULA
RNGC 5223	13 32 04.	+ 34 57		14.5	GALAXY
RRS 1.041	13 32 05.	+ 30 55		18.65	BLUE OBJECT
HN 1070	13 32 05.	- 26 52		14.	NEBULA
IC 4289	13 32 05.	- 26 53			NONSTELLAR OBJECT
SC 1329-2829.4	13 32 05.	- 28 44 49.	18		NEBULA

OBJECT NAME	RIGHT ASCEN.	DECLINATION	DIAM.	MAGN.	TYPE OF OBJECT
ZWG 045.028	13 32 06.	+ 04 23		14.9	GALAXY
UGC 08552	13 32 06.	+ 04 23	60	14.9	GALAXY SBb
MCG+01-35-008	13 32 06.	+ 04 23	48	14.9	GALAXY
ZC 1332.1+2421	13 32 06.	+ 24 21	2220		CLUSTEF OF GALAXIES
ZC 1332.1+3148	13 32 06.	+ 31 48	1140		CLUSTER OF GALAXIES
ZWG 190.025	13 32 06.	+ 34 57		14.4	GALAXY
UGC 08553	13 32 06.	+ 34 57	102	14.4	GALAXY E
UGC 08554	13 32 06.	+ 37 27	96	16.5	GALAXY Sb-c
MCG+06-30-039	13 32 06.	+ 37 29	48	15.	GALAXY
MCG+10-19-090	13 32 06.	+ 58 07	45	16.	GALAXY
MCG-04-32-041	13 32 06.	- 26 53	15	15.	GALAXY
REIN 5.317	13 32 06.00	+ 08 56 54.2			NEBULA
REIN 5.318	13 32 06.27	+ 08 58 32.4			NEBULA
SC 1329-3055.5	13 32 07.	- 31 10 55.	12		NEBULA
RNGC 5213	13 32 08.	+ 04 23		15.0	GALAXY
ZL 142	13 32 08.	+ 20 27 54.		22.4	ULTRAFAINT BLUE STAF
SC 1329-3106.3	13 32 08.	- 31 21 43.	6		NEBULA
REIN 5.319	13 32 08.43	+ 09 02 58.7			NEBULA
REIZ 3507	13 32 09.	+ 34 58	24	14.0	GALAXY
REIZ 3508	13 32 09.	+ 35 30	66	13.8	GALAXY
ZL 143	13 32 10.	+ 20 31 06.		21.2	ULTRAFAINT BLUF STAR
RRS 1.042	13 32 11.	+ 31 14		17.73	BLUE OBJECT
RNGC 5215	13 32 11.	- 33 14		15.0	GALAXY
VV 018B	13 32 12.	+ 04 23	12	18.	INTERACTING GALAXY
VV 018A	13 32 12.	+ 04 23	60	14.	INTEFACTING GALAXY
ZWG 073.037	13 32 12.	+ 09 36		14.3	GALAXY
UGC 08555	13 32 12.	+ 09 36	84	14.3	GALAXY Sc
MCG+02-35-004	13 32 12.	+ 09 36	96	14.3	GALAXY
MCG+06-30-040	13 32 12.	+ 34 57	24	14.	GALAXY
ZC 1332.2+4134	13 32 12.	+ 41 34	1010		CLUSTER OF GALAXIES
MRK 264	13 32 12.	+ 52 09	12	17.	GALAXY WITH UV CONTINUUM
MCG+10-19-091	13 32 12.	+ 57 01	39	16.	GALAXY
SC 1329-3108.7	13 32 12.	- 31 24 07.	12		NEBULA
MCG-05-32-040	13 32 12.	- 33 14	48	15.	GALAXY
BC RSII13	13 32 12.0	+ 27 41 00.		17.94	QUASI-STELLAR OBJECT
REIN 5.320	13 32 12.14	+ 09 34 47.8			NEBULA
LB 00685	13 32 13.	+ 55 16 24.		15.6	FAINT BLUE STAR
REIN 5.321	13 32 13.48	+ 09 33 41.5			NEBULA
REIN 5.322	13 32 13.97	+ 09 35 33.4			NEBULA
IC 0900	13 32 14.	+ 09 35 43.			NONSTELLAR OBJECT
REIZ 3506	13 32 14.	+ 28 13	12	15.0	GALAXY
HN 1686	13 32 14.9	- 01 09 18.	18		NEBULA
ZL 144	13 32 16.	+ 20 26 42.		21.4	ULTRAFAINT BLUF STAR
RNGC 5228	13 32 16.	+ 35 02		14.5	GALAXY
REIN 5.323	13 32 16.41	+ 09 40 26.2			NEBULA
REIN 5.324	13 32 16.79	+ 09 43 53.6			NEBULA
BC 4C55.27	13 32 17.	+ 55 16			QUASI-STELLAR OBJECT
ZWG 017.028	13 32 18.	+ 01 58		15.4	GALAXY
ZWG 073.038	13 32 18.	+ 11 44		15.7	GALAXY
ZWG 190.026	13 32 18.	+ 35 02		14.5	GALAXY
UGC 08556	13 32 18.	+ 35 02	60	14.5	GALAXY F-S0
MRK 458	13 32 18.	+ 40 22	7	16.	GALAXY WITH UV CONTINUUM
MCG+10-19-092	13 32 18.	+ 61 59	36	16.	GALAXY
UGC 08557	13 32 18.	+ 78 32	66	16.0	GALAXY Sa-b
MCG-05-32-042	13 32 18.	- 29 51	30	15.	GALAXY
SC 1329-3055.1	13 32 18.	- 31 10 31.	12		NEBULA
SC 1329-3258.4	13 32 18.	- 33 13 49.	60		NEBULA
MCG-05-32-041	13 32 18.	- 33 15	54	15.	GALAXY
MCG-06-30-012	13 32 18.	- 35 00	90	14.5	GALAXY
ZL 145	13 32 19.	+ 20 31 30.		20.8	ULTRAFAINT BLUE STAR
REIZ 3500	13 32 19.	- 07 50	18	14.9	GALAXY
REIZ 3511	13 32 21.	+ 35 03	18	14.4	GALAXY
MCG+06-30-041	13 32 21.	+ 35 30	30	15.	GALAXY
MCG+06-30-042	13 32 21.	+ 36 18 30.	42	15.	GALAXY
MCG+07-28-033	13 32 21.	+ 38 32	42	16.	GALAXY
IC 0899	13 32 22.	- 07 50 05.			NONSTELLAR OBJECT
ZL 146	13 32 23.	+ 20 36 00.		18.4	ULTRAFAINT BLUE STAR
SC 1329-3055.7	13 32 23.	- 31 11 07.	6		NEBULA
TON-N 0719	13 32 24.	+ 26 12		15.5	BLUE STAR
MCG+06-30-044	13 32 24.	+ 33 11 30.	15	15.	GALAXY
ZWG 190.027	13 32 24.	+ 33 13		15.4	GALAXY
MCG+06-30-043	13 32 24.	+ 35 02	42	13.5	GALAXY
ZWG 294.045	13 32 24.	+ 61 59		15.5	GALAXY
MCG-03-35-005	13 32 24.	- 15 36	42	15.	GALAXY
MCG-05-32-043	13 32 24.	- 33 40	42	15.	GALAXY
ZL 147	13 32 27.	+ 20 29 30.		21.6	ULTRAFAINT BLUF STAR
RNGC 5222	13 32 28.	+ 14 00		14.0	GALAXY
RNGC 5221	13 32 28.	+ 14 05		14.5	GALAXY
SC 1329-3054.9	13 32 28.	- 31 10 19.	12		NEBULA
SC 1329-3201.7	13 32 28.	- 32 17 07.	30		NEBULA
ZWG 073.039	13 32 30.	+ 14 00		14.1	GALAXY
UGC 08558	13 32 30.	+ 14 00	96	14.1	GALAXY E+SPIRL
KARA.72 383A	13 32 30.	+ 14 00	90	14.1	PART OF DOUBLE GALAXY
MCG+02-35-005	13 32 30.	+ 14 00	24	14.1	GALAXY
VV 315B	13 32 30.	+ 14 00 00.	180	15.	INTERACTING GALAXY
ZWG 073.040	13 32 30.	+ 14 05		14.5	GALAXY
8ZW 1332+14.1	13 32 30.	+ 14 05		14.5	COMPACT GALAXY
UGC 08559	13 32 30.	+ 14 05	180	14.5	GALAXY S(b)
MCG+02-35-006	13 32 30.	+ 14 05	96	14.5	GALAXY
VV 315C	13 32 30.	+ 14 06 30.		14.	INTERACTING GALAXY
VV 315A	13 32 30.	+ 14 06 30.	60	14.	INTERACTING GALAXY
ZC 1332.5+1740	13 32 30.	+ 17 40	1080		CLUSTEF OF GALAXIES
ZC 1332.5+2556	13 32 30.	+ 25 56	670		CLUSTER OF GALAXIES
STOCK 46	13 32 30.	+ 34 08			BLUE KNOT NEAR ELLIP GLXY
ZC 1332.5+3901	13 32 30.	+ 39 01	1010		CLUSTER OF GALAXIES
ZWG 246.014	13 32 30.	+ 46 25		15.6	GALAXY
MCG+13-10-008	13 32 30.	+ 78 32	78	16.	GALAXY
MCG-05-32-044	13 32 30.	- 27 44	72	15.	GALAXY
ARP 288	13 32 31.	+ 14 04			PECULIAR GALAXY
MCG+04-32-014	13 32 33.	+ 22 50	30	15.5	GALAXY
MCG+04-32-013	13 32 33.	+ 26 28 30.	18	15.0	GALAXY
MCG-06-30-013	13 32 33.	- 33 56	72	14.	GALAXY
RNGC 5226	13 32 34.	+ 14 10			GALAXY
REIZ 3513	13 32 34.	+ 34 18	30	15.2	GALAXY
HN 1071	13 32 35.	- 27 46	60		NEBULA
ZWG 045.029	13 32 36.	+ 03 54		15.5	GALAXY
ZWG 045.030	13 32 36.	+ 06 44		15.0	GALAXY
MCG+01-35-009	13 32 36.	+ 06 44	36	15.0	GALAXY
KARA.72 383B	13 32 36.	+ 14 00	42		PART OF DOUBLF GALAXY
ZWG 131.012	13 32 36.	+ 26 28		15.0	GALAXY
ZWG 161.081	13 32 36.	+ 31 39		15.1	GALAXY
UGC 08560	13 32 36.	+ 31 39	78	15.1	GALAXY Sb
MCG+06-30-045	13 32 36.	+ 34 17 30.	18	15.	GALAXY
ZWG 190.028	13 32 36.	+ 34 18		13.8	GALAXY
UGC 08561	13 32 36.	+ 34 18	66	13.8	GALAXY Sc
MCG+08-25-020	13 32 36.	+ 46 24	15	16.	GALAXY
MCG+09-22-081	13 32 36.	+ 50 42 30.	18	17.	GALAXY
MCG+09-22-082	13 32 36.	+ 51 51	90	13.7	GALAXY

OBJECT NAME	RIGHT ASCEN.	DECLINATION	DIAM.	MAGN.	TYPE OF OBJECT
ZWG 294.046	13 32 36.	+ 56 44		15.3	GALAXY
MCG+10-19-093	13 32 36.	+ 60 30	39	14.	GALAXY
ZWG 294.047	13 32 36.	+ 61 57		15.6	GALAXY
IC 4290	13 32 36.	- 27 47			NONSTELLAR OBJECT
SVEN 413	13 32 36.	- 29 50 54.	12	15.4	GALAXY
HN 1687	13 32 36.9	- 02 45 24.	18		NEBULA
REIZ 3510	13 32 38.	+ 03 55	18	14.9	GALAXY
RNGC 5224	13 32 39.	+ 06 44		15.0	GALAXY
ARP 183	13 32 41.	+ 31 39			PECULIAR GALAXY
ZWG 073.041	13 32 42.	+ 10 56		15.2	GALAXY
UGC 08562	13 32 42.	+ 10 56	60	15.2	GALAXY DBL SYS
KARA.72 385A	13 32 42.	+ 10 56	30	15.2	PART OF DOUBLE GALAXY
MCG+02-35-007	13 32 42.	+ 10 56	24	15.2	GALAXY
ZWG 073.042	13 32 42.	+ 10 57		14.9	GALAXY
UGC 08563	13 32 42.	+ 10 57	60	14.9	GALAXY DBL SYS
KARA.72 385B	13 32 42.	+ 10 57	36	14.9	PART OF DOUBLE GALAXY
MCG+02-35-008	13 32 42.	+ 10 57	24	14.9	GALAXY
MCG+05-32-039	13 32 42.	+ 31 39	60	15.1	GALAXY
MCG+06-30-046	13 32 42.	+ 34 18	66	13.	GALAXY
MCG+07-28-034	13 32 42.	+ 38 42	78	14.5	GALAXY
ZWG 218.024	13 32 42.	+ 38 43		14.7	GALAXY
UGC 08564	13 32 42.	+ 38 43	102	14.7	GALAXY S0
ZWG 271.052	13 32 42.	+ 51 53		14.2	GALAXY
RNGC 5238	13 32 42.	+ 51 53		14.0	GALAXY
UGC 08565	13 32 42.	+ 51 53	120	14.2	GALAXY
KARA.72 384B	13 32 42.	+ 51 53	36		PART OF DOUBLE GALAXY
KARA.72 384A	13 32 42.	+ 51 53	48	14.2	PART OF DOUBLE GALAXY
MCG+10-19-094	13 32 42.	+ 56 42	15	15.	GALAXY
MCG-02-35-004	13 32 42.	- 10 37	36	15.5	GALAXY
MCG-02-35-003	13 32 42.	- 10 38	36	15.5	GALAXY
MCG-06-30-014	13 32 42.	- 33 55	60	16.	GALAXY
RRS 1.043	13 32 44.	+ 30 03		19.80	BLUE OBJECT
VV 211B	13 32 45.	+ 10 57	48	15.	INTERACTING GALAXY
VV 211A	13 32 45.	+ 10 57	27	15.	INTERACTING GALAXY
REIZ 3514	13 32 46.	+ 34 18	60	14.2	GALAXY
ZWG 017.029	13 32 48.	+ 01 40		14.6	GALAXY
UGC 08566	13 32 48.	+ 01 40	96	14.6	GALAXY SBb
MCG+00-35-010	13 32 48.	+ 01 40	120	14.6	GALAXY
FEIG 084	13 32 48.	+ 13 43		11.9	FAINT BLUE STAR
MRK 459	13 32 48.	+ 34 18	9	17.	GALAXY WITH UV CONTINUUM
1ZW 063	13 32 48.	+ 35 44			COMPACT GALAXY
ZC 1332.8+5043	13 32 43.	+ 50 43	2080		CLUSTER OF GALAXIES
ZWG 294.048	13 32 48.	+ 60 31		15.7	GALAXY
RNGC 5227	13 32 50.	+ 01 40		14.5	GALAXY
RS 16	13 32 50.2	+ 27 55 24.		18.67	BLUE STELLAR OBJECT
REIZ 3512	13 32 51.	+ 01 41	48	14.1	GALAXY
RNGC 5233	13 32 52.	+ 34 56		15.0	GALAXY
RRS 1.044	13 32 53.	+ 30 58		17.15	BLUE OBJECT
ZWG 131.013	13 32 54.	+ 25 17		15.6	GALAXY
MCG+05-32-040	13 32 54.	+ 26 40	48	15.5	GALAXY
UGC 08567	13 32 54.	+ 27 54	66	16.5	GALAXY
TON-N 0165	13 32 54.	+ 28 09		14.9	BLUE STAR
ZWG 190.029	13 32 54.	+ 34 56		14.8	GALAXY
UGC 08568	13 32 54.	+ 34 56	84	14.8	GALAXY Sa-b
MCG-04-32-042	13 32 54.	- 23 49	120	13.	GALAXY
MCG-05-32-045	13 32 54.	- 30 37	60	15.	GALAXY
ZL 148	13 32 55.	+ 20 22 54.		17.1	ULTRAFAINT BLUF STAR
ARC 1764	13 32 55.	+ 60 11		17.2	RICH CLUSTER OF GALAXIES
RRS 1.045	13 32 56.	+ 30 00		19.70	BLUE OBJECT
MCG+08-25-021	13 32 57.	+ 46 01	60	16.	GALAXY
SC 1330-3021.9	13 32 57.	- 30 37 18.	60		NEBULA
RS 17	13 32 57.1	+ 28 08 00.		15.68	BLUE STELLAR OBJECT
HN 1072	13 32 59.	- 27 25		13.5	NEBULA
IC 4292	13 32 59.	- 27 26			NONSTELLAR OBJECT
VDB.66G 178	13 33	+ 46 11	70		DWARF GALAXY
VDB.66G 177	13 33	+ 46 26	70		DWARF GALAXY
LB 09981	13 33	- 87 03		14.8	FAINT BLUE STAR
ZWG 045.031	13 33 00.	+ 06 27		15.4	GALAXY
UGC 08569	13 33 00.	+ 06 27	60	15.4	GALAXY Sa-b
MCG+01-35-010	13 33 00.	+ 06 27	48	15.4	GALAXY
ZWG 161.082	13 33 00.	+ 26 41		15.5	GALAXY
UGC 08570	13 33 00.	+ 26 41	66	15.5	GALAXY
IC 4297	13 33 00.	+ 26 41 46.			NONSTELLAR OBJECT
MCG+06-30-047	13 33 00.	+ 34 56	60	13.5	GALAXY
ZWG 218.025	13 33 00.	+ 41 27		15.7	GALAXY
ZWG 246.015	13 33 00.	+ 46 02		15.4	GALAXY
REIZ 3522	13 33 00.	+ 51 53	108	13.7	GALAXY
ZC 1333.0+6012	13 33 00.	+ 60 12	1410		CLUSTEF OF GALAXIES
ZWG 294.049	13 33 00.	+ 62 15		14.1	GALAXY
UGC 08571	13 33 00.	+ 62 15	72	14.1	GALAXY S
MCG+10-19-095	13 33 00.	+ 62 15	72	14.	GALAXY
KEEN 5216A	13 33 00.	+ 62 16		14.	GALAXY
MCG+14-06-027	13 33 00.	+ 86 13	84	16.	GALAXY
MCG-05-32-047	13 33 00.	- 30 07	42	15.	GALAXY
MCG-05-32-046	13 33 00.	- 33 13	84	15.	GALAXY
SC 1330-2952.1	13 33 01.	- 30 07 30.	30		NEBULA
REIZ 3517	13 33 02.	+ 28 10	24	14.6	GALAXY
RRS 1.046	13 33 02.	+ 31 27		17.46	BLUF OBJECT
RNGC 5216A	13 33 02.	+ 62 15		14.0	GALAXY
REIZ 3518	13 33 03.	+ 34 56	30	14.3	GALAXY
FATH 2.140	13 33 03.	- 00 40			NEBULA
SC 1330-2954.8	13 33 03.	- 30 10 11.	18		NEBULA
RNGC 5230	13 33 04.	+ 13 56		13.0	GALAXY
RNGC 5220	13 33 04.	- 33 12			GALAXY
SN 1970P	13 33 05.	+ 13 55		17.2	SUPFRNOVA
SC 1330-3256.7	13 33 05.	- 33 12 06.	18		NEBULA
HN 1688	13 33 05.3	- 00 39 59.	18		NEBULA
ZWG 045.032	13 33 06.	+ 03 32		15.7	GALAXY
ZWG 045.033	13 33 06.	+ 07 48		15.3	GALAXY
UGC 08572	13 33 06.	+ 07 48	102	15.3	GALAXY Sb
ZWG 073.043	13 33 06.	+ 13 56		13.4	GALAXY
8ZW 1333+13.9	13 33 06.	+ 13 56		18.3	COMPACT GALAXY
UGC 08573	13 33 06.	+ 13 56	126	13.4	GALAXY Sc
MCG+02-35-009	13 33 06.	+ 13 56	132	13.4	GALAXY
ZC 1333.1+1605	13 33 06.	+ 16 05	810		CLUSTEF OF GALAXIES
ZC 1333.1+2320	13 33 06.	+ 23 20	1080		CLUSTER OF GALAXIES
ZWG 161.083	13 33 06.	+ 28 10		15.5	GALAXY
ZWG 017.030	13 33 06.	- 00 40		15.5	GALAXY
MCG-01-35-002	13 33 06.	- 04 40	72	14.5	GALAXY
MCG-04-32-043	13 33 06.	- 23 00	36	15.	GALAXY
MCG-06-30-015	13 33 06.	- 34 02	36	15.	GALAXY
HOLM 529B	13 33 08.	+ 03 16	18	15.3	PART OF MULTIPLE GALAXY
REIZ 3515	13 33 09.	+ 00 39	24	16.3	GALAXY
RRS 1.047	13 33 09.	+ 30 33		16.73	BLUE OBJECT
MCG+06-30-048	13 33 09.	+ 33 40	9	16.	GALAXY
IC 4300	13 33 09.	+ 33 40 23.			NONSTELLAR OBJECT
ARC 1763	13 33 09.	+ 41 14		17.7	RICH CLUSTER OF GALAXIES
ZWG 045.034	13 33 12.	+ 03 15		14.7	GALAXY

OBJECT NAME	RIGHT ASCEN.	DECLINATION	DIAM.	MAGN.	TYPE OF OBJECT
UGC 08574	13 33 12.	+ 03 15	78	14.7	GALAXY SBa
MCG+01-35-011	13 33 12.	+ 03 15	60	14.7	GALAXY
ZWG 073.044	13 33 12.	+ 09 14		15.5	GALAXY
UGC 08575	13 33 12.	+ 09 14	156	15.5	GALAXY DWF:IRR
MCG+02-35-010	13 33 12.	+ 09 14	156	15.5	GALAXY
UGC 08576	13 33 12.	+ 13 35	78	16.0	GALAXY S
ARC 1762	13 33 12.	+ 23 22		17.6	RICH CLUSTER OF GALAXIES
ZC 1333.2+3105	13 33 12.	+ 31 05	740		CLUSTER OF GALAXIES
ZWG 190.030	13 33 12.	+ 33 58		15.7	GALAXY
MCG+06-30-049	13 33 12.	+ 33 58 30.	18	15.5	GALAXY
UGC 08577	13 33 12.	+ 45 00	78	16.5	GALAXY Sc-IRR
MCG+08-25-022	13 33 12.	+ 45 00	90	16.	GALAXY
ZC 1333.2+4515	13 33 12.	+ 45 15	1280		CLUSTER OF GALAXIES
MCG-05-32-048	13 33 12.	- 32 46	48	15.	GALAXY
OCL 0906	13 33 12.	- 61 54	240	10.5	OPEN STAR CLUSTER
RNGC 5231	13 33 14.	+ 03 15		14.5	GALAXY
REIZ 3516	13 33 14.	+ 03 15	30	14.0	GALAXY
HOLM 529A	13 33 14.	+ 03 16	36	13.8	PART OF MULTIPLE GALAXY
IC 0901	13 33 14.	+ 13 34 16.			NONSTELLAR OBJECT
SC 1330-3230.3	13 33 15.	- 32 45 41.	30		NEBULA
RS 18	13 33 15.1	+ 27 06 48.		16.94	BLUE STELLAR OBJECT
REIN 5.325	13 33 16.22	+ 09 13 23.7			NEBULA
ZC 1333.3+0358	13 33 18.	+ 03 58	1750		CLUSTER OF GALAXIES
ZWG 073.045	13 33 18.	+ 13 35		15.2	GALAXY
MCG+02-35-011	13 33 18.	+ 13 35	90	15.2	GALAXY
TON-N 0166	13 33 18.	+ 25 41		14.5	BLUE STAR
ZWG 131.014	13 33 18.	+ 26 08		15.6	GALAXY
MCG+04-32-015	13 33 18.	+ 26 08	54	15.6	GALAXY
ZWG 161.084	13 33 18.	+ 29 29		15.1	GALAXY
HARO 38	13 33 18.	+ 29 29			BLUE EMISSION-LINE GALAXY
UGC 08578	13 33 18.	+ 29 29	66	15.1	GALAXY
MCG+05-32-041	13 33 18.	+ 29 29	54	15.1	GALAXY
RRS 1.043	13 33 18.	+ 29 47		17.82	BLUE OBJECT
UGC 08579	13 33 18.	+ 33 33	60	16.0	GALAXY S
IC 4301	13 33 18.	+ 33 37 47.			NONSTELLAR OBJECT
UGC 08580	13 33 18.	+ 33 44	78	16.5	GALAXY Sc
RS 19	13 33 18.4	+ 26 59 12.		15.99	BLUE STELLAR OBJECT
IC 4302	13 33 20.	+ 33 44 05.			NONSTELLAR OBJECT
REIN 5.327	13 33 20.39	+ 09 13 43.4			NEBULA
MCG+06-30-050	13 33 21.	+ 33 37	48	15.	GALAXY
MCG-04-32-044	13 33 21.	- 25 37	24	14.	GALAXY
RNGC 5234	13 33 21.	- 49 38			GALAXY
REIN 5.328	13 33 21.99	+ 09 01 45.2			NEBULA
RRS 1.049	13 33 22.	+ 28 42		18.45	BLUE OBJECT
HN 1073	13 33 22.	- 25 38		13.	NEBULA
RNGC 5232	13 33 23.	- 08 14		13.0	GALAXY
IC 4293	13 33 23.	- 25 39			NONSTELLAR OBJECT
UGC 08581	13 33 24.	+ 02 00	66	16.0	GALAXY DBL SYS
ZWG 073.046	13 33 24.	+ 13 41		15.4	GALAXY
ZC 1333.4+1935	13 33 24.	+ 19 35	1010		CLUSTER OF GALAXIES
ZC 1333.4+2431	13 33 24.	+ 24 31	670		CLUSTER OF GALAXIES
MCG+05-32-042	13 33 24.	+ 27 39 30.	48	15.7	GALAXY
ZWG 161.085	13 33 24.	+ 27 40		15.7	GALAXY
MCG+06-30-051	13 33 24.	+ 33 44	78	15.	GALAXY
ZC 1333.4+5847	13 33 24.	+ 58 47	2220		CLUSTER OF GALAXIES
ZWG 353.020	13 33 24.	+ 76 01		15.4	GALAXY
MCG-01-35-003	13 33 24.	- 08 14	78	13.	GALAXY
SC 1330-3029.9	13 33 25.	- 30 45 17.	18		NEBULA
REIN 5.329	13 33 25.88	+ 09 02 44.6			NEBULA
HN 1689	13 33 26.9	+ 01 59 20.	60		NEBULA
REIZ 3521	13 33 27.	+ 02 00	48	15.1	GALAXY
MCG-02-35-005	13 33 27.	- 14 39	60	15.	GALAXY
SC 1330-3115.8	13 33 28.	- 31 31 11.	24		NEBULA
SC 1330-3117.6	13 33 28.	- 31 32 59.	24		NEBULA
HN 1690	13 33 28.2	+ 00 30 32.	18		NEBULA
HN 1691	13 33 29.6	- 01 20 22.	24		NEBULA
ZWG 017.032	13 33 30.	+ 01 59		15.2	GALAXY
ZWG 045.035	13 33 30.	+ 06 47		15.7	GALAXY
ZWG 045.036	13 33 30.	+ 06 50		14.9	GALAXY
UGC 08582	13 33 30.	+ 06 50	78	14.9	GALAXY SB
MCG+01-35-012	13 33 30.	+ 06 50	72	14.9	GALAXY
ZWG 073.047	13 33 30.	+ 13 36		15.6	GALAXY
MCG+02-35-012	13 33 30.	+ 13 36	36	15.6	GALAXY
FAI 1333+14	13 33 30.	+ 14 24			COMPACT GALAXY
TON-N 0720	13 33 30.	+ 24 41		14.0	BLUE STAR
TON-N 0721	13 33 30.	+ 25 19		15.3	BLUE STAR
TON-N 0722	13 33 30.	+ 27 11		15.2	BLUE STAR
ZWG 190.031	13 33 30.	+ 35 15		14.6	GALAXY
UGC 08583	13 33 30.	+ 35 15	72	14.6	GALAXY SBa-b
MCG+09-22-083	13 33 30.	+ 51 26	48	17.	GALAXY
ZWG 294.050	13 33 30.	+ 56 40		15.4	GALAXY
ZWG 017.031	13 33 30.	- 01 20		15.4	GALAXY
MCG-01-35-004	13 33 30.	- 04 43	54	15.5	GALAXY
RRS 1.050	13 33 31.	+ 31 01		18.13	BLUE OBJECT
REIZ 3520	13 33 31.	- 08 14	30	13.6	GALAXY
REIN 5.330	13 33 31.03	+ 08 53 42.5			NEBULA
HN 1692	13 33 31.3	+ 01 35 26.	12		NEBULA
REIZ 3519	13 33 32.	- 08 30	18	15.4	GALAXY
RNGC 5235	13 33 33.	+ 06 50		15.0	GALAXY
REIZ 3524	13 33 33.	+ 35 49	72	14.2	GALAXY
RS 20	13 33 34.0	+ 26 25 06.		18.29	BLUE STELLAR OBJECT
REIZ 3526	13 33 35.	+ 56 41	30	14.7	GALAXY
ZWG 045.037	13 33 36.	+ 03 45		15.5	GALAXY
MCG+01-35-013	13 33 36.	+ 03 45	108	15.5	GALAXY
ZC 1333.6+1200	13 33 36.	+ 12 00	810		CLUSTER OF GALAXIES
ZWG 161.086	13 33 36.	+ 26 58		15.6	GALAXY
ZWG 161.087	13 33 36.	+ 28 07		15.7	GALAXY
MCG+06-30-052	13 33 36.	+ 35 15 30.	66	14.	GALAXY
MCG+06-30-053	13 33 36.	+ 36 07	18	16.	GALAXY
MCG+09-22-084	13 33 36.	+ 56 39	18	15.	GALAXY
FATH 2.141	13 33 36.	- 00 47	24		NEBULA
ZC 1333.6-0321	13 33 36.	- 03 21	2550		CLUSTER OF GALAXIES
FATH 2.142	13 33 38.	- 00 15	19		NEBULA
IC 4291	13 33 38.	- 61 48 09.			NONSTELLAR OBJECT
HN 1693	13 33 38.0	- 00 46 58.	18		NEBULA
MCG+07-28-035	13 33 39.	+ 40 31	48	15.	GALAXY
FATH 2.143	13 33 39.	- 00 47	19		NEBULA
HN 1694	13 33 39.2	- 00 15 52.	12		NEBULA
HN 1695	13 33 40.4	- 00 47 28.	24		NEBULA
RNGC 5240	13 33 41.	+ 35 50		14.0	GALAXY
ZWG 045.038	13 33 42.	+ 05 00		15.4	GALAXY
ZWG 073.048	13 33 42.	+ 09 31		15.5	GALAXY
FAI 1333+09	13 33 42.	+ 09 44			COMPACT GALAXY
ZWG 073.049	13 33 42.	+ 10 44		15.3	GALAXY
UGC 08585	13 33 42.	+ 10 44	78	15.3	GALAXY SBc
MCG+02-35-013	13 33 42.	+ 10 44	72	15.3	GALAXY
ZWG 131.015	13 33 42.	+ 22 54		15.7	GALAXY
MCG+06-30-055	13 33 42.	+ 33 40	63	14.5	GALAXY

OBJECT NAME	RIGHT ASCEN.	DECLINATION	DIAM.	MAGN.	TYPE OF OBJECT
ZWG 190.032	13 33 42.	+ 33 41		15.0	GALAXY
UGC 08586	13 33 42.	+ 33 41	78	15.0	GALAXY Sa-b
IC 4304	13 33 42.	+ 33 41 06.			NONSTELLAR OBJECT
MCG+06-30-054	13 33 42.	+ 33 43	18	14.5	GALAXY
IC 4305	13 33 42.	+ 33 43 42.			NONSTELLAR OBJECT
ZWG 190.033	13 33 42.	+ 33 44		15.1	GALAXY
ZWG 190.034	13 33 42.	+ 35 50		14.1	GALAXY
UGC 08587	13 33 42.	+ 35 50	126	14.1	GALAXY SBc
MCG+06-30-056	13 33 42.	+ 35 51	108	13.	GALAXY
ZWG 218.026	13 33 42.	+ 40 32		15.5	GALAXY
ZC 1333.7+4117	13 33 42.	+ 41 17	1550		CLUSTER OF GALAXIES
ZWG 246.016	13 33 42.	+ 46 11		15.3	GALAXY
UGC 08588	13 33 42.	+ 46 11	84	15.3	GALAXY DWRF SP
MCG+08-25-023	13 33 42.	+ 46 11	72	15.	GALAXY
ZWG 017.033	13 33 42.	- 00 47		15.4	GALAXY
UGC 08584	13 33 42.	- 00 47	72	15.4	GALAXY TRP SYS
MCG+00-35-012	13 33 42.	- 00 47	24	15.4	GALAXY
MCG+00-35-011	13 33 42.	- 00 47	18		GALAXY
MCG-01-35-005	13 33 42.	- 08 14	42	16.	GALAXY
MCG-03-35-006	13 33 42.	- 17 00	42	15.5	GALAXY
HN 1074	13 33 42.	- 28 31		14.	NEBULA
IC 4294	13 33 42.	- 28 32			NONSTELLAR OBJECT
MCG-05-32-049	13 33 42.	- 28 48	48	15.	GALAXY
HN 1075	13 33 42.	- 28 50	30	14.	NEBULA
IC 4295	13 33 42.	- 28 51			NONSTELLAR OBJECT
VHA 149	13 33 42.	- 61 48	240		OPEN STAR CLUSTER
REIN 5.331	13 33 42.50	+ 09 14 23.1			NEBULA
SC 1330-2843.8	13 33 43.	- 28 59 10.	60		NEBULA
SC 1331+4649.5	13 33 47.	+ 46 34 08.	42		NEBULA
LB 02694	13 33 47.	+ 52 27 00.		16.7	FAINT BLUE STAR
LB 00250	13 33 47.	+ 56 11 12.		18.0	FAINT BLUE STAR
HN 1696	13 33 47.7	- 00 13 10.	18		NEBULA
ZWG 045.039	13 33 48.	+ 03 35		14.9	GALAXY
MCG+01-35-014	13 33 48.	+ 03 45	48	14.9	GALAXY
ZWG 073.050	13 33 48.	+ 11 30		15.6	GALAXY
MCG+06-30-057	13 33 48.	+ 35 04	30	15.5	GALAXY
ZC 1333.8+5931	13 33 48.	+ 59 31	4370		CLUSTER OF GALAXIES
FATH 2.144	13 33 48.	- 00 12	27		NEBULA
IC 4296	13 33 48.	- 33 43	54	11.9	GALAXY E1
RRS 1.051	13 33 49.	+ 29 20		19.21	BLUE OBJECT
SC 1331-2958.8	13 33 49.	- 30 14 10.	18		NEBULA
REIZ 3523	13 33 50.	+ 04 15	18	14.2	GALAXY
MCG-06-30-016	13 33 51.	- 33 42	180	13.1	GALAXY
RS 21	13 33 51.3	+ 27 26 12.		18.08	BLUE STELLAR OBJECT
RS 22	13 33 51.8	+ 27 25 30.		17.56	BLUE STELLAR OBJECT
RRS 1.052	13 33 52.	+ 30 53		18.79	BLUE OBJECT
IC 4298	13 33 52.	- 26 19			NONSTELLAR OBJECT
HN 1076	13 33 53.	- 26 18			NEBULA
SC 1331-2954.4	13 33 53.	- 30 09 46.	18		NEBULA
ZWG 045.040	13 33 54.	+ 07 38		14.7	GALAXY
UGC 08589	13 33 54.	+ 07 38	126	14.7	GALAXY SBb
MCG+01-35-015	13 33 54.	+ 07 38	96	14.7	GALAXY
ZWG 045.041	13 33 54.	+ 08 04		15.6	GALAXY
ZC 1333.9+1140	13 33 54.	+ 11 40	1810		CLUSTER OF GALAXIES
ZC 1333.9+2553	13 33 54.	+ 25 53	1140		CLUSTER OF GALAXIES
UGC 08590	13 33 54.	+ 37 13	90	17.	GALAXY Sc
ZWG 317.004	13 33 54.	+ 66 23		15.7	GALAXY
ZWG 316.021	13 33 54.	+ 66 23		15.7	GALAXY
MCG-04-32-045	13 33 54.	- 26 19	84	14.	GALAXY
BRON 14	13 33 54.4	+ 28 40 16.		18.54	BLUE OBJECT
RS 23	13 33 54.4	+ 28 40 18.		18.55	BLUE STELLAR OBJECT
RRS 1.053	13 33 55.	+ 28 40		18.57	BLUE OBJECT
SHB 271	13 33 55.	+ 28 40 18.		18.7	QUASI-STELLAR OBJECT
RRS 1.054	13 33 56.	+ 29 31		17.69	BLUE OBJECT
RNGC 5239	13 33 57.	+ 07 38		14.5	GALAXY
IC 0902	13 33 57.	+ 50 12 14.			NONSTELLAR OBJECT
REIN 5.333	13 33 57.75	+ 09 13 40.9			NEBULA
RNGC 5241	13 33 59.	- 08 08		15.0	GALAXY
VDB.66G 179	13 34	+ 07 56	130		DWARF GALAXY
SER 105.02	13 34	- 49 37	1200	14.	LOOSE GROUP OF GALAXIES
LB 09876	13 34	- 83 46		13.5	FAINT BLUE STAR
ZWG 045.042	13 34 00.	+ 08 28		15.6	GALAXY
ZWG 102.021	13 34 00.	+ 16 20		15.7	GALAXY
UGC 08591	13 34 00.	+ 16 20	78	15.7	GALAXY S
ZC 1334.0+2936	13 34 00.	+ 29 36	3630		CLUSTER OF GALAXIES
ZC 1334.0+3116	13 34 00.	+ 31 16	610		CLUSTER OF GALAXIES
ZWG 190.035	13 34 00.	+ 33 40		15.7	GALAXY
MCG+06-30-058	13 34 00.	+ 33 40	24	15.	GALAXY
MCG+07-28-036	13 34 00.	+ 38 35	90	13.5	GALAXY
ZWG 218.027	13 34 00.	+ 38 36		14.0	GALAXY
RNGC 5243	13 34 00.	+ 38 36		14.0	GALAXY
UGC 08592	13 34 00.	+ 38 36	90	14.0	GALAXY S
ZWG 246.017	13 34 00.	+ 50 13		14.7	GALAXY
UGC 08593	13 34 00.	+ 50 13	132	14.7	GALAXY Sb
MCG+08-25-024	13 34 00.	+ 50 13	132	14.	GALAXY
MCG+09-22-085	13 34 00.	+ 51 29	36	13.7	GALAXY
ZWG 271.053	13 34 00.	+ 51 30		14.0	GALAXY
RNGC 5250	13 34 00.	+ 51 30		14.0	GALAXY
UGC 08594	13 34 00.	+ 51 30	60	14.0	GALAXY S0
MCG-01-35-006	13 34 00.	- 08 09	54	15.	GALAXY
MCG-03-35-007	13 34 00.	- 16 43	30	15.	GALAXY
MCG-06-30-017	13 34 00.	- 33 48	96	13.5	GALAXY
IC 4299	13 34 02.	- 33 49 18.			NONSTELLAR OBJECT
IC 4306	13 34 03.	+ 33 40 43.			NONSTELLAR OBJECT
LB 00686	13 34 04.	+ 53 49 12.		16.2	FAINT BLUE STAR
KARA.73 47	13 34 05.	- 37 12	67		DWARF GALAXY
RNGC 5219	13 34 05.	- 45 38			GALAXY
ZC 1334.1+1040	13 34 06.	+ 10 40	2490		CLUSTER OF GALAXIES
ZC 1334.1+1608	13 34 06.	+ 16 08	1080		CLUSTER OF GALAXIES
ZC 1334.1+3438	13 34 06.	+ 34 38	610		CLUSTER OF GALAXIES
REIZ 3528	13 34 06.	+ 38 37	78	13.8	GALAXY
ZWG 353.021	13 34 06.	+ 75 17		15.7	GALAXY
UGC 08595	13 34 06.	+ 75 17	66	15.7	GALAXY Sa-b
KARA.72 386A	13 34 06.	+ 75 17	48	15.7	PART OF DOUBLE GALAXY
ZWG 017.034	13 34 06.	- 03 15		15.6	GALAXY
MCG-03-35-009	13 34 06.	- 16 41	48	14.5	GALAXY
HELF 060	13 34 06.	- 18 03			NEBULA
MCG-03-35-008	13 34 06.	- 18 04	30	15.5	GALAXY
REIN 5.334	13 34 06.51	+ 08 58 31.7			NEBULA
REIZ 3525	13 34 07.	- 08 22	48	15.2	GALAXY
REIN 2.195	13 34 07.59	+ 51 29 23.1			NEBULA
HN 1697	13 34 08.4	- 03 14 51.	12		NEBULA
ZWG 045.043	13 34 12.	+ 06 45		15.1	GALAXY
UGC 08596	13 34 12.	+ 06 45	66	15.1	GALAXY S
ZWG 073.051	13 34 12.	+ 08 48		15.6	GALAXY
MCG+05-32-043	13 34 12.	+ 27 30	30	15.2	GALAXY
MCG+08-25-025	13 34 12.	+ 46 27	72	15.	GALAXY
ZWG 246.018	13 34 12.	+ 46 28		15.2	GALAXY

OBJECT NAME	RIGHT ASCEN.	DECLINATION	DIAM.	MAGN.	TYPE OF OBJECT
UGC 08597	13 34 12.	+ 46 28	108	15.2	GALAXY SBc
ZC 1334.2+5350	13 34 12.	+ 53 50	540		CLUSTER OF GALAXIES
MCG+10-19-096	13 34 12.	+ 59 27 30.	30	16.	GALAXY
MCG-05-32-050	13 34 12.	- 29 36	630	9.	GALAXY
SN 1950B	13 34 12.	- 29 37		14.5	SUPERNOVA
SVEN 414	13 34 13.	- 29 36	660	10.3	GALAXY
REIN 5.336	13 34 14.36	+ 08 48 04.9			NEBULA
HN 1698	13 34 14.5	+ 01 24 15.	18		NEBULA
ARC 1767	13 34 15.	+ 59 29		15.7	RICH CLUSTER OF GALAXIES
SN 1957D	13 34 16.	- 29 35		15.0	SUPERNOVA
KEEL 673	13 34 16.6	- 17 19 44.		16.	NEBULA
IC 4307	13 34 17.	+ 27 29 44.			NONSTELLAR OBJECT
ZWG 017.035	13 34 18.	+ 01 25		15.3	GALAXY
ZWG 045.044	13 34 18.	+ 03 46		15.5	GALAXY
ZWG 045.045	13 34 18.	+ 06 06		15.7	GALAXY
ZC 1334.3+1259	13 34 18.	+ 12 59	400		CLUSTER OF GALAXIES
ZWG 102.022	13 34 18.	+ 20 27		14.8	GALAXY
UGC 08598	13 34 18.	+ 20 27	102	14.8	GALAXY Sb-c
KARA.73B 0593	13 34 18.	+ 20 27	102	14.8	ISOLATED GALAXY S
ZWG 161.088	13 34 18.	+ 27 30		15.2	GALAXY
UGC 08599	13 34 18.	+ 75 03	60	16.0	GALAXY Sc
MCG-01-35-007	13 34 18.	- 07 58	42	14.	GALAXY
RNGC 5236	13 34 18.	- 29 36		8.5	GALAXY
SN 1968L	13 34 18.	- 29 37		11.9	SUPERNOVA
HN 1699	13 34 18.9	+ 01 26 21.	18		NEBULA
RS 24	13 34 20.8	+ 26 58 47.		17.16	BLUE STELLAR OBJECT
KEEL 674	13 34 20.8	- 17 37 25.		16.	NEBULA
MCG+11-17-006	13 34 21.	+ 66 34	78	16.	GALAXY
SC 1331-3157.1	13 34 22.	- 32 12 27.	30		NEBULA
ZWG 017.036	13 34 24.	+ 01 27		15.4	GALAXY
ZWG 045.046	13 34 24.	+ 03 44		15.6	GALAXY
ARC 1765	13 34 24.	+ 10 42		17.8	RICH CLUSTER OF GALAXIES
ZWG 102.023	13 34 24.	+ 17 43		15.0	GALAXY
KARA.72 387A	13 34 24.	+ 17 43	42	15.0	PART OF DOUBLE GALAXY
ZWG 102.024	13 34 24.	+ 17 44		15.0	GALAXY
KARA.72 387B	13 34 24.	+ 17 44	42	15.0	PART OF DOUBLE GALAXY
MCG+03-35-010	13 34 24.	+ 20 29	108	14.8	GALAXY
TON-N 0723	13 34 24.	+ 27 40		15.4	BLUE STAR
ZWG 190.036	13 34 24.	+ 35 00		15.4	GALAXY
UGC 08600	13 34 24.	+ 35 00	102	15.4	GALAXY Sb
UGC 08601	13 34 24.	+ 48 00	66	17.	GALAXY DWRF SP
ZC 1334.4+5428	13 34 24.	+ 54 28	2080		CLUSTER OF GALAXIES
ZC 1334.4+6420	13 34 24.	+ 64 20	1880		CLUSTER OF GALAXIES
HN 1077	13 34 24.	- 28 24		13.5	NEBULA
IC 4303	13 34 24.	- 28 25			NONSTELLAR OBJECT
SN 1923A	13 34 24.	- 29 36		14.0	SUPERNOVA
VHA 150	13 34 24.	- 63 05	150		OPEN STAR CLUSTER
REIZ 3527	13 34 25.	- 07 57	18	13.8	GALAXY
HN 1700	13 34 26.2	+ 00 28 03.	18		NEBULA
MCG-02-35-006	13 34 27.	- 11 34	78	15.	GALAXY
KEEL 675	13 34 29.8	- 17 24 54.		18.	NEBULA
ZWG 017.037	13 34 30.	+ 00 29		15.5	GALAXY
UGC 08644	13 34 30.	+ 07 39	66	16.0	GALAXY
ZC 1334.5+1531	13 34 30.	+ 15 31	610		CLUSTER OF GALAXIES
ZWG 102.025	13 34 30.	+ 16 13		15.2	GALAXY
MCG+03-35-011	13 34 30.	+ 16 13	42	15.2	GALAXY
ZWG 102.026	13 34 30.	+ 16 51		15.6	GALAXY
KARA.73B 0594	13 34 30.	+ 16 51	24	15.6	ISOLATED GALAXY S
ZWG 102.027	13 34 30.	+ 17 13		15.4	GALAXY
MCG+03-35-012	13 34 30.	+ 17 13	48	15.4	GALAXY
MCG+03-35-013	13 34 30.	+ 17 39	60	15.0	GALAXY
UGC 08602	13 34 30.	+ 32 21	72	18.	GALAXY DWARF
MCG+06-30-059	13 34 30.	+ 35 00	90	14.5	GALAXY
MCG+08-25-026	13 34 30.	+ 44 50	54	16.	GALAXY
ZWG 246.019	13 34 30.	+ 44 51		15.3	GALAXY
UGC 08603	13 34 30.	+ 44 51	66	15.3	GALAXY
MCG+08-25-027	13 34 30.	+ 47 59 30.	42	17.	GALAXY
REIZ 3530	13 34 30.	+ 51 31	60	13.7	GALAXY
ZC 1334.5+6317	13 34 30.	+ 63 17	2550		CLUSTER OF GALAXIES
ZWG 317.005	13 34 30.	+ 66 33		15.6	GALAXY
UGC 08604	13 34 30.	+ 66 33	96	15.6	GALAXY Sc/SBc
MCG-02-35-007	13 34 30.	- 14 49 30.	30	15.5	GALAXY
MCG-05-32-051	13 34 30.	- 28 23	24	15.5	GALAXY
RNGC 5242	13 34 32.	+ 03 02			NON-EXISTENT OBJECT
REIN 5.337	13 34 32.49	+ 09 16 08.6			NEBULA
MCG+03-35-014	13 34 33.	+ 17 40	18	15.0	GALAXY
IC 4308	13 34 34.	+ 32 58 45.			NONSTELLAR OBJECT
RNGC 5262	13 34 34.	+ 75 18		15.0	GALAXY
HN 1701	13 34 34.2	+ 00 36 34.	12		NEBULA
KEEL 676	13 34 34.3	- 17 26 13.		18.	NEBULA
KEEL 677	13 34 34.4	- 17 28 08.		18.	NEBULA
ZWG 017.038	13 34 36.	+ 00 37		15.7	GALAXY
ZWG 073.052	13 34 36.	+ 09 17		15.7	GALAXY
RRS 3.021	13 34 36.	+ 30 39		18.79	BLUE OBJECT
UGC 08605	13 34 36.	+ 32 21	108	18.	GALAXY DWRF SP
MCG+09-22-086	13 34 36.	+ 55 41	30	16.	GALAXY
LB 02695	13 34 36.	+ 56 36 36.		16.7	FAINT BLUE STAR
ZWG 353.022	13 34 36.	+ 75 18		15.0	GALAXY
UGC 08606	13 34 36.	+ 75 18	84	15.0	GALAXY E-S0
KARA.72 386B	13 34 36.	+ 75 18	66	15.0	PART OF DOUBLE GALAXY
MCG-02-35-008	13 34 36.	- 14 50	42	15.	GALAXY
MCG-05-32-052	13 34 36.	- 32 46	48	14.5	GALAXY
AGU 48	13 34 36.	- 42 35 42.	54	13.5	PECULIAR GALAXY
BC PSII23	13 34 36.1	+ 28 35 42.		18.74	QUASI-STELLAR OBJECT
BRON 15	13 34 36.4	+ 28 29 25.		17.48	BLUE OBJECT
RS 25	13 34 36.4	+ 28 29 35.		17.25	BLUE STELLAR OBJECT
RS 26	13 34 36.9	+ 26 58 53.		17.12	BLUE STELLAR OBJECT
RRS 1.055	13 34 37.	+ 28 29		17.19	BLUE OBJECT
SVEN 415	13 34 37.	- 29 34 51.	12	17.0	GALAXY
SC 1331-3230.3	13 34 37.	- 32 45 39.	60		NEBULA
RNGC 5237	13 34 37.	- 42 36			GALAXY
REIN 5.338	13 34 39.66	+ 09 04 00.1			NEBULA
REIN 5.339	13 34 39.82	+ 09 04 31.5			NEBULA
RRS 1.056	13 34 40.	+ 30 54		18.77	BLUE OBJECT
REIZ 3529	13 34 40.	+ 33 50	36	14.6	GALAXY
BC M1334+119	13 34 41.4	+ 11 55 29.		18.	QUASI-STELLAR OBJECT
SHB 272	13 34 41.4	+ 11 55 29.		18.	QUASI-STELLAR OBJECT
HN 1702	13 34 41.6	- 01 54 02.	18		NEBULA
ZWG 045.047	13 34 42.	+ 03 25		15.6	GALAXY
FEIG 085	13 34 42.	+ 08 35		15.0	FAINT BLUE STAR
ZC 1334.7+1501	13 34 42.	+ 15 01	2290		CLUSTER OF GALAXIES
ZWG 161.089	13 34 42.	+ 32 01		15.7	GALAXY
UGC 08608	13 34 42.	+ 32 01	90	15.7	GALAXY Sc
UGC 08609	13 34 42.	+ 33 48	66	16.0	GALAXY Sc/SBc
MCG+06-30-060	13 34 42.	+ 33 48 30.	48	15.	GALAXY
MCG+08-25-028	13 34 42.	+ 44 50	60	16.	GALAXY
ZWG 017.039	13 34 42.	- 01 55		15.3	GALAXY
UGC 08607	13 34 42.	- 01 55	60	15.3	GALAXY S0?
MCG-05-32-053	13 34 42.	- 33 34	42	15.	GALAXY
MCG-07-28-005	13 34 42.	- 42 36	66	14.	GALAXY
SVEN 416	13 34 43.	- 29 35	12	16.5	GALAXY
HN 1703	13 34 45.6	- 03 19 50.	12		NEBULA
HN 1704	13 34 47.2	- 01 29 32.	18		NEBULA
ZC 1334.8+0249	13 34 48.	+ 02 49	2490		CLUSTER OF GALAXIES
ZWG 045.048	13 34 48.	+ 04 09		15.3	GALAXY
ZWG 045.049	13 34 48.	+ 06 44		15.4	GALAXY
UGC 08610	13 34 48.	+ 06 44	60	15.4	GALAXY Sa
MCG+01-35-016	13 34 48.	+ 06 44	60	15.4	GALAXY
ZC 1334.8+1730	13 34 48.	+ 17 30	1950		CLUSTER OF GALAXIES
RRS 1.057	13 34 48.	+ 31 35		16.69	BLUE OBJECT
ZC 1334.8+3241	13 34 48.	+ 32 41	870		CLUSTER OF GALAXIES
MCG+07-28-037	13 34 48.	+ 38 50	45	16.	GALAXY
MCG+08-25-029	13 34 48.	+ 45 08	90	15.	GALAXY
ZWG 246.020	13 34 48.	+ 45 09		15.5	GALAXY
UGC 08611	13 34 48.	+ 45 09	96	15.5	GALAXY Sc
ZWG 017.041	13 34 48.	- 01 30		15.7	GALAXY
ZWG 017.040	13 34 48.	- 03 19		15.7	GALAXY
HELW 216	13 34 49.	- 31 30 56.			NEBULA
ARC 1766	13 34 50.	+ 32 44		17.5	RICH CLUSTER OF GALAXIES
VV 006C	13 34 51.	+ 06 41	12	17.	INTERACTING GALAXY
VV 006B	13 34 51.	+ 06 41	12	17.	INTERACTING GALAXY
VV 006A	13 34 51.	+ 06 41	36	15.	INTERACTING GALAXY
VV 006	13 34 51.	+ 06 41	90		INTERACTING GALAXY
REIN 5.342	13 34 51.39	+ 08 52 14.7			NEBULA
RRS 1.058	13 34 53.	+ 31 46		18.29	BLUE OBJECT
ZWG 045.050	13 34 54.	+ 04 21		14.8	GALAXY
MCG+01-35-017	13 34 54.	+ 04 21	60	14.8	GALAXY
ZWG 045.051	13 34 54.	+ 06 41		15.3	GALAXY
UGC 08613	13 34 54.	+ 06 41	90	15.3	GALAXY SB
MCG+01-35-018	13 34 54.	+ 06 41	72	15.3	GALAXY
ZWG 045.052	13 34 54.	+ 06 46		15.3	GALAXY
MCG+01-35-019	13 34 54.	+ 06 46	36	15.3	GALAXY
ZWG 045.053	13 34 54.	+ 07 54		15.4	GALAXY
UGC 08614	13 34 54.	+ 07 54	240	15.4	GALAXY DWRF IR
MCG+01-35-020	13 34 54.	+ 07 54	180	15.4	GALAXY
ZWG 073.053	13 34 54.	+ 09 17		15.5	GALAXY
MCG+02-35-014	13 34 54.	+ 09 17	36	15.5	GALAXY
ZC 1334.9+1718	13 34 54.	+ 17 18	1280		CLUSTER OF GALAXIES
1ZW 065	13 34 54.	+ 45 31			COMPACT GALAXY
MCG+09-22-087	13 34 54.	+ 55 40	36	16.	GALAXY
MCG+10-19-097	13 34 54.	+ 62 19	27	16.	GALAXY
MCG-03-35-010	13 34 54.	- 16 22 30.	36	15.5	GALAXY
RS 27	13 34 54.6	+ 27 27 35.		18.82	BLUE STELLAR OBJECT
REIN 5.343	13 34 55.25	+ 09 16 38.7			NEBULA
REIN 5.344	13 34 55.96	+ 08 54 57.5			NEBULA
RNGC 5245	13 34 56.	+ 04 09			GALAXY
RNGC 5246	13 34 56.	+ 04 21		15.0	GALAXY
RNGC 5248	13 34 57.	+ 09 09		11.0	GALAXY
RRS 1.059	13 34 57.	+ 29 55		18.44	BLUE OBJECT
SC 1332-3107.1	13 34 57.	- 31 22 26.	18		NEBULA
LB 09877	13 35	- 85 03		14.7	FAINT BLUE STAR
ZWG 073.054	13 35 00.	+ 09 08		11.4	GALAXY
UGC 08616	13 35 00.	+ 09 08	420	11.4	GALAXY Sc
MCG+02-35-015	13 35 00.	+ 09 08	360	11.4	GALAXY
FAI 13350+09	13 35 00.	+ 09 15			COMPACT GALAXY
MCG+05-32-044	13 35 00.	+ 27 40	36	14.7	GALAXY
UGC 08615	13 35 00.	+ 86 16	96	16.0	GALAXY Sc
MCG-02-35-009	13 35 00.	- 14 51	48	15.5	GALAXY
REIZ 3531	13 35 02.	+ 27 41	30	14.0	GALAXY
REIN 5.345	13 35 02.62	+ 09 08 20.8			NEBULA
REIN 5.346	13 35 03.19	+ 09 08 58.9			NEBULA
REIN 5.347	13 35 05.89	+ 09 07 28.4			NEBULA
ZWG 045.054	13 35 06.	+ 03 43		15.5	GALAXY
ARP 326	13 35 06.	+ 06 41			PECULIAR GALAXY
ARP 023	13 35 06.	+ 06 41			PECULIAR GALAXY
ZWG 161.090	13 35 06.	+ 27 41		14.7	GALAXY
1ZW 066	13 35 06.	+ 53 29			COMPACT GALAXY
ZWG 017.042	13 35 06.	- 01 25		15.6	GALAXY
HELW 217	13 35 06.	- 31 42 56.			NEBULA
SC 1332-3127.7	13 35 06.	- 31 43 02.	24		NEBULA
MCG-05-32-054	13 35 06.	- 33 17	48	15.	GALAXY
REIN 5.348	13 35 06.24	+ 09 07 59.1			NEBULA
BC PKS1335+023	13 35 07.	+ 02 22 06.		18.	QUASI-STELLAR OBJECT
SC 1332-3025.5	13 35 07.	- 30 40 50.	90		NEBULA
REIN 5.349	13 35 07.25	+ 09 06 21.9			NEBULA
SHB 273	13 35 07.3	+ 02 22 06.		18.	QUASI-STELLAR OBJECT
HN 1705	13 35 07.4	- 01 25 25.	12		NEBULA
RNGC 5251	13 35 08.	+ 27 41		14.5	GALAXY
SC 1332-3259.4	13 35 09.	- 33 14 44.	48		NEBULA
RNGC 5249	13 35 11.	+ 16 14		14.5	GALAXY
ZWG 045.055	13 35 12.	+ 05 30		15.0	GALAXY
UGC 08617	13 35 12.	+ 05 30	60	15.0	GALAXY SB
MCG+01-35-021	13 35 12.	+ 05 30	48	15.0	GALAXY
ZWG 102.028	13 35 12.	+ 16 14		14.5	GALAXY
UGC 08618	13 35 12.	+ 16 14	96	14.5	GALAXY S0?
MCG+03-35-015	13 35 12.	+ 16 14	96	14.5	GALAXY
MCG+07-28-039	13 35 12.	+ 38 51	30	16.	GALAXY
MCG+07-28-038	13 35 12.	+ 39 12	48	15.	GALAXY
MCG-05-32-056	13 35 12.	- 30 42	90	15.	GALAXY
MCG-05-32-055	13 35 12.	- 33 38	36	15.	GALAXY
SC 1332-3003.6	13 35 14.	- 30 18 56.	36		NEBULA
RS 28	13 35 14.2	+ 26 15 29.		18.15	BLUE STELLAR OBJECT
RS 29	13 35 14.9	+ 27 16 23.		18.24	BLUE STELLAR OBJECT
MCG+07-28-040	13 35 15.	+ 38 51	54	15.	GALAXY
RNGC 5247	13 35 15.	- 17 37		11.5	GALAXY
MCG-07-28-006	13 35 15.	- 39 69	60	15.	GALAXY
RRS 1.060	13 35 16.	+ 30 40		18.45	BLUE OBJECT
HOLM 530B	13 35 17.	+ 29 03	66	16.1	PART OF MULTIPLE GALAXY
ZC 1335.3+1944	13 35 18.	+ 19 44	1410		CLUSTER OF GALAXIES
ZWG 161.091	13 35 18.	+ 29 04		15.6	GALAXY
MCG+05-32-045	13 35 18.	+ 29 04	33	15.6	GALAXY
ZC 1335.3+3010	13 35 18.	+ 30 10	810		CLUSTER OF GALAXIES
ZWG 218.028	13 35 18.	+ 38 53		15.1	GALAXY
UGC 08619	13 35 18.	+ 38 53	84	15.1	GALAXY S
ZWG 218.029	13 35 18.	+ 39 14		15.3	GALAXY
UGC 08620	13 35 18.	+ 39 14	66	15.3	GALAXY Sa-b
ZC 1335.3+7900	13 35 18.	+ 79 00	270		CLUSTER OF GALAXIES
REIN 5.350	13 35 18.86	+ 08 57 52.7			NEBULA
HOLM 530A	13 35 19.	+ 29 04	30	15.0	PART OF MULTIPLE GALAXY
RS 30	13 35 20.1	+ 26 41 11.		19.08	BLUE STELLAR OBJECT
REIN 5.351	13 35 21.49	+ 09 04 44.2			NEBULA
HN 1706	13 35 21.5	- 01 28 25.	30		NEBULA
REIZ 3537	13 35 22.	+ 57 22	36	14.5	GALAXY
HELW 218	13 35 22.	- 31 09 31.			NEBULA
ZC 1335.4+1552	13 35 24.	+ 15 52	1410		CLUSTER OF GALAXIES
ZWG 161.092	13 35 24.	+ 28 15		15.7	GALAXY

OBJECT NAME	RIGHT ASCEN.	DECLINATION	DIAM.	MAGN.	TYPE OF OBJECT
MCG+06-30-061	13 35 24.	+ 36 14 30.	24	17.	GALAXY
MCG+07-28-041	13 35 24.	+ 39 24	48	14.	GALAXY
SVEN 417	13 35 24.	- 17 37	300	11.2	GALAXY
MCG-03-35-011	13 35 24.	- 17 38	300	11.5	GALAXY
MCG-05-32-057	13 35 24.	- 31 10	48	15.	GALAXY
RS 31	13 35 24.8	+ 27 16 47.		18.58	BLUE STELLAR OBJECT
RNGC 5255	13 35 28.	+ 57 22		15.5	GALAXY
ZWG 161.093	13 35 30.	+ 28 05		15.6	GALAXY
ZWG 218.030	13 35 30.	+ 39 25		14.2	GALAXY
UGC 08621	13 35 30.	+ 39 25	48	14.2	GALAXY S
ZWG 294.051	13 35 30.	+ 57 22		15.3	GALAXY
SHB 274	13 35 31.3	- 06 11 57.		17.7	QUASI-STELLAR OBJECT
BC ESH13-011	13 35 31.34	- 06 11 57.4		17.68	QUASI-STELLAR OBJECT
REIZ 3532	13 35 32.	- 09 32	72	13.4	GALAXY
HN 1707	13 35 33.8	+ 00 16 17.			NEBULA
PATH 2.145	13 35 34.	+ 00 17	14		NEBULA
RNGC 5244	13 35 35.	- 45 37			GALAXY
ZWG 017.043	13 35 36.	+ 00 16		15.6	GALAXY
ZC 1335.6+2117	13 35 36.	+ 21 17	4770		CLUSTER OF GALAXIES
TON-N 0724	13 35 36.	+ 31 06		16.8	BLUE STAR
RRS 1.061	13 35 36.	+ 31 15		17.13	BLUE OBJECT
ZC 1335.6+4206	13 35 36.	+ 42 06	1140		CLUSTER OF GALAXIES
MCG+10-19-098	13 35 36.	+ 57 21	42	14.5	GALAXY
MCG+10-19-099	13 35 36.	+ 60 31	18	17.	GALAXY
MCG-02-35-010	13 35 36.	- 09 32	120	13.5	GALAXY
MCG-03-35-012	13 35 36.	- 20 26	48	15.	GALAXY
HELW 219	13 35 36.	- 31 00 13.			NEBULA
KEEL 678	13 35 37.0	- 17 45 16.		17.	NEBULA
SC 1332-3054.4	13 35 40.	- 31 09 43.	60		NEBULA
BROF 16	13 35 41.0	+ 28 32 17.		18.91	BLUE OBJECT
HN 1708	13 35 41.1	- 00 09 00.	12		NEBULA
ZWG 045.056	13 35 42.	+ 04 47		14.5	GALAXY
UGC 08622	13 35 42.	+ 04 47	108	14.5	GALAXY S0
MCG+01-35-022	13 35 42.	+ 04 47	60	14.5	GALAXY
ZC 1335.7+0540	13 35 42.	+ 05 40	940		CLUSTER OF GALAXIES
UGC 08623	13 35 42.	+ 06 43	84	16.0	GALAXY
MCG+08-25-030	13 35 42.	+ 48 29	12	16.	GALAXY
MCG+10-19-100	13 35 42.	+ 60 31	27	16.	GALAXY
MCG-07-28-007	13 35 42.	- 45 36	78	13.5	GALAXY
ACK 309+06.1	13 35 42.	- 55 52			PLANETARY NEBULA
REIZ 3533	13 35 43.	+ 04 49	42	14.1	GALAXY
SC 1332-3255.7	13 35 43.	- 33 11 01.	48		NEBULA
RNGC 5252	13 35 44.	+ 04 47		14.5	GALAXY
ZWG 017.044	13 35 46.	+ 01 44		15.3	GALAXY
ZWG 045.057	13 35 48.	+ 07 48		15.7	GALAXY
ZWG 073.055	13 35 48.	+ 11 30		15.7	GALAXY
ZWG 102.029	13 35 48.	+ 18 39		15.2	GALAXY
UGC 08624	13 35 48.	+ 18 39	60	15.2	GALAXY Sb-c
ZWG 190.037	13 35 48.	+ 32 48		15.7	GALAXY
LB 02696	13 35 48.	+ 51 28 54.		16.7	FAINT BLUE STAR
1ZW 064	13 35 48.	+ 51 53			COMPACT GALAXY
MCG+10-19-101	13 35 48.	+ 61 17	18	16.	GALAXY
MCG-02-35-011	13 35 48.	- 10 26 30.	15	15.	GALAXY
SC 1332-3023.0	13 35 48.	- 30 38 19.	60		NEBULA
RRS 1.062	13 35 49.	+ 29 25		18.39	BLUE OBJECT
SVEN 418	13 35 49.	- 29 23 48.	18	16.1	GALAXY
REIZ 3534	13 35 52.	+ 00 02	18	14.5	GALAXY
RRS 1.063	13 35 52.	+ 30 09		17.12	BLUE OBJECT
IC 0903	13 35 53.	+ 00 01	19		NEBULA
PATH 2.146	13 35 53.	+ 00 01	19		NEBULA
PATH 1.631	13 35 53.	- 00 52	14		NEBULA
ZWG 017.045	13 35 54.	+ 00 01		14.7	GALAXY
UGC 08625	13 35 54.	+ 00 01	126	14.7	GALAXY Sb
MCG+00-35-013	13 35 54.	+ 00 01	114	14.7	GALAXY
ZWG 017.046	13 35 54.	+ 00 45		15.7	GALAXY
ZWG 045.058	13 35 54.	+ 02 46		15.7	GALAXY
ZWG 045.059	13 35 54.	+ 07 08		15.0	GALAXY
UGC 08626	13 35 54.	+ 07 08	90	15.0	GALAXY SBb
MCG+01-35-023	13 35 54.	+ 07 08	90	15.0	GALAXY
FAI 13359+09	13 35 54.	+ 09 48			COMPACT GALAXY
ZC 1335.9+1640	13 35 54.	+ 16 40	1410		CLUSTER OF GALAXIES
MCG+05-32-046	13 35 54.	+ 27 00	33	15.6	GALAXY
MCG+06-30-062	13 35 54.	+ 33 03 30.	66	14.5	GALAXY
ZWG 190.038	13 35 54.	+ 33 05		15.0	GALAXY
UGC 08627	13 35 54.	+ 33 05	66	15.0	GALAXY Sb-c
ZWG 295.001	13 35 54.	+ 61 17		15.2	GALAXY
ZWG 294.052	13 35 54.	+ 61 17		15.2	GALAXY
MCG-04-32-046	13 35 54.	- 23 54	48	14.	GALAXY
PK307-04.1	13 35 54.	- 67 07 43.	25	12.2	PLANETARY NEBULA
HN 1078	13 35 55.	- 29 25		14.	NEBULA
IC 4309	13 35 55.	- 29 26			NONSTELLAR OBJECT
BROF 33	13 35 56.2	+ 29 01 26.		19.45	BLUE OBJECT
REIZ 3535	13 35 57.	+ 00 48	42	14.8	GALAXY
RRS 1.064	13 35 57.	+ 30 41		18.08	BLUE OBJECT
REIZ 3540	13 35 58.	+ 33 04	30	14.3	GALAXY
ABC 1768	13 35 58.	- 13 42		17.2	RICH CLUSTER OF GALAXIES
VDB.66G 180	13 36	- 09 34	70		DWARF GALAXY
ZWG 017.047	13 36 00.	+ 00 47		15.1	GALAXY
UGC 08628	13 36 00.	+ 00 47	66	15.1	GALAXY Sa?
MCG+00-35-014	13 36 00.	+ 00 47	60	15.1	GALAXY
IC 0904	13 36 00.	+ 00 47 13.			NONSTELLAR OBJECT
ZWG 073.056	13 36 00.	+ 08 42		15.5	GALAXY
UGC 08629	13 36 00.	+ 08 42	84	15.5	GALAXY DWARF
MCG+05-32-047	13 36 00.	+ 26 58 30.	18	15.0	GALAXY
ZWG 161.094	13 36 00.	+ 27 01		15.6	GALAXY
ZWG 161.095	13 36 00.	+ 28 02		15.5	GALAXY
MRK 265	13 36 00.	+ 28 02	15	16.	GALAXY WITH UV CONTINUUM
MCG-03-35-013	13 36 00.	- 20 14	36	16.	GALAXY
SHB 275	13 36 03.	+ 26 29 06.		18.9	QUASI-STELLAR OBJECT
IC 4313	13 36 03.	+ 27 00 19.			NONSTELLAR OBJECT
RRS 1.066	13 36 03.	+ 29 30		16.70	BLUE OBJECT
RRS 1.065	13 36 03.	+ 29 37		18.88	BLUE OBJECT
RS 32	13 36 03.1	+ 26 29 04.		17.97	BLUE STELLAR OBJECT
BRON 17	13 36 04.3	+ 28 31 38.		19.56	BLUE OBJECT
REIZ 3536	13 36 05.	- 03 09	18	14.6	GALAXY
ZC 1336.1+1535	13 36 06.	+ 15 35	610		CLUSTER OF GALAXIES
ZWG 161.096	13 36 06.	+ 27 00		15.0	GALAXY
TON-N 0725	13 36 06.	+ 29 31		16.2	BLUE STAR
FEIG 086	13 36 06.	+ 29 37		9.3	FAINT BLUE STAR
ZWG 190.039	13 36 06.	+ 33 23		14.3	GALAXY
UGC 08630	13 36 06.	+ 33 23	108	14.3	GALAXY PECULR
ZC 1336.1+4010	13 36 06.	+ 40 10	1080		CLUSTER OF GALAXIES
IC 4314	13 36 07.	+ 26 59 20.			NONSTELLAR OBJECT
HOLM 531B	13 36 08.	+ 26 22	18	15.3	PART OF MULTIPLE GALAXY
REIZ 3538	13 36 09.	+ 00 48	30	16.3	GALAXY
MCG+06-30-063	13 36 09.	+ 33 21 30.	93	14.	GALAXY
REIZ 3541	13 36 10.	+ 33 22	90	13.4	GALAXY
RNGC 5256	13 36 11.	+ 48 32		14.0	GALAXY
HN 1709	13 36 11.5	- 00 10 00.	30		NEBULA
ZWG 017.049	13 36 12.	+ 00 48		15.6	GALAXY
UGC 08631	13 36 12.	+ 00 48	72	15.6	GALAXY
ZWG 017.050	13 36 12.	+ 02 12		15.7	GALAXY
ZWG 045.060	13 36 12.	+ 04 42		15.4	GALAXY
ZWG 073.057	13 36 12.	+ 12 42		15.6	GALAXY
MCG+04-34-001	13 36 12.	+ 23 35 30.	15	15.0	GALAXY
ZWG 131.016	13 36 12.	+ 26 21		15.7	GALAXY
MCG+04-32-016	13 36 12.	+ 26 22 30.	36	15.7	GALAXY
ZWG 161.097	13 36 12.	+ 31 15		15.7	GALAXY
1ZW 067	13 36 12.	+ 48 32			COMPACT GALAXY
ZWG 246.021	13 36 12.	+ 48 32		14.1	GALAXY
MRK 266	13 36 12.	+ 48 32	25	14.	GALAXY WITH UV CONTINUUM
UGC 08632	13 36 12.	+ 48 32	72	14.1	GALAXY PECULR
KARA.72 388B	13 36 12.	+ 48 32	24		PART OF DOUBLE GALAXY
KARA.72 388A	13 36 12.	+ 48 32	24	14.1	PART OF DOUBLE GALAXY
MCG+08-25-031	13 36 12.	+ 48 32	72	13.6	GALAXY
ZWG 017.048	13 36 12.	- 00 10		15.6	GALAXY
REIZ 3539	13 36 13.	+ 04 43	30	15.2	GALAXY
PATH 2.147	13 36 13.	- 00 11	27		NEBULA
HOLM 531A	13 36 14.	+ 26 21	48	14.7	PART OF MULTIPLE GALAXY
HN 1710	13 36 14.5	- 01 26 53.	24		NEBULA
REIZ 3545	13 36 15.	+ 48 33	24	13.6	GALAXY
HN 1079	13 36 16.	- 25 35	18	13.	NEBULA
IC 4310	13 36 16.	- 25 36			NONSTELLAR OBJECT
ZWG 017.052	13 36 18.	+ 01 00		15.6	GALAXY
MCG+10-19-102	13 36 18.	+ 60 32	24	15.	GALAXY
ZWG 017.051	13 36 18.	- 01 27		15.7	GALAXY
MCG-04-32-047	13 36 18.	- 25 36	72	14.	GALAXY
HN 1711	13 36 20.7	+ 01 00 25.	18		NEBULA
MCG+06-30-064	13 36 21.	+ 32 32	30	15.	GALAXY
PATH 1.632	13 36 21.	- 00 33			NEBULA
ZWG 102.030	13 36 24.	+ 15 00		15.4	GALAXY
UGC 08633	13 36 24.	+ 15 00	78	15.4	GALAXY Sb-c
ZWG 102.031	13 36 24.	+ 19 13		15.3	GALAXY
ZWG 131.017	13 36 24.	+ 24 45		15.7	GALAXY
MCG+05-32-048	13 36 24.	+ 26 34 30.	42	15.6	GALAXY
ZWG 161.098	13 36 24.	+ 26 36		15.6	GALAXY
ZWG 161.099	13 36 24.	+ 28 00		15.7	GALAXY
ZC 1336.4+2812	13 36 24.	+ 28 12	3830		CLUSTER OF GALAXIES
ZWG 161.100	13 36 24.	+ 30 44		15.6	GALAXY
ZWG 161.101	13 36 24.	+ 31 31		15.1	GALAXY
REIZ 3542	13 36 27.	+ 25 17	48	15.8	GALAXY
RRS 1.067	13 36 27.	+ 28 51		18.62	BLUE OBJECT
RRS 1.068	13 36 27.	+ 29 17		18.14	BLUE OBJECT
HN 1712	13 36 27.4	- 00 20 59.	12		NEBULA
ZWG 017.053	13 36 30.	+ 02 25		15.7	GALAXY
UGC 08634	13 36 30.	+ 02 25	96	15.7	GALAXY S?
FAI 1336+08	13 36 30.	+ 08 46			COMPACT GALAXY
ZC 1336.5+2230	13 36 30.	+ 22 30	740		CLUSTER OF GALAXIES
REIZ 3544	13 36 30.	+ 30 44	18	14.4	GALAXY
MCG+05-32-049	13 36 30.	+ 31 31	18	15.1	GALAXY
ZWG 294.053	13 36 30.	+ 56 47		15.5	GALAXY
MCG+10-19-103	13 36 30.	+ 57 35	36	16.	GALAXY
ZWG 295.002	13 36 30.	+ 60 32		15.1	GALAXY
ZWG 294.054	13 36 30.	+ 60 32		15.1	GALAXY
MCG-05-32-058	13 36 30.	- 30 32	60	14.	GALAXY
HELW 220	13 36 30.	- 31 25 05.			NEBULA
MCG-05-32-059	13 36 30.	- 31 59	24	15.	GALAXY
MRSL 308+00/1	13 36 30.	- 61 30	540		HII REGION
RRS 1.069	13 36 31.	+ 30 53		18.39	BLUE OBJECT
SC 1333-3142.9	13 36 31.	- 31 58 12.	18		NEBULA
RRS 1.070	13 36 32.	+ 30 20		19.75	BLUE OBJECT
SC 1333-3016.1	13 36 32.	- 30 31 23.	48		NEBULA
HELW 221	13 36 32.	- 31 16 23.			NEBULA
REIZ 3547	13 36 35.	+ 31 37	24	13.9	GALAXY
8ZW 1336+12.3	13 36 36.	+ 12 15		17.8	COMPACT GALAXY
ZC 1336.6+1251	13 36 36.	+ 12 51	470		CLUSTER OF GALAXIES
ZWG 161.102	13 36 36.	+ 31 38		15.0	GALAXY
RS 33	13 36 37.1	+ 28 03 52.		19.00	BLUE STELLAR OBJECT
HN 1713	13 36 38.4	- 00 51 53.	24		NEBULA
MCG+04-32-017	13 36 39.	+ 25 15	36	15.5	GALAXY
MCG+07-28-042	13 36 39.	+ 44 09	30	17.	GALAXY
REIZ 3556	13 36 39.	+ 57 17	24	15.0	GALAXY
PATH 2.148	13 36 39.	- 00 52	16		NEBULA
SC 1333-3135.7	13 36 40.	- 31 50 59.	18		NEBULA
ZWG 045.061	13 36 42.	+ 04 52		14.9	GALAXY
UGC 08635	13 36 42.	+ 04 52	60	14.9	GALAXY Sb-c
MCG+01-35-024	13 36 42.	+ 04 52	48	14.9	GALAXY
ZWG 073.058	13 36 42.	+ 09 09		15.7	GALAXY
ZC 1336.7+1957	13 36 42.	+ 19 57	610		CLUSTER OF GALAXIES
MCG+08-25-032	13 36 42.	+ 45 36	36	16.	GALAXY
REIZ 3555	13 36 42.	+ 55 45	30	15.0	GALAXY
MCG+10-19-104	13 36 42.	+ 57 33	18	16.	GALAXY
ZC 1336.7+6653	13 36 42.	+ 66 53	2490		CLUSTER OF GALAXIES
ZWG 017.054	13 36 42.	- 00 52		15.3	GALAXY
REIZ 3543	13 36 43.	+ 04 53	36	14.4	GALAXY
VV 108	13 36 45.	+ 04 53	66	14.	INTERACTING GALAXY
BC BSII32	13 36 45.1	+ 26 24 30.		18.91	QUASI-STELLAR OBJECT
REIZ 3549	13 36 46.	+ 32 24	24	14.9	GALAXY
ZWG 073.059	13 36 48.	+ 12 58		15.1	GALAXY
MCG+02-35-016	13 36 48.	+ 12 58	42	15.1	GALAXY
REIZ 3548	13 36 48.	+ 29 13	24	15.0	GALAXY
MCG+05-32-050	13 36 48.	+ 31 37	30	15.0	GALAXY
ZWG 161.103	13 36 48.	+ 32 25		15.7	GALAXY
ZC 1336.8+3407	13 36 48.	+ 34 07	1210		CLUSTER OF GALAXIES
MCG+08-25-033	13 36 48.	+ 48 07 30.	30	16.	GALAXY
MCG+09-22-088	13 36 48.	+ 55 43	30	16.	GALAXY
ZWG 295.003	13 36 48.	+ 57 34		15.7	GALAXY
ZWG 294.055	13 36 48.	+ 57 34		15.7	GALAXY
MCG+10-19-105	13 36 48.	+ 62 00	18	16.	GALAXY
MCG+05-32-051	13 36 51.	+ 29 12 30.	66	15.4	GALAXY
MCG-04-32-048	13 36 51.	- 22 15	42	15.	GALAXY
RNGC 5254	13 36 52.	- 11 13		13.0	GALAXY
ZWG 045.062	13 36 54.	+ 04 45		15.5	GALAXY
8ZW 1336+11.5	13 36 54.	+ 11 32		17.0	COMPACT GALAXY
ZC 1336.9+1749	13 36 54.	+ 17 49	2220		CLUSTER OF GALAXIES
VV 133A	13 36 54.	+ 25 02	24	17.	INTERACTING GALAXY
VV 133	13 36 54.	+ 25 02	66	16.	INTERACTING GALAXY
MCG+04-32-018	13 36 54.	+ 25 02	66	14.8	GALAXY
ZWG 161.104	13 36 54.	+ 29 13		15.4	GALAXY
UGC 08636	13 36 54.	+ 29 13	60	15.4	GALAXY Sb-c
RRS 1.071	13 36 54.	+ 30 56		18.89	BLUE OBJECT
TON-N 0726	13 36 54.	+ 31 53		14.8	BLUE STAR
ZWG 218.031	13 36 54.	+ 43 49		15.4	GALAXY
MCG+09-22-089	13 36 54.	+ 51 40	84	14.	GALAXY
VHA 151	13 36 54.	- 61 28	180		STAR CLSTR IN NEBULOSITY
ABC 1769	13 36 55.	+ 28 02		17.2	RICH CLUSTER OF GALAXIES

OBJECT NAME	RIGHT ASCEN.	DECLINATION	DIAM.	MAGN.	TYPE OF OBJECT
RS 34	13 36 56.0	+ 27 23 52.		18.15	BLUE STELLAR OBJECT
REIZ 3546	13 36 57.	- 11 14	132	13.3	GALAXY
RS 35	13 36 58.1	+ 27 43 04.		18.59	BLUE STELLAR OBJECT
ZWG 045.063	13 37 00.	+ 02 38		15.6	GALAXY
ZWG 045.064	13 37 00.	+ 06 25		15.2	GALAXY
UGC 08637	13 37 00.	+ 06 25	66	15.2	GALAXY Sc
MCG+01-35-025	13 37 00.	+ 06 25	78	15.2	GALAXY
ZWG 131.018	13 37 00.	+ 25 01		14.8	GALAXY
UGC 08638	13 37 00.	+ 25 01	78	14.8	GALAXY IRR
ZC 1337.0+3840	13 37 00.	+ 38 40	1480		CLUSTER OF GALAXIES
MCG+07-28-043	13 37 00.	+ 43 50	36	16.	GALAXY
MCG+07-28-044	13 37 00.	+ 44 02	42	17.	GALAXY
ZWG 271.054	13 37 00.	+ 51 42		15.5	GALAXY
UGC 08639	13 37 00.	+ 51 42	96	15.5	GALAXY DWARF
ZC 1337.0+5413	13 37 00.	+ 54 13	1140		CLUSTER OF GALAXIES
MCG+10-20-001	13 37 00.	+ 57 12	15	16.	GALAXY
ZWG 295.004	13 37 00.	+ 57 15		15.3	GALAXY
ZWG 294.056	13 37 00.	+ 57 15		15.3	GALAXY
KARA.73B 0595	13 37 00.	+ 61 45	66	15.2	ISOLATED GALAXY S
ZWG 295.005	13 37 00.	+ 62 00		15.7	GALAXY
ZWG 294.057	13 37 00.	+ 62 00		15.7	GALAXY
KARA.73B 0596	13 37 00.	+ 62 00	18	15.7	ISOLATED GALAXY S0
MCG-02-35-012	13 37 00.	- 11 14	156	13.	GALAXY
HN 1080	13 37 01.	- 50 47			NEBULA
IC 4311	13 37 01.	- 50 48			NONSTELLAR OBJECT
BRON 18	13 37 01.1	+ 28 54 20.		18.97	BLUE OBJECT
SC 1334-3135.0	13 37 02.	- 31 50 17.	30		NEBULA
REIZ 3551	13 37 03.	+ 25 02	72	13.8	GALAXY
RRS 1.072	13 37 03.	+ 29 11		19.76	BLUE OBJECT
HOLM 533B	13 37 03.	+ 31 15	12	15.9	PART OF MULTIPLE GALAXY
BRON 19	13 37 03.0	+ 28 28 55.		19.36	BLUE OBJECT
RNGC 5259	13 37 04.	+ 31 15		15.0	GALAXY
SN 1972E	13 37 04.	- 31 26		8.5	SUPERNOVA
REIZ 3553	13 37 05.	+ 31 15	12	16.1	GALAXY
HOLM 533A	13 37 05.	+ 31 15	18	15.1	PART OF MULTIPLE GALAXY
REIZ 3554	13 37 05.	+ 31 28	42	15.1	GALAXY
RNGC 5253	13 37 05.	- 31 23		11.0	GALAXY
HARO 10	13 37 05.	- 31 23			BLUE EMISSION-LINE GALAXY
ZWG 161.105	13 37 06.	+ 31 15		15.2	GALAXY
ZWG 161.106	13 37 06.	+ 31 28		15.3	GALAXY
UGC 08640	13 37 06.	+ 31 28	72	15.3	GALAXY SB:a-b
ZWG 218.032	13 37 06.	+ 44 30		15.7	GALAXY
SC 1335+4709.3	13 37 06.	+ 46 54 02.	18		NEBULA
MCG-05-32-060	13 37 06.	- 31 23	270	11.	GALAXY
MCG-05-32-061	13 37 06.	- 32 21	48	15.	GALAXY
SN 1895B	13 37 07.	- 31 24		8.0	SUPERNOVA
SC 1334-3205.1	13 37 09.	- 32 20 22.	60		NEBULA
REIZ 3558	13 37 11.	+ 31 15	12	14.7	GALAXY
ZC 1337.2+1016	13 37 12.	+ 10 16	3360		CLUSTER OF GALAXIES
ZC 1337.2+1458	13 37 12.	+ 14 58	810		CLUSTER OF GALAXIES
ZC 1337.2+2621	13 37 12.	+ 26 21	1480		CLUSTER OF GALAXIES
ZWG 161.107	13 37 12.	+ 28 11		15.3	GALAXY
ZWG 161.108	13 37 12.	+ 29 37		15.7	GALAXY
MCG+05-32-052	13 37 12.	+ 31 14	15	15.2	GALAXY
ZWG 161.109	13 37 12.	+ 31 34		15.5	GALAXY
ZC 1337.2+5041	13 37 12.	+ 50 41	1210		CLUSTER OF GALAXIES
RS 36	13 37 12.4	+ 28 02 46.		17.83	BLUE STELLAR OBJECT
REIZ 3550	13 37 13.	+ 06 25	48	14.8	GALAXY
RRS 1.073	13 37 15.	+ 30 54		18.44	BLUE OBJECT
MCG+09-22-090	13 37 15.	+ 51 17 30.	78	14.	GALAXY
HN 1082	13 37 16.	- 25 13	78		NEBULA
IC 4315	13 37 16.	- 25 14			NONSTELLAR OBJECT
RNGC 5266A	13 37 16.	- 48 05			GALAXY
PATH 2.149	13 37 17.	- 00 05	8		NEBULA
HN 1714	13 37 17.0	- 00 04 34.	6		NEBULA
VV C55B	13 37 18.	+ 01 05	108	12.9	INTERACTING GALAXY
ZWG 102.032	13 37 18.	+ 17 04		15.6	GALAXY
ZWG 161.110	13 37 18.	+ 30 15		15.7	GALAXY
RRS 1.074	13 37 18.	+ 31 04		15.43	BLUE OBJECT
MCG+05-32-053	13 37 18.	+ 31 27	63	15.3	GALAXY
MCG+05-32-054	13 37 18.	+ 31 33	30	15.5	GALAXY
MCG+10-20-002	13 37 18.	+ 61 46	24	15.	GALAXY
PATH 1.634	13 37 18.	- 00 22	14		NEBULA
PATH 1.633	13 37 18.	- 00 57	14		NEBULA
MCG-04-32-049	13 37 18.	- 25 13 30.	78	15.5	GALAXY
RNGC 5257	13 37 19.	+ 01 06		13.5	GALAXY
HN 1081	13 37 19.	- 50 49		17.	NEBULA
IC 4312	13 37 19.	- 50 50			NONSTELLAR OBJECT
LB 00251	13 37 20.	+ 56 09 24.		14.4	FAINT BLUE STAR
REIZ 3552	13 37 21.	+ 01 06	60	13.4	GALAXY
HOLM 532A	13 37 21.	+ 01 06	36	12.9	PART OF MULTIPLE GALAXY
MCG+05-32-055	13 37 21.	+ 30 15	48	15.7	GALAXY
DV.56 N5266A	13 37 21.	- 48 06	216		SAc GALAXY
ARP 240	13 37 23.	+ 01 05			PECULIAR GALAXY
RRS 1.075	13 37 23.	+ 29 45		18.11	BLUE OBJECT
ZWG 017.055	13 37 24.	+ 01 05		13.7	GALAXY
UGC 08641	13 37 24.	+ 01 05	108	13.7	GALAXY S
VV 055A	13 37 24.	+ 01 05	66	13.3	INTERACTING GALAXY
KARA.72 389A	13 37 24.	+ 01 05	90	13.7	PART OF DOUBLE GALAXY
MCG+00-35-015	13 37 24.	+ 01 05	90	13.7	GALAXY
ZC 1337.4+1640	13 37 24.	+ 16 40	3160		CLUSTER OF GALAXIES
ZC 1337.4+2233	13 37 24.	+ 22 33	610		CLUSTER OF GALAXIES
TON-N 0727	13 37 24.	+ 25 10		15.6	BLUE STAR
ZWG 161.111	13 37 24.	+ 28 02		14.9	GALAXY
MCG+05-32-056	13 37 24.	+ 28 02	30	14.9	GALAXY
UGC 08642	13 37 24.	+ 46 16	90	16.5	GALAXY Sc
MCG+08-25-034	13 37 24.	+ 46 16	90	16.	GALAXY
ZWG 271.055	13 37 24.	+ 51 19		15.3	GALAXY
UGC 08643	13 37 24.	+ 51 19	78	15.3	GALAXY SB?b
BRON 20	13 37 24.5	+ 28 42 28.		19.14	BLUE OBJECT
RNGC 5258	13 37 25.	+ 01 05		14.0	GALAXY
HOLM 532B	13 37 26.	+ 01 06	66	13.3	PART OF MULTIPLE GALAXY
REIZ 3557	13 37 27.	+ 01 05	72	13.6	GALAXY
MCG+04-32-019	13 37 27.	+ 22 54	48	15.5	GALAXY
MCG+06-30-065	13 37 27.	+ 35 16 30.	18	15.5	GALAXY
RS 37	13 37 27.3	+ 26 10 46.		17.50	BLUE STELLAR OBJECT
IC 0907	13 37 29.	+ 50 58 25.			NONSTELLAR OBJECT
ZWG 017.056	13 37 30.	+ 01 05		13.8	GALAXY
UGC 08645	13 37 30.	+ 01 05	102	13.8	GALAXY S
KARA.72 389B	13 37 30.	+ 01 05	90	13.8	PART OF DOUBLE GALAXY
MCG+00-35-016	13 37 30.	+ 01 05	78	13.8	GALAXY
ZWG 045.065	13 37 30.	+ 04 56		15.0	GALAXY
MCG+01-35-026	13 37 30.	+ 04 56	36	15.0	GALAXY
ZC 1337.5+1854	13 37 30.	+ 18 54	3160		CLUSTER OF GALAXIES
MCG+05-32-057	13 37 30.	+ 30 22	54	15.7	GALAXY
ZWG 161.112	13 37 30.	+ 30 23		15.7	GALAXY
UGC 08646	13 37 30.	+ 30 23	60	15.7	GALAXY SBa-b
UGC 08647	13 37 30.	+ 31 33	66	16.5	GALAXY IRR
MCG+06-30-066	13 37 30.	+ 33 55	24	15.	GALAXY
MCG+06-30-067	13 37 30.	+ 33 56	24	15.	GALAXY
ZC 1337.5+3920	13 37 30.	+ 39 20	1280		CLUSTER OF GALAXIES
ZWG 218.033	13 37 30.	+ 43 18		15.2	GALAXY
MRK 267	13 37 30.	+ 43 18	20	15.	GALAXY WITH UV CONTINUUM
MCG+07-28-045	13 37 30.	+ 43 18	30	16.	GALAXY
MCG+08-25-035	13 37 30.	+ 46 49	24	16.	GALAXY
MCG-05-32-062	13 37 30.	- 28 37	24	15.5	GALAXY
HN 1083	13 37 30.	- 28 38		14.5	NEBULA
BC PKS1337-013	13 37 30.4	- 01 22 36.		18.5	QUASI-STELLAR OBJECT
SHB 276	13 37 30.4	- 01 22 38.		18.5	QUASI-STELLAR OBJECT
REIZ 3559	13 37 31.	+ 04 56	18	15.0	GALAXY
IC 4316	13 37 31.	- 28 39			NONSTELLAR OBJECT
HN 1715	13 37 31.4	- 02 37 51.	18		NEBULA
ZC 1337.6+0324	13 37 36.	+ 03 24	2220		CLUSTER OF GALAXIES
MCG+05-32-058	13 37 36.	+ 28 38 30.	96	14.0	GALAXY
ZWG 161.113	13 37 36.	+ 28 40		14.0	GALAXY
UGC 08648	13 37 36.	+ 28 40	96	14.0	GALAXY S
MCG+05-32-059	13 37 36.	+ 31 32 30.	66	15.	GALAXY
ZC 1337.6+5331	13 37 36.	+ 53 31	2080		CLUSTER OF GALAXIES
ZWG 295.006	13 37 36.	+ 61 45		15.2	GALAXY
ZWG 294.058	13 37 36.	+ 61 45		15.2	GALAXY
UGC 08649	13 37 36.	+ 61 45	78	15.2	GALAXY E
ZWG 017.057	13 37 36.	- 02 38		15.5	GALAXY
MCG-04-32-050	13 37 36.	- 23 36 30.	90	13.5	GALAXY
MCG-05-32-063	13 37 36.	- 33 25	72	15.	GALAXY
REIZ 3560	13 37 37.	+ 28 39	72	13.2	GALAXY
RNGC 5260	13 37 37.	- 23 36		13.0	GALAXY
REIN 7.101	13 37 37.13	+ 28 39 06.2			NEBULA
RNGC 5263	13 37 39.	+ 28 40		14.0	GALAXY
BRON 34	13 37 39.4	+ 29 53 22.		19.82	BLUE OBJECT
BRON 35	13 37 39.9	+ 29 53 05.		18.92	BLUE OBJECT
IC 0905	13 37 41.	+ 23 23 54.			NONSTELLAR OBJECT
ZWG 045.066	13 37 42.	+ 02 43		14.9	GALAXY
UGC 08650	13 37 42.	+ 02 43	114	14.9	GALAXY Sb
MCG+01-35-027	13 37 42.	+ 02 43	108	14.9	GALAXY
ZWG 045.067	13 37 42.	+ 05 19		15.3	GALAXY
ZC 1337.7+0732	13 37 42.	+ 07 32	2220		CLUSTER OF GALAXIES
ZWG 102.033	13 37 42.	+ 15 11		15.6	GALAXY
ZWG 131.019	13 37 42.	+ 23 24		15.0	GALAXY
MCG+04-32-020	13 37 42.	+ 23 25	48	15.0	GALAXY
MCG+05-32-060	13 37 42.	+ 30 21	18	15.4	GALAXY
ZWG 161.114	13 37 42.	+ 30 22		15.4	GALAXY
MCG+07-28-046	13 37 42.	+ 40 59	120	14.5	GALAXY
ZC 1337.7+5530	13 37 42.	+ 55 30	1140		CLUSTER OF GALAXIES
RNGC 5261	13 37 44.	+ 05 19		15.5	GALAXY
MCG+04-32-021	13 37 45.	+ 23 37	48	15.7	GALAXY
ZWG 131.020	13 37 48.	+ 23 35		15.7	GALAXY
STOCK 47	13 37 48.	+ 27 21			BLUE KNOT NEAR ELLIP GLXY
ZWG 161.115	13 37 48.	+ 29 24		15.7	GALAXY
ZC 1337.8+3753	13 37 48.	+ 37 53	940		CLUSTER OF GALAXIES
MCG+07-28-047	13 37 48.	+ 40 40	33	15.	GALAXY
ZWG 218.034	13 37 48.	+ 41 00		15.3	GALAXY
UGC 08651	13 37 48.	+ 41 00	180	15.3	GALAXY DWRF IR
MCG+08-25-036	13 37 48.	+ 47 48	60	17.	GALAXY
ZC 1337.8+7151	13 37 48.	+ 71 51	1080		CLUSTER OF GALAXIES
IC 0906	13 37 49.	+ 23 35 37.			NONSTELLAR OBJECT
RRS 1.076	13 37 49.	+ 29 22		16.30	BLUE OBJECT
MCG-04-32-051	13 37 51.	- 21 40	42	15.	GALAXY
SC 1335-3209.1	13 37 52.	- 32 24 21.	30		NEBULA
ZWG 045.068	13 37 54.	+ 05 01		14.9	GALAXY
MCG+01-35-028	13 37 54.	+ 05 01	48	14.9	GALAXY
ZWG 190.040	13 37 54.	+ 37 07		14.9	GALAXY
RNGC 5265	13 37 54.	+ 37 07		15.0	GALAXY
ARC 1770	13 37 54.	+ 41 33		17.6	RICH CLUSTER OF GALAXIES
PK318+41.1	13 37 54.	- 19 38	478	13.0	PLANETARY NEBULA
MCG-05-32-064	13 37 54.	- 32 25	36	15.	GALAXY
REIZ 3561	13 37 55.	+ 05 02	21	14.4	GALAXY
SC 1335-3147.7	13 37 55.	- 32 02 57.	30		NEBULA
MCG+04-32-022	13 37 57.	+ 22 08	39	15.7	GALAXY
MCG+06-30-068	13 37 57.	+ 37 08	33	13.5	GALAXY
REIZ 3563	13 37 58.	+ 32 24	30	14.7	GALAXY
RRS 1.077	13 37 59.	+ 31 09		18.01	BLUE OBJECT
VDB.66G 181	13 38	+ 40 56	100		DWARF GALAXY
SER 105.01	13 38	- 48 02	2280	13.	LOOSE GROUP OF 5 GALAXIES
LB 09878	13 38	- 80 43		13.5	FAINT BLUE STAR
ZWG 017.058	13 38 00.	+ 00 15		15.7	GALAXY
ZWG 131.021	13 38 00.	+ 22 09		15.7	GALAXY
KARA.73B 0597	13 38 00.	+ 22 09	48	15.7	ISOLATED GALAXY S
ZWG 161.116	13 38 00.	+ 26 37		15.6	GALAXY
UGC 08652	13 38 00.	+ 26 37	66	15.6	GALAXY S
TON-N 0167	13 38 00.	+ 30 28		16.0	BLUE STAR
ZWG 161.117	13 38 00.	+ 31 39		15.7	GALAXY
UGC 08653	13 38 00.	+ 31 39	72	15.7	GALAXY Sb-c
ZWG 161.118	13 38 00.	+ 32 24		15.7	GALAXY
MCG+07-28-048	13 38 00.	+ 39 06	36	15.5	GALAXY
ZWG 218.035	13 38 00.	+ 39 07		15.2	GALAXY
UGC 08654	13 38 00.	+ 76 15	72	16.0	GALAXY Sc
MCG-01-35-008	13 38 00.	- 07 29	90	13.5	GALAXY
REIZ 3564	13 38 01.	+ 37 07	30	14.4	GALAXY
REIZ 3566	13 38 02.	+ 57 54	15	15.2	GALAXY
RS 38	13 38 03.9	+ 27 56 09.		17.71	BLUE STELLAR OBJECT
RRS 1.078	13 38 04.	+ 29 09		17.33	BLUE OBJECT
MCG+05-32-061	13 38 06.	+ 31 37 30.	66	15.7	GALAXY
FEIG 087	13 38 06.	+ 61 07		12.8	FAINT BLUE STAR
REIZ 3562	13 38 07.	- 07 30	54	14.2	GALAXY
ARC 1777	13 38 10.	+ 71 52		17.6	RICH CLUSTER OF GALAXIES
RRS 1.079	13 38 12.	+ 30 27		17.30	BLUE OBJECT
MCG+10-20-003	13 38 12.	+ 57 22	18	16.	GALAXY
BRON 36	13 38 15.4	+ 29 38 27.		18.50	BLUE OBJECT
ZWG 045.069	13 38 18.	+ 05 40		15.6	GALAXY
ZC 1338.3+2611	13 38 18.	+ 26 11	400		CLUSTER OF GALAXIES
VV 195B	13 38 18.	+ 26 25	18	16.	INTERACTING GALAXY
VV 195A	13 38 18.	+ 26 25	18	16.	INTERACTING GALAXY
VV 195	13 38 18.	+ 26 25	54		INTERACTING GALAXY
8ZW 1338+10.3	13 38 24.	+ 10 16		16.0	COMPACT GALAXY
ZC 1338.4+2420	13 38 24.	+ 24 20	6250		CLUSTER OF GALAXIES
ZWG 218.036	13 38 24.	+ 39 02		14.3	GALAXY
RNGC 5267	13 38 24.	+ 39 02		14.5	GALAXY
UGC 08655	13 38 24.	+ 39 02	96	14.3	GALAXY SBb
MCG+07-28-049	13 38 24.	+ 39 02	81	14.	GALAXY
ZC 1338.4+4132	13 38 24.	+ 41 32	1480		CLUSTER OF GALAXIES
ZWG 218.037	13 38 24.	+ 43 15		15.4	GALAXY
UGC 08656	13 38 24.	+ 43 15	60	15.4	GALAXY S
MCG+07-28-050	13 38 24.	+ 43 15	48	15.	GALAXY
MCG+10-20-004	13 38 24.	+ 57 27	27	16.	GALAXY
MCG-02-35-013	13 38 24.	- 10 30	42	15.5	GALAXY
BRON 21	13 38 26.7	+ 28 46 40.		19.36	BLUE OBJECT

OBJECT NAME	RIGHT ASCEN.	DECLINATION	DIAM.	MAGN.	TYPE OF OBJECT
IC 0909	13 38 27.	+ 24 45 27.			NONSTELLAR OBJECT
RRS 1.080	13 38 29.	+ 30 04		17.81	BLUE OBJECT
REIZ 3567	13 38 29.	+ 39 03	60	13.9	GALAXY
ZWG 045.070	13 38 30.	+ 05 21		15.3	GALAXY
UGC 08657	13 38 30.	+ 05 21	96	15.3	GALAXY Sb
MCG+01-35-029	13 38 30.	+ 05 21	84	15.3	GALAXY
ZWG 073.060	13 38 30.	+ 11 57		15.6	GALAXY
ZC 1338.5+2224	13 38 30.	+ 22 24	1280		CLUSTER OF GALAXIES
MCG+04-32-023	13 38 30.	+ 24 42 30.	24	14.9	GALAXY
ZWG 131.022	13 38 30.	+ 24 44		14.9	GALAXY
ZWG 131.023	13 38 30.	+ 26 09		15.2	GALAXY
MCG+04-32-024	13 38 30.	+ 26 10	48	15.2	GALAXY
MCG+10-20-005	13 38 30.	+ 57 02	12	16.	GALAXY
MCG+13-10-009	13 38 30.	+ 79 44	18	14.	GALAXY
BRON 01	13 38 30.2	+ 27 48 25.		18.99	BLUE OBJECT
HELW 222	13 38 34.	- 31 45 02.			NEBULA
RNGC 5295	13 38 35.	+ 79 43		15.0	GALAXY
BRON 37	13 38 35.1	+ 29 37 32.		19.37	BLUE OBJECT
ZC 1338.6+0445	13 38 36.	+ 04 45	1080		CLUSTER OF GALAXIES
RRS 1.081	13 38 36.	+ 28 50		17.10	BLUE OBJECT
ZC 1338.6+3722	13 38 36.	+ 37 22	1610		CLUSTER OF GALAXIES
ZC 1338.6+4010	13 38 36.	+ 40 10	2550		CLUSTER OF GALAXIES
MCG+10-20-006	13 38 36.	+ 61 05	10	16.	GALAXY
ZWG 353.023	13 38 36.	+ 79 43		14.9	GALAXY
MCG-01-35-009	13 38 36.	- 04 04	48	15.	GALAXY
MCG-05-32-065	13 38 36.	- 31 45	36	15.	GALAXY
REIZ 3565	13 38 37.	+ 05 22	18	15.8	GALAXY
SC 1335-3129.9	13 38 37.	- 31 45 08.	48		NEBULA
HN 1716	13 38 40.0	+ 02 02 05.	24		NEBULA
RNGC 5264	13 38 41.	- 29 40		13.0	GALAXY
ZWG 017.059	13 38 42.	+ 02 02		15.2	GALAXY
ZWG 073.061	13 38 42.	+ 13 01		15.5	GALAXY
MCG+04-32-025	13 38 42.	+ 23 32	30	15.1	GALAXY
ZWG 272.001	13 38 42.	+ 54 35		14.4	GALAXY
ZWG 271.056	13 38 42.	+ 54 35		14.4	GALAXY
UGC 08658	13 38 42.	+ 54 35	162	14.4	GALAXY SBb/Sc
MCG+09-22-091	13 38 42.	+ 54 35	150	13.2	GALAXY
UGC 08659	13 38 42.	+ 55 41	66	17.	GALAXY DWARF
MCG-05-32-066	13 38 42.	- 29 40	96	13.	GALAXY
IC 0908	13 38 44.	- 04 05 51.			NONSTELLAR OBJECT
MCG+08-25-037	13 38 45.	+ 48 25	48	16.	GALAXY
HN 1717	13 38 45.1	- 02 16 49.	12		NEBULA
RS 39	13 38 46.5	+ 27 05 39.		18.17	BLUE STELLAR OBJECT
BRON 22	13 38 46.8	+ 28 53 35.		19.26	BLUE OBJECT
IC 0910	13 38 47.	+ 23 32 22.			NONSTELLAR OBJECT
RRS 1.082	13 38 47.	+ 30 22		18.74	BLUE OBJECT
HO 5	13 38 47.	+ 54 35	210	13.23	Sc GALAXY
BRON 38	13 38 47.1	+ 29 32 19.		19.38	BLUE OBJECT
ZWG 045.071	13 38 48.	+ 05 17		15.4	GALAXY
ZWG 073.062	13 38 48.	+ 14 01		15.5	GALAXY
ZWG 131.024	13 38 48.	+ 23 32		15.1	GALAXY
MCG+04-32-026	13 38 48.	+ 24 45	30	15.	GALAXY
ZWG 218.038	13 38 48.	+ 42 41		15.1	GALAXY
MCG-03-35-014	13 38 48.	- 19 07	42	16.	GALAXY
RRS 1.083	13 38 49.	+ 29 33		18.93	BLUE OBJECT
BRON 23	13 38 49.0	+ 28 54 47.		18.78	BLUE OBJECT
RRS 1.085	13 38 50.	+ 28 55		18.75	BLUE OBJECT
RRS 1.084	13 38 50.	+ 31 35		18.59	BLUE OBJECT
RS 40	13 38 50.8	+ 28 54 45.		18.23	BLUE STELLAR OBJECT
ZC 1338.9+2034	13 38 54.	+ 20 34	1280		CLUSTER OF GALAXIES
RRS 3.001	13 38 54.	+ 30 19		19.90	BLUE OBJECT
ZWG 161.119	13 38 54.	+ 30 38		15.3	GALAXY
MRK 268	13 38 54.	+ 30 38	12	15.	GALAXY WITH UV CONTINUUM
KW 22	13 38 54.	+ 30 33	36		SEYFERT GALAXY
VVI 65	13 38 54.	+ 30 38	28	15.60	SEYFERT GALAXY
ZC 1338.9-0042	13 38 54.	- 00 42	2150		CLUSTER OF GALAXIES
ZWG 017.060	13 38 54.	- 03 27		15.7	GALAXY
BRON 39	13 38 54.0	+ 29 52 30.		20.00	BLUE OBJECT
REIZ 3572	13 38 55.	+ 54 35	108	14.1	GALAXY
REIZ 3570	13 38 57.	+ 34 02	60	14.9	GALAXY
VDB.66G 242	13 39	- 29 38	100		DWARF GALAXY
LB 09879	13 39	- 84 49		15.5	FAINT BLUE STAR
ZWG 073.063	13 39 00.	+ 14 06		15.7	GALAXY
UGC 08660	13 39 00.	+ 14 06	66	15.7	GALAXY Sc
ZC 1339.0+1536	13 39 00.	+ 15 36	1880		CLUSTER OF GALAXIES
MCG+04-32-027	13 39 00.	+ 23 30	24	15.6	GALAXY
UGC 08661	13 39 00.	+ 24 45	84	16.0	GALAXY SBa?
UGC 08662	13 39 00.	+ 34 00	84	16.0	GALAXY Sc
MCG+06-30-069	13 39 00.	+ 34 00	78	15.	GALAXY
MCG+06-30-070	13 39 00.	+ 37 17	42	15.	GALAXY
ZC 1339.0+3938	13 39 00.	+ 39 38	1140		CLUSTER OF GALAXIES
ZC 1339.0+4236	13 39 00.	+ 42 36	1480		CLUSTER OF GALAXIES
ZC 1339.0+4806	13 39 00.	+ 48 06	870		CLUSTER OF GALAXIES
ZC 1339.0+8035	13 39 00.	+ 80 35	4570		CLUSTER OF GALAXIES
SC 1336-3204.3	13 39 00.	- 32 19 31.	30		NEBULA
MCG-06-30-018	13 39 00.	- 38 04	36	15.	GALAXY
BRON 40	13 39 00.9	+ 29 18 22.		19.21	BLUE OBJECT
ARC 1774	13 39 01.	+ 40 16		17.6	RICH CLUSTER OF GALAXIES
HN 1718	13 39 02.1	- 00 10 12.	6		NEBULA
MCG+04-32-028	13 39 03.	+ 23 30 30.	36	15.6	GALAXY
RS 41	13 39 03.9	+ 27 24 39.		17.28	BLUE STELLAR OBJECT
IC 0911	13 39 04.	+ 23 29 59.			NONSTELLAR OBJECT
ZWG 045.072	13 39 06.	+ 04 30		15.5	GALAXY
ZWG 045.073	13 39 06.	+ 05 17		14.6	GALAXY
UGC 08663	13 39 06.	+ 05 17	66	14.6	GALAXY E-S0
MCG+01-35-030	13 39 06.	+ 05 17	36	14.6	GALAXY
ZWG 073.064	13 39 06.	+ 09 03		15.7	GALAXY
ZWG 102.034	13 39 06.	+ 17 43		15.2	GALAXY
ZWG 131.025	13 39 06.	+ 23 25		15.6	GALAXY
UGC 08664	13 39 06.	+ 23 25	60	15.6	GALAXY SBb
MCG+04-32-029	13 39 06.	+ 23 25 30.	36	15.6	GALAXY
ZWG 131.026	13 39 06.	+ 23 30		15.6	GALAXY
HOLM 534B	13 39 06.	+ 23 30	24	14.9	PART OF MULTIPLE GALAXY
UGC 08665	13 39 06.	+ 23 30	84	15.6	GALAXY DBL SYS
MCG+06-30-071	13 39 06.	+ 37 11 30.	30	15.	GALAXY
ZC 1339.1+5147	13 39 06.	+ 51 47	1410		CLUSTER OF GALAXIES
ZC 1339.1+5815	13 39 06.	+ 58 15	1680		CLUSTER OF GALAXIES
7ZW 522	13 39 06.	+ 60 43			COMPACT GALAXY
IC 0912	13 39 08.	+ 23 29 59.			NONSTELLAR OBJECT
IC 0913	13 39 09.	+ 23 25 23.			NONSTELLAR OBJECT
HOLM 534A	13 39 10.	+ 23 30	48	14.4	PART OF MULTIPLE GALAXY
ZC 1339.2+0323	13 39 12.	+ 03 23	340		CLUSTER OF GALAXIES
ZWG 073.065	13 39 12.	+ 12 53		15.6	GALAXY
PK309+00.1	13 39 12.	- 61 07 52.	10		PLANETARY NEBULA
REIZ 3569	13 39 13.	+ 05 17	12	14.6	GALAXY
SC 1336-3048.9	13 39 13.	- 31 04 07.	24		NEBULA
REIZ 3568	13 39 14.	+ 02 20	54	15.0	GALAXY
ARC 1776	13 39 14.	+ 58 17		17.2	RICH CLUSTER OF GALAXIES

OBJECT NAME	RIGHT ASCEN.	DECLINATION	DIAM.	MAGN.	TYPE OF OBJECT
BRON 02	13 39 14.9	+ 27 36 50.		18.90	BLUE OBJECT
MCG+07-28-051	13 39 15.	+ 43 33	48	16.	GALAXY
MCG+09-22-092	13 39 15.	+ 52 21	42	15.	GALAXY
RS 42	13 39 16.3	+ 27 16 57.		16.10	BLUE STELLAR OBJECT
ZWG 045.074	13 39 18.	+ 07 52		15.7	GALAXY
8ZW 1339+10.4	13 39 18.	+ 10 25		19.3	COMPACT GALAXY
ZWG 131.027	13 39 18.	+ 23 26		15.2	GALAXY
MCG+04-32-030	13 39 18.	+ 23 27	36	15.2	GALAXY
MCG+09-22-093	13 39 18.	+ 56 27	24	16.	GALAXY
MRK 269	13 39 18.	+ 66 06	8	17.5	GALAXY WITH UV CONTINUUM
MCG-05-32-067	13 39 18.	- 30 31	42	14.5	GALAXY
IC 0914	13 39 19.	+ 23 26 48.			NONSTELLAR OBJECT
SC 1336-3015.4	13 39 19.	- 30 30 36.	48		NEBULA
RNGC 5271	13 39 21.	+ 30 23		15.5	GALAXY
REIZ 3575	13 39 23.	+ 30 22	12	14.7	GALAXY
ARC 1771	13 39 23.	- 26 02		16.8	RICH CLUSTER OF GALAXIES
ZWG 017.061	13 39 24.	+ 02 19		15.5	GALAXY
UGC 08666	13 39 24.	+ 02 19	90	15.5	GALAXY CHAIN
MCG+00-35-017	13 39 24.	+ 02 19	42		GALAXY
ZC 1339.4+1025	13 39 24.	+ 10 25	1010		CLUSTER OF GALAXIES
MCG+05-32-064	13 39 24.	+ 26 36	60	15.3	GALAXY
MCG+05-32-063	13 39 24.	+ 26 36	60	15.3	GALAXY
MCG+05-32-062	13 39 24.	+ 27 15	45	15.0	GALAXY
ZWG 161.120	13 39 24.	+ 30 23		15.4	GALAXY
ZC 1339.4+3229	13 39 24.	+ 32 29	670		CLUSTER OF GALAXIES
UGC 08667	13 39 24.	+ 38 44	72	16.5	GALAXY
MCG+09-22-095	13 39 24.	+ 52 33	42	16.	GALAXY
MCG+09-22-094	13 39 24.	+ 55 56	30	16.	GALAXY
MRK 270	13 39 24.	+ 67 56	15	14.5	GALAXY WITH UV CONTINUUM
KW 23	13 39 24.	+ 67 56	45		SEYFERT GALAXY
VVI 66	13 39 24.	+ 67 56	45	14.97	SEYFERT GALAXY
MCG+11-17-007	13 39 24.	+ 67 56	45	16.	GALAXY
MCG-05-32-068	13 39 24.	- 30 47	60	15.	GALAXY
SC 1336-3031.9	13 39 24.	- 30 47 06.	60		NEBULA
REIN 7.102	13 39 25.20	+ 30 22 38.0			NEBULA
ARC 1772	13 39 26.	- 10 51		17.0	RICH CLUSTER OF GALAXIES
HN 1719	13 39 26.6	+ 00 50 42.	12		NEBULA
VV 170C	13 39 27.	+ 02 21	18	17.	INTERACTING GALAXY
VV 170B	13 39 27.	+ 02 21	18	17.	INTERACTING GALAXY
VV 170A	13 39 27.	+ 02 21	24	16.5	INTERACTING GALAXY
MCG+05-32-065	13 39 27.	+ 30 21 30.	48	15.4	GALAXY
RNGC 5268	13 39 28.	- 13 36			NON-EXISTENT OBJECT
ZWG 017.062	13 39 30.	+ 00 51		15.6	GALAXY
ZC 1339.5+0233	13 39 30.	+ 02 33	2420		CLUSTER OF GALAXIES
ZWG 102.035	13 39 30.	+ 18 24		15.4	GALAXY
UGC 08668	13 39 30.	+ 18 24	66	15.4	GALAXY Sc
ZWG 161.121	13 39 30.	+ 26 38		15.3	GALAXY
UGC 08669	13 39 30.	+ 26 38	60	15.3	GALAXY DBL SYS
ZWG 161.122	13 39 30.	+ 27 16		15.0	GALAXY
IC 4317	13 39 30.	+ 27 21 06.			NONSTELLAR OBJECT
MCG+07-28-052	13 39 30.	+ 38 44	60	16.	GALAXY
ZWG 218.039	13 39 30.	+ 41 08		15.5	GALAXY
UGC 08670	13 39 30.	+ 41 08	66	15.5	GALAXY Sc
ZWG 272.002	13 39 30.	+ 55 54		14.1	GALAXY
ZWG 271.057	13 39 30.	+ 55 54		14.1	GALAXY
UGC 08671	13 39 30.	+ 55 54	33	14.1	GALAXY
MCG+09-22-096	13 39 30.	+ 55 54	36	15.	GALAXY
BRON 03	13 39 32.2	+ 27 39 32.		18.48	BLUE OBJECT
MCG+07-28-053	13 39 33.	+ 41 07	72	15.	GALAXY
BRON 04	13 39 34.3	+ 27 24 40.		18.72	BLUE OBJECT
MCG+03-35-016	13 39 36.	+ 18 25	72	15.4	GALAXY
1ZW 068	13 39 36.	+ 53 46			COMPACT GALAXY
MCG+09-22-098	13 39 36.	+ 56 20	24	16.	GALAXY
MCG+09-22-097	13 39 36.	+ 56 26	36	16.	GALAXY
ZWG 317.006	13 39 36.	+ 67 55		14.3	GALAXY
RNGC 5283	13 39 36.	+ 67 55		14.5	GALAXY
UGC 08672	13 39 36.	+ 67 55	66	14.3	GALAXY S0?
REIZ 3571	13 39 36.	- 04 50	54	14.7	GALAXY
ARC 1773	13 39 37.	+ 02 30		15.6	RICH CLUSTER OF GALAXIES
REIZ 3574	13 39 37.	+ 04 31	18	14.9	GALAXY
ARC 1775	13 39 37.	+ 26 37		15.7	RICH CLUSTER OF GALAXIES
REIZ 3573	13 39 38.	+ 02 29	42	16.0	GALAXY
RRS 4.001	13 39 38.	+ 27 14		18.18	BLUE OBJECT
RRS 4.002	13 39 40.	+ 26 44		17.88	BLUE OBJECT
VV 019A	13 39 40.	+ 55 55	66	14.	INTERACTING GALAXY
RS 43	13 39 40.5	+ 26 51 15.		18.24	BLUE STELLAR OBJECT
ZWG 017.063	13 39 42.	+ 02 06		14.9	GALAXY
MCG+00-35-018	13 39 42.	+ 02 06	42	14.9	GALAXY
ZWG 045.075	13 39 42.	+ 04 30		14.7	GALAXY
UGC 08673	13 39 42.	+ 04 30	72	14.7	GALAXY SBb
MCG+01-35-031	13 39 42.	+ 04 30	60	14.7	GALAXY
MRK 067	13 39 42.	+ 30 46	10	16.5	GALAXY WITH UV CONTINUUM
ZC 1339.7+5506	13 39 42.	+ 55 06	1950		CLUSTER OF GALAXIES
MCG+09-22-100	13 39 42.	+ 56 22	24	16.	GALAXY
MCG+09-22-099	13 39 42.	+ 56 26	36	16.	GALAXY
ZC 1339.7+7548	13 39 42.	+ 75 48	1680		CLUSTER OF GALAXIES
8ZW 1339-31.8	13 39 42.	- 31 51		15.6	COMPACT GALAXY
REIN 7.103	13 39 42.48	+ 30 02 01.7			NEBULA
HN 1720	13 39 42.5	+ 00 43 37.	12		NEBULA
HN 1721	13 39 42.8	+ 02 06 37.	24		NEBULA
RNGC 5270	13 39 44.	+ 04 30		14.5	GALAXY
REIZ 3576	13 39 45.	+ 02 06	24	14.4	GALAXY
RRS 4.003	13 39 45.	+ 27 14		19.50	BLUE OBJECT
ARP 239	13 39 45.	+ 55 55			PECULIAR GALAXY
RNGC 5278	13 39 46.	+ 55 55		13.5	GALAXY
BRON 24	13 39 46.4	+ 28 49 02.		19.53	BLUE OBJECT
ZWG 045.076	13 39 48.	+ 04 52		15.7	GALAXY
ZC 1339.8+1617	13 39 48.	+ 16 17	870		CLUSTER OF GALAXIES
ZWG 161.123	13 39 48.	+ 32 17		15.7	GALAXY
UGC 08674	13 39 48.	+ 32 17	66	15.7	GALAXY SB
ZWG 190.041	13 39 48.	+ 35 54		12.7	GALAXY
UGC 08675	13 39 48.	+ 35 54	168	12.7	GALAXY S0
KARA.72 391A	13 39 48.	+ 35 54	192	12.7	PART OF DOUBLE GALAXY
MCG+06-30-072	13 39 48.	+ 35 54	42	12.	GALAXY
ZWG 190.042	13 39 48.	+ 37 17		14.6	GALAXY
UGC 08676	13 39 48.	+ 43 09	66	17.	GALAXY
MCG+08-25-038	13 39 48.	+ 48 09	30	17.	GALAXY
1ZW 069	13 39 48.	+ 55 55			COMPACT GALAXY
ZWG 272.003	13 39 48.	+ 55 56		13.6	GALAXY
ZWG 271.058	13 39 48.	+ 55 56		13.6	GALAXY
UGC 08678	13 39 48.	+ 55 56	60	13.6	GALAXY DBL SYS
UGC 08677	13 39 48.	+ 55 56	60	13.6	GALAXY DBL SYS
KARA.72 390A	13 39 48.	+ 55 56	48	13.6	PART OF DOUBLE GALAXY
MRK 271	13 39 48.	+ 55 57	25	14.	GALAXY WITH UV CONTINUUM
MCG+09-22-101	13 39 48.	+ 55 57	60	13.6	GALAXY
ZC 1339.8+6328	13 39 48.	+ 63 28	2080		CLUSTER OF GALAXIES
BRON 05	13 39 49.2	+ 27 23 29.		18.22	BLUE OBJECT
RRS 4.004	13 39 50.	+ 27 25			BLUE OBJECT

OBJECT NAME	RIGHT ASCEN.	DECLINATION	DIAM.	MAGN.	TYPE OF OBJECT
RNGC 5266	13 39 51.	- 47 56		12.5	GALAXY
RRS 1.086	13 39 52.	+ 31 05		17.82	BLUE OBJECT
RNGC 5279	13 39 52.	+ 55 55		15.0	GALAXY
REIZ 3585	13 39 52.	+ 55 55	42	13.6	GALAXY
BRON 06	13 39 52.7	+ 27 35 32.		16.33	BLUE OBJECT
RRS 4.005	13 39 53.	+ 27 01		18.94	BLUE OBJECT
RNGC 5273	13 39 53.	+ 35 54		12.5	GALAXY
REIN 2.196	13 39 53.12	+ 28 37 44.2			NEBULA
ZWG 045.077	13 39 54.	+ 04 06		15.6	GALAXY
ZC 1339.9+1327	13 39 54.	+ 13 27	340		CLUSTER OF GALAXIES
MCG+03-35-017	13 39 54.	+ 14 59	60	15.3	GALAXY
GCL 025	13 39 54.	+ 28 38	1116	7.21	GLOBULAR STAR CLUSTER
ZC 1339.9+3030	13 39 54.	+ 30 30	7530		CLUSTER OF GALAXIES
MCG+06-30-073	13 39 54.	+ 37 18	48	14.5	GALAXY
MRK 272	13 39 54.	+ 43 02	11	16.5	GALAXY WITH UV CONTINUUM
VV 019B	13 39 54.	+ 55 55	24	15.	INTERACTING GALAXY
KARA.72 390B	13 39 54.	+ 55 56	30		PART OF DOUBLE GALAXY
MCG+09-22-102	13 39 54.	+ 55 57	60	15.	GALAXY
ZC 1339.9+5646	13 39 54.	+ 56 46	870		CLUSTER OF GALAXIES
REIZ 3578	13 39 55.	+ 35 54	72	13.1	GALAXY
HOLM 535A	13 39 55.	+ 35 55	108	12.6	PART OF MULTIPLE GALAXY
BRON 41	13 39 55.0	+ 29 49 47.		18.75	BLUE OBJECT
RS 44	13 39 56.9	+ 27 53 14.		17.42	BLUE STELLAR OBJECT
RNGC 5272	13 39 57.	+ 28 38		7.0	GLOBULAR CLUSTER
REIZ 3577	13 39 58.	+ 32 17	42	14.5	GALAXY
RS 45	13 39 59.0	+ 26 14 26.		16.17	BLUE STELLAR OBJECT
VDB.66G 182	13 40	+ 39 51	70		DWARF GALAXY
ZWG 045.078	13 40 00.	+ 03 19		15.7	GALAXY
ZWG 102.036	13 40 00.	+ 15 00		15.3	GALAXY
MCG+05-32-066	13 40 00.	+ 30 06	24	15.7	GALAXY
UGC 08679	13 40 00.	+ 38 47	84	17.	GALAXY DWRF SP
ZC 1340.0+5218	13 40 00.	+ 52 18	1080		CLUSTER OF GALAXIES
BRON 07	13 40 02.9	+ 27 41 27.		19.36	BLUE OBJECT
RNGC 5275	13 40 03.	+ 30 05		15.5	GALAXY
RNGC 5274	13 40 03.	+ 30 06		15.5	GALAXY
MCG+07-28-054	13 40 03.	+ 38 46	78	17.	GALAXY
RRS 1.087	13 40 04.	+ 30 37		18.55	BLUE OBJECT
REIZ 3579	13 40 05.	+ 30 05	6	15.5	GALAXY
REIZ 3580	13 40 05.	+ 30 06	18	14.8	GALAXY
RRS 1.088	13 40 05.	+ 30 49		17.91	BLUE OBJECT
BRON 25	13 40 05.0	+ 28 49 27.		19.38	BLUE OBJECT
BRON 08	13 40 05.6	+ 27 35 45.		17.79	BLUE OBJECT
RS 46	13 40 05.7	+ 27 35 20.		17.98	BLUE STELLAR OBJECT
MCG+05-32-068	13 40 06.	+ 26 44	42	15.4	GALAXY
RRS 4.006	13 40 06.	+ 27 36		17.74	BLUE OBJECT
TON-N 0728	13 40 06.	+ 30 01		16.3	BLUE STAR
MCG+05-32-067	13 40 06.	+ 30 04	30	15.4	GALAXY
ZWG 161.124	13 40 06.	+ 30 05		15.4	GALAXY
ZWG 161.125	13 40 06.	+ 30 06		15.7	GALAXY
ZWG 190.043	13 40 06.	+ 35 53		14.6	GALAXY
UGC 08680	13 40 06.	+ 35 53	66	14.6	GALAXY SBa
KARA.72 391B	13 40 06.	+ 35 53	66	14.6	PART OF DOUBLE GALAXY
MCG+13-10-010	13 40 06.	+ 74 57	48	16.	GALAXY
REIN 7.104	13 40 06.12	+ 30 05 57.6			NEBULA
REIN 7.105	13 40 06.26	+ 30 04 36.2			NEBULA
REIZ 3583	13 40 07.	+ 35 53	30	14.5	GALAXY
BRON 42	13 40 08.9	+ 29 18 51.		20.09	BLUE OBJECT
RRS 3.002	13 40 09.	+ 29 19			BLUE OBJECT
MCG+06-30-074	13 40 09.	+ 35 52	42	14.	GALAXY
HOLM 535B	13 40 09.	+ 35 53	36	14.3	PART OF MULTIPLE GALAXY
RRS 1.090	13 40 11.	+ 28 54		19.06	BLUE OBJECT
RRS 1.089	13 40 11.	+ 29 19		19.70	BLUE OBJECT
REIZ 3581	13 40 11.	+ 30 08	24	15.4	GALAXY
REIZ 3582	13 40 11.	+ 30 30	18	14.6	GALAXY
RNGC 5276	13 40 11.	+ 35 53		14.5	GALAXY
ZWG 161.126	13 40 12.	+ 26 45		15.4	GALAXY
ZWG 161.127	13 40 12.	+ 31 47		15.4	GALAXY
MCG-03-35-015	13 40 12.	- 18 35	9	15.5	GALAXY
BRON 43	13 40 12.6	+ 29 11 34.		19.38	BLUE OBJECT
IC 0916	13 40 13.	+ 24 43 39.			NONSTELLAR OBJECT
BRON 26	13 40 13.0	+ 28 52 32.		17.76	BLUE OBJECT
RRS 1.091	13 40 14.	+ 28 53		17.50	BLUE OBJECT
REIN 7.106	13 40 14.19	+ 30 30 06.2			NEBULA
REIN 7.107	13 40 14.75	+ 30 07 05.0			NEBULA
RRS 4.007	13 40 16.	+ 26 52		17.56	BLUE OBJECT
REIZ 3584	13 40 17.	+ 30 13	12	14.9	GALAXY
RS 47	13 40 17.1	+ 26 29 08.		17.18	BLUE STELLAR OBJECT
ZWG 131.028	13 40 18.	+ 24 43		15.4	GALAXY
MCG+04-32-031	13 40 18.	+ 24 43 30.	30	15.4	GALAXY
HOLM 536B	13 40 18.	+ 30 07	30	15.5	PART OF MULTIPLE GALAXY
ZWG 161.128	13 40 18.	+ 30 30		15.5	GALAXY
ZWG 190.044	13 40 18.	+ 35 16		15.2	GALAXY
UGC 08681	13 40 18.	+ 35 16	60	15.2	GALAXY Sa-b
MCG+07-28-055	13 40 18.	+ 39 54	60	16.	GALAXY
MCG-06-30-019	13 40 18.	- 38 10	36	15.5	GALAXY
BRON 27	13 40 19.8	+ 28 49 43.		19.18	BLUE OBJECT
RRS 3.003	13 40 21.	+ 29 52		18.80	BLUE OBJECT
RNGC 5277	13 40 21.	+ 30 13		15.5	GALAXY
BRON 44	13 40 21.0	+ 29 51 40.		18.88	BLUE OBJECT
BRON 28	13 40 21.1	+ 28 48 01.		19.80	BLUE OBJECT
REIN 7.108	13 40 21.18	+ 30 12 22.0			NEBULA
HOLM 536A	13 40 23.	+ 30 06	60	15.4	PART OF MULTIPLE GALAXY
ZWG 045.079	13 40 24.	+ 03 30		15.7	GALAXY
FAI 1340+12	13 40 24.	+ 12 04			COMPACT GALAXY
MCG+05-32-070	13 40 24.	+ 30 05 30.	72	15.4	GALAXY
UGC 08682	13 40 24.	+ 30 06	72	16.0	GALAXY S
ZWG 161.129	13 40 24.	+ 30 13		15.4	GALAXY
ZC 1340.4+3100	13 40 24.	+ 31 00	400		CLUSTER OF GALAXIES
MCG+05-32-069	13 40 24.	+ 31 46	36	15.4	GALAXY
MCG+06-30-075	13 40 24.	+ 35 16	48	14.	GALAXY
ZWG 218.040	13 40 24.	+ 39 55		15.7	GALAXY
UGC 08683	13 40 24.	+ 39 55	138	15.7	GALAXY DWRF IR
MCG+10-20-007	13 40 24.	+ 56 43	24	16.	GALAXY
MCG+10-20-008	13 40 24.	+ 61 02	102	13.	GALAXY
BRON 45	13 40 24.7	+ 29 36 54.		19.50	BLUE OBJECT
ZWG 131.029	13 40 30.	+ 22 24		15.2	GALAXY
MCG+04-32-032	13 40 30.	+ 22 24	36	15.2	GALAXY
ZWG 161.130	13 40 30.	+ 29 58		15.7	GALAXY
MCG+05-32-071	13 40 30.	+ 29 58	36	15.7	GALAXY
ZC 1340.5+5435	13 40 30.	+ 54 35	1610		CLUSTER OF GALAXIES
MCG+10-20-009	13 40 30.	+ 61 24	36	16.	GALAXY
MCG-05-32-069	13 40 30.	- 28 42	48	15.	GALAXY
BC 3CR288.1	13 40 30.29	+ 60 36 48.0		18.12	QUASI-STELLAR OBJECT
SHE 277	13 40 30.4	+ 60 36 55.		18.1	QUASI-STELLAR OBJECT
HN 1084	13 40 31.	- 28 43		14.	NEBULA
IC 4318	13 40 31.	- 28 43			NONSTELLAR OBJECT
RS 48	13 40 31.1	+ 26 12 20.		15.89	BLUE STELLAR OBJECT
REIN 7.109	13 40 31.29	+ 29 57 24.0			NEBULA
RNGC 5280	13 40 33.	+ 30 07		15.0	GALAXY
SC 1337-2951.7	13 40 34.	- 30 06 52.	36		NEBULA
BRON 46	13 40 35.3	+ 29 48 48.		19.01	BLUE OBJECT
BRON 09	13 40 35.6	+ 27 48 14.		19.00	BLUE OBJECT
ZC 1340.6+1328	13 40 36.	+ 13 28	340		CLUSTER OF GALAXIES
ZWG 102.037	13 40 36.	+ 16 57		15.4	GALAXY
ZWG 102.038	13 40 36.	+ 18 08		15.5	GALAXY
ZWG 161.131	13 40 36.	+ 30 07		15.1	GALAXY
MCG+05-32-072	13 40 36.	+ 30 07	18	15.1	GALAXY
ZWG 295.007	13 40 36.	+ 61 02		14.5	GALAXY
ZWG 294.059	13 40 36.	+ 61 02		14.5	GALAXY
UGC 08684	13 40 36.	+ 61 02	102	14.5	GALAXY Sc
7ZW 523	13 40 36.	+ 62 04			COMPACT GALAXY
MCG+10-20-010	13 40 36.	+ 62 06	39	16.	GALAXY
ZC 1340.6-0234	13 40 36.	- 02 34	1810		CLUSTER OF GALAXIES
MCG-05-32-071	13 40 36.	- 29 33	72	14.5	GALAXY
MCG-05-32-070	13 40 36.	- 30 07	24	15.5	GALAXY
HN 1085	13 40 37.	- 29 33	90	13.5	NEBULA
IC 4319	13 40 38.	- 29 33			NONSTELLAR OBJECT
REIN 7.110	13 40 38.44	+ 30 07 12.5			NEBULA
RRS 4.008	13 40 39.	+ 27 10		17.76	BLUE OBJECT
MCG+05-32-073	13 40 39.	+ 30 07 30.	60	15.1	GALAXY
RS 49	13 40 40.4	+ 28 09 32.		18.33	BLUE STELLAR OBJECT
RS 50	13 40 41.0	+ 28 02 26.		18.76	BLUE STELLAR OBJECT
MCG+03-35-018	13 40 42.	+ 16 58	48	15.4	GALAXY
RRS 3.004	13 40 42.	+ 28 59		16.79	BLUE OBJECT
MCG+10-20-011	13 40 42.	+ 60 22	39	15.	GALAXY
REIZ 3593	13 40 42.	+ 61 02	90	14.2	GALAXY
MCG-06-30-020	13 40 42.	- 37 56		14.	GALAXY
BRON 29	13 40 42.1	+ 28 59 15.		16.85	BLUE OBJECT
BRON 47	13 40 44.8	+ 29 04 52.		18.66	BLUE OBJECT
REIZ 3586	13 40 45.	+ 13 41	12	15.2	GALAXY
REIZ 3587	13 40 45.	+ 13 42	12	15.0	GALAXY
BRON 10	13 40 46.3	+ 27 21 41.		19.00	BLUE OBJECT
RRS 4.009	13 40 47.	+ 26 07		19.20	BLUE OBJECT
REIZ 3589	13 40 47.	+ 30 36	42	14.2	GALAXY
IC 0915	13 40 47.	- 17 04 45.			NONSTELLAR OBJECT
SC 1337-3224.9	13 40 47.	- 32 40 04.	72		NEBULA
BC B21340+31	13 40 47.0	+ 31 58 53.0		20.0	QUASI-STELLAR OBJECT
ZC 1340.8+0419	13 40 48.	+ 04 19	2760		CLUSTER OF GALAXIES
ZWG 102.039	13 40 48.	+ 14 37		15.6	GALAXY
ZC 1340.8+1942	13 40 48.	+ 19 42	1480		CLUSTER OF GALAXIES
RRS 4.010	13 40 48.	+ 27 07		18.64	BLUE OBJECT
MCG+05-32-074	13 40 51.	+ 30 35	72	14.5	GALAXY
REIZ 3588	13 40 52.	+ 12 03	24	15.2	GALAXY
REIN 7.111	13 40 52.02	+ 30 35 21.2			NEBULA
IC 0917	13 40 53.	+ 55 53 12.			NONSTELLAR OBJECT
8ZW 1340+18.2	13 40 54.	+ 18 12		16.0	COMPACT GALAXY
ZWG 161.132	13 40 54.	+ 30 36		14.5	GALAXY
UGC 08685	13 40 54.	+ 30 36	78	14.5	GALAXY SBb
ZC 1340.9+3238	13 40 54.	+ 32 38	870		CLUSTER OF GALAXIES
ZC 1340.9+3827	13 40 54.	+ 38 27	940		CLUSTER OF GALAXIES
ZWG 272.004	13 40 54.	+ 55 46		15.6	GALAXY
ZWG 271.059	13 40 54.	+ 55 46		15.6	GALAXY
ACK 309+01.1	13 40 54.	- 60 35			PLANETARY NEBULA
BRON 53	13 40 54.1	+ 30 07 45.		18.54	BLUE OBJECT
REIZ 3590	13 40 55.	+ 27 42	42	13.9	GALAXY
HOLM 537A	13 40 55.	+ 27 43	36	13.0	PART OF MULTIPLE GALAXY
IC 0918	13 40 56.	+ 55 50 42.			NONSTELLAR OBJECT
MCG+08-25-039	13 40 57.	+ 49 19	12	16.	GALAXY
RS 51	13 40 57.9	+ 27 09 32.		18.73	BLUE STELLAR OBJECT
RRS 4.011	13 40 58.	+ 26 03		19.52	BLUE OBJECT
HOLM 537B	13 40 58.	+ 27 42	18	15.1	PART OF MULTIPLE GALAXY
IC 0919	13 40 59.	+ 55 50 13.			NONSTELLAR OBJECT
8ZW 1341+16.9	13 41 00.	+ 16 53		17.2	COMPACT GALAXY
ZC 1341.0+2820	13 41 00.	+ 28 20	940		CLUSTER OF GALAXIES
ZC 1341.0+3150	13 41 00.	+ 31 50	740		CLUSTER OF GALAXIES
MCG+09-23-001	13 41 00.	+ 53 00	30	16.	GALAXY
ZC 1341.0+5930	13 41 00.	+ 59 30	23720		CLUSTER OF GALAXIES
ZWG 366.005	13 41 00.	+ 80 57		15.7	GALAXY
ZWG 365.014	13 41 00.	+ 80 57		15.7	GALAXY
ZWG 370.010	13 41 00.	+ 87 13		15.5	GALAXY
ZWG 370.004	13 41 00.	+ 87 13		15.5	GALAXY
MCG-04-32-052	13 41 00.	- 27 07	48	16.	GALAXY
RNGC 5282	13 41 04.	+ 30 20		15.0	GALAXY
ZWG 045.080	13 41 06.	+ 04 08		15.0	GALAXY
UGC 08686	13 41 06.	+ 04 08	102	15.0	GALAXY S
MCG+01-35-032	13 41 06.	+ 04 08	90	15.0	GALAXY
ZWG 102.040	13 41 06.	+ 18 18		14.9	GALAXY
8ZW 1341+18.3	13 41 06.	+ 18 18		14.9	COMPACT GALAXY
MCG+05-32-075	13 41 06.	+ 30 19	18	15.0	GALAXY
ZWG 161.133	13 41 06.	+ 30 20		15.0	GALAXY
UGC 08687	13 41 06.	+ 30 20	96	15.0	GALAXY
MCG-06-30-021	13 41 06.	- 38 13	48	15.5	GALAXY
RRS 3.005	13 41 07.	+ 29 54		19.05	BLUE OBJECT
REIN 7.112	13 41 07.85	+ 30 19 13.8			NEBULA
BRON 48	13 41 08.4	+ 29 54 07.		19.61	BLUE OBJECT
ARC 1779	13 41 11.	+ 47 52		17.6	RICH CLUSTER OF GALAXIES
ZWG 017.064	13 41 12.	+ 00 43		15.6	GALAXY
ZWG 045.081	13 41 12.	+ 04 08		15.5	GALAXY
MCG+01-35-033	13 41 12.	+ 04 08	24	15.5	GALAXY
ZWG 102.041	13 41 12.	+ 19 49		15.5	GALAXY
RRS 4.012	13 41 12.	+ 26 29		19.21	BLUE OBJECT
ZC 1341.2+4022	13 41 12.	+ 40 22	1340		CLUSTER OF GALAXIES
MCG+10-20-012	13 41 12.	+ 61 07	39	16.	GALAXY
ZC 1341.2+7902	13 41 12.	+ 79 02	1340		CLUSTER OF GALAXIES
REIZ 3591	13 41 14.	+ 04 09	78	14.4	GALAXY
RRS 3.006	13 41 14.	+ 29 47		18.47	BLUE OBJECT
MCG+03-35-019	13 41 15.	+ 18 19	54	14.9	GALAXY
MCG-03-35-016	13 41 15.	- 19 30	36	15.	GALAXY
RNGC 5269	13 41 15.	- 62 39			NON-EXISTENT OBJECT
ZC 1341.3+1000	13 41 18.	+ 10 00	810		CLUSTER OF GALAXIES
ZWG 102.042	13 41 18.	+ 18 16		15.5	GALAXY
TON-N 0729	13 41 18.	+ 27 05		15.2	BLUE STAR
ZC 1341.3+3943	13 41 18.	+ 39 43	1950		CLUSTER OF GALAXIES
ZC 1341.3+4752	13 41 18.	+ 47 52	1480		CLUSTER OF GALAXIES
ZC 1341.3+5333	13 41 18.	+ 53 33	810		CLUSTER OF GALAXIES
HN 1086	13 41 18.	- 26 59		13.5	NEBULA
IC 4320	13 41 18.	- 26 59			NONSTELLAR OBJECT
MCG-04-32-053	13 41 18.	- 26 59 30.	36	14.5	GALAXY
IC 0921	13 41 19.	+ 55 55 08.			NONSTELLAR OBJECT
RS 52	13 41 20.4	+ 27 23 20.		16.42	BLUE STELLAR OBJECT
REIZ 3592	13 41 21.	+ 02 08	24	15.3	GALAXY
PK307-09.1	13 41 22.	- 71 13 47.	5		PLANETARY NEBULA
RS 53	13 41 22.3	+ 26 11 56.		17.39	BLUE STELLAR OBJECT
HN 1723	13 41 22.9	+ 00 42 34.	12		NEBULA
SC 1338-3119.6	13 41 23.	- 31 34 45.	24		NEBULA
ZWG 045.082	13 41 24.	+ 08 01		15.7	GALAXY

OBJECT NAME	RIGHT ASCEN.	DECLINATION	DIAM.	MAGN.	TYPE OF OBJECT
ZWG 073.066	13 41 24.	+ 13 29		15.7	GALAXY
ZWG 102.043	13 41 24.	+ 19 06		15.7	GALAXY
ZWG 131.030	13 41 24.	+ 25 38		15.5	GALAXY
TON-N 0730	13 41 24.	+ 25 55		15.4	BLUE STAR
ZWG 218.041	13 41 24.	+ 43 43		14.8	GALAXY
UGC 08688	13 41 24.	+ 43 43	66	14.8	GALAXY PECULR
IC 0922	13 41 24.	+ 55 51 20.	8		NEBULA
ZWG 295.008	13 41 24.	+ 61 07		15.6	GALAXY
ZC 1341.4+6601	13 41 24.	+ 66 01	2080		CLUSTER OF GALAXIES
MCG-05-32-072	13 41 24.	- 27 45	30	15.5	GALAXY
SC 1338-3020.3	13 41 24.	- 30 35 27.	30		NEBULA
BRON 11	13 41 24.0	+ 27 30 25.		19.10	BLUE OBJECT
IC 4322	13 41 25.	+ 25 39 19.			NONSTELLAR OBJECT
IC 0923	13 41 26.	+ 55 52 14.			NONSTELLAR OBJECT
PRA 5	13 41 26.9	+ 25 57 18.		16.	NONSTELLAR BLUE OBJECT
ZWG 045.083	13 41 30.	+ 08 17		15.4	GALAXY
ZWG 102.044	13 41 30.	+ 18 34		15.6	GALAXY
MCG+07-28-056	13 41 30.	+ 43 43	48	14.	GALAXY
MCG+10-20-013	13 41 30.	+ 61 11	27	17.	GALAXY
BRON 30	13 41 30.1	+ 28 50 08.		19.29	BLUE OBJECT
HN 1722	13 41 30.4	- 46 54 56.	36		NEBULA
BRON 31	13 41 30.6	+ 28 38 54.		19.17	BLUE OBJECT
IC 0925	13 41 31.	+ 55 51 14.			NONSTELLAR OBJECT
RRS 4.013	13 41 32.	+ 28 17		17.22	BLUE OBJECT
ARC 1783	13 41 32.	+ 55 51		16.3	RICH CLUSTER OF GALAXIES
ARC 1778	13 41 32.	- 10 54		16.6	RICH CLUSTER OF GALAXIES
RRS 4.014	13 41 34.	+ 28 39		18.80	BLUE OBJECT
RRS 4.015	13 41 34.	+ 28 50		19.53	BLUE OBJECT
RRS 4.016	13 41 35.	+ 26 08		18.60	BLUE OBJECT
RRS 3.007	13 41 35.	+ 29 56		19.13	BLUE OBJECT
ZWG 045.084	13 41 36.	+ 04 09		15.2	GALAXY
ZC 1341.6+0714	13 41 36.	+ 07 14	1210		CLUSTER OF GALAXIES
FAI 1341+09	13 41 36.	+ 09 45			COMPACT GALAXY
ZWG 102.045	13 41 36.	+ 18 05		15.5	GALAXY
ZC 1341.6+2344	13 41 36.	+ 23 44	670		CLUSTER OF GALAXIES
ZC 1341.6+2614	13 41 36.	+ 26 14	8400		CLUSTER OF GALAXIES
ZC 1341.6+2641	13 41 36.	+ 26 41	8400		CLUSTER OF GALAXIES
MCG+07-28-057	13 41 36.	+ 41 57	30	15.	GALAXY
ZC 1341.6+5524	13 41 36.	+ 55 24	740		CLUSTER OF GALAXIES
MCG-05-32-073	13 41 36.	- 29 53	30	15.	GALAXY
BRON 49	13 41 37.0	+ 29 56 11.		19.33	BLUE OBJECT
HN 1724	13 41 37.1	+ 00 00 40.	12		NEBULA
MCG+03-35-020	13 41 39.	+ 18 06	48	15.5	GALAXY
IC 0926	13 41 40.	+ 55 53 45.			NONSTELLAR OBJECT
SC 1338-2938.6	13 41 40.	- 29 53 44.	30		NEBULA
ZWG 073.067	13 41 42.	+ 12 23		15.5	GALAXY
MCG+04-32-033	13 41 42.	+ 25 55 30.	27	15.4	GALAXY
ZWG 161.134	13 41 42.	+ 30 16		15.6	GALAXY
MCG+05-32-076	13 41 42.	+ 30 16 30.	36	15.6	GALAXY
ZC 1341.7+3257	13 41 42.	+ 32 57	740		CLUSTER OF GALAXIES
ZC 1341.7+3805	13 41 42.	+ 38 05	940		CLUSTER OF GALAXIES
MCG+10-20-014	13 41 42.	+ 62 06	42	16.	GALAXY
BRON 50	13 41 43.1	+ 29 44 45.		19.10	BLUE OBJECT
REIZ 3595	13 41 44.	+ 25 54	24	14.5	GALAXY
HN 1087	13 41 44.	- 29 53		15.	NEBULA
IC 4321	13 41 44.	- 29 53			NONSTELLAR OBJECT
BRON 12	13 41 46.5	+ 27 41 40.		19.60	BLUE OBJECT
ZWG 045.085	13 41 48.	+ 06 02		14.7	GALAXY
UGC 08689	13 41 48.	+ 06 02	96	14.7	GALAXY S0
KARA.72 392A	13 41 48.	+ 06 02	90	14.7	PART OF DOUBLE GALAXY
MCG+01-35-034	13 41 48.	+ 06 02	60	14.7	GALAXY
ZWG 102.046	13 41 48.	+ 14 44		15.4	GALAXY
ZC 1341.8+1556	13 41 48.	+ 15 56	540		CLUSTER OF GALAXIES
8ZW 1341+19.4	13 41 48.	+ 19 22		17.5	COMPACT GALAXY
ZWG 131.031	13 41 48.	+ 25 55		15.4	GALAXY
ZWG 295.009	13 41 48.	+ 62 03		15.6	GALAXY
REIZ 3594	13 41 49.	+ 06 02	12	14.4	GALAXY
RRS 3.008	13 41 50.	+ 29 20		17.65	BLUE OBJECT
IC 0928	13 41 50.	+ 55 52 09.	8		NEBULA
IC 0929	13 41 52.	+ 55 54 16.			NONSTELLAR OBJECT
SC 1339-2933.7	13 41 52.	- 29 48 50.	30		NEBULA
ZWG 017.065	13 41 54.	+ 02 21		15.5	GALAXY
ZC 1341.9+0310	13 41 54.	+ 03 10	2690		CLUSTER OF GALAXIES
ZWG 045.086	13 41 54.	+ 05 22		15.6	GALAXY
ZC 1341.9+1226	13 41 54.	+ 12 26	1340		CLUSTER OF GALAXIES
ZWG 073.068	13 41 54.	+ 13 25		15.6	GALAXY
ZWG 102.047	13 41 54.	+ 14 35		15.5	GALAXY
ZC 1341.9+1444	13 41 54.	+ 14 44	870		CLUSTER OF GALAXIES
ZWG 102.048	13 41 54.	+ 19 50		15.7	GALAXY
VV 163C	13 41 54.	+ 20 39	12	16.5	INTERACTING GALAXY
VV 163B	13 41 54.	+ 20 39	12	16.5	INTERACTING GALAXY
VV 163A	13 41 54.	+ 20 39	12	16.	INTERACTING GALAXY
VV 163	13 41 54.	+ 20 39	78	16.	INTERACTING GALAXY
ZC 1341.9+5550	13 41 54.	+ 55 50	2080		CLUSTER OF GALAXIES
RRS 4.018	13 41 55.	+ 27 56		19.02	BLUE OBJECT
RNGC 5285	13 41 56.	+ 02 21		15.5	GALAXY
IC 0931	13 41 56.	+ 55 52 15.	5		NEBULA
IC 0930	13 41 56.	+ 55 55 46.			NONSTELLAR OBJECT
MCG+04-32-034	13 41 57.	+ 20 39	72	15.0	GALAXY
RRS 3.009	13 41 57.	+ 29 19			BLUE OBJECT
REIN 7.113	13 41 57.18	+ 30 03 17.0			NEBULA
IC 0932	13 41 58.	+ 55 53 16.			NONSTELLAR OBJECT
REIZ 3597	13 41 59.	+ 30 04	6	15.8	GALAXY
LB 09880	13 42	- 84 18		14.0	FAINT BLUE STAR
ZWG 045.087	13 42 00.	+ 05 02		15.3	GALAXY
UGC 08690	13 42 00.	+ 05 02	66	15.3	GALAXY Sc
MCG+01-35-035	13 42 00.	+ 05 02	60	15.3	GALAXY
ZWG 131.032	13 42 00.	+ 20 38		15.2	GALAXY
ZWG 131.033	13 42 00.	+ 20 40		15.0	GALAXY
UGC 08691	13 42 00.	+ 20 40	66	15.0	GALAXY S
TON-N 0731	13 42 00.	+ 30 05		15.7	BLUE STAR
ZWG 336.015	13 42 00.	+ 72 34		15.0	GALAXY
MCG+12-13-008	13 42 00.	+ 72 34 30.	45	16.	GALAXY
MCG+13-10-011	13 42 00.	+ 78 30	36	17.	GALAXY
MCG-06-30-022	13 42 00.	- 37 20	24	15.	GALAXY
HOLM 538C	13 42 02.	+ 30 10	12	15.9	PART OF MULTIPLE GALAXY
HOLM 538B	13 42 04.	+ 30 08	24	15.8	PART OF MULTIPLE GALAXY
BRON 13	13 42 04.0	+ 27 25 12.		17.79	BLUE OBJECT
REIN 7.114	13 42 04.52	+ 30 10 13.8			NEBULA
REIZ 3596	13 42 05.	+ 20 35	84	14.5	GALAXY
RRS 4.019	13 42 05.	+ 27 25		17.69	BLUE OBJECT
REIZ 3598	13 42 05.	+ 30 04	6	15.8	GALAXY
REIZ 3599	13 42 05.	+ 30 09	12	16.0	GALAXY
HOLM 538A	13 42 05.	+ 30 09	36	15.8	PART OF MULTIPLE GALAXY
REIZ 3600	13 42 05.	+ 30 10	12	15.5	GALAXY
REIZ 3601	13 42 05.	+ 30 11	6	15.8	GALAXY
HN 1725	13 42 05.5	+ 00 49 11.	18		NEBULA
ZWG 045.088	13 42 06.	+ 06 06		15.7	GALAXY
KARA.72 392B	13 42 06.	+ 06 06	42	15.7	PART OF DOUBLE GALAXY
ZWG 102.049	13 42 06.	+ 20 28		15.4	GALAXY
ZC 1342.1+2033	13 42 06.	+ 20 33	2420		CLUSTER OF GALAXIES
MCG+05-32-078	13 42 06.	+ 30 07	48	15.7	GALAXY
ZWG 161.135	13 42 06.	+ 30 08		15.7	GALAXY
MCG+05-32-077	13 42 06.	+ 30 08 30.	66	15.6	GALAXY
ZWG 161.136	13 42 06.	+ 30 10		15.6	GALAXY
UGC 08692	13 42 06.	+ 30 10	72	15.6	GALAXY Sa-b
ZC 1342.1+3658	13 42 06.	+ 36 58	1140		CLUSTER OF GALAXIES
ZC 1342.1+3921	13 42 06.	+ 39 21	400		CLUSTER OF GALAXIES
MRK 460	13 42 06.	+ 40 16	7	16.	GALAXY WITH UV CONTINUUM
MCG-06-30-023	13 42 06.	- 35 54	42	15.	GALAXY
ARC 1780	13 42 07.	+ 03 08		16.6	RICH CLUSTER OF GALAXIES
RRS 4.020	13 42 07.	+ 26 29		18.91	BLUE OBJECT
REIN 7.115	13 42 07.00	+ 30 07 53.3			NEBULA
REIN 7.116	13 42 07.79	+ 30 09 16.0			NEBULA
HN 1726	13 42 07.9	+ 00 33 23.	12		NEBULA
REIN 7.117	13 42 08.46	+ 30 03 16.1			NEBULA
IC 0934	13 42 11.	+ 55 52 17.			NONSTELLAR OBJECT
ZWG 045.089	13 42 12.	+ 05 11		15.7	GALAXY
ZC 1342.2+1149	13 42 12.	+ 11 49	3020		CLUSTER OF GALAXIES
ZC 1342.2+1635	13 42 12.	+ 16 35	2490		CLUSTER OF GALAXIES
ZWG 102.050	13 42 12.	+ 16 53		15.5	GALAXY
MCG+03-35-021	13 42 12.	+ 20 28	48	15.4	GALAXY
ZWG 190.045	13 42 12.	+ 35 26		14.5	GALAXY
UGC 08693	13 42 12.	+ 35 26	78	14.5	GALAXY S
IC 0935	13 42 12.	+ 55 51 17.			NONSTELLAR OBJECT
IC 0936	13 42 12.	+ 55 52 17.			NONSTELLAR OBJECT
MCG+10-20-015	13 42 12.	+ 58 22	24	16.	GALAXY
ARC 1781	13 42 15.	+ 30 06		15.4	RICH CLUSTER OF GALAXIES
MCG+06-30-076	13 42 15.	+ 35 26	66	14.	GALAXY
BRON 32	13 42 16.8	+ 28 42 22.		19.13	BLUE OBJECT
RRS 4.017	13 42 17.	+ 27 42			BLUE OBJECT
RRS 4.021	13 42 17.	+ 26 48		18.00	BLUE OBJECT
ZWG 045.090	13 42 18.	+ 04 16		15.4	GALAXY
ZWG 102.051	13 42 18.	+ 15 46		15.7	GALAXY
TON-N 0732	13 42 18.	+ 27 06		15.5	BLUE STAR
ZWG 162.001	13 42 18.	+ 30 35		15.5	GALAXY
ZWG 161.137	13 42 18.	+ 30 35		15.5	GALAXY
ZC 1342.3+6437	13 42 18.	+ 64 37	1550		CLUSTER OF GALAXIES
HN 1088	13 42 19.	- 28 24	36		NEBULA
IC 4323	13 42 19.	- 28 24			NONSTELLAR OBJECT
SC 1339-2254.6	13 42 20.	- 23 09 42.	120		NEBULA
ZC 1342.4+0016	13 42 24.	+ 00 16	2760		CLUSTER OF GALAXIES
8ZW 1342+10.8	13 42 24.	+ 10 51		18.9	COMPACT GALAXY
ZC 1342.4+1726	13 42 24.	+ 17 26	3090		CLUSTER OF GALAXIES
ZWG 161.138	13 42 24.	+ 27 24		15.7	GALAXY
ZWG 162.002	13 42 24.	+ 31 28		15.7	GALAXY
ZWG 161.139	13 42 24.	+ 31 28		15.7	GALAXY
UGC 08694	13 42 24.	+ 31 28	60	15.7	GALAXY Sa-b
ZWG 190.046	13 42 24.	+ 37 25		15.6	GALAXY
KARA.72 393A	13 42 24.	+ 37 25	36	15.6	PART OF DOUBLE GALAXY
1ZW 070	13 42 24.	+ 37 28			COMPACT GALAXY
ZC 1342.4+3826	13 42 24.	+ 38 26	1080		CLUSTER OF GALAXIES
MCG-02-35-014	13 42 24.	- 14 15	36	16.	GALAXY
MCG-03-35-017	13 42 24.	- 15 36	48	15.	GALAXY
HOLM 539A	13 42 26.	+ 25 42	18	14.6	PART OF MULTIPLE GALAXY
REIZ 3602	13 42 27.	+ 01 50	18	15.0	GALAXY
HOLM 539B	13 42 29.	+ 25 43	24	15.0	PART OF MULTIPLE GALAXY
HN 1727	13 42 29.8	+ 00 00 24.	12		NEBULA
ZWG 162.003	13 42 30.	+ 31 52		15.6	GALAXY
ZWG 161.140	13 42 30.	+ 31 52		15.6	GALAXY
TON-N 0733	13 42 30.	+ 33 19		15.9	BLUE STAR
ZC 1342.5+4100	13 42 30.	+ 41 00	2490		CLUSTER OF GALAXIES
MCG+09-23-002	13 42 30.	+ 56 07	42	16.	GALAXY
MCG+10-20-016	13 42 30.	+ 58 10	27	15.	GALAXY
ZC 1342.5+7841	13 42 30.	+ 78 41	7860		CLUSTER OF GALAXIES
MCG-05-32-074	13 42 30.	- 29 46	15	15.5	GALAXY
ARC 1782	13 42 32.	+ 13 57		17.5	RICH CLUSTER OF GALAXIES
ARC 1785	13 42 33.	+ 38 24		17.2	RICH CLUSTER OF GALAXIES
BRON 54	13 42 35.9	+ 30 00 30.		18.59	BLUE OBJECT
REIN 7.118	13 42 35.98	+ 30 01 26.8			NEBULA
ZWG 017.066	13 42 36.	+ 00 22		15.5	GALAXY
ZWG 073.069	13 42 36.	+ 08 44		15.5	GALAXY
MCG+02-35-017	13 42 36.	+ 08 44	36	15.5	GALAXY
ZWG 162.004	13 42 36.	+ 31 19		15.6	GALAXY
ZWG 161.141	13 42 36.	+ 31 19		15.6	GALAXY
ZWG 190.047	13 42 36.	+ 32 47		15.2	GALAXY
REIZ 3604	13 42 36.	+ 37 23	54	14.6	GALAXY
ZWG 190.048	13 42 36.	+ 37 25		15.2	GALAXY
UGC 08695	13 42 36.	+ 37 25	66	15.2	GALAXY S
KARA.72 393B	13 42 36.	+ 37 25	66	15.2	PART OF DOUBLE GALAXY
MCG-05-32-075	13 42 36.	- 29 58	24	15.5	GALAXY
HN 1728	13 42 37.3	+ 00 22 00.	12		NEBULA
RRS 4.022	13 42 38.	+ 26 57		19.42	BLUE OBJECT
HN 1089	13 42 38.	- 29 59		13.5	NEBULA
IC 4324	13 42 38.	- 29 59			NONSTELLAR OBJECT
MCG+06-30-077	13 42 39.	+ 37 25	54	14.5	GALAXY
RNGC 5287	13 42 40.	+ 30 03		16.0	GALAXY
IC 0937	13 42 40.	+ 55 53 48.			NONSTELLAR OBJECT
ZWG 073.070	13 42 42.	+ 12 27		15.6	GALAXY
MCG+05-32-079	13 42 42.	+ 30 03	30	16.5	GALAXY
MCG+10-20-017	13 42 42.	+ 61 29	24	16.	GALAXY
MCG+10-20-018	13 42 42.	+ 61 46	24	16.	GALAXY
MCG-04-32-054	13 42 42.	- 25 08	30	15.5	GALAXY
MIL 27	13 42 42.	- 60 13 54.	378		SUPERNOVA REMNANT
IC 0938	13 42 43.	+ 55 52 18.			NONSTELLAR OBJECT
REIN 7.119	13 42 43.45	+ 30 03 57.3			NEBULA
IC 0920	13 42 44.	- 12 19 26.			NONSTELLAR OBJECT
BRON 51	13 42 47.1	+ 29 03 15.		19.27	BLUE OBJECT
ZWG 045.091	13 42 48.	+ 04 03		15.4	GALAXY
ZC 1342.8+1033	13 42 48.	+ 10 33	870		CLUSTER OF GALAXIES
ZWG 131.034	13 42 48.	+ 23 19		15.3	GALAXY
ZC 1342.8+5427	13 42 48.	+ 54 27	1010		CLUSTER OF GALAXIES
1ZW 071	13 42 48.	+ 56 08			COMPACT GALAXY
ZWG 272.005	13 42 48.	+ 56 08		15.0	GALAXY
ZWG 271.060	13 42 48.	+ 56 08		15.0	GALAXY
MRK 273	13 42 48.	+ 56 08	20	14.5	GALAXY WITH UV CONTINUUM
KW 24	13 42 48.	+ 56 08	70		SEYFERT GALAXY
VVI 67	13 42 48.	+ 56 08	66	15.39	SEYFERT GALAXY
UGC 08696	13 42 48.	+ 56 08	72	15.0	GALAXY PECULR
MCG+09-23-004	13 42 48.	+ 56 09	66	15.	GALAXY
MCG+09-23-003	13 42 48.	+ 56 15	24	15.	GALAXY
MCG-02-35-015	13 42 48.	- 12 18 30.	24	15.	GALAXY
MCG+04-32-036	13 42 51.	+ 23 28 30.	54	14.7	GALAXY
MCG+04-32-035	13 42 51.	+ 23 34	42	15.5	GALAXY
ZWG 045.092	13 42 54.	+ 04 01		15.6	GALAXY
ZC 1342.9+1401	13 42 54.	+ 14 01	2350		CLUSTER OF GALAXIES

OBJECT NAME	RIGHT ASCEN.	DECLINATION	DIAM.	MAGN.	TYPE OF OBJECT
ZWG 102.052	13 42 54.	+ 19 50		15.4	GALAXY
ZWG 131.035	13 42 54.	+ 23 28		14.7	GALAXY
UGC 08697	13 42 54.	+ 23 28	78	14.7	GALAXY S0
ZWG 131.036	13 42 54.	+ 23 34		15.5	GALAXY
MRK 068	13 42 54.	+ 27 23	12	17.	GALAXY WITH UV CONTINUUM
TON-N 0734	13 42 54.	+ 31 07		15.5	BLUE STAR
ZWG 190.049	13 42 54.	+ 35 27		15.2	GALAXY
UGC 08698	13 42 54.	+ 35 27	66	15.2	GALAXY S
MCG+10-20-019	13 42 54.	+ 61 40	18	15.	GALAXY
MCG-01-35-010	13 42 54.	- 05 43	90	14.	GALAXY
IC 0933	13 42 55.	+ 23 29 12.			NONSTELLAR OBJECT
RRS 4.023	13 42 56.	+ 27 41		19.41	BLUE OBJECT
RRS 3.010	13 42 56.	+ 29 21			BLUE OBJECT
REIZ 3605	13 42 56.	+ 41 45	120	13.5	GALAXY
IC 0924	13 42 56.	- 12 12 19.			NONSTELLAR OBJECT
BRON 52	13 42 56.4	+ 29 21 05.		18.64	BLUE OBJECT
MCG+06-30-078	13 42 57.	+ 35 28	48	14.	GALAXY
HOLM 540B	13 42 58.	- 05 44	18	13.6	PART OF MULTIPLE GALAXY
ZC 1343.0+1436	13 43 00.	+ 14 36	940		CLUSTER OF GALAXIES
ZWG 102.053	13 43 00.	+ 16 33		15.5	GALAXY
ZWG 218.042	13 43 00.	+ 41 45		13.5	GALAXY
UGC 08699	13 43 00.	+ 41 45	114	13.5	GALAXY Sa-b
MCG+10-20-020	13 43 00.	+ 59 33	30	16.	GALAXY
ZWG 366.006	13 43 00.	+ 83 45		15.5	GALAXY
ZWG 365.015	13 43 00.	+ 83 45		15.5	GALAXY
MCG+14-06-028	13 43 00.	+ 83 46	30	15.	GALAXY
ZC 1343.0-0036	13 43 00.	- 00 36	1750		CLUSTER OF GALAXIES
REIZ 3603	13 43 00.	- 05 44	30	13.9	GALAXY
HOLM 540A	13 43 00.	- 05 44	42	13.6	PART OF MULTIPLE GALAXY
GCL 026	13 43 00.	- 51 07	816	9.5	GLOBULAR STAR CLUSTER
RNGC 5286	13 43 01.	- 51 07		9.5	GLOBULAR CLUSTER
REIN 7.120	13 43 01.24	+ 41 45 13.4			NEBULA
RNGC 5289	13 43 02.	+ 41 45		13.5	GALAXY
HOLM 540C	13 43 02.	- 05 43	30	13.9	PART OF MULTIPLE GALAXY
RRS 3.011	13 43 03.	+ 29 17		16.48	BLUE OBJECT
MCG+07-28-058	13 43 03.	+ 41 45	78	13.5	GALAXY
ARC 1784	13 43 04.	+ 06 00		17.3	RICH CLUSTER OF GALAXIES
RRS 4.024	13 43 04.	+ 26 24		19.10	BLUE OBJECT
ZC 1343.1+0601	13 43 06.	+ 06 01	940		CLUSTER OF GALAXIES
ZWG 218.043	13 43 06.	+ 41 58		13.0	GALAXY
UGC 08700	13 43 06.	+ 41 58	228	13.0	GALAXY Sb-c
MCG+07-28-059	13 43 06.	+ 42 59	42	16.	GALAXY
MCG+10-20-021	13 43 06.	+ 59 40	45	15.	GALAXY
MCG+10-20-022	13 43 06.	+ 61 16	51	16.	GALAXY
MCG+10-20-023	13 43 06.	+ 61 31	30	15.	GALAXY
OCL 0911	13 43 06.	- 62 39	660	8.4	OPEN STAR CLUSTER
VHA 152	13 43 06.	- 62 39	180		OPEN STAR CLUSTER
REIZ 3606	13 43 08.	+ 41 58	162	13.2	GALAXY
RNGC 5281	13 43 09.	- 62 39		8.0	OPEN CLUSTER
IC 0927	13 43 11.	- 12 12 55.			NONSTELLAR OBJECT
ZC 1343.2+0825	13 43 12.	+ 08 25	1010		CLUSTER OF GALAXIES
FAI 1343+14	13 43 12.	+ 14 30			COMPACT GALAXY
MCG+07-28-060	13 43 12.	+ 41 38	42	16.	GALAXY
MCG+09-23-005	13 43 12.	+ 51 15	36	16.	GALAXY
MCG+09-23-006	13 43 12.	+ 51 19	36	16.	GALAXY
REIN 2.197	13 43 12.19	+ 41 57 48.0			NEBULA
RNGC 5284	13 43 13.	- 58 56			NON-EXISTENT OBJECT
RNGC 5290	13 43 14.	+ 41 58		13.0	GALAXY
MCG+07-28-061	13 43 15.	+ 41 57 30.	228	12.5	GALAXY
ZWG 045.093	13 43 18.	+ 07 15		15.7	GALAXY
ZWG 045.094	13 43 18.	+ 07 46		15.4	GALAXY
ZWG 132.001	13 43 18.	+ 22 18		15.7	GALAXY
ZC 1343.3+2245	13 43 18.	+ 22 45	4100		CLUSTER OF GALAXIES
ARC 1788	13 43 18.	+ 54 01		17.7	RICH CLUSTER OF GALAXIES
ZWG 295.010	13 43 18.	+ 61 30		15.0	GALAXY
RRS 3.012	13 43 20.	+ 31 22			BLUE OBJECT
MCG+03-35-022	13 43 21.	+ 15 46	36	15.6	GALAXY
MCG+04-32-037	13 43 21.	+ 22 19	84	15.2	GALAXY
REIZ 3609	13 43 21.	+ 55 32	30	14.3	GALAXY
RNGC 5294	13 43 22.	+ 55 32		15.0	GALAXY
ZC 1343.4+1529	13 43 24.	+ 15 29	870		CLUSTER OF GALAXIES
ZWG 102.054	13 43 24.	+ 15 47		15.6	GALAXY
ZWG 132.002	13 43 24.	+ 22 20		15.2	GALAXY
UGC 08701	13 43 24.	+ 22 20	108	15.2	GALAXY S
RRS 4.025	13 43 24.	+ 28 00		19.18	BLUE OBJECT
ZWG 190.050	13 43 24.	+ 35 52		15.0	GALAXY
MCG+06-30-079	13 43 24.	+ 35 52		14.5	GALAXY
ZC 1343.4+4513	13 43 24.	+ 45 13	1410		CLUSTER OF GALAXIES
MCG+08-25-040	13 43 24.	+ 48 11	60	15.	GALAXY
ZWG 272.006	13 43 24.	+ 55 32		15.2	GALAXY
ZWG 271.061	13 43 24.	+ 55 32		15.2	GALAXY
MCG+10-20-024	13 43 24.	+ 61 42	30	16.	GALAXY
ZWG 353.024	13 43 24.	+ 78 56		15.6	GALAXY
MRSL 309-00/1	13 43 24.	- 62 20	300		HII REGION
RRS 4.026	13 43 25.	+ 26 40			BLUE OBJECT
RRS 4.027	13 43 26.	+ 26 40			BLUE OBJECT
MCG+05-32-080	13 43 27.	+ 27 00	42	15.5	GALAXY
ARC 1786	13 43 27.	+ 45 14		17.6	RICH CLUSTER OF GALAXIES
RRS 4.028	13 43 28.	+ 28 34			BLUE OBJECT
REIZ 3613	13 43 28.	+ 61 32	30	14.3	GALAXY
ZC 1343.5+0926	13 43 30.	+ 09 26	470		CLUSTER OF GALAXIES
ZC 1343.5+2151	13 43 30.	+ 21 51	810		CLUSTER OF GALAXIES
ZWG 162.005	13 43 30.	+ 27 01		15.5	GALAXY
TON-N 0735	13 43 30.	+ 28 35		15.5	BLUE STAR
ZC 1343.5+2839	13 43 30.	+ 28 39	870		CLUSTER OF GALAXIES
TON-N 0736	13 43 30.	+ 29 20		14.0	BLUE STAR
ZC 1343.5+3906	13 43 30.	+ 39 06	1210		CLUSTER OF GALAXIES
1ZW 072	13 43 30.	+ 46 14			COMPACT GALAXY
ZWG 246.022	13 43 30.	+ 48 11		15.1	GALAXY
UGC 08702	13 43 30.	+ 48 11	102	15.1	GALAXY Sc
MIL 28	13 43 33.	- 60 08 18.	426		SUPERNOVA REMNANT
RRS 3.013	13 43 35.	+ 29 21		18.16	BLUE OBJECT
ZWG 073.071	13 43 36.	+ 09 32		15.1	GALAXY
ZC 1343.6+1253	13 43 36.	+ 12 53	610		CLUSTER OF GALAXIES
ZC 1343.6+2005	13 43 36.	+ 20 05	1280		CLUSTER OF GALAXIES
TON-N 0737	13 43 36.	+ 28 59		15.2	BLUE STAR
ZWG 190.051	13 43 36.	+ 35 07		15.6	GALAXY
ZC 1343.6+3955	13 43 36.	+ 39 55	1280		CLUSTER OF GALAXIES
ZC 1343.6+5530	13 43 36.	+ 55 30	940		CLUSTER OF GALAXIES
MCG+10-20-025	13 43 36.	+ 61 30	15	16.	GALAXY
ZC 1343.6-0213	13 43 36.	- 02 13	1140		CLUSTER OF GALAXIES
MCG-04-33-001	13 43 36.	- 23 51	24	15.5	GALAXY
REIZ 3607	13 43 37.	+ 27 02	30	14.7	GALAXY
RRS 3.014	13 43 40.	+ 28 47		18.68	BLUE OBJECT
ARC 1787	13 43 40.	+ 37 03		17.8	RICH CLUSTER OF GALAXIES
RRS 4.029	13 43 41.	+ 27 56		18.98	BLUE OBJECT
SC 1340-3022.4	13 43 41.	- 30 37 28.	48		NEBULA
ZWG 132.003	13 43 42.	+ 22 22		15.3	GALAXY
ZC 1343.7+3700	13 43 42.	+ 37 00	670		CLUSTER OF GALAXIES
ZC 1343.7+5303	13 43 42.	+ 53 03	1610		CLUSTER OF GALAXIES
ZWG 017.067	13 43 42.	- 03 08		15.6	GALAXY
MCG+00-35-019	13 43 42.	- 03 08	42	15.6	GALAXY
MCG-05-33-001	13 43 42.	- 30 38	60	16.	GALAXY
REIZ 3608	13 43 43.	+ 35 06	18	14.8	GALAXY
RRS 4.030	13 43 44.	+ 26 43		18.79	BLUE OBJECT
MCG+04-32-038	13 43 45.	+ 22 08 30.	42	15.1	GALAXY
ZC 1343.8+0936	13 43 48.	+ 09 36	1810		CLUSTER OF GALAXIES
ZWG 132.004	13 43 48.	+ 22 09		15.1	GALAXY
UGC 08703	13 43 48.	+ 22 09	66	15.1	GALAXY S
MRK 069	13 43 48.	+ 29 53	8	16.5	GALAXY WITH UV CONTINUUM
KW 11	13 43 48.	+ 29 53	11		SEYFERT GALAXY
VVI 68	13 43 48.	+ 29 53	8	16.5	SEYFERT GALAXY
ZC 1343.8+3100	13 43 48.	+ 31 00	1880		CLUSTER OF GALAXIES
ZC 1343.8+3528	13 43 48.	+ 35 28	400		CLUSTER OF GALAXIES
MCG+10-20-026	13 43 48.	+ 56 51	51	15.	GALAXY
ZWG 295.011	13 43 43.	+ 56 52		14.8	GALAXY
UGC 08704	13 43 48.	+ 56 52	72	14.8	GALAXY Sa
ZWG 017.068	13 43 48.	- 03 11		14.8	GALAXY
MCG+00-35-020	13 43 48.	- 03 11	48	14.8	GALAXY
MCG-01-35-011	13 43 48.	- 09 23	60	15.	GALAXY
MCG-06-30-024	13 43 48.	- 37 40	36	15.	GALAXY
ZWG 132.005	13 43 54.	+ 23 18		15.3	GALAXY
RRS 4.031	13 43 59.	+ 28 29		18.38	BLUE OBJECT
RRS 3.015	13 43 59.	+ 31 30			BLUE OBJECT
ZWG 045.095	13 44 00.	+ 04 07		15.6	GALAXY
ZWG 045.096	13 44 00.	+ 05 27		15.6	GALAXY
ZC 1344.0+0727	13 44 00.	+ 07 27	2550		CLUSTER OF GALAXIES
ZWG 102.055	13 44 00.	+ 18 13		15.3	GALAXY
ZC 1344.0+2555	13 44 00.	+ 25 55	1010		CLUSTER OF GALAXIES
RRS 3.016	13 44 00.	+ 31 43			BLUE OBJECT
REIZ 3612	13 44 00.	+ 36 48	30	14.6	GALAXY
ZWG 190.052	13 44 00.	+ 36 50		15.0	GALAXY
MCG+06-30-080	13 44 00.	+ 36 50	42	14.5	GALAXY
ARC 1789	13 44 00.	+ 39 54		17.8	RICH CLUSTER OF GALAXIES
ZWG 366.007	13 44 00.	+ 80 51		15.4	GALAXY
ZWG 365.016	13 44 00.	+ 80 51		15.4	GALAXY
RRS 3.017	13 44 03.	+ 28 41		17.09	BLUE OBJECT
MCG+03-35-023	13 44 06.	+ 18 13	42	15.3	GALAXY
ZC 1344.1+3955	13 44 06.	+ 39 55	810		CLUSTER OF GALAXIES
ZC 1344.1+7031	13 44 06.	+ 70 31	1080		CLUSTER OF GALAXIES
REIZ 3614	13 44 07.	+ 42 29	60	14.1	GALAXY
MAI 087	13 44 08.	+ 65 44	33		DWARF SPHEROIDAL GALAXY
RNGC 5296	13 44 09.	+ 44 06		15.0	GALAXY
REIZ 3611	13 44 10.	+ 21 07	48	14.7	GALAXY
RRS 4.032	13 44 10.	+ 26 49			BLUE OBJECT
RRS 3.018	13 44 11.	+ 29 51		18.38	BLUE OBJECT
ZC 1344.2+1237	13 44 12.	+ 12 37	1340		CLUSTER OF GALAXIES
ZWG 132.006	13 44 12.	+ 21 06		14.8	GALAXY
UGC 08705	13 44 12.	+ 21 06	72	14.8	GALAXY Sc
KARA.73B 0598	13 44 12.	+ 21 06	60	14.8	ISOLATED GALAXY S
ZWG 132.007	13 44 12.	+ 22 56		15.5	GALAXY
ZWG 132.008	13 44 12.	+ 23 20		15.4	GALAXY
UGC 08706	13 44 12.	+ 23 21	60	16.0	GALAXY Sb-c
ZWG 218.044	13 44 12.	+ 44 05		15.0	GALAXY
KARA.72 394A	13 44 12.	+ 44 05	54	15.0	PART OF DOUBLE GALAXY
ZWG 272.007	13 44 12.	+ 55 58		15.4	GALAXY
UGC 08707	13 44 12.	+ 55 58	84	15.4	GALAXY Sb
MCG+09-23-007	13 44 12.	+ 55 58	54	14.	GALAXY
ZC 1344.2+7703	13 44 12.	+ 77 03	540		CLUSTER OF GALAXIES
RNGC 5297	13 44 15.	+ 44 07		12.5	GALAXY
REIZ 3617	13 44 16.	+ 38 49	54	15.4	GALAXY
RRS 4.033	13 44 17.	+ 26 42		18.75	BLUE OBJECT
REIZ 3618	13 44 17.	+ 44 05	168	12.8	GALAXY
ZWG 045.097	13 44 18.	+ 04 00		15.3	GALAXY
ZWG 045.098	13 44 18.	+ 07 38		15.2	GALAXY
UGC 08708	13 44 18.	+ 07 38	96	15.2	GALAXY IRR
ZC 1344.3+0855	13 44 18.	+ 08 55	1080		CLUSTER OF GALAXIES
MCG+04-33-001	13 44 18.	+ 21 05	54	14.8	GALAXY
ZWG 132.009	13 44 18.	+ 22 56		15.2	GALAXY
RRS 3.019	13 44 18.	+ 28 32		14.92	BLUE OBJECT
TON-N 0168	13 44 18.	+ 28 33		14.8	BLUE STAR
ZWG 218.045	13 44 18.	+ 44 07		12.3	GALAXY
UGC 08709	13 44 18.	+ 44 07	348	12.3	GALAXY Sc
KARA.72 394B	13 44 18.	+ 44 07	324	12.3	PART OF DOUBLE GALAXY
MCG+07-28-062	13 44 18.	+ 44 07	42	14.5	GALAXY
RRS 4.034	13 44 19.	+ 27 06		18.31	BLUE OBJECT
SC 1341-2312.0	13 44 19.	- 23 27 03.	48		NEBULA
REIZ 3610	13 44 20.	+ 04 00	18	15.0	GALAXY
RNGC 5301	13 44 23.	+ 46 21		12.5	GALAXY
ARC 1790	13 44 23.	+ 54 17		17.7	RICH CLUSTER OF GALAXIES
ZWG 017.069	13 44 24.	+ 02 21		15.6	GALAXY
ZWG 073.072	13 44 24.	+ 11 53		15.1	GALAXY
KARA.72 395A	13 44 24.	+ 11 53	42	15.1	PART OF DOUBLE GALAXY
ZWG 102.056	13 44 24.	+ 14 40		15.5	GALAXY
ZWG 102.057	13 44 24.	+ 16 31		14.3	GALAXY
UGC 08710	13 44 24.	+ 16 31	102	14.3	GALAXY Sc
TON-N 0738	13 44 24.	+ 27 01		15.3	BLUE STAR
ZC 1344.4+4136	13 44 24.	+ 41 36	540		CLUSTER OF GALAXIES
MCG+07-28-063	13 44 24.	+ 44 08	348	12.5	GALAXY
MCG+08-25-041	13 44 24.	+ 46 21	240	13.0	GALAXY
ZWG 246.023	13 44 24.	+ 46 22		13.0	GALAXY
UGC 08711	13 44 24.	+ 46 22	252	13.0	GALAXY Sc
ZC 1344.4+5403	13 44 24.	+ 54 03	2490		CLUSTER OF GALAXIES
MCG-04-33-002	13 44 24.	- 24 10	36	15.	GALAXY
MCG+03-35-024	13 44 27.	+ 16 31	114	14.3	GALAXY
RRS 4.035	13 44 28.	+ 26 24		18.85	BLUE OBJECT
RRS 3.020	13 44 29.	+ 28 53		17.50	BLUE OBJECT
RNGC 5291	13 44 29.	- 30 08		15.0	GALAXY
SC 1341-3006.9	13 44 29.	- 30 21 57.	18		NEBULA
ZWG 073.073	13 44 30.	+ 11 52		15.4	GALAXY
KARA.72 395B	13 44 30.	+ 11 53	18	15.4	PART OF DOUBLE GALAXY
RNGC 5293	13 44 30.	+ 16 31		14.5	GALAXY
ZWG 102.058	13 44 30.	+ 17 57		14.7	GALAXY
8ZW 1344+17.9	13 44 30.	+ 17 57		14.7	COMPACT GALAXY
MCG-05-33-003	13 44 30.	- 28 04	60	17.	GALAXY
MCG-05-33-002	13 44 30.	- 32 38	60	17.	GALAXY
REIZ 3616	13 44 33.	+ 11 49	30	15.3	GALAXY
ZWG 017.070	13 44 36.	+ 02 09		15.6	GALAXY
ZWG 045.099	13 44 36.	+ 03 54		15.4	GALAXY
ZWG 073.074	13 44 36.	+ 11 21		15.5	GALAXY
ZC 1344.6+1834	13 44 36.	+ 18 34	1340		CLUSTER OF GALAXIES
ZC 1344.6+1939	13 44 36.	+ 19 39	1610		CLUSTER OF GALAXIES
MRK 274	13 44 36.	+ 29 55	5	17.5	GALAXY WITH UV CONTINUUM
ZC 1344.6+3015	13 44 36.	+ 30 15	1010		CLUSTER OF GALAXIES
ZC 1344.6+4545	13 44 36.	+ 45 45	3020		CLUSTER OF GALAXIES
7ZW 524	13 44 36.	+ 70 20			COMPACT GALAXY

OBJECT NAME	RIGHT ASCEN.	DECLINATION	DIAM.	MAGN.	TYPE OF OBJECT
ZWG 336.016	13 44 36.	+ 70 20		15.6	GALAXY
MCG-05-33-004	13 44 36.	- 29 34	12	15.5	GALAXY
MCG-05-33-006	13 44 36.	- 30 09	42	15.	GALAXY
MCG-05-33-005	13 44 36.	- 30 10	24	16.	GALAXY
SC 1341-3222.1	13 44 36.	- 32 37 09.	96		NEBULA
REIZ 3615	13 44 38.	+ 03 54	18	16.0	GALAXY
ZWG 132.010	13 44 42.	+ 25 14		15.5	GALAXY
UGC 08712	13 44 42.	+ 50 43	72	17.	GALAXY Sc
MCG+09-23-008	13 44 42.	+ 50 43	30	16.	GALAXY
SHAH 129	13 44 42.	+ 74 20	60	17.5	GROUP OF COMPACT GALAXIES
ZC 1344.7-0232	13 44 42.	- 02 32	470		CLUSTER OF GALAXIES
REIZ 3621	13 44 44.	+ 25 15	24	14.6	GALAXY
REIZ 3622	13 44 44.	+ 34 07	90	13.8	GALAXY
HOLM 541B	13 44 44.	+ 34 08	78	14.0	PART OF MULTIPLE GALAXY
RRS 3.022	13 44 45.	+ 31 34		18.58	BLUE OBJECT
VV 317A	13 44 45.	+ 34 07	78	14.5	INTERACTING GALAXY
VV 317B	13 44 45.	+ 34 07 30.	108	14.5	INTERACTING GALAXY
MCG+07-28-064	13 44 45.	+ 42 55	30	15.	GALAXY
SC 1341-3026.3	13 44 46.	- 30 41 20.	72		NEBULA
RNGC 5292	13 44 46.	- 30 42		14.0	GALAXY
ZWG 045.100	13 44 48.	+ 04 01		15.4	GALAXY
ZWG 045.101	13 44 48.	+ 07 50		15.5	GALAXY
ZC 1344.8+2140	13 44 48.	+ 21 40	740		CLUSTER OF GALAXIES
RRS 3.023	13 44 48.	+ 29 50			BLUE OBJECT
MCG+06-30-081	13 44 48.	+ 34 07 30.	90	14.	GALAXY
ZWG 190.053	13 44 48.	+ 34 09		15.5	GALAXY
UGC 08713	13 44 48.	+ 34 09	120	15.5	GALAXY Sc
KARA.72 396A	13 44 48.	+ 34 09	96	15.5	PART OF DOUBLE GALAXY
ZWG 272.008	13 44 48.	+ 55 20		15.7	GALAXY
MCG+10-20-027	13 44 48.	+ 60 37	57	15.	GALAXY
UGC 08714	13 44 48.	+ 60 39	102	16.5	GALAXY DWRF IR
KARA.68 228	13 44 48.	- 12 16	27		DWARF GALAXY
MCG-05-33-007	13 44 48.	- 29 11	48	16.	GALAXY
MCG-05-33-008	13 44 48.	- 30 42	60	14.	GALAXY
REIZ 3620	13 44 49.	+ 16 41	48	15.4	GALAXY
IC 4325	13 44 49.	- 29 11			NONSTELLAR OBJECT
REIZ 3619	13 44 50.	+ 02 26	9	16.0	GALAXY
HOLM 541A	13 44 50.	+ 34 07	60	13.6	PART OF MULTIPLE GALAXY
REIZ 3623	13 44 50.	+ 34 08	60	13.5	GALAXY
HN 1090	13 44 50.	- 29 11		14.	NEBULA
RRS 4.036	13 44 51.	+ 27 15		19.51	BLUE OBJECT
MCG+06-30-083	13 44 51.	+ 34 06 30.	72	14.	GALAXY
MCG+06-30-082	13 44 51.	+ 35 32	48	17.	GALAXY
RRS 3.024	13 44 52.	+ 29 47			BLUE OBJECT
RRS 4.037	13 44 53.	+ 28 19			BLUE OBJECT
ZC 1344.9+1121	13 44 54.	+ 11 21	870		CLUSTER OF GALAXIES
ZWG 132.011	13 44 54.	+ 22 05		15.7	GALAXY
ZC 1344.9+2436	13 44 54.	+ 24 36	3630		CLUSTER OF GALAXIES
ZWG 162.006	13 44 54.	+ 29 46		15.6	GALAXY
ZC 1344.9+3143	13 44 54.	+ 31 43	1080		CLUSTER OF GALAXIES
ZWG 190.054	13 44 54.	+ 34 08		14.8	GALAXY
UGC 08715	13 44 54.	+ 34 08	84	14.8	GALAXY SBc
KARA.72 396B	13 44 54.	+ 34 08	72	14.8	PART OF DOUBLE GALAXY
ZC 1344.9+4204	13 44 54.	+ 42 04	810		CLUSTER OF GALAXIES
MCG+10-20-028	13 44 54.	+ 60 04	30	16.	GALAXY
UGC 08716	13 44 54.	+ 60 35	72	17.	GALAXY
RRS 4.038	13 44 56.	+ 28 15		19.50	BLUE OBJECT
REIZ 3626	13 44 56.	+ 34 19	24	14.7	GALAXY
MCG+06-30-084	13 44 57.	+ 34 19	48	15.	GALAXY
REIZ 3629	13 44 57.	+ 46 24	180	12.8	GALAXY
RRS 3.025	13 44 59.	+ 29 41		18.80	BLUE OBJECT
LB 09982	13 45	- 88 38		13.6	FAINT BLUE STAR
RRS 3.026	13 45 00.	+ 29 57			BLUE OBJECT
RRS 3.027	13 45 00.	+ 30 26		18.40	BLUE OBJECT
ZWG 162.007	13 45 00.	+ 30 35		14.8	GALAXY
UGC 08717	13 45 00.	+ 30 35	78	14.8	GALAXY Sa
ZWG 336.017	13 45 00.	+ 70 35		14.6	GALAXY
MCG+13-10-012	13 45 00.	+ 77 04	84	14.	GALAXY
OCL 0909	13 45 00.	- 65 50	900	9.2	OPEN STAR CLUSTER
RNGC 5314	13 45 01.	+ 70 35		14.5	GALAXY
SC 1342-3157.8	13 45 01.	- 32 12 50.	72		NEBULA
REIZ 3628	13 45 02.	+ 34 23	30	13.8	GALAXY
REIZ 3627	13 45 04.	+ 30 34	60	13.7	GALAXY
RNGC 5323	13 45 05.	+ 77 05		14.5	GALAXY
ZWG 045.102	13 45 06.	+ 03 35		15.5	GALAXY
ZWG 073.075	13 45 06.	+ 10 01		15.4	GALAXY
MCG+06-30-085	13 45 06.	+ 34 22	42	14.	GALAXY
ZWG 190.055	13 45 06.	+ 34 24		14.5	GALAXY
MRK 461	13 45 06.	+ 34 24	24	15.5	GALAXY WITH UV CONTINUUM
UGC 08718	13 45 06.	+ 34 24	48	14.5	GALAXY S
MCG+10-20-029	13 45 06.	+ 61 13	156	12.2	GALAXY
MCG+12-13-009	13 45 06.	+ 70 35	39	15.	GALAXY
ZWG 353.025	13 45 06.	+ 77 05		14.3	GALAXY
UGC 08719	13 45 06.	+ 77 05	90	14.3	GALAXY Sa-b
MCG-04-33-003	13 45 06.	- 25 42	24	15.5	GALAXY
OCL 0910	13 45 06.	- 64 26	420	11.8	OPEN STAR CLUSTER
VHA 153	13 45 06.	- 64 26	180		OPEN STAR CLUSTER
RNGC 5288	13 45 06.	- 64 26		12.0	OPEN CLUSTER
RRS 3.028	13 45 07.	+ 31 10		18.25	BLUE OBJECT
RRS 4.039	13 45 09.	+ 27 19			BLUE OBJECT
MCG+05-33-001	13 45 09.	+ 30 36	60	14.8	GALAXY
SC 1342-3018.6	13 45 09.	- 30 33 38.	30		NEBULA
ZWG 045.103	13 45 12.	+ 03 39		14.7	GALAXY
MCG+01-35-036	13 45 12.	+ 03 39	60	14.7	GALAXY
MCG+10-20-030	13 45 12.	+ 61 13	12	16.	GALAXY
IC 0939	13 45 13.	+ 03 38 19.			NONSTELLAR OBJECT
REIZ 3625	13 45 14.	+ 03 40	12	14.8	GALAXY
SC 1342-3001.8	13 45 14.	- 30 16 49.	72		NEBULA
RRS 4.040	13 45 16.	+ 26 08		19.00	BLUE OBJECT
RNGC 5308	13 45 16.	+ 61 13		13.0	GALAXY
SC 1342-2214.3	13 45 16.	- 22 29 19.	18		NEBULA
SC 1342-3015.3	13 45 17.	- 30 30 19.	6		NEBULA
ZWG 073.076	13 45 18.	+ 11 21		15.6	GALAXY
ZWG 102.059	13 45 18.	+ 17 14		15.2	GALAXY
ZWG 102.060	13 45 18.	+ 17 58		15.2	GALAXY
UGC 08720	13 45 18.	+ 17 58	60	15.2	GALAXY S
ZC 1345.3+4120	13 45 18.	+ 41 20	870		CLUSTER OF GALAXIES
ZC 1345.3+5744	13 45 18.	+ 57 44	4030		CLUSTER OF GALAXIES
SC 1342-2956.7	13 45 18.	- 30 11 43.	54		NEBULA
RRS 3.029	13 45 19.	+ 29 15		19.06	BLUE OBJECT
REIZ 3624	13 45 19.	- 06 28	12	15.4	GALAXY
REIN 2.198	13 45 20.71	+ 61 13 20.8			NEBULA
MCG+03-35-025	13 45 21.	+ 17 58	66	14.7	GALAXY
RRS 3.030	13 45 21.	+ 28 48		16.34	BLUE OBJECT
RRS 4.041	13 45 22.	+ 27 10		18.56	BLUE OBJECT
RRS 3.031	13 45 22.	+ 30 28		16.36	BLUE OBJECT
RNGC 5298	13 45 22.	- 30 11		14.0	GALAXY
SC 1342-3144.3	13 45 23.	- 31 59 19.	30		NEBULA

OBJECT NAME	RIGHT ASCEN.	DECLINATION	DIAM.	MAGN.	TYPE OF OBJECT
ZWG 017.071	13 45 24.	+ 02 09		15.7	GALAXY
ZWG 045.104	13 45 24.	+ 03 42		15.3	GALAXY
ZC 1345.4+1956	13 45 24.	+ 19 56	610		CLUSTER OF GALAXIES
ZWG 132.012	13 45 24.	+ 22 22		15.6	GALAXY
TON-N 0739	13 45 24.	+ 30 28		15.6	BLUE STAR
UGC 08721	13 45 24.	+ 40 36	72	16.0	GALAXY
MCG+07-28-065	13 45 24.	+ 40 36	66	15.5	GALAXY
ZWG 272.009	13 45 24.	+ 55 49		15.5	GALAXY
ZWG 295.012	13 45 24.	+ 61 13		12.5	GALAXY
UGC 08722	13 45 24.	+ 61 13	180	12.5	GALAXY S0-a
MCG-05-33-010	13 45 24.	- 30 12	36	15.	GALAXY
MCG-05-33-009	13 45 24.	- 30 17	36	16.	GALAXY
SC 1342-3116.6	13 45 25.	- 31 31 37.	18		NEBULA
RRS 3.032	13 45 26.	+ 29 12			BLUE OBJECT
HOLM 542A	13 45 28.	+ 38 33	42	12.5	PART OF MULTIPLE GALAXY
REIZ 3639	13 45 28.	+ 61 14	174	12.5	GALAXY
IC 0940	13 45 29.	+ 03 40 38.			NONSTELLAR OBJECT
HOLM 542B	13 45 29.	+ 38 30	36	14.7	PART OF MULTIPLE GALAXY
ZWG 045.105	13 45 30.	+ 05 14		15.7	GALAXY
SN 1954P	13 45 30.	+ 09 36		19.5	SUPERNOVA
ZC 1345.5+1645	13 45 30.	+ 16 45	1480		CLUSTER OF GALAXIES
ZC 1345.5+3309	13 45 30.	+ 33 09	670		CLUSTER OF GALAXIES
MCG+07-28-066	13 45 30.	+ 38 29	39	15.	GALAXY
MCG+07-28-067	13 45 30.	+ 38 32	45	14.	GALAXY
MCG-05-33-011	13 45 30.	- 30 17	30	15.5	GALAXY
RNGC 5303B	13 45 31.	+ 38 29		15.5	GALAXY
ARC 1803	13 45 31.	+ 71 06		17.5	RICH CLUSTER OF GALAXIES
REIZ 3638	13 45 32.	+ 59 17	24	15.2	GALAXY
HN 1091	13 45 32.	- 29 23		14.	NEBULA
IC 4326	13 45 32.	- 29 23			NONSTELLAR OBJECT
REIZ 3632	13 45 34.	+ 38 30	36	14.7	GALAXY
REIZ 3633	13 45 34.	+ 38 33	48	13.1	GALAXY
ZWG 045.106	13 45 36.	+ 03 05		15.6	GALAXY
ZWG 045.107	13 45 36.	+ 07 38		14.7	GALAXY
KARA.72 398A	13 45 36.	+ 07 38	48	14.7	PART OF DOUBLE GALAXY
MCG+01-35-037	13 45 36.	+ 07 38	18	14.7	GALAXY
ZWG 073.077	13 45 36.	+ 12 19		15.3	GALAXY
ZC 1345.6+1812	13 45 36.	+ 18 12	1410		CLUSTER OF GALAXIES
UGC 08723	13 45 36.	+ 25 10	66	16.0	GALAXY S
ZC 1345.6+3105	13 45 36.	+ 31 05	810		CLUSTER OF GALAXIES
ZWG 190.056	13 45 36.	+ 37 59		15.4	GALAXY
UGC 08724	13 45 36.	+ 37 59	72	15.4	GALAXY SB
ZWG 218.046	13 45 36.	+ 38 30		15.3	GALAXY
KARA.72 397B	13 45 36.	+ 38 30	48	15.3	PART OF DOUBLE GALAXY
ZWG 218.047	13 45 36.	+ 38 33		12.9	GALAXY
UGC 08725	13 45 36.	+ 38 33	60	12.9	GALAXY PECULR
KARA.72 397A	13 45 36.	+ 38 33	72	12.9	PART OF DOUBLE GALAXY
ZWG 218.048	13 45 36.	+ 40 44		15.4	GALAXY
UGC 08726	13 45 36.	+ 40 44	138	15.4	GALAXY Sc
MCG+07-28-068	13 45 36.	+ 40 44	126	14.5	GALAXY
MCG-05-33-012	13 45 36.	- 29 22	36	15.	GALAXY
MCG-05-33-013	13 45 36.	- 32 00	36	15.	GALAXY
RNGC 5303A	13 45 37.	+ 38 33		13.0	GALAXY
RRS 4.042	13 45 38.	+ 26 11		18.05	BLUE OBJECT
VV 306B	13 45 39.	+ 07 38 30.		15.	INTERACTING GALAXY
RRS 4.043	13 45 39.	+ 26 00		18.42	BLUE OBJECT
RRS 4.044	13 45 40.	+ 27 40		18.14	BLUE OBJECT
REIZ 3634	13 45 41.	+ 38 01	48	14.9	GALAXY
ZWG 045.108	13 45 42.	+ 04 12		13.7	GALAXY
UGC 08727	13 45 42.	+ 04 12	228	13.7	GALAXY Sc
MCG+01-35-038	13 45 42.	+ 04 12	228	13.7	GALAXY
ZWG 045.109	13 45 42.	+ 07 38		14.9	GALAXY
UGC 08728	13 45 42.	+ 07 38	90	14.9	GALAXY SB
VV 306C	13 45 42.	+ 07 38	15	16.	INTERACTING GALAXY
KARA.72 398B	13 45 42.	+ 07 38	78	14.9	PART OF DOUBLE GALAXY
MCG+01-35-039	13 45 42.	+ 07 38	72	14.9	GALAXY
ARC 1792	13 45 42.	+ 31 05		17.8	RICH CLUSTER OF GALAXIES
ZWG 190.057	13 45 42.	+ 38 04		14.7	GALAXY
UGC 08729	13 45 42.	+ 38 04	96	14.7	GALAXY SBb
ZC 1345.7+4808	13 45 42.	+ 48 08	1410		CLUSTER OF GALAXIES
MCG-02-35-016	13 45 42.	- 13 18	48	15.5	GALAXY
RNGC 5305	13 45 43.	+ 38 04		14.5	GALAXY
RNGC 5300	13 45 44.	+ 04 12		12.0	GALAXY
VV 306A	13 45 45.	+ 07 38 30.	84	15.	INTERACTING GALAXY
MCG+06-30-086	13 45 45.	+ 38 00	36	14.5	GALAXY
REIZ 3636	13 45 46.	+ 38 05	42	14.6	GALAXY
RRS 4.045	13 45 47.	+ 28 02			BLUE OBJECT
SC 1342-3018.7	13 45 47.	- 30 33 42.	72		NEBULA
REIZ 3630	13 45 48.	+ 07 37	9	16.0	GALAXY
REIZ 3631	13 45 48.	+ 07 38	36	14.7	GALAXY
ZWG 102.061	13 45 48.	+ 15 40		15.0	GALAXY
8ZW 1345+15.7	13 45 48.	+ 15 40		15.0	COMPACT GALAXY
MCG+03-35-026	13 45 48.	+ 15 40	30	15.0	GALAXY
RRS 3.033	13 45 48.	+ 28 42		18.45	BLUE OBJECT
MCG+06-30-087	13 45 48.	+ 38 04 30.	90	13.5	GALAXY
MCG+08-25-042	13 45 43.	+ 48 02	9	16.	GALAXY
MCG+10-20-031	13 45 48.	+ 56 51	18	15.	GALAXY
MCG-05-33-015	13 45 48.	- 30 11	60	14.5	GALAXY
MCG-05-33-014	13 45 48.	- 30 34	48	15.	GALAXY
DVDV 4	13 45 48.	- 35 48	156		GALAXY
RRS 3.034	13 45 49.	+ 28 55			BLUE OBJECT
RRS 4.046	13 45 51.	+ 28 17		17.96	BLUE OBJECT
HN 1092	13 45 51.	- 29 59		12.5	NEBULA
RRS 3.035	13 45 52.	+ 31 01		19.13	BLUE OBJECT
IC 4327	13 45 53.1	- 29 58 15.			GALAXY
ZC 1345.9+1526	13 45 54.	+ 15 26	1080		CLUSTER OF GALAXIES
ZWG 102.062	13 45 54.	+ 18 36		15.3	GALAXY
UGC 08730	13 45 54.	+ 25 02	72	16.0	GALAXY SB0
ZWG 132.013	13 45 54.	+ 25 21		15.6	GALAXY
MCG+04-33-002	13 45 54.	+ 25 23	18	15.6	GALAXY
ZWG 132.014	13 45 54.	+ 25 59		15.4	GALAXY
ZC 1345.9+3614	13 45 54.	+ 36 14	1010		CLUSTER OF GALAXIES
ZWG 190.058	13 45 54.	+ 36 22		15.7	GALAXY
UGC 08731	13 45 54.	+ 36 22	96	15.7	GALAXY S
ZWG 295.013	13 45 54.	+ 56 53		15.3	GALAXY
MCG-05-33-016	13 45 54.	- 29 57	36	16.	GALAXY
IC 0942	13 45 55.	+ 56 52 11.			NONSTELLAR OBJECT
BC 4C58.27	13 45 55.79	+ 58 27 37.3		17.5	QUASI-STELLAR OBJECT
SHB 278	13 45 55.8	+ 58 28 12.		17.5	QUASI-STELLAR OBJECT
HOLM 543A	13 45 56.	+ 25 23	12	14.7	PART OF MULTIPLE GALAXY
SC 1343-3039.4	13 45 56.	- 30 54 24.	72		NEBULA
VV 190A	13 45 57.	+ 25 59	30	15.0	INTERACTING GALAXY
MCG+06-30-088	13 45 57.	+ 36 22	66	15.	GALAXY
HOLM 543B	13 45 58.	+ 25 24	24	14.8	PART OF MULTIPLE GALAXY
RNGC 5302	13 45 58.	- 30 14		14.0	GALAXY
REIZ 3635	13 46 00.	+ 18 32	60	15.2	GALAXY
ZWG 132.015	13 46 00.	+ 25 55		15.5	GALAXY
VV 191B	13 46 00.	+ 25 56 30.	15	15.	INTERACTING GALAXY

OBJECT NAME	RIGHT ASCEN.	DECLINATION	DIAM.	MAGN.	TYPE OF OBJECT
VV 191A	13 46 00.	+ 25 56 30.	15	15.	INTERACTING GALAXY
REIZ 3637	13 46 00.	+ 36 21	36	15.8	GALAXY
MCG+10-20-032	13 46 00.	+ 61 34	42	16.	GALAXY
HOLM 544B	13 46 01.	+ 25 56	12	15.3	PART OF MULTIPLE GALAXY
HOLM 544A	13 46 01.	+ 25 56	30	15.0	PART OF MULTIPLE GALAXY
RRS 3.036	13 46 02.	+ 30 56		18.55	BLUE OBJECT
ARC 1793	13 46 04.	+ 32 32		16.4	RICH CLUSTER OF GALAXIES
MCG+04-33-004	13 46 06.	+ 25 56	9	15.4	GALAXY
MCG+04-33-003	13 46 06.	+ 25 56	24	15.4	GALAXY
RRS 3.037	13 46 06.	+ 30 35		17.86	BLUE OBJECT
ZC 1346.1+3230	13 46 06.	+ 32 30	1880		CLUSTER OF GALAXIES
ZWG 336.018	13 46 06.	+ 72 18		15.1	GALAXY
UGC 08732	13 46 06.	+ 72 18	66	15.1	GALAXY Sb
MCG+12-13-010	13 46 06.	+ 72 19	51	15.	GALAXY
MCG-01-35-012	13 46 06.	- 07 23	72	15.	GALAXY
ARC 1791	13 46 06.	- 25 13		17.0	RICH CLUSTER OF GALAXIES
MCG-05-33-018	13 46 06.	- 30 16	60	14.5	GALAXY
MCG-05-33-017	13 46 06.	- 30 55	48	15.	GALAXY
LB 00687	13 46 07.	+ 26 30 18.		17.8	FAINT BLUE STAR
RRS 4.047	13 46 07.	+ 27 21			BLUE OBJECT
RRS 3.038	13 46 10.	+ 28 56			BLUE OBJECT
SC 1343-2220.8	13 46 11.	- 22 35 47.	24		NEBULA
ZWG 132.016	13 46 12.	+ 24 15		15.2	GALAXY
MCG+04-33-005	13 46 12.	+ 25 54	24	15.5	GALAXY
MCG+04-33-006	13 46 12.	+ 25 54 30.	30	15.5	GALAXY
ZC 1346.2+2814	13 46 12.	+ 28 14	5650		CLUSTER OF GALAXIES
VVI 69	13 46 12.	- 30 03	90	12.79	SEYFERT GALAXY
RRS 4.048	13 46 13.	+ 28 17		18.76	BLUE OBJECT
RRS 3.039	13 46 13.	+ 31 34		18.64	BLUE OBJECT
RRS 3.040	13 46 14.	+ 28 58		18.79	BLUE OBJECT
HN 1093	13 46 14.	- 29 42		14.	NEBULA
IC 4328	13 46 14.	- 29 42			NONSTELLAR OBJECT
SC 1343-3056.0	13 46 14.	- 31 11 00.	18		NEBULA
KW 60	13 46 14.4	- 30 02 59.	70		SEYFERT GALAXY
IC 4329	13 46 14.4	- 30 02 59.	90	12.0	GALAXY S0
IC 0941	13 46 15.	+ 24 14 59.			NONSTELLAR OBJECT
RRS 3.041	13 46 15.	+ 29 10		18.12	BLUE OBJECT
SC 1343-3039.9	13 46 15.	- 30 54 54.	72		NEBULA
RRS 4.049	13 46 16.	+ 28 05		18.46	BLUE OBJECT
ARC 1798	13 46 16.	+ 57 51		17.6	RICH CLUSTER OF GALAXIES
ZWG 045.110	13 46 18.	+ 06 36		15.7	GALAXY
ZWG 073.078	13 46 18.	+ 12 12		15.6	GALAXY
8ZW 1346+12.8	13 46 18.	+ 12 48		15.9	COMPACT GALAXY
TON-N 0740	13 46 18.	+ 24 11		15.4	BLUE STAR
ZWG 132.017	13 46 18.	+ 26 19		15.7	GALAXY
ZWG 162.008	13 46 18.	+ 28 29		15.3	GALAXY
ZWG 162.009	13 46 18.	+ 31 43		15.5	GALAXY
MRK 275	13 46 18.	+ 31 43	15	16.	GALAXY WITH UV CONTINUUM
MCG+05-33-002	13 46 18.	+ 31 43	48	15.5	GALAXY
ZWG 218.049	13 46 18.	+ 43 57		15.7	GALAXY
MCG+10-20-033	13 46 18.	+ 62 05	60	15.	GALAXY
MCG-01-35-013	13 46 18.	- 06 56	120	15.	GALAXY
REIZ 3642	13 46 19.	+ 26 20	24	14.6	GALAXY
MCG-05-33-019	13 46 21.	- 30 03	12	14.5	GALAXY
RRS 4.050	13 46 22.	+ 25 50			BLUE OBJECT
RRS 3.042	13 46 22.	+ 28 58		17.01	BLUE OBJECT
MCG+04-33-007	13 46 24.	+ 24 15	36	15.2	GALAXY
MCG+05-33-003	13 46 24.	+ 28 29	42	15.3	GALAXY
MCG+05-33-004	13 46 24.	+ 28 30	24	15.3	GALAXY
RRS 3.043	13 46 24.	+ 29 10			BLUE OBJECT
ZC 1346.4+3124	13 46 24.	+ 31 24	870		CLUSTER OF GALAXIES
ZWG 190.059	13 46 24.	+ 37 21		14.9	GALAXY
KARA.73B 0599	13 46 24.	+ 37 21	54	14.9	ISOLATED GALAXY S0
MCG+07-28-069	13 46 24.	+ 43 34	36	16.5	GALAXY
MCG+09-23-009	13 46 24.	+ 55 02	36	14.	GALAXY
ZWG 272.010	13 46 24.	+ 55 03		15.2	GALAXY
MCG+12-13-011	13 46 24.	+ 72 55 30.	51	17.	GALAXY
SC 1343-2924.4	13 46 24.	- 29 29 23.	36		NEBULA
MCG-05-33-020	13 46 24.	- 30 55 30.	48	15.	GALAXY
MCG-06-30-025	13 46 24.	- 35 49	240	12.	GALAXY
RRS 4.051	13 46 25.	+ 28 01		18.46	BLUE OBJECT
HN 1094	13 46 25.	- 28 04	60	13.5	NEBULA
IC 4330	13 46 25.	- 28 04			NONSTELLAR OBJECT
REIZ 3641	13 46 26.	+ 03 11	60	15.0	GALAXY
RRS 3.044	13 46 28.	+ 28 57		17.80	BLUE OBJECT
RNGC 5306	13 46 29.	- 06 58		14.5	GALAXY
8ZW 1346+10.4	13 46 30.	+ 10 26		18.5	COMPACT GALAXY
RRS 4.052	13 46 30.	+ 26 38		18.51	BLUE OBJECT
ZWG 162.010	13 46 30.	+ 26 50		15.5	GALAXY
MCG+05-33-005	13 46 30.	+ 26 50	48	15.5	GALAXY
MCG+06-30-089	13 46 30.	+ 37 21	18	14.5	GALAXY
ZWG 218.050	13 46 30.	+ 43 39		14.6	GALAXY
UGC 08733	13 46 30.	+ 43 39	162	14.6	GALAXY SBc
ZWG 295.014	13 46 30.	+ 62 03		15.2	GALAXY
UGC 08734	13 46 30.	+ 62 03	66	15.2	GALAXY S0-a
MCG-01-35-015	13 46 30.	- 06 57	36	16.	GALAXY
MCG-01-35-014	13 46 30.	- 06 58	30	14.5	GALAXY
RRS 3.045	13 46 31.	+ 28 53		17.19	BLUE OBJECT
REIZ 3645	13 46 31.	+ 41 56	30	14.3	GALAXY
REIZ 3640	13 46 31.	- 06 58	12	13.7	GALAXY
RRS 3.046	13 46 32.	+ 29 40		17.98	BLUE OBJECT
SC 1343-3234.8	13 46 33.	- 32 49 47.	60		NEBULA
RRS 3.047	13 46 35.	+ 29 15		17.70	BLUE OBJECT
ZWG 045.111	13 46 36.	+ 03 10		15.7	GALAXY
ZC 1346.6+1320	13 46 36.	+ 13 20	2150		CLUSTER OF GALAXIES
ZWG 102.063	13 46 36.	+ 17 29		15.5	GALAXY
ZWG 218.051	13 46 36.	+ 41 58		15.6	GALAXY
ZC 1346.6+4250	13 46 36.	+ 42 50	2620		CLUSTER OF GALAXIES
MCG+07-28-070	13 46 36.	+ 43 40	132	14.	GALAXY
ZWG 017.072	13 46 36.	- 02 11		15.6	GALAXY
VV 135C	13 46 36.	- 06 58	42	16.	INTERACTING GALAXY
VV 135B	13 46 36.	- 06 58	15	16.	INTERACTING GALAXY
VV 135A	13 46 36.	- 06 58	60	14.	INTERACTING GALAXY
VV 135	13 46 36.	- 06 58	132		INTERACTING GALAXY
MCG-05-33-021	13 46 36.	- 30 04	48	15.	GALAXY
REIZ 3644	13 46 37.	+ 26 51	18	14.6	GALAXY
RRS 3.048	13 46 37.	+ 30 04		19.01	BLUE OBJECT
RRS 4.053	13 46 38.	+ 26 14		19.10	BLUE OBJECT
REIN 7.121	13 46 38.55	+ 41 57 28.7			NEBULA
MCG+07-28-071	13 46 39.	+ 41 57	30	15.	GALAXY
RRS 3.049	13 46 40.	+ 29 08		18.32	BLUE OBJECT
ARC 1795	13 46 41.	+ 26 51		16.0	RICH CLUSTER OF GALAXIES
MCG+03-35-027	13 46 42.	+ 17 28	48	15.5	GALAXY
ZWG 317.007	13 46 42.	+ 67 33		15.4	GALAXY
ZC 1346.7+7105	13 46 42.	+ 71 05	1880		CLUSTER OF GALAXIES
MCG-03-35-018	13 46 42.	- 17 51	24	15.5	GALAXY
REIZ 3643	13 46 43.	+ 04 42	24	13.6	GALAXY
RRS 3.050	13 46 45.	+ 28 57		17.69	BLUE OBJECT
RRS 3.051	13 46 45.	+ 29 59			BLUE OBJECT
MCG+07-28-072	13 46 45.	+ 40 13	150	14.	GALAXY
ZC 1346.8+1457	13 46 48.	+ 14 57	870		CLUSTER OF GALAXIES
ZWG 218.052	13 46 48.	+ 40 14		13.7	GALAXY
UGC 08735	13 46 48.	+ 40 14	162	13.7	GALAXY S0-a
MCG+09-23-010	13 46 48.	+ 56 21	30	16.	GALAXY
REIZ 3654	13 46 48.	+ 60 02	36	15.5	GALAXY
MCG+11-17-008	13 46 48.	+ 68 21	126	14.	GALAXY
ZWG 336.019	13 46 48.	+ 72 17		15.7	GALAXY
IC 0945	13 46 48.	+ 72 18 11.			NONSTELLAR OBJECT
REIN 7.122	13 46 48.04	+ 40 14 02.1			NEBULA
RNGC 5311	13 46 50.	+ 40 14		13.5	GALAXY
REIZ 3649	13 46 50.	+ 40 14	24	13.7	GALAXY
REIZ 3648	13 46 51.	+ 39 45	60	13.7	GALAXY
ZWG 073.079	13 46 54.	+ 08 45		14.9	GALAXY
MCG+02-35-018	13 46 54.	+ 08 45	48	14.9	GALAXY
ZC 1346.9+2655	13 46 54.	+ 26 55	3020		CLUSTER OF GALAXIES
RRS 3.052	13 46 54.	+ 30 00			BLUE OBJECT
ZC 1346.9+3931	13 46 54.	+ 39 31	870		CLUSTER OF GALAXIES
MCG+07-28-073	13 46 54.	+ 39 44	72	14.	GALAXY
ZWG 218.053	13 46 54.	+ 39 45		14.3	GALAXY
UGC 08736	13 46 54.	+ 39 45	84	14.3	GALAXY S
REIZ 3656	13 46 54.	+ 62 27	60	14.8	GALAXY
ZWG 317.008	13 46 54.	+ 68 20		14.8	GALAXY
UGC 08737	13 46 54.	+ 68 20	144	14.8	GALAXY Sb-c
ZC 1346.9-0124	13 46 54.	- 01 24	2420		CLUSTER OF GALAXIES
REIN 7.123	13 46 56.11	+ 39 44 47.9			NEBULA
RRS 3.053	13 46 57.	+ 31 21		18.04	BLUE OBJECT
HOLM 545D	13 46 57.	+ 35 32	24	14.9	PART OF MULTIPLE GALAXY
RRS 4.054	13 46 58.	+ 27 05		19.33	BLUE OBJECT
RRS 3.054	13 46 58.	+ 29 54		18.02	BLUE OBJECT
RRS 3.055	13 46 59.	+ 29 05		18.59	BLUE OBJECT
RNGC 5299	13 46 59.	- 59 42			NON-EXISTENT OBJECT
SER 108.01	13 47	- 48 20	4320	15.	RICH GROUP OF 14 GALAXIES
ZC 1347.0+0200	13 47 00.	+ 02 00	1550		CLUSTER OF GALAXIES
ZWG 073.080	13 47 00.	+ 14 07		15.5	GALAXY
KARA.73B 0600	13 47 00.	+ 14 07	24	15.5	ISOLATED GALAXY S
ZC 1347.0+1701	13 47 00.	+ 17 01	1950		CLUSTER OF GALAXIES
ZWG 162.011	13 47 00.	+ 28 22		15.6	GALAXY
UGC 08738	13 47 00.	+ 28 22	66	15.6	GALAXY E-S
ZWG 162.012	13 47 00.	+ 29 51		15.7	GALAXY
RRS 3.056	13 47 00.	+ 30 06			BLUE OBJECT
ZWG 190.060	13 47 00.	+ 35 30		14.7	GALAXY
REIZ 3651	13 47 00.	+ 35 30	60	14.1	GALAXY
HOLM 545A	13 47 00.	+ 35 30	108	13.5	PART OF MULTIPLE GALAXY
UGC 08739	13 47 00.	+ 35 30	120	14.7	GALAXY S
HOLM 545C	13 47 00.	+ 35 32	18	14.6	PART OF MULTIPLE GALAXY
ZC 1347.0+4928	13 47 00.	+ 49 28	1340		CLUSTER OF GALAXIES
MCG+08-25-043	13 47 00.	+ 49 40	360	16.	GALAXY
MCG+12-13-012	13 47 00.	+ 72 18	27	16.	GALAXY
RRS 3.057	13 47 01.	+ 30 01		17.88	BLUE OBJECT
IC 4331	13 47 04.	+ 25 24 02.			NONSTELLAR OBJECT
RNGC 5304	13 47 04.	- 30 19		15.0	GALAXY
ZWG 045.112	13 47 06.	+ 04 29		14.4	GALAXY
UGC 08740	13 47 06.	+ 04 29	114	14.4	GALAXY Sa/SBb
MCG+01-35-040	13 47 06.	+ 04 29	72	14.4	GALAXY
ZWG 073.081	13 47 06.	+ 11 20		15.6	GALAXY
ZC 1347.1+2515	13 47 06.	+ 25 15	940		CLUSTER OF GALAXIES
TON-N 0169	13 47 06.	+ 27 52		16.5	BLUE STAR
REIZ 3650	13 47 06.	+ 28 16	18	15.8	GALAXY
MCG+06-30-090	13 47 06.	+ 35 30	114	13.5	GALAXY
ZC 1347.1+3544	13 47 06.	+ 35 44	470		CLUSTER OF GALAXIES
REIZ 3647	13 47 07.	+ 04 29	48	14.3	GALAXY
RRS 4.055	13 47 09.	+ 28 12		19.42	BLUE OBJECT
MCG+05-33-006	13 47 09.	+ 28 22	51	15.6	GALAXY
ARC 1804	13 47 09.	+ 49 30		17.6	RICH CLUSTER OF GALAXIES
ARC 1799	13 47 11.	+ 35 42		17.2	RICH CLUSTER OF GALAXIES
REIZ 3652	13 47 12.	+ 28 15	12	15.9	GALAXY
1ZW 073	13 47 12.	+ 35 43			COMPACT GALAXY
ZC 1347.2+5739	13 47 12.	+ 57 39	1210		CLUSTER OF GALAXIES
MCG+10-20-034	13 47 12.	+ 60 05	45	14.	GALAXY
MCG-04-33-004	13 47 12.	- 22 01	30	15.	GALAXY
ARC 1797	13 47 13.	+ 25 16		17.0	RICH CLUSTER OF GALAXIES
HN 0019	13 47 15.	+ 00 19			NEBULA
REIZ 3646	13 47 15.	- 09 35	48	14.0	GALAXY
RNGC 5309	13 47 15.	- 15 31			NON-EXISTENT OBJECT
KLEM 27	13 47 16.	- 30 32	3600	11.	CLUSTER OF 30 GALAXIES
REIZ 3661	13 47 17.	+ 60 05	48	14.8	GALAXY
ZWG 162.013	13 47 18.	+ 29 13		15.7	GALAXY
REIZ 3653	13 47 18.	+ 35 30	24	15.8	GALAXY
ZWG 295.015	13 47 18.	+ 60 05		15.3	GALAXY
UGC 08741	13 47 18.	+ 60 05	72	15.3	GALAXY SB
ZC 1347.3+6709	13 47 18.	+ 67 09	1080		CLUSTER OF GALAXIES
RRS 3.058	13 47 19.	+ 29 38		18.34	BLUE OBJECT
ARC 1800	13 47 21.	+ 28 20		15.4	RICH CLUSTER OF GALAXIES
RRS 3.059	13 47 21.	+ 29 43		17.60	BLUE OBJECT
HOLM 545B	13 47 21.	+ 35 30	30	14.2	PART OF MULTIPLE GALAXY
MCG-05-33-022	13 47 21.	- 30 20	24	15.	GALAXY
ARC 1805	13 47 22.	+ 57 38		17.8	RICH CLUSTER OF GALAXIES
ZWG 017.073	13 47 24.	+ 02 19		15.6	GALAXY
TON-N 0170	13 47 24.	+ 25 23		15.1	BLUE STAR
MCG+05-33-007	13 47 24.	+ 29 14 30.	30	15.7	GALAXY
ZC 1347.4+3029	13 47 24.	+ 30 29	670		CLUSTER OF GALAXIES
MCG+06-30-091	13 47 24.	+ 35 29	24	15.	GALAXY
ZC 1347.4+3855	13 47 24.	+ 38 55	2690		CLUSTER OF GALAXIES
UGC 08742	13 47 24.	+ 39 10	96	17.	GALAXY DWRF IR
ZC 1347.4+4605	13 47 24.	+ 46 05	1340		CLUSTER OF GALAXIES
UGC 08743	13 47 24.	+ 58 55	78	16.0	GALAXY SBc
MCG+10-20-035	13 47 24.	+ 60 26	84	11.0	GALAXY
ARC 1794	13 47 25.	- 26 05		17.0	RICH CLUSTER OF GALAXIES
REIN 7.124B	13 47 25.99	+ 39 54 16.1			NEBULA
RRS 3.060	13 47 26.	+ 30 25		18.83	BLUE OBJECT
REIN 7.124A	13 47 26.03	+ 39 54 16.1			NEBULA
REIZ 3655	13 47 27.	+ 39 54	18	14.3	GALAXY
ZC 1347.5+1815	13 47 30.	+ 18 15	8400		CLUSTER OF GALAXIES
ZC 1347.5+2006	13 47 30.	+ 20 06	670		CLUSTER OF GALAXIES
ZC 1347.5+2405	13 47 30.	+ 24 05	670		CLUSTER OF GALAXIES
ZWG 132.018	13 47 30.	+ 25 26		15.3	GALAXY
RRS 4.056	13 47 30.	+ 26 49			BLUE OBJECT
RRS 3.061	13 47 30.	+ 29 12		19.58	BLUE OBJECT
ZWG 162.014	13 47 30.	+ 25 56		15.4	GALAXY
MCG+08-25-044	13 47 30.	+ 46 48	18	17.	GALAXY
MCG+10-20-036	13 47 30.	+ 57 10	15	15.	GALAXY
ZWG 295.016	13 47 30.	+ 57 11		15.4	GALAXY
MCG+10-20-037	13 47 30.	+ 58 54	78	15.	GALAXY
MCG+10-20-038	13 47 30.	+ 61 47	36	16.	GALAXY
ZC 1347.5+6632	13 47 30.	+ 66 32	610		CLUSTER OF GALAXIES
MCG-03-35-019	13 47 30.	- 19 32 30.	18	15.	GALAXY

OBJECT NAME	RIGHT ASCEN.	DECLINATION	DIAM.	MAGN.	TYPE OF OBJECT
RRS 3.062	13 47 31.	+ 30 16			BLUE OBJECT
IC 4335	13 47 31.	+ 33 55			NONSTELLAR OBJECT
IC 4332	13 47 32.	+ 25 26 09.			NONSTELLAR OBJECT
RRS 4.057	13 47 32.	+ 26 08		18.42	BLUE OBJECT
RRS 3.063	13 47 32.	+ 29 13			BLUE OBJECT
IC 4334	13 47 32.	+ 29 55 10.	4		PLANETARY NEBULA
MCG+05-33-008	13 47 33.	+ 29 58	24	15.4	GALAXY
MCG+07-28-074	13 47 33.	+ 40 13	108	13.	GALAXY
ARC 1796	13 47 33.	- 11 40		17.2	RICH CLUSTER OF GALAXIES
RRS 3.064	13 47 34.	+ 30 45		17.49	BLUE OBJECT
RNGC 5322	13 47 34.	+ 60 26		11.5	GALAXY
REIZ 3663	13 47 34.	+ 60 27	108	12.0	GALAXY
RNGC 5312	13 47 35.	+ 33 52		15.0	GALAXY
REIN 2.199	13 47 35.15	+ 60 26 19.7			NEBULA
REIN 2.200	13 47 35.32	+ 60 25 57.5			NEBULA
8ZW 1347+15.5	13 47 36.	+ 15 28		17.9	COMPACT GALAXY
ZWG 102.064	13 47 36.	+ 19 37		15.7	GALAXY
MCG+04-33-008	13 47 36.	+ 25 26	48	15.3	GALAXY
MCG+06-30-092	13 47 36.	+ 33 50	42	14.5	GALAXY
ZWG 190.061	13 47 36.	+ 33 52		14.8	GALAXY
ZWG 218.054	13 47 36.	+ 40 14		12.4	GALAXY
UGC 08744	13 47 36.	+ 40 14	114	12.4	GALAXY S
ZWG 295.017	13 47 36.	+ 60 26		11.3	GALAXY
UGC 08745	13 47 36.	+ 60 26	360	11.3	GALAXY E
REIN 7.125	13 47 36.43	+ 40 13 59.5			NEBULA
RRS 4.058	13 47 37.	+ 27 29			BLUE OBJECT
RRS 4.059	13 47 37.	+ 28 22		18.73	BLUE OBJECT
REIZ 3657	13 47 37.	+ 33 52	36	14.2	GALAXY
RRS 3.065	13 47 38.	+ 28 51			BLUE OBJECT
RNGC 5313	13 47 38.	+ 40 14		13.0	GALAXY
REIZ 3660	13 47 38.	+ 40 14	90	13.0	GALAXY
RRS 4.060	13 47 41.	+ 26 40			BLUE OBJECT
TON-N 0741	13 47 42.	+ 25 45		15.5	BLUE STAR
ZWG 295.018	13 47 42.	+ 61 46		15.7	GALAXY
MCG-06-30-026	13 47 42.	- 36 30	30	15.	GALAXY
MCG-06-30-027	13 47 42.	- 38 13		15.	GALAXY
SER 108.02	13 47 42.	- 48 47			PEC. GALAXY
RRS 3.066	13 47 43.	+ 30 16		19.85	BLUE OBJECT
RRS 4.061	13 47 46.	+ 27 41		18.44	BLUE OBJECT
ZWG 045.113	13 47 48.	+ 03 15		15.7	GALAXY
ZWG 045.114	13 47 48.	+ 08 03		15.6	GALAXY
RRS 3.067	13 47 48.	+ 31 05		17.94	BLUE OBJECT
ZWG 336.020	13 47 48.	+ 69 04		15.3	GALAXY
RRS 3.068	13 47 49.	+ 30 01			BLUE OBJECT
BIGO 532	13 47 50.	+ 41 04			NEBULA
RRS 3.069	13 47 54.	+ 29 25		18.52	BLUE OBJECT
MCG+10-20-039	13 47 54.	+ 60 16	24	16.	GALAXY
RNGC 5307	13 47 54.	- 50 58		12.0	PLANETARY NEBULA
PK312+10.1	13 47 54.	- 51 03 34.	15	12.1	PLANETARY NEBULA
ZWG 045.115	13 48 00.	+ 03 26		14.8	GALAXY
MCG+01-35-041	13 48 00.	+ 03 26	36	14.8	GALAXY
IC 0943	13 48 00.	+ 03 26 10.			NONSTELLAR OBJECT
ZC 1348.0+0535	13 48 00.	+ 05 35	610		CLUSTER OF GALAXIES
ZWG 102.065	13 48 00.	+ 18 24		15.4	GALAXY
UGC 08746	13 48 00.	+ 18 24	60	15.4	GALAXY S
MCG+03-35-028	13 48 00.	+ 18 24	66	15.4	GALAXY
ZWG 218.055	13 48 00.	+ 38 28		15.4	GALAXY
MCG+06-30-093	13 48 00.	+ 38 28	24	15.	GALAXY
ZC 1348.0+5024	13 48 00.	+ 50 24	810		CLUSTER OF GALAXIES
MCG+12-13-013	13 48 00.	+ 72 50	30	16.	GALAXY
ZWG 353.026	13 48 00.	+ 74 30		15.1	GALAXY
ZWG 336.021	13 48 00.	+ 74 30		15.1	GALAXY
UGC 08747	13 48 00.	+ 74 30	102	15.1	GALAXY Sc
REIZ 3659	13 48 01.	+ 04 30	9	15.2	GALAXY
REIZ 3658	13 48 02.	+ 03 26	24	14.7	GALAXY
RRS 4.062	13 48 03.	+ 26 16		18.74	BLUE OBJECT
ARC 1801	13 48 04.	+ 05 35		17.8	RICH CLUSTER OF GALAXIES
LB 00688	13 48 04.	+ 27 08 06.		17.0	FAINT BLUE STAR
ZWG 102.066	13 48 06.	+ 16 20		15.2	GALAXY
ZWG 102.067	13 48 06.	+ 19 32		15.1	GALAXY
TON-N 0171	13 48 06.	+ 24 50		15.4	BLUE STAR
RRS 4.063	13 48 06.	+ 26 14		19.06	BLUE OBJECT
RRS 3.070	13 48 06.	+ 29 14		18.40	BLUE OBJECT
RRS 3.071	13 48 06.	+ 29 31		18.21	BLUE OBJECT
ZWG 190.062	13 48 06.	+ 35 43		15.3	GALAXY
MCG+12-13-014	13 48 06.	+ 72 43	21	16.	GALAXY
ZWG 336.022	13 48 06.	+ 72 53		15.2	GALAXY
REIZ 3664	13 48 07.	+ 41 35	60	13.3	GALAXY
LB 00689	13 48 08.	+ 26 48 24.		17.6	FAINT BLUE STAR
RNGC 5340	13 48 08.	+ 72 53		15.0	GALAXY
MCG+06-30-094	13 48 09.	+ 35 42	27	15.	GALAXY
MCG+07-28-075	13 48 09.	+ 43 00	36	15.	GALAXY
ZWG 045.116	13 48 12.	+ 02 33		15.1	GALAXY
MCG+01-35-042	13 48 12.	+ 02 33	24	15.1	GALAXY
ZC 1348.2+1830	13 48 12.	+ 18 30	1080		CLUSTER OF GALAXIES
MCG+03-35-029	13 48 12.	+ 19 32	27	15.1	GALAXY
MCG+04-33-009	13 48 12.	+ 25 11	30	14.9	GALAXY
KARA.72 399A	13 48 12.	+ 25 12	36	14.9	PART OF DOUBLE GALAXY
MCG+04-33-010	13 48 12.	+ 25 12	24	14.9	GALAXY
ZWG 132.019	13 48 12.	+ 25 13		14.9	GALAXY
KARA.72 399B	13 48 12.	+ 25 13	36		PART OF DOUBLE GALAXY
ZWG 162.015	13 48 12.	+ 28 24		14.9	GALAXY
UGC 08748	13 48 12.	+ 28 24	60	14.9	GALAXY S
ZWG 162.016	13 48 12.	+ 29 37		15.3	GALAXY
MCG+07-28-076	13 48 12.	+ 41 36 30.	216	13.	GALAXY
ZWG 219.001	13 48 12.	+ 41 37		13.1	GALAXY
ZWG 218.056	13 48 12.	+ 41 37		13.1	GALAXY
UGC 08749	13 48 12.	+ 41 37	216	13.1	GALAXY Sc
ZWG 219.002	13 48 12.	+ 42 59		15.5	GALAXY
ZWG 218.057	13 48 12.	+ 42 59		15.5	GALAXY
ZC 1348.2+4342	13 48 12.	+ 43 42	940		CLUSTER OF GALAXIES
MCG+12-13-015	13 48 12.	+ 68 40	24	17.	GALAXY
HOLM 546B	13 48 13.	+ 25 13	12	14.8	PART OF MULTIPLE GALAXY
HOLM 546A	13 48 13.	+ 25 14	12	14.7	PART OF MULTIPLE GALAXY
REIN 7.126	13 48 13.72	+ 41 36 48.5			NEBULA
MAI 088	13 48 14.	+ 63 03	47		DWARF SPHEROIDAL GALAXY
RNGC 5320	13 48 15.	+ 41 37		13.0	GALAXY
SC 1345-3017.1	13 48 16.	- 30 32 02.	54		NEBULA
RNGC 5319	13 48 17.	+ 33 57		15.0	GALAXY
RNGC 5318B	13 48 17.	+ 33 57			GALAXY
RNGC 5318A	13 48 17.	+ 33 57		13.5	GALAXY
ZWG 045.117	13 48 18.	+ 02 34		15.3	GALAXY
UGC 08750	13 48 18.	+ 02 34	78	15.3	GALAXY S+COMP
MCG+01-35-043	13 48 18.	+ 02 34	72	15.3	GALAXY
RRS 3.072	13 48 18.	+ 30 01		18.69	BLUE OBJECT
ZWG 190.063	13 48 18.	+ 33 57		13.5	GALAXY
UGC 08751	13 48 18.	+ 33 57	96	13.5	GALAXY S0?
MCG+06-30-095	13 48 18.	+ 33 57 30.	36	15.	GALAXY
REIZ 3665	13 48 18.	+ 35 23	42	15.7	GALAXY
UGC 08752	13 48 18.	+ 35 23	60	16.0	GALAXY Sb-c
MCG-05-33-023	13 48 18.	- 30 02	36	15.5	GALAXY
RRS 4.064	13 48 21.	+ 26 30		19.22	BLUE OBJECT
MCG+05-33-009	13 48 21.	+ 28 24 30.	48	14.9	GALAXY
RRS 3.073	13 48 21.	+ 29 22			BLUE OBJECT
MCG+06-30-096	13 48 21.	+ 33 56	30	13.5	GALAXY
MCG+06-30-097	13 48 21.	+ 33 56 30.	6	17.	GALAXY
RRS 3.074	13 48 22.	+ 30 27		17.55	BLUE OBJECT
HOLM 547B	13 48 23.	+ 25 27	12	14.8	PART OF MULTIPLE GALAXY
RRS 3.075	13 48 23.	+ 28 48		18.08	BLUE OBJECT
RRS 3.076	13 48 23.	+ 31 01		19.15	BLUE OBJECT
HOLM 548B	13 48 23.	+ 33 59	24	14.8	PART OF MULTIPLE GALAXY
ZWG 102.068	13 48 24.	+ 15 14		15.5	GALAXY
ZWG 132.020	13 48 24.	+ 25 25		15.4	GALAXY
UGC 08753	13 48 24.	+ 25 25	66	15.4	GALAXY Sb-c
MCG+04-33-011	13 48 24.	+ 25 26	60	15.4	GALAXY
ZWG 190.064	13 48 24.	+ 35 17		14.8	GALAXY
UGC 08754	13 48 24.	+ 35 17	60	14.8	GALAXY E-S0
MCG+06-30-098	13 48 24.	+ 35 21 30.	60	15.	GALAXY
MCG+06-30-099	13 48 24.	+ 35 51	30	17.	GALAXY
ZC 1348.4+6935	13 48 24.	+ 69 35	1080		CLUSTER OF GALAXIES
MCG-06-30-028	13 48 24.	- 37 24	78	14.5	GALAXY
RNGC 5310	13 48 25.	+ 00 08			GALAXY
HOLM 547A	13 48 25.	+ 25 26	48	14.4	PART OF MULTIPLE GALAXY
REIZ 3666	13 48 25.	+ 33 57	36	13.7	GALAXY
HOLM 548A	13 48 25.	+ 33 57	24	13.8	PART OF MULTIPLE GALAXY
RRS 3.077	13 48 26.	+ 29 29		17.37	BLUE OBJECT
SC 1345-3039.2	13 48 26.	- 30 54 07.	84		NEBULA
MCG+06-30-100	13 48 27.	+ 35 16	30	15.	GALAXY
HN 1729	13 48 27.8	+ 00 05 47.	18		NEBULA
RRS 4.065	13 48 28.	+ 25 43		17.57	BLUE OBJECT
RRS 4.066	13 48 28.	+ 26 57		19.07	BLUE OBJECT
ARC 1806	13 48 29.	+ 21 34		18.0	RICH CLUSTER OF GALAXIES
RNGC 5321	13 48 29.	+ 33 53		15.5	GALAXY
ZWG 017.074	13 48 30.	+ 00 05		15.3	GALAXY
KARA.73B 0601	13 48 30.	+ 00 05	72	15.3	ISOLATED GALAXY S
ZC 1348.5+1341	13 48 30.	+ 13 41	740		CLUSTER OF GALAXIES
UGC 08755	13 48 30.	+ 17 18	66	16.0	GALAXY Sb
MCG+06-30-101	13 48 30.	+ 33 51	36	15.	GALAXY
ZWG 190.065	13 48 30.	+ 33 53		15.3	GALAXY
MCG+06-30-102	13 48 30.	+ 35 50	18	16.	GALAXY
MCG+07-28-077	13 48 30.	+ 39 56	72	14.5	GALAXY
MCG+07-28-078	13 48 30.	+ 40 31	36	15.	GALAXY
ZC 1348.5+4234	13 48 30.	+ 42 34	870		CLUSTER OF GALAXIES
ZWG 219.003	13 48 30.	+ 42 47		14.5	GALAXY
ZWG 218.058	13 48 30.	+ 42 47		14.5	GALAXY
UGC 08756	13 48 30.	+ 42 47	114	14.5	GALAXY S0-a
MCG+07-28-079	13 48 30.	+ 42 47	90	14.5	GALAXY
MCG+08-25-045	13 48 30.	+ 49 39	30	16.	GALAXY
UGC 08757	13 48 30.	+ 56 17	60	16.0	GALAXY
MCG+12-13-016	13 48 30.	+ 74 29	84	15.	GALAXY
MCG-04-32-005	13 48 30.	- 22 49	24	15.	GALAXY
MCG-05-33-024	13 48 30.	- 30 32	36	15.	GALAXY
REIZ 3667	13 48 31.	+ 33 54	18	14.9	GALAXY
REIZ 3662	13 48 32.	- 07 47	24	13.8	GALAXY
RRS 3.078	13 48 33.	+ 28 43		17.34	BLUE OBJECT
MCG+08-25-046	13 48 33.	+ 45 05	36	16.	GALAXY
FATH 1.635	13 48 33.	+ 45 05	19		NEBULA
REIN 7.127	13 48 33.97	+ 40 31 34.6			NEBULA
RRS 3.079	13 48 35.	+ 30 54			BLUE OBJECT
REIN 7.128	13 48 35.53	+ 39 57 11.5			NEBULA
ZWG 045.118	13 48 36.	+ 04 36		15.5	GALAXY
ZWG 073.082	13 48 36.	+ 08 37		15.6	GALAXY
ZWG 073.083	13 48 36.	+ 14 01		15.4	GALAXY
ZC 1348.6+1850	13 48 36.	+ 18 50	400		CLUSTER OF GALAXIES
ZC 1349.6+2135	13 48 36.	+ 21 35	670		CLUSTER OF GALAXIES
ZWG 132.021	13 48 36.	+ 21 47		15.4	GALAXY
UGC 08758	13 48 36.	+ 21 47	72	15.4	GALAXY Sa
ZWG 132.022	13 48 36.	+ 22 14		15.5	GALAXY
UGC 08759	13 48 36.	+ 22 14	72	15.5	GALAXY S0
ZWG 190.066	13 48 36.	+ 38 15		15.4	GALAXY
UGC 08760	13 48 36.	+ 38 15	138	15.4	GALAXY DWRF IR
MCG+06-30-103	13 48 36.	+ 38 28	27	16.	GALAXY
MCG+07-28-081	13 48 36.	+ 38 30	30	16.	GALAXY
MCG+07-28-080	13 48 36.	+ 38 32	45	15.	GALAXY
MCG+07-28-082	13 48 36.	+ 39 48	138	13.5	GALAXY
ZWG 219.004	13 48 36.	+ 39 57		14.6	GALAXY
ZWG 218.059	13 48 36.	+ 39 57		14.6	GALAXY
UGC 08761	13 48 36.	+ 39 57	96	14.6	GALAXY SBb
IC 4336	13 48 36.	+ 39 58			NONSTELLAR OBJECT
ZWG 219.005	13 48 36.	+ 40 31		15.2	GALAXY
ZWG 218.060	13 48 36.	+ 40 31		15.2	GALAXY
ZC 1348.6+4201	13 48 36.	+ 42 01	740		CLUSTER OF GALAXIES
ZC 1348.6+4957	13 48 36.	+ 49 57	1480		CLUSTER OF GALAXIES
MCG+09-23-011	13 48 36.	+ 56 16	18	15.	GALAXY
MCG-05-33-025	13 48 36.	- 30 54	60	15.	GALAXY
RNGC 5325B	13 48 37.	+ 38 30		16.0	GALAXY
RRS 3.080	13 48 38.	+ 29 43		18.05	BLUE OBJECT
RRS 3.081	13 48 38.	+ 30 49			BLUE OBJECT
REIZ 3668	13 48 38.	+ 39 57	42	13.8	GALAXY
REIZ 3669	13 48 38.	+ 40 31	18	14.9	GALAXY
ARC 1802	13 48 38.	- 26 28		17.0	RICH CLUSTER OF GALAXIES
RNGC 5317	13 48 39.	+ 05 14			NON-EXISTENT OBJECT
RRS 4.067	13 48 40.	+ 26 11		17.58	BLUE OBJECT
LB 00690	13 48 40.	+ 28 32 54.		17.4	FAINT BLUE STAR
RRS 3.082	13 48 40.	+ 30 17		18.68	BLUE OBJECT
PK309-04.1	13 48 41.	- 66 08	25		PLANETARY NEBULA
ZC 1348.7+0249	13 48 42.	+ 02 49	4910		CLUSTER OF GALAXIES
ZWG 073.084	13 48 42.	+ 08 44		15.5	GALAXY
ZC 1348.7+1241	13 48 42.	+ 12 41	2690		CLUSTER OF GALAXIES
UGC 08762	13 48 42.	+ 24 19	102	17.	GALAXY
ZWG 132.023	13 48 42.	+ 25 20		14.6	GALAXY
UGC 08763	13 48 42.	+ 25 20	72	14.6	GALAXY DBL SYS
RRS 3.083	13 48 42.	+ 28 33			BLUE OBJECT
TON-N 0742	13 48 42.	+ 30 47		15.5	BLUE STAR
ZC 1348.7+3109	13 48 42.	+ 31 09	610		CLUSTER OF GALAXIES
ZWG 219.006	13 48 42.	+ 39 49		12.9	GALAXY
ZWG 218.061	13 48 42.	+ 39 49		12.9	GALAXY
UGC 08764	13 48 42.	+ 39 49	132	12.9	GALAXY Sa
ARC 1811	13 48 42.	+ 71 17		17.5	RICH CLUSTER OF GALAXIES
REIN 7.129	13 48 42.68	+ 39 49 19.3			NEBULA
RRS 3.084	13 48 43.	+ 30 30		19.41	BLUE OBJECT
HOLM 550B	13 48 43.	+ 38 30	36	14.6	PART OF MULTIPLE GALAXY
RNGC 5326	13 48 44.	+ 39 49		13.5	GALAXY
REIZ 3672	13 48 44.	+ 39 49	54	13.2	GALAXY
REIZ 3671	13 48 45.	+ 38 30	54	14.3	GALAXY
HOLM 550A	13 48 45.	+ 38 32	60	14.1	PART OF MULTIPLE GALAXY

OBJECT NAME	RIGHT ASCEN.	DECLINATION	DIAM.	MAGN.	TYPE OF OBJECT
REIZ 3670	13 48 46.	+ 38 13	96	15.2	GALAXY
RRS 4.068	13 48 47.	+ 26 30			BLUE OBJECT
RRS 4.069	13 48 47.	+ 26 55		18.62	BLUE OBJECT
ZC 1348.8+1525	13 48 48.	+ 15 25	740		CLUSTER OF GALAXIES
SN 1954X	13 48 48.	+ 17 53		18.5	SUPERNOVA
ZWG 132.024	13 48 48.	+ 24 15		15.3	GALAXY
MCG+04-33-012	13 48 48.	+ 24 19	72	16.	GALAXY
ZWG 162.017	13 48 48.	+ 29 49		15.7	GALAXY
MCG+06-30-104	13 48 48.	+ 37 11 30.	24	15.	GALAXY
ZWG 190.067	13 48 48.	+ 37 12		15.1	GALAXY
MCG+06-30-105	13 48 48.	+ 38 16 30.	126	14.	GALAXY
ZWG 219.007	13 48 48.	+ 38 31		15.1	GALAXY
ZWG 218.062	13 48 48.	+ 38 31		15.1	GALAXY
MCG+08-25-047	13 48 48.	+ 49 35	60	16.	GALAXY
ZWG 336.023	13 48 48.	+ 70 58		15.4	GALAXY
RNGC 5325A	13 48 49.	+ 38 31		15.0	GALAXY
RRS 3.085	13 48 51.	+ 30 10		18.25	BLUE OBJECT
IC 0954	13 48 53.	+ 71 25 48.			NONSTELLAR OBJECT
ZWG 132.025	13 48 54.	+ 21 47		15.2	GALAXY
7ZW 525	13 48 54.	+ 71 17			COMPACT GALAXY
ZWG 336.024	13 48 54.	+ 71 25		14.5	GALAXY
UGC 08765	13 48 54.	+ 71 25	72	14.5	GALAXY PECULR
RRS 4.070	13 48 58.	+ 25 56			BLUE OBJECT
REIZ 3673	13 48 58.	+ 29 49	24	14.9	GALAXY
VDB.66G 183	13 49	+ 38 17	130		DWARF GALAXY
LB 09881	13 49	- 82 21		13.3	FAINT BLUE STAR
ZC 1349.0+4525	13 49 00.	+ 45 25	1410		CLUSTER OF GALAXIES
7ZW 526	13 49 00.	+ 70 31			COMPACT GALAXY
MCG+12-13-017	13 49 00.	+ 70 59	12	16.	GALAXY
7ZW 527	13 49 00.	+ 71 24			COMPACT GALAXY
MCG+12-13-018	13 49 00.	+ 71 25	30	15.	GALAXY
RRS 4.071	13 49 01.	+ 26 30			BLUE OBJECT
HOLM 549A	13 49 02.	+ 14 20	60	13.6	PART OF MULTIPLE GALAXY
IC 0944	13 49 04.	+ 14 20 57.			NONSTELLAR OBJECT
HOLM 549B	13 49 04.	+ 14 21	36	15.4	PART OF MULTIPLE GALAXY
RRS 3.086	13 49 05.	+ 31 30		18.70	BLUE OBJECT
8ZW 1349+12.9	13 49 06.	+ 12 54		16.5	COMPACT GALAXY
ZWG 073.085	13 49 06.	+ 14 20		14.7	GALAXY
UGC 08766	13 49 06.	+ 14 20	102	14.7	GALAXY Sa
KARA.72 400A	13 49 06.	+ 14 20	114	14.7	PART OF DOUBLE GALAXY
MCG+02-35-019	13 49 06.	+ 14 20	96	14.7	GALAXY
ZWG 073.086	13 49 06.	+ 14 21		15.4	GALAXY
MCG+02-35-020	13 49 06.	+ 14 21	36	15.4	GALAXY
KARA.72 400B	13 49 06.	+ 14 21	42	15.4	PART OF DOUBLE GALAXY
ZC 1349.1+2414	13 49 06.	+ 24 14	2150		CLUSTER OF GALAXIES
RRS 4.072	13 49 06.	+ 28 08		17.66	BLUE OBJECT
ZWG 219.008	13 49 06.	+ 43 47		15.6	GALAXY
ZWG 218.063	13 49 06.	+ 43 47		15.6	GALAXY
UGC 08767	13 49 06.	+ 43 47	60	15.6	GALAXY Sc
ZWG 336.025	13 49 06.	+ 68 49		15.5	GALAXY
RRS 3.087	13 49 09.	+ 31 28		18.66	BLUE OBJECT
ZC 1349.2+0940	13 49 12.	+ 09 40	5380		CLUSTER OF GALAXIES
MCG+07-29-001	13 49 12.	+ 43 48	60	15.	GALAXY
ZC 1349.2+5847	13 49 12.	+ 58 47	4910		CLUSTER OF GALAXIES
MCG+12-13-019	13 49 12.	+ 68 47 30.	39	16.	GALAXY
MCG+12-13-020	13 49 12.	+ 71 02	18	17.	GALAXY
ZWG 336.026	13 49 12.	+ 74 12		15.4	GALAXY
RNGC 5344	13 49 13.	+ 74 12		15.5	GALAXY
MCG+07-29-002	13 49 15.	+ 40 27	42	15.	GALAXY
ZWG 045.119	13 49 18.	+ 03 42		15.3	GALAXY
MCG+06-30-106	13 49 18.	+ 34 47 30.	18	15.	GALAXY
ZWG 219.009	13 49 18.	+ 40 27		15.4	GALAXY
ZWG 218.064	13 49 18.	+ 40 27		15.4	GALAXY
MRK 462	13 49 18.	+ 40 27	9	16.	GALAXY WITH UV CONTINUUM
MCG+10-20-040	13 49 18.	+ 62 07	60	15.	GALAXY
ZWG 017.075	13 49 18.	- 01 52		15.7	GALAXY
RRS 3.088	13 49 19.	+ 28 34		17.02	BLUE OBJECT
RRS 3.089	13 49 21.	+ 29 14		16.35	BLUE OBJECT
RRS 4.073	13 49 22.	+ 27 27		18.69	BLUE OBJECT
ZWG 017.076	13 49 24.	+ 00 47		15.6	GALAXY
ZWG 017.077	13 49 24.	+ 02 05		15.1	GALAXY
ZWG 102.069	13 49 24.	+ 17 11		15.7	GALAXY
ZWG 162.018	13 49 24.	+ 29 35		15.4	GALAXY
ZC 1349.4+2945	13 49 24.	+ 29 45	4500		CLUSTER OF GALAXIES
ZWG 162.019	13 49 24.	+ 32 21		15.6	GALAXY
ZWG 219.010	13 49 24.	+ 44 03		15.2	GALAXY
ZWG 218.065	13 49 24.	+ 44 03		15.2	GALAXY
MRK 276	13 49 24.	+ 44 05	15	15.	GALAXY WITH UV CONTINUUM
RNGC 5324	13 49 24.	- 05 48		12.5	GALAXY
MCG-01-35-016	13 49 24.	- 05 48	132	12.	GALAXY
RRS 4.074	13 49 25.	+ 26 13		17.86	BLUE OBJECT
REIZ 3674	13 49 25.	- 05 48	72	12.5	GALAXY
REIZ 3675	13 49 26.	+ 03 42	24	15.2	GALAXY
MCG-02-35-017	13 49 27.	- 10 21 30.	30	17.	GALAXY
ZWG 017.080	13 49 30.	+ 02 15		15.1	GALAXY
ZWG 045.120	13 49 30.	+ 03 37		15.6	GALAXY
ZWG 073.087	13 49 30.	+ 09 46		15.3	GALAXY
UGC 08769	13 49 30.	+ 09 46	78	15.3	GALAXY E-S0
ZWG 295.019	13 49 30.	+ 62 05		15.5	GALAXY
UGC 08770	13 49 30.	+ 62 05	60	15.5	GALAXY S0
ZWG 017.079	13 49 30.	- 01 53		15.3	GALAXY
ZWG 017.078	13 49 30.	- 01 58		14.2	GALAXY
UGC 08768	13 49 30.	- 01 58	126	14.2	GALAXY SBb
MCG+00-35-021	13 49 30.	- 01 58	108	14.2	GALAXY
MCG-02-35-018	13 49 30.	- 10 22	24	15.	GALAXY
MCG-06-30-029	13 49 30.	- 33 53 30.	36	15.5	GALAXY
RNGC 5327	13 49 31.	- 01 58		14.0	GALAXY
RRS 3.090	13 49 34.	+ 28 42		18.39	BLUE OBJECT
ZWG 045.121	13 49 36.	+ 02 34		14.4	GALAXY
UGC 08771	13 49 36.	+ 02 34	96	14.4	GALAXY E
MCG+01-35-044	13 49 36.	+ 02 34	60	14.4	GALAXY
ZC 1349.6+1317	13 49 36.	+ 13 17	2020		CLUSTER OF GALAXIES
ZWG 073.088	13 49 36.	+ 14 12		15.6	GALAXY
ZC 1349.6+1510	13 49 36.	+ 15 10	670		CLUSTER OF GALAXIES
ZC 1349.6+2447	13 49 36.	+ 24 47	610		CLUSTER OF GALAXIES
ZC 1349.6+4035	13 49 36.	+ 40 35	810		CLUSTER OF GALAXIES
MCG+10-20-041	13 49 36.	+ 60 05 30.	66	14.0	GALAXY
ZC 1349.6+6416	13 49 36.	+ 64 16	3290		CLUSTER OF GALAXIES
MCG-02-35-019	13 49 36.	- 10 21 30.	36	16.	GALAXY
MCG-02-35-020	13 49 36.	- 10 22	30	16.	GALAXY
RNGC 5329	13 49 38.	+ 02 34		14.5	GALAXY
REIZ 3676	13 49 38.	+ 02 34	12	14.5	GALAXY
ZWG 017.081	13 49 42.	+ 02 20		15.7	GALAXY
VV 253B	13 49 42.	+ 02 20	30	15.	INTERACTING GALAXY
VV 253A	13 49 42.	+ 02 20	42	15.	INTERACTING GALAXY
ZWG 045.122	13 49 42.	+ 02 36		15.7	GALAXY
ZWG 045.123	13 49 42.	+ 02 39		15.7	GALAXY
ARC 1808	13 49 42.	+ 09 40		17.2	RICH CLUSTER OF GALAXIES

OBJECT NAME	RIGHT ASCEN.	DECLINATION	DIAM.	MAGN.	TYPE OF OBJECT
ZWG 073.089	13 49 42.	+ 14 21		14.5	GALAXY
UGC 08772	13 49 42.	+ 14 21	60	14.5	GALAXY Sa
MCG+02-35-021	13 49 42.	+ 14 21	60	14.5	GALAXY
MCG+03-35-030	13 49 42.	+ 17 12	24	14.1	GALAXY
ZWG 102.070	13 49 42.	+ 17 13		14.1	GALAXY
RNGC 5332	13 49 42.	+ 17 13		14.0	GALAXY
UGC 08773	13 49 42.	+ 17 13	60	14.1	GALAXY E-S0
ZWG 162.020	13 49 42.	+ 29 18		15.6	GALAXY
MCG+05-33-010	13 49 42.	+ 29 19	78	15.6	GALAXY
MCG+08-25-048	13 49 42.	+ 46 37	9	16.	GALAXY
MCG-05-33-026	13 49 42.	- 27 40	24	15.5	GALAXY
ARC 1807	13 49 43.	- 09 30		18.0	RICH CLUSTER OF GALAXIES
REIZ 3678	13 49 44.	+ 02 21	18	14.8	GALAXY
REIZ 3677	13 49 44.	+ 02 21	18	14.7	GALAXY
IC 0946	13 49 44.	+ 14 22 11.			NONSTELLAR OBJECT
IC 4337	13 49 46.	+ 14 31 59.			NONSTELLAR OBJECT
RNGC 5342	13 49 46.	+ 60 07		14.5	GALAXY
REIZ 3688	13 49 46.	+ 60 07	12	14.0	GALAXY
REIN 2.201A	13 49 46.16	+ 60 06 39.7			NEBULA
REIN 2.201B	13 49 46.38	+ 60 06 39.2			NEBULA
ZWG 017.082	13 49 48.	+ 02 20		14.3	GALAXY
UGC 08774	13 49 48.	+ 02 20	66	14.3	GALAXY DBL SYS
KARA.72 401A	13 49 48.	+ 02 20	48	14.3	PART OF DOUBLE GALAXY
MCG+00-35-022	13 49 48.	+ 02 20	66	14.3	GALAXY
ZWG 045.124	13 49 48.	+ 02 35		15.6	GALAXY
FAI 1349+10	13 49 48.	+ 10 13			COMPACT GALAXY
FAI 1349+14	13 49 48.	+ 14 02			COMPACT GALAXY
ZWG 073.090	13 49 48.	+ 14 15		15.5	GALAXY
ZWG 102.071	13 49 48.	+ 14 31		15.2	GALAXY
MCG+03-35-031	13 49 48.	+ 17 31	42	15.4	GALAXY
ZWG 102.072	13 49 48.	+ 17 32		15.4	GALAXY
KARA.72 401B	13 49 48.	+ 20 21	42		PART OF DOUBLE GALAXY
ZWG 272.011	13 49 48.	+ 51 15		14.4	GALAXY
UGC 08775	13 49 48.	+ 51 15	84	14.4	GALAXY Sc
ZWG 295.020	13 49 48.	+ 60 07		14.4	GALAXY
UGC 08776	13 49 48.	+ 60 07	60	14.4	GALAXY S0
ZC 1349.8-0046	13 49 48.	- 00 46	1550		CLUSTER OF GALAXIES
RNGC 5331	13 49 50.	+ 02 20		14.5	GALAXY
REIZ 3679	13 49 50.	+ 02 36	12	15.8	GALAXY
RRS 4.075	13 49 50.	+ 26 51		18.98	BLUE OBJECT
HOLM 551B	13 49 51.	+ 13 49	36	15.6	PART OF MULTIPLE GALAXY
REIZ 3680	13 49 51.	+ 21 48	48	15.2	GALAXY
IC 0949	13 49 51.	+ 22 47 12.			NONSTELLAR OBJECT
MCG+04-33-014	13 49 51.	+ 25 19 30.	12	14.6	GALAXY
MCG+04-33-013	13 49 51.	+ 25 20	36	14.6	GALAXY
KLEM 28	13 49 51.	- 28 12	900	11.	GROUP OF 7 GALAXIES
ZWG 045.125	13 49 54.	+ 08 03		15.2	GALAXY
ZWG 073.091	13 49 54.	+ 13 49		14.9	GALAXY
MCG+02-35-022	13 49 54.	+ 13 49	48	14.9	GALAXY
MCG+03-35-032	13 49 54.	+ 14 43	60	15.2	GALAXY
ZWG 132.026	13 49 54.	+ 22 46		15.4	GALAXY
UGC 08777	13 49 54.	+ 22 46	72	15.4	GALAXY S-IRR
TON-N 0172	13 49 54.	+ 30 29		15.0	BLUE STAR
ZWG 191.001	13 49 54.	+ 38 19		14.8	GALAXY
ZWG 190.068	13 49 54.	+ 38 19		14.8	GALAXY
UGC 08778	13 49 54.	+ 38 19	78	14.8	GALAXY S
MCG-01-35-017	13 49 54.	- 07 37	42	15.	GALAXY
HOLM 551A	13 49 56.	+ 13 49	54	13.7	PART OF MULTIPLE GALAXY
RRS 3.091	13 49 56.	+ 30 28		16.50	BLUE OBJECT
HOLM 553A	13 49 57.	+ 31 43	36	14.4	PART OF MULTIPLE GALAXY
REIZ 3683	13 49 57.	+ 38 18	48	14.0	GALAXY
IC 0951	13 49 57.	+ 51 13 13.			NONSTELLAR OBJECT
SHB 279	13 49 58.	+ 02 47 42.		19.	QUASI-STELLAR OBJECT
BC LW1349+02	13 49 58.	+ 02 48			QUASI-STELLAR OBJECT
IC 0948	13 49 59.	+ 14 21 42.			NONSTELLAR OBJECT
RNGC 5328	13 49 59.	- 28 14		13.0	GALAXY
LB 09882	13 50	- 83 06		12.7	FAINT BLUE STAR
ZWG 017.083	13 50 00.	+ 02 11		15.2	GALAXY
ZWG 045.126	13 50 00.	+ 08 06		15.7	GALAXY
FAI 13500+14	13 50 00.	+ 14 20			COMPACT GALAXY
ZWG 073.092	13 50 00.	+ 14 20		14.4	GALAXY
UGC 08779	13 50 00.	+ 14 20	72	14.4	GALAXY E
MCG+02-35-023	13 50 00.	+ 14 20	60	14.4	GALAXY
ZWG 102.073	13 50 00.	+ 14 44		15.2	GALAXY
UGC 08780	13 50 00.	+ 14 44	84	15.2	GALAXY
REIZ 3681	13 50 00.	+ 17 56	30	16.2	GALAXY
ZWG 132.027	13 50 00.	+ 21 47		14.5	GALAXY
UGC 08781	13 50 00.	+ 21 47	102	14.5	GALAXY SBb
ZWG 132.028	13 50 00.	+ 21 50		15.5	GALAXY
ZC 1350.0+2646	13 50 00.	+ 26 46	1880		CLUSTER OF GALAXIES
ZWG 162.021	13 50 00.	+ 31 42		15.6	GALAXY
UGC 08782	13 50 00.	+ 31 42	66	15.6	GALAXY
ZC 1350.0+3148	13 50 00.	+ 31 48	1080		CLUSTER OF GALAXIES
MCG+06-31-001	13 50 00.	+ 38 19 30.	48	14.5	GALAXY
MCG+08-25-049	13 50 00.	+ 48 22	18	18.	GALAXY
MCG+08-25-050	13 50 00.	+ 48 22 30.	18	18.	GALAXY
ACK 310+03.1	13 50 00.	- 58 43			PLANETARY NEBULA
IC 0950	13 50 01.	+ 14 45 12.			NONSTELLAR OBJECT
RRS 4.076	13 50 02.	+ 28 05		18.56	BLUE OBJECT
MCG+05-33-011	13 50 03.	+ 27 57	24	15.5	GALAXY
IC 0947	13 50 04.	+ 01 04 18.			NONSTELLAR OBJECT
RRS 3.092	13 50 04.	+ 30 57		18.70	BLUE OBJECT
RNGC 5336	13 50 04.	+ 43 29		13.5	GALAXY
REIZ 3686	13 50 04.	+ 43 30	42	13.9	GALAXY
RRS 3.093	13 50 05.	+ 29 55		18.87	BLUE OBJECT
HOLM 553B	13 50 05.	+ 31 43	12	15.5	PART OF MULTIPLE GALAXY
ZWG 017.085	13 50 06.	+ 01 03		14.2	GALAXY
UGC 08784	13 50 06.	+ 01 03	102	14.2	GALAXY E-S0
MCG+00-35-023	13 50 06.	+ 01 03	48	14.2	GALAXY
ZWG 045.127	13 50 06.	+ 03 43		15.6	GALAXY
ZC 1350.1+0524	13 50 06.	+ 05 24	1810		CLUSTER OF GALAXIES
MCG+04-33-015	13 50 06.	+ 22 46	66	15.4	GALAXY
ZWG 219.011	13 50 06.	+ 43 29		13.6	GALAXY
ZWG 218.066	13 50 06.	+ 43 29		13.6	GALAXY
UGC 08785	13 50 06.	+ 43 29	96	13.6	GALAXY Sc
MCG+07-29-003	13 50 06.	+ 43 30	78	13.	GALAXY
MCG+09-23-012	13 50 06.	+ 51 11 30.	60	13.	GALAXY
ZWG 017.084	13 50 06.	- 01 46		15.4	GALAXY
UGC 08783	13 50 06.	- 01 46	60	15.4	GALAXY Sb
MCG-05-33-027	13 50 06.	- 29 41	72	15.	GALAXY
IC 4338	13 50 07.	- 00 52 54.			SAME AS NGC 5334
MCG+05-33-012	13 50 09.	+ 31 43	51	15.6	GALAXY
RRS 3.094	13 50 11.	+ 28 37		17.82	BLUE OBJECT
RRS 3.095	13 50 11.	+ 31 38		18.44	BLUE OBJECT
ZWG 017.086	13 50 12.	+ 00 22		15.4	GALAXY
MCG+04-33-016	13 50 12.	+ 21 46 30.	90	14.5	GALAXY
MCG+07-29-004	13 50 12.	+ 39 55	96	13.5	GALAXY
MCG-05-33-028	13 50 12.	- 28 14	60	14.5	GALAXY

OBJECT NAME	RIGHT ASCEN.	DECLINATION	DIAM.	MAGN.	TYPE OF OBJECT
BRS 4.077	13 50 13.	+ 26 46		18.22	BLUE OBJECT
HOLM 552A	13 50 14.	+ 02 30	90	14.3	PART OF MULTIPLE GALAXY
REIZ 3687	13 50 14.	+ 39 56	54	13.4	GALAXY
BRS 4.078	13 50 15.	+ 26 21		18.18	BLUE OBJECT
REIN 7.130	13 50 15.49	+ 39 55 59.8			NEBULA
BRS 4.079	13 50 17.	+ 26 16			BLUE OBJECT
RNGC 5330	13 50 17.	- 28 13		15.0	GALAXY
ZWG 045.128	13 50 18.	+ 02 30		15.2	GALAXY
UGC 08787	13 50 18.	+ 02 30	114	15.2	GALAXY Sb-c
MCG+01-35-045	13 50 18.	+ 02 30	78	15.2	GALAXY
FAI 13503+14	13 50 18.	+ 14 16			COMPACT GALAXY
ZWG 132.029	13 50 18.	+ 21 53		15.7	GALAXY
ZWG 132.030	13 50 18.	+ 25 00		14.9	GALAXY
UGC 08788	13 50 18.	+ 25 00	78	14.9	GALAXY S0-a
ZWG 219.012	13 50 18.	+ 39 56		13.4	GALAXY
UGC 08789	13 50 18.	+ 39 56	108	13.4	GALAXY S
ZWG 017.087	13 50 18.	- 01 39		15.1	GALAXY
UGC 08786	13 50 18.	- 01 39	114	15.1	GALAXY Sa-b
MCG-04-33-006	13 50 18.	- 24 27	36	15.	GALAXY
MCG-05-33-028A	13 50 18.	- 28 13	12	15.	GALAXY
RNGC 5334	13 50 19.	- 00 52		12.0	GALAXY
REIZ 3682	13 50 20.	+ 02 30	72	15.0	GALAXY
RNGC 5337	13 50 20.	+ 39 56		13.5	GALAXY
BRS 4.080	13 50 23.	+ 27 03		18.84	BLUE OBJECT
ZWG 045.129	13 50 24.	+ 03 04		14.5	GALAXY
UGC 08791	13 50 24.	+ 03 04	156	14.5	GALAXY SBb
MCG+01-35-046	13 50 24.	+ 03 04	120	14.5	GALAXY
BZW 1350+12.3	13 50 24.	+ 12 18		17.4	COMPACT GALAXY
MCG+04-33-017	13 50 24.	+ 24 58 30.	48	14.9	GALAXY
ZWG 132.031	13 50 24.	+ 25 17		15.3	GALAXY
ZWG 191.002	13 50 24.	+ 38 04		14.1	GALAXY
ZWG 190.069	13 50 24.	+ 38 04		14.1	GALAXY
UGC 08792	13 50 24.	+ 38 04	84	14.1	GALAXY S
MCG+06-31-002	13 50 24.	+ 38 04	72	14.	GALAXY
MCG+07-29-005	13 50 24.	+ 38 57	84	15.	GALAXY
ZWG 317.009	13 50 24.	+ 64 37		15.7	GALAXY
7ZW 528	13 50 24.	+ 64 38			COMPACT GALAXY
ZWG 017.088	13 50 24.	- 00 52		13.7	GALAXY
UGC 08790	13 50 24.	- 00 52	270	13.7	GALAXY SBc
MCG+00-35-024	13 50 24.	- 00 52	240	13.7	GALAXY
MCG-03-35-020	13 50 24.	- 16 42	60	15.	GALAXY
OCL 0913	13 50 24.	- 61 37	1860	9.1	OPEN STAR CLUSTER
VHA 154	13 50 24.	- 61 37	480		OPEN STAR CLUSTER
HOLM 552B	13 50 25.	+ 02 30	54	14.8	PART OF MULTIPLE GALAXY
RNGC 5341	13 50 25.	+ 38 04		14.0	GALAXY
REIZ 3685	13 50 26.	+ 02 30	60	15.1	GALAXY
RNGC 5335	13 50 26.	+ 03 04		14.5	GALAXY
RNGC 5316	13 50 26.	- 61 37		9.0	OPEN CLUSTER
ABC 1810	13 50 27.	+ 36 31		17.2	RICH CLUSTER OF GALAXIES
BRS 3.096	13 50 29.	+ 29 39		19.52	BLUE OBJECT
ZWG 045.130	13 50 30.	+ 02 30		15.4	GALAXY
MCG+01-35-047	13 50 30.	+ 02 30	60	15.4	GALAXY
ZWG 102.074	13 50 30.	+ 15 12		15.6	GALAXY
ZWG 132.032	13 50 30.	+ 20 37		15.6	GALAXY
UGC 08793	13 50 30.	+ 38 56	114	16.0	GALAXY Sc
MRK 277	13 50 30.	+ 64 37	7	16.5	GALAXY WITH UV CONTINUUM
MCG+11-17-009	13 50 30.	+ 64 37	39	16.	GALAXY
MCG-05-33-029	13 50 30.	- 28 10	42	16.	GALAXY
REIZ 3684	13 50 32.	- 07 42	72	13.2	GALAXY
BRS 4.081	13 50 34.	+ 27 06		19.12	BLUE OBJECT
REIZ 3690	13 50 34.	+ 38 01	48	13.9	GALAXY
MCG+03-35-033	13 50 36.	+ 16 04	36	14.9	GALAXY
ZWG 102.075	13 50 36.	+ 16 05		14.9	GALAXY
ZC 1350.6+2053	13 50 36.	+ 20 53	270		CLUSTER OF GALAXIES
ZWG 132.033	13 50 36.	+ 21 10		15.4	GALAXY
UGC 08794	13 50 36.	+ 21 10	126	15.4	GALAXY Sb
ZC 1350.6+3239	13 50 36.	+ 32 39	670		CLUSTER OF GALAXIES
ZC 1350.6+3630	13 50 36.	+ 36 30	1010		CLUSTER OF GALAXIES
ZWG 191.003	13 50 36.	+ 37 45		15.5	GALAXY
ZWG 190.070	13 50 36.	+ 37 45		15.5	GALAXY
UGC 08795	13 50 36.	+ 37 45	84	15.5	GALAXY Sc
ZWG 246.024	13 50 36.	+ 46 36		15.7	GALAXY
MCG-03-35-021	13 50 36.	- 19 52 30.	9	15.5	GALAXY
REIZ 3697	13 50 37.	+ 61 23	42	15.2	GALAXY
BRS 4.082	13 50 38.	+ 26 12		18.58	BLUE OBJECT
MCG+06-31-003	13 50 39.	+ 37 44 30.	66	14.5	GALAXY
BRS 4.083	13 50 41.	+ 28 08			BLUE OBJECT
ZWG 017.089	13 50 42.	+ 00 40		15.5	GALAXY
ZWG 045.131	13 50 42.	+ 05 13		15.7	GALAXY
UGC 08796	13 50 42.	+ 05 13	60	15.7	GALAXY Sc
ZC 1350.7+1142	13 50 42.	+ 11 42	1280		CLUSTER OF GALAXIES
ZWG 102.076	13 50 42.	+ 14 54		15.0	GALAXY
MCG+03-35-034	13 50 42.	+ 14 54	30	15.0	GALAXY
ZWG 102.077	13 50 42.	+ 17 35		14.8	GALAXY
BZW 1350+17.6	13 50 42.	+ 17 35		14.8	COMPACT GALAXY
MCG+03-35-035	13 50 42.	+ 17 35	18	14.8	GALAXY
MCG+04-33-018	13 50 42.	+ 21 09	54	15.4	GALAXY
ZWG 132.034	13 50 42.	+ 24 48		15.6	GALAXY
UGC 08797	13 50 42.	+ 24 48	66	15.6	GALAXY S
ZWG 162.022	13 50 42.	+ 32 03		15.4	GALAXY
UGC 08798	13 50 42.	+ 44 04	108	16.0	GALAXY
MCG+08-25-051	13 50 42.	+ 46 36	12	16.	GALAXY
MCG+10-20-042	13 50 42.	+ 59 55	24	16.	GALAXY
7ZW 529	13 50 42.	+ 62 40			COMPACT GALAXY
ZWG 317.010	13 50 42.	+ 67 48		15.5	GALAXY
BRS 4.084	13 50 44.	+ 28 09		18.80	BLUE OBJECT
ABC 1809	13 50 46.	+ 05 25		15.8	RICH CLUSTER OF GALAXIES
BRS 3.097	13 50 47.	+ 30 19			BLUE OBJECT
ZWG 017.090	13 50 48.	+ 02 17		15.4	GALAXY
UGC 08799	13 50 48.	+ 06 00	78	16.5	GALAXY DWARF
ZWG 132.035	13 50 48.	+ 24 37		15.5	GALAXY
1ZW 074	13 50 48.	+ 35 57			COMPACT GALAXY
ZWG 191.004	13 50 48.	+ 37 37		15.7	GALAXY
ZWG 190.071	13 50 48.	+ 37 37		15.7	GALAXY
MCG+07-29-006	13 50 48.	+ 44 05	84	16.	GALAXY
MCG-06-31-001	13 50 48.	- 33 42	30	16.	GALAXY
BRS 3.098	13 50 49.	+ 28 33		18.50	BLUE OBJECT
REIZ 3705	13 50 49.	+ 61 25	60	14.7	GALAXY
BRS 4.085	13 50 51.	+ 26 51			BLUE OBJECT
ZWG 045.132	13 50 54.	+ 05 27		14.3	GALAXY
UGC 08800	13 50 54.	+ 05 27	168	14.3	GALAXY SB:0
MCG+01-35-048	13 50 54.	+ 05 27	120	14.3	GALAXY
BIGO 533	13 50 54.	+ 05 33			NEBULA
ZC 1350.9+0709	13 50 54.	+ 07 09	1210		CLUSTER OF GALAXIES
ZC 1350.9+2142	13 50 54.	+ 21 42	5710		CLUSTER OF GALAXIES
TON-N 0173	13 50 54.	+ 26 10		16.1	BLUE STAR
MCG+06-31-004	13 50 54.	+ 35 57	60	17.	GALAXY
ABC 1812	13 50 54.	+ 37 58		17.0	RICH CLUSTER OF GALAXIES

OBJECT NAME	RIGHT ASCEN.	DECLINATION	DIAM.	MAGN.	TYPE OF OBJECT
MCG+07-29-007	13 50 54.	+ 39 49	108	14.	GALAXY
ZWG 219.013	13 50 54.	+ 39 58		15.6	GALAXY
ZC 1350.9+4638	13 50 54.	+ 46 38	3630		CLUSTER OF GALAXIES
PK309-04.2	13 50 54.	- 66 18	5	13.0	PLANETARY NEBULA
REIN 7.131A	13 50 54.44	+ 39 49 35.9			NEBULA
REIN 7.131B	13 50 54.48	+ 39 49 35.9			NEBULA
REIZ 3689	13 50 55.	+ 05 26	15	14.7	GALAXY
BRS 3.099	13 50 55.	+ 30 40			BLUE OBJECT
RNGC 5315	13 50 55.	- 66 18		13.0	PLANETARY NEBULA
REIZ 3692	13 50 56.	+ 39 49	48	15.0	GALAXY
RNGC 5338	13 50 57.	+ 05 27		14.5	GALAXY
SN 1954Q	13 50 59.	+ 09 58		18.5	SUPERNOVA
ZWG 073.093	13 51 00.	+ 11 35		15.5	GALAXY
MCG+04-33-019	13 51 00.	+ 21 30	48	15.5	GALAXY
ZWG 132.036	13 51 00.	+ 23 17		15.7	GALAXY
KARA.73B 0602	13 51 00.	+ 23 17	24	15.7	ISOLATED GALAXY E
ZC 1351.0+2731	13 51 00.	+ 27 31	670		CLUSTER OF GALAXIES
ZWG 191.005	13 51 00.	+ 35 50		15.6	GALAXY
UGC 08802	13 51 00.	+ 35 58	78	17.	GALAXY Sc
ZWG 191.006	13 51 00.	+ 38 08		15.1	GALAXY
ZWG 190.072	13 51 00.	+ 38 08		15.1	GALAXY
HOLM 554B	13 51 00.	+ 38 08	54	14.4	PART OF MULTIPLE GALAXY
UGC 08803	13 51 00.	+ 38 08	108	15.1	GALAXY SBb
MCG+06-31-005	13 51 00.	+ 38 08	102	15.	GALAXY
ZWG 219.014	13 51 00.	+ 39 49		14.9	GALAXY
UGC 08804	13 51 00.	+ 39 49	144	14.9	GALAXY Sc
ZWG 017.092	13 51 00.	- 00 49		15.1	GALAXY
ZWG 017.091	13 51 00.	- 00 58		15.0	GALAXY
UGC 08801	13 51 00.	- 00 58	78	15.0	GALAXY S
MCG+00-35-025	13 51 00.	- 00 58	78	15.0	GALAXY
BRS 4.086	13 51 01.	+ 26 28		18.38	BLUE OBJECT
RNGC 5349	13 51 01.	+ 38 08		15.0	GALAXY
RNGC 5346	13 51 02.	+ 39 49		15.0	GALAXY
REIZ 3704	13 51 02.	+ 54 51	30	14.4	GALAXY
ZWG 017.093	13 51 06.	+ 00 18		15.1	GALAXY
ZC 1351.1+1345	13 51 06.	+ 13 45	1140		CLUSTER OF GALAXIES
BRS 4.087	13 51 06.	+ 28 37		16.82	BLUE OBJECT
RNGC 5347	13 51 06.	+ 33 44		13.0	GALAXY
ZWG 191.007	13 51 06.	+ 33 45		13.3	GALAXY
UGC 08805	13 51 06.	+ 33 45	102	13.3	GALAXY SBa-b
ZWG 219.015	13 51 06.	+ 38 28		15.0	GALAXY
UGC 08806	13 51 06.	+ 38 28	120	15.0	GALAXY Sb
ZWG 219.016	13 51 06.	+ 41 02		15.4	GALAXY
UGC 08807	13 51 06.	+ 41 02	60	15.4	GALAXY DBL SYS
MCG+07-29-008	13 51 06.	+ 41 02	63	15.	GALAXY
7ZW 530	13 51 06.	+ 76 24			COMPACT GALAXY
REIZ 3694	13 51 07.	+ 33 44	78	13.5	GALAXY
BRS 4.088	13 51 08.	+ 28 36		18.59	BLUE OBJECT
MCG+06-31-007	13 51 09.	+ 33 42 30.	90	12.5	GALAXY
REIZ 3696	13 51 09.	+ 38 07	48	14.4	GALAXY
MCG+06-31-006	13 51 09.	+ 38 29	114	14.5	GALAXY
IC 0952	13 51 10.	+ 03 35 52.			NONSTELLAR OBJECT
RNGC 5339	13 51 11.	- 07 41		12.0	GALAXY
ZWG 045.133	13 51 12.	+ 03 37		14.9	GALAXY
UGC 08808	13 51 12.	+ 03 37	84	14.9	GALAXY SBb-c
MCG+01-35-049	13 51 12.	+ 03 37	72	14.9	GALAXY
ZC 1351.2+1544	13 51 12.	+ 15 44	870		CLUSTER OF GALAXIES
ZWG 132.037	13 51 12.	+ 25 07		15.5	GALAXY
MCG+07-29-009	13 51 12.	+ 40 36	180	12.5	GALAXY
MCG-01-35-018	13 51 12.	- 07 41	90	12.	GALAXY
REIZ 3699	13 51 13.	+ 40 37	72	13.2	GALAXY
HOLM 555C	13 51 13.	+ 40 37	102	13.8	PART OF MULTIPLE GALAXY
REIZ 3691	13 51 14.	+ 03 37	36	14.9	GALAXY
REIN 7.132A	13 51 14.78	+ 40 36 34.2			NEBULA
REIN 7.132B	13 51 14.78	+ 40 36 34.8			NEBULA
BRS 4.089	13 51 15.	+ 26 47			BLUE OBJECT
BRS 4.090	13 51 15.	+ 27 01			BLUE OBJECT
BRS 3.100	13 51 15.	+ 29 08		18.74	BLUE OBJECT
REIZ 3700	13 51 15.	+ 38 10	12	13.4	GALAXY
HOLM 554A	13 51 15.	+ 38 10	168	12.6	PART OF MULTIPLE GALAXY
ABC 1813	13 51 16.	+ 35 47		16.0	RICH CLUSTER OF GALAXIES
BRS 3.101	13 51 17.	+ 29 44		17.64	BLUE OBJECT
ZWG 132.038	13 51 18.	+ 25 04		15.7	GALAXY
ZC 1351.3+3156	13 51 18.	+ 31 56	1080		CLUSTER OF GALAXIES
ZC 1351.3+3333	13 51 18.	+ 33 33	4170		CLUSTER OF GALAXIES
ZWG 191.008	13 51 18.	+ 38 10		13.1	GALAXY
ZWG 190.073	13 51 18.	+ 38 10		13.1	GALAXY
UGC 08809	13 51 18.	+ 38 10	180	13.1	GALAXY Sb
MCG+06-31-008	13 51 18.	+ 38 10	168	12.5	GALAXY
MCG+07-29-010	13 51 18.	+ 40 32	60	13.	GALAXY
MCG+07-29-011	13 51 18.	+ 40 33	42	13.	GALAXY
ZWG 219.017	13 51 18.	+ 40 36		12.4	GALAXY
UGC 08810	13 51 18.	+ 40 36	204	12.4	GALAXY SBb-c
ZC 1351.3+5524	13 51 18.	+ 55 24	1080		CLUSTER OF GALAXIES
ZWG 336.027	13 51 18.	+ 72 58		14.8	GALAXY
UGC 08811	13 51 18.	+ 72 58	102	14.8	GALAXY SBb
MCG-04-33-007	13 51 18.	- 27 23	90	14.	GALAXY
IC 4339	13 51 19.	+ 37 47 05.			NONSTELLAR OBJECT
RNGC 5351	13 51 19.	+ 38 10		13.0	GALAXY
REIZ 3702	13 51 19.	+ 40 32	60	12.7	GALAXY
HOLM 555B	13 51 19.	+ 40 32	48	12.6	PART OF MULTIPLE GALAXY
REIZ 3703	13 51 19.	+ 40 33	30	13.4	GALAXY
HOLM 555A	13 51 19.	+ 40 33	24	13.6	PART OF MULTIPLE GALAXY
REIN 7.133	13 51 19.80	+ 40 32 55.0			NEBULA
REIN 7.134	13 51 19.85	+ 40 31 43.9			NEBULA
RNGC 5333	13 51 20.	- 48 16			UNVERIFIED SOUTHERN OBJECT
RNGC 5353	13 51 21.	+ 40 32		12.5	GALAXY
RNGC 5354	13 51 21.	+ 40 33		12.5	GALAXY
RNGC 5350	13 51 21.	+ 40 36		12.5	GALAXY
RNGC 5342	13 51 23.	- 07 20		14.0	GALAXY
ZWG 191.009	13 51 24.	+ 36 23		14.2	GALAXY
UGC 08812	13 51 24.	+ 36 23	72	14.4	GALAXY E-S0
ZWG 191.010	13 51 24.	+ 37 38		15.0	GALAXY
MCG+06-31-009	13 51 24.	+ 37 38	42	14.5	GALAXY
ZWG 191.011	13 51 24.	+ 37 46		14.9	GALAXY
MCG+06-31-010	13 51 24.	+ 37 46	45	14.	GALAXY
ZWG 219.018	13 51 24.	+ 40 31		11.8	GALAXY
UGC 08813	13 51 24.	+ 40 31	168	11.8	GALAXY S0
ZWG 219.019	13 51 24.	+ 40 32		12.3	GALAXY
UGC 08814	13 51 24.	+ 40 32	150	12.3	GALAXY S0
PATH 1.636	13 51 24.	+ 45 18	5		NEBULA
UGC 08815	13 51 24.	+ 68 42	66	16.5	GALAXY Sb-c
MCG-01-35-020	13 51 24.	- 04 34	90	14.5	GALAXY
MCG-01-35-019	13 51 24.	- 07 20	36	14.	GALAXY
SC 1348-2239.0	13 51 24.	- 22 53 49.	48		NEBULA
LP 00691	13 51 25.	+ 26 06 06.		17.7	FAINT BLUE STAR
BRS 4.091	13 51 25.	+ 26 07			BLUE OBJECT
RNGC 5352	13 51 25.	+ 36 23		14.0	GALAXY

OBJECT NAME	RIGHT ASCEN.	DECLINATION	DIAM.	MAGN.	TYPE OF OBJECT
IC 4340	13 51 25.	+ 37 38 54.			NONSTELLAR OBJECT
IC 4341	13 51 25.	+ 37 47 00.			NONSTELLAR OBJECT
RRS 4.092	13 51 27.	+ 28 02		18.22	BLUE OBJECT
REIZ 3707	13 51 29.	+ 36 23	18	14.2	GALAXY
ZWG 045.134	13 51 30.	+ 04 12		15.4	GALAXY
UGC 08816	13 51 30.	+ 04 12	78	15.4	GALAXY Sc
ZC 1351.5+0414	13 51 30.	+ 04 14	1880		CLUSTER OF GALAXIES
ZC 1351.5+1509	13 51 30.	+ 15 09	1550		CLUSTER OF GALAXIES
MCG+05-33-013	13 51 30.	+ 28 51 30.	48	15.5	GALAXY
MCG+06-31-013	13 51 30.	+ 33 26	30	16.	GALAXY
MCG+06-31-012	13 51 30.	+ 33 27	18	15.5	GALAXY
ZWG 191.012	13 51 30.	+ 33 28		15.3	GALAXY
UGC 08817	13 51 30.	+ 33 28	96	15.3	GALAXY TRP SYS
MCG+06-31-011	13 51 30.	+ 36 22	27	14.5	GALAXY
ZC 1351.5+4103	13 51 30.	+ 41 03	470		CLUSTER OF GALAXIES
MRK 278	13 51 30.	+ 72 58	50	15.	GALAXY WITH UV CONTINUUM
MCG+12-13-021	13 51 30.	+ 72 58	84	15.	GALAXY
REIZ 3695	13 51 31.	+ 04 14	42	14.9	GALAXY
REIZ 3693	13 51 32.	- 07 21	18	13.3	GALAXY
MCG+06-31-016	13 51 33.	+ 33 26 30.	12	18.	GALAXY
MCG+06-31-015	13 51 33.	+ 33 26 30.	24	15.5	GALAXY
MCG+06-31-014	13 51 33.	+ 35 02	24	16.	GALAXY
RRS 4.093	13 51 34.	+ 27 49		18.27	BLUE OBJECT
PK311+03.1	13 51 34.	- 58 11	10		PLANETARY NEBULA
RRS 3.102	13 51 35.	+ 29 43		17.89	BLUE OBJECT
ZWG 045.135	13 51 36.	+ 05 11		15.6	GALAXY
ZWG 045.136	13 51 36.	+ 05 36		15.2	GALAXY
UGC 08818	13 51 36.	+ 05 36	66	15.2	GALAXY Sb-c
MCG+01-35-050	13 51 36.	+ 05 36	48	15.2	GALAXY
FAI 1351+09	13 51 36.	+ 09 18			COMPACT GALAXY
ZWG 073.094	13 51 36.	+ 09 39		15.6	GALAXY
ZWG 103.001	13 51 36.	+ 17 24		15.6	GALAXY
ZWG 102.078	13 51 36.	+ 17 24		15.6	GALAXY
MCG+03-35-036	13 51 36.	+ 17 24	24	15.6	GALAXY
ZC 1351.6+1947	13 51 36.	+ 19 47	2350		CLUSTER OF GALAXIES
ZC 1351.6+3957	13 51 36.	+ 39 57	1140		CLUSTER OF GALAXIES
ZWG 219.020	13 51 36.	+ 40 35		14.0	GALAXY
UGC 08819	13 51 36.	+ 40 35	72	14.0	GALAXY S0?
MCG+07-29-012	13 51 36.	+ 40 35	27	14.	GALAXY
MCG+10-20-043	13 51 36.	+ 61 55	57	15.	GALAXY
MCG-04-33-009	13 51 36.	- 26 39	60	15.	GALAXY
MCG-04-33-008	13 51 36.	- 26 40	60	14.	GALAXY
REIZ 3698	13 51 37.	+ 05 36	24	15.2	GALAXY
REIZ 3709	13 51 37.	+ 33 49	12	14.8	GALAXY
REIZ 3710	13 51 37.	+ 40 35	12	14.5	GALAXY
VV 202A	13 51 38.	+ 33 26 30.	9	15.5	INTERACTING GALAXY
REIN 7.135	13 51 38.87	+ 40 35 03.3			NEBULA
RNGC 5355	13 51 39.	+ 40 35		14.0	GALAXY
HOLM 555D	13 51 39.	+ 40 35	18	14.8	PART OF MULTIPLE GALAXY
SC 1348-2244.0	13 51 39.	- 22 58 48.	30		NEBULA
SC 1348-2340.8	13 51 39.	- 23 55 36.	30		NEBULA
VV 351A	13 51 39.	- 26 20	72	15.	INTERACTING GALAXY
VV 202B	13 51 41.	+ 33 26 00.	9	15.5	INTERACTING GALAXY
ZWG 045.137	13 51 42.	+ 05 29		14.5	GALAXY
UGC 08821	13 51 42.	+ 05 29	210	14.5	GALAXY Sb
MCG+01-35-051	13 51 42.	+ 05 29	216	14.5	GALAXY
ZWG 073.095	13 51 42.	+ 13 50		14.7	GALAXY
MCG+02-35-024	13 51 42.	+ 13 50	60	14.7	GALAXY
ARC 1814	13 51 42.	+ 15 10		17.2	RICH CLUSTER OF GALAXIES
MRK 662	13 51 42.	+ 23 39	9	15.5	GALAXY WITH UV CONTINUUM
ZC 1351.7+2852	13 51 42.	+ 28 52	1480		CLUSTER OF GALAXIES
ZC 1351.7+4320	13 51 42.	+ 43 20	1210		CLUSTER OF GALAXIES
ZWG 017.094	13 51 42.	- 01 11		13.8	GALAXY
UGC 08820	13 51 42.	- 01 11	120	13.8	GALAXY Sa
MCG+00-35-026	13 51 42.	- 01 11	66	13.8	GALAXY
MCG-04-33-010	13 51 42.	- 26 20	72	14.5	GALAXY
REIZ 3701	13 51 43.	+ 05 28	150	13.9	GALAXY
RNGC 5345	13 51 43.	- 01 11		14.0	GALAXY
REIZ 3706	13 51 44.	+ 13 49	48	14.6	GALAXY
RNGC 5348	13 51 45.	+ 05 28		14.5	GALAXY
VV 351B	13 51 45.	- 26 21	54	15.5	INTERACTING GALAXY
ZC 1351.8+1056	13 51 48.	+ 10 56	1340		CLUSTER OF GALAXIES
ZWG 162.023	13 51 48.	+ 31 05		15.2	GALAXY
ZWG 162.024	13 51 48.	+ 31 08		15.5	GALAXY
MCG+07-29-013	13 51 48.	+ 40 32	48	14.5	GALAXY
ZC 1351.8+5617	13 51 48.	+ 56 17	1080		CLUSTER OF GALAXIES
ZWG 295.021	13 51 48.	+ 61 54		15.6	GALAXY
UGC 08822	13 51 48.	+ 61 54	78	15.6	GALAXY S
ZWG 336.028	13 51 48.	+ 69 34		14.5	GALAXY
UGC 08823	13 51 48.	+ 69 34	54	14.5	GALAXY S0
7ZW 531	13 51 48.	+ 78 30			COMPACT GALAXY
ZC 1351.8-0048	13 51 48.	- 00 48	1750		CLUSTER OF GALAXIES
ZWG 017.095	13 51 48.	- 01 18		15.7	GALAXY
MCG-04-33-011	13 51 48.	- 26 19	60	15.	GALAXY
REIZ 3708	13 51 49.	+ 14 44	60	14.8	GALAXY
RRS 4.094	13 51 51.	+ 28 20		19.11	BLUE OBJECT
MCG+06-31-017	13 51 51.	+ 37 37 30.	42	15.	GALAXY
MCG+06-31-018	13 51 51.	+ 37 40	24	15.	GALAXY
MCG-04-33-012	13 51 51.	- 26 17	24	15.	GALAXY
ARC 1815	13 51 53.	+ 04 17		17.8	RICH CLUSTER OF GALAXIES
REIN 7.136	13 51 53.59	+ 40 31 20.7			NEBULA
ZWG 074.001	13 51 54.	+ 08 39		15.7	GALAXY
ZWG 073.096	13 51 54.	+ 08 39		15.7	GALAXY
FAI 1351+12	13 51 54.	+ 12 56			COMPACT GALAXY
ZC 1351.9+1754	13 51 54.	+ 17 54	940		CLUSTER OF GALAXIES
MCG+04-33-020	13 51 54.	+ 22 13	42	15.5	GALAXY
TON-N 0174	13 51 54.	+ 29 10		14.9	BLUE STAR
ZWG 162.025	13 51 54.	+ 31 20		15.7	GALAXY
UGC 08824	13 51 54.	+ 31 20	66	15.7	GALAXY PECULR
MCG+06-31-019	13 51 54.	+ 33 48 30.	24	15.	GALAXY
1ZW 075	13 51 54.	+ 33 50			COMPACT GALAXY
ZWG 191.013	13 51 54.	+ 33 50		14.4	GALAXY
UGC 08825	13 51 54.	+ 33 50	42	14.4	GALAXY COMPACT
ZWG 191.014	13 51 54.	+ 37 38		15.6	GALAXY
ZWG 219.021	13 51 54.	+ 39 57		15.7	GALAXY
MCG+07-29-014	13 51 54.	+ 40 13	36	15.	GALAXY
ZWG 219.022	13 51 54.	+ 40 31		14.6	GALAXY
UGC 08826	13 51 54.	+ 40 31	78	14.6	GALAXY S0-a
ZC 1351.9+6027	13 51 54.	+ 60 27	1010		CLUSTER OF GALAXIES
ZC 1351.9+7143	13 51 54.	+ 71 43	1280		CLUSTER OF GALAXIES
RRS 4.095	13 51 57.	+ 26 53		18.30	BLUE OBJECT
MCG+05-33-014	13 51 57.	+ 31 06	30	15.2	GALAXY
MCG+05-33-015	13 51 57.	+ 31 09	36	15.5	GALAXY
RNGC 5358	13 51 57.	+ 40 31		14.5	GALAXY
REIN 7.137	13 51 57.01	+ 40 13 51.9			NEBULA
VDB.66G 184	13 52	+ 18 02	130		DWARF GALAXY
VDB.66G 185	13 52	+ 54 07	170		DWARF GALAXY
ZWG 103.002	13 52 00.	+ 14 55		15.7	GALAXY
ZWG 102.079	13 52 00.	+ 14 55		15.7	GALAXY
MCG+03-35-037	13 52 00.	+ 15 16	27	14.1	GALAXY
ZC 1352.0+1636	13 52 00.	+ 16 36	1550		CLUSTER OF GALAXIES
ZWG 132.039	13 52 00.	+ 25 23		15.4	GALAXY
ZC 1352.0+3107	13 52 00.	+ 31 07	3630		CLUSTER OF GALAXIES
MCG+05-33-016	13 52 00.	+ 31 21	54	15.7	GALAXY
ZC 1352.0+3212	13 52 00.	+ 32 12	740		CLUSTER OF GALAXIES
MCG+06-31-020	13 52 00.	+ 34 25	36	16.	GALAXY
ZWG 219.023	13 52 00.	+ 40 13		15.3	GALAXY
MCG+12-13-023	13 52 00.	+ 68 48		16.	GALAXY
MCG+12-13-022	13 52 00.	+ 69 33	45	16.	GALAXY
RNGC 5385	13 52 00.	+ 76 25			NON-EXISTENT OBJECT
MCG-02-35-021	13 52 00.	- 13 31 30.	30	16.	GALAXY
MCG-03-35-022	13 52 00.	- 19 25	42	16.	GALAXY
ARC 1817	13 52 01.	+ 28 50		17.2	RICH CLUSTER OF GALAXIES
IC 4342	13 52 02.	+ 25 23 50.			NONSTELLAR OBJECT
ARC 1818	13 52 04.	+ 27 09		17.2	RICH CLUSTER OF GALAXIES
SN 1954Y	13 52 05.	+ 15 17		19.4	SUPERNOVA
IC 0953	13 52 05.	- 30 06 30.			NONSTELLAR OBJECT
ZWG 018.001	13 52 06.	+ 00 21		15.7	GALAXY
FAI 1352+10	13 52 06.	+ 10 59			COMPACT GALAXY
FAI 13521+11	13 52 06.	+ 11 05			COMPACT GALAXY
ZWG 103.003	13 52 06.	+ 15 17		14.1	GALAXY
ZWG 102.080	13 52 06.	+ 15 17		14.1	GALAXY
UGC 08827	13 52 06.	+ 15 17	66	14.1	GALAXY SB0
ZWG 132.040	13 52 06.	+ 22 05		15.4	GALAXY
UGC 08828	13 52 06.	+ 22 05	72	15.4	GALAXY S?
ZWG 132.041	13 52 06.	+ 24 00		15.3	GALAXY
KARA.73B 0603	13 52 06.	+ 24 00	54	15.3	ISOLATED GALAXY S
MCG+04-33-021	13 52 06.	+ 25 22	45	15.4	GALAXY
ZWG 162.026	13 52 06.	+ 31 10		15.4	GALAXY
MCG+06-31-021	13 52 06.	+ 33 09	66	14.5	GALAXY
ZWG 191.015	13 52 06.	+ 33 11		14.8	GALAXY
MRK 663	13 52 06.	+ 33 11	18	16.	GALAXY WITH UV CONTINUUM
UGC 08829	13 52 06.	+ 33 11	66	14.8	GALAXY SBa
ZWG 219.024	13 52 06.	+ 44 27		15.4	GALAXY
MRK 279	13 52 06.	+ 69 33	15	15.	GALAXY WITH UV CONTINUUM
KW 25	13 52 06.	+ 69 33	43		SEYFERT GALAXY
VVI 70	13 52 06.	+ 69 33	15	15.37	SEYFERT GALAXY
MCG+12-13-024	13 52 06.	+ 69 33	12	17.	GALAXY
ZWG 336.029	13 52 06.	+ 72 46		15.6	GALAXY
RRS 4.096	13 52 07.	+ 27 49		18.24	BLUE OBJECT
FATH 1.637	13 52 08.	+ 44 27	14		NEBULA
RRS 4.097	13 52 10.	+ 28 05			BLUE OBJECT
LB 00692	13 52 11.	+ 25 49 42.		17.1	FAINT BLUE STAR
MCG+04-33-022	13 52 12.	+ 22 04	72	15.4	GALAXY
MCG+04-33-023	13 52 12.	+ 24 05	42	15.3	GALAXY
ZWG 191.016	13 52 12.	+ 36 28		15.6	GALAXY
UGC 08830	13 52 12.	+ 36 28	66	15.6	GALAXY S
ZWG 336.030	13 52 12.	+ 69 33		15.7	GALAXY
MCG-05-33-030	13 52 12.	- 28 06	15	15.	GALAXY
MCG+09-23-013	13 52 15.	+ 54 42	42	16.	GALAXY
LB 00693	13 52 16.	+ 29 56 42.		17.3	FAINT BLUE STAR
IC 0956	13 52 17.	+ 20 57 32.			NONSTELLAR OBJECT
HN 1730	13 52 17.3	- 44 14 06.	36		NEBULA
ZWG 074.002	13 52 18.	+ 11 36		15.7	GALAXY
ZWG 074.003	13 52 18.	+ 13 07		15.6	GALAXY
FAI 1352+13	13 52 18.	+ 13 22			COMPACT GALAXY
ZWG 132.042	13 52 18.	+ 20 58		15.7	GALAXY
TON-N 0743	13 52 18.	+ 24 24		15.6	BLUE STAR
TON-N 0744	13 52 18.	+ 26 49		15.5	BLUE STAR
RRS 3.103	13 52 18.	+ 29 57			BLUE OBJECT
MCG+05-33-017	13 52 18.	+ 31 10	54	15.4	GALAXY
ZC 1352.3+5225	13 52 18.	+ 52 25	1210		CLUSTER OF GALAXIES
MCG+10-20-044	13 52 18.	+ 60 55	42	13.9	GALAXY
MCG+12-13-025	13 52 18.	+ 69 32 30.	33	17.	GALAXY
MCG-04-33-013	13 52 18.	- 26 32	90	14.	GALAXY
MAI 089	13 52 19.	+ 60 22	47		DWARF SPHEROIDAL GALAXY
SC 1349-2202.9	13 52 19.	- 22 17 41.	24		NEBULA
RRS 4.098	13 52 21.	+ 28 28		19.14	BLUE OBJECT
MCG+06-31-022	13 52 21.	+ 36 27	63	15.5	GALAXY
RRS 3.104	13 52 23.	+ 30 32		18.03	BLUE OBJECT
ZWG 046.001	13 52 24.	+ 05 35		14.1	GALAXY
UGC 08831	13 52 24.	+ 05 35	180	14.1	GALAXY Sb
MCG+01-35-052	13 52 24.	+ 05 35	168	14.1	GALAXY
FAI 13524+11.0	13 52 24.	+ 11 02			COMPACT GALAXY
FAI 13524+11.1	13 52 24.	+ 11 05			COMPACT GALAXY
ZWG 132.043	13 52 24.	+ 22 05		15.5	GALAXY
MCG+07-29-015	13 52 24.	+ 38 41	45	15.	GALAXY
REIZ 3711	13 52 25.	+ 05 35	132	13.4	GALAXY
FAI 13525+11	13 52 30.	+ 11 03			COMPACT GALAXY
ZWG 074.004	13 52 30.	+ 11 36		15.5	GALAXY
ZWG 103.004	13 52 30.	+ 20 00		15.2	GALAXY
ZWG 102.081	13 52 30.	+ 20 00		15.2	GALAXY
ZC 1352.5+2438	13 52 30.	+ 24 38	1950		CLUSTER OF GALAXIES
ZWG 219.025	13 52 30.	+ 38 41		14.7	GALAXY
KARA.68 229	13 52 30.	+ 60 16	54		DWARF GALAXY
ZWG 295.022	13 52 30.	+ 60 55		14.3	GALAXY
RNGC 5370	13 52 30.	+ 60 55		14.5	GALAXY
UGC 08832	13 52 30.	+ 60 55	84	14.3	GALAXY SB0
ZWG 018.002	13 52 30.	- 01 44		15.6	GALAXY
MCG-04-33-014	13 52 30.	- 23 00	36	16.	GALAXY
SC 1349-2252.9	13 52 30.	- 23 07 41.	36		NEBULA
REIZ 3717	13 52 31.	+ 54 47	36	14.5	GALAXY
RNGC 5361	13 52 32.	+ 38 41		14.5	GALAXY
REIN 2.202A	13 52 32.75	+ 60 55 22.5			NEBULA
REIN 2.202B	13 52 32.76	+ 60 55 23.1			NEBULA
RNGC 5356	13 52 33.	+ 05 35		14.0	GALAXY
RRS 4.099	13 52 34.	+ 26 28		19.22	BLUE OBJECT
ZC 1352.6+0906	13 52 36.	+ 09 06	1280		CLUSTER OF GALAXIES
FEIG 088	13 52 36.	+ 11 58		15.0	FAINT BLUE STAR
ZC 1352.6+1741	13 52 36.	+ 17 41	670		CLUSTER OF GALAXIES
ZWG 132.044	13 52 36.	+ 25 21		15.6	GALAXY
IC 4343	13 52 36.	+ 25 21 58.			NONSTELLAR OBJECT
TON-N 0175	13 52 36.	+ 25 48		16.7	BLUE STAR
UGC 08833	13 52 36.	+ 36 05	66	16.5	GALAXY IRR
ZC 1352.6+3636	13 52 36.	+ 36 36	1140		CLUSTER OF GALAXIES
ZWG 272.012	13 52 36.	+ 54 35		13.8	GALAXY
UGC 08834	13 52 36.	+ 54 35	54	13.8	GALAXY Sa
REIZ 3720	13 52 37.	+ 60 56	24	13.9	GALAXY
LB 00694	13 52 39.	+ 29 57 30.		17.1	FAINT BLUE STAR
MCG+06-31-023	13 52 39.	+ 36 04 30.	48	15.	GALAXY
RNGC 5368	13 52 41.	+ 54 35		14.0	GALAXY
ZWG 018.003	13 52 42.	+ 02 09		15.7	GALAXY
ZWG 074.005	13 52 42.	+ 10 17		15.5	GALAXY
ZC 1352.7+1400	13 52 42.	+ 14 00	740		CLUSTER OF GALAXIES
ZWG 103.005	13 52 42.	+ 18 29		15.6	GALAXY
MCG+04-33-024	13 52 42.	+ 25 20 30.	39	15.6	GALAXY

OBJECT NAME	RIGHT ASCEN.	DECLINATION	DIAM.	MAGN.	TYPE OF OBJECT
RRS 4.100	13 52 42.	+ 26 25			BLUE OBJECT
ZC 1352.7+2935	13 52 42.	+ 29 35	940		CLUSTER OF GALAXIES
MCG+09-23-014	13 52 42.	+ 54 34	60	13.8	GALAXY
MCG-06-31-002	13 52 42.	- 33 39	36	15.	GALAXY
REIZ 3713	13 52 43.	+ 33 45	90	13.7	GALAXY
REIZ 3719	13 52 43.	+ 54 35	36	13.8	GALAXY
RNGC 5362	13 52 45.	+ 41 34		13.0	GALAXY
LB 00695	13 52 47.	+ 25 30 30.		16.0	FAINT BLUE STAR
RRS 3.105	13 52 47.	+ 29 41		17.75	BLUE OBJECT
REIN 7.138	13 52 47.67	+ 41 33 31.5			NEBULA
ZWG 074.006	13 52 48.	+ 09 15		15.6	GALAXY
ZWG 103.006	13 52 48.	+ 19 48		15.6	GALAXY
ZWG 191.017	13 52 48.	+ 33 09		15.6	GALAXY
REIZ 3716	13 52 48.	+ 41 30	72	13.3	GALAXY
ZWG 219.026	13 52 48.	+ 41 33		13.2	GALAXY
UGC 08835	13 52 48.	+ 41 33	144	13.2	GALAXY S
MCG+07-29-016	13 52 48.	+ 41 33	114	13.	GALAXY
MCG+10-20-045	13 52 48.	+ 58 38	42	15.	GALAXY
ZWG 295.023	13 52 48.	+ 58 39		15.7	GALAXY
UGC 08836	13 52 48.	+ 58 39	72	15.7	GALAXY SB
7ZW 532	13 52 48.	+ 68 48			COMPACT GALAXY
VV 099B	13 52 48.	- 05 43	24	16.	INTERACTING GALAXY
VV 099A	13 52 48.	- 05 43	30	15.	INTERACTING GALAXY
VV 099	13 52 48.	- 05 43	42	15.	INTERACTING GALAXY
MCG-01-35-021	13 52 48.	- 05 43	36	14.5	GALAXY
IC 0955	13 52 51.	- 30 01 03.			NONSTELLAR OBJECT
REIZ 3727	13 52 53.	+ 58 54	24	13.7	GALAXY
ZWG 046.002	13 52 54.	+ 06 51		15.1	GALAXY
ZC 1352.9+2010	13 52 54.	+ 20 10	670		CLUSTER OF GALAXIES
ZWG 132.045	13 52 54.	+ 25 15		15.5	GALAXY
HOLM 556B	13 52 54.	+ 25 16	30	14.4	PART OF MULTIPLE GALAXY
IC 4344	13 52 54.	+ 25 16 47.			NONSTELLAR OBJECT
ZWG 132.046	13 52 54.	+ 25 17		14.7	GALAXY
ZC 1352.9+2625	13 52 54.	+ 26 25	1810		CLUSTER OF GALAXIES
RRS 4.101	13 52 54.	+ 27 18		19.82	BLUE OBJECT
ZWG 162.027	13 52 54.	+ 31 31		15.5	GALAXY
MCG+06-31-024	13 52 54.	+ 33 07 30.	36	16.	GALAXY
ZC 1352.9+3856	13 52 54.	+ 38 56	14850		CLUSTER OF GALAXIES
ZWG 246.025	13 52 54.	+ 47 12		15.6	GALAXY
ZWG 272.013	13 52 54.	+ 54 09		14.2	GALAXY
UGC 08837	13 52 54.	+ 54 09	300	14.2	GALAXY IRR
MCG-01-35-022	13 52 54.	- 05 45	66	14.5	GALAXY
ARC 1819	13 52 55.	+ 24 37		17.2	RICH CLUSTER OF GALAXIES
HOLM 556A	13 52 55.	+ 25 18	30	14.2	PART OF MULTIPLE GALAXY
IC 4345	13 52 55.	+ 25 18 35.			NONSTELLAR OBJECT
REIZ 3715	13 52 55.	+ 33 10	30	14.9	GALAXY
REIZ 3714	13 52 56.	+ 31 32	24	14.2	GALAXY
ARC 1816	13 52 56.	- 26 07		17.6	RICH CLUSTER OF GALAXIES
SC 1350-2112.9	13 52 57.	- 21 27 40.	18		NEBULA
HO 4	13 52 58.	+ 54 08	420	12.95	S GALAXY
RRS 3.106	13 52 59.	+ 30 10		18.34	BLUE OBJECT
ZWG 046.003	13 53 00.	+ 05 14		14.9	GALAXY
UGC 08838	13 53 00.	+ 05 14	108	14.9	GALAXY PECULR
MCG+01-36-001	13 53 00.	+ 05 14	84	14.9	GALAXY
ZC 1353.0+1559	13 53 00.	+ 15 59	1140		CLUSTER OF GALAXIES
ZWG 103.007	13 53 00.	+ 18 01		15.7	GALAXY
UGC 08839	13 53 00.	+ 18 01	198	15.7	GALAXY DWRF IR
MCG+03-36-001	13 53 00.	+ 18 03	240	15.7	GALAXY
MCG+04-33-026	13 53 00.	+ 25 14	36	15.5	GALAXY
MCG+04-33-025	13 53 00.	+ 25 16	15	14.7	GALAXY
MCG+05-33-018	13 53 00.	+ 31 32	36	15.5	GALAXY
MCG+06-31-025	13 53 00.	+ 33 05	45	15.	GALAXY
ZWG 191.018	13 53 00.	+ 33 07		15.5	GALAXY
ZWG 219.027	13 53 00.	+ 40 10		15.3	GALAXY
UGC 08840	13 53 00.	+ 40 10	66	15.3	GALAXY
MCG+07-29-017	13 53 00.	+ 40 10	36	16.	GALAXY
ZWG 219.028	13 53 00.	+ 40 24		15.1	GALAXY
UGC 08841	13 53 00.	+ 40 24	102	15.1	GALAXY SBb
MCG+07-29-018	13 53 00.	+ 40 25	96	14.5	GALAXY
ZWG 246.026	13 53 00.	+ 44 50		15.5	GALAXY
MCG+09-23-016	13 53 00.	+ 52 48	30	16.	GALAXY
MCG+09-23-015	13 53 00.	+ 52 48	24	16.	GALAXY
MCG+09-23-017	13 53 00.	+ 54 09	180	12.9	GALAXY
MCG+10-20-046	13 53 00.	+ 58 54	39	13.7	GALAXY
VV 100D	13 53 00.	- 05 45	3	19.	INTERACTING GALAXY
VV 100C	13 53 00.	- 05 45	6	18.	INTERACTING GALAXY
VV 100B	13 53 00.	- 05 45	18	16.	INTERACTING GALAXY
VV 100A	13 53 00.	- 05 45	30	16.	INTERACTING GALAXY
VV 100	13 53 00.	- 05 45	120	15.	INTERACTING GALAXY
MCG-05-33-031	13 53 00.	- 33 30	30	15.	GALAXY
REIN 7.139B	13 53 00.91	+ 40 24 45.5			NEBULA
HOLM 557B	13 53 01.	+ 05 14	120	13.2	PART OF MULTIPLE GALAXY
REIZ 3712	13 53 01.	+ 05 15	60	14.7	GALAXY
REIZ 3718	13 53 01.	+ 33 06	42	14.7	GALAXY
REIN 7.139A	13 53 01.06	+ 40 24 45.5			NEBULA
RNGC 5372	13 53 03.	+ 58 55		13.5	GALAXY
RNGC 5357	13 53 04.	- 30 06		14.0	GALAXY
RRS 3.107	13 53 05.	+ 30 48		18.86	BLUE OBJECT
ZWG 046.004	13 53 06.	+ 06 02		15.3	GALAXY
MCG+03-36-002	13 53 06.	+ 15 00	36	15.7	GALAXY
ZWG 132.047	13 53 06.	+ 25 18		15.3	GALAXY
UGC 08842	13 53 06.	+ 25 18	84	15.3	GALAXY
LB 00696	13 53 06.	+ 26 38 00.		17.0	FAINT BLUE STAR
MCG+05-33-019	13 53 06.	+ 27 02	24	15.	GALAXY
ZC 1353.1+3010	13 53 06.	+ 30 10	670		CLUSTER OF GALAXIES
ZWG 295.024	13 53 06.	+ 58 55		13.7	GALAXY
UGC 08843	13 53 06.	+ 58 55	42	13.7	GALAXY PECULR
RRS 4.102	13 53 08.	+ 28 39		18.55	BLUE OBJECT
RNGC 5360	13 53 09.	+ 05 14		15.0	GALAXY
RRS 3.108	13 53 11.	+ 29 46			BLUE OBJECT
ZWG 046.005	13 53 12.	+ 08 27		15.2	GALAXY
ZC 1353.2+1123	13 53 12.	+ 11 23	670		CLUSTER OF GALAXIES
8ZW 1353+12.5	13 53 12.	+ 12 28		15.4	COMPACT GALAXY
FAI 1353+14	13 53 12.	+ 14 27			COMPACT GALAXY
ZWG 103.008	13 53 12.	+ 14 59		15.7	GALAXY
ZWG 103.009	13 53 12.	+ 16 12		15.7	GALAXY
ZC 1353.2+2508	13 53 12.	+ 25 08	8530		CLUSTER OF GALAXIES
ZWG 132.048	13 53 12.	+ 25 17		15.7	GALAXY
MCG+05-33-020	13 53 12.	+ 27 02	60	15.5	GALAXY
ZC 1353.2+2711	13 53 12.	+ 27 11	3290		CLUSTER OF GALAXIES
ZC 1353.2+3207	13 53 12.	+ 32 07	940		CLUSTER OF GALAXIES
ZC 1353.2+3250	13 53 12.	+ 32 50	870		CLUSTER OF GALAXIES
ZC 1353.2-0210	13 53 12.	- 02 10	870		CLUSTER OF GALAXIES
VV 281D	13 53 15.	+ 25 18 30.	18	17.	INTERACTING GALAXY
VV 281C	13 53 15.	+ 25 18 30.	9	17.	INTERACTING GALAXY
VV 281B	13 53 15.	+ 25 18 30.	30	15.	INTERACTING GALAXY
VV 281A	13 53 15.	+ 25 18 30.	30	15.	INTERACTING GALAXY
VV 281	13 53 15.	+ 25 18 30.	72	14.4	INTERACTING GALAXY
REIZ 3724	13 53 15.	+ 37 51	78	14.2	GALAXY
IC 0957	13 53 15.	- 29 59 38.			NONSTELLAR OBJECT
IC 0958	13 53 16.	+ 05 16 41.			MAY NOT EXIST
RNGC 5365A	13 53 16.	- 43 45			GALAXY
MCG+04-33-027	13 53 18.	+ 25 18	72	15.3	GALAXY
ZWG 132.049	13 53 18.	+ 25 23		15.4	GALAXY
ZC 1353.3+6848	13 53 18.	+ 68 48	1280		CLUSTER OF GALAXIES
MCG-05-33-032	13 53 18.	- 30 05	18	14.5	GALAXY
DV.56 N5365A	13 53 18.	- 43 45	138		S GALAXY
RRS 4.103	13 53 19.	+ 26 44		18.24	BLUE OBJECT
RRS 4.104	13 53 20.	+ 27 54		18.81	BLUE OBJECT
MCG+04-33-028	13 53 21.	+ 25 18	12	15.3	GALAXY
IC 4346	13 53 21.	+ 25 23 48.			NONSTELLAR OBJECT
SC 1350-2315.4	13 53 21.	- 23 30 09.	12		NEBULA
SC 1350-2316.6	13 53 23.	- 23 31 21.	18		NEBULA
ZWG 103.010	13 53 24.	+ 14 39		15.6	GALAXY
ZWG 103.011	13 53 24.	+ 15 17		15.5	GALAXY
ZWG 132.050	13 53 24.	+ 25 26		15.7	GALAXY
ZWG 353.027	13 53 24.	+ 79 03		15.7	GALAXY
UGC 08845	13 53 24.	+ 79 03	60	15.7	GALAXY SBb
MCG+13-10-013	13 53 24.	+ 79 05	51	16.	GALAXY
ZWG 018.004	13 53 24.	- 01 01		15.1	GALAXY
UGC 08844	13 53 24.	- 01 01	108	15.1	GALAXY SB:c
MCG+00-36-001	13 53 24.	- 01 01	90	15.1	GALAXY
MCG-04-33-015	13 53 24.	- 25 55	24	15.5	GALAXY
MCG-06-31-003	13 53 24.	- 34 18	66	15.	GALAXY
REIZ 3722	13 53 25.	+ 25 25	30	14.6	GALAXY
REIZ 3723	13 53 25.	+ 25 28	24	15.3	GALAXY
IC 4348	13 53 26.	+ 25 26 49.			NONSTELLAR OBJECT
IC 4349	13 53 28.	+ 25 23 43.			NONSTELLAR OBJECT
REIZ 3726	13 53 29.	+ 27 29	72	14.3	GALAXY
ZWG 046.006	13 53 30.	+ 08 27		15.3	GALAXY
ZC 1353.5+1223	13 53 30.	+ 12 23	400		CLUSTER OF GALAXIES
ZWG 103.012	13 53 30.	+ 19 00		15.7	GALAXY
MCG+04-33-029	13 53 30.	+ 25 22	48	15.4	GALAXY
ZWG 132.051	13 53 30.	+ 25 23		15.3	GALAXY
ZWG 132.052	13 53 30.	+ 26 05		15.5	GALAXY
MRK 664	13 53 30.	+ 26 05	20	16.	GALAXY WITH UV CONTINUUM
IC 0961	13 53 30.	+ 26 05 37.			NONSTELLAR OBJECT
MRK 070	13 53 30.	+ 26 56	12	16.	GALAXY WITH UV CONTINUUM
TON-N 0745	13 53 30.	+ 28 13		14.7	BLUE STAR
MCG+07-29-019	13 53 30.	+ 39 57 30.	72	15.	GALAXY
ZWG 219.029	13 53 30.	+ 40 42		11.5	GALAXY
UGC 08846	13 53 30.	+ 40 42	270	11.5	GALAXY SBb
MCG+10-20-047	13 53 30.	+ 59 44	144	13.0	GALAXY
MCG-04-33-017	13 53 30.	- 22 23	36	15.	GALAXY
MCG-04-33-016	13 53 30.	- 27 30	60	15.	GALAXY
MCG-06-31-004	13 53 30.	- 33 52	36	15.	GALAXY
RNGC 5371	13 53 33.	+ 40 42		11.5	GALAXY
REIN 7.140	13 53 33.72	+ 40 42 22.1			NEBULA
RRS 4.105	13 53 34.	+ 27 43		19.02	BLUE OBJECT
RNGC 5376	13 53 34.	+ 59 45		13.0	GALAXY
HN 1731	13 53 34.5	- 43 45 51.	60		NEBULA
ZWG 046.007	13 53 36.	+ 05 30		11.4	GALAXY
UGC 08847	13 53 36.	+ 05 30	330	11.4	GALAXY
MCG+01-36-002	13 53 36.	+ 05 30	90	11.4	GALAXY
ZWG 046.008	13 53 36.	+ 05 54		15.7	GALAXY
ZC 1353.6+1156	13 53 36.	+ 11 56	1610		CLUSTER OF GALAXIES
ZWG 074.007	13 53 36.	+ 13 45		14.4	GALAXY
UGC 08848	13 53 36.	+ 13 45	108	14.4	GALAXY Sa
MCG+02-36-001	13 53 36.	+ 13 45	96	14.4	GALAXY
UGC 08849	13 53 36.	+ 14 45	90	14.8	GALAXY DBL SYS
ZWG 103.013	13 53 36.	+ 17 45		14.8	GALAXY
KARA.72 402A	13 53 36.	+ 17 45	36	14.8	PART OF DOUBLE GALAXY
MCG+03-36-003	13 53 36.	+ 17 45	48	14.8	GALAXY
MCG+03-36-004	13 53 36.	+ 17 45 30.	48	14.8	GALAXY
VV 335B	13 53 36.	+ 17 46	45	15.5	INTERACTING GALAXY
VV 335A	13 53 36.	+ 17 46	54	15.5	INTERACTING GALAXY
KARA.72 402B	13 53 36.	+ 17 46	42		PART OF DOUBLE GALAXY
IC 0960	13 53 36.	+ 17 46 13.			NONSTELLAR OBJECT
ZWG 103.014	13 53 36.	+ 18 37		14.8	GALAXY
MRK 463	13 53 36.	+ 18 37	25	16.	GALAXY WITH UV CONTINUUM
UGC 08850	13 53 36.	+ 18 37	60	14.8	GALAXY PECULR
MCG+03-36-005	13 53 36.	+ 18 37	27	14.8	GALAXY
MCG+04-33-030	13 53 36.	+ 25 25 30.	18	15.7	GALAXY
MCG+04-33-031	13 53 36.	+ 26 04	36	15.5	GALAXY
ZC 1353.6+3107	13 53 36.	+ 31 07	1480		CLUSTER OF GALAXIES
UGC 08851	13 53 36.	+ 39 57	96	16.5	GALAXY
MCG+07-29-020	13 53 36.	+ 40 42	240	10.	GALAXY
REIZ 3729	13 53 36.	+ 40 44	228	12.0	GALAXY
ZWG 295.025	13 53 36.	+ 59 45		12.9	GALAXY
UGC 08852	13 53 36.	+ 59 45	102	12.9	GALAXY Sa-b
ZWG 336.031	13 53 36.	+ 71 31		15.3	GALAXY
HOLM 557A	13 53 37.	+ 05 16	330	11.4	PART OF MULTIPLE GALAXY
REIZ 3721	13 53 37.	+ 05 30	120	11.5	GALAXY
REIZ 3730	13 53 37.	+ 54 50	96	14.6	GALAXY
REIN 2.203	13 53 37.21	+ 59 45 04.0			NEBULA
RNGC 5363	13 53 39.	+ 05 30		11.5	GALAXY
MCG+04-33-032	13 53 39.	+ 25 27	48	15.3	GALAXY
REIZ 3732	13 53 39.	+ 59 45	84	12.9	GALAXY
IC 0959	13 53 40.	+ 13 45 31.			NONSTELLAR OBJECT
ZWG 046.009	13 53 42.	+ 05 16		13.2	GALAXY
UGC 08853	13 53 42.	+ 05 16	432	13.2	GALAXY Sb/Sc
MCG+01-36-003	13 53 42.	+ 05 16	390	13.2	GALAXY
ZC 1353.7+0553	13 53 42.	+ 05 53	7530		CLUSTER OF GALAXIES
ZWG 103.015	13 53 42.	+ 20 00		15.7	GALAXY
TON-N 0176	13 53 42.	+ 26 17		16.8	BLUE STAR
ZWG 191.019	13 53 42.	+ 37 26		14.9	GALAXY
UGC 08854	13 53 42.	+ 37 26	66	14.9	GALAXY S0
MCG+10-20-048	13 53 42.	+ 59 18	30	16.	GALAXY
MCG+10-20-049	13 53 42.	+ 59 58 30.	120	12.9	GALAXY
ZWG 018.005	13 53 42.	- 01 18		15.7	GALAXY
MCG-07-29-001	13 53 42.	- 43 45 30.	144	13.	GALAXY
REIZ 3725	13 53 43.	+ 05 16	270	12.1	GALAXY
RNGC 5364	13 53 45.	+ 05 16		11.0	GALAXY
MCG+06-31-026	13 53 45.	+ 37 26 30.	48	15.	GALAXY
ARC 1820	13 53 46.	+ 11 02		17.6	RICH CLUSTER OF GALAXIES
ARC 1821	13 53 46.	+ 31 04		17.5	RICH CLUSTER OF GALAXIES
ZWG 018.007	13 53 48.	+ 00 00		14.7	GALAXY
KARA.72 403A	13 53 48.	+ 00 00	48	14.7	PART OF DOUBLE GALAXY
ZC 1353.8+0354	13 53 48.	+ 03 54	340		CLUSTER OF GALAXIES
ZWG 046.010	13 53 48.	+ 04 50		15.7	GALAXY
ZWG 074.008	13 53 48.	+ 13 03		15.6	GALAXY
ZWG 132.053	13 53 48.	+ 24 44		15.5	GALAXY
UGC 08855	13 53 48.	+ 24 44	60	15.5	GALAXY S
ZWG 132.054	13 53 48.	+ 25 16		15.6	GALAXY
ZWG 162.028	13 53 48.	+ 29 15		15.7	GALAXY
ZWG 162.029	13 53 48.	+ 29 55		15.4	GALAXY

OBJECT NAME	RIGHT ASCEN.	DECLINATION	DIAM.	MAGN.	TYPE OF OBJECT
ZWG 162.030	13 53 48.	+ 30 20		15.6	GALAXY
UGC 08856	13 53 48.	+ 30 20	270	15.6	GALAXY DBL SYS
MCG+10-20-050	13 53 48.	+ 59 51	84	15.	GALAXY
MCG+00-36-002	13 53 48.	- 00 00	30	14.7	GALAXY
ZWG 018.006	13 53 48.	- 01 42		15.7	GALAXY
MCG-05-33-033	13 53 48.	- 32 24	72	15.	GALAXY
RNGC 5366	13 53 49.	+ 00 00		14.5	GALAXY
REIZ 3728	13 53 49.	+ 25 28	18	15.2	GALAXY
MCG+05-33-021	13 53 51.	+ 30 20	42	15.6	GALAXY
RNGC 5379	13 53 53.	+ 59 59		14.0	GALAXY
HOLM 561B	13 53 53.	+ 59 59	60	14.3	PART OF MULTIPLE GALAXY
ZWG 046.011	13 53 54.	+ 04 38		15.3	GALAXY
UGC 08857	13 53 54.	+ 04 38	60	15.3	GALAXY Sa
ZC 1353.9+1110	13 53 54.	+ 11 10	1140		CLUSTER OF GALAXIES
ZWG 103.016	13 53 54.	+ 18 32		15.5	GALAXY
ZWG 132.055	13 53 54.	+ 25 25		15.7	GALAXY
ZC 1353.9+2539	13 53 54.	+ 25 39	1550		CLUSTER OF GALAXIES
ZWG 162.031	13 53 54.	+ 28 47		15.4	GALAXY
MCG+05-33-023	13 53 54.	+ 29 55	42	15.4	GALAXY
MCG+05-33-022	13 53 54.	+ 30 21	48	15.6	GALAXY
MCG+07-29-021	13 53 54.	+ 38 32	90	14.	GALAXY
MRK 464	13 53 54.	+ 38 48	6	16.5	GALAXY WITH UV CONTINUUM
ZWG 219.030	13 53 54.	+ 42 47		15.7	GALAXY
ZWG 018.008	13 53 54.	- 00 03		15.6	GALAXY
KARA.72 403B	13 53 54.	- 00 03	54	15.6	PART OF DOUBLE GALAXY
MCG+00-36-003	13 53 54.	- 00 03	48	15.6	GALAXY
HOLM 558B	13 53 55.	+ 30 19	30	15.0	PART OF MULTIPLE GALAXY
HOLM 558A	13 53 57.	+ 30 20	30	15.0	PART OF MULTIPLE GALAXY
REIZ 3736	13 53 57.	+ 59 59	60	14.3	GALAXY
HN 0014	13 53 58.	+ 00 02			NEBULA
ZC 1354.0+1228	13 54 00.	+ 12 28	870		CLUSTER OF GALAXIES
ZC 1354.0+1834	13 54 00.	+ 18 34	3970		CLUSTER OF GALAXIES
ZWG 162.032	13 54 00.	+ 28 46		15.7	GALAXY
ZWG 219.031	13 54 00.	+ 38 32		15.3	GALAXY
UGC 08858	13 54 00.	+ 38 32	96	15.3	GALAXY S-IRR
ZWG 295.026	13 54 00.	+ 59 59		14.1	GALAXY
UGC 08860	13 54 00.	+ 59 59	120	14.1	GALAXY S0
VHA 155	13 54 00.	- 59 20	660		OPEN STAR CLUSTER
VV 158D	13 54 03.	+ 28 46 30.	30	15.	INTERACTING GALAXY
MCG+05-33-024	13 54 03.	+ 28 47	18	15.4	GALAXY
ZWG 074.009	13 54 06.	+ 10 06		15.7	GALAXY
ZWG 074.010	13 54 06.	+ 10 27		15.6	GALAXY
UGC 08861	13 54 06.	+ 10 27	78	15.6	GALAXY
ZWG 074.011	13 54 06.	+ 12 31		15.7	GALAXY
ZWG 162.033	13 54 06.	+ 28 46		15.6	GALAXY
TON-N 0746	13 54 06.	+ 29 04		16.9	BLUE STAR
ZC 1354.1+3142	13 54 06.	+ 31 42	1550		CLUSTER OF GALAXIES
LB 00697	13 54 10.	+ 26 55 12.		16.7	FAINT BLUE STAR
ZWG 046.012	13 54 12.	+ 05 46		15.4	GALAXY
TON-N 0177	13 54 12.	+ 25 17		15.6	BLUE STAR
MCG+05-33-024A	13 54 12.	+ 28 46 30.	18	15.7	GALAXY
MCG+10-20-051	13 54 12.	+ 59 58	210	12.5	GALAXY
ZC 1354.2+7254	13 54 12.	+ 72 54	670		CLUSTER OF GALAXIES
MCG+13-10-014	13 54 12.	+ 78 28	102	14.	GALAXY
RNGC 5369	13 54 12.	- 05 15			GALAXY
RNGC 5452	13 54 13.	+ 78 28		14.0	GALAXY
ZWG 103.017	13 54 18.	+ 20 24		15.6	GALAXY
UGC 08862	13 54 18.	+ 20 24	60	15.6	GALAXY Sb-c
ZWG 162.034	13 54 18.	+ 28 45		15.5	GALAXY
ZWG 246.027	13 54 18.	+ 47 29		12.5	GALAXY
UGC 08863	13 54 18.	+ 47 29	270	12.5	GALAXY SBa
KARA.73B 0604	13 54 18.	+ 47 29	264	12.5	ISOLATED GALAXY S
MCG+08-25-052	13 54 18.	+ 47 30	258	12.0	GALAXY
7ZW 533	13 54 18.	+ 64 57			COMPACT GALAXY
ZC 1354.3+6501	13 54 18.	+ 65 01	1810		CLUSTER OF GALAXIES
REIN 2.204	13 54 18.14	+ 47 28 45.3			NEBULA
RNGC 5377	13 54 19.	+ 47 29		12.0	GALAXY
MCG+05-33-025	13 54 21.	+ 28 46	24	15.6	GALAXY
MCG-04-33-018	13 54 21.	- 24 33	9	15.5	GALAXY
REIZ 3731	13 54 22.	+ 20 25	54	16.4	GALAXY
ZWG 046.013	13 54 24.	+ 05 24		15.7	GALAXY
MCG+03-36-006	13 54 24.	+ 20 26	60	15.6	GALAXY
VV 158C	13 54 24.	+ 28 46 00.	21	15.	INTERACTING GALAXY
VV 158B	13 54 24.	+ 28 46 00.	30	15.	INTERACTING GALAXY
TON-N 0747	13 54 24.	+ 28 55		15.1	BLUE STAR
ZC 1354.4+3234	13 54 24.	+ 32 34	1140		CLUSTER OF GALAXIES
MCG-04-33-019	13 54 24.	- 25 00	60	14.	GALAXY
REIZ 3746	13 54 26.	+ 59 59	120	13.0	GALAXY
HOLM 561A	13 54 26.	+ 59 59	102	13.3	PART OF MULTIPLE GALAXY
MCG+05-33-026	13 54 27.	+ 28 45	30	15.5	GALAXY
RNGC 5375	13 54 28.	+ 29 25		13.0	GALAXY
REIN 2.205A	13 54 28.25	+ 59 59 09.1			NEBULA
REIN 2.205B	13 54 28.32	+ 59 59 09.2			NEBULA
RNGC 5389	13 54 29.	+ 59 59		13.0	GALAXY
IC 4350	13 54 29.	- 24 58 45.			NONSTELLAR OBJECT
ZWG 074.012	13 54 30.	+ 11 06		15.3	GALAXY
ZWG 074.013	13 54 30.	+ 14 23		15.1	GALAXY
ZWG 103.018	13 54 30.	+ 15 56		14.9	GALAXY
MCG+03-36-007	13 54 30.	+ 15 57	24	14.9	GALAXY
ZWG 103.019	13 54 30.	+ 20 22		15.4	GALAXY
UGC 08864	13 54 30.	+ 23 28	78	16.0	GALAXY Sa
ZWG 132.056	13 54 30.	+ 26 09		15.6	GALAXY
ZWG 162.035	13 54 30.	+ 29 25		13.2	GALAXY
UGC 08865	13 54 30.	+ 29 25	222	13.2	GALAXY SBb
KARA.73B 0605	13 54 30.	+ 29 25	204	13.2	ISOLATED GALAXY S
ZWG 295.027	13 54 30.	+ 59 59		13.2	GALAXY
UGC 08866	13 54 30.	+ 59 59	270	13.2	GALAXY S0
ZWG 353.028	13 54 30.	+ 78 28		14.2	GALAXY
UGC 08867	13 54 30.	+ 78 28	138	14.2	GALAXY Sc/SBc
VV 158A	13 54 33.	+ 28 45 00.	30	15.	INTERACTING GALAXY
REIZ 3735	13 54 34.	+ 29 24	42	13.0	GALAXY
SC 1351-2414.8	13 54 34.	- 24 29 30.	18		NEBULA
IC 4347	13 54 35.	- 39 46 09.			NONSTELLAR OBJECT
ZWG 046.014	13 54 36.	+ 05 30		15.3	GALAXY
FAI 1354+13	13 54 36.	+ 13 55			COMPACT GALAXY
1ZW 076	13 54 36.	+ 43 50			COMPACT GALAXY
ZC 1354.6+6132	13 54 36.	+ 61 32	870		CLUSTER OF GALAXIES
MCG-04-33-020	13 54 36.	- 26 56	36	15.	GALAXY
HN 1732	13 54 36.9	- 46 12 07.	18		NEBULA
REIZ 3733	13 54 37.	+ 05 31	9	15.4	GALAXY
SC 1351-5331.8	13 54 37.	- 53 46 31.	36		NEBULA
RNGC 5373	13 54 39.	+ 05 30		15.5	GALAXY
MCG+05-33-027	13 54 39.	+ 29 25	180	13.2	GALAXY
REIZ 3739	13 54 39.	+ 38 02	60	14.0	GALAXY
MIL 29	13 54 41.	- 61 56 36.	450		SUPERNOVA REMNANT
ZWG 074.014	13 54 42.	+ 12 15		14.9	GALAXY
MCG+02-36-002	13 54 42.	+ 12 15	48	14.9	GALAXY
ZWG 074.015	13 54 42.	+ 12 16		14.0	GALAXY

OBJECT NAME	RIGHT ASCEN.	DECLINATION	DIAM.	MAGN.	TYPE OF OBJECT
UGC 08868	13 54 42.	+ 12 16	48	14.0	GALAXY
MCG+02-36-003	13 54 42.	+ 12 16	48	14.0	GALAXY
ZWG 103.020	13 54 42.	+ 14 54		15.4	GALAXY
ZWG 103.021	13 54 42.	+ 19 47		15.6	GALAXY
ZWG 132.057	13 54 42.	+ 22 15		15.6	GALAXY
ZWG 191.020	13 54 42.	+ 38 02		13.8	GALAXY
UGC 08869	13 54 42.	+ 38 02	144	13.8	GALAXY SBa
MCG+06-31-027	13 54 42.	+ 38 03	186	12.5	GALAXY
ZC 1354.7-0123	13 54 42.	- 01 23	1410		CLUSTER OF GALAXIES
RNGC 5367	13 54 42.	- 39 44			DIFFUSE NEBULA
BC PKS1354+19	13 54 42.06	+ 19 33 43.5		16.02	QUASI-STELLAR OBJECT
SHB 280	13 54 42.3	+ 19 33 41.		16.0	QUASI-STELLAR OBJECT
RNGC 5378	13 54 44.	+ 38 03		14.0	GALAXY
REIZ 3734	13 54 45.	+ 12 13	18	15.1	GALAXY
REIZ 3742	13 54 45.	+ 37 51	24	13.9	GALAXY
ABC 1823	13 54 45.	+ 45 10		17.5	RICH CLUSTER OF GALAXIES
LB 00698	13 54 46.	+ 26 12 36.		16.0	FAINT BLUE STAR
RNGC 5365	13 54 46.	- 43 42		13.0	GALAXY
PK311+02.1	13 54 46.	- 58 40	10		PLANETARY NEBULA
ZWG 074.016	13 54 48.	+ 12 14		15.2	GALAXY
MCG+02-36-004	13 54 48.	+ 12 14	36	15.2	GALAXY
ZWG 191.021	13 54 48.	+ 37 51		13.5	GALAXY
UGC 08870	13 54 48.	+ 37 51	120	13.5	GALAXY S0
MCG+06-31-028	13 54 48.	+ 37 51	90	13.	GALAXY
ZC 1354.8+5257	13 54 48.	+ 52 57	3290		CLUSTER OF GALAXIES
ZC 1354.8+5700	13 54 48.	+ 57 00	3020		CLUSTER OF GALAXIES
MCG-06-31-005	13 54 48.	- 33 58	60	13.5	GALAXY
MCG-07-29-002	13 54 48.	- 43 41	72	13.0	GALAXY
BC OP291	13 54 48.39	+ 25 52 05.5		18.5	QUASI-STELLAR OBJECT
RNGC 5380	13 54 50.	+ 37 51		13.0	GALAXY
IC 0962	13 54 51.	+ 12 17 29.			NONSTELLAR OBJECT
SC 1352-2405.3	13 54 52.	- 24 20 00.	48		NEBULA
REIZ 3737	13 54 54.	+ 06 10	18	15.1	GALAXY
ZWG 046.015	13 54 54.	+ 06 12		15.5	GALAXY
UGC 08871	13 54 54.	+ 06 12	66	15.5	GALAXY
ZWG 103.022	13 54 54.	+ 15 41		14.5	GALAXY
UGC 08872	13 54 54.	+ 15 41	84	14.5	GALAXY S0?
MCG+03-36-008	13 54 54.	+ 15 42	78	14.5	GALAXY
ZWG 132.058	13 54 54.	+ 24 29		15.2	GALAXY
UGC 08873	13 54 54.	+ 24 29	90	15.2	GALAXY S
HOLM 559B	13 54 54.	+ 24 32	30	14.8	PART OF MULTIPLE GALAXY
ZWG 162.036	13 54 54.	+ 29 02		15.2	GALAXY
MRK 280	13 54 54.	+ 29 02	12	15.	GALAXY WITH UV CONTINUUM
ZWG 219.032	13 54 54.	+ 43 35		15.0	GALAXY
REIZ 3741	13 54 55.	+ 24 30	60	15.0	GALAXY
IC 4353	13 54 55.	+ 38 01			NONSTELLAR OBJECT
FATH 1.638	13 54 57.	+ 44 43	27		NEBULA
RNGC 5383	13 54 58.	+ 42 06		12.5	GALAXY
SHB 281	13 54 59.	+ 25 44		18.5	QUASI-STELLAR OBJECT
REIZ 3750	13 54 59.	+ 42 05	180	12.5	GALAXY
ZWG 046.016	13 55 00.	+ 06 21		13.7	GALAXY
REIZ 3738	13 55 00.	+ 06 21	72	13.2	GALAXY
UGC 08874	13 55 00.	+ 06 21	108	13.7	GALAXY Sb
MCG+01-36-004	13 55 00.	+ 06 21	96	13.7	GALAXY
ZWG 074.017	13 55 00.	+ 10 12		15.6	GALAXY
ZC 1355.0+1422	13 55 00.	+ 14 22	1480		CLUSTER OF GALAXIES
ZWG 103.023	13 55 00.	+ 17 38		15.2	GALAXY
MCG+04-33-033	13 55 00.	+ 24 29	90	15.2	GALAXY
MCG+05-33-029	13 55 00.	+ 29 02	24	15.2	GALAXY
MCG+05-33-028	13 55 00.	+ 29 02 30.	30	15.2	GALAXY
MCG+07-29-022	13 55 00.	+ 42 02 30.	60	16.	GALAXY
ZWG 219.033	13 55 00.	+ 42 05		12.5	GALAXY
MRK 281	13 55 00.	+ 42 05	100	13.	GALAXY WITH UV CONTINUUM
UGC 08875	13 55 00.	+ 42 05	210	12.5	GALAXY SBb
MCG+07-29-023	13 55 00.	+ 42 06	150	12.5	GALAXY
MCG+07-29-024	13 55 00.	+ 43 37	16	15.	GALAXY
ZC 1355.0+4510	13 55 00.	+ 45 10	2150		CLUSTER OF GALAXIES
ZWG 246.028	13 55 00.	+ 46 13		14.3	GALAXY
UGC 08876	13 55 00.	+ 46 13	60	14.3	GALAXY S0-a
MCG+08-25-053	13 55 00.	+ 46 13	72	15.	GALAXY
RNGC 5391	13 55 00.	+ 46 31		16.0	GALAXY
MCG-06-31-006	13 55 00.	- 33 45	36	15.	GALAXY
REIN 2.206A	13 55 00.40	+ 42 05 24.3			NEBULA
REIN 2.206B	13 55 00.44	+ 42 05 24.8			NEBULA
HOLM 559A	13 55 01.	+ 24 31	78	13.9	PART OF MULTIPLE GALAXY
LB 00699	13 55 02.	+ 25 00 00.		16.7	FAINT BLUE STAR
IC 4351	13 55 02.2	- 29 04 18.	420	12.0	GALAXY SA(s)
RNGC 5374	13 55 03.	+ 06 20		13.5	GALAXY
MCG+03-36-009	13 55 03.	+ 17 40	42	15.2	GALAXY
MCG+08-25-054	13 55 03.	+ 46 31	54	16.	GALAXY
MCG+11-17-010	13 55 03.	+ 64 54	39	16.	GALAXY
REIZ 3740	13 55 05.	+ 07 26	12	16.	GALAXY
ZC 1355.1+2237	13 55 06.	+ 22 37	5240		CLUSTER OF GALAXIES
STOCK 48	13 55 06.	+ 29 02			BLUE KNOT NEAR ELLIP GLXY
MCG+06-31-029	13 55 06.	+ 34 45	36	15.	GALAXY
UGC 08877	13 55 06.	+ 42 02	78	16.5	GALAXY
REIN 2.207	13 55 06.19	+ 42 05 47.3			NEBULA
IC 0963	13 55 07.	+ 11 39 12.			NONSTELLAR OBJECT
REIZ 3743	13 55 11.	+ 07 37	30	14.7	GALAXY
REIZ 3744	13 55 11.	+ 07 38	24	16.0	GALAXY
SC 1352-2411.3	13 55 11.	- 24 25 59.	42		NEBULA
MCG+09-23-018	13 55 12.	+ 54 07	36	16.	GALAXY
ZC 1355.2+6747	13 55 12.	+ 67 47	2690		CLUSTER OF GALAXIES
MCG-04-33-021	13 55 12.	- 25 46	42	15.	GALAXY
RNGC 5390	13 55 15.	+ 40 40			NON-EXISTENT OBJECT
IC 4352	13 55 15.	- 34 17 31.			NONSTELLAR OBJECT
HOLM 560A	13 55 17.	+ 07 38	42	13.4	PART OF MULTIPLE GALAXY
REIZ 3745	13 55 17.	+ 07 40	18	15.9	GALAXY
LB 00700	13 55 17.	+ 28 15 00.		16.8	FAINT BLUE STAR
ZWG 046.017	13 55 18.	+ 07 39		14.8	GALAXY
MCG+01-36-005	13 55 18.	+ 07 39	48	14.8	GALAXY
ZC 1355.3+0743	13 55 18.	+ 07 43	1210		CLUSTER OF GALAXIES
FAI 13553+09	13 55 18.	+ 09 51			COMPACT GALAXY
ZWG 074.018	13 55 18.	+ 10 08		14.5	GALAXY
UGC 08878	13 55 18.	+ 10 08	66	14.5	GALAXY SBb
MCG+02-36-005	13 55 18.	+ 10 08	60	14.5	GALAXY
ZWG 103.024	13 55 18.	+ 17 45		15.3	GALAXY
MCG+03-36-010	13 55 18.	+ 17 45 30.	18	15.3	GALAXY
IC 0964	13 55 18.	+ 17 45 31.			NONSTELLAR OBJECT
ZC 1355.3+4649	13 55 18.	+ 46 49	1140		CLUSTER OF GALAXIES
MCG-05-33-034	13 55 18.	- 29 03	240	14.	GALAXY
HOLM 560B	13 55 22.	+ 07 39	30	14.8	PART OF MULTIPLE GALAXY
RNGC 5365B	13 55 22.	- 43 44			GALAXY
REIZ 3747	13 55 23.	+ 07 24	60	14.9	GALAXY
REIZ 3748	13 55 23.	+ 07 38	24	16.1	GALAXY
REIZ 3749	13 55 23.	+ 07 46	30	15.1	GALAXY
ZWG 103.025	13 55 24.	+ 17 30		15.6	GALAXY
ZWG 103.026	13 55 24.	+ 17 45		15.2	GALAXY

OBJECT NAME	RIGHT ASCEN.	DECLINATION	DIAM.	MAGN.	TYPE OF OBJECT
MCG+03-36-011	13 55 24.	+ 17 45 30.	48	15.2	GALAXY
ZC 1355.4+2059	13 55 24.	+ 20 59	2350		CLUSTER OF GALAXIES
ZWG 132.059	13 55 24.	+ 26 00		15.7	GALAXY
UGC 08879	13 55 24.	+ 26 00	96	15.7	GALAXY Sc
UGC 08880	13 55 24.	+ 50 40	72	16.0	GALAXY Sa
DV.56 N5265B	13 55 24.	- 43 44	93		S GALAXY
HOLM 560C	13 55 25.	+ 07 40	42	14.5	PART OF MULTIPLE GALAXY
IC 0965	13 55 25.	+ 17 45 32.			NONSTELLAR OBJECT
ARC 1824	13 55 27.	+ 27 05		17.2	RICH CLUSTER OF GALAXIES
REIZ 3755	13 55 27.	+ 30 44	24	14.7	GALAXY
ZWG 046.018	13 55 30.	+ 07 25		15.2	GALAXY
UGC 08881	13 55 30.	+ 07 25	72	15.2	GALAXY Sc
MCG+01-36-005A	13 55 30.	+ 07 25	72	15.2	GALAXY
HOLM 560D	13 55 30.	+ 07 38	48	15.2	PART OF MULTIPLE GALAXY
ZWG 046.019	13 55 30.	+ 07 41		15.3	GALAXY
ZWG 074.019	13 55 30.	+ 10 58		15.6	GALAXY
MCG+06-31-030	13 55 30.	+ 36 35	18	16.	GALAXY
MCG+09-23-019	13 55 30.	+ 50 40	60	16.	GALAXY
UGC 08882	13 55 30.	+ 54 21	66	16.0	GALAXY
MCG+09-23-020	13 55 30.	+ 54 21	30	14.	GALAXY
ZC 1355.5+6408	13 55 30.	+ 64 08	1080		CLUSTER OF GALAXIES
MCG-03-36-001	13 55 30.	- 15 31	15	15.5	GALAXY
MCG-06-31-007	13 55 30.	- 34 16	96	13.	GALAXY
ARC 1826	13 55 34.	+ 30 50		17.5	RICH CLUSTER OF GALAXIES
HN 1733	13 55 34.2	- 43 43 23.	48		NEBULA
ARC 1825	13 55 35.	+ 20 54		15.7	RICH CLUSTER OF GALAXIES
REIZ 3751	13 55 36.	+ 06 45	24	13.4	GALAXY
ZWG 103.027	13 55 36.	+ 15 32		14.3	GALAXY
UGC 08883	13 55 36.	+ 15 32	60	14.3	GALAXY S
ZWG 103.028	13 55 36.	+ 19 55		15.7	GALAXY
TON-N 0178	13 55 36.	+ 28 01		14.5	BLUE STAR
ZC 1355.6+3052	13 55 36.	+ 30 52	1550		CLUSTER OF GALAXIES
ZC 1355.6+5541	13 55 36.	+ 55 41	1950		CLUSTER OF GALAXIES
ZWG 018.010	13 55 36.	- 00 09		14.8	GALAXY
MCG+00-36-004	13 55 36.	- 00 09	36	14.8	GALAXY
ZWG 018.009	13 55 36.	- 01 08		15.5	GALAXY
8ZW 1355-01.8	13 55 36.	- 01 46		19.9	COMPACT GALAXY
MCG-01-36-001	13 55 36.	- 03 50	42	15.	GALAXY
MCG-07-29-003	13 55 36.	- 43 43	72	14.	GALAXY
RNGC 5382	13 55 39.	+ 06 30		14.0	GALAXY
RNGC 5384	13 55 39.	+ 06 46		14.0	GALAXY
RNGC 5359	13 55 40.	- 70 10			NON-EXISTENT OBJECT
REIZ 3754	13 55 41.	+ 07 27	36	15.5	GALAXY
ZWG 018.011	13 55 42.	+ 01 23		15.2	GALAXY
ZWG 046.020	13 55 42.	+ 04 14		15.7	GALAXY
ZWG 046.021	13 55 42.	+ 05 39		15.0	GALAXY
UGC 08884	13 55 42.	+ 05 39	96	15.0	GALAXY S0
MCG+01-36-006	13 55 42.	+ 05 39	36	15.0	GALAXY
REIZ 3752	13 55 42.	+ 05 40	18	14.6	GALAXY
ZWG 046.022	13 55 42.	+ 06 30		14.0	GALAXY
UGC 08885	13 55 42.	+ 06 30	108	14.0	GALAXY S0
MCG+01-36-007	13 55 42.	+ 06 30	30	14.0	GALAXY
REIZ 3753	13 55 42.	+ 06 40	12	14.1	GALAXY
ZWG 046.023	13 55 42.	+ 06 46		14.0	GALAXY
UGC 08836	13 55 42.	+ 06 46	108	14.0	GALAXY S0
MCG+01-36-008	13 55 42.	+ 06 46	36	14.0	GALAXY
FAI 13557+09	13 55 42.	+ 09 51			COMPACT GALAXY
MCG+03-36-012	13 55 42.	+ 15 33	60	14.3	GALAXY
ZWG 103.029	13 55 42.	+ 17 26		15.5	GALAXY
ZWG 132.060	13 55 42.	+ 20 39		15.2	GALAXY
UGC 08887	13 55 42.	+ 20 39	72	15.2	GALAXY Sa-b
UGC 08838	13 55 42.	+ 20 52	108	16.0	GALAXY QUD SYS
ZWG 132.061	13 55 42.	+ 22 02		15.4	GALAXY
UGC 08889	13 55 42.	+ 22 02	102	15.4	GALAXY S
ZC 1355.7+4010	13 55 42.	+ 40 10	470		CLUSTER OF GALAXIES
MRSL 310+00/1	13 55 42.	- 61 12	300		HII REGION
IC 0966	13 55 43.	+ 05 38 20.			NONSTELLAR OBJECT
ARC 1822	13 55 44.	- 25 09		17.6	RICH CLUSTER OF GALAXIES
RNGC 5386	13 55 45.	+ 06 35		13.5	GALAXY
REIZ 3760	13 55 45.	+ 20 39	54	15.6	GALAXY
REIZ 3756	13 55 43.	+ 06 30	30	13.3	GALAXY
ZWG 046.024	13 55 48.	+ 06 35		13.7	GALAXY
REIZ 3757	13 55 48.	+ 06 35	24	14.1	GALAXY
UGC 08890	13 55 48.	+ 06 35	66	13.7	GALAXY S0-a
MCG+01-36-010	13 55 48.	+ 06 35	60	13.7	GALAXY
ZWG 046.025	13 55 48.	+ 07 29		15.1	GALAXY
MCG+01-36-009	13 55 48.	+ 07 29	48	15.1	GALAXY
ZWG 074.020	13 55 48.	+ 08 30		15.1	GALAXY
8ZW 1355+08.5	13 55 48.	+ 08 30		15.1	COMPACT GALAXY
FAI 1355+08	13 55 48.	+ 08 31			COMPACT GALAXY
MCG+03-36-013	13 55 48.	+ 17 27	33	15.5	GALAXY
ZWG 103.030	13 55 48.	+ 17 46		15.7	GALAXY
MCG+04-33-034	13 55 48.	+ 20 38	54	15.2	GALAXY
ZWG 162.037	13 55 48.	+ 28 40		15.0	GALAXY
MCG+05-33-030	13 55 48.	+ 28 40	24	15.0	GALAXY
IC 4355	13 55 49.	+ 28 39 04.			NONSTELLAR OBJECT
ARC 1827	13 55 51.	+ 21 57		16.6	RICH CLUSTER OF GALAXIES
MCG+04-33-035	13 55 51.	+ 22 01	72	15.4	GALAXY
ARC 1830	13 55 51.	+ 47 39		17.5	RICH CLUSTER OF GALAXIES
MCG-02-36-001	13 55 51.	- 12 23	60	15.	GALAXY
LB 00701	13 55 52.	+ 26 27 00.		17.4	FAINT BLUE STAR
IC 4354	13 55 53.	- 12 20			NONSTELLAR OBJECT
HN 0383	13 55 53.	- 12 21			NEBULA
REIZ 3758	13 55 54.	+ 06 18	72	13.8	GALAXY
ZWG 046.026	13 55 54.	+ 06 19		14.8	GALAXY
UGC 08891	13 55 54.	+ 06 19	96	14.8	GALAXY Sb-c
MCG+01-36-011	13 55 54.	+ 06 19	96	14.8	GALAXY
REIZ 3759	13 55 54.	+ 06 41	18	14.7	GALAXY
ZWG 103.031	13 55 54.	+ 14 41		14.7	GALAXY
MCG+03-36-014	13 55 54.	+ 17 47	36	15.7	GALAXY
ZC 1355.9+1840	13 55 54.	+ 18 40	1080		CLUSTER OF GALAXIES
MCG+06-31-031	13 55 54.	+ 32 52	24	15.5	GALAXY
ZWG 191.022	13 55 54.	+ 32 53		15.7	GALAXY
ZC 1355.9+3827	13 55 54.	+ 38 27	1210		CLUSTER OF GALAXIES
REIZ 3762	13 55 55.	+ 32 52	30	14.6	GALAXY
IC 0967	13 55 56.	+ 14 42 04.			NONSTELLAR OBJECT
RNGC 5387	13 55 57.	+ 06 19		15.0	GALAXY
BC PKS1355-41	13 55 57.27	- 41 38 19.3		16.5	QUASI-STELLAR OBJECT
SHB 282	13 55 57.3	- 41 38 19.		16.	QUASI-STELLAR OBJECT
REIZ 3761	13 55 59.	+ 07 27	60	14.8	GALAXY
ZC 1356.0+1327	13 56 00.	+ 13 27	1080		CLUSTER OF GALAXIES
MCG+03-36-015	13 56 00.	+ 14 42	42	14.7	GALAXY
ZWG 103.032	13 56 00.	+ 17 38		15.2	GALAXY
ZWG 132.062	13 56 00.	+ 23 08		15.1	GALAXY
ZWG 132.063	13 56 00.	+ 23 10		15.3	GALAXY
MCG+09-23-021	13 56 00.	+ 52 12	30	16.	GALAXY
MCG+10-20-052	13 56 00.	+ 57 14	114	15.	GALAXY
ZWG 295.028	13 56 00.	+ 57 15		15.6	GALAXY

OBJECT NAME	RIGHT ASCEN.	DECLINATION	DIAM.	MAGN.	TYPE OF OBJECT
UGC 08892	13 56 00.	+ 57 15	144	15.6	GALAXY IRR
ZWG 336.032	13 56 00.	+ 71 00		15.0	GALAXY
RNGC 5415	13 56 00.	+ 71 00		15.0	GALAXY
ARC 1828	13 56 01.	+ 18 38		16.6	RICH CLUSTER OF GALAXIES
LB 00702	13 56 03.	+ 29 19 30.		17.4	FAINT BLUE STAR
MCG+06-31-032	13 56 03.	+ 36 53	48	15.	GALAXY
ARC 1829	13 56 03.	+ 38 32		17.8	RICH CLUSTER OF GALAXIES
SC 1353-2315.8	13 56 03.	- 23 30 27.	12		NEBULA
HARO 39	13 56 04.	+ 25 47			BLUE EMISSION-LINE GALAXY
FAI 13561+11	13 56 06.	+ 11 10			COMPACT GALAXY
ZWG 074.021	13 56 06.	+ 12 12		15.4	GALAXY
ZWG 074.022	13 56 06.	+ 13 43		15.5	GALAXY
ZWG 132.064	13 56 06.	+ 24 23		15.7	GALAXY
ZWG 132.065	13 56 06.	+ 25 47		15.2	GALAXY
ZC 1356.1+2654	13 56 06.	+ 26 54	1140		CLUSTER OF GALAXIES
ZWG 191.023	13 56 06.	+ 36 54		15.7	GALAXY
UGC 08893	13 56 06.	+ 36 54	84	15.7	GALAXY Sc/SBc
UGC 08894	13 56 06.	+ 63 42	132	16.5	GALAXY DWRF SP
MCG+03-36-016	13 56 09.	+ 17 38	60	15.2	GALAXY
REIZ 3771	13 56 11.	+ 61 10	48	15.0	GALAXY
ZWG 046.027	13 56 12.	+ 02 42		15.6	GALAXY
UGC 08895	13 56 12.	+ 02 42	78	15.6	GALAXY Sc
ZWG 046.028	13 56 12.	+ 03 05		15.2	GALAXY
ZWG 046.029	13 56 12.	+ 07 28		15.0	GALAXY
UGC 08896	13 56 12.	+ 07 28	90	15.0	GALAXY S
MCG+01-36-012	13 56 12.	+ 07 28	84	15.0	GALAXY
ZWG 074.023	13 56 12.	+ 12 49		15.5	GALAXY
UGC 08897	13 56 12.	+ 12 49	60	15.5	GALAXY Sc
ZWG 074.024	13 56 12.	+ 13 42		15.3	GALAXY
FAI 1356+14	13 56 12.	+ 14 24			COMPACT GALAXY
MCG+04-33-037	13 56 12.	+ 23 07	30	15.1	GALAXY
MCG+04-33-036	13 56 12.	+ 25 46	36	15.2	GALAXY
ZC 1356.2+5136	13 56 12.	+ 51 36	1480		CLUSTER OF GALAXIES
MRSL 310-00/1	13 56 12.	- 62 24	1080		HII REGION
REIZ 3764	13 56 13.	+ 32 54	24	15.0	GALAXY
RNGC 5388	13 56 15.	- 13 55			NON-EXISTENT OBJECT
REIZ 3763	13 56 17.	+ 07 29	54	15.4	GALAXY
LB 00703	13 56 17.	+ 26 47 00.		16.8	FAINT BLUE STAR
FAI 13563+08	13 56 18.	+ 08 57			COMPACT GALAXY
ZWG 103.033	13 56 18.	+ 14 59		15.6	GALAXY
ZWG 162.038	13 56 18.	+ 29 06		15.5	GALAXY
ZWG 219.034	13 56 18.	+ 42 51		15.6	GALAXY
MCG+09-23-022	13 56 18.	+ 53 15	24	16.	GALAXY
MCG+10-20-053	13 56 18.	+ 61 00	27	16.	GALAXY
ZWG 317.011	13 56 18.	+ 67 43		15.7	GALAXY
HOLM 563B	13 56 23.	+ 37 42	42	13.6	PART OF MULTIPLE GALAXY
ZC 1356.4+0813	13 56 24.	+ 08 13	2150		CLUSTER OF GALAXIES
FAI 13564+08	13 56 24.	+ 08 52			COMPACT GALAXY
ZC 1356.4+1045	13 56 24.	+ 10 45	1140		CLUSTER OF GALAXIES
ZC 1356.4+2912	13 56 24.	+ 29 12	270		CLUSTER OF GALAXIES
VV 048B	13 56 24.	+ 37 41	78	13.6	INTERACTING GALAXY
MCG+06-31-033	13 56 24.	+ 37 41	90	13.	GALAXY
ZWG 191.024	13 56 24.	+ 37 42		13.7	GALAXY
ZCG 1356+37.7	13 56 24.	+ 37 42		13.7	COMPACT GALAXY
UGC 08898	13 56 24.	+ 37 42	114	13.7	GALAXY S
KARA.72 404A	13 56 24.	+ 37 42	96	13.7	PART OF DOUBLE GALAXY
MCG+11-17-011	13 56 24.	+ 63 39 30.	78	16.	GALAXY
MCG-03-36-002	13 56 24.	- 18 50	48	15.	GALAXY
MCG-07-29-004	13 56 24.	- 39 50	84	14.5	GALAXY
RNGC 5394	13 56 26.	+ 37 42		13.5	GALAXY
REIZ 3768	13 56 27.	+ 37 40	132	13.1	GALAXY
HOLM 563A	13 56 27.	+ 37 40	126	12.5	PART OF MULTIPLE GALAXY
REIZ 3767	13 56 27.	+ 37 42	30	13.9	GALAXY
SC 1353-2243.6	13 56 28.	- 22 58 14.	60		NEBULA
SC 1353-2244.4	13 56 28.	- 22 59 02.	18		NEBULA
HOLM 562A	13 56 29.	+ 07 30	60	14.1	PART OF MULTIPLE GALAXY
REIZ 3766	13 56 29.	+ 34 46	18	14.3	GALAXY
HN 0382	13 56 29.	- 84 02			NEBULA
IC 4333	13 56 29.	- 84 02			NONSTELLAR OBJECT
ZWG 046.030	13 56 30.	+ 07 30		15.4	GALAXY
FAI 13565+08	13 56 30.	+ 08 58			COMPACT GALAXY
ZWG 103.034	13 56 30.	+ 15 52		15.0	GALAXY
MCG+03-36-017	13 56 30.	+ 15 53	48	15.0	GALAXY
ZWG 132.066	13 56 30.	+ 22 14		15.6	GALAXY
UGC 08899	13 56 30.	+ 22 14	78	15.6	GALAXY Sb-c
ZWG 191.025	13 56 30.	+ 34 46		15.6	GALAXY
MCG+06-31-035	13 56 30.	+ 34 46	48	15.	GALAXY
MCG+06-31-036	13 56 30.	+ 35 19 30.	48	15.	GALAXY
VV 048A	13 56 30.	+ 37 39	168	12.5	INTERACTING GALAXY
ZWG 191.026	13 56 30.	+ 37 40		12.6	GALAXY
UGC 08900	13 56 30.	+ 37 40	180	12.6	GALAXY Sb
KARA.72 404B	13 56 30.	+ 37 40	162	12.6	PART OF DOUBLE GALAXY
MCG+06-31-034	13 56 30.	+ 37 40	180	11.	GALAXY
MCG+10-20-054	13 56 30.	+ 60 03	84	14.0	GALAXY
ZWG 317.012	13 56 30.	+ 65 10		14.4	GALAXY
RNGC 5413	13 56 30.	+ 65 10		14.5	GALAXY
UGC 08901	13 56 30.	+ 65 10	72	14.4	GALAXY E
MCG+11-17-012	13 56 30.	+ 65 10	51	14.	GALAXY
ZWG 018.012	13 56 30.	- 02 09		15.4	GALAXY
HOLM 562B	13 56 31.	+ 07 29	18	15.5	PART OF MULTIPLE GALAXY
RNGC 5395	13 56 32.	+ 37 40		13.0	GALAXY
MIL 30	13 56 32.	- 62 00 48.	660		SUPERNOVA REMNANT
ARP 084	13 56 33.	+ 37 41			PECULIAR GALAXY
REIZ 3770	13 56 35.	+ 35 18	30	14.9	GALAXY
REIZ 3769	13 56 35.	+ 35 18	18	16.2	GALAXY
RNGC 5402	13 56 35.	+ 60 03		14.5	GALAXY
BC 4C58.29	13 56 35.81	+ 58 06 37.8		13.	QUASI-STELLAR OBJECT
ZWG 103.035	13 56 36.	+ 15 48		14.5	GALAXY
UGC 08902	13 56 36.	+ 15 48	78	14.5	GALAXY Sb
TON-N 0179	13 56 36.	+ 24 15		14.0	BLUE STAR
2ZW 068	13 56 36.	+ 37 40			COMPACT GALAXY
1ZW 077	13 56 36.	+ 37 40			COMPACT GALAXY
ZCG 1356+37.8	13 56 36.	+ 37 45		16.2	COMPACT GALAXY
IC 4356	13 56 36.	+ 37 45 01.			NONSTELLAR OBJECT
ZWG 247.001	13 56 36.	+ 46 30		15.7	GALAXY
ZWG 246.029	13 56 36.	+ 46 30		15.7	GALAXY
ZWG 295.029	13 56 36.	+ 60 03		14.6	GALAXY
UGC 08903	13 56 36.	+ 60 03	72	14.6	GALAXY
SC 1353-2251.5	13 56 36.	- 23 06 08.	60		NEBULA
OCL 0914	13 56 36.	- 61 57	180		OPEN STAR CLUSTER
SHB 283	13 56 36.4	+ 58 04 12.		16.	QUASI-STELLAR OBJECT
REIZ 3773	13 56 37.	+ 60 05	60	14.0	GALAXY
REIZ 3772	13 56 38.	+ 56 56	30	15.1	GALAXY
RNGC 5412	13 56 39.	+ 73 50		14.5	GALAXY
FAI 1356+09	13 56 42.	+ 09 22			COMPACT GALAXY
MCG+03-36-019	13 56 42.	+ 15 48	66	14.5	GALAXY
MCG+03-36-018	13 56 42.	+ 16 05	42	16.	GALAXY
UGC 08904	13 56 42.	+ 26 21	108	16.0	GALAXY

OBJECT NAME	RIGHT ASCEN.	DECLINATION	DIAM.	MAGN.	TYPE OF OBJECT
MCG+06-31-037	13 56 42.	+ 35 30	30	15.5	GALAXY
ZC 1356.7+4930	13 56 42.	+ 49 30	2960		CLUSTER OF GALAXIES
ZWG 336.033	13 56 42.	+ 73 50		14.7	GALAXY
UGC 08905	13 56 42.	+ 73 50	72	14.7	GALAXY E-SO
MCG+03-36-021	13 56 45.	+ 15 48	30	17.	GALAXY
MCG+03-36-020	13 56 45.	+ 15 49	24	17.	GALAXY
RNGC 5396	13 56 46.	+ 29 21			NON-EXISTENT OBJECT
SC 1353-2236.6	13 56 46.	- 22 51 14.	48		NEBULA
REIZ 3765	13 56 48.	+ 05 48	54	14.6	GALAXY
ZWG 074.025	13 56 48.	+ 10 23		15.3	GALAXY
FAI 13568+11	13 56 48.	+ 11 09			COMPACT GALAXY
ZWG 074.026	13 56 48.	+ 13 43		15.6	GALAXY
ZWG 103.036	13 56 48.	+ 17 04		15.7	GALAXY
MCG+05-33-032	13 56 48.	+ 28 15 30.	36	15.6	GALAXY
ZWG 162.039	13 56 48.	+ 28 16		15.6	GALAXY
ZWG 162.040	13 56 48.	+ 28 18		15.6	GALAXY
MCG+05-33-031	13 56 48.	+ 28 18	27	15.6	GALAXY
TON-N 0748	13 56 48.	+ 29 20		14.1	BLUE STAR
ZC 1356.8+3006	13 56 48.	+ 30 06	1340		CLUSTER OF GALAXIES
ZC 1356.8+3108	13 56 48.	+ 31 08	870		CLUSTER OF GALAXIES
MCG+10-20-055	13 56 48.	+ 59 23	30	16.	GALAXY
ZWG 018.013	13 56 48.	- 02 58		15.1	GALAXY
MCG+00-36-005	13 56 48.	- 02 58	18	15.1	GALAXY
MCG-04-33-022	13 56 48.	- 24 08	15	15.	GALAXY
BIGO 534	13 56 51.	+ 38 26			NEBULA
ARC 1834	13 56 51.	+ 49 47		17.2	RICH CLUSTER OF GALAXIES
HOLM 564A	13 56 54.	+ 05 46	72	14.3	PART OF MULTIPLE GALAXY
ZWG 046.031	13 56 54.	+ 05 47		15.2	GALAXY
UGC 08906	13 56 54.	+ 05 47	90	15.2	GALAXY Sc
MCG+01-36-013	13 56 54.	+ 05 47	84	15.2	GALAXY
ZWG 103.037	13 56 54.	+ 15 12		15.5	GALAXY
MCG+03-36-022	13 56 54.	+ 15 13	48	15.5	GALAXY
MRK 332	13 56 54.	+ 20 29	35	15.5	GALAXY WITH UV CONTINUUM
ZWG 162.041	13 56 54.	+ 28 13		15.7	GALAXY
MCG+05-33-033	13 56 54.	+ 28 13 30.	60	15.7	GALAXY
MCG+10-20-056	13 56 54.	+ 61 02	66	13.	GALAXY
ARC 1831	13 56 55.	+ 28 14		15.4	RICH CLUSTER OF GALAXIES
SC 1354-2351.8	13 56 57.	- 24 06 25.	24		NEBULA
REIZ 3776	13 56 58.	+ 55 36	66	15.3	GALAXY
HOLM 564B	13 56 59.	+ 05 47	18	14.5	PART OF MULTIPLE GALAXY
ARC 1832	13 56 59.	+ 29 47		17.2	RICH CLUSTER OF GALAXIES
REIZ 3779	13 56 59.	+ 61 03	42	14.8	GALAXY
ZWG 074.027	13 57 00.	+ 13 02		15.2	GALAXY
UGC 08907	13 57 00.	+ 13 02	96	15.2	GALAXY
ZWG 103.038	13 57 00.	+ 15 25		15.4	GALAXY
MCG+03-36-023	13 57 00.	+ 15 25	60	15.4	GALAXY
ZC 1357.0+1948	13 57 00.	+ 19 48	870		CLUSTER OF GALAXIES
MCG+06-31-038	13 57 00.	+ 34 18	39	15.	GALAXY
UGC 08908	13 57 00.	+ 43 21	66	16.0	GALAXY Sb-c
UGC 08859	13 57 00.	+ 59 53	96	16.5	GALAXY
ZWG 295.030	13 57 00.	+ 61 01		14.7	GALAXY
UGC 08909	13 57 00.	+ 61 01	108	14.7	GALAXY Sc/SBc
MCG+10-20-057	13 57 00.	+ 62 07	15	16.	GALAXY
LB 00704	13 57 01.	+ 27 30 48.		17.7	FAINT BLUE STAR
MCG+07-29-025	13 57 03.	+ 43 21	48	16.	GALAXY
RNGC 5381	13 57 03.	- 59 19			OPEN CLUSTER
UGC 08910	13 57 06.	+ 17 43	60	16.0	GALAXY
MCG+03-36-024	13 57 06.	+ 17 43 30.	48	17.	GALAXY
ZC 1357.1+2836	13 57 06.	+ 28 36	8800		CLUSTER OF GALAXIES
ZWG 162.042	13 57 06.	+ 30 19		15.6	GALAXY
ZWG 018.014	13 57 06.	- 02 49		15.7	GALAXY
MCG+00-36-006	13 57 06.	- 02 49	36	15.7	GALAXY
OCL 0915	13 57 06.	- 59 19	840	12.	OPEN STAR CLUSTER
VHA 156	13 57 06.	- 59 19	360		OPEN STAR CLUSTER
SC 1353-5314.4	13 57 08.	- 53 29 02.	12		NEBULA
MCG+05-33-034	13 57 09.	+ 30 19	42	15.6	GALAXY
ZWG 018.015	13 57 12.	+ 01 19		15.3	GALAXY
KARA.73B 0606	13 57 12.	+ 01 19	48	15.3	ISOLATED GALAXY S
ZC 1357.2+2947	13 57 12.	+ 29 47	1140		CLUSTER OF GALAXIES
ZWG 162.043	13 57 12.	+ 30 20		15.5	GALAXY
ZWG 074.028	13 57 18.	+ 12 59		15.2	GALAXY
ZWG 162.044	13 57 18.	+ 28 19		15.3	GALAXY
UGC 08911	13 57 18.	+ 28 19	72	15.3	GALAXY Sc
ZC 1357.3+2845	13 57 18.	+ 28 45	400		CLUSTER OF GALAXIES
MCG+05-33-035	13 57 18.	+ 30 20	27	15.5	GALAXY
ZWG 191.027	13 57 18.	+ 35 01		14.7	GALAXY
UGC 08912	13 57 18.	+ 35 01	78	14.7	GALAXY S
UGC 08913	13 57 18.	+ 39 02	90	16.0	GALAXY Sc
MCG+07-29-026	13 57 18.	+ 39 02	66	15.	GALAXY
MCG+10-20-058	13 57 18.	+ 58 38	12	16.	GALAXY
RNGC 5399	13 57 19.	+ 35 01		14.5	GALAXY
REIZ 3775	13 57 19.	+ 39 12	36	14.0	GALAXY
REIZ 3783	13 57 22.	+ 55 23	30	14.4	GALAXY
REIZ 3774	13 57 23.	+ 35 01	60	13.6	GALAXY
ZWG 074.029	13 57 24.	+ 12 18		15.7	GALAXY
ZC 1357.4+1430	13 57 24.	+ 14 30	1210		CLUSTER OF GALAXIES
MCG+05-33-036	13 57 24.	+ 28 18	72	15.3	GALAXY
MCG+06-31-039	13 57 24.	+ 35 00	66	14.	GALAXY
UGC 08914	13 57 24.	+ 52 36	60	17.	GALAXY DWARF
ZC 1357.4+5601	13 57 24.	+ 56 01	870		CLUSTER OF GALAXIES
REIZ 3777	13 57 25.	+ 39 01	60	14.5	GALAXY
ZC 1357.5+1855	13 57 30.	+ 18 55	1680		CLUSTER OF GALAXIES
UGC 08915	13 57 30.	+ 26 28	66	17.	GALAXY DWARF
ZC 1357.5+2913	13 57 30.	+ 29 13	470		CLUSTER OF GALAXIES
ZWG 191.028	13 57 30.	+ 36 29		14.6	GALAXY
UGC 08916	13 57 30.	+ 36 29	96	14.6	GALAXY Sa
RNGC 5401	13 57 31.	+ 36 29		14.5	GALAXY
BIGO 535	13 57 31.	+ 48 41			NEBULA
REIZ 3786	13 57 31.	+ 59 54	18	15.9	GALAXY
ARC 1848	13 57 31.	+ 74 22		17.7	RICH CLUSTER OF GALAXIES
SC 1354-2159.7	13 57 31.	- 22 14 18.	42		NEBULA
REIZ 3778	13 57 33.	+ 36 29	60	14.0	GALAXY
ZWG 046.032	13 57 36.	+ 03 40		15.6	GALAXY
ZWG 074.030	13 57 36.	+ 12 21		15.2	GALAXY
ZWG 103.039	13 57 36.	+ 16 09		14.8	GALAXY
ZC 1357.6+3244	13 57 36.	+ 32 44	4970		CLUSTER OF GALAXIES
MCG+06-31-040	13 57 36.	+ 36 29	90	14.	GALAXY
ZWG 219.035	13 57 36.	+ 40 37		15.1	GALAXY
UGC 08917	13 57 36.	+ 40 37	108	15.1	GALAXY Sc
MCG+07-29-027	13 57 36.	+ 40 37 30.	39	15.5	GALAXY
SC 1354-2054.4	13 57 36.	- 21 09 00.	30		NEBULA
SC 1354-2055.3	13 57 36.	- 21 09 54.	24		NEBULA
HOLM 565A	13 57 38.	+ 38 26	72	14.2	PART OF MULTIPLE GALAXY
SC 1354-2233.7	13 57 39.	- 22 48 18.	36		NEBULA
HN 1734	13 57 39.0	- 45 10 48.	42		NEBULA
ARC 1833	13 57 40.	+ 04 52		17.0	RICH CLUSTER OF GALAXIES
RNGC 5393	13 57 40.	- 28 36		15.0	GALAXY
ZC 1357.7+0451	13 57 42.	+ 04 51	1080		CLUSTER OF GALAXIES
ZWG 074.031	13 57 42.	+ 09 13		14.8	GALAXY
UGC 08918	13 57 42.	+ 09 13	102	14.8	GALAXY S
MCG+02-36-006	13 57 42.	+ 09 13	72	14.8	GALAXY
MCG+03-36-025	13 57 42.	+ 16 10	36	14.8	GALAXY
ZC 1357.7+2118	13 57 42.	+ 21 18	810		CLUSTER OF GALAXIES
LB 00705	13 57 42.	+ 23 59 30.		15.6	FAINT BLUE STAR
MCG+05-33-037	13 57 42.	+ 28 45	36	16.	GALAXY
VV 310B	13 57 42.	+ 38 26	30	14.6	INTERACTING GALAXY
VV 310A	13 57 42.	+ 38 26	180	14.2	INTERACTING GALAXY
MCG+06-31-041	13 57 42.	+ 38 26	180	13.	GALAXY
MCG+07-29-028	13 57 42.	+ 40 37	90	14.5	GALAXY
ZWG 018.016	13 57 42.	- 00 43		15.5	GALAXY
MCG-04-33-023	13 57 42.	- 23 11	48	15.5	GALAXY
MCG-05-33-035	13 57 42.	- 28 36	24	15.	GALAXY
MCG-07-29-005	13 57 42.	- 45 11	102	13.0	GALAXY
MCG-04-33-025	13 57 43.	- 23 03	18	16.	GALAXY
MCG-04-33-024	13 57 43.	- 23 03 30.	24	15.5	GALAXY
REIN 7.141B	13 57 43.33	+ 38 25 27.7			NEBULA
REIN 7.141A	13 57 43.62	+ 38 25 27.7			NEBULA
RNGC 5403	13 57 44.	+ 38 25		15.0	GALAXY
REIZ 3782	13 57 44.	+ 38 26	48	14.0	GALAXY
HOLM 565B	13 57 44.	+ 38 28	18	14.6	PART OF MULTIPLE GALAXY
MCG-04-33-026	13 57 44.	- 23 03 30.	36	15.5	GALAXY
ZWG 046.033	13 57 48.	+ 05 17		15.2	GALAXY
ZWG 074.032	13 57 48.	+ 09 33		15.6	GALAXY
ZWG 074.033	13 57 48.	+ 13 12		15.2	GALAXY
KARA.72 405A	13 57 48.	+ 13 12	48	15.2	PART OF DOUBLE GALAXY
MCG+02-36-007	13 57 48.	+ 13 12	48	15.2	GALAXY
VV 239A	13 57 48.	+ 13 13	54	15.	INTERACTING GALAXY
ZWG 103.040	13 57 48.	+ 16 57		15.7	GALAXY
ZWG 191.029	13 57 48.	+ 38 25		14.9	GALAXY
UGC 08919	13 57 48.	+ 38 25	192	14.9	GALAXY Sb
MCG+06-31-042	13 57 48.	+ 38 27	24	15.	GALAXY
ZWG 272.014	13 57 48.	+ 56 17		15.6	GALAXY
ZWG 018.017	13 57 48.	- 02 05		15.5	GALAXY
REIN 7.142	13 57 49.56	+ 38 26 34.9			NEBULA
REIZ 3784	13 57 50.	+ 38 26	30	14.2	GALAXY
RNGC 5403A	13 57 50.	+ 38 27		15.0	GALAXY
FAI 1357+13	13 57 54.	+ 13 05			COMPACT GALAXY
ZWG 074.034	13 57 54.	+ 13 12		15.5	GALAXY
UGC 08920	13 57 54.	+ 13 12	72	15.5	GALAXY S(c)
KARA.72 405B	13 57 54.	+ 13 12	54	15.5	PART OF DOUBLE GALAXY
MCG+02-36-008	13 57 54.	+ 13 12	48	15.5	GALAXY
VV 239B	13 57 54.	+ 13 13	66	17.	INTERACTING GALAXY
ZWG 191.030	13 57 54.	+ 38 26		15.2	GALAXY
UGC 08921	13 57 54.	+ 61 32	60	16.5	GALAXY PECULR?
ZWG 018.018	13 57 54.	- 02 35		15.5	GALAXY
LB 09883	13 58	- 85 00		12.5	FAINT BLUE STAR
LB 09884	13 58	- 86 32		12.5	FAINT BLUE STAR
LB 09885	13 58	- 86 52		13.3	FAINT BLUE STAR
ZWG 074.035	13 58 00.	+ 08 55		15.6	GALAXY
ZWG 074.036	13 58 00.	+ 09 09		15.4	GALAXY
ZWG 074.037	13 58 00.	+ 12 33		15.6	GALAXY
FAI 13580+13	13 58 00.	+ 13 12			COMPACT GALAXY
ZWG 103.041	13 58 00.	+ 18 05		15.7	GALAXY
ZC 1358.0+2313	13 58 00.	+ 23 13	940		CLUSTER OF GALAXIES
MCG+07-29-029	13 58 00.	+ 38 54	60	14.5	GALAXY
ZWG 219.036	13 58 00.	+ 38 55		15.3	GALAXY
UGC 08922	13 58 00.	+ 38 55	66	15.3	GALAXY SBa
ZC 1358.0+6846	13 58 00.	+ 68 46	1880		CLUSTER OF GALAXIES
ZWG 336.034	13 58 00.	+ 69 39		15.2	GALAXY
ZWG 336.035	13 58 00.	+ 70 59		15.6	GALAXY
ZC 1358.0+7228	13 58 00.	+ 72 28	1010		CLUSTER OF GALAXIES
MCG+12-13-026	13 58 00.	+ 72 44	36	16.	GALAXY
ZC 1358.0+8421	13 58 00.	+ 84 21	2150		CLUSTER OF GALAXIES
ZWG 018.020	13 58 00.	- 02 37		14.5	GALAXY
RNGC 5400	13 58 00.	- 02 37		14.5	GALAXY
MKW 05	13 58 00.	- 02 37		14.5	POOR GALAXY CLUSTER
MCG+00-36-008	13 58 00.	- 02 37	15	14.5	GALAXY
ZWG 018.019	13 58 00.	- 02 40		14.8	GALAXY
MCG+00-36-007	13 58 00.	- 02 40	24	14.8	GALAXY
REIZ 3785	13 58 01.	+ 38 54	30	14.5	GALAXY
IC 0968	13 58 02.	- 02 42			NONSTELLAR OBJECT
REIN 7.143	13 58 02.30	+ 38 54 56.1			NEBULA
MCG+07-29-030	13 58 03.	+ 38 43	48	14.5	GALAXY
REIZ 3781	13 58 05.	- 02 37	42	14.2	GALAXY
REIZ 3780	13 58 05.	- 02 40	60	14.6	GALAXY
SC 1355-2147.1	13 58 05.	- 22 01 41.	42		NEBULA
HOLM 566B	13 58 06.	+ 14 05	30	14.9	PART OF MULTIPLE GALAXY
ZC 1358.1+2658	13 58 06.	+ 26 58	1410		CLUSTER OF GALAXIES
ZWG 162.045	13 58 06.	+ 28 01		15.5	GALAXY
ZWG 219.037	13 58 06.	+ 38 44		15.0	GALAXY
UGC 08923	13 58 06.	+ 38 44	72	15.0	GALAXY SO-a
ZC 1358.1+4117	13 58 06.	+ 41 17	1010		CLUSTER OF GALAXIES
ZWG 272.015	13 58 06.	+ 55 01		15.3	GALAXY
MCG+09-23-023	13 58 06.	+ 55 01	42	15.	GALAXY
ZC 1358.1+5939	13 58 06.	+ 59 39	810		CLUSTER OF GALAXIES
ZC 1358.1+6245	13 58 06.	+ 62 45	940		CLUSTER OF GALAXIES
ZWG 336.036	13 58 06.	+ 72 44		15.7	GALAXY
ZWG 018.021	13 58 06.	- 00 16		15.4	GALAXY
8ZW 1358-01.7	13 58 06.	- 01 40		16.9	COMPACT GALAXY
HOLM 566A	13 58 07.	+ 14 04	30	14.3	PART OF MULTIPLE GALAXY
REIZ 3788	13 58 07.	+ 38 44	42	14.2	GALAXY
REIN 7.144	13 58 08.50	+ 38 44 43.6			NEBULA
ARC 1838	13 58 09.	+ 41 18		18.0	RICH CLUSTER OF GALAXIES
REIZ 3793	13 58 11.	+ 55 01	30	14.4	GALAXY
ZWG 018.022	13 58 12.	+ 02 16		14.7	GALAXY
UGC 08924	13 58 12.	+ 02 16	102	14.7	GALAXY Sc
MCG+00-36-009	13 58 12.	+ 02 16	78	14.7	GALAXY
ZWG 132.067	13 58 12.	+ 20 56		15.4	GALAXY
ZC 1358.2+2420	13 58 12.	+ 24 20	870		CLUSTER OF GALAXIES
ZWG 132.068	13 58 12.	+ 26 14		15.7	GALAXY
MCG+07-29-031	13 58 12.	+ 39 08	114	13.	GALAXY
ZWG 219.038	13 58 12.	+ 39 09		13.1	GALAXY
UGC 08925	13 58 12.	+ 39 09	126	13.1	GALAXY SBb
MCG+10-20-059	13 58 12.	+ 59 30	15	15.	GALAXY
ZC 1358.2+7502	13 58 12.	+ 75 02	1680		CLUSTER OF GALAXIES
REIZ 3789	13 58 13.	+ 39 10	60	13.7	GALAXY
REIN 7.145	13 58 13.42	+ 39 09 26.6			NEBULA
RNGC 5398	13 58 14.	- 32 50		14.0	GALAXY
RNGC 5397	13 58 14.	- 33 43			GALAXY
MCG+05-33-038	13 58 15.	+ 28 54 30.	42	15.	GALAXY
RNGC 5406	13 58 15.	+ 39 09		13.0	GALAXY
ZWG 018.023	13 58 18.	+ 02 20		15.6	GALAXY
ZWG 046.034	13 58 18.	+ 06 43		15.5	GALAXY
FAI 13583+13	13 58 18.	+ 13 07			COMPACT GALAXY
ZWG 103.042	13 58 18.	+ 15 10		15.7	GALAXY
MCG+03-36-026	13 58 18.	+ 15 10	48	15.7	GALAXY

OBJECT NAME	RIGHT ASCEN.	DECLINATION	DIAM.	MAGN.	TYPE OF OBJECT
ZWG 132.069	13 58 18.	+ 25 33		15.7	GALAXY
REIZ 3787	13 58 18.	+ 25 33	24	15.7	GALAXY
KARA.73B 0607	13 58 18.	+ 25 33	36	15.7	ISOLATED GALAXY S
ZC 1358.3+2847	13 58 18.	+ 28 47	940		CLUSTER OF GALAXIES
MCG+07-29-032	13 58 18.	+ 38 44	27	14.5	GALAXY
ZWG 219.039	13 58 18.	+ 38 45		15.4	GALAXY
ZC 1358.3+4633	13 58 18.	+ 46 33	1210		CLUSTER OF GALAXIES
MCG+10-20-060	13 58 18.	+ 58 36	12	16.	GALAXY
ZC 1358.3+6326	13 58 18.	+ 63 26	340		CLUSTER OF GALAXIES
FAI 1358+08	13 58 24.	+ 08 32			COMPACT GALAXY
FAI 1358+10	13 58 24.	+ 10 53			COMPACT GALAXY
ZC 1358.4+1542	13 58 24.	+ 15 42	1080		CLUSTER OF GALAXIES
ZWG 295.031	13 58 24.	+ 59 30		15.5	GALAXY
MCG-05-33-036	13 58 24.	- 30 04	150	14.	GALAXY
REIZ 3790	13 58 25.	+ 32 07	30	14.4	GALAXY
ZC 1358.5+0305	13 58 30.	+ 03 05	740		CLUSTER OF GALAXIES
ARC 1835	13 58 30.	+ 03 06		17.6	RICH CLUSTER OF GALAXIES
ZWG 103.043	13 58 30.	+ 15 58		15.7	GALAXY
ZC 1358.5+2249	13 58 30.	+ 22 49	1410		CLUSTER OF GALAXIES
ZWG 162.046	13 58 30.	+ 30 19		15.1	GALAXY
MCG+05-33-039	13 58 30.	+ 30 19	42	15.1	GALAXY
ZWG 162.047	13 58 30.	+ 32 08		14.9	GALAXY
UGC 08926	13 58 30.	+ 32 08	72	14.9	GALAXY S
MCG-05-33-037	13 58 30.	- 32 49	150	14.	GALAXY
RNGC 5404	13 58 31.	+ 00 19			NON-EXISTENT OBJECT
IC 4357	13 58 31.	+ 32 08 26.			NONSTELLAR OBJECT
REIZ 3792	13 58 32.	+ 30 17	42	13.8	GALAXY
HE 0017	13 58 33.	+ 00 19			NEBULA
ARC 1851	13 58 33.	+ 72 22		17.2	RICH CLUSTER OF GALAXIES
FAI 1358+09	13 58 36.	+ 09 20			COMPACT GALAXY
ZC 1358.6+3053	13 58 36.	+ 30 53	2890		CLUSTER OF GALAXIES
MCG+05-33-040	13 58 36.	+ 32 09	60	14.9	GALAXY
MCG-04-33-027	13 58 36.	- 22 20	48	15.	GALAXY
REIZ 3794	13 58 37.	+ 39 24	12	14.4	GALAXY
RNGC 5405	13 58 39.	+ 07 57		14.5	GALAXY
REIN 7.146	13 58 39.83	+ 39 24 28.2			NEBULA
RNGC 5410	13 58 40.	+ 41 14		14.0	GALAXY
REIZ 3791	13 58 41.	+ 17 19	54	15.6	GALAXY
ZWG 046.035	13 58 42.	+ 07 44		15.3	GALAXY
UGC 08927	13 58 42.	+ 07 44	66	15.3	GALAXY Sb-c
ZWG 046.036	13 58 42.	+ 07 57		14.5	GALAXY
UGC 08928	13 58 42.	+ 07 57	54	14.5	GALAXY S
MCG+01-36-014	13 58 42.	+ 07 57	48	14.5	GALAXY
ZWG 074.038	13 58 42.	+ 08 51		15.3	GALAXY
ZC 1358.7+1521	13 58 42.	+ 15 21	16330		CLUSTER OF GALAXIES
ZWG 132.070	13 58 42.	+ 21 28		15.1	GALAXY
UGC 08929	13 58 42.	+ 21 28	60	15.1	GALAXY DBL SYS
ZWG 162.048	13 58 42.	+ 29 48		15.3	GALAXY
ZC 1358.7+3017	13 58 42.	+ 30 17	1810		CLUSTER OF GALAXIES
MCG+07-29-033	13 58 42.	+ 39 22 30.	21	14.5	GALAXY
ZWG 219.040	13 58 42.	+ 39 23		14.5	GALAXY
UGC 08930	13 58 42.	+ 39 23	84	14.5	GALAXY
ZWG 219.041	13 58 42.	+ 41 14		14.1	GALAXY
UGC 08931	13 58 42.	+ 41 14	96	14.1	GALAXY SB
KARA.72 406A	13 58 42.	+ 41 14	90	14.1	PART OF DOUBLE GALAXY
ZWG 018.024	13 58 42.	- 02 15		15.6	GALAXY
MCG-04-33-029	13 58 42.	- 22 53	42	14.5	GALAXY
MCG-04-33-028	13 58 42.	- 25 57	84	14.5	GALAXY
REIN 7.147	13 58 43.72	+ 39 23 53.4			NEBULA
FATH 1.639	13 58 45.	+ 29 48	19		NEBULA
RNGC 5407	13 58 45.	+ 39 23		14.5	GALAXY
VV 256A	13 58 45.	+ 41 13	90	14.	INTERACTING GALAXY
REIZ 3795	13 58 47.	+ 41 14	48	13.4	GALAXY
ZWG 046.037	13 58 48.	+ 03 14		15.7	GALAXY
ZWG 074.039	13 58 48.	+ 10 22		15.3	GALAXY
FAI 1358+12	13 58 48.	+ 12 20			COMPACT GALAXY
ZWG 074.040	13 58 48.	+ 13 39		15.7	GALAXY
ZWG 103.044	13 58 48.	+ 16 00		15.4	GALAXY
VV 277C	13 58 48.	+ 21 28	9	18.	INTERACTING GALAXY
VV 277B	13 58 48.	+ 21 28	12	17.	INTERACTING GALAXY
VV 277A	13 58 48.	+ 21 28	24	16.	INTERACTING GALAXY
VV 277	13 58 48.	+ 21 28	54	16.	INTERACTING GALAXY
LB 00706	13 58 48.	+ 27 11 00.		17.1	FAINT BLUE STAR
TON-N 0749	13 58 48.	+ 28 42		15.6	BLUE STAR
ZWG 162.049	13 58 48.	+ 29 46		15.1	GALAXY
MCG+05-33-041	13 58 48.	+ 29 46	60	15.1	GALAXY
MRK 333	13 58 48.	+ 31 09	20	16.	GALAXY WITH UV CONTINUUM
MCG+07-29-034	13 58 48.	+ 41 13	78	14.	GALAXY
VV 256B	13 58 48.	+ 41 14	66	15.5	INTERACTING GALAXY
ZWG 219.042	13 58 48.	+ 41 15		15.4	GALAXY
UGC 08932	13 58 48.	+ 41 15	66	15.4	GALAXY IRR
KARA.72 406B	13 58 48.	+ 41 15	48	15.4	PART OF DOUBLE GALAXY
ZWG 247.002	13 58 48.	+ 48 41		14.3	GALAXY
UGC 08933	13 58 48.	+ 48 41	114	14.3	GALAXY Sc
ZWG 018.026	13 58 48.	- 00 23		15.2	GALAXY
ZWG 018.025	13 58 48.	- 02 28		15.7	GALAXY
MCG+00-36-010	13 58 48.	- 02 28	42	15.7	GALAXY
RNGC 5392	13 58 48.	- 02 58		15.0	GALAXY
MCG-04-33-030	13 58 48.	- 25 00	60	15.5	GALAXY
FATH 1.640	13 58 49.	+ 29 46	27		NEBULA
RNGC 5425	13 58 50.	+ 48 41		14.5	GALAXY
REIZ 3796	13 58 50.	+ 48 42	60	13.4	GALAXY
REIN 7.148	13 58 50.03	+ 41 13 46.2			NEBULA
MCG+07-29-035	13 58 51.	+ 41 14	30	15.5	GALAXY
MCG+08-26-001	13 58 51.	+ 48 41	120	14.	GALAXY
ZWG 074.041	13 58 54.	+ 10 43		15.2	GALAXY
UGC 08934	13 58 54.	+ 10 43	60	15.2	GALAXY Sc
MCG+03-36-027	13 58 54.	+ 16 01	48	15.4	GALAXY
MCG+04-33-038	13 58 54.	+ 21 27 30.	36	15.1	GALAXY
ZC 1358.9+2735	13 58 54.	+ 27 35	670		CLUSTER OF GALAXIES
ZWG 162.050	13 58 54.	+ 32 05		15.6	GALAXY
MCG+07-29-036	13 58 54.	+ 43 06	24	16.	GALAXY
ZWG 272.016	13 58 54.	+ 55 24		13.1	GALAXY
UGC 08935	13 58 54.	+ 55 24	210	13.1	GALAXY S0-a
MCG+09-23-024	13 58 54.	+ 55 25	180	13.0	GALAXY
MCG-02-36-002	13 58 54.	- 11 22	15	14.5	GALAXY
RNGC 5422	13 58 55.	+ 55 24		13.0	GALAXY
REIN 2.208	13 58 56.74	+ 55 24 19.9			NEBULA
REIZ 3797	13 58 58.	+ 55 24	120	13.0	GALAXY
HOLM 567A	13 58 58.	+ 55 24	90	14.0	PART OF MULTIPLE GALAXY
ARC 1836	13 58 58.	- 11 23		15.7	RICH CLUSTER OF GALAXIES
ZWG 018.027	13 59 00.	+ 01 47		15.6	GALAXY
3ZW 1359+01.9	13 59 00.	+ 01 53		17.6	COMPACT GALAXY
ZWG 074.042	13 59 00.	+ 09 25		15.7	GALAXY
ZC 1359.0+2208	13 59 00.	+ 22 08	2420		CLUSTER OF GALAXIES
TON-N 0750	13 59 00.	+ 23 12		15.2	BLUE STAR
TON-N 0180	13 59 00.	+ 27 55		15.0	BLUE STAR
UGC 08936	13 59 00.	+ 49 38	84	17.	GALAXY DWARF
MCG+10-20-061	13 59 00.	+ 59 03	45	16.	GALAXY
MCG+10-20-062	13 59 00.	+ 59 33	126	12.8	GALAXY
MCG+10-20-063	13 59 00.	+ 59 52	18	16.	GALAXY
ZC 1359.0+8323	13 59 00.	+ 83 23	2490		CLUSTER OF GALAXIES
VV 254B	13 59 03.	+ 23 11 30.	66	14.	INTERACTING GALAXY
VV 254A	13 59 03.	+ 23 11 30.	66	14.	INTERACTING GALAXY
REIN 2.209	13 59 03.06	+ 59 34 46.7			NEBULA
REIZ 3803	13 59 04.	+ 55 26	12	15.2	GALAXY
ARC 1837	13 59 04.	- 10 56		15.7	RICH CLUSTER OF GALAXIES
HOLM 567B	13 59 05.	+ 55 26	18	15.0	PART OF MULTIPLE GALAXY
RNGC 5430	13 59 05.	+ 59 34		13.0	GALAXY
6ZW 1359+08.4	13 59 06.	+ 08 25		16.5	COMPACT GALAXY
ZC 1359.1+3337	13 59 06.	+ 33 37	400		CLUSTER OF GALAXIES
ZWG 191.031	13 59 06.	+ 38 07		15.5	GALAXY
MCG+06-31-043	13 59 06.	+ 38 08	36	15.	GALAXY
ZWG 295.032	13 59 06.	+ 59 33		12.7	GALAXY
UGC 08937	13 59 06.	+ 59 33	138	12.7	GALAXY SBb
REIZ 3805	13 59 07.	+ 59 34	60	13.0	GALAXY
HOLM 569A	13 59 07.	+ 59 34	54	14.0	PART OF MULTIPLE GALAXY
REIN 2.210	13 59 08.86	+ 59 34 11.8			NEBULA
HOLM 569B	13 59 09.	+ 59 34	12	15.0	PART OF MULTIPLE GALAXY
SC 1356-2503.2	13 59 09.	- 25 17 45.	36		NEBULA
IC 0969	13 59 10.	- 03 56 27.			NONSTELLAR OBJECT
REIN 2.211	13 59 10.30	+ 59 33 55.3			NEBULA
LB 00707	13 59 11.	+ 28 34 54.		16.0	FAINT BLUE STAR
ZC 1359.2+0826	13 59 12.	+ 08 26	540		CLUSTER OF GALAXIES
ZWG 074.043	13 59 12.	+ 10 07		15.2	GALAXY
ZWG 103.045	13 59 12.	+ 19 38		15.6	GALAXY
ZWG 162.051	13 59 12.	+ 28 32		15.5	GALAXY
ZWG 191.032	13 59 12.	+ 37 02		15.2	GALAXY
MRK 465	13 59 12.	+ 37 02	10	16.	GALAXY WITH UV CONTINUUM
ZWG 219.043	13 59 12.	+ 43 49		15.7	GALAXY
ZWG 018.028	13 59 12.	- 03 15		15.3	GALAXY
MCG-04-33-031	13 59 12.	- 22 07	42	15.	GALAXY
MCG-04-33-032	13 59 12.	- 25 17	96	14.5	GALAXY
REIZ 3809	13 59 13.	+ 59 33	12	15.1	GALAXY
SC 1356-2510.5	13 59 13.	- 25 25 03.	30		NEBULA
RNGC 5409	13 59 16.	+ 09 44		14.5	GALAXY
ZWG 074.044	13 59 18.	+ 09 44		14.4	GALAXY
UGC 08938	13 59 18.	+ 09 44	102	14.4	GALAXY Sb/SBb
MCG+02-36-009	13 59 18.	+ 09 44	120	14.4	GALAXY
BIGO 536	13 59 18.	+ 09 49			NEBULA
FAI 1359+13	13 59 18.	+ 13 25			COMPACT GALAXY
ZC 1359.3+3147	13 59 18.	+ 31 47	810		CLUSTER OF GALAXIES
MCG+06-31-044	13 59 18.	+ 37 02 30.	30	15.	GALAXY
MCG-04-33-033	13 59 18.	- 25 24	48	15.	GALAXY
MRSL 311+01/1	13 59 18.	- 60 11	236		HII REGION
LB 00708	13 59 21.	+ 26 48 00.		16.9	FAINT BLUE STAR
ARC 1840	13 59 21.	+ 30 49		17.2	RICH CLUSTER OF GALAXIES
ZWG 074.045	13 59 24.	+ 09 02		14.7	GALAXY
MCG+02-36-010	13 59 24.	+ 09 02	36	14.7	GALAXY
ZWG 074.046	13 59 24.	+ 09 53		15.5	GALAXY
ZC 1359.4+2815	13 59 24.	+ 28 15	400		CLUSTER OF GALAXIES
ZC 1359.4+3821	13 59 24.	+ 38 21	810		CLUSTER OF GALAXIES
ZWG 272.017	13 59 24.	+ 52 08		15.4	GALAXY
RNGC 5411	13 59 28.	+ 09 11		14.5	GALAXY
HOLM 568C	13 59 28.	+ 34 04	9	14.9	PART OF MULTIPLE GALAXY
REIZ 3802	13 59 29.	+ 34 04	66	14.3	GALAXY
HOLM 568A	13 59 29.	+ 34 04	54	14.5	PART OF MULTIPLE GALAXY
REIZ 3812	13 59 29.	+ 54 39	24	14.3	GALAXY
KEEL 679	13 59 29.3	+ 54 39 34.			NEBULA
REIN 4.266	13 59 29.73	+ 54 39 32.6			NEBULA
ZWG 074.047	13 59 30.	+ 09 11		14.6	GALAXY
UGC 08940	13 59 30.	+ 09 11	84	14.6	GALAXY E-S0
MCG+02-36-011	13 59 30.	+ 09 11	60	14.6	GALAXY
ZWG 074.048	13 59 30.	+ 09 47		15.6	GALAXY
ZWG 074.049	13 59 30.	+ 12 37		15.6	GALAXY
MCG+02-36-012	13 59 30.	+ 12 37	24	15.6	GALAXY
ZC 1359.5+2650	13 59 30.	+ 26 50	1210		CLUSTER OF GALAXIES
ZWG 162.052	13 59 30.	+ 28 48		15.5	GALAXY
1ZW 078	13 59 30.	+ 34 04			COMPACT GALAXY
ZWG 191.033	13 59 30.	+ 34 04		14.3	GALAXY
RNGC 5421B	13 59 30.	+ 34 04		17.0	GALAXY
RNGC 5421A	13 59 30.	+ 34 04		14.5	GALAXY
MRK 665	13 59 30.	+ 34 04	20	16.5	GALAXY WITH UV CONTINUUM
HOLM 568B	13 59 30.	+ 34 04	12	14.9	PART OF MULTIPLE GALAXY
UGC 08941	13 59 30.	+ 34 04	66	14.3	GALAXY DBL SYS
VV 120E	13 59 30.	+ 34 04	30	18.	INTERACTING GALAXY
VV 120D	13 59 30.	+ 34 04	30	18.	INTERACTING GALAXY
VV 120C	13 59 30.	+ 34 04	18	14.9	INTERACTING GALAXY
VV 120B	13 59 30.	+ 34 04	18	14.9	INTERACTING GALAXY
VV 120A	13 59 30.	+ 34 04	42	14.5	INTERACTING GALAXY
VV 120	13 59 30.	+ 34 04	108		INTERACTING GALAXY
KARA.72 407B	13 59 30.	+ 34 04	30		PART OF DOUBLE GALAXY
KARA.72 407A	13 59 30.	+ 34 04	42	14.3	PART OF DOUBLE GALAXY
ZWG 219.044	13 59 30.	+ 38 55		14.7	GALAXY
MCG+07-29-037	13 59 30.	+ 38 55	54	15.	GALAXY
ZWG 272.018	13 59 30.	+ 54 40		14.9	GALAXY
MCG+09-23-025	13 59 30.	+ 54 40	36	15.	GALAXY
ZWG 018.030	13 59 30.	- 01 07		14.8	GALAXY
UGC 08939	13 59 30.	- 01 07	84	14.8	GALAXY Sb
MCG+00-36-012	13 59 30.	- 01 07	60	14.8	GALAXY
ZWG 018.029	13 59 30.	- 01 09		15.3	GALAXY
MCG+00-36-011	13 59 30.	- 01 09	30	15.3	GALAXY
ARP 111	13 59 32.	+ 34 04			PECULIAR GALAXY
MCG+06-31-046	13 59 33.	+ 34 02	24	17.	GALAXY
MCG+06-31-045	13 59 33.	+ 34 03 30.	66	14.	GALAXY
RNGC 5414	13 59 34.	+ 10 11		14.0	GALAXY
ZWG 018.032	13 59 36.	+ 00 00		15.4	GALAXY
KARA.73B 0608	13 59 36.	+ 00 00	54	15.4	ISOLATED GALAXY S
ZWG 074.050	13 59 36.	+ 10 11		13.8	GALAXY
UGC 08942	13 59 36.	+ 10 11	60	13.8	GALAXY PECULR
MCG+02-36-013	13 59 36.	+ 10 11	60	13.8	GALAXY
FAI 1359+10	13 59 36.	+ 10 51			COMPACT GALAXY
ZWG 103.046	13 59 36.	+ 17 38		15.6	GALAXY
ZWG 103.047	13 59 36.	+ 18 39		15.4	GALAXY
ZC 1359.6+2727	13 59 36.	+ 27 27	1680		CLUSTER OF GALAXIES
ZWG 191.034	13 59 36.	+ 32 42		15.7	GALAXY
ZWG 191.035	13 59 36.	+ 34 55		15.3	GALAXY
ZWG 018.031	13 59 36.	- 01 08		15.6	GALAXY
MCG+00-36-013	13 59 36.	- 01 08	42	15.6	GALAXY
MCG-01-36-002	13 59 36.	- 06 05	42	16.	GALAXY
MCG-05-33-038	13 59 36.	- 33 09	72	15.	GALAXY
ARC 1841	13 59 37.	+ 23 33		17.5	RICH CLUSTER OF GALAXIES
MCG+06-31-047	13 59 39.	+ 34 54	36	14.	GALAXY
RNGC 5417	13 59 40.	+ 08 17		14.0	GALAXY
RNGC 5416	13 59 40.	+ 09 41		13.5	GALAXY
ZWG 046.038	13 59 42.	+ 04 49		15.2	GALAXY

OBJECT NAME	RIGHT ASCEN.	DECLINATION	DIAM.	MAGN.	TYPE OF OBJECT
ZWG 046.039	13 59 42.	+ 08 17		13.8	GALAXY
UGC 08943	13 59 42.	+ 08 17	102	13.8	GALAXY Sa
MCG+01-36-015	13 59 42.	+ 08 17	90	13.8	GALAXY
ZWG 074.051	13 59 42.	+ 09 18		15.7	GALAXY
ZWG 074.052	13 59 42.	+ 09 41		13.6	GALAXY
UGC 08944	13 59 42.	+ 09 41	90	13.6	GALAXY Sc
MCG+02-36-014	13 59 42.	+ 09 41	84	13.6	GALAXY
ZWG 074.053	13 59 42.	+ 09 49		15.7	GALAXY
FAI 1359+11	13 59 42.	+ 11 00			COMPACT GALAXY
ZC 1359.7+2415	13 59 42.	+ 24 15	610		CLUSTER OF GALAXIES
ZWG 191.036	13 59 42.	+ 37 15		14.6	GALAXY
MRK 466	13 59 42.	+ 37 15	20	15.5	GALAXY WITH UV CONTINUUM
UGC 08945	13 59 42.	+ 37 15	66	14.6	GALAXY S0-a
MCG+06-31-048	13 59 42.	+ 37 16	72	14.	GALAXY
RNGC 5418	13 59 45.	+ 07 56		14.5	GALAXY
REIZ 3801	13 59 46.	+ 09 22	48	14.5	GALAXY
REIZ 3798	13 59 46.	+ 09 41	24	13.5	GALAXY
REIZ 3800	13 59 47.	+ 07 55	54	14.4	GALAXY
REIZ 3799	13 59 47.	+ 08 16	60	13.5	GALAXY
REIZ 3804	13 59 47.	+ 26 18	30	14.7	GALAXY
ZWG 046.040	13 59 48.	+ 07 56		14.4	GALAXY
UGC 08946	13 59 48.	+ 07 56	66	14.4	GALAXY SB
MCG+01-36-016	13 59 48.	+ 07 56	72	14.4	GALAXY
ZC 1359.8+2455	13 59 48.	+ 24 55	3020		CLUSTER OF GALAXIES
ZWG 132.071	13 59 48.	+ 26 18		15.0	GALAXY
REIZ 3808	13 59 48.	+ 32 41	18	14.5	GALAXY
ARC 1842	13 59 49.	+ 18 02		17.2	RICH CLUSTER OF GALAXIES
LB 00709	13 59 50.	+ 27 14 36.		16.7	FAINT BLUE STAR
MCG+04-33-039	13 59 51.	+ 26 18	48	15.0	GALAXY
LB 00710	13 59 53.	+ 28 05 36.		17.0	FAINT BLUE STAR
ZWG 132.072	13 59 54.	+ 25 34		15.5	GALAXY
REIZ 3807	13 59 54.	+ 25 34	24	15.5	GALAXY
ARC 1839	13 59 57.	- 04 37		17.4	RICH CLUSTER OF GALAXIES
ZWG 074.054	14 00 00.	+ 10 39		15.4	GALAXY
ARC 1843	14 00 00.	+ 13 51		17.8	RICH CLUSTER OF GALAXIES
ZC 1400.0+1801	14 00 00.	+ 18 01	1880		CLUSTER OF GALAXIES
ZWG 191.037	14 00 00.	+ 32 41		15.6	GALAXY
MCG+08-26-002	14 00 00.	+ 46 32	66	15.	GALAXY
ZWG 247.003	14 00 00.	+ 46 33		14.6	GALAXY
UGC 08947	14 00 00.	+ 46 33	78	14.6	GALAXY S
KARA.73B 0609	14 00 00.	+ 46 33	66	14.6	ISOLATED GALAXY S
RNGC 5439	14 00 01.	+ 46 33		14.5	GALAXY
HOLM 572B	14 00 02.	+ 32 41	12	15.2	PART OF MULTIPLE GALAXY
REIZ 3806	14 00 04.	+ 09 37	36	14.0	GALAXY
ZC 1400.1+0134	14 00 06.	+ 01 34	740		CLUSTER OF GALAXIES
ZWG 018.033	14 00 06.	+ 02 08		15.7	GALAXY
ZWG 074.055	14 00 06.	+ 09 19		15.0	GALAXY
UGC 08948	14 00 06.	+ 09 19	90	15.0	GALAXY SBb
MCG+02-36-015	14 00 06.	+ 09 19	96	15.0	GALAXY
ZWG 074.056	14 00 06.	+ 09 22		15.6	GALAXY
ZWG 103.048	14 00 06.	+ 14 46		15.6	GALAXY
KARA.72 408B	14 00 06.	+ 14 46	48	15.6	PART OF DOUBLE GALAXY
ZWG 103.049	14 00 06.	+ 14 47		14.7	GALAXY
UGC 08949	14 00 06.	+ 14 47	72	14.7	GALAXY S0-a
KARA.72 408A	14 00 06.	+ 14 47	72	14.7	PART OF DOUBLE GALAXY
ZWG 103.050	14 00 06.	+ 18 48		15.7	GALAXY
ZC 1400.1+2335	14 00 06.	+ 23 35	2890		CLUSTER OF GALAXIES
HOLM 572A	14 00 06.	+ 32 41	24	14.3	PART OF MULTIPLE GALAXY
MCG+06-31-049	14 00 06.	+ 32 41	48	15.	GALAXY
HOLM 570B	14 00 08.	+ 14 59	54	14.2	PART OF MULTIPLE GALAXY
REIZ 3810	14 00 10.	+ 09 41	15	14.7	GALAXY
REIZ 3811	14 00 10.	+ 09 43	24	14.1	GALAXY
ARC 1844	14 00 10.	+ 10 45		17.2	RICH CLUSTER OF GALAXIES
IC 0970	14 00 10.	+ 14 46 32.			NONSTELLAR OBJECT
REIZ 3816	14 00 11.	+ 33 38	48	13.9	GALAXY
ZWG 074.057	14 00 12.	+ 09 25		15.7	GALAXY
UGC 08950	14 00 12.	+ 09 25	72	15.7	GALAXY Sc
MCG+02-36-016	14 00 12.	+ 09 35	24	15.5	GALAXY
ZWG 074.058	14 00 12.	+ 09 36		15.5	GALAXY
MCG+03-36-029	14 00 12.	+ 14 46	48	15.6	GALAXY
MCG+03-36-028	14 00 12.	+ 14 47	66	14.7	GALAXY
ZC 1400.2+2037	14 00 12.	+ 20 37	1010		CLUSTER OF GALAXIES
REIZ 3815	14 00 12.	+ 32 42	24	14.5	GALAXY
ZC 1400.2+4241	14 00 12.	+ 42 41	1010		CLUSTER OF GALAXIES
HOLM 570A	14 00 13.	+ 15 03	180	13.3	PART OF MULTIPLE GALAXY
HOLM 571B	14 00 15.	+ 09 34	24	15.3	PART OF MULTIPLE GALAXY
MCG+07-29-038	14 00 15.	+ 38 56	36	16.	GALAXY
RNGC 5423	14 00 16.	+ 09 35		14.0	GALAXY
SC 1357-2200.3	14 00 16.	- 22 14 48.	18		NEBULA
RNGC 5408	14 00 16.	- 41 10			GALAXY
FATH 1.641	14 00 17.	+ 29 37	5		NEBULA
UGC 08951	14 00 18.	+ 09 01	66	15.6	GALAXY Sc
ZWG 074.059	14 00 18.	+ 09 35		13.9	GALAXY
UGC 08952	14 00 18.	+ 09 35	90	13.9	GALAXY E-S0
MCG+02-36-017	14 00 18.	+ 09 35	24	13.9	GALAXY
ZWG 103.051	14 00 18.	+ 18 45		15.2	GALAXY
UGC 08953	14 00 18.	+ 18 45	78	15.2	GALAXY S
MCG+03-36-030	14 00 18.	+ 18 48	72	15.7	GALAXY
FATH 1.642	14 00 18.	+ 28 57	8		NEBULA
ZWG 191.038	14 00 18.	+ 32 45		14.0	GALAXY
RNGC 5433	14 00 18.	+ 32 45		14.0	GALAXY
UGC 08954	14 00 18.	+ 32 45	102	14.0	GALAXY S-IRR
ZWG 219.045	14 00 18.	+ 38 57		15.7	GALAXY
ZWG 272.019	14 00 18.	+ 52 05		15.7	GALAXY
MCG-05-33-039	14 00 18.	- 33 06	18	15.5	GALAXY
ACK 317+19.1	14 00 18.	- 41 09			PLANETARY NEBULA
MCG-07-29-006	14 00 18.	- 41 09	132	13.	GALAXY
REIZ 3813	14 00 22.	+ 09 15	42	14.5	GALAXY
RNGC 5469	14 00 22.	+ 09 34		15.5	GALAXY
HOLM 571A	14 00 22.	+ 09 35	24	14.5	PART OF MULTIPLE GALAXY
ZWG 074.060	14 00 24.	+ 09 02		15.6	GALAXY
ZWG 074.061	14 00 24.	+ 09 09		15.6	GALAXY
ZWG 074.062	14 00 24.	+ 09 36		15.5	GALAXY
MCG+02-36-018	14 00 24.	+ 09 36	24	15.5	GALAXY
ZC 1400.4+0949	14 00 24.	+ 09 49	8060		CLUSTER OF GALAXIES
MCG+06-31-050	14 00 24.	+ 32 44	90	14.	GALAXY
REIZ 3820	14 00 24.	+ 32 46	84	13.2	GALAXY
HOLM 574A	14 00 24.	+ 32 46	84	13.3	PART OF MULTIPLE GALAXY
ZWG 191.039	14 00 24.	+ 35 06		15.5	GALAXY
UGC 08955	14 00 24.	+ 35 06	72	15.5	GALAXY S
MCG+09-23-026	14 00 24.	+ 56 04	114	13.7	GALAXY
MCG-03-36-003	14 00 24.	- 18 12 30.	36	15.	GALAXY
MCG-05-33-040	14 00 24.	- 31 05	72	15.5	GALAXY
ARC 1847	14 00 25.	+ 22 58		17.6	RICH CLUSTER OF GALAXIES
REIZ 3833	14 00 26.	+ 56 04	108	13.7	GALAXY
REIZ 3814	14 00 28.	+ 09 12	42	14.6	GALAXY
HOLM 571C	14 00 28.	+ 09 36	30	15.6	PART OF MULTIPLE GALAXY
RNGC 5424	14 00 28.	+ 09 40		14.5	GALAXY

OBJECT NAME	RIGHT ASCEN.	DECLINATION	DIAM.	MAGN.	TYPE OF OBJECT
REIZ 3823	14 00 28.	+ 35 07	54	13.9	GALAXY
ZWG 074.063	14 00 30.	+ 09 40		14.3	GALAXY
UGC 08956	14 00 30.	+ 09 40	96	14.3	GALAXY S0
MKW 12	14 00 30.	+ 09 40		14.3	POOR GALAXY CLUSTER
MCG+02-36-019	14 00 30.	+ 09 40	30	14.3	GALAXY
UGC 08957	14 00 30.	+ 24 04	66	16.5	GALAXY DWARF
REIZ 3822	14 00 30.	+ 32 47	12	15.5	GALAXY
MCG+06-31-051	14 00 30.	+ 35 05	66	14.5	GALAXY
REIZ 3824	14 00 30.	+ 39 26	54	14.5	GALAXY
ZWG 272.020	14 00 30.	+ 56 03		13.2	GALAXY
UGC 08958	14 00 30.	+ 56 03	198	13.2	GALAXY S
UGC 08959	14 00 30.	+ 69 08	108	16.5	GALAXY DWRF SP
MCG+13-10-015	14 00 30.	+ 79 06	84	15.	GALAXY
MCG-04-33-034	14 00 30.	- 22 19	72	14.	GALAXY
HOLM 574B	14 00 32.	+ 32 48	18	15.3	PART OF MULTIPLE GALAXY
REIZ 3837	14 00 32.	+ 56 05	12	15.1	GALAXY
HOLM 578A	14 00 32.	+ 56 05	90	13.7	PART OF MULTIPLE GALAXY
MCG+07-29-039	14 00 33.	+ 39 23	90	14.	GALAXY
RNGC 5431	14 00 34.	+ 09 37		15.0	GALAXY
ZWG 046.041	14 00 36.	+ 04 13		15.6	GALAXY
ZWG 074.064	14 00 36.	+ 09 12		15.1	GALAXY
ZWG 074.065	14 00 36.	+ 09 37		14.8	GALAXY
MCG+02-36-020	14 00 36.	+ 09 37	42	14.8	GALAXY
MCG+07-29-041	14 00 36.	+ 39 20 30.	6	15.5	GALAXY
MCG+07-29-040	14 00 36.	+ 39 21	60	15.	GALAXY
ZWG 219.046	14 00 36.	+ 39 24		15.2	GALAXY
UGC 08960	14 00 36.	+ 39 24	102	15.2	GALAXY S
KARA.72 409A	14 00 36.	+ 39 24	108	15.2	PART OF DOUBLE GALAXY
1ZW 079	14 00 36.	+ 51 57			COMPACT GALAXY
MCG+10-20-064	14 00 36.	+ 61 02	18	17.	GALAXY
MCG+11-17-013	14 00 36.	+ 63 32	42	16.	GALAXY
ZWG 018.034	14 00 36.	- 00 44		15.5	GALAXY
REIN 7.149	14 00 36.88	+ 39 24 31.1			NEBULA
ARC 1845	14 00 37.	- 08 53		17.2	RICH CLUSTER OF GALAXIES
HOLM 578B	14 00 38.	+ 56 07	12	15.0	PART OF MULTIPLE GALAXY
REIZ 3819	14 00 40.	+ 09 37	18	15.0	GALAXY
REIN 4.267B	14 00 41.34	+ 54 34 12.9			NEBULA
REIN 4.267A	14 00 41.54	+ 54 34 12.9			NEBULA
ZWG 074.066	14 00 42.	+ 09 09		15.0	GALAXY
MCG+02-36-021	14 00 42.	+ 09 09	48	15.0	GALAXY
8ZW 1400+12.0	14 00 42.	+ 11 58		15.3	COMPACT GALAXY
ZWG 074.067	14 00 42.	+ 12 24		15.7	GALAXY
ZC 1400.7+1552	14 00 42.	+ 15 52	3020		CLUSTER OF GALAXIES
ZC 1400.7+2255	14 00 42.	+ 22 55	1080		CLUSTER OF GALAXIES
ZWG 162.053	14 00 42.	+ 28 16		14.7	GALAXY
UGC 08961	14 00 42.	+ 28 16	90	14.7	GALAXY S
ZWG 219.047	14 00 42.	+ 39 22		15.7	GALAXY
UGC 08962	14 00 42.	+ 39 22	78	15.7	GALAXY SB
KARA.72 409B	14 00 42.	+ 39 22	78	15.7	PART OF DOUBLE GALAXY
MCG+09-23-027	14 00 42.	+ 51 26	36	16.	GALAXY
MCG+10-20-065	14 00 42.	+ 59 39	36	16.	GALAXY
MCG-02-36-003	14 00 42.	- 14 44	24	15.5	GALAXY
MCG-05-33-041	14 00 42.	- 33 22	30	15.5	GALAXY
REIN 4.268	14 00 42.92	+ 54 30 41.7			NEBULA
RNGC 5450	14 00 43.	+ 54 28			NON-EXISTENT OBJECT
RNGC 5447	14 00 43.	+ 54 31			NEBULA IN EXTERNAL GALAXY
RNGC 5449	14 00 43.	+ 54 33			NON-EXISTENT OBJECT
REIN 4.269	14 00 43.62	+ 54 31 18.1			NEBULA
RNGC 5419	14 00 44.	- 33 44			GALAXY
REIN 4.270	14 00 44.10	+ 54 35 54.6			NEBULA
MCG+05-33-042	14 00 45.	+ 28 16 30.	96	14.7	GALAXY
ARP 271	14 00 45.	- 05 49			PECULIAR GALAXY
REIN 4.271	14 00 45.16	+ 54 28 26.1			NEBULA
REIZ 3821	14 00 46.	+ 09 31	24	14.6	GALAXY
HOLM 576A	14 00 46.	+ 35 01	120	13.9	PART OF MULTIPLE GALAXY
RNGC 5427	14 00 47.	- 05 47		12.0	GALAXY
RNGC 5429	14 00 47.	- 05 48			NON-EXISTENT OBJECT
RNGC 5428	14 00 47.	- 05 48			NON-EXISTENT OBJECT
RNGC 5426	14 00 47.	- 05 49		12.5	GALAXY
MCG+03-36-031	14 00 48.	+ 16 15	42	17.	GALAXY
ZC 1400.8+2201	14 00 48.	+ 22 01	1080		CLUSTER OF GALAXIES
ZC 1400.8+2416	14 00 48.	+ 24 16	670		CLUSTER OF GALAXIES
ZWG 191.040	14 00 48.	+ 35 00		13.4	GALAXY
UGC 08963	14 00 48.	+ 35 00	192	13.4	GALAXY Sa
MCG+06-31-052	14 00 48.	+ 35 00	180	13.	GALAXY
MCG+12-13-027	14 00 48.	+ 69 06	102	17.	GALAXY
ZWG 336.037	14 00 48.	+ 72 02		15.7	GALAXY
ZWG 353.029	14 00 48.	+ 79 05		15.7	GALAXY
UGC 08964	14 00 48.	+ 79 05	90	15.7	GALAXY
ZWG 018.035	14 00 48.	- 00 18		14.7	GALAXY
MCG+00-36-014	14 00 48.	- 00 18	24	14.7	GALAXY
VV 021B	14 00 48.	- 05 47	132	12.5	INTERACTING GALAXY
VV 021A	14 00 48.	- 05 49	132	11.6	INTERACTING GALAXY
HOLM 573B	14 00 48.	- 05 50	126	12.5	PART OF MULTIPLE GALAXY
MCG-02-36-004	14 00 48.	- 09 55	72	15.	GALAXY
MCG-04-33-035	14 00 48.	- 25 10	30	15.	GALAXY
RNGC 5440	14 00 49.	+ 35 00		13.5	GALAXY
REIZ 3818	14 00 49.	- 05 47	138	11.8	GALAXY
HOLM 573A	14 00 49.	- 05 47	120	11.6	PART OF MULTIPLE GALAXY
REIZ 3817	14 00 49.	- 05 49	132	12.5	GALAXY
RNGC 5434A	14 00 52.	+ 09 41		14.5	GALAXY
REIZ 3829	14 00 52.	+ 28 17	54	14.2	GALAXY
REIZ 3832	14 00 52.	+ 35 01	102	13.8	GALAXY
REIN 4.272	14 00 52.54	+ 54 29 14.5			NEBULA
ZWG 018.036	14 00 54.	+ 00 58		15.2	GALAXY
ZWG 046.042	14 00 54.	+ 02 45		15.5	GALAXY
ZWG 074.068	14 00 54.	+ 09 41		14.3	GALAXY
UGC 08965	14 00 54.	+ 09 41	108	14.3	GALAXY Sc
KARA.72 410A	14 00 54.	+ 09 41	90	14.3	PART OF DOUBLE GALAXY
MCG+02-36-022	14 00 54.	+ 09 41	108	14.3	GALAXY
ZWG 103.052	14 00 54.	+ 16 52		15.6	GALAXY
ZC 1400.9+2102	14 00 54.	+ 21 02	670		CLUSTER OF GALAXIES
ZC 1400.9+2237	14 00 54.	+ 22 37	940		CLUSTER OF GALAXIES
MCG+08-26-003	14 00 54.	+ 49 24	210	12.2	GALAXY
RNGC 5451	14 00 55.	+ 54 36			NON-EXISTENT OBJECT
IC 4358	14 00 55.	- 09 54 38.			NONSTELLAR OBJECT
ARC 1846	14 00 57.	- 25 09		17.6	RICH CLUSTER OF GALAXIES
REIN 4.273	14 00 57.42	+ 54 36 55.0			NEBULA
REIZ 3825	14 00 58.	+ 09 42	42	14.2	GALAXY
HOLM 575A	14 00 58.	+ 09 42	66	13.6	PART OF MULTIPLE GALAXY
RNGC 5434B	14 00 58.	+ 09 43		14.5	GALAXY
REIZ 3826	14 00 58.	+ 09 43	42	14.3	GALAXY
REIZ 3827	14 00 58.	+ 09 53	30	14.2	GALAXY
HOLM 576B	14 00 58.	+ 34 57	36	14.9	PART OF MULTIPLE GALAXY
RNGC 5432	14 00 59.	- 05 43			NON-EXISTENT OBJECT
8ZW 1401+01.9	14 01 00.	+ 01 54		15.2	COMPACT GALAXY
ZWG 046.043	14 01 00.	+ 07 01		15.3	GALAXY
UGC 08966	14 01 00.	+ 07 01	66	15.3	GALAXY Sc

OBJECT NAME	RIGHT ASCEN.	DECLINATION	DIAM.	MAGN.	TYPE OF OBJECT
ZWG 074.069	14 01 00.	+ 09 28		14.8	GALAXY
MCG+02-36-023	14 01 00.	+ 09 28	30	14.8	GALAXY
ZWG 074.070	14 01 00.	+ 09 43		14.7	GALAXY
UGC 08967	14 01 00.	+ 09 43	108	14.7	GALAXY Sb-c
KARA.72 410B	14 01 00.	+ 09 43	102	14.7	PART OF DOUBLE GALAXY
MCG+02-36-024	14 01 00.	+ 09 43	108	14.7	GALAXY
ZWG 103.053	14 01 00.	+ 18 13		15.7	GALAXY
MCG+06-31-053	14 01 00.	+ 34 55	27	15.	GALAXY
UGC 08968	14 01 00.	+ 35 36	66	16.5	GALAXY S?
MCG+07-29-042	14 01 00.	+ 39 00 30.	36	16.	GALAXY
ZWG 247.004	14 01 00.	+ 49 25		12.7	GALAXY
REIZ 3841	14 01 00.	+ 49 25	168	12.6	GALAXY
UGC 08969	14 01 00.	+ 49 25	258	12.7	GALAXY SBb
UGC 08970	14 01 00.	+ 58 13	96	16.0	GALAXY
ZWG 336.038	14 01 00.	+ 69 43		15.5	GALAXY
ZWG 018.037	14 01 00.	- 01 54		15.2	GALAXY
MCG-01-36-003	14 01 00.	- 05 47	120	12.	GALAXY
MCG-01-36-004	14 01 00.	- 05 50	132	13.	GALAXY
REIN 4.274	14 01 00.94	+ 54 36 34.3			NEBULA
RNGC 5441	14 01 01.	+ 34 55		15.5	GALAXY
HOLM 575B	14 01 02.	+ 09 43	60	14.0	PART OF MULTIPLE GALAXY
REIZ 3828	14 01 04.	+ 09 55	18	14.1	GALAXY
REIN 4.275	14 01 04.62	+ 54 36 33.7			NEBULA
ZWG 018.038	14 01 06.	+ 01 04		15.0	GALAXY
MCG+00-36-015	14 01 06.	+ 01 04	24	15.0	GALAXY
ZWG 162.054	14 01 06.	+ 30 06		15.6	GALAXY
MCG+07-29-043	14 01 06.	+ 38 45	60	14.	GALAXY
SN 1909A	14 01 06.	+ 54 42		12.1	SUPERNOVA
MCG+12-13-028	14 01 06.	+ 70 12	51	16.	GALAXY
ZC 1401.1+7448	14 01 06.	+ 74 48	340		CLUSTER OF GALAXIES
MCG-02-36-005	14 01 06.	- 09 55	132	13.	GALAXY
REIN 4.276	14 01 08.36	+ 54 30 49.6			NEBULA
REIZ 3830	14 01 09.	+ 11 37	66	14.7	GALAXY
REIZ 3831	14 01 09.	+ 11 38	15	15.4	GALAXY
HOLM 577A	14 01 09.	+ 11 38	48	14.1	PART OF MULTIPLE GALAXY
HOLM 577B	14 01 09.	+ 11 39	48	14.1	PART OF MULTIPLE GALAXY
RNGC 5436	14 01 10.	+ 09 49		15.0	GALAXY
ARC 1849	14 01 10.	+ 15 40		16.6	RICH CLUSTER OF GALAXIES
REIZ 3838	14 01 10.	+ 34 56	18	14.9	GALAXY
IC 0971	14 01 10.	- 09 53 49.			NONSTELLAR OBJECT
REIN 7.150	14 01 10.36	+ 38 46 11.7			NEBULA
REIN 4.277	14 01 10.86	+ 54 32 49.3			NEBULA
SN 1951B	14 01 11.	+ 11 37		18.2	SUPERNOVA
RNGC 5435	14 01 11.	- 05 40			NON-EXISTENT OBJECT
ZWG 046.044	14 01 12.	+ 06 42		15.6	GALAXY
ZC 1401.2+0921	14 01 12.	+ 09 21	940		CLUSTER OF GALAXIES
ZWG 074.071	14 01 12.	+ 09 49		14.9	GALAXY
UGC 08971	14 01 12.	+ 09 49	72	14.9	GALAXY S0-a
MCG+02-36-025	14 01 12.	+ 09 49	72	14.9	GALAXY
ZWG 074.072	14 01 12.	+ 11 37		15.1	GALAXY
UGC 08972	14 01 12.	+ 11 37	84	15.1	GALAXY Sb
KARA.72 411B	14 01 12.	+ 11 37	60	15.1	PART OF DOUBLE GALAXY
MCG+02-36-026	14 01 12.	+ 11 37	60	15.1	GALAXY
ZWG 074.073	14 01 12.	+ 11 38		15.1	GALAXY
UGC 08973	14 01 12.	+ 11 38	72	15.1	GALAXY Sb
KARA.72 411A	14 01 12.	+ 11 38	60	15.1	PART OF DOUBLE GALAXY
MCG+02-36-027	14 01 12.	+ 11 38	72	15.1	GALAXY
MCG+03-36-032	14 01 12.	+ 15 30	48	16.	GALAXY
REIZ 3836	14 01 12.	+ 16 29	24	15.5	GALAXY
LB 00711	14 01 12.	+ 24 26 24.		18.0	FAINT BLUE STAR
ZC 1401.2+2721	14 01 12.	+ 27 21	940		CLUSTER OF GALAXIES
MCG+06-31-054	14 01 12.	+ 35 22	42	12.	GALAXY
ZWG 191.041	14 01 12.	+ 35 23		12.8	GALAXY
UGC 08974	14 01 12.	+ 35 23	156	12.8	GALAXY E
ZWG 219.048	14 01 12.	+ 38 46		14.2	GALAXY
UGC 08975	14 01 12.	+ 38 46	66	14.2	GALAXY S
MCG+10-20-066	14 01 12.	+ 58 12	66	15.	GALAXY
MRK 282	14 01 12.	+ 69 45	12	16.	GALAXY WITH UV CONTINUUM
MCG-04-33-036	14 01 12.	- 25 58	21	14.5	GALAXY
RNGC 5444	14 01 13.	+ 35 22		13.0	GALAXY
REIZ 3840	14 01 13.	+ 38 47	36	13.5	GALAXY
RNGC 5455	14 01 13.	+ 54 26			NEBULA IN EXTERNAL GALAXY
RNGC 5453	14 01 13.	+ 54 32			NON-EXISTENT OBJECT
RNGC 5420	14 01 15.	- 14 23		13.0	GALAXY
MCG-02-36-006	14 01 15.	- 14 23	66	13.5	GALAXY
REIN 4.278	14 01 15.37	+ 54 28 47.6			NEBULA
RNGC 5437	14 01 16.	+ 09 45		15.0	GALAXY
REIZ 3834	14 01 16.	+ 09 47	18	14.3	GALAXY
REIZ 3835	14 01 16.	+ 09 50	24	14.1	GALAXY
RNGC 5438	14 01 16.	+ 09 51		14.5	GALAXY
REIZ 3839	14 01 16.	+ 35 23	36	13.2	GALAXY
REIN 4.279	14 01 16.52	+ 54 34 23.0			NEBULA
ARC 1850	14 01 17.	+ 09 22		17.5	RICH CLUSTER OF GALAXIES
ZWG 074.074	14 01 18.	+ 09 45		15.1	GALAXY
MCG+02-36-028	14 01 18.	+ 09 45	48	15.1	GALAXY
ZWG 074.075	14 01 18.	+ 09 51		14.7	GALAXY
MCG+02-36-029	14 01 18.	+ 09 51	18	14.7	GALAXY
ZWG 103.054	14 01 18.	+ 16 02		15.6	GALAXY
MCG+03-36-033	14 01 18.	+ 16 02	30	15.6	GALAXY
ZC 1401.3+2125	14 01 18.	+ 21 25	870		CLUSTER OF GALAXIES
ZC 1401.3+2611	14 01 18.	+ 26 11	870		CLUSTER OF GALAXIES
ZWG 191.042	14 01 18.	+ 35 16		14.1	GALAXY
UGC 08976	14 01 18.	+ 35 16	102	14.1	GALAXY S0?
MRK 283	14 01 18.	+ 41 50	7	17.	GALAXY WITH UV CONTINUUM
MCG+10-20-067	14 01 18.	+ 61 59	36	14.	GALAXY
RNGC 5445	14 01 19.	+ 35 16		14.0	GALAXY
IC 4365	14 01 20.	+ 09 46			NONSTELLAR OBJECT
LB 00712	14 01 20.	+ 28 56 06.		17.2	FAINT BLUE STAR
MCG+06-31-055	14 01 21.	+ 35 15 30.	48	14.	GALAXY
REIN 4.280	14 01 22.63	+ 54 41 55.8			NEBULA
LB 00713	14 01 24.	+ 02 25 06.		16.8	FAINT BLUE STAR
ZWG 103.055	14 01 24.	+ 15 06		15.1	GALAXY
MCG+03-36-034	14 01 24.	+ 15 06	18	15.1	GALAXY
ZWG 103.056	14 01 24.	+ 15 38		15.6	GALAXY
UGC 08977	14 01 24.	+ 15 38	72	15.6	GALAXY S
MCG+03-36-035	14 01 24.	+ 15 38	84	15.6	GALAXY
ZWG 103.057	14 01 24.	+ 15 58		15.5	GALAXY
UGC 08978	14 01 24.	+ 15 58	66	15.5	GALAXY S
MCG+03-36-036	14 01 24.	+ 15 58	54	15.5	GALAXY
MCG+07-29-044	14 01 24.	+ 39 17	54	14.5	GALAXY
VV 344A	14 01 24.	+ 54 35	1320	8.5	INTERACTING GALAXY
ARP 026	14 01 24.	+ 54 35			PECULIAR GALAXY
MCG+09-23-028	14 01 24.	+ 54 36	2400	8.2	GALAXY
MCG-02-36-007	14 01 24.	- 09 32	36	15.	GALAXY
MCG-04-33-037	14 01 24.	- 22 22	30	15.5	GALAXY
MRSL 311-00/2	14 01 24.	- 61 50	7200		HII REGION
RNGC 5458	14 01 25.	+ 54 30			NON-EXISTENT OBJECT
SN 1970G	14 01 26.	+ 54 29		11.5	SUPERNOVA

OBJECT NAME	RIGHT ASCEN.	DECLINATION	DIAM.	MAGN.	TYPE OF OBJECT
REIN 4.281A	14 01 26.67	+ 54 32 15.7			NEBULA
REIN 4.281B	14 01 26.67	+ 54 32 16.1			NEBULA
REIN 2.212	14 01 26.96	+ 54 35 16.2			NEBULA
REIZ 3843	14 01 27.	+ 36 00	54	13.1	GALAXY
HN 0384	14 01 27.	- 09 31			NEBULA
IC 4361	14 01 27.	- 09 31			NONSTELLAR OBJECT
MCG-02-36-008	14 01 27.	- 10 31	54	15.	GALAXY
REIN 4.282	14 01 27.92	+ 54 30 05.1			NEBULA
REIZ 3842	14 01 28.	+ 35 16	60	14.1	GALAXY
REIZ 3847	14 01 28.	+ 54 36	450	9.8	GALAXY
HN 0385	14 01 28.	- 11 10			NEBULA
IC 4360	14 01 28.	- 11 10			NONSTELLAR OBJECT
REIN 2.213	14 01 28.79	+ 54 36 34.0			NEBULA
ZWG 018.039	14 01 30.	+ 01 09		15.3	GALAXY
UGC 08979	14 01 30.	+ 06 43	78	16.0	GALAXY Sb-c
FAI 1401+10	14 01 30.	+ 10 17			COMPACT GALAXY
ZC 1401.5+1846	14 01 30.	+ 18 46	810		CLUSTER OF GALAXIES
ZC 1401.5+2359	14 01 30.	+ 23 59	540		CLUSTER OF GALAXIES
ZWG 162.055	14 01 30.	+ 32 14		15.7	GALAXY
ZWG 191.043	14 01 30.	+ 33 48		15.5	GALAXY
MRK 666	14 01 30.	+ 33 48	10	16.5	GALAXY WITH UV CONTINUUM
ZWG 219.049	14 01 30.	+ 39 17		14.4	GALAXY
UGC 08980	14 01 30.	+ 39 17	66	14.4	GALAXY SBb
REIZ 3845	14 01 30.	+ 39 19	30	13.5	GALAXY
ZWG 272.021	14 01 30.	+ 54 35		8.7	GALAXY
UGC 08981	14 01 30.	+ 54 35	1680	8.7	GALAXY Sc
KARA.73B 0610	14 01 30.	+ 54 35	1542	8.7	ISOLATED GALAXY S
MCG+10-20-068	14 01 30.	+ 59 40	102	15.	GALAXY
ZWG 295.033	14 01 30.	+ 61 59		15.5	GALAXY
UGC 08982	14 01 30.	+ 61 59	84	15.5	GALAXY PECULR
MCG-04-33-038	14 01 30.	- 25 23	72	14.	GALAXY
REIN 4.283	14 01 30.55	+ 54 41 51.3			NEBULA
REIN 7.151	14 01 30.75	+ 39 17 33.1			NEBULA
FATH 1.643	14 01 31.	+ 29 36	19		NEBULA
RNGC 5457	14 01 31.	+ 54 36		8.5	GALAXY
RNGC 5443	14 01 32.	+ 56 03		13.0	GALAXY
HN 0386	14 01 33.	- 09 24			NEBULA
IC 4363	14 01 33.	- 09 24			NONSTELLAR OBJECT
HN 0387	14 01 33.	- 09 45			NEBULA
ARC 1852	14 01 34.	+ 16 00		16.6	RICH CLUSTER OF GALAXIES
IC 4364	14 01 34.	- 09 45			NONSTELLAR OBJECT
UGC 08983	14 01 36.	+ 12 14	66	16.0	GALAXY Sc
ZWG 103.058	14 01 36.	+ 15 54		15.7	GALAXY
ZC 1401.6+3237	14 01 36.	+ 32 37	740		CLUSTER OF GALAXIES
ZWG 191.044	14 01 36.	+ 35 59		14.2	GALAXY
UGC 08984	14 01 36.	+ 35 59	78	14.2	GALAXY S
MCG+06-31-056	14 01 36.	+ 35 59	48	14.5	GALAXY
UGC 08985	14 01 36.	+ 59 42	90	16.0	GALAXY
MCG+10-20-069	14 01 36.	+ 61 13	78	15.	GALAXY
MCG-02-36-009	14 01 36.	- 09 45 30.	36	15.5	GALAXY
MCG-02-36-010	14 01 36.	- 15 00	78	15.	GALAXY
REIN 4.284	14 01 36.94	+ 54 29 53.4			NEBULA
REIN 4.285	14 01 38.07	+ 54 28 55.8			NEBULA
LB 00714	14 01 39.	+ 28 08 48.		15.6	FAINT BLUE STAR
REIN 4.286	14 01 40.02	+ 54 34 16.3			NEBULA
IC 0972	14 01 41.	- 16 59 06.			PLANETARY NEBULA
ZWG 046.045	14 01 42.	+ 04 20		15.2	GALAXY
UGC 08986	14 01 42.	+ 04 20	78	15.2	GALAXY S0?
MCG+01-36-017	14 01 42.	+ 04 20	78	15.2	GALAXY
ZWG 074.076	14 01 42.	+ 12 57		15.2	GALAXY
ZWG 074.077	14 01 42.	+ 13 18		15.2	GALAXY
ZWG 018.040	14 01 42.	- 01 24		15.7	GALAXY
REIN 4.287	14 01 47.79	+ 54 37 35.9			NEBULA
ZWG 046.046	14 01 48.	+ 04 57		15.5	GALAXY
ZWG 074.078	14 01 48.	+ 12 15		15.1	GALAXY
ZWG 103.059	14 01 48.	+ 16 34		15.7	GALAXY
UGC 08987	14 01 48.	+ 16 34	222	15.7	GALAXY
ZC 1401.8+2513	14 01 48.	+ 25 13	670		CLUSTER OF GALAXIES
ZWG 295.034	14 01 48.	+ 61 13		15.6	GALAXY
UGC 08988	14 01 48.	+ 61 13	96	15.6	GALAXY Sc
MCG+11-17-014	14 01 48.	+ 66 54	36	16.	GALAXY
SHAH 130	14 01 48.	+ 67 50	120	18.3	GROUP OF COMPACT GALAXIES
MCG-04-33-039	14 01 48.	- 24 35	66	14.	GALAXY
SC 1359-2421.1	14 01 49.	- 24 35 33.	66		NEBULA
REIN 4.288	14 01 49.51	+ 54 40 37.7			NEBULA
PK310-02.1	14 01 50.	- 64 27	20		PLANETARY NEBULA
MCG+06-31-057	14 01 51.	+ 33 33 30.	15	16.	GALAXY
ZWG 074.079	14 01 54.	+ 11 59		15.3	GALAXY
UGC 08989	14 01 54.	+ 11 59	66	15.3	GALAXY S0-a
ZC 1401.9+3218	14 01 54.	+ 32 18	610		CLUSTER OF GALAXIES
ZWG 191.045	14 01 54.	+ 33 33		15.3	GALAXY
MCG+06-31-058	14 01 54.	+ 33 33	24	15.5	GALAXY
ZWG 191.046	14 01 54.	+ 33 35		15.0	GALAXY
UGC 08990	14 01 54.	+ 33 35	90	15.0	GALAXY Sa
ZC 1401.9+3829	14 01 54.	+ 38 29	540		CLUSTER OF GALAXIES
MCG-01-36-005	14 01 54.	- 07 42	42	15.	GALAXY
MCG-02-36-011	14 01 54.	- 09 59	30	15.	GALAXY
IC 4369	14 01 55.	+ 33 33 33.			NONSTELLAR OBJECT
RNGC 5461	14 01 55.	+ 54 33			NEBULA IN EXTERNAL GALAXY
REIN 4.289	14 01 55.80	+ 54 33 24.6			NEBULA
RNGC 5448	14 01 57.	+ 49 25		12.5	GALAXY
SC 1359-2420.9	14 01 58.	- 24 35 21.	120		NEBULA
FATH 1.644	14 01 59.	+ 29 54	5		NEBULA
HOLM 579A	14 01 59.	+ 33 33	18	14.3	PART OF MULTIPLE GALAXY
IC 4370	14 01 59.	+ 33 35 04.			NONSTELLAR OBJECT
LB 09983	14 02	- 88 24		12.5	FAINT BLUE STAR
8ZW 1402+01.1	14 02 00.	+ 01 06		17.3	COMPACT GALAXY
ZWG 103.060	14 02 00.	+ 16 32		14.6	GALAXY
MCG+03-36-037	14 02 00.	+ 16 34	156	15.7	GALAXY
TON-N 0181	14 02 00.	+ 25 08		13.9	BLUE STAR
ZC 1402.0+2704	14 02 00.	+ 27 04	3020		CLUSTER OF GALAXIES
ZC 1402.0+3105	14 02 00.	+ 31 05	3290		CLUSTER OF GALAXIES
ZWG 191.047	14 02 00.	+ 33 25		15.4	GALAXY
MCG+06-31-061	14 02 00.	+ 33 32	18	14.5	GALAXY
IC 4371	14 02 00.	+ 33 32 40.			NONSTELLAR OBJECT
MCG+06-31-059	14 02 00.	+ 33 34	54	15.	GALAXY
MCG+06-31-060	14 02 00.	+ 33 34 30.	18	15.	GALAXY
MCG+09-23-029	14 02 00.	+ 56 37	18	17.	GALAXY
ZWG 317.013	14 02 00.	+ 66 53		15.7	GALAXY
ZWG 317.014	14 02 00.	+ 67 25		15.5	GALAXY
MCG-04-33-040	14 02 00.	- 24 35	120	15.	GALAXY
HN 0391	14 02 03.	- 09 43			NEBULA
REIZ 3846	14 02 04.	+ 09 06	72	15.0	GALAXY
IC 4368	14 02 04.	- 09 43			NONSTELLAR OBJECT
8ZW 1402+01.7	14 02 06.	+ 01 42		17.8	COMPACT GALAXY
ZWG 074.080	14 02 06.	+ 11 13		15.5	GALAXY
ZWG 103.061	14 02 06.	+ 15 43		15.2	GALAXY
UGC 08991	14 02 06.	+ 15 43	72	15.2	GALAXY

OBJECT NAME	RIGHT ASCEN.	DECLINATION	DIAM.	MAGN.	TYPE OF OBJECT
MCG+03-36-038	14 02 06.	+ 16 32 30.	24	14.6	GALAXY
MCG+06-31-063	14 02 06.	+ 33 24	18	16.	GALAXY
ZWG 191.048	14 02 06.	+ 36 02		15.4	GALAXY
MCG+06-31-062	14 02 06.	+ 36 02	24	15.	GALAXY
ARC 1854	14 02 07.	+ 31 20		17.5	RICH CLUSTER OF GALAXIES
RNGC 5462	14 02 07.	+ 54 36			NEBULA IN EXTERNAL GALAXY
SVEN 419	14 02 07.	- 05 19	18	14.9	GALAXY
REIN 4.290	14 02 07.87	+ 54 36 16.1			NEBULA
HOLM 579B	14 02 09.	+ 33 34	48	14.8	PART OF MULTIPLE GALAXY
HELW 473	14 02 09.	- 05 19 32.			NEBULA
REIZ 3850	14 02 10.	+ 35 22	48	15.0	GALAXY
ZWG 074.081	14 02 12.	+ 11 53		14.6	GALAXY
MCG+02-36-030	14 02 12.	+ 11 53	42	14.6	GALAXY
MCG+03-36-039	14 02 12.	+ 15 42 30.	66	15.2	GALAXY
ZC 1402.2+1613	14 02 12.	+ 16 13	810		CLUSTER OF GALAXIES
ZWG 103.062	14 02 12.	+ 19 53		15.7	GALAXY
ZC 1402.2+2246	14 02 12.	+ 22 46	1080		CLUSTER OF GALAXIES
MCG+06-31-065	14 02 12.	+ 33 33	48	15.	GALAXY
MCG+06-31-064	14 02 12.	+ 33 33 30.	18	18.	GALAXY
ZWG 191.049	14 02 12.	+ 34 12		15.7	GALAXY
ZWG 336.039	14 02 12.	+ 73 19		15.6	GALAXY
UGC 08992	14 02 12.	+ 73 19	60	15.6	GALAXY SBb
MCG+12-13-029	14 02 12.	+ 73 20	51	16.	GALAXY
MCG-02-36-012	14 02 12.	- 14 02	60	14.5	GALAXY
MCG-05-33-042	14 02 12.	- 33 32	72	15.	GALAXY
LB 00715	14 02 13.	+ 26 29 00.		17.4	FAINT BLUE STAR
PK326+42.1	14 02 13.	- 17 01 58.	47	14.9	PLANETARY NEBULA
HN 0388	14 02 13.	- 45 02			NEBULA
IC 4359	14 02 13.	- 45 02			NONSTELLAR OBJECT
HN 0390	14 02 14.	- 33 31			NEBULA
IC 4366	14 02 14.	- 33 31			NONSTELLAR OBJECT
MCG+03-36-040	14 02 15.	+ 15 42 30.	24	15.2	GALAXY
LB 00716	14 02 15.	+ 24 50 30.		16.1	FAINT BLUE STAR
REIZ 3853	14 02 15.	+ 35 48	30	13.8	GALAXY
MCG+06-31-066	14 02 15.	+ 35 51	36	15.	GALAXY
MCG-02-36-013	14 02 15.	- 14 37	54	15.	GALAXY
HN 0389	14 02 15.	- 41 34			NEBULA
IC 4362	14 02 15.	- 41 34			NONSTELLAR OBJECT
RNGC 5442	14 02 16.	- 09 29		14.5	GALAXY
SN 1950F	14 02 17.	+ 09 04		16.2	SUPERNOVA
ZWG 074.082	14 02 18.	+ 08 35		15.4	GALAXY
ZWG 074.083	14 02 18.	+ 09 04		14.9	GALAXY
UGC 08995	14 02 18.	+ 09 04	138	14.9	GALAXY
MCG+02-36-031	14 02 18.	+ 09 04	120	14.9	GALAXY
ZWG 103.063	14 02 18.	+ 14 30		15.5	GALAXY
ZWG 074.084	14 02 18.	+ 14 30		15.5	GALAXY
UGC 08996	14 02 18.	+ 14 30	102	15.5	GALAXY Sc
KARA.72 412A	14 02 18.	+ 14 30	96	15.5	PART OF DOUBLE GALAXY
MCG+03-36-041	14 02 18.	+ 14 31	102	15.5	GALAXY
ZWG 103.064	14 02 18.	+ 14 37		14.4	GALAXY
RNGC 5454	14 02 18.	+ 14 37		14.5	GALAXY
UGC 08997	14 02 18.	+ 14 37	96	14.4	GALAXY S0
KARA.72 412B	14 02 18.	+ 14 37	102	14.4	PART OF DOUBLE GALAXY
UGC 08998	14 02 18.	+ 26 02	60	16.0	GALAXY Sc
ZWG 162.056	14 02 18.	+ 28 09		15.3	GALAXY
ZWG 162.057	14 02 18.	+ 29 26		15.5	GALAXY
UGC 08999	14 02 18.	+ 29 26	60	15.5	GALAXY Sc
ZC 1402.3+5925	14 02 18.	+ 59 25	1480		CLUSTER OF GALAXIES
ZWG 018.042	14 02 18.	- 00 22		14.7	GALAXY
UGC 08994	14 02 18.	- 00 22	66	14.7	GALAXY Sb/SBc
KARA.72 413B	14 02 18.	- 00 22	60	14.7	PART OF DOUBLE GALAXY
MCG+00-36-017	14 02 18.	- 00 22	66	14.7	GALAXY
ZWG 018.041	14 02 18.	- 00 24		14.8	GALAXY
UGC 08993	14 02 18.	- 00 24	66	14.8	GALAXY SBa
KARA.72 413A	14 02 18.	- 00 24	72	14.8	PART OF DOUBLE GALAXY
MCG+00-36-016	14 02 18.	- 00 24	60	14.8	GALAXY
MCG-01-36-006	14 02 18.	- 09 29	60	14.5	GALAXY
MCG-07-29-007	14 02 18.	- 41 35	78	14.	GALAXY
HOLM 580B	14 02 19.	+ 29 25	42	15.1	PART OF MULTIPLE GALAXY
MCG+03-36-042	14 02 21.	+ 14 37	90	14.4	GALAXY
FATH 1.645	14 02 21.	+ 29 24	19		NEBULA
HOLM 580A	14 02 21.	+ 29 27	36	14.4	PART OF MULTIPLE GALAXY
REIZ 3852	14 02 22.	+ 28 09	36	15.2	GALAXY
REIN 4.291	14 02 22.17	+ 54 35 27.3			NEBULA
FATH 1.646	14 02 23.	+ 29 26	11		NEBULA
ZWG 074.085	14 02 24.	+ 11 02		15.1	GALAXY
UGC 09001	14 02 24.	+ 11 02	66	15.1	GALAXY SB
UGC 09000	14 02 24.	+ 11 02	66	15.1	GALAXY S
MCG+02-36-032	14 02 24.	+ 11 02	60	15.1	GALAXY
ZWG 074.086	14 02 24.	+ 11 38		14.8	GALAXY
MCG+02-36-033	14 02 24.	+ 11 38	36	14.8	GALAXY
ZWG 074.087	14 02 24.	+ 12 57		15.2	GALAXY
VV 328B	14 02 24.	+ 12 57	18	15.	INTERACTING GALAXY
KARA.72 414B	14 02 24.	+ 12 57	30	15.2	PART OF DOUBLE GALAXY
MCG+02-36-034	14 02 24.	+ 12 57	84	15.2	GALAXY
ZWG 074.088	14 02 24.	+ 12 58		15.3	GALAXY
UGC 09002	14 02 24.	+ 12 58	78	15.3	GALAXY SB
VV 328A	14 02 24.	+ 12 58	108	15.	INTERACTING GALAXY
KARA.72 414A	14 02 24.	+ 12 58	66	15.3	PART OF DOUBLE GALAXY
MCG+02-36-035	14 02 24.	+ 12 58	18	15.3	GALAXY
ZWG 103.065	14 02 24.	+ 16 31		15.7	GALAXY
ZC 1402.4+2045	14 02 24.	+ 20 45	610		CLUSTER OF GALAXIES
ZC 1402.4+2205	14 02 24.	+ 22 05	2550		CLUSTER OF GALAXIES
ZWG 162.058	14 02 24.	+ 28 50		15.4	GALAXY
MCG+05-33-043	14 02 24.	+ 28 50	33	15.4	GALAXY
MCG+05-33-044	14 02 24.	+ 29 25 30.	54	15.5	GALAXY
MCG+05-33-045	14 02 24.	+ 29 27	60	15.5	GALAXY
ZWG 191.050	14 02 24.	+ 35 47		15.4	GALAXY
UGC 09003	14 02 24.	+ 35 47	84	15.4	GALAXY S0
MCG+06-31-067	14 02 24.	+ 35 47	30	14.5	GALAXY
ZC 1402.4+7212	14 02 24.	+ 72 12	1410		CLUSTER OF GALAXIES
FATH 1.647	14 02 25.	+ 28 49	14		NEBULA
REIN 4.292	14 02 26.16	+ 54 39 38.2			NEBULA
REIZ 3849	14 02 27.	+ 11 03	42	14.8	GALAXY
MCG-03-36-004	14 02 27.	- 17 16	36	15.	GALAXY
REIZ 3861	14 02 28.	+ 50 57	72	12.7	GALAXY
REIZ 3863	14 02 28.	+ 57 46	90	13.6	GALAXY
RNGC 5456	14 02 29.	+ 12 07		14.0	GALAXY
RNGC 5459	14 02 29.	+ 13 22		14.5	GALAXY
ZWG 074.089	14 02 30.	+ 12 07		14.2	GALAXY
UGC 09004	14 02 30.	+ 12 07	90	14.2	GALAXY S0
MCG+02-36-036	14 02 30.	+ 12 07	23	14.2	GALAXY
ZWG 074.090	14 02 30.	+ 13 22		14.5	GALAXY
UGC 09005	14 02 30.	+ 13 22	78	14.5	GALAXY S0
MCG+02-36-037	14 02 30.	+ 13 22	18	14.5	GALAXY
MCG+03-36-043	14 02 30.	+ 16 31	36	15.7	GALAXY
ZWG 132.073	14 02 30.	+ 21 53		14.9	GALAXY
MRK 667	14 02 30.	+ 21 53	13	15.	GALAXY WITH UV CONTINUUM

OBJECT NAME	RIGHT ASCEN.	DECLINATION	DIAM.	MAGN.	TYPE OF OBJECT
MCG+04-33-040	14 02 30.	+ 21 53	15	14.9	GALAXY
ZC 1402.5+2743	14 02 30.	+ 27 43	1140		CLUSTER OF GALAXIES
ZWG 162.059	14 02 30.	+ 30 59		15.5	GALAXY
MCG+07-29-045	14 02 30.	+ 38 46	36	16.5	GALAXY
ZC 1402.5+4217	14 02 30.	+ 42 17	1080		CLUSTER OF GALAXIES
MCG-03-36-005	14 02 30.	- 17 15	36	16.	GALAXY
OCL 0912	14 02 30.	- 67 20	1680	10.	OPEN STAR CLUSTER
REIN 4.293	14 02 31.13	+ 54 40 02.0			NEBULA
REIZ 3851	14 02 32.	+ 12 07	30	14.0	GALAXY
REIZ 3854	14 02 34.	+ 19 05	24	15.2	GALAXY
REIZ 3862	14 02 34.	+ 50 57	42	13.3	GALAXY
REIZ 3848	14 02 35.	- 03 08	24	14.1	GALAXY
ZWG 018.044	14 02 36.	+ 00 10		15.3	GALAXY
UGC 09006	14 02 36.	+ 00 10	78	15.3	GALAXY S
KARA.73B 0611	14 02 36.	+ 00 10	66	15.3	ISOLATED GALAXY S
ZWG 074.091	14 02 36.	+ 09 35		15.6	GALAXY
UGC 09007	14 02 36.	+ 09 35	60	15.6	GALAXY
FAI 1402+11	14 02 36.	+ 11 15			COMPACT GALAXY
ZWG 074.092	14 02 36.	+ 11 15		15.3	GALAXY
UGC 09008	14 02 36.	+ 11 15	66	15.3	GALAXY Sc
ZWG 103.066	14 02 36.	+ 16 49		14.9	GALAXY
MCG+03-36-044	14 02 36.	+ 16 49	72	14.9	GALAXY
ZC 1402.6+2118	14 02 36.	+ 21 18	1340		CLUSTER OF GALAXIES
MCG+05-33-046	14 02 36.	+ 30 59	51	15.5	GALAXY
ZWG 162.060	14 02 36.	+ 31 18		15.4	GALAXY
ZC 1402.6+3227	14 02 36.	+ 32 27	740		CLUSTER OF GALAXIES
ZWG 018.043	14 02 36.	- 03 08		14.6	GALAXY
MCG+00-36-018	14 02 36.	- 03 08	30	14.6	GALAXY
MCG-04-33-041	14 02 36.	- 26 21	36	14.5	GALAXY
REIZ 3857	14 02 37.	+ 31 01	48	14.7	GALAXY
RNGC 5446	14 02 40.	+ 09 51			GALAXY
IC 4367	14 02 41.	- 38 58 09.			NONSTELLAR OBJECT
8ZW 1402+09.5	14 02 42.	+ 09 32		17.3	COMPACT GALAXY
ZWG 103.067	14 02 42.	+ 16 00		15.7	GALAXY
UGC 09009	14 02 42.	+ 16 00	60	15.7	GALAXY S
MCG+03-36-045	14 02 42.	+ 16 01	60	15.7	GALAXY
ZC 1402.7+2138	14 02 42.	+ 21 38	740		CLUSTER OF GALAXIES
STOCK 49	14 02 42.	+ 42 01			BLUE KNOT NEAR ELLIP GLXY
MCG+09-23-030	14 02 42.	+ 54 38	54	15.	GALAXY
MCG+10-20-070	14 02 42.	+ 60 16	39	16.	GALAXY
MCG+11-17-015	14 02 42.	+ 64 56	12	17.	GALAXY
REIN 4.294	14 02 43.91	+ 54 38 07.2			NEBULA
ZWG 074.093	14 02 48.	+ 09 52		15.5	GALAXY
MCG+02-36-038	14 02 48.	+ 10 38	24	16.	GALAXY
ZWG 074.094	14 02 48.	+ 10 39		15.1	GALAXY
ZC 1402.8+1755	14 02 48.	+ 17 55	670		CLUSTER OF GALAXIES
TON-N 0182	14 02 48.	+ 26 11		15.4	BLUE STAR
REIZ 3860	14 02 48.	+ 33 02	36	14.8	GALAXY
ZWG 191.051	14 02 48.	+ 35 51		15.7	GALAXY
ZC 1402.8+3954	14 02 48.	+ 39 54	3020		CLUSTER OF GALAXIES
ZC 1402.8+4719	14 02 48.	+ 47 19	1410		CLUSTER OF GALAXIES
ZC 1402.8+5823	14 02 48.	+ 58 23	2290		CLUSTER OF GALAXIES
RNGC 5471	14 02 49.	+ 54 38			NEBULA IN EXTERNAL GALAXY
REIZ 3858	14 02 50.	+ 21 15	54	15.1	GALAXY
ARC 1853	14 02 50.	- 19 32		17.3	RICH CLUSTER OF GALAXIES
REIZ 3855	14 02 52.	+ 09 11	48	15.1	GALAXY
REIZ 3856	14 02 52.	+ 09 13	12	15.0	GALAXY
ZWG 132.074	14 02 54.	+ 21 17		15.6	GALAXY
MCG+05-33-047	14 02 54.	+ 29 39 30.	48	15.5	GALAXY
ZC 1402.9+6653	14 02 54.	+ 66 53	4230		CLUSTER OF GALAXIES
ZWG 018.045	14 02 54.	- 00 35		15.4	GALAXY
REIZ 3859	14 02 56.	+ 21 17	48	14.8	GALAXY
MCG-02-36-014	14 02 57.	- 13 22	24	16.	GALAXY
FATH 1.648	14 02 58.	+ 29 39	16		NEBULA
REIN 2.214	14 02 58.70	+ 55 07 53.6			NEBULA
VDB.66G 186	14 03	+ 54 43	70		DWARF GALAXY
ZWG 018.047	14 03 00.	+ 01 47		15.7	GALAXY
8ZW 1403+01.8	14 03 00.	+ 01 47		15.7	COMPACT GALAXY
FAI 14030+09	14 03 00.	+ 09 20			COMPACT GALAXY
ZWG 074.095	14 03 00.	+ 12 41		15.6	GALAXY
ZC 1403.0+2313	14 03 00.	+ 23 13	1210		CLUSTER OF GALAXIES
ZC 1403.0+2429	14 03 00.	+ 24 29	2890		CLUSTER OF GALAXIES
MCG+05-33-048	14 03 00.	+ 29 40	24	15.5	GALAXY
ZWG 162.061	14 03 00.	+ 31 02		15.7	GALAXY
UGC 09010	14 03 00.	+ 31 02	60	15.7	GALAXY Sc
ZWG 272.022	14 03 00.	+ 55 08		12.5	GALAXY
UGC 09011	14 03 00.	+ 55 08	132	12.5	GALAXY S0
MCG+09-23-031	14 03 00.	+ 55 09	60	12.4	GALAXY
ZWG 018.046	14 03 00.	- 01 07		15.6	GALAXY
REIN 2.215	14 03 00.90	+ 55 08 08.6			NEBULA
FATH 1.649	14 03 01.	+ 29 40	19		NEBULA
RNGC 5473	14 03 02.	+ 55 08		13.0	GALAXY
REIZ 3869	14 03 03.	+ 55 09	90	12.8	GALAXY
HN 0392	14 03 04.	- 10 39			NEBULA
IC 4372	14 03 04.	- 10 39			NONSTELLAR OBJECT
ZWG 074.096	14 03 06.	+ 09 09		15.3	GALAXY
ZWG 074.097	14 03 06.	+ 12 06		15.7	GALAXY
ZC 1403.1+3004	14 03 06.	+ 30 04	1080		CLUSTER OF GALAXIES
MCG+05-33-049	14 03 06.	+ 31 03	42	15.7	GALAXY
ARC 1855	14 03 06.	+ 47 20		17.6	RICH CLUSTER OF GALAXIES
ZC 1403.1+4912	14 03 06.	+ 49 12	1210		CLUSTER OF GALAXIES
MCG+05-33-050	14 03 09.	+ 30 59	30	15.5	GALAXY
HOLM 581A	14 03 10.	+ 09 10	48	14.7	PART OF MULTIPLE GALAXY
HOLM 581B	14 03 11.	+ 09 12	12	15.0	PART OF MULTIPLE GALAXY
RNGC 5466	14 03 11.	+ 28 46		10.5	GLOBULAR CLUSTER
REIZ 3871	14 03 11.	+ 53 54	252	11.6	GALAXY
ZWG 074.098	14 03 12.	+ 13 53		15.6	GALAXY
GCL 027	14 03 12.	+ 28 46	552	10.39	GLOBULAR STAR CLUSTER
UGC 10691	14 03 12.	+ 30 28	60	15.5	GALAXY S
ZWG 162.062	14 03 12.	+ 31 00		15.5	GALAXY
UGC 09012	14 03 12.	+ 31 00	66	15.5	GALAXY S0
MCG+05-33-051	14 03 12.	+ 31 00	60	15.5	GALAXY
ZWG 272.023	14 03 12.	+ 53 54		11.9	GALAXY
UGC 09013	14 03 12.	+ 53 54	390	11.9	GALAXY S(c)
VV 344B	14 03 12.	+ 53 54	270	12.1	INTERACTING GALAXY
MCG-02-36-015	14 03 12.	- 11 50	42	15.	GALAXY
MCG-02-36-016	14 03 12.	- 15 04	48	15.5	GALAXY
MCG+09-23-032	14 03 15.	+ 53 53	180	11.2	GALAXY
REIZ 3864	14 03 16.	+ 19 16	42	15.0	GALAXY
REIZ 3867	14 03 16.	+ 35 10	60	13.8	GALAXY
ZWG 074.099	14 03 18.	+ 09 45		15.7	GALAXY
ZC 1403.3+1413	14 03 18.	+ 14 13	1480		CLUSTER OF GALAXIES
MCG+06-31-068	14 03 18.	+ 35 09	66	14.5	GALAXY
ZWG 191.052	14 03 18.	+ 35 10		15.7	GALAXY
UGC 09014	14 03 18.	+ 35 10	72	15.7	GALAXY S
MCG+08-26-004	14 03 18.	+ 46 22	30	16.	GALAXY
RNGC 5474	14 03 19.	+ 53 54		11.5	GALAXY
HOLM 582A	14 03 22.	+ 09 39	30	14.1	PART OF MULTIPLE GALAXY

OBJECT NAME	RIGHT ASCEN.	DECLINATION	DIAM.	MAGN.	TYPE OF OBJECT
FATH 1.650	14 03 22.	+ 29 40	5		NEBULA
HOLM 582E	14 03 24.	+ 09 40	18	15.5	PART OF MULTIPLE GALAXY
ZWG 132.075	14 03 24.	+ 25 28		15.3	GALAXY
MCG+09-23-033	14 03 24.	+ 56 00	102	13.4	GALAXY
IC 4373	14 03 25.	+ 25 28 15.			NONSTELLAR OBJECT
FATH 1.651	14 03 25.	+ 29 21	5		NEBULA
REIZ 3866	14 03 28.	+ 09 39	18	15.6	GALAXY
REIZ 3865	14 03 28.	+ 09 39	30	13.8	GALAXY
ZWG 046.047	14 03 30.	+ 02 35		15.6	GALAXY
FAI 14035+09	14 03 30.	+ 09 16			COMPACT GALAXY
ZWG 074.100	14 03 30.	+ 09 16		15.2	GALAXY
UGC 09015	14 03 30.	+ 09 16	72	15.2	GALAXY S
ZWG 074.101	14 03 30.	+ 13 01		14.7	GALAXY
FAI 1403+13	14 03 30.	+ 13 42			COMPACT GALAXY
TON-N 0751	14 03 30.	+ 24 39		15.0	BLUE STAR
ZWG 272.024	14 03 30.	+ 55 59		13.4	GALAXY
UGC 09016	14 03 30.	+ 55 59	126	13.4	GALAXY Sa?
ZWG 018.048	14 03 30.	- 00 04		15.3	GALAXY
MCG-04-33-042	14 03 30.	- 25 33	12	15.5	GALAXY
REIZ 3877	14 03 31.	+ 55 59	120	13.4	GALAXY
RNGC 5475	14 03 33.	+ 55 59		13.5	GALAXY
ARC 1859	14 03 34.	+ 60 21		17.8	RICH CLUSTER OF GALAXIES
ZWG 046.048	14 03 36.	+ 02 33		15.1	GALAXY
8ZW 1403+02.5	14 03 36.	+ 02 33		15.1	COMPACT GALAXY
MCG+01-36-018	14 03 36.	+ 02 33	36	15.1	GALAXY
FEIG 089	14 03 36.	+ 07 02		14.2	FAINT BLUE STAR
MCG+02-36-039	14 03 36.	+ 13 02	36	14.7	GALAXY
8ZW 1403+14.5	14 03 36.	+ 14 29		17.8	COMPACT GALAXY
ZWG 219.050	14 03 36.	+ 41 05		15.7	GALAXY
MCG-03-36-006	14 03 36.	- 16 24	42	15.	GALAXY
KEEL 680	14 03 36.5	+ 54 30 28.			NEBULA
SVEN 420	14 03 37.	- 05 17	24	16.2	GALAXY
HOLM 583A	14 03 38.	+ 13 02	30	14.1	PART OF MULTIPLE GALAXY
RNGC 5463A	14 03 40.	+ 09 36		14.0	GALAXY
HELW 474	14 03 40.	- 05 17 17.			NEBULA
HOLM 583B	14 03 41.	+ 13 02	18	15.5	PART OF MULTIPLE GALAXY
REIN 4.296	14 03 41.56	+ 54 41 10.4			NEBULA
ZWG 074.102	14 03 42.	+ 09 36		14.1	GALAXY
UGC 09017	14 03 42.	+ 09 36	72	14.1	GALAXY S
MCG+02-36-040	14 03 42.	+ 09 36	72	14.1	GALAXY
ZWG 074.103	14 03 42.	+ 12 01		15.1	GALAXY
ZWG 074.104	14 03 42.	+ 12 45		15.6	GALAXY
MCG+10-20-071	14 03 42.	+ 58 39	24	16.	GALAXY
MCG-05-33-043	14 03 42.	- 30 57	60	16.	GALAXY
SVEN 421	14 03 43.	- 05 28 53.	60	14.7	GALAXY
ARC 1856	14 03 45.	+ 25 22		17.2	RICH CLUSTER OF GALAXIES
REIZ 3879	14 03 45.	+ 54 43	90	13.9	GALAXY
HELW 475	14 03 45.	- 05 28 47.			NEBULA
HOLM 584A	14 03 46.	+ 09 36	36	13.9	PART OF MULTIPLE GALAXY
FATH 1.652	14 03 47.	+ 29 27	8		NEBULA
ZWG 046.049	14 03 48.	+ 04 57		15.4	GALAXY
HOLM 584B	14 03 48.	+ 09 37	30	15.4	PART OF MULTIPLE GALAXY
ZWG 103.068	14 03 48.	+ 14 56		15.2	GALAXY
ZC 1403.8+2524	14 03 48.	+ 25 24	1880		CLUSTER OF GALAXIES
ZC 1403.8+2841	14 03 48.	+ 28 41	540		CLUSTER OF GALAXIES
TON-N 0183	14 03 48.	+ 31 40		14.0	BLUE STAR
MCG+07-29-046	14 03 48.	+ 43 47	36	15.	GALAXY
ZWG 272.025	14 03 48.	+ 54 42		14.5	GALAXY
UGC 09018	14 03 48.	+ 54 42	114	14.5	GALAXY Sc-IRR
MCG+09-23-034	14 03 48.	+ 54 42	72	13.	GALAXY
RNGC 5465	14 03 48.	- 05 17			NON-EXISTENT OBJECT
HOLM 585E	14 03 48.	- 05 18	48	15.4	PART OF MULTIPLE GALAXY
KEEL 681	14 03 48.3	+ 54 41 56.			NEBULA
REIN 4.297	14 03 48.61	+ 54 41 51.6			NEBULA
RNGC 5477	14 03 49.	+ 54 42		14.5	GALAXY
REIZ 3868	14 03 49.	- 05 16	18	14.8	GALAXY
SVEN 422	14 03 49.	- 05 29	18	16.4	GALAXY
REIZ 3873	14 03 51.	+ 20 24	30	14.9	GALAXY
REIZ 3875	14 03 51.	+ 36 04	54	13.7	GALAXY
IC 0973	14 03 52.	- 05 15			MAY NOT EXIST
FATH 1.653	14 03 53.	+ 29 09	8		NEBULA
ZWG 103.069	14 03 54.	+ 20 22		15.4	GALAXY
UGC 09019	14 03 54.	+ 20 22	60	15.4	GALAXY Sb-c
MCG+03-36-046	14 03 54.	+ 20 24	48	15.4	GALAXY
RNGC 5467	14 03 54.	- 05 15			NON-EXISTENT OBJECT
RNGC 5470	14 03 57.	+ 06 16		14.5	GALAXY
IC 0974	14 03 57.	- 05 17			MAY NOT EXIST
ZWG 046.050	14 04 00.	+ 06 16		14.5	GALAXY
REIZ 3872	14 04 00.	+ 06 16	156	14.1	GALAXY
UGC 09020	14 04 00.	+ 06 16	156	14.5	GALAXY Sb
MCG+01-36-019	14 04 00.	+ 06 16	150	14.5	GALAXY
FAI 1404+09	14 04 00.	+ 09 23			COMPACT GALAXY
ZWG 103.070	14 04 00.	+ 16 43		14.7	GALAXY
MCG+03-36-047	14 04 00.	+ 16 43	48	14.7	GALAXY
ZC 1404.0+2013	14 04 00.	+ 20 13	940		CLUSTER OF GALAXIES
ZC 1404.0+2650	14 04 00.	+ 26 50	740		CLUSTER OF GALAXIES
FATH 1.654	14 04 00.	+ 29 55	8		NEBULA
ZC 1404.0+2956	14 04 00.	+ 29 56	610		CLUSTER OF GALAXIES
ZC 1404.0+5443	14 04 00.	+ 54 43	1080		CLUSTER OF GALAXIES
ZWG 336.040	14 04 00.	+ 69 20		15.6	GALAXY
8ZW 1404-02.0	14 04 00.	- 02 02		18.1	COMPACT GALAXY
8ZW 1404-03.3	14 04 00.	- 03 20		17.1	COMPACT GALAXY
RNGC 5468	14 04 00.	- 05 14		12.5	GALAXY
HOLM 585A	14 04 01.	- 05 12	126	12.4	PART OF MULTIPLE GALAXY
SVEN 423	14 04 01.	- 05 12 52.	150	12.4	GALAXY
REIZ 3870	14 04 01.	- 05 13	120	12.7	GALAXY
ZWG 046.051	14 04 06.	+ 03 59		15.3	GALAXY
ZWG 074.105	14 04 06.	+ 12 57		15.7	GALAXY
UGC 09021	14 04 06.	+ 12 57	72	15.7	GALAXY Sc
TON-N 0752	14 04 06.	+ 25 24		16.1	BLUE STAR
ZC 1404.1+3029	14 04 06.	+ 30 29	1210		CLUSTER OF GALAXIES
ZWG 191.053	14 04 06.	+ 36 01		15.1	GALAXY
UGC 09022	14 04 06.	+ 36 01	108	15.1	GALAXY
MCG+06-31-069	14 04 06.	+ 36 02	90	13.5	GALAXY
ZWG 272.026	14 04 06.	+ 53 22		15.4	GALAXY
SN 1968J	14 04 06.	+ 53 22		16.6	SUPERNOVA
MRK 284	14 04 06.	+ 69 23	15	16.	GALAXY WITH UV CONTINUUM
MCG-01-36-007	14 04 06.	- 05 12	126	12.	GALAXY
REIZ 3874	14 04 07.	+ 04 00	24	15.3	GALAXY
LB 00717	14 04 09.	+ 27 46 00.		17.3	FAINT BLUE STAR
RNGC 5464	14 04 09.	- 29 46		14.0	GALAXY
ZC 1404.2+2334	14 04 12.	+ 23 34	870		CLUSTER OF GALAXIES
ZC 1404.2+2855	14 04 12.	+ 28 55	2620		CLUSTER OF GALAXIES
REIZ 3881	14 04 12.	+ 59 19	42	14.5	GALAXY
ZC 1404.2+6010	14 04 12.	+ 60 10	1610		CLUSTER OF GALAXIES
MCG-05-33-045	14 04 12.	- 29 46	48	14.5	GALAXY
MCG-05-33-044	14 04 12.	- 30 57	42	16.	GALAXY
LB 00718	14 04 15.	+ 03 02 54.		17.7	FAINT BLUE STAR
ARC 1865	14 04 16.	+ 58 55		17.8	RICH CLUSTER OF GALAXIES
ZWG 018.050	14 04 18.	+ 01 05		15.7	GALAXY
8ZW 1404+01.1	14 04 18.	+ 01 05		15.7	COMPACT GALAXY
ZC 1404.3+1944	14 04 18.	+ 19 44	1550		CLUSTER OF GALAXIES
ZC 1404.3+4241	14 04 18.	+ 42 41	740		CLUSTER OF GALAXIES
ZWG 018.049	14 04 18.	- 01 18		15.2	GALAXY
RNGC 5472	14 04 18.	- 05 14		15.0	GALAXY
SVEN 424	14 04 19.	- 05 13	42	14.7	GALAXY
HOLM 585B	14 04 23.	- 05 14	60	14.8	PART OF MULTIPLE GALAXY
FAI 1404+08	14 04 24.	+ 08 28			COMPACT GALAXY
ZWG 074.106	14 04 24.	+ 09 34		15.4	GALAXY
UGC 09023	14 04 24.	+ 09 34	72	15.4	GALAXY Sc
FAI 1404+11	14 04 24.	+ 11 12			COMPACT GALAXY
ZWG 103.071	14 04 24.	+ 15 16		15.5	GALAXY
UGC 09024	14 04 24.	+ 22 16	138	16.0	GALAXY S
ZWG 219.051	14 04 24.	+ 39 44		15.3	GALAXY
MCG+11-17-016	14 04 24.	+ 65 58	30	18.	GALAXY
MCG+11-17-017	14 04 24.	+ 66 03	24	17.	GALAXY
ZWG 018.051	14 04 24.	- 01 21		15.1	GALAXY
MCG-01-36-008	14 04 24.	- 05 12	72	15.	GALAXY
MCG-03-36-007	14 04 24.	- 18 53 30.	48	15.5	GALAXY
MCG-04-33-043	14 04 24.	- 26 36	48	15.	GALAXY
MCG-04-33-044	14 04 24.	- 26 56	42	15.5	GALAXY
MCG-04-33-044	14 04 24.	- 26 56	42	15.5	GALAXY
MCG-05-33-046	14 04 24.	- 28 42	72	15.5	GALAXY
OCL 0925	14 04 24.	- 48 05	3000	6.3	OPEN STAR CLUSTER
REIZ 3876	14 04 25.	- 05 13	30	14.7	GALAXY
RNGC 5480	14 04 28.	+ 50 58		13.0	GALAXY
ZWG 046.052	14 04 30.	+ 08 29		15.5	GALAXY
8ZW 1404+08.5	14 04 30.	+ 08 29		15.5	COMPACT GALAXY
ZWG 074.107	14 04 30.	+ 10 42		14.8	GALAXY
VV 103	14 04 30.	+ 10 42	42	15.	INTERACTING GALAXY
MCG+02-36-041	14 04 30.	+ 10 42	48	14.8	GALAXY
ZWG 074.108	14 04 30.	+ 12 48		15.3	GALAXY
UGC 09025	14 04 30.	+ 12 48	66	15.3	GALAXY Sb/SBc
ZWG 074.109	14 04 30.	+ 13 14		15.6	GALAXY
8ZW 1404+13.2	14 04 30.	+ 13 14		15.6	COMPACT GALAXY
KARA.72 415A	14 04 30.	+ 13 14	42	15.6	PART OF DOUBLE GALAXY
ZC 1404.5+2737	14 04 30.	+ 27 37	2150		CLUSTER OF GALAXIES
UGC 09026	14 04 30.	+ 50 39	102	12.6	GALAXY Sc?
ZWG 272.027	14 04 30.	+ 50 59		12.6	GALAXY
KARA.72 416A	14 04 30.	+ 50 59	102	12.6	PART OF DOUBLE GALAXY
MCG+11-17-018	14 04 30.	+ 65 57	18	18.	GALAXY
8ZW 1404-02.3	14 04 30.	- 02 19		18.6	COMPACT GALAXY
MCG-04-33-045	14 04 30.	- 21 47	48	17.	GALAXY
RNGC 5460	14 04 30.	- 48 05		6.0	OPEN CLUSTER
HOLM 586A	14 04 33.	+ 10 42	30	14.5	PART OF MULTIPLE GALAXY
REIZ 3880	14 04 33.	+ 35 40	30	14.8	GALAXY
HOLM 588A	14 04 33.	+ 50 57	72	12.1	PART OF MULTIPLE GALAXY
REIZ 3878	14 04 34.	- 01 25	18	14.5	GALAXY
HOLM 586B	14 04 35.	+ 10 44	18	15.5	PART OF MULTIPLE GALAXY
ZWG 074.110	14 04 36.	+ 10 53		15.4	GALAXY
UGC 09027	14 04 36.	+ 10 53	60	15.4	GALAXY S
ZWG 074.111	14 04 36.	+ 13 14		15.3	GALAXY
8ZW 1404+13.2	14 04 36.	+ 13 14		15.3	COMPACT GALAXY
KARA.72 415B	14 04 36.	+ 13 14	42	15.3	PART OF DOUBLE GALAXY
ZWG 191.054	14 04 36.	+ 34 00		15.6	GALAXY
MCG+09-23-035	14 04 36.	+ 50 57	90	12.1	GALAXY
MCG+11-17-019	14 04 36.	+ 65 56	27	16.	GALAXY
ZWG 317.015	14 04 36.	+ 65 57		15.7	GALAXY
ZC 1404.6+6843	14 04 36.	+ 68 43	1880		CLUSTER OF GALAXIES
ZWG 018.052	14 04 36.	- 00 47		15.6	GALAXY
MCG+06-31-070	14 04 39.	+ 34 00	48	15.	GALAXY
RNGC 5479	14 04 40.	+ 65 55		15.0	GALAXY
IC 4374	14 04 41.	- 26 46 18.			NONSTELLAR OBJECT
ZWG 074.112	14 04 42.	+ 12 50		14.8	GALAXY
MCG+02-36-042	14 04 42.	+ 12 50	48	14.8	GALAXY
ZWG 103.072	14 04 42.	+ 15 33		15.6	GALAXY
ZC 1404.7+2404	14 04 42.	+ 24 04	870		CLUSTER OF GALAXIES
ZC 1404.7+2616	14 04 42.	+ 26 16	2350		CLUSTER OF GALAXIES
REIZ 3885	14 04 42.	+ 48 52	30	15.3	GALAXY
ZWG 317.016	14 04 42.	+ 65 55		15.2	GALAXY
UGC 09028	14 04 42.	+ 67 49	60	16.5	GALAXY
MCG-04-33-046	14 04 42.	- 26 47	18	14.5	GALAXY
SVEN 425	14 04 43.	- 04 53 50.	36	15.9	GALAXY
MCG+03-36-048	14 04 45.	+ 15 33	24	15.6	GALAXY
IC 0975	14 04 45.	+ 15 35 20.			NONSTELLAR OBJECT
RNGC 5481	14 04 47.	+ 50 58		13.5	GALAXY
ZC 1404.8+1254	14 04 48.	+ 12 54	3020		CLUSTER OF GALAXIES
ZC 1404.8+2802	14 04 48.	+ 28 02	940		CLUSTER OF GALAXIES
MRK 668	14 04 48.	+ 28 41	11	16.	GALAXY WITH UV CONTINUUM
TON-N 0184	14 04 48.	+ 29 54		16.5	BLUE STAR
ZC 1404.8+2958	14 04 48.	+ 29 58	740		CLUSTER OF GALAXIES
REIZ 3887	14 04 48.	+ 48 55	18	15.0	GALAXY
MCG-04-33-047	14 04 48.	- 24 52	21	14.5	GALAXY
HOLM 588B	14 04 53.	+ 50 57	36	13.2	PART OF MULTIPLE GALAXY
ZWG 074.113	14 04 54.	+ 13 07		15.4	GALAXY
ZWG 103.073	14 04 54.	+ 15 19		14.6	GALAXY
ZWG 103.074	14 04 54.	+ 15 24		15.0	GALAXY
MCG+03-36-049	14 04 54.	+ 15 24	15	15.0	GALAXY
ZWG 272.028	14 04 54.	+ 50 59		13.5	GALAXY
UGC 09029	14 04 54.	+ 50 59	108	13.5	GALAXY E
KARA.72 416B	14 04 54.	+ 50 59	114	13.5	PART OF DOUBLE GALAXY
MCG+11-17-020	14 04 54.	+ 67 45	39	17.	GALAXY
MCG-04-33-048	14 04 54.	- 26 19	36	15.5	GALAXY
MCG+03-36-050	14 04 57.	+ 15 19	48	14.6	GALAXY
REIZ 3892	14 04 57.	+ 60 16	72	14.7	GALAXY
ZWG 046.053	14 05 00.	+ 07 25		15.5	GALAXY
ZWG 103.075	14 05 00.	+ 15 09		15.6	GALAXY
MCG+03-36-051	14 05 00.	+ 15 10	24	15.6	GALAXY
MCG+09-23-036	14 05 00.	+ 50 57	48	13.2	GALAXY
MCG+10-20-072	14 05 00.	+ 60 18	15	16.	GALAXY
MCG-05-33-047	14 05 00.	- 31 48	60	15.	GALAXY
MRSL 311-00/1	14 05 00.	- 61 50	3600		HII REGION
REIZ 3883	14 05 02.	+ 20 59	30	15.3	GALAXY
RNGC 5488	14 05 02.	- 33 04			GALAXY
REIZ 3882	14 05 03.	+ 20 18	18	15.1	GALAXY
ZC 1405.1+1844	14 05 06.	+ 18 44	870		CLUSTER OF GALAXIES
ZWG 272.029	14 05 06.	+ 55 16		15.6	GALAXY
ZC 1405.1+6957	14 05 06.	+ 69 57	870		CLUSTER OF GALAXIES
ZWG 018.053	14 05 06.	- 00 43		15.7	GALAXY
MCG-04-33-049	14 05 06.	- 26 31	48	15.5	GALAXY
MCG-05-33-048	14 05 06.	- 33 05	84	14.5	GALAXY
REIZ 3890	14 05 08.	+ 55 13	6	15.7	GALAXY
RNGC 5484	14 05 08.	+ 55 16		15.5	GALAXY
REIZ 3891	14 05 08.	+ 55 17	18	15.0	GALAXY
EN 0393	14 05 08.	- 33 04			NEBULA
IC 4375	14 05 08.	- 33 04			NONSTELLAR OBJECT

OBJECT NAME	RIGHT ASCEN.	DECLINATION	DIAM.	MAGN.	TYPE OF OBJECT
HOLM 587A	14 05 09.	+ 10 59	30	15.6	PART OF MULTIPLE GALAXY
ARC 1860	14 05 11.	+ 14 13		17.2	RICH CLUSTER OF GALAXIES
ARC 1861	14 05 11.	+ 28 04		17.2	RICH CLUSTER OF GALAXIES
ZWG 046.054	14 05 12.	+ 07 27		15.3	GALAXY
ZWG 074.114	14 05 12.	+ 09 56		15.6	GALAXY
UGC 09030	14 05 12.	+ 09 56	66	15.6	GALAXY Sb-c
HOLM 587B	14 05 12.	+ 10 58	24	15.7	PART OF MULTIPLE GALAXY
ZWG 103.076	14 05 12.	+ 15 06		15.5	GALAXY
UGC 09031	14 05 12.	+ 15 06	66	15.5	GALAXY S
MCG+03-36-053	14 05 12.	+ 15 06	60	15.5	GALAXY
ZWG 103.077	14 05 12.	+ 16 15		15.4	GALAXY
REIZ 3884	14 05 12.	+ 16 16	24	15.1	GALAXY
MCG+03-36-052	14 05 12.	+ 16 21	48	15.	GALAXY
ZC 1405.2+2320	14 05 12.	+ 23 20	610		CLUSTER OF GALAXIES
LB 00719	14 05 12.	+ 25 36 36.		17.7	FAINT BLUE STAR
ZC 1405.2+3847	14 05 12.	+ 38 47	1610		CLUSTER OF GALAXIES
ZWG 018.054	14 05 12.	- 01 25		15.6	GALAXY
8ZW 1405+01.0	14 05 18.	+ 00 59		19.5	COMPACT GALAXY
ZC 1405.3+1413	14 05 18.	+ 14 13	2290		CLUSTER OF GALAXIES
ARC 1863	14 05 18.	+ 27 31		17.2	RICH CLUSTER OF GALAXIES
ZWG 219.052	14 05 18.	+ 39 59		15.3	GALAXY
ARC 1858	14 05 20.	- 04 05		17.6	RICH CLUSTER OF GALAXIES
FEIG 090	14 05 24.	+ 05 22		13.9	FAINT BLUE STAR
ZWG 046.055	14 05 24.	+ 07 17		15.5	GALAXY
8ZW 1405+10.1	14 05 24.	+ 10 05		17.6	COMPACT GALAXY
TON-N 0185	14 05 24.	+ 24 09		15.4	BLUE STAR
TON-N 0186	14 05 24.	+ 29 49		16.1	BLUE STAR
MCG+09-23-037	14 05 24.	+ 55 16	48	12.6	GALAXY
ZWG 295.035	14 05 24.	+ 56 47		15.0	GALAXY
UGC 09032	14 05 24.	+ 56 47	60	15.0	GALAXY Sb-c
MCG+10-20-073	14 05 24.	+ 56 47	54	14.	GALAXY
ZC 1405.4+6022	14 05 24.	+ 60 22	2150		CLUSTER OF GALAXIES
8ZW 1405-00.6	14 05 24.	- 00 35		17.5	COMPACT GALAXY
RNGC 5485	14 05 26.	+ 55 14		13.0	GALAXY
REIZ 3889	14 05 28.	+ 08 52	78	14.1	GALAXY
REIN 2.216	14 05 28.07	+ 55 14 19.7			NEBULA
8ZW 1405+01.0	14 05 30.	+ 00 59		18.9	COMPACT GALAXY
ZC 1405.5+0654	14 05 30.	+ 06 54	5710		CLUSTER OF GALAXIES
FAI 1405+10	14 05 30.	+ 10 04			COMPACT GALAXY
ZWG 132.076	14 05 30.	+ 23 20		15.3	GALAXY
LB 00720	14 05 30.	+ 23 49 12.		16.2	FAINT BLUE STAR
1ZW 080	14 05 30.	+ 50 43			COMPACT GALAXY
ZWG 272.030	14 05 30.	+ 55 15		12.4	GALAXY
UGC 09033	14 05 30.	+ 55 15	162	12.4	GALAXY S0
REIZ 3886	14 05 31.	- 05 51	54	13.3	GALAXY
REIZ 3901	14 05 32.	+ 55 15	36	13.0	GALAXY
ARC 1857	14 05 33.	- 24 32		17.6	RICH CLUSTER OF GALAXIES
REIZ 3888	14 05 34.	- 01 27	24	14.3	GALAXY
RNGC 5476	14 05 35.	- 05 51		14.0	GALAXY
ZC 1405.6+2815	14 05 36.	+ 28 15	1340		CLUSTER OF GALAXIES
TON-N 0187	14 05 36.	+ 31 19		14.6	BLUE STAR
ZC 1405.6+6202	14 05 36.	+ 62 02	5440		CLUSTER OF GALAXIES
ZWG 018.055	14 05 36.	- 01 28		14.7	GALAXY
UGC 09034	14 05 36.	- 01 28	72	14.7	GALAXY Sb
MCG+00-36-019	14 05 36.	- 01 28	54	14.7	GALAXY
MCG-01-36-009	14 05 36.	- 05 51	60	14.	GALAXY
RNGC 5478	14 05 37.	- 01 28		14.5	GALAXY
ARC 1862	14 05 38.	+ 06 48		17.2	RICH CLUSTER OF GALAXIES
PATH 1.655	14 05 38.	+ 29 08	16		NEBULA
HOLM 589B	14 05 39.	+ 06 04	12	15.1	PART OF MULTIPLE GALAXY
MCG+05-33-052	14 05 39.	+ 30 06	90	15.2	GALAXY
ZC 1405.7+0538	14 05 42.	+ 05 38	1880		CLUSTER OF GALAXIES
ZWG 046.056	14 05 42.	+ 06 02		15.2	GALAXY
HOLM 589A	14 05 42.	+ 06 04	18	14.4	PART OF MULTIPLE GALAXY
ZC 1405.7+2501	14 05 42.	+ 25 01	940		CLUSTER OF GALAXIES
ZWG 162.063	14 05 42.	+ 30 07		15.2	GALAXY
UGC 09035	14 05 42.	+ 30 07	90	15.2	GALAXY SBb
KARA.73B 0612	14 05 42.	+ 30 07	66	15.2	ISOLATED GALAXY S
ZWG 272.031	14 05 42.	+ 55 20		14.0	GALAXY
UGC 09036	14 05 42.	+ 55 20	90	14.0	GALAXY Sc-IRR
MCG+09-23-038	14 05 42.	+ 55 22	90	13.7	GALAXY
MCG-01-36-010	14 05 42.	- 08 51	12	15.5	GALAXY
REIZ 3904	14 05 43.	+ 55 21	120	13.7	GALAXY
REIZ 3896	14 05 44.	+ 30 05	42	14.0	GALAXY
RNGC 5486	14 05 44.	+ 55 20		14.0	GALAXY
ARC 1864	14 05 45.	+ 05 42		17.0	RICH CLUSTER OF GALAXIES
ZWG 103.078	14 05 48.	+ 16 25		15.5	GALAXY
ZC 1405.8+1900	14 05 48.	+ 19 00	1080		CLUSTER OF GALAXIES
ZC 1405.8+2955	14 05 48.	+ 29 55	610		CLUSTER OF GALAXIES
ZC 1405.8+3030	14 05 48.	+ 30 30	1010		CLUSTER OF GALAXIES
ZWG 018.056	14 05 48.	- 00 21		15.6	GALAXY
REIZ 3895	14 05 50.	+ 21 03	12	15.4	GALAXY
REIZ 3894	14 05 53.	+ 16 25	30	14.9	GALAXY
ZWG 018.057	14 05 54.	+ 00 06		15.2	GALAXY
MCG+09-23-039	14 05 54.	+ 55 02	30	14.	GALAXY
REIZ 3906	14 05 56.	+ 55 02	18	14.7	GALAXY
HOLM 590B	14 05 58.	+ 05 56	36	15.1	PART OF MULTIPLE GALAXY
RNGC 5482	14 05 58.	+ 09 10		14.0	GALAXY
REIZ 3893	14 05 59.	+ 07 17	60	14.0	GALAXY
HOLM 591A	14 05 59.	+ 07 18	54	13.4	PART OF MULTIPLE GALAXY
ZWG 046.057	14 06 00.	+ 03 15		15.3	GALAXY
ZWG 046.058	14 06 00.	+ 03 39		15.4	GALAXY
ZWG 046.059	14 06 00.	+ 05 54		15.4	GALAXY
HOLM 590A	14 06 00.	+ 05 54	36	14.8	PART OF MULTIPLE GALAXY
ZWG 046.060	14 06 00.	+ 07 18		14.5	GALAXY
UGC 09037	14 06 00.	+ 07 18	108	14.5	GALAXY Sc
MCG+01-36-020	14 06 00.	+ 07 18	72	14.5	GALAXY
ZWG 074.115	14 06 00.	+ 09 10		14.2	GALAXY
UGC 09038	14 06 00.	+ 09 10	72	14.2	GALAXY S0
MCG+02-36-043	14 06 00.	+ 09 10	30	14.2	GALAXY
ZWG 074.116	14 06 00.	+ 13 50		15.4	GALAXY
ZWG 103.079	14 06 00.	+ 15 13		15.7	GALAXY
ZWG 103.080	14 06 00.	+ 16 10		15.2	GALAXY
MCG+03-36-054	14 06 00.	+ 17 10	30	15.5	GALAXY
ZC 1406.0+2828	14 06 00.	+ 28 28	940		CLUSTER OF GALAXIES
ZC 1406.0+4415	14 06 00.	+ 44 15	670		CLUSTER OF GALAXIES
ZWG 272.032	14 06 00.	+ 55 02		14.7	GALAXY
ZWG 337.001	14 06 00.	+ 72 08		15.5	GALAXY
ZWG 336.041	14 06 00.	+ 72 08		15.5	GALAXY
UGC 09039	14 06 00.	+ 72 23	96	16.0	GALAXY Sc
HOLM 591B	14 06 03.	+ 07 18	18	15.4	PART OF MULTIPLE GALAXY
MCG+06-31-071	14 06 03.	+ 35 58 30.	48	15.	GALAXY
IC 0976	14 06 05.	- 00 54 17.			NONSTELLAR OBJECT
ZWG 074.117	14 06 06.	+ 09 17		15.4	GALAXY
ZWG 074.118	14 06 06.	+ 09 27		15.5	GALAXY
ZWG 074.119	14 06 06.	+ 12 03		14.9	GALAXY
UGC 09041	14 06 06.	+ 12 03	72	14.9	GALAXY SBa
MCG+02-36-044	14 06 06.	+ 12 03	72	14.9	GALAXY
ZWG 074.120	14 06 06.	+ 12 11		15.7	GALAXY
MCG+03-36-055	14 06 06.	+ 16 10	24	15.2	GALAXY
ZC 1406.1+2343	14 06 06.	+ 23 43	2020		CLUSTER OF GALAXIES
TON-N 0188	14 06 06.	+ 24 25		16.2	BLUE STAR
UGC 09042	14 06 06.	+ 35 58	60	16.5	GALAXY
ZWG 191.055	14 06 06.	+ 37 07		15.5	GALAXY
ZWG 018.059	14 06 06.	- 00 55		14.1	GALAXY
UGC 09040	14 06 06.	- 00 55	90	14.1	GALAXY S0-a
MCG+00-36-020	14 06 06.	- 00 55	78	14.1	GALAXY
ZWG 018.058	14 06 06.	- 02 46		15.3	GALAXY
KARA.72 417A	14 06 06.	- 02 46	36	15.3	PART OF DOUBLE GALAXY
MCG-04-33-050	14 06 06.	- 26 42	84	15.	GALAXY
REIZ 3899	14 06 10.	+ 09 10	24	14.0	GALAXY
REIZ 3898	14 06 10.	- 00 55	18	14.0	GALAXY
REIZ 3897	14 06 11.	- 02 47	30	14.6	GALAXY
ZC 1406.2+0809	14 06 12.	+ 08 09	1340		CLUSTER OF GALAXIES
ZWG 103.081	14 06 12.	+ 15 56		15.7	GALAXY
ZC 1406.2+2546	14 06 12.	+ 25 46	1140		CLUSTER OF GALAXIES
ZC 1406.2+5905	14 06 12.	+ 59 05	1210		CLUSTER OF GALAXIES
REIZ 3903	14 06 16.	+ 17 48	18	15.5	GALAXY
REIZ 3902	14 06 17.	+ 16 20	36	14.6	GALAXY
ZWG 074.121	14 06 18.	+ 09 07		15.6	GALAXY
ZWG 103.082	14 06 18.	+ 16 20		14.9	GALAXY
UGC 09043	14 06 18.	+ 16 20	66	14.9	GALAXY S0
MCG+03-36-056	14 06 18.	+ 16 20	84	14.9	GALAXY
MCG+12-13-030	14 06 18.	+ 72 22 30.	102	16.	GALAXY
ZWG 337.002	14 06 18.	+ 72 25		15.2	GALAXY
ZWG 336.042	14 06 18.	+ 72 25		15.2	GALAXY
REIZ 3909	14 06 20.	+ 60 29	18	14.0	GALAXY
IC 0978	14 06 23.	- 02 44 10.			NONSTELLAR OBJECT
REIZ 3900	14 06 23.	- 02 46	12	15.5	GALAXY
FAI 14064+10	14 06 24.	+ 10 28			COMPACT GALAXY
1ZW 081	14 06 24.	+ 49 05			COMPACT GALAXY
VVI 71	14 06 24.	+ 49 05	12	16.	SEYFERT GALAXY
ZC 1406.4+5513	14 06 24.	+ 55 13	5440		CLUSTER OF GALAXIES
MCG+10-20-074	14 06 24.	+ 60 28	42	14.	GALAXY
MCG+10-20-075	14 06 24.	+ 60 34	39	16.	GALAXY
MCG+11-17-021	14 06 24.	+ 65 53	30	16.	GALAXY
MCG+12-13-031	14 06 24.	+ 72 26	39	15.	GALAXY
ZWG 018.060	14 06 24.	- 02 44		15.0	GALAXY
KARA.72 417B	14 06 24.	- 02 44	48	15.0	PART OF DOUBLE GALAXY
MCG+00-36-021	14 06 24.	- 02 44	36	15.0	GALAXY
ZC 1406.4-0334	14 06 24.	- 03 34	4640		CLUSTER OF GALAXIES
ZWG 103.083	14 06 30.	+ 16 40		15.3	GALAXY
HOLM 592A	14 06 30.	+ 24 27	18	15.3	PART OF MULTIPLE GALAXY
ZWG 162.064	14 06 30.	+ 27 30		15.7	GALAXY
ZC 1406.5+3150	14 06 30.	+ 31 50	740		CLUSTER OF GALAXIES
ZWG 272.033	14 06 30.	+ 54 07		15.6	GALAXY
ZWG 317.017	14 06 30.	+ 65 51		15.6	GALAXY
7ZW 534	14 06 30.	+ 73 08			COMPACT GALAXY
ZWG 337.003	14 06 30.	+ 73 09		15.6	GALAXY
ZWG 336.043	14 06 30.	+ 73 09		15.6	GALAXY
MCG-04-33-051	14 06 30.	- 26 47	36	15.5	GALAXY
MCG+03-36-057	14 06 33.	+ 16 40	36	15.3	GALAXY
HOLM 592B	14 06 33.	+ 24 26	24	15.5	PART OF MULTIPLE GALAXY
REIZ 3908	14 06 34.	+ 33 45	84	14.0	GALAXY
REIZ 3905	14 06 35.	+ 16 41	30	14.9	GALAXY
FAI 14066+10	14 06 36.	+ 10 26			COMPACT GALAXY
MCG+03-36-058	14 06 36.	+ 14 32	66	14.6	GALAXY
ZWG 103.084	14 06 36.	+ 14 33		14.6	GALAXY
UGC 09044	14 06 36.	+ 14 33	66	14.6	GALAXY S
ZWG 103.085	14 06 36.	+ 15 00		15.4	GALAXY
ZWG 191.056	14 06 36.	+ 35 51		15.5	GALAXY
UGC 09045	14 06 36.	+ 35 51	72	15.5	GALAXY
UGC 09046	14 06 36.	+ 48 13	72	17.	GALAXY DWARF
ZWG 272.034	14 06 36.	+ 55 32		15.5	GALAXY
UGC 09047	14 06 36.	+ 55 32	72	15.5	GALAXY SBb
MCG+13-10-016	14 06 36.	+ 76 09	51	15.	GALAXY
MCG-03-36-008	14 06 36.	- 17 38 30.	9	15.5	GALAXY
REIZ 3907	14 06 39.	+ 19 26	9	15.6	GALAXY
MCG+06-31-072	14 06 39.	+ 33 45	102	14.5	GALAXY
MCG+06-31-073	14 06 39.	+ 35 50 30.	42	14.5	GALAXY
ZWG 103.086	14 06 42.	+ 18 00		15.1	GALAXY
MCG+03-36-059	14 06 42.	+ 18 00	48	15.1	GALAXY
ZC 1406.7+1956	14 06 42.	+ 19 56	2080		CLUSTER OF GALAXIES
ZWG 191.057	14 06 42.	+ 33 46		15.6	GALAXY
UGC 09048	14 06 42.	+ 33 46	102	15.6	GALAXY Sb-c
KARA.73B 0613	14 06 42.	+ 33 46	102	15.6	ISOLATED GALAXY S
ZWG 219.053	14 06 42.	+ 38 46		15.5	GALAXY
MCG+09-23-041	14 06 42.	+ 55 10	48	15.	GALAXY
MCG+09-23-040	14 06 42.	+ 55 33	42	15.	GALAXY
ZWG 295.036	14 06 42.	+ 60 28		15.6	GALAXY
UGC 09049	14 06 42.	+ 60 28	66	15.6	GALAXY Sb
ZC 1406.7+6307	14 06 42.	+ 63 07	1080		CLUSTER OF GALAXIES
ZC 1406.7+6543	14 06 42.	+ 65 43	270		CLUSTER OF GALAXIES
MCG-03-36-009	14 06 45.	- 17 37 30.	15	15.5	GALAXY
ZWG 103.087	14 06 48.	+ 14 48		15.7	GALAXY
MCG+03-36-060	14 06 48.	+ 14 48	36	15.7	GALAXY
ZWG 103.088	14 06 48.	+ 18 40		15.7	GALAXY
ZC 1406.8+2756	14 06 48.	+ 27 56	740		CLUSTER OF GALAXIES
TON-N 0189	14 06 48.	+ 29 16		15.8	BLUE STAR
UGC 09050	14 06 48.	+ 51 22	78	17.	GALAXY DWARF
UGC 09051	14 06 48.	+ 55 09	78	16.5	GALAXY
ARC 1867	14 06 53.	+ 31 24		17.5	RICH CLUSTER OF GALAXIES
ZWG 074.122	14 06 54.	+ 09 08		15.4	GALAXY
IC 0979	14 06 54.	+ 15 04 47.			NONSTELLAR OBJECT
HOLM 593A	14 06 54.	+ 24 24	18	15.2	PART OF MULTIPLE GALAXY
HOLM 593B	14 06 54.	+ 24 25	24	15.7	PART OF MULTIPLE GALAXY
FEIG 091	14 06 54.	+ 59 55		13.9	FAINT BLUE STAR
ZWG 337.004	14 06 54.	+ 73 16		15.5	GALAXY
ZWG 336.044	14 06 54.	+ 73 16		15.5	GALAXY
ZWG 018.061	14 06 54.	- 01 40		15.6	GALAXY
ZWG 103.089	14 07 00.	+ 19 46		15.7	GALAXY
ZC 1407.0+3118	14 07 00.	+ 31 18	1080		CLUSTER OF GALAXIES
MCG+06-31-074	14 07 00.	+ 35 36	36	15.5	GALAXY
MCG+12-13-032	14 07 00.	+ 73 18	33	16.	GALAXY
ZWG 353.030	14 07 00.	+ 76 10		15.4	GALAXY
UGC 09052	14 07 00.	+ 76 10	66	15.4	GALAXY Sc
ARC 1868	14 07 05.	+ 27 52		17.5	RICH CLUSTER OF GALAXIES
FAI 14071+12	14 07 06.	+ 12 56			COMPACT GALAXY
ZWG 103.090	14 07 06.	+ 15 04		14.5	GALAXY
UGC 09053	14 07 06.	+ 15 04	66	14.5	GALAXY SBa-b
MCG+03-36-061	14 07 06.	+ 15 04	60	14.5	GALAXY
ZC 1407.1+4017	14 07 06.	+ 40 17	1550		CLUSTER OF GALAXIES
ZWG 272.035	14 07 06.	+ 53 31		15.6	GALAXY
ZC 1407.1+6635	14 07 06.	+ 66 35	470		CLUSTER OF GALAXIES
ZWG 018.062	14 07 06.	- 00 25		15.6	GALAXY
8ZW 1407-00.6	14 07 06.	- 00 38		19.5	COMPACT GALAXY

OBJECT NAME	RIGHT ASCEN.	DECLINATION	DIAM.	MAGN.	TYPE OF OBJECT
HARO 40	14 07 07.	+ 26 36			BLUE EMISSION-LINE GALAXY
SVEN 426	14 07 07.	- 05 17 45.	36	15.7	GALAXY
ARC 1869	14 07 08.	+ 29 42		17.2	RICH CLUSTER OF GALAXIES
PK 1735	14 07 08.8	- 46 22 28.	18		NEBULA
HN 0036	14 07 10.	+ 08 18			NEBULA
MCG+04-33-041	14 07 12.	+ 26 27	24	17.	GALAXY
ZWG 191.058	14 07 12.	+ 35 22		15.7	GALAXY
ZC 1407.2+3715	14 07 12.	+ 37 15	1410		CLUSTER OF GALAXIES
ZWG 247.005	14 07 12.	+ 47 00		15.5	GALAXY
MCG+10-20-076	14 07 12.	+ 60 32	27	17.	GALAXY
LB 00271	14 07 14.	- 14 09 42.		15.8	FAINT BLUE STAR
RNGC 5487	14 07 16.	+ 08 19		14.5	GALAXY
REIZ 3912	14 07 16.	+ 17 45	18	15.4	GALAXY
HOLM 594B	14 07 18.	+ 08 17	30	15.3	PART OF MULTIPLE GALAXY
ZWG 046.061	14 07 18.	+ 08 19		14.6	GALAXY
MCG+01-36-021	14 07 18.	+ 08 19	48	14.6	GALAXY
ZWG 074.123	14 07 18.	+ 10 06		15.3	GALAXY
UGC 09054	14 07 18.	+ 10 06	60	15.3	GALAXY SB
ZC 1407.3+1107	14 07 18.	+ 11 07	3700		CLUSTER OF GALAXIES
ZWG 074.124	14 07 18.	+ 11 49		15.1	GALAXY
MCG-07-29-008	14 07 18.	- 43 06	180	12.4	GALAXY
ARC 1866	14 07 20.	+ 06 58		17.2	RICH CLUSTER OF GALAXIES
RNGC 5483	14 07 21.	- 43 05		12.0	GALAXY
HOLM 594A	14 07 22.	+ 08 19	36	14.1	PART OF MULTIPLE GALAXY
SVEN 427	14 07 22.	- 01 00	24	14.9	GALAXY
REIZ 3911	14 07 23.	+ 07 10	90	14.2	GALAXY
ZWG 046.062	14 07 24.	+ 04 49		15.1	GALAXY
ZWG 074.125	14 07 24.	+ 11 12		15.3	GALAXY
MCG+05-33-053	14 07 24.	+ 26 34	30	15.7	GALAXY
ZC 1407.4+2730	14 07 24.	+ 27 30	2220		CLUSTER OF GALAXIES
ZC 1407.4+2936	14 07 24.	+ 29 36	1410		CLUSTER OF GALAXIES
ZWG 272.036	14 07 24.	+ 54 04		15.4	GALAXY
ZWG 018.063	14 07 24.	- 01 00		14.8	GALAXY
MCG+00-36-022	14 07 24.	- 01 00	24	14.8	GALAXY
REIZ 3910	14 07 25.	+ 04 49	24	15.2	GALAXY
MCG+03-36-062	14 07 27.	+ 15 06	60	15.0	GALAXY
HOLM 595A	14 07 28.	+ 17 47	36	14.1	PART OF MULTIPLE GALAXY
REIZ 3913	14 07 28.	+ 17 50	12	15.3	GALAXY
ZWG 103.091	14 07 30.	+ 15 06		15.0	GALAXY
UGC 09055	14 07 30.	+ 15 06	60	15.0	GALAXY SBb
ZWG 103.092	14 07 30.	+ 15 25		14.8	GALAXY
MCG+03-36-063	14 07 30.	+ 15 25	48	14.8	GALAXY
ZWG 103.093	14 07 30.	+ 17 49		15.7	GALAXY
MCG+03-36-064	14 07 30.	+ 17 50	15	15.7	GALAXY
ZC 1407.5+2020	14 07 30.	+ 20 20	1210		CLUSTER OF GALAXIES
ZC 1407.5+3216	14 07 30.	+ 32 16	3360		CLUSTER OF GALAXIES
ZC 1407.5+3456	14 07 30.	+ 34 56	1010		CLUSTER OF GALAXIES
ZWG 247.006	14 07 30.	+ 49 16		14.4	GALAXY
UGC 09056	14 07 30.	+ 49 16	66	14.4	GALAXY PECULR
MCG+08-26-005	14 07 30.	+ 49 16	60	15.	GALAXY
MCG+09-23-042	14 07 30.	+ 54 03	42	15.	GALAXY
MCG+09-23-043	14 07 30.	+ 56 38	12	16.	GALAXY
1ZW 082	14 07 30.	+ 56 39			COMPACT GALAXY
8ZW 1407-01.3	14 07 30.	- 01 18		19.5	COMPACT GALAXY
RNGC 5490A	14 07 31.	+ 17 49			GALAXY
ARP 079	14 07 33.	+ 17 52			PECULIAR GALAXY
REIZ 3915	14 07 34.	+ 17 47	24	13.8	GALAXY
REIZ 3916	14 07 34.	+ 17 55	18	14.9	GALAXY
HOLM 595B	14 07 35.	+ 17 47	24	15.1	PART OF MULTIPLE GALAXY
ZWG 074.126	14 07 36.	+ 10 24		15.1	GALAXY
FAI 14076+12	14 07 36.	+ 12 14			COMPACT GALAXY
ZWG 103.094	14 07 36.	+ 17 38		15.6	GALAXY
ZWG 103.095	14 07 36.	+ 17 47		13.4	GALAXY
UGC 09058	14 07 36.	+ 17 47	150	13.4	GALAXY E
MCG+03-36-065	14 07 36.	+ 17 47	120	13.4	GALAXY
ZWG 103.096	14 07 36.	+ 17 56		14.6	GALAXY
UGC 09059	14 07 36.	+ 17 56	78	14.6	GALAXY S0
MCG+03-36-066	14 07 36.	+ 17 56	66	14.6	GALAXY
ZWG 162.065	14 07 36.	+ 26 34		15.7	GALAXY
UGC 09060	14 07 36.	+ 55 31	60	16.0	GALAXY Sc
MCG+09-23-044	14 07 36.	+ 55 32	42	15.	GALAXY
ZC 1407.6+6329	14 07 36.	+ 63 29	870		CLUSTER OF GALAXIES
ZWG 018.064	14 07 36.	- 02 20		14.4	GALAXY
UGC 09057	14 07 36.	- 02 20	180	14.4	GALAXY
MCG+00-36-023	14 07 36.	- 02 20	180	14.4	GALAXY
ZC 1407.6-0250	14 07 36.	- 02 50	1080		CLUSTER OF GALAXIES
IC 0982	14 07 36.7	+ 17 55 53.			GALAXY
RNGC 5490B	14 07 37.	+ 17 47			GALAXY
RNGC 5490	14 07 37.	+ 17 47		13.5	GALAXY
ARP 117	14 07 39.	+ 17 55			PECULIAR GALAXY
ARC 1872	14 07 39.	+ 62 12		17.2	RICH CLUSTER OF GALAXIES
REIZ 3918	14 07 40.	+ 17 39	18	15.1	GALAXY
REIZ 3919	14 07 40.	+ 17 47	24	15.2	GALAXY
REIZ 3920	14 07 40.	+ 17 51	18	14.7	GALAXY
REIZ 3921	14 07 40.	+ 17 57	18	14.8	GALAXY
REIZ 3925	14 07 41.	+ 49 18	48	13.9	GALAXY
IC 0983	14 07 41.8	+ 17 58 12.	240	15.0	GALAXY SB(r)
ZWG 074.127	14 07 42.	+ 12 10		15.4	GALAXY
ZWG 103.097	14 07 42.	+ 17 47		15.7	GALAXY
ZWG 103.098	14 07 42.	+ 17 58		14.3	GALAXY
UGC 09061	14 07 42.	+ 17 58	360	14.3	GALAXY SBa/SBb
ZWG 103.099	14 07 42.	+ 18 36		14.5	GALAXY
UGC 09062	14 07 42.	+ 18 36	114	14.5	GALAXY Sb
TON-N 0753	14 07 42.	+ 26 57		14.8	BLUE STAR
ZC 1407.7+3803	14 07 42.	+ 38 03	1080		CLUSTER OF GALAXIES
MCG-01-36-011	14 07 42.	- 06 34	36	15.	GALAXY
RNGC 5490C	14 07 43.	+ 17 51		15.0	GALAXY
IC 0980	14 07 44.	- 07 06 29.			NONSTELLAR OBJECT
LB 00721	14 07 45.	+ 00 32 12.		16.3	FAINT BLUE STAR
REIZ 3922	14 07 46.	+ 18 36	120	13.8	GALAXY
HOLM 596A	14 07 46.	+ 18 36	126	13.9	PART OF MULTIPLE GALAXY
IC 0984	14 07 46.	+ 18 36 50.			NONSTELLAR OBJECT
UGC 09063	14 07 48.	+ 05 49	78	16.0	GALAXY
ZWG 103.100	14 07 48.	+ 17 51		15.2	GALAXY
ZC 1407.8+2154	14 07 48.	+ 21 54	5240		CLUSTER OF GALAXIES
ZC 1407.8+3032	14 07 48.	+ 30 32	1810		CLUSTER OF GALAXIES
MCG+10-20-077	14 07 48.	+ 60 39	18	16.	GALAXY
ZWG 018.065	14 07 48.	- 00 36		15.1	GALAXY
REIZ 3914	14 07 48.	- 03 54	24	15.5	GALAXY
MCG-01-36-012	14 07 48.	- 07 56	36	15.	GALAXY
HOLM 596B	14 07 51.	+ 18 36	9	15.4	PART OF MULTIPLE GALAXY
RNGC 5503	14 07 51.	+ 60 39		16.0	GALAXY
REIZ 3923	14 07 52.	+ 18 36	6	15.6	GALAXY
IC 0981	14 07 52.	- 03 56 16.			NONSTELLAR OBJECT
8ZW 1407+00.7	14 07 54.	+ 00 41		18.8	COMPACT GALAXY
8ZW 1407+02.2	14 07 54.	+ 02 13		19.5	COMPACT GALAXY
MCG+06-31-075	14 07 54.	+ 37 47	27	14.	GALAXY
REIZ 3917	14 07 54.	- 03 56	36	14.7	GALAXY
HN 0394	14 07 54.	- 30 34			NEBULA
IC 4376	14 07 54.	- 30 35			THREE STARS
IC 4380	14 07 55.	+ 37 47 10.			NONSTELLAR OBJECT
LB 09886	14 08	- 82 42		13.3	FAINT BLUE STAR
ZWG 103.101	14 08 00.	+ 16 20		14.7	GALAXY
ZWG 103.102	14 08 00.	+ 16 25		15.7	GALAXY
ZWG 103.103	14 08 00.	+ 16 35		15.6	GALAXY
UGC 09064	14 08 00.	+ 16 35	72	15.6	GALAXY Sb
ZWG 191.059	14 08 00.	+ 37 47		15.1	GALAXY
ZC 1408.0+5353	14 08 00.	+ 53 53	3360		CLUSTER OF GALAXIES
MRK 285	14 08 00.	+ 71 56	10	16.	GALAXY WITH UV CONTINUUM
MCG+14-07-001	14 08 00.	+ 83 43	18	16.	GALAXY
MCG+03-36-072	14 08 03.	+ 16 21	30	14.7	GALAXY
MCG+03-36-073	14 08 03.	+ 16 25	48	15.7	GALAXY
RNGC 5502	14 08 03.	+ 60 40			NON-EXISTENT OBJECT
ZWG 018.066	14 08 06.	+ 01 42		15.6	GALAXY
ZC 1408.1+0206	14 08 06.	+ 02 06	1610		CLUSTER OF GALAXIES
FAI 1408+13	14 08 06.	+ 13 40			COMPACT GALAXY
ZWG 103.104	14 08 06.	+ 15 41		15.7	GALAXY
MCG+08-26-006	14 08 06.	+ 46 40	72	16.	GALAXY
MCG+09-23-045	14 08 06.	+ 55 10	24	17.	GALAXY
REIZ 3924	14 08 10.	+ 18 02	18	14.8	GALAXY
LB 00272	14 08 10.	- 15 46 54.		15.1	FAINT BLUE STAR
ZWG 074.128	14 08 12.	+ 09 14		14.7	GALAXY
MCG+02-36-045	14 08 12.	+ 09 14	36	14.7	GALAXY
ZWG 103.105	14 08 12.	+ 17 32		15.5	GALAXY
ZWG 103.106	14 08 12.	+ 19 51		13.7	GALAXY
UGC 09065	14 08 12.	+ 19 51	108	13.7	GALAXY S
UGC 09066	14 08 12.	+ 46 41	72	16.5	GALAXY Sc
ZWG 295.037	14 08 12.	+ 57 03		15.4	GALAXY
MCG+10-20-078	14 08 12.	+ 57 03	39	15.	GALAXY
RNGC 5492	14 08 14.	+ 19 51		13.5	GALAXY
REIZ 3927	14 08 15.	+ 19 51	84	13.5	GALAXY
FAI 1408+12	14 08 18.	+ 12 13			COMPACT GALAXY
ZWG 074.129	14 08 18.	+ 13 47		15.2	GALAXY
ZWG 103.107	14 08 18.	+ 15 27		14.7	GALAXY
UGC 09067	14 08 18.	+ 15 27	96	14.7	GALAXY Sa-b
MCG+03-36-074	14 08 18.	+ 19 52	168	13.7	GALAXY
ZC 1408.3+2115	14 08 18.	+ 21 15	1480		CLUSTER OF GALAXIES
ZWG 132.077	14 08 18.	+ 22 11		15.7	GALAXY
KARA.73B 0614	14 08 18.	+ 22 11	36	15.7	ISOLATED GALAXY S0
ZC 1408.3+2317	14 08 18.	+ 23 17	1340		CLUSTER OF GALAXIES
MCG+07-29-047	14 08 18.	+ 39 07 30.	24	17.	GALAXY
MCG+08-26-007	14 08 18.	+ 48 32	90	15.	GALAXY
UGC 09068	14 08 18.	+ 48 33	102	16.0	GALAXY
MCG+09-23-046	14 08 18.	+ 54 27 30.	120	14.	GALAXY
ZWG 272.037	14 08 18.	+ 55 07		15.1	GALAXY
MCG+09-23-047	14 08 18.	+ 55 08	36	15.	GALAXY
ARC 1870	14 08 20.	+ 06 54		17.2	RICH CLUSTER OF GALAXIES
HOLM 597C	14 08 21.	+ 06 36	12	15.4	PART OF MULTIPLE GALAXY
MCG+03-36-075	14 08 21.	+ 14 36	30	16.	GALAXY
MCG+03-36-076	14 08 21.	+ 15 26	84	14.7	GALAXY
RNGC 5500	14 08 21.	+ 48 47		14.5	GALAXY
RNGC 5497	14 08 22.	+ 39 07		15.0	GALAXY
REIZ 3926	14 08 23.	+ 06 37	60	13.7	GALAXY
HOLM 597A	14 08 23.	+ 06 37	48	13.7	PART OF MULTIPLE GALAXY
HOLM 597B	14 08 23.	+ 06 38	18	15.1	PART OF MULTIPLE GALAXY
ZWG 018.067	14 08 24.	+ 01 16		15.5	GALAXY
ZWG 103.108	14 08 24.	+ 18 06		15.7	GALAXY
ZC 1408.4+1915	14 08 24.	+ 19 15	2220		CLUSTER OF GALAXIES
ZC 1408.4+3006	14 08 24.	+ 30 06	940		CLUSTER OF GALAXIES
ZWG 219.054	14 08 24.	+ 39 07		15.1	GALAXY
UGC 09069	14 08 24.	+ 39 07	84	15.1	GALAXY SBb
MCG+08-26-008	14 08 24.	+ 48 46	30	13.9	GALAXY
ZWG 247.007	14 08 24.	+ 48 47		14.5	GALAXY
UGC 09070	14 08 24.	+ 48 47	60	14.5	GALAXY E
ZWG 272.038	14 08 24.	+ 54 27		15.1	GALAXY
UGC 09071	14 08 24.	+ 54 27	132	15.1	GALAXY Sc
MCG+09-23-048	14 08 24.	+ 55 10	15	17.	GALAXY
ZWG 295.038	14 08 24.	+ 58 16		15.7	GALAXY
MCG+10-20-079	14 08 24.	+ 58 17	24	15.	GALAXY
ZC 1408.4+6352	14 08 24.	+ 63 52	1680		CLUSTER OF GALAXIES
REIZ 3935	14 08 26.	+ 54 27	72	14.4	GALAXY
RNGC 5491A	14 08 27.	+ 06 36		14.0	GALAXY
RNGC 5491B	14 08 27.	+ 06 37			GALAXY
MCG+07-29-048	14 08 27.	+ 39 07 30.	72	14.5	GALAXY
REIZ 3928	14 08 28.	+ 09 15	18	14.7	GALAXY
REIZ 3933	14 08 29.	+ 48 48	18	13.9	GALAXY
8ZW 1408+00.8	14 08 30.	+ 00 50		17.8	COMPACT GALAXY
ZC 1408.5+0056	14 08 30.	+ 00 56	1950		CLUSTER OF GALAXIES
ZWG 046.063	14 08 30.	+ 06 36		13.9	GALAXY
UGC 09072	14 08 30.	+ 06 36	96	13.9	GALAXY S
MCG+01-36-022	14 08 30.	+ 06 36	72	13.9	GALAXY
FAI 1408+14	14 08 30.	+ 14 23			COMPACT GALAXY
ZC 1408.5+1758	14 08 30.	+ 17 58	540		CLUSTER OF GALAXIES
ZC 1408.5+2640	14 08 30.	+ 26 40	1140		CLUSTER OF GALAXIES
MCG+09-23-049	14 08 30.	+ 51 18	36	16.	GALAXY
MCG+10-20-080	14 08 30.	+ 59 30	12	16.	GALAXY
MCG-04-33-052	14 08 30.	- 24 47	48	15.	GALAXY
PK315+09.1	14 08 33.	- 51 12	5		PLANETARY NEBULA
ZWG 018.068	14 08 36.	+ 00 48		15.3	GALAXY
ZWG 018.069	14 08 36.	+ 01 42		15.2	GALAXY
ZC 1408.6+0321	14 08 36.	+ 03 21	1550		CLUSTER OF GALAXIES
ZWG 103.109	14 08 36.	+ 16 55		15.5	GALAXY
ZWG 163.001	14 08 36.	+ 27 03		15.5	GALAXY
ZWG 162.066	14 08 36.	+ 27 03		15.5	GALAXY
ZC 1408.6+2931	14 08 36.	+ 29 31	2290		CLUSTER OF GALAXIES
MCG+06-31-076	14 08 36.	+ 36 08 30.	36	13.	GALAXY
ZC 1408.6+5956	14 08 36.	+ 59 56	1550		CLUSTER OF GALAXIES
HOLM 598A	14 08 41.	+ 25 44	42	14.4	PART OF MULTIPLE GALAXY
ZWG 018.070	14 08 42.	+ 00 43		15.7	GALAXY
ZWG 018.071	14 08 42.	+ 01 30		15.4	GALAXY
8ZW 1408+01.5	14 08 42.	+ 01 33		16.6	COMPACT GALAXY
IC 4381	14 08 42.	+ 25 43 06.			NONSTELLAR OBJECT
ZWG 133.001	14 08 42.	+ 25 44		14.8	GALAXY
ZWG 132.078	14 08 42.	+ 25 44		14.8	GALAXY
UGC 09073	14 08 42.	+ 25 44	102	14.8	GALAXY Sc
ZWG 191.060	14 08 42.	+ 36 08		14.5	GALAXY
UGC 09074	14 08 42.	+ 36 08	72	14.5	GALAXY
ZC 1408.7+4330	14 08 42.	+ 43 30	1680		CLUSTER OF GALAXIES
8ZW 1408-03.3	14 08 42.	- 03 16		17.7	COMPACT GALAXY
SVEN 428	14 08 42.	- 30 16 42.	18	15.1	GALAXY
RNGC 5499	14 08 44.	+ 36 08		14.5	GALAXY
REIZ 3932	14 08 44.	+ 36 09	36	13.8	GALAXY
MCG+04-33-042	14 08 45.	+ 25 44	180	14.8	GALAXY
HOLM 598B	14 08 46.	+ 25 46	18	14.9	PART OF MULTIPLE GALAXY
RNGC 5498	14 08 46.	+ 25 56		15.0	GALAXY
REIZ 3934	14 08 46.	+ 40 00	30	14.9	GALAXY

OBJECT NAME	RIGHT ASCEN.	DECLINATION	DIAM.	MAGN.	TYPE OF OBJECT
IC 4382	14 08 47.	+ 25 44 31.			NONSTELLAR OBJECT
ZWG 074.130	14 08 48.	+ 11 37		15.7	GALAXY
ZWG 103.110	14 08 48.	+ 19 25		15.5	GALAXY
ZWG 133.002	14 08 48.	+ 25 45		15.4	GALAXY
ZWG 132.079	14 08 48.	+ 25 45		15.4	GALAXY
ZWG 133.003	14 08 48.	+ 25 56		15.0	GALAXY
ZWG 132.080	14 08 48.	+ 25 56		15.0	GALAXY
UGC 09075	14 08 48.	+ 25 56	78	15.0	GALAXY E-S0
ZWG 163.002	14 08 48.	+ 27 04		15.7	GALAXY
ZWG 162.067	14 08 48.	+ 27 04		15.7	GALAXY
UGC 09076	14 08 48.	+ 38 59	78	16.0	GALAXY S
MCG+07-29-049	14 08 48.	+ 38 59 30.	42	17.	GALAXY
ZWG 219.055	14 08 48.	+ 40 00		14.9	GALAXY
UGC 09077	14 08 48.	+ 40 00	60	14.9	GALAXY S
MCG+07-29-050	14 08 48.	+ 40 00	45	15.	GALAXY
ZC 1408.8+7820	14 08 48.	+ 78 20	870		CLUSTER OF GALAXIES
LB 00273	14 08 48.	- 14 01 48.		15.7	FAINT BLUE STAR
RNGC 5489	14 08 49.	- 45 51			GALAXY
LB 00722	14 08 50.	+ 24 16 48.		17.3	FAINT BLUE STAR
REIN 7.152	14 08 51.90	+ 40 00 27.7			NEBULA
REIZ 3931	14 08 52.	+ 17 45	12	14.9	GALAXY
REIZ 3941	14 08 52.	+ 56 15	18	15.6	GALAXY
REIN 2.217	14 08 52.53	- 04 48 33.7			NEBULA
ZWG 018.073	14 08 54.	+ 01 31		14.8	GALAXY
MCG+00-36-025	14 08 54.	+ 01 31	15	14.8	GALAXY
IC 0986	14 08 54.	+ 01 33 54.			NONSTELLAR OBJECT
ZWG 103.111	14 08 54.	+ 17 44		14.6	GALAXY
UGC 09078	14 08 54.	+ 17 44	90	14.6	GALAXY SBa
MCG+03-36-077	14 08 54.	+ 17 45	90	14.6	GALAXY
MCG+04-33-043	14 08 54.	+ 25 56	48	15.0	GALAXY
MCG+10-20-081	14 08 54.	+ 59 29	15	16.	GALAXY
ZWG 018.072	14 08 54.	- 03 00		14.8	GALAXY
MCG+00-36-024	14 08 54.	- 03 00	30	14.8	GALAXY
RNGC 5493	14 08 54.	- 04 49		13.0	GALAXY
ARC 1877	14 08 55.	+ 60 01		17.8	RICH CLUSTER OF GALAXIES
REIZ 3929	14 08 55.	- 04 48	54	12.5	GALAXY
IC 0985	14 08 58.	- 02 59 12.			NONSTELLAR OBJECT
REIZ 3930	14 08 59.	- 02 59	12	15.0	GALAXY
SN 1955I	14 09 00.	+ 01 23		19.3	SUPERNOVA
ZWG 074.131	14 09 00.	+ 08 36		15.1	GALAXY
TON-N 0190	14 09 00.	+ 28 34		14.8	BLUE STAR
SHAH 010	14 09 00.	+ 46 30	240	17.5	GROUP OF COMPACT GALAXIES
ZWG 295.039	14 09 00.	+ 59 35		15.7	GALAXY
UGC 09080	14 09 00.	+ 59 35	78	15.7	GALAXY Sc
MCG+10-20-082	14 09 00.	+ 59 36	78	15.	GALAXY
ZC 1409.0+7850	14 09 00.	+ 78 50	1080		CLUSTER OF GALAXIES
ZWG 018.074	14 09 00.	- 00 55		13.4	GALAXY
UGC 09079	14 09 00.	- 00 55	270	13.4	GALAXY Sc
MCG+00-36-026	14 09 00.	- 00 55	240	13.4	GALAXY
KARA.73B 0615	14 09 00.	- 00 55	360	13.4	ISOLATED GALAXY S
MCG-01-36-013	14 09 00.	- 04 47	90	13.5	GALAXY
ARC 1879	14 09 04.	+ 63 50		17.8	RICH CLUSTER OF GALAXIES
SVEN 429	14 09 04.	- 00 55	210	12.5	GALAXY
ZWG 074.132	14 09 06.	+ 09 03		15.4	GALAXY
ZWG 103.112	14 09 06.	+ 19 24		15.4	GALAXY
SHAH 011	14 09 06.	+ 44 58	168	17.	GROUP OF COMPACT GALAXIES
RNGC 5496	14 09 07.	- 00 55		12.5	GALAXY
MCG+03-36-078	14 09 09.	+ 19 25	72	15.4	GALAXY
HN 0395	14 09 09.	- 34 01			NEBULA
HN 0396	14 09 09.	- 34 02			NEBULA
IC 0987	14 09 10.	+ 19 25 14.			NONSTELLAR OBJECT
IC 4378	14 09 10.	- 34 02			NONSTELLAR OBJECT
IC 4379	14 09 10.	- 34 03			NONSTELLAR OBJECT
ZWG 018.075	14 09 12.	+ 00 57		15.6	GALAXY
SN 1955J	14 09 12.	+ 00 57		17.9	SUPERNOVA
ZC 1409.2+0517	14 09 12.	+ 05 17	1810		CLUSTER OF GALAXIES
ZWG 133.004	14 09 12.	+ 24 16		15.7	GALAXY
ZC 1409.2+4355	14 09 12.	+ 43 55	540		CLUSTER OF GALAXIES
ZWG 247.008	14 09 12.	+ 49 12		15.7	GALAXY
REIZ 3936	14 09 15.	+ 19 25	24	15.1	GALAXY
HOLM 599A	14 09 15.	+ 20 10	36	15.8	PART OF MULTIPLE GALAXY
ZWG 018.076	14 09 18.	+ 00 11		15.5	GALAXY
HOLM 599B	14 09 18.	+ 20 10	36	15.9	PART OF MULTIPLE GALAXY
MCG+06-31-077	14 09 18.	+ 36 07 30.	30	15.	GALAXY
ZC 1409.3+5751	14 09 18.	+ 57 51	3360		CLUSTER OF GALAXIES
MCG-02-36-017	14 09 18.	- 09 50	48	14.5	GALAXY
ZWG 133.005	14 09 24.	+ 20 46		15.6	GALAXY
ZWG 191.061	14 09 24.	+ 36 07		15.6	GALAXY
ZC 1409.4+7758	14 09 24.	+ 77 58	1280		CLUSTER OF GALAXIES
MCG-05-34-001	14 09 24.	- 30 24	120	12.5	GALAXY
REIZ 3938	14 09 26.	+ 20 46	24	15.2	GALAXY
REIZ 3939	14 09 26.	+ 21 37	42	14.3	GALAXY
REIZ 3937	14 09 27.	+ 19 45	12	16.1	GALAXY
ARC 1873	14 09 27.	+ 28 23		16.3	RICH CLUSTER OF GALAXIES
RNGC 5494	14 09 27.	- 30 26		12.5	GALAXY
VV 070	14 09 30.	+ 07 55 30.	60	14.	INTERACTING GALAXY
ZWG 074.133	14 09 30.	+ 10 37		15.4	GALAXY
ZC 1409.5+1414	14 09 30.	+ 14 14	3360		CLUSTER OF GALAXIES
ZWG 133.006	14 09 30.	+ 21 37		15.0	GALAXY
MCG+07-29-051	14 09 30.	+ 39 52	24	14.5	GALAXY
ZWG 219.056	14 09 30.	+ 39 53		13.9	GALAXY
MRK 669	14 09 30.	+ 39 53	27	15.5	GALAXY WITH UV CONTINUUM
UGC 09081	14 09 30.	+ 39 53	102	13.9	GALAXY Sa
SVEN 431	14 09 30.	- 29 59 40.	48	14.4	GALAXY
SVEN 430	14 09 30.	- 30 24	90	12.9	GALAXY
ARC 1874	14 09 31.	+ 29 59		17.4	RICH CLUSTER OF GALAXIES
REIZ 3940	14 09 33.	+ 19 45	18	15.5	GALAXY
REIZ 3949	14 09 34.	+ 39 52	12	14.0	GALAXY
RNGC 5495	14 09 34.	- 26 53		13.0	GALAXY
REIN 7.153	14 09 34.93	+ 39 52 37.5			NEBULA
8ZW 1409+00.2	14 09 36.	+ 00 12		19.4	COMPACT GALAXY
MCG+04-33-044	14 09 36.	+ 21 36	48	15.0	GALAXY
ZC 1409.6+2247	14 09 36.	+ 22 47	610		CLUSTER OF GALAXIES
ZWG 163.003	14 09 36.	+ 27 21		15.0	GALAXY
UGC 09082	14 09 36.	+ 27 21	66	15.0	GALAXY S0-a
ZWG 163.004	14 09 36.	+ 32 09		15.4	GALAXY
ZWG 162.068	14 09 36.	+ 32 09		15.4	GALAXY
ZC 1409.6+4440	14 09 36.	+ 44 40	740		CLUSTER OF GALAXIES
ZWG 272.039	14 09 36.	+ 50 27		15.1	GALAXY
ZWG 247.009	14 09 36.	+ 50 27		15.1	GALAXY
UGC 09083	14 09 36.	+ 50 27	60	15.1	GALAXY
VV 125	14 09 36.	+ 50 27 30.	78	14.	INTERACTING GALAXY
MCG+09-23-051	14 09 36.	+ 55 27	42	15.	GALAXY
ZWG 272.040	14 09 36.	+ 55 32		15.7	GALAXY
MCG+09-23-050	14 09 36.	+ 55 46	48	16.	GALAXY
7ZW 535	14 09 36.	+ 62 58			COMPACT GALAXY
MCG-04-34-001	14 09 36.	- 26 53	72	13.5	GALAXY
MCG+05-33-054	14 09 39.	+ 27 20	36	15.0	GALAXY
MCG+08-26-009	14 09 39.	+ 50 26 30.	72	15.	GALAXY
IC 4384	14 09 40.	+ 27 20 53.			NONSTELLAR OBJECT
SVEN 432	14 09 40.	- 00 22 39.	24	15.4	GALAXY
ZWG 074.134	14 09 42.	+ 08 54		14.9	GALAXY
UGC 09084	14 09 42.	+ 08 54	66	14.9	GALAXY Sb
MCG+02-36-046	14 09 42.	+ 08 54	60	14.9	GALAXY
ZC 1409.7+2238	14 09 42.	+ 22 38	340		CLUSTER OF GALAXIES
ZWG 133.007	14 09 42.	+ 25 36		15.7	GALAXY
MCG+09-23-052	14 09 42.	+ 55 22	30	15.	GALAXY
MCG+12-13-033	14 09 42.	+ 72 35	27	16.	GALAXY
ZWG 018.077	14 09 42.	- 00 23		15.3	GALAXY
ARC 1871	14 09 43.	- 13 17		17.1	RICH CLUSTER OF GALAXIES
MCG+06-31-078	14 09 45.	+ 38 26	60	14.	GALAXY
REIZ 3942	14 09 46.	+ 08 54	72	14.6	GALAXY
HOLM 600A	14 09 46.	+ 08 54	96	14.2	PART OF MULTIPLE GALAXY
REIZ 3944	14 09 47.	+ 16 06	30	14.6	GALAXY
HMS 1.22	14 09 48.	+ 52 35			GALAXY
ZWG 018.078	14 09 48.	+ 01 30		15.1	GALAXY
MCG+00-36-027	14 09 48.	+ 01 30	42	15.1	GALAXY
ZWG 074.135	14 09 48.	+ 08 45		15.2	GALAXY
RNGC 5504B	14 09 48.	+ 16 05		15.5	GALAXY
MCG+03-36-079	14 09 48.	+ 16 05 30.	36	15.4	GALAXY
ZWG 103.113	14 09 48.	+ 16 06		15.4	GALAXY
ZC 1409.8+1753	14 09 48.	+ 17 53	870		CLUSTER OF GALAXIES
ZC 1409.8+3655	14 09 48.	+ 36 55	1140		CLUSTER OF GALAXIES
MCG+09-23-053	14 09 48.	+ 52 35	48	18.	GALAXY
ZWG 272.041	14 09 48.	+ 55 21		15.3	GALAXY
ZC 1409.8+5805	14 09 48.	+ 58 05	610		CLUSTER OF GALAXIES
7ZW 536	14 09 48.	+ 63 26			COMPACT GALAXY
ZWG 337.005	14 09 48.	+ 72 34		15.7	GALAXY
ZWG 336.045	14 09 48.	+ 72 34		15.7	GALAXY
RNGC 5501	14 09 50.	+ 01 30		15.0	GALAXY
HOLM 601B	14 09 50.	+ 16 07	30	14.1	PART OF MULTIPLE GALAXY
IC 4383	14 09 51.	+ 16 05			NONSTELLAR OBJECT
HOLM 602B	14 09 51.	+ 18 33	18	14.7	PART OF MULTIPLE GALAXY
MCG+03-36-080	14 09 51.	+ 18 35	36	16.	GALAXY
HOLM 603B	14 09 51.	+ 34 27	12	13.9	PART OF MULTIPLE GALAXY
HOLM 603A	14 09 51.	+ 34 27	18	13.6	PART OF MULTIPLE GALAXY
REIZ 3943	14 09 52.	+ 09 05	24	15.4	GALAXY
REIZ 3948	14 09 52.	+ 18 31	30	14.3	GALAXY
REIZ 3946	14 09 53.	+ 16 05	60	13.4	GALAXY
HOLM 601A	14 09 53.	+ 16 05	66	12.9	PART OF MULTIPLE GALAXY
REIZ 3947	14 09 53.	+ 16 07	48	14.9	GALAXY
HOLM 601C	14 09 53.	+ 16 07	60	14.4	PART OF MULTIPLE GALAXY
ZWG 103.114	14 09 54.	+ 16 04		13.9	GALAXY
RNGC 5504A	14 09 54.	+ 16 04		14.0	GALAXY
UGC 09085	14 09 54.	+ 16 04	84	13.9	GALAXY Sb/SBc
MCG+03-36-081	14 09 54.	+ 16 04	72	13.9	GALAXY
RNGC 5504C	14 09 54.	+ 16 06		15.5	GALAXY
MCG+03-36-082	14 09 54.	+ 16 06	66	15.5	GALAXY
ZWG 103.115	14 09 54.	+ 16 07		15.5	GALAXY
UGC 09086	14 09 54.	+ 16 07	66	15.5	GALAXY Sc-IRR
ZWG 103.116	14 09 54.	+ 18 32		14.9	GALAXY
UGC 09087	14 09 54.	+ 18 32	66	14.9	GALAXY S0?
ZWG 103.117	14 09 54.	+ 18 54		15.7	GALAXY
ZWG 191.062	14 09 54.	+ 38 25		14.8	GALAXY
UGC 09088	14 09 54.	+ 38 25	150	14.8	GALAXY S0-a
KARA.73B 0616	14 09 54.	+ 38 25	168	14.8	ISOLATED GALAXY S
MCG+08-26-010	14 09 54.	+ 48 11	42	16.	GALAXY
ZC 1409.9+5405	14 09 54.	+ 54 05	1340		CLUSTER OF GALAXIES
8ZW 1409-00.6	14 09 54.	- 00 35		17.3	COMPACT GALAXY
MCG+03-36-083	14 09 57.	+ 18 33	66	14.9	GALAXY
REIZ 3945	14 09 58.	+ 08 52	30	14.6	GALAXY
HOLM 602A	14 09 58.	+ 18 32	48	14.2	PART OF MULTIPLE GALAXY
REIZ 3951	14 09 58.	+ 18 38	12	14.8	GALAXY
HOLM 600C	14 09 59.	+ 08 52	18	15.4	PART OF MULTIPLE GALAXY
ZWG 074.136	14 10 00.	+ 08 53		15.1	GALAXY
ZWG 074.137	14 10 00.	+ 11 29		14.9	GALAXY
UGC 09089	14 10 00.	+ 11 29	66	14.9	GALAXY S
MCG+02-36-047	14 10 00.	+ 11 29	60	14.9	GALAXY
ZC 1410.0+1538	14 10 00.	+ 15 38	870		CLUSTER OF GALAXIES
ZWG 103.118	14 10 00.	+ 18 39		14.9	GALAXY
UGC 09090	14 10 00.	+ 18 39	60	14.9	GALAXY E-S0
ZC 1410.0+2412	14 10 00.	+ 24 12	740		CLUSTER OF GALAXIES
ZC 1410.0+2431	14 10 00.	+ 24 31	1080		CLUSTER OF GALAXIES
ZC 1410.0+2509	14 10 00.	+ 25 09	5440		CLUSTER OF GALAXIES
ZWG 163.005	14 10 00.	+ 29 26		15.6	GALAXY
KARA.73B 0617	14 10 00.	+ 29 26	24	15.6	ISOLATED GALAXY S
UGC 09091	14 10 00.	+ 30 09	60	16.5	GALAXY Sc
REIZ 3956	14 10 00.	+ 48 12	12	14.8	GALAXY
FEIG 092	14 10 00.	+ 50 21		11.2	FAINT BLUE STAR
MCG+09-23-054	14 10 00.	+ 54 17	30	17.	GALAXY
ZWG 018.079	14 10 00.	- 00 47		15.6	GALAXY
REIZ 3950	14 10 01.	+ 13 32	42	13.8	GALAXY
MCG+03-36-084	14 10 03.	+ 16 05	36	18.	GALAXY
HOLM 600B	14 10 04.	+ 08 53	60	14.7	PART OF MULTIPLE GALAXY
RNGC 5509	14 10 04.	+ 24 53			GALAXY
RNGC 5505	14 10 05.	+ 13 32		14.0	GALAXY
ARC 1875	14 10 05.	+ 14 16		17.6	RICH CLUSTER OF GALAXIES
ZWG 018.080	14 10 06.	+ 01 59		15.7	GALAXY
ZWG 074.138	14 10 06.	+ 13 32		14.1	GALAXY
UGC 09092	14 10 06.	+ 13 32	60	14.1	GALAXY SBa
MCG+02-36-048	14 10 06.	+ 13 32	60	14.1	GALAXY
MCG+03-36-085	14 10 06.	+ 18 40	27	14.9	GALAXY
ZC 1410.1+2820	14 10 06.	+ 28 20	4440		CLUSTER OF GALAXIES
ZC 1410.1+4206	14 10 06.	+ 42 06	1010		CLUSTER OF GALAXIES
REIZ 3955	14 10 09.	+ 40 40	36	14.7	GALAXY
RNGC 5508	14 10 10.	+ 24 53		15.5	GALAXY
ZWG 074.139	14 10 12.	+ 12 16		15.7	GALAXY
UGC 09093	14 10 12.	+ 12 16	60	15.7	GALAXY Sc
ZWG 133.008	14 10 12.	+ 21 50		15.7	GALAXY
MCG+04-34-001	14 10 12.	+ 21 50 30.	36	15.7	GALAXY
MCG+04-34-002	14 10 12.	+ 24 52	90	15.3	GALAXY
ZWG 133.009	14 10 12.	+ 24 53		15.3	GALAXY
UGC 09094	14 10 12.	+ 24 53	126	15.2	GALAXY S0
MCG+09-23-055	14 10 12.	+ 54 11	24	16.	GALAXY
ZWG 353.031	14 10 12.	+ 78 50		14.5	GALAXY
UGC 09095	14 10 12.	+ 78 50	60	14.5	GALAXY DBL SYS
REIZ 3953	14 10 13.	+ 21 49	48	14.9	GALAXY
RNGC 5547	14 10 14.	+ 78 50		14.5	GALAXY
ZWG 133.010	14 10 18.	+ 20 38		15.2	GALAXY
MCG+04-34-003	14 10 18.	+ 20 38	54	15.2	GALAXY
ZC 1410.3+2056	14 10 18.	+ 20 56	1480		CLUSTER OF GALAXIES
ZC 1410.3+3030	14 10 18.	+ 30 30	4030		CLUSTER OF GALAXIES
MRK 467	14 10 18.	+ 34 47	14	16.5	GALAXY WITH UV CONTINUUM
ZWG 272.042	14 10 18.	+ 55 38		15.5	GALAXY
7ZW 537	14 10 18.	+ 63 28			COMPACT GALAXY
MCG-07-29-009	14 10 18.	- 45 11	132	14.5	GALAXY

OBJECT NAME	RIGHT ASCEN.	DECLINATION	DIAM.	MAGN.	TYPE OF OBJECT
ARC 1884	14 10 21.	+ 61 27		17.8	RICH CLUSTER OF GALAXIES
HN 1736	14 10 21.0	- 45 11 08.	60		NEBULA
REIZ 3952	14 10 22.	+ 09 21	18	15.5	GALAXY
PK312-02.1	14 10 22.	- 63 11 40.	10		PLANETARY NEBULA
ZC 1410.4+1500	14 10 24.	+ 15 00	1140		CLUSTER OF GALAXIES
ZWG 163.006	14 10 24.	+ 31 05		15.3	GALAXY
RNGC 5512	14 10 24.	+ 31 05		15.5	GALAXY
SVEN 433	14 10 24.	- 30 14 38.	30	15.1	GALAXY
HOLM 609A	14 10 27.	+ 45 56	24	14.2	PART OF MULTIPLE GALAXY
RNGC 5515	14 10 28.	+ 39 32		13.5	GALAXY
RNGC 5520	14 10 29.	+ 50 36		13.5	GALAXY
8ZW 1410+01.3	14 10 30.	+ 01 15		16.5	COMPACT GALAXY
ZWG 074.140	14 10 30.	+ 10 09		15.0	GALAXY
MCG+02-36-049	14 10 30.	+ 10 09	36	15.0	GALAXY
TON-N 0754	14 10 30.	+ 29 27		14.3	BLUE STAR
ZWG 219.057	14 10 30.	+ 39 32		13.7	GALAXY
UGC 09096	14 10 30.	+ 39 32	90	13.7	GALAXY Sa-b
MCG+07-29-052	14 10 30.	+ 39 32	84	14.	GALAXY
ZWG 247.010	14 10 30.	+ 44 42		15.6	GALAXY
MCG+08-26-011	14 10 30.	+ 45 54 30.	27	14.2	GALAXY
KARA.72 418A	14 10 30.	+ 45 55	36	14.1	PART OF DOUBLE GALAXY
HOLM 609B	14 10 30.	+ 45 56	18	14.3	PART OF MULTIPLE GALAXY
ZWG 272.043	14 10 30.	+ 50 36		13.3	GALAXY
UGC 09097	14 10 30.	+ 50 36	108	13.3	GALAXY Sb
MCG+09-23-056	14 10 30.	+ 55 28	42	16.	GALAXY
8ZW 1410-00.9	14 10 30.	- 00 52		19.6	COMPACT GALAXY
MCG-05-34-002	14 10 30.	- 29 21	120	15.	GALAXY
HOLM 605B	14 10 32.	+ 10 09	9	15.5	PART OF MULTIPLE GALAXY
REIZ 3969	14 10 32.	+ 50 35	72	13.2	GALAXY
MCG+08-26-013	14 10 33.	+ 50 35	132	13.2	GALAXY
HOLM 605A	14 10 33.	+ 10 09	30	14.5	PART OF MULTIPLE GALAXY
MCG+08-26-012	14 10 33.	+ 45 54 30.	48	14.3	GALAXY
RNGC 5511B	14 10 34.	+ 08 52		15.5	GALAXY
REIZ 3962	14 10 34.	+ 33 29	24	14.7	GALAXY
REIZ 3965	14 10 34.	+ 39 33	24	13.8	GALAXY
REIN 7.154	14 10 34.23	+ 39 32 37.1			NEBULA
ZWG 046.064	14 10 36.	+ 03 56		15.3	GALAXY
ZWG 074.141	14 10 36.	+ 08 52		15.5	GALAXY
8ZW 1410+08.9	14 10 36.	+ 08 52		15.5	COMPACT GALAXY
MCG+02-36-050	14 10 36.	+ 08 52	24	15.5	GALAXY
MCG+04-34-004	14 10 36.	+ 20 38	48		GALAXY
ZWG 247.011	14 10 36.	+ 45 55		14.1	GALAXY
UGC 09098	14 10 36.	+ 45 55	72	14.1	GALAXY DBL SYS
KARA.72 418B	14 10 36.	+ 45 55	42		PART OF DOUBLE GALAXY
MCG+08-26-014	14 10 36.	+ 47 28	36	16.	GALAXY
HOLM 606B	14 10 37.	+ 08 50	24	14.8	PART OF MULTIPLE GALAXY
ARC 1878	14 10 37.	+ 29 27		17.5	RICH CLUSTER OF GALAXIES
MCG+08-26-015	14 10 39.	+ 46 25	60	15.	GALAXY
RNGC 5511A	14 10 40.	+ 08 51		15.0	GALAXY
REIZ 3959	14 10 40.	+ 08 51	24	15.2	GALAXY
HOLM 606A	14 10 40.	+ 08 51	48	14.1	PART OF MULTIPLE GALAXY
HOLM 607B	14 10 40.	+ 20 39	12	15.2	PART OF MULTIPLE GALAXY
REIZ 3954	14 10 41.	- 02 58	126	13.3	GALAXY
HOLM 604A	14 10 41.	- 02 58	126	13.3	PART OF MULTIPLE GALAXY
ZWG 074.142	14 10 42.	+ 08 51		15.0	GALAXY
8ZW 1410+08.8	14 10 42.	+ 08 51		15.0	COMPACT GALAXY
MCG+02-36-051	14 10 42.	+ 08 51	36	15.0	GALAXY
MCG+04-34-005	14 10 42.	+ 20 39	36	14.1	GALAXY
ZC 1410.7+2929	14 10 42.	+ 29 29	810		CLUSTER OF GALAXIES
MCG+06-31-079	14 10 42.	+ 35 56	60	14.	GALAXY
ZWG 247.012	14 10 42.	+ 46 27		15.0	GALAXY
RNGC 5506	14 10 42.	- 02 58		13.5	GALAXY
ZWG 018.081	14 10 42.	- 02 59		13.6	GALAXY
KARA.72 419A	14 10 42.	- 02 59	150	13.6	PART OF DOUBLE GALAXY
MCG+00-36-028	14 10 42.	- 02 59	144	13.6	GALAXY
REIZ 3960	14 10 43.	+ 13 15	78	14.9	GALAXY
HOLM 607A	14 10 44.	+ 20 40	54	13.9	PART OF MULTIPLE GALAXY
REIZ 3967	14 10 44.	+ 35 57	42	13.7	GALAXY
REIZ 3966	14 10 45.	+ 34 56	60	14.4	GALAXY
REIZ 3958	14 10 46.	- 00 21	150	12.9	GALAXY
HOLM 504B	14 10 46.	- 02 54	60	13.9	PART OF MULTIPLE GALAXY
REIZ 3957	14 10 47.	- 02 55	90	13.6	GALAXY
PK308-12.1	14 10 47.	- 73 59 17.	35		PLANETARY NEBULA
ZC 1410.8+0710	14 10 48.	+ 07 10	2420		CLUSTER OF GALAXIES
VV 299B	14 10 48.	+ 08 51	24	14.8	INTERACTING GALAXY
VV 299A	14 10 48.	+ 08 51	42	14.1	INTERACTING GALAXY
ZWG 074.143	14 10 48.	+ 09 05		14.9	GALAXY
MCG+02-36-052	14 10 48.	+ 09 05	60	14.9	GALAXY
ZWG 074.144	14 10 48.	+ 13 14		15.0	GALAXY
MCG+02-36-053	14 10 48.	+ 13 14	60	15.0	GALAXY
ZWG 103.119	14 10 48.	+ 19 17		15.5	GALAXY
ZC 1410.8+2005	14 10 48.	+ 20 05	3630		CLUSTER OF GALAXIES
ZWG 133.011	14 10 48.	+ 20 40		14.1	GALAXY
UGC 09099	14 10 48.	+ 20 40	138	14.1	GALAXY S0
ZWG 191.063	14 10 48.	+ 35 56		15.0	GALAXY
UGC 09100	14 10 48.	+ 35 56	72	15.0	GALAXY S0-a
ZWG 018.082	14 10 48.	- 02 55		13.7	GALAXY
RNGC 5507	14 10 48.	- 02 55		13.5	GALAXY
KARA.72 419B	14 10 48.	- 02 55	90	13.7	PART OF DOUBLE GALAXY
MCG+00-36-029	14 10 48.	- 02 55	84	13.7	GALAXY
RNGC 5510	14 10 49.	- 17 45		14.0	GALAXY
RNGC 5513	14 10 50.	+ 20 40		14.0	GALAXY
REIZ 3964	14 10 50.	+ 20 40	72	13.6	GALAXY
RNGC 5517	14 10 50.	+ 35 56		15.0	GALAXY
ARC 1876	14 10 50.	- 13 51		17.1	RICH CLUSTER OF GALAXIES
PATE 1.656	14 10 51.	- 15 04	11		NEBULA
MCG-03-36-010	14 10 51.	- 17 45	72	14.	GALAXY
REIZ 3961	14 10 52.	+ 08 44	24	14.5	GALAXY
SN 1971P	14 10 52.	- 32 22		17.3	SUPERNOVA
HOLM 608B	14 10 53.	+ 09 12	30	15.4	PART OF MULTIPLE GALAXY
ARC 1830	14 10 53.	+ 22 38		17.2	RICH CLUSTER OF GALAXIES
ZWG 018.083	14 10 54.	+ 00 25		15.6	GALAXY
ZC 1410.9+2240	14 10 54.	+ 22 40	1010		CLUSTER OF GALAXIES
REIZ 3963	14 10 58.	+ 09 13	42	14.4	GALAXY
HOLM 608A	14 10 58.	+ 09 13	60	14.2	PART OF MULTIPLE GALAXY
ZWG 074.145	14 11 00.	+ 09 14		15.0	GALAXY
MCG+02-36-054	14 11 00.	+ 09 14	60	15.0	GALAXY
ZC 1411.0+2133	14 11 00.	+ 21 33	1480		CLUSTER OF GALAXIES
ZC 1411.0+2334	14 11 00.	+ 23 34	2490		CLUSTER OF GALAXIES
ZWG 163.007	14 11 00.	+ 27 15		14.7	GALAXY
UGC 09101	14 11 00.	+ 27 15	126	14.7	GALAXY Sb
ZWG 219.058	14 11 00.	+ 42 14		15.5	GALAXY
MCG+07-29-053	14 11 00.	+ 42 14	36	17.	GALAXY
ZWG 046.065	14 11 06.	+ 06 24		15.7	GALAXY
MCG+02-36-055	14 11 06.	+ 09 13	12	15.4	GALAXY
ZWG 103.120	14 11 06.	+ 16 21		15.7	GALAXY
MCG+03-36-086	14 11 06.	+ 16 21	48	15.7	GALAXY
ZWG 103.121	14 11 06.	+ 17 59		15.6	GALAXY

OBJECT NAME	RIGHT ASCEN.	DECLINATION	DIAM.	MAGN.	TYPE OF OBJECT
KARA.73B 0618	14 11 06.	+ 17 59	42	15.6	ISOLATED GALAXY S
ZWG 133.012	14 11 06.	+ 24 46		15.5	GALAXY
TON-N 0755	14 11 06.	+ 27 12		14.3	BLUE STAR
MCG+05-34-001	14 11 06.	+ 27 13	120	14.7	GALAXY
ZWG 317.018	14 11 06.	+ 64 36		15.6	GALAXY
ZC 1411.1+6538	14 11 06.	+ 65 38	3090		CLUSTER OF GALAXIES
KARA.68 230	14 11 06.	- 01 56	54		DWARF GALAXY
HOLM 612B	14 11 08.	+ 16 22	18	15.2	PART OF MULTIPLE GALAXY
RNGC 5514	14 11 10.	+ 07 54		14.5	GALAXY
HOLM 611B	14 11 10.	+ 12 26	12	15.4	PART OF MULTIPLE GALAXY
REIZ 3968	14 11 11.	+ 07 54	60	13.7	GALAXY
HOLM 612A	14 11 11.	+ 16 22	30	14.5	PART OF MULTIPLE GALAXY
ZWG 046.066	14 11 12.	+ 07 54		14.5	GALAXY
UGC 09102	14 11 12.	+ 07 54	138	14.5	GALAXY DBL SYS
KARA.72 420B	14 11 12.	+ 07 54	48		PART OF DOUBLE GALAXY
KARA.720A	14 11 12.	Q 07 54	60	14.5	PART OF DOUBLE GALAXY
MCG+01-36-023	14 11 12.	+ 07 54	120	14.5	GALAXY
VV 223C	14 11 12.	+ 08 26	78	16.	INTERACTING GALAXY
VV 223B	14 11 12.	+ 08 26	24	15.	INTERACTING GALAXY
VV 223A	14 11 12.	+ 08 26	24	15.	INTERACTING GALAXY
UGC 09103	14 11 12.	+ 08 27	72	16.5	GALAXY S
HOLM 610C	14 11 12.	+ 08 28	24	15.4	PART OF MULTIPLE GALAXY
ZWG 074.146	14 11 12.	+ 12 44		14.4	GALAXY
UGC 09104	14 11 12.	+ 12 44	60	14.4	GALAXY S
MCG+02-36-056	14 11 12.	+ 12 44	60	14.4	GALAXY
ZWG 337.006	14 11 12.	+ 72 31		15.1	GALAXY
ZWG 336.046	14 11 12.	+ 72 31		15.1	GALAXY
MCG+12-13-034	14 11 12.	+ 72 31	51	15.	GALAXY
HOLM 611A	14 11 14.	+ 12 26	30	15.4	PART OF MULTIPLE GALAXY
HOLM 610A	14 11 16.	+ 08 28	18	14.7	PART OF MULTIPLE GALAXY
HOLM 610B	14 11 17.	+ 08 28	12	14.7	PART OF MULTIPLE GALAXY
IC 4404	14 11 17.	+ 78 52			NONSTELLAR OBJECT
ZWG 018.084	14 11 18.	+ 01 57		15.6	GALAXY
ZWG 046.067	14 11 18.	+ 08 27		14.7	GALAXY
8ZW 1411+08.4	14 11 18.	+ 08 27		14.7	COMPACT GALAXY
MCG+01-36-024	14 11 18.	+ 08 27	30	14.7	GALAXY
ZC 1411.2+3510	14 11 18.	+ 35 10	1550		CLUSTER OF GALAXIES
ZWG 018.085	14 11 24.	+ 02 03		15.6	GALAXY
MCG+02-36-057	14 11 24.	+ 08 28	27	15.	GALAXY
MCG+04-34-006	14 11 24.	+ 21 05	24	15.0	GALAXY
ZWG 163.008	14 11 24.	+ 29 18		15.7	GALAXY
KARA.73B 0619	14 11 24.	+ 29 18	24	15.7	ISOLATED GALAXY S
ZWG 219.059	14 11 24.	+ 44 05		15.6	GALAXY
ZWG 018.086	14 11 30.	+ 02 04		15.5	GALAXY
ZWG 074.147	14 11 30.	+ 10 02		15.7	GALAXY
ZWG 103.122	14 11 30.	+ 15 44		15.3	GALAXY
MCG+03-36-087	14 11 30.	+ 15 44	48	15.3	GALAXY
ZWG 133.013	14 11 30.	+ 21 05		15.0	GALAXY
MCG+04-34-007	14 11 30.	+ 21 13	60	15.4	GALAXY
ZWG 133.014	14 11 30.	+ 21 14		15.4	GALAXY
TON-N 0756	14 11 30.	+ 25 54		15.8	BLUE STAR
ZC 1411.5+2635	14 11 30.	+ 26 35	3160		CLUSTER OF GALAXIES
ZWG 163.009	14 11 30.	+ 29 40		14.8	GALAXY
MCG+10-20-083	14 11 30.	+ 58 04	39	16.	GALAXY
RNGC 5518	14 11 32.	+ 21 05		15.0	GALAXY
REIZ 3970	14 11 32.	+ 21 05	18	14.5	GALAXY
MCG+05-34-002	14 11 33.	+ 29 17 30.	36	15.7	GALAXY
SVEN 434	14 11 34.	- 00 38	30	15.1	GALAXY
ZWG 046.068	14 11 36.	+ 03 27		15.2	GALAXY
ZC 1411.6+2406	14 11 36.	+ 24 06	810		CLUSTER OF GALAXIES
ZC 1411.6+2631	14 11 36.	+ 26 31	940		CLUSTER OF GALAXIES
MCG+05-34-003	14 11 36.	+ 29 39 30.	42	14.8	GALAXY
ZWG 219.060	14 11 36.	+ 44 06		15.1	GALAXY
UGC 09105	14 11 36.	+ 44 06	72	15.1	GALAXY S
ZWG 272.044	14 11 36.	+ 54 03		15.7	GALAXY
ZWG 018.087	14 11 36.	- 03 16		15.7	GALAXY
HN 1095	14 11 36.	- 42 04	12		NEBULA
IC 4385	14 11 36.	- 42 04			NONSTELLAR OBJECT
MCG+07-29-054	14 11 39.	+ 44 07	66	14.5	GALAXY
8ZW 1411+01.3	14 11 42.	+ 01 19		19.4	COMPACT GALAXY
ZC 1411.7+1859	14 11 42.	+ 18 59	1950		CLUSTER OF GALAXIES
ZWG 133.015	14 11 42.	+ 26 09		15.7	GALAXY
UGC 09106	14 11 42.	+ 26 09	66	15.7	GALAXY Sb-c
ZWG 163.010	14 11 42.	+ 31 48		15.5	GALAXY
UGC 09107	14 11 42.	+ 31 48	72	15.5	GALAXY SBb
ZWG 191.064	14 11 42.	+ 37 29		15.3	GALAXY
KARA.73B 0620	14 11 42.	+ 37 29	36	15.3	ISOLATED GALAXY S
MCG+09-23-057	14 11 42.	+ 54 55	36	16.	GALAXY
ZWG 018.088	14 11 42.	- 00 39		15.2	GALAXY
ARC 1881	14 11 43.	+ 07 05		17.0	RICH CLUSTER OF GALAXIES
MCG-02-36-018	14 11 45.	- 10 28	48	15.	GALAXY
ARC 1885	14 11 46.	+ 43 54		17.0	RICH CLUSTER OF GALAXIES
ZWG 018.089	14 11 48.	+ 00 02		15.4	GALAXY
ZWG 046.069	14 11 48.	+ 03 12		15.6	GALAXY
UGC 09108	14 11 48.	+ 03 12	72	15.6	GALAXY TRP SYS
ZWG 074.148	14 11 48.	+ 10 49		15.2	GALAXY
MCG+02-36-058	14 11 48.	+ 10 49	30	15.2	GALAXY
ZWG 074.149	14 11 48.	+ 12 34		15.5	GALAXY
UGC 09109	14 11 48.	+ 12 34	60	15.5	GALAXY DBL SYS
ZWG 103.123	14 11 48.	+ 15 51		14.0	GALAXY
UGC 09110	14 11 48.	+ 15 51	138	14.0	GALAXY SBb
MCG+03-36-088	14 11 48.	+ 15 51	126	14.0	GALAXY
ZWG 133.016	14 11 48.	+ 21 16		15.5	GALAXY
MCG+08-26-016	14 11 48.	+ 47 52	60	15.	GALAXY
REIZ 3973	14 11 50.	+ 21 16	36	14.7	GALAXY
HOLM 613A	14 11 51.	+ 10 49	30	15.2	PART OF MULTIPLE GALAXY
MCG+05-34-004	14 11 51.	+ 31 48	66	15.5	GALAXY
RNGC 5519	14 11 52.	+ 07 45		14.5	GALAXY
REIZ 3971	14 11 53.	+ 07 40	18	14.1	GALAXY
REIZ 3972	14 11 53.	+ 15 51	60	13.6	GALAXY
HOLM 614A	14 11 53.	+ 15 51	108	13.5	PART OF MULTIPLE GALAXY
ZWG 046.070	14 11 54.	+ 07 45		14.6	GALAXY
UGC 09111	14 11 54.	+ 07 45	126	14.6	GALAXY Sa
ZWG 247.013	14 11 54.	+ 47 53		15.0	GALAXY
UGC 09112	14 11 54.	+ 47 53	72	15.0	GALAXY S
ZWG 018.090	14 11 54.	- 03 04		15.4	GALAXY
MCG-07-29-011	14 11 54.	- 43 43	90	15.5	GALAXY
MCG-07-29-010	14 11 54.	- 43 45	30	17.	GALAXY
HN 1096	14 11 56.	- 43 43	12		NEBULA
IC 4386	14 11 56.	- 43 43			NONSTELLAR OBJECT
HN 1097	14 11 56.	- 43 45	12		NEBULA
IC 4387	14 11 56.	- 43 45			NONSTELLAR OBJECT
HOLM 613B	14 11 58.	+ 10 49	30	15.5	PART OF MULTIPLE GALAXY
REIZ 3974	14 11 59.	+ 15 51	30	15.4	GALAXY
ZWG 018.091	14 12 00.	+ 01 58		15.5	GALAXY
8ZW 1412+02.0	14 12 00.	+ 01 58		15.5	COMPACT GALAXY
IC 0988	14 12 00.	+ 03 24 49.			NONSTELLAR OBJECT
ZWG 046.071	14 12 00.	+ 03 25		14.8	GALAXY

OBJECT NAME	RIGHT ASCEN.	DECLINATION	DIAM.	MAGN.	TYPE OF OBJECT
ZWG 103.124	14 12 00.	+ 15 50		15.3	GALAXY
ZWG 133.017	14 12 00.	+ 24 53		15.6	GALAXY
MRK 670	14 12 00.	+ 26 59	15	15.5	GALAXY WITH UV CONTINUUM
ZC 1412.0+2717	14 12 00.	+ 27 17	1550		CLUSTER OF GALAXIES
MCG+09-23-058	14 12 00.	+ 54 52 30.	42	16.	GALAXY
MCG+10-20-085	14 12 00.	+ 57 59 30.	96	13.9	GALAXY
MCG+10-20-084	14 12 00.	+ 58 00	39	15.	GALAXY
7ZW 538	14 12 00.	+ 68 55			COMPACT GALAXY
MCG+14-07-002	14 12 00.	+ 84 20	24	17.	GALAXY
REIZ 3976	14 12 01.	+ 36 42	48	14.7	GALAXY
LB 00723	14 12 02.	+ 00 25 48.		16.5	FAINT BLUE STAR
HN 0397	14 12 02.	- 75 24			NEBULA
IC 4377	14 12 02.	- 75 25			NONSTELLAR OBJECT
ARC 1882	14 12 04.	- 00 07		17.2	RICH CLUSTER OF GALAXIES
ZC 1412.1+3248	14 12 06.	+ 32 48	740		CLUSTER OF GALAXIES
ZWG 191.065	14 12 06.	+ 35 39		15.7	GALAXY
UGC 09113	14 12 06.	+ 35 39	138	15.7	GALAXY S
MCG+06-31-080	14 12 06.	+ 35 39	114	14.5	GALAXY
ZWG 018.092	14 12 06.	- 00 03		15.4	GALAXY
HOLM 614B	14 12 08.	+ 15 50	30	14.5	PART OF MULTIPLE GALAXY
REIZ 3991	14 12 11.	+ 58 00	132	13.9	GALAXY
ZC 1412.2+0000	14 12 12.	+ 00 00	4030		CLUSTER OF GALAXIES
ZC 1412.2+2227	14 12 12.	+ 22 27	540		CLUSTER OF GALAXIES
ZC 1412.2+4149	14 12 12.	+ 41 49	1080		CLUSTER OF GALAXIES
KARA.72 421A	14 12 12.	+ 58 00	54	14.2	PART OF DOUBLE GALAXY
MCG+10-20-087	14 12 12.	+ 58 23	18	16.	GALAXY
MCG+10-20-086	14 12 12.	+ 60 34	36	16.	GALAXY
SER 111.01	14 12 12.	- 43 43	300	14.	FRAGMENTED SB GALAXIES
8ZW 1412+00.2	14 12 18.	+ 00 13		18.4	COMPACT GALAXY
ZWG 046.072	14 12 18.	+ 03 21		14.4	GALAXY
UGC 09114	14 12 18.	+ 03 21	78	14.4	GALAXY E
MCG+01-36-027	14 12 18.	+ 03 21	60	14.4	GALAXY
REIZ 3979	14 12 18.	+ 22 38	42	14.6	GALAXY
ZWG 191.066	14 12 18.	+ 36 04		15.6	GALAXY
ZWG 191.067	14 12 18.	+ 36 38		15.3	GALAXY
MCG+06-31-081	14 12 18.	+ 36 38	42	14.	GALAXY
ZC 1412.3+4105	14 12 18.	+ 41 05	470		CLUSTER OF GALAXIES
ZWG 295.040	14 12 18.	+ 58 00		14.2	GALAXY
RNGC 5526	14 12 18.	+ 58 00		14.0	GALAXY
UGC 09115	14 12 18.	+ 58 00	120	14.2	GALAXY Sb-c
KARA.72 421B	14 12 18.	+ 58 00	108		PART OF DOUBLE GALAXY
IC 0989	14 12 20.	+ 03 22 09.			NONSTELLAR OBJECT
REIZ 3978	14 12 20.	+ 21 09	18	15.7	GALAXY
RNGC 5524	14 12 21.	+ 36 38		15.5	GALAXY
ZWG 046.073	14 12 24.	+ 05 08		15.7	GALAXY
REIZ 3975	14 12 24.	+ 05 09	18	16.1	GALAXY
FAI 1412+09	14 12 24.	+ 09 13			COMPACT GALAXY
8ZW 1412+14.4	14 12 24.	+ 14 22		17.0	COMPACT GALAXY
REIZ 3977	14 12 24.	+ 15 21	72	14.0	GALAXY
MCG+03-36-089	14 12 24.	+ 15 22	114	14.1	GALAXY
ZWG 103.125	14 12 24.	+ 15 23		14.1	GALAXY
RNGC 5522	14 12 24.	+ 15 23		14.0	GALAXY
UGC 09116	14 12 24.	+ 15 23	114	14.1	GALAXY Sb
ZC 1412.4+7428	14 12 24.	+ 74 28	1080		CLUSTER OF GALAXIES
ZWG 018.093	14 12 24.	- 00 34		15.4	GALAXY
REIZ 3986	14 12 25.	+ 36 38	60	14.4	GALAXY
LB 00724	14 12 27.	- 12 21 18.		17.5	FAINT BLUE STAR
ARC 1886	14 12 28.	+ 27 22		17.2	RICH CLUSTER OF GALAXIES
LB 00725	14 12 28.	- 10 55 36.		17.2	FAINT BLUE STAR
ZWG 074.150	14 12 30.	+ 14 21		14.3	GALAXY
UGC 09117	14 12 30.	+ 14 21	120	14.3	GALAXY Sa
MCG+02-36-059	14 12 30.	+ 14 21	84	14.3	GALAXY
ZC 1412.5+1430	14 12 30.	+ 14 30	1010		CLUSTER OF GALAXIES
ZWG 018.094	14 12 30.	- 01 19		15.6	GALAXY
HOLM 618B	14 12 33.	+ 34 58	36	13.8	PART OF MULTIPLE GALAXY
RNGC 5523	14 12 34.	+ 25 33		12.5	GALAXY
REIZ 3985	14 12 34.	+ 25 34	108	14.3	GALAXY
REIZ 3993	14 12 35.	+ 58 00	24	14.9	GALAXY
RNGC 5516	14 12 35.	- 47 53			GALAXY
UGC 09118	14 12 36.	+ 03 20	72	16.0	GALAXY Sa-b
ZWG 046.074	14 12 36.	+ 05 10		15.3	GALAXY
REIZ 3981	14 12 36.	+ 05 10	12	16.1	GALAXY
ZWG 103.126	14 12 36.	+ 18 45		15.7	GALAXY
MCG+03-36-090	14 12 36.	+ 18 45	12	15.7	GALAXY
TON-N 0191	14 12 36.	+ 24 22		16.1	BLUE STAR
ZWG 133.018	14 12 36.	+ 25 33		13.4	GALAXY
UGC 09119	14 12 36.	+ 25 33	282	13.4	GALAXY Sc
MCG+04-34-008	14 12 36.	+ 25 33	270	13.4	GALAXY
KARA.73B 0621	14 12 36.	+ 25 33	300	13.4	ISOLATED GALAXY S
7ZW 539	14 12 36.	+ 63 47			COMPACT GALAXY
8ZW 1412-00.3	14 12 36.	- 00 15		19.0	COMPACT GALAXY
REIZ 3984	14 12 37.	+ 21 16	48	15.2	GALAXY
ARC 1893	14 12 37.	+ 74 34		17.5	RICH CLUSTER OF GALAXIES
LB 00274	14 12 37.	- 16 10 00.		15.1	FAINT BLUE STAR
HOLM 618A	14 12 38.	+ 34 58	18	13.7	PART OF MULTIPLE GALAXY
HOLM 615B	14 12 40.	- 04 10	24	15.7	PART OF MULTIPLE GALAXY
ZWG 046.075	14 12 42.	+ 05 03		14.6	GALAXY
UGC 09120	14 12 42.	+ 05 03	72	14.6	GALAXY Sc
MCG+01-36-028	14 12 42.	+ 05 03	66	14.6	GALAXY
REIZ 3982	14 12 42.	+ 05 04	42	14.3	GALAXY
HOLM 616A	14 12 42.	+ 05 04	60	13.5	PART OF MULTIPLE GALAXY
ZWG 046.076	14 12 42.	+ 05 06		15.3	GALAXY
HOLM 616B	14 12 42.	+ 05 06	78	14.3	PART OF MULTIPLE GALAXY
MCG+01-36-029	14 12 42.	+ 05 06	48	15.3	GALAXY
TON-N 0192	14 12 42.	+ 27 41		15.3	BLUE STAR
ZWG 219.061	14 12 42.	+ 44 12		15.7	GALAXY
ZWG 018.095	14 12 42.	- 02 12		15.5	GALAXY
REIZ 3980	14 12 42.	- 04 08	42	15.2	GALAXY
HOLM 615A	14 12 42.	- 04 08	72	14.1	PART OF MULTIPLE GALAXY
REIZ 3994	14 12 43.	+ 57 24	15	15.7	GALAXY
HOLM 617B	14 12 45.	+ 15 56	48	14.5	PART OF MULTIPLE GALAXY
MCG+06-31-082	14 12 45.	+ 37 01	30	15.	GALAXY
REIZ 3987	14 12 47.	+ 15 56	30	14.9	GALAXY
REIZ 3988	14 12 47.	+ 15 58	78	14.2	GALAXY
HOLM 617A	14 12 47.	+ 15 58	72	13.8	PART OF MULTIPLE GALAXY
ZWG 074.151	14 12 48.	+ 11 31		15.4	GALAXY
ZWG 103.127	14 12 48.	+ 15 58		14.8	GALAXY
UGC 09121	14 12 48.	+ 15 58	102	14.8	GALAXY Sb-c
MCG+03-36-091	14 12 48.	+ 15 58	72	14.8	GALAXY
ZWG 103.128	14 12 48.	+ 16 06		15.2	GALAXY
MCG+03-36-092	14 12 48.	+ 16 06	24	15.2	GALAXY
ZWG 103.129	14 12 48.	+ 18 21		14.7	GALAXY
MCG+03-36-093	14 12 48.	+ 18 21	27	14.7	GALAXY
ZC 1412.8+2245	14 12 48.	+ 22 45	3020		CLUSTER OF GALAXIES
ZWG 163.011	14 12 48.	+ 30 50		15.5	GALAXY
ZWG 191.068	14 12 48.	+ 37 00		15.4	GALAXY
ZWG 018.096	14 12 48.	- 02 53		15.0	GALAXY
8ZW 1412-02.9	14 12 48.	- 02 53		15.0	COMPACT GALAXY

OBJECT NAME	RIGHT ASCEN.	DECLINATION	DIAM.	MAGN.	TYPE OF OBJECT
MCG+00-36-030	14 12 48.	- 02 53	42	15.0	GALAXY
REIZ 3983	14 12 49.	+ 04 39	18	16.0	GALAXY
ARC 1883	14 12 51.	- 23 02		17.1	RICH CLUSTER OF GALAXIES
REIZ 3989	14 12 53.	+ 16 07	18	14.7	GALAXY
LB 00726	14 12 54.	+ 02 28 30.		17.2	FAINT BLUE STAR
ZWG 046.077	14 12 54.	+ 04 38		14.3	GALAXY
UGC 09122	14 12 54.	+ 04 38	42	14.3	GALAXY
MCG+01-36-030	14 12 54.	+ 04 38	30	14.3	GALAXY
ZWG 046.078	14 12 54.	+ 08 15		15.6	GALAXY
ZWG 103.130	14 12 54.	+ 17 07		15.4	GALAXY
MCG+03-36-094	14 12 54.	+ 17 07	18	15.4	GALAXY
UGC 09123	14 12 54.	+ 36 42	72	16.5	GALAXY
ZC 1412.9+5506	14 12 54.	+ 55 06	1550		CLUSTER OF GALAXIES
7ZW 540	14 12 54.	+ 63 46			COMPACT GALAXY
RNGC 5521	14 12 57.	+ 04 38		14.5	GALAXY
MCG+06-31-083	14 12 57.	+ 36 40	60	14.5	GALAXY
HOLM 619A	14 12 58.	+ 17 07	12	15.3	PART OF MULTIPLE GALAXY
HOLM 619B	14 12 58.	+ 17 08	12	15.6	PART OF MULTIPLE GALAXY
REIZ 3990	14 12 58.	+ 18 22	24	14.3	GALAXY
VDB.66G 188	14 13	+ 16 48	70		DWARF GALAXY
VDB.66G 187	14 13	+ 23 13	70		DWARF GALAXY
ZC 1413.0+4100	14 13 00.	+ 41 00	2550		CLUSTER OF GALAXIES
MCG+10-20-088	14 13 00.	+ 60 17	24	16.	GALAXY
ZC 1413.0+7129	14 13 00.	+ 71 29	1340		CLUSTER OF GALAXIES
ZWG 353.032	14 13 00.	+ 80 16		15.6	GALAXY
ZC 1413.1+1729	14 13 06.	+ 17 29	940		CLUSTER OF GALAXIES
ZWG 103.131	14 13 06.	+ 18 45		15.7	GALAXY
ZC 1413.1+2313	14 13 06.	+ 23 13	610		CLUSTER OF GALAXIES
ZC 1413.1+2414	14 13 06.	+ 24 14	740		CLUSTER OF GALAXIES
MCG+06-31-084	14 13 06.	+ 34 34	42	14.5	GALAXY
MCG+08-26-017	14 13 06.	+ 45 48	66	15.	GALAXY
ZC 1413.1+5649	14 13 06.	+ 56 49	2080		CLUSTER OF GALAXIES
MCG-05-34-003	14 13 06.	- 31 32	48	15.	GALAXY
HN 0398	14 13 08.	- 31 32			NEBULA
IC 4388	14 13 08.	- 31 32			NONSTELLAR OBJECT
MCG+03-36-095	14 13 09.	+ 18 45	36	15.7	GALAXY
MCG+03-36-096	14 13 12.	+ 14 30	84	14.0	GALAXY
ZWG 103.132	14 13 12.	+ 14 31		14.0	GALAXY
RNGC 5525	14 13 12.	+ 14 31		14.0	GALAXY
UGC 09124	14 13 12.	+ 14 31	78	14.1	GALAXY S0
MRK 671	14 13 12.	+ 34 46	8	16.5	GALAXY WITH UV CONTINUUM
MCG+06-31-085A	14 13 12.	+ 36 25 30.	48	16.	GALAXY
ZWG 247.014	14 13 12.	+ 45 50		14.8	GALAXY
UGC 09125	14 13 12.	+ 45 50	72	14.8	GALAXY S
MCG+10-20-089	14 13 12.	+ 60 12	30	16.	GALAXY
MCG+10-20-090	14 13 12.	+ 60 14	27	15.	GALAXY
ZC 1413.2+6310	14 13 12.	+ 63 10	4570		CLUSTER OF GALAXIES
HN 0399	14 13 12.	- 12 53			NEBULA
IC 4392	14 13 12.	- 12 53			NONSTELLAR OBJECT
MCG-05-34-004	14 13 12.	- 30 03	18	15.	GALAXY
RNGC 5527	14 13 15.	+ 36 25		16.0	GALAXY
ZWG 074.152	14 13 18.	+ 10 40		15.3	GALAXY
ZWG 103.133	14 13 18.	+ 16 26		15.4	GALAXY
MCG+03-36-097	14 13 18.	+ 16 26	36	15.4	GALAXY
UGC 09126	14 13 18.	+ 16 46	108	16.0	GALAXY DWARF
MCG+03-36-098	14 13 18.	+ 16 46	108	15.	GALAXY
ZC 1413.3+2105	14 13 18.	+ 21 05	2420		CLUSTER OF GALAXIES
RNGC 5540	14 13 21.	+ 60 14		15.0	GALAXY
ZWG 191.069	14 13 24.	+ 36 27		12.9	GALAXY
UGC 09127	14 13 24.	+ 36 27	372	12.9	GALAXY Sc
MCG+06-31-085	14 13 24.	+ 36 27	360	12.	GALAXY
ZWG 295.041	14 13 24.	+ 60 14		14.9	GALAXY
ZC 1413.4+6122	14 13 24.	+ 61 22	1610		CLUSTER OF GALAXIES
MCG-05-34-005	14 13 24.	- 31 27	30	15.	GALAXY
TON-N 0193	14 13 30.	+ 23 09		15.9	BLUE STAR
ZWG 191.070	14 13 30.	+ 36 36		15.4	GALAXY
MCG+06-31-086	14 13 30.	+ 36 36	18	15.5	GALAXY
ZWG 247.015	14 13 30.	+ 49 33		14.9	GALAXY
MRK 672	14 13 30.	+ 49 33	15	14.5	GALAXY WITH UV CONTINUUM
ZWG 018.097	14 13 30.	- 01 01		15.5	GALAXY
REIZ 3997	14 13 31.	+ 36 27	252	13.4	GALAXY
HN 0400	14 13 32.	- 31 28			NEBULA
IC 4391	14 13 32.	- 31 28			NONSTELLAR OBJECT
RNGC 5529	14 13 33.	+ 36 27		13.0	GALAXY
ZWG 133.019	14 13 36.	+ 23 17		15.3	GALAXY
UGC 09128	14 13 36.	+ 23 17	108	15.3	GALAXY DWRF IR
MCG+04-34-009	14 13 36.	+ 23 17 30.	90	15.3	GALAXY
ZWG 191.071	14 13 36.	+ 36 24		15.7	GALAXY
MCG+06-31-085B	14 13 36.	+ 36 29		17.5	GALAXY
REIZ 3992	14 13 36.	- 04 09	24	15.3	GALAXY
MCG+06-31-087	14 13 39.	+ 36 24 30.	30	15.	GALAXY
ARC 1895	14 13 40.	+ 71 28		17.7	RICH CLUSTER OF GALAXIES
HN 1098	14 13 40.	- 40 19	12		NEBULA
IC 4389	14 13 40.	- 40 19			NONSTELLAR OBJECT
ZC 1413.7+3709	14 13 42.	+ 37 09	200		CLUSTER OF GALAXIES
MCG+07-29-056	14 13 42.	+ 40 01	36	15.	GALAXY
UGC 09129	14 13 42.	+ 40 20	66	16.0	GALAXY Sc-c
MCG+07-29-055	14 13 42.	+ 40 20	60	16.	GALAXY
ZWG 219.062	14 13 42.	+ 41 13		15.5	GALAXY
MRK 468	14 13 42.	+ 41 13	12	15.5	GALAXY WITH UV CONTINUUM
ZWG 247.016	14 13 42.	+ 50 25		15.6	GALAXY
8ZW 1413-01.2	14 13 42.	- 01 13		17.6	COMPACT GALAXY
RNGC 5528	14 13 46.	+ 08 31		15.0	GALAXY
IC 0990	14 13 47.	+ 40 02			NONSTELLAR OBJECT
8ZW 1413+01.8	14 13 48.	+ 01 50		16.7	COMPACT GALAXY
ZC 1413.8+0207	14 13 48.	+ 02 07	4640		CLUSTER OF GALAXIES
ZWG 074.153	14 13 48.	+ 08 31		14.8	GALAXY
MCG+02-36-060	14 13 48.	+ 08 31	48	14.8	GALAXY
UGC 09130	14 13 48.	+ 23 15	60	16.0	GALAXY DBL SYS
MCG+04-34-010	14 13 48.	+ 23 15 30.	30		GALAXY
ZWG 163.012	14 13 48.	+ 32 21		15.4	GALAXY
ZWG 219.063	14 13 48.	+ 40 01		15.4	GALAXY
UGC 09131	14 13 48.	+ 63 57	102	16.5	GALAXY Sc
MCG-07-29-012	14 13 48.	- 44 45	96	14.	GALAXY
HOLM 620B	14 13 51.	+ 08 33	9	15.4	PART OF MULTIPLE GALAXY
REIZ 3995	14 13 52.	+ 08 32	48	14.4	GALAXY
HOLM 620A	14 13 52.	+ 08 32	60	14.5	PART OF MULTIPLE GALAXY
REIZ 3996	14 13 52.	+ 08 33	12	15.3	GALAXY
HN 0401	14 13 52.	- 44 45			NEBULA
IC 4390	14 13 52.	- 44 45			NONSTELLAR OBJECT
ZC 1413.9+0548	14 13 54.	+ 05 48	740		CLUSTER OF GALAXIES
HOLM 621A	14 13 54.	+ 23 15	54	14.7	PART OF MULTIPLE GALAXY
ZC 1413.9+2330	14 13 54.	+ 23 30	1340		CLUSTER OF GALAXIES
ZWG 133.020	14 13 54.	+ 26 25		15.7	GALAXY
ZWG 163.013	14 13 54.	+ 26 42		15.5	GALAXY
UGC 09132	14 13 54.	+ 33 57	60	16.5	GALAXY Sc
MCG+06-31-088	14 13 54.	+ 33 57	42	16.5	GALAXY
ZC 1413.9+5806	14 13 54.	+ 58 06	1480		CLUSTER OF GALAXIES

OBJECT NAME	RIGHT ASCEN.	DECLINATION	DIAM.	MAGN.	TYPE OF OBJECT
ZC 1413.9+5826	14 13 54.	+ 58 26	400		CLUSTER OF GALAXIES
ZWG 018.098	14 13 54.	- 00 31		15.3	GALAXY
ZWG 018.099	14 14 00.	+ 01 50		15.7	GALAXY
ZC 1414.0+2156	14 14 00.	+ 21 56	1610		CLUSTER OF GALAXIES
ZWG 133.021	14 14 00.	+ 25 47		15.4	GALAXY
TON-N 0757	14 14 00.	+ 28 51		15.2	BLUE STAR
ZWG 191.072	14 14 00.	+ 35 34		13.0	GALAXY
UGC 09133	14 14 00.	+ 35 34	222	13.0	GALAXY Sb
MCG+06-31-089	14 14 00.	+ 35 34	150	12.	GALAXY
MRSL 313+01/1	14 14 00.	- 59 00	9000		HII REGION
HOLM 621B	14 14 01.	+ 23 14	18	15.4	PART OF MULTIPLE GALAXY
REIZ 3999	14 14 01.	+ 35 35	54	12.8	GALAXY
RNGC 5533	14 14 02.	+ 35 34		12.5	GALAXY
MCG+05-34-005	14 14 03.	+ 26 40 30.	63	15.5	GALAXY
ARC 1887	14 14 05.	+ 17 49		17.2	RICH CLUSTER OF GALAXIES
ZWG 018.100	14 14 06.	+ 02 16		15.6	GALAXY
ZWG 074.154	14 14 06.	+ 10 13		15.5	GALAXY
UGC 09134	14 14 06.	+ 10 13	90	15.5	GALAXY Sc
ZC 1414.1+1806	14 14 06.	+ 18 06	540		CLUSTER OF GALAXIES
ZC 1414.1+2932	14 14 06.	+ 29 32	2820		CLUSTER OF GALAXIES
ZC 1414.1+3758	14 14 06.	+ 37 58	1140		CLUSTER OF GALAXIES
ZWG 272.045	14 14 06.	+ 54 16		15.5	GALAXY
ZC 1414.2+1750	14 14 12.	+ 17 50	940		CLUSTER OF GALAXIES
TON-N 0758	14 14 12.	+ 27 33		16.5	BLUE STAR
ZC 1414.2+3059	14 14 12.	+ 30 59	940		CLUSTER OF GALAXIES
8ZW 1414-01.1	14 14 12.	- 01 07		17.5	COMPACT GALAXY
REIZ 3998	14 14 14.	+ 11 07	9	14.8	GALAXY
MCG+07-29-057	14 14 15.	+ 39 43	48	14.5	GALAXY
RNGC 5531	14 14 17.	+ 11 07		14.5	GALAXY
RNGC 5536	14 14 17.	+ 39 44		14.5	GALAXY
ZWG 074.155	14 14 18.	+ 11 07		14.7	GALAXY
MCG+02-36-061	14 14 18.	+ 11 07	15	14.7	GALAXY
ZWG 133.022	14 14 18.	+ 24 44		15.5	GALAXY
UGC 09135	14 14 18.	+ 24 44	72	15.5	GALAXY S
MCG+04-34-011	14 14 18.	+ 24 44	60	15.5	GALAXY
ZWG 219.064	14 14 18.	+ 39 44		14.5	GALAXY
UGC 09136	14 14 18.	+ 39 44	60	14.5	GALAXY SBa
MCG+07-29-058	14 14 18.	+ 39 52	36	15.	GALAXY
REIZ 4003	14 14 21.	+ 39 43	18	14.1	GALAXY
RNGC 5532B	14 14 23.	+ 11 02		16.0	GALAXY
RNGC 5532A	14 14 23.	+ 11 02		13.5	GALAXY
RNGC 5541	14 14 23.	+ 39 49		13.5	GALAXY
ZWG 018.101	14 14 24.	+ 01 52		15.3	GALAXY
8ZW 1414+01.9	14 14 24.	+ 01 52		15.3	COMPACT GALAXY
ZWG 018.102	14 14 24.	+ 02 24		15.6	GALAXY
FAI 1414+09	14 14 24.	+ 09 54			COMPACT GALAXY
ZWG 074.156	14 14 24.	+ 11 02		13.3	GALAXY
UGC 09137	14 14 24.	+ 11 02	120	13.3	GALAXY S0
MCG+02-36-062	14 14 24.	+ 11 02	30	13.3	GALAXY
ZWG 133.023	14 14 24.	+ 23 14		15.5	GALAXY
UGC 09138	14 14 24.	+ 23 14	114	15.5	GALAXY Sc
ZWG 219.065	14 14 24.	+ 39 49		13.4	GALAXY
UGC 09139	14 14 24.	+ 39 49	54	13.4	GALAXY S?
MCG+07-29-059	14 14 24.	+ 39 49	42	14.	GALAXY
REIZ 4000	14 14 26.	+ 11 02	12	15.0	GALAXY
HOLM 622B	14 14 26.	+ 11 02	18	15.0	PART OF MULTIPLE GALAXY
REIZ 4001	14 14 26.	+ 11 03	18	13.8	GALAXY
HOLM 622A	14 14 26.	+ 11 03	48	13.4	PART OF MULTIPLE GALAXY
MCG+04-34-012	14 14 27.	+ 23 15	102	15.5	GALAXY
REIZ 4004	14 14 27.	+ 39 49	36	13.8	GALAXY
ARC 1889	14 14 28.	+ 30 57		17.3	RICH CLUSTER OF GALAXIES
ZWG 018.103	14 14 30.	+ 00 04		15.7	GALAXY
ZWG 018.104	14 14 30.	+ 00 42		15.7	GALAXY
MCG+00-36-031	14 14 30.	+ 00 42	36	15.7	GALAXY
ZWG 074.157	14 14 30.	+ 09 21		15.4	GALAXY
TON-N 0759	14 14 30.	+ 26 52		16.4	BLUE STAR
IC 4394	14 14 30.	+ 39 56			NONSTELLAR OBJECT
7ZW 541	14 14 30.	+ 63 01			COMPACT GALAXY
REIZ 4013	14 14 34.	+ 58 03	60	14.4	GALAXY
SC 1411-1709.9	14 14 34.	- 17 23 51.	30		NEBULA
ZWG 046.079	14 14 36.	+ 07 30		15.4	GALAXY
ZC 1414.6+1355	14 14 36.	+ 13 55	1880		CLUSTER OF GALAXIES
ZWG 103.134	14 14 36.	+ 17 19		15.5	GALAXY
ZC 1414.6+3417	14 14 36.	+ 34 17	400		CLUSTER OF GALAXIES
REIZ 4005	14 14 36.	+ 36 48	12	14.7	GALAXY
OCL 0917	14 14 36.	- 58 44	3180	11.	OPEN STAR CLUSTER
HOLM 624B	14 14 39.	+ 22 40	18	15.0	PART OF MULTIPLE GALAXY
ARC 1888	14 14 40.	+ 14 06		17.5	RICH CLUSTER OF GALAXIES
8ZW 1414+00.8	14 14 42.	+ 00 47		19.8	COMPACT GALAXY
ZWG 018.105	14 14 42.	+ 01 58		15.7	GALAXY
ZWG 103.135	14 14 42.	+ 18 24		15.5	GALAXY
HOLM 624A	14 14 42.	+ 22 40	30	14.9	PART OF MULTIPLE GALAXY
7ZW 542	14 14 42.	+ 63 08			COMPACT GALAXY
PK316+08.1	14 14 47.	- 51 56 48.	25	10.1	PLANETARY NEBULA
ZWG 018.106	14 14 48.	+ 00 15		15.5	GALAXY
8ZW 1414+01.9	14 14 48.	+ 01 53		17.9	COMPACT GALAXY
MCG-04-34-002	14 14 48.	- 23 58	30	15.	GALAXY
MCG-05-34-006	14 14 48.	- 31 07	132	15.	GALAXY
RNGC 5544	14 14 51.	+ 36 48		14.0	GALAXY
ZWG 018.107	14 14 54.	+ 00 43		15.6	GALAXY
ZWG 074.158	14 14 54.	+ 09 08		15.4	GALAXY
ZWG 163.014	14 14 54.	+ 30 28		15.7	GALAXY
VV 210B	14 14 54.	+ 36 48	60	14.	INTERACTING GALAXY
VV 210A	14 14 54.	+ 36 48	72	14.	INTERACTING GALAXY
KARA.72 422A	14 14 54.	+ 36 48	60	13.2	PART OF DOUBLE GALAXY
MCG+06-31-090	14 14 54.	+ 36 48	24	14.	GALAXY
MCG+10-20-091	14 14 54.	+ 58 01	72	14.	GALAXY
REIZ 4002	14 14 54.	- 03 57	18	14.7	GALAXY
IC 0995	14 14 55.	+ 58 02 29.			NONSTELLAR OBJECT
PK312-01.1	14 14 55.	- 62 53 13.	25		PLANETARY NEBULA
HOLM 623A	14 14 56.	- 07 11	42	14.3	PART OF MULTIPLE GALAXY
RNGC 5545	14 14 57.	+ 36 48		14.0	GALAXY
IC 4393	14 14 57.	- 31 07			NONSTELLAR OBJECT
ARP 199	14 14 58.	+ 36 48			PECULIAR GALAXY
HOLM 623B	14 14 58.	- 07 11	30	14.9	PART OF MULTIPLE GALAXY
RNGC 5534	14 14 59.	- 07 11		13.5	GALAXY
ZWG 018.108	14 15 00.	+ 00 35		15.7	GALAXY
UGC 09140	14 15 00.	+ 01 21	66	16.5	GALAXY Sc
ZWG 018.109	14 15 00.	+ 02 02		15.2	GALAXY
ZWG 046.080	14 15 00.	+ 04 47		14.9	GALAXY
MCG+01-36-031	14 15 00.	+ 04 47	36	14.9	GALAXY
HOLM 625A	14 15 00.	+ 04 48	54	14.0	PART OF MULTIPLE GALAXY
ZWG 046.081	14 15 00.	+ 08 17		15.5	GALAXY
ZC 1415.0+2014	14 15 00.	+ 20 14	670		CLUSTER OF GALAXIES
ZC 1415.0+2058	14 15 00.	+ 20 58	610		CLUSTER OF GALAXIES
ZWG 163.015	14 15 00.	+ 27 05		15.1	GALAXY
UGC 09141	14 15 00.	+ 27 05	90	15.1	GALAXY S
ZC 1415.0+2819	14 15 00.	+ 28 19	2420		CLUSTER OF GALAXIES

OBJECT NAME	RIGHT ASCEN.	DECLINATION	DIAM.	MAGN.	TYPE OF OBJECT
ZC 1415.0+3430	14 15 00.	+ 34 30	1080		CLUSTER OF GALAXIES
MCG+06-31-092	14 15 00.	+ 35 38	36	15.	GALAXY
ZWG 191.073	14 15 00.	+ 36 48		13.2	GALAXY
REIZ 4012	14 15 00.	+ 36 48	90	13.3	GALAXY
UGC 09143	14 15 00.	+ 36 48	66	13.2	GALAXY Sb-c
UGC 09142	14 15 00.	+ 36 48	66	13.2	GALAXY Sa
KARA.72 422B	14 15 00.	+ 36 48	60		PART OF DOUBLE GALAXY
MCG+06-31-091	14 15 00.	+ 36 48 30.	51	14.	GALAXY
UGC 09144	14 15 00.	+ 42 55	90	16.0	GALAXY Sc
MCG+07-29-061	14 15 00.	+ 42 55	54	17.	GALAXY
ZWG 219.066	14 15 00.	+ 43 44		15.6	GALAXY
MCG+07-29-060	14 15 00.	+ 43 44	45	16.	GALAXY
ZWG 295.042	14 15 00.	+ 58 02		14.5	GALAXY
UGC 09145	14 15 00.	+ 58 02	84	14.5	GALAXY Sc-IRR
MCG+14-07-003	14 15 00.	+ 86 36	27	17.	GALAXY
ZWG 370.011	14 15 00.	+ 86 37		15.7	GALAXY
MCG-02-36-019	14 15 00.	- 13 39	84	13.	GALAXY
MCG-04-34-003	14 15 00.	- 23 58	42	15.5	GALAXY
HOLM 625B	14 15 03.	+ 04 49	72	14.9	PART OF MULTIPLE GALAXY
FATH 1.657	14 15 03.	+ 14 41	14		NEBULA
FATH 1.658	14 15 03.	+ 14 50	8		NEBULA
RNGC 5537	14 15 04.	+ 07 17		15.0	GALAXY
REIZ 4006	14 15 04.	+ 08 26	12	15.2	GALAXY
REIZ 4007	14 15 05.	+ 07 17	42	14.6	GALAXY
REIZ 4008	14 15 05.	+ 07 42	30	15.0	GALAXY
ARC 1890	14 15 05.	+ 08 25		15.5	RICH CLUSTER OF GALAXIES
ZWG 018.110	14 15 06.	+ 00 33		15.7	GALAXY
ZWG 018.111	14 15 06.	+ 02 16		15.3	GALAXY
8ZW 1415+02.3	14 15 06.	+ 02 16		15.3	COMPACT GALAXY
MKW 06	14 15 06.	+ 02 16		15.3	POOR GALAXY CLUSTER
MCG+00-36-032	14 15 06.	+ 02 16	30	15.3	GALAXY
ZC 1415.1+0621	14 15 06.	+ 06 21	1480		CLUSTER OF GALAXIES
ZWG 046.082	14 15 06.	+ 07 17		15.1	GALAXY
MCG+01-36-032	14 15 06.	+ 07 17	54	15.1	GALAXY
FAI 1415+09	14 15 06.	+ 09 05			COMPACT GALAXY
ZC 1415.1+1645	14 15 06.	+ 16 45	1610		CLUSTER OF GALAXIES
ZWG 133.024	14 15 06.	+ 25 17		15.3	GALAXY
IC 4395	14 15 06.	+ 27 04 58.			NONSTELLAR OBJECT
MCG+05-34-006	14 15 06.	+ 30 27 30.	33	15.7	GALAXY
IC 0991	14 15 06.	- 13 38 28.			NONSTELLAR OBJECT
HN 0402	14 15 08.	- 31 07			NEBULA
REIZ 4011	14 15 09.	+ 27 05	18	14.8	GALAXY
RNGC 5538	14 15 10.	+ 07 43		15.5	GALAXY
RNGC 5539	14 15 10.	+ 08 25		15.5	GALAXY
RNGC 5535	14 15 10.	+ 08 25		15.5	GALAXY
REIZ 4009	14 15 10.	+ 08 25	24	14.2	GALAXY
ZWG 046.083	14 15 12.	+ 07 43		15.4	GALAXY
ZWG 046.084	14 15 12.	+ 08 25		15.4	GALAXY
MCG+01-36-033	14 15 12.	+ 08 25	24	15.4	GALAXY
ZWG 074.159	14 15 12.	+ 13 11		15.1	GALAXY
MCG+05-34-007	14 15 12.	+ 27 04	48	15.1	GALAXY
ZWG 163.016	14 15 12.	+ 29 02		15.6	GALAXY
MCG-01-36-014	14 15 12.	- 07 10	54	14.	GALAXY
ZC 1415.3+0038	14 15 18.	+ 00 38	3700		CLUSTER OF GALAXIES
ZWG 018.112	14 15 18.	+ 00 45		15.5	GALAXY
FEIG 093	14 15 18.	+ 13 16		15.0	FAINT BLUE STAR
ZC 1415.3+2333	14 15 18.	+ 23 33	2350		CLUSTER OF GALAXIES
ZC 1415.3+2500	14 15 18.	+ 25 00	2150		CLUSTER OF GALAXIES
IC 4396	14 15 18.	+ 29 01 47.			NONSTELLAR OBJECT
UGC 09146	14 15 18.	+ 56 07	60	17.	GALAXY DWARF
MCG-07-29-013	14 15 18.	- 43 09	198	12.3	GALAXY
ARC 1891	14 15 20.	+ 28 16		17.0	RICH CLUSTER OF GALAXIES
RNGC 5542	14 15 22.	+ 07 47		15.0	GALAXY
REIZ 4010	14 15 23.	+ 07 48	24	14.1	GALAXY
LP 00727	14 15 24.	+ 03 20 48.		17.6	FAINT BLUE STAR
ZWG 046.085	14 15 24.	+ 07 47		15.0	GALAXY
MCG+01-36-034	14 15 24.	+ 07 47	24	15.0	GALAXY
FAI 1415+12	14 15 24.	+ 12 55			COMPACT GALAXY
MCG+05-34-008	14 15 24.	+ 29 01	36	15.6	GALAXY
MCG+05-34-009	14 15 24.	+ 29 02	12		GALAXY
MCG-04-34-004	14 15 24.	- 27 12	24	15.5	GALAXY
RNGC 5530	14 15 26.	- 43 09		12.0	GALAXY
IC 0996	14 15 27.	+ 57 52 32.			NONSTELLAR OBJECT
ZWG 046.086	14 15 30.	+ 05 18		15.5	GALAXY
ZC 1415.5+1936	14 15 30.	+ 19 36	670		CLUSTER OF GALAXIES
ZC 1415.5+4130	14 15 30.	+ 41 30	1480		CLUSTER OF GALAXIES
ZC 1415.5+5024	14 15 30.	+ 50 24	1080		CLUSTER OF GALAXIES
RNGC 5543	14 15 34.	+ 07 53		15.5	GALAXY
REIZ 4014	14 15 34.	+ 07 54	30	14.3	GALAXY
ZWG 046.087	14 15 36.	+ 07 02		15.3	GALAXY
ZWG 046.088	14 15 36.	+ 07 53		15.3	GALAXY
ZWG 163.017	14 15 36.	+ 26 59		15.6	GALAXY
MCG+10-20-092	14 15 36.	+ 57 50	78	15.	GALAXY
MCG+10-20-093	14 15 36.	+ 58 56	45	15.	GALAXY
ZC 1415.6+6537	14 15 36.	+ 65 37	1810		CLUSTER OF GALAXIES
ZWG 018.113	14 15 36.	- 02 53		15.6	GALAXY
MCG-04-34-005	14 15 36.	- 27 10	24	15.	GALAXY
MCG-04-34-006	14 15 36.	- 27 11	48	16.	GALAXY
ARC 1891	14 15 38.	+ 28 17		17.0	RICH CLUSTER OF GALAXIES
RNGC 5546	14 15 40.	+ 07 48		14.0	GALAXY
RNGC 5548	14 15 40.	+ 25 22		13.5	GALAXY
REIZ 4016	14 15 40.	+ 25 22	18	12.9	GALAXY
IC 0992	14 15 41.	+ 01 05 41.			NONSTELLAR OBJECT
REIZ 4015	14 15 41.	+ 07 48	42	13.1	GALAXY
BC M1415+172	14 15 41.4	+ 17 16 59.		18.	QUASI-STELLAR OBJECT
SHB 284	14 15 41.4	+ 17 16 59.		18.	QUASI-STELLAR OBJECT
ZWG 018.114	14 15 42.	+ 01 07		14.9	GALAXY
UGC 09147	14 15 42.	+ 01 07	96	14.9	GALAXY Sb-c
MCG+00-36-033	14 15 42.	+ 01 07	66	14.9	GALAXY
ZWG 046.089	14 15 42.	+ 07 48		14.1	GALAXY
UGC 09148	14 15 42.	+ 07 48	108	14.1	GALAXY E
MCG+01-36-035	14 15 42.	+ 07 48	24	14.1	GALAXY
ZC 1415.7+1520	14 15 42.	+ 15 20	1610		CLUSTER OF GALAXIES
ZWG 133.025	14 15 42.	+ 25 22		13.1	GALAXY
VVI 72	14 15 42.	+ 25 22	110	14.48	SEYFERT GALAXY
UGC 09149	14 15 42.	+ 25 22	102	13.1	GALAXY Sa
MCG+04-34-013	14 15 42.	+ 25 22	108	13.1	GALAXY
ZWG 163.018	14 15 42.	+ 26 39		14.2	GALAXY
MRK 673	14 15 42.	+ 26 39	26	15.	GALAXY WITH UV CONTINUUM
UGC 09150	14 15 42.	+ 26 39	66	14.2	GALAXY S
MCG+05-34-010	14 15 42.	+ 26 58 30.	60	15.6	GALAXY
ZWG 163.019	14 15 42.	+ 29 28		15.6	GALAXY
UGC 09151	14 15 42.	+ 58 56	84	16.0	GALAXY
IC 0993	14 15 43.	+ 11 30 00.			NONSTELLAR OBJECT
IC 4397	14 15 43.	+ 26 38 36.			NONSTELLAR OBJECT
REIZ 4017	14 15 45.	+ 26 39	18	14.0	GALAXY
MCG+05-34-011	14 15 45.	+ 29 27	36	15.6	GALAXY
ARC 1894	14 15 45.	+ 43 37		17.0	RICH CLUSTER OF GALAXIES

OBJECT NAME	RIGHT ASCEN.	DECLINATION	DIAM.	MAGN.	TYPE OF OBJECT
IC 0994	14 15 47.	+ 11 26 24.			NONSTELLAR OBJECT
ZWG 046.090	14 15 48.	+ 07 34		15.4	GALAXY
ZWG 075.001	14 15 48.	+ 11 27		15.4	GALAXY
ZWG 074.160	14 15 48.	+ 11 27		15.4	GALAXY
MCG+02-36-063	14 15 48.	+ 11 27	15	15.4	GALAXY
MCG+05-34-012	14 15 48.	+ 26 37	66	14.2	GALAXY
ZWG 163.020	14 15 48.	+ 29 06		15.3	GALAXY
ZWG 295.043	14 15 48.	+ 57 51		14.6	GALAXY
UGC 09152	14 15 48.	+ 57 51	84	14.6	GALAXY Sb-c
ZWG 295.044	14 15 48.	+ 58 58		15.5	GALAXY
ZC 1415.8+6635	14 15 48.	+ 66 35	1480		CLUSTER OF GALAXIES
IC 4398	14 15 50.	+ 29 05 49.			NONSTELLAR OBJECT
RNGC 5561	14 15 50.	+ 58 58		15.5	GALAXY
LB 00728	14 15 51.	+ 02 57 48.		17.4	FAINT BLUE STAR
HOLM 626B	14 15 52.	+ 11 27	24	15.1	PART OF MULTIPLE GALAXY
ZWG 046.091	14 15 54.	+ 08 20		15.5	GALAXY
ZWG 075.002	14 15 54.	+ 11 25		14.8	GALAXY
ZWG 074.161	14 15 54.	+ 11 25		14.8	GALAXY
UGC 09153	14 15 54.	+ 11 25	84	14.8	GALAXY Sa
MCG+02-36-064	14 15 54.	+ 11 25	72	14.8	GALAXY
ZWG 133.026	14 15 54.	+ 26 15		15.6	GALAXY
HOLM 626A	14 15 56.	+ 11 26	54	14.8	PART OF MULTIPLE GALAXY
MCG+05-34-013	14 15 57.	+ 29 05	30	15.3	GALAXY
FATH 1.659	14 15 59.	+ 14 52	8		NEBULA
ZWG 019.001	14 16 00.	+ 00 23		15.6	GALAXY
8ZW 1416+02.9	14 16 00.	+ 02 56		18.1	COMPACT GALAXY
ZC 1416.0+0752	14 16 00.	+ 07 52	5380		CLUSTER OF GALAXIES
FAI 14160+11	14 16 00.	+ 11 52			COMPACT GALAXY
ZWG 075.003	14 16 00.	+ 13 06		14.2	GALAXY
ZWG 074.162	14 16 00.	+ 13 06		14.2	GALAXY
RNGC 5550	14 16 00.	+ 13 06		14.0	GALAXY
UGC 09154	14 16 00.	+ 13 06	78	14.2	GALAXY S
MCG+02-36-065	14 16 00.	+ 13 06	72	14.2	GALAXY
ZC 1416.0+2558	14 16 00.	+ 25 58	870		CLUSTER OF GALAXIES
ZWG 163.021	14 16 00.	+ 29 12		15.6	GALAXY
UGC 09155	14 16 00.	+ 29 12	78	15.6	GALAXY Sc
ZC 1416.0+3802	14 16 00.	+ 38 02	2150		CLUSTER OF GALAXIES
ZC 1416.0+5505	14 16 00.	+ 55 05	1280		CLUSTER OF GALAXIES
MCG-04-34-007	14 16 00.	- 27 12	30	15.	GALAXY
RNGC 5549	14 16 04.	+ 07 36		14.0	GALAXY
REIZ 4025	14 16 04.	+ 31 52	36	14.1	GALAXY
ZWG 047.001	14 16 06.	+ 07 36		14.2	GALAXY
UGC 09156	14 16 06.	+ 07 36	108	14.2	GALAXY S0
MCG+01-36-036	14 16 06.	+ 07 36	84	14.2	GALAXY
FAI 14161+10	14 16 06.	+ 10 40			COMPACT GALAXY
MCG+03-36-099	14 16 06.	+ 14 30	66	15.4	GALAXY
FATH 1.660	14 16 06.	+ 15 24	14		NEBULA
ZC 1416.1+1628	14 16 06.	+ 16 28	870		CLUSTER OF GALAXIES
ZC 1416.1+1916	14 16 06.	+ 19 16	2020		CLUSTER OF GALAXIES
ZWG 133.027	14 16 06.	+ 22 02		15.7	GALAXY
REIZ 4021	14 16 06.	+ 22 28	24	14.9	GALAXY
ZWG 133.028	14 16 06.	+ 25 44		15.7	GALAXY
ZWG 163.022	14 16 06.	+ 26 37		15.5	GALAXY
UGC 09157	14 16 06.	+ 26 37	72	15.5	GALAXY Sb
ZWG 163.023	14 16 06.	+ 31 53		14.9	GALAXY
UGC 09158	14 16 06.	+ 31 53	84	14.9	GALAXY S
KARA.73B 0622	14 16 06.	+ 31 53	60	14.9	ISOLATED GALAXY S
MRK 469	14 16 06.	+ 34 35	8	17.	GALAXY WITH UV CONTINUUM
MCG-03-36-011	14 16 06.	- 18 54	30	15.	GALAXY
REIZ 4019	14 16 07.	+ 13 07	60	14.2	GALAXY
REIZ 4024	14 16 07.	+ 29 12	60	14.3	GALAXY
IC 4403	14 16 07.	+ 31 53 20.			NONSTELLAR OBJECT
IC 0977	14 16 07.	- 02 45 17.			NONSTELLAR OBJECT
IC 4399	14 16 08.	+ 26 36 50.			NONSTELLAR OBJECT
REIZ 4023	14 16 09.	+ 26 37	6	15.7	GALAXY
FATH 1.661	14 16 10.	+ 15 14	19		NEBULA
REIZ 4018	14 16 11.	+ 07 36	60	13.9	GALAXY
RNGC 5553	14 16 11.	+ 26 31		15.0	GALAXY
ZWG 047.002	14 16 12.	+ 06 04		15.6	GALAXY
ZC 1416.2+0904	14 16 12.	+ 09 04	670		CLUSTER OF GALAXIES
ZC 1416.2+1000	14 16 12.	+ 10 00	1080		CLUSTER OF GALAXIES
FAI 14162+10	14 16 12.	+ 10 03			COMPACT GALAXY
FAI 14162+11	14 16 12.	+ 11 53			COMPACT GALAXY
FAI 14162+13.0	14 16 12.	+ 13 00			COMPACT GALAXY
ZWG 075.004	14 16 12.	+ 13 12		15.5	GALAXY
FAI 14162+13.2	14 16 12.	+ 13 12			COMPACT GALAXY
ZWG 104.001	14 16 12.	+ 14 29		15.4	GALAXY
ZWG 103.136	14 16 12.	+ 14 29		15.4	GALAXY
ZWG 075.005	14 16 12.	+ 14 29		15.4	GALAXY
UGC 09159	14 16 12.	+ 14 29	66	15.4	GALAXY Sb-c
ZC 1416.2+2525	14 16 12.	+ 25 25	610		CLUSTER OF GALAXIES
ZWG 163.024	14 16 12.	+ 26 31		14.8	GALAXY
UGC 09160	14 16 12.	+ 26 31	78	14.8	GALAXY Sa
MCG+05-34-017	14 16 12.	+ 26 31	78	14.8	GALAXY
MCG+05-34-014	14 16 12.	+ 26 35 30.	60	15.5	GALAXY
MCG+05-34-015	14 16 12.	+ 29 11 30.	48	15.6	GALAXY
ZWG 163.025	14 16 12.	+ 30 05		15.6	GALAXY
MCG+05-34-016	14 16 12.	+ 31 53 30.	63	14.9	GALAXY
ZC 1416.2+3606	14 16 12.	+ 36 06	8800		CLUSTER OF GALAXIES
MCG-04-34-008	14 16 12.	- 27 11	24	15.5	GALAXY
REIZ 4022	14 16 13.	+ 22 02	30	15.1	GALAXY
RNGC 5555	14 16 13.	- 18 54		15.0	GALAXY
REIZ 4027	14 16 15.	+ 26 30	30	15.0	GALAXY
RNGC 5557	14 16 15.	+ 36 43		13.0	GALAXY
ZWG 019.002	14 16 18.	+ 00 08		15.5	GALAXY
REIZ 4020	14 16 18.	+ 05 40	12	14.8	GALAXY
ZWG 163.026	14 16 18.	+ 29 24		15.4	GALAXY
ZWG 191.074	14 16 18.	+ 36 43		12.2	GALAXY
REIZ 4032	14 16 18.	+ 36 43	48	12.6	GALAXY
UGC 09161	14 16 18.	+ 36 43	180	12.2	GALAXY E
ZC 1416.3+4331	14 16 18.	+ 43 31	3760		CLUSTER OF GALAXIES
8ZW 1416-02.2	14 16 18.	- 02 10		17.7	COMPACT GALAXY
SER 111.02	14 16 18.	- 45 01	70	16.	PEC. COMET-LIKE OBJECT
RNGC 5551	14 16 21.	+ 05 40		15.0	GALAXY
MCG+06-31-093	14 16 21.	+ 36 43	60	11.	GALAXY
HOLM 627B	14 16 23.	+ 13 14	18	15.4	PART OF MULTIPLE GALAXY
HOLM 628C	14 16 23.	+ 25 09	18	15.4	PART OF MULTIPLE GALAXY
ZWG 047.003	14 16 24.	+ 05 40		15.1	GALAXY
MCG+01-36-037	14 16 24.	+ 05 40	24	15.1	GALAXY
ZWG 075.006	14 16 24.	+ 11 05		15.4	GALAXY
UGC 09162	14 16 24.	+ 11 05	72	15.4	GALAXY SBb
ZWG 075.007	14 16 24.	+ 11 31		15.6	GALAXY
UGC 09163	14 16 24.	+ 11 31	78	15.6	GALAXY Sc
FAI 14164+13	14 16 24.	+ 13 13			COMPACT GALAXY
ZWG 133.029	14 16 24.	+ 22 03		15.4	GALAXY
MRK 674	14 16 24.	+ 22 03	15	16.5	GALAXY WITH UV CONTINUUM
UGC 09164	14 16 24.	+ 22 03	66	15.4	GALAXY PECULR
MCG+04-34-014	14 16 24.	+ 22 03	66	15.4	GALAXY
ZC 1416.4+2357	14 16 24.	+ 23 57	540		CLUSTER OF GALAXIES
MCG+05-34-018	14 16 24.	+ 29 23 30.	36	15.4	GALAXY
ZC 1416.4+4915	14 16 24.	+ 49 15	1480		CLUSTER OF GALAXIES
ZC 1416.4+5531	14 16 24.	+ 55 31	1410		CLUSTER OF GALAXIES
REIZ 4026	14 16 25.	+ 13 08	24	14.4	GALAXY
HOLM 627A	14 16 25.	+ 13 13	24	15.2	PART OF MULTIPLE GALAXY
REIZ 4029	14 16 25.	+ 22 03	36	14.8	GALAXY
RNGC 5552	14 16 28.	+ 07 16		15.0	GALAXY
HOLM 628A	14 16 28.	+ 25 10	72	14.4	PART OF MULTIPLE GALAXY
ZWG 047.004	14 16 30.	+ 07 16		15.2	GALAXY
FAI 1416+09	14 16 30.	+ 09 03			COMPACT GALAXY
ZC 1416.5+1850	14 16 30.	+ 18 50	3090		CLUSTER OF GALAXIES
ZWG 133.030	14 16 30.	+ 25 10		15.3	GALAXY
UGC 09165	14 16 30.	+ 25 10	90	15.3	GALAXY S
MCG+06-31-094	14 16 30.	+ 37 15 30.	36	16.	GALAXY
FATH 1.662	14 16 31.	+ 15 24	8		NEBULA
REIZ 4033	14 16 34.	+ 25 11	48	15.6	GALAXY
REIZ 4028	14 16 35.	+ 07 15	12	15.2	GALAXY
8ZW 1416+06.7	14 16 36.	+ 06 43		16.8	COMPACT GALAXY
ZWG 047.005	14 16 36.	+ 08 09		15.5	GALAXY
MCG+04-34-015	14 16 36.	+ 25 10	60	15.3	GALAXY
ZWG 163.027	14 16 36.	+ 26 50		15.4	GALAXY
MCG-04-34-010	14 16 36.	- 26 26	150	14.	GALAXY
MCG-04-34-009	14 16 36.	- 27 10	66	14.5	GALAXY
ARC 1896	14 16 38.	+ 38 02		17.2	RICH CLUSTER OF GALAXIES
SHB 285	14 16 38.8	+ 06 42 20.		16.8	QUASI-STELLAR OBJECT
BC 3CR298	14 16 38.80	+ 06 42 20.6		16.79	QUASI-STELLAR OBJECT
REIZ 4031	14 16 41.	+ 07 14	12	15.3	GALAXY
ZC 1416.7+2112	14 16 42.	+ 21 12	1010		CLUSTER OF GALAXIES
ZWG 133.031	14 16 42.	+ 25 10		15.7	GALAXY
HOLM 628B	14 16 42.	+ 25 10	18	15.4	PART OF MULTIPLE GALAXY
MCG+09-23-059	14 16 42.	+ 54 58	30	16.	GALAXY
MCG-04-34-011	14 16 42.	- 27 10	36	15.	GALAXY
RN 0412	14 16 43.	- 36 41			NEBULA
RNGC 5554	14 16 46.	+ 07 15		15.0	GALAXY
ZWG 019.003	14 16 48.	+ 01 24		15.6	GALAXY
8ZW 1416+01.9	14 16 48.	+ 01 52		16.8	COMPACT GALAXY
ZWG 047.006	14 16 48.	+ 07 15		15.2	GALAXY
FAI 14168+09	14 16 48.	+ 09 55			COMPACT GALAXY
FAI 14168+12	14 16 48.	+ 12 07			COMPACT GALAXY
ZC 1416.8+2110	14 16 48.	+ 21 10	270		CLUSTER OF GALAXIES
MCG+04-34-016	14 16 48.	+ 25 10	39	15.7	GALAXY
KW 56	14 16 48.	+ 25 15	80		SEYFERT GALAXY
ZWG 353.033	14 16 48.	+ 77 38		15.6	GALAXY
IC 4401	14 16 48.	- 04 15 27.			NONSTELLAR OBJECT
REIZ 4030	14 16 48.	- 04 16	66	14.5	GALAXY
MCG-01-36-015	14 16 48.	- 04 17	60	15.	GALAXY
RNGC 5559	14 16 52.	+ 25 02		15.0	GALAXY
FAI 14169+09	14 16 54.	+ 09 00			COMPACT GALAXY
ZC 1416.9+2326	14 16 54.	+ 23 26	400		CLUSTER OF GALAXIES
MCG+04-34-017	14 16 54.	+ 25 01	78	15.0	GALAXY
ZWG 133.032	14 16 54.	+ 25 02		15.0	GALAXY
UGC 09166	14 16 54.	+ 25 02	96	15.0	GALAXY SBb
ZWG 163.028	14 16 54.	+ 26 32		14.9	GALAXY
KARA.72 423A	14 16 54.	+ 26 32	96	14.9	PART OF DOUBLE GALAXY
MCG+08-26-018	14 16 54.	+ 46 10	30	16.	GALAXY
HOLM 629A	14 16 57.	+ 26 31	24	14.0	PART OF MULTIPLE GALAXY
RNGC 5570	14 16 58.	+ 07 40			GALAXY
REIZ 4036	14 16 58.	+ 25 02	54	15.4	GALAXY
REIZ 4034	14 16 59.	+ 07 41	42	15.0	GALAXY
VDB.66G 243	14 17	- 29 01	130		DWARF GALAXY
LB 09887	14 17	- 81 47		14.5	FAINT BLUE STAR
ZWG 047.007	14 17 00.	+ 07 41		15.4	GALAXY
ZWG 047.008	14 17 00.	+ 08 08		15.7	GALAXY
ZWG 163.029	14 17 00.	+ 26 32		15.7	GALAXY
KARA.72 423B	14 17 00.	+ 26 32	48	15.7	PART OF DOUBLE GALAXY
MCG+14-07-004	14 17 00.	+ 84 02	51	16.	GALAXY
ZC 1417.0-0344	14 17 00.	- 03 44	3020		CLUSTER OF GALAXIES
MRSL 313-00/1	14 17 00.	- 61 10	1500		HII REGION
IC 4405	14 17 01.	+ 26 31 42.			NONSTELLAR OBJECT
HOLM 629B	14 17 02.	+ 26 31	18	15.0	PART OF MULTIPLE GALAXY
REIZ 4038	14 17 03.	+ 26 32	12	14.7	GALAXY
REIZ 4035	14 17 05.	+ 07 04	15	15.8	GALAXY
ZWG 047.009	14 17 06.	+ 06 22		15.5	GALAXY
ZWG 191.075	14 17 06.	+ 35 22		15.0	GALAXY
ZWG 191.076	14 17 06.	+ 35 33		15.4	GALAXY
MRK 675	14 17 06.	+ 36 35	13	16.5	GALAXY WITH UV CONTINUUM
MRK 676	14 17 06.	+ 40 06	14	16.	GALAXY WITH UV CONTINUUM
MCG+05-34-019	14 17 09.	+ 26 30	54	14.9	GALAXY
RNGC 5567	14 17 09.	+ 35 22		15.0	GALAXY
MCG+06-31-095	14 17 09.	+ 35 33 30.	15	15.	GALAXY
FAI 14172+11	14 17 12.	+ 11 53			COMPACT GALAXY
ZWG 104.002	14 17 12.	+ 17 52		15.3	GALAXY
UGC 09167	14 17 12.	+ 17 52	90	15.3	GALAXY Sa
ZWG 104.003	14 17 12.	+ 18 06		14.5	GALAXY
UGC 09168	14 17 12.	+ 18 06	48	14.5	GALAXY S0?
MCG+03-37-001	14 17 12.	+ 18 07	36	14.5	GALAXY
TON-N 0760	14 17 12.	+ 28 01		15.1	BLUE STAR
TON-N 0761	14 17 12.	+ 30 01		16.8	BLUE STAR
ZWG 191.077	14 17 12.	+ 35 19		15.7	GALAXY
MCG+06-31-096	14 17 12.	+ 35 21	42	14.5	GALAXY
ZWG 192.001	14 17 12.	+ 37 14		15.7	GALAXY
ZWG 191.078	14 17 12.	+ 37 14		15.7	GALAXY
MCG+06-31-097	14 17 12.	+ 37 15	30	16.	GALAXY
ZWG 220.001	14 17 12.	+ 44 22		15.7	GALAXY
ZWG 219.067	14 17 12.	+ 44 22		15.7	GALAXY
ZWG 247.017	14 17 12.	+ 47 59		15.6	GALAXY
FATH 1.663	14 17 13.	+ 15 30	14		NEBULA
IC 0999	14 17 13.	+ 18 06 06.			NONSTELLAR OBJECT
REIZ 4042	14 17 13.	+ 35 19	48	14.5	GALAXY
REIZ 4041	14 17 13.	+ 35 21	90	13.6	GALAXY
MCG+05-34-020	14 17 15.	+ 26 30 30.	30	15.7	GALAXY
RNGC 5568	14 17 15.	+ 35 19		15.5	GALAXY
MCG+06-31-098	14 17 15.	+ 35 19	30	14.5	GALAXY
RNGC 5558	14 17 16.	+ 07 16			NON-EXISTENT OBJECT
ZWG 075.008	14 17 18.	+ 09 36		14.7	GALAXY
UGC 09169	14 17 18.	+ 09 36	270	14.7	GALAXY IRR
MCG+02-37-001	14 17 18.	+ 09 36	240	14.7	GALAXY
ZWG 075.009	14 17 18.	+ 11 51		15.4	GALAXY
MCG+03-37-002	14 17 18.	+ 17 52 30.	60	15.3	GALAXY
ZWG 104.004	14 17 18.	+ 18 05		14.4	GALAXY
UGC 09170	14 17 18.	+ 18 05	48	14.4	GALAXY S0
MCG+03-37-003	14 17 18.	+ 18 05 30.	54	14.4	GALAXY
ZC 1417.3+1837	14 17 18.	+ 18 37	610		CLUSTER OF GALAXIES
ZWG 104.005	14 17 18.	+ 20 10		15.4	GALAXY
KARA.73B 0623	14 17 18.	+ 20 10	18	15.4	ISOLATED GALAXY E
ZC 1417.3+2408	14 17 18.	+ 24 08	2550		CLUSTER OF GALAXIES
MCG-01-37-001	14 17 18.	- 04 13	60	14.5	GALAXY

OBJECT NAME	RIGHT ASCEN.	DECLINATION	DIAM.	MAGN.	TYPE OF OBJECT
MCG-05-34-007	14 17 18.	- 29 31	24	15.	GALAXY
SN 19540	14 17 19.	+ 13 34		17.6	SUPERNOVA
IC 0997	14 17 19.	- 04 15 00.			NONSTELLAR OBJECT
PK315+05.1	14 17 20.	- 55 14 15.	10		PLANETARY NEBULA
REIZ 4039	14 17 21.	+ 09 36	120	14.5	GALAXY
FATH 1.664	14 17 21.	+ 15 10	19		NEBULA
IC 1000	14 17 21.	+ 18 04 55.			NONSTELLAR OBJECT
UGC 09171	14 17 24.	+ 18 04	60	16.0	GALAXY Sb
ZC 1417.4+3936	14 17 24.	+ 39 36	1880		CLUSTER OF GALAXIES
MCG+08-26-019	14 17 24.	+ 48 31	48	16.	GALAXY
REIZ 4037	14 17 24.	- 04 13	42	13.6	GALAXY
FATH 1.665	14 17 25.	+ 15 00	11		NEBULA
REIZ 4047	14 17 25.	+ 35 23	66	14.3	GALAXY
RNGC 5571	14 17 27.	+ 35 23			NON-EXISTENT OBJECT
RNGC 5572	14 17 27.	+ 36 22		15.0	GALAXY
ZWG 047.010	14 17 30.	+ 04 13		13.7	GALAXY
UGC 09172	14 17 30.	+ 04 13	240	13.7	GALAXY SBb
MCG+01-37-001	14 17 30.	+ 04 13	192	13.7	GALAXY
FAI 1417+09	14 17 30.	+ 09 06			COMPACT GALAXY
ZWG 075.010	14 17 30.	+ 12 12		15.7	GALAXY
ZC 1417.5+2542	14 17 30.	+ 25 42	610		CLUSTER OF GALAXIES
TON-N 0762	14 17 30.	+ 28 34		17.0	BLUE STAR
ZC 1417.5+3041	14 17 30.	+ 30 41	3490		CLUSTER OF GALAXIES
REIZ 4048	14 17 30.	+ 36 21	24	14.4	GALAXY
ZWG 192.002	14 17 30.	+ 36 22		15.2	GALAXY
ZWG 191.079	14 17 30.	+ 36 22		15.2	GALAXY
MRK 677	14 17 30.	+ 36 22	7	17.	GALAXY WITH UV CONTINUUM
UGC 09173	14 17 30.	+ 36 22	72	15.2	GALAXY Sb
MCG+06-31-099	14 17 30.	+ 36 22	48	14.	GALAXY
REIZ 4049	14 17 30.	+ 36 24	84	14.6	GALAXY
ZWG 219.068	14 17 30.	+ 40 33		15.4	GALAXY
MCG+08-26-020	14 17 30.	+ 46 02	54	15.	GALAXY
ZWG 247.018	14 17 30.	+ 46 05		15.4	GALAXY
ZWG 247.019	14 17 30.	+ 46 06		15.6	GALAXY
ZC 1417.5-0239	14 17 30.	- 02 39	1750		CLUSTER OF GALAXIES
MCG-05-34-008	14 17 30.	- 33 09	36	15.5	GALAXY
REIZ 4040	14 17 31.	+ 04 13	120	13.4	GALAXY
REIZ 4046	14 17 31.	+ 29 07	48	15.2	GALAXY
HOLM 630B	14 17 33.	+ 04 14	66	13.0	PART OF MULTIPLE GALAXY
MCG+03-37-004	14 17 33.	+ 18 05	60	16.	GALAXY
RNGC 5556	14 17 33.	- 29 01		12.5	GALAXY
FAI 14176+09	14 17 36.	+ 09 39			COMPACT GALAXY
FAI 14176+12	14 17 36.	+ 12 52			COMPACT GALAXY
MCG+05-34-021	14 17 36.	+ 29 05 30.	18		GALAXY
MCG+06-31-100	14 17 36.	+ 38 29 30.	48	17.5	GALAXY
ZC 1417.6+4024	14 17 36.	+ 40 24	740		CLUSTER OF GALAXIES
MCG-05-34-009	14 17 36.	- 29 00	180	14.5	GALAXY
FATH 1.666	14 17 37.	+ 14 53	14		NEBULA
IC 0998	14 17 37.	- 04 13 59.			NONSTELLAR OBJECT
FATH 1.667	14 17 38.	+ 15 18	5		NEBULA
MCG+05-34-022	14 17 38.	+ 29 05 30.	24		GALAXY
IC 1005	14 17 38.	+ 71 49 33.			NONSTELLAR OBJECT
RNGC 5560	14 17 39.	+ 04 13		13.5	GALAXY
RNGC 5563	14 17 40.	+ 07 17		15.5	GALAXY
REIZ 4043	14 17 41.	+ 07 16	12	15.3	GALAXY
REIZ 4044	14 17 41.	+ 07 17	9	15.6	GALAXY
RNGC 5562	14 17 41.	+ 10 29		14.5	GALAXY
ZWG 047.011	14 17 42.	+ 07 17		15.3	GALAXY
ZWG 075.011	14 17 42.	+ 10 29		14.5	GALAXY
UGC 09174	14 17 42.	+ 10 29	42	14.5	GALAXY
MCG+02-37-002	14 17 42.	+ 10 29	48	14.5	GALAXY
MCG+09-23-060	14 17 42.	+ 52 09	36	17.	GALAXY
MCG+09-23-061	14 17 42.	+ 52 22	30	15.	GALAXY
7ZW 543	14 17 42.	+ 67 25			COMPACT GALAXY
HOLM 631B	14 17 45.	+ 10 01	18	15.4	PART OF MULTIPLE GALAXY
HOLM 631A	14 17 45.	+ 10 01	18	15.4	PART OF MULTIPLE GALAXY
RNGC 5565	14 17 46.	+ 07 14			NON-EXISTENT OBJECT
RNGC 5564	14 17 46.	+ 07 15			NON-EXISTENT OBJECT
ZC 1417.8+0150	14 17 48.	+ 01 50	870		CLUSTER OF GALAXIES
ZWG 047.012	14 17 48.	+ 04 09		12.0	GALAXY
UGC 09175	14 17 48.	+ 04 09	372	12.0	GALAXY SBa
MCG+01-37-002	14 17 48.	+ 04 09	420	12.0	GALAXY
ARP 286	14 17 48.	+ 04 12			PECULIAR GALAXY
ZWG 075.012	14 17 48.	+ 09 09		15.4	GALAXY
ZC 1417.8+1600	14 17 48.	+ 16 00	670		CLUSTER OF GALAXIES
TON-N 0194	14 17 48.	+ 25 42		13.6	BLUE STAR
ZWG 273.001	14 17 48.	+ 54 19		15.7	GALAXY
ZWG 272.046	14 17 48.	+ 54 19		15.7	GALAXY
MCG-01-37-002	14 17 48.	- 04 07	60	15.5	GALAXY
HOLM 630A	14 17 49.	+ 04 10	102	12.1	PART OF MULTIPLE GALAXY
REIZ 4045	14 17 49.	+ 04 11	210	12.0	GALAXY
FATH 1.668	14 17 49.	+ 14 54	14		NEBULA
RNGC 5566	14 17 51.	+ 04 10		11.5	GALAXY
FAI 14179+09	14 17 54.	+ 09 10			COMPACT GALAXY
ZWG 075.013	14 17 54.	+ 11 37		15.7	GALAXY
FAI 14179+13	14 17 54.	+ 13 13			COMPACT GALAXY
ZWG 247.020	14 17 54.	+ 49 28		15.4	GALAXY
MCG+09-23-062	14 17 54.	+ 52 20	12	16.	GALAXY
MCG+10-20-094	14 17 54.	+ 56 57	324	11.2	GALAXY
7ZW 544	14 17 54.	+ 58 06			COMPACT GALAXY
7ZW 545	14 17 54.	+ 67 29			COMPACT GALAXY
MCG-04-34-012	14 17 54.	- 27 10	60	15.5	GALAXY
HN 1099	14 17 54.	- 46 04	174		NEBULA
IC 4402	14 17 54.	- 46 04			NONSTELLAR OBJECT
FATH 1.669	14 17 55.	+ 15 34	8		NEBULA
REIZ 4051	14 17 58.	+ 07 44	12	15.4	GALAXY
ZWG 047.013	14 18 00.	+ 04 12		14.9	GALAXY
UGC 09176	14 18 00.	+ 04 12	120	14.9	GALAXY Sc
MCG+01-37-003	14 18 00.	+ 04 12	90	14.9	GALAXY
ZWG 075.014	14 18 00.	+ 10 40		14.9	GALAXY
UGC 09177	14 18 00.	+ 10 40	84	14.9	GALAXY Sc
MCG+02-37-003	14 18 00.	+ 10 40	84	14.9	GALAXY
ZWG 104.006	14 18 00.	+ 17 59		15.5	GALAXY
ZC 1418.0+2139	14 18 00.	+ 21 39	6120		CLUSTER OF GALAXIES
TON-N 0763	14 18 00.	+ 26 34		15.4	BLUE STAR
MCG+06-32-001	14 18 00.	+ 35 52	48	16.	GALAXY
ZC 1418.0+4145	14 18 00.	+ 41 45	1550		CLUSTER OF GALAXIES
ARP 045	14 18 00.	+ 52 06			PECULIAR GALAXY
MCG+09-23-063	14 18 00.	+ 52 19	24	16.	GALAXY
ZC 1418.0+5510	14 18 00.	+ 55 10	940		CLUSTER OF GALAXIES
REIZ 4050	14 18 01.	+ 04 13	60	14.6	GALAXY
HOLM 630C	14 18 01.	+ 04 13	42	14.7	PART OF MULTIPLE GALAXY
REIZ 4052	14 18 02.	+ 10 40	30	15.2	GALAXY
RNGC 5569	14 18 03.	+ 04 13		15.0	GALAXY
REIZ 4064	14 18 05.	+ 56 57	150	12.3	GALAXY
MCG+03-37-005	14 18 06.	+ 18 00	24	15.5	GALAXY
MCG+09-23-064	14 18 06.	+ 52 07	60	15.	GALAXY
ZWG 273.002	14 18 06.	+ 52 08		15.5	GALAXY
ZWG 272.047	14 18 06.	+ 52 08		15.5	GALAXY
UGC 09178	14 18 06.	+ 52 08	108	15.5	GALAXY DBL SYS
ZC 1418.1+7102	14 18 06.	+ 71 02	1340		CLUSTER OF GALAXIES
RNGC 5573	14 18 10.	+ 07 08		15.0	GALAXY
IC 1001	14 18 11.	+ 05 21 22.			NONSTELLAR OBJECT
REIZ 4053	14 18 11.	+ 07 08	12	15.3	GALAXY
ZWG 047.014	14 18 12.	+ 05 39		15.0	GALAXY
ZWG 047.015	14 18 12.	+ 05 43		15.2	GALAXY
ZWG 047.016	14 18 12.	+ 07 08		15.0	GALAXY
ZWG 075.015	14 18 12.	+ 10 05		15.5	GALAXY
FAI 14182+11	14 18 12.	+ 11 11			COMPACT GALAXY
FATH 1.670	14 18 12.	+ 15 40	19		NEBULA
MCG+04-34-018	14 18 12.	+ 20 40	36	15.3	GALAXY
ZC 1418.2+2808	14 18 12.	+ 28 08	1950		CLUSTER OF GALAXIES
ZWG 247.021	14 18 12.	+ 45 23		15.7	GALAXY
ZWG 295.045	14 18 12.	+ 56 57		11.7	GALAXY
UGC 09179	14 18 12.	+ 56 57	366	11.7	GALAXY Sc
KARA.73B 0624	14 18 12.	+ 56 57	372	11.7	ISOLATED GALAXY S
RNGC 5585	14 18 12.	+ 56 58		11.5	GALAXY
MCG+10-20-095	14 18 12.	+ 58 35	36	16.	GALAXY
REIN 2.218	14 18 13.06	+ 56 57 29.7			NEBULA
IC 1002	14 18 14.	+ 05 24 52.			NONSTELLAR OBJECT
VV 002C	14 18 15.	+ 52 08	9	17.	INTERACTING GALAXY
VV 002B	14 18 15.	+ 52 08	24	16.	INTERACTING GALAXY
VV 002A	14 18 15.	+ 52 08	48	15.	INTERACTING GALAXY
SC 1415-1538.3	14 18 16.	- 15 52 06.	18		NEBULA
HOLM 633B	14 18 17.	+ 22 05	42	15.0	PART OF MULTIPLE GALAXY
ZWG 047.017	14 18 18.	+ 04 33		15.1	GALAXY
MCG+01-37-004	14 18 18.	+ 04 33	30	15.0	GALAXY
8ZW 1418+05.7	14 18 18.	+ 05 41		16.0	COMPACT GALAXY
MCG+01-37-005	14 18 18.	+ 07 08	72	15.1	GALAXY
8ZW 1418+08.1	14 18 18.	+ 08 07		16.4	COMPACT GALAXY
ZWG 075.016	14 18 18.	+ 09 13		15.4	GALAXY
ZC 1418.3+2000	14 18 18.	+ 20 00	400		CLUSTER OF GALAXIES
ZWG 133.033	14 18 18.	+ 20 41		15.3	GALAXY
ARC 1898	14 18 18.	+ 25 24		17.0	RICH CLUSTER OF GALAXIES
ZWG 163.030	14 18 18.	+ 27 05		15.0	GALAXY
ZWG 192.003	14 18 18.	+ 35 25		14.7	GALAXY
ZWG 191.080	14 18 18.	+ 35 25		14.7	GALAXY
UGC 09180	14 18 18.	+ 35 25	114	14.7	GALAXY Sc
ZC 1418.3+5410	14 18 18.	+ 54 10	2690		CLUSTER OF GALAXIES
ZC 1418.3+6016	14 18 18.	+ 60 16	3630		CLUSTER OF GALAXIES
MCG+10-20-096	14 18 18.	+ 61 43	42	17.	GALAXY
REIZ 4058	14 18 19.	+ 20 40	36	14.1	GALAXY
REIZ 4062	14 18 19.	+ 35 25	90	13.4	GALAXY
ARC 1897	14 18 20.	- 08 33		17.0	RICH CLUSTER OF GALAXIES
ARP 069	14 18 21.	+ 35 24			PECULIAR GALAXY
RNGC 5580	14 18 21.	+ 35 25		14.5	GALAXY
RNGC 5579	14 18 21.	+ 35 25		14.5	GALAXY
VV 142A	14 18 21.	+ 35 25			INTERACTING GALAXY
VV 142	14 18 21.	+ 35 25	108	15.	INTERACTING GALAXY
HOLM 632B	14 18 23.	+ 03 28	54	13.2	PART OF MULTIPLE GALAXY
IC 1003	14 18 23.	+ 05 40 47.			NONSTELLAR OBJECT
REIZ 4055	14 18 23.	+ 06 27	12	14.9	GALAXY
ZWG 047.018	14 18 24.	+ 03 28		13.4	GALAXY
UGC 09181	14 18 24.	+ 03 28	66	13.4	GALAXY S0?
MCG+01-37-006	14 18 24.	+ 03 28	72	13.4	GALAXY
ZWG 047.019	14 18 24.	+ 04 05		15.6	GALAXY
MCG+04-34-019	14 18 24.	+ 22 09	102	14.7	GALAXY
ZWG 133.034	14 18 24.	+ 22 10		14.7	GALAXY
UGC 09182	14 18 24.	+ 22 10	156	14.7	GALAXY Sc
KARA.73B 0625	14 18 24.	+ 22 10	174	14.7	ISOLATED GALAXY S
TON-N 0764	14 18 24.	+ 27 49		15.0	BLUE STAR
TON-N 0765	14 18 24.	+ 28 17		14.6	BLUE STAR
MCG+06-32-002	14 18 24.	+ 35 25	114	13.	GALAXY
7ZW 546	14 18 24.	+ 67 34			COMPACT GALAXY
MCG-04-34-013	14 18 24.	- 27 05	36	15.	GALAXY
MCG-05-34-010	14 18 24.	- 29 02	72	15.5	GALAXY
REIZ 4054	14 18 25.	+ 03 28	54	13.1	GALAXY
RNGC 5574	14 18 26.	+ 03 28		13.5	GALAXY
RNGC 5575	14 18 27.	+ 06 26		14.5	GALAXY
REIZ 4057	14 18 29.	+ 06 25	24	14.2	GALAXY
ZWG 047.020	14 18 30.	+ 03 30		12.3	GALAXY
UGC 09183	14 18 30.	+ 03 30	198	12.3	GALAXY E
MCG+01-37-007	14 18 30.	+ 03 30	48	12.3	GALAXY
ZWG 047.021	14 18 30.	+ 06 26		14.5	GALAXY
UGC 09184	14 18 30.	+ 06 26	72	14.5	GALAXY S0
MCG+01-37-008	14 18 30.	+ 06 26	36	14.5	GALAXY
ZC 1418.5+1344	14 18 30.	+ 13 44	340		CLUSTER OF GALAXIES
ZWG 104.007	14 18 30.	+ 15 13		15.0	GALAXY
UGC 09185	14 18 30.	+ 15 13	66	15.0	GALAXY S
MCG+03-37-006	14 18 30.	+ 15 13	45	15.0	GALAXY
FATH 1.671	14 18 30.	+ 15 13	33		NEBULA
ZC 1418.5+1630	14 18 30.	+ 16 30	870		CLUSTER OF GALAXIES
ZC 1418.5+2045	14 18 30.	+ 20 45	740		CLUSTER OF GALAXIES
REIZ 4061	14 18 30.	+ 22 10	132	14.2	GALAXY
HOLM 633A	14 18 30.	+ 22 10	150	13.0	PART OF MULTIPLE GALAXY
ZC 1418.5+2531	14 18 30.	+ 25 31	2080		CLUSTER OF GALAXIES
ZC 1418.5+2536	14 18 30.	+ 25 36	340		CLUSTER OF GALAXIES
ZWG 133.035	14 18 30.	+ 26 11		15.7	GALAXY
MCG+05-34-023	14 18 30.	+ 27 04 30.	54	15.0	GALAXY
ZC 1418.5+3545	14 18 30.	+ 35 45	2820		CLUSTER OF GALAXIES
MCG+07-29-062	14 18 30.	+ 40 32	21	15.	GALAXY
SC 1415-1635.6	14 18 30.	- 16 49 24.	18		NEBULA
REIZ 4056	14 18 31.	+ 03 30	72	11.7	GALAXY
HOLM 632A	14 18 31.	+ 03 30	36	12.0	PART OF MULTIPLE GALAXY
IC 1004	14 18 31.	+ 17 53 00.			NONSTELLAR OBJECT
HOLM 633C	14 18 31.	+ 22 09	18	15.3	PART OF MULTIPLE GALAXY
REIZ 4063	14 18 31.	+ 35 26	24	14.7	GALAXY
RNGC 5576	14 18 32.	+ 03 30		12.0	GALAXY
ZC 1418.6+1006	14 18 36.	+ 10 06	1280		CLUSTER OF GALAXIES
ZWG 133.036	14 18 36.	+ 23 50		15.6	GALAXY
UGC 09186	14 18 36.	+ 23 50	66	15.6	GALAXY Sa
MCG+04-34-020	14 18 36.	+ 23 50	48	15.6	GALAXY
ZC 1418.6+2637	14 18 36.	+ 26 37	470		CLUSTER OF GALAXIES
TON-N 0195	14 18 36.	+ 29 39		14.9	BLUE STAR
VV 142B	14 18 36.	+ 35 26			INTERACTING GALAXY
MCG+07-29-063	14 18 36.	+ 39 54	36	13.	GALAXY
ZWG 220.002	14 18 36.	+ 40 33		15.5	GALAXY
ZWG 219.069	14 18 36.	+ 40 33		15.5	GALAXY
ZWG 019.004	14 18 36.	- 00 09		15.6	GALAXY
SC 1415-1731.7	14 18 37.	- 17 45 29.	12		NEBULA
IC 4400	14 18 38.	- 60 21			NONSTELLAR OBJECT
HOLM 634A	14 18 39.	+ 17 55	24	15.4	PART OF MULTIPLE GALAXY
REIZ 4069	14 18 39.	+ 48 24	30	14.5	GALAXY
HOLM 634B	14 18 40.	+ 17 56	18	15.7	PART OF MULTIPLE GALAXY
RNGC 5582	14 18 41.	+ 39 56		13.0	GALAXY
RNGC 5607	14 18 41.	+ 71 50		14.0	GALAXY

OBJECT NAME	RIGHT ASCEN.	DECLINATION	DIAM.	MAGN.	TYPE OF OBJECT
ZWG 047.022	14 18 42.	+ 03 40		13.6	GALAXY
UGC 09187	14 18 42.	+ 03 40	192	13.6	GALAXY Sb
MCG+01-37-009	14 18 42.	+ 03 40	192	13.6	GALAXY
ZWG 104.008	14 18 42.	+ 17 30		15.7	GALAXY
MCG+03-37-007	14 18 42.	+ 17 56	18	15.4	GALAXY
MCG+06-32-003	14 18 42.	+ 34 48	24	17.	GALAXY
ZWG 220.003	14 18 42.	+ 39 56		13.0	GALAXY
ZWG 219.070	14 18 42.	+ 39 56		13.0	GALAXY
UGC 09188	14 18 42.	+ 39 56	168	13.0	GALAXY E
ZWG 337.007	14 18 42.	+ 71 50		13.9	GALAXY
UGC 09189	14 18 42.	+ 71 50	54	13.9	GALAXY PECULR
REIZ 4060	14 18 43.	+ 03 40	168	13.0	GALAXY
IC 4407	14 18 43.	- 05 46			NONSTELLAR OBJECT
RNGC 5577	14 18 44.	+ 03 40		13.5	GALAXY
REIZ 4066	14 18 44.	+ 39 55	54	13.4	GALAXY
RNGC 5578	14 18 45.	+ 06 26			NON-EXISTENT OBJECT
MCG-01-37-003	14 18 45.	- 03 32	60	15.	GALAXY
ZWG 047.023	14 18 48.	+ 02 31		15.3	GALAXY
FAI 14188+12	14 18 48.	+ 12 13			COMPACT GALAXY
FAI 14188+14	14 18 48.	+ 14 11			COMPACT GALAXY
ZC 1418.8+4502	14 18 48.	+ 45 02	3700		CLUSTER OF GALAXIES
1ZW 083	14 18 48.	+ 51 44			COMPACT GALAXY
7ZW 547	14 18 48.	+ 71 49			COMPACT GALAXY
MRK 286	14 18 48.	+ 71 50	20	14.5	GALAXY WITH UV CONTINUUM
MCG+12-14-001	14 18 48.	+ 71 50	51	15.	GALAXY
ZWG 019.005	14 18 48.	- 02 01		15.7	GALAXY
REIZ 4059	14 18 48.	- 03 33	36	14.6	GALAXY
RNGC 5581	14 18 52.	+ 23 43		15.0	GALAXY
ZWG 047.024	14 18 54.	+ 05 17		15.3	GALAXY
UGC 09190	14 18 54.	+ 05 17	78	15.3	GALAXY Sb-c
ZWG 075.017	14 18 54.	+ 11 30		15.5	GALAXY
FAI 14189+12	14 18 54.	+ 12 44			COMPACT GALAXY
ZWG 133.037	14 18 54.	+ 22 16		15.7	GALAXY
ZWG 133.038	14 18 54.	+ 23 43		15.1	GALAXY
TON-N 0196	14 18 54.	+ 25 51		15.4	BLUE STAR
ZWG 220.004	14 18 54.	+ 40 21		15.5	GALAXY
ZWG 219.071	14 18 54.	+ 40 21		15.5	GALAXY
ZWG 337.008	14 18 54.	+ 72 46		15.4	GALAXY
SC 1416-1548.9	14 18 55.	- 16 02 41.	42		NEBULA
MCG+04-34-021	14 18 57.	+ 23 42	24	15.1	GALAXY
ARC 1899	14 18 59.	+ 17 56		16.0	RICH CLUSTER OF GALAXIES
ARC 1900	14 18 59.	+ 36 16		17.2	RICH CLUSTER OF GALAXIES
ZC 1419.0+0115	14 19 00.	+ 01 15	3560		CLUSTER OF GALAXIES
ZC 1419.0+2206	14 19 00.	+ 22 06	610		CLUSTER OF GALAXIES
ZC 1419.0+2300	14 19 00.	+ 23 00	1010		CLUSTER OF GALAXIES
ZC 1419.0+2349	14 19 00.	+ 23 49	540		CLUSTER OF GALAXIES
HOLM 635B	14 19 00.	+ 30 12	36	15.5	PART OF MULTIPLE GALAXY
ZWG 163.031	14 19 00.	+ 30 13		15.0	GALAXY
REIZ 4067	14 19 00.	+ 30 13	12	15.7	GALAXY
HOLM 635A	14 19 00.	+ 30 13	30	14.1	PART OF MULTIPLE GALAXY
UGC 09191	14 19 00.	+ 30 13	60	15.0	GALAXY S
REIZ 4068	14 19 00.	+ 30 14	24	14.3	GALAXY
MCG+07-30-001	14 19 00.	+ 40 20	30	15.5	GALAXY
ZC 1419.0+4221	14 19 00.	+ 42 21	400		CLUSTER OF GALAXIES
SHAH 015	14 19 00.	+ 44 47	138	16.5	GROUP OF COMPACT GALAXIES
MCG+09-24-001	14 19 00.	+ 51 22	72	15.	GALAXY
UGC 09192	14 19 00.	+ 51 23	66	16.5	GALAXY IRR
ZWG 353.034	14 19 00.	+ 80 20		15.3	GALAXY
ARC 1901	14 19 02.	+ 40 34		16.6	RICH CLUSTER OF GALAXIES
IC 4408	14 19 03.	+ 30 14 09.			NONSTELLAR OBJECT
REIZ 4065	14 19 04.	+ 17 35	24	15.7	GALAXY
ZWG 133.039	14 19 06.	+ 26 21		15.7	GALAXY
ZWG 163.032	14 19 06.	+ 27 21		15.6	GALAXY
MCG+05-34-024	14 19 06.	+ 30 12 30.	60	15.0	GALAXY
MCG+06-32-004	14 19 06.	+ 35 31	42	16.	GALAXY
UGC 09193	14 19 06.	+ 36 58	60	17.	GALAXY DWARF
MCG+07-30-002	14 19 06.	+ 39 42	48	15.5	GALAXY
ZWG 220.005	14 19 06.	+ 39 44		15.6	GALAXY
ZWG 219.072	14 19 06.	+ 39 44		15.6	GALAXY
UGC 09194	14 19 06.	+ 39 44	78	15.6	GALAXY S
SHAH 074	14 19 06.	+ 43 18	252	17.4	GROUP OF COMPACT GALAXIES
REIZ 4077	14 19 06.	+ 53 50	42	14.5	GALAXY
REIZ 4073	14 19 09.	+ 48 04	24	14.5	GALAXY
HN 0403	14 19 10.	- 43 56			NEBULA
ZC 1419.2+1324	14 19 12.	+ 13 24	810		CLUSTER OF GALAXIES
ZWG 133.040	14 19 12.	+ 24 10		15.5	GALAXY
UGC 09195	14 19 12.	+ 24 10	72	15.5	GALAXY Sb
ZC 1419.2+2653	14 19 12.	+ 26 53	940		CLUSTER OF GALAXIES
ZC 1419.2+3616	14 19 12.	+ 36 16	1210		CLUSTER OF GALAXIES
REIZ 4070	14 19 13.	+ 20 55	36	14.5	GALAXY
MCG+04-34-022	14 19 15.	+ 24 10	66	15.5	GALAXY
MCG+05-34-025	14 19 15.	+ 27 20	36	15.6	GALAXY
PK319+15.1	14 19 15.2	- 43 54 55.	100	10.6	PLANETARY NEBULA
IC 4406	14 19 15.5	- 43 55 27.	20	10.6	PLANETARY NEBULA
HN 0094	14 19 16.	- 43 55			NEBULA
ZWG 019.006	14 19 18.	+ 02 15		15.2	GALAXY
8ZW 1419+02.3	14 19 18.	+ 02 15		15.2	COMPACT GALAXY
ZWG 075.018	14 19 18.	+ 13 27		14.2	GALAXY
RNGC 5583	14 19 18.	+ 13 27		14.0	GALAXY
UGC 09196	14 19 18.	+ 13 27	48	14.2	GALAXY
MCG+02-37-004	14 19 18.	+ 13 27	48	14.2	GALAXY
ZWG 133.041	14 19 18.	+ 24 31		15.7	GALAXY
ZWG 163.033	14 19 18.	+ 31 49		15.1	GALAXY
MCG+05-34-026	14 19 18.	+ 31 49	42	15.1	GALAXY
ZWG 192.004	14 19 18.	+ 35 29		14.3	GALAXY
UGC 09197	14 19 18.	+ 35 29	78	14.3	GALAXY SBa
REIZ 4072	14 19 18.	+ 35 30	36	13.8	GALAXY
UGC 09198	14 19 18.	+ 36 40	60	16.5	GALAXY Sb
MCG-04-34-014	14 19 18.	- 26 38	54	15.	GALAXY
SC 1416-1613.8	14 19 20.	- 16 27 34.	6		NEBULA
BC B21419+31	14 19 20.0	+ 31 32 36.5		20.0	QUASI-STELLAR OBJECT
RNGC 5588	14 19 21.	+ 35 21			NON-EXISTENT OBJECT
RNGC 5589	14 19 21.	+ 35 29		14.5	GALAXY
MCG+06-32-055	14 19 21.	+ 35 30	48	13.	GALAXY
IC 4409	14 19 22.	+ 31 48 29.			NONSTELLAR OBJECT
ZWG 047.025	14 19 24.	+ 04 11		15.1	GALAXY
ZWG 047.026	14 19 24.	+ 08 07		15.4	GALAXY
FAI 14194+09	14 19 24.	+ 09 25			COMPACT GALAXY
MCG+03-37-008	14 19 24.	+ 17 59	12	17.	GALAXY
ZWG 192.005	14 19 24.	+ 38 20		15.1	GALAXY
MCG+10-20-097	14 19 24.	+ 58 53	24	16.	GALAXY
ZWG 019.007	14 19 24.	- 00 19		15.7	GALAXY
FAI 14195+09	14 19 30.	+ 09 23			COMPACT GALAXY
ZWG 075.019	14 19 30.	+ 11 20		15.3	GALAXY
UGC 09199	14 19 30.	+ 11 20	84	15.3	GALAXY S
ZC 1419.5+1158	14 19 30.	+ 11 58	810		CLUSTER OF GALAXIES
ZWG 192.006	14 19 30.	+ 35 25		13.6	GALAXY
UGC 09200	14 19 30.	+ 35 25	120	13.6	GALAXY S0
REIZ 4076	14 19 31.	+ 35 26	48	13.7	GALAXY
SC 1416-1430.9	14 19 32.	- 14 44 39.	18		NEBULA
RNGC 5590	14 19 33.	+ 35 25		13.5	GALAXY
ZWG 133.042	14 19 36.	+ 23 45		15.6	GALAXY
MRK 678	14 19 36.	+ 23 45	20	15.5	GALAXY WITH UV CONTINUUM
ZWG 133.043	14 19 36.	+ 24 20		15.6	GALAXY
MCG+06-32-006	14 19 36.	+ 35 26	30	13.	GALAXY
MCG+10-20-098	14 19 36.	+ 61 50	27	16.	GALAXY
ZC 1419.6+6255	14 19 36.	+ 62 55	1410		CLUSTER OF GALAXIES
REIZ 4075	14 19 37.	+ 28 41	36	15.2	GALAXY
SC 1416-1444.1	14 19 41.	- 14 57 51.	36		NEBULA
SC 1416-1713.5	14 19 41.	- 17 27 15.	18		NEBULA
RNGC 5586	14 19 42.	+ 13 24			NON-EXISTENT OBJECT
ZC 1419.7+2128	14 19 42.	+ 21 28	540		CLUSTER OF GALAXIES
TON-N 0766	14 19 42.	+ 24 53		16.0	BLUE STAR
MCG+06-32-007	14 19 42.	+ 34 41	30	17.	GALAXY
SC 1416-1504.3	14 19 43.	- 15 18 03.	12		NEBULA
SC 1416-1612.7	14 19 43.	- 16 26 27.	12		NEBULA
ARC 1902	14 19 44.	+ 37 32		17.2	RICH CLUSTER OF GALAXIES
HN 1100	14 19 46.	+ 17 37	12		NEBULA
IC 4410	14 19 46.	+ 17 37			NONSTELLAR OBJECT
REIZ 4071	14 19 46.	- 00 10	180	12.4	GALAXY
REIZ 4074	14 19 48.	+ 14 08	60	14.1	GALAXY
ZWG 075.020	14 19 48.	+ 14 09		14.0	GALAXY
RNGC 5587	14 19 48.	+ 14 09		14.0	GALAXY
UGC 09202	14 19 48.	+ 14 09	162	14.0	GALAXY S0-a
MCG+02-37-005	14 19 48.	+ 14 09	156	14.0	GALAXY
8ZW 1419+19.1	14 19 48.	+ 19 06		19.0	COMPACT GALAXY
TON-N 0767	14 19 48.	+ 26 49		16.9	BLUE STAR
ZC 1419.8+3201	14 19 48.	+ 32 01	870		CLUSTER OF GALAXIES
ZC 1419.8+4232	14 19 48.	+ 42 32	940		CLUSTER OF GALAXIES
MCG+08-26-021	14 19 48.	+ 49 46	48	16.	GALAXY
ZWG 273.003	14 19 48.	+ 50 39		15.3	GALAXY
ZWG 019.008	14 19 48.	- 00 09		12.8	GALAXY
UGC 09201	14 19 48.	- 00 09	210	12.8	GALAXY Sc
MCG+00-37-001	14 19 48.	- 00 09	180	12.8	GALAXY
KARA.73B 0626	14 19 48.	- 00 09	216	12.8	ISOLATED GALAXY S
RNGC 5584	14 19 49.	- 00 10		12.0	GALAXY
HOLM 636B	14 19 50.	+ 17 38	30	15.4	PART OF MULTIPLE GALAXY
REIZ 4078	14 19 51.	+ 17 37	36	14.8	GALAXY
HOLM 636A	14 19 51.	+ 17 37	30	14.4	PART OF MULTIPLE GALAXY
MCG+06-32-008	14 19 51.	+ 35 02	48	15.	GALAXY
MCG+08-26-022	14 19 51.	+ 50 37	60	15.	GALAXY
REIZ 4085	14 19 53.	+ 50 39	42	15.4	GALAXY
ZWG 019.009	14 19 54.	+ 00 17		15.4	GALAXY
8ZW 1419+00.3	14 19 54.	+ 00 17		15.4	COMPACT GALAXY
ZWG 104.009	14 19 54.	+ 17 37		15.6	GALAXY
MCG+03-37-009	14 19 54.	+ 17 37	36	15.6	GALAXY
REIZ 4080	14 19 54.	+ 22 04	42	15.3	GALAXY
MCG+07-30-003	14 19 54.	+ 40 11	63	14.	GALAXY
ZWG 220.006	14 19 54.	+ 40 13		15.0	GALAXY
UGC 09203	14 19 54.	+ 40 13	60	15.0	GALAXY SBb
PATR 1.672	14 19 56.	+ 15 08	14		NEBULA
SC 1417-1452.1	14 19 57.	- 15 05 50.	12		NEBULA
SC 1417-1702.2	14 19 59.	- 17 15 56.	12		NEBULA
VDB.66G 189	14 20	+ 45 39	70		DWARF GALAXY
ZWG 075.021	14 20 00.	+ 11 32		15.5	GALAXY
ZWG 075.022	14 20 00.	+ 13 38		15.7	GALAXY
ZWG 104.010	14 20 00.	+ 15 18		14.2	GALAXY
UGC 09206	14 20 00.	+ 15 18	48	14.2	GALAXY PECULR
SN 1955K	14 20 00.	+ 15 18		18.4	SUPERNOVA
ZWG 133.044	14 20 00.	+ 24 16		15.7	GALAXY
ZC 1420.0+3734	14 20 00.	+ 37 34	2420		CLUSTER OF GALAXIES
1ZW 084	14 20 00.	+ 46 57			COMPACT GALAXY
ZWG 296.001	14 20 00.	+ 61 48		15.7	GALAXY
ZWG 295.046	14 20 00.	+ 61 48		15.7	GALAXY
7ZW 548	14 20 00.	+ 76 00			COMPACT GALAXY
UGC 09204	14 20 00.	+ 82 52	78	16.5	GALAXY Sc
UGC 09205	14 20 00.	+ 86 03	72	17.	GALAXY
HOLM 637A	14 20 03.	+ 09 30	36	15.3	PART OF MULTIPLE GALAXY
SC 1417-1622.0	14 20 05.	- 16 35 44.	12		NEBULA
ZWG 047.027	14 20 06.	+ 06 24		15.5	GALAXY
MCG+03-37-010	14 20 06.	+ 15 18	36	14.2	GALAXY
PATR 1.673	14 20 06.	+ 15 18	27		NEBULA
MCG+06-32-009	14 20 06.	+ 36 01	12	15.5	GALAXY
MCG+08-26-023	14 20 06.	+ 48 42 30.	6	17.	GALAXY
REIZ 4079	14 20 08.	+ 01 56	18	14.0	GALAXY
REIZ 4081	14 20 09.	+ 17 37	18	15.5	GALAXY
HOLM 637B	14 20 11.	+ 09 30	24	15.4	PART OF MULTIPLE GALAXY
ZWG 075.023	14 20 12.	+ 13 57		14.5	GALAXY
RNGC 5591	14 20 12.	+ 13 57		14.5	GALAXY
UGC 09207	14 20 12.	+ 13 57	84	14.5	GALAXY DBL SYS
KARA.72 424B	14 20 12.	+ 13 57	42		PART OF DOUBLE GALAXY
KARA.72 424A	14 20 12.	+ 13 57	42	14.5	PART OF DOUBLE GALAXY
MCG+02-37-006	14 20 12.	+ 13 57	84	14.5	GALAXY
FAI 14202+13	14 20 12.	+ 13 57			COMPACT GALAXY
ZC 1420.2+3417	14 20 12.	+ 34 17	1280		CLUSTER OF GALAXIES
ZC 1420.2+4827	14 20 12.	+ 48 27	6120		CLUSTER OF GALAXIES
ZC 1420.2+4952	14 20 12.	+ 49 52	940		CLUSTER OF GALAXIES
REIZ 4082	14 20 13.	+ 20 25	36	15.1	GALAXY
SC 1417-1713.9	14 20 16.	- 17 27 37.	12		NEBULA
SC 1417-1739.2	14 20 16.	- 17 52 55.	6		NEBULA
ARC 1904	14 20 17.	+ 48 48		15.6	RICH CLUSTER OF GALAXIES
REIZ 4084	14 20 18.	+ 29 11	30	15.1	GALAXY
ZC 1420.3+4113	14 20 18.	+ 41 13	740		CLUSTER OF GALAXIES
OCL 0916	14 20 18.	- 61 10	720		OPEN STAR CLUSTER
VHA 157	14 20 18.	- 61 10	600		OPEN STAR CLUSTER
SHB 286	14 20 19.2	+ 32 36 54.		17.5	QUASI-STELLAR OBJECT
BC 0Q334	14 20 20.66	+ 32 36 40.2		17.5	QUASI-STELLAR OBJECT
RNGC 5596	14 20 22.	+ 37 20		14.5	GALAXY
REIZ 4087	14 20 22.	+ 37 21	24	14.3	GALAXY
8ZW 1420+01.0	14 20 24.	+ 01 02		17.4	COMPACT GALAXY
ZWG 047.028	14 20 24.	+ 06 24		15.6	GALAXY
ZWG 192.007	14 20 24.	+ 37 20		14.5	GALAXY
MRK 470	14 20 24.	+ 37 20	24	15.5	GALAXY WITH UV CONTINUUM
UGC 09208	14 20 24.	+ 37 20	72	14.5	GALAXY S0
1ZW 085	14 20 24.	+ 48 44			COMPACT GALAXY
MCG+08-26-024	14 20 24.	+ 48 47	42	16.	GALAXY
ZWG 247.022	14 20 24.	+ 50 04		15.6	GALAXY
REIZ 4091	14 20 24.	+ 50 06	18	14.8	GALAXY
REIZ 4089	14 20 25.	+ 40 32	24	14.1	GALAXY
MCG+06-32-010	14 20 27.	+ 37 21	54	14.5	GALAXY
MCG+08-26-026	14 20 27.	+ 48 42	6	17.	GALAXY
MCG+08-26-025	14 20 27.	+ 48 42	12	17.	GALAXY
REIZ 4095	14 20 29.	+ 50 45	24	13.7	GALAXY
MCG+07-30-004	14 20 30.	+ 40 32	72	14.	GALAXY
ZWG 220.007	14 20 30.	+ 40 33		14.3	GALAXY
RNGC 5598	14 20 30.	+ 40 33		14.5	GALAXY

OBJECT NAME	RIGHT ASCEN.	DECLINATION	DIAM.	MAGN.	TYPE OF OBJECT
UGC 09209	14 20 30.	+ 40 33	90	14.3	GALAXY S0
MCG+08-26-027	14 20 30.	+ 48 52	39	16.	GALAXY
MCG+08-26-028	14 20 30.	+ 50 04	48	15.	GALAXY
ZWG 273.004	14 20 30.	+ 50 45		13.5	GALAXY
UGC 09210	14 20 30.	+ 50 45	84	13.5	GALAXY Sa
MCG-01-37-004	14 20 30.	- 04 31	30	15.	GALAXY
REIZ 4086	14 20 31.	+ 20 23	24	15.3	GALAXY
ARC 1907	14 20 31.	+ 49 57		17.2	RICH CLUSTER OF GALAXIES
RNGC 5602	14 20 31.	+ 50 45		13.5	GALAXY
ZWG 019.010	14 20 36.	+ 01 54		15.2	GALAXY
ZWG 047.029	14 20 36.	+ 06 23		15.6	GALAXY
FAI 14206+09	14 20 36.	+ 09 07			COMPACT GALAXY
8ZW 1420+15.1	14 20 36.	+ 15 09		17.7	COMPACT GALAXY
ZC 1420.6+2036	14 20 36.	+ 20 36	340		CLUSTER OF GALAXIES
ZC 1420.6+2259	14 20 36.	+ 22 59	540		CLUSTER OF GALAXIES
ZWG 133.045	14 20 36.	+ 24 01		15.3	GALAXY
ZWG 220.008	14 20 36.	+ 39 49		15.5	GALAXY
MCG+08-26-029	14 20 36.	+ 45 35	72	15.	GALAXY
ZWG 247.023	14 20 36.	+ 45 37		15.7	GALAXY
UGC 09211	14 20 36.	+ 45 37	150	15.7	GALAXY DWRF IR
MCG+09-24-002	14 20 36.	+ 50 44	90	13.7	GALAXY
MCG-02-37-001	14 20 36.	- 10 37	6	15.5	GALAXY
REIZ 4083	14 20 38.	+ 01 55	24	14.6	GALAXY
IC 1006	14 20 38.	+ 24 01 04.			NONSTELLAR OBJECT
ARC 1903	14 20 38.	+ 27 36		17.0	RICH CLUSTER OF GALAXIES
MCG+06-32-011	14 20 39.	+ 34 28 30.	45	17.	GALAXY
REIZ 4097	14 20 39.	+ 47 51	36	14.3	GALAXY
ZWG 019.011	14 20 42.	+ 01 12		15.3	GALAXY
ZWG 104.011	14 20 42.	+ 18 30		14.7	GALAXY
UGC 09212	14 20 42.	+ 18 30	78	14.7	GALAXY S0
MCG+04-34-023	14 20 42.	+ 24 01	60	15.3	GALAXY
ZC 1420.7+2735	14 20 42.	+ 27 35	1140		CLUSTER OF GALAXIES
ZWG 192.008	14 20 42.	+ 38 13		15.0	GALAXY
UGC 09213	14 20 42.	+ 38 13	120	15.0	GALAXY Sb
MCG+07-30-005	14 20 42.	+ 39 47	36	15.5	GALAXY
ZC 1420.7+4025	14 20 42.	+ 40 25	4170		CLUSTER OF GALAXIES
ZWG 247.024	14 20 42.	+ 47 50		15.7	GALAXY
REIZ 4094	14 20 45.	+ 38 13	30	14.1	GALAXY
MCG+06-32-013	14 20 45.	+ 38 15	102	14.5	GALAXY
MCG+06-32-012	14 20 45.	+ 38 16	24	16.	GALAXY
MCG+03-37-011	14 20 48.	+ 18 31	18	14.7	GALAXY
ZC 1420.8+2401	14 20 48.	+ 24 01	540		CLUSTER OF GALAXIES
ZWG 163.034	14 20 48.	+ 28 50		15.7	GALAXY
ZWG 192.009	14 20 48.	+ 33 04		14.5	GALAXY
MRK 471	14 20 48.	+ 33 04	24	15.5	GALAXY WITH UV CONTINUUM
KW 40	14 20 48.	+ 33 04	50		SEYFERT GALAXY
UGC 09214	14 20 48.	+ 33 04	54	14.5	GALAXY SBa
MCG+06-32-014	14 20 48.	+ 33 04	48	14.5	GALAXY
REIZ 4090	14 20 49.	+ 20 57	18	14.5	GALAXY
REIZ 4096	14 20 49.	+ 40 32	30	15.0	GALAXY
SC 1418-1619.1	14 20 50.	- 16 32 48.	6		NEBULA
REIZ 4093	14 20 51.	+ 33 04	18	14.3	GALAXY
ZWG 019.012	14 20 54.	+ 01 57		13.6	GALAXY
8ZW 1420+01.9	14 20 54.	+ 01 57		13.6	COMPACT GALAXY
UGC 09215	14 20 54.	+ 01 57	150	13.6	GALAXY SBc
MCG+00-37-002	14 20 54.	+ 01 57	120	13.6	GALAXY
FAI 14209+08	14 20 54.	+ 08 52			COMPACT GALAXY
ZWG 104.012	14 20 54.	+ 15 00		15.5	GALAXY
ZWG 133.046	14 20 54.	+ 26 29		15.1	GALAXY
MCG+06-32-015	14 20 54.	+ 37 45 30.	12	16.	GALAXY
ZWG 220.009	14 20 54.	+ 40 32		15.6	GALAXY
RNGC 5601	14 20 54.	+ 40 32		15.5	GALAXY
ZWG 247.025	14 20 54.	+ 46 46		15.5	GALAXY
MCG+09-24-003	14 20 54.	+ 56 29	60	16.	GALAXY
KARA.68 231	14 20 54.	- 10 52	47		DWARF GALAXY
SC 1418-1619.4	14 20 54.	- 16 33 06.	6		NEBULA
MCG-05-34-011	14 20 54.	- 28 27	72	14.5	GALAXY
IC 4413	14 20 55.	+ 37 46 30.			NONSTELLAR OBJECT
REIZ 4088	14 20 56.	+ 01 58	60	13.5	GALAXY
IC 4412	14 20 56.	+ 26 29 53.			SAME AS NGC 5594
REIZ 4092	14 20 56.	+ 26 30	12	14.9	GALAXY
MCG+06-32-016	14 20 57.	+ 37 22	27	16.	GALAXY
MCG+07-30-006	14 20 57.	+ 40 31	42	15.5	GALAXY
MCG-02-37-002	14 20 57.	- 13 06 30.	60	15.	GALAXY
RNGC 5592	14 20 57.	- 28 27		14.0	GALAXY
SC 1418-1626.2	14 20 58.	- 16 39 54.	12		NEBULA
RNGC 5594	14 20 59.	+ 26 29		15.0	GALAXY
ZC 1421.0+0022	14 21 00.	+ 00 22	1950		CLUSTER OF GALAXIES
FAI 14210+09	14 21 00.	+ 09 11			COMPACT GALAXY
FAI 14210+11	14 21 00.	+ 11 32			COMPACT GALAXY
ZWG 104.013	14 21 00.	+ 14 28		15.3	GALAXY
ZWG 075.024	14 21 00.	+ 14 28		15.3	GALAXY
MCG+03-37-012	14 21 00.	+ 14 30	30	15.3	GALAXY
FEIG 094	14 21 00.	+ 21 03		12.0	FAINT BLUE STAR
MCG+04-34-024	14 21 00.	+ 26 29	60	15.1	GALAXY
ZC 1421.0+3224	14 21 00.	+ 32 24	1410		CLUSTER OF GALAXIES
ZWG 192.010	14 21 00.	+ 38 21		15.6	GALAXY
MCG+06-32-017	14 21 00.	+ 38 23	30	14.5	GALAXY
RNGC 5603B	14 21 00.	+ 40 35		15.0	GALAXY
1ZW 086	14 21 00.	+ 40 36			COMPACT GALAXY
MCG+07-30-007	14 21 00.	+ 40 38	78	14.	GALAXY
ZWG 220.010	14 21 00.	+ 40 39		15.1	GALAXY
UGC 09216	14 21 00.	+ 40 39	120	15.1	GALAXY Sc
ZC 1421.0+5846	14 21 00.	+ 58 46	1080		CLUSTER OF GALAXIES
MCG+10-21-001	14 21 00.	+ 60 10	15	16.	GALAXY
ZC 1421.0+6704	14 21 00.	+ 67 04	610		CLUSTER OF GALAXIES
ZWG 366.008	14 21 00.	+ 83 39		15.6	GALAXY
ZWG 365.017	14 21 00.	+ 83 39		15.6	GALAXY
MCG+14-07-005	14 21 00.	+ 83 40	27	16.	GALAXY
ZWG 019.013	14 21 00.	- 02 17		15.6	GALAXY
MCG-01-37-005	14 21 00.	- 05 47	72	15.	GALAXY
SC 1418-1737.0	14 21 00.	- 17 50 42.			NEBULA
REIZ 4100	14 21 01.	+ 40 36	24	14.2	GALAXY
HOLM 641B	14 21 02.	+ 40 40	66	14.4	PART OF MULTIPLE GALAXY
REIZ 4099	14 21 03.	+ 38 21	42	14.7	GALAXY
MCG+07-30-008	14 21 03.	+ 40 35	66	14.	GALAXY
BC M1421+12.2	14 21 04.69	+ 12 13 26.7		18.	QUASI-STELLAR OBJECT
SHB 287	14 21 04.7	+ 12 13 26.		17.	QUASI-STELLAR OBJECT
SC 1418-1622.0	14 21 05.	- 16 35 41.	12		NEBULA
FAI 14211+09	14 21 06.	+ 09 41			COMPACT GALAXY
FAI 14211+14	14 21 06.	+ 14 31			COMPACT GALAXY
TON-N 0768	14 21 06.	+ 24 39		15.1	BLUE STAR
ZWG 220.011	14 21 06.	+ 40 36		14.0	GALAXY
RNGC 5603A	14 21 06.	+ 40 36		14.0	GALAXY
UGC 09217	14 21 06.	+ 40 36	84	14.0	GALAXY S0
ZC 1421.1+5113	14 21 06.	+ 51 13	870		CLUSTER OF GALAXIES
ZWG 296.002	14 21 06.	+ 61 55		15.5	GALAXY
ZWG 295.047	14 21 06.	+ 61 55		15.5	GALAXY
ZC 1421.1+7241	14 21 06.	+ 72 41	870		CLUSTER OF GALAXIES
HOLM 641A	14 21 07.	+ 40 37	48	13.9	PART OF MULTIPLE GALAXY
ZWG 019.014	14 21 12.	+ 02 18		15.7	GALAXY
8ZW 1421+06.8	14 21 12.	+ 06 50		17.3	COMPACT GALAXY
FAI 14212+13	14 21 12.	+ 13 01			COMPACT GALAXY
ZC 1421.2+1641	14 21 12.	+ 16 41	610		CLUSTER OF GALAXIES
ZC 1421.2+2944	14 21 12.	+ 29 44	4230		CLUSTER OF GALAXIES
ZWG 296.003	14 21 12.	+ 60 10		15.3	GALAXY
ZWG 295.048	14 21 12.	+ 60 10		15.3	GALAXY
SC 1418-1608.3	14 21 13.	- 16 21 59.	12		NEBULA
SHB 288	14 21 14.6	- 38 13 26.		18.	QUASI-STELLAR OBJECT
RNGC 5599	14 21 15.	+ 06 48		14.5	GALAXY
HOLM 639C	14 21 17.	+ 06 49	18	15.4	PART OF MULTIPLE GALAXY
HOLM 639B	14 21 17.	+ 06 50	18	15.2	PART OF MULTIPLE GALAXY
ZWG 047.030	14 21 18.	+ 06 48		14.7	GALAXY
UGC 09218	14 21 18.	+ 06 48	90	14.7	GALAXY Sb
MCG+01-37-010	14 21 18.	+ 06 48	84	14.7	GALAXY
ZWG 075.025	14 21 18.	+ 09 28		15.4	GALAXY
FAI 14213+10	14 21 18.	+ 10 05			COMPACT GALAXY
FAI 14213+11	14 21 18.	+ 11 34			COMPACT GALAXY
FAI 14213+12	14 21 18.	+ 12 41			COMPACT GALAXY
ARC 1905	14 21 18.	+ 16 43		18.0	RICH CLUSTER OF GALAXIES
ZWG 104.014	14 21 18.	+ 17 14		15.7	GALAXY
MRK 679	14 21 18.	+ 33 05	6	17.	GALAXY WITH UV CONTINUUM
ZWG 220.012	14 21 18.	+ 42 00		14.3	GALAXY
UGC 09219	14 21 18.	+ 42 00	180	14.3	GALAXY IRR
KARA.73B 0627	14 21 18.	+ 42 00	156	14.3	ISOLATED GALAXY S
RNGC 5608	14 21 19.	+ 42 00		14.5	GALAXY
MCG+07-30-009	14 21 21.	+ 42 00	144	13.5	GALAXY
REIZ 4098	14 21 23.	+ 06 48	72	13.8	GALAXY
HOLM 639A	14 21 23.	+ 06 48	42	13.7	PART OF MULTIPLE GALAXY
ARC 1906	14 21 23.	+ 17 40		16.6	RICH CLUSTER OF GALAXIES
REIZ 4102	14 21 23.	+ 42 00	120	13.8	GALAXY
FAI 14214+11	14 21 24.	+ 11 26			COMPACT GALAXY
FAI 14214+12	14 21 24.	+ 12 59			COMPACT GALAXY
MCG+03-37-013	14 21 24.	+ 14 51	66	11.9	GALAXY
RNGC 5600	14 21 24.	+ 14 52		13.0	GALAXY
ZC 1421.4+1846	14 21 24.	+ 18 46	1210		CLUSTER OF GALAXIES
ZWG 133.047	14 21 24.	+ 26 03		15.6	GALAXY
ZWG 163.035	14 21 24.	+ 28 35		15.2	GALAXY
ZWG 337.009	14 21 24.	+ 72 05		15.6	GALAXY
MCG-03-37-001	14 21 24.	- 16 30	108	12.5	GALAXY
MCG-05-34-012	14 21 24.	- 28 24	36	16.	GALAXY
HOLM 640B	14 21 25.	+ 20 54	18	15.3	PART OF MULTIPLE GALAXY
HOLM 640A	14 21 25.	+ 20 54	18	15.2	PART OF MULTIPLE GALAXY
ZWG 104.015	14 21 30.	+ 14 52		11.9	GALAXY
8ZW 1421+14.9	14 21 30.	+ 14 52		11.9	COMPACT GALAXY
UGC 09220	14 21 30.	+ 14 52	84	11.9	GALAXY S
ZC 1421.5+2503	14 21 30.	+ 25 03	2350		CLUSTER OF GALAXIES
TON-N 0197	14 21 30.	+ 31 50		14.6	BLUE STAR
REIZ 4107	14 21 30.	+ 53 16	72	14.8	GALAXY
REIZ 4101	14 21 31.	+ 28 34	30	14.0	GALAXY
RNGC 5595	14 21 31.	- 16 30		13.0	GALAXY
IC 4414	14 21 32.	+ 28 34 20.			NONSTELLAR OBJECT
MCG+05-34-027	14 21 33.	+ 28 33	30	15.2	GALAXY
SC 1418-1555.0	14 21 33.	- 16 08 40.	54		NEBULA
HOLM 638A	14 21 33.	- 16 30	120	12.0	PART OF MULTIPLE GALAXY
ZWG 047.031	14 21 36.	+ 06 25		15.4	GALAXY
ZWG 047.032	14 21 36.	+ 07 59		15.6	GALAXY
8ZW 1421+08.0	14 21 36.	+ 07 59		15.6	COMPACT GALAXY
FAI 14216+12	14 21 36.	+ 12 42			COMPACT GALAXY
ZWG 192.011	14 21 36.	+ 34 14		14.4	GALAXY
UGC 09221	14 21 36.	+ 34 14	54	14.4	GALAXY S
KARA.72 425A	14 21 36.	+ 34 14	54	14.4	PART OF DOUBLE GALAXY
MCG+06-32-018	14 21 36.	+ 34 14	42	14.5	GALAXY
ZWG 019.015	14 21 36.	- 00 27		15.7	GALAXY
SC 1418-1604.3	14 21 36.	- 16 17 58.	42		NEBULA
MCG-03-37-002	14 21 36.	- 16 32	96	13.	GALAXY
ARC 1908	14 21 39.	+ 26 40		17.1	RICH CLUSTER OF GALAXIES
ZWG 075.026	14 21 42.	+ 12 28		15.6	GALAXY
FAI 14217+12	14 21 42.	+ 12 29			COMPACT GALAXY
ZC 1421.7+2642	14 21 42.	+ 26 42	940		CLUSTER OF GALAXIES
ZWG 163.036	14 21 42.	+ 29 13		15.5	GALAXY
MCG+06-32-019	14 21 42.	+ 34 14 30.	42	14.5	GALAXY
ZWG 192.012	14 21 42.	+ 34 15		15.1	GALAXY
UGC 09222	14 21 42.	+ 34 15	60	15.1	GALAXY S
KARA.72 425B	14 21 42.	+ 34 15	60	15.1	PART OF DOUBLE GALAXY
ZWG 220.013	14 21 42.	+ 40 15		14.7	GALAXY
UGC 09223	14 21 42.	+ 40 15	66	14.7	GALAXY S0
RNGC 5620	14 21 42.	+ 69 48		15.0	GALAXY
MCG+12-14-002	14 21 42.	+ 69 48	18	15.	GALAXY
REIZ 4104	14 21 43.	+ 35 04	42	13.2	GALAXY
RNGC 5597	14 21 43.	- 16 33		13.0	GALAXY
RNGC 5609	14 21 45.	+ 35 04			GALAXY
ARC 1909	14 21 47.	+ 25 10		16.8	RICH CLUSTER OF GALAXIES
HOLM 638B	14 21 47.	- 16 33	96	13.7	PART OF MULTIPLE GALAXY
TON-N 0769	14 21 48.	+ 29 04		15.8	BLUE STAR
MCG+05-34-028	14 21 48.	+ 29 12	24	15.5	GALAXY
UGC 09224	14 21 48.	+ 34 57	60	17.	GALAXY Sc
MCG+07-30-010	14 21 48.	+ 40 14	36	14.5	GALAXY
ZWG 337.010	14 21 48.	+ 69 50		15.1	GALAXY
ZC 1421.8-0127	14 21 48.	- 01 27	1810		CLUSTER OF GALAXIES
SC 1419-1348.9	14 21 48.	- 14 02 33.	36		NEBULA
SC 1419-1439.2	14 21 53.	- 14 52 51.	12		NEBULA
ZWG 075.027	14 21 54.	+ 08 30		15.4	GALAXY
ZWG 047.033	14 21 54.	+ 08 30		15.4	GALAXY
UGC 09225	14 21 54.	+ 08 30	66	15.4	GALAXY
ZWG 133.048	14 21 54.	+ 24 08		15.6	GALAXY
KARA.73B 0628	14 21 54.	+ 24 08	36	15.6	ISOLATED GALAXY S
ZWG 163.037	14 21 54.	+ 26 52		15.7	GALAXY
ZWG 163.038	14 21 54.	+ 27 56		15.4	GALAXY
REIZ 4106	14 21 56.	+ 33 16	42	13.5	GALAXY
MCG+05-34-029	14 21 57.	+ 26 50	42	15.7	GALAXY
VDB.66G 190	14 22	+ 44 44	70		DWARF GALAXY
VDB.66G 191	14 22	+ 56 29	100		DWARF GALAXY
FAI 14220+09	14 22 00.	+ 09 27			COMPACT GALAXY
ZC 1422.0+1732	14 22 00.	+ 17 32	7800		CLUSTER OF GALAXIES
ZC 1422.0+2710	14 22 00.	+ 27 10	2820		CLUSTER OF GALAXIES
MCG+05-34-030	14 22 00.	+ 27 55	54	15.4	GALAXY
ZWG 163.039	14 22 00.	+ 29 52		15.7	GALAXY
ZWG 192.013	14 22 00.	+ 33 15		13.5	GALAXY
UGC 09227	14 22 00.	+ 33 15	90	13.5	GALAXY S0
MCG+06-32-020	14 22 00.	+ 33 16	45	14.	GALAXY
VV 077A	14 22 00.	+ 35 03	120	12.5	INTERACTING GALAXY
VV 077C	14 22 00.	+ 35 03 30.	24	15.	INTERACTING GALAXY
VV 077B	14 22 00.	+ 35 03 30.	66	15.	INTERACTING GALAXY
ZWG 192.014	14 22 00.	+ 35 05		12.6	GALAXY
UGC 09226	14 22 00.	+ 35 05	168	12.6	GALAXY Sa

OBJECT NAME	RIGHT ASCEN.	DECLINATION	DIAM.	MAGN.	TYPE OF OBJECT
MCG+06-32-022	14 22 00.	+ 35 05	120	12.	GALAXY
MCG+06-32-023	14 22 00.	+ 35 06	12	15.	GALAXY
UGC 09228	14 22 00.	+ 35 07	60	16.0	GALAXY S0-a
MCG+06-32-021	14 22 00.	+ 35 07	24	15.	GALAXY
ZC 1422.0+4911	14 22 00.	+ 49 11	1480		CLUSTER OF GALAXIES
MCG+09-24-004	14 22 00.	+ 54 10	48	15.	GALAXY
ZC 1422.0+6126	14 22 00.	+ 61 26	1480		CLUSTER OF GALAXIES
ZWG 353.035	14 22 00.	+ 80 20		15.4	GALAXY
RNGC 5640	14 22 00.	+ 80 20		15.5	GALAXY
HN 0404	14 22 00.	- 34 48			NEBULA
IC 4411	14 22 00.	- 34 48			NONSTELLAR OBJECT
RNGC 5611	14 22 02.	+ 33 15		13.5	GALAXY
IC 4415	14 22 03.	+ 16 51			NONSTELLAR OBJECT
ARP 178	14 22 03.	+ 35 04			PECULIAR GALAXY
RNGC 5614	14 22 03.	+ 35 05		13.0	GALAXY
RNGC 5615	14 22 03.	+ 35 06		15.0	GALAXY
RNGC 5613	14 22 03.	+ 35 07		15.0	GALAXY
HN 1101	14 22 04.	+ 16 51	12		NEBULA
RNGC 5610	14 22 04.	+ 24 50		14.5	GALAXY
ARC 1910	14 22 05.	+ 25 27		17.8	RICH CLUSTER OF GALAXIES
IC 4416	14 22 05.	+ 29 51 29.			NONSTELLAR OBJECT
ZWG 019.017	14 22 06.	+ 01 24		14.9	GALAXY
UGC 09229	14 22 06.	+ 01 24	60	14.9	GALAXY Sb
MCG+00-37-004	14 22 06.	+ 01 24	48	14.9	GALAXY
ZWG 047.034	14 22 06.	+ 04 46		15.0	GALAXY
8ZW 1422+04.8	14 22 06.	+ 04 46		15.0	COMPACT GALAXY
FAI 14221+11	14 22 06.	+ 11 16			COMPACT GALAXY
FAI 14221+13	14 22 06.	+ 13 54			COMPACT GALAXY
FAI 14221+14	14 22 06.	+ 14 01			COMPACT GALAXY
ZWG 133.049	14 22 06.	+ 24 50		14.5	GALAXY
UGC 09230	14 22 06.	+ 24 50	132	14.5	GALAXY SBa
MCG+04-34-025	14 22 06.	+ 24 50	114	14.5	GALAXY
ZC 1422.1+2526	14 22 06.	+ 25 26	670		CLUSTER OF GALAXIES
ZWG 163.040	14 22 06.	+ 26 55		15.7	GALAXY
ZWG 019.016	14 22 06.	- 02 59		13.8	GALAXY
RNGC 5604	14 22 06.	- 02 59		14.0	GALAXY
REIZ 4103	14 22 06.	- 02 59	66	13.4	GALAXY
MCG+00-37-003	14 22 06.	- 02 59	90	13.8	GALAXY
IC 1007	14 22 07.	+ 04 46 34.			NONSTELLAR OBJECT
MCG+06-32-024	14 22 09.	+ 35 22	36	15.	GALAXY
REIZ 4105	14 22 10.	+ 16 49	18	15.6	GALAXY
RNGC 5616	14 22 10.	+ 36 40		15.0	GALAXY
REIZ 4108	14 22 11.	+ 36 41	108	13.7	GALAXY
MCG+05-34-031	14 22 12.	+ 29 51	18	15.7	GALAXY
ZWG 192.015	14 22 12.	+ 36 40		14.8	GALAXY
UGC 09231	14 22 12.	+ 36 40	150	14.8	GALAXY Sb-c
8ZW 1422-02.0	14 22 12.	- 01 59		17.8	COMPACT GALAXY
MCG+06-32-025	14 22 15.	+ 35 28	24	16.	GALAXY
RNGC 5605	14 22 15.	- 12 57		13.0	GALAXY
SC 1419-1706.0	14 22 16.	- 17 19 38.	30		NEBULA
MCG+05-34-032	14 22 18.	+ 26 53	48	15.7	GALAXY
UGC 09232	14 22 18.	+ 33 10	66	16.5	GALAXY
MCG+06-32-027	14 22 18.	+ 33 10	48	16.	GALAXY
MCG+06-32-026	14 22 18.	+ 36 41	114	14.	GALAXY
UGC 10832	14 22 18.	+ 62 13	78	14.6	GALAXY SBc
ZWG 353.036	14 22 18.	+ 75 03		15.6	GALAXY
ARC 1911	14 22 22.	+ 39 11		17.2	RICH CLUSTER OF GALAXIES
RNGC 5593	14 22 22.	- 54 35			OPEN CLUSTER
HOLM 642B	14 22 23.	+ 19 05	18	15.2	PART OF MULTIPLE GALAXY
FAI 14224+13	14 22 24.	+ 13 54			COMPACT GALAXY
ZWG 104.016	14 22 24.	+ 16 23		15.6	GALAXY
ZWG 163.041	14 22 24.	+ 26 51		15.6	GALAXY
REIZ 4110	14 22 24.	+ 28 55	24	15.3	GALAXY
ZWG 019.018	14 22 24.	- 02 51		15.4	GALAXY
8ZW 1422-02.8	14 22 24.	- 02 51		15.4	COMPACT GALAXY
MCG-02-37-003	14 22 24.	- 12 56	84	12.5	GALAXY
OCL 0926	14 22 24.	- 54 35	480		OPEN STAR CLUSTER
HOLM 642A	14 22 26.	+ 19 05	18	15.2	PART OF MULTIPLE GALAXY
HN 1102	14 22 28.	+ 17 16	12		NEBULA
8ZW 1422+03.9	14 22 30.	+ 03 57		18.1	COMPACT GALAXY
ZWG 104.017	14 22 30.	+ 17 15		15.5	GALAXY
MCG+05-34-033	14 22 30.	+ 26 50	27	15.6	GALAXY
ZWG 163.042	14 22 30.	+ 29 41		15.6	GALAXY
ZWG 192.016	14 22 30.	+ 35 30		15.4	GALAXY
UGC 09233	14 22 30.	+ 35 30	78	15.4	GALAXY S
MCG+03-37-014	14 22 33.	+ 17 15 30.	12	15.5	GALAXY
MCG+06-32-028	14 22 33.	+ 35 30	66	14.5	GALAXY
SC 1419-1428.6	14 22 33.	- 14 42 13.	12		NEBULA
REIZ 4109	14 22 34.	+ 17 15	30	15.0	GALAXY
SC 1419-1430.7	14 22 35.	- 14 44 19.	12		NEBULA
ZWG 019.020	14 22 36.	+ 02 14		15.6	GALAXY
ZC 1422.6+2257	14 22 36.	+ 22 57	1210		CLUSTER OF GALAXIES
ZWG 133.050	14 22 36.	+ 26 22		15.7	GALAXY
UGC 09234	14 22 36.	+ 26 22	126	15.7	GALAXY S
TON-N 0198	14 22 36.	+ 27 05		14.9	BLUE STAR
ZWG 163.043	14 22 36.	+ 28 07		15.6	GALAXY
ZWG 192.017	14 22 36.	+ 35 29		15.4	GALAXY
UGC 09235	14 22 36.	+ 35 29	66	15.4	GALAXY SBa
ZC 1422.6+3908	14 22 36.	+ 39 08	1880		CLUSTER OF GALAXIES
ZWG 019.019	14 22 36.	- 00 54		15.6	GALAXY
SHB 289	14 22 37.5	+ 20 13 49.		17.9	QUASI-STELLAR OBJECT
BC PKS1422+20	14 22 38.	+ 20 14 04.		17.65	QUASI-STELLAR OBJECT
ZC 1422.7+0040	14 22 42.	+ 00 40	1410		CLUSTER OF GALAXIES
FAI 14227+09	14 22 42.	+ 09 07			COMPACT GALAXY
ZWG 133.051	14 22 42.	+ 25 15		15.4	GALAXY
UGC 09236	14 22 42.	+ 25 15	84	15.4	GALAXY SB?b-c
MCG+04-34-026	14 22 42.	+ 25 16	66	15.4	GALAXY
ZWG 163.044	14 22 42.	+ 28 03		15.5	GALAXY
UGC 09237	14 22 42.	+ 28 03	60	15.5	GALAXY S
MCG+05-34-034	14 22 42.	+ 28 05	54	15.6	GALAXY
REIZ 4111	14 22 42.	+ 29 31	18	14.9	GALAXY
ZC 1422.7+3241	14 22 42.	+ 32 41	870		CLUSTER OF GALAXIES
MCG+06-32-029	14 22 42.	+ 35 29 30.	48	15.	GALAXY
UGC 09238	14 22 42.	+ 35 30	78	16.0	GALAXY Sc
MCG+08-26-030	14 22 42.	+ 44 42 30.	60	14.	GALAXY
SC 1419-1430.0	14 22 42.	- 14 43 37.	18		NEBULA
HELW 031	14 22 44.	- 14 44			NEBULA
ZWG 075.028	14 22 48.	+ 09 02		15.6	GALAXY
ZC 1422.8+1220	14 22 48.	+ 12 20	870		CLUSTER OF GALAXIES
ZWG 075.029	14 22 48.	+ 13 58		15.3	GALAXY
UGC 09239	14 22 48.	+ 13 58	84	15.3	GALAXY Sb-c
ZWG 163.045	14 22 48.	+ 27 59		15.6	GALAXY
MCG+05-34-035	14 22 48.	+ 28 01	45	15.5	GALAXY
MCG+06-32-030	14 22 48.	+ 35 30 30.	57	15.	GALAXY
ZWG 220.014	14 22 48.	+ 38 28		15.2	GALAXY
ZWG 192.018	14 22 48.	+ 38 28		15.2	GALAXY
1ZW 087	14 22 48.	+ 44 45			COMPACT GALAXY
ZWG 247.026	14 22 48.	+ 44 45		13.9	GALAXY
UGC 09240	14 22 48.	+ 44 45	120	13.9	GALAXY IRR
IC 1008	14 22 50.	+ 28 33 38.			NONSTELLAR OBJECT
MCG+05-34-036	14 22 51.	+ 27 57	42	15.6	GALAXY
REIZ 4113	14 22 51.	+ 38 28	15	14.6	GALAXY
MCG+06-32-031	14 22 54.	+ 38 29	48	14.5	GALAXY
ZC 1422.9+4600	14 22 54.	+ 46 00	1410		CLUSTER OF GALAXIES
ZC 1422.9+6636	14 22 54.	+ 66 36	1480		CLUSTER OF GALAXIES
REIZ 4115	14 22 55.	+ 44 45	66	13.4	GALAXY
FAI 14230+09.2	14 23 00.	+ 09 14			COMPACT GALAXY
FAI 14230+09.8	14 23 00.	+ 09 48			COMPACT GALAXY
ZWG 104.018	14 23 00.	+ 17 30		15.5	GALAXY
8ZW 1423+18.2	14 23 00.	+ 18 13		16.3	COMPACT GALAXY
ZC 1423.0+2213	14 23 00.	+ 22 13	4030		CLUSTER OF GALAXIES
MCG+05-34-037	14 23 00.	+ 28 04	48	15.7	GALAXY
ZWG 163.046	14 23 00.	+ 28 05		15.7	GALAXY
ZWG 337.011	14 23 00.	+ 68 35		15.6	GALAXY
MCG+12-14-003	14 23 00.	+ 68 35	33	16.	GALAXY
KARA.73F 0629	14 23 00.	+ 68 35	42	15.6	ISOLATED GALAXY S
ZWG 019.022	14 23 00.	- 00 54		15.4	GALAXY
ZWG 019.021	14 23 00.	- 02 48		15.3	GALAXY
REIZ 4112	14 23 02.	+ 19 41	18	15.2	GALAXY
IC 4418	14 23 03.	+ 25 44 57.			NONSTELLAR OBJECT
MCG-01-37-006	14 23 03.	- 05 11	66	14.	GALAXY
ZWG 019.023	14 23 06.	+ 01 49		15.6	GALAXY
8ZW 1423+03.2	14 23 06.	+ 03 13		16.5	COMPACT GALAXY
FAI 14231+11	14 23 06.	+ 11 24			COMPACT GALAXY
ZWG 163.047	14 23 06.	+ 26 42		15.5	GALAXY
ZWG 273.005	14 23 06.	+ 55 55		15.5	GALAXY
MCG-04-34-015	14 23 06.	- 23 45	72	15.5	GALAXY
MCG+06-32-032	14 23 09.	+ 36 46	42	16.	GALAXY
FAI 14232+11	14 23 12.	+ 11 57			COMPACT GALAXY
FAI 14232+12	14 23 12.	+ 12 42			COMPACT GALAXY
FAI 14232+14	14 23 12.	+ 14 03			COMPACT GALAXY
ZC 1423.2+2214	14 23 12.	+ 22 14	340		CLUSTER OF GALAXIES
ZC 1423.2+2245	14 23 12.	+ 22 45	610		CLUSTER OF GALAXIES
ZWG 133.052	14 23 12.	+ 25 45		15.1	GALAXY
ZWG 192.019	14 23 12.	+ 32 42		14.2	GALAXY
UGC 09241	14 23 12.	+ 32 42	33	14.2	GALAXY PECULR
MCG+06-32-033	14 23 15.	+ 32 41 30.	24	14.5	GALAXY
ZC 1423.3+2135	14 23 18.	+ 21 35	1340		CLUSTER OF GALAXIES
ZWG 220.015	14 23 18.	+ 39 45		14.8	GALAXY
UGC 09242	14 23 18.	+ 39 45	324	14.8	GALAXY Sc
ZWG 019.024	14 23 18.	- 02 07		15.7	GALAXY
REIZ 4114	14 23 19.	+ 21 08	30	15.2	GALAXY
HOLM 643A	14 23 19.	+ 39 45	300	13.0	PART OF MULTIPLE GALAXY
ZWG 047.035	14 23 24.	+ 03 05		15.3	GALAXY
ZWG 047.036	14 23 24.	+ 04 52		15.2	GALAXY
8ZW 1423+05.2	14 23 24.	+ 05 10		16.6	COMPACT GALAXY
ZWG 047.037	14 23 24.	+ 06 40		15.2	GALAXY
FAI 14234+09.4	14 23 24.	+ 09 27			COMPACT GALAXY
FAI 14234+09.6	14 23 24.	+ 09 39			COMPACT GALAXY
FAI 14234+10.2	14 23 24.	+ 10 13			COMPACT GALAXY
FAI 14234+10.8	14 23 24.	+ 10 51			COMPACT GALAXY
FAI 14234+13	14 23 24.	+ 13 52			COMPACT GALAXY
ZC 1423.4+2319	14 23 24.	+ 23 19	610		CLUSTER OF GALAXIES
MCG+04-34-027	14 23 24.	+ 25 35	42	15.5	GALAXY
ZWG 133.053	14 23 24.	+ 25 36		15.5	GALAXY
IC 4420	14 23 24.	+ 25 36 10.			NONSTELLAR OBJECT
ZC 1423.4+2700	14 23 24.	+ 27 00	1010		CLUSTER OF GALAXIES
MCG+05-34-038	14 23 24.	+ 28 28	24		GALAXY
ZWG 192.020	14 23 24.	+ 34 04		15.6	GALAXY
UGC 09243	14 23 24.	+ 34 04	90	15.6	GALAXY Sc
MCG+07-30-011	14 23 24.	+ 39 44	288	13.5	GALAXY
1ZW 088	14 23 24.	+ 39 46			COMPACT GALAXY
SHAH 014	14 23 24.	+ 47 28	54	16.	GROUP OF COMPACT GALAXIES
ARC 1912	14 23 26.	+ 26 58		17.0	RICH CLUSTER OF GALAXIES
IC 4419	14 23 27.	+ 16 51			NONSTELLAR OBJECT
HN 1103	14 23 28.	+ 16 51	12		NEBULA
FAI 14235+08	14 23 30.	+ 08 44			COMPACT GALAXY
MCG+06-32-034	14 23 30.	+ 34 05	90	14.5	GALAXY
SC 1420-1421.8	14 23 31.	- 14 35 23.	18		NEBULA
HOLM 643B	14 23 32.	+ 39 48	24	15.1	PART OF MULTIPLE GALAXY
HELW 032	14 23 32.	- 14 30			NEBULA
BC 4C24.31	14 23 34.6	+ 24 17 28.		17.2	QUASI-STELLAR OBJECT
8ZW 1423+01.1	14 23 36.	+ 01 05		15.6	COMPACT GALAXY
ZWG 047.038	14 23 36.	+ 05 27		14.8	GALAXY
UGC 09244	14 23 36.	+ 05 27	78	14.8	GALAXY SBb
MCG+01-37-011	14 23 36.	+ 05 27	84	14.8	GALAXY
ZWG 047.039	14 23 36.	+ 08 20		15.3	GALAXY
FAI 14236+12	14 23 36.	+ 12 48			COMPACT GALAXY
FAI 14236+13	14 23 36.	+ 13 55			COMPACT GALAXY
MCG-02-37-004	14 23 36.	- 11 40	30	15.	GALAXY
SHB 290	14 23 37.	+ 24 17 28.		18.	QUASI-STELLAR OBJECT
FAI 14237+08	14 23 42.	+ 08 58			COMPACT GALAXY
FAI 14237+10	14 23 42.	+ 10 19			COMPACT GALAXY
FAI 14237+12	14 23 42.	+ 12 47			COMPACT GALAXY
FAI 14237+14	14 23 42.	+ 14 20			COMPACT GALAXY
ZWG 104.019	14 23 42.	+ 16 45		15.5	GALAXY
TON-N 0770	14 23 42.	+ 26 19		14.6	BLUE STAR
ZWG 163.048	14 23 42.	+ 30 42		15.3	GALAXY
MCG-02-37-005	14 23 42.	- 14 20	60	16.	GALAXY
HOLM 644A	14 23 46.	+ 30 41	12	14.8	PART OF MULTIPLE GALAXY
HOLM 644B	14 23 47.	+ 30 41	9	15.1	PART OF MULTIPLE GALAXY
FAI 14238+10	14 23 48.	+ 10 23			COMPACT GALAXY
FAI 14238+11	14 23 48.	+ 11 08			COMPACT GALAXY
ZWG 104.020	14 23 48.	+ 19 37		15.4	GALAXY
ZC 1423.8+2550	14 23 48.	+ 25 50	740		CLUSTER OF GALAXIES
MRK 680	14 23 48.	+ 28 31	9	16.5	GALAXY WITH UV CONTINUUM
ZWG 296.004	14 23 48.	+ 56 33		15.4	GALAXY
UGC 09245	14 23 48.	+ 56 33	126	15.4	GALAXY SBc
MCG+09-24-005	14 23 48.	+ 56 33	120	15.	GALAXY
IC 4422	14 23 49.	+ 30 41 49.			NONSTELLAR OBJECT
REIZ 4116	14 23 50.	+ 19 36	24	14.8	GALAXY
SC 1421-1533.4	14 23 50.	- 15 46 58.	12		NEBULA
MCG+05-34-039	14 23 51.	+ 30 41 30.	12	15.3	GALAXY
ARC 1918	14 23 53.	+ 63 23		17.5	RICH CLUSTER OF GALAXIES
ZWG 075.030	14 23 54.	+ 12 34		15.3	GALAXY
IC 1009	14 23 54.	+ 12 35 00.			NONSTELLAR OBJECT
ZWG 104.021	14 23 54.	+ 17 01		15.7	GALAXY
MCG+03-37-015	14 23 54.	+ 19 37	30	15.4	GALAXY
ZC 1423.9+3804	14 23 54.	+ 38 04	2150		CLUSTER OF GALAXIES
ZWG 047.040	14 24 00.	+ 02 58		15.3	GALAXY
ZWG 047.041	14 24 00.	+ 06 12		15.1	GALAXY
UGC 09246	14 24 00.	+ 06 12	78	15.1	GALAXY S
ZC 1424.0+1043	14 24 00.	+ 10 43	1680		CLUSTER OF GALAXIES
ZWG 075.031	14 24 00.	+ 11 23		15.5	GALAXY
MCG+02-37-007	14 24 00.	+ 11 23	30	15.5	GALAXY
FAI 14240+11	14 24 00.	+ 11 25			COMPACT GALAXY

OBJECT NAME	RIGHT ASCEN.	DECLINATION	DIAM.	MAGN.	TYPE OF OBJECT
ZWG 104.022	14 24 00.	+ 17 01		15.6	GALAXY
MRK 681	14 24 00.	+ 23 09	10	16.5	GALAXY WITH UV CONTINUUM
ZWG 133.054	14 24 00.	+ 25 14		15.6	GALAXY
ZC 1424.0+2613	14 24 00.	+ 26 13	12230		CLUSTER OF GALAXIES
ARC 1914	14 24 00.	+ 38 03		17.2	RICH CLUSTER OF GALAXIES
ZWG 220.016	14 24 00.	+ 39 00		15.5	GALAXY
IC 4423	14 24 03.	+ 26 27 43.			NONSTELLAR OBJECT
MCG+07-30-012	14 24 03.	+ 38 59	51	15.	GALAXY
RNGC 5606	14 24 04.	- 59 25		10.0	OPEN CLUSTER
FAI 14241+13	14 24 06.	+ 13 07			COMPACT GALAXY
ZWG 104.023	14 24 06.	+ 17 02		15.7	GALAXY
ZC 1424.1+2104	14 24 06.	+ 21 04	540		CLUSTER OF GALAXIES
ZC 1424.1+2224	14 24 06.	+ 22 24	610		CLUSTER OF GALAXIES
ZWG 133.055	14 24 06.	+ 26 28		15.5	GALAXY
UGC 09247	14 24 06.	+ 26 28	66	15.5	GALAXY S
ZWG 019.025	14 24 06.	- 03 12		15.4	GALAXY
OCL 0922	14 24 06.	- 59 25	600	10.0	OPEN STAR CLUSTER
VHA 158	14 24 06.	- 59 25	150		OPEN STAR CLUSTER
MCG+04-34-028	14 24 09.	+ 26 28	66	15.5	GALAXY
SC 1421-1444.4	14 24 09.	- 14 57 57.	6		NEBULA
ZC 1424.2+0106	14 24 12.	+ 01 06	1550		CLUSTER OF GALAXIES
FAI 14242+10	14 24 12.	+ 10 25			COMPACT GALAXY
ZWG 075.032	14 24 12.	+ 11 37		15.5	GALAXY
ZWG 104.024	14 24 12.	+ 15 37		15.3	GALAXY
ZWG 104.025	14 24 12.	+ 16 58		15.6	GALAXY
ZWG 104.026	14 24 12.	+ 17 02		15.7	GALAXY
ZWG 133.056	14 24 12.	+ 21 47		15.6	GALAXY
ZWG 133.057	14 24 12.	+ 22 03		15.4	GALAXY
8ZW 1424+04.8	14 24 18.	+ 04 47		17.1	COMPACT GALAXY
ZWG 104.027	14 24 18.	+ 16 55		15.7	GALAXY
ZC 1424.3+3227	14 24 18.	+ 32 27	1010		CLUSTER OF GALAXIES
UGC 09851	14 24 18.	+ 41 51	66	13.0	GALAXY E-SO
MCG+08-26-031	14 24 18.	+ 49 44	54	15.	GALAXY
ZWG 247.027	14 24 18.	+ 49 45		15.7	GALAXY
REIZ 4122	14 24 19.	+ 48 46	54	14.3	GALAXY
REIZ 4117	14 24 22.	+ 16 54	60	15.7	GALAXY
FAI 14244+09	14 24 24.	+ 09 25			COMPACT GALAXY
ZC 1424.4+2001	14 24 24.	+ 20 01	940		CLUSTER OF GALAXIES
TON-N 0771	14 24 24.	+ 26 52		15.2	BLUE STAR
TON-N 0772	14 24 24.	+ 28 26		14.7	BLUE STAR
ZWG 247.028	14 24 24.	+ 48 47		14.2	GALAXY
RNGC 5622	14 24 24.	+ 48 47		14.0	GALAXY
UGC 09248	14 24 24.	+ 48 47	108	14.2	GALAXY Sb
KARA.73B 0630	14 24 24.	+ 48 47	102	14.2	ISOLATED GALAXY S
ZC 1424.4+4906	14 24 24.	+ 49 06	1680		CLUSTER OF GALAXIES
ZC 1424.4+5545	14 24 24.	+ 55 45	4500		CLUSTER OF GALAXIES
MCG+13-10-017	14 24 24.	+ 75 23	60	16.	GALAXY
HELW 033	14 24 26.	- 14 44			NEBULA
MCG+08-26-032	14 24 27.	+ 48 46	102	14.3	GALAXY
ARC 1913	14 24 29.	+ 16 54		16.0	RICH CLUSTER OF GALAXIES
ARC 1915	14 24 29.	+ 32 24		17.6	RICH CLUSTER OF GALAXIES
ZWG 075.033	14 24 30.	+ 08 54		15.1	GALAXY
UGC 09249	14 24 30.	+ 08 54	138	15.1	GALAXY
FAI 14245+13	14 24 30.	+ 13 21			COMPACT GALAXY
ZWG 163.049	14 24 30.	+ 27 25		15.3	GALAXY
VV 152A	14 24 30.	+ 45 05	24	16.	INTERACTING GALAXY
VV 152	14 24 30.	+ 45 05	78	16.	INTERACTING GALAXY
ZWG 353.037	14 24 30.	+ 75 24		15.7	GALAXY
ZC 1424.5-0127	14 24 30.	- 01 27	2220		CLUSTER OF GALAXIES
MCG+08-26-033	14 24 33.	+ 45 04	66	16.	GALAXY
SC 1421-1429.9	14 24 33.	- 14 43 26.	12		NEBULA
IC 4425	14 24 34.	+ 27 24 46.			NONSTELLAR OBJECT
ZWG 075.034	14 24 36.	+ 13 07		15.1	GALAXY
FAI 14246+14	14 24 36.	+ 14 28			COMPACT GALAXY
ZC 1424.6+1911	14 24 36.	+ 19 11	1880		CLUSTER OF GALAXIES
ZWG 133.058	14 24 36.	+ 25 15		15.6	GALAXY
ZC 1424.6+2632	14 24 36.	+ 26 32	740		CLUSTER OF GALAXIES
ZWG 247.029	14 24 36.	+ 45 05		15.7	GALAXY
UGC 09251	14 24 36.	+ 45 05	60	15.7	GALAXY Sc
ZC 1424.6-0017	14 24 36.	- 00 17	670		CLUSTER OF GALAXIES
ZWG 019.026	14 24 36.	- 02 03		14.8	GALAXY
RNGC 5618	14 24 36.	- 02 03		15.0	GALAXY
UGC 09250	14 24 36.	- 02 03	108	14.8	GALAXY SBc
MCG+00-37-005	14 24 36.	- 02 03	78	14.8	GALAXY
HOLM 645A	14 24 42.	+ 05 01	150	13.4	PART OF MULTIPLE GALAXY
ZWG 047.042	14 24 42.	+ 05 21		15.7	GALAXY
UGC 09252	14 24 42.	+ 05 21	66	15.7	GALAXY IRR
FAI 14247+13	14 24 42.	+ 13 00			COMPACT GALAXY
ZWG 163.050	14 24 42.	+ 27 06		15.4	GALAXY
ZWG 163.051	14 24 42.	+ 31 45		14.7	GALAXY
UGC 09253	14 24 42.	+ 31 45	114	14.7	GALAXY Sb-c
TON-N 0199	14 24 42.	+ 33 11		14.1	BLUE STAR
REIZ 4118	14 24 45.	+ 01 12	48	14.7	GALAXY
IC 4427	14 24 45.	+ 27 05 17.			NONSTELLAR OBJECT
ZWG 019.028	14 24 48.	+ 01 15		14.8	GALAXY
UGC 09254	14 24 48.	+ 01 15	138	14.8	GALAXY SBb
MCG+00-37-006	14 24 48.	+ 01 15	66	14.8	GALAXY
ZWG 047.043	14 24 48.	+ 03 02		15.4	GALAXY
ZWG 047.044	14 24 48.	+ 05 01		14.0	GALAXY
UGC 09255	14 24 48.	+ 05 01	144	14.0	GALAXY Sb
MCG+01-37-012	14 24 48.	+ 05 01	120	14.0	GALAXY
REIZ 4119	14 24 48.	+ 05 01	132	13.6	GALAXY
ZWG 047.045	14 24 48.	+ 08 07		15.7	GALAXY
ZWG 075.035	14 24 48.	+ 11 34		15.0	GALAXY
MCG+02-37-008	14 24 48.	+ 11 34	48	15.0	GALAXY
TON-N 0200	14 24 48.	+ 24 31		14.7	BLUE STAR
MCG+05-34-040	14 24 48.	+ 26 40	36		GALAXY
MCG+05-34-041	14 24 48.	+ 27 04	42	15.4	GALAXY
TON-N 0201	14 24 48.	+ 27 39		14.9	BLUE STAR
ZWG 273.006	14 24 48.	+ 51 50		14.1	GALAXY
UGC 09256	14 24 48.	+ 51 50	66	14.1	GALAXY S
ZC 1424.8+6331	14 24 48.	+ 63 31	3970		CLUSTER OF GALAXIES
ZWG 019.027	14 24 49.	- 01 28		15.7	GALAXY
IC 1010	14 24 49.	+ 01 15 10.			NONSTELLAR OBJECT
HOLM 646B	14 24 50.	+ 31 45	18	15.8	PART OF MULTIPLE GALAXY
RNGC 5624	14 24 50.	+ 51 50		14.0	GALAXY
RNGC 5619A	14 24 51.	+ 05 01		14.0	GALAXY
REIZ 4124	14 24 51.	+ 31 44	96	13.9	GALAXY
HOLM 646A	14 24 51.	+ 31 44	84	13.6	PART OF MULTIPLE GALAXY
ZWG 047.046	14 24 54.	+ 03 47		15.1	GALAXY
HOLM 645C	14 24 54.	+ 05 00	18	15.3	PART OF MULTIPLE GALAXY
FAI 14249+11	14 24 54.	+ 11 58			COMPACT GALAXY
ZC 1424.9+1244	14 24 54.	+ 12 44	2350		CLUSTER OF GALAXIES
ZWG 104.028	14 24 54.	+ 17 03		15.6	GALAXY
MCG+03-37-016	14 24 54.	+ 17 03	24	15.6	GALAXY
ZC 1424.9+2707	14 24 54.	+ 27 07	1010		CLUSTER OF GALAXIES
ZWG 163.052	14 24 54.	+ 31 10		14.8	GALAXY
UGC 09257	14 24 54.	+ 31 10	78	14.8	GALAXY S

OBJECT NAME	RIGHT ASCEN.	DECLINATION	DIAM.	MAGN.	TYPE OF OBJECT
MCG+05-34-042	14 24 54.	+ 31 44	108	14.7	GALAXY
ZC 1424.9+4941	14 24 54.	+ 49 41	1280		CLUSTER OF GALAXIES
MCG+09-24-006	14 24 54.	+ 51 49	72	14.	GALAXY
ZC 1424.9+7340	14 24 54.	+ 73 40	870		CLUSTER OF GALAXIES
HOLM 645B	14 24 56.	+ 05 02	18	14.3	PART OF MULTIPLE GALAXY
SHB 291	14 24 56.0	- 11 50 25.		19.	QUASI-STELLAR OBJECT
REIZ 4130	14 24 57.	+ 54 11	30	14.5	GALAXY
REIZ 4133	14 24 57.	+ 56 48	30	12.5	GALAXY
HN 1104	14 24 58.	+ 17 02	12		NEBULA
REIZ 4123	14 24 58.	+ 17 02	42	15.1	GALAXY
IC 4426	14 24 58.	+ 17 03			NONSTELLAR OBJECT
ZWG 047.047	14 25 00.	+ 05 00		15.3	GALAXY
UGC 09258	14 25 00.	+ 05 00	60	15.3	GALAXY Sb-c
MCG+01-37-013	14 25 00.	+ 05 00	48	15.3	GALAXY
REIZ 4120	14 25 00.	+ 05 00	15	15.0	GALAXY
ZWG 047.048	14 25 00.	+ 05 02		14.8	GALAXY
REIZ 4121	14 25 00.	+ 05 02	18	15.4	GALAXY
MCG+01-37-014	14 25 00.	+ 05 02	48	14.8	GALAXY
ZWG 075.036	14 25 00.	+ 11 16		15.2	GALAXY
UGC 09259	14 25 00.	+ 11 16	66	15.2	GALAXY S
ZC 1425.0+2520	14 25 00.	+ 25 20	870		CLUSTER OF GALAXIES
MCG+04-34-029	14 25 00.	+ 25 42 30.	60	15.4	GALAXY
ZWG 192.021	14 25 00.	+ 33 28		13.7	GALAXY
UGC 09260	14 25 00.	+ 33 28	108	13.7	GALAXY E
REIZ 4128	14 25 00.	+ 40 10	42	14.2	GALAXY
ZWG 220.017	14 25 00.	+ 40 11		14.8	GALAXY
RNGC 5625	14 25 00.	+ 40 11		15.0	GALAXY
VV 024B	14 25 00.	+ 40 11	18	17.	INTERACTING GALAXY
VV 024A	14 25 00.	+ 40 11	66	14.	INTERACTING GALAXY
MCG+10-21-002	14 25 00.	+ 56 48	39	12.6	GALAXY
ZWG 296.005	14 25 00.	+ 56 49		12.4	GALAXY
UGC 09261	14 25 00.	+ 56 49	120	12.4	GALAXY S0/Sa
ZWG 019.030	14 25 00.	- 00 04		15.7	GALAXY
ZWG 019.029	14 25 00.	- 00 42		15.2	GALAXY
REIN 2.219	14 25 00.21	+ 56 48 24.4			NEBULA
RNGC 5631	14 25 01.	+ 56 48		12.5	GALAXY
IC 4431	14 25 02.	+ 31 09 30.			NONSTELLAR OBJECT
RNGC 5623	14 25 02.	+ 33 28		13.5	GALAXY
REIZ 4127	14 25 02.	+ 33 28	36	13.6	GALAXY
RNGC 5619B	14 25 03.	+ 05 00		15.5	GALAXY
IC 4424	14 25 03.	+ 05 04			NONSTELLAR OBJECT
MCG+05-34-043	14 25 03.	+ 31 10	66	14.8	GALAXY
MCG+06-32-035	14 25 03.	+ 33 28	24	13.	GALAXY
HN 1105	14 25 04.	+ 16 25	6		NEBULA
IC 4428	14 25 04.	+ 16 26			NONSTELLAR OBJECT
REIZ 4126	14 25 04.	+ 31 09	18	14.8	GALAXY
IC 1012	14 25 04.	+ 31 12 12.			NONSTELLAR OBJECT
IC 4421	14 25 04.	- 37 21 45.			NONSTELLAR OBJECT
HOLM 647A	14 25 05.	+ 36 08	60	14.0	PART OF MULTIPLE GALAXY
ZWG 047.049	14 25 06.	+ 03 10		15.1	GALAXY
ZWG 047.050	14 25 06.	+ 08 21		15.2	GALAXY
ZWG 075.037	14 25 06.	+ 11 33		14.9	GALAXY
MCG+02-37-009	14 25 06.	+ 11 33	24	14.9	GALAXY
FAI 14251+11	14 25 06.	+ 11 36			COMPACT GALAXY
ZWG 075.038	14 25 06.	+ 13 26		15.5	GALAXY
FAI 14251+14	14 25 06.	+ 14 31			COMPACT GALAXY
UGC 09262	14 25 06.	+ 36 08	78	16.0	GALAXY S
MCG+07-30-013	14 25 06.	+ 40 10	45	14.5	GALAXY
ZC 1425.1+4241	14 25 06.	+ 42 41	1340		CLUSTER OF GALAXIES
ZWG 019.031	14 25 06.	- 01 22		15.6	GALAXY
HOLM 647B	14 25 07.	+ 36 09	24	14.8	PART OF MULTIPLE GALAXY
MCG+06-32-036	14 25 09.	+ 36 09	84	14.5	GALAXY
SC 1422-1333.1	14 25 10.	- 13 46 37.	18		NEBULA
HOLM 648B	14 25 11.	+ 30 11	24	15.7	PART OF MULTIPLE GALAXY
ZWG 047.051	14 25 12.	+ 06 16		15.5	GALAXY
UGC 09264	14 25 12.	+ 06 16	60	15.5	GALAXY Sc
FAI 14252+09	14 25 12.	+ 09 32			COMPACT GALAXY
FAI 14252+11	14 25 12.	+ 11 57			COMPACT GALAXY
FAI 14252+14	14 25 12.	+ 14 30			COMPACT GALAXY
ZWG 104.029	14 25 12.	+ 17 07		15.4	GALAXY
ZWG 133.059	14 25 12.	+ 25 44		15.4	GALAXY
UGC 09265	14 25 12.	+ 25 44	60	15.4	GALAXY S
ZWG 163.053	14 25 12.	+ 30 10		15.4	GALAXY
UGC 09266	14 25 12.	+ 30 10	60	15.4	GALAXY SBc-IRR
ZWG 019.032	14 25 12.	- 00 23		15.6	GALAXY
UGC 09263	14 25 12.	- 00 23	66	15.6	GALAXY
MCG+00-37-007	14 25 12.	- 00 23	42	15.6	GALAXY
SC 1422-1444.2	14 25 12.	- 14 57 43.	30		NEBULA
MCG-05-34-013	14 25 12.	- 33 06	36	16.	GALAXY
HN 1106	14 25 15.	+ 17 06	12		NEBULA
IC 4429	14 25 16.	+ 17 07			NONSTELLAR OBJECT
HOLM 648A	14 25 17.	+ 30 10	48	14.4	PART OF MULTIPLE GALAXY
ZWG 047.052	14 25 18.	+ 05 07		15.4	GALAXY
ZWG 047.053	14 25 18.	+ 06 49		15.3	GALAXY
FAI 14253+11	14 25 18.	+ 11 58			COMPACT GALAXY
FAI 14253+12	14 25 18.	+ 12 59			COMPACT GALAXY
FAI 14253+14.1	14 25 18.	+ 14 08			COMPACT GALAXY
FAI 14253+14.5	14 25 18.	+ 14 31			COMPACT GALAXY
MCG+03-37-017	14 25 18.	+ 17 06	36	15.4	GALAXY
ZC 1425.3+2142	14 25 18.	+ 21 42	540		CLUSTER OF GALAXIES
TON-N 0202	14 25 18.	+ 26 46		15.1	BLUE STAR
SHB 292	14 25 18.	+ 26 46		15.7	QUASI-STELLAR OBJECT
ZC 1425.3+2728	14 25 18.	+ 27 28	940		CLUSTER OF GALAXIES
ZC 1425.3+2751	14 25 18.	+ 27 51	1410		CLUSTER OF GALAXIES
ZWG 192.022	14 25 18.	+ 37 41		15.3	GALAXY
MCG+05-34-044	14 25 21.	+ 30 10	60	15.4	GALAXY
REIZ 4125	14 25 22.	+ 08 28	42	13.0	GALAXY
RNGC 5621	14 25 22.	+ 08 29			NON-EXISTENT OBJECT
IC 4435	14 25 22.	+ 37 41 38.			NONSTELLAR OBJECT
REIZ 4129	14 25 23.	+ 30 05	24	14.6	GALAXY
ZWG 047.054	14 25 24.	+ 07 59		15.7	GALAXY
ZWG 075.039	14 25 24.	+ 11 47		14.9	GALAXY
UGC 09267	14 25 24.	+ 11 47	66	14.9	GALAXY SBb
MCG+02-37-010	14 25 24.	+ 11 47	72	14.9	GALAXY
FAI 14254+12	14 25 24.	+ 12 12			COMPACT GALAXY
ZWG 075.040	14 25 24.	+ 13 08		15.6	GALAXY
FAI 14254+13.1	14 25 24.	+ 13 09			COMPACT GALAXY
FAI 14254+13.6	14 25 24.	+ 13 40			COMPACT GALAXY
UGC 09268	14 25 24.	+ 32 22	78	17.	GALAXY DWRF SP
MCG+06-32-037	14 25 24.	+ 37 42	30	14.5	GALAXY
UGC 09269	14 25 24.	+ 50 48	78	16.5	GALAXY Sc
IC 4433	14 25 27.	+ 16 26			NONSTELLAR OBJECT
IC 4437	14 25 27.	+ 41 43			NONSTELLAR OBJECT
HN 1107	14 25 28.	+ 16 25	12		NEBULA
REIZ 4135	14 25 28.	+ 46 31	48	14.2	GALAXY
FAI 14255+12	14 25 30.	+ 12 30			COMPACT GALAXY
FAI 14255+13.3	14 25 30.	+ 13 23			COMPACT GALAXY
FAI 14255+13.8	14 25 30.	+ 13 53			COMPACT GALAXY

OBJECT NAME	RIGHT ASCEN.	DECLINATION	DIAM.	MAGN.	TYPE OF OBJECT
PAI 14255+13.9	14 25 30.	+ 13 57			COMPACT GALAXY
ZWG 104.030	14 25 30.	+ 16 26		15.3	GALAXY
ZC 1425.5+2253	14 25 30.	+ 22 53	1480		CLUSTER OF GALAXIES
MCG+04-34-030	14 25 30.	+ 26 02	36	15.4	GALAXY
MCG+09-24-007	14 25 30.	+ 55 47	90	15.	GALAXY
ZWG 019.033	14 25 30.	- 01 38		15.7	GALAXY
IC 1011	14 25 33.	+ 01 13 55.			NONSTELLAR OBJECT
IC 4434	14 25 33.	+ 16 27			NONSTELLAR OBJECT
MCG+08-26-034	14 25 33.	+ 46 20 30.	120	12.9	GALAXY
HN 1108	14 25 34.	+ 16 27	12		NEBULA
RNGC 5633	14 25 34.	+ 46 22		13.0	GALAXY
ZWG 019.036	14 25 36.	+ 01 13		14.7	GALAXY
MCG+00-37-008	14 25 36.	+ 01 13	24	14.7	GALAXY
8ZW 1425+01.4	14 25 36.	+ 01 27		15.3	COMPACT GALAXY
ZC 1425.6+1126	14 25 36.	+ 11 26	940		CLUSTER OF GALAXIES
ZWG 075.041	14 25 36.	+ 11 55		15.1	GALAXY
ZC 1425.6+2018	14 25 36.	+ 20 18	670		CLUSTER OF GALAXIES
MCG+04-34-031	14 25 36.	+ 21 32	78	14.7	GALAXY
ZWG 133.060	14 25 36.	+ 26 04		15.4	GALAXY
ZC 1425.6+3305	14 25 36.	+ 33 05	4370		CLUSTER OF GALAXIES
ZWG 192.023	14 25 36.	+ 33 37		15.6	GALAXY
ZWG 192.024	14 25 36.	+ 34 25		15.4	GALAXY
ZWG 220.018	14 25 36.	+ 41 29		13.6	GALAXY
UGC 09270	14 25 36.	+ 41 29	150	13.6	GALAXY Sc-IRR
1ZW 089	14 25 36.	+ 46 22			COMPACT GALAXY
ZWG 247.030	14 25 36.	+ 46 22		12.9	GALAXY
UGC 09271	14 25 36.	+ 46 22	150	12.9	GALAXY Sb
KARA.73B 0631	14 25 36.	+ 46 22	132	12.9	ISOLATED GALAXY S
7ZW 549	14 25 36.	+ 61 25			COMPACT GALAXY
UGC 09272	14 25 36.	+ 61 26	60	16.0	GALAXY Sc
ZWG 019.035	14 25 36.	- 01 27		15.3	GALAXY
ZWG 019.034	14 25 36.	- 01 58		15.6	GALAXY
REIN 2.220	14 25 36.71	+ 46 22 12.8			NEBULA
IC 1013	14 25 37.	+ 26 03 15.			NONSTELLAR OBJECT
RNGC 5630	14 25 37.	+ 41 29		13.5	GALAXY
REIZ 4139	14 25 40.	+ 46 23	48	13.0	GALAXY
REIZ 4137	14 25 41.	+ 41 28	18	14.5	GALAXY
REIZ 4138	14 25 41.	+ 41 29	108	13.3	GALAXY
HOLM 649A	14 25 41.	+ 41 29	72	12.7	PART OF MULTIPLE GALAXY
ZWG 075.042	14 25 42.	+ 13 46		15.1	GALAXY
UGC 09273	14 25 42.	+ 13 46	66	15.1	GALAXY IRR
ARC 1917	14 25 42.	+ 20 15		18.0	RICH CLUSTER OF GALAXIES
ZWG 133.061	14 25 42.	+ 21 32		14.7	GALAXY
REIZ 4132	14 25 42.	+ 21 32	60	13.8	GALAXY
UGC 09274	14 25 42.	+ 21 32	84	14.7	GALAXY SB
MCG+05-34-045	14 25 42.	+ 26 42	48	15.2	GALAXY
ZWG 163.054	14 25 42.	+ 26 44		15.2	GALAXY
MCG+07-30-014	14 25 42.	+ 41 28	120	13.	GALAXY
ARC 1920	14 25 42.	+ 56 00		17.0	RICH CLUSTER OF GALAXIES
ZWG 019.037	14 25 42.	- 01 44		15.5	GALAXY
RNGC 5632	14 25 43.	+ 00 02			GALAXY
HOLM 649B	14 25 43.	+ 41 27	18	14.0	PART OF MULTIPLE GALAXY
REIZ 4134	14 25 44.	+ 33 25	24	15.5	GALAXY
ARC 1916	14 25 44.	- 07 57		17.8	RICH CLUSTER OF GALAXIES
IC 4436	14 25 45.	+ 26 43 09.			NONSTELLAR OBJECT
ZWG 047.055	14 25 48.	+ 04 57		15.6	GALAXY
ZWG 047.056	14 25 48.	+ 05 16		15.1	GALAXY
UGC 10862	14 25 48.	+ 07 28	198	15.0	GALAXY SBc
ZC 1425.8+1915	14 25 48.	+ 19 15	610		CLUSTER OF GALAXIES
MCG+04-34-032	14 25 48.	+ 26 05	48	14.9	GALAXY
ZWG 163.055	14 25 48.	+ 28 10		15.7	GALAXY
ZC 1425.8+2945	14 25 48.	+ 29 45	1010		CLUSTER OF GALAXIES
SHAH 075	14 25 48.	+ 39 01	426	17.3	GROUP OF COMPACT GALAXIES
ZWG 019.039	14 25 48.	- 00 00		15.5	GALAXY
ZWG 019.038	14 25 48.	- 03 23		14.6	GALAXY
MCG+00-37-009	14 25 48.	- 03 23	60	14.6	GALAXY
REIZ 4131	14 25 49.	+ 11 52	60	13.6	GALAXY
MCG+05-34-046	14 25 51.	+ 28 09	30	15.7	GALAXY
ZWG 075.043	14 25 54.	+ 11 35		15.4	GALAXY
ZWG 075.044	14 25 54.	+ 11 38		14.9	GALAXY
MCG+02-37-011	14 25 54.	+ 11 38	60	14.9	GALAXY
ZWG 075.045	14 25 54.	+ 14 00		14.1	GALAXY
UGC 09275	14 25 54.	+ 14 00	168	14.1	GALAXY
MCG+02-37-012	14 25 54.	+ 14 00	156	14.1	GALAXY
1ZW 090	14 25 54.	+ 15 38			COMPACT GALAXY
ZWG 104.031	14 25 54.	+ 15 38		15.2	GALAXY
8ZW 1425+15.6	14 25 54.	+ 15 38		15.2	COMPACT GALAXY
MCG+03-37-018	14 25 54.	+ 15 38	36	15.2	GALAXY
ZC 1425.9+1858	14 25 54.	+ 18 58	870		CLUSTER OF GALAXIES
MCG+04-34-034	14 25 54.	+ 26 03	36	14.2	GALAXY
ZWG 133.062	14 25 54.	+ 26 05		14.9	GALAXY
UGC 09276	14 25 54.	+ 26 05	66	14.9	GALAXY S0?
IC 1017	14 25 54.	+ 26 05 10.			NONSTELLAR OBJECT
MCG+04-34-033	14 25 54.	+ 26 09 30.	24	15.3	GALAXY
ZWG 163.056	14 25 54.	+ 27 10		15.5	GALAXY
IC 1014	14 25 55.	+ 14 00 03.			NONSTELLAR OBJECT
IC 1015	14 25 57.	+ 15 38 34.			NONSTELLAR OBJECT
MCG+05-34-047	14 25 57.	+ 27 08	48	15.5	GALAXY
IC 1018	14 25 59.	+ 26 02 52.			NONSTELLAR OBJECT
ZWG 047.057	14 26 00.	+ 03 29		15.0	GALAXY
UGC 09277	14 26 00.	+ 03 29	90	15.0	GALAXY Sb
MCG+01-37-015	14 26 00.	+ 03 29	84	15.0	GALAXY
ZWG 104.032	14 26 00.	+ 17 49		15.7	GALAXY
ZWG 104.033	14 26 00.	+ 18 09		14.5	GALAXY
UGC 09278	14 26 00.	+ 18 09	66	14.5	GALAXY E
ZC 1426.0+2039	14 26 00.	+ 20 39	470		CLUSTER OF GALAXIES
ZC 1426.0+2407	14 26 00.	+ 24 07	870		CLUSTER OF GALAXIES
ZWG 133.063	14 26 00.	+ 26 03		15.6	GALAXY
IC 1019	14 26 00.	+ 26 09 52.			NONSTELLAR OBJECT
ZWG 133.064	14 26 00.	+ 26 10		15.3	GALAXY
UGC 09279	14 26 00.	+ 34 02	60	16.0	GALAXY Sc
ZWG 019.040	14 26 00.	- 03 07		14.9	GALAXY
MCG+00-37-010	14 26 00.	- 03 07	48	14.9	GALAXY
OCL 0919	14 26 00.	- 60 30	1680	8.5	OPEN STAR CLUSTER
VHA 159	14 26 00.	- 60 30	480		OPEN STAR CLUSTER
RNGC 5628	14 26 02.	+ 18 09		14.5	GALAXY
RNGC 5617	14 26 02.	- 60 30		8.5	OPEN CLUSTER
RNGC 5627	14 26 05.	+ 11 36		14.5	GALAXY
RNGC 5629	14 26 05.	+ 26 04		14.0	GALAXY
ZWG 019.041	14 26 06.	+ 00 46		15.2	GALAXY
ZWG 075.046	14 26 06.	+ 11 36		14.7	GALAXY
UGC 09280	14 26 06.	+ 11 36	108	14.7	GALAXY S0
MCG+03-37-019	14 26 06.	+ 18 09	54	14.5	GALAXY
ZWG 133.065	14 26 06.	+ 26 04		14.2	GALAXY
UGC 09281	14 26 06.	+ 26 04	138	14.2	GALAXY S0
BC TON202	14 26 06.	+ 26 36		16.00	QUASI-STELLAR OBJECT
ZC 1426.1+3938	14 26 06.	+ 39 38	1280		CLUSTER OF GALAXIES
ZC 1426.1+6505	14 26 06.	+ 65 05	2080		CLUSTER OF GALAXIES
SN 1954W	14 26 07.	+ 00 46		19.4	SUPERNOVA
REIZ 4136	14 26 07.	+ 11 37	60	14.2	GALAXY
ZWG 019.042	14 26 12.	+ 00 55		15.7	GALAXY
ZWG 075.047	14 26 12.	+ 10 26		15.3	GALAXY
ZWG 075.048	14 26 12.	+ 11 36		15.3	GALAXY
MCG+02-37-014	14 26 12.	+ 11 36	39	15.3	GALAXY
ZWG 104.034	14 26 12.	+ 17 33		15.7	GALAXY
MCG+05-34-048	14 26 12.	+ 27 29	24	15.3	GALAXY
MCG+09-24-008	14 26 12.	+ 52 47	42	16.	GALAXY
HARO 41	14 26 13.	+ 27 29			BLUE EMISSION-LINE GALAXY
REIZ 4141	14 26 13.	+ 33 29	36	14.0	GALAXY
HN 1109	14 26 15.	+ 17 34	12		NEBULA
MCG+03-37-020	14 26 15.	+ 17 34	24	15.7	GALAXY
IC 4438	14 26 15.	+ 17 34			NONSTELLAR OBJECT
ZWG 104.035	14 26 18.	+ 17 15		15.7	GALAXY
ZWG 104.036	14 26 18.	+ 17 48		15.6	GALAXY
FEIG 095	14 26 18.	+ 21 20		13.0	FAINT BLUE STAR
ZWG 133.066	14 26 18.	+ 21 34		15.7	GALAXY
UGC 09282	14 26 18.	+ 21 34	96	15.7	GALAXY DWARF
ZWG 163.057	14 26 18.	+ 27 29		15.3	GALAXY
ZWG 163.058	14 26 18.	+ 27 38		13.9	GALAXY
RNGC 5635	14 26 18.	+ 27 38		14.0	GALAXY
UGC 09283	14 26 18.	+ 27 38	150	13.9	GALAXY S
ZWG 192.025	14 26 18.	+ 32 37		14.8	GALAXY
ZWG 192.026	14 26 18.	+ 33 28		14.9	GALAXY
UGC 09284	14 26 18.	+ 33 28	96	14.9	GALAXY Sa?
ZWG 192.027	14 26 18.	+ 35 37		15.3	GALAXY
MCG-05-34-014	14 26 18.	- 33 15	48	14.	GALAXY
REIZ 4140	14 26 19.	+ 27 38	96	13.2	GALAXY
IC 1016	14 26 20.	+ 05 02 59.			NONSTELLAR OBJECT
HN 1110	14 26 21.	+ 17 16	16		NEBULA
IC 4439	14 26 21.	+ 17 16			NONSTELLAR OBJECT
MCG+05-34-049	14 26 21.	+ 27 36 30.	138	13.9	GALAXY
MCG+06-32-039	14 26 21.	+ 32 37	48	14.	GALAXY
MCG+06-32-038	14 26 21.	+ 33 28	90	14.	GALAXY
MCG+06-32-040	14 26 21.	+ 35 37 30.	48	14.5	GALAXY
ZC 1426.4+1132	14 26 24.	+ 11 32	5710		CLUSTER OF GALAXIES
ZC 1426.4+2030	14 26 24.	+ 20 30	1810		CLUSTER OF GALAXIES
ZC 1426.4+2216	14 26 24.	+ 22 16	540		CLUSTER OF GALAXIES
ZWG 133.067	14 26 24.	+ 25 56		15.7	GALAXY
ZWG 163.059	14 26 24.	+ 30 51		15.5	GALAXY
ZWG 192.028	14 26 24.	+ 36 42		15.7	GALAXY
MCG+12-14-004	14 26 24.	+ 70 06	33	16.	GALAXY
ZWG 337.012	14 26 24.	+ 70 08		14.8	GALAXY
HN 0405	14 26 24.	- 33 23			NEBULA
IC 4430	14 26 24.	- 33 23			NONSTELLAR OBJECT
RNGC 5639B	14 26 25.	+ 30 38		17.0	GALAXY
MCG+06-32-041	14 26 27.	+ 36 43	42	16.	GALAXY
HOLM 651A	14 26 28.	+ 30 39	72	12.4	PART OF MULTIPLE GALAXY
REIZ 4143	14 26 28.	+ 30 51	24	15.0	GALAXY
ZWG 047.058	14 26 30.	+ 03 22		15.3	GALAXY
UGC 09285	14 26 30.	+ 03 22	78	15.3	GALAXY Sc?
MCG+01-37-016	14 26 30.	+ 03 22	84	15.3	GALAXY
ZWG 047.059	14 26 30.	+ 04 54		15.2	GALAXY
ZC 1426.5+1020	14 26 30.	+ 10 20	2290		CLUSTER OF GALAXIES
ZWG 075.050	14 26 30.	+ 11 25		15.2	GALAXY
UGC 09286	14 26 30.	+ 11 25	66	15.2	GALAXY SBb
MCG+02-37-015	14 26 30.	+ 11 25	72	15.2	GALAXY
MCG+03-37-021	14 26 30.	+ 16 56	48	16.5	GALAXY
MCG+04-34-035	14 26 30.	+ 26 15	72	15.2	GALAXY
ZC 1426.5+2640	14 26 30.	+ 26 40	810		CLUSTER OF GALAXIES
VV C15C	14 26 30.	+ 29 09 30.	9	18.	INTERACTING GALAXY
VV 015B	14 26 30.	+ 29 09 30.	9	18.	INTERACTING GALAXY
VV 015A	14 26 30.	+ 29 09 30.	72	16.	INTERACTING GALAXY
TON-N 0773	14 26 30.	+ 29 11		16.5	BLUE STAR
ZWG 163.060	14 26 30.	+ 29 12		15.2	GALAXY
UGC 09287	14 26 30.	+ 29 12	66	15.2	GALAXY SBa
ZC 1426.5-0300	14 26 30.	- 03 00	1210		CLUSTER OF GALAXIES
MCG-04-34-016	14 26 30.	- 22 42	72	14.5	GALAXY
BC B21426+29	14 26 31.8	+ 29 32 22.5		18.1	QUASI-STELLAR OBJECT
IC 4442	14 26 32.	+ 29 09 31.			NONSTELLAR OBJECT
HN 1111	14 26 33.	+ 17 33		15.	NEBULA
IC 4440	14 26 33.	+ 17 33			NONSTELLAR OBJECT
REIZ 4144	14 26 34.	+ 30 38	72	13.7	GALAXY
HOLM 652A	14 26 35.	+ 29 12	18	14.8	PART OF MULTIPLE GALAXY
ZWG 047.060	14 26 36.	+ 08 04		15.1	GALAXY
ZWG 075.049	14 26 36.	+ 13 40		15.5	GALAXY
ZWG 075.051	14 26 36.	+ 14 05		14.4	GALAXY
UGC 09288	14 26 36.	+ 14 05	72	14.4	GALAXY S0
MCG+02-37-015A	14 26 36.	+ 14 05	60	14.4	GALAXY
ZWG 104.037	14 26 36.	+ 17 32		15.6	GALAXY
MCG+04-34-037	14 26 36.	+ 23 23 30.	42	14.6	GALAXY
MCG+04-34-036	14 26 36.	+ 25 45	72	15.1	GALAXY
IC 1020	14 26 36.	+ 26 14 25.			NONSTELLAR OBJECT
ZWG 133.068	14 26 36.	+ 26 15		15.2	GALAXY
UGC 09289	14 26 36.	+ 26 15	72	15.2	GALAXY S0
MRK 682	14 26 36.	+ 27 28	15	15.5	GALAXY WITH UV CONTINUUM
MCG+05-34-050	14 26 36.	+ 29 10	60	15.2	GALAXY
ZWG 163.061	14 26 36.	+ 30 38		14.6	GALAXY
UGC 09290	14 26 36.	+ 30 38	84	14.6	GALAXY Sc
MCG+06-32-042	14 26 36.	+ 34 36	42	16.	GALAXY
MCG+07-30-015	14 26 36.	+ 39 12	150	14.	GALAXY
ZWG 220.019	14 26 36.	+ 39 13		14.0	GALAXY
UGC 09291	14 26 36.	+ 39 13	168	14.0	GALAXY Sc
ZWG 247.031	14 26 36.	+ 49 46		15.6	GALAXY
MCG+08-26-035	14 26 36.	+ 49 46	21	16.	GALAXY
MCG+12-14-005	14 26 36.	+ 70 08	24	16.	GALAXY
RNGC 5639A	14 26 37.	+ 30 38		14.5	GALAXY
REIZ 4147	14 26 37.	+ 39 14	90	13.8	GALAXY
MCG+03-37-022	14 26 39.	+ 17 33	36	15.6	GALAXY
MCG+05-34-051	14 26 39.	+ 30 38	84	14.6	GALAXY
HOLM 651B	14 26 39.	+ 30 39	24	15.5	PART OF MULTIPLE GALAXY
HOLM 650A	14 26 40.	+ 08 04	30	14.0	PART OF MULTIPLE GALAXY
RNGC 5637	14 26 40.	+ 23 25		14.5	GALAXY
HOLM 652B	14 26 40.	+ 29 12	18	15.7	PART OF MULTIPLE GALAXY
SC 1423-1329.7	14 26 40.	- 13 43 09.	12		NEBULA
REIZ 4149	14 26 41.	+ 49 47	30	14.7	GALAXY
ZWG 047.061	14 26 42.	+ 02 30		15.2	GALAXY
ZWG 019.043	14 26 42.	+ 02 30		15.2	GALAXY
UGC 09292	14 26 42.	+ 02 30	96	15.2	GALAXY S
MCG+00-37-011	14 26 42.	+ 02 30	84	15.2	GALAXY
HOLM 650B	14 26 42.	+ 08 05	24	14.7	PART OF MULTIPLE GALAXY
MCG+04-34-038	14 26 42.	+ 20 50 30.	66	15.2	GALAXY
ZWG 133.069	14 26 42.	+ 23 25		14.6	GALAXY
UGC 09293	14 26 42.	+ 23 25	60	14.6	GALAXY S
ZWG 133.070	14 26 42.	+ 25 46		15.1	GALAXY
UGC 09294	14 26 42.	+ 25 46	78	15.1	GALAXY SBb
ZWG 296.006	14 26 42.	+ 57 25		15.4	GALAXY

OBJECT NAME	RIGHT ASCEN.	DECLINATION	DIAM.	MAGN.	TYPE OF OBJECT
7ZW 550	14 26 42.	+ 58 24			COMPACT GALAXY
MCG+10-21-003	14 26 42.	+ 60 51	42	16.	GALAXY
ZWG 337.013	14 26 42.	+ 70 10		14.5	GALAXY
UGC 09295	14 26 42.	+ 70 10	21	14.5	GALAXY PECULR
8ZW 1426-02.9	14 26 42.	- 02 55		19.4	COMPACT GALAXY
MCG-01-37-007	14 26 42.	- 04 48	60	15.	GALAXY
MCG-05-34-015	14 26 42.	- 29 32	24	15.	GALAXY
HN 0406	14 26 42.	- 39 18			NEBULA
IC 4432	14 26 42.	- 39 18			NONSTELLAR OBJECT
RNGC 5626	14 26 44.	- 29 32		15.0	GALAXY
HN 0011	14 26 46.	- 00 13			NEBULA
MCG+02-37-013	14 26 48.	+ 11 36	84	14.7	GALAXY
ZWG 075.052	14 26 48.	+ 11 55		15.4	GALAXY
ZWG 075.053	14 26 48.	+ 13 04		15.6	GALAXY
KARA.73B 0632	14 26 48.	+ 13 04	60	15.6	ISOLATED GALAXY S
ZC 1426.8+2105	14 26 48.	+ 21 05	1080		CLUSTER OF GALAXIES
ZC 1426.8+2947	14 26 48.	+ 29 47	16600		CLUSTER OF GALAXIES
MCG+12-14-006	14 26 48.	+ 69 55	102	15.	GALAXY
REIZ 4142	14 26 52.	- 00 13	6	14.7	GALAXY
8ZW 1426+02.6	14 26 54.	+ 02 35		17.7	COMPACT GALAXY
ZWG 133.071	14 26 54.	+ 20 53		15.2	GALAXY
UGC 09296	14 26 54.	+ 20 53	72	15.2	GALAXY SBa
MCG+04-34-039	14 26 54.	+ 22 02	48	15.5	GALAXY
ZWG 163.062	14 26 54.	+ 30 18		15.5	GALAXY
ZC 1426.9+3200	14 26 54.	+ 32 00	1140		CLUSTER OF GALAXIES
ZWG 192.029	14 26 54.	+ 36 10		15.5	GALAXY
MRK 472	14 26 54.	+ 36 10	20	15.5	GALAXY WITH UV CONTINUUM
ZC 1426.9+5818	14 26 54.	+ 58 18	1750		CLUSTER OF GALAXIES
ZWG 296.007	14 26 54.	+ 61 25		15.3	GALAXY
MCG+11-18-001	14 26 54.	+ 63 09	36	16.	GALAXY
ZWG 337.014	14 26 54.	+ 69 56		14.4	GALAXY
UGC 09297	14 26 54.	+ 69 56	102	14.4	GALAXY SBb
7ZW 551	14 26 54.	+ 70 10			COMPACT GALAXY
VHA 160	14 26 54.	- 60 40	180		OPEN STAR CLUSTER
OCL 0921	14 26 54.	- 60 46	132	12.	OPEN STAR CLUSTER
ARC 1925	14 26 55.	+ 57 05		17.2	RICH CLUSTER OF GALAXIES
RNGC 5671	14 26 56.	+ 69 56		14.5	GALAXY
IC 4443	14 26 57.	+ 16 25			NONSTELLAR OBJECT
HN 1112	14 26 58.	+ 16 25	12		NEBULA
IC 4446	14 26 58.	+ 37 41 04.			NONSTELLAR OBJECT
RNGC 5634	14 26 59.	- 05 45		11.0	GLOBULAR CLUSTER
REIN 2.221	14 26 59.43	- 05 45 18.5			NEBULA
VDB.66G 192	14 27	+ 44 39	70		DWARF GALAXY
ZWG 019.044	14 27 00.	+ 00 11		14.7	GALAXY
UGC 09299	14 27 00.	+ 00 11	102	14.7	GALAXY Sc
MCG+00-37-012	14 27 00.	+ 00 11	102	14.7	GALAXY
IC 1021	14 27 00.	+ 20 52 57.			NONSTELLAR OBJECT
ZC 1427.0+2317	14 27 00.	+ 23 17	870		CLUSTER OF GALAXIES
ZWG 133.072	14 27 00.	+ 24 19		15.7	GALAXY
ZWG 163.063	14 27 00.	+ 29 03		13.6	GALAXY
UGC 09300	14 27 00.	+ 29 03	168	13.6	GALAXY SBb
MCG+05-34-052	14 27 00.	+ 30 14 30.	30	14.3	GALAXY
ZWG 163.064	14 27 00.	+ 30 15		14.3	GALAXY
UGC 09301	14 27 00.	+ 30 15	180	14.3	GALAXY E
MCG+05-34-053	14 27 00.	+ 30 18	27	15.5	GALAXY
ZWG 163.065	14 27 00.	+ 32 00		15.6	GALAXY
UGC 09302	14 27 00.	+ 32 00	66	15.6	GALAXY S-IRR
UGC 09303	14 27 00.	+ 33 07	60	16.0	GALAXY S
MCG+06-32-043	14 27 00.	+ 33 07	66	15.	GALAXY
MCG+06-32-044	14 27 00.	+ 37 41 30.	48	15.	GALAXY
UGC 09298	14 27 00.	+ 84 42	72	16.5	GALAXY DWARF
GCL 028	14 27 00.	- 05 45	222	10.8	GLOBULAR STAR CLUSTER
ARC 1921	14 27 01.	+ 23 19		17.2	RICH CLUSTER OF GALAXIES
RNGC 5642	14 27 01.	+ 30 15		14.5	GALAXY
REIZ 4148	14 27 04.	+ 30 15	30	14.7	GALAXY
ARC 1951	14 27 05.	+ 83 32		17.7	RICH CLUSTER OF GALAXIES
ZWG 019.045	14 27 06.	+ 00 35		15.7	GALAXY
ZWG 047.062	14 27 06.	+ 03 29		14.6	GALAXY
UGC 09304	14 27 06.	+ 03 29	90	14.6	GALAXY SBa
MCG+01-37-017	14 27 06.	+ 03 29	72	14.6	GALAXY
HOLM 653B	14 27 06.	+ 03 31	72	13.8	PART OF MULTIPLE GALAXY
UGC 09305	14 27 06.	+ 04 16	66	16.0	GALAXY Sc-IRR
ZWG 133.073	14 27 06.	+ 22 04		15.5	GALAXY
RNGC 5641	14 27 06.	+ 29 03		13.0	GALAXY
ZWG 163.066	14 27 06.	+ 31 03		14.7	GALAXY
UGC 09306	14 27 06.	+ 31 03	78	14.7	GALAXY S0-a
MCG+05-34-054	14 27 06.	+ 32 01	66	15.6	GALAXY
ZWG 220.020	14 27 06.	+ 42 03		15.3	GALAXY
UGC 09307	14 27 06.	+ 42 03	60	15.3	GALAXY S
REIZ 4146	14 27 07.	+ 03 28	54	12.7	GALAXY
HOLM 653A	14 27 07.	+ 03 29	48	12.4	PART OF MULTIPLE GALAXY
REIZ 4145	14 27 07.	+ 03 30	60	13.9	GALAXY
RNGC 5638	14 27 08.	+ 03 27		12.5	GALAXY
RNGC 5636	14 27 08.	+ 03 29		14.5	GALAXY
MCG+05-34-055	14 27 09.	+ 29 02	132	13.6	GALAXY
REIZ 4153	14 27 10.	+ 31 03	12	14.3	GALAXY
ZWG 019.046	14 27 12.	+ 00 03		15.7	GALAXY
ZWG 047.063	14 27 12.	+ 03 27		12.5	GALAXY
UGC 09308	14 27 12.	+ 03 27	150	12.5	GALAXY E
MCG+01-37-018	14 27 12.	+ 03 27	54	12.5	GALAXY
ZWG 047.064	14 27 12.	+ 07 28		15.5	GALAXY
ZWG 075.054	14 27 12.	+ 10 48		15.3	GALAXY
UGC 09309	14 27 12.	+ 10 48	60	15.3	GALAXY Sb-c
ZC 1427.2+2047	14 27 12.	+ 20 47	670		CLUSTER OF GALAXIES
ZWG 133.074	14 27 12.	+ 23 32		15.6	GALAXY
REIZ 4151	14 27 12.	+ 28 12	12	14.7	GALAXY
REIZ 4152	14 27 12.	+ 28 56	30	15.1	GALAXY
MCG+05-34-056	14 27 12.	+ 31 03 30.	75	14.7	GALAXY
MCG+07-30-016	14 27 12.	+ 42 03	48	15.	GALAXY
MCG-07-30-001	14 27 12.	- 43 19	42	15.	GALAXY
ARC 1919	14 27 14.	+ 05 37		17.6	RICH CLUSTER OF GALAXIES
MCG+08-26-036	14 27 15.	+ 44 53	48	16.	GALAXY
ZC 1427.3+0540	14 27 18.	+ 05 40	1080		CLUSTER OF GALAXIES
ZWG 075.055	14 27 18.	+ 11 59		15.5	GALAXY
ZC 1427.3+7803	14 27 18.	+ 78 03	1080		CLUSTER OF GALAXIES
REIZ 4154	14 27 23.	+ 29 12	15	15.8	GALAXY
ZC 1427.4+1146	14 27 24.	+ 11 46	1810		CLUSTER OF GALAXIES
ZWG 075.056	14 27 24.	+ 14 01		15.2	GALAXY
ZWG 133.075	14 27 24.	+ 26 17		15.5	GALAXY
ZWG 247.032	14 27 24.	+ 44 55		15.2	GALAXY
REIZ 4156	14 27 24.	+ 44 55	24	14.1	GALAXY
ZC 1427.4+5713	14 27 24.	+ 57 13	4230		CLUSTER OF GALAXIES
OCL 0920	14 27 24.	- 60 57	1500	10.0	OPEN STAR CLUSTER
VHA 161	14 27 24.	- 60 57	300		OPEN STAR CLUSTER
IC 4447	14 27 27.	+ 31 02 41.			NONSTELLAR OBJECT
MCG+08-26-037	14 27 27.	+ 45 15	72	15.	GALAXY
SC 1424-1418.9	14 27 27.	- 14 32 19.	12		NEBULA
RNGC 5646	14 27 28.	+ 35 40		15.0	GALAXY

OBJECT NAME	RIGHT ASCEN.	DECLINATION	DIAM.	MAGN.	TYPE OF OBJECT
ZWG 019.047	14 27 30.	+ 01 15		15.6	GALAXY
ZWG 047.065	14 27 30.	+ 03 26		15.3	GALAXY
UGC 09310	14 27 30.	+ 03 26	132	15.3	GALAXY
ZWG 047.066	14 27 30.	+ 03 59		15.3	GALAXY
UGC 09311	14 27 30.	+ 03 59	72	15.3	GALAXY S
ZWG 047.067	14 27 30.	+ 07 28		15.5	GALAXY
ZWG 192.030	14 27 30.	+ 35 40		15.2	GALAXY
UGC 09312	14 27 30.	+ 35 40	96	15.2	GALAXY SB:b
MCG+06-32-045	14 27 30.	+ 35 41 30.	90	14.5	GALAXY
ZWG 220.021	14 27 30.	+ 38 53		15.5	GALAXY
UGC 09313	14 27 30.	+ 38 53	66	15.5	GALAXY Sb-c
ZWG 247.033	14 27 30.	+ 45 18		15.1	GALAXY
UGC 09314	14 27 30.	+ 45 18	78	15.1	GALAXY S
ZC 1427.5+4856	14 27 30.	+ 48 56	1210		CLUSTER OF GALAXIES
REIZ 4150	14 27 31.	+ 04 00	72	14.1	GALAXY
IC 1022	14 27 31.	+ 04 00 52.			NONSTELLAR OBJECT
REIZ 4155	14 27 31.	+ 33 42	72	14.1	GALAXY
MCG+07-30-017	14 27 33.	+ 38 52	60	16.	GALAXY
HELW 223	14 27 33.	- 06 12 07.			NEBULA
REIZ 4161	14 27 35.	+ 45 18	72	14.4	GALAXY
REIZ 4162	14 27 35.	+ 45 20	48	15.1	GALAXY
ZC 1427.6+1511	14 27 36.	+ 15 11	2420		CLUSTER OF GALAXIES
ZWG 273.007	14 27 36.	+ 52 53		15.4	GALAXY
ZC 1427.6+7026	14 27 36.	+ 70 26	2760		CLUSTER OF GALAXIES
ARC 1922	14 27 41.	+ 20 49		17.6	RICH CLUSTER OF GALAXIES
REIZ 4158	14 27 41.	+ 35 48	48	14.7	GALAXY
ZC 1427.7+1939	14 27 42.	+ 19 39	2960		CLUSTER OF GALAXIES
ARC 1923	14 27 42.	+ 19 51		17.2	RICH CLUSTER OF GALAXIES
MCG+04-34-040	14 27 42.	+ 23 13	18	15.3	GALAXY
UGC 09315	14 27 42.	+ 25 47	60	15.4	GALAXY S
TON-N 0203	14 27 42.	+ 28 16		15.0	BLUE STAR
ZC 1427.7+2846	14 27 42.	+ 28 46	670		CLUSTER OF GALAXIES
ZWG 192.031	14 27 42.	+ 35 47		15.4	GALAXY
ZWG 019.048	14 27 42.	- 01 35		15.7	GALAXY
REIZ 4160	14 27 43.	+ 38 54	48	14.9	GALAXY
MCG+06-32-046	14 27 45.	+ 35 48	42	14.5	GALAXY
REIZ 4157	14 27 47.	+ 29 04	78	13.7	GALAXY
ZWG 047.068	14 27 48.	+ 03 17		15.7	GALAXY
8ZW 1427+04.9	14 27 48.	+ 04 53		15.6	COMPACT GALAXY
ZWG 104.036	14 27 48.	+ 15 06		15.5	GALAXY
MCG+04-34-041	14 27 48.	+ 23 16	54	15.1	GALAXY
ZC 1427.8+2443	14 27 48.	+ 24 43	2420		CLUSTER OF GALAXIES
MCG+06-32-048	14 27 48.	+ 34 50	42	15.	GALAXY
MCG+06-32-047	14 27 48.	+ 35 48	36	15.	GALAXY
MCG+08-26-038	14 27 48.	+ 49 50	42	16.	GALAXY
MCG-04-34-017	14 27 48.	- 27 13	36	15.	GALAXY
HOLM 655B	14 27 49.	+ 23 15	18	14.0	PART OF MULTIPLE GALAXY
REIZ 4165	14 27 51.	+ 37 06	48	14.7	GALAXY
RNGC 5654	14 27 52.	+ 36 35		14.0	GALAXY
HELW 224	14 27 52.	- 05 26 06.			NEBULA
ZC 1427.9+1925	14 27 54.	+ 19 25	810		CLUSTER OF GALAXIES
ZWG 133.076	14 27 54.	+ 21 59		15.6	GALAXY
UGC 09316	14 27 54.	+ 21 59	108	15.6	GALAXY S0?
ZWG 133.077	14 27 54.	+ 23 15		15.3	GALAXY
MRK 683	14 27 54.	+ 23 15	18	15.	GALAXY WITH UV CONTINUUM
HOLM 655C	14 27 54.	+ 23 17	36	14.8	PART OF MULTIPLE GALAXY
ZWG 163.067	14 27 54.	+ 27 45		15.1	GALAXY
UGC 09317	14 27 54.	+ 27 45	78	15.1	GALAXY SBc
ZWG 163.068	14 27 54.	+ 31 26		12.7	GALAXY
UGC 09318	14 27 54.	+ 31 26	114	12.7	GALAXY S
ZWG 192.032	14 27 54.	+ 36 35		14.1	GALAXY
UGC 09319	14 27 54.	+ 36 35	96	14.1	GALAXY PECULR
ZWG 192.033	14 27 54.	+ 37 05		15.5	GALAXY
UGC 09320	14 27 54.	+ 37 05	72	15.5	GALAXY
MCG+06-32-049	14 27 54.	+ 37 07	66	14.5	GALAXY
ZC 1427.9+5454	14 27 54.	+ 54 54	2020		CLUSTER OF GALAXIES
HOLM 655A	14 27 58.	+ 23 17	72	13.6	PART OF MULTIPLE GALAXY
REIZ 4168	14 27 58.	+ 36 30	48	15.0	GALAXY
REIZ 4169	14 27 58.	+ 36 35	24	13.9	GALAXY
RNGC 5644	14 27 59.	+ 12 09		14.0	GALAXY
ZWG 019.049	14 28 00.	+ 02 03		15.5	GALAXY
ZWG 075.057	14 28 00.	+ 12 09		14.1	GALAXY
UGC 09321	14 28 00.	+ 12 09	108	14.1	GALAXY S0
MCG+02-37-016	14 28 00.	+ 12 09	72	14.1	GALAXY
FEIG 096	14 28 00.	+ 21 30		13.3	FAINT BLUE STAR
ZWG 133.078	14 28 00.	+ 23 17		15.1	GALAXY
UGC 09322	14 28 00.	+ 23 17	60	15.1	GALAXY Sb/SBc
MCG+05-34-057	14 28 00.	+ 27 44	72	15.1	GALAXY
ZC 1428.0+3225	14 28 00.	+ 32 25	1080		CLUSTER OF GALAXIES
MCG+06-32-050	14 28 00.	+ 36 35	90	14.	GALAXY
UGC 09323	14 28 00.	+ 40 09	60	16.0	GALAXY Sb-c
MCG+07-30-018	14 28 00.	+ 44 39	120	15.	GALAXY
ZWG 247.034	14 28 00.	+ 44 40		15.7	GALAXY
REIZ 4172	14 28 00.	+ 44 40	30	15.0	GALAXY
UGC 09324	14 28 00.	+ 44 40	156	15.7	GALAXY
REIZ 4175	14 28 00.	+ 48 51	42	14.2	GALAXY
ZWG 247.035	14 28 00.	+ 49 50		12.2	GALAXY
UGC 09325	14 28 00.	+ 49 50	192	12.2	GALAXY Sc
MCG+08-26-039	14 28 00.	+ 49 50	144	12.3	GALAXY
ZC 1428.0+8334	14 28 00.	+ 83 34	1140		CLUSTER OF GALAXIES
MCG+14-07-006	14 28 00.	+ 84 42	51	16.	GALAXY
REIZ 4164	14 28 01.	+ 27 46	42	14.4	GALAXY
RNGC 5653	14 28 01.	+ 31 26		13.5	GALAXY
HOLM 654B	14 28 02.	+ 14 16	18	14.4	PART OF MULTIPLE GALAXY
MCG+05-34-058	14 28 03.	+ 31 26 30.	96	12.7	GALAXY
REIZ 4167	14 28 03.	+ 31 27	24	12.7	GALAXY
MCG+07-30-019	14 28 03.	+ 40 09	60	16.	GALAXY
REIZ 4177	14 28 04.	+ 49 51	84	12.7	GALAXY
HOLM 654A	14 28 05.	+ 14 15	54	13.6	PART OF MULTIPLE GALAXY
ZWG 047.069	14 28 06.	+ 03 38		15.6	GALAXY
ZC 1428.1+2050	14 28 06.	+ 20 50	810		CLUSTER OF GALAXIES
RNGC 5660	14 28 07.	+ 49 51		12.5	GALAXY
MAI 090	14 28 09.	+ 59 06	40		DWARF SPHEROIDAL GALAXY
RNGC 5645	14 28 10.	+ 07 30		13.0	GALAXY
REIZ 4159	14 28 10.	+ 07 30	108	13.0	GALAXY
HN 1737	14 28 10.3	+ 02 05 37.	18		NEBULA
RNGC 5647	14 28 11.	+ 12 06		15.5	GALAXY
REIZ 4163	14 28 11.	+ 14 14	54	13.4	GALAXY
IC 4441	14 28 11.	- 43 17 55.			NONSTELLAR OBJECT
ZWG 019.050	14 28 12.	+ 00 28		15.5	GALAXY
UGC 09327	14 28 12.	+ 00 28	78	15.5	GALAXY S
UGC 09326	14 28 12.	+ 00 28	60	15.5	GALAXY
ZC 1428.2+0029	14 28 12.	+ 00 29	3760		CLUSTER OF GALAXIES
ZWG 019.051	14 28 12.	+ 02 05		14.8	GALAXY
MCG+00-37-013	14 28 12.	+ 02 05	36	14.6	GALAXY
ZWG 047.070	14 28 12.	+ 07 30		12.8	GALAXY
UGC 09328	14 28 12.	+ 07 30	192	12.8	GALAXY
MCG+01-37-019	14 28 12.	+ 07 30	132	12.8	GALAXY

OBJECT NAME	RIGHT ASCEN.	DECLINATION	DIAM.	MAGN.	TYPE OF OBJECT
ZWG 075.058	14 28 12.	+ 12 06		15.3	GALAXY
UGC 09329	14 28 12.	+ 12 06	78	15.3	GALAXY Sa
MCG+02-37-018	14 28 12.	+ 12 06	12	16.	GALAXY
MCG+02-37-017	14 28 12.	+ 12 06	60	15.3	GALAXY
ZWG 075.059	14 28 12.	+ 14 15		14.1	GALAXY
RNGC 5648	14 28 12.	+ 14 15		14.0	GALAXY
UGC 09330	14 28 12.	+ 14 15	60	14.1	GALAXY S
MCG+02-37-019	14 28 12.	+ 14 15	72	14.1	GALAXY
MCG+04-34-042	14 28 12.	+ 25 43	42	15.3	GALAXY
ZC 1428.2+3054	14 28 12.	+ 30 54	1010		CLUSTER OF GALAXIES
IC 1027	14 28 12.	+ 54 10 23.			NONSTELLAR OBJECT
ZWG 273.008	14 28 12.	+ 54 12		15.4	GALAXY
UGC 09331	14 28 12.	+ 54 12	60	15.4	GALAXY
MCG+09-24-009	14 28 12.	+ 54 12	54	15.	GALAXY
7ZW 552	14 28 12.	+ 57 30			COMPACT GALAXY
RNGC 5612	14 28 12.	- 78 11		13.5	GALAXY
RNGC 5656	14 28 16.	+ 35 32		12.5	GALAXY
ARC 1926	14 28 17.	+ 24 53		16.6	RICH CLUSTER OF GALAXIES
REIZ 4176	14 28 17.	+ 35 33	54	13.3	GALAXY
ZWG 047.071	14 28 18.	+ 03 28		15.7	GALAXY
ZWG 133.079	14 28 18.	+ 25 44		15.3	GALAXY
MCG+06-32-051	14 28 18.	+ 33 10	42	15.	GALAXY
UGC 09332	14 28 18.	+ 35 22	108	12.7	GALAXY Sa-b
ZWG 192.034	14 28 18.	+ 35 32		12.7	GALAXY
ZWG 220.022	14 28 18.	+ 40 57		15.7	GALAXY
ZWG 247.036	14 28 18.	+ 45 46		15.2	GALAXY
ZC 1428.3-0135	14 28 18.	- 01 35	2490		CLUSTER OF GALAXIES
SER 114.02	14 28 18.	- 44 43	70	16.	INTERACTING GALAXIES
REIZ 4174	14 28 19.	+ 33 09	18	15.5	GALAXY
MCG+06-32-052	14 28 19.	+ 33 09	24	15.	GALAXY
HOLM 657A	14 28 19.	+ 33 10	24	15.4	PART OF MULTIPLE GALAXY
HOLM 657B	14 28 20.	+ 33 09	24	15.5	PART OF MULTIPLE GALAXY
RNGC 5650	14 28 21.	+ 06 13			NON-EXISTENT OBJECT
HOLM 656B	14 28 21.	+ 06 13	9	14.8	PART OF MULTIPLE GALAXY
MCG+05-34-059	14 28 21.	+ 30 59	30		GALAXY
ARC 1933	14 28 22.	+ 70 22		17.3	RICH CLUSTER OF GALAXIES
REIZ 4166	14 28 23.	+ 06 13	9	15.2	GALAXY
ZWG 075.060	14 28 24.	+ 14 11		14.0	GALAXY
RNGC 5649	14 28 24.	+ 14 11		14.0	GALAXY
UGC 09333	14 28 24.	+ 14 11	54	14.0	GALAXY Sc
MCG+02-37-020	14 28 24.	+ 14 11	72	14.0	GALAXY
MCG+03-37-023	14 28 24.	+ 14 42	42	15.3	GALAXY
ZWG 104.039	14 28 24.	+ 14 44		15.3	GALAXY
MCG+06-32-053	14 28 24.	+ 35 33	120	13.	GALAXY
REIZ 4179	14 28 24.	+ 40 14	48	15.1	GALAXY
MCG+08-26-040	14 28 24.	+ 50 17	30	16.	GALAXY
MCG-05-34-016	14 28 24.	- 28 32	72	15.	GALAXY
RNGC 5652	14 28 27.	+ 06 12		14.0	GALAXY
REIZ 4170	14 28 29.	+ 06 13	66	13.5	GALAXY
HOLM 656A	14 28 29.	+ 06 13	72	13.2	PART OF MULTIPLE GALAXY
HN 0407	14 28 29.	- 43 11			NEBULA
HN 1738	14 28 29.7	- 00 41 35.	18		NEBULA
ZWG 019.053	14 28 30.	+ 01 27		15.4	GALAXY
8ZW 1428+02.6	14 28 30.	+ 02 37		18.1	COMPACT GALAXY
ZWG 047.072	14 28 30.	+ 06 12		13.8	GALAXY
8ZW 1428+06.2	14 28 30.	+ 06 12		18.5	COMPACT GALAXY
UGC 09334	14 28 30.	+ 06 12	126	13.8	GALAXY Sb
MCG+01-37-020	14 28 30.	+ 06 12	120	13.8	GALAXY
ZC 1428.5+2822	14 28 30.	+ 28 22	870		CLUSTER OF GALAXIES
ZWG 163.069	14 28 30.	+ 29 24		14.4	GALAXY
UGC 09335	14 28 30.	+ 29 24	120	14.4	GALAXY SB:b
ZWG 220.023	14 28 30.	+ 40 15		15.5	GALAXY
REIZ 4180	14 28 30.	+ 40 15	30	14.9	GALAXY
UGC 09336	14 28 30.	+ 40 15	72	15.5	GALAXY Sa-b
ZWG 220.024	14 28 30.	+ 41 02		15.6	GALAXY
UGC 09337	14 28 30.	+ 41 02	60	15.6	GALAXY S
ZWG 247.037	14 28 30.	+ 48 17		15.7	GALAXY
MCG+14-07-007	14 28 30.	+ 83 32 30.	15	16.	GALAXY
ZWG 019.052	14 28 30.	- 00 43		15.3	GALAXY
SER 114.01	14 28 30.	- 43 12	80	13.	HIGH SURF. BRGHTNSS GLXY
IC 4444	14 28 30.	- 43 12	90	12.2	GALAXY S
RNGC 5657	14 28 31.	+ 29 24		14.5	GALAXY
HN 1739	14 28 31.2	+ 01 27 02.	18		NEBULA
MCG+07-30-020	14 28 33.	+ 40 13	60	15.	GALAXY
REIZ 4173	14 28 35.	+ 14 08	30	13.3	GALAXY
RNGC 5655	14 28 36.	+ 14 13			GALAXY
ZWG 104.040	14 28 36.	+ 14 39		15.6	GALAXY
ZC 1428.6+2114	14 28 36.	+ 21 14	610		CLUSTER OF GALAXIES
MCG+04-34-043	14 28 36.	+ 25 42	48	15.3	GALAXY
ZC 1428.6+2629	14 28 36.	+ 26 29	670		CLUSTER OF GALAXIES
TON-N 0204	14 28 36.	+ 28 42		15.8	BLUE STAR
MCG+05-34-060	14 28 36.	+ 29 24	108	14.4	GALAXY
ZC 1428.6+3146	14 28 36.	+ 31 46	540		CLUSTER OF GALAXIES
MCG+07-30-021	14 28 36.	+ 41 02	54	15.	GALAXY
ZC 1428.6+4817	14 28 36.	+ 48 17	2350		CLUSTER OF GALAXIES
ZC 1428.6+6654	14 28 36.	+ 66 54	870		CLUSTER OF GALAXIES
MCG-04-34-018	14 28 36.	- 22 09	54	15.	GALAXY
MCG-07-30-002	14 28 36.	- 43 11	90	12.2	GALAXY
REIZ 4185	14 28 38.	+ 47 01	60	12.9	GALAXY
HN 1113	14 28 39.	- 45 49	48		NEBULA
IC 4445	14 28 39.	- 45 49			NONSTELLAR OBJECT
HN 0012	14 28 40.	- 00 06			NEBULA
REIZ 4171	14 28 40.	- 00 06	9	14.6	GALAXY
ARC 1924	14 28 40.	- 22 10		17.0	RICH CLUSTER OF GALAXIES
ZWG 047.073	14 28 42.	+ 05 31		15.2	GALAXY
UGC 09338	14 28 42.	+ 05 31	72	15.2	GALAXY Sb
MCG+04-34-044	14 28 42.	+ 25 33	96	15.0	GALAXY
ZC 1428.7+2555	14 28 42.	+ 25 55	2420		CLUSTER OF GALAXIES
MCG-04-34-019	14 28 42.	- 25 09	36	14.5	GALAXY
RNGC 5651	14 28 43.	- 00 42			NON-EXISTENT OBJECT
REIZ 4181	14 28 47.	+ 29 25	42	13.8	GALAXY
ZWG 047.074	14 28 48.	+ 03 42		15.5	GALAXY
ZWG 047.075	14 28 48.	+ 08 10		14.9	GALAXY
UGC 09339	14 28 48.	+ 08 10	66	14.9	GALAXY S
MCG+01-37-021	14 28 48.	+ 08 10	60	14.9	GALAXY
ZWG 133.080	14 28 48.	+ 23 16		15.7	GALAXY
ZWG 133.081	14 28 48.	+ 25 43		15.3	GALAXY
UGC 09340	14 28 48.	+ 25 43	60	15.3	GALAXY SBc
ZWG 163.070	14 28 48.	+ 28 30		15.1	GALAXY
MRK 684	14 28 48.	+ 28 30	13	15.	GALAXY WITH UV CONTINUUM
ARC 1927	14 28 51.	+ 25 54		16.0	RICH CLUSTER OF GALAXIES
RNGC 5659	14 28 53.	+ 25 35		15.0	GALAXY
ZWG 047.076	14 28 54.	+ 03 13		14.0	GALAXY
UGC 09341	14 28 54.	+ 03 13	96	14.0	GALAXY S0?
MCG+01-37-022	14 28 54.	+ 03 13	72	14.0	GALAXY
8ZW 1428+03.6	14 28 54.	+ 03 37		15.8	COMPACT GALAXY
ZWG 133.082	14 28 54.	+ 25 35		15.0	GALAXY
UGC 09342	14 28 54.	+ 25 35	102	15.0	GALAXY Sb

OBJECT NAME	RIGHT ASCEN.	DECLINATION	DIAM.	MAGN.	TYPE OF OBJECT
MCG+05-34-061	14 28 54.	+ 27 26	36	15.2	GALAXY
ZWG 163.071	14 28 54.	+ 27 27		15.2	GALAXY
MRK 685	14 28 54.	+ 27 27	15	15.	GALAXY WITH UV CONTINUUM
HARO 42	14 28 54.	+ 27 28			BLUE EMISSION-LINE GALAXY
MCG+05-34-062	14 28 54.	+ 28 30	39	15.1	GALAXY
MCG+12-14-007	14 28 54.	+ 71 19 30.	33	17.	GALAXY
ZWG 019.054	14 28 54.	- 03 05		15.2	GALAXY
MCG+00-37-014	14 28 54.	- 03 05	24	15.2	GALAXY
REIZ 4178	14 28 55.	+ 03 14	36	13.3	GALAXY
REIZ 4182	14 28 56.	+ 25 37	60	14.2	GALAXY
IC 1024	14 28 58.	+ 03 13 59.			NONSTELLAR OBJECT
HN 1114	14 28 58.	+ 15 28	12		NEBULA
IC 4449	14 28 58.	+ 15 28			NONSTELLAR OBJECT
LB 09888	14 29	- 86 30		13.5	FAINT BLUE STAR
ZWG 047.077	14 29 00.	+ 07 04		15.4	GALAXY
ZWG 047.078	14 29 00.	+ 07 17		15.3	GALAXY
IC 1025	14 29 00.	+ 07 18 11.			NONSTELLAR OBJECT
ZC 1429.0+2310	14 29 00.	+ 23 10	1340		CLUSTER OF GALAXIES
ZC 1429.0+2922	14 29 00.	+ 29 22	670		CLUSTER OF GALAXIES
MCG+06-32-054	14 29 00.	+ 35 44 30.	66	15.	GALAXY
ZWG 220.025	14 29 00.	+ 40 56		15.7	GALAXY
UGC 09343	14 29 00.	+ 40 56	72	15.7	GALAXY S
MCG+07-30-022	14 29 00.	+ 40 56	60	15.	GALAXY
ZWG 296.008	14 29 00.	+ 59 42		13.1	GALAXY
RNGC 5667	14 29 00.	+ 59 42		13.0	GALAXY
UGC 09344	14 29 00.	+ 59 42	108	13.1	GALAXY PECULR
MCG+10-21-004	14 29 00.	+ 59 42	96	13.	GALAXY
ZWG 354.001	14 29 00.	+ 78 39		15.6	GALAXY
ZWG 353.038	14 29 00.	+ 78 39		15.6	GALAXY
ZWG 354.002	14 29 00.	+ 79 13		15.6	GALAXY
ZWG 353.039	14 29 00.	+ 79 13		15.6	GALAXY
MCG+13-10-018	14 29 00.	+ 79 13	51	16.	GALAXY
ZWG 366.009	14 29 00.	+ 83 31		15.3	GALAXY
IC 4470	14 29 01.	+ 79 07			OPEN CLUSTER
REIZ 4184	14 29 02.	+ 25 29	72	14.2	GALAXY
8ZW 1429+01.2	14 29 06.	+ 01 12		17.0	COMPACT GALAXY
REIZ 4187	14 29 06.	+ 28 32	18	14.4	GALAXY
ZWG 163.072	14 29 06.	+ 31 00		15.5	GALAXY
IC 1026	14 29 06.	+ 31 26 37.			SINGLE STAR
ZC 1429.1+4915	14 29 06.	+ 49 15	1210		CLUSTER OF GALAXIES
ZWG 019.055	14 29 06.	- 01 47		15.7	GALAXY
MCG+00-37-015	14 29 06.	- 01 47	30	15.7	GALAXY
ZWG 047.079	14 29 12.	+ 06 22		15.4	GALAXY
UGC 09345	14 29 12.	+ 06 22	72	15.4	GALAXY Sc
MCG+06-32-055	14 29 12.	+ 35 33	30	16.	GALAXY
8ZW 1429-02.0	14 29 12.	- 01 59		15.4	COMPACT GALAXY
TON-N 0774	14 29 18.	+ 28 31		14.5	BLUE STAR
ZWG 192.035	14 29 18.	+ 33 27		14.6	GALAXY
MCG+06-32-056	14 29 18.	+ 35 51	42	15.	GALAXY
ZWG 273.009	14 29 18.	+ 55 28		15.6	GALAXY
ZWG 354.003	14 29 18.	+ 79 06		15.0	GALAXY
ZWG 353.040	14 29 18.	+ 79 06		15.0	GALAXY
MCG+13-10-019	14 29 18.	+ 79 07	39	16.	GALAXY
HN 1740	14 29 18.5	+ 00 08 52.	60		NEBULA
HOLM 658B	14 29 21.	+ 06 29	18	14.9	PART OF MULTIPLE GALAXY
MCG+06-32-057	14 29 21.	+ 33 27	48	15.	GALAXY
HN 0013	14 29 22.	- 00 09			NEBULA
REIZ 4183	14 29 22.	- 00 09	15	14.7	GALAXY
REIZ 4186	14 29 23.	+ 06 29	60	13.6	GALAXY
HOLM 658A	14 29 23.	+ 06 29	54	13.1	PART OF MULTIPLE GALAXY
IC 1023	14 29 23.	- 35 34 43.			NONSTELLAR OBJECT
RNGC 5643	14 29 23.	- 43 59		11.5	GALAXY
ZC 1429.4+0536	14 29 24.	+ 05 36	1550		CLUSTER OF GALAXIES
ZWG 047.080	14 29 24.	+ 07 32		15.6	GALAXY
ZWG 075.061	14 29 24.	+ 09 09		15.3	GALAXY
REIZ 4190	14 29 24.	+ 39 30	18	14.6	GALAXY
ZC 1429.4+4540	14 29 24.	+ 45 40	2420		CLUSTER OF GALAXIES
MCG+12-14-008	14 29 24.	+ 69 26	57	17.	GALAXY
ZWG 019.056	14 29 24.	- 01 46		15.6	GALAXY
MCG-05-34-017	14 29 24.	- 28 09	48	15.	GALAXY
VVI 73	14 29 24.	- 43 59	300	11.36	SEYFERT GALAXY
REIZ 4189	14 29 25.	+ 33 28	24	14.3	GALAXY
REIZ 4195	14 29 25.	+ 51 14	60	14.0	GALAXY
RNGC 5661	14 29 27.	+ 06 28		14.0	GALAXY
ZWG 047.081	14 29 30.	+ 06 28		14.2	GALAXY
8ZW 1429+06.5	14 29 30.	+ 06 28		17.5	COMPACT GALAXY
UGC 09346	14 29 30.	+ 06 28	102	14.2	GALAXY SBb
MCG+01-37-023	14 29 30.	+ 06 28	84	14.2	GALAXY
ZWG 075.062	14 29 30.	+ 08 52		15.3	GALAXY
MCG+07-30-023	14 29 30.	+ 39 27 30.	42	15.	GALAXY
ZWG 220.026	14 29 30.	+ 39 29		15.4	GALAXY
ZC 1429.5+4705	14 29 30.	+ 47 05	3760		CLUSTER OF GALAXIES
ZWG 247.038	14 29 30.	+ 47 43		15.5	GALAXY
SMI 05	14 29 30.	- 60 37	720		FAINT NEBULOSITY
PK315-00.1	14 29 31.	- 60 36 36.	30		PLANETARY NEBULA
ZWG 047.082	14 29 36.	+ 06 24		15.3	GALAXY
8ZW 1429+06.4	14 29 36.	+ 06 24		15.3	COMPACT GALAXY
ZWG 104.041	14 29 36.	+ 16 59		15.5	GALAXY
ZWG 133.083	14 29 36.	+ 22 10		15.7	GALAXY
ZC 1429.6+2947	14 29 36.	+ 29 47	810		CLUSTER OF GALAXIES
REIZ 4193	14 29 36.	+ 39 51	42	15.4	GALAXY
MCG-07-30-003	14 29 36.	- 43 56	210	11.4	GALAXY
ARC 1928	14 29 39.	+ 04 46		17.6	RICH CLUSTER OF GALAXIES
ARC 1929	14 29 44.	+ 29 46		17.3	RICH CLUSTER OF GALAXIES
MCG+08-26-041	14 29 45.	+ 50 11	132	13.8	GALAXY
REIZ 4192	14 29 47.	+ 29 12	15	15.3	GALAXY
ZC 1429.8+2931	14 29 48.	+ 29 31	740		CLUSTER OF GALAXIES
ZWG 192.036	14 29 48.	+ 33 51		15.2	GALAXY
REIZ 4194	14 29 48.	+ 33 52	42	14.1	GALAXY
ZWG 248.001	14 29 48.	+ 50 10		14.0	GALAXY
ZWG 247.039	14 29 48.	+ 50 10		14.0	GALAXY
UGC 09347	14 29 48.	+ 50 10	156	14.0	GALAXY Sc
RNGC 5658	14 29 49.	+ 00 30		14.5	GALAXY
REIZ 4188	14 29 49.	+ 03 09	24	14.6	GALAXY
RNGC 5673	14 29 50.	+ 50 11		14.0	GALAXY
MCG+06-32-058A	14 29 51.	+ 33 51 30.	48	15.	GALAXY
MCG+06-32-058	14 29 51.	+ 33 51 30.	18	16.	GALAXY
REIZ 4202	14 29 51.	+ 50 10	120	13.8	GALAXY
HOLM 659A	14 29 53.	+ 29 11	24	15.1	PART OF MULTIPLE GALAXY
ZWG 019.057	14 29 54.	+ 00 30		14.6	GALAXY
UGC 09348	14 29 54.	+ 00 30	108	14.6	GALAXY S
MCG+00-37-016	14 29 54.	+ 00 30	90	14.6	GALAXY
ZC 1429.9+0336	14 29 54.	+ 03 36	3020		CLUSTER OF GALAXIES
ZWG 075.063	14 29 54.	+ 10 09		15.3	GALAXY
ZWG 163.073	14 29 54.	+ 28 46		15.4	GALAXY
UGC 09349	14 29 54.	+ 28 46	72	15.4	GALAXY S
MCG+05-34-063	14 29 54.	+ 29 11	24		GALAXY
ZWG 163.074	14 29 54.	+ 31 48		15.5	GALAXY

OBJECT NAME	RIGHT ASCEN.	DECLINATION	DIAM.	MAGN.	TYPE OF OBJECT
ZWG 192.037	14 29 54.	+ 36 31		15.3	GALAXY
UGC 09350	14 29 54.	+ 36 31	60	15.3	GALAXY S
MCG+06-32-059	14 29 54.	+ 36 31 30.	36	16.	GALAXY
VV 262B	14 29 54.	+ 36 32	36	16.	INTERACTING GALAXY
VV 262A	14 29 54.	+ 36 32	21	16.	INTERACTING GALAXY
VV 262	14 29 54.	+ 36 32	54		INTERACTING GALAXY
ZC 1429.9+5256	14 29 54.	+ 52 56	10010		CLUSTER OF GALAXIES
SC 1427-1625.1	14 29 54.	- 16 38 25.	12		NEBULA
OCL 0923	14 29 54.	- 61 10	420		OPEN STAR CLUSTER
HN 1741	14 29 54.9	+ 00 30 29.	60		NEBULA
ARP 049	14 29 55.	+ 08 17			PECULIAR GALAXY
REIZ 4203	14 29 55.	+ 51 31	90	15.3	GALAXY
RNGC 5665	14 29 58.	+ 08 18		13.0	GALAXY
REIZ 4191	14 29 58.	+ 08 18	48	12.5	GALAXY
HOLM 659B	14 29 58.	+ 29 11	18	15.9	PART OF MULTIPLE GALAXY
REIZ 4198	14 29 58.	+ 36 32	60	15.2	GALAXY
HOLM 660A	14 29 58.	+ 36 32	90	14.9	PART OF MULTIPLE GALAXY
HOLM 660A	14 29 58.	+ 36 32	60	14.4	PART OF MULTIPLE GALAXY
REIZ 4196	14 29 59.	+ 28 47	36	14.1	GALAXY
LB 09889	14 30	- 83 25		13.3	FAINT BLUE STAR
ZWG 047.083	14 30 00.	+ 05 47		15.7	GALAXY
UGC 09351	14 30 00.	+ 05 54	66	17.	GALAXY
ZWG 047.084	14 30 00.	+ 08 18		12.6	GALAXY
UGC 09352	14 30 00.	+ 08 18	138	12.6	GALAXY S(c)
MCG+01-37-024	14 30 00.	+ 08 18	120	12.6	GALAXY
MCG+05-34-064	14 30 00.	+ 26 32	42	15.7	GALAXY
ZWG 163.075	14 30 00.	+ 26 33		15.7	GALAXY
IC 4450	14 30 00.	+ 28 45 23.			NONSTELLAR OBJECT
MCG+05-34-065	14 30 00.	+ 28 46	72	15.4	GALAXY
MCG+06-32-060	14 30 00.	+ 36 31 30.	54	15.	GALAXY
MCG+05-34-066	14 30 03.	+ 31 48 30.	30	15.5	GALAXY
REIZ 4199	14 30 04.	+ 36 32	48	14.9	GALAXY
SC 1427-1421.6	14 30 05.	- 14 34 54.	6		NEBULA
SC 1427-1654.2	14 30 05.	- 17 07 30.	12		NEBULA
ZC 1430.1+1209	14 30 06.	+ 12 09	1680		CLUSTER OF GALAXIES
ZWG 133.084	14 30 06.	+ 23 30		15.7	GALAXY
MCG+04-34-045	14 30 06.	+ 24 01	36		GALAXY
ARC 1931	14 30 06.	+ 44 30		17.2	RICH CLUSTER OF GALAXIES
ZWG 019.058	14 30 12.	+ 02 05		15.5	GALAXY
ZWG 047.085	14 30 12.	+ 03 08		14.9	GALAXY
MCG+01-37-025	14 30 12.	+ 03 08	48	14.9	GALAXY
ZWG 133.085	14 30 12.	+ 26 24		15.2	GALAXY
ZWG 163.076	14 30 12.	+ 27 39		14.9	GALAXY
KARA.73B 0633	14 30 12.	+ 27 39	24	14.9	ISOLATED GALAXY E
ZWG 273.010	14 30 12.	+ 56 10		15.4	GALAXY
MCG-02-37-006	14 30 12.	- 12 45	90	14.5	GALAXY
REIZ 4197	14 30 14.	+ 10 08	120	12.4	GALAXY
HN 1742	14 30 14.4	- 01 05 06.	12		NEBULA
IC 4452	14 30 15.	+ 27 37 42.			NONSTELLAR OBJECT
MCG+09-24-010	14 30 15.	+ 56 11	42	15.	GALAXY
RNGC 5669	14 30 17.	+ 10 07		12.5	GALAXY
ZWG 019.059	14 30 18.	+ 01 44		15.5	GALAXY
ZWG 075.064	14 30 18.	+ 10 06		13.2	GALAXY
UGC 09353	14 30 18.	+ 10 06	270	13.2	GALAXY Sc
MCG+02-37-021	14 30 18.	+ 10 06	240	13.2	GALAXY
ZC 1430.3+2006	14 30 18.	+ 20 06	1750		CLUSTER OF GALAXIES
REIZ 4201	14 30 18.	+ 27 39	30	14.3	GALAXY
MCG+06-32-061	14 30 18.	+ 33 48	24	16.	GALAXY
ARC 1932	14 30 22.	+ 47 20		17.2	RICH CLUSTER OF GALAXIES
ZWG 104.042	14 30 24.	+ 14 49		15.2	GALAXY
8ZW 1430+14.8	14 30 24.	+ 14 49		15.2	COMPACT GALAXY
ZWG 163.077	14 30 24.	+ 31 53		14.5	GALAXY
UGC 09354	14 30 24.	+ 31 53	54	14.5	GALAXY PECULR?
ZWG 247.040	14 30 24.	+ 46 08		15.3	GALAXY
MCG+13-10-020	14 30 24.	+ 79 27	66	15.	GALAXY
ZWG 354.004	14 30 24.	+ 79 28		14.9	GALAXY
ZWG 353.041	14 30 24.	+ 79 28		14.9	GALAXY
UGC 09355	14 30 24.	+ 79 28	72	14.9	GALAXY S
ARC 1930	14 30 28.	+ 31 51		17.0	RICH CLUSTER OF GALAXIES
ZWG 075.065	14 30 30.	+ 11 49		14.3	GALAXY
UGC 09356	14 30 30.	+ 11 49	102	14.3	GALAXY S
MCG+02-37-022	14 30 30.	+ 11 49	96	14.3	GALAXY
SHAH 076	14 30 30.	+ 41 30	90	18.0	GROUP OF COMPACT GALAXIES
MCG+10-21-005	14 30 30.	+ 58 08	162	12.1	GALAXY
MCG+10-21-006	14 30 30.	+ 58 09	10	16.	GALAXY
MCG-02-37-007	14 30 30.	- 15 12	15	15.	GALAXY
RNGC 5672	14 30 32.	+ 31 53		14.5	GALAXY
REIZ 4207	14 30 32.	+ 31 53	30	13.6	GALAXY
MCG+05-34-067	14 30 33.	+ 31 52	6		GALAXY
MCG+05-34-068	14 30 33.	+ 31 53 30.	42	14.5	GALAXY
RNGC 5675	14 30 34.	+ 36 31		14.0	GALAXY
RNGC 5678	14 30 34.	+ 58 08		12.5	GALAXY
REIZ 4214	14 30 34.	+ 58 09	150	12.2	GALAXY
RNGC 5712	14 30 34.	+ 79 05		15.5	GALAXY
IC 1028	14 30 35.	+ 42 03 51.			NONSTELLAR OBJECT
ZWG 047.086	14 30 36.	+ 04 10		14.9	GALAXY
MCG+01-37-026	14 30 36.	+ 04 10	60	14.9	GALAXY
ZWG 133.086	14 30 36.	+ 24 30		15.6	GALAXY
MCG+05-34-070	14 30 36.	+ 31 47	9		GALAXY
MCG+05-34-069	14 30 36.	+ 31 49	5		GALAXY
ZC 1430.6+3150	14 30 36.	+ 31 50	2550		CLUSTER OF GALAXIES
ZWG 192.038	14 30 36.	+ 36 31		14.0	GALAXY
UGC 09357	14 30 36.	+ 36 31	174	14.0	GALAXY S
MCG+06-32-062	14 30 36.	+ 36 32	180	13.	GALAXY
ZC 1430.6+4959	14 30 36.	+ 49 59	870		CLUSTER OF GALAXIES
MCG+08-26-042	14 30 36.	+ 50 08	150	13.	GALAXY
ZWG 296.009	14 30 36.	+ 58 09		12.1	GALAXY
UGC 09358	14 30 36.	+ 58 09	210	12.1	GALAXY Sb
KARA.73B 0634	14 30 36.	+ 58 09	222	12.1	ISOLATED GALAXY S
7ZW 553	14 30 36.	+ 79 05			COMPACT GALAXY
ZWG 354.005	14 30 36.	+ 79 05		15.3	GALAXY
ZWG 353.042	14 30 36.	+ 79 05		15.3	GALAXY
REIZ 4200	14 30 37.	+ 04 11	42	14.2	GALAXY
REIZ 4210	14 30 39.	+ 36 31	90	13.5	GALAXY
RNGC 5666	14 30 41.	+ 10 43		13.5	GALAXY
ZWG 047.087	14 30 42.	+ 02 52		15.5	GALAXY
ZWG 075.066	14 30 42.	+ 10 43		13.5	GALAXY
UGC 09360	14 30 42.	+ 10 43	54	13.5	GALAXY VY CMPT
MCG+02-37-023	14 30 42.	+ 10 43	60	13.5	GALAXY
TON-N 0775	14 30 42.	+ 24 49		16.0	BLUE STAR
ZWG 192.039	14 30 42.	+ 36 55		15.5	GALAXY
ZWG 248.002	14 30 42.	+ 50 07		13.7	GALAXY
ZWG 247.041	14 30 42.	+ 50 07		13.7	GALAXY
UGC 09361	14 30 42.	+ 50 07	168	13.7	GALAXY Sb
7ZW 554	14 30 42.	+ 59 28			COMPACT GALAXY
ZWG 019.060	14 30 42.	- 00 56		15.6	GALAXY
UGC 09359	14 30 42.	- 00 56	90	15.6	GALAXY
REIZ 4205	14 30 44.	+ 10 44	48	13.4	GALAXY

OBJECT NAME	RIGHT ASCEN.	DECLINATION	DIAM.	MAGN.	TYPE OF OBJECT
MCG+06-32-063	14 30 45.	+ 36 56	30	14.5	GALAXY
IC 1029	14 30 46.	+ 50 08		13.33	GALAXY S
ZWG 019.061	14 30 48.	+ 01 02		15.6	GALAXY
ZWG 047.088	14 30 48.	+ 04 07		14.9	GALAXY
UGC 09362	14 30 48.	+ 04 07	66	14.9	GALAXY Sb-c
MCG+01-37-027	14 30 48.	+ 04 07	48	14.9	GALAXY
ZWG 047.089	14 30 48.	+ 06 59		15.4	GALAXY
ZWG 075.067	14 30 48.	+ 09 18		15.4	GALAXY
ZWG 075.068	14 30 48.	+ 09 53		15.7	GALAXY
ZC 1430.8+1246	14 30 48.	+ 12 46	1550		CLUSTER OF GALAXIES
ZWG 075.069	14 30 48.	+ 14 02		15.5	GALAXY
KARA.73B 0635	14 30 48.	+ 14 02	60	15.5	ISOLATED GALAXY S
ZC 1430.8+2040	14 30 48.	+ 20 40	2150		CLUSTER OF GALAXIES
ZWG 296.010	14 30 48.	+ 57 05		15.6	GALAXY
MRK 473	14 30 48.	+ 57 05	12	16.	GALAXY WITH UV CONTINUUM
MCG+12-14-009	14 30 48.	+ 71 00	57	17.	GALAXY
MCG+13-10-021	14 30 48.	+ 79 05	12	16.	GALAXY
ZC 1430.8+7959	14 30 48.	+ 79 59	1010		CLUSTER OF GALAXIES
REIZ 4204	14 30 49.	+ 04 08	42	13.9	GALAXY
SN 1952G	14 30 53.	+ 04 40		17.9	SUPERNOVA
ZWG 047.090	14 30 54.	+ 04 40		12.7	GALAXY
REIZ 4206	14 30 54.	+ 04 40	60	12.8	GALAXY
UGC 09363	14 30 54.	+ 04 40	210	12.7	GALAXY Sc
SN 1954B	14 30 54.	+ 04 40		12.3	SUPERNOVA
MCG+01-37-028	14 30 54.	+ 04 40	180	12.7	GALAXY
ZWG 047.091	14 30 54.	+ 07 05		15.0	GALAXY
UGC 09364	14 30 54.	+ 07 05	108	15.0	GALAXY
MCG+01-37-029	14 30 54.	+ 07 05	72	15.0	GALAXY
ZC 1430.9+1518	14 30 54.	+ 15 18	1280		CLUSTER OF GALAXIES
ZC 1430.9+2214	14 30 54.	+ 22 14	540		CLUSTER OF GALAXIES
TON-N 0776	14 30 54.	+ 28 21		14.5	BLUE STAR
ZC 1430.9+2928	14 30 54.	+ 29 28	810		CLUSTER OF GALAXIES
ZC 1430.9+2955	14 30 54.	+ 29 55	740		CLUSTER OF GALAXIES
HN 1115	14 30 56.	+ 17 56	6		NEBULA
RNGC 5668	14 30 57.	+ 04 40		12.0	GALAXY
IC 4454	14 30 57.	+ 17 56			NONSTELLAR OBJECT
REIZ 4215	14 30 57.	+ 50 07	180	13.3	GALAXY
HMS 1431+3146	14 31	+ 31 46			BOOTES GALAXY CLUSTER
ZWG 047.092	14 31 00.	+ 02 51		15.5	GALAXY
ZWG 047.093	14 31 00.	+ 03 54		15.0	GALAXY
UGC 09365	14 31 00.	+ 03 54	66	15.0	GALAXY Sb-c
MCG+01-37-030	14 31 00.	+ 03 54	48	15.0	GALAXY
ZWG 075.070	14 31 00.	+ 09 28		15.4	GALAXY
ZC 1431.0+2338	14 31 00.	+ 23 38	2080		CLUSTER OF GALAXIES
TON-N 0205	14 31 00.	+ 25 43		16.3	BLUE STAR
ZC 1431.0+2940	14 31 00.	+ 29 40	810		CLUSTER OF GALAXIES
MCG+06-32-064	14 31 00.	+ 34 58	42	15.5	GALAXY
ZC 1431.0+3459	14 31 00.	+ 34 59	870		CLUSTER OF GALAXIES
ZWG 248.003	14 31 00.	+ 49 40		11.7	GALAXY
ZWG 247.042	14 31 00.	+ 49 40		11.7	GALAXY
UGC 09366	14 31 00.	+ 49 40	240	11.7	GALAXY Sc
MCG+08-26-043	14 31 00.	+ 49 40	204	11.7	GALAXY
MCG-02-37-008	14 31 00.	- 14 24	36	15.	GALAXY
REIZ 4208	14 31 01.	+ 03 55	30	14.6	GALAXY
RNGC 5676	14 31 01.	+ 49 41		12.0	GALAXY
RNGC 5664	14 31 02.	- 14 23		15.0	GALAXY
REIZ 4216	14 31 03.	+ 49 41	120	11.9	GALAXY
REIZ 4209	14 31 04.	+ 08 27	60	14.5	GALAXY
REIZ 4213	14 31 05.	+ 40 29	60	14.8	GALAXY
ZWG 104.043	14 31 06.	+ 20 13		15.0	GALAXY
MCG+03-37-024	14 31 06.	+ 20 13	42	15.0	GALAXY
ZWG 133.087	14 31 06.	+ 26 25		15.6	GALAXY
ZWG 163.078	14 31 06.	+ 28 55		15.5	GALAXY
ZC 1431.1+6601	14 31 06.	+ 66 01	340		CLUSTER OF GALAXIES
RNGC 5663	14 31 07.	- 16 20		15.0	GALAXY
MCG-03-37-003	14 31 09.	- 16 20 30.	18	15.5	GALAXY
ZC 1431.2+3137	14 31 12.	+ 31 37	670		CLUSTER OF GALAXIES
ZWG 192.040	14 31 12.	+ 34 57		15.6	GALAXY
UGC 09367	14 31 12.	+ 34 57	60	15.6	GALAXY S0-a
MCG+06-32-065	14 31 12.	+ 34 57	48	15.	GALAXY
ZWG 220.027	14 31 12.	+ 41 32		15.6	GALAXY
MCG-02-37-009	14 31 12.	- 13 03	48	15.	GALAXY
SC 1428-1648.1	14 31 12.	- 17 01 21.	6		NEBULA
ARC 1934	14 31 14.	+ 29 41		17.4	RICH CLUSTER OF GALAXIES
ZWG 047.094	14 31 18.	+ 04 10		15.2	GALAXY
MCG+07-30-024	14 31 18.	+ 41 31	39	15.	GALAXY
ZWG 220.028	14 31 18.	+ 41 52		14.8	GALAXY
UGC 09368	14 31 19.	+ 41 52	72	14.8	GALAXY S
REIZ 4211	14 31 19.	+ 04 11	24	14.5	GALAXY
RNGC 5674	14 31 21.	+ 05 40		13.5	GALAXY
MCG+07-30-025	14 31 21.	+ 41 51	66	14.5	GALAXY
ZWG 047.095	14 31 24.	+ 03 54		15.5	GALAXY
ZWG 047.096	14 31 24.	+ 05 40		13.7	GALAXY
8ZW 1431+05.7	14 31 24.	+ 05 40		13.7	COMPACT GALAXY
UGC 09369	14 31 24.	+ 05 40	72	13.7	GALAXY Sb/SBc
MCG+01-37-031	14 31 24.	+ 05 40	60	13.7	GALAXY
REIZ 4212	14 31 24.	+ 05 41	54	13.5	GALAXY
8ZW 1431+19.1	14 31 24.	+ 19 05		17.8	COMPACT GALAXY
ZWG 163.079	14 31 24.	+ 28 16		15.3	GALAXY
MCG+06-32-066	14 31 24.	+ 34 47	24	15.	GALAXY
ZWG 248.004	14 31 24.	+ 46 00		15.7	GALAXY
MCG+08-26-044	14 31 24.	+ 46 00	18	16.	GALAXY
UGC 09370	14 31 24.	+ 53 14	60	16.0	GALAXY Sb
ZWG 273.011	14 31 24.	+ 55 57		15.7	GALAXY
ZWG 019.062	14 31 24.	- 02 51		15.7	GALAXY
MCG-01-37-008	14 31 24.	- 07 29	60	15.	GALAXY
ZWG 047.097	14 31 30.	+ 03 58		15.4	GALAXY
ZWG 047.098	14 31 30.	+ 04 00		15.0	GALAXY
UGC 09371	14 31 30.	+ 04 00	60	15.0	GALAXY DBL SYS
MKW 07	14 31 30.	+ 04 00		15.0	POOR GALAXY CLUSTER
MCG+01-37-032	14 31 30.	+ 04 00	30	15.0	GALAXY
ZWG 047.099	14 31 30.	+ 05 48		15.6	GALAXY
MCG+05-34-071	14 31 30.	+ 28 15	15	15.3	GALAXY
TON-N 0206	14 31 30.	+ 29 48		15.5	BLUE STAR
ZWG 192.041	14 31 30.	+ 36 10		15.5	GALAXY
UGC 09372	14 31 30.	+ 36 10	60	15.5	GALAXY SBb
MCG+09-24-011	14 31 30.	+ 53 13	60	14.	GALAXY
MCG+09-24-012	14 31 30.	+ 53 28	60	14.	GALAXY
ZWG 273.012	14 31 30.	+ 53 30		15.4	GALAXY
MCG+09-24-013	14 31 30.	+ 53 30	48	15.	GALAXY
RNGC 5662	14 31 30.	- 56 21		7.5	OPEN CLUSTER
HN 1743	14 31 32.2	+ 00 50 10.	18		NEBULA
HN 0408	14 31 33.	- 36 04			NEBULA
IC 4451	14 31 33.	- 36 04			NONSTELLAR OBJECT
IC 4453	14 31 34.	- 27 18 02.			NONSTELLAR OBJECT
IC 1030	14 31 35.	+ 31 54 49.			NONSTELLAR OBJECT
ZWG 019.063	14 31 36.	+ 00 49		15.3	GALAXY
ZWG 047.100	14 31 36.	+ 03 58		15.4	GALAXY

OBJECT NAME	RIGHT ASCEN.	DECLINATION	DIAM.	MAGN.	TYPE OF OBJECT
ZC 1431.6+0724	14 31 36.	+ 07 24	1550		CLUSTER OF GALAXIES
ZC 1431.6+1902	14 31 36.	+ 19 02	2550		CLUSTER OF GALAXIES
ZWG 192.042	14 31 36.	+ 35 31		15.7	GALAXY
MCG+06-32-067	14 31 36.	+ 36 11	48	14.5	GALAXY
ZC 1431.6+3614	14 31 36.	+ 36 14	1210		CLUSTER OF GALAXIES
MCG+07-30-026	14 31 36.	+ 41 00	60	16.	GALAXY
UGC 09373	14 31 36.	+ 41 01	60	16.0	GALAXY S
MCG+09-24-014	14 31 36.	+ 53 11	18	16.	GALAXY
ZWG 273.013	14 31 36.	+ 53 28		15.0	GALAXY
MCG+09-24-015	14 31 36.	+ 56 30	36	16.	GALAXY
MCG+10-21-007	14 31 36.	+ 58 48	27	16.	GALAXY
MCG-04-34-020	14 31 36.	- 27 18	36	14.	GALAXY
MCG-05-34-018	14 31 36.	- 27 45	48	15.	GALAXY
OCL 0928	14 31 36.	- 56 20	1860	8.2	OPEN STAR CLUSTER
VHA 162	14 31 36.	- 56 20	1200		OPEN STAR CLUSTER
MCG+06-32-069	14 31 39.	+ 35 12	42	15.	GALAXY
MCG+06-32-068	14 31 39.	+ 35 32	15	15.	GALAXY
REIZ 4217	14 31 40.	+ 36 11	30	14.7	GALAXY
ZWG 075.071	14 31 42.	+ 09 15		15.6	GALAXY
ZWG 075.072	14 31 42.	+ 10 26		15.4	GALAXY
UGC 09374	14 31 42.	+ 10 26	66	15.4	GALAXY S
ZWG 075.073	14 31 42.	+ 12 54		15.5	GALAXY
ZWG 075.074	14 31 42.	+ 12 56		15.2	GALAXY
UGC 09375	14 31 42.	+ 28 38	60	16.0	GALAXY Sb
ZWG 163.080	14 31 42.	+ 29 21		15.7	GALAXY
HN 0410	14 31 44.	- 14 25			NEBULA
IC 4455	14 31 44.	- 14 25			NONSTELLAR OBJECT
HN 1116	14 31 46.	+ 16 24	18		NEBULA
IC 4456	14 31 46.	+ 16 24			NONSTELLAR OBJECT
HOLM 661A	14 31 47.	+ 40 18	54	13.9	PART OF MULTIPLE GALAXY
REIZ 4219	14 31 47.	+ 40 19	42	14.2	GALAXY
ZWG 047.101	14 31 48.	+ 03 53		15.3	GALAXY
ZWG 075.075	14 31 48.	+ 09 12		15.5	GALAXY
MCG+04-34-046	14 31 48.	+ 25 41	48	14.8	GALAXY
ZC 1431.8+3235	14 31 48.	+ 32 35	1950		CLUSTER OF GALAXIES
ZWG 220.029	14 31 48.	+ 40 17		15.0	GALAXY
UGC 09376	14 31 48.	+ 40 17	102	15.0	GALAXY S
KARA.72 426A	14 31 48.	+ 40 17	84	15.0	PART OF DOUBLE GALAXY
ZWG 220.030	14 31 48.	+ 40 18		14.8	GALAXY
KARA.72 426B	14 31 48.	+ 40 18	84	14.8	PART OF DOUBLE GALAXY
UGC 09377	14 31 48.	+ 44 18	60	16.5	GALAXY Sc
HOLM 661B	14 31 49.	+ 40 19	36	14.1	PART OF MULTIPLE GALAXY
MCG+07-30-027	14 31 51.	+ 40 16 30.	90	14.	GALAXY
REIZ 4221	14 31 53.	+ 40 25	42	15.4	GALAXY
ZWG 047.102	14 31 54.	+ 03 00		15.7	GALAXY
ZC 1431.9+1420	14 31 54.	+ 14 20	3360		CLUSTER OF GALAXIES
ZC 1431.9+2019	14 31 54.	+ 20 19	940		CLUSTER OF GALAXIES
ZC 1431.9+2204	14 31 54.	+ 22 04	740		CLUSTER OF GALAXIES
ZWG 163.081	14 31 54.	+ 28 10		15.2	GALAXY
MCG+07-30-028	14 31 54.	+ 40 17	36	14.5	GALAXY
ZC 1431.9+6020	14 31 54.	+ 60 20	9010		CLUSTER OF GALAXIES
REIZ 4218	14 31 56.	+ 25 41	48	14.1	GALAXY
RNGC 5677	14 31 59.	+ 25 41		15.0	GALAXY
VDB.66G 193	14 32	+ 58 45	70		DWARF GALAXY
ZWG 133.088	14 32 00.	+ 25 41		14.8	GALAXY
UGC 09378	14 32 00.	+ 25 41	60	14.8	GALAXY S
MCG+05-34-072	14 32 00.	+ 28 09	24	15.2	GALAXY
ZWG 220.031	14 32 00.	+ 40 28		15.6	GALAXY
UGC 09379	14 32 00.	+ 40 28	102	15.6	GALAXY Sc
ZC 1432.0+5054	14 32 00.	+ 50 54	810		CLUSTER OF GALAXIES
ZC 1432.0+5450	14 32 00.	+ 54 50	2220		CLUSTER OF GALAXIES
MCG+10-21-008	14 32 00.	+ 57 55	24	16.	GALAXY
ZWG 296.011	14 32 00.	+ 57 56		15.7	GALAXY
KARA.73B 0636	14 32 00.	+ 57 56	36	15.7	ISOLATED GALAXY S0
MCG+07-30-029	14 32 03.	+ 40 27	72	14.5	GALAXY
ZWG 047.103	14 32 06.	+ 03 05		15.6	GALAXY
ZWG 047.104	14 32 06.	+ 04 29		15.3	GALAXY
8ZW 1432+04.5	14 32 06.	+ 04 29		15.3	COMPACT GALAXY
UGC 09380	14 32 06.	+ 04 29	126	15.3	GALAXY DWRF IR
ZWG 075.076	14 32 06.	+ 10 00		15.3	GALAXY
ZWG 104.044	14 32 06.	+ 19 58		15.6	GALAXY
ZC 1432.1+5000	14 32 06.	+ 50 00	870		CLUSTER OF GALAXIES
OCL 0924	14 32 06.	- 59 44	510	13.	OPEN STAR CLUSTER
HN 1117	14 32 08.	+ 18 26	12		NEBULA
IC 4457	14 32 08.	+ 18 26			NONSTELLAR OBJECT
ZWG 019.064	14 32 12.	+ 01 05		15.5	GALAXY
ZWG 047.105	14 32 12.	+ 03 30		15.3	GALAXY
TON-N 0777	14 32 12.	+ 28 04		14.5	BLUE STAR
ZWG 163.082	14 32 12.	+ 31 46		15.3	GALAXY
MCG+05-34-073	14 32 12.	+ 31 46	36	15.3	GALAXY
ZWG 192.043	14 32 12.	+ 37 24		15.7	GALAXY
ZC 1432.2+3919	14 32 12.	+ 39 19	1480		CLUSTER OF GALAXIES
MCG+08-27-001	14 32 15.	+ 49 40	48	16.	GALAXY
ZWG 047.106	14 32 18.	+ 03 33		15.2	GALAXY
ZWG 047.107	14 32 18.	+ 03 52		15.3	GALAXY
ZWG 047.108	14 32 18.	+ 08 23		14.9	GALAXY
MCG+01-37-033	14 32 18.	+ 08 23	48	14.9	GALAXY
ZWG 075.077	14 32 18.	+ 10 03		15.3	GALAXY
ZWG 075.078	14 32 18.	+ 10 15		15.4	GALAXY
ZWG 075.079	14 32 18.	+ 13 30		15.7	GALAXY
MCG+03-37-025	14 32 18.	+ 19 58	54	15.6	GALAXY
ZC 1432.3+2854	14 32 18.	+ 28 54	1010		CLUSTER OF GALAXIES
ZWG 163.083	14 32 18.	+ 31 11		15.3	GALAXY
MCG+09-24-016	14 32 18.	+ 56 43	36	15.	GALAXY
ZWG 296.012	14 32 18.	+ 56 45		15.6	GALAXY
MCG-01-37-009	14 32 18.	- 08 40	84	16.	GALAXY
REIZ 4223	14 32 21.	+ 31 11	24	15.2	GALAXY
HOLM 662A	14 32 21.	+ 31 11	36	14.5	PART OF MULTIPLE GALAXY
REIZ 4220	14 32 22.	+ 08 22	54	14.6	GALAXY
RNGC 5670	14 32 22.	- 45 43			NON-EXISTENT OBJECT
HOLM 662B	14 32 23.	+ 31 11	12	15.2	PART OF MULTIPLE GALAXY
ZC 1432.4+3017	14 32 24.	+ 30 17	2760		CLUSTER OF GALAXIES
ZWG 163.084	14 32 24.	+ 30 30		15.0	GALAXY
MCG+05-34-075	14 32 24.	+ 30 30	42		GALAXY
MCG+05-34-074	14 32 24.	+ 31 11	30	15.3	GALAXY
IC 4459	14 32 24.	+ 31 11 17.			NONSTELLAR OBJECT
MCG+10-21-009	14 32 24.	+ 60 17	51	17.	GALAXY
SC 1429-1543.9	14 32 27.	- 15 57 06.	30		NEBULA
KARA.72 427A	14 32 30.	+ 05 34	48	14.2	PART OF DOUBLE GALAXY
ARP 274	14 32 30.	+ 05 34			PECULIAR GALAXY
UGC 09381	14 32 30.	+ 36 30	90	17.	GALAXY DWRF SP
ZC 1432.5+4325	14 32 30.	+ 43 25	3760		CLUSTER OF GALAXIES
MCG+15-01-015	14 32 30.	+ 87 23	24	17.	GALAXY
IC 4460	14 32 32.	+ 30 28 36.			NONSTELLAR OBJECT
RNGC 5679C	14 32 33.	+ 05 34		16.0	GALAXY
RNGC 5679B	14 32 33.	+ 05 34		14.5	GALAXY
RNGC 5679A	14 32 33.	+ 05 34		14.0	GALAXY
ZWG 047.109	14 32 36.	+ 03 45		15.2	GALAXY
UGC 09382	14 32 36.	+ 03 45	78	15.2	GALAXY Sa-b
ZWG 047.110	14 32 36.	+ 05 34		14.2	GALAXY
UGC 09383	14 32 36.	+ 05 34	72	14.2	GALAXY Sb
KARA.72 427B	14 32 36.	+ 05 34	78		PART OF DOUBLE GALAXY
MCG+01-37-036	14 32 36.	+ 05 34	9	14.2	GALAXY
MCG+01-37-035	14 32 36.	+ 05 34	72	14.2	GALAXY
MCG+01-37-034	14 32 36.	+ 05 34	42	14.2	GALAXY
REIZ 4222	14 32 36.	+ 05 35	72	14.7	GALAXY
ZWG 075.080	14 32 36.	+ 12 57		15.6	GALAXY
TON-N 0207	14 32 36.	+ 25 04		14.6	BLUE STAR
ZWG 248.005	14 32 36.	+ 48 15		15.4	GALAXY
MCG+09-24-017	14 32 36.	+ 54 20	60	15.	GALAXY
ZWG 104.045	14 32 42.	+ 16 20		15.7	GALAXY
MCG+05-34-076	14 32 42.	+ 26 43 30.	21	15.3	GALAXY
ZC 1432.7+5023	14 32 42.	+ 50 23	940		CLUSTER OF GALAXIES
MCG-02-37-010	14 32 42.	- 13 31	72	15.	GALAXY
MCG+05-34-077	14 32 45.	+ 26 44 30.	72	15.3	GALAXY
REIZ 4229	14 32 47.	+ 48 12	60	14.6	GALAXY
HN 1744	14 32 47.6	+ 00 41 25.	18		NEBULA
ZWG 019.065	14 32 48.	+ 00 40		15.5	GALAXY
MCG+05-34-078	14 32 48.	+ 26 44	12	15.3	GALAXY
VV 303B	14 32 48.	+ 26 44 30.	12	16.5	INTERACTING GALAXY
VV 303A	14 32 48.	+ 26 44 30.	36	15.	INTERACTING GALAXY
IC 4461	14 32 48.	+ 26 44 37.			NONSTELLAR OBJECT
ZWG 163.085	14 32 48.	+ 26 46		15.3	GALAXY
UGC 09384	14 32 48.	+ 26 46	60	15.3	GALAXY DBL SYS
ZC 1432.8+3645	14 32 48.	+ 36 45	1140		CLUSTER OF GALAXIES
ZWG 248.006	14 32 48.	+ 48 11		15.6	GALAXY
IC 1031	14 32 48.	+ 48 15 14.			NONSTELLAR OBJECT
MCG+09-24-018	14 32 48.	+ 51 28	60	16.	GALAXY
ZC 1432.8+6940	14 32 48.	+ 69 40	2620		CLUSTER OF GALAXIES
IC 4462	14 32 50.	+ 26 45 19.			NONSTELLAR OBJECT
REIZ 4230	14 32 52.	+ 48 53	72	14.0	GALAXY
HOLM 663A	14 32 52.	+ 48 54	60	13.7	PART OF MULTIPLE GALAXY
HN 1745	14 32 52.9	+ 00 33 37.	18		NEBULA
ZWG 019.066	14 32 54.	+ 00 32		14.9	GALAXY
MCG+00-37-017	14 32 54.	+ 00 32	15	14.9	GALAXY
ZWG 019.067	14 32 54.	+ 02 04		15.7	GALAXY
MCG+01-37-037	14 32 54.	+ 05 29	60	17.5	GALAXY
UGC 09385	14 32 54.	+ 05 30	96	17.	GALAXY DWARF
ZWG 075.081	14 32 54.	+ 09 41		15.2	GALAXY
ZC 1432.9+2826	14 32 54.	+ 28 26	1680		CLUSTER OF GALAXIES
TON-N 0208	14 32 54.	+ 29 01		14.9	BLUE STAR
ZWG 220.032	14 32 54.	+ 40 58		14.9	GALAXY
UGC 09386	14 32 54.	+ 40 58	72	14.9	GALAXY SBa-b
ZWG 248.007	14 32 54.	+ 48 09		15.3	GALAXY
IC 1032	14 32 54.	+ 48 10 33.			NONSTELLAR OBJECT
1ZW 091	14 32 54.	+ 48 11			COMPACT GALAXY
ARC 1936	14 32 54.	+ 55 03		17.0	RICH CLUSTER OF GALAXIES
ARP 095	14 32 55.	+ 26 44			PECULIAR GALAXY
REIZ 4224	14 32 57.	+ 00 29	72	14.0	GALAXY
REIZ 4228	14 32 58.	+ 40 56	54	13.9	GALAXY
REIZ 4231	14 32 58.	+ 48 53	12	14.9	GALAXY
IC 1033	14 32 59.	+ 48 09 15.			NONSTELLAR OBJECT
HOLM 663B	14 32 59.	+ 48 54	18	14.2	PART OF MULTIPLE GALAXY
VDB.66G 194	14 33	+ 57 25	70		DWARF GALAXY
ZWG 047.111	14 33 00.	+ 03 54		15.7	GALAXY
ZC 1433.0+2053	14 33 00.	+ 20 53	610		CLUSTER OF GALAXIES
ZWG 133.089	14 33 00.	+ 23 11		15.5	GALAXY
UGC 09387	14 23 00.	+ 23 11	60	15.5	GALAXY S0
TON-N 0209	14 33 00.	+ 24 00		12.7	BLUE STAR
MCG+07-30-030	14 33 00.	+ 40 57 30.	66	14.	GALAXY
ZWG 248.008	14 33 00.	+ 48 53		15.1	GALAXY
UGC 09388	14 33 00.	+ 48 53	120	15.1	GALAXY SBb
MCG+08-27-002	14 33 00.	+ 48 53	96	13.7	GALAXY
ZC 1433.0+5830	14 33 00.	+ 58 30	1210		CLUSTER OF GALAXIES
MCG+10-21-010	14 33 00.	+ 59 53	36	15.	GALAXY
ZC 1433.0+6752	14 33 00.	+ 67 52	540		CLUSTER OF GALAXIES
RNGC 5682	14 33 01.	+ 48 53		15.0	GALAXY
ARC 1937	14 33 02.	+ 58 30		17.2	RICH CLUSTER OF GALAXIES
KARA.68 232	14 33 06.	+ 02 53	27		DWARF GALAXY
ZWG 075.082	14 33 06.	+ 13 07		14.6	GALAXY
UGC 09389	14 33 06.	+ 13 07	132	14.6	GALAXY SBb
KARA.72 428A	14 33 06.	+ 13 07	150	14.6	PART OF DOUBLE GALAXY
MCG+02-37-024	14 33 06.	+ 13 07	144	14.6	GALAXY
ZC 1433.1+2415	14 33 06.	+ 24 15	2350		CLUSTER OF GALAXIES
ZWG 248.009	14 33 06.	+ 48 52		15.5	GALAXY
MRK 474	14 33 06.	+ 48 52	12	16.5	GALAXY WITH UV CONTINUUM
KW 41	14 33 06.	+ 48 52	22		SEYFERT GALAXY
MCG+08-27-003	14 33 06.	+ 48 53	24	14.2	GALAXY
MCG+09-24-019	14 33 06.	+ 54 42	24	15.	GALAXY
ARC 1935	14 33 06.	- 19 06		17.5	RICH CLUSTER OF GALAXIES
RNGC 5683	14 33 07.	+ 48 53		15.5	GALAXY
REIZ 4225	14 33 09.	+ 00 13	18	15.0	GALAXY
REIZ 4226	14 33 09.	+ 00 15	42	14.4	GALAXY
REIZ 4227	14 33 09.	+ 08 31	42	13.7	GALAXY
ZWG 019.068	14 33 12.	+ 00 11		15.4	GALAXY
VV 146C	14 33 12.	+ 13 23 30.	18	15.	INTERACTING GALAXY
VV 146B	14 33 12.	+ 13 23 30.	18	15.	INTERACTING GALAXY
VV 146A	14 33 12.	+ 13 23 30.	114	14.	INTERACTING GALAXY
VV 146	14 33 12.	+ 13 23 30.	120	14.	INTERACTING GALAXY
ZC 1433.2+1609	14 33 12.	+ 16 09	340		CLUSTER OF GALAXIES
UGC 09390	14 33 12.	+ 22 37	60	17.	GALAXY
ZWG 133.090	14 33 12.	+ 22 57		15.5	GALAXY
ZWG 133.091	14 33 12.	+ 23 45		15.7	GALAXY
ZC 1433.2+3116	14 33 12.	+ 31 16	810		CLUSTER OF GALAXIES
MCG+06-32-070	14 33 12.	+ 35 20	24	14.5	GALAXY
MCG+09-24-020	14 33 12.	+ 54 42	120	12.8	GALAXY
MCG+10-21-011	14 33 12.	+ 59 33	90	15.	GALAXY
ZWG 296.013	14 33 12.	+ 59 34		15.5	GALAXY
UGC 09391	14 33 12.	+ 59 34	108	15.5	GALAXY
RNGC 5680	14 33 13.	+ 00 11		15.5	GALAXY
LB 00275	14 33 13.	- 27 00 42.		15.0	FAINT BLUE STAR
MCG+09-24-021	14 33 15.	+ 55 23	42	16.	GALAXY
RNGC 5681	14 33 16.	+ 08 31		14.5	GALAXY
REIZ 4235	14 33 17.	+ 54 42	24	12.8	GALAXY
UGC 09392	14 33 18.	+ 03 15	84	17.	GALAXY
ZWG 075.083	14 33 18.	+ 08 31		14.3	GALAXY
UGC 09393	14 33 18.	+ 08 31	54	14.3	GALAXY S
MCG+02-37-025	14 33 18.	+ 08 31	60	14.3	GALAXY
ZWG 075.084	14 33 18.	+ 13 23		14.9	GALAXY
UGC 09394	14 33 18.	+ 13 23	138	14.9	GALAXY Sc
KARA.72 428B	14 33 18.	+ 13 23	132	14.9	PART OF DOUBLE GALAXY
MCG+02-37-026	14 33 18.	+ 13 23	132	14.9	GALAXY
ZWG 273.014	14 33 18.	+ 54 42		13.3	GALAXY
RNGC 5687	14 33 18.	+ 54 42		13.0	GALAXY
UGC 09395	14 33 18.	+ 54 42	162	13.3	GALAXY S0
KARA.73B 0637	14 33 18.	+ 54 42	168	13.3	ISOLATED GALAXY S

OBJECT NAME	RIGHT ASCEN.	DECLINATION	DIAM.	MAGN.	TYPE OF OBJECT
MCG+04-34-047	14 33 21.	+ 24 56	66	15.0	GALAXY
8ZW 1433+03.0	14 33 24.	+ 03 03		15.9	COMPACT GALAXY
ZWG 104.046	14 33 24.	+ 16 15		15.4	GALAXY
ZC 1433.4+5355	14 33 24.	+ 53 55	2490		CLUSTER OF GALAXIES
ZWG 019.069	14 33 24.	- 03 02		15.7	GALAXY
REIZ 4232	14 33 26.	+ 37 03	24	14.4	GALAXY
HN 1118	14 33 28.	+ 16 14	12		NEBULA
IC 4463	14 33 28.	+ 16 14			NONSTELLAR OBJECT
ZWG 104.047	14 33 30.	+ 16 07		15.6	GALAXY
ZWG 133.092	14 33 30.	+ 24 56		15.0	GALAXY
UGC 09396	14 33 30.	+ 24 56	60	15.0	GALAXY SBa
ZWG 192.044	14 33 30.	+ 37 01		15.3	GALAXY
UGC 09397	14 33 30.	+ 37 01	96	15.3	GALAXY S0
MCG+06-32-071	14 33 30.	+ 37 03	102	14.5	GALAXY
IC 4465	14 33 33.	+ 15 47			NONSTELLAR OBJECT
HN 1119	14 33 34.	+ 15 47	12		NEBULA
ZC 1433.6+0001	14 33 36.	+ 00 01	810		CLUSTER OF GALAXIES
MCG+04-34-048	14 33 36.	+ 21 58	84	14.6	GALAXY
ZC 1433.6+2739	14 33 36.	+ 27 39	740		CLUSTER OF GALAXIES
MCG+06-32-072	14 33 36.	+ 35 18	24	16.	GALAXY
BC M1433+177	14 33 36.0	+ 17 42 39.		18.	QUASI-STELLAR OBJECT
SHB 293	14 33 36.1	+ 17 42 36.		18.	QUASI-STELLAR OBJECT
HN 1746	14 33 37.3	- 01 48 51.	18		NEBULA
LYNG 10	14 33 38.	- 60 45 18.	2160		OB CONCENTRATION
MCG+05-34-079	14 33 39.	+ 28 20 30.	30		GALAXY
ZWG 075.085	14 33 42.	+ 09 21		15.4	GALAXY
ZWG 192.045	14 33 42.	+ 35 17		15.7	GALAXY
ZWG 248.010	14 33 42.	+ 48 57		12.7	GALAXY
UGC 09399	14 33 42.	+ 48 57	252	12.7	GALAXY SBa
MCG+08-27-004	14 33 42.	+ 48 58	150	12.9	GALAXY
MCG+10-21-012	14 33 42.	+ 59 37	39	17.	GALAXY
ZC 1433.7-0050	14 33 42.	- 00 50	810		CLUSTER OF GALAXIES
UGC 09398	14 33 42.	- 01 00	66	15.7	GALAXY SBb-c
ZWG 019.070	14 33 42.	- 01 11		15.7	GALAXY
MCG+00-37-018	14 33 42.	- 01 11	36	15.7	GALAXY
8ZW 1433-01.8	14 33 42.	- 01 46		17.3	COMPACT GALAXY
ZC 1433.7-0239	14 33 42.	- 02 39	940		CLUSTER OF GALAXIES
RNGC 5689	14 33 43.	+ 48 58		12.5	GALAXY
SC 1430-1726.5	14 33 43.	- 17 39 39.	18		NEBULA
REIN 2.222	14 33 43.64	+ 48 57 31.6			NEBULA
REIZ 4234	14 33 45.	+ 36 45	24	14.1	GALAXY
MCG+06-32-073	14 33 45.	+ 36 46 30.	72	13.5	GALAXY
REIZ 4239	14 33 46.	+ 48 57	90	12.5	GALAXY
RNGC 5684	14 33 47.	+ 36 45		14.0	GALAXY
MAI 091	14 33 47.	+ 57 26	53		DWARF SPHEROIDAL GALAXY
ZWG 047.112	14 33 48.	+ 05 33		14.9	GALAXY
UGC 09400	14 33 48.	+ 05 33	66	14.9	GALAXY S0
MCG+01-37-038	14 33 48.	+ 05 33	36	14.9	GALAXY
ZWG 133.093	14 33 48.	+ 22 01		14.6	GALAXY
UGC 09401	14 33 48.	+ 22 01	102	14.6	GALAXY S
FEIG 097	14 33 48.	+ 30 20		12.0	FAINT BLUE STAR
ZWG 192.046	14 33 48.	+ 36 45		14.2	GALAXY
UGC 09402	14 33 48.	+ 36 45	126	14.2	GALAXY S0
MCG+10-21-013	14 33 48.	+ 57 28	78	15.	GALAXY
REIZ 4236	14 33 50.	+ 36 56	18	14.5	GALAXY
MCG+06-32-074	14 33 51.	+ 36 57	27	14.5	GALAXY
ARC 1940	14 33 52.	+ 55 22		17.0	RICH CLUSTER OF GALAXIES
ZWG 075.086	14 33 54.	+ 08 57		15.7	GALAXY
ZWG 163.086	14 33 54.	+ 28 40		15.4	GALAXY
ZWG 192.047	14 33 54.	+ 36 56		15.3	GALAXY
MCG+08-27-005	14 33 54.	+ 45 12 30.	30	17.	GALAXY
HARO 43	14 33 57.	+ 28 40			BLUE EMISSION-LINE GALAXY
REIZ 4238	14 33 57.	+ 36 43	21	14.6	GALAXY
MCG+06-32-075	14 33 57.	+ 36 44	21	14.5	GALAXY
RNGC 5686	14 33 59.	+ 36 43		15.0	GALAXY
ZC 1434.0+1755	14 34 00.	+ 17 55	4700		CLUSTER OF GALAXIES
ZWG 133.094	14 34 00.	+ 23 48		15.6	GALAXY
TON-N 0778	14 34 00.	+ 24 30		15.7	BLUE STAR
ZC 1434.0+2625	14 34 00.	+ 26 25	870		CLUSTER OF GALAXIES
MCG+05-34-080	14 34 00.	+ 28 39	27	15.4	GALAXY
ZWG 163.087	14 34 00.	+ 30 07		14.9	GALAXY
UGC 09403	14 34 00.	+ 30 07	66	14.9	GALAXY (F)
ZWG 192.048	14 34 00.	+ 36 43		15.2	GALAXY
UGC 09404	14 34 00.	+ 56 02	60	16.5	GALAXY SBc
UGC 09405	14 34 00.	+ 57 28	150	17.	GALAXY DWARF
ZWG 296.014	14 34 00.	+ 57 52		15.6	GALAXY
MCG+10-21-014	14 34 00.	+ 57 52	27	16.	GALAXY
ZC 1434.0+6003	14 34 00.	+ 60 03	1140		CLUSTER OF GALAXIES
RNGC 5685	14 34 01.	+ 30 07		15.0	GALAXY
HN 0411	14 34 01.	- 39 15			NEBULA
IC 4458	14 34 01.	- 39 15			NONSTELLAR OBJECT
REIZ 4233	14 34 03.	+ 08 50	12	15.7	GALAXY
REIZ 4237	14 34 04.	+ 30 08	18	14.5	GALAXY
ZWG 047.113	14 34 06.	+ 03 24		15.4	GALAXY
ZWG 075.087	14 34 06.	+ 10 09		15.3	GALAXY
ZWG 075.088	14 34 06.	+ 11 37		15.5	GALAXY
MCG+05-34-081	14 34 06.	+ 30 07	60	14.9	GALAXY
ZC 1434.1+6338	14 34 06.	+ 63 38	1750		CLUSTER OF GALAXIES
HN 1747	14 34 06.6	- 01 53 26.	18		NEBULA
ZWG 047.114	14 34 12.	+ 02 36		15.5	GALAXY
8ZW 1434+02.6	14 34 12.	+ 02 36		15.5	COMPACT GALAXY
ZWG 075.089	14 34 12.	+ 10 09		15.4	GALAXY
ZWG 075.090	14 34 12.	+ 11 47		15.6	GALAXY
ZWG 075.091	14 34 12.	+ 12 09		15.4	GALAXY
ZC 1434.2+2847	14 34 12.	+ 28 47	870		CLUSTER OF GALAXIES
ZWG 047.115	14 34 18.	+ 08 20		15.5	GALAXY
ZC 1434.3+7419	14 34 18.	+ 74 19	1880		CLUSTER OF GALAXIES
ZWG 047.116	14 34 24.	+ 06 10		15.2	GALAXY
ZWG 047.117	14 34 24.	+ 06 21		15.6	GALAXY
ZWG 133.095	14 34 24.	+ 21 18		15.6	GALAXY
ZC 1434.4+2451	14 34 24.	+ 24 51	2080		CLUSTER OF GALAXIES
ZC 1434.4+2802	14 34 24.	+ 28 02	610		CLUSTER OF GALAXIES
ZC 1434.4+3143	14 34 24.	+ 31 43	3020		CLUSTER OF GALAXIES
ZC 1434.4+3620	14 34 24.	+ 36 20	1140		CLUSTER OF GALAXIES
ZWG 248.011	14 34 24.	+ 48 47		14.5	GALAXY
UGC 09406	14 34 24.	+ 48 47	102	14.5	GALAXY SBc
MCG+08-27-006	14 34 24.	+ 48 48	120	13.9	GALAXY
IC 4466	14 34 25.	+ 18 33			NONSTELLAR OBJECT
RNGC 5693	14 34 25.	+ 48 48		14.5	GALAXY
HN 0409	14 34 25.	- 78 36			NEBULA
HN 1120	14 34 26.	+ 18 33	12		NEBULA
IC 4448	14 34 26.	- 78 36	54		GALAXY SB(s)
ZWG 104.048	14 34 30.	+ 18 34		15.5	GALAXY
TON-N 0210	14 34 30.	+ 28 57		14.8	BLUE STAR
IC 4471	14 34 30.	+ 41 53			SAME AS NGC 5697
ZC 1434.5+5532	14 34 30.	+ 55 32	3630		CLUSTER OF GALAXIES
IC 4467	14 34 31.	+ 18 35			NONSTELLAR OBJECT
HN 1121	14 34 32.	+ 18 35	12		NEBULA
REIZ 4242	14 34 32.	+ 37 07	30	14.8	GALAXY
REIZ 4243	14 34 32.	+ 41 53	36	13.8	GALAXY
REIZ 4247	14 34 32.	+ 53 00	30	14.6	GALAXY
REIZ 4246	14 34 34.	+ 48 45	60	13.9	GALAXY
REIZ 4241	14 34 35.	+ 28 59	30	15.0	GALAXY
ZWG 133.096	14 34 36.	+ 21 31		15.7	GALAXY
ZWG 164.001	14 34 36.	+ 29 58		15.7	GALAXY
ZWG 163.088	14 34 36.	+ 29 58		15.7	GALAXY
MCG+07-30-031	14 34 36.	+ 41 53	48	14.5	GALAXY
ZWG 220.033	14 34 36.	+ 41 54		14.6	GALAXY
UGC 09407	14 34 36.	+ 41 54	66	14.6	GALAXY PECULR
MCG+10-21-015	14 34 36.	+ 59 28	39	16.	GALAXY
LB 00276	14 34 37.	- 24 47 54.		14.8	FAINT BLUE STAR
RNGC 5697	14 34 38.	+ 41 54		14.5	GALAXY
REIZ 4240	14 34 39.	+ 08 52	30	14.9	GALAXY
ZL 149	14 34 40.	+ 30 27 12.		20.7	ULTRAFAINT BLUE STAR
HN 1748	14 34 40.0	+ 00 39 42.	12		NEBULA
ZWG 019.071	14 34 42.	+ 01 24		15.6	GALAXY
8ZW 1434+01.4	14 34 42.	+ 01 24		15.6	COMPACT GALAXY
ZWG 220.034	14 34 42.	+ 40 10		15.0	GALAXY
MCG+07-30-032	14 34 42.	+ 41 16	54	15.	GALAXY
UGC 09408	14 34 42.	+ 41 18	66	16.0	GALAXY S?
ZWG 220.035	14 34 42.	+ 41 22		15.6	GALAXY
UGC 09409	14 34 42.	+ 41 22	60	15.6	GALAXY S
MCG+10-21-016	14 34 42.	+ 60 42 30.	39	16.	GALAXY
SER 113.01	14 34 42.	- 78 35	60	15.	PEC. RING-GALAXY
MCG+07-30-033	14 34 45.	+ 41 21	48	15.	GALAXY
IC 4464	14 34 46.	- 36 40			NONSTELLAR OBJECT
ZWG 075.092	14 34 48.	+ 08 51		15.3	GALAXY
UGC 09410	14 34 48.	+ 08 51	84	15.3	GALAXY Sb-c
ZWG 075.093	14 34 48.	+ 10 13		14.9	GALAXY
UGC 09411	14 34 48.	+ 10 13	60	14.9	GALAXY Sa-b
MCG+02-37-027	14 34 48.	+ 10 13	48	14.9	GALAXY
ZWG 104.049	14 34 48.	+ 14 53		15.6	GALAXY
ZC 1434.8+1919	14 34 48.	+ 19 19	2080		CLUSTER OF GALAXIES
ZWG 104.050	14 34 48.	+ 19 38		15.7	GALAXY
ZC 1434.8+2109	14 34 48.	+ 21 09	540		CLUSTER OF GALAXIES
ZWG 134.001	14 34 48.	+ 25 12		15.7	GALAXY
ZWG 133.097	14 34 48.	+ 25 12		15.7	GALAXY
MCG+06-32-076	14 34 48.	+ 34 30	24	15.	GALAXY
MCG+07-30-034	14 34 48.	+ 40 08	15	15.	GALAXY
MCG+10-21-017	14 34 48.	+ 62 38	54	15.	GALAXY
MCG+12-14-010	14 34 48.	+ 69 26	39	17.	GALAXY
ZL 150	14 34 49.	+ 30 17 36.		21.5	ULTRAFAINT BLUE STAR
ZL 151	14 34 50.	+ 30 30 48.		20.6	ULTRAFAINT BLUE STAR
MCG+05-34-082	14 34 51.	+ 28 58	36		GALAXY
MCG+07-30-035	14 34 51.	+ 40 10	36	15.5	GALAXY
IC 1034	14 34 52.	+ 14 52 52.			NONSTELLAR OBJECT
8ZW 1434+04.8	14 34 54.	+ 04 49		17.8	COMPACT GALAXY
ZWG 075.094	14 34 54.	+ 10 12		15.1	GALAXY
ZWG 134.002	14 34 54.	+ 25 28		15.6	GALAXY
ZWG 133.098	14 34 54.	+ 25 28		15.6	GALAXY
ZC 1434.9+3341	14 34 54.	+ 33 41	1480		CLUSTER OF GALAXIES
ZWG 296.015	14 34 54.	+ 59 01		14.3	GALAXY
UGC 09412	14 34 54.	+ 59 01	42	14.3	GALAXY VY CMPT
ZWG 354.006	14 34 54.	+ 78 06		14.8	GALAXY
ZWG 353.043	14 34 54.	+ 78 06		14.8	GALAXY
UGC 09413	14 34 54.	+ 78 06	60	14.8	GALAXY Sb-c
KARA.72 429A	14 34 54.	+ 78 06	60	14.8	PART OF DOUBLE GALAXY
MCG+13-11-001	14 34 54.	+ 78 06	48	14.	GALAXY
HELW 034	14 34 57.	- 14 59			NEBULA
SHB 294	14 35	+ 17			QUASI-STELLAR OBJECT
ZWG 104.051	14 35 00.	+ 18 28		15.6	GALAXY
UGC 09414	14 35 00.	+ 18 28	96	15.6	GALAXY Sc
MCG+03-37-026	14 35 00.	+ 18 28	96	15.6	GALAXY
ZC 1435.0+2055	14 35 00.	+ 20 55	740		CLUSTER OF GALAXIES
ZC 1435.0+2925	14 35 00.	+ 29 25	1010		CLUSTER OF GALAXIES
ZWG 220.036	14 35 00.	+ 42 02		14.1	GALAXY
UGC 09415	14 35 00.	+ 42 02	120	14.1	GALAXY Sb-c
ZC 1435.0+5203	14 35 00.	+ 52 03	2350		CLUSTER OF GALAXIES
MCG+10-21-018	14 35 00.	+ 60 42	12	16.	GALAXY
ZC 1435.0+8028	14 35 00.	+ 80 28	740		CLUSTER OF GALAXIES
ACK 322+14.1	14 35 00.	- 44 00			PLANETARY NEBULA
IC 4469	14 35 01.	+ 18 27			NONSTELLAR OBJECT
HN 1122	14 35 02.	+ 18 27	72		NEBULA
RNGC 5696	14 35 02.	+ 42 02		14.0	GALAXY
ARC 1939	14 35 03.	+ 25 03		16.6	RICH CLUSTER OF GALAXIES
MCG+07-30-036	14 35 03.	+ 42 02	120	13.	GALAXY
ZL 152	14 35 05.	+ 30 22 48.		17.5	ULTRAFAINT BLUE STAR
ZWG 047.118	14 35 06.	+ 08 03		15.6	GALAXY
ZC 1435.1+1651	14 35 06.	+ 16 51	340		CLUSTER OF GALAXIES
ZWG 248.012	14 35 06.	+ 47 00		15.7	GALAXY
MCG+09-24-022	14 35 06.	+ 51 41	108	15.	GALAXY
MCG+10-21-019	14 35 06.	+ 58 00	18	16.	GALAXY
ZWG 337.015	14 35 06.	+ 72 40		15.6	GALAXY
7ZW 555	14 35 06.	+ 77 05			COMPACT GALAXY
MCG+13-11-002	14 35 06.	+ 78 10	39	15.	GALAXY
ZC 1435.1-0009	14 35 06.	- 00 09	1610		CLUSTER OF GALAXIES
REIZ 4248	14 35 08.	+ 42 00	60	13.5	GALAXY
REIN 2.223	14 35 09.21	+ 02 30 25.5			NEBULA
ARC 1938	14 35 10.	- 00 04		17.2	RICH CLUSTER OF GALAXIES
REIN 2.224	14 35 11.35	+ 02 29 50.0			NEBULA
ZWG 019.072	14 35 12.	+ 02 28		13.1	GALAXY
UGC 09416	14 35 12.	+ 02 28	216	13.1	GALAXY Sc
MCG+00-37-019	14 35 12.	+ 02 28	210	13.1	GALAXY
KARA.73B 0638	14 35 12.	+ 02 28	204	13.1	ISOLATED GALAXY S
ZWG 047.119	14 35 12.	+ 02 30		13.1	GALAXY
UGC 09417	14 35 12.	+ 22 11	72	16.5	GALAXY S
ZWG 134.003	14 35 12.	+ 25 59		15.2	GALAXY
ZWG 133.099	14 35 12.	+ 25 59		15.2	GALAXY
UGC 09418	14 35 12.	+ 25 59	60	15.2	GALAXY SBa
MCG+04-34-049	14 35 12.	+ 25 59	60	15.2	GALAXY
TON-N 0779	14 35 12.	+ 27 38		15.0	BLUE STAR
ZWG 220.037	14 35 12.	+ 38 40		14.0	GALAXY
RNGC 5698	14 35 12.	+ 38 40		14.0	GALAXY
UGC 09419	14 35 12.	+ 38 40	114	14.0	GALAXY SBb
MCG+07-30-037	14 35 12.	+ 39 51	30	16.	GALAXY
MCG+08-27-007	14 35 12.	+ 48 46 30.	60	14.2	GALAXY
MCG-01-37-010	14 35 12.	- 08 27	150	14.	GALAXY
RNGC 5690	14 35 14.	+ 02 30		13.0	GALAXY
REIZ 4244	14 35 14.	+ 02 30	84	13.2	GALAXY
RNGC 5695	14 35 17.	+ 36 46		14.0	GALAXY
ZWG 047.120	14 35 18.	+ 04 01		15.5	GALAXY
ZWG 047.121	14 35 18.	+ 05 13		15.3	GALAXY
ZWG 047.122	14 35 18.	+ 06 58		15.5	GALAXY
KARA.73B 0639	14 35 18.	+ 06 58	42	15.5	ISOLATED GALAXY S
8ZW 1435+07.9	14 35 18.	+ 07 57		17.5	COMPACT GALAXY
ZWG 192.049	14 35 18.	+ 36 46		13.9	GALAXY

OBJECT NAME	RIGHT ASCEN.	DECLINATION	DIAM.	MAGN.	TYPE OF OBJECT
MRK 686	14 35 18.	+ 36 46	45	14.5	GALAXY WITH UV CONTINUUM
UGC 09421	14 35 18.	+ 36 46	96	13.9	GALAXY S
MCG+06-32-077	14 35 18.	+ 36 47	90	13.5	GALAXY
MCG+07-30-038	14 35 18.	+ 38 38	108	14.	GALAXY
REIZ 4250	14 35 18.	+ 38 40	96	13.9	GALAXY
REIZ 4251	14 35 18.	+ 39 10	42	14.9	GALAXY
ZWG 220.038	14 35 18.	+ 43 55		15.4	GALAXY
UGC 09422	14 35 18.	+ 43 55	120	15.4	GALAXY Sc
ZWG 248.013	14 35 18.	+ 48 45		15.2	GALAXY
UGC 09423	14 35 18.	+ 48 45	60	15.2	GALAXY SB
ZWG 354.007	14 35 18.	+ 78 11		14.9	GALAXY
ZWG 353.044	14 35 18.	+ 78 11		14.9	GALAXY
KARA.72 429B	14 35 18.	+ 78 11	66	14.9	PART OF DOUBLE GALAXY
ZWG 019.073	14 35 18.	- 00 12		12.9	GALAXY
UGC 09420	14 35 18.	- 00 12	126	12.9	GALAXY S
MCG+00-37-020	14 35 18.	- 00 12	96	12.9	GALAXY
RNGC 5700	14 35 19.	+ 48 45		15.0	GALAXY
RNGC 5691	14 35 19.	- 00 11		13.5	GALAXY
REIZ 4249	14 35 20.	+ 36 46	42	13.4	GALAXY
MCG+07-30-039	14 35 21.	+ 43 55	102	14.5	GALAXY
REIZ 4253	14 35 22.	+ 48 43	30	14.2	GALAXY
REIZ 4245	14 35 22.	- 00 11	42	12.7	GALAXY
ARC 1941	14 35 23.	+ 30 43		17.5	RICH CLUSTER OF GALAXIES
ZC 1435.4+2545	14 35 24.	+ 25 45	610		CLUSTER OF GALAXIES
ZL 153	14 35 24.	+ 30 26 06.		20.2	ULTRAFAINT BLUE STAR
ZC 1435.4+4649	14 35 24.	+ 46 49	1080		CLUSTER OF GALAXIES
7ZW 556	14 35 24.	+ 73 44			COMPACT GALAXY
ZL 154	14 35 26.	+ 30 30 54.		22.2	ULTRAFAINT BLUE STAR
ARC 1943	14 35 29.	+ 30 27		17.6	RICH CLUSTER OF GALAXIES
MCG+03-37-027	14 35 30.	+ 16 03	27	15.2	GALAXY
ZWG 104.052	14 35 30.	+ 16 05		15.2	GALAXY
MCG+04-34-050	14 35 30.	+ 21 36	72	15.	GALAXY
2ZW 069	14 35 30.	+ 29 02			COMPACT GALAXY
ZC 1435.5+3037	14 35 30.	+ 30 37	1010		CLUSTER OF GALAXIES
ZC 1435.5+4538	14 35 30.	+ 45 38	3700		CLUSTER OF GALAXIES
MRSL 315-02/1	14 35 30.	- 62 27	180		HII REGION
ARP 241	14 35 32.	+ 30 40			PECULIAR GALAXY
HN 1123	14 35 34.	+ 16 04	12		NEBULA
IC 4473	14 35 34.	+ 16 04			NONSTELLAR OBJECT
ZWG 075.095	14 35 36.	+ 10 22		15.4	GALAXY
ZWG 075.096	14 35 36.	+ 10 23		15.4	GALAXY
ZWG 104.053	14 35 36.	+ 14 40		15.4	GALAXY
ZWG 134.004	14 35 36.	+ 21 35		15.6	GALAXY
UGC 09424	14 35 36.	+ 21 35	66	15.6	GALAXY Sc
ZWG 134.005	14 35 36.	+ 22 11		15.2	GALAXY
ZC 1435.6+3020	14 35 36.	+ 30 20	740		CLUSTER OF GALAXIES
VV 264B	14 35 36.	+ 30 40	30	16.	INTERACTING GALAXY
VV 264A	14 35 36.	+ 30 40	30	16.	INTERACTING GALAXY
VV 264	14 35 36.	+ 30 40	48		INTERACTING GALAXY
ZWG 164.002	14 35 36.	+ 30 42		15.0	GALAXY
ZWG 163.089	14 35 36.	+ 30 42		15.0	GALAXY
UGC 09425	14 35 36.	+ 30 42	60	15.0	GALAXY DBL SYS
ZWG 337.016	14 35 36.	+ 73 45		15.3	GALAXY
MCG+13-11-003	14 35 36.	+ 80 26	39	14.	GALAXY
ZL 155	14 35 37.	+ 30 32 54.		19.8	ULTRAFAINT BLUE STAR
REIZ 4252	14 35 37.	+ 38 09	18	14.7	GALAXY
HOLM 664A	14 35 39.	+ 09 33	18	14.3	PART OF MULTIPLE GALAXY
MCG-04-35-001	14 35 39.	- 22 09	84	14.	GALAXY
HN 0413	14 35 39.	- 22 10			NEBULA
IC 4468	14 35 39.	- 22 10			NONSTELLAR OBJECT
HOLM 664B	14 35 40.	+ 09 33	12	14.7	PART OF MULTIPLE GALAXY
REIZ 4254	14 35 40.	+ 40 36	12	14.6	GALAXY
ARC 1944	14 35 41.	+ 30 38		17.2	RICH CLUSTER OF GALAXIES
ZWG 075.097	14 35 42.	+ 09 33		15.2	GALAXY
MCG+02-37-028	14 35 42.	+ 09 33	42	15.2	GALAXY
UGC 09426	14 35 42.	+ 48 49	120	17.	GALAXY DWARF
MCG+08-27-008	14 35 42.	+ 48 50	36	17.	GALAXY
ZWG 019.074	14 35 42.	- 00 09		15.4	GALAXY
HN 1749	14 35 42.3	- 00 07 51.	12		NEBULA
IC 1035	14 35 44.	+ 09 33 08.			NONSTELLAR OBJECT
MCG+05-34-083	14 35 45.	+ 30 41	60	15.0	GALAXY
RNGC 5707	14 35 46.	+ 51 47		13.5	GALAXY
REIZ 4259	14 35 46.	+ 51 47	120	13.1	GALAXY
ZL 156	14 35 47.	+ 30 18 54.		20.0	ULTRAFAINT BLUE STAR
ZWG 047.123	14 35 48.	+ 03 37		13.3	GALAXY
UGC 09427	14 35 48.	+ 03 37	60	13.3	GALAXY PECULR
MCG+01-37-039	14 35 48.	+ 03 37	48	13.3	GALAXY
ZC 1435.8+0355	14 35 48.	+ 03 55	810		CLUSTER OF GALAXIES
ZWG 075.098	14 35 48.	+ 11 47		15.3	GALAXY
ZWG 075.099	14 35 48.	+ 12 15		15.7	GALAXY
ZWG 104.054	14 35 48.	+ 17 40		15.7	GALAXY
ZC 1435.8+3944	14 35 48.	+ 39 44	1140		CLUSTER OF GALAXIES
MCG+09-24-023	14 35 48.	+ 51 47	150	13.1	GALAXY
RNGC 5692	14 35 50.	+ 03 37		13.5	GALAXY
ZL 157	14 35 50.	+ 30 25 30.		18.5	ULTRAFAINT BLUE STAR
HN 1750	14 35 52.9	- 02 45 09.	18		NEBULA
ZC 1435.9+2552	14 35 54.	+ 25 52	540		CLUSTER OF GALAXIES
REIZ 4257	14 35 54.	+ 38 35	21	14.9	GALAXY
REIZ 4256	14 35 54.	+ 38 35	18	14.8	GALAXY
ZC 1435.9+4030	14 35 54.	+ 40 30	1010		CLUSTER OF GALAXIES
ZC 1435.9+4417	14 35 54.	+ 44 17	1340		CLUSTER OF GALAXIES
MCG+08-27-009	14 35 54.	+ 45 24	48	17.	GALAXY
ZC 1435.9+4845	14 35 54.	+ 48 45	1410		CLUSTER OF GALAXIES
ZWG 273.015	14 35 54.	+ 51 48		13.3	GALAXY
UGC 09428	14 35 54.	+ 51 48	162	13.3	GALAXY Sa-b
MCG+09-24-024	14 35 54.	+ 51 49	54	15.	GALAXY
RNGC 5704	14 35 55.	+ 40 43			NON-EXISTENT OBJECT
ZL 158	14 35 56.	+ 30 36 30.		18.0	ULTRAFAINT BLUE STAR
REIZ 4255	14 35 56.	+ 37 14	36	14.8	GALAXY
ARC 1946	14 35 56.	+ 40 26		17.6	RICH CLUSTER OF GALAXIES
ARC 1948	14 35 57.	+ 48 49		17.2	RICH CLUSTER OF GALAXIES
VDB.66G 195	14 36	- 08 25	130		DWARF GALAXY
LB 09984	14 36	- 89 24		13.8	FAINT BLUE STAR
ZC 1436.0+0926	14 36 00.	+ 09 26	11090		CLUSTER OF GALAXIES
ZWG 104.055	14 36 00.	+ 17 50		15.7	GALAXY
ZWG 104.056	14 36 00.	+ 18 20		15.6	GALAXY
ZWG 192.050	14 36 00.	+ 37 13		15.7	GALAXY
MCG+07-30-040	14 36 00.	+ 38 32	33	16.	GALAXY
MCG+07-30-041	14 36 00.	+ 38 33	15	16.	GALAXY
ZWG 220.039	14 36 00.	+ 38 35		15.6	GALAXY
ZWG 220.040	14 36 00.	+ 40 19		14.8	GALAXY
UGC 09429	14 36 00.	+ 40 19	72	14.8	GALAXY S
ZWG 220.041	14 36 00.	+ 41 34		15.7	GALAXY
ZC 1436.0+5940	14 36 00.	+ 59 40	1210		CLUSTER OF GALAXIES
MCG+10-21-020	14 36 00.	+ 60 01	39	16.	GALAXY
MCG-02-37-011	14 36 00.	- 12 56 30.	36	17.	GALAXY
IC 1036	14 36 03.	+ 18 19 58.			NONSTELLAR OBJECT
MCG+03-37-028	14 36 03.	+ 18 23	48	15.0	GALAXY
MCG+07-30-042	14 36 03.	+ 40 18	66	14.5	GALAXY
MCG+07-30-043	14 36 03.	+ 41 32	45	14.5	GALAXY
ARC 1942	14 36 04.	+ 03 53		17.5	RICH CLUSTER OF GALAXIES
ARC 1947	14 36 04.	+ 39 39		17.5	RICH CLUSTER OF GALAXIES
REIZ 4258	14 36 04.	+ 40 19	60	13.6	GALAXY
ZL 159	14 36 05.	+ 30 27 06.		19.3	ULTRAFAINT BLUE STAR
ZWG 019.075	14 36 06.	+ 00 10		15.6	GALAXY
ZC 1436.1+1531	14 36 06.	+ 15 31	2760		CLUSTER OF GALAXIES
ZWG 104.057	14 36 06.	+ 18 24		15.0	GALAXY
IC 1037	14 36 06.	+ 18 24 22.			NONSTELLAR OBJECT
ZWG 134.006	14 36 06.	+ 23 33		15.0	GALAXY
IC 4474	14 36 06.	+ 23 38 40.			NONSTELLAR OBJECT
MCG+06-32-078	14 36 06.	+ 36 40 30.	24	15.	GALAXY
MCG+06-32-079	14 36 06.	+ 37 14	45	15.	GALAXY
MCG+08-27-010	14 36 06.	+ 48 25	42	16.	GALAXY
ZWG 273.016	14 36 06.	+ 52 57		15.6	GALAXY
RNGC 5699	14 36 07.	+ 29 42			NON-EXISTENT OBJECT
IC 4475	14 36 08.	+ 23 33 11.			NONSTELLAR OBJECT
HN 1751	14 36 09.3	+ 00 12 52.	18		NEBULA
ZWG 075.100	14 36 12.	+ 09 45		15.4	GALAXY
ZC 1436.2+1141	14 36 12.	+ 11 41	470		CLUSTER OF GALAXIES
FEIG 098	14 36 12.	+ 27 43		11.7	FAINT BLUE STAR
REIZ 4263	14 36 16.	+ 40 40	60	13.3	GALAXY
RNGC 5688	14 36 16.	- 44 49			GALAXY
ZWG 047.124	14 36 18.	+ 07 50		15.7	GALAXY
ZWG 164.003	14 36 18.	+ 28 40		15.5	GALAXY
MCG+07-30-044	14 36 18.	+ 40 39 30.	90	13.5	GALAXY
ZWG 220.042	14 36 18.	+ 40 40		13.9	GALAXY
UGC 09430	14 36 18.	+ 40 40	102	13.9	GALAXY S-IRR
RNGC 5703	14 36 19.	+ 29 42			NON-EXISTENT OBJECT
RNGC 5708	14 36 19.	+ 40 40		14.0	GALAXY
REIZ 4267	14 36 19.	+ 46 52	168	13.4	GALAXY
REIZ 4261	14 36 23.	+ 28 41	18	15.1	GALAXY
ZWG 075.101	14 36 24.	+ 10 20		15.5	GALAXY
ZWG 164.004	14 36 24.	+ 30 41		15.7	GALAXY
ZC 1436.4+3252	14 36 24.	+ 32 52	3630		CLUSTER OF GALAXIES
ZC 1436.4+3852	14 36 24.	+ 38 52	1480		CLUSTER OF GALAXIES
ZWG 248.014	14 36 24.	+ 46 51		14.2	GALAXY
RNGC 5714	14 36 24.	+ 46 51		14.0	GALAXY
UGC 09431	14 36 24.	+ 46 51	180	14.2	GALAXY Sc
MCG+08-27-011	14 36 24.	+ 46 51	210	13.4	GALAXY
MCG-07-30-004	14 36 24.	- 44 47	240	12.	GALAXY
IC 4477	14 36 25.	+ 28 40 18.			NONSTELLAR OBJECT
REIZ 4269	14 36 25.	+ 46 51	15	14.7	GALAXY
REIZ 4270	14 36 25.	+ 46 57	24	14.4	GALAXY
RNGC 5706	14 36 26.	+ 30 41		15.5	GALAXY
ZWG 047.125	14 36 30.	+ 03 09		15.4	GALAXY
UGC 09432	14 36 30.	+ 03 09	96	15.4	GALAXY DWARF
MCG+01-37-040	14 36 30.	+ 03 09	60	15.4	GALAXY
ZWG 075.102	14 36 30.	+ 09 37		15.6	GALAXY
ZWG 164.005	14 36 30.	+ 28 43		14.8	GALAXY
UGC 09433	14 36 30.	+ 28 43	90	14.8	GALAXY Sc
ZC 1436.5+2914	14 36 30.	+ 29 14	810		CLUSTER OF GALAXIES
MCG+09-24-025	14 36 30.	+ 51 02	120	13.	GALAXY
MCG-04-35-002	14 36 30.	- 25 16 30.	24	15.5	GALAXY
HOLM 665B	14 36 31.	+ 30 40	18	15.1	PART OF MULTIPLE GALAXY
RNGC 5702	14 36 33.	+ 20 43		14.5	GALAXY
MCG+04-35-002	14 36 33.	+ 20 45	24	14.5	GALAXY
MCG+05-35-001	14 36 33.	+ 28 43	72	14.8	GALAXY
REIZ 4265	14 36 33.	+ 30 41	12	14.8	GALAXY
MCG+05-35-002	14 36 33.	+ 30 41	18	15.7	GALAXY
REIZ 4264	14 36 35.	+ 28 43	36	14.4	GALAXY
IC 4479	14 36 35.	+ 28 43 13.			NONSTELLAR OBJECT
MCG+03-37-029	14 36 36.	+ 16 11	30	15.6	GALAXY
ZWG 104.058	14 36 36.	+ 16 13		15.6	GALAXY
ZWG 134.007	14 36 36.	+ 20 43		14.5	GALAXY
UGC 09434	14 36 36.	+ 20 43	66	14.5	GALAXY S0
ZC 1436.6+2250	14 36 36.	+ 22 50	810		CLUSTER OF GALAXIES
ZWG 164.006	14 36 36.	+ 30 39		14.5	GALAXY
UGC 09435	14 36 36.	+ 30 39	96	14.5	GALAXY SBa
ZC 1436.6+5442	14 36 36.	+ 54 42	810		CLUSTER OF GALAXIES
ZC 1436.6-0138	14 36 36.	- 01 38	1550		CLUSTER OF GALAXIES
RNGC 5709	14 36 38.	+ 30 39		14.5	GALAXY
RNGC 5701	14 36 39.	+ 05 35		12.0	GALAXY
REIZ 4266	14 36 39.	+ 30 39	84	13.6	GALAXY
HOLM 665A	14 36 39.	+ 30 39	96	12.8	PART OF MULTIPLE GALAXY
MCG+05-35-003	14 36 39.	+ 30 40	96	14.5	GALAXY
WRAY 19.39	14 36 40.3	- 62 27 26.			DIFFUSE NEBULA
REIZ 4260	14 36 41.	+ 05 35	120	12.7	GALAXY
ZWG 047.126	14 36 42.	+ 02 56		14.9	GALAXY
MCG+01-37-041	14 36 42.	+ 02 56	42	14.9	GALAXY
8ZW 1436+03.9	14 36 42.	+ 03 54		17.9	COMPACT GALAXY
ZWG 047.127	14 36 42.	+ 05 35		12.9	GALAXY
UGC 09436	14 36 42.	+ 05 35	276	12.9	GALAXY SB0
MCG+01-37-042	14 36 42.	+ 05 35	240	12.9	GALAXY
MRK 287	14 36 42.	+ 73 50	12	16.	GALAXY WITH UV CONTINUUM
MCG-05-35-001	14 36 42.	- 32 27	36	15.	GALAXY
REIZ 4272	14 36 43.	+ 46 54	18	14.3	GALAXY
RNGC 5720	14 36 45.	+ 51 04		14.5	GALAXY
RNGC 5694	14 36 45.	- 26 19		11.0	GLOBULAR CLUSTER
ZWG 047.128	14 36 48.	+ 03 35		15.3	GALAXY
ZWG 104.059	14 36 48.	+ 16 06		15.3	GALAXY
UGC 09437	14 36 48.	+ 18 55	60	16.0	GALAXY S
ZWG 134.008	14 36 48.	+ 21 36		15.6	GALAXY
MRK 687	14 36 48.	+ 21 36	13	16.5	GALAXY WITH UV CONTINUUM
ZWG 164.007	14 36 48.	+ 31 01		15.5	GALAXY
ZWG 164.008	14 36 48.	+ 31 10		15.3	GALAXY
UGC 09438	14 36 48.	+ 31 10	66	15.3	GALAXY S0
ZC 1436.8+3249	14 36 48.	+ 32 49	740		CLUSTER OF GALAXIES
MCG+08-27-012	14 36 48.	+ 46 52 30.	42	14.3	GALAXY
ZWG 248.015	14 36 48.	+ 46 53		15.4	GALAXY
RNGC 5717	14 36 48.	+ 46 53		15.5	GALAXY
ZWG 273.017	14 36 48.	+ 51 04		14.7	GALAXY
UGC 09439	14 36 48.	+ 51 04	132	14.7	GALAXY Sb
REIZ 4262	14 36 49.	+ 02 56	18	14.5	GALAXY
REIZ 4274	14 36 49.	+ 46 49	48	14.4	GALAXY
HN 1124	14 36 50.	- 44 08	120		NEBULA
HN 1125	14 36 51.	+ 16 05		15.	NEBULA
MCG+03-37-030	14 36 51.	+ 18 55	54	16.	GALAXY
MCG+05-35-004	14 36 51.	+ 31 02	42	15.5	GALAXY
MCG+05-35-005	14 36 51.	+ 31 11	30	15.3	GALAXY
IC 4472	14 36 51.	- 44 08			NONSTELLAR OBJECT
IC 4478	14 36 52.	+ 16 05			NONSTELLAR OBJECT
REIZ 4278	14 36 52.	+ 51 52	54	13.9	GALAXY
8ZW 1436+03.9	14 36 54.	+ 03 54		16.8	COMPACT GALAXY
ZWG 104.060	14 36 54.	+ 20 15		14.3	GALAXY
UGC 09440	14 36 54.	+ 20 15	84	14.5	GALAXY E
ZC 1436.9+2158	14 36 54.	+ 21 58	470		CLUSTER OF GALAXIES

OBJECT NAME	RIGHT ASCEN.	DECLINATION	DIAM.	MAGN.	TYPE OF OBJECT
MRK 475	14 36 54.	+ 37 01	10	17.	GALAXY WITH UV CONTINUUM
MCG+07-30-045	14 36 54.	+ 41 12	48	14.5	GALAXY
ZWG 220.043	14 36 54.	+ 41 14		14.6	GALAXY
MRK 476	14 36 54.	+ 41 14	50	16.	GALAXY WITH UV CONTINUUM
UGC 09441	14 36 54.	+ 41 14	84	14.6	GALAXY Sa
UGC 09442	14 36 54.	+ 52 50	60	16.0	GALAXY SBb
ZC 1436.9+6625	14 36 54.	+ 66 25	5240		CLUSTER OF GALAXIES
RNGC 5710	14 36 57.	+ 20 15		14.5	GALAXY
MCG+01-36-025	14 37 00.	+ 06 25	18	15.2	GALAXY
ZWG 075.103	14 37 00.	+ 09 27		15.4	GALAXY
UGC 09443	14 37 00.	+ 09 27	90	15.4	GALAXY S
ZC 1437.0+1119	14 37 00.	+ 11 19	1140		CLUSTER OF GALAXIES
ZWG 075.104	14 37 00.	+ 12 08		15.3	GALAXY
MCG+03-37-031	14 37 00.	+ 17 12	60	14.8	GALAXY
ZWG 104.061	14 37 00.	+ 17 14		14.8	GALAXY
UGC 09444	14 37 00.	+ 17 14	78	14.8	GALAXY S0?
8ZW 1437+18.0	14 37 00.	+ 17 58		19.1	COMPACT GALAXY
ZWG 104.062	14 37 00.	+ 20 12		15.1	GALAXY
UGC 09445	14 37 00.	+ 20 12	72	15.1	GALAXY S
MCG+03-37-032	14 37 00.	+ 20 15	60	14.3	GALAXY
ZC 1437.0+2409	14 37 00.	+ 24 09	2150		CLUSTER OF GALAXIES
MCG+09-24-026	14 37 00.	+ 51 31	72	14.	GALAXY
ZWG 273.018	14 37 00.	+ 51 34		15.5	GALAXY
ZWG 273.019	14 37 00.	+ 52 56		15.7	GALAXY
ZWG 273.020	14 37 00.	+ 54 30		15.1	GALAXY
KARA.73B 0640	14 37 00.	+ 54 30	18	15.1	ISOLATED GALAXY E
REIZ 4277	14 37 01.	+ 46 54	21	14.4	GALAXY
PK319+06.1	14 37 01.	- 52 21 53.	25		PLANETARY NEBULA
IC 1038	14 37 02.	+ 12 08 26.			NONSTELLAR OBJECT
RNGC 5711	14 37 03.	+ 20 12		15.0	GALAXY
MCG-07-30-005	14 37 03.	- 44 05	150	14.	GALAXY
ZWG 047.129	14 37 06.	+ 03 22		14.9	GALAXY
MCG+01-37-043	14 37 06.	+ 03 22	48	14.9	GALAXY
ZWG 047.130	14 37 06.	+ 04 08		15.6	GALAXY
ZC 1437.1+1235	14 37 06.	+ 12 35	1680		CLUSTER OF GALAXIES
MCG+03-37-033	14 37 06.	+ 20 12	60	15.1	GALAXY
ZWG 192.051	14 37 06.	+ 38 20		15.6	GALAXY
MCG+06-32-080	14 37 06.	+ 38 22	27	15.5	GALAXY
UGC 09446	14 37 06.	+ 38 28	72	17.	GALAXY Sc
MCG+08-27-013	14 37 06.	+ 46 52 30.	9	14.4	GALAXY
ZWG 248.016	14 37 06.	+ 46 53		15.3	GALAXY
RNGC 5723	14 37 06.	+ 46 53		15.5	GALAXY
RNGC 5721	14 37 06.	+ 46 53		15.5	GALAXY
REIZ 4279	14 37 06.	+ 47 01	24	14.9	GALAXY
ZWG 273.021	14 37 06.	+ 52 30		15.7	GALAXY
MCG+09-24-027	14 37 06.	+ 52 30	66	15.	GALAXY
ZC 1437.1+5346	14 37 06.	+ 53 46	1010		CLUSTER OF GALAXIES
MCG+11-18-002	14 37 06.	+ 64 37	42	16.	GALAXY
ZWG 318.001	14 37 06.	+ 66 33		14.8	GALAXY
MCG+12-14-011	14 37 06.	+ 69 12 30.	45	15.	GALAXY
ZWG 337.017	14 37 06.	+ 69 15		14.7	GALAXY
IC 4476	14 37 06.	- 16 01 35.			NONSTELLAR OBJECT
REIZ 4268	14 37 07.	+ 03 22	18	14.4	GALAXY
IC 1046	14 37 07.	+ 69 14 16.			NONSTELLAR OBJECT
MCG+06-32-081	14 37 09.	+ 38 29	66	16.	GALAXY
FEIG 099	14 37 12.	+ 19 39		10.0	FAINT BLUE STAR
RNGC 5722	14 37 12.	+ 46 52		15.0	GALAXY
MCG+08-27-014	14 37 12.	+ 46 54	15	15.	GALAXY
ZWG 273.022	14 37 12.	+ 51 22		15.3	GALAXY
UGC 09448	14 37 12.	+ 51 22	102	15.3	GALAXY Sb
SHAH 131	14 37 12.	+ 62 58	114	17.0	GROUP OF COMPACT GALAXIES
MCG+11-18-003	14 37 12.	+ 64 21 30.	27	16.	GALAXY
ZWG 019.076	14 37 12.	- 00 32		14.5	GALAXY
UGC 09447	14 37 12.	- 00 32	174	14.5	GALAXY SBc
MCG+00-37-021	14 37 12.	- 00 32	156	14.5	GALAXY
MCG+08-27-015	14 37 15.	+ 46 56	36	17.	GALAXY
MCG+09-24-028	14 37 15.	+ 51 20	120	14.	GALAXY
REIZ 4286	14 37 15.	+ 52 10	78	14.3	GALAXY
REIZ 4271	14 37 16.	- 00 29	180	13.7	GALAXY
FATH 2.150	14 37 17.	- 00 30	41		NEBULA
RNGC 5724	14 37 18.	+ 46 56		17.0	GALAXY
ZC 1437.3+5701	14 37 18.	+ 57 01	2290		CLUSTER OF GALAXIES
ZC 1437.3+6436	14 37 18.	+ 64 36	1140		CLUSTER OF GALAXIES
MCG+11-18-004	14 37 18.	+ 66 32	45	15.	GALAXY
ARC 1945	14 37 18.	- 22 05		17.6	RICH CLUSTER OF GALAXIES
RNGC 5705	14 37 19.	- 00 30		14.5	GALAXY
MCG+06-32-082	14 37 21.	+ 36 46	27	15.	GALAXY
MCG+08-27-016	14 37 21.	+ 46 56	6	18.	GALAXY
ZC 1437.4+1148	14 37 24.	+ 11 48	400		CLUSTER OF GALAXIES
8ZW 1437+15.9	14 37 24.	+ 15 56		18.4	COMPACT GALAXY
ZWG 104.063	14 37 24.	+ 18 43		15.4	GALAXY
FEIG 100	14 37 24.	+ 19 21		11.6	FAINT BLUE STAR
ZWG 134.009	14 37 24.	+ 24 45		15.4	GALAXY
UGC 09449	14 37 24.	+ 24 45	66	15.4	GALAXY Sb-c
ZC 1437.4+2912	14 37 24.	+ 29 12	540		CLUSTER OF GALAXIES
ZC 1437.4+3707	14 37 24.	+ 37 07	340		CLUSTER OF GALAXIES
ZC 1437.4+3725	14 37 24.	+ 37 25	2290		CLUSTER OF GALAXIES
ZC 1437.4+6225	14 37 24.	+ 62 25	3160		CLUSTER OF GALAXIES
MCG-04-35-003	14 37 24.	- 25 34 30.	150	14.	GALAXY
HN 1126	14 37 25.	+ 18 42	24		NEBULA
IC 4480	14 37 25.	+ 18 42			NONSTELLAR OBJECT
ARP 171	14 37 30.	+ 03 41			PECULIAR GALAXY
REIZ 4273	14 37 30.	+ 04 25	48	14.5	GALAXY
MCG+01-36-026	14 37 30.	+ 04 56	36	15.0	GALAXY
MCG+03-37-034	14 37 30.	+ 18 42	30	15.4	GALAXY
UGC 09450	14 37 30.	+ 23 37	60	17.	GALAXY Sc-IRR
7ZW 557	14 37 30.	+ 73 13			COMPACT GALAXY
SMI 06	14 37 30.	- 61 59	360		FAINT NEBULOSITY
REIZ 4276	14 37 31.	+ 18 29	24	14.3	GALAXY
UGC 09452	14 37 36.	+ 54 08	84	16.5	GALAXY DWRF SP
UGC 09453	14 37 36.	+ 79 07	78	16.0	GALAXY Sb
ZWG 019.077	14 37 36.	- 00 05		11.7	GALAXY
8ZW 1437-00.1	14 37 36.	- 00 05		11.7	COMPACT GALAXY
UGC 09451	14 37 36.	- 00 05	198	11.7	GALAXY Sc
MCG+00-37-022	14 37 36.	- 00 05	138	11.7	GALAXY
RNGC 5713	14 37 37.	- 00 05		12.0	GALAXY
FATH 2.151	14 37 39.	- 00 04	54		NEBULA
REIZ 4275	14 37 40.	- 00 05	120	11.6	GALAXY
IC 4481	14 37 41.	+ 19 21			NONSTELLAR OBJECT
ZWG 075.105	14 37 42.	+ 13 35		15.5	GALAXY
MCG+09-24-029	14 37 42.	+ 54 07	120	15.	GALAXY
7ZW 558	14 37 42.	+ 69 18			COMPACT GALAXY
ARC 1949	14 37 43.	+ 18 21		17.2	RICH CLUSTER OF GALAXIES
REIZ 4281	14 37 43.	+ 18 32	36	14.2	GALAXY
MIL 32	14 37 43.	- 59 47	1020		SUPERNOVA REMNANT
HN 1127	14 37 45.	+ 16 21	12		NEBULA
ZWG 104.064	14 37 48.	+ 19 10		15.5	GALAXY
ZC 1437.8+5850	14 37 48.	+ 58 50	1410		CLUSTER OF GALAXIES
7ZW 559	14 37 48.	+ 80 26			COMPACT GALAXY
MCG-05-35-002	14 37 48.	- 32 11	48	15.	GALAXY
HN 1128	14 37 49.	+ 19 09	24		NEBULA
IC 4482	14 37 49.	+ 19 09			NONSTELLAR OBJECT
ZWG 047.131	14 37 54.	+ 03 49		15.7	GALAXY
ZWG 047.132	14 37 54.	+ 06 31		14.9	GALAXY
UGC 09454	14 37 54.	+ 06 31	84	14.9	GALAXY Sc
MCG+01-37-044	14 37 54.	+ 06 31	48	14.9	GALAXY
ZWG 075.106	14 37 54.	+ 09 41		15.3	GALAXY
ZC 1437.9+1825	14 37 54.	+ 18 25	1210		CLUSTER OF GALAXIES
TON-N 0780	14 37 54.	+ 23 20		14.6	BLUE STAR
ZWG 134.010	14 37 54.	+ 26 07		15.6	GALAXY
ZC 1437.9+2900	14 37 54.	+ 29 00	610		CLUSTER OF GALAXIES
ZWG 164.009	14 37 54.	+ 31 43		15.2	GALAXY
REIZ 4280	14 37 55.	+ 03 37	12	15.1	GALAXY
REIZ 4285	14 37 55.	+ 18 32	18	13.9	GALAXY
HN 1752	14 37 55.7	+ 00 48 03.	18		NEBULA
IC 1040	14 37 57.	+ 09 41 07.			NONSTELLAR OBJECT
HN 1129	14 37 57.	+ 16 53	36		NEBULA
IC 4483	14 37 57.	+ 16 53			NONSTELLAR OBJECT
MCG+07-30-046	14 37 57.	+ 42 57	102	14.	GALAXY
IC 1039	14 37 58.	+ 03 37 07.			NONSTELLAR OBJECT
ZWG 019.078	14 38 00.	+ 00 46		15.5	GALAXY
MCG+00-37-023	14 38 00.	+ 00 46	36	15.5	GALAXY
ZWG 047.133	14 38 00.	+ 03 38		15.6	GALAXY
ZC 1438.0+1313	14 38 00.	+ 13 13	1140		CLUSTER OF GALAXIES
ZWG 104.065	14 38 00.	+ 15 03		15.3	GALAXY
MCG+03-37-035	14 38 00.	+ 16 52	90	15.0	GALAXY
ZWG 104.066	14 38 00.	+ 16 54		15.0	GALAXY
UGC 09455	14 38 00.	+ 16 54	96	15.0	GALAXY Sb
ZC 1438.0+2136	14 38 00.	+ 21 36	740		CLUSTER OF GALAXIES
ZC 1438.0+2940	14 38 00.	+ 29 40	1280		CLUSTER OF GALAXIES
REIZ 4292	14 38 00.	+ 42 57	108	13.7	GALAXY
ZWG 220.044	14 38 00.	+ 42 58		14.7	GALAXY
HOLM 667A	14 38 00.	+ 42 58	120	13.7	PART OF MULTIPLE GALAXY
UGC 09456	14 38 00.	+ 42 58	114	14.7	GALAXY IRR
KARA.72 430A	14 38 00.	+ 42 58	120	14.7	PART OF DOUBLE GALAXY
MCG+08-27-017	14 38 00.	+ 47 22 30.	24	16.	GALAXY
REIZ 4284	14 38 03.	+ 08 58	12	15.2	GALAXY
RNGC 5731	14 38 03.	+ 42 58		14.5	GALAXY
ZWG 047.134	14 38 06.	+ 03 35		15.1	GALAXY
MCG+01-37-045	14 38 06.	+ 03 35	36	15.1	GALAXY
ZWG 047.135	14 38 06.	+ 03 40		14.9	GALAXY
UGC 09457	14 38 06.	+ 03 40	72	14.9	GALAXY S0?
KARA.72 431A	14 38 06.	+ 03 40	60	14.9	PART OF DOUBLE GALAXY
MCG+01-37-046	14 38 06.	+ 03 40	48	14.9	GALAXY
ZWG 134.011	14 38 06.	+ 24 56		15.5	GALAXY
UGC 09458	14 38 06.	+ 24 56	60	15.5	GALAXY S0-a
IC 1041	14 38 07.	+ 03 33 43.			NONSTELLAR OBJECT
REIZ 4282	14 38 07.	+ 03 34	18	14.3	GALAXY
REIZ 4283	14 38 07.	+ 03 39	30	14.2	GALAXY
IC 1042	14 38 07.	+ 03 39 19.			NONSTELLAR OBJECT
ARC 1950	14 38 09.	+ 13 17		17.2	RICH CLUSTER OF GALAXIES
ZWG 047.136	14 38 12.	+ 03 28		15.7	GALAXY
ZWG 047.137	14 38 12.	+ 03 40		14.6	GALAXY
UGC 09459	14 38 12.	+ 03 40	96	14.6	GALAXY E-S0
MKW 08	14 38 12.	+ 03 40		14.6	POOR GALAXY CLUSTER
KARA.72 431B	14 38 12.	+ 03 40	84	14.6	PART OF DOUBLE GALAXY
MCG+01-37-047	14 38 12.	+ 03 40	72	14.6	GALAXY
ZWG 075.107	14 38 12.	+ 09 19		15.5	GALAXY
ZC 1438.2+1414	14 38 12.	+ 14 14	4300		CLUSTER OF GALAXIES
ZWG 164.010	14 38 12.	+ 28 53		15.7	GALAXY
REIZ 4294	14 38 12.	+ 42 59	42	13.9	GALAXY
IC 1043	14 38 13.	+ 03 33 38.			NONSTELLAR OBJECT
REIZ 4287	14 38 13.	+ 03 34	9	15.0	GALAXY
REIZ 4288	14 38 13.	+ 03 41	36	13.6	GALAXY
RNGC 5718	14 38 15.	+ 03 40		14.5	GALAXY
MCG+07-30-047	14 38 15.	+ 42 59	90	14.	GALAXY
RNGC 5730	14 38 15.	+ 43 00		14.0	GALAXY
HOLM 666A	14 38 16.	+ 14 52	42	14.3	PART OF MULTIPLE GALAXY
HOLM 666B	14 38 16.	+ 14 53	18	14.4	PART OF MULTIPLE GALAXY
HOLM 667B	14 38 16.	+ 43 01	54	13.9	PART OF MULTIPLE GALAXY
ZWG 047.138	14 38 18.	+ 03 19		15.6	GALAXY
ZWG 075.108	14 38 18.	+ 08 40		15.2	GALAXY
MCG+03-37-036	14 38 18.	+ 14 52	24	16.	GALAXY
ZWG 134.012	14 38 18.	+ 22 07		15.5	GALAXY
ZWG 220.045	14 38 18.	+ 43 00		14.0	GALAXY
UGC 09460	14 38 18.	+ 43 00	102	14.0	GALAXY S
KARA.72 430B	14 38 18.	+ 43 00	114	14.0	PART OF DOUBLE GALAXY
ZWG 296.016	14 38 18.	+ 62 13		14.8	GALAXY
UGC 09461	14 38 18.	+ 62 13	72	14.8	GALAXY SBa
KARA.73B 0641	14 38 18.	+ 62 13	48	14.8	ISOLATED GALAXY S
ZC 1438.3+7707	14 38 18.	+ 77 07	5310		CLUSTER OF GALAXIES
8ZW 1438-02.4	14 38 18.	- 02 23		17.3	COMPACT GALAXY
MCG-01-37-011	14 38 18.	- 08 42	60	14.	GALAXY
MCG-03-37-004	14 38 18.	- 17 14	96	13.	GALAXY
RNGC 5716	14 38 19.	- 17 15		13.0	GALAXY
REIZ 4291	14 38 21.	+ 08 39	24	14.7	GALAXY
IC 4485	14 38 21.	+ 28 52 34.			NONSTELLAR OBJECT
RNGC 5727	14 38 22.	+ 34 12		14.5	GALAXY
REIZ 4289	14 38 22.	- 00 04	90	13.3	GALAXY
REIZ 4293	14 38 23.	+ 34 12	84	13.6	GALAXY
FATH 2.152	14 38 23.	- 00 06	33		NEBULA
ZWG 047.139	14 38 24.	+ 03 21		15.2	GALAXY
UGC 09463	14 38 24.	+ 03 21	66	15.2	GALAXY S0-a
ZWG 047.140	14 38 24.	+ 03 37		15.5	GALAXY
ZWG 047.141	14 38 24.	+ 03 44		15.4	GALAXY
ZC 1438.4+0405	14 38 24.	+ 04 05	7590		CLUSTER OF GALAXIES
FEIG 101	14 38 24.	+ 18 07		11.5	FAINT BLUE STAR
ZC 1438.4+2457	14 38 24.	+ 24 57	4100		CLUSTER OF GALAXIES
ZWG 164.011	14 38 24.	+ 31 35		14.9	GALAXY
UGC 09464	14 38 24.	+ 31 35	60	14.9	GALAXY S0-a
ZWG 192.052	14 38 24.	+ 34 12		14.6	GALAXY
UGC 09465	14 38 24.	+ 34 12	138	14.6	GALAXY
MCG+06-32-083	14 38 24.	+ 34 12	135	13.	GALAXY
KARA.73B 0642	14 38 24.	+ 34 12	138	14.6	ISOLATED GALAXY S
MCG+10-21-021	14 38 24.	+ 62 13	42	14.	GALAXY
ZWG 019.079	14 38 24.	- 00 07		13.8	GALAXY
UGC 09462	14 38 24.	- 00 07	204	13.8	GALAXY Sb
MCG+00-37-024	14 38 24.	- 00 07	180	13.8	GALAXY
RNGC 5719	14 38 25.	- 00 06		14.0	GALAXY
REIZ 4290	14 38 26.	+ 02 24	30	13.9	GALAXY
ZWG 019.080	14 38 30.	+ 02 22		14.5	GALAXY
UGC 09466	14 38 30.	+ 02 22	66	14.5	GALAXY SBc?
MCG+00-37-025	14 38 30.	+ 02 22	66	14.5	GALAXY
ZWG 104.067	14 38 30.	+ 14 33		15.4	GALAXY
ZWG 134.013	14 38 30.	+ 21 25		15.5	GALAXY
MCG+05-35-006	14 38 30.	+ 31 36	54	14.9	GALAXY

OBJECT NAME	RIGHT ASCEN.	DECLINATION	DIAM.	MAGN.	TYPE OF OBJECT
MCG-05-35-003	14 38 30.	- 33 07	54	15.	GALAXY
RNGC 5725	14 38 32.	+ 02 22		14.5	GALAXY
REIZ 4298	14 38 35.	+ 38 51	54	14.1	GALAXY
ZWG 047.142	14 38 36.	+ 06 16		15.5	GALAXY
ZWG 220.046	14 38 36.	+ 38 51		14.4	GALAXY
RNGC 5732	14 38 36.	+ 38 51		14.5	GALAXY
UGC 09467	14 38 36.	+ 38 51	90	14.4	GALAXY Sb-c
MCG+07-30-048	14 38 42.	+ 38 49	72	14.5	GALAXY
REIZ 4295	14 38 44.	+ 18 10	15	14.4	GALAXY
REIZ 4296	14 38 44.	+ 18 11	18	14.1	GALAXY
UGC 09468	14 38 48.	+ 10 15	60	16.0	GALAXY S
ZWG 075.109	14 38 48.	+ 10 51		15.6	GALAXY
ZC 1438.8+2557	14 38 49.	+ 25 57	1280		CLUSTER OF GALAXIES
ZC 1438.8+3741	14 38 48.	+ 37 41	740		CLUSTER OF GALAXIES
7ZW 560	14 38 48.	+ 60 53			COMPACT GALAXY
IC 1045	14 38 51.	+ 42 57 20.			NONSTELLAR OBJECT
MCG-02-37-012	14 38 51.	- 10 49	48	16.	GALAXY
ZWG 075.110	14 38 54.	+ 11 01		15.2	GALAXY
TON-N 0781	14 38 54.	+ 25 59		14.2	BLUE STAR
ZC 1438.9+2845	14 38 54.	+ 28 45	940		CLUSTER OF GALAXIES
ARC 1952	14 38 55.	+ 28 51		18.0	RICH CLUSTER OF GALAXIES
REIZ 4297	14 38 56.	+ 18 16	36	14.3	GALAXY
REIZ 4301	14 38 58.	+ 29 09	24	15.0	GALAXY
LB 09890	14 39	- 82 34		14.3	FAINT BLUE STAR
ZWG 047.143	14 39 00.	+ 03 49		15.6	GALAXY
ZC 1439.0+2107	14 39 00.	+ 21 07	1010		CLUSTER OF GALAXIES
ZWG 134.014	14 39 00.	+ 22 12		15.7	GALAXY
IC 1044	14 39 02.	+ 09 39 12.			NONSTELLAR OBJECT
ZWG 075.111	14 39 06.	+ 09 39		15.0	GALAXY
MCG+02-37-029	14 39 06.	+ 09 39	60	15.0	GALAXY
ZWG 273.023	14 39 06.	+ 53 44		15.2	GALAXY
MRK 477	14 39 06.	+ 53 44	17	16.	GALAXY WITH UV CONTINUUM
ZWG 019.081	14 39 06.	- 01 37		15.7	GALAXY
UGC 09469	14 39 06.	- 01 37	96	15.7	GALAXY DWARF
MCG+00-37-026	14 39 06.	- 01 37	60	15.7	GALAXY
VMT 14	14 39 08.	- 62 15	1860		SUPERNOVA REMNANT
MIL 31	14 39 08.	- 62 15	2400		SUPERNOVA REMNANT
REIZ 4300	14 39 11.	+ 06 11	12	15.2	GALAXY
ZWG 019.082	14 39 12.	+ 00 52		14.9	GALAXY
UGC 09470	14 39 12.	+ 00 52	72	14.9	GALAXY
MCG+00-37-027	14 39 12.	+ 00 52	60	14.9	GALAXY
ZWG 047.144	14 39 12.	+ 06 10		14.9	GALAXY
UGC 09471	14 39 12.	+ 06 10	72	14.9	GALAXY SBb
MCG+01-37-048	14 39 12.	+ 06 10	84	14.9	GALAXY
ZC 1439.2+1623	14 39 12.	+ 16 23	670		CLUSTER OF GALAXIES
TON-N 0211	14 39 12.	+ 32 38		15.1	BLUE STAR
1ZW 092	14 39 12.	+ 53 44			COMPACT GALAXY
VVI 74	14 39 12.	+ 53 44	25	14.4	SEYFERT GALAXY
REIZ 4299	14 39 15.	+ 00 54	60	14.2	GALAXY
RNGC 5729	14 39 16.	- 08 50		13.0	GALAXY
HN 1753	14 39 16.2	+ 00 54 07.	60		NEBULA
ZC 1439.3+1213	14 39 18.	+ 12 13	3360		CLUSTER OF GALAXIES
ZWG 105.001	14 39 18.	+ 18 46		15.2	GALAXY
ZWG 104.068	14 39 18.	+ 18 46		15.2	GALAXY
ZWG 134.015	14 39 18.	+ 23 09		15.3	GALAXY
ZWG 164.012	14 39 18.	+ 31 51		15.3	GALAXY
MCG+10-21-022	14 39 18.	+ 57 06	30	16.	GALAXY
MCG-01-37-012	14 39 18.	- 08 50	138	13.	GALAXY
RNGC 5726	14 39 18.	- 18 14			NON-EXISTENT OBJECT
ZWG 105.002	14 39 24.	+ 19 28		15.5	GALAXY
ZWG 104.069	14 39 24.	+ 19 28		15.5	GALAXY
ZWG 248.017	14 39 24.	+ 44 42		14.8	GALAXY
KARA.72 432A	14 39 24.	+ 44 42	48	14.8	PART OF DOUBLE GALAXY
UGC 09472	14 39 24.	- 01 13	66	16.5	GALAXY
LB 00277	14 39 26.	- 27 36 24.		15.5	FAINT BLUE STAR
REIZ 4302	14 39 27.	+ 08 40	24	15.2	GALAXY
REIZ 4304	14 39 29.	+ 39 05	24	13.9	GALAXY
ZWG 047.145	14 39 30.	+ 03 36		15.5	GALAXY
ZWG 075.112	14 39 30.	+ 09 36		15.3	GALAXY
ZWG 105.003	14 39 30.	+ 18 47		15.7	GALAXY
ZWG 104.070	14 39 30.	+ 18 47		15.7	GALAXY
ZC 1439.5+2225	14 39 30.	+ 22 25	2620		CLUSTER OF GALAXIES
ZC 1439.5+3145	14 39 30.	+ 31 45	740		CLUSTER OF GALAXIES
ZWG 220.047	14 39 30.	+ 39 04		13.8	GALAXY
UGC 09473	14 39 30.	+ 39 04	84	13.8	GALAXY S0
MCG+08-27-018	14 39 30.	+ 44 41	30	15.	GALAXY
1ZW 093	14 39 30.	+ 44 42			COMPACT GALAXY
HN 1130	14 39 31.	+ 18 47	12		NEBULA
IC 4486	14 39 31.	+ 18 47			NONSTELLAR OBJECT
ARC 1953	14 39 32.	+ 13 30		16.6	RICH CLUSTER OF GALAXIES
MCG-03-37-005	14 39 33.	- 17 01 30.	120	12.5	GALAXY
HN 1755	14 39 33.0	- 00 35 23.	12		NEBULA
ZWG 047.146	14 39 36.	+ 03 32		15.5	GALAXY
ZWG 075.113	14 39 36.	+ 08 41		15.0	GALAXY
UGC 09474	14 39 36.	+ 08 41	66	15.0	GALAXY Sc
MCG+02-37-030	14 39 36.	+ 08 41	60	15.0	GALAXY
ZWG 075.114	14 39 36.	+ 12 17		15.7	GALAXY
UGC 09475	14 39 36.	+ 12 17	66	15.7	GALAXY Sc
ZC 1439.6+1331	14 39 36.	+ 13 31	1140		CLUSTER OF GALAXIES
MCG+03-37-037	14 39 36.	+ 18 46	42	15.7	GALAXY
MCG+06-32-085	14 39 36.	+ 36 52	45	15.	GALAXY
ZWG 192.053	14 39 36.	+ 36 57		15.7	GALAXY
MCG+06-32-084	14 39 36.	+ 36 57	30	15.	GALAXY
MCG+07-30-049	14 39 36.	+ 39 02	27	14.5	GALAXY
VVI 75	14 39 36.	- 17 03	120	12.50	SEYFERT GALAXY
GCL 02	14 39 36.	- 26 32	138	10.87	GLOBULAR STAR CLUSTER
RNGC 5728	14 39 37.	- 17 03		12.5	GALAXY
REIZ 4308	14 39 39.	+ 44 44	60	13.8	GALAXY
ZWG 075.115	14 39 42.	+ 09 36		15.7	GALAXY
ZWG 076.001	14 39 42.	+ 13 10		15.6	GALAXY
ZWG 075.116	14 39 42.	+ 13 10		15.6	GALAXY
MCG+08-27-019	14 39 42.	+ 44 43	60	13.	GALAXY
ZWG 248.018	14 39 42.	+ 44 44		13.9	GALAXY
UGC 09476	14 39 42.	+ 44 44	96	13.9	GALAXY Sc
KARA.72 432B	14 39 42.	+ 44 44	78	13.9	PART OF DOUBLE GALAXY
OCL 0929	14 39 42.	- 57 20	900	9.8	OPEN STAR CLUSTER
VHA 163	14 39 42.	- 57 20	600		OPEN STAR CLUSTER
VHE 62	14 39 42.	- 58 32	96		REFLECTION NEBULA
HN 1131	14 39 43.	+ 18 50	6		NEBULA
IC 4487	14 39 43.	+ 18 50			NONSTELLAR OBJECT
RNGC 5715	14 39 46.	- 57 20		10.0	OPEN CLUSTER
REIZ 4307	14 39 47.	+ 38 38	12	15.4	GALAXY
ZWG 075.117	14 39 48.	+ 08 53		15.4	GALAXY
ZWG 075.118	14 39 48.	+ 09 13		15.7	GALAXY
KARA.72 433A	14 39 48.	+ 09 13	42	15.7	PART OF DOUBLE GALAXY
MCG+02-37-031	14 39 48.	+ 09 13	30	15.7	GALAXY
MCG+07-30-050	14 39 48.	+ 40 14	30	16.	GALAXY
MCG+10-21-023	14 39 48.	+ 59 30	27	16.	GALAXY
ZWG 354.008	14 39 48.	+ 78 23		15.2	GALAXY
ZWG 353.045	14 39 48.	+ 78 23		15.2	GALAXY
HOLM 668B	14 39 53.	+ 09 12	48	14.8	PART OF MULTIPLE GALAXY
ZWG 019.083	14 39 54.	+ 01 05		15.5	GALAXY
8ZW 1439+01.1	14 39 54.	+ 01 05		15.5	COMPACT GALAXY
ZWG 047.147	14 39 54.	+ 03 26		15.2	GALAXY
ZWG 076.002	14 39 54.	+ 09 13		15.3	GALAXY
ZWG 075.119	14 39 54.	+ 09 13		15.3	GALAXY
KARA.72 433B	14 39 54.	+ 09 13	54	15.3	PART OF DOUBLE GALAXY
MCG+02-37-032	14 39 54.	+ 09 13	60	15.3	GALAXY
ZC 1439.9+3000	14 39 54.	+ 30 00	940		CLUSTER OF GALAXIES
ZC 1439.9+3415	14 39 54.	+ 34 15	3090		CLUSTER OF GALAXIES
UGC 09477	14 39 54.	+ 59 32	72	17.	GALAXY DWARF
MCG-02-37-013	14 39 54.	- 09 55	54	15.5	GALAXY
VHA 164	14 39 54.	- 66 11	1800		OPEN STAR CLUSTER
ARC 1954	14 39 55.	+ 28 45		17.6	RICH CLUSTER OF GALAXIES
HN 1756	14 39 56.9	+ 01 42 51.	24		NEBULA
HOLM 668A	14 39 57.	+ 09 12	30	14.6	PART OF MULTIPLE GALAXY
ZWG 020.001	14 40 00.	+ 01 42		15.1	GALAXY
ZWG 019.084	14 40 00.	+ 01 42		15.1	GALAXY
ZWG 047.148	14 40 00.	+ 04 38		15.2	GALAXY
REIZ 4303	14 40 00.	+ 04 38	54	14.5	GALAXY
UGC 09479	14 40 00.	+ 04 38	90	15.2	GALAXY Sb
MCG+01-37-049	14 40 00.	+ 04 38	90	15.2	GALAXY
ZWG 105.004	14 40 00.	+ 19 24		15.7	GALAXY
ZWG 104.071	14 40 00.	+ 19 24		15.7	GALAXY
ZWG 134.016	14 40 00.	+ 25 43		15.0	GALAXY
ZWG 134.017	14 40 00.	+ 26 27		15.5	GALAXY
UGC 09478	14 40 00.	+ 85 31	66	17.	GALAXY Sc-IRR
MCG+14-07-008	14 40 00.	+ 85 32 30.	39	17.	GALAXY
IC 1047	14 40 02.	+ 19 24 06.			NONSTELLAR OBJECT
8ZW 1440+03.8	14 40 06.	+ 03 47		17.2	COMPACT GALAXY
ZWG 048.001	14 40 06.	+ 04 11		15.4	GALAXY
ZWG 048.002	14 40 06.	+ 05 34		15.5	GALAXY
MCG+01-37-050	14 40 06.	+ 05 34	36	15.5	GALAXY
MCG+03-37-038	14 40 06.	+ 19 23	36	15.7	GALAXY
ZWG 134.018	14 40 06.	+ 22 33		14.8	GALAXY
UGC 09480	14 40 06.	+ 22 33	60	14.8	GALAXY SB0-a
MCG+04-35-003	14 40 06.	+ 22 35	51	14.8	GALAXY
ZC 1440.1+2840	14 40 06.	+ 28 40	810		CLUSTER OF GALAXIES
MRK 478	14 40 06.	+ 35 39	10	15.	GALAXY WITH UV CONTINUUM
KW 42	14 40 06.	+ 35 39	16		SEYFERT GALAXY
ZWG 220.048	14 40 06.	+ 44 06		15.7	GALAXY
MCG-03-37-006	14 40 06.	- 18 14	15	15.	GALAXY
VVI 76	14 40 06.	- 35 39	10	15.	SEYFERT GALAXY
ACK 318+03.1	14 40 06.	- 56 05			PLANETARY NEBULA
HN 1754	14 40 08.7	- 44 27 10.	42		NEBULA
HN 1757	14 40 08.9	+ 01 32 21.	12		NEBULA
REIZ 4305	14 40 10.	- 00 08	60	13.9	GALAXY
ZWG 134.019	14 40 12.	+ 22 34		15.4	GALAXY
MCG+04-35-004	14 40 12.	+ 22 36	36	15.4	GALAXY
MCG+04-35-005	14 40 12.	+ 25 45	42	15.0	GALAXY
ZWG 164.013	14 40 12.	+ 28 56		13.8	GALAXY
UGC 09481	14 40 12.	+ 28 56	168	13.8	GALAXY SBb
ZC 1440.2+2935	14 40 12.	+ 29 35	1210		CLUSTER OF GALAXIES
MCG+07-30-051	14 40 12.	+ 41 51	42	16.	GALAXY
ZWG 354.009	14 40 12.	+ 75 15		15.7	GALAXY
ZWG 020.002	14 40 12.	- 00 09		14.6	GALAXY
MCG+00-38-001	14 40 12.	- 00 09	48	14.6	GALAXY
HN 1758	14 40 12.2	+ 00 52 15.	30		NEBULA
RNGC 5735	14 40 13.	+ 28 56		14.0	GALAXY
RNGC 5733	14 40 13.	- 00 08		14.5	GALAXY
FATH 2.153	14 40 13.	- 00 09	41		NEBULA
REIZ 4306	14 40 15.	+ 00 52	90	14.9	GALAXY
ZWG 020.003	14 40 18.	+ 00 53		15.4	GALAXY
UGC 09482	14 40 18.	+ 00 53	72	15.4	GALAXY Sc
ZC 1440.3+0128	14 40 18.	+ 01 28	8940		CLUSTER OF GALAXIES
ZWG 048.003	14 40 18.	+ 06 53		15.7	GALAXY
ZWG 076.003	14 40 18.	+ 08 56		15.3	GALAXY
ZWG 076.004	14 40 18.	+ 09 02		15.6	GALAXY
ZWG 164.014	14 40 18.	+ 28 52		15.7	GALAXY
ZC 1440.3+3243	14 40 18.	+ 32 43	1680		CLUSTER OF GALAXIES
MCG+10-21-024	14 40 18.	+ 57 08	15	16.	GALAXY
OCL 0252	14 40 18.	+ 69 47			OPEN STAR CLUSTER
ZC 1440.3-0036	14 40 18.	- 00 36	2150		CLUSTER OF GALAXIES
MCG+05-35-007	14 40 21.	+ 28 56	120	13.8	GALAXY
REIZ 4312	14 40 22.	+ 28 56	60	13.8	GALAXY
ZWG 048.004	14 40 24.	+ 05 06		14.0	GALAXY
REIZ 4309	14 40 24.	+ 05 06	108	13.3	GALAXY
UGC 09483	14 40 24.	+ 05 06	156	14.0	GALAXY S
MCG+01-37-051	14 40 24.	+ 05 06	120	14.0	GALAXY
ZWG 048.005	14 40 24.	+ 05 21		15.3	GALAXY
ZWG 048.006	14 40 24.	+ 06 57		15.6	GALAXY
ZWG 076.005	14 40 24.	+ 08 58		15.3	GALAXY
ZWG 105.005	14 40 24.	+ 15 07		15.7	GALAXY
MCG+06-32-086	14 40 27.	+ 35 34	24	16.	GALAXY
IC 1048	14 40 29.	+ 05 06 13.			NONSTELLAR OBJECT
HOLM 670A	14 40 30.	+ 05 06	120	12.5	PART OF MULTIPLE GALAXY
ZC 1440.5+3155	14 40 30.	+ 31 55	940		CLUSTER OF GALAXIES
MCG+06-32-087	14 40 30.	+ 33 57	36	15.	GALAXY
ZWG 192.054	14 40 30.	+ 37 40		15.6	GALAXY
UGC 09484	14 40 30.	+ 42 42	60	17.	GALAXY Sc
MCG+10-21-025	14 40 30.	+ 62 32	12	17.	GALAXY
MCG-02-37-014	14 40 30.	- 10 10	24	15.	GALAXY
HN 1132	14 40 31.	+ 18 49	6		NEBULA
IC 4488	14 40 31.	+ 18 49			NONSTELLAR OBJECT
HOLM 671B	14 40 33.	+ 05 13	30	14.6	PART OF MULTIPLE GALAXY
HOLM 669B	14 40 33.	- 15 43	18	14.5	PART OF MULTIPLE GALAXY
HOLM 669A	14 40 34.	- 15 44	54	12.2	PART OF MULTIPLE GALAXY
HN 1759	14 40 34.4	+ 01 18 52.	18		NEBULA
ZWG 048.007	14 40 36.	+ 04 59		15.4	GALAXY
REIZ 4310	14 40 36.	+ 04 59	90	14.9	GALAXY
UGC 09485	14 40 36.	+ 04 59	96	15.4	GALAXY
HOLM 671A	14 40 36.	+ 05 14	24	14.6	PART OF MULTIPLE GALAXY
IC 4492	14 40 36.	+ 37 37 22.			NONSTELLAR OBJECT
MCG+06-32-088	14 40 36.	+ 37 40	12	16.	GALAXY
ZWG 220.049	14 40 36.	+ 42 03		13.7	GALAXY
UGC 09486	14 40 36.	+ 42 03	180	13.7	GALAXY S0/Sa
MCG+07-30-052	14 40 36.	+ 42 03	84	13.	GALAXY
ZC 1440.6+5056	14 40 36.	+ 50 56	1080		CLUSTER OF GALAXIES
ZC 1440.6+6202	14 40 36.	+ 62 02	1550		CLUSTER OF GALAXIES
MCG-04-35-004	14 40 36.	- 24 15	90	14.5	GALAXY
LB 00278	14 40 36.	- 27 59 48.		12.5	FAINT BLUE STAR
RNGC 5739	14 40 39.	+ 42 03		12.5	GALAXY
MCG-04-35-005	14 40 39.	- 23 13 30.	36	14.5	GALAXY
HOLM 670B	14 40 40.	+ 05 06	30	15.0	PART OF MULTIPLE GALAXY
ZWG 134.020	14 40 42.	+ 21 38		15.7	GALAXY
UGC 09487	14 40 42.	+ 21 38	84	15.7	GALAXY Sa-b

OBJECT NAME	RIGHT ASCEN.	DECLINATION	DIAM.	MAGN.	TYPE OF OBJECT
KARA.73B 0643	14 40 42.	+ 21 38	66	15.7	ISOLATED GALAXY S
ZWG 134.021	14 40 42.	+ 25 13		15.6	GALAXY
MCG+09-24-030	14 40 42.	+ 55 48	60	16.	GALAXY
MCG-04-35-006	14 40 42.	- 24 15	60	15.5	GALAXY
REIZ 4318	14 40 43.	+ 42 03	120	13.1	GALAXY
HN 1760	14 40 43.1	- 01 02 31.	12		NEBULA
REIZ 4311	14 40 44.	+ 02 24	18	14.3	GALAXY
ZWG 105.006	14 40 48.	+ 16 40		15.6	GALAXY
ZWG 105.007	14 40 48.	+ 19 05		14.6	GALAXY
UGC 09488	14 40 48.	+ 19 05	84	14.6	GALAXY SBb
ZWG 105.008	14 40 48.	+ 19 43		15.2	GALAXY
BZW 1440+19.7	14 40 48.	+ 19 43		16.6	COMPACT GALAXY
ZC 1440.8+2054	14 40 48.	+ 20 54	740		CLUSTER OF GALAXIES
UGC 09489	14 40 48.	+ 25 23	66	16.0	GALAXY DBL SYS
ZWG 192.055	14 40 48.	+ 38 05		15.7	GALAXY
MCG+06-32-089	14 40 48.	+ 38 05 30.	48	15.	GALAXY
SMI 07	14 40 48.	- 59 39	360		FAINT NEBULOSITY
ARC 1956	14 40 49.	+ 31 53		17.6	RICH CLUSTER OF GALAXIES
HN 1761	14 40 49.3	+ 01 18 41.	12		NEBULA
RNGC 5737	14 40 51.	+ 19 05		14.5	GALAXY
LB 00279	14 40 53.	+ 00 59 42.		15.7	FAINT BLUE STAR
ZWG 048.008	14 40 54.	+ 04 35		15.2	GALAXY
REIZ 4313	14 40 54.	+ 04 35	12	14.8	GALAXY
MCG+03-37-039	14 40 54.	+ 19 04	72	14.6	GALAXY
MCG+11-18-005	14 40 54.	+ 63 31 30.	30	16.	GALAXY
HN 1133	14 40 55.	+ 18 44	6		NEBULA
IC 4489	14 40 55.	+ 18 44			NONSTELLAR OBJECT
REIZ 4314	14 40 55.	+ 19 06	24	14.2	GALAXY
HN 1762	14 40 56.4	+ 00 24 05.	18		NEBULA
ZWG 076.006	14 41 00.	+ 10 58		15.5	GALAXY
ZC 1441.0+1614	14 41 00.	+ 16 14	1880		CLUSTER OF GALAXIES
ZWG 134.022	14 41 00.	+ 21 00		15.7	GALAXY
ZC 1441.0+3112	14 41 00.	+ 31 12	810		CLUSTER OF GALAXIES
MCG+08-27-020	14 41 00.	+ 46 02 30.	30	17.	GALAXY
MCG-01-38-001	14 41 00.	- 05 00	12	15.	GALAXY
ARC 1957	14 41 02.	+ 31 26		17.8	RICH CLUSTER OF GALAXIES
ZWG 076.007	14 41 06.	+ 11 25		14.9	GALAXY
RNGC 5736	14 41 06.	+ 11 25		15.0	GALAXY
MCG+02-38-001	14 41 06.	+ 11 25	60	14.9	GALAXY
ZC 1441.1+3142	14 41 06.	+ 31 42	870		CLUSTER OF GALAXIES
ZWG 220.050	14 41 06.	+ 41 43		15.7	GALAXY
MCG+09-24-031	14 41 06.	+ 55 45	60	16.	GALAXY
REIZ 4315	14 41 07.	+ 11 25	60	14.7	GALAXY
HARO 44	14 41 07.	+ 28 31			BLUE EMISSION-LINE GALAXY
ARC 1958	14 41 08.	+ 31 12		17.8	RICH CLUSTER OF GALAXIES
MCG-03-38-001	14 41 09.	- 18 16	48	14.5	GALAXY
ZWG 076.008	14 41 12.	+ 11 22		15.6	GALAXY
UGC 09490	14 41 12.	+ 11 22	84	15.6	GALAXY Sc
ZC 1441.2+1909	14 41 12.	+ 19 09	6450		CLUSTER OF GALAXIES
ZC 1441.2+2801	14 41 12.	+ 28 01	1080		CLUSTER OF GALAXIES
MCG+05-35-008	14 41 12.	+ 28 30	24		GALAXY
REIZ 4319	14 41 13.	+ 19 05	12	15.4	GALAXY
ZWG 076.009	14 41 18.	+ 11 26		15.2	GALAXY
ZC 1441.3+2526	14 41 18.	+ 25 26	2220		CLUSTER OF GALAXIES
MCG+10-21-026	14 41 18.	+ 61 24	24	16.	GALAXY
HOLM 673A	14 41 19.	+ 72 51	24	14.8	PART OF MULTIPLE GALAXY
REIZ 4316	14 41 20.	+ 01 42	24	14.4	GALAXY
REIZ 4317	14 41 20.	+ 01 49	24	14.1	GALAXY
MCG-02-38-001	14 41 21.	- 12 44	90	14.5	GALAXY
MCG-03-38-002	14 41 21.	- 17 57	54	15.	GALAXY
ARC 1955	14 41 22.	- 04 19		17.8	RICH CLUSTER OF GALAXIES
ARC 1962	14 41 23.	+ 55 26		17.2	RICH CLUSTER OF GALAXIES
HOLM 673B	14 41 23.	+ 72 49	30	15.2	PART OF MULTIPLE GALAXY
ZWG 020.004	14 41 24.	+ 01 49		14.7	GALAXY
MCG+00-38-002	14 41 24.	+ 01 49	48	14.7	GALAXY
ZWG 076.010	14 41 24.	+ 11 49		15.2	GALAXY
ZC 1441.4+1757	14 41 24.	+ 17 57	810		CLUSTER OF GALAXIES
ZWG 134.023	14 41 24.	+ 23 14		15.1	GALAXY
FEIG 102	14 41 24.	+ 51 56		16.0	FAINT BLUE STAR
RNGC 5738	14 41 26.	+ 01 49		14.5	GALAXY
ZWG 076.011	14 41 30.	+ 09 44		15.4	GALAXY
ZWG 105.009	14 41 30.	+ 16 41		15.3	GALAXY
MCG+04-35-006	14 41 30.	+ 23 16	48	15.1	GALAXY
MCG+10-21-027	14 41 30.	+ 61 56	36	16.	GALAXY
MCG-05-35-004	14 41 30.	- 30 56	48	15.5	GALAXY
REIZ 4320	14 41 32.	+ 10 06	54	14.4	GALAXY
HN 1763	14 41 35.1	+ 01 23 43.	24		NEBULA
ZWG 020.005	14 41 36.	+ 01 25		15.4	GALAXY
ZWG 048.009	14 41 36.	+ 06 14		15.7	GALAXY
TON-N 0212	14 41 36.	+ 32 54		14.5	BLUE STAR
MCG-04-35-007	14 41 36.	- 22 40	24	15.5	GALAXY
ARC 1959	14 41 37.	+ 17 56		17.4	RICH CLUSTER OF GALAXIES
REIZ 4324	14 41 37.	+ 36 56	42	14.7	GALAXY
MCG-02-38-002	14 41 39.	- 11 37	36	15.	GALAXY
ZWG 020.006	14 41 42.	+ 02 20		15.7	GALAXY
ZWG 048.010	14 41 42.	+ 04 26		15.3	GALAXY
REIZ 4321	14 41 42.	+ 04 26	9	15.1	GALAXY
UGC 09491	14 41 42.	+ 04 26	78	15.3	GALAXY S
MCG+01-38-001	14 41 42.	+ 04 26	72	15.3	GALAXY
ZWG 048.011	14 41 42.	+ 08 10		15.0	GALAXY
UGC 09492	14 41 42.	+ 08 10	108	15.0	GALAXY Sa
MCG+01-38-002	14 41 42.	+ 08 10	48	15.0	GALAXY
ZWG 076.012	14 41 42.	+ 09 30		15.3	GALAXY
MCG+08-27-021	14 41 42.	+ 48 25	36	16.	GALAXY
ZC 1441.7+5542	14 41 42.	+ 55 42	1080		CLUSTER OF GALAXIES
HN 1764	14 41 42.5	+ 00 15 02.	30		NEBULA
HN 0415	14 41 44.	- 13 31			NEBULA
IC 4491	14 41 44.	- 13 31			NONSTELLAR OBJECT
MCG+09-24-032	14 41 45.	+ 52 50	36	16.	GALAXY
ZWG 020.007	14 41 48.	+ 00 15		15.3	GALAXY
MCG+03-38-001	14 41 48.	+ 15 45	18	15.6	GALAXY
ZWG 105.010	14 41 48.	+ 15 47		15.6	GALAXY
ZWG 105.011	14 41 48.	+ 18 14		15.1	GALAXY
IC 1050	14 41 48.	+ 18 14 15.			NONSTELLAR OBJECT
ZWG 134.024	14 41 48.	+ 26 14		15.7	GALAXY
TON-N 0782	14 41 48.	+ 28 44		15.5	BLUE STAR
MCG+06-32-090	14 41 48.	+ 33 36	18	15.	GALAXY
ZWG 248.019	14 41 48.	+ 49 36		15.5	GALAXY
REIZ 4322	14 41 50.	+ 01 54	90	12.6	GALAXY
IC 1051	14 41 52.	+ 19 14 09.			NONSTELLAR OBJECT
HN 1765	14 41 53.7	- 02 13 28.	18		NEBULA
ZWG 020.008	14 41 54.	+ 01 54		13.2	GALAXY
UGC 09493	14 41 54.	+ 01 54	192	13.2	GALAXY Sb
KARA.72 434A	14 41 54.	+ 01 54	180	13.2	PART OF DOUBLE GALAXY
MCG+00-38-003	14 41 54.	+ 01 54	132	13.2	GALAXY
ZWG 048.012	14 41 54.	+ 04 57		15.6	GALAXY
REIZ 4323	14 41 54.	+ 12 20	42	13.8	GALAXY
MCG+03-38-002	14 41 54.	+ 18 12	60	15.1	GALAXY
ZWG 105.012	14 41 54.	+ 19 14		15.4	GALAXY
ZWG 134.025	14 41 54.	+ 20 50		15.3	GALAXY
UGC 09494	14 41 54.	+ 20 50	66	15.3	GALAXY SB:c
MCG+04-35-007	14 41 54.	+ 20 50	66	15.3	GALAXY
IC 4496	14 41 54.	+ 33 36 16.			NONSTELLAR OBJECT
ZWG 192.056	14 41 54.	+ 33 37		15.0	GALAXY
UGC 09495	14 41 54.	+ 41 27	84	16.0	GALAXY Sc
IZW 094	14 41 54.	+ 49 35			COMPACT GALAXY
ZC 1441.9+5806	14 41 54.	+ 58 06	1140		CLUSTER OF GALAXIES
IC 1052	14 41 55.	+ 20 49 27.			NONSTELLAR OBJECT
RNGC 5740	14 41 56.	+ 01 53		12.5	GALAXY
IC 4493	14 41 57.	+ 12 20			SAME AS NGC 5747
REIZ 4327	14 41 57.	+ 29 18	60	14.9	GALAXY
HN 1766	14 41 58.5	+ 01 43 08.	18		NEBULA
VDB.66G 196	14 42	+ 08 06	70		DWARF GALAXY
LB 09891	14 42	- 82 52		14.4	FAINT BLUE STAR
LB 09892	14 42	- 86 13		11.8	FAINT BLUE STAR
ZWG 020.010	14 42 00.	+ 01 46		15.7	GALAXY
ZWG 076.013	14 42 00.	+ 12 21		14.4	GALAXY
RNGC 5747	14 42 00.	+ 12 21		14.5	GALAXY
UGC 09496	14 42 00.	+ 12 21	60	14.4	GALAXY S
KARA.72 435B	14 42 00.	+ 12 21	36		PART OF DOUBLE GALAXY
KARA.72 435A	14 42 00.	+ 12 21	24	14.4	PART OF DOUBLE GALAXY
MCG+02-38-002	14 42 00.	+ 12 21	60	14.4	GALAXY
ZWG 105.013	14 42 00.	+ 15 45		15.2	GALAXY
ZWG 134.026	14 42 00.	+ 22 58		15.6	GALAXY
ZWG 134.027	14 42 00.	+ 23 46		15.7	GALAXY
IC 4495	14 42 00.	+ 23 46 22.			NONSTELLAR OBJECT
ZC 1442.0+2414	14 42 00.	+ 24 14	20970		CLUSTER OF GALAXIES
MCG+07-30-053	14 42 00.	+ 41 26	60	15.	GALAXY
ZWG 220.051	14 42 00.	+ 43 47		15.4	GALAXY
KARA.73B 0644	14 42 00.	+ 43 47	54	15.4	ISOLATED GALAXY S
MCG+07-30-054	14 42 00.	+ 43 48	42	15.5	GALAXY
7ZW 561	14 42 00.	+ 57 39			COMPACT GALAXY
MCG+10-21-028	14 42 00.	+ 59 05	27	16.	GALAXY
ZWG 020.009	14 42 00.	- 03 02		15.5	GALAXY
MCG+00-38-004	14 42 00.	- 03 02	72	15.5	GALAXY
MCG-02-38-003	14 42 00.	- 09 27	48	15.	GALAXY
ZWG 076.014	14 42 06.	+ 09 10		15.4	GALAXY
ZWG 076.015	14 42 06.	+ 12 28		15.6	GALAXY
MCG+03-38-004	14 42 06.	+ 15 39	36	15.5	GALAXY
ZWG 105.014	14 42 06.	+ 15 42		15.5	GALAXY
MCG+03-38-003	14 42 06.	+ 15 43	60	15.2	GALAXY
ZWG 164.015	14 42 06.	+ 28 45		15.1	GALAXY
MCG+05-35-009	14 42 06.	+ 28 45	36	15.1	GALAXY
ACK 333+32.1	14 42 06.	- 23 35			PLANETARY NEBULA
MCG-04-35-008	14 42 06.	- 23 35	30	15.	GALAXY
REIZ 4325	14 42 07.	+ 11 09	54	15.3	GALAXY
HN 1134	14 42 09.	+ 15 44	12		NEBULA
IC 4494	14 42 09.	+ 15 44			NONSTELLAR OBJECT
ARC 1960	14 42 11.	+ 19 34		16.6	RICH CLUSTER OF GALAXIES
IC 4497	14 42 11.	+ 28 45 47.			NONSTELLAR OBJECT
LB 00280	14 42 11.	- 24 36 12.		15.6	FAINT BLUE STAR
MCG+03-38-005	14 42 12.	+ 15 47	27	16.5	GALAXY
ZC 1442.2+2539	14 42 12.	+ 25 39	610		CLUSTER OF GALAXIES
UGC 09497	14 42 12.	+ 42 50	66	16.0	GALAXY PECULR
MCG+09-24-033	14 42 12.	+ 53 38	90	13.	GALAXY
MCG+11-18-006	14 42 12.	+ 65 31	24	16.	GALAXY
ZC 1442.2+7842	14 42 12.	+ 78 42	1140		CLUSTER OF GALAXIES
RNGC 5734	14 42 17.	- 20 40		14.0	GALAXY
RNGC 5743	14 42 17.	- 20 42		13.0	GALAXY
IC 4490	14 42 17.	- 35 57 47.			NONSTELLAR OBJECT
ZC 1442.3+1436	14 42 18.	+ 14 36	1550		CLUSTER OF GALAXIES
MCG+07-30-055	14 42 18.	+ 42 50	60	15.5	GALAXY
ZWG 273.024	14 42 18.	+ 53 37		13.9	GALAXY
UGC 09498	14 42 18.	+ 53 37	90	13.9	GALAXY Sc
VV 098D	14 42 18.	- 13 44	12	16.	INTERACTING GALAXY
VV 098C	14 42 18.	- 13 44	12	15.	INTERACTING GALAXY
VV 098B	14 42 18.	- 13 44	24	15.	INTERACTING GALAXY
VV 098A	14 42 18.	- 13 44	18	15.	INTERACTING GALAXY
VV 098	14 42 18.	- 13 44	102	14.	INTERACTING GALAXY
RNGC 5744	14 42 18.	- 18 16			NON-EXISTENT OBJECT
MCG-03-38-003	14 42 18.	- 20 40	36	14.	GALAXY
REIZ 4328	14 42 19.	+ 11 19	24	15.6	GALAXY
RNGC 5751	14 42 19.	+ 53 37		14.0	GALAXY
HN 1767	14 42 19.4	+ 00 33 51.	30		NEBULA
REIZ 4326	14 42 20.	+ 02 10	420	11.8	GALAXY
RNGC 5745	14 42 20.	- 13 44		14.0	GALAXY
MCG-02-38-004	14 42 21.	- 13 44 30.	36	14.5	GALAXY
MCG-03-38-004	14 42 21.	- 20 42 30.	60	13.5	GALAXY
ZWG 134.028	14 42 24.	+ 22 39		15.4	GALAXY
ZC 1442.4+3059	14 42 24.	+ 30 59	810		CLUSTER OF GALAXIES
ZC 1442.4+3125	14 42 24.	+ 31 25	1340		CLUSTER OF GALAXIES
RNGC 5746	14 42 26.	+ 02 01		11.5	GALAXY
ARC 1961	14 42 26.	+ 31 25		17.8	RICH CLUSTER OF GALAXIES
BC B1442+117	14 42 26.2	+ 11 44 34.		18.	QUASI-STELLAR OBJECT
SHB 295	14 42 26.2	+ 11 44 34.		18.	QUASI-STELLAR OBJECT
HN 1768	14 42 27.0	+ 01 36 52.	18		NEBULA
HN 1769	14 42 27.7	+ 00 44 16.	36		NEBULA
ZWG 020.011	14 42 30.	+ 00 44		15.5	GALAXY
ZWG 020.012	14 42 30.	+ 02 10		12.3	GALAXY
UGC 09499	14 42 30.	+ 02 10	444	12.3	GALAXY Sb
KARA.72 434B	14 42 30.	+ 02 10	384	12.3	PART OF DOUBLE GALAXY
MCG+00-38-005	14 42 30.	+ 02 10	420	12.3	GALAXY
ZWG 048.013	14 42 30.	+ 04 43		15.5	GALAXY
ZC 1442.5+3015	14 42 30.	+ 30 15	3970		CLUSTER OF GALAXIES
ZWG 192.057	14 42 30.	+ 37 41		15.6	GALAXY
MCG+07-30-056	14 42 30.	+ 41 00	21	17.	GALAXY
IZW 095	14 42 30.	+ 50 48			COMPACT GALAXY
REIZ 4329	14 42 33.	+ 00 44	72	15.0	GALAXY
ZC 1442.6+2229	14 42 36.	+ 22 29	1280		CLUSTER OF GALAXIES
IC 4500	14 42 36.	+ 37 40 50.			NONSTELLAR OBJECT
MCG+06-32-091	14 42 36.	+ 37 42	18	15.	GALAXY
REIZ 4330	14 42 37.	+ 11 04	60	15.5	GALAXY
MCG+07-30-057	14 42 39.	+ 41 00 30.	30	16.	GALAXY
HN 1771	14 42 41.9	+ 00 03 10.	18		NEBULA
ZWG 048.014	14 42 42.	+ 03 36		15.6	GALAXY
ZWG 076.016	14 42 42.	+ 09 29		15.2	GALAXY
ZWG 164.016	14 42 42.	+ 26 30		15.1	GALAXY
ZWG 164.017	14 42 42.	+ 30 27		15.7	GALAXY
ZC 1442.7+3611	14 42 42.	+ 36 11	2420		CLUSTER OF GALAXIES
MCG+10-21-029	14 42 42.	+ 62 25	51	15.	GALAXY
MCG-02-38-005	14 42 42.	- 10 45	30	15.	GALAXY
MCG-03-38-005	14 42 42.	- 20 29	84	16.	GALAXY
HN 0414	14 42 42.	- 73 06			NEBULA
IC 4484	14 42 42.	- 73 06			NONSTELLAR OBJECT
ARC 1963	14 42 43.	+ 31 42		17.7	RICH CLUSTER OF GALAXIES
ARC 1966	14 42 45.	+ 59 07		18.0	RICH CLUSTER OF GALAXIES

OBJECT NAME	RIGHT ASCEN.	DECLINATION	DIAM.	MAGN.	TYPE OF OBJECT
RNGC 5748	14 42 46.	+ 22 07		15.5	GALAXY
UGC 09500	14 42 48.	+ 08 05	180	18.	GALAXY DWRF SP
ZWG 105.015	14 42 48.	+ 17 58		15.3	GALAXY
MCG+03-38-006	14 42 48.	+ 19 37 30.	15	15.3	GALAXY
ZWG 105.016	14 42 48.	+ 19 39		15.3	GALAXY
ZWG 134.029	14 42 48.	+ 22 07		15.4	GALAXY
ZC 1442.8+2625	14 42 48.	+ 26 25	940		CLUSTER OF GALAXIES
MCG+05-35-010	14 42 48.	+ 26 30	24	15.1	GALAXY
IC 4498	14 42 48.	+ 26 30 38.			NONSTELLAR OBJECT
UGC 09501	14 42 48.	+ 62 27	60	16.5	GALAXY Sa-b
VV 164D	14 42 48.	- 20 31 30.	3	18.	INTERACTING GALAXY
VV 164C	14 42 48.	- 20 31 30.	9	17.	INTERACTING GALAXY
VV 164B	14 42 48.	- 20 31 30.	9	17.	INTERACTING GALAXY
VV 164A	14 42 48.	- 20 31 30.	6	17.	INTERACTING GALAXY
VV 164	14 42 48.	- 20 31 30.	90		INTERACTING GALAXY
REIZ 4338	14 42 50.	+ 57 16	120	14.1	GALAXY
ARC 1969	14 42 50.	+ 63 55		17.7	RICH CLUSTER OF GALAXIES
SHB 296	14 42 50.4	+ 10 11 11.		17.8	QUASI-STELLAR OBJECT
BC 0Q172	14 42 50.55	+ 10 11 13.0			QUASI-STELLAR OBJECT
ARP 064	14 42 53.	+ 19 41			PECULIAR GALAXY
ZWG 105.017	14 42 54.	+ 16 02		15.6	GALAXY
ZWG 105.018	14 42 54.	+ 17 36		15.4	GALAXY
ZC 1442.9+3144	14 42 54.	+ 31 44	540		CLUSTER OF GALAXIES
MCG+07-30-058	14 42 54.	+ 41 01	42	16.	GALAXY
ZC 1442.9+5401	14 42 54.	+ 54 01	810		CLUSTER OF GALAXIES
ZC 1442.9+5511	14 42 54.	+ 55 11	610		CLUSTER OF GALAXIES
7ZW 562	14 42 54.	+ 73 07			COMPACT GALAXY
MCG-02-38-006	14 42 57.	- 11 26	15	15.5	GALAXY
UGC 09502	14 43 00.	+ 41 06	60	16.5	GALAXY Sc
ZC 1443.0+6359	14 43 00.	+ 63 59	940		CLUSTER OF GALAXIES
7ZW 563	14 43 00.	+ 72 08			COMPACT GALAXY
MCG-02-38-007	14 43 00.	- 11 36	48	14.5	GALAXY
RNGC 5742	14 43 03.	- 11 36		14.0	GALAXY
ZWG 076.017	14 43 06.	+ 10 07		15.7	GALAXY
MCG+03-38-007	14 43 06.	+ 19 39 30.	90	15.1	GALAXY
ZWG 105.019	14 43 06.	+ 19 41		15.1	GALAXY
UGC 09503	14 43 06.	+ 19 41	90	15.1	GALAXY Sb
ZC 1443.1+3014	14 43 06.	+ 30 14	1080		CLUSTER OF GALAXIES
MCG+06-32-092	14 43 06.	+ 33 13 30.	30	16.	GALAXY
ZC 1443.1+3639	14 43 06.	+ 36 39	740		CLUSTER OF GALAXIES
MCG+07-30-059	14 43 06.	+ 41 05	60	16.	GALAXY
MCG+10-21-030	14 43 06.	+ 58 24	36	17.	GALAXY
ZC 1443.1+7500	14 43 06.	+ 75 00	1410		CLUSTER OF GALAXIES
HN 1772	14 43 06.0	- 02 26 55.	18		NEBULA
HOLM 672B	14 43 09.	+ 17 32	18	14.9	PART OF MULTIPLE GALAXY
HOLM 674C	14 43 10.	+ 38 58	24	14.8	PART OF MULTIPLE GALAXY
ZWG 076.018	14 43 12.	+ 13 57		15.7	GALAXY
ZC 1443.2+2356	14 43 12.	+ 23 56	2490		CLUSTER OF GALAXIES
ZWG 164.018	14 43 12.	+ 30 38		15.0	GALAXY
MCG+05-35-011	14 43 12.	+ 30 38	36	15.0	GALAXY
ZWG 248.020	14 43 12.	+ 49 55		15.4	GALAXY
HOLM 672A	14 43 14.	+ 17 31	24	13.5	PART OF MULTIPLE GALAXY
ARC 1965	14 43 14.	+ 36 41		17.6	RICH CLUSTER OF GALAXIES
HOLM 674D	14 43 14.	+ 39 03	48	15.3	PART OF MULTIPLE GALAXY
HN 1770	14 42 14.7	- 43 42 19.	18		NEBULA
IC 4502	14 43 15.	+ 37 30 17.			NONSTELLAR OBJECT
REIZ 4333	14 43 16.	+ 38 56	30	14.9	GALAXY
HOLM 674A	14 43 16.	+ 38 58	60	13.7	PART OF MULTIPLE GALAXY
REIZ 4334	14 43 16.	+ 39 00	12	15.7	GALAXY
ZWG 105.020	14 43 18.	+ 17 10		15.1	GALAXY
MCG+03-38-008	14 43 18.	+ 17 10	30	15.1	GALAXY
ZWG 164.019	14 43 18.	+ 27 15		15.4	GALAXY
TON-N 0213	14 43 18.	+ 29 33		14.6	BLUE STAR
ZWG 164.020	14 43 18.	+ 31 39		15.3	GALAXY
UGC 09504	14 43 18.	+ 31 39	78	15.3	GALAXY DBL SYS
MCG+07-30-060	14 43 18.	+ 38 55	30	15.	GALAXY
ZWG 220.052	14 43 18.	+ 38 56		14.1	GALAXY
UGC 09505	14 43 18.	+ 38 56	120	14.1	GALAXY SBb
1ZW 096	14 43 18.	+ 51 35			COMPACT GALAXY
VVI 77	14 43 18.	+ 51 35	48	16.	SEYFERT GALAXY
ARC 1974	14 43 18.	+ 75 03		17.5	RICH CLUSTER OF GALAXIES
MCG-02-38-008	14 43 18.	- 11 42	24	15.	GALAXY
MCG-03-38-006	14 43 18.	- 17 09	96	14.5	GALAXY
REIZ 4331	14 43 19.	+ 11 05	48	15.4	GALAXY
RNGC 5752	14 43 19.	+ 38 55		14.0	GALAXY
ARP 297	14 43 19.	+ 38 57			PECULIAR GALAXY
HOLM 674B	14 43 20.	+ 39 01	48	14.2	PART OF MULTIPLE GALAXY
REIZ 4337	14 43 20.	+ 40 57	30	15.0	GALAXY
MCG+05-35-013	14 43 21.	+ 31 39 30.	84	15.3	GALAXY
MCG+05-35-012	14 43 21.	+ 31 40	84	15.3	GALAXY
RNGC 5741	14 43 21.	- 11 42		15.0	GALAXY
REIZ 4335	14 43 22.	+ 38 57	42	14.0	GALAXY
REIZ 4336	14 43 22.	+ 38 59	12	14.5	GALAXY
SN 1954M	14 43 24.	+ 10 23		18.0	SUPERNOVA
ZC 1443.4+1617	14 43 24.	+ 16 17	1750		CLUSTER OF GALAXIES
IC 1053	14 43 24.	+ 17 09 47.			NONSTELLAR OBJECT
UGC 09506	14 43 24.	+ 31 38	60	18.	GALAXY DWRF IR
MCG+07-30-061	14 43 24.	+ 38 55	120	13.	GALAXY
ZWG 220.053	14 43 24.	+ 38 59		15.1	GALAXY
UGC 09507	14 43 24.	+ 38 59	102	15.1	GALAXY SB
MCG+07-30-062	14 43 24.	+ 38 59 30.	24	15.	GALAXY
UGC 09508	14 43 24.	+ 49 18	78	16.0	GALAXY IRR
MCG+08-27-022	14 43 24.	+ 49 19	36	16.	GALAXY
RNGC 5753	14 43 25.	+ 38 55		15.0	GALAXY
RNGC 5754	14 43 25.	+ 38 59		14.0	GALAXY
MCG+07-30-063	14 43 27.	+ 38 58	42	14.5	GALAXY
ZWG 076.019	14 43 30.	+ 08 43		15.4	GALAXY
UGC 09509	14 43 30.	+ 08 43	72	15.4	GALAXY DBL SYS
MCG+02-38-003	14 43 30.	+ 08 43	66	15.4	GALAXY
MCG+02-38-004	14 43 30.	+ 08 44	18	15.4	GALAXY
UGC 09510	14 43 30.	+ 32 50	66	16.0	GALAXY Sb-c
MCG+06-32-093	14 43 30.	+ 34 54 30.	18	16.	GALAXY
UGC 09511	14 43 30.	+ 51 35	78	16.5	GALAXY PECULR
RNGC 5755	14 43 31.	+ 38 58		15.0	GALAXY
REIZ 4340	14 43 34.	+ 47 38	12	15.0	GALAXY
ZWG 076.020	14 43 36.	+ 09 09		15.6	GALAXY
ZWG 076.021	14 43 36.	+ 11 01		15.4	GALAXY
ZWG 105.021	14 43 36.	+ 16 52		15.7	GALAXY
ZC 1443.6+2139	14 43 36.	+ 21 39	2890		CLUSTER OF GALAXIES
ZWG 164.021	14 43 36.	+ 28 44		15.4	GALAXY
MCG+06-32-094	14 43 36.	+ 32 49 30.	90	15.	GALAXY
MCG-02-38-009	14 43 36.	- 11 18 30.	48	15.	GALAXY
RNGC 5750	14 43 37.	- 00 01		12.5	GALAXY
HN 1773	14 43 39.3	+ 01 04 43.	12		NEBULA
HOLM 675A	14 43 40.	+ 28 44	30	14.5	PART OF MULTIPLE GALAXY
REIZ 4332	14 43 40.	- 00 01	72	12.4	GALAXY
ZC 1443.7+0413	14 43 42.	+ 04 13	940		CLUSTER OF GALAXIES
ZWG 076.022	14 43 42.	+ 13 14		15.4	GALAXY
UGC 09513	14 43 42.	+ 13 14	66	15.4	GALAXY
ZC 1443.7+2727	14 43 42.	+ 27 27	810		CLUSTER OF GALAXIES
ZC 1443.7+4217	14 43 42.	+ 42 17	1140		CLUSTER OF GALAXIES
ZWG 020.013	14 43 42.	- 00 01		13.1	GALAXY
UGC 09512	14 43 42.	- 00 01	192	13.1	GALAXY SBa
MCG+00-38-006	14 43 42.	- 00 01	114	13.1	GALAXY
HOLM 675B	14 43 44.	+ 28 45	30	15.8	PART OF MULTIPLE GALAXY
VV 109A	14 43 45.	+ 08 41	66	16.	INTERACTING GALAXY
REIZ 4339	14 43 46.	+ 28 43	24	15.1	GALAXY
VV 109B	14 43 48.	+ 08 42	18	17.	INTERACTING GALAXY
ZWG 076.023	14 43 48.	+ 10 37		15.4	GALAXY
ZWG 076.024	14 43 48.	+ 10 45		15.5	GALAXY
ZC 1443.8+2521	14 43 48.	+ 25 21	940		CLUSTER OF GALAXIES
ZWG 134.030	14 43 48.	+ 25 48		15.4	GALAXY
MCG+05-35-014	14 43 48.	+ 28 44	24	15.4	GALAXY
ZC 1443.8+3114	14 43 48.	+ 31 14	670		CLUSTER OF GALAXIES
ZC 1443.8+3504	14 43 48.	+ 35 04	3360		CLUSTER OF GALAXIES
ZC 1443.8+4043	14 43 48.	+ 40 43	1550		CLUSTER OF GALAXIES
MCG+09-24-034	14 43 48.	+ 54 40	36	15.	GALAXY
MCG+06-32-095	14 43 51.	+ 36 21	36	16.	GALAXY
MCG-03-38-007	14 43 51.	- 18 18 30.	30	14.	GALAXY
ZWG 048.015	14 43 54.	+ 06 45		15.4	GALAXY
IC 1054	14 43 58.	+ 01 28 31.			NONSTELLAR OBJECT
ZWG 020.014	14 44 00.	+ 01 29		15.2	GALAXY
UGC 09514	14 44 00.	+ 01 29	72	15.2	GALAXY S0
ZWG 076.025	14 44 00.	+ 11 42		14.9	GALAXY
MCG+02-38-005	14 44 00.	+ 11 42	24	14.9	GALAXY
ZWG 076.026	14 44 00.	+ 13 14		15.4	GALAXY
UGC 09515	14 44 00.	+ 13 14	96	15.4	GALAXY S
ZC 1444.0+1459	14 44 00.	+ 14 59	400		CLUSTER OF GALAXIES
ZWG 105.022	14 44 00.	+ 19 29		15.7	GALAXY
ZWG 164.022	14 44 00.	+ 30 10		15.7	GALAXY
MCG+08-27-023	14 44 00.	+ 50 36 30.	84	14.	GALAXY
MCG+09-24-035	14 44 00.	+ 51 47 30.	48	14.	GALAXY
IC 1056	14 44 03.	+ 50 36 29.			NONSTELLAR OBJECT
ARC 1964	14 44 04.	- 08 34		16.9	RICH CLUSTER OF GALAXIES
HN 1774	14 44 04.6	- 01 18 22.	18		NEBULA
REIZ 4342	14 44 05.	+ 27 25	24	14.9	GALAXY
ZWG 076.027	14 44 06.	+ 13 48		15.7	GALAXY
MCG+03-38-009	14 44 06.	+ 19 28	30	15.7	GALAXY
ZWG 273.025	14 44 06.	+ 50 38		14.4	GALAXY
UGC 09516	14 44 06.	+ 50 38	120	14.4	GALAXY Sb
KARA.73B 0645	14 44 06.	+ 50 39	120	14.4	ISOLATED GALAXY
ZWG 273.026	14 44 06.	+ 51 48		14.7	GALAXY
ZC 1444.1+6926	14 44 06.	+ 69 26	540		CLUSTER OF GALAXIES
ZWG 076.028	14 44 12.	+ 10 57		15.4	GALAXY
ZWG 105.023	14 44 12.	+ 16 21		14.9	GALAXY
ZC 1444.2+3820	14 44 12.	+ 38 20	1340		CLUSTER OF GALAXIES
ZWG 020.015	14 44 12.	- 01 45		15.7	GALAXY
MCG-05-35-005	14 44 12.	- 27 31	60	15.5	GALAXY
HN 1775	14 44 13.7	- 01 44 33.	12		NEBULA
MCG+06-32-096	14 44 15.	+ 32 58	24	14.5	GALAXY
ZWG 076.029	14 44 18.	+ 10 23		15.7	GALAXY
ZWG 076.030	14 44 18.	+ 12 49		15.0	GALAXY
UGC 09517	14 44 18.	+ 12 49	60	15.0	GALAXY Sb
MCG+02-38-006	14 44 18.	+ 12 49	66	15.0	GALAXY
ZWG 076.031	14 44 18.	+ 13 53		15.2	GALAXY
MCG+03-38-010	14 44 18.	+ 16 19	48	14.9	GALAXY
MCG+03-38-011	14 44 18.	+ 16 28 30.	42	15.2	GALAXY
ZWG 105.024	14 44 18.	+ 19 30		15.4	GALAXY
ZWG 192.058	14 44 18.	+ 32 59		14.5	GALAXY
UGC 09518	14 44 18.	+ 32 59	84	14.5	GALAXY E
MCG+06-32-097	14 44 13.	+ 34 34 30.	36	14.5	GALAXY
ZWG 192.059	14 44 18.	+ 34 35		14.4	GALAXY
UGC 09519	14 44 18.	+ 34 35	48	14.4	GALAXY S0
HN 1776	14 44 18.7	+ 02 03 09.	18		NEBULA
REIZ 4341	14 44 19.	+ 11 10	36	14.3	GALAXY
IC 1057	14 44 20.	+ 50 34 54.			NONSTELLAR OBJECT
HN 1135	14 44 21.	+ 16 20	12		NEBULA
IC 4503	14 44 21.	+ 16 20			NONSTELLAR OBJECT
MCG+06-32-098	14 44 21.	+ 35 50	12	16.	GALAXY
ZWG 076.032	14 44 24.	+ 11 47		15.0	GALAXY
MCG+02-38-007	14 44 24.	+ 11 47	24	15.0	GALAXY
ZC 1444.4+1800	14 44 24.	+ 18 00	1340		CLUSTER OF GALAXIES
ZWG 134.031	14 44 24.	+ 20 49		15.7	GALAXY
ZWG 134.032	14 44 24.	+ 21 16		15.3	GALAXY
ZC 1444.4+2644	14 44 24.	+ 26 44	810		CLUSTER OF GALAXIES
ZWG 164.023	14 44 24.	+ 31 55		15.4	GALAXY
ZWG 192.060	14 44 24.	+ 32 51		15.6	GALAXY
ZC 1444.4+6418	14 44 24.	+ 64 18	270		CLUSTER OF GALAXIES
HN 1777	14 44 26.2	+ 02 02 21.	18		NEBULA
MCG+06-32-099	14 44 27.	+ 33 36 30.	60	14.5	GALAXY
MCG-02-38-010	14 44 27.	- 13 07	72	15.	GALAXY
IC 4505	14 44 29.	+ 33 36 53.			NONSTELLAR OBJECT
HN 1778	14 44 29.4	+ 00 40 57.	24		NEBULA
ZWG 076.033	14 44 30.	+ 11 50		15.0	GALAXY
MCG+02-38-008	14 44 30.	+ 11 50	24	15.0	GALAXY
ZWG 076.034	14 44 30.	+ 14 03		15.1	GALAXY
IC 4504	14 44 30.	+ 31 54 53.			NONSTELLAR OBJECT
ZWG 192.061	14 44 30.	+ 33 37		15.1	GALAXY
UGC 09520	14 44 30.	+ 33 37	66	15.1	GALAXY E-S0
MCG-05-35-006	14 44 30.	- 30 26	24	16.	GALAXY
HN 0416	14 44 33.	- 22 11			NEBULA
IC 4501	14 44 33.	- 22 11			NONSTELLAR OBJECT
HELW 035	14 44 34.	- 16 07			NEBULA
ZWG 048.016	14 44 36.	+ 03 49		15.4	GALAXY
ZWG 076.035	14 44 36.	+ 09 53		15.7	GALAXY
ZC 1444.6+1001	14 44 36.	+ 10 01	810		CLUSTER OF GALAXIES
ZWG 076.036	14 44 36.	+ 11 48		14.7	GALAXY
UGC 09521	14 44 36.	+ 11 48	78	14.7	GALAXY SBa
MCG+02-38-009	14 44 36.	+ 11 48	66	14.7	GALAXY
ZC 1444.6+2041	14 44 36.	+ 20 41	2490		CLUSTER OF GALAXIES
ZWG 192.062	14 44 36.	+ 33 36		15.5	GALAXY
IC 4506	14 44 36.	+ 33 36 24.			NONSTELLAR OBJECT
UGC 09522	14 44 36.	+ 54 53	66	16.0	GALAXY Sa-b
MCG-03-38-008	14 44 36.	- 17 15	180	13.5	GALAXY
MCG-04-35-009	14 44 36.	- 22 12	54	14.	GALAXY
ZWG 076.037	14 44 42.	+ 11 12		15.3	GALAXY
ZWG 076.038	14 44 42.	+ 11 48		14.7	GALAXY
UGC 09523	14 44 42.	+ 11 48	60	14.7	GALAXY S0-a
MCG+02-38-010	14 44 42.	+ 11 48	48	14.7	GALAXY
ZWG 076.039	14 44 42.	+ 13 52		15.0	GALAXY
UGC 09524	14 44 42.	+ 13 52	60	15.0	GALAXY E-S0
MCG+02-38-011	14 44 42.	+ 13 52	60	15.0	GALAXY
ZC 1444.7+3205	14 44 42.	+ 32 05	740		CLUSTER OF GALAXIES
ARC 1968	14 44 42.	+ 32 05		17.8	RICH CLUSTER OF GALAXIES
ZC 1444.7+3757	14 44 42.	+ 37 57	810		CLUSTER OF GALAXIES
ZWG 020.016	14 44 42.	- 00 45		15.5	GALAXY

OBJECT NAME	RIGHT ASCEN.	DECLINATION	DIAM.	MAGN.	TYPE OF OBJECT
MCG+00-38-007	14 44 42.	- 00 45	42	15.5	GALAXY
IC 1055	14 44 42.	- 13 30 21.			NONSTELLAR OBJECT
MCG-02-38-011	14 44 42.	- 13 31	120	13.5	GALAXY
MCG-03-38-009	14 44 42.	- 19 33 30.	120	14.	GALAXY
MCG-03-38-010	14 44 42.	- 20 16 30.	36	15.	GALAXY
MCG-04-35-010	14 44 42.	- 22 03 30.	60	14.5	GALAXY
RNGC 5758	14 44 43.	+ 13 52		15.0	GALAXY
HOLM 677A	14 44 44.	- 13 31	120	12.6	PART OF MULTIPLE GALAXY
HOLM 676A	14 44 45.	- 14 39	108	12.7	PART OF MULTIPLE GALAXY
HN 1779	14 44 45.3	- 00 44 14.	24		NEBULA
ZWG 076.040	14 44 48.	+ 11 48		15.5	GALAXY
ZWG 076.041	14 44 48.	+ 12 40		15.2	GALAXY
ZWG 076.042	14 44 48.	+ 13 37		15.4	GALAXY
ZWG 076.043	14 44 48.	+ 13 52		15.2	GALAXY
ZWG 105.025	14 44 48.	+ 16 30		15.2	GALAXY
ZC 1444.8+2503	14 44 48.	+ 25 03	610		CLUSTER OF GALAXIES
ZC 1444.8+6406	14 44 48.	+ 64 06	2820		CLUSTER OF GALAXIES
HOLM 677C	14 44 48.	- 13 33	24	14.6	PART OF MULTIPLE GALAXY
HOLM 677B	14 44 48.	- 13 34	24	14.5	PART OF MULTIPLE GALAXY
MCG-02-38-012	14 44 48.	- 14 39	96	13.	GALAXY
MCG-03-38-013	14 44 48.	- 17 52	84	14.	GALAXY
MCG-03-38-012	14 44 48.	- 19 11	54	15.	GALAXY
MCG-03-38-011	14 44 48.	- 20 11 30.	30	15.5	GALAXY
ARC 1967	14 44 49.	+ 10 00		17.8	RICH CLUSTER OF GALAXIES
MCG+07-30-064	14 44 51.	+ 42 31	15	16.	GALAXY
ARC 1975	14 44 51.	+ 69 15		17.1	RICH CLUSTER OF GALAXIES
HOLM 676B	14 44 51.	- 14 38	78	15.1	PART OF MULTIPLE GALAXY
ARP 047	14 44 53.	+ 19 04			PECULIAR GALAXY
HN 1780	14 44 53.2	- 01 25 26.	24		NEBULA
ZWG 020.018	14 44 54.	+ 02 11		15.5	GALAXY
ZWG 048.017	14 44 54.	+ 07 23		15.5	GALAXY
ZWG 076.044	14 44 54.	+ 13 40		14.9	GALAXY
UGC 09525	14 44 54.	+ 13 40	90	14.9	GALAXY DBL SYS
MCG+02-38-012	14 44 54.	+ 13 40	18	14.9	GALAXY
MCG+03-38-012	14 44 54.	+ 16 27	54	16.	GALAXY
ZWG 105.026	14 44 54.	+ 19 04		15.4	GALAXY
ZC 1444.9+3322	14 44 54.	+ 33 22	2020		CLUSTER OF GALAXIES
ZC 1444.9+5356	14 44 54.	+ 53 56	940		CLUSTER OF GALAXIES
MCG+09-24-036	14 44 54.	+ 54 51	72	15.	GALAXY
MCG+09-24-037	14 44 54.	+ 55 00	36	16.	GALAXY
ZWG 020.017	14 44 54.	- 01 26		15.6	GALAXY
MCG-03-38-014	14 44 54.	- 18 53 30.	108	12.5	GALAXY
MCG-05-35-007	14 44 54.	- 30 07	18	15.5	GALAXY
RNGC 5759	14 44 55.	+ 13 40		15.0	GALAXY
RNGC 5756	14 44 55.	- 14 39		13.0	GALAXY
MCG+03-38-013	14 44 57.	+ 16 27	18	16.	GALAXY
ZC 1445.0+1458	14 45 00.	+ 14 58	1880		CLUSTER OF GALAXIES
MCG+03-38-014	14 45 00.	+ 19 02 30.	45	15.4	GALAXY
UGC 09527	14 45 00.	+ 25 04	60	16.0	GALAXY Sc-IRR
ZC 1445.0+2730	14 45 00.	+ 27 30	610		CLUSTER OF GALAXIES
ZC 1445.0+3723	14 45 00.	+ 37 23	1140		CLUSTER OF GALAXIES
ZC 1445.0+4259	14 45 00.	+ 42 59	1080		CLUSTER OF GALAXIES
ZC 1445.0+5520	14 45 00.	+ 55 20	1140		CLUSTER OF GALAXIES
ZWG 273.027	14 45 00.	+ 55 23		15.7	GALAXY
ZC 1445.0+6209	14 45 00.	+ 62 09	1080		CLUSTER OF GALAXIES
ZC 1445.0+6513	14 45 00.	+ 65 13	1080		CLUSTER OF GALAXIES
MCG+12-14-012	14 45 00.	+ 69 07 30.	102	15.	GALAXY
ZWG 337.018	14 45 00.	+ 70 33		15.7	GALAXY
ZC 1445.0+7216	14 45 00.	+ 72 16	740		CLUSTER OF GALAXIES
ZWG 337.019	14 45 00.	+ 73 01		15.3	GALAXY
UGC 09526	14 45 00.	- 00 18	66	17.	GALAXY Sc
MCG-02-38-013	14 45 00.	- 14 56	48	15.	GALAXY
MCG-02-38-014	14 45 00.	- 14 57	12	17.	GALAXY
RNGC 5757	14 45 00.	- 18 53		13.0	GALAXY
HELW 036	14 45 03.	- 14 56			NEBULA
ZWG 076.045	14 45 06.	+ 09 27		15.5	GALAXY
ZWG 076.046	14 45 06.	+ 09 53		15.6	GALAXY
ZWG 076.047	14 45 06.	+ 12 59		15.5	GALAXY
ZWG 076.048	14 45 06.	+ 13 14		15.3	GALAXY
ZWG 134.033	14 45 06.	+ 24 09		15.1	GALAXY
MCG+09-24-039	14 45 06.	+ 55 23	42	16.	GALAXY
MCG+09-24-038	14 45 06.	+ 56 28	42	15.	GALAXY
ZC 1445.1+6129	14 45 06.	+ 61 29	2080		CLUSTER OF GALAXIES
MCG-03-38-015	14 45 06.	- 18 56	72	15.	GALAXY
MCG+04-35-008	14 45 09.	+ 24 10	48	15.1	GALAXY
ZWG 020.019	14 45 12.	+ 01 15		15.6	GALAXY
ZWG 076.049	14 45 12.	+ 10 37		15.4	GALAXY
ZC 1445.2+1356	14 45 12.	+ 13 56	3700		CLUSTER OF GALAXIES
ZWG 164.024	14 45 12.	+ 30 07		15.7	GALAXY
UGC 09528	14 45 12.	+ 30 07	60	15.7	GALAXY S
MCG+05-35-015	14 45 12.	+ 30 46 30.	42		GALAXY
ZWG 337.020	14 45 12.	+ 69 10		15.2	GALAXY
UGC 09529	14 45 12.	+ 69 10	96	15.2	GALAXY Sb
MCG-02-38-015	14 45 12.	- 14 04 30.	120	13.	GALAXY
VHE 63	14 45 12.	- 65 02	78		REFLECTION NEBULA
BSO 12	14 45 13.8	+ 42 58 10.		17.55	BLUE STELLAR OBJECT
HN 1781	14 45 17.3	+ 01 15 24.	18		NEBULA
ZWG 076.050	14 45 18.	+ 09 44		15.1	GALAXY
ZWG 076.051	14 45 18.	+ 09 52		15.0	GALAXY
UGC 09530	14 45 18.	+ 09 52	60	15.0	GALAXY S
MCG+02-38-013	14 45 18.	+ 09 52	42	15.0	GALAXY
ZWG 105.027	14 45 18.	+ 18 40		15.7	GALAXY
ZWG 105.028	14 45 18.	+ 18 43		14.3	GALAXY
UGC 09531	14 45 18.	+ 18 43	102	14.3	GALAXY Sa
REIZ 4346	14 45 18.	+ 26 23	48	14.4	GALAXY
MCG-03-38-016	14 45 18.	- 20 20	42	15.5	GALAXY
OCL 0930	14 45 18.	- 54 19	960	9.0	OPEN STAR CLUSTER
VHA 165	14 45 18.	- 54 19	480		OPEN STAR CLUSTER
RNGC 5749	14 45 18.	- 54 19		9.0	OPEN CLUSTER
REIZ 4343	14 45 19.	+ 18 43	9	15.4	GALAXY
ARC 1970	14 45 20.	+ 13 35		17.2	RICH CLUSTER OF GALAXIES
RNGC 5760	14 45 21.	+ 18 43		14.5	GALAXY
ZC 1445.4+1125	14 45 24.	+ 11 25	4700		CLUSTER OF GALAXIES
ZC 1445.4+1334	14 45 24.	+ 13 34	1280		CLUSTER OF GALAXIES
MCG+03-38-016	14 45 24.	+ 18 39	45	15.7	GALAXY
MCG+03-38-015	14 45 24.	+ 18 42	90	14.3	GALAXY
ZC 1445.4+2338	14 45 24.	+ 23 38	870		CLUSTER OF GALAXIES
ZC 1445.4+4031	14 45 24.	+ 40 31	1080		CLUSTER OF GALAXIES
HN 1136	14 45 25.	+ 18 39	6		NEBULA
REIZ 4344	14 45 25.	+ 18 39	18	15.5	GALAXY
IC 4507	14 45 25.	+ 18 39			NONSTELLAR OBJECT
REIZ 4345	14 45 25.	+ 18 43	48	13.8	GALAXY
ZWG 105.029	14 45 30.	+ 20 18		15.6	GALAXY
ZC 1445.5+4726	14 45 30.	+ 47 26	1410		CLUSTER OF GALAXIES
ARC 1971	14 45 35.	+ 15 01		17.2	RICH CLUSTER OF GALAXIES
ARP 328	14 45 35.	+ 19 16			PECULIAR GALAXY
ZC 1445.6+0351	14 45 36.	+ 03 51	1410		CLUSTER OF GALAXIES
ZWG 076.052	14 45 36.	+ 09 38		15.4	GALAXY
ZWG 076.053	14 45 36.	+ 13 53		15.4	GALAXY
ZWG 105.030	14 45 36.	+ 19 16		15.1	GALAXY
UGC 09532	14 45 36.	+ 19 16	102	15.1	GALAXY MLT SYS
MCG+03-38-017	14 45 36.	+ 19 16	24	15.1	GALAXY
MCG+03-38-021	14 45 37.	+ 19 15	27	15.1	GALAXY
MCG+03-38-020	14 45 37.	+ 19 15	12	15.1	GALAXY
MCG+03-38-019	14 45 37.	+ 19 15	4	18.	GALAXY
MCG+03-38-018	14 45 37.	+ 19 15	12	18.	GALAXY
MCG+03-38-022	14 45 38.	+ 19 14	15	15.1	GALAXY
HELW 037	14 45 40.	- 16 06			NEBULA
VV 165B	14 45 42.	+ 19 17	18	16.	INTERACTING GALAXY
VV 165A	14 45 42.	+ 19 18	19	16.	INTERACTING GALAXY
ZWG 164.025	14 45 42.	+ 31 59		15.3	GALAXY
IC 4508	14 45 44.	+ 31 58 48.			NONSTELLAR OBJECT
VV 165G	14 45 45.	+ 19 16 30.	6	19.	INTERACTING GALAXY
VV 165F	14 45 45.	+ 19 16 30.	9	17.5	INTERACTING GALAXY
VV 165E	14 45 45.	+ 19 16 30.	9	17.	INTERACTING GALAXY
VV 165D	14 45 45.	+ 19 16 30.	24	15.	INTERACTING GALAXY
ZWG 076.054	14 45 48.	+ 14 21		15.7	GALAXY
ZWG 164.026	14 45 48.	+ 30 07		15.6	GALAXY
MCG+09-24-040	14 45 48.	+ 53 02	48	16.	GALAXY
LB 00729	14 45 49.	+ 25 23 42.		17.7	FAINT BLUE STAR
ZWG 076.055	14 45 54.	+ 09 20		15.3	GALAXY
ZC 1445.9+2213	14 45 54.	+ 22 13	2020		CLUSTER OF GALAXIES
ZC 1445.9+2608	14 45 54.	+ 26 08	870		CLUSTER OF GALAXIES
MCG+10-21-031	14 45 54.	+ 62 46	36	16.	GALAXY
LB 09985	14 46	- 87 23		15.2	FAINT BLUE STAR
ZC 1446.0+1352	14 46 00.	+ 13 52	870		CLUSTER OF GALAXIES
ZWG 076.056	14 46 00.	+ 14 10		15.2	GALAXY
UGC 09533	14 46 00.	+ 14 10	60	15.2	GALAXY S
KARA.72 436A	14 46 00.	+ 14 11	42	15.3	PART OF DOUBLE GALAXY
ZWG 076.057	14 46 00.	+ 14 12		15.3	GALAXY
MCG+03-38-023	14 46 00.	+ 18 33	66	15.5	GALAXY
ZWG 105.031	14 46 00.	+ 18 34		15.5	GALAXY
ZC 1446.0+2123	14 46 00.	+ 21 23	810		CLUSTER OF GALAXIES
ZC 1446.0+2413	14 46 00.	+ 24 13	1210		CLUSTER OF GALAXIES
ZC 1446.0+2731	14 46 00.	+ 27 31	670		CLUSTER OF GALAXIES
TON-N 0214	14 46 00.	+ 28 38		14.5	BLUE STAR
ZC 1446.0+2959	14 46 00.	+ 29 59	940		CLUSTER OF GALAXIES
ZWG 164.027	14 46 00.	+ 31 55		15.6	GALAXY
ZWG 164.028	14 46 00.	+ 31 57		15.7	GALAXY
ZC 1446.0+5412	14 46 00.	+ 54 12	1140		CLUSTER OF GALAXIES
ZWG 318.002	14 46 00.	+ 62 48		15.7	GALAXY
KARA.73B 0646	14 46 00.	+ 62 48	36	15.7	ISOLATED GALAXY S
MCG+11-18-007	14 46 00.	+ 63 37	39	16.	GALAXY
ZWG 318.003	14 46 00.	+ 63 38		15.6	GALAXY
KARA.73B 0647	14 46 00.	+ 63 38	48	15.6	ISOLATED GALAXY S
ZC 1446.0+6920	14 46 00.	+ 69 20	870		CLUSTER OF GALAXIES
7ZW 564	14 46 00.	+ 69 47			COMPACT GALAXY
ZWG 354.010	14 46 00.	+ 76 59		15.6	GALAXY
MCG-01-38-002	14 46 00.	- 04 30	84	15.	GALAXY
ZWG 048.018	14 46 06.	+ 07 02		15.3	GALAXY
HOLM 679B	14 46 06.	+ 07 02	18	15.0	PART OF MULTIPLE GALAXY
SN 1952H	14 46 06.	+ 07 02		18.9	SUPERNOVA
ZWG 076.058	14 46 06.	+ 10 29		15.2	GALAXY
KARA.72 436B	14 46 06.	+ 14 10	72	15.2	PART OF DOUBLE GALAXY
REIZ 4348	14 46 07.	+ 57 11	48	13.7	GALAXY
ARC 1972	14 46 08.	+ 24 09		17.2	RICH CLUSTER OF GALAXIES
ARC 1973	14 46 08.	+ 27 39		17.6	RICH CLUSTER OF GALAXIES
HOLM 679A	14 46 10.	+ 07 01	48	14.4	PART OF MULTIPLE GALAXY
ZWG 076.059	14 46 12.	+ 11 29		15.7	GALAXY
ZWG 076.060	14 46 12.	+ 11 32		15.3	GALAXY
UGC 09534	14 46 12.	+ 11 32	60	15.3	GALAXY E?
ZWG 076.061	14 46 12.	+ 12 02		15.5	GALAXY
ZC 1446.2+2518	14 46 12.	+ 25 18	2960		CLUSTER OF GALAXIES
HOLM 678A	14 46 12.	- 03 31	72	13.5	PART OF MULTIPLE GALAXY
MCG-03-38-017	14 46 12.	- 20 40	72	14.5	GALAXY
HOLM 678B	14 46 14.	- 03 32	36	14.3	PART OF MULTIPLE GALAXY
RNGC 5761	14 46 17.	- 20 10		14.0	GALAXY
ZWG 048.019	14 46 18.	+ 08 19		15.2	GALAXY
KARA.73B 0648	14 46 18.	+ 08 19	24	15.2	ISOLATED GALAXY S
ZWG 076.062	14 46 18.	+ 08 59		15.3	GALAXY
ZWG 076.063	14 46 18.	+ 12 40		14.3	GALAXY
RNGC 5762	14 46 18.	+ 12 40		14.5	GALAXY
UGC 09535	14 46 18.	+ 12 40	120	14.3	GALAXY S
MCG+02-38-014	14 46 18.	+ 12 40	78	14.3	GALAXY
ZWG 164.029	14 46 18.	+ 32 00		14.7	GALAXY
UGC 09536	14 46 18.	+ 32 00	60	14.7	GALAXY S
MCG+08-27-024	14 46 18.	+ 48 18	24	16.	GALAXY
ZC 1446.3-0322	14 46 18.	- 03 22	5710		CLUSTER OF GALAXIES
MCG-01-38-003	14 46 18.	- 03 31	48	15.	GALAXY
MCG-03-38-018	14 46 18.	- 20 10	60	14.	GALAXY
IC 4509	14 46 20.	+ 32 00 21.			NONSTELLAR OBJECT
REIZ 4347	14 46 20.	+ 48 08	15	14.5	GALAXY
MCG+05-35-016	14 46 21.	+ 32 00 30.	48	14.7	GALAXY
MCG+03-38-024	14 46 24.	+ 18 29	48	15.6	GALAXY
ZWG 105.032	14 46 24.	+ 18 30		15.6	GALAXY
LB 00730	14 46 24.	+ 23 22 18.		17.1	FAINT BLUE STAR
ZC 1446.4+2747	14 46 24.	+ 27 47	870		CLUSTER OF GALAXIES
TON-N 0783	14 46 24.	+ 31 36		15.7	BLUE STAR
ZWG 193.001	14 46 24.	+ 35 12		14.9	GALAXY
ZWG 192.063	14 46 24.	+ 35 12		14.9	GALAXY
UGC 09537	14 46 24.	+ 35 12	150	14.9	GALAXY Sb
ZWG 248.021	14 46 24.	+ 48 17		15.3	GALAXY
ZWG 020.020	14 46 24.	- 01 53		15.5	GALAXY
MCG+06-33-001	14 46 27.	+ 35 12	132	14.	GALAXY
ZWG 048.020	14 46 30.	+ 02 38		15.7	GALAXY
MCG+01-38-003	14 46 30.	+ 02 38	30	15.7	GALAXY
ZC 1446.5+0308	14 46 30.	+ 03 08	1280		CLUSTER OF GALAXIES
ZC 1446.5+0516	14 46 30.	+ 05 16	1480		CLUSTER OF GALAXIES
ZWG 105.033	14 46 30.	+ 16 22		15.6	GALAXY
ZC 1446.5+3041	14 46 30.	+ 30 41	470		CLUSTER OF GALAXIES
ZWG 220.054	14 46 30.	+ 41 50		15.7	GALAXY
KARA.73B 0649	14 46 30.	+ 41 50	36	15.7	ISOLATED GALAXY S
ZC 1446.5-0214	14 46 30.	- 02 14	1410		CLUSTER OF GALAXIES
FATH 1.674	14 46 32.	+ 44 38	35		NEBULA
ZWG 076.064	14 46 36.	+ 12 41		15.3	GALAXY
RNGC 5763	14 46 36.	+ 12 41		15.5	GALAXY
MCG+03-38-025	14 46 36.	+ 15 47	42	16.5	GALAXY
MCG+03-38-026	14 46 36.	+ 16 20	36	15.6	GALAXY
ZC 1446.6+1809	14 46 36.	+ 18 09	3230		CLUSTER OF GALAXIES
ZL 160	14 46 41.	+ 26 26 30.		21.8	ULTRAFAINT BLUE STAR
ZWG 076.065	14 46 42.	+ 11 11		15.3	GALAXY
ZC 1446.7+1506	14 46 42.	+ 15 06	1080		CLUSTER OF GALAXIES
ZWG 105.034	14 46 42.	+ 16 56		15.1	GALAXY
MCG+03-38-027	14 46 42.	+ 18 18	30	15.5	GALAXY
ZWG 134.034	14 46 42.	+ 22 20		15.7	GALAXY
ARP 261	14 46 43.	- 09 57			PECULIAR GALAXY

OBJECT NAME	RIGHT ASCEN.	DECLINATION	DIAM.	MAGN.	TYPE OF OBJECT
ZL 161	14 46 44.	+ 26 16 24.		21.2	ULTRAFAINT BLUE STAR
ZL 162	14 46 45.	+ 26 21 00.		20.2	ULTRAFAINT BLUE STAR
ZWG 076.066	14 46 48.	+ 11 20		15.3	GALAXY
MCG+03-38-028	14 46 48.	+ 16 55	42	15.1	GALAXY
ZWG 105.035	14 46 48.	+ 17 14		14.8	GALAXY
UGC 09538	14 46 48.	+ 17 14	96	14.8	GALAXY S
ZC 1446.8+1954	14 46 48.	+ 19 54	2290		CLUSTER OF GALAXIES
ZWG 134.035	14 46 48.	+ 21 32		15.4	GALAXY
ZWG 134.036	14 46 48.	+ 22 24		15.0	GALAXY
UGC 09539	14 46 48.	+ 22 25	78	15.0	GALAXY Sa
ZC 1446.8+3046	14 46 48.	+ 30 46	3900		CLUSTER OF GALAXIES
UGC 09540	14 46 48.	+ 34 55	60	17.	GALAXY TRP SYS
ZC 1446.8+4255	14 46 48.	+ 42 55	1080		CLUSTER OF GALAXIES
ZC 1446.8+5942	14 46 48.	+ 59 42	810		CLUSTER OF GALAXIES
ZL 163	14 46 50.	+ 26 08 06.		21.7	ULTRAFAINT BLUE STAR
MCG+04-35-009	14 46 51.	+ 22 25	60	15.0	GALAXY
VV 140D	14 46 51.	- 09 57	12	17.	INTERACTING GALAXY
VV 140C	14 46 51.	- 09 57	12	16.	INTERACTING GALAXY
VV 140B	14 46 51.	- 09 57	42	15.	INTERACTING GALAXY
VV 140A	14 46 51.	- 09 57	42	14.	INTERACTING GALAXY
VV 140	14 46 51.	- 09 57	168		INTERACTING GALAXY
MCG-02-38-016	14 46 51.	- 09 57	180	15.5	GALAXY
ZL 164	14 46 52.	+ 26 13 06.		22.5	ULTRAFAINT BLUE STAR
FATH 1.675	14 46 52.	+ 44 39	33		NEBULA
HELW 038	14 46 52.	- 15 46			NEBULA
ZWG 105.036	14 46 54.	+ 16 51		15.4	GALAXY
MCG+03-38-029	14 46 54.	+ 17 12	54	14.8	GALAXY
IC 1058	14 46 54.	+ 17 14 35.			NONSTELLAR OBJECT
ZC 1446.9+2855	14 46 54.	+ 28 55	2690		CLUSTER OF GALAXIES
ZWG 164.030	14 46 54.	+ 29 57		15.7	GALAXY
UGC 09541	14 46 54.	+ 29 57	84	15.7	GALAXY Sb-c
ZWG 193.002	14 46 54.	+ 33 24		15.7	GALAXY
MCG-02-38-017	14 46 54.	- 09 58	180	14.5	GALAXY
ZL 165	14 46 59.	+ 26 06 18.		18.6	ULTRAFAINT BLUE STAR
VDB.66G 197	14 47	- 09 54	130		DWARF GALAXY
ZWG 105.037	14 47 00.	+ 16 56		15.7	GALAXY
ZL 166	14 47 00.	+ 26 27 18.		20.8	ULTRAFAINT BLUE STAR
MCG+05-35-017	14 47 00.	+ 27 58	12	15.0	GALAXY
MCG+07-30-065	14 47 00.	+ 38 57	48	15.	GALAXY
ZC 1447.0+5313	14 47 00.	+ 53 13	1140		CLUSTER OF GALAXIES
LB 00281	14 47 01.	- 00 20 12.		15.5	FAINT BLUE STAR
ZL 167	14 47 04.	+ 26 24 00.		22.6	ULTRAFAINT BLUE STAR
ZWG 076.067	14 47 06.	+ 09 16		15.6	GALAXY
ZC 1447.1+2329	14 47 06.	+ 23 29	1680		CLUSTER OF GALAXIES
ZWG 164.031	14 47 06.	+ 27 59		15.0	GALAXY
ZC 1447.1+3355	14 47 06.	+ 33 55	2550		CLUSTER OF GALAXIES
ZWG 220.055	14 47 06.	+ 42 40		15.3	GALAXY
UGC 09542	14 47 06.	+ 42 40	120	15.3	GALAXY Sc
MCG+07-30-066	14 47 06.	+ 42 40	72	14.5	GALAXY
ZL 168	14 47 07.	+ 26 07 42.		19.4	ULTRAFAINT BLUE STAR
MCG+03-38-030	14 47 09.	+ 16 44	60	15.5	GALAXY
REIZ 4350	14 47 11.	+ 42 40	42	14.8	GALAXY
SN 1954M	14 47 12.	+ 10 29		17.5	SUPERNOVA
ZWG 076.068	14 47 12.	+ 11 27		15.3	GALAXY
MCG+02-38-015	14 47 12.	+ 11 27	18	15.3	GALAXY
ZWG 076.069	14 47 12.	+ 14 00		15.4	GALAXY
ZWG 105.038	14 47 12.	+ 16 52		15.5	GALAXY
ZWG 105.039	14 47 12.	+ 17 24		15.7	GALAXY
ZC 1447.2+2619	14 47 12.	+ 26 19	670		CLUSTER OF GALAXIES
ZWG 164.032	14 47 12.	+ 28 05		15.5	GALAXY
ZC 1447.2+3238	14 47 12.	+ 32 38	740		CLUSTER OF GALAXIES
ZWG 220.056	14 47 12.	+ 38 37		15.4	GALAXY
UGC 09543	14 47 12.	+ 38 37	90	15.4	GALAXY Sa-b
ZC 1447.2+4010	14 47 12.	+ 40 10	940		CLUSTER OF GALAXIES
OCL 0933	14 47 12.	- 52 03	480		OPEN STAR CLUSTER
VHA 166	14 47 12.	- 52 03	180		OPEN STAR CLUSTER
MCG+05-35-018	14 47 15.	+ 27 58 30.	30	15.0	GALAXY
MCG+07-30-067	14 47 15.	+ 38 36	66	15.	GALAXY
MCG+08-27-025	14 47 15.	+ 50 10	24	16.	GALAXY
ZC 1447.3+1451	14 47 18.	+ 14 51	670		CLUSTER OF GALAXIES
MCG+03-38-031	14 47 18.	+ 18 53	30	15.6	GALAXY
ZWG 105.040	14 47 18.	+ 18 54		15.6	GALAXY
ZWG 134.037	14 47 18.	+ 25 35		14.8	GALAXY
UGC 09544	14 47 18.	+ 25 35	96	14.8	GALAXY S
UGC 09545	14 47 18.	+ 50 09	90	16.0	GALAXY Sc
REIZ 4351	14 47 20.	+ 48 12	30	14.8	GALAXY
ZL 169	14 47 21.	+ 26 27 12.		22.0	ULTRAFAINT BLUE STAR
REIZ 4352	14 47 21.	+ 51 02	60	14.3	GALAXY
LB 00731	14 47 23.	+ 24 56 42.		16.0	FAINT BLUE STAR
ZWG 020.021	14 47 24.	+ 00 45		15.4	GALAXY
UGC 09546	14 47 24.	+ 23 47	60	17.	GALAXY Sc
REIZ 4349	14 47 24.	+ 25 36	18	14.8	GALAXY
ZC 1447.4+2603	14 47 24.	+ 26 03	540		CLUSTER OF GALAXIES
MCG+10-21-032	14 47 24.	+ 57 58	30	16.	GALAXY
HN 1782	14 47 24.8	+ 00 46 12.	18		NEBULA
ZL 170	14 47 27.	+ 26 18 00.		22.8	ULTRAFAINT BLUE STAR
ZWG 048.021	14 47 30.	+ 05 56		15.4	GALAXY
ZC 1447.5+1418	14 47 30.	+ 14 18	740		CLUSTER OF GALAXIES
ZC 1447.5+2036	14 47 30.	+ 20 36	540		CLUSTER OF GALAXIES
MCG+04-35-010	14 47 30.	+ 25 35 30.	60	14.8	GALAXY
HN 1783	14 47 31.1	+ 00 16 54.	24		NEBULA
ZL 171	14 47 35.	+ 26 12 00.		22.5	ULTRAFAINT BLUE STAR
ZWG 193.003	14 47 36.	+ 35 54		15.7	GALAXY
ZC 1447.6+6727	14 47 36.	+ 67 27	1280		CLUSTER OF GALAXIES
REIZ 4354	14 47 38.	+ 48 02	18	14.5	GALAXY
HN 1784	14 47 38.8	+ 00 45 19.	12		NEBULA
FATH 1.676	14 47 41.	+ 45 04	8		NEBULA
MCG+03-38-032	14 47 42.	+ 17 14	42	15.7	GALAXY
ZWG 105.041	14 47 42.	+ 17 15		15.7	GALAXY
ZC 1447.7+2107	14 47 42.	+ 21 07	1410		CLUSTER OF GALAXIES
IC 4512	14 47 43.	+ 27 54 22.			NONSTELLAR OBJECT
ZL 172	14 47 44.	+ 26 16 12.		20.7	ULTRAFAINT BLUE STAR
MCG-03-38-019	14 47 45.	- 17 57	108	13.5	GALAXY
ZWG 048.022	14 47 48.	+ 06 43		15.2	GALAXY
ZWG 134.038	14 47 48.	+ 21 45		15.7	GALAXY
MCG+09-24-041	14 47 48.	+ 55 35	48	16.	GALAXY
HN 0417	14 47 50.	- 20 31			NEBULA
IC 4510	14 47 50.	- 20 32			NONSTELLAR OBJECT
ZL 173	14 47 51.	+ 26 12 24.		22.2	ULTRAFAINT BLUE STAR
REIZ 4355	14 47 51.	+ 47 36	36	14.0	GALAXY
HOLM 681A	14 47 51.	+ 47 36	36	13.7	PART OF MULTIPLE GALAXY
ZWG 076.070	14 47 54.	+ 10 33		15.6	GALAXY
UGC 09547	14 47 54.	+ 10 33	84	15.6	GALAXY Sb
MCG+03-38-033	14 47 54.	+ 16 54	27	15.3	GALAXY
ZWG 105.042	14 47 54.	+ 16 55		15.3	GALAXY
UGC 09548	14 47 54.	+ 16 55	60	15.3	GALAXY Sa-b
ZWG 248.022	14 47 54.	+ 47 35		15.1	GALAXY
UGC 09549	14 47 54.	+ 47 35	60	15.1	GALAXY SB:a-b
ZL 174	14 47 56.	+ 26 26 24.		20.0	ULTRAFAINT BLUE STAR
RNGC 5767	14 47 56.	+ 47 35		15.0	GALAXY
IC 1065	14 47 56.	+ 63 27 58.			NONSTELLAR OBJECT
ZL 175	14 47 57.	+ 26 15 06.		22.8	ULTRAFAINT BLUE STAR
HMS 1.24	14 48 00	+ 26 23			E3 GALAXY
ZC 1448.0+1632	14 48 00.	+ 16 32	1210		CLUSTER OF GALAXIES
ZWG 105.043	14 48 00.	+ 16 57		15.2	GALAXY
UGC 09550	14 48 00.	+ 16 57	78	15.2	GALAXY S
ZC 1448.0+1911	14 48 00.	+ 19 11	540		CLUSTER OF GALAXIES
ZWG 134.039	14 48 00.	+ 24 31		15.6	GALAXY
HMS 1.23	14 48 00.	+ 26 23			E0 GALAXY
MCG+04-35-012	14 48 00.	+ 26 23	3		GALAXY
MCG+04-35-011	14 48 00.	+ 26 23	3		GALAXY
ZL 176	14 48 00.	+ 26 26 42.		18.7	ULTRAFAINT BLUE STAR
ZWG 164.033	14 48 00.	+ 29 58		15.6	GALAXY
MCG+11-18-008	14 48 00.	+ 63 27	45	16.	GALAXY
HOLM 681B	14 48 02.	+ 47 37	18	14.1	PART OF MULTIPLE GALAXY
MCG+04-35-013	14 48 03.	+ 24 32	30	15.6	GALAXY
REIZ 4356	14 48 03.	+ 47 37	24	14.4	GALAXY
ZWG 076.071	14 48 06.	+ 09 51		15.6	GALAXY
ZWG 076.072	14 48 06.	+ 10 00		15.4	GALAXY
MCG+03-38-034	14 48 06.	+ 16 55	72	15.2	GALAXY
ZC 1448.1+3144	14 48 06.	+ 31 44	670		CLUSTER OF GALAXIES
ZWG 020.024	14 48 06.	- 00 40		15.6	GALAXY
ZWG 020.023	14 48 06.	- 01 31		15.3	GALAXY
ZWG 020.022	14 48 06.	- 01 35		15.6	GALAXY
UGC 09551	14 48 06.	- 01 35	60	15.6	GALAXY
MCG+00-38-008	14 48 06.	- 01 35	54	15.6	GALAXY
VV 322C	14 48 06.	- 13 19	18	18.	INTERACTING GALAXY
VV 322B	14 48 06.	- 13 19	60	16.	INTERACTING GALAXY
VV 322A	14 48 06.	- 13 19	30	16.	INTERACTING GALAXY
HN 1785	14 48 06.7	- 01 34 04.	24		NEBULA
ARC 1976	14 48 07.	+ 21 10		16.9	RICH CLUSTER OF GALAXIES
IC 1059	14 48 08.	- 00 40 02.			NONSTELLAR OBJECT
HN 1786	14 48 08.3	- 01 30 28.	18		NEBULA
MCG-02-38-018	14 48 09.	- 13 19	42	15.5	GALAXY
ZC 1448.2+1543	14 48 12.	+ 15 43	1750		CLUSTER OF GALAXIES
ZWG 134.040	14 48 12.	+ 26 13		15.7	GALAXY
ZWG 134.041	14 48 12.	+ 26 17		15.6	GALAXY
UGC 09552	14 48 12.	+ 26 17	66	15.6	GALAXY Sa-b
ZWG 220.057	14 48 12.	+ 40 39		15.6	GALAXY
ZWG 313.004	14 48 12.	+ 63 29		15.0	GALAXY
UGC 09553	14 48 12.	+ 63 29	60	15.0	GALAXY SB0
ZC 1448.2+6445	14 48 12.	+ 64 45	3290		CLUSTER OF GALAXIES
MCG-02-38-019	14 48 12.	- 13 20 30.	36	15.5	GALAXY
MCG-03-38-020	14 48 12.	- 18 17 30.	54	14.	GALAXY
MCG-04-35-011	14 48 12.	- 26 05 30.	24	15.	GALAXY
VV 130C	14 48 15.	- 20 15	12	17.	INTERACTING GALAXY
VV 130B	14 48 15.	- 20 15	12	16.	INTERACTING GALAXY
VV 130A	14 48 15.	- 20 15	30	15.	INTERACTING GALAXY
VV 130	14 48 15.	- 20 15	42	14.	INTERACTING GALAXY
REIZ 4353	14 48 17.	+ 05 19	60	14.2	GALAXY
ZC 1448.3+0530	14 48 18.	+ 05 30	940		CLUSTER OF GALAXIES
ZWG 076.073	14 48 18.	+ 09 44		15.5	GALAXY
ZWG 134.042	14 48 18.	+ 25 01		15.7	GALAXY
MCG+09-24-042	14 48 18.	+ 55 09	36	16.	GALAXY
MCG-04-35-012	14 48 18.	- 26 25	72	15.	GALAXY
RNGC 5765B	14 48 21.	+ 05 19		14.5	GALAXY
RNGC 5765A	14 48 21.	+ 05 19		14.5	GALAXY
HN 1787	14 48 22.3	- 00 40 58.	18		NEBULA
HOLM 680B	14 48 23.	+ 10 20	54	15.0	PART OF MULTIPLE GALAXY
ARC 1978	14 48 23.	+ 14 48		16.8	RICH CLUSTER OF GALAXIES
ZWG 020.025	14 48 24.	+ 00 09		15.5	GALAXY
ZWG 048.023	14 48 24.	+ 05 09		15.3	GALAXY
ZWG 048.024	14 48 24.	+ 05 19		14.6	GALAXY
UGC 09554	14 48 24.	+ 05 19	66	14.6	GALAXY DBL SYS
KARA.72 437B	14 48 24.	+ 05 19	42		PART OF DOUBLE GALAXY
KARA.72 437A	14 48 24.	+ 05 19	48	14.6	PART OF DOUBLE GALAXY
MCG+01-38-005	14 48 24.	+ 05 19	36	14.6	GALAXY
MCG+01-38-004	14 48 24.	+ 05 19	48	14.6	GALAXY
ZWG 048.025	14 48 24.	+ 08 19		15.3	GALAXY
KARA.73B 0650	14 48 24.	+ 08 19	24	15.3	ISOLATED GALAXY S
ZWG 076.074	14 48 24.	+ 10 19		15.7	GALAXY
UGC 09555	14 48 24.	+ 10 19	102	15.7	GALAXY TRP SYS
MCG+02-38-018	14 48 24.	+ 10 19	12	15.7	GALAXY
MCG+02-38-017	14 48 24.	+ 10 19	15	15.7	GALAXY
MCG+02-38-016	14 48 24.	+ 10 20	9	15.7	GALAXY
ZC 1448.4+1454	14 48 24.	+ 14 54	2080		CLUSTER OF GALAXIES
ZWG 134.043	14 48 24.	+ 22 56		15.7	GALAXY
ZC 1448.4+3009	14 48 24.	+ 30 09	670		CLUSTER OF GALAXIES
MCG+09-24-043	14 48 24.	+ 52 37 30.	24	15.	GALAXY
MCG+10-21-033	14 48 24.	+ 60 36	51	15.	GALAXY
ZC 1448.4+7817	14 48 24.	+ 78 17	540		CLUSTER OF GALAXIES
MCG-03-38-021	14 48 24.	- 20 14 30.	96	14.5	GALAXY
HOLM 680A	14 48 25.	+ 10 19	30	13.9	PART OF MULTIPLE GALAXY
REIZ 4358	14 48 26.	+ 47 57	24	14.7	GALAXY
ZWG 048.026	14 48 30.	+ 06 49		15.5	GALAXY
ZC 1448.5+2040	14 48 30.	+ 20 40	870		CLUSTER OF GALAXIES
ZWG 296.017	14 48 30.	+ 60 35		15.3	GALAXY
UGC 09556	14 48 30.	+ 60 35	120	15.3	GALAXY SB?c
KARA.73B 0651	14 48 30.	+ 60 35	102	15.3	ISOLATED GALAXY S
MCG+11-18-009	14 48 30.	+ 63 29	24	17.	GALAXY
ACK 315-04.1	14 48 30.	- 63 50			PLANETARY NEBULA
ZWG 048.027	14 48 36.	+ 05 02		15.4	GALAXY
ZC 1448.6+2000	14 48 36.	+ 20 00	810		CLUSTER OF GALAXIES
7ZW 565	14 48 36.	+ 57 30			COMPACT GALAXY
MRSL 315-04/1	14 48 36.	- 64 08	1320		HII REGION
LB 00282	14 48 39.	+ 01 34 36.		15.8	FAINT BLUE STAR
MCG+03-38-035	14 48 39.	+ 16 17	48	15.5	GALAXY
ZWG 076.075	14 48 42.	+ 12 12		15.7	GALAXY
MCG+03-38-036	14 48 42.	+ 16 39	30	16.	GALAXY
ZC 1448.7+1651	14 48 42.	+ 16 51	7390		CLUSTER OF GALAXIES
ZWG 164.034	14 48 42.	+ 27 46		15.3	GALAXY
UGC 09557	14 48 42.	+ 27 46	60	15.3	GALAXY SBa-b
MCG+05-35-019	14 48 42.	+ 27 46	42	15.3	GALAXY
ZC 1448.7+5735	14 48 42.	+ 57 35	2150		CLUSTER OF GALAXIES
IC 4514	14 48 45.	+ 27 49 03.			NONSTELLAR OBJECT
MCG+07-30-068	14 48 45.	+ 42 57	66	14.5	GALAXY
HOLM 683A	14 48 46.	+ 42 57	54	13.6	PART OF MULTIPLE GALAXY
REIZ 4359	14 48 46.	+ 42 58	72	13.8	GALAXY
REIZ 4360	14 48 46.	+ 43 00	18	14.6	GALAXY
ZWG 105.044	14 48 48.	+ 17 24		15.0	GALAXY
UGC 09558	14 48 48.	+ 17 24	66	15.0	GALAXY Sc
ZWG 221.001	14 48 48.	+ 42 56		14.6	GALAXY
ZWG 220.058	14 48 48.	+ 42 56		14.6	GALAXY
UGC 09559	14 48 48.	+ 42 56	72	14.6	GALAXY S
ZC 1448.8+4654	14 48 48.	+ 46 54	4700		CLUSTER OF GALAXIES
HOLM 683B	14 48 50.	+ 42 59	24	14.4	PART OF MULTIPLE GALAXY

OBJECT NAME	RIGHT ASCEN.	DECLINATION	DIAM.	MAGN.	TYPE OF OBJECT
MCG+03-38-037	14 48 51.	+ 17 23	66	15.0	GALAXY
MCG+03-38-038	14 48 51.	+ 18 46	42	15.5	GALAXY
MCG-04-35-013	14 48 51.	- 24 22	36	15.	GALAXY
HN 1137	14 48 53.	- 40 17	36		NEBULA
ZWG 048.028	14 48 54.	+ 07 00		15.1	GALAXY
MCG+01-38-006	14 48 54.	+ 07 00	39	15.1	GALAXY
ZWG 105.045	14 48 54.	+ 15 39		15.6	GALAXY
ZWG 105.046	14 48 54.	+ 16 54		15.4	GALAXY
ARC 1979	14 48 54.	+ 31 29		17.2	RICH CLUSTER OF GALAXIES
ZC 1448.9+3456	14 48 54.	+ 34 56	3970		CLUSTER OF GALAXIES
2ZW 070	14 48 54.	+ 35 45			COMPACT GALAXY
ZWG 193.004	14 48 54.	+ 35 47		14.5	GALAXY
UGC 09560	14 48 54.	+ 35 47	48	14.5	GALAXY PECULR
KARA.72 438A	14 48 54.	+ 35 47	60	14.5	PART OF DOUBLE GALAXY
MCG+06-33-002	14 48 54.	+ 35 47	42	15.	GALAXY
ZWG 221.002	14 48 54.	+ 42 58		15.5	GALAXY
ZWG 220.059	14 48 54.	+ 42 58		15.5	GALAXY
MCG-02-38-020	14 48 54.	- 13 23	24	15.	GALAXY
IC 4511	14 48 54.	- 40 18			NONSTELLAR OBJECT
HOLM 682B	14 48 56.	+ 07 01	24	14.9	PART OF MULTIPLE GALAXY
IC 1061	14 48 56.	+ 18 57 39.			NONSTELLAR OBJECT
MCG+03-38-039	14 48 57.	+ 16 32 30.	30	16.	GALAXY
VV 324B	14 48 57.	+ 35 48	36	15.	INTERACTING GALAXY
REIZ 4357	14 48 58.	+ 07 01	36	14.8	GALAXY
HOLM 682A	14 48 58.	+ 07 01	42	14.3	PART OF MULTIPLE GALAXY
IC 1062	14 48 59.	+ 18 53 22.			NONSTELLAR OBJECT
ZWG 048.029	14 49 00.	+ 06 16		15.2	GALAXY
ZWG 048.030	14 49 00.	+ 06 19		15.5	GALAXY
KARA.72 439A	14 49 00.	+ 09 31	18	14.9	PART OF DOUBLE GALAXY
ZWG 076.076	14 49 00.	+ 09 32		14.9	GALAXY
UGC 09561	14 49 00.	+ 09 32	90	14.9	GALAXY
KARA.72 439B	14 49 00.	+ 09 32	36		PART OF DOUBLE GALAXY
MCG+02-38-020	14 49 00.	+ 09 32	15	14.9	GALAXY
MCG+02-38-019	14 49 00.	+ 09 32	54	14.9	GALAXY
ZWG 076.077	14 49 00.	+ 09 48		15.2	GALAXY
MCG+03-38-040	14 49 00.	+ 15 37	48	15.6	GALAXY
ZWG 105.047	14 49 00.	+ 18 53		15.3	GALAXY
MCG+03-38-041	14 49 00.	+ 18 53	18	15.3	GALAXY
ZC 1449.0+2200	14 49 00.	+ 22 00	810		CLUSTER OF GALAXIES
MCG+05-35-020	14 49 00.	+ 30 54	48	15.6	GALAXY
ZC 1449.0+3130	14 49 00.	+ 31 30	870		CLUSTER OF GALAXIES
ZWG 273.028	14 49 00.	+ 51 55		15.6	GALAXY
ZC 1449.0+6806	14 49 00.	+ 68 06	1880		CLUSTER OF GALAXIES
ARP 173	14 49 02.	+ 09 34			PECULIAR GALAXY
ZWG 048.031	14 49 06.	+ 02 55		15.3	GALAXY
ZWG 105.048	14 49 06.	+ 17 43		15.7	GALAXY
ZC 1449.1+2252	14 49 06.	+ 22 52	870		CLUSTER OF GALAXIES
ZC 1449.1+2757	14 49 06.	+ 27 57	810		CLUSTER OF GALAXIES
ZWG 164.035	14 49 06.	+ 28 44		15.7	GALAXY
ZWG 164.036	14 49 06.	+ 30 54		15.6	GALAXY
ZWG 193.005	14 49 06.	+ 37 42		15.6	GALAXY
ZC 1449.1-0227	14 49 06.	- 02 27	1610		CLUSTER OF GALAXIES
ARC 1982	14 49 07.	+ 30 56		16.6	RICH CLUSTER OF GALAXIES
IC 4515	14 49 08.	+ 37 48 42.			NONSTELLAR OBJECT
MCG+06-33-003	14 49 09.	+ 37 43	24	15.	GALAXY
HELW 039	14 49 10.	- 15 29			NEBULA
VV 296B	14 49 12.	+ 09 31	84	15.	INTERACTING GALAXY
ZWG 076.078	14 49 12.	+ 10 56		15.2	GALAXY
ZWG 134.044	14 49 12.	+ 26 01		15.5	GALAXY
ZC 1449.2+2626	14 49 12.	+ 26 26	870		CLUSTER OF GALAXIES
ZC 1449.2+3035	14 49 12.	+ 30 35	610		CLUSTER OF GALAXIES
ZWG 193.006	14 49 12.	+ 35 45		14.2	GALAXY
UGC 09562	14 49 12.	+ 35 45	66	14.2	GALAXY PECULR
KARA.72 438B	14 49 12.	+ 35 45	84	14.2	PART OF DOUBLE GALAXY
MCG+06-33-004	14 49 12.	+ 35 45	48	14.	GALAXY
ZWG 273.029	14 49 12.	+ 54 37		15.1	GALAXY
UGC 09563	14 49 12.	+ 54 37	96	15.1	GALAXY S0
MCG-01-38-004	14 49 12.	- 07 01	60	14.5	GALAXY
IC 1060	14 49 12.	- 07 02 03.			NONSTELLAR OBJECT
BC PKS1449-012	14 49 12.1	- 01 15 15.			QUASI-STELLAR OBJECT
SHB 297	14 49 12.1	- 01 15 15.		18.	QUASI-STELLAR OBJECT
VV 296A	14 49 15.	+ 09 32	33	14.	INTERACTING GALAXY
MCG+09-24-044	14 49 15.	+ 54 39	48	14.	GALAXY
ARC 1980	14 49 16.	+ 22 52		17.2	RICH CLUSTER OF GALAXIES
FATH 1.677	14 49 16.	+ 45 15	27		NEBULA
ZWG 048.032	14 49 18.	+ 02 44		15.6	GALAXY
ZWG 076.079	14 49 18.	+ 08 46		15.4	GALAXY
FEIG 103	14 49 18.	+ 11 52		13.0	FAINT BLUE STAR
ZC 1449.3+3241	14 49 18.	+ 32 41	1340		CLUSTER OF GALAXIES
2ZW 071	14 49 18.	+ 35 45			COMPACT GALAXY
VV 324A	14 49 18.	+ 35 46	60	14.	INTERACTING GALAXY
REIZ 4362	14 49 18.	+ 54 37	9	14.7	GALAXY
MCG-02-38-021	14 49 18.	- 15 30 30.	24	15.	GALAXY
ARC 1977	14 49 18.	- 24 18		17.0	RICH CLUSTER OF GALAXIES
ZWG 048.033	14 49 24.	+ 03 34		15.3	GALAXY
ZWG 048.034	14 49 24.	+ 05 32		15.4	GALAXY
ZC 1449.4+2008	14 49 24.	+ 20 08	810		CLUSTER OF GALAXIES
TON-N 0215	14 49 24.	+ 25 35		16.9	BLUE STAR
HN 0419	14 49 26.	- 20 31			NEBULA
IC 4513	14 49 26.	- 20 31			NONSTELLAR OBJECT
IC 1069	14 49 30.	+ 54 34 52.			NONSTELLAR OBJECT
HN 1788	14 49 32.6	+ 01 54 18.	36		NEBULA
MCG-02-38-022	14 49 33.	- 10 32	72	14.	GALAXY
ZC 1449.6+4738	14 49 36.	+ 47 38	1010		CLUSTER OF GALAXIES
ZC 1449.6+6909	14 49 36.	+ 69 09	610		CLUSTER OF GALAXIES
ZWG 020.027	14 49 36.	- 00 03		15.4	GALAXY
ZWG 020.026	14 49 36.	- 02 20		14.2	GALAXY
RNGC 5768	14 49 36.	- 02 20		13.0	GALAXY
UGC 09564	14 49 36.	- 02 20	120	14.2	GALAXY Sc
MCG+00-38-009	14 49 36.	- 02 20	90	14.2	GALAXY
KARA.73B 0652	14 49 36.	- 02 20	114	14.2	ISOLATED GALAXY S
HN 1789	14 49 37.5	- 00 03 54.	18		NEBULA
HN 1790	14 49 40.9	- 02 37 18.	42		NEBULA
IC 1063	14 49 41.	+ 04 53 12.			NONSTELLAR OBJECT
ZWG 048.035	14 49 42.	+ 04 43		15.7	GALAXY
IC 1064	14 49 42.	+ 04 52 00.			NONSTELLAR OBJECT
ZWG 048.036	14 49 42.	+ 04 53		14.8	GALAXY
UGC 09565	14 49 42.	+ 04 53	66	14.8	GALAXY SBb
MCG+01-38-007	14 49 42.	+ 04 53	60	14.8	GALAXY
ZWG 048.037	14 49 42.	+ 07 57		15.6	GALAXY
ZWG 221.003	14 49 42.	+ 40 49		13.9	GALAXY
ZWG 220.060	14 49 42.	+ 40 49		13.9	GALAXY
UGC 09566	14 49 42.	+ 40 49	138	13.9	GALAXY Sb
KARA.73B 0653	14 49 42.	+ 40 49	132	13.9	ISOLATED GALAXY S
HN 1791	14 49 42.2	- 00 06 06.	12		NEBULA
RNGC 5772	14 49 45.	+ 40 48		14.0	GALAXY
MCG+07-31-001	14 49 45.	+ 40 50	120	13.5	GALAXY
ZWG 020.028	14 49 48.	+ 02 07		15.7	GALAXY

OBJECT NAME	RIGHT ASCEN.	DECLINATION	DIAM.	MAGN.	TYPE OF OBJECT
ZWG 048.038	14 49 48.	+ 03 55		15.7	GALAXY
ZC 1449.8+0822	14 49 48.	+ 08 22	1410		CLUSTER OF GALAXIES
ZWG 105.049	14 49 48.	+ 20 16		15.7	GALAXY
FEIG 104	14 49 48.	+ 55 49		13.8	FAINT BLUE STAR
MCG-04-35-014	14 49 48.	- 24 36	24	15.5	GALAXY
REIZ 4363	14 49 49.	+ 40 49	42	13.6	GALAXY
ARC 2010	14 49 50.	+ 81 22		17.5	RICH CLUSTER OF GALAXIES
FATH 1.678	14 49 51.	+ 44 43	16		NEBULA
ZWG 076.080	14 49 54.	+ 12 30		15.2	GALAXY
ZC 1449.9+2541	14 49 54.	+ 25 41	2350		CLUSTER OF GALAXIES
ZC 1449.9+2813	14 49 54.	+ 28 13	1610		CLUSTER OF GALAXIES
ZWG 221.004	14 49 54.	+ 43 51		14.9	GALAXY
ZWG 220.061	14 49 54.	+ 43 51		14.9	GALAXY
UGC 09567	14 49 54.	+ 43 51	72	14.9	GALAXY IRR
REIZ 4367	14 49 56.	+ 43 51	48	13.9	GALAXY
ZWG 048.039	14 50 00.	+ 03 20		15.4	GALAXY
ZWG 048.040	14 50 00.	+ 03 37		15.7	GALAXY
ZWG 048.041	14 50 00.	+ 04 45		15.4	GALAXY
MCG+03-38-042	14 50 00.	+ 17 19	36	15.5	GALAXY
ZWG 105.050	14 50 00.	+ 17 20		15.5	GALAXY
ZWG 134.045	14 50 00.	+ 20 33		15.5	GALAXY
ZC 1450.0+2040	14 50 00.	+ 20 40	810		CLUSTER OF GALAXIES
ZWG 164.037	14 50 00.	+ 30 03		14.6	GALAXY
MCG+05-35-021	14 50 00.	+ 30 03	18	14.6	GALAXY
MCG+07-31-002	14 50 00.	+ 43 52	48	14.	GALAXY
ZWG 296.018	14 50 00.	+ 59 10		14.2	GALAXY
UGC 09568	14 50 00.	+ 59 10	210	14.2	GALAXY Sb
MCG+10-21-034	14 50 00.	+ 59 10	162	12.	GALAXY
MCG+11-18-010	14 50 00.	+ 67 34	18	17.	GALAXY
ZWG 354.011	14 50 00.	+ 79 58		15.2	GALAXY
ZWG 353.046	14 50 00.	+ 79 58		15.2	GALAXY
ZWG 020.029	14 50 00.	- 03 21		15.2	GALAXY
MCG+00-38-010	14 50 00.	- 03 21	192	15.2	GALAXY
OCL 0934	14 50 00.	- 52 29	240	12.8	OPEN STAR CLUSTER
VHA 167	14 50 00.	- 52 29	150		OPEN STAR CLUSTER
REIZ 4372	14 50 01.	+ 48 04	36	14.6	GALAXY
RNGC 5771	14 50 02.	+ 30 03		14.5	GALAXY
REIZ 4368	14 50 02.	+ 44 01	12	14.6	GALAXY
RNGC 5764	14 50 02.	- 52 29		12.5	OPEN CLUSTER
RNGC 5777	14 50 03.	+ 59 10		14.0	GALAXY
ZWG 048.042	14 50 06.	+ 06 08		15.6	GALAXY
ZWG 048.043	14 50 06.	+ 06 28		15.1	GALAXY
ZWG 076.081	14 50 06.	+ 10 07		15.7	GALAXY
ZWG 105.051	14 50 06.	+ 15 14		15.7	GALAXY
MCG+03-38-043	14 50 06.	+ 16 38	30	15.5	GALAXY
ZWG 105.052	14 50 06.	+ 17 16		15.6	GALAXY
ZWG 221.005	14 50 06.	+ 43 56		14.4	GALAXY
ZWG 220.062	14 50 06.	+ 43 56		14.4	GALAXY
UGC 09569	14 50 06.	+ 43 56	102	14.4	GALAXY SBc
ZC 1450.1+5109	14 50 06.	+ 51 09	3430		CLUSTER OF GALAXIES
MCG+09-24-045	14 50 06.	+ 53 39	60	16.	GALAXY
MCG+09-24-046	14 50 06.	+ 53 40	12	17.	GALAXY
LB 00732	14 50 08.	+ 23 00 00.		16.8	FAINT BLUE STAR
REIZ 4364	14 50 08.	+ 30 03	36	14.3	GALAXY
REIZ 4371	14 50 08.	+ 43 56	48	14.0	GALAXY
MCG+07-31-003	14 50 09.	+ 43 56	72	14.	GALAXY
RNGC 5769	14 50 10.	+ 08 08		15.0	GALAXY
HELW 040	14 50 10.	- 15 40			NEBULA
ZWG 048.044	14 50 12.	+ 03 25		15.2	GALAXY
ZWG 048.045	14 50 12.	+ 07 07		15.3	GALAXY
ZWG 048.046	14 50 12.	+ 08 01		15.5	GALAXY
ZWG 048.047	14 50 12.	+ 08 08		14.9	GALAXY
MCG+01-38-008	14 50 12.	+ 08 08	18	14.9	GALAXY
ZWG 076.082	14 50 12.	+ 12 16		15.7	GALAXY
ARC 1984	14 50 12.	+ 28 09		17.2	RICH CLUSTER OF GALAXIES
ZC 1450.2+3811	14 50 12.	+ 38 11	2150		CLUSTER OF GALAXIES
MCG+10-21-035	14 50 12.	+ 59 08	27	15.	GALAXY
UGC 09570	14 50 12.	+ 59 09	90	16.5	GALAXY DWRF SP
MCG-01-38-005	14 50 12.	- 03 35	42	15.5	GALAXY
MCG+09-24-047	14 50 15.	+ 51 29	72	14.	GALAXY
MCG-03-38-022	14 50 15.	- 15 41 30.	60	15.5	GALAXY
REIZ 4361	14 50 16.	+ 07 06	30	14.7	GALAXY
HELW 041	14 50 16.	- 15 39			NEBULA
ZC 1450.3+1601	14 50 18.	+ 16 01	1140		CLUSTER OF GALAXIES
ZWG 105.053	14 50 18.	+ 17 06		15.4	GALAXY
ZWG 134.046	14 50 18.	+ 22 12		15.7	GALAXY
ZWG 164.038	14 50 18.	+ 30 00		14.5	GALAXY
UGC 09571	14 50 18.	+ 30 00	66	14.5	GALAXY S
BSO 13	14 50 18.	+ 45 45 23.		17.29	BLUE STELLAR OBJECT
ZWG 318.005	14 50 18.	+ 66 13		15.3	GALAXY
ZC 1450.3+6958	14 50 18.	+ 69 58	540		CLUSTER OF GALAXIES
MCG-03-38-023	14 50 18.	- 19 33	96	15.	GALAXY
MCG-03-38-024	14 50 18.	- 21 12	42	14.5	GALAXY
RNGC 5773	14 50 20.	+ 30 00		14.5	GALAXY
MCG+05-35-022	14 50 21.	+ 30 00	42	14.5	GALAXY
IC 1074	14 50 21.	+ 51 28 26.			NONSTELLAR OBJECT
RNGC 5766	14 50 22.	- 21 10		14.0	GALAXY
ZWG 048.048	14 50 24.	+ 08 09		15.3	GALAXY
ZWG 273.030	14 50 24.	+ 51 29		15.2	GALAXY
UGC 09572	14 50 24.	+ 51 29	60	15.2	GALAXY S
BSO 14	14 50 24.5	+ 43 15 00.		14.60	BLUE STELLAR OBJECT
ARC 1983	14 50 26.	+ 16 57		15.4	RICH CLUSTER OF GALAXIES
REIZ 4370	14 50 26.	+ 30 00	27	14.4	GALAXY
HOLM 684B	14 50 29.	+ 03 31	42	13.6	PART OF MULTIPLE GALAXY
ZWG 048.049	14 50 30.	+ 03 30		14.2	GALAXY
UGC 09573	14 50 30.	+ 03 30	96	14.2	GALAXY S
MCG+01-38-009	14 50 30.	+ 03 30	60	14.2	GALAXY
ZWG 076.083	14 50 30.	+ 10 05		15.3	GALAXY
ZWG 105.054	14 50 30.	+ 16 54		15.6	GALAXY
ZWG 105.055	14 50 30.	+ 16 56		15.7	GALAXY
ZWG 134.047	14 50 30.	+ 20 32		15.7	GALAXY
ZC 1450.5+2230	14 50 30.	+ 22 30	2150		CLUSTER OF GALAXIES
SHAK 077	14 50 30.	+ 39 53	138	17.5	GROUP OF COMPACT GALAXIES
MCG-01-38-006	14 50 30.	- 06 48	54	15.	GALAXY
ARC 1981	14 50 30.	- 24 11		17.0	RICH CLUSTER OF GALAXIES
HOLM 684A	14 50 31.	+ 03 33	48	12.9	PART OF MULTIPLE GALAXY
IC 1066	14 50 32.	+ 03 29 41.			NONSTELLAR OBJECT
IC 1067	14 50 34.4	+ 03 32 09.	114		GALAXY SB(s)b
ZWG 048.050	14 50 36.	+ 03 32		13.6	GALAXY
UGC 09574	14 50 36.	+ 03 32	132	13.6	GALAXY SBb
MCG+01-38-010	14 50 36.	+ 03 32	108	13.6	GALAXY
ZWG 076.084	14 50 36.	+ 12 57		15.2	GALAXY
ZWG 105.056	14 50 36.	+ 16 53		15.5	GALAXY
MCG+03-38-044	14 50 36.	+ 16 53	90	15.5	GALAXY
MCG+03-38-045	14 50 36.	+ 16 55	60	15.5	GALAXY
ZWG 296.019	14 50 36.	+ 60 56		15.4	GALAXY
ZC 1450.6+7159	14 50 36.	+ 71 59	1410		CLUSTER OF GALAXIES
REIZ 4365	14 50 37.	+ 03 30	54	13.2	GALAXY

OBJECT NAME	RIGHT ASCEN.	DECLINATION	DIAM.	MAGN.	TYPE OF OBJECT
REIZ 4366	14 50 37.	+ 03 32	48	13.4	GALAXY
ZC 1450.7+1517	14 50 42.	+ 15 17	2220		CLUSTER OF GALAXIES
ZC 1450.7+3140	14 50 42.	+ 31 40	2820		CLUSTER OF GALAXIES
MCG+08-27-026	14 50 42.	+ 49 13	48	17.	GALAXY
MCG+09-24-048	14 50 42.	+ 56 08	24	15.	GALAXY
ZC 1450.7+6034	14 50 42.	+ 60 34	1480		CLUSTER OF GALAXIES
MCG+10-21-036	14 50 42.	+ 60 57	39	15.	GALAXY
SHAH 132	14 50 42.	+ 64 55	54	17.4	GROUP OF COMPACT GALAXIES
RNGC 5779	14 50 47.	+ 56 06		15.5	GALAXY
ZWG 048.051	14 50 48.	+ 03 10		15.5	GALAXY
ZWG 048.052	14 50 48.	+ 04 09		13.3	GALAXY
UGC 09575	14 50 48.	+ 04 09	96	13.3	GALAXY SB0
MCG+01-38-011	14 50 48.	+ 04 09	84	13.3	GALAXY
REIZ 4369	14 50 48.	+ 04 11	42	15.7	GALAXY
ARC 1985	14 50 48.	+ 06 08		17.2	RICH CLUSTER OF GALAXIES
ZWG 048.053	14 50 48.	+ 08 02		15.4	GALAXY
ZWG 076.085	14 50 48.	+ 09 44		15.6	GALAXY
ZC 1450.8+1427	14 50 48.	+ 14 27	1680		CLUSTER OF GALAXIES
ZWG 105.057	14 50 48.	+ 16 53		15.5	GALAXY
ZWG 105.058	14 50 48.	+ 17 15		15.7	GALAXY
ZC 1450.8+2204	14 50 48.	+ 22 04	1210		CLUSTER OF GALAXIES
ZWG 134.048	14 50 48.	+ 24 22		15.7	GALAXY
ZWG 273.031	14 50 48.	+ 56 06		15.7	GALAXY
ZWG 337.021	14 50 48.	+ 74 01		15.4	GALAXY
SHAH 161	14 50 48.	+ 78 53	104		GROUP OF COMPACT GALAXIES
MCG-01-38-007	14 50 48.	- 04 28	84	15.	GALAXY
RNGC 5770	14 50 51.	+ 04 09		13.5	GALAXY
ARC 1987	14 50 51.	+ 32 35		17.6	RICH CLUSTER OF GALAXIES
ARC 1986	14 50 53.	+ 22 07		16.9	RICH CLUSTER OF GALAXIES
ZWG 048.054	14 50 54.	+ 03 28		15.7	GALAXY
ZWG 048.055	14 50 54.	+ 08 25		15.1	GALAXY
ZWG 105.059	14 50 54.	+ 16 50		15.7	GALAXY
ZC 1450.9+3233	14 50 54.	+ 32 33	1080		CLUSTER OF GALAXIES
HN 1792	14 50 59.7	+ 00 13 04.	18		NEBULA
ZWG 020.030	14 51 00.	+ 00 12		15.3	GALAXY
ZWG 048.056	14 51 00.	+ 03 16		14.9	GALAXY
MCG+01-38-012	14 51 00.	+ 03 16	12	14.9	GALAXY
REIZ 4374	14 51 00.	+ 04 10	36	15.3	GALAXY
ZC 1451.0+2144	14 51 00.	+ 21 44	2760		CLUSTER OF GALAXIES
ZWG 164.039	14 51 00.	+ 27 36		15.6	GALAXY
ZC 1451.0+4135	14 51 00.	+ 41 35	2080		CLUSTER OF GALAXIES
MRK 288	14 51 00.	+ 74 02	30	15.	GALAXY WITH UV CONTINUUM
MCG+12-14-013	14 51 00.	+ 74 02	39	16.	GALAXY
REIZ 4373	14 51 01.	+ 03 29	54	14.7	GALAXY
IC 1068	14 51 02.	+ 03 16 19.			NONSTELLAR OBJECT
REIZ 4376	14 51 04.	+ 07 09	18	15.3	GALAXY
ZWG 048.057	14 51 06.	+ 03 47		13.9	GALAXY
UGC 09576	14 51 06.	+ 03 47	204	13.9	GALAXY Sc
KARA.72 440A	14 51 06.	+ 03 47	210	13.9	PART OF DOUBLE GALAXY
MCG+01-38-013	14 51 06.	+ 03 47	168	13.9	GALAXY
ZC 1451.1+2057	14 51 06.	+ 20 57	1140		CLUSTER OF GALAXIES
ZWG 193.007	14 51 06.	+ 37 07		15.6	GALAXY
REIZ 4375	14 51 07.	+ 03 17	18	14.4	GALAXY
MCG+06-33-005	14 51 09.	+ 37 08 30.	18	15.	GALAXY
HOLM 685B	14 51 10.	+ 03 48	60	13.5	PART OF MULTIPLE GALAXY
ZWG 076.086	14 51 12.	+ 09 41		15.7	GALAXY
ZWG 105.060	14 51 12.	+ 18 18		15.7	GALAXY
UGC 09577	14 51 12.	+ 18 18	60	15.7	GALAXY SBc
MCG+03-38-046	14 51 12.	+ 18 18	48	15.7	GALAXY
ZC 1451.2+3224	14 51 12.	+ 32 24	470		CLUSTER OF GALAXIES
ZC 1451.2+7313	14 51 12.	+ 73 13	270		CLUSTER OF GALAXIES
REIZ 4377	14 51 13.	+ 03 48	60	13.7	GALAXY
ARC 1988	14 51 13.	+ 21 00		17.2	RICH CLUSTER OF GALAXIES
RNGC 5774	14 51 15.	+ 03 47		14.0	GALAXY
MCG+06-33-006	14 51 15.	+ 37 08	12	17.	GALAXY
ZC 1451.3+0000	14 51 18.	+ 00 00	1610		CLUSTER OF GALAXIES
ZC 1451.3+0506	14 51 18.	+ 05 06	1550		CLUSTER OF GALAXIES
ZWG 076.087	14 51 18.	+ 10 48		15.6	GALAXY
ZC 1451.3+2815	14 51 18.	+ 28 15	610		CLUSTER OF GALAXIES
BC PKS1451-375	14 51 18.25	- 37 35 22.9		17.	QUASI-STELLAR OBJECT
SHB 298	14 51 18.3	- 37 35 22.		17.	QUASI-STELLAR OBJECT
ZWG 048.058	14 51 24.	+ 03 13		15.4	GALAXY
ZWG 048.059	14 51 24.	+ 03 41		15.4	GALAXY
IC 1070	14 51 24.	+ 03 41 09.			NONSTELLAR OBJECT
ZC 1451.4+0555	14 51 24.	+ 05 55	2550		CLUSTER OF GALAXIES
ZWG 076.088	14 51 24.	+ 10 20		15.7	GALAXY
ZWG 105.061	14 51 24.	+ 17 39		15.6	GALAXY
ZWG 105.062	14 51 24.	+ 20 19		15.4	GALAXY
UGC 09578	14 51 24.	+ 20 19	72	15.4	GALAXY SBb
ZC 1451.4+4241	14 51 24.	+ 42 41	1410		CLUSTER OF GALAXIES
ZWG 248.023	14 51 24.	+ 46 41		15.7	GALAXY
KARA.72 441A	14 51 24.	+ 46 41	48	15.7	PART OF DOUBLE GALAXY
MCG+08-27-027	14 51 24.	+ 46 41	48	15.	GALAXY
ZC 1451.4+5335	14 51 24.	+ 53 35	1550		CLUSTER OF GALAXIES
HOLM 685A	14 51 25.	+ 03 45	210	12.1	PART OF MULTIPLE GALAXY
REIZ 4378	14 51 25.	+ 03 46	228	12.2	GALAXY
RNGC 5775	14 51 27.	+ 03 45		12.5	GALAXY
MCG+03-38-047	14 51 27.	+ 20 19	60	15.4	GALAXY
SHB 299	14 51 27.9	+ 09 46 33.		18.5	QUASI-STELLAR OBJECT
FATH 1.679	14 51 28.	+ 45 23	27		NEBULA
ZWG 048.060	14 51 30.	+ 03 45		13.0	GALAXY
UGC 09579	14 51 30.	+ 03 45	252	13.0	GALAXY Sc
KARA.72 440B	14 51 30.	+ 03 45	246	13.0	PART OF DOUBLE GALAXY
MCG+01-38-014	14 51 30.	+ 03 45	240	13.0	GALAXY
ZWG 105.063	14 51 30.	+ 15 24		15.6	GALAXY
ZC 1451.5+2709	14 51 30.	+ 27 09	870		CLUSTER OF GALAXIES
ZWG 248.024	14 51 30.	+ 45 23		15.6	GALAXY
MCG+10-21-037	14 51 30.	+ 60 32	39	16.	GALAXY
7ZW 566	14 51 30.	+ 63 17			COMPACT GALAXY
ARC 1995	14 51 32.	+ 58 16		18.4	RICH CLUSTER OF GALAXIES
IC 4516	14 51 35.	+ 16 35 11.			NONSTELLAR OBJECT
ARC 1990	14 51 35.	+ 28 18		17.2	RICH CLUSTER OF GALAXIES
ZWG 048.061	14 51 36.	+ 03 24		15.6	GALAXY
ZWG 076.089	14 51 36.	+ 10 18		15.3	GALAXY
UGC 09580	14 51 36.	+ 10 18	42	15.3	GALAXY DBL SYS
KARA.72 442A	14 51 36.	+ 10 18	48	15.3	PART OF DOUBLE GALAXY
ZC 1451.6+1456	14 51 36.	+ 14 56	940		CLUSTER OF GALAXIES
ZC 1451.6+1855	14 51 36.	+ 18 55	5780		CLUSTER OF GALAXIES
ZC 1451.6+2144	14 51 36.	+ 21 44	540		CLUSTER OF GALAXIES
ZC 1451.6+2758	14 51 36.	+ 27 58	1140		CLUSTER OF GALAXIES
ZC 1451.6+4342	14 51 36.	+ 43 42	1340		CLUSTER OF GALAXIES
ZWG 248.025	14 51 36.	+ 46 41		15.6	GALAXY
KARA.72 441B	14 51 36.	+ 46 41	42	15.6	PART OF DOUBLE GALAXY
MCG+09-24-049	14 51 36.	+ 52 16	30	14.	GALAXY
7ZW 567	14 51 36.	+ 60 20			COMPACT GALAXY
ZWG 354.012	14 51 36.	+ 74 44		15.4	GALAXY
7ZW 568	14 51 36.	+ 76 52			COMPACT GALAXY
IC 4499	14 51 36.	- 82 02	186	11.5	GLOBULAR CLUSTER
RNGC 5788	14 51 37.	+ 52 15		15.5	GALAXY
IC 1071	14 51 39.	+ 04 55 11.			NONSTELLAR OBJECT
REIZ 4379	14 51 40.	+ 07 16	24	14.7	GALAXY
ZWG 048.062	14 51 42.	+ 04 57		14.4	GALAXY
UGC 09582	14 51 42.	+ 04 57	72	14.4	GALAXY S0
MCG+01-38-015	14 51 42.	+ 04 57	48	14.4	GALAXY
ZWG 048.063	14 51 42.	+ 05 00		15.3	GALAXY
IC 1072	14 51 42.	+ 05 02 47.			NONSTELLAR OBJECT
ZWG 048.064	14 51 42.	+ 05 03		15.1	GALAXY
MCG+01-38-016	14 51 42.	+ 05 03	30	15.1	GALAXY
ZWG 076.090	14 51 42.	+ 10 17		15.3	GALAXY
UGC 09581	14 51 42.	+ 10 17	54	15.3	GALAXY DBL SYS
KARA.72 442B	14 51 42.	+ 10 17	42	15.3	PART OF DOUBLE GALAXY
ZWG 076.091	14 51 42.	+ 12 42		15.2	GALAXY
ZC 1451.7+2842	14 51 42.	+ 28 42	870		CLUSTER OF GALAXIES
ZWG 221.006	14 51 42.	+ 42 38		15.7	GALAXY
UGC 09583	14 51 42.	+ 42 38	84	15.7	GALAXY S
ZC 1451.7+4526	14 51 42.	+ 45 26	1080		CLUSTER OF GALAXIES
MCG+08-27-028	14 51 42.	+ 46 41	30	15.	GALAXY
ZWG 273.032	14 51 42.	+ 52 15		15.6	GALAXY
IC 1073	14 51 44.	+ 04 59 53.			NONSTELLAR OBJECT
ZWG 048.065	14 51 48.	+ 04 29		14.8	GALAXY
MCG+01-38-017	14 51 48.	+ 04 29	15	14.8	GALAXY
ZWG 048.066	14 51 48.	+ 04 43		15.3	GALAXY
UGC 09584	14 51 48.	+ 04 43	66	15.3	GALAXY Sb
MCG+07-31-004	14 51 48.	+ 42 21	36	15.5	GALAXY
ARC 1992	14 51 48.	+ 45 25		17.8	RICH CLUSTER OF GALAXIES
MCG+08-27-029	14 51 48.	+ 49 20	42	16.	GALAXY
MCG+09-24-050	14 51 48.	+ 52 17 30.	180	12.	GALAXY
MCG-03-38-025	14 51 51.	- 17 13	108	14.	GALAXY
REIZ 4380	14 51 52.	+ 07 12	18	14.8	GALAXY
REIZ 4384	14 51 53.	+ 52 17	42	14.2	GALAXY
UGC 09585	14 51 54.	+ 42 45	66	16.0	GALAXY SBb-c
ZWG 273.033	14 51 54.	+ 52 17		14.0	GALAXY
UGC 09586	14 51 54.	+ 52 17	168	14.0	GALAXY SBc
MCG+09-24-051	14 51 54.	+ 56 23	72	16.	GALAXY
ZC 1451.9+6838	14 51 54.	+ 68 38	1010		CLUSTER OF GALAXIES
ARC 1989	14 51 55.	+ 05 55		17.2	RICH CLUSTER OF GALAXIES
RNGC 5783	14 51 55.	+ 52 17		14.0	GALAXY
ZWG 048.067	14 52 00.	+ 03 10		14.7	GALAXY
MCG+01-38-018	14 52 00.	+ 03 10	48	14.7	GALAXY
ZWG 048.068	14 52 00.	+ 03 29		15.3	GALAXY
MCG+03-38-049	14 52 00.	+ 16 32	24	14.9	GALAXY
ZWG 105.064	14 52 00.	+ 16 34		14.9	GALAXY
UGC 09587	14 52 00.	+ 16 34	96	14.9	GALAXY E
ZWG 105.065	14 52 00.	+ 17 55		15.7	GALAXY
MCG+03-38-048	14 52 00.	+ 18 45	24	16.	GALAXY
ZC 1452.0+2114	14 52 00.	+ 21 14	540		CLUSTER OF GALAXIES
ZC 1452.0+2821	14 52 00.	+ 28 21	1280		CLUSTER OF GALAXIES
ZWG 164.040	14 52 00.	+ 30 25		15.3	GALAXY
UGC 09588	14 52 00.	+ 30 25	78	15.3	GALAXY DBL SYS
MCG+05-35-023	14 52 00.	+ 30 25	54	15.3	GALAXY
ZC 1452.0+3642	14 52 00.	+ 36 42	1410		CLUSTER OF GALAXIES
MCG+07-31-005	14 52 00.	+ 42 48	48	15.	GALAXY
MCG+09-24-052	14 52 00.	+ 54 22 30.	72	17.	GALAXY
UGC 09589	14 52 00.	+ 54 24	72	17.	GALAXY SB
7ZW 569	14 52 00.	+ 67 45			COMPACT GALAXY
MRSL 315-05/1	14 52 00.	- 64 51	900		HII REGION
REIZ 4381	14 52 01.	+ 03 30	15	14.3	GALAXY
RNGC 5776	14 52 02.	+ 03 10		14.5	GALAXY
HELW 042	14 52 05.	- 16 04			NEBULA
ZWG 076.092	14 52 06.	+ 10 31		15.2	GALAXY
KARA.72 443A	14 52 06.	+ 10 31	36	15.2	PART OF DOUBLE GALAXY
ZWG 105.066	14 52 06.	+ 18 50		15.6	GALAXY
UGC 09590	14 52 06.	+ 18 50	78	15.6	GALAXY E?
ZWG 221.007	14 52 06.	+ 38 51		15.4	GALAXY
7ZW 570	14 52 06.	+ 67 41			COMPACT GALAXY
REIZ 4382	14 52 07.	+ 03 11	15	14.4	GALAXY
RNGC 5778	14 52 09.	+ 18 50		15.5	GALAXY
REIZ 4383	14 52 10.	+ 07 15	18	14.8	GALAXY
ARC 1991	14 52 10.	+ 18 51		15.4	RICH CLUSTER OF GALAXIES
KARA.72 443B	14 52 12.	+ 10 30	36	15.7	PART OF DOUBLE GALAXY
ZWG 076.093	14 52 12.	+ 10 31		15.7	GALAXY
MCG+03-38-050	14 52 12.	+ 18 50	60	15.6	GALAXY
ZWG 134.049	14 52 12.	+ 23 39		15.6	GALAXY
MCG+05-35-024	14 52 12.	+ 29 08	48	14.7	GALAXY
ZWG 164.041	14 52 12.	+ 29 09		14.7	GALAXY
ZC 1452.2+5735	14 52 12.	+ 57 35	1010		CLUSTER OF GALAXIES
MCG+11-18-011	14 52 12.	+ 63 51	15	17.	GALAXY
BSO 15	14 52 13.	+ 44 18 32.		17.83	BLUE STELLAR OBJECT
RNGC 5780	14 52 14.	+ 29 09		14.5	GALAXY
SHB 300	14 52 15.	+ 29 54		18.5	QUASI-STELLAR OBJECT
ZWG 020.031	14 52 18.	+ 00 50		15.5	GALAXY
ZWG 105.067	14 52 18.	+ 20 14		15.5	GALAXY
MCG+03-38-051	14 52 18.	+ 20 14	36	15.5	GALAXY
ZWG 134.050	14 52 18.	+ 23 50		15.5	GALAXY
ZWG 134.051	14 52 18.	+ 24 13		15.4	GALAXY
ZWG 164.042	14 52 18.	+ 29 16		15.6	GALAXY
ZWG 318.006	14 52 18.	+ 63 53		15.5	GALAXY
MCG+11-18-012	14 52 18.	+ 64 00	24	17.	GALAXY
MCG+11-18-013	14 52 18.	+ 64 01	45	16.	GALAXY
ZWG 318.007	14 52 18.	+ 64 04		15.6	GALAXY
UGC 09591	14 52 18.	+ 64 04	78	15.6	GALAXY Sb-c
VV 137	14 52 18.	- 19 28 30.	78	14.	INTERACTING GALAXY
MCG-04-35-015	14 52 18.	- 23 47	48	15.5	GALAXY
RNGC 5785	14 52 19.	+ 52 21			NON-EXISTENT OBJECT
REIZ 4385	14 52 21.	+ 42 46	24	13.5	GALAXY
ARC 2002	14 52 21.	+ 68 35		17.7	RICH CLUSTER OF GALAXIES
IC 4517	14 52 22.	+ 23 51 16.			NONSTELLAR OBJECT
RNGC 5784	14 52 22.	+ 42 45		13.5	GALAXY
ZWG 105.068	14 52 24.	+ 18 46		15.7	GALAXY
ZWG 134.052	14 52 24.	+ 21 01		15.6	GALAXY
ZWG 134.053	14 52 24.	+ 25 55		15.6	GALAXY
ZWG 221.008	14 52 24.	+ 39 00		15.6	GALAXY
ZWG 221.009	14 52 24.	+ 42 45		13.7	GALAXY
UGC 09592	14 52 24.	+ 42 45	114	13.7	GALAXY S0
SHAH 162	14 52 24.	+ 85 09	66		GROUP OF COMPACT GALAXIES
MCG-03-38-026	14 52 24.	- 19 29	84	14.5	GALAXY
MCG-04-35-016	14 52 24.	- 25 16 30.	72	15.	GALAXY
BC OQ287	14 52 25.23	+ 30 08 06.9		18.5	QUASI-STELLAR OBJECT
REIN 2.225	14 52 25.32	+ 42 45 37.6			NEBULA
MCG+03-38-052	14 52 27.	+ 18 45	60	15.7	GALAXY
ZWG 105.069	14 52 30.	+ 18 18		14.9	GALAXY
UGC 09593	14 52 30.	+ 18 18	78	14.9	GALAXY SBb
KARA.72 444A	14 52 30.	+ 18 18	72	14.9	PART OF DOUBLE GALAXY
MCG+03-38-053	14 52 30.	+ 18 18	72	14.9	GALAXY
ZWG 134.054	14 52 30.	+ 24 17		15.2	GALAXY
UGC 09594	14 52 30.	+ 24 17	66	15.2	GALAXY SBc

OBJECT NAME	RIGHT ASCEN.	DECLINATION	DIAM.	MAGN.	TYPE OF OBJECT
ZC 1452.5+2631	14 52 30.	+ 26 31	940		CLUSTER OF GALAXIES
ZWG 193.008	14 52 30.	+ 37 17		15.5	GALAXY
MCG+07-31-006	14 52 30.	+ 42 47 30.	84	14.	GALAXY
IC 1075	14 52 31.	+ 18 19 34.			NONSTELLAR OBJECT
ZWG 048.069	14 52 36.	+ 02 55		15.4	GALAXY
MCG+01-38-019	14 52 36.	+ 02 55	48	15.4	GALAXY
ZC 1452.6+0430	14 52 36.	+ 04 30	1210		CLUSTER OF GALAXIES
ZWG 076.094	14 52 36.	+ 11 54		15.1	GALAXY
RNGC 5782	14 52 36.	+ 11 54		15.0	GALAXY
MCG+02-38-021	14 52 36.	+ 11 54	48	15.1	GALAXY
MCG+03-38-054	14 52 36.	+ 17 48	36	15.4	GALAXY
ZWG 105.070	14 52 36.	+ 17 49		15.4	GALAXY
ZWG 105.071	14 52 36.	+ 18 14		13.9	GALAXY
MRK 479	14 52 36.	+ 18 14	30	15.	GALAXY WITH UV CONTINUUM
UGC 09595	14 52 36.	+ 18 14	72	13.9	GALAXY
KARA.72 444B	14 52 36.	+ 18 14	66	13.9	PART OF DOUBLE GALAXY
MCG+04-35-014	14 52 36.	+ 24 18	60	15.2	GALAXY
ZWG 134.055	14 52 36.	+ 26 00		15.3	GALAXY
UGC 09596	14 52 36.	+ 26 00	60	15.3	GALAXY Sb-c
ZWG 318.008	14 52 36.	+ 64 48		15.1	GALAXY
ZC 1452.6+6953	14 52 36.	+ 69 53	870		CLUSTER OF GALAXIES
ARC 1999	14 52 38.	+ 54 32		15.7	RICH CLUSTER OF GALAXIES
MCG+03-38-055	14 52 39.	+ 18 14	42	13.9	GALAXY
IC 1076	14 52 41.	+ 18 14 47.			NONSTELLAR OBJECT
ZWG 076.095	14 52 42.	+ 09 48		15.5	GALAXY
ZWG 105.072	14 52 42.	+ 17 48		15.7	GALAXY
MCG+04-35-015	14 52 42.	+ 26 00 30.	48	15.3	GALAXY
MCG+09-24-053	14 52 42.	+ 53 28	24	16.	GALAXY
MCG+11-18-014	14 52 42.	+ 64 48	8	17.	GALAXY
ZWG 354.013	14 52 42.	+ 79 49		15.4	GALAXY
ZWG 353.047	14 52 42.	+ 79 49		15.4	GALAXY
MCG-01-38-008	14 52 42.	- 05 49	66	15.	GALAXY
MCG-03-38-027	14 52 42.	- 21 24 30.	36	16.	GALAXY
GCL 030	14 52 42.	- 82 01	372	11.6	GLOBULAR STAR CLUSTER
BSO 16	14 52 44.	+ 47 32 16.		16.17	BLUE STELLAR OBJECT
IC 4519	14 52 46.	+ 37 37 37.			NONSTELLAR OBJECT
ZWG 076.096	14 52 48.	+ 10 04		15.7	GALAXY
ZWG 105.073	14 52 48.	+ 18 19		15.6	GALAXY
ZWG 134.056	14 52 48.	+ 25 34		15.7	GALAXY
UGC 09597	14 52 48.	+ 31 02	66	17.	GALAXY DWRF SP
MCG+05-35-025	14 52 48.	+ 31 02 30.	60		GALAXY
ZWG 193.009	14 52 48.	+ 37 37		15.3	GALAXY
MCG+06-33-007	14 52 48.	+ 37 37	36	15.	GALAXY
1ZW 097	14 52 48.	+ 42 13			COMPACT GALAXY
ZWG 221.010	14 52 48.	+ 42 13		14.8	GALAXY
MCG+11-18-015	14 52 48.	+ 64 46	24	16.	GALAXY
HN 0418	14 52 50.	- 82 01			NEBULA
ZWG 048.070	14 52 54.	+ 05 30		15.2	GALAXY
ZWG 105.074	14 52 54.	+ 16 43		14.8	GALAXY
ZWG 105.075	14 52 54.	+ 16 45		15.5	GALAXY
ZWG 221.011	14 52 54.	+ 40 59		15.0	GALAXY
MCG+07-31-007	14 52 54.	+ 41 00	36	15.	GALAXY
ZWG 273.034	14 52 54.	+ 53 25		15.5	GALAXY
MCG+09-24-054	14 52 54.	+ 53 26	36	15.	GALAXY
REIZ 4386	14 52 56.	+ 53 26	12	14.7	GALAXY
MCG+03-38-056	14 52 57.	+ 16 43	30	15.5	GALAXY
RNGC 5808	14 52 59.	+ 73 13			NON-EXISTENT OBJECT
MCG+03-38-057	14 53 00.	+ 16 41	21	14.8	GALAXY
ZWG 105.076	14 53 00.	+ 19 02		15.6	GALAXY
ZWG 193.010	14 53 00.	+ 33 55		15.0	GALAXY
ZC 1453.0+5429	14 53 00.	+ 54 29	2350		CLUSTER OF GALAXIES
MCG+09-24-055	14 53 00.	+ 55 17	42	16.	GALAXY
MCG+09-24-056	14 53 00.	+ 55 20	24	16.	GALAXY
ZWG 337.022	14 53 00.	+ 70 40		15.5	GALAXY
KARA.73B 0654	14 53 00.	+ 70 40	48	15.5	ISOLATED GALAXY S
OCL 0927	14 53 00.	- 62 21	510	14.	OPEN STAR CLUSTER
MCG+03-38-058	14 53 03.	+ 19 01	36	15.6	GALAXY
IC 4520	14 53 03.	+ 33 55 26.			NONSTELLAR OBJECT
ZC 1453.1+0202	14 53 06.	+ 02 02	1080		CLUSTER OF GALAXIES
ZWG 076.097	14 53 06.	+ 09 01		15.1	GALAXY
FEIG 105	14 53 06.	+ 12 57		13.9	FAINT BLUE STAR
MCG+06-33-008	14 53 06.	+ 33 55	24	15.	GALAXY
MCG+09-24-057	14 53 06.	+ 53 30	42	16.	GALAXY
ZWG 020.032	14 53 06.	- 00 49		15.6	GALAXY
KLEM 29	14 53 06.	- 37 23	1200	14.	CLUSTER OF 15 GALAXIES
ARC 2000	14 53 07.	+ 54 41		16.6	RICH CLUSTER OF GALAXIES
HN 1793	14 53 07.4	- 00 47 38.	12		NEBULA
SHB 301	14 53 12.2	- 10 56 39.		17.4	QUASI-STELLAR OBJECT
BC MSH14-121	14 53 12.22	- 10 56 39.9		17.37	QUASI-STELLAR OBJECT
ZWG 048.071	14 53 18.	+ 06 38		15.6	GALAXY
ZWG 076.098	14 53 18.	+ 09 47		15.2	GALAXY
ZC 1453.3+2221	14 53 18.	+ 22 21	810		CLUSTER OF GALAXIES
ZC 1453.3+3849	14 53 18.	+ 38 49	1080		CLUSTER OF GALAXIES
ZWG 221.012	14 53 18.	+ 44 00		15.0	GALAXY
UGC 09598	14 53 18.	+ 44 00	96	15.0	GALAXY Sc
KARA.73B 0655	14 53 18.	+ 44 00	102	15.0	ISOLATED GALAXY S
REIZ 4387	14 53 21.	+ 42 42	21	13.6	GALAXY
RNGC 5787	14 53 22.	+ 42 42		14.0	GALAXY
ZC 1453.4+2742	14 53 24.	+ 27 42	1340		CLUSTER OF GALAXIES
MCG+06-33-009	14 53 24.	+ 33 02	24	15.	GALAXY
1ZW 098	14 53 24.	+ 42 42			COMPACT GALAXY
ZWG 221.013	14 53 24.	+ 42 42		14.1	GALAXY
UGC 09599	14 53 24.	+ 42 42	66	14.1	GALAXY
UGC 09600	14 53 24.	+ 51 15	60	17.	GALAXY Sb
REIN 2.226	14 53 24.28	+ 42 42 29.2			NEBULA
ARC 1993	14 53 25.	+ 02 04		17.5	RICH CLUSTER OF GALAXIES
HN 1794	14 53 26.9	- 01 11 01.	24		NEBULA
VV 274B	14 53 27.	+ 33 01 30.	21	16.	INTERACTING GALAXY
VV 274A	14 53 27.	+ 33 01 30.	21	16.	INTERACTING GALAXY
VV 274	14 53 27.	+ 33 01 30.	36	15.	INTERACTING GALAXY
MCG+06-33-011	14 53 27.	+ 33 01 30.	24	17.5	GALAXY
MCG+06-33-010	14 53 27.	+ 33 01 30.	11	16.5	GALAXY
ZWG 048.072	14 53 30.	+ 02 39		15.1	GALAXY
MCG+01-38-020	14 53 30.	+ 02 39	48	15.1	GALAXY
ZWG 076.099	14 53 30.	+ 12 04		14.9	GALAXY
UGC 09602	14 53 30.	+ 12 04	60	14.9	GALAXY S0
KARA.72 445A	14 53 30.	+ 12 04	78	14.9	PART OF DOUBLE GALAXY
MCG+02-38-022	14 53 30.	+ 12 04	30	14.9	GALAXY
UGC 09603	14 53 30.	+ 33 01	72	16.0	GALAXY DBL SYS
VV 275B	14 53 30.	+ 33 02	30	18.	INTERACTING GALAXY
VV 275A	14 53 30.	+ 33 02	30	17.5	INTERACTING GALAXY
UGC 09604	14 53 30.	+ 37 38	66	16.5	GALAXY Sc
MCG+06-33-012	14 53 30.	+ 37 38	66	16.	GALAXY
MCG+07-31-008	14 53 30.	+ 42 44	48	14.	GALAXY
MCG+07-31-009	14 53 30.	+ 44 03	84	14.5	GALAXY
UGC 09605	14 53 30.	+ 48 33	60	16.0	GALAXY Sc
MCG+08-27-030	14 53 30.	+ 48 34	66	16.	GALAXY
ZWG 020.033	14 53 30.	- 01 12		14.6	GALAXY
UGC 09601	14 53 30.	- 01 12	90	14.6	GALAXY
MCG+00-38-011	14 53 30.	- 01 12	78	14.6	GALAXY
ZWG 076.100	14 53 36.	+ 12 04		15.3	GALAXY
MCG+02-38-023	14 53 36.	+ 12 04	48	15.3	GALAXY
KARA.72 445B	14 53 36.	+ 12 05	30	15.3	PART OF DOUBLE GALAXY
ZWG 105.077	14 53 36.	+ 18 15		15.7	GALAXY
ZC 1453.6+2345	14 53 36.	+ 23 45	810		CLUSTER OF GALAXIES
MCG+04-35-016	14 53 36.	+ 24 48	24		GALAXY
UGC 09606	14 53 36.	+ 24 55	78	16.0	GALAXY Sb
ZWG 221.014	14 53 36.	+ 41 27		15.2	GALAXY
MCG+07-31-010	14 53 36.	+ 41 29	15	15.	GALAXY
ARC 1994	14 53 36.	- 05 38		17.7	RICH CLUSTER OF GALAXIES
ARP 177	14 53 37.	+ 24 48			PECULIAR GALAXY
ARC 1997	14 53 38.	+ 20 17		17.0	RICH CLUSTER OF GALAXIES
MCG+04-35-017	14 53 39.	+ 24 48	12		GALAXY
MCG-01-38-009	14 53 42.	- 05 40	54	15.5	GALAXY
MCG-04-35-017	14 53 42.	- 24 18 30.	60	14.5	GALAXY
MCG+02-38-024	14 53 48.	+ 09 41	30	15.	GALAXY
ZWG 076.101	14 53 48.	+ 12 31		15.6	GALAXY
ZWG 105.078	14 53 48.	+ 19 57		15.7	GALAXY
MCG+10-21-038	14 53 48.	+ 62 27	15	16.	GALAXY
UGC 09607	14 53 48.	+ 62 29	72	16.0	GALAXY
7ZW 571	14 53 48.	+ 66 30			COMPACT GALAXY
RNGC 5781	14 53 48.	- 17 03		14.0	GALAXY
SC 1451-1604.7	14 53 51.	- 16 16 53.	18		NEBULA
MCG-03-38-029	14 53 51.	- 16 18	36	15.	GALAXY
MCG-03-38-028	14 53 51.	- 17 03	54	14.5	GALAXY
ZC 1453.9+2015	14 53 54.	+ 20 15	1280		CLUSTER OF GALAXIES
ZWG 134.057	14 53 54.	+ 20 44		15.3	GALAXY
ZC 1453.9+2401	14 53 54.	+ 24 01	1010		CLUSTER OF GALAXIES
ZC 1453.9+2901	14 53 54.	+ 29 01	2150		CLUSTER OF GALAXIES
MCG+08-27-031	14 53 54.	+ 45 44	60	15.	GALAXY
ZWG 248.026	14 53 54.	+ 45 45		15.5	GALAXY
REIZ 4389	14 53 57.	+ 49 47	12	15.7	GALAXY
REIZ 4388	14 53 58.	+ 20 44	24	14.6	GALAXY
ZC 1454.0+0514	14 54 00.	+ 05 14	1080		CLUSTER OF GALAXIES
ZWG 076.102	14 54 00.	+ 09 33		14.8	GALAXY
UGC 09608	14 54 00.	+ 09 33	72	14.8	GALAXY Sa
MCG+02-38-025	14 54 00.	+ 09 33	48	14.8	GALAXY
ZC 1454.0+1721	14 54 00.	+ 17 21	870		CLUSTER OF GALAXIES
ZC 1454.0+2206	14 54 00.	+ 22 06	1080		CLUSTER OF GALAXIES
TON-N 0784	14 54 00.	+ 28 03		15.5	BLUE STAR
ZWG 337.023	14 54 00.	+ 73 20		14.3	GALAXY
RNGC 5819	14 54 00.	+ 73 20		14.5	GALAXY
UGC 09609	14 54 00.	+ 73 20	66	14.3	GALAXY SBb
ZWG 020.034	14 54 00.	- 01 55		15.1	GALAXY
BC PKS1454-06	14 54 02.58	- 06 05 40.4		18.03	QUASI-STELLAR OBJECT
SHB 302	14 54 02.7	- 06 05 45.		18.0	QUASI-STELLAR OBJECT
IC 1078	14 54 04.	+ 09 32 48.			NONSTELLAR OBJECT
RNGC 5804	14 54 04.	+ 49 51			GALAXY
RNGC 5794	14 54 05.	+ 49 56		14.5	GALAXY
VV 026B	14 54 06.	+ 09 33 30.	6	19.	INTERACTING GALAXY
VV 026A	14 54 06.	+ 09 33 30.	60	15.	INTERACTING GALAXY
ZWG 248.027	14 54 06.	+ 49 56		14.5	GALAXY
UGC 09610	14 54 06.	+ 49 56	66	14.5	GALAXY
MCG-05-35-008	14 54 06.	- 31 38	48	15.	GALAXY
IC 1079	14 54 11.	+ 09 33 43.			NONSTELLAR OBJECT
ZWG 048.073	14 54 12.	+ 06 40		15.2	GALAXY
MCG+01-38-021	14 54 12.	+ 06 40	30	15.2	GALAXY
ZWG 076.103	14 54 12.	+ 09 34		14.8	GALAXY
UGC 09611	14 54 12.	+ 09 34	96	14.8	GALAXY E
MCG+02-38-026	14 54 12.	+ 09 34	84	14.8	GALAXY
ZC 1454.2+2537	14 54 12.	+ 25 37	2890		CLUSTER OF GALAXIES
ZWG 248.028	14 54 12.	+ 45 37		15.2	GALAXY
UGC 09612	14 54 12.	+ 45 37	90	15.2	GALAXY Sc/SBc
MCG+08-27-032	14 54 12.	+ 49 56	24	14.1	GALAXY
7ZW 572	14 54 12.	+ 71 43			COMPACT GALAXY
LB 00733	14 54 14.	+ 25 41 24.		17.5	FAINT BLUE STAR
MCG+08-27-033	14 54 15.	+ 49 19	42	16.	GALAXY
RNGC 5805	14 54 17.	+ 49 55			GALAXY
ZWG 048.074	14 54 18.	+ 05 47		15.2	GALAXY
UGC 09613	14 54 18.	+ 05 47	66	15.2	GALAXY S
ZC 1454.3+0915	14 54 18.	+ 09 15	4700		CLUSTER OF GALAXIES
UGC 09614	14 54 18.	+ 09 41	96	16.5	GALAXY DWARF
ZC 1454.3+2813	14 54 18.	+ 28 13	810		CLUSTER OF GALAXIES
MCG+07-31-011	14 54 18.	+ 38 54	42	15.	GALAXY
MCG+08-27-034	14 54 18.	+ 45 37	72	15.	GALAXY
REIZ 4392	14 54 21.	+ 49 50	12	15.7	GALAXY
REIZ 4393	14 54 21.	+ 49 55	18	14.1	GALAXY
HN 1138	14 54 22.	- 42 56	60		NEBULA
IC 4518	14 54 22.	- 42 56			NONSTELLAR OBJECT
UGC 09615	14 54 24.	+ 03 25	60	13.9	GALAXY
MCG+02-38-027	14 54 24.	+ 09 42	108	15.5	GALAXY
ZWG 105.079	14 54 24.	+ 14 36		15.4	GALAXY
ZWG 164.043	14 54 24.	+ 30 25		13.9	GALAXY
ZWG 020.035	14 54 24.	- 01 56		15.2	GALAXY
VHE 64B	14 54 24.	- 59 34	30		REFLECTION NEBULA
ARC 1998	14 54 25.	+ 01 43		17.5	RICH CLUSTER OF GALAXIES
RNGC 5789	14 54 27.	+ 30 25		14.0	GALAXY
MCG+05-35-026	14 54 27.	+ 30 27	48	13.9	GALAXY
MCG-07-31-001	14 54 27.	- 42 56	48	15.	GALAXY
MCG-07-31-002	14 54 27.	- 42 56 30.	60	17.	GALAXY
ZWG 048.075	14 54 30.	+ 02 55		15.7	GALAXY
ZC 1454.5+0601	14 54 30.	+ 06 01	540		CLUSTER OF GALAXIES
ZC 1454.5+0656	14 54 30.	+ 06 56	810		CLUSTER OF GALAXIES
ZWG 076.104	14 54 30.	+ 09 28		14.8	GALAXY
UGC 09616	14 54 30.	+ 09 28	66	14.8	GALAXY Sa-b
MCG+02-38-028	14 54 30.	+ 09 28	60	14.8	GALAXY
ZWG 076.105	14 54 30.	+ 09 57		15.3	GALAXY
MCG-03-38-030	14 54 30.	- 19 02	60	13.5	GALAXY
ARC 1996	14 54 30.	- 23 43		17.4	RICH CLUSTER OF GALAXIES
MCG-07-31-003	14 54 30.	- 42 56	90	14.5	GALAXY
AGU 50	14 54 30.	- 42 56 06.		12.5	2 INTERACTING GALAXIES
REIZ 4390	14 54 31.	+ 30 25	60	13.5	GALAXY
IC 1077	14 54 32.	- 19 00 41.			NONSTELLAR OBJECT
BC PKS1454-034	14 54 32.8	- 03 27 48.		19.	QUASI-STELLAR OBJECT
IC 1077	14 54 33.	- 19 01 47.			NON-STELLAR OBJECT
ZWG 020.036	14 54 36.	+ 02 13		15.6	GALAXY
ZC 1454.6+2221	14 54 36.	+ 22 21	4840		CLUSTER OF GALAXIES
TON-N 0785	14 54 36.	+ 30 18		15.7	BLUE STAR
ZWG 248.029	14 54 36.	+ 49 36		14.7	GALAXY
UGC 09617	14 54 36.	+ 49 36	96	14.7	GALAXY S
MCG+08-27-035	14 54 36.	+ 49 37	96	14.	GALAXY
MCG-03-38-031	14 54 36.	- 18 17	108	14.5	GALAXY
MCG-03-38-032	14 54 36.	- 21 28	48	16.	GALAXY
VHE 64A	14 54 36.	- 59 28	30		REFLECTION NEBULA
RNGC 5795	14 54 40.	+ 49 35		14.5	GALAXY
RNGC 5797	14 54 41.	+ 49 54		13.5	GALAXY

OBJECT NAME	RIGHT ASCEN.	DECLINATION	DIAM.	MAGN.	TYPE OF OBJECT
ZC 1454.7+2255	14 54 42.	+ 22 55	870		CLUSTER OF GALAXIES
ZWG 134.058	14 54 42.	+ 24 48		14.3	GALAXY
UGC 09618	14 54 42.	+ 24 48	96	14.3	GALAXY DBL SYS
KARA.72 446A	14 54 42.	+ 24 48	42	14.3	PART OF DOUBLE GALAXY
KARA.72 446B	14 54 42.	+ 24 49	48		PART OF DOUBLE GALAXY
ZWG 248.030	14 54 42.	+ 49 54		13.6	GALAXY
UGC 09619	14 54 42.	+ 49 54	90	13.6	GALAXY S0-a
MCG+08-27-036	14 54 42.	+ 49 54	66	13.8	GALAXY
ZC 1454.7+6112	14 54 42.	+ 61 12	1410		CLUSTER OF GALAXIES
ZWG 318.009	14 54 42.	+ 64 07		15.3	GALAXY
IC 1083	14 54 42.	+ 68 37 36.			NONSTELLAR OBJECT
RNGC 5807	14 54 43.	+ 64 07		15.5	GALAXY
REIZ 4396	14 54 45.	+ 49 53	18	13.8	GALAXY
ZWG 076.106	14 54 48.	+ 12 02		15.5	GALAXY
ZWG 105.080	14 54 48.	+ 17 23		15.7	GALAXY
ZWG 105.081	14 54 48.	+ 19 54		14.7	GALAXY
UGC 09620	14 54 48.	+ 19 54	78	14.7	GALAXY S
KARA.72 447A	14 54 48.	+ 19 54	66	14.7	PART OF DOUBLE GALAXY
ZC 1454.8+2233	14 54 48.	+ 22 33	810		CLUSTER OF GALAXIES
VV 340B	14 54 48.	+ 24 49 30.	36	15.	INTERACTING GALAXY
VV 340A	14 54 48.	+ 24 49 30.	42	15.	INTERACTING GALAXY
VV 340	14 54 48.	+ 24 49 30.	90		INTERACTING GALAXY
MCG+04-35-018	14 54 48.	+ 24 50	36	14.3	GALAXY
MCG+04-35-019	14 54 48.	+ 24 50 30.	48	14.3	GALAXY
ZWG 164.044	14 54 48.	+ 31 41		15.5	GALAXY
ZC 1454.8+3719	14 54 48.	+ 37 19	2420		CLUSTER OF GALAXIES
7ZW 573	14 54 48.	+ 68 42			COMPACT GALAXY
7ZW 574	14 54 48.	+ 77 41			COMPACT GALAXY
ARP 302	14 54 49.	+ 24 49			PECULIAR GALAXY
ARC 2001	14 54 51.	+ 22 58		17.2	RICH CLUSTER OF GALAXIES
MCG+05-35-027	14 54 51.	+ 31 42	39	15.5	GALAXY
REIZ 4398	14 54 51.	+ 49 35	54	14.1	GALAXY
ZWG 076.107	14 54 54.	+ 11 52		15.0	GALAXY
MCG+02-38-029	14 54 54.	+ 11 52	42	15.0	GALAXY
ZWG 076.108	14 54 54.	+ 14 29		15.4	GALAXY
MCG+03-38-059	14 54 54.	+ 19 53 30.	66	14.7	GALAXY
ZWG 164.045	14 54 54.	+ 31 45		15.6	GALAXY
MCG+11-18-016	14 54 54.	+ 64 05	18	16.	GALAXY
MCG-03-38-033	14 54 54.	- 19 06	54	15.	GALAXY
BSO 17	14 54 54.8	+ 45 29 36.		18.63	BLUE STELLAR OBJECT
SC 1452-1320.4	14 54 58.	- 13 32 32.	24		NEBULA
REIZ 4391	14 54 59.	+ 19 54	18	14.2	GALAXY
LB 09893	14 55	- 84 03		14.1	FAINT BLUE STAR
ZWG 020.037	14 55 00.	+ 01 46		15.7	GALAXY
ZWG 076.109	14 55 00.	+ 14 29		15.3	GALAXY
ZWG 105.082	14 55 00.	+ 19 52		14.0	GALAXY
UGC 09622	14 55 00.	+ 19 52	78	14.0	GALAXY S
KARA.72 447B	14 55 00.	+ 19 52	72	14.0	PART OF DOUBLE GALAXY
MCG+03-38-060	14 55 00.	+ 19 52	66	14.0	GALAXY
ZWG 221.015	14 55 00.	+ 38 56		14.4	GALAXY
UGC 09623	14 55 00.	+ 38 56	78	14.4	GALAXY S0
MCG+12-14-014	14 55 00.	+ 68 36	39	16.	GALAXY
ZWG 337.024	14 55 00.	+ 68 37		15.2	GALAXY
ZWG 366.010	14 55 00.	+ 82 45		15.4	GALAXY
UGC 09621	14 55 00.	+ 82 45	72	15.4	GALAXY Sc
MCG-02-38-023	14 55 00.	- 13 32	36	14.5	GALAXY
HN 1795	14 55 04.9	- 47 28 09.	42		NEBULA
RNGC 5790	14 55 05.	+ 08 29		15.0	GALAXY
REIZ 4394	14 55 05.	+ 19 52	60	14.0	GALAXY
ZWG 076.110	14 55 06.	+ 08 29		15.2	GALAXY
ZWG 048.076	14 55 06.	+ 08 29		15.1	GALAXY
UGC 09624	14 55 06.	+ 08 29	72	15.1	GALAXY S0/Sa
MCG+01-38-022	14 55 06.	+ 08 29	48	15.1	GALAXY
ZWG 076.111	14 55 06.	+ 08 35		15.2	GALAXY
ZWG 076.112	14 55 12.	+ 09 21		15.0	GALAXY
MCG+02-38-030	14 55 12.	+ 09 21	42	15.0	GALAXY
ZWG 076.113	14 55 12.	+ 12 19		15.7	GALAXY
ZWG 134.059	14 55 12.	+ 22 09		15.6	GALAXY
ZWG 164.046	14 55 12.	+ 27 37		15.5	GALAXY
MCG+10-21-039	14 55 12.	+ 59 48	36	16.	GALAXY
MCG-03-38-034	14 55 12.	- 19 13	108	16.	GALAXY
REIZ 4395	14 55 13.	+ 17 27	30	14.2	GALAXY
ZWG 048.077	14 55 18.	+ 06 50		14.7	GALAXY
UGC 09625	14 55 18.	+ 06 50	102	14.7	GALAXY SB:a-b
MCG+01-38-023	14 55 18.	+ 06 50	84	14.7	GALAXY
ZWG 105.083	14 55 18.	+ 15 28		15.6	GALAXY
ZC 1455.3+1549	14 55 18.	+ 15 49	940		CLUSTER OF GALAXIES
IC 1080	14 55 22.	- 06 31 30.			NONSTELLAR OBJECT
BSO 18	14 55 23.5	+ 43 30 20.		17.67	BLUE STELLAR OBJECT
ZWG 048.078	14 55 24.	+ 03 58		15.6	GALAXY
ZWG 248.031	14 55 24.	+ 48 50		15.6	GALAXY
UGC 09626	14 55 24.	+ 48 50	66	15.6	GALAXY SBb
MCG+08-27-037	14 55 24.	+ 48 50	60	15.	GALAXY
MCG+08-27-039	14 55 24.	+ 49 09 30.	15	15.	GALAXY
ZWG 248.032	14 55 24.	+ 49 52		14.0	GALAXY
UGC 09627	14 55 24.	+ 49 52	78	14.0	GALAXY SBb
MCG+08-27-038	14 55 24.	+ 49 52 30.	72	13.9	GALAXY
MCG-01-38-010	14 55 24.	- 06 30	24	14.5	GALAXY
MCG+05-35-028	14 55 27.	+ 30 10	84	13.5	GALAXY
REIZ 4402	14 55 27.	+ 49 51	18	15.3	GALAXY
REIZ 4401	14 55 27.	+ 49 53	42	13.9	GALAXY
ZWG 164.047	14 55 30.	+ 30 10		13.5	GALAXY
UGC 09628	14 55 30.	+ 30 10	84	13.5	GALAXY IRR
MCG-01-38-011	14 55 30.	- 06 14	60	15.	GALAXY
REIZ 4397	14 55 31.	+ 17 21	60	14.5	GALAXY
RNGC 5798	14 55 33.	+ 30 10		13.5	GALAXY
ZWG 048.079	14 55 36.	+ 02 38		15.5	GALAXY
ZWG 048.080	14 55 36.	+ 03 37		15.3	GALAXY
ZWG 076.114	14 55 36.	+ 11 13		15.6	GALAXY
ZWG 076.115	14 55 36.	+ 14 14		15.7	GALAXY
ZC 1455.6+1759	14 55 36.	+ 17 59	1080		CLUSTER OF GALAXIES
ZWG 221.016	14 55 36.	+ 41 11		15.6	GALAXY
MCG+07-31-012	14 55 36.	+ 41 12	45	15.	GALAXY
ZWG 274.001	14 55 36.	+ 52 32		14.6	GALAXY
ZWG 273.035	14 55 36.	+ 52 32		14.6	GALAXY
UGC 09629	14 55 36.	+ 52 32	84	14.6	GALAXY Sa
MCG+09-24-058	14 55 36.	+ 52 33	90	14.	GALAXY
REIZ 4399	14 55 37.	+ 17 20	30	14.3	GALAXY
REIZ 4400	14 55 37.	+ 30 10	66	13.4	GALAXY
ZWG 048.081	14 55 42.	+ 03 30		15.5	GALAXY
ZWG 048.082	14 55 42.	+ 04 18		15.6	GALAXY
ZWG 076.116	14 55 42.	+ 11 10		15.5	GALAXY
ZC 1455.7+1136	14 55 42.	+ 11 36	2080		CLUSTER OF GALAXIES
ZWG 076.117	14 55 42.	+ 11 43		15.6	GALAXY
ZWG 164.048	14 55 42.	+ 26 30		15.6	GALAXY
MCG+08-27-040	14 55 42.	+ 46 02	54	15.	GALAXY
ZWG 248.033	14 55 42.	+ 46 03		15.1	GALAXY
UGC 09630	14 55 42.	+ 53 04	66	17.	GALAXY DWARF

OBJECT NAME	RIGHT ASCEN.	DECLINATION	DIAM.	MAGN.	TYPE OF OBJECT
ZC 1455.7+7336	14 55 42.	+ 73 36	670		CLUSTER OF GALAXIES
MCG-07-31-004	14 55 42.	- 41 50	150	12.	GALAXY
REIZ 4403	14 55 46.	+ 46 03	48	14.9	GALAXY
RNGC 5786	14 55 46.	- 41 49			GALAXY
ZC 1455.8+0314	14 55 48.	+ 03 14	1080		CLUSTER OF GALAXIES
ZWG 020.038	14 55 48.	- 00 55		13.5	GALAXY
UGC 09631	14 55 48.	- 00 55	492	13.5	GALAXY SBb
MCG+00-38-012	14 55 48.	- 00 55	420	13.5	GALAXY
MCG-01-38-012	14 55 49.	- 06 35	78	14.5	GALAXY
RNGC 5792	14 55 49.	- 00 53		12.0	GALAXY
MCG-03-38-035	14 55 54.	- 19 05	48	13.	GALAXY
RNGC 5791	14 55 59.	- 19 04		13.0	GALAXY
BC M1456+09.2	14 56	+ 09 12		18.	QUASI-STELLAR OBJECT
ZWG 048.083	14 56 00.	+ 03 43		15.2	GALAXY
ZWG 048.084	14 56 00.	+ 08 25		15.0	GALAXY
MCG+01-38-024	14 56 00.	+ 08 25	15	15.0	GALAXY
ZC 1456.0+2339	14 56 00.	+ 23 39	2220		CLUSTER OF GALAXIES
MCG+06-33-013	14 56 00.	+ 34 22	18	16.	GALAXY
ZC 1456.0+6415	14 56 00.	+ 64 15	1480		CLUSTER OF GALAXIES
MCG+11-18-017	14 56 00.	+ 66 07	51	16.	GALAXY
IC 1081	14 56 04.	- 19 02 27.			NONSTELLAR OBJECT
MCG-03-38-036	14 56 06.	- 19 03 30.	78	14.5	GALAXY
ZWG 048.085	14 56 12.	+ 06 59		15.4	GALAXY
ZWG 048.086	14 56 12.	+ 08 11		15.1	GALAXY
MCG+01-38-025	14 56 12.	+ 08 11	30	15.1	GALAXY
ZWG 134.060	14 56 12.	+ 24 09		15.5	GALAXY
TON-N 0786	14 56 12.	+ 29 36		14.5	BLUE STAR
ZC 1456.2+4901	14 56 12.	+ 49 01	8400		CLUSTER OF GALAXIES
ZWG 020.039	14 56 18.	+ 02 13		14.9	GALAXY
MCG+00-38-013	14 56 18.	+ 02 13	36	14.9	GALAXY
ZWG 076.118	14 56 18.	+ 11 49		15.3	GALAXY
MCG+13-11-004	14 56 18.	+ 79 30	66	16.	GALAXY
MCG+04-35-020	14 56 21.	+ 24 10	24	15.5	GALAXY
ARC 2004	14 56 23.	+ 25 08		17.0	RICH CLUSTER OF GALAXIES
ZWG 048.087	14 56 24.	+ 07 12		15.2	GALAXY
ZC 1456.4+2503	14 56 24.	+ 25 03	1140		CLUSTER OF GALAXIES
ZWG 274.002	14 56 24.	+ 53 59		15.3	GALAXY
ZWG 273.036	14 56 24.	+ 53 59		15.3	GALAXY
UGC 09632	14 56 24.	+ 53 59	90	15.3	GALAXY Sc
MCG+09-24-059	14 56 24.	+ 54 00	72	14.	GALAXY
IC 1082	14 56 26.	+ 07 12 49.			NONSTELLAR OBJECT
ARC 2003	14 56 26.	+ 19 39		17.2	RICH CLUSTER OF GALAXIES
ZWG 076.119	14 56 30.	+ 11 51		15.4	GALAXY
MCG+03-38-061	14 56 30.	+ 14 37	42	15.7	GALAXY
ZWG 105.084	14 56 30.	+ 14 39		15.7	GALAXY
ZC 1456.5+1654	14 56 30.	+ 16 54	400		CLUSTER OF GALAXIES
ZC 1456.5+1841	14 56 30.	+ 18 41	2550		CLUSTER OF GALAXIES
ZC 1456.5+1940	14 56 30.	+ 19 40	610		CLUSTER OF GALAXIES
ZC 1456.5+3810	14 56 30.	+ 38 10	1080		CLUSTER OF GALAXIES
ZWG 354.014	14 56 30.	+ 79 30		15.6	GALAXY
KARA.72 448A	14 56 30.	+ 79 30	54	15.6	PART OF DOUBLE GALAXY
MCG+14-07-009	14 56 30.	+ 81 01 30.	24	17.	GALAXY
MCG-01-38-013	14 56 30.	- 04 05	42	15.5	GALAXY
MCG-03-38-037	14 56 30.	- 19 51	54	16.5	GALAXY
ARC 2005	14 56 35.	+ 28 01		16.0	RICH CLUSTER OF GALAXIES
ZC 1456.6+0055	14 56 36.	+ 00 55	1550		CLUSTER OF GALAXIES
UGC 09633	14 56 36.	+ 08 53	90	16.0	GALAXY DBL SYS
ZWG 105.085	14 56 36.	+ 16 36		15.5	GALAXY
ZWG 105.086	14 56 36.	+ 20 15		15.5	GALAXY
UGC 09634	14 56 36.	+ 20 15	60	15.5	GALAXY SBb
MCG+03-38-062	14 56 36.	+ 20 15	66	15.5	GALAXY
ZC 1456.6+3810	14 56 36.	+ 38 10	340		CLUSTER OF GALAXIES
RNGC 5796	14 56 36.	- 16 26		13.0	GALAXY
MCG-03-38-039	14 56 36.	- 16 26	42	13.	GALAXY
RNGC 5793	14 56 36.	- 16 30		14.0	GALAXY
MCG-03-38-038	14 56 36.	- 16 30	84	14.5	GALAXY
MCG+05-35-029	14 56 39.	+ 26 56	30	15.4	GALAXY
ZC 1456.7+1744	14 56 42.	+ 17 44	670		CLUSTER OF GALAXIES
MCG+03-38-063	14 56 42.	+ 19 46	60	15.7	GALAXY
ZWG 105.087	14 56 42.	+ 19 47		15.7	GALAXY
UGC 09635	14 56 42.	+ 19 47	66	15.7	GALAXY Sa-b
ZWG 164.049	14 56 42.	+ 26 56		15.4	GALAXY
ZC 1456.7+2803	14 56 42.	+ 28 03	2420		CLUSTER OF GALAXIES
ZC 1456.7+4210	14 56 42.	+ 42 10	940		CLUSTER OF GALAXIES
ZWG 248.034	14 56 42.	+ 48 41		15.5	GALAXY
MCG+08-27-041	14 56 42.	+ 48 42	30	15.	GALAXY
MCG+09-24-060	14 56 42.	+ 53 27 30.	30	15.	GALAXY
MCG+08-27-042	14 56 45.	+ 45 05	66	15.	GALAXY
MCG+08-27-043	14 56 45.	+ 48 42	48	15.	GALAXY
MCG-03-38-040	14 56 45.	- 18 33	90	15.	GALAXY
ZWG 076.120	14 56 48.	+ 13 37		15.5	GALAXY
UGC 09636	14 56 48.	+ 13 37	66	15.5	GALAXY S
ZC 1456.8+1945	14 56 48.	+ 19 45	1610		CLUSTER OF GALAXIES
UGC 09637	14 56 48.	+ 41 58	84	16.0	GALAXY IRR
ZWG 221.017	14 56 48.	+ 44 04		15.3	GALAXY
ZWG 248.035	14 56 48.	+ 48 40		15.7	GALAXY
ZWG 274.003	14 56 48.	+ 53 26		15.6	GALAXY
ZWG 273.037	14 56 48.	+ 53 26		15.6	GALAXY
ZC 1456.8+5814	14 56 48.	+ 58 14	540		CLUSTER OF GALAXIES
ZWG 296.020	14 56 48.	+ 59 04		15.5	GALAXY
UGC 09638	14 56 48.	+ 59 04	120	15.5	GALAXY DWRF IR
7ZW 575	14 56 48.	+ 75 28			COMPACT GALAXY
RNGC 5817	14 56 48.	- 15 59		15.0	GALAXY
MCG-03-38-041	14 56 48.	- 15 59 30.	12	15.	GALAXY
REIZ 4406	14 56 49.	+ 53 26	9	14.9	GALAXY
ZC 1456.9+0332	14 56 54.	+ 03 32	470		CLUSTER OF GALAXIES
ARC 2007	14 56 54.	+ 38 12		16.9	RICH CLUSTER OF GALAXIES
MCG+07-31-013	14 56 54.	+ 44 06	33	14.	GALAXY
ZWG 248.036	14 56 54.	+ 45 05		14.6	GALAXY
UGC 09639	14 56 54.	+ 45 09	60	14.6	GALAXY Sa-b
1ZW 099	14 56 54.	+ 53 26			COMPACT GALAXY
SHB 303	14 56 56.9	+ 09 16 08.		18.	QUASI-STELLAR OBJECT
BSO 19	14 56 57.	+ 45 48 22.		17.84	BLUE STELLAR OBJECT
ZC 1457.0+0752	14 57 00.	+ 07 52	1080		CLUSTER OF GALAXIES
ZWG 134.061	14 57 00.	+ 24 13		15.1	GALAXY
ZWG 248.037	14 57 00.	+ 48 40		15.6	GALAXY
MCG+10-21-040	14 57 00.	+ 59 03	60	15.	GALAXY
ZC 1457.0+6045	14 57 00.	+ 60 45	1680		CLUSTER OF GALAXIES
ZWG 354.015	14 57 00.	+ 79 31		15.5	GALAXY
KARA.72 448B	14 57 00.	+ 79 31	48	15.5	PART OF DOUBLE GALAXY
MCG+13-11-005	14 57 00.	+ 79 31	39	16.	GALAXY
MCG-01-38-014	14 57 00.	- 06 46	96	15.	GALAXY
VHE 65A	14 57 00.	- 63 05			REFLECTION NEBULA
ZC 1457.1+1556	14 57 06.	+ 15 56	1010		CLUSTER OF GALAXIES
MCG+03-38-064	14 57 06.	+ 16 49	30	14.5	GALAXY
ZWG 105.088	14 57 06.	+ 16 50		14.5	GALAXY
UGC 09640	14 57 06.	+ 16 50	108	14.5	GALAXY E
ZC 1457.1+2427	14 57 06.	+ 24 27	870		CLUSTER OF GALAXIES

OBJECT NAME	RIGHT ASCEN.	DECLINATION	DIAM.	MAGN.	TYPE OF OBJECT
ZC 1457.1+2946	14 57 06.	+ 29 46	3020		CLUSTER OF GALAXIES
ZWG 221.018	14 57 06.	+ 41 38		14.9	GALAXY
UGC 09641	14 57 06.	+ 41 38	90	14.9	GALAXY S
ZWG 221.019	14 57 06.	+ 43 35		15.3	GALAXY
MCG+08-27-044	14 57 06.	+ 48 13	36	17.	GALAXY
ZWG 274.004	14 57 06.	+ 54 05		13.0	GALAXY
ZWG 273.038	14 57 06.	+ 54 05		13.0	GALAXY
UGC 09642	14 57 06.	+ 54 05	132	13.0	GALAXY E-S0
MCG+09-25-001	14 57 06.	+ 54 06	60	13.1	GALAXY
HOLM 686B	14 57 09.	+ 13 26	42	15.0	PART OF MULTIPLE GALAXY
RNGC 5820	14 57 09.	+ 54 05		13.0	GALAXY
HOLM 686A	14 57 10.	+ 13 25	72	14.8	PART OF MULTIPLE GALAXY
RNGC 5818	14 57 11.	+ 50 01		15.0	GALAXY
REIZ 4410	14 57 11.	+ 54 05	30	13.1	GALAXY
MCG+04-35-021	14 57 12.	+ 24 13	15	15.1	GALAXY
ZWG 134.062	14 57 12.	+ 25 46		15.3	GALAXY
MCG+07-31-014	14 57 12.	+ 41 40	18	15.	GALAXY
ZWG 248.038	14 57 12.	+ 48 36		15.4	GALAXY
ZWG 248.039	14 57 12.	+ 50 01		15.0	GALAXY
UGC 09643	14 57 12.	+ 50 01	72	15.0	GALAXY S0
ARP 136	14 57 12.	+ 54 05			PECULIAR GALAXY
MCG+07-31-015	14 57 15.	+ 43 38	36	15.	GALAXY
MCG+08-27-045	14 57 15.	+ 48 37 30.	18	16.	GALAXY
MCG+08-27-046	14 57 15.	+ 50 02	24	14.3	GALAXY
REIZ 4405	14 57 16.	+ 13 25	24	15.0	GALAXY
REIZ 4404	14 57 17.	+ 13 24	48	14.6	GALAXY
IC 4521	14 57 17.	+ 25 46 01.			NONSTELLAR OBJECT
ZWG 076.121	14 57 18.	+ 13 25		15.0	GALAXY
KARA.72 449B	14 57 18.	+ 13 25	42	15.0	PART OF DOUBLE GALAXY
MCG+02-38-031	14 57 18.	+ 13 25	30	15.0	GALAXY
ZWG 076.122	14 57 18.	+ 13 26		15.3	GALAXY
KARA.72 449A	14 57 18.	+ 13 26	48	15.3	PART OF DOUBLE GALAXY
MCG+02-38-032	14 57 18.	+ 13 26	36	15.3	GALAXY
ZC 1457.3+1628	14 57 18.	+ 16 28	3160		CLUSTER OF GALAXIES
ZWG 164.050	14 57 18.	+ 27 19		14.9	GALAXY
UGC 09644	14 57 18.	+ 27 19	84	14.9	GALAXY SBa
ZC 1457.3+3634	14 57 18.	+ 36 34	3020		CLUSTER OF GALAXIES
ZC 1457.3+5241	14 57 18.	+ 52 41	610		CLUSTER OF GALAXIES
ARC 2006	14 57 20.	+ 19 52		17.2	RICH CLUSTER OF GALAXIES
REIZ 4409	14 57 20.	+ 50 03	12	14.3	GALAXY
MCG+05-35-030	14 57 21.	+ 27 18	66	14.9	GALAXY
MCG-03-38-042	14 57 21.	- 16 11	72	14.5	GALAXY
ZWG 020.041	14 57 24.	+ 02 05		12.9	GALAXY
UGC 09645	14 57 24.	+ 02 05	210	12.9	GALAXY Sb
MCG+00-38-014	14 57 24.	+ 02 05	180	12.9	GALAXY
MCG+04-35-022	14 57 24.	+ 25 47	27	15.3	GALAXY
ZC 1457.4+3032	14 57 24.	+ 30 32	740		CLUSTER OF GALAXIES
ZC 1457.4+4232	14 57 24.	+ 42 32	1010		CLUSTER OF GALAXIES
BSO 20	14 57 24.	+ 44 00 48.			BLUE STELLAR OBJECT
ZWG 020.040	14 57 24.	- 00 55		15.5	GALAXY
MCG-03-38-043	14 57 24.	- 18 23	48	14.5	GALAXY
RNGC 5821A	14 57 27.	+ 54 07		15.0	GALAXY
REIZ 4411	14 57 29.	+ 54 08	24	14.5	GALAXY
HOLM 687A	14 57 29.	+ 54 08	30	14.3	PART OF MULTIPLE GALAXY
ZWG 134.063	14 57 30.	+ 21 10		15.7	GALAXY
TON-N 0216	14 57 30.	+ 23 40		15.2	BLUE STAR
ZWG 164.051	14 57 30.	+ 26 49		15.5	GALAXY
UGC 09646	14 57 30.	+ 27 31	66	16.0	GALAXY Sc
ZWG 164.052	14 57 30.	+ 29 08		15.7	GALAXY
ZWG 193.011	14 57 30.	+ 33 02		15.7	GALAXY
UGC 09647	14 57 30.	+ 33 02	102	15.7	GALAXY DWRF SP
ZWG 274.005	14 57 30.	+ 54 07		14.9	GALAXY
ZWG 273.039	14 57 30.	+ 54 07		14.9	GALAXY
UGC 09648	14 57 30.	+ 54 07	102	14.9	GALAXY S
MCG+09-25-002	14 57 30.	+ 54 08	120	14.3	GALAXY
ZC 1457.5+5415	14 57 30.	+ 54 15	13370		CLUSTER OF GALAXIES
RNGC 5806	14 57 32.	+ 02 05		12.5	GALAXY
MCG+06-33-014	14 57 33.	+ 33 01 30.	66	14.5	GALAXY
HOLM 687B	14 57 34.	+ 54 07	30	14.5	PART OF MULTIPLE GALAXY
SC 1454-1438.8	14 57 35.	- 14 50 48.	24		NEBULA
ZC 1457.6+1916	14 57 36.	+ 19 16	1340		CLUSTER OF GALAXIES
ZC 1457.6+2316	14 57 36.	+ 23 16	740		CLUSTER OF GALAXIES
ZC 1457.6+4404	14 57 36.	+ 44 04	2220		CLUSTER OF GALAXIES
MCG+08-27-047	14 57 36.	+ 48 32	36	16.	GALAXY
ZC 1457.6+6020	14 57 36.	+ 60 20	610		CLUSTER OF GALAXIES
ZWG 337.025	14 57 36.	+ 71 53		13.3	GALAXY
RNGC 5832	14 57 36.	+ 71 53		13.5	GALAXY
UGC 09649	14 57 36.	+ 71 53	222	13.3	GALAXY SBc?
KARA.73B 0656	14 57 36.	+ 71 53	204	13.3	ISOLATED GALAXY S
MCG+14-07-010	14 57 36.	+ 81 02	18	16.	GALAXY
ARC 2013	14 57 37.	+ 60 42		18.0	RICH CLUSTER OF GALAXIES
RNGC 5801	14 57 37.	- 13 42			GALAXY
ZWG 020.042	14 57 42.	+ 02 29		15.5	GALAXY
ZWG 164.053	14 57 42.	+ 27 32		15.7	GALAXY
MCG+10-21-041	14 57 42.	+ 61 26	42	16.	GALAXY
ZC 1457.7+7016	14 57 42.	+ 70 16	1280		CLUSTER OF GALAXIES
MCG+12-14-015	14 57 42.	+ 71 53	222	13.1	GALAXY
MCG-02-38-024	14 57 42.	- 13 22	120	16.	GALAXY
RNGC 5815	14 57 42.	- 16 38		14.0	GALAXY
MCG-03-38-044	14 57 42.	- 16 38 30.	42	14.5	GALAXY
RNGC 5803	14 57 43.	- 13 42			GALAXY
RNGC 5802	14 57 43.	- 13 43			GALAXY
REIZ 4408	14 57 46.	+ 13 23	12	15.9	GALAXY
REIZ 4407	14 57 47.	+ 13 22	36	15.4	GALAXY
ZWG 048.088	14 57 48.	+ 02 30		15.4	GALAXY
ZWG 076.123	14 57 48.	+ 10 08		15.5	GALAXY
ZWG 076.124	14 57 48.	+ 14 24		15.5	GALAXY
ZC 1457.8+1524	14 57 48.	+ 15 24	2890		CLUSTER OF GALAXIES
ZC 1457.8+2407	14 57 48.	+ 24 07	1340		CLUSTER OF GALAXIES
ARC 2008	14 57 50.	+ 23 20		17.5	RICH CLUSTER OF GALAXIES
MCG+06-33-015	14 57 51.	+ 33 00	24	17.	GALAXY
ZWG 020.043	14 57 54.	+ 01 48		14.8	GALAXY
KARA.72 450B	14 57 54.	+ 01 48	18		PART OF DOUBLE GALAXY
KARA.72 450A	14 57 54.	+ 01 48	24	14.8	PART OF DOUBLE GALAXY
MCG+00-38-015	14 57 54.	+ 01 48	42	14.8	GALAXY
ZWG 076.125	14 57 54.	+ 11 32		15.3	GALAXY
ZWG 076.126	14 57 54.	+ 12 40		15.7	GALAXY
ZC 1457.9+1405	14 57 54.	+ 14 05	1410		CLUSTER OF GALAXIES
TON-N 0217	14 57 54.	+ 24 50		15.6	BLUE STAR
ZWG 248.040	14 57 54.	+ 49 31		15.2	GALAXY
RNGC 5811	14 57 56.	+ 01 49		15.0	GALAXY
RNGC 5821B	14 57 58.	+ 54 06		14.5	GALAXY
ARC 2009	14 57 59.	+ 21 34		17.2	RICH CLUSTER OF GALAXIES
ZWG 048.089	14 58 00.	+ 06 39		15.3	GALAXY
ZWG 076.127	14 58 00.	+ 14 20		15.5	GALAXY
ZC 1458.0+3733	14 58 00.	+ 37 33	2350		CLUSTER OF GALAXIES
MCG+07-31-016	14 58 00.	+ 44 30	36	17.	GALAXY
ZC 1458.0+4544	14 58 00.	+ 45 44	940		CLUSTER OF GALAXIES

OBJECT NAME	RIGHT ASCEN.	DECLINATION	DIAM.	MAGN.	TYPE OF OBJECT
UGC 09651	14 58 00.	+ 49 43	66	16.0	GALAXY SBc
MCG+09-25-003	14 58 00.	+ 54 06	54	14.5	GALAXY
MCG+14-07-011	14 58 00.	+ 81 01	27	16.	GALAXY
ZWG 366.011	14 58 00.	+ 83 48		14.1	GALAXY
UGC 09650	14 58 00.	+ 83 48	96	14.1	GALAXY Sa
KARA.72 451A	14 58 00.	+ 83 48	84	14.1	PART OF DOUBLE GALAXY
MCG+14-07-012	14 58 00.	+ 83 48	66	15.	GALAXY
MCG-01-38-015	14 58 00.	- 03 36	18	15.5	GALAXY
ZC 1458.1+2135	14 58 06.	+ 21 35	1210		CLUSTER OF GALAXIES
RNGC 5809	14 58 07.	- 13 58		14.0	GALAXY
MCG-02-38-025	14 58 09.	- 13 58 30.	60	14.5	GALAXY
RNGC 5812	14 58 10.	- 07 16		13.0	GALAXY
SC 1455-1500.8	14 58 11.	- 15 12 46.	60		NEBULA
ZWG 076.128	14 58 12.	+ 11 44		15.3	GALAXY
ZWG 105.089	14 58 12.	+ 16 22		15.1	GALAXY
UGC 09652	14 58 12.	+ 16 22	72	15.1	GALAXY SBa
ZWG 105.090	14 58 12.	+ 17 45		15.7	GALAXY
ZC 1458.2+4121	14 58 12.	+ 41 21	1750		CLUSTER OF GALAXIES
ZC 1458.2+4956	14 58 12.	+ 49 56	1480		CLUSTER OF GALAXIES
ARC 2011	14 58 12.	+ 49 57		17.2	RICH CLUSTER OF GALAXIES
ZWG 296.021	14 58 12.	+ 61 14		15.7	GALAXY
UGC 09653	14 58 12.	+ 61 14	72	15.7	GALAXY SBb
MCG+10-21-042	14 58 12.	+ 61 14	42	16.	GALAXY
MCG+03-38-065	14 58 15.	+ 16 21	66	15.1	GALAXY
ARC 2015	14 58 15.	+ 56 11		17.6	RICH CLUSTER OF GALAXIES
MCG-02-38-026	14 58 15.	- 15 12	36	17.	GALAXY
REIN 2.227	14 58 15.91	- 07 15 38.0			NEBULA
ZWG 248.041	14 58 18.	+ 49 35		15.7	GALAXY
MCG-01-38-016	14 58 18.	- 07 14	45	12.	GALAXY
MCG-02-38-027	14 58 18.	- 14 56	150	15.	GALAXY
MCG-02-38-028	14 58 18.	- 15 13 30.	54	15.	GALAXY
ZC 1458.4+2301	14 58 24.	+ 23 01	870		CLUSTER OF GALAXIES
MCG+08-27-048	14 58 24.	+ 46 38	48	16.	GALAXY
MCG+09-25-004	14 58 24.	+ 54 49	30	16.	GALAXY
RNGC 5826	14 58 24.	+ 55 43			NON-EXISTENT OBJECT
ZC 1458.4+5613	14 58 24.	+ 56 13	1140		CLUSTER OF GALAXIES
RNGC 5800	14 58 25.	- 51 43			NON-EXISTENT OBJECT
ZWG 020.044	14 58 30.	+ 00 53		15.6	GALAXY
ZWG 076.129	14 58 30.	+ 11 43		15.2	GALAXY
UGC 09654	14 58 30.	+ 11 43	66	15.2	GALAXY Sa-b
ZWG 105.091	14 58 30.	+ 17 09		15.3	GALAXY
MCG+03-38-066	14 58 30.	+ 19 48	33	16.	GALAXY
ZWG 134.064	14 58 30.	+ 23 48		15.7	GALAXY
ZWG 221.020	14 58 30.	+ 43 28		15.4	GALAXY
IC 1084	14 58 35.	- 07 16 37.			NONSTELLAR OBJECT
ZWG 020.045	14 58 36.	+ 01 53		12.5	GALAXY
UGC 09655	14 58 36.	+ 01 53	240	12.5	GALAXY E
MCG+00-38-016	14 58 36.	+ 01 53	48	12.5	GALAXY
MCG+03-38-068	14 58 36.	+ 17 08	24	15.3	GALAXY
MCG+03-38-067	14 58 36.	+ 19 56	45	16.	GALAXY
ZC 1458.6+4735	14 58 36.	+ 47 35	3560		CLUSTER OF GALAXIES
MCG-01-38-018	14 58 36.	- 04 18	30	15.	GALAXY
MCG-01-38-017	14 58 36.	- 07 15	24	14.5	GALAXY
BSO 21	14 58 36.5	+ 42 17 29.		13.88	BLUE STELLAR OBJECT
SC 1455-1614.8	14 58 37.	- 16 26 45.	24		NEBULA
RNGC 5813	14 58 38.	+ 01 54		12.0	GALAXY
HOLM 688A	14 58 38.	+ 01 54	138	13.4	PART OF MULTIPLE GALAXY
BSO 22	14 58 40.5	+ 42 17 56.		18.17	BLUE STELLAR OBJECT
MCG+07-31-018	14 58 42.	+ 43 30	24	15.5	GALAXY
MCG+07-31-017	14 58 42.	+ 43 30	30	15.5	GALAXY
MCG-01-38-019	14 58 42.	- 04 15	30	15.	GALAXY
MCG-02-38-029	14 58 45.	- 14 07	36	15.	GALAXY
ZWG 020.046	14 58 48.	+ 01 49		14.7	GALAXY
MCG+00-38-017	14 58 48.	+ 01 49	48	14.7	GALAXY
HOLM 688B	14 58 48.	+ 01 50	18	14.7	PART OF MULTIPLE GALAXY
ZC 1458.8+1614	14 58 48.	+ 16 14	740		CLUSTER OF GALAXIES
ZWG 134.065	14 58 48.	+ 23 40		15.6	GALAXY
MCG+08-27-049	14 58 48.	+ 49 23	60	16.	GALAXY
MCG-03-38-045	14 58 48.	- 19 07	72	14.5	GALAXY
RNGC 5814	14 58 50.	+ 01 49		14.5	GALAXY
ZWG 076.130	14 58 54.	+ 09 21		15.2	GALAXY
UGC 09656	14 58 54.	+ 09 21	66	15.2	GALAXY S
ZWG 076.131	14 58 54.	+ 12 51		15.6	GALAXY
ZWG 105.092	14 58 54.	+ 16 58		15.7	GALAXY
ZC 1458.9+3218	14 58 54.	+ 32 18	3230		CLUSTER OF GALAXIES
REIZ 4412	14 58 56.	+ 33 44	24	15.4	GALAXY
SHB 304	14 58 56.6	+ 71 52 11.		16.8	QUASI-STELLAR OBJECT
BC 3CR309.1	14 58 56.70	+ 71 52 10.		16C78	QUASI-STELLAR OBJECT
REIZ 4417	14 58 58.	+ 48 33	12	15.6	GALAXY
RNGC 5828A	14 58 59.	+ 50 11		14.5	GALAXY
VDB.66G 198	14 59	+ 52 52	70		DWARF GALAXY
ZC 1459.0+1945	14 59 00.	+ 19 45	2420		CLUSTER OF GALAXIES
ZWG 248.042	14 59 00.	+ 48 33		15.5	GALAXY
HOLM 689B	14 59 00.	+ 48 33	12	14.4	PART OF MULTIPLE GALAXY
UGC 09657	14 59 00.	+ 48 33	90	15.5	GALAXY SB?c
MCG+08-27-050	14 59 00.	+ 48 33	60	13.7	GALAXY
ZWG 248.043	14 59 00.	+ 50 11		14.3	GALAXY
UGC 09658	14 59 00.	+ 50 11	36	14.3	GALAXY S
REIZ 4419	14 59 01.	+ 50 23	12	15.6	GALAXY
RNGC 5828B	14 59 05.	+ 50 11		14.5	GALAXY
ZWG 020.047	14 59 06.	+ 02 21		15.6	GALAXY
ZWG 076.132	14 59 06.	+ 09 09		15.4	GALAXY
ZC 1459.1+4319	14 59 06.	+ 43 19	1080		CLUSTER OF GALAXIES
HOLM 690B	14 59 06.	+ 50 10	30	14.5	PART OF MULTIPLE GALAXY
MCG+08-27-052	14 59 06.	+ 50 11	18	14.5	GALAXY
MCG+08-27-051	14 59 06.	+ 50 12	36	13.5	GALAXY
REIZ 4421	14 59 07.	+ 50 11	12	15.4	GALAXY
HOLM 690A	14 59 07.	+ 50 11	36	13.5	PART OF MULTIPLE GALAXY
REIZ 4422	14 59 07.	+ 50 12	27	14.2	GALAXY
REIZ 4415	14 59 08.	+ 33 51	42	15.0	GALAXY
REIZ 4420	14 59 10.	+ 48 33	78	14.5	GALAXY
HOLM 689A	14 59 10.	+ 48 33	102	13.7	PART OF MULTIPLE GALAXY
ZWG 105.093	14 59 12.	+ 16 55		15.4	GALAXY
VHE 65B	14 59 12.	- 63 11			REFLECTION NEBULA
ZWG 076.133	14 59 18.	+ 10 37		15.2	GALAXY
ZWG 105.094	14 59 18.	+ 16 04		15.7	GALAXY
ZC 1459.3+2453	14 59 18.	+ 24 53	2350		CLUSTER OF GALAXIES
MCG+08-27-053	14 59 18.	+ 44 53	48	15.	GALAXY
RNGC 5816	14 59 18.	- 15 56			NON-EXISTENT OBJECT
REIZ 4413	14 59 22.	+ 13 46	30	15.4	GALAXY
REIZ 4423	14 59 23.	+ 44 53	48	13.5	GALAXY
ZWG 048.090	14 59 24.	+ 05 08		15.3	GALAXY
UGC 09659	14 59 24.	+ 05 08	66	15.3	GALAXY Sc
MCG+03-38-070	14 59 24.	+ 15 17	21	15.3	GALAXY
MCG+03-38-069	14 59 24.	+ 16 02	27	15.7	GALAXY
ZC 1459.4+4240	14 59 24.	+ 42 40	2290		CLUSTER OF GALAXIES
ZWG 248.044	14 59 24.	+ 44 54		14.3	GALAXY
UGC 09660	14 59 24.	+ 44 54	48	14.3	GALAXY PECULR?

OBJECT NAME	RIGHT ASCEN.	DECLINATION	DIAM.	MAGN.	TYPE OF OBJECT
KARA.73B 0657	14 59 24.	+ 44 54	48	14.3	ISOLATED GALAXY
REIZ 4414	14 59 28.	+ 13 44	9	15.8	GALAXY
ZWG 020.048	14 59 30.	+ 02 01		15.0	GALAXY
UGC 09661	14 59 30.	+ 02 01	84	15.0	GALAXY
MCG+00-38-018	14 59 30.	+ 02 01	66	15.0	GALAXY
ZWG 105.095	14 59 30.	+ 15 16		15.3	GALAXY
ARC 2012	14 59 31.	+ 16 43		16.6	RICH CLUSTER OF GALAXIES
ARC 2018	14 59 32.	+ 47 28		16.6	RICH CLUSTER OF GALAXIES
MCG+03-38-071	14 59 33.	+ 17 09	54	15.	GALAXY
MCG+03-38-072	14 59 36.	+ 15 15	24	15.3	GALAXY
ZC 1459.6+1651	14 59 36.	+ 16 51	1550		CLUSTER OF GALAXIES
ZC 1459.6+1803	14 59 36.	+ 18 03	1810		CLUSTER OF GALAXIES
ZWG 134.066	14 59 36.	+ 26 10		13.7	GALAXY
UGC 09662	14 59 36.	+ 26 10	72	13.7	GALAXY S
ZWG 274.006	14 59 36.	+ 52 48		15.2	GALAXY
UGC 09663	14 59 36.	+ 52 48	90	15.2	GALAXY DWARF
7ZW 576	14 59 36.	+ 74 05			COMPACT GALAXY
RNGC 5827	14 59 37.	+ 26 10		13.5	GALAXY
REIZ 4418	14 59 38.	+ 22 22	60	13.6	GALAXY
ZWG 020.049	14 59 42.	+ 02 13		15.3	GALAXY
ZWG 076.134	14 59 42.	+ 11 02		15.6	GALAXY
REIZ 4416	14 59 42.	+ 11 02	24	15.1	GALAXY
ZWG 134.067	14 59 42.	+ 21 32		15.2	GALAXY
ZWG 193.012	14 59 42.	+ 38 15		15.6	GALAXY
MCG+06-33-016	14 59 42.	+ 38 16	12	16.	GALAXY
ZC 1459.7+7227	14 59 42.	+ 72 27	670		CLUSTER OF GALAXIES
ZWG 337.026	14 59 42.	+ 74 05		14.9	GALAXY
RNGC 5836	14 59 42.	+ 74 05		15.0	GALAXY
UGC 09664	14 59 42.	+ 74 05	78	14.9	GALAXY SBb
RNGC 5825	14 59 45.	+ 18 54			NON-EXISTENT OBJECT
MCG+09-25-005	14 59 45.	+ 52 47	102	14.	GALAXY
ZWG 048.091	14 59 48.	+ 06 33		15.5	GALAXY
ZWG 076.135	14 59 48.	+ 11 42		15.6	GALAXY
ZC 1459.8+2043	14 59 48.	+ 20 43	9070		CLUSTER OF GALAXIES
MCG+04-35-023	14 59 48.	+ 21 30	36	15.2	GALAXY
ZWG 248.045	14 59 48.	+ 48 31		14.9	GALAXY
UGC 09665	14 59 48.	+ 48 31	96	14.9	GALAXY Sb-c
MCG+08-27-054	14 59 48.	+ 48 31	108	13.2	GALAXY
ZWG 248.046	14 59 48.	+ 49 18		15.3	GALAXY
MCG+08-27-055	14 59 48.	+ 49 19	42	15.	GALAXY
REIZ 4426	14 59 48.	+ 50 29	18	15.2	GALAXY
MCG+10-21-043	14 59 48.	+ 60 25	51	15.	GALAXY
UGC 09666	14 59 48.	+ 60 27	72	16.0	GALAXY S
RNGC 5810	14 59 48.	- 17 41		14.0	GALAXY
MCG-03-38-046	14 59 48.	- 17 41 30.	48	14.	GALAXY
SMI 09	14 59 48.	- 57 28	360		FAINT NEBULOSITY
REIZ 4424	14 59 52.	+ 48 28	24	15.7	GALAXY
REIZ 4425	14 59 52.	+ 48 31	114	13.9	GALAXY
HOLM 691A	14 59 52.	+ 48 31	108	13.2	PART OF MULTIPLE GALAXY
IC 4524	14 59 53.	+ 25 47 39.			NONSTELLAR OBJECT
HOLM 691B	14 59 53.	+ 48 28	24	15.0	PART OF MULTIPLE GALAXY
ZWG 048.092	14 59 54.	+ 05 51		14.6	GALAXY
UGC 09667	14 59 54.	+ 05 51	126	14.6	GALAXY S0-a
MCG+01-38-026	14 59 54.	+ 05 51	36	14.6	GALAXY
ZWG 076.136	14 59 54.	+ 12 07		15.0	GALAXY
MCG+02-38-033	14 59 54.	+ 12 07	48	15.0	GALAXY
ZWG 134.068	14 59 54.	+ 25 47		15.6	GALAXY
MCG+04-35-024	14 59 54.	+ 26 10	66	13.7	GALAXY
IC 4528	14 59 54.	+ 49 18			NONSTELLAR OBJECT
ZC 1459.9+6410	14 59 54.	+ 64 10	340		CLUSTER OF GALAXIES
ZWG 048.093	15 00 00.	+ 05 45		15.5	GALAXY
ZWG 076.137	15 00 00.	+ 11 24		15.3	GALAXY
MCG+03-38-073	15 00 00.	+ 16 22 30.	21	15.7	GALAXY
ZWG 105.096	15 00 00.	+ 16 24		15.7	GALAXY
ZC 1500.0+1930	15 00 00.	+ 19 30	540		CLUSTER OF GALAXIES
ZC 1500.0+2420	15 00 00.	+ 24 20	870		CLUSTER OF GALAXIES
ZC 1500.0+5228	15 00 00.	+ 52 28	940		CLUSTER OF GALAXIES
MCG+12-14-016	15 00 00.	+ 74 05	66	15.	GALAXY
ZWG 366.012	15 00 00.	+ 83 44		13.8	GALAXY
UGC 09668	15 00 00.	+ 83 44	96	13.8	GALAXY
KARA.72 451B	15 00 00.	+ 83 44	96	13.8	PART OF DOUBLE GALAXY
ZWG 367.001	15 00 00.	+ 85 13		15.7	GALAXY
ZWG 366.013	15 00 00.	+ 85 13		15.7	GALAXY
ZWG 365.018	15 00 00.	+ 85 13		15.7	GALAXY
UGC 09669	15 00 00.	+ 85 13	78	15.7	GALAXY S?
MCG+14-07-013	15 00 00.	+ 85 13	12	16.	GALAXY
MIL 38	15 00 00.	- 41 45	1800		SUPERNOVA REMNANT
RNGC 5830	15 00 03.	+ 48 04		15.0	GALAXY
ZC 1500.1+1947	15 00 06.	+ 19 47	340		CLUSTER OF GALAXIES
ZC 1500.1+2325	15 00 06.	+ 23 25	1410		CLUSTER OF GALAXIES
ZWG 165.001	15 00 06.	+ 32 19		15.4	GALAXY
ZWG 164.054	15 00 06.	+ 32 19		15.4	GALAXY
ZWG 248.047	15 00 06.	+ 48 04		15.2	GALAXY
UGC 09670	15 00 06.	+ 48 04	66	15.2	GALAXY Sb
MCG+08-27-056	15 00 06.	+ 48 05	42	15.	GALAXY
PK318-02.1	15 00 08.	- 60 41 26.	30		PLANETARY NEBULA
IC 4525	15 00 11.	+ 25 49 52.			NONSTELLAR OBJECT
REIZ 4428	15 00 11.	+ 48 04	12	14.6	GALAXY
ZC 1500.2+1710	15 00 12.	+ 17 10	810		CLUSTER OF GALAXIES
ZWG 134.069	15 00 12.	+ 25 50		15.7	GALAXY
MCG+07-31-019	15 00 12.	+ 42 13	36	15.	GALAXY
ZWG 221.021	15 00 12.	+ 42 23		15.6	GALAXY
UGC 09671	15 00 12.	+ 42 23	90	15.6	GALAXY Sc
REIZ 4427	15 00 15.	+ 42 23	60	14.8	GALAXY
SC 1457-1253.2	15 00 15.	- 13 05 03.	60		NEBULA
ARC 2016	15 00 16.	+ 11 24		17.6	RICH CLUSTER OF GALAXIES
ZC 1500.3+1845	15 00 18.	+ 18 45	2890		CLUSTER OF GALAXIES
ZC 1500.3+2715	15 00 18.	+ 27 15	3090		CLUSTER OF GALAXIES
MCG-02-38-030	15 00 18.	- 13 05	48	14.5	GALAXY
ARC 2017	15 00 19.	+ 23 22		16.6	RICH CLUSTER OF GALAXIES
ZC 1500.4+1126	15 00 24.	+ 11 26	1680		CLUSTER OF GALAXIES
MCG+03-38-075	15 00 24.	+ 17 00	54	15.3	GALAXY
ZWG 105.097	15 00 24.	+ 17 01		15.3	GALAXY
MCG+03-38-074	15 00 24.	+ 17 26	72	15.0	GALAXY
ZWG 105.098	15 00 24.	+ 17 27		15.0	GALAXY
UGC 09672	15 00 24.	+ 20 00	96	17.	GALAXY
ZC 1500.4+2246	15 00 24.	+ 22 46	4230		CLUSTER OF GALAXIES
ZWG 134.070	15 00 24.	+ 23 31		14.6	GALAXY
UGC 09673	15 00 24.	+ 23 31	120	14.6	GALAXY Sc
MCG+04-35-025	15 00 24.	+ 25 50	24	15.7	GALAXY
MCG+11-18-018	15 00 24.	+ 65 40	39	16.	GALAXY
ZWG 318.010	15 00 24.	+ 65 42		15.7	GALAXY
ZWG 020.050	15 00 24.	- 02 24		15.4	GALAXY
IC 1085	15 00 25.	+ 17 25 41.			NONSTELLAR OBJECT
IC 4526	15 00 25.	+ 23 33 30.			NONSTELLAR OBJECT
VV 007C	15 00 27.	+ 23 31	15	17.	INTERACTING GALAXY
VV 007B	15 00 27.	+ 23 31	18	17.	INTERACTING GALAXY
VV 007A	15 00 27.	+ 23 31	108	14.	INTERACTING GALAXY
RNGC 5799	15 00 29.	- 72 14			GALAXY
ZWG 048.094	15 00 30.	+ 07 25		15.7	GALAXY
MCG+03-38-076	15 00 30.	+ 17 05 30.	60	15.2	GALAXY
ZWG 105.099	15 00 30.	+ 17 07		15.2	GALAXY
MCG+04-35-026	15 00 30.	+ 23 32	24	14.6	GALAXY
MCG+07-31-020	15 00 30.	+ 42 08	36	16.	GALAXY
7ZW 577	15 00 30.	+ 75 15			COMPACT GALAXY
ARP 042	15 00 31.	+ 23 32			PECULIAR GALAXY
MCG+04-35-027	15 00 33.	+ 23 31 30.	84	14.6	GALAXY
MCG+06-33-017	15 00 33.	+ 33 22	42	15.	GALAXY
REIZ 4431	15 00 33.	+ 42 07	48	14.8	GALAXY
ZC 1500.6+2559	15 00 36.	+ 25 59	6590		CLUSTER OF GALAXIES
RNGC 5835	15 00 40.	+ 49 04		15.5	GALAXY
ZWG 076.138	15 00 42.	+ 11 24		15.2	GALAXY
HOLM 692A	15 00 42.	+ 11 24	24	14.7	PART OF MULTIPLE GALAXY
MCG+02-38-035	15 00 42.	+ 11 24	18	15.2	GALAXY
MCG+02-38-034	15 00 42.	+ 11 24	30	15.2	GALAXY
ZC 1500.7+1623	15 00 42.	+ 16 23	1240		CLUSTER OF GALAXIES
ZWG 193.013	15 00 42.	+ 38 08		15.5	GALAXY
ZWG 248.048	15 00 42.	+ 49 04		15.7	GALAXY
UGC 09674	15 00 42.	+ 49 04	78	15.7	GALAXY Sa
MCG+08-27-057	15 00 42.	+ 49 05	60	16.	GALAXY
ZWG 318.011	15 00 42.	+ 62 42		15.6	GALAXY
HOLM 692B	15 00 44.	+ 11 24	18	15.1	PART OF MULTIPLE GALAXY
MCG-04-35-018	15 00 45.	- 25 41 30.	24	15.	GALAXY
ARC 2019	15 00 47.	+ 27 23		16.3	RICH CLUSTER OF GALAXIES
ARC 2014	15 00 47.	- 15 51		17.5	RICH CLUSTER OF GALAXIES
MCG+06-33-018	15 00 48.	+ 38 10	30	17.	GALAXY
ZWG 193.014	15 00 48.	+ 38 14		15.6	GALAXY
MCG+10-21-044	15 00 48.	+ 62 40	18	16.	GALAXY
RNGC 5824	15 00 52.	- 32 53		10.0	GLOBULAR CLUSTER
ZC 1500.9+1222	15 00 54.	+ 12 22	1140		CLUSTER OF GALAXIES
ZC 1500.9+1833	15 00 54.	+ 18 33	470		CLUSTER OF GALAXIES
ZWG 134.071	15 00 54.	+ 21 44		15.7	GALAXY
GCL 031	15 00 54.	- 32 52	222	10.08	GLOBULAR STAR CLUSTER
ZWG 048.095	15 01 00.	+ 04 53		15.0	GALAXY
MCG+01-38-027	15 01 00.	+ 04 53	24	15.0	GALAXY
ZWG 048.096	15 01 00.	+ 07 32		15.3	GALAXY
ZWG 076.139	15 01 00.	+ 10 51		15.3	GALAXY
UGC 09675	15 01 00.	+ 10 51	72	15.3	GALAXY SB
ZWG 134.072	15 01 00.	+ 21 48		15.6	GALAXY
ZWG 165.002	15 01 00.	+ 31 23		15.7	GALAXY
ZWG 164.055	15 01 00.	+ 31 23		15.7	GALAXY
MCG+07-31-022	15 01 00.	+ 43 28	24	15.	GALAXY
MCG+07-31-021	15 01 00.	+ 43 30	42	15.	GALAXY
ZC 1501.0+5536	15 01 00.	+ 55 36	1410		CLUSTER OF GALAXIES
7ZW 578	15 01 00.	+ 67 13			COMPACT GALAXY
MCG+14-07-014	15 01 00.	+ 83 44	27	15.	GALAXY
ZC 1501.0+8734	15 01 00.	+ 87 34	5110		CLUSTER OF GALAXIES
MRSL 320+00/2	15 01 00.	- 57 19	120		HII REGION
REIZ 4429	15 01 01.	+ 10 51	30	15.2	GALAXY
MCG+03-38-077	15 01 05.	+ 17 17	24	15.4	GALAXY
ZWG 020.051	15 01 06.	+ 02 25		15.7	GALAXY
ZWG 076.140	15 01 06.	+ 10 48		15.3	GALAXY
ZWG 076.141	15 01 06.	+ 11 01		15.6	GALAXY
ZWG 105.100	15 01 06.	+ 15 00		15.6	GALAXY
ZWG 105.101	15 01 06.	+ 17 18		15.4	GALAXY
ZC 1501.1+1734	15 01 06.	+ 17 34	1880		CLUSTER OF GALAXIES
ZC 1501.1+2308	15 01 06.	+ 23 08	1010		CLUSTER OF GALAXIES
ZC 1501.1+4613	15 01 06.	+ 46 13	1010		CLUSTER OF GALAXIES
REIZ 4430	15 01 07.	+ 10 48	18	15.0	GALAXY
MCG+03-38-078	15 01 07.	+ 17 16	30	15.7	GALAXY
IC 1086	15 01 08.	+ 17 16 57.			NONSTELLAR OBJECT
ZWG 105.102	15 01 12.	+ 17 17		15.7	GALAXY
ZC 1501.2+2008	15 01 12.	+ 20 08	670		CLUSTER OF GALAXIES
LB 09389	15 01 12.	+ 26 44		16.0	FAINT BLUE STAR
ZWG 274.007	15 01 12.	+ 53 43		15.5	GALAXY
ZWG 020.052	15 01 12.	- 03 06		15.2	GALAXY
MCG+00-38-019	15 01 12.	- 03 06	108	15.2	GALAXY
ZWG 165.003	15 01 18.	+ 28 00		15.7	GALAXY
ZWG 164.056	15 01 18.	+ 28 00		15.7	GALAXY
UGC 09676	15 01 18.	+ 28 00	66	15.7	GALAXY SBc
MCG+09-25-007	15 01 18.	+ 53 42	24	16.	GALAXY
MCG+09-25-006	15 01 18.	+ 53 42	15	16.	GALAXY
1ZW 100	15 01 18.	+ 53 43			COMPACT GALAXY
ARC 2021	15 01 19.	+ 23 13		17.0	RICH CLUSTER OF GALAXIES
ARC 2020	15 01 21.	+ 08 07		16.0	RICH CLUSTER OF GALAXIES
MCG-02-38-031	15 01 21.	- 10 32	72	15.	GALAXY
ZC 1501.4+1928	15 01 24.	+ 19 28	1410		CLUSTER OF GALAXIES
UGC 09677	15 01 24.	+ 22 53	84	16.0	GALAXY Sb
ZWG 135.001	15 01 24.	+ 22 54		15.7	GALAXY
ZC 1501.4+2341	15 01 24.	+ 23 41	540		CLUSTER OF GALAXIES
MCG+11-18-019	15 01 24.	+ 67 53	39	16.	GALAXY
ZC 1501.4+7822	15 01 24.	+ 78 22	610		CLUSTER OF GALAXIES
LB 09390	15 01 30.	+ 22 36		18.0	FAINT BLUE STAR
RNGC 5829	15 01 30.	+ 23 32		14.5	GALAXY
ZWG 165.004	15 01 30.	+ 28 40		15.7	GALAXY
ZWG 164.057	15 01 30.	+ 28 40		15.7	GALAXY
ZC 1501.5+3550	15 01 30.	+ 35 50	7120		CLUSTER OF GALAXIES
ZWG 248.049	15 01 30.	+ 50 02		15.6	GALAXY
ACK 320+00.1	15 01 30.	- 51 19			PLANETARY NEBULA
OCL 0937	15 01 30.	- 54 09	3300	6.5	OPEN STAR CLUSTER
VHA 168	15 01 30.	- 54 09	1800		OPEN STAR CLUSTER
KEEL 682	15 01 31.8	+ 55 47 16.		18.	NEBULA
RNGC 5822	15 01 34.	- 54 09		6.5	OPEN CLUSTER
IC 4530	15 01 35.	+ 26 16 48.			NONSTELLAR OBJECT
ZWG 020.054	15 01 36.	+ 01 24		13.1	GALAXY
UGC 09678	15 01 36.	+ 01 24	144	13.1	GALAXY E
MCG+00-38-020	15 01 36.	+ 01 24	36	13.1	GALAXY
LB 09392	15 01 36.	+ 25 22		17.3	FAINT BLUE STAR
ZWG 135.002	15 01 36.	+ 26 17		15.3	GALAXY
UGC 09679	15 01 36.	+ 26 17	60	15.3	GALAXY S
MCG+04-36-001	15 01 36.	+ 26 18	60	15.3	GALAXY
LB 09391	15 01 36.	+ 26 36		16.8	FAINT BLUE STAR
ZWG 337.027	15 01 36.	+ 68 28		15.4	GALAXY
ZWG 318.012	15 01 36.	+ 68 28		15.4	GALAXY
ZWG 020.053	15 01 36.	- 01 50		15.6	GALAXY
MCG-01-38-020	15 01 36.	- 07 00	48	13.5	GALAXY
PK321+02.1	15 01 36.	- 54 59 19.	5		PLANETARY NEBULA
ACK 320+00.3	15 01 36.	- 60 37			PLANETARY NEBULA
RNGC 5831	15 01 38.	+ 01 25		13.0	GALAXY
MCG+11-18-020	15 01 39.	+ 68 27	18	16.	GALAXY
UGC 09680	15 01 42.	+ 18 51	78	16.5	GALAXY
FATH 1.680	15 01 42.	+ 29 30	22		NEBULA
UGC 09681	15 01 42.	+ 42 19	90	16.5	GALAXY Sc
MCG+08-27-058	15 01 42.	+ 50 02	48	16.	GALAXY
MCG-02-38-032	15 01 42.	- 14 15	108	15.	GALAXY
ZWG 048.097	15 01 48.	+ 03 30		15.7	GALAXY

OBJECT NAME	RIGHT ASCEN.	DECLINATION	DIAM.	MAGN.	TYPE OF OBJECT
KARA.73B 0658	15 01 48.	+ 03 30	42	15.7	ISOLATED GALAXY S
ZWG 048.098	15 01 48.	+ 07 40		15.6	GALAXY
ZC 1501.8+1810	15 01 48.	+ 18 10	1340		CLUSTER OF GALAXIES
LB 09393	15 01 48.	+ 21 14		16.1	FAINT BLUE STAR
ZC 1501.8+3803	15 01 48.	+ 38 03	3970		CLUSTER OF GALAXIES
MCG+07-31-023	15 01 48.	+ 42 20	66	15.	GALAXY
MCG-01-38-021	15 01 48.	- 04 22	36	15.	GALAXY
IC 4523	15 01 48.	- 43 19			NONSTELLAR OBJECT
HW 1139	15 01 48.	- 43 20		14.	NEBULA
MCG-07-31-005	15 01 48.	- 43 20	90	14.	GALAXY
PATH 1.681	15 01 49.	+ 29 07	8		NEBULA
ABC 2032	15 01 49.	+ 78 01		17.9	RICH CLUSTER OF GALAXIES
LB 09394	15 01 54.	+ 22 13		17.0	FAINT BLUE STAR
ZWG 165.005	15 01 54.	+ 28 43		15.5	GALAXY
1ZW 101	15 01 54.	+ 42 53			COMPACT GALAXY
ZWG 221.022	15 01 54.	+ 42 53		15.1	GALAXY
ZWG 020.055	15 01 54.	- 00 40		15.3	GALAXY
UGC 09682	15 01 54.	- 00 40	120	15.3	GALAXY
MCG+00-38-021	15 01 54.	- 00 40	114	15.3	GALAXY
REIZ 4435	15 01 56.	+ 42 19	42	14.0	GALAXY
RNGC 5823	15 01 56.	- 55 24		8.5	OPEN CLUSTER
KEEL 683	15 01 56.0	+ 55 46 36.		18.	NEBULA
KEEL 684	15 01 57.0	+ 55 53 07.		17.	NEBULA
LB 00734	15 01 59.	+ 28 59 00.		17.4	FAINT BLUE STAR
ZWG 048.099	15 02 00.	+ 02 32		15.3	GALAXY
LB 09395	15 02 00.	+ 25 58		16.8	FAINT BLUE STAR
ZC 1502.0+2841	15 02 00.	+ 28 41	3970		CLUSTER OF GALAXIES
ZWG 221.023	15 02 00.	+ 42 18		14.8	GALAXY
UGC 09684	15 02 00.	+ 42 18	102	14.8	GALAXY SBa-b
MCG+07-31-024	15 02 00.	+ 42 20	90	14.5	GALAXY
MCG+07-31-025	15 02 00.	+ 42 55	18	15.	GALAXY
ZWG 337.028	15 02 00.	+ 70 05		14.9	GALAXY
ZWG 366.014	15 02 00.	+ 81 37		14.8	GALAXY
UGC 09683	15 02 00.	+ 81 37	66	14.8	GALAXY SBa
OCL 0936	15 02 00.	- 55 14	1380	8.7	OPEN STAR CLUSTER
VHA 169	15 02 00.	- 55 24	660		OPEN STAR CLUSTER
SHB 305	15 02 00.2	+ 10 41 18.		18.5	QUASI-STELLAR OBJECT
BC OR103	15 02 00.3	+ 10 41 12.		17.	QUASI-STELLAR OBJECT
PK318-02.2	15 02 01.	- 61 09 48.	45		PLANETARY NEBULA
ZWG 076.142	15 02 06.	+ 11 41		15.4	GALAXY
ZWG 105.103	15 02 06.	+ 17 30		14.7	GALAXY
ZWG 105.104	15 02 06.	+ 17 52		15.3	GALAXY
MCG+03-38-079	15 02 06.	+ 17 52 30.	36	15.5	GALAXY
ZWG 105.105	15 02 06.	+ 17 53		15.5	GALAXY
LB 09397	15 02 06.	+ 20 44		16.8	FAINT BLUE STAR
LB 09396	15 02 06.	+ 24 03		15.6	FAINT BLUE STAR
ZWG 165.006	15 02 06.	+ 28 42		15.3	GALAXY
MCG+05-36-001	15 02 06.	+ 28 42	18	15.5	GALAXY
ZWG 165.007	15 02 06.	+ 28 46		15.6	GALAXY
MCG+08-27-059	15 02 06.	+ 48 47	60	16.	GALAXY
MCG+03-38-080	15 02 07.	+ 17 51 30.	24	15.3	GALAXY
MCG+03-38-081	15 02 09.	+ 17 30	48	14.7	GALAXY
RNGC 5840	15 02 09.	+ 29 42			NON-EXISTENT OBJECT
PK321+02.2	15 02 09.	- 53 47 46.	50		PLANETARY NEBULA
REIZ 4432	15 02 10.	+ 13 43	30	15.7	GALAXY
ZWG 048.100	15 02 12.	+ 06 33		15.7	GALAXY
ZWG 076.143	15 02 12.	+ 10 32		15.4	GALAXY
LB 09400	15 02 12.	+ 23 01		16.5	FAINT BLUE STAR
ZWG 135.003	15 02 12.	+ 23 36		15.6	GALAXY
LB 09399	15 02 12.	+ 24 32		17.1	FAINT BLUE STAR
LB 09398	15 02 12.	+ 25 08		16.7	FAINT BLUE STAR
ZWG 165.008	15 02 12.	+ 28 35		15.7	GALAXY
ZWG 165.009	15 02 12.	+ 28 40		15.7	GALAXY
MCG+05-36-002	15 02 12.	+ 28 41	24	15.3	GALAXY
ZC 1502.2+2925	15 02 12.	+ 29 25	870		CLUSTER OF GALAXIES
ZC 1502.2+3024	15 02 12.	+ 30 24	6050		CLUSTER OF GALAXIES
ZWG 248.050	15 02 12.	+ 48 44		15.0	GALAXY
MCG+08-27-060	15 02 12.	+ 48 45	42	15.	GALAXY
BSO 23	15 02 13.5	+ 44 21 14.			BLUE STELLAR OBJECT
IC 4531	15 02 14.	+ 23 35 16.			NONSTELLAR OBJECT
MCG+03-38-083	15 02 15.	+ 17 52 30.	48	15.5	GALAXY
MCG+03-38-082	15 02 15.	+ 19 41	30	15.5	GALAXY
ABC 2022	15 02 15.	+ 28 38		15.6	RICH CLUSTER OF GALAXIES
REIZ 4436	15 02 15.	+ 48 45	12	14.5	GALAXY
REIZ 4433	15 02 17.	+ 12 49	48	14.0	GALAXY
ZWG 048.101	15 02 18.	+ 05 10		15.6	GALAXY
ZWG 048.102	15 02 18.	+ 06 32		15.6	GALAXY
ZWG 048.103	15 02 18.	+ 08 22		15.1	GALAXY
UGC 09685	15 02 18.	+ 08 22	60	15.1	GALAXY Sc
ZWG 076.144	15 02 18.	+ 12 50		14.5	GALAXY
UGC 09686	15 02 18.	+ 12 50	60	14.5	GALAXY S
MCG+02-38-036	15 02 18.	+ 12 50	60	14.5	GALAXY
MCG+03-38-084	15 02 18.	+ 16 19 30.	30	15.7	GALAXY
ZWG 105.106	15 02 18.	+ 16 20		15.7	GALAXY
IC 4532	15 02 18.	+ 23 24 22.			NONSTELLAR OBJECT
ZC 1502.3+2350	15 02 18.	+ 23 50	670		CLUSTER OF GALAXIES
ZWG 165.010	15 02 18.	+ 28 00		14.9	GALAXY
UGC 09687	15 02 18.	+ 28 00	60	14.9	GALAXY Sa
ZC 1502.3+2814	15 02 18.	+ 28 14	270		CLUSTER OF GALAXIES
MCG+05-36-003	15 02 18.	+ 28 45 30.	18	15.6	GALAXY
LB 00735	15 02 18.	+ 29 11 54.		17.1	FAINT BLUE STAR
ZC 1502.3+7701	15 02 18.	+ 77 01	1340		CLUSTER OF GALAXIES
RNGC 5837	15 02 19.	+ 12 50		14.5	GALAXY
LB 00736	15 02 21.	+ 30 26 54.		17.0	FAINT BLUE STAR
REIZ 4434	15 02 22.	+ 13 40	18	14.9	GALAXY
IC 4533	15 02 22.	+ 27 58 59.			NONSTELLAR OBJECT
PATH 2.154	15 02 22.	+ 29 50	5		NEBULA
IC 4527	15 02 22.	- 42 15			NONSTELLAR OBJECT
HW 1140	15 02 23.	- 42 16	36		NEBULA
ZC 1502.4+0516	15 02 24.	+ 05 16	940		CLUSTER OF GALAXIES
ZC 1502.4+1614	15 02 24.	+ 16 14	1080		CLUSTER OF GALAXIES
ZWG 105.107	15 02 24.	+ 16 32		15.6	GALAXY
ZWG 105.108	15 02 24.	+ 19 14		15.6	GALAXY
LB 09402	15 02 24.	+ 25 35		16.4	FAINT BLUE STAR
ZWG 135.004	15 02 24.	+ 26 05		15.5	GALAXY
MCG+04-36-002	15 02 24.	+ 26 05 30.	30	15.5	GALAXY
LB 09401	15 02 24.	+ 26 38		17.0	FAINT BLUE STAR
ZC 1502.4+4219	15 02 24.	+ 42 19	340		CLUSTER OF GALAXIES
ZC 1502.4+5155	15 02 24.	+ 51 55	1210		CLUSTER OF GALAXIES
ZC 1502.4+6420	15 02 24.	+ 64 20	1480		CLUSTER OF GALAXIES
MCG-07-31-006	15 02 24.	- 42 16	78	15.	GALAXY
KEEL 685	15 02 26.6	+ 55 40 10.		17.	NEBULA
KEEL 686	15 02 27.4	+ 55 49 15.		17.	NEBULA
ZC 1502.5+0240	15 02 30.	+ 02 40	1080		CLUSTER OF GALAXIES
ZWG 048.104	15 02 30.	+ 06 07		15.7	GALAXY
ZC 1502.5+1908	15 02 30.	+ 19 08	400		CLUSTER OF GALAXIES
LB 09403	15 02 30.	+ 20 54		17.0	FAINT BLUE STAR
ZC 1502.5+2328	15 02 30.	+ 23 28	1010		CLUSTER OF GALAXIES

OBJECT NAME	RIGHT ASCEN.	DECLINATION	DIAM.	MAGN.	TYPE OF OBJECT
ZWG 165.011	15 02 30.	+ 27 11		15.3	GALAXY
MCG+05-36-004	15 02 30.	+ 27 58 30.	60	14.9	GALAXY
MCG+06-33-019	15 02 30.	+ 36 05 30.	57	15.5	GALAXY
ZWG 193.015	15 02 30.	+ 36 09		14.9	GALAXY
ZC 1502.5+4408	15 02 30.	+ 44 08	1680		CLUSTER OF GALAXIES
MCG+09-25-008	15 02 30.	+ 50 56	42	16.	GALAXY
MCG+09-25-009	15 02 30.	+ 52 27	60	16.	GALAXY
REIZ 4440	15 02 30.	+ 53 05	18	14.4	GALAXY
ZWG 274.008	15 02 30.	+ 53 08		14.8	GALAXY
UGC 09688	15 02 30.	+ 53 08	78	14.8	GALAXY S0-a
RNGC 5842	15 02 34.	+ 21 16		15.0	GALAXY
ZWG 048.105	15 02 36.	+ 05 59		14.9	GALAXY
MCG+01-38-028	15 02 36.	+ 05 59	54	14.9	GALAXY
ZWG 048.106	15 02 36.	+ 06 04		15.3	GALAXY
ZWG 048.107	15 02 36.	+ 08 02		15.3	GALAXY
UGC 09689	15 02 36.	+ 08 02	60	15.3	GALAXY Sb-c
ZWG 105.109	15 02 36.	+ 16 30		15.7	GALAXY
ZWG 105.110	15 02 36.	+ 16 32		15.7	GALAXY
ZWG 135.005	15 02 36.	+ 21 16		15.2	GALAXY
LB 09404	15 02 36.	+ 21 48		16.8	FAINT BLUE STAR
MCG+05-36-005	15 02 36.	+ 27 09	30	15.3	GALAXY
MCG+06-33-020	15 02 36.	+ 36 10	18	15.	GALAXY
MCG+09-25-010	15 02 36.	+ 53 07	90	14.	GALAXY
MCG+11-18-021	15 02 36.	+ 65 23	24	17.	GALAXY
REIZ 4437	15 02 41.	+ 40 33	36	14.4	GALAXY
MCG+04-36-003	15 02 42.	+ 21 14	21	15.2	GALAXY
ZWG 135.006	15 02 42.	+ 23 26		15.6	GALAXY
PATH 1.682	15 02 42.	+ 29 31	22		NEBULA
UGC 09690	15 02 42.	+ 38 13	66	16.5	GALAXY Sb-c
ZWG 221.024	15 02 42.	+ 40 34		15.1	GALAXY
UGC 09691	15 02 42.	+ 40 34	72	15.1	GALAXY SBb
MCG+07-31-026	15 02 42.	+ 40 35	72	14.5	GALAXY
MCG+08-27-061	15 02 42.	+ 49 35	30	16.	GALAXY
MCG+08-27-062	15 02 42.	+ 49 37	42	15.	GALAXY
ZWG 274.009	15 02 42.	+ 52 42		15.4	GALAXY
MRSL 320+00/1	15 02 42.	- 57 36	180		HII REGION
LB 00737	15 02 43.	+ 28 42 18.		17.2	FAINT BLUE STAR
REIZ 4441	15 02 43.	+ 49 36	15	14.4	GALAXY
REIZ 4442	15 02 44.	+ 54 54	12	14.7	GALAXY
VV 204C	15 02 45.	+ 26 12	6	19.	INTERACTING GALAXY
VV 204B	15 02 45.	+ 26 12	12	16.5	INTERACTING GALAXY
VV 204A	15 02 45.	+ 26 12	12	16.5	INTERACTING GALAXY
VV 204	15 02 45.	+ 26 12	60		INTERACTING GALAXY
ZWG 020.057	15 02 48.	+ 02 16		12.1	GALAXY
WEED 9	15 02 48.	+ 02 16		19.0	VERY BLUE STELLAR OBJECT
UGC 09692	15 02 48.	+ 02 16	228	12.1	GALAXY S0
MCG+00-38-022	15 02 48.	+ 02 16	180	12.1	GALAXY
ZWG 048.108	15 02 48.	+ 05 15		15.7	GALAXY
ZWG 076.145	15 02 48.	+ 08 59		15.3	GALAXY
LB 09406	15 02 48.	+ 20 55		16.7	FAINT BLUE STAR
LB 09405	15 02 48.	+ 26 09		16.8	FAINT BLUE STAR
MCG+04-36-004	15 02 48.	+ 26 12	24		GALAXY
ZC 1502.8+2613	15 02 48.	+ 26 13	2420		CLUSTER OF GALAXIES
ZWG 165.012	15 02 48.	+ 31 19		15.6	GALAXY
ZWG 249.001	15 02 48.	+ 49 34		15.2	GALAXY
ZWG 248.051	15 02 48.	+ 49 34		15.2	GALAXY
ZWG 249.002	15 02 48.	+ 49 36		14.8	GALAXY
ZWG 248.052	15 02 48.	+ 49 36		14.8	GALAXY
ZWG 020.056	15 02 48.	- 01 51		15.3	GALAXY
MCG+04-36-005	15 02 51.	+ 26 12	18		GALAXY
ZWG 020.058	15 02 54.	+ 01 48		13.9	GALAXY
UGC 09693	15 02 54.	+ 01 48	90	13.9	GALAXY S0
MCG+00-38-023	15 02 54.	+ 01 48	18	13.9	GALAXY
ZWG 048.109	15 02 54.	+ 04 41		15.3	GALAXY
UGC 09694	15 02 54.	+ 04 41	66	15.3	GALAXY Sb
ZC 1502.9+1316	15 02 54.	+ 13 16	810		CLUSTER OF GALAXIES
LB 09407	15 02 54.	+ 21 13		16.9	FAINT BLUE STAR
MCG+04-36-006	15 02 54.	+ 26 12	21		GALAXY
ZWG 165.013	15 02 54.	+ 31 52		15.7	GALAXY
RNGC 5839	15 02 56.	+ 01 50		14.0	GALAXY
RNGC 5838	15 02 56.	+ 02 18		12.0	GALAXY
PK327+13.1	15 02 56.	- 42 48 41.	5		PLANETARY NEBULA
SHB 306	15 02 58.	+ 60 12 36.		18.	QUASI-STELLAR OBJECT
BC 3C311	15 02 58.	+ 60 13		18.	QUASI-STELLAR OBJECT
IC 4529	15 02 59.	- 43 02			NONSTELLAR OBJECT
ZWG 048.110	15 03 00.	+ 03 17		15.3	GALAXY
ZWG 048.111	15 03 00.	+ 06 24		15.5	GALAXY
ZWG 105.111	15 03 00.	+ 14 37		15.1	GALAXY
UGC 09695	15 03 00.	+ 14 37	60	15.1	GALAXY SBb
ZWG 105.112	15 03 00.	+ 16 35		15.5	GALAXY
ZWG 105.113	15 03 00.	+ 16 36		15.7	GALAXY
LB 09408	15 03 00.	+ 23 10		11.5	FAINT BLUE STAR
MCG+14-07-015	15 03 00.	+ 81 35	18	15.	GALAXY
MCG-01-38-022	15 03 00.	- 06 37	30	15.	GALAXY
MCG-02-38-033	15 03 00.	- 14 46	16	15.5	GALAXY
HW 1141	15 03 00.	- 43 03		14.	NEBULA
RNGC 5841	15 03 02.	+ 02 11			NON-EXISTENT OBJECT
ZWG 076.146	15 03 06.	+ 08 42		14.5	GALAXY
UGC 09696	15 03 06.	+ 08 42	54	14.5	GALAXY Sa
MCG+02-38-037	15 03 06.	+ 08 42	42	14.5	GALAXY
MCG+03-38-085	15 03 06.	+ 14 36	48	15.1	GALAXY
LB 09409	15 03 06.	+ 23 31		17.1	FAINT BLUE STAR
ZWG 165.014	15 03 06.	+ 28 32		15.1	GALAXY
MCG+09-25-011	15 03 06.	+ 51 46	24	16.	GALAXY
ZWG 274.010	15 03 06.	+ 54 01		15.6	GALAXY
ZWG 297.001	15 03 06.	+ 59 19		15.7	GALAXY
ZWG 296.022	15 03 06.	+ 59 19		15.7	GALAXY
MCG-07-31-007	15 03 06.	- 43 03	60	15.	GALAXY
PATH 1.683	15 03 07.	+ 28 58	11		NEBULA
REIZ 4443	15 03 11.	+ 47 52	18	15.2	GALAXY
ZWG 048.112	15 03 12.	+ 05 18		15.3	GALAXY
MCG+01-38-029	15 03 12.	+ 05 18	12	15.3	GALAXY
ZWG 048.113	15 03 12.	+ 05 58		15.5	GALAXY
LB 09411	15 03 12.	+ 22 51		16.4	FAINT BLUE STAR
ZWG 135.007	15 03 12.	+ 23 53		15.7	GALAXY
UGC 09698	15 03 12.	+ 23 53	66	15.7	GALAXY PECULR
LB 09410	15 03 12.	+ 25 26		17.5	FAINT BLUE STAR
MCG+09-25-012	15 03 12.	+ 54 01	42	15.	GALAXY
ZWG 274.011	15 03 12.	+ 54 54		15.5	GALAXY
MCG+09-25-013	15 03 12.	+ 54 54	60	15.	GALAXY
UGC 09697	15 03 12.	- 00 32	66	16.0	GALAXY
LB 00738	15 03 16.	+ 30 02 18.		17.8	FAINT BLUE STAR
ABC 2023	15 03 17.	+ 03 04		17.2	RICH CLUSTER OF GALAXIES
REIZ 4438	15 03 17.	+ 12 55	12	15.2	GALAXY
REIZ 4439	15 03 17.	+ 12 56	12	15.3	GALAXY
LB 00739	15 03 17.	+ 28 21 06.		17.6	FAINT BLUE STAR
ZWG 048.114	15 03 18.	+ 06 01		15.3	GALAXY
MCG+04-36-007	15 03 18.	+ 23 52	60	15.7	GALAXY

OBJECT NAME	RIGHT ASCEN.	DECLINATION	DIAM.	MAGN.	TYPE OF OBJECT
MCG+05-36-006	15 03 18.	+ 28 30	24	15.1	GALAXY
ZWG 249.003	15 03 18.	+ 48 51		15.7	GALAXY
ZWG 248.053	15 03 18.	+ 48 51		15.7	GALAXY
MCG+08-27-063	15 03 18.	+ 48 51	36	16.	GALAXY
MCG-07-31-008	15 03 18.	- 41 07 30.	36	14.	GALAXY
RNGC 5834	15 03 22.	- 32 55			NON-EXISTENT OBJECT
ZWG 076.147	15 03 24.	+ 09 43		15.4	GALAXY
UGC 09699	15 03 24.	+ 09 43	60	15.4	GALAXY Sc
ZWG 105.114	15 03 24.	+ 14 45		15.6	GALAXY
LB 09413	15 03 24.	+ 23 56		16.7	FAINT BLUE STAR
MCG+04-36-008	15 03 24.	+ 25 17	27		GALAXY
ZWG 135.008	15 03 24.	+ 25 38		15.7	GALAXY
LB 09412	15 03 24.	+ 26 15		17.0	FAINT BLUE STAR
ZWG 165.015	15 03 24.	+ 31 06		15.6	GALAXY
ACK 320+00.2	15 03 24.	- 57 37			PLANETARY NEBULA
FATH 1.684	15 03 28.	+ 29 28	14		NEBULA
LB 00740	15 03 29.	- 11 24 06.		16.8	FAINT BLUE STAR
ZWG 020.059	15 03 30.	+ 01 48		13.8	GALAXY
UGC 09700	15 03 30.	+ 01 48	42	13.8	GALAXY E
MCG+00-38-024	15 03 30.	+ 01 48	24	13.8	GALAXY
ZWG 048.115	15 03 30.	+ 03 54		15.2	GALAXY
ZWG 165.016	15 03 30.	+ 31 21		15.3	GALAXY
UGC 09701	15 03 30.	+ 31 21	60	15.3	GALAXY Sc
MCG+07-31-028	15 03 30.	+ 40 09	36	15.	GALAXY
MCG+07-31-027	15 03 30.	+ 40 10	18	16.	GALAXY
RNGC 5845	15 03 32.	+ 01 50		14.0	GALAXY
HOLM 693A	15 03 35.	+ 12 56	24	15.1	PART OF MULTIPLE GALAXY
ZWG 048.116	15 03 36.	+ 03 54		15.3	GALAXY
ZWG 076.148	15 03 36.	+ 08 59		15.4	GALAXY
ZWG 076.149	15 03 36.	+ 12 55		15.2	GALAXY
HOLM 693B	15 03 36.	+ 12 55	24	15.3	PART OF MULTIPLE GALAXY
KARA.72 452B	15 03 36.	+ 12 55	30	15.2	PART OF DOUBLE GALAXY
MCG+02-38-038	15 03 36.	+ 12 55	21	15.2	GALAXY
ZWG 076.150	15 03 36.	+ 12 56		15.2	GALAXY
KARA.72 452A	15 03 36.	+ 12 56	30	15.2	PART OF DOUBLE GALAXY
MCG+02-38-039	15 03 36.	+ 12 56	21	15.2	GALAXY
ZC 1503.6+2321	15 03 36.	+ 23 21	610		CLUSTER OF GALAXIES
ZWG 221.025	15 03 36.	+ 40 08		15.5	GALAXY
ZWG 221.026	15 03 36.	+ 43 18		15.7	GALAXY
HOLM 696B	15 03 36.	+ 46 46	12	15.1	PART OF MULTIPLE GALAXY
ZWG 274.012	15 03 36.	+ 51 22		15.7	GALAXY
UGC 09702	15 03 36.	+ 51 22	72	15.7	GALAXY Sb-c
ZC 1503.6+6030	15 03 36.	+ 60 30	1680		CLUSTER OF GALAXIES
REIZ 4444	15 03 37.	+ 46 47	12	15.6	GALAXY
MCG+05-36-007	15 03 39.	+ 31 20 30.	54	15.3	GALAXY
SC 1500-1344.5	15 03 39.	- 13 56 11.	12		NEBULA
ZWG 048.117	15 03 42.	+ 03 52		15.5	GALAXY
ZWG 135.009	15 03 42.	+ 21 02		15.4	GALAXY
MCG+07-31-029	15 03 42.	+ 43 18	36	15.	GALAXY
ZWG 249.004	15 03 42.	+ 46 46		15.0	GALAXY
ZWG 248.054	15 03 42.	+ 46 46		15.0	GALAXY
UGC 09703	15 03 42.	+ 46 46	72	15.0	GALAXY S
MCG+08-27-064	15 03 42.	+ 46 46	72	13.8	GALAXY
MCG+10-21-045	15 03 42.	+ 57 29	45	16.	GALAXY
MCG-02-38-034	15 03 42.	- 09 43	36	15.	GALAXY
REIZ 4447	15 03 43.	+ 46 45	42	14.4	GALAXY
HOLM 696A	15 03 43.	+ 46 45	48	13.8	PART OF MULTIPLE GALAXY
KEEL 687	15 03 47.0	+ 19 29 16.			NEBULA
ZWG 048.118	15 03 48.	+ 03 20		15.6	GALAXY
ZWG 048.119	15 03 48.	+ 06 01		15.4	GALAXY
ZC 1503.8+0853	15 03 48.	+ 08 53	7730		CLUSTER OF GALAXIES
ZWG 077.001	15 03 48.	+ 10 32		15.5	GALAXY
ZWG 076.151	15 03 48.	+ 10 32		15.5	GALAXY
MCG+04-36-009	15 03 48.	+ 21 01	36		GALAXY
LB 09415	15 03 48.	+ 23 20		16.3	FAINT BLUE STAR
LB 09414	15 03 48.	+ 25 55		16.5	FAINT BLUE STAR
ZWG 165.017	15 03 48.	+ 31 53		15.7	GALAXY
ZC 1503.8+4716	15 03 48.	+ 47 16	1010		CLUSTER OF GALAXIES
ZWG 249.005	15 03 48.	+ 49 27		15.3	GALAXY
ZWG 248.055	15 03 48.	+ 49 27		15.3	GALAXY
MCG+09-25-014	15 03 48.	+ 51 20	90	15.	GALAXY
UGC 09704	15 03 48.	+ 57 30	60	16.5	GALAXY Sc
ZWG 020.060	15 03 48.	- 01 34		15.7	GALAXY
KEEL 688	15 03 48.5	+ 19 59 35.			NEBULA
ARC 2024	15 03 49.	+ 47 20		18.0	RICH CLUSTER OF GALAXIES
RNGC 5866A	15 03 49.	+ 56 01			GALAXY
RNGC 5847	15 03 52.	+ 06 35		15.0	GALAXY
KEEL 689	15 03 53.5	+ 19 51 21.			NEBULA
KEEL 690	15 03 53.8	+ 56 01 14.		16.	NEBULA
ZWG 048.120	15 03 54.	+ 06 35		15.1	GALAXY
MCG+01-38-030	15 03 54.	+ 06 35	36	15.1	GALAXY
MCG+04-36-010	15 03 54.	+ 24 34	30	15.6	GALAXY
ZWG 135.010	15 03 54.	+ 24 35		15.6	GALAXY
ZWG 135.011	15 03 54.	+ 25 58		15.0	GALAXY
UGC 09705	15 03 54.	+ 25 58	60	15.0	GALAXY S0-a
FATH 1.685	15 03 54.	+ 29 27	8		NEBULA
MCG+07-31-030	15 03 54.	+ 39 42 30.	90	14.	GALAXY
IC 1090	15 03 54.	+ 42 52			NONSTELLAR OBJECT
MCG+07-31-031	15 03 54.	+ 44 36	54	16.	GALAXY
ZWG 318.013	15 03 54.	+ 62 55		15.5	GALAXY
HOLM 694B	15 03 56.	+ 01 46	12	14.3	PART OF MULTIPLE GALAXY
RNGC 5846A	15 03 56.	+ 01 47			GALAXY
HOLM 694A	15 03 56.	+ 01 47	66	12.1	PART OF MULTIPLE GALAXY
RNGC 5846	15 03 56.	+ 01 48		11.5	GALAXY
RNGC 5853	15 03 57.	+ 39 43		15.0	GALAXY
ZWG 020.061	15 04 00.	+ 01 46		11.9	GALAXY
UGC 09706	15 04 00.	+ 01 46	270	11.9	GALAXY E
MCG+00-38-025	15 04 00.	+ 01 46	54	11.9	GALAXY
MCG+00-38-026	15 04 00.	+ 01 47	15	14.1	GALAXY
ZWG 049.001	15 04 00.	+ 05 25		15.2	GALAXY
ZWG 077.002	15 04 00.	+ 09 14		15.5	GALAXY
ZWG 076.152	15 04 00.	+ 09 14		15.5	GALAXY
ZC 1504.0+1915	15 04 00.	+ 19 15	2760		CLUSTER OF GALAXIES
ZWG 135.012	15 04 00.	+ 25 48		15.7	GALAXY
MCG+04-36-011	15 04 00.	+ 25 58	24	15.0	GALAXY
REIZ 4449	15 04 00.	+ 39 42	42	14.2	GALAXY
ZWG 221.027	15 04 00.	+ 39 43		14.8	GALAXY
UGC 09707	15 04 00.	+ 39 43	90	14.8	GALAXY SB
MCG+07-31-032	15 04 00.	+ 42 26	48	16.	GALAXY
MCG+11-18-022	15 04 00.	+ 67 11	18	17.	GALAXY
HMS 1.25	15 04 009	+ 01 47			E2 GALAXY
REIZ 4448	15 04 01.	+ 38 27	30	14.7	GALAXY
FATH 1.686	15 04 04.	+ 30 06	11		NEBULA
BC PKS1504-164	15 04 04.0	- 16 26 32.		18.	QUASI-STELLAR OBJECT
REIZ 4452	15 04 05.	+ 40 25	27	15.0	GALAXY
ZWG 021.001	15 04 06.	+ 02 12		14.8	GALAXY
MCG+00-39-001	15 04 06.	+ 02 12	48	14.8	GALAXY
ZWG 077.003	15 04 06.	+ 09 38		15.2	GALAXY
ZWG 076.153	15 04 06.	+ 09 38		15.2	GALAXY
UGC 09708	15 04 06.	+ 09 38	66	15.2	GALAXY Sb
KARA.72 453A	15 04 06.	+ 09 38	48	15.2	PART OF DOUBLE GALAXY
MCG+02-38-040	15 04 06.	+ 09 38	72	15.2	GALAXY
LB 09418	15 04 06.	+ 24 41		15.8	FAINT BLUE STAR
LB 09417	15 04 06.	+ 24 46		16.4	FAINT BLUE STAR
LB 09416	15 04 06.	+ 25 58		16.5	FAINT BLUE STAR
FATH 1.687	15 04 06.	+ 29 36	14		NEBULA
UGC 09709	15 04 06.	+ 42 25	72	17.	GALAXY DWRF SP
MCG+10-21-046	15 04 06.	+ 58 15	39	15.	GALAXY
MCG-02-38-035	15 04 06.	- 14 24	48	15.	GALAXY
RNGC 5849	15 04 07.	- 14 24		15.0	GALAXY
RNGC 5848	15 04 08.	+ 02 12		15.0	GALAXY
IC 1087	15 04 10.	+ 03 57 07.			NONSTELLAR OBJECT
ZWG 021.002	15 04 12.	+ 01 11		15.3	GALAXY
ZWG 049.002	15 04 12.	+ 03 58		15.1	GALAXY
UGC 09710	15 04 12.	+ 03 58	60	15.1	GALAXY Sb
MCG+01-38-032	15 04 12.	+ 03 58	72	15.1	GALAXY
MCG+01-38-031	15 04 12.	+ 03 58	36	15.1	GALAXY
2ZW 072	15 04 12.	+ 04 16			COMPACT GALAXY
ZWG 049.003	15 04 12.	+ 05 31		15.5	GALAXY
ZWG 077.004	15 04 12.	+ 09 37		15.3	GALAXY
ZWG 076.154	15 04 12.	+ 09 37		15.3	GALAXY
UGC 09711	15 04 12.	+ 09 37	78	15.3	GALAXY Sa
KARA.72 453B	15 04 12.	+ 09 37	96	15.3	PART OF DOUBLE GALAXY
MCG+02-38-041	15 04 12.	+ 09 37	72	15.3	GALAXY
ZC 1504.2+1345	15 04 12.	+ 13 45	1880		CLUSTER OF GALAXIES
LB 09421	15 04 12.	+ 22 12		18.2	FAINT BLUE STAR
LB 09420	15 04 12.	+ 22 36		15.5	FAINT BLUE STAR
LB 09419	15 04 12.	+ 26 02		17.1	FAINT BLUE STAR
TON-N 0787	15 04 12.	+ 29 28		15.0	BLUE STAR
ZWG 274.013	15 04 12.	+ 51 28		15.4	GALAXY
IC 1088	15 04 14.	+ 03 58 01.			NONSTELLAR OBJECT
ZWG 021.003	15 04 18.	+ 00 41		15.6	GALAXY
ZWG 135.013	15 04 18.	+ 21 54		15.1	GALAXY
MCG+04-36-012	15 04 18.	+ 21 54	36	15.1	GALAXY
ZC 1504.3+4043	15 04 18.	+ 40 43	870		CLUSTER OF GALAXIES
ZWG 297.002	15 04 18.	+ 58 16		15.6	GALAXY
ZWG 296.023	15 04 18.	+ 58 16		15.6	GALAXY
KARA.73B 0659	15 04 18.	+ 58 16	48	15.6	ISOLATED GALAXY S
7ZW 579	15 04 18.	+ 60 51			COMPACT GALAXY
SC 1501-1315.8	15 04 21.	- 13 27 27.	42		NEBULA
KEEL 692	15 04 21.4	+ 56 09 07.		17.	NEBULA
FATH 1.688	15 04 22.	+ 29 55	22		NEBULA
REIZ 4445	15 04 23.	+ 12 49	12	15.7	GALAXY
HOLM 695B	15 04 23.	+ 12 50	18	15.6	PART OF MULTIPLE GALAXY
REIZ 4446	15 04 23.	+ 12 51	54	14.2	GALAXY
HOLM 695A	15 04 23.	+ 12 52	54	14.1	PART OF MULTIPLE GALAXY
ZWG 021.005	15 04 24.	+ 00 22		15.4	GALAXY
ZWG 077.005	15 04 24.	+ 12 46		15.2	GALAXY
UGC 09712	15 04 24.	+ 12 46	90	15.2	GALAXY Sc
ZWG 077.006	15 04 24.	+ 12 53		14.9	GALAXY
MCG+02-38-042	15 04 24.	+ 12 53	48	14.9	GALAXY
ZWG 077.007	15 04 24.	+ 13 03		15.3	GALAXY
MCG+02-38-043	15 04 24.	+ 13 03	36	15.3	GALAXY
ZWG 135.014	15 04 24.	+ 23 50		14.7	GALAXY
UGC 09713	15 04 24.	+ 23 50	90	14.7	GALAXY SB0
LB 09425	15 04 24.	+ 24 27		17.2	FAINT BLUE STAR
LB 09424	15 04 24.	+ 24 33		15.6	FAINT BLUE STAR
LB 09423	15 04 24.	+ 24 42		16.2	FAINT BLUE STAR
LB 09422	15 04 24.	+ 25 56		16.6	FAINT BLUE STAR
7ZW 580	15 04 24.	+ 62 08			COMPACT GALAXY
ZWG 021.004	15 04 24.	- 00 10		15.6	GALAXY
MCG-02-38-036	15 04 24.	- 09 47	36	15.	GALAXY
HOLM 697C	15 04 27.	+ 13 02	30	15.4	PART OF MULTIPLE GALAXY
ZWG 077.008	15 04 30.	+ 13 04		14.9	GALAXY
UGC 09714	15 04 30.	+ 13 04	66	14.9	GALAXY
MCG+02-38-044	15 04 30.	+ 13 04	60	14.9	GALAXY
ZWG 135.015	15 04 30.	+ 20 42		15.5	GALAXY
MCG+04-36-013	15 04 30.	+ 23 49 30.	72	14.7	GALAXY
LB 09427	15 04 30.	+ 24 14		16.7	FAINT BLUE STAR
LB 09426	15 04 30.	+ 26 45		18.0	FAINT BLUE STAR
ZWG 165.018	15 04 30.	+ 31 56		15.7	GALAXY
RNGC 5851	15 04 31.	+ 13 04		15.0	GALAXY
FATH 1.689	15 04 32.	+ 29 44	16		NEBULA
REIZ 4451	15 04 34.	+ 13 02	24	14.2	GALAXY
REIZ 4450	15 04 34.	+ 13 02	60	14.2	GALAXY
HOLM 697A	15 04 34.	+ 13 02	66	14.3	PART OF MULTIPLE GALAXY
BSO 24	15 04 35.	+ 42 04 37.			BLUE STELLAR OBJECT
ZWG 077.009	15 04 36.	+ 09 16		15.3	GALAXY
ZWG 077.010	15 04 36.	+ 13 03		14.7	GALAXY
MCG+02-38-045	15 04 36.	+ 13 03	60	14.7	GALAXY
ZC 1504.6+1639	15 04 36.	+ 16 39	4570		CLUSTER OF GALAXIES
ZWG 106.001	15 04 36.	+ 18 51		15.3	GALAXY
ZWG 105.115	15 04 36.	+ 18 51		15.3	GALAXY
LB 09428	15 04 36.	+ 26 44		16.4	FAINT BLUE STAR
ZWG 165.019	15 04 36.	+ 26 45		15.7	GALAXY
ZWG 274.014	15 04 36.	+ 53 37		15.5	GALAXY
ZWG 274.015	15 04 36.	+ 55 46		15.7	GALAXY
HOLM 697B	15 04 37.	+ 13 01	60	14.3	PART OF MULTIPLE GALAXY
RNGC 5852	15 04 37.	+ 13 03		14.5	GALAXY
RNGC 5862	15 04 37.	+ 55 46		15.5	GALAXY
ARC 2047	15 04 37.	+ 78 19		17.9	RICH CLUSTER OF GALAXIES
RNGC 5850	15 04 38.	+ 01 44		12.5	GALAXY
MCG+04-36-014	15 04 39.	+ 20 40	42	15.5	GALAXY
REIZ 4453	15 04 40.	+ 13 02	12	15.5	GALAXY
KEEL 693	15 04 40.9	+ 55 45 58.		13.	NEBULA
REIZ 4459	15 04 41.	+ 55 47	12	14.7	GALAXY
ZWG 021.006	15 04 42.	+ 01 44		13.6	GALAXY
UGC 09715	15 04 42.	+ 01 44	300	13.6	GALAXY SBb
MCG+00-39-002	15 04 42.	+ 01 44	240	13.6	GALAXY
ZWG 049.004	15 04 42.	+ 07 05		15.7	GALAXY
ZWG 077.011	15 04 42.	+ 09 51		15.2	GALAXY
ZC 1504.7+3104	15 04 42.	+ 31 04	740		CLUSTER OF GALAXIES
UGC 09716	15 04 42.	+ 37 29	60	16.0	GALAXY Sb-c
MCG+06-33-021	15 04 42.	+ 37 29	48	15.5	GALAXY
1ZW 102	15 04 42.	+ 42 50			COMPACT GALAXY
ZWG 221.028	15 04 42.	+ 42 50		14.2	GALAXY
MRK 480	15 04 42.	+ 42 50	20	14.5	GALAXY WITH UV CONTINUUM
RNGC 5860	15 04 42.	+ 42 50		14.0	GALAXY
UGC 09717	15 04 42.	+ 42 50	60	14.2	GALAXY DBL SYS
KARA.72 454B	15 04 42.	+ 42 50	30		PART OF DOUBLE GALAXY
KARA.72 454A	15 04 42.	+ 42 50	36	14.2	PART OF DOUBLE GALAXY
MCG+08-27-065	15 04 42.	+ 46 32 30.	15	17.	GALAXY
MCG+09-25-015	15 04 42.	+ 53 36	48	15.	GALAXY
SMI 08	15 04 42.	- 61 33	180		FAINT NEBULOSITY
ARC 2025	15 04 43.	+ 34 40		17.2	RICH CLUSTER OF GALAXIES
MCG+08-27-066	15 04 45.	+ 46 49	36	16.	GALAXY

OBJECT NAME	RIGHT ASCEN.	DECLINATION	DIAM.	MAGN.	TYPE OF OBJECT
REIZ 4457	15 04 46.	+ 40 46	30	14.6	GALAXY
ZWG 049.005	15 04 48.	+ 07 04		15.5	GALAXY
ZWG 077.012	15 04 48.	+ 10 48		15.3	GALAXY
ZWG 135.016	15 04 48.	+ 21 25		15.2	GALAXY
UGC 09718	15 04 48.	+ 21 25	84	15.2	GALAXY Sb
LB 09432	15 04 48.	+ 21 34		17.1	FAINT BLUE STAR
LB 09431	15 04 48.	+ 23 10		17.7	FAINT BLUE STAR
LB 09430	15 04 48.	+ 23 58		16.0	FAINT BLUE STAR
LB 09429	15 04 48.	+ 25 18		17.0	FAINT BLUE STAR
ZWG 135.017	15 04 48.	+ 25 57		15.3	GALAXY
TON-N 0218	15 04 48.	+ 27 03		14.9	BLUE STAR
ZC 1504.8+3239	15 04 48.	+ 32 39	940		CLUSTER OF GALAXIES
ZWG 221.029	15 04 48.	+ 40 46		15.7	GALAXY
UGC 09719	15 04 48.	+ 40 46	60	15.7	GALAXY COMPACT
MCG+07-31-033	15 04 48.	+ 42 50	36	14.5	GALAXY
REIZ 4458	15 04 48.	+ 46 48	9	14.9	GALAXY
ZWG 249.006	15 04 48.	+ 46 50		15.6	GALAXY
ZWG 248.056	15 04 48.	+ 46 50		15.6	GALAXY
RNGC 5843	15 04 49.	- 36 08			GALAXY
MCG+04-36-015	15 04 51.	+ 21 23	66	15.2	GALAXY
MCG+04-36-016	15 04 51.	+ 25 56	48	15.3	GALAXY
KEEL 695	15 04 53.5	+ 55 49 13.		16.	NEBULA
ZWG 021.007	15 04 54.	+ 00 50		15.5	GALAXY
ZWG 049.006	15 04 54.	+ 03 23		15.6	GALAXY
ZWG 049.007	15 04 54.	+ 07 19		15.4	GALAXY
ZWG 077.013	15 04 54.	+ 10 08		15.5	GALAXY
8ZW 1504+10.5	15 04 54.	+ 10 30		15.7	COMPACT GALAXY
REIZ 4454	15 04 54.	+ 10 47	24	14.9	GALAXY
ZWG 135.018	15 04 54.	+ 20 40		15.3	GALAXY
ZC 1504.9+3442	15 04 54.	+ 34 42	1010		CLUSTER OF GALAXIES
ZC 1504.9+7223	15 04 54.	+ 72 23	1610		CLUSTER OF GALAXIES
ZC 1504.9+7820	15 04 54.	+ 78 20	340		CLUSTER OF GALAXIES
IC 1089	15 04 57.	+ 07 17 36.			NONSTELLAR OBJECT
REIZ 4455	15 04 59.	+ 12 46	42	15.0	GALAXY
ZWG 077.014	15 05 00.	+ 12 46		15.3	GALAXY
UGC 09721	15 05 00.	+ 12 46	60	15.3	GALAXY
ZC 1505.0+1506	15 05 00.	+ 15 06	870		CLUSTER OF GALAXIES
ZWG 106.002	15 05 00.	+ 16 26		15.6	GALAXY
ZWG 106.003	15 05 00.	+ 16 34		15.7	GALAXY
MCG+03-39-001	15 05 00.	+ 16 34	36	15.7	GALAXY
MCG+04-36-017	15 05 00.	+ 20 38	54	15.3	GALAXY
LB 09433	15 05 00.	+ 24 08		16.9	FAINT BLUE STAR
LB 00741	15 05 00.	+ 28 49 36.		17.7	FAINT BLUE STAR
UGC 09722	15 05 00.	+ 52 06	66	16.0	GALAXY Sc
MCG+11-18-023	15 05 00.	+ 66 58	36	17.	GALAXY
MCG+11-18-024	15 05 00.	+ 67 49 30.	27	18.	GALAXY
7ZW 581	15 05 00.	+ 74 10			COMPACT GALAXY
7ZW 582	15 05 00.	+ 82 06			COMPACT GALAXY
UGC 09720	15 05 00.	+ 85 12	72	16.5	GALAXY
ZC 1505.0+8601	15 05 00.	+ 86 01	2080		CLUSTER OF GALAXIES
RNGC 5867	15 05 01.	+ 55 56			NON-EXISTENT OBJECT
PATH 1.690	15 05 02.	+ 29 40	5		NEBULA
RNGC 5856	15 05 03.	+ 18 38			NON-EXISTENT OBJECT
IC 1098	15 05 04.	+ 55 48			SINGLE STAR
KEEL 691	15 05 05.6	+ 19 38 20.			NEBULA
ZWG 021.008	15 05 06.	+ 00 04		15.7	GALAXY
ZWG 021.009	15 05 06.	+ 02 13		15.7	GALAXY
MCG+00-39-003	15 05 06.	+ 02 13	18	15.7	GALAXY
ZWG 077.015	15 05 06.	+ 09 34		15.1	GALAXY
LB 09434	15 05 06.	+ 26 45		16.8	FAINT BLUE STAR
PATH 1.691	15 05 06.	+ 29 40	5		NEBULA
MCG+09-25-016	15 05 06.	+ 55 40	30	14.	GALAXY
ZWG 274.016	15 05 06.	+ 55 57		11.1	GALAXY
UGC 09723	15 05 06.	+ 55 57	390	11.1	GALAXY SO
MCG+09-25-017	15 05 06.	+ 55 57	168	10.9	GALAXY
RNGC 5866	15 05 07.	+ 55 57		11.5	GALAXY
MCG+09-25-018	15 05 09.	+ 52 05	60	14.	GALAXY
RNGC 5857	15 05 10.	+ 19 47		13.5	GALAXY
SN 1955M	15 05 10.	+ 19 47		14.5	SUPERNOVA
REIZ 4462	15 05 10.	+ 55 57	120	11.4	GALAXY
KEEL 698	15 05 10.7	+ 55 39 30.		17.	NEBULA
REIN 2.228A	15 05 10.89	+ 19 47 22.6			NEBULA
IC 1092	15 05 11.	+ 09 33 13.			NONSTELLAR OBJECT
APC 2037	15 05 11.	+ 72 24		17.5	RICH CLUSTER OF GALAXIES
REIN 2.228B	15 05 11.10	+ 19 47 18.3			NEBULA
ZWG 049.008	15 05 12.	+ 07 05		15.6	GALAXY
MCG+03-39-002	15 05 12.	+ 14 45	66	14.9	GALAXY
ZWG 106.004	15 05 12.	+ 16 23		15.6	GALAXY
MCG+03-39-003	15 05 12.	+ 16 23 30.	36	15.6	GALAXY
ZWG 106.005	15 05 12.	+ 19 47		13.6	GALAXY
UGC 09724	15 05 12.	+ 19 47	78	13.6	GALAXY Sa
KARA.72 455A	15 05 12.	+ 19 47	72	13.6	PART OF DOUBLE GALAXY
MCG+03-39-004	15 05 12.	+ 19 48	72	13.6	GALAXY
LB 09436	15 05 12.	+ 21 54		17.6	FAINT BLUE STAR
LB 09435	15 05 12.	+ 22 16		16.2	FAINT BLUE STAR
ZWG 274.017	15 05 12.	+ 55 40		15.3	GALAXY
UGC 09725	15 05 12.	+ 55 40	72	15.3	GALAXY SO?
MCG+13-11-006	15 05 12.	+ 77 50	66	15.	GALAXY
ZC 1505.2-0142	15 05 12.	- 01 42	6180		CLUSTER OF GALAXIES
SN 1950H	15 05 13.	+ 19 47		18.1	SUPERNOVA
RNGC 5870	15 05 13.	+ 55 40		15.5	GALAXY
PATH 1.692	15 05 14.	+ 29 39	16		NEBULA
IC 1100	15 05 14.	+ 63 10 57.			NONSTELLAR OBJECT
REIZ 4456	15 05 15.	+ 14 42	30	15.1	GALAXY
IC 1093	15 05 16.	+ 14 42 44.			NONSTELLAR OBJECT
RNGC 5859	15 05 16.	+ 19 46		13.0	GALAXY
SC 1502-1724.7	15 05 16.	- 17 36 18.	18		NEBULA
KEEL 699	15 05 16.7	+ 55 48 06.		17.	NEBULA
ZWG 021.011	15 05 18.	+ 01 29		15.6	GALAXY
ZWG 049.009	15 05 18.	+ 02 45		13.1	GALAXY
UGC 09726	15 05 18.	+ 02 45	138	13.1	GALAXY SO
MCG+01-39-001	15 05 18.	+ 02 45	120	13.1	GALAXY
ZWG 049.010	15 05 18.	+ 04 10		15.5	GALAXY
ZWG 106.006	15 05 18.	+ 14 44		14.9	GALAXY
UGC 09727	15 05 18.	+ 14 44	60	14.9	GALAXY Sb/SBc
8ZW 1505+14.8	15 05 18.	+ 14 48		18.0	COMPACT GALAXY
ZWG 106.007	15 05 18.	+ 19 46		13.1	GALAXY
UGC 09728	15 05 18.	+ 19 46	174	13.1	GALAXY SBb
KARA.72 455B	15 05 18.	+ 19 46	162	13.1	PART OF DOUBLE GALAXY
ZWG 135.019	15 05 18.	+ 20 39		15.5	GALAXY
MCG+11-18-025	15 05 18.	+ 63 08 30.	39	14.	GALAXY
ZWG 318.014	15 05 18.	+ 63 12		14.1	GALAXY
UGC 09729	15 05 18.	+ 63 12	60	14.1	GALAXY S
ZWG 354.016	15 05 18.	+ 77 50		15.7	GALAXY
UGC 09730	15 05 18.	+ 77 50	108	15.7	GALAXY SBc
KARA.73B 0660	15 05 18.	+ 77 50	72	15.7	ISOLATED GALAXY S
ZWG 021.010	15 05 18.	- 01 17		15.5	GALAXY
REIN 2.229A	15 05 18.35	+ 19 46 28.0			NEBULA

OBJECT NAME	RIGHT ASCEN.	DECLINATION	DIAM.	MAGN.	TYPE OF OBJECT
REIN 2.229B	15 05 18.62	+ 19 46 27.5			NEBULA
RNGC 5854	15 05 20.	+ 02 46		13.0	GALAXY
RNGC 5855	15 05 21.	+ 04 10		15.5	GALAXY
MCG+03-39-005	15 05 21.	+ 19 47	180	13.1	GALAXY
IC 1094	15 05 22.	+ 14 47 15.			NONSTELLAR OBJECT
REIZ 4460	15 05 22.	+ 40 50	36	14.7	GALAXY
REIZ 4463	15 05 23.	+ 55 40	48	14.7	GALAXY
ZWG 106.008	15 05 24.	+ 14 49		15.2	GALAXY
MCG+03-39-006	15 05 24.	+ 14 49	30	15.2	GALAXY
LB 09438	15 05 24.	+ 23 21		17.0	FAINT BLUE STAR
LB 09437	15 05 24.	+ 23 48		16.4	FAINT BLUE STAR
ZWG 221.030	15 05 24.	+ 40 51		15.6	GALAXY
ZC 1505.4+4250	15 05 24.	+ 42 50	2820		CLUSTER OF GALAXIES
MCG+09-25-019	15 05 24.	+ 56 30	24	16.	GALAXY
MCG+09-25-020	15 05 24.	+ 56 31	12	17.	GALAXY
7ZW 583	15 05 24.	+ 60 46			COMPACT GALAXY
LB 00742	15 05 27.	+ 29 00 30.		16.8	FAINT BLUE STAR
ZWG 021.012	15 05 30.	+ 01 25		14.9	GALAXY
MCG+00-39-004	15 05 30.	+ 01 25	24	14.9	GALAXY
REIZ 4461	15 05 30.	+ 39 38	15	14.8	GALAXY
MCG+09-25-021	15 05 30.	+ 56 41	72	14.	GALAXY
MCG-02-39-001	15 05 30.	- 10 57	60	14.5	GALAXY
IC 1091	15 05 30.9	- 10 57 07.			NONSTELLAR OBJECT
SVEN 435	15 05 31.	- 10 57	24	14.6	GALAXY
ARC 2027	15 05 32.	+ 42 58		17.7	RICH CLUSTER OF GALAXIES
REIZ 4465	15 05 32.	+ 56 41	48	14.0	GALAXY
LB 00743	15 05 34.	+ 28 15 24.		15.2	FAINT BLUE STAR
HOLM 698A	15 05 35.	- 08 52	42	13.5	PART OF MULTIPLE GALAXY
ZWG 021.013	15 05 36.	+ 01 50		15.6	GALAXY
ZWG 049.011	15 05 36.	+ 02 37		15.3	GALAXY
ZWG 077.016	15 05 36.	+ 14 04		15.3	GALAXY
LB 09439	15 05 36.	+ 23 14		16.7	FAINT BLUE STAR
ZC 1505.6+2635	15 05 36.	+ 26 35	610		CLUSTER OF GALAXIES
ZWG 165.020	15 05 36.	+ 27 39		15.7	GALAXY
ZWG 297.003	15 05 36.	+ 56 44		15.0	GALAXY
UGC 09731	15 05 36.	+ 56 44	72	15.0	GALAXY Sc
SN 1940C	15 05 36.	+ 56 44		16.3	SUPERNOVA
LB 00744	15 05 38.	+ 14 16 30.		17.8	FAINT BLUE STAR
PATH 1.693	15 05 38.	+ 29 01	11		NEBULA
HOLM 698B	15 05 40.	- 08 53	18	14.3	PART OF MULTIPLE GALAXY
LB 00745	15 05 41.	+ 29 43 24.		16.0	FAINT BLUE STAR
ZWG 021.014	15 05 42.	+ 01 26		14.7	GALAXY
UGC 09732	15 05 42.	+ 01 26	78	14.7	GALAXY SBb
MCG+00-39-005	15 05 42.	+ 01 26	84	14.7	GALAXY
ZWG 021.015	15 05 42.	+ 01 47		15.7	GALAXY
ZWG 049.012	15 05 42.	+ 07 01		15.4	GALAXY
ZWG 049.013	15 05 42.	+ 08 15		15.4	GALAXY
ACK 359+29.1	15 05 42.	- 23 03			PLANETARY NEBULA
MRSL 321+02/1	15 05 42.	- 55 23	168		HII REGION
WRAY 19.40	15 05 42.6	- 55 21 53.			DIFFUSE NEBULA
KEEL 694	15 05 45.3	+ 19 44 55.			NEBULA
IC 1099	15 05 46.	+ 56 40 58.			NONSTELLAR OBJECT
PATH 1.694	15 05 47.	+ 28 54	14		NEBULA
ZWG 077.017	15 05 48.	+ 10 29		15.0	GALAXY
MCG+02-39-001	15 05 48.	+ 10 29	30	15.0	GALAXY
MCG+05-36-008	15 05 48.	+ 27 37 30.	42	15.7	GALAXY
ZWG 165.021	15 05 48.	+ 28 10		15.5	GALAXY
PATH 1.695	15 05 48.	+ 29 46	5		NEBULA
IC 4522	15 05 51.	- 75 39			NONSTELLAR OBJECT
KEEL 696	15 05 51.0	+ 19 54 05.			NEBULA
HN 0420	15 05 52.	- 75 40			NEBULA
ZWG 106.009	15 05 54.	+ 20 22		15.2	GALAXY
KEEL 703	15 05 54.8	+ 55 34 34.		18.	NEBULA
BC PKS1505+01	15 05 56.	+ 01 13 24.		17.5	QUASI-STELLAR OBJECT
PATH 1.696	15 05 56.	+ 29 49	5		NEBULA
IC 4534	15 05 57.	+ 23 53 43.			NONSTELLAR OBJECT
KEEL 697	15 05 57.0	+ 19 55 33.			SPIRAL NEBULA
REIZ 4464	15 05 58.	+ 40 54	24	15.1	GALAXY
APC 2026	15 05 58.	- 00 05		16.7	RICH CLUSTER OF GALAXIES
SN 1950I	15 05 59.	+ 19 23		20.2	SUPERNOVA
PATH 1.697	15 05 59.	+ 29 49	8		NEBULA
ZWG 049.014	15 06 00.	+ 05 52		15.4	GALAXY
ZWG 106.010	15 06 00.	+ 19 23		15.6	GALAXY
MCG+03-39-007	15 06 00.	+ 20 24	48	15.2	GALAXY
MCG+04-36-018	15 06 00.	+ 20 38 30.	42		GALAXY
ZWG 135.020	15 06 00.	+ 20 41		15.7	GALAXY
LB 09440	15 06 00.	+ 22 03		16.8	FAINT BLUE STAR
VV 059B	15 06 00.	+ 34 33	15	16.5	INTERACTING GALAXY
VV 059A	15 06 00.	+ 34 33	21	16.	INTERACTING GALAXY
VV 059	15 06 00.	+ 34 33	42		INTERACTING GALAXY
ZWG 354.017	15 06 00.	+ 80 25		15.4	GALAXY
UGC 09733	15 06 00.	+ 81 23	72	17.	GALAXY DWARF
SC 1503-1248.2	15 06 01.	- 12 59 46.	36		NEBULA
RNGC 5858	15 06 02.	- 11 01		14.0	GALAXY
MCG+11-18-026	15 06 03.	+ 66 22	78	16.	GALAXY
IC 1096	15 06 04.	+ 19 22 31.			NONSTELLAR OBJECT
PATH 1.698	15 06 04.	+ 29 49	8		NEBULA
ZWG 077.018	15 06 06.	+ 10 23		15.5	GALAXY
ZWG 106.011	15 06 06.	+ 19 24		15.1	GALAXY
MCG+06-33-022	15 06 06.	+ 34 34	42	15.	GALAXY
ZWG 193.016	15 06 06.	+ 34 35		15.6	GALAXY
ZWG 274.018	15 06 06.	+ 51 40		15.6	GALAXY
ZWG 318.015	15 06 06.	+ 66 21		15.5	GALAXY
UGC 09734	15 06 06.	+ 66 21	90	15.5	GALAXY
KARA.73B 0661	15 06 06.	+ 66 21	72	15.5	ISOLATED GALAXY S
ZWG 337.029	15 06 06.	+ 68 33		15.7	GALAXY
MCG-02-39-002	15 06 06.	- 11 01	30	14.	GALAXY
PATH 1.699	15 06 07.	+ 29 58	16		NEBULA
SVEN 436	15 06 07.	- 11 01	12	12.9	GALAXY
LB 00283	15 06 07.	- 14 43 30.		13.2	FAINT BLUE STAR
IC 1095	15 06 09.	+ 14 12 07.			NONSTELLAR OBJECT
MCG+03-39-008	15 06 09.	+ 19 24	30	15.6	GALAXY
MCG+04-36-019	15 06 09.	+ 20 39	18		GALAXY
LB 00746	15 06 10.	+ 14 59 42.		17.3	FAINT BLUE STAR
REIZ 4466	15 06 10.	+ 47 40	6	15.4	GALAXY
ZWG 077.019	15 06 12.	+ 13 52		15.2	GALAXY
8ZW 1506+13.9	15 06 12.	+ 13 52		15.2	COMPACT GALAXY
MCG+02-39-003	15 06 12.	+ 13 52	9	15.2	GALAXY
MCG+02-39-002	15 06 12.	+ 13 52	30	15.2	GALAXY
ZWG 106.012	15 06 12.	+ 19 22		14.7	GALAXY
UGC 09735	15 06 12.	+ 19 22	66	14.7	GALAXY S
MCG+03-39-009	15 06 12.	+ 19 25	42	15.1	GALAXY
MCG+04-36-020	15 06 12.	+ 20 40 30.	48	15.7	GALAXY
ZC 1506.2+2205	15 06 12.	+ 22 05	610		CLUSTER OF GALAXIES
ZWG 135.021	15 06 12.	+ 22 08		15.3	GALAXY
ZC 1506.2+3615	15 06 12.	+ 36 15	3230		CLUSTER OF GALAXIES
ZC 1506.2+4046	15 06 12.	+ 40 46	1480		CLUSTER OF GALAXIES
MCG+11-18-027	15 06 12.	+ 66 07	27	17.	GALAXY

OBJECT NAME	RIGHT ASCEN.	DECLINATION	DIAM.	MAGN.	TYPE OF OBJECT
PATH 1.700	15 06 13.	+ 30 05	8		NEBULA
KEEL 700	15 06 13.1	+ 19 43 08.			NEBULA
IC 1097	15 06 14.	+ 19 22 02.			NONSTELLAR OBJECT
PATH 1.701	15 06 14.	+ 29 49	19		NEBULA
PATH 1.702	15 06 16.	+ 29 46	14		NEBULA
ZWG 021.016	15 06 18.	+ 00 22		15.5	GALAXY
MCG+03-39-010	15 06 18.	+ 19 23	66	14.7	GALAXY
LB 09441	15 06 18.	+ 22 12		16.5	FAINT BLUE STAR
ZC 1506.3+2703	15 06 18.	+ 27 03	4300		CLUSTER OF GALAXIES
ZC 1506.3+2937	15 06 18.	+ 29 37	1610		CLUSTER OF GALAXIES
MCG+06-33-023	15 06 18.	+ 37 35 30.	7	17.	GALAXY
MCG+09-25-022	15 06 18.	+ 51 37 30.	54	15.	GALAXY
MCG+09-25-023	15 06 18.	+ 55 22 30.	72	15.	GALAXY
ZWG 318.016	15 06 18.	+ 66 07		15.4	GALAXY
RNGC 5861	15 06 20.	- 11 08		12.5	GALAXY
MCG+04-36-021	15 06 21.	+ 22 07 30.	30	15.3	GALAXY
ZC 1506.4+4343	15 06 24.	+ 43 43	1340		CLUSTER OF GALAXIES
REIZ 4468	15 06 24.	+ 52 42	9	14.5	GALAXY
ZWG 274.019	15 06 24.	+ 52 43		15.0	GALAXY
ZWG 274.020	15 06 24.	+ 54 56		14.1	GALAXY
UGC 09736	15 06 24.	+ 54 56	156	14.1	GALAXY Sc
MCG+09-25-024	15 06 24.	+ 54 57	120	11.	GALAXY
ZWG 274.021	15 06 24.	+ 55 22		14.7	GALAXY
UGC 09737	15 06 24.	+ 55 22	78	14.7	GALAXY S
ZWG 021.017	15 06 24.	- 00 01		15.7	GALAXY
SN 1971D	15 06 26.	- 11 08		15.5	SUPERNOVA
PK317-05.1	15 06 27.	- 64 29 16.	50		PLANETARY NEBULA
BSO 25	15 06 28.	+ 45 16 59.			BLUE STELLAR OBJECT
RNGC 5844	15 06 28.	- 64 29			UNVERIFIED SOUTHERN OBJECT
ZWG 021.018	15 06 30.	+ 00 01		15.4	GALAXY
ZWG 077.020	15 06 30.	+ 11 00		15.4	GALAXY
MCG+03-39-011	15 06 30.	+ 19 08	36	15.5	GALAXY
ZWG 106.013	15 06 30.	+ 20 16		15.7	GALAXY
LB 00747	15 06 30.	+ 31 18 42.		17.5	FAINT BLUE STAR
ZWG 193.017	15 06 30.	+ 37 51		15.7	GALAXY
MCG+06-33-024	15 06 30.	+ 37 52	15	16.	GALAXY
ZC 1506.5+3850	15 06 30.	+ 38 50	1010		CLUSTER OF GALAXIES
RNGC 5874	15 06 30.	+ 54 57		14.0	GALAXY
ZC 1506.5+5921	15 06 30.	+ 59 21	1280		CLUSTER OF GALAXIES
ZC 1506.5+6931	15 06 30.	+ 69 31	1340		CLUSTER OF GALAXIES
ABC 2041	15 06 31.	+ 69 04		17.0	RICH CLUSTER OF GALAXIES
SVEN 437	15 06 31.	- 11 08	162	12.5	GALAXY
ZWG 106.014	15 06 36.	+ 20 15		15.6	GALAXY
LB 09442	15 06 36.	+ 26 06		16.3	FAINT BLUE STAR
REIZ 4469	15 06 36.	+ 52 30	12	14.5	GALAXY
ZWG 274.022	15 06 36.	+ 52 31		15.3	GALAXY
KARA.68 233	15 06 36.	+ 56 25	34		DWARF GALAXY
MCG-02-39-003	15 06 36.	- 11 08	168	12.	GALAXY
RNGC 5633	15 06 37.	- 72 40			GALAXY
KEEL 701	15 06 40.1	+ 19 28 52.			SPIRAL NEBULA
ZWG 077.021	15 06 42.	+ 09 15		15.5	GALAXY
ZWG 106.015	15 06 42.	+ 19 28		15.7	GALAXY
UGC 09738	15 06 42.	+ 19 28	60	15.7	GALAXY Sc
MCG+03-39-012	15 06 42.	+ 19 30	36	15.7	GALAXY
ZWG 135.022	15 06 42.	+ 20 38		15.6	GALAXY
LB 09443	15 06 42.	+ 22 56		16.6	FAINT BLUE STAR
ZWG 135.023	15 06 42.	+ 25 55		15.6	GALAXY
UGC 09739	15 06 42.	+ 25 55	60	15.6	GALAXY
KARA.73B 0662	15 06 42.	+ 25 55	60	15.6	ISOLATED GALAXY S
ZWG 165.022	15 06 42.	+ 30 22		15.5	GALAXY
MCG+09-25-025	15 06 42.	+ 52 30	42	16.	GALAXY
MCG-02-39-004	15 06 42.	- 10 30	72	15.	GALAXY
MRSL 319-01/1	15 06 42.	- 60 03	18600		HII REGION
REIZ 4467	15 06 44.	+ 37 40	12	15.4	GALAXY
KEEL 702	15 06 44.5	+ 19 38 54.			NEBULA
IC 4535	15 06 46.	+ 37 45 37.			NONSTELLAR OBJECT
8ZW 1506+13.5	15 06 48.	+ 13 28		17.8	COMPACT GALAXY
8ZW 1506+13.5	15 06 48.	+ 13 28		17.3	COMPACT GALAXY
LB 09445	15 06 48.	+ 25 45		16.7	FAINT BLUE STAR
ZWG 165.023	15 06 48.	+ 26 43		14.8	GALAXY
LB 09444	15 06 48.	+ 26 46		16.0	FAINT BLUE STAR
ZC 1506.8+3032	15 06 48.	+ 30 32	2550		CLUSTER OF GALAXIES
ZWG 249.007	15 06 48.	+ 49 52		15.4	GALAXY
MCG+04-36-022	15 06 51.	+ 20 36	30	15.6	GALAXY
ZWG 077.022	15 06 54.	+ 09 15		15.7	GALAXY
ZWG 077.023	15 06 54.	+ 11 22		15.7	GALAXY
LB 09447	15 06 54.	+ 20 52		17.3	FAINT BLUE STAR
LB 09446	15 06 54.	+ 25 42		17.1	FAINT BLUE STAR
ZC 1506.9+2640	15 06 54.	+ 26 40	610		CLUSTER OF GALAXIES
ZC 1506.9+3025	15 06 54.	+ 30 25	740		CLUSTER OF GALAXIES
MCG+08-28-001	15 06 54.	+ 49 50	21	16.	GALAXY
SVEN 438	15 06 55.	- 11 24	42	15.4	GALAXY
KEEN 5875A	15 07	+ 52 29	54		Sa GALAXY
ZWG 049.015	15 07 00.	+ 03 15		12.9	GALAXY
UGC 09740	15 07 00.	+ 03 15	150	12.9	GALAXY SB0
MCG+01-39-002	15 07 00.	+ 03 15	120	12.9	GALAXY
ZWG 049.016	15 07 00.	+ 04 53		15.6	GALAXY
ZWG 049.017	15 07 00.	+ 05 04		15.5	GALAXY
LB 09448	15 07 00.	+ 22 02		16.9	FAINT BLUE STAR
MCG+05-36-009	15 07 00.	+ 26 40 30.	18	14.8	GALAXY
LB 00748	15 07 00.	+ 29 45 24.		16.6	FAINT BLUE STAR
ZWG 221.031	15 07 00.	+ 39 30		15.6	GALAXY
MCG+07-31-034	15 07 00.	+ 39 30	36	15.5	GALAXY
ZC 1507.0+4443	15 07 00.	+ 44 43	1210		CLUSTER OF GALAXIES
ZWG 274.023	15 07 00.	+ 52 29		14.1	GALAXY
UGC 09741	15 07 00.	+ 52 29	27	14.1	GALAXY PECULR
MCG+09-25-026	15 07 00.	+ 52 29	36	14.	GALAXY
REIZ 4470	15 07 00.	+ 52 29	36	14.0	GALAXY
ZC 1507.0+8440	15 07 00.	+ 84 40	2080		CLUSTER OF GALAXIES
ZWG 021.019	15 07 00.	- 00 35		15.5	GALAXY
HOLM 699C	15 07 02.	+ 00 42	30	14.3	PART OF MULTIPLE GALAXY
RNGC 5864	15 07 03.	+ 03 15		13.0	GALAXY
ABC 2028	15 07 03.	+ 07 43		15.7	RICH CLUSTER OF GALAXIES
RNGC 5875A	15 07 03.	+ 52 29		14.0	GALAXY
ZWG 021.020	15 07 06.	+ 01 34		15.2	GALAXY
LB 09449	15 07 06.	+ 24 31		16.9	FAINT BLUE STAR
ZWG 165.024	15 07 06.	+ 28 11		15.7	GALAXY
ZWG 274.024	15 07 06.	+ 52 39		15.5	GALAXY
KEEL 704	15 07 06.5	+ 56 09 17.		17.	NEBULA
HOLM 699A	15 07 09.	+ 00 41	30	12.6	PART OF MULTIPLE GALAXY
LB 00749	15 07 11.	+ 00 35 54.		16.2	FAINT BLUE STAR
LB 09450	15 07 12.	+ 21 44		16.9	FAINT BLUE STAR
ZWG 165.025	15 07 12.	+ 27 37		15.6	GALAXY
ZC 1507.2+4310	15 07 12.	+ 43 10	1610		CLUSTER OF GALAXIES
REIZ 4472	15 07 12.	+ 52 39	15	14.5	GALAXY
HOLM 699B	15 07 13.	+ 00 39	18	13.9	PART OF MULTIPLE GALAXY
KEEL 705	15 07 14.2	+ 55 43 37.		18.	NEBULA
ZWG 021.022	15 07 18.	+ 00 39		13.5	GALAXY

OBJECT NAME	RIGHT ASCEN.	DECLINATION	DIAM.	MAGN.	TYPE OF OBJECT
UGC 09742	15 07 18.	+ 00 39	144	13.5	GALAXY S0
KARA.72 456B	15 07 18.	+ 00 39	150	13.5	PART OF DOUBLE GALAXY
MCG+00-39-006	15 07 18.	+ 00 39	102	13.5	GALAXY
ZWG 021.023	15 07 18.	+ 00 43		15.2	GALAXY
UGC 09743	15 07 18.	+ 00 43	66	15.2	GALAXY S0
KARA.72 456A	15 07 18.	+ 00 43	54	15.2	PART OF DOUBLE GALAXY
MCG+00-39-007	15 07 18.	+ 00 43	42	15.2	GALAXY
ZWG 274.025	15 07 18.	+ 55 47		15.6	GALAXY
ZWG 021.021	15 07 18.	- 00 11		15.6	GALAXY
RNGC 5865	15 07 19.	+ 00 39		13.5	GALAXY
RNGC 5869	15 07 19.	+ 00 42			NON-EXISTENT OBJECT
RNGC 5868	15 07 19.	+ 00 43		15.0	GALAXY
LB 00750	15 07 20.	+ 29 33 42.		16.9	FAINT BLUE STAR
ZWG 021.024	15 07 24.	+ 01 35		15.4	GALAXY
ZWG 021.025	15 07 24.	+ 01 58		15.3	GALAXY
UGC 09744	15 07 24.	+ 01 58	66	15.3	GALAXY SBa-b
ZC 1507.4+1228	15 07 24.	+ 12 28	1680		CLUSTER OF GALAXIES
ZC 1507.4+1320	15 07 24.	+ 13 20	1750		CLUSTER OF GALAXIES
LB 09453	15 07 24.	+ 21 06		16.8	FAINT BLUE STAR
LB 09452	15 07 24.	+ 21 26		16.1	FAINT BLUE STAR
ZC 1507.4+2149	15 07 24.	+ 21 49	670		CLUSTER OF GALAXIES
LB 09451	15 07 24.	+ 22 02		14.8	FAINT BLUE STAR
ZWG 165.026	15 07 24.	+ 31 01		15.6	GALAXY
MCG+08-28-002	15 07 24.	+ 48 17	36	18.	GALAXY
ZWG 274.026	15 07 24.	+ 52 38		15.7	GALAXY
3ZW 070	15 07 24.	+ 55 48			COMPACT GALAXY
ZWG 077.024	15 07 30.	+ 09 35		15.3	GALAXY
ZC 1507.5+1157	15 07 30.	+ 11 57	1340		CLUSTER OF GALAXIES
ZWG 077.025	15 07 30.	+ 13 27		15.2	GALAXY
ZC 1507.5+2319	15 07 30.	+ 23 19	610		CLUSTER OF GALAXIES
REIZ 4474	15 07 30.	+ 52 38	9	14.8	GALAXY
7ZW 584	15 07 30.	+ 69 07			COMPACT GALAXY
SER 122.01	15 07 30.	- 78 41	150		COMPACT GROUP OF GALAXIES
RNGC 5871	15 07 31.	+ 00 43			NON-EXISTENT OBJECT
LB 00751	15 07 32.	+ 29 20 00.		17.2	FAINT BLUE STAR
MCG+12-14-017	15 07 36.	+ 69 47 30.	36	15.	GALAXY
ZWG 021.026	15 07 36.	- 02 04		15.5	GALAXY
MCG-07-31-009	15 07 36.	- 40 47	60	14.5	GALAXY
LB 00284	15 07 37.	- 12 19 42.		15.8	FAINT BLUE STAR
RNGC 5875	15 07 39.	+ 52 43		13.5	GALAXY
ZWG 049.018	15 07 42.	+ 07 49		15.4	GALAXY
ZWG 077.026	15 07 42.	+ 09 56		15.2	GALAXY
LB 09454	15 07 42.	+ 23 52		17.0	FAINT BLUE STAR
ZC 1507.7+3727	15 07 42.	+ 37 27	1080		CLUSTER OF GALAXIES
REIZ 4475	15 07 42.	+ 52 43	90	13.4	GALAXY
MCG+09-25-027	15 07 42.	+ 52 43	120	13.4	GALAXY
ZWG 274.027	15 07 42.	+ 52 44		13.4	GALAXY
UGC 09745	15 07 42.	+ 52 44	156	13.4	GALAXY Sb
ZWG 021.027	15 07 48.	+ 00 13		15.6	GALAXY
ZWG 021.028	15 07 48.	+ 02 07		15.0	GALAXY
UGC 09746	15 07 48.	+ 02 07	84	15.0	GALAXY Sb-c
MCG+00-39-008	15 07 48.	+ 02 07	84	15.0	GALAXY
ZWG 049.019	15 07 48.	+ 07 05		15.6	GALAXY
ZWG 049.020	15 07 48.	+ 07 54		15.0	GALAXY
MCG+01-39-003	15 07 48.	+ 07 54	42	15.0	GALAXY
LB 09455	15 07 48.	+ 25 02		16.9	FAINT BLUE STAR
MCG+08-28-003	15 07 48.	+ 46 40	18	15.	GALAXY
LB 09458	15 07 54.	+ 22 56		17.8	FAINT BLUE STAR
LB 09457	15 07 54.	+ 25 44		16.6	FAINT BLUE STAR
LB 09456	15 07 54.	+ 26 10		16.4	FAINT BLUE STAR
ZC 1507.9+3656	15 07 54.	+ 36 56	1550		CLUSTER OF GALAXIES
ZWG 221.032	15 07 54.	+ 42 54		15.5	GALAXY
ZWG 249.008	15 07 54.	+ 46 42		15.4	GALAXY
MCG+11-18-028	15 07 54.	+ 64 20	24	17.	GALAXY
MCG+11-18-029	15 07 54.	+ 65 07	90	15.	GALAXY
ZWG 021.029	15 07 54.	- 01 59		14.9	GALAXY
MCG+00-39-009	15 07 54.	- 01 59	24	14.9	GALAXY
MCG-04-36-001	15 07 54.	- 21 50	48	15.	GALAXY
KEEL 706	15 07 58.8	+ 55 53 51.		18.	NEBULA
RNGC 5863	15 07 59.	- 18 15		14.0	GALAXY
VDB.66G 199	15 08	+ 67 28	1340		DWARF GALAXY
LB 00752	15 08 00.	+ 02 24 54.		16.2	FAINT BLUE STAR
ZWG 049.021	15 08 00.	+ 06 41		15.7	GALAXY
ZWG 049.022	15 08 00.	+ 07 37		15.6	GALAXY
ZWG 106.016	15 08 00.	+ 19 32		15.5	GALAXY
LB 09460	15 08 00.	+ 20 42		14.7	FAINT BLUE STAR
LB 09459	15 08 00.	+ 24 34		16.7	FAINT BLUE STAR
ZC 1508.0+4706	15 08 00.	+ 47 06	740		CLUSTER OF GALAXIES
ZWG 274.028	15 08 00.	+ 54 41		13.9	GALAXY
UGC 09747	15 08 00.	+ 54 41	168	13.9	GALAXY SBa
ZWG 318.017	15 08 00.	+ 68 24		15.7	GALAXY
MCG-03-39-001	15 08 03.	- 18 15	60	14.	GALAXY
REIZ 4471	15 08 04.	+ 13 46	30	15.0	GALAXY
ZWG 077.027	15 08 06.	+ 08 48		15.4	GALAXY
ZC 1508.1+1558	15 08 06.	+ 15 58	1810		CLUSTER OF GALAXIES
LB 09461	15 08 06.	+ 24 47		17.1	FAINT BLUE STAR
ZWG 165.027	15 08 06.	+ 32 05		14.8	GALAXY
ZC 1508.1+5227	15 08 06.	+ 52 27	1480		CLUSTER OF GALAXIES
RNGC 5876	15 08 06.	+ 54 42		14.0	GALAXY
7ZW 585	15 08 06.	+ 61 25			COMPACT GALAXY
ZWG 354.018	15 08 06.	+ 76 15		15.1	GALAXY
UGC 09748	15 08 06.	+ 76 15	78	15.1	GALAXY SB0
KARA.72 457A	15 08 06.	+ 76 15	72	15.1	PART OF DOUBLE GALAXY
REIZ 4485	15 08 07.	+ 54 43	12	15.5	GALAXY
MCG+05-36-010	15 08 09.	+ 32 05	30	14.8	GALAXY
MCG+09-25-028	15 08 09.	+ 54 42	120	12.	GALAXY
REIZ 4473	15 08 10.	+ 13 39	24	15.1	GALAXY
LB 00753	15 08 11.	+ 12 53 06.		16.7	FAINT BLUE STAR
PK321+01.1	15 08 11.	- 55 28 48.	30		PLANETARY NEBULA
ZWG 077.028	15 08 12.	+ 13 39		15.3	GALAXY
ZWG 106.017	15 08 12.	+ 17 57		15.7	GALAXY
LB 09464	15 08 12.	+ 24 18		16.6	FAINT BLUE STAR
TON-N 0219	15 08 12.	+ 25 50		14.9	BLUE STAR
LB 09463	15 08 12.	+ 25 51		15.7	FAINT BLUE STAR
LB 09462	15 08 12.	+ 26 17		15.8	FAINT BLUE STAR
ZC 1508.2+3341	15 08 12.	+ 33 41	3360		CLUSTER OF GALAXIES
MCG+11-18-030	15 08 12.	+ 67 18			GALAXY
ZWG 319.001	15 08 12.	+ 67 23		13.6	GALAXY
ZWG 318.018	15 08 12.	+ 67 23		13.6	GALAXY
UGC 09749	15 08 12.	+ 67 23	2400	13.6	GALAXY DWRF EL
KARA.73B 0663	15 08 12.	+ 67 23	1344	13.6	ISOLATED GALAXY E
ZWG 021.030	15 08 12.	- 00 10		15.6	GALAXY
MCG-02-39-005	15 08 12.	- 11 18	24	14.5	GALAXY
HN 0030	15 08 13.	- 11 16			NEBULA
SVEN 439	15 08 13.	- 11 17	12	15.1	GALAXY
ABC 2034	15 08 14.	+ 33 43		16.9	RICH CLUSTER OF GALAXIES
RNGC 5872	15 08 14.	- 11 18		14.0	GALAXY
BC PKS1508-05	15 08 14.95	- 05 31 49.1			QUASI-STELLAR OBJECT

OBJECT NAME	RIGHT ASCEN.	DECLINATION	DIAM.	MAGN.	TYPE OF OBJECT
REIZ 4477	15 08 15.	+ 37 14	30	15.2	GALAXY
SHB 307	15 08 15.0	- 05 31 49.		17.	QUASI-STELLAR OBJECT
REIZ 4476	15 08 16.	+ 36 13	30	15.0	GALAXY
ZWG 021.031	15 08 18.	+ 01 52		15.6	GALAXY
ZWG 077.029	15 08 18.	+ 09 55		15.6	GALAXY
LB 00754	15 08 18.	+ 28 57 00.		17.5	FAINT BLUE STAR
MCG+13-11-007	15 08 18.	+ 76 15	51	15.	GALAXY
ZWG 354.019	15 08 18.	+ 76 21		15.3	GALAXY
UGC 09750	15 08 18.	+ 76 21	96	15.3	GALAXY Sb
KARA.72 457B	15 08 18.	+ 76 21	90	15.3	PART OF DOUBLE GALAXY
SC 1505-1647.8	15 08 21.	- 16 59 14.	6		NEBULA
SC 1505-1647.8	15 08 21.	- 16 59 14.	6		NEBULA
SN 1954C	15 08 23.	+ 57 12		14.9	SUPERNOVA
ZWG 021.032	15 08 24.	+ 01 37		15.6	GALAXY
UGC 09751	15 08 24.	+ 01 37	66	15.6	GALAXY Sc
ZWG 049.023	15 08 24.	+ 05 56		15.4	GALAXY
UGC 09752	15 08 24.	+ 05 56	84	15.4	GALAXY E-SO
ZWG 049.024	15 08 24.	+ 08 11		15.6	GALAXY
ZWG 106.018	15 08 24.	+ 17 53		15.7	GALAXY
LB 09468	15 08 24.	+ 23 39		15.5	FAINT BLUE STAR
LB 09467	15 08 24.	+ 24 13		16.2	FAINT BLUE STAR
LB 09466	15 08 24.	+ 24 14		16.5	FAINT BLUE STAR
TON-N 0220	15 08 24.	+ 25 41		15.6	BLUE STAR
LB 09465	15 08 24.	+ 25 53		16.9	FAINT BLUE STAR
ZC 1508.4+3200	15 08 24.	+ 32 00	3360		CLUSTER OF GALAXIES
ZWG 297.004	15 08 24.	+ 57 12		11.9	GALAXY
REIZ 4487	15 08 24.	+ 57 12	210	12.4	GALAXY
UGC 09753	15 08 24.	+ 57 12	288	11.9	GALAXY Sb
RNGC 5879	15 08 27.	+ 57 11		12.0	GALAXY
IC 1101	15 08 28.	+ 05 57 08.			NONSTELLAR OBJECT
REIZ 4480	15 08 28.	+ 36 12	24	15.5	GALAXY
REIN 2.230	15 08 28.89	+ 57 11 21.1			NEBULA
ABC 2029	15 08 30.	+ 05 57		16.0	RICH CLUSTER OF GALAXIES
ZWG 077.030	15 08 30.	+ 11 09		15.7	GALAXY
MCG+10-22-001	15 08 30.	+ 57 10	192	11.9	GALAXY
MCG+13-11-008	15 08 30.	+ 76 20	84	15.	GALAXY
MCG+14-07-016	15 08 30.	+ 82 02	24	17.	GALAXY
SVEN 440	15 08 31.	- 11 04	60	14.1	GALAXY
LB 00755	15 08 35.	+ 27 46 24.		17.2	FAINT BLUE STAR
ZWG 049.025	15 08 36.	+ 04 29		15.3	GALAXY
UGC 09754	15 08 36.	+ 04 29	72	15.3	GALAXY S(b)
ZWG 049.026	15 08 36.	+ 05 43		15.7	GALAXY
ZWG 049.027	15 08 36.	+ 06 53		15.4	GALAXY
ZC 1508.6+2105	15 08 36.	+ 21 05	670		CLUSTER OF GALAXIES
ZWG 135.024	15 08 36.	+ 21 07		15.7	GALAXY
ZC 1508.6+2522	15 08 36.	+ 25 22	4170		CLUSTER OF GALAXIES
LB 09470	15 08 36.	+ 25 22		16.2	FAINT BLUE STAR
LB 09469	15 08 36.	+ 25 26		16.0	FAINT BLUE STAR
ZWG 249.009	15 08 36.	+ 45 22		15.6	GALAXY
ABC 2030	15 08 39.	+ 00 06		16.9	RICH CLUSTER OF GALAXIES
IC 1102	15 08 39.	+ 04 27 21.			NONSTELLAR OBJECT
LB 00756	15 08 40.	+ 11 48 42.		17.3	FAINT BLUE STAR
REIZ 4484	15 08 40.	+ 36 07	36	14.6	GALAXY
MCG+04-36-023	15 08 42.	+ 21 06	30	15.7	GALAXY
LB 09471	15 08 42.	+ 22 52		16.4	FAINT BLUE STAR
ZWG 221.033	15 08 42.	+ 41 15		15.5	GALAXY
MCG+08-28-004	15 08 42.	+ 45 20	48	15.	GALAXY
SMI 10	15 08 42.	- 56 04	360		FAINT NEBULOSITY
REIZ 4481	15 08 44.	+ 32 28	18	16.0	GALAXY
REIZ 4482	15 08 44.	+ 32 30	12	15.3	GALAXY
REIZ 4483	15 08 44.	+ 32 33	18	16.0	GALAXY
MCG+07-31-035	15 08 45.	+ 41 13 30.	36	15.	GALAXY
SC 1506-1327.2	15 08 46.	- 13 38 37.	12		NEBULA
SC 1506-1327.6	15 08 46.	- 13 39 01.	12		NEBULA
ZWG 021.033	15 08 48.	+ 01 45		15.6	GALAXY
ZWG 021.034	15 08 48.	+ 01 52		15.7	GALAXY
ZWG 049.028	15 08 48.	+ 04 50		15.7	GALAXY
ZWG 077.031	15 08 48.	+ 10 38		14.8	GALAXY
REIZ 4478	15 08 48.	+ 10 38	48	14.8	GALAXY
UGC 09755	15 08 48.	+ 10 38	66	14.8	GALAXY Sc
KARA.72 458B	15 08 48.	+ 10 38	60	14.8	PART OF DOUBLE GALAXY
MCG+02-39-004	15 08 48.	+ 10 38	72	14.8	GALAXY
ZWG 077.032	15 08 48.	+ 10 39		15.3	GALAXY
REIZ 4479	15 08 48.	+ 10 39	24	15.0	GALAXY
KARA.72 458A	15 08 48.	+ 10 39	48	15.3	PART OF DOUBLE GALAXY
MCG+02-39-005	15 08 48.	+ 10 39	48	15.3	GALAXY
LB 09474	15 08 48.	+ 20 54		15.6	FAINT BLUE STAR
ZWG 135.025	15 08 48.	+ 23 48		15.5	GALAXY
LB 09473	15 08 48.	+ 23 48		15.8	FAINT BLUE STAR
LB 09472	15 08 48.	+ 26 33		16.5	FAINT BLUE STAR
ZC 1508.8+4054	15 08 48.	+ 40 54	12030		CLUSTER OF GALAXIES
MCG+09-25-029	15 08 48.	+ 56 33	60	15.	GALAXY
MCG+04-36-024	15 08 51.	+ 23 47	54	15.5	GALAXY
ZC 1508.9+0026	15 08 54.	+ 00 26	2350		CLUSTER OF GALAXIES
ZWG 021.035	15 08 54.	+ 01 46		15.5	GALAXY
ZWG 049.029	15 08 54.	+ 06 32		15.5	GALAXY
UGC 09756	15 08 54.	+ 06 32	72	15.5	GALAXY DBL SYS
ZWG 077.033	15 08 54.	+ 09 25		15.7	GALAXY
ZWG 077.034	15 08 54.	+ 11 35		15.6	GALAXY
ZWG 077.035	15 08 54.	+ 11 36		15.5	GALAXY
ZC 1508.9+6000	15 08 54.	+ 60 00	1340		CLUSTER OF GALAXIES
ZWG 337.030	15 08 54.	+ 70 16		15.6	GALAXY
HOLM 700A	15 08 56.	+ 01 58	24	14.8	PART OF MULTIPLE GALAXY
HOLM 700B	15 08 57.	+ 01 59	24	15.3	PART OF MULTIPLE GALAXY
ABC 2033	15 08 59.	+ 06 33		15.7	RICH CLUSTER OF GALAXIES
MIL 40	15 09	- 40 00	16200		SUPERNOVA REMNANT
ZWG 021.037	15 09 00.	+ 01 57		15.3	GALAXY
ZWG 049.030	15 09 00.	+ 04 42		15.5	GALAXY
ZWG 049.031	15 09 00.	+ 05 24		15.3	GALAXY
ZWG 049.032	15 09 00.	+ 05 26		15.3	GALAXY
SN 1963B	15 09 00.	+ 05 26		17.0	SUPERNOVA
ZWG 049.033	15 09 00.	+ 07 26		15.1	GALAXY
LB 09475	15 09 00.	+ 24 02		16.9	FAINT BLUE STAR
ZC 1509.0+4216	15 09 00.	+ 42 16	1010		CLUSTER OF GALAXIES
ZC 1509.0+4320	15 09 00.	+ 43 20	1080		CLUSTER OF GALAXIES
ZWG 021.036	15 09 00.	- 00 05		15.4	GALAXY
MCG-04-36-002	15 09 00.	- 23 51 30.	48	15.	GALAXY
LB 00757	15 09 03.	+ 31 15 18.		17.2	FAINT BLUE STAR
REIZ 4486	15 09 03.	+ 36 24	24	15.2	GALAXY
ZWG 049.034	15 09 06.	+ 04 39		15.7	GALAXY
ZWG 049.035	15 09 06.	+ 05 35		15.0	GALAXY
UGC 09757	15 09 06.	+ 05 35	78	15.0	GALAXY Sb
MCG+01-39-004	15 09 06.	+ 05 35	72	15.0	GALAXY
ZWG 077.036	15 09 06.	+ 09 40		15.3	GALAXY
ZWG 077.037	15 09 06.	+ 09 48		15.0	GALAXY
MCG+02-39-006	15 09 06.	+ 09 48	48	15.0	GALAXY
8ZW 1509+09.9	15 09 06.	+ 09 54		17.9	COMPACT GALAXY
ZWG 077.038	15 09 06.	+ 12 31		15.7	GALAXY
ZWG 077.039	15 09 06.	+ 13 40		15.0	GALAXY
UGC 09758	15 09 06.	+ 13 40	78	15.0	GALAXY SO
MCG+02-39-007	15 09 06.	+ 13 40	42	15.0	GALAXY
MCG+06-33-025	15 09 06.	+ 37 13	9	17.	GALAXY
7ZW 586	15 09 06.	+ 69 35			COMPACT GALAXY
ZC 1509.1-0257	15 09 06.	- 02 57	810		CLUSTER OF GALAXIES
BSO 26	15 09 07.6	+ 42 43 05.			BLUE STELLAR OBJECT
ABC 2036	15 09 09.	+ 18 15		16.0	RICH CLUSTER OF GALAXIES
LB 00758	15 09 10.	+ 30 21 30.		15.8	FAINT BLUE STAR
ABC 2031	15 09 10.	- 11 01		17.1	RICH CLUSTER OF GALAXIES
ZWG 021.038	15 09 12.	+ 00 50		15.5	GALAXY
ZWG 021.039	15 09 12.	+ 01 33		15.6	GALAXY
ZWG 049.036	15 09 12.	+ 03 51		15.7	GALAXY
ZWG 049.037	15 09 12.	+ 04 40		15.6	GALAXY
ZWG 049.038	15 09 12.	+ 07 18		15.5	GALAXY
LB 09478	15 09 12.	+ 20 37		16.1	FAINT BLUE STAR
LB 09477	15 09 12.	+ 24 44		17.3	FAINT BLUE STAR
LB 09476	15 09 12.	+ 25 26		16.8	FAINT BLUE STAR
MCG+09-25-030	15 09 12.	+ 55 33	120	14.	GALAXY
ZC 1509.3+1434	15 09 18.	+ 14 34	1610		CLUSTER OF GALAXIES
ZWG 106.019	15 09 18.	+ 18 10		15.5	GALAXY
ZWG 274.029	15 09 18.	+ 55 32		15.4	GALAXY
UGC 09759	15 09 18.	+ 55 32	120	15.4	GALAXY S
REIZ 4488	15 09 20.	+ 37 15	21	15.0	GALAXY
MCG+06-33-026	15 09 21.	+ 37 14	18	16.	GALAXY
BSO 27	15 09 21.3	+ 42 46 00.			BLUE STELLAR OBJECT
IC 1103	15 09 22.	+ 19 23 32.			NONSTELLAR OBJECT
ZC 1509.4+0745	15 09 24.	+ 07 45	4030		CLUSTER OF GALAXIES
ZWG 077.040	15 09 24.	+ 12 25		15.7	GALAXY
LB 09482	15 09 24.	+ 21 14		16.6	FAINT BLUE STAR
LB 09481	15 09 24.	+ 24 46		18.3	FAINT BLUE STAR
LB 09480	15 09 24.	+ 25 18		17.0	FAINT BLUE STAR
LB 09479	15 09 24.	+ 25 48		17.0	FAINT BLUE STAR
ABC 2042	15 09 24.	+ 36 44		17.2	RICH CLUSTER OF GALAXIES
ZC 1509.4+3645	15 09 24.	+ 36 45	1280		CLUSTER OF GALAXIES
REIZ 4490	15 09 24.	+ 49 47	30	15.2	GALAXY
ZWG 274.030	15 09 24.	+ 52 39		15.3	GALAXY
REIZ 4492	15 09 29.	+ 52 39	42	14.3	GALAXY
ZWG 021.041	15 09 30.	+ 01 52		14.9	GALAXY
UGC 09760	15 09 30.	+ 01 52	180	14.9	GALAXY Sc
MCG+00-39-011	15 09 30.	+ 01 52	138	14.9	GALAXY
ZWG 021.042	15 09 30.	+ 02 00		15.6	GALAXY
ZWG 049.039	15 09 30.	+ 05 45		15.2	GALAXY
LB 09483	15 09 30.	+ 22 52		17.7	FAINT BLUE STAR
ZC 1509.5+2805	15 09 30.	+ 28 05	4370		CLUSTER OF GALAXIES
ZWG 221.034	15 09 30.	+ 41 17		15.2	GALAXY
ZWG 249.010	15 09 30.	+ 46 20		14.1	GALAXY
UGC 09761	15 09 30.	+ 46 20	78	14.1	GALAXY SBa
MCG+08-28-005	15 09 30.	+ 46 20	60	14.	GALAXY
ZWG 274.031	15 09 30.	+ 52 15		15.2	GALAXY
REIZ 4491	15 09 30.	+ 52 15	27	14.3	GALAXY
MCG+09-25-031	15 09 30.	+ 52 39	60	14.	GALAXY
ZWG 274.032	15 09 30.	+ 52 45		15.6	GALAXY
ZC 1509.5+7351	15 09 30.	+ 73 51	1140		CLUSTER OF GALAXIES
ZC 1509.5-0118	15 09 30.	- 01 18	1880		CLUSTER OF GALAXIES
ZWG 021.040	15 09 30.	- 03 03		15.0	GALAXY
MCG+00-39-010	15 09 30.	- 03 03	30	15.0	GALAXY
VMT 15	15 09 30.	- 58 46	580		SUPERNOVA REMNANT
ABC 2035	15 09 31.	- 05 53		17.1	RICH CLUSTER OF GALAXIES
ZC 1509.6+1303	15 09 36.	+ 13 03	2220		CLUSTER OF GALAXIES
ZC 1509.6+1500	15 09 36.	+ 15 00	810		CLUSTER OF GALAXIES
ZWG 135.026	15 09 36.	+ 23 20		15.4	GALAXY
1ZW 103	15 09 36.	+ 47 50			COMPACT GALAXY
MCG+09-25-032	15 09 36.	+ 52 14	54	15.	GALAXY
MCG+10-22-002	15 09 36.	+ 61 20	18	16.	GALAXY
MCG+13-11-009	15 09 36.	+ 75 22	27	16.	GALAXY
REIZ 4489	15 09 39.	+ 41 16	24	14.2	GALAXY
MIL 33	15 09 39.	- 58 49	480		SUPERNOVA REMNANT
ZWG 021.043	15 09 42.	+ 02 11		15.6	GALAXY
ZWG 049.040	15 09 42.	+ 05 26		15.4	GALAXY
ZC 1509.7+1349	15 09 42.	+ 13 49	1750		CLUSTER OF GALAXIES
LB 09484	15 09 42.	+ 22 06		16.5	FAINT BLUE STAR
UGC 09762	15 09 42.	+ 32 51	66	17.	GALAXY DWRF SP
ZC 1509.7+3502	15 09 42.	+ 35 02	3970		CLUSTER OF GALAXIES
RNGC 5873	15 09 42.	- 37 55		13.5	PLANETARY NEBULA
PK331+16.1	15 09 42.	- 37 55	10	13.3	PLANETARY NEBULA
LB 00759	15 09 44.	- 00 43 42.		15.6	FAINT BLUE STAR
BSO 28	15 09 45.	+ 45 28 33.			BLUE STELLAR OBJECT
ZWG 021.045	15 09 48.	+ 02 13		15.6	GALAXY
ZWG 077.041	15 09 48.	+ 11 07		15.4	GALAXY
LB 09490	15 09 48.	+ 20 54		16.3	FAINT BLUE STAR
ZWG 135.027	15 09 48.	+ 21 29		15.6	GALAXY
UGC 09763	15 09 48.	+ 21 29	120	15.6	GALAXY S?
LB 09489	15 09 48.	+ 22 42		16.2	FAINT BLUE STAR
LB 09488	15 09 48.	+ 24 45		16.7	FAINT BLUE STAR
LB 09487	15 09 48.	+ 24 57		17.0	FAINT BLUE STAR
LB 09486	15 09 48.	+ 26 06		17.2	FAINT BLUE STAR
ZC 1509.8+2617	15 09 48.	+ 26 17	870		CLUSTER OF GALAXIES
LB 09485	15 09 48.	+ 26 27		16.7	FAINT BLUE STAR
ZWG 319.002	15 09 48.	+ 65 05		15.7	GALAXY
ZWG 318.019	15 09 48.	+ 65 05		15.7	GALAXY
UGC 09764	15 09 48.	+ 65 05	156	15.7	GALAXY
ZWG 021.044	15 09 48.	- 02 50		15.6	GALAXY
MCG+04-36-025	15 09 51.	+ 21 27	72	15.6	GALAXY
LB 00760	15 09 52.	+ 28 10 36.		17.3	FAINT BLUE STAR
SHB 308	15 09 52.5	+ 15 51 39.		18.	QUASI-STELLAR OBJECT
BC M1509+159	15 09 52.5	+ 15 51 42.		17.	QUASI-STELLAR OBJECT
ZWG 021.046	15 09 54.	+ 01 40		15.5	GALAXY
ZWG 049.041	15 09 54.	+ 07 55		15.7	GALAXY
ZWG 077.042	15 09 54.	+ 09 32		15.4	GALAXY
ZWG 165.028	15 09 54.	+ 28 37		15.7	GALAXY
RNGC 5881	15 09 54.	+ 65 03		15.0	GALAXY
ABC 2038	15 09 56.	+ 15 25		17.2	RICH CLUSTER OF GALAXIES
REIZ 4494	15 09 57.	+ 57 50	96	14.2	GALAXY
ZWG 021.047	15 10 00.	+ 01 41		15.3	GALAXY
ZC 1510.0+0315	15 10 00.	+ 03 15	21100		CLUSTER OF GALAXIES
ZC 1510.0+0957	15 10 00.	+ 09 57	2350		CLUSTER OF GALAXIES
ZC 1510.0+1753	15 10 00.	+ 17 53	5040		CLUSTER OF GALAXIES
LB 09494	15 10 00.	+ 20 38		16.2	FAINT BLUE STAR
LB 09493	15 10 00.	+ 23 21		16.4	FAINT BLUE STAR
LB 09492	15 10 00.	+ 23 28		18.7	FAINT BLUE STAR
LB 09491	15 10 00.	+ 24 20		16.7	FAINT BLUE STAR
ZWG 193.018	15 10 00.	+ 36 58		15.7	GALAXY
ZC 1510.0+3707	15 10 00.	+ 37 07	1080		CLUSTER OF GALAXIES
MCG+10-22-003	15 10 00.	+ 61 19	78	14.	GALAXY
ZWG 318.020	15 10 00.	+ 63 13		15.7	GALAXY
KARA.73B 0664	15 10 00.	+ 63 13	24	15.7	ISOLATED GALAXY E
REIZ 4493	15 10 03.	+ 36 50	18	15.1	GALAXY

OBJECT NAME	RIGHT ASCEN.	DECLINATION	DIAM.	MAGN.	TYPE OF OBJECT
ZWG 021.048	15 10 06.	+ 02 07		15.3	GALAXY
ZWG 049.042	15 10 06.	+ 07 57		15.5	GALAXY
ZWG 077.043	15 10 06.	+ 09 02		15.7	GALAXY
ZWG 077.044	15 10 06.	+ 13 00		15.3	GALAXY
8ZW 1510+13.0	15 10 06.	+ 13 00		15.3	COMPACT GALAXY
ZC 1510.1+1521	15 10 06.	+ 15 21	1810		CLUSTER OF GALAXIES
ZWG 106.020	15 10 06.	+ 18 50		15.2	GALAXY
KARA.73B 0665	15 10 06.	+ 18 50	48	15.2	ISOLATED GALAXY S
LB 09495	15 10 06.	+ 25 42		17.7	FAINT BLUE STAR
MCG+06-33-027	15 10 06.	+ 36 50	12	16.	GALAXY
MCG+08-28-006	15 10 06.	+ 46 26	24	16.	GALAXY
ZWG 249.011	15 10 06.	+ 46 27		15.4	GALAXY
ZC 1510.1+5339	15 10 06.	+ 53 39	1210		CLUSTER OF GALAXIES
SHB 309	15 10 08.9	- 08 54 48.		16.5	QUASI-STELLAR OBJECT
BC PKS1510-08	15 10 08.97	- 08 54 48.3		16.52	QUASI-STELLAR OBJECT
LB 00761	15 10 10.	+ 27 46 12.		17.3	FAINT BLUE STAR
ABC 2039	15 10 11.	+ 09 59		17.2	RICH CLUSTER OF GALAXIES
RNGC 5877	15 10 11.	- 04 44			NON-EXISTENT OBJECT
ZWG 021.049	15 10 12.	+ 01 38		15.2	GALAXY
MCG+03-39-013	15 10 12.	+ 18 51	42	15.2	GALAXY
ZWG 135.028	15 10 12.	+ 20 50		15.3	GALAXY
UGC 09765	15 10 12.	+ 20 50	72	15.3	GALAXY Sa-b
ZC 1510.2+2253	15 10 12.	+ 22 53	1550		CLUSTER OF GALAXIES
LB 09500	15 10 12.	+ 23 45		16.8	FAINT BLUE STAR
LB 09499	15 10 12.	+ 24 50		17.1	FAINT BLUE STAR
LB 09498	15 10 12.	+ 25 38		17.6	FAINT BLUE STAR
LB 09497	15 10 12.	+ 25 49		18.1	FAINT BLUE STAR
LB 09496	15 10 12.	+ 25 52		16.9	FAINT BLUE STAR
ZWG 165.029	15 10 12.	+ 30 42		15.6	GALAXY
ZWG 297.005	15 10 12.	+ 61 19		15.3	GALAXY
UGC 09766	15 10 12.	+ 61 19	96		GALAXY SPb
IC 1104	15 10 13.	- 04 54			NONSTELLAR OBJECT
ABC 2040	15 10 15.	+ 07 37		15.7	RICH CLUSTER OF GALAXIES
UGC 09767	15 10 18.	+ 07 37	66	16.0	GALAXY
MCG+03-39-014	15 10 18.	+ 19 22	48	15.	GALAXY
MCG+04-36-026	15 10 18.	+ 20 49	60	15.3	GALAXY
LB 09502	15 10 18.	+ 23 46		17.7	FAINT BLUE STAR
LB 09501	15 10 18.	+ 25 02		16.6	FAINT BLUE STAR
TON-N 0221	15 10 18.	+ 28 02		15.2	BLUE STAR
ZWG 049.043	15 10 24.	+ 04 42		15.3	GALAXY
ABC 2043	15 10 24.	+ 16 37		17.2	RICH CLUSTER OF GALAXIES
LB 09506	15 10 24.	+ 21 42		16.6	FAINT BLUE STAR
LB 09505	15 10 24.	+ 22 13		16.8	FAINT BLUE STAR
ZC 1510.4+2229	15 10 24.	+ 22 29	470		CLUSTER OF GALAXIES
LB 09504	15 10 24.	+ 25 54		17.5	FAINT BLUE STAR
LB 09503	15 10 24.	+ 26 21		16.5	FAINT BLUE STAR
ZC 1510.4+3859	15 10 24.	+ 38 59	1210		CLUSTER OF GALAXIES
MCG+10-22-004	15 10 24.	+ 59 59	168	13.	GALAXY
LB 00762	15 10 26.	+ 29 15 48.		16.0	FAINT BLUE STAR
ABC 2044	15 10 28.	+ 14 30		17.6	RICH CLUSTER OF GALAXIES
ZWG 021.050	15 10 30.	+ 02 12		15.7	GALAXY
ZWG 049.044	15 10 30.	+ 03 20		14.8	GALAXY
MCG+01-39-005	15 10 30.	+ 03 20	18	14.8	GALAXY
ZWG 221.035	15 10 30.	+ 40 45		15.2	GALAXY
7ZW 587	15 10 30.	+ 61 08			COMPACT GALAXY
MCG-03-39-002	15 10 30.	- 17 57	120	14.	GALAXY
IC 4536	15 10 33.	- 17 56 48.			NONSTELLAR OBJECT
ZWG 049.045	15 10 36.	+ 04 41		15.5	GALAXY
ZWG 049.046	15 10 36.	+ 07 29		15.2	GALAXY
8ZW 1510+09.6	15 10 36.	+ 09 38		18.2	COMPACT GALAXY
LB 09508	15 10 36.	+ 25 42		17.5	FAINT BLUE STAR
LB 09507	15 10 36.	+ 26 21		15.6	FAINT BLUE STAR
MCG+07-31-036	15 10 36.	+ 40 44	30	15.	GALAXY
ZWG 297.006	15 10 36.	+ 60 00		13.2	GALAXY
UGC 09768	15 10 36.	+ 60 00	210	13.2	GALAXY SB?c-IE
ZC 1510.6+6347	15 10 36.	+ 63 47	2690		CLUSTER OF GALAXIES
ZWG 337.031	15 10 36.	+ 71 12		15.5	GALAXY
MCG-03-39-003	15 10 36.	- 20 30	216	14.	GALAXY
RNGC 5894	15 10 38.	+ 60 00		13.0	GALAXY
ABC 2046	15 10 40.	+ 35 02		16.9	RICH CLUSTER OF GALAXIES
ZWG 049.047	15 10 42.	+ 03 26		15.6	GALAXY
ZWG 049.048	15 10 42.	+ 04 28		15.4	GALAXY
KARA.72 459A	15 10 42.	+ 04 28	36	16.4	PART OF DOUBLE GALAXY
MCG+01-39-006	15 10 42.	+ 04 28	36	15.4	GALAXY
ZWG 049.049	15 10 42.	+ 04 40		15.4	GALAXY
ZWG 049.050	15 10 42.	+ 05 32		15.3	GALAXY
ZWG 049.051	15 10 42.	+ 08 03		15.3	GALAXY
ZC 1510.7+1427	15 10 42.	+ 14 27	1010		CLUSTER OF GALAXIES
LB 09510	15 10 42.	+ 24 52		16.4	FAINT BLUE STAR
LB 09509	15 10 42.	+ 26 05		17.7	FAINT BLUE STAR
TON-N 0222	15 10 42.	+ 26 18		14.9	BLUE STAR
ZWG 274.033	15 10 42.	+ 55 59		15.7	GALAXY
UGC 09769	15 10 42.	+ 55 59	180	15.7	GALAXY
ZC 1510.7+6733	15 10 42.	+ 67 33	1210		CLUSTER OF GALAXIES
MCG+12-14-018	15 10 42.	+ 71 11	21	16.	GALAXY
LB 00763	15 10 46.	+ 27 45 00.		16.8	FAINT BLUE STAR
ZWG 049.052	15 10 48.	+ 02 42		15.4	GALAXY
ZWG 049.053	15 10 48.	+ 04 28		14.8	GALAXY
KARA.72 459B	15 10 48.	+ 04 28	42	14.8	PART OF DOUBLE GALAXY
MCG+01-39-007	15 10 48.	+ 04 28	48	14.8	GALAXY
ZWG 049.054	15 10 48.	+ 04 40		15.5	GALAXY
ZWG 049.055	15 10 49.	+ 04 56		15.7	GALAXY
ZWG 049.056	15 10 48.	+ 05 13		15.7	GALAXY
ZWG 049.057	15 10 48.	+ 07 25		15.5	GALAXY
ZWG 049.058	15 10 48.	+ 08 14		15.4	GALAXY
LB 09512	15 10 48.	+ 21 31		16.1	FAINT BLUE STAR
LB 09511	15 10 48.	+ 22 04		15.7	FAINT BLUE STAR
ZC 1510.8+4328	15 10 48.	+ 43 28	810		CLUSTER OF GALAXIES
ZC 1510.8+4345	15 10 48.	+ 43 45	740		CLUSTER OF GALAXIES
MCG+09-25-033	15 10 48.	+ 50 44	54	15.	GALAXY
MCG+09-25-034	15 10 48.	+ 55 59	108	14.	GALAXY
MCG-04-36-003	15 10 48.	- 23 28 30.	24	15.5	GALAXY
IC 1105	15 10 49.	+ 04 26 57.			NONSTELLAR OBJECT
REIZ 4495	15 10 50.	+ 41 25	18	14.3	GALAXY
RNGC 5866B	15 10 50.	+ 55 58		15.5	GALAXY
MAI 092	15 10 52.	+ 57 11	47		DWARF SPHEROIDAL GALAXY
LB 00764	15 10 53.	+ 29 35 18.		17.2	FAINT BLUE STAR
RNGC 5886	15 10 53.	+ 41 25		15.0	GALAXY
LB 00765	15 10 54.	+ 60 14 00.		16.7	FAINT BLUE STAR
ZWG 049.059	15 10 54.	+ 06 48		15.3	GALAXY
ZWG 077.045	15 10 54.	+ 09 29		15.6	GALAXY
LB 09514	15 10 54.	+ 23 54		16.3	FAINT BLUE STAR
LB 09513	15 10 54.	+ 25 14		16.9	FAINT BLUE STAR
ZWG 221.036	15 10 54.	+ 41 25		15.1	GALAXY
KEEF 5866B	15 10 54.	+ 55 58	132		Sc GALAXY
MCG-02-39-006	15 10 54.	- 14 06	180	12.5	GALAXY
RNGC 5892	15 10 54.	- 15 18		13.0	GALAXY
MCG-02-39-007	15 10 54.	- 15 18	180	13.	GALAXY

OBJECT NAME	RIGHT ASCEN.	DECLINATION	DIAM.	MAGN.	TYPE OF OBJECT
MRSL 323+04/1	15 10 54.	- 52 48	60		HII REGION
VHE 66	15 10 54.	- 62 35	96		REFLECTION NEBULA
ZWG 049.060	15 11 00.	+ 05 50		15.6	GALAXY
ZC 1511.0+1912	15 11 00.	+ 19 12	3490		CLUSTER OF GALAXIES
LB 09515	15 11 00.	+ 26 24		16.8	FAINT BLUE STAR
LB 00766	15 11 00.	+ 30 52 24.		16.8	FAINT BLUE STAR
ZWG 165.030	15 11 00.	+ 32 16		15.7	GALAXY
MCG+07-31-037	15 11 00.	+ 40 43	36	17.	GALAXY
ZWG 319.003	15 11 00.	+ 64 10		15.6	GALAXY
ZWG 318.021	15 11 00.	+ 64 10		15.6	GALAXY
7ZW 588	15 11 00.	+ 64 11			COMPACT GALAXY
SC 1508-1506.0	15 11 00.	- 15 17 18.	96		NEBULA
RNGC 5878	15 11 01.	- 14 04		12.5	GALAXY
FATH 1.703	15 11 01.	- 15 16	136		NEBULA
RNGC 5884	15 11 05.	+ 32 03			NON-EXISTENT OBJECT
ZWG 049.061	15 11 06.	+ 04 15		15.7	GALAXY
ZWG 049.062	15 11 06.	+ 05 38		15.5	GALAXY
ZC 1511.1+1140	15 11 06.	+ 11 40	1140		CLUSTER OF GALAXIES
ZC 1511.1+2102	15 11 06.	+ 21 02	2890		CLUSTER OF GALAXIES
ZWG 165.031	15 11 06.	+ 27 05		15.6	GALAXY
ZC 1511.1+3735	15 11 06.	+ 37 35	1210		CLUSTER OF GALAXIES
ZC 1511.1+5924	15 11 06.	+ 59 24	1280		CLUSTER OF GALAXIES
MCG+12-14-019	15 11 06.	+ 71 25 30.	24	17.	GALAXY
FATH 1.704	15 11 08.	+ 14 59	14		NEBULA
MCG-02-39-008	15 11 09.	- 12 55	36	15.	GALAXY
SC 1508-1244.2	15 11 09.	- 12 55 29.	18		NEBULA
RNGC 5888	15 11 11.	+ 41 27		14.5	GALAXY
IC 1110	15 11 11.	+ 67 33 09.			NONSTELLAR OBJECT
ZWG 021.051	15 11 12.	+ 02 12		15.6	GALAXY
ZWG 049.063	15 11 12.	+ 04 16		15.6	GALAXY
MCG+01-39-008	15 11 12.	+ 04 16	36	15.6	GALAXY
ZWG 049.064	15 11 12.	+ 08 17		15.4	GALAXY
LB 09518	15 11 12.	+ 22 38		17.0	FAINT BLUE STAR
ZWG 135.029	15 11 12.	+ 25 23		15.6	GALAXY
UGC 09770	15 11 12.	+ 25 23	78	15.6	GALAXY Sb
TON-N 0223	15 11 12.	+ 26 33		16.6	BLUE STAR
LB 09517	15 11 12.	+ 26 34		16.0	FAINT BLUE STAR
LB 09516	15 11 12.	+ 26 36		16.2	FAINT BLUE STAR
SHAH 078	15 11 12.	+ 40 21	72	16.8	GROUP OF COMPACT GALAXIES
ZWG 221.037	15 11 12.	+ 41 27		14.3	GALAXY
UGC 09771	15 11 12.	+ 41 27	90	14.3	GALAXY SBb
ZWG 354.020	15 11 12.	+ 75 17		15.7	GALAXY
REIZ 4496	15 11 14.	+ 41 27	60	13.9	GALAXY
MCG+06-33-028	15 11 15.	+ 34 19	30	16.	GALAXY
MCG+07-31-038	15 11 15.	+ 41 28	66	14.	GALAXY
LB 00767	15 11 17.	+ 12 53 00.		17.5	FAINT BLUE STAR
RNGC 5889	15 11 17.	+ 41 27		15.5	GALAXY
ZWG 021.052	15 11 18.	+ 02 16		15.6	GALAXY
ZWG 021.053	15 11 18.	+ 02 18		15.7	GALAXY
LB 09520	15 11 18.	+ 22 21		16.8	FAINT BLUE STAR
LB 09519	15 11 18.	+ 25 34		17.0	FAINT BLUE STAR
TON-N 0224	15 11 18.	+ 26 32		15.5	BLUE STAR
ZC 1511.3+2922	15 11 18.	+ 29 22	740		CLUSTER OF GALAXIES
ZWG 165.032	15 11 18.	+ 31 08		15.4	GALAXY
UGC 09772	15 11 18.	+ 34 21	60	16.5	GALAXY SBc
ZWG 221.038	15 11 18.	+ 40 44		15.0	GALAXY
ZWG 274.034	15 11 18.	+ 51 37		15.3	GALAXY
MIL 34	15 11 20.	- 59 07	870		SUPERNOVA REMNANT
MCG+04-36-027	15 11 21.	+ 25 22 30.	66	15.6	GALAXY
REIZ 4497	15 11 21.	+ 40 43	18	14.3	GALAXY
MCG+07-31-039	15 11 21.	+ 40 43	30	15.	GALAXY
ZWG 049.065	15 11 24.	+ 04 38		15.6	GALAXY
ZC 1511.4+1549	15 11 24.	+ 15 49	2490		CLUSTER OF GALAXIES
LB 09522	15 11 24.	+ 25 52		18.2	FAINT BLUE STAR
LB 09521	15 11 24.	+ 26 00		17.7	FAINT BLUE STAR
MCG+07-31-040	15 11 24.	+ 40 42	42	15.	GALAXY
ZWG 221.039	15 11 24.	+ 40 43		15.3	GALAXY
MCG+09-25-036	15 11 24.	+ 51 34	36	15.	GALAXY
MCG+09-25-035	15 11 24.	+ 51 34	42	17.	GALAXY
MCG-02-39-009	15 11 24.	- 14 10 30.	24	15.	GALAXY
IC 1106	15 11 27.	+ 04 53 49.			NONSTELLAR OBJECT
LB 00768	15 11 27.	+ 12 09 24.		17.5	FAINT BLUE STAR
REIZ 4498	15 11 27.	+ 40 42	12	14.3	GALAXY
ZWG 021.054	15 11 30.	+ 01 55		15.7	GALAXY
ZWG 049.066	15 11 30.	+ 04 54		15.2	GALAXY
LB 09523	15 11 30.	+ 22 48		17.0	FAINT BLUE STAR
ZWG 165.033	15 11 30.	+ 27 05		15.7	GALAXY
ZWG 319.004	15 11 30.	+ 67 33		14.9	GALAXY
ZWG 318.022	15 11 30.	+ 67 33		14.9	GALAXY
UGC 09773	15 11 30.	+ 67 33	78	14.9	GALAXY Sa
KARA.73B 0666	15 11 30.	+ 67 33	96	14.9	ISOLATED GALAXY S
OCL 0932	15 11 30.	- 58 53	270	12.	OPEN STAR CLUSTER
VHA 170	15 11 30.	- 58 53	90		OPEN STAR CLUSTER
ZWG 049.067	15 11 36.	+ 04 24		15.6	GALAXY
LB 09526	15 11 36.	+ 23 00		16.8	FAINT BLUE STAR
LB 09525	15 11 36.	+ 25 07		16.3	FAINT BLUE STAR
LB 09524	15 11 36.	+ 26 35		17.0	FAINT BLUE STAR
ZWG 165.034	15 11 36.	+ 28 41		15.7	GALAXY
ZWG 221.040	15 11 36.	+ 42 31		15.4	GALAXY
REIZ 4500	15 11 36.	+ 42 32	30	14.2	GALAXY
ZWG 249.012	15 11 36.	+ 47 18		15.7	GALAXY
KARA.73B 0667	15 11 36.	+ 47 18	54	15.7	ISOLATED GALAXY S
MCG+11-19-001	15 11 36.	+ 67 32	72	15.	GALAXY
MCG-02-39-010	15 11 36.	- 14 29	60	15.5	GALAXY
RNGC 5880	15 11 37.	- 14 29		15.0	GALAXY
REIZ 4499	15 11 38.	+ 41 31	42	14.7	GALAXY
ABC 2045	15 11 38.	- 02 35		17.1	RICH CLUSTER OF GALAXIES
IC 1107	15 11 40.	+ 04 54 02.			NONSTELLAR OBJECT
HELW 476	15 11 40.	- 14 28 04.			NEBULA
ZWG 049.068	15 11 42.	+ 04 54		15.5	GALAXY
ZC 1511.7+2406	15 11 42.	+ 24 06	3230		CLUSTER OF GALAXIES
LB 09527	15 11 42.	+ 24 39		16.7	FAINT BLUE STAR
ZWG 221.041	15 11 42.	+ 42 09		14.1	GALAXY
UGC 09774	15 11 42.	+ 42 09	84	14.1	GALAXY SBb
MCG+07-31-042	15 11 42.	+ 42 09	60	14.	GALAXY
MCG+07-31-041	15 11 42.	+ 42 31	45	15.	GALAXY
ZC 1511.7+4342	15 11 42.	+ 43 42	2080		CLUSTER OF GALAXIES
MCG+08-28-007	15 11 42.	+ 47 16	30	16.	GALAXY
MCG+10-22-005	15 11 42.	+ 57 09	42	16.	GALAXY
MRSL 320-01/1	15 11 42.	- 59 09	3060		HII REGION
HOLM 701B	15 11 45.	+ 42 10	48	13.6	PART OF MULTIPLE GALAXY
LB 00769	15 11 47.	+ 00 58 00.		16.0	FAINT BLUE STAR
ZWG 077.046	15 11 48.	+ 12 16		15.3	GALAXY
ZWG 077.047	15 11 48.	+ 14 05		15.4	GALAXY
ZWG 106.021	15 11 48.	+ 20 10		15.5	GALAXY
UGC 09775	15 11 48.	+ 20 10	66	15.5	GALAXY Sa
MCG+03-39-015	15 11 48.	+ 20 11	36		GALAXY
LB 09528	15 11 48.	+ 21 44		16.3	FAINT BLUE STAR

OBJECT NAME	RIGHT ASCEN.	DECLINATION	DIAM.	MAGN.	TYPE OF OBJECT
RNGC 5893	15 11 48.	+ 42 09		14.0	GALAXY
UGC 09776	15 11 48.	+ 57 10	60	17.	GALAXY DWRF IR
REIZ 4501	15 11 49.	+ 42 08	54	13.7	GALAXY
FATH 1.705	15 11 50.	+ 14 04	19		NEBULA
BSO 29	15 11 53.	+ 44 46 05.		17.36	BLUE STELLAR OBJECT
ZWG 049.069	15 11 54.	+ 08 30		15.7	GALAXY
MCG+03-39-016	15 11 54.	+ 20 11 30.	54	15.5	GALAXY
LB 09530	15 11 54.	+ 25 46		17.6	FAINT BLUE STAR
LB 09529	15 11 54.	+ 25 46		17.1	FAINT BLUE STAR
ZWG 135.030	15 11 54.	+ 25 53		15.7	GALAXY
ZC 1511.9+3227	15 11 54.	+ 32 27	740		CLUSTER OF GALAXIES
7ZW 589	15 11 54.	+ 58 21			COMPACT GALAXY
REIZ 4503	15 11 55.	+ 42 11	72	14.9	GALAXY
REIZ 4504	15 11 55.	+ 42 12	48	14.6	GALAXY
HOLM 701A	15 11 55.	+ 42 12	66	14.9	PART OF MULTIPLE GALAXY
REIZ 4502	15 11 56.	+ 41 32	48	15.1	GALAXY
IC 1114	15 11 56.	+ 75 38			NONSTELLAR OBJECT
ZWG 049.070	15 12 00.	+ 05 38		15.6	GALAXY
ZWG 049.071	15 12 00.	+ 08 05		15.6	GALAXY
ZWG 077.048	15 12 00.	+ 10 08		15.6	GALAXY
ZWG 135.031	15 12 00.	+ 20 40		15.0	GALAXY
UGC 09777	15 12 00.	+ 20 40	102	15.0	GALAXY S
LB 09531	15 12 00.	+ 23 26		16.2	FAINT BLUE STAR
ZWG 221.042	15 12 00.	+ 42 11		15.5	GALAXY
RNGC 5896	15 12 00.	+ 42 11		15.5	GALAXY
RNGC 5895	15 12 00.	+ 42 11		15.5	GALAXY
MCG+07-31-044	15 12 00.	+ 42 12	9	16.	GALAXY
MCG+07-31-043	15 12 00.	+ 42 12	48	15.	GALAXY
HOLM 701C	15 12 00.	+ 42 13	48	14.4	PART OF MULTIPLE GALAXY
ZC 1512.0+5209	15 12 00.	+ 52 09	1080		CLUSTER OF GALAXIES
ZC 1512.0-0247	15 12 00.	- 02 47	2620		CLUSTER OF GALAXIES
HOLM 703C	15 12 02.	+ 75 36	12	14.3	PART OF MULTIPLE GALAXY
LB 00771	15 12 03.	+ 15 44 48.		17.6	FAINT BLUE STAR
LB 00770	15 12 03.	+ 27 07 00.		16.8	FAINT BLUE STAR
ZWG 021.055	15 12 06.	+ 01 59		15.5	GALAXY
MCG+04-36-028	15 12 06.	+ 20 39	90	15.0	GALAXY
LB 09532	15 12 06.	+ 20 45		14.5	FAINT BLUE STAR
ZWG 135.032	15 12 06.	+ 22 29		15.3	GALAXY
ZWG 135.033	15 12 06.	+ 23 05		15.6	GALAXY
ZC 1512.1+3210	15 12 06.	+ 32 10	470		CLUSTER OF GALAXIES
ZC 1512.1+3627	15 12 06.	+ 36 27	2350		CLUSTER OF GALAXIES
ZWG 354.021	15 12 06.	+ 75 20		14.7	GALAXY
RNGC 5909	15 12 06.	+ 75 20		14.5	GALAXY
UGC 09778	15 12 06.	+ 75 20	66	14.7	GALAXY S
KARA.72 460A	15 12 06.	+ 75 20	66	14.7	PART OF DOUBLE GALAXY
HOLM 703B	15 12 10.	+ 75 34	60	14.0	PART OF MULTIPLE GALAXY
ZWG 021.056	15 12 12.	+ 01 20		15.2	GALAXY
UGC 09779	15 12 12.	+ 01 20	78	15.2	GALAXY S0-a
MCG+00-39-012	15 12 12.	+ 01 20	42	15.2	GALAXY
ZWG 077.049	15 12 12.	+ 08 32		15.0	GALAXY
MCG+02-39-008	15 12 12.	+ 08 32	48	15.0	GALAXY
ZWG 135.034	15 12 12.	+ 20 59		15.4	GALAXY
MCG+04-36-029	15 12 12.	+ 22 27	24	15.3	GALAXY
LB 09533	15 12 12.	+ 26 14		16.6	FAINT BLUE STAR
ZWG 249.013	15 12 12.	+ 44 47		15.5	GALAXY
UGC 09780	15 12 12.	+ 44 47	126	15.5	GALAXY Sc
KARA.73B 0668	15 12 12.	+ 44 47	132	15.5	ISOLATED GALAXY S
MCG+10-22-006	15 12 12.	+ 58 41	27	15.	GALAXY
ZWG 319.005	15 12 12.	+ 64 54		15.4	GALAXY
ZWG 318.023	15 12 12.	+ 64 54		15.4	GALAXY
ZWG 319.006	15 12 12.	+ 66 15		15.6	GALAXY
ZWG 318.024	15 12 12.	+ 66 15		15.6	GALAXY
LB 00772	15 12 13.	+ 27 48 30.		17.5	FAINT BLUE STAR
SC 1509-1338.5	15 12 13.	- 13 49 44.	54		NEBULA
RNGC 5887	15 12 14.	+ 01 20		15.0	GALAXY
MCG+08-28-008	15 12 15.	+ 47 41 30.	36	16.	GALAXY
HELW 477	15 12 15.	- 14 23 26.			NEBULA
MCG-04-36-004	15 12 15.	- 22 37 30.	42	14.5	GALAXY
ZWG 049.072	15 12 18.	+ 05 43		15.7	GALAXY
UGC 09781	15 12 18.	+ 05 43	84	15.7	GALAXY Sb-c
MCG+01-39-009	15 12 18.	+ 05 43	60	15.7	GALAXY
TON-N 0768	15 12 18.	+ 24 21		14.3	BLUE STAR
ZWG 135.035	15 12 18.	+ 25 57		15.7	GALAXY
ZC 1512.3+3115	15 12 18.	+ 31 15	4030		CLUSTER OF GALAXIES
MCG+08-28-009	15 12 13.	+ 44 45	120	14.	GALAXY
ZWG 297.007	15 12 18.	+ 58 42		15.0	GALAXY
MCG+10-22-007	15 12 18.	+ 59 07	10	17.	GALAXY
MCG+11-19-002	15 12 18.	+ 62 55	42	16.	GALAXY
MCG+11-19-003	15 12 18.	+ 67 08 30.	30	17.	GALAXY
ZWG 354.022	15 12 18.	+ 75 20		14.6	GALAXY
RNGC 5912	15 12 18.	+ 75 20		14.5	GALAXY
KARA.72 460B	15 12 18.	+ 75 20	60	14.6	PART OF DOUBLE GALAXY
MCG+13-11-010	15 12 18.	+ 75 34	66	14.0	GALAXY
MCG-02-39-011	15 12 18.	- 13 50	48	15.	GALAXY
MCG-02-39-012	15 12 18.	- 14 24 30.	18	15.5	GALAXY
LB 00773	15 12 19.	- 00 18 24.		16.3	FAINT BLUE STAR
RNGC 5883	15 12 19.	- 14 24		15.0	GALAXY
RNGC 5885	15 12 21.	- 09 52		12.0	GALAXY
MCG-02-39-013	15 12 21.	- 09 54	240	12.	GALAXY
HN 0031	15 12 22.	- 14 26			NEBULA
FATH 1.707	15 12 23.	+ 14 32	19		NEBULA
UGC 09782	15 12 24.	+ 01 23	60	16.5	GALAXY Sc
ZC 1512.4+1646	15 12 24.	+ 16 46	540		CLUSTER OF GALAXIES
LB 09534	15 12 24.	+ 24 49		17.9	FAINT BLUE STAR
ZWG 165.035	15 12 24.	+ 26 47		15.3	GALAXY
SVEK 441	15 12 24.	- 09 53	180	12.9	GALAXY
MCG-02-39-014	15 12 24.	- 14 27	36	15.	GALAXY
HOLM 703A	15 12 25.	+ 75 34	18	13.8	PART OF MULTIPLE GALAXY
ZWG 049.073	15 12 30.	+ 06 46		15.6	GALAXY
LB 00774	15 12 30.	+ 31 13 30.		15.9	FAINT BLUE STAR
UGC 09783	15 12 30.	+ 46 32	60	16.5	GALAXY
MCG+13-11-011	15 12 30.	+ 75 34	27	13.8	GALAXY
MRSL 321-00/1	15 12 30.	- 58 01	660		HII REGION
FATH 1.706	15 12 32.	- 15 29	11		NEBULA
LB 09536	15 12 36.	+ 22 22		16.4	FAINT BLUE STAR
LB 09535	15 12 36.	+ 23 46		16.1	FAINT BLUE STAR
MCG+05-36-011	15 12 36.	+ 26 45 30.	42	15.3	GALAXY
SN 1972A	15 12 36.	+ 27 50		17.0	SUPERNOVA
LB 00775	15 12 36.	+ 28 10 36.		15.9	FAINT BLUE STAR
MCG+09-25-037	15 12 36.	+ 55 43	60	15.	GALAXY
UGC 09784	15 12 36.	+ 62 52	84	16.5	GALAXY DWARF
OCL 0935	15 12 36.	- 58 08	180	15.	OPEN STAR CLUSTER
LB 00776	15 12 39.	+ 31 33 54.		16.6	FAINT BLUE STAR
SN 1973B	15 12 40.	+ 02 55		15.0	SUPERNOVA
REIZ 4505	15 12 41.	+ 12 13	24	15.7	GALAXY
ARC 2049	15 12 41.	+ 31 59		17.0	RICH CLUSTER OF GALAXIES
ZC 1512.7+0434	15 12 42.	+ 04 34	1550		CLUSTER OF GALAXIES
ZWG 077.050	15 12 42.	+ 12 13		15.4	GALAXY
8ZW 1512+12.5	15 12 42.	+ 12 31		17.4	COMPACT GALAXY
LB 09537	15 12 42.	+ 24 00		17.1	FAINT BLUE STAR
ZC 1512.7+5451	15 12 42.	+ 54 51	1680		CLUSTER OF GALAXIES
MCG+12-11-012	15 12 42.	+ 75 30	33	14.3	GALAXY
HELW 478	15 12 43.	- 14 18 13.			NEBULA
MCG+08-28-010	15 12 45.	+ 46 30	60	16.	GALAXY
MCG+08-28-011	15 12 45.	+ 50 31	60	14.1	GALAXY
BC 4C37.43	15 12 46.	+ 37 02 30.			QUASI-STELLAR OBJECT
SHB 310	15 12 46.9	+ 37 01 54.		15.5	QUASI-STELLAR OBJECT
LB 00777	15 12 47.	+ 30 37 12.		15.7	FAINT BLUE STAR
ZWG 049.074	15 12 48.	+ 05 23		15.4	GALAXY
UGC 09785	15 12 48.	+ 05 23	60	15.4	GALAXY Sa
ZWG 049.075	15 12 48.	+ 05 46		15.3	GALAXY
LB 09538	15 12 48.	+ 23 44		15.8	FAINT BLUE STAR
ZWG 135.036	15 12 48.	+ 26 05		15.6	GALAXY
ZC 1512.8+2929	15 12 48.	+ 29 29	610		CLUSTER OF GALAXIES
ZC 1512.8+3044	15 12 48.	+ 30 44	740		CLUSTER OF GALAXIES
ZWG 274.035	15 12 48.	+ 50 33		15.4	GALAXY
MCG-07-31-010	15 12 48.	- 43 50	78	14.	GALAXY
OCL 0931	15 12 48.	- 59 28	120	13.	OPEN STAR CLUSTER
VHA 171	15 12 48.	- 59 28	150		OPEN STAR CLUSTER
MRSL 317-06/1	15 12 48.	- 65 07	1020		HII REGION
LB 00778	15 12 49.	+ 30 58 42.		15.9	FAINT BLUE STAR
RNGC 5902	15 12 49.	+ 50 33		15.5	GALAXY
ARC 2048	15 12 50.	+ 04 35		16.0	RICH CLUSTER OF GALAXIES
ZWG 049.076	15 12 54.	+ 04 52		15.3	GALAXY
ZC 1512.9+2459	15 12 54.	+ 24 59	810		CLUSTER OF GALAXIES
FATH 1.708	15 12 55.	- 15 46	11		NEBULA
HMS 1513+0433	15 13	+ 04 33			CLUSTER OF GALAXIES
ZWG 021.057	15 13 00.	+ 01 57		15.5	GALAXY
MCG+00-39-013	15 13 00.	+ 01 57	30	15.5	GALAXY
ZWG 021.058	15 13 00.	+ 02 25		15.1	GALAXY
MCG+00-39-014	15 13 00.	+ 02 25	36	15.1	GALAXY
ZWG 049.077	15 13 00.	+ 06 34		14.9	GALAXY
MCG+01-39-010	15 13 00.	+ 06 34	24	14.9	GALAXY
ZWG 049.078	15 13 00.	+ 07 40		15.6	GALAXY
LB 09540	15 13 00.	+ 23 02		16.0	FAINT BLUE STAR
ZWG 135.037	15 13 00.	+ 25 11		15.7	GALAXY
LB 09539	15 13 00.	+ 26 19		16.4	FAINT BLUE STAR
IC 1111	15 13 00.	+ 54 42 39.			NONSTELLAR OBJECT
7ZW 590	15 13 00.	+ 63 05			COMPACT GALAXY
LB 00779	15 13 03.	+ 01 46 18.		17.7	FAINT BLUE STAR
REIZ 4511	15 13 03.	+ 50 28	24	14.1	GALAXY
LB 00780	15 13 04.	+ 14 15 54.		15.9	FAINT BLUE STAR
ZWG 021.059	15 13 06.	+ 00 09		15.6	GALAXY
LB 00781	15 13 06.	+ 00 44 48.		17.4	FAINT BLUE STAR
MCG+01-39-011	15 13 06.	+ 04 33	9	17.	GALAXY
LB 09542	15 13 06.	+ 21 56		16.3	FAINT BLUE STAR
LB 09541	15 13 06.	+ 23 17		16.6	FAINT BLUE STAR
MCG+04-36-030	15 13 06.	+ 25 10	54	15.7	GALAXY
ZC 1513.1+4130	15 13 06.	+ 41 30	1410		CLUSTER OF GALAXIES
MCG+12-14-020	15 13 06.	+ 71 23	66	16.	GALAXY
ZWG 337.032	15 13 06.	+ 71 25		15.6	GALAXY
UGC 09786	15 13 06.	+ 71 25	90	15.6	GALAXY Sb-c
REIZ 4512	15 13 08.	+ 50 49	48	14.4	GALAXY
REIZ 4506	15 13 10.	+ 39 48	15	15.1	GALAXY
ZWG 021.060	15 13 12.	+ 01 37		15.4	GALAXY
UGC 09787	15 13 12.	+ 01 37	60	15.4	GALAXY
MCG+00-39-015	15 13 12.	+ 01 37	42	15.4	GALAXY
ZWG 049.079	15 13 12.	+ 08 29		15.4	GALAXY
UGC 09788	15 13 12.	+ 08 29	60	15.4	GALAXY IRR?
ZWG 135.038	15 13 12.	+ 25 38		15.2	GALAXY
ZC 1513.2+3203	15 13 12.	+ 32 03	2080		CLUSTER OF GALAXIES
ZWG 221.043	15 13 12.	+ 42 14		12.6	GALAXY
RNGC 5899	15 13 12.	+ 42 14		12.5	GALAXY
UGC 09789	15 13 12.	+ 42 14	168	12.6	GALAXY SBc
MCG+07-31-045	15 13 12.	+ 42 14	210	12.	GALAXY
ZWG 221.044	15 13 12.	+ 42 24		15.0	GALAXY
UGC 09790	15 13 12.	+ 42 24	90	15.0	GALAXY Sb
ARC 2054	15 13 12.	+ 54 59		17.6	RICH CLUSTER OF GALAXIES
MCG+07-31-046	15 13 15.	+ 42 23	72	14.	GALAXY
HOLM 702B	15 13 15.	+ 42 25	24	15.2	PART OF MULTIPLE GALAXY
MCG+07-31-047	15 13 15.	+ 43 21	12	14.5	GALAXY
LB 00782	15 13 17.	+ 31 07 30.		16.5	FAINT BLUE STAR
MAI 093	15 13 17.	+ 63 02	67		DWARF SPHEROIDAL GALAXY
REIZ 4508	15 13 18.	+ 42 14	132	12.5	GALAXY
REIZ 4509	15 13 18.	+ 42 23	72	14.0	GALAXY
RNGC 5901	15 13 18.	+ 42 24			NON-EXISTENT OBJECT
RNGC 5900	15 13 18.	+ 42 24		15.0	GALAXY
REIZ 4510	15 13 18.	+ 42 24	12	15.7	GALAXY
HOLM 702A	15 13 18.	+ 42 24	108	13.6	PART OF MULTIPLE GALAXY
ZWG 221.045	15 13 18.	+ 43 20		15.2	GALAXY
MCG+10-22-008	15 13 18.	+ 59 01	54	14.	GALAXY
ZWG 297.008	15 13 18.	+ 59 02		15.5	GALAXY
UGC 09791	15 13 18.	+ 59 02	66	15.5	GALAXY SBb
FATH 1.709	15 13 21.	+ 15 09	14		NEBULA
FATH 1.711	15 13 23.	+ 59 02	41		NEBULA
ZWG 049.080	15 13 24.	+ 03 13		15.7	GALAXY
ZWG 077.051	15 13 24.	+ 12 20		15.5	GALAXY
ZWG 077.052	15 13 24.	+ 12 24		15.5	GALAXY
ZC 1513.4+1646	15 13 24.	+ 16 46	2420		CLUSTER OF GALAXIES
LB 09544	15 13 24.	+ 23 56		16.8	FAINT BLUE STAR
LB 09543	15 13 24.	+ 25 12		17.2	FAINT BLUE STAR
MCG+04-36-031	15 13 24.	+ 25 37	39	15.2	GALAXY
ZC 1513.4+4351	15 13 24.	+ 43 51	470		CLUSTER OF GALAXIES
IC 1108	15 13 24.	- 45 28			NONSTELLAR OBJECT
PK327+10.1	15 13 25.	- 45 27 47.	10	10.5	PLANETARY NEBULA
REIZ 4513	15 13 29.	+ 49 30	60	14.8	GALAXY
ZWG 021.061	15 13 30.	+ 00 03		15.1	GALAXY
UGC 09792	15 13 30.	+ 00 03	900	15.1	GALAXY DWARF
MCG+00-39-016	15 13 30.	+ 00 03	600	15.1	GALAXY
GCL 032	15 13 30.	+ 00 04	618		GLOBULAR STAR CLUSTER
ZWG 106.022	15 13 30.	+ 18 34		15.7	GALAXY
UGC 09793	15 13 30.	+ 18 34	72	15.7	GALAXY Sb
ZWG 106.023	15 13 30.	+ 20 12		15.6	GALAXY
LB 09545	15 13 30.	+ 24 32		17.6	FAINT BLUE STAR
REIZ 4514	15 13 30.	+ 56 39	18	14.8	GALAXY
MCG-02-39-015	15 13 30.	- 11 19	36	14.	GALAXY
RNGC 5882	15 13 30.	- 45 27		10.5	PLANETARY NEBULA
RNGC 5891	15 13 32.	- 11 19		14.0	GALAXY
8ZW 1513+10.6	15 13 36.	+ 10 35		19.0	COMPACT GALAXY
LB 00783	15 13 36.	+ 12 08 54.		18.2	FAINT BLUE STAR
ZC 1513.6+1611	15 13 36.	+ 16 11	1140		CLUSTER OF GALAXIES
LB 09547	15 13 36.	+ 25 08		17.4	FAINT BLUE STAR
LB 09546	15 13 36.	+ 26 02		16.8	FAINT BLUE STAR
ZC 1513.6+3731	15 13 36.	+ 37 31	940		CLUSTER OF GALAXIES
ZC 1513.7+0015	15 13 42.	+ 00 15	810		CLUSTER OF GALAXIES
1ZW 104	15 13 42.	+ 06 11			COMPACT GALAXY

OBJECT NAME	RIGHT ASCEN.	DECLINATION	DIAM.	MAGN.	TYPE OF OBJECT
ZC 1513.7+1822	15 13 42.	+ 18 22	940		CLUSTER OF GALAXIES
ZC 1513.7+4038	15 13 42.	+ 40 38	810		CLUSTER OF GALAXIES
ZC 1513.7-0051	15 13 42.	- 00 51	2550		CLUSTER OF GALAXIES
ARC 2050	15 13 45.	+ 00 18		17.1	RICH CLUSTER OF GALAXIES
ZWG 049.081	15 13 48.	+ 06 59		15.5	GALAXY
ZWG 049.082	15 13 48.	+ 07 20		15.3	GALAXY
ZWG 077.053	15 13 48.	+ 08 52		15.5	GALAXY
REIZ 4507	15 13 48.	+ 10 31	54	13.9	GALAXY
ZWG 077.054	15 13 48.	+ 10 42		14.3	GALAXY
8ZW 1513+10.7	15 13 48.	+ 10 42		14.3	COMPACT GALAXY
UGC 09794	15 13 48.	+ 10 42	180	14.3	GALAXY SBc-IRR
MCG+02-39-009	15 13 48.	+ 10 42	180	14.3	GALAXY
ZWG 077.055	15 13 48.	+ 12 55		15.2	GALAXY
ZWG 106.024	15 13 48.	+ 17 20		15.3	GALAXY
ZC 1513.8+5020	15 13 48.	+ 50 20	940		CLUSTER OF GALAXIES
MCG+10-22-009	15 13 48.	+ 61 24	57	15.	GALAXY
ARC 2058	15 13 49.	+ 72 03		17.9	RICH CLUSTER OF GALAXIES
LB 00784	15 13 50.	+ 13 48 42.		17.3	FAINT BLUE STAR
ZWG 021.062	15 13 54.	+ 00 53		15.2	GALAXY
ZWG 049.083	15 13 54.	+ 06 03		15.5	GALAXY
ZWG 049.084	15 13 54.	+ 08 18		15.5	GALAXY
LB 09549	15 13 54.	+ 21 30		16.6	FAINT BLUE STAR
LB 09548	15 13 54.	+ 22 12		17.0	FAINT BLUE STAR
ZC 1513.9+2949	15 13 54.	+ 29 49	870		CLUSTER OF GALAXIES
MCG+08-28-012	15 13 54.	+ 48 19	30	16.	GALAXY
MCG+03-39-017	15 13 57.	+ 17 22	36	15.3	GALAXY
SN 19630	15 13 58.	+ 55 42		16.0	SUPERNOVA
ZWG 049.085	15 14 00.	+ 05 14		15.6	GALAXY
ZWG 049.086	15 14 00.	+ 07 56		15.3	GALAXY
UGC 09795	15 14 00.	+ 07 56	72	15.3	GALAXY Sa-b
ZC 1514.0+1422	15 14 00.	+ 14 22	1080		CLUSTER OF GALAXIES
LB 09550	15 14 00.	+ 22 07		17.1	FAINT BLUE STAR
ZC 1514.0+2231	15 14 00.	+ 22 31	870		CLUSTER OF GALAXIES
TON-N 0225	15 14 00.	+ 27 50		15.3	BLUE STAR
UGC 09796	15 14 00.	+ 43 22	96	16.0	GALAXY PECULR
ZWG 274.036	15 14 00.	+ 55 42		13.6	GALAXY
UGC 09797	15 14 00.	+ 55 42	282	13.6	GALAXY SBb
ZWG 297.009	15 14 00.	+ 61 23		14.8	GALAXY
ZC 1514.0+7159	15 14 00.	+ 71 59	740		CLUSTER OF GALAXIES
RNGC 5905	15 14 02.	+ 55 42		12.5	GALAXY
LB 00785	15 14 02.	+ 59 40 30.		17.7	FAINT BLUE STAR
REIZ 4515	15 14 03.	+ 50 34	24	14.8	GALAXY
ZWG 021.063	15 14 06.	+ 00 25		15.6	GALAXY
LB 09551	15 14 06.	+ 21 24		16.8	FAINT BLUE STAR
ZWG 135.039	15 14 06.	+ 25 02		15.1	GALAXY
MCG+04-36-032	15 14 06.	+ 25 02	42	15.1	GALAXY
2ZW 073	15 14 06.	+ 43 21			COMPACT GALAXY
MCG+07-31-048	15 14 06.	+ 43 21	72	17.	GALAXY
MCG+09-25-038	15 14 06.	+ 55 42	192	13.1	GALAXY
7ZW 591	15 14 05.	+ 61 23			COMPACT GALAXY
MCG-02-39-016	15 14 06.	- 13 14 30.	60	15.	GALAXY
LB 00786	15 14 07.	+ 28 55 12.		17.7	FAINT BLUE STAR
REIZ 4516	15 14 08.	+ 55 42	90	13.1	GALAXY
ARC 2051	15 14 11.	- 00 46		17.4	RICH CLUSTER OF GALAXIES
ZWG 021.064	15 14 12.	+ 01 26		15.5	GALAXY
MCG+00-39-017	15 14 12.	+ 01 26	18	15.5	GALAXY
ZWG 049.087	15 14 12.	+ 07 35		15.4	GALAXY
UGC 09798	15 14 12.	+ 07 35	66	15.4	GALAXY S0-a
ZWG 077.056	15 14 12.	+ 12 55		15.3	GALAXY
ZWG 106.025	15 14 12.	+ 17 40		15.6	GALAXY
LB 09552	15 14 12.	+ 24 38		16.3	FAINT BLUE STAR
ZC 1514.2+4232	15 14 12.	+ 42 32	1340		CLUSTER OF GALAXIES
ZWG 221.046	15 14 12.	+ 43 20		15.7	GALAXY
MCG+07-31-049	15 14 15.	+ 43 20 30.	30	15.5	GALAXY
ARC 2052	15 14 16.	+ 07 12		15.0	RICH CLUSTER OF GALAXIES
LB 00787	15 14 17.	+ 28 17 36.		16.8	FAINT BLUE STAR
ZWG 021.065	15 14 18.	+ 00 24		15.7	GALAXY
ZWG 049.088	15 14 18.	+ 05 05		15.6	GALAXY
ZWG 049.089	15 14 18.	+ 07 11		15.5	GALAXY
ZWG 049.090	15 14 18.	+ 07 12		14.8	GALAXY
UGC 09799	15 14 18.	+ 07 12	114	14.8	GALAXY E
MCG+01-39-012	15 14 16.	+ 07 12	18	14.8	GALAXY
ZWG 049.091	15 14 18.	+ 07 14		15.6	GALAXY
ZC 1514.3+1904	15 14 18.	+ 19 04	2290		CLUSTER OF GALAXIES
ZWG 106.026	15 14 18.	+ 19 16		15.1	GALAXY
MRK 688	15 14 18.	+ 19 16	20	16.	GALAXY WITH UV CONTINUUM
MCG+03-39-018	15 14 18.	+ 19 16	42	15.1	GALAXY
ZC 1514.3+2345	15 14 18.	+ 23 45	1410		CLUSTER OF GALAXIES
ZC 1514.3+3837	15 14 18.	+ 38 37	2020		CLUSTER OF GALAXIES
ZWG 106.027	15 14 24.	+ 20 20		15.7	GALAXY
LB 09555	15 14 24.	+ 23 08		15.2	FAINT BLUE STAR
LB 09554	15 14 24.	+ 24 18		17.7	FAINT BLUE STAR
LB 09553	15 14 24.	+ 25 27		16.9	FAINT BLUE STAR
TON-N 0226	15 14 24.	+ 29 41		15.8	BLUE STAR
ZC 1514.4+5847	15 14 24.	+ 58 47	1410		CLUSTER OF GALAXIES
MCG-04-36-005	15 14 24.	- 22 07 30.	72	14.5	GALAXY
RNGC 5897	15 14 28.	- 20 49		9.5	GLOBULAR CLUSTER
ZWG 049.092	15 14 30.	+ 07 07		15.3	GALAXY
ZWG 049.093	15 14 30.	+ 07 26		15.6	GALAXY
ZWG 077.057	15 14 30.	+ 09 04		15.5	GALAXY
ZC 1514.5+1545	15 14 30.	+ 15 45	2350		CLUSTER OF GALAXIES
LB 00788	15 14 30.	+ 30 46 12.		15.5	FAINT BLUE STAR
HOLM 704B	15 14 30.	+ 56 31	30	14.3	PART OF MULTIPLE GALAXY
HOLM 704A	15 14 30.	+ 56 31	720	11.6	PART OF MULTIPLE GALAXY
LB 00789	15 14 31.	+ 12 06 06.		16.0	FAINT BLUE STAR
FATH 1.710	15 14 32.	- 15 13	8		NEBULA
RNGC 5907	15 14 33.	+ 56 30		11.5	GALAXY
RNGC 5906	15 14 33.	+ 56 31			NON-EXISTENT OBJECT
IC 1109	15 14 34.	+ 05 25 13.			NONSTELLAR OBJECT
ZWG 049.094	15 14 36.	+ 05 26		15.5	GALAXY
ZWG 077.058	15 14 36.	+ 11 13		15.5	GALAXY
8ZW 1514+13.3	15 14 36.	+ 13 17		15.2	COMPACT GALAXY
ZC 1514.6+2047	15 14 36.	+ 20 47	670		CLUSTER OF GALAXIES
LB 09557	15 14 36.	+ 22 03		16.7	FAINT BLUE STAR
LB 09556	15 14 36.	+ 26 48		17.5	FAINT BLUE STAR
ZC 1514.6+2837	15 14 36.	+ 28 37	3430		CLUSTER OF GALAXIES
ZC 1514.6+3619	15 14 36.	+ 36 19	1140		CLUSTER OF GALAXIES
MCG+08-28-013	15 14 36.	+ 47 14	42	17.	GALAXY
ZWG 274.037	15 14 36.	+ 54 40		15.5	GALAXY
UGC 09800	15 14 36.	+ 54 40	78	15.5	GALAXY Sb
MCG+09-25-039	15 14 36.	+ 54 40	72	15.	GALAXY
ZWG 297.010	15 14 36.	+ 56 30		11.4	GALAXY
ZWG 274.038	15 14 36.	+ 56 30		11.4	GALAXY
REIZ 4521	15 14 36.	+ 56 30	120	11.6	GALAXY
UGC 09801	15 14 36.	+ 56 30	768	11.4	GALAXY Sc
ZWG 021.066	15 14 36.	- 03 01		15.3	GALAXY
GCL 033	15 14 36.	- 20 50	522	9.61	GLOBULAR STAR CLUSTER
MRSL 320-02/1	15 14 36.	- 59 46	450		HII REGION
LB 00790	15 14 37.	+ 59 05 36.		17.5	FAINT BLUE STAR
ARC 2053	15 14 40.	- 00 30		17.4	RICH CLUSTER OF GALAXIES
SN 1940A	15 14 41.	+ 56 25		14.3	SUPERNOVA
ZWG 049.095	15 14 42.	+ 03 31		15.7	GALAXY
ZWG 049.096	15 14 42.	+ 07 04		15.3	GALAXY
ZWG 077.059	15 14 42.	+ 13 17		15.2	GALAXY
LB 09558	15 14 42.	+ 23 28		16.8	FAINT BLUE STAR
TON-N 0227	15 14 42.	+ 29 21		16.1	BLUE STAR
MCG+07-31-050	15 14 42.	+ 41 12 30.	30	16.	GALAXY
UGC 09802	15 14 42.	+ 43 09	78	16.0	GALAXY Sc
MCG+09-25-040	15 14 42.	+ 56 30	660	11.6	GALAXY
ZC 1514.7-0031	15 14 42.	- 00 31	1550		CLUSTER OF GALAXIES
ZWG 021.067	15 14 48.	+ 00 24		15.7	GALAXY
ZWG 049.097	15 14 48.	+ 07 12		14.9	GALAXY
MCG+01-39-013	15 14 48.	+ 07 12	18	14.9	GALAXY
ZWG 077.060	15 14 48.	+ 12 04		15.4	GALAXY
LB 09559	15 14 48.	+ 24 13		16.8	FAINT BLUE STAR
ZWG 221.047	15 14 48.	+ 41 13		15.6	GALAXY
ZC 1514.8+4408	15 14 48.	+ 44 08	2820		CLUSTER OF GALAXIES
BSO 30	15 14 49.	+ 43 55 50.		17.29	BLUE STELLAR OBJECT
REIZ 4517	15 14 50.	+ 41 13	18	14.9	GALAXY
ZWG 021.069	15 14 54.	+ 01 11		15.5	GALAXY
ZWG 049.098	15 14 54.	+ 04 23		15.7	GALAXY
ZWG 049.099	15 14 54.	+ 05 41		15.4	GALAXY
ZWG 049.100	15 14 54.	+ 07 32		15.5	GALAXY
ZWG 049.101	15 14 54.	+ 08 05		15.0	GALAXY
MCG+01-39-014	15 14 54.	+ 08 05	72	15.0	GALAXY
1ZW 105	15 14 54.	+ 40 05			COMPACT GALAXY
MCG+07-31-051	15 14 54.	+ 43 09	48	16.	GALAXY
ZC 1514.9+6058	15 14 54.	+ 60 58	340		CLUSTER OF GALAXIES
ZWG 021.068	15 14 54.	- 02 50		15.7	GALAXY
IC 4537	15 14 55.	+ 02 13 45.			NONSTELLAR OBJECT
LB 00791	15 14 56.	+ 00 01 48.		16.1	FAINT BLUE STAR
ZWG 021.070	15 15 00.	+ 02 13		15.6	GALAXY
ZWG 049.102	15 15 00.	+ 08 03		15.3	GALAXY
ZWG 077.061	15 15 00.	+ 13 21		15.5	GALAXY
8ZW 1515+13.3	15 15 00.	+ 13 21		15.5	COMPACT GALAXY
LB 09562	15 15 00.	+ 20 41		15.7	FAINT BLUE STAR
LB 09561	15 15 00.	+ 21 15		13.6	FAINT BLUE STAR
HOAG 1	15 15 00.	+ 21 46	45	15.	COMPACT GALAXY WITH RING
LB 09560	15 15 00.	+ 25 50		17.2	FAINT BLUE STAR
MCG+06-34-001	15 15 00.	+ 38 26	30	15.	GALAXY
MCG+07-31-052	15 15 00.	+ 39 52	30	15.	GALAXY
MCG+12-14-021	15 15 00.	+ 69 28	51	16.	GALAXY
LB 00285	15 15 02.	- 13 50 18.		14.5	FAINT BLUE STAR
SN 1961E	15 15 05.	+ 00 14		17.0	SUPERNOVA
RNGC 5890	15 15 05.	- 17 25		14.0	GALAXY
ZWG 021.071	15 15 06.	+ 01 32		15.4	GALAXY
LB 09563	15 15 06.	+ 22 15		16.5	FAINT BLUE STAR
LB 00792	15 15 06.	+ 30 46 54.		17.1	FAINT BLUE STAR
ZC 1515.1+3337	15 15 06.	+ 33 37	3430		CLUSTER OF GALAXIES
ZWG 221.048	15 15 06.	+ 39 53		15.1	GALAXY
MCG-03-39-004	15 15 06.	- 17 25 30.	30	14.	GALAXY
REIZ 4520	15 15 07.	+ 41 54	30	14.6	GALAXY
REIZ 4519	15 15 10.	+ 39 53	36	13.9	GALAXY
ZWG 049.103	15 15 12.	+ 06 50		15.7	GALAXY
ZWG 049.104	15 15 12.	+ 07 21		15.5	GALAXY
SN 1950J	15 15 12.	+ 19 57		20.3	SUPERNOVA
LB 09568	15 15 12.	+ 23 38		17.4	FAINT BLUE STAR
LB 09567	15 15 12.	+ 23 47		17.0	FAINT BLUE STAR
LB 09566	15 15 12.	+ 24 54		16.9	FAINT BLUE STAR
LB 09565	15 15 12.	+ 25 02		16.5	FAINT BLUE STAR
LB 09564	15 15 12.	+ 25 33		16.0	FAINT BLUE STAR
UGC 09803	15 15 12.	+ 29 36	60	18.	GALAXY DWARF
ZWG 021.072	15 15 12.	- 00 36		15.6	GALAXY
MCG+00-39-018	15 15 12.	- 00 36	24	15.6	GALAXY
LB 00793	15 15 13.	+ 61 01 42.		16.3	FAINT BLUE STAR
RNGC 5898	15 15 14.	- 23 56		13.0	GALAXY
IC 1112	15 15 17.	+ 07 24 41.			NONSTELLAR OBJECT
ZWG 049.105	15 15 18.	+ 04 20		14.9	GALAXY
UGC 09804	15 15 18.	+ 04 20	66	14.9	GALAXY DBL SYS
MCG+01-39-015	15 15 18.	+ 04 20	60	14.9	GALAXY
8ZW 1515+09.6	15 15 18.	+ 09 39		18.1	COMPACT GALAXY
8ZW 1515+10.1	15 15 18.	+ 10 08		17.9	COMPACT GALAXY
ZWG 106.028	15 15 18.	+ 17 40		15.7	GALAXY
LB 09572	15 15 18.	+ 23 01		18.3	FAINT BLUE STAR
LB 09571	15 15 18.	+ 24 08		16.6	FAINT BLUE STAR
LB 09570	15 15 18.	+ 26 08		16.8	FAINT BLUE STAR
LB 09569	15 15 18.	+ 26 36		15.7	FAINT BLUE STAR
MCG+07-31-053	15 15 18.	+ 41 07	30	15.	GALAXY
ZWG 274.039	15 15 18.	+ 55 35		13.5	GALAXY
UGC 09805	15 15 18.	+ 55 35	180	13.5	GALAXY Sb
ZWG 337.033	15 15 18.	+ 69 30		15.4	GALAXY
ZC 1515.3+7618	15 15 18.	+ 76 18	1080		CLUSTER OF GALAXIES
ZWG 021.073	15 15 18.	- 00 37		15.7	GALAXY
MCG+00-39-019	15 15 18.	- 00 37	30	15.7	GALAXY
MCG-04-36-006	15 15 18.	- 23 56	36	12.	GALAXY
REIZ 4526	15 15 20.	+ 41 07	18	14.8	GALAXY
LB 00794	15 15 22.	+ 27 14 42.		17.3	FAINT BLUE STAR
BSO 31	15 15 22.	+ 44 07 41.			BLUE STELLAR OBJECT
ZWG 021.074	15 15 24.	+ 01 00		15.7	GALAXY
ZWG 049.106	15 15 24.	+ 05 17		15.6	GALAXY
ZWG 049.107	15 15 24.	+ 07 24		15.3	GALAXY
ZWG 049.108	15 15 24.	+ 08 25		15.5	GALAXY
LB 09577	15 15 24.	+ 22 59		16.4	FAINT BLUE STAR
LB 09576	15 15 24.	+ 23 08		16.8	FAINT BLUE STAR
LB 09575	15 15 24.	+ 23 10		15.7	FAINT BLUE STAR
LB 09574	15 15 24.	+ 23 50		17.2	FAINT BLUE STAR
LB 09573	15 15 24.	+ 25 51		16.2	FAINT BLUE STAR
ZWG 221.049	15 15 24.	+ 41 08		15.5	GALAXY
MCG+09-25-041	15 15 24.	+ 55 36	180	13.0	GALAXY
RNGC 5908	15 15 26.	+ 55 36		13.0	GALAXY
REIZ 4528	15 15 26.	+ 55 36	180	12.8	GALAXY
LB 00795	15 15 27.	+ 02 33 30.		15.9	FAINT BLUE STAR
ZWG 049.109	15 15 30.	+ 07 25		15.5	GALAXY
UGC 09806	15 15 30.	+ 38 50	60	16.5	GALAXY Sc
ZWG 337.034	15 15 30.	+ 69 28		15.3	GALAXY
MRSL 322+00/1	15 15 30.	- 56 30	480		HII REGION
ZWG 077.062	15 15 36.	+ 11 05		15.1	GALAXY
UGC 09807	15 15 36.	+ 11 05	84	15.1	GALAXY Sa-b
ZWG 077.063	15 15 36.	+ 13 09		15.2	GALAXY
ZWG 106.029	15 15 36.	+ 16 30		15.7	GALAXY
MCG+03-39-019	15 15 36.	+ 16 30	48	15.7	GALAXY
ZWG 106.030	15 15 36.	+ 20 20		15.6	GALAXY
LB 09579	15 15 36.	+ 21 49		16.6	FAINT BLUE STAR
LB 09578	15 15 36.	+ 22 01		16.7	FAINT BLUE STAR
MCG+07-31-054	15 15 36.	+ 43 09	36	17.	GALAXY
MCG-04-36-007	15 15 36.	- 23 57	24	15.	GALAXY

OBJECT NAME	RIGHT ASCEN.	DECLINATION	DIAM.	MAGN.	TYPE OF OBJECT
RNGC 5903	15 15 38.	- 23 51		13.0	GALAXY
REIZ 4518	15 15 39.	+ 13 42	42	14.4	GALAXY
GCL 034	15 15 42.	+ 02 16	1200	7.04	GLOBULAR STAR CLUSTER
ZWG 049.110	15 15 42.	+ 05 30		14.7	GALAXY
MCG+01-39-016	15 15 42.	+ 05 30	36	14.7	GALAXY
ZWG 049.111	15 15 42.	+ 08 26		15.3	GALAXY
ZWG 077.064	15 15 42.	+ 13 42		14.9	GALAXY
MCG+02-39-010	15 15 42.	+ 13 42	42	14.9	GALAXY
1ZW 106	15 15 42.	+ 44 20			COMPACT GALAXY
ZWG 337.035	15 15 42.	+ 69 31		15.7	GALAXY
MCG-04-36-008	15 15 42.	- 23 54	36	12.	GALAXY
LB 00796	15 15 43.	+ 28 00 12.		16.8	FAINT BLUE STAR
HOLM 705A	15 15 45.	+ 14 01	24	14.7	PART OF MULTIPLE GALAXY
FATH 1.713	15 15 45.	+ 59 01	27		NEBULA
REIZ 4522	15 15 46.	+ 12 35	18	15.7	GALAXY
FATH 1.714	15 15 46.	+ 59 36			NEBULA
ZWG 077.065	15 15 48.	+ 14 00		15.2	GALAXY
8ZW 1515+14.0	15 15 48.	+ 14 00		15.2	COMPACT GALAXY
UGC 09808	15 15 48.	+ 14 00	84	15.2	GALAXY Sb
MCG+02-39-011	15 15 48.	+ 14 00	72	15.2	GALAXY
ZWG 106.031	15 15 48.	+ 14 46		15.4	GALAXY
LB 09584	15 15 48.	+ 21 49		16.8	FAINT BLUE STAR
LB 09583	15 15 48.	+ 22 24		18.4	FAINT BLUE STAR
LB 09582	15 15 48.	+ 23 44		16.3	FAINT BLUE STAR
LB 09581	15 15 48.	+ 24 34		16.8	FAINT BLUE STAR
LB 09580	15 15 48.	+ 25 38		17.1	FAINT BLUE STAR
REIZ 4527	15 15 48.	+ 38 24	12	14.7	GALAXY
HOLM 705B	15 15 50.	+ 14 02	18	14.9	PART OF MULTIPLE GALAXY
REIZ 4524	15 15 51.	+ 14 01	24	14.8	GALAXY
REIZ 4525	15 15 51.	+ 14 02	12	15.0	GALAXY
REIZ 4523	15 15 52.	+ 12 41	24	15.2	GALAXY
ZWG 049.112	15 15 54.	+ 05 03		15.4	GALAXY
ZWG 077.066	15 15 54.	+ 12 40		15.2	GALAXY
MCG+02-39-012	15 15 54.	+ 12 40	36	15.2	GALAXY
IC 1113	15 15 54.	+ 12 41 09.			NONSTELLAR OBJECT
LB 09586	15 15 54.	+ 24 14		15.9	FAINT BLUE STAR
LB 09585	15 15 54.	+ 26 32		16.8	FAINT BLUE STAR
ZWG 165.036	15 15 54.	+ 30 52		14.7	GALAXY
UGC 09809	15 15 54.	+ 30 52	78	14.7	GALAXY SBc
LBN 1091	15 16	- 29 40	4920		BRIGHT NEBULA
ZWG 049.113	15 16 00.	+ 04 51		15.7	GALAXY
ZWG 049.114	15 16 00.	+ 05 24		15.4	GALAXY
ZWG 077.067	15 16 00.	+ 12 22		15.4	GALAXY
ZWG 077.068	15 16 00.	+ 13 09		15.1	GALAXY
MCG+02-39-013	15 16 00.	+ 13 09	42	15.1	GALAXY
LB 09588	15 16 00.	+ 20 40		16.6	FAINT BLUE STAR
LB 09587	15 16 00.	+ 26 29		17.9	FAINT BLUE STAR
MCG+05-36-012	15 16 00.	+ 30 51 30.	90	14.7	GALAXY
ZC 1516.0+5336	15 16 00.	+ 53 36	1680		CLUSTER OF GALAXIES
MCG+10-22-010	15 16 00.	+ 56 50 30.	27	15.	GALAXY
MCG-07-31-011	15 16 00.	- 41 04	108	14.5	GALAXY
LB 00797	15 16 01.	+ 11 15 36.		17.4	FAINT BLUE STAR
RNGC 5904	15 16 02.	+ 02 16		7.0	GLOBULAR CLUSTER
MCG-04-36-009	15 16 03.	- 23 38 30.	90	14.	GALAXY
REIZ 4531	15 16 04.	+ 56 50	18	14.6	GALAXY
ZWG 077.069	15 16 06.	+ 13 07		14.9	GALAXY
MCG+02-39-013A	15 16 06.	+ 13 07	60	14.9	GALAXY
ZWG 135.040	15 16 06.	+ 21 25		15.4	GALAXY
ZWG 249.014	15 16 06.	+ 49 16		15.7	GALAXY
MCG+08-28-014	15 16 06.	+ 49 46	36	16.	GALAXY
ZWG 297.011	15 16 06.	+ 56 51		15.1	GALAXY
ZWG 049.115	15 16 12.	+ 06 05		15.2	GALAXY
UGC 09810	15 16 12.	+ 08 01	66	16.0	GALAXY S
LB 09590	15 16 12.	+ 21 51		15.4	FAINT BLUE STAR
LB 09589	15 16 12.	+ 25 32		16.2	FAINT BLUE STAR
ZWG 194.001	15 16 12.	+ 38 24		15.4	GALAXY
MCG+06-34-002	15 16 12.	+ 38 24	27	15.	GALAXY
1ZW 107	15 16 12.	+ 42 55			COMPACT GALAXY
ZWG 221.050	15 16 12.	+ 42 55		14.9	GALAXY
UGC 09811	15 16 12.	+ 65 22	60	16.0	GALAXY S
MCG+11-19-004	15 16 12.	+ 65 26	42	16.	GALAXY
ARC 2055	15 16 17.	+ 06 24		16.0	RICH CLUSTER OF GALAXIES
MCG+04-36-033	15 16 18.	+ 21 23	30	15.4	GALAXY
LB 09591	15 16 18.	+ 21 32		16.4	FAINT BLUE STAR
ZC 1516.3+3641	15 16 18.	+ 36 41	1080		CLUSTER OF GALAXIES
MCG+07-31-054A	15 16 18.	+ 43 03	42	15.	GALAXY
FATH 1.712	15 16 18.	- 15 49	11		NEBULA
MCG-04-36-010	15 16 18.	- 24 15	60	15.	GALAXY
LB 00798	15 16 20.	+ 30 06 36.		17.2	FAINT BLUE STAR
MCG-03-39-005	15 16 21.	- 15 50	30	15.	GALAXY
ZWG 049.116	15 16 24.	+ 04 17		15.4	GALAXY
ZWG 049.117	15 16 24.	+ 05 03		14.9	GALAXY
MCG+01-39-017	15 16 24.	+ 05 03	48	14.9	GALAXY
ZWG 049.118	15 16 24.	+ 06 37		15.4	GALAXY
ZC 1516.4+1459	15 16 24.	+ 14 59	1140		CLUSTER OF GALAXIES
LB 09593	15 16 24.	+ 21 10		15.7	FAINT BLUE STAR
ZWG 135.041	15 16 24.	+ 22 16		15.6	GALAXY
ZWG 135.042	15 16 24.	+ 23 47		15.7	GALAXY
LB 09592	15 16 24.	+ 24 02		16.0	FAINT BLUE STAR
ZC 1516.4+5406	15 16 24.	+ 54 06	1610		CLUSTER OF GALAXIES
7ZW 592	15 16 24.	+ 66 47			COMPACT GALAXY
MCG-02-39-017	15 16 24.	- 14 22	36	15.5	GALAXY
LB 00799	15 16 27.	+ 31 03 12.		17.1	FAINT BLUE STAR
ZWG 021.075	15 16 30.	+ 01 13		15.7	GALAXY
ZWG 049.119	15 16 30.	+ 03 10		15.4	GALAXY
ZWG 049.120	15 16 30.	+ 04 42		15.0	GALAXY
MCG+01-39-018	15 16 30.	+ 04 42	15	15.0	GALAXY
ZWG 049.121	15 16 30.	+ 05 23		15.3	GALAXY
ZC 1516.5+2221	15 16 30.	+ 22 21	3760		CLUSTER OF GALAXIES
LB 09595	15 16 30.	+ 25 50		16.7	FAINT BLUE STAR
LB 09594	15 16 30.	+ 26 16		16.5	FAINT BLUE STAR
MCG+06-34-003	15 16 30.	+ 32 34 30.	24	15.	GALAXY
FATH 1.715	15 16 30.	+ 59 16	14		NEBULA
IC 4539	15 16 31.	+ 32 33 21.			NONSTELLAR OBJECT
LB 00800	15 16 36.	+ 01 44 30.		17.4	FAINT BLUE STAR
ZWG 049.122	15 16 36.	+ 03 32		15.7	GALAXY
ZWG 049.123	15 16 36.	+ 04 30		15.6	GALAXY
3ZW 071	15 16 36.	+ 28 30			COMPACT GALAXY
ZC 1516.6+3559	15 16 36.	+ 35 59	2220		CLUSTER OF GALAXIES
ZWG 049.124	15 16 42.	+ 05 41		15.5	GALAXY
ZWG 049.125	15 16 42.	+ 08 19		15.5	GALAXY
ZWG 077.070	15 16 42.	+ 09 57		15.5	GALAXY
KARA.72 461A	15 16 42.	+ 09 57	36	15.5	PART OF DOUBLE GALAXY
ZWG 077.071	15 16 42.	+ 09 59		15.5	GALAXY
8ZW 1516+10.0	15 16 42.	+ 09 59		15.5	COMPACT GALAXY
UGC 09812	15 16 42.	+ 09 59	60	15.5	GALAXY Sc
KARA.72 461B	15 16 42.	+ 09 59	54	15.5	PART OF DOUBLE GALAXY
ZWG 135.043	15 16 42.	+ 21 00		15.6	GALAXY
UGC 09813	15 16 42.	+ 21 00	66	15.6	GALAXY SBb
LB 09596	15 16 42.	+ 23 12		16.0	FAINT BLUE STAR
LB 00801	15 16 47.	+ 57 19 12.		17.4	FAINT BLUE STAR
ZWG 049.126	15 16 48.	+ 08 03		15.6	GALAXY
ZWG 049.127	15 16 48.	+ 08 11		15.7	GALAXY
MCG+04-36-034	15 16 48.	+ 20 58	60	15.6	GALAXY
LB 09601	15 16 48.	+ 22 10		17.2	FAINT BLUE STAR
LB 09600	15 16 48.	+ 22 12		16.9	FAINT BLUE STAR
LB 09599	15 16 48.	+ 22 20		17.0	FAINT BLUE STAR
LB 09598	15 16 48.	+ 25 05		15.5	FAINT BLUE STAR
LB 09597	15 16 48.	+ 25 14		16.6	FAINT BLUE STAR
ZWG 077.072	15 16 54.	+ 13 19		15.5	GALAXY
LB 09603	15 16 54.	+ 21 15		15.8	FAINT BLUE STAR
LB 09602	15 16 54.	+ 25 31		17.1	FAINT BLUE STAR
ZWG 222.001	15 16 54.	+ 42 02		14.9	GALAXY
ZWG 221.051	15 16 54.	+ 42 02		14.9	GALAXY
RNGC 5914A	15 16 54.	+ 42 02		15.0	GALAXY
REIZ 4532	15 16 54.	+ 42 02	21	14.2	GALAXY
MCG+07-31-055	15 16 54.	+ 42 02	42	14.5	GALAXY
HOLM 706A	15 16 54.	+ 42 03	24	14.1	PART OF MULTIPLE GALAXY
RNGC 5914B	15 16 54.	+ 42 04		17.0	GALAXY
REIZ 4533	15 16 54.	+ 42 04	36	15.0	GALAXY
MCG+07-31-056	15 16 54.	+ 42 04	24	17.	GALAXY
LB 00802	15 16 56.	+ 11 48 24.		17.3	FAINT BLUE STAR
HOLM 706B	15 16 56.	+ 42 05	30	14.9	PART OF MULTIPLE GALAXY
LB 00803	15 16 57.	+ 57 44 36.		17.1	FAINT BLUE STAR
MCG-04-36-011	15 16 57.	- 23 45 30.	48	15.	GALAXY
LB 09894	15 17	- 83 37		14.6	FAINT BLUE STAR
ZWG 049.128	15 17 00.	+ 05 26		15.5	GALAXY
ZWG 077.073	15 17 00.	+ 11 14		14.7	GALAXY
REIZ 4529	15 17 00.	+ 11 14	60	14.4	GALAXY
UGC 09814	15 17 00.	+ 11 14	66	14.7	GALAXY Sc-IRR
MCG+02-39-014	15 17 00.	+ 11 14	72	14.7	GALAXY
LB 09606	15 17 00.	+ 22 58		16.6	FAINT BLUE STAR
LB 09605	15 17 00.	+ 24 00		16.8	FAINT BLUE STAR
ZC 1517.0+2519	15 17 00.	+ 25 19	610		CLUSTER OF GALAXIES
LB 09604	15 17 00.	+ 25 45		16.6	FAINT BLUE STAR
TON-N 0228	15 17 00.	+ 26 26		15.0	BLUE STAR
REIZ 4530	15 17 04.	+ 13 19	36	14.9	GALAXY
RNGC 5910	15 17 05.	+ 21 05		15.0	GALAXY
ZWG 077.074	15 17 06.	+ 10 09		15.1	GALAXY
ZWG 135.044	15 17 06.	+ 20 44		15.7	GALAXY
ZWG 135.045	15 17 06.	+ 21 05		14.9	GALAXY
LB 09608	15 17 06.	+ 23 00		17.8	FAINT BLUE STAR
LB 09607	15 17 06.	+ 25 56		18.0	FAINT BLUE STAR
ARC 2056	15 17 06.	+ 28 27		16.9	RICH CLUSTER OF GALAXIES
MCG+10-22-011	15 17 06.	+ 58 17	30	16.	GALAXY
ZC 1517.1-0303	15 17 06.	- 03 03	2150		CLUSTER OF GALAXIES
ZWG 021.076	15 17 12.	+ 01 03		15.7	GALAXY
MCG+00-39-020	15 17 12.	+ 01 03	30	15.7	GALAXY
LB 09614	15 17 12.	+ 20 41		16.1	FAINT BLUE STAR
MCG+04-36-036	15 17 12.	+ 21 03	9	14.9	GALAXY
MCG+04-36-035	15 17 12.	+ 21 03	18	14.9	GALAXY
LB 09613	15 17 12.	+ 23 34		17.8	FAINT BLUE STAR
LB 09612	15 17 12.	+ 23 58		16.4	FAINT BLUE STAR
LB 09611	15 17 12.	+ 24 40		17.1	FAINT BLUE STAR
LB 09610	15 17 12.	+ 25 49		16.4	FAINT BLUE STAR
ZWG 135.046	15 17 12.	+ 25 58		15.4	GALAXY
LB 09609	15 17 12.	+ 26 04		16.2	FAINT BLUE STAR
VV 139D	15 17 18.	+ 21 03	6	18.5	INTERACTING GALAXY
VV 139C	15 17 18.	+ 21 03	9	16.	INTERACTING GALAXY
VV 139B	15 17 18.	+ 21 03	15	16.	INTERACTING GALAXY
VV 139A	15 17 18.	+ 21 03	21	15.	INTERACTING GALAXY
VV 139	15 17 18.	+ 21 03	60		INTERACTING GALAXY
MCG+04-36-037	15 17 18.	+ 25 58	18	15.4	GALAXY
ZC 1517.3+2910	15 17 18.	+ 29 10	3020		CLUSTER OF GALAXIES
ZC 1517.3+5745	15 17 18.	+ 57 45	940		CLUSTER OF GALAXIES
MCG-05-36-001	15 17 18.	- 31 15	48	15.	GALAXY
LB 00804	15 17 21.	+ 30 11 42.		16.7	FAINT BLUE STAR
ZWG 049.129	15 17 24.	+ 05 18		15.6	GALAXY
ZWG 049.130	15 17 24.	+ 05 59		15.6	GALAXY
ZWG 049.131	15 17 24.	+ 08 17		15.4	GALAXY
ZC 1517.4+1906	15 17 24.	+ 19 06	810		CLUSTER OF GALAXIES
LB 09616	15 17 24.	+ 20 49		16.5	FAINT BLUE STAR
ZC 1517.4+2338	15 17 24.	+ 23 38	870		CLUSTER OF GALAXIES
LB 09615	15 17 24.	+ 24 35		16.8	FAINT BLUE STAR
ZC 1517.5+0220	15 17 30.	+ 02 20	1080		CLUSTER OF GALAXIES
ZWG 049.132	15 17 30.	+ 05 36		15.3	GALAXY
ZC 1517.5+1949	15 17 30.	+ 19 49	3160		CLUSTER OF GALAXIES
LB 09617	15 17 30.	+ 23 38		16.2	FAINT BLUE STAR
MCG+08-28-015	15 17 30.	+ 45 04	90	15.	GALAXY
ZWG 249.015	15 17 30.	+ 45 06		15.7	GALAXY
UGC 09815	15 17 30.	+ 45 06	96	15.7	GALAXY
MCG+09-25-042	15 17 30.	+ 50 42 30.	60	16.	GALAXY
7ZW 593	15 17 30.	+ 56 41			COMPACT GALAXY
UGC 09816	15 17 30.	+ 59 40	90	16.0	GALAXY IRR
MCG-04-36-012	15 17 30.	- 24 17 30.	36	15.5	GALAXY
IC 4540	15 17 31.	+ 01 58 00.			NONSTELLAR OBJECT
LB 00805	15 17 34.	+ 02 19 30.		16.2	FAINT BLUE STAR
LB 09619	15 17 36.	+ 21 14		13.3	FAINT BLUE STAR
LB 09618	15 17 36.	+ 23 42		17.4	FAINT BLUE STAR
MCG+10-22-012	15 17 36.	+ 59 39	72	14.	GALAXY
SMI 12	15 17 36.	- 53 18	720		FAINT NEBULOSITY
RNGC 5918	15 17 40.	+ 46 04		14.0	GALAXY
REIZ 4534	15 17 41.	+ 11 22	36	14.3	GALAXY
REIZ 4536	15 17 41.	+ 46 03	108	13.8	GALAXY
FATH 1.716	15 17 41.	+ 59 38	73		NEBULA
ZWG 077.075	15 17 42.	+ 12 03		15.4	GALAXY
MCG+02-39-015	15 17 42.	+ 12 03	54	15.4	GALAXY
MCG+08-28-016	15 17 42.	+ 45 57	48	17.	GALAXY
MCG+08-28-017	15 17 42.	+ 46 02 30.	120	13.8	GALAXY
ZWG 249.016	15 17 42.	+ 46 04		14.0	GALAXY
UGC 09817	15 17 42.	+ 46 04	114	14.0	GALAXY Sc
7ZW 594	15 17 42.	+ 59 15			COMPACT GALAXY
FATH 1.717	15 17 45.	+ 59 09	19		NEBULA
LB 00286	15 17 46.	+ 58 12 06.		14.4	FAINT BLUE STAR
ARC 2068	15 17 46.	+ 71 40		17.7	RICH CLUSTER OF GALAXIES
LB 00806	15 17 47.	+ 00 01 18.		16.4	FAINT BLUE STAR
ZWG 049.133	15 17 48.	+ 03 42		14.7	GALAXY
MCG+01-39-019	15 17 48.	+ 03 42	36	14.7	GALAXY
ZWG 165.037	15 17 48.	+ 31 33		15.2	GALAXY
ZWG 021.077	15 17 48.	- 00 04		15.3	GALAXY
RNGC 5911	15 17 51.	+ 03 42		14.5	GALAXY
ZWG 049.134	15 17 54.	+ 03 43		15.7	GALAXY
ZWG 077.076	15 17 54.	+ 12 37		15.5	GALAXY
MCG+08-28-018	15 17 54.	+ 47 24	24	16.	GALAXY
MCG-01-39-001	15 17 54.	- 08 22	18	15.5	GALAXY
MCG-03-39-006	15 17 54.	- 18 12	48	15.5	GALAXY

OBJECT NAME	RIGHT ASCEN.	DECLINATION	DIAM.	MAGN.	TYPE OF OBJECT
LB 00807	15 17 55.	+ 11 29 54.		17.3	FAINT BLUE STAR
BC M1517+176	15 17 57.8	+ 17 36 46.		18.	QUASI-STELLAR OBJECT
SHB 311	15 17 57.8	+ 17 36 46.		18.	QUASI-STELLAR OBJECT
ZWG 049.135	15 18 00.	+ 03 37		15.4	GALAXY
ZC 1518.0+1628	15 18 00.	+ 16 28	740		CLUSTER OF GALAXIES
LB 09621	15 18 00.	+ 24 04		16.5	FAINT BLUE STAR
LB 09620	15 18 00.	+ 25 46		17.3	FAINT BLUE STAR
MCG+04-36-038	15 18 00.	+ 25 53 30.	15	15.2	GALAXY
ZWG 135.047	15 18 00.	+ 25 55		15.2	GALAXY
ZC 1518.0+3210	15 18 00.	+ 32 10	540		CLUSTER OF GALAXIES
7ZW 595	15 18 00.	+ 59 17			COMPACT GALAXY
ZWG 366.015	15 18 00.	+ 82 08		15.6	GALAXY
ZWG 021.078	15 18 06.	+ 02 00		15.4	GALAXY
ZWG 049.136	15 18 06.	+ 08 08		15.5	GALAXY
LB 09623	15 18 06.	+ 22 07		17.8	FAINT BLUE STAR
LB 09622	15 18 06.	+ 25 09		16.2	FAINT BLUE STAR
ZC 1518.1+2520	15 18 06.	+ 25 20	670		CLUSTER OF GALAXIES
ZC 1518.1+5501	15 18 06.	+ 55 01	940		CLUSTER OF GALAXIES
IC 4538	15 18 06.	- 23 30 41.			NONSTELLAR OBJECT
ARC 2059	15 18 10.	+ 29 01		17.0	RICH CLUSTER OF GALAXIES
LB 00808	15 18 11.	+ 01 59 30.		17.2	FAINT BLUE STAR
REIZ 4535	15 18 11.	+ 12 11	18	15.3	GALAXY
ZC 1518.2+0205	15 18 12.	+ 02 05	3970		CLUSTER OF GALAXIES
ZWG 077.077	15 18 12.	+ 09 10		15.6	GALAXY
KARA.72 462A	15 18 12.	+ 09 10	48	15.6	PART OF DOUBLE GALAXY
ZWG 077.078	15 18 12.	+ 12 10		15.2	GALAXY
LB 09626	15 18 12.	+ 21 56		16.5	FAINT BLUE STAR
LB 09625	15 18 12.	+ 23 04		15.8	FAINT BLUE STAR
LB 09624	15 18 12.	+ 23 41		18.0	FAINT BLUE STAR
ZC 1518.2+3040	15 18 12.	+ 30 40	3230		CLUSTER OF GALAXIES
ARC 2057	15 18 15.	- 10 30		17.1	RICH CLUSTER OF GALAXIES
ZWG 077.079	15 18 18.	+ 09 07		15.7	GALAXY
KARA.72 462B	15 18 18.	+ 09 07	42	15.7	PART OF DOUBLE GALAXY
ZWG 021.079	15 18 18.	- 02 25		14.6	GALAXY
RNGC 5913	15 18 18.	- 02 25		14.5	GALAXY
UGC 09818	15 18 18.	- 02 25	150	14.6	GALAXY SB
MCG+00-39-021	15 18 18.	- 02 25	78	14.6	GALAXY
KARA.73B 0669	15 18 18.	- 02 25	96	14.6	ISOLATED GALAXY S
MCG-04-36-013	15 18 18.	- 23 29	138	13.	GALAXY
REIZ 4537	15 18 21.	+ 31 16	42	15.5	GALAXY
LB 09628	15 18 24.	+ 24 42		16.4	FAINT BLUE STAR
ZWG 135.048	15 18 24.	+ 25 55		15.6	GALAXY
LB 09627	15 18 24.	+ 26 36		16.7	FAINT BLUE STAR
MCG-02-39-018	15 18 24.	- 12 56	60	15.	GALAXY
MCG+04-36-039	15 18 27.	+ 25 55	48	15.6	GALAXY
MCG+06-34-004	15 18 27.	+ 33 03	24	15.	GALAXY
HELW 479	15 18 28.	- 12 55 12.			NEBULA
ZWG 077.080	15 18 30.	+ 08 35		15.3	GALAXY
ZWG 135.049	15 18 30.	+ 25 52		15.7	GALAXY
ZC 1518.5+2814	15 18 30.	+ 28 14	2490		CLUSTER OF GALAXIES
MCG+06-34-005	15 18 30.	+ 33 02	42	15.5	GALAXY
1ZW 108	15 18 30.	+ 51 31			COMPACT GALAXY
MCG+12-14-022	15 18 30.	+ 70 20	39	16.	GALAXY
UGC 09819	15 18 30.	+ 73 07	84	16.0	GALAXY PECULR
RNGC 5916A	15 18 31.	- 12 55		15.0	GALAXY
LB 09630	15 18 36.	+ 23 34		17.3	FAINT BLUE STAR
LB 09629	15 18 36.	+ 24 16		16.1	FAINT BLUE STAR
ZC 1518.6+3327	15 18 36.	+ 33 27	3760		CLUSTER OF GALAXIES
ZC 1518.6+4353	15 18 36.	+ 43 53	1210		CLUSTER OF GALAXIES
ZWG 338.001	15 18 36.	+ 70 23		15.7	GALAXY
ZWG 337.036	15 18 36.	+ 70 23		15.7	GALAXY
7ZW 596	15 18 36.	+ 73 06			COMPACT GALAXY
MCG-03-39-008	15 18 36.	- 18 03 30.	48	15.5	GALAXY
MCG-03-39-007	15 18 36.	- 18 43	48	16.	GALAXY
PK323+02.1	15 18 37.	- 53 57 27.	25		PLANETARY NEBULA
LB 00809	15 18 39.	+ 27 23 18.		16.0	FAINT BLUE STAR
LB 09631	15 18 42.	+ 26 15		16.8	FAINT BLUE STAR
REIZ 4540	15 18 44.	+ 36 16	60	14.0	GALAXY
LB 00810	15 18 45.	+ 30 03 06.		16.1	FAINT BLUE STAR
MCG-02-39-019	15 18 45.	- 12 55	72	13.	GALAXY
RNGC 5917	15 18 46.	- 07 12		14.5	GALAXY
ZWG 049.137	15 18 48.	+ 07 33		15.6	GALAXY
ZC 1518.8+0747	15 18 48.	+ 07 47	3900		CLUSTER OF GALAXIES
LB 09636	15 18 48.	+ 22 46		16.7	FAINT BLUE STAR
LB 09635	15 18 48.	+ 23 02		15.4	FAINT BLUE STAR
LB 09634	15 18 48.	+ 25 16		17.5	FAINT BLUE STAR
LB 09633	15 18 48.	+ 26 42		16.1	FAINT BLUE STAR
LB 09632	15 18 48.	+ 26 45		17.0	FAINT BLUE STAR
MCG-01-39-002	15 18 48.	- 07 12	60	14.5	GALAXY
MCG-01-39-003	15 18 48.	- 07 16	84	15.	GALAXY
MCG-02-39-021	15 18 48.	- 11 55	48	14.5	GALAXY
MCG-02-39-020	15 18 48.	- 13 00	108	14.	GALAXY
RNGC 5915	15 18 49.	- 12 55		12.5	GALAXY
ARP 254	15 18 50.	- 07 11			PECULIAR GALAXY
LB 00811	15 18 51.	+ 03 00 42.		16.5	FAINT BLUE STAR
REIZ 4538	15 18 51.	+ 14 25	12	15.0	GALAXY
REIZ 4539	15 18 51.	+ 30 40	30	13.6	GALAXY
FATH 1.718	15 18 53.	+ 59 18	8		NEBULA
ZWG 049.138	15 18 54.	+ 04 31		15.6	GALAXY
ZWG 077.081	15 18 54.	+ 13 06		15.6	GALAXY
ZWG 077.082	15 18 54.	+ 14 25		15.3	GALAXY
ZWG 106.032	15 18 54.	+ 15 29		15.3	GALAXY
MCG+03-39-020	15 18 54.	+ 15 30	60	15.3	GALAXY
FATH 1.719	15 18 54.	+ 59 36	19		NEBULA
MCG+11-19-005	15 18 54.	+ 65 45 30.	30	16.	GALAXY
ZWG 319.007	15 18 54.	+ 65 46		15.7	GALAXY
MCG+12-14-023	15 18 54.	+ 73 09	30	17.	GALAXY
RNGC 5916	15 18 55.	- 12 59		14.0	GALAXY
LB 00812	15 18 58.	- 00 20 06.		15.8	FAINT BLUE STAR
3ZW 072	15 19 00.	+ 02 58			COMPACT GALAXY
ZWG 049.139	15 19 00.	+ 06 40		15.4	GALAXY
LB 09638	15 19 00.	+ 21 48		16.0	FAINT BLUE STAR
LB 09637	15 19 00.	+ 25 26		16.5	FAINT BLUE STAR
ZWG 165.038	15 19 00.	+ 26 31		15.6	GALAXY
SN 1962N	15 19 00.	+ 26 31		17.0	SUPERNOVA
MCG+05-36-013	15 19 00.	+ 26 32	48	15.6	GALAXY
ZWG 165.039	15 19 00.	+ 30 50		15.5	GALAXY
MCG+08-28-019	15 19 00.	+ 49 17	42	15.	GALAXY
ZC 1519.0+6351	15 19 00.	+ 63 51	540		CLUSTER OF GALAXIES
MCG+11-19-006	15 19 00.	+ 65 47	30	16.	GALAXY
LB 00813	15 19 02.	+ 11 53 30.		16.6	FAINT BLUE STAR
RNGC 5919	15 19 05.	+ 07 54		15.5	GALAXY
ZWG 021.080	15 19 06.	+ 00 38		15.1	GALAXY
ZWG 049.140	15 19 06.	+ 05 25		15.5	GALAXY
ZWG 049.141	15 19 06.	+ 07 46		15.5	GALAXY
ZWG 049.142	15 19 06.	+ 07 54		15.5	GALAXY
ZWG 106.033	15 19 06.	+ 15 19		15.1	GALAXY
MCG+03-39-021	15 19 06.	+ 15 19	48	15.1	GALAXY

OBJECT NAME	RIGHT ASCEN.	DECLINATION	DIAM.	MAGN.	TYPE OF OBJECT
LB 09643	15 19 06.	+ 22 24		16.2	FAINT BLUE STAR
LB 09642	15 19 06.	+ 24 26		16.6	FAINT BLUE STAR
LB 09641	15 19 06.	+ 24 37		16.0	FAINT BLUE STAR
LB 09640	15 19 06.	+ 25 02		16.3	FAINT BLUE STAR
LB 09639	15 19 06.	+ 26 24		16.4	FAINT BLUE STAR
1ZW 109	15 19 06.	+ 49 17			COMPACT GALAXY
ZWG 249.017	15 19 06.	+ 49 17		15.4	GALAXY
FATH 1.720	15 19 06.	+ 60 23	8		NEBULA
ZWG 319.008	15 19 06.	+ 67 41		15.1	GALAXY
KARA.73B 0670	15 19 06.	+ 67 41	36	15.1	ISOLATED GALAXY E
MCG-03-39-009	15 19 06.	- 17 34	24	15.	GALAXY
LB 00814	15 19 09.	+ 28 26 36.		17.2	FAINT BLUE STAR
REIZ 4541	15 19 09.	+ 30 50	12	15.2	GALAXY
ZWG 049.143	15 19 12.	+ 03 54		15.4	GALAXY
ZWG 049.144	15 19 12.	+ 07 54		15.7	GALAXY
MCG+01-39-020	15 19 12.	+ 07 54	180		GALAXY
ZWG 077.083	15 19 12.	+ 11 26		15.4	GALAXY
ZWG 106.034	15 19 12.	+ 14 34		15.7	GALAXY
ZWG 135.050	15 19 12.	+ 21 23		15.4	GALAXY
LB 09646	15 19 12.	+ 23 02		16.2	FAINT BLUE STAR
LB 09645	15 19 12.	+ 26 03		18.1	FAINT BLUE STAR
LB 09644	15 19 12.	+ 26 18		16.5	FAINT BLUE STAR
ZWG 165.040	15 19 12.	+ 28 44		15.3	GALAXY
UGC 09820	15 19 12.	+ 28 44	72	15.3	GALAXY S0-a
ARC 2061	15 19 12.	+ 30 50		15.7	RICH CLUSTER OF GALAXIES
ZWG 165.041	15 19 12.	+ 30 51		15.7	GALAXY
ZWG 249.018	15 19 12.	+ 45 55		15.1	GALAXY
7ZW 597	15 19 12.	+ 62 32			COMPACT GALAXY
VV 225B	15 19 15.	+ 03 25	72	14.	INTERACTING GALAXY
VV 225A	15 19 15.	+ 03 25	66	14.	INTERACTING GALAXY
VV 225	15 19 15.	+ 03 25	96		INTERACTING GALAXY
REIZ 4542	15 19 15.	+ 30 51	24	15.3	GALAXY
ZWG 077.084	15 19 18.	+ 08 36		15.1	GALAXY
UGC 09821	15 19 18.	+ 08 36	90	15.1	GALAXY SBb
MCG+02-39-016	15 19 18.	+ 08 36	90	15.1	GALAXY
ZC 1519.3+1220	15 19 18.	+ 12 20	340		CLUSTER OF GALAXIES
ZC 1519.3+2631	15 19 18.	+ 26 31	3020		CLUSTER OF GALAXIES
MCG+08-28-020	15 19 18.	+ 48 49	18	15.	GALAXY
LB 00815	15 19 20.	- 00 45 06.		17.1	FAINT BLUE STAR
MCG+05-36-014	15 19 21.	+ 28 44	27	15.3	GALAXY
ARC 2062	15 19 21.	+ 32 16		16.6	RICH CLUSTER OF GALAXIES
FATH 1.721	15 19 21.	+ 59 22	27		NEBULA
RNGC 5920	15 19 23.	+ 07 53		15.5	GALAXY
ZWG 049.145	15 19 24.	+ 07 53		15.5	GALAXY
UGC 09822	15 19 24.	+ 07 53	66	15.5	GALAXY S0
MKW 03S	15 19 24.	+ 07 53		15.5	POOR GALAXY CLUSTER
ZC 1519.4+2213	15 19 24.	+ 22 13	1750		CLUSTER OF GALAXIES
LB 09650	15 19 24.	+ 22 17		16.7	FAINT BLUE STAR
LB 09649	15 19 24.	+ 23 27		17.5	FAINT BLUE STAR
LB 09648	15 19 24.	+ 24 32		8.0	FAINT BLUE STAR
LB 09647	15 19 24.	+ 24 40		16.9	FAINT BLUE STAR
ZWG 135.051	15 19 24.	+ 26 01		15.4	GALAXY
ZC 1519.4+2610	15 19 24.	+ 26 10	9610		CLUSTER OF GALAXIES
RNGC 5922	15 19 24.	+ 41 49			NON-EXISTENT OBJECT
REIZ 4544	15 19 24.	+ 41 50	18	15.4	GALAXY
ZWG 222.002	15 19 24.	+ 41 54		14.7	GALAXY
ZWG 221.052	15 19 24.	+ 41 54		14.7	GALAXY
RNGC 5923	15 19 24.	+ 41 54		14.5	GALAXY
REIZ 4545	15 19 24.	+ 41 54	66	14.0	GALAXY
UGC 09823	15 19 24.	+ 41 54	132	14.7	GALAXY Sb/SBc
MCG+07-32-001	15 19 24.	+ 41 54	84	14.	GALAXY
ARC 2064	15 19 25.	+ 48 49		16.6	RICH CLUSTER OF GALAXIES
HOLM 707B	15 19 26.	+ 41 51	30	15.2	PART OF MULTIPLE GALAXY
PK342+27.1	15 19 26.	- 23 27	8		PLANETARY NEBULA
REIN 2.231	15 19 27.41	+ 05 14 55.9			NEBULA
RNGC 5921	15 19 28.	+ 05 15		12.0	GALAXY
LB 00816	15 19 28.	+ 29 31 00.		16.4	FAINT BLUE STAR
ZWG 049.146	15 19 30.	+ 05 15		12.7	GALAXY
UGC 09824	15 19 30.	+ 05 15	318	12.7	GALAXY SBb
MCG+01-39-021	15 19 30.	+ 05 15	270	12.7	GALAXY
ZWG 077.085	15 19 30.	+ 08 37		14.9	GALAXY
MCG+02-39-017	15 19 30.	+ 08 37	72	14.9	GALAXY
ZWG 077.086	15 19 30.	+ 10 45		15.5	GALAXY
ZWG 077.087	15 19 30.	+ 13 26		15.4	GALAXY
SN 1950K	15 19 30.	+ 17 37		20.0	SUPERNOVA
LB 09652	15 19 30.	+ 21 58		16.2	FAINT BLUE STAR
ZWG 135.052	15 19 30.	+ 23 25		15.4	GALAXY
UGC 09825	15 19 30.	+ 23 25	66	15.4	GALAXY Sc
ZC 1519.5+2401	15 19 30.	+ 24 01	3970		CLUSTER OF GALAXIES
LB 09651	15 19 30.	+ 26 36		16.8	FAINT BLUE STAR
HOLM 707A	15 19 30.	+ 41 54	78	13.6	PART OF MULTIPLE GALAXY
ZC 1519.5+4618	15 19 30.	+ 46 18	1080		CLUSTER OF GALAXIES
ZC 1519.5+7020	15 19 30.	+ 70 20	870		CLUSTER OF GALAXIES
IC 1116	15 19 35.	+ 08 36 55.			NONSTELLAR OBJECT
ZC 1519.6+0140	15 19 36.	+ 01 40	810		CLUSTER OF GALAXIES
ZWG 049.147	15 19 36.	+ 03 56		15.3	GALAXY
ZWG 049.148	15 19 36.	+ 06 02		15.2	GALAXY
KARA.72 463A	15 19 36.	+ 06 02	42	15.2	PART OF DOUBLE GALAXY
LB 09654	15 19 36.	+ 21 46		16.9	FAINT BLUE STAR
MCG+04-36-040	15 19 36.	+ 23 24	60	15.4	GALAXY
LB 09653	15 19 36.	+ 25 37		16.7	FAINT BLUE STAR
ZWG 165.042	15 19 36.	+ 26 30		15.7	GALAXY
MCG+07-32-002	15 19 36.	+ 39 23	60	15.	GALAXY
MCG+09-25-043	15 19 36.	+ 50 50	48	16.	GALAXY
ZWG 274.040	15 19 36.	+ 50 52		15.5	GALAXY
ARC 2060	15 19 36.	- 12 00		17.7	RICH CLUSTER OF GALAXIES
MCG+09-25-044	15 19 39.	+ 50 49 30.	18	17.	GALAXY
HELW 480	15 19 39.	- 12 41 38.			NEBULA
ZWG 049.149	15 19 42.	+ 06 01		15.5	GALAXY
KARA.72 463B	15 19 42.	+ 06 01	30	15.5	PART OF DOUBLE GALAXY
ZC 1519.7+2107	15 19 42.	+ 21 07	5240		CLUSTER OF GALAXIES
LB 09655	15 19 42.	+ 24 19		16.0	FAINT BLUE STAR
ZWG 222.003	15 19 42.	+ 39 22		15.3	GALAXY
UGC 09826	15 19 42.	+ 39 22	66	15.3	GALAXY SBc
KARA.73B 0671	15 19 42.	+ 39 22	72	15.3	ISOLATED GALAXY S
1ZW 110	15 19 42.	+ 42 06			COMPACT GALAXY
LB 00817	15 19 43.	+ 12 12 30.		17.1	FAINT BLUE STAR
IC 1115	15 19 44.	- 04 16 59.			MAY NOT EXIST
PK322-00.1	15 19 47.	- 56 58 22.	10		PLANETARY NEBULA
ZWG 077.088	15 19 48.	+ 14 06		15.0	GALAXY
MCG+02-39-018	15 19 48.	+ 14 06	42	15.0	GALAXY
LB 09656	15 19 48.	+ 24 24		15.7	FAINT BLUE STAR
ZWG 249.019	15 19 48.	+ 48 35		15.7	GALAXY
ZC 1519.8+6700	15 19 48.	+ 67 00	400		CLUSTER OF GALAXIES
REIZ 4543	15 19 51.	+ 14 06	18	15.0	GALAXY
RNGC 5924	15 19 53.	+ 31 24		15.5	GALAXY
PK324+02.1	15 19 53.	- 53 40 39.	25		PLANETARY NEBULA
ZWG 049.150	15 19 54.	+ 05 34		15.2	GALAXY

OBJECT NAME	RIGHT ASCEN.	DECLINATION	DIAM.	MAGN.	TYPE OF OBJECT
SN 1963H	15 19 54.	+ 05 34		18.5	SUPERNOVA
ZWG 077.089	15 19 54.	+ 08 46		15.7	GALAXY
ZWG 077.090	15 19 54.	+ 08 50		15.7	GALAXY
LB 09661	15 19 54.	+ 23 24		16.5	FAINT BLUE STAR
LB 09660	15 19 54.	+ 23 36		16.2	FAINT BLUE STAR
LB 09659	15 19 54.	+ 24 52		17.8	FAINT BLUE STAR
LB 09658	15 19 54.	+ 25 30		18.0	FAINT BLUE STAR
LB 09657	15 19 54.	+ 25 46		16.3	FAINT BLUE STAR
ZC 1519.9+2947	15 19 54.	+ 29 47	2490		CLUSTER OF GALAXIES
ZWG 165.043	15 19 54.	+ 31 24		15.3	GALAXY
MCG+05-36-015	15 19 54.	+ 31 24	36	15.3	GALAXY
MCG+12-14-024	15 19 54.	+ 69 17 30.	24	18.	GALAXY
ZL 177	15 19 57.	+ 27 53 18.		22.1	ULTRAFAINT BLUE STAR
MCG+12-14-025	15 19 57.	+ 69 17 30.	27	15.	GALAXY
HMS 1520+2754	15 20	+ 27 54			COR BOR GALAXY CLUSTER
ZWG 077.091	15 20 00.	+ 08 32		15.7	GALAXY
LB 09662	15 20 00.	+ 20 56		16.8	FAINT BLUE STAR
ZWG 165.044	15 20 00.	+ 26 35		15.5	GALAXY
ZC 1520.0+2748	15 20 00.	+ 27 48	2760		CLUSTER OF GALAXIES
MCG+05-36-016	15 20 00.	+ 27 51	6		GALAXY
ZWG 194.002	15 20 00.	+ 33 20		15.1	GALAXY
ZWG 338.002	15 20 00.	+ 69 19		15.7	GALAXY
ZL 178	15 20 01.	+ 27 43 48.		18.8	ULTRAFAINT BLUE STAR
ARC 2075	15 20 04.	+ 74 12		17.1	RICH CLUSTER OF GALAXIES
MCG+06-34-007	15 20 06.	+ 33 20	15	16.	GALAXY
MCG+06-34-006	15 20 06.	+ 33 20	42	15.	GALAXY
IC 4542	15 20 06.	+ 33 20 48.			NONSTELLAR OBJECT
ZWG 222.004	15 20 06.	+ 41 18		15.2	GALAXY
UGC 09827	15 20 06.	+ 41 18	60	15.2	GALAXY
MCG+07-32-003	15 20 06.	+ 41 19	48	14.5	GALAXY
ZL 179	15 20 08.	+ 27 42 24.		18.7	ULTRAFAINT BLUE STAR
ZWG 106.035	15 20 12.	+ 19 26		15.7	GALAXY
UGC 09828	15 20 12.	+ 19 26	102	15.7	GALAXY Sc
KARA.73B 0672	15 20 12.	+ 19 26	102	15.7	ISOLATED GALAXY S
LB 09665	15 20 12.	+ 21 28		15.7	FAINT BLUE STAR
LB 09664	15 20 12.	+ 26 01		17.4	FAINT BLUE STAR
LB 09663	15 20 12.	+ 26 14		16.9	FAINT BLUE STAR
MCG+05-36-017	15 20 12.	+ 27 52	12		GALAXY
ZL 180	15 20 12.	+ 28 00 12.		20.8	ULTRAFAINT BLUE STAR
MCG+05-36-018	15 20 12.	+ 28 01	4		GALAXY
LB 00818	15 20 12.	+ 29 11 30.		17.3	FAINT BLUE STAR
ZL 181	15 20 14.	+ 27 44 36.		22.3	ULTRAFAINT BLUE STAR
MCG+03-39-022	15 20 15.	+ 19 27	120	15.7	GALAXY
REIZ 4546	15 20 15.	+ 30 47	9	16.0	GALAXY
MCG+08-28-021	15 20 15.	+ 48 26	30	16.	GALAXY
MCG+09-25-045	15 20 15.	+ 50 41	42	15.	GALAXY
LB 00237	15 20 15.	+ 60 18 00.		14.9	FAINT BLUE STAR
ZWG 077.092	15 20 18.	+ 08 54		15.5	GALAXY
ZC 1520.3+2244	15 20 18.	+ 22 44	2420		CLUSTER OF GALAXIES
MCG+05-36-020	15 20 18.	+ 27 54	18		GALAXY
MCG+05-36-019	15 20 18.	+ 27 55	9		GALAXY
SN 1963F	15 20 18.	+ 28 01		17.6	SUPERNOVA
MCG+08-28-022	15 20 18.	+ 48 47	36	16.	GALAXY
1ZW 111	15 20 18.	+ 53 23			COMPACT GALAXY
7ZW 598	15 20 18.	+ 78 05			COMPACT GALAXY
ZC 1520.3-0018	15 20 18.	- 00 18	1410		CLUSTER OF GALAXIES
ZWG 021.082	15 20 18.	- 00 44		15.6	GALAXY
ZWG 021.081	15 20 18.	- 01 11		15.7	GALAXY
MCG-01-39-004	15 20 18.	- 07 09	12	16.	GALAXY
LB 00819	15 20 19.	+ 12 51 36.		17.7	FAINT BLUE STAR
ZL 183	15 20 20.	+ 27 54 06.		19.6	ULTRAFAINT BLUE STAR
ZL 184	15 20 21.	+ 28 00 06.		18.6	ULTRAFAINT BLUE STAR
LB 00820	15 20 21.	+ 28 18 06.		17.4	FAINT BLUE STAR
ZWG 021.084	15 20 24.	+ 00 16		15.6	GALAXY
MCG+00-39-022	15 20 24.	+ 00 16	24	15.6	GALAXY
LB 00821	15 20 24.	+ 00 28 36.		15.9	FAINT BLUE STAR
ZWG 049.151	15 20 24.	+ 08 16		15.4	GALAXY
ZWG 077.093	15 20 24.	+ 08 56		15.6	GALAXY
MCG+05-36-021	15 20 24.	+ 27 52	6		GALAXY
ZWG 021.083	15 20 24.	- 01 13		15.6	GALAXY
ZL 185	15 20 28.	+ 27 46 42.		22.0	ULTRAFAINT BLUE STAR
ZC 1520.5+0429	15 20 30.	+ 04 29	1210		CLUSTER OF GALAXIES
ZWG 049.152	15 20 30.	+ 04 42		15.5	GALAXY
UGC 09830	15 20 30.	+ 04 42	90	15.5	GALAXY Sc
ZWG 049.153	15 20 30.	+ 08 14		15.5	GALAXY
ZWG 077.094	15 20 30.	+ 08 32		15.7	GALAXY
ZWG 077.095	15 20 30.	+ 08 42		15.6	GALAXY
MCG+02-39-019	15 20 30.	+ 08 42	12	15.6	GALAXY
MCG+05-36-022	15 20 30.	+ 27 53	6		GALAXY
ZC 1520.5+3116	15 20 30.	+ 31 16	3230		CLUSTER OF GALAXIES
ZWG 319.009	15 20 30.	+ 65 42		15.6	GALAXY
MCG+12-14-026	15 20 30.	+ 73 54	12	17.	GALAXY
ZWG 021.085	15 20 30.	- 01 10		15.3	GALAXY
UGC 09829	15 20 30.	- 01 10	138	15.3	GALAXY S
MCG+00-39-023	15 20 30.	- 01 10	120	15.3	GALAXY
MCG+03-39-023	15 20 33.	+ 15 10	48	15.0	GALAXY
ZL 186	15 20 33.	+ 27 47 48.		17.6	ULTRAFAINT BLUE STAR
ZWG 077.096	15 20 36.	+ 08 46		15.7	GALAXY
ZWG 077.097	15 20 36.	+ 08 47		15.0	GALAXY
MCG+02-39-020	15 20 36.	+ 08 47	48	15.0	GALAXY
ZWG 077.098	15 20 36.	+ 08 57		15.5	GALAXY
ZWG 106.036	15 20 36.	+ 15 11		15.0	GALAXY
ZC 1520.6+1516	15 20 36.	+ 15 16	740		CLUSTER OF GALAXIES
LB 09666	15 20 36.	+ 25 46		16.3	FAINT BLUE STAR
ZL 187	15 20 36.	+ 27 39 18.		19.3	ULTRAFAINT BLUE STAR
MCG+05-36-024	15 20 36.	+ 27 51	18		GALAXY
MCG+05-36-023	15 20 36.	+ 27 51	9		GALAXY
ARC 2065	15 20 36.	+ 27 54		15.6	RICH CLUSTER OF GALAXIES
ZC 1520.6+4350	15 20 36.	+ 43 50	740		CLUSTER OF GALAXIES
ZWG 338.003	15 20 36.	+ 73 54		15.5	GALAXY
ZWG 337.037	15 20 36.	+ 73 54		15.5	GALAXY
ZWG 021.086	15 20 36.	- 03 20		15.5	GALAXY
MCG+00-39-024	15 20 36.	- 03 20	24	15.5	GALAXY
ARC 2063	15 20 37.	+ 08 49		15.1	RICH CLUSTER OF GALAXIES
ZWG 077.099	15 20 42.	+ 08 41		15.6	GALAXY
ZWG 077.100	15 20 42.	+ 08 43		15.6	GALAXY
MCG+02-39-021	15 20 42.	+ 08 43	30	15.6	GALAXY
ZWG 077.101	15 20 42.	+ 11 25		15.4	GALAXY
MCG+05-36-025	15 20 42.	+ 29 56	42	14.9	GALAXY
ZWG 165.045	15 20 42.	+ 29 57		14.9	GALAXY
UGC 09831	15 20 42.	+ 29 57	60	14.9	GALAXY Sa
SN 1962B	15 20 42.	+ 29 57		17.0	SUPERNOVA
MCG-01-39-005	15 20 42.	- 03 56	84	14.	GALAXY
LB 00822	15 20 44.	+ 60 47 06.		16.0	FAINT BLUE STAR
LB 00823	15 20 45.	+ 30 31 48.		16.0	FAINT BLUE STAR
ZL 189	15 20 46.	+ 27 48 54.		20.4	ULTRAFAINT BLUE STAR
ZL 188	15 20 46.	+ 27 56 54.		22.6	ULTRAFAINT BLUE STAR
ZL 190	15 20 47.	+ 28 04 30.		20.5	ULTRAFAINT BLUE STAR
ZWG 077.102	15 20 48.	+ 08 44		15.5	GALAXY
ZWG 077.103	15 20 48.	+ 08 45		15.5	GALAXY
MCG+02-39-022	15 20 48.	+ 08 45	24	15.5	GALAXY
ZWG 077.104	15 20 48.	+ 08 49		15.6	GALAXY
MCG+02-39-023	15 20 48.	+ 08 49	12	15.6	GALAXY
ZWG 077.105	15 20 48.	+ 08 50		15.5	GALAXY
MCG+02-39-024	15 20 48.	+ 08 50	12	15.5	GALAXY
ZWG 077.106	15 20 48.	+ 08 56		15.7	GALAXY
LB 09670	15 20 48.	+ 24 26		16.6	FAINT BLUE STAR
LB 09669	15 20 48.	+ 25 28		15.3	FAINT BLUE STAR
LB 09668	15 20 48.	+ 25 38		16.6	FAINT BLUE STAR
LB 09667	15 20 48.	+ 25 46		16.2	FAINT BLUE STAR
ZC 1520.8+3207	15 20 48.	+ 32 07	5040		CLUSTER OF GALAXIES
ZL 191	15 20 51.	+ 27 55 30.		20.1	ULTRAFAINT BLUE STAR
ZWG 077.107	15 20 54.	+ 08 48		15.5	GALAXY
MCG+02-39-025	15 20 54.	+ 08 48	12	15.5	GALAXY
ZWG 077.108	15 20 54.	+ 08 54		15.6	GALAXY
ZC 1520.9+1045	15 20 54.	+ 10 45	1550		CLUSTER OF GALAXIES
LB 09671	15 20 54.	+ 26 22		16.7	FAINT BLUE STAR
ZC 1520.9+7407	15 20 54.	+ 74 07	1340		CLUSTER OF GALAXIES
ZWG 021.088	15 20 54.	- 01 00		15.6	GALAXY
ZWG 021.087	15 20 54.	- 01 17		15.6	GALAXY
ZL 192	15 20 56.	+ 27 55 24.		18.0	ULTRAFAINT BLUE STAR
MCG+02-39-027	15 21 00.	+ 12 51	12	16.	GALAXY
ZWG 077.109	15 21 00.	+ 12 52		14.8	GALAXY
8ZW 1521+12.9	15 21 00.	+ 12 52		16.2	COMPACT GALAXY
MCG+02-39-026	15 21 00.	+ 12 52	48	14.8	GALAXY
ZL 193	15 21 00.	+ 27 52 18.		22.7	ULTRAFAINT BLUE STAR
ZC 1521.0+2835	15 21 00.	+ 28 35	1480		CLUSTER OF GALAXIES
ZC 1521.0+4844	15 21 00.	+ 48 44	8940		CLUSTER OF GALAXIES
ZWG 021.089	15 21 00.	- 02 57		15.6	GALAXY
SMI 11	15 21 00.	- 58 41	360		FAINT NEBULOSITY
RNGC 5926	15 21 01.	+ 12 52		15.0	GALAXY
HOLM 708B	15 21 01.	+ 12 53	12	14.9	PART OF MULTIPLE GALAXY
HOLM 708A	15 21 04.	+ 12 54	18	13.6	PART OF MULTIPLE GALAXY
ZL 194	15 21 05.	+ 27 49 54.		20.8	ULTRAFAINT BLUE STAR
ZWG 077.110	15 21 06.	+ 09 32		14.9	GALAXY
KARA.72 464A	15 21 06.	+ 09 32	48	14.9	PART OF DOUBLE GALAXY
MCG+02-39-028	15 21 06.	+ 09 32	12	14.9	GALAXY
ZWG 077.111	15 21 06.	+ 13 44		15.5	GALAXY
ZC 1521.1+2029	15 21 06.	+ 20 29	470		CLUSTER OF GALAXIES
LB 09674	15 21 06.	+ 21 38		16.9	FAINT BLUE STAR
LB 09673	15 21 06.	+ 25 09		17.0	FAINT BLUE STAR
LB 09672	15 21 06.	+ 26 12		16.4	FAINT BLUE STAR
LB 00824	15 21 06.	+ 29 42 48.		17.6	FAINT BLUE STAR
TON-N 0229	15 21 06.	+ 31 04		15.2	BLUE STAR
ZWG 319.010	15 21 06.	+ 65 11		15.6	GALAXY
KARA.73B 0673	15 21 06.	+ 65 11	24	15.6	ISOLATED GALAXY S
REIZ 4548	15 21 08.	+ 14 20	18	15.1	GALAXY
REIZ 4547	15 21 09.	+ 13 44	60	15.0	GALAXY
ARC 2067	15 21 11.	+ 31 06		15.7	RICH CLUSTER OF GALAXIES
ZC 1521.2+0851	15 21 12.	+ 08 51	6050		CLUSTER OF GALAXIES
ZWG 077.112	15 21 12.	+ 08 57		15.5	GALAXY
ZWG 077.113	15 21 12.	+ 09 10		15.7	GALAXY
ZWG 077.114	15 21 12.	+ 14 17		15.7	GALAXY
ZC 1521.2+2014	15 21 12.	+ 20 14	610		CLUSTER OF GALAXIES
LB 09676	15 21 12.	+ 22 53		16.1	FAINT BLUE STAR
LB 09675	15 21 12.	+ 25 22		17.5	FAINT BLUE STAR
ZWG 165.046	15 21 12.	+ 32 15		15.7	GALAXY
ZWG 338.004	15 21 12.	+ 68 28		15.7	GALAXY
ZWG 319.011	15 21 12.	+ 68 28		15.7	GALAXY
PK325+04.2	15 21 14.	- 51 39 14.	5		PLANETARY NEBULA
ZL 182	15 21 17.	+ 27 51 54.		21.8	ULTRAFAINT BLUE STAR
ZWG 049.154	15 21 18.	+ 05 57		15.3	GALAXY
ZWG 077.115	15 21 18.	+ 09 06		15.7	GALAXY
ZWG 077.116	15 21 18.	+ 09 34		15.4	GALAXY
KARA.72 464B	15 21 18.	+ 09 34	42	15.4	PART OF DOUBLE GALAXY
MCG+04-36-041	15 21 18.	+ 20 36 30.	36		GALAXY
LB 09677	15 21 18.	+ 22 34		16.8	FAINT BLUE STAR
3ZW 1521+14.0	15 21 24.	+ 14 03		18.0	COMPACT GALAXY
LB 09678	15 21 24.	+ 25 04		16.4	FAINT BLUE STAR
MCG+08-28-023	15 21 24.	+ 46 14	42	16.	GALAXY
MCG+12-15-004	15 21 24.	+ 71 05	45	16.	GALAXY
MCG+12-15-001	15 21 24.	+ 71 05	12	16.	GALAXY
LB 00825	15 21 25.	+ 61 15 24.		17.8	FAINT BLUE STAR
ARC 2066	15 21 26.	+ 01 14		17.2	RICH CLUSTER OF GALAXIES
LB 00826	15 21 29.	+ 12 06 54.		16.9	FAINT BLUE STAR
ZC 1521.5+0115	15 21 30.	+ 01 15	810		CLUSTER OF GALAXIES
LB 00827	15 21 30.	+ 03 00 36.		16.2	FAINT BLUE STAR
UGC 09832	15 21 30.	+ 14 03	66	16.5	GALAXY S
TON-N 0230	15 21 30.	+ 27 50		16.2	BLUE STAR
ZWG 165.047	15 21 30.	+ 31 47		15.6	GALAXY
7ZW 599	15 21 30.	+ 66 49			COMPACT GALAXY
ZWG 338.005	15 21 30.	+ 69 05		15.3	GALAXY
PK325+04.1	15 21 30.	- 51 09 13.	5		PLANETARY NEBULA
REIZ 4549	15 21 34.	+ 12 54	18	13.8	GALAXY
ZWG 135.053	15 21 36.	+ 23 44		15.7	GALAXY
UGC 09833	15 21 36.	+ 23 44	72	15.7	GALAXY
KARA.73B 0674	15 21 36.	+ 23 44	60	15.7	ISOLATED GALAXY S
ZC 1521.6+2819	15 21 36.	+ 28 19	740		CLUSTER OF GALAXIES
MCG+12-15-002	15 21 36.	+ 69 04	18	16.	GALAXY
MCG+04-36-042	15 21 42.	+ 23 43	66	15.7	GALAXY
LB 09679	15 21 42.	+ 24 52		16.2	FAINT BLUE STAR
TON-N 0789	15 21 42.	+ 26 58		14.7	BLUE STAR
ZC 1521.7+6140	15 21 42.	+ 61 40	2960		CLUSTER OF GALAXIES
SHAH 163	15 21 42.	+ 75 14	42		GROUP OF COMPACT GALAXIES
REIZ 4550	15 21 45.	+ 13 46	48	14.9	GALAXY
MCG+09-25-046	15 21 45.	+ 55 52	30	16.	GALAXY
ZWG 049.155	15 21 48.	+ 05 01		15.4	GALAXY
ZWG 077.117	15 21 48.	+ 08 44		15.6	GALAXY
ZWG 077.118	15 21 48.	+ 13 45		15.4	GALAXY
LB 09680	15 21 48.	+ 23 42		16.4	FAINT BLUE STAR
ZWG 165.048	15 21 48.	+ 26 44		15.6	GALAXY
ZWG 194.003	15 21 48.	+ 37 59		15.7	GALAXY
ZC 1521.8+4410	15 21 48.	+ 44 10	3020		CLUSTER OF GALAXIES
MCG-03-39-010	15 21 48.	- 21 12 30.	48	15.5	GALAXY
REIZ 4552	15 21 50.	+ 26 42	36	14.9	GALAXY
MCG+04-36-044	15 21 51.	+ 20 56	30	15.4	GALAXY
MCG+04-36-043	15 21 51.	+ 20 56	24	15.4	GALAXY
PK325+03.1	15 21 51.	- 52 40 11.	10		PLANETARY NEBULA
ZC 1521.9+0045	15 21 54.	+ 00 45	400		CLUSTER OF GALAXIES
ZWG 049.156	15 21 54.	+ 03 48		15.7	GALAXY
8ZW 1521+11.2	15 21 54.	+ 11 10		16.9	COMPACT GALAXY
ZWG 135.054	15 21 54.	+ 20 57		15.4	GALAXY
LB 09681	15 21 54.	+ 22 25		16.2	FAINT BLUE STAR
ZWG 165.049	15 21 54.	+ 31 23		15.7	GALAXY
ARC 2069	15 21 55.	+ 30 05		16.6	RICH CLUSTER OF GALAXIES
ARC 2074	15 21 55.	+ 63 57		17.1	RICH CLUSTER OF GALAXIES

OBJECT NAME	RIGHT ASCEN.	DECLINATION	DIAM.	MAGN.	TYPE OF OBJECT
REIZ 4553	15 21 56.	+ 31 23	30	15.3	GALAXY
REIZ 4554	15 21 56.	+ 31 26	18	15.7	GALAXY
MCG+05-36-026	15 21 57.	+ 31 23	24	15.7	GALAXY
ZC 1522.0+1439	15 22 00.	+ 14 39	940		CLUSTER OF GALAXIES
ZWG 106.037	15 22 00.	+ 15 40		15.6	GALAXY
ZC 1522.0+2500	15 22 00.	+ 25 00	2890		CLUSTER OF GALAXIES
TON-N 0231	15 22 00.	+ 30 40		15.1	BLUE STAR
ZC 1522.0+5604	15 22 00.	+ 56 04	940		CLUSTER OF GALAXIES
REIZ 4551	15 22 01.	+ 15 37	24	15.2	GALAXY
IC 1117	15 22 04.	+ 15 37 28.			NONSTELLAR OBJECT
LB 00828	15 22 05.	+ 29 05 06.		17.0	FAINT BLUE STAR
ZWG 077.119	15 22 06.	+ 08 34		15.5	GALAXY
ZC 1522.1+1659	15 22 06.	+ 16 59	1140		CLUSTER OF GALAXIES
LB 09682	15 22 06.	+ 25 06		16.9	FAINT BLUE STAR
TON-N 0232	15 22 06.	+ 33 08		16.1	BLUE STAR
ZWG 194.004	15 22 06.	+ 38 18		15.3	GALAXY
UGC 09834	15 22 06.	+ 38 19	60	15.3	GALAXY S0-a
MCG+06-34-008	15 22 06.	+ 38 20	48	14.	GALAXY
ZC 1522.1+6635	15 22 06.	+ 66 35	1080		CLUSTER OF GALAXIES
ZWG 135.055	15 22 12.	+ 20 56		15.5	GALAXY
LB 09686	15 22 12.	+ 22 18		17.2	FAINT BLUE STAR
LB 09685	15 22 12.	+ 25 41		15.8	FAINT BLUE STAR
LB 09684	15 22 12.	+ 25 52		16.9	FAINT BLUE STAR
ZC 1522.2+2611	15 22 12.	+ 26 11	740		CLUSTER OF GALAXIES
LB 09683	15 22 12.	+ 26 26		16.5	FAINT BLUE STAR
ZWG 165.050	15 22 12.	+ 26 37		15.0	GALAXY
MCG+05-36-027	15 22 12.	+ 26 37	12	15.0	GALAXY
ZWG 165.051	15 22 12.	+ 30 08		15.6	GALAXY
MCG+05-36-028	15 22 12.	+ 30 08	60	15.6	GALAXY
ZWG 249.020	15 22 12.	+ 46 47		14.8	GALAXY
UGC 09835	15 22 12.	+ 46 47	66	14.8	GALAXY Sb
IC 4543	15 22 13.	+ 13 39 29.			PLANETARY NEBULA
LB 00288	15 22 13.	+ 58 07 06.		15.3	FAINT BLUE STAR
MCG+08-28-024	15 22 15.	+ 46 46	60	15.	GALAXY
MCG+09-25-047	15 22 15.	+ 56 22	54	16.	GALAXY
ZWG 021.090	15 22 18.	+ 01 15		15.6	GALAXY
LB 09687	15 22 18.	+ 24 18		17.7	FAINT BLUE STAR
ZC 1522.3+3005	15 22 18.	+ 30 05	2420		CLUSTER OF GALAXIES
ZWG 165.052	15 22 18.	+ 30 27		15.4	GALAXY
UGC 09836	15 22 18.	+ 30 27	60	15.4	GALAXY Sa-b
ABC 2070	15 22 18.	+ 35 26		17.5	RICH CLUSTER OF GALAXIES
ZC 1522.3+5456	15 22 18.	+ 54 56	1610		CLUSTER OF GALAXIES
REIZ 4556	15 22 20.	+ 26 46	12	15.2	GALAXY
REIZ 4557	15 22 22.	+ 38 49	18	15.0	GALAXY
BC M1522+155	15 22 22.4	+ 15 31 48.		17.	QUASI-STELLAR OBJECT
SHB 312	15 22 22.4	+ 15 31 48.		17.	QUASI-STELLAR OBJECT
ZWG 077.120	15 22 24.	+ 08 37		15.3	GALAXY
LB 09692	15 22 24.	+ 21 02		13.2	FAINT BLUE STAR
LB 09691	15 22 24.	+ 22 58		15.2	FAINT BLUE STAR
LB 09690	15 22 24.	+ 24 12		16.5	FAINT BLUE STAR
LB 09689	15 22 24.	+ 25 42		16.4	FAINT BLUE STAR
LB 09688	15 22 24.	+ 26 03		16.1	FAINT BLUE STAR
ZC 1522.4+3527	15 22 24.	+ 35 27	1410		CLUSTER OF GALAXIES
7ZW 600	15 22 24.	+ 61 27			COMPACT GALAXY
LB 00829	15 22 27.	+ 14 34 54.		16.2	FAINT BLUE STAR
SN 1951E	15 22 30.	+ 08 30		20.0	SUPERNOVA
LB 09693	15 22 30.	+ 22 51		17.0	FAINT BLUE STAR
TON-N 0790	15 22 30.	+ 27 48		16.0	BLUE STAR
ZWG 222.005	15 22 30.	+ 42 34		15.4	GALAXY
MCG+07-32-004	15 22 30.	+ 42 35	30	15.	GALAXY
ZC 1522.5+6004	15 22 30.	+ 60 04	670		CLUSTER OF GALAXIES
REIZ 4555	15 22 33.	+ 13 37	12	14.9	GALAXY
ZWG 049.157	15 22 36.	+ 05 00		15.6	GALAXY
ZWG 077.121	15 22 36.	+ 08 54		15.6	GALAXY
LB 09697	15 22 36.	+ 22 06		16.4	FAINT BLUE STAR
ZC 1522.6+2235	15 22 36.	+ 22 35	870		CLUSTER OF GALAXIES
LB 09696	15 22 36.	+ 25 41		15.4	FAINT BLUE STAR
LB 09695	15 22 36.	+ 25 54		17.5	FAINT BLUE STAR
LB 09694	15 22 36.	+ 26 34		17.3	FAINT BLUE STAR
ZC 1522.6+3054	15 22 36.	+ 30 54	740		CLUSTER OF GALAXIES
MCG+10-22-013	15 22 36.	+ 58 13	102	13.	GALAXY
ZWG 297.012	15 22 36.	+ 58 14		14.6	GALAXY
UGC 09837	15 22 36.	+ 58 14	120	14.6	GALAXY Sc
IC 1118	15 22 38.	+ 13 37 19.			NONSTELLAR OBJECT
HOLM 709A	15 22 39.	+ 13 37	12	14.8	PART OF MULTIPLE GALAXY
ZWG 077.122	15 22 42.	+ 13 37		15.2	GALAXY
MCG+02-39-029	15 22 42.	+ 13 37	48	15.2	GALAXY
LB 09699	15 22 42.	+ 23 00		16.3	FAINT BLUE STAR
LB 09698	15 22 42.	+ 26 14		16.8	FAINT BLUE STAR
ZWG 021.091	15 22 42.	- 01 39		15.4	GALAXY
UGC 09839	15 22 42.	- 01 39	66	15.4	GALAXY S
LB 00830	15 22 43.	+ 61 20 42.		16.5	FAINT BLUE STAR
HOLM 709B	15 22 44.	+ 13 37	18	15.2	PART OF MULTIPLE GALAXY
LB 00831	15 22 44.	+ 14 36 00.		17.8	FAINT BLUE STAR
ZWG 049.158	15 22 48.	+ 07 20		15.3	GALAXY
UGC 09838	15 22 48.	+ 07 20	84	15.3	GALAXY Sa-b
ZWG 077.123	15 22 48.	+ 13 04		15.4	GALAXY
ZWG 106.038	15 22 48.	+ 20 00		15.3	GALAXY
MCG+03-39-024	15 22 48.	+ 20 01	36	15.3	GALAXY
LB 09700	15 22 48.	+ 23 42		16.0	FAINT BLUE STAR
ZWG 021.092	15 22 48.	- 03 24		15.7	GALAXY
ABC 2071	15 22 49.	+ 37 12		17.2	RICH CLUSTER OF GALAXIES
ZWG 077.124	15 22 54.	+ 09 55		15.3	GALAXY
LB 09703	15 22 54.	+ 22 34		16.6	FAINT BLUE STAR
LB 09702	15 22 54.	+ 22 48		15.4	FAINT BLUE STAR
LB 09701	15 22 54.	+ 25 30		16.0	FAINT BLUE STAR
7ZW 601	15 22 54.	+ 79 41			COMPACT GALAXY
ZWG 354.023	15 22 54.	+ 79 41		15.1	GALAXY
REIZ 4558	15 22 57.	+ 13 54	18	14.9	GALAXY
ZWG 049.159	15 23 00.	+ 04 59		15.5	GALAXY
ZWG 049.160	15 23 00.	+ 07 16		15.5	GALAXY
ZWG 077.125	15 23 00.	+ 13 54		15.2	GALAXY
ZWG 106.039	15 23 00.	+ 17 30		15.7	GALAXY
LB 09704	15 23 00.	+ 21 56		16.5	FAINT BLUE STAR
ZC 1523.0+3716	15 23 00.	+ 37 16	2350		CLUSTER OF GALAXIES
UGC 09840	15 23 00.	+ 42 22	66	16.0	GALAXY Sc-IRR
MCG+07-32-005	15 23 00.	+ 42 22	42	15.	GALAXY
ZWG 274.041	15 23 00.	+ 56 22		15.6	GALAXY
MIL 35	15 23 03.	- 57 55 36.	156		SUPERNOVA REMNANT
LB 09706	15 23 06.	+ 20 49		15.0	FAINT BLUE STAR
LB 09705	15 23 06.	+ 21 42		15.8	FAINT BLUE STAR
TON-N 0233	15 23 06.	+ 32 16		15.4	BLUE STAR
MCG+12-15-003	15 23 06.	+ 71 04	27	16.	GALAXY
ZWG 021.093	15 23 06.	- 03 29		15.3	GALAXY
MCG+00-39-025	15 23 06.	- 03 29	30	15.3	GALAXY
SN 1971E	15 23 07.	+ 26 33		17.5	SUPERNOVA
LB 00832	15 23 08.	+ 00 31 54.		16.3	FAINT BLUE STAR
IC 1119	15 23 08.	- 03 28 57.			NONSTELLAR OBJECT

OBJECT NAME	RIGHT ASCEN.	DECLINATION	DIAM.	MAGN.	TYPE OF OBJECT
LB 00289	15 23 10.	+ 56 49 36.		14.1	FAINT BLUE STAR
ZWG 049.161	15 23 12.	+ 04 59		15.7	GALAXY
UGC 09841	15 23 12.	+ 18 22	150	14.6	GALAXY Sb-c
ZWG 106.040	15 23 12.	+ 18 27		14.6	GALAXY
KARA.72 465A	15 23 12.	+ 18 27	132	14.6	PART OF DOUBLE GALAXY
LB 09707	15 23 12.	+ 21 26		16.9	FAINT BLUE STAR
MCG+09-25-048	15 23 12.	+ 54 56	30	16.	GALAXY
ZC 1523.2+7132	15 23 12.	+ 71 32	1210		CLUSTER OF GALAXIES
MCG-04-36-014	15 23 12.	- 22 07 30.	120	13.	GALAXY
MRSL 319-06/1	15 23 12.	- 63 59	36		HII REGION
ABC 2077	15 23 14.	+ 62 14		17.7	RICH CLUSTER OF GALAXIES
MCG+06-34-009	15 23 15.	+ 38 10	90	14.5	GALAXY
REIZ 4559	15 23 16.	+ 13 05	24	14.9	GALAXY
REIZ 4561	15 23 16.	+ 18 27	96	14.3	GALAXY
ZWG 021.094	15 23 18.	+ 01 12		15.7	GALAXY
ZC 1523.3+1235	15 23 18.	+ 12 35	1680		CLUSTER OF GALAXIES
REIZ 4560	15 23 18.	+ 16 30	36	15.3	GALAXY
MCG+03-39-025	15 23 18.	+ 18 28	144	14.6	GALAXY
VV 227	15 23 18.	+ 20 55 30.	72	14.5	INTERACTING GALAXY
ZWG 194.005	15 23 18.	+ 38 08		15.4	GALAXY
UGC 09842	15 23 18.	+ 38 08	102	15.4	GALAXY SBb
MCG+08-28-025	15 23 18.	+ 50 29	24	16.	GALAXY
ZWG 135.056	15 23 24.	+ 20 58		15.5	GALAXY
UGC 09843	15 23 24.	+ 20 58	66	15.5	GALAXY S
LB 09711	15 23 24.	+ 20 59		16.8	FAINT BLUE STAR
LB 09710	15 23 24.	+ 23 02		16.7	FAINT BLUE STAR
LB 09709	15 23 24.	+ 25 31		17.7	FAINT BLUE STAR
LB 09708	15 23 24.	+ 26 36		17.1	FAINT BLUE STAR
ZWG 165.053	15 23 24.	+ 26 44		15.1	GALAXY
SCH 88	15 23 24.	+ 26 44		15.1	PECULIAR GALAXY
ZC 1523.4+2839	15 23 24.	+ 28 39	870		CLUSTER OF GALAXIES
MCG+08-28-026	15 23 24.	+ 48 37	48	16.	GALAXY
ZWG 249.021	15 23 24.	+ 48 38		15.5	GALAXY
ZWG 338.006	15 23 24.	+ 68 50		15.5	GALAXY
REIZ 4562	15 23 26.	+ 26 43	18	14.5	GALAXY
MCG+04-36-045	15 23 27.	+ 20 57	66	15.5	GALAXY
MCG+05-36-029	15 23 27.	+ 26 42	18	15.1	GALAXY
ZWG 049.162	15 23 30.	+ 08 00		14.8	GALAXY
UGC 09844	15 23 30.	+ 08 00	96	14.8	GALAXY SBb
MCG+01-39-022	15 23 30.	+ 08 00	48	14.8	GALAXY
MCG+05-36-030	15 23 30.	+ 26 42 30.	30	15.6	GALAXY
ZC 1523.5+4555	15 23 30.	+ 45 55	940		CLUSTER OF GALAXIES
MCG+08-28-027	15 23 30.	+ 48 21	36	16.	GALAXY
ZC 1523.5+6211	15 23 30.	+ 62 11	1950		CLUSTER OF GALAXIES
ABC 2073	15 23 34.	+ 28 36		16.9	RICH CLUSTER OF GALAXIES
LB 00833	15 23 35.	+ 00 26 36.		16.5	FAINT BLUE STAR
ZC 1523.6+1822	15 23 36.	+ 18 22	1810		CLUSTER OF GALAXIES
LB 09713	15 23 36.	+ 20 44		16.0	FAINT BLUE STAR
ZWG 135.057	15 23 36.	+ 25 16		15.6	GALAXY
KARA.73B 0675	15 23 36.	+ 25 16	36	15.6	ISOLATED GALAXY S
ZWG 165.054	15 23 36.	+ 29 46		15.2	GALAXY
ZC 1523.6+3123	15 23 36.	+ 31 23	670		CLUSTER OF GALAXIES
MCG+08-28-028	15 23 36.	+ 47 49	42	15.	GALAXY
ABC 2072	15 23 37.	+ 18 24		17.0	RICH CLUSTER OF GALAXIES
RNGC 5928	15 23 40.	+ 18 15		14.0	GALAXY
ZC 1523.7+0305	15 23 42.	+ 03 05	1010		CLUSTER OF GALAXIES
ZWG 077.126	15 23 42.	+ 09 04		15.7	GALAXY
ZWG 077.127	15 23 42.	+ 09 23		15.3	GALAXY
UGC 09845	15 23 42.	+ 09 23	96	15.3	GALAXY Sc
ZWG 106.041	15 23 42.	+ 16 30		14.7	GALAXY
UGC 09846	15 23 42.	+ 16 30	66	14.7	GALAXY Sb/Sc
MCG+03-39-026	15 23 42.	+ 16 30	66	14.7	GALAXY
ZWG 106.042	15 23 42.	+ 18 15		13.8	GALAXY
UGC 09847	15 23 42.	+ 18 15	132	13.8	GALAXY S0
KARA.72 465B	15 23 42.	+ 18 15	132	13.8	PART OF DOUBLE GALAXY
LB 09714	15 23 42.	+ 26 42		16.3	FAINT BLUE STAR
ZC 1523.7+3108	15 23 42.	+ 31 08	670		CLUSTER OF GALAXIES
MCG+10-22-014	15 23 42.	+ 62 31	27	16.	GALAXY
ZC 1523.7+6405	15 23 42.	+ 64 05	3830		CLUSTER OF GALAXIES
ZWG 319.012	15 23 42.	+ 67 20		15.3	GALAXY
UGC 09848	15 23 42.	+ 72 00	60	16.0	GALAXY Sb-c
MCG+03-39-027	15 23 45.	+ 18 15	36	13.8	GALAXY
MCG+08-28-029	15 23 45.	+ 48 36	12	16.	GALAXY
REIZ 4564	15 23 47.	+ 18 15	21	13.7	GALAXY
REIZ 4563	15 23 48.	+ 16 30	42	14.1	GALAXY
LB 09718	15 23 48.	+ 23 02		15.6	FAINT BLUE STAR
LB 09717	15 23 48.	+ 23 08		15.8	FAINT BLUE STAR
LB 09716	15 23 48.	+ 23 44		17.3	FAINT BLUE STAR
LB 09715	15 23 48.	+ 25 48		17.6	FAINT BLUE STAR
TON-N 0234	15 23 48.	+ 28 36		16.8	BLUE STAR
7ZW 602	15 23 48.	+ 62 31			COMPACT GALAXY
ZWG 319.013	15 23 48.	+ 62 31		15.7	GALAXY
MCG+12-15-005	15 23 48.	+ 69 56	18	15.	GALAXY
ZWG 338.007	15 23 48.	+ 69 57		15.5	GALAXY
ZWG 021.095	15 23 48.	- 03 23		15.4	GALAXY
ZWG 049.163	15 23 54.	+ 04 00		15.5	GALAXY
ZWG 077.128	15 23 54.	+ 08 45		15.4	GALAXY
ZWG 274.042	15 23 54.	+ 53 35		15.7	GALAXY
UGC 09849	15 23 54.	+ 53 35	102	15.7	GALAXY Sc
MCG+09-25-049	15 23 54.	+ 53 35	108	14.	GALAXY
OCL 0938	15 23 54.	- 54 21	2520	8.5	OPEN STAR CLUSTER
VHA 172	15 23 54.	- 54 21	1200		OPEN STAR CLUSTER
IC 1120	15 23 55.	+ 19 02 16.			NONSTELLAR OBJECT
RNGC 5925	15 23 55.	- 54 21		8.5	OPEN CLUSTER
REIZ 4568	15 23 59.	+ 53 35	84	14.1	GALAXY
ZWG 077.129	15 24 00.	+ 09 34		15.4	GALAXY
ZC 1524.0+1700	15 24 00.	+ 17 00	400		CLUSTER OF GALAXIES
ZWG 222.006	15 24 00.	+ 38 56		15.7	GALAXY
SHAH 028	15 24 00.	+ 55 04	54	18.	GROUP OF COMPACT GALAXIES
MCG+12-15-006	15 24 00.	+ 71 04	24	16.	GALAXY
MCG+14-07-017	15 24 00.	+ 82 22	8	17.	GALAXY
ZWG 021.096	15 24 06.	+ 00 45		15.6	GALAXY
ZWG 077.130	15 24 06.	+ 08 46		15.2	GALAXY
ZWG 106.043	15 24 06.	+ 17 35		15.7	GALAXY
ZC 1524.1+1904	15 24 06.	+ 19 04	2690		CLUSTER OF GALAXIES
LB 09723	15 24 06.	+ 23 57		18.0	FAINT BLUE STAR
LB 09722	15 24 06.	+ 24 18		16.7	FAINT BLUE STAR
ZC 1524.1+2444	15 24 06.	+ 24 44	1080		CLUSTER OF GALAXIES
LB 09721	15 24 06.	+ 25 41		17.6	FAINT BLUE STAR
LB 09720	15 24 06.	+ 25 51		17.7	FAINT BLUE STAR
LB 09719	15 24 06.	+ 26 19		16.8	FAINT BLUE STAR
ZWG 249.022	15 24 06.	+ 48 40		15.7	GALAXY
UGC 09850	15 24 06.	+ 48 40	78	15.7	GALAXY SBb
MCG+08-28-030	15 24 06.	+ 48 40	78	15.	GALAXY
7ZW 603	15 24 06.	+ 60 25			COMPACT GALAXY
ZC 1524.2+0101	15 24 12.	+ 01 01	670		CLUSTER OF GALAXIES
ZC 1524.2+2016	15 24 12.	+ 20 16	1140		CLUSTER OF GALAXIES
LB 09725	15 24 12.	+ 25 34		17.8	FAINT BLUE STAR

OBJECT NAME	RIGHT ASCEN.	DECLINATION	DIAM.	MAGN.	TYPE OF OBJECT
LB 09724	15 24 12.	+ 25 59		17.6	FAINT BLUE STAR
ZC 1524.2+2825	15 24 12.	+ 28 25	1410		CLUSTER OF GALAXIES
TON-N 0791	15 24 12.	+ 29 29		14.9	BLUE STAR
MCG+07-32-006	15 24 15.	+ 41 50 30.	42	14.1	GALAXY
ARC 2086	15 24 16.	+ 72 33		17.7	RICH CLUSTER OF GALAXIES
REIZ 4566	15 24 17.	+ 41 50	36	14.4	GALAXY
1ZW 112	15 24 18.	+ 41 51			COMPACT GALAXY
ZWG 222.007	15 24 18.	+ 41 51		13.0	GALAXY
RNGC 5930	15 24 18.	+ 41 51		13.0	GALAXY
RNGC 5929	15 24 19.	+ 41 51		14.0	GALAXY
UGC 09852	15 24 18.	+ 41 51	132	13.0	GALAXY Sa
KARA.72 466A	15 24 18.	+ 41 51	54	13.0	PART OF DOUBLE GALAXY
MCG+07-32-007	15 24 18.	+ 41 51	90	13.5	GALAXY
LB 00290	15 24 18.	+ 58 08 42.		15.1	FAINT BLUE STAR
ZC 1524.3+7233	15 24 18.	+ 72 33	1210		CLUSTER OF GALAXIES
HOLM 710B	15 24 21.	+ 41 51	30	14.1	PART OF MULTIPLE GALAXY
ARC 2076	15 24 21.	+ 44 19		17.8	RICH CLUSTER OF GALAXIES
BC B1524+102	15 24 21.6	+ 10 09 32.		19.	QUASI-STELLAR OBJECT
SHB 313	15 24 21.6	+ 10 09 32.		18.	QUASI-STELLAR OBJECT
REIZ 4567	15 24 23.	+ 41 51	48	13.9	GALAXY
HOLM 710A	15 24 23.	+ 41 51	36	13.6	PART OF MULTIPLE GALAXY
ZC 1524.4+1447	15 24 24.	+ 14 47	540		CLUSTER OF GALAXIES
ZC 1524.4+1656	15 24 24.	+ 16 56	270		CLUSTER OF GALAXIES
LB 09712	15 24 24.	+ 20 42		14.6	FAINT BLUE STAR
ZC 1524.4+3100	15 24 24.	+ 31 00	740		CLUSTER OF GALAXIES
KARA.72 466B	15 24 24.	+ 41 51	96		PART OF DOUBLE GALAXY
ZWG 274.043	15 24 24.	+ 52 37		15.1	GALAXY
UGC 09853	15 24 24.	+ 52 37	108	15.1	GALAXY Sb
MCG+09-25-050	15 24 24.	+ 52 37	120	14.	GALAXY
KARA.73B 0676	15 24 24.	+ 52 37	168	15.1	ISOLATED GALAXY S
MCG+09-25-051	15 24 24.	+ 54 33	36	16.	GALAXY
MCG+11-19-007	15 24 24.	+ 66 27	84	15.	GALAXY
ZWG 338.008	15 24 24.	+ 68 55		13.7	GALAXY
UGC 09854	15 24 24.	+ 68 55	54	13.7	GALAXY S
MCG+12-15-007	15 24 24.	+ 68 55	57	14.	GALAXY
RNGC 5927	15 24 24.	- 50 29		9.5	GLOBULAR CLUSTER
GCL 035	15 24 24.	- 50 30	720	9.7	GLOBULAR STAR CLUSTER
VHA 173	15 24 24.	- 50 30	240		GLOBULAR STAR CLUSTER
RNGC 5939	15 24 26.	+ 68 55		13.5	GALAXY
ZC 1524.5+1006	15 24 30.	+ 10 06	2690		CLUSTER OF GALAXIES
ZWG 106.044	15 24 30.	+ 15 28		15.3	GALAXY
MCG+03-39-028	15 24 30.	+ 15 28	42	15.3	GALAXY
ZWG 106.045	15 24 30.	+ 16 44		15.4	GALAXY
ZC 1524.5+3046	15 24 30.	+ 30 46	740		CLUSTER OF GALAXIES
ZC 1524.5+6055	15 24 30.	+ 60 55	740		CLUSTER OF GALAXIES
ZWG 319.014	15 24 30.	+ 66 25		15.0	GALAXY
UGC 09855	15 24 30.	+ 66 25	96	15.0	GALAXY IRR
LB 00834	15 24 34.	+ 57 35 00.		17.3	FAINT BLUE STAR
LB 09727	15 24 36.	+ 23 01		17.0	FAINT BLUE STAR
LB 09726	15 24 36.	+ 25 11		16.4	FAINT BLUE STAR
ZWG 222.008	15 24 36.	+ 41 28		15.5	GALAXY
UGC 09856	15 24 36.	+ 41 28	150	15.5	GALAXY Sc
MCG+12-15-008	15 24 36.	+ 69 33	39	16.	GALAXY
REIZ 4565	15 24 37.	+ 15 26	42	14.7	GALAXY
ARP 090	15 24 41.	+ 41 50			PECULIAR GALAXY
ZWG 077.131	15 24 42.	+ 13 21		15.5	GALAXY
ZWG 165.055	15 24 42.	+ 29 14		15.7	GALAXY
MCG+07-32-008	15 24 42.	+ 41 27 30.	114	14.5	GALAXY
UGC 09857	15 24 42.	+ 41 55	114	16.0	GALAXY IRR
ZWG 222.009	15 24 42.	+ 42 32		15.0	GALAXY
MCG+07-32-009	15 24 45.	+ 41 55	72	15.	GALAXY
REIZ 4572	15 24 47.	+ 41 53	60	14.9	GALAXY
ZWG 049.164	15 24 48.	+ 06 40		15.5	GALAXY
LB 09728	15 24 48.	+ 25 04		16.6	FAINT BLUE STAR
ZC 1524.8+2850	15 24 48.	+ 28 50	9010		CLUSTER OF GALAXIES
ZWG 222.010	15 24 48.	+ 40 44		14.0	GALAXY
UGC 09858	15 24 48.	+ 40 44	270	14.0	GALAXY SBb/Sc
MCG+07-32-010	15 24 48.	+ 40 45	252	14.	GALAXY
REIZ 4571	15 24 48.	+ 41 27	48	14.2	GALAXY
ARC 2087	15 24 48.	+ 71 46		17.6	RICH CLUSTER OF GALAXIES
ZWG 021.097	15 24 48.	- 02 44		15.2	GALAXY
KARA.73B 0677	15 24 48.	- 02 44	36	15.2	ISOLATED GALAXY S
REIZ 4569	15 24 49.	+ 26 41	30	14.8	GALAXY
SN 1963A	15 24 53.	+ 26 37		18.1	SUPERNOVA
MCG+05-36-032	15 24 54.	+ 26 37	24		GALAXY
ZWG 165.056	15 24 54.	+ 26 53		15.2	GALAXY
UGC 09859	15 24 54.	+ 26 53	60	15.2	GALAXY COMPACT
ZWG 165.057	15 24 54.	+ 29 01		15.2	GALAXY
MCG+05-36-031	15 24 54.	+ 29 01	36	15.2	GALAXY
IC 4546	15 24 54.	+ 29 01 11.			NONSTELLAR OBJECT
ZC 1524.9+3022	15 24 54.	+ 30 22	670		CLUSTER OF GALAXIES
ZWG 021.098	15 24 54.	- 00 17		15.7	GALAXY
REIZ 4573	15 24 55.	+ 40 44	192	13.3	GALAXY
ZWG 021.099	15 25 00.	+ 00 27		15.3	GALAXY
LB 09729	15 25 00.	+ 22 58		16.7	FAINT BLUE STAR
MCG+08-28-031	15 25 03.	+ 50 30	60	15.	GALAXY
LB 09730	15 25 06.	+ 25 30		14.8	FAINT BLUE STAR
REIZ 4570	15 25 08.	+ 26 31	54	14.7	GALAXY
IC 4547	15 25 10.	+ 28 57 30.			NONSTELLAR OBJECT
HN 0421	15 25 10.	- 70 25			NEBULA
IC 4541	15 25 11.	- 70 24			NONSTELLAR OBJECT
LB 09732	15 25 12.	+ 23 44		16.3	FAINT BLUE STAR
LB 09731	15 25 12.	+ 25 16		16.8	FAINT BLUE STAR
ZWG 165.058	15 25 12.	+ 28 58		15.4	GALAXY
ZWG 165.059	15 25 12.	+ 31 08		15.6	GALAXY
UGC 09860	15 25 12.	+ 31 08	66	15.6	GALAXY Sc
MCG+08-28-032	15 25 12.	+ 46 36	48	16.	GALAXY
ZWG 249.023	15 25 12.	+ 48 47		15.2	GALAXY
MCG+08-28-033	15 25 12.	+ 48 47	42	14.	GALAXY
RNGC 5932	15 25 13.	+ 48 47		15.0	GALAXY
IC 1121	15 25 17.	+ 06 59 23.			NONSTELLAR OBJECT
BC B21525+31	15 25 17.4	+ 31 25 46.5		19.1	QUASI-STELLAR OBJECT
ZWG 049.165	15 25 18.	+ 05 03		15.7	GALAXY
ZWG 049.166	15 25 18.	+ 06 59		15.6	GALAXY
3ZW 073	15 25 18.	+ 28 30			COMPACT GALAXY
IC 4548	15 25 19.	+ 29 01 07.			NONSTELLAR OBJECT
ZWG 049.167	15 25 24.	+ 05 01		15.7	GALAXY
LB 09737	15 25 24.	+ 22 09		16.7	FAINT BLUE STAR
LB 09736	15 25 24.	+ 22 46		16.5	FAINT BLUE STAR
LB 09735	15 25 24.	+ 23 30		16.9	FAINT BLUE STAR
LB 09734	15 25 24.	+ 24 14		17.0	FAINT BLUE STAR
LB 09733	15 25 24.	+ 25 46		16.0	FAINT BLUE STAR
ZWG 165.060	15 25 24.	+ 28 34		15.7	GALAXY
ZWG 249.024	15 25 24.	+ 48 47		15.7	GALAXY
MCG+08-28-033	15 25 24.	+ 48 47	16	15.	GALAXY
RNGC 5933	15 25 25.	+ 48 47		15.5	GALAXY
ZWG 049.168	15 25 30.	+ 07 50		15.3	GALAXY
LB 09738	15 25 30.	+ 22 18		14.8	FAINT BLUE STAR
TON-N 0235	15 25 30.	+ 25 46		15.3	BLUE STAR
MCG+10-22-015	15 25 30.	+ 61 08	36	16.	GALAXY
LB 00835	15 25 32.	+ 59 53 00.		17.9	FAINT BLUE STAR
ZC 1525.6+0347	15 25 36.	+ 03 47	4370		CLUSTER OF GALAXIES
TON-N 0792	15 25 36.	+ 26 06		15.0	BLUE STAR
LB 09739	15 25 36.	+ 26 26		16.7	FAINT BLUE STAR
ZWG 165.061	15 25 36.	+ 28 24		15.5	GALAXY
3ZW 074	15 25 36.	+ 28 25			COMPACT GALAXY
LB 00836	15 25 41.	+ 00 42 24.		16.3	FAINT BLUE STAR
ZWG 049.169	15 25 42.	+ 05 58		15.6	GALAXY
ZWG 077.132	15 25 42.	+ 12 10		15.7	GALAXY
KARA.73B 0678	15 25 42.	+ 12 10	24	15.7	ISOLATED GALAXY S0
ZWG 165.062	15 25 42.	+ 29 05		15.4	GALAXY
UGC 09861	15 25 42.	+ 29 05	90	15.4	GALAXY DBL SYS
MCG+05-36-033	15 25 42.	+ 29 06	66	15.4	GALAXY
MCG+10-22-016	15 25 42.	+ 59 26	18	16.	GALAXY
ARC 2080	15 25 47.	+ 41 54		17.3	RICH CLUSTER OF GALAXIES
ZWG 021.100	15 25 48.	+ 00 02		15.5	GALAXY
LB 09744	15 25 48.	+ 22 02		17.0	FAINT BLUE STAR
LB 09743	15 25 48.	+ 22 44		16.0	FAINT BLUE STAR
LB 09742	15 25 48.	+ 24 16		18.0	FAINT BLUE STAR
LB 09741	15 25 48.	+ 25 38		16.6	FAINT BLUE STAR
LB 09740	15 25 48.	+ 26 00		16.8	FAINT BLUE STAR
ZC 1525.8+3053	15 25 48.	+ 30 53	400		CLUSTER OF GALAXIES
ZWG 319.015	15 25 48.	+ 66 15		15.4	GALAXY
HN 0071	15 25 48.	- 50 25			NEBULA
IC 4544	15 25 48.	- 50 25			NONSTELLAR OBJECT
LB 00837	15 25 49.	+ 57 16 18.		17.5	FAINT BLUE STAR
ZWG 106.046	15 25 54.	+ 15 33		15.5	GALAXY
LB 09745	15 25 54.	+ 21 50		17.0	FAINT BLUE STAR
ZC 1525.9+5422	15 25 54.	+ 54 22	1140		CLUSTER OF GALAXIES
REIZ 4574	15 25 57.	+ 13 52	18	15.0	GALAXY
ZWG 077.133	15 26 00.	+ 13 51		15.3	GALAXY
KARA.73B 0679	15 26 00.	+ 13 51	24	15.3	ISOLATED GALAXY E
SN 1967I	15 26 00.	+ 28 49		18.5	SUPERNOVA
SN 1962G	15 26 00.	+ 29 11		19.0	SUPERNOVA
MCG+05-36-034	15 26 00.	+ 29 11	12		GALAXY
ZWG 194.006	15 26 00.	+ 35 09		15.7	GALAXY
MCG+06-34-010	15 26 00.	+ 35 09	36	15.	GALAXY
ZC 1526.0+4153	15 26 00.	+ 41 53	1810		CLUSTER OF GALAXIES
ARC 2079	15 26 03.	+ 29 03		15.4	RICH CLUSTER OF GALAXIES
ZWG 135.058	15 26 06.	+ 24 42		15.4	GALAXY
LB 09746	15 26 06.	+ 24 44		16.5	FAINT BLUE STAR
ZWG 249.025	15 26 06.	+ 47 12		15.6	GALAXY
8ZW 1526+10.8	15 26 12.	+ 10 45		18.0	COMPACT GALAXY
LB 09748	15 26 12.	+ 25 16		17.0	FAINT BLUE STAR
LB 09747	15 26 12.	+ 25 58		16.4	FAINT BLUE STAR
ZC 1526.2+2902	15 26 12.	+ 29 02	610		CLUSTER OF GALAXIES
MCG+08-28-035	15 26 12.	+ 47 11	30	16.	GALAXY
REIZ 4576	15 26 13.	+ 31 41	54	15.2	GALAXY
ZWG 021.101	15 26 18.	+ 00 56		15.5	GALAXY
ZWG 049.170	15 26 18.	+ 07 32		15.3	GALAXY
ZWG 049.171	15 26 18.	+ 07 34		15.2	GALAXY
ZWG 049.172	15 26 18.	+ 07 39		15.6	GALAXY
LB 09750	15 26 24.	+ 22 10		16.3	FAINT BLUE STAR
LB 09749	15 26 24.	+ 24 28		17.1	FAINT BLUE STAR
ZC 1526.4+2607	15 26 24.	+ 26 07	870		CLUSTER OF GALAXIES
ZWG 222.011	15 26 24.	+ 43 05		14.5	GALAXY
UGC 09862	15 26 24.	+ 43 05	78	14.5	GALAXY
MCG+07-32-011	15 26 24.	+ 43 07 30.	36	15.	GALAXY
ZWG 222.012	15 26 24.	+ 43 11		15.7	GALAXY
MCG+07-32-012	15 26 24.	+ 43 12	45	15.	GALAXY
7ZW 604	15 26 24.	+ 58 02			COMPACT GALAXY
RNGC 5934	15 26 26.	+ 43 05		14.5	GALAXY
REIZ 4575	15 26 27.	+ 13 37	24	15.6	GALAXY
ZWG 049.173	15 26 30.	+ 07 46		15.4	GALAXY
ZWG 049.174	15 26 30.	+ 07 55		15.5	GALAXY
ZWG 077.134	15 26 30.	+ 13 36		15.4	GALAXY
ZWG 222.013	15 26 30.	+ 43 06		15.1	GALAXY
1ZW 113	15 26 30.	+ 43 07			COMPACT GALAXY
MCG+07-32-013	15 26 30.	+ 43 08	39	15.5	GALAXY
MCG+09-25-052	15 26 30.	+ 51 51	60	15.	GALAXY
MCG+09-25-053	15 26 30.	+ 54 27 30.	60	15.	GALAXY
ZWG 274.044	15 26 30.	+ 55 36		15.6	GALAXY
MRK 481	15 26 30.	+ 55 36	8	15.5	GALAXY WITH UV CONTINUUM
ZC 1526.5+6228	15 26 30.	+ 62 28	1280		CLUSTER OF GALAXIES
RNGC 5935	15 26 32.	+ 43 06		15.0	GALAXY
LB 09751	15 26 36.	+ 25 10		16.8	FAINT BLUE STAR
ZWG 136.001	15 26 36.	+ 25 55		15.7	GALAXY
ZWG 135.059	15 26 36.	+ 25 55		15.7	GALAXY
TON-N 0236	15 26 36.	+ 28 37		16.3	BLUE STAR
ZC 1526.6+7121	15 26 36.	+ 71 21	940		CLUSTER OF GALAXIES
ZWG 021.102	15 26 42.	+ 00 53		15.4	GALAXY
MCG+09-25-054	15 26 42.	+ 55 44	42	16.	GALAXY
REIZ 4577	15 26 43.	+ 15 26	24	15.1	GALAXY
LB 00838	15 26 46.	+ 57 58 48.		18.0	FAINT BLUE STAR
ZWG 049.175	15 26 48.	+ 05 15		15.2	GALAXY
ZC 1526.8+1147	15 26 48.	+ 11 47	1680		CLUSTER OF GALAXIES
LB 09755	15 26 48.	+ 24 14		16.4	FAINT BLUE STAR
LB 09754	15 26 48.	+ 25 32		17.2	FAINT BLUE STAR
LB 09753	15 26 48.	+ 25 54		16.9	FAINT BLUE STAR
LB 09752	15 26 48.	+ 26 31		16.8	FAINT BLUE STAR
ZWG 274.045	15 26 48.	+ 55 43		14.8	GALAXY
MRK 482	15 26 48.	+ 55 43	24	15.	GALAXY WITH UV CONTINUUM
SHAH 133	15 26 48.	+ 57 10	54	18.0	GROUP OF COMPACT GALAXIES
MCG+13-11-013	15 26 48.	+ 75 13	27	16.	GALAXY
ZWG 049.176	15 26 54.	+ 03 46		15.6	GALAXY
UGC 09863	15 26 54.	+ 03 46	60	15.6	GALAXY Sa
ZWG 077.135	15 26 54.	+ 09 17		15.3	GALAXY
KARA.73B 0680	15 26 54.	+ 09 17	48	15.3	ISOLATED GALAXY S
ZC 1526.9+1129	15 26 54.	+ 11 29	670		CLUSTER OF GALAXIES
ZWG 136.002	15 26 54.	+ 25 38		15.1	GALAXY
ZWG 135.060	15 26 54.	+ 25 38		15.1	GALAXY
SCH 89	15 26 54.	+ 25 38		15.1	PECULIAR GALAXY
ZC 1526.9+3105	15 26 54.	+ 31 05	610		CLUSTER OF GALAXIES
MCG+04-36-046	15 26 57.	+ 25 38 30.	48	15.1	GALAXY
IC 1122	15 26 59.	+ 07 45 34.			NONSTELLAR OBJECT
ZWG 049.177	15 27 00.	+ 02 55		15.7	GALAXY
ZWG 049.178	15 27 00.	+ 03 38		15.1	GALAXY
ZWG 049.179	15 27 00.	+ 05 16		15.3	GALAXY
UGC 09864	15 27 00.	+ 05 16	102	15.3	GALAXY S0?
ZC 1527.0+0750	15 27 00.	+ 07 50	870		CLUSTER OF GALAXIES
ZC 1527.0+3040	15 27 00.	+ 30 40	3760		CLUSTER OF GALAXIES
LB 00839	15 27 02.	- 15 23 18.		17.1	FAINT BLUE STAR
MCG-03-39-011	15 27 03.	- 18 28	108	14.5	GALAXY
RNGC 5931	15 27 05.	+ 07 45		15.0	GALAXY
ZWG 049.180	15 27 06.	+ 07 45		15.0	GALAXY
MCG+01-39-023	15 27 06.	+ 07 45	12	15.0	GALAXY

OBJECT NAME	RIGHT ASCEN.	DECLINATION	DIAM.	MAGN.	TYPE OF OBJECT
ZC 1527.1+1955	15 27 06.	+ 19 55	740		CLUSTER OF GALAXIES
ZWG 136.003	15 27 06.	+ 20 55		15.3	GALAXY
LB 09756	15 27 06.	+ 25 32		16.6	FAINT BLUE STAR
ZWG 166.001	15 27 06.	+ 29 49		15.7	GALAXY
ZWG 165.063	15 27 06.	+ 29 49		15.7	GALAXY
TON-N 0237	15 27 06.	+ 30 02		15.3	BLUE STAR
MCG+10-22-017	15 27 06.	+ 61 57	42	16.	GALAXY
ZC 1527.1+6959	15 27 06.	+ 69 59	3090		CLUSTER OF GALAXIES
ZWG 021.103	15 27 06.	- 01 45		15.5	GALAXY
ARC 2078	15 27 08.	- 12 51		17.5	RICH CLUSTER OF GALAXIES
MCG+05-37-001	15 27 09.	+ 29 48 30.	21	15.7	GALAXY
IC 1123	15 27 11.	+ 43 03			NONSTELLAR OBJECT
ZWG 049.181	15 27 12.	+ 02 56		15.7	GALAXY
ZWG 049.182	15 27 12.	+ 08 13		15.4	GALAXY
ZWG 077.136	15 27 12.	+ 12 25		15.7	GALAXY
KARA.73B 0681	15 27 12.	+ 12 25	48	15.7	ISOLATED GALAXY S
ZC 1527.2+1553	15 27 12.	+ 15 53	810		CLUSTER OF GALAXIES
ZC 1527.2+1839	15 27 12.	+ 18 39	670		CLUSTER OF GALAXIES
LB 09760	15 27 12.	+ 23 13		16.8	FAINT BLUE STAR
LB 09759	15 27 12.	+ 24 14		17.6	FAINT BLUE STAR
LB 09758	15 27 12.	+ 24 46		16.3	FAINT BLUE STAR
LB 09757	15 27 12.	+ 26 18		16.2	FAINT BLUE STAR
MCG+06-34-011	15 27 12.	+ 33 00	30	15.	GALAXY
MCG+13-11-014	15 27 12.	+ 75 15	51	15.	GALAXY
ZWG 354.024	15 27 12.	+ 75 16		15.7	GALAXY
UGC 09865	15 27 12.	+ 75 16	66	15.7	GALAXY Sb-c
IC 4549	15 27 14.	+ 32 59 14.			NONSTELLAR OBJECT
RNGC 5949	15 27 15.	+ 64 56		13.0	GALAXY
ZWG 049.183	15 27 18.	+ 03 41		14.9	GALAXY
MCG+01-39-024	15 27 18.	+ 03 41	30	14.9	GALAXY
ZC 1527.3+2323	15 27 18.	+ 23 23	870		CLUSTER OF GALAXIES
ZWG 136.004	15 27 18.	+ 26 10		15.4	GALAXY
ZWG 135.061	15 27 18.	+ 26 10		15.4	GALAXY
SCH 90	15 27 18.	+ 26 10		15.4	PECULIAR GALAXY
TON-N 0793	15 27 18.	+ 27 37		15.5	BLUE STAR
ZWG 194.007	15 27 18.	+ 33 00		15.5	GALAXY
MCG+07-32-014	15 27 18.	+ 39 47 30.	30	15.	GALAXY
MCG+11-19-008	15 27 18.	+ 64 56	132	12.9	GALAXY
ARC 2083	15 27 22.	+ 30 55		16.9	RICH CLUSTER OF GALAXIES
REIZ 4580	15 27 23.	+ 41 38	54	14.7	GALAXY
ZWG 049.184	15 27 24.	+ 05 11		15.3	GALAXY
LB 09763	15 27 24.	+ 20 50		16.6	FAINT BLUE STAR
LB 09762	15 27 24.	+ 26 00		18.2	FAINT BLUE STAR
LB 09761	15 27 24.	+ 26 38		16.6	FAINT BLUE STAR
ZWG 166.002	15 27 24.	+ 28 29		15.6	GALAXY
ZWG 165.064	15 27 24.	+ 28 29		15.6	GALAXY
ZWG 222.014	15 27 24.	+ 39 48		14.7	GALAXY
ZWG 319.016	15 27 24.	+ 64 55		12.7	GALAXY
UGC 09866	15 27 24.	+ 64 55	138	12.7	GALAXY Sc
KARA.73B 0682	15 27 24.	+ 64 55	132	12.7	ISOLATED GALAXY S
ZC 1527.5+2833	15 27 30.	+ 28 33	740		CLUSTER OF GALAXIES
ZC 1527.5+3134	15 27 30.	+ 31 34	610		CLUSTER OF GALAXIES
VHA 174	15 27 30.	- 55 05	360		OPEN STAR CLUSTER
REIZ 4578	15 27 32.	+ 07 44	12	15.2	GALAXY
ZC 1527.6+2159	15 27 36.	+ 21 59	610		CLUSTER OF GALAXIES
7ZW 605	15 27 36.	+ 57 47			COMPACT GALAXY
REIZ 4579	15 27 39.	+ 13 10	60	12.9	GALAXY
ZWG 049.185	15 27 42.	+ 06 01		15.4	GALAXY
ZWG 078.001	15 27 42.	+ 13 10		13.0	GALAXY
ZWG 077.137	15 27 42.	+ 13 10		13.0	GALAXY
UGC 09867	15 27 42.	+ 13 10	78	13.0	GALAXY SBb
MCG+02-39-030	15 27 42.	+ 13 10	84	13.0	GALAXY
LB 09765	15 27 42.	+ 21 58		16.8	FAINT BLUE STAR
LB 09764	15 27 42.	+ 25 31		16.4	FAINT BLUE STAR
ZWG 166.003	15 27 42.	+ 30 39		15.1	GALAXY
ZWG 165.065	15 27 42.	+ 30 39		15.1	GALAXY
MCG+05-37-002	15 27 42.	+ 30 39	15	15.1	GALAXY
RNGC 5936	15 27 43.	+ 13 10		13.0	GALAXY
LB 00291	15 27 45.	+ 59 50 24.		14.7	FAINT BLUE STAR
ARC 2081	15 27 47.	- 10 50		17.7	RICH CLUSTER OF GALAXIES
ZC 1527.8+2006	15 27 48.	+ 20 06	670		CLUSTER OF GALAXIES
LB 09766	15 27 48.	+ 21 24		16.5	FAINT BLUE STAR
TON-N 0794	15 27 48.	+ 29 25		15.8	BLUE STAR
ZC 1527.8+3055	15 27 48.	+ 30 55	810		CLUSTER OF GALAXIES
ZC 1527.8+4411	15 27 48.	+ 44 11	1410		CLUSTER OF GALAXIES
MCG+09-25-055	15 27 48.	+ 53 02 30.	15	16.	GALAXY
UGC 09868	15 27 48.	+ 53 03	60	16.5	GALAXY
7ZW 606	15 27 48.	+ 59 21			COMPACT GALAXY
IC 1124	15 27 50.	+ 23 48 16.			NONSTELLAR OBJECT
LB 00840	15 27 50.	+ 61 11 54.		16.2	FAINT BLUE STAR
MCG+04-37-001	15 27 51.	+ 23 50	45	14.5	GALAXY
REIZ 4582	15 27 52.	+ 34 15	18	14.2	GALAXY
8ZW 1527+12.1	15 27 54.	+ 12 05		17.5	COMPACT GALAXY
8ZW 1527+13.0	15 27 54.	+ 13 02		17.8	COMPACT GALAXY
ZWG 136.005	15 27 54.	+ 23 48		14.5	GALAXY
UGC 09869	15 27 54.	+ 23 48	54	14.5	GALAXY S
ZWG 222.015	15 27 54.	+ 42 50		15.2	GALAXY
MCG+07-32-015	15 27 54.	+ 42 50	30	15.	GALAXY
MCG+10-22-018	15 27 54.	+ 62 02	18	16.	GALAXY
ZWG 021.104	15 27 54.	- 02 06		15.2	GALAXY
REIZ 4581	15 27 58.	+ 29 17	9	15.4	GALAXY
LB 09767	15 28 00.	+ 23 56		9.0	FAINT BLUE STAR
TON-N 0795	15 28 00.	+ 24 01		16.2	BLUE STAR
ZWG 222.016	15 28 00.	+ 42 57		14.6	GALAXY
UGC 09870	15 28 00.	+ 42 57	72	14.6	GALAXY S0?
MCG+07-32-016	15 28 00.	+ 42 58	78	14.	GALAXY
ZWG 222.017	15 28 00.	+ 43 05		14.1	GALAXY
UGC 09871	15 28 00.	+ 43 05	198	14.1	GALAXY SBa
MCG+07-32-017	15 28 00.	+ 43 06	84	13.	GALAXY
MCG+12-15-009	15 28 00.	+ 68 53	66	16.	GALAXY
UGC 09872	15 28 00.	+ 68 54	66	16.0	GALAXY Sc
LB 00841	15 28 02.	+ 00 24 18.		16.0	FAINT BLUE STAR
RNGC 5943	15 28 02.	+ 42 57		14.5	GALAXY
RNGC 5945	15 28 02.	+ 43 05		14.0	GALAXY
HN 1796	15 28 02.5	- 02 02 39.	30		NEBULA
ZWG 050.001	15 28 06.	+ 06 50		15.6	GALAXY
8ZW 1528+08.7	15 28 06.	+ 08 40		17.6	COMPACT GALAXY
LB 09768	15 28 06.	+ 25 06		16.5	FAINT BLUE STAR
ZWG 166.004	15 28 06.	+ 27 18		15.4	GALAXY
REIZ 4584	15 28 06.	+ 27 19	9	15.2	GALAXY
ZC 1528.1+3525	15 28 06.	+ 35 25	1340		CLUSTER OF GALAXIES
ZWG 222.018	15 28 06.	+ 42 48		15.6	GALAXY
UGC 09873	15 28 06.	+ 42 48	96	15.6	GALAXY Sc
MCG+07-32-018	15 28 06.	+ 42 49	72	15.	GALAXY
7ZW 607	15 28 06.	+ 57 02			COMPACT GALAXY
ZWG 022.001	15 28 06.	- 02 02		15.3	GALAXY
REIZ 4583	15 28 07.	+ 27 16	6	15.3	GALAXY
ARC 2082	15 28 09.	+ 03 37		17.0	RICH CLUSTER OF GALAXIES

OBJECT NAME	RIGHT ASCEN.	DECLINATION	DIAM.	MAGN.	TYPE OF OBJECT
REIN 2.232	15 28 09.97	- 02 39 35.6			NEBULA
REIZ 4586	15 28 10.	+ 29 24	12	15.3	GALAXY
ARC 2084	15 28 11.	+ 35 28		17.3	RICH CLUSTER OF GALAXIES
ZWG 050.002	15 28 12.	+ 03 08		15.6	GALAXY
LB 09769	15 28 12.	+ 22 12		16.4	FAINT BLUE STAR
ZWG 166.005	15 28 12.	+ 27 16		15.3	GALAXY
ZC 1528.2+4259	15 28 12.	+ 42 59	2220		CLUSTER OF GALAXIES
ZWG 022.003	15 28 12.	- 00 11		15.6	GALAXY
KARA.73B 0683	15 28 12.	- 00 11	36	15.6	ISOLATED GALAXY S
ZWG 022.002	15 28 12.	- 02 39		13.1	GALAXY
RNGC 5937	15 28 12.	- 02 39		13.0	GALAXY
MCG+00-40-001	15 28 12.	- 02 39	90	13.1	GALAXY
ZWG 249.026	15 28 18.	+ 46 05		15.3	GALAXY
REIZ 4585	15 28 19.	+ 21 23	24	14.9	GALAXY
ZC 1528.4+0049	15 28 24.	+ 00 49	11630		CLUSTER OF GALAXIES
ZWG 022.004	15 28 24.	+ 01 09		15.5	GALAXY
ZC 1528.4+1841	15 28 24.	+ 18 41	2020		CLUSTER OF GALAXIES
ZC 1528.4+1920	15 28 24.	+ 19 20	1480		CLUSTER OF GALAXIES
ZWG 022.005	15 28 30.	+ 00 39		15.7	GALAXY
ZWG 050.003	15 28 30.	+ 04 31		15.5	GALAXY
ZWG 354.025	15 28 30.	+ 77 20		14.6	GALAXY
UGC 09874	15 28 30.	+ 77 20	84	14.6	GALAXY E
KARA.73B 0684	15 28 30.	+ 77 20	42	14.6	ISOLATED GALAXY S0
ZWG 050.004	15 28 36.	+ 07 10		15.5	GALAXY
UGC 09875	15 28 36.	+ 23 14	102	17.	GALAXY DWRF EL
MCG+04-37-002	15 28 36.	+ 23 16	48		GALAXY
TON-N 0238	15 28 36.	+ 29 44		15.1	BLUE STAR
ZWG 050.005	15 28 42.	+ 06 36		15.5	GALAXY
ZC 1528.7+2202	15 28 42.	+ 22 02	610		CLUSTER OF GALAXIES
ZC 1528.7+2333	15 28 42.	+ 23 33	940		CLUSTER OF GALAXIES
ZC 1528.7+3147	15 28 42.	+ 31 47	610		CLUSTER OF GALAXIES
MRK 483	15 28 42.	+ 34 06	10	17.	GALAXY WITH UV CONTINUUM
ZC 1528.7+5019	15 28 42.	+ 50 19	3630		CLUSTER OF GALAXIES
MCG-01-39-006	15 28 42.	- 04 58	36	15.	GALAXY
RNGC 5940	15 28 47.	+ 07 38		14.5	GALAXY
ZWG 050.006	15 28 48.	+ 04 30		15.4	GALAXY
ZWG 050.007	15 28 48.	+ 07 38		14.3	GALAXY
UGC 09876	15 28 48.	+ 07 38	48	14.3	GALAXY SBa-b
MCG+01-39-025	15 28 48.	+ 07 38	48	14.3	GALAXY
ZC 1528.8+1010	15 28 48.	+ 10 10	810		CLUSTER OF GALAXIES
MCG+07-32-019	15 28 48.	+ 42 53	66	14.	GALAXY
ARC 2090	15 28 51.	+ 61 44		17.1	RICH CLUSTER OF GALAXIES
ZC 1528.9+2940	15 28 54.	+ 29 40	610		CLUSTER OF GALAXIES
ZWG 222.019	15 28 54.	+ 42 53		14.8	GALAXY
UGC 09877	15 28 54.	+ 42 53	78	14.8	GALAXY SBb
MCG-04-37-001	15 28 54.	- 24 37	66	15.5	GALAXY
RNGC 5947	15 28 56.	+ 42 53		15.0	GALAXY
ZWG 050.008	15 29 00.	+ 06 05		15.5	GALAXY
ZC 1529.0+3220	15 29 00.	+ 32 20	1280		CLUSTER OF GALAXIES
ZWG 194.008	15 29 00.	+ 36 58		15.6	GALAXY
UGC 09879	15 29 00.	+ 36 58	66	15.6	GALAXY Sa-b
UGC 09880	15 29 00.	+ 47 30	72	17.	GALAXY DWARF
ZWG 297.013	15 29 00.	+ 57 00		15.7	GALAXY
KARA.73B 0685	15 29 00.	+ 57 00	24	15.7	ISOLATED GALAXY E
ZWG 319.017	15 29 00.	+ 67 48		15.5	GALAXY
UGC 09881	15 29 00.	+ 79 58	72	16.0	GALAXY Sb
ZWG 366.016	15 29 00.	+ 83 15		15.2	GALAXY
UGC 09878	15 29 00.	+ 83 15	60	15.2	GALAXY S
MCG+14-07-018	15 29 00.	+ 83 16	45	16.	GALAXY
RNGC 5941	15 29 05.	+ 07 30		15.0	GALAXY
ZWG 050.009	15 29 05.	+ 07 30		15.2	GALAXY
MCG+01-40-001	15 29 06.	+ 07 30	12	15.2	GALAXY
8ZW 1529+13.9	15 29 06.	+ 13 54		18.2	COMPACT GALAXY
MCG+06-34-012	15 29 06.	+ 36 59 30.	48	15.	GALAXY
HN 1797	15 29 07.8	- 01 12 18.	24		NEBULA
RNGC 5942	15 29 11.	+ 07 28		15.5	GALAXY
ZWG 050.010	15 29 12.	+ 07 28		15.4	GALAXY
MCG+01-40-002	15 29 12.	+ 07 28	9	15.4	GALAXY
ZWG 050.011	15 29 12.	+ 07 31		15.1	GALAXY
MCG+01-40-003	15 29 12.	+ 07 31	12	15.1	GALAXY
ZWG 050.012	15 29 12.	+ 07 39		15.7	GALAXY
ZC 1529.2+3425	15 29 12.	+ 34 25	540		CLUSTER OF GALAXIES
ARC 2088	15 29 12.	+ 39 02		17.5	RICH CLUSTER OF GALAXIES
MCG+10-22-019	15 29 12.	+ 59 40	27	17.	GALAXY
ZWG 022.006	15 29 12.	- 01 12		15.7	GALAXY
ARC 2085	15 29 14.	+ 07 32		17.7	RICH CLUSTER OF GALAXIES
RNGC 5944	15 29 17.	+ 07 29		15.0	GALAXY
ZWG 050.013	15 29 18.	+ 07 29		15.2	GALAXY
MCG+01-40-004	15 29 18.	+ 07 29	36	15.2	GALAXY
ZWG 107.001	15 29 18.	+ 15 45		15.3	GALAXY
MCG+03-40-001	15 29 18.	+ 15 45	42	15.3	GALAXY
ZWG 022.007	15 29 24.	+ 00 21		15.7	GALAXY
ZWG 078.002	15 29 24.	+ 09 38		15.5	GALAXY
ZWG 078.003	15 29 24.	+ 10 01		15.7	GALAXY
LB 00842	15 29 28.	- 13 36 00.		16.8	FAINT BLUE STAR
ZWG 136.006	15 29 30.	+ 25 24		15.3	GALAXY
SCH 91	15 29 30.	+ 25 24		15.3	PECULIAR GALAXY
UGC 09882	15 29 30.	+ 40 27	72	16.0	GALAXY Sc
MCG+07-32-020	15 29 30.	+ 40 27	63	15.	GALAXY
MCG+08-28-036	15 29 30.	+ 47 12 30.	48	15.	GALAXY
OCL 0941	15 29 30.	- 55 03	360		OPEN STAR CLUSTER
ZWG 050.014	15 29 36.	+ 04 55		15.4	GALAXY
ZWG 050.015	15 29 36.	+ 07 38		15.4	GALAXY
UGC 09883	15 29 36.	+ 47 13	60	16.0	GALAXY Sb
MCG+12-15-010	15 29 36.	+ 70 44	51	16.	GALAXY
LB 00292	15 29 39.	- 12 23 42.		14.9	FAINT BLUE STAR
ZWG 078.004	15 29 42.	+ 13 26		15.6	GALAXY
MCG+06-34-013	15 29 42.	+ 35 28	30	15.	GALAXY
ZWG 222.020	15 29 42.	+ 40 36		14.8	GALAXY
RNGC 5950	15 29 42.	+ 40 36		15.0	GALAXY
REIZ 4588	15 29 42.	+ 40 36	60	14.9	GALAXY
UGC 09884	15 29 42.	+ 40 36	96	14.8	GALAXY Sb
MCG+07-32-021	15 29 42.	+ 40 36	90	14.	GALAXY
ZWG 274.046	15 29 42.	+ 54 51		15.1	GALAXY
MRK 484	15 29 42.	+ 54 51	12	15.5	GALAXY WITH UV CONTINUUM
ZWG 022.008	15 29 42.	- 02 39		15.7	GALAXY
MCG+00-40-002	15 29 42.	- 02 39	78	15.7	GALAXY
MCG+09-25-056	15 29 45.	+ 55 55	54	16.	GALAXY
ZWG 050.016	15 29 48.	+ 03 40		15.5	GALAXY
ZWG 050.017	15 29 48.	+ 05 51		15.4	GALAXY
ZC 1529.8+3903	15 29 48.	+ 39 03	2690		CLUSTER OF GALAXIES
UGC 09885	15 29 48.	+ 70 43	60	16.0	GALAXY Sc
ZWG 078.005	15 29 54.	+ 09 38		15.5	GALAXY
ZWG 136.007	15 29 54.	+ 25 44		15.6	GALAXY
REIZ 4587	15 29 54.	+ 32 20	36	14.6	GALAXY
1ZW 114	15 29 54.	+ 41 02			COMPACT GALAXY
ZWG 274.047	15 29 54.	+ 55 54		15.7	GALAXY
LB 00843	15 29 58.	- 00 52 12.		17.6	FAINT BLUE STAR

OBJECT NAME	RIGHT ASCEN.	DECLINATION	DIAM.	MAGN.	TYPE OF OBJECT
ZWG 050.018	15 30 00.	+ 04 51		15.3	GALAXY
UGC 09886	15 30 00.	+ 04 51	66	15.3	GALAXY E-SO
MKW 09	15 30 00.	+ 04 51		15.3	POOR GALAXY CLUSTER
ZWG 050.019	15 30 00.	+ 04 57		15.4	GALAXY
ZWG 050.020	15 30 00.	+ 05 16		15.4	GALAXY
ZWG 107.002	15 30 00.	+ 18 50		15.5	GALAXY
MCG-02-40-001	15 30 00.	- 15 08	36	15.	GALAXY
OCL 0944	15 30 00.	- 53 26	180		OPEN STAR CLUSTER
LB 00844	15 30 04.	+ 60 21 00.		16.5	FAINT BLUE STAR
REIZ 4589	15 30 05.	+ 32 57	24	14.9	GALAXY
MCG+02-40-002	15 30 06.	+ 18 51	48	15.5	GALAXY
ZC 1530.1+3231	15 30 06.	+ 32 31	670		CLUSTER OF GALAXIES
PK322-02.1	15 30 09.	- 58 59 29.	55	12.5	PLANETARY NEBULA
ZWG 050.021	15 30 12.	+ 04 55		15.1	GALAXY
UGC 09887	15 30 12.	+ 04 55	78	15.1	GALAXY S
MCG+01-40-005	15 30 12.	+ 04 55	60	15.1	GALAXY
ZWG 078.006	15 30 12.	+ 10 30		15.5	GALAXY
ZWG 078.007	15 30 12.	+ 10 37		15.6	GALAXY
MCG+02-40-001	15 30 12.	+ 10 37	30	15.6	GALAXY
ZWG 249.027	15 30 12.	+ 45 36		15.5	GALAXY
SC 1526-5848.9	15 30 17.	- 58 59 11.	30		NEBULA
ZWG 022.009	15 30 18.	+ 00 38		15.6	GALAXY
ZWG 050.022	15 30 18.	+ 08 26		15.4	GALAXY
ZWG 136.008	15 30 18.	+ 25 39		15.7	GALAXY
MRK 435	15 30 18.	+ 51 56	20	15.	GALAXY WITH UV CONTINUUM
ZWG 022.010	15 30 24.	+ 00 26		15.5	GALAXY
ZWG 022.011	15 30 24.	+ 00 52		15.7	GALAXY
HOLM 712B	15 30 24.	+ 10 38	30	14.7	PART OF MULTIPLE GALAXY
HOLM 712A	15 30 24.	+ 10 38	36	14.8	PART OF MULTIPLE GALAXY
ZC 1530.4+2240	15 30 24.	+ 22 40	810		CLUSTER OF GALAXIES
ZWG 222.021	15 30 24.	+ 42 23		15.7	GALAXY
RNGC 5948	15 30 27.	+ 04 09			NON-EXISTENT OBJECT
LB 00845	15 30 28.	- 00 22 36.		17.0	FAINT BLUE STAR
IC 1125	15 30 29.	- 01 27 06.			NONSTELLAR OBJECT
ZC 1530.5+1900	15 30 30.	+ 19 00	670		CLUSTER OF GALAXIES
ZWG 022.012	15 30 30.	- 01 28		14.5	GALAXY
UGC 09888	15 30 30.	- 01 28	102	14.5	GALAXY S-IRR
KARA.72 467B	15 30 30.	- 01 28	48		PART OF DOUBLE GALAXY
KARA.72 467A	15 30 30.	- 01 28	36	14.5	PART OF DOUBLE GALAXY
MCG+00-40-003	15 30 30.	- 01 28	102	14.5	GALAXY
LB 00846	15 30 31.	+ 00 35 30.		16.2	FAINT BLUE STAR
HN 1798	15 30 31.9	+ 00 42 47.	18		NEBULA
ABC 2089	15 30 34.	+ 28 12		15.8	RICH CLUSTER OF GALAXIES
ZWG 022.013	15 30 36.	+ 00 44		15.4	GALAXY
UGC 09889	15 30 36.	+ 00 44	66	15.4	GALAXY S
ZWG 136.009	15 30 36.	+ 23 20		15.4	GALAXY
MCG+07-32-022	15 30 36.	+ 42 00	30	15.	GALAXY
MCG-01-40-001	15 30 36.	- 08 32	72	14.5	GALAXY
HOLM 711A	15 30 41.	- 08 31	66	13.5	PART OF MULTIPLE GALAXY
ZWG 050.023	15 30 42.	+ 06 40		15.6	GALAXY
ZWG 166.006	15 30 42.	+ 28 32		15.7	GALAXY
ZWG 222.022	15 30 42.	+ 41 59		15.6	GALAXY
UGC 09890	15 30 42.	+ 42 10	78	16.0	GALAXY Sc
MCG+07-32-023	15 30 45.	+ 42 10	60	15.	GALAXY
MCG+08-28-037	15 30 45.	+ 48 34	36	15.	GALAXY
LB 00847	15 30 45.	- 16 14 36.		14.4	FAINT BLUE STAR
ZWG 050.024	15 30 48.	+ 04 38		15.6	GALAXY
ZWG 136.010	15 30 48.	+ 25 44		15.7	GALAXY
ABC 2098	15 30 53.	+ 69 42		17.7	RICH CLUSTER OF GALAXIES
LB 00848	15 30 54.	+ 01 26 24.		16.1	FAINT BLUE STAR
ZC 1530.9+0454	15 30 54.	+ 04 54	4370		CLUSTER OF GALAXIES
MCG+07-32-024	15 30 54.	+ 40 23	48	15.	GALAXY
ZWG 222.023	15 30 54.	+ 40 24		15.6	GALAXY
UGC 09891	15 30 54.	+ 40 24	66	15.6	GALAXY S
SHAH 079	15 30 54.	+ 43 14	60	17.3	GROUP OF COMPACT GALAXIES
ZC 1530.9+4316	15 30 54.	+ 43 16	1080		CLUSTER OF GALAXIES
ZC 1530.9+5400	15 30 54.	+ 54 00	1210		CLUSTER OF GALAXIES
HOLM 711B	15 30 54.	- 08 33	54	14.3	PART OF MULTIPLE GALAXY
MCG-01-40-003	15 30 54.	- 08 36	30	15.5	GALAXY
SHB 314	15 30 54.3	+ 13 42 28.		19.	QUASI-STELLAR OBJECT
LB 00849	15 30 55.	- 01 57 00.		16.1	FAINT BLUE STAR
ZC 1531.0+1408	15 31 00.	+ 14 08	1210		CLUSTER OF GALAXIES
ZC 1531.0+2357	15 31 00.	+ 23 57	870		CLUSTER OF GALAXIES
ZC 1531.0+3230	15 31 00.	+ 32 30	740		CLUSTER OF GALAXIES
MCG+07-32-025	15 31 00.	+ 41 21	84	14.5	GALAXY
ZWG 222.024	15 31 00.	+ 41 22		15.3	GALAXY
UGC 09892	15 31 00.	+ 41 22	102	15.3	GALAXY Sb
ZC 1531.0+5156	15 31 00.	+ 51 56	810		CLUSTER OF GALAXIES
7ZW 608	15 31 00.	+ 59 04			COMPACT GALAXY
ZWG 022.014	15 31 00.	- 02 48		15.5	GALAXY
TON-N 0239	15 31 06.	+ 27 42		14.8	BLUE STAR
ZWG 194.009	15 31 06.	+ 36 38		15.5	GALAXY
IC 4550	15 31 08.	- 50 29 08.			NONSTELLAR OBJECT
MCG+06-34-014	15 31 09.	+ 35 14 30.	30	15.	GALAXY
ZWG 078.008	15 31 12.	+ 08 36		15.5	GALAXY
TON-N 0240	15 31 12.	+ 28 28		14.8	BLUE STAR
ZWG 166.007	15 31 12.	+ 31 29		15.6	GALAXY
ZC 1531.2+3446	15 31 12.	+ 34 46	1810		CLUSTER OF GALAXIES
ZWG 249.028	15 31 12.	+ 44 42		15.5	GALAXY
MCG+08-28-038	15 31 15.	+ 46 37	72	15.	GALAXY
ZWG 050.025	15 31 18.	+ 05 28		15.4	GALAXY
ZWG 249.029	15 31 18.	+ 46 37		15.2	GALAXY
UGC 09893	15 31 18.	+ 46 37	72	15.2	GALAXY PECULR
KARA.73B 0686	15 31 18.	+ 46 37	66	15.2	ISOLATED GALAXY S
1ZW 115	15 31 18.	+ 46 38			COMPACT GALAXY
ZWG 297.014	15 31 18.	+ 58 03		15.5	GALAXY
MCG-01-40-002	15 31 18.	- 08 39	12	15.5	GALAXY
HOLM 713A	15 31 19.	+ 15 10	102	13.1	PART OF MULTIPLE GALAXY
ABC 2092	15 31 20.	+ 31 20		15.7	RICH CLUSTER OF GALAXIES
REIN 2.233	15 31 23.69	+ 15 10 27.2			NEBULA
ZWG 050.026	15 31 24.	+ 05 27		15.3	GALAXY
UGC 09894	15 31 24.	+ 05 27	60	15.3	GALAXY SBa
ZWG 078.009	15 31 24.	+ 10 57		15.5	GALAXY
ZWG 107.003	15 31 24.	+ 15 10		13.8	GALAXY
UGC 09895	15 31 24.	+ 15 10	210	13.8	GALAXY Sc
MCG+03-40-003	15 31 24.	+ 15 11	228	13.8	GALAXY
ZWG 136.011	15 31 24.	+ 21 39		15.6	GALAXY
ZC 1531.4+3127	15 31 24.	+ 31 27	3090		CLUSTER OF GALAXIES
MCG+07-32-026	15 31 24.	+ 41 14	36	15.	GALAXY
ZWG 222.025	15 31 24.	+ 41 15		14.9	GALAXY
ZWG 249.030	15 31 24.	+ 44 42		15.3	GALAXY
ZC 1531.4+5901	15 31 24.	+ 59 01	3560		CLUSTER OF GALAXIES
ZWG 319.018	15 31 24.	+ 67 44		14.4	GALAXY
UGC 09896	15 31 24.	+ 67 44	96	14.4	GALAXY Sc
MCG+11-19-009	15 31 24.	+ 67 44	66	15.	GALAXY
HOLM 713B	15 31 25.	+ 15 09	24	15.1	PART OF MULTIPLE GALAXY
REIZ 4590	15 31 25.	+ 15 10	108	13.5	GALAXY
IC 4551	15 31 26.	+ 06 10 49.			SAME AS NGC 5964
RNGC 5951	15 31 26.	+ 15 10		14.0	GALAXY
LB 00850	15 31 30.	+ 01 25 12.		17.4	FAINT BLUE STAR
ZWG 078.010	15 31 30.	+ 11 11		15.0	GALAXY
UGC 09897	15 31 30.	+ 11 11	66	15.0	GALAXY SBa
MCG+02-40-002	15 31 30.	+ 11 11	42	15.0	GALAXY
ZWG 136.012	15 31 30.	+ 24 34		15.3	GALAXY
MCG+06-34-015	15 31 30.	+ 38 07	42	15.5	GALAXY
MCG+14-07-019	15 31 30.	+ 82 52	30	17.	GALAXY
REIZ 4591	15 31 31.	+ 15 09	18	15.4	GALAXY
REIZ 4592	15 31 31.	+ 21 19	18	14.2	GALAXY
MCG+04-37-003	15 31 33.	+ 21 18 30.	24	14.9	GALAXY
ZWG 022.015	15 31 36.	+ 02 22		15.7	GALAXY
ZWG 136.013	15 31 36.	+ 21 11		15.2	GALAXY
ZWG 136.014	15 31 36.	+ 26 18		15.3	GALAXY
MCG+08-28-039	15 31 36.	+ 45 51	48	16.	GALAXY
ABC 2105	15 31 36.	+ 74 16		17.2	RICH CLUSTER OF GALAXIES
IC 1129	15 31 37.	+ 68 25 20.			NONSTELLAR OBJECT
ZWG 050.027	15 31 42.	+ 07 16		15.4	GALAXY
UGC 09898	15 31 42.	+ 16 05	60	16.0	GALAXY S
ZWG 136.015	15 31 42.	+ 21 19		14.9	GALAXY
MRK 289	15 31 42.	+ 58 04	11	16.5	GALAXY WITH UV CONTINUUM
ZWG 319.019	15 31 42.	+ 68 25		13.7	GALAXY
UGC 09899	15 31 42.	+ 68 25	72	13.7	GALAXY Sc
MCG+11-19-010	15 31 42.	+ 68 26	57	14.	GALAXY
LB 00851	15 31 45.	- 16 25 30.		16.3	FAINT BLUE STAR
LB 00852	15 31 47.	+ 59 31 54.		18.0	FAINT BLUE STAR
RNGC 5946	15 31 47.	- 50 30		11.0	GLOBULAR CLUSTER
RNGC 5938	15 31 47.	- 66 42			UNVERIFIED SOUTHERN OBJECT
MCG+03-40-004	15 31 48.	+ 14 33	78	15.2	GALAXY
ZWG 107.004	15 31 48.	+ 14 34		15.2	GALAXY
UGC 09900	15 31 48.	+ 14 34	66	15.2	GALAXY
ZWG 107.005	15 31 48.	+ 18 49		15.7	GALAXY
ZWG 107.006	15 31 48.	+ 19 44		15.7	GALAXY
ZC 1531.8+2744	15 31 48.	+ 27 44	870		CLUSTER OF GALAXIES
REIZ 4594	15 31 48.	+ 31 53	30	15.2	GALAXY
MCG+09-25-057	15 31 48.	+ 54 26	54	16.	GALAXY
GCL 036	15 31 48.	- 50 29	156	11.0	GLOBULAR STAR CLUSTER
VHA 175	15 31 48.	- 50 29	90		GLOBULAR STAR CLUSTER
LB 00853	15 31 49.	- 12 37 18.		15.5	FAINT BLUE STAR
REIZ 4593	15 31 50.	+ 14 34	24	15.3	GALAXY
ABC 2091	15 31 51.	+ 10 25		17.5	RICH CLUSTER OF GALAXIES
MCG+04-37-004	15 31 51.	+ 23 07	45	14.6	GALAXY
ZWG 078.011	15 31 54.	+ 09 35		15.6	GALAXY
ZWG 107.007	15 31 54.	+ 16 40		15.6	GALAXY
ZWG 136.016	15 31 54.	+ 23 05		14.6	GALAXY
ZWG 166.008	15 31 54.	+ 27 30		15.5	GALAXY
PK315-13.1	15 31 54.	- 71 44 49.	6		PLANETARY NEBULA
HN 1799	15 31 58.9	+ 01 34 04.	24		NEBULA
KHAV 231	15 32	- 36 04	9250		DARK NEBULA
ZC 1532.0+1023	15 32 00.	+ 10 23	1340		CLUSTER OF GALAXIES
TON-N 0796	15 32 00.	+ 29 28		14.8	BLUE STAR
ZWG 194.010	15 32 00.	+ 32 59		15.4	GALAXY
MCG+06-34-016	15 32 00.	+ 32 59	30	15.	GALAXY
LB 00854	15 32 01.	+ 01 12 18.		16.1	FAINT BLUE STAR
LB 00855	15 32 03.	- 00 37 24.		16.7	FAINT BLUE STAR
ZWG 078.012	15 32 06.	+ 08 31		15.4	GALAXY
ZWG 078.013	15 32 06.	+ 12 26		15.2	GALAXY
UGC 09901	15 32 06.	+ 12 26	102	15.2	GALAXY Sb-c
ZC 1532.1+3708	15 32 06.	+ 37 08	1410		CLUSTER OF GALAXIES
HOLM 714B	15 32 10.	+ 15 22	18	13.2	PART OF MULTIPLE GALAXY
LB 00856	15 32 10.	- 00 19 24.		15.4	FAINT BLUE STAR
REIN 2.234	15 32 11.90	+ 15 21 21.0			NEBULA
ZWG 050.028	15 32 12.	+ 06 28		15.3	GALAXY
UGC 09902	15 32 12.	+ 15 17	60	17.	GALAXY DWARF
ZWG 107.008	15 32 12.	+ 15 21		12.7	GALAXY
UGC 09904	15 32 12.	+ 15 21	66	13.7	GALAXY S(c)
UGC 09903	15 32 12.	+ 15 21	108	13.3	GALAXY S0
VV 244A	15 32 12.	+ 15 21	36	13.1	INTERACTING GALAXY
KARA.72 468A	15 32 12.	+ 15 21	102	13.3	PART OF DOUBLE GALAXY
MCG+03-40-005	15 32 12.	+ 15 21 30.	120	12.7	GALAXY
7ZW 609	15 32 12.	+ 58 55			COMPACT GALAXY
HOLM 714A	15 32 13.	+ 15 22	54	13.1	PART OF MULTIPLE GALAXY
REIN 2.235	15 32 13.22	+ 15 21 35.3			NEBULA
RNGC 5953	15 32 15.	+ 15 22		13.5	GALAXY
LB 00857	15 32 15.	+ 57 19 12.		18.1	FAINT BLUE STAR
REIN 2.236	15 32 15.96	+ 15 21 59.4			NEBULA
ARP 091	15 32 16.	+ 15 22			PECULIAR GALAXY
ABC 2102	15 32 17.	+ 70 21		17.7	RICH CLUSTER OF GALAXIES
ZWG 078.014	15 32 18.	+ 08 30		15.6	GALAXY
ZWG 050.029	15 32 18.	+ 08 30		15.6	GALAXY
UGC 09905	15 32 18.	+ 08 30	60	15.6	GALAXY S(c)
ZWG 078.015	15 32 18.	+ 10 27		15.6	GALAXY
VV 244B	15 32 18.	+ 15 22	60	13.2	INTERACTING GALAXY
KARA.72 468B	15 32 18.	+ 15 22	90	13.7	PART OF DOUBLE GALAXY
MCG+03-40-006	15 32 18.	+ 15 22	66	12.7	GALAXY
MCG+09-25-058	15 32 18.	+ 56 44	150	12.	GALAXY
ZWG 297.015	15 32 18.	+ 56 45		13.0	GALAXY
RNGC 5963	15 32 18.	+ 56 45		13.0	GALAXY
UGC 09906	15 32 18.	+ 56 45	240	13.0	GALAXY S
KARA.72 469A	15 32 18.	+ 56 45	222	13.0	PART OF DOUBLE GALAXY
REIZ 4597	15 32 19.	+ 15 22	6	15.2	GALAXY
REIZ 4596	15 32 19.	+ 15 22	48	13.2	GALAXY
REIZ 4595	15 32 19.	+ 15 22	18	13.2	GALAXY
RNGC 5954	15 32 21.	+ 15 22		13.5	GALAXY
RNGC 5952	15 32 22.	+ 05 07		15.5	GALAXY
ABC 2093	15 32 23.	+ 37 13		17.2	RICH CLUSTER OF GALAXIES
ZWG 050.030	15 32 24.	+ 05 07		15.5	GALAXY
MCG+07-32-027	15 32 24.	+ 43 25	39	15.5	GALAXY
IC 4552	15 32 29.	+ 04 51 55.			NONSTELLAR OBJECT
ZWG 078.016	15 32 30.	+ 09 45		15.3	GALAXY
UGC 09907	15 32 30.	+ 09 45	66	15.3	GALAXY S
IC 1126	15 32 32.	+ 05 10			SAME AS NGC 5952
ZWG 078.017	15 32 36.	+ 11 55		13.3	GALAXY
UGC 09908	15 32 36.	+ 11 55	102	13.3	GALAXY Sc?
MCG+02-40-003	15 32 36.	+ 11 55	96	13.3	GALAXY
ZC 1532.6+2505	15 32 36.	+ 25 05	2960		CLUSTER OF GALAXIES
LB 00858	15 32 36.	- 00 32 12.		17.2	FAINT BLUE STAR
LB 00859	15 32 36.	- 15 50 24.		17.0	FAINT BLUE STAR
RNGC 5956	15 32 37.	+ 11 55		13.5	GALAXY
REIZ 4601	15 32 38.	+ 30 56	18	15.2	GALAXY
RNGC 5955	15 32 40.	+ 05 14		15.0	GALAXY
REIZ 4599	15 32 40.	+ 28 49	18	14.6	GALAXY
RNGC 5958	15 32 40.	+ 28 50		13.0	GALAXY
REIZ 4600	15 32 40.	+ 28 51	24	13.7	GALAXY
ZWG 050.031	15 32 42.	+ 05 14		15.0	GALAXY
MCG+01-40-006	15 32 42.	+ 05 14	48	15.0	GALAXY
ZC 1532.7+2405	15 32 42.	+ 24 05	670		CLUSTER OF GALAXIES
ZWG 166.009	15 32 42.	+ 28 50		13.2	GALAXY

OBJECT NAME	RIGHT ASCEN.	DECLINATION	DIAM.	MAGN.	TYPE OF OBJECT
UGC 09909	15 32 42.	+ 28 50	66	13.2	GALAXY S
MCG+05-37-003	15 32 42.	+ 28 50	60	13.2	GALAXY
ZWG 166.010	15 32 42.	+ 31 14		15.5	GALAXY
UGC 09910	15 32 42.	+ 31 14	108	15.5	GALAXY
UGC 09911	15 32 42.	+ 41 19	78	16.0	GALAXY Sc
MCG+10-22-020	15 32 42.	+ 56 50	276	11.	GALAXY
ZC 1532.7+7020	15 32 42.	+ 70 20	1480		CLUSTER OF GALAXIES
MRSL 324-00/1	15 32 42.	- 56 40	360		HII REGION
ARP 220	15 32 44.	+ 23 39			PECULIAR GALAXY
MCG+07-32-028	15 32 45.	+ 41 18	66	15.	GALAXY
IC 4553	15 32 46.	+ 23 39 29.			NONSTELLAR OBJECT
IC 4557	15 32 47.	+ 39 53 13.			NONSTELLAR OBJECT
ZWG 050.032	15 32 48.	+ 05 16		15.3	GALAXY
MCG+01-40-006A	15 32 48.	+ 05 16	36	15.3	GALAXY
ZWG 050.033	15 32 48.	+ 07 18		15.4	GALAXY
REIZ 4598	15 32 48.	+ 16 42	30	14.3	GALAXY
ZWG 107.009	15 32 48.	+ 16 43		15.0	GALAXY
UGC 09912	15 32 48.	+ 16 43	102	15.0	GALAXY
VV 132A	15 32 48.	+ 16 43	420	15.	INTERACTING GALAXY
VV 132	15 32 48.	+ 16 43	66	14.5	INTERACTING GALAXY
ZWG 136.017	15 32 48.	+ 23 40		14.4	GALAXY
UGC 09913	15 32 48.	+ 23 40	120	14.4	GALAXY PECULR
KARA.72 470B	15 32 48.	+ 23 40	36		PART OF DOUBLE GALAXY
KARA.72 470A	15 32 48.	+ 23 40	30	14.4	PART OF DOUBLE GALAXY
MCG+04-37-005	15 32 48.	+ 23 40	72	14.4	GALAXY
ZWG 136.018	15 32 48.	+ 24 17		15.6	GALAXY
ZWG 222.026	15 32 48.	+ 39 54		15.7	GALAXY
ZWG 297.016	15 32 48.	+ 56 52		13.4	GALAXY
RNGC 5965	15 32 48.	+ 56 52		13.5	GALAXY
UGC 09914	15 32 48.	+ 56 52	360	13.4	GALAXY Sb
KARA.72 469B	15 32 48.	+ 56 52	270	13.4	PART OF DOUBLE GALAXY
MCG+10-22-021	15 32 48.	+ 57 31	57	15.	GALAXY
ZWG 297.017	15 32 48.	+ 57 32		15.6	GALAXY
ZC 1532.8-0307	15 32 48.	- 03 07	2350		CLUSTER OF GALAXIES
REIZ 4603	15 32 49.	+ 39 53	30	14.9	GALAXY
REIN 2.237	15 32 52.61	+ 16 42 55.4			NEBULA
MCG+03-40-007	15 32 54.	+ 16 43	96	15.0	GALAXY
MCG+03-40-008	15 32 54.	+ 17 02	42		GALAXY
IC 4554	15 32 54.	+ 23 38 00.			NONSTELLAR OBJECT
ZC 1532.9+2912	15 32 54.	+ 29 12	3700		CLUSTER OF GALAXIES
7ZW 610	15 32 54.	+ 60 09			COMPACT GALAXY
LB 00293	15 32 58.	- 16 19 42.		16.0	FAINT BLUE STAR
LB 09895	15 33	- 80 42		13.8	FAINT BLUE STAR
ZWG 050.034	15 33 00.	+ 06 28		15.5	GALAXY
ZWG 078.018	15 33 00.	+ 12 13		13.3	GALAXY
UGC 09915	15 33 00.	+ 12 13	168	13.3	GALAXY SBb
MCG+02-40-004	15 33 00.	+ 12 13	210	13.3	GALAXY
TON-N 0241	15 33 00.	+ 23 58		15.3	BLUE STAR
ZWG 136.019	15 33 00.	+ 24 15		15.7	GALAXY
UGC 09916	15 33 00.	+ 24 15	60	15.7	GALAXY S
ARC 2095	15 33 00.	+ 40 42		17.8	RICH CLUSTER OF GALAXIES
7ZW 611	15 33 00.	+ 57 28			COMPACT GALAXY
MRSL 323-01/1	15 33 00.	- 57 12	480		HII REGION
RNGC 5957	15 33 01.	+ 12 13		13.5	GALAXY
MCG+04-37-006	15 33 03.	+ 21 00	54	15.2	GALAXY
ZWG 050.035	15 33 06.	+ 02 35		15.6	GALAXY
KARA.73B 0687	15 33 06.	+ 02 35	42	15.6	ISOLATED GALAXY S
ZWG 136.020	15 33 06.	+ 21 00		15.2	GALAXY
UGC 09917	15 33 06.	+ 21 00	66	15.2	GALAXY SB
MCG+05-37-004	15 33 06.	+ 27 29 30.	48	15.2	GALAXY
ZWG 166.011	15 33 06.	+ 27 30		15.2	GALAXY
ZWG 166.012	15 33 06.	+ 27 31		15.7	GALAXY
ZC 1533.1+5303	15 33 06.	+ 53 03	670		CLUSTER OF GALAXIES
MCG+06-34-017	15 33 09.	+ 35 20 30.	30	15.5	GALAXY
RNGC 5961	15 33 11.	+ 31 01		14.0	GALAXY
ZWG 078.019	15 33 12.	+ 13 57		15.7	GALAXY
ZWG 136.021	15 33 12.	+ 25 28		14.8	GALAXY
ZWG 166.013	15 33 12.	+ 31 01		14.0	GALAXY
UGC 09918	15 33 12.	+ 31 01	54	14.0	GALAXY
MCG+10-22-022	15 33 12.	+ 59 53	39	15.	GALAXY
REIZ 4602	15 33 13.	+ 21 02	24	14.5	GALAXY
REIZ 4605	15 33 13.	+ 31 02	36	13.8	GALAXY
HOLM 715A	15 33 13.	+ 31 03	48	13.6	PART OF MULTIPLE GALAXY
REIZ 4604	15 33 14.	+ 30 57	90	14.7	GALAXY
HOLM 715B	15 33 15.	+ 30 59	72	14.9	PART OF MULTIPLE GALAXY
IC 4556	15 33 16.	+ 25 28 20.			NONSTELLAR OBJECT
ZWG 050.036	15 33 18.	+ 08 07		15.4	GALAXY
ZWG 078.020	15 33 18.	+ 12 46		15.1	GALAXY
UGC 09919	15 33 18.	+ 12 46	102	15.1	GALAXY Sc
ZWG 166.014	15 33 18.	+ 30 58		15.1	GALAXY
UGC 09920	15 33 18.	+ 30 58	84	15.1	GALAXY Sb
MCG+05-37-005	15 33 18.	+ 31 03 30.	45	14.0	GALAXY
HOLM 715C	15 33 20.	+ 31 03	36	15.0	PART OF MULTIPLE GALAXY
MCG+04-37-007	15 33 21.	+ 25 28 30.	21	14.8	GALAXY
MCG+05-37-006	15 33 21.	+ 31 00	72	15.1	GALAXY
LB 00294	15 33 21.	+ 60 32 42.		14.3	FAINT BLUE STAR
ZWG 050.037	15 33 24.	+ 06 14		15.3	GALAXY
ZWG 078.021	15 33 24.	+ 09 02		15.5	GALAXY
ZWG 078.022	15 33 24.	+ 09 06		15.7	GALAXY
ZWG 078.023	15 33 24.	+ 14 00		15.7	GALAXY
ZWG 136.022	15 33 24.	+ 21 40		15.7	GALAXY
ZC 1533.4+2415	15 33 24.	+ 24 15	2020		CLUSTER OF GALAXIES
1ZW 116	15 33 24.	+ 46 59			COMPACT GALAXY
LB 00860	15 33 24.	+ 58 22 42.		16.5	FAINT BLUE STAR
ARC 2096	15 33 28.	+ 37 31		17.0	RICH CLUSTER OF GALAXIES
ZWG 078.024	15 33 30.	+ 08 54		15.4	GALAXY
ZWG 107.010	15 33 30.	+ 14 41		14.7	GALAXY
ZWG 166.015	15 33 30.	+ 26 30		15.6	GALAXY
ZWG 136.023	15 33 30.	+ 26 30		15.6	GALAXY
UGC 09921	15 33 30.	+ 26 30	66	15.6	GALAXY S
ZC 1533.5+2724	15 33 30.	+ 27 24	5040		CLUSTER OF GALAXIES
ZWG 166.016	15 33 30.	+ 31 42		15.7	GALAXY
ZC 1533.5+4151	15 33 30.	+ 41 51	1280		CLUSTER OF GALAXIES
7ZW 612	15 33 30.	+ 57 48			COMPACT GALAXY
ZC 1533.5+6347	15 33 30.	+ 63 47	1550		CLUSTER OF GALAXIES
REIZ 4606	15 33 32.	+ 30 55	12	15.5	GALAXY
ARC 2097	15 33 33.	+ 39 46		17.8	RICH CLUSTER OF GALAXIES
MCG+09-25-059	15 33 33.	+ 56 37	18	16.	GALAXY
MCG+03-40-009	15 33 36.	+ 14 40	27	14.7	GALAXY
ZWG 136.024	15 33 36.	+ 25 30		15.6	GALAXY
ZWG 166.017	15 33 36.	+ 27 17		15.6	GALAXY
ZC 1533.6+3732	15 33 36.	+ 37 32	1080		CLUSTER OF GALAXIES
REIZ 4607	15 33 36.	+ 40 40	18	15.3	GALAXY
IC 4558	15 33 39.	+ 25 31 17.			NONSTELLAR OBJECT
MCG+06-34-018	15 33 39.	+ 36 25	42	15.	GALAXY
LB 00861	15 33 42.	+ 00 49 36.		14.9	FAINT BLUE STAR
ZWG 078.025	15 33 42.	+ 09 54		15.4	GALAXY
ZWG 136.025	15 33 42.	+ 25 20		15.6	GALAXY
ZWG 136.026	15 33 42.	+ 25 30		15.1	GALAXY
TON-N 0242	15 33 42.	+ 32 03		15.3	BLUE STAR
ZC 1533.7+3218	15 33 42.	+ 32 18	940		CLUSTER OF GALAXIES
ZC 1533.7+4355	15 33 42.	+ 43 55	740		CLUSTER OF GALAXIES
ZWG 297.018	15 33 42.	+ 56 38		15.4	GALAXY
RNGC 5969	15 33 42.	+ 56 38		15.5	GALAXY
IC 1127	15 33 44.	+ 23 38 41.			NONSTELLAR OBJECT
RNGC 5960	15 33 46.	+ 05 50		15.0	GALAXY
IC 4559	15 33 47.	+ 25 30 59.			NONSTELLAR OBJECT
ZWG 050.038	15 33 48.	+ 05 50		15.1	GALAXY
MCG+01-40-007	15 33 48.	+ 05 50	36	15.1	GALAXY
ZWG 050.039	15 33 48.	+ 06 05		15.3	GALAXY
ZC 1533.8+1430	15 33 48.	+ 14 30	1340		CLUSTER OF GALAXIES
MCG+04-37-008	15 33 48.	+ 25 30 30.	15	15.1	GALAXY
ZWG 249.031	15 33 48.	+ 46 55		15.7	GALAXY
ZWG 022.016	15 33 48.	- 03 06		15.7	GALAXY
MCG-03-40-001	15 33 48.	- 18 38	72	15.	GALAXY
LB 00295	15 33 54.	+ 01 45 48.		15.6	FAINT BLUE STAR
ZWG 136.027	15 33 54.	+ 25 34		15.7	GALAXY
MCG+07-32-029	15 33 54.	+ 40 15	30	16.	GALAXY
IC 4562	15 33 56.	+ 43 36			NONSTELLAR OBJECT
ARC 2099	15 33 56.	+ 43 55		17.8	RICH CLUSTER OF GALAXIES
PK323-02.1	15 33 59.	- 58 34 46.	25		PLANETARY NEBULA
HMS 1534+3749	15 34	+ 37 49			CLUSTER OF GALAXIES
ZWG 078.026	15 34 00.	+ 12 17		15.7	GALAXY
MCG+04-37-009	15 34 00.	+ 25 34	24	15.7	GALAXY
1ZW 117	15 34 00.	+ 38 50			COMPACT GALAXY
ZWG 222.027	15 34 00.	+ 38 50		14.4	GALAXY
UGC 09922	15 34 00.	+ 38 50	54	14.4	GALAXY DBL SYS
MCG+07-32-030	15 34 00.	+ 38 50	30	15.	GALAXY
MCG+07-32-031	15 34 00.	+ 38 50 30.	12	16.	GALAXY
MCG+07-32-032	15 34 00.	+ 39 56	30	14.	GALAXY
ZWG 222.028	15 34 00.	+ 39 57		13.9	GALAXY
RNGC 5966	15 34 00.	+ 39 57		14.0	GALAXY
UGC 09923	15 34 00.	+ 39 57	108	13.9	GALAXY E
IC 4560	15 34 00.	+ 40 00 27.			NONSTELLAR OBJECT
ZC 1534.0+4222	15 34 00.	+ 42 22	14580		CLUSTER OF GALAXIES
ZWG 249.032	15 34 00.	+ 45 00		14.4	GALAXY
UGC 09924	15 34 00.	+ 45 00	48	14.4	GALAXY S(B?)
ZC 1534.0+6150	15 34 00.	+ 61 50	4570		CLUSTER OF GALAXIES
MCG+14-07-020	15 34 00.	+ 82 35	9	16.	GALAXY
ZWG 366.017	15 34 00.	+ 82 46		15.6	GALAXY
ZC 1534.0-0153	15 34 00.	- 01 53	2220		CLUSTER OF GALAXIES
REIZ 4612	15 34 01.	+ 40 00	12	15.3	GALAXY
ARC 2094	15 34 01.	- 01 52		16.7	RICH CLUSTER OF GALAXIES
REIZ 4611	15 34 02.	+ 38 50	48	14.1	GALAXY
LB 00862	15 34 03.	- 01 27 18.		16.1	FAINT BLUE STAR
REIZ 4610	15 34 04.	+ 28 49	30	15.1	GALAXY
ZWG 050.040	15 34 06.	+ 03 31		15.4	GALAXY
KARA.73B 0688	15 34 06.	+ 03 31	36	15.4	ISOLATED GALAXY S
ZWG 078.027	15 34 06.	+ 12 20		15.6	GALAXY
ZWG 166.018	15 34 06.	+ 27 29		15.4	GALAXY
REIZ 4616	15 34 07.	+ 39 56	24	14.0	GALAXY
RNGC 5962	15 34 09.	+ 16 46		12.5	GALAXY
MCG+05-37-007	15 34 09.	+ 27 28	30	15.4	GALAXY
REIZ 4608	15 34 11.	+ 16 45	12	15.0	GALAXY
REIZ 4609	15 34 11.	+ 16 46	90	12.5	GALAXY
HOLM 716A	15 34 11.	+ 16 46	108	12.4	PART OF MULTIPLE GALAXY
IC 4563	15 34 11.	+ 40 01 22.			NONSTELLAR OBJECT
IC 1139	15 34 11.	+ 82 45 59.			NONSTELLAR OBJECT
ZWG 050.041	15 34 12.	+ 03 58		15.5	GALAXY
ZWG 107.011	15 34 12.	+ 16 36		15.4	GALAXY
UGC 09925	15 34 12.	+ 16 36	84	15.4	GALAXY Sc
ZWG 107.012	15 34 12.	+ 16 46		12.2	GALAXY
UGC 09926	15 34 12.	+ 16 46	168	12.2	GALAXY Sc
ZWG 136.028	15 34 12.	+ 22 40		14.9	GALAXY
UGC 09927	15 34 12.	+ 22 40	60	14.9	GALAXY SB0
ZWG 222.029	15 34 12.	+ 40 00		15.1	GALAXY
MCG+07-32-033	15 34 12.	+ 40 00	42	14.	GALAXY
1ZW 118	15 34 12.	+ 43 40			COMPACT GALAXY
MCG+07-32-034	15 34 12.	+ 43 40	60	14.	GALAXY
MCG+09-26-001	15 34 12.	+ 56 14	36	17.	GALAXY
ZWG 319.020	15 34 12.	+ 67 59		15.5	GALAXY
HOLM 716B	15 34 13.	+ 16 45	30	14.9	PART OF MULTIPLE GALAXY
REIZ 4617	15 34 13.	+ 40 01	18	14.6	GALAXY
REIN 2.238	15 34 14.14	+ 16 46 19.0			NEBULA
MCG+03-40-010	15 34 15.	+ 16 36	78	15.4	GALAXY
MCG+03-40-011	15 34 15.	+ 16 46	180	12.2	GALAXY
ZWG 050.042	15 34 18.	+ 08 00		15.6	GALAXY
ZWG 078.028	15 34 18.	+ 09 11		15.5	GALAXY
ZC 1534.3+1015	15 34 18.	+ 10 15	1210		CLUSTER OF GALAXIES
ZWG 107.013	15 34 18.	+ 17 30		15.0	GALAXY
KARA.73B 0689	15 34 18.	+ 17 30	48	15.0	ISOLATED GALAXY S
MCG+04-37-010	15 34 18.	+ 22 40	30	14.9	GALAXY
ZWG 166.019	15 34 18.	+ 30 51		15.2	GALAXY
MRK 689	15 34 18.	+ 30 51	15	15.5	GALAXY WITH UV CONTINUUM
ZWG 222.030	15 34 18.	+ 43 39		13.8	GALAXY
UGC 09928	15 34 18.	+ 43 39	66	13.8	GALAXY E?
ZWG 249.033	15 34 18.	+ 46 54		15.6	GALAXY
MCG+09-26-002	15 34 18.	+ 56 37	90	14.	GALAXY
MCG-04-37-003	15 34 18.	- 26 18	36	15.5	GALAXY
MRSL 330+07/1	15 34 18.	- 46 07	2400		HII REGION
REIZ 4614	15 34 19.	+ 30 53	30	15.6	GALAXY
REIZ 4615	15 34 19.	+ 31 17	30	14.8	GALAXY
FATH 1.722	15 34 21.	+ 00 04	8		NEBULA
MCG+05-37-008	15 34 21.	+ 30 51	27	15.2	GALAXY
ZWG 050.043	15 34 24.	+ 04 55		15.5	GALAXY
MCG+03-40-012	15 34 24.	+ 17 30	36	15.0	GALAXY
ZWG 136.029	15 34 24.	+ 25 18		15.4	GALAXY
ZC 1534.4+2553	15 34 24.	+ 25 53	4030		CLUSTER OF GALAXIES
ZC 1534.4+3100	15 34 24.	+ 31 00	6120		CLUSTER OF GALAXIES
ZWG 166.020	15 34 24.	+ 31 17		15.2	GALAXY
MCG+06-34-020	15 34 24.	+ 37 42			GALAXY
MCG+06-34-019	15 34 24.	+ 37 48			GALAXY
ZC 1534.4+3749	15 34 24.	+ 37 49	1680		CLUSTER OF GALAXIES
ZWG 222.031	15 34 24.	+ 43 40		15.3	GALAXY
MCG+08-28-040	15 34 24.	+ 46 55	48	16.	GALAXY
ZCG 1534+47.3	15 34 24.	+ 47 11		16.5	COMPACT GALAXY
ZWG 297.019	15 34 24.	+ 56 39		14.9	GALAXY
RNGC 5971	15 34 24.	+ 56 39		15.0	GALAXY
UGC 09929	15 34 24.	+ 56 39	102	14.9	GALAXY Sa
HN 0422	15 34 24.	- 81 29			NEBULA
IC 4545	15 34 25.	- 81 30			NONSTELLAR OBJECT
MCG+05-37-009	15 34 27.	+ 31 18 30.	48	15.2	GALAXY
REIZ 4618	15 34 27.	+ 38 16	48	15.3	GALAXY
LB 00863	15 34 29.	- 14 04 12.		17.0	FAINT BLUE STAR
HN 1800	15 34 29.0	- 01 40 35.	18		NEBULA
ZWG 050.044	15 34 30.	+ 05 13		15.6	GALAXY

OBJECT NAME	RIGHT ASCEN.	DECLINATION	DIAM.	MAGN.	TYPE OF OBJECT
ZCG 1534+47.2	15 34 30.	+ 47 09		16.3	COMPACT GALAXY
MCG+14-07-021	15 34 30.	+ 82 35	30	16.	GALAXY
ARC 2100	15 34 33.	+ 37 49		17.0	RICH CLUSTER OF GALAXIES
RNGC 5959	15 34 35.	- 16 27		14.0	GALAXY
ZWG 136.030	15 34 36.	+ 25 35		15.2	GALAXY
1ZW 119	15 34 36.	+ 54 28			COMPACT GALAXY
MCG-03-40-002	15 34 36.	- 16 27	30	14.5	GALAXY
REIZ 4613	15 34 37.	+ 15 05	18	15.8	GALAXY
IC 4561	15 34 40.	+ 25 34 29.			NONSTELLAR OBJECT
ZWG 078.029	15 34 42.	+ 10 50		15.7	GALAXY
ZC 1534.7+3753	15 34 42.	+ 37 53	4500		CLUSTER OF GALAXIES
MCG+07-32-035	15 34 42.	+ 41 58	30	15.	GALAXY
REIZ 4619	15 34 42.	+ 43 40	72	13.7	GALAXY
MCG+07-32-036	15 34 42.	+ 43 41	66	14.	GALAXY
MRK 290	15 34 42.	+ 58 05	10	15.	GALAXY WITH UV CONTINUUM
KW 26	15 34 42.	+ 58 05	14		SEYFERT GALAXY
VVI 78	15 34 42.	+ 58 05	10	15.56	SEYFERT GALAXY
MCG-03-40-003	15 34 42.	- 16 26 30.	24	15.	GALAXY
MCG+04-37-011	15 34 45.	+ 25 35	18	15.2	GALAXY
MCG+08-28-041	15 34 45.	+ 49 44	42	16.	GALAXY
MCG+09-26-003	15 34 45.	+ 53 43	24	16.	GALAXY
ZC 1534.8+2435	15 34 48.	+ 24 35	940		CLUSTER OF GALAXIES
ZC 1534.8+3300	15 34 48.	+ 33 00	2960		CLUSTER OF GALAXIES
MCG+06-34-021	15 34 48.	+ 37 51			GALAXY
ZWG 222.032	15 34 48.	+ 41 58		15.3	GALAXY
ZWG 222.033	15 34 48.	+ 43 41		14.4	GALAXY
UGC 09930	15 34 48.	+ 43 41	90	14.4	GALAXY S
ZCG 1534+47.1	15 34 48.	+ 47 08		18.0	COMPACT GALAXY
ZWG 249.034	15 34 48.	+ 48 00		15.2	GALAXY
KARA.73B 0690	15 34 48.	+ 48 00	42	15.2	ISOLATED GALAXY S0
7ZW 613	15 34 48.	+ 58 58			COMPACT GALAXY
ZC 1534.8+7420	15 34 48.	+ 74 20	270		CLUSTER OF GALAXIES
REIZ 4620	15 34 49.	+ 43 15	48	14.0	GALAXY
MCG+07-32-037	15 34 51.	+ 43 35	48	14.5	GALAXY
IC 4564	15 34 52.	+ 43 41			NONSTELLAR OBJECT
LB 00864	15 34 53.	+ 58 53 54.		17.0	FAINT BLUE STAR
ZWG 136.031	15 34 54.	+ 25 43		15.4	GALAXY
ZWG 166.021	15 34 54.	+ 29 59		15.7	GALAXY
ZWG 222.034	15 34 54.	+ 43 35		14.8	GALAXY
REIZ 4621	15 34 54.	+ 43 35	30	14.4	GALAXY
UGC 09931	15 34 54.	+ 43 35	60	14.8	GALAXY
MCG-01-40-004	15 34 54.	- 08 34	54	15.	GALAXY
LB 00296	15 34 55.	- 12 48 48.		13.7	FAINT BLUE STAR
RNGC 5976A	15 34 59.	+ 59 44		15.5	GALAXY
VDB.66C 200	15 35	+ 44 21	70		DWARF GALAXY
KHAV 232	15 35	- 34 46	3610		DARK NEBULA
ZWG 050.045	15 35 00.	+ 06 27		15.7	GALAXY
ZWG 078.030	15 35 00.	+ 10 47		15.6	GALAXY
MCG+04-37-012	15 35 00.	+ 25 44	48	15.4	GALAXY
ZWG 166.022	15 35 00.	+ 29 30		15.3	GALAXY
REIZ 4622	15 35 00.	+ 43 41	48	13.9	GALAXY
MCG+07-32-038	15 35 00.	+ 43 42 30.	90	14.	GALAXY
ZWG 222.035	15 35 00.	+ 43 43		14.3	GALAXY
UGC 09933	15 35 00.	+ 43 43	114	14.3	GALAXY Sb
MCG+07-32-039	15 35 00.	+ 44 25	78	15.	GALAXY
1ZW 120	15 35 00.	+ 47 28			COMPACT GALAXY
ZWG 297.020	15 35 00.	+ 59 44		15.4	GALAXY
UGC 09934	15 35 00.	+ 59 44	72	15.4	GALAXY SBa-b
MCG+10-22-023	15 35 00.	+ 59 44	45	14.	GALAXY
ZWG 366.018	15 35 00.	+ 82 38		14.7	GALAXY
UGC 09932	15 35 00.	+ 82 38	90	14.7	GALAXY E
MCG+14-07-022	15 35 00.	+ 82 38	18	15.	GALAXY
LB 00865	15 35 04.	+ 00 54 06.		17.2	FAINT BLUE STAR
RNGC 5964	15 35 04.	+ 06 08		14.0	GALAXY
IC 4565	15 35 04.	+ 43 34			NONSTELLAR OBJECT
ZWG 050.046	15 35 06.	+ 05 06		15.6	GALAXY
ZWG 050.047	15 35 06.	+ 06 09		14.2	GALAXY
UGC 09935	15 35 06.	+ 06 09	270	14.2	GALAXY SBc
MCG+01-40-008	15 35 06.	+ 06 09	222	14.2	GALAXY
KARA.73B 0691	15 35 06.	+ 06 09	264	14.2	ISOLATED GALAXY S
ZWG 078.031	15 35 06.	+ 13 19		15.4	GALAXY
MCG+04-37-013	15 35 06.	+ 20 42	60	14.7	GALAXY
ZWG 222.036	15 35 06.	+ 44 24		15.5	GALAXY
UGC 09936	15 35 06.	+ 44 24	102	15.5	GALAXY DWRF SP
KEEN 5976A	15 35 06.	+ 59 43			GALAXY
ZWG 022.017	15 35 06.	- 01 36		15.3	GALAXY
LB 00866	15 35 07.	- 16 52 12.		16.0	FAINT BLUE STAR
HN 1801	15 35 07.2	- 01 35 27.	12		NEBULA
LB 00867	15 35 10.	- 00 06 12.		16.9	FAINT BLUE STAR
ZWG 050.048	15 35 12.	+ 03 11		15.4	GALAXY
ZWG 078.032	15 35 12.	+ 09 29		15.5	GALAXY
ZWG 136.032	15 35 12.	+ 20 43		14.7	GALAXY
UGC 09937	15 35 12.	+ 20 43	72	14.7	GALAXY S0-a
ZWG 136.033	15 35 12.	+ 21 55		15.3	GALAXY
ZC 1535.2+2507	15 35 12.	+ 25 07	870		CLUSTER OF GALAXIES
UGC 09938	15 35 12.	+ 30 15	90	17.	GALAXY DWRF SP
ZC 1535.2+7220	15 35 12.	+ 72 20	1080		CLUSTER OF GALAXIES
LB 00868	15 35 15.	+ 02 08 12.		18.1	FAINT BLUE STAR
HN 1802	15 35 16.9	- 01 34 20.	12		NEBULA
ZWG 050.049	15 35 18.	+ 03 36		15.7	GALAXY
ZC 1535.3+1210	15 35 18.	+ 12 10	1080		CLUSTER OF GALAXIES
ZWG 107.014	15 35 18.	+ 16 20		15.7	GALAXY
ZWG 022.018	15 35 18.	- 01 35		15.5	GALAXY
UGC 09939	15 35 18.	- 01 35	84	15.5	GALAXY (S0)
MCG+00-40-004	15 35 18.	- 01 35	18	15.5	GALAXY
LB 00297	15 35 20.	+ 00 04 36.		15.3	FAINT BLUE STAR
IC 1128	15 35 21.	- 01 22 47.			NONSTELLAR OBJECT
MCG+03-40-013	15 35 24.	+ 16 20	66	15.7	GALAXY
ZWG 107.015	15 35 24.	+ 17 25		15.5	GALAXY
ZWG 194.011	15 35 24.	+ 33 50		15.5	GALAXY
MCG+06-34-022	15 35 24.	+ 33 50	42	14.5	GALAXY
SHAH 080	15 35 24.	+ 41 48	36	17.5	GROUP OF COMPACT GALAXIES
ZWG 275.001	15 35 24.	+ 54 07		15.7	GALAXY
ZWG 274.048	15 35 24.	+ 54 07		15.7	GALAXY
MCG+09-26-004	15 35 24.	+ 54 10	36	16.	GALAXY
ZCG 1535+54	15 35 24.	+ 54 41		17.8	COMPACT GALAXY
1ZW 121	15 35 24.	+ 54 43			COMPACT GALAXY
VVI 79	15 35 24.	+ 54 43		15.2	SEYFERT GALAXY
ZC 1535.4+6803	15 35 24.	+ 68 03	1140		CLUSTER OF GALAXIES
IC 1130	15 35 26.	+ 17 24			NONSTELLAR OBJECT
LB 00869	15 35 29.	- 01 21 18.		15.8	FAINT BLUE STAR
LB 00870	15 35 29.	- 15 13 36.		17.0	FAINT BLUE STAR
ZC 1535.5+1229	15 35 30.	+ 12 29	1010		CLUSTER OF GALAXIES
ZWG 107.016	15 35 30.	+ 16 29		15.6	GALAXY
MCG+03-40-014	15 35 30.	+ 17 25	36	15.5	GALAXY
ZC 1535.5+2408	15 35 30.	+ 24 08	610		CLUSTER OF GALAXIES
ZWG 222.037	15 35 30.	+ 43 29		13.5	GALAXY
UGC 09940	15 35 30.	+ 43 28	96	13.5	GALAXY S(c?)
MCG+07-32-040	15 35 30.	+ 43 28	78	14.	GALAXY
MRK 486	15 35 30.	+ 54 42	8	15.	GALAXY WITH UV CONTINUUM
KW 43	15 35 30.	+ 54 42	25		SEYFERT GALAXY
7ZW 614	15 35 30.	+ 56 53			COMPACT GALAXY
KARA.68 234	15 35 30.	- 01 00	34		DWARF GALAXY
VVI 80	15 35 30.	- 54 42	8	15.	SEYFERT GALAXY
REIZ 4623	15 35 31.	+ 43 27	54	13.6	GALAXY
IC 4566	15 35 31.	+ 43 42			NONSTELLAR OBJECT
MCG+03-40-015	15 35 36.	+ 16 28	66	15.6	GALAXY
ZWG 136.034	15 35 36.	+ 22 35		14.6	GALAXY
MCG+04-37-014	15 35 36.	+ 22 35 30.	36	14.6	GALAXY
MCG+06-34-023	15 35 36.	+ 36 25	9	15.	GALAXY
MCG+10-22-024	15 35 36.	+ 57 05	36	16.	GALAXY
ZWG 297.021	15 35 36.	+ 57 06		15.7	GALAXY
ZWG 297.022	15 35 36.	+ 59 33		15.7	GALAXY
MCG+10-22-025	15 35 36.	+ 59 34	39	15.9	GALAXY
IC 1143	15 35 37.	+ 82 37 07.			NONSTELLAR OBJECT
CED 124	15 35 40.	- 29 37			DIFFUSE GALACTIC NEBULA
ARC 2101	15 35 41.	+ 12 27		17.2	RICH CLUSTER OF GALAXIES
ZWG 050.050	15 35 42.	+ 04 57		15.2	GALAXY
TON-N 0797	15 35 42.	+ 29 18		15.2	BLUE STAR
ZWG 249.035	15 35 42.	+ 45 11		15.6	GALAXY
7ZW 615	15 35 42.	+ 59 33			COMPACT GALAXY
ARC 2115	15 35 46.	+ 70 13		17.8	RICH CLUSTER OF GALAXIES
RNGC 5976	15 35 47.	+ 59 34		15.5	GALAXY
IC 4567	15 35 48.	+ 43 26			NONSTELLAR OBJECT
1ZW 122	15 35 48.	+ 49 34			COMPACT GALAXY
1ZW 123	15 35 48.	+ 55 25			COMPACT GALAXY
ZWG 275.002	15 35 48.	+ 55 25		15.2	GALAXY
ZWG 274.049	15 35 48.	+ 55 25		15.2	GALAXY
MRK 487	15 35 48.	+ 55 25	10	16.	GALAXY WITH UV CONTINUUM
ZWG 022.019	15 35 48.	- 00 18		15.7	GALAXY
VHA 176	15 35 48.	- 49 54	180		OPEN STAR CLUSTER
PATH 1.723	15 35 49.	- 00 18	8		NEBULA
LB 00298	15 35 53.	- 14 18 24.		15.0	FAINT BLUE STAR
ZWG 050.051	15 35 54.	+ 03 19		15.6	GALAXY
ZC 1535.9+1640	15 35 54.	+ 16 40	540		CLUSTER OF GALAXIES
ZWG 136.035	15 35 54.	+ 26 25		15.5	GALAXY
ZWG 136.036	15 35 54.	+ 26 26		15.2	GALAXY
MCG+10-22-026	15 35 54.	+ 57 44	18	16.	GALAXY
UGC 09941	15 36 00.	+ 13 07	90	15.7	GALAXY DWARF
ZWG 078.033	15 36 00.	+ 13 07		15.7	GALAXY
MCG+02-40-005	15 36 00.	+ 13 07	90	15.7	GALAXY
UGC 09942	15 36 00.	+ 71 35	60	16.0	GALAXY Sa
ZWG 355.001	15 36 00.	+ 80 12		15.6	GALAXY
ZWG 354.026	15 36 00.	+ 80 12		15.6	GALAXY
MCG+14-07-023	15 36 00.	+ 82 24	51	15.	GALAXY
ZWG 022.020	15 36 00.	- 01 24		15.7	GALAXY
MCG-04-37-002	15 36 00.	- 22 30	72	14.5	GALAXY
ZWG 050.052	15 36 06.	+ 02 53		15.3	GALAXY
ZWG 050.053	15 36 06.	+ 06 49		15.7	GALAXY
ZWG 078.034	15 36 06.	+ 12 21		12.2	GALAXY
UGC 09943	15 36 06.	+ 12 21	174	12.2	GALAXY SBc
MCG+02-40-006	15 36 06.	+ 12 21	192	12.2	GALAXY
MCG+02-40-007	15 36 06.	+ 12 23	36	16.	GALAXY
ZWG 022.022	15 36 06.	- 03 07		15.6	GALAXY
ZWG 022.021	15 36 06.	- 03 13		15.3	GALAXY
RNGC 5970	15 36 07.	+ 12 21		12.5	GALAXY
ZWG 050.054	15 36 12.	+ 06 54		15.7	GALAXY
ZWG 078.035	15 36 12.	+ 10 25		15.5	GALAXY
LB 00871	15 36 16.	+ 58 20 18.		17.8	FAINT BLUE STAR
ZWG 022.023	15 36 18.	+ 01 38		15.6	GALAXY
ZWG 050.055	15 36 18.	+ 07 04		15.3	GALAXY
ZWG 194.012	15 36 18.	+ 33 20		15.2	GALAXY
ZC 1536.3+3941	15 36 18.	+ 39 41	2020		CLUSTER OF GALAXIES
ZWG 338.009	15 36 18.	+ 73 37		14.9	GALAXY
UGC 09944	15 36 18.	+ 73 37	90	14.9	GALAXY S
KARA.73B 0692	15 36 18.	+ 73 37	102	14.9	ISOLATED GALAXY S
PATH 1.724	15 36 21.	+ 00 02	5		NEBULA
ZWG 050.056	15 36 24.	+ 04 45		14.4	GALAXY
UGC 09945	15 36 24.	+ 04 45	78	14.4	GALAXY Sc
MCG+01-40-009	15 36 24.	+ 04 45	78	14.4	GALAXY
ZWG 136.037	15 36 24.	+ 26 22		15.6	GALAXY
ZC 1536.4+6546	15 36 24.	+ 65 46	1080		CLUSTER OF GALAXIES
MCG-01-40-004A	15 36 24.	- 06 43	3	16.	GALAXY
HOLM 717B	15 36 27.	+ 12 13	36	14.2	PART OF MULTIPLE GALAXY
HOLM 717A	15 36 28.	+ 12 14	48	14.0	PART OF MULTIPLE GALAXY
IC 1131	15 36 29.6	+ 12 14 33.			NONSTELLAR OBJECT
ZWG 050.057	15 36 30.	+ 08 15		15.7	GALAXY
ZWG 078.036	15 36 30.	+ 12 14		14.8	GALAXY
MCG+02-40-008	15 36 30.	+ 12 14	30	14.8	GALAXY
ZC 1536.5+2147	15 36 30.	+ 21 47	8330		CLUSTER OF GALAXIES
ZWG 222.038	15 36 30.	+ 42 38		15.7	GALAXY
KARA.73B 0693	15 36 30.	+ 42 38	24	15.7	ISOLATED GALAXY S
ZC 1536.5+4636	15 36 30.	+ 46 36	2820		CLUSTER OF GALAXIES
RNGC 5972	15 36 34.	+ 17 12		15.0	GALAXY
REIZ 4624	15 36 35.	+ 17 11	18	14.5	GALAXY
ZWG 022.024	15 36 36.	+ 02 08		15.6	GALAXY
ZWG 050.058	15 36 36.	+ 05 44		14.9	GALAXY
MCG+01-40-010	15 36 36.	+ 05 44	18	14.9	GALAXY
ZWG 050.059	15 36 36.	+ 06 00		15.2	GALAXY
ZWG 050.060	15 36 36.	+ 07 54		15.3	GALAXY
ZWG 107.017	15 36 36.	+ 16 37		15.3	GALAXY
MCG+03-40-016	15 36 36.	+ 17 11	60	14.8	GALAXY
ZWG 107.018	15 36 36.	+ 17 12		14.8	GALAXY
UGC 09946	15 36 36.	+ 17 12	66	14.8	GALAXY S0-a
ZC 1536.6+4600	15 36 36.	+ 46 00	1010		CLUSTER OF GALAXIES
ZC 1536.6+7015	15 36 36.	+ 70 15	470		CLUSTER OF GALAXIES
HN 1803	15 36 37.0	+ 02 08 15.	18		NEBULA
LB 00299	15 36 39.	+ 02 10 42.		15.2	FAINT BLUE STAR
ZWG 050.061	15 36 42.	+ 08 15		15.7	GALAXY
ZWG 078.037	15 36 42.	+ 10 58		15.5	GALAXY
ZWG 136.038	15 36 42.	+ 21 54		15.7	GALAXY
ZC 1536.7+2355	15 36 42.	+ 23 55	1610		CLUSTER OF GALAXIES
ZWG 166.023	15 36 42.	+ 29 35		15.5	GALAXY
MCG+10-22-027	15 36 42.	+ 59 33	156	13.9	GALAXY
MCG-01-40-005	15 36 42.	- 06 41	66	14.5	GALAXY
MAI 094	15 36 47.	+ 59 11	40		DWARF SPHEROIDAL GALAXY
HOLM 719C	15 36 47.	+ 59 33	132	12.5	PART OF MULTIPLE GALAXY
ZWG 050.062	15 36 48.	+ 05 59		15.5	GALAXY
ZC 1536.8+2132	15 36 48.	+ 21 32	940		CLUSTER OF GALAXIES
ZWG 136.039	15 36 48.	+ 23 57		15.5	GALAXY
ZWG 166.024	15 36 48.	+ 29 35		15.6	GALAXY
ZWG 194.013	15 36 48.	+ 37 07		14.7	GALAXY
MCG+06-34-024	15 36 48.	+ 37 07 30.	33	14.5	GALAXY
MCG+07-32-041	15 36 48.	+ 41 10	54	17.	GALAXY
UGC 09947	15 36 48.	+ 41 12	90	16.5	GALAXY
ZWG 297.023	15 36 48.	+ 59 33		14.2	GALAXY

OBJECT NAME	RIGHT ASCEN.	DECLINATION	DIAM.	MAGN.	TYPE OF OBJECT
UGC 09948	15 36 48.	+ 59 33	168	14.2	GALAXY Sb-c
RNGC 5968	15 36 51.	- 30 24		13.0	GALAXY
REIN 7.155	15 36 51.85	+ 59 33 12.5			NEBULA
REIZ 4625	15 36 53.	+ 17 23	12	15.2	GALAXY
REIZ 4625	15 36 53.	+ 17 23	12	15.2	GALAXY
RNGC 5981	15 36 53.	+ 59 33		14.0	GALAXY
ZWG 078.038	15 36 54.	+ 14 19		15.5	GALAXY
UGC 09949	15 36 54.	+ 14 19	60	15.5	GALAXY Sc
LB 00872	15 36 54.	- 01 34 06.		16.0	FAINT BLUE STAR
ZWG 022.025	15 37 00.	+ 01 25		15.7	GALAXY
MCG+00-40-005	15 37 00.	+ 01 25	36	15.7	GALAXY
ZWG 078.039	15 37 00.	+ 14 08		15.6	GALAXY
MCG+03-40-017	15 37 00.	+ 15 32	90	15.7	GALAXY
ZWG 107.019	15 37 00.	+ 15 33		15.7	GALAXY
UGC 09951	15 37 00.	+ 15 33	84	15.7	GALAXY Sc
ZWG 166.025	15 37 00.	+ 31 55		14.3	GALAXY
RNGC 5974	15 37 00.	+ 31 55		14.5	GALAXY
UGC 09952	15 37 00.	+ 31 55	39	14.3	GALAXY
ZWG 166.026	15 37 00.	+ 32 25		15.6	GALAXY
ZC 1537.0+3345	15 37 00.	+ 33 45	1140		CLUSTER OF GALAXIES
ZC 1537.0+5328	15 37 00.	+ 53 28	1340		CLUSTER OF GALAXIES
MCG+10-22-028	15 37 00.	+ 57 46	27	16.	GALAXY
7ZW 616	15 37 00.	+ 62 36			COMPACT GALAXY
ZWG 366.019	15 37 00.	+ 82 25		15.2	GALAXY
UGC 09950	15 37 00.	+ 82 25	72	15.2	GALAXY Sb-c
MCG-05-37-001	15 37 00.	- 30 24	102	13.	GALAXY
MRSL 326+00/3	15 37 00.	- 54 00	1200		HII REGION
ARC 2105	15 37 01.	+ 33 50		17.2	RICH CLUSTER OF GALAXIES
REIZ 4628	15 37 05.	+ 40 35	60	14.6	GALAXY
ZWG 050.063	15 37 06.	+ 03 22		15.0	GALAXY
UGC 09953	15 37 06.	+ 03 22	84	15.0	GALAXY SBc
MCG+01-40-011	15 37 06.	+ 03 22	60	15.0	GALAXY
ZWG 136.040	15 37 06.	+ 24 37		15.1	GALAXY
UGC 09954	15 37 06.	+ 24 37	66	15.1	GALAXY Sa-b
ZWG 166.027	15 37 06.	+ 31 59		15.6	GALAXY
UGC 09955	15 37 06.	+ 79 48	72	16.5	GALAXY S
MCG+05-37-010	15 37 09.	+ 31 57	36	14.3	GALAXY
ZWG 107.020	15 37 12.	+ 17 35		15.0	GALAXY
ZWG 136.041	15 37 12.	+ 21 55		15.6	GALAXY
MCG+04-37-015	15 37 12.	+ 24 36	24	15.1	GALAXY
LDN 1778	15 37 12.	- 07 00	1140		DARK NEBULA
MCG+03-40-018	15 37 15.	+ 17 36	27	15.0	GALAXY
REIZ 4626	15 37 16.	+ 17 36	24	14.5	GALAXY
ZWG 078.040	15 37 18.	+ 10 23		15.4	GALAXY
UGC 09956	15 37 18.	+ 10 23	60	15.4	GALAXY SBb
ZWG 078.041	15 37 18.	+ 14 12		15.7	GALAXY
ZWG 078.042	15 37 18.	+ 14 16		15.6	GALAXY
MCG+03-40-019	15 37 18.	+ 15 11	30		GALAXY
ZWG 136.042	15 37 18.	+ 25 06		15.1	GALAXY
MCG+06-34-025	15 37 18.	+ 32 33	30	15.	GALAXY
MCG+07-32-042	15 37 18.	+ 44 02 30.	84	14.5	GALAXY
ZC 1537.3+6435	15 37 18.	+ 64 35	2290		CLUSTER OF GALAXIES
ARC 2103	15 37 19.	- 02 01		17.1	RICH CLUSTER OF GALAXIES
MCG+03-40-020	15 37 24.	+ 17 37	42		GALAXY
REIZ 4627	15 37 24.	+ 21 57	18	15.2	GALAXY
ZWG 136.043	15 37 24.	+ 23 22		15.7	GALAXY
MCG+04-37-017	15 37 24.	+ 25 05 30.	30	15.1	GALAXY
MCG+04-37-016	15 37 24.	+ 25 06	24	15.1	GALAXY
UGC 09957	15 37 24.	+ 40 19	60	17.	GALAXY Sc
ZWG 022.026	15 37 24.	- 00 45		15.7	GALAXY
ARC 2104	15 37 27.	- 03 09		17.4	RICH CLUSTER OF GALAXIES
LB 00873	15 37 29.	- 15 52 42.		15.9	FAINT BLUE STAR
ZC 1537.5+1044	15 37 30.	+ 10 44	810		CLUSTER OF GALAXIES
ZWG 078.043	15 37 30.	+ 14 20		14.9	GALAXY
MCG+02-40-009	15 37 30.	+ 14 20	36	14.9	GALAXY
ZC 1537.5+1813	15 37 30.	+ 18 13	3430		CLUSTER OF GALAXIES
MCG+04-37-018	15 37 30.	+ 21 56	72	15.0	GALAXY
ZWG 136.044	15 37 30.	+ 21 57		15.0	GALAXY
UGC 09958	15 37 30.	+ 21 57	72	15.0	GALAXY E-S0
ZWG 136.045	15 37 30.	+ 25 54		15.4	GALAXY
ZWG 222.039	15 37 30.	+ 44 02		14.6	GALAXY
UGC 09959	15 37 30.	+ 44 02	120	14.6	GALAXY SBb
ZC 1537.5-0308	15 37 30.	- 03 08	1410		CLUSTER OF GALAXIES
MCG-03-40-004	15 37 30.	- 18 18 30.	42	15.5	GALAXY
RNGC 5973	15 37 33.	- 08 27			GALAXY
ARC 2107	15 37 34.	+ 21 56		15.7	RICH CLUSTER OF GALAXIES
RNGC 5982	15 37 35.	+ 59 31		12.5	GALAXY
ZWG 022.027	15 37 36.	+ 02 05		15.5	GALAXY
UGC 09960	15 37 36.	+ 02 05	96	15.5	GALAXY Sa
KARA.73B 0694	15 37 36.	+ 02 05	102	15.5	ISOLATED GALAXY S
TON-N 0243	15 37 36.	+ 27 15		16.3	BLUE STAR
ZC 1537.6+3630	15 37 36.	+ 36 30	810		CLUSTER OF GALAXIES
ZWG 297.024	15 37 36.	+ 59 31		12.4	GALAXY
UGC 09961	15 37 36.	+ 59 31	198	12.4	GALAXY E
MCG+10-22-029	15 37 36.	+ 59 31	66	12.4	GALAXY
LB 00874	15 37 36.	- 15 14 00.		16.3	FAINT BLUE STAR
HOLM 719A	15 37 37.	+ 59 31	30	12.4	PART OF MULTIPLE GALAXY
REIN 2.239	15 37 38.38	+ 59 31 01.9			NEBULA
IC 1132	15 37 41.	+ 20 49 24.			NONSTELLAR OBJECT
ARC 2111	15 37 41.	+ 34 34		17.8	RICH CLUSTER OF GALAXIES
ZWG 050.064	15 37 42.	+ 04 22		15.6	GALAXY
ZWG 078.044	15 37 42.	+ 08 40		15.6	GALAXY
ZWG 078.045	15 37 42.	+ 14 07		15.3	GALAXY
UGC 09962	15 37 42.	+ 14 07	60	15.3	GALAXY IRR
MCG+04-37-019	15 37 42.	+ 21 36 30.	63	14.7	GALAXY
ZC 1537.7+3436	15 37 42.	+ 34 36	1280		CLUSTER OF GALAXIES
ARC 2110	15 37 44.	+ 30 52		17.0	RICH CLUSTER OF GALAXIES
MCG+06-34-026	15 37 45.	+ 34 11 30.	27	15.	GALAXY
REIN 7.156B	15 37 47.88	+ 59 33 26.8			NEBULA
REIN 7.156A	15 37 47.88	+ 59 33 29.9			NEBULA
ZWG 078.046	15 37 48.	+ 08 37		15.6	GALAXY
ARC 2108	15 37 48.	+ 18 03		15.7	RICH CLUSTER OF GALAXIES
MCG+04-37-020	15 37 48.	+ 20 50	66	14.4	GALAXY
ZWG 136.046	15 37 48.	+ 21 38		14.7	GALAXY
RNGC 5975	15 37 48.	+ 21 38		14.5	GALAXY
REIZ 4629	15 37 48.	+ 21 38	30	14.0	GALAXY
UGC 09963	15 37 48.	+ 21 38	66	14.7	GALAXY S
ZWG 136.047	15 37 48.	+ 21 59		15.5	GALAXY
ZWG 136.048	15 37 48.	+ 22 02		15.5	GALAXY
LB 00300	15 37 49.	- 16 24 06.		14.2	FAINT BLUE STAR
MCG+03-40-021	15 37 51.	+ 16 13	12		GALAXY
ZC 1537.9+0052	15 37 54.	+ 00 52	2890		CLUSTER OF GALAXIES
ZWG 050.065	15 37 54.	+ 07 27		15.2	GALAXY
UGC 09964	15 37 54.	+ 07 27	78	15.2	GALAXY SB0/SBa
ZWG 136.049	15 37 54.	+ 20 51		14.4	GALAXY
UGC 09965	15 37 54.	+ 20 51	78	14.4	GALAXY Sc
MCG+04-37-021	15 37 54.	+ 21 39 30.	45	15.2	GALAXY
REIZ 4631	15 37 54.	+ 21 41	18	15.3	GALAXY

OBJECT NAME	RIGHT ASCEN.	DECLINATION	DIAM.	MAGN.	TYPE OF OBJECT
TON-N 0244	15 37 54.	+ 28 22		14.8	BLUE STAR
ZC 1537.9+3224	15 37 54.	+ 32 24	940		CLUSTER OF GALAXIES
REIZ 4630	15 37 55.	+ 20 50	48	14.0	GALAXY
LB 00875	15 37 57.	+ 59 10 24.		18.2	FAINT BLUE STAR
HOLM 718B	15 37 59.	+ 21 41	24	15.2	PART OF MULTIPLE GALAXY
LBN 1122	15 38	- 07 00	2700		BRIGHT NEBULA
ZC 1538.0+0317	15 38 00.	+ 03 17	3020		CLUSTER OF GALAXIES
ZC 1538.0+0610	15 38 00.	+ 06 10	1550		CLUSTER OF GALAXIES
ZWG 050.066	15 38 00.	+ 07 38		15.6	GALAXY
ZWG 107.021	15 38 00.	+ 16 08		15.5	GALAXY
HOLM 718A	15 38 00.	+ 21 40	30	14.5	PART OF MULTIPLE GALAXY
ZWG 136.050	15 38 00.	+ 21 41		15.2	GALAXY
ZWG 166.028	15 38 00.	+ 28 19		15.3	GALAXY
ARC 2112	15 38 00.	+ 36 27		17.8	RICH CLUSTER OF GALAXIES
ZWG 367.002	15 38 00.	+ 83 15		15.6	GALAXY
ZWG 366.020	15 38 00.	+ 83 15		15.6	GALAXY
UGC 09966	15 38 00.	+ 83 15	66	15.6	GALAXY SBa
MCG+14-07-024	15 38 00.	+ 83 15	39	16.	GALAXY
LB 00876	15 38 00.	- 00 13 00.		15.2	FAINT BLUE STAR
LDN 1780	15 38 00.	- 07 00	1800		DARK NEBULA
PK324-01.1	15 38 02.	- 56 27	10		PLANETARY NEBULA
IC 4568	15 38 04.	+ 28 18 39.			NONSTELLAR OBJECT
MCG+03-40-022	15 38 06.	+ 16 09	12	15.5	GALAXY
ZC 1538.1+1643	15 38 06.	+ 16 43	870		CLUSTER OF GALAXIES
ZC 1538.1+3055	15 38 06.	+ 30 55	1550		CLUSTER OF GALAXIES
7ZW 617	15 38 06.	+ 61 38			COMPACT GALAXY
ZWG 338.010	15 38 06.	+ 73 25		15.5	GALAXY
OCL 0942	15 38 06.	- 56 28	360		OPEN STAR CLUSTER
PATH 1.725	15 38 09.	- 00 01	41		NEBULA
RNGC 5977	15 38 10.	+ 17 17		15.0	GALAXY
ARC 2109	15 38 11.	+ 06 11		17.4	RICH CLUSTER OF GALAXIES
ZWG 050.067	15 38 12.	+ 07 36		15.4	GALAXY
ZWG 107.022	15 38 12.	+ 16 52		15.6	GALAXY
ZWG 107.023	15 38 12.	+ 17 17		15.1	GALAXY
UGC 09967	15 38 12.	+ 17 17	66	15.1	GALAXY SB0
ZC 1538.2+6959	15 38 12.	+ 69 59	2220		CLUSTER OF GALAXIES
MCG+07-32-043	15 38 15.	+ 44 29	45	15.	GALAXY
MCG+03-40-023	15 38 18.	+ 17 18	66	15.1	GALAXY
LB 00301	15 38 22.	+ 01 21 30.		15.8	FAINT BLUE STAR
ZWG 050.068	15 38 24.	+ 05 18		15.5	GALAXY
MCG+04-37-022	15 38 24.	+ 20 43	48	15.3	GALAXY
ZWG 136.051	15 38 24.	+ 24 43		15.6	GALAXY
ZWG 166.029	15 38 24.	+ 26 31		15.7	GALAXY
TON-N 0245	15 38 24.	+ 26 57		13.2	BLUE STAR
ZWG 166.030	15 38 24.	+ 28 31		15.2	GALAXY
ZWG 166.031	15 38 24.	+ 28 41		15.6	GALAXY
MCG+08-28-042	15 38 24.	+ 46 06	60	15.	GALAXY
ZWG 338.011	15 38 24.	+ 72 43		15.4	GALAXY
VHA 177	15 38 24.	- 56 34	360		OPEN STAR CLUSTER
ZWG 078.047	15 38 30.	+ 09 55		14.6	GALAXY
MCG+02-40-010	15 38 30.	+ 09 55	48	14.6	GALAXY
KARA.73B 0695	15 38 30.	+ 09 55	66	14.6	ISOLATED GALAXY S
ZC 1538.5+1609	15 38 30.	+ 16 09	3230		CLUSTER OF GALAXIES
ZWG 136.052	15 38 30.	+ 20 44		15.3	GALAXY
MCG+05-37-012	15 38 30.	+ 28 31	24	15.2	GALAXY
MCG+05-37-011	15 38 30.	+ 28 40	24	15.6	GALAXY
ZWG 194.014	15 38 30.	+ 35 05		15.7	GALAXY
MRK 488	15 38 30.	+ 35 05	14	16.5	GALAXY WITH UV CONTINUUM
ZWG 250.001	15 38 30.	+ 46 07		15.6	GALAXY
MCG+14-07-025	15 38 30.	+ 83 10	30	17.	GALAXY
MCG+03-40-024	15 38 33.	+ 17 19	36		GALAXY
RNGC 5985	15 38 35.	+ 59 30		12.0	GALAXY
ZWG 136.053	15 38 36.	+ 21 28		15.7	GALAXY
MCG+09-26-005	15 38 36.	+ 51 16	72	15.	GALAXY
HOLM 719B	15 38 36.	+ 59 29	240	11.4	PART OF MULTIPLE GALAXY
MCG+10-22-030	15 38 36.	+ 59 29	324	11.9	GALAXY
ZWG 297.025	15 38 36.	+ 59 30		12.0	GALAXY
UGC 09969	15 38 36.	+ 59 30	348	12.0	GALAXY Sb
ZWG 022.028	15 38 36.	- 01 33		15.2	GALAXY
UGC 09968	15 38 36.	- 01 33	84	15.2	GALAXY Sb
MCG+00-40-006	15 38 36.	- 01 33	48	15.2	GALAXY
REIN 2.240	15 38 36.03	+ 59 29 31.9			NEBULA
REIZ 4633	15 38 39.	+ 38 18	48	14.5	GALAXY
MCG+05-37-013	15 38 42.	+ 28 27	30	15.0	GALAXY
IC 4569	15 38 42.	+ 28 27 07.			NONSTELLAR OBJECT
ZWG 166.032	15 38 42.	+ 28 28		15.0	GALAXY
ZWG 166.033	15 38 42.	+ 32 10		15.6	GALAXY
MCG+06-34-027	15 38 42.	+ 35 56	30	15.	GALAXY
ARC 2114	15 38 45.	+ 43 11		17.8	RICH CLUSTER OF GALAXIES
RNGC 5987	15 38 45.	+ 58 15		13.5	GALAXY
ZWG 050.069	15 38 48.	+ 04 14		15.5	GALAXY
UGC 09970	15 38 48.	+ 51 15	72	16.5	GALAXY Sc
MCG+10-22-032	15 38 48.	+ 58 14	234	12.	GALAXY
ZWG 297.026	15 38 48.	+ 58 15		13.3	GALAXY
UGC 09971	15 38 48.	+ 58 15	306	13.3	GALAXY Sb
ZWG 297.027	15 38 48.	+ 59 59		15.6	GALAXY
MCG+10-22-031	15 38 48.	+ 59 59	57	15.	GALAXY
UGC 09972	15 38 48.	+ 59 59	60	15.6	GALAXY SBa
ZC 1538.8+7623	15 38 48.	+ 76 23	1010		CLUSTER OF GALAXIES
REIN 7.158	15 38 49.41	+ 59 58 31.0			NEBULA
MCG+03-40-025	15 38 51.	+ 15 44	66	14.8	GALAXY
REIN 7.157	15 38 51.77	+ 58 14 24.3			NEBULA
IC 1133	15 38 52.	+ 15 44 01.			NONSTELLAR OBJECT
ZWG 107.024	15 38 54.	+ 15 44		14.8	GALAXY
UGC 09973	15 38 54.	+ 15 44	78	14.8	GALAXY Sc
MCG+06-34-028	15 38 54.	+ 32 56	48	14.5	GALAXY
ZWG 194.015	15 38 54.	+ 32 57		15.3	GALAXY
ARC 2127	15 38 54.	+ 76 26		17.9	RICH CLUSTER OF GALAXIES
HOLM 720B	15 38 59.	+ 15 56	36	14.5	PART OF MULTIPLE GALAXY
ARC 2121	15 38 59.	+ 69 58		17.4	RICH CLUSTER OF GALAXIES
VDB.66G 201	15 39	+ 00 37	70		DWARF GALAXY
ZWG 078.048	15 39 00.	+ 12 08		15.7	GALAXY
REIZ 4632	15 39 00.	+ 15 56	18	15.6	GALAXY
ZC 1539.0+2114	15 39 00.	+ 21 14	540		CLUSTER OF GALAXIES
ZWG 166.034	15 39 00.	+ 32 23		15.7	GALAXY
ZWG 194.016	15 39 00.	+ 32 50		15.6	GALAXY
ZWG 222.040	15 39 00.	+ 43 55		15.5	GALAXY
ZC 1539.0+5155	15 39 00.	+ 51 55	1480		CLUSTER OF GALAXIES
LB 00877	15 39 05.	- 01 44 12.		15.8	FAINT BLUE STAR
HOLM 720A	15 39 06.	+ 15 56	114	12.9	PART OF MULTIPLE GALAXY
ZWG 107.025	15 39 06.	+ 15 57		13.3	GALAXY
UGC 09974	15 39 06.	+ 15 57	114	13.3	GALAXY S
ZWG 136.054	15 39 06.	+ 22 11		15.5	GALAXY
ZC 1539.1+2820	15 39 06.	+ 23 20	6380		CLUSTER OF GALAXIES
MCG+09-26-006	15 39 06.	+ 53 41	54	15.	GALAXY
MCG+09-26-007	15 39 06.	+ 53 42	6	20.	GALAXY
ARC 2113	15 39 07.	+ 04 50		17.1	RICH CLUSTER OF GALAXIES
RNGC 5980	15 39 09.	+ 15 57		13.5	GALAXY

OBJECT NAME	RIGHT ASCEN.	DECLINATION	DIAM.	MAGN.	TYPE OF OBJECT
MCG+03-40-026	15 39 09.	+ 15 57	114	13.3	GALAXY
MCG+09-26-008	15 39 09.	+ 53 40 30.	15	17.	GALAXY
ZC 1539.2+1448	15 39 12.	+ 14 48	540		CLUSTER OF GALAXIES
REIZ 4634	15 39 12.	+ 15 56	90	13.4	GALAXY
ZWG 166.035	15 39 12.	+ 28 24		15.1	GALAXY
UGC 09975	15 39 12.	+ 28 24	60	15.1	GALAXY Sc
MCG+09-26-009	15 39 12.	+ 53 40	9	18.	GALAXY
LB 00878	15 39 16.	- 00 58 36.		17.2	FAINT BLUE STAR
ZWG 050.070	15 39 18.	+ 06 46		15.6	GALAXY
ZWG 166.036	15 39 18.	+ 28 09		15.2	GALAXY
IC 4570	15 39 18.	+ 28 23 23.			NONSTELLAR OBJECT
MCG+05-37-014	15 39 18.	+ 28 23 30.	42	15.1	GALAXY
MCG+05-37-014	15 39 18.	+ 28 23 30.	42	15.1	GALAXY
MCG+10-22-033	15 39 18.	+ 59 46	27	16.	GALAXY
REIN 7.159A	15 39 19.42	+ 59 30 26.9			NEBULA
REIN 7.159B	15 39 19.42	+ 59 30 27.4			NEBULA
HN 1804	15 39 22.3	- 01 09 42.	18		NEBULA
ZWG 050.071	15 39 24.	+ 06 00		15.5	GALAXY
UGC 09976	15 39 24.	+ 06 00	66	15.5	GALAXY SBb
ZWG 050.072	15 39 24.	+ 08 02		15.4	GALAXY
ZWG 078.049	15 39 24.	+ 10 13		15.4	GALAXY
TON-N 0246	15 39 24.	+ 25 32		14.9	BLUE STAR
MCG+05-37-015	15 39 24.	+ 28 08 30.	18	15.2	GALAXY
MCG+05-37-015	15 39 24.	+ 28 08 30.	18	15.2	GALAXY
ZC 1539.4+3517	15 39 24.	+ 35 17	9010		CLUSTER OF GALAXIES
HN 1805	15 39 27.4	+ 00 51 49.	120		NEBULA
ZWG 022.029	15 39 30.	+ 00 52		14.7	GALAXY
UGC 09977	15 39 30.	+ 00 52	234	14.7	GALAXY Sc
MCG+00-40-007	15 39 30.	+ 00 52	240	14.7	GALAXY
ZWG 136.055	15 39 30.	+ 25 16		15.7	GALAXY
MCG+09-26-010	15 39 30.	+ 55 20	18	16.	GALAXY
LB 00879	15 39 32.	- 02 02 54.		16.0	FAINT BLUE STAR
LB 00880	15 39 35.	+ 60 26 48.		18.0	FAINT BLUE STAR
ZWG 050.073	15 39 36.	+ 02 50		15.5	GALAXY
ZWG 107.026	15 39 36.	+ 17 00		15.3	GALAXY
MCG+03-40-027	15 39 36.	+ 17 00	42	15.3	GALAXY
MCG+03-40-028	15 39 36.	+ 19 15	48		GALAXY
ZWG 136.056	15 39 36.	+ 23 22		15.5	GALAXY
ZC 1539.6+6130	15 39 36.	+ 61 30	270		CLUSTER OF GALAXIES
ZC 1539.6+6623	15 39 36.	+ 66 23	1280		CLUSTER OF GALAXIES
MCG-02-40-002	15 39 36.	- 13 04 30.	42	14.5	GALAXY
RNGC 5978	15 39 37.	- 13 04		14.0	GALAXY
ZWG 050.074	15 39 42.	+ 06 49		15.2	GALAXY
UGC 09978	15 39 42.	+ 06 49	78	15.2	GALAXY Sc
ZWG 022.030	15 39 42.	- 01 04		15.7	GALAXY
KARA.73B 0696	15 39 42.	- 01 04	42	15.7	ISOLATED GALAXY S
LB 00881	15 39 42.	- 01 44 48.		15.3	FAINT BLUE STAR
MRSL 326+00/2	15 39 42.	- 53 47	180		HII REGION
ZWG 022.031	15 39 48.	+ 00 37		15.7	GALAXY
UGC 09979	15 39 48.	+ 00 37	90	15.7	GALAXY DWRF IR
MCG+00-40-008	15 39 48.	+ 00 37	72	15.7	GALAXY
ZWG 050.075	15 39 48.	+ 02 37		15.2	GALAXY
MCG+05-37-016	15 39 48.	+ 28 17	54	15.0	GALAXY
ZWG 166.037	15 39 48.	+ 28 18		15.0	GALAXY
MCG+09-26-011	15 39 48.	+ 55 09	24	17.	GALAXY
IC 4572	15 39 51.	+ 28 17 38.			NONSTELLAR OBJECT
ZWG 050.076	15 39 54.	+ 07 59		15.7	GALAXY
ZWG 166.038	15 39 54.	+ 28 25		15.7	GALAXY
IC 4574	15 39 55.	+ 28 24 09.			NONSTELLAR OBJECT
IC 4573	15 39 59.	+ 24 56 45.			NONSTELLAR OBJECT
LB 09986	15 40	- 87 56		13.6	FAINT BLUE STAR
ZWG 022.032	15 40 00.	+ 02 10		15.2	GALAXY
UGC 09980	15 40 00.	+ 02 10	66	15.2	GALAXY SB0
KARA.73B 0697	15 40 00.	+ 02 10	60	15.2	ISOLATED GALAXY S
ZWG 050.077	15 40 00.	+ 07 32		15.6	GALAXY
SN 1950L	15 40 00.	+ 22 07		19.9	SUPERNOVA
ZC 1540.0+2230	15 40 00.	+ 22 30	1680		CLUSTER OF GALAXIES
ZWG 136.057	15 40 00.	+ 23 58		15.6	GALAXY
ZC 1540.0+3653	15 40 00.	+ 36 53	740		CLUSTER OF GALAXIES
ZC 1540.0+6240	15 40 00.	+ 62 40	1410		CLUSTER OF GALAXIES
HN 1806	15 40 01.9	+ 02 09 51.	18		NEBULA
ZWG 078.050	15 40 06.	+ 12 26		15.7	GALAXY
ZWG 136.058	15 40 06.	+ 23 59		15.5	GALAXY
ZWG 166.039	15 40 06.	+ 27 46		15.6	GALAXY
TON-N 0247	15 40 06.	+ 30 37		14.9	BLUE STAR
MCG+12-15-011	15 40 06.	+ 70 55	120	15.	GALAXY
IC 4575	15 40 08.	+ 23 57 16.			NONSTELLAR OBJECT
MCG+08-29-003	15 40 12.	+ 45 42	54	16.	GALAXY
ZWG 250.002	15 40 12.	+ 45 43		15.5	GALAXY
UGC 09981	15 40 12.	+ 45 43	60	15.5	GALAXY S0
MCG+08-29-002	15 40 12.	+ 45 44	48	16.	GALAXY
MCG+08-29-001	15 40 12.	+ 48 42	36	16.	GALAXY
ZWG 338.012	15 40 12.	+ 70 55		15.3	GALAXY
UGC 09982	15 40 12.	+ 70 55	72	15.3	GALAXY Sc
RNGC 5983	15 40 17.	+ 08 24		15.0	GALAXY
LB 00882	15 40 17.	- 00 29 36.		15.9	FAINT BLUE STAR
ZWG 050.078	15 40 18.	+ 07 58		15.5	GALAXY
ZC 1540.3+0803	15 40 18.	+ 08 03	3020		CLUSTER OF GALAXIES
ZWG 050.079	15 40 18.	+ 08 24		15.1	GALAXY
UGC 09983	15 40 18.	+ 08 24	60	15.1	GALAXY E
MCG+01-40-012	15 40 18.	+ 08 24	15	15.1	GALAXY
ZC 1540.3+5557	15 40 18.	+ 55 57	8740		CLUSTER OF GALAXIES
LB 00883	15 40 19.	+ 02 05 06.		15.0	FAINT BLUE STAR
MCG+07-32-044	15 40 21.	+ 40 08	45	14.5	GALAXY
ZWG 107.027	15 40 24.	+ 17 38		15.7	GALAXY
ZWG 136.059	15 40 24.	+ 23 50		14.8	GALAXY
IC 4576	15 40 24.	+ 23 50 29.			NONSTELLAR OBJECT
MCG+08-29-004	15 40 24.	+ 48 42	30	15.	GALAXY
MCG+10-22-034	15 40 24.	+ 59 55	51	13.	GALAXY
MCG-02-40-003	15 40 24.	- 12 24 30.	48	15.	GALAXY
ARC 2125	15 40 27.	+ 66 29		17.6	RICH CLUSTER OF GALAXIES
LB 00884	15 40 28.	- 14 01 30.		16.6	FAINT BLUE STAR
REIZ 4635	15 40 29.	+ 40 08	36	14.5	GALAXY
LB 00885	15 40 29.	+ 59 16 18.		18.2	FAINT BLUE STAR
LB 00886	15 40 29.	- 15 13 36.		16.3	FAINT BLUE STAR
ZWG 078.051	15 40 30.	+ 09 23		15.6	GALAXY
ZWG 136.060	15 40 30.	+ 23 02		15.3	GALAXY
UGC 09984	15 40 30.	+ 23 02	60	15.3	GALAXY SBb
ZWG 222.041	15 40 30.	+ 40 09		14.7	GALAXY
ZWG 250.003	15 40 30.	+ 48 43		15.6	GALAXY
ZWG 297.028	15 40 30.	+ 59 55		13.6	GALAXY
RNGC 5989	15 40 30.	+ 59 55		13.5	GALAXY
UGC 09985	15 40 30.	+ 59 55	60	13.6	GALAXY Sc
ZWG 338.013	15 40 30.	+ 68 58		15.6	GALAXY
UGC 09986	15 40 30.	+ 68 58	60	15.6	GALAXY Sa-b
DV.56 N5967A	15 40 30.	- 75 38	84		S GALAXY
RNGC 5967A	15 40 31.	- 75 38			GALAXY
IC 4577	15 40 32.	+ 24 56 18.			NONSTELLAR OBJECT
MCG+04-37-023	15 40 33.	+ 23 01 30.	60	15.3	GALAXY
ARC 2116	15 40 33.	+ 43 00		17.8	RICH CLUSTER OF GALAXIES
BC M1540+111	15 40 33.4	+ 11 03 58.		20.	QUASI-STELLAR OBJECT
S4B 315	15 40 33.4	+ 11 03 58.		20.	QUASI-STELLAR OBJECT
REIN 7.160B	15 40 33.64	+ 59 54 46.9			NEBULA
REIN 7.160A	15 40 33.64	+ 59 54 49.1			NEBULA
ZWG 022.033	15 40 36.	+ 01 28		15.6	GALAXY
ZWG 050.080	15 40 36.	+ 06 02		15.4	GALAXY
ZWG 078.052	15 40 36.	+ 14 23		13.5	GALAXY
UGC 09987	15 40 36.	+ 14 23	174	13.5	GALAXY SBc
MCG+02-40-011	15 40 36.	+ 14 23	180	13.5	GALAXY
ZWG 136.061	15 40 36.	+ 23 57		15.4	GALAXY
ZWG 136.062	15 40 36.	+ 25 08		15.1	GALAXY
UGC 09988	15 40 36.	+ 41 48	60	17.	GALAXY Sc
ARC 2117	15 40 36.	+ 43 56		17.8	RICH CLUSTER OF GALAXIES
ZC 1540.6+4815	15 40 36.	+ 48 15	810		CLUSTER OF GALAXIES
RNGC 5984	15 40 38.	+ 14 23		13.0	GALAXY
MCG+04-37-024	15 40 39.	+ 25 08	45	15.1	GALAXY
LB 00887	15 40 39.	- 00 11 18.		17.2	FAINT BLUE STAR
IC 4579	15 40 40.	+ 23 55 07.			NONSTELLAR OBJECT
ZWG 136.063	15 40 42.	+ 23 56		15.5	GALAXY
MCG+08-29-005	15 40 42.	+ 48 22 30.	30	15.	GALAXY
MCG+12-15-012	15 40 42.	+ 69 58	45	16.	GALAXY
MCG-07-32-001	15 40 42.	- 41 05	72	15.5	GALAXY
LB 00888	15 40 43.	- 00 53 06.		15.8	FAINT BLUE STAR
LB 00889	15 40 47.	- 13 25 12.		16.4	FAINT BLUE STAR
ZC 1540.8+0554	15 40 48.	+ 05 54	2420		CLUSTER OF GALAXIES
ZWG 107.028	15 40 48.	+ 15 55		15.4	GALAXY
KARA.73B 0698	15 40 48.	+ 15 55	24	15.4	ISOLATED GALAXY S
MCG+03-40-029	15 40 48.	+ 15 56	27	15.4	GALAXY
ZC 1540.8+2320	15 40 48.	+ 23 20	940		CLUSTER OF GALAXIES
ZWG 166.040	15 40 48.	+ 29 05		15.7	GALAXY
ARC 2118	15 40 48.	+ 41 59		17.1	RICH CLUSTER OF GALAXIES
ZWG 250.004	15 40 49.	+ 48 22		15.7	GALAXY
LB 00890	15 40 51.	+ 76 25 00.		18.1	FAINT BLUE STAR
ZWG 078.053	15 40 54.	+ 09 53		15.3	GALAXY
UGC 09989	15 40 54.	+ 09 53	60	15.3	GALAXY S?
ZC 1540.9+4200	15 40 54.	+ 42 00	1410		CLUSTER OF GALAXIES
ZC 1540.9+4758	15 40 54.	+ 47 58	6050		CLUSTER OF GALAXIES
ZC 1540.9+6427	15 40 54.	+ 64 27	1140		CLUSTER OF GALAXIES
B 228	15 41	- 34 21	14400		DARK OBJECT
LB 09896	15 41	- 82 38		14.3	FAINT BLUE STAR
ZWG 050.081	15 41 00.	+ 04 57		15.3	GALAXY
UGC 09990	15 41 00.	+ 04 57	84	15.3	GALAXY Sc
ZWG 050.082	15 41 00.	+ 08 04		15.5	GALAXY
ZWG 078.054	15 41 00.	+ 09 09		15.5	GALAXY
ZWG 078.055	15 41 00.	+ 11 40		15.5	GALAXY
MCG+03-40-030	15 41 00.	+ 14 35	102	15.1	GALAXY
ZWG 107.029	15 41 00.	+ 14 36		15.1	GALAXY
UGC 09991	15 41 00.	+ 14 36	102	15.1	GALAXY Sc
ZWG 366.021	15 41 00.	+ 82 32		15.7	GALAXY
MRSL 326+00/1	15 41 00.	- 54 03	720		HII REGION
ZWG 050.083	15 41 06.	+ 05 01		15.1	GALAXY
ZWG 166.041	15 41 06.	+ 28 31		15.4	GALAXY
ZC 1541.1+6313	15 41 06.	+ 63 13	1280		CLUSTER OF GALAXIES
IC 4580	15 41 11.	+ 28 31 47.			NONSTELLAR OBJECT
ZWG 050.084	15 41 12.	+ 08 19		15.7	GALAXY
ZWG 107.030	15 41 12.	+ 17 28		15.5	GALAXY
MCG+03-40-031	15 41 12.	+ 17 28	27	15.5	GALAXY
LB 00891	15 41 14.	+ 75 47 30.		17.6	FAINT BLUE STAR
HN 0423	15 41 14.	- 77 31			NEBULA
IC 4555	15 41 14.	- 77 31			NONSTELLAR OBJECT
ZC 1541.3+1825	15 41 18.	+ 18 25	2020		CLUSTER OF GALAXIES
TON-N 0798	15 41 18.	+ 30 30		16.5	BLUE STAR
ZWG 136.064	15 41 24.	+ 22 55		15.1	GALAXY
ZWG 166.042	15 41 24.	+ 29 57		15.7	GALAXY
ZWG 319.021	15 41 24.	+ 67 25		15.7	GALAXY
UGC 09992	15 41 24.	+ 67 25	108	15.5	GALAXY DWRF IR
KARA.73B 0699	15 41 24.	+ 67 25	102	15.7	ISOLATED GALAXY IR
MCG+11-19-011	15 41 24.	+ 67 26	84	15.	GALAXY
ZC 1541.4-0141	15 41 24.	- 01 41	2760		CLUSTER OF GALAXIES
LB 00892	15 41 25.	- 01 15 18.		16.5	FAINT BLUE STAR
MCG+04-37-025	15 41 27.	+ 22 53 30.	24	15.1	GALAXY
LB 00893	15 41 29.	+ 60 03 06.		16.8	FAINT BLUE STAR
ZWG 050.085	15 41 30.	+ 06 37		15.5	GALAXY
ZWG 050.086	15 41 30.	+ 08 00		15.5	GALAXY
MCG+14-07-026	15 41 30.	+ 82 32	6	16.	GALAXY
WRAY 19.41	15 41 32.9	- 46 08 59.			DIFFUSE NEBULA
ZWG 050.087	15 41 36.	+ 04 56		15.4	GALAXY
ZWG 050.088	15 41 36.	+ 08 26		15.6	GALAXY
ZC 1541.6+2441	15 41 36.	+ 24 41	810		CLUSTER OF GALAXIES
ZWG 166.043	15 41 36.	+ 28 41		15.5	GALAXY
TON-N 0799	15 41 36.	+ 30 32		15.6	BLUE STAR
MCG+08-29-006	15 41 36.	+ 47 55	60	15.	GALAXY
7ZW 124	15 41 36.	+ 51 24			COMPACT GALAXY
MCG+05-37-017	15 41 39.	+ 28 34	30	15.1	GALAXY
MCG+05-37-018	15 41 39.	+ 28 40	30	15.5	GALAXY
UGC 09993	15 41 42.	+ 08 00	66	16.5	GALAXY
ZWG 107.031	15 41 42.	+ 19 22		15.5	GALAXY
ZWG 166.044	15 41 42.	+ 28 35		15.1	GALAXY
UGC 09994	15 41 42.	+ 33 27	66	16.5	GALAXY Sc
ZWG 222.042	15 41 42.	+ 44 16		15.5	GALAXY
MCG+07-32-045	15 41 42.	+ 44 17 30.	30	15.5	GALAXY
ZWG 250.005	15 41 42.	+ 47 54		15.4	GALAXY
UGC 09995	15 41 42.	+ 47 54	60	15.4	GALAXY Sa-b
MCG+03-40-032	15 41 45.	+ 19 24	12	15.5	GALAXY
LB 00302	15 41 46.	- 00 58 12.		14.6	FAINT BLUE STAR
ZWG 078.056	15 41 48.	+ 11 25		15.7	GALAXY
ZC 1541.8+1804	15 41 48.	+ 18 04	810		CLUSTER OF GALAXIES
RNGC 5967	15 41 50.	- 75 30		13.0	GALAXY
LB 00303	15 41 52.	- 00 17 36.		14.9	FAINT BLUE STAR
ZWG 050.089	15 41 54.	+ 08 27		15.6	GALAXY
ZWG 166.045	15 41 54.	+ 27 52		15.3	GALAXY
MCG+05-37-019	15 41 54.	+ 28 25	48	15.3	GALAXY
ZWG 166.046	15 41 54.	+ 28 26		15.3	GALAXY
ZWG 166.047	15 41 54.	+ 28 41		15.6	GALAXY
MCG+06-34-029	15 41 54.	+ 34 39	9	16.	GALAXY
ZWG 222.043	15 41 54.	+ 44 04		15.6	GALAXY
ZC 1541.9+7054	15 41 54.	+ 70 54	940		CLUSTER OF GALAXIES
IC 4581	15 41 58.	+ 28 27 04.			NONSTELLAR OBJECT
ARC 2120	15 41 58.	+ 34 40		17.8	RICH CLUSTER OF GALAXIES
LB 09897	15 42	- 85 19		14.0	FAINT BLUE STAR
MCG+03-40-033	15 42 00.	+ 18 58	27		GALAXY
ZWG 166.048	15 42 00.	+ 27 37		15.4	GALAXY
MCG+07-32-046	15 42 00.	+ 43 57	36	15.	GALAXY
MCG+07-32-047	15 42 00.	+ 43 58	48	15.5	GALAXY
7ZW 618	15 42 00.	+ 78 25			COMPACT GALAXY
ZWG 078.057	15 42 06.	+ 11 42		15.3	GALAXY

OBJECT NAME	RIGHT ASCEN.	DECLINATION	DIAM.	MAGN.	TYPE OF OBJECT
UGC 09996	15 42 06.	+ 11 42	102	15.3	GALAXY Sb
ZC 1542.1+3620	15 42 06.	+ 36 20	940		CLUSTER OF GALAXIES
ZWG 222.044	15 42 06.	+ 43 56		15.3	GALAXY
ZWG 222.045	15 42 06.	+ 43 57		15.6	GALAXY
UGC 09997	15 42 06.	+ 43 57	60	15.6	GALAXY Sb-c
ZWG 078.058	15 42 12.	+ 10 27		15.3	GALAXY
RNGC 5988	15 42 12.	+ 10 27		15.5	GALAXY
UGC 09998	15 42 12.	+ 10 27	78	15.3	GALAXY Sc
MCG+02-40-012	15 42 12.	+ 10 27	72	15.3	GALAXY
ZWG 136.065	15 42 12.	+ 25 29		14.7	GALAXY
UGC 09999	15 42 12.	+ 25 29	84	14.7	GALAXY S0
MCG+04-37-026	15 42 12.	+ 25 29	21	14.7	GALAXY
MCG+07-32-048	15 42 12.	+ 44 30	84	15.	GALAXY
ZC 1542.2+6837	15 42 12.	+ 68 37	340		CLUSTER OF GALAXIES
MRSL 332+07/1	15 42 12.	- 44 49	1800		HII REGION
ZWG 050.090	15 42 18.	+ 04 06		15.7	GALAXY
UGC 10000	15 42 18.	+ 04 06	72	15.7	GALAXY Sc
TON-N 0800	15 42 18.	+ 27 54		16.1	BLUE STAR
ZWG 222.046	15 42 18.	+ 44 28		15.4	GALAXY
UGC 10001	15 42 18.	+ 44 28	90	15.4	GALAXY S
ZWG 275.003	15 42 18.	+ 53 44		15.3	GALAXY
MCG+09-26-012	15 42 18.	+ 53 45	48	15.	GALAXY
HN 1807	15 42 18.7	+ 00 53 17.	48		NEBULA
MCG+06-35-001	15 42 21.	+ 34 52	18	15.	GALAXY
ZWG 078.059	15 42 24.	+ 09 10		15.5	GALAXY
ZWG 078.060	15 42 24.	+ 11 39		15.2	GALAXY
TON-N 0801	15 42 24.	+ 28 37		16.4	BLUE STAR
PK322-05.1	15 42 27.	- 61 03 47.	25		PLANETARY NEBULA
RNGC 5979	15 42 29.	- 61 02			PLANETARY NEBULA
ZWG 050.091	15 42 30.	+ 02 37		15.2	GALAXY
ZC 1542.5+0939	15 42 30.	+ 09 39	1280		CLUSTER OF GALAXIES
ZC 1542.5+3006	15 42 30.	+ 30 06	740		CLUSTER OF GALAXIES
MCG+07-32-049	15 42 30.	+ 41 15	45	14.5	GALAXY
MCG+08-29-007	15 42 30.	+ 47 28	60	16.	GALAXY
UGC 10002	15 42 30.	+ 57 25	72	16.0	GALAXY SB?
7ZW 619	15 42 30.	+ 67 33			COMPACT GALAXY
PK319-09.1	15 42 30.	- 66 19 28.	10	15.0	PLANETARY NEBULA
ARC 2119	15 42 34.	+ 09 41		17.2	RICH CLUSTER OF GALAXIES
ARC 2122	15 42 35.	+ 36 18		16.6	RICH CLUSTER OF GALAXIES
ZWG 222.047	15 42 36.	+ 41 15		14.2	GALAXY
MRK 489	15 42 36.	+ 41 15	12	14.5	GALAXY WITH UV CONTINUUM
UGC 10003	15 42 36.	+ 41 15	60	14.2	GALAXY S
KARA.72 471A	15 42 36.	+ 41 15	54	14.2	PART OF DOUBLE GALAXY
UGC 10004	15 42 36.	+ 47 26	66	17.	GALAXY Sc
LB 00304	15 42 37.	+ 01 41 54.		15.4	FAINT BLUE STAR
RNGC 5992	15 42 37.	+ 41 15		14.0	GALAXY
MCG+03-40-034	15 42 39.	+ 17 07	24	15.2	GALAXY
REIZ 4636	15 42 39.	+ 41 17	12	14.2	GALAXY
IC 1134	15 42 40.	+ 17 08 01.			NONSTELLAR OBJECT
REIN 7.161	15 42 41.77	+ 57 23 21.0			NEBULA
ZWG 022.034	15 42 42.	+ 00 55		15.7	GALAXY
UGC 10005	15 42 42.	+ 00 55	96	15.7	GALAXY Sc?
MCG+00-40-009	15 42 42.	+ 00 55	48	15.7	GALAXY
KARA.73B 0700	15 42 42.	+ 00 55	30	15.7	ISOLATED GALAXY S
MCG+03-40-035	15 42 42.	+ 17 04	30		GALAXY
ZWG 107.032	15 42 42.	+ 17 07		15.2	GALAXY
ZWG 136.066	15 42 42.	+ 23 01		15.5	GALAXY
MCG+04-37-027	15 42 42.	+ 23 01	42	15.5	GALAXY
UGC 10006	15 42 42.	+ 32 36	66	16.0	GALAXY Sb-c
ZWG 222.048	15 42 42.	+ 41 17		13.9	GALAXY
UGC 10007	15 42 42.	+ 41 17	72	13.9	GALAXY SB:b
KARA.72 471B	15 42 42.	+ 41 17	66	13.9	PART OF DOUBLE GALAXY
MCG+07-32-050	15 42 42.	+ 41 17	72	14.	GALAXY
MCG+10-22-035	15 42 42.	+ 57 23	45	16.	GALAXY
RNGC 5993	15 42 43.	+ 41 17		14.0	GALAXY
REIZ 4638	15 42 45.	+ 41 18	24	14.1	GALAXY
RNGC 5986	15 42 46.	- 37 37		8.5	GLOBULAR CLUSTER
ZWG 050.092	15 42 48.	+ 05 19		15.5	GALAXY
ZWG 222.049	15 42 48.	+ 38 35		15.6	GALAXY
KARA.73B 0701	15 42 48.	+ 38 35	12	15.6	ISOLATED GALAXY S0
ZC 1542.8+5142	15 42 48.	+ 51 42	2290		CLUSTER OF GALAXIES
GCL 037	15 42 48.	- 37 37	300	8.72	GLOBULAR STAR CLUSTER
BC 4C37.45	15 42 53.6	+ 37 22 41.		17.7	QUASI-STELLAR OBJECT
ZC 1542.9+2325	15 42 54.	+ 23 25	1950		CLUSTER OF GALAXIES
SHB 316	15 42 54.	+ 37 22 41.		18.	QUASI-STELLAR OBJECT
ZWG 250.006	15 42 54.	+ 47 10		15.3	GALAXY
MCG+08-29-008	15 42 54.	+ 47 10	54	15.	GALAXY
MCG+06-35-002	15 42 57.	+ 36 40	18	16.	GALAXY
LB 00305	15 42 58.	+ 00 18 48.		15.1	FAINT BLUE STAR
KHAV 233	15 43	- 34 27	7020		DARK NEBULA
LB 09898	15 43	- 82 37		14.5	FAINT BLUE STAR
ZWG 107.033	15 43 00.	+ 17 29		15.6	GALAXY
ZC 1543.0+2447	15 43 00.	+ 24 47	670		CLUSTER OF GALAXIES
ZWG 166.049	15 43 00.	+ 26 40		15.4	GALAXY
MCG+06-35-003	15 43 00.	+ 36 18	48	14.5	GALAXY
ZWG 367.003	15 43 00.	+ 82 30		15.7	GALAXY
ZWG 366.022	15 43 00.	+ 82 30		15.7	GALAXY
UGC 10008	15 43 00.	+ 82 30	66	15.7	GALAXY Sc
ARC 2124	15 43 05.	+ 36 14		15.6	RICH CLUSTER OF GALAXIES
UGC 10009	15 43 06.	+ 04 19	66	17.	GALAXY DWARF
ZC 1543.1+1203	15 43 06.	+ 12 03	2350		CLUSTER OF GALAXIES
ZWG 136.067	15 43 06.	+ 24 47		14.7	GALAXY
ZWG 222.050	15 43 06.	+ 43 40		15.0	GALAXY
ZWG 250.007	15 43 06.	+ 46 14		15.4	GALAXY
UGC 10010	15 43 06.	+ 46 14	96	15.4	GALAXY IRR
MCG+12-15-013	15 43 06.	+ 70 58	18	15.	GALAXY
ZWG 338.014	15 43 06.	+ 71 00		15.5	GALAXY
ZWG 022.035	15 43 06.	- 01 58		15.6	GALAXY
MRSL 327+00/1	15 43 06.	- 53 48	1080		HII REGION
VHE 67	15 43 06.	- 56 53	30		REFLECTION NEBULA
RNGC 5991	15 43 08.	+ 24 47		14.5	GALAXY
MCG+07-32-051	15 43 09.	+ 43 41	36	15.	GALAXY
LB 00894	15 43 10.	+ 01 27 42.		17.9	FAINT BLUE STAR
ZWG 050.093	15 43 12.	+ 02 34		15.0	GALAXY
MCG+01-40-013	15 43 12.	+ 02 34	48	15.0	GALAXY
ZWG 136.068	15 43 12.	+ 24 40		15.3	GALAXY
MCG+04-37-028	15 43 12.	+ 24 47	24	14.7	GALAXY
ZWG 136.069	15 43 12.	+ 25 36		15.4	GALAXY
UGC 10011	15 43 12.	+ 25 36	66	15.4	GALAXY Sb-c
UGC 10012	15 43 12.	+ 36 15	84	16.0	GALAXY E
MCG+08-29-009	15 43 12.	+ 46 14	84	15.	GALAXY
ZWG 275.004	15 43 12.	+ 55 49		14.9	GALAXY
UGC 10013	15 43 12.	+ 55 49	60	14.9	GALAXY SB:a-b
MCG+09-26-013	15 43 12.	+ 55 50	54	15.	GALAXY
MCG+04-37-029	15 43 15.	+ 25 35 30.	60	15.4	GALAXY
IC 1135	15 43 16.	+ 17 50 47.			NONSTELLAR OBJECT
REIZ 4637	15 43 16.	+ 17 51	24	14.7	GALAXY
ZWG 050.094	15 43 18.	+ 07 17		15.3	GALAXY

OBJECT NAME	RIGHT ASCEN.	DECLINATION	DIAM.	MAGN.	TYPE OF OBJECT
ZWG 078.061	15 43 18.	+ 09 13		15.3	GALAXY
ZWG 107.034	15 43 18.	+ 17 51		15.1	GALAXY
MCG+03-40-036	15 43 18.	+ 17 51	36	15.1	GALAXY
MCG+04-37-030	15 43 18.	+ 24 39	30	15.3	GALAXY
ZWG 166.050	15 43 18.	+ 26 46		15.5	GALAXY
7ZW 620	15 43 18.	+ 58 03			COMPACT GALAXY
7ZW 621	15 43 18.	+ 62 40			COMPACT GALAXY
ZWG 078.062	15 43 24.	+ 09 14		15.4	GALAXY
ZWG 078.063	15 43 24.	+ 12 39		15.7	GALAXY
UGC 10014	15 43 24.	+ 12 39	78	15.7	GALAXY DWARF
MCG+02-40-013	15 43 24.	+ 12 39	72	15.7	GALAXY
UGC 10015	15 43 24.	+ 21 11	66	16.5	GALAXY Sc-IRR
UGC 10017	15 43 24.	+ 21 35	72	17.	GALAXY DWARF
ZC 1543.4+2247	15 43 24.	+ 22 47	470		CLUSTER OF GALAXIES
UGC 10016	15 43 24.	+ 26 46	66	15.5	GALAXY Sa-b
ZWG 275.005	15 43 24.	+ 52 51		15.5	GALAXY
ZWG 319.022	15 43 24.	+ 67 55		15.0	GALAXY
UGC 10018	15 43 24.	+ 67 55	66	15.0	GALAXY SBb
ZWG 050.095	15 43 30.	+ 02 51		15.4	GALAXY
ZWG 050.096	15 43 30.	+ 05 18		15.3	GALAXY
MCG+05-37-020	15 43 30.	+ 28 14	72	14.9	GALAXY
ZWG 166.051	15 43 30.	+ 30 18		15.4	GALAXY
MCG+11-19-012	15 43 30.	+ 67 55	72	15.	GALAXY
MCG+14-07-027	15 43 30.	+ 82 29	51	16.	GALAXY
SMI 13	15 43 30.	- 53 50	1080		FAINT NEBULOSITY
UGC 10019	15 43 30.	+ 30 18	84	15.4	GALAXY Sb
REIZ 4639	15 43 31.	+ 20 44	36	14.2	GALAXY
ARC 2132	15 43 31.	+ 70 22		16.9	RICH CLUSTER OF GALAXIES
IC 4582	15 43 32.	+ 28 13 56.			NONSTELLAR OBJECT
MCG+05-37-021	15 43 33.	+ 30 19	51	15.4	GALAXY
ZWG 050.097	15 43 36.	+ 03 22		15.3	GALAXY
MCG+03-40-037	15 43 36.	+ 17 41	42		GALAXY
MCG+04-37-031	15 43 36.	+ 20 41 30.	120	14.5	GALAXY
ZWG 136.070	15 43 36.	+ 20 44		14.5	GALAXY
UGC 10020	15 43 36.	+ 20 44	138	14.5	GALAXY Sc
ZWG 136.071	15 43 36.	+ 23 02		15.5	GALAXY
ZWG 166.052	15 43 36.	+ 28 15		14.9	GALAXY
UGC 10021	15 43 36.	+ 28 15	84	14.9	GALAXY S
MCG+09-26-014	15 43 36.	+ 54 42 30.	30	17.	GALAXY
UGC 10022	15 43 36.	+ 57 32	90	16.5	GALAXY Sc
ZC 1543.6+6328	15 43 36.	+ 63 28	1610		CLUSTER OF GALAXIES
ZWG 050.098	15 43 42.	+ 02 36		15.7	GALAXY
ZWG 050.099	15 43 42.	+ 04 46		15.4	GALAXY
ZWG 050.100	15 43 42.	+ 07 03		15.2	GALAXY
UGC 10023	15 43 42.	+ 07 03	66	15.2	GALAXY IRR
ZWG 078.064	15 43 42.	+ 08 57		15.2	GALAXY
ZC 1543.7+1105	15 43 42.	+ 11 05	870		CLUSTER OF GALAXIES
ZC 1543.7+4139	15 43 42.	+ 41 39	1950		CLUSTER OF GALAXIES
MCG+10-22-036	15 43 42.	+ 57 30	66	16.	GALAXY
MCG-01-40-006	15 43 42.	- 04 32	42	16.	GALAXY
MCG-05-37-002	15 43 42.	- 28 35	72	14.5	GALAXY
LB 00895	15 43 43.	+ 76 46 48.		18.0	FAINT BLUE STAR
REIN 7.162	15 43 46.43	+ 57 31 02.7			NEBULA
ZC 1543.8+0105	15 43 48.	+ 01 05	2080		CLUSTER OF GALAXIES
ZWG 050.101	15 43 48.	+ 02 34		13.1	GALAXY
UGC 10024	15 43 48.	+ 02 34	102	13.1	GALAXY Sa?
MCG+01-40-014	15 43 48.	+ 02 34	84	13.1	GALAXY
ZWG 166.053	15 43 48.	+ 30 21		15.6	GALAXY
MCG+05-37-022	15 43 48.	+ 30 22	36	15.6	GALAXY
MCG+06-35-004	15 43 48.	+ 35 51	30	16.	GALAXY
MCG+08-29-010	15 43 48.	+ 49 27 30.	36	16.	GALAXY
RNGC 5990	15 43 50.	+ 02 34		13.0	GALAXY
UGC 10025	15 43 54.	+ 02 59	60	17.	GALAXY
ZWG 078.065	15 43 54.	+ 10 55		15.3	GALAXY
UGC 10026	15 43 54.	+ 10 55	60	15.3	GALAXY Sc
ZC 1543.9+1549	15 43 54.	+ 15 49	2080		CLUSTER OF GALAXIES
ZC 1543.9+2215	15 43 54.	+ 22 15	610		CLUSTER OF GALAXIES
TON-N 0802	15 43 54.	+ 27 17		16.3	BLUE STAR
ZC 1543.9+2808	15 43 54.	+ 28 08	3900		CLUSTER OF GALAXIES
ZWG 250.008	15 43 54.	+ 49 27		15.7	GALAXY
ZC 1543.9+5340	15 43 54.	+ 53 40	2290		CLUSTER OF GALAXIES
LB 00896	15 43 54.	- 00 12 36.		16.1	FAINT BLUE STAR
LB 00897	15 43 57.	+ 00 03 18.		17.3	FAINT BLUE STAR
UGC 10027	15 44 00.	+ 04 35	60	16.0	GALAXY Sb-c
ZWG 136.072	15 44 00.	+ 21 14		15.4	GALAXY
REIZ 4640	15 44 00.	+ 21 14	18	15.2	GALAXY
TON-N 0803	15 44 00.	+ 25 18		13.9	BLUE STAR
MCG+07-32-052	15 44 00.	+ 44 25	48	16.	GALAXY
MCG+10-22-037	15 44 00.	+ 57 33	45	16.	GALAXY
MCG+11-19-013	15 44 00.	+ 67 00	12	16.	GALAXY
SHAH 134	15 44 00.	+ 67 25	108	17.5	GROUP OF COMPACT GALAXIES
MCG+15-01-016	15 44 00.	+ 87 23 30.	24	17.	GALAXY
BC M1544+173	15 44 01.3	+ 17 23 14.		18.	QUASI-STELLAR OBJECT
SHB 317	15 44 01.3	+ 17 23 14.		18.	QUASI-STELLAR OBJECT
ZWG 078.066	15 44 06.	+ 12 30		15.7	GALAXY
REIZ 4641	15 44 06.	+ 21 15	9	15.1	GALAXY
ZWG 136.073	15 44 06.	+ 22 32		15.6	GALAXY
ZWG 222.051	15 44 06.	+ 44 24		15.7	GALAXY
UGC 10028	15 44 06.	+ 44 24	60	15.7	GALAXY
ZC 1544.1+5600	15 44 06.	+ 56 00	1550		CLUSTER OF GALAXIES
HN 0424	15 44 08.	- 67 09			NEBULA
IC 4571	15 44 08.	- 67 09			NONSTELLAR OBJECT
LB 00898	15 44 10.	+ 00 54 30.		15.3	FAINT BLUE STAR
ZWG 078.067	15 44 12.	+ 12 18		15.5	GALAXY
VV 016A	15 44 12.	+ 18 03	87	12.5	INTERACTING GALAXY
ZWG 136.074	15 44 12.	+ 23 58		15.1	GALAXY
MCG+04-37-032	15 44 12.	+ 23 58	48	15.1	GALAXY
REIN 7.163	15 44 12.10	+ 57 34 02.6			NEBULA
IC 4583	15 44 15.	+ 23 57 18.			NONSTELLAR OBJECT
ZWG 166.054	15 44 18.	+ 30 25		15.2	GALAXY
ZC 1544.3+3412	15 44 18.	+ 34 12	1880		CLUSTER OF GALAXIES
MCG+12-15-014	15 44 18.	+ 71 20	36	16.	GALAXY
ZWG 050.102	15 44 24.	+ 06 03		15.1	GALAXY
UGC 10029	15 44 24.	+ 06 03	78	15.1	GALAXY S0
MRK 690	15 44 24.	+ 18 53	8	16.5	GALAXY WITH UV CONTINUUM
ZC 1544.4+2543	15 44 24.	+ 25 43	7530		CLUSTER OF GALAXIES
ARC 2123	15 44 24.	- 08 02		17.5	RICH CLUSTER OF GALAXIES
LB 00899	15 44 25.	+ 00 57 00.		15.2	FAINT BLUE STAR
LB 00900	15 44 27.	+ 00 07 54.		16.3	FAINT BLUE STAR
ZWG 078.068	15 44 30.	+ 11 56		15.3	GALAXY
VV 016B	15 44 30.	+ 18 02	48	13.1	INTERACTING GALAXY
ZWG 107.035	15 44 30.	+ 18 13		15.6	GALAXY
ZWG 166.055	15 44 30.	+ 32 16		15.1	GALAXY
SHAH 022	15 44 30.	+ 55 17	144	16.5	GROUP OF COMPACT GALAXIES
UGC 10031	15 44 30.	+ 61 45	102	17.	GALAXY DWRF SP
ZWG 022.036	15 44 30.	- 00 50		15.4	GALAXY
UGC 10030	15 44 30.	- 00 50	84	15.4	GALAXY SBa-b
MCG+00-40-010	15 44 30.	- 00 50	90	15.4	GALAXY

OBJECT NAME	RIGHT ASCEN.	DECLINATION	DIAM.	MAGN.	TYPE OF OBJECT
MCG+06-35-005	15 44 33.	+ 33 23	48	14.5	GALAXY
RNGC 5994	15 44 34.	+ 18 01		15.0	GALAXY
KARA.72 472A	15 44 36.	+ 18 01	48	13.2	PART OF DOUBLE GALAXY
MCG+03-40-038	15 44 36.	+ 18 01 30.	24	13.2	GALAXY
ARC 2126	15 44 36.	+ 26 08		17.6	RICH CLUSTER OF GALAXIES
MCG+05-37-023	15 44 36.	+ 32 17 30.	42	15.1	GALAXY
ZWG 195.001	15 44 36.	+ 33 22		15.5	GALAXY
KARA.73B 0702	15 44 36.	+ 33 22	54	15.5	ISOLATED GALAXY S
ZWG 338.015	15 44 36.	+ 72 35		15.2	GALAXY
UGC 10032	15 44 36.	+ 72 35	84	15.2	GALAXY Sb-c
MCG+12-15-015	15 44 36.	+ 72 35	102	15.	GALAXY
RNGC 5996	15 44 40.	+ 18 02		13.0	GALAXY
REIZ 4642	15 44 40.	+ 18 02	12	14.9	GALAXY
ARC 2134	15 44 40.	+ 71 10		17.5	RICH CLUSTER OF GALAXIES
HOLM 721B	15 44 41.	+ 18 02	12	15.1	PART OF MULTIPLE GALAXY
ZWG 078.069	15 44 42.	+ 08 57		15.6	GALAXY
ZWG 107.036	15 44 42.	+ 18 02		13.2	GALAXY
MRK 691	15 44 42.	+ 18 02	10	15.	GALAXY WITH UV CONTINUUM
UGC 10033	15 44 42.	+ 18 02	228	13.2	GALAXY SB+COMP
KARA.72 472B	15 44 42.	+ 18 02	102		PART OF DOUBLE GALAXY
MCG+03-40-039	15 44 42.	+ 18 02	96	13.2	GALAXY
ZWG 136.075	15 44 42.	+ 22 19		15.7	GALAXY
KARA.73B 0703	15 44 42.	+ 22 19	18	15.7	ISOLATED GALAXY E
ZC 1544.7+2610	15 44 42.	+ 26 10	1140		CLUSTER OF GALAXIES
ZWG 166.056	15 44 42.	+ 31 10		15.3	GALAXY
UGC 10034	15 44 42.	+ 31 10	72	15.3	GALAXY SBb/Sb
REIZ 4643	15 44 46.	+ 18 03	90	13.2	GALAXY
HOLM 721A	15 44 46.	+ 18 03	78	12.5	PART OF MULTIPLE GALAXY
ZWG 050.103	15 44 48.	+ 03 05		15.6	GALAXY
ZWG 050.104	15 44 48.	+ 06 04		15.2	GALAXY
ZWG 078.070	15 44 48.	+ 09 44		15.7	GALAXY
ARP 072	15 44 48.	+ 18 04			PECULIAR GALAXY
ZC 1544.8+1850	15 44 48.	+ 18 50	2420		CLUSTER OF GALAXIES
MCG+05-37-024	15 44 48.	+ 31 10	78	15.3	GALAXY
ZWG 250.009	15 44 48.	+ 45 17		15.6	GALAXY
MAI 095	15 44 49.	+ 61 48	40		DWARF SPHEROIDAL GALAXY
REIZ 4644	15 44 50.	+ 42 05	24	15.1	GALAXY
ZWG 078.071	15 44 54.	+ 10 56		15.5	GALAXY
ZC 1544.9+2549	15 44 54.	+ 25 49	610		CLUSTER OF GALAXIES
ZWG 250.010	15 44 54.	+ 46 09		15.0	GALAXY
MRK 490	15 44 54.	+ 46 09	12	15.5	GALAXY WITH UV CONTINUUM
IC 1136	15 44 57.	- 01 24 10.			NONSTELLAR OBJECT
RNGC 5997	15 44 59.	+ 08 29		15.5	GALAXY
ZWG 050.105	15 45 00.	+ 08 29		15.5	GALAXY
ZWG 078.072	15 45 00.	+ 10 26		15.6	GALAXY
KARA.73B 0704	15 45 00.	+ 10 26	42	15.6	ISOLATED GALAXY S
ZC 1545.0+1044	15 45 00.	+ 10 44	610		CLUSTER OF GALAXIES
ZWG 107.037	15 45 00.	+ 16 55		15.3	GALAXY
MCG+03-40-040	15 45 00.	+ 16 55	27	15.3	GALAXY
ZWG 136.076	15 45 00.	+ 23 03		15.7	GALAXY
TON-N 0248	15 45 00.	+ 23 28		15.8	BLUE STAR
ZC 1545.0+5515	15 45 00.	+ 55 15	1080		CLUSTER OF GALAXIES
ZWG 022.037	15 45 00.	- 01 24		15.4	GALAXY
ZC 1545.1+2104	15 45 06.	+ 21 04	540		CLUSTER OF GALAXIES
ZWG 136.077	15 45 06.	+ 21 42		15.5	GALAXY
ZC 1545.2+1259	15 45 12.	+ 12 59	1080		CLUSTER OF GALAXIES
ZWG 078.073	15 45 12.	+ 14 13		15.7	GALAXY
MCG+06-35-006	15 45 12.	+ 35 50	24	16.	GALAXY
ZWG 250.011	15 45 12.	+ 44 59		15.7	GALAXY
REIZ 4646	15 45 13.	+ 42 08	24	15.5	GALAXY
MCG-04-37-005	15 45 15.	- 24 44	36	15.	GALAXY
ZWG 136.078	15 45 18.	+ 21 53		15.2	GALAXY
MCG+10-22-038	15 45 18.	+ 59 21	42	15.	GALAXY
MCG-04-37-004	15 45 18.	- 25 13 30.	36	15.	GALAXY
ZWG 297.029	15 45 24.	+ 59 22		15.7	GALAXY
ZC 1545.4+7153	15 45 24.	+ 71 53	740		CLUSTER OF GALAXIES
MIL 37	15 45 24.	- 53 39 54.	600		SUPERNOVA REMNANT
ZWG 136.079	15 45 30.	+ 26 13		14.9	GALAXY
UGC 10035	15 45 30.	+ 26 13	66	14.9	GALAXY PECULR?
ZWG 166.057	15 45 30.	+ 32 10		15.5	GALAXY
ZWG 022.038	15 45 30.	- 02 00		15.5	GALAXY
SHB 318	15 45 31.2	+ 21 01 34.		16.7	QUASI-STELLAR OBJECT
BC 3CR323.1	15 45 32.29	+ 21 01 42.1		16.59	QUASI-STELLAR OBJECT
ZWG 050.106	15 45 36.	+ 06 05		15.4	GALAXY
MCG+04-37-033	15 45 36.	+ 26 13	30	14.9	GALAXY
MCG+05-37-025	15 45 36.	+ 32 11	36	15.5	GALAXY
SHAH 020	15 45 36.	+ 55 09	150	17.	GROUP OF COMPACT GALAXIES
RNGC 5995	15 45 36.	- 13 37		14.0	GALAXY
MCG-02-40-004	15 45 36.	- 13 37	42	14.5	GALAXY
LB 00901	15 45 37.	+ 61 13 18.		17.6	FAINT BLUE STAR
LB 00902	15 45 38.	+ 18 55 54.		15.9	FAINT BLUE STAR
MCG+05-37-026	15 45 39.	+ 28 46	12	17.	GALAXY
RNGC 6001	15 45 40.	+ 28 48		14.5	GALAXY
MCG+03-40-041	15 45 42.	+ 17 58	36	15.5	GALAXY
TON-N 0249	15 45 42.	+ 24 31		15.4	BLUE STAR
MCG+05-37-027	15 45 42.	+ 28 47	60	14.4	GALAXY
ZWG 166.058	15 45 42.	+ 28 48		14.4	GALAXY
UGC 10036	15 45 42.	+ 28 48	66	14.4	GALAXY Sc
MCG+13-11-015	15 45 42.	+ 77 50	51	16.	GALAXY
REIZ 4645	15 45 43.	+ 20 29	9	15.4	GALAXY
LB 00903	15 45 44.	+ 61 53 00.		17.5	FAINT BLUE STAR
RNGC 6002	15 45 46.	+ 28 47			NON-EXISTENT OBJECT
ZWG 078.074	15 45 48.	+ 09 17		15.6	GALAXY
ZWG 078.075	15 45 48.	+ 11 26		15.2	GALAXY
UGC 10037	15 45 48.	+ 11 26	72	15.2	GALAXY SBb
MCG+03-40-042	15 45 48.	+ 17 57	15	15.5	GALAXY
ZWG 107.038	15 45 48.	+ 17 58		15.5	GALAXY
ZC 1545.8+2017	15 45 48.	+ 20 17	1410		CLUSTER OF GALAXIES
ZWG 136.080	15 45 48.	+ 25 53		15.7	GALAXY
1ZW 125	15 45 48.	+ 54 30			COMPACT GALAXY
ZWG 338.016	15 45 48.	+ 70 53		15.6	GALAXY
ZC 1545.9+1125	15 45 54.	+ 11 25	1680		CLUSTER OF GALAXIES
ZC 1545.9+3325	15 45 54.	+ 33 25	2690		CLUSTER OF GALAXIES
1ZW 126	15 45 54.	+ 37 21			COMPACT GALAXY
ZWG 195.002	15 45 54.	+ 37 21		15.2	GALAXY
KARA.73B 0705	15 45 54.	+ 37 21	36	15.2	ISOLATED GALAXY E
VDB.66G 203	15 46	+ 81 58	70		DWARF GALAXY
LBN 0023	15 46	- 01 48	1140		BRIGHT NEBULA
LB 09899	15 46	- 82 06		13.5	FAINT BLUE STAR
ZWG 050.107	15 46 00.	+ 07 17		15.3	GALAXY
ZC 1546.0+0853	15 46 00.	+ 08 53	5710		CLUSTER OF GALAXIES
ZWG 078.076	15 46 00.	+ 10 58		15.6	GALAXY
ZWG 136.081	15 46 00.	+ 21 12		15.2	GALAXY
ZC 1546.0+3646	15 46 00.	+ 36 46	1080		CLUSTER OF GALAXIES
ZC 1546.0+6722	15 46 00.	+ 67 22	8670		CLUSTER OF GALAXIES
MCG+14-07-028	15 46 00.	+ 82 01	12	17.	GALAXY
ARC 2130	15 46 03.	+ 36 43		17.8	RICH CLUSTER OF GALAXIES
ZWG 022.039	15 46 06.	+ 00 25		15.7	GALAXY
ZC 1546.1+1918	15 46 06.	+ 19 18	470		CLUSTER OF GALAXIES
ZWG 136.082	15 46 06.	+ 26 21		15.3	GALAXY
UGC 10038	15 46 06.	+ 26 21	60	15.3	GALAXY S0-a
HN 1808	15 46 08.4	+ 00 24 31.	18		NEBULA
IC 1138	15 46 10.	+ 26 20 55.			NONSTELLAR OBJECT
ZWG 078.077	15 46 12.	+ 11 48		15.4	GALAXY
UGC 10039	15 46 12.	+ 11 48	66	15.4	GALAXY S
ARC 2129	15 46 13.	+ 20 12		17.8	RICH CLUSTER OF GALAXIES
MCG+09-26-015	15 46 15.	+ 55 20	48	17.	GALAXY
ZWG 136.083	15 46 18.	+ 23 08		15.7	GALAXY
MCG+08-29-011	15 46 18.	+ 45 07	60	16.	GALAXY
ARC 2128	15 46 21.	- 02 55		16.5	RICH CLUSTER OF GALAXIES
RNGC 5998	15 46 22.	- 28 26			NON-EXISTENT OBJECT
ZWG 078.078	15 46 24.	+ 08 56		15.3	GALAXY
ZWG 107.039	15 46 24.	+ 18 01		15.3	GALAXY
UGC 10040	15 46 24.	+ 18 01	60	15.3	GALAXY S
ZC 1546.4+3610	15 46 24.	+ 36 10	1750		CLUSTER OF GALAXIES
ZWG 250.012	15 46 24.	+ 45 09		15.7	GALAXY
MCG+04-37-034	15 46 27.	+ 22 00	144	15.4	GALAXY
HOLM 722A	15 46 28.	+ 05 22	42	15.0	PART OF MULTIPLE GALAXY
IC 1137	15 46 29.	+ 08 44 13.			NONSTELLAR OBJECT
REIZ 4647	15 46 29.	+ 22 02	72	14.4	GALAXY
ZWG 050.108	15 46 30.	+ 05 20		15.0	GALAXY
UGC 10041	15 46 30.	+ 05 20	198	15.0	GALAXY
MCG+01-40-015	15 46 30.	+ 05 20	180	15.0	GALAXY
HOLM 722B	15 46 30.	+ 05 21	18	15.2	PART OF MULTIPLE GALAXY
ZWG 050.109	15 46 30.	+ 07 22		14.8	GALAXY
UGC 10042	15 46 30.	+ 07 22	96	14.8	GALAXY Sc
MCG+01-40-016	15 46 30.	+ 07 22	84	14.8	GALAXY
ZWG 078.079	15 46 30.	+ 09 00		15.4	GALAXY
MCG+03-40-043	15 46 30.	+ 18 01	51	15.3	GALAXY
ZWG 136.084	15 46 30.	+ 22 02		15.4	GALAXY
UGC 10043	15 46 30.	+ 22 02	144	15.4	GALAXY Sb/Sc
ZWG 136.085	15 46 30.	+ 25 33		15.2	GALAXY
MCG+09-26-016	15 46 30.	+ 55 18	24	17.	GALAXY
MCG+11-19-014	15 46 30.	+ 68 27 30.	27	17.	GALAXY
IC 1145	15 46 35.	+ 72 36 37.			NONSTELLAR OBJECT
UGC 10044	15 46 36.	+ 18 15	78	16.0	GALAXY Sc
MCG+04-37-035	15 46 36.	+ 21 59	48		GALAXY
MCG+04-37-036	15 46 36.	+ 25 33 30.	30	15.2	GALAXY
MCG+07-32-053	15 46 36.	+ 42 08	42	15.5	GALAXY
MCG+04-37-038	15 46 39.	+ 21 55	18		GALAXY
ARC 2131	15 46 40.	+ 36 13		17.5	RICH CLUSTER OF GALAXIES
ZWG 050.110	15 46 42.	+ 04 40		15.6	GALAXY
ZWG 107.040	15 46 42.	+ 18 06		15.7	GALAXY
MCG+03-40-044	15 46 42.	+ 18 15	72		GALAXY
OCL 0943	15 46 42.	- 57 31	960	7.9	OPEN STAR CLUSTER
RNGC 6000	15 46 45.	- 29 15		13.0	GALAXY
HOLM 723B	15 46 46.	+ 18 38	9	15.3	PART OF MULTIPLE GALAXY
ZWG 022.040	15 46 48.	+ 00 19		15.5	GALAXY
UGC 10045	15 46 48.	+ 00 19	72	15.5	GALAXY S
ZWG 166.059	15 46 48.	+ 28 54		15.6	GALAXY
MCG-05-37-003	15 46 48.	- 29 15	90	13.	GALAXY
HOLM 723A	15 46 51.	+ 18 39	36	14.6	PART OF MULTIPLE GALAXY
IC 1140	15 46 53.	+ 19 13			NONSTELLAR OBJECT
ZWG 022.041	15 46 54.	+ 00 22		15.1	GALAXY
UGC 10046	15 46 54.	+ 00 22	102	15.1	GALAXY DBL SYS
MCG+00-40-011	15 46 54.	+ 00 22	66	15.1	GALAXY
ZWG 107.041	15 46 54.	+ 17 58		15.3	GALAXY
ZWG 166.060	15 46 54.	+ 27 46		15.3	GALAXY
MCG+09-26-017	15 46 54.	+ 52 51	36	16.	GALAXY
ZWG 275.006	15 46 54.	+ 55 52		15.1	GALAXY
MCG-03-40-005	15 46 54.	- 20 56	72	15.	GALAXY
BC PKS1546+027	15 46 58.4	+ 02 46 07.		18.	QUASI-STELLAR OBJECT
SHB 319	15 46 58.4	+ 02 46 07.		18.	QUASI-STELLAR OBJECT
RNGC 6011	15 46 59.	+ 72 19		14.5	GALAXY
MCG+03-40-046	15 47 00.	+ 16 26	48		GALAXY
MCG+03-40-045	15 47 00.	+ 17 58	48	15.3	GALAXY
ZWG 107.042	15 47 00.	+ 18 43		15.5	GALAXY
ZWG 136.086	15 47 00.	+ 21 51		15.7	GALAXY
MCG+04-37-037	15 47 00.	+ 21 57	66	15.1	GALAXY
MCG+09-26-018	15 47 00.	+ 55 53	36	15.	GALAXY
7ZW 622	15 47 00.	+ 59 56			COMPACT GALAXY
ZWG 338.017	15 47 00.	+ 72 19		14.5	GALAXY
UGC 10047	15 47 00.	+ 72 19	114	14.5	GALAXY Sb
MCG+12-15-016	15 47 00.	+ 72 20	132	15.	GALAXY
RNGC 6003	15 47 05.	+ 19 11		14.5	GALAXY
REIZ 4648	15 47 05.	+ 21 59	78	14.1	GALAXY
ZWG 078.080	15 47 06.	+ 12 05		15.2	GALAXY
MCG+03-40-047	15 47 06.	+ 18 43	18	15.5	GALAXY
ZWG 107.043	15 47 06.	+ 19 11		14.4	GALAXY
UGC 10048	15 47 06.	+ 19 11	66	14.4	GALAXY S0?
ZWG 136.087	15 47 06.	+ 21 59		15.1	GALAXY
UGC 10049	15 47 06.	+ 21 59	78	15.1	GALAXY CHAIN
TON-N 0250	15 47 06.	+ 25 43		15.4	BLUE STAR
ZC 1547.1+5805	15 47 06.	+ 58 05	810		CLUSTER OF GALAXIES
MCG+11-19-015	15 47 06.	+ 68 31 30.	27	17.	GALAXY
IC 1142	15 47 11.	+ 18 18 25.			NONSTELLAR OBJECT
MCG+03-40-048	15 47 12.	+ 19 11	27	14.4	GALAXY
MCG+04-37-039	15 47 12.	+ 21 09	36	15.1	GALAXY
ZWG 136.088	15 47 12.	+ 21 13		15.1	GALAXY
ZWG 338.018	15 47 12.	+ 68 32		14.8	GALAXY
MCG-01-40-007	15 47 12.	- 07 20	21	16.	GALAXY
IC 1141	15 47 17.	+ 12 33 01.			NONSTELLAR OBJECT
ZWG 136.089	15 47 18.	+ 20 49		15.7	GALAXY
ZWG 319.023	15 47 18.	+ 68 13		15.6	GALAXY
HN 0425	15 47 19.	- 74 40			NEBULA
IC 4578	15 47 19.	- 74 40			NONSTELLAR OBJECT
ZWG 107.044	15 47 24.	+ 18 41		14.6	GALAXY
UGC 10050	15 47 24.	+ 18 41	102	14.6	GALAXY S0
ZWG 275.007	15 47 24.	+ 52 30		15.6	GALAXY
PK329+02.1	15 47 24.	- 51 20 57.	80	13.6	PLANETARY NEBULA
HN 1809	15 47 24.7	+ 00 20 35.	18		NEBULA
HOLM 724A	15 47 27.	+ 18 41	36	14.6	PART OF MULTIPLE GALAXY
ZWG 022.043	15 47 30.	+ 00 20		15.4	GALAXY
ZWG 078.081	15 47 30.	+ 12 33		14.5	GALAXY
UGC 10051	15 47 30.	+ 12 33	30	14.5	GALAXY
MCG+02-40-014	15 47 30.	+ 12 33	36	14.5	GALAXY
MCG+03-40-049	15 47 30.	+ 18 41	102	14.6	GALAXY
ZC 1547.5+3405	15 47 30.	+ 34 05	1080		CLUSTER OF GALAXIES
MCG+08-29-012	15 47 30.	+ 44 50	22	15.	GALAXY
ZWG 022.042	15 47 30.	- 02 41		15.7	GALAXY
KARA.73B 0706	15 47 30.	- 02 41	54	15.7	ISOLATED GALAXY S
HOLM 724B	15 47 31.	+ 18 40	18	15.4	PART OF MULTIPLE GALAXY
FATH 1.726	15 47 31.	+ 44 52	14		NEBULA
REIZ 4649	15 47 33.	+ 18 40	6	15.0	GALAXY
REIZ 4650	15 47 33.	+ 18 41	18	14.0	GALAXY
MCG+04-37-040	15 47 36.	+ 20 56	27	14.4	GALAXY

OBJECT NAME	RIGHT ASCEN.	DECLINATION	DIAM.	MAGN.	TYPE OF OBJECT
ZC 1547.6+2555	15 47 36.	+ 25 55	400		CLUSTER OF GALAXIES
ZWG 250.013	15 47 36.	+ 44 51		15.5	GALAXY
MCG+04-37-041	15 47 39.	+ 20 56	30	14.4	GALAXY
ZWG 050.111	15 47 42.	+ 07 20		15.5	GALAXY
KARA.72 473A	15 47 42.	+ 20 57	42	14.4	PART OF DOUBLE GALAXY
ZC 1547.7+2219	15 47 42.	+ 22 19	540		CLUSTER OF GALAXIES
ZWG 166.061	15 47 42.	+ 27 17		15.7	GALAXY
ZWG 166.062	15 47 42.	+ 28 42		15.4	GALAXY
ZC 1547.7+2941	15 47 42.	+ 29 41	810		CLUSTER OF GALAXIES
ZC 1547.7+3215	15 47 42.	+ 32 15	470		CLUSTER OF GALAXIES
ACK 330+04.1	15 47 42.	- 48 17			PLANETARY NEBULA
MRSL 327-00/2	15 47 42.	- 54 36	360		HII REGION
PK330+04.1	15 47 43.	- 48 36			PLANETARY NEBULA
KEEL 560	15 47 43.5	+ 41 20 06.		18.	NEBULA
HOLM 725A	15 47 48.	+ 20 57	24	14.0	PART OF MULTIPLE GALAXY
KARA.72 473B	15 47 48.	+ 20 57	36		PART OF DOUBLE GALAXY
ZWG 136.090	15 47 48.	+ 20 58		14.4	GALAXY
UGC 10052	15 47 48.	+ 20 58	66	14.4	GALAXY DBL SYS
MCG+07-32-054	15 47 48.	+ 42 11	18	14.5	GALAXY
7ZW 623	15 47 48.	+ 69 37			COMPACT GALAXY
VV 291A	15 47 48.	+ 69 37 30.	24	15.	INTERACTING GALAXY
MCG+12-15-017	15 47 48.	+ 69 37 30.	72	16.	GALAXY
HOLM 725B	15 47 50.	+ 20 57	30	14.3	PART OF MULTIPLE GALAXY
ZWG 022.044	15 47 54.	+ 00 42		15.7	GALAXY
ZWG 050.112	15 47 54.	+ 04 37		15.4	GALAXY
ZC 1547.9+1425	15 47 54.	+ 14 25	340		CLUSTER OF GALAXIES
ZWG 136.091	15 47 54.	+ 24 57		15.0	GALAXY
ZWG 223.001	15 47 54.	+ 42 11		15.0	GALAXY
ZWG 222.052	15 47 54.	+ 42 11		15.0	GALAXY
ZWG 338.019	15 47 54.	+ 69 38		15.2	GALAXY
UGC 10053	15 47 54.	+ 69 38	66	15.2	GALAXY
REIZ 4652	15 47 55.	+ 42 12	9	15.0	GALAXY
ARC 2133	15 47 57.	+ 36 27		17.8	RICH CLUSTER OF GALAXIES
PK322-06.1	15 47 59.	- 62 21 50.	10		PLANETARY NEBULA
LBN 0019	15 48	- 02 40	6900		BRIGHT NEBULA
LBN 0018	15 48	- 02 45	1140		BRIGHT NEBULA
LBN 0011	15 48	- 03 54	900		BRIGHT NEBULA
LBN 0008	15 48	- 05 50	1680		BRIGHT NEBULA
ZWG 050.113	15 48 00.	+ 06 56		15.1	GALAXY
ZC 1548.0+1111	15 48 00.	+ 11 11	470		CLUSTER OF GALAXIES
ZC 1548.0+1914	15 48 00.	+ 19 14	1610		CLUSTER OF GALAXIES
MCG+04-37-042	15 48 00.	+ 24 57	18	15.0	GALAXY
ZWG 136.092	15 48 00.	+ 25 13		15.3	GALAXY
ZWG 319.024	15 48 00.	+ 68 21		15.2	GALAXY
MCG+11-19-016	15 48 00.	+ 68 21	39	16.	GALAXY
MCG+12-15-018	15 48 00.	+ 69 37	21	16.	GALAXY
ZWG 366.023	15 48 00.	+ 81 59		14.9	GALAXY
UGC 10054	15 48 00.	+ 81 59	192	14.9	GALAXY SB IV
MCG-01-40-008	15 48 00.	- 07 25	24	15.	GALAXY
DG 126	15 48 00.	- 25 36	780		REFLECTION NEBULA
CED 125A	15 48 00.	- 25 36	4020		DIFFUSE GALACTIC NEBULA
ACK 329+01.1	15 48 00.	- 51 23			PLANETARY NEBULA
REIZ 4653	15 48 01.	+ 42 12	6	15.3	GALAXY
RNGC 6004	15 48 05.	+ 19 05		13.5	GALAXY
ARP 109	15 48 05.	+ 69 35			PECULIAR GALAXY
ZWG 107.045	15 48 06.	+ 18 17		15.1	GALAXY
UGC 10055	15 48 06.	+ 18 17	90	15.1	GALAXY Sc
MCG+03-40-050	15 48 06.	+ 18 17	96	15.1	GALAXY
ZWG 107.046	15 48 06.	+ 19 05		13.4	GALAXY
UGC 10056	15 48 06.	+ 19 05	108	13.4	GALAXY SBc
MCG+03-40-051	15 48 06.	+ 19 05	114	13.4	GALAXY
MCG+04-37-043	15 48 06.	+ 25 13	30	15.3	GALAXY
ZC 1548.1+3735	15 48 06.	+ 37 35	3970		CLUSTER OF GALAXIES
ZWG 223.002	15 48 06.	+ 42 08		15.7	GALAXY
ZWG 222.053	15 48 06.	+ 42 08		15.7	GALAXY
MCG+11-19-017	15 48 06.	+ 68 22	78	15.	GALAXY
ZWG 338.020	15 48 06.	+ 69 37		15.1	GALAXY
VV 291B	15 48 06.	+ 69 37	18	15.5	INTERACTING GALAXY
MCG-01-40-009	15 48 06.	- 07 32	36	16.	GALAXY
REIZ 4651	15 48 08.	+ 19 06	60	13.6	GALAXY
ZWG 022.045	15 48 12.	+ 01 53		15.7	GALAXY
MCG+00-40-012	15 48 12.	+ 01 53	42	15.7	GALAXY
KARA.73B 0707	15 48 12.	+ 01 53	42	15.7	ISOLATED GALAXY S
ZWG 166.063	15 48 12.	+ 28 57		15.5	GALAXY
ZWG 319.025	15 48 12.	+ 68 22		15.3	GALAXY
UGC 10057	15 48 12.	+ 68 22	96	15.3	GALAXY SBb
OCL 0946	15 48 12.	- 56 19	960	9.2	OPEN STAR CLUSTER
VHA 178	15 48 12.	- 56 19	480		OPEN STAR CLUSTER
REIZ 4655	15 48 13.	+ 42 15	30	15.2	GALAXY
IC 1146	15 48 13.	+ 69 33 21.			NONSTELLAR OBJECT
PNGC 5999	15 48 13.	- 56 20		9.0	OPEN CLUSTER
ZC 1548.3+1604	15 48 18.	+ 16 04	2350		CLUSTER OF GALAXIES
ZWG 223.003	15 48 18.	+ 42 07		15.5	GALAXY
ZWG 222.054	15 48 18.	+ 42 07		15.5	GALAXY
ZWG 223.004	15 48 18.	+ 42 14		15.6	GALAXY
ZWG 222.055	15 48 18.	+ 42 14		15.6	GALAXY
UGC 10058	15 48 18.	+ 26 05	72	17.	GALAXY DWRF SP
REIZ 4654	15 48 19.	+ 38 17	18	14.9	GALAXY
SHB 320	15 48 21.2	+ 11 29 47.		17.	QUASI-STELLAR OBJECT
BC 4C11.50A	15 48 21.4	+ 11 29 48.		17.	QUASI-STELLAR OBJECT
SHB 321	15 48 21.5	+ 11 29 46.		18.5	QUASI-STELLAR OBJECT
BC 4C11.50B	15 48 21.8	+ 11 29 48.		19.	QUASI-STELLAR OBJECT
LB 00904	15 48 24.	+ 17 07 18.		17.0	FAINT BLUE STAR
ZWG 136.093	15 48 24.	+ 20 33		14.9	GALAXY
KARA.72 474A	15 48 24.	+ 20 33	30	14.9	PART OF DOUBLE GALAXY
KARA.72 474B	15 48 24.	+ 20 34	30		PART OF DOUBLE GALAXY
ZC 1548.4+2615	15 48 24.	+ 26 15	610		CLUSTER OF GALAXIES
ZWG 223.005	15 48 24.	+ 42 27		15.3	GALAXY
ZWG 222.056	15 48 24.	+ 42 27		15.3	GALAXY
ZWG 319.026	15 48 24.	+ 68 15		15.2	GALAXY
MCG+11-19-018	15 48 24.	+ 68 15	45	16.	GALAXY
MCG+12-15-019	15 48 24.	+ 69 32	51	14.	GALAXY
ZWG 338.021	15 48 24.	+ 69 33		14.7	GALAXY
MIL 36	15 48 26.	- 56 02 48.	660		SUPERNOVA REMNANT
MCG+04-37-044	15 48 27.	+ 20 33	48	14.9	GALAXY
ZWG 078.082	15 48 30.	+ 13 26		15.6	GALAXY
ZC 1548.5+2109	15 48 30.	+ 21 09	670		CLUSTER OF GALAXIES
ZWG 166.064	15 48 30.	+ 28 47		15.3	GALAXY
ZC 1548.5+7029	15 48 30.	+ 70 29	1410		CLUSTER OF GALAXIES
MCG+14-07-029	15 48 30.	+ 81 58 30.	132	15.	GALAXY
UGC 10059	15 48 36.	+ 22 24	72	15.7	GALAXY Sc
ZWG 078.083	15 48 36.	+ 10 12		15.7	GALAXY
ZC 1548.6+2136	15 48 36.	+ 21 36	7590		CLUSTER OF GALAXIES
ZWG 136.094	15 48 36.	+ 22 24		15.7	GALAXY
ZC 1548.6+2855	15 48 36.	+ 28 55	4500		CLUSTER OF GALAXIES
ZC 1548.6+3637	15 48 36.	+ 36 37	2350		CLUSTER OF GALAXIES
ZC 1548.6+6107	15 48 36.	+ 61 07	1280		CLUSTER OF GALAXIES
MCG+04-37-045	15 48 39.	+ 22 22	66	15.7	GALAXY

OBJECT NAME	RIGHT ASCEN.	DECLINATION	DIAM.	MAGN.	TYPE OF OBJECT
ZWG 078.084	15 48 42.	+ 09 53		15.5	GALAXY
ZC 1548.7+5903	15 48 42.	+ 59 03	810		CLUSTER OF GALAXIES
ZC 1548.7+7140	15 48 42.	+ 71 40	810		CLUSTER OF GALAXIES
ZWG 078.085	15 48 48.	+ 10 27		15.6	GALAXY
UGC 10061	15 48 48.	+ 16 27	138	16.5	GALAXY DWRF IR
ZWG 107.047	15 48 48.	+ 20 20		15.7	GALAXY
UGC 10060	15 48 48.	+ 20 20	60	15.7	GALAXY SBb
ZWG 136.095	15 48 48.	+ 23 10		15.5	GALAXY
MCG+12-15-020	15 48 48.	+ 69 36	33	16.	GALAXY
ZWG 078.086	15 48 54.	+ 11 57		15.5	GALAXY
LB 00905	15 48 54.	+ 18 18 18.		18.0	FAINT BLUE STAR
MCG+03-40-053	15 48 54.	+ 18 42 30.	30	15.2	GALAXY
ZWG 107.048	15 48 54.	+ 18 43		15.2	GALAXY
MCG+03-40-052	15 48 54.	+ 20 22	90	15.7	GALAXY
MCG+04-37-046	15 48 54.	+ 24 06 30.	42	15.5	GALAXY
ZWG 136.096	15 48 54.	+ 24 07		15.5	GALAXY
ZWG 338.022	15 48 54.	+ 69 49		15.3	GALAXY
MRSL 326-01/1	15 48 54.	- 55 06	120		HII REGION
VDB.66G 202	15 49	+ 16 28	130		DWARF GALAXY
LBN 0015	15 49	- 03 36	960		BRIGHT NEBULA
MCG+04-37-047	15 49 00.	+ 22 04 30.	84	15.2	GALAXY
ZWG 136.097	15 49 00.	+ 22 06		15.2	GALAXY
UGC 10062	15 49 00.	+ 22 06	90	15.2	GALAXY Sc
UGC 10063	15 49 00.	+ 25 53	72	16.5	GALAXY
TON-N 0804	15 49 00.	+ 29 55		15.2	BLUE STAR
ZWG 166.065	15 49 00.	+ 30 41		15.3	GALAXY
MCG+06-35-008	15 49 00.	+ 34 01 30.	36	15.	GALAXY
MCG+06-35-007	15 49 00.	+ 34 16	42	15.	GALAXY
ZC 1549.0+5315	15 49 00.	+ 53 15	3160		CLUSTER OF GALAXIES
MCG+12-15-021	15 49 00.	+ 69 48	33	15.	GALAXY
MCG+12-15-022	15 49 00.	+ 71 34 30.	36	16.	GALAXY
ZWG 354.027	15 49 00.	+ 74 41		15.5	GALAXY
KARA.73B 0708	15 49 00.	+ 74 41	36	15.5	ISOLATED GALAXY S
ZWG 078.087	15 49 06.	+ 12 01		15.7	GALAXY
ZWG 136.098	15 49 06.	+ 25 51		15.0	GALAXY
UGC 10064	15 49 06.	+ 25 51	60	15.0	GALAXY S0
7ZW 624	15 49 06.	+ 61 14			COMPACT GALAXY
ZWG 338.023	15 49 06.	+ 71 35		15.6	GALAXY
ZWG 050.114	15 49 12.	+ 06 42		15.6	GALAXY
MCG+03-40-055	15 49 12.	+ 18 32 30.	18		GALAXY
MCG+03-40-054	15 49 12.	+ 18 33	18		GALAXY
ZWG 136.099	15 49 12.	+ 21 07		15.6	GALAXY
MCG+04-37-048	15 49 12.	+ 25 50 30.	36	15.0	GALAXY
ZC 1549.2+2635	15 49 12.	+ 26 35	1280		CLUSTER OF GALAXIES
ZC 1549.2+3856	15 49 12.	+ 38 56	810		CLUSTER OF GALAXIES
MCG+08-29-013	15 49 12.	+ 50 37 30.	36	15.	GALAXY
7ZW 625	15 49 12.	+ 58 53			COMPACT GALAXY
MCG+12-15-024	15 49 12.	+ 71 24	9	16.	GALAXY
MCG+12-15-023	15 49 12.	+ 71 24	6	16.	GALAXY
ZWG 078.088	15 49 18.	+ 12 08		15.7	GALAXY
ZWG 078.089	15 49 18.	+ 13 20		15.6	GALAXY
ZWG 136.100	15 49 18.	+ 24 35		15.4	GALAXY
UGC 10065	15 49 18.	+ 37 10	60	15.1	GALAXY Sb
ZWG 275.008	15 49 18.	+ 50 38		15.1	GALAXY
ZWG 338.024	15 49 18.	+ 71 24		15.4	GALAXY
ZWG 050.115	15 49 24.	+ 04 29		15.7	GALAXY
ZWG 078.090	15 49 24.	+ 12 54		15.1	GALAXY
MCG+02-40-015	15 49 24.	+ 12 54	54	15.1	GALAXY
ZWG 136.101	15 49 24.	+ 23 57		15.4	GALAXY
UGC 10066	15 49 24.	+ 23 57	72	15.4	GALAXY S
MCG+06-35-009	15 49 24.	+ 34 44	24	17.	GALAXY
ZWG 275.009	15 49 30.	+ 55 46		15.4	GALAXY
ZWG 022.046	15 49 30.	- 01 06		15.6	GALAXY
UGC 10067	15 49 30.	- 01 06	72	15.6	GALAXY Sc
ZWG 050.116	15 49 36.	+ 04 40		15.5	GALAXY
ZC 1549.6+2417	15 49 36.	+ 24 17	5850		CLUSTER OF GALAXIES
MCG+05-37-028	15 49 36.	+ 28 38 30.	36	15.6	GALAXY
ZWG 166.066	15 49 36.	+ 28 39		15.6	GALAXY
MCG+07-33-001	15 49 36.	+ 43 34	36	15.	GALAXY
MCG+08-29-014	15 49 36.	+ 47 24	72	14.	GALAXY
MCG+09-26-019	15 49 36.	+ 51 56	60	15.	GALAXY
FATH 1.727	15 49 41.	+ 45 02	16		NEBULA
ZWG 078.091	15 49 42.	+ 13 03		15.0	GALAXY
UGC 10068	15 49 42.	+ 13 03	60	15.0	GALAXY S0
MCG+02-40-016	15 49 42.	+ 13 03	18	15.0	GALAXY
ZWG 136.102	15 49 42.	+ 26 08		15.3	GALAXY
ZWG 223.006	15 49 42.	+ 43 33		14.4	GALAXY
MRK 491	15 49 42.	+ 43 33	16	16.	GALAXY WITH UV CONTINUUM
UGC 10069	15 49 42.	+ 43 33	42	14.4	GALAXY E-S0
ZWG 250.014	15 49 42.	+ 47 23		13.6	GALAXY
UGC 10070	15 49 42.	+ 47 23	84	13.6	GALAXY S
ZWG 275.010	15 49 42.	+ 51 57		15.4	GALAXY
UGC 10071	15 49 42.	+ 51 57	66	15.4	GALAXY S
ZWG 338.025	15 49 42.	+ 71 20		15.3	GALAXY
MCG+12-15-025	15 49 42.	+ 71 24	48	15.	GALAXY
IC 1144	15 49 43.	+ 43 35 45.			NONSTELLAR OBJECT
ZWG 107.049	15 49 48.	+ 20 17		15.5	GALAXY
ZWG 136.103	15 49 48.	+ 23 54		15.7	GALAXY
ZWG 136.104	15 49 48.	+ 24 29		15.5	GALAXY
ZC 1549.8+6436	15 49 48.	+ 64 36	1140		CLUSTER OF GALAXIES
ZWG 338.026	15 49 48.	+ 71 25		15.4	GALAXY
UGC 10072	15 49 48.	+ 71 25	114	15.4	GALAXY SBb/Sc
MCG+12-15-026	15 49 48.	+ 71 25	84	15.	GALAXY
ACK 331+03.1	15 49 48.	- 48 34			PLANETARY NEBULA
LB 00906	15 49 50.	+ 20 23 06.		16.6	FAINT BLUE STAR
MCG+06-35-010	15 49 51.	+ 36 24 30.	24	17.	GALAXY
ZWG 050.117	15 49 54.	+ 02 36		15.3	GALAXY
KARA.73B 0709	15 49 54.	+ 02 36	42	15.3	ISOLATED GALAXY S
MCG+04-37-049	15 49 54.	+ 26 08 30.	60	15.3	GALAXY
MCG+08-29-015	15 49 54.	+ 47 48	45	16.	GALAXY
IC 1147	15 49 55.	+ 69 43 26.			NONSTELLAR OBJECT
FATH 1.728	15 49 57.	+ 44 23	8		NEBULA
LBN 1097	15 50	- 23 20	2700		BRIGHT NEBULA
LBN 1092	15 50	- 25 50	4800		BRIGHT NEBULA
KHAV 234	15 50	- 34 21	3880		DARK NEBULA
LB 09900	15 50	- 82 34		15.1	FAINT BLUE STAR
ZC 1550.0+1132	15 50 00.	+ 11 32	740		CLUSTER OF GALAXIES
ZWG 107.050	15 50 00.	+ 16 36		15.4	GALAXY
ZC 1550.0+2740	15 50 00.	+ 27 40	6520		CLUSTER OF GALAXIES
1ZW 127	15 50 00.	+ 45 43			COMPACT GALAXY
MCG+11-19-023	15 50 00.	+ 67 17	12	17.	GALAXY
LDN 0169	15 50 00.	- 03 10	3720		DARK NEBULA
ZC 1550.1+2019	15 50 06.	+ 20 19	810		CLUSTER OF GALAXIES
ZWG 136.105	15 50 06.	+ 23 30		15.4	GALAXY
ZWG 136.106	15 50 06.	+ 24 46		14.7	GALAXY
UGC 10073	15 50 06.	+ 24 46	72	14.7	GALAXY SBb/Sc
MCG+09-26-020	15 50 06.	+ 52 11	30	16.	GALAXY
MCG+09-26-021	15 50 06.	+ 52 15	48	16.	GALAXY

OBJECT NAME	RIGHT ASCEN.	DECLINATION	DIAM.	MAGN.	TYPE OF OBJECT
MCG+12-15-027	15 50 06.	+ 69 42	24	15.	GALAXY
LB 00907	15 50 08.	+ 18 41 42.		16.7	FAINT BLUE STAR
ZWG 136.107	15 50 12.	+ 22 34		15.7	GALAXY
MCG+06-35-011	15 50 12.	+ 34 17 30.	12	15.5	GALAXY
MCG+09-26-022	15 50 12.	+ 53 33	36	17.	GALAXY
ZC 1550.2+5550	15 50 12.	+ 55 50	2960		CLUSTER OF GALAXIES
MCG+11-19-019	15 50 12.	+ 67 19	27	16.	GALAXY
MCG+12-15-028	15 50 12.	+ 71 39	33	16.	GALAXY
MCG+12-15-029	15 50 12.	+ 71 40	33	16.	GALAXY
MCG+04-37-051	15 50 15.	+ 23 29	42	15.4	GALAXY
MCG+04-37-050	15 50 15.	+ 24 46	72	14.7	GALAXY
MCG+06-35-012	15 50 15.	+ 34 15	24	15.	GALAXY
MCG+07-33-002	15 50 15.	+ 39 57	30	16.	GALAXY
ZC 1550.3+5239	15 50 18.	+ 52 39	540		CLUSTER OF GALAXIES
ZWG 319.027	15 50 18.	+ 67 20		15.3	GALAXY
ZWG 338.027	15 50 18.	+ 69 43		15.5	GALAXY
ZWG 022.047	15 50 18.	- 00 23		15.4	GALAXY
LB 00908	15 50 23.	+ 20 36 06.		16.5	FAINT BLUE STAR
ZWG 078.092	15 50 24.	+ 10 25		15.7	GALAXY
ZC 1550.4+1243	15 50 24.	+ 12 43	5580		CLUSTER OF GALAXIES
MCG+07-33-003	15 50 24.	+ 42 53	45	15.5	GALAXY
ZWG 298.001	15 50 24.	+ 61 13		15.2	GALAXY
ZWG 297.030	15 50 24.	+ 61 13		15.2	GALAXY
MCG+10-23-001	15 50 24.	+ 61 14	30	16.	GALAXY
MCG+10-23-002	15 50 24.	+ 61 19	36	16.	GALAXY
MCG+10-23-003	15 50 24.	+ 62 28	240	11.6	GALAXY
ZWG 107.051	15 50 30.	+ 20 22		15.7	GALAXY
MCG+03-40-056	15 50 30.	+ 20 23	48	15.7	GALAXY
ZWG 136.108	15 50 30.	+ 21 56		15.5	GALAXY
ZC 1550.5+2310	15 50 30.	+ 23 10	2420		CLUSTER OF GALAXIES
ZWG 298.002	15 50 30.	+ 61 18		15.6	GALAXY
ZWG 297.031	15 50 30.	+ 61 18		15.6	GALAXY
MCG+12-15-030	15 50 30.	+ 70 01	36	16.	GALAXY
ZWG 050.118	15 50 36.	+ 08 10		15.7	GALAXY
ZWG 078.093	15 50 36.	+ 12 10		15.3	GALAXY
ZWG 136.109	15 50 36.	+ 24 32		15.2	GALAXY
UGC 10074	15 50 36.	+ 24 32	66	15.2	GALAXY Sc
UGC 10075	15 50 36.	+ 32 27	420	11.7	GALAXY Sc
ZC 1550.6+4540	15 50 36.	+ 45 40	1880		CLUSTER OF GALAXIES
ZWG 319.028	15 50 36.	+ 62 27		11.7	GALAXY
ZWG 298.003	15 50 36.	+ 62 27		11.7	GALAXY
KARA.73B 0710	15 50 36.	+ 62 27	342	11.7	ISOLATED GALAXY S
MCG+11-19-021	15 50 36.	+ 64 15	39	15.	GALAXY
MCG+11-19-020	15 50 36.	+ 67 20	12	17.	GALAXY
DG 127	15 50 36.	- 25 11	660		REFLECTION NEBULA
CED 125B	15 50 36.	- 25 11	2040		DIFFUSE GALACTIC NEBULA
RNGC 6006	15 50 37.	+ 12 10		15.5	GALAXY
MCG+04-37-052	15 50 39.	+ 21 13	84	14.2	GALAXY
ZWG 136.110	15 50 42.	+ 21 16		14.2	GALAXY
RNGC 6008A	15 50 42.	+ 21 16		14.0	GALAXY
HOLM 726A	15 50 42.	+ 21 16	66	13.6	PART OF MULTIPLE GALAXY
UGC 10076	15 50 42.	+ 21 16	90	14.2	GALAXY SBb
MCG+04-37-053	15 50 42.	+ 24 32	48	15.2	GALAXY
RNGC 6015	15 50 43.	+ 62 28		12.0	GALAXY
ZC 1550.8+1029	15 50 48.	+ 10 29	340		CLUSTER OF GALAXIES
ZWG 078.094	15 50 48.	+ 13 23		15.0	GALAXY
UGC 10077	15 50 48.	+ 13 23	66	15.0	GALAXY Sb-c
MCG+02-40-017	15 50 48.	+ 13 23	60	15.0	GALAXY
ZWG 136.111	15 50 48.	+ 20 48		15.4	GALAXY
REIZ 4656	15 50 48.	+ 42 12	24	15.0	GALAXY
ZWG 275.011	15 50 48.	+ 54 49		15.6	GALAXY
MCG+11-19-022	15 50 48.	+ 67 18	27	14.	GALAXY
ZWG 319.029	15 50 48.	+ 67 19		14.6	GALAXY
UGC 10078	15 50 48.	+ 67 19	84	14.6	GALAXY E
PK326-01.1	15 50 52.	- 55 21	10		PLANETARY NEBULA
CED 125C	15 50 53.	- 24 24	2820		DIFFUSE GALACTIC NEBULA
ZWG 050.119	15 50 54.	+ 07 01		15.6	GALAXY
HOLM 726B	15 50 54.	+ 21 14	36	14.4	PART OF MULTIPLE GALAXY
ZWG 136.112	15 50 54.	+ 21 15		15.0	GALAXY
RNGC 6008B	15 50 54.	+ 21 15		15.0	GALAXY
MCG+04-37-054	15 50 54.	+ 21 15	18	15.0	GALAXY
SHAH 025	15 50 54.	+ 56 36	72	18.5	GROUP OF COMPACT GALAXIES
MCG+10-23-004	15 50 54.	+ 56 50	30	16.	GALAXY
ZWG 319.030	15 50 54.	+ 64 17		15.4	GALAXY
KARA.73B 0711	15 50 54.	+ 64 17	48	15.4	ISOLATED GALAXY IR
DG 128	15 50 54.	- 24 24	660		REFLECTION NEBULA
REIN 7.164	15 50 54.86	+ 56 51 39.7			NEBULA
LBN 0010	15 51	- 04 33	2700		BRIGHT NEBULA
ZWG 078.095	15 51 00.	+ 12 06		14.1	GALAXY
UGC 10079	15 51 00.	+ 12 06	102	14.1	GALAXY Sc
MCG+02-40-018	15 51 00.	+ 12 06	108	14.1	GALAXY
ZC 1551.0+1755	15 51 00.	+ 17 55	1550		CLUSTER OF GALAXIES
TON-N 0805	15 51 00.	+ 29 02		15.4	BLUE STAR
ZWG 250.015	15 51 00.	+ 46 29		15.7	GALAXY
MCG+12-15-031	15 51 00.	+ 71 41	27	16.	GALAXY
LDN 0134	15 51 00.	- 04 30	1860		DARK NEBULA
RNGC 6007	15 51 01.	+ 12 06		14.0	GALAXY
ZWG 078.096	15 51 06.	+ 12 13		15.4	GALAXY
ZWG 223.007	15 51 06.	+ 40 47		14.6	GALAXY
UGC 10080	15 51 06.	+ 40 47	90	14.6	GALAXY SBb
MCG+10-23-005A	15 51 06.	+ 59 24	18	17.	GALAXY
MCG+10-23-005	15 51 06.	+ 59 24	18	17.	GALAXY
SMI 14	15 51 06.	- 51 13	180		FAINT NEBULOSITY
RNGC 6009	15 51 07.	+ 12 13		15.5	GALAXY
RNGC 6013	15 51 08.	+ 40 47		14.5	GALAXY
REIZ 4657	15 51 09.	+ 40 47	72	13.8	GALAXY
LB 00909	15 51 10.	+ 74 27 54.		17.1	FAINT BLUE STAR
ZC 1551.2+2436	15 51 12.	+ 24 36	1410		CLUSTER OF GALAXIES
MCG+07-33-004	15 51 12.	+ 40 48	72	14.	GALAXY
ZWG 223.008	15 51 12.	+ 42 42		15.6	GALAXY
ZWG 298.004	15 51 12.	+ 59 24		15.6	GALAXY
ZC 1551.2+7633	15 51 12.	+ 76 33	670		CLUSTER OF GALAXIES
ZWG 108.001	15 51 18.	+ 18 45		15.6	GALAXY
ZWG 107.052	15 51 18.	+ 18 45		15.6	GALAXY
MCG+03-40-057	15 51 18.	+ 18 45	48	15.6	GALAXY
VV 311B	15 51 18.	+ 18 46	30	18.	INTERACTING GALAXY
VV 311A	15 51 18.	+ 18 46	48	15.	INTERACTING GALAXY
ZWG 136.113	15 51 18.	+ 23 44		15.7	GALAXY
ZC 1551.3+3013	15 51 18.	+ 30 13	3830		CLUSTER OF GALAXIES
LDN 0184	15 51 18.	- 03 00	1980		DARK NEBULA
LDN 0183	15 51 18.	- 03 00	1980		DARK NEBULA
PK320-09.1	15 51 22.	- 66 01	25	9.9	PLANETARY NEBULA
ZC 1551.4+1208	15 51 24.	+ 12 08	340		CLUSTER OF GALAXIES
ZWG 078.097	15 51 24.	+ 12 31		15.6	GALAXY
ZC 1551.4+1539	15 51 24.	+ 15 39	1080		CLUSTER OF GALAXIES
ARP 218	15 51 24.	+ 18 46			PECULIAR GALAXY
FATH 1.729	15 51 24.	+ 45 00	11		NEBULA
MCG+11-19-024	15 51 24.	+ 67 20	51	17.	GALAXY

OBJECT NAME	RIGHT ASCEN.	DECLINATION	DIAM.	MAGN.	TYPE OF OBJECT
ACK 329+01.2	15 51 24.	- 51 15			PLANETARY NEBULA
ZC 1551.5+5044	15 51 30.	+ 50 44	870		CLUSTER OF GALAXIES
MCG+12-15-032	15 51 30.	+ 71 11 30.	51	16.	GALAXY
MCG+12-15-033	15 51 33.	+ 71 12	9	16.	GALAXY
ARC 2135	15 51 34.	+ 50 44		17.6	RICH CLUSTER OF GALAXIES
ZC 1551.6+0645	15 51 36.	+ 06 45	810		CLUSTER OF GALAXIES
MRK 692	15 51 36.	+ 23 18	9	17.	GALAXY WITH UV CONTINUUM
ZWG 223.009	15 51 36.	+ 39 32		15.6	GALAXY
MRSL 327-00/1	15 51 36.	- 54 30	840		HII REGION
MCG+07-33-005	15 51 42.	+ 39 30	48	15.5	GALAXY
ZWG 319.031	15 51 42.	+ 64 59		15.7	GALAXY
RNGC 6019	15 51 44.	+ 64 59		15.5	GALAXY
MIL 39	15 51 47.	- 53 08 24.	222		SUPERNOVA REMNANT
RNGC 6005	15 51 47.	- 57 18		11.5	OPEN CLUSTER
WRAY 19.42	15 51 47.4	- 54 31 40.			DIFFUSE NEBULA
ZWG 022.048	15 51 48.	+ 00 40		13.3	GALAXY
UGC 10081	15 51 48.	+ 00 40	126	13.3	GALAXY Sa
MCG+00-40-013	15 51 48.	+ 00 40	114	13.3	GALAXY
ZWG 079.001	15 51 48.	+ 12 01		15.7	GALAXY
ZWG 078.098	15 51 48.	+ 12 01		15.7	GALAXY
ZWG 108.002	15 51 48.	+ 18 47		15.1	GALAXY
ZWG 107.053	15 51 48.	+ 18 47		15.1	GALAXY
ZC 1551.8+2625	15 51 48.	+ 26 25	1080		CLUSTER OF GALAXIES
ZWG 223.010	15 51 43.	+ 41 43		15.3	GALAXY
MCG+07-33-006	15 51 48.	+ 41 43	12	15.	GALAXY
OCL 0945	15 51 48.	- 57 17	540	11.8	OPEN STAR CLUSTER
VHA 179	15 51 48.	- 57 17	240		OPEN STAR CLUSTER
RNGC 6010	15 51 49.	+ 00 40		13.5	GALAXY
LB 00910	15 51 49.	+ 57 48 42.		17.5	FAINT BLUE STAR
LB 00911	15 51 50.	+ 17 30 06.		17.1	FAINT BLUE STAR
MCG+03-40-058	15 51 51.	+ 18 47	48	15.1	GALAXY
ZWG 079.002	15 51 54.	+ 12 16		15.5	GALAXY
ZWG 078.099	15 51 54.	+ 12 16		15.5	GALAXY
UGC 10082	15 51 54.	+ 12 16	66	15.5	GALAXY SB+COMP
ZWG 108.003	15 51 54.	+ 14 44		13.1	GALAXY
ZWG 107.054	15 51 54.	+ 14 44		13.1	GALAXY
UGC 10083	15 51 54.	+ 14 44	126	13.1	GALAXY SBa
KARA.73B 0712	15 51 54.	+ 14 44	108	13.1	ISOLATED GALAXY S
MCG+03-40-059	15 51 54.	+ 14 45	120	13.1	GALAXY
ZWG 108.004	15 51 54.	+ 19 15		15.0	GALAXY
ZWG 107.055	15 51 54.	+ 19 15		15.0	GALAXY
MCG+03-40-060	15 51 54.	+ 19 15	33	15.0	GALAXY
ZWG 136.114	15 51 54.	+ 23 17		14.8	GALAXY
MRK 693	15 51 54.	+ 23 17	33	15.	GALAXY WITH UV CONTINUUM
ZC 1551.9+5114	15 51 54.	+ 51 14	1340		CLUSTER OF GALAXIES
ZWG 275.012	15 51 54.	+ 54 18		15.7	GALAXY
7ZW 626	15 51 54.	+ 61 21			COMPACT GALAXY
ZC 1551.9+6210	15 51 54.	+ 62 10	940		CLUSTER OF GALAXIES
MCG+12-15-034	15 51 54.	+ 71 41	12	16.	GALAXY
HUB C22	15 51 54.	- 29 33			DIFFUSE NEBULA
RNGC 6012	15 51 57.	+ 14 44		13.0	GALAXY
MCG+04-37-055	15 51 57.	+ 23 16	45	14.8	GALAXY
ARC 2137	15 51 57.	+ 62 11		17.1	RICH CLUSTER OF GALAXIES
LB 00912	15 51 58.	+ 20 29 36.		16.3	FAINT BLUE STAR
LBN 1098	15 52	- 23 10	1500		BRIGHT NEBULA
ZWG 051.001	15 52 00.	+ 07 40		15.6	GALAXY
ZWG 050.120	15 52 00.	+ 07 40		15.6	GALAXY
KARA.73B 0713	15 52 00.	+ 07 40	36	15.6	ISOLATED GALAXY S
MCG+09-26-024	15 52 00.	+ 52 13	60	15.	GALAXY
ZWG 275.013	15 52 00.	+ 54 52		15.4	GALAXY
MCG+09-26-023	15 52 00.	+ 54 56	36	17.	GALAXY
ZWG 338.028	15 52 00.	+ 69 01		15.7	GALAXY
MCG+13-11-016	15 52 00.	+ 77 55	33	16.	GALAXY
ZWG 108.005	15 52 06.	+ 18 47		14.8	GALAXY
ZWG 107.056	15 52 06.	+ 18 47		14.8	GALAXY
UGC 10084	15 52 06.	+ 18 47	72	14.8	GALAXY S0?
ZWG 167.001	15 52 06.	+ 30 18		15.3	GALAXY
ZWG 166.067	15 52 06.	+ 30 18		15.3	GALAXY
MCG+03-40-061	15 52 09.	+ 18 47	78	14.8	GALAXY
ZWG 108.006	15 52 12.	+ 18 40		14.7	GALAXY
ZWG 107.057	15 52 12.	+ 18 40		14.7	GALAXY
UGC 10085	15 52 12.	+ 18 40	60	14.7	GALAXY Sc
MCG+03-40-062	15 52 12.	+ 18 40	66	14.7	GALAXY
MCG+06-35-013	15 52 12.	+ 37 21	24	16.	GALAXY
MCG+08-29-016	15 52 12.	+ 50 06	48	16.	GALAXY
DG 129	15 52 12.	- 25 37	5400		REFLECTION NEBULA
ARC 2136	15 52 13.	+ 51 16		17.6	RICH CLUSTER OF GALAXIES
LB 00913	15 52 16.	+ 18 11 06.		17.6	FAINT BLUE STAR
ZWG 108.007	15 52 18.	+ 16 45		14.5	GALAXY
ZWG 107.058	15 52 18.	+ 16 45		14.5	GALAXY
UGC 10086	15 52 18.	+ 16 45	42	14.5	GALAXY
ZWG 250.016	15 52 18.	+ 50 05		15.4	GALAXY
ZWG 051.002	15 52 24.	+ 04 08		15.7	GALAXY
ZWG 079.003	15 52 24.	+ 11 01		15.7	GALAXY
MCG+03-40-063	15 52 24.	+ 16 45	30	14.5	GALAXY
ZWG 223.011	15 52 24.	+ 41 45		15.1	GALAXY
UGC 10087	15 52 24.	+ 41 45	66	15.1	GALAXY Sc
MCG+07-33-007	15 52 24.	+ 41 46	60	14.5	GALAXY
MCG+09-26-025	15 52 24.	+ 53 38	60	16.	GALAXY
MCG+11-19-025	15 52 24.	+ 66 04	39	17.	GALAXY
ZC 1552.4+7044	15 52 24.	+ 70 44	1340		CLUSTER OF GALAXIES
IC 4586	15 52 25.	+ 06 10 10.			SAME AS NGC 6014
ZC 1552.5+3435	15 52 30.	+ 34 35	8270		CLUSTER OF GALAXIES
REIZ 4658	15 52 30.	+ 39 07	48	15.2	GALAXY
MCG+11-19-026	15 52 30.	+ 65 02 30.	39	15.	GALAXY
7ZW 627	15 52 30.	+ 74 20			COMPACT GALAXY
IC 1154	15 52 31.	+ 70 31 02.			NONSTELLAR OBJECT
ZWG 051.003	15 52 36.	+ 05 56		15.6	GALAXY
ZWG 079.004	15 52 36.	+ 11 06		15.5	GALAXY
ZWG 079.005	15 52 36.	+ 11 13		15.6	GALAXY
ZWG 108.008	15 52 36.	+ 19 03		15.4	GALAXY
ZWG 107.059	15 52 36.	+ 19 03		15.4	GALAXY
ZC 1552.6+2355	15 52 36.	+ 23 55	1950		CLUSTER OF GALAXIES
ZC 1552.6+3230	15 52 36.	+ 32 30	610		CLUSTER OF GALAXIES
ZC 1552.6+6044	15 52 36.	+ 60 44	3560		CLUSTER OF GALAXIES
ZWG 319.032	15 52 36.	+ 65 04		15.1	GALAXY
ZWG 338.029	15 52 36.	+ 70 32		14.8	GALAXY
UGC 10088	15 52 36.	+ 70 32	108	14.8	GALAXY E
RNGC 6024	15 52 38.	+ 65 04		15.0	GALAXY
ZC 1552.7+1112	15 52 42.	+ 11 12	6520		CLUSTER OF GALAXIES
ZWG 136.115	15 52 42.	+ 21 17		15.4	GALAXY
UGC 10089	15 52 42.	+ 21 17	66	15.4	GALAXY Sc
ZWG 167.002	15 52 42.	+ 28 00		15.5	GALAXY
ZWG 166.068	15 52 42.	+ 28 00		15.5	GALAXY
ARC 2144	15 52 42.	+ 70 40		17.5	RICH CLUSTER OF GALAXIES
MCG-04-37-006	15 52 45.	- 24 32 30.	30	15.	GALAXY
ZWG 051.004	15 52 48.	+ 05 02		15.7	GALAXY
ZWG 079.006	15 52 48.	+ 10 40		15.5	GALAXY

OBJECT NAME	RIGHT ASCEN.	DECLINATION	DIAM.	MAGN.	TYPE OF OBJECT
REIZ 4659	15 52 48.	+ 39 05	36	15.0	GALAXY
MCG+07-33-008	15 52 51.	+ 39 53 30.	27	16.5	GALAXY
ARC 2139	15 52 53.	+ 65 32		17.4	RICH CLUSTER OF GALAXIES
ZC 1552.9+2419	15 52 54.	+ 24 19	740		CLUSTER OF GALAXIES
MCG+07-33-009	15 52 54.	+ 39 55	24	16.	GALAXY
MCG+08-29-017	15 52 54.	+ 45 30	9	18.	GALAXY
MCG+11-19-027	15 52 54.	+ 67 38	39	17.	GALAXY
MCG+12-15-035	15 52 54.	+ 70 32	24	14.	GALAXY
ZWG 023.001	15 52 54.	- 00 40		15.1	GALAXY
MCG+00-41-001	15 52 54.	- 00 40	48	15.1	GALAXY
LBN 1093	15 53	- 25 50	3000		BRIGHT NEBULA
KHAV 235	15 53	- 37 57	3450		DARK NEBULA
LB 09901	15 53	- 85 47		15.0	FAINT BLUE STAR
ZWG 051.005	15 53 00.	+ 06 37		15.6	GALAXY
MRK 291	15 53 00.	+ 19 20	13	15.	GALAXY WITH UV CONTINUUM
KW 27	15 53 00.	+ 19 20	35		SEYFERT GALAXY
VVI 81	15 53 00.	+ 19 20	13	15.	SEYFERT GAL1XY
ZWG 167.003	15 53 00.	+ 31 00		15.7	GALAXY
ZWG 166.069	15 53 00.	+ 31 00		15.7	GALAXY
MCG+06-35-014	15 53 00.	+ 36 22 30.	24	16.	GALAXY
ZC 1553.0+6538	15 53 00.	+ 65 38	2290		CLUSTER OF GALAXIES
ZWG 338.030	15 53 00.	+ 69 02		15.6	GALAXY
MCG+12-15-036	15 53 00.	+ 69 02	51	16.	GALAXY
LB 00915	15 53 00.	+ 73 32 42.		17.4	FAINT BLUE STAR
LB 00914	15 53 00.	+ 75 13 06.		16.8	FAINT BLUE STAR
OCL 0940	15 53 00.	- 59 19	2700	9.	OPEN STAR CLUSTER
REIZ 4660	15 53 05.	+ 39 10	18	15.0	GALAXY
1ZW 128	15 53 06.	+ 45 29			COMPACT GALAXY
ZWG 250.017	15 53 06.	+ 45 29		15.2	GALAXY
MCG+08-29-019	15 53 06.	+ 45 29	21	15.5	GALAXY
MCG+08-29-018	15 53 06.	+ 45 29	30	15.	GALAXY
MCG+10-23-006	15 53 06.	+ 58 22	36	16.	GALAXY
UGC 10090	15 53 06.	+ 61 13	60	17.	GALAXY DWRF SP
ZC 1553.1+6316	15 53 06.	+ 63 16	1280		CLUSTER OF GALAXIES
FATH 1.730	15 53 14.	+ 44 42	5		NEBULA
ZWG 051.006	15 53 18.	+ 04 59		15.5	GALAXY
MCG+01-41-001	15 53 18.	+ 04 59	24	15.5	GALAXY
ZWG 108.009	15 52 18.	+ 19 33		15.7	GALAXY
ZC 1553.2+2518	15 53 18.	+ 25 18	670		CLUSTER OF GALAXIES
TON-N 0806	15 53 18.	+ 33 04		16.0	BLUE STAR
MCG+07-33-010	15 53 18.	+ 39 53	24	16.5	GALAXY
ZWG 223.012	15 53 18.	+ 41 43		14.9	GALAXY
MCG+09-26-026	15 53 18.	+ 56 12	48	16.	GALAXY
RNGC 6014	15 53 22.	+ 06 05		14.0	GALAXY
ZWG 023.002	15 53 24.	+ 00 45		15.5	GALAXY
ZWG 051.007	15 53 24.	+ 06 05		13.8	GALAXY
UGC 10091	15 53 24.	+ 06 05	120	13.8	GALAXY S0
MCG+01-41-002	15 53 24.	+ 06 05	72	13.8	GALAXY
ZWG 079.007	15 53 24.	+ 08 50		15.7	GALAXY
ZWG 079.008	15 53 24.	+ 11 46		15.2	GALAXY
ZWG 108.010	15 53 24.	+ 18 25		14.7	GALAXY
UGC 10092	15 53 24.	+ 18 25	66	14.7	GALAXY IRR
ZC 1553.4+2820	15 53 24.	+ 28 20	670		CLUSTER OF GALAXIES
ZC 1553.4+3614	15 53 24.	+ 36 14	4500		CLUSTER OF GALAXIES
MCG+07-33-011	15 53 24.	+ 41 43	60	14.5	GALAXY
MCG+08-29-020	15 53 24.	+ 45 37	42	16.	GALAXY
MCG+09-26-027	15 53 24.	+ 53 18	30	15.	GALAXY
MCG+10-23-007	15 53 24.	+ 61 32	30	16.	GALAXY
MCG+11-19-028	15 53 24.	+ 64 57	51	16.	GALAXY
ZC 1553.4+7104	15 53 24.	+ 71 04	810		CLUSTER OF GALAXIES
ZWG 079.009	15 53 30.	+ 12 07		15.2	GALAXY
MCG+06-35-015	15 53 30.	+ 34 20	30	15.	GALAXY
ZWG 250.018	15 53 30.	+ 45 38		15.1	GALAXY
ZWG 275.014	15 53 30.	+ 53 18		15.7	GALAXY
ZC 1553.5+6513	15 53 30.	+ 65 13	670		CLUSTER OF GALAXIES
MCG+03-41-001	15 53 33.	+ 18 26	66	14.7	GALAXY
ZWG 108.011	15 53 36.	+ 17 18		15.0	GALAXY
UGC 10093	15 53 36.	+ 17 18	96	15.0	GALAXY SBc/Sc
ZWG 137.001	15 53 36.	+ 24 38		15.5	GALAXY
UGC 10094	15 53 36.	+ 24 38	60	15.5	GALAXY Sb-c
KARA.72 475A	15 53 36.	+ 24 38	60	15.5	PART OF DOUBLE GALAXY
REIZ 4661	15 53 36.	+ 42 28	18	14.8	GALAXY
ZWG 298.005	15 53 36.	+ 61 31		15.7	GALAXY
MCG+03-41-002	15 53 42.	+ 17 18	90	15.0	GALAXY
MRK 292	15 53 42.	+ 19 01	12	16.5	GALAXY WITH UV CONTINUUM
REIZ 4662	15 53 42.	+ 42 05	36	14.6	GALAXY
MCG+08-29-021	15 53 42.	+ 48 06	42	16.	GALAXY
7ZW 628	15 53 42.	+ 63 11			COMPACT GALAXY
CED 126	15 53 47.	- 29 04			DIFFUSE GALACTIC NEBULA
ZWG 079.010	15 53 48.	+ 09 55		15.7	GALAXY
ZWG 223.013	15 53 48.	+ 42 05		15.4	GALAXY
UGC 10095	15 53 48.	+ 45 32	84	16.5	GALAXY DWRF IR
MCG+08-29-022	15 53 48.	+ 45 33	66	15.	GALAXY
ZWG 250.019	15 53 48.	+ 48 06		15.6	GALAXY
MCG-01-41-001	15 53 48.	- 08 47	24	17.	GALAXY
RNGC 6016	15 53 52.	+ 27 07		15.0	GALAXY
ZC 1553.9+1559	15 53 54.	+ 15 59	1010		CLUSTER OF GALAXIES
ZWG 137.002	15 53 54.	+ 21 32		15.6	GALAXY
MCG+05-38-001	15 53 54.	+ 27 04 30.	54	15.1	GALAXY
ZWG 167.004	15 53 54.	+ 27 07		15.1	GALAXY
UGC 10096	15 53 54.	+ 27 07	66	15.1	GALAXY Sc
REIZ 4663	15 53 54.	+ 38 53	60	14.8	GALAXY
MCG+07-33-012	15 53 54.	+ 42 06	39	16.	GALAXY
MCG+07-33-013	15 53 54.	+ 42 46	27	16.5	GALAXY
ZWG 250.020	15 53 54.	+ 47 01		15.6	GALAXY
ZWG 275.015	15 53 54.	+ 52 20		15.3	GALAXY
ZC 1553.9+6154	15 53 54.	+ 61 54	1480		CLUSTER OF GALAXIES
ZC 1553.9+7333	15 53 54.	+ 73 33	1280		CLUSTER OF GALAXIES
VHE 68	15 53 54.	- 53 50	66		REFLECTION NEBULA
ZWG 079.011	15 54 00.	+ 09 13		15.5	GALAXY
ZC 1554.0+2418	15 54 00.	+ 24 18	810		CLUSTER OF GALAXIES
ZWG 137.003	15 54 00.	+ 24 35		15.4	GALAXY
KARA.72 475B	15 54 00.	+ 24 35	72	15.4	PART OF DOUBLE GALAXY
MCG+05-38-002	15 54 00.	+ 30 30	45	15.3	GALAXY
ZWG 167.005	15 54 00.	+ 30 33		15.3	GALAXY
MCG+09-26-028	15 54 00.	+ 54 40	30	16.	GALAXY
MCG+14-07-030	15 54 00.	+ 84 27	36	17.	GALAXY
LB 00916	15 54 02.	+ 17 26 42.		16.4	FAINT BLUE STAR
ZWG 023.003	15 54 06.	+ 02 14		15.4	GALAXY
ZWG 108.012	15 54 06.	+ 16 40		14.7	GALAXY
ZC 1554.1+2145	15 54 06.	+ 21 45	540		CLUSTER OF GALAXIES
ZWG 223.014	15 54 06.	+ 39 55		15.4	GALAXY
REIZ 4664	15 54 06.	+ 42 03	30	15.0	GALAXY
MCG+07-33-014	15 54 06.	+ 42 46	30	15.5	GALAXY
MCG+03-41-003	15 54 12.	+ 16 40	66	14.7	GALAXY
ZC 1554.2+2239	15 54 12.	+ 22 39	810		CLUSTER OF GALAXIES
ZWG 250.021	15 54 12.	+ 48 00		14.1	GALAXY
UGC 10097	15 54 12.	+ 48 00	84	14.1	GALAXY S0
MCG+08-29-023	15 54 12.	+ 48 00	36	14.	GALAXY
MCG+09-26-029	15 54 12.	+ 53 11	12	16.	GALAXY
PK327-01.2	15 54 14.	- 55 33	5		PLANETARY NEBULA
ZWG 108.013	15 54 18.	+ 20 11		14.9	GALAXY
ZWG 250.022	15 54 18.	+ 49 40		15.3	GALAXY
KARA.73B 0714	15 54 18.	+ 49 40	36	15.3	ISOLATED GALAXY S
ZWG 079.012	15 54 24.	+ 14 08		15.6	GALAXY
MCG+03-41-004	15 54 24.	+ 20 13	42	14.9	GALAXY
MCG+04-38-001	15 54 24.	+ 21 24 30.	24	14.7	GALAXY
ZWG 137.004	15 54 24.	+ 21 27		14.7	GALAXY
ZC 1554.4+2551	15 54 24.	+ 25 51	2960		CLUSTER OF GALAXIES
MCG+05-38-003	15 54 24.	+ 28 31	36	15.5	GALAXY
MCG+07-33-015	15 54 24.	+ 42 16	30	15.	GALAXY
MCG-01-41-002	15 54 24.	- 06 02	36	17.	GALAXY
ZC 1554.5+1429	15 54 30.	+ 14 29	870		CLUSTER OF GALAXIES
ZC 1554.5+2444	15 54 30.	+ 24 44	1280		CLUSTER OF GALAXIES
ZC 1554.5+3955	15 54 30.	+ 39 55	3230		CLUSTER OF GALAXIES
ZWG 079.013	15 54 36.	+ 14 16		15.3	GALAXY
ZWG 250.023	15 54 36.	+ 48 14		14.9	GALAXY
ZWG 319.033	15 54 36.	+ 63 05		15.7	GALAXY
7ZW 629	15 54 36.	+ 67 10			COMPACT GALAXY
ZWG 023.004	15 54 36.	- 01 24		15.4	GALAXY
KARA.73B 0715	15 54 36.	- 01 24	54	15.4	ISOLATED GALAXY S
ZWG 108.014	15 54 42.	+ 18 47		15.5	GALAXY
TON-N 0807	15 54 42.	+ 32 13		15.1	BLUE STAR
ZWG 223.015	15 54 42.	+ 42 14		15.7	GALAXY
ZWG 223.016	15 54 42.	+ 42 32		15.5	GALAXY
ZC 1554.7+4301	15 54 42.	+ 43 01	940		CLUSTER OF GALAXIES
IC 1148	15 54 43.	+ 22 33 09.			SAME AS NGC 6020
ARC 2138	15 54 45.	+ 33 54		17.5	RICH CLUSTER OF GALAXIES
RNGC 6017	15 54 46.	+ 06 08		14.0	GALAXY
REIN 2.241	15 54 47.77	+ 06 08 29.2			NEBULA
ZWG 051.008	15 54 48.	+ 06 08		13.8	GALAXY
UGC 10098	15 54 48.	+ 06 08	48	13.8	GALAXY
MCG+01-41-003	15 54 48.	+ 06 08	48	13.8	GALAXY
ZC 1554.8+1613	15 54 48.	+ 16 13	870		CLUSTER OF GALAXIES
LB 00917	15 54 48.	+ 16 41 48.		17.5	FAINT BLUE STAR
ZWG 108.015	15 54 48.	+ 18 19		15.2	GALAXY
ZWG 223.017	15 54 48.	+ 42 02		15.0	GALAXY
REIZ 4665	15 54 48.	+ 42 02	24	14.6	GALAXY
ZC 1554.8+5106	15 54 48.	+ 51 06	1010		CLUSTER OF GALAXIES
MCG+09-26-031	15 54 48.	+ 53 10	30	17.	GALAXY
ASS 65	15 54 48.	- 54 21			OB ASSOCIATION NOR OB1
MCG+04-38-002	15 54 54.	+ 22 32 30.	30	14.5	GALAXY
ZWG 223.018	15 54 54.	+ 42 01		14.3	GALAXY
UGC 10099	15 54 54.	+ 42 01	24	14.3	GALAXY COMPACT
MCG+07-33-016	15 54 54.	+ 42 02	30	14.5	GALAXY
LYNG 11	15 54 58.	- 54 13 36.	5040		OB CONCENTRATION
ZWG 023.005	15 55 00.	+ 01 16		15.5	GALAXY
ZWG 137.005	15 55 00.	+ 22 33		14.5	GALAXY
UGC 10100	15 55 00.	+ 22 33	102	14.5	GALAXY E
ZC 1555.0+3235	15 55 00.	+ 32 35	610		CLUSTER OF GALAXIES
ZC 1555.0+3310	15 55 00.	+ 33 10	1880		CLUSTER OF GALAXIES
1ZW 129	15 55 00.	+ 42 01			COMPACT GALAXY
ZWG 367.004	15 55 00.	+ 81 55		15.3	GALAXY
ZWG 366.024	15 55 00.	+ 81 55		15.3	GALAXY
RNGC 6020	15 55 01.	+ 22 33		14.5	GALAXY
IC 1152	15 55 02.	+ 48 14 03.			NONSTELLAR OBJECT
PK325-04.1	15 55 02.	- 58 14 35.	25		PLANETARY NEBULA
ZC 1555.1+2212	15 55 06.	+ 22 12	2220		CLUSTER OF GALAXIES
MCG+04-38-003	15 55 06.	+ 25 57	48	15.3	GALAXY
ZWG 137.006	15 55 06.	+ 25 58		15.3	GALAXY
ZC 1555.1+4146	15 55 06.	+ 41 46	6790		CLUSTER OF GALAXIES
ARC 2140	15 55 06.	+ 48 48		17.5	RICH CLUSTER OF GALAXIES
RNGC 6018	15 55 09.	+ 16 00		14.5	GALAXY
RNGC 6021	15 55 09.	+ 16 05		14.0	GALAXY
ZWG 108.016	15 55 12.	+ 16 00		14.6	GALAXY
UGC 10101	15 55 12.	+ 16 00	96	14.6	GALAXY S0
ZWG 108.017	15 55 12.	+ 16 05		14.1	GALAXY
UGC 10102	15 55 12.	+ 16 05	96	14.1	GALAXY E
MCG+08-29-025	15 55 12.	+ 46 28	36	16.	GALAXY
ZWG 250.024	15 55 12.	+ 46 29		15.3	GALAXY
MCG+08-29-024	15 55 12.	+ 48 14	60	14.	GALAXY
ZC 1555.2-0032	15 55 12.	- 00 32	1610		CLUSTER OF GALAXIES
IC 1164	15 55 14.	+ 70 43			NONSTELLAR OBJECT
MCG+03-41-006	15 55 15.	+ 16 01	90	14.6	GALAXY
MCG+03-41-005	15 55 15.	+ 16 06	96	14.1	GALAXY
IC 1153	15 55 17.	+ 48 18 53.			NONSTELLAR OBJECT
ZWG 051.009	15 55 18.	+ 04 24		15.6	GALAXY
ZWG 108.018	15 55 18.	+ 18 10		15.3	GALAXY
ZC 1555.3+4529	15 55 18.	+ 45 29	810		CLUSTER OF GALAXIES
ZWG 250.025	15 55 18.	+ 48 14		14.4	GALAXY
UGC 10103	15 55 18.	+ 48 14	66	14.4	GALAXY E
MCG+10-23-008	15 55 18.	+ 58 51	42	16.	GALAXY
HN 0426	15 55 18.	- 66 14			NEBULA
IC 4584	15 55 18.	- 66 14			NONSTELLAR OBJECT
ZWG 108.019	15 55 24.	+ 16 21		15.7	GALAXY
MCG+03-41-007	15 55 24.	+ 18 11	24	15.3	GALAXY
MCG+05-38-004	15 55 24.	+ 30 11	180	14.9	GALAXY
ZWG 167.006	15 55 24.	+ 30 12		14.9	GALAXY
UGC 10104	15 55 24.	+ 30 12	168	14.9	GALAXY Sc
KARA.73B 0716	15 55 24.	+ 30 12	132	14.9	ISOLATED GALAXY S
UGC 10105	15 55 24.	+ 58 51	78	17.	GALAXY DWARF
HN 0427	15 55 24.	- 66 11			NEBULA
IC 4585	15 55 24.	- 66 11			NONSTELLAR OBJECT
RNGC 6022	15 55 28.	+ 16 25		15.0	GALAXY
RNGC 6023	15 55 28.	+ 16 27		14.5	GALAXY
ZWG 079.014	15 55 30.	+ 09 33		15.6	GALAXY
MCG+03-41-008	15 55 30.	+ 16 21	42	15.7	GALAXY
ZWG 108.020	15 55 30.	+ 16 25		15.2	GALAXY
ZWG 108.021	15 55 30.	+ 16 27		14.7	GALAXY
UGC 10106	15 55 30.	+ 16 27	72	14.7	GALAXY E
MCG+08-29-026	15 55 30.	+ 48 18	36	14.	GALAXY
MCG+14-07-031	15 55 30.	+ 81 55 30.	18	16.	GALAXY
MAI 096	15 55 31.	+ 58 54	53		DWARF SPHEROIDAL GALAXY
ZWG 051.010	15 55 36.	+ 06 55		15.6	GALAXY
MCG+03-41-009	15 55 36.	+ 16 25	42	15.2	GALAXY
MCG+03-41-010	15 55 36.	+ 16 27	30	14.7	GALAXY
SN 1969I	15 55 36.	+ 19 37		17.0	SUPERNOVA
TON-N 0808	15 55 36.	+ 31 28		16.7	BLUE STAR
ZC 1555.6+3735	15 55 36.	+ 37 35	2150		CLUSTER OF GALAXIES
ZWG 223.019	15 55 36.	+ 42 02		15.7	GALAXY
ZWG 250.026	15 55 36.	+ 48 18		13.6	GALAXY
UGC 10107	15 55 36.	+ 48 18	72	13.6	GALAXY S0
MCG+08-29-027	15 55 36.	+ 48 19	36	15.	GALAXY
ACK 343+15.1	15 55 36.	- 31 52			PLANETARY NEBULA
IC 1149	15 55 39.	+ 12 13 20.			NONSTELLAR OBJECT
ARC 2146	15 55 41.	+ 66 30		17.7	RICH CLUSTER OF GALAXIES

OBJECT NAME	RIGHT ASCEN.	DECLINATION	DIAM.	MAGN.	TYPE OF OBJECT
ZWG 051.011	15 55 42.	+ 07 13		15.7	GALAXY
KARA.73B 0717	15 55 42.	+ 07 13	24	15.7	ISOLATED GALAXY S
ZWG 108.022	15 55 42.	+ 15 05		15.6	GALAXY
ZWG 108.023	15 55 42.	+ 16 29		15.3	GALAXY
ZC 1555.7+3535	15 55 42.	+ 35 35	1010		CLUSTER OF GALAXIES
MCG+07-33-017	15 55 42.	+ 42 02	30	16.5	GALAXY
ZWG 250.027	15 55 42.	+ 48 19		15.1	GALAXY
MCG+10-23-009	15 55 42.	+ 59 56	60	15.	GALAXY
MCG+05-38-005	15 55 45.	+ 28 52 30.	48	15.5	GALAXY
MCG+09-26-030	15 55 45.	+ 51 38	60	17.	GALAXY
ZWG 051.012	15 55 48.	+ 03 05		15.5	GALAXY
ZWG 051.013	15 55 48.	+ 04 35		15.4	GALAXY
ZWG 079.015	15 55 48.	+ 12 13		14.1	GALAXY
UGC 10108	15 55 48.	+ 12 13	78	14.1	GALAXY Sb-c
MCG+02-41-001	15 55 48.	+ 12 13	72	14.1	GALAXY
MCG+03-41-011	15 55 48.	+ 15 07	48	15.6	GALAXY
ZWG 108.024	15 55 48.	+ 15 27		15.6	GALAXY
MCG+03-41-012	15 55 48.	+ 16 29	48	15.3	GALAXY
ZWG 250.028	15 55 48.	+ 47 18		14.9	GALAXY
UGC 10109	15 55 48.	+ 47 18	84	14.9	GALAXY Sb/SBb
MCG+08-29-028	15 55 48.	+ 47 18	90	15.	GALAXY
1ZW 130	15 55 48.	+ 51 40			COMPACT GALAXY
MCG+10-23-010	15 55 48.	+ 60 04	30	16.	GALAXY
DG 130	15 55 48.	- 25 59	2640		REFLECTION NEBULA
MRSL 347+20/1	15 55 48.	- 25 59	9000		HII REGION
MRSL 328-00/1	15 55 48.	- 53 35	240		HII REGION
ACK 328-00.1	15 55 48.	- 53 37			PLANETARY NEBULA
CED 125D	15 55 49.	- 25 59			DIFFUSE GALACTIC NEBULA
ARC 2141	15 55 52.	+ 35 36		17.2	RICH CLUSTER OF GALAXIES
VDB.66N 099	15 55 52.	- 26 00	12000		REFLECTION NEBULA
UGC 10110	15 55 54.	+ 28 57	72	16.0	GALAXY SO?
TON-N 0809	15 55 54.	+ 31 47		15.0	BLUE STAR
UGC 10111	15 56 00.	+ 13 27	108	16.0	GALAXY PECULR?
IC 1150	15 56 00.	+ 16 00 35.			NONSTELLAR OBJECT
ZWG 108.025	15 56 00.	+ 18 14		15.6	GALAXY
ZWG 108.026	15 56 00.	+ 20 06		15.2	GALAXY
UGC 10112	15 56 00.	+ 20 06	60	15.2	GALAXY S0-a
ZC 1556.0+2310	15 56 00.	+ 23 10	470		CLUSTER OF GALAXIES
ZWG 250.029	15 56 00.	+ 48 24		15.6	GALAXY
MCG+09-26-032	15 56 00.	+ 54 50	15	15.	GALAXY
ZWG 366.025	15 56 00.	+ 80 15		15.7	GALAXY
ZWG 355.002	15 56 00.	+ 80 15		15.7	GALAXY
ZWG 354.028	15 56 00.	+ 80 15		15.7	GALAXY
MCG+14-07-032	15 56 00.	+ 81 59	12	17.	GALAXY
PK327-02.1	15 56 01.	- 55 47 11.	5		PLANETARY NEBULA
ZWG 108.027	15 56 06.	+ 18 11		15.7	GALAXY
ZC 1556.1+2141	15 56 06.	+ 21 41	1340		CLUSTER OF GALAXIES
ZWG 137.007	15 56 06.	+ 22 49		15.7	GALAXY
ZC 1556.1+2829	15 56 06.	+ 28 29	5510		CLUSTER OF GALAXIES
1ZW 131	15 56 06.	+ 41 52			COMPACT GALAXY
ZWG 275.016	15 56 06.	+ 54 49		15.6	GALAXY
ARC 2143	15 56 10.	+ 37 32		16.6	RICH CLUSTER OF GALAXIES
ZWG 108.028	15 56 12.	+ 17 35		13.4	GALAXY
UGC 10113	15 56 12.	+ 17 35	156	13.4	GALAXY SBc
MCG+03-41-013	15 56 12.	+ 18 14	36	15.7	GALAXY
MCG+03-41-014	15 56 12.	+ 20 07	60		GALAXY
ZC 1556.2+2725	15 56 12.	+ 27 25	3090		CLUSTER OF GALAXIES
TON-N 0251	15 56 12.	+ 30 19		15.0	BLUE STAR
ZWG 223.020	15 56 12.	+ 40 10		15.2	GALAXY
ZWG 223.021	15 56 12.	+ 41 55		15.1	GALAXY
REIZ 4666	15 56 12.	+ 41 56	60	14.3	GALAXY
ZWG 223.022	15 56 12.	+ 42 04		15.0	GALAXY
MCG+07-33-018	15 56 12.	+ 42 05	33	14.5	GALAXY
MCG+09-26-033	15 56 12.	+ 53 30	36	16.	GALAXY
MCG+10-23-011	15 56 12.	+ 60 07	42	16.	GALAXY
ARC 2142	15 56 14.	+ 27 22		16.0	RICH CLUSTER OF GALAXIES
REIF 7.165	15 56 15.89	+ 60 07 02.7			NEBULA
IC 1151	15 56 16.5	+ 17 35 04.		13.4	GALAXY SB(sr)bc
SC 1552-5803.3	15 56 17.	- 58 12 01.	18		NEBULA
MCG+03-41-015	15 56 18.	+ 17 35	156	13.4	GALAXY
ZWG 108.029	15 56 18.	+ 20 19		15.7	GALAXY
MCG+07-33-019	15 56 18.	+ 40 10	18	15.	GALAXY
7ZW 630	15 56 18.	+ 77 30			COMPACT GALAXY
ZWG 051.014	15 56 24.	+ 08 19		15.5	GALAXY
ZWG 079.016	15 56 24.	+ 14 20		15.4	GALAXY
MCG+06-35-016	15 56 24.	+ 37 04	48	14.5	GALAXY
ZC 1556.4+6023	15 56 24.	+ 60 23	1410		CLUSTER OF GALAXIES
ZWG 319.034	15 56 24.	+ 63 59		15.3	GALAXY
MCG+11-19-030	15 56 24.	+ 64 01 30.	24	16.	GALAXY
ZWG 319.035	15 56 24.	+ 67 07		15.6	GALAXY
MCG+11-19-029	15 56 24.	+ 67 07 30.	27	16.	GALAXY
MCG+12-15-037	15 56 24.	+ 71 09	21	16.	GALAXY
ZC 1556.5+0245	15 56 30.	+ 02 45	3700		CLUSTER OF GALAXIES
ZWG 079.017	15 56 30.	+ 13 02		15.7	GALAXY
ZWG 108.030	15 56 30.	+ 15 05		15.1	GALAXY
ZWG 167.007	15 56 30.	+ 27 07		15.7	GALAXY
ZWG 167.008	15 56 30.	+ 27 45		15.2	GALAXY
UGC 10114	15 56 30.	+ 37 01	78	16.0	GALAXY Sc
ZWG 319.036	15 56 30.	+ 64 04		13.9	GALAXY
UGC 10115	15 56 30.	+ 64 04	84	13.9	GALAXY E
MCG+11-19-031	15 56 30.	+ 64 05	30	16.	GALAXY
ZC 1556.5+6446	15 56 30.	+ 64 46	1550		CLUSTER OF GALAXIES
MCG+03-41-016	15 56 33.	+ 15 06	30	15.1	GALAXY
ZWG 108.031	15 56 36.	+ 15 06		15.6	GALAXY
MCG+03-41-017	15 56 36.	+ 18 30	30		GALAXY
LF 00918	15 56 36.	+ 19 31 06.		16.2	FAINT BLUE STAR
ZWG 167.009	15 56 36.	+ 26 58		14.9	GALAXY
MRK 492	15 56 36.	+ 26 58	25	15.5	GALAXY WITH UV CONTINUUM
TON-N 0810	15 56 36.	+ 28 37		16.5	BLUE STAR
ZC 1556.6+5635	15 56 36.	+ 56 35	1080		CLUSTER OF GALAXIES
OCL 0950	15 56 36.	- 53 23	960	11.9	OPEN STAR CLUSTER
VHA 180	15 56 36.	- 53 23	360		OPEN STAR CLUSTER
MCG+03-41-018	15 56 39.	+ 15 07	36	15.6	GALAXY
ZWG 108.032	15 56 42.	+ 19 53		15.6	GALAXY
ZWG 137.008	15 56 42.	+ 25 02		15.6	GALAXY
MCG+04-38-004	15 56 42.	+ 26 17	48	15.5	GALAXY
MCG+05-38-006	15 56 42.	+ 28 55 30.	36	14.9	GALAXY
ZC 1556.7+3257	15 56 42.	+ 32 57	1010		CLUSTER OF GALAXIES
ZWG 338.031	15 56 42.	+ 68 32		15.6	GALAXY
ZWG 320.001	15 56 42.	+ 68 32		15.6	GALAXY
KARA.73B 0718	15 56 42.	+ 68 32	18	15.6	ISOLATED GALAXY E
ZWG 108.033	15 56 48.	+ 15 03		15.4	GALAXY
MCG+03-41-019	15 56 48.	+ 15 04	36	15.4	GALAXY
ZWG 137.009	15 56 48.	+ 26 16		15.5	GALAXY
ZC 1556.8+2641	15 56 48.	+ 26 41	4770		CLUSTER OF GALAXIES
ZWG 298.006	15 56 54.	+ 58 18		15.1	GALAXY
LBN 1095	15 57	- 25 30	3600		BRIGHT NEBULA
LBN 1094	15 57	- 26 20	1920		BRIGHT NEBULA
ZWG 108.034	15 57 00.	+ 15 18		15.3	GALAXY
RNGC 6027E	15 57 00.	+ 20 51		16.5	GALAXY
RNGC 6027C	15 57 00.	+ 20 52		15.5	GALAXY
RNGC 6027B	15 57 00.	+ 20 52		15.0	GALAXY
MCG+04-38-006	15 57 00.	+ 20 52	18	13.4	GALAXY
MCG+04-38-005	15 57 00.	+ 20 52 30.	18	13.4	GALAXY
RNGC 6027D	15 57 00.	+ 20 54		17.0	GALAXY
RNGC 6027A	15 57 00.	+ 20 54		14.5	GALAXY
VV 115E	15 57 00.	+ 20 54 30.	42	16.9	INTERACTING GALAXY
VV 115D	15 57 00.	+ 20 54 30.	16	16.8	INTERACTING GALAXY
VV 115C	15 57 00.	+ 20 54 30.	42	16.7	INTERACTING GALAXY
VV 115E	15 57 00.	+ 20 54 30.	36	15.4	INTERACTING GALAXY
VV 115A	15 57 00.	+ 20 54 30.	60	15.1	INTERACTING GALAXY
VV 115	15 57 00.	+ 20 54 30.	126		INTERACTING GALAXY
7ZW 631	15 57 00.	+ 20 55			COMPACT GALAXY
ZWG 137.010	15 57 00.	+ 20 55		13.4	GALAXY
UGC 10116	15 57 00.	+ 20 55	114	13.4	GALAXY GROUP
ZWG 223.023	15 57 00.	+ 42 04		15.1	GALAXY
MCG+13-11-017	15 57 00.	+ 79 08	45	14.1	GALAXY
MCG+13-11-018	15 57 00.	+ 79 20	33	16.	GALAXY
MCG+04-38-007	15 57 02.	+ 20 51 30.	48	13.4	GALAXY
ZWG 051.015	15 57 06.	+ 06 28		15.7	GALAXY
ZWG 079.018	15 57 06.	+ 09 58		15.2	GALAXY
ZC 1557.1+1448	15 57 06.	+ 14 48	340		CLUSTER OF GALAXIES
RNGC 6027F	15 57 06.	+ 20 52		17.0	GALAXY
MCG+04-38-009	15 57 06.	+ 20 52	9	13.4	GALAXY
MCG+04-38-008	15 57 06.	+ 20 52 30.	24	13.4	GALAXY
ZC 1557.1+4021	15 57 06.	+ 40 21	940		CLUSTER OF GALAXIES
MCG+07-33-020	15 57 06.	+ 42 04	27	15.	GALAXY
PK327-01.1	15 57 06.	- 54 57	25		PLANETARY NEBULA
MCG+04-38-010	15 57 08.	+ 20 52 30.	36	13.4	GALAXY
ZWG 051.016	15 57 12.	+ 02 51		14.8	GALAXY
MCG+01-41-004	15 57 12.	+ 02 51	36	14.8	GALAXY
ZWG 108.035	15 57 12.	+ 20 21		15.7	GALAXY
ZWG 137.011	15 57 12.	+ 21 45		15.4	GALAXY
UGC 10117	15 57 12.	+ 21 45	72	15.4	GALAXY Sa-b
MCG+05-38-007	15 57 12.	+ 28 28	48	15.7	GALAXY
ZWG 167.010	15 57 12.	+ 28 32		15.7	GALAXY
MCG+06-35-017	15 57 12.	+ 35 11	66	14.	GALAXY
ZWG 250.030	15 57 12.	+ 48 49		14.3	GALAXY
UGC 10118	15 57 12.	+ 48 49	66	14.3	GALAXY E-S0
MCG+08-29-029	15 57 12.	+ 48 49	60	15.	GALAXY
ZWG 319.037	15 57 12.	+ 64 09		15.5	GALAXY
MCG+11-19-032	15 57 12.	+ 64 11	27	15.	GALAXY
RNGC 6068A	15 57 12.	+ 79 08		14.5	GALAXY
ZWG 355.003	15 57 12.	+ 79 18		15.5	GALAXY
ZWG 354.029	15 57 12.	+ 79 18		15.5	GALAXY
UGC 10119	15 57 12.	+ 79 18	72	15.5	GALAXY S?
MCG+04-38-011	15 57 15.	+ 21 43 30.	66	15.4	GALAXY
ZWG 051.017	15 57 18.	+ 02 48		15.2	GALAXY
ZWG 195.003	15 57 18.	+ 35 10		14.9	GALAXY
MRK 493	15 57 18.	+ 35 10	10	16.	GALAXY WITH UV CONTINUUM
UGC 10120	15 57 18.	+ 35 10	78	14.9	GALAXY SBb
KARA.73B 0719	15 57 18.	+ 35 10	66	14.9	ISOLATED GALAXY S
MCG+06-35-018	15 57 18.	+ 35 53	30	15.5	GALAXY
ZWG 355.004	15 57 18.	+ 79 07		14.7	GALAXY
ZWG 354.030	15 57 18.	+ 79 07		14.7	GALAXY
KARA.72 476A	15 57 18.	+ 79 07	54	14.7	PART OF DOUBLE GALAXY
MRSL 008+36/1	15 57 18.	- 01 29	2100		HII REGION
CED 125E	15 57 21.	- 22 29			DIFFUSE GALACTIC NEBULA
ZWG 051.018	15 57 24.	+ 04 07		15.6	GALAXY
MRK 293	15 57 24.	+ 18 52	8	17.5	GALAXY WITH UV CONTINUUM
ZWG 250.031	15 57 24.	+ 44 54		15.5	GALAXY
ZC 1557.4+6235	15 57 24.	+ 62 35	3020		CLUSTER OF GALAXIES
MRSL 349+22/1	15 57 24.	- 22 49	14400		HII REGION
ZWG 108.036	15 57 30.	+ 18 56		14.6	GALAXY
UGC 10121	15 57 30.	+ 18 56	78	14.6	GALAXY SBb-c
MCG+03-41-020	15 57 30.	+ 18 56	48	14.6	GALAXY
ZWG 250.032	15 57 30.	+ 46 10		15.6	GALAXY
ZWG 250.033	15 57 30.	+ 48 40		15.5	GALAXY
MCG+08-29-030	15 57 30.	+ 48 40	60	16.	GALAXY
MCG+09-26-035	15 57 30.	+ 51 25	60	14.	GALAXY
MCG+09-26-034	15 57 30.	+ 52 55	54	17.	GALAXY
ZC 1557.5+7036	15 57 30.	+ 70 36	1410		CLUSTER OF GALAXIES
MRSL 328-01/1	15 57 30.	- 54 01	10800		HII REGION
HOLM 727B	15 57 31.	+ 79 07	30	14.1	PART OF MULTIPLE GALAXY
ZWG 051.019	15 57 36.	+ 05 01		15.7	GALAXY
ZWG 108.037	15 57 36.	+ 15 44		15.6	GALAXY
ZC 1557.6+2218	15 57 36.	+ 22 18	740		CLUSTER OF GALAXIES
MCG+06-35-019	15 57 36.	+ 33 35 30.	30	17.	GALAXY
MCG+06-35-020	15 57 39.	+ 36 50	24	17.	GALAXY
ARC 2150	15 57 40.	+ 71 33		17.7	RICH CLUSTER OF GALAXIES
ZWG 223.024	15 57 42.	+ 39 58		15.5	GALAXY
UGC 10122	15 57 42.	+ 39 58	78	15.5	GALAXY Sa
MCG+07-33-021	15 57 42.	+ 43 04	30	16.	GALAXY
ZWG 275.017	15 57 42.	+ 51 28		15.0	GALAXY
UGC 10123	15 57 42.	+ 51 28	84	15.0	GALAXY Sa-b
MCG+11-19-033	15 57 42.	+ 64 04 30.	30	16.	GALAXY
MCG-03-41-001	15 57 42.	- 20 17	36	15.	GALAXY
FATH 1.731	15 57 43.	+ 29 34	8		NEBULA
IC 4587	15 57 46.	+ 26 04 54.			NONSTELLAR OBJECT
ZWG 137.012	15 57 48.	+ 26 05		15.5	GALAXY
MCG+07-33-022	15 57 48.	+ 39 57 30.	72	15.	GALAXY
ZWG 338.032	15 57 48.	+ 70 50		13.6	GALAXY
RNGC 6048	15 57 48.	+ 70 50		13.5	GALAXY
UGC 10124	15 57 48.	+ 70 50	132	13.6	GALAXY E
MCG+12-15-038	15 57 48.	+ 70 50	51	14.	GALAXY
RNGC 6068	15 57 48.	+ 79 08		13.5	GALAXY
MCG+13-11-019	15 57 48.	+ 79 09	54	13.3	GALAXY
ZWG 195.004	15 57 54.	+ 37 10		15.1	GALAXY
MCG+06-35-021	15 57 54.	+ 37 12	36	14.	GALAXY
MCG+03-41-021	15 57 57.	+ 18 32	9		GALAXY
LBN 0037	15 58	- 01 30	1500		BRIGHT NEBULA
LBN 0036	15 58	- 01 30	2400		BRIGHT NEBULA
SIV 10	15 58	- 20 11	50400		FAINT H EMISSION REGION
LBN 1099	15 58	- 23 00	6900		BRIGHT NEBULA
ZWG 023.007	15 58 00.	+ 00 39		15.3	GALAXY
ZWG 108.038	15 58 00.	+ 15 54		15.6	GALAXY
ZWG 108.039	15 58 00.	+ 16 17		15.6	GALAXY
ZC 1558.0+5021	15 58 00.	+ 50 21	1080		CLUSTER OF GALAXIES
MCG+09-26-037	15 58 00.	+ 53 25	48	16.	GALAXY
MCG+09-26-036	15 58 00.	+ 53 30	36	17.	GALAXY
MCG+10-23-012	15 58 00.	+ 59 32	12	16.	GALAXY
ZWG 355.005	15 58 00.	+ 79 08		13.3	GALAXY
ZWG 354.031	15 58 00.	+ 79 08		13.3	GALAXY
UGC 10126	15 58 00.	+ 79 08	66	13.3	GALAXY SB
KARA.72 476B	15 58 00.	+ 79 08	78	13.3	PART OF DOUBLE GALAXY
ZWG 367.005	15 58 00.	+ 81 58		15.5	GALAXY

OBJECT NAME	RIGHT ASCEN.	DECLINATION	DIAM.	MAGN.	TYPE OF OBJECT
ZWG 366.026	15 58 00.	+ 81 58		15.5	GALAXY
UGC 10125	15 58 00.	+ 84 00	84	17.	GALAXY DWARF
ZWG 023.006	15 58 00.	- 00 01		15.6	GALAXY
KARA.73B 0720	15 58 00.	- 00 01	60	15.6	ISOLATED GALAXY S
MCG+03-41-022	15 58 03.	+ 18 32	30		GALAXY
PK341+13.1	15 58 05.	- 34 23 32.	55		PLANETARY NEBULA
RNGC 6026	15 58 05.	- 34 25			PLANETARY NEBULA
SC 1554-5546.8	15 58 05.	- 55 55 24.	18		NEBULA
VV 156	15 58 06.	+ 18 32	30	16.	INTERACTING GALAXY
ZWG 108.040	15 58 06.	+ 19 35		15.7	GALAXY
PATH 1.732	15 58 11.	+ 29 33	5		NEBULA
ZC 1558.2+0024	15 58 12.	+ 00 24	870		CLUSTER OF GALAXIES
ZWG 108.041	15 58 12.	+ 16 46		15.7	GALAXY
ZWG 137.013	15 58 12.	+ 21 00		14.2	GALAXY
UGC 10127	15 58 12.	+ 21 00	96	14.2	GALAXY Sb
ZWG 223.025	15 58 12.	+ 41 37		15.2	GALAXY
KARA.72 477A	15 58 12.	+ 41 37	48	15.2	PART OF DOUBLE GALAXY
MCG+10-23-013	15 58 12.	+ 58 00	30	16.	GALAXY
HOLM 727A	15 58 13.	+ 79 07	54	13.3	PART OF MULTIPLE GALAXY
MCG+07-33-023	15 58 15.	+ 41 26	42	15.	GALAXY
MCG-01-41-003	15 58 15.	- 06 28	24	18.	GALAXY
MCG+03-41-023	15 58 18.	+ 15 49	48	14.9	GALAXY
ZWG 108.042	15 58 18.	+ 15 50		14.9	GALAXY
IC 1155	15 58 18.	+ 15 50			GALAXY
MCG+04-38-012	15 58 18.	+ 20 58	96	14.2	GALAXY
ZWG 223.026	15 58 18.	+ 41 37		15.3	GALAXY
KARA.72 477B	15 58 18.	+ 41 37	36	15.3	PART OF DOUBLE GALAXY
MCG+07-33-024	15 58 21.	+ 41 36	27	15.	GALAXY
ABC 2145	15 58 22.	+ 33 25		16.6	RICH CLUSTER OF GALAXIES
MCG+03-41-024	15 58 24.	+ 16 49	30		GALAXY
ZWG 108.043	15 58 24.	+ 16 51		15.4	GALAXY
ZWG 108.044	15 58 24.	+ 19 52		14.9	GALAXY
MCG+03-41-025	15 58 24.	+ 19 52	48	14.9	GALAXY
IC 1156	15 58 24.	+ 19 52		14.9	GALAXY
ZWG 167.011	15 58 24.	+ 30 31		15.2	GALAXY
MRK 494	15 58 24.	+ 30 31	8	16.5	GALAXY WITH UV CONTINUUM
UGC 10128	15 58 24.	+ 30 31	72	15.2	GALAXY DBL SYS
MCG+09-26-038	15 58 24.	+ 52 53	42	16.	GALAXY
MCG+03-41-026	15 58 27.	+ 16 51	42	15.4	GALAXY
ZWG 108.045	15 58 30.	+ 15 17		15.4	GALAXY
ZWG 108.046	15 58 30.	+ 18 13		15.5	GALAXY
MCG+05-38-009	15 58 30.	+ 30 30 30.	15	15.2	GALAXY
MCG+05-38-008	15 58 30.	+ 30 31	24	15.2	GALAXY
ZWG 079.019	15 58 36.	+ 12 48		14.8	GALAXY
MCG+02-41-002	15 58 36.	+ 12 48	36	14.8	GALAXY
ZC 1558.6+1306	15 58 36.	+ 13 06	340		CLUSTER OF GALAXIES
ZWG 108.047	15 58 36.	+ 15 40		15.7	GALAXY
ZWG 108.048	15 58 36.	+ 16 28		15.0	GALAXY
MCG+03-41-028	15 58 36.	+ 16 28 30.	30	15.0	GALAXY
MCG+03-41-029	15 58 36.	+ 16 29	42	15.0	GALAXY
ZWG 108.049	15 58 36.	+ 17 41		15.4	GALAXY
ZWG 108.050	15 58 36.	+ 19 04		15.6	GALAXY
SN 1956A	15 58 36.	+ 19 52		19.0	SUPERNOVA
MCG+03-41-027	15 58 36.	+ 19 52	18		GALAXY
ZWG 250.034	15 58 36.	+ 46 55		15.4	GALAXY
MCG+09-26-039	15 58 36.	+ 51 56	60	15.	GALAXY
ZC 1558.6+5410	15 58 36.	+ 54 10	1140		CLUSTER OF GALAXIES
IC 1157	15 58 38.	+ 15 29 41.			NONSTELLAR OBJECT
MCG+03-41-030	15 58 39.	+ 17 41	42	15.4	GALAXY
ZWG 108.051	15 58 42.	+ 15 38		15.7	GALAXY
MCG+03-41-031	15 58 42.	+ 15 39	36	15.7	GALAXY
ZWG 108.052	15 58 42.	+ 18 45		15.7	GALAXY
2ZW 074	15 58 42.	+ 18 58			COMPACT GALAXY
MCG+05-38-010	15 58 42.	+ 28 11	30	15.5	GALAXY
ZWG 167.012	15 58 42.	+ 28 14		15.5	GALAXY
1ZW 132	15 58 42.	+ 51 58			COMPACT GALAXY
MCG+10-23-014	15 58 42.	+ 59 37	18	16.	GALAXY
ZWG 338.033	15 58 42.	+ 70 33		15.6	GALAXY
IC 1159	15 58 44.	+ 15 33 05.			NONSTELLAR OBJECT
IC 1160	15 58 46.	+ 15 37 30.			NONSTELLAR OBJECT
ZWG 079.020	15 58 48.	+ 12 05		15.7	GALAXY
MCG+03-41-032	15 58 48.	+ 15 37 30.	21	15.7	GALAXY
ZWG 108.053	15 58 48.	+ 19 35		15.0	GALAXY
MCG+07-33-025	15 58 48.	+ 44 05	42	14.5	GALAXY
ZWG 275.018	15 58 48.	+ 51 58		15.7	GALAXY
UGC 10129	15 58 48.	+ 51 58	72	15.7	GALAXY PECULR
ZWG 298.007	15 58 48.	+ 59 37		15.0	GALAXY
ZWG 319.038	15 58 48.	+ 65 23		15.3	GALAXY
MCG+12-15-039	15 58 48.	+ 70 33	36	15.	GALAXY
ACK 336+08.1	15 58 48.	- 41 25			PLANETARY NEBULA
ZWG 079.021	15 58 54.	+ 08 58		15.1	GALAXY
UGC 10130	15 58 54.	+ 08 58	84	15.1	GALAXY Sa
UGC 10131	15 58 54.	+ 14 14	66	16.0	GALAXY Sb
ZWG 223.027	15 58 54.	+ 44 04		15.2	GALAXY
UGC 10132	15 58 54.	+ 78 37	72	16.0	GALAXY SBb/Sc
MCG+03-41-033	15 58 57.	+ 19 36	48	15.0	GALAXY
IC 1161	15 58 59.	+ 15 47 31.			NONSTELLAR OBJECT
ZWG 023.008	15 59 00.	+ 01 51		14.4	GALAXY
UGC 10133	15 59 00.	+ 01 51	168	14.4	GALAXY Sc/SBc
MCG+00-41-002	15 59 00.	+ 01 51	138	14.4	GALAXY
ZWG 108.054	15 59 00.	+ 15 47		15.2	GALAXY
MCG+03-41-036	15 59 00.	+ 15 47	36	15.2	GALAXY
ZWG 108.055	15 59 00.	+ 16 27		15.2	GALAXY
UGC 10134	15 59 00.	+ 16 27	60	15.2	GALAXY SBb
ZWG 108.056	15 59 00.	+ 17 49		15.2	GALAXY
MCG+03-41-034	15 59 00.	+ 17 50	42	15.2	GALAXY
ZWG 108.057	15 59 00.	+ 17 54		15.6	GALAXY
MCG+03-41-035	15 59 00.	+ 17 55	24	15.6	GALAXY
ZC 1559.0+2320	15 59 00.	+ 23 20	540		CLUSTER OF GALAXIES
ZWG 167.013	15 59 00.	+ 28 51		15.4	GALAXY
ZC 1559.0+3819	15 59 00.	+ 38 19	2490		CLUSTER OF GALAXIES
ZC 1559.0+5353	15 59 00.	+ 53 53	6320		CLUSTER OF GALAXIES
ZWG 355.006	15 59 00.	+ 78 37		15.7	GALAXY
ZWG 354.032	15 59 00.	+ 78 37		15.7	GALAXY
MCG+13-11-020	15 59 00.	+ 78 47	66	16.	GALAXY
MCG+14-07-033	15 59 00.	+ 81 56	15	16.	GALAXY
IC 1162	15 59 01.2	+ 17 49 01.		15.2	GALAXY SBc
MCG+03-41-037	15 59 03.	+ 16 26 30.	66	15.2	GALAXY
ZWG 108.058	15 59 06.	+ 16 21		15.3	GALAXY
ZWG 108.059	15 59 06.	+ 16 49		15.2	GALAXY
MCG+03-41-038	15 59 06.	+ 16 49	42	15.2	GALAXY
ZWG 275.019	15 59 06.	+ 53 03		15.2	GALAXY
MCG+10-23-015	15 59 06.	+ 58 30	36	14.	GALAXY
IC 1158	15 59 07.	+ 01 51 07.			NONSTELLAR OBJECT
MCG+06-35-022	15 59 09.	+ 36 45	48	15.	GALAXY
RNGC 6028	15 59 11.	+ 19 29		15.0	GALAXY
IC 1187	15 59 11.	+ 70 42			NONSTELLAR OBJECT
REIN 7.166	15 59 11.59	+ 58 31 30.3			NEBULA

OBJECT NAME	RIGHT ASCEN.	DECLINATION	DIAM.	MAGN.	TYPE OF OBJECT
ZWG 079.022	15 59 12.	+ 12 30		14.7	GALAXY
MCG+02-41-003	15 59 12.	+ 12 30	48	14.7	GALAXY
ZWG 108.060	15 59 12.	+ 15 38		15.3	GALAXY
MCG+03-41-039	15 59 12.	+ 15 38	12	15.3	GALAXY
IC 1163	15 59 12.	+ 15 38 09.			NONSTELLAR OBJECT
ZWG 108.061	15 59 12.	+ 16 53		15.4	GALAXY
MCG+03-41-040	15 59 12.	+ 16 54	27	15.4	GALAXY
ZWG 108.062	15 59 12.	+ 17 23		15.5	GALAXY
MCG+03-41-041	15 59 12.	+ 17 23	24	15.5	GALAXY
MCG+03-41-042	15 59 12.	+ 18 10	48		GALAXY
ZWG 108.063	15 59 12.	+ 19 29		14.8	GALAXY
UGC 10135	15 59 12.	+ 19 29	78	14.8	GALAXY S0
ZC 1559.2+3234	15 59 12.	+ 32 34	740		CLUSTER OF GALAXIES
ZWG 195.005	15 59 12.	+ 36 44		15.6	GALAXY
UGC 10136	15 59 12.	+ 36 44	60	15.6	GALAXY S
ZWG 250.035	15 59 12.	+ 48 42		15.0	GALAXY
MCG+08-29-031	15 59 12.	+ 48 42	48	15.	GALAXY
ZWG 298.008	15 59 12.	+ 58 32		14.8	GALAXY
ZWG 023.009	15 59 12.	- 00 40		15.5	GALAXY
PATH 1.733	15 59 16.	+ 29 58	27		NEBULA
ZWG 108.064	15 59 18.	+ 16 34		15.6	GALAXY
1ZW 133	15 59 18.	+ 19 29			COMPACT GALAXY
MCG+03-41-043	15 59 18.	+ 19 30	84	14.8	GALAXY
ZWG 137.014	15 59 18.	+ 22 34		15.1	GALAXY
ZWG 137.015	15 59 18.	+ 22 56		15.6	GALAXY
MCG+08-29-032	15 59 18.	+ 45 33	30	16.	GALAXY
ZC 1559.3+4615	15 59 18.	+ 46 15	2420		CLUSTER OF GALAXIES
MCG+04-38-014	15 59 21.	+ 22 34	60	15.1	GALAXY
MCG+04-38-013	15 59 21.	+ 22 56	48	15.6	GALAXY
RNGC 6025	15 59 22.	- 60 22		6.0	OPEN CLUSTER
ZWG 051.020	15 59 24.	+ 08 17		15.1	GALAXY
UGC 10137	15 59 24.	+ 08 17	78	15.1	GALAXY Sc
MCG+01-41-005	15 59 24.	+ 08 17	60	15.1	GALAXY
ZWG 250.036	15 59 24.	+ 45 33		15.5	GALAXY
MCG+09-26-040	15 59 24.	+ 55 08	18	16.	GALAXY
MCG+09-26-041	15 59 24.	+ 55 15	36	16.	GALAXY
ZWG 338.034	15 59 24.	+ 70 42		15.7	GALAXY
OCL 0939	15 59 24.	- 60 22	1680	6.1	OPEN STAR CLUSTER
VHA 181	15 59 24.	- 60 22	780		OPEN STAR CLUSTER
ZWG 023.010	15 59 30.	+ 01 38		15.7	GALAXY
ZC 1559.5+1321	15 59 30.	+ 13 21	1010		CLUSTER OF GALAXIES
ZWG 137.016	15 59 30.	+ 20 34		15.4	GALAXY
ZWG 137.017	15 59 30.	+ 21 30		15.2	GALAXY
UGC 10138	15 59 30.	+ 21 30	96	15.2	GALAXY Sa
MCG+06-35-023	15 59 30.	+ 35 58 30.	48	15.	GALAXY
MCG+12-15-040	15 59 30.	+ 70 42 30.	18	16.	GALAXY
RNGC 6030	15 59 35.	+ 18 06		14.5	GALAXY
ZWG 079.023	15 59 36.	+ 12 42		15.6	GALAXY
ZWG 079.024	15 59 36.	+ 13 32		15.5	GALAXY
VV 159F	15 59 36.	+ 16 03 00.	36	16.	INTERACTING GALAXY
VV 159E	15 59 36.	+ 16 03 00.	12	17.	INTERACTING GALAXY
VV 159D	15 59 36.	+ 16 03 00.	9	16.	INTERACTING GALAXY
VV 159C	15 59 36.	+ 16 03 00.	27	14.5	INTERACTING GALAXY
MCG+03-41-045	15 59 36.	+ 17 47	30		GALAXY
ZWG 108.065	15 59 36.	+ 18 06		14.5	GALAXY
UGC 10139	15 59 36.	+ 18 06	84	14.5	GALAXY S0-a
MCG+03-41-044	15 59 36.	+ 18 06	42	14.5	GALAXY
UGC 10140	15 59 36.	+ 18 53	60	18.	GALAXY DWARF
ZWG 167.014	15 59 36.	+ 31 52		15.5	GALAXY
UGC 10141	15 59 36.	+ 35 57	60	16.0	GALAXY S
RNGC 6029	15 59 38.	+ 12 42		15.5	GALAXY
MCG-04-38-001	15 59 39.	- 25 40 30.	48	16.	GALAXY
ZWG 051.021	15 59 42.	+ 07 14		15.0	GALAXY
MCG+01-41-006	15 59 42.	+ 07 14	54	15.0	GALAXY
ZWG 108.066	15 59 42.	+ 16 35		15.7	GALAXY
ZWG 338.035	15 59 42.	+ 70 29		15.4	GALAXY
UGC 10142	15 59 42.	+ 70 29	102	15.4	GALAXY SBa
MCG+12-15-041	15 59 42.	+ 70 29	57	16.	GALAXY
SHB 322	15 59 44.8	+ 06 23 20.		17.5	QUASI-STELLAR OBJECT
MCG+03-41-046	15 59 48.	+ 15 49	12		GALAXY
ZWG 108.067	15 59 48.	+ 15 50		14.6	GALAXY
MRK 694	15 59 48.	+ 16 34	17	16.	GALAXY WITH UV CONTINUUM
ZWG 108.068	15 59 48.	+ 17 13		14.9	GALAXY
ZWG 108.069	15 59 48.	+ 18 57		15.2	GALAXY
MRK 294	15 59 48.	+ 18 57	20	15.	GALAXY WITH UV CONTINUUM
ZWG 167.015	15 59 48.	+ 32 02		15.7	GALAXY
KARA.72 478A	15 59 48.	+ 32 02	48	15.7	PART OF DOUBLE GALAXY
ZWG 223.028	15 59 48.	+ 42 31		15.7	GALAXY
MCG+07-33-026	15 59 48.	+ 42 31	39	15.	GALAXY
IC 1165	15 59 51.	+ 15 50 31.			NONSTELLAR OBJECT
MCG+03-41-047	15 59 51.	+ 17 13	36	14.9	GALAXY
SC 1557+1609.9	15 59 52.	+ 16 01 28.	12		NEBULA
SC 1557+1612.8	15 59 53.	+ 16 04 22.	18		NEBULA
ARP 324	15 59 53.	+ 16 05			PECULIAR GALAXY
ZWG 051.022	15 59 54.	+ 04 22		15.7	GALAXY
ZWG 051.023	15 59 54.	+ 06 55		15.6	GALAXY
MCG+03-41-049	15 59 54.	+ 15 50	24	14.6	GALAXY
MCG+03-41-048	15 59 54.	+ 15 50	30	14.6	GALAXY
VV 092B	15 59 54.	+ 15 51	6	19.	INTERACTING GALAXY
VV 092A	15 59 54.	+ 15 51	24	18.	INTERACTING GALAXY
VV 092	15 59 54.	+ 15 51	42	18.	INTERACTING GALAXY
MCG+03-41-050	15 59 54.	+ 15 51	18		GALAXY
VV 090B	15 59 54.	+ 15 54	15	16.5	INTERACTING GALAXY
VV 090A	15 59 54.	+ 15 54	18	16.	INTERACTING GALAXY
VV 090	15 59 54.	+ 15 54	54		INTERACTING GALAXY
ZWG 108.070	15 59 54.	+ 16 02		15.5	GALAXY
3ZW 075	15 59 54.	+ 16 03			COMPACT GALAXY
ZWG 108.071	15 59 54.	+ 16 04		15.3	GALAXY
ZWG 108.072	15 59 54.	+ 16 34		15.4	GALAXY
ZWG 167.016	15 59 54.	+ 32 01		15.4	GALAXY
KARA.72 478B	15 59 54.	+ 32 01	48	15.4	PART OF DOUBLE GALAXY
SC 1557+1610.8	15 59 56.	+ 16 02 22.	12		NEBULA
MCG+03-41-051	15 59 57.	+ 16 03	48	15.5	GALAXY
MCG+03-41-052	15 59 57.	+ 16 05	21	15.3	GALAXY
ZWG 108.073	16 00 00.	+ 16 06		14.9	GALAXY
UGC 10143	16 00 00.	+ 16 06	138	14.9	GALAXY E
ZWG 108.074	16 00 00.	+ 16 17		15.6	GALAXY
ZWG 108.075	16 00 00.	+ 16 29		14.6	GALAXY
UGC 10144	16 00 00.	+ 16 29	102	14.6	GALAXY E
ZWG 108.076	16 00 00.	+ 16 30		15.7	GALAXY
MCG+03-41-053	16 00 00.	+ 16 34	27	15.4	GALAXY
ZWG 137.018	16 00 00.	+ 26 28		15.3	GALAXY
MCG+09-26-042	16 00 00.	+ 52 29	48	16.	GALAXY
MCG+09-26-044	16 00 00.	+ 53 22 30.	48	17.	GALAXY
MCG+09-26-043	16 00 00.	+ 53 25	30	16.	GALAXY
MCG+09-26-045	16 00 00.	+ 54 00	60	16.	GALAXY
MCG+10-23-016	16 00 00.	+ 58 03	36	16.	GALAXY
MCG+13-11-021	16 00 00.	+ 78 46	30	17.	GALAXY

OBJECT NAME	RIGHT ASCEN.	DECLINATION	DIAM.	MAGN.	TYPE OF OBJECT
MCG-03-41-002	16 00 00.	- 15 52	36	16.	GALAXY
SC 1557+1615.0	16 00 01.	+ 16 06 34.	18		NEBULA
ABC 2147	16 00 02.	+ 16 03		13.8	RICH CLUSTER OF GALAXIES
MCG+03-41-054	16 00 03.	+ 16 07	78	14.9	GALAXY
MCG+03-41-055	16 00 03.	+ 16 30	15	14.6	GALAXY
IC 1166	16 00 05.	+ 26 26 04.			NONSTELLAR OBJECT
ZWG 051.024	16 00 06.	+ 03 07		15.7	GALAXY
VV 159B	16 00 06.	+ 16 05 00.	24	14.5	INTERACTING GALAXY
VV 159A	16 00 06.	+ 16 07 42.	30	14.	INTERACTING GALAXY
MCG+03-41-056	16 00 06.	+ 15 29	54	15.7	GALAXY
ZWG 195.006	16 00 06.	+ 36 50		15.5	GALAXY
MCG+06-35-024	16 00 06.	+ 36 51	45	15.	GALAXY
MCG+09-26-046	16 00 06.	+ 54 02 30.	15	16.	GALAXY
MCG+09-26-047	16 00 06.	+ 54 05	15	16.	GALAXY
MCG-04-38-002	16 00 06.	- 22 24	36	15.	GALAXY
ZWG 108.077	16 00 12.	+ 14 42		15.7	GALAXY
ZC 1600.2+6135	16 00 12.	+ 61 35	3430		CLUSTER OF GALAXIES
MCG+08-29-033	16 00 15.	+ 48 57	48	16.	GALAXY
ZWG 079.025	16 00 18.	+ 09 01		15.6	GALAXY
ZWG 079.026	16 00 18.	+ 12 18		15.7	GALAXY
ZWG 079.027	16 00 18.	+ 14 04		15.5	GALAXY
ZWG 137.019	16 00 18.	+ 21 16		14.8	GALAXY
MCG+08-29-034	16 00 18.	+ 45 29	54	17.	GALAXY
ZWG 250.037	16 00 18.	+ 47 31		15.5	GALAXY
ZWG 250.038	16 00 18.	+ 48 58		15.7	GALAXY
ZWG 108.078	16 00 24.	+ 16 15		15.5	GALAXY
ZC 1600.4+1925	16 00 24.	+ 19 25	22710		CLUSTER OF GALAXIES
MCG+04-38-015	16 00 24.	+ 21 13 30.	18	14.8	GALAXY
TON-N 0811	16 00 24.	+ 30 32		15.1	BLUE STAR
MCG+08-29-035	16 00 24.	+ 47 14	21	16.	GALAXY
MCG+09-26-048	16 00 24.	+ 54 20	42	17.	GALAXY
MCG-04-38-003	16 00 24.	- 22 12	36	15.	GALAXY
ABC 2149	16 00 26.	+ 54 01		16.1	RICH CLUSTER OF GALAXIES
ZWG 079.028	16 00 30.	+ 12 46		15.7	GALAXY
MRK 695	16 00 30.	+ 16 05	8	16.	GALAXY WITH UV CONTINUUM
ZWG 195.007	16 00 30.	+ 36 15		15.3	GALAXY
MCG+06-35-025	16 00 30.	+ 36 16	30	14.5	GALAXY
ZWG 250.039	16 00 30.	+ 47 14		15.2	GALAXY
UGC 10145	16 00 30.	+ 47 14	60	15.2	GALAXY E
ZWG 051.025	16 00 36.	+ 05 15		15.6	GALAXY
UGC 10146	16 00 36.	+ 05 15	60	15.6	GALAXY
MCG+01-41-007	16 00 36.	+ 05 15	36	15.6	GALAXY
ZWG 108.079	16 00 36.	+ 16 42		15.4	GALAXY
MCG+03-41-057	16 00 36.	+ 16 42	48	15.4	GALAXY
ZWG 108.080	16 00 36.	+ 19 56		15.7	GALAXY
ZWG 137.020	16 00 36.	+ 25 23		15.5	GALAXY
ZWG 223.029	16 00 36.	+ 43 02		14.7	GALAXY
MCG+07-33-027	16 00 36.	+ 43 03	48	15.	GALAXY
ZWG 250.040	16 00 36.	+ 48 48		15.3	GALAXY
ZC 1600.7+0322	16 00 42.	+ 03 22	1080		CLUSTER OF GALAXIES
MCG+03-41-058	16 00 42.	+ 15 58 30.	36	15.1	GALAXY
ZWG 108.081	16 00 42.	+ 15 59		15.1	GALAXY
RNGC 6038	16 00 47.	+ 37 29		14.5	GALAXY
ZWG 051.026	16 00 48.	+ 05 47		14.7	GALAXY
UGC 10147	16 00 48.	+ 05 47	72	14.7	GALAXY DBL SYS
MCG+01-41-008	16 00 48.	+ 05 47	30	14.7	GALAXY
ZWG 137.021	16 00 48.	+ 21 06		15.0	GALAXY
RNGC 6032	16 00 48.	+ 21 06		15.0	GALAXY
UGC 10148	16 00 48.	+ 21 06	102	15.0	GALAXY SB
TON-N 0252	16 00 48.	+ 30 46		15.6	BLUE STAR
ZWG 195.008	16 00 48.	+ 37 29		14.4	GALAXY
UGC 10149	16 00 48.	+ 37 29	72	14.4	GALAXY Sc
MCG+06-35-026	16 00 48.	+ 37 31	66	13.	GALAXY
ZC 1600.8+3804	16 00 48.	+ 38 04	5170		CLUSTER OF GALAXIES
MCG+08-29-036	16 00 48.	+ 47 22	48	16.	GALAXY
ZWG 250.041	16 00 48.	+ 49 20		15.6	GALAXY
UGC 10150	16 00 48.	+ 49 20	60	15.6	GALAXY SBc
MCG+08-29-037	16 00 48.	+ 49 20	60	16.	GALAXY
MCG+10-23-017	16 00 48.	+ 59 01	9	16.	GALAXY
ZC 1600.8+6508	16 00 48.	+ 65 08	1880		CLUSTER OF GALAXIES
ZWG 338.036	16 00 48.	+ 69 52		15.7	GALAXY
SN 1968K	16 00 49.	+ 17 19		17.5	SUPERNOVA
MCG+04-38-016	16 00 51.	+ 21 03 30.	90	15.0	GALAXY
MCG+06-35-027	16 00 51.	+ 33 16	15	15.	GALAXY
MCG+03-41-059	16 00 54.	+ 15 59	27		GALAXY
ZC 1600.9+2528	16 00 54.	+ 25 28	3900		CLUSTER OF GALAXIES
ZWG 167.017	16 00 54.	+ 27 09		14.8	GALAXY
UGC 10151	16 00 54.	+ 27 09	60	14.8	GALAXY Sc
TON-N 0812	16 00 54.	+ 31 30		14.5	BLUE STAR
ZWG 167.018	16 00 54.	+ 32 17		15.0	GALAXY
ZWG 223.030	16 00 54.	+ 41 20		14.5	GALAXY
UGC 10152	16 00 54.	+ 41 20	60	14.5	GALAXY DBL SYS
KARA.72 479B	16 00 54.	+ 41 20	24		PART OF DOUBLE GALAXY
KARA.72 479A	16 00 54.	+ 41 20	36	14.5	PART OF DOUBLE GALAXY
MCG+07-33-028	16 00 54.	+ 41 25	30	17.	GALAXY
MCG+03-41-060	16 00 57.	+ 16 32	36	15.5	GALAXY
MCG+05-38-011	16 00 57.	+ 27 05 30.	60	14.8	GALAXY
MCG+05-38-012	16 00 57.	+ 32 17	24	15.0	GALAXY
LB 09902	16 01	- 84 39		14.5	FAINT BLUE STAR
ZC 1601.0+0305	16 01 00.	+ 03 05	1750		CLUSTER OF GALAXIES
ZWG 108.082	16 01 00.	+ 16 32		15.5	GALAXY
SC 1558+1804.9	16 01 00.	+ 17 56 32.	12		NEBULA
ZWG 195.009	16 01 00.	+ 33 16		15.7	GALAXY
MCG+07-33-030	16 01 00.	+ 41 19	30	14.5	GALAXY
MCG+07-33-029	16 01 00.	+ 41 19	24	15.	GALAXY
ZC 1601.0+5039	16 01 00.	+ 50 39	940		CLUSTER OF GALAXIES
MCG+12-15-042	16 01 00.	+ 69 02	51	16.	GALAXY
OCL 0955	16 01 00.	- 51 47	300		OPEN STAR CLUSTER
VDA 182	16 01 00.	- 51 49	240		OPEN STAR CLUSTER
ZWG 137.022	16 01 06.	+ 20 47		15.2	GALAXY
UGC 10153	16 01 06.	+ 20 47	66	15.2	GALAXY SBb
ZWG 137.023	16 01 06.	+ 23 09		15.7	GALAXY
ZC 1601.1+2534	16 01 06.	+ 25 34	870		CLUSTER OF GALAXIES
MCG+10-23-018	16 01 06.	+ 62 37	6	16.	GALAXY
ZWG 320.002	16 01 06.	+ 68 24		15.5	GALAXY
ZWG 319.039	16 01 06.	+ 68 24		15.5	GALAXY
ZC 1601.1+7318	16 01 06.	+ 73 18	1210		CLUSTER OF GALAXIES
MCG+04-38-017	16 01 09.	+ 20 45	78	15.2	GALAXY
RNGC 6034	16 01 10.	+ 17 20		15.0	GALAXY
ABC 2148	16 01 11.	+ 25 36		15.4	RICH CLUSTER OF GALAXIES
ZWG 079.029	16 01 12.	+ 14 03		15.3	GALAXY
ZWG 108.083	16 01 12.	+ 16 28		14.8	GALAXY
ZWG 108.084	16 01 12.	+ 17 20		15.2	GALAXY
ZWG 108.085	16 01 12.	+ 19 18		15.5	GALAXY
MRK 296	16 01 12.	+ 19 18	25	15.5	GALAXY WITH UV CONTINUUM
MRK 295	16 01 12.	+ 19 19	8	16.	GALAXY WITH UV CONTINUUM
ZWG 137.024	16 01 12.	+ 21 02		14.7	GALAXY
RNGC 6035	16 01 12.	+ 21 02		14.5	GALAXY
UGC 10154	16 01 12.	+ 21 02	66	14.7	GALAXY Sc
ZWG 223.031	16 01 12.	+ 39 46		14.5	GALAXY
UGC 10155	16 01 12.	+ 39 46	60	14.5	GALAXY Sc
MCG+12-15-043	16 01 12.	+ 70 45	54	15.	GALAXY
MCG+03-41-061	16 01 15.	+ 16 27 30.	36	14.8	GALAXY
MCG+04-38-018	16 01 15.	+ 21 00	48	14.7	GALAXY
ABC 2156	16 01 16.	+ 73 20		17.5	RICH CLUSTER OF GALAXIES
SC 1559+1745.0	16 01 17.	+ 17 36 39.	12		NEBULA
SC 1559+1535.2	16 01 18.	+ 15 26 51.	18		NEBULA
MCG+03-41-062	16 01 18.	+ 17 21	60	15.2	GALAXY
ZWG 108.086	16 01 18.	+ 20 25		15.6	GALAXY
ZC 1601.3+4220	16 01 18.	+ 42 20	1210		CLUSTER OF GALAXIES
ZWG 250.042	16 01 18.	+ 47 21		14.6	GALAXY
UGC 10156	16 01 18.	+ 47 21	96	14.6	GALAXY DBL SYS
MCG+08-29-039	16 01 18.	+ 47 21	18	16.	GALAXY
MCG+08-29-038	16 01 18.	+ 47 21	54	15.	GALAXY
ZWG 338.037	16 01 19.	+ 70 45		15.0	GALAXY
RNGC 6071	16 01 18.	+ 70 45		15.0	GALAXY
UGC 10157	16 01 18.	+ 70 45	60	15.0	GALAXY SBb
MCG+07-33-031	16 01 21.	+ 39 46	60	14.5	GALAXY
ZWG 108.087	16 01 24.	+ 16 28		15.5	GALAXY
SC 1559+1535.8	16 01 29.	+ 15 27 28.	12		NEBULA
ZWG 051.027	16 01 30.	+ 08 29		15.5	GALAXY
ZWG 108.088	16 01 30.	+ 17 23		15.4	GALAXY
7ZW 632	16 01 30.	+ 62 37			COMPACT GALAXY
ZWG 338.038	16 01 30.	+ 71 38		15.1	GALAXY
ZC 1601.5-0201	16 01 30.	- 02 01	2490		CLUSTER OF GALAXIES
IC 1167	16 01 33.	+ 15 05 12.			NONSTELLAR OBJECT
MCG+03-41-063	16 01 33.	+ 17 23	42	15.4	GALAXY
ZWG 051.028	16 01 36.	+ 02 45		15.6	GALAXY
ZWG 079.030	16 01 36.	+ 11 52		15.2	GALAXY
UGC 10158	16 01 36.	+ 11 52	66	15.2	GALAXY S0
ZWG 108.089	16 01 36.	+ 14 46		15.7	GALAXY
MCG+03-41-064	16 01 36.	+ 14 46	48	15.7	GALAXY
ZWG 108.090	16 01 36.	+ 15 02		15.6	GALAXY
IC 1168	16 01 36.	+ 15 02 37.			NONSTELLAR OBJECT
MCG+03-41-065	16 01 36.	+ 15 05	30		GALAXY
ZWG 108.091	16 01 36.	+ 16 08		15.6	GALAXY
ZWG 108.092	16 01 36.	+ 16 30		15.6	GALAXY
7ZW 633	16 01 36.	+ 66 02			COMPACT GALAXY
MCG+12-15-044	16 01 36.	+ 71 40	33	16.	GALAXY
MCG+03-41-066	16 01 39.	+ 15 02	30	15.6	GALAXY
SC 1559+1615.4	16 01 40.	+ 16 07 04.	18		NEBULA
ZC 1601.7+0855	16 01 42.	+ 08 55	1210		CLUSTER OF GALAXIES
SC 1559+1615.4	16 01 42.	+ 16 07 05.	18		NEBULA
MCG+08-29-040	16 01 42.	+ 19 34	48	16.	GALAXY
ZC 1601.7+2331	16 01 42.	+ 23 31	400		CLUSTER OF GALAXIES
ZWG 338.039	16 01 42.	+ 70 32		15.7	GALAXY
MCG+12-15-045	16 01 42.	+ 70 32	27	14.	GALAXY
MCG+04-38-019	16 01 45.	+ 25 08	84	15.2	GALAXY
ZWG 079.031	16 01 48.	+ 11 20		15.5	GALAXY
ZWG 108.093	16 01 48.	+ 17 25		15.7	GALAXY
MCG+03-41-067	16 01 48.	+ 17 26	24	15.7	GALAXY
UGC 10160	16 01 48.	+ 25 10	90	15.2	GALAXY SBa
ZWG 137.025	16 01 48.	+ 25 13		15.2	GALAXY
MCG+05-38-013	16 01 48.	+ 26 38 30.	18	15.5	GALAXY
ZWG 167.019	16 01 48.	+ 26 41		15.5	GALAXY
7ZW 634	16 01 48.	+ 58 00			COMPACT GALAXY
MCG+12-15-046	16 01 49.	+ 71 34	90	16.	GALAXY
ZWG 023.011	16 01 48.	- 02 00		15.3	GALAXY
RNGC 6033	16 01 48.	- 02 00		15.5	GALAXY
UGC 10159	16 01 48.	- 02 00	66	15.3	GALAXY
MCG+00-41-003	16 01 48.	- 02 00	48	15.3	GALAXY
ZWG 051.029	16 01 54.	+ 05 45		15.7	GALAXY
ZWG 079.032	16 01 54.	+ 09 24		15.6	GALAXY
ZWG 079.033	16 01 54.	+ 13 53		14.1	GALAXY
UGC 10161	16 01 54.	+ 13 53	60	14.1	GALAXY S
MCG+02-41-004	16 01 54.	+ 13 53	60	14.1	GALAXY
SC 1559+1556.9	16 01 54.	+ 15 48 35.	12		NEBULA
SC 1559+1556.9	16 01 54.	+ 15 48 35.	12		NEBULA
ZWG 137.026	16 01 54.	+ 26 05		15.4	GALAXY
ZWG 223.032	16 01 54.	+ 41 17		15.6	GALAXY
ZWG 320.003	16 01 54.	+ 64 29		15.5	GALAXY
ZWG 319.040	16 01 54.	+ 64 29		15.5	GALAXY
ZWG 338.040	16 01 54.	+ 71 32		15.4	GALAXY
UGC 10162	16 01 54.	+ 71 32	78	15.4	GALAXY SBb
RNGC 6037	16 01 57.	+ 03 56		15.0	GALAXY
RNGC 6036	16 01 57.	+ 04 00		14.0	GALAXY
MCG+04-38-020	16 01 57.	+ 24 30	27		GALAXY
MCG+07-33-032	16 01 57.	+ 41 17	48	16.	GALAXY
IC 1169	16 01 58.	+ 13 53 39.			NONSTELLAR OBJECT
LB 00919	16 01 58.	+ 73 58 00.		17.0	FAINT BLUE STAR
ZWG 051.030	16 02 00.	+ 03 03		15.6	GALAXY
ZWG 051.031	16 02 00.	+ 03 56		15.2	GALAXY
KARA.72 480A	16 02 00.	+ 03 56	36	15.2	PART OF DOUBLE GALAXY
MCG+01-41-009	16 02 00.	+ 03 56	36	15.2	GALAXY
ZWG 051.032	16 02 00.	+ 04 00		13.9	GALAXY
UGC 10163	16 02 00.	+ 04 00	66	13.9	GALAXY S0-a
KARA.72 480B	16 02 00.	+ 04 00	78	13.9	PART OF DOUBLE GALAXY
MCG+01-41-010	16 02 00.	+ 04 00	66	13.9	GALAXY
MCG+08-29-041	16 02 00.	+ 49 29	48	15.	GALAXY
MCG+08-29-042	16 02 00.	+ 49 35	36	16.	GALAXY
7ZW 635	16 02 00.	+ 58 03			COMPACT GALAXY
ABC 2154	16 02 03.	+ 65 14		17.6	RICH CLUSTER OF GALAXIES
RNGC 6040	16 02 05.	+ 17 53		14.5	GALAXY
RNGC 6039	16 02 05.	+ 17 53		14.5	GALAXY
ZWG 079.034	16 02 06.	+ 11 21		15.4	GALAXY
ZWG 079.035	16 02 06.	+ 12 14		15.6	GALAXY
ZWG 108.094	16 02 06.	+ 14 55		15.5	GALAXY
UGC 10164	16 02 06.	+ 14 55	66	15.5	GALAXY E?
ZWG 108.095	16 02 06.	+ 16 50		15.6	GALAXY
MCG+03-41-068	16 02 06.	+ 16 50	48	15.6	GALAXY
ZWG 108.096	16 02 06.	+ 17 53		14.6	GALAXY
UGC 10165	16 02 06.	+ 17 53	108	14.6	GALAXY DBL SYS
ZWG 137.027	16 02 06.	+ 21 55		15.7	GALAXY
MRK 495	16 02 06.	+ 26 12	7	15.5	GALAXY WITH UV CONTINUUM
ZC 1602.1+2615	16 02 06.	+ 26 15	610		CLUSTER OF GALAXIES
ZWG 223.033	16 02 06.	+ 40 07		14.8	GALAXY
UGC 10166	16 02 06.	+ 40 07	66	14.8	GALAXY SB
ZWG 223.034	16 02 06.	+ 41 39		15.4	GALAXY
UGC 10167	16 02 06.	+ 41 39	90	15.4	GALAXY S0-a
MCG+07-33-033	16 02 06.	+ 41 39	66	15.	GALAXY
ZWG 250.043	16 02 06.	+ 49 28		14.3	GALAXY
UGC 10168	16 02 06.	+ 49 28	96	14.3	GALAXY SB0
MCG+09-26-049	16 02 06.	+ 52 23	36	17.	GALAXY
SC 1559+1809.6	16 02 07.	+ 18 01 18.	12		NEBULA
SC 1559+1801.8	16 02 10.	+ 17 53 30.	60		NEBULA
SC 1559+1816.2	16 02 10.	+ 18 07 54.	6		NEBULA

OBJECT NAME	RIGHT ASCEN.	DECLINATION	DIAM.	MAGN.	TYPE OF OBJECT
ARP 101	16 02 11.	+ 15 00			PECULIAR GALAXY
ZWG 051.033	16 02 12.	+ 02 39		15.4	GALAXY
ZWG 051.034	16 02 12.	+ 05 19		15.7	GALAXY
ZWG 079.036	16 02 12.	+ 10 05		15.7	GALAXY
ZWG 079.037	16 02 12.	+ 12 19		15.2	GALAXY
ZWG 079.038	16 02 12.	+ 12 23		15.4	GALAXY
MCG+03-41-069	16 02 12.	+ 14 55	54		GALAXY
VV 318A	16 02 12.	+ 14 55 30.	66	15.5	INTERACTING GALAXY
ZWG 108.097	16 02 12.	+ 14 57		15.2	GALAXY
UGC 10169	16 02 12.	+ 14 57	150	15.2	GALAXY
HELW 043	16 02 12.	+ 14 58			NEBULA
ZWG 108.098	16 02 12.	+ 17 36		15.7	GALAXY
ARP 122	16 02 12.	+ 17 49			PECULIAR GALAXY
VV 212B	16 02 12.	+ 17 50	15	16.	INTERACTING GALAXY
VV 212A	16 02 12.	+ 17 50	90	15.	INTERACTING GALAXY
VV 212	16 02 12.	+ 17 50	102		INTERACTING GALAXY
MCG+07-33-034	16 02 12.	+ 40 06	42	14.5	GALAXY
SHAH 008	16 02 12.	+ 52 30	36	17.5	GROUP OF COMPACT GALAXIES
MCG+03-41-070	16 02 15.	+ 14 57	96	15.2	GALAXY
VV 318B	16 02 15.	+ 14 58 00.		15.5	INTERACTING GALAXY
MCG+03-41-071	16 02 15.	+ 17 02	36	15.3	GALAXY
MCG+03-41-072	16 02 15.	+ 17 36	36	15.7	GALAXY
MCG+03-41-074	16 02 15.	+ 17 53 30.	72	14.6	GALAXY
MCG+03-41-073	16 02 15.	+ 17 53 30.	36	14.6	GALAXY
IC 1170	16 02 16.6	+ 17 51 28.	16		GALAXY SB0
RNGC 6041	16 02 17.	+ 17 51		15.0	GALAXY
MCG+03-41-075	16 02 18.	+ 16 36 30.	54	15.5	GALAXY
ZWG 108.099	16 02 18.	+ 16 37		15.5	GALAXY
ZWG 108.100	16 02 18.	+ 17 01		15.3	GALAXY
ZWG 108.101	16 02 18.	+ 17 51		14.9	GALAXY
UGC 10170	16 02 18.	+ 17 51	72	14.9	GALAXY E
UGC 10171	16 02 18.	+ 43 02	72	17.	GALAXY Sc
ZC 1602.3+5917	16 02 18.	+ 59 17	1080		CLUSTER OF GALAXIES
MCG+12-15-047	16 02 18.	+ 70 34	21	15.	GALAXY
HUB C23	16 02 20.	- 18 39			DIFFUSE NEBULA
MCG+03-41-076	16 02 21.	+ 16 36	27	15.5	GALAXY
VV 213B	16 02 21.	+ 17 51	15	16.	INTERACTING GALAXY
VV 213A	16 02 21.	+ 17 51	21	15.	INTERACTING GALAXY
VV 213	16 02 21.	+ 17 51	78		INTERACTING GALAXY
SC 1600+1800.0	16 02 21.	+ 17 51 43.	18		NEBULA
BC PKS1602-002	16 02 21.3	- 00 10 55.		17.5	QUASI-STELLAR OBJECT
RNGC 6042	16 02 23.	+ 17 50		15.5	GALAXY
SC 1600+1758.8	16 02 23.	+ 17 50 31.	18		NEBULA
ZWG 079.039	16 02 24.	+ 08 37		15.5	GALAXY
MCG+03-41-077	16 02 24.	+ 16 33	30	15.7	GALAXY
ZWG 108.102	16 02 24.	+ 16 34		15.7	GALAXY
ZWG 108.103	16 02 24.	+ 16 40		15.6	GALAXY
ZWG 108.104	16 02 24.	+ 17 50		15.6	GALAXY
MCG+03-41-078	16 02 24.	+ 17 52	72	14.9	GALAXY
ZWG 137.028	16 02 24.	+ 25 19		15.4	GALAXY
ZC 1602.4+2734	16 02 24.	+ 27 34	670		CLUSTER OF GALAXIES
MCG+05-38-014	16 02 24.	+ 30 49 30.	72	15.7	GALAXY
ZWG 338.041	16 02 24.	+ 70 34		14.9	GALAXY
MCG+03-41-079	16 02 27.	+ 17 51	36	15.6	GALAXY
ZWG 079.040	16 02 30.	+ 14 15		15.2	GALAXY
ZWG 079.041	16 02 30.	+ 14 25		15.2	GALAXY
UGC 10172	16 02 30.	+ 14 25	72	15.2	GALAXY Sb-c
ZWG 108.105	16 02 30.	+ 15 52		15.6	GALAXY
ZWG 108.106	16 02 30.	+ 16 43		15.1	GALAXY
ZWG 108.107	16 02 30.	+ 17 01		15.7	GALAXY
ZWG 108.108	16 02 30.	+ 17 35		15.7	GALAXY
MCG+03-41-080	16 02 30.	+ 17 35	42	15.7	GALAXY
ZC 1602.5+2344	16 02 30.	+ 23 44	740		CLUSTER OF GALAXIES
ZWG 137.029	16 02 30.	+ 25 51		15.7	GALAXY
MCG+06-35-028	16 02 30.	+ 33 27 30.	24	16.	GALAXY
ZWG 320.004	16 02 30.	+ 66 37		15.4	GALAXY
ZWG 319.041	16 02 30.	+ 66 37		15.4	GALAXY
ZWG 320.005	16 02 30.	+ 68 20		15.6	GALAXY
ZWG 319.042	16 02 30.	+ 68 20		15.6	GALAXY
UGC 10173	16 02 30.	+ 68 20	60	15.6	GALAXY S
MCG+03-41-081	16 02 33.	+ 17 01	36	15.7	GALAXY
SC 1600+1652.1	16 02 35.	+ 16 43 50.	18		NEBULA
ZWG 051.035	16 02 36.	+ 02 58		15.7	GALAXY
ZWG 079.042	16 02 36.	+ 14 15		15.4	GALAXY
MCG+03-41-082	16 02 36.	+ 15 51	36	15.6	GALAXY
MCG+03-41-083	16 02 36.	+ 16 43	36	15.1	GALAXY
UGC 10174	16 02 36.	+ 23 28	66	16.5	GALAXY Sa-b
ZWG 167.020	16 02 36.	+ 30 51		15.7	GALAXY
UGC 10175	16 02 36.	+ 30 51	66	15.7	GALAXY IRR
ZWG 223.035	16 02 36.	+ 42 09		15.7	GALAXY
ZWG 275.020	16 02 36.	+ 53 53		15.6	GALAXY
MCG+11-20-001	16 02 36.	+ 66 39	51	15.	GALAXY
MCG+11-20-002	16 02 36.	+ 68 21	57	16.	GALAXY
OCL 0947	16 02 36.	- 56 47	528	12.	OPEN STAR CLUSTER
IC 1171	16 02 37.1	+ 18 06 49.			SINGLE STAR
SC 1600+1803.2	16 02 41.	+ 17 54 56.	12		NEBULA
RNGC 6043	16 02 41.	+ 17 55		15.5	GALAXY
RNGC 6044	16 02 41.	+ 18 00		15.5	GALAXY
RNGC 6046	16 02 41.	+ 19 29			NON-EXISTENT OBJECT
ZWG 079.043	16 02 42.	+ 13 50		15.2	GALAXY
UGC 10176	16 02 42.	+ 13 50	90	15.2	GALAXY Sc
MCG+02-41-005	16 02 42.	+ 13 50	90	15.2	GALAXY
ZWG 108.109	16 02 42.	+ 17 55		15.4	GALAXY
ZWG 108.110	16 02 42.	+ 18 00		15.3	GALAXY
IC 1172	16 02 42.	+ 18 00		15.3	GALAXY
MCG+03-41-084	16 02 42.	+ 18 02	24		GALAXY
ZWG 167.021	16 02 42.	+ 28 14		15.6	GALAXY
MRK 696	16 02 42.	+ 28 14	12	16.	GALAXY WITH UV CONTINUUM
ZWG 195.010	16 02 42.	+ 37 17		15.7	GALAXY
MCG+07-33-035	16 02 42.	+ 42 09	30	16.	GALAXY
PK064+48.1	16 02 43.28	+ 40 49 06.6	40	13.3	PLANETARY NEBULA
MCG+04-38-021	16 02 45.	+ 24 03 30.	27	14.9	GALAXY
MCG+03-41-085	16 02 48.	+ 17 43	24		GALAXY
ZWG 108.111	16 02 48.	+ 17 52		15.4	GALAXY
ZWG 108.112	16 02 48.	+ 17 54		14.8	GALAXY
UGC 10177	16 02 48.	+ 17 54	66	14.8	GALAXY Sb
MCG+03-41-086	16 02 48.	+ 17 54	24	15.4	GALAXY
ZWG 137.030	16 02 48.	+ 24 04		14.9	GALAXY
UGC 10178	16 02 48.	+ 24 04	126	14.9	GALAXY E
ZC 1602.8+2913	16 02 48.	+ 29 13	940		CLUSTER OF GALAXIES
UGC 10179	16 02 48.	+ 31 07	66	16.5	GALAXY COMPACT
MCG+06-35-030	16 02 48.	+ 34 45 30.	12	16.	GALAXY
MCG+06-35-029	16 02 48.	+ 34 45 30.	24	15.	GALAXY
SN 1962C	16 02 49.	+ 17 43		18.0	SUPERNOVA
RNGC 6051	16 02 50.	+ 24 04		15.0	GALAXY
RNGC 6058	16 02 50.	+ 40 49		13.5	PLANETARY NEBULA
SC 1600+1802.0	16 02 52.	+ 17 53 45.	84		NEBULA
RNGC 6047	16 02 53.	+ 17 52		15.5	GALAXY
RNGC 6045	16 02 53.	+ 17 53		15.0	GALAXY
MRSL 011+36/1	16 02 54.	+ 00 32	2700		HII REGION
ZC 1602.9+0235	16 02 54.	+ 02 35	810		CLUSTER OF GALAXIES
ZWG 051.036	16 02 54.	+ 05 08		15.7	GALAXY
ZWG 051.037	16 02 54.	+ 07 13		15.5	GALAXY
ZWG 079.044	16 02 54.	+ 14 28		15.5	GALAXY
ZWG 108.113	16 02 54.	+ 17 33		15.6	GALAXY
UGC 10180	16 02 54.	+ 17 33	66	15.6	GALAXY S
ARP 071	16 02 54.	+ 17 49			PECULIAR GALAXY
MCG+03-41-087	16 02 54.	+ 17 52	54	15.4	GALAXY
MCG+03-41-088	16 02 54.	+ 17 54	72	14.8	GALAXY
ZWG 137.031	16 02 54.	+ 23 48		15.7	GALAXY
ZWG 167.022	16 02 54.	+ 28 18		15.5	GALAXY
TON-N 0253	16 02 54.	+ 33 17		15.7	BLUE STAR
ZWG 195.011	16 02 54.	+ 34 45		15.4	GALAXY
MCG+03-41-089	16 02 57.	+ 17 34	60	15.6	GALAXY
SC 1600+1759.6	16 02 57.	+ 17 51 21.	18		NEBULA
IC 4588	16 02 57.	+ 24 02 53.			NONSTELLAR OBJECT
MCG+06-35-031	16 02 57.	+ 34 44	24	15.	GALAXY
IC 1173	16 02 57.1	+ 17 33 24.	35		GALAXY SAB(s)c
ABC 2151	16 02 58.	+ 17 53		13.8	RICH CLUSTER OF GALAXIES
MCG+03-41-090	16 03 00.	+ 14 47	60	15.2	GALAXY
ZWG 108.114	16 03 00.	+ 15 53		15.7	GALAXY
ZC 1603.0+1639	16 03 00.	+ 16 39	1880		CLUSTER OF GALAXIES
MCG+04-38-022	16 03 00.	+ 20 39	45	14.1	GALAXY
RNGC 6052	16 03 00.	+ 20 40		13.5	GALAXY
ZWG 137.032	16 03 00.	+ 20 41		14.1	GALAXY
MRK 297	16 03 00.	+ 20 41	40	14.	GALAXY WITH UV CONTINUUM
UGC 10182	16 03 00.	+ 20 41	54	14.1	GALAXY DBL SYS
MCG+04-38-023	16 03 00.	+ 25 04 30.	45	15.4	GALAXY
ZWG 355.007	16 03 00.	+ 80 00		15.3	GALAXY
ZWG 354.033	16 03 00.	+ 80 00		15.3	GALAXY
UGC 10183	16 03 00.	+ 80 00	72	15.3	GALAXY SB0-a
MCG+13-11-022	16 03 00.	+ 80 00	54	16.	GALAXY
ZWG 367.006	16 03 00.	+ 81 50		15.3	GALAXY
ZWG 366.027	16 03 00.	+ 81 50		15.3	GALAXY
UGC 10181	16 03 00.	+ 81 50	78	15.3	GALAXY Sa
MCG+14-07-034	16 03 00.	+ 81 50	54	17.	GALAXY
IC 1175	16 03 03.	+ 18 16			MAY NOT EXIST
RNGC 6050	16 03 05.	+ 17 54		15.0	GALAXY
IC 1177	16 03 05.1	+ 18 26 57.			GALAXY SA(s)0
ZWG 051.038	16 03 06.	+ 03 11		15.6	GALAXY
UGC 10184	16 03 06.	+ 03 11	60	15.6	GALAXY Sa
ZC 1603.1+0738	16 03 06.	+ 07 38	1340		CLUSTER OF GALAXIES
ZWG 108.115	16 03 06.	+ 14 46		15.2	GALAXY
ZWG 108.116	16 03 06.	+ 15 10		14.5	GALAXY
UGC 10185	16 03 06.	+ 15 10	60	14.5	GALAXY Sa
MCG+03-41-091	16 03 06.	+ 15 10	60	14.5	GALAXY
ZWG 108.117	16 03 06.	+ 16 20		15.3	GALAXY
ZWG 108.118	16 03 06.	+ 17 54		14.9	GALAXY
UGC 10186	16 03 06.	+ 17 54	60	14.9	GALAXY DBL SYS
KARA.72 481B	16 03 06.	+ 17 54	48		PART OF DOUBLE GALAXY
KARA.72 481A	16 03 06.	+ 17 54	42	14.9	PART OF DOUBLE GALAXY
ARP 272	16 03 06.	+ 17 55			PECULIAR GALAXY
VV 086	16 03 06.	+ 20 42	48	13.	INTERACTING GALAXY
ZWG 137.033	16 03 06.	+ 25 05		15.4	GALAXY
ZWG 195.012	16 03 06.	+ 34 44		15.3	GALAXY
MCG+11-20-003	16 03 06.	+ 66 32	39	17.	GALAXY
ABC 2152	16 03 07.	+ 16 35		13.8	RICH CLUSTER OF GALAXIES
SC 1600+1801.9	16 03 07.	+ 17 53 40.	60		NEBULA
SC 1600+1808.0	16 03 07.	+ 17 59 46.	12		NEBULA
ARP 209	16 03 07.	+ 20 41			PECULIAR GALAXY
IC 1179	16 03 07.2	+ 17 53 15.	36	15.8	GALAXY SAB(rs)
IC 1174	16 03 08.	+ 15 09 41.			NONSTELLAR OBJECT
MCG+03-41-094	16 03 09.	+ 16 24	12	15.4	GALAXY
MCG+03-41-093	16 03 09.	+ 17 54	30	14.9	GALAXY
MCG+03-41-092	16 03 09.	+ 17 54 30.	48	14.9	GALAXY
SC 1600+1758.3	16 03 10.	+ 17 50 04.	30		NEBULA
RNGC 6054	16 03 11.	+ 17 55		15.5	GALAXY
IC 1180	16 03 11.	+ 18 15			MAY NOT EXIST
ZWG 051.039	16 03 12.	+ 06 22		15.7	GALAXY
MCG+03-41-095	16 03 12.	+ 16 34	60	15.4	GALAXY
ZWG 108.119	16 03 12.	+ 16 35		15.4	GALAXY
UGC 10187	16 03 12.	+ 16 35	72	15.4	GALAXY DBL SYS
ZWG 108.120	16 03 12.	+ 17 44		15.0	GALAXY
UGC 10189	16 03 12.	+ 17 44	66	15.0	GALAXY DBL SYS
UGC 10188	16 03 12.	+ 17 44	72	15.0	GALAXY DBL SYS
UGC 10190	16 03 12.	+ 17 51	60	16.0	GALAXY Sc
ZWG 108.121	16 03 12.	+ 17 55		15.7	GALAXY
SC 1600+1811.1	16 03 12.	+ 18 02 52.	12		NEBULA
MCG+03-41-096	16 03 12.	+ 18 03	42		GALAXY
ZWG 108.122	16 03 12.	+ 18 06		15.1	GALAXY
IC 1176	16 03 12.	+ 18 06		15.1	GALAXY;MAY BE NGC6056
ZWG 108.123	16 03 12.	+ 18 17		15.4	GALAXY
UGC 10191	16 03 12.	+ 18 17	84	15.4	GALAXY S0
ZWG 108.124	16 03 12.	+ 20 04		15.2	GALAXY
ZWG 137.034	16 03 12.	+ 23 54		15.7	GALAXY
ZWG 167.023	16 03 12.	+ 30 36		15.7	GALAXY
ZWG 167.024	16 03 12.	+ 31 53		15.6	GALAXY
MCG+08-29-043	16 03 12.	+ 49 34	48	17.	GALAXY
REIF 4.298	16 03 12.57	+ 08 13 48.8			NEBULA
SC 1600+1837.2	16 03 13.	+ 18 28 59.	18		NEBULA
SC 1601+1814.4	16 03 16.	+ 18 06 11.	30		NEBULA
RNGC 6049	16 03 17.	+ 08 14			NON-EXISTENT OBJECT
SC 1601+1752.6	16 03 17.	+ 17 44 23.	12		NEBULA
RNGC 6056	16 03 17.	+ 18 06		15.0	GALAXY
SC 1601+1826.2	16 03 17.	+ 18 17 59.	12		NEBULA
RNGC 6055	16 03 17.	+ 18 18		15.5	GALAXY
IC 1181A	16 03 17.5	+ 17 43 59.			SAME AS IC 1178
IC 1178	16 03 17.6	+ 17 44 04.	21	15.0	GALAXY SA0
ZWG 051.040	16 03 18.	+ 02 55		15.7	GALAXY
ZWG 079.045	16 03 18.	+ 10 10		15.0	GALAXY
MCG+02-41-006	16 03 18.	+ 10 10	48	15.0	GALAXY
ZWG 108.125	16 03 18.	+ 16 40		15.6	GALAXY
MCG+03-41-098	16 03 18.	+ 17 43 30.	36	15.0	GALAXY
MCG+03-41-097	16 03 18.	+ 17 44	42	15.0	GALAXY
VV 194B	16 03 18.	+ 17 45	15	16.	INTERACTING GALAXY
VV 194A	16 03 18.	+ 17 45	15	15.	INTERACTING GALAXY
VV 194	16 03 18.	+ 17 45	120		INTERACTING GALAXY
ARP 172	16 03 18.	+ 17 46			PECULIAR GALAXY
VV 091B	16 03 18.	+ 17 54	9	17.5	INTERACTING GALAXY
VV 091A	16 03 18.	+ 17 54	21	17.	INTERACTING GALAXY
VV 091	16 03 18.	+ 17 54	36		INTERACTING GALAXY
MCG+03-41-099	16 03 18.	+ 17 54 30.	45	15.7	GALAXY
ZWG 108.126	16 03 18.	+ 17 56		15.2	GALAXY
MRK 298	16 03 18.	+ 17 56	20	15.	GALAXY WITH UV CONTINUUM
KW 28	16 03 18.	+ 17 56	42		SEYFERT GALAXY
VVI 82	16 03 18.	+ 17 56	20	16.20	SEYFERT GALAXY

OBJECT NAME	RIGHT ASCEN.	DECLINATION	DIAM.	MAGN.	TYPE OF OBJECT
UGC 10192	16 03 18.	+ 17 56	96	15.2	GALAXY PECULR
MCG+03-41-100	16 03 18.	+ 18 06 30.	48	15.1	GALAXY
MCG+03-41-101	16 03 18.	+ 18 18 30.	72	15.4	GALAXY
ZWG 108.127	16 03 18.	+ 18 25		15.6	GALAXY
MCG+03-41-102	16 03 18.	+ 20 05	36	15.2	GALAXY
ZWG 250.044	16 03 18.	+ 49 35		15.4	GALAXY
MCG+08-29-044	16 03 18.	+ 49 36	42	16.	GALAXY
IC 1181B	16 03 18.3	+ 17 43 30.			SAME AS IC 1181
IC 1181	16 03 18.8	+ 17 43 35.			GALAXY SAB(s)
SC 1601+1648.4	16 03 20.	+ 16 40 11.	12		NEBULA
SC 1601+1802.8	16 03 21.	+ 17 54 35.	18		NEBULA
SC 1601+1804.6	16 03 21.	+ 17 56 23.	18		NEBULA
IC 1182	16 03 21.6	+ 17 46 06.	22	15.2	GALAXY S0
SC 1601+1802.6	16 03 22.	+ 17 54 23.	18		NEBULA
SC 1601+1833.2	16 03 22.	+ 18 24 59.	30		NEBULA
IC 1183	16 03 22.9	+ 17 54 02.	20	15.6	GALAXY S0
RNGC 6053	16 03 23.	+ 18 12		15.5	GALAXY
RNGC 6057	16 03 23.	+ 18 18		15.5	GALAXY
MCG+03-41-103	16 03 24.	+ 17 54 30.	27	15.6	GALAXY
ZWG 108.128	16 03 24.	+ 17 55		15.6	GALAXY
VV 220B	16 03 24.	+ 17 55	30	15.	INTERACTING GALAXY
VV 220A	16 03 24.	+ 17 55	42	15.	INTERACTING GALAXY
MCG+03-41-104	16 03 24.	+ 17 56 30.	30	15.2	GALAXY
SC 1601+1820.0	16 03 24.	+ 18 11 47.	18		NEBULA
ZWG 108.129	16 03 24.	+ 18 12		15.7	GALAXY
MCG+03-41-105	16 03 24.	+ 18 12	27	15.7	GALAXY
ZWG 108.130	16 03 24.	+ 18 18		15.7	GALAXY
SC 1601+1826.6	16 03 24.	+ 18 18 23.	12		NEBULA
MCG+03-41-106	16 03 24.	+ 18 19	30	15.7	GALAXY
MCG+03-41-107	16 03 24.	+ 18 25	30	15.6	GALAXY
ZWG 137.035	16 03 24.	+ 22 20		15.7	GALAXY
MCG+08-29-045	16 03 24.	+ 45 34	48	16.	GALAXY
ZC 1603.4+5247	16 03 24.	+ 52 47	1010		CLUSTER OF GALAXIES
IC 1184	16 03 27.3	+ 17 55 20.	6	15.6	GALAXY E0
IC 1186	16 03 28.9	+ 17 29 50.	28		GALAXY SA(s)
IC 1185	16 03 29.3	+ 17 51 05.	23	15.1	GALAXY SA(s)
ZWG 108.131	16 03 30.	+ 15 55		15.5	GALAXY
MCG+03-41-108	16 03 30.	+ 15 55 30.	48	15.5	GALAXY
SC 1601+1604.6	16 03 30.	+ 15 56 23.	12		NEBULA
ZWG 108.132	16 03 30.	+ 16 20		15.5	GALAXY
UGC 10193	16 03 30.	+ 16 20	66	15.5	GALAXY S+COMP
MCG+03-41-109	16 03 30.	+ 16 20	72	15.5	GALAXY
MRK 299	16 03 30.	+ 17 26	10	17.5	GALAXY WITH UV CONTINUUM
ZWG 108.133	16 03 30.	+ 17 29		15.4	GALAXY
MCG+03-41-111	16 03 30.	+ 17 30	42	15.4	GALAXY
ZWG 108.134	16 03 30.	+ 17 51		15.1	GALAXY
MCG+03-41-110	16 03 30.	+ 17 51	24	15.1	GALAXY
ZWG 108.135	16 03 30.	+ 18 09		15.7	GALAXY
MCG+03-41-112	16 03 30.	+ 18 10	30	15.7	GALAXY
MCG+04-38-024	16 03 30.	+ 22 18	48	15.7	GALAXY
MCG+06-35-032	16 03 30.	+ 32 47	36	15.	GALAXY
ZWG 250.045	16 03 30.	+ 45 35		15.4	GALAXY
MCG+09-26-050	16 03 30.	+ 53 20	24	17.	GALAXY
MCG+09-26-051	16 03 30.	+ 54 10	36	16.	GALAXY
ZC 1603.5+5642	16 03 30.	+ 56 42	810		CLUSTER OF GALAXIES
ZC 1603.5+6337	16 03 30.	+ 63 37	1010		CLUSTER OF GALAXIES
UGC 10194	16 03 30.	+ 63 51	108	16.0	GALAXY
MCG+11-20-004	16 03 30.	+ 63 51	90	16.	GALAXY
SC 1601+1817.7	16 03 31.	+ 18 09 30.	12		NEBULA
ZWG 079.046	16 03 36.	+ 08 38		15.6	GALAXY
ZWG 108.136	16 03 36.	+ 18 21		15.7	GALAXY
UGC 10195	16 03 36.	+ 18 21	102	15.7	GALAXY Sb
ZWG 108.137	16 03 36.	+ 18 40		15.7	GALAXY
ZWG 195.013	16 03 36.	+ 32 47		15.6	GALAXY
SC 1601+1829.9	16 03 38.	+ 18 21 42.	30		NEBULA
MCG+03-41-113	16 03 39.	+ 18 21 30.	72	15.7	GALAXY
SC 1601+1602.8	16 03 40.	+ 15 54 36.	12		NEBULA
SC 1601+1822.2	16 03 40.	+ 18 14 00.	12		NEBULA
ZC 1603.7+0006	16 03 42.	+ 00 06	7800		CLUSTER OF GALAXIES
MCG+04-38-025	16 03 42.	+ 21 35 30.	120	14.3	GALAXY
ZWG 137.036	16 03 42.	+ 21 38		14.3	GALAXY
UGC 10196	16 03 42.	+ 21 38	132	14.3	GALAXY Sc
OCL 0951	16 03 42.	- 53 56	540	12.2	OPEN STAR CLUSTER
VHA 183	16 03 42.	- 53 56	180		OPEN STAR CLUSTER
RNGC 6060	16 03 43.	+ 21 38		14.5	GALAXY
SC 1601+1651.4	16 03 44.	+ 16 43 12.	24		NEBULA
SC 1601+1821.8	16 03 45.	+ 18 13 37.	12		NEBULA
SC 1601+1828.3	16 03 45.	+ 18 20 07.	36		NEBULA
RNGC 6031	16 03 45.	- 53 55		12.0	OPEN CLUSTER
ZWG 023.012	16 03 48.	+ 02 18		15.3	GALAXY
SN 1964C	16 03 48.	+ 17 35		17.3	SUPERNOVA
ZWG 108.138	16 03 48.	+ 18 15		15.7	GALAXY
MCG+03-41-114	16 03 48.	+ 18 15	24	15.7	GALAXY
ZWG 108.139	16 03 48.	+ 18 20		15.7	GALAXY
MCG+03-41-115	16 03 48.	+ 18 20	66	15.7	GALAXY
MCG+03-41-116	16 03 48.	+ 18 48 30.	36	15.7	GALAXY
ZWG 108.140	16 03 48.	+ 18 49		15.7	GALAXY
ZWG 137.037	16 03 48.	+ 21 30		15.4	GALAXY
MCG+07-33-036	16 03 48.	+ 40 50	30	16.	GALAXY
ZWG 223.036	16 03 48.	+ 42 45		15.1	GALAXY
MCG+07-33-037	16 03 48.	+ 42 45 30.	42	15.	GALAXY
IC 1188	16 03 51.6	+ 17 35 32.	32		DOUBLE GALAXY
LB 00920	16 03 53.	+ 40 33 42.		15.3	FAINT BLUE STAR
SN 1967G	16 03 54.	+ 18 28		19.	SUPERNOVA
ZWG 108.141	16 03 54.	+ 18 45		15.6	GALAXY
MCG+03-41-117	16 03 54.	+ 18 45	24	15.6	GALAXY
MCG+04-38-026	16 03 54.	+ 20 53 30.	60	15.2	GALAXY
MCG+04-38-027	16 03 54.	+ 20 54 30.	60	15.2	GALAXY
ZWG 137.038	16 03 54.	+ 20 55		15.2	GALAXY
UGC 10198	16 03 54.	+ 20 55	84	15.2	GALAXY
UGC 10197	16 03 54.	+ 20 55	84	15.2	GALAXY
MCG+07-33-038	16 03 54.	+ 44 21	36	16.	GALAXY
ZWG 320.006	16 03 54.	+ 65 40		15.2	GALAXY
ZWG 319.043	16 03 54.	+ 65 40		15.2	GALAXY
SC 1601+1818.5	16 03 55.	+ 18 10 19.	12		NEBULA
SC 1601+1813.8	16 03 59.	+ 18 05 37.	12		NEBULA
RNGC 6061	16 03 59.	+ 18 23		15.0	GALAXY
LBN 0045	16 04	+ 00 30	1500		BRIGHT NEBULA
LBN 0044	16 04	+ 00 30	3600		BRIGHT NEBULA
LB 09903	16 04	- 82 49		15.3	FAINT BLUE STAR
ZWG 051.041	16 04 00.	+ 04 25		15.4	GALAXY
ZC 1604.0+0707	16 04 00.	+ 07 07	1480		CLUSTER OF GALAXIES
ZWG 108.142	16 04 00.	+ 16 34		15.6	GALAXY
ZWG 108.143	16 04 00.	+ 16 40		15.7	GALAXY
ZWG 108.144	16 04 00.	+ 18 19		15.5	GALAXY
MRK 300	16 04 00.	+ 18 19	20	15.5	GALAXY WITH UV CONTINUUM
MCG+03-41-119	16 04 00.	+ 18 19	30	15.5	GALAXY
ZWG 108.145	16 04 00.	+ 18 23		15.0	GALAXY

OBJECT NAME	RIGHT ASCEN.	DECLINATION	DIAM.	MAGN.	TYPE OF OBJECT
UGC 10199	16 04 00.	+ 18 23	72	15.0	GALAXY S0
MCG+03-41-118	16 04 00.	+ 18 23	60	15.0	GALAXY
ZWG 108.146	16 04 00.	+ 18 33		15.7	GALAXY
MCG+03-41-120	16 04 00.	+ 18 33 30.	39	15.7	GALAXY
VV 327A	16 04 00.	+ 20 55	60	15.	INTERACTING GALAXY
ZC 1604.0+2717	16 04 00.	+ 27 17	3430		CLUSTER OF GALAXIES
ZWG 223.037	16 04 00.	+ 41 27		15.0	GALAXY
KARA.72 482B	16 04 00.	+ 41 27	24	15.0	PART OF DOUBLE GALAXY
MCG+07-33-040	16 04 00.	+ 41 28	36	15.5	GALAXY
ZWG 223.038	16 04 00.	+ 41 29		13.6	GALAXY
UGC 10200	16 04 00.	+ 41 29	48	13.6	GALAXY
KARA.72 482A	16 04 00.	+ 41 29	48	13.6	PART OF DOUBLE GALAXY
MCG+07-33-039	16 04 00.	+ 41 29	36	14.	GALAXY
ZWG 223.039	16 04 00.	+ 44 20		15.6	GALAXY
ZC 1604.0+5544	16 04 00.	+ 55 44	1080		CLUSTER OF GALAXIES
MCG+12-15-048	16 04 00.	+ 69 48	12	17.	GALAXY
ZWG 366.028	16 04 00.	+ 80 22		15.7	GALAXY
ZWG 355.008	16 04 00.	+ 80 22		15.7	GALAXY
ZWG 023.013	16 04 00.	- 02 14		15.7	GALAXY
SC 1601+1813.0	16 04 01.	+ 18 04 49.	12		NEBULA
VV 214B	16 04 03.	+ 15 49	6	19.	INTERACTING GALAXY
VV 214A	16 04 03.	+ 15 49	6	19.	INTERACTING GALAXY
MCG+03-41-121	16 04 03.	+ 16 34	42	15.6	GALAXY
VV 327C	16 04 03.	+ 20 54	12	18.5	INTERACTING GALAXY
VV 327B	16 04 03.	+ 20 54	60	15.	INTERACTING GALAXY
SC 1601+1649.2	16 04 04.	+ 16 41 02.	30		NEBULA
IC 1190	16 04 04.2	+ 18 22 03.			MAY NOT EXIST
IC 1189	16 04 04.6	+ 18 18 54.	30	15.5	GALAXY SB(s)
ZWG 051.042	16 04 06.	+ 03 16		15.3	GALAXY
ZC 1604.1+1407	16 04 06.	+ 14 07	2490		CLUSTER OF GALAXIES
ZWG 108.147	16 04 06.	+ 15 49		14.3	GALAXY
UGC 10201	16 04 06.	+ 15 49	66	14.3	GALAXY DBL SYS
KARA.72 483A	16 04 06.	+ 15 49	54	14.3	PART OF DOUBLE GALAXY
KARA.72 483B	16 04 06.	+ 15 50	36		PART OF DOUBLE GALAXY
MCG+03-41-122	16 04 06.	+ 19 54 30.	24	16.	GALAXY
ZWG 108.148	16 04 06.	+ 19 55		14.4	GALAXY
RNGC 6062	16 04 06.	+ 19 55		14.5	GALAXY
UGC 10202	16 04 06.	+ 19 55	66	14.4	GALAXY SBb
ZC 1604.1+5445	16 04 06.	+ 54 45	1410		CLUSTER OF GALAXIES
MCG+03-41-123	16 04 09.	+ 15 49	60	14.3	GALAXY
MCG+03-41-124	16 04 09.	+ 15 49 30.	24	14.3	GALAXY
HOLM 728B	16 04 09.	+ 19 54	30	14.4	PART OF MULTIPLE GALAXY
MCG+03-41-125	16 04 09.	+ 19 55 30.	72	14.4	GALAXY
ZWG 051.043	16 04 12.	+ 06 43		15.3	GALAXY
VV 215B	16 04 12.	+ 15 50	42	15.	INTERACTING GALAXY
VV 215A	16 04 12.	+ 15 50	42	15.	INTERACTING GALAXY
HOLM 728A	16 04 12.	+ 19 55	60	13.5	PART OF MULTIPLE GALAXY
MCG+10-23-019	16 04 12.	+ 61 42	24	16.	GALAXY
IC 1191	16 04 14.4	+ 18 24 02.	27		DOUBLE GALAXY
SC 1602+1832.7	16 04 16.	+ 18 24 33.	30		NEBULA
IC 1193	16 04 17.1	+ 17 50 43.	24		GALAXY SB0
ZWG 051.044	16 04 18.	+ 03 55		15.1	GALAXY
UGC 10203	16 04 18.	+ 03 55	60	15.1	GALAXY S0
MCG+01-41-011	16 04 18.	+ 03 55	60	15.1	GALAXY
ZWG 108.149	16 04 18.	+ 18 01		15.7	GALAXY
ZWG 195.014	16 04 18.	+ 36 07		15.7	GALAXY
MCG+12-15-049	16 04 18.	+ 69 36	21	17.	GALAXY
IC 1192	16 04 18.3	+ 17 54 26.	41		GALAXY SB(s)0
SC 1602+1803.0	16 04 19.	+ 17 54 51.	30		NEBULA
SC 1602+1759.6	16 04 21.	+ 17 51 27.	18		NEBULA
SC 1602+1809.9	16 04 21.	+ 18 01 45.	12		NEBULA
SN 1963Q	16 04 23.	+ 17 19		17.0	SUPERNOVA
IC 1194A	16 04 23.7	+ 17 55 04.	18		GALAXY SA(s)0
ZWG 079.047	16 04 24.	+ 12 01		15.2	GALAXY
ZWG 108.150	16 04 24.	+ 16 27		14.3	GALAXY
UGC 10204	16 04 24.	+ 16 27	108	14.3	GALAXY S0?
MCG+03-41-127	16 04 24.	+ 16 27	54	14.3	GALAXY
ZWG 108.151	16 04 24.	+ 17 19		15.4	GALAXY
MCG+03-41-126	16 04 24.	+ 17 19	30	15.4	GALAXY
ZWG 108.152	16 04 24.	+ 17 54		15.5	GALAXY
MCG+03-41-128	16 04 24.	+ 17 55	24	15.5	GALAXY
SC 1602+1803.2	16 04 24.	+ 17 55 03.	18		NEBULA
ZC 1604.4+6113	16 04 24.	+ 61 13	470		CLUSTER OF GALAXIES
ZWG 338.042	16 04 24.	+ 69 36		15.3	GALAXY
IC 1194B	16 04 24.7	+ 17 53 32.	20	15.5	GALAXY E
IC 1195	16 04 25.2	+ 17 19 31.	30	15.4	GALAXY Sb
RNGC 6059	16 04 28.	- 06 17			NON-EXISTENT OBJECT
ABC 2153	16 04 30.	+ 14 13		16.9	RICH CLUSTER OF GALAXIES
MCG+03-41-129	16 04 30.	+ 14 55	30	15.4	GALAXY
ZWG 108.153	16 04 30.	+ 14 56		15.4	GALAXY
ZWG 108.154	16 04 30.	+ 17 38		15.7	GALAXY
MCG+07-33-041	16 04 30.	+ 40 31	27	17.	GALAXY
MCG+09-26-052	16 04 30.	+ 55 45	54	15.	GALAXY
SC 1602+1744.4	16 04 32.	+ 17 36 15.	18		NEBULA
ARP 188	16 04 33.	+ 55 40			PECULIAR GALAXY
ZC 1604.6+1149	16 04 36.	+ 11 49	1140		CLUSTER OF GALAXIES
ZC 1604.6+2250	16 04 36.	+ 22 50	740		CLUSTER OF GALAXIES
ZWG 167.025	16 04 36.	+ 28 00		15.4	GALAXY
ZWG 167.026	16 04 36.	+ 30 14		14.4	GALAXY
UGC 10205	16 04 36.	+ 30 14	102	14.4	GALAXY Sa
MCG+06-35-033	16 04 36.	+ 36 49 30.	63	15.	GALAXY
ZC 1604.6+3807	16 04 36.	+ 38 07	1010		CLUSTER OF GALAXIES
ZWG 275.021	16 04 36.	+ 55 30		15.5	GALAXY
VV 010C	16 04 36.	+ 55 39	54	17.	INTERACTING GALAXY
VV 010B	16 04 36.	+ 55 39	12	17.	INTERACTING GALAXY
VV 010A	16 04 36.	+ 55 39	54	15.	INTERACTING GALAXY
VV 010	16 04 36.	+ 55 39	264		INTERACTING GALAXY
ZWG 338.043	16 04 36.	+ 69 50		13.9	GALAXY
UGC 10206	16 04 36.	+ 69 50	84	13.9	GALAXY E
SC 1602+1557.2	16 04 37.	+ 15 49 04.	24		NEBULA
MCG+05-38-015	16 04 39.	+ 32 02	90	15.3	GALAXY
RNGC 6079	16 04 39.	+ 69 50		14.0	GALAXY
ZWG 108.155	16 04 42.	+ 15 44		15.3	GALAXY
ZC 1604.7+2758	16 04 42.	+ 27 58	1340		CLUSTER OF GALAXIES
MCG+05-38-017	16 04 42.	+ 30 14	72	14.4	GALAXY
MCG+05-38-016	16 04 42.	+ 32 00 30.	48	15.3	GALAXY
ZWG 167.027	16 04 42.	+ 32 01		15.3	GALAXY
UGC 10208	16 04 42.	+ 32 01	72	15.3	GALAXY
UGC 10207	16 04 42.	+ 32 01	114	15.3	GALAXY SBb
UGC 10209	16 04 42.	+ 36 48	66	16.0	GALAXY Sb
ZWG 275.022	16 04 42.	+ 55 24		15.7	GALAXY
MCG+09-26-053	16 04 42.	+ 55 24	42	15.	GALAXY
MCG+09-26-054	16 04 42.	+ 55 29	12	16.	GALAXY
SC 1602+1652.4	16 04 44.	+ 16 44 16.	12		NEBULA
MCG+03-41-130	16 04 45.	+ 15 43	48	15.3	GALAXY
IC 4589	16 04 45.	- 06 15			NONSTELLAR OBJECT
RNGC 6063	16 04 47.	+ 08 06		14.0	GALAXY
SC 1602+1817.7	16 04 47.	+ 18 09 34.	24		NEBULA

OBJECT NAME	RIGHT ASCEN.	DECLINATION	DIAM.	MAGN.	TYPE OF OBJECT
ZWG 051.045	16 04 48.	+ 08 06		14.1	GALAXY
UGC 10210	16 04 48.	+ 08 06	114	14.1	GALAXY Sc
MCG+01-41-012	16 04 48.	+ 08 06	90	14.1	GALAXY
ZWG 079.048	16 04 48.	+ 14 24		15.2	GALAXY
RNGC 6064	16 04 48.	+ 20 41			NON-EXISTENT OBJECT
ZC 1604.8+2407	16 04 48.	+ 24 07	1080		CLUSTER OF GALAXIES
MCG+06-35-034	16 04 48.	+ 37 15	30	15.	GALAXY
VV 029B	16 04 48.	+ 55 33	90	16.	INTERACTING GALAXY
VV 029A	16 04 48.	+ 55 33	66	14.	INTERACTING GALAXY
VV 029	16 04 48.	+ 55 33	210		INTERACTING GALAXY
MCG+09-26-055	16 04 48.	+ 55 47 30.	30	17.	GALAXY
MCG+12-15-050	16 04 48.	+ 69 48	33	14.	GALAXY
ZWG 051.046	16 04 54.	+ 03 42		15.5	GALAXY
ZWG 079.049	16 04 54.	+ 10 29		15.6	GALAXY
KARA.72 484A	16 04 54.	+ 10 29	60	15.6	PART OF DOUBLE GALAXY
ZWG 137.039	16 04 54.	+ 22 12		15.0	GALAXY
UGC 10211	16 04 54.	+ 22 12	72	15.0	GALAXY Sa-b
UGC 10212	16 04 54.	+ 77 45	90	16.0	GALAXY Sb-c
MCG+13-11-023	16 04 54.	+ 77 45	102	16.	GALAXY
SC 1602+1651.2	16 04 55.	+ 16 43 05.	30		NEBULA
MCG+03-41-131	16 04 57.	+ 18 36	45	16.	GALAXY
MCG+03-41-132	16 04 57.	+ 18 47	30	16.	GALAXY
MCG+04-38-028	16 04 57.	+ 22 10	66	15.0	GALAXY
LB 00921	16 04 58.	+ 77 36 36.		17.8	FAINT BLUE STAR
ZWG 079.050	16 05 00.	+ 10 33		15.0	GALAXY
UGC 10213	16 05 00.	+ 10 33	66	15.0	GALAXY S
KARA.72 484B	16 05 00.	+ 10 33	66	15.0	PART OF DOUBLE GALAXY
MCG+02-41-007	16 05 00.	+ 10 33	72	15.0	GALAXY
ZWG 275.023	16 05 00.	+ 55 33		15.0	GALAXY
UGC 10214	16 05 00.	+ 55 33	246	15.0	GALAXY SB
MCG+09-26-056	16 05 00.	+ 55 33	198	14.	GALAXY
ZC 1605.0+7212	16 05 00.	+ 72 12	2080		CLUSTER OF GALAXIES
SC 1602+1608.4	16 05 04.	+ 16 00 17.	18		NEBULA
ZWG 051.047	16 05 06.	+ 02 38		15.7	GALAXY
ZC 1605.1+1250	16 05 06.	+ 12 50	1480		CLUSTER OF GALAXIES
ZWG 079.051	16 05 06.	+ 14 01		15.0	GALAXY
MCG+02-41-008	16 05 06.	+ 14 01	36	15.0	GALAXY
ZWG 167.028	16 05 06.	+ 30 39		15.4	GALAXY
RNGC 6065	16 05 08.	+ 14 00		15.0	GALAXY
SC 1602+1755.5	16 05 10.	+ 17 47 24.	18		NEBULA
PK331+00.1	16 05 11.	- 50 54 06.	25		PLANETARY NEBULA
ZWG 051.048	16 05 12.	+ 03 06		15.5	GALAXY
ZWG 079.052	16 05 12.	+ 13 58		15.5	GALAXY
ZWG 108.156	16 05 12.	+ 19 41		15.7	GALAXY
UGC 10215	16 05 12.	+ 19 41	60	15.7	GALAXY S
ARC 2155	16 05 12.	+ 25 09		17.5	RICH CLUSTER OF GALAXIES
ZC 1605.2+2818	16 05 12.	+ 28 18	400		CLUSTER OF GALAXIES
ZC 1605.2+3216	16 05 12.	+ 32 16	470		CLUSTER OF GALAXIES
MCG+03-41-133	16 05 15.	+ 19 41	30	15.7	GALAXY
ZWG 023.014	16 05 18.	+ 01 32		15.1	GALAXY
ZWG 079.053	16 05 18.	+ 13 51		15.2	GALAXY
UGC 10216	16 05 18.	+ 13 51	60	15.2	GALAXY S
ZWG 079.054	16 05 18.	+ 14 04		15.2	GALAXY
SC 1603+1754.2	16 05 18.	+ 17 46 06.	12		NEBULA
ZC 1605.3+2505	16 05 18.	+ 25 05	940		CLUSTER OF GALAXIES
TON-N 0813	16 05 18.	+ 29 43		15.5	BLUE STAR
RNGC 6066	16 05 21.	+ 14 04		15.0	GALAXY
ZC 1605.4+5247	16 05 24.	+ 52 47	1080		CLUSTER OF GALAXIES
SC 1603+1845.0	16 05 25.	+ 18 36 55.	12		NEBULA
ZC 1605.5+0421	16 05 30.	+ 04 21	1550		CLUSTER OF GALAXIES
MCG+04-38-029	16 05 30.	+ 22 27	60		GALAXY
UGC 10217	16 05 30.	+ 22 29	60	16.0	GALAXY SBb
ZWG 223.040	16 05 30.	+ 41 32		15.5	GALAXY
IC 1196	16 05 35.	+ 10 54 22.			NONSTELLAR OBJECT
ZC 1605.6+1004	16 05 36.	+ 10 04	870		CLUSTER OF GALAXIES
ZWG 079.055	16 05 36.	+ 10 55		14.8	GALAXY
UGC 10218	16 05 36.	+ 10 55	66	14.8	GALAXY Sa
MCG+02-41-009	16 05 36.	+ 10 55	72	14.8	GALAXY
TON-N 0254	16 05 36.	+ 31 47		15.3	BLUE STAR
MCG+07-33-042	16 05 36.	+ 41 32	30	15.	GALAXY
MCG+08-29-046	16 05 36.	+ 47 52	36	16.	GALAXY
SC 1603+1833.7	16 05 40.	+ 18 25 38.	12		NEBULA
PATH 1.734	16 05 40.	+ 74 45	5		NEBULA
ZWG 051.049	16 05 42.	+ 03 15		15.6	GALAXY
ZWG 079.056	16 05 42.	+ 14 02		15.6	GALAXY
MCG+06-35-035	16 05 42.	+ 34 14	30	16.	GALAXY
ZWG 223.041	16 05 42.	+ 41 50		15.6	GALAXY
MCG+08-29-047	16 05 42.	+ 48 42 30.	21	16.	GALAXY
IC 1201	16 05 47.	+ 69 44 38.			NONSTELLAR OBJECT
ZWG 051.050	16 05 48.	+ 03 20		15.7	GALAXY
ZWG 051.051	16 05 48.	+ 05 51		15.6	GALAXY
ZWG 195.015	16 05 48.	+ 36 37		15.7	GALAXY
ZWG 250.046	16 05 48.	+ 48 42		15.7	GALAXY
MCG+12-15-051	16 05 48.	+ 69 45	78	15.	GALAXY
MCG-01-41-004	16 05 48.	- 05 32	60	15.5	GALAXY
ARC 2161	16 05 49.	+ 71 40		17.5	RICH CLUSTER OF GALAXIES
IC 1197	16 05 53.	+ 07 41 11.			NONSTELLAR OBJECT
ZWG 051.052	16 05 54.	+ 07 40		14.7	GALAXY
UGC 10219	16 05 54.	+ 07 40	168	14.7	GALAXY Sc
MCG+01-41-013	16 05 54.	+ 07 40	180	14.7	GALAXY
ZWG 108.157	16 05 54.	+ 16 54		15.7	GALAXY
MCG+03-41-134	16 05 54.	+ 16 54	42	15.7	GALAXY
UGC 10220	16 05 54.	+ 20 20	60	16.0	GALAXY SBc
MCG+03-41-135	16 05 54.	+ 20 20	42	15.5	GALAXY
ZC 1605.9+2445	16 05 54.	+ 24 45	1010		CLUSTER OF GALAXIES
ZWG 223.042	16 05 54.	+ 39 03		15.5	GALAXY
ZWG 338.044	16 05 54.	+ 69 45		15.6	GALAXY
UGC 10221	16 05 54.	+ 69 45	66	15.6	GALAXY Sb
SC 1603+1701.4	16 05 55.	+ 16 53 21.	18		NEBULA
LB 00922	16 05 55.	+ 38 34 36.		15.7	FAINT BLUE STAR
RNGC 6069	16 05 55.	+ 39 03		15.5	GALAXY
ZWG 051.053	16 06 00.	+ 07 35		15.4	GALAXY
ZWG 137.040	16 06 00.	+ 25 36		15.7	GALAXY
MCG+04-38-030	16 06 00.	+ 25 36	36	15.7	GALAXY
MCG+07-33-043	16 06 00.	+ 39 03	42	16.	GALAXY
ZWG 223.043	16 06 00.	+ 39 19		15.1	GALAXY
KARA.73B 0721	16 06 00.	+ 39 19	18	15.1	ISOLATED GALAXY E
MCG+08-29-048	16 06 00.	+ 45 12	36	16.	GALAXY
ARC 2157	16 06 00.	+ 48 00		17.7	RICH CLUSTER OF GALAXIES
ZC 1606.0+4801	16 06 00.	+ 48 01	1010		CLUSTER OF GALAXIES
MCG+10-23-020	16 06 00.	+ 62 40	54	16.	GALAXY
UGC 10223	16 06 00.	+ 76 58	60	16.0	GALAXY SBb
7ZW 636	16 06 00.	+ 82 01			COMPACT GALAXY
UGC 10222	16 06 00.	+ 82 01	27	14.0	GALAXY DBL SYS
SC 1603+1850.4	16 06 02.	+ 18 42 21.	18		NEBULA
ZWG 079.057	16 06 06.	+ 09 44		14.9	GALAXY
MCG+02-41-010	16 06 06.	+ 09 44	36	14.9	GALAXY
KARA.73B 0722	16 06 06.	+ 09 44	48	14.9	ISOLATED GALAXY S0

OBJECT NAME	RIGHT ASCEN.	DECLINATION	DIAM.	MAGN.	TYPE OF OBJECT
ZC 1606.1+6602	16 06 06.	+ 66 02	1980		CLUSTER OF GALAXIES
ZC 1606.2+0530	16 06 12.	+ 05 30	1750		CLUSTER OF GALAXIES
ZWG 051.054	16 06 12.	+ 07 00		15.7	GALAXY
ZWG 137.041	16 06 12.	+ 23 37		15.3	GALAXY
IC 1198	16 06 15.	+ 12 28 32.			NONSTELLAR OBJECT
MCG+04-38-032	16 06 15.	+ 23 35 30.	12	15.3	GALAXY
MCG+04-38-031	16 06 15.	+ 23 36	48	15.3	GALAXY
ZWG 079.058	16 06 18.	+ 12 08		15.5	GALAXY
ZWG 079.059	16 06 18.	+ 12 28		14.9	GALAXY
MCG+02-41-011	16 06 18.	+ 12 28	36	14.9	GALAXY
ZWG 167.029	16 06 18.	+ 28 37		15.7	GALAXY
MCG+05-38-018	16 06 18.	+ 28 37	18	15.7	GALAXY
MCG+09-26-057	16 06 18.	+ 54 05	48	16.	GALAXY
IC 4590	16 06 20.	+ 28 36 53.			NONSTELLAR OBJECT
SHB 323	16 06 23.4	+ 10 36 59.		18.0	QUASI-STELLAR OBJECT
ZWG 051.055	16 06 24.	+ 05 45		15.5	GALAXY
ZWG 079.060	16 06 24.	+ 13 04		15.4	GALAXY
ZWG 137.042	16 06 24.	+ 21 13		15.5	GALAXY
ZC 1606.4+3404	16 06 24.	+ 34 04	3490		CLUSTER OF GALAXIES
MCG+06-35-036	16 06 24.	+ 36 14	30	15.	GALAXY
SC 1604+1646.1	16 06 26.	+ 16 38 05.	12		NEBULA
ZWG 023.015	16 06 30.	+ 01 10		15.5	GALAXY
MCG+03-41-136	16 06 30.	+ 16 53	48	15.4	GALAXY
ZWG 108.158	16 06 30.	+ 16 54		15.4	GALAXY
TON-N 0255	16 06 30.	+ 27 21		16.0	BLUE STAR
ZC 1606.5+3033	16 06 30.	+ 30 33	1080		CLUSTER OF GALAXIES
ZWG 195.016	16 06 30.	+ 36 13		15.6	GALAXY
ZWG 275.024	16 06 30.	+ 54 06		15.6	GALAXY
SC 1604+1700.5	16 06 31.	+ 16 52 29.	72		NEBULA
ZWG 137.043	16 06 36.	+ 22 11		15.5	GALAXY
UGC 10224	16 06 36.	+ 22 11	66	15.5	GALAXY Sb-c
ARC 2158	16 06 37.	+ 43 09		16.8	RICH CLUSTER OF GALAXIES
PK328-02.1	16 06 41.	- 54 49			PLANETARY NEBULA
MCG+04-38-033	16 06 42.	+ 22 09	54	15.5	GALAXY
ZWG 079.061	16 06 48.	+ 12 06		15.6	GALAXY
ZWG 079.062	16 06 48.	+ 12 13		15.2	GALAXY
ZC 1606.8+1434	16 06 48.	+ 14 34	340		CLUSTER OF GALAXIES
MCG-01-41-005	16 06 48.	- 04 21	12	16.	GALAXY
MRSL 333+01/1	16 06 48.	- 48 55	2520		HII REGION
WRAY 19.43	16 06 52.2	- 48 58 31.			DIFFUSE NEBULA
SC 1604+1620.1	16 06 53.	+ 16 12 06.	18		NEBULA
ZWG 079.063	16 06 54.	+ 08 53		14.8	GALAXY
UGC 10225	16 06 54.	+ 08 53	108	14.8	GALAXY
MCG+02-41-012	16 06 54.	+ 08 53	84	14.8	GALAXY
ZWG 137.044	16 06 54.	+ 22 00		15.6	GALAXY
SC 1604+1642.7	16 06 58.	+ 16 34 43.	18		NEBULA
ZWG 137.045	16 07 00.	+ 25 00		15.5	GALAXY
MCG+10-23-021	16 07 00.	+ 60 00	48	16.	GALAXY
ZC 1607.0+6132	16 07 00.	+ 61 32	540		CLUSTER OF GALAXIES
ZC 1607.0+6736	16 07 00.	+ 67 36	1950		CLUSTER OF GALAXIES
MCG-01-41-006	16 07 00.	- 04 30	36	15.5	GALAXY
OCL 0949	16 07 00.	- 55 10	180		OPEN STAR CLUSTER
VHA 184	16 07 00.	- 55 10	150		OPEN STAR CLUSTER
ZWG 051.056	16 07 06.	+ 08 24		15.6	GALAXY
MCG+06-35-038	16 07 06.	+ 35 55 30.	60	14.	GALAXY
MCG+06-35-037	16 07 06.	+ 36 45	135	14.	GALAXY
ZC 1607.1+4830	16 07 06.	+ 48 30	1080		CLUSTER OF GALAXIES
MCG+12-15-052	16 07 06.	+ 72 38	36	14.	GALAXY
MCG+04-38-034	16 07 09.	+ 25 50	36	15.2	GALAXY
RNGC 6094	16 07 11.	+ 72 38		14.5	GALAXY
ZWG 079.064	16 07 12.	+ 09 05		15.4	GALAXY
MCG+03-41-137	16 07 12.	+ 18 23	36	15.	GALAXY
ZWG 137.046	16 07 12.	+ 25 50		15.2	GALAXY
ZWG 195.017	16 07 12.	+ 35 55		15.4	GALAXY
UGC 10226	16 07 12.	+ 35 55	66	15.4	GALAXY Sb
KARA.73B 0723	16 07 12.	+ 35 55	66	15.4	ISOLATED GALAXY S
ZWG 195.018	16 07 12.	+ 36 45		15.6	GALAXY
UGC 10227	16 07 12.	+ 36 45	132	15.6	GALAXY SB?c
KARA.73B 0724	16 07 12.	+ 36 45	132	15.6	ISOLATED GALAXY S
ZWG 338.045	16 07 12.	+ 72 38		14.6	GALAXY
UGC 10228	16 07 12.	+ 72 38	120	14.6	GALAXY S0
SC 1604+1646.2	16 07 13.	+ 16 38 14.	18		NEBULA
IC 1204	16 07 16.	+ 70 04			NONSTELLAR OBJECT
UGC 10229	16 07 18.	+ 00 00	66	17.	GALAXY DWRF IR
ZC 1607.3+2943	16 07 18.	+ 29 43	2620		CLUSTER OF GALAXIES
MCG+06-35-039	16 07 18.	+ 34 38	24	16.	GALAXY
MCG+10-23-022	16 07 18.	+ 57 57	36	16.	GALAXY
ZWG 298.009	16 07 18.	+ 57 58		15.6	GALAXY
MCG+10-23-023	16 07 18.	+ 58 04	42	15.	GALAXY
ZWG 298.010	16 07 18.	+ 58 05		15.4	GALAXY
SC 1605+1830.4	16 07 19.	+ 18 22 26.	18		NEBULA
HOLM 729A	16 07 21.	+ 00 51	180	12.1	PART OF MULTIPLE GALAXY
MCG+06-35-040	16 07 21.	+ 33 07	21	15.	GALAXY
ZWG 023.017	16 07 24.	+ 00 50		13.0	GALAXY
UGC 10230	16 07 24.	+ 00 50	240	13.0	GALAXY Sc
MCG+00-41-004	16 07 24.	+ 00 50	198	13.0	GALAXY
ZWG 079.065	16 07 24.	+ 13 34		15.5	GALAXY
ZWG 137.047	16 07 24.	+ 20 33		15.6	GALAXY
SHAH 029	16 07 24.	+ 52 34	54	18.	GROUP OF COMPACT GALAXIES
MCG+10-23-024	16 07 24.	+ 58 58	36	16.	GALAXY
UGC 10231	16 07 24.	+ 64 28	96	16.0	GALAXY S
ZWG 338.046	16 07 24.	+ 70 04		15.5	GALAXY
MCG+12-15-053	16 07 24.	+ 70 04	39	15.	GALAXY
MCG+13-12-001	16 07 24.	+ 79 45 30.	51	15.	GALAXY
ZWG 023.016	16 07 24.	- 02 32		15.7	GALAXY
REIN 2.242	16 07 25.87	+ 00 50 20.5			NEBULA
RNGC 6070	16 07 26.	+ 00 50		12.5	GALAXY
SC 1605+1901.7	16 07 26.	+ 18 53 45.	36		NEBULA
ZWG 079.066	16 07 30.	+ 11 23		15.7	GALAXY
ZWG 079.067	16 07 30.	+ 13 39		15.6	GALAXY
ZWG 195.019	16 07 30.	+ 33 08		15.7	GALAXY
ZWG 223.044	16 07 30.	+ 41 52		15.5	GALAXY
HOLM 729B	16 07 31.	+ 00 54	54	14.9	PART OF MULTIPLE GALAXY
RNGC 6070A	16 07 32.	+ 00 53		15.5	GALAXY
MCG+03-41-138	16 07 33.	+ 20 18 30.	90	15.7	GALAXY
HOLM 729C	16 07 34.	+ 00 55	30	14.9	PART OF MULTIPLE GALAXY
SC 1605+1551.2	16 07 34.	+ 15 43 15.	18		NEBULA
ZWG 023.018	16 07 36.	+ 00 53		15.6	GALAXY
HOLM 730B	16 07 36.	+ 01 12	18	15.4	PART OF MULTIPLE GALAXY
ZWG 051.057	16 07 36.	+ 06 42		15.6	GALAXY
ZWG 108.159	16 07 36.	+ 20 18		15.7	GALAXY
UGC 10232	16 07 36.	+ 20 18	96	15.7	GALAXY Sc
ZWG 137.048	16 07 36.	+ 25 20		15.6	GALAXY
MCG+05-38-019	16 07 36.	+ 28 10	30	14.7	GALAXY
ZWG 167.030	16 07 36.	+ 28 11		14.7	GALAXY
MCG+07-33-044	16 07 36.	+ 40 16	30	16.	GALAXY
MCG+07-33-045	16 07 36.	+ 41 52	36	16.	GALAXY
1ZW 134	16 07 36.	+ 41 53			COMPACT GALAXY

OBJECT NAME	RIGHT ASCEN.	DECLINATION	DIAM.	MAGN.	TYPE OF OBJECT
ZC 1607.6+4303	16 07 36.	+ 43 03	1010		CLUSTER OF GALAXIES
ZWG 250.047	16 07 36.	+ 46 47		15.7	GALAXY
ZC 1607.6+6055	16 07 36.	+ 60 55	1880		CLUSTER OF GALAXIES
RNGC 6070B	16 07 38.	+ 00 54		15.5	GALAXY
HOLM 730A	16 07 38.	+ 01 12	24	15.1	PART OF MULTIPLE GALAXY
ZWG 023.019	16 07 42.	+ 01 10		15.1	GALAXY
MCG+00-41-005	16 07 42.	+ 01 10	78	15.1	GALAXY
MCG+04-38-035	16 07 42.	+ 22 44	90	15.7	GALAXY
ZWG 137.049	16 07 42.	+ 22 45		15.7	GALAXY
UGC 10233	16 07 42.	+ 22 45	90	15.7	GALAXY Sc
MCG+05-38-020	16 07 42.	+ 30 16	30	15.6	GALAXY
ZWG 167.031	16 07 42.	+ 30 35		15.4	GALAXY
UGC 10234	16 07 42.	+ 30 35	78	15.4	GALAXY S0
TON-N 0814	16 07 42.	+ 31 22		16.0	BLUE STAR
MCG+09-26-058	16 07 42.	+ 51 42 30.	30	16.	GALAXY
MPSL 332+00/1	16 07 42.	- 50 13	2520		HII REGION
ZWG 167.032	16 07 48.	+ 30 16		15.6	GALAXY
ZWG 275.025	16 07 48.	+ 51 45		15.7	GALAXY
KARA.73B 0725	16 07 48.	+ 51 45	36	15.7	ISOLATED GALAXY S
SC 1605+1908.3	16 07 50.	+ 19 00 22.	18		NEBULA
RNGC 6073	16 07 52.	+ 16 50		14.5	GALAXY
MCG+03-41-139	16 07 54.	+ 16 49	60	14.5	GALAXY
ZWG 108.160	16 07 54.	+ 16 50		14.5	GALAXY
UGC 10235	16 07 54.	+ 16 50	72	14.5	GALAXY Sc
MCG+04-38-036	16 07 54.	+ 22 46	90	15.6	GALAXY
ZWG 137.050	16 07 54.	+ 22 47		15.6	GALAXY
UGC 10236	16 07 54.	+ 22 47	90	15.6	GALAXY Sb
MCG+07-33-046	16 07 54.	+ 40 11	27	17.	GALAXY
SC 1605+1552.6	16 07 58.	+ 15 44 40.	18		NEBULA
HOLM 731A	16 07 58.	+ 16 49	66	13.5	PART OF MULTIPLE GALAXY
RNGC 6091	16 07 58.	+ 70 03		14.5	GALAXY
LBN 0105	16 08	+ 22 00	2400		BRIGHT NEBULA
LBN 0017	16 08	- 06 58	1200		BRIGHT NEBULA
ZWG 051.058	16 08 00.	+ 04 31		15.4	GALAXY
ZWG 108.161	16 08 00.	+ 15 45		15.5	GALAXY
ZC 1608.0+2826	16 08 00.	+ 28 26	870		CLUSTER OF GALAXIES
MCG+10-23-025	16 08 00.	+ 58 13	18	16.	GALAXY
ZWG 338.047	16 08 00.	+ 70 03		14.7	GALAXY
MCG+12-15-054	16 08 00.	+ 70 03	24	15.	GALAXY
ZWG 355.009	16 08 00.	+ 79 45		15.2	GALAXY
ZWG 354.034	16 08 00.	+ 79 45		15.2	GALAXY
UGC 10237	16 08 00.	+ 79 45	78	15.2	GALAXY Sc
ZWG 023.020	16 08 00.	- 01 36		15.6	GALAXY
VHE 69	16 08 00.	- 51 08	96		REFLECTION NEBULA
HOLM 731B	16 08 01.	+ 16 50	18	15.0	PART OF MULTIPLE GALAXY
ZWG 051.059	16 08 06.	+ 04 53		15.5	GALAXY
ZWG 079.068	16 08 06.	+ 12 26		15.1	GALAXY
UGC 10238	16 08 06.	+ 12 26	96	15.1	GALAXY SBb
ZWG 079.069	16 08 06.	+ 12 57		15.1	GALAXY
UGC 10239	16 08 06.	+ 12 57	60	15.1	GALAXY SBa
KARA.73B 0726	16 08 06.	+ 12 57	66	15.1	ISOLATED GALAXY S
UGC 10240	16 08 06.	+ 32 15	60	16.0	GALAXY Sa-b
ZWG 223.045	16 08 06.	+ 42 27		15.3	GALAXY
UGC 10241	16 08 06.	+ 42 27	66	15.3	GALAXY Sb-c
MCG+07-33-047	16 08 06.	+ 42 28	48	15.	GALAXY
IC 1199	16 08 11.	+ 10 10 15.			NONSTELLAR OBJECT
BC M1608+114	16 08 11.6	+ 11 23 19.		19.	QUASI-STELLAR OBJECT
SHB 324	16 08 11.6	+ 11 23 19.		19.	QUASI-STELLAR OBJECT
ZWG 023.021	16 08 12.	+ 02 14		15.4	GALAXY
ZWG 079.070	16 08 12.	+ 10 10		14.6	GALAXY
UGC 10242	16 08 12.	+ 10 10	78	14.6	GALAXY S
MCG+02-41-013	16 08 12.	+ 10 10	84	14.6	GALAXY
ZWG 079.071	16 08 12.	+ 13 48		15.6	GALAXY
ZWG 108.162	16 08 12.	+ 20 05		15.4	GALAXY
UGC 10243	16 08 12.	+ 20 05	60	15.4	GALAXY SBc
MCG+10-23-026	16 08 12.	+ 58 01	15	16.	GALAXY
SC 1606+1913.2	16 08 14.	+ 19 05 18.	42		NEBULA
MCG+03-41-140	16 08 18.	+ 20 05	60	15.4	GALAXY
ZWG 223.046	16 08 18.	+ 43 15		15.4	GALAXY
UGC 10244	16 08 18.	+ 43 15	72	15.4	GALAXY S
MCG+07-33-048	16 08 18.	+ 43 15	48	16.	GALAXY
MCG+05-38-021	16 08 21.	+ 31 55 30.	51	15.3	GALAXY
SN 1965F	16 08 23.	+ 19 11		18.0	SUPERNOVA
MCG+05-38-022	16 08 24.	+ 27 36	60	16.	GALAXY
UGC 10245	16 08 24.	+ 49 17	60	16.0	GALAXY Sb-c
MCG+08-29-049	16 08 24.	+ 49 18	72	16.	GALAXY
MPSL 005+30/1	16 08 24.	- 06 57	1800		HII REGION
SC 1606+1919.1	16 08 27.	+ 19 11 13.	24		NEBULA
UGC 10246	16 08 30.	+ 27 38	72	16.0	GALAXY Sa-b
ZC 1608.5+3044	16 08 30.	+ 30 44	21370		CLUSTER OF GALAXIES
MCG+10-23-027	16 08 30.	+ 60 12	48	15.	GALAXY
ZWG 298.011	16 08 30.	+ 60 13		15.6	GALAXY
UGC 10247	16 08 30.	+ 60 13	60	15.6	GALAXY
SC 1606+1719.1	16 08 35.	+ 17 11 13.	18		NEBULA
REIN 7.167	16 08 35.33	+ 60 12 57.6			NEBULA
ZWG 079.072	16 08 36.	+ 12 04		15.5	GALAXY
ZWG 079.073	16 08 36.	+ 13 53		14.8	GALAXY
MCG+02-41-014	16 08 36.	+ 13 53	60	14.8	GALAXY
ZWG 108.163	16 08 36.	+ 17 11		15.0	GALAXY
MCG+03-41-141	16 08 36.	+ 17 11	48	15.0	GALAXY
ZWG 108.164	16 08 36.	+ 18 06		15.4	GALAXY
MCG+03-41-142	16 08 36.	+ 18 06	42	15.4	GALAXY
UGC 10248	16 08 36.	+ 29 39	60	16.0	GALAXY Sb
ZWG 167.033	16 08 36.	+ 31 55		15.3	GALAXY
ZWG 223.047	16 08 36.	+ 41 59		15.2	GALAXY
MCG+07-33-049	16 08 36.	+ 42 00	30	15.	GALAXY
SHAH 023	16 08 36.	+ 52 23	48	18.	GROUP OF COMPACT GALAXIES
ZWG 298.012	16 08 36.	+ 58 16		15.6	GALAXY
SC 1606+1812.4	16 08 37.	+ 18 04 31.	36		NEBULA
SC 1606+1921.4	16 08 37.	+ 19 13 31.	24		NEBULA
LB 00923	16 08 40.	+ 40 20 12.		16.5	FAINT BLUE STAR
ZWG 137.051	16 08 42.	+ 22 45		15.6	GALAXY
MCG+08-29-050	16 08 42.	+ 47 12	48	16.	GALAXY
MCG+10-23-028	16 08 42.	+ 58 15	24	16.	GALAXY
GCL 038	16 08 48.	+ 15 05	504		GLOBULAR STAR CLUSTER
MRK 301	16 08 48.	+ 18 12	7	17.5	GALAXY WITH UV CONTINUUM
ZWG 108.165	16 08 48.	+ 18 38		15.5	GALAXY
MCG+06-35-041	16 08 48.	+ 37 00	42	15.	GALAXY
ZWG 079.074	16 08 54.	+ 13 59		15.2	GALAXY
UGC 10249	16 08 54.	+ 13 59	72	15.2	GALAXY SBa-b
ZC 1608.9+1705	16 08 54.	+ 17 05	1950		CLUSTER OF GALAXIES
SC 1606+1845.0	16 08 54.	+ 18 37 08.	18		NEBULA
ZWG 137.052	16 08 54.	+ 26 28		15.7	GALAXY
MCG+04-38-037	16 08 54.	+ 26 28	60	15.7	GALAXY
UGC 10250	16 08 54.	+ 26 30	60	17.	GALAXY S
TON-N 0815	16 08 54.	+ 31 31		15.7	BLUE STAR
UGC 10251	16 08 54.	+ 56 44	72	16.0	GALAXY Sb
MCG+09-26-059	16 08 54.	+ 56 45 30.	72	16.	GALAXY

OBJECT NAME	RIGHT ASCEN.	DECLINATION	DIAM.	MAGN.	TYPE OF OBJECT
ZWG 320.007	16 08 54.	+ 64 31		15.7	GALAXY
REIN 7.168	16 08 55.20	+ 56 45 28.8			NEBULA
LBN 1096	16 09	- 27 48	660		BRIGHT NEBULA
ZWG 079.075	16 09 00.	+ 14 23		15.3	GALAXY
MCG+02-41-016	16 09 00.	+ 14 23	15	16.	GALAXY
MCG+02-41-015	16 09 00.	+ 14 23	12	15.3	GALAXY
MPSL 037+44/1	16 09 00.	+ 22 00	4500		HII REGION
ZWG 137.053	16 09 00.	+ 24 20		15.4	GALAXY
ZWG 223.048	16 09 00.	+ 41 16		15.3	GALAXY
UGC 10252	16 09 00.	+ 41 16	72	15.3	GALAXY Sa-b
MCG+07-33-050	16 09 00.	+ 41 17	54	14.5	GALAXY
ZC 1609.0+5345	16 09 00.	+ 53 45	2420		CLUSTER OF GALAXIES
ZC 1609.0+6411	16 09 00.	+ 64 11	6920		CLUSTER OF GALAXIES
ZWG 320.008	16 09 00.	+ 64 30		15.6	GALAXY
ZC 1609.0+8212	16 09 00.	+ 82 12	11760		CLUSTER OF GALAXIES
OCL 0956	16 09 00.	- 52 15	330	13.	OPEN STAR CLUSTER
VHA 185	16 09 00.	- 52 15	180		OPEN STAR CLUSTER
RNGC 6074	16 09 03.	+ 14 23		15.5	GALAXY
MCG+06-35-042	16 09 03.	+ 36 07	60	15.	GALAXY
VDB.66N 100	16 09 03.	- 19 20	684		REFLECTION NEBULA
RNGC 6076	16 09 04.	+ 27 00		15.5	GALAXY
RNGC 6077	16 09 04.	+ 27 04		15.0	GALAXY
IC 4592	16 09 05.	- 19 20	4800		DIFFUSE NEBULA
ZWG 167.034	16 09 06.	+ 27 00		15.4	GALAXY
UGC 10253	16 09 06.	+ 27 00	72	15.4	GALAXY DBL SYS
ZWG 167.035	16 09 06.	+ 27 04		14.9	GALAXY
UGC 10254	16 09 06.	+ 27 04	84	14.9	GALAXY E
ZWG 196.001	16 09 06.	+ 36 07		15.4	GALAXY
ZWG 195.020	16 09 06.	+ 36 07		15.4	GALAXY
UGC 10255	16 09 06.	+ 36 07	66	15.4	GALAXY Sb
ZWG 223.049	16 09 06.	+ 42 00		14.9	GALAXY
CED 128	16 09 06.	- 19 20	11700		DIFFUSE GALACTIC NEBULA
REIN 7.169	16 09 06.62	+ 58 35 09.9			NEBULA
SC 1606+1756.7	16 09 10.	+ 17 48 51.	12		NEBULA
CED 127	16 09 11.	- 27 48	1380		DIFFUSE GALACTIC NEBULA
ZWG 137.054	16 09 12.	+ 22 30		15.5	GALAXY
ZWG 137.055	16 09 12.	+ 24 05		15.3	GALAXY
MCG+07-33-051	16 09 12.	+ 42 02	42	15.	GALAXY
7ZW 637	16 09 12.	+ 64 06			COMPACT GALAXY
DG 132	16 09 12.	- 19 20	8400		REFLECTION NEBULA
DG 131	16 09 12.	- 27 48	660		REFLECTION NEBULA
IC 4594	16 09 13.	+ 23 47 30.			NONSTELLAR OBJECT
IC 4591	16 09 13.	- 27 48			NONSTELLAR OBJECT
RNGC 6075	16 09 14.	+ 24 05		15.5	GALAXY
MCG+05-38-023	16 09 15.	+ 26 59	54	15.4	GALAXY
MCG+05-38-024	16 09 15.	+ 27 02 30.	48	14.9	GALAXY
ARC 2159	16 09 17.	+ 17 07		15.9	RICH CLUSTER OF GALAXIES
ZWG 137.056	16 09 18.	+ 23 56		15.5	GALAXY
UGC 10256	16 09 18.	+ 23 56	78	15.5	GALAXY Sb
MCG+04-38-038	16 09 18.	+ 24 05 30.	48	15.3	GALAXY
ZC 1609.3+3148	16 09 18.	+ 31 48	340		CLUSTER OF GALAXIES
ZWG 196.002	16 09 18.	+ 38 23		15.0	GALAXY
ZWG 195.021	16 09 18.	+ 38 23		15.0	GALAXY
UGC 10257	16 09 18.	+ 38 23	108	15.0	GALAXY Sb-c
ZC 1609.3+4245	16 09 18.	+ 42 45	1080		CLUSTER OF GALAXIES
VHE 70	16 09 18.	- 50 16	30		REFLECTION NEBULA
GCL 0953	16 09 18.	- 54 05	1920	6.7	OPEN STAR CLUSTER
VHA 186	16 09 18.	- 54 05	600		OPEN STAR CLUSTER
RNGC 6067	16 09 20.	- 54 05		6.5	OPEN CLUSTER
HH 0108	16 09 21.	+ 12 12			NEBULA
MCG+05-38-025	16 09 21.	+ 29 29 30.	60	15.7	GALAXY
MCG+06-35-043	16 09 21.	+ 38 24	114	14.	GALAXY
SC 1607+1909.6	16 09 22.	+ 19 01 46.	36		NEBULA
IC 4593	16 09 23.6	+ 12 11 57.3	16	11.1	PLANETARY NEBULA
MCG+04-38-039	16 09 24.	+ 23 56	66	15.5	GALAXY
ZWG 167.036	16 09 24.	+ 29 29		15.7	GALAXY
UGC 10258	16 09 24.	+ 29 29	72	15.7	GALAXY Sa-b
ZC 1609.4+3006	16 09 24.	+ 30 06	670		CLUSTER OF GALAXIES
MCG+06-35-045	16 09 24.	+ 36 04 30.	12	16.	GALAXY
MCG+06-35-044	16 09 24.	+ 36 04 30.	36	15.	GALAXY
ZWG 196.003	16 09 24.	+ 36 05		15.3	GALAXY
ZWG 195.022	16 09 24.	+ 36 05		15.3	GALAXY
ARC 2160	16 09 29.	+ 30 04		17.9	RICH CLUSTER OF GALAXIES
ZWG 276.001	16 09 30.	+ 56 23		15.6	GALAXY
ZWG 275.026	16 09 30.	+ 56 23		15.6	GALAXY
SC 1607+1700.5	16 09 32.	+ 16 52 41.	18		NEBULA
ZWG 108.166	16 09 36.	+ 16 54		15.6	GALAXY
ZC 1609.6+2107	16 09 36.	+ 21 07	2550		CLUSTER OF GALAXIES
MCG+05-38-026	16 09 36.	+ 30 27 30.	42	15.6	GALAXY
ZWG 250.048	16 09 36.	+ 49 55		15.2	GALAXY
ZC 1609.6+5219	16 09 36.	+ 52 19	1550		CLUSTER OF GALAXIES
MCG+09-26-060	16 09 36.	+ 56 25	24	16.	GALAXY
OCL 0957	16 09 36.	- 51 47	240	13.	OPEN STAR CLUSTER
VHA 187	16 09 36.	- 51 47	120		OPEN STAR CLUSTER
ARC 2164	16 09 38.	+ 60 32		17.5	RICH CLUSTER OF GALAXIES
HOLM 732B	16 09 40.	+ 57 37	12	14.1	PART OF MULTIPLE GALAXY
RNGC 6072	16 09 40.	- 36 07		14.0	PLANETARY NEBULA
SC 1607+1702.2	16 09 41.	+ 16 54 23.	12		NEBULA
PK342+10.1	16 09 41.62	- 36 06 08.9	70	14.1	PLANETARY NEBULA
REIN 7.170A	16 09 41.89	+ 57 35 44.1			NEBULA
REIN 7.170B	16 09 41.91	+ 57 35 43.4			NEBULA
ZWG 051.060	16 09 42.	+ 03 06		15.6	GALAXY
ZWG 079.076	16 09 42.	+ 14 20		14.6	GALAXY
MCG+02-41-018	16 09 42.	+ 14 20	9	15.	GALAXY
MCG+02-41-017	16 09 42.	+ 14 20	18	14.6	GALAXY
UGC 10259	16 09 42.	+ 29 53	108	16.0	GALAXY CHAIN
MCG+08-29-051	16 09 42.	+ 47 45	36	16.	GALAXY
MCG+09-26-061	16 09 42.	+ 52 34	18	15.	GALAXY
MCG+10-23-030	16 09 42.	+ 57 35	36	14.1	GALAXY
MCG+10-23-029	16 09 42.	+ 57 35	24	13.8	GALAXY
ZWG 298.013	16 09 42.	+ 57 36		14.7	GALAXY
RNGC 6086B	16 09 42.	+ 57 36		14.5	GALAXY
RNGC 6088A	16 09 42.	+ 57 36		14.5	GALAXY
KARA.72 485B	16 09 42.	+ 57 36	42		PART OF DOUBLE GALAXY
KARA.72 485A	16 09 42.	+ 57 36	36	14.7	PART OF DOUBLE GALAXY
HOLM 732A	16 09 42.	+ 57 37	42	13.8	PART OF MULTIPLE GALAXY
YM 01	16 09 42.	- 45 28	96		SYMMETRIC GALACTIC NEBULA
REIN 7.171A	16 09 43.62	+ 57 35 28.5			NEBULA
REIN 7.171B	16 09 43.70	+ 57 35 28.2			NEBULA
RNGC 6078	16 09 45.	+ 14 20		14.5	GALAXY
MCG+09-26-062	16 09 45.	+ 51 18	36	16.	GALAXY
ZWG 051.061	16 09 48.	+ 04 44		15.1	GALAXY
MCG+01-41-014	16 09 48.	+ 04 44	24	15.1	GALAXY
ZWG 137.057	16 09 48.	+ 21 03		15.4	GALAXY
UGC 10260	16 09 48.	+ 21 03	66	15.4	GALAXY S
MCG+05-38-028	16 09 48.	+ 29 18 30.	48	15.6	GALAXY
SN 1961J	16 09 48.	+ 29 42		16.0	SUPERNOVA
MCG+05-38-027	16 09 48.	+ 29 42	24	17.	GALAXY

OBJECT NAME	RIGHT ASCEN.	DECLINATION	DIAM.	MAGN.	TYPE OF OBJECT
ZWG 167.037	16 09 48.	+ 30 28		15.6	GALAXY
ZC 1609.8+3204	16 09 48.	+ 32 04	670		CLUSTER OF GALAXIES
ZWG 275.027	16 09 48.	+ 52 34		15.7	GALAXY
UGC 10261	16 09 48.	+ 52 35	72	15.7	GALAXY E-SO
ZC 1609.8+6031	16 09 48.	+ 60 31	1680		CLUSTER OF GALAXIES
ZC 1609.8+6717	16 09 48.	+ 67 17	540		CLUSTER OF GALAXIES
SC 1607+1700.6	16 09 50.	+ 16 52 48.	12		NEBULA
MCG+05-38-029	16 09 51.	+ 29 59	48	14.8	GALAXY
ZWG 167.038	16 09 54.	+ 29 19		15.6	GALAXY
ZWG 167.039	16 09 54.	+ 29 58		14.8	GALAXY
UGC 10262	16 09 54.	+ 29 58	78	14.8	GALAXY SO
ZWG 196.090	16 09 54.	+ 32 15		15.6	GALAXY
ZWG 167.040	16 09 54.	+ 32 15		15.6	GALAXY
MCG+05-38-030	16 09 54.	+ 32 15	24	15.6	GALAXY
MCG+06-35-046	16 09 54.	+ 37 37	42	15.	GALAXY
MCG-04-38-004	16 09 54.	- 21 33	36	15.	GALAXY
PK025+40.1	16 09 55.	+ 12 12	43	11.4	PLANETARY NEBULA
PK327-04.1	16 09 57.	- 56 51 40.	25		PLANETARY NEBULA
LBN 0013	16 10	- 07 55	1500		BRIGHT NEBULA
LBN 1113	16 10	- 19 10	3000		BRIGHT NEBULA
MCG+09-26-063	16 10 00.	+ 54 05	36	16.	GALAXY
ZC 1610.0+5543	16 10 00.	+ 55 43	1140		CLUSTER OF GALAXIES
MCG+10-23-031	16 10 00.	+ 58 47	36	16.	GALAXY
MCG+10-23-032	16 10 00.	+ 60 45	42	16.	GALAXY
UGC 10263	16 10 00.	+ 86 20	96	16.5	GALAXY DWRF SP
MCG+05-38-031	16 10 03.	+ 28 38	30	15.5	GALAXY
ZWG 051.062	16 10 06.	+ 03 08		15.7	GALAXY
ZWG 137.058	16 10 06.	+ 23 08		15.6	GALAXY
ZWG 137.059	16 10 06.	+ 25 41		15.7	GALAXY
ZWG 167.041	16 10 06.	+ 28 38		15.5	GALAXY
ZWG 167.042	16 10 06.	+ 29 43		15.6	GALAXY
ZWG 275.028	16 10 06.	+ 52 35		15.5	GALAXY
ZWG 051.063	16 10 12.	+ 06 13		15.6	GALAXY
ZWG 137.060	16 10 12.	+ 26 07		15.6	GALAXY
ZWG 223.050	16 10 12.	+ 43 50		15.5	GALAXY
MCG+10-23-033	16 10 12.	+ 61 22	18	14.	GALAXY
SC 1607+1627.0	16 10 13.	+ 16 19 13.	18		NEBULA
PK343+11.1	16 10 13.3	- 34 28 03.	3		PLANETARY NEBULA
MCG+04-38-040	16 10 15.	+ 23 07	36	15.6	GALAXY
MCG+09-26-064	16 10 15.	+ 52 35	78	15.	GALAXY
ZWG 023.022	16 10 18.	+ 00 01		14.9	GALAXY
UGC 10264	16 10 18.	+ 00 01	78	14.9	GALAXY S
MCG+00-41-006	16 10 18.	+ 00 01	66	14.9	GALAXY
ZWG 051.064	16 10 18.	+ 04 43		15.3	GALAXY
MRK 697	16 10 18.	+ 29 13	11	16.5	GALAXY WITH UV CONTINUUM
ZC 1610.3+4955	16 10 18.	+ 49 55	17940		CLUSTER OF GALAXIES
MCG+09-26-065	16 10 18.	+ 56 03	48	16.	GALAXY
ZWG 298.014	16 10 18.	+ 61 24		14.5	GALAXY
UGC 10265	16 10 18.	+ 61 24	120	14.5	GALAXY E-SO
RNGC 6095	16 10 19.	+ 61 24		14.5	GALAXY
RNGC 6090	16 10 22.	+ 52 35		14.0	GALAXY
ZWG 051.065	16 10 24.	+ 06 11		15.5	GALAXY
ZWG 079.077	16 10 24.	+ 11 17		15.3	GALAXY
ZWG 137.061	16 10 24.	+ 25 37		15.5	GALAXY
MCG+05-38-032	16 10 24.	+ 29 46	24	15.5	GALAXY
UGC 10266	16 10 24.	+ 49 02	78	17.	GALAXY DWRF SP
1ZW 135	16 10 24.	+ 52 35			COMPACT GALAXY
ZWG 276.002	16 10 24.	+ 52 35		14.0	GALAXY
ZWG 275.029	16 10 24.	+ 52 35		14.0	GALAXY
MRK 496	16 10 24.	+ 52 35	10	16.	GALAXY WITH UV CONTINUUM
UGC 10267	16 10 24.	+ 52 35	168	14.0	GALAXY DBL SYS
KARA.72 486B	16 10 24.	+ 52 35	24		PART OF DOUBLE GALAXY
KARA.72 486A	16 10 24.	+ 52 35	18	14.0	PART OF DOUBLE GALAXY
ZWG 320.009	16 10 24.	+ 67 58		15.1	GALAXY
KARA.73B 0727	16 10 24.	+ 67 58	24	15.1	ISOLATED GALAXY SO
PK329-02.1	16 10 25.	- 54 40	5		PLANETARY NEBULA
REIN 7.174	16 10 25.89	+ 61 23 41.2			NEBULA
SC 1608+1708.0	16 10 27.	+ 17 00 14.	18		NEBULA
ZWG 023.023	16 10 30.	+ 02 18		14.1	GALAXY
UGC 10268	16 10 30.	+ 02 18	78	14.1	GALAXY Sa
KARA.72 487A	16 10 30.	+ 02 18	54	14.1	PART OF DOUBLE GALAXY
MCG+00-41-007	16 10 30.	+ 02 18	48	14.1	GALAXY
KARA.72 487B	16 10 30.	+ 02 19	30		PART OF DOUBLE GALAXY
ZWG 051.066	16 10 30.	+ 04 28		15.2	GALAXY
ZWG 137.062	16 10 30.	+ 23 07		15.7	GALAXY
ZWG 167.043	16 10 30.	+ 28 27		15.6	GALAXY
ZWG 167.044	16 10 30.	+ 29 30		14.5	GALAXY
RNGC 6085	16 10 30.	+ 29 30		14.5	GALAXY
UGC 10269	16 10 30.	+ 29 30	102	14.5	GALAXY Sa
MCG+05-38-034	16 10 30.	+ 29 30	90	14.5	GALAXY
MCG+05-38-035	16 10 30.	+ 29 37 30.	72	14.8	GALAXY
ZWG 167.045	16 10 30.	+ 29 38		14.8	GALAXY
RNGC 6086	16 10 30.	+ 29 38		15.0	GALAXY
UGC 10270	16 10 30.	+ 29 38	132	14.8	GALAXY E
ARC 2162	16 10 30.	+ 29 40		13.7	RICH CLUSTER OF GALAXIES
MCG+05-38-033	16 10 30.	+ 32 08	66	15.5	GALAXY
ZC 1610.5+4135	16 10 30.	+ 41 35	2690		CLUSTER OF GALAXIES
MCG+08-29-052	16 10 30.	+ 49 02	30	17.	GALAXY
MCG+10-23-036	16 10 30.	+ 57 30	30	16.	GALAXY
MCG+10-23-034	16 10 30.	+ 58 54	78	14.	GALAXY
ZWG 298.015	16 10 30.	+ 58 55		15.0	GALAXY
UGC 10271	16 10 30.	+ 58 55	78	15.0	GALAXY SB
MCG+10-23-035	16 10 30.	+ 61 21	15	16.	GALAXY
REIN 7.172A	16 10 30.79	+ 57 30 37.1			NEBULA
REIN 7.172B	16 10 30.82	+ 57 30 36.5			NEBULA
RNGC 6080	16 10 32.	+ 02 18		14.0	GALAXY
IC 1202	16 10 32.	+ 09 59 50.			NONSTELLAR OBJECT
MIL 41	16 10 32.	- 50 51			SUPERNOVA REMNANT
PK329-02.2	16 10 33.	- 54 49 30.	45	12.6	PLANETARY NEBULA
SC 1606-5441.7	16 10 33.	- 54 49 31.	30		NEBULA
REIN 7.173	16 10 34.96	+ 58 54 49.0			NEBULA
ZWG 051.067	16 10 36.	+ 07 17		15.6	GALAXY
ZWG 079.078	16 10 36.	+ 10 00		14.4	GALAXY
RNGC 6081	16 10 36.	+ 10 00		14.5	GALAXY
UGC 10272	16 10 36.	+ 10 00	102	14.4	GALAXY SO
MCG+02-41-019	16 10 36.	+ 10 00	96	14.4	GALAXY
ZWG 079.079	16 10 36.	+ 14 24		15.6	GALAXY
MRK 302	16 10 36.	+ 15 55	7	17.5	GALAXY WITH UV CONTINUUM
MCG+04-38-041	16 10 36.	+ 23 06	36	15.7	GALAXY
ZWG 167.046	16 10 36.	+ 29 47		15.5	GALAXY
RNGC 6092	16 10 41.	+ 28 06		15.0	GALAXY
ZC 1610.7+1110	16 10 42.	+ 11 10	2690		CLUSTER OF GALAXIES
ZWG 167.047	16 10 42.	+ 28 25		15.2	GALAXY
UGC 10273	16 10 42.	+ 28 25	90	15.2	GALAXY PECULR
MCG+05-38-036	16 10 42.	+ 28 25	84	15.2	GALAXY
MCG+06-36-001	16 10 42.	+ 33 09	36	15.	GALAXY
ZWG 196.091	16 10 42.	+ 33 10		15.0	GALAXY
ZWG 196.004	16 10 42.	+ 33 10		15.0	GALAXY

OBJECT NAME	RIGHT ASCEN.	DECLINATION	DIAM.	MAGN.	TYPE OF OBJECT
UGC 10274	16 10 42.	+ 38 23	84	16.0	GALAXY Sc
MCG+10-23-037	16 10 42.	+ 57 15	60	16.	GALAXY
MCG+10-23-038	16 10 42.	+ 60 42	42	15.	GALAXY
MRSL 004+29/1	16 10 42.	- 08 15	3000		HII REGION
SC 1608+1533.0	16 10 43.	+ 15 25 15.	6		NEBULA
MCG+05-38-037	16 10 45.	+ 28 07	24	15.0	GALAXY
RNGC 6089	16 10 45.	+ 33 10		15.0	GALAXY
SC 1608+1530.4	16 10 48.	+ 15 22 39.	12		NEBULA
ZWG 167.048	16 10 48.	+ 28 07		15.0	GALAXY
UGC 10275	16 10 48.	+ 28 07	90	15.0	GALAXY E
ZWG 167.049	16 10 48.	+ 32 07		15.6	GALAXY
UGC 10276	16 10 48.	+ 32 07	84	15.6	GALAXY Sb-c
MCG+08-29-053	16 10 48.	+ 49 32	60	15.	GALAXY
UGC 10277	16 10 48.	+ 57 15	60	16.0	GALAXY S
ZWG 298.016	16 10 48.	+ 60 17		15.3	GALAXY
MCG+10-23-039	16 10 48.	+ 60 17	30	15.	GALAXY
ZWG 023.024	16 10 48.	- 00 46		15.4	GALAXY
MCG+00-41-008	16 10 48.	- 00 46	48	15.4	GALAXY
KARA.73B 0728	16 10 48.	- 00 46	60	15.4	ISOLATED GALAXY IR
SC 1608+1914.3	16 10 50.	+ 19 06 34.	12		NEBULA
REIN 7.175A	16 10 50.99	+ 57 15 14.0			NEBULA
MCG+05-38-038	16 10 51.	+ 30 58	42	15.5	GALAXY
MCG+05-38-039	16 10 51.	+ 31 03		15.5	GALAXY
REIN 7.175B	16 10 51.02	+ 57 15 14.0			NEBULA
SC 1608+1908.5	16 10 52.	+ 19 00 46.	24		NEBULA
ZWG 051.068	16 10 54.	+ 07 10		15.5	GALAXY
ZWG 079.080	16 10 54.	+ 14 18		15.2	GALAXY
MCG+02-41-020	16 10 54.	+ 14 18	36	15.2	GALAXY
ZWG 251.001	16 10 54.	+ 49 31		14.8	GALAXY
ZWG 250.049	16 10 54.	+ 49 31		14.8	GALAXY
UGC 10278	16 10 54.	+ 49 31	60	14.8	GALAXY Sa-b
ZWG 298.017	16 10 54.	+ 60 43		14.5	GALAXY
UGC 10279	16 10 54.	+ 60 43	60	14.5	GALAXY DBL SYS
KARA.72 488B	16 10 54.	+ 60 43	42		PART OF DOUBLE GALAXY
KARA.72 488A	16 10 54.	+ 60 43	30	14.5	PART OF DOUBLE GALAXY
ARC 2171	16 10 54.	+ 71 48		17.9	RICH CLUSTER OF GALAXIES
REIN 7.176	16 10 55.50	+ 60 17 30.1			NEBULA
REIN 7.177	16 10 56.72	+ 60 42 37.9			NEBULA
RNGC 6083	16 10 57.	+ 14 18		15.0	GALAXY
REIN 7.178	16 10 58.90	+ 60 42 28.4			NEBULA
LBN 0106	16 11	+ 21 50	1500		BRIGHT NEBULA
UGC 10281	16 11 00.	+ 17 20	120	18.	GALAXY DWARF
ZWG 167.050	16 11 00.	+ 31 01		15.6	GALAXY
MCG+06-36-002	16 11 00.	+ 32 37	66	14.5	GALAXY
ZWG 196.092	16 11 00.	+ 32 38		15.3	GALAXY
UGC 10282	16 11 00.	+ 32 38	60	15.3	GALAXY Sc
MCG+10-23-040	16 11 00.	+ 61 18	42	15.	GALAXY
ZWG 367.007	16 11 00.	+ 81 26		15.0	GALAXY
ZWG 366.029	16 11 00.	+ 81 26		15.0	GALAXY
UGC 10280	16 11 00.	+ 81 26	96	15.0	GALAXY Sb/SBb
MCG+14-07-035	16 11 00.	+ 86 20	39	17.	GALAXY
HOLM 734B	16 11 03.	+ 60 41	18	14.1	PART OF MULTIPLE GALAXY
ARC 2166	16 11 04.	+ 56 06		17.1	RICH CLUSTER OF GALAXIES
ZWG 167.051	16 11 06.	+ 30 57		15.5	GALAXY
UGC 10283	16 11 06.	+ 30 57	108	15.5	GALAXY E
MCG+05-38-040	16 11 06.	+ 31 03	9	15.6	GALAXY
ZC 1611.1+4213	16 11 06.	+ 42 13	810		CLUSTER OF GALAXIES
MCG+10-23-041	16 11 06.	+ 58 01	24	16.	GALAXY
HOLM 734A	16 11 06.	+ 60 41	24	14.0	PART OF MULTIPLE GALAXY
ZWG 298.018	16 11 06.	+ 61 18		15.4	GALAXY
UGC 10284	16 11 06.	+ 61 18	60	15.4	GALAXY SO-a
MCG+05-38-041	16 11 09.	+ 31 03 30.	36	15.7	GALAXY
ZWG 023.025	16 11 12.	+ 00 39		15.5	GALAXY
ZWG 137.063	16 11 12.	+ 22 02		15.4	GALAXY
ZWG 167.052	16 11 12.	+ 31 01		15.7	GALAXY
UGC 10285	16 11 12.	+ 31 01	90	15.7	GALAXY E+COMP
HOLM 733A	16 11 12.	+ 37 24	30	15.0	PART OF MULTIPLE GALAXY
ZWG 196.005	16 11 12.	+ 37 25		15.5	GALAXY
ZWG 298.019	16 11 12.	+ 58 01		15.7	GALAXY
REIN 7.179	16 11 14.18	+ 61 18 11.2			NEBULA
MCG+06-36-003	16 11 15.	+ 37 25 30.	48	15.	GALAXY
HUB C24	16 11 15.	- 19 18			DIFFUSE NEBULA
LB 00306	16 11 16.	- 16 54 00.		16.0	FAINT BLUE STAR
ZWG 196.006	16 11 18.	+ 36 43		15.7	GALAXY
ZWG 223.051	16 11 18.	+ 39 22		15.4	GALAXY
HOLM 733B	16 11 20.	+ 37 25	30	15.5	PART OF MULTIPLE GALAXY
MCG+06-36-004	16 11 21.	+ 36 43 30.	42	15.	GALAXY
ZWG 079.081	16 11 24.	+ 13 12		15.4	GALAXY
ZWG 196.093	16 11 24.	+ 32 25		15.7	GALAXY
ZWG 167.053	16 11 24.	+ 32 25		15.7	GALAXY
UGC 10286	16 11 24.	+ 32 25	72	15.7	GALAXY Sc
MCG+09-26-066	16 11 24.	+ 54 06	54	16.	GALAXY
MCG+09-26-067	16 11 24.	+ 54 45	60	17.	GALAXY
ACK 330-02.2	16 11 24.	- 53 44			PLANETARY NEBULA
PK326-06.1	16 11 26.	- 59 46 14.	5		PLANETARY NEBULA
LB 03541	16 11 28.	- 24 44 18.		17.8	FAINT BLUE STAR
ZC 1611.5+2855	16 11 30.	+ 28 55	470		CLUSTER OF GALAXIES
ZWG 051.069	16 11 36.	+ 02 45		15.6	GALAXY
ZWG 108.167	16 11 36.	+ 15 03		15.7	GALAXY
ZWG 137.064	16 11 36.	+ 23 03		15.5	GALAXY
ZC 1611.6+3717	16 11 36.	+ 37 17	12370		CLUSTER OF GALAXIES
ZC 1611.6+7740	16 11 36.	+ 77 40	1410		CLUSTER OF GALAXIES
PK333+01.1	16 11 36.	- 49 05 47.	5		PLANETARY NEBULA
KARA.72 490A	16 11 42.	+ 04 06	24	14.7	PART OF DOUBLE GALAXY
ZWG 051.070	16 11 42.	+ 06 16		15.5	GALAXY
KARA.72 489A	16 11 42.	+ 06 16	30	15.5	PART OF DOUBLE GALAXY
ZWG 079.082	16 11 42.	+ 14 16		15.4	GALAXY
ZWG 079.083	16 11 42.	+ 14 24		14.7	GALAXY
UGC 10287	16 11 42.	+ 14 24	78	14.7	GALAXY SBb
MCG+02-41-021	16 11 42.	+ 14 24	84	14.7	GALAXY
ZWG 196.007	16 11 42.	+ 34 25		15.7	GALAXY
LB 00924	16 11 45.	+ 38 51 00.		15.8	FAINT BLUE STAR
B 040	16 11 45.	- 18 51	900		DARK OBJECT
BC DA406	16 11 47.93	+ 34 20 20.9		17.5	QUASI-STELLAR OBJECT
ZWG 051.071	16 11 48.	+ 04 07		14.7	GALAXY
KARA.72 490B	16 11 48.	+ 04 07	36		PART OF DOUBLE GALAXY
MCG+01-41-015	16 11 48.	+ 04 07	24	14.7	GALAXY
ZWG 051.072	16 11 48.	+ 06 17		15.4	GALAXY
KARA.72 489B	16 11 48.	+ 06 17	36	15.4	PART OF DOUBLE GALAXY
ZWG 137.065	16 11 48.	+ 20 35		15.6	GALAXY
ZC 1611.8+2056	16 11 48.	+ 20 56	940		CLUSTER OF GALAXIES
UGC 10289	16 11 48.	+ 48 20	72	16.5	GALAXY Sc
ZWG 023.026	16 11 48.	- 00 05		14.7	GALAXY
UGC 10288	16 11 48.	- 00 05	318	14.7	GALAXY Sc
MCG+00-41-009	16 11 48.	- 00 05	300	14.7	GALAXY
MIL 42	16 11 48.	- 50 33 12.	1320		SUPERNOVA REMNANT
SHB 325	16 11 48.0	+ 34 20 19.		17.5	QUASI-STELLAR OBJECT
SC 1609+1510.1	16 11 49.	+ 15 02 25.	18		NEBULA

OBJECT NAME	RIGHT ASCEN.	DECLINATION	DIAM.	MAGN.	TYPE OF OBJECT
ZWG 079.084	16 11 54.	+ 09 39		14.6	GALAXY
MCG+02-41-022	16 11 54.	+ 09 39	36	14.6	GALAXY
ZC 1611.9+2653	16 11 54.	+ 26 53	1340		CLUSTER OF GALAXIES
LB 00307	16 11 54.	- 14 53 36.		15.6	FAINT BLUE STAR
ASS 68	16 11 54.	- 25 48			OB ASSOCIATION SCO OB2
ARC 2168	16 11 56.	+ 54 17		16.5	RICH CLUSTER OF GALAXIES
LB 03542	16 11 57.	- 27 10 36.		18.2	FAINT BLUE STAR
IC 1205	16 11 58.	+ 09 38 54.			NONSTELLAR OBJECT
RNGC 6084	16 11 59.	+ 17 53		15.5	GALAXY
LBN 1114	16 12	- 19 10	9300		BRIGHT NEBULA
ZWG 023.027	16 12 00.	+ 00 56		15.0	GALAXY
UGC 10290	16 12 00.	+ 00 56	120	15.0	GALAXY DWRF SP
MCG+00-41-010	16 12 00.	+ 00 56	102	15.0	GALAXY
ZWG 051.073	16 12 00.	+ 03 20		15.2	GALAXY
ZWG 051.074	16 12 00.	+ 04 20		15.6	GALAXY
SC 1609+1800.2	16 12 00.	+ 17 52 33.	18		NEBULA
ZWG 108.168	16 12 00.	+ 17 53		15.4	GALAXY
UGC 10291	16 12 00.	+ 17 53	66	15.4	GALAXY Sa
MCG+03-41-143	16 12 00.	+ 17 53	60	15.4	GALAXY
SC 1609+1932.3	16 12 00.	+ 19 24 39.	24		NEBULA
ZC 1612.0+2438	16 12 00.	+ 24 38	3830		CLUSTER OF GALAXIES
TON-N 0256	16 12 00.	+ 26 13		15.2	BLUE STAR
ZWG 167.054	16 12 00.	+ 28 38		15.4	GALAXY
ZC 1612.0+2955	16 12 00.	+ 29 55	740		CLUSTER OF GALAXIES
ZWG 196.008	16 12 00.	+ 38 17		15.2	GALAXY
MCG+14-07-036	16 12 00.	+ 81 27	66	16.	GALAXY
MCG+14-08-001	16 12 00.	+ 86 20	48	16.	GALAXY
MCG-03-41-003	16 12 00.	- 20 41	30	15.5	GALAXY
ZWG 079.085	16 12 06.	+ 10 38		15.1	GALAXY
ZWG 167.055	16 12 06.	+ 29 59		15.3	GALAXY
ZWG 196.094	16 12 06.	+ 33 08		15.7	GALAXY
ZWG 196.009	16 12 06.	+ 33 08		15.7	GALAXY
MCG+06-36-005	16 12 06.	+ 38 18	24	15.	GALAXY
ZC 1612.1+4320	16 12 06.	+ 43 20	1140		CLUSTER OF GALAXIES
1ZW 136	16 12 06.	+ 51 13			COMPACT GALAXY
ZC 1612.1+5425	16 12 06.	+ 54 25	1880		CLUSTER OF GALAXIES
NAB 1612+26	16 12 07.	+ 26 40 15.		17.3	QUASI-STELLAR OBJECT
SHB 326	16 12 07.	+ 26 40 15.		17.3	QUASI-STELLAR OBJECT
ARC 2165	16 12 07.	+ 26 48		17.4	RICH CLUSTER OF GALAXIES
SHB 327	16 12 08.7	+ 26 13		15.4	QUASI-STELLAR OBJECT
BC TON256	16 12 09.	+ 26 12 14.		16.06	QUASI-STELLAR OBJECT
ZWG 023.028	16 12 12.	+ 02 15		15.7	GALAXY
KARA.73B 0729	16 12 12.	+ 02 15	36	15.7	ISOLATED GALAXY S
ZC 1612.2+2328	16 12 12.	+ 23 28	400		CLUSTER OF GALAXIES
ZWG 196.010	16 12 12.	+ 35 00		15.3	GALAXY
ZC 1612.2+6125	16 12 12.	+ 61 25	610		CLUSTER OF GALAXIES
MCG+10-23-042	16 12 12.	+ 61 59	42	16.	GALAXY
ZWG 320.010	16 12 12.	+ 63 28		15.6	GALAXY
MCG+06-36-006	16 12 15.	+ 35 00	24	15.	GALAXY
IC 1203	16 12 18.	- 22 13 02.			NONSTELLAR OBJECT
RNGC 6082	16 12 23.	- 34 07			NON-EXISTENT OBJECT
ZC 1612.4+0901	16 12 24.	+ 09 01	1340		CLUSTER OF GALAXIES
ZC 1612.4+2927	16 12 24.	+ 29 27	610		CLUSTER OF GALAXIES
ZWG 167.056	16 12 24.	+ 30 06		15.5	GALAXY
ZC 1612.4+4358	16 12 24.	+ 43 58	4230		CLUSTER OF GALAXIES
ZWG 251.002	16 12 24.	+ 48 25		15.5	GALAXY
ZWG 250.050	16 12 24.	+ 48 25		15.5	GALAXY
ZC 1612.4+6018	16 12 24.	+ 60 18	270		CLUSTER OF GALAXIES
MCG+10-23-043	16 12 24.	+ 62 14	30	16.	GALAXY
SC 1610+1712.6	16 12 26.	+ 17 04 58.	18		NEBULA
RNGC 6097	16 12 28.	+ 35 14		15.0	GALAXY
ARC 2167	16 12 28.	+ 38 33		17.6	RICH CLUSTER OF GALAXIES
ZWG 196.011	16 12 30.	+ 35 14		14.9	GALAXY
LYNG 12	16 12 30.	- 51 57 48.	2520		OB CONCENTRATION
ARC 2181	16 12 32.	+ 77 38		17.7	RICH CLUSTER OF GALAXIES
MCG+06-36-007	16 12 36.	+ 35 14	60	15.5	GALAXY
ZC 1612.6+3832	16 12 36.	+ 38 32	870		CLUSTER OF GALAXIES
UGC 10292	16 12 36.	+ 56 25	66	16.0	GALAXY Sb
MCG+05-38-042	16 12 39.	+ 27 23	30	15.5	GALAXY
HOLM 735A	16 12 40.	+ 26 40	24	14.9	PART OF MULTIPLE GALAXY
RNGC 6096	16 12 40.	+ 26 42		15.5	GALAXY
ARC 2169	16 12 40.	+ 49 16		15.9	RICH CLUSTER OF GALAXIES
ZC 1612.7+2624	16 12 42.	+ 26 24	1550		CLUSTER OF GALAXIES
MCG+05-38-044	16 12 42.	+ 26 40	48	15.3	GALAXY
ZWG 167.057	16 12 42.	+ 26 42		15.3	GALAXY
MCG+05-38-043	16 12 42.	+ 31 13	48	15.5	GALAXY
ZWG 196.012	16 12 42.	+ 34 21		15.2	GALAXY
ZC 1612.7+4645	16 12 42.	+ 46 45	1210		CLUSTER OF GALAXIES
MCG+09-27-001	16 12 42.	+ 56 25	54	16.	GALAXY
MCG+10-23-044	16 12 42.	+ 61 20	24	16.	GALAXY
ZWG 338.048	16 12 42.	+ 68 55		15.6	GALAXY
ZWG 320.011	16 12 42.	+ 68 55		15.6	GALAXY
HUB C26	16 12 43.	- 23 25			DIFFUSE NEBULA
HOLM 735B	16 12 44.	+ 26 40	24	15.3	PART OF MULTIPLE GALAXY
ZWG 079.086	16 12 48.	+ 11 24		15.7	GALAXY
KARA.72 491A	16 12 48.	+ 11 24	36	15.7	PART OF DOUBLE GALAXY
MCG+04-38-042	16 12 48.	+ 22 03	48	15.3	GALAXY
ZWG 137.066	16 12 48.	+ 22 04		15.3	GALAXY
ZWG 276.003	16 12 48.	+ 52 32		15.5	GALAXY
MCG+10-23-045	16 12 48.	+ 61 09	36	16.	GALAXY
IC 1206	16 12 51.	+ 11 25 24.			NONSTELLAR OBJECT
ARC 2163	16 12 52.	- 06 01		17.5	RICH CLUSTER OF GALAXIES
ZWG 079.087	16 12 54.	+ 11 25		14.8	GALAXY
UGC 10293	16 12 54.	+ 11 25	66	14.8	GALAXY Sa-b
KARA.72 491B	16 12 54.	+ 11 25	66	14.8	PART OF DOUBLE GALAXY
MCG+02-41-023	16 12 54.	+ 11 25	72	14.8	GALAXY
ZC 1612.9+2120	16 12 54.	+ 21 20	670		CLUSTER OF GALAXIES
ZWG 196.013	16 12 54.	+ 36 40		15.7	GALAXY
ZWG 298.020	16 12 54.	+ 61 09		15.7	GALAXY
UGC 10294	16 12 54.	+ 65 33	120	16.5	GALAXY DWRF SP
FATH 1.735	16 12 56.	+ 14 57	14		NEBULA
KHAV 236	16 13	- 24 32	1570		DARK NEBULA
ZWG 023.029	16 13 00.	+ 01 39		15.2	GALAXY
ZC 1613.0+2755	16 13 00.	+ 27 55	400		CLUSTER OF GALAXIES
ZWG 167.058	16 13 00.	+ 31 12		15.5	GALAXY
UGC 10295	16 13 00.	+ 31 12	72	15.5	GALAXY S
ZWG 196.095	16 13 00.	+ 33 11		15.7	GALAXY
ZWG 196.014	16 13 00.	+ 33 11		15.7	GALAXY
MCG+07-33-052	16 13 00.	+ 39 39	60	16.5	GALAXY
LDN 1721	16 13 00.	- 19 00	2160		DARK NEBULA
LDN 1678	16 13 00.	- 24 30	1500		DARK NEBULA
HN 0428	16 13 04.	- 22 30			NEBULA
IC 4596	16 13 05.	- 22 30			NONSTELLAR OBJECT
ZC 1613.1+2225	16 13 06.	+ 22 25	540		CLUSTER OF GALAXIES
UGC 10296	16 13 06.	+ 38 38	84	15.7	GALAXY Sc
LB 00925	16 13 06.	+ 39 16 48.		15.7	FAINT BLUE STAR
ZWG 223.052	16 13 05.	+ 39 38		15.7	GALAXY
MCG+10-23-046	16 13 06.	+ 62 19	24	16.	GALAXY
MCG+11-20-005	16 13 06.	+ 63 50	54	15.	GALAXY
ZC 1613.1+7145	16 13 06.	+ 71 45	2490		CLUSTER OF GALAXIES
SHAH 164	16 13 06.	+ 85 25	198		GROUP OF COMPACT GALAXIES
HUB C25	16 13 08.	- 24 48			DIFFUSE NEBULA
ZWG 108.169	16 13 12.	+ 19 01		15.1	GALAXY
UGC 10297	16 13 12.	+ 19 01	126	15.1	GALAXY S
MCG+03-41-144	16 13 12.	+ 19 01	126	15.1	GALAXY
VV 192B	16 13 12.	+ 19 35	30	14.	INTERACTING GALAXY
MCG+09-27-002	16 13 12.	+ 53 43	24	16.	GALAXY
ZWG 320.012	16 13 12.	+ 63 50		14.5	GALAXY
UGC 10298	16 13 12.	+ 63 50	72	14.5	GALAXY S
MCG-04-38-005	16 13 12.	- 22 31 30.	48	15.5	GALAXY
LB 00193	16 13 16.	- 14 51 36.		15.7	FAINT BLUE STAR
SC 1611+1908.5	16 13 17.	+ 19 00 56.	66		NEBULA
VV 192A	16 13 18.	+ 19 34 30.	30	14.	INTERACTING GALAXY
ZC 1613.3+2105	16 13 18.	+ 21 05	610		CLUSTER OF GALAXIES
KARA.72 492B	16 13 18.	+ 26 44	42	15.7	PART OF DOUBLE GALAXY
MCG+05-38-045	16 13 18.	+ 26 44	30	15.6	GALAXY
ZWG 167.059	16 13 18.	+ 26 45		14.9	GALAXY
KARA.72 492A	16 13 18.	+ 26 45	60	15.6	PART OF DOUBLE GALAXY
SHAH 135	16 13 18.	+ 64 30	144	16.8	GROUP OF COMPACT GALAXIES
PK330-02.1	16 13 19.	- 53 24 44.	20		PLANETARY NEBULA
MCG+03-41-145	16 13 21.	+ 19 33 30.	66	14.6	GALAXY
MCG+04-38-043	16 13 21.	+ 26 15 30.	42	15.1	GALAXY
MCG+05-38-046	16 13 21.	+ 26 43 30.	27	15.7	GALAXY
SC 1609-5136.2	16 13 23.	- 51 43 49.	12		NEBULA
ZC 1613.4+1239	16 13 24.	+ 12 39	1680		CLUSTER OF GALAXIES
MCG+03-41-146	16 13 24.	+ 19 33	60	14.6	GALAXY
KARA.72 493B	16 13 24.	+ 19 34	42		PART OF DOUBLE GALAXY
ZWG 108.170	16 13 24.	+ 19 35		14.6	GALAXY
RNGC 6099	16 13 24.	+ 19 35		14.5	GALAXY
RNGC 6098	16 13 24.	+ 19 35		14.5	GALAXY
UGC 10299	16 13 24.	+ 19 35	96	14.6	GALAXY DBL SYS
KARA.72 493A	16 13 24.	+ 19 35	48	14.6	PART OF DOUBLE GALAXY
ZWG 137.067	16 13 24.	+ 26 14		15.1	GALAXY
LB 03543	16 13 25.	- 25 10 18.		19.0	FAINT BLUE STAR
PK331-01.1	16 13 25.	- 51 51 37.	35		PLANETARY NEBULA
ZWG 108.171	16 13 30.	+ 19 46		15.6	GALAXY
ZWG 196.015	16 13 30.	+ 34 21		15.6	GALAXY
MCG+09-27-003	16 13 30.	+ 52 41	30	16.	GALAXY
SC 1611+1951.7	16 13 32.	+ 19 44 09.	12		NEBULA
MCG+05-38-047	16 13 33.	+ 28 17	48	15.4	GALAXY
RNGC 6102	16 13 35.	+ 28 17		15.5	GALAXY
ZWG 167.060	16 13 36.	+ 28 17		15.4	GALAXY
UGC 10300	16 13 36.	+ 28 17	90	15.4	GALAXY S
ZWG 167.061	16 13 36.	+ 31 26		15.7	GALAXY
UGC 10301	16 13 36.	+ 31 26	84	15.7	GALAXY SB:b
ZWG 251.003	16 13 36.	+ 46 54		15.0	GALAXY
MCG+08-30-001	16 13 36.	+ 46 54	54	15.	GALAXY
ZWG 320.013	16 13 36.	+ 63 23		15.7	GALAXY
MCG+11-20-006	16 13 36.	+ 63 23	45	16.	GALAXY
FATH 1.736	16 13 38.	+ 15 25	16		NEBULA
SC 1611+1921.0	16 13 39.	+ 19 13 27.	18		NEBULA
FATH 1.737	16 13 40.	+ 15 17	11		NEBULA
SC 1611+1702.0	16 13 40.	+ 16 54 27.	12		NEBULA
SC 1611+1523.8	16 13 41.	+ 15 16 15.	30		NEBULA
MCG+05-38-048	16 13 42.	+ 31 33	48	15.2	GALAXY
ZWG 167.062	16 13 42.	+ 32 05		14.4	GALAXY
UGC 10302	16 13 42.	+ 32 05	48	14.4	GALAXY
7ZW 638	16 13 42.	+ 63 23			COMPACT GALAXY
FATH 1.738	16 13 44.	+ 15 22	14		NEBULA
RNGC 6103	16 13 44.	+ 32 05		14.5	GALAXY
MIL 43	16 13 44.	- 50 56 00.	474		SUPERNOVA REMNANT
FATH 1.739	16 13 46.	+ 15 22	16		NEBULA
LB 00926	16 13 46.	+ 72 53 00.		15.9	FAINT BLUE STAR
WRAY 15.44	16 13 47.1	- 50 59 01.			DIFFUSE NEBULA
ZC 1613.8+0913	16 13 48.	+ 09 13	1340		CLUSTER OF GALAXIES
ZWG 137.068	16 13 48.	+ 22 24		15.7	GALAXY
ZC 1613.8+2737	16 13 48.	+ 27 37	1750		CLUSTER OF GALAXIES
MCG+05-38-049	16 13 48.	+ 32 07 30.	42	14.4	GALAXY
ZWG 276.004	16 13 48.	+ 56 00		15.2	GALAXY
ZWG 276.005	16 13 48.	+ 56 31		15.6	GALAXY
ZC 1613.8+5632	16 13 48.	+ 56 32	13240		CLUSTER OF GALAXIES
MCG+10-23-047	16 13 48.	+ 61 01	36	16.	GALAXY
MCG+10-23-048	16 13 48.	+ 62 39	96	14.	GALAXY
IC 1210	16 13 48.	+ 62 40 00.			NONSTELLAR OBJECT
ZWG 320.014	16 13 48.	+ 63 23		15.2	GALAXY
MCG+11-20-007	16 13 48.	+ 63 23	30	15.	GALAXY
REIN 7.180	16 13 48.99	+ 56 31 28.8			NEBULA
LB 00927	16 13 50.	- 00 22 54.		17.9	FAINT BLUE STAR
LB 00188	16 13 52.	+ 15 19 18.		17.1	FAINT BLUE STAR
RNGC 6111	16 13 53.	+ 62 40		14.0	GALAXY
BIGO 537	16 13 53.	+ 62 41			NEBULA
ZWG 167.063	16 13 54.	+ 31 31		15.2	GALAXY
UGC 10303	16 13 54.	+ 31 31	60	15.2	GALAXY S
ZWG 196.016	16 13 54.	+ 36 39		15.3	GALAXY
MCG+09-27-004	16 13 54.	+ 55 59	48	14.	GALAXY
ZWG 298.021	16 13 54.	+ 62 40		13.8	GALAXY
UGC 10304	16 13 54.	+ 62 40	96	13.8	GALAXY Sa-b
VMT 16	16 13 54.	- 50 55 48.	570		SUPERNOVA REMNANT
ARC 2176	16 13 56.	+ 71 34		17.1	RICH CLUSTER OF GALAXIES
IC 1208	16 13 59.	+ 36 39 05.			NONSTELLAR OBJECT
VDB.66G 204	16 14	+ 47 11	100		DWARF GALAXY
B 229	16 14	- 27 11	2700		DARK OBJECT
KHAV 237	16 14	- 37 38	12920		DARK NEBULA
ZC 1614.0+0708	16 14 00.	+ 07 08	1340		CLUSTER OF GALAXIES
UGC 10305	16 14 00.	+ 31 50	60	16.0	GALAXY S
ZWG 196.017	16 14 00.	+ 37 07		15.5	GALAXY
MCG+10-23-049	16 14 00.	+ 60 15	60	16.	GALAXY
ZWG 320.015	16 14 00.	+ 65 07		14.8	GALAXY
MCG+11-20-008	16 14 00.	+ 65 07	54	15.	GALAXY
ZWG 023.030	16 14 00.	- 00 29		15.5	GALAXY
BRSL 331-01/1	16 14 00.	- 51 48	720		HII REGION
LB 03544	16 14 02.	- 24 30 42.		19.1	FAINT BLUE STAR
REIN 2.243	16 14 03.43	- 22 51 13.6			NEBULA
FATH 1.740	16 14 06.	+ 14 38	5		NEBULA
ZWG 196.096	16 14 06.	+ 32 58		15.6	GALAXY
ZWG 196.018	16 14 06.	+ 32 58		15.6	GALAXY
GCL 039	16 14 06.	- 22 52	516	8.93	GLOBULAR STAR CLUSTER
RNGC 6093	16 14 07.	- 22 52		8.5	GLOBULAR CLUSTER
MCG+06-36-008	16 14 09.	+ 37 08	15	15.	GALAXY
FATH 1.741	16 14 10.	+ 15 00	16		NEBULA
HOLM 736A	16 14 11.	+ 38 03	24	15.4	PART OF MULTIPLE GALAXY
ZWG 023.031	16 14 12.	+ 00 21		14.8	GALAXY
UGC 10306	16 14 12.	+ 00 21	144	14.8	GALAXY Sb
MCG+00-41-011	16 14 12.	+ 00 21	120	14.8	GALAXY
ZC 1614.2+1310	16 14 12.	+ 13 10	1340		CLUSTER OF GALAXIES
ZWG 196.019	16 14 12.	+ 38 02		15.5	GALAXY

OBJECT NAME	RIGHT ASCEN.	DECLINATION	DIAM.	MAGN.	TYPE OF OBJECT
7ZW 639	16 14 12.	+ 57 50			COMPACT GALAXY
MCG+10-23-050	16 14 12.	+ 61 33	54	16.	GALAXY
MCG+10-23-051	16 14 12.	+ 61 54	18	16.	GALAXY
HOLM 736B	16 14 13.	+ 38 04	24	16.1	PART OF MULTIPLE GALAXY
SC 1612+2000.6	16 14 14.	+ 19 53 05.	12		NEBULA
SC 1611+1507.3	16 14 16.	+ 14 59 47.	18		NEBULA
FATH 1.742	16 14 18.	+ 14 44	14		NEBULA
ZWG 167.064	16 14 18.	+ 27 23		15.1	GALAXY
MCG+05-38-050	16 14 18.	+ 27 23	24	15.1	GALAXY
ZWG 298.022	16 14 18.	+ 61 54		14.9	GALAXY
MCG+06-36-009	16 14 21.	+ 38 04	33	15.	GALAXY
FATH 1.743	16 14 22.	+ 14 46	41		NEBULA
LB 00189	16 14 22.	+ 14 46 42.		16.2	FAINT BLUE STAR
LB 00194	16 14 22.	- 14 40 30.		15.5	FAINT BLUE STAR
FATH 1.744	16 14 23.	+ 14 50	19		NEBULA
ZWG 023.032	16 14 24.	+ 00 56		14.7	GALAXY
UGC 10307	16 14 24.	+ 00 56	96	14.7	GALAXY SO
MCG+00-41-012	16 14 24.	+ 00 56	102	14.7	GALAXY
MCG+06-36-010	16 14 24.	+ 38 05	21	16.	GALAXY
UGC 10308	16 14 24.	+ 52 22	60	16.5	GALAXY Sc
MCG+10-23-052	16 14 24.	+ 60 04	60	16.	GALAXY
OCL 0952	16 14 24.	- 55 00	1230	6.9	OPEN STAR CLUSTER
FATH 1.745	16 14 25.	+ 14 56	33		NEBULA
RNGC 6100	16 14 26.	+ 00 56		14.5	GALAXY
LB 00190	16 14 27.	+ 15 17 18.		15.3	FAINT BLUE STAR
MCG-02-41-001	16 14 27.	- 11 36	96	15.	GALAXY
FATH 1.746	16 14 29.	+ 15 21	11		NEBULA
MCG+03-41-148	16 14 30.	+ 19 35	33	15.7	GALAXY
MCG+03-41-147	16 14 30.	+ 19 38	36	15.5	GALAXY
ZWG 223.053	16 14 30.	+ 42 31		15.2	GALAXY
ARP 002	16 14 30.	+ 47 10			PECULIAR GALAXY
MCG+09-27-005	16 14 30.	+ 52 22 30.	54	16.	GALAXY
ZWG 276.006	16 14 30.	+ 54 47		15.1	GALAXY
1ZW 137	16 14 30.	+ 54 48			COMPACT GALAXY
LB 00308	16 14 30.	- 01 32 06.		16.1	FAINT BLUE STAR
SC 1612+2005.4	16 14 31.	+ 19 57 55.	12		NEBULA
LB 00191	16 14 32.	+ 14 49 30.		16.3	FAINT BLUE STAR
HN 1142	16 14 32.	- 34 15		14.	NEBULA
IC 4597	16 14 32.	- 34 15			NONSTELLAR OBJECT
FATH 1.747	16 14 34.	+ 14 41	3		NEBULA
FATH 1.748	16 14 34.	+ 15 05	14		NEBULA
SC 1612+1957.6	16 14 34.	+ 19 50 07.	36		NEBULA
FATH 1.749	16 14 35.	+ 15 18	14		NEBULA
ZWG 108.172	16 14 36.	+ 19 36		15.7	GALAXY
SC 1612+1944.3	16 14 36.	+ 19 36 49.	36		NEBULA
ZWG 108.173	16 14 36.	+ 19 39		15.5	GALAXY
ZWG 137.069	16 14 36.	+ 21 04		15.7	GALAXY
ZC 1614.6+2315	16 14 36.	+ 23 15	1750		CLUSTER OF GALAXIES
ZC 1614.6+2701	16 14 36.	+ 27 01	2420		CLUSTER OF GALAXIES
7ZW 640	16 14 36.	+ 62 50			COMPACT GALAXY
BC PKS1614-09	16 14 36.	- 09 55 18.		18.	QUASI-STELLAR OBJECT
SC 1612+1941.5	16 14 37.	+ 19 34 01.	24		NEBULA
RNGC 6104	16 14 41.	+ 35 50		14.0	GALAXY
ZWG 051.075	16 14 42.	+ 04 40		15.7	GALAXY
KARA.73B 0730	16 14 42.	+ 04 40	24	15.7	ISOLATED GALAXY S
ZWG 196.020	16 14 42.	+ 35 50		14.1	GALAXY
UGC 10309	16 14 42.	+ 35 50	54	14.1	GALAXY PECULR
MCG+06-36-011	16 14 42.	+ 35 50	48	14.	GALAXY
1ZW 138	16 14 42.	+ 51 04			COMPACT GALAXY
OCL 0948	16 14 42.	- 57 47	1740	6.0	OPEN STAR CLUSTER
VHA 188	16 14 42.	- 57 47	900		OPEN STAR CLUSTER
RNGC 6087	16 14 43.	- 57 47		6.0	OPEN CLUSTER
ABC 2170	16 14 45.	+ 23 19		15.9	RICH CLUSTER OF GALAXIES
LB 00192	16 14 48.	+ 15 13 42.		15.9	FAINT BLUE STAR
ZC 1614.8+4230	16 14 48.	+ 42 30	1280		CLUSTER OF GALAXIES
ZWG 251.004	16 14 48.	+ 47 10		14.9	GALAXY
UGC 10310	16 14 48.	+ 47 10	192	14.9	GALAXY DWRF IR
MCG+08-30-002	16 14 48.	+ 47 10	150	14.	GALAXY
ZWG 251.005	16 14 48.	+ 48 06		15.6	GALAXY
FATH 1.750	16 14 50.	+ 15 01	3		NEBULA
SC 1612+1919.9	16 14 51.	+ 19 12 26.	18		NEBULA
MCG+07-33-053	16 14 54.	+ 40 27	42	16.	GALAXY
ZWG 223.054	16 14 54.	+ 40 28		15.7	GALAXY
MCG+08-30-003	16 14 54.	+ 48 07	51	16.	GALAXY
MCG+10-23-053	16 14 54.	+ 60 30	60	15.	GALAXY
KHAV 238	16 15	- 26 13	3700		DARK NEBULA
FATH 1.751	16 15 00.	+ 15 31	11		NEBULA
ZWG 196.097	16 15 00.	+ 33 03		15.7	GALAXY
ZWG 196.021	16 15 00.	+ 33 03		15.7	GALAXY
ZWG 196.022	16 15 00.	+ 35 50		15.4	GALAXY
MCG+06-36-012	16 15 00.	+ 35 50	30	15.	GALAXY
ZWG 298.023	16 15 00.	+ 60 30		15.7	GALAXY
ZWG 320.016	16 15 00.	+ 64 21		15.2	GALAXY
ZWG 367.008	16 15 00.	+ 81 26		15.4	GALAXY
LDN 1675	16 15 00.	- 26 00	4260		DARK NEBULA
IC 4598	16 15 03.	- 31 20			NONSTELLAR OBJECT
IC 1212	16 15 04.	+ 64 20 58.			NONSTELLAR OBJECT
HN 1143	16 15 04.	- 31 19	60		NEBULA
REIN 7.181	16 15 05.24	+ 60 30 16.5			NEBULA
MCG-01-41-007	16 15 06.	- 06 41	60	15.5	GALAXY
ABC 2172	16 15 07.	+ 42 32		17.1	RICH CLUSTER OF GALAXIES
FATH 1.752	16 15 08.	+ 15 02	11		NEBULA
RNGC 6105	16 15 10.	+ 35 00		15.5	GALAXY
FATH 1.753	16 15 11.	+ 15 16	14		NEBULA
ZC 1615.2+1524	16 15 12.	+ 15 24	2490		CLUSTER OF GALAXIES
ZWG 196.023	16 15 12.	+ 35 00		15.3	GALAXY
MCG+09-27-006	16 15 12.	+ 52 03	30	16.	GALAXY
7ZW 641	16 15 12.	+ 61 27			COMPACT GALAXY
7ZW 642	16 15 12.	+ 66 34			COMPACT GALAXY
HN 0430	16 15 17.	- 22 39			NEBULA
IC 4600	16 15 17.	- 22 40			NONSTELLAR OBJECT
ZWG 079.088	16 15 18.	+ 10 30		15.3	GALAXY
KARA.73B 0731	16 15 18.	+ 10 30	36	15.3	ISOLATED GALAXY
ZWG 079.089	16 15 18.	+ 13 37		15.2	GALAXY
MCG+06-36-013	16 15 18.	+ 35 00	33	15.	GALAXY
ZC 1615.3+4811	16 15 18.	+ 48 11	1950		CLUSTER OF GALAXIES
MCG+10-23-054	16 15 18.	+ 60 32 30.	42	15.	GALAXY
BC PKS1615+029	16 15 18.8	+ 02 53 58.		18.	QUASI-STELLAR OBJECT
FATH 1.754	16 15 20.	+ 15 09	5		NEBULA
RNGC 6107	16 15 22.	+ 35 01		14.5	GALAXY
ZWG 079.090	16 15 24.	+ 08 42		15.7	GALAXY
ZWG 196.024	16 15 24.	+ 35 01		14.7	GALAXY
UGC 10311	16 15 24.	+ 35 01	90	14.7	GALAXY (E)
ZC 1615.4+4137	16 15 24.	+ 41 37	3630		CLUSTER OF GALAXIES
MRK 497	16 15 24.	+ 52 09	8	16.	GALAXY WITH UV CONTINUUM
ZWG 298.024	16 15 24.	+ 60 33		15.5	GALAXY
SC 1613+1542.2	16 15 25.	+ 15 34 46.	12		NEBULA
MCG+06-36-014	16 15 27.	+ 35 01 30.	18	14.5	GALAXY

OBJECT NAME	RIGHT ASCEN.	DECLINATION	DIAM.	MAGN.	TYPE OF OBJECT
SC 1613+1541.6	16 15 28.	+ 15 34 10.	6		NEBULA
RNGC 6108	16 15 28.	+ 35 15		15.5	GALAXY
SC 1613+1542.7	16 15 30.	+ 15 35 16.	12		NEBULA
ZWG 137.070	16 15 30.	+ 20 49		15.6	GALAXY
ZWG 167.065	16 15 30.	+ 31 19		15.1	GALAXY
UGC 10312	16 15 30.	+ 31 19	102	15.1	GALAXY S
MCG+05-38-051	16 15 30.	+ 31 20	60	15.1	GALAXY
ZWG 196.025	16 15 30.	+ 35 15		15.4	GALAXY
MCG+07-33-054	16 15 30.	+ 40 13	42	16.	GALAXY
ABC 2174	16 15 30.	+ 61 13		17.1	RICH CLUSTER OF GALAXIES
IC 1215	16 15 30.	+ 68 31 46.			NONSTELLAR OBJECT
REIN 7.182	16 15 30.30	+ 60 33 03.3			NEBULA
LB 00195	16 15 35.	- 14 41 06.		15.8	FAINT BLUE STAR
HN 0429	16 15 35.	- 70 02			NEBULA
IC 4595	16 15 35.	- 70 02			NONSTELLAR OBJECT
ZWG 079.091	16 15 36.	+ 14 05		15.5	GALAXY
MCG+04-38-044	16 15 36.	+ 20 47	36	15.6	GALAXY
ZC 1615.6+2508	16 15 36.	+ 25 08	470		CLUSTER OF GALAXIES
ZWG 167.066	16 15 36.	+ 31 28		15.7	GALAXY
UGC 10313	16 15 36.	+ 31 43	66	17.	GALAXY
MCG+06-36-015	16 15 36.	+ 35 15 30.	42	14.5	GALAXY
MCG+09-27-007	16 15 36.	+ 52 07 30.	36	16.	GALAXY
1ZW 139	16 15 36.	+ 53 08			COMPACT GALAXY
ZWG 276.007	16 15 36.	+ 53 08		13.8	GALAXY
UGC 10314	16 15 36.	+ 53 08	72	13.8	GALAXY (E)
KARA.73B 0732	16 15 36.	+ 53 08	60	13.8	ISOLATED GALAXY E
ZWG 338.049	16 15 36.	+ 68 32		14.0	GALAXY
ZWG 320.017	16 15 36.	+ 68 32		14.0	GALAXY
UGC 10315	16 15 36.	+ 68 32	78	14.0	GALAXY SB
MRSL 332-00/1	16 15 36.	- 50 41	1800		HII REGION
LB 03545	16 15 37.	- 25 42 48.		20.4	FAINT BLUE STAR
IC 1211	16 15 38.	+ 53 07 57.			NONSTELLAR OBJECT
RNGC 6109	16 15 40.	+ 35 07		15.0	GALAXY
ZWG 109.001	16 15 42.	+ 20 49		15.7	GALAXY
ZWG 137.071	16 15 42.	+ 23 04		15.0	GALAXY
KARA.73B 0733	16 15 42.	+ 23 04	54	15.0	ISOLATED GALAXY SO
ZWG 196.026	16 15 42.	+ 35 07		14.9	GALAXY
UGC 10316	16 15 42.	+ 35 07	66	14.9	GALAXY COMPACT
MCG+07-33-055	16 15 42.	+ 39 00	66	17.	GALAXY
MCG+09-27-008	16 15 42.	+ 51 04	15	16.	GALAXY
SEAH 012	16 15 42.	+ 53 40	96	17.5	GROUP OF COMPACT GALAXIES
ZC 1615.7+5700	16 15 42.	+ 57 00	1410		CLUSTER OF GALAXIES
LB 00309	16 15 43.	- 03 01 24.		14.2	FAINT BLUE STAR
MCG+09-27-009	16 15 45.	+ 53 08	36	14.	GALAXY
RNGC 6110	16 15 46.	+ 35 13		15.5	GALAXY
ZWG 080.001	16 15 48.	+ 11 17		15.7	GALAXY
ZWG 079.092	16 15 48.	+ 11 17		15.7	GALAXY
MCG+04-38-045	16 15 48.	+ 23 04	48	15.0	GALAXY
ZC 1615.8+3505	16 15 48.	+ 35 05	7120		CLUSTER OF GALAXIES
MCG+06-36-016	16 15 48.	+ 35 08	60	14.5	GALAXY
ZWG 196.027	16 15 48.	+ 35 13		15.6	GALAXY
UGC 10317	16 15 48.	+ 39 03	90	16.5	GALAXY S
MCG+09-27-011	16 15 48.	+ 51 02	36	16.	GALAXY
MCG+09-27-010	16 15 48.	+ 52 03	30	17.	GALAXY
ZWG 320.018	16 15 48.	+ 64 13		15.5	GALAXY
UGC 10318	16 15 48.	+ 64 13	60	15.5	GALAXY S
IC 4599	16 15 51.	- 42 09			NONSTELLAR OBJECT
IC 1214	16 15 52.	+ 66 05 58.			NONSTELLAR OBJECT
HN 1144	16 15 52.	- 42 08		15.	NEBULA
ZC 1615.9+0919	16 15 54.	+ 09 19	940		CLUSTER OF GALAXIES
ZWG 080.002	16 15 54.	+ 12 50		15.4	GALAXY
ZWG 079.093	16 15 54.	+ 12 50		15.4	GALAXY
UGC 10319	16 15 54.	+ 12 55	72	16.0	GALAXY COMPACT
ZWG 137.072	16 15 54.	+ 21 12		15.4	GALAXY
UGC 10320	16 15 54.	+ 21 12	84	15.4	GALAXY Sa-b
VV 129E	16 15 54.	+ 21 39 30.	6	17.	INTERACTING GALAXY
VV 129D	16 15 54.	+ 21 39 30.	18	16.	INTERACTING GALAXY
VV 129C	16 15 54.	+ 21 39 30.	12	17.	INTERACTING GALAXY
VV 129B	16 15 54.	+ 21 39 30.	12	16.	INTERACTING GALAXY
VV 129A	16 15 54.	+ 21 39 30.	21	15.	INTERACTING GALAXY
VV 129	16 15 54.	+ 21 39 30.	96		INTERACTING GALAXY
ZWG 137.073	16 15 54.	+ 21 41		15.5	GALAXY
UGC 10321	16 15 54.	+ 21 41	102	15.5	GALAXY GROUP
ZWG 137.074	16 15 54.	+ 22 20		14.9	GALAXY
UGC 10322	16 15 54.	+ 22 20	72	14.9	GALAXY S
MCG+08-30-004	16 15 54.	+ 50 07	48	16.	GALAXY
ZC 1615.9+6110	16 15 54.	+ 61 10	2420		CLUSTER OF GALAXIES
ZWG 320.019	16 15 54.	+ 66 05		15.0	GALAXY
UGC 10323	16 15 54.	+ 66 05	66	15.0	GALAXY S0-a
UGC 10324	16 15 54.	+ 66 58	90	16.0	GALAXY
PK338+05.1	16 15 54.	- 42 08 45.	25		PLANETARY NEBULA
OCL 0962	16 15 54.	- 50 06	60		OPEN STAR CLUSTER
MRSL 331-02/1	16 15 54.	- 52 52	5400		HII REGION
OCL 0954	16 15 54.	- 54 52	1800		OPEN STAR CLUSTER
IC 1216	16 15 56.	+ 68 28 25.			NONSTELLAR OBJECT
MCG+04-38-046	16 15 57.	+ 21 39 30.	24	15.5	GALAXY
LB 00928	16 15 58.	- 00 09 24.		16.9	FAINT BLUE STAR
VDB.66G 205	16 16	+ 64 01	70		DWARF GALAXY
KHAV 240	16 16	- 23 31	6890		DARK NEBULA
KHAV 239	16 16	- 35 13	5060		DARK NEBULA
MCG+04-38-049	16 16 00.	+ 21 10	78	15.4	GALAXY
MCG+04-38-047	16 16 00.	+ 21 40	30	15.5	GALAXY
MCG+04-38-048	16 16 00.	+ 22 20	66	14.9	GALAXY
ZC 1616.0+2814	16 16 00.	+ 28 14	540		CLUSTER OF GALAXIES
ZWG 251.006	16 16 00.	+ 46 13		14.5	GALAXY
UGC 10325	16 16 00.	+ 46 13		14.5	GALAXY DBL SYS
KARA.72 494A	16 16 00.	+ 46 13	36	14.5	PART OF DOUBLE GALAXY
ZWG 251.007	16 16 00.	+ 50 45		15.6	GALAXY
ZWG 320.020	16 16 00.	+ 65 28		15.3	GALAXY
MCG+11-20-009	16 16 00.	+ 66 06	39	15.	GALAXY
ZWG 338.050	16 16 00.	+ 68 29		14.9	GALAXY
ZWG 320.021	16 16 00.	+ 68 29		14.9	GALAXY
UGC 10326	16 16 00.	+ 68 29	66	14.9	GALAXY Sc
MCG+11-20-010	16 16 00.	+ 68 30	51	15.	GALAXY
MCG+11-20-009A	16 16 00.	+ 68 32	57	14.	GALAXY
LDN 1717	16 16 00.	- 20 00	2400		DARK NEBULA
VDB.66R 101	16 16 00.	- 20 04	684		REFLECTION NEBULA
MRSL 332-01/3	16 16 00.	- 51 44	1440		HII REGION
RNGC 6112	16 16 04.	+ 35 14		15.0	GALAXY
ZWG 196.028	16 16 06.	+ 35 14		14.8	GALAXY
MCG+06-36-017	16 16 06.	+ 35 14	30	15.	GALAXY
ZWG 196.029	16 16 06.	+ 35 23		15.7	GALAXY
KARA.72 494B	16 16 06.	+ 46 12	18		PART OF DOUBLE GALAXY
CED 129A	16 16 07.	- 20 06	900		DIFFUSE GALACTIC NEBULA
FATH 1.755	16 16 11.	+ 14 25	14		NEBULA
LB 00929	16 16 11.	+ 42 02 00.		16.1	FAINT BLUE STAR
MCG+06-36-018	16 16 12.	+ 35 23 30.	42	16.	GALAXY
ZWG 251.008	16 16 12.	+ 50 21		15.5	GALAXY

OBJECT NAME	RIGHT ASCEN.	DECLINATION	DIAM.	MAGN.	TYPE OF OBJECT
7ZW 643	16 16 12.	+ 74 20			COMPACT GALAXY
ZWG 339.001	16 16 12.	+ 74 20		15.2	GALAXY
ZWG 338.051	16 16 12.	+ 74 20		15.2	GALAXY
DG 133	16 16 12.	- 20 07	300		REFLECTION NEBULA
LB 00930	16 16 13.	+ 41 06 00.		16.3	FAINT BLUE STAR
IC 1217	16 16 13.	+ 69 47 52.			NONSTELLAR OBJECT
FATH 1.757	16 16 13.	+ 74 21	16		NEBULA
ZC 1616.3+1344	16 16 18.	+ 13 44	740		CLUSTER OF GALAXIES
MCG+03-41-149	16 16 18.	+ 15 40	30	15.1	GALAXY
ZWG 137.075	16 16 18.	+ 22 17		15.7	GALAXY
UGC 10327	16 16 18.	+ 22 17	66	15.7	GALAXY Sb-c
ZWG 167.067	16 16 18.	+ 31 43		15.7	GALAXY
MCG+05-38-052	16 16 18.	+ 31 43	36	15.7	GALAXY
MCG+10-23-055	16 16 18.	+ 59 54	60	15.	GALAXY
IC 1207	16 16 18.	- 29 31 23.			NONSTELLAR OBJECT
LB 00310	16 16 20.	- 13 49 00.		16.0	FAINT BLUE STAR
SG 3.105	16 16 20.	- 25 12	720		DIFFUSE EMISSION NEBULA
REIN 2.244	16 16 20.44	+ 07 32 23.2			NEBULA
REIN 2.245	16 16 21.55	+ 07 31 49.9			NEBULA
IC 1209	16 16 23.	+ 15 40 43.			NONSTELLAR OBJECT
IC 1209	16 16 23.	+ 15 41	16		FAINT NEBULA
RNGC 6106	16 16 23.	+ 07 32		13.0	GALAXY
FATH 2.155	16 16 23.	+ 15 41	16		NEBULA
ZWG 052.001	16 16 24.	+ 07 32		13.4	GALAXY
UGC 10328	16 16 24.	+ 07 32	150	13.4	GALAXY Sc
MCG+01-41-016	16 16 24.	+ 07 32	144	13.4	GALAXY
FATH 1.756	16 16 24.	+ 14 40	11		NEBULA
ZWG 109.002	16 16 24.	+ 15 40		15.1	GALAXY
ZWG 108.174	16 16 24.	+ 15 40		15.1	GALAXY
UGC 10329	16 16 24.	+ 15 40	96	15.1	GALAXY S0
SC 1614+1548.2	16 16 24.	+ 15 40 50.	12		NEBULA
SC 1614+1604.5	16 16 24.	+ 15 57 08.	24		NEBULA
ZWG 137.076	16 16 24.	+ 22 05		15.3	GALAXY
MCG+04-38-050	16 16 24.	+ 22 16	60	15.7	GALAXY
ZWG 223.055	16 16 24.	+ 40 12		15.4	GALAXY
UGC 10330	16 16 24.	+ 40 12	66	15.4	GALAXY SBb
MCG+07-33-056	16 16 24.	+ 40 13	72	15.	GALAXY
MCG+10-23-056	16 16 24.	+ 59 26	108	14.	GALAXY
ZWG 299.025	16 16 24.	+ 59 27		14.8	GALAXY
UGC 10331	16 16 24.	+ 59 27	96	14.8	GALAXY S
MCG+10-23-057	16 16 24.	+ 59 30	42	16.	GALAXY
RNGC 6114	16 16 28.	+ 35 18		15.5	GALAXY
REIN 7.183	16 16 28.49	+ 59 26 30.2			NEBULA
REIN 7.184	16 16 28.59	+ 59 30 29.2			NEBULA
ZWG 196.030	16 16 30.	+ 35 18		15.3	GALAXY
ZWG 276.008	16 16 30.	+ 52 20		15.0	GALAXY
MCG+10-23-058	16 16 30.	+ 58 27	30	16.	GALAXY
ZWG 298.026	16 16 30.	+ 58 28		15.4	GALAXY
MCG+10-23-059	16 16 30.	+ 59 25	30	16.	GALAXY
ZWG 298.027	16 16 30.	+ 59 27		15.7	GALAXY
MCG+10-23-060	16 16 30.	+ 62 03	48	15.	GALAXY
ZWG 338.052	16 16 30.	+ 71 05		15.7	GALAXY
MCG+12-15-055	16 16 30.	+ 71 06	48	16.	GALAXY
MRSL 332-01/4	16 16 30.	- 51 23	5040		HII REGION
MCG+06-36-019	16 16 33.	+ 35 18	45	15.	GALAXY
RNGC 6123	16 16 34.	+ 62 04		14.5	GALAXY
REIN 7.186	16 16 35.34	+ 59 26 15.9			NEBULA
ZWG 137.077	16 16 36.	+ 21 40		15.5	GALAXY
ZWG 137.078	16 16 36.	+ 25 46		15.5	GALAXY
UGC 10332	16 16 36.	+ 29 57	78	16.5	GALAXY DBL SYS
ZWG 298.028	16 16 36.	+ 62 04		14.4	GALAXY
UGC 10333	16 16 36.	+ 62 04	54	14.4	GALAXY S0-a
KARA.73B 0734	16 16 36.	+ 62 04	48	14.4	ISOLATED GALAXY S
7ZW 870	16 16 36.	+ 64 13			COMPACT GALAXY
ZWG 320.022	16 16 36.	+ 66 20		14.6	GALAXY
REIN 7.185	16 16 37.03	+ 58 27 31.2			NEBULA
HOLM 738B	16 16 39.	+ 59 28	24	13.9	PART OF MULTIPLE GALAXY
LB 03546	16 16 39.	- 25 51 48.		19.3	FAINT BLUE STAR
HOLM 738A	16 16 40.	+ 59 27	78	13.1	PART OF MULTIPLE GALAXY
ZWG 196.031	16 16 42.	+ 35 37		15.2	GALAXY
MCG+06-36-020	16 16 42.	+ 35 37	30	14.5	GALAXY
MCG+09-27-012	16 16 42.	+ 52 20	48	15.	GALAXY
ZWG 320.023	16 16 42.	+ 63 58		15.4	GALAXY
UGC 10334	16 16 42.	+ 63 58	96	15.4	GALAXY IRR
IC 1218	16 16 42.	+ 68 20 12.			NONSTELLAR OBJECT
REIN 7.187	16 16 44.19	+ 58 12 03.1			NEBULA
SC 1614+1600.7	16 16 46.	+ 15 53 21.	72		NEBULA
HOLM 738C	16 16 47.	+ 59 27	18	14.1	PART OF MULTIPLE GALAXY
LB 00931	16 16 47.	+ 76 53 54.		16.6	FAINT BLUE STAR
ZWG 024.001	16 16 48.	+ 01 50		15.4	GALAXY
ZWG 080.003	16 16 48.	+ 14 12		15.5	GALAXY
ZWG 080.004	16 16 48.	+ 14 15		14.8	GALAXY
MCG+02-41-024	16 16 48.	+ 14 15	48	14.8	GALAXY
ZC 1616.8+2331	16 16 48.	+ 23 31	2220		CLUSTER OF GALAXIES
MCG+11-20-011	16 16 48.	+ 68 20	51	15.	GALAXY
OCL 0966	16 16 48.	- 48 26	360		OPEN STAR CLUSTER
VHA 189	16 16 48.	- 48 26	360		OPEN STAR CLUSTER
RNGC 6113	16 16 51.	+ 14 15		15.0	GALAXY
FATH 2.156	16 16 51.	+ 14 15	27		NEBULA
MCG+02-41-025	16 16 51.	+ 14 16	18	16.	GALAXY
LB 03547	16 16 53.	- 25 00 48.		18.8	FAINT BLUE STAR
ZWG 109.003	16 16 54.	+ 16 36		15.5	GALAXY
UGC 10335	16 16 54.	+ 16 36	72	15.5	GALAXY E-S0
MCG+03-41-150	16 16 54.	+ 16 36	72	15.5	GALAXY
MCG+07-33-057	16 16 54.	+ 43 12	42	17.	GALAXY
ZWG 276.009	16 16 54.	+ 54 31		15.5	GALAXY
ZC 1616.9-0309	16 16 54.	- 03 09	2290		CLUSTER OF GALAXIES
LB 00932	16 16 56.	- 00 17 18.		17.2	FAINT BLUE STAR
SC 1614+1644.0	16 16 57.	+ 16 36 40.	12		NEBULA
RNGC 6116	16 16 58.	+ 35 17		15.5	GALAXY
LB 00933	16 16 59.	- 00 59 42.		15.8	FAINT BLUE STAR
LBN 1105	16 17	- 25 10	480		BRIGHT NEBULA
LBN 1101	16 17	- 25 30	600		BRIGHT NEBULA
ZWG 196.032	16 17 00.	+ 35 17		15.3	GALAXY
UGC 10336	16 17 00.	+ 35 17	60	15.3	GALAXY Sb
MCG+06-36-021	16 17 00.	+ 35 17	66	14.5	GALAXY
ZWG 223.056	16 17 00.	+ 39 27		15.6	GALAXY
KARA.73B 0735	16 17 00.	+ 39 27	42	15.6	ISOLATED GALAXY S
MCG+08-30-005	16 17 00.	+ 50 23	36	16.	GALAXY
MRK 498	16 17 00.	+ 52 58	7	17.	GALAXY WITH UV CONTINUUM
1ZW 140	16 17 00.	+ 52 59			COMPACT GALAXY
MCG+09-27-013	16 17 00.	+ 54 32	36	15.	GALAXY
LDN 1719	16 17 00.	- 20 00	3120		DARK NEBULA
SC 1614+1556.7	16 17 01.	+ 15 49 22.	12		NEBULA
SG 3.106	16 17 01.	- 24 57	720		DIFFUSE EMISSION NEBULA
SC 1614+1555.5	16 17 03.	+ 15 48 10.	18		NEBULA
ABC 2185	16 17 05.	+ 70 43		17.4	RICH CLUSTER OF GALAXIES
ZWG 052.002	16 17 06.	+ 07 24		14.9	GALAXY
UGC 10337	16 17 06.	+ 07 24	72	14.9	GALAXY Sb
MCG+01-42-001	16 17 06.	+ 07 24	72	14.9	GALAXY
MCG+10-23-061	16 17 06.	+ 61 47	9	16.	GALAXY
VDB.66N 102	16 17 07.	- 19 55	852		REFLECTION NEBULA
CED 129B	16 17 07.	- 19 55	1320		DIFFUSE GALACTIC NEBULA
SC 1614+1806.9	16 17 08.	+ 17 59 35.	12		NEBULA
SC 1614+1753.2	16 17 10.	+ 17 45 53.	12		NEBULA
ZWG 137.079	16 17 12.	+ 22 16		15.6	GALAXY
ZWG 196.033	16 17 12.	+ 35 10		15.7	GALAXY
DG 134	16 17 12.	- 19 56	300		REFLECTION NEBULA
LB 03548	16 17 13.	- 26 20 00.		19.0	FAINT BLUE STAR
SC 1615+1901.3	16 17 15.	+ 18 53 59.	18		NEBULA
SC 1615+1909.8	16 17 15.	+ 19 02 29.	12		NEBULA
SC 1615+1843.5	16 17 17.	+ 18 36 11.	30		NEBULA
ZWG 024.002	16 17 18.	+ 02 05		15.2	GALAXY
ZWG 052.003	16 17 18.	+ 03 14		15.3	GALAXY
ZWG 052.004	16 17 18.	+ 05 17		15.1	GALAXY
ZWG 109.004	16 17 18.	+ 15 23		15.4	GALAXY
FAI 1617+17	16 17 18.	+ 17 44			COMPACT GALAXY
MCG+03-42-001	16 17 18.	+ 18 35	27	15.5	GALAXY
ZWG 109.005	16 17 18.	+ 18 36		15.5	GALAXY
SC 1615+1640.8	16 17 22.	+ 16 33 30.	12		NEBULA
SC 1615+1849.5	16 17 22.	+ 18 42 12.	18		NEBULA
SC 1615+1639.9	16 17 23.	+ 16 32 36.	12		NEBULA
ZWG 080.005	16 17 24.	+ 13 22		15.6	GALAXY
ZWG 109.006	16 17 24.	+ 18 42		15.7	GALAXY
ZC 1617.4+2755	16 17 24.	+ 27 55	1010		CLUSTER OF GALAXIES
ZWG 196.034	16 17 24.	+ 34 40		15.4	GALAXY
HN 0431	16 17 25.	- 19 57			NEBULA
VDB.66N 103	16 17 26.	- 19 59	540		REFLECTION NEBULA
REIN 7.188A	16 17 28.91	+ 57 43 34.5			NEBULA
REIN 7.188B	16 17 28.92	+ 57 43 34.9			NEBULA
ZC 1617.5+0901	16 17 30.	+ 09 01	1340		CLUSTER OF GALAXIES
ZWG 196.035	16 17 30.	+ 35 42		15.7	GALAXY
HOLM 737C	16 17 30.	+ 37 10	24	15.8	PART OF MULTIPLE GALAXY
HOLM 737A	16 17 30.	+ 37 12	42	14.3	PART OF MULTIPLE GALAXY
ZWG 196.036	16 17 30.	+ 37 13		14.7	GALAXY
RNGC 6117A	16 17 30.	+ 37 13		14.5	GALAXY
UGC 10338	16 17 30.	+ 37 13	78	14.7	GALAXY Sc
MCG+06-36-022	16 17 30.	+ 37 14	66	14.	GALAXY
HOLM 737D	16 17 31.	+ 37 11	18	16.3	PART OF MULTIPLE GALAXY
MCG+06-36-023	16 17 33.	+ 35 42	42	15.	GALAXY
HOLM 737B	16 17 34.	+ 37 10	18	15.7	PART OF MULTIPLE GALAXY
ZWG 024.003	16 17 36.	+ 02 08		14.3	GALAXY
UGC 10339	16 17 36.	+ 02 08	42	14.3	GALAXY
MCG+00-42-001	16 17 36.	+ 02 08	36	14.3	GALAXY
ZWG 080.006	16 17 36.	+ 10 37		15.2	GALAXY
ZWG 196.037	16 17 36.	+ 36 24		15.6	GALAXY
UGC 10340	16 17 36.	+ 36 24	66	15.6	GALAXY S
MCG+06-36-025	16 17 36.	+ 36 26	66	15.5	GALAXY
RNGC 6117B	16 17 36.	+ 37 13		15.0	GALAXY
MCG+06-36-024	16 17 36.	+ 37 13	33	15.5	GALAXY
LB 00934	16 17 36.	- 00 19 48.		17.7	FAINT BLUE STAR
CED 129C	16 17 37.	- 20 00	1320		DIFFUSE GALACTIC NEBULA
PK338+05.2	16 17 40.	- 42 16	10		PLANETARY NEBULA
FATH 1.758	16 17 41.	+ 75 15	27		NEBULA
ZWG 167.068	16 17 42.	+ 31 24		15.3	GALAXY
MCG+10-23-062	16 17 42.	+ 60 26	24	16.	GALAXY
UGC 10341	16 17 42.	+ 60 55	72	16.0	GALAXY Sb-c
ZC 1617.7+6507	16 17 42.	+ 65 07	870		CLUSTER OF GALAXIES
LB 00311	16 17 42.	- 01 39 00.		15.7	FAINT BLUE STAR
DG 135	16 17 42.	- 20 01	1320		REFLECTION NEBULA
IC 4601	16 17 43.	- 19 57	1560		REFLECTION NEBULAE
SC 1615+1940.7	16 17 45.	+ 19 33 25.	12		NEBULA
REIN 7.189	16 17 46.01	+ 57 42 53.1			NEBULA
ZWG 024.004	16 17 48.	+ 01 20		15.3	GALAXY
ZWG 024.005	16 17 48.	+ 01 52		15.5	GALAXY
ZWG 052.005	16 17 48.	+ 06 24		15.7	GALAXY
SC 1615+1905.8	16 17 48.	+ 18 58 31.	12		NEBULA
ZC 1617.8+2511	16 17 48.	+ 25 11	810		CLUSTER OF GALAXIES
ZWG 196.038	16 17 48.	+ 34 27		15.5	GALAXY
ZWG 196.039	16 17 48.	+ 36 12		15.3	GALAXY
UGC 10342	16 17 48.	+ 36 12	60	15.3	GALAXY SBb
ZC 1617.8+3739	16 17 48.	+ 37 39	1480		CLUSTER OF GALAXIES
MCG+07-33-058	16 17 49.	+ 41 46	24	16.	GALAXY
MCG+10-23-063	16 17 48.	+ 60 55	36	15.	GALAXY
ABC 2173	16 17 52.	+ 09 01		17.1	RICH CLUSTER OF GALAXIES
MCG+06-36-027	16 17 54.	+ 36 13	54	15.	GALAXY
ZWG 196.040	16 17 54.	+ 37 56		15.4	GALAXY
MCG+06-36-026	16 17 54.	+ 37 56 30.	45	15.	GALAXY
RNGC 6119	16 17 55.	+ 37 56		15.5	GALAXY
RNGC 6125	16 17 55.	+ 57 44			NON-EXISTENT OBJECT
REIN 7.190A	16 17 56.77	+ 58 06 07.0			NEBULA
REIN 7.190B	16 17 56.78	+ 58 06 06.1			NEBULA
REIN 7.191	16 17 57.24	+ 58 05 46.4			NEBULA
LBN 1115	16 18	- 20 00	900		BRIGHT NEBULA
LBN 1106	16 18	- 25 10	720		BRIGHT NEBULA
LBN 1104	16 18	- 25 30	3300		BRIGHT NEBULA
KHAV 241	16 18	- 37 31	9050		DARK NEBULA
1ZW 141	16 18 00.	+ 37 53			COMPACT GALAXY
ZWG 196.041	16 18 00.	+ 37 54		14.3	GALAXY
UGC 10343	16 18 00.	+ 37 54	30	14.3	GALAXY PECULR
ZWG 196.042	16 18 00.	+ 38 21		15.4	GALAXY
UGC 10344	16 18 00.	+ 38 21	84	15.4	GALAXY S
MCG+06-36-028	16 18 00.	+ 38 22	42	15.	GALAXY
MCG+08-30-006	16 18 00.	+ 46 47 30.	30	17.	GALAXY
MCG+10-23-064	16 18 00.	+ 56 51	24	16.	GALAXY
LB 00312	16 18 00.	- 03 45 06.		16.0	FAINT BLUE STAR
LDN 1680	16 18 00.	- 24 00	6840		DARK NEBULA
LDN 1676	16 18 00.	- 26 00	6840		DARK NEBULA
LDN 1672	16 18 00.	- 28 00	3480		DARK NEBULA
RNGC 6120	16 18 01.	+ 37 54		14.5	GALAXY
HOLM 739B	16 18 03.	+ 37 53	18	15.8	PART OF MULTIPLE GALAXY
MCG+06-36-029	16 18 03.	+ 37 55	27	14.5	GALAXY
HOLM 739A	16 18 05.	+ 37 53	24	14.1	PART OF MULTIPLE GALAXY
ZWG 052.006	16 18 06.	+ 07 28		15.3	GALAXY
ZWG 109.007	16 18 06.	+ 18 35		15.5	GALAXY
MCG+13-12-002	16 18 06.	+ 75 15	51	16.	GALAXY
MRSL 351+17/1	16 18 06.	- 25 28	4800		HII REGION
VDB.66N 104	16 18 06.	- 25 29	3360		REFLECTION NEBULA
DG 136	16 18 06.	- 25 29	900		REFLECTION NEBULA
SHB 328	16 18 07.3	+ 17 43 29.		16.4	QUASI-STELLAR OBJECT
BC 3CR334	16 18 07.40	+ 17 43 30.5		16.41	QUASI-STELLAR OBJECT
CED 130	16 18 08.	- 25 28	7800		DIFFUSE GALACTIC NEBULA
ZWG 024.006	16 18 12.	+ 00 38		15.5	GALAXY
1ZW 142	16 18 12.	+ 58 06			COMPACT GALAXY
ZWG 298.029	16 18 12.	+ 58 06		13.0	GALAXY
UGC 10345	16 18 12.	+ 58 06	84	13.0	GALAXY E

OBJECT NAME	RIGHT ASCEN.	DECLINATION	DIAM.	MAGN.	TYPE OF OBJECT
MCG+10-23-065	16 18 12.	+ 58 06	42	13.	GALAXY
ZC 1618.2+6839	16 18 12.	+ 68 39	1140		CLUSTER OF GALAXIES
RNGC 6127	16 18 14.	+ 58 06		13.0	GALAXY
REIN 7.192	16 18 14.82	+ 58 06 11.9			NEBULA
UGC 10346	16 18 18.	+ 37 42	72	17.	GALAXY Sc
MCG+09-27-014	16 18 18.	+ 52 36	30	16.	GALAXY
MRSL 333-00/1	16 18 18.	- 50 41	720		HII REGION
PK331-02.1	16 18 18.	- 53 34 33.	5		PLANETARY NEBULA
RNGC 6128	16 18 20.	+ 58 07			NON-EXISTENT OBJECT
MCG+06-36-030	16 18 21.	+ 37 56	18	17.	GALAXY
ARC 2175	16 18 22.	+ 30 02		16.2	RICH CLUSTER OF GALAXIES
SC 1616+1716.2	16 18 23.	+ 17 08 58.	18		NEBULA
ZC 1618.4+3000	16 18 24.	+ 30 00	2550		CLUSTER OF GALAXIES
MCG+06-36-031	16 18 24.	+ 35 16	30	15.5	GALAXY
RNGC 6122	16 18 25.	+ 37 56		15.0	GALAXY
SC 1616+1741.8	16 18 26.	+ 17 34 34.	12		NEBULA
MCG+06-36-032	16 18 27.	+ 37 56	42	15.5	GALAXY
ZWG 080.007	16 18 30.	+ 14 50		15.5	GALAXY
ZWG 109.008	16 18 30.	+ 18 41		15.7	GALAXY
MCG+03-42-002	16 18 30.	+ 18 41	27	15.7	GALAXY
ZC 1618.5+5346	16 18 30.	+ 53 46	3230		CLUSTER OF GALAXIES
ZWG 298.030	16 18 30.	+ 57 45		14.2	GALAXY
UGC 10347	16 18 30.	+ 57 45	66	14.2	GALAXY SB:b-c
PK013+32.1	16 18 30.	- 00 10			PLANETARY NEBULA
RNGC 6130	16 18 31.	+ 57 45		14.0	GALAXY
REIN 7.193	16 18 35.24	+ 57 44 00.6			NEBULA
ZWG 024.007	16 18 36.	+ 02 48		15.6	GALAXY
ZWG 052.007	16 18 36.	+ 04 19		15.4	GALAXY
ZC 1618.6+4603	16 18 36.	+ 46 03	1550		CLUSTER OF GALAXIES
ZC 1618.6+4718	16 18 36.	+ 47 18	1010		CLUSTER OF GALAXIES
MCG+09-27-015	16 18 36.	+ 52 36	18	16.	GALAXY
MCG+09-27-016	16 18 36.	+ 52 40	15	16.	GALAXY
MCG+10-23-066	16 18 36.	+ 57 43	60	13.	GALAXY
RNGC 6135	16 18 36.	+ 65 02			NON-EXISTENT OBJECT
ARC 2179	16 18 37.	+ 42 32		17.1	RICH CLUSTER OF GALAXIES
ARC 2180	16 18 39.	+ 47 47		17.1	RICH CLUSTER OF GALAXIES
UGC 10348	16 18 42.	+ 56 20	60	16.0	GALAXY Sb
MCG+09-27-017	16 18 45.	+ 51 40	48	16.	GALAXY
ZWG 080.008	16 18 48.	+ 13 15		15.2	GALAXY
ZWG 137.080	16 18 48.	+ 21 17		15.4	GALAXY
ZC 1618.9+1407	16 18 54.	+ 14 07	940		CLUSTER OF GALAXIES
ZWG 137.081	16 18 54.	+ 21 28		15.5	GALAXY
PATH 1.759	16 18 54.	+ 74 29	14		NEBULA
MCG+12-15-056	16 18 54.	+ 74 31	33	17.	GALAXY
SC 1616+1711.8	16 18 56.	+ 17 04 36.	54		NEBULA
ARC 2177	16 18 56.	+ 25 52		17.5	RICH CLUSTER OF GALAXIES
LB 00935	16 18 58.	- 04 08 18.		15.4	FAINT BLUE STAR
KHAV 242	16 19	- 28 19	2370		DARK NEBULA
KHAV 243	16 19	- 32 37	10240		DARK NEBULA
ZC 1619.0+2420	16 19 00.	+ 24 20	5980		CLUSTER OF GALAXIES
ZC 1619.0+4246	16 19 00.	+ 42 46	2490		CLUSTER OF GALAXIES
ZWG 251.009	16 19 00.	+ 50 03		15.4	GALAXY
ZWG 339.002	16 19 00.	+ 74 30		15.6	GALAXY
MCG+14-08-002	16 19 00.	+ 81 10 30.	30	17.	GALAXY
LB 03549	16 19 01.	- 23 58 30.		18.5	FAINT BLUE STAR
SC 1616+1650.9	16 19 04.	+ 16 43 42.	12		NEBULA
RNGC 6133	16 19 05.	+ 56 48			NON-EXISTENT OBJECT
ZWG 080.009	16 19 06.	+ 14 38		15.3	GALAXY
ZC 1619.1+2554	16 19 06.	+ 25 54	740		CLUSTER OF GALAXIES
1ZW 143	16 19 06.	+ 34 53			COMPACT GALAXY
ZWG 196.043	16 19 06.	+ 34 53		15.6	GALAXY
HOLM 740B	16 19 09.	+ 40 00	36	15.1	PART OF MULTIPLE GALAXY
HOLM 740C	16 19 09.	+ 40 03	36	15.3	PART OF MULTIPLE GALAXY
LB 00936	16 19 09.	+ 74 45 06.		17.2	FAINT BLUE STAR
ZWG 080.010	16 19 12.	+ 14 35		15.7	GALAXY
ZWG 080.011	16 19 12.	+ 14 51		15.6	GALAXY
FAI 16192+14	16 19 12.	+ 14 53			COMPACT GALAXY
MCG+07-34-001	16 19 12.	+ 40 12	72	14.	GALAXY
ZWG 224.001	16 19 12.	+ 40 14		15.2	GALAXY
UGC 10349	16 19 12.	+ 40 14	90	15.2	GALAXY SBa-b
SHAE 004	16 19 12.	+ 61 50	54	17.	GROUP OF COMPACT GALAXIES
SC 1616+1458.6	16 19 13.	+ 14 51 25.	18		NEBULA
SN 1953B	16 19 13.	+ 40 14		17.5	SUPERNOVA
HDB C27	16 19 14.	- 24 45			DIFFUSE NEBULA
ZC 1619.3+2445	16 19 18.	+ 24 45	740		CLUSTER OF GALAXIES
ZWG 196.098	16 19 18.	+ 32 28		15.5	GALAXY
ZWG 168.001	16 19 18.	+ 32 28		15.5	GALAXY
ZWG 167.069	16 19 18.	+ 32 28		15.5	GALAXY
ZWG 196.045	16 19 18.	+ 34 53		15.4	GALAXY
ZCG 1619+34	16 19 18.	+ 34 53		15.4	COMPACT GALAXY
ZWG 196.044	16 19 18.	+ 36 10		15.4	GALAXY
ZC 1619.3+6538	16 19 18.	+ 65 38	5040		CLUSTER OF GALAXIES
ZWG 338.053	16 19 18.	+ 68 56		15.7	GALAXY
ZWG 320.024	16 19 18.	+ 68 56		15.7	GALAXY
ZWG 024.008	16 19 18.	- 02 10		13.2	GALAXY
RNGC 6118	16 19 18.	- 02 10		12.0	GALAXY
UGC 10350	16 19 18.	- 02 10	300	13.2	GALAXY Sc
MCG+00-42-002	16 19 18.	- 02 10	294	13.2	GALAXY
KARA.73B 0736	16 19 18.	- 02 10	294	13.2	ISOLATED GALAXY S
HOLM 740A	16 19 19.	+ 40 03	24	14.3	PART OF MULTIPLE GALAXY
PK327-06.1	16 19 19.	- 58 11 58.	5		PLANETARY NEBULA
MCG+06-36-033	16 19 21.	+ 34 53	15	15.	GALAXY
B 041	16 19 22.	- 19 31	2700		DARK OBJECT
ARC 2178	16 19 23.	+ 24 46		17.1	RICH CLUSTER OF GALAXIES
ZWG 080.012	16 19 24.	+ 10 03		15.5	GALAXY
ZWG 168.002	16 19 24.	+ 28 46		14.9	GALAXY
ZWG 167.070	16 19 24.	+ 28 46		14.9	GALAXY
UGC 10351	16 19 24.	+ 28 46	60	14.9	GALAXY S-IRR
MCG+05-39-001	16 19 24.	+ 28 46	42	14.9	GALAXY
MCG+06-36-034	16 19 24.	+ 36 12	48	15.	GALAXY
MCG+07-34-002	16 19 24.	+ 40 01	30	15.	GALAXY
ZWG 224.002	16 19 24.	+ 40 03		15.6	GALAXY
IC 1213	16 19 26.	- 01 24 27.			NONSTELLAR OBJECT
FAI 1619+19	16 19 30.	+ 19 05			COMPACT GALAXY
ZC 1619.5+4445	16 19 30.	+ 44 45	1010		CLUSTER OF GALAXIES
LB 00937	16 19 30.	- 01 46 54.		16.3	FAINT BLUE STAR
ARC 2189	16 19 31.	+ 72 34		17.1	RICH CLUSTER OF GALAXIES
SC 1617+1911.5	16 19 32.	+ 19 04 20.	12		NEBULA
MCG+04-39-001	16 19 36.	+ 25 00	39		GALAXY
1ZW 144	16 19 36.	+ 36 29			COMPACT GALAXY
ZWG 196.046	16 19 36.	+ 36 30		14.5	GALAXY
RNGC 6126	16 19 36.	+ 36 30		14.5	GALAXY
UGC 10353	16 19 36.	+ 36 30	60	14.5	GALAXY COMPACT
ZWG 024.009	16 19 36.	- 01 24		13.9	GALAXY
UGC 10352	16 19 36.	- 01 24	60	13.9	GALAXY E
MCG+00-42-003	16 19 36.	- 01 24	24	13.9	GALAXY
SC 1617+1928.9	16 19 41.	+ 19 21 45.	90		NEBULA
ZWG 080.013	16 19 42.	+ 10 00		15.5	GALAXY
FAI 16197+14	16 19 42.	+ 14 55			COMPACT GALAXY
ZC 1619.7+2429	16 19 42.	+ 24 29	670		CLUSTER OF GALAXIES
ZWG 138.001	16 19 42.	+ 25 56		15.7	GALAXY
MCG+06-36-035	16 19 42.	+ 36 30	66	14.5	GALAXY
3ZW 076	16 19 42.	+ 40 19			COMPACT GALAXY
ZWG 224.003	16 19 42.	+ 40 56		15.7	GALAXY
UGC 10354	16 19 42.	+ 40 56	78	15.7	GALAXY Sc/SBc
MCG+07-34-003	16 19 42.	+ 40 56	66	14.5	GALAXY
OCL 0958	16 19 42.	- 51 53	300	9.	OPEN STAR CLUSTER
ARC 2184	16 19 45.	+ 50 19		15.9	RICH CLUSTER OF GALAXIES
LB 00938	16 19 46.	+ 38 16 18.		16.4	FAINT BLUE STAR
RNGC 6136	16 19 46.	+ 56 05		15.5	GALAXY
SC 1617+1709.6	16 19 48.	+ 17 02 27.	12		NEBULA
ZC 1619.8+1712	16 19 48.	+ 17 12	1140		CLUSTER OF GALAXIES
ZWG 196.047	16 19 48.	+ 38 07		15.6	GALAXY
HOLM 741B	16 19 48.	+ 38 07	18	15.7	PART OF MULTIPLE GALAXY
MCG+06-36-036	16 19 48.	+ 38 09	30	15.5	GALAXY
MCG+09-27-018	16 19 48.	+ 53 40	42	15.	GALAXY
ZWG 276.010	16 19 48.	+ 56 05		15.5	GALAXY
ZC 1619.8+5959	16 19 48.	+ 59 59	1480		CLUSTER OF GALAXIES
OCL 0959	16 19 48.	- 51 46	102	12.	OPEN STAR CLUSTER
HOLM 741C	16 19 51.	+ 38 07	12	15.8	PART OF MULTIPLE GALAXY
ARC 2183	16 19 53.	+ 42 50		17.1	RICH CLUSTER OF GALAXIES
ZC 1619.9+2403	16 19 54.	+ 24 03	670		CLUSTER OF GALAXIES
ZWG 138.002	16 19 54.	+ 25 44		15.7	GALAXY
ZC 1619.9+2558	16 19 54.	+ 25 58	6720		CLUSTER OF GALAXIES
RNGC 6101	16 19 57.	- 72 06		10.0	GLOBULAR CLUSTER
HOLM 741A	16 19 58.	+ 38 06	18	14.7	PART OF MULTIPLE GALAXY
ZWG 080.014	16 20 00.	+ 13 57		15.7	GALAXY
UGC 10355	16 20 00.	+ 13 57	66	15.7	GALAXY Sc
ZWG 196.048	16 20 00.	+ 38 06		14.7	GALAXY
MCG+06-36-037	16 20 00.	+ 38 08	36	15.	GALAXY
MCG+09-27-019	16 20 00.	+ 56 07	48	15.	GALAXY
MCG+11-20-012	16 20 00.	+ 65 31	270	12.	GALAXY
MCG+12-15-057	16 20 00.	+ 74 00	24	18.	GALAXY
MCG+14-08-003	16 20 00.	+ 81 11	39	17.	GALAXY
LDN 1687	16 20 00.	- 23 30	7200		DARK NEBULA
GCL 040	16 20 00.	- 72 05	876	10.2	GLOBULAR STAR CLUSTER
RNGC 6129	16 20 01.	+ 38 06		14.5	GALAXY
HOLM 742A	16 20 02.	+ 39 04	60	12.8	PART OF MULTIPLE GALAXY
ZWG 080.015	16 20 06.	+ 10 37		15.7	GALAXY
ZC 1620.1+3343	16 20 06.	+ 33 43	1410		CLUSTER OF GALAXIES
MCG+09-27-020	16 20 06.	+ 51 40	30	16.	GALAXY
PK346+12.1	16 20 06.	- 31 38	154	15.7	PLANETARY NEBULA
OCL 0957	16 20 06.	- 48 48	240	15.	OPEN STAR CLUSTER
VHA 190	16 20 06.	- 48 48	90		OPEN STAR CLUSTER
WRAY 19.45	16 20 07.4	- 31 36 35.		10.0	STAR-NEBULA ASSOCIATION
HOLM 742B	16 20 10.	+ 39 03	18	14.8	PART OF MULTIPLE GALAXY
MCG+07-34-004	16 20 12.	+ 39 01	60	14.	GALAXY
ZWG 224.004	16 20 12.	+ 39 03		14.2	GALAXY
UGC 10356	16 20 12.	+ 39 03	66	14.2	GALAXY Sc/SBc
ZWG 251.010	16 20 12.	+ 49 33		15.0	GALAXY
MCG+10-23-067	16 20 12.	+ 62 32	60	16.	GALAXY
MRSL 332-01/2	16 20 12.	- 51 24	1200		HII REGION
RNGC 6131B	16 20 14.	+ 39 01		15.0	GALAXY
RNGC 6131A	16 20 14.	+ 39 03		14.0	GALAXY
MCG+08-30-007	16 20 15.	+ 49 33	36	16.	GALAXY
PK336+01.1	16 20 16.1	- 46 35 22.	14		PLANETARY NEBULA
ZC 1620.3+0130	16 20 18.	+ 01 30	940		CLUSTER OF GALAXIES
ZWG 224.005	16 20 18.	+ 40 34		15.1	GALAXY
UGC 10357	16 20 18.	+ 40 34	78	15.1	GALAXY
MCG+09-27-021	16 20 18.	+ 56 09	42	16.	GALAXY
PATH 1.760	16 20 20.	+ 74 56	8		NEBULA
PK330-03.1	16 20 20.	- 54 29	25		PLANETARY NEBULA
MCG+07-34-005	16 20 21.	+ 39 00	27	15.	GALAXY
ZC 1620.4+1747	16 20 24.	+ 17 47	2290		CLUSTER OF GALAXIES
FAI 16204+20	16 20 24.	+ 20 43			COMPACT GALAXY
TON-N 0816	16 20 24.	+ 26 01		14.8	BLUE STAR
MCG+07-34-006	16 20 24.	+ 40 32	48	15.	GALAXY
LB 03550	16 20 27.	- 22 55 06.		18.2	FAINT BLUE STAR
ZWG 052.008	16 20 30.	+ 03 45		15.6	GALAXY
ZWG 052.009	16 20 30.	+ 06 41		15.7	GALAXY
ZC 1620.5+0716	16 20 30.	+ 07 16	1550		CLUSTER OF GALAXIES
ZWG 080.016	16 20 30.	+ 11 29		15.6	GALAXY
FAI 16205+20	16 20 30.	+ 20 42			COMPACT GALAXY
ZWG 276.011	16 20 30.	+ 55 12		13.9	GALAXY
UGC 10358	16 20 30.	+ 55 12	60	13.9	GALAXY S
ZWG 024.010	16 20 30.	- 00 25		15.7	GALAXY
KARA.73B 0737	16 20 30.	- 00 25	36	15.7	ISOLATED GALAXY S
SC 1618+1422.6	16 20 31.	+ 14 15 30.	12		NEBULA
RNGC 6121	16 20 34.	- 26 24		7.5	GLOBULAR CLUSTER
RNGC 6115	16 20 34.	- 51 51			NON-EXISTENT OBJECT
ZWG 080.017	16 20 36.	+ 14 42		15.1	GALAXY
MCG+02-42-001	16 20 36.	+ 14 42	48	15.1	GALAXY
1ZW 145	16 20 36.	+ 35 46			COMPACT GALAXY
ZWG 196.049	16 20 36.	+ 35 46		15.3	GALAXY
ZWG 251.011	16 20 36.	+ 49 37		15.5	GALAXY
MCG+10-23-068	16 20 36.	+ 58 48	36	15.	GALAXY
ZWG 298.031	16 20 36.	+ 58 49		15.4	GALAXY
ZWG 320.025	16 20 36.	+ 65 30		12.6	GALAXY
UGC 10359	16 20 36.	+ 65 30	480	12.6	GALAXY SBc
GCL 041	16 20 36.	- 26 24	1368	7.41	GLOBULAR STAR CLUSTER
LB 00313	16 20 37.	- 03 40 24.		15.7	FAINT BLUE STAR
HOLM 743B	16 20 38.	+ 27 25	12	15.5	PART OF MULTIPLE GALAXY
HOLM 743A	16 20 38.	+ 27 25	18	15.3	PART OF MULTIPLE GALAXY
RNGC 6140	16 20 38.	+ 65 30		12.5	GALAXY
RNGC 6143	16 20 39.	+ 55 11		14.0	GALAXY
PK331-02.2	16 20 41.	- 53 15 31.	25		PLANETARY NEBULA
ZWG 024.011	16 20 42.	+ 01 53		15.1	GALAXY
MCG+00-42-004	16 20 42.	+ 01 53	15	15.1	GALAXY
FAI 16207+20	16 20 42.	+ 20 15			COMPACT GALAXY
MCG+08-30-008	16 20 42.	+ 49 39	39	16.	GALAXY
MCG+09-27-023	16 20 42.	+ 54 47	30	16.	GALAXY
MCG+09-27-022	16 20 42.	+ 54 48	30	16.	GALAXY
MCG+09-27-024	16 20 42.	+ 55 12 30.	60	13.	GALAXY
OCL 0960	16 20 42.	- 51 51	204	11.	OPEN STAR CLUSTER
REIN 7.194A	16 20 43.07	+ 58 48 52.4			NEBULA
REIN 7.194B	16 20 43.17	+ 58 48 52.4			NEBULA
ARC 2182	16 20 46.	+ 14 32		17.4	RICH CLUSTER OF GALAXIES
FAI 16208+20	16 20 48.	+ 20 39			COMPACT GALAXY
ZWG 052.010	16 20 54.	+ 06 46		15.6	GALAXY
ZWG 109.009	16 20 54.	+ 17 03		15.1	GALAXY
UGC 10360	16 20 54.	+ 17 03	66	15.1	GALAXY S0
ZWG 109.010	16 20 54.	+ 19 06		15.3	GALAXY
FAI 16209+20	16 20 54.	+ 20 38			COMPACT GALAXY
ZWG 196.050	16 20 54.	+ 38 29		14.9	GALAXY
ZC 1620.9+6715	16 20 54.	+ 67 15	670		CLUSTER OF GALAXIES
SC 1618+1915.0	16 20 58.	+ 19 07 56.	18		NEBULA

OBJECT NAME	RIGHT ASCEN.	DECLINATION	DIAM.	MAGN.	TYPE OF OBJECT
REIN 6.102	16 20 58.95	+ 41 05 40.8			NEBULA
KHAV 244	16 21	- 25 07	3880		DARK NEBULA
MCG+03-42-003	16 21 00.	+ 17 02	60	15.1	GALAXY
ZWG 138.003	16 21 00.	+ 26 39		15.5	GALAXY
ZWG 196.051	16 21 00.	+ 37 40		15.0	GALAXY
ZWG 224.006	16 21 00.	+ 39 42		15.7	GALAXY
ZWG 224.007	16 21 00.	+ 43 35		15.3	GALAXY
ZWG 251.012	16 21 00.	+ 50 29		15.7	GALAXY
MCG+08-30-009	16 21 00.	+ 50 30	54	16.	GALAXY
MCG+09-27-025	16 21 00.	+ 55 10	36	16.	GALAXY
MCG+10-23-069	16 21 00.	+ 57 22	24	15.	GALAXY
7ZW 644	16 21 00.	+ 86 10			COMPACT GALAXY
MRSL 332-01/1	16 21 00.	- 51 57	4680		HII REGION
IC 4602	16 21 04.	+ 12 52 10.			NONSTELLAR OBJECT
ZWG 052.011	16 21 06.	+ 07 58		15.4	GALAXY
MCG+07-34-007	16 21 06.	+ 43 34	42	15.	GALAXY
MCG+09-27-026	16 21 06.	+ 51 15	48	15.	GALAXY
ZWG 298.032	16 21 06.	+ 57 24		15.2	GALAXY
UGC 10361	16 21 06.	+ 57 24	84	15.2	GALAXY S0
REIN 6.103	16 21 08.64	+ 41 06 15.9			NEBULA
REIN 6.104	16 21 11.38	+ 41 04 31.1			NEBULA
ZWG 196.052	16 21 12.	+ 38 04		15.5	GALAXY
ZWG 224.008	16 21 12.	+ 39 55		15.4	GALAXY
UGC 10362	16 21 12.	+ 39 55	96	15.4	GALAXY SBb
1ZW 146	16 21 12.	+ 57 23			COMPACT GALAXY
MCG+12-15-058	16 21 12.	+ 73 57	51	17.	GALAXY
HOLM 744B	16 21 13.	+ 38 05	24	15.2	PART OF MULTIPLE GALAXY
PK334+00.1	16 21 14.	- 48 36 45.			PLANETARY NEBULA
MCG+07-34-008	16 21 15.	+ 39 52 30.	66	14.5	GALAXY
HOLM 744A	16 21 16.	+ 38 03	24	13.9	PART OF MULTIPLE GALAXY
ZWG 080.018	16 21 18.	+ 09 49		15.6	GALAXY
ZWG 080.019	16 21 18.	+ 09 52		15.2	GALAXY
ZWG 080.020	16 21 18.	+ 11 54		14.8	GALAXY
UGC 10363	16 21 18.	+ 11 54	96	14.8	GALAXY Sa-b
MCG+02-42-002	16 21 18.	+ 11 54	84	14.8	GALAXY
ZC 1621.3+2347	16 21 18.	+ 23 47	810		CLUSTER OF GALAXIES
TON-N 0257	16 21 18.	+ 24 51		15.1	BLUE STAR
ZWG 196.054	16 21 18.	+ 35 57		15.3	GALAXY
ZWG 196.053	16 21 18.	+ 38 02		14.1	GALAXY
UGC 10364	16 21 18.	+ 38 02	126	14.1	GALAXY E
MCG+06-36-038	16 21 18.	+ 38 05 30.	24	15.	GALAXY
MCG+07-34-009	16 21 18.	+ 41 37	36	17.	GALAXY
DG 137	16 21 18.	- 25 35	60		REFLECTION NEBULA
RNGC 6137A	16 21 19.	+ 38 02		14.0	GALAXY
RNGC 6137B	16 21 19.	+ 38 05		15.0	GALAXY
RNGC 6132	16 21 20.	+ 11 54		15.0	GALAXY
MCG+06-36-039	16 21 21.	+ 38 04	42	14.	GALAXY
SS 56	16 21 21.	- 25 35			DIFFUSE GALACTIC NEBULA
REIN 6.105	16 21 23.89	+ 41 04 53.6			NEBULA
ZWG 080.021	16 21 24.	+ 09 54		15.3	GALAXY
MCG+06-36-040	16 21 24.	+ 35 57 30.	42	15.	GALAXY
SN 1964I	16 21 25.	+ 41 20		18.0	SUPERNOVA
REIN 6.106	16 21 25.46	+ 40 58 25.4			NEBULA
REIN 6.107	16 21 29.56	+ 41 04 33.6			NEBULA
ZWG 196.055	16 21 30.	+ 37 19		15.6	GALAXY
MCG+07-34-010	16 21 30.	+ 39 24	36	15.	GALAXY
ZWG 276.012	16 21 30.	+ 56 42		15.5	GALAXY
MCG+12-15-059	16 21 30.	+ 74 00	51	15.	GALAXY
HOLM 745B	16 21 31.	+ 39 58	24	15.4	PART OF MULTIPLE GALAXY
HOLM 745A	16 21 31.	+ 40 00	60	13.5	PART OF MULTIPLE GALAXY
MCG+09-27-027	16 21 33.	+ 56 42	42	15.	GALAXY
REIN 6.109	16 21 34.55	+ 41 05 14.6			NEBULA
UGC 10365	16 21 36.	+ 04 49	60	16.0	GALAXY S
ZWG 138.004	16 21 36.	+ 24 35		15.4	GALAXY
ZWG 168.003	16 21 36.	+ 30 48		15.7	GALAXY
ZWG 196.056	16 21 36.	+ 37 22		14.8	GALAXY
RNGC 6142	16 21 36.	+ 37 22		15.0	GALAXY
UGC 10366	16 21 36.	+ 37 22	114	14.8	GALAXY Sb
MCG+06-36-041	16 21 36.	+ 37 23 30.	108	14.	GALAXY
MCG+07-34-011	16 21 36.	+ 40 00	72	13.5	GALAXY
ZWG 224.009	16 21 36.	+ 40 02		14.6	GALAXY
UGC 10367	16 21 36.	+ 40 02	78	14.6	GALAXY SBb
ZWG 339.003	16 21 36.	+ 73 58		14.8	GALAXY
ZWG 338.054	16 21 36.	+ 73 58		14.8	GALAXY
UGC 10368	16 21 36.	+ 73 58	78	14.8	GALAXY E-S0
LB 00939	16 21 38.	- 00 47 06.		16.8	FAINT BLUE STAR
REIN 6.110	16 21 39.42	+ 40 58 40.6			NEBULA
ZWG 080.022	16 21 42.	+ 13 50		15.5	GALAXY
SC 1619+1534.0	16 21 42.	+ 15 26 59.	18		NEBULA
FAI 1621+20	16 21 42.	+ 20 48			COMPACT GALAXY
MCG+09-27-028	16 21 42.	+ 54 02	36	15.	GALAXY
UGC 10369	16 21 42.	+ 67 25	72	16.5	GALAXY Sc
MCG+11-20-013	16 21 42.	+ 67 25	72	16.	GALAXY
MCG+12-15-060	16 21 42.	+ 73 55	21	17.	GALAXY
REIN 6.111	16 21 42.77	+ 41 05 51.3			NEBULA
REIN 6.112	16 21 44.94	+ 41 08 41.5			NEBULA
MCG+07-34-012	16 21 45.	+ 41 44	42	15.	GALAXY
REIN 6.113	16 21 47.33	+ 41 00 01.1			NEBULA
ZWG 080.023	16 21 48.	+ 09 19		15.4	GALAXY
ZWG 138.005	16 21 48.	+ 21 55		15.5	GALAXY
MCG+04-39-002	16 21 48.	+ 22 29	48	15.0	GALAXY
ZWG 138.006	16 21 48.	+ 22 30		15.0	GALAXY
ZWG 196.057	16 21 48.	+ 35 07		15.1	GALAXY
ZWG 224.010	16 21 48.	+ 39 15		14.9	GALAXY
MCG+07-34-013	16 21 48.	+ 39 16	30	15.	GALAXY
ZWG 224.011	16 21 48.	+ 41 46		15.3	GALAXY
MCG+07-34-014	16 21 48.	+ 41 47 30.	42	14.5	GALAXY
MCG+09-27-029	16 21 48.	+ 52 20	36	16.	GALAXY
MCG+07-34-015	16 21 51.	+ 39 12	48	14.	GALAXY
ZWG 224.012	16 21 54.	+ 39 20		15.7	GALAXY
SN 1964G	16 21 54.	+ 39 20		16.0	SUPERNOVA
ZWG 224.013	16 21 54.	+ 41 50		15.0	GALAXY
HN 1810	16 21 56.1	- 16 52 44.			NEBULA
LBN 0030	16 22	- 07 20	4500		BRIGHT NEBULA
KHAV 246	16 22	- 16 49	5010		DARK NEBULA
KHAV 247	16 22	- 20 49	6050		DARK NEBULA
B 042	16 22	- 23 20			DARK OBJECT
KHAV 245	16 22	- 34 31	4940		DARK NEBULA
ZWG 109.011	16 22 00.	+ 20 17		15.2	GALAXY
MCG+03-42-004	16 22 00.	+ 20 18 30.	48	15.2	GALAXY
ZWG 251.013	16 22 00.	+ 50 35		15.6	GALAXY
MRK 698	16 22 00.	+ 52 39	12	16.5	GALAXY WITH UV CONTINUUM
1ZW 147	16 22 00.	+ 54 16			COMPACT GALAXY
ZWG 338.055	16 22 00.	+ 68 40		15.7	GALAXY
ZWG 320.026	16 22 00.	+ 68 40		15.7	GALAXY
LB 03551	16 22 00.	- 22 30 42.		14.3	FAINT BLUE STAR
MCG+07-34-016	16 22 03.	+ 39 16	21	15.	GALAXY
ZWG 138.007	16 22 06.	+ 25 55		15.6	GALAXY
ZWG 196.058	16 22 06.	+ 36 20		15.2	GALAXY
KARA.72 495A	16 22 06.	+ 36 20	48	15.2	PART OF DOUBLE GALAXY
3ZW 077	16 22 06.	+ 41 12			COMPACT GALAXY
ZWG 224.014	16 22 06.	+ 41 12		15.4	GALAXY
VVI 83	16 22 06.	+ 41 12	2	15.4	SEYFERT GALAXY
MRK 699	16 22 06.	+ 41 12	12	16.5	GALAXY WITH UV CONTINUUM
ZWG 251.014	16 22 06.	+ 45 51		15.3	GALAXY
UGC 10370	16 22 06.	+ 45 51	66	15.3	GALAXY Sa
ZC 1622.1+4802	16 22 06.	+ 48 02	1080		CLUSTER OF GALAXIES
ZWG 276.013	16 22 06.	+ 53 13		15.7	GALAXY
ZWG 276.014	16 22 06.	+ 55 52		15.6	GALAXY
ZWG 320.027	16 22 06.	+ 64 35		15.5	GALAXY
RNGC 6124	16 22 11.	- 40 35		6.5	OPEN CLUSTER
ZWG 109.012	16 22 12.	+ 19 35		15.4	GALAXY
UGC 10371	16 22 12.	+ 19 35	84	15.4	GALAXY S0
KARA.72 496A	16 22 12.	+ 19 35	54	15.4	PART OF DOUBLE GALAXY
MCG+03-42-005	16 22 12.	+ 19 36	54	15.4	GALAXY
ZWG 168.004	16 22 12.	+ 30 17		15.4	GALAXY
UGC 10372	16 22 12.	+ 30 17	66	15.4	GALAXY Sb
MCG+09-27-030	16 22 12.	+ 55 55	18	16.	GALAXY
OCL 0990	16 22 12.	- 40 33	2580	6.3	OPEN STAR CLUSTER
IC 1219	16 22 14.	+ 19 35 43.			NONSTELLAR OBJECT
LB 03552	16 22 15.	- 22 26 30.		18.0	FAINT BLUE STAR
HOLM 746A	16 22 17.	+ 41 02	18	15.0	PART OF MULTIPLE GALAXY
UGC 10373	16 22 18.	+ 07 17	66	15.3	GALAXY Sc
ZWG 052.012	16 22 18.	+ 07 25		15.3	GALAXY
MCG+05-39-002	16 22 18.	+ 30 16	48	15.4	GALAXY
ZWG 168.005	16 22 18.	+ 31 59		15.6	GALAXY
ZWG 196.059	16 22 18.	+ 36 20		15.7	GALAXY
KARA.72 495B	16 22 18.	+ 36 20	36	15.7	PART OF DOUBLE GALAXY
ZWG 276.015	16 22 18.	+ 51 06		14.6	GALAXY
UGC 10374	16 22 18.	+ 51 06	78	14.6	GALAXY SBa
MCG+09-27-031	16 22 18.	+ 55 53	60	14.	GALAXY
ISS 1042	16 22 18.	- 42 09	115		STELLAR RING
REIN 6.114	16 22 18.35	+ 41 01 59.9			NEBULA
IC 4603	16 22 19.	- 24 20			DIFFUSE NEBULA
SC 1620+1643.0	16 22 20.	+ 16 36 01.	18		NEBULA
HOLM 746B	16 22 22.	+ 41 02	12	15.1	PART OF MULTIPLE GALAXY
REIN 6.115	16 22 22.12	+ 41 02 05.5			NEBULA
LB 00940	16 22 23.	- 03 48 36.		17.9	FAINT BLUE STAR
ZWG 080.024	16 22 24.	+ 09 41		15.3	GALAXY
KARA.72 497A	16 22 24.	+ 09 41	72	15.3	PART OF DOUBLE GALAXY
ZWG 080.025	16 22 24.	+ 09 43		15.5	GALAXY
UGC 10375	16 22 24.	+ 09 43	84	15.5	GALAXY Sb
KARA.72 497B	16 22 24.	+ 09 43	72	15.5	PART OF DOUBLE GALAXY
ZWG 109.013	16 22 24.	+ 19 36		15.4	GALAXY
KARA.72 496B	16 22 24.	+ 19 36	42	15.4	PART OF DOUBLE GALAXY
MCG+03-42-006	16 22 24.	+ 19 37	42	15.4	GALAXY
ZWG 138.008	16 22 24.	+ 21 19		15.6	GALAXY
UGC 10376	16 22 24.	+ 65 33	90	16.5	GALAXY DWRF SP
DG 138	16 22 24.	- 24 21	1800		REFLECTION NEBULA
CED 131A	16 22 25.	- 24 21			DIFFUSE GALACTIC NEBULA
ZWG 052.013	16 22 30.	+ 06 11		15.0	GALAXY
MCG+01-42-002	16 22 30.	+ 06 11	36	15.0	GALAXY
ZWG 080.026	16 22 30.	+ 13 48		15.5	GALAXY
ZWG 168.006	16 22 30.	+ 28 25		15.7	GALAXY
ZWG 196.060	16 22 30.	+ 38 43		15.6	GALAXY
MCG+09-27-032	16 22 30.	+ 51 04	72	14.	GALAXY
BC 3CR336	16 22 32.45	+ 23 52 00.7		17.47	QUASI-STELLAR OBJECT
SHB 329	16 22 32.5	+ 23 52 00.		17.5	QUASI-STELLAR OBJECT
ARC 2187	16 22 34.	+ 41 23		17.1	RICH CLUSTER OF GALAXIES
VDB.66N 106	16 22 34.	- 23 19	3720		REFLECTION NEBULA
IC 4604	16 22 35.	- 23 19 55.	7200		DIFFUSE NEBULA
ZWG 109.014	16 22 36.	+ 18 30		15.3	GALAXY
ZWG 138.009	16 22 36.	+ 24 04		15.7	GALAXY
MCG+07-34-017	16 22 36.	+ 39 17	48	16.	GALAXY
ZWG 224.015	16 22 36.	+ 39 20		15.6	GALAXY
DG 139	16 22 36.	- 23 20	4260		REFLECTION NEBULA
CED 131B	16 22 36.	- 23 20	7440		DIFFUSE GALACTIC NEBULA
VDB.66N 105	16 22 38.	- 24 21	1680		REFLECTION NEBULA
REIN 6.116	16 22 39.86	+ 41 05 41.4			NEBULA
MCG+03-42-007	16 22 42.	+ 18 30	30	15.3	GALAXY
ZWG 138.010	16 22 42.	+ 21 44		15.7	GALAXY
ZWG 298.033	16 22 42.	+ 56 58		15.7	GALAXY
REIN 6.117	16 22 46.78	+ 41 03 11.6			NEBULA
REIN 6.118	16 22 46.96	+ 41 04 16.9			NEBULA
ZWG 080.027	16 22 48.	+ 11 59		15.7	GALAXY
MCG+02-42-003	16 22 48.	+ 11 59	60	15.7	GALAXY
ZWG 168.007	16 22 48.	+ 30 53		15.6	GALAXY
ZC 1622.8+5443	16 22 48.	+ 54 43	3900		CLUSTER OF GALAXIES
LDN 1686	16 22 48.	- 24 13	480		DARK NEBULA
REIN 6.119	16 22 48.33	+ 41 09 25.9			NEBULA
LB 00941	16 22 53.	- 02 17 06.		15.5	FAINT BLUE STAR
UGC 10377	16 22 54.	+ 23 11	84	17.	GALAXY PECULR?
MCG+07-34-018	16 22 54.	+ 39 12	48	15.5	GALAXY
SC 1620+1620.7	16 22 59.	+ 16 13 46.	12		NEBULA
REIN 6.120	16 22 59.06	+ 41 08 02.6			NEBULA
LBN 1119	16 23	- 19 30	6300		BRIGHT NEBULA
LBN 1112	16 23	- 23 20	4200		BRIGHT NEBULA
LBN 1111	16 23	- 23 20	2520		BRIGHT NEBULA
LBN 1109	16 23	- 24 20	720		BRIGHT NEBULA
LB 09987	16 23	- 87 04		13.3	FAINT BLUE STAR
MCG+06-36-042	16 23 00.	+ 37 38 30.	24	15.	GALAXY
RDS 002	16 23 00.	+ 39 41 06.	26		S GALAXY IN ARC 2199
1ZW 148	16 23 00.	+ 41 08			COMPACT GALAXY
MCG+10-23-070	16 23 00.	+ 58 29	54	16.	GALAXY
ZWG 298.034	16 23 00.	+ 58 30		15.5	GALAXY
ZWG 367.009	16 23 00.	+ 81 10		15.5	GALAXY
UGC 10378	16 23 00.	+ 84 08	60	17.	GALAXY Sc
LDN 1688	16 23 00.	- 24 00	3540		DARK NEBULA
OCL 1011	16 23 00.	- 26 07	30600	1.1	OPEN STAR CLUSTER
SHB 330	16 23 00.8	+ 26 57 18.		17.5	QUASI-STELLAR OBJECT
ARC 2186	16 23 01.	+ 28 40		17.1	RICH CLUSTER OF GALAXIES
ZWG 080.028	16 23 06.	+ 09 54		15.6	GALAXY
MCG+07-34-019	16 23 06.	+ 39 49	54	14.5	GALAXY
RDS 001	16 23 06.	+ 39 53 06.	51		SB GALAXY IN ARC 2199
ZC 1623.1+6201	16 23 06.	+ 62 01	540		CLUSTER OF GALAXIES
LB 00942	16 23 07.	+ 77 43 12.		16.9	FAINT BLUE STAR
LB 00943	16 23 08.	+ 73 13 42.		17.6	FAINT BLUE STAR
BC 4C26.48	16 23 10.4	+ 26 57 22.		18.1	QUASI-STELLAR OBJECT
ZC 1623.2+2404	16 23 12.	+ 24 04	810		CLUSTER OF GALAXIES
MCG+09-27-033	16 23 12.	+ 54 37	30	16.	GALAXY
REIN 6.121	16 23 13.46	+ 41 09 53.0			NEBULA
ARC 2188	16 23 16.	+ 33 52		17.1	RICH CLUSTER OF GALAXIES
RNGC 6138	16 23 16.	+ 41 08		15.0	GALAXY
REIN 6.122	16 23 16.27	+ 40 41 16.5			NEBULA
ZWG 168.008	16 23 18.	+ 30 27		15.3	GALAXY
RDS 004	16 23 18.	+ 39 24 00.	15		S0 GALAXY IN ARC 2199

OBJECT NAME	RIGHT ASCEN.	DECLINATION	DIAM.	MAGN.	TYPE OF OBJECT
ZWG 224.016	16 23 18.	+ 39 39		15.6	GALAXY
RDS 003	16 23 18.	+ 39 52 36.	41		S GALAXY IN ARC 2199
MCG+07-34-020	16 23 18.	+ 41 08	48	15.	GALAXY
REIN 6.123	16 23 21.88	+ 41 03 35.2			NEBULA
RNGC 6147	16 23 22.	+ 41 00		17.0	GALAXY
RNGC 6145	16 23 22.	+ 41 02		15.0	GALAXY
HOLM 747A	16 23 23.	+ 41 04	24	14.6	PART OF MULTIPLE GALAXY
MCG+03-42-008	16 23 24.	+ 16 33	18	14.9	GALAXY
ZWG 109.015	16 23 24.	+ 16 35		14.9	GALAXY
ZC 1623.4+2235	16 23 24.	+ 22 35	940		CLUSTER OF GALAXIES
ZWG 168.009	16 23 24.	+ 32 35		15.5	GALAXY
MCG+07-34-022	16 23 24.	+ 41 00 30.	12	17.	GALAXY
MCG+07-34-021	16 23 24.	+ 41 02	45	14.5	GALAXY
ZWG 224.017	16 23 24.	+ 41 04		15.1	GALAXY
1ZW 149	16 23 24.	+ 56 01			COMPACT GALAXY
ZWG 276.016	16 23 24.	+ 56 01		15.0	GALAXY
ZWG 276.017	16 23 24.	+ 56 06		15.7	GALAXY
7ZW 645	16 23 24.	+ 60 32			COMPACT GALAXY
REIN 6.124	16 23 24.68	+ 41 09 09.8			NEBULA
REIN 6.125	16 23 24.96	+ 41 02 26.1			NEBULA
SC 1621+1641.3	16 23 26.	+ 16 34 24.	12		NEBULA
HOLM 747B	16 23 26.	+ 41 03	18	14.7	PART OF MULTIPLE GALAXY
MCG+07-34-023	16 23 27.	+ 41 01	24	16.	GALAXY
REIN 6.126	16 23 27.41	+ 41 05 32.3			NEBULA
REIN 6.127	16 23 27.71	+ 41 08 14.0			NEBULA
RNGC 6141	16 23 28.	+ 41 01		16.0	GALAXY
REIN 6.128	16 23 29.74	+ 41 00 20.3			NEBULA
ZWG 052.014	16 23 30.	+ 04 21		15.7	GALAXY
ZC 1623.5+1408	16 23 30.	+ 14 08	5040		CLUSTER OF GALAXIES
ZWG 224.018	16 23 30.	+ 41 01		13.8	GALAXY
UGC 10379	16 23 30.	+ 41 01	90	13.8	GALAXY E
MCG+09-27-035	16 23 30.	+ 53 14	60	16.	GALAXY
1ZW 150	16 23 30.	+ 54 17			COMPACT GALAXY
MCG+09-27-034	16 23 30.	+ 56 08	15		GALAXY
LDN 1692	16 23 30.	- 23 50	300		DARK NEBULA
REIN 6.129	16 23 30.49	+ 40 39 46.2			NEBULA
REIN 6.130	16 23 32.09	+ 40 57 18.6			NEBULA
MCG+03-42-009	16 23 33.	+ 16 41	102	15.5	GALAXY
MCG+07-34-024	16 23 33.	+ 40 59	30	14.	GALAXY
SC 1621+1648.2	16 23 34.	+ 16 41 18.	36		NEBULA
RNGC 6146	16 23 34.	+ 40 59		14.0	GALAXY
ZWG 052.015	16 23 36.	+ 03 00		15.4	GALAXY
ZWG 109.016	16 23 36.	+ 16 41		15.5	GALAXY
UGC 10380	16 23 36.	+ 16 41	102	15.5	GALAXY Sb
ZC 1623.6+2820	16 23 36.	+ 28 20	2490		CLUSTER OF GALAXIES
MCG+08-30-010	16 23 36.	+ 50 25	36	15.	GALAXY
ZWG 298.035	16 23 36.	+ 57 48		15.6	GALAXY
DG 140	16 23 36.	- 24 17	120		REFLECTION NEBULA
SS 57	16 23 37.	- 24 17			DIFFUSE GALACTIC NEBULA
REIN 6.131	16 23 40.42	+ 41 00 19.5			NEBULA
RDS 005	16 23 42.	+ 39 40 36.	22		SB GALAXY IN ARC 2199
RDS 006	16 23 42.	+ 40 00 06.	55		S GALAXY IN ARC 2199
ZC 1623.7+4145	16 23 42.	+ 41 45	3090		CLUSTER OF GALAXIES
ZWG 251.015	16 23 42.	+ 50 25		15.6	GALAXY
MCG+09-27-036	16 23 42.	+ 56 45	24	16.	GALAXY
MCG+10-23-071	16 23 42.	+ 57 47	36	16.	GALAXY
ARC 2203	16 23 42.	+ 73 28		17.1	RICH CLUSTER OF GALAXIES
MCG+07-34-025	16 23 45.	+ 39 57	66	14.	GALAXY
REIN 6.133	16 23 45.58	+ 41 02 05.7			NEBULA
REIN 6.132	16 23 45.92	+ 40 42 02.7			NEBULA
REIN 6.134	16 23 47.06	+ 40 43 48.8			NEBULA
ZWG 196.061	16 23 48.	+ 36 05		15.7	GALAXY
ZWG 224.019	16 23 48.	+ 39 59		14.9	GALAXY
UGC 10381	16 23 48.	+ 39 59	78	14.9	GALAXY S0-a
MCG+10-23-073	16 23 48.	+ 58 02	36	16.	GALAXY
ZWG 298.036	16 23 48.	+ 58 03		15.7	GALAXY
MCG+10-23-072	16 23 48.	+ 58 11	48	15.	GALAXY
ZWG 298.037	16 23 48.	+ 58 13		15.6	GALAXY
LDN 1690	16 23 48.	- 23 56	540		DARK NEBULA
REIN 6.135	16 23 50.72	+ 40 40 38.2			NEBULA
REIN 6.136	16 23 52.32	+ 40 46 58.7			NEBULA
ZWG 080.029	16 23 54.	+ 13 17		15.3	GALAXY
MCG+02-42-004	16 23 54.	+ 13 17	48	15.3	GALAXY
SC 1621+1950.4	16 23 54.	+ 19 43 32.	12		NEBULA
ZWG 109.017	16 23 54.	+ 19 44		15.7	GALAXY
RDS 008	16 23 54.	+ 39 43 48.	30		E GALAXY IN ARC 2199
RDS 007	16 23 54.	+ 39 56 24.	22		SO GALAXY IN ARC 2199
PK331-03.1	16 23 54.	- 53 54 30.	5		PLANETARY NEBULA
REIN 6.137	16 23 54.53	+ 40 56 23.4			NEBULA
LBN 0020	16 24	- 09 30	2100		BRIGHT NEBULA
B 230	16 24	- 16 40	3600		DARK OBJECT
ZWG 109.018	16 24 00.	+ 19 40		15.6	GALAXY
MCG+07-34-026	16 24 00.	+ 39 40	42	15.	GALAXY
ZWG 224.020	16 24 00.	+ 39 43		15.4	GALAXY
MCG+12-15-061	16 24 00.	+ 72 11 30.	39	16.	GALAXY
MCG+14-08-004	16 24 00.	+ 81 09	30	16.	GALAXY
MCG+15-01-017	16 24 00.	+ 87 57	10	17.	GALAXY
LDN 1681	16 24 00.	- 24 40	1800		DARK NEBULA
ISS 1043	16 24 00.	- 44 59	78		STELLAR RING
OCL 0968	16 24 00.	- 49 02	1440	9.1	OPEN STAR CLUSTER
VHA 191	16 24 00.	- 49 02	360		OPEN STAR CLUSTER
REIN 6.138	16 24 00.20	+ 40 30 36.4			NEBULA
SC 1621+1632.9	16 24 02.	+ 16 26 02.	30		NEBULA
RNGC 6134	16 24 02.	- 49 04		9.0	OPEN CLUSTER
RNGC 6150B	16 24 03.	+ 40 33		15.0	GALAXY
ARC 2194	16 24 03.	+ 57 18		16.9	RICH CLUSTER OF GALAXIES
REIN 6.139A	16 24 03.21	+ 40 35 17.6			NEBULA
REIN 6.139B	16 24 03.23	+ 40 35 15.9			NEBULA
REIN 6.140	16 24 04.34	+ 41 21 40.2			NEBULA
1ZW 151	16 24 06.	+ 37 31			COMPACT GALAXY
RDS 009	16 24 06.	+ 39 35 12.	10		E GALAXY IN ARC 2199
MCG+07-34-028	16 24 06.	+ 40 26	24	15.	GALAXY
MCG+07-34-027	16 24 06.	+ 40 33 30.	27	15.5	GALAXY
HOLM 748B	16 24 06.	+ 40 34	18	15.2	PART OF MULTIPLE GALAXY
MCG+08-30-011	16 24 06.	+ 49 22	48	16.	GALAXY
MCG+09-27-037	16 24 06.	+ 51 46	36	16.	GALAXY
MCG+09-27-038	16 24 06.	+ 51 58	48	16.	GALAXY
MCG+10-23-074	16 24 06.	+ 57 47	30	16.	GALAXY
7ZW 646	16 24 06.	+ 68 19			COMPACT GALAXY
GCL 042	16 24 06.	- 25 56	666	10.85	GLOBULAR STAR CLUSTER
REIN 6.141	16 24 07.70	+ 40 27 26.9			NEBULA
REIN 6.142A	16 24 08.64	+ 40 36 02.0			NEBULA
REIN 6.142B	16 24 08.65	+ 40 36 02.0			NEBULA
RNGC 6150A	16 24 09.	+ 40 36		15.0	GALAXY
RNGC 6144	16 24 10.	- 25 56		11.0	GLOBULAR CLUSTER
HOLM 748A	16 24 11.	+ 40 35	24	14.7	PART OF MULTIPLE GALAXY
ZWG 168.010	16 24 12.	+ 32 52		15.4	GALAXY
ZC 1624.2+3403	16 24 12.	+ 34 03	3290		CLUSTER OF GALAXIES
RDS 011	16 24 12.	+ 39 33 36.	19		SO GALAXY IN ARC 2199
RDS 012	16 24 12.	+ 39 49 06.	18		SB GALAXY IN ARC 2199
RDS 010	16 24 12.	+ 40 01 54.	23		SO GALAXY IN ARC 2199
ZWG 224.021	16 24 12.	+ 40 28		15.3	GALAXY
MCG+07-34-029	16 24 12.	+ 40 34 30.	30	14.5	GALAXY
ZWG 224.022	16 24 12.	+ 40 36		14.9	GALAXY
ZWG 251.016	16 24 12.	+ 49 57		14.0	GALAXY
UGC 10382	16 24 12.	+ 49 57	132	14.0	GALAXY SBa
ZC 1624.2+7237	16 24 12.	+ 72 37	1010		CLUSTER OF GALAXIES
ZC 1624.2+7340	16 24 12.	+ 73 40	940		CLUSTER OF GALAXIES
RNGC 6154	16 24 13.	+ 49 57		14.0	GALAXY
ZWG 080.030	16 24 18.	+ 13 47		15.6	GALAXY
RDS 015	16 24 18.	+ 39 15 42.	10		SO GALAXY IN ARC 2199
RDS 014	16 24 18.	+ 39 21 12.	17		SO GALAXY IN ARC 2199
RDS 013	16 24 18.	+ 39 44 18.	18		SO GALAXY IN ARC 2199
ZWG 320.028	16 24 18.	+ 64 37		15.3	GALAXY
UGC 10383	16 24 18.	+ 64 37	96	15.3	GALAXY
MCG+11-20-014	16 24 18.	+ 64 38	78	15.	GALAXY
LB 03553	16 24 19.	- 23 58 36.		18.4	FAINT BLUE STAR
RNGC 6139	16 24 19.	- 38 44		10.5	GLOBULAR CLUSTER
REIN 6.144	16 24 22.72	+ 41 21 25.6			NEBULA
REIN 6.143B	16 24 23.56	+ 40 35 39.0			NEBULA
REIN 6.143A	16 24 23.67	+ 40 35 39.0			NEBULA
ZWG 052.016	16 24 24.	+ 03 20		15.5	GALAXY
ZWG 052.017	16 24 24.	+ 05 58		15.5	GALAXY
KARA.73B 0738	16 24 24.	+ 05 58	36	15.5	ISOLATED GALAXY S0
ZWG 080.031	16 24 24.	+ 11 41		15.0	GALAXY
UGC 10384	16 24 24.	+ 11 41	84	15.0	GALAXY S
MCG+02-42-005	16 24 24.	+ 11 41	72	15.0	GALAXY
RDS 021	16 24 24.	+ 39 27 54.	12		SO GALAXY IN ARC 2199
ARC 2190	16 24 24.	+ 43 49		17.7	RICH CLUSTER OF GALAXIES
MCG+07-34-030	16 24 24.	+ 44 03	30	15.5	GALAXY
ZWG 251.017	16 24 24.	+ 48 12		15.2	GALAXY
MCG+08-30-012	16 24 24.	+ 49 57 30.	120	14.	GALAXY
GCL 043	16 24 24.	- 38 43	156	10.4	GLOBULAR STAR CLUSTER
ZWG 052.018	16 24 30.	+ 03 17		15.6	GALAXY
ZWG 080.032	16 24 30.	+ 10 28		15.7	GALAXY
RDS 018	16 24 30.	+ 39 12 30.	10		SO GALAXY IN ARC 2199
RDS 022	16 24 30.	+ 39 25 48.	13		E GALAXY IN ARC 2199
RDS 019	16 24 30.	+ 39 31 12.	25		SO GALAXY IN ARC 2199
RDS 016	16 24 30.	+ 39 47 24.	21		SO GALAXY IN ARC 2199
7ZW 647	16 24 30.	+ 66 23			COMPACT GALAXY
OCL 0963	16 24 30.	- 51 24	486	10.	OPEN STAR CLUSTER
REIN 6.146	16 24 30.27	+ 41 02 33.4			NEBULA
REIN 6.145	16 24 30.93	+ 40 36 00.2			NEBULA
ZWG 052.019	16 24 36.	+ 07 40		15.2	GALAXY
ZWG 080.033	16 24 36.	+ 09 10		15.3	GALAXY
ZWG 196.062	16 24 36.	+ 35 47		15.0	GALAXY
MCG+06-36-043	16 24 36.	+ 35 47	36	15.	GALAXY
RDS 023	16 24 36.	+ 39 11 24.	13		E GALAXY IN ARC 2199
ZWG 224.023	16 24 36.	+ 40 05		15.2	GALAXY
RDS 020	16 24 36.	+ 40 06 18.	13		PEC GALAXY IN ARC 2199
LB 00944	16 24 36.	- 03 40 54.		16.4	FAINT BLUE STAR
LB 00945	16 24 37.	+ 40 46 06.		16.8	FAINT BLUE STAR
RNGC 6157	16 24 40.	+ 55 28		15.5	GALAXY
REIN 6.147	16 24 41.43	+ 41 01 21.5			NEBULA
ZWG 196.063	16 24 42.	+ 33 38		15.7	GALAXY
ZWG 196.064	16 24 42.	+ 35 10		15.7	GALAXY
RDS 026	16 24 42.	+ 39 18 06.	15		SO GALAXY IN ARC 2199
RDS 025	16 24 42.	+ 39 29 36.	14		S GALAXY IN ARC 2199
RDS 024	16 24 42.	+ 39 33 36.	11		SB GALAXY IN ARC 2199
ZWG 224.024	16 24 42.	+ 41 01		15.7	GALAXY
ZWG 251.018	16 24 42.	+ 48 28		13.0	GALAXY
RNGC 6155	16 24 42.	+ 48 28		13.0	GALAXY
UGC 10385	16 24 42.	+ 48 28	84	13.0	GALAXY S
ZWG 276.018	16 24 42.	+ 55 28		15.5	GALAXY
ZC 1624.7+5948	16 24 42.	+ 59 48	1140		CLUSTER OF GALAXIES
REIN 6.148	16 24 43.53	+ 41 22 21.1			NEBULA
MCG+07-34-031	16 24 45.	+ 41 00	42	15.	GALAXY
MCG+09-27-039	16 24 45.	+ 55 30	18	16.	GALAXY
LB 00314	16 24 46.	- 23 28 42.		16.1	FAINT BLUE STAR
ZWG 024.012	16 24 48.	+ 02 01		15.2	GALAXY
UGC 10386	16 24 48.	+ 02 01	84	15.2	GALAXY Sc
ZWG 052.020	16 24 48.	+ 03 11		15.1	GALAXY
MCG+01-42-003	16 24 48.	+ 03 11	48	15.1	GALAXY
ZWG 080.034	16 24 48.	+ 13 05		15.3	GALAXY
UGC 10387	16 24 48.	+ 13 05	60	15.3	GALAXY SBc
SC 1622+1635.7	16 24 48.	+ 16 28 53.	78		NEBULA
MCG+03-42-010	16 24 48.	+ 16 29	90	15.1	GALAXY
ZWG 109.019	16 24 48.	+ 16 30		15.1	GALAXY
UGC 10388	16 24 48.	+ 16 30	102	15.1	GALAXY Sa
ZWG 196.065	16 24 48.	+ 33 12		15.6	GALAXY
RDS 027	16 24 48.	+ 39 14 30.	32		S GALAXY IN ARC 2199
RDS 028	16 24 48.	+ 40 00 36.	10		E GALAXY IN ARC 2199
REIN 6.149	16 24 50.08	+ 41 01 02.6			NEBULA
ZWG 080.035	16 24 54.	+ 14 26		15.3	GALAXY
RDS 029	16 24 54.	+ 39 58 30.	24		S GALAXY IN ARC 2199
RDS 031	16 24 54.	+ 40 18 42.	18		S GALAXY IN ARC 2199
ZC 1624.9+4249	16 24 54.	+ 42 49	1340		CLUSTER OF GALAXIES
MCG+07-34-032	16 24 54.	+ 44 12 30.	30	15.5	GALAXY
RNGC 6148	16 24 57.	+ 24 11			GALAXY
ARC 2192	16 24 58.	+ 42 47		17.1	RICH CLUSTER OF GALAXIES
REIN 6.151	16 24 58.34	+ 41 12 26.6			NEBULA
REIN 6.150	16 24 59.15	+ 40 17 33.3			NEBULA
KHAV 249	16 25	- 24 25	2230		DARK NEBULA
KHAV 248	16 25	- 36 01	5860		DARK NEBULA
LB 09988	16 25	- 89 43		12.7	FAINT BLUE STAR
ZWG 080.036	16 25 00.	+ 14 12		15.4	GALAXY
ZWG 168.011	16 25 00.	+ 28 50		15.7	GALAXY
RDS 032	16 25 00.	+ 39 10 06.	26		S GALAXY IN ARC 2199
UGC 10389	16 25 00.	+ 39 14	66	16.0	GALAXY SB
RDS 030	16 25 00.	+ 39 44 48.	19		SB GALAXY IN ARC 2199
RDS 033	16 25 00.	+ 39 58 30.	33		S GALAXY IN ARC 2199
MCG+07-34-033	16 25 00.	+ 40 34	45	14.	GALAXY
ZWG 224.025	16 25 00.	+ 40 35		15.0	GALAXY
UGC 10390	16 25 00.	+ 44 06	66	16.0	GALAXY S
MCG+07-34-034	16 25 00.	+ 44 13	18	17.	GALAXY
MCG+08-30-013	16 25 00.	+ 48 28	78	13.	GALAXY
MCG+12-15-062	16 25 00.	+ 71 15	33	17.	GALAXY
LDN 0027	16 25 00.	- 15 00	7500		DARK NEBULA
LDN 1781	16 25 00.	- 17 00	4380		DARK NEBULA
REIN 6.152	16 25 02.82	+ 40 57 07.0			NEBULA
REIN 6.153	16 25 04.36	+ 40 47 40.2			NEBULA
ZWG 109.020	16 25 06.	+ 16 06		15.7	GALAXY
KARA.73B 0739	16 25 06.	+ 16 06	42	15.7	ISOLATED GALAXY S
ZWG 138.011	16 25 06.	+ 21 45		15.5	GALAXY
ZWG 138.012	16 25 06.	+ 22 08		15.7	GALAXY
ZWG 138.013	16 25 06.	+ 22 10		15.6	GALAXY

OBJECT NAME	RIGHT ASCEN.	DECLINATION	DIAM.	MAGN.	TYPE OF OBJECT
ZWG 196.066	16 25 06.	+ 35 35		15.6	GALAXY
3ZW 078	16 25 06.	+ 40 48			COMPACT GALAXY
MCG+08-30-014	16 25 06.	+ 45 43	48	16.	GALAXY
ZC 1625.1+6025	16 25 06.	+ 60 25	2020		CLUSTER OF GALAXIES
OCL 0969	16 25 06.	- 49 01	240		OPEN STAR CLUSTER
REIN 6.154	16 25 07.48	+ 40 55 07.9			NEBULA
SC 1622+1613.1	16 25 08.	+ 16 06 18.	30		NEBULA
ARC 2195	16 25 10.	+ 48 39		17.4	RICH CLUSTER OF GALAXIES
ZWG 052.021	16 25 12.	+ 05 12		15.7	GALAXY
MCG+01-42-004	16 25 12.	+ 05 12	30	15.7	GALAXY
ZWG 109.021	16 25 12.	+ 19 42		14.8	GALAXY
RNGC 6149	16 25 12.	+ 19 42		15.0	GALAXY
UGC 10391	16 25 12.	+ 19 42	78	14.8	GALAXY S0
MCG+03-42-011	16 25 12.	+ 19 42	48	14.8	GALAXY
SC 1623+1949.0	16 25 12.	+ 19 42 13.	12		NEBULA
ZWG 138.014	16 25 12.	+ 21 43		15.0	GALAXY
RDS 035	16 25 12.	+ 39 25 36.	30		SO GALAXY IN ARC 2199
RDS 034	16 25 12.	+ 39 29 12.	15		SB GALAXY IN ARC 2199
UGC 10392	16 25 12.	+ 45 42	60	16.0	GALAXY Sb
UGC 10393	16 25 12.	+ 66 24	60	16.5	GALAXY SBb-c
MCG+12-15-063	16 25 12.	+ 73 28	33	16.	GALAXY
REIN 6.155	16 25 12.58	+ 40 20 26.0			NEBULA
REIN 6.156A	16 25 13.06	+ 41 21 55.1			NEBULA
REIN 6.156B	16 25 13.06	+ 41 21 55.4			NEBULA
PK327-07.1	16 25 14.	- 58 35 53.	25		PLANETARY NEBULA
MCG+04-39-003	16 25 15.	+ 21 42	48	15.0	GALAXY
MCG+07-34-035	16 25 15.	+ 41 20	48	13.5	GALAXY
ARC 2209	16 25 17.	+ 73 41		17.9	RICH CLUSTER OF GALAXIES
REIN 6.157	16 25 17.90	+ 40 51 09.3			NEBULA
RDS 037	16 25 18.	+ 39 32 48.	14		SO GALAXY IN ARC 2199
RDS 038	16 25 18.	+ 39 38 30.	22		E GALAXY IN ARC 2199
ZWG 224.026	16 25 18.	+ 41 22		15.2	GALAXY
SN 1968T	16 25 18.	+ 41 22		18.0	SUPERNOVA
ZWG 339.004	16 25 18.	+ 70 19		15.5	GALAXY
ZWG 338.056	16 25 18.	+ 70 19		15.5	GALAXY
UGC 10395	16 25 18.	+ 70 19	78	15.5	GALAXY Sa/SBa
ZWG 024.013	16 25 18.	- 01 40		15.3	GALAXY
UGC 10394	16 25 18.	- 01 40	95	15.3	GALAXY Sc
TER 03	16 25 22.89	- 35 14 37.0	200		STAR CLUSTER
ZWG 080.037	16 25 24.	+ 13 33		15.3	GALAXY
RDS 036	16 25 24.	+ 39 03 48.	15		SO GALAXY IN ARC 2199
RDS 041	16 25 24.	+ 39 16 18.	16		S GALAXY IN ARC 2199
ZWG 224.027	16 25 24.	+ 39 38		15.3	GALAXY
RDS 039	16 25 24.	+ 39 42 00.	20		SO GALAXY IN ARC 2199
MCG+08-30-015	16 25 24.	+ 47 33	60	16.	GALAXY
MCG+12-15-064	16 25 24.	+ 70 18	51	16.	GALAXY
ZWG 339.005	16 25 24.	+ 73 26		15.2	GALAXY
ZWG 338.057	16 25 24.	+ 73 26		15.2	GALAXY
YM 05	16 25 26.	- 35 12	420		SYMMETRIC GALACTIC NEBULA
ZWG 052.022	16 25 30.	+ 08 36		15.7	GALAXY
PK047+42.1	16 25 30.	+ 28 01	174	13.7	PLANETARY NEBULA
ZWG 168.012	16 25 30.	+ 29 55		15.4	GALAXY
RDS 043	16 25 30.	+ 39 12 18.	37		S GALAXY IN ARC 2199
RDS 044	16 25 30.	+ 39 21 30.	23		SO GALAXY IN ARC 2199
RDS 040	16 25 30.	+ 39 25 54.	19		SO GALAXY IN ARC 2199
RDS 042	16 25 30.	+ 39 28 48.	41		SO GALAXY IN ARC 2199
ZC 1625.5+4006	16 25 30.	+ 40 06	10950		CLUSTER OF GALAXIES
MCG+12-15-065	16 25 30.	+ 71 14	51	16.	GALAXY
REIN 6.158	16 25 34.14	+ 40 21 22.5			NEBULA
ARP 066	16 25 35.	+ 51 39			PECULIAR GALAXY
FAI 1625+20	16 25 36.	+ 20 10			COMPACT GALAXY
ZC 1625.6+2600	16 25 36.	+ 26 00	870		CLUSTER OF GALAXIES
ZC 1625.6+3006	16 25 36.	+ 30 06	470		CLUSTER OF GALAXIES
OCL 0128	16 25 36.	+ 38 11	1620		OPEN STAR CLUSTER
RDS 045	16 25 36.	+ 38 58 30.	26		SB GALAXY IN ARC 2199
ZWG 224.028	16 25 36.	+ 39 13		14.8	GALAXY
ZC 1625.6+4217	16 25 36.	+ 42 17	1080		CLUSTER OF GALAXIES
ZWG 251.019	16 25 36.	+ 49 38		14.7	GALAXY
ZWG 276.019	16 25 36.	+ 51 41		14.7	GALAXY
UGC 10396	16 25 36.	+ 51 41	72	14.7	GALAXY Sb
ARC 2191	16 25 37.	+ 25 59		17.7	RICH CLUSTER OF GALAXIES
MCG+07-34-036	16 25 42.	+ 39 11	45	15.	GALAXY
MCG-01-42-001	16 25 42.	- 09 00	60	15.	GALAXY
MRSL 335-00/1	16 25 42.	- 49 11	5400		HII REGION
MCG+07-34-037	16 25 45.	+ 38 56	36	15.5	GALAXY
ARC 2196	16 25 45.	+ 41 36		16.9	RICH CLUSTER OF GALAXIES
MCG+08-30-016	16 25 45.	+ 49 40	36	16.	GALAXY
MCG+09-27-040	16 25 45.	+ 51 40	66	14.	GALAXY
ZC 1625.8+2153	16 25 48.	+ 21 53	1410		CLUSTER OF GALAXIES
ZWG 196.067	16 25 48.	+ 38 58		15.5	GALAXY
RDS 046	16 25 48.	+ 39 00 18.	15		S GALAXY IN ARC 2199
RDS 047	16 25 48.	+ 39 12 48.	12		PEC GALAXY IN ARC 2199
RDS 048	16 25 48.	+ 39 14 48.	12		E GALAXY IN ARC 2199
RDS 049	16 25 48.	+ 39 22 54.	36		SB GALAXY IN ARC 2199
MCG+07-34-038	16 25 48.	+ 42 46	18	15.	GALAXY
ZWG 224.029	16 25 48.	+ 42 47		15.2	GALAXY
RNGC 6159	16 25 48.	+ 42 47		15.0	GALAXY
UGC 10397	16 25 48.	+ 42 47	66	15.2	GALAXY E-SO
ZC 1625.8+5538	16 25 48.	+ 55 38	1140		CLUSTER OF GALAXIES
LDN 1696	16 25 48.	- 24 13	720		DARK NEBULA
REIN 6.159	16 25 49.69	+ 40 07 44.6			NEBULA
UGC 10398	16 25 54.	+ 17 45	60	16.0	GALAXY
ZWG 138.015	16 25 54.	+ 21 46		15.6	GALAXY
ZC 1625.9+3016	16 25 54.	+ 30 16	870		CLUSTER OF GALAXIES
RDS 051	16 25 54.	+ 39 04 12.	18		I GALAXY IN ARC 2199
RDS 052	16 25 54.	+ 39 28 36.	41		E GALAXY IN ARC 2199
RDS 054	16 25 54.	+ 39 28 42.	16		S GALAXY IN ARC 2199
RDS 053	16 25 54.	+ 39 29 24.	17		SO GALAXY IN ARC 2199
RDS 050	16 25 54.	+ 39 41 36.	19		E GALAXY IN ARC 2199
RDS 057	16 25 54.	+ 40 15 36.	15		S GALAXY IN ARC 2199
RDS 058	16 25 54.	+ 40 19 54.	15		SO GALAXY IN ARC 2199
RDS 055	16 25 54.	+ 40 21 00.	13		PEC GALAXY IN ARC 2199
ZWG 224.030	16 25 54.	+ 40 47		15.7	GALAXY
UGC 10399	16 25 54.	+ 48 14	72	16.0	GALAXY PECULR
MCG+08-30-017	16 25 54.	+ 48 15	60	15.	GALAXY
MCG+09-27-041	16 25 54.	+ 52 27	42	17.	GALAXY
HOLM 749A	16 25 55.	+ 39 30	18	14.7	PART OF MULTIPLE GALAXY
HOLM 749B	16 25 55.	+ 39 31	12	15.8	PART OF MULTIPLE GALAXY
ARC 2201	16 25 56.	+ 55 34		17.1	RICH CLUSTER OF GALAXIES
PK332-03.1	16 25 56.	- 53 16 42.	25		PLANETARY NEBULA
REIN 6.160	16 25 56.27	+ 40 46 48.1			NEBULA
HOLM 749C	16 25 57.	+ 39 30	24	15.6	PART OF MULTIPLE GALAXY
MCG+07-34-039	16 25 57.	+ 40 46	27	15.5	GALAXY
RNGC 6160	16 25 58.	+ 41 02		15.0	GALAXY
REIN 6.161	16 25 58.40	+ 41 01 41.6			NEBULA
ARC 2193	16 25 59.	+ 21 53		17.1	RICH CLUSTER OF GALAXIES
LBN 0022	16 26	- 09 25	1800		BRIGHT NEBULA
LBN 0021	16 26	- 09 30	720		BRIGHT NEBULA
LBN 1108	16 26	- 26 20	9000		BRIGHT NEBULA
LBN 1107	16 26	- 26 30	3000		BRIGHT NEBULA
LBN 1100	16 26	- 28 10	1200		BRIGHT NEBULA
ZWG 052.023	16 26 00.	+ 08 38		15.5	GALAXY
ZWG 196.068	16 26 00.	+ 36 15		15.5	GALAXY
MCG+06-36-044	16 26 00.	+ 36 16 30.	36	16.	GALAXY
1ZW 152	16 26 00.	+ 38 31			COMPACT GALAXY
ZWG 196.069	16 26 00.	+ 38 31		15.4	GALAXY
RDS 061	16 26 00.	+ 39 23 06.	22		E GALAXY IN ARC 2199
MCG+07-34-041	16 26 00.	+ 39 27 30.	18	15.	GALAXY
ZWG 224.031	16 26 00.	+ 39 30		15.5	GALAXY
RDS 056	16 26 00.	+ 39 39 00.	15		E GALAXY IN ARC 2199
RDS 060	16 26 00.	+ 40 16 42.	18		S GALAXY IN ARC 2199
RDS 059	16 26 00.	+ 40 19 36.	36		SB GALAXY IN ARC 2199
ZWG 224.032	16 26 00.	+ 41 02		14.8	GALAXY
UGC 10400	16 26 00.	+ 41 02	138	14.8	GALAXY E
MCG+07-34-040	16 26 00.	+ 42 43	18	16.	GALAXY
ZWG 224.033	16 26 00.	+ 42 45		15.5	GALAXY
UGC 10401	16 26 00.	+ 57 00	84	17.	GALAXY DWRF SP
LDN 0001	16 26 00.	- 16 00	900		DARK NEBULA
LDN 0001	16 26 00.	- 16 00	900		DARK NEBULA
LDN 1683	16 26 00.	- 25 00	6300		DARK NEBULA
REIN 6.162	16 26 00.84	+ 41 02 12.8			NEBULA
PK331-03.2	16 26 01.	- 54 03 17.	35		PLANETARY NEBULA
RNGC 6158	16 26 02.	+ 39 30		15.5	GALAXY
REIN 6.163	16 26 02.04	+ 41 02 46.3			NEBULA
REIN 6.164	16 26 02.26	+ 41 02 28.4			NEBULA
REIN 6.165	16 26 02.78	+ 40 49 01.7			NEBULA
HOLM 750B	16 26 03.	+ 40 23	12	15.4	PART OF MULTIPLE GALAXY
MCG+07-34-042	16 26 03.	+ 41 00 30.	21	14.	GALAXY
REIN 6.166	16 26 03.39	+ 41 03 35.1			NEBULA
VDB.66N 107	16 26 04.	- 26 20	14040		REFLECTION NEBULA
HOLM 750A	16 26 05.	+ 40 23	42	14.9	PART OF MULTIPLE GALAXY
REIN 6.167	16 26 05.48	+ 41 10 56.6			NEBULA
RDS 064	16 26 06.	+ 39 22 54.	11		E GALAXY IN ARC 2199
RDS 063	16 26 06.	+ 39 50 06.	8		S GALAXY IN ARC 2199
REIN 6.168	16 26 07.14	+ 40 24 25.7			NEBULA
REIN 6.169B	16 26 08.13	+ 41 03 04.5			NEBULA
REIN 6.169A	16 26 08.16	+ 41 03 04.5			NEBULA
REIN 6.172	16 26 09.94	+ 41 02 50.4			NEBULA
REIN 6.171	16 26 09.97	+ 40 57 58.5			NEBULA
REIN 6.170	16 26 10.02	+ 40 24 35.8			NEBULA
ZWG 052.024	16 26 12.	+ 03 56		15.7	GALAXY
ZC 1626.2+2045	16 26 12.	+ 20 45	7390		CLUSTER OF GALAXIES
MCG+07-34-043	16 26 12.	+ 39 20	30	15.	GALAXY
MCG+07-34-044	16 26 12.	+ 39 21	18	15.5	GALAXY
RDS 066	16 26 12.	+ 39 21 36.	15		SO GALAXY IN ARC 2199
ZWG 224.034	16 26 12.	+ 39 22		15.4	GALAXY
RDS 065	16 26 12.	+ 39 24 24.	22		SB GALAXY IN ARC 2199
RDS 062	16 26 12.	+ 39 39 42.	26		SO GALAXY IN ARC 2199
RDS 067	16 26 12.	+ 39 42 48.	14		SBO GALAXY IN ARC 2199
RNGC 6170	16 26 12.	+ 59 41			NON-EXISTENT OBJECT
MCG+07-34-045	16 26 15.	+ 39 23	24	15.5	GALAXY
ZWG 080.038	16 26 18.	+ 12 52		15.2	GALAXY
UGC 10402	16 26 18.	+ 12 52	78	15.2	GALAXY DBL SYS
MCG+02-42-006	16 26 18.	+ 12 52	72	15.2	GALAXY
RDS 068	16 26 18.	+ 39 08 12.	12		SO GALAXY IN ARC 2199
RDS 069	16 26 18.	+ 39 38 06.	12		E GALAXY IN ARC 2199
ZWG 276.020	16 26 18.	+ 54 47		15.3	GALAXY
KARA.73B 0740	16 26 18.	+ 54 47	36	15.3	ISOLATED GALAXY S
MCG+09-27-042	16 26 18.	+ 54 49	36	16.	GALAXY
REIN 6.173	16 26 18.14	+ 40 36 24.0			NEBULA
RNGC 6161	16 26 21.	+ 32 55		15.5	GALAXY
CED 132	16 26 22.	- 26 20	7560		DIFFUSE GALACTIC NEBULA
REIN 6.175	16 26 22.02	+ 40 38 59.3			NEBULA
ZWG 080.039	16 26 24.	+ 10 32		15.3	GALAXY
MCG+05-39-003	16 26 24.	+ 28 04	60	16.	GALAXY
ZWG 168.013	16 26 24.	+ 32 55		15.6	GALAXY
MCG+06-36-046	16 26 24.	+ 32 55	42	15.	GALAXY
MCG+06-36-047	16 26 24.	+ 32 57	48	14.5	GALAXY
MCG+06-36-045	16 26 24.	+ 33 16 30.	30	15.	GALAXY
ZWG 196.070	16 26 24.	+ 33 17		15.6	GALAXY
RDS 071	16 26 24.	+ 39 18 42.	18		S GALAXY IN ARC 2199
7ZW 648	16 26 24.	+ 79 33			COMPACT GALAXY
LB 03554	16 26 24.	- 22 32 06.		15.0	FAINT BLUE STAR
RNGC 6162	16 26 27.	+ 32 58		15.0	GALAXY
ARC 2197	16 26 28.	+ 41 01		13.9	RICH CLUSTER OF GALAXIES
ZWG 138.016	16 26 30.	+ 24 40		15.7	GALAXY
MCG+04-39-004	16 26 30.	+ 24 40	18	15.7	GALAXY
MCG+06-36-048	16 26 30.	+ 32 56 30.	15	15.	GALAXY
ZWG 168.014	16 26 30.	+ 32 58		15.2	GALAXY
UGC 10403	16 26 30.	+ 32 58	66	15.2	GALAXY S0
RDS 072	16 26 30.	+ 38 54 12.	17		E GALAXY IN ARC 2199
RDS 073	16 26 30.	+ 39 38 36.	14		S GALAXY IN ARC 2199
MCG+07-34-046	16 26 30.	+ 39 54	78	14.	GALAXY
ZWG 224.035	16 26 30.	+ 39 56		15.5	GALAXY
UGC 10404	16 26 30.	+ 39 56	90	15.5	GALAXY SB
RDS 070	16 26 30.	+ 39 56 30.	43		SB GALAXY IN ARC 2199
RDS 075	16 26 30.	+ 40 13 06.	18		SO GALAXY IN ARC 2199
RDS 074	16 26 30.	+ 40 24 42.	31		SB GALAXY IN ARC 2199
ZWG 224.036	16 26 30.	+ 40 25		15.5	GALAXY
ARC 2198	16 26 30.	+ 43 56		17.7	RICH CLUSTER OF GALAXIES
REIN 6.177	16 26 32.27	+ 41 12 51.4			NEBULA
REIN 6.176	16 26 32.78	+ 40 25 22.7			NEBULA
RNGC 6163	16 26 33.	+ 32 58		15.5	GALAXY
MCG+10-23-075	16 26 33.	+ 62 45 30.	48	15.	GALAXY
ZWG 080.040	16 26 36.	+ 13 04		15.7	GALAXY
SC 1624+1538.2	16 26 36.	+ 15 31 30.	12		NEBULA
ZWG 109.022	16 26 36.	+ 18 00		15.2	GALAXY
UGC 10405	16 26 36.	+ 18 00	102	15.2	GALAXY Sc
KARA.73B 0741	16 26 36.	+ 18 00	78	15.2	ISOLATED GALAXY S
MCG+04-39-005	16 26 36.	+ 21 34	30	15.3	GALAXY
ZWG 138.017	16 26 36.	+ 21 36		15.3	GALAXY
ZC 1626.6+2325	16 26 36.	+ 23 25	940		CLUSTER OF GALAXIES
ZWG 168.015	16 26 36.	+ 32 58		15.4	GALAXY
ZC 1626.6+3326	16 26 36.	+ 33 26	5240		CLUSTER OF GALAXIES
ZWG 196.071	16 26 36.	+ 38 55		15.1	GALAXY
RDS 076	16 26 36.	+ 39 40 30.	20		E GALAXY IN ARC 2199
RDS 077	16 26 36.	+ 39 42 48.	14		SO GALAXY IN ARC 2199
RDS 078	16 26 36.	+ 40 04 06.	16		E GALAXY IN ARC 2199
MCG+07-34-047	16 26 36.	+ 40 24	36	14.5	GALAXY
DG 141	16 26 36.	- 26 20	7560		REFLECTION NEBULA
REIN 6.178	16 26 36.10	+ 41 11 24.7			NEBULA
MCG+03-42-012	16 26 39.	+ 18 00	108	15.2	GALAXY
HOLM 751D	16 26 40.	+ 39 41	18	15.1	PART OF MULTIPLE GALAXY
ZWG 052.025	16 26 42.	+ 08 12		15.5	GALAXY
ZWG 052.026	16 26 42.	+ 08 14		15.3	GALAXY
ZWG 138.018	16 26 42.	+ 25 35		15.4	GALAXY

OBJECT NAME	RIGHT ASCEN.	DECLINATION	DIAM.	MAGN.	TYPE OF OBJECT
RDS 079	16 26 42.	+ 39 12 42.	12		SO GALAXY IN ARC 2199
RDS 081	16 26 42.	+ 39 12 48.	27		S GALAXY IN ARC 2199
RDS 091	16 26 42.	+ 39 30 48.	13		SO GALAXY IN ARC 2199
RDS 083	16 26 42.	+ 39 37 06.	29		SO GALAXY IN ARC 2199
MCG+07-34-048	16 26 42.	+ 39 40	24	15.	GALAXY
MCG+07-34-049	16 26 42.	+ 39 42	15	16.	GALAXY
RDS 084	16 26 42.	+ 39 49 30.	42		SO GALAXY IN ARC 2199
RDS 080	16 26 42.	+ 39 51 48.	23		SO GALAXY IN ARC 2199
ZWG 224.037	16 26 42.	+ 41 17		15.3	GALAXY
ZC 1626.7+4355	16 26 42.	+ 43 55	940		CLUSTER OF GALAXIES
MCG+10-23-076	16 26 42.	+ 59 40	18	15.	GALAXY
ZWG 298.038	16 26 42.	+ 59 41		14.8	GALAXY
RNGC 6176	16 26 42.	+ 59 41		15.0	GALAXY
ZWG 298.039	16 26 42.	+ 62 46		15.5	GALAXY
KARA.72 498A	16 26 42.	+ 62 46	48	15.5	PART OF DOUBLE GALAXY
MCG+13-12-003	16 26 42.	+ 78 54	33	15.	GALAXY
LDN 1674	16 26 42.	- 28 40	360		DARK NEBULA
REIN 6.179	16 26 44.87	+ 40 58 02.6			NEBULA
RNGC 6166C	16 26 45.	+ 39 41			GALAXY
REIN 6.180A	16 26 45.41	+ 41 16 41.3			NEBULA
REIN 6.180B	16 26 45.41	+ 41 16 42.7			NEBULA
LB 00946	16 26 47.	+ 77 35 30.		17.4	FAINT BLUE STAR
REIN 6.181	16 26 47.60	+ 41 16 11.6			NEBULA
ZWG 052.027	16 26 48.	+ 03 13		15.7	GALAXY
UGC 10406	16 26 48.	+ 03 13	102	15.7	GALAXY Sb-c
RDS 082	16 26 48.	+ 39 30 36.	17		SO GALAXY IN ARC 2199
RDS 087	16 26 48.	+ 39 33 12.	13		E GALAXY IN ARC 2199
MCG+07-34-050	16 26 48.	+ 39 37	18	16.	GALAXY
HOLM 751B	16 26 48.	+ 39 38	18	14.9	PART OF MULTIPLE GALAXY
RDS 086	16 26 48.	+ 39 38 18.	14		E GALAXY IN ARC 2199
RDS 085	16 26 48.	+ 39 39 00.	15		SO GALAXY IN ARC 2199
RDS 088	16 26 48.	+ 39 48 30.	27		S GALAXY IN ARC 2199
MCG+07-34-051	16 26 48.	+ 41 15	42	15.5	GALAXY
ZWG 224.038	16 26 48.	+ 41 20		14.3	GALAXY
UGC 10407	16 26 48.	+ 41 20	33	14.3	GALAXY CMPT GP
ZWG 276.021	16 26 48.	+ 52 30		15.5	GALAXY
UGC 10408	16 26 48.	+ 56 08	84	16.0	GALAXY Sb
REIN 6.182	16 26 48.37	+ 41 19 41.5			NEBULA
REIN 6.183	16 26 48.78	+ 41 17 16.2			NEBULA
IC 1220	16 26 49.	+ 08 33 57.			NONSTELLAR OBJECT
RNGC 6166A	16 26 51.	+ 39 38			GALAXY
ARC 2199	16 26 51.	+ 39 38		13.9	RICH CLUSTER OF GALAXIES
MCG+07-34-054	16 26 51.	+ 39 38 30.	18	16.	GALAXY
RNGC 6166D	16 26 51.	+ 39 40			GALAXY
MCG+07-34-052	16 26 51.	+ 41 14 30.	15	15.5	GALAXY
MCG+07-34-053	16 26 51.	+ 41 18	30	14.5	GALAXY
PK337+01.1	16 26 51.7	- 45 56 12.	5		PLANETARY NEBULA
REIN 6.184	16 26 52.83	+ 41 16 59.4			NEBULA
ZWG 052.028	16 26 54.	+ 08 30		15.6	GALAXY
MCG+07-34-057	16 26 54.	+ 39 09	24	15.	GALAXY
MCG+07-34-056	16 26 54.	+ 39 36 30.	24	16.	GALAXY
MCG+07-34-055	16 26 54.	+ 39 37 30.	15	16.	GALAXY
MCG+07-34-060	16 26 54.	+ 39 38 30.	36	13.5	GALAXY
MCG+07-34-060	16 26 54.	+ 39 38 30.	36	13.5	GALAXY
RDS 089	16 26 54.	+ 39 39 42.	56		D GALAXY IN ARC 2199
ZWG 224.039	16 26 54.	+ 39 40		13.9	GALAXY
UGC 10409	16 26 54.	+ 39 40	138	13.9	GALAXY E
MCG+07-34-061	16 26 54.	+ 40 11 30.	36	14.5	GALAXY
RDS 092	16 26 54.	+ 40 12 48.	32		SB GALAXY IN ARC 2199
HOLM 752A	16 26 54.	+ 40 13	24	14.3	PART OF MULTIPLE GALAXY
KARA.72 499A	16 26 54.	+ 40 14	48	14.9	PART OF DOUBLE GALAXY
MCG+07-34-059	16 26 54.	+ 41 11 30.	12	16.	GALAXY
MCG+07-34-058	16 26 54.	+ 41 15	24	16.	GALAXY
MCG+09-27-043	16 26 54.	+ 56 10	90	15.	GALAXY
DG 142	16 26 54.	- 25 03	1740		REFLECTION NEBULA
HOLM 751A	16 26 55.	+ 39 40	30	13.6	PART OF MULTIPLE GALAXY
HOLM 752C	16 26 55.	+ 40 13	12	15.2	PART OF MULTIPLE GALAXY
HOLM 751E	16 26 56.	+ 39 38	18	15.3	PART OF MULTIPLE GALAXY
CED 133	16 26 56.	- 25 03			DIFFUSE GALACTIC NEBULA
RNGC 6166	16 26 57.	+ 39 40		14.0	GALAXY
MCG+07-34-062	16 26 57.	+ 40 11 30.	30	15.	GALAXY
HOLM 752B	16 26 57.	+ 40 13	18	14.8	PART OF MULTIPLE GALAXY
MCG+07-34-063	16 26 57.	+ 41 13	12	16.	GALAXY
IC 4605	16 26 57.	- 25 02 42.	6000		DIFFUSE NEBULA
REIN 6.185	16 26 57.91	+ 40 13 56.6			NEBULA
REIN 6.187A	16 26 58.03	+ 41 16 21.4			NEBULA
REIN 6.187B	16 26 58.06	+ 41 16 21.4			NEBULA
REIN 6.186	16 26 58.20	+ 40 13 35.5			NEBULA
REIN 6.189	16 26 59.56	+ 41 13 12.0			NEBULA
B 043	16 27	- 19 40			DARK OBJECT
LBN 1002	16 27	- 27 30	2700		BRIGHT NEBULA
KHAV 252	16 27	- 28 07	2070		DARK NEBULA
KHAV 250	16 27	- 29 13	1950		DARK NEBULA
KHAV 251	16 27	- 33 25	4020		DARK NEBULA
ZWG 080.041	16 27 00.	+ 11 18		15.6	GALAXY
ZWG 109.023	16 27 00.	+ 20 28		15.6	GALAXY
MCG+04-39-006	16 27 00.	+ 25 48	42	15.3	GALAXY
ZWG 138.019	16 27 00.	+ 25 49		15.3	GALAXY
UGC 10410	16 27 00.	+ 25 49	66	15.3	GALAXY E
RDS 101	16 27 00.	+ 39 08 48.	27		SO GALAXY IN ARC 2199
MCG+07-34-067	16 27 00.	+ 39 31 30.	12	15.	GALAXY
RDS 094	16 27 00.	+ 39 32 18.	14		SO GALAXY IN ARC 2199
MCG+07-34-069	16 27 00.	+ 39 33	36	15.	GALAXY
RDS 095	16 27 00.	+ 39 33 48.	38		SO GALAXY IN ARC 2199
MCG+07-34-064	16 27 00.	+ 39 34	12	15.	GALAXY
MCG+07-34-065	16 27 00.	+ 39 34 30.	18	16.5	GALAXY
RDS 096	16 27 00.	+ 39 34 36.	35		SB GALAXY IN ARC 2199
ZWG 224.040	16 27 00.	+ 39 35		15.4	GALAXY
RDS 097	16 27 00.	+ 39 35 30.	15		SO GALAXY IN ARC 2199
MCG+07-34-066	16 27 00.	+ 39 36 30.	12	16.	GALAXY
MCG+07-34-070	16 27 00.	+ 39 37	9	16.	GALAXY
RDS 098	16 27 00.	+ 39 37 30.	17		E GALAXY IN ARC 2199
1ZW 153	16 27 00.	+ 39 38			COMPACT GALAXY
RDS 090	16 27 00.	+ 39 38 24.	16		SB GALAXY IN ARC 2199
RDS 099	16 27 00.	+ 39 43 30.	16		SO GALAXY IN ARC 2199
MCG+07-34-068	16 27 00.	+ 39 44	9	16.5	GALAXY
RDS 093	16 27 00.	+ 40 13 18.	24		SB GALAXY IN ARC 2199
ZWG 224.041	16 27 00.	+ 40 14		14.9	GALAXY
KARA.72 499B	16 27 00.	+ 40 14	36		PART OF DOUBLE GALAXY
ZWG 224.042	16 27 00.	+ 41 14		15.6	GALAXY
ZWG 224.043	16 27 00.	+ 41 15		15.5	GALAXY
ZC 1627.0+4708	16 27 00.	+ 47 08	3700		CLUSTER OF GALAXIES
ZWG 251.020	16 27 00.	+ 49 10		15.7	GALAXY
MCG+09-27-045	16 27 00.	+ 52 30	36	15.	GALAXY
MCG+09-27-044	16 27 00.	+ 54 20	42	16.	GALAXY
MCG+10-23-077	16 27 00.	+ 62 47	60	15.	GALAXY
ZC 1627.0+7101	16 27 00.	+ 71 01	610		CLUSTER OF GALAXIES
LDN 1752	16 27 00.	- 19 30	5400		DARK NEBULA

OBJECT NAME	RIGHT ASCEN.	DECLINATION	DIAM.	MAGN.	TYPE OF OBJECT
LDN 1673	16 27 00.	- 29 20	1080		DARK NEBULA
REIN 6.188	16 27 00.58	+ 40 13 59.3			NEBULA
REIN 6.190	16 27 01.72	+ 41 14 45.0			NEBULA
MCG+07-34-071	16 27 03.	+ 39 42	9	16.5	GALAXY
MCG+07-34-072	16 27 03.	+ 39 45	12	16.5	GALAXY
REIN 6.191	16 27 03.73	+ 41 17 42.7			NEBULA
RDS 108	16 27 06.	+ 39 25 36.	48		S GALAXY IN ARC 2199
RDS 105	16 27 06.	+ 39 39 36.	18		PEC GALAXY IN ARC 2199
MCG+07-34-074	16 27 06.	+ 39 42 30.	24	16.	GALAXY
MCG+07-34-073	16 27 06.	+ 39 44 30.	12	16.	GALAXY
RDS 103	16 27 06.	+ 39 51 00.	15		SO GALAXY IN ARC 2199
MCG+07-34-075	16 27 06.	+ 39 55	36	15.	GALAXY
RDS 102	16 27 06.	+ 39 55 30.	27		S GALAXY IN ARC 2199
ZWG 224.044	16 27 06.	+ 39 57		15.5	GALAXY
MCG+07-34-076	16 27 09.	+ 39 39	24	15.	GALAXY
VDB.66N 108	16 27 09.	- 25 00	1680		REFLECTION NEBULA
HOLM 751C	16 27 10.	+ 39 41	12	15.0	PART OF MULTIPLE GALAXY
ZWG 052.029	16 27 12.	+ 06 25		15.5	GALAXY
ZWG 052.030	16 27 12.	+ 08 34		15.0	GALAXY
MCG+01-42-005	16 27 12.	+ 08 34	48	15.0	GALAXY
ZWG 138.020	16 27 12.	+ 24 58		15.6	GALAXY
RDS 111	16 27 12.	+ 39 16 12.	29		SO GALAXY IN ARC 2199
RDS 100	16 27 12.	+ 39 17 30.	13		SO GALAXY IN ARC 2199
RDS 110	16 27 12.	+ 39 23 36.	15		SO GALAXY IN ARC 2199
ZWG 224.045	16 27 12.	+ 39 41		15.7	GALAXY
RDS 106	16 27 12.	+ 39 45 00.	18		SO GALAXY IN ARC 2199
RDS 109	16 27 12.	+ 39 47 30.	16		E GALAXY IN ARC 2199
RDS 104	16 27 12.	+ 39 48 30.	12		SBO GALAXY IN ARC 2199
RDS 107	16 27 12.	+ 39 54 30.	10		E GALAXY IN ARC 2199
3ZW 079	16 27 12.	+ 40 58			COMPACT GALAXY
ZCG 1627+42	16 27 12.	+ 42 54		17.4	COMPACT GALAXY
ZWG 299.001	16 27 12.	+ 62 47		15.3	GALAXY
ZWG 298.040	16 27 12.	+ 62 47		15.3	GALAXY
UGC 10411	16 27 12.	+ 62 47	84	15.3	GALAXY Sa-b
KARA.72 498B	16 27 12.	+ 62 47	60	15.3	PART OF DOUBLE GALAXY
REIN 6.192	16 27 13.94	+ 40 58 30.9			NEBULA
RNGC 6166B	16 27 15.	+ 39 40			GALAXY
MCG+07-34-078	16 27 15.	+ 40 57	30	16.	GALAXY
MCG+07-34-077	16 27 15.	+ 42 54	36	15.	GALAXY
MCG+03-42-013	16 27 18.	+ 15 45	54	15.0	GALAXY
ZWG 109.024	16 27 18.	+ 15 46		15.0	GALAXY
UGC 10412	16 27 18.	+ 15 46	60	15.0	GALAXY S0
SC 1625+1553.7	16 27 18.	+ 15 47 03.	18		NEBULA
MCG+04-39-007	16 27 18.	+ 21 24	120	15.2	GALAXY
ZWG 138.021	16 27 18.	+ 21 27		15.2	GALAXY
UGC 10413	16 27 18.	+ 21 27	150	15.2	GALAXY Sc
4ZW 062	16 27 18.	+ 39 23			COMPACT GALAXY
RDS 112	16 27 18.	+ 39 35 06.	17		SO GALAXY IN ARC 2199
3ZW 080	16 27 18.	+ 40 38			COMPACT GALAXY
3ZW 081	16 27 18.	+ 42 56			COMPACT GALAXY
ZWG 251.021	16 27 18.	+ 48 52		15.6	GALAXY
REIN 6.194	16 27 19.68	+ 41 16 15.3			NEBULA
REIN 6.193	16 27 20.70	+ 40 38 36.9			NEBULA
REIN 6.195	16 27 20.71	+ 41 23 33.9			NEBULA
ARC 2207	16 27 22.	+ 65 32		17.6	RICH CLUSTER OF GALAXIES
ZWG 052.031	16 27 24.	+ 08 45		15.4	GALAXY
UGC 10414	16 27 24.	+ 08 45	60	15.4	GALAXY SB:0-a
ZWG 080.042	16 27 24.	+ 11 57		15.6	GALAXY
ZC 1627.4+2815	16 27 24.	+ 28 15	1140		CLUSTER OF GALAXIES
ZWG 168.016	16 27 24.	+ 29 40		15.5	GALAXY
MCG+07-34-079	16 27 24.	+ 39 35	24	15.5	GALAXY
RDS 114	16 27 24.	+ 39 46 06.	31		SP GALAXY IN ARC 2199
RDS 113	16 27 24.	+ 39 54 30.	10		E GALAXY IN ARC 2199
MCG+07-34-080	16 27 24.	+ 41 22	42	14.5	GALAXY
ZWG 224.046	16 27 24.	+ 41 24		14.8	GALAXY
UGC 10415	16 27 24.	+ 41 24	60	14.8	GALAXY Sb/SBb
MCG+08-30-018	16 27 24.	+ 48 42 30.	48	16.	GALAXY
REIN 6.196	16 27 24.38	+ 41 15 39.0			NEBULA
ARC 2200	16 27 25.	+ 28 17		17.1	RICH CLUSTER OF GALAXIES
REIN 6.197	16 27 26.58	+ 41 18 21.7			NEBULA
REIN 6.198	16 27 27.55	+ 41 22 20.9			NEBULA
REIN 6.199	16 27 28.85	+ 40 59 40.1			NEBULA
RDS 115	16 27 30.	+ 39 29 54.	38		SO GALAXY IN ARC 2199
RDS 116	16 27 30.	+ 39 31 48.	27		SBO GALAXY IN ARC 2199
MCG+07-34-081	16 27 30.	+ 40 56	48	15.5	GALAXY
ZWG 251.022	16 27 30.	+ 48 50		15.5	GALAXY
ZC 1627.5+6543	16 27 30.	+ 65 43	2290		CLUSTER OF GALAXIES
LB 00315	16 27 30.	+ 73 59 30.		15.7	FAINT BLUE STAR
SC 1625+1851.9	16 27 31.	+ 18 45 16.	6		NEBULA
SN 1966C	16 27 31.	+ 41 06		16.8	SUPERNOVA
REIN 6.200A	16 27 31.12	+ 40 57 54.4			NEBULA
REIN 6.200B	16 27 31.20	+ 40 57 54.4			NEBULA
REIN 6.201	16 27 32.95	+ 41 15 35.9			NEBULA
REIN 6.202	16 27 33.69	+ 41 19 47.5			NEBULA
FAI 1627+17	16 27 36.	+ 17 12			COMPACT GALAXY
ZWG 138.022	16 27 36.	+ 23 27		15.7	GALAXY
RDS 126	16 27 36.	+ 38 54 48.	11		SB GALAXY IN ARC 2199
RDS 117	16 27 36.	+ 39 43 24.	6		PEC GALAXY IN ARC 2199
RDS 119	16 27 36.	+ 40 20 42.	25		S GALAXY IN ARC 2199
RDS 118	16 27 36.	+ 40 21 54.	40		SB GALAXY IN ARC 2199
3ZW 082	16 27 36.	+ 40 59			COMPACT GALAXY
ZWG 109.025	16 27 42.	+ 20 20		15.5	GALAXY
RDS 131	16 27 42.	+ 39 52 00.	28		SO GALAXY IN ARC 2199
RDS 120	16 27 42.	+ 39 56 06.	28		SO GALAXY IN ARC 2199
ZWG 224.047	16 27 42.	+ 40 59		15.5	GALAXY
ZWG 276.022	16 27 42.	+ 52 03		15.7	GALAXY
REIN 6.203	16 27 43.78	+ 40 58 57.9			NEBULA
MCG+07-34-082	16 27 45.	+ 40 57	30	15.5	GALAXY
REIN 6.204	16 27 46.37	+ 40 49 47.3			NEBULA
LB 00947	16 27 47.	- 02 28 06.		16.6	FAINT BLUE STAR
ZWG 138.023	16 27 48.	+ 24 32		15.5	GALAXY
MCG+04-39-008	16 27 48.	+ 24 32	24	15.5	GALAXY
RDS 123	16 27 48.	+ 40 00 18.	16		E GALAXY IN ARC 2199
RDS 122	16 27 48.	+ 40 04 00.	11		PEC GALAXY IN ARC 2199
LDN 1704	16 27 48.	- 23 40	360		DARK NEBULA
VHE 71	16 27 48.	- 38 17	96		REFLECTION NEBULA
REIN 6.205B	16 27 48.60	+ 40 48 12.0			NEBULA
REIN 6.205A	16 27 48.76	+ 40 48 12.0			NEBULA
REIN 6.206	16 27 50.69	+ 41 03 40.6			NEBULA
HN 1811	16 27 51.3	- 01 55 25.	18		NEBULA
ZWG 052.032	16 27 54.	+ 08 44		15.2	GALAXY
UGC 10416	16 27 54.	+ 08 44	66	15.2	GALAXY Sb/SBc
MCG+01-42-006	16 27 54.	+ 08 44	60	15.2	GALAXY
ZWG 138.024	16 27 54.	+ 23 27		15.6	GALAXY
RDS 124	16 27 54.	+ 40 19 42.	32		S GALAXY IN ARC 2199
MCG+09-27-046	16 27 54.	+ 52 01	24	16.	GALAXY
ARC 2202	16 27 56.	+ 48 56		17.1	RICH CLUSTER OF GALAXIES
MCG+04-39-009	16 27 57.	+ 21 55	42	15.4	GALAXY

OBJECT NAME	RIGHT ASCEN.	DECLINATION	DIAM.	MAGN.	TYPE OF OBJECT
RNGC 6153	16 27 59.	- 40 08		11.5	PLANETARY NEBULA
LBN 0012	16 28	- 11 40	2520		BRIGHT NEBULA
KHAV 254	16 28	- 22 07	3070		DARK NEBULA
LBN 1110	16 28	- 25 00	1320		BRIGHT NEBULA
CED 134	16 28	- 25 57			DIFFUSE GALACTIC NEBULA
LB 09904	16 28	- 80 55		14.0	FAINT BLUE STAR
IC 4606	16 28 .	- 25 57			NONSTELLAR OBJECT
ZWG 052.033	16 28 00.	+ 03 41		15.7	GALAXY
ZWG 138.025	16 28 00.	+ 21 58		15.4	GALAXY
ZC 1628.0+2438	16 28 00.	+ 24 38	3830		CLUSTER OF GALAXIES
MCG+04-39-010	16 28 00.	+ 24 42	78		GALAXY
ZC 1628.0+3346	16 28 00.	+ 33 46	1210		CLUSTER OF GALAXIES
RDS 133	16 28 00.	+ 39 19 12.	12		SO GALAXY IN ARC 2199
RDS 132	16 28 00.	+ 39 48 06.	32		SO GALAXY IN ARC 2199
RDS 128	16 28 00.	+ 40 07 06.	18		S GALAXY IN ARC 2199
RDS 125	16 28 00.	+ 40 20 48.	15		SBO GALAXY IN ARC 2199
UGC 10417	16 28 00.	+ 40 49	66	16.5	GALAXY Sb-c
ZWG 251.023	16 28 00.	+ 48 58		15.7	GALAXY
MCG+10-23-078	16 28 00.	+ 60 20	48	16.	GALAXY
7ZW 649	16 28 00.	+ 75 00			COMPACT GALAXY
UGC 10418	16 28 00.	+ 75 00	15	14.2	GALAXY COMPACT
MCG+14-08-005	16 28 00.	+ 82 30	39	16.	GALAXY
ZWG 367.010	16 28 00.	+ 82 31		15.7	GALAXY
LDN 1684	16 28 00.	- 25 20	1140		DARK NEBULA
RNGC 6173	16 28 04.	+ 40 55		14.0	GALAXY
HOLM 753A	16 28 04.	+ 40 55	24	14.2	PART OF MULTIPLE GALAXY
RNGC 6174	16 28 04.	+ 40 57		15.0	GALAXY
REIN 6.207	16 28 04.62	+ 40 55 09.4			NEBULA
ZWG 052.034	16 28 06.	+ 08 05		15.3	GALAXY
MCG+01-42-007	16 28 06.	+ 08 05	30	15.3	GALAXY
ZWG 138.026	16 28 06.	+ 23 10		15.6	GALAXY
ZC 1628.1+2644	16 28 06.	+ 26 44	540		CLUSTER OF GALAXIES
UGC 10419	16 28 06.	+ 27 50	90	17.	GALAXY DWARF
RDS 137	16 28 06.	+ 38 58 06.	13		SO GALAXY IN ARC 2199
RDS 134	16 28 06.	+ 39 51 36.	62		E GALAXY IN ARC 2199
ZWG 224.048	16 28 06.	+ 39 53		15.6	GALAXY
UGC 10420	16 28 06.	+ 39 53	108	15.6	GALAXY SBb
RDS 130	16 28 06.	+ 39 55 30.	21		SB GALAXY IN ARC 2199
RDS 129	16 28 06.	+ 40 05 30.	21		SO GALAXY IN ARC 2199
RDS 127	16 28 06.	+ 40 19 18.	28		SBO GALAXY IN ARC 2199
MCG+07-34-083	16 28 06.	+ 40 53 30.	30	13.5	GALAXY
ZWG 224.049	16 28 06.	+ 40 55		14.0	GALAXY
UGC 10421	16 28 06.	+ 40 55	132	14.0	GALAXY E
MCG+07-34-085	16 28 06.	+ 40 57	24	15.5	GALAXY
4ZW 063	16 28 06.	+ 40 59			COMPACT GALAXY
MCG+07-34-084	16 28 06.	+ 41 51 30.	30	16.	GALAXY
REIN 6.208	16 28 06.32	+ 40 58 56.0			NEBULA
REIN 6.209	16 28 06.87	+ 40 57 00.1			NEBULA
REIN 6.210	16 28 06.99	+ 40 58 50.9			NEBULA
MCG+07-34-086	16 28 09.	+ 39 50 30.	96	14.	GALAXY
IC 4607	16 28 11.	+ 24 40 09.			NONSTELLAR OBJECT
HOLM 753B	16 28 11.	+ 40 55	18	15.1	PART OF MULTIPLE GALAXY
REIN 6.211	16 28 11.23	+ 40 54 46.9			NEBULA
REIN 6.212	16 28 11.85	+ 40 55 56.9			NEBULA
ZWG 052.035	16 28 12.	+ 03 37		15.5	GALAXY
ZWG 052.036	16 28 12.	+ 06 08		15.5	GALAXY
ZWG 138.027	16 28 12.	+ 24 50		15.4	GALAXY
RDS 136	16 28 12.	+ 39 37 30.	24		SO GALAXY IN ARC 2199
RDS 135	16 28 12.	+ 40 02 24.	22		E GALAXY IN ARC 2199
3ZW 083	16 28 12.	+ 40 57			COMPACT GALAXY
REIN 6.213	16 28 13.79	+ 40 44 20.5			NEBULA
PK341+05.1	16 28 15.	- 40 08	28	11.5	PLANETARY NEBULA
RNGC 6175	16 28 16.	+ 40 45		15.0	GALAXY
REIN 6.214	16 28 17.11	+ 40 44 14.2			NEBULA
MCG+07-34-087	16 28 18.	+ 40 42 30.	72	14.	GALAXY
ZWG 224.050	16 28 18.	+ 40 45		15.0	GALAXY
UGC 10422	16 28 18.	+ 40 45	72	15.0	GALAXY DBL SYS
MCG+07-34-088	16 28 18.	+ 41 41	78	16.	GALAXY
UGC 10423	16 28 18.	+ 41 43	66	16.0	GALAXY
MCG+08-30-019	16 28 18.	+ 50 17	36	16.	GALAXY
MCG+09-27-047	16 28 18.	+ 55 37	24	15.	GALAXY
ZC 1628.3+5840	16 28 18.	+ 58 40	1810		CLUSTER OF GALAXIES
ZWG 052.037	16 28 24.	+ 04 11		14.9	GALAXY
MCG+01-42-008	16 28 24.	+ 04 11	36	14.9	GALAXY
ZWG 138.028	16 28 24.	+ 25 17		15.5	GALAXY
RDS 141	16 28 24.	+ 39 09 30.	10		S GALAXY IN ARC 2199
RDS 139	16 28 24.	+ 39 23 54.	17		SBO GALAXY IN ARC 2199
RDS 138	16 28 24.	+ 40 02 36.	16		SB GALAXY IN ARC 2199
1ZW 154	16 28 24.	+ 52 00			COMPACT GALAXY
ZWG 276.023	16 28 24.	+ 55 35		15.7	GALAXY
MCG+11-20-015	16 28 24.	+ 64 18	96	15.	GALAXY
LB 03555	16 28 24.	- 25 50 06.		19.2	FAINT BLUE STAR
RNGC 6182	16 28 28.	+ 55 37		14.5	GALAXY
ZC 1628.5+0035	16 28 30.	+ 00 35	1610		CLUSTER OF GALAXIES
MCG+03-42-014	16 28 30.	+ 20 28	42	15.1	GALAXY
ZC 1628.5+3540	16 28 30.	+ 35 40	1480		CLUSTER OF GALAXIES
RDS 140	16 28 30.	+ 38 58 00.	17		PEC GALAXY IN ARC 2199
RDS 143	16 28 30.	+ 39 58 36.	42		SO GALAXY IN ARC 2199
ZWG 276.024	16 28 30.	+ 55 37		14.6	GALAXY
UGC 10424	16 28 30.	+ 55 37	114	14.6	GALAXY Sa
ZWG 320.029	16 28 30.	+ 64 19		15.4	GALAXY
UGC 10425	16 28 30.	+ 64 19	108	15.4	GALAXY Sb
MIL 44	16 28 30.	- 47 16 00.			SUPERNOVA REMNANT
SC 1626+1628.0	16 28 35.	+ 16 21 26.	18		NEBULA
ZWG 052.038	16 28 36.	+ 05 16		15.2	GALAXY
ZWG 109.026	16 28 36.	+ 16 21		15.5	GALAXY
UGC 10426	16 28 36.	+ 16 21	66	15.5	GALAXY S
MCG+03-42-015	16 28 36.	+ 16 21	54	15.5	GALAXY
ZWG 109.027	16 28 36.	+ 20 27		15.1	GALAXY
ZWG 168.017	16 28 36.	+ 27 35		15.5	GALAXY
RDS 142	16 28 36.	+ 39 23 48.	17		SO GALAXY IN ARC 2199
RDS 144	16 28 36.	+ 39 53 48.	13		PEC GALAXY IN ARC 2199
ZWG 224.051	16 28 36.	+ 40 43		15.6	GALAXY
MCG+09-27-048	16 28 36.	+ 55 39	48	14.	GALAXY
ZWG 339.006	16 28 36.	+ 69 49		15.0	GALAXY
MCG-05-39-001	16 28 36.	- 27 58	24	16.	GALAXY
SS 58	16 28 37.	- 24 19			DIFFUSE GALACTIC NEBULA
REIN 6.215	16 28 37.21	+ 40 42 22.4			NEBULA
MCG+04-39-011	16 28 39.	+ 26 11	36	15.6	GALAXY
MCG+07-34-089	16 28 39.	+ 40 41	27	15.	GALAXY
REIN 6.216A	16 28 39.34	+ 41 12 30.7			NEBULA
REIN 6.216B	16 28 39.34	+ 41 12 34.0			NEBULA
ARC 2208	16 28 40.	+ 58 38		17.1	RICH CLUSTER OF GALAXIES
REIN 6.217	16 28 40.45	+ 40 49 05.0			NEBULA
REIN 6.218	16 28 41.77	+ 41 01 54.5			NEBULA
FAI 1628+19	16 28 42.	+ 19 04			COMPACT GALAXY
ZWG 138.029	16 28 42.	+ 26 11		15.6	GALAXY
ZWG 168.018	16 28 42.	+ 30 36		15.7	GALAXY
MCG+06-36-049	16 28 42.	+ 35 09 30.	90	13.	GALAXY
RDS 151	16 28 42.	+ 39 18 30.	34		E GALAXY IN ARC 2199
RDS 147	16 28 42.	+ 39 28 54.	28		SO GALAXY IN ARC 2199
RDS 146	16 28 42.	+ 39 43 18.	43		S GALAXY IN ARC 2199
RDS 145	16 28 42.	+ 39 55 30.	15		S GALAXY IN ARC 2199
ZWG 224.052	16 28 42.	+ 40 47		15.7	GALAXY
MCG+07-34-090	16 28 42.	+ 41 00	30	14.5	GALAXY
ZWG 224.053	16 28 42.	+ 41 02		15.7	GALAXY
MCG+07-34-091	16 28 42.	+ 41 11	72	15.	GALAXY
ZWG 224.054	16 28 42.	+ 41 13		15.4	GALAXY
UGC 10427	16 28 42.	+ 41 13	90	15.4	GALAXY SBc
DG 143	16 28 42.	- 24 19	300		REFLECTION NEBULA
REIN 6.219	16 28 46.39	+ 40 49 55.3			NEBULA
RNGC 6177	16 28 47.	+ 35 10		15.0	GALAXY
ZWG 138.030	16 28 48.	+ 24 40		15.7	GALAXY
ZWG 168.019	16 28 48.	+ 29 31		15.4	GALAXY
ZWG 196.072	16 28 48.	+ 35 10		14.8	GALAXY
UGC 10428	16 28 48.	+ 35 10	102	14.8	GALAXY SBb
ZWG 224.055	16 28 48.	+ 39 44		15.6	GALAXY
RDS 149	16 28 48.	+ 39 55 48.	32		S GALAXY IN ARC 2199
RDS 148	16 28 48.	+ 40 15 12.	25		SO GALAXY IN ARC 2199
MCG+07-34-092	16 28 48.	+ 40 48	54	15.	GALAXY
ZWG 224.056	16 28 48.	+ 40 50		15.7	GALAXY
MCG+07-34-093	16 28 48.	+ 41 34	66	14.5	GALAXY
ZWG 298.041	16 28 48.	+ 57 19		15.7	GALAXY
OCL 0961	16 28 48.	- 52 31	1800	8.6	OPEN STAR CLUSTER
REIN 6.220A	16 28 48.15	+ 40 54 59.1			NEBULA
REIN 6.220B	16 28 48.31	+ 40 54 59.1			NEBULA
SC 1626+1701.5	16 28 49.	+ 16 54 57.	12		NEBULA
REIN 6.221	16 28 49.09	+ 40 35 26.4			NEBULA
RNGC 6152	16 28 51.	- 52 31		8.0	OPEN CLUSTER
RNGC 6180	16 28 52.	+ 40 40		15.0	GALAXY
RNGC 6179	16 28 53.	+ 35 13		15.5	GALAXY
REIN 6.222	16 28 53.12	+ 40 38 44.8			NEBULA
ZWG 196.073	16 28 54.	+ 35 13		15.7	GALAXY
RDS 153	16 28 54.	+ 39 43 06.	39		SO GALAXY IN ARC 2199
MCG+07-34-094	16 28 54.	+ 39 54	48	15.	GALAXY
ZWG 224.057	16 28 54.	+ 39 57		15.3	GALAXY
UGC 10429	16 28 54.	+ 39 57	66	15.3	GALAXY
MCG+07-34-095	16 28 54.	+ 40 37	21	15.	GALAXY
ZWG 224.058	16 28 54.	+ 40 40		15.2	GALAXY
ZWG 224.059	16 28 54.	+ 41 36		15.4	GALAXY
UGC 10430	16 28 54.	+ 41 36	66	15.4	GALAXY SBb
ZC 1628.9+5815	16 28 54.	+ 58 15	1080		CLUSTER OF GALAXIES
ZWG 024.014	16 28 54.	- 01 56		15.6	GALAXY
KARA.73B 0742	16 28 54.	- 01 56	30	15.6	ISOLATED GALAXY E
REIN 6.223	16 28 54.46	+ 41 13 13.4			NEBULA
MCG+07-34-096	16 28 57.	+ 41 40	18	16.	GALAXY
BC 4C36.28	16 28 57.7	+ 36 19 36.		17.	QUASI-STELLAR OBJECT
SHB 331	16 28 58.	+ 36 19 36.		17.	QUASI-STELLAR OBJECT
SC 1626+1727.1	16 28 59.	+ 17 20 34.	18		NEBULA
REIN 6.224B	16 28 59.36	+ 41 18 57.1			NEBULA
REIN 6.224A	16 28 59.36	+ 41 18 57.4			NEBULA
KHAV 255	16 29	- 19 24	4320		DARK NEBULA
KHAV 253	16 29	- 36 31	2370		DARK NEBULA
ZWG 138.031	16 29 00.	+ 23 56		15.5	GALAXY
ZC 1629.0+3436	16 29 00.	+ 34 36	1610		CLUSTER OF GALAXIES
MCG+07-34-099	16 29 00.	+ 39 16	30	16.5	GALAXY
RDS 152	16 29 00.	+ 39 17 48.	53		SO GALAXY IN ARC 2199
RDS 154	16 29 00.	+ 40 04 06.	10		E GALAXY IN ARC 2199
RDS 150	16 29 00.	+ 40 17 48.	26		E GALAXY IN ARC 2199
MCG+07-34-097	16 29 00.	+ 40 36 30.	24	15.5	GALAXY
MCG+07-34-098	16 29 00.	+ 41 17 30.	72	16.	GALAXY
UGC 10432	16 29 00.	+ 41 19	90	16.0	GALAXY Sb
ZC 1629.0+4128	16 29 00.	+ 41 28	2080		CLUSTER OF GALAXIES
ZC 1629.0+5516	16 29 00.	+ 55 16	2820		CLUSTER OF GALAXIES
ZC 1629.0+6045	16 29 00.	+ 60 45	1410		CLUSTER OF GALAXIES
ZWG 339.007	16 29 00.	+ 74 03		15.4	GALAXY
ZWG 338.058	16 29 00.	+ 74 03		15.4	GALAXY
UGC 10431	16 29 00.	+ 81 55	60	16.5	GALAXY Sc
LDN 1757	16 29 00.	- 19 30	960		DARK NEBULA
LDN 1689	16 29 00.	- 24 59	300		DARK NEBULA
REIN 6.225A	16 29 01.66	+ 40 38 09.0			NEBULA
REIN 6.225B	16 29 01.79	+ 40 38 09.0			NEBULA
ZWG 052.039	16 29 06.	+ 05 10		15.5	GALAXY
MCG+03-42-016	16 29 06.	+ 20 17 30.	78	14.7	GALAXY
ZC 1629.1+2140	16 29 06.	+ 21 40	3490		CLUSTER OF GALAXIES
MCG+07-34-100	16 29 06.	+ 39 17	60	15.	GALAXY
ZWG 224.060	16 29 06.	+ 39 20		15.7	GALAXY
UGC 10433	16 29 06.	+ 39 20	90	15.7	GALAXY S
RDS 156	16 29 06.	+ 39 56 36.	34		SO GALAXY IN ARC 2199
RDS 155	16 29 06.	+ 40 04 48.	40		S GALAXY IN ARC 2199
3ZW 089	16 29 06.	+ 40 54			COMPACT GALAXY
MCG+07-34-101	16 29 06.	+ 44 37	30	16.	GALAXY
ZWG 355.010	16 29 06.	+ 75 59		15.6	GALAXY
MIL 46	16 29 06.	- 46 29 30.	660		SUPERNOVA REMNANT
REIN 6.226	16 29 08.83	+ 40 39 33.4			NEBULA
HOLM 754B	16 29 10.	+ 39 57	12	14.7	PART OF MULTIPLE GALAXY
ZWG 109.028	16 29 12.	+ 20 17		14.7	GALAXY
UGC 10434	16 29 12.	+ 20 17	96	14.7	GALAXY Sc-IRR
MCG+04-39-012	16 29 12.	+ 22 47	48	15.3	GALAXY
ZWG 138.032	16 29 12.	+ 22 48		15.3	GALAXY
UGC 10435	16 29 12.	+ 22 48	66	15.3	GALAXY SBb
KARA.73B 0743	16 29 12.	+ 22 48	60	15.3	ISOLATED GALAXY S
HOLM 754A	16 29 12.	+ 39 56	18	14.7	PART OF MULTIPLE GALAXY
RDS 161	16 29 12.	+ 39 57 00.	42		SB GALAXY IN ARC 2199
ZWG 224.061	16 29 12.	+ 39 58		15.5	GALAXY
ZC 1629.2+4328	16 29 12.	+ 43 28	1080		CLUSTER OF GALAXIES
ZWG 224.062	16 29 12.	+ 44 38		15.6	GALAXY
RNGC 6168	16 29 13.	+ 20 18		14.5	GALAXY
MCG+07-34-102	16 29 15.	+ 39 55	36	15.	GALAXY
ZWG 080.043	16 29 18.	+ 13 44		15.6	GALAXY
ZWG 080.044	16 29 18.	+ 13 58		15.6	GALAXY
RDS 157	16 29 18.	+ 39 18 00.	16		S GALAXY IN ARC 2199
RDS 158	16 29 18.	+ 39 39 30.	56		SO GALAXY IN ARC 2199
RDS 159	16 29 18.	+ 39 53 24.	42		SO GALAXY IN ARC 2199
ZWG 224.063	16 29 18.	+ 41 02		15.6	GALAXY
REIN 6.227	16 29 18.80	+ 41 02 12.5			NEBULA
REIN 6.228	16 29 23.37	+ 40 54 16.9			NEBULA
REIN 6.229B	16 29 23.99	+ 41 15 42.5			NEBULA
REIN 6.229A	16 29 23.99	+ 41 15 43.0			NEBULA
RDS 160	16 29 24.	+ 39 32 18.	67		S GALAXY IN ARC 2199
MCG+07-34-104	16 29 24.	+ 39 51 30.	54	14.5	GALAXY
ZWG 224.064	16 29 24.	+ 39 54		14.8	GALAXY
MCG+07-34-105	16 29 24.	+ 39 54 30.	42	15.	GALAXY
ZWG 224.065	16 29 24.	+ 39 57		15.6	GALAXY
3ZW 085	16 29 24.	+ 41 01			COMPACT GALAXY
MCG+07-34-103	16 29 24.	+ 41 14	72	14.	GALAXY

OBJECT NAME	RIGHT ASCEN.	DECLINATION	DIAM.	MAGN.	TYPE OF OBJECT
ZWG 224.066	16 29 24.	+ 41 16		14.8	GALAXY
UGC 10436	16 29 24.	+ 41 16	84	14.8	GALAXY Sc
ARC 2206	16 29 25.	+ 43 27		17.7	RICH CLUSTER OF GALAXIES
MCG+03-42-017	16 29 27.	+ 20 51 30.	42	15.6	GALAXY
REIN 6.230	16 29 29.72	+ 40 56 23.2			NEBULA
ZWG 109.029	16 29 30.	+ 18 45		15.7	GALAXY
MCG+03-42-018	16 29 30.	+ 18 46	27	15.7	GALAXY
ZWG 109.030	16 29 30.	+ 20 55		15.6	GALAXY
ZWG 138.033	16 29 30.	+ 22 25		15.2	GALAXY
RDS 162	16 29 30.	+ 39 13 42.	24		E GALAXY IN ARC 2199
RDS 163	16 29 30.	+ 39 15 06.	39		SO GALAXY IN ARC 2199
RDS 168	16 29 30.	+ 39 50 24.	28		E GALAXY IN ARC 2199
RDS 165	16 29 30.	+ 39 53 00.	19		E GALAXY IN ARC 2199
RDS 166	16 29 30.	+ 39 58 12.	18		S GALAXY IN ARC 2199
RDS 167	16 29 30.	+ 40 13 42.	22		E GALAXY IN ARC 2199
3ZW 086	16 29 30.	+ 40 37			COMPACT GALAXY
MCG+07-34-106	16 29 30.	+ 43 26	132	14.5	GALAXY
RNGC 6172	16 29 30.	- 01 23			NON-EXISTENT OBJECT
LB 03556	16 29 30.	- 22 50 18.		17.8	FAINT BLUE STAR
SC 1627+1850.7	16 29 32.	+ 18 44 12.	12		NEBULA
LB 03557	16 29 32.	- 24 16 24.		16.2	FAINT BLUE STAR
ZWG 052.040	16 29 36.	+ 07 52		15.6	GALAXY
ZC 1629.6+2111	16 29 36.	+ 21 11	810		CLUSTER OF GALAXIES
ZWG 138.034	16 29 36.	+ 21 45		15.5	GALAXY
ZC 1629.6+2535	16 29 36.	+ 25 35	2020		CLUSTER OF GALAXIES
MCG+07-34-107	16 29 36.	+ 39 13	24	16.	GALAXY
ZWG 224.067	16 29 36.	+ 39 16		15.5	GALAXY
RDS 169	16 29 36.	+ 39 53 12.	21		SBO GALAXY IN ARC 2199
ZWG 224.068	16 29 36.	+ 43 27		15.4	GALAXY
UGC 10437	16 29 36.	+ 43 27	132	15.4	GALAXY S
KARA.73B 0744	16 29 36.	+ 43 27	120	15.4	ISOLATED GALAXY S
MCG+07-34-108	16 29 39.	+ 39 14	45	16.	GALAXY
ZWG 024.015	16 29 42.	+ 02 25		15.7	GALAXY
RDS 171	16 29 42.	+ 40 06 06.	18		SB GALAXY IN ARC 2199
ZC 1629.7+5027	16 29 42.	+ 50 27	3700		CLUSTER OF GALAXIES
ZWG 320.030	16 29 42.	+ 67 30		15.4	GALAXY
KARA.73B 0745	16 29 42.	+ 67 30	42	15.4	ISOLATED GALAXY S
RNGC 6171	16 29 42.	- 12 56		10.0	GLOBULAR CLUSTER
GCL 044	16 29 42.	- 12 57	768	10.10	GLOBULAR STAR CLUSTER
REIN 6.231	16 29 44.40	+ 40 37 44.5			NEBULA
ZWG 052.041	16 29 48.	+ 03 57		15.4	GALAXY
KARA.73B 0746	16 29 48.	+ 03 57	24	15.4	ISOLATED GALAXY E
ZC 1629.8+1028	16 29 48.	+ 10 28	2020		CLUSTER OF GALAXIES
ZWG 080.045	16 29 48.	+ 13 44		15.4	GALAXY
ZWG 224.069	16 29 48.	+ 40 00		15.6	GALAXY
ZWG 251.024	16 29 48.	+ 50 08		15.5	GALAXY
MCG+11-20-016	16 29 48.	+ 67 30	42	15.	GALAXY
MCG+03-42-019	16 29 51.	+ 20 49	48	15.	GALAXY
RNGC 6184	16 29 52.	+ 40 41		15.0	GALAXY
REIN 6.232	16 29 53.92	+ 40 40 15.0			NEBULA
TON-N 0258	16 29 54.	+ 24 55		16.0	BLUE STAR
RDS 170	16 29 54.	+ 40 00 06.	31		SO GALAXY IN ARC 2199
ZWG 224.070	16 29 54.	+ 40 41		15.1	GALAXY
MCG+08-30-020	16 29 54.	+ 46 25	36	15.	GALAXY
MCG+07-34-109	16 29 57.	+ 40 39	30	15.	GALAXY
LBN 0032	16 30	- 08 30	2520		BRIGHT NEBULA
LBN 0024	16 30	- 10 00	21600		BRIGHT NEBULA
LBN 0016	16 30	- 11 10	5400		BRIGHT NEBULA
KHAV 256	16 30	- 23 48	3070		DARK NEBULA
LBN 1103	16 30	- 28 00	12600		BRIGHT NEBULA
ZWG 224.071	16 30 00.	+ 41 37		15.6	GALAXY
UGC 10438	16 30 00.	+ 81 32	60	16.0	GALAXY S0-a
LDN 1770	16 30 00.	- 18 30	3660		DARK NEBULA
LDN 1709	16 30 00.	- 23 40	1260		DARK NEBULA
CED 135A	16 30 00.	- 48 00			DIFFUSE GALACTIC NEBULA
ZWG 109.031	16 30 06.	+ 19 56		12.7	GALAXY
UGC 10439	16 30 06.	+ 19 56	168	12.7	GALAXY Sc
MCG+03-42-020	16 30 06.	+ 19 56	150	12.7	GALAXY
SN 1926B	16 30 06.	+ 19 57		14.8	SUPERNOVA
ZWG 138.035	16 30 06.	+ 21 29		15.4	GALAXY
ZWG 168.020	16 30 06.	+ 32 10		15.7	GALAXY
KARA.73B 0747	16 30 06.	+ 32 10	24	15.7	ISOLATED GALAXY S
RDS 172	16 30 06.	+ 39 11 54.	25		SO GALAXY IN ARC 2199
MCG+07-34-110	16 30 06.	+ 41 34	36	15.	GALAXY
ZC 1630.1+4227	16 30 06.	+ 42 27	3830		CLUSTER OF GALAXIES
ZC 1630.1+4728	16 30 06.	+ 47 28	1280		CLUSTER OF GALAXIES
ZC 1630.1+7435	16 30 06.	+ 74 35	670		CLUSTER OF GALAXIES
MCG+04-39-013	16 30 09.	+ 21 26	39	15.4	GALAXY
PK336-00.1	16 30 09.5	- 48 00 23.	420		PLANETARY NEBULA
WRAY 19.46	16 30 10.2	- 48 00 36.		7.5	STAR-NEBULA ASSOCIATION
SC 1627+1854.9	16 30 11.	+ 18 48 27.	6		NEBULA
RNGC 6181	16 30 12.	+ 19 56		13.0	GALAXY
ARC 2205	16 30 13.	+ 13 00		16.5	RICH CLUSTER OF GALAXIES
YM 02	16 30 14.	- 48 10	192		SYMMETRIC GALACTIC NEBULA
RNGC 6165	16 30 15.	- 48 00			PLANETARY NEBULA
RNGC 6164	16 30 15.	- 48 00			PLANETARY NEBULA
ARC 2204	16 30 16.	+ 05 42		17.1	RICH CLUSTER OF GALAXIES
MCG+07-34-111	16 30 18.	+ 42 47 30.	27	15.5	GALAXY
CED 135B	16 30 18.	- 48 02			DIFFUSE GALACTIC NEBULA
1ZW 155	16 30 24.	+ 38 05			COMPACT GALAXY
3ZW 087	16 30 24.	+ 42 18			COMPACT GALAXY
MCG+08-30-021	16 30 24.	+ 50 21	36	16.	GALAXY
RNGC 6156	16 30 24.	- 60 30			GALAXY
LB 00948	16 30 29.	+ 74 30 42.		17.0	FAINT BLUE STAR
PK335-01.1	16 30 29.	- 49 14 55.	25		PLANETARY NEBULA
ZC 1630.5+0535	16 30 30.	+ 05 35	1610		CLUSTER OF GALAXIES
UGC 10440	16 30 30.	+ 19 32	72	16.0	GALAXY Sc
MCG+03-42-021	16 30 30.	+ 19 32	66	16.	GALAXY
ZWG 138.036	16 30 30.	+ 26 28		15.7	GALAXY
ZWG 196.074	16 30 30.	+ 37 20		15.7	GALAXY
MCG+13-12-004	16 30 30.	+ 78 59	30	15.	GALAXY
OCL 0984	16 30 30.	- 43 57	420	6.6	OPEN STAR CLUSTER
RNGC 6169	16 30 31.	- 43 57		5.5	OPEN CLUSTER
REIN 6.233	16 30 31.92	+ 40 42 39.2			NEBULA
ZWG 080.046	16 30 36.	+ 11 50		15.2	GALAXY
ZC 1630.6+1253	16 30 36.	+ 12 53	1950		CLUSTER OF GALAXIES
ZC 1630.6+2710	16 30 36.	+ 27 10	1140		CLUSTER OF GALAXIES
MCG+05-39-004	16 30 36.	+ 27 33	36	17.	GALAXY
RDS 173	16 30 36.	+ 39 36 30.	24		S GALAXY IN ARC 2199
MCG+10-23-079	16 30 36.	+ 57 48	24	15.	GALAXY
ZWG 299.002	16 30 36.	+ 57 49		15.1	GALAXY
ZWG 298.042	16 30 36.	+ 57 49		15.1	GALAXY
MCG+11-20-017	16 30 36.	+ 64 41	42	16.	GALAXY
ZWG 339.008	16 30 36.	+ 74 08		15.4	GALAXY
ZWG 338.059	16 30 36.	+ 74 08		15.4	GALAXY
ZWG 024.016	16 30 36.	- 00 07		15.3	GALAXY
UGC 10441	16 30 36.	- 00 07	60	15.3	GALAXY Sc
OCL 0971	16 30 36.	- 49 30	1680	6.9	OPEN STAR CLUSTER

OBJECT NAME	RIGHT ASCEN.	DECLINATION	DIAM.	MAGN.	TYPE OF OBJECT
RNGC 6167	16 30 37.	- 49 30		6.5	OPEN CLUSTER
HN 1812	16 30 37.7	- 00 06 38.	36		NEBULA
RNGC 6187	16 30 38.	+ 57 49		15.0	GALAXY
REIN 6.234	16 30 39.22	+ 40 40 46.7			NEBULA
VHA 192	16 30 42.	- 49 30	360		OPEN STAR CLUSTER
PK346+08.1	16 30 47.	- 34 59 22.	10		PLANETARY NEBULA
ZWG 299.003	16 30 48.	+ 59 45		13.3	GALAXY
ZWG 298.043	16 30 48.	+ 59 45		13.3	GALAXY
RNGC 6189	16 30 48.	+ 59 45		13.5	GALAXY
UGC 10442	16 30 48.	+ 59 45	108	13.3	GALAXY Sc
MCG+10-23-081	16 30 48.	+ 59 45	120	13.	GALAXY
MCG+10-23-080	16 30 48.	+ 62 13	48	16.	GALAXY
ZC 1630.8+6430	16 30 48.	+ 64 30	2290		CLUSTER OF GALAXIES
IC 4609	16 30 52.	+ 22 53 46.			NONSTELLAR OBJECT
MCG+04-39-014	16 30 54.	+ 22 52 30.	42	15.4	GALAXY
ZWG 138.037	16 30 54.	+ 22 54		15.4	GALAXY
KHAV 257	16 31	- 12 12	6230		DARK NEBULA
LBN 0007	16 31	- 15 40	1800		BRIGHT NEBULA
KHAV 258	16 31	- 22 18	3070		DARK NEBULA
ZWG 196.075	16 31 00.	+ 37 27		15.2	GALAXY
MCG+06-36-050	16 31 00.	+ 37 28	48	15.	GALAXY
MCG+08-30-022	16 31 00.	+ 49 57	36	16.	GALAXY
MCG+08-30-023	16 31 00.	+ 50 18	36	16.	GALAXY
MCG+09-27-049	16 31 00.	+ 52 07 30.	24	16.	GALAXY
ZC 1631.0+7216	16 31 00.	+ 72 16	940		CLUSTER OF GALAXIES
ZC 1631.1+2031	16 31 06.	+ 20 31	1340		CLUSTER OF GALAXIES
RNGC 6190	16 31 10.	+ 58 33		13.0	GALAXY
RNGC 6191	16 31 10.	+ 58 54			NON-EXISTENT OBJECT
1ZW 156	16 31 12.	+ 35 01			COMPACT GALAXY
ZWG 299.004	16 31 12.	+ 58 33		13.2	GALAXY
ZWG 298.044	16 31 12.	+ 58 33		13.2	GALAXY
UGC 10443	16 31 12.	+ 58 33	102	13.2	GALAXY Sc
MCG+10-23-082	16 31 12.	+ 58 33	78	14.	GALAXY
MCG+12-16-001	16 31 12.	+ 74 30	51	16.	GALAXY
ZWG 196.076	16 31 18.	+ 35 01		15.5	GALAXY
MCG+06-36-051	16 31 18.	+ 35 01	27	15.	GALAXY
RNGC 6185	16 31 23.	+ 35 26		14.5	GALAXY
ZWG 052.042	16 31 24.	+ 07 19		15.4	GALAXY
ZWG 196.077	16 31 24.	+ 35 26		14.5	GALAXY
UGC 10444	16 31 24.	+ 35 26	72	14.5	GALAXY
MCG+06-36-052	16 31 24.	+ 35 26	72	14.	GALAXY
MCG+09-27-050	16 31 24.	+ 52 07	30	15.	GALAXY
ZWG 339.009	16 31 24.	+ 74 28		15.7	GALAXY
OCL 0965	16 31 24.	- 50 52	390	10.5	OPEN STAR CLUSTER
PK332-04.1	16 31 24.	- 53 44	5		PLANETARY NEBULA
ZC 1631.5+3613	16 31 30.	+ 36 13	670		CLUSTER OF GALAXIES
MCG+13-12-005	16 31 30.	+ 80 29	102	15.	GALAXY
LDN 0083	16 31 30.	- 14 10	960		DARK NEBULA
ZC 1631.6+2439	16 31 36.	+ 24 39	940		CLUSTER OF GALAXIES
ZWG 251.025	16 31 36.	+ 50 30		14.9	GALAXY
ZC 1631.6+5243	16 31 36.	+ 52 43	1010		CLUSTER OF GALAXIES
LDN 0043	16 31 36.	- 15 44	1020		DARK NEBULA
VHA 193	16 31 36.	- 50 52	360		OPEN STAR CLUSTER
SC 1629+1940.7	16 31 37.	+ 19 34 21.	6		NEBULA
SC 1629+1940.8	16 31 38.	+ 19 34 27.	12		NEBULA
HN 1814	16 31 38.3	+ 01 56 08.	18		NEBULA
HN 1813	16 31 38.8	- 02 34 04.	18		NEBULA
MCG+08-30-024	16 31 42.	+ 50 30	60	15.	GALAXY
OCL 0974	16 31 42.	- 48 12	204	12.	OPEN STAR CLUSTER
LB 03558	16 31 43.	- 23 55 36.		18.1	FAINT BLUE STAR
ZWG 052.043	16 31 48.	+ 08 25		15.7	GALAXY
ZC 1631.8+2556	16 31 48.	+ 25 56	1750		CLUSTER OF GALAXIES
TON-N 0817	16 31 48.	+ 26 38		15.0	BLUE STAR
MCG+05-39-005	16 31 48.	+ 29 04 30.	168	14.2	GALAXY
ZWG 168.021	16 31 48.	+ 29 06		14.2	GALAXY
UGC 10445	16 31 48.	+ 29 06	174	14.2	GALAXY Sc
KARA.73B 0748	16 31 48.	+ 29 06	156	14.2	ISOLATED GALAXY S
MCG+06-36-053	16 31 48.	+ 33 25	30	15.	GALAXY
ZWG 276.025	16 31 48.	+ 53 18		15.3	GALAXY
ZC 1631.8+5605	16 31 48.	+ 56 05	1140		CLUSTER OF GALAXIES
7ZW 650	16 31 48.	+ 64 17			COMPACT GALAXY
UGC 10446	16 31 48.	+ 79 36	108	16.0	GALAXY Sb-c
MCG+03-42-022	16 31 54.	+ 19 45	36	15.5	GALAXY
ZWG 196.078	16 31 54.	+ 33 25		15.1	GALAXY
ZWG 251.026	16 31 54.	+ 50 30		15.2	GALAXY
ZC 1631.9+6021	16 31 54.	+ 60 21	2620		CLUSTER OF GALAXIES
ZWG 339.010	16 31 54.	+ 69 50		15.7	GALAXY
IC 4610	16 31 55.	+ 39 21 56.			NONSTELLAR OBJECT
IC 4611	16 31 58.	+ 39 26 27.			NONSTELLAR OBJECT
ZWG 024.017	16 32 00.	+ 02 06		15.3	GALAXY
SC 1629+1950.7	16 32 00.	+ 19 44 22.	18		NEBULA
ZWG 109.032	16 32 00.	+ 19 45		15.5	GALAXY
ZC 1632.0+2508	16 32 00.	+ 25 08	2220		CLUSTER OF GALAXIES
ZWG 196.079	16 32 00.	+ 35 59		15.7	GALAXY
KARA.73B 0749	16 32 00.	+ 35 59	36	15.7	ISOLATED GALAXY S
MCG+07-34-112	16 32 00.	+ 39 15	27	16.	GALAXY
ZC 1632.0+4105	16 32 00.	+ 41 05	1680		CLUSTER OF GALAXIES
MCG+08-30-025	16 32 00.	+ 50 30	48	15.	GALAXY
ZWG 251.027	16 32 00.	+ 50 31		15.4	GALAXY
MCG+08-30-026	16 32 00.	+ 50 31	36	15.	GALAXY
MCG+10-23-083	16 32 00.	+ 62 47	48	15.	GALAXY
MCG+11-20-018	16 32 00.	+ 66 10	42	16.	GALAXY
MCG+13-12-006	16 32 00.	+ 79 37	102	16.	GALAXY
MCG+13-12-006	16 32 00.	+ 79 37	102	16.	GALAXY
ZWG 355.011	16 32 00.	+ 80 28		15.5	GALAXY
UGC 10447	16 32 00.	+ 80 28	156	15.5	GALAXY Sb
LDN 1739	16 32 00.	- 21 55	1200		DARK NEBULA
MCG+10-23-085	16 32 03.	+ 60 10	54	15.	GALAXY
ARC 2212	16 32 04.	+ 49 23		16.9	RICH CLUSTER OF GALAXIES
IC 4612	16 32 05.	+ 39 22 16.			NONSTELLAR OBJECT
MCG+07-34-113	16 32 06.	+ 39 20	48	14.5	GALAXY
1ZW 157	16 32 06.	+ 39 22			COMPACT GALAXY
ZWG 224.072	16 32 06.	+ 39 22		14.6	GALAXY
MCG+10-23-084	16 32 06.	+ 59 49	15	16.	GALAXY
DG 144	16 32 06.	- 15 50	60		REFLECTION NEBULA
RNGC 6178	16 32 06.	- 45 31		7.0	OPEN CLUSTER
OCL 0980	16 32 06.	- 45 32	1740	7.4	OPEN STAR CLUSTER
SC 1629+1712.2	16 32 07.	+ 17 05 53.	12		NEBULA
SS 59	16 32 09.	- 15 49			DIFFUSE GALACTIC NEBULA
MIL 45	16 32 10.	- 47 30	540		SUPERNOVA REMNANT
ZWG 138.038	16 32 12.	+ 21 39		14.2	GALAXY
UGC 10448	16 32 12.	+ 21 39	108	14.2	GALAXY SBa
ZC 1632.2+2428	16 32 12.	+ 24 28	540		CLUSTER OF GALAXIES
ZWG 299.005	16 32 12.	+ 62 46		15.5	GALAXY
ZWG 298.045	16 32 12.	+ 62 46		15.5	GALAXY
UGC 10449	16 32 12.	+ 62 46	90	15.5	GALAXY
KARA.73B 0750	16 32 12.	+ 62 46	66	15.5	ISOLATED GALAXY S
RNGC 6186	16 32 13.	+ 21 39		14.0	GALAXY
ARC 2210	16 32 16.	+ 05 36		17.1	RICH CLUSTER OF GALAXIES

OBJECT NAME	RIGHT ASCEN.	DECLINATION	DIAM.	MAGN.	TYPE OF OBJECT
ZWG 052.044	16 32 18.	+ 06 50		15.7	GALAXY
ZWG 052.045	16 32 18.	+ 07 34		15.5	GALAXY
ZWG 052.046	16 32 18.	+ 08 46		15.4	GALAXY
MCG+04-39-015	16 32 18.	+ 21 36	84	14.2	GALAXY
ZC 1632.3+2416	16 32 18.	+ 24 16	670		CLUSTER OF GALAXIES
UGC 10450	16 32 18.	+ 36 17	66	16.5	GALAXY Sc
ZWG 196.080	16 32 18.	+ 37 26		15.6	GALAXY
MCG+08-30-027	16 32 18.	+ 50 29	42	15.	GALAXY
ZC 1632.3+5213	16 32 18.	+ 52 13	1080		CLUSTER OF GALAXIES
MCG+10-23-086	16 32 18.	+ 61 05	24	16.	GALAXY
RNGC 6151	16 32 18.	- 73 09			UNVERIFIED SOUTHERN OBJECT
BC 4C39.46	16 32 19.7	+ 39 06 16.		18.	QUASI-STELLAR OBJECT
SHB 332	16 32 20.	+ 39 06 16.		18.	QUASI-STELLAR OBJECT
ABC 2211	16 32 21.	+ 41 03		17.4	RICH CLUSTER OF GALAXIES
UGC 10451	16 32 24.	+ 30 04	60	17.	GALAXY
ZWG 224.073	16 32 24.	+ 41 27		15.7	GALAXY
ZWG 251.028	16 32 24.	+ 48 05		15.6	GALAXY
MCG+08-30-028	16 32 24.	+ 49 56	42	16.	GALAXY
UGC 10452	16 32 24.	+ 67 52	72	16.5	GALAXY Sb-c
DG 145	16 32 24.	- 15 46	60		REFLECTION NEBULA
SS 60	16 32 27.	- 15 45			DIFFUSE GALACTIC NEBULA
SC 1630+1718.8	16 32 29.	+ 17 12 30.	12		NEBULA
ZC 1632.5+1649	16 32 30.	+ 16 49	1480		CLUSTER OF GALAXIES
ZC 1632.5+2205	16 32 30.	+ 22 05	1340		CLUSTER OF GALAXIES
ZC 1632.5+2240	16 32 30.	+ 22 40	940		CLUSTER OF GALAXIES
LDN 0145	16 32 30.	- 12 10	1500		DARK NEBULA
ZWG 052.047	16 32 42.	+ 07 13		15.6	GALAXY
ZWG 080.047	16 32 42.	+ 13 13		15.5	GALAXY
MCG+08-30-029	16 32 42.	+ 45 55	36	16.	GALAXY
ZC 1632.8+1450	16 32 48.	+ 14 50	340		CLUSTER OF GALAXIES
ZWG 109.033	16 32 48.	+ 20 41		15.6	GALAXY
UGC 10453	16 32 48.	+ 20 41	90	15.6	GALAXY Sc
MCG+04-39-016	16 32 48.	+ 23 17	24	15.4	GALAXY
ZWG 138.039	16 32 48.	+ 23 19		15.4	GALAXY
ZWG 276.026	16 32 48.	+ 56 13		15.0	GALAXY
UGC 10454	16 32 48.	+ 56 13	78	15.0	GALAXY Sa-b
ZWG 052.048	16 32 54.	+ 08 28		15.5	GALAXY
ZWG 138.040	16 32 54.	+ 25 46		15.6	GALAXY
UGC 10455	16 32 54.	+ 25 46	66	15.6	GALAXY SBb
ZWG 138.041	16 32 54.	+ 26 21		15.6	GALAXY
MCG+09-27-051	16 32 54.	+ 56 16	60	14.	GALAXY
PK342+05.1	16 32 58.	- 39 45 33.	10		PLANETARY NEBULA
LBN 0035	16 33	- 08 55	900		BRIGHT NEBULA
KHAV 261	16 33	- 20 42	2370		DARK NEBULA
KHAV 259	16 33	- 21 48	1950		DARK NEBULA
ZC 1633.0+0614	16 33 00.	+ 06 14	3160		CLUSTER OF GALAXIES
ZWG 276.027	16 33 00.	+ 53 03		15.1	GALAXY
UGC 10456	16 33 00.	+ 53 03	96	15.1	GALAXY Sb
LDN 0238	16 33 00.	- 08 40	900		DARK NEBULA
LDN 0098	16 33 00.	- 13 51	420		DARK NEBULA
ZWG 024.018	16 33 06.	+ 00 25		14.9	GALAXY
MCG+00-42-005	16 33 06.	+ 00 25	36	14.9	GALAXY
ZWG 109.034	16 33 06.	+ 17 52		15.5	GALAXY
MCG+08-30-030	16 33 06.	+ 46 29	78	14.	GALAXY
MCG+09-27-052	16 33 06.	+ 53 02 30.	96	14.	GALAXY
MCG+10-24-001	16 33 06.	+ 61 40	9	17.	GALAXY
HN 1815	16 33 07.0	+ 00 26 14.	12		NEBULA
ZC 1633.2+2641	16 33 12.	+ 26 41	2420		CLUSTER OF GALAXIES
ZC 1633.2+2835	16 33 12.	+ 28 35	3020		CLUSTER OF GALAXIES
1ZW 158	16 33 12.	+ 40 18			COMPACT GALAXY
MCG+08-30-031	16 33 12.	+ 45 22	60	15.	GALAXY
ZWG 251.029	16 33 12.	+ 45 25		15.3	GALAXY
UGC 10457	16 33 12.	+ 45 25	60	15.3	GALAXY Sc
ZWG 251.030	16 33 12.	+ 46 30		14.7	GALAXY
UGC 10458	16 33 12.	+ 46 30	78	14.7	GALAXY Sc
KARA.72 500A	16 33 12.	+ 46 30	72	14.7	PART OF DOUBLE GALAXY
ZWG 276.028	16 33 12.	+ 55 49		15.7	GALAXY
MCG-01-42-002	16 33 12.	- 05 00	84	15.	GALAXY
LDN 1740	16 33 12.	- 22 07	1140		DARK NEBULA
MCG+03-42-023	16 33 15.	+ 19 07 30.	27	16.	GALAXY
IC 1221	16 33 17.	+ 46 30 50.			NONSTELLAR OBJECT
FAI 1633+20	16 33 18.	+ 20 17			COMPACT GALAXY
LB 03559	16 33 18.	- 24 00 18.		18.7	FAINT BLUE STAR
MRSL 336-00/1	16 33 18.	- 48 27	7500		HII REGION
SC 1631+1848.9	16 33 23.	+ 18 42 40.	6		NEBULA
SC 1631+1913.0	16 33 23.	+ 19 06 46.	12		NEBULA
ZC 1633.4+4256	16 33 24.	+ 42 56	1080		CLUSTER OF GALAXIES
7ZW 651	16 33 24.	+ 64 00			COMPACT GALAXY
LB 00949	16 33 24.	+ 75 19 18.		17.4	FAINT BLUE STAR
LB 03560	16 33 26.	- 24 53 42.		18.9	FAINT BLUE STAR
ZWG 138.042	16 33 30.	+ 25 11		15.6	GALAXY
ZC 1633.5+3106	16 33 30.	+ 31 06	870		CLUSTER OF GALAXIES
ZWG 168.022	16 33 30.	+ 31 07		15.7	GALAXY
UGC 10459	16 33 30.	+ 41 06	96	16.0	GALAXY Sc
ZC 1633.5+4916	16 33 30.	+ 49 16	3700		CLUSTER OF GALAXIES
UGC 10460	16 33 30.	+ 57 26	72	16.0	GALAXY Sc
SHB 333	16 33 30.6	+ 38 14 10.		17.	QUASI-STELLAR OBJECT
ZWG 052.049	16 33 36.	+ 06 52		15.7	GALAXY
ZWG 196.081	16 33 36.	+ 35 56		15.5	GALAXY
KARA.73B 0751	16 33 36.	+ 35 56	30	15.5	ISOLATED GALAXY S
ZWG 251.031	16 33 36.	+ 46 19		14.6	GALAXY
UGC 10461	16 33 36.	+ 46 19	102	14.6	GALAXY SBb
KARA.72 500B	16 33 36.	+ 46 19	102	14.6	PART OF DOUBLE GALAXY
IC 1222	16 33 36.	+ 46 19 52.			NONSTELLAR OBJECT
MCG+10-24-002	16 33 36.	+ 57 26	78	15.	GALAXY
ZC 1633.6+6748	16 33 36.	+ 67 48	2150		CLUSTER OF GALAXIES
LDN 0105	16 33 36.	- 13 51	120		DARK NEBULA
ARP 073	16 33 37.	+ 46 19			PECULIAR GALAXY
PK331-05.1	16 33 40.	- 55 36 32.	5	12.59	PLANETARY NEBULA
SC 1631+1816.6	16 33 42.	+ 18 10 23.	6		NEBULA
ZWG 168.023	16 33 42.	+ 27 36		15.3	GALAXY
MCG+05-39-006	16 33 42.	+ 27 36	30	15.3	GALAXY
MCG+08-30-032	16 33 42.	+ 46 18	102	14.	GALAXY
ZWG 355.012	16 33 42.	+ 79 03		15.3	GALAXY
LB 00316	16 33 44.	+ 00 44 24.		14.1	FAINT BLUE STAR
UGC 10462	16 33 48.	+ 05 29	60	16.5	GALAXY S
ZWG 299.006	16 33 48.	+ 62 05		15.7	GALAXY
ZWG 298.046	16 33 48.	+ 62 05		15.7	GALAXY
OCL 0025	16 33 48.	- 08 51	1440		OPEN STAR CLUSTER
ZWG 080.048	16 33 54.	+ 10 28		15.3	GALAXY
UGC 10463	16 33 54.	+ 10 28	60	15.3	GALAXY SBb
ZWG 080.049	16 33 54.	+ 10 57		15.5	GALAXY
LDN 0106	16 33 54.	- 13 51	120		DARK NEBULA
LBN 0039	16 34	- 07 45	2520		BRIGHT NEBULA
LBN 0006	16 34	- 16 30	2940		BRIGHT NEBULA
LBN 1123	16 34	- 18 30	10800		BRIGHT NEBULA
KHAV 260	16 34	- 32 36	6470		DARK NEBULA
LB 09989	16 34	- 87 43		14.7	FAINT BLUE STAR

OBJECT NAME	RIGHT ASCEN.	DECLINATION	DIAM.	MAGN.	TYPE OF OBJECT
ZWG 168.024	16 34 00.	+ 31 55		15.6	GALAXY
MCG+07-34-114	16 34 00.	+ 42 32	30	16.	GALAXY
1ZW 159	16 34 00.	+ 52 20			COMPACT GALAXY
ZWG 276.029	16 34 00.	+ 52 20		15.6	GALAXY
MCG+09-27-053	16 34 00.	+ 52 53	48	16.	GALAXY
ZWG 367.011	16 34 00.	+ 83 00		15.5	GALAXY
ZWG 366.030	16 34 00.	+ 83 00		15.5	GALAXY
UGC 10464	16 34 00.	+ 83 00	78	15.5	GALAXY DBL SYS
MRSL 351+12/1	16 34 00.	- 28 00	10800		HII REGION
SHB 334	16 34 01.1	+ 62 51 42.		21.	QUASI-STELLAR OBJECT
BC 3CR343	16 34 01.12	+ 62 51 42.4		21.	QUASI-STELLAR OBJECT
IC 1223	16 34 02.	+ 49 21 56.			NONSTELLAR OBJECT
ZWG 138.043	16 34 06.	+ 26 01		15.6	GALAXY
ZC 1634.1+5014	16 34 06.	+ 50 14	1340		CLUSTER OF GALAXIES
MCG+09-27-054	16 34 06.	+ 54 26	36	15.	GALAXY
ZWG 276.030	16 34 06.	+ 55 44		15.5	GALAXY
7ZW 652	16 34 06.	+ 60 15			COMPACT GALAXY
ZWG 024.019	16 34 12.	+ 01 45		14.7	GALAXY
UGC 10465	16 34 12.	+ 01 45	66	14.7	GALAXY Sb-c
MCG+00-42-006	16 34 12.	+ 01 45	66	14.7	GALAXY
1ZW 160	16 34 12.	+ 55 44			COMPACT GALAXY
ZWG 355.013	16 34 12.	+ 76 51		14.9	GALAXY
UGC 10466	16 34 12.	+ 76 51	72	14.9	GALAXY Sb/SBb
HN 1816	16 34 12.7	+ 01 47 06.	24		NEBULA
LB 00196	16 34 17.	- 00 39 06.		15.6	FAINT BLUE STAR
ZC 1634.3+0951	16 34 18.	+ 09 51	1680		CLUSTER OF GALAXIES
FAI 1634+20	16 34 18.	+ 20 08			COMPACT GALAXY
ZWG 251.032	16 34 18.	+ 49 19		15.4	GALAXY
MCG+13-12-007	16 34 18.	+ 76 50	60	15.	GALAXY
ISS 1044	16 34 18.	- 41 08	261		STELLAR RING
LB 03561	16 34 21.	- 24 35 48.		19.8	FAINT BLUE STAR
SHB 335	16 34 21.4	+ 26 54 18.		17.8	QUASI-STELLAR OBJECT
SC 1632+1736.6	16 34 22.	+ 17 30 26.	12		NEBULA
ZC 1634.4+4412	16 34 24.	+ 44 12	4700		CLUSTER OF GALAXIES
MCG+08-30-033	16 34 24.	+ 49 19	48	15.	GALAXY
MRSL 006+23/1	16 34 24.	- 10 28	28800		HII REGION
OCL 0979	16 34 24.	- 46 13	240		OPEN STAR CLUSTER
MCG+07-34-115	16 34 30.	+ 44 42 30.	60	15.	GALAXY
MCG+10-24-003	16 34 30.	+ 57 35	42	14.	GALAXY
ZWG 299.007	16 34 30.	+ 57 37		14.8	GALAXY
UGC 10467	16 34 30.	+ 57 37	60	14.8	GALAXY E
ZC 1634.5+7101	16 34 30.	+ 71 01	670		CLUSTER OF GALAXIES
MCG+14-08-006	16 34 30.	+ 83 01	18	17.	GALAXY
LDN 0156	16 34 30.	- 12 10	1440		DARK NEBULA
ABC 2216	16 34 31.	+ 67 46		17.4	RICH CLUSTER OF GALAXIES
RNGC 6198	16 34 32.	+ 57 37		15.0	GALAXY
SC 1632+1851.9	16 34 34.	+ 18 45 45.	12		NEBULA
BC 4C26.49	16 34 34.4	+ 26 54 11.		17.7	QUASI-STELLAR OBJECT
ZWG 052.050	16 34 36.	+ 08 54		15.7	GALAXY
ISS 1045	16 34 36.	- 44 32	167		STELLAR RING
HN 1817	16 34 38.8	+ 00 27 14.	18		NEBULA
MCG+07-34-116	16 34 39.	+ 44 10	24	15.5	GALAXY
ZWG 138.044	16 34 42.	+ 25 51		15.7	GALAXY
ZWG 138.045	16 34 42.	+ 25 54		15.7	GALAXY
ZWG 224.074	16 34 42.	+ 44 10		15.4	GALAXY
MCG+07-34-117	16 34 42.	+ 44 13	42	15.	GALAXY
UGC 10468	16 34 42.	+ 44 43	66	16.0	GALAXY Sc
HN 1818	16 34 42.3	- 00 08 58.	36		NEBULA
MCG+06-36-054	16 34 45.	+ 36 18	21	14.5	GALAXY
MCG+08-30-034	16 34 45.	+ 48 40	60	16.	GALAXY
ARP 185	16 34 47.	+ 78 18			PECULIAR GALAXY
ZWG 138.046	16 34 48.	+ 21 10		15.5	GALAXY
ZWG 196.082	16 34 48.	+ 36 17		14.6	GALAXY
RNGC 6194	16 34 48.	+ 36 17		14.5	GALAXY
ZWG 224.075	16 34 48.	+ 39 08		14.7	GALAXY
UGC 10469	16 34 48.	+ 39 08	96	14.7	GALAXY Sb
ZWG 224.076	16 34 48.	+ 44 14		15.3	GALAXY
ZWG 251.033	16 34 48.	+ 50 30		15.5	GALAXY
MCG+08-30-035	16 34 48.	+ 50 30	60	15.	GALAXY
MCG+10-24-004	16 34 48.	+ 58 18	30	17.	GALAXY
LB 03562	16 34 48.	- 23 28 00.		18.4	FAINT BLUE STAR
REIN 2.246	16 34 49.52	+ 39 07 43.4			NEBULA
RNGC 6195	16 34 50.	+ 39 08		14.5	GALAXY
PK338+01.1	16 34 50.	- 45 16 52.	5		PLANETARY NEBULA
MCG+07-34-118	16 34 51.	+ 39 06	90	13.5	GALAXY
ZC 1634.9+0216	16 34 54.	+ 02 16	1080		CLUSTER OF GALAXIES
ZC 1634.9+4115	16 34 54.	+ 41 15	1680		CLUSTER OF GALAXIES
MCG+10-24-005	16 34 54.	+ 60 25	18	17.	GALAXY
ZWG 355.014	16 34 54.	+ 78 18		12.1	GALAXY
UGC 10470	16 34 54.	+ 78 18	216	12.1	GALAXY Sb
LB 00197	16 34 54.	- 00 30 42.		14.8	FAINT BLUE STAR
ABC 2213	16 34 56.	+ 41 23		17.7	RICH CLUSTER OF GALAXIES
SIV 09	16 35	- 28 40	36000		FAINT H EMISSION REGION
KHAV 262	16 35	- 37 36	3960		DARK NEBULA
ZC 1635.0+4613	16 35 00.	+ 46 13	1810		CLUSTER OF GALAXIES
MCG+09-27-055	16 35 00.	+ 54 23	24	16.	GALAXY
ZWG 276.031	16 35 00.	+ 56 08		15.5	GALAXY
MCG+13-12-008	16 35 00.	+ 78 18	204	11.8	GALAXY
ZWG 367.012	16 35 00.	+ 81 40		15.2	GALAXY
UGC 10471	16 35 00.	+ 81 40	120	15.2	GALAXY
MCG+14-08-007	16 35 00.	+ 83 00	36	17.	GALAXY
UGC 10472	16 35 00.	+ 84 39	60	16.5	GALAXY Sc
LDN 1712	16 35 00.	- 24 20	5040		DARK NEBULA
VHE 72B	16 35 00.	- 48 47			REFLECTION NEBULA
RNGC 6217	16 35 01.	+ 78 18		12.5	GALAXY
SHB 336	16 35 05.8	+ 11 03 45.		17.	QUASI-STELLAR OBJECT
ZWG 196.083	16 35 06.	+ 36 31		15.0	GALAXY
UGC 10473	16 35 06.	+ 36 31	96	15.0	GALAXY SBa
MCG+06-36-055	16 35 06.	+ 36 31	72	14.5	GALAXY
MCG+09-27-056	16 35 06.	+ 56 10	54	14.	GALAXY
B 231	16 35 06.	- 35 19	1800		DARK OBJECT
ZWG 080.050	16 35 12.	+ 11 50		15.6	GALAXY
ZC 1635.2+1500	16 35 12.	+ 15 00	470		CLUSTER OF GALAXIES
MCG+03-42-024	16 35 12.	+ 18 02	24	15.5	GALAXY
ZWG 196.084	16 35 12.	+ 36 33		15.5	GALAXY
HN 1819	16 35 13.8	- 01 16 02.			NEBULA
SC 1633+1807.8	16 35 17.	+ 18 01 41.	24		NEBULA
ZWG 052.051	16 35 18.	+ 07 22		15.6	GALAXY
ZWG 109.035	16 35 18.	+ 18 02		15.5	GALAXY
ZWG 024.020	16 35 18.	- 01 16		15.1	GALAXY
UGC 10474	16 35 18.	- 01 16	66	15.1	GALAXY S
MCG+00-42-007	16 35 18.	- 01 16	48	15.1	GALAXY
MCG-01-42-003	16 35 18.	- 04 40	96	15.5	GALAXY
LB 03563	16 35 20.	- 24 00 42.		18.3	FAINT BLUE STAR
MIL 47	16 35 22.	- 46 51 54.	300		SUPERNOVA REMNANT
ZC 1635.4+6618	16 35 24.	+ 66 18	2550		CLUSTER OF GALAXIES
VHE 72C	16 35 24.	- 48 50			REFLECTION NEBULA
BC M1635+119	16 35 26.1	+ 11 55 41.		17.	QUASI-STELLAR OBJECT

OBJECT NAME	RIGHT ASCEN.	DECLINATION	DIAM.	MAGN.	TYPE OF OBJECT
BIGO 538	16 35 27.	+ 36 11			NEBULA
IC 4613	16 35 27.8	+ 36 04 43.			GALAXY S
ZWG 052.052	16 35 30.	+ 06 22		15.3	GALAXY
ZWG 052.053	16 35 30.	+ 07 40		15.7	GALAXY
UGC 10475	16 35 30.	+ 07 40	66	15.7	GALAXY Sb-c
ZC 1635.5+2113	16 35 30.	+ 21 13	1140		CLUSTER OF GALAXIES
ZC 1635.5+2608	16 35 30.	+ 26 08	10620		CLUSTER OF GALAXIES
MCG+14-08-008	16 35 30.	+ 81 38	24	18.	GALAXY
ISS 1046	16 35 30.	- 40 02	910		STELLAR RING
ZWG 052.054	16 35 36.	+ 03 48		15.7	GALAXY
ZWG 052.055	16 35 36.	+ 07 43		15.6	GALAXY
RNGC 6202	16 35 36.	+ 62 03			NON-EXISTENT OBJECT
ZWG 320.031	16 35 36.	+ 68 37		15.7	GALAXY
UGC 10476	16 35 36.	+ 68 37	90	15.7	GALAXY Sa
MCG+12-16-002	16 35 36.	+ 68 37	84	16.	GALAXY
ZWG 355.015	16 35 36.	+ 77 12		15.2	GALAXY
MCG+07-34-119	16 35 39.	+ 44 30	36	15.	GALAXY
ZC 1635.7+3800	16 35 42.	+ 38 00	200		CLUSTER OF GALAXIES
ZWG 276.032	16 35 42.	+ 55 50		15.2	GALAXY
ABC 2218	16 35 45.	+ 66 20		17.7	RICH CLUSTER OF GALAXIES
ZWG 052.056	16 35 48.	+ 06 17		15.3	GALAXY
FAI 1635+20	16 35 48.	+ 20 13			COMPACT GALAXY
ZWG 109.036	16 35 48.	+ 20 13		15.7	GALAXY
KARA.73B 0752	16 35 48.	+ 20 13	42	15.7	ISOLATED GALAXY S
ZWG 196.085	16 35 48.	+ 37 22		15.4	GALAXY
UGC 10477	16 35 48.	+ 37 22	108	15.4	GALAXY S
KARA.73B 0753	16 35 48.	+ 37 22	102	15.4	ISOLATED GALAXY S
MCG+06-36-056	16 35 48.	+ 37 23	78	14.5	GALAXY
MCG+07-34-120	16 35 48.	+ 44 26	15	15.5	GALAXY
ZWG 224.077	16 35 48.	+ 44 31		15.4	GALAXY
MCG+10-24-006	16 35 48.	+ 60 27	15	17.	GALAXY
UGC 10478	16 35 48.	+ 68 31	102	16.0	GALAXY Sb-c
ASS 66	16 35 48.	- 46 40	21600		OB ASSOCIATION ARA OB1
ZC 1635.9+1608	16 35 54.	+ 16 08	1550		CLUSTER OF GALAXIES
ZWG 168.025	16 35 54.	+ 27 12		15.7	GALAXY
ZC 1635.9+2939	16 35 54.	+ 29 39	3760		CLUSTER OF GALAXIES
MCG+06-36-057	16 35 54.	+ 36 12	24	15.	GALAXY
MCG+07-34-121	16 35 54.	+ 40 57	66	15.5	GALAXY
ZWG 224.078	16 35 54.	+ 44 27		15.5	GALAXY
MCG+07-34-122	16 35 57.	+ 44 13 30.	60	15.	GALAXY
CED 136A	16 35 57.	- 48 55			DIFFUSE GALACTIC NEBULA
IC 4614	16 35 59.1	+ 36 12 42.			GALAXY SA(r)
KHAV 263	16 36	- 20 06	4320		DARK NEBULA
ZWG 138.047	16 36 00.	+ 25 23		15.5	GALAXY
ZWG 196.086	16 36 00.	+ 32 58		15.4	GALAXY
ZWG 168.026	16 36 00.	+ 32 58		15.4	GALAXY
RNGC 6197	16 36 00.	+ 36 09		14.0	GALAXY
MCG+06-36-058	16 36 00.	+ 36 10	24	14.5	GALAXY
ZWG 196.087	16 36 00.	+ 36 13		15.3	GALAXY
UGC 10479	16 36 00.	+ 40 59	66	16.0	GALAXY
ZWG 224.079	16 36 00.	+ 44 14		14.9	GALAXY
UGC 10480	16 36 00.	+ 44 14	72	14.9	GALAXY E?
MCG+07-34-123	16 36 00.	+ 44 26	21	15.5	GALAXY
1ZW 161	16 36 00.	+ 44 27			COMPACT GALAXY
MCG+08-30-037	16 36 00.	+ 46 56	30	16.	GALAXY
MCG+08-30-036	16 36 00.	+ 46 56	48	16.	GALAXY
MCG+08-30-038	16 36 00.	+ 47 20	48	16.	GALAXY
UGC 10481	16 36 00.	+ 55 00	60	16.5	GALAXY
MCG+09-27-057	16 36 00.	+ 55 01	12	16.	GALAXY
ZWG 320.032	16 36 00.	+ 66 07		15.2	GALAXY
MCG+11-20-020	16 36 00.	+ 66 07 30.	30	16.	GALAXY
MCG+11-20-019	16 36 00.	+ 66 20	60	15.	GALAXY
MCG+11-20-021	16 36 00.	+ 68 32	84	16.	GALAXY
MCG+14-08-009	16 36 00.	+ 81 38	15	16.	GALAXY
7ZW 653	16 36 00.	+ 85 36			COMPACT GALAXY
MFSL 336-01/1	16 36 00.	- 48 51	2400		HII REGION
ABC 2214	16 36 01.	+ 38 00		17.6	RICH CLUSTER OF GALAXIES
IC 4615	16 36 05.	+ 36 11			SAME AS NGC 6196
ZWG 138.048	16 36 06.	+ 26 25		15.7	GALAXY
ZWG 196.088	16 36 06.	+ 36 10		14.2	GALAXY
RNGC 6196	16 36 06.	+ 36 10			GALAXY
UGC 10482	16 36 06.	+ 36 10	84	14.2	GALAXY S0
3ZW 088	16 36 06.	+ 42 37			COMPACT GALAXY
ZWG 320.033	16 36 06.	+ 66 20		14.7	GALAXY
UGC 10483	16 36 06.	+ 76 09	78	16.0	GALAXY S
LB 00317	16 36 07.	+ 00 39 18.		15.2	FAINT BLUE STAR
LB 00950	16 36 07.	+ 46 53 48.		16.4	FAINT BLUE STAR
PK345+06.1	16 36 08.	- 36 28 43.	5		PLANETARY NEBULA
MCG+06-36-059	16 36 09.	+ 36 05 30.	33	15.5	GALAXY
IC 4616	16 36 11.7	+ 36 05 44.			GALAXY S0
ZWG 138.049	16 36 12.	+ 26 22		15.6	GALAXY
UGC 10484	16 36 12.	+ 26 22	72	15.6	GALAXY Sb
ZWG 196.089	16 36 12.	+ 36 05		15.4	GALAXY
RNGC 6199	16 36 12.	+ 36 05		15.5	GALAXY
BIGO 539	16 36 12.	+ 36 06			NEBULA
UGC 10485	16 36 12.	+ 39 21	90	16.0	GALAXY Sb
MCG+08-30-039	16 36 12.	+ 49 34	24	15.	GALAXY
ZWG 251.034	16 36 12.	+ 50 26		13.9	GALAXY
UGC 10486	16 36 12.	+ 50 26	66	13.9	GALAXY E-S0
MCG+10-24-007	16 36 12.	+ 58 16	30	16.	GALAXY
RNGC 6183	16 36 14.	- 69 17			GALAXY
WRAY 19.47	16 36 15.5	- 48 46 02.			DIFFUSE NEBULA
ZWG 251.035	16 36 18.	+ 49 32		15.7	GALAXY
UGC 10487	16 36 18.	+ 49 32	84	15.7	GALAXY Sa-b
MCG+08-30-040	16 36 18.	+ 50 27	36	14.	GALAXY
ZWG 276.033	16 36 18.	+ 51 46		15.7	GALAXY
ZWG 339.011	16 36 18.	+ 71 47		15.5	GALAXY
VHE 72A	16 36 18.	- 48 42	30		REFLECTION NEBULA
ZC 1636.4+3758	16 36 24.	+ 37 58	870		CLUSTER OF GALAXIES
ZC 1636.4+4729	16 36 24.	+ 47 29	2550		CLUSTER OF GALAXIES
MCG+08-30-041	16 36 24.	+ 49 32 30.	66	15.	GALAXY
MCG+09-27-058	16 36 24.	+ 51 45	48	14.	GALAXY
ZC 1636.4+7400	16 36 24.	+ 74 00	870		CLUSTER OF GALAXIES
ISS 1047	16 36 24.	- 40 13	113		STELLAR RING
SC 1634+1618.4	16 36 25.	+ 16 12 22.	6		NEBULA
LB 00318	16 36 26.	- 01 36 36.		15.0	FAINT BLUE STAR
SC 1634+1926.1	16 36 27.	+ 19 20 04.	18		NEBULA
ZC 1636.5+2807	16 36 30.	+ 28 07	2820		CLUSTER OF GALAXIES
ZWG 224.080	16 36 30.	+ 40 01		15.4	GALAXY
UGC 10488	16 36 30.	+ 40 01	90	15.4	GALAXY Sa-b
UGC 10489	16 36 30.	+ 62 53	90	16.5	GALAXY
MCG+03-42-025	16 36 33.	+ 17 27	60	15.3	GALAXY
ARP 125	16 36 33.	+ 42 01			PECULIAR GALAXY
ZC 1636.6+1419	16 36 36.	+ 14 19	610		CLUSTER OF GALAXIES
ZWG 109.037	16 36 36.	+ 17 27		15.3	GALAXY
UGC 10490	16 36 36.	+ 17 27	66	15.3	GALAXY Sc
KARA.73B 0754	16 36 36.	+ 17 27	60	15.3	ISOLATED GALAXY S
MCG+07-34-125	16 36 36.	+ 40 13	42	15.	GALAXY
ZWG 224.081	16 36 36.	+ 40 15		15.5	GALAXY
MCG+07-34-126	16 36 36.	+ 40 15	18	16.	GALAXY
MCG+07-34-127	16 36 36.	+ 42 01	60	15.	GALAXY
1ZW 162	16 36 36.	+ 42 03			COMPACT GALAXY
ZC 1636.6+4809	16 36 36.	+ 48 09	1280		CLUSTER OF GALAXIES
MCG+10-24-008	16 36 36.	+ 62 49	78	15.	GALAXY
ABC 2215	16 36 39.	+ 48 10		17.1	RICH CLUSTER OF GALAXIES
ZWG 168.027	16 36 42.	+ 32 39		15.4	GALAXY
ZWG 224.082	16 36 42.	+ 42 02		15.5	GALAXY
UGC 10491	16 36 42.	+ 42 02	66	15.5	GALAXY DBL SYS
SC 1634+1732.0	16 36 45.	+ 17 25 59.	36		NEBULA
ZWG 024.021	16 36 48.	- 02 20		15.5	GALAXY
UGC 10492	16 36 48.	- 02 20	102	15.5	GALAXY Sa
MCG+00-42-008	16 36 48.	- 02 20	36	15.5	GALAXY
KARA.73B 0755	16 36 48.	- 02 20	66	15.5	ISOLATED GALAXY S
OCL 0988	16 36 48.	- 43 16	2580	8.7	OPEN STAR CLUSTER
VHA 194	16 36 48.	- 43 16	360		OPEN STAR CLUSTER
HH 1820	16 36 48.0	- 02 18 25.	12		NEBULA
RNGC 6192	16 36 50.	- 43 17		8.5	OPEN CLUSTER
ZC 1636.9+3151	16 36 54.	+ 31 51	540		CLUSTER OF GALAXIES
ZWG 276.034	16 36 54.	+ 55 31		14.9	GALAXY
UGC 10493	16 36 54.	+ 55 31	72	14.9	GALAXY (E)
ZWG 320.034	16 36 54.	+ 67 44		15.5	GALAXY
UGC 10494	16 36 54.	+ 67 44	90	15.5	GALAXY SB?b
MCG+11-20-022	16 36 54.	+ 67 44	60	15.	GALAXY
KARA.73B 0756	16 36 54.	+ 67 44	90	15.5	ISOLATED GALAXY S
KHAV 266	16 37	- 21 06	2630		DARK NEBULA
B 044	16 37	- 23 59			DARK OBJECT
KHAV 264	16 37	- 24 06	9120		DARK NEBULA
ZWG 052.057	16 37 00.	+ 07 23		15.5	GALAXY
UGC 10495	16 37 00.	+ 07 23	60	15.5	GALAXY S
ZWG 080.051	16 37 00.	+ 11 19		15.1	GALAXY
MCG+02-42-007	16 37 00.	+ 11 19	48	15.1	GALAXY
MCG+08-30-042	16 37 00.	+ 45 42 30.	48	16.	GALAXY
MCG+08-30-043	16 37 00.	+ 48 24	36	16.	GALAXY
MCG+09-27-059	16 37 00.	+ 55 32 30.	18	14.	GALAXY
LDN 0121	16 37 00.	- 13 55	480		DARK NEBULA
LDN 1744	16 37 00.	- 22 10	1260		DARK NEBULA
IC 1225	16 37 01.	+ 67 44 03.			NONSTELLAR OBJECT
UGC 10496	16 37 06.	+ 72 06	90	16.0	GALAXY
SCHO 0264	16 37 07.	- 33 10 48.	870		ISOLATED DARK CLOUD
ZWG 251.036	16 37 12.	+ 49 14		14.8	GALAXY
MCG+10-24-009	16 37 12.	+ 60 07	30	17.	GALAXY
MCG+04-39-017	16 37 18.	+ 21 22	48	15.1	GALAXY
ZWG 138.050	16 37 18.	+ 21 25		15.1	GALAXY
KARA.73B 0757	16 37 18.	+ 21 25	54	15.1	ISOLATED GALAXY S
TON-N 0259	16 37 18.	+ 25 27		14.8	BLUE STAR
ZWG 224.083	16 37 18.	+ 39 23		15.5	GALAXY
MCG+10-24-010	16 37 18.	+ 60 52	18	17.	GALAXY
ZWG 339.012	16 37 18.	+ 72 30		14.4	GALAXY
UGC 10497	16 37 18.	+ 72 30	72	14.4	GALAXY S
UGC 10498	16 37 24.	+ 29 28	60	17.	GALAXY Sb-c
MCG+09-27-060	16 37 24.	+ 55 01	36	16.	GALAXY
MCG+12-16-003	16 37 24.	+ 72 07	45	17.	GALAXY
SC 1635+1616.8	16 37 25.	+ 16 10 50.	6		NEBULA
HH 1821	16 37 26.2	- 01 05 53.	12		NEBULA
ZC 1637.5+1510	16 37 30.	+ 15 10	1140		CLUSTER OF GALAXIES
ZWG 138.051	16 37 30.	+ 23 36		15.6	GALAXY
MCG+10-24-011	16 37 30.	+ 60 54	15	17.	GALAXY
ZWG 320.035	16 37 30.	+ 63 22		15.1	GALAXY
MCG+12-16-004	16 37 30.	+ 72 30	66	14.	GALAXY
MCG+14-08-010	16 37 30.	+ 82 37 30.	24	15.	GALAXY
MCG+14-08-011	16 37 30.	+ 82 40	39	16.	GALAXY
OCL 0975	16 37 30.	- 48 40	1200	5.7	OPEN STAR CLUSTER
VHA 195	16 37 30.	- 48 40	660		STAR CLSTR IN NEBULOSITY
CED 136B	16 37 32.	- 48 40	1140		DIFFUSE GALACTIC NEBULA
ZC 1637.6+2208	16 37 36.	+ 22 08	940		CLUSTER OF GALAXIES
ZC 1637.6+2341	16 37 36.	+ 23 41	1480		CLUSTER OF GALAXIES
ZC 1637.6+2455	16 37 36.	+ 24 55	1080		CLUSTER OF GALAXIES
7ZW 654	16 37 36.	+ 60 58			COMPACT GALAXY
ZWG 339.013	16 37 36.	+ 72 40		15.6	GALAXY
ISS 1048	16 37 36.	- 45 24	61		STELLAR RING
RNGC 6193	16 37 38.	- 48 40		5.5	OPEN CLUSTER
HH 1822	16 37 40.7	- 02 36 16.	18		NEBULA
ZWG 224.084	16 37 42.	+ 42 17		15.7	GALAXY
ZWG 275.035	16 37 42.	+ 51 38		15.4	GALAXY
SC 1635+1604.5	16 37 47.	+ 15 58 34.	18		NEBULA
ZWG 109.038	16 37 48.	+ 15 59		14.8	GALAXY
MCG+03-42-026	16 37 48.	+ 15 59	27	14.8	GALAXY
MCG+09-27-061	16 37 48.	+ 51 36	36	16.	GALAXY
PK339+00.1	16 37 53.	- 45 08	10		PLANETARY NEBULA
MCG+08-30-044	16 37 54.	+ 46 00	30	16.	GALAXY
MCG+10-24-012	16 37 54.	+ 58 54	36	16.	GALAXY
SC 1635+1926.8	16 37 55.	+ 19 20 52.	6		NEBULA
RNGC 6252	16 37 58.	+ 82 42		15.0	GALAXY
LBN 0005	16 38	- 17 30	1080		BRIGHT NEBULA
KHAV 265	16 38	- 32 54	1950		DARK NEBULA
UGC 10499	16 38 00.	+ 49 08	60	16.5	GALAXY S
ZWG 299.008	16 38 00.	+ 57 50		14.1	GALAXY
UGC 10500	16 38 00.	+ 57 50	96	14.1	GALAXY S0/Sa
MCG+10-24-013	16 38 00.	+ 57 50	78	13.	GALAXY
ZWG 367.013	16 38 00.	+ 82 39		14.0	GALAXY
UGC 10501	16 38 00.	+ 82 39	168	14.0	GALAXY E
ZWG 367.014	16 38 00.	+ 82 42		15.1	GALAXY
RNGC 6251	16 38 02.	+ 82 39		14.0	GALAXY
RNGC 6188	16 38 02.	- 48 00			DIFFUSE NEBULA
MCG+04-39-018	16 38 03.	+ 23 05 30.	54	15.4	GALAXY
RNGC 6201	16 38 03.	+ 23 51		15.5	GALAXY
MCG+07-34-128	16 38 03.	+ 43 22	27	15.5	GALAXY
ZWG 138.052	16 38 06.	+ 23 07		15.4	GALAXY
ZWG 138.053	16 38 06.	+ 23 51		15.5	GALAXY
ZWG 138.054	16 38 06.	+ 26 11		15.5	GALAXY
ZWG 197.001	16 38 06.	+ 37 16		15.2	GALAXY
MCG+06-37-001	16 38 06.	+ 37 16 30.	24	15.	GALAXY
ZC 1638.1+3913	16 38 06.	+ 39 13	1140		CLUSTER OF GALAXIES
ZWG 224.085	16 38 06.	+ 43 22		15.6	GALAXY
LB 00951	16 38 08.	+ 45 19 12.		17.2	FAINT BLUE STAR
ZWG 080.052	16 38 12.	+ 09 06		15.7	GALAXY
MCG+10-24-014	16 38 12.	+ 58 30	42	16.	GALAXY
ZWG 339.014	16 38 12.	+ 72 28		13.1	GALAXY
UGC 10502	16 38 12.	+ 72 28	156	13.1	GALAXY Sc
MCG+12-16-005	16 38 12.	+ 72 28	132	13.	GALAXY
RNGC 6203	16 38 15.	+ 23 52		15.5	GALAXY
ZWG 052.058	16 38 18.	+ 08 43		15.7	GALAXY
ZWG 080.053	16 38 18.	+ 14 27		15.7	GALAXY
KARA.73B 0758	16 38 18.	+ 14 27	24	15.7	ISOLATED GALAXY S
MCG+04-39-019	16 38 18.	+ 23 50	42	15.3	GALAXY
ZWG 138.055	16 38 18.	+ 23 52		15.3	GALAXY

OBJECT NAME	RIGHT ASCEN.	DECLINATION	DIAM.	MAGN.	TYPE OF OBJECT
ZWG 197.002	16 38 18.	+ 37 17		14.8	GALAXY
MCG+06-37-002	16 38 18.	+ 37 17	42	14.5	GALAXY
ZC 1638.3+4307	16 38 18.	+ 43 07	810		CLUSTER OF GALAXIES
UGC 10503	16 38 18.	+ 71 37	108	16.0	GALAXY IRR
ABC 2226	16 38 20.	+ 67 10		17.6	RICH CLUSTER OF GALAXIES
SC 1634-5122.5	16 38 23.	- 51 28 27.	12		NEBULA
MCG+07-34-129	16 38 24.	+ 43 47	30	16.	GALAXY
ZC 1638.4+6038	16 38 24.	+ 60 38	20560		CLUSTER OF GALAXIES
OCL 0983	16 38 24.	- 46 05	504	13.	OPEN STAR CLUSTER
ZWG 197.003	16 38 30.	+ 33 46		14.9	GALAXY
UGC 10504	16 38 30.	+ 33 46	66	14.9	GALAXY Sa-b
ZC 1638.5+4922	16 38 30.	+ 49 22	540		CLUSTER OF GALAXIES
MCG+10-24-015	16 38 30.	+ 59 30	12	16.	GALAXY
MCG+11-20-023	16 38 30.	+ 66 05 30.	39	16.	GALAXY
MCG+12-16-006	16 38 30.	+ 71 36	84	16.	GALAXY
LDN 0033	16 38 30.	- 17 20	2220		DARK NEBULA
MCG+06-37-003	16 38 33.	+ 33 45 30.	48	15.	GALAXY
ABC 2220	16 38 33.	+ 53 52		17.5	RICH CLUSTER OF GALAXIES
LB 00952	16 38 33.	+ 81 03 18.		17.3	FAINT BLUE STAR
ZC 1638.6+1307	16 38 36.	+ 13 07	540		CLUSTER OF GALAXIES
ACK 061+41.1	16 38 36.	+ 38 48			PLANETARY NEBULA
MCG+08-30-045	16 38 36.	+ 47 51	36	16.	GALAXY
PK352+11.1	16 38 37.	- 27 53			PLANETARY NEBULA
ZWG 024.022	16 38 42.	+ 02 43		15.2	GALAXY
KARA.73B 0759	16 38 42.	+ 02 43	48	15.2	ISOLATED GALAXY S
ZC 1638.7+4643	16 38 42.	+ 46 43	3090		CLUSTER OF GALAXIES
MCG+10-24-016	16 38 42.	+ 62 34	24	16.	GALAXY
ZWG 355.016	16 38 42.	+ 78 53		15.7	GALAXY
VDB.66N 109	16 38 42.	- 17 41	2160		REFLECTION NEBULA
ABC 2217	16 38 43.	+ 28 01		17.1	RICH CLUSTER OF GALAXIES
SC 1636+2001.7	16 38 44.	+ 19 55 50.	12		NEBULA
ABC 2225	16 38 44.	+ 55 52		17.5	RICH CLUSTER OF GALAXIES
MCG+07-34-130	16 38 45.	+ 39 23 30.	48	14.5	GALAXY
ZWG 138.056	16 38 48.	+ 23 45		15.4	GALAXY
ZWG 168.028	16 38 48.	+ 32 20		15.7	GALAXY
ZWG 197.004	16 38 48.	+ 34 01		15.7	GALAXY
MCG+07-34-131	16 38 48.	+ 39 19 30.	42	15.	GALAXY
ZWG 224.086	16 38 48.	+ 39 20		15.6	GALAXY
ZWG 224.087	16 38 48.	+ 39 25		15.1	GALAXY
MCG+10-24-017	16 38 48.	+ 59 51	24	17.	GALAXY
SHB 337	16 38 48.2	+ 39 52 30.		18.5	QUASI-STELLAR OBJECT
BC FRA0512	16 38 48.29	+ 39 52 30.3		18.5	QUASI-STELLAR OBJECT
ZWG 080.054	16 38 54.	+ 13 30		15.6	GALAXY
ZC 1638.9+2802	16 38 54.	+ 28 02	1480		CLUSTER OF GALAXIES
1ZW 163	16 38 54.	+ 46 48			COMPACT GALAXY
LB 00319	16 38 56.	- 00 07 36.		15.5	FAINT BLUE STAR
ABC 2219	16 38 57.	+ 46 48		17.4	RICH CLUSTER OF GALAXIES
ZWG 052.059	16 39 00.	+ 05 50		15.7	GALAXY
ZWG 052.060	16 39 00.	+ 07 30		15.6	GALAXY
ZWG 080.055	16 39 00.	+ 09 00		15.3	GALAXY
MCG+09-27-062	16 39 00.	+ 55 13	54	16.	GALAXY
ZC 1639.0+5553	16 39 00.	+ 55 53	1080		CLUSTER OF GALAXIES
ZWG 024.023	16 39 00.	- 00 02		15.7	GALAXY
UGC 10505	16 39 00.	- 00 02	66	15.7	GALAXY S
LDN 1796	16 39 00.	- 19 00	2340		DARK NEBULA
LDN 1782	16 39 00.	- 19 40	840		DARK NEBULA
HN 1823	16 39 04.7	+ 01 00 50.	12		NEBULA
ZWG 024.024	16 39 06.	+ 01 00		15.6	GALAXY
ZWG 052.061	16 39 06.	+ 03 14		15.6	GALAXY
KARA.73B 0760	16 39 06.	+ 03 14	48	15.6	ISOLATED GALAXY S
ZWG 138.057	16 39 06.	+ 22 22		15.6	GALAXY
ZWG 197.005	16 39 06.	+ 33 52		15.4	GALAXY
PK344+04.1	16 39 09.	- 38 48 55.	10	12.	PLANETARY NEBULA
MCG+06-37-004	16 39 12.	+ 33 51	45	15.	GALAXY
MCG+11-20-024	16 39 12.	+ 66 08	54	14.	GALAXY
IC 1227	16 39 15.	+ 58 43			MAY NOT EXIST
HN 1824	16 39 15.3	- 01 47 03.	60		NEBULA
RNGC 6206	16 39 17.	+ 58 44		14.5	GALAXY
ZWG 080.056	16 39 18.	+ 09 31		15.5	GALAXY
ZWG 080.057	16 39 18.	+ 09 33		15.6	GALAXY
MCG+06-37-005	16 39 18.	+ 33 21	36	16.	GALAXY
MCG+10-24-018	16 39 18.	+ 58 43	24	14.	GALAXY
ZWG 299.009	16 39 18.	+ 58 44		14.5	GALAXY
UGC 10506	16 39 18.	+ 58 44	42	14.5	GALAXY (S0)
MCG+10-24-019	16 39 18.	+ 58 46	18	16.	GALAXY
MCG+10-24-020	16 39 18.	+ 62 08	72	15.	GALAXY
ZWG 320.036	16 39 18.	+ 66 07		14.3	GALAXY
RNGC 6214	16 39 18.	+ 66 07		14.5	GALAXY
UGC 10507	16 39 18.	+ 66 07	66	14.3	GALAXY S
MCG-01-42-004	16 39 18.	- 04 58	90	15.	GALAXY
ZWG 024.025	16 39 24.	+ 01 48		15.5	GALAXY
ZWG 138.058	16 39 24.	+ 26 47		15.7	GALAXY
ZC 1639.4+4253	16 39 24.	+ 42 53	1340		CLUSTER OF GALAXIES
ZWG 299.010	16 39 24.	+ 62 05		15.7	GALAXY
UGC 10508	16 39 24.	+ 62 05	72	15.7	GALAXY Sa-b
ZWG 355.017	16 39 24.	+ 78 14		15.1	GALAXY
UGC 10509	16 39 24.	+ 78 14	60	15.1	GALAXY SBb
MCG+07-34-132	16 39 27.	+ 39 49 30.	24	15.5	GALAXY
MCG+07-34-124	16 39 27.	+ 39 59 30.	66	14.5	GALAXY
ABC 2221	16 39 29.	+ 43 22		17.7	RICH CLUSTER OF GALAXIES
ZC 1639.5+2207	16 39 30.	+ 22 07	1140		CLUSTER OF GALAXIES
ZWG 197.006	16 39 30.	+ 33 27		15.2	GALAXY
ZC 1639.5+3729	16 39 30.	+ 37 29	1080		CLUSTER OF GALAXIES
MCG+07-34-133	16 39 30.	+ 39 49	27	15.5	GALAXY
ZWG 224.088	16 39 30.	+ 39 52		15.4	GALAXY
ZWG 224.089	16 39 30.	+ 40 31		15.7	GALAXY
ZWG 251.037	16 39 30.	+ 48 59		15.7	GALAXY
HN 0432	16 39 30.	- 77 24			NEBULA
IC 4608	16 39 30.	- 77 24			NONSTELLAR OBJECT
ABC 2222	16 39 31.	+ 42 54		17.7	RICH CLUSTER OF GALAXIES
MCG+07-34-134	16 39 33.	+ 39 21 30.	48	16.	GALAXY
ZWG 052.062	16 39 36.	+ 07 20		15.7	GALAXY
ZC 1639.6+4320	16 39 36.	+ 43 20	1340		CLUSTER OF GALAXIES
ZWG 251.038	16 39 36.	+ 46 06		15.4	GALAXY
ZWG 251.039	16 39 36.	+ 50 36		15.4	GALAXY
ZWG 299.011	16 39 36.	+ 58 13		14.9	GALAXY
UGC 10510	16 39 36.	+ 58 13	96	14.9	GALAXY Sc
MCG+13-12-009	16 39 36.	+ 78 14	51	15.	GALAXY
PK339+00.2	16 39 36.	- 45 54 23.	10		PLANETARY NEBULA
IC 1226	16 39 41.	+ 46 06 19.			NONSTELLAR OBJECT
MCG+08-30-046	16 39 42.	+ 50 36 30.	42	16.	GALAXY
MCG+10-24-021	16 39 42.	+ 58 11	54	14.	GALAXY
ZWG 052.063	16 39 54.	+ 07 12		15.6	GALAXY
GCL 045	16 39 54.	+ 36 33	2538	6.79	GLOBULAR STAR CLUSTER
RNGC 6205	16 39 54.	+ 36 33		7.0	GLOBULAR CLUSTER
ZC 1639.9+3805	16 39 54.	+ 38 05	1340		CLUSTER OF GALAXIES
MCG+07-34-135	16 39 54.	+ 40 15	27	15.	GALAXY
ZWG 224.090	16 39 54.	+ 40 16		15.1	GALAXY
SIV 08	16 40	- 41 18	25200		FAINT H EMISSION REGION
ZWG 110.001	16 40 00.	+ 19 15		15.7	GALAXY
ZWG 109.039	16 40 00.	+ 19 15		15.7	GALAXY
ZC 1640.0+3032	16 40 00.	+ 30 32	2690		CLUSTER OF GALAXIES
MCG+06-37-006	16 40 00.	+ 37 50	36	16.	GALAXY
ZWG 224.091	16 40 00.	+ 39 45		15.2	GALAXY
MCG+09-27-063	16 40 00.	+ 51 17	48	15.	GALAXY
ZWG 276.036	16 40 00.	+ 51 18		15.2	GALAXY
UGC 10511	16 40 00.	+ 51 18	60	15.2	GALAXY S0
LDN 1729	16 40 00.	- 24 00	3840		DARK NEBULA
PK326-10.1	16 40 02.	- 62 32 13.	5	12.43	PLANETARY NEBULA
MCG+07-34-136	16 40 03.	+ 39 43	30	14.5	GALAXY
MCG+07-34-137	16 40 06.	+ 40 03	18	15.	GALAXY
ZWG 224.092	16 40 06.	+ 40 05		14.8	GALAXY
UGC 10512	16 40 06.	+ 40 05	78	14.8	GALAXY E
ZC 1640.1+4932	16 40 06.	+ 49 32	2490		CLUSTER OF GALAXIES
ZWG 339.015	16 40 06.	+ 72 18		15.7	GALAXY
SCHO 0265	16 40 06.	- 41 14 42.	2700		ISOLATED DARK CLOUD
LB 00953	16 40 12.	+ 45 46 18.		15.9	FAINT BLUE STAR
MCG+09-27-064	16 40 12.	+ 51 14	30	16.	GALAXY
MCG+10-24-022	16 40 12.	+ 61 13	6	17.	GALAXY
MCG+10-24-023	16 40 12.	+ 61 24	42	16.	GALAXY
UGC 10513	16 40 18.	+ 00 44	72	16.0	GALAXY Sc
RAI 1640+19	16 40 18.	+ 19 15			COMPACT GALAXY
ZWG 138.059	16 40 18.	+ 25 10		15.4	GALAXY
UGC 10514	16 40 18.	+ 25 10	108	15.4	GALAXY S-IRR
KARA.73B 0761	16 40 18.	+ 25 10	108	15.4	ISOLATED GALAXY S
ZWG 299.012	16 40 18.	+ 59 26		15.1	GALAXY
UGC 10515	16 40 18.	+ 59 26	60	15.1	GALAXY Sb
MCG+10-24-024	16 40 18.	+ 59 26	42	15.	GALAXY
MCG+10-24-025	16 40 18.	+ 60 54	8	17.	GALAXY
ZWG 299.013	16 40 19.	+ 61 23		15.6	GALAXY
ZC 1640.3+6704	16 40 18.	+ 67 04	2080		CLUSTER OF GALAXIES
B 232	16 40 18.	- 39 44	600		DARK OBJECT
IC 4617	16 40 20.1	+ 36 46 42.			GALAXY Sb
MCG+04-39-020	16 40 21.	+ 25 10	102	15.4	GALAXY
MCG+07-34-138	16 40 24.	+ 40 21	48	14.5	GALAXY
ZWG 224.093	16 40 24.	+ 40 22		15.5	GALAXY
ZC 1640.4+4400	16 40 24.	+ 44 00	740		CLUSTER OF GALAXIES
MCG+10-24-026	16 40 24.	+ 59 11	42	16.	GALAXY
ZC 1640.4+7230	16 40 24.	+ 72 30	670		CLUSTER OF GALAXIES
SC 1636+1949.4	16 40 29.	+ 19 43 39.	6		NEBULA
ZWG 110.002	16 40 30.	+ 20 27		15.7	GALAXY
ZWG 109.040	16 40 30.	+ 20 27		15.7	GALAXY
KARA.73B 0762	16 40 30.	+ 20 27	36	15.7	ISOLATED GALAXY S
MCG+07-34-139	16 40 30.	+ 39 31 30.	27	15.	GALAXY
ZWG 224.094	16 40 30.	+ 39 33		15.7	GALAXY
MCG+10-24-027	16 40 30.	+ 57 52 30.	78	12.	GALAXY
ZWG 299.014	16 40 30.	+ 57 54		13.8	GALAXY
UGC 10516	16 40 30.	+ 57 54	102	13.8	GALAXY S0
MCG+10-24-028	16 40 30.	+ 61 27	84	14.	GALAXY
MCG+10-24-029	16 40 30.	+ 62 31	60	13.	GALAXY
MCG+11-20-025	16 40 30.	+ 65 52	60	15.	GALAXY
ZWG 025.001	16 40 30.	- 00 34		15.2	GALAXY
MCG+00-43-001	16 40 30.	- 00 34	48	15.2	GALAXY
OCL 0978	16 40 30.	- 47 23	960	8.0	OPEN STAR CLUSTER
ABC 2223	16 40 32.	+ 27 32		16.5	RICH CLUSTER OF GALAXIES
RNGC 6211	16 40 33.	+ 57 53		14.0	GALAXY
RNGC 6200	16 40 33.	- 47 23		8.0	OPEN CLUSTER
ZWG 025.002	16 40 36.	+ 02 11		15.2	GALAXY
KARA.73B 0763	16 40 36.	+ 02 11	36	15.2	ISOLATED GALAXY S0
7ZW 655	16 40 36.	+ 57 53			COMPACT GALAXY
ZWG 299.015	16 40 36.	+ 61 25		13.7	GALAXY
UGC 10517	16 40 36.	+ 61 25	96	13.7	GALAXY Sa
ZWG 299.016	16 40 36.	+ 62 29		14.7	GALAXY
UGC 10518	16 40 36.	+ 62 29	66	14.7	GALAXY Sb/Sc
HOLM 755B	16 40 37.	+ 61 23	18	15.5	PART OF MULTIPLE GALAXY
RNGC 6213	16 40 39.	+ 57 56		15.0	GALAXY
ZWG 053.001	16 40 42.	+ 08 37		15.5	GALAXY
MCG+03-42-027	16 40 42.	+ 19 21	42	15.5	GALAXY
MCG+10-24-030	16 40 42.	+ 57 55	36	14.	GALAXY
ZWG 299.017	16 40 42.	+ 57 56		15.2	GALAXY
MCG+10-24-031	16 40 42.	+ 60 55	12	17.	GALAXY
ZC 1640.7+6426	16 40 42.	+ 64 26	3490		CLUSTER OF GALAXIES
ZWG 320.037	16 40 42.	+ 65 52		15.7	GALAXY
UGC 10519	16 40 42.	+ 65 52	66	15.7	GALAXY Sa-b
OCL 0976	16 40 42.	- 47 28	240		OPEN STAR CLUSTER
IC 1224	16 40 45.	+ 19 20 16.			NONSTELLAR OBJECT
SCHO 0266	16 40 46.	- 35 12 30.	2420		ISOLATED DARK CLOUD
ZWG 081.001	16 40 48.	+ 10 01		15.3	GALAXY
KARA.73B 0764	16 40 48.	+ 10 01	54	15.3	ISOLATED GALAXY S
ZWG 110.003	16 40 48.	+ 19 21		15.5	GALAXY
ZC 1640.8+3921	16 40 48.	+ 39 21	940		CLUSTER OF GALAXIES
ZWG 251.040	16 40 48.	+ 50 00		15.7	GALAXY
ZWG 276.037	16 40 48.	+ 51 38		15.6	GALAXY
MCG+10-24-032	16 40 48.	+ 57 57	36	15.	GALAXY
MCG+10-24-033	16 40 48.	+ 61 42	42	16.	GALAXY
SCHO 0267	16 40 48.	- 37 10 48.	1430		ISOLATED DARK CLOUD
HOLM 755A	16 40 50.	+ 61 25	84	14.2	PART OF MULTIPLE GALAXY
ZWG 299.018	16 40 54.	+ 57 58		15.7	GALAXY
BRSL 336-02/1	16 40 54.	- 49 09	2700		HII REGION
PK000+17.1	16 40 59.	- 18 51 41.	5		PLANETARY NEBULA
KHAV 267	16 41	- 27 24	2960		DARK NEBULA
KHAV 268	16 41	- 32 24	5360		DARK NEBULA
KHAV 271	16 41	- 35 12	5750		DARK NEBULA
KHAV 269	16 41	- 36 18	3350		DARK NEBULA
KHAV 270	16 41	- 37 00	3790		DARK NEBULA
ZC 1641.0+1329	16 41 00.	+ 13 29	940		CLUSTER OF GALAXIES
ZC 1641.0+2135	16 41 00.	+ 21 35	1080		CLUSTER OF GALAXIES
MCG+09-27-065	16 41 00.	+ 54 48	30	16.	GALAXY
LDN 1755	16 41 00.	- 21 50	3300		DARK NEBULA
ABC 2224	16 41 05.	+ 13 27		17.4	RICH CLUSTER OF GALAXIES
ZWG 025.003	16 41 06.	+ 00 58		15.6	GALAXY
KARA.73B 0765	16 41 06.	+ 00 58	66	15.6	ISOLATED GALAXY S
ZC 1641.1+2737	16 41 06.	+ 27 37	2150		CLUSTER OF GALAXIES
ZC 1641.1+4349	16 41 06.	+ 43 49	1340		CLUSTER OF GALAXIES
ZC 1641.1+4729	16 41 06.	+ 47 29	1880		CLUSTER OF GALAXIES
ZWG 299.019	16 41 06.	+ 58 01		15.6	GALAXY
ZWG 168.029	16 41 12.	+ 32 46		15.6	GALAXY
UGC 10520	16 41 12.	+ 48 01	66	16.0	GALAXY Sc
MCG+10-24-034	16 41 12.	+ 57 59	36	16.	GALAXY
MCG+10-24-035	16 41 12.	+ 61 15	9	18.	GALAXY
ZC 1641.2+6817	16 41 12.	+ 68 17	1080		CLUSTER OF GALAXIES
MCG+06-37-007	16 41 15.	+ 36 54	120	11.	GALAXY
MCG+08-30-047	16 41 15.	+ 48 02	66	16.	GALAXY
SHB 338	16 41 17.6	+ 39 54 11.		16.0	QUASI-STELLAR OBJECT
BC 3CR345	16 41 17.70	+ 39 54 11.1		15.96	QUASI-STELLAR OBJECT
ZWG 197.007	16 41 18.	+ 36 55		11.9	GALAXY

OBJECT NAME	RIGHT ASCEN.	DECLINATION	DIAM.	MAGN.	TYPE OF OBJECT
UGC 10521	16 41 18.	+ 36 55	198	11.9	GALAXY S
KARA.73B 0766	16 41 18.	+ 36 55	186	11.9	ISOLATED GALAXY S
ZC 1641.3+3723	16 41 18.	+ 37 23	1680		CLUSTER OF GALAXIES
MCG+07-34-140	16 41 18.	+ 44 05	48	15.5	GALAXY
MCG+09-27-066	16 41 18.	+ 52 55	24	16.	GALAXY
MCG+10-24-036	16 41 18.	+ 61 57	60	14.	GALAXY
B 044a	16 41 18.	- 40 15	300		DARK OBJECT
RNGC 6207	16 41 19.	+ 36 56		12.5	GALAXY
ZWG 110.004	16 41 24.	+ 20 19		15.7	GALAXY
ZWG 138.060	16 41 24.	+ 21 47		15.6	GALAXY
MCG+10-24-037	16 41 24.	+ 61 15	9	18.	GALAXY
B 233	16 41 27.	- 35 19	3300		DARK OBJECT
HN 1826	16 41 28.1	+ 01 35 12.	12		NEBULA
MCG+07-34-141	16 41 30.	+ 42 16	60	15.	GALAXY
ZWG 299.020	16 41 30.	+ 61 55		15.5	GALAXY
UGC 10522	16 41 30.	+ 61 55	60	15.5	GALAXY Sc
ZWG 053.002	16 41 36.	+ 07 32		15.4	GALAXY
ZWG 138.061	16 41 36.	+ 24 19		15.7	GALAXY
ZWG 224.095	16 41 36.	+ 42 16		15.3	GALAXY
UGC 10523	16 41 36.	+ 42 16	78	15.3	GALAXY S
ZWG 251.041	16 41 36.	+ 45 56		15.3	GALAXY
MCG+08-30-048	16 41 36.	+ 45 56	24	16.	GALAXY
ZWG 276.038	16 41 36.	+ 51 40		15.4	GALAXY
ZWG 276.039	16 41 36.	+ 52 20		15.7	GALAXY
7ZW 656	16 41 36.	+ 62 59			COMPACT GALAXY
MCG+07-34-142	16 41 39.	+ 39 52	30	15.	GALAXY
RNGC 6212	16 41 39.	+ 39 54		15.0	GALAXY
ZWG 224.096	16 41 42.	+ 39 54		15.0	GALAXY
MCG+09-27-067	16 41 42.	+ 52 20	12	16.	GALAXY
MCG+10-24-038	16 41 42.	+ 61 11	24	16.	GALAXY
MCG+11-20-026	16 41 42.	+ 65 40	96	15.	GALAXY
ZC 1641.7+6541	16 41 42.	+ 65 41	1280		CLUSTER OF GALAXIES
ISS 1003	16 41 42.	- 35 45	109		STELLAR RING
SC 1637-5118.3	16 41 42.	- 51 24 02.	6		NEBULA
ARC 2227	16 41 46.	+ 51 22		17.4	RICH CLUSTER OF GALAXIES
ZWG 138.062	16 41 48.	+ 25 05		15.5	GALAXY
ZWG 320.038	16 41 49.	+ 65 40		14.5	GALAXY
UGC 10524	16 41 48.	+ 65 40	96	14.5	GALAXY SBb
ZC 1641.9+1129	16 41 54.	+ 11 29	940		CLUSTER OF GALAXIES
ZWG 081.002	16 41 54.	+ 14 23		15.6	GALAXY
ZWG 138.063	16 41 54.	+ 23 29		15.7	GALAXY
UGC 10525	16 41 54.	+ 23 29	60	15.7	GALAXY Sb
KARA.73B 0767	16 41 54.	+ 23 29	66	15.7	ISOLATED GALAXY S
ZC 1641.9+5303	16 41 54.	+ 53 03	940		CLUSTER OF GALAXIES
IC 1228	16 41 55.	+ 65 40 44.			NONSTELLAR OBJECT
KHAV 273	16 42	- 21 06	5590		DARK NEBULA
KHAV 272	16 42	- 33 42	1770		DARK NEBULA
ZWG 110.005	16 42 00.	+ 17 51		15.2	GALAXY
ZC 1642.0+4310	16 42 00.	+ 43 10	740		CLUSTER OF GALAXIES
ZWG 367.015	16 42 00.	+ 82 07		15.1	GALAXY
LDN 1765	16 42 00.	- 21 10	1140		DARK NEBULA
HN 1145	16 42 01.	+ 17 50			NEBULA
IC 4619	16 42 02.	+ 17 51			NONSTELLAR OBJECT
ZC 1642.1+2314	16 42 06.	+ 23 14	1340		CLUSTER OF GALAXIES
OCL 0973	16 42 06.	- 50 42	360		OPEN STAR CLUSTER
ZWG 053.003	16 42 12.	+ 08 15		15.4	GALAXY
ZWG 081.003	16 42 12.	+ 14 20		15.7	GALAXY
MCG+10-24-039	16 42 12.	+ 62 07	36	16.	GALAXY
OCL 0977	16 42 12.	- 47 40	240		OPEN STAR CLUSTER
LB 00954	16 42 14.	+ 43 52 24.		17.5	FAINT BLUE STAR
ZWG 110.006	16 42 18.	+ 20 01		15.4	GALAXY
KARA.73B 0768	16 42 18.	+ 20 01	24	15.4	ISOLATED GALAXY E
ZWG 138.064	16 42 18.	+ 25 13		15.5	GALAXY
UGC 10526	16 42 18.	+ 25 13	60	15.5	GALAXY Sb-c
VHE 73E	16 42 18.	- 41 08			REFLECTION NEBULA
ZC 1642.4+1428	16 42 24.	+ 14 28	2350		CLUSTER OF GALAXIES
ZWG 138.065	16 42 24.	+ 26 09		15.7	GALAXY
RNGC 6210	16 42 27.	+ 23 53		9.5	PLANETARY NEBULA
TON-N 0260	16 42 30.	+ 24 27		16.0	BLUE STAR
ZWG 138.066	16 42 30.	+ 26 45		15.5	GALAXY
MCG+07-34-143	16 42 30.	+ 39 51	45	15.	GALAXY
ZWG 224.097	16 42 30.	+ 39 53		15.1	GALAXY
ZC 1642.5+4544	16 42 30.	+ 45 44	1080		CLUSTER OF GALAXIES
7ZW 657	16 42 30.	+ 61 40			COMPACT GALAXY
ZWG 299.021	16 42 30.	+ 61 40		13.1	GALAXY
RNGC 6223	16 42 30.	+ 61 40		13.0	GALAXY
UGC 10527	16 42 30.	+ 61 40	210	13.1	GALAXY PECULR
MCG+10-24-040	16 42 30.	+ 61 42	60	11.	GALAXY
MCG-01-43-001	16 42 30.	- 07 05	60	16.	GALAXY
VHE 73C	16 42 30.	- 41 07	30		REFLECTION NEBULA
MCG+05-39-007	16 42 33.	+ 32 18 30.	48	15.4	GALAXY
MCG+04-39-021	16 42 36.	+ 22 35	120	13.5	GALAXY
ZWG 138.067	16 42 36.	+ 22 37		13.5	GALAXY
UGC 10528	16 42 36.	+ 22 37	144	13.5	GALAXY S0
KARA.73B 0769	16 42 36.	+ 22 37	126	13.5	ISOLATED GALAXY S
ZWG 168.030	16 42 36.	+ 32 18		15.4	GALAXY
UGC 10529	16 42 36.	+ 32 18	72	15.4	GALAXY Sb-c
ZWG 168.031	16 42 36.	+ 32 40		15.6	GALAXY
UGC 10530	16 42 36.	+ 49 36	72	17.	GALAXY S
ARC 2229	16 42 37.	+ 65 43		17.7	RICH CLUSTER OF GALAXIES
ZWG 138.068	16 42 42.	+ 21 52		15.7	GALAXY
MCG+07-34-144	16 42 42.	+ 43 49	48	15.	GALAXY
MCG+10-24-041	16 42 42.	+ 57 08	18	17.	GALAXY
RNGC 6204	16 42 45.	- 46 56		8.5	OPEN CLUSTER
HN 1827	16 42 45.6	+ 01 49 59.			NEBULA
ZWG 197.008	16 42 48.	+ 33 34		15.7	GALAXY
ZWG 224.098	16 42 48.	+ 39 01		15.6	GALAXY
MCG+10-24-042	16 42 48.	+ 62 04	42	16.	GALAXY
MCG+10-24-043	16 42 48.	+ 62 05	48	13.	GALAXY
VHE 73D	16 42 48.	- 41 06			REFLECTION NEBULA
OCL 0982	16 42 48.	- 46 56	2340	8.7	OPEN STAR CLUSTER
VHA 196	16 42 48.	- 46 56	240		STAR CLSTR IN NEBULOSITY
LB 00955	16 42 49.	+ 45 29 06.		16.5	FAINT BLUE STAR
UGC 10531	16 42 54.	+ 43 50	60	16.5	GALAXY Sb-c
ZWG 299.022	16 42 54.	+ 62 04		13.8	GALAXY
UGC 10532	16 42 54.	+ 62 04	48	13.8	GALAXY PECULR
VHE 73A	16 42 54.	- 41 09	96		REFLECTION NEBULA
RNGC 6226	16 42 55.	+ 62 04		14.0	GALAXY
SC 1640+1612.9	16 42 58.	+ 16 07 19.	12		NEBULA
SCHO 0268	16 42 59.	- 39 00 06.	1760		ISOLATED DARK CLOUD
B 045	16 43	- 21 30	7200		DARK OBJECT
KHAV 274	16 43	- 30 12	1570		DARK NEBULA
ZWG 053.004	16 43 00.	+ 08 56		15.6	GALAXY
MCG+01-43-001	16 43 00.	+ 08 56	30	15.6	GALAXY
ZWG 138.069	16 43 00.	+ 22 14		15.7	GALAXY
ZWG 276.040	16 43 00.	+ 51 37		15.7	GALAXY
VHE 73B	16 43 00.	- 41 08	12		REFLECTION NEBULA
OCL 0981	16 43 00.	- 47 01	90		OPEN STAR CLUSTER

OBJECT NAME	RIGHT ASCEN.	DECLINATION	DIAM.	MAGN.	TYPE OF OBJECT
ZC 1643.1+3717	16 43 06.	+ 37 17	1550		CLUSTER OF GALAXIES
ZWG 224.099	16 43 06.	+ 40 05		15.6	GALAXY
MCG+09-27-068	16 43 06.	+ 51 38	48	14.	GALAXY
ISS 1049	16 43 06.	- 40 17	99		STELLAR RING
SC 1640+1822.2	16 43 12.	+ 18 16 38.	18		NEBULA
ZWG 110.007	16 43 12.	+ 18 18		15.3	GALAXY
UGC 10533	16 43 12.	+ 18 18	60	15.3	GALAXY S
ZWG 168.032	16 43 12.	+ 31 23		15.5	GALAXY
MCG+12-16-007	16 43 12.	+ 70 43	84	14.	GALAXY
VHA 197	16 43 12.	- 45 44	180		OPEN STAR CLUSTER
SER 125.01	16 43 12.	- 57 25	1000		INTERACTING GALAXIES
SC 1641+1822.8	16 43 13.	+ 18 17 14.	48		NEBULA
B 234	16 43 16.	- 30 23	1800		DARK OBJECT
HN 1825	16 43 17.6	- 60 03 38.	60		NEBULA
MCG+03-43-001	16 43 18.	+ 18 17 30.	15	15.3	GALAXY
MCG+03-43-002	16 43 18.	+ 18 18	45	15.3	GALAXY
PK345+04.1	16 43 18.	- 38 32		13.6	PLANETARY NEBULA
ZWG 053.005	16 43 24.	+ 05 26		15.5	GALAXY
UGC 10534	16 43 24.	+ 58 58	60	17.	GALAXY Sc
LB 00956	16 43 26.	+ 46 47 42.		16.6	FAINT BLUE STAR
ZWG 168.033	16 43 30.	+ 27 04		15.6	GALAXY
LB 00957	16 43 30.	+ 44 53 24.		17.6	FAINT BLUE STAR
MCG+10-24-044	16 43 30.	+ 58 56	60	17.	GALAXY
MCG+10-24-045	16 43 30.	+ 61 27	30	16.	GALAXY
MCG+10-24-046	16 43 30.	+ 62 12	120	14.	GALAXY
LDN 0008	16 43 30.	- 19 10	1140		DARK NEBULA
ISS 1050	16 43 30.	- 39 40	81		STELLAR RING
B 235	16 43 31.	- 44 23	420		DARK OBJECT
ZWG 053.006	16 43 36.	+ 06 34		15.2	GALAXY
UGC 10535	16 43 36.	+ 06 34	72	15.2	GALAXY Sa-b
OCL 0129	16 43 36.	+ 38 22	1020		OPEN STAR CLUSTER
MCG+09-27-069	16 43 36.	+ 51 36	48	15.	GALAXY
MCG+10-24-047	16 43 36.	+ 61 27	6	16.	GALAXY
ZWG 299.023	16 43 36.	+ 62 10		15.0	GALAXY
UGC 10536	16 43 36.	+ 62 10	150	15.0	GALAXY Sb
LDN 0148	16 43 36.	- 14 05	420		DARK NEBULA
ISS 1004	16 43 36.	- 36 09	206		STELLAR RING
ZC 1643.7+3106	16 43 42.	+ 31 06	3560		CLUSTER OF GALAXIES
MCG+08-30-049	16 43 42.	+ 47 09	15	16.	GALAXY
MCG+09-27-071	16 43 42.	+ 51 21	18	16.	GALAXY
MCG+09-27-070	16 43 42.	+ 52 52	36	16.	GALAXY
7ZW 658	16 43 42.	+ 58 03			COMPACT GALAXY
ZWG 339.016	16 43 42.	+ 70 44		13.5	GALAXY
UGC 10537	16 43 42.	+ 70 44	108	13.5	GALAXY SBa
RNGC 6232	16 43 43.	+ 70 44		13.5	GALAXY
IC 1229	16 43 44.	+ 51 22 50.			NONSTELLAR OBJECT
ZWG 168.034	16 43 48.	+ 27 37		15.5	GALAXY
ZWG 251.042	16 43 48.	+ 50 54		15.6	GALAXY
MCG+09-27-074	16 43 48.	+ 50 55	12	16.	GALAXY
MCG+09-27-073	16 43 49.	+ 51 21	24	16.	GALAXY
ZWG 276.041	16 43 48.	+ 51 22		15.5	GALAXY
UGC 10538	16 43 48.	+ 51 22	60	15.5	GALAXY
MCG+09-27-072	16 43 48.	+ 51 24	24	16.	GALAXY
ZWG 339.017	16 43 48.	+ 73 45		15.0	GALAXY
ZWG 081.004	16 43 54.	+ 09 08		15.2	GALAXY
RNGC 6219	16 43 54.	+ 09 08		15.0	GALAXY
MCG+02-43-001	16 43 54.	+ 09 08	36	15.2	GALAXY
ZWG 110.008	16 43 54.	+ 18 18		15.4	GALAXY
ZWG 138.070	16 43 54.	+ 22 04		15.6	GALAXY
SC 1641+1822.8	16 43 56.	+ 18 17 17.	18		NEBULA
KHAV 227	16 44	- 12 17	5840		DARK NEBULA
KHAV 276	16 44	- 33 12	2840		DARK NEBULA
KHAV 275	16 44	- 38 12	5480		DARK NEBULA
ZWG 168.035	16 44 00.	+ 29 03		15.7	GALAXY
UGC 10539	16 44 00.	+ 58 10	60	16.5	GALAXY Sb-c
ZWG 367.016	16 44 00.	+ 81 47		15.6	GALAXY
LDN 0022	16 44 00.	- 18 50	2100		DARK NEBULA
LDN 1799	16 44 00.	- 19 35	480		DARK NEBULA
IC 1230	16 44 02.	+ 51 18 53.			NONSTELLAR OBJECT
ZWG 053.007	16 44 06.	+ 06 39		15.6	GALAXY
ZWG 138.071	16 44 06.	+ 21 58		15.7	GALAXY
MCG+10-24-048	16 44 06.	+ 58 08	60	16.	GALAXY
IC 1200	16 44 08.	+ 69 53			NONSTELLAR OBJECT
SCHO 0269	16 44 08.	- 33 20 00.	960		ISOLATED DARK CLOUD
UGC 10540	16 44 12.	+ 30 04	66	16.5	GALAXY Sb
MCG+07-34-145	16 44 12.	+ 43 57	39	16.	GALAXY
SHAH 013	16 44 12.	+ 53 48	102	17.5	GROUP OF COMPACT GALAXIES
ZWG 224.100	16 44 18.	+ 43 57		15.7	GALAXY
1ZW 164	16 44 18.	+ 44 44			COMPACT GALAXY
MCG+09-27-075	16 44 18.	+ 52 01	36	16.	GALAXY
SCHO 0270	16 44 19.	- 37 53 48.	1040		ISOLATED DARK CLOUD
MCG+07-34-146	16 44 24.	+ 42 32	30	15.	GALAXY
ZWG 276.042	16 44 24.	+ 52 01		15.6	GALAXY
TON-N 0261	16 44 30.	+ 26 43		15.8	BLUE STAR
ZWG 169.001	16 44 30.	+ 27 25		15.5	GALAXY
ZWG 168.036	16 44 30.	+ 27 25		15.5	GALAXY
MCG+10-24-049	16 44 30.	+ 61 57	84	15.	GALAXY
ZWG 339.018	16 44 30.	+ 74 47		15.1	GALAXY
KARA.73B 0770	16 44 30.	+ 74 47	30	15.1	ISOLATED GALAXY E
HN 1829	16 44 33.7	+ 01 01 07.	19		NEBULA
ZWG 299.024	16 44 36.	+ 61 55		15.7	GALAXY
UGC 10542	16 44 36.	+ 61 55	96	15.7	GALAXY Sc
ZWG 025.004	16 44 36.	- 00 11		15.5	GALAXY
UGC 10541	16 44 36.	- 00 11	120	15.5	GALAXY Sa?
GCL 046	16 44 36.	- 01 52	1290	7.95	GLOBULAR STAR CLUSTER
RNGC 6218	16 44 36.	- 01 52		8.0	GLOBULAR CLUSTER
MCG-05-39-002	16 44 36.	- 31 26	24	15.	GALAXY
OCL 0985	16 44 36.	- 45 45	120		OPEN STAR CLUSTER
RNGC 6220	16 44 37.	- 00 11		15.5	GALAXY
ZWG 138.072	16 44 42.	+ 22 59		15.6	GALAXY
KARA.73B 0771	16 44 42.	+ 22 59	24	15.6	ISOLATED GALAXY E
ZWG 169.002	16 44 42.	+ 27 02		15.5	GALAXY
ZWG 168.037	16 44 42.	+ 27 02		15.5	GALAXY
UGC 10543	16 44 42.	+ 27 02	60	15.5	GALAXY SBa
MCG+10-24-050	16 44 42.	+ 62 31	24	16.	GALAXY
MCG+05-39-008	16 44 45.	+ 27 00	42	15.5	GALAXY
ZWG 138.073	16 44 48.	+ 22 36		15.6	GALAXY
MCG+09-27-076	16 44 48.	+ 51 53	36	17.	GALAXY
LB 03564	16 44 50.	- 23 45 54.		17.6	FAINT BLUE STAR
MCG+06-37-009	16 44 51.	+ 36 10	54	14.5	GALAXY
ZWG 197.009	16 44 54.	+ 36 10		15.4	GALAXY
UGC 10544	16 44 54.	+ 36 10	60	15.4	GALAXY Sa
MCG+07-34-147	16 44 54.	+ 43 06	30	15.	GALAXY
ZC 1644.9+4411	16 44 54.	+ 44 11	1410		CLUSTER OF GALAXIES
MCG+09-27-077	16 44 54.	+ 51 57 30.	42	16.	GALAXY
LBN 0406	16 45	+ 60 20	4500		BRIGHT NEBULA
LBN 0415	16 45	+ 61 30	6900		BRIGHT NEBULA
KHAV 278	16 45	- 13 53	3610		DARK NEBULA

OBJECT NAME	RIGHT ASCEN.	DECLINATION	DIAM.	MAGN.	TYPE OF OBJECT
KHAV 279	16 45	- 19 11	4650		DARK NEBULA
LBN 1118	16 45	- 24 20	9900		BRIGHT NEBULA
B 236	16 45	- 29 43			DARK OBJECT
ZWG 197.010	16 45 00.	+ 34 30		15.1	GALAXY
UGC 10545	16 45 00.	+ 34 30	60	15.1	GALAXY Sb-c
ZWG 224.101	16 45 00.	+ 43 06		15.5	GALAXY
MCG+12-16-008	16 45 00.	+ 70 51	168	13.	GALAXY
ZWG 339.019	16 45 00.	+ 70 53		12.7	GALAXY
UGC 10546	16 45 00.	+ 70 53	180	12.7	GALAXY Sc/SBc
ZWG 025.005	16 45 00.	- 01 20		15.6	GALAXY
LDN 0260	16 45 00.	- 09 30	1080		DARK NEBULA
LDN 0255	16 45 00.	- 09 50	720		DARK NEBULA
LDN 0204	16 45 00.	- 12 00	1620		DARK NEBULA
LDN 0191	16 45 00.	- 12 40	1560		DARK NEBULA
LDN 0158	16 45 00.	- 14 00	960		DARK NEBULA
VHE 74	16 45 00.	- 48 00	60		REFLECTION NEBULA
RNGC 6236	16 45 02.	+ 70 53		12.5	GALAXY
ZC 1645.1+0620	16 45 06.	+ 06 20	3630		CLUSTER OF GALAXIES
ZWG 053.008	16 45 06.	+ 08 51		15.7	GALAXY
ZC 1645.1+2933	16 45 06.	+ 29 33	3230		CLUSTER OF GALAXIES
ZWG 197.011	16 45 06.	+ 34 15		14.7	GALAXY
UGC 10547	16 45 06.	+ 34 15	90	14.7	GALAXY SBb
MCG+06-37-010	16 45 06.	+ 34 15	66	14.	GALAXY
MCG+06-37-009	16 45 06.	+ 34 30	60	14.5	GALAXY
ZC 1645.1+5540	16 45 06.	+ 55 40	940		CLUSTER OF GALAXIES
ZWG 299.025	16 45 06.	+ 59 44		14.7	GALAXY
UGC 10548	16 45 06.	+ 59 44	96	14.7	GALAXY SBb
ZWG 053.009	16 45 12.	+ 06 44		15.7	GALAXY
MCG+10-24-051	16 45 12.	+ 59 42	84	14.	GALAXY
MCG+13-12-010	16 45 12.	+ 75 22	30	17.	GALAXY
ZWG 139.001	16 45 18.	+ 21 13		15.7	GALAXY
UGC 10549	16 45 18.	+ 21 13	66	15.7	GALAXY IRR
MCG+07-34-148	16 45 18.	+ 40 13	66	16.	GALAXY
ZWG 320.039	16 45 18.	+ 68 01		15.0	GALAXY
KARA.72 501A	16 45 18.	+ 68 01	66	15.0	PART OF DOUBLE GALAXY
UGC 10550	16 45 24.	+ 40 13	84	16.0	GALAXY Sc
MCG+09-27-078	16 45 24.	+ 51 56	36	17.	GALAXY
MCG+09-27-079	16 45 24.	+ 54 17 30.	42	16.	GALAXY
ZC 1645.4+6100	16 45 24.	+ 61 00	1880		CLUSTER OF GALAXIES
MCG+11-20-027	16 45 24.	+ 68 01	39	16.	GALAXY
ZWG 320.040	16 45 24.	+ 68 02		14.9	GALAXY
KARA.72 501B	16 45 24.	+ 68 02	66	14.9	PART OF DOUBLE GALAXY
MCG+07-34-149	16 45 30.	+ 40 18 30.	60	14.5	GALAXY
MCG+10-24-052	16 45 30.	+ 62 20	48	15.	GALAXY
ZWG 299.026	16 45 30.	+ 62 40		15.7	GALAXY
UGC 10551	16 45 30.	+ 63 35	60	16.0	GALAXY Sc
MCG+11-20-028	16 45 30.	+ 68 02	36	16.	GALAXY
ZWG 355.018	16 45 30.	+ 78 29		15.4	GALAXY
UGC 10552	16 45 30.	+ 78 29	66	15.4	GALAXY
LDN 0190	16 45 30.	- 12 50	600		DARK NEBULA
OCL 0964	16 45 30.	- 53 44	1980	9.6	OPEN STAR CLUSTER
VHA 198	16 45 30.	- 53 44	720		OPEN STAR CLUSTER
RNGC 6208	16 45 30.	- 53 44		9.0	OPEN CLUSTER
ARC 2231	16 45 31.	+ 56 36		17.9	RICH CLUSTER OF GALAXIES
REIN 2.247	16 45 34.38	+ 47 36 56.1			NEBULA
PK347+05.1	16 45 34.4	- 35 41 56.	5		PLANETARY NEBULA
ZWG 224.102	16 45 36.	+ 40 20		15.0	GALAXY
UGC 10553	16 45 36.	+ 40 20	66	15.0	GALAXY SB:a-b
GCL 047	16 45 36.	+ 47 37	228	10.26	GLOBULAR STAR CLUSTER
RNGC 6229	16 45 36.	+ 47 37		10.5	GLOBULAR CLUSTER
ZC 1645.6+5634	16 45 36.	+ 56 34	740		CLUSTER OF GALAXIES
ZWG 299.027	16 45 36.	+ 62 19		14.9	GALAXY
PK359+15.1	16 45 36.	- 20 56	30	16.8	PLANETARY NEBULA
ZC 1645.7+2254	16 45 42.	+ 22 54	1550		CLUSTER OF GALAXIES
TON-N 0262	16 45 42.	+ 26 45		15.1	BLUE STAR
OCL 0995	16 45 42.	- 43 21	480		OPEN STAR CLUSTER
B 237	16 45 45.	- 29 53			DARK OBJECT
SCHO 0271	16 45 46.	- 33 40 48.	890		ISOLATED DARK CLOUD
RNGC 6224	16 45 47.	+ 06 24		15.0	GALAXY
ARC 2230	16 45 47.	+ 48 42		16.8	RICH CLUSTER OF GALAXIES
ZWG 053.010	16 45 48.	+ 06 24		15.0	GALAXY
UGC 10555	16 45 48.	+ 06 24	96	15.0	GALAXY E-S0
MCG+01-43-002	16 45 48.	+ 06 24	36	15.0	GALAXY
ARC 2228	16 45 48.	+ 30 02		16.9	RICH CLUSTER OF GALAXIES
ZC 1645.8+4835	16 45 48.	+ 48 35	2890		CLUSTER OF GALAXIES
MCG+10-24-053	16 45 48.	+ 62 32	48	16.	GALAXY
MCG+13-12-011	16 45 48.	+ 75 44	51	17.	GALAXY
ZWG 025.006	16 45 48.	- 01 32		15.7	GALAXY
UGC 10554	16 45 48.	- 01 32	90	15.7	GALAXY Sc
RNGC 6216	16 45 48.	- 44 38			NON-EXISTENT OBJECT
OCL 0989	16 45 48.	- 44 39	2340	10.2	OPEN STAR CLUSTER
VHA 199	16 45 48.	- 44 39	270		OPEN STAR CLUSTER
ARC 2232	16 45 49.	+ 61 49		18.1	RICH CLUSTER OF GALAXIES
RNGC 6225	16 45 53.	+ 06 19		15.0	GALAXY
ZWG 053.011	16 45 54.	+ 06 19		15.0	GALAXY
UGC 10556	16 45 54.	+ 06 19	72	15.0	GALAXY E-S0
MCG+01-43-003	16 45 54.	+ 06 19	30	15.0	GALAXY
ZWG 081.005	16 45 54.	+ 13 59		15.3	GALAXY
MCG+02-43-002	16 45 54.	+ 13 59	60	15.3	GALAXY
ZWG 110.009	16 45 54.	+ 18 07		15.6	GALAXY
MCG+09-27-080	16 45 54.	+ 52 22	30	16.	GALAXY
ZWG 320.041	16 45 54.	+ 65 37		15.7	GALAXY
RNGC 6245	16 45 56.	+ 70 54			GALAXY
SC 1643+1811.2	16 45 58.	+ 18 05 49.	18		NEBULA
RNGC 6228	16 45 59.	+ 26 18		15.5	GALAXY
KHAV 280	16 46	- 17 47	3470		DARK NEBULA
UGC 10557	16 46 00.	+ 13 58	66	16.5	GALAXY SBb-c
MCG+03-43-003	16 46 00.	+ 18 07	30	15.6	GALAXY
ZWG 139.002	16 46 00.	+ 25 00		15.7	GALAXY
ZWG 139.003	16 46 00.	+ 26 18		15.3	GALAXY
UGC 10558	16 46 00.	+ 26 18	66	15.3	GALAXY S
MCG+04-40-001	16 46 00.	+ 26 18	66	15.3	GALAXY
MCG+08-30-050	16 46 00.	+ 50 39	30	16.	GALAXY
MCG+09-27-081	16 46 00.	+ 53 40	36	16.	GALAXY
MCG+09-27-082	16 46 00.	+ 54 02 30.	42	16.	GALAXY
MCG+10-24-054	16 46 00.	+ 59 10	60	15.	GALAXY
ZWG 299.028	16 46 00.	+ 59 12		15.4	GALAXY
UGC 10559	16 46 00.	+ 59 12	66	15.4	GALAXY
MCG+10-24-055	16 46 00.	+ 62 54	138	13.	GALAXY
ZWG 320.042	16 46 00.	+ 63 30		15.6	GALAXY
ZC 1646.0+8236	16 46 00.	+ 82 36	1340		CLUSTER OF GALAXIES
LDN 0234	16 46 00.	- 11 00	4800		DARK NEBULA
LDN 1748	16 46 00.	- 23 20	1500		DARK NEBULA
LDN 1745	16 46 00.	- 23 50	2040		DARK NEBULA
ISS 1051	16 46 00.	- 39 42	155		STELLAR RING
FATH 1.761	16 46 04.	+ 45 14	5		NEBULA
ZWG 299.029	16 46 06.	+ 58 32		13.7	GALAXY
UGC 10560	16 46 06.	+ 58 32	144	13.7	GALAXY Sc

OBJECT NAME	RIGHT ASCEN.	DECLINATION	DIAM.	MAGN.	TYPE OF OBJECT
KARA.73B 0772	16 46 06.	+ 58 32	132	13.7	ISOLATED GALAXY S
ZWG 299.030	16 46 06.	+ 62 54		15.2	GALAXY
UGC 10561	16 46 06.	+ 62 54	150	15.2	GALAXY Sb
ZWG 355.019	16 46 06.	+ 75 47		15.0	GALAXY
UGC 10562	16 46 06.	+ 75 47	84	15.0	GALAXY
PK344+03.1	16 46 06.	- 19 15			PLANETARY NEBULA
IC 1231	16 46 10.	+ 58 31 35.			NONSTELLAR OBJECT
ZWG 169.003	16 46 12.	+ 30 10		15.7	GALAXY
ZC 1646.2+4016	16 46 12.	+ 40 16	1680		CLUSTER OF GALAXIES
MCG+10-24-056	16 46 12.	+ 58 30	120	13.	GALAXY
MCG+09-27-083	16 46 15.	+ 53 00	36	16.	GALAXY
HN 1146	16 46 17.	+ 19 23			NEBULA
ZWG 110.010	16 46 18.	+ 19 23		15.6	GALAXY
ZWG 139.004	16 46 18.	+ 22 42		15.7	GALAXY
ZWG 320.043	16 46 18.	+ 65 38		15.3	GALAXY
MCG+13-12-012	16 46 18.	+ 75 45	39	16.	GALAXY
LDN 0162	16 46 18.	- 14 10	1380		DARK NEBULA
ISS 1005	16 46 18.	- 37 36	91		STELLAR RING
IC 4620	16 46 20.	+ 18 22			NONSTELLAR OBJECT
HN 1828	16 46 22.9	- 61 43 44.			NEBULA
MCG+03-43-004	16 46 24.	+ 19 24	27	15.6	GALAXY
ZC 1646.4+4659	16 46 24.	+ 46 59	1610		CLUSTER OF GALAXIES
ZWG 276.043	16 46 24.	+ 53 57		15.7	GALAXY
VHA 200	16 46 24.	- 44 08	210		OPEN STAR CLUSTER
LB 00320	16 46 27.	+ 11 53 00.		14.2	FAINT BLUE STAR
FATH 1.762	16 46 27.	+ 45 53	14		NEBULA
ZC 1646.5+1945	16 46 30.	+ 19 45	1410		CLUSTER OF GALAXIES
ZWG 197.012	16 46 30.	+ 35 55		15.3	GALAXY
ZWG 197.013	16 46 30.	+ 35 59		15.3	GALAXY
MCG+09-27-084	16 46 30.	+ 53 56	24	16.	GALAXY
MCG+09-27-085	16 46 30.	+ 53 57 30.	24	15.	GALAXY
ZWG 277.001	16 46 30.	+ 56 50		15.4	GALAXY
ZWG 276.044	16 46 30.	+ 56 50		15.4	GALAXY
LB 00321	16 46 34.	+ 13 37 42.		14.5	FAINT BLUE STAR
HN 1830	16 46 35.0	- 60 43 25.	12		NEBULA
ZC 1646.6+6148	16 46 36.	+ 61 48	470		CLUSTER OF GALAXIES
ZWG 299.031	16 46 36.	+ 62 14		14.4	GALAXY
UGC 10563	16 46 36.	+ 62 14	30	14.4	GALAXY PECULR
ZWG 355.020	16 46 36.	+ 77 56		15.5	GALAXY
RNGC 6238	16 46 37.	+ 62 14		14.5	GALAXY
HOLM 756B	16 46 38.	+ 62 16	18	14.9	PART OF MULTIPLE GALAXY
HOLM 756A	16 46 39.	+ 62 16	24	14.4	PART OF MULTIPLE GALAXY
MCG+09-27-086	16 46 42.	+ 52 38	36	16.	GALAXY
MCG+10-24-057	16 46 42.	+ 62 15	42	15.	GALAXY
RNGC 6215	16 46 44.	- 58 55		12.0	GALAXY
ZWG 224.103	16 46 45.	+ 42 02		15.6	GALAXY
ZWG 339.020	16 46 48.	+ 70 28		14.1	GALAXY
UGC 10564	16 46 48.	+ 70 28	198	14.1	GALAXY SBc
TON-N 0263	16 46 54.	+ 25 03		15.0	BLUE STAR
MCG+06-37-011	16 46 54.	+ 36 00	42	14.5	GALAXY
ZWG 197.014	16 46 54.	+ 36 01		15.5	GALAXY
MCG+07-34-150	16 46 54.	+ 41 40	33	15.	GALAXY
UGC 10565	16 46 54.	+ 48 49	60	17.	GALAXY DWARF
1ZW 165	16 46 54.	+ 49 42			COMPACT GALAXY
MCG+12-16-009	16 46 54.	+ 70 26	186	13.	GALAXY
RNGC 6237	16 46 55.	+ 70 29			GALAXY
HN 1831	16 46 59.6	- 59 09 17.	18		NEBULA
IC 1232	16 47	+ 46 10 18.			NONSTELLAR OBJECT
KHAV 282	16 47	- 23 35	2960		DARK NEBULA
KHAV 281	16 47	- 31 29	5690		DARK NEBULA
1ZW 166	16 47 00.	+ 48 47			COMPACT GALAXY
ZWG 252.001	16 47 00.	+ 48 47		14.6	GALAXY
ZWG 251.043	16 47 00.	+ 48 47		14.6	GALAXY
MRK 499	16 47 00.	+ 48 47	12	15.	GALAXY WITH UV CONTINUUM
MCG+08-31-001	16 47 00.	+ 48 50	60	17.	GALAXY
ZC 1647.0+5943	16 47 00.	+ 59 43	1080		CLUSTER OF GALAXIES
LDN 0163	16 47 00.	- 14 15	720		DARK NEBULA
LDN 0152	16 47 00.	- 14 30	4140		DARK NEBULA
LDN 0110	16 47 00.	- 16 00	2400		DARK NEBULA
LDN 0063	16 47 00.	- 18 00	780		DARK NEBULA
LDN 0062	16 47 00.	- 18 00	5340		DARK NEBULA
LDN 0047	16 47 00.	- 18 30	2100		DARK NEBULA
LDN 0031	16 47 00.	- 19 00	600		DARK NEBULA
PK345+03.1	16 47 00.	- 39 03		14.6	PLANETARY NEBULA
PK351+09.1	16 47 04.	- 30 13			PLANETARY NEBULA
ZWG 053.012	16 47 06.	+ 06 06		15.3	GALAXY
TON-N 0264	16 47 06.	+ 25 16		14.2	BLUE STAR
ZWG 169.004	16 47 06.	+ 29 51		15.3	GALAXY
UGC 10566	16 47 06.	+ 36 16	120	16.0	GALAXY S
RNGC 6222	16 47 06.	- 44 39		10.0	OPEN CLUSTER
ZWG 081.006	16 47 12.	+ 09 52		15.7	GALAXY
MCG+06-37-012	16 47 12.	+ 36 17	114	15.	GALAXY
MRK 500	16 47 12.	+ 48 48	18	16.5	GALAXY WITH UV CONTINUUM
MCG+09-27-087	16 47 12.	+ 51 05	30	16.	GALAXY
MCG+10-24-058	16 47 12.	+ 60 55	30	15.	GALAXY
7ZW 659	16 47 12.	+ 67 46			COMPACT GALAXY
PK344+02.1	16 47 12.	- 39 58		14.8	PLANETARY NEBULA
ZWG 053.013	16 47 18.	+ 05 00		15.5	GALAXY
ZWG 139.005	16 47 18.	+ 23 01		15.6	GALAXY
ZWG 197.015	16 47 18.	+ 36 18		15.2	GALAXY
UGC 10567	16 47 18.	+ 36 18	66	15.2	GALAXY
MCG+06-37-013	16 47 18.	+ 36 19	30	14.5	GALAXY
MCG+09-27-088	16 47 18.	+ 54 06	36	15.	GALAXY
MCG+10-24-059	16 47 18.	+ 62 18	90	14.	GALAXY
SCHO 0272	16 47 20.	- 35 07 18.	1750		ISOLATED DARK CLOUD
FATH 1.763	16 47 22.	+ 45 07	5		NEBULA
ZC 1647.4+3801	16 47 24.	+ 38 01	1880		CLUSTER OF GALAXIES
ZC 1647.4+4908	16 47 24.	+ 49 08	1010		CLUSTER OF GALAXIES
MCG+10-24-060	16 47 24.	+ 61 33	30	16.	GALAXY
ZWG 110.011	16 47 30.	+ 17 57		15.1	GALAXY
MCG+03-43-005	16 47 30.	+ 17 57	42	15.1	GALAXY
MCG+09-27-089	16 47 30.	+ 54 04	42	15.	GALAXY
ZWG 299.032	16 47 30.	+ 62 17		14.3	GALAXY
UGC 10568	16 47 30.	+ 62 17	102	14.3	GALAXY SB:a
SCHO 0273	16 47 30.	- 39 20 12.	930		ISOLATED DARK CLOUD
RNGC 6244	16 47 32.	+ 62 17		14.5	GALAXY
ZWG 081.007	16 47 36.	+ 09 31		15.4	GALAXY
UGC 10569	16 47 36.	+ 09 31	114	15.4	GALAXY Sb
MCG+02-43-003	16 47 36.	+ 09 31	60	15.3	GALAXY
ZWG 110.012	16 47 36.	+ 16 43		15.7	GALAXY
MCG+03-43-006	16 47 36.	+ 16 43	30	15.7	GALAXY
ZWG 169.005	16 47 36.	+ 27 30		15.6	GALAXY
ZWG 197.016	16 47 36.	+ 36 08		15.7	GALAXY
ZC 1647.6+3730	16 47 36.	+ 37 30	1080		CLUSTER OF GALAXIES
ZC 1647.6+5337	16 47 36.	+ 53 37	6720		CLUSTER OF GALAXIES
UGC 10570	16 47 42.	+ 59 38	78	16.0	GALAXY Sc
ARP 330	16 47 45.	+ 53 29			PECULIAR GALAXY
RNGC 6247	16 47 46.	+ 63 04		13.5	GALAXY

OBJECT NAME	RIGHT ASCEN.	DECLINATION	DIAM.	MAGN.	TYPE OF OBJECT
ZWG 252.002	16 47 48.	+ 48 43		15.4	GALAXY
UGC 10571	16 47 48.	+ 48 43	78	15.4	GALAXY SB:c
MCG+10-24-061	16 47 48.	+ 59 35	60	15.	GALAXY
ZWG 320.044	16 47 48.	+ 63 04		13.5	GALAXY
UGC 10572	16 47 48.	+ 63 04	66	13.5	GALAXY PECULR
MCG+12-16-010	16 47 48.	+ 68 54	27	16.	GALAXY
MCG+08-31-002	16 47 54.	+ 48 42 30.	84	15.	GALAXY
MCG+09-27-090	16 47 54.	+ 53 05	54	16.	GALAXY
IC 1233	16 47 58.	+ 63 14 03.			NONSTELLAR OBJECT
KEEN 6246A	16 48	+ 55 47		13.5	GALAXY
KHAV 284	16 48	- 15 17	4700		DARK NEBULA
KHAV 283	16 48	- 35 17	4470		DARK NEBULA
ZWG 169.006	16 48 00.	+ 28 15		15.7	GALAXY
MCG+05-40-001	16 48 00.	+ 28 15	39	15.7	GALAXY
MCG+08-31-003	16 48 00.	+ 45 32 30.	21	15.2	GALAXY
ZWG 252.003	16 48 00.	+ 45 33		15.4	GALAXY
1ZW 167	16 48 00.	+ 53 30			COMPACT GALAXY
ZWG 277.002	16 48 00.	+ 53 31		15.5	GALAXY
ZWG 276.045	16 48 00.	+ 53 31		15.5	GALAXY
SHAH 016	16 48 00.	+ 53 31	300	14.	GROUP OF COMPACT GALAXIES
ZWG 355.021	16 48 00.	+ 75 22		15.7	GALAXY
LDN 0141	16 48 00.	- 15 10	1080		DARK NEBULA
LDN 0137	16 48 00.	- 15 20	3840		DARK NEBULA
LDN 1747	16 48 00.	- 23 45	2100		DARK NEBULA
RNGC 6233	16 48 03.	+ 23 40		15.0	GALAXY
HOLM 757B	16 48 03.	+ 45 33	12	15.5	PART OF MULTIPLE GALAXY
HOLM 757A	16 48 05.	+ 45 33	12	15.2	PART OF MULTIPLE GALAXY
ZWG 139.006	16 48 06.	+ 23 31		15.6	GALAXY
ZWG 139.007	16 48 06.	+ 23 40		14.9	GALAXY
UGC 10573	16 48 06.	+ 23 40	96	14.9	GALAXY S0
UGC 10574	16 48 06.	+ 27 07	66	16.0	GALAXY Sa-b
MCG+08-31-003A	16 48 06.	+ 45 34 30.	18	16.	GALAXY
MCG+09-27-091	16 48 06.	+ 53 27	15	16.	GALAXY
ZWG 277.003	16 48 06.	+ 55 11		15.6	GALAXY
ZWG 276.046	16 48 06.	+ 55 11		15.6	GALAXY
MCG+10-24-062	16 48 06.	+ 62 01	9	16.	GALAXY
RNGC 6227	16 48 09.	- 41 08			NON-EXISTENT OBJECT
RNGC 6230	16 48 10.	+ 04 41		15.5	GALAXY
ZWG 053.014	16 48 12.	+ 04 41		15.5	GALAXY
MCG+01-43-005	16 48 12.	+ 04 41	24	15.5	GALAXY
ZC 1648.2+1203	16 48 12.	+ 12 03	1880		CLUSTER OF GALAXIES
MCG+04-40-002	16 48 12.	+ 23 39	18	14.9	GALAXY
ZWG 225.001	16 48 12.	+ 40 31		15.7	GALAXY
ZWG 224.104	16 48 12.	+ 40 31		15.7	GALAXY
ARP 103	16 48 12.	+ 45 32			PECULIAR GALAXY
HMS 1.26	16 48 12.	+ 45 33			SO GALAXY
MCG+09-27-093	16 48 12.	+ 53 27 30.	6	19.	GALAXY
MCG+09-27-092	16 48 12.	+ 53 28	18	16.	GALAXY
MCG+09-27-094	16 48 12.	+ 53 29	18	16.	GALAXY
MCG+09-27-096	16 48 12.	+ 53 30	12	17.	GALAXY
MCG+09-27-095	16 48 12.	+ 53 31	15	17.	GALAXY
RNGC 6215A	16 48 14.	- 58 50			GALAXY
ZWG 081.008	16 48 18.	+ 12 05		15.7	GALAXY
MCG+02-43-005	16 48 18.	+ 12 05	24	15.7	GALAXY
MCG+02-43-004	16 48 18.	+ 12 05	48	15.7	GALAXY
ZC 1648.3+2632	16 48 18.	+ 26 32	2760		CLUSTER OF GALAXIES
MCG+08-31-004	16 48 18.	+ 46 47 30.	60	16.	GALAXY
MCG+08-31-006	16 48 18.	+ 46 48	18	15.7	GALAXY
MCG+08-31-005	16 48 18.	+ 46 48	24	15.4	GALAXY
DV.56 N6215A	16 48 18.	- 58 51	75		S GALAXY
HOLM 758A	16 48 20.	+ 46 48	9	15.4	PART OF MULTIPLE GALAXY
HOLM 758B	16 48 22.	+ 46 48	9	15.7	PART OF MULTIPLE GALAXY
UGC 10575	16 48 24.	+ 04 41	60	15.5	GALAXY E?
ZWG 053.015	16 48 24.	+ 08 53		15.0	GALAXY
UGC 10576	16 48 24.	+ 08 53	66	15.0	GALAXY Sa-b
MCG+01-43-004	16 48 24.	+ 08 53	36	15.0	GALAXY
LB 00958	16 48 24.	+ 46 30 36.		15.9	FAINT BLUE STAR
ZWG 252.004	16 48 24.	+ 46 48		15.6	GALAXY
VV 197B	16 48 24.	+ 46 49	12	16.5	INTERACTING GALAXY
VV 197B	16 48 24.	+ 46 49	42	16.	INTERACTING GALAXY
VV 197	16 48 24.	+ 46 49	48		INTERACTING GALAXY
IC 4621	16 48 26.	+ 08 52 54.			NONSTELLAR OBJECT
ARP 312	16 48 26.	+ 46 47			PECULIAR GALAXY
MCG+07-35-001	16 48 30.	+ 42 49	138	13.5	GALAXY
ZWG 225.002	16 48 30.	+ 42 50		12.9	GALAXY
ZWG 224.105	16 48 30.	+ 42 50		12.9	GALAXY
UGC 10577	16 48 30.	+ 42 50	150	12.9	GALAXY SB
ZC 1648.5+4440	16 48 30.	+ 44 40	1010		CLUSTER OF GALAXIES
RNGC 6239	16 48 31.	+ 42 49		13.0	GALAXY
RNGC 6221	16 48 32.	- 59 07		11.5	GALAXY
ZWG 299.033	16 48 36.	+ 59 00		15.5	GALAXY
MCG+10-24-063	16 48 36.	+ 62 00	42	16.	GALAXY
ZC 1648.6-0307	16 48 36.	- 03 07	6250		CLUSTER OF GALAXIES
RNGC 6241	16 48 40.	+ 45 30		15.0	GALAXY
FATH 2.157	16 48 41.	+ 45 30	16		NEBULA
MCG+01-43-006	16 48 42.	+ 05 04 30.	15	18.5	GALAXY
ZWG 252.005	16 48 42.	+ 45 30		14.8	GALAXY
MCG+08-31-007	16 48 42.	+ 45 30	42	15.	GALAXY
MCG+10-24-064	16 48 42.	+ 58 59	36	15.	GALAXY
SCHO 0274	16 48 42.	- 38 38 30.	400		ISOLATED DARK CLOUD
MRSL 343+01/1	16 48 42.	- 41 09	18060		HII REGION
LB 00959	16 48 45.	+ 46 59 24.		17.6	FAINT BLUE STAR
ZWG 277.004	16 48 48.	+ 54 02		14.4	GALAXY
ZWG 276.047	16 48 48.	+ 54 02		14.4	GALAXY
UGC 10579	16 48 48.	+ 54 02	72	14.4	GALAXY E
ZC 1648.8+5715	16 48 48.	+ 57 15	1140		CLUSTER OF GALAXIES
ZWG 025.008	16 48 48.	- 02 23		15.7	GALAXY
UGC 10578	16 48 48.	- 02 23	72	15.7	GALAXY
MCG+00-43-002	16 48 48.	- 02 23	48	15.7	GALAXY
ZWG 025.007	16 48 48.	- 02 44		15.6	GALAXY
CED 137A	16 48 48.	- 41 46			DIFFUSE GALACTIC NEBULA
ZWG 053.016	16 48 54.	+ 07 57		15.6	GALAXY
ZC 1648.9+5110	16 48 54.	+ 51 10	810		CLUSTER OF GALAXIES
MCG+09-27-097	16 48 54.	+ 54 03	30	14.	GALAXY
ZWG 277.005	16 48 54.	+ 55 38		14.2	GALAXY
ZWG 276.048	16 48 54.	+ 55 38		14.2	GALAXY
RNGC 6246A	16 48 54.	+ 55 38			GALAXY
UGC 10580	16 48 54.	+ 55 38	90	14.2	GALAXY SB?b
MCG+10-24-065	16 48 54.	+ 61 49	36	16.	GALAXY
MCG-01-43-002	16 48 54.	- 03 01	120	13.	GALAXY
IC 4623	16 48 57.	+ 22 36 35.			NONSTELLAR OBJECT
SCHO 0275	16 48 58.	- 41 39 36.	410		ISOLATED DARK CLOUD
LB 09905	16 49	- 84 11		13.7	FAINT BLUE STAR
ZWG 139.008	16 49 00.	+ 22 37		15.4	GALAXY
MCG+09-27-098	16 49 00.	+ 55 40	90	14.	GALAXY
MCG+10-24-066	16 49 00.	+ 62 02	30	16.	GALAXY
UGC 10581	16 49 00.	+ 82 43	60	17.	GALAXY
LDN 0244	16 49 00.	- 11 10	1140		DARK NEBULA
LDN 1750	16 49 00.	- 23 50	2100		DARK NEBULA
MCG+04-40-003	16 49 03.	+ 22 36	36	15.4	GALAXY
UGC 10582	16 49 06.	+ 24 02	66	16.5	GALAXY Sb-c
ZWG 277.006	16 49 06.	+ 53 39		15.5	GALAXY
ZWG 276.049	16 49 06.	+ 53 39		15.5	GALAXY
UGC 10583	16 49 06.	+ 53 39	72	15.5	GALAXY SBa
MCG+09-27-099	16 49 06.	+ 53 41	36	16.	GALAXY
MCG+10-24-067	16 49 06.	+ 61 51	12	16.	GALAXY
ZWG 025.009	16 49 06.	- 02 32		14.9	GALAXY
KARA.72 502A	16 49 06.	- 02 32	42	14.9	PART OF DOUBLE GALAXY
MCG+00-43-003	16 49 06.	- 02 32	48	14.9	GALAXY
KARA.72 502B	16 49 06.	- 02 33	48		PART OF DOUBLE GALAXY
LDN 1759	16 49 06.	- 23 00	660		DARK NEBULA
SCHO 0276	16 49 06.	- 38 47 54.	410		ISOLATED DARK CLOUD
ISS 1052	16 49 06.	- 40 14	180		STELLAR RING
RNGC 6209	16 49 08.	- 72 28			GALAXY
LB 00322	16 49 09.	+ 13 35 00.		15.7	FAINT BLUE STAR
RNGC 6246	16 49 11.	+ 55 28		14.0	GALAXY
ZWG 053.017	16 49 12.	+ 04 47		15.6	GALAXY
ZWG 169.007	16 49 12.	+ 30 45		15.6	GALAXY
ZWG 225.003	16 49 12.	+ 39 41		15.7	GALAXY
MCG+09-27-100	16 49 12.	+ 53 36	48	16.	GALAXY
ZWG 277.007	16 49 12.	+ 55 28		14.7	GALAXY
ZWG 276.050	16 49 12.	+ 55 28		14.7	GALAXY
UGC 10584	16 49 12.	+ 55 28	150	14.7	GALAXY Sc/SBc
MCG+12-16-010A	16 49 12.	+ 68 49	84	15.	GALAXY
ZC 1649.2-0016	16 49 12.	- 00 16	2620		CLUSTER OF GALAXIES
ISS 1053	16 49 12.	- 40 15	233		STELLAR RING
MCG+09-27-101	16 49 15.	+ 55 30	120	13.	GALAXY
HN 0434	16 49 16.	- 16 09			NEBULA
IC 4622	16 49 16.	- 16 09			MAY NOT EXIST
ZWG 110.013	16 49 18.	+ 17 32		15.7	GALAXY
ZWG 110.014	16 49 18.	+ 19 30		14.8	GALAXY
UGC 10585	16 49 18.	+ 19 30	72	14.8	GALAXY S0-a
MCG+03-43-007	16 49 18.	+ 19 30 30.	45	14.8	GALAXY
ZC 1649.3+2322	16 49 18.	+ 23 22	810		CLUSTER OF GALAXIES
MCG+08-31-008	16 49 18.	+ 45 28	102	15.	GALAXY
ZWG 252.006	16 49 18.	+ 45 29		15.1	GALAXY
UGC 10586	16 49 18.	+ 45 29	90	15.1	GALAXY Sb
ZWG 321.001	16 49 18.	+ 68 50		15.4	GALAXY
ZWG 320.045	16 49 18.	+ 68 50		15.4	GALAXY
UGC 10587	16 49 18.	+ 68 50	84	15.4	GALAXY Sa-b
IC 4624	16 49 19.	+ 17 31			NONSTELLAR OBJECT
FATH 1.764	16 49 19.	+ 45 29	5		NEBULA
LB 00323	16 49 20.	+ 15 03 00.		15.6	FAINT BLUE STAR
HN 1147	16 49 20.	+ 17 31			NEBULA
B 238	16 49 21.	- 23 03	780		DARK OBJECT
CED 137B	16 49 22.	- 40 23	2040		DIFFUSE GALACTIC NEBULA
UGC 10588	16 49 24.	+ 21 58	90	16.5	GALAXY SB:b-c
ZWG 277.008	16 49 24.	+ 55 55		15.4	GALAXY
ZWG 276.051	16 49 24.	+ 55 55		15.4	GALAXY
UGC 10589	16 49 24.	+ 55 55	108	15.4	GALAXY PECULR
ZWG 025.010	16 49 24.	- 02 34		15.6	GALAXY
FATH 1.765	16 49 25.	+ 45 29	5		NEBULA
ARP 208	16 49 26.	+ 47 19			PECULIAR GALAXY
RNGC 6234	16 49 28.	+ 04 28		15.5	GALAXY
ZWG 053.018	16 49 30.	+ 04 28		15.3	GALAXY
MCG+01-43-007	16 49 30.	+ 04 28	18	15.3	GALAXY
ZWG 139.009	16 49 30.	+ 23 48		15.7	GALAXY
ZWG 139.010	16 49 30.	+ 24 03		15.4	GALAXY
MCG+09-27-102	16 49 30.	+ 55 57	120	14.	GALAXY
SCHO 0277	16 49 30.	- 41 46 18.	290		ISOLATED DARK CLOUD
MCG+08-31-009	16 49 33.	+ 47 17 30.	42	17.	GALAXY
SCHO 0278	16 49 34.	- 41 52 48.	260		ISOLATED DARK CLOUD
ZWG 252.007	16 49 36.	+ 47 18		15.2	GALAXY
ZWG 277.009	16 49 36.	+ 53 45		15.4	GALAXY
ZWG 276.052	16 49 36.	+ 53 45		15.4	GALAXY
MCG+08-31-011	16 49 42.	+ 47 19	24	16.	GALAXY
MCG+08-31-010	16 49 42.	+ 47 19	24	16.	GALAXY
MCG+09-27-103	16 49 42.	+ 53 42	18	16.	GALAXY
MCG+09-27-104	16 49 45.	+ 53 45	18	16.	GALAXY
LB 00960	16 49 47.	+ 45 03 48.		17.7	FAINT BLUE STAR
MCG+10-24-068	16 49 48.	+ 59 59	36	17.	GALAXY
ABC 2234	16 49 49.	+ 56 31		17.7	RICH CLUSTER OF GALAXIES
PK341-00.1	16 49 49.	- 44 48	10		PLANETARY NEBULA
PK325-12.1	16 49 50.	- 64 09	10		PLANETARY NEBULA
YM 03	16 49 53.	- 45 07	204		SYMMETRIC GALACTIC NEBULA
ZC 1649.9+2343	16 49 54.	+ 23 43	7730		CLUSTER OF GALAXIES
ZC 1649.9+3320	16 49 54.	+ 33 20	4640		CLUSTER OF GALAXIES
VV 271B	16 49 54.	+ 47 19	18	16.	INTERACTING GALAXY
VV 271A	16 49 54.	+ 47 19	24	15.	INTERACTING GALAXY
VV 271	16 49 54.	+ 47 19	36		INTERACTING GALAXY
PK342+00.1	16 49 58.7	- 42 34 28.	18		PLANETARY NEBULA
LBN 0412	16 50	+ 60 50	1560		BRIGHT NEBULA
KHAV 285	16 50	- 37 29	3070		DARK NEBULA
ZC 1650.0+0705	16 50 00.	+ 07 05	1280		CLUSTER OF GALAXIES
ZC 1650.0+5155	16 50 00.	+ 51 55	1410		CLUSTER OF GALAXIES
ZC 1650.0+5633	16 50 00.	+ 56 33	1010		CLUSTER OF GALAXIES
MCG+10-24-069	16 50 00.	+ 59 48	60	13.	GALAXY
LDN 1677	16 50 00.	- 32 00	12900		DARK NEBULA
ASS 67	16 50 00.	- 41 52	10800		OB ASSOCIATION SCO OB1
MRSL 340-00/1	16 50 00.	- 45 00	420		HII REGION
SCHO 0279	16 50 04.	- 32 38 30.	380		ISOLATED DARK CLOUD
ZWG 139.011	16 50 06.	+ 23 23		15.7	GALAXY
ZWG 139.012	16 50 06.	+ 23 57		15.5	GALAXY
ZWG 225.004	16 50 06.	+ 41 45		15.5	GALAXY
MCG+10-24-070	16 50 06.	+ 60 46	18	17.	GALAXY
MRSL 348+05/1	16 50 06.	- 35 41	3000		HII REGION
HOLM 759B	16 50 08.	+ 23 23	18	15.2	PART OF MULTIPLE GALAXY
LB 00961	16 50 11.	+ 82 11 18.		16.5	FAINT BLUE STAR
VMT 17	16 50 11.	- 43 30 18.	1200		SUPERNOVA REMNANT
MTL 48	16 50 11.	- 43 30 18.	1800		SUPERNOVA REMNANT
HOLM 759A	16 50 12.	+ 23 23	18	15.1	PART OF MULTIPLE GALAXY
ZWG 299.034	16 50 12.	+ 59 49		14.1	GALAXY
UGC 10590	16 50 12.	+ 59 49	54	14.1	GALAXY SB(c)
ISS 1054	16 50 12.	- 39 37	65		STELLAR RING
SCHO 0280	16 50 14.	- 36 16 48.	510		ISOLATED DARK CLOUD
ZWG 139.013	16 50 18.	+ 23 25		15.1	GALAXY
UGC 10591	16 50 18.	+ 23 25	90	15.1	GALAXY Sa
ZC 1650.3+7234	16 50 18.	+ 72 34	1680		CLUSTER OF GALAXIES
YM 04	16 50 19.	- 45 11	276		SYMMETRIC GALACTIC NEBULA
RNGC 6243	16 50 21.	+ 23 25		15.0	GALAXY
MCG+04-40-004	16 50 24.	+ 23 24	42	15.1	GALAXY
ZC 1650.4+2328	16 50 24.	+ 23 28	540		CLUSTER OF GALAXIES
ZWG 169.008	16 50 24.	+ 28 13		15.7	GALAXY
GCL 048	16 50 24.	- 22 05	114	10.8	GLOBULAR STAR CLUSTER
RNGC 6235	16 50 24.	- 22 05		11.0	GLOBULAR CLUSTER
ISS 1055	16 50 24.	- 44 02	89		STELLAR RING

OBJECT NAME	RIGHT ASCEN.	DECLINATION	DIAM.	MAGN.	TYPE OF OBJECT
PK351+07.1	16 50 24.0	- 31 35 35.			PLANETARY NEBULA
IC 4625	16 50 25.	+ 02 31 14.			SAME AS NGC 6240
REIN 2.248	16 50 25.44	- 22 05 34.0			NEBULA
HOLM 760A	16 50 26.	+ 30 48	30	14.9	PART OF MULTIPLE GALAXY
SCHO 0281	16 50 28.	- 39 28 12.	330		ISOLATED DARK CLOUD
HN 0433	16 50 29.	- 76 54			NEBULA
IC 4618	16 50 29.	- 76 54			NONSTELLAR OBJECT
ZWG 025.011	16 50 30.	+ 02 29		14.7	GALAXY
UGC 10592	16 50 30.	+ 02 29	132	14.7	GALAXY PECULR
MCG+00-43-004	16 50 30.	+ 02 29	102	14.7	GALAXY
MCG+05-40-002	16 50 30.	+ 30 47 30.	30	15.5	GALAXY
ZWG 169.009	16 50 30.	+ 30 48		15.5	GALAXY
ZC 1650.5+6547	16 50 30.	+ 65 47	1410		CLUSTER OF GALAXIES
OCL 0997	16 50 30.	- 41 43	1980	8.5	OPEN STAR CLUSTER
VHA 201	16 50 30.	- 41 43	480		OPEN STAR CLUSTER
RNGC 6240	16 50 33.	+ 02 29		14.5	GALAXY
HOLM 760B	16 50 33.	+ 30 49	24	15.2	PART OF MULTIPLE GALAXY
ZWG 110.015	16 50 36.	+ 17 31		15.4	GALAXY
ZWG 139.014	16 50 36.	+ 23 20		15.4	GALAXY
MCG+05-40-003	16 50 36.	+ 30 48	36	15.5	GALAXY
SCHO 0282	16 50 40.	- 41 43 18.	380		ISOLATED DARK CLOUD
MCG+04-40-005	16 50 42.	+ 23 18 30.	12	15.4	GALAXY
ZC 1650.7+4453	16 50 42.	+ 44 53	870		CLUSTER OF GALAXIES
ZWG 252.008	16 50 42.	+ 49 24		15.7	GALAXY
MCG+12-16-011	16 50 42.	+ 69 06	51	16.	GALAXY
ZWG 339.021	16 50 42.	+ 69 08		15.6	GALAXY
RNGC 6231	16 50 45.	- 41 43		8.5	OPEN CLUSTER
ZWG 225.005	16 50 48.	+ 40 24		15.6	GALAXY
MCG+08-31-012	16 50 48.	+ 49 23	36	16.	GALAXY
ZC 1650.8+5947	16 50 48.	+ 59 47	740		CLUSTER OF GALAXIES
PK345+03.2	16 50 49.2	- 38 40 13.			PLANETARY NEBULA
IC 4626	16 50 50.	+ 02 24			MAY NOT EXIST
ARC 2233	16 50 52.	+ 43 15		17.4	RICH CLUSTER OF GALAXIES
ZWG 025.012	16 50 54.	+ 02 40		15.7	GALAXY
ZWG 053.019	16 50 54.	+ 04 19		15.4	GALAXY
MCG+04-40-006	16 50 54.	+ 23 01 30.	30		GALAXY
ARC 2236	16 50 58.	+ 71 34		17.1	RICH CLUSTER OF GALAXIES
KHAV 287	16 51	- 16 29	3250		DARK NEBULA
B 239	16 51	- 31 03	900		DARK OBJECT
KHAV 286	16 51	- 32 59	2630		DARK NEBULA
ZWG 110.016	16 51 00.	+ 20 17		15.7	GALAXY
ZWG 169.010	16 51 00.	+ 30 35		15.2	GALAXY
MCG+05-40-004	16 51 00.	+ 30 35	42	15.2	GALAXY
ZC 1651.0+3135	16 51 00.	+ 31 35	3970		CLUSTER OF GALAXIES
ZWG 225.006	16 51 00.	+ 39 27		15.0	GALAXY
ZC 1651.0+6200	16 51 00.	+ 62 00	540		CLUSTER OF GALAXIES
MCG+14-08-012	16 51 00.	+ 81 30	27	17.	GALAXY
LDN 0115	16 51 00.	- 16 35	1500		DARK NEBULA
HOLM 761A	16 51 02.	+ 30 35	36	14.7	PART OF MULTIPLE GALAXY
HOLM 761B	16 51 04.	+ 30 35	18	15.4	PART OF MULTIPLE GALAXY
ZC 1651.1+2634	16 51 06.	+ 26 34	1340		CLUSTER OF GALAXIES
MCG+12-16-012	16 51 06.	+ 69 49	54	16.	GALAXY
VHA 209	16 51 06.	- 44 36	600		OPEN STAR CLUSTER
LB 00324	16 51 11.	+ 15 47 12.		14.0	FAINT BLUE STAR
ZWG 139.015	16 51 12.	+ 23 32		15.6	GALAXY
ZWG 277.010	16 51 18.	+ 55 59		15.2	GALAXY
ZWG 276.053	16 51 18.	+ 55 59		15.2	GALAXY
UGC 10593	16 51 18.	+ 55 59	90	15.2	GALAXY Sa-b
LB 00962	16 51 22.	+ 46 40 00.		17.6	FAINT BLUE STAR
ZWG 110.017	16 51 24.	+ 20 00		15.5	GALAXY
ZC 1651.4+4930	16 51 24.	+ 49 30	1550		CLUSTER OF GALAXIES
MCG+09-28-001	16 51 24.	+ 55 58	90	15.	GALAXY
MCG+10-24-071	16 51 24.	+ 59 50	6	17.	GALAXY
ZWG 320.046	16 51 24.	+ 63 47		14.8	GALAXY
RNGC 6260	16 51 24.	+ 63 47		15.0	GALAXY
IC 4627	16 51 26.	- 07 33 16.			NONSTELLAR OBJECT
MCG+11-20-029	16 51 27.	+ 63 47	42	14.	GALAXY
ZWG 277.011	16 51 30.	+ 51 10		15.3	GALAXY
UGC 10594	16 51 30.	+ 51 10	66	15.3	GALAXY S-IRR
KARA.73B 0773	16 51 30.	+ 51 10	78	15.3	ISOLATED GALAXY S
MCG+10-24-072	16 51 30.	+ 59 49	3	17.	GALAXY
MRSL 341-01/1	16 51 30.	- 45 00	300		HII REGION
LB 00963	16 51 32.	+ 47 59 36.		16.5	FAINT BLUE STAR
ZWG 277.012	16 51 36.	+ 54 55		15.7	GALAXY
MCG+09-28-002	16 51 36.	+ 55 55	36	15.	GALAXY
ZWG 320.047	16 51 36.	+ 63 12		14.9	GALAXY
ZWG 321.002	16 51 36.	+ 65 17		15.4	GALAXY
ZWG 320.048	16 51 36.	+ 65 17		15.4	GALAXY
7ZW 660	16 51 36.	+ 66 16			COMPACT GALAXY
OCL 0992	16 51 36.	- 45 14	120		OPEN STAR CLUSTER
IC 1235	16 51 37.	+ 63 12			NONSTELLAR OBJECT
RNGC 6258	16 51 40.	+ 60 37		14.5	GALAXY
ZWG 081.009	16 51 42.	+ 10 39		15.6	GALAXY
ZWG 110.018	16 51 42.	+ 20 40		15.6	GALAXY
ZWG 277.013	16 51 42.	+ 53 03		15.6	GALAXY
ZWG 299.035	16 51 42.	+ 60 37		14.5	GALAXY
UGC 10595	16 51 42.	+ 60 37	54	14.5	GALAXY E
MCG+12-16-012A	16 51 42.	+ 69 00	33	15.	GALAXY
LB 00964	16 51 46.	+ 48 09 12.		17.2	FAINT BLUE STAR
ZWG 053.020	16 51 48.	+ 03 35		15.6	GALAXY
KARA.73B 0774	16 51 48.	+ 03 35	42	15.6	ISOLATED GALAXY S
MCG+09-28-003	16 51 48.	+ 51 06	72	16.	GALAXY
MCG+10-24-073	16 51 48.	+ 60 37	24	13.	GALAXY
MCG+11-20-030	16 51 48.	+ 65 17	36	16.	GALAXY
ZWG 339.022	16 51 48.	+ 69 01		14.7	GALAXY
VHA 202	16 51 48.	- 40 52	180		STAR CLSTR IN NEBULOSITY
MPSL 344+01/1	16 51 48.	- 40 59	3300		HII REGION
MCG+09-28-004	16 51 54.	+ 53 01	42	15.	GALAXY
ZWG 320.049	16 51 54.	+ 64 01		15.7	GALAXY
SHAH 165	16 51 54.	+ 81 58	78		GROUP OF COMPACT GALAXIES
SCHO 0283	16 51 55.	- 32 01 12.	890		ISOLATED DARK CLOUD
IC 1234	16 51 56.	+ 56 59			NONSTELLAR OBJECT
ARC 2247	16 51 58.	+ 81 39		15.3	RICH CLUSTER OF GALAXIES
KHAV 291	16 52	- 21 59	6850		DARK NEBULA
KHAV 288	16 52	- 33 41	3170		DARK NEBULA
UGC 10596	16 52 00.	+ 10 30	66	16.0	GALAXY SB
ZWG 139.016	16 52 00.	+ 23 30		15.7	GALAXY
UGC 10597	16 52 00.	+ 23 30	66	15.7	GALAXY Sb
ZWG 277.014	16 52 00.	+ 54 55		15.2	GALAXY
LDN 0132	16 52 00.	- 16 10	900		DARK NEBULA
LDN 0118	16 52 00.	- 16 40	900		DARK NEBULA
OCL 0998	16 52 00.	- 40 45	6300	4.0	OPEN STAR CLUSTER
VHA 203	16 52 00.	- 40 45	240		OPEN STAR CLUSTER
LB 00965	16 52 06.	+ 46 06 12.		17.6	FAINT BLUE STAR
MCG+09-28-005	16 52 06.	+ 55 55	42	15.	GALAXY
7ZW 661	16 52 06.	+ 57 26			COMPACT GALAXY
LDN 0129	16 52 06.	- 16 15	360		DARK NEBULA
LB 00966	16 52 10.	+ 30 41 42.		15.4	FAINT BLUE STAR
RNGC 6242	16 52 11.	- 39 25		8.0	OPEN CLUSTER
ZWG 169.011	16 52 12.	+ 27 15		15.7	GALAXY
UGC 10598	16 52 12.	+ 27 15	66	15.7	GALAXY S
MCG+05-40-005	16 52 12.	+ 27 15	60	15.7	GALAXY
ZC 1652.2+2920	16 52 12.	+ 29 20	1550		CLUSTER OF GALAXIES
ZWG 225.007	16 52 12.	+ 39 50		13.7	GALAXY
MRK 501	16 52 12.	+ 39 50	25	13.5	GALAXY WITH UV CONTINUUM
KW 44	16 52 12.	+ 39 50	60		SEYFERT GALAXY
UGC 10599	16 52 12.	+ 39 50	72	13.7	GALAXY EX CMPT
MCG+07-35-002	16 52 12.	+ 39 50	21	14.5	GALAXY
MCG+10-24-074	16 52 12.	+ 60 38	24	16.	GALAXY
OCL 1001	16 52 12.	- 39 25	1680	8.3	OPEN STAR CLUSTER
VHA 204	16 52 12.	- 39 25	360		OPEN STAR CLUSTER
LB 00967	16 52 16.	+ 47 01 54.		17.7	FAINT BLUE STAR
SCHO 0284	16 52 16.	- 31 03 18.	330		ISOLATED DARK CLOUD
ZWG 139.017	16 52 18.	+ 26 38		15.7	GALAXY
LDN 0122	16 52 18.	- 15 40	540		DARK NEBULA
LB 00968	16 52 20.	+ 28 29 54.		16.3	FAINT BLUE STAR
ZC 1652.5+2245	16 52 30.	+ 22 45	740		CLUSTER OF GALAXIES
ZWG 139.018	16 52 30.	+ 24 08		15.5	GALAXY
ZWG 197.017	16 52 30.	+ 36 05		15.7	GALAXY
UGC 10600	16 52 30.	+ 36 05	66	15.7	GALAXY S
MCG+09-28-006	16 52 30.	+ 54 39	60	16.	GALAXY
UGC 10601	16 52 30.	+ 54 41	96	16.0	GALAXY Sb
MCG+14-08-013	16 52 30.	+ 81 39	36	17.	GALAXY
ZC 1652.6+3714	16 52 36.	+ 37 14	1750		CLUSTER OF GALAXIES
ZC 1652.6+4005	16 52 36.	+ 40 05	1950		CLUSTER OF GALAXIES
7ZW 662	16 52 36.	+ 68 12			COMPACT GALAXY
VHA 205	16 52 36.	- 40 34	240		STAR CLSTR IN NEBULOSITY
RNGC 6262	16 52 38.	+ 57 00			NON-EXISTENT OBJECT
PK353+08.1	16 52 39.6	- 29 47 03.	5	13.0	PLANETARY NEBULA
ZWG 225.008	16 52 42.	+ 41 25		14.9	GALAXY
UGC 10602	16 52 42.	+ 41 25	60	14.9	GALAXY SBb-c
LB 00969	16 52 45.	+ 44 32 36.		17.4	FAINT BLUE STAR
UGC 10603	16 52 48.	+ 22 14	72	16.0	GALAXY Sb-c
MCG+07-35-003	16 52 48.	+ 41 25	48	14.	GALAXY
LB 00970	16 52 48.	+ 46 40 24.		16.4	FAINT BLUE STAR
ZWG 321.003	16 52 48.	+ 64 33		15.6	GALAXY
ZWG 320.050	16 52 48.	+ 64 33		15.6	GALAXY
SHAH 136	16 52 48.	+ 65 59	174	17.5	GROUP OF COMPACT GALAXIES
MCG+12-16-013	16 52 48.	+ 74 38	27	16.	GALAXY
ZWG 339.023	16 52 48.	+ 74 39		15.3	GALAXY
SCHO 0285	16 52 49.	- 40 55 30.	380		ISOLATED DARK CLOUD
MCG+06-37-014	16 52 51.	+ 36 35	198	12.5	GALAXY
ZWG 277.015	16 52 54.	+ 51 58		15.4	GALAXY
7ZW 663	16 52 54.	+ 64 32			COMPACT GALAXY
HOLM 762B	16 52 55.	+ 11 57	24	12.3	PART OF MULTIPLE GALAXY
HN 1148	16 52 58.	- 40 23			NEBULA
IC 4628	16 52 58.	- 40 23			HII REGION
VDB.66G 206	16 53	+ 53 11	70		DWARF GALAXY
KHAV 289	16 53	- 23 11	2230		DARK NEBULA
KHAV 290	16 53	- 24 29	3700		DARK NEBULA
ZWG 139.019	16 53 00.	+ 23 06		15.7	GALAXY
ZWG 197.018	16 53 00.	+ 36 35		13.8	GALAXY
ZCG 1653+36	16 53 00.	+ 36 35		13.8	COMPACT GALAXY
UGC 10606	16 53 00.	+ 36 35	210	13.8	GALAXY SBc
KARA.73B 0775	16 53 00.	+ 36 35	234	13.8	ISOLATED GALAXY S
ZWG 252.009	16 53 00.	+ 50 33		15.6	GALAXY
ZWG 367.017	16 53 00.	+ 81 56		15.7	GALAXY
UGC 10604	16 53 00.	+ 81 56	108	15.7	GALAXY S
UGC 10605	16 53 00.	+ 83 23	90	17.	GALAXY DWARF
LDN 1763	16 53 00.	- 23 20	2520		DARK NEBULA
LDN 1761	16 53 00.	- 23 30	2520		DARK NEBULA
RNGC 6255	16 53 01.	+ 36 25		14.0	GALAXY
HOLM 762A	16 53 02.	+ 11 58	48	12.2	PART OF MULTIPLE GALAXY
MCG+04-40-007	16 53 06.	+ 26 43 30.	60	14.8	GALAXY
ZWG 139.020	16 53 06.	+ 26 44		14.8	GALAXY
UGC 10607	16 53 06.	+ 26 44	78	14.8	GALAXY PECULR
1ZW 168	16 53 06.	+ 50 33			COMPACT GALAXY
ZC 1653.1+5240	16 53 06.	+ 52 40	1080		CLUSTER OF GALAXIES
MRK 502	16 53 06.	+ 64 12	12	15.5	GALAXY WITH UV CONTINUUM
VHE 75	16 53 06.	- 40 11	78		REFLECTION NEBULA
IC 4630	16 53 07.	+ 26 44 19.			NONSTELLAR OBJECT
ARC 2235	16 53 15.	+ 40 07		17.1	RICH CLUSTER OF GALAXIES
HN 0435	16 53 17.	- 16 38			NEBULA
IC 4629	16 53 17.	- 16 38			NONSTELLAR OBJECT
ZC 1653.3+3800	16 53 18.	+ 38 00	2550		CLUSTER OF GALAXIES
MCG+09-28-007	16 53 18.	+ 53 10	72	15.	GALAXY
UGC 10608	16 53 18.	+ 53 13	78	16.5	GALAXY
UGC 10609	16 53 18.	+ 69 58	72	16.5	GALAXY DWRF SP
MCG+06-37-015	16 53 24.	+ 36 59 30.	60	15.	GALAXY
ZWG 225.009	16 53 24.	+ 43 08		15.0	GALAXY
UGC 10610	16 53 24.	+ 43 08	126	15.0	GALAXY DBL SYS
ZC 1653.4+6610	16 53 24.	+ 66 10	540		CLUSTER OF GALAXIES
ISS 1006	16 53 24.	- 36 54	552		STELLAR RING
MCG+07-35-004	16 53 27.	+ 43 08	120	15.	GALAXY
UGC 10611	16 53 30.	+ 36 58	66	16.0	GALAXY S
MCG+07-35-005	16 53 30.	+ 43 08	36	15.5	GALAXY
VV 289B	16 53 30.	+ 43 17 30.	12	16.	INTERACTING GALAXY
VV 289A	16 53 30.	+ 43 18	30	15.5	INTERACTING GALAXY
ZC 1653.5+6124	16 53 30.	+ 61 24	2760		CLUSTER OF GALAXIES
ZWG 339.024	16 53 30.	+ 70 31		15.6	GALAXY
ZC 1653.5+7132	16 53 30.	+ 71 32	1680		CLUSTER OF GALAXIES
MCG+14-08-014	16 53 30.	+ 81 55	57	16.	GALAXY
LDN 0146	16 53 30.	- 15 55	1020		DARK NEBULA
OCL 0999	16 53 30.	- 40 35	3600	8.6	OPEN STAR CLUSTER
MCG+12-16-014	16 53 36.	+ 70 29	51	16.	GALAXY
SCHO 0286	16 53 38.	- 36 46 12.	710		ISOLATED DARK CLOUD
ZWG 139.021	16 53 42.	+ 26 20		15.6	GALAXY
ZC 1653.7+5858	16 53 42.	+ 58 58	2350		CLUSTER OF GALAXIES
ZWG 225.010	16 53 48.	+ 44 24		14.8	GALAXY
MCG+07-35-006	16 53 48.	+ 44 24	45	15.	GALAXY
SCHO 0287	16 53 50.	- 37 31 00.	570		ISOLATED DARK CLOUD
PK344+00.1	16 53 53.0	- 41 33 22.	6		PLANETARY NEBULA
ZC 1653.9+2315	16 53 54.	+ 23 15	1010		CLUSTER OF GALAXIES
ZC 1653.9+7856	16 53 54.	+ 78 56	8400		CLUSTER OF GALAXIES
RNGC 6249	16 53 59.	- 44 42		9.5	OPEN CLUSTER
KHAV 293	16 54	- 22 35	1150		DARK NEBULA
KHAV 292	16 54	- 27 41	4470		DARK NEBULA
KHAV 294	16 54	- 36 47	6120		DARK NEBULA
ZWG 053.021	16 54 00.	+ 06 01		15.3	GALAXY
KARA.73B 0776	16 54 00.	+ 06 01	24	15.3	ISOLATED GALAXY S0
ZC 1654.0+3738	16 54 00.	+ 37 38	470		CLUSTER OF GALAXIES
MCG+10-24-075	16 54 00.	+ 58 57	9	17.	GALAXY
ZC 1654.0+6015	16 54 00.	+ 60 15	1410		CLUSTER OF GALAXIES
ARC 2240	16 54 00.	+ 66 50		17.4	RICH CLUSTER OF GALAXIES
UGC 10612	16 54 00.	+ 80 26	66	16.0	GALAXY
ZWG 367.018	16 54 00.	+ 81 42		15.6	GALAXY

OBJECT NAME	RIGHT ASCEN.	DECLINATION	DIAM.	MAGN.	TYPE OF OBJECT
LDN 1777	16 54 00.	- 22 40	900		DARK NEBULA
LDN 1775	16 54 00.	- 22 45	360		DARK NEBULA
OCL 0994	16 54 00.	- 44 42	2160	9.4	OPEN STAR CLUSTER
ZWG 139.022	16 54 06.	+ 21 04		15.7	GALAXY
ZC 1654.1+2527	16 54 06.	+ 25 27	740		CLUSTER OF GALAXIES
ZWG 225.011	16 54 06.	+ 39 23		15.7	GALAXY
UGC 10613	16 54 06.	+ 53 28	60	16.0	GALAXY Sa-b
UGC 10614	16 54 06.	+ 65 23	90	16.0	GALAXY
PK347+03.1	16 54 06.	- 37 02		16.	PLANETARY NEBULA
MCG+07-35-007	16 54 09.	+ 39 22	51	15.	GALAXY
MCG+09-28-008	16 54 12.	+ 53 26	60	16.	GALAXY
ZWG 321.004	16 54 12.	+ 65 22		15.7	GALAXY
ZWG 320.051	16 54 12.	+ 65 22		15.7	GALAXY
KARA.73B 0777	16 54 12.	+ 65 22	66	15.7	ISOLATED GALAXY S
7ZW 664	16 54 12.	+ 67 34			COMPACT GALAXY
ZC 1654.2+7030	16 54 12.	+ 70 30	1880		CLUSTER OF GALAXIES
B 046	16 54 12.	- 22 39			DARK OBJECT
MRSL 344+01/2	16 54 12.	- 40 26	2400		HII REGION
LB 00971	16 54 16.	+ 28 16 54.		16.2	FAINT BLUE STAR
RNGC 6250	16 54 16.	- 45 43		8.0	OPEN CLUSTER
SCHO 0288	16 54 17.	- 29 32 06.	380		ISOLATED DARK CLOUD
ZWG 197.019	16 54 18.	+ 34 55		15.6	GALAXY
UGC 10615	16 54 18.	+ 34 55	66	15.6	GALAXY SBc
UGC 10616	16 54 18.	+ 58 47	66	17.	GALAXY
7ZW 665	16 54 18.	+ 70 10			COMPACT GALAXY
OCL 0991	16 54 18.	- 45 43	930	8.0	OPEN STAR CLUSTER
VHA 206	16 54 18.	- 45 43	660		STAR CLSTR IN NEBULOSITY
HN 1832	16 54 18.8	- 61 12 41.	12		NEBULA
ARC 2237	16 54 20.	+ 55 20		17.7	RICH CLUSTER OF GALAXIES
MCG+06-37-016	16 54 21.	+ 34 56	54	15.	GALAXY
MCG+11-21-001	16 54 21.	+ 65 22	57	15.	GALAXY
RNGC 6257	16 54 22.	+ 39 43		15.5	GALAXY
ARC 2239	16 54 23.	+ 59 00		17.1	RICH CLUSTER OF GALAXIES
ZWG 169.012	16 54 24.	+ 29 25		15.6	GALAXY
ZWG 225.012	16 54 24.	+ 39 43		15.6	GALAXY
ZWG 252.010	16 54 24.	+ 47 36		15.6	GALAXY
ZWG 321.005	16 54 24.	+ 66 33		15.6	GALAXY
ZWG 320.052	16 54 24.	+ 66 33		15.6	GALAXY
RNGC 6254	16 54 29.	- 04 02		7.5	GLOBULAR CLUSTER
MCG+05-40-006	16 54 30.	+ 28 03	63	15.2	GALAXY
ZWG 169.013	16 54 30.	+ 28 04		15.2	GALAXY
RNGC 6261	16 54 30.	+ 28 04		15.0	GALAXY
UGC 10617	16 54 30.	+ 28 04	90	15.2	GALAXY S0-a
MCG+08-31-013	16 54 30.	+ 47 35	24	16.	GALAXY
7ZW 666	16 54 30.	+ 63 32			COMPACT GALAXY
MCG+11-21-002	16 54 30.	+ 66 31	39	16.	GALAXY
ZWG 321.006	16 54 30.	+ 66 34		15.3	GALAXY
ZWG 320.053	16 54 30.	+ 66 34		15.3	GALAXY
ZC 1654.5+6646	16 54 30.	+ 66 46	1750		CLUSTER OF GALAXIES
GCL 049	16 54 30.	- 04 02	972	7.64	GLOBULAR STAR CLUSTER
MRSL 345+01/1	16 54 30.	- 39 43	3000		HII REGION
MCG+03-43-008	16 54 33.	+ 18 38	42	16.	GALAXY
ZWG 053.022	16 54 36.	+ 05 27		15.5	GALAXY
ZC 1654.6+1841	16 54 36.	+ 18 41	3830		CLUSTER OF GALAXIES
ZWG 139.023	16 54 36.	+ 26 55		15.5	GALAXY
ZC 1654.6+3535	16 54 36.	+ 35 35	2150		CLUSTER OF GALAXIES
HN 1833	16 54 36.3	- 60 08 27.	18		NEBULA
MCG+05-40-007	16 54 39.	+ 27 21 30.	42	15.5	GALAXY
MCG+11-21-003	16 54 39.	+ 66 32	39	16.	GALAXY
MCG+05-40-008	16 54 42.	+ 27 53 30.	36	15.1	GALAXY
ZWG 169.014	16 54 42.	+ 27 54		15.1	GALAXY
RNGC 6263	16 54 42.	+ 27 54		15.0	GALAXY
UGC 10618	16 54 42.	+ 27 54	66	15.1	GALAXY E
ZWG 197.020	16 54 42.	+ 38 26		15.5	GALAXY
VHE 76	16 54 42.	- 45 39	78		REFLECTION NEBULA
MCG+07-35-008	16 54 45.	+ 39 20	30	16.	GALAXY
MCG+11-21-004	16 54 45.	+ 66 34	39	16.	GALAXY
MCG+08-31-014	16 54 48.	+ 50 20	36	16.	GALAXY
UGC 10619	16 54 48.	+ 54 30	90	16.5	GALAXY Sc
MCG+09-28-009	16 54 54.	+ 54 29	78	16.	GALAXY
SCHO 0289	16 54 57.	- 29 13 48.	490		ISOLATED DARK CLOUD
RNGC 6275	16 54 59.	+ 63 20		15.0	GALAXY
KHAV 296	16 55	- 14 17	11130		DARK NEBULA
KHAV 295	16 55	- 34 41	2630		DARK NEBULA
LB 09906	16 55	- 80 39		15.5	FAINT BLUE STAR
MCG+10-24-076	16 55 00.	+ 58 59	30	16.	GALAXY
ZWG 321.007	16 55 00.	+ 63 20		15.1	GALAXY
ZWG 320.054	16 55 00.	+ 63 20		15.1	GALAXY
MRK 503	16 55 00.	+ 63 20	14	16.5	GALAXY WITH UV CONTINUUM
SCHO 0290	16 55 00.	- 34 50 48.	630		ISOLATED DARK CLOUD
7ZW 667	16 55 06.	+ 63 18			COMPACT GALAXY
OCL 0972	16 55 06.	- 52 38	600	10.2	OPEN STAR CLUSTER
VHA 207	16 55 06.	- 52 38	360		OPEN STAR CLUSTER
RNGC 6253	16 55 07.	- 52 38		10.0	OPEN CLUSTER
UGC 10620	16 55 12.	+ 31 18	60	16.0	GALAXY S
LB 00972	16 55 12.	+ 45 13 48.		16.3	FAINT BLUE STAR
MCG+09-28-010	16 55 12.	+ 55 04	90	14.	GALAXY
ZWG 277.016	16 55 12.	+ 55 06		14.7	GALAXY
UGC 10621	16 55 12.	+ 55 06	156	14.7	GALAXY Sb
MCG+12-16-015	16 55 12.	+ 70 30	84	16.	GALAXY
ZWG 339.025	16 55 12.	+ 70 31		15.3	GALAXY
UGC 10622	16 55 12.	+ 70 31	84	15.3	GALAXY Sc
SCHO 0291	16 55 13.	- 32 56 42.	320		ISOLATED DARK CLOUD
IC 1237	16 55 14.	+ 55 07 42.		14.02	GALAXY SBc
MCG+05-40-009	16 55 18.	+ 27 55	39	15.5	GALAXY
ZWG 169.015	16 55 18.	+ 27 56		15.5	GALAXY
RNGC 6264	16 55 18.	+ 27 56		15.5	GALAXY
HOLM 763A	16 55 18.	+ 27 56	36	14.8	PART OF MULTIPLE GALAXY
ZWG 169.016	16 55 18.	+ 28 16		15.5	GALAXY
ZWG 225.013	16 55 18.	+ 42 16		15.7	GALAXY
VHE 79A	16 55 18.	- 46 03	288		REFLECTION NEBULA
HOLM 763B	16 55 20.	+ 27 56	9	15.6	PART OF MULTIPLE GALAXY
MCG+05-40-010	16 55 21.	+ 28 15	24	15.5	GALAXY
ZWG 025.013	16 55 24.	+ 02 33		15.7	GALAXY
UGC 10623	16 55 24.	+ 02 33	60	15.7	GALAXY Sc
MCG+00-43-005	16 55 24.	+ 02 33	60	15.7	GALAXY
ZWG 139.024	16 55 24.	+ 22 40		15.3	GALAXY
1ZW 169	16 55 24.	+ 34 06			COMPACT GALAXY
LB 00973	16 55 26.	+ 29 39 12.		17.3	FAINT BLUE STAR
MCG+04-40-008	16 55 27.	+ 22 39	42	15.3	GALAXY
ZWG 169.017	16 55 30.	+ 27 55		15.4	GALAXY
RNGC 6265	16 55 30.	+ 27 55		15.5	GALAXY
UGC 10624	16 55 30.	+ 27 55	60	15.4	GALAXY S0
MCG+05-40-011	16 55 30.	+ 27 55	24	15.4	GALAXY
ZC 1655.5+3723	16 55 30.	+ 37 23	2150		CLUSTER OF GALAXIES
MCG+09-28-011	16 55 30.	+ 55 03	48	16.	GALAXY
VHE 77	16 55 30.	- 40 32	30		REFLECTION NEBULA
MCG-07-35-001	16 55 30.	- 42 37	72	13.	GALAXY
VHE 80	16 55 30.	- 42 38	96		REFLECTION NEBULA
SER 125.02	16 55 30.	- 58 41	100		PEC. GALAXY
SCHO 0292	16 55 35.	- 34 20 54.	300		ISOLATED DARK CLOUD
MCG+10-24-077	16 55 36.	+ 58 06	42	14.	GALAXY
SCHO 0293	16 55 38.	- 35 11 36.	500		ISOLATED DARK CLOUD
SCHO 0294	16 55 40.	- 30 57 12.	420		ISOLATED DARK CLOUD
PK336-05.1	16 55 40.	- 51 37 45.	25		PLANETARY NEBULA
ZC 1655.7+3038	16 55 42.	+ 30 38	1210		CLUSTER OF GALAXIES
UGC 10625	16 55 42.	+ 38 45	84	16.5	GALAXY Sc
ZWG 299.036	16 55 42.	+ 58 08		15.4	GALAXY
UGC 10626	16 55 42.	+ 58 08	84	15.4	GALAXY S
YM 06	16 55 45.	- 37 05	180		SYMMETRIC GALACTIC NEBULA
PK321-16.1	16 55 47.	- 70 01 58.	10		PLANETARY NEBULA
MCG+03-43-009	16 55 48.	+ 20 57	30	15.	GALAXY
ZC 1655.8+3130	16 55 48.	+ 31 30	1210		CLUSTER OF GALAXIES
ZWG 225.014	16 55 48.	+ 40 49		15.4	GALAXY
ZWG 252.011	16 55 48.	+ 46 56		15.6	GALAXY
UGC 10627	16 55 48.	+ 46 56	102	15.6	GALAXY Sc
KARA.73B 0778	16 55 48.	+ 46 56	114	15.6	ISOLATED GALAXY S
7ZW 668	16 55 48.	+ 60 43			COMPACT GALAXY
ZC 1655.8+6844	16 55 48.	+ 68 44	13640		CLUSTER OF GALAXIES
VHE 78	16 55 48.	- 40 05	84		REFLECTION NEBULA
MCG+05-40-012	16 55 54.	+ 27 55	30	14.4	GALAXY
MCG+05-40-013	16 55 54.	+ 28 11 30.	36	15.6	GALAXY
ZWG 169.018	16 55 54.	+ 28 13		15.6	GALAXY
MCG+07-35-009	16 55 54.	+ 40 48 30.	51	15.	GALAXY
ZWG 225.015	16 55 54.	+ 41 40		15.4	GALAXY
MCG+08-31-015	16 55 54.	+ 46 56	96	15.	GALAXY
ARC 2238	16 55 55.	+ 37 18		17.4	RICH CLUSTER OF GALAXIES
B 240	16 55 58.	- 35 17	1200		DARK OBJECT
BC N1656+05.3	16 56	+ 05 18		16.	QUASI-STELLAR OBJECT
KHAV 297	16 56	- 35 17	3450		DARK NEBULA
ZWG 139.025	16 56 00.	+ 23 03		14.0	GALAXY
UGC 10628	16 56 00.	+ 23 03	84	14.0	GALAXY SBc
ZWG 169.019	16 56 00.	+ 27 56		14.4	GALAXY
RNGC 6269	16 56 00.	+ 27 56		14.5	GALAXY
UGC 10629	16 56 00.	+ 27 56	132	14.4	GALAXY E
UGC 10630	16 56 00.	+ 32 03	60	16.0	GALAXY Sa-b
ZC 1656.0+3211	16 56 00.	+ 32 11	5310		CLUSTER OF GALAXIES
UGC 10631	16 56 00.	+ 36 30	60	15.4	GALAXY S
MCG+06-37-017	16 56 00.	+ 36 30	48	15.	GALAXY
ZWG 225.016	16 56 00.	+ 40 47		14.8	GALAXY
ZWG 225.017	16 56 00.	+ 40 48		14.9	GALAXY
MCG+07-35-010	16 56 00.	+ 41 40	39	15.5	GALAXY
ZC 1656.0+5637	16 56 00.	+ 56 37	940		CLUSTER OF GALAXIES
MCG+11-21-005	16 56 00.	+ 67 32	15	16.	GALAXY
ZWG 355.022	16 56 00.	+ 79 09		15.5	GALAXY
UGC 10632	16 56 00.	+ 79 09	102	15.5	GALAXY Sc
ZWG 367.019	16 56 00.	+ 81 40		15.5	GALAXY
ZC 1656.0-0145	16 56 00.	- 01 45	4500		CLUSTER OF GALAXIES
LDN 0280	16 56 00.	- 10 00	1980		DARK NEBULA
LDN 0278	16 56 00.	- 10 10	4320		DARK NEBULA
LDN 1779	16 56 00.	- 22 50	1200		DARK NEBULA
RNGC 6267	16 56 03.	+ 23 03		14.0	GALAXY
SHB 339	16 56 05.6	+ 05 19 46.		17.5	QUASI-STELLAR OBJECT
MCG+04-40-009	16 56 06.	+ 23 02	66	14.0	GALAXY
HOLM 764B	16 56 06.	+ 23 04	30	15.7	PART OF MULTIPLE GALAXY
HOLM 764A	16 56 06.	+ 23 05	60	12.7	PART OF MULTIPLE GALAXY
ZWG 225.018	16 56 06.	+ 39 18		15.7	GALAXY
MCG+07-35-012	16 56 06.	+ 40 46 30.	30	15.	GALAXY
MCG+07-35-011	16 56 06.	+ 40 47 30.	30	15.	GALAXY
ZC 1656.1+5421	16 56 06.	+ 54 21	2690		CLUSTER OF GALAXIES
ZWG 169.020	16 56 12.	+ 29 59		15.1	GALAXY
MCG+05-40-014	16 56 12.	+ 29 59	39	15.1	GALAXY
ZWG 225.019	16 56 12.	+ 39 20		15.6	GALAXY
MCG+13-12-013	16 56 12.	+ 79 10	102	15.	GALAXY
ISS 1007	16 56 12.	- 34 07	82		STELLAR RING
MRSL 349+04/1	16 56 12.	- 35 16	2400		HII REGION
VHA 208	16 56 12.	- 37 02	90		OPEN STAR CLUSTER
SCHO 0295	16 56 13.	- 34 33 12.	600		ISOLATED DARK CLOUD
RNGC 6256	16 56 13.	- 37 02			GLOBULAR CLUSTER
LB 00974	16 56 15.	+ 27 06 24.		15.9	FAINT BLUE STAR
B 241	16 56 16.	- 30 07	1080		DARK OBJECT
UGC 10634	16 56 18.	+ 15 17	60	17.	GALAXY S
ZWG 110.019	16 56 18.	+ 20 07		14.6	GALAXY
UGC 10633	16 56 18.	+ 20 07	72	14.6	GALAXY SBc
ZC 1656.3+2619	16 56 18.	+ 26 19	1750		CLUSTER OF GALAXIES
2ZW 075	16 56 18.	+ 38 17			COMPACT GALAXY
UGC 10635	16 56 18.	+ 38 17	42	13.5	GALAXY COMPACT
ZWG 225.020	16 56 18.	+ 40 27		15.6	GALAXY
LB 00975	16 56 18.	+ 45 04 30.		17.5	FAINT BLUE STAR
SCHO 0296	16 56 20.	- 30 37 24.	380		ISOLATED DARK CLOUD
IC 1236	16 56 23.	+ 20 07 55.			NONSTELLAR OBJECT
IC 4632	16 56 23.	+ 23 00			NONSTELLAR OBJECT
MCG+03-43-010	16 56 24.	+ 20 07	48	14.6	GALAXY
ZWG 139.026	16 56 24.	+ 25 27		15.7	GALAXY
ZWG 277.017	16 56 30.	+ 52 29		15.5	GALAXY
KARA.73B 0779	16 56 30.	+ 52 29	36	15.5	ISOLATED GALAXY S
ARC 2242	16 56 30.	+ 54 30		16.9	RICH CLUSTER OF GALAXIES
MCG+10-24-078	16 56 30.	+ 58 32	6	17.	GALAXY
MCG+14-08-015	16 56 30.	+ 81 39	12	16.	GALAXY
SCHO 0298	16 56 32.	- 31 33 36.	510		ISOLATED DARK CLOUD
LB 00976	16 56 33.	+ 45 41 54.		17.8	FAINT BLUE STAR
SCHO 0299	16 56 34.	- 30 11 24.	700		ISOLATED DARK CLOUD
SCHO 0297	16 56 34.	- 34 09 12.	310		ISOLATED DARK CLOUD
ZWG 197.021	16 56 36.	+ 33 07		15.6	GALAXY
ZC 1656.6+5016	16 56 36.	+ 50 16	1210		CLUSTER OF GALAXIES
LDN 1792	16 56 36.	- 22 31	480		DARK NEBULA
LDN 1791	16 56 36.	- 22 31	480		DARK NEBULA
ZWG 025.014	16 56 42.	+ 01 27		15.4	GALAXY
RNGC 6270	16 56 42.	+ 27 42		15.0	GALAXY
MCG+05-40-015	16 56 42.	+ 27 42	36	15.5	GALAXY
UGC 10636	16 56 42.	+ 70 31	66	16.5	GALAXY DWRF IR
B 047	16 56 42.	- 22 34	900		DARK OBJECT
SCHO 0300	16 56 44.	- 37 02 30.	1030		ISOLATED DARK CLOUD
ZC 1656.8+2558	16 56 48.	+ 25 58	810		CLUSTER OF GALAXIES
MCG+05-40-016	16 56 48.	+ 28 01	48	15.6	GALAXY
ZWG 169.021	16 56 48.	+ 28 03		15.6	GALAXY
RNGC 6271	16 56 48.	+ 28 03		15.5	GALAXY
MCG+05-40-017	16 56 48.	+ 28 21 30.	54	16.	GALAXY
UGC 10637	16 56 48.	+ 32 02	66	16.5	GALAXY S
7ZW 669	16 56 48.	+ 68 28			COMPACT GALAXY
SHAH 166	16 56 48.	+ 81 42	474		GROUP OF COMPACT GALAXIES
LDN 1784	16 56 48.	- 22 50	660		DARK NEBULA
ZWG 169.022	16 56 54.	+ 28 00		15.5	GALAXY
RNGC 6272	16 56 54.	+ 28 00		15.5	GALAXY
MCG+12-16-016	16 56 54.	+ 70 30	51	16.	GALAXY
SCHO 0301	16 56 54.	- 34 20 30.	440		ISOLATED DARK CLOUD

OBJECT NAME	RIGHT ASCEN.	DECLINATION	DIAM.	MAGN.	TYPE OF OBJECT
ZL 195	16 56 59.	- 00 30 36.		18.8	ULTRAFAINT BLUE STAR
KHAV 299	16 57	- 32 11	5120		DARK NEBULA
KHAV 298	16 57	- 33 29	3250		DARK NEBULA
MCG+06-37-018	16 57 00.	+ 36 36	48	15.	GALAXY
MCG+08-31-016	16 57 00.	+ 47 12	42	16.	GALAXY
ZC 1657.0+5830	16 57 00.	+ 58 30	3090		CLUSTER OF GALAXIES
ZWG 367.020	16 57 00.	+ 81 05		15.6	GALAXY
UGC 10638	16 57 00.	+ 81 40	480		GALAXY CHAIN
LDN 1787	16 57 00.	- 22 40	1200		DARK NEBULA
LDN 1785	16 57 00.	- 22 50	660		DARK NEBULA
LB 00977	16 57 04.	+ 30 00 54.		17.6	FAINT BLUE STAR
ZL 196	16 57 05.	- 00 32 48.		19.4	ULTRAFAINT BLUE STAR
RNGC 6259	16 57 05.	- 44 36		8.5	OPEN CLUSTER
ZWG 169.023	16 57 06.	+ 29 05		15.3	GALAXY
UGC 10639	16 57 06.	+ 29 05	60	15.3	GALAXY SBb
ZC 1657.1+6817	16 57 06.	+ 68 17	870		CLUSTER OF GALAXIES
SCHO 0302	16 57 06.	- 29 00 06.	400		ISOLATED DARK CLOUD
OCL 0996	16 57 06.	- 44 36	6180	8.6	OPEN STAR CLUSTER
MCG+05-40-018	16 57 09.	+ 29 03 30.	48	15.3	GALAXY
LB 00978	16 57 11.	+ 47 53 48.		17.5	FAINT BLUE STAR
ZL 197	16 57 11.	- 00 26 48.		21.2	ULTRAFAINT BLUE STAR
ZWG 025.015	16 57 12.	+ 02 34		15.3	GALAXY
UGC 10640	16 57 12.	+ 02 34	60	15.3	GALAXY SB
ZWG 110.020	16 57 12.	+ 20 55		15.7	GALAXY
KARA.73B 0780	16 57 12.	+ 20 55	54	15.7	ISOLATED GALAXY S
ZWG 197.022	16 57 12.	+ 36 35		15.4	GALAXY
MCG+07-35-013	16 57 12.	+ 40 11	48	15.	GALAXY
ZWG 225.021	16 57 12.	+ 40 12		14.9	GALAXY
UGC 10641	16 57 12.	+ 58 59	90	16.5	GALAXY Sc
MCG+03-43-011	16 57 15.	+ 20 55	30	15.7	GALAXY
MCG+10-24-079	16 57 15.	+ 58 57	72	15.	GALAXY
ZL 198	16 57 16.	- 00 28 00.		20.0	ULTRAFAINT BLUE STAR
ZL 199	16 57 17.	- 00 26 12.		19.7	ULTRAFAINT BLUE STAR
ZWG 252.012	16 57 18.	+ 50 34		15.6	GALAXY
GCL 050	16 57 18.	- 00 28			GLOBULAR STAR CLUSTER
ZWG 025.016	16 57 18.	- 00 28		15.7	GALAXY
UGC 10642	16 57 18.	- 00 28	1200	15.7	GALAXY DWF SYS
MCG+00-43-006	16 57 18.	- 00 28	240	15.7	GALAXY
KARA.73B 0781	16 57 18.	- 00 28	48	15.7	ISOLATED GALAXY IR
SCHO 0303	16 57 21.	- 40 46 36.	460		ISOLATED DARK CLOUD
SHB 340	16 57 22.0	+ 26 36 42.		18.	QUASI-STELLAR OBJECT
BC 4C26.51	16 57 22.34	+ 26 34 04.5		18.	QUASI-STELLAR OBJECT
SCHO 0305	16 57 23.	- 31 25 54.	360		ISOLATED DARK CLOUD
ZWG 169.024	16 57 24.	+ 30 01		14.5	GALAXY
UGC 10643	16 57 24.	+ 30 01	60	14.5	GALAXY DBL SYS
KARA.72 503B	16 57 24.	+ 30 01	42		PART OF DOUBLE GALAXY
KARA.72 503A	16 57 24.	+ 30 01	42	14.5	PART OF DOUBLE GALAXY
MCG+05-40-020	16 57 24.	+ 30 01	36	15.5	GALAXY
MCG+05-40-019	16 57 24.	+ 30 01	27	14.5	GALAXY
FATE 1.766	16 57 24.	+ 30 01	14		NEBULA
ZC 1657.4+6221	16 57 24.	+ 62 21	670		CLUSTER OF GALAXIES
ZC 1657.4+7355	16 57 24.	+ 73 55	340		CLUSTER OF GALAXIES
SCHO 0304	16 57 24.	- 31 57 54.	920		ISOLATED DARK CLOUD
RNGC 6274	16 57 26.	+ 30 01		14.5	GALAXY
ZL 200	16 57 28.	- 00 33 30.		20.6	ULTRAFAINT BLUE STAR
ZWG 277.018	16 57 30.	+ 54 56		14.8	GALAXY
UGC 10644	16 57 30.	+ 54 56	60	14.8	GALAXY
MCG+14-08-016	16 57 30.	+ 81 40	8	16.	GALAXY
MCG+05-40-021	16 57 33.	+ 32 26	30	15.5	GALAXY
ZWG 197.023	16 57 36.	+ 38 56		15.7	GALAXY
ZWG 252.013	16 57 36.	+ 47 19		14.9	GALAXY
RNGC 6279	16 57 36.	+ 47 18		15.0	GALAXY
UGC 10645	16 57 36.	+ 47 18	60	14.9	GALAXY SB0
7ZW 670	16 57 36.	+ 57 36			COMPACT GALAXY
ZWG 299.037	16 57 36.	+ 59 03		14.6	GALAXY
MCG+11-21-006	16 57 36.	+ 68 32	27	15.	GALAXY
IC 4636	16 57 37.	+ 47 16			NONSTELLAR OBJECT
RNGC 6285	16 57 37.	+ 59 03		14.5	GALAXY
RNGC 6288	16 57 41.	+ 68 31		15.5	GALAXY
SCHO 0306	16 57 41.	- 41 40 54.	1110		ISOLATED DARK CLOUD
ZWG 081.010	16 57 42.	+ 11 28		15.7	GALAXY
ZWG 225.022	16 57 42.	+ 41 07		15.5	GALAXY
MCG+08-31-017	16 57 42.	+ 47 18	60	15.	GALAXY
MCG+09-28-012	16 57 42.	+ 54 56	60	15.	GALAXY
MCG+10-24-080	16 57 42.	+ 57 09	30	14.	GALAXY
MCG+10-24-081	16 57 42.	+ 59 02	60	14.	GALAXY
ZWG 299.038	16 57 42.	+ 59 12		15.0	GALAXY
UGC 10646	16 57 42.	+ 59 12	84	15.0	GALAXY E
MCG+10-24-082	16 57 42.	+ 59 12	60	15.	GALAXY
MCG+10-24-083	16 57 42.	+ 59 50	54	14.	GALAXY
ZWG 321.008	16 57 42.	+ 68 31		15.3	GALAXY
ZWG 320.055	16 57 42.	+ 68 31		15.3	GALAXY
ABC 2241	16 57 46.	+ 32 37		15.6	RICH CLUSTER OF GALAXIES
PK337-05.1	16 57 46.	- 50 18 31.	10		PLANETARY NEBULA
ARP 293	16 57 47.	+ 59 00			PECULIAR GALAXY
PK349+04.1	16 57 47.7	- 34 45 17.	5		PLANETARY NEBULA
MRSL 347+02/2	16 57 48.	+ 38 08	1500		HII REGION
ZWG 299.039	16 57 48.	+ 57 12		14.6	GALAXY
MCG+10-24-084	16 57 48.	+ 59 00	66	14.	GALAXY
ZWG 299.040	16 57 48.	+ 59 02		14.2	GALAXY
UGC 10647	16 57 48.	+ 59 02	90	14.2	GALAXY
ZWG 299.041	16 57 48.	+ 59 50		15.3	GALAXY
UGC 10648	16 57 48.	+ 59 50	84	15.3	GALAXY SBb
ZL 201	16 57 48.	- 00 22 48.		21.0	ULTRAFAINT BLUE STAR
ISS 1056	16 57 48.	- 42 20	883		STELLAR RING
RNGC 6286	16 57 49.	+ 59 02		14.0	GALAXY
ZL 202	16 57 50.	- 00 17 06.		20.6	ULTRAFAINT BLUE STAR
ZC 1657.9+6248	16 57 54.	+ 62 48	2690		CLUSTER OF GALAXIES
MCG+11-21-007	16 57 54.	+ 68 35	18	15.	GALAXY
FATH 1.767	16 57 55.	+ 30 01	16		NEBULA
RNGC 6289	16 57 59.	+ 68 35		15.5	GALAXY
KHAV 300	16 58	- 23 05	2230		DARK NEBULA
B 048	16 58	- 40 37	2400		DARK OBJECT
ZWG 110.021	16 58 00.	+ 17 14		15.7	GALAXY
UGC 10649	16 58 00.	+ 23 00	60	17.	GALAXY Sc
1ZW 170	16 58 00.	+ 51 55			COMPACT GALAXY
ZWG 321.009	16 58 00.	+ 68 35		15.5	GALAXY
ZWG 320.056	16 58 00.	+ 68 35		15.5	GALAXY
ZWG 367.021	16 58 00.	+ 81 44		15.4	GALAXY
LDN 0317	16 58 00.	- 08 00	5100		DARK NEBULA
LDN 0295	16 58 00.	- 09 10	1800		DARK NEBULA
SCHO 0307	16 58 00.	- 32 49 30.	300		ISOLATED DARK CLOUD
LB 00979	16 58 01.	+ 28 54 42.		14.3	FAINT BLUE STAR
ZWG 139.027	16 58 06.	+ 23 11		15.3	GALAXY
UGC 10650	16 58 06.	+ 23 11	102	15.3	GALAXY IRR
ZWG 169.025	16 58 06.	+ 29 55		15.7	GALAXY
MCG-01-43-003	16 58 06.	- 05 07	60	16.	GALAXY
GCL 051	16 58 06.	- 30 03	840	8.16	GLOBULAR STAR CLUSTER

OBJECT NAME	RIGHT ASCEN.	DECLINATION	DIAM.	MAGN.	TYPE OF OBJECT
VHA 210	16 58 06.	- 30 03	480		GLOBULAR STAR CLUSTER
ISS 1008	16 58 06.	- 36 48	116		STELLAR RING
RNGC 6266	16 58 07.	- 30 03		8.0	GLOBULAR CLUSTER
RNGC 6276	16 58 09.	+ 23 11			GALAXY
RNGC 6283	16 58 10.	+ 50 00		13.5	GALAXY
ZC 1658.2+2229	16 58 12.	+ 22 29	1140		CLUSTER OF GALAXIES
MCG+06-37-020	16 58 12.	+ 37 55 30.	24	15.	GALAXY
MCG+06-37-019	16 58 12.	+ 38 55	42	14.5	GALAXY
MCG+07-35-014	16 58 12.	+ 42 36	60	15.	GALAXY
ZWG 225.023	16 58 12.	+ 42 37		15.6	GALAXY
UGC 10651	16 58 12.	+ 42 37	78	15.6	GALAXY S
ZWG 225.024	16 58 12.	+ 43 40		15.7	GALAXY
ZWG 252.014	16 58 12.	+ 50 00		13.7	GALAXY
UGC 10652	16 58 12.	+ 50 00	72	13.7	GALAXY S
YM 08	16 58 12.	- 35 57	126		SYMMETRIC GALACTIC NEBULA
ISS 1009	16 58 12.	- 36 49	53		STELLAR RING
MCG+05-40-022	16 58 15.	+ 27 28 30.	48	15.4	GALAXY
MCG+08-31-018	16 58 15.	+ 49 59	60	14.	GALAXY
LB 00980	16 58 17.	+ 45 50 12.		16.7	FAINT BLUE STAR
ZWG 025.017	16 58 18.	+ 02 09		15.4	GALAXY
ZWG 081.011	16 58 18.	+ 10 21		15.4	GALAXY
TON-N 0265	16 58 18.	+ 24 20		14.9	BLUE STAR
ZWG 169.026	16 58 18.	+ 27 40		15.4	GALAXY
UGC 10653	16 58 18.	+ 27 40	90	15.4	GALAXY SBb
ZWG 197.024	16 58 18.	+ 38 53		15.4	GALAXY
SCHO 0308	16 58 18.	- 30 44 36.	260		ISOLATED DARK CLOUD
SCHO 0309	16 58 18.	- 34 34 48.	640		ISOLATED DARK CLOUD
ISS 1057	16 58 18.	- 43 54	173		STELLAR RING
SCHO 0310	16 58 21.	- 30 47 00.	350		ISOLATED DARK CLOUD
LB 00981	16 58 26.	+ 13 22 18.		14.8	FAINT BLUE STAR
IC 1238	16 58 27.	+ 23 10			NONSTELLAR OBJECT
ZC 1658.5+0136	16 58 30.	+ 01 36	3090		CLUSTER OF GALAXIES
ZC 1658.5+2459	16 58 30.	+ 24 59	7800		CLUSTER OF GALAXIES
MCG+14-08-017	16 58 30.	+ 81 42	15	16.	GALAXY
MCG+14-08-018	16 58 30.	+ 81 43	12	17.	GALAXY
LDN 1800	16 58 30.	- 22 10	480		DARK NEBULA
FATH 1.768	16 58 32.	+ 29 33	19		NEBULA
SCHO 0311	16 58 32.	- 30 21 36.	340		ISOLATED DARK CLOUD
IC 4634	16 58 33.6	- 21 45 12.	20	11.1	PLANETARY NEBULA
SCHO 0312	16 58 34.	- 36 07 06.	1330		ISOLATED DARK CLOUD
RNGC 6268	16 58 34.	- 39 39		9.5	OPEN CLUSTER
PK000+12.1	16 58 35.	- 21 44	22	12.3	PLANETARY NEBULA
HN 0072	16 58 35.	- 21 45			NEBULA
ZWG 139.028	16 58 36.	+ 23 07		15.2	GALAXY
HOLM 765B	16 58 36.	+ 23 08	18	14.7	PART OF MULTIPLE GALAXY
ZWG 169.027	16 58 36.	+ 29 48		15.5	GALAXY
UGC 10654	16 58 36.	+ 29 48	66	15.5	GALAXY S0-a
ZC 1658.6+3730	16 58 36.	+ 37 30	2290		CLUSTER OF GALAXIES
ZWG 277.019	16 58 36.	+ 56 08		15.4	GALAXY
UGC 10655	16 58 36.	+ 56 08	60	15.4	GALAXY Sb-c
MCG+09-28-013	16 58 36.	+ 56 09	72	15.	GALAXY
SN 1971M	16 58 36.	+ 66 16		19.0	SUPERNOVA
IC 1239	16 58 38.	+ 23 06			NONSTELLAR OBJECT
RNGC 6277	16 58 39.	+ 23 07		15.0	GALAXY
FATH 2.158	16 58 39.	+ 29 47	11		NEBULA
HOLM 765C	16 58 40.	+ 23 08	12	15.5	PART OF MULTIPLE GALAXY
LB 00982	16 58 40.	+ 27 19 06.		15.8	FAINT BLUE STAR
ZWG 053.023	16 58 42.	+ 06 56		15.4	GALAXY
ZWG 139.029	16 58 42.	+ 23 05		13.8	GALAXY
UGC 10656	16 58 42.	+ 23 05	126	13.8	GALAXY S0
HOLM 765A	16 58 42.	+ 23 06	30	13.6	PART OF MULTIPLE GALAXY
MCG+04-40-010	16 58 42.	+ 23 06	21	15.2	GALAXY
MCG+05-40-023	16 58 42.	+ 27 53	30	15.5	GALAXY
ZWG 169.028	16 58 42.	+ 32 45		15.5	GALAXY
UGC 10657	16 58 42.	+ 39 13	60	16.5	GALAXY Sa-b
SCHO 0313	16 58 42.	- 30 17 36.	540		ISOLATED DARK CLOUD
VHA 211	16 58 42.	- 41 00	240		OPEN STAR CLUSTER
RNGC 6278	16 58 45.	+ 23 05		14.0	GALAXY
ZWG 110.022	16 58 48.	+ 20 15		15.6	GALAXY
MCG+04-40-011	16 58 48.	+ 23 04	90	13.8	GALAXY
MCG+05-40-024	16 58 48.	+ 27 47	24	15.5	GALAXY
ZWG 169.029	16 58 48.	+ 29 54		15.2	GALAXY
ZWG 321.010	16 58 48.	+ 68 37		15.7	GALAXY
ZWG 320.057	16 58 48.	+ 68 37		15.7	GALAXY
RNGC 6282	16 58 50.	+ 29 54		15.0	GALAXY
FATH 2.159	16 58 51.	+ 29 53	27		NEBULA
SCHO 0314	16 58 52.	- 33 32 24.	990		ISOLATED DARK CLOUD
ZWG 053.024	16 58 54.	+ 05 21		15.5	GALAXY
ZWG 110.023	16 58 54.	+ 20 45		15.3	GALAXY
UGC 10658	16 58 54.	+ 28 05	78	15.	GALAXY S0-a
OCL 1002	16 58 54.	- 39 40	1020	9.6	OPEN STAR CLUSTER
VHA 212	16 58 54.	- 39 40	270		OPEN STAR CLUSTER
MCG+05-40-025	16 58 57.	+ 29 29	24	15.5	GALAXY
KHAV 304	16 59	- 21 10	1150		DARK NEBULA
KHAV 301	16 59	- 28 28	3170		DARK NEBULA
KHAV 303	16 59	- 34 22	1950		DARK NEBULA
KHAV 302	16 59	- 36 16	3960		DARK NEBULA
LB 09990	16 59	- 87 01		15.6	FAINT BLUE STAR
ZWG 053.025	16 59 00.	+ 07 00		15.2	GALAXY
UGC 10659	16 59 00.	+ 07 00	78	15.2	GALAXY S
LDN 1797	16 59 00.	- 22 30	1560		DARK NEBULA
SCHO 0315	16 59 01.	- 36 33 00.	680		ISOLATED DARK CLOUD
MCG+03-43-012	16 59 03.	+ 20 45	24	15.3	GALAXY
PK351+05.1	16 59 03.2	- 33 05 49.	5		PLANETARY NEBULA
UGC 10660	16 59 06.	+ 25 56	60	16.0	GALAXY S
MCG+05-40-026	16 59 06.	+ 29 28 30.	18	16.	GALAXY
ZWG 197.025	16 59 06.	+ 37 10		15.5	GALAXY
LB 00983	16 59 10.	+ 11 07 36.		16.1	FAINT BLUE STAR
FATH 1.769	16 59 12.	+ 29 28	14		NEBULA
MRK 504	16 59 12.	+ 29 29	18	17.	GALAXY WITH UV CONTINUUM
KW 45	16 59 12.	+ 29 29	19		SEYFERT GALAXY
MCG+08-31-019	16 59 12.	+ 47 19 30.	42	16.	GALAXY
ZWG 252.015	16 59 12.	+ 47 20		14.7	GALAXY
MCG+14-08-019	16 59 12.	+ 81 03	27	16.	GALAXY
ZWG 025.018	16 59 12.	- 01 51		15.6	GALAXY
KARA.73B 0782	16 59 12.	- 01 51	48	15.6	ISOLATED GALAXY S
MCG+06-37-021	16 59 18.	+ 33 33	42	17.	GALAXY
ZWG 197.026	16 59 18.	+ 33 35		15.7	GALAXY
ZWG 225.025	16 59 18.	+ 39 39		15.1	GALAXY
LDN 1802	16 59 18.	- 22 09	480		DARK NEBULA
IC 4638	16 59 23.	+ 33 38			NONSTELLAR OBJECT
B 049	16 59 23.	- 33 12			DARK OBJECT
ZC 1659.4+2354	16 59 24.	+ 23 54	1010		CLUSTER OF GALAXIES
MCG+05-40-027	16 59 24.	+ 27 48	48	16.	GALAXY
MCG+07-35-015	16 59 24.	+ 39 37 30.	48	15.	GALAXY
7ZW 671	16 59 24.	+ 64 41			COMPACT GALAXY
SCHO 0316	16 59 26.	- 34 21 12.	420		ISOLATED DARK CLOUD
RNGC 6273	16 59 27.	- 26 11		8.5	GLOBULAR CLUSTER

OBJECT NAME	RIGHT ASCEN.	DECLINATION	DIAM.	MAGN.	TYPE OF OBJECT
SCHO 0317	16 59 28.	- 32 26 48.	330		ISOLATED DARK CLOUD
RNGC 6280	16 59 29.	+ 06 44		15.5	GALAXY
SCHO 0318	16 59 29.	- 33 33 30.	330		ISOLATED DARK CLOUD
ZWG 053.026	16 59 30.	+ 06 44		15.5	GALAXY
MCG+01-43-008	16 59 30.	+ 06 44	24	15.5	GALAXY
MCG+05-40-029	16 59 30.	+ 30 15	18	15.2	GALAXY
MCG+05-40-028	16 59 30.	+ 30 24	90	15.3	GALAXY
LDN 0004	16 59 30.	- 22 08	240		DARK NEBULA
GCL 052	16 59 30.	- 26 12	558		GLOBULAR STAR CLUSTER
LDN 1679	16 59 30.	- 32 50	420		DARK NEBULA
PATH 1.771	16 59 35.	+ 30 14	11		NEBULA
PATH 1.770	16 59 35.	+ 30 23	11		NEBULA
UGC 10661	16 59 36.	+ 28 00	60	16.5	GALAXY Sb
ZWG 169.030	16 59 36.	+ 30 15		15.2	GALAXY
UGC 10662	16 59 36.	+ 30 15	108	15.2	GALAXY E
ZWG 169.031	16 59 36.	+ 30 24		15.3	GALAXY
UGC 10663	16 59 36.	+ 30 24	84	15.3	GALAXY SBb
1ZW 171	16 59 36.	+ 53 26			COMPACT GALAXY
7ZW 672	16 59 36.	+ 68 29			COMPACT GALAXY
B 050	16 59 36.	- 34 19			DARK OBJECT
SCHO 0319	16 59 37.	- 33 15 06.	330		ISOLATED DARK CLOUD
MCG+09-28-014	16 59 42.	+ 53 24	60	16.	GALAXY
LB 00984	16 59 47.	+ 12 19 18.		16.9	FAINT BLUE STAR
2ZW 076	16 59 48.	+ 33 38			COMPACT GALAXY
ZWG 277.020	16 59 48.	+ 55 10		15.5	GALAXY
MCG+10-24-085	16 59 48.	+ 59 00	36	15.	GALAXY
ISS 1010	16 59 48.	- 36 54	80		STELLAR RING
SCHO 0320	16 59 50.	- 31 57 18.	540		ISOLATED DARK CLOUD
MCG+05-40-030	16 59 51.	+ 28 27	12	15.7	GALAXY
MCG+05-40-031	16 59 51.	+ 28 29	18	15.4	GALAXY
ARC 2248	16 59 53.	+ 77 05		15.5	RICH CLUSTER OF GALAXIES
ZWG 081.012	16 59 54.	+ 10 45		15.6	GALAXY
ZWG 110.024	16 59 54.	+ 17 15		15.7	GALAXY
ZWG 169.032	16 59 54.	+ 28 28		15.7	GALAXY
ZWG 169.033	16 59 54.	+ 28 30		15.4	GALAXY
MCG+09-28-015	16 59 54.	+ 55 10	42	15.	GALAXY
KHAV 311	17 00	- 17 04	9780		DARK NEBULA
ZC 1700.0+2433	17 00 00.	+ 24 33	940		CLUSTER OF GALAXIES
LDN 1695	17 00 00.	- 30 30	2940		DARK NEBULA
ARC 2243	17 00 02.	+ 35 09		17.1	RICH CLUSTER OF GALAXIES
SCHO 0321	17 00 03.	- 31 40 18.	910		ISOLATED DARK CLOUD
MCG+07-35-016	17 00 06.	+ 41 17 30.	78	15.	GALAXY
ZWG 225.026	17 00 06.	+ 41 18		15.4	GALAXY
UGC 10664	17 00 06.	+ 41 18	90	15.4	GALAXY Sc
ZWG 299.042	17 00 06.	+ 59 02		14.8	GALAXY
ZWG 299.043	17 00 06.	+ 59 04		14.3	GALAXY
UGC 10665	17 00 06.	+ 59 04	72	14.3	GALAXY SBa
VHE 81	17 00 06.	- 51 01	384		REFLECTION NEBULA
RNGC 6291	17 00 07.	+ 59 02		15.0	GALAXY
RNGC 6290	17 00 07.	+ 59 04		14.5	GALAXY
SCHO 0322	17 00 07.	- 35 13 00.	400		ISOLATED DARK CLOUD
SC 1656-5051.6	17 00 07.	- 50 56 03.			NEBULA
LB 00985	17 00 09.	+ 11 58 18.		16.4	FAINT BLUE STAR
PATH 1.772	17 00 09.	+ 29 45	5		NEBULA
ZWG 252.016	17 00 12.	+ 48 30		15.0	GALAXY
MCG+10-24-087	17 00 12.	+ 58 45	48	15.	GALAXY
MCG+10-24-086	17 00 12.	+ 59 00	30	15.	GALAXY
MCG+10-24-088	17 00 12.	+ 59 02	60	14.	GALAXY
ZC 1700.2+5941	17 00 12.	+ 59 41	3700		CLUSTER OF GALAXIES
MAI 097	17 00 16.	+ 70 23	74		DWARF SPHEROIDAL GALAXY
SCHO 0323	17 00 16.	- 38 01 36.	520		ISOLATED DARK CLOUD
PATH 1.773	17 00 17.	+ 29 37	5		NEBULA
ZC 1700.3+1556	17 00 18.	+ 15 56	1550		CLUSTER OF GALAXIES
MCG+05-40-032	17 00 18.	+ 30 48 30.	54	17.	GALAXY
UGC 10666	17 00 18.	+ 30 49	66	16.0	GALAXY S
ZWG 299.044	17 00 18.	+ 58 46		15.6	GALAXY
UGC 10667	17 00 24.	+ 21 38	66	16.0	GALAXY S
MCG+05-40-033	17 00 24.	+ 30 46 30.	60	15.3	GALAXY
ZWG 169.034	17 00 24.	+ 30 47		15.3	GALAXY
UGC 10668	17 00 24.	+ 30 47	66	15.3	GALAXY S0
MCG+10-24-089	17 00 24.	+ 60 16	30	16.	GALAXY
IC 1240	17 00 24.	+ 61 07			NONSTELLAR OBJECT
ARC 2246	17 00 24.	+ 64 18		17.6	RICH CLUSTER OF GALAXIES
LB 00986	17 00 25.	+ 30 34 36.		17.4	FAINT BLUE STAR
SCHO 0324	17 00 25.	- 36 23 42.	560		ISOLATED DARK CLOUD
ZC 1700.5+3322	17 00 30.	+ 33 22	2890		CLUSTER OF GALAXIES
MCG+10-24-090	17 00 30.	+ 59 17	18	16.	GALAXY
MCG+10-24-091	17 00 30.	+ 60 19	24	15.	GALAXY
ZWG 299.045	17 00 30.	+ 60 20		15.1	GALAXY
ZC 1700.5+7042	17 00 30.	+ 70 42	1810		CLUSTER OF GALAXIES
MCG+14-08-020	17 00 30.	+ 84 25	15	17.	GALAXY
PK351+04.1	17 00 30.4	- 33 25 35.	10		PLANETARY NEBULA
LB 00198	17 00 36.	+ 29 57 00.		17.3	FAINT BLUE STAR
ZC 1700.6+3516	17 00 36.	+ 35 16	2150		CLUSTER OF GALAXIES
CED 138	17 00 36.	- 37 46			DIFFUSE GALACTIC NEBULA
LB 00199	17 00 40.	+ 29 59 54.		16.8	FAINT BLUE STAR
ZWG 081.013	17 00 42.	+ 09 18		15.7	GALAXY
ZC 1700.7+3806	17 00 42.	+ 38 06	2420		CLUSTER OF GALAXIES
UGC 10669	17 00 42.	+ 70 22	84	17.	GALAXY DWARF
MRSL 347+02/1	17 00 42.	- 37 52	2700		HII REGION
HN 1834	17 00 47.8	- 43 45 36.	18		NEBULA
ZWG 110.025	17 00 48.	+ 16 22		15.6	GALAXY
IC 4639	17 00 48.	+ 22 58 35.			NONSTELLAR OBJECT
ZWG 225.027	17 00 48.	+ 44 55		15.0	GALAXY
LB 00987	17 00 48.	+ 80 16 06.		16.3	FAINT BLUE STAR
ISS 1011	17 00 48.	- 35 39	140		STELLAR RING
LB 00988	17 00 49.	+ 31 29 36.		16.0	FAINT BLUE STAR
ARC 2244	17 00 53.	+ 34 08		16.6	RICH CLUSTER OF GALAXIES
ARC 2245	17 00 54.	+ 33 37		16.5	RICH CLUSTER OF GALAXIES
ZWG 225.028	17 00 54.	+ 41 51		15.6	GALAXY
MCG+07-35-017	17 00 54.	+ 44 56	42	15.	GALAXY
MCG+11-21-008	17 00 54.	+ 68 11 30.	42	16.	GALAXY
KHAV 309	17 01	+ 01 26	3610		DARK NEBULA
KHAV 307	17 01	- 22 04	2840		DARK NEBULA
KHAV 310	17 01	- 25 58	2370		DARK NEBULA
KHAV 305	17 01	- 35 40	2730		DARK NEBULA
ZWG 081.014	17 01 00.	+ 12 18		15.4	GALAXY
MCG+06-37-022	17 01 00.	+ 36 44	78	15.	GALAXY
MCG+07-35-018	17 01 00.	+ 40 33	12	15.5	GALAXY
ZWG 225.029	17 01 00.	+ 43 20		15.5	GALAXY
ZWG 321.011	17 01 00.	+ 63 46		14.2	GALAXY
UGC 10670	17 01 00.	+ 63 46	102	14.2	GALAXY S
7ZW 673	17 01 00.	+ 83 57			COMPACT GALAXY
LDN 0013	17 01 00.	- 22 06	420		DARK NEBULA
SCHO 0325	17 01 02.	- 34 09 18.	380		ISOLATED DARK CLOUD
LB 00989	17 01 05.	+ 29 22 06.		17.1	FAINT BLUE STAR
SCHO 0326	17 01 05.	- 23 44 00.	380		ISOLATED DARK CLOUD
PK353+06.2	17 01 05.9	- 30 49 20.	10		PLANETARY NEBULA

OBJECT NAME	RIGHT ASCEN.	DECLINATION	DIAM.	MAGN.	TYPE OF OBJECT
1ZW 172	17 01 06.	+ 33 08			COMPACT GALAXY
UGC 10671	17 01 06.	+ 36 44	90	16.0	GALAXY Sb
MCG+09-28-016	17 01 06.	+ 54 06	42	16.	GALAXY
PATH 1.774	17 01 07.	+ 29 57	8		NEBULA
IC 1241	17 01 07.	+ 63 46 31.			NONSTELLAR OBJECT
ZWG 053.027	17 01 12.	+ 04 19		15.7	GALAXY
ZWG 139.030	17 01 12.	+ 25 00		15.1	GALAXY
UGC 10672	17 01 12.	+ 25 00	126	15.1	GALAXY S0-a
UGC 10673	17 01 12.	+ 29 56	102	16.0	GALAXY
MCG+09-28-017	17 01 12.	+ 56 16	72	17.	GALAXY
SCHO 0327	17 01 15.	- 35 01 06.	300		ISOLATED DARK CLOUD
HOLM 766B	17 01 17.	+ 31 43	18	14.8	PART OF MULTIPLE GALAXY
ZWG 081.015	17 01 18.	+ 09 22		15.5	GALAXY
UGC 10674	17 01 18.	+ 09 22	102	15.5	GALAXY S
HOLM 766A	17 01 18.	+ 31 43	18	14.7	PART OF MULTIPLE GALAXY
ZC 1701.3+3406	17 01 18.	+ 34 06	2690		CLUSTER OF GALAXIES
ZWG 225.030	17 01 18.	+ 40 36		14.7	GALAXY
PK350+04.1	17 01 18.9	- 33 55 13.	6	12.5	PLANETARY NEBULA
MCG+04-40-012	17 01 21.	+ 25 00	96	15.1	GALAXY
ZC 1701.4+2830	17 01 24.	+ 28 30	10890		CLUSTER OF GALAXIES
ZWG 169.035	17 01 24.	+ 31 31		15.4	GALAXY
MRK 700	17 01 24.	+ 31 31	12	15.5	GALAXY WITH UV CONTINUUM
UGC 10675	17 01 24.	+ 31 31	150	15.4	GALAXY DBL SYS
MCG+05-40-034	17 01 24.	+ 31 31	36	15.4	GALAXY
MCG+07-35-019	17 01 24.	+ 40 35	30	15.	GALAXY
KARA.68 235	17 01 24.	+ 70 20	74		DWARF GALAXY
LDN 0015	17 01 24.	- 22 06	840		DARK NEBULA
GCL 053	17 01 24.	- 24 41	342	10.61	GLOBULAR STAR CLUSTER
RNGC 6281	17 01 24.	- 37 49		8.5	OPEN CLUSTER
OCL 1003	17 01 24.	- 37 50	3600	8.6	OPEN STAR CLUSTER
VHA 213	17 01 24.	- 37 50	360		OPEN STAR CLUSTER
RNGC 6284	17 01 28.	- 24 41		10.5	GLOBULAR CLUSTER
ZWG 225.031	17 01 30.	+ 40 43		15.1	GALAXY
7ZW 674	17 01 30.	+ 61 02			COMPACT GALAXY
ZC 1701.5+6427	17 01 30.	+ 64 27	1750		CLUSTER OF GALAXIES
SCHO 0328	17 01 32.	- 35 50 48.	860		ISOLATED DARK CLOUD
HN 0095	17 01 35.	- 40 48			NEBULA
MCG+06-37-023	17 01 36.	+ 36 08	36	15.	GALAXY
MCG+07-35-020	17 01 36.	+ 40 42	33	14.5	GALAXY
ZWG 225.032	17 01 36.	+ 44 14		15.7	GALAXY
MCG+08-31-020	17 01 36.	+ 46 26	48	16.	GALAXY
7ZW 675	17 01 36.	+ 67 09			COMPACT GALAXY
OCL 0986	17 01 36.	- 48 07	900		OPEN STAR CLUSTER
MCG+07-35-021	17 01 39.	+ 44 14	36	16.	GALAXY
IC 4637	17 01 39.2	- 40 48 52.	2	13.6	PLANETARY NEBULA
PK345+00.1	17 01 40.8	- 40 49 05.	21	13.6	PLANETARY NEBULA
ZWG 053.028	17 01 42.	+ 06 51		15.6	GALAXY
KARA.73B 0783	17 01 42.	+ 06 51	42	15.6	ISOLATED GALAXY S
ZWG 169.036	17 01 42.	+ 31 26		15.4	GALAXY
MCG+06-37-024	17 01 42.	+ 36 30	36	14.5	GALAXY
ZWG 252.017	17 01 42.	+ 46 27		15.3	GALAXY
B 051	17 01 44.	- 22 12	1200		DARK OBJECT
MCG+08-31-021	17 01 45.	+ 46 25	51	15.	GALAXY
UGC 10676	17 01 48.	+ 26 35	120	18.	GALAXY DWARF
ZWG 197.027	17 01 48.	+ 36 29		14.5	GALAXY
UGC 10677	17 01 48.	+ 36 29	54	14.5	GALAXY PECULR
ZWG 299.046	17 01 48.	+ 61 14		15.5	GALAXY
PK347+01.1	17 01 48.3	- 37 48 44.		14.8	PLANETARY NEBULA
B 242	17 01 52.	- 32 22	1800		DARK OBJECT
SCHO 0329	17 01 53.	- 32 23 54.	800		ISOLATED DARK CLOUD
MCG+04-40-013	17 01 54.	+ 24 14	12	15.1	GALAXY
ZWG 139.031	17 01 54.	+ 24 15		15.1	GALAXY
UGC 10678	17 01 54.	+ 24 15	78	15.1	GALAXY S0
ZWG 139.032	17 01 54.	+ 25 07		15.4	GALAXY
ZC 1701.9+2555	17 01 54.	+ 25 55	1210		CLUSTER OF GALAXIES
MCG+05-40-035	17 01 54.	+ 31 34	90	15.5	GALAXY
LB 00990	17 01 56.	+ 30 51 12.		15.2	FAINT BLUE STAR
SCHO 0330	17 01 56.	- 34 18 36.	340		ISOLATED DARK CLOUD
MCG+04-40-015	17 01 57.	+ 25 07	30	15.4	GALAXY
MCG+04-40-014	17 01 57.	+ 25 08	36		GALAXY
KHAV 306	17 02	- 22 28	2070		DARK NEBULA
KHAV 308	17 02	- 33 28	3350		DARK NEBULA
B 053	17 02	- 33 31	1800		DARK OBJECT
LB 09907	17 02	- 83 49		14.0	FAINT BLUE STAR
ZWG 025.019	17 02 00.	+ 01 24		15.4	GALAXY
ZWG 081.016	17 02 00.	+ 09 16		15.7	GALAXY
ZWG 081.017	17 02 00.	+ 13 58		15.7	GALAXY
ZWG 169.037	17 02 00.	+ 31 34		15.5	GALAXY
UGC 10679	17 02 00.	+ 31 34	96	15.5	GALAXY Sb
ZWG 277.021	17 02 00.	+ 51 28		15.3	GALAXY
UGC 10680	17 02 00.	+ 51 28	72	15.3	GALAXY SB
PK343-01.1	17 02 00.	- 43 52			PLANETARY NEBULA
PK353+06.1	17 02 01.9	- 30 28 15.	8		PLANETARY NEBULA
MCG+08-31-022	17 02 03.	+ 45 30	84	15.	GALAXY
ZWG 053.029	17 02 06.	+ 04 20		15.6	GALAXY
ZWG 252.018	17 02 06.	+ 45 32		15.7	GALAXY
UGC 10681	17 02 06.	+ 45 32	78	15.7	GALAXY Sc
MCG+09-28-018	17 02 06.	+ 51 34	72	15.	GALAXY
ZC 1702.1+6528	17 02 06.	+ 65 28	1480		CLUSTER OF GALAXIES
GCL 054	17 02 06.	- 22 38	348	11.24	GLOBULAR STAR CLUSTER
RNGC 6287	17 02 06.	- 22 38		11.0	GLOBULAR CLUSTER
BC 4C29.50	17 02 10.9	+ 29 51 05.		19.14	QUASI-STELLAR OBJECT
LB 00991	17 02 11.	+ 28 58 36.		16.2	FAINT BLUE STAR
SHB 341	17 02 11.	+ 29 51 00.		19.1	QUASI-STELLAR OBJECT
VHE 82	17 02 12.	- 39 55	30		REFLECTION NEBULA
LB 00200	17 02 14.	+ 29 39 24.		16.9	FAINT BLUE STAR
PK352+05.1	17 02 15.6	- 32 28 05.	5		PLANETARY NEBULA
PK336-06.1	17 02 17.	- 52 26 04.	10		PLANETARY NEBULA
ZC 1702.3+2055	17 02 18.	+ 20 55	540		CLUSTER OF GALAXIES
ZWG 169.038	17 02 18.	+ 31 33		15.7	GALAXY
UGC 10682	17 02 18.	+ 60 26	66	16.0	GALAXY S
VHA 214	17 02 18.	- 36 36	180		STAR CLSTR IN NEBULOSITY
HN 1835	17 02 23.7	- 62 01 07.	36		NEBULA
ZWG 053.030	17 02 24.	+ 04 32		15.4	GALAXY
MCG+05-40-036	17 02 24.	+ 28 31	36	15.3	GALAXY
ZWG 169.039	17 02 24.	+ 28 32		15.3	GALAXY
7ZW 676	17 02 24.	+ 59 47			COMPACT GALAXY
ZWG 299.047	17 02 24.	+ 61 07		14.4	GALAXY
RNGC 6292	17 02 24.	+ 61 07		14.5	GALAXY
UGC 10684	17 02 24.	+ 61 07	108	14.4	GALAXY Sb-c
UGC 10683	17 02 24.	- 01 29	96	16.0	GALAXY DBL SYS
ISS 1012	17 02 24.	- 35 57	76		STELLAR RING
SCHO 0331	17 02 26.	- 31 22 48.	420		ISOLATED DARK CLOUD
RNGC 6295	17 02 28.	+ 60 24		15.0	GALAXY
ZWG 277.022	17 02 30.	+ 55 13		14.9	GALAXY
MCG+09-28-019	17 02 30.	+ 55 13	60	15.	GALAXY
MCG+10-24-092	17 02 30.	+ 60 24	48	15.	GALAXY
MCG+10-24-093	17 02 30.	+ 61 07	78	13.	GALAXY

OBJECT NAME	RIGHT ASCEN.	DECLINATION	DIAM.	MAGN.	TYPE OF OBJECT
MRSL 348+02/1	17 02 30.	- 37 23	2700		HII REGION
LB 00992	17 02 33.	+ 29 20 36.		16.6	FAINT BLUE STAR
ZWG 081.018	17 02 36.	+ 10 00		15.1	GALAXY
MCG+02-43-006	17 02 36.	+ 10 00	36	15.1	GALAXY
KARA.73B 0784	17 02 36.	+ 10 00	36	15.1	ISOLATED GALAXY S
ZWG 081.019	17 02 36.	+ 13 00		15.2	GALAXY
UGC 10685	17 02 36.	+ 13 00	78	15.2	GALAXY S
KARA.73B 0785	17 02 36.	+ 13 00	132	15.2	ISOLATED GALAXY S
ZWG 299.048	17 02 36.	+ 61 32		14.9	GALAXY
ZWG 225.033	17 02 42.	+ 42 00		15.5	GALAXY
UGC 10686	17 02 42.	+ 42 00	72	15.5	GALAXY
ZWG 299.049	17 02 42.	+ 59 48		15.4	GALAXY
UGC 10687	17 02 42.	+ 59 48	60	15.4	GALAXY SBc
PK358+09.1	17 02 42.	- 25 20			PLANETARY NEBULA
MCG+07-35-022	17 02 45.	+ 42 00	54	15.	GALAXY
PK010+18.1	17 02 45.	- 10 02	11		PLANETARY NEBULA
PK342-02.1	17 02 45.9	- 44 09 13.	25		PLANETARY NEBULA
ZWG 139.033	17 02 48.	+ 23 15		15.2	GALAXY
MCG+06-37-025	17 02 48.	+ 34 38	78	14.5	GALAXY
ZC 1702.8+3725	17 02 48.	+ 37 25	1810		CLUSTER OF GALAXIES
ZWG 197.028	17 02 48.	+ 38 01		15.7	GALAXY
MCG+10-24-094	17 02 48.	+ 59 48	60	15.	GALAXY
PK010+18.2	17 02 52.7	- 10 04 26.	70		PLANETARY NEBULA
SCHO 0332	17 02 53.	- 30 41 30.	570		ISOLATED DARK CLOUD
MCG+04-40-016	17 02 54.	+ 23 14	30	15.2	GALAXY
ZWG 197.029	17 02 54.	+ 34 38		15.5	GALAXY
UGC 10688	17 02 54.	+ 34 38	78	15.5	GALAXY Sc
ZC 1702.9+3510	17 02 54.	+ 35 10	11160		CLUSTER OF GALAXIES
SIV 07	17 03	- 46 35	5400		FAINT H EMISSION REGION
FATH 1.775	17 03 00.	+ 29 23	14		NEBULA
ZWG 197.030	17 03 00.	+ 38 17		15.5	GALAXY
ZWG 277.023	17 03 00.	+ 56 05		15.7	GALAXY
UGC 10689	17 03 00.	+ 56 05	60	15.7	GALAXY Sa-b
MCG+09-28-020	17 03 00.	+ 56 06	72	16.	GALAXY
ZWG 299.050	17 03 00.	+ 62 06		14.4	GALAXY
UGC 10690	17 03 00.	+ 62 06	42	14.4	GALAXY S0
LDN 1731	17 03 00.	- 28 00	4620		DARK NEBULA
SCHO 0333	17 03 01.	- 31 44 00.	700		ISOLATED DARK CLOUD
RNGC 6297	17 03 02.	+ 62 06		14.5	GALAXY
ZC 1703.1+7104	17 03 06.	+ 71 04	1280		CLUSTER OF GALAXIES
LB 00993	17 03 11.	+ 80 54 24.		17.0	FAINT BLUE STAR
ZWG 169.040	17 03 12.	+ 30 28		15.5	GALAXY
MCG+05-40-037	17 03 12.	+ 30 28 30.	42	15.5	GALAXY
ZWG 197.031	17 03 12.	+ 38 05		15.3	GALAXY
ZWG 321.012	17 03 12.	+ 64 13		15.2	GALAXY
B 054	17 03 16.	- 34 11	300		DARK OBJECT
ZWG 139.034	17 03 18.	+ 23 14		15.4	GALAXY
UGC 10692	17 03 18.	+ 23 14	132	15.4	GALAXY Sb
ZC 1703.3+2827	17 03 18.	+ 28 27	2490		CLUSTER OF GALAXIES
HOLM 767B	17 03 18.	+ 30 30	12	15.0	PART OF MULTIPLE GALAXY
MCG+07-35-023	17 03 18.	+ 41 55	30	14.	GALAXY
ZWG 225.034	17 03 18.	+ 41 56		14.2	GALAXY
UGC 10693	17 03 18.	+ 41 56	108	14.2	GALAXY E
UGC 10694	17 03 18.	+ 43 18	60	17.	GALAXY Sc
ZWG 225.035	17 03 18.	+ 44 57		15.7	GALAXY
HOLM 767A	17 03 20.	+ 30 29	30	14.8	PART OF MULTIPLE GALAXY
SCHO 0334	17 03 22.	- 34 13 18.	310		ISOLATED DARK CLOUD
ZWG 081.020	17 03 24.	+ 11 41		15.3	GALAXY
ZC 1703.4+1816	17 03 24.	+ 18 16	3490		CLUSTER OF GALAXIES
MCG+04-40-017	17 03 24.	+ 23 13	90	15.4	GALAXY
MCG+09-28-021	17 03 24.	+ 56 07	30	16.	GALAXY
LB 03101	17 03 24.	- 59 06		13.4	FAINT BLUE STAR
PK344-01.1	17 03 25.7	- 42 37 13.	14		PLANETARY NEBULA
SCHO 0335	17 03 28.	- 36 22 12.	760		ISOLATED DARK CLOUD
ZWG 111.001	17 03 30.	+ 18 10		15.5	GALAXY
ZWG 110.026	17 03 30.	+ 18 10		15.5	GALAXY
MCG+03-43-013	17 03 30.	+ 18 28	45	15.5	GALAXY
ZWG 111.002	17 03 30.	+ 18 29		15.5	GALAXY
ZWG 110.027	17 03 30.	+ 18 29		15.5	GALAXY
ZC 1703.5+2200	17 03 30.	+ 22 00	4770		CLUSTER OF GALAXIES
ZWG 225.036	17 03 30.	+ 43 06		14.9	GALAXY
UGC 10695	17 03 30.	+ 43 06	84	14.9	GALAXY E
MCG+14-08-021	17 03 30.	+ 84 48	36	16.	GALAXY
IC 4631	17 03 30.	- 77 32			NONSTELLAR OBJECT
HN 0436	17 03 31.	- 77 32			NEBULA
MCG+07-35-024	17 03 33.	+ 43 07	84	14.	GALAXY
SCHO 0336	17 03 35.	- 34 01 36.	350		ISOLATED DARK CLOUD
ZWG 139.035	17 03 36.	+ 25 36		15.6	GALAXY
UGC 10696	17 03 36.	+ 25 36	66	15.6	GALAXY S
MRSL 344-01/1	17 03 36.	- 42 43	54600		HII REGION
SC 1701+2001.7	17 03 37.	+ 19 57 33.	30		NEBULA
ZWG 053.031	17 03 42.	+ 07 51		15.1	GALAXY
MCG+01-43-009	17 03 42.	+ 07 51	36	15.1	GALAXY
LB 00994	17 03 42.	+ 29 24 18.		17.4	FAINT BLUE STAR
ZWG 197.032	17 03 42.	+ 35 42		15.6	GALAXY
MCG+07-35-025	17 03 42.	+ 41 53	15	15.5	GALAXY
ZWG 225.037	17 03 42.	+ 41 54		15.6	GALAXY
ZWG 339.026	17 03 42.	+ 72 57		15.7	GALAXY
UGC 10697	17 03 42.	+ 72 57	72	15.7	GALAXY S
ISS 1013	17 03 42.	- 34 00	530		STELLAR RING
SCHO 0337	17 03 43.	- 30 50 24.	590		ISOLATED DARK CLOUD
ZWG 111.003	17 03 48.	+ 18 15		15.6	GALAXY
ZWG 110.028	17 03 48.	+ 18 15		15.6	GALAXY
UGC 10698	17 03 48.	+ 32 33	84	17.	GALAXY S
ZC 1703.8-0129	17 03 48.	- 01 29	1410		CLUSTER OF GALAXIES
ZWG 082.001	17 03 54.	+ 10 28		14.5	GALAXY
ZWG 081.021	17 03 54.	+ 10 28		14.5	GALAXY
UGC 10699	17 03 54.	+ 10 28	36	14.5	GALAXY SB
MCG+02-43-007	17 03 54.	+ 10 28	36	14.5	GALAXY
KARA.73B 0786	17 03 54.	+ 10 28	42	14.5	ISOLATED GALAXY S
ZWG 225.038	17 03 54.	+ 41 49		15.7	GALAXY
SCHO 0338	17 03 57.	- 31 51 18.	380		ISOLATED DARK CLOUD
KHAV 315	17 04	- 07 16	19030		DARK NEBULA
KHAV 313	17 04	- 21 16	2500		DARK NEBULA
KHAV 312	17 04	- 35 58	4610		DARK NEBULA
ZWG 082.002	17 04 00.	+ 13 40		15.7	GALAXY
ZWG 081.022	17 04 00.	+ 13 40		15.7	GALAXY
KARA.73B 0787	17 04 00.	+ 13 40	12	15.7	ISOLATED GALAXY IR
UGC 10700	17 04 00.	+ 25 53	66	16.0	GALAXY S
ZWG 277.024	17 04 00.	+ 51 41		15.6	GALAXY
MCG+09-28-022	17 04 00.	+ 52 46	72	16.	GALAXY
UGC 10701	17 04 00.	+ 52 48	66	16.0	GALAXY Sb
MCG+10-24-095	17 04 00.	+ 58 15	36	16.	GALAXY
SCHO 0339	17 04 01.	- 30 27 30.	380		ISOLATED DARK CLOUD
SC 1700-5115.6	17 04 01.	- 51 19 47.	18		NEBULA
RC 3CR351	17 04 03.39	+ 60 48 31.3		15.28	QUASI-STELLAR OBJECT
STB 342	17 04 03.6	+ 60 48 29.		15.3	QUASI-STELLAR OBJECT
PK351+03.1	17 04 04.1	- 34 01 19.			PLANETARY NEBULA
ZC 1704.1+3159	17 04 06.	+ 31 59	1010		CLUSTER OF GALAXIES
ZC 1704.1+3700	17 04 06.	+ 37 00	1210		CLUSTER OF GALAXIES
ZWG 054.001	17 04 12.	+ 04 57		15.4	GALAXY
KARA.73B 0788	17 04 12.	+ 04 57	66	15.4	ISOLATED GALAXY S
ZWG 082.003	17 04 12.	+ 14 13		15.5	GALAXY
ZWG 081.023	17 04 12.	+ 14 13		15.5	GALAXY
TON-N 0266	17 04 12.	+ 25 47		14.7	BLUE STAR
ZWG 277.025	17 04 12.	+ 56 16		15.6	GALAXY
MCG+10-24-096	17 04 12.	+ 58 20	30	16.	GALAXY
FATH 1.776	17 04 15.	+ 29 59	16		NEBULA
ZWG 026.001	17 04 18.	+ 02 18		15.2	GALAXY
MCG+00-44-001	17 04 18.	+ 02 18	36	15.2	GALAXY
MCG+07-35-026	17 04 18.	+ 39 56	45	14.5	GALAXY
ZWG 225.039	17 04 18.	+ 39 57		15.5	GALAXY
MCG+09-28-023	17 04 18.	+ 56 17	42	16.	GALAXY
B 055	17 04 19.	- 31 56	960		DARK OBJECT
ZWG 111.004	17 04 24.	+ 18 10		15.7	GALAXY
ZWG 110.029	17 04 24.	+ 18 10		15.7	GALAXY
MCG+10-24-097	17 04 24.	+ 62 21	36	14.	GALAXY
SCHO 0340	17 04 29.	- 32 03 48.	530		ISOLATED DARK CLOUD
ZWG 139.036	17 04 30.	+ 24 50		15.4	GALAXY
UGC 10702	17 04 30.	+ 24 50	78	15.4	GALAXY DISTRBD
MCG+06-37-026	17 04 30.	+ 38 28	30	15.	GALAXY
RNGC 6299	17 04 33.	+ 62 31		15.0	GALAXY
MCG+04-40-018	17 04 36.	+ 24 50	36	15.4	GALAXY
ZWG 198.001	17 04 36.	+ 38 27		15.5	GALAXY
ZWG 197.033	17 04 36.	+ 38 27		15.5	GALAXY
UGC 10703	17 04 36.	+ 38 27	66	15.5	GALAXY Sc
ZWG 299.051	17 04 36.	+ 62 31		15.0	GALAXY
ZWG 355.023	17 04 36.	+ 79 06		15.6	GALAXY
UGC 10704	17 04 36.	+ 79 06	66	15.6	GALAXY S
MCG+07-35-027	17 04 39.	+ 41 06 30.	24	17.	GALAXY
ZWG 198.002	17 04 42.	+ 33 45		15.5	GALAXY
ZWG 197.034	17 04 42.	+ 33 45		15.5	GALAXY
7ZW 677	17 04 42.	+ 71 30			COMPACT GALAXY
MCG+06-37-028	17 04 45.	+ 35 42 30.	36	16.	GALAXY
MCG+06-37-027	17 04 45.	+ 38 27	66	14.	GALAXY
SCHO 0341	17 04 46.	- 33 48 30.	420		ISOLATED DARK CLOUD
LB 00995	17 04 47.	+ 27 59 24.		16.0	FAINT BLUE STAR
SCHO 0342	17 04 47.	- 31 37 48.	330		ISOLATED DARK CLOUD
ZWG 082.004	17 04 48.	+ 11 29		15.2	GALAXY
UGC 10705	17 04 48.	+ 11 29	60	15.2	GALAXY SBa
ZWG 111.005	17 04 48.	+ 15 43		15.6	GALAXY
KARA.73B 0789	17 04 48.	+ 15 43	36	15.6	ISOLATED GALAXY S
ZWG 198.003	17 04 48.	+ 38 25		15.1	GALAXY
ZWG 197.035	17 04 48.	+ 38 25		15.1	GALAXY
UGC 10706	17 04 48.	+ 38 25	72	15.1	GALAXY Sb-c
MCG+07-35-028	17 04 48.	+ 41 06 30.	60	14.	GALAXY
ZWG 225.040	17 04 48.	+ 41 07		15.6	GALAXY
UGC 10707	17 04 48.	+ 41 07	66	15.6	GALAXY Sc
MCG+13-12-014	17 04 48.	+ 79 08	108	15.	GALAXY
MCG+05-40-038	17 04 54.	+ 30 20	48	14.7	GALAXY
ZC 1704.9+3056	17 04 54.	+ 30 56	7060		CLUSTER OF GALAXIES
1ZW 173	17 04 54.	+ 34 08			COMPACT GALAXY
ZWG 198.004	17 04 54.	+ 35 56		14.7	GALAXY
ZWG 197.036	17 04 54.	+ 35 56		14.7	GALAXY
MCG+06-37-029	17 04 54.	+ 35 56	36	15.	GALAXY
SCHO 0343	17 04 58.	- 33 09 06.	460		ISOLATED DARK CLOUD
KHAV 314	17 05	- 25 28	3700		DARK NEBULA
ZC 1705.0+2545	17 05 00.	+ 25 45	1550		CLUSTER OF GALAXIES
ZWG 169.041	17 05 00.	+ 30 20		14.7	GALAXY
MCG+06-37-030	17 05 00.	+ 35 42	66	14.5	GALAXY
ZWG 198.005	17 05 00.	+ 35 43		14.8	GALAXY
ZWG 197.037	17 05 00.	+ 35 43		14.8	GALAXY
UGC 10708	17 05 00.	+ 35 43	66	14.8	GALAXY S
MCG+08-31-023	17 05 00.	+ 48 00	30	16.	GALAXY
LDN 0017	17 05 00.	- 22 40	2580		DARK NEBULA
LDN 1682	17 05 00.	- 32 02	600		DARK NEBULA
SCHO 0344	17 05 05.	- 23 09 24.	670		ISOLATED DARK CLOUD
MCG+05-40-039	17 05 06.	+ 31 29	36	15.6	GALAXY
ZWG 225.041	17 05 06.	+ 43 09		15.6	GALAXY
ZWG 252.019	17 05 06.	+ 48 01		15.5	GALAXY
MCG-01-44-001	17 05 06.	- 04 51	36	16.	GALAXY
LDN 0011	17 05 06.	- 22 50	420		DARK NEBULA
MRSL 348+01/1	17 05 06.	- 38 02	3600		HII REGION
ZWG 082.005	17 05 12.	+ 11 31		15.3	GALAXY
ZC 1705.2+2624	17 05 12.	+ 26 24	3560		CLUSTER OF GALAXIES
ZWG 169.042	17 05 12.	+ 31 29		15.6	GALAXY
UGC 10709	17 05 12.	+ 31 29	60	15.6	GALAXY SB
ZWG 198.006	17 05 12.	+ 34 18		15.7	GALAXY
ZWG 197.038	17 05 12.	+ 34 18		15.7	GALAXY
ZWG 225.042	17 05 12.	+ 42 30		14.8	GALAXY
ZWG 225.043	17 05 12.	+ 43 04		15.3	GALAXY
MCG+07-35-029	17 05 12.	+ 43 04	39	15.	GALAXY
MCG+12-16-017	17 05 12.	+ 68 54	30	15.	GALAXY
MCG+07-35-030	17 05 15.	+ 42 29	21	16.	GALAXY
LB 00996	17 05 17.	+ 27 58 06.		16.1	FAINT BLUE STAR
ZWG 082.006	17 05 18.	+ 09 20		15.6	GALAXY
ZC 1705.3+2055	17 05 18.	+ 20 55	810		CLUSTER OF GALAXIES
ZWG 198.007	17 05 18.	+ 34 20		15.4	GALAXY
ZWG 197.039	17 05 18.	+ 34 20		15.4	GALAXY
ZWG 225.044	17 05 18.	+ 43 11		15.2	GALAXY
UGC 10710	17 05 18.	+ 43 11	96	15.2	GALAXY Sb
MCG+07-35-031	17 05 18.	+ 43 11	90	14.	GALAXY
ZC 1705.3+6025	17 05 18.	+ 60 25	1680		CLUSTER OF GALAXIES
ZWG 321.013	17 05 18.	+ 68 53		15.1	GALAXY
UGC 10711	17 05 18.	+ 68 53	84	15.1	GALAXY F
ISS 1014	17 05 18.	- 34 56	636		STELLAR RING
RNGC 6303	17 05 19.	+ 68 53		15.0	GALAXY
MCG+06-37-031	17 05 21.	+ 34 18	30	15.	GALAXY
B 057	17 05 21.	- 22 46	300		DARK OBJECT
ZWG 169.043	17 05 24.	+ 31 30		15.5	GALAXY
UGC 10712	17 05 24.	+ 31 30	60	15.5	GALAXY Sa
MCG+05-40-040	17 05 24.	+ 31 30	54	15.5	GALAXY
MCG+06-37-032	17 05 24.	+ 34 20	24	15.	GALAXY
ZWG 198.008	17 05 24.	+ 34 38		15.7	GALAXY
ZWG 197.040	17 05 24.	+ 34 38		15.7	GALAXY
ZWG 339.027	17 05 24.	+ 72 31		13.8	GALAXY
UGC 10713	17 05 24.	+ 72 31	120	13.8	GALAXY Sb
LB 00997	17 05 26.	+ 27 30 36.		16.0	FAINT BLUE STAR
MCG+05-40-041	17 05 27.	+ 30 17	96	15.5	GALAXY
SC 1701-5106.3	17 05 29.	- 51 10 22.	30		NEBULA
ZWG 169.044	17 05 30.	+ 30 17		15.5	GALAXY
UGC 10714	17 05 30.	+ 30 17	102	15.5	GALAXY Sb-c
ZWG 169.045	17 05 30.	+ 31 11		15.3	GALAXY
ZWG 169.046	17 05 30.	+ 31 18		15.0	GALAXY
UGC 10715	17 05 30.	+ 31 18	66	15.0	GALAXY SBa
MCG+05-40-042	17 05 30.	+ 31 19	60	15.0	GALAXY

OBJECT NAME	RIGHT ASCEN.	DECLINATION	DIAM.	MAGN.	TYPE OF OBJECT
ZWG 225.045	17 05 30.	+ 39 05		15.5	GALAXY
ZWG 277.026	17 05 30.	+ 52 12		15.5	GALAXY
MCG+12-16-018	17 05 30.	+ 72 28	102	14.	GALAXY
LDN 0020	17 05 30.	- 22 40	840		DARK NEBULA
B 056	17 05 34.	- 32 02	180		DARK OBJECT
ZWG 299.052	17 05 36.	+ 60 46		15.7	GALAXY
HN 1836	17 05 40.4	- 61 51 17.	48		NEBULA
7ZW 678	17 05 42.	+ 72 12			COMPACT GALAXY
SCHO 0346	17 05 45.	- 30 55 54.	390		ISOLATED DARK CLOUD
SCHO 0345	17 05 45.	- 31 34 18.	340		ISOLATED DARK CLOUD
SCHO 0347	17 05 47.	- 22 44 24.	860		ISOLATED DARK CLOUD
ZWG 026.002	17 05 48.	+ 01 01		15.5	GALAXY
UGC 10716	17 05 48.	+ 30 23	78	16.0	GALAXY Sc
MCG+06-37-033	17 05 48.	+ 36 30	24	16.	GALAXY
ZWG 225.046	17 05 48.	+ 39 14		15.5	GALAXY
ZC 1705.8+6434	17 05 48.	+ 64 34	1410		CLUSTER OF GALAXIES
7ZW 679	17 05 48.	+ 69 14			COMPACT GALAXY
ZC 1705.8+7417	17 05 48.	+ 74 17	1280		CLUSTER OF GALAXIES
SCHO 0348	17 05 48.	- 32 07 12.	260		ISOLATED DARK CLOUD
ZWG 082.007	17 05 54.	+ 11 55		15.5	GALAXY
ISS 1015	17 05 54.	- 34 58	364		STELLAR RING
KHAV 319	17 06	- 11 40	15460		DARK NEBULA
KHAV 317	17 06	- 24 04	6070		DARK NEBULA
KHAV 316	17 06	- 29 28	2070		DARK NEBULA
SIV 11	17 06	- 30 48	21600		FAINT H EMISSION REGION
ZWG 082.008	17 06 00.	+ 10 09		15.6	GALAXY
ZWG 139.037	17 06 00.	+ 26 26		15.5	GALAXY
UGC 10717	17 06 00.	+ 26 26	60	15.5	GALAXY Sc
LDN 0018	17 06 00.	- 22 50	360		DARK NEBULA
LDN 0006	17 06 00.	- 23 10	900		DARK NEBULA
LDN 1685	17 06 00.	- 32 05	240		DARK NEBULA
PK345-00.1	17 06 02.4	- 41 32 17.			PLANETARY NEBULA
IC 4633	17 06 10.	- 77 27			NONSTELLAR OBJECT
ZWG 054.002	17 06 12.	+ 04 06		14.9	GALAXY
UGC 10718	17 06 12.	+ 04 06	60	14.9	GALAXY S
MCG+01-44-001	17 06 12.	+ 04 06	54	14.9	GALAXY
ZWG 082.009	17 06 12.	+ 11 57		15.6	GALAXY
ZWG 198.009	17 06 12.	+ 34 53		15.2	GALAXY
7ZW 680	17 06 12.	+ 66 59			COMPACT GALAXY
MCG+13-12-015	17 06 12.	+ 78 42 30.	21	16.	GALAXY
HN 0437	17 06 12.	- 77 27			NEBULA
IC 1242	17 06 14.	+ 04 06 32.			NONSTELLAR OBJECT
RNGC 6296	17 06 15.	+ 03 57		14.0	GALAXY
LB 00998	17 06 17.	+ 31 29 42.		18.0	FAINT BLUE STAR
ZWG 054.003	17 06 18.	+ 03 57		14.2	GALAXY
UGC 10719	17 06 18.	+ 03 57	60	14.2	GALAXY Sb-c
MCG+01-44-002	17 06 18.	+ 03 57	42	14.2	GALAXY
ZWG 082.010	17 06 18.	+ 09 40		15.5	GALAXY
UGC 10720	17 06 18.	+ 09 40	66	15.5	GALAXY Sc
ZWG 139.038	17 06 18.	+ 25 35		14.2	GALAXY
UGC 10721	17 06 18.	+ 25 35	72	14.2	GALAXY Sc?
MCG+06-38-001	17 06 18.	+ 36 24	24	15.	GALAXY
ZWG 198.010	17 06 18.	+ 36 25		15.7	GALAXY
ZWG 355.024	17 06 18.	+ 78 42		15.4	GALAXY
ISS 1016	17 06 18.	- 33 58	149		STELLAR RING
RNGC 6331	17 06 19.	+ 78 42		15.5	GALAXY
MCG+04-40-019	17 06 24.	+ 25 35 30.	66	14.2	GALAXY
LB 01000	17 06 28.	+ 12 33 42.		17.2	FAINT BLUE STAR
LB 00999	17 06 28.	+ 30 05 24.		15.8	FAINT BLUE STAR
ZWG 139.039	17 06 30.	+ 26 41		15.3	GALAXY
B 243	17 06 35.	- 29 31	1500		DARK OBJECT
ZWG 054.004	17 06 36.	+ 08 28		15.7	GALAXY
KARA.73B 0790	17 06 36.	+ 08 28	48	15.7	ISOLATED GALAXY S
ZWG 082.011	17 06 36.	+ 11 32		15.0	GALAXY
MCG+02-44-001	17 06 36.	+ 11 32	42	15.0	GALAXY
ZC 1706.6+2554	17 06 36.	+ 25 54	670		CLUSTER OF GALAXIES
MCG+04-40-020	17 06 36.	+ 26 41	18	15.3	GALAXY
ZWG 169.047	17 06 36.	+ 31 52		14.8	GALAXY
ABC 2256	17 06 36.	+ 78 47		15.3	RICH CLUSTER OF GALAXIES
MCG+05-40-043	17 06 39.	+ 31 52	42	14.8	GALAXY
LB 01001	17 06 41.	+ 31 10 24.		16.3	FAINT BLUE STAR
ZWG 082.012	17 06 42.	+ 09 29		15.4	GALAXY
MCG+07-35-032	17 06 45.	+ 43 01	14	15.	GALAXY
ZWG 082.013	17 06 48.	+ 09 19		15.5	GALAXY
ZWG 169.048	17 06 48.	+ 32 37		15.6	GALAXY
MCG+07-35-033	17 06 48.	+ 40 51	21	14.5	GALAXY
ZWG 225.047	17 06 48.	+ 40 52		14.7	GALAXY
ZWG 225.048	17 06 48.	+ 43 01		15.4	GALAXY
UGC 10722	17 06 48.	+ 43 01	72	15.4	GALAXY (E)
1ZW 174	17 06 48.	+ 53 45			COMPACT GALAXY
MCG+10-24-098	17 06 54.	+ 60 48	60	14.0	GALAXY
SCHO 0349	17 06 55.	- 29 26 30.	1350		ISOLATED DARK CLOUD
PK345-01.1	17 06 55.4	- 41 49 06.	9		PLANETARY NEBULA
IC 4643	17 06 58.	+ 42 24 01.			SAME AS NGC 6301
RNGC 6306	17 06 59.	+ 60 48		14.5	GALAXY
RNGC 6307	17 06 59.	+ 60 49		14.0	GALAXY
KHAV 322	17 07	- 20 52	1570		DARK NEBULA
KHAV 318	17 07	- 22 28	2500		DARK NEBULA
KHAV 320	17 07	- 35 16	5190		DARK NEBULA
LB 09908	17 07	- 84 10		15.4	FAINT BLUE STAR
ZWG 082.014	17 07 00.	+ 09 22		15.7	GALAXY
ZWG 225.049	17 07 00.	+ 42 24		14.6	GALAXY
UGC 10723	17 07 00.	+ 42 24	150	14.6	GALAXY Sc
MCG+07-35-034	17 07 00.	+ 42 24	126	13.	GALAXY
ZWG 299.053	17 07 00.	+ 60 47		14.3	GALAXY
UGC 10724	17 07 00.	+ 60 47	60	14.3	GALAXY S
KARA.72 504A	17 07 00.	+ 60 47	66	14.3	PART OF DOUBLE GALAXY
MCG+10-24-099	17 07 00.	+ 60 50	72	13.2	GALAXY
MCG+13-12-016	17 07 00.	+ 75 28	54	15.	GALAXY
ZWG 355.025	17 07 00.	+ 75 29		13.5	GALAXY
UGC 10725	17 07 00.	+ 75 29	60	13.5	GALAXY S
ZWG 355.026	17 07 00.	+ 78 42		15.6	GALAXY
UGC 10726	17 07 00.	+ 78 42	84	15.6	GALAXY E
MCG+13-12-017	17 07 00.	+ 78 42 30.	33	16.	GALAXY
LDN 0009	17 07 00.	- 23 10	1020		DARK NEBULA
LDN 1736	17 07 00.	- 28 20	780		DARK NEBULA
LDN 1716	17 07 00.	- 29 40	1980		DARK NEBULA
LDN 1708	17 07 00.	- 30 20	840		DARK NEBULA
LDN 1694	17 07 00.	- 31 40	2940		DARK NEBULA
RNGC 6301	17 07 01.	+ 42 24		14.5	GALAXY
RNGC 6324	17 07 01.	+ 75 29		13.5	GALAXY
HOLM 769B	17 07 03.	+ 60 48	18	14.0	PART OF MULTIPLE GALAXY
ZC 1707.1+1450	17 07 06.	+ 14 50	870		CLUSTER OF GALAXIES
ZWG 299.054	17 07 06.	+ 60 48		14.0	GALAXY
UGC 10727	17 07 06.	+ 60 48	84	14.0	GALAXY S0-a
KARA.72 504B	17 07 06.	+ 60 48	90	14.0	PART OF DOUBLE GALAXY
GCL 055	17 07 06.	- 26 30	372	9.38	GLOBULAR STAR CLUSTER
VHA 215	17 07 06.	- 26 30	180		GLOBULAR STAR CLUSTER
HOLM 769A	17 07 07.	+ 60 49	42	13.2	PART OF MULTIPLE GALAXY
RNGC 6294	17 07 09.	- 26 30			NON-EXISTENT OBJECT
RNGC 6293	17 07 09.	- 26 30		9.5	GLOBULAR CLUSTER
ZWG 139.040	17 07 12.	+ 22 17		15.6	GALAXY
UGC 10728	17 07 12.	+ 22 17	96	15.6	GALAXY Sb
ZWG 198.011	17 07 12.	+ 34 58		15.7	GALAXY
ZWG 225.050	17 07 12.	+ 42 21		15.5	GALAXY
ZWG 355.027	17 07 12.	+ 78 43		15.5	GALAXY
ISS 1058	17 07 12.	- 39 21	121		STELLAR RING
ZC 1707.3+1435	17 07 18.	+ 14 35	810		CLUSTER OF GALAXIES
MCG+05-40-044	17 07 18.	+ 31 40 30.	72	15.4	GALAXY
ZWG 252.020	17 07 18.	+ 47 16		15.6	GALAXY
RNGC 6310	17 07 18.	+ 61 03		14.0	GALAXY
MCG+10-24-100	17 07 18.	+ 61 04	120	12.	GALAXY
MCG+13-12-018	17 07 18.	+ 78 43	54	16.	GALAXY
HOLM 768B	17 07 20.	+ 37 06	54	14.3	PART OF MULTIPLE GALAXY
HOLM 768A	17 07 20.	+ 37 06	18	13.8	PART OF MULTIPLE GALAXY
ZWG 169.049	17 07 24.	+ 31 40		15.4	GALAXY
UGC 10729	17 07 24.	+ 31 40	84	15.4	GALAXY S-IRR
ZWG 299.055	17 07 24.	+ 61 03		13.8	GALAXY
UGC 10730	17 07 24.	+ 61 03	114	13.8	GALAXY S
SCHO 0350	17 07 24.	- 34 26 48.	440		ISOLATED DARK CLOUD
SCHO 0351	17 07 26.	- 23 10 48.	500		ISOLATED DARK CLOUD
IC 1254	17 07 27.	+ 72 28 46.			NON-STELLAR OBJECT
ZC 1707.5+3208	17 07 30.	+ 32 08	1550		CLUSTER OF GALAXIES
VHE 88C	17 07 30.	- 45 51			REFLECTION NEBULA
VHE 88B	17 07 30.	- 45 56	48		REFLECTION NEBULA
VHE 88A	17 07 30.	- 46 04	30		REFLECTION NEBULA
PK357+07.1	17 07 35.	- 27 05 02.	10	12.9	PLANETARY NEBULA
ZWG 111.006	17 07 36.	+ 19 51		15.6	GALAXY
ZC 1707.6+4045	17 07 35.	+ 40 45	20030		CLUSTER OF GALAXIES
ZWG 225.051	17 07 36.	+ 41 00		15.2	GALAXY
HN 0096	17 07 37.	- 55 20			NEBULA
IC 4642	17 07 37.	- 55 20	2	13.4	PLANETARY NEBULA
PK334-09.1	17 07 37.	- 55 20 14.	30	14.4	PLANETARY NEBULA
ZWG 198.012	17 07 42.	+ 34 29		15.3	GALAXY
ZWG 225.052	17 07 42.	+ 40 41		15.5	GALAXY
ZWG 321.014	17 07 42.	+ 63 43		13.1	GALAXY
UGC 10731	17 07 42.	+ 63 43	66	13.1	GALAXY S
KARA.72 505B	17 07 42.	+ 63 43	24		PART OF DOUBLE GALAXY
KARA.72 505A	17 07 42.	+ 63 43	42	13.1	PART OF DOUBLE GALAXY
ACK 341-04.1	17 07 42.	- 47 21			PLANETARY NEBULA
B 244	17 07 43.	- 28 21	1800		DARK OBJECT
B 058	17 07 43.	- 40 21	1800		DARK OBJECT
ZWG 198.013	17 07 48.	+ 34 10		15.4	GALAXY
7ZW 681	17 07 48.	+ 63 42			COMPACT GALAXY
MCG+13-12-019	17 07 48.	+ 78 43	9	17.	GALAXY
LB 03102	17 07 48.	- 58 47		13.5	FAINT BLUE STAR
SCHO 0352	17 07 53.	- 28 22 12.	1130		ISOLATED DARK CLOUD
ZWG 139.041	17 07 54.	+ 21 43		15.5	GALAXY
ZC 1707.9+3426	17 07 54.	+ 34 26	2220		CLUSTER OF GALAXIES
ABC 2249	17 07 57.	+ 34 32		15.4	RICH CLUSTER OF GALAXIES
KHAV 325	17 08	- 21 40	2230		DARK NEBULA
KHAV 323	17 08	- 27 22	3540		DARK NEBULA
KHAV 321	17 08	- 29 04	1570		DARK NEBULA
LB 09909	17 08	- 81 33		14.4	FAINT BLUE STAR
LB 09910	17 08	- 85 56		13.7	FAINT BLUE STAR
MCG+06-38-002	17 08 00.	+ 36 27 30.	48	15.	GALAXY
ZWG 198.014	17 08 00.	+ 36 28		15.4	GALAXY
UGC 10732	17 08 00.	+ 36 28	60	15.4	GALAXY Sc
ZC 1708.0+5825	17 08 00.	+ 58 25	3560		CLUSTER OF GALAXIES
MCG+10-24-101	17 08 00.	+ 62 06	42	15.	GALAXY
LDN 0032	17 08 00.	- 22 35	1320		DARK NEBULA
LDN 1790	17 08 00.	- 24 30	4200		DARK NEBULA
LDN 1746	17 08 00.	- 27 20	2220		DARK NEBULA
LDN 1737	17 08 00.	- 28 25	1740		DARK NEBULA
LDN 1728	17 08 00.	- 28 58	540		DARK NEBULA
LDN 1713	17 08 00.	- 30 08	300		DARK NEBULA
LDN 1711	17 08 00.	- 30 10	840		DARK NEBULA
IC 4635	17 08 03.	- 77 24			NONSTELLAR OBJECT
HN 0438	17 08 05.	- 77 24			NEBULA
ZWG 169.050	17 08 12.	+ 28 39		15.5	GALAXY
MCG+05-40-046	17 08 12.	+ 28 39	36	15.5	GALAXY
ZWG 169.051	17 08 12.	+ 32 33		15.1	GALAXY
UGC 10733	17 08 12.	+ 32 33	84	15.1	GALAXY E
MCG+05-40-045	17 08 12.	+ 32 34	36	15.1	GALAXY
ZCG 1708+38	17 08 12.	+ 38 55		17.0	COMPACT GALAXY
ZWG 225.053	17 08 12.	+ 40 40		15.2	GALAXY
ZWG 355.028	17 08 12.	+ 77 11		15.6	GALAXY
UGC 10734	17 08 12.	+ 77 11	78	15.6	GALAXY COMPACT
IC 1243	17 08 14.	+ 10 51 06.			MAY NOT EXIST
MCG+05-40-047	17 08 15.	+ 30 22	48	15.1	GALAXY
B 059	17 08 16.	- 27 26	3600		DARK OBJECT
OCL 0093	17 08 18.	+ 15 36	1200		OPEN STAR CLUSTER
ZWG 169.052	17 08 18.	+ 30 22		15.1	GALAXY
UGC 10735	17 08 18.	+ 32 20	78	16.5	GALAXY Sb-c
MCG+07-35-035	17 08 18.	+ 40 40	42	15.	GALAXY
MCG+07-35-036	17 08 18.	+ 41 17	30	15.	GALAXY
ZWG 299.056	17 08 18.	+ 62 06		15.6	GALAXY
7ZW 682	17 08 18.	+ 64 39			COMPACT GALAXY
MCG+12-16-019	17 08 18.	+ 69 28	33	16.	GALAXY
MCG+12-16-020	17 08 18.	+ 69 31	168	15.	GALAXY
MCG+13-12-020	17 08 18.	+ 78 42	24	17.	GALAXY
LDN 0038	17 08 18.	- 22 25	360		DARK NEBULA
ZWG 169.053	17 08 24.	+ 32 17		15.6	GALAXY
ZC 1708.4+3749	17 08 24.	+ 37 49	1550		CLUSTER OF GALAXIES
ZWG 339.028	17 08 24.	+ 69 33		15.0	GALAXY
UGC 10736	17 08 24.	+ 69 33	210	15.0	GALAXY
SCHO 0353	17 08 26.	- 22 25 42.	730		ISOLATED DARK CLOUD
ZWG 111.007	17 08 30.	+ 20 19		15.3	GALAXY
1ZW 175	17 08 30.	+ 39 53			COMPACT GALAXY
ZWG 225.054	17 08 30.	+ 39 05		15.5	GALAXY
ZWG 225.055	17 08 30.	+ 42 26		15.0	GALAXY
MCG+07-35-037	17 08 30.	+ 42 26 30.	39	14.5	GALAXY
ZWG 252.021	17 08 30.	+ 49 22		15.4	GALAXY
UGC 10737	17 08 30.	+ 49 22	66	15.4	GALAXY SB:a-b
PK358+07.1	17 08 33.9	- 25 40 05.	5		PLANETARY NEBULA
SCHO 0354	17 08 34.	- 29 26 18.	530		ISOLATED DARK CLOUD
ZWG 054.005	17 08 36.	+ 05 55		15.5	GALAXY
UGC 10738	17 08 36.	+ 05 55	120	15.5	GALAXY Sb-c
LDN 0039	17 08 36.	- 22 26	480		DARK NEBULA
YM 07	17 08 37.	- 28 21	420		SYMMETRIC GALACTIC NEBULA
SCHO 0355	17 08 40.	- 31 47 30.	780		ISOLATED DARK CLOUD
IC 1244	17 08 42.	+ 36 19 19.			NONSTELLAR OBJECT
MCG+08-31-024	17 08 42.	+ 49 21	72	15.	GALAXY
ISS 1017	17 08 42.	- 37 40	112		STELLAR RING
B 245	17 08 45.	- 29 21	480		DARK OBJECT
ZWG 054.006	17 08 48.	+ 05 28		15.7	GALAXY

OBJECT NAME	RIGHT ASCEN.	DECLINATION	DIAM.	MAGN.	TYPE OF OBJECT
ZWG 111.008	17 08 48.	+ 19 33		15.7	GALAXY
ZWG 111.009	17 08 48.	+ 20 41		15.7	GALAXY
ZWG 198.015	17 08 48.	+ 36 21		14.7	GALAXY
UGC 10739	17 08 48.	+ 36 21	72	14.7	GALAXY
ZWG 225.056	17 08 48.	+ 41 40		15.7	GALAXY
7ZW 683	17 08 48.	+ 64 19			COMPACT GALAXY
MCG+06-38-003	17 08 51.	+ 36 21	48	14.5	GALAXY
B 060	17 08 51.	- 22 23	780		DARK OBJECT
SCHO 0356	17 08 52.	- 31 15 00.	590		ISOLATED DARK CLOUD
RNGC 6317	17 08 53.	+ 62 57		16.0	GALAXY
ZWG 139.042	17 08 54.	+ 21 09		15.7	GALAXY
ZWG 225.057	17 08 54.	+ 40 00		15.1	GALAXY
ZWG 225.058	17 08 54.	+ 43 26		15.5	GALAXY
ZWG 277.027	17 08 54.	+ 55 43		15.5	GALAXY
MCG+11-21-009	17 08 54.	+ 62 57	57	16.	GALAXY
MRSL 348+00/2	17 08 54.	- 28 26	420		HII REGION
SCHO 0357	17 08 55.	- 32 38 00.	300		ISOLATED DARK CLOUD
WRAY 19.48	17 08 55.0	- 38 25 44.		10.0	STAR-NEBULA ASSOCIATION
PK352+03.1	17 08 56.4	- 32 34 17.			PLANETARY NEBULA
MCG+07-35-038	17 08 57.	+ 39 59	42	15.	GALAXY
LB 01002	17 08 58.	+ 14 32 24.		17.8	FAINT BLUE STAR
KHAV 326	17 09	- 19 22	4740		DARK NEBULA
KHAV 324	17 09	- 22 34	1350		DARK NEBULA
ZWG 225.059	17 09 00.	+ 41 43		14.9	GALAXY
RNGC 6311	17 09 00.	+ 41 43		15.0	GALAXY
UGC 10741	17 09 00.	+ 41 43	72	14.9	GALAXY E
ZC 1709.0+4243	17 09 00.	+ 42 43	870		CLUSTER OF GALAXIES
ZWG 252.022	17 09 00.	+ 48 23		14.8	GALAXY
UGC 10742	17 09 00.	+ 48 23	78	14.8	GALAXY Sa-b
7ZW 684	17 09 00.	+ 66 29			COMPACT GALAXY
ZWG 370.005	17 09 00.	+ 86 40		15.2	GALAXY
ZWG 367.022	17 09 00.	+ 86 40		15.2	GALAXY
UGC 10740	17 09 00.	+ 86 40	84	15.2	GALAXY SB0
LDN 0060	17 09 00.	- 21 55	720		DARK NEBULA
LDN 1743	17 09 00.	- 28 00	2880		DARK NEBULA
LDN 1742	17 09 00.	- 28 20	720		DARK NEBULA
LDN 1727	17 09 00.	- 29 10	180		DARK NEBULA
LDN 1726	17 09 00.	- 29 10	1200		DARK NEBULA
LDN 1715	17 09 00.	- 30 10	180		DARK NEBULA
LDN 1714	17 09 00.	- 30 15	300		DARK NEBULA
LDN 1706	17 09 00.	- 30 50	3900		DARK NEBULA
LDN 1700	17 09 00.	- 32 30	2940		DARK NEBULA
B 246	17 09 01.	- 22 36	900		DARK OBJECT
RNGC 6313	17 09 02.	+ 48 23		15.0	GALAXY
ZWG 054.007	17 09 06.	+ 08 03		14.8	GALAXY
UGC 10743	17 09 06.	+ 08 03	78	14.8	GALAXY Sa?
MCG+01-44-003	17 09 06.	+ 08 03	72	14.8	GALAXY
KARA.73B 0791	17 09 06.	+ 08 03	72	14.8	ISOLATED GALAXY S
ZC 1709.1+2459	17 09 06.	+ 24 59	4300		CLUSTER OF GALAXIES
MCG+07-35-039	17 09 06.	+ 41 43	24	14.	GALAXY
MCG+08-31-025	17 09 06.	+ 48 22	84	15.	GALAXY
ARC 2250	17 09 09.	+ 39 46		16.5	RICH CLUSTER OF GALAXIES
SCHO 0358	17 09 10.	- 22 36 42.	420		ISOLATED DARK CLOUD
RNGC 6319	17 09 11.	+ 63 03		14.5	GALAXY
ZWG 225.060	17 09 12.	+ 42 21		15.3	GALAXY
MCG+07-35-040	17 09 12.	+ 42 21	36	15.	GALAXY
ZWG 321.015	17 09 12.	+ 63 03		14.4	GALAXY
UGC 10744	17 09 12.	+ 63 03	66	14.4	GALAXY
RNGC 6312	17 09 13.	+ 42 21		15.5	GALAXY
MCG+08-31-026	17 09 18.	+ 45 52 30.	72	16.	GALAXY
LDN 1733	17 09 18.	- 28 57	540		DARK NEBULA
MRSL 348+00/1	17 09 18.	- 38 46	5400		HII REGION
VHE 83	17 09 18.	- 42 40	48		REFLECTION NEBULA
ZWG 198.016	17 09 24.	+ 38 04		15.6	GALAXY
UGC 10746	17 09 24.	+ 45 56	66	16.5	GALAXY Sc
UGC 10745	17 09 24.	+ 61 28	60	16.5	GALAXY
HN 1837	17 09 28.8	- 61 45 07.	42		NEBULA
ZWG 198.017	17 09 30.	+ 38 39		15.7	GALAXY
MCG+14-08-022	17 09 30.	+ 86 40	27	16.	GALAXY
LB 01003	17 09 33.	+ 29 23 12.		17.3	FAINT BLUE STAR
MCG+08-31-027	17 09 36.	+ 46 27	36	16.	GALAXY
ZWG 252.023	17 09 36.	+ 46 28		15.6	GALAXY
MCG+11-21-010	17 09 36.	+ 63 02	27	15.	GALAXY
LDN 005[illegible]	17 09 36.	- 22 17	720		DARK NEBULA
ISS 0921	17 09 36.	- 33 02	119		STELLAR RING
ZWG 111.010	17 09 42.	+ 18 54		15.5	GALAXY
KARA.73B 0792	17 09 42.	+ 18 54	24	15.5	ISOLATED GALAXY E
SCHO 0359	17 09 42.	- 28 53 00.	320		ISOLATED DARK CLOUD
VHE 84	17 09 42.	- 38 44	30		REFLECTION NEBULA
LB 01004	17 09 43.	+ 28 40 54.		15.9	FAINT BLUE STAR
ZWG 111.011	17 09 48.	+ 20 19		15.4	GALAXY
KARA.73B 0793	17 09 48.	+ 20 19	48	15.4	ISOLATED GALAXY S
ZWG 139.043	17 09 48.	+ 23 26		14.4	GALAXY
UGC 10747	17 09 48.	+ 23 26	84	14.4	GALAXY Sc/SBc
ZWG 169.054	17 09 48.	+ 28 23		15.6	GALAXY
RNGC 6308	17 09 51.	+ 23 26		14.5	GALAXY
HUB E19	17 09 51.	- 38 22			DIFFUSE NEBULA
B 250	17 09 53.	- 28 21	900		DARK OBJECT
MCG+04-40-021	17 09 54.	+ 23 27	60	14.4	GALAXY
MCG+09-28-024	17 09 54.	+ 55 51	36	16.	GALAXY
B 248	17 09 54.	- 28 56	600		DARK OBJECT
B 249	17 09 55.	- 29 06	300		DARK OBJECT
B 247	17 09 55.	- 30 12	240		DARK OBJECT
KHAV 328	17 10	- 28 22	3630		DARK NEBULA
KHAV 327	17 10	- 28 58	1150		DARK NEBULA
KHAV 329	17 10	- 33 46	2230		DARK NEBULA
ZWG 252.024	17 10 00.	+ 50 02		15.4	GALAXY
MCG+09-28-026	17 10 00.	+ 56 30	24	17.	GALAXY
MCG+09-28-025	17 10 00.	+ 56 30	30	17.	GALAXY
LDN 0065	17 10 00.	- 21 50	1200		DARK NEBULA
LDN 0046	17 10 00.	- 22 30	2340		DARK NEBULA
LDN 0030	17 10 00.	- 23 00	1860		DARK NEBULA
LDN 1707	17 10 00.	- 30 55	660		DARK NEBULA
LDN 1693	17 10 00.	- 32 10	480		DARK NEBULA
LDN 1697	17 10 00.	- 32 30	660		DARK NEBULA
SCHO 0360	17 10 05.	- 29 00 06.	520		ISOLATED DARK CLOUD
ZWG 111.012	17 10 06.	+ 16 37		15.5	GALAXY
ZWG 198.018	17 10 06.	+ 37 37		15.2	GALAXY
ZWG 225.061	17 10 06.	+ 42 45		15.1	GALAXY
MCG+09-28-027	17 10 06.	+ 55 20	42	16.	GALAXY
ZWG 277.028	17 10 06.	+ 56 48		15.6	GALAXY
ZWG 299.057	17 10 06.	+ 57 00		15.2	GALAXY
ISS 0922	17 10 06.	- 31 36	145		STELLAR RING
ISS 0923	17 10 06.	- 31 37	184		STELLAR RING
LB 01005	17 10 09.	+ 30 40 00.		16.4	FAINT BLUE STAR
MCG+06-38-004	17 10 09.	+ 37 37	24	15.	GALAXY
MCG+07-35-041	17 10 09.	+ 42 46	39	15.	GALAXY
UGC 10748	17 10 12.	+ 28 29	78	17.	GALAXY TRP SYS

OBJECT NAME	RIGHT ASCEN.	DECLINATION	DIAM.	MAGN.	TYPE OF OBJECT
MCG+05-40-048	17 10 12.	+ 30 14	63	15.2	GALAXY
MCG+10-24-102	17 10 12.	+ 57 00	36	16.	GALAXY
SCHO 0361	17 10 14.	- 28 25 12.	970		ISOLATED DARK CLOUD
PK018+20.1	17 10 14.4	- 03 12 25.		13.4	PLANETARY NEBULA
ZWG 170.001	17 10 18.	+ 30 14		15.2	GALAXY
ZWG 169.055	17 10 18.	+ 30 14		15.2	GALAXY
UGC 10749	17 10 16.	+ 30 14	66	15.2	GALAXY SBb
ZWG 252.025	17 10 18.	+ 49 37		15.5	GALAXY
UGC 10750	17 10 18.	+ 49 37	78	15.5	GALAXY Sb-0
UGC 10751	17 10 18.	+ 54 26	60	16.5	GALAXY Sc
MCG+09-28-028	17 10 18.	+ 55 20	15	17.	GALAXY
PK349+01.1	17 10 21.92	- 37 02 40.7	240	12.8	PLANETARY NEBULA
CED 139	17 10 23.	- 37 03	120		DIFFUSE GALACTIC NEBULA
ZC 1710.4+2552	17 10 24.	+ 25 52	1280		CLUSTER OF GALAXIES
MCG+08-31-028	17 10 24.	+ 49 37	66	16.	GALAXY
ZC 1710.4+6401	17 10 24.	+ 64 01	7260		CLUSTER OF GALAXIES
RNGC 6302	17 10 25.	- 37 03			PLANETARY NEBULA
SCHO 0362	17 10 29.	- 32 00 36.	470		ISOLATED DARK CLOUD
ZWG 139.044	17 10 30.	+ 23 20		14.3	GALAXY
UGC 10752	17 10 30.	+ 23 20	108	14.3	GALAXY Sa
ZC 1710.5+2453	17 10 30.	+ 24 53	1140		CLUSTER OF GALAXIES
ARC 2251	17 10 30.	+ 24 53		17.6	RICH CLUSTER OF GALAXIES
ZWG 252.026	17 10 30.	+ 46 51		15.7	GALAXY
7ZW 685	17 10 30.	+ 63 43			COMPACT GALAXY
ISS 1018	17 10 30.	- 34 31	114		STELLAR RING
RNGC 6314	17 10 33.	+ 23 20		14.5	GALAXY
LB 00325	17 10 33.	- 10 16 00.		16.1	FAINT BLUE STAR
ZWG 139.045	17 10 36.	+ 23 17		15.4	GALAXY
MCG+04-40-022	17 10 36.	+ 23 20	78	14.3	GALAXY
ZWG 170.002	17 10 36.	+ 30 58		15.6	GALAXY
ZWG 169.056	17 10 36.	+ 30 58		15.6	GALAXY
ZWG 198.019	17 10 36.	+ 35 30		15.3	GALAXY
ZWG 225.062	17 10 36.	+ 39 48		15.6	GALAXY
MCG+07-35-042	17 10 36.	+ 39 48	30	15.	GALAXY
MCG+08-31-029	17 10 36.	+ 46 35	36	16.	GALAXY
MCG+08-31-030	17 10 36.	+ 46 50	36	16.	GALAXY
7ZW 686	17 10 36.	+ 64 31			COMPACT GALAXY
RNGC 6315	17 10 39.	+ 23 17		15.5	GALAXY
HN 1838	17 10 40.4	- 61 19 20.	18		NEBULA
ZWG 198.020	17 10 42.	+ 35 57		15.1	GALAXY
ZC 1710.7+5603	17 10 42.	+ 56 03	1210		CLUSTER OF GALAXIES
MCG+10-24-103	17 10 42.	+ 57 29	18	18.	GALAXY
7ZW 687	17 10 42.	+ 64 05			COMPACT GALAXY
ISS 0924	17 10 42.	- 31 48	131		STELLAR RING
MCG+04-40-023	17 10 45.	+ 23 18	30	15.4	GALAXY
MCG+06-38-005	17 10 45.	+ 35 56	42	15.	GALAXY
MCG+08-31-031	17 10 45.	+ 46 33	60	16.	GALAXY
IC 1245	17 10 47.	+ 38 05 42.			NONSTELLAR OBJECT
MIL 50	17 10 47.	- 38 07 42.	480		SUPERNOVA REMNANT
ZWG 139.046	17 10 48.	+ 21 22		15.4	GALAXY
MCG+06-38-006	17 10 48.	+ 34 44	51	15.	GALAXY
ZWG 198.021	17 10 48.	+ 34 45		15.5	GALAXY
UGC 10753	17 10 48.	+ 34 45	60	15.5	GALAXY Sc
UGC 10754	17 10 48.	+ 46 36	72	16.0	GALAXY S
MCG+08-31-032	17 10 48.	+ 47 49 30.	42	16.	GALAXY
LB 01006	17 10 48.	+ 58 59 54.		17.6	FAINT BLUE STAR
7ZW 688	17 10 48.	+ 63 42			COMPACT GALAXY
LB 03103	17 10 48.	- 59 23		14.8	FAINT BLUE STAR
B 251	17 10 51.	- 20 06	1200		DARK OBJECT
SCHO 0363	17 10 53.	- 20 06 42.	490		ISOLATED DARK CLOUD
PK354+04.1	17 10 53.6	- 31 16 16.	5		PLANETARY NEBULA
ZWG 082.015	17 10 54.	+ 13 06		15.2	GALAXY
ZWG 198.022	17 10 54.	+ 38 05		15.0	GALAXY
UGC 10755	17 10 54.	+ 38 05	102	15.0	GALAXY S0
MCG+06-38-007	17 10 54.	+ 38 05	48	14.	GALAXY
ZWG 225.063	17 10 54.	+ 41 02		15.7	GALAXY
MCG+07-35-043	17 10 54.	+ 41 02	48	16.	GALAXY
MCG+08-31-033	17 10 54.	+ 47 49	42	16.	GALAXY
ZWG 252.027	17 10 54.	+ 47 50		15.5	GALAXY
MCG+10-24-104	17 10 54.	+ 57 38	15	16.	GALAXY
SCHO 0364	17 10 56.	- 28 21 00.	400		ISOLATED DARK CLOUD
LB 01007	17 10 58.	+ 31 12 00.		17.2	FAINT BLUE STAR
KHAV 330	17 11	- 25 52	2500		DARK NEBULA
ZWG 198.023	17 11 00.	+ 35 09		15.4	GALAXY
MCG+10-24-105	17 11 00.	+ 57 09	30	17.	GALAXY
ZWG 299.058	17 11 00.	+ 57 40		15.2	GALAXY
ZWG 299.059	17 11 00.	+ 60 03		14.6	GALAXY
UGC 10756	17 11 00.	+ 60 03	84	14.6	GALAXY SBc
MCG+10-24-106	17 11 00.	+ 60 03	72	14.	GALAXY
IC 1248	17 11 00.	+ 60 04		14.06	GALAXY
7ZW 689	17 11 00.	+ 64 14			COMPACT GALAXY
ZWG 339.029	17 11 00.	+ 72 28		14.4	GALAXY
UGC 10757	17 11 00.	+ 72 28	102	14.4	GALAXY Sc
LDN 0071	17 11 00.	- 21 40	1500		DARK NEBULA
LDN 0067	17 11 00.	- 21 50	1200		DARK NEBULA
ISS 0925	17 11 00.	- 29 32	97		STELLAR RING
ISS 1019	17 11 00.	- 34 30	52		STELLAR RING
ZWG 140.001	17 11 06.	+ 26 32		15.6	GALAXY
ZWG 139.047	17 11 06.	+ 26 32		15.6	GALAXY
ZWG 225.064	17 11 06.	+ 40 49		15.6	GALAXY
ZWG 277.029	17 11 06.	+ 53 11		15.3	GALAXY
UGC 10758	17 11 06.	+ 53 11	66	15.3	GALAXY S
KARA.73B 0794	17 11 06.	+ 53 11	54	15.3	ISOLATED GALAXY S
MCG+12-16-021	17 11 06.	+ 72 27	72	14.	GALAXY
LB 01008	17 11 12.	+ 12 44 30.		17.4	FAINT BLUE STAR
ZWG 111.013	17 11 12.	+ 17 57		15.7	GALAXY
ZWG 225.065	17 11 12.	+ 42 52		14.5	GALAXY
UGC 10759	17 11 12.	+ 42 52	60	14.5	GALAXY COMPACT
ZWG 225.066	17 11 12.	+ 44 13		15.3	GALAXY
UGC 10760	17 11 12.	+ 44 13	84	15.3	GALAXY S
MCG+09-28-029	17 11 12.	+ 53 10	36	14.	GALAXY
ZC 1711.2+6147	17 11 12.	+ 61 47	870		CLUSTER OF GALAXIES
IC 1251	17 11 12.	+ 72 28 45.			NONSTELLAR OBJECT
7ZW 690	17 11 12.	+ 76 55			COMPACT GALAXY
RNGC 6309	17 11 12.	- 12 51		11.5	PLANETARY NEBULA
MIL 49	17 11 12.	- 38 26 42.	486		SUPERNOVA REMNANT
HN 1839	17 11 14.9	- 61 20 35.	18		NEBULA
MCG+07-35-044	17 11 15.	+ 40 19 30.	72	14.	GALAXY
MCG+07-35-045	17 11 15.	+ 42 53	18	14.	GALAXY
MCG+07-35-046	17 11 15.	+ 44 13 30.	60	16.	GALAXY
PK009+14.1	17 11 15.87	- 12 51 11.7	56	11.6	PLANETARY NEBULA
RNGC 6320	17 11 17.	+ 40 20		15.0	GALAXY
ZWG 111.014	17 11 18.	+ 16 06		15.1	GALAXY
MCG+03-44-001	17 11 18.	+ 16 06	27	15.1	GALAXY
ZWG 225.067	17 11 18.	+ 40 20		14.9	GALAXY
UGC 10761	17 11 18.	+ 40 20	78	14.9	GALAXY Sb-c
SHAH 137	17 11 18.	+ 67 37	162	17.5	GROUP OF COMPACT GALAXIES
MCG+12-16-022	17 11 18.	+ 69 38	51	17.	GALAXY

OBJECT NAME	RIGHT ASCEN.	DECLINATION	DIAM.	MAGN.	TYPE OF OBJECT
ZWG 339.030	17 11 18.	+ 69 41		15.5	GALAXY
MCG+12-16-023	17 11 18.	+ 72 20	57	12.0	GALAXY
ZWG 339.031	17 11 18.	+ 72 21		11.9	GALAXY
UGC 10762	17 11 18.	+ 72 21	222	11.9	GALAXY Sa
ISS 1020	17 11 18.	- 33 24	142		STELLAR RING
RNGC 6340	17 11 20.	+ 72 22		12.5	GALAXY
IC 1255	17 11 21.	+ 78 41 54.			NONSTELLAR OBJECT
UGC 10763	17 11 24.	+ 40 32	90	16.5	GALAXY DISTRBD
7ZW 691	17 11 24.	+ 63 41			COMPACT GALAXY
7ZW 692	17 11 24.	+ 64 25			COMPACT GALAXY
GCL 056	17 11 24.	- 29 24	480	9.83	GLOBULAR STAR CLUSTER
VHA 216	17 11 24.	- 29 24	240		GLOBULAR STAR CLUSTER
ASS 69	17 11 24.	- 33 07			OB ASSOCIATION SCO OB4
ISS 1021	17 11 24.	- 34 41	77		STELLAR RING
RNGC 6304	17 11 25.	- 29 24		10.0	GLOBULAR CLUSTER
PK352+03.2	17 11 26.1	- 33 21 24.	4		PLANETARY NEBULA
LB 01009	17 11 27.	+ 30 15 36.		17.2	FAINT BLUE STAR
1ZW 176	17 11 30.	+ 38 08			COMPACT GALAXY
ZWG 198.024	17 11 30.	+ 38 08		15.6	GALAXY
ZWG 225.068	17 11 30.	+ 42 27		15.4	GALAXY
MCG+07-35-047	17 11 30.	+ 42 27	36	15.	GALAXY
ZWG 225.069	17 11 30.	+ 42 55		15.6	GALAXY
ZWG 225.070	17 11 30.	+ 43 45		15.4	GALAXY
SCHO 0365	17 11 30.	- 30 52 18.	1140		ISOLATED DARK CLOUD
ZWG 198.025	17 11 36.	+ 37 53		15.4	GALAXY
ZC 1711.6+6245	17 11 36.	+ 62 45	2290		CLUSTER OF GALAXIES
7ZW 693	17 11 36.	+ 67 18			COMPACT GALAXY
ZWG 225.071	17 11 42.	+ 43 50		14.8	GALAXY
UGC 10764	17 11 42.	+ 43 50	66	14.8	GALAXY Sa-b
MCG+07-35-048	17 11 42.	+ 43 51	63	14.5	GALAXY
ISS 0926	17 11 42.	- 27 50	149		STELLAR RING
RNGC 6323	17 11 45.	+ 43 50		15.0	GALAXY
ZWG 225.072	17 11 48.	+ 42 57		15.0	GALAXY
UGC 10765	17 11 48.	+ 42 57	66	15.0	GALAXY Sc
MCG+07-35-049	17 11 48.	+ 42 58	60	14.	GALAXY
UGC 10766	17 11 48.	+ 74 29	60	17.	GALAXY Sc
ISS 0927	17 11 48.	- 29 36	338		STELLAR RING
LB 01010	17 11 49.	+ 60 32 48.		18.0	FAINT BLUE STAR
ZC 1711.9+1221	17 11 54.	+ 12 21	1610		CLUSTER OF GALAXIES
ISS 0928	17 11 54.	- 29 38	119		STELLAR RING
KHAV 335	17 12	- 20 52	1150		DARK NEBULA
KHAV 331	17 12	- 22 28	5060		DARK NEBULA
B 253	17 12	- 22 30	3600		DARK OBJECT
KHAV 332	17 12	- 36 46	2990		DARK NEBULA
ZWG 225.073	17 12 00.	+ 41 01		15.4	GALAXY
7ZW 694	17 12 00.	+ 64 05			COMPACT GALAXY
UGC 10767	17 12 00.	+ 73 40	66	16.0	GALAXY Sb-c
LDN 0079	17 12 00.	- 21 25	2220		DARK NEBULA
LDN 0056	17 12 00.	- 22 30	2760		DARK NEBULA
LDN 1718	17 12 00.	- 30 11	240		DARK NEBULA
LDN 1698	17 12 00.	- 32 10	780		DARK NEBULA
LDN 1691	17 12 00.	- 32 30	4020		DARK NEBULA
B 061	17 12 02.	- 20 25			DARK OBJECT
MCG+07-35-050	17 12 03.	+ 41 01 30.	54	15.	GALAXY
B 252	17 12 04.	- 32 05	1200		DARK OBJECT
PK354+04.2	17 12 04.7	- 31 30 42.			PLANETARY NEBULA
IC 1246	17 12 06.	+ 20 18			NONSTELLAR OBJECT
ZC 1712.1+6001	17 12 06.	+ 60 01	940		CLUSTER OF GALAXIES
ZC 1712.1+7740	17 12 06.	+ 77 40	5440		CLUSTER OF GALAXIES
ISS 0849	17 12 06.	- 22 09	70		STELLAR RING
ISS 0929	17 12 06.	- 29 32	200		STELLAR RING
SCHO 0366	17 12 09.	- 28 56 06.	540		ISOLATED DARK CLOUD
ARC 2255	17 12 11.	+ 64 09		15.3	RICH CLUSTER OF GALAXIES
ZC 1712.2+7135	17 12 12.	+ 71 35	1140		CLUSTER OF GALAXIES
LDN 0111	17 12 12.	- 20 18	300		DARK NEBULA
SCHO 0367	17 12 12.	- 24 58 00.	430		ISOLATED DARK CLOUD
ISS 0930	17 12 12.	- 32 45	119		STELLAR RING
ARP 032	17 12 13.	+ 59 23			PECULIAR GALAXY
SCHO 0368	17 12 14.	- 20 25 48.	410		ISOLATED DARK CLOUD
MCG+03-44-002	17 12 15.	+ 20 23	66	14.5	GALAXY
RNGC 6300	17 12 16.	- 62 46		11.5	GALAXY
ZWG 111.015	17 12 18.	+ 20 22		14.5	GALAXY
UGC 10768	17 12 18.	+ 20 22	66	14.5	GALAXY SBb
ZWG 140.002	17 12 18.	+ 23 07		15.5	GALAXY
ZWG 252.028	17 12 18.	+ 49 11		15.3	GALAXY
LDN 1699	17 12 18.	- 32 08	720		DARK NEBULA
RNGC 6321	17 12 19.	+ 20 22		14.5	GALAXY
ZWG 054.008	17 12 24.	+ 08 26		15.6	GALAXY
ZWG 140.003	17 12 24.	+ 23 06		15.1	GALAXY
MCG+04-41-001	17 12 24.	+ 23 06	21	15.1	GALAXY
ZWG 225.074	17 12 24.	+ 43 42		15.7	GALAXY
MCG+08-31-034	17 12 24.	+ 49 10	48	16.	GALAXY
ZWG 277.030	17 12 24.	+ 51 15		15.7	GALAXY
ZWG 299.060	17 12 24.	+ 57 45		15.7	GALAXY
7ZW 695	17 12 24.	+ 63 46			COMPACT GALAXY
ZWG 339.032	17 12 24.	+ 72 27		14.7	GALAXY
UGC 10769	17 12 24.	+ 72 27	114	14.7	GALAXY S?
ISS 1022	17 12 24.	- 38 58	92		STELLAR RING
MRSL 347-01/1	17 12 24.	- 40 19	6480		HII REGION
RNGC 6327	17 12 26.	+ 43 42		15.5	GALAXY
ZWG 054.009	17 12 30.	+ 06 16		15.4	GALAXY
ZWG 225.075	17 12 30.	+ 39 34		15.0	GALAXY
ZWG 299.061	17 12 30.	+ 59 23		14.3	GALAXY
UGC 10770	17 12 30.	+ 59 23	102	14.3	GALAXY DBL SYS
KARA.72 506A	17 12 30.	+ 59 23	60	14.3	PART OF DOUBLE GALAXY
MCG+12-16-024	17 12 30.	+ 72 26	84	15.	GALAXY
ZC 1712.6+0616	17 12 36.	+ 06 16	5710		CLUSTER OF GALAXIES
ZWG 054.010	17 12 36.	+ 08 29		15.6	GALAXY
KARA.72 506B	17 12 36.	+ 59 24	48		PART OF DOUBLE GALAXY
VHA 217	17 12 36.	- 40 46	240		OPEN STAR CLUSTER
MCG+07-35-051	17 12 39.	+ 43 45	96	14.	GALAXY
ARC 2252	17 12 39.	+ 49 27		17.5	RICH CLUSTER OF GALAXIES
MCG+10-24-107	17 12 39.	+ 59 23	90	14.	GALAXY
ZWG 054.011	17 12 42.	+ 08 30		15.4	GALAXY
MCG+05-40-001	17 12 42.	+ 28 11	30	17.	GALAXY
ZWG 170.003	17 12 42.	+ 30 47		15.5	GALAXY
MCG+07-35-052	17 12 42.	+ 39 07 30.	39	15.	GALAXY
ZWG 225.076	17 12 42.	+ 39 08		15.6	GALAXY
ZWG 225.077	17 12 42.	+ 43 45		14.3	GALAXY
UGC 10771	17 12 42.	+ 43 45	66	14.3	GALAXY E
VHA 218	17 12 42.	- 39 22	300		OPEN STAR CLUSTER
IC 1254	17 12 43.	+ 72 29 27.			NONSTELLAR OBJECT
RNGC 6329	17 12 44.	+ 43 45		14.5	GALAXY
ZWG 054.012	17 12 48.	+ 06 30		15.7	GALAXY
ZWG 054.013	17 12 48.	+ 08 28		15.2	GALAXY
MCG+08-31-036	17 12 48.	+ 46 00	42	16.	GALAXY
MCG+08-31-035	17 12 48.	+ 46 00	24	16.	GALAXY
7ZW 696	17 12 48.	+ 64 08			COMPACT GALAXY

OBJECT NAME	RIGHT ASCEN.	DECLINATION	DIAM.	MAGN.	TYPE OF OBJECT
ISS 1023	17 12 48.	- 33 22	156		STELLAR RING
ZWG 252.029	17 12 54.	+ 45 49		15.3	GALAXY
ZC 1712.9+4913	17 12 54.	+ 49 13	2960		CLUSTER OF GALAXIES
ZWG 339.033	17 12 54.	+ 72 39		15.3	GALAXY
KHAV 336	17 13	- 21 21	3540		DARK NEBULA
B 063	17 13	- 21 25	1140		DARK OBJECT
KHAV 334	17 13	- 24 04	4470		DARK NEBULA
KHAV 333	17 13	- 27 16	3250		DARK NEBULA
ZC 1713.0+1306	17 13 00.	+ 13 06	1340		CLUSTER OF GALAXIES
ZWG 170.004	17 13 00.	+ 29 13		15.4	GALAXY
UGC 10772	17 13 00.	+ 29 13	66	15.4	GALAXY S
ZWG 225.078	17 13 00.	+ 41 28		15.5	GALAXY
MCG+07-35-053	17 13 00.	+ 41 28	36	15.	GALAXY
ZWG 252.030	17 13 00.	+ 46 02		15.7	GALAXY
MCG+13-12-021	17 13 00.	+ 77 35	39	16.	GALAXY
LDN 0391	17 13 00.	- 03 00	2520		DARK NEBULA
LDN 0104	17 13 00.	- 20 45	540		DARK NEBULA
LDN 0100	17 13 00.	- 20 50	1080		DARK NEBULA
LDN 0016	17 13 00.	- 24 00	1380		DARK NEBULA
LDN 1756	17 13 00.	- 27 20	2160		DARK NEBULA
LDN 1722	17 13 00.	- 30 00	11280		DARK NEBULA
LDN 1720	17 13 00.	- 30 10	1200		DARK NEBULA
LDN 1702	17 13 00.	- 31 45	960		DARK NEBULA
LDN 1701	17 13 00.	- 32 03	300		DARK NEBULA
MCG+05-40-002	17 13 03.	+ 29 14	63	15.4	GALAXY
MCG+05-40-003	17 13 03.	+ 29 15	18	15.5	GALAXY
ZWG 198.026	17 13 06.	+ 35 34		14.9	GALAXY
ZWG 225.079	17 13 06.	+ 42 55		15.5	GALAXY
ZWG 225.080	17 13 06.	+ 43 10		15.7	GALAXY
ZWG 225.081	17 13 06.	+ 43 59		15.5	GALAXY
MCG+12-16-025	17 13 06.	+ 72 40	66	17.	GALAXY
ISS 0931	17 13 06.	- 30 33	118		STELLAR RING
VDB.66N 110	17 13 07.	- 20 59	72		REFLECTION NEBULA
IC 1249	17 13 08.	+ 35 35 53.			NONSTELLAR OBJECT
ZWG 170.005	17 13 12.	+ 29 19		15.7	GALAXY
ZWG 252.031	17 13 12.	+ 48 27		15.4	GALAXY
LDN 0005	17 13 12.	- 24 22	420		DARK NEBULA
IC 4645	17 13 13.	+ 43 09			NONSTELLAR OBJECT
RUB E20	17 13 13.	- 34 07			DIFFUSE NEBULA
B 062	17 13 14.	- 20 49	1140		DARK OBJECT
MCG+10-24-108	17 13 15.	+ 58 52	36	14.	GALAXY
SC 1709-5045.1	17 13 17.	- 50 48 37.	18		NEBULA
ZWG 111.016	17 13 18.	+ 16 22		15.1	GALAXY
MCG+03-44-003	17 13 18.	+ 16 22 30.	48	15.1	GALAXY
ZWG 111.017	17 13 19.	+ 20 06		15.6	GALAXY
ARC 2253	17 13 24.	+ 38 43		16.5	RICH CLUSTER OF GALAXIES
ZWG 225.082	17 13 24.	+ 43 43		14.6	GALAXY
UGC 10773	17 13 24.	+ 43 43	78	14.6	GALAXY Sa
ZWG 299.062	17 13 24.	+ 58 52		15.3	GALAXY
UGC 10774	17 13 24.	+ 58 52	60	15.3	GALAXY S
KARA.73B 0795	17 13 24.	+ 58 52	42	15.3	ISOLATED GALAXY S
SCHO 0369	17 13 24.	- 20 55 18.	690		ISOLATED DARK CLOUD
GCL 057	17 13 24.	- 28 05	432	10.10	GLOBULAR STAR CLUSTER
VHA 219	17 13 24.	- 28 05	180		GLOBULAR STAR CLUSTER
ISS 0932	17 13 24.	- 28 10	737		STELLAR RING
RNGC 6332	17 13 26.	+ 43 43		14.5	GALAXY
RNGC 6316	17 13 26.	- 28 05		10.0	GLOBULAR CLUSTER
MCG+07-35-054	17 13 27.	+ 43 43 30.	66	14.	GALAXY
ZWG 170.006	17 13 30.	+ 30 57		15.2	GALAXY
ZWG 225.083	17 13 30.	+ 43 06		15.7	GALAXY
IC 1247	17 13 34.	- 12 44			MAY NOT EXIST
ZWG 054.014	17 13 36.	+ 06 54		15.3	GALAXY
UGC 10775	17 13 36.	+ 06 54	78	15.3	GALAXY SBa
ZWG 111.018	17 13 36.	+ 18 45		15.7	GALAXY
ZC 1713.6+3845	17 13 36.	+ 38 45	1950		CLUSTER OF GALAXIES
ZWG 277.031	17 13 36.	+ 52 27		15.5	GALAXY
IC 1250	17 13 36.	+ 57 28 39.			NONSTELLAR OBJECT
MCG+05-40-004	17 13 39.	+ 30 57	30	15.2	GALAXY
ARC 2265	17 13 39.	+ 77 30		17.4	RICH CLUSTER OF GALAXIES
ZC 1713.7+2027	17 13 42.	+ 20 27	1810		CLUSTER OF GALAXIES
MCG+06-38-008	17 13 42.	+ 37 52 30.	24	15.5	GALAXY
ZWG 225.084	17 13 42.	+ 43 20		15.1	GALAXY
MCG+07-35-055	17 13 42.	+ 43 21	36	15.5	GALAXY
ZWG 252.032	17 13 42.	+ 47 44		15.3	GALAXY
ZWG 299.063	17 13 42.	+ 57 22		15.7	GALAXY
MCG+10-24-109	17 13 42.	+ 57 52	30	16.	GALAXY
7ZW 697	17 13 42.	+ 72 28			COMPACT GALAXY
7ZW 698	17 13 42.	+ 72 34			COMPACT GALAXY
ISS 0933	17 13 42.	- 31 07	149		STELLAR RING
SC 1709-5141.9	17 13 42.	- 51 45 24.	12		NEBULA
RNGC 6305	17 13 42.	- 59 06			UNVERIFIED SOUTHERN OBJECT
SCHO 0370	17 13 44.	- 30 21 06.	480		ISOLATED DARK CLOUD
MCG+08-31-037	17 13 45.	+ 47 42 30.	36	16.	GALAXY
ZC 1713.8+2144	17 13 48.	+ 21 44	3360		CLUSTER OF GALAXIES
ZWG 170.007	17 13 48.	+ 29 27		15.3	GALAXY
UGC 10776	17 13 48.	+ 29 27	102	15.3	GALAXY SBb
UGC 10777	17 13 48.	+ 37 52	66	16.0	GALAXY SBb
ZWG 225.085	17 13 48.	+ 39 15		15.6	GALAXY
ZWG 225.086	17 13 48.	+ 42 08		14.6	GALAXY
MCG+09-28-030	17 13 48.	+ 52 25	60	15.	GALAXY
MCG+10-24-110	17 13 48.	+ 57 54	30	18.	GALAXY
LDN 0139	17 13 48.	- 19 40	1320		DARK NEBULA
ISS 0934	17 13 48.	- 32 07	305		STELLAR RING
SCHO 0371	17 13 49.	- 29 56 24.	490		ISOLATED DARK CLOUD
RNGC 6330	17 13 50.	+ 29 27		15.5	GALAXY
MCG+05-40-005	17 13 51.	+ 29 29	72	15.3	GALAXY
MCG+07-35-056	17 13 51.	+ 42 07 30.	21	14.5	GALAXY
ZWG 054.015	17 13 54.	+ 06 37		14.7	GALAXY
UGC 10778	17 13 54.	+ 06 37	60	14.7	GALAXY E-S0?
MCG+01-44-004	17 13 54.	+ 06 37	48	14.7	GALAXY
ZWG 225.087	17 13 54.	+ 43 49		15.5	GALAXY
ZWG 252.033	17 13 54.	+ 48 30		15.7	GALAXY
MRSL 350+00/2	17 13 54.	- 36 19	1020		HII REGION
SCHO 0372	17 13 56.	- 29 28 48.	840		ISOLATED DARK CLOUD
LB 01011	17 13 58.	+ 59 41 24.		17.7	FAINT BLUE STAR
HN 1840	17 13 59.9	- 60 21 35.	24		NEBULA
KHAV 337	17 14	- 18 33	2840		DARK NEBULA
ZWG 026.003	17 14 00.	+ 00 55		15.7	GALAXY
MCG+00-44-002	17 14 00.	+ 00 55	30	15.7	GALAXY
ZWG 054.016	17 14 00.	+ 06 29		14.8	GALAXY
UGC 10779	17 14 00.	+ 06 29	120	14.8	GALAXY Sb
MCG+01-44-005	17 14 00.	+ 06 29	96	14.8	GALAXY
ZWG 198.027	17 14 00.	+ 38 56		15.6	GALAXY
ZC 1714.0+7209	17 14 00.	+ 72 09	1680		CLUSTER OF GALAXIES
SCHO 0373	17 14 04.	- 29 06 12.	590		ISOLATED DARK CLOUD
MCG+10-24-111	17 14 06.	+ 61 42	24	16.	GALAXY
ISS 0935	17 14 06.	- 28 39	173		STELLAR RING
PK356+05.1	17 14 06.	- 28 56		13.5	PLANETARY NEBULA

OBJECT NAME	RIGHT ASCEN.	DECLINATION	DIAM.	MAGN.	TYPE OF OBJECT
MCG+06-38-009	17 14 09.	+ 36 27 30.	36	16.	GALAXY
UGC 10780	17 14 12.	+ 07 22	72	16.5	GALAXY Sc
ZWG 198.028	17 14 12.	+ 36 26		15.6	GALAXY
MCG+10-24-112	17 14 12.	+ 57 21	18	17.	GALAXY
ZC 1714.2+5840	17 14 12.	+ 58 40	1080		CLUSTER OF GALAXIES
LDN 0023	17 14 12.	- 23 55	180		DARK NEBULA
MRSL 352+02/1	17 14 12.	- 34 06	1800		HII REGION
RNGC 6318	17 14 16.	- 39 24		12.0	OPEN CLUSTER
LB 01012	17 14 17.	+ 31 11 24.		16.5	FAINT BLUE STAR
ARC 2260	17 14 17.	+ 72 12		17.1	RICH CLUSTER OF GALAXIES
ZWG 054.017	17 14 18.	+ 08 15		15.3	GALAXY
ZC 1714.3+1939	17 14 18.	+ 19 39	740		CLUSTER OF GALAXIES
ZC 1714.3+2611	17 14 18.	+ 26 11	540		CLUSTER OF GALAXIES
ZWG 170.008	17 14 18.	+ 31 51		15.1	GALAXY
UGC 10781	17 14 18.	+ 31 51	60	15.1	GALAXY S
UGC 10782	17 14 18.	+ 36 26	72	16.0	GALAXY Sb-c
MCG+06-38-010	17 14 18.	+ 36 26 30.	24	15.5	GALAXY
MCG+08-31-038	17 14 18.	+ 49 08	30	16.	GALAXY
MCG+10-24-113	17 14 18.	+ 57 48	42	16.	GALAXY
LDN 0173	17 14 18.	- 18 20	540		DARK NEBULA
OCL 1004	17 14 18.	- 39 24	780	12.0	OPEN STAR CLUSTER
B 064	17 14 23.	- 18 26			DARK OBJECT
ZWG 277.032	17 14 24.	+ 52 22		15.7	GALAXY
7ZW 699	17 14 24.	+ 63 04			COMPACT GALAXY
MCG+05-40-006	17 14 27.	+ 31 53	54	15.1	GALAXY
RNGC 6346	17 14 28.	+ 57 23		15.5	GALAXY
RNGC 6345	17 14 29.	+ 57 25		15.5	GALAXY
RNGC 6338	17 14 29.	+ 57 29		14.0	GALAXY
ZWG 170.009	17 14 30.	+ 29 30		15.6	GALAXY
UGC 10783	17 14 30.	+ 29 30	78	15.6	GALAXY S
ZWG 198.029	17 14 30.	+ 37 10		15.7	GALAXY
ZWG 299.064	17 14 30.	+ 57 23		15.3	GALAXY
MCG+10-24-114	17 14 30.	+ 57 24 30.	12	15.	GALAXY
ZWG 299.065	17 14 30.	+ 57 25		15.4	GALAXY
MCG+10-24-115	17 14 30.	+ 57 25	36	16.	GALAXY
MCG+10-24-116	17 14 30.	+ 57 28	30	14.	GALAXY
ZWG 299.066	17 14 30.	+ 57 29		14.2	GALAXY
UGC 10784	17 14 30.	+ 57 29	90	14.2	GALAXY S0
MCG+10-24-117	17 14 30.	+ 57 29 30.	15	16.	GALAXY
7ZW 700	17 14 30.	+ 57 30			COMPACT GALAXY
ZWG 299.067	17 14 30.	+ 57 30		15.6	GALAXY
MCG+10-24-118	17 14 30.	+ 57 42	36	17.	GALAXY
7ZW 701	17 14 30.	+ 64 04			COMPACT GALAXY
LDN 0024	17 14 30.	- 23 58	240		DARK NEBULA
IC 4648	17 14 34.	+ 45 55			OPEN CLUSTER
ZWG 140.004	17 14 36.	+ 25 40		15.7	GALAXY
ZWG 225.088	17 14 36.	+ 44 07		15.4	GALAXY
1ZW 177	17 14 36.	+ 52 54			COMPACT GALAXY
MCG+10-24-119	17 14 36.	+ 59 38	30	17.	GALAXY
MIL 51	17 14 37.	- 37 23 18.	156		SUPERNOVA REMNANT
BIGO 540	17 14 40.	+ 43 55			NEBULA
KARA.72 507A	17 14 42.	+ 21 41	36	14.6	PART OF DOUBLE GALAXY
ZC 1714.7+3630	17 14 42.	+ 36 30	3290		CLUSTER OF GALAXIES
ZWG 198.030	17 14 42.	+ 37 23		14.9	GALAXY
UGC 10785	17 14 42.	+ 37 23	90	14.9	GALAXY S
MCG+06-38-011	17 14 42.	+ 37 24	84	14.5	GALAXY
ZWG 225.089	17 14 42.	+ 43 52		14.5	GALAXY
UGC 10786	17 14 42.	+ 43 52	66	14.5	GALAXY SBa
MCG+07-35-057	17 14 42.	+ 43 53	48	14.5	GALAXY
ZWG 277.033	17 14 42.	+ 52 54		14.7	GALAXY
KARA.73B 0796	17 14 42.	+ 52 54	66	14.7	ISOLATED GALAXY S
ISS 0936	17 14 42.	- 29 52	191		STELLAR RING
SCHO 0374	17 14 42.	- 32 50 36.	500		ISOLATED DARK CLOUD
RNGC 6336	17 14 45.	+ 43 52		14.5	GALAXY
ZWG 054.018	17 14 48.	+ 07 48		15.3	GALAXY
ZWG 140.005	17 14 48.	+ 21 40		14.6	GALAXY
UGC 10787	17 14 48.	+ 21 40	66	14.6	GALAXY SB
KARA.72 507B	17 14 48.	+ 21 40	66		PART OF DOUBLE GALAXY
MCG+04-41-002	17 14 48.	+ 21 40	60	14.6	GALAXY
ZWG 225.090	17 14 48.	+ 44 07		15.7	GALAXY
MCG+09-28-031	17 14 48.	+ 52 53	36	15.	GALAXY
ZWG 225.091	17 14 54.	+ 43 24		15.3	GALAXY
MCG+07-35-058	17 14 54.	+ 43 24 30.	42	15.5	GALAXY
ZWG 252.034	17 14 54.	+ 47 21		15.5	GALAXY
ZWG 300.001	17 14 54.	+ 57 26		15.7	GALAXY
ZWG 299.068	17 14 54.	+ 57 26		15.7	GALAXY
UGC 10788	17 14 54.	+ 57 26	66	15.7	GALAXY Sa-b
GCL 058	17 14 54.	- 23 42	294	12.66	GLOBULAR STAR CLUSTER
SCHO 0375	17 14 54.	- 29 51 12.	380		ISOLATED DARK CLOUD
OCL 1000	17 14 54.	- 42 54	960	7.0	OPEN STAR CLUSTER
VHA 220	17 14 54.	- 42 54	360		OPEN STAR CLUSTER
RNGC 6322	17 14 55.	- 42 54		6.5	OPEN CLUSTER
PK348-00.1	17 14 55.8	- 39 16 10.			PLANETARY NEBULA
MCG+10-24-120	17 14 57.	+ 57 25 30.	60	16.	GALAXY
IC 1252	17 14 58.	+ 57 26			NONSTELLAR OBJECT
RNGC 6325	17 14 59.	- 23 42		12.5	GLOBULAR CLUSTER
KHAV 338	17 15	- 28 39	1350		DARK NEBULA
DDKS	17 15	- 78 40	36000		REFLECTION NEBULA
ZWG 054.019	17 15 00.	+ 07 45		15.4	GALAXY
ZWG 170.010	17 15 00.	+ 31 35		15.7	GALAXY
ZWG 170.011	17 15 00.	+ 31 39		15.6	GALAXY
BIGO 541	17 15 00.	+ 57 23			NEBULA
IC 4650	17 15 00.	+ 57 23			NONSTELLAR OBJECT
IC 4649	17 15 00.	+ 57 26			NONSTELLAR OBJECT
ZC 1715.0+5903	17 15 00.	+ 59 03	3760		CLUSTER OF GALAXIES
ZC 1715.0+8507	17 15 00.	+ 85 07	1280		CLUSTER OF GALAXIES
LDN 0371	17 15 00.	- 06 00	4080		DARK NEBULA
SCHO 0376	17 15 00.	- 31 21 00.	920		ISOLATED DARK CLOUD
MRSL 350+00/1	17 15 00.	- 37 09	480		HII REGION
SG 3.107	17 15 02.	- 23 31	60		DIFFUSE EMISSION NEBULA
ZWG 054.020	17 15 06.	+ 07 43		15.2	GALAXY
ISS 0850	17 15 06.	- 22 05	149		STELLAR RING
SCHO 0377	17 15 06.	- 29 13 12.	500		ISOLATED DARK CLOUD
VHA 221	17 15 12.	- 32 18	600		OPEN STAR CLUSTER
MRSL 348-00/1	17 15 12.	- 38 38	3600		HII REGION
SCHO 0378	17 15 13.	- 18 36 00.	1430		ISOLATED DARK CLOUD
SCHO 0379	17 15 14.	- 28 35 48.	650		ISOLATED DARK CLOUD
MCG+10-24-121	17 15 15.	+ 61 26	36	15.	GALAXY
ZWG 054.021	17 15 18.	+ 07 39		15.7	GALAXY
UGC 10789	17 15 18.	+ 07 39	102	15.7	GALAXY DBL SYS
ZWG 339.034	17 15 18.	+ 71 16		15.1	GALAXY
ZWG 300.002	17 15 24.	+ 61 25		15.3	GALAXY
ZWG 299.069	17 15 24.	+ 61 25		15.3	GALAXY
LDN 0175	17 15 24.	- 18 28	480		DARK NEBULA
VHA 222	17 15 24.	- 38 14	150		STAR CLSTR IN NEBULOSITY
MCG+07-35-059	17 15 27.	+ 40 54	150	13.	GALAXY
ZC 1715.5+1944	17 15 30.	+ 19 44	810		CLUSTER OF GALAXIES
ZC 1715.5+2555	17 15 30.	+ 25 55	940		CLUSTER OF GALAXIES
ZWG 225.092	17 15 30.	+ 40 54		13.7	GALAXY
RNGC 6339	17 15 30.	+ 40 54		13.5	GALAXY
UGC 10790	17 15 30.	+ 40 54	198	13.7	GALAXY SBc
ZC 1715.5+4229	17 15 30.	+ 42 29	810		CLUSTER OF GALAXIES
MCG+12-16-026	17 15 30.	+ 72 25	60	15.	GALAXY
ZWG 339.035	17 15 30.	+ 72 27		15.5	GALAXY
UGC 10791	17 15 30.	+ 72 27	138	15.5	GALAXY DWRF SP
LDN 0089	17 15 30.	- 21 39	240		DARK NEBULA
SCHO 0380	17 15 30.	- 22 47 00.	370		ISOLATED DARK CLOUD
LDN 1767	17 15 30.	- 26 42	300		DARK NEBULA
LDN 1703	17 15 30.	- 32 03	360		DARK NEBULA
ARC 2254	17 15 31.	+ 19 47		17.7	RICH CLUSTER OF GALAXIES
REIF 2.251	17 15 35.31	+ 43 11 18.0			NEBULA
ZWG 140.006	17 15 36.	+ 24 15		15.6	GALAXY
ZWG 225.093	17 15 36.	+ 38 57		15.5	GALAXY
ZWG 198.031	17 15 36.	+ 38 57		15.5	GALAXY
MCG+06-38-012	17 15 36.	+ 38 57	30	15.	GALAXY
MCG+07-35-060	17 15 36.	+ 41 07 30.	60	14.5	GALAXY
GCL 059	17 15 36.	+ 43 12	738	7.30	GLOBULAR STAR CLUSTER
ZWG 252.035	17 15 36.	+ 45 35		15.7	GALAXY
UGC 10792	17 15 36.	+ 75 17	120	17.	GALAXY DWRF IR
LDN 0087	17 15 36.	- 21 45	120		DARK NEBULA
VHE 85A	17 15 36.	- 35 38			REFLECTION NEBULA
RNGC 6344	17 15 37.	+ 42 31			GALAXY
HN 0439	17 15 37.	- 80 01			NEBULA
PK354+03.1	17 15 37.8	- 31 36 00.		14.1	PLANETARY NEBULA
RNGC 6341	17 15 38.	+ 43 12		7.5	GLOBULAR CLUSTER
IC 4640	17 15 38.	- 80 01			NONSTELLAR OBJECT
MCG+07-35-061	17 15 39.	+ 40 38	33	15.	GALAXY
MCG+07-35-062	17 15 39.	+ 40 55 30.	48	15.5	GALAXY
HN 1841	17 15 40.1	- 60 06 22.	24		NEBULA
ZWG 225.094	17 15 42.	+ 40 39		15.4	GALAXY
ZWG 225.095	17 15 42.	+ 41 07		14.7	GALAXY
RNGC 6343	17 15 42.	+ 41 07		14.5	GALAXY
MCG+10-24-122	17 15 42.	+ 57 55	36	16.	GALAXY
ISS 0851	17 15 42.	- 22 26	123		STELLAR RING
ISS 1059	17 15 42.	- 43 03	426		STELLAR RING
SCHO 0381	17 15 44.	- 19 26 30.	490		ISOLATED DARK CLOUD
HN 0440	17 15 46.	- 80 06			NEBULA
IC 4641	17 15 47.	- 80 06			NONSTELLAR OBJECT
MCG+10-24-123	17 15 48.	+ 58 28	42	15.	GALAXY
ZWG 300.003	17 15 48.	+ 58 29		15.7	GALAXY
ZWG 299.070	17 15 48.	+ 58 29		15.7	GALAXY
PK342-04.1	17 15 49.	- 45 50	35		PLANETARY NEBULA
PK355+03.1	17 15 50.6	- 30 51 09.			PLANETARY NEBULA
LB 01013	17 15 51.	+ 30 29 36.		17.7	FAINT BLUE STAR
LB 01014	17 15 52.	+ 31 31 54.		14.9	FAINT BLUE STAR
ZWG 170.012	17 15 54.	+ 29 42		15.7	GALAXY
ZC 1715.9+3218	17 15 54.	+ 32 18	4230		CLUSTER OF GALAXIES
SCHO 0382	17 15 55.	- 21 45 18.	450		ISOLATED DARK CLOUD
MCG+05-40-007	17 15 57.	+ 29 43	48	15.7	GALAXY
LBN 0088	17 16	+ 06 10	600		BRIGHT NEBULA
KHAV 341	17 16	- 01 51	17080		DARK NEBULA
KHAV 339	17 16	- 25 03	2960		DARK NEBULA
ZWG 082.016	17 16 00.	+ 11 00		15.7	GALAXY
UGC 10793	17 16 00.	+ 11 00	60	15.7	GALAXY Sb-c
UGC 10794	17 16 00.	+ 40 30	66	16.5	GALAXY Sc-IRR
LB 01015	17 16 00.	+ 61 10 24.		16.7	FAINT BLUE STAR
7ZW 702	17 16 00.	+ 64 12			COMPACT GALAXY
ZWG 356.001	17 16 00.	+ 80 14		15.4	GALAXY
ZWG 355.029	17 16 00.	+ 80 14		15.4	GALAXY
LDN 0387	17 16 00.	- 04 00	3000		DARK NEBULA
LDN 0099	17 16 00.	- 21 20	1200		DARK NEBULA
LDN 0092	17 16 00.	- 21 40	240		DARK NEBULA
LDN 0077	17 16 00.	- 22 12	240		DARK NEBULA
LDN 0028	17 16 00.	- 24 00	6000		DARK NEBULA
LDN 1725	17 16 00.	- 30 20	720		DARK NEBULA
SCHO 0383	17 16 01.	- 19 41 36.	340		ISOLATED DARK CLOUD
ZWG 198.032	17 16 06.	+ 33 23		14.6	GALAXY
ZWG 225.096	17 16 06.	+ 39 33		15.4	GALAXY
ZWG 225.097	17 16 06.	+ 40 45		15.5	GALAXY
PK355+03.3	17 16 06.	- 31 08		13.3	PLANETARY NEBULA
ARC 2257	17 16 08.	+ 32 39		17.1	RICH CLUSTER OF GALAXIES
PK359+06.1	17 16 08.4	- 25 14 14.	11		PLANETARY NEBULA
UGC 10795	17 16 12.	+ 30 58	96	17.	GALAXY DWARF
SM2.01	17 16 12.	+ 44 40		16.5	FAINT BLUE OBJECT
MCG+10-24-124	17 16 12.	+ 62 00	66	15.	GALAXY
GCL 060	17 16 12.	- 18 28	474	8.92	GLOBULAR STAR CLUSTER
SCHO 0384	17 16 12.	- 30 17 42.	700		ISOLATED DARK CLOUD
MRSL 351+00/6	17 16 12.	- 35 40	1500		HII REGION
LB 03104	17 16 12.	- 56 49		14.3	FAINT BLUE STAR
RNGC 6333	17 16 14.	- 18 28		9.0	GLOBULAR CLUSTER
REIN 2.249	17 16 15.95	- 18 27 53.8			NEBULA
SCHO 0385	17 16 16.	- 32 05 36.	550		ISOLATED DARK CLOUD
SM1.1	17 16 18.	+ 44 41		18.2	BLUE GALAXY
ZWG 252.036	17 16 18.	+ 48 41		15.1	GALAXY
ZWG 300.004	17 16 18.	+ 61 58		14.8	GALAXY
ZWG 299.071	17 16 18.	+ 61 58		14.8	GALAXY
UGC 10796	17 16 18.	+ 61 58	120	14.8	GALAXY SBb
MAI 098	17 16 18.	+ 75 25	47		DWARF SPHEROIDAL GALAXY
SCHO 0386	17 16 18.	- 21 54 12.	300		ISOLATED DARK CLOUD
ISS 1060	17 16 18.	- 41 43	136		STELLAR RING
REIF 2.250	17 16 19.85	- 18 28 08.4			NEBULA
WRAY 19.49	17 16 23.9	- 36 02 57.		10.0	STAR-NEBULA ASSOCIATION
ZWG 170.013	17 16 24.	+ 28 12		15.6	GALAXY
VHE 86	17 16 24.	- 36 02			REFLECTION NEBULA
DG 146	17 16 30.	+ 06 09	660		REFLECTION NEBULA
ZWG 054.022	17 16 30.	+ 08 29		15.0	GALAXY
UGC 10797	17 16 30.	+ 08 29	66	15.0	GALAXY SB:C
MCG+01-44-006	17 16 30.	+ 08 29	72	15.0	GALAXY
ZWG 198.033	17 16 30.	+ 38 53		15.7	GALAXY
B 065	17 16 30.	- 26 39	720		DARK OBJECT
MRSL 351+00/2	17 16 30.	- 36 03	420		HII REGION
VDB.66N 111	17 16 34.	+ 06 07	744		REFLECTION NEBULA
LB 01016	17 16 34.	+ 81 23 24.		16.4	FAINT BLUE STAR
ZWG 140.007	17 16 36.	+ 21 35		15.4	GALAXY
SM2.26	17 16 36.	+ 42 12		18.6	FAINT BLUE OBJECT
1ZW 178	17 16 36.	+ 48 31			COMPACT GALAXY
ZWG 252.037	17 16 36.	+ 48 31		15.2	GALAXY
LB 03105	17 16 36.	- 59 43		14.5	FAINT BLUE STAR
ZWG 111.019	17 16 42.	+ 19 34		15.4	GALAXY
UGC 10798	17 16 42.	+ 19 34	60	15.4	GALAXY Sb-c
KARA.73B 0797	17 16 42.	+ 19 34	60	15.4	ISOLATED GALAXY S
MCG+04-41-003	17 16 42.	+ 21 34	42	15.4	GALAXY
MCG+07-35-063	17 16 42.	+ 41 42	36	15.	GALAXY
LDN 1772	17 16 42.	- 26 35	480		DARK NEBULA
MRSL 351+00/4	17 16 42.	- 35 55	120		HII REGION
VHE 85B	17 16 42.	- 35 57			REFLECTION NEBULA

OBJECT NAME	RIGHT ASCEN.	DECLINATION	DIAM.	MAGN.	TYPE OF OBJECT
WRAY 19.50	17 16 43.6	- 35 52 03.			DIFFUSE NEBULA
PK338-08.1	17 16 44.	- 51 41	16	12.2	PLANETARY NEBULA
ZWG 225.098	17 16 48.	+ 41 43		15.6	GALAXY
RNGC 6348	17 16 48.	+ 41 43		15.5	GALAXY
SM2.02	17 16 48.	+ 42 52		18.0	FAINT BLUE OBJECT
SM2.03	17 16 48.	+ 43 09		17.6	FAINT BLUE OBJECT
ZWG 252.038	17 16 48.	+ 50 17		15.1	GALAXY
MCG+08-31-039	17 16 48.	+ 50 17	30	16.	GALAXY
ISS 0937	17 16 48.	- 27 34	59		STELLAR RING
MRSL 351+00/5	17 16 48.	- 35 42	720		HII REGION
SCHO 0387	17 16 50.	- 22 06 18.	600		ISOLATED DARK CLOUD
SCHO 0388	17 16 50.	- 22 42 18.	550		ISOLATED DARK CLOUD
B 066	17 16 50.	- 26 51	480		DARK OBJECT
RNGC 6326	17 16 50.	- 51 42		12.0	PLANETARY NEBULA
ZWG 170.014	17 16 54.	+ 29 55		15.7	GALAXY
UGC 10799	17 16 54.	+ 29 55	72	15.7	GALAXY Sc
ZWG 198.034	17 16 54.	+ 36 03		15.5	GALAXY
SM2.04	17 16 54.	+ 43 58		18.7	FAINT BLUE OBJECT
7ZW 703	17 16 54.	+ 64 19			COMPACT GALAXY
SCHO 0389	17 16 57.	- 19 54 06.	300		ISOLATED DARK CLOUD
ARC 2258	17 16 59.	+ 31 47		17.7	RICH CLUSTER OF GALAXIES
VDB.66G 207	17 17	+ 14 29	70		DWARF GALAXY
KHAV 340	17 17	- 31 45	2960		DARK NEBULA
SM2.05	17 17 00.	+ 42 14		18.8	FAINT BLUE OBJECT
SM1.2	17 17 00.	+ 44 30		18.7	BLUE GALAXY
7ZW 704	17 17 00.	+ 61 12			COMPACT GALAXY
MCG+12-16-027	17 17 00.	+ 69 25	45	16.	GALAXY
LDN 0095	17 17 00.	- 21 38	360		DARK NEBULA
LDN 0094	17 17 00.	- 21 43	780		DARK NEBULA
LDN 0072	17 17 00.	- 22 30	2220		DARK NEBULA
LDN 0044	17 17 00.	- 23 40	720		DARK NEBULA
LDN 0037	17 17 00.	- 23 50	480		DARK NEBULA
LDN 1768	17 17 00.	- 26 49	420		DARK NEBULA
LDN 1705	17 17 00.	- 32 05	720		DARK NEBULA
MCG+07-35-064	17 17 03.	+ 41 45	24	14.	GALAXY
MIN.46 06	17 17 03.	- 36 01			DIFFUSE NEBULA
CED 140	17 17 03.	- 36 01			DIFFUSE GALACTIC NEBULA
RNGC 6248	17 17 05.	+ 69 28		15.5	GALAXY
ZWG 140.008	17 17 06.	+ 21 55		15.7	GALAXY
ZC 1717.1+2729	17 17 06.	+ 27 29	740		CLUSTER OF GALAXIES
ZWG 226.001	17 17 06.	+ 41 45		14.3	GALAXY
ZWG 225.099	17 17 06.	+ 41 45		14.3	GALAXY
RNGC 6350	17 17 06.	+ 41 45		14.5	GALAXY
UGC 10800	17 17 06.	+ 41 45	60	14.3	GALAXY S0
SM2.06	17 17 06.	+ 42 27		18.5	FAINT BLUE OBJECT
SM2.07	17 17 06.	+ 43 16		16.6	FAINT BLUE OBJECT
ZWG 339.036	17 17 06.	+ 69 28		15.7	GALAXY
VHE 85C	17 17 06.	- 35 47			REFLECTION NEBULA
MRSL 348-01/1	17 17 06.	- 38 54	120		HII REGION
SCHO 0390	17 17 10.	- 31 44 06.	1030		ISOLATED DARK CLOUD
WRAY 19.51	17 17 11.0	- 35 48 25.			DIFFUSE NEBULA
ZWG 082.017	17 17 12.	+ 11 20		15.6	GALAXY
ZWG 111.020	17 17 12.	+ 20 00		15.7	GALAXY
UGC 10801	17 17 12.	+ 32 19	96	16.0	GALAXY Sc
SM2.08	17 17 12.	+ 42 02		18.2	FAINT BLUE OBJECT
LB 01017	17 17 12.	+ 60 22 30.		17.9	FAINT BLUE STAR
MCG+10-25-001	17 17 12.	+ 61 51	42	13.8	GALAXY
MCG+12-16-028	17 17 12.	+ 73 28	84	14.	GALAXY
VHA 223	17 17 12.	- 35 50	300		STAR CLSTR IN NEBULOSITY
RNGC 6334	17 17 13.	- 36 01			DIFFUSE NEBULA
MCG+06-38-013	17 17 15.	+ 34 15	7	16.	GALAXY
B 254	17 17 17.	- 30 05	3600		DARK OBJECT
ZWG 082.018	17 17 18.	+ 11 16		15.5	GALAXY
UGC 10802	17 17 18.	+ 11 16	60	15.5	GALAXY S0
MCG+06-38-014	17 17 18.	+ 34 14 30.	18	16.	GALAXY
ZWG 198.035	17 17 18.	+ 34 15		15.3	GALAXY
ZWG 198.036	17 17 18.	+ 36 06		15.1	GALAXY
ZWG 339.037	17 17 18.	+ 73 29		12.9	GALAXY
UGC 10803	17 17 18.	+ 73 29	78	12.9	GALAXY
KARA.73B 0798	17 17 18.	+ 73 29	66	12.9	ISOLATED GALAXY E
RNGC 6335	17 17 18.	- 30 05			NON-EXISTENT OBJECT
MRSL 349-00/1	17 17 18.	- 37 58	3600		HII REGION
RNGC 6349	17 17 19.	+ 36 06		15.0	GALAXY
LB 00326	17 17 19.	+ 58 17 06.		14.8	FAINT BLUE STAR
MCG+05-40-008	17 17 21.	+ 32 20	72	17.	GALAXY
PK356+04.1	17 17 23.1	- 28 57 40.	10		PLANETARY NEBULA
MCG+06-38-015	17 17 24.	+ 34 14 30.	12	15.5	GALAXY
MCG+06-38-016	17 17 24.	+ 36 06	42	15.	GALAXY
SM2.09	17 17 24.	+ 42 19		18.9	FAINT BLUE OBJECT
SM2.10	17 17 24.	+ 43 01		16.0	FAINT BLUE OBJECT
SM2.11	17 17 24.	+ 44 47		18.1	FAINT BLUE OBJECT
MCG+09-28-032	17 17 24.	+ 55 30	48	16.	GALAXY
ZWG 300.005	17 17 24.	+ 61 50		13.6	GALAXY
ZWG 299.072	17 17 24.	+ 61 50		13.6	GALAXY
UGC 10804	17 17 24.	+ 61 50	60	13.6	GALAXY S0
CED 141	17 17 24.	- 30 06			DIFFUSE GALACTIC NEBULA
RNGC 6351	17 17 25.	+ 36 06		16.0	GALAXY
RNGC 6359	17 17 26.	+ 61 50		13.5	GALAXY
MCG+06-38-018	17 17 27.	+ 34 15	18	15.	GALAXY
MCG+06-38-017	17 17 27.	+ 36 06	12	16.	GALAXY
WRAY 19.52	17 17 28.9	- 35 49 03.		10.0	STAR-NEBULA ASSOCIATION
ZC 1717.5+5752	17 17 30.	+ 57 52	1950		CLUSTER OF GALAXIES
LDN 0088	17 17 30.	- 22 00	660		DARK NEBULA
LDN 1710	17 17 30.	- 31 54	1320		DARK NEBULA
MRSL 351+00/3	17 17 30.	- 35 50	360		HII REGION
WRAY 19.53	17 17 30.9	- 36 01 46.		9.3	STAR-NEBULA ASSOCIATION
B 255	17 17 33.	- 23 25	300		DARK OBJECT
IC 1253	17 17 35.	+ 18 42 49.			NONSTELLAR OBJECT
ZWG 082.019	17 17 36.	+ 14 27		15.3	GALAXY
UGC 10805	17 17 36.	+ 14 27	108	15.3	GALAXY DWRF SP
MCG+02-44-002	17 17 36.	+ 14 27	96	15.3	GALAXY
KARA.73B 0799	17 17 36.	+ 14 27	102	15.3	ISOLATED GALAXY S
MCG+03-44-004	17 17 36.	+ 16 42	72	14.6	GALAXY
SM2.12	17 17 36.	+ 42 04		17.7	FAINT BLUE OBJECT
ZWG 252.039	17 17 36.	+ 49 55		13.9	GALAXY
UGC 10806	17 17 36.	+ 49 55	144	13.9	GALAXY
LDN 0059	17 17 36.	- 23 30	300		DARK NEBULA
ISS 0938	17 17 36.	- 31 56	49		STELLAR RING
MRSL 351+00/1	17 17 36.	- 36 02	480		HII REGION
RNGC 6347	17 17 41.	+ 16 43		14.5	GALAXY
ZWG 111.021	17 17 42.	+ 16 43		14.6	GALAXY
UGC 10807	17 17 42.	+ 16 43	72	14.6	GALAXY SBb
KARA.73B 0800	17 17 42.	+ 16 43	72	14.6	ISOLATED GALAXY S
UGC 10808	17 17 42.	+ 28 22	66	17.	GALAXY DWRF SP
SM2.13	17 17 42.	+ 44 25		18.6	FAINT BLUE OBJECT
MCG+08-31-040	17 17 42.	+ 49 55	108	14.	GALAXY
SCHO 0391	17 17 42.	- 29 15 30.	500		ISOLATED DARK CLOUD
SCHO 0392	17 17 43.	- 23 26 54.	310		ISOLATED DARK CLOUD

OBJECT NAME	RIGHT ASCEN.	DECLINATION	DIAM.	MAGN.	TYPE OF OBJECT
LB 01018	17 17 46.	+ 58 02 00.		17.7	FAINT BLUE STAR
ZWG 140.009	17 17 48.	+ 24 58		15.7	GALAXY
UGC 10809	17 17 48.	+ 24 58	84	15.7	GALAXY DBL SYS
MCG+09-28-033	17 17 48.	+ 52 39	54	15.	GALAXY
ZWG 277.034	17 17 48.	+ 52 40		15.1	GALAXY
UGC 10810	17 17 48.	+ 52 40	60	15.1	GALAXY
MCG+10-25-002	17 17 48.	+ 58 10 30.	84	15.	GALAXY
ZWG 300.006	17 17 48.	+ 58 12		15.1	GALAXY
UGC 10811	17 17 48.	+ 58 12	108	15.1	GALAXY SB?a-b
VV 089B	17 17 48.	+ 59 22	48	15.	INTERACTING GALAXY
VV 089A	17 17 48.	+ 59 22	48	14.	INTERACTING GALAXY
VV 089	17 17 48.	+ 59 22	108	14.	INTERACTING GALAXY
MCG+10-25-003	17 17 48.	+ 60 39	24	16.	GALAXY
RNGC 6358	17 17 50.	+ 52 40		15.0	GALAXY
B 067	17 17 50.	- 26 49			DARK OBJECT
MCG+07-35-065	17 17 51.	+ 40 59	66	16.	GALAXY
ZWG 226.002	17 17 54.	+ 40 59		15.0	GALAXY
ZWG 225.100	17 17 54.	+ 40 59		15.0	GALAXY
UGC 10812	17 17 54.	+ 40 59	72	15.0	GALAXY SBb
SM2.14	17 17 54.	+ 44 02		19.0	FAINT BLUE OBJECT
ZWG 252.040	17 17 54.	+ 49 01		15.5	GALAXY
KARA.72 508A	17 17 54.	+ 49 01	54	15.5	PART OF DOUBLE GALAXY
1ZW 179	17 17 54.	+ 51 32			COMPACT GALAXY
ZC 1717.9+5636	17 17 54.	+ 56 36	2350		CLUSTER OF GALAXIES
ZWG 300.007	17 17 54.	+ 60 38		15.5	GALAXY
ISS 0939	17 17 54.	- 29 15	228		STELLAR RING
PK356+04.3	17 17 54.	- 29 19		14.0	PLANETARY NEBULA
PK356+04.2	17 17 54.3	- 29 00 03.	25		PLANETARY NEBULA
SCHO 0393	17 17 56.	- 25 33 36.	770		ISOLATED DARK CLOUD
SCHO 0394	17 17 57.	- 26 15 54.	600		ISOLATED DARK CLOUD
KHAV 347	17 18	- 09 39	13020		DARK NEBULA
KHAV 345	17 18	- 25 45	1950		DARK NEBULA
KHAV 343	17 18	- 27 03	1570		DARK NEBULA
KHAV 344	17 18	- 29 45	1150		DARK NEBULA
KHAV 342	17 18	- 30 09	3250		DARK NEBULA
ZWG 082.020	17 18 00.	+ 11 22		15.7	GALAXY
ZWG 140.010	17 18 00.	+ 23 53		14.5	GALAXY
UGC 10813	17 18 00.	+ 23 53	54	14.5	GALAXY
MCG+08-31-041	17 18 00.	+ 49 00	12	16.	GALAXY
ZWG 252.041	17 18 00.	+ 49 05		14.9	GALAXY
UGC 10814	17 18 00.	+ 49 05	228	14.9	GALAXY S
KARA.72 508B	17 18 00.	+ 49 05	72	14.9	PART OF DOUBLE GALAXY
ZWG 252.042	17 18 00.	+ 50 00		15.6	GALAXY
7ZW 705	17 18 00.	+ 57 25			COMPACT GALAXY
ZWG 300.008	17 18 00.	+ 57 26		15.7	GALAXY
ZWG 300.009	17 18 00.	+ 60 40		13.9	GALAXY
RNGC 6361	17 18 00.	+ 60 40		14.0	GALAXY
UGC 10815	17 18 00.	+ 60 40	138	13.9	GALAXY Sb
MCG+10-25-004	17 18 00.	+ 60 40	126	13.	GALAXY
7ZW 706	17 18 00.	+ 62 53			COMPACT GALAXY
7ZW 707	17 18 00.	+ 68 47			COMPACT GALAXY
LDN 0086	17 18 00.	- 22 10	960		DARK NEBULA
LDN 1773	17 18 00.	- 26 45	720		DARK NEBULA
MRSL 343-04/1	17 18 00.	- 45 00	19800		HII REGION
PK358+05.1	17 18 04.1	- 27 08 42.	25		PLANETARY NEBULA
ZWG 111.022	17 18 06.	+ 19 22		15.3	GALAXY
MCG+04-41-004	17 18 06.	+ 23 53	21	14.5	GALAXY
MCG+08-31-042	17 18 06.	+ 49 03	36	17.	GALAXY
MCG+08-31-043	17 18 06.	+ 49 03 30.	48	15.	GALAXY
UGC 10816	17 18 06.	+ 49 59	66	16.5	GALAXY Sc
UGC 10817	17 18 06.	+ 61 17	66	16.5	GALAXY Sc
ZC 1718.1-0108	17 18 06.	- 01 08	3900		CLUSTER OF GALAXIES
ISS 0940	17 18 06.	- 29 47	163		STELLAR RING
ISS 0941	17 18 06.	- 30 28	130		STELLAR RING
ARC 2259	17 18 10.	+ 27 43		17.1	RICH CLUSTER OF GALAXIES
ARP 102	17 18 11.	+ 49 06			PECULIAR GALAXY
MCG+08-31-044	17 18 12.	+ 47 43	45	15.	GALAXY
ZWG 252.043	17 18 12.	+ 47 45		15.2	GALAXY
KARA.73B 0801	17 18 12.	+ 47 45	42	15.2	ISOLATED GALAXY S
ZC 1718.2+4757	17 18 12.	+ 47 57	940		CLUSTER OF GALAXIES
ZC 1718.2+5403	17 18 12.	+ 54 03	1340		CLUSTER OF GALAXIES
VHE 87	17 18 12.	- 44 05	60		REFLECTION NEBULA
REIN 2.252	17 18 13.05	- 19 32 22.6			NEBULA
RNGC 6342	17 18 14.	- 19 32		11.5	GLOBULAR CLUSTER
SCHO 0395	17 18 14.	- 19 51 54.	460		ISOLATED DARK CLOUD
HN 1842	17 18 14.4	- 59 27 41.	42		NEBULA
ARP 124	17 18 15.	+ 60 39			PECULIAR GALAXY
PK002+08.1	17 18 17.0	- 22 15 40.	4		PLANETARY NEBULA
ZWG 111.023	17 18 18.	+ 17 49		15.6	GALAXY
ZC 1718.3+3538	17 18 18.	+ 35 38	1480		CLUSTER OF GALAXIES
ZWG 226.003	17 18 18.	+ 39 12		15.6	GALAXY
ZWG 225.101	17 18 18.	+ 39 12		15.6	GALAXY
SM2.15	17 18 18.	+ 42 25		16.0	FAINT BLUE OBJECT
GCL 061	17 18 18.	- 19 32	282	11.35	GLOBULAR STAR CLUSTER
HN 0441	17 18 20.	- 73 53			NEBULA
IC 4644	17 18 20.	- 73 53			NONSTELLAR OBJECT
PK355+03.2	17 18 20.0	- 30 18 01.	10	10.	PLANETARY NEBULA
ZC 1718.4+2740	17 18 24.	+ 27 40	1080		CLUSTER OF GALAXIES
7ZW 708	17 18 24.	+ 68 18			COMPACT GALAXY
ZWG 226.004	17 18 30.	+ 40 43		15.6	GALAXY
ZWG 225.102	17 18 30.	+ 40 43		15.6	GALAXY
SM2.16	17 18 30.	+ 44 42		18.4	FAINT BLUE OBJECT
MCG+08-31-045	17 18 30.	+ 46 17	30	16.	GALAXY
MCG+10-25-005	17 18 30.	+ 61 36	30	17.	GALAXY
LDN 0103	17 18 30.	- 21 40	360		DARK NEBULA
ZWG 054.023	17 18 36.	+ 06 27		15.7	GALAXY
MCG+07-36-001	17 18 36.	+ 39 18	24	14.5	GALAXY
MCG+07-36-002	17 18 36.	+ 40 54	30	15.	GALAXY
ZC 1718.6+5229	17 18 36.	+ 52 29	9480		CLUSTER OF GALAXIES
MCG+09-28-034	17 18 36.	+ 56 42	30	16.	GALAXY
ZWG 140.011	17 18 42.	+ 24 56		15.4	GALAXY
ZWG 226.005	17 18 42.	+ 39 19		14.6	GALAXY
SM2.17	17 18 42.	+ 41 50		18.0	FAINT BLUE OBJECT
SM2.18	17 18 42.	+ 43 25		17.4	FAINT BLUE OBJECT
ZWG 277.035	17 18 42.	+ 56 40		15.7	GALAXY
ISS 0721	17 18 42.	- 21 05	1039		STELLAR RING
SCHO 0396	17 18 45.	- 29 57 12.	910		ISOLATED DARK CLOUD
RNGC 6328	17 18 45.	- 64 57			NON-EXISTENT OBJECT
RNGC 6337	17 18 47.	- 38 26			PLANETARY NEBULA
ZWG 140.012	17 18 48.	+ 24 03		15.6	GALAXY
SM2.19	17 18 48.	+ 44 46		18.2	FAINT BLUE OBJECT
LDN 0101	17 18 48.	- 21 45	180		DARK NEBULA
PK349-01.1	17 18 49.5	- 38 26 11.	60		PLANETARY NEBULA
ZWG 111.024	17 18 54.	+ 16 06		15.4	GALAXY
SM2.20	17 18 54.	+ 44 25		15.4	FAINT BLUE OBJECT
ZC 1718.9+6724	17 18 54.	+ 67 24	740		CLUSTER OF GALAXIES
HN 1843	17 18 54.8	- 60 27 44.	24		NEBULA
LB 01019	17 18 55.	+ 59 36 54.		17.5	FAINT BLUE STAR

OBJECT NAME	RIGHT ASCEN.	DECLINATION	DIAM.	MAGN.	TYPE OF OBJECT
RNGC 6353	17 18 58.	+ 15 44			NON-EXISTENT OBJECT
VDB.66G 208	17 19	+ 57 59	540		DWARF GALAXY
KHAV 351	17 19	- 06 45	8570		DARK NEBULA
KHAV 350	17 19	- 22 03	1950		DARK NEBULA
KHAV 346	17 19	- 28 39	5590		DARK NEBULA
KHAV 348	17 19	- 34 39	4020		DARK NEBULA
KHAV 349	17 19	- 36 03	3250		DARK NEBULA
MCG+10-25-006	17 19 00.	+ 61 22	60	16.	GALAXY
7ZW 709	17 19 00.	+ 65 47			COMPACT GALAXY
ZC 1719.0+7150	17 19 00.	+ 71 50	1080		CLUSTER OF GALAXIES
LDN 0324	17 19 00.	- 10 20	9780		DARK NEBULA
LDN 1741	17 19 00.	- 30 00	16980		DARK NEBULA
B 256	17 19 02.	- 28 47	3000		DARK OBJECT
B 259	17 19 05.	- 19 15	1800		DARK OBJECT
SM2.21	17 19 06.	+ 41 47		17.8	FAINT BLUE OBJECT
SM2.22	17 19 06.	+ 41 59		18.1	FAINT BLUE OBJECT
SM2.23	17 19 06.	+ 44 38		18.1	FAINT BLUE OBJECT
LB 01020	17 19 08.	+ 59 18 36.		16.4	FAINT BLUE STAR
SM2.24	17 19 12.	+ 42 04		18.4	FAINT BLUE OBJECT
ZWG 253.001	17 19 12.	+ 50 25		15.3	GALAXY
ZWG 252.044	17 19 12.	+ 50 25		15.3	GALAXY
MCG+10-25-007	17 19 12.	+ 60 12	48	15.	GALAXY
ZWG 300.010	17 19 12.	+ 61 22		15.7	GALAXY
UGC 10813	17 19 12.	+ 61 22	78	15.7	GALAXY
ZWG 026.004	17 19 12.	- 00 45		15.3	GALAXY
ISS 0942	17 19 12.	- 26 53	127		STELLAR RING
MCG+08-31-046	17 19 18.	+ 50 24	36	16.	GALAXY
ZWG 300.011	17 19 18.	+ 60 12		15.3	GALAXY
UGC 10819	17 19 18.	+ 60 12	66	15.3	GALAXY Sa-b
ISS 0943	17 19 18.	- 29 11	107		STELLAR RING
PK358+05.2	17 19 20.8	- 27 05 52.	10		PLANETARY NEBULA
UGC 10820	17 19 24.	+ 42 13	96	16.5	GALAXY Sc
SM2.25	17 19 24.	+ 42 43		18.0	FAINT BLUE OBJECT
MCG+09-28-035	17 19 24.	+ 52 40	72	15.	GALAXY
ZWG 277.036	17 19 24.	+ 52 42		15.5	GALAXY
UGC 10821	17 19 24.	+ 52 42	72	15.5	GALAXY SBb
ZWG 300.012	17 19 24.	+ 57 58		12.4	GALAXY
UGC 10822	17 19 24.	+ 57 58	3000	12.4	GALAXY DWRF EL
KARA.73B 0802	17 19 24.	+ 57 58	1680	12.4	ISOLATED GALAXY E
MCG+10-25-008	17 19 24.	+ 57 58 30.			GALAXY
ZC 1719.4+6343	17 19 24.	+ 63 43	2820		CLUSTER OF GALAXIES
B 257	17 19 26.	- 35 35	900		DARK OBJECT
MCG+07-36-003	17 19 27.	+ 42 12	90	15.5	GALAXY
SCHO 0397	17 19 29.	- 23 47 54.	240		ISOLATED DARK CLOUD
ZWG 111.025	17 19 30.	+ 16 01		15.6	GALAXY
ZWG 140.013	17 19 30.	+ 22 53		15.7	GALAXY
UGC 10823	17 19 30.	+ 22 53	66	15.7	GALAXY SB
ZC 1719.5+2946	17 19 30.	+ 29 46	1010		CLUSTER OF GALAXIES
ZWG 277.037	17 19 30.	+ 51 57		15.4	GALAXY
LB 00327	17 19 30.	+ 58 18 06.		14.5	FAINT BLUE STAR
LDN 0102	17 19 30.	- 21 50	720		DARK NEBULA
B 067a	17 19 30.	- 21 51	780		DARK OBJECT
LDN 1774	17 19 30.	- 26 58	660		DARK NEBULA
HN 1149	17 19 33.	- 59 58	60		NEBULA
IC 4646	17 19 33.	- 59 58			NONSTELLAR OBJECT
B 068	17 19 34.	- 23 45	1200		DARK OBJECT
SM2.27	17 19 36.	+ 42 36		18.3	FAINT BLUE OBJECT
MCG+09-28-036	17 19 36.	+ 51 56	15	16.	GALAXY
SCHO 0398	17 19 36.	- 21 55 48.	620		ISOLATED DARK CLOUD
LDN 0057	17 19 36.	- 23 41	180		DARK NEBULA
PK354+02.1	17 19 36.	- 32 11		13.7	PLANETARY NEBULA
ISS 1061	17 19 36.	- 43 38	171		STELLAR RING
LB 03106	17 19 36.	- 55 56		14.2	FAINT BLUE STAR
B 258	17 19 39.	- 34 41	2400		DARK OBJECT
HN 1844	17 19 41.9	- 59 20 11.	90		NEBULA
MRK 505	17 19 42.	+ 39 45	12	15.5	GALAXY WITH UV CONTINUUM
SM2.28	17 19 42.	+ 44 06		18.4	FAINT BLUE OBJECT
ISS 0944	17 19 42.	- 27 50	130		STELLAR RING
ISS 0946	17 19 42.	- 28 34	81		STELLAR RING
ISS 0945	17 19 42.	- 29 20	215		STELLAR RING
ZL 203	17 19 46.	+ 32 15 24.		20.6	ULTRAFAINT BLUE STAR
ZWG 026.005	17 19 48.	+ 02 03		15.1	GALAXY
UGC 10824	17 19 48.	+ 02 03	66	15.1	GALAXY S
MCG+00-44-003	17 19 48.	+ 02 03	48	15.1	GALAXY
ZWG 111.026	17 19 48.	+ 18 37		15.2	GALAXY
ZC 1719.8+3431	17 19 48.	+ 34 31	3900		CLUSTER OF GALAXIES
SN 1972G	17 19 48.	+ 65 58		18.0	SUPERNOVA
SCHO 0399	17 19 52.	- 23 56 12.	260		ISOLATED DARK CLOUD
B 069	17 19 53.	- 23 53	240		DARK OBJECT
ZWG 054.024	17 19 54.	+ 06 40		15.3	GALAXY
KARA.72 509A	17 19 54.	+ 06 40	60	15.3	PART OF DOUBLE GALAXY
ZWG 170.015	17 19 54.	+ 30 06		15.6	GALAXY
ZWG 226.006	17 19 54.	+ 43 56		15.1	GALAXY
MCG+10-25-009	17 19 54.	+ 57 08	24	17.	GALAXY
ARC 2271	17 19 56.	+ 78 05		15.7	RICH CLUSTER OF GALAXIES
MCG+10-25-010	17 19 57.	+ 57 07 30.	30	17.	GALAXY
ZL 204	17 19 58.	+ 32 16 18.		20.0	ULTRAFAINT BLUE STAR
B 071	17 19 59.	- 23 58	75		DARK OBJECT
LB 09911	17 20	- 82 01		15.3	FAINT BLUE STAR
LB 09912	17 20	- 82 11		15.2	FAINT BLUE STAR
ZWG 026.006	17 20 00.	+ 01 00		15.4	GALAXY
UGC 10825	17 20 00.	+ 01 00	72	15.4	GALAXY S
ZWG 054.025	17 20 00.	+ 06 42		15.6	GALAXY
KARA.72 509B	17 20 00.	+ 06 42	48	15.6	PART OF DOUBLE GALAXY
MCG+07-36-004	17 20 00.	+ 43 56	27	15.	GALAXY
MCG+08-31-047	17 20 00.	+ 49 15	48	16.	GALAXY
ZWG 253.002	17 20 00.	+ 49 17		15.5	GALAXY
ZWG 252.045	17 20 00.	+ 49 17		15.5	GALAXY
ZWG 300.013	17 20 00.	+ 57 09		15.7	GALAXY
ZWG 300.014	17 20 00.	+ 62 38		14.7	GALAXY
LDN 0287	17 20 00.	- 13 20	1560		DARK NEBULA
ISS 0722	17 20 00.	- 17 02	132		STELLAR RING
LDN 0177	17 20 00.	- 19 10	720		DARK NEBULA
LDN 0055	17 20 00.	- 23 50	180		DARK NEBULA
LDN 1749	17 20 00.	- 29 00	1380		DARK NEBULA
COU 001	17 20 00.	- 34 20	2100		EMISSION NEBULA
OCL 1008	17 20 00.	- 37 53	540	10.	OPEN STAR CLUSTER
ZWG 170.016	17 20 06.	+ 30 44		15.4	GALAXY
ZWG 170.017	17 20 06.	+ 30 49		15.7	GALAXY
ZWG 226.007	17 20 06.	+ 44 31		15.5	GALAXY
KARA.73B 0803	17 20 06.	+ 44 31	36	15.5	ISOLATED GALAXY S
ISS 0723	17 20 06.	- 15 56	97		STELLAR RING
LDN 0053	17 20 06.	- 23 52	120		DARK NEBULA
ISS 0947	17 20 06.	- 28 53	137		STELLAR RING
MRSL 355+02/1	17 20 06.	- 31 23	2400		HII REGION
ZL 205	17 20 07.	+ 32 17 18.		19.2	ULTRAFAINT BLUE STAR
ZL 206	17 20 11.	+ 32 09 24.		21.8	ULTRAFAINT BLUE STAR
ZWG 054.026	17 20 12.	+ 08 24		15.5	GALAXY

OBJECT NAME	RIGHT ASCEN.	DECLINATION	DIAM.	MAGN.	TYPE OF OBJECT
ZWG 111.027	17 20 12.	+ 15 40		15.7	GALAXY
ZWG 170.018	17 20 12.	+ 30 45		15.6	GALAXY
ZWG 198.037	17 20 12.	+ 33 43		15.6	GALAXY
ISS 0948	17 20 12.	- 26 58	85		STELLAR RING
PK357+04.1	17 20 14.7	- 28 56 19.	5		PLANETARY NEBULA
ZWG 111.028	17 20 18.	+ 18 50		15.5	GALAXY
MCG+05-40-009	17 20 18.	+ 30 45 30.	42	15.6	GALAXY
ZWG 198.038	17 20 18.	+ 35 23		15.6	GALAXY
MCG+09-28-037	17 20 18.	+ 54 21	42	16.	GALAXY
ZL 207	17 20 20.	+ 32 11 54.		19.8	ULTRAFAINT BLUE STAR
ZL 208	17 20 21.	+ 32 21 54.		19.0	ULTRAFAINT BLUE STAR
ZWG 082.021	17 20 24.	+ 12 54		15.7	GALAXY
ZWG 082.022	17 20 24.	+ 14 37		15.6	GALAXY
KARA.73B 0804	17 20 24.	+ 14 37	66	15.6	ISOLATED GALAXY S
ZWG 111.029	17 20 24.	+ 15 43		15.7	GALAXY
LB 01021	17 20 24.	+ 29 27 00.		17.2	FAINT BLUE STAR
ZC 1720.4+3320	17 20 24.	+ 33 20	2420		CLUSTER OF GALAXIES
MCG+06-38-019	17 20 24.	+ 35 23	42	15.	GALAXY
ZWG 277.038	17 20 24.	+ 54 20		15.7	GALAXY
ZC 1720.4+6700	17 20 24.	+ 67 00	740		CLUSTER OF GALAXIES
MRSL 351-00/1	17 20 24.	- 36 15	600		HII REGION
ZL 209	17 20 26.	+ 32 00 00.		18.8	ULTRAFAINT BLUE STAR
BON 1	17 20 27.	+ 24 48		17.7	VARIABLE GALAXY
MCG+05-40-010	17 20 27.	+ 30 46 30.	48	15.6	GALAXY
ZL 210	17 20 29.	+ 32 13 06.		19.8	ULTRAFAINT BLUE STAR
B 070	17 20 29.	- 24 00	240		DARK OBJECT
ZWG 111.030	17 20 30.	+ 19 28		15.6	GALAXY
ZC 1720.5+2345	17 20 30.	+ 23 45	4640		CLUSTER OF GALAXIES
8ZW 1720+24.8	17 20 30.	+ 24 48		17.7	COMPACT GALAXY
ZL 211	17 20 30.	+ 32 01 48.		17.3	ULTRAFAINT BLUE STAR
LDN 0091	17 20 30.	- 22 25	480		DARK NEBULA
LDN 0054	17 20 30.	- 23 55	180		DARK NEBULA
B 072	17 20 33.	- 23 35			DARK OBJECT
LB 01022	17 20 36.	+ 30 20 36.		17.6	FAINT BLUE STAR
ZWG 356.002	17 20 36.	+ 78 04		15.7	GALAXY
ZWG 355.030	17 20 36.	+ 78 04		15.7	GALAXY
ZL 212	17 20 39.	+ 31 59 42.		20.2	ULTRAFAINT BLUE STAR
ARC 2261	17 20 39.	+ 32 12		17.4	RICH CLUSTER OF GALAXIES
REIN 2.253	17 20 39.86	- 17 46 04.7			NEBULA
SCHO 0400	17 20 40.	- 24 01 06.	300		ISOLATED DARK CLOUD
ZWG 170.019	17 20 42.	+ 30 55		15.4	GALAXY
KARA.72 510A	17 20 42.	+ 30 55	48	15.4	PART OF DOUBLE GALAXY
ZWG 170.020	17 20 42.	+ 30 56		15.3	GALAXY
MRK 506	17 20 42.	+ 30 56	24	15.5	GALAXY WITH UV CONTINUUM
KW 46	17 20 42.	+ 30 56	34		SEYFERT GALAXY
KARA.72 510B	17 20 42.	+ 30 56	66	15.3	PART OF DOUBLE GALAXY
GCL 062	17 20 42.	- 17 46	278	9.68	GLOBULAR STAR CLUSTER
SCHO 0402	17 20 42.	- 20 00 12.	520		ISOLATED DARK CLOUD
SCHO 0401	17 20 42.	- 23 32 00.	710		ISOLATED DARK CLOUD
VVI 84	17 20 42.	- 30 56	24	15.3	SEYFERT GAL1XY
ACK 331-13.1	17 20 42.	- 59 30			PLANETARY NEBULA
MCG+09-28-038	17 20 45.	+ 51 35	42	16.	GALAXY
RNGC 6356	17 20 45.	- 17 46		9.5	GLOBULAR CLUSTER
SCHO 0403	17 20 45.	- 22 54 30.	620		ISOLATED DARK CLOUD
ZL 213	17 20 47.	+ 32 02 54.		19.4	ULTRAFAINT BLUE STAR
ZWG 082.023	17 20 48.	+ 12 44		14.2	GALAXY
UGC 10826	17 20 48.	+ 12 44	66	14.2	GALAXY S
MCG+02-44-003	17 20 48.	+ 12 44	60	14.2	GALAXY
ZWG 082.024	17 20 48.	+ 12 57		15.6	GALAXY
MCG+05-40-011	17 20 48.	+ 30 56	24	15.4	GALAXY
ZWG 198.039	17 20 48.	+ 38 32		15.5	GALAXY
SCHO 0404	17 20 48.	- 24 18 36.	900		ISOLATED DARK CLOUD
PK355+02.1	17 20 48.	- 30 59		13.2	PLANETARY NEBULA
OCL 0987	17 20 48.	- 49 54	1440	8.2	OPEN STAR CLUSTER
VHA 224	17 20 48.	- 49 54	900		OPEN STAR CLUSTER
MCG+05-40-012	17 20 51.	+ 30 57	45	15.3	GALAXY
ZWG 170.021	17 20 54.	+ 30 05		15.7	GALAXY
SG 2.108	17 20 54.	- 26 17	60		DIFFUSE EMISSION NEBULA
OCL 1036	17 20 54.	- 26 18	78	12.3	OPEN STAR CLUSTER
GCL 063	17 20 54.	- 26 18	366	9.6	GLOBULAR STAR CLUSTER
PK359+05.1	17 20 55.6	- 25 56 40.	5		PLANETARY NEBULA
ZL 214	17 20 57.	+ 20 14 36.		19.7	ULTRAFAINT BLUE STAR
RNGC 6355	17 20 57.	- 26 18		9.5	GLOBULAR CLUSTER
ZL 215	17 20 58.	+ 32 09 06.		20.0	ULTRAFAINT BLUE STAR
KHAV 355	17 21	- 19 03	6560		DARK NEBULA
KHAV 352	17 21	- 23 39			DARK NEBULA
IC 4651	17 21	- 49 55	840	6.6	OPEN CLUSTER
ZWG 170.022	17 21 00.	+ 31 23		15.2	GALAXY
ZC 1721.0+5425	17 21 00.	+ 54 25	1480		CLUSTER OF GALAXIES
MCG+13-12-022	17 21 00.	+ 78 03	66	16.	GALAXY
LDN 0179	17 21 00.	- 19 15	840		DARK NEBULA
ISS 0724	17 21 00.	- 19 18	271		STELLAR RING
LDN 0138	17 21 00.	- 20 50	4080		DARK NEBULA
LDN 0066	17 21 00.	- 23 30	360		DARK NEBULA
ZL 216	17 21 04.	+ 32 09 06.		21.0	ULTRAFAINT BLUE STAR
B 073	17 21 05.	- 24 14	60		DARK OBJECT
ZWG 082.025	17 21 06.	+ 14 31		15.6	GALAXY
ZWG 140.014	17 21 06.	+ 23 43		15.3	GALAXY
MCG+07-36-005	17 21 06.	+ 41 09	24	14.5	GALAXY
ZWG 226.008	17 21 06.	+ 41 10		14.5	GALAXY
RNGC 6363	17 21 06.	+ 41 10		14.5	GALAXY
UGC 10827	17 21 06.	+ 41 10	66	14.5	GALAXY E
SCHO 0405	17 21 11.	- 29 24 00.	330		ISOLATED DARK CLOUD
RNGC 6354	17 21 11.	- 38 30			NON-EXISTENT OBJECT
MCG+04-41-005	17 21 12.	+ 23 43	30	15.3	GALAXY
ZC 1721.2+2345	17 21 12.	+ 23 45	540		CLUSTER OF GALAXIES
ZC 1721.2+2354	17 21 12.	+ 23 54	400		CLUSTER OF GALAXIES
ARC 2263	17 21 12.	+ 26 59		16.9	RICH CLUSTER OF GALAXIES
1ZW 180	17 21 12.	+ 52 27			COMPACT GALAXY
ISS 0725	17 21 12.	- 17 44	91		STELLAR RING
PK355+02.3	17 21 12.0	- 31 40 37.		13.5	PLANETARY NEBULA
ZL 217	17 21 13.	+ 32 06 24.		18.5	ULTRAFAINT BLUE STAR
ARC 2262	17 21 15.	+ 23 48		17.4	RICH CLUSTER OF GALAXIES
ZWG 140.015	17 21 18.	+ 23 41		14.9	GALAXY
UGC 10828	17 21 18.	+ 23 41	84	14.9	GALAXY S0?+CMP
MCG+08-31-001	17 21 18.	+ 48 12 30.	60	16.	GALAXY
MCG+08-32-002	17 21 18.	+ 48 32	66	16.	GALAXY
ZWG 300.015	17 21 18.	+ 59 10		14.9	GALAXY
OCL 1007	17 21 18.	- 38 57	780	11.8	OPEN STAR CLUSTER
VHA 225	17 21 18.	- 38 57	360		OPEN STAR CLUSTER
GCL 064	17 21 18.	- 48 25	534	9.1	GLOBULAR STAR CLUSTER
VHA 226	17 21 18.	- 48 25	300		GLOBULAR STAR CLUSTER
ZL 218	17 21 19.	+ 32 06 54.		20.0	ULTRAFAINT BLUE STAR
LB 00328	17 21 21.	+ 29 26 42.		15.8	FAINT BLUE STAR
SCHO 0406	17 21 21.	- 29 40 54.	340		ISOLATED DARK CLOUD
RNGC 6357	17 21 21.	- 34 07			DIFFUSE NEBULA
SC 1717-5117.4	17 21 22.	- 51 20 21.			NEBULA
PK357+03.1	17 21 23.6	- 29 21 38.	7		PLANETARY NEBULA

OBJECT NAME	RIGHT ASCEN.	DECLINATION	DIAM.	MAGN.	TYPE OF OBJECT
MCG+03-44-005	17 21 24.	+ 15 59	42	15.4	GALAXY
MCG+04-41-006	17 21 24.	+ 23 41	60	14.9	GALAXY
ZWG 140.016	17 21 24.	+ 23 43		15.5	GALAXY
ZC 1721.4+3523	17 21 24.	+ 35 23	1280		CLUSTER OF GALAXIES
1ZW 181	17 21 24.	+ 38 00			COMPACT GALAXY
ZWG 253.003	17 21 24.	+ 48 15		15.3	GALAXY
CED 142	17 21 24.	- 34 09	240		DIFFUSE GALACTIC NEBULA
LB 00329	17 21 27.	+ 60 15 24.		15.5	FAINT BLUE STAR
PK355+02.2	17 21 27.8	- 30 49 19.		14.1	PLANETARY NEBULA
ZWG 026.007	17 21 30.	+ 01 41		15.5	GALAXY
MCG+03-44-006	17 21 30.	+ 16 00	24	15.4	GALAXY
ZWG 170.023	17 21 30.	+ 27 30		15.6	GALAXY
ZWG 170.024	17 21 30.	+ 32 27		15.2	GALAXY
ZC 1721.5+3600	17 21 30.	+ 36 00	2150		CLUSTER OF GALAXIES
MCG+10-25-011	17 21 30.	+ 59 09	24	15.	GALAXY
ISS 0949	17 21 30.	- 29 05	173		STELLAR RING
VHE 89	17 21 30.	- 34 07	30		REFLECTION NEBULA
VHA 227	17 21 30.	- 34 08	90		STAR CLSTR IN NEBULOSITY
SCHO 0407	17 21 31.	- 20 35 00.	400		ISOLATED DARK CLOUD
ZWG 111.031	17 21 36.	+ 16 01		15.4	GALAXY
MCG+10-25-012	17 21 36.	+ 59 07	54	16.	GALAXY
ISS 0950	17 21 36.	- 29 48	155		STELLAR RING
OCL 0993	17 21 36.	- 48 25	180	10.8	OPEN STAR CLUSTER
RNGC 6352	17 21 36.	- 48 26		9.0	GLOBULAR CLUSTER
SCHO 0408	17 21 41.	- 29 38 18.	480		ISOLATED DARK CLOUD
ZWG 082.026	17 21 42.	+ 12 38		15.4	GALAXY
ZWG 226.009	17 21 42.	+ 43 07		15.7	GALAXY
ISS 0951	17 21 42.	- 30 34	59		STELLAR RING
PK358+04.1	17 21 43.2	- 28 03 15.	6		PLANETARY NEBULA
B 260	17 21 44.	- 25 34	720		DARK OBJECT
IC 1256	17 21 46.	+ 26 31 04.			NONSTELLAR OBJECT
SCHO 0409	17 21 46.	- 25 32 24.	470		ISOLATED DARK CLOUD
PK003+07.1	17 21 46.2	- 21 31 04.	4		PLANETARY NEBULA
ZWG 140.017	17 21 48.	+ 26 31		14.5	GALAXY
UGC 10829	17 21 48.	+ 26 31	102	14.5	GALAXY Sb
MCG+04-41-007	17 21 48.	+ 26 31	90	14.5	GALAXY
KARA.73B 0805	17 21 48.	+ 26 31	96	14.5	ISOLATED GALAXY S
ARC 2264	17 21 48.	+ 29 14		17.7	RICH CLUSTER OF GALAXIES
ISS 1062	17 21 48.	- 43 11	113		STELLAR RING
ZWG 140.018	17 21 54.	+ 21 11		15.6	GALAXY
KARA.73B 0806	17 21 54.	+ 21 11	48	15.6	ISOLATED GALAXY S
ZWG 170.025	17 21 54.	+ 30 36		15.7	GALAXY
ZWG 170.026	17 21 54.	+ 30 45		15.6	GALAXY
PK356+03.1	17 21 54.6	- 29 42 38.		13.7	PLANETARY NEBULA
HN 1845	17 21 55.5	- 60 30 07.	12		NEBULA
KHAV 358	17 22	- 21 15	3170		DARK NEBULA
KHAV 353	17 22	- 22 27	1770		DARK NEBULA
B 262	17 22	- 22 34	1800		DARK OBJECT
B 076	17 22	- 24 22	1800		DARK OBJECT
KHAV 354	17 22	- 29 15	3070		DARK NEBULA
MCG+09-28-039	17 22 00.	+ 53 47	60	17.	GALAXY
MCG+10-25-013	17 22 00.	+ 59 53	36	16.	GALAXY
ZWG 300.016	17 22 00.	+ 59 55		15.6	GALAXY
MCG+10-25-014	17 22 00.	+ 60 03 30.	48	15.	GALAXY
ZWG 300.017	17 22 00.	+ 60 04		15.6	GALAXY
7ZW 710	17 22 00.	+ 77 06			COMPACT GALAXY
ZWG 355.031	17 22 00.	+ 77 06		15.5	GALAXY
LDN 0131	17 22 00.	- 21 10	2400		DARK NEBULA
LDN 0117	17 22 00.	- 21 40	1680		DARK NEBULA
LDN 0081	17 22 00.	- 22 58	360		DARK NEBULA
LDN 0070	17 22 00.	- 23 30	360		DARK NEBULA
LDN 0050	17 22 00.	- 24 20	540		DARK NEBULA
LDN 0014	17 22 00.	- 25 30	540		DARK NEBULA
MRSL 353+00/1	17 22 00.	- 34 13	2100		HII REGION
OCL 1016	17 22 00.	- 34 18	240	10.	OPEN STAR CLUSTER
B 261	17 22 02.	- 22 59	840		DARK OBJECT
B 074	17 22 05.	- 24 09	300		DARK OBJECT
MCG+10-25-015	17 22 06.	+ 60 30	60	15.	GALAXY
PK356+02.1	17 22 06.	- 30 38		14.0	PLANETARY NEBULA
IC 4652	17 22 08.	- 59 40			NONSTELLAR OBJECT
HN 1150	17 22 08.	- 59 41		15.	NEBULA
MCG+10-25-016	17 22 09.	+ 60 10	24	16.	GALAXY
SCHO 0410	17 22 09.	- 23 01 06.	530		ISOLATED DARK CLOUD
IC 4417	17 22 12.	+ 17 36			NONSTELLAR OBJECT
1ZW 182	17 22 12.	+ 52 03			COMPACT GALAXY
MCG+10-25-017	17 22 12.	+ 60 10	30	15.	GALAXY
ZWG 300.018	17 22 12.	+ 60 11		15.5	GALAXY
ZWG 300.019	17 22 12.	+ 60 30		14.4	GALAXY
UGC 10830	17 22 12.	+ 60 30	72	14.4	GALAXY SPIRAL?
ISS 0952	17 22 12.	- 28 33	268		STELLAR RING
CED 143	17 22 12.	- 29 57			DIFFUSE GALACTIC NEBULA
RNGC 6360	17 22 12.	- 29 58			NON-EXISTENT OBJECT
ISS 0953	17 22 12.	- 31 12	113		STELLAR RING
ISS 1063	17 22 12.	- 41 38	134		STELLAR RING
SCHO 0411	17 22 14.	- 24 06 54.	580		ISOLATED DARK CLOUD
HN 1846	17 22 14.5	- 59 53 18.	24		NEBULA
RNGC 6365	17 22 16.	+ 62 13		14.5	GALAXY
B 075	17 22 17.	- 21 59			DARK OBJECT
ZWG 140.019	17 22 18.	+ 25 00		15.2	GALAXY
UGC 10831	17 22 18.	+ 25 00	84	15.2	GALAXY Sc
KARA.72 512A	17 22 18.	+ 25 00	60	15.2	PART OF DOUBLE GALAXY
ZWG 198.040	17 22 18.	+ 33 53		15.2	GALAXY
MCG+07-36-006	17 22 18.	+ 41 49	36	15.	GALAXY
ZWG 300.020	17 22 18.	+ 62 13		14.6	GALAXY
UGC 10833	17 22 18.	+ 62 13	66	14.6	GALAXY Sc-IRR
KARA.72 511B	17 22 18.	+ 62 13	60		PART OF DOUBLE GALAXY
KARA.72 511A	17 22 18.	+ 62 13	60	14.6	PART OF DOUBLE GALAXY
MCG+10-25-019	17 22 18.	+ 62 13	60	15.	GALAXY
MCG+10-25-018	17 22 18.	+ 62 13	66	15.	GALAXY
ISS 0852	17 22 18.	- 25 22	858		STELLAR RING
ARP 030	17 22 19.	+ 62 12			PECULIAR GALAXY
IC 4653	17 22 19.	- 60 51			NONSTELLAR OBJECT
LB 00330	17 22 20.	+ 30 36 12.		16.0	FAINT BLUE STAR
EN 0442	17 22 20.	- 60 52			NEBULA
MCG+03-44-007	17 22 21.	+ 20 27	60	15.3	GALAXY
ARC 2267	17 22 23.	+ 61 05		17.9	RICH CLUSTER OF GALAXIES
ZWG 111.032	17 22 24.	+ 20 26		15.3	GALAXY
UGC 10834	17 22 24.	+ 20 26	60	15.3	GALAXY Sc
MCG+04-41-008	17 22 24.	+ 25 00	66	15.2	GALAXY
HOLM 770A	17 22 27.	+ 25 01	60	14.5	PART OF MULTIPLE GALAXY
MCG+10-25-020	17 22 27.	+ 57 01	30	14.	GALAXY
RNGC 6370	17 22 28.	+ 57 02		14.0	GALAXY
ZWG 170.027	17 22 30.	+ 29 26		14.4	GALAXY
UGC 10835	17 22 30.	+ 29 26	96	14.4	GALAXY S0
MCG+09-28-040	17 22 30.	+ 53 33	60	16.	GALAXY
ZWG 300.021	17 22 30.	+ 57 02		14.2	GALAXY
UGC 10836	17 22 30.	+ 57 02	84	14.2	GALAXY E
ZC 1722.5+5840	17 22 30.	+ 58 40	1880		CLUSTER OF GALAXIES
LDN 1793	17 22 30.	- 26 30	4200		DARK NEBULA
RNGC 6364	17 22 32.	+ 29 26		14.5	GALAXY
MCG+05-40-013	17 22 33.	+ 29 26	24	14.4	GALAXY
PK345-04.1	17 22 35.	- 44 08 52.	5		PLANETARY NEBULA
ZWG 140.020	17 22 36.	+ 25 00		15.0	GALAXY
UGC 10837	17 22 36.	+ 25 00	120	15.0	GALAXY SBb
KARA.72 512B	17 22 36.	+ 25 00	108	15.0	PART OF DOUBLE GALAXY
ZWG 253.004	17 22 36.	+ 48 31		15.4	GALAXY
UGC 10838	17 22 36.	+ 48 31	72	15.4	GALAXY Sb
MCG+09-28-041	17 22 36.	+ 52 03	60	15.	GALAXY
ZWG 277.039	17 22 36.	+ 52 04		14.8	GALAXY
UGC 10839	17 22 36.	+ 52 04	60	14.8	GALAXY S
ISS 0954	17 22 36.	- 27 52	106		STELLAR RING
PK359+04.1	17 22 36.9	- 26 55 12.		13.4	PLANETARY NEBULA
PK359+05.2	17 22 37.2	- 26 09 18.	18		PLANETARY NEBULA
SCHO 0412	17 22 39.	- 22 17 54.	800		ISOLATED DARK CLOUD
ZWG 026.008	17 22 42.	+ 02 07		15.6	GALAXY
ZWG 140.021	17 22 42.	+ 23 47		14.9	GALAXY
UGC 10840	17 22 42.	+ 23 47	72	14.9	GALAXY E
MCG+04-41-009	17 22 42.	+ 25 00	108	15.0	GALAXY
ZWG 226.010	17 22 42.	+ 43 30		15.0	GALAXY
MCG+07-36-007	17 22 42.	+ 43 30	42	15.	GALAXY
VV 232B	17 22 42.	+ 62 12 30.	78	15.	INTERACTING GALAXY
VV 232A	17 22 42.	+ 62 12 30.	60	15.	INTERACTING GALAXY
ZWG 321.016	17 22 42.	+ 68 44		15.6	GALAXY
7ZW 711	17 22 42.	+ 68 45			COMPACT GALAXY
ISS 0955	17 22 42.	- 27 37	173		STELLAR RING
MCG+04-41-010	17 22 45.	+ 23 47	48	14.9	GALAXY
HOLM 770B	17 22 45.	+ 25 01	60	14.6	PART OF MULTIPLE GALAXY
ZWG 082.027	17 22 48.	+ 10 42		15.6	GALAXY
UGC 10841	17 22 48.	+ 10 42	66	15.6	GALAXY Sc-IRR
ZWG 111.033	17 22 48.	+ 15 54		15.7	GALAXY
ZWG 170.028	17 22 48.	+ 23 07		15.7	GALAXY
UGC 10842	17 22 48.	+ 28 07	78	15.7	GALAXY SBb
ZC 1722.8+3120	17 22 48.	+ 31 20	6720		CLUSTER OF GALAXIES
ZC 1722.8+6104	17 22 48.	+ 61 04	810		CLUSTER OF GALAXIES
ZWG 321.017	17 22 48.	+ 63 58		15.5	GALAXY
KARA.73B 0807	17 22 48.	+ 63 58	30	15.5	ISOLATED GALAXY S0
ISS 0726	17 22 48.	- 20 53	192		STELLAR RING
HN 1847	17 22 48.4	- 61 58 04.	12		NEBULA
PK357+03.2	17 22 49.0	- 29 19 14.	4		PLANETARY NEBULA
ARC 2266	17 22 51.	+ 32 10		17.5	RICH CLUSTER OF GALAXIES
MCG+05-40-014	17 22 54.	+ 28 07	63	15.7	GALAXY
ZC 1722.9+3206	17 22 54.	+ 32 06	870		CLUSTER OF GALAXIES
MCG+08-32-003	17 22 54.	+ 45 00	66	14.	GALAXY
ZWG 277.040	17 22 54.	+ 52 00		15.5	GALAXY
UGC 10843	17 22 54.	+ 52 00	66	15.5	GALAXY Sa-b
MCG+10-25-021	17 22 54.	+ 57 02 30.	30	17.	GALAXY
LB 00331	17 22 57.	+ 60 17 30.		15.3	FAINT BLUE STAR
KHAV 356	17 23	- 23 03	1350		DARK NEBULA
KHAV 357	17 23	- 38 27	4340		DARK NEBULA
ZWG 082.028	17 23 00.	+ 10 20		15.4	GALAXY
UGC 10844	17 23 00.	+ 10 20	60	15.4	GALAXY S0
ZWG 253.005	17 23 00.	+ 45 00		14.4	GALAXY
ZWG 226.011	17 23 00.	+ 45 00		14.4	GALAXY
UGC 10845	17 23 00.	+ 45 00	66	14.4	GALAXY Sc
MCG+09-28-042	17 23 00.	+ 52 00	72	16.	GALAXY
MCG+09-28-043	17 23 00.	+ 53 30	18	15.	GALAXY
LDN 0107	17 23 00.	- 22 15	600		DARK NEBULA
LDN 0085	17 23 00.	- 22 58	720		DARK NEBULA
PK352+00.1	17 23 04.3	- 34 59 07.	9		PLANETARY NEBULA
ZWG 253.006	17 23 06.	+ 49 59		15.7	GALAXY
UGC 10846	17 23 06.	+ 49 59	60	15.7	GALAXY Sb
MCG+09-28-044	17 23 06.	+ 53 23	30	16.	GALAXY
ZWG 277.041	17 23 06.	+ 53 30		15.7	GALAXY
LB 03107	17 23 06.	- 56 09		14.0	FAINT BLUE STAR
LB 00332	17 23 07.	+ 58 35 54.		16.0	FAINT BLUE STAR
ZWG 140.022	17 23 12.	+ 23 21		15.3	GALAXY
UGC 10847	17 23 12.	+ 23 21	66	15.3	GALAXY Sc-IRR
ZWG 253.007	17 23 12.	+ 45 33		15.5	GALAXY
KARA.72 513A	17 23 12.	+ 45 33	30	15.5	PART OF DOUBLE GALAXY
MCG+08-32-004	17 23 12.	+ 50 00	66	16.	GALAXY
ISS 0956	17 23 12.	- 27 52	72		STELLAR RING
MCG+04-41-011	17 23 15.	+ 23 22	48	15.3	GALAXY
ZWG 140.023	17 23 18.	+ 26 47		15.6	GALAXY
MCG+08-32-005	17 23 18.	+ 45 30	42	15.	GALAXY
ZWG 253.008	17 23 18.	+ 48 23		15.5	GALAXY
LDN 1794	17 23 18.	- 26 36	300		DARK NEBULA
IC 4660	17 23 22.	+ 75 54 26.			NONSTELLAR OBJECT
B 263	17 23 22.	- 42 44	1800		DARK OBJECT
ZWG 198.041	17 23 24.	+ 37 48		15.0	GALAXY
ZWG 253.009	17 23 24.	+ 45 30		14.7	GALAXY
KARA.72 513B	17 23 24.	+ 45 30	54	14.7	PART OF DOUBLE GALAXY
ZWG 253.010	17 23 24.	+ 45 59		15.5	GALAXY
MCG+10-25-022	17 23 24.	+ 59 06	30	17.	GALAXY
ZWG 355.032	17 23 24.	+ 75 54		14.3	GALAXY
UGC 10848	17 23 24.	+ 75 54	84	14.3	GALAXY S
VHE 90	17 23 24.	- 35 51	48		REFLECTION NEBULA
SCHO 0413	17 23 25.	- 31 49 48.	860		ISOLATED DARK CLOUD
RNGC 6367	17 23 27.	+ 37 48		15.0	GALAXY
MCG+06-38-020	17 23 27.	+ 37 48 30.	45	14.5	GALAXY
MCG+10-25-023	17 23 27.	+ 59 02 30.	60	14.	GALAXY
ZWG 111.034	17 23 30.	+ 20 49		15.1	GALAXY
UGC 10849	17 23 30.	+ 20 49	60	15.1	GALAXY SB
MCG+07-36-008	17 23 30.	+ 41 58	30	15.	GALAXY
ZWG 226.012	17 23 30.	+ 41 59		15.3	GALAXY
ZWG 226.013	17 23 30.	+ 42 03		15.5	GALAXY
ZWG 300.022	17 23 30.	+ 59 03		14.5	GALAXY
UGC 10850	17 23 30.	+ 59 03	96	14.5	GALAXY Sc/SBc
RNGC 6373	17 23 32.	+ 59 03		14.5	GALAXY
MCG+07-36-009	17 23 33.	+ 42 01 30.	24	16.	GALAXY
MCG+07-36-010	17 23 33.	+ 42 02 30.	36	16.	GALAXY
LB 01023	17 23 35.	+ 62 16 48.		17.5	FAINT BLUE STAR
SCHO 0414	17 23 35.	- 20 46 12.	420		ISOLATED DARK CLOUD
1ZW 183	17 23 36.	+ 46 04			COMPACT GALAXY
ZWG 300.023	17 23 36.	+ 56 56		15.5	GALAXY
ZWG 277.042	17 23 36.	+ 56 56		15.5	GALAXY
MCG+09-28-045	17 23 36.	+ 56 57	42	15.	GALAXY
MCG+12-16-029	17 23 36.	+ 74 27	9	17.	GALAXY
UGC 10851	17 23 42.	+ 49 23	60	17.	GALAXY Sc
MCG+12-16-030	17 23 42.	+ 74 27	9	17.	GALAXY
MCG+13-12-023	17 23 42.	+ 75 54	60	15.	GALAXY
ISS 0957	17 23 42.	- 27 30	157		STELLAR RING
ZC 1723.8+7722	17 23 48.	+ 77 22	1480		CLUSTER OF GALAXIES
PK358+04.2	17 23 48.	- 27 41		14.2	PLANETARY NEBULA
PK357+03.4	17 23 49.2	- 29 13 00.	8		PLANETARY NEBULA
ZWG 082.029	17 23 54.	+ 11 21		15.7	GALAXY
UGC 10852	17 23 54.	+ 11 21	66	15.7	GALAXY

OBJECT NAME	RIGHT ASCEN.	DECLINATION	DIAM.	MAGN.	TYPE OF OBJECT
MCG+10-25-024	17 23 54.	+ 58 15	60	15.	GALAXY
MCG+12-16-031	17 23 54.	+ 74 24	9	17.	GALAXY
ZWG 340.001	17 23 54.	+ 74 25		15.3	GALAXY
ZWG 339.038	17 23 54.	+ 74 25		15.3	GALAXY
UGC 10853	17 23 54.	+ 74 25	96	15.3	GALAXY E+2COMP
KHAV 362	17 24	- 14 09	11060		DARK NEBULA
KHAV 360	17 24	- 26 51	3450		DARK NEBULA
KHAV 359	17 24	- 31 57	4390		DARK NEBULA
ZWG 111.035	17 24 00.	+ 19 19		15.6	GALAXY
UGC 10854	17 24 00.	+ 58 15	72	16.0	GALAXY Sc
ZWG 339.039	17 24 00.	+ 71 18		14.9	GALAXY
KARA.72 514A	17 24 00.	+ 71 18	36	14.9	PART OF DOUBLE GALAXY
MCG+12-16-032B	17 24 00.	+ 71 18		16.	GALAXY
MCG+12-16-032A	17 24 00.	+ 71 18		16.	GALAXY
MCG+12-16-033	17 24 00.	+ 74 24	18	16.	GALAXY
LDN 0149	17 24 00.	- 20 50	2880		DARK NEBULA
LB 01024	17 24 01.	+ 59 33 48.		17.5	FAINT BLUE STAR
PK342-06.1	17 24 05.	- 46 53 01.	10		PLANETARY NEBULA
OCL 0104	17 24 06.	+ 24 14	840		OPEN STAR CLUSTER
KARA.72 514B	17 24 06.	+ 71 18	36		PART OF DOUBLE GALAXY
MCG+12-16-034	17 24 06.	+ 74 24	21	17.	GALAXY
ISS 0958	17 24 06.	- 27 22	128		STELLAR RING
SCHO 0415	17 24 08.	- 21 08 06.	840		ISOLATED DARK CLOUD
B 264	17 24 08.	- 25 29	600		DARK OBJECT
IC 1261	17 24 09.	+ 71 19 24.			NONSTELLAR OBJECT
PK358+03.1	17 24 10.6	- 28 25 22.	4		PLANETARY NEBULA
ZWG 226.014	17 24 12.	+ 44 30		15.6	GALAXY
1ZW 184	17 24 12.	+ 45 41			COMPACT GALAXY
1ZW 185	17 24 12.	+ 51 31			COMPACT GALAXY
PK357+03.5	17 24 13.4	- 29 18 44.		14.0	PLANETARY NEBULA
SCHO 0416	17 24 16.	- 30 18 06.	480		ISOLATED DARK CLOUD
ZWG 082.030	17 24 18.	+ 13 58		15.7	GALAXY
KARA.72 515A	17 24 18.	+ 13 58	30	15.7	PART OF DOUBLE GALAXY
ZWG 253.011	17 24 18.	+ 45 34		15.6	GALAXY
ZC 1724.3+6725	17 24 18.	+ 67 25	1280		CLUSTER OF GALAXIES
PK357+03.3	17 24 18.	- 29 01		14.0	PLANETARY NEBULA
VHA 228	17 24 18.	- 30 45	90		OPEN STAR CLUSTER
TER 02	17 24 20.3	- 30 45 40.2	30		STAR CLUSTER
LB 00333	17 24 23.	+ 59 00 00.		13.7	FAINT BLUE STAR
PK358+03.2	17 24 23.2	- 28 28 39.	5		PLANETARY NEBULA
ZWG 082.031	17 24 24.	+ 13 57		15.5	GALAXY
KARA.72 515B	17 24 24.	+ 13 57	36	15.5	PART OF DOUBLE GALAXY
ZWG 226.015	17 24 24.	+ 39 22		15.2	GALAXY
SAAK 1	17 24 24.	+ 41 38		15.5	EXTR BLUE OBJ NEAR D GLXY
OCL 0051	17 24 24.	- 07 03			OPEN STAR CLUSTER
IC 1257	17 24 27.	- 07 03 10.			NONSTELLAR OBJECT
HN 1848	17 24 28.7	- 62 24 21.	12		NEBULA
SCHO 0417	17 24 29.	- 25 05 18.	550		ISOLATED DARK CLOUD
ZWG 054.027	17 24 30.	+ 07 10		15.6	GALAXY
ZWG 226.016	17 24 30.	+ 39 58		15.5	GALAXY
CED 144	17 24 30.	- 07 03			DIFFUSE GALACTIC NEBULA
OCL 1005	17 24 30.	- 40 44	120	12.	OPEN STAR CLUSTER
B 265	17 24 32.	- 25 09	1080		DARK OBJECT
ZWG 111.036	17 24 36.	+ 17 22		15.6	GALAXY
ZWG 140.024	17 24 36.	+ 26 52		15.7	GALAXY
ZC 1724.6+2703	17 24 36.	+ 27 03	3290		CLUSTER OF GALAXIES
ZWG 198.042	17 24 36.	+ 35 18		15.2	GALAXY
MCG+06-38-021	17 24 36.	+ 35 18 30.	42	15.	GALAXY
SAAK 1A	17 24 36.	+ 41 37		18.2	EXTR BLUE OBJ NEAR D GLXY
SAAK 1B	17 24 36.	+ 41 39		18.2	EXTR BLUE OBJ NEAR D GLXY
7ZW 712	17 24 36.	+ 58 52			COMPACT GALAXY
ZWG 300.024	17 24 36.	+ 58 52		14.1	GALAXY
UGC 10855	17 24 36.	+ 58 52	60	14.1	GALAXY DBL SYS
KARA.72 516A	17 24 36.	+ 58 52	36	14.1	PART OF DOUBLE GALAXY
MCG+10-25-025	17 24 36.	+ 58 52	30	15.	GALAXY
SC 1720-5250.9	17 24 36.	- 52 53 37.			NEBULA
RNGC 6377	17 24 38.	+ 58 52		14.0	GALAXY
RNGC 6376	17 24 38.	+ 58 52		14.0	GALAXY
KARA.72 516B	17 24 42.	+ 58 52	36		PART OF DOUBLE GALAXY
MCG+10-25-026	17 24 42.	+ 58 52	30	15.	GALAXY
7ZW 713	17 24 42.	+ 60 03			COMPACT GALAXY
ZWG 300.025	17 24 42.	+ 60 04		15.5	GALAXY
MCG+10-25-027	17 24 42.	+ 60 04	30	16.	GALAXY
ZC 1724.7+6956	17 24 42.	+ 69 56	870		CLUSTER OF GALAXIES
MCG+12-16-035	17 24 42.	+ 71 16	12	17.	GALAXY
ZWG 082.032	17 24 48.	+ 11 35		13.7	GALAXY
UGC 10856	17 24 48.	+ 11 35	240	13.7	GALAXY Sb
MCG+02-44-004	17 24 48.	+ 11 35	216	13.7	GALAXY
ZWG 111.037	17 24 48.	+ 16 27		15.5	GALAXY
ZC 1724.8+3921	17 24 48.	+ 39 21	2420		CLUSTER OF GALAXIES
MCG+10-25-028	17 24 48.	+ 59 35	30	17.	GALAXY
MCG+12-16-036	17 24 48.	+ 71 13	21	16.	GALAXY
ISS 1064	17 24 48.	- 44 59	400		STELLAR RING
SCHO 0418	17 24 49.	- 25 18 00.	320		ISOLATED DARK CLOUD
RNGC 6368	17 24 50.	+ 11 35		13.5	GALAXY
ZWG 140.025	17 24 54.	+ 24 38		15.7	GALAXY
MCG+05-40-015	17 24 54.	+ 28 57	27	15.5	GALAXY
ZC 1724.9+3401	17 24 54.	+ 34 01	1410		CLUSTER OF GALAXIES
FATH 1.777	17 24 54.	+ 59 36	33		NEBULA
ZC 1724.9+6547	17 24 54.	+ 65 47	1410		CLUSTER OF GALAXIES
PK001+05.1	17 24 58.0	- 24 22 58.	7		PLANETARY NEBULA
HN 1849	17 24 59.6	- 62 25 31.	18		NEBULA
B 077	17 25	- 23 49	3600		DARK OBJECT
KHAV 361	17 25	- 24 03			DARK NEBULA
UGC 10857	17 25 00.	+ 53 37	78	16.5	GALAXY
MCG+10-25-029	17 25 00.	+ 59 29	60	17.	GALAXY
UGC 10858	17 25 00.	+ 59 30	60	16.5	GALAXY Sc
ZC 1725.0+6310	17 25 00.	+ 63 10	2290		CLUSTER OF GALAXIES
LDN 0069	17 25 00.	- 24 00	6000		DARK NEBULA
LDN 1764	17 25 00.	- 28 30	5940		DARK NEBULA
FATH 1.778	17 25 02.	+ 59 30	54		NEBULA
SCHO 0419	17 25 03.	- 29 41 18.	320		ISOLATED DARK CLOUD
RNGC 6366	17 25 04.	- 05 02		12.0	GLOBULAR CLUSTER
ZWG 170.029	17 25 06.	+ 29 31		15.5	GALAXY
1ZW 186	17 25 06.	+ 52 28			COMPACT GALAXY
ZWG 278.001	17 25 06.	+ 56 40		15.6	GALAXY
ZWG 277.043	17 25 06.	+ 56 40		15.6	GALAXY
MCG+10-25-030	17 25 06.	+ 59 39	78	17.	GALAXY
ZWG 300.026	17 25 06.	+ 59 40		15.6	GALAXY
UGC 10859	17 25 06.	+ 59 40	72	15.6	GALAXY SBb-c
GCL 065	17 25 06.	- 05 02	480	12.1	GLOBULAR STAR CLUSTER
PK352-00.1	17 25 07.3	- 35 05 06.	13		PLANETARY NEBULA
B 266	17 25 08.	- 20 54	1800		DARK OBJECT
FATH 1.779	17 25 10.	+ 59 40	19		NEBULA
ZWG 140.026	17 25 12.	+ 21 10		15.3	GALAXY
ZWG 140.027	17 25 12.	+ 26 32		15.2	GALAXY
RNGC 6371	17 25 12.	+ 26 32		15.0	GALAXY
LB 00334	17 25 12.	+ 29 44 18.		14.5	FAINT BLUE STAR
ZWG 170.030	17 25 12.	+ 31 54		15.7	GALAXY
MCG+10-25-031	17 25 12.	+ 60 14	30	17.	GALAXY
FATH 1.780	17 25 12.	+ 60 22	8		NEBULA
ZWG 339.040	17 25 12.	+ 72 12		15.2	GALAXY
SCHO 0420	17 25 13.	- 32 17 54.	1090		ISOLATED DARK CLOUD
SCHO 0421	17 25 15.	- 20 46 48.	830		ISOLATED DARK CLOUD
PK358+03.5	17 25 17.1	- 28 36 09.	10	12.6	PLANETARY NEBULA
ISS 0959	17 25 18.	- 27 02	146		STELLAR RING
MCG+04-41-012	17 25 18.	+ 26 31	42	15.2	GALAXY
ZWG 170.031	17 25 18.	+ 31 46		15.7	GALAXY
ZWG 198.043	17 25 18.	+ 37 53		15.4	GALAXY
MCG+12-16-037	17 25 18.	+ 72 10	30	15.	GALAXY
OCL 1032	17 25 18.	- 29 27	1080	9.8	OPEN STAR CLUSTER
LB 03108	17 25 18.	- 56 10		13.4	FAINT BLUE STAR
LB 01026	17 25 20.	+ 61 00 42.		16.6	FAINT BLUE STAR
LB 01025	17 25 20.	+ 80 56 48.		17.2	FAINT BLUE STAR
MCG+06-38-022	17 25 21.	+ 37 54	18	15.	GALAXY
ZC 1725.4+6611	17 25 24.	+ 66 11	6520		CLUSTER OF GALAXIES
B 267	17 25 27.	- 25 11	300		DARK OBJECT
ZWG 054.028	17 25 30.	+ 06 31		15.5	GALAXY
UGC 10860	17 25 30.	+ 06 31	66	15.5	GALAXY S0-a
ZWG 111.038	17 25 30.	+ 15 20		15.7	GALAXY
ZWG 140.028	17 25 30.	+ 26 30		14.1	GALAXY
RNGC 6372	17 25 30.	+ 26 30		14.0	GALAXY
UGC 10861	17 25 30.	+ 26 30	108	14.1	GALAXY S
MCG+04-41-013	17 25 30.	+ 26 30	96	14.1	GALAXY
7ZW 714	17 25 30.	+ 59 21			COMPACT GALAXY
LDN 0007	17 25 30.	- 26 15	720		DARK NEBULA
PK358+03.3	17 25 30.	- 28 24		14.0	PLANETARY NEBULA
PK001+05.2	17 25 33.3	- 24 48 45.	6		PLANETARY NEBULA
MCG+10-25-032	17 25 36.	+ 58 37 30.	30	15.	GALAXY
LDN 0036	17 25 36.	- 25 10	240		DARK NEBULA
ACK 346-04.1	17 25 36.	- 42 30			PLANETARY NEBULA
PK357+02.5	17 25 38.2	- 30 05 23.	7		PLANETARY NEBULA
PK357+02.1	17 25 40.4	- 29 04 33.	9		PLANETARY NEBULA
ZWG 140.029	17 25 42.	+ 26 53		15.4	GALAXY
ZWG 140.030	17 25 42.	+ 26 55		15.5	GALAXY
MCG+04-41-014	17 25 42.	+ 26 55	30	15.5	GALAXY
ZWG 170.032	17 25 42.	+ 32 20		15.3	GALAXY
MCG+07-36-011	17 25 42.	+ 41 10	39	15.	GALAXY
ZWG 226.017	17 25 42.	+ 41 12		15.6	GALAXY
ZWG 300.027	17 25 42.	+ 58 39		15.7	GALAXY
MCG+10-25-033	17 25 42.	+ 62 11	48	15.	GALAXY
ZWG 339.041	17 25 42.	+ 68 53		15.5	GALAXY
PK011+11.1	17 25 44.	- 13 23 56.	10		PLANETARY NEBULA
ZWG 054.029	17 25 48.	+ 07 28		15.0	GALAXY
MCG+01-44-007	17 25 48.	+ 07 28	180	15.0	GALAXY
KARA.73B 0808	17 25 48.	+ 07 28	126	15.0	ISOLATED GALAXY S
ZWG 226.018	17 25 48.	+ 44 27		15.3	GALAXY
ZC 1725.8+7841	17 25 48.	+ 78 41	2690		CLUSTER OF GALAXIES
SCHO 0422	17 25 48.	- 25 15 48.	420		ISOLATED DARK CLOUD
ABC 2268	17 25 50.	+ 55 23		17.7	RICH CLUSTER OF GALAXIES
HN 1850	17 25 53.1	- 60 01 32.	12		NEBULA
PK357+02.2	17 25 53.5	- 29 11 58.			PLANETARY NEBULA
ZWG 226.019	17 25 54.	+ 44 58		15.5	GALAXY
ZWG 300.028	17 25 54.	+ 62 11		15.5	GALAXY
UGC 10862	17 25 54.	+ 62 11	90	15.5	GALAXY S
MRSL 355+01/1	17 25 54.	- 31 30	2580		HII REGION
PK357+02.3	17 25 54.9	- 29 41 01.		13.5	PLANETARY NEBULA
HN 1851	17 25 55.9	- 28 22 47.			NEBULA
SCHO 0423	17 25 59.	- 29 46 12.	270		ISOLATED DARK CLOUD
KHAV 363	17 26	- 22 33			DARK NEBULA
ZWG 082.033	17 26 00.	+ 11 24		15.4	GALAXY
LB 00335	17 26 00.	+ 58 41 18.		14.6	FAINT BLUE STAR
MCG+14-08-023	17 26 00.	+ 81 57	27	16.	GALAXY
LDN 0136	17 26 00.	- 21 40	1560		DARK NEBULA
LDN 0021	17 26 00.	- 25 54	600		DARK NEBULA
LDN 0019	17 26 00.	- 26 00	3240		DARK NEBULA
LDN 1732	17 26 00.	- 31 30	3840		DARK NEBULA
PK006+08.1	17 26 00.7	- 19 13 31.	10		PLANETARY NEBULA
ZWG 082.034	17 26 06.	+ 14 13		14.3	GALAXY
UGC 10864	17 26 06.	+ 14 13	66	14.3	GALAXY S0
MCG+02-44-005	17 26 06.	+ 14 13	60	14.3	GALAXY
ZWG 340.002	17 26 06.	+ 72 56		15.4	GALAXY
ZWG 339.042	17 26 06.	+ 72 56		15.4	GALAXY
UGC 10865	17 26 06.	+ 72 56	78	15.4	GALAXY SBb/Sb
SCHO 0424	17 26 08.	- 21 04 30.	320		ISOLATED DARK CLOUD
LB 00336	17 26 10.	+ 28 02 48.		14.9	FAINT BLUE STAR
ZWG 140.031	17 26 12.	+ 23 40		15.7	GALAXY
MCG+04-41-015	17 26 12.	+ 23 40	18	15.7	GALAXY
ZWG 170.033	17 26 12.	+ 32 01		15.6	GALAXY
MCG+10-25-034	17 26 12.	+ 59 46	42	16.	GALAXY
MCG+12-16-038	17 26 12.	+ 72 55	39	16.	GALAXY
PK009+10.1	17 26 12.	- 15 11	21	17.2	PLANETARY NEBULA
OCL 1006	17 26 12.	- 40 28	840	11.	OPEN STAR CLUSTER
FATH 1.781	17 26 16.	+ 59 46	5		NEBULA
PK000+04.2	17 26 16.7	- 26 23 45.	5		PLANETARY NEBULA
RNGC 6369	17 26 17.	- 23 44		14.0	PLANETARY NEBULA
PK002+05.1	17 26 17.5	- 23 43 15.	68	10.4	PLANETARY NEBULA
REIN 2.254	17 26 17.51	- 23 43 10.6			NEBULA
ABC 2274	17 26 18.	+ 77 29		17.2	RICH CLUSTER OF GALAXIES
PK000+04.1	17 26 20.3	- 25 46 46.			PLANETARY NEBULA
ABC 2270	17 26 21.	+ 55 13		17.7	RICH CLUSTER OF GALAXIES
ZWG 253.012	17 26 24.	+ 48 10		15.6	GALAXY
ZC 1726.4+5515	17 26 24.	+ 55 15	1210		CLUSTER OF GALAXIES
MCG+10-25-035	17 26 24.	+ 58 32	48	14.	GALAXY
MCG+10-25-036	17 26 24.	+ 58 42	30	16.	GALAXY
ACK 344-06.1	17 26 24.	- 45 20			PLANETARY NEBULA
IC 1258	17 26 25.	+ 58 32 13.			NONSTELLAR OBJECT
ZWG 140.032	17 26 30.	+ 26 35		15.4	GALAXY
UGC 10866	17 26 30.	+ 26 35	66	15.4	GALAXY Sc
ZWG 300.029	17 26 30.	+ 58 33		14.4	GALAXY
UGC 10867	17 26 30.	+ 58 33	60	14.4	GALAXY S0-a
PK358+03.7	17 26 30.7	- 28 38 03.	5		PLANETARY NEBULA
PK357+02.4	17 26 31.5	- 29 30 30.	25		PLANETARY NEBULA
ABC 2269	17 26 32.	+ 49 12		17.9	RICH CLUSTER OF GALAXIES
FATH 1.782	17 26 34.	+ 60 04	14		NEBULA
IC 1259	17 26 35.	+ 58 34 45.			NONSTELLAR OBJECT
RNGC 6381	17 26 35.	+ 60 03		13.5	GALAXY
ZWG 054.030	17 26 36.	+ 06 22		15.2	GALAXY
UGC 10868	17 26 36.	+ 06 22	126	15.2	GALAXY DBL SYS
MCG+01-44-008	17 26 36.	+ 06 22	84	15.2	GALAXY
ZWG 111.039	17 26 36.	+ 17 44		15.5	GALAXY
MCG+04-41-016	17 26 36.	+ 26 35	60	15.4	GALAXY
ARP 311	17 26 36.	+ 58 34			PECULIAR GALAXY
ARP 310	17 26 36.	+ 58 34			PECULIAR GALAXY
ZWG 300.030	17 26 36.	+ 58 35		14.0	GALAXY
UGC 10869	17 26 36.	+ 58 35	72	14.0	GALAXY DBL SYS

OBJECT NAME	RIGHT ASCEN.	DECLINATION	DIAM.	MAGN.	TYPE OF OBJECT
VV 101B	17 26 36.	+ 58 35	36	15.	INTERACTING GALAXY
VV 101A	17 26 36.	+ 58 35	48	15.	INTERACTING GALAXY
VV 101	17 26 36.	+ 58 35	36	14.	INTERACTING GALAXY
KARA.72 517B	17 26 36.	+ 58 35	60		PART OF DOUBLE GALAXY
KARA.72 517A	17 26 36.	+ 58 35	42	14.0	PART OF DOUBLE GALAXY
MCG+10-25-037A	17 26 36.	+ 58 35	42	15.	GALAXY
MCG+10-25-037	17 26 36.	+ 58 35	18	15.	GALAXY
MCG+10-25-038	17 26 36.	+ 60 03	60	13.	GALAXY
ZWG 300.031	17 26 36.	+ 60 04		15.6	GALAXY
UGC 10870	17 26 36.	+ 60 04	66	15.6	GALAXY S
KARA.72 518A	17 26 36.	+ 60 04	72	15.6	PART OF DOUBLE GALAXY
MCG+10-25-039	17 26 36.	+ 60 04	60	16.	GALAXY
GCL 066	17 26 36.	- 67 01	600	9.1	GLOBULAR STAR CLUSTER
RNGC 6362	17 26 37.	- 67 01		8.5	GLOBULAR CLUSTER
MCG+10-25-040	17 26 39.	+ 58 31	15	16.	GALAXY
IC 1260	17 26 40.	+ 58 31 21.			NONSTELLAR OBJECT
FATH 2.160	17 26 40.	+ 60 03	54		NEBULA
ZWG 054.031	17 26 42.	+ 08 52		15.6	GALAXY
ZWG 140.033	17 26 42.	+ 24 26		15.7	GALAXY
ZC 1726.7+4913	17 26 42.	+ 49 13	940		CLUSTER OF GALAXIES
ZWG 300.032	17 26 42.	+ 58 32		15.7	GALAXY
ZWG 300.033	17 26 42.	+ 59 42		15.2	GALAXY
ZWG 300.034	17 26 42.	+ 60 03		13.6	GALAXY
UGC 10871	17 26 42.	+ 60 03	84	13.6	GALAXY Sc
KARA.72 518B	17 26 42.	+ 60 03	90	13.6	PART OF DOUBLE GALAXY
UGC 10872	17 26 42.	+ 61 14	72	16.5	GALAXY
ISS 0960	17 26 42.	- 28 35	105		STELLAR RING
FATH 1.783	17 26 43.	+ 59 41	5		NEBULA
FATH 1.784	17 26 44.	+ 59 40	22		NEBULA
ZWG 082.035	17 26 48.	+ 14 20		15.4	GALAXY
ZC 1726.8+1458	17 26 48.	+ 14 58	2690		CLUSTER OF GALAXIES
ZWG 140.034	17 26 48.	+ 25 51		15.6	GALAXY
MCG+07-36-012	17 26 48.	+ 41 09	36	15.5	GALAXY
MCG+10-25-041	17 26 48.	+ 61 13	66	16.	GALAXY
PK356+01.1	17 26 48.	- 30 35		14.0	PLANETARY NEBULA
MCG+10-25-043	17 25 51.	+ 59 40	30	16.	GALAXY
MCG+10-25-042	17 26 51.	+ 59 40	42	16.	GALAXY
PK358+03.4	17 26 53.7	- 27 57 00.	8		PLANETARY NEBULA
ZWG 140.035	17 26 54.	+ 23 23		15.7	GALAXY
ZC 1726.9+6132	17 26 54.	+ 61 32	810		CLUSTER OF GALAXIES
ZWG 340.003	17 26 54.	+ 73 18		15.7	GALAXY
ZWG 339.043	17 26 54.	+ 73 18		15.7	GALAXY
MCG+03-44-008	17 26 57.	+ 16 11	90	15.2	GALAXY
RNGC 6382	17 26 58.	+ 56 54		15.0	GALAXY
KHAV 364	17 27	- 34 50	7980		DARK NEBULA
LB 09913	17 27	- 83 57		13.5	FAINT BLUE STAR
LB 09914	17 27	- 86 38		13.2	FAINT BLUE STAR
ZWG 054.032	17 27 00.	+ 05 07		15.7	GALAXY
ZWG 111.040	17 27 00.	+ 16 12		15.2	GALAXY
UGC 10873	17 27 00.	+ 16 12	96	15.2	GALAXY SB?
KARA.72 519A	17 27 00.	+ 16 12	78	15.2	PART OF DOUBLE GALAXY
ZWG 111.041	17 27 00.	+ 17 35		15.6	GALAXY
ZWG 140.036	17 27 00.	+ 22 32		15.5	GALAXY
MCG+04-41-017	17 27 00.	+ 22 32	36	15.5	GALAXY
UGC 10874	17 27 00.	+ 29 21	84	16.0	GALAXY Sc
ZWG 277.044	17 27 00.	+ 56 54		15.2	GALAXY
COU 003	17 27 00.	- 31 35	3000		EMISSION NEBULA
COU 002	17 27 00.	- 29 00	12000		EMISSION NEBULA
MCG+03-44-009	17 27 03.	+ 16 14	24	14.5	GALAXY
RNGC 6375	17 27 05.	+ 16 14		14.5	GALAXY
ZWG 111.042	17 27 06.	+ 16 14		14.5	GALAXY
UGC 10875	17 27 05.	+ 16 14	102	14.5	GALAXY E
KARA.72 519B	17 27 06.	+ 16 14	84	14.5	PART OF DOUBLE GALAXY
2ZW 077	17 27 06.	+ 50 15			COMPACT GALAXY
1ZW 187	17 27 06.	+ 50 15			COMPACT GALAXY
ZWG 253.013	17 27 06.	+ 50 20		15.7	GALAXY
MCG+08-32-006	17 27 06.	+ 50 21	15	16.	GALAXY
MCG+09-29-001	17 27 06.	+ 56 55	48	15.	GALAXY
MCG+12-16-039	17 27 06.	+ 71 07	132	13.	GALAXY
ZWG 339.044	17 27 06.	+ 71 08		12.8	GALAXY
UGC 10876	17 27 06.	+ 71 08	162	12.8	GALAXY Sc
RNGC 6395	17 27 08.	+ 71 08		13.0	GALAXY
PK358+02.2	17 27 10.8	- 29 07 56.		14.0	PLANETARY NEBULA
RNGC 6385	17 27 11.	+ 57 35		14.0	GALAXY
ZWG 140.037	17 27 12.	+ 24 55		14.8	GALAXY
ZC 1727.2+3427	17 27 12.	+ 34 27	1280		CLUSTER OF GALAXIES
1ZW 188	17 27 12.	+ 45 35			COMPACT GALAXY
MCG+10-25-044	17 27 12.	+ 57 33	72	14.	GALAXY
ZWG 300.035	17 27 12.	+ 57 35		14.2	GALAXY
UGC 10877	17 27 12.	+ 57 35	90	14.2	GALAXY SBa
MCG+10-25-045	17 27 12.	+ 60 08	24	16.	GALAXY
7ZW 715	17 27 12.	+ 74 33			COMPACT GALAXY
ZC 1727.2+7643	17 27 12.	+ 76 43	6320		CLUSTER OF GALAXIES
LB 01027	17 27 15.	+ 82 16 30.		16.9	FAINT BLUE STAR
FATH 1.785	17 27 16.	+ 60 08	8		NEBULA
MCG+04-41-018	17 27 18.	+ 24 55	48	14.8	GALAXY
ZWG 300.036	17 27 18.	+ 60 09		15.5	GALAXY
SHAH 138	17 27 18.	+ 64 42	192	17.0	GROUP OF COMPACT GALAXIES
FATH 1.786	17 27 20.	+ 60 01	19		NEBULA
SCHO 0425	17 27 23.	- 19 49 00.	840		ISOLATED DARK CLOUD
LDN 0058	17 27 24.	- 24 53	240		DARK NEBULA
OCL 1010	17 27 24.	- 37 03	660	9.2	OPEN STAR CLUSTER
TER 04	17 27 24.42	- 31 33 29.4	20		STAR CLUSTER
7ZW 716	17 27 30.	+ 67 06			COMPACT GALAXY
MRSL 355+00/3	17 27 30.	- 32 13	1800		HII REGION
IC 4647	17 27 33.	- 80 07			NONSTELLAR OBJECT
HN 0443	17 27 34.	- 80 08			NEBULA
PK358+03.6	17 27 34.8	- 28 01 51.	5		PLANETARY NEBULA
RNGC 6387	17 27 35.	+ 57 36		15.0	GALAXY
MCG+09-29-002	17 27 36.	+ 52 49	42	17.	GALAXY
1ZW 189	17 27 36.	+ 57 35			COMPACT GALAXY
ZWG 300.037	17 27 36.	+ 57 36		15.0	GALAXY
PK004+06.1	17 27 36.7	- 21 26	5		PLANETARY NEBULA
PK357+02.7	17 27 39.0	- 30 15 00.		14.0	PLANETARY NEBULA
PK359+03.2	17 27 39.1	- 27 03 45.	10	13.7	PLANETARY NEBULA
SCHO 0426	17 27 40.	- 18 48 24.	310		ISOLATED DARK CLOUD
VMT 18	17 27 41.	- 21 26 36.	64		SUPERNOVA REMNANT
ZWG 140.038	17 27 42.	+ 23 22		15.4	GALAXY
UGC 10878	17 27 42.	+ 23 22	66	15.4	GALAXY PECULR
KARA.73B 0809	17 27 42.	+ 23 22	60	15.4	ISOLATED GALAXY S
MCG+07-36-013	17 27 42.	+ 40 16	30	15.	GALAXY
ZWG 226.020	17 27 42.	+ 40 18		15.7	GALAXY
LB 01028	17 27 42.	+ 46 50 36.		17.2	FAINT BLUE STAR
MCG+09-29-003	17 27 42.	+ 52 18	36	16.	GALAXY
ZWG 278.002	17 27 42.	+ 52 46		14.9	GALAXY
ZWG 277.045	17 27 42.	+ 52 46		14.9	GALAXY
MCG+10-25-046	17 27 42.	+ 57 10	30	15.	GALAXY
FATH 1.787	17 27 42.	+ 60 03	5		NEBULA
ZC 1727.7+6424	17 27 42.	+ 64 24	2490		CLUSTER OF GALAXIES
MIL 55	17 27 43.	- 21 26 00.	132		SUPERNOVA REMNANT
MCG+04-41-019	17 27 45.	+ 23 22	42	15.4	GALAXY
RNGC 6386	17 27 45.	+ 52 46		15.0	GALAXY
PK356+01.2	17 27 45.2	- 30 58 53.			PLANETARY NEBULA
ZWG 140.039	17 27 48.	+ 24 55		14.7	GALAXY
UGC 10879	17 27 48.	+ 24 55	60	14.7	GALAXY Sc
MCG+09-29-004	17 27 48.	+ 52 47	54	15.	GALAXY
ZWG 300.038	17 27 48.	+ 57 13		15.4	GALAXY
MCG+10-25-047	17 27 48.	+ 60 07 30.	90	14.	GALAXY
ZWG 300.039	17 27 48.	+ 60 59		15.5	GALAXY
UGC 10880	17 27 48.	+ 60 59	60	15.5	GALAXY Sc-IRR
MCG+10-25-048	17 27 48.	+ 60 59	48	15.	GALAXY
ISS 0961	17 27 48.	- 27 44	82		STELLAR RING
PK359+03.1	17 27 49.3	- 26 56 57.		13.9	PLANETARY NEBULA
MCG+10-25-049	17 27 51.	+ 58 53	24	15.	GALAXY
RNGC 6390	17 27 53.	+ 60 08		14.5	GALAXY
FATH 2.161	17 27 53.	+ 60 08	68		NEBULA
MCG+04-41-020	17 27 54.	+ 24 55	54	14.7	GALAXY
MCG+08-32-007	17 27 54.	+ 50 38	42	16.	GALAXY
ZWG 300.040	17 27 54.	+ 60 08		14.5	GALAXY
UGC 10881	17 27 54.	+ 60 08	102	14.5	GALAXY Sb-c
GCL 067	17 27 54.	- 29 57	78		GLOBULAR STAR CLUSTER
VHA 229	17 27 54.	- 29 57	60		GLOBULAR STAR CLUSTER
MRSL 357+02/1	17 27 54.	- 29 57	120		HII REGION
LB 03109	17 27 54.	- 56 24		14.4	FAINT BLUE STAR
PK357+02.6	17 27 55.8	- 30 08 15.			PLANETARY NEBULA
PK363+03.8	17 27 59.5	- 28 12 48.	25	13.7	PLANETARY NEBULA
KHAV 368	17 28	- 11 44	12040		DARK NEBULA
KHAV 369	17 28	- 15 32	10060		DARK NEBULA
KHAV 366	17 28	- 18 26	5300		DARK NEBULA
B 268	17 28	- 20 29	6300		DARK OBJECT
KHAV 367	17 28	- 20 32	6530		DARK NEBULA
MCG+07-36-014	17 28 00.	+ 40 17	30	15.	GALAXY
ZWG 226.021	17 28 00.	+ 40 19		15.6	GALAXY
MCG+09-29-005	17 28 00.	+ 52 13	60	16.	GALAXY
MCG+10-25-050	17 28 00.	+ 58 33 30.	36	16.	GALAXY
UGC 10882	17 28 00.	+ 58 35	66	16.0	GALAXY Sc
ZWG 321.018	17 28 00.	+ 68 26		15.5	GALAXY
ZC 1728.0+7104	17 28 00.	+ 71 04	670		CLUSTER OF GALAXIES
7ZW 717	17 28 00.	+ 77 21			COMPACT GALAXY
LDN 0220	17 28 00.	- 18 26	300		DARK NEBULA
LDN 0206	17 28 00.	- 19 00	7620		DARK NEBULA
LDN 0192	17 28 00.	- 19 45	2100		DARK NEBULA
SCHO 0427	17 28 00.	- 21 27 36.	800		ISOLATED DARK CLOUD
LDN 0128	17 28 00.	- 22 10	3600		DARK NEBULA
LDN 0073	17 28 00.	- 24 10	2100		DARK NEBULA
LDN 0045	17 28 00.	- 25 20	2340		DARK NEBULA
ISS 0962	17 28 00.	- 29 57	142		STELLAR RING
LDN 1751	17 28 00.	- 30 00	4500		DARK NEBULA
LDN 1723	17 28 00.	- 32 10	240		DARK NEBULA
OCL 1017	17 28 00.	- 34 03	300	9.9	OPEN STAR CLUSTER
OCL 1012	17 28 00.	- 36 49	900		OPEN STAR CLUSTER
VHA 230	17 28 00.	- 36 49	421		OPEN STAR CLUSTER
PK358+02.1	17 28 04.7	- 28 21 59.			PLANETARY NEBULA
ZWG 112.001	17 28 06.	+ 19 47		15.6	GALAXY
ZWG 111.043	17 28 06.	+ 19 47		15.6	GALAXY
ZWG 226.022	17 28 06.	+ 43 38		15.4	GALAXY
ZWG 278.003	17 28 06.	+ 52 15		15.2	GALAXY
UGC 10883	17 28 06.	+ 52 15	60	15.2	GALAXY S0-a
ZWG 300.041	17 28 06.	+ 58 54		15.0	GALAXY
ISS 0963	17 28 06.	- 27 12	91		STELLAR RING
MRSL 346-05/1	17 28 06.	- 42 57	2100		HII REGION
LB 01029	17 28 07.	+ 58 08 36.		16.5	FAINT BLUE STAR
RNGC 6391	17 28 08.	+ 58 54		15.0	GALAXY
MCG+07-36-015	17 28 09.	+ 42 28	24	17.	GALAXY
RNGC 6378	17 28 11.	+ 06 19		15.0	GALAXY
ZWG 055.001	17 28 12.	+ 06 19		15.1	GALAXY
UGC 10884	17 28 12.	+ 06 19	84	15.1	GALAXY S
MCG+01-44-009	17 28 12.	+ 06 19	72	15.1	GALAXY
ZWG 055.002	17 28 12.	+ 06 26		15.2	GALAXY
MCG+07-36-016	17 28 12.	+ 42 29	45	16.	GALAXY
MAI 099	17 28 12.	+ 59 15	40		DWARF SPHEROIDAL GALAXY
ISS 0964	17 28 12.	- 27 46	51		STELLAR RING
MCG+03-44-010	17 28 18.	+ 16 19	60	14.6	GALAXY
ZWG 198.044	17 28 18.	+ 35 24		15.3	GALAXY
UGC 10885	17 28 18.	+ 35 24	60	15.3	GALAXY Sb
LDN 0061	17 28 18.	- 24 53	360		DARK NEBULA
SCHO 0428	17 28 20.	- 18 31 54.	470		ISOLATED DARK CLOUD
RNGC 6379	17 28 23.	+ 16 19		14.5	GALAXY
ZWG 112.002	17 28 24.	+ 16 19		14.6	GALAXY
ZWG 111.044	17 28 24.	+ 16 19		14.6	GALAXY
UGC 10886	17 28 24.	+ 16 19	72	14.6	GALAXY Sc
ZWG 140.040	17 28 24.	+ 23 40		15.3	GALAXY
ZWG 300.042	17 28 24.	+ 60 07		15.6	GALAXY
MCG+10-25-051	17 28 24.	+ 60 07 30.	54	16.	GALAXY
ZWG 356.003	17 28 24.	+ 77 45		15.4	GALAXY
ZWG 355.033	17 28 24.	+ 77 45		15.4	GALAXY
UGC 10887	17 28 24.	+ 77 45	72	15.4	GALAXY Sc
MCG+06-38-023	17 28 27.	+ 35 24	48	14.5	GALAXY
ZC 1728.5+4353	17 28 30.	+ 43 53	7530		CLUSTER OF GALAXIES
LDN 1724	17 28 30.	- 32 11	180		DARK NEBULA
MIN.48 02	17 28 34.	- 28 37			DIFFUSE NEBULA
ZWG 112.003	17 28 36.	+ 19 48		15.4	GALAXY
ZWG 111.045	17 28 36.	+ 19 48		15.4	GALAXY
ZWG 170.034	17 28 36.	+ 29 21		15.2	GALAXY
ISS 0965	17 28 36.	- 28 18	67		STELLAR RING
ISS 0966	17 28 36.	- 31 04	90		STELLAR RING
VHA 231	17 28 36.	- 31 52	240		OPEN STAR CLUSTER
HN 1852	17 28 37.2	- 59 51 33.	18		NEBULA
ZWG 198.045	17 28 42.	+ 37 02		15.6	GALAXY
ZWG 339.045	17 28 42.	+ 70 45		15.7	GALAXY
MCG+13-12-024	17 28 48.	+ 77 46	78	16.	GALAXY
PK016+13.1	17 28 48.	- 08 17	60	17.8	PLANETARY NEBULA
MRSL 355+00/2	17 28 48.	- 32 23	11400		HII REGION
MCG+08-32-008	17 28 51.	+ 47 27	12	16.	GALAXY
LDN 1735	17 28 54.	- 31 41	180		DARK NEBULA
PK350-02.1	17 28 56.8	- 37 55 15.	25		PLANETARY NEBULA
RNGC 6374	17 28 58.	- 32 34		10.0	OPEN CLUSTER
B 078	17 29	- 25 34	10800		DARK OBJECT
KHAV 365	17 29	- 26 02	9680		DARK NEBULA
LB 09915	17 29	- 81 58		15.3	FAINT BLUE STAR
MCG+09-29-006	17 29 00.	+ 56 51	42	17.	GALAXY
LDN 0185	17 29 00.	- 20 15	1320		DARK NEBULA
LDN 0178	17 29 00.	- 20 30	3300		DARK NEBULA
LDN 1754	17 29 00.	- 30 00	2280		DARK NEBULA
OCL 1022	17 29 00.	- 32 34	150	10.0	OPEN STAR CLUSTER
PK359+03.3	17 29 01.0	- 27 02 36.	8		PLANETARY NEBULA

OBJECT NAME	RIGHT ASCEN.	DECLINATION	DIAM.	MAGN.	TYPE OF OBJECT
ZWG 140.041	17 29 06.	+ 23 37		15.4	GALAXY
MCG+10-25-052	17 29 06.	+ 58 25	60	16.	GALAXY
LDN 0052	17 29 06.	- 25 19	300		DARK NEBULA
LB C1030	17 29 08.	+ 79 40 06.		17.4	FAINT BLUE STAR
ZWG 300.043	17 29 12.	+ 58 27		15.7	GALAXY
B 269	17 29 12.	- 22 44	3600		DARK OBJECT
LDN 1734	17 29 12.	- 31 50	300		DARK NEBULA
PK348-04.1	17 29 16.5	- 40 56 23.			PLANETARY NEBULA
PK358+02.3	17 29 17.3	- 29 03 01.		13.4	PLANETARY NEBULA
ZC 1729.3+4039	17 29 18.	+ 40 39	5440		CLUSTER OF GALAXIES
FATH 1.788	17 29 18.	+ 59 23	8		NEBULA
SCHO 0429	17 29 18.	- 19 39 06.	950		ISOLATED DARK CLOUD
SCHO 0430	17 29 21.	- 30 56 18.	730		ISOLATED DARK CLOUD
SCHO 0432	17 29 23.	- 33 09 30.	650		ISOLATED DARK CLOUD
PK349-03.1	17 29 23.4	- 39 50 15.			PLANETARY NEBULA
ZWG 055.003	17 29 24.	+ 06 31		15.5	GALAXY
ZWG 055.004	17 29 24.	+ 06 32		15.5	GALAXY
ZWG 278.004	17 29 24.	+ 51 41		15.7	GALAXY
FATH 1.790	17 29 24.	+ 60 23	14		NEBULA
MCG+10-25-053	17 29 24.	+ 60 23 30.	66	15.	GALAXY
ZWG 300.044	17 29 24.	+ 60 24		14.2	GALAXY
UGC 10888	17 29 24.	+ 60 24	66	14.2	GALAXY Sb
7ZW 718	17 29 24.	+ 65 59			COMPACT GALAXY
MCG+13-12-025	17 29 24.	+ 77 41	51	15.	GALAXY
ISS 0727	17 29 24.	- 15 25	114		STELLAR RING
LDN 0025	17 29 24.	- 26 09	300		DARK NEBULA
FATH 1.789	17 29 26.	+ 59 34	16		NEBULA
RNGC 6394	17 29 28.	+ 59 34		16.0	GALAXY
ZWG 055.005	17 29 30.	+ 06 30		15.4	GALAXY
ZWG 055.006	17 29 30.	+ 06 34		15.4	GALAXY
ZWG 198.046	17 29 30.	+ 36 28		15.7	GALAXY
MCG+10-25-054	17 29 30.	+ 59 34	27	16.	GALAXY
ISS 0967	17 29 30.	- 29 14	110		STELLAR RING
SCHO 0431	17 29 34.	- 32 17 30.	450		ISOLATED DARK CLOUD
PK357+01.1	17 29 34.9	- 29 58 08.	3		PLANETARY NEBULA
SCHO 0433	17 29 35.	- 22 18 54.	300		ISOLATED DARK CLOUD
ZWG 2[illegible].014	17 29 36.	+ 50 55		15.3	GALAXY
ZWG 2[illegible] 005	17 29 36.	+ 51 29		15.5	GALAXY
MCG+1[illegible]-055	17 29 36.	+ 59 41	72	15.	GALAXY
ACK 3[illegible]1.1	17 29 36.	- 36 42			PLANETARY NEBULA
RNGC 6[illegible]	17 29 40.	+ 59 41		15.5	GALAXY
SCHO 0[illegible]	17 29 40.	- 22 05 12.	230		ISOLATED DARK CLOUD
FATH 1.[illegible]	17 29 41.	+ 59 40	33		NEBULA
MCG+08-[illegible]09	17 29 42.	+ 50 55	48	16.	GALAXY
ZWG 278.[illegible]	17 29 42.	+ 51 30		15.1	GALAXY
ZWG 300.[illegible]	17 29 42.	+ 59 41		15.4	GALAXY
UGC 10889	17 29 42.	+ 59 41	90	15.4	GALAXY SBb
LDN 1738	17 29 42.	- 31 41	180		DARK NEBULA
MCG+07-36-017	17 29 45.	+ 40 15	42	15.	GALAXY
HN 1853	17 29 45.8	- 60 50 10.	18		NEBULA
B 270	17 29 46.	- 19 34	660		DARK OBJECT
ZWG 170.035	17 29 48.	+ 32 16		15.6	GALAXY
UGC 10890	17 29 48.	+ 32 16	114	15.6	GALAXY Sc
KARA.73B 0810	17 29 48.	+ 32 16	102	15.6	ISOLATED GALAXY S
ZWG 278.007	17 29 48.	+ 56 23		15.4	GALAXY
ZWG 278.008	17 29 48.	+ 56 40		15.6	GALAXY
ZWG 277.046	17 29 48.	+ 56 40		15.6	GALAXY
MCG+10-25-056	17 29 48.	+ 57 00	30	17.	GALAXY
CED 146	17 29 48.	- 17 29			DIFFUSE GALACTIC NEBULA
IC 4657	17 29 48.	- 17 29 24.			MAY NOT EXIST
SCHO 0436	17 29 48.	- 18 44 36.	800		ISOLATED DARK CLOUD
SCHO 0437	17 29 48.	- 21 59 48.	250		ISOLATED DARK CLOUD
SCHO 0435	17 29 48.	- 32 29 42.	440		ISOLATED DARK CLOUD
ZWG 170.036	17 29 54.	+ 28 27		15.5	GALAXY
MCG+09-29-007	17 29 54.	+ 56 24	36	16.	GALAXY
SCHO 0438	17 29 56.	- 18 34 48.	300		ISOLATED DARK CLOUD
REIN 2.255A	17 29 58.87	+ 07 05 40.3			NEBULA
REIN 2.255B	17 29 58.91	+ 07 05 41.9			NEBULA
RNGC 6384	17 29 59.	+ 07 06		12.0	GALAXY
KHAV 371	17 30	+ 02 58	16190		DARK NEBULA
LBN 1116	17 30	- 32 10	7200		BRIGHT NEBULA
B 271	17 30	- 34 13	7200		DARK OBJECT
ZWG 055.007	17 30 00.	+ 07 06		13.2	GALAXY
UGC 10891	17 30 00.	+ 07 06	420	13.2	GALAXY Sb
MCG+01-45-001	17 30 00.	+ 07 06	276	13.2	GALAXY
LDN 0225	17 30 00.	- 18 30	1200		DARK NEBULA
LDN 0182	17 30 00.	- 20 30	4260		DARK NEBULA
LDN 0168	17 30 00.	- 21 00	4140		DARK NEBULA
LDN 0042	17 30 00.	- 25 40	4860		DARK NEBULA
LDN 1776	17 30 00.	- 28 30	6300		DARK NEBULA
LDN 1730	17 30 00.	- 32 10	5220		DARK NEBULA
HN 0445	17 30 01.	- 60 41			NEBULA
PK359+02.2	17 30 01.3	- 28 19 21.	8		PLANETARY NEBULA
SN 1971L	17 30 02.	+ 07 06		12.8	SUPERNOVA
LB C1031	17 30 02.	+ 78 36 42.		17.6	FAINT BLUE STAR
IC 4655	17 30 02.	- 60 41			NONSTELLAR OBJECT
ZWG 055.008	17 30 06.	+ 03 41		15.6	GALAXY
SCHO 0442	17 30 06.	- 18 16 24.	460		ISOLATED DARK CLOUD
LB 03110	17 30 06.	- 56 33		14.0	FAINT BLUE STAR
MIL 52	17 30 09.	- 32 52 42.	570		SUPERNOVA REMNANT
SCHO 0443	17 30 10.	- 18 17 00.	540		ISOLATED DARK CLOUD
SCHO 0439	17 30 10.	- 22 57 36.	550		ISOLATED DARK CLOUD
7ZW 719	17 30 12.	+ 65 56			COMPACT GALAXY
SCHO 0440	17 30 12.	- 32 09 24.	420		ISOLATED DARK CLOUD
CED 145	17 30 12.	- 37 04	600		DIFFUSE GALACTIC NEBULA
BC MRA0530	17 30 13.43	- 13 02 46.2		18.	QUASI-STELLAR OBJECT
SHB 343	17 30 13.5	- 13 02 11.		18.	QUASI-STELLAR OBJECT
FATH 1.792	17 30 14.	+ 60 08	14		NEBULA
SCHO 0441	17 30 14.	- 32 25 12.	380		ISOLATED DARK CLOUD
ZWG 140.042	17 30 18.	+ 26 20		15.7	GALAXY
ZWG 226.023	17 30 18.	+ 41 31		15.7	GALAXY
UGC 10892	17 30 18.	+ 74 18	108	16.0	GALAXY
FATH 1.793	17 30 22.	+ 60 25	8		NEBULA
ZWG 083.001	17 30 24.	+ 13 20		15.6	GALAXY
MCG+03-45-001	17 30 24.	+ 16 25	156	13.6	GALAXY
ZC 1730.4+5829	17 30 24.	+ 58 29	17470		CLUSTER OF GALAXIES
ZWG 321.019	17 30 24.	+ 67 22		15.0	GALAXY
KARA.73B 0811	17 30 24.	+ 67 22	36	15.0	ISOLATED GALAXY E
SCHO 0444	17 30 27.	- 14 45 36.	440		ISOLATED DARK CLOUD
SCHO 0445	17 30 27.	- 19 30 12.	620		ISOLATED DARK CLOUD
RNGC 6389	17 30 29.	+ 16 26		13.5	GALAXY
ZWG 027.001	17 30 30.	+ 01 45		15.5	GALAXY
ZWG 112.004	17 30 30.	+ 15 37		15.7	GALAXY
ZWG 112.005	17 30 30.	+ 16 26		13.6	GALAXY
UGC 10893	17 30 30.	+ 16 26	192	13.6	GALAXY Sb
KARA.73B 0812	17 30 30.	+ 16 26	168	13.6	ISOLATED GALAXY S
ZWG 112.006	17 30 30.	+ 19 34		15.6	GALAXY
ZC 1730.5+2526	17 30 30.	+ 25 26	2550		CLUSTER OF GALAXIES
ZWG 253.015	17 30 30.	+ 45 46		15.7	GALAXY
ZWG 278.009	17 30 30.	+ 56 46		15.5	GALAXY
MCG+10-25-057	17 30 30.	+ 60 00	18	17.	GALAXY
FATH 1.794	17 30 30.	+ 60 00	8		NEBULA
MCG+12-16-040	17 30 30.	+ 74 17	102	16.	GALAXY
LB 03111	17 30 30.	- 58 09		13.8	FAINT BLUE STAR
PK359+02.1	17 30 34.2	- 28 05 20.	10	13.1	PLANETARY NEBULA
FATH 1.795	17 30 35.	+ 59 32	41		NEBULA
ZC 1730.6+3405	17 30 36.	+ 34 05	1950		CLUSTER OF GALAXIES
ZWG 300.046	17 30 36.	+ 59 32		15.6	GALAXY
PK004+06.2	17 30 37.4	- 21 44 23.	9		PLANETARY NEBULA
MCG+10-25-058	17 30 42.	+ 59 31	48	16.	GALAXY
ISS 0728	17 30 42.	- 15 53	154		STELLAR RING
ISS 0729	17 30 42.	- 17 07	175		STELLAR RING
ISS 0853	17 30 42.	- 21 36	173		STELLAR RING
ISS 0854	17 30 42.	- 22 19	149		STELLAR RING
ISS 0968	17 30 42.	- 27 14	112		STELLAR RING
ISS 1024	17 30 42.	- 39 13	478		STELLAR RING
VHR 91	17 30 42.	- 39 21	30		REFLECTION NEBULA
HN 0444	17 30 47.	- 74 20			NEBULA
MCG+07-36-018	17 30 45.	+ 40 53	36	16.	GALAXY
ZC 1730.8+6733	17 30 48.	+ 67 33	1340		CLUSTER OF GALAXIES
SHAH 167	17 30 48.	+ 74 59	78		GROUP OF COMPACT GALAXIES
MCG+13-12-026	17 30 48.	+ 75 45	114	12.1	GALAXY
IC 4654	17 30 48.	- 74 21			NONSTELLAR OBJECT
ARP 038	17 30 49.	+ 75 46			PECULIAR GALAXY
FATH 1.796	17 30 51.	+ 59 33	11		NEBULA
ZC 1730.9+7304	17 30 54.	+ 73 04	13570		CLUSTER OF GALAXIES
ISS 1025	17 30 54.	- 36 48	100		STELLAR RING
KHAV 372	17 31	- 01 08	3610		DARK NEBULA
KHAV 370	17 31	- 31 38	2960		DARK NEBULA
ZWG 170.037	17 31 00.	+ 27 37		15.7	GALAXY
UGC 10894	17 31 00.	+ 27 37	96	15.7	GALAXY S
MCG+10-25-059	17 31 00.	+ 59 38	60	15.	GALAXY
ZWG 340.004	17 31 00.	+ 71 22		15.4	GALAXY
ZWG 339.046	17 31 00.	+ 71 22		15.4	GALAXY
ISS 0855	17 31 00.	- 22 00	197		STELLAR RING
ISS 0969	17 31 00.	- 28 17	61		STELLAR RING
GCL 068	17 31 00.	- 39 03			GLOBULAR STAR CLUSTER
PK358+01.2	17 31 05.6	- 29 27 12.	5	13.6	PLANETARY NEBULA
ZC 1731.1+3745	17 31 06.	+ 37 45	1080		CLUSTER OF GALAXIES
ISS 0730	17 31 06.	- 20 53	91		STELLAR RING
ISS 0970	17 31 06.	- 29 11	64		STELLAR RING
SCHO 0446	17 31 07.	- 17 59 12.	550		ISOLATED DARK CLOUD
FATH 2.162	17 31 08.	+ 59 39	19		NEBULA
RNGC 6399	17 31 10.	+ 59 40		15.0	GALAXY
SCHO 0448	17 31 10.	- 31 25 30.	680		ISOLATED DARK CLOUD
SCHO 0447	17 31 11.	- 18 00 00.	500		ISOLATED DARK CLOUD
ZWG 112.007	17 31 12.	+ 20 10		15.7	GALAXY
ZWG 140.043	17 31 12.	+ 26 50		15.6	GALAXY
MCG+09-29-008	17 31 12.	+ 51 55	36	16.	GALAXY
UGC 10895	17 31 12.	+ 59 30	66	16.0	GALAXY SB:b-c
ZWG 300.047	17 31 12.	+ 59 40		14.8	GALAXY
UGC 10896	17 31 12.	+ 59 40	72	14.8	GALAXY S0-a
ZWG 356.004	17 31 12.	+ 75 45		12.4	GALAXY
ZWG 355.034	17 31 12.	+ 75 45		12.4	GALAXY
UGC 10897	17 31 12.	+ 75 45	144	12.4	GALAXY Sc
KARA.73B 0813	17 31 12.	+ 75 45	150	12.4	ISOLATED GALAXY S
SCHO 0449	17 31 14.	- 16 33 00.	240		ISOLATED DARK CLOUD
ABC 2272	17 31 16.	+ 40 39		16.5	RICH CLUSTER OF GALAXIES
IC 4659	17 31 17.	- 17 53 42.			NONSTELLAR OBJECT
UGC 10898	17 31 18.	+ 51 56	66	16.5	GALAXY S
MCG+09-29-009	17 31 18.	+ 52 00	36	16.	GALAXY
MCG+10-25-060	17 31 18.	+ 59 29	60	16.	GALAXY
7ZW 720	17 31 18.	+ 60 00			COMPACT GALAXY
ZWG 027.002	17 31 18.	- 00 34		15.6	GALAXY
ISS 0856	17 31 18.	- 25 08	103		STELLAR RING
ISS 0971	17 31 18.	- 27 35	95		STELLAR RING
CED 148	17 31 19.	- 17 54			DIFFUSE GALACTIC NEBULA
PK006+07.1	17 31 20.5	- 19 07 23.	10		PLANETARY NEBULA
MCG+03-45-002	17 31 21.	+ 20 50	132	15.0	GALAXY
MCG+09-29-010	17 31 21.	+ 56 54 30.	30	16.	GALAXY
FATH 1.797	17 31 22.	+ 59 29	54		NEBULA
RNGC 6363	17 31 22.	- 32 33		5.5	OPEN CLUSTER
CED 147	17 31 22.	- 32 33			DIFFUSE GALACTIC NEBULA
RNGC 6412	17 31 23.	+ 75 44		12.5	GALAXY
VHA 232	17 31 24.	- 32 32	300		STAR CLSTR IN NEBULOSITY
PK003+05.1	17 31 25.0	- 22 51 22.	5		PLANETARY NEBULA
SCHO 0450	17 31 28.	- 16 27 12.	230		ISOLATED DARK CLOUD
ZWG 112.008	17 31 30.	+ 20 48		15.0	GALAXY
UGC 10899	17 31 30.	+ 20 48	108	15.0	GALAXY S
ZWG 198.047	17 31 30.	+ 34 12		15.5	GALAXY
MCG+07-36-019	17 31 30.	+ 39 19	30	15.5	GALAXY
ZWG 226.024	17 31 30.	+ 39 21		15.7	GALAXY
KARA.73B 0814	17 31 30.	+ 39 21	36	15.7	ISOLATED GALAXY S
ZWG 226.025	17 31 30.	+ 43 47		14.9	GALAXY
UGC 10900	17 31 30.	+ 43 47	90	14.9	GALAXY E
ZC 1731.5+7948	17 31 30.	+ 79 48	2820		CLUSTER OF GALAXIES
ISS 0857	17 31 30.	- 24 31	134		STELLAR RING
ISS 0858	17 31 30.	- 27 01	79		STELLAR RING
OCL 1026	17 31 30.	- 32 32	360	5.5	OPEN STAR CLUSTER
MCG+07-36-020	17 31 33.	+ 43 47	21	14.5	GALAXY
IC 1262	17 31 33.	+ 43 48 24.			NONSTELLAR OBJECT
UGC 10901	17 31 36.	+ 05 30	72	17.	GALAXY
ZWG 198.048	17 31 36.	+ 36 46		15.5	GALAXY
ZWG 226.026	17 31 36.	+ 43 51		14.8	GALAXY
UGC 10902	17 31 36.	+ 43 51	102	14.8	GALAXY SB
ZC 1731.6+5348	17 31 36.	+ 53 48	5780		CLUSTER OF GALAXIES
7ZW 721	17 31 36.	+ 59 54			COMPACT GALAXY
ZWG 300.048	17 31 36.	+ 59 55		15.7	GALAXY
MCG+10-25-061	17 31 36.	+ 59 58	24	16.	GALAXY
ZWG 300.049	17 31 36.	+ 59 59		15.0	GALAXY
MCG+10-25-062	17 31 36.	+ 60 06	42	17.	GALAXY
ISS 0972	17 31 36.	- 27 14	64		STELLAR RING
IC 1263	17 31 38.	+ 43 51 25.			NONSTELLAR OBJECT
FATH 1.798	17 31 38.	+ 59 58	8		NEBULA
MCG+07-36-021	17 31 39.	+ 43 51	96	14.	GALAXY
SCHO 0451	17 31 41.	- 19 07 42.	300		ISOLATED DARK CLOUD
UGC 10903	17 31 42.	+ 23 54	60	17.	GALAXY DWRF IR
ZWG 198.049	17 31 42.	+ 36 34		15.6	GALAXY
ISS 0859	17 31 42.	- 26 58	195		STELLAR RING
SCHO 0452	17 31 42.	- 32 22 24.	450		ISOLATED DARK CLOUD
LB 01032	17 31 43.	+ 58 32 36.		16.0	FAINT BLUE STAR
HN 1151	17 31 44.	- 59 33		15.	NEBULA
FATH 1.799	17 31 45.	+ 60 07	19		NEBULA
SCHO 0453	17 31 46.	- 32 01 24.	390		ISOLATED DARK CLOUD
PK000+03.1	17 31 47.8	- 26 34 01.	6		PLANETARY NEBULA
ZC 1731.8+3315	17 31 48.	+ 33 15	1550		CLUSTER OF GALAXIES

OBJECT NAME	RIGHT ASCEN.	DECLINATION	DIAM.	MAGN.	TYPE OF OBJECT
MCG+07-36-022	17 31 48.	+ 43 39	72	15.	GALAXY
ZWG 226.027	17 31 48.	+ 43 40		15.6	GALAXY
UGC 10904	17 31 48.	+ 43 40	84	15.6	GALAXY SB0?
IC 1264	17 31 48.	+ 43 40 56.			NONSTELLAR OBJECT
OCL 1019	17 31 48.	- 34 16	420	13.	OPEN STAR CLUSTER
SCHO 0454	17 31 49.	- 20 57 30.	280		ISOLATED DARK CLOUD
SCHO 0455	17 31 52.	- 22 04 36.	390		ISOLATED DARK CLOUD
RNGC 6380	17 31 52.	- 39 02			GLOBULAR CLUSTER
HN 1854	17 31 52.3	- 58 13 30.	24		NEBULA
ZWG 198.050	17 31 54.	+ 36 35		15.7	GALAXY
ZWG 226.028	17 31 54.	+ 43 57		15.5	GALAXY
MCG+09-29-011	17 31 54.	+ 56 55	36	17.	GALAXY
VHA 233	17 31 54.	- 39 03	90		GLOBULAR STAR CLUSTER
MCG+07-36-023	17 31 57.	+ 43 56	42	15.5	GALAXY
PK343-07.1	17 31 58.	- 46 57 52.	5		PLANETARY NEBULA
KHAV 373	17 32	- 01 44	3330		DARK NEBULA
KHAV 375	17 32	- 16 02	2570		DARK NEBULA
LB 09991	17 32	- 88 17		14.4	FAINT BLUE STAR
ZWG 140.044	17 32 00.	+ 23 33		15.5	GALAXY
ZWG 140.045	17 32 00.	+ 25 22		14.6	GALAXY
UGC 10905	17 32 00.	+ 25 22	90	14.6	GALAXY S0-a
MCG+04-41-021	17 32 00.	+ 25 22 30.	72	14.6	GALAXY
MCG+07-36-024	17 32 00.	+ 43 39	18	15.5	GALAXY
MCG+07-36-025	17 32 00.	+ 44 07 30.	36	16.	GALAXY
MCG+12-16-041	17 32 00.	+ 74 24	51	17.	GALAXY
ZWG 340.005	17 32 00.	+ 74 25		15.6	GALAXY
ZWG 339.047	17 32 00.	+ 74 25		15.6	GALAXY
RNGC 6414	17 32 00.	+ 74 25		15.5	GALAXY
UGC 10906	17 32 00.	+ 74 25	72	15.6	GALAXY
ISS 0860	17 32 00.	- 24 47	114		STELLAR RING
VHA 234	17 32 00.	- 44 42	210		GLOBULAR STAR CLUSTER
SCHO 0456	17 32 02.	- 19 20 48.	440		ISOLATED DARK CLOUD
LB 01033	17 32 03.	+ 60 18 54.		16.2	FAINT BLUE STAR
PK358+01.1	17 32 03.5	- 29 01 16.			PLANETARY NEBULA
SCHO 0457	17 32 04.	- 22 02 00.	320		ISOLATED DARK CLOUD
ZC 1732.1+4500	17 32 06.	+ 45 00	670		CLUSTER OF GALAXIES
MCG+10-25-063	17 32 06.	+ 60 28	48	17.	GALAXY
PK359+02.4	17 32 06.3	- 28 05 08.		13.8	PLANETARY NEBULA
PK358+01.3	17 32 10.6	- 29 43 26.	5	16.	PLANETARY NEBULA
SCHO 0458	17 32 11.	- 19 58 12.	330		ISOLATED DARK CLOUD
ZWG 055.009	17 32 12.	+ 06 16		15.7	GALAXY
ZWG 356.005	17 32 12.	+ 77 26		15.3	GALAXY
ZWG 355.035	17 32 12.	+ 77 26		15.3	GALAXY
UGC 10907	17 32 12.	+ 77 26	72	15.3	GALAXY Sb-c
MRSL 355+00/1	17 32 12.	- 32 23	5400		HII REGION
SCHO 0459	17 32 13.	- 20 10 36.	240		ISOLATED DARK CLOUD
PK007+07.1	17 32 14.2	- 18 32 26.	9		PLANETARY NEBULA
ABC 2273	17 32 15.	+ 42 26		17.4	RICH CLUSTER OF GALAXIES
PK341-09.1	17 32 15.	- 49 23	5		PLANETARY NEBULA
PATH 1.800	17 32 21.	+ 59 44	16		NEBULA
PK005+06.1	17 32 22.3	- 20 55 28.	12		PLANETARY NEBULA
MCG+08-32-010	17 32 24.	+ 50 24	66	16.	GALAXY
UGC 10908	17 32 24.	+ 50 25	114	16.0	GALAXY SBc/IRR
MCG+13-12-027	17 32 24.	+ 77 26	66	15.	GALAXY
ISS 0861	17 32 24.	- 21 35	305		STELLAR RING
ISS 0862	17 32 24.	- 26 56	77		STELLAR RING
ZWG 140.046	17 32 30.	+ 25 36		15.5	GALAXY
UGC 10909	17 32 30.	+ 25 36	66	15.5	GALAXY Sc
ZWG 226.029	17 32 30.	+ 43 42		15.7	GALAXY
VV 268C	17 32 30.	+ 50 24	3	19.	INTERACTING GALAXY
VV 268B	17 32 30.	+ 50 24	21	17.	INTERACTING GALAXY
VV 268A	17 32 30.	+ 50 24	24	17.	INTERACTING GALAXY
VV 268	17 32 30.	+ 50 24	72		INTERACTING GALAXY
ISS 0863	17 32 30.	- 25 12	80		STELLAR RING
ISS 0864	17 32 30.	- 26 57	110		STELLAR RING
SCHO 0460	17 32 34.	- 32 01 48.	460		ISOLATED DARK CLOUD
RNGC 6388	17 32 34.	- 44 43		8.5	GLOBULAR CLUSTER
TER 01	17 32 35.	- 30 26 18.	120		STAR CLUSTER
MCG+12-16-042	17 32 36.	+ 71 51	51	16.	GALAXY
SCHO 0461	17 32 36.	- 19 09 42.	650		ISOLATED DARK CLOUD
ISS 0865	17 32 36.	- 23 49	114		STELLAR RING
ISS 0866	17 32 36.	- 24 41	52		STELLAR RING
GCL 069	17 32 36.	- 30 26			GLOBULAR STAR CLUSTER
VHA 235	17 32 36.	- 30 27	60		GLOBULAR STAR CLUSTER
GCL 070	17 32 36.	- 44 42	408	8.7	GLOBULAR STAR CLUSTER
PK359+02.3	17 32 39.6	- 27 41 28.		13.9	PLANETARY NEBULA
LB 01034	17 32 42.	+ 62 05 12.		17.2	FAINT BLUE STAR
MCG+12-16-043	17 32 42.	+ 69 20	66	16.	GALAXY
ZWG 340.006	17 32 42.	+ 71 53		15.2	GALAXY
ZWG 339.048	17 32 42.	+ 71 53		15.2	GALAXY
ISS 0731	17 32 42.	- 16 53	118		STELLAR RING
ISS 0867	17 32 42.	- 23 37	109		STELLAR RING
ISS 0973	17 32 42.	- 32 53	150		STELLAR RING
GCL 071	17 32 42.	- 38 30			GLOBULAR STAR CLUSTER
VHA 236	17 32 42.	- 38 30	90		GLOBULAR STAR CLUSTER
SCHO 0462	17 32 44.	- 19 11 48.	830		ISOLATED DARK CLOUD
IC 4658	17 32 44.	- 59 33			NONSTELLAR OBJECT
SCHO 0463	17 32 45.	- 33 26 54.	650		ISOLATED DARK CLOUD
ZWG 055.010	17 32 48.	+ 06 12		14.8	GALAXY
MCG+01-45-002	17 32 48.	+ 06 12	36	14.8	GALAXY
ZWG 112.009	17 32 48.	+ 20 37		15.5	GALAXY
KARA.73B 0815	17 32 48.	+ 20 37	42	15.5	ISOLATED GALAXY S
ISS 0868	17 32 48.	- 23 06	215		STELLAR RING
ISS 0869	17 32 48.	- 23 07	97		STELLAR RING
ISS 0870	17 32 48.	- 23 40	185		STELLAR RING
ISS 0871	17 32 48.	- 24 02	52		STELLAR RING
LB 01035	17 32 53.	+ 44 13 06.		17.1	FAINT BLUE STAR
IC 4656	17 32 53.	- 63 42			NONSTELLAR OBJECT
UGC 10910	17 32 54.	+ 33 57	78	17.	GALAXY DWARF
UGC 10911	17 32 54.	+ 69 23	60	16.0	GALAXY Sc
MCG+12-16-044	17 32 54.	+ 71 30	18	16.	GALAXY
ISS 0872	17 32 54.	- 25 12	65		STELLAR RING
OCL 1021	17 32 54.	- 33 27	1020	9.1	OPEN STAR CLUSTER
VHA 237	17 32 54.	- 33 27	420		STAR CLSTR IN NEBULOSITY
HN 0446	17 32 54.	- 63 42			NEBULA
PK002+04.1	17 32 54.8	- 24 23 37.		13.2	PLANETARY NEBULA
PATH 1.801	17 32 57.	+ 59 31	11		NEBULA
ABC 2277	17 32 58.	+ 70 56		17.5	RICH CLUSTER OF GALAXIES
SCHO 0464	17 32 59.	- 20 51 18.	340		ISOLATED DARK CLOUD
KHAV 374	17 33	- 22 44	3170		DARK NEBULA
LBN 1117	17 33	- 32 20	3300		BRIGHT NEBULA
MCG+10-25-064	17 33 00.	+ 59 30	24	16.	GALAXY
7ZW 722	17 33 00.	+ 59 37			COMPACT GALAXY
MCG+12-16-045	17 33 00.	+ 68 45	33	16.	GALAXY
LDN 0261	17 33 00.	- 17 10	2100		DARK NEBULA
LDN 0193	17 33 00.	- 20 30	1260		DARK NEBULA
ISS 0873	17 33 00.	- 21 33	164		STELLAR RING
LDN 0125	17 33 00.	- 23 00	3960		DARK NEBULA
LDN 0084	17 33 00.	- 24 30	2280		DARK NEBULA
COU 004	17 33 00.	- 32 40	7200		EMISSION NEBULA
PK350-03.1	17 33 01.7	- 39 20 08.	23		PLANETARY NEBULA
SCHO 0465	17 33 04.	- 20 51 36.	310		ISOLATED DARK CLOUD
ZWG 083.002	17 33 06.	+ 09 28		15.4	GALAXY
ZWG 253.016	17 33 06.	+ 47 15		15.5	GALAXY
SCHO 0466	17 33 09.	- 19 56 12.	360		ISOLATED DARK CLOUD
LB 01036	17 33 10.	+ 44 34 06.		18.0	FAINT BLUE STAR
ZC 1733.2+5644	17 33 12.	+ 56 44	1080		CLUSTER OF GALAXIES
MCG+12-16-046	17 33 12.	+ 71 27	12	16.	GALAXY
SCHO 0467	17 33 16.	- 19 21 54.	530		ISOLATED DARK CLOUD
SCHO 0468	17 33 17.	- 21 41 30.	270		ISOLATED DARK CLOUD
ZWG 140.047	17 33 18.	+ 21 51		15.7	GALAXY
KARA.72 520A	17 33 18.	+ 21 51	42	15.7	PART OF DOUBLE GALAXY
ZWG 253.017	17 33 18.	+ 50 01		14.8	GALAXY
ZWG 253.018	17 33 18.	+ 50 17		15.1	GALAXY
ISS 0874	17 33 18.	- 24 45	190		STELLAR RING
ISS 0875	17 33 18.	- 27 07	231		STELLAR RING
PK005+05.1	17 33 22.7	- 21 29 23.	7		PLANETARY NEBULA
SCHO 0469	17 33 23.	- 20 33 30.	290		ISOLATED DARK CLOUD
ZWG 055.011	17 33 24.	+ 07 00		15.3	GALAXY
ZWG 112.010	17 33 24.	+ 20 49		14.8	GALAXY
MCG+03-45-003	17 33 24.	+ 20 49	36	14.8	GALAXY
ZWG 140.048	17 33 24.	+ 25 22		15.3	GALAXY
ZWG 226.030	17 33 24.	+ 43 25		15.7	GALAXY
ZWG 340.007	17 33 24.	+ 73 54		15.5	GALAXY
ZWG 339.049	17 33 24.	+ 73 54		15.5	GALAXY
ISS 1026	17 33 24.	- 37 50	94		STELLAR RING
LB 03112	17 33 24.	- 57 27		14.9	FAINT BLUE STAR
LB 01037	17 33 29.	+ 45 40 30.		15.4	FAINT BLUE STAR
SCHO 0470	17 33 29.	- 20 21 54.	380		ISOLATED DARK CLOUD
ZWG 140.049	17 33 30.	+ 21 50		15.6	GALAXY
KARA.72 520B	17 33 30.	+ 21 50	36	15.6	PART OF DOUBLE GALAXY
ISS 0876	17 33 30.	- 20 53	140		STELLAR RING
ISS 0877	17 33 30.	- 24 43	314		STELLAR RING
ISS 0878	17 33 30.	- 26 48	137		STELLAR RING
OCL 1029	17 33 30.	- 32 27	1020	9.5	OPEN STAR CLUSTER
VHA 238	17 33 30.	- 32 27	360		OPEN STAR CLUSTER
ZWG 140.050	17 33 36.	+ 24 49		15.4	GALAXY
ZC 1733.6+3611	17 33 36.	+ 36 11	1080		CLUSTER OF GALAXIES
ZC 1733.6+6440	17 33 36.	+ 64 40	940		CLUSTER OF GALAXIES
MCG+12-16-047	17 33 36.	+ 70 51	12	16.	GALAXY
ZWG 083.003	17 33 42.	+ 13 15		15.7	GALAXY
ZWG 253.019	17 33 42.	+ 50 47		15.4	GALAXY
MCG+10-25-065	17 33 42.	+ 58 15	54	17.	GALAXY
7ZW 723	17 33 42.	+ 68 35			COMPACT GALAXY
7ZW 724	17 33 42.	+ 72 11			COMPACT GALAXY
ISS 0974	17 33 42.	- 29 06	65		STELLAR RING
PK358+01.4	17 33 47.	- 29 38		16.	PLANETARY NEBULA
ZWG 340.008	17 33 48.	+ 70 48		15.6	GALAXY
ZWG 339.050	17 33 48.	+ 70 48		15.6	GALAXY
LB 01038	17 33 51.	+ 60 19 00.		16.0	FAINT BLUE STAR
ZC 1733.9+6407	17 33 54.	+ 64 07	1750		CLUSTER OF GALAXIES
SCHO 0471	17 33 55.	- 21 00 36.	600		ISOLATED DARK CLOUD
B 272	17 34	- 23 23	2700		DARK OBJECT
KHAV 376	17 34	- 27 50	3960		DARK NEBULA
ZWG 083.004	17 34 00.	+ 10 03		15.5	GALAXY
KARA.73B 0816	17 34 00.	+ 10 03	48	15.5	ISOLATED GALAXY P
MCG+10-25-066	17 34 00.	+ 58 58	18	16.	GALAXY
ZWG 300.050	17 34 00.	+ 58 59		15.6	GALAXY
MCG+12-16-048	17 34 00.	+ 70 47	33	16.	GALAXY
ZC 1734.0+8557	17 34 00.	+ 85 57	1480		CLUSTER OF GALAXIES
LDN 0164	17 34 00.	- 21 50	960		DARK NEBULA
LDN 0076	17 34 00.	- 25 00	2940		DARK NEBULA
ISS 0879	17 34 00.	- 26 55	257		STELLAR RING
ZWG 112.011	17 34 06.	+ 19 00		15.4	GALAXY
ZWG 140.051	17 34 06.	+ 22 54		15.6	GALAXY
ZWG 140.052	17 34 06.	+ 22 56		15.5	GALAXY
SCHO 0472	17 34 09.	- 21 03 48.	520		ISOLATED DARK CLOUD
SCHO 0473	17 34 09.	- 23 15 30.	790		ISOLATED DARK CLOUD
SCHO 0474	17 34 10.	- 23 37 12.	540		ISOLATED DARK CLOUD
ZWG 083.005	17 34 12.	+ 09 04		15.7	GALAXY
ZWG 340.009	17 34 12.	+ 70 52		15.7	GALAXY
ZWG 339.051	17 34 12.	+ 70 52		15.7	GALAXY
UGC 10912	17 34 12.	+ 70 52	138	15.7	GALAXY S
SCHO 0475	17 34 12.	- 21 52 36.	530		ISOLATED DARK CLOUD
ZWG 112.012	17 34 18.	+ 15 20		15.6	GALAXY
UGC 10913	17 34 18.	+ 15 20	66	15.6	GALAXY SBc
KARA.73B 0817	17 34 18.	+ 15 20	66	15.6	ISOLATED GALAXY S
ZWG 140.053	17 34 18.	+ 23 50		15.5	GALAXY
KARA.73B 0818	17 34 18.	+ 23 50	36	15.5	ISOLATED GALAXY S
7ZW 725	17 34 18.	+ 62 30			COMPACT GALAXY
ISS 0732	17 34 18.	- 15 55	81		STELLAR RING
ISS 1027	17 34 18.	- 35 20	64		STELLAR RING
OCL 1015	17 34 18.	- 36 14	504	11.	OPEN STAR CLUSTER
ISS 1065	17 34 18.	- 41 44	170		STELLAR RING
MCG+03-45-004	17 34 21.	+ 15 18	78	15.6	GALAXY
ZWG 300.051	17 34 24.	+ 62 30		15.6	GALAXY
ISS 0733	17 34 24.	- 18 07	320		STELLAR RING
PK007+06.1	17 34 25.7	- 18 44 56.	8		PLANETARY NEBULA
SCHO 0476	17 34 26.	- 16 37 18.	280		ISOLATED DARK CLOUD
B 079	17 34 26.	- 19 35	1800		DARK OBJECT
PK356-00.1	17 34 28.7	- 32 13 33.			PLANETARY NEBULA
ZWG 083.006	17 34 30.	+ 13 11		15.6	GALAXY
ZWG 112.013	17 34 30.	+ 17 14		15.4	GALAXY
ZC 1734.5+3942	17 34 30.	+ 39 42	1210		CLUSTER OF GALAXIES
7ZW 726	17 34 30.	+ 63 05			COMPACT GALAXY
7ZW 726	17 34 30.	+ 66 05			COMPACT GALAXY
ISS 1028	17 34 30.	- 35 32	136		STELLAR RING
SCHO 0477	17 34 32.	- 19 16 12.	550		ISOLATED DARK CLOUD
MCG+07-36-026	17 34 36.	+ 42 59	30	16.	GALAXY
ZWG 321.020	17 34 36.	+ 67 04		15.6	GALAXY
ZWG 321.021	17 34 36.	+ 68 02		15.6	GALAXY
ZWG 340.010	17 34 36.	+ 73 00		15.5	GALAXY
ZWG 339.052	17 34 36.	+ 73 00		15.5	GALAXY
ISS 0880	17 34 36.	- 24 39	77		STELLAR RING
ISS 0975	17 34 36.	- 26 59	67		STELLAR RING
ISS 0976	17 34 36.	- 27 02	58		STELLAR RING
RNGC 6410	17 34 37.	+ 60 51			NON-EXISTENT OBJECT
ABC 2276	17 34 38.	+ 64 04		17.4	RICH CLUSTER OF GALAXIES
ZWG 112.014	17 34 42.	+ 17 45		15.6	GALAXY
ZWG 140.054	17 34 42.	+ 21 09		15.3	GALAXY
KARA.73B 0819	17 34 42.	+ 21 09	42	15.3	ISOLATED GALAXY PEC
ZWG 226.031	17 34 42.	+ 40 10		15.4	GALAXY
UGC 10914	17 34 42.	+ 40 10	60	15.4	GALAXY Sc
SCHO 0478	17 34 42.	- 19 36 00.	500		ISOLATED DARK CLOUD
B 080	17 34 43.	- 21 15	180		DARK OBJECT
UGC 10915	17 34 48.	+ 24 57	72	17.	GALAXY

OBJECT NAME	RIGHT ASCEN.	DECLINATION	DIAM.	MAGN.	TYPE OF OBJECT
ZWG 321.022	17 34 48.	+ 63 26		15.4	GALAXY
OCL 1018	17 34 48.	- 34 58	600	11.1	OPEN STAR CLUSTER
VHA 239	17 34 48.	- 34 58	240		STAR CLSTR IN NEBULOSITY
OCL 1013	17 34 48.	- 37 32	1500	8.0	OPEN STAR CLUSTER
RNGC 6396	17 34 50.	- 34 58		11.0	OPEN CLUSTER
LB 01039	17 34 52.	+ 45 57 24.		17.1	FAINT BLUE STAR
ZC 1734.9+4215	17 34 54.	+ 42 15	1680		CLUSTER OF GALAXIES
SCHO 0479	17 34 54.	- 22 37 12.	530		ISOLATED DARK CLOUD
ISS 0881	17 34 54.	- 26 32	51		STELLAR RING
ABC 2275	17 34 57.	+ 53 13		16.5	RICH CLUSTER OF GALAXIES
REIN 2.256	17 34 58.66	- 03 13 04.1			NEBULA
RNGC 6402	17 34 59.	- 03 15		9.5	GLOBULAR CLUSTER
KHAV 379	17 35	- 05 32	12870		DARK NEBULA
KHAV 380	17 35	- 16 32	2660		DARK NEBULA
KHAV 381	17 35	- 19 50	7290		DARK NEBULA
KHAV 378	17 35	- 23 20	2500		DARK NEBULA
KHAV 377	17 35	- 25 44			DARK NEBULA
MCG+06-39-001	17 35 00.	+ 36 00	30	15.	GALAXY
MCG+10-25-067	17 35 00.	+ 59 56	30	16.	GALAXY
ZWG 300.052	17 35 00.	+ 60 50		13.2	GALAXY
UGC 10916	17 35 00.	+ 60 50	138	13.2	GALAXY E
MCG+10-25-068	17 35 00.	+ 60 50	45	13.	GALAXY
GCL 072	17 35 00.	- 03 13	402	9.44	GLOBULAR STAR CLUSTER
LDN 0216	17 35 00.	- 19 38	420		DARK NEBULA
LDN 0209	17 35 00.	- 19 50	420		DARK NEBULA
ISS 0734	17 35 00.	- 20 56	73		STELLAR RING
LDN 0172	17 35 00.	- 21 35	600		DARK NEBULA
LDN 0155	17 35 00.	- 22 15	5100		DARK NEBULA
LDN 0074	17 35 00.	- 25 10	780		DARK NEBULA
ISS 0977	17 35 00.	- 28 00	140		STELLAR RING
SCHO 0481	17 35 00.	- 33 15 00.	1090		ISOLATED DARK CLOUD
SCHO 0480	17 35 00.	- 24 37 36.	570		ISOLATED DARK CLOUD
RNGC 6411	17 35 01.	+ 60 50		13.0	GALAXY
SCHO 0483	17 35 01.	- 22 38 12.	420		ISOLATED DARK CLOUD
SCHO 0482	17 35 01.	- 23 17 54.	370		ISOLATED DARK CLOUD
B 274	17 35 02.	- 22 42	1080		DARK OBJECT
ZWG 083.007	17 35 06.	+ 13 02		15.3	GALAXY
MCG+07-36-027	17 35 06.	+ 42 06	120	14.	GALAXY
ZWG 226.032	17 35 06.	+ 42 07		14.3	GALAXY
UGC 10917	17 35 06.	+ 42 07	126	14.3	GALAXY Sa-b
MCG+08-32-011	17 35 06.	+ 49 39	9	16.	GALAXY
ISS 0882	17 35 06.	- 26 28	118		STELLAR RING
SCHO 0484	17 35 08.	- 22 23 00.	490		ISOLATED DARK CLOUD
IC 1265	17 35 09.	+ 42 08 04.			NONSTELLAR OBJECT
ZWG 083.008	17 35 12.	+ 11 10		14.8	GALAXY
UGC 10918	17 35 12.	+ 11 10	78	14.8	GALAXY E
MCG+02-45-001	17 35 12.	+ 11 10	15	14.8	GALAXY
MCG+03-45-005	17 35 12.	+ 19 36	42	15.0	GALAXY
ZWG 140.055	17 35 12.	+ 26 16		15.2	GALAXY
B 273	17 35 12.	- 33 19	900		DARK OBJECT
PK007+06.2	17 35 14.1	- 19 35 55.	7		PLANETARY NEBULA
ZWG 112.015	17 35 18.	+ 17 34		14.5	GALAXY
UGC 10919	17 35 18.	+ 17 34	84	14.5	GALAXY S
MCG+03-45-006	17 35 18.	+ 17 34	72	14.5	GALAXY
ZWG 112.016	17 35 18.	+ 19 35		15.0	GALAXY
ISS 0283	17 35 18.	- 27 04	157		STELLAR RING
SCHO 0485	17 35 19.	- 21 22 00.	450		ISOLATED DARK CLOUD
SCHO 0486	17 35 23.	- 23 46 42.	230		ISOLATED DARK CLOUD
ZWG 253.020	17 35 24.	+ 50 48		14.8	GALAXY
RNGC 6409	17 35 24.	+ 50 48		15.0	GALAXY
SCHO 0487	17 35 24.	- 16 36 36.	300		ISOLATED DARK CLOUD
SCHO 0488	17 35 27.	- 21 11 42.	350		ISOLATED DARK CLOUD
PK004+04.1	17 35 29.5	- 22 06 58.	5		PLANETARY NEBULA
ZWG 083.009	17 35 30.	+ 11 17		15.6	GALAXY
ZWG 083.010	17 35 30.	+ 12 58		15.4	GALAXY
UGC 10920	17 35 30.	+ 49 55	60	17.	GALAXY
MCG+11-21-011	17 35 30.	+ 68 02	15	16.	GALAXY
ZWG 321.023	17 35 30.	+ 68 07		15.5	GALAXY
LDN 0202	17 35 30.	- 20 20	1380		DARK NEBULA
LDN 0113	17 35 30.	- 23 50	780		DARK NEBULA
B 081	17 35 30.	- 23 54			DARK OBJECT
PK359+01.1	17 35 30.	- 28 41		14.0	PLANETARY NEBULA
CED 149	17 35 33.	- 23 53	90		DIFFUSE GALACTIC NEBULA
REIN 2.257	17 35 33.83	- 23 52 52.6			NEBULA
B 082	17 35 34.	- 23 45	480		DARK OBJECT
RNGC 6401	17 35 34.	- 23 53			GLOBULAR CLUSTER
REIN 2.258	17 35 34.81	- 23 53 02.5			NEBULA
ISS 0735	17 35 36.	- 17 16	188		STELLAR RING
SCHO 0489	17 35 36.	- 21 11 00.	270		ISOLATED DARK CLOUD
GCL 073	17 35 36.	- 23 52	60	8.0	GLOBULAR STAR CLUSTER
PK346-06.1	17 35 38.	- 44 08	5		PLANETARY NEBULA
SCHO 0490	17 35 39.	- 23 45 00.	290		ISOLATED DARK CLOUD
SCHO 0492	17 35 40.	- 24 09 54.	390		ISOLATED DARK CLOUD
SCHO 0491	17 35 40.	- 24 21 00.	350		ISOLATED DARK CLOUD
UGC 10921	17 35 42.	+ 17 24	60	17.	GALAXY DWARF
ISS 0736	17 35 42.	- 21 09	70		STELLAR RING
LB 01040	17 35 43.	+ 80 02 54.		15.3	FAINT BLUE STAR
B 275	17 35 43.	- 32 18	780		DARK OBJECT
ZWG 083.011	17 35 48.	+ 13 55		15.2	GALAXY
ZC 1735.8+4248	17 35 48.	+ 42 48	1010		CLUSTER OF GALAXIES
UGC 10922	17 35 48.	+ 60 19	84	16.0	GALAXY Sc
ISS 0884	17 35 48.	- 21 48	110		STELLAR RING
LB 03991	17 35 49.	+ 07 41 54.		19.4	FAINT BLUE STAR
SCHO 0493	17 35 49.	- 21 46 18.	430		ISOLATED DARK CLOUD
B 083	17 35 55.	- 24 09	420		DARK OBJECT
KHAV 382	17 36	- 24 08	810		DARK NEBULA
KHAV 383	17 36	- 30 38	3960		DARK NEBULA
ZWG 083.012	17 36 00.	+ 13 50		15.5	GALAXY
MCG+04-41-022	17 36 00.	+ 22 42 30.	36	15.3	GALAXY
ZWG 140.056	17 36 00.	+ 22 44		15.3	GALAXY
ZWG 278.010	17 36 00.	+ 54 15		15.4	GALAXY
ZWG 278.011	17 36 00.	+ 55 02		15.3	GALAXY
MCG+10-25-069	17 36 00.	+ 60 18	30	16.	GALAXY
7ZW 728	17 36 00.	+ 67 22			COMPACT GALAXY
MCG+11-21-012	17 36 00.	+ 68 13	57	16.	GALAXY
7ZW 729	17 36 00.	+ 86 47			COMPACT GALAXY
ZWG 370.006	17 36 00.	+ 86 47		14.3	GALAXY
ZWG 367.023	17 36 00.	+ 86 47		14.3	GALAXY
UGC 10923	17 36 00.	+ 86 47	84	14.3	GALAXY DBL SYS
LDN 0283	17 36 00.	- 16 00	8880		DARK NEBULA
ISS 0737	17 36 00.	- 17 36	82		STELLAR RING
SCHO 0494	17 36 00.	- 18 25 30.	390		ISOLATED DARK CLOUD
LDN 0174	17 36 00.	- 21 40	540		DARK NEBULA
LDN 0109	17 36 00.	- 24 05	300		DARK NEBULA
SCHO 0495	17 36 00.	- 24 07 42.	420		ISOLATED DARK CLOUD
LDN 0080	17 36 00.	- 25 05	240		DARK NEBULA
LB 03992	17 36 02.	+ 07 35 12.		14.0	FAINT BLUE STAR
PK008+06.1	17 36 02.	- 18 16 09.	10	11.	PLANETARY NEBULA
ABC 2294	17 36 03.	+ 85 55		17.7	RICH CLUSTER OF GALAXIES
SCHO 0497	17 36 05.	- 21 16 00.	270		ISOLATED DARK CLOUD
SCHO 0496	17 36 05.	- 21 25 42.	300		ISOLATED DARK CLOUD
ZWG 033.013	17 36 06.	+ 09 43		15.5	GALAXY
KARA.73B 0820	17 36 06.	+ 09 43	36	15.5	ISOLATED GALAXY E
MCG+07-36-028	17 36 06.	+ 43 32	39	15.	GALAXY
ZWG 226.033	17 36 06.	+ 43 33		15.7	GALAXY
KARA.73B 0821	17 36 06.	+ 43 33	42	15.7	ISOLATED GALAXY S
ZWG 253.021	17 36 06.	+ 46 20		15.2	GALAXY
MCG+09-29-012	17 36 06.	+ 54 14	24	17.	GALAXY
MCG+09-29-013	17 36 06.	+ 54 15	30	16.	GALAXY
MCG+09-29-014	17 36 06.	+ 55 01	54	15.	GALAXY
MCG+11-21-013	17 36 06.	+ 69 06	39	16.	GALAXY
ZWG 321.024	17 36 06.	+ 68 10		15.5	GALAXY
RNGC 6419	17 36 06.	+ 68 10		15.5	GALAXY
UGC 10924	17 36 06.	+ 68 10	66	15.5	GALAXY Sa
RNGC 6406	17 36 07.	+ 18 52			NON-EXISTENT OBJECT
LB 03993	17 36 08.	+ 07 40 30.		19.9	FAINT BLUE STAR
SCHO 0498	17 36 09.	- 19 06 54.	310		ISOLATED DARK CLOUD
SCHO 0499	17 36 10.	- 24 37 00.	440		ISOLATED DARK CLOUD
ZWG 083.014	17 36 12.	+ 11 24		15.5	GALAXY
ZWG 083.015	17 36 12.	+ 11 48		15.4	GALAXY
ZWG 170.038	17 36 12.	+ 28 29		15.3	GALAXY
ZWG 253.022	17 36 12.	+ 48 59		15.6	GALAXY
KARA.73B 0822	17 36 12.	+ 48 59	42	15.6	ISOLATED GALAXY S
MCG+10-25-070	17 36 12.	+ 59 13	36	16.	GALAXY
7ZW 730	17 36 12.	+ 67 47			COMPACT GALAXY
ISS 0885	17 36 12.	- 26 32	81		STELLAR RING
PK359+01.2	17 36 12.	- 28 46		13.7	PLANETARY NEBULA
ISS 1029	17 36 12.	- 35 22	98		STELLAR RING
RNGC 6404	17 36 15.	- 33 13		10.5	OPEN CLUSTER
ZWG 083.016	17 36 18.	+ 12 46		15.5	GALAXY
ZWG 140.057	17 36 18.	+ 24 59		15.1	GALAXY
UGC 10926	17 36 18.	+ 24 59	72	15.1	GALAXY SB
UGC 10925	17 36 18.	+ 34 58	60	16.0	GALAXY S
ZC 1736.2+3846	17 36 18.	+ 38 46	870		CLUSTER OF GALAXIES
ZWG 253.023	17 36 18.	+ 46 25		15.4	GALAXY
ZWG 321.025	17 36 18.	+ 68 03		15.5	GALAXY
RNGC 6420	17 36 18.	+ 68 03		15.5	GALAXY
ISS 0738	17 36 18.	- 16 36	53		STELLAR RING
OCL 1024	17 36 18.	- 33 13	840	10.6	OPEN STAR CLUSTER
VHA 240	17 36 18.	- 33 13	300		STAR CLSTR IN NEBULOSITY
SCHO 0500	17 36 20.	- 21 19 30.	300		ISOLATED DARK CLOUD
MCG+06-39-002	17 36 21.	+ 34 59	45	15.5	GALAXY
SCHO 0501	17 36 23.	- 21 18 00.	340		ISOLATED DARK CLOUD
UGC 10927	17 36 24.	+ 34 57	66	16.0	GALAXY S0
MCG+06-39-003	17 36 24.	+ 34 58	24	15.5	GALAXY
ZC 1736.4+6439	17 36 24.	+ 64 39	870		CLUSTER OF GALAXIES
MCG+11-21-014	17 36 24.	+ 68 07	12	16.	GALAXY
VHA 241	17 36 24.	- 36 56	360		OPEN STAR CLUSTER
LB 03994	17 36 26.	+ 07 58 12.		18.3	FAINT BLUE STAR
MCG+04-41-023	17 36 27.	+ 24 59 30.	60	15.1	GALAXY
ZWG 055.012	17 36 30.	+ 06 14		15.5	GALAXY
MCG+03-45-007	17 36 30.	+ 18 55	96	14.0	GALAXY
ZWG 140.058	17 36 30.	+ 21 43		15.2	GALAXY
UGC 10928	17 36 30.	+ 21 43	96	15.2	GALAXY Sb
ZWG 140.059	17 36 30.	+ 23 57		15.5	GALAXY
ZWG 199.001	17 36 30.	+ 35 33		14.9	GALAXY
UGC 10929	17 36 30.	+ 35 33	72	14.9	GALAXY Sa-b
ZWG 321.026	17 36 30.	+ 68 04		15.1	GALAXY
RNGC 6422	17 36 30.	+ 68 04		15.0	GALAXY
MCG+11-21-015	17 36 30.	+ 68 06 30.	27	15.	GALAXY
ZC 1736.5+7242	17 36 30.	+ 72 42	1010		CLUSTER OF GALAXIES
LDN 0219	17 36 30.	- 19 45	1140		DARK NEBULA
LDN 0090	17 36 30.	- 24 50	720		DARK NEBULA
ISS 1066	17 36 30.	- 43 49	187		STELLAR RING
LB 03995	17 36 31.	+ 07 30 00.		12.0	FAINT BLUE STAR
SCHO 0502	17 36 32.	- 16 52 00.	450		ISOLATED DARK CLOUD
MCG+06-39-004	17 36 33.	+ 35 34	66	14.	GALAXY
LB 03996	17 36 35.	+ 08 06 54.		17.3	FAINT BLUE STAR
SCHO 0503	17 36 35.	- 24 22 48.	610		ISOLATED DARK CLOUD
ZWG 083.017	17 36 36.	+ 13 01		15.3	GALAXY
ZWG 112.017	17 36 36.	+ 18 54		14.0	GALAXY
UGC 10930	17 36 36.	+ 18 54	96	14.0	GALAXY SBa
MCG+04-41-024	17 36 36.	+ 21 42	72	15.2	GALAXY
UGC 10931	17 36 36.	+ 60 28	72	16.5	GALAXY Sc
ZWG 340.011	17 36 36.	+ 70 01		14.5	GALAXY
UGC 10932	17 36 36.	+ 70 01	54	14.5	GALAXY COMPACT
MCG+12-17-001	17 36 36.	+ 70 01	21	15.	GALAXY
ISS 0739	17 36 36.	- 18 40	79		STELLAR RING
RNGC 6408	17 36 37.	+ 18 54		14.0	GALAXY
RNGC 6424	17 36 38.	+ 70 01		15.0	GALAXY
LB 03997	17 36 40.	+ 08 08 06.		14.6	FAINT BLUE STAR
LB 01041	17 36 40.	+ 45 18 24.		16.7	FAINT BLUE STAR
B 276	17 36 41.	- 19 48	2700		DARK OBJECT
MCG+10-25-071	17 36 42.	+ 60 28	60	17.	GALAXY
RNGC 6405	17 36 46.	- 32 11		4.5	OPEN CLUSTER
RNGC 6397	17 36 46.	- 53 39		7.5	GLOBULAR CLUSTER
ZWG 112.018	17 36 49.	+ 19 13		15.4	GALAXY
ZWG 141.001	17 36 48.	+ 26 25		15.6	GALAXY
ZWG 140.060	17 36 48.	+ 26 25		15.6	GALAXY
ZWG 278.012	17 36 48.	+ 51 16		15.7	GALAXY
ZWG 278.013	17 36 48.	+ 54 05		15.7	GALAXY
ISS 0740	17 36 48.	- 19 49	68		STELLAR RING
OCL 1030	17 36 48.	- 32 11	2700	5.8	OPEN STAR CLUSTER
VHA 242	17 36 48.	- 32 11	900		OPEN STAR CLUSTER
GCL 074	17 36 48.	- 53 39	1200	7.3	GLOBULAR STAR CLUSTER
LB 03113	17 36 48.	- 57 39		14.0	FAINT BLUE STAR
LB 03998	17 36 49.	+ 07 50 42.		16.0	FAINT BLUE STAR
LB 03999	17 36 51.	+ 08 06 24.		18.3	FAINT BLUE STAR
SCHO 0504	17 36 52.	- 23 51 54.	360		ISOLATED DARK CLOUD
ZWG 226.034	17 36 54.	+ 39 15		15.4	GALAXY
UGC 10933	17 36 54.	+ 39 15	96	15.4	GALAXY PECULR
ZWG 253.024	17 36 54.	+ 48 10		15.7	GALAXY
MCG+10-25-072	17 36 54.	+ 57 08	30	16.	GALAXY
ZWG 321.027	17 36 54.	+ 68 11		15.6	GALAXY
RNGC 6423	17 36 54.	+ 68 11		15.5	GALAXY
MCG+11-21-016	17 36 54.	+ 68 14	15	16.	GALAXY
PK005+05.2	17 36 55.3	- 21 12 32.	20		PLANETARY NEBULA
KHAV 384	17 37	- 28 56	4390		DARK NEBULA
KHAV 385	17 37	- 36 38	8910		DARK NEBULA
ZWG 199.002	17 37 00.	+ 38 45		15.4	GALAXY
MCG+14-08-025	17 37 00.	+ 86 46	30	17.	GALAXY
MCG+14-08-024	17 37 00.	+ 86 46 30.	66	15.	GALAXY
ISS 0741	17 37 00.	- 20 18	46		STELLAR RING
LDN 0153	17 37 00.	- 22 35	2880		DARK NEBULA
LDN 0114	17 37 00.	- 24 00	2820		DARK NEBULA
LDN 0064	17 37 00.	- 26 00	4020		DARK NEBULA

OBJECT NAME	RIGHT ASCEN.	DECLINATION	DIAM.	MAGN.	TYPE OF OBJECT
ISS 1030	17 37 00.	- 35 59	161		STELLAR RING
SCHO 0505	17 37 01.	- 16 54 48.	420		ISOLATED DARK CLOUD
PK003+03.1	17 37 03.5	- 24 24 11.	4		PLANETARY NEBULA
MIL 53	17 37 05.	- 30 56 54.	264		SUPERNOVA REMNANT
PK001+02.1	17 37 05.0	- 26 42 44.	10		PLANETARY NEBULA
MCG+06-39-005	17 37 06.	+ 33 27	36	15.	GALAXY
MCG+10-25-073	17 37 06.	+ 57 08	18	17.	GALAXY
SCHO 0507	17 37 09.	- 21 33 12.	320		ISOLATED DARK CLOUD
SCHO 0506	17 37 09.	- 33 59 12.	590		ISOLATED DARK CLOUD
ZWG 300.053	17 37 12.	+ 57 10		15.7	GALAXY
SCHO 0508	17 37 12.	- 17 07 00.	350		ISOLATED DARK CLOUD
LDN 0229	17 37 12.	- 19 29	180		DARK NEBULA
ISS 0979	17 37 12.	- 27 23	74		STELLAR RING
ISS 0978	17 37 12.	- 27 23	79		STELLAR RING
SCHO 0509	17 37 15.	- 24 48 06.	1050		ISOLATED DARK CLOUD
PK005+04.1	17 37 16.9	- 22 17 45.	7		PLANETARY NEBULA
ZWG 226.035	17 37 18.	+ 44 05		15.7	GALAXY
ZWG 300.054	17 37 18.	+ 58 45		14.7	GALAXY
ZWG 340.012	17 37 18.	+ 70 57		15.3	GALAXY
LB 04000	17 37 20.	+ 07 57 30.		17.5	FAINT BLUE STAR
RNGC 6418	17 37 20.	+ 58 45		14.5	GALAXY
SCHO 0510	17 37 20.	- 22 04 24.	420		ISOLATED DARK CLOUD
PK001+01.1	17 37 20.4	- 26 59 15.	47		PLANETARY NEBULA
LB 04001	17 37 21.	+ 07 43 36.		15.2	FAINT BLUE STAR
PK344-08.1	17 37 21.	- 47 01			PLANETARY NEBULA
SCHO 0511	17 37 23.	- 20 27 00.	330		ISOLATED DARK CLOUD
ZWG 027.003	17 37 24.	+ 02 51		15.3	GALAXY
ZC 1737.4+4016	17 37 24.	+ 40 16	3290		CLUSTER OF GALAXIES
MCG+10-25-074	17 37 24.	+ 58 44	30	16.	GALAXY
7ZW 731	17 37 24.	+ 76 35			COMPACT GALAXY
LDN 0231	17 37 24.	- 19 30	120		DARK NEBULA
RNGC 6400	17 37 24.	- 36 54		9.0	OPEN CLUSTER
OCL 1014	17 37 24.	- 36 55	900	8.9	OPEN STAR CLUSTER
SCHO 0512	17 37 25.	- 20 42 00.	350		ISOLATED DARK CLOUD
LB 04002	17 37 28.	+ 07 58 24.		13.0	FAINT BLUE STAR
SCHO 0513	17 37 28.	- 21 56 00.	460		ISOLATED DARK CLOUD
SCHO 0514	17 37 29.	- 21 47 00.	610		ISOLATED DARK CLOUD
ZWG 253.025	17 37 30.	+ 49 48		15.7	GALAXY
MCG+12-17-002	17 37 30.	+ 72 05	132	14.	GALAXY
LDN 0144	17 37 30.	- 23 00	660		DARK NEBULA
SCHO 0515	17 37 30.	- 24 31 00.	370		ISOLATED DARK CLOUD
ISS 0980	17 37 30.	- 27 14	142		STELLAR RING
LB 04003	17 37 31.	+ 07 54 06.		17.2	FAINT BLUE STAR
SCHO 0516	17 37 32.	- 21 23 48.	350		ISOLATED DARK CLOUD
B 277	17 37 33.	- 23 03	1080		DARK OBJECT
ZWG 112.019	17 37 36.	+ 16 54		15.4	GALAXY
ZWG 112.020	17 37 36.	+ 19 49		15.3	GALAXY
KARA.73B 0823	17 37 36.	+ 19 49	36	15.3	ISOLATED GALAXY E
MCG+03-45-008	17 37 36.	+ 19 50	36	15.3	GALAXY
ZWG 340.013	17 37 36.	+ 72 07		13.2	GALAXY
ZWG 339.053	17 37 36.	+ 72 07		13.2	GALAXY
UGC 10934	17 37 36.	+ 72 07	144	13.2	GALAXY SB:b-c
LDN 0226	17 37 36.	- 19 38	180		DARK NEBULA
ISS 1031	17 37 36.	- 34 44	365		STELLAR RING
SCHO 0517	17 37 37.	- 24 17 12.	460		ISOLATED DARK CLOUD
LB 01042	17 37 38.	+ 59 52 06.		17.7	FAINT BLUE STAR
RNGC 6434	17 37 38.	+ 72 07		13.0	GALAXY
MCG+03-45-009	17 37 39.	+ 16 54	36	15.4	GALAXY
SCHO 0518	17 37 40.	- 23 01 24.	510		ISOLATED DARK CLOUD
ZWG 199.003	17 37 42.	+ 33 55		15.7	GALAXY
MCG+10-25-075	17 37 42.	+ 57 15	24	17.	GALAXY
SCHO 0519	17 37 42.	- 20 14 00.	350		ISOLATED DARK CLOUD
SCHO 0520	17 37 45.	- 23 02 18.	500		ISOLATED DARK CLOUD
SCHO 0521	17 37 47.	- 21 45 42.	640		ISOLATED DARK CLOUD
ZWG 199.004	17 37 48.	+ 34 46		15.6	GALAXY
UGC 10935	17 37 48.	+ 57 17	72	16.0	GALAXY SO
MCG+10-25-076	17 37 48.	+ 61 03	60	15.	GALAXY
ZWG 300.055	17 37 48.	+ 61 04		15.1	GALAXY
UGC 10936	17 37 48.	+ 61 04	102	15.1	GALAXY S
SCHO 0522	17 37 48.	- 17 15	410		ISOLATED DARK CLOUD
LDN 0228	17 37 48.	- 19 36	180		DARK NEBULA
LDN 0223	17 37 48.	- 19 41	120		DARK NEBULA
SCHO 0523	17 37 48.	- 21 22 48.	610		ISOLATED DARK CLOUD
RNGC 6392	17 37 55.	- 69 44			UNVERIFIED SOUTHERN OBJECT
IC 1267	17 37 59.	+ 59 24 44.			NONSTELLAR OBJECT
KHAV 390	17 38	- 17 20	1860		DARK NEBULA
KHAV 386	17 38	- 22 50	2230		DARK NEBULA
LB 09916	17 38	- 81 00		14.2	FAINT BLUE STAR
LB 09917	17 38	- 81 47		14.7	FAINT BLUE STAR
LB 09918	17 38	- 83 34		13.2	FAINT BLUE STAR
ZC 1738.0+3516	17 38 00.	+ 35 16	6180		CLUSTER OF GALAXIES
ZWG 300.056	17 38 00.	+ 59 25		14.5	GALAXY
UGC 10937	17 38 00.	+ 59 25	90	14.5	GALAXY SBb
MCG+10-25-077	17 38 00.	+ 59 25	90	14.	GALAXY
LB 01043	17 38 00.	+ 60 29 30.		17.2	FAINT BLUE STAR
SCHO 0524	17 38 05.	- 18 44 12.	340		ISOLATED DARK CLOUD
RNGC 6398	17 38 05.	- 61 39			NON-EXISTENT OBJECT
OCL 1009	17 38 06.	- 40 05	1440	7.6	OPEN STAR CLUSTER
LB 04004	17 38 10.	+ 07 57 24.		14.0	FAINT BLUE STAR
SCHO 0525	17 38 11.	- 23 13 54.	240		ISOLATED DARK CLOUD
ZC 1738.2+2415	17 38 12.	+ 24 15	1080		CLUSTER OF GALAXIES
ZWG 199.005	17 38 12.	+ 33 49		14.9	GALAXY
ZC 1738.2+4340	17 38 12.	+ 43 40	1550		CLUSTER OF GALAXIES
ZWG 253.026	17 38 12.	+ 48 22		14.9	GALAXY
MCG+10-25-078	17 38 12.	+ 59 21	18	16.	GALAXY
LDN 0222	17 38 12.	- 19 45	120		DARK NEBULA
SCHO 0527	17 38 15.	- 17 05 36.	380		ISOLATED DARK CLOUD
SCHO 0528	17 38 15.	- 17 31 00.	480		ISOLATED DARK CLOUD
SCHO 0526	17 38 15.	- 21 43 30.	300		ISOLATED DARK CLOUD
ZWG 226.036	17 38 18.	+ 43 20		15.2	GALAXY
MCG+08-32-012	17 38 18.	+ 49 37	84	15.	GALAXY
ZWG 253.027	17 38 18.	+ 49 38		15.0	GALAXY
UGC 10938	17 38 18.	+ 49 38	84	15.0	GALAXY SO
ZC 1738.3+6452	17 38 18.	+ 64 52	1010		CLUSTER OF GALAXIES
SCHO 0529	17 38 18.	- 20 32 42.	320		ISOLATED DARK CLOUD
SCHO 0530	17 38 20.	- 21 34 48.	290		ISOLATED DARK CLOUD
SCHO 0531	17 38 20.	- 22 01 30.	330		ISOLATED DARK CLOUD
RNGC 6413	17 38 21.	+ 12 39			NON-EXISTENT OBJECT
MCG+07-36-029	17 38 21.	+ 43 18 30.	18	15.	GALAXY
ZWG 055.013	17 38 24.	+ 07 04		15.6	GALAXY
LB 04005	17 38 24.	+ 07 27 00.		18.7	FAINT BLUE STAR
ZWG 199.006	17 38 24.	+ 37 42		15.7	GALAXY
ISS 1032	17 38 24.	- 34 40	221		STELLAR RING
LB 04006	17 38 28.	+ 07 28 48.		19.8	FAINT BLUE STAR
UGC 10939	17 38 30.	+ 19 32	60	16.0	GALAXY SBb-c
MCG+06-39-006	17 38 30.	+ 37 43 30.	30	15.	GALAXY
7ZW 732	17 38 30.	+ 68 04			COMPACT GALAXY
ZWG 340.014	17 38 30.	+ 70 00		15.5	GALAXY

OBJECT NAME	RIGHT ASCEN.	DECLINATION	DIAM.	MAGN.	TYPE OF OBJECT
RNGC 6403	17 38 35.	- 61 39			NON-EXISTENT OBJECT
PK005+04.2	17 38 35.7	- 22 11 36.	7		PLANETARY NEBULA
MCG+03-45-010	17 38 36.	+ 18 52	42	15.2	GALAXY
MCG+06-39-008	17 38 36.	+ 35 25	6	16.	GALAXY
MCG+06-39-007	17 38 36.	+ 35 25	30	16.	GALAXY
ARC 2278	17 38 36.	+ 39 56		17.5	RICH CLUSTER OF GALAXIES
ZWG 321.028	17 38 36.	+ 68 03		15.7	GALAXY
SCHO 0532	17 38 36.	- 23 17 24.	240		ISOLATED DARK CLOUD
ISS 1033	17 38 36.	- 35 20	114		STELLAR RING
LB 01044	17 38 41.	+ 44 17 42.		16.6	FAINT BLUE STAR
ZWG 112.021	17 38 42.	+ 18 51		15.2	GALAXY
SCHO 0533	17 38 43.	- 17 21 12.	340		ISOLATED DARK CLOUD
ZWG 199.007	17 38 48.	+ 35 40		15.4	GALAXY
MCG+06-39-010	17 38 48.	+ 35 40	18	15.	GALAXY
MCG+06-39-009	17 38 48.	+ 35 40	18	15.	GALAXY
ZC 1738.8+3956	17 38 49.	+ 39 56	540		CLUSTER OF GALAXIES
1ZW 190	17 38 48.	+ 49 28			COMPACT GALAXY
ISS 1034	17 38 48.	- 34 57	233		STELLAR RING
PK003+02.1	17 38 48.4	- 24 40 42.	7	13.6	PLANETARY NEBULA
SCHO 0534	17 38 49.	- 17 40 00.	440		ISOLATED DARK CLOUD
LB 04007	17 38 53.	+ 07 47 00.		18.7	FAINT BLUE STAR
SCHO 0535	17 38 53.	- 19 24 54.	600		ISOLATED DARK CLOUD
PK003+03.2	17 38 53.6	- 24 09 51.			PLANETARY NEBULA
ZWG 141.002	17 38 54.	+ 26 10		15.6	GALAXY
ZWG 321.029	17 38 54.	+ 63 59		15.3	GALAXY
UGC 10940	17 38 54.	+ 63 59	60	15.3	GALAXY
SCHO 0536	17 38 58.	- 17 31 06.	840		ISOLATED DARK CLOUD
SCHO 0537	17 38 58.	- 19 50 00.	150		ISOLATED DARK CLOUD
KHAV 389	17 39	- 23 50	2230		DARK NEBULA
KHAV 388	17 39	- 27 02	4540		DARK NEBULA
KHAV 387	17 39	- 30 02	4940		DARK NEBULA
LB 09919	17 39	- 81 11		15.1	FAINT BLUE STAR
ZWG 112.022	17 39 00.	+ 17 23		15.4	GALAXY
UGC 10941	17 39 00.	+ 17 23	72	15.4	GALAXY S
ZC 1739.0+1726	17 39 00.	+ 17 26	3830		CLUSTER OF GALAXIES
MCG+07-36-030	17 39 00.	+ 41 18	42	15.	GALAXY
ZWG 226.037	17 39 00.	+ 41 20		14.8	GALAXY
KARA.73B 0824	17 39 00.	+ 41 20	42	14.8	ISOLATED GALAXY E
1ZW 191	17 39 00.	+ 47 46			COMPACT GALAXY
ZWG 278.014	17 39 00.	+ 51 04		15.7	GALAXY
UGC 10942	17 39 00.	+ 51 04	66	15.7	GALAXY SB
KARA.72 521B	17 39 00.	+ 51 04	60	15.7	PART OF DOUBLE GALAXY
ZWG 278.015	17 39 00.	+ 51 06		15.4	GALAXY
KARA.72 521A	17 39 00.	+ 51 06	48	15.4	PART OF DOUBLE GALAXY
MCG+11-21-017	17 39 00.	+ 68 09	27	16.	GALAXY
LDN 0154	17 39 00.	- 22 50	2820		DARK NEBULA
LDN 1769	17 39 00.	- 30 00	14340		DARK NEBULA
SCHO 0538	17 39 01.	- 24 28 00.	270		ISOLATED DARK CLOUD
LB 01045	17 39 02.	+ 60 09 12.		17.7	FAINT BLUE STAR
MCG+03-45-011	17 39 03.	+ 17 24	45	15.4	GALAXY
ZWG 141.003	17 39 06.	+ 23 04		15.5	GALAXY
MCG+09-29-015	17 39 06.	+ 51 02	60	15.	GALAXY
MCG+09-29-016	17 39 06.	+ 51 04 30.	48	15.	GALAXY
MCG+10-25-079	17 39 06.	+ 59 18	30	16.	GALAXY
ZWG 300.057	17 39 06.	+ 59 19		15.7	GALAXY
PK358-00.1	17 39 06.0	- 30 25 38.6		17.40	PLANETARY NEBULA
LB 04008	17 39 07.	+ 07 49 00.		19.3	FAINT BLUE STAR
1ZW 192	17 39 12.	+ 38 45			COMPACT GALAXY
MCG+08-32-013	17 39 12.	+ 46 37	54	16.	GALAXY
ACK 010+07.1	17 39 12.	- 15 55			PLANETARY NEBULA
SCHO 0539	17 39 16.	- 21 08 00.	650		ISOLATED DARK CLOUD
SCHO 0542	17 39 17.	- 17 30 12.	470		ISOLATED DARK CLOUD
SCHO 0541	17 39 17.	- 24 14 54.	260		ISOLATED DARK CLOUD
SCHO 0540	17 39 17.	- 34 22 36.	650		ISOLATED DARK CLOUD
ZWG 199.008	17 39 18.	+ 38 44		15.0	GALAXY
MCG+06-39-011	17 39 21.	+ 38 47	24	15.	GALAXY
B 278	17 39 23.	- 32 18	900		DARK OBJECT
ZWG 278.016	17 39 24.	+ 51 14		15.4	GALAXY
PK350-05.1	17 39 25.3	- 39 35 02.	8		PLANETARY NEBULA
ZC 1739.5+3354	17 39 30.	+ 33 54	1080		CLUSTER OF GALAXIES
MCG+09-29-017	17 39 30.	+ 51 11	54	14.	GALAXY
ISS 0742	17 39 30.	- 15 42	145		STELLAR RING
ZWG 083.018	17 39 36.	+ 11 13		15.7	GALAXY
ZWG 112.023	17 39 36.	+ 20 09		15.2	GALAXY
MCG+08-32-014	17 39 36.	+ 45 09	96	15.	GALAXY
MCG+03-45-012	17 39 39.	+ 18 20	60	15.2	GALAXY
SCHO 0543	17 39 39.	- 21 24 00.	350		ISOLATED DARK CLOUD
RNGC 6417	17 39 40.	+ 23 42		14.5	GALAXY
RNGC 6435	17 39 41.	+ 62 39		15.0	GALAXY
ZWG 055.014	17 39 42.	+ 03 12		15.5	GALAXY
UGC 10943	17 39 42.	+ 03 12	96	15.5	GALAXY S
ZWG 112.024	17 39 42.	+ 18 19		15.2	GALAXY
UGC 10944	17 39 42.	+ 18 19	60	15.2	GALAXY E
ZWG 141.004	17 39 42.	+ 23 42		14.4	GALAXY
UGC 10945	17 39 42.	+ 23 42	96	14.4	GALAXY SBb
ZWG 171.001	17 39 42.	+ 29 13		15.6	GALAXY
KARA.73B 0825	17 39 42.	+ 29 13	42	15.6	ISOLATED GALAXY S
ZWG 253.028	17 39 42.	+ 45 10		14.9	GALAXY
UGC 10946	17 39 42.	+ 45 10	90	14.9	GALAXY SBb
LB 01046	17 39 42.	+ 45 33 00.		16.1	FAINT BLUE STAR
ZWG 300.058	17 39 42.	+ 62 39		14.9	GALAXY
UGC 10947	17 39 42.	+ 62 39	66	14.9	GALAXY COMPACT
MRSL 001+01/1	17 39 42.	- 26 18	2100		HII REGION
SCHO 0544	17 39 44.	- 21 43 00.	250		ISOLATED DARK CLOUD
MCG+04-42-001	17 39 48.	+ 23 43	78	14.4	GALAXY
MCG+10-25-080	17 39 48.	+ 62 40	24	15.	GALAXY
SCHO 0545	17 39 48.	- 21 10 42.	410		ISOLATED DARK CLOUD
LB 01047	17 39 49.	+ 47 18 30.		17.4	FAINT BLUE STAR
SCHO 0546	17 39 51.	- 34 47 00.	830		ISOLATED DARK CLOUD
LB 01048	17 39 53.	+ 80 22 42.		15.6	FAINT BLUE STAR
ZWG 112.025	17 39 54.	+ 18 02		15.7	GALAXY
ZWG 199.009	17 39 54.	+ 36 00		15.5	GALAXY
7ZW 733	17 39 54.	+ 68 29			COMPACT GALAXY
MCG+12-17-003	17 39 54.	+ 68 50	45	16.	GALAXY
ZWG 340.015	17 39 54.	+ 73 32		15.7	GALAXY
ZWG 339.054	17 39 54.	+ 73 32		15.7	GALAXY
RNGC 6461	17 39 54.	+ 73 32		15.5	GALAXY
SCHO 0547	17 39 55.	- 37 25 42.	1020		ISOLATED DARK CLOUD
SCHO 0548	17 39 56.	- 21 41 54.	430		ISOLATED DARK CLOUD
KHAV 391	17 40	- 34 44	3960		DARK NEBULA
LB 09920	17 40	- 85 52		13.5	FAINT BLUE STAR
ZWG 027.004	17 40 00.	+ 00 14		15.5	GALAXY
UGC 10948	17 40 00.	+ 00 14	120	15.5	GALAXY S
MCG+00-45-001	17 40 00.	+ 00 14	54	15.5	GALAXY
ZWG 141.005	17 40 00.	+ 25 39		14.8	GALAXY
MCG+04-42-002	17 40 00.	+ 25 39	30	14.8	GALAXY
MCG+14-08-026	17 40 00.	+ 86 51	12	17.	GALAXY
LDN 0290	17 40 00.	- 16 00	4620		DARK NEBULA

OBJECT NAME	RIGHT ASCEN.	DECLINATION	DIAM.	MAGN.	TYPE OF OBJECT
LDN 0189	17 40 00.	- 21 40	3120		DARK NEBULA
LDN 0142	17 40 00.	- 23 30	4920		DARK NEBULA
LDN 0126	17 40 00.	- 24 00	2400		DARK NEBULA
ISS 0886	17 40 00.	- 26 28	113		STELLAR RING
COU 005	17 40 00.	- 27 00	10800		EMISSION NEBULA
SCHO 0549	17 40 11.	- 21 46 24.	260		ISOLATED DARK CLOUD
ZWG 112.026	17 40 12.	+ 17 01		15.5	GALAXY
MCG+03-45-013	17 40 12.	+ 17 02	36	15.5	GALAXY
SCHO 0550	17 40 13.	- 21 39 48.	240		ISOLATED DARK CLOUD
SCHO 0551	17 40 14.	- 24 35 00.	300		ISOLATED DARK CLOUD
SCHO 0552	17 40 16.	- 33 39 00.	970		ISOLATED DARK CLOUD
ZWG 356.006	17 40 18.	+ 74 52		14.9	GALAXY
ZWG 340.016	17 40 18.	+ 74 52		14.9	GALAXY
UGC 10949	17 40 18.	+ 74 52	60	14.9	GALAXY E
ISS 0887	17 40 18.	- 21 53	69		STELLAR RING
LB 01049	17 40 19.	+ 47 16 42.		17.4	FAINT BLUE STAR
RNGC 6407	17 40 20.	- 60 43			UNVERIFIED SOUTHRN OBJECT
LB 01050	17 40 21.	+ 48 07 48.		17.7	FAINT BLUE STAR
SCHO 0553	17 40 23.	- 22 11 36.	500		ISOLATED DARK CLOUD
MCG+03-45-014	17 40 24.	+ 19 32	60	15.6	GALAXY
ZWG 300.059	17 40 24.	+ 62 37		15.2	GALAXY
PK045+24.1	17 40 25.	+ 21 28	49		PLANETARY NEBULA
SCHO 0554	17 40 28.	- 17 34 00.	1240		ISOLATED DARK CLOUD
PK006+04.1	17 40 28.9	- 21 03 33.	4		PLANETARY NEBULA
ZWG 112.027	17 40 30.	+ 19 30		15.6	GALAXY
ZC 1740.5+2330	17 40 30.	+ 23 30	1280		CLUSTER OF GALAXIES
MCG+10-25-081	17 40 30.	+ 62 38	36	15.	GALAXY
SCHO 0555	17 40 35.	- 35 20 36.	310		ISOLATED DARK CLOUD
ZWG 112.028	17 40 36.	+ 16 14		15.1	GALAXY
UGC 10950	17 40 35.	+ 16 14	96	15.1	GALAXY DBL SYS
MCG+03-45-015	17 40 36.	+ 16 14	42	15.1	GALAXY
SCHO 0556	17 40 36.	+ 20 21 48.	630		ISOLATED DARK CLOUD
MCG+07-36-031	17 40 36.	+ 44 18	30	15.	GALAXY
ZWG 226.038	17 40 36.	+ 44 20		15.6	GALAXY
ZWG 300.060	17 40 36.	+ 60 29		14.9	GALAXY
RNGC 6436	17 40 36.	+ 60 29		15.0	GALAXY
UGC 10951	17 40 36.	+ 60 29	90	14.9	GALAXY Sc
ZWG 322.001	17 40 36.	+ 68 29		15.6	GALAXY
ZWG 321.030	17 40 36.	+ 68 29		15.6	GALAXY
7ZW 734	17 40 36.	+ 68 30			COMPACT GALAXY
GCL 075	17 40 36.	- 26 12	108		GLOBULAR STAR CLUSTER
MCG+03-45-016	17 40 39.	+ 16 13 30.	42	15.1	GALAXY
MCG+10-25-082	17 40 39.	+ 60 29	72	14.	GALAXY
LB 01051	17 40 41.	+ 60 31 00.		15.8	FAINT BLUE STAR
ZWG 112.029	17 40 42.	+ 19 01		15.7	GALAXY
MCG+07-36-032	17 40 42.	+ 39 19	36	14.5	GALAXY
SCHO 0557	17 40 42.	- 17 28 00.	380		ISOLATED DARK CLOUD
ZWG 226.039	17 40 48.	+ 39 21		15.3	GALAXY
ZWG 027.005	17 40 48.	- 01 40		15.4	GALAXY
MCG+00-45-002	17 40 48.	- 01 40	36	15.4	GALAXY
SCHO 0558	17 40 49.	- 21 10 00.	480		ISOLATED DARK CLOUD
SCHO 0559	17 40 53.	- 17 25 42.	410		ISOLATED DARK CLOUD
ZWG 112.030	17 40 54.	+ 19 02		15.5	GALAXY
OCL 1020	17 40 54.	- 34 51	204	13.	OPEN STAR CLUSTER
VHA 243	17 40 54.	- 34 51	360		STAR CLSTR IN NEBULOSITY
PK355-02.2	17 40 54.6	- 34 16 17.	10		PLANETARY NEBULA
SCHO 0560	17 40 57.	- 23 34 54.	270		ISOLATED DARK CLOUD
KHAV 392	17 41	- 10 01	5170		DARK NEBULA
B 279	17 41	- 22 32	3600		DARK OBJECT
ZWG 055.015	17 41 00.	+ 07 15		15.7	GALAXY
ZC 1741.0+3337	17 41 00.	+ 33 37	1480		CLUSTER OF GALAXIES
ZWG 253.029	17 41 00.	+ 50 30		15.6	GALAXY
MCG+08-32-015	17 41 00.	+ 50 30	42	16.	GALAXY
MCG+12-17-004	17 41 00.	+ 74 03	39	16.	GALAXY
LDN 0276	17 41 00.	- 17 20	840		DARK NEBULA
LDN 0186	17 41 00.	- 22 00	2220		DARK NEBULA
PK355-02.1	17 41 01.6	- 34 05 25.	10		PLANETARY NEBULA
RNGC 6415	17 41 02.	- 34 59			NON-EXISTENT OBJECT
RNGC 6416	17 41 04.	- 32 20		8.5	OPEN CLUSTER
SCHO 0561	17 41 05.	- 19 15 12.	420		ISOLATED DARK CLOUD
ZC 1741.1+5450	17 41 06.	+ 54 50	870		CLUSTER OF GALAXIES
OCL 1031	17 41 06.	- 32 20	2220	8.9	OPEN STAR CLUSTER
SG 3.109	17 41 07.	- 26 21	3000		DIFFUSE EMISSION NEBULA
MCG+11-21-018	17 41 09.	+ 66 43	54	15.	GALAXY
SCHO 0562	17 41 11.	- 17 46 00.	320		ISOLATED DARK CLOUD
ZWG 278.017	17 41 12.	+ 56 10		15.6	GALAXY
UGC 10952	17 41 12.	+ 56 10	78	15.6	GALAXY SB
ZC 1741.2+6438	17 41 12.	+ 64 38	2150		CLUSTER OF GALAXIES
ZWG 321.031	17 41 12.	+ 66 42		15.3	GALAXY
UGC 10953	17 41 12.	+ 66 42	66	15.3	GALAXY SBa-b
ZWG 340.017	17 41 12.	+ 74 03		15.6	GALAXY
ZWG 339.055	17 41 12.	+ 74 03		15.6	GALAXY
UGC 10954	17 41 12.	+ 74 03	60	15.6	GALAXY Sa-b
SCHO 0563	17 41 12.	- 23 26 54.	350		ISOLATED DARK CLOUD
ISS 1067	17 41 12.	- 41 15	470		STELLAR RING
SCHO 0564	17 41 13.	- 19 44 06.	470		ISOLATED DARK CLOUD
SCHO 0565	17 41 13.	- 19 44 30.	630		ISOLATED DARK CLOUD
LB 01052	17 41 16.	+ 44 31 42.		15.9	FAINT BLUE STAR
ZWG 278.018	17 41 18.	+ 56 00		15.6	GALAXY
UGC 10955	17 41 18.	+ 56 00	60	15.6	GALAXY Sc
MCG+09-29-018	17 41 18.	+ 56 10	48	16.	GALAXY
ISS 0888	17 41 18.	- 23 45	240		STELLAR RING
CED 150	17 41 18.	- 33 43	420		DIFFUSE GALACTIC NEBULA
OCL 1027	17 41 18.	- 33 44	360	10.9	OPEN STAR CLUSTER
ZWG 055.016	17 41 24.	+ 04 34		15.7	GALAXY
UGC 10956	17 41 24.	+ 04 34	90	15.7	GALAXY Sb-c
MCG+01-45-003	17 41 24.	+ 04 34	48	15.7	GALAXY
ZWG 112.031	17 41 24.	+ 20 45		15.7	GALAXY
MCG+09-29-019	17 41 24.	+ 56 00	54	15.	GALAXY
7ZW 735	17 41 24.	+ 67 56			COMPACT GALAXY
ISS 0889	17 41 24.	- 26 00	80		STELLAR RING
MIL 54	17 41 24.	- 29 24 00.	402		SUPERNOVA REMNANT
ARC 2279	17 41 30.	+ 24 46		17.3	RICH CLUSTER OF GALAXIES
ZC 1741.5+4945	17 41 30.	+ 49 45	1340		CLUSTER OF GALAXIES
ZWG 278.019	17 41 30.	+ 55 50		15.7	GALAXY
LDN 0273	17 41 30.	- 17 35	300		DARK NEBULA
RNGC 6427	17 41 35.	+ 25 31		14.5	GALAXY
ZC 1741.6+2453	17 41 36.	+ 24 53	2350		CLUSTER OF GALAXIES
ZWG 141.006	17 41 36.	+ 25 31		14.6	GALAXY
UGC 10957	17 41 36.	+ 25 31	90	14.6	GALAXY E-S0
MCG+04-42-003	17 41 36.	+ 25 31	78	14.6	GALAXY
ZWG 226.040	17 41 36.	+ 39 11		15.5	GALAXY
ZC 1741.6+5836	17 41 36.	+ 58 36	1550		CLUSTER OF GALAXIES
MCG+12-17-005	17 41 36.	+ 68 58	33	16.	GALAXY
MRSL 002+01/1	17 41 36.	- 25 59	2700		HII REGION
SCHO 0566	17 41 38.	- 23 25 48.	270		ISOLATED DARK CLOUD
PK352-04.1	17 41 40.7	- 38 07 37.	6		PLANETARY NEBULA
MCG+06-39-012	17 41 42.	+ 33 05 30.	30	15.	GALAXY

OBJECT NAME	RIGHT ASCEN.	DECLINATION	DIAM.	MAGN.	TYPE OF OBJECT
ZWG 226.041	17 41 42.	+ 39 15		15.5	GALAXY
MRSL 360+00/1	17 41 42.	- 28 49	480		HII REGION
LB 03871	17 41 45.	+ 05 41 24.		20.2	FAINT BLUE STAR
HN 0097	17 41 46.	- 44 53			NEBULA
PK346-08.1	17 41 46.	- 44 54	14	13.1	PLANETARY NEBULA
IC 4663	17 41 46.	- 44 54			NONSTELLAR OBJECT
LB 01053	17 41 47.	+ 46 48 00.		16.0	FAINT BLUE STAR
PK351-04.1	17 41 47.	- 38 16 13.			PLANETARY NEBULA
LB 03872	17 41 48.	+ 05 48 18.		19.7	FAINT BLUE STAR
ZWG 199.010	17 41 48.	+ 33 05		15.7	GALAXY
MCG+06-39-013	17 41 48.	+ 33 06 30.	15	15.5	GALAXY
UGC 10958	17 41 48.	+ 66 53	90	16.5	GALAXY DWRF SP
ACK 011+07.1	17 41 48.	- 15 44			PLANETARY NEBULA
SCHO 0567	17 41 50.	- 19 43 18.	300		ISOLATED DARK CLOUD
SCHO 0568	17 41 50.	- 37 59 12.	820		ISOLATED DARK CLOUD
SCHO 0569	17 41 52.	- 19 27 24.	280		ISOLATED DARK CLOUD
RNGC 6428	17 41 53.	+ 25 34			NON-EXISTENT OBJECT
ZWG 199.011	17 41 54.	+ 33 06		15.7	GALAXY
MCG+06-39-014	17 41 54.	+ 33 06	30	15.	GALAXY
LB 03873	17 41 55.	+ 06 01 24.		18.2	FAINT BLUE STAR
PK345-08.1	17 41 55.	- 46 04	10		PLANETARY NEBULA
IC 1266	17 41 55.	- 46 05			NONSTELLAR OBJECT
SCHO 0570	17 41 56.	- 19 34 06.	270		ISOLATED DARK CLOUD
BC 4C27.38	17 41 57.8	+ 27 54 02.		17.7	QUASI-STELLAR OBJECT
SHB 344	17 41 58.	+ 27 54 02.		19.	QUASI-STELLAR OBJECT
B 280	17 41 58.	- 20 42	3600		DARK OBJECT
SCHO 0571	17 41 58.	- 23 16 36.	260		ISOLATED DARK CLOUD
RNGC 6429	17 41 59.	+ 25 23		14.5	GALAXY
KHAV 395	17 42	- 16 25	4540		DARK NEBULA
KHAV 393	17 42	- 22 25	4880		DARK NEBULA
LBN 0001	17 42	- 28 50	240		BRIGHT NEBULA
KHAV 394	17 42	- 38 01	2500		DARK NEBULA
ZWG 141.007	17 42 00.	+ 25 23		14.3	GALAXY
UGC 10960	17 42 00.	+ 25 23	138	14.3	GALAXY SBa
MCG+04-42-004	17 42 00.	+ 25 23	78	14.3	GALAXY
ZWG 199.012	17 42 00.	+ 34 36		15.6	GALAXY
KARA.73B 0826	17 42 00.	+ 34 36	18	15.5	ISOLATED GALAXY E
MCG+10-25-083	17 42 00.	+ 58 48	60	15.	GALAXY
7ZW 736	17 42 00.	+ 66 33			COMPACT GALAXY
7ZW 737	17 42 00.	+ 67 59			COMPACT GALAXY
UGC 10959	17 42 00.	+ 84 40	72	16.5	GALAXY
LDN 0393	17 42 00.	- 07 10	18120		DARK NEBULA
LDN 0258	17 42 00.	- 18 50	1680		DARK NEBULA
LDN 0246	17 42 00.	- 19 30	12840		DARK NEBULA
LDN 0207	17 42 00.	- 21 00	5700		DARK NEBULA
LDN 0160	17 42 00.	- 23 10	900		DARK NEBULA
LDN 0130	17 42 00.	- 24 10	2160		DARK NEBULA
SCHO 0572	17 42 01.	- 38 14 30.	430		ISOLATED DARK CLOUD
PK351-05.1	17 42 04.	- 38 38 14.	10		PLANETARY NEBULA
LB 03874	17 42 05.	+ 05 31 42.		16.5	FAINT BLUE STAR
ZC 1742.1+3306	17 42 06.	+ 33 06	2290		CLUSTER OF GALAXIES
ZWG 300.061	17 42 06.	+ 58 50		15.7	GALAXY
UGC 10961	17 42 06.	+ 58 50	66	15.7	GALAXY Sc
LDN 0233	17 42 06.	- 20 00	120		DARK NEBULA
IC 4662	17 42 06.	- 64 39	144	11.7	GALAXY IRR
SCHO 0573	17 42 10.	- 33 55 54.	670		ISOLATED DARK CLOUD
SCHO 0574	17 42 11.	- 23 40 18.	240		ISOLATED DARK CLOUD
LB 03876	17 42 12.	+ 05 40 42.		17.4	FAINT BLUE STAR
LB 03875	17 42 12.	+ 06 02 30.		17.0	FAINT BLUE STAR
ZC 1742.2+2345	17 42 12.	+ 23 45	6720		CLUSTER OF GALAXIES
ZWG 199.013	17 42 12.	+ 36 48		14.1	GALAXY
UGC 10962	17 42 12.	+ 36 48	138	14.1	GALAXY Sb
ZWG 199.014	17 42 12.	+ 36 52		15.2	GALAXY
ZWG 253.030	17 42 12.	+ 49 07		15.4	GALAXY
SCHO 0575	17 42 12.	- 23 41 48.	320		ISOLATED DARK CLOUD
RNGC 6433	17 42 14.	+ 36 48		14.0	GALAXY
PK328-17.1	17 42 15.	- 64 37 20.	10		PLANETARY NEBULA
LB 03877	17 42 16.	+ 05 17 42.		16.8	FAINT BLUE STAR
LB 03878	17 42 17.	+ 05 21 54.		20.2	FAINT BLUE STAR
RNGC 6431	17 42 17.	+ 25 32			NON-EXISTENT OBJECT
LB 03879	17 42 18.	+ 05 50 48.		16.3	FAINT BLUE STAR
MCG+06-39-015	17 42 18.	+ 36 49	114	13.	GALAXY
ZWG 322.002	17 42 18.	+ 67 30		15.4	GALAXY
ZWG 321.032	17 42 18.	+ 67 30		15.4	GALAXY
ISS 0743	17 42 18.	- 19 37	804		STELLAR RING
ISS 0890	17 42 18.	- 23 09	62		STELLAR RING
RNGC 6426	17 42 21.	+ 03 12		12.5	GLOBULAR CLUSTER
B 083a	17 42 21.	- 19 59	300		DARK OBJECT
SCHO 0576	17 42 21.	- 38 40 36.	520		ISOLATED DARK CLOUD
REIN 2.259	17 42 21.82	+ 03 10 59.5			NEBULA
SCHO 0577	17 42 22.	- 20 00 36.	340		ISOLATED DARK CLOUD
REIN 2.260	17 42 22.05	+ 03 11 44.9			NEBULA
SCHO 0578	17 42 23.	- 19 42 48.	270		ISOLATED DARK CLOUD
OCL 0081	17 42 24.	+ 03 11	138	12.9	OPEN STAR CLUSTER
GCL 076	17 42 24.	+ 03 11	1260	12.33	GLOBULAR STAR CLUSTER
RNGC 6430	17 42 24.	+ 18 11			NON-EXISTENT OBJECT
MCG+07-36-033	17 42 24.	+ 40 52	27	15.5	GALAXY
ZWG 322.003	17 42 24.	+ 68 22		15.3	GALAXY
ZWG 321.033	17 42 24.	+ 68 22		15.3	GALAXY
UGC 10963	17 42 24.	+ 68 22	108	15.3	GALAXY SBb
REIN 2.261	17 42 25.11	+ 03 11 22.6			NEBULA
PK355-02.4	17 42 25.6	- 34 29 47.	5		PLANETARY NEBULA
REIN 2.262	17 42 26.34	+ 03 11 58.3			NEBULA
SCHO 0579	17 42 27.	- 23 12 06.	290		ISOLATED DARK CLOUD
RNGC 6421	17 42 27.	- 33 40			NON-EXISTENT OBJECT
SCHO 0580	17 42 27.	- 38 23 12.	410		ISOLATED DARK CLOUD
REIN 2.263	17 42 27.79	+ 03 11 26.9			NEBULA
ZWG 112.032	17 42 30.	+ 19 05		15.7	GALAXY
KARA.72 522A	17 42 30.	+ 19 05	24	15.7	PART OF DOUBLE GALAXY
1ZW 193	17 42 30.	+ 38 05			COMPACT GALAXY
ZWG 226.042	17 42 30.	+ 40 53		15.0	GALAXY
ZWG 253.031	17 42 30.	+ 48 03		15.1	GALAXY
MCG+11-21-020	17 42 30.	+ 67 31 30.	45	15.	GALAXY
7ZW 738	17 42 30.	+ 68 23			COMPACT GALAXY
MCG+11-21-019	17 42 30.	+ 68 23 30.	78	15.	GALAXY
LDN 0277	17 42 30.	- 17 28	240		DARK NEBULA
MRSL 003+02/1	17 42 30.	- 24 54	420		HII REGION
SCHO 0581	17 42 31.	- 21 50 00.	530		ISOLATED DARK CLOUD
SCHO 0582	17 42 32.	- 23 28 30.	260		ISOLATED DARK CLOUD
PK006+04.2	17 42 32.4	- 20 56 52.	4		PLANETARY NEBULA
PK002+01.1	17 42 33.6	- 25 38 49.	4		PLANETARY NEBULA
RNGC 6456	17 42 34.	+ 67 36		15.5	GALAXY
PK005+03.1	17 42 34.8	- 23 01 17.	15		PLANETARY NEBULA
ZWG 083.019	17 42 36.	+ 14 13		15.4	GALAXY
ZWG 112.033	17 42 36.	+ 19 04		15.7	GALAXY
KARA.72 522B	17 42 36.	+ 19 04	18	15.7	PART OF DOUBLE GALAXY
ZWG 199.015	17 42 35.	+ 38 05		15.5	GALAXY
ZWG 322.004	17 42 36.	+ 67 36		15.7	GALAXY

OBJECT NAME	RIGHT ASCEN.	DECLINATION	DIAM.	MAGN.	TYPE OF OBJECT
ZWG 321.034	17 42 36.	+ 67 36		15.7	GALAXY
SCHO 0583	17 42 36.	- 23 51 00.	260		ISOLATED DARK CLOUD
MCG+06-39-016	17 42 39.	+ 38 07	21	15.	GALAXY
SCHO 0584	17 42 39.	- 36 08 36.	850		ISOLATED DARK CLOUD
ZWG 112.034	17 42 42.	+ 19 33		15.5	GALAXY
ISS 0744	17 42 42.	- 15 04	107		STELLAR RING
LB 03880	17 42 43.	+ 05 52 48.		20.5	FAINT BLUE STAR
PK358-00.2	17 42 45.0	- 30 10 52.	5	10.6	PLANETARY NEBULA
ARC 2280	17 42 47.	+ 63 47		17.9	RICH CLUSTER OF GALAXIES
PK355-02.3	17 42 47.4	- 34 02 37.	5		PLANETARY NEBULA
ARC 2281	17 42 48.	+ 64 41		17.6	RICH CLUSTER OF GALAXIES
ZWG 322.005	17 42 48.	+ 66 30		14.7	GALAXY
ZWG 321.035	17 42 48.	+ 66 30		14.7	GALAXY
RNGC 6457	17 42 48.	+ 66 30		14.5	GALAXY
UGC 10964	17 42 48.	+ 66 30	78	14.7	GALAXY E-SO
ISS 0745	17 42 48.	- 18 16	81		STELLAR RING
PK358-01.1	17 42 48.8	- 31 02 30.	14	16.	PLANETARY NEBULA
SCHO 0585	17 42 49.	- 23 39 00.	260		ISOLATED DARK CLOUD
PK001+00.1	17 42 50.4	- 26 56 59.1		15.69	PLANETARY NEBULA
RNGC 6449	17 42 52.	+ 56 49		14.5	GALAXY
SCHO 0587	17 42 52.	- 34 26 36.	380		ISOLATED DARK CLOUD
SCHO 0586	17 42 52.	- 34 39 24.	520		ISOLATED DARK CLOUD
ZWG 083.020	17 42 54.	+ 14 50		15.7	GALAXY
MCG+08-32-016	17 42 54.	+ 50 34	24	17.	GALAXY
ZWG 278.020	17 42 54.	+ 56 49		14.6	GALAXY
UGC 10965	17 42 54.	+ 56 49	66	14.6	GALAXY Sc
MCG+09-29-020	17 42 54.	+ 56 49 30.	60	13.	GALAXY
ZWG 300.062	17 42 54.	+ 62 50		15.7	GALAXY
MCG+11-21-021	17 42 54.	+ 66 30	27	15.	GALAXY
ISS 0746	17 42 54.	- 18 18	63		STELLAR RING
ISS 0747	17 42 54.	- 19 36	350		STELLAR RING
ISS 0748	17 42 54.	- 21 08	64		STELLAR RING
SCHO 0588	17 42 55.	- 19 55 48.	340		ISOLATED DARK CLOUD
SCHO 0590	17 42 56.	- 23 42 00.	200		ISOLATED DARK CLOUD
SCHO 0589	17 42 56.	- 34 16 00.	440		ISOLATED DARK CLOUD
LB 03881	17 42 59.	+ 05 21 54.		19.8	FAINT BLUE STAR
PK051+25.1	17 42 59.	+ 27 21	43		PLANETARY NEBULA
LBN 0009	17 43	- 26 10	300		BRIGHT NEBULA
LBN 1124	17 43	- 29 20	720		BRIGHT NEBULA
MCG+03-45-018	17 43 00.	+ 15 30	30	17.	GALAXY
MCG+03-45-017	17 43 00.	+ 15 31	18	16.	GALAXY
ZWG 112.035	17 43 00.	+ 18 09		14.8	GALAXY
UGC 10966	17 43 00.	+ 18 09	108	14.8	GALAXY Sa-b
ZWG 112.036	17 43 00.	+ 20 35		15.6	GALAXY
ZWG 253.032	17 43 00.	+ 50 34		15.5	GALAXY
MCG+08-32-017	17 43 00.	+ 50 34	42	16.	GALAXY
ZWG 340.018	17 43 00.	+ 69 43		15.4	GALAXY
LDN 0352	17 43 00.	- 12 00	9300		DARK NEBULA
LDN 0082	17 43 00.	- 26 00	6060		DARK NEBULA
COU 006	17 43 00.	- 29 20	1500		EMISSION NEBULA
SCHO 0591	17 43 00.	- 35 32 48.	490		ISOLATED DARK CLOUD
LB 03882	17 43 03.	+ 05 34 00.		19.5	FAINT BLUE STAR
SCHO 0592	17 43 03.	- 37 02 36.	830		ISOLATED DARK CLOUD
SCHO 0593	17 43 05.	- 23 17 06.	250		ISOLATED DARK CLOUD
ZC 1743.1+6344	17 43 06.	+ 63 44	1410		CLUSTER OF GALAXIES
ISS 0891	17 43 06.	- 24 03	64		STELLAR RING
LB 03883	17 43 07.	+ 05 32 24.		19.3	FAINT BLUE STAR
ZWG 253.033	17 43 12.	+ 48 07		14.8	GALAXY
UGC 10967	17 43 12.	+ 48 07	78	14.8	GALAXY Sa-b
MCG+08-32-018	17 43 12.	+ 48 07	72	15.	GALAXY
OCL 1039	17 43 12.	- 29 17	240	10.5	OPEN STAR CLUSTER
VHA 244	17 43 12.	- 29 17	300		STAR CLSTR IN NEBULOSITY
VHA 245	17 43 12.	- 29 41	120		OPEN STAR CLUSTER
ISS 0981	17 43 12.	- 33 00	99		STELLAR RING
RNGC 6443	17 43 14.	+ 48 07		15.0	GALAXY
SCHO 0594	17 43 14.	- 21 09 00.	300		ISOLATED DARK CLOUD
LB 03884	17 43 16.	+ 06 03 18.		15.2	FAINT BLUE STAR
RNGC 6448	17 43 16.	+ 53 34			NON-EXISTENT OBJECT
LB 03885	17 43 17.	+ 05 51 12.		15.5	FAINT BLUE STAR
PK008+05.1	17 43 17.	- 18 39			PLANETARY NEBULA
LB 03886	17 43 18.	+ 05 36 36.		17.6	FAINT BLUE STAR
MCG+03-45-019	17 43 18.	+ 18 11	120	14.8	GALAXY
UGC 10968	17 43 18.	+ 58 01	60	16.0	GALAXY Sb-c
SCHO 0595	17 43 22.	- 17 14 00.	590		ISOLATED DARK CLOUD
PK007+04.1	17 43 23.	- 20 13	10	14.7	PLANETARY NEBULA
MIN.47 09	17 43 23.	- 29 22			DIFFUSE NEBULA
ZWG 171.002	17 43 24.	+ 28 38		15.7	GALAXY
MCG+09-29-021	17 43 24.	+ 51 37	42	16.	GALAXY
MCG+10-25-084	17 43 24.	+ 58 00	60	16.	GALAXY
MRSL 359-00/1	17 43 24.	- 29 17	720		HII REGION
B 084	17 43 27.	- 20 14	1320		DARK OBJECT
PK356-02.2	17 43 28.2	- 33 07 30.	5		PLANETARY NEBULA
LB 03887	17 43 29.	+ 05 35 42.		17.2	FAINT BLUE STAR
SCHO 0596	17 43 29.	- 19 14 00.	2370		ISOLATED DARK CLOUD
ZWG 141.008	17 43 30.	+ 22 15		15.7	GALAXY
MCG+09-29-022	17 43 30.	+ 53 56	42	16.	GALAXY
ZWG 278.021	17 43 30.	+ 55 51		15.4	GALAXY
ZWG 322.006	17 43 30.	+ 68 46		15.4	GALAXY
ZWG 321.036	17 43 30.	+ 68 46		15.4	GALAXY
LDN 0235	17 43 30.	- 20 10	780		DARK NEBULA
ISS 0892	17 43 30.	- 22 41	99		STELLAR RING
LB 03888	17 43 32.	+ 06 05 24.		17.4	FAINT BLUE STAR
ARC 2287	17 43 32.	+ 79 36		17.5	RICH CLUSTER OF GALAXIES
ZWG 199.016	17 43 36.	+ 38 55		15.6	GALAXY
UGC 10969	17 43 36.	+ 38 55	90	15.6	GALAXY S
MCG+09-29-023	17 43 36.	+ 55 50	48	15.	GALAXY
ISS 0749	17 43 36.	- 18 22	97		STELLAR RING
LB 03889	17 43 38.	+ 05 40 06.		17.5	FAINT BLUE STAR
ARC 2282	17 43 38.	+ 71 50		17.4	RICH CLUSTER OF GALAXIES
MCG+06-39-017	17 43 39.	+ 38 57	90	14.5	GALAXY
LB 01054	17 43 39.	+ 43 43 42.		15.8	FAINT BLUE STAR
MCG+12-17-006	17 43 39.	+ 68 48	18	16.	GALAXY
RNGC 6463	17 43 40.	+ 67 37		15.0	GALAXY
RNGC 6425	17 43 41.	- 31 30		9.0	OPEN CLUSTER
ZWG 141.009	17 43 42.	+ 24 23		15.7	GALAXY
UGC 10970	17 43 42.	+ 24 23	78	15.7	GALAXY Sa-b
ZWG 322.007	17 43 42.	+ 67 37		15.2	GALAXY
ZWG 321.037	17 43 42.	+ 67 37		15.2	GALAXY
MCG+11-21-022	17 43 42.	+ 67 38 30.	18	16.	GALAXY
OCL 1033	17 43 42.	- 31 31	1260	9.2	OPEN STAR CLUSTER
VHA 246	17 43 42.	- 31 31	540		OPEN STAR CLUSTER
B 281	17 43 44.	- 23 42			DARK OBJECT
MCG+09-29-024	17 43 45.	+ 56 31	36	16.	GALAXY
SCHO 0597	17 43 45.	- 23 45 42.	330		ISOLATED DARK CLOUD
LB 03890	17 43 46.	+ 05 47 48.		19.3	FAINT BLUE STAR
LB 01055	17 43 47.	+ 46 17 12.		16.3	FAINT BLUE STAR
OCL 0085	17 43 48.	+ 05 44	3600	5.4	OPEN STAR CLUSTER
ZC 1743.8+5528	17 43 48.	+ 55 28	4570		CLUSTER OF GALAXIES

OBJECT NAME	RIGHT ASCEN.	DECLINATION	DIAM.	MAGN.	TYPE OF OBJECT
ZWG 278.022	17 43 48.	+ 56 30		15.3	GALAXY
PK000-00.2	17 43 51.8	- 28 59 55.2		14.03	PLANETARY NEBULA
LB 01056	17 43 53.	+ 45 58 30.		17.1	FAINT BLUE STAR
ZWG 226.043	17 43 54.	+ 40 31		15.6	GALAXY
ZWG 278.023	17 43 54.	+ 55 21		15.0	GALAXY
UGC 10971	17 43 54.	+ 55 21	66	15.0	GALAXY Sa
MCG+09-29-025	17 43 54.	+ 55 22	66	15.	GALAXY
SCHO 0598	17 43 56.	- 17 47 12.	740		ISOLATED DARK CLOUD
SCHO 0599	17 43 57.	- 36 47 12.	350		ISOLATED DARK CLOUD
SCHO 0600	17 43 58.	- 35 25 24.	410		ISOLATED DARK CLOUD
KHAV 401	17 44	+ 02 35	17240		DARK NEBULA
KHAV 400	17 44	- 06 07	3940		DARK NEBULA
KHAV 399	17 44	- 14 01	6230		DARK NEBULA
KHAV 398	17 44	- 18 25	5760		DARK NEBULA
KHAV 397	17 44	- 19 07	4380		DARK NEBULA
KHAV 396	17 44	- 25 07	7370		DARK NEBULA
LBN 0003	17 44	- 28 45	120		BRIGHT NEBULA
LBN 1125	17 44	- 29 12	120		BRIGHT NEBULA
ZWG 171.003	17 44 00.	+ 32 54		15.3	GALAXY
MCG+05-42-001	17 44 00.	+ 32 55	18	15.3	GALAXY
ZWG 278.024	17 44 00.	+ 55 43		14.6	GALAXY
MCG+09-29-026	17 44 00.	+ 55 43	54	13.	GALAXY
ZC 1744.0+5625	17 44 00.	+ 56 25	1410		CLUSTER OF GALAXIES
LDN 0170	17 44 00.	- 23 00	2760		DARK NEBULA
SCHO 0601	17 44 00.	- 35 43 36.	450		ISOLATED DARK CLOUD
SCHO 0602	17 44 00.	- 36 26 42.	600		ISOLATED DARK CLOUD
PK000-00.1	17 44 00.0	- 28 45 22.5		16.83	PLANETARY NEBULA
PK011+06.1	17 44 01.	- 16 16 20.	6		PLANETARY NEBULA
LB 03891	17 44 02.	+ 06 06 06.		17.0	FAINT BLUE STAR
RNGC 6454	17 44 02.	+ 55 43		14.5	GALAXY
RNGC 6477	17 44 04.	+ 67 39			GALAXY
IC 4665	17 44 06.	+ 05 44	3000		OPEN CLUSTER
ZWG 112.037	17 44 06.	+ 18 38		15.5	GALAXY
MCG+03-45-020	17 44 06.	+ 18 39	42	15.5	GALAXY
7ZW 739	17 44 06.	+ 64 24			COMPACT GALAXY
OCL 1037	17 44 06.	- 29 35	516	12.	OPEN STAR CLUSTER
LB 03892	17 44 07.	+ 05 56 06.		14.6	FAINT BLUE STAR
LB 03893	17 44 08.	+ 06 00 42.		16.6	FAINT BLUE STAR
PK356-02.1	17 44 08.6	- 32 49 03.	7		PLANETARY NEBULA
HN 0448	17 44 10.	- 63 14			NEBULA
LB 03896	17 44 11.	+ 05 24 24.		18.0	FAINT BLUE STAR
LB 03895	17 44 11.	+ 05 51 36.		15.7	FAINT BLUE STAR
LB 03894	17 44 11.	+ 05 58 18.		15.4	FAINT BLUE STAR
ZWG 199.017	17 44 12.	+ 37 17		15.7	GALAXY
KARA.73B 0827	17 44 12.	+ 37 17	18	15.7	ISOLATED GALAXY E
MCG+10-25-085	17 44 12.	+ 61 56	21	16.	GALAXY
MCG+12-17-007	17 44 12.	+ 70 43	33	17.	GALAXY
LDN 0425	17 44 12.	- 04 40	480		DARK NEBULA
IC 4664	17 44 13.	- 63 15			NONSTELLAR OBJECT
LB 03897	17 44 16.	+ 05 49 06.		16.1	FAINT BLUE STAR
RNGC 6462	17 44 16.	+ 61 55		14.5	GALAXY
RNGC 6471	17 44 16.	+ 67 36		15.5	GALAXY
RNGC 6470	17 44 16.	+ 67 36		15.5	GALAXY
SCHO 0603	17 44 16.	- 21 08 00.	300		ISOLATED DARK CLOUD
MIN.48 03	17 44 16.	- 28 42			DIFFUSE NEBULA
LB 03898	17 44 18.	+ 05 18 06.		16.3	FAINT BLUE STAR
ZWG 141.010	17 44 18.	+ 26 34		14.5	GALAXY
UGC 10972	17 44 18.	+ 26 34	156	14.5	GALAXY Sc
KARA.73B 0828	17 44 18.	+ 26 34	162	14.5	ISOLATED GALAXY S
ZWG 199.018	17 44 18.	+ 35 35		15.5	GALAXY
KARA.72 523A	17 44 18.	+ 35 35	36	15.5	PART OF DOUBLE GALAXY
ZWG 226.044	17 44 18.	+ 39 43		15.2	GALAXY
MCG+09-29-027	17 44 18.	+ 55 37 30.	48	16.	GALAXY
7ZW 740	17 44 18.	+ 61 55			COMPACT GALAXY
ZWG 300.063	17 44 18.	+ 61 55		14.7	GALAXY
KARA.73B 0829	17 44 18.	+ 61 55	24	14.7	ISOLATED GALAXY E
ZWG 322.008	17 44 18.	+ 67 36		15.4	GALAXY
ZWG 321.038	17 44 18.	+ 67 36		15.4	GALAXY
UGC 10973	17 44 18.	+ 67 36	84	15.4	GALAXY Sc
MRSL 360-00/1	17 44 18.	- 29 13	180		HII REGION
RNGC 6446	17 44 19.	+ 35 35		15.5	GALAXY
LB 03900	17 44 20.	+ 05 38 30.		18.8	FAINT BLUE STAR
LB 03899	17 44 20.	+ 05 46 12.		17.2	FAINT BLUE STAR
MCG+11-21-024	17 44 21.	+ 67 37 30.	12	16.	GALAXY
MCG+11-21-023	17 44 21.	+ 67 37 30.	78	16.	GALAXY
SCHO 0604	17 44 21.	- 34 25 30.	550		ISOLATED DARK CLOUD
RNGC 6472	17 44 22.	+ 67 38		15.0	GALAXY
RNGC 6432	17 44 22.	- 24 52			NON-EXISTENT OBJECT
MCG+04-42-005	17 44 24.	+ 26 35	138	14.5	GALAXY
1ZW 194	17 44 24.	+ 39 43			COMPACT GALAXY
ZWG 322.009	17 44 24.	+ 67 38		15.2	GALAXY
ZWG 321.039	17 44 24.	+ 67 38		15.2	GALAXY
UGC 10974	17 44 24.	+ 67 38	72	15.2	GALAXY SBb
MCG+11-21-025	17 44 24.	+ 67 39 30.	57	16.	GALAXY
OCL 1034	17 44 24.	- 30 06	234	14.	OPEN STAR CLUSTER
VHA 247	17 44 24.	- 30 06	180		OPEN STAR CLUSTER
SCHO 0605	17 44 25.	- 33 46 06.	830		ISOLATED DARK CLOUD
MCG+06-39-018	17 44 27.	+ 35 35	18	15.	GALAXY
LB 01057	17 44 29.	+ 46 05 54.		16.2	FAINT BLUE STAR
ZWG 199.019	17 44 30.	+ 35 35		13.8	GALAXY
UGC 10975	17 44 30.	+ 35 35	102	13.8	GALAXY SB?
KARA.72 523B	17 44 30.	+ 35 35	78	13.8	PART OF DOUBLE GALAXY
ZC 1744.5+3846	17 44 30.	+ 38 46	11290		CLUSTER OF GALAXIES
PK355-03.1	17 44 30.4	- 34 07 04.	10		PLANETARY NEBULA
LB 03901	17 44 31.	+ 06 07 36.		14.3	FAINT BLUE STAR
RNGC 6447	17 44 31.	+ 35 35		14.0	GALAXY
HN 0447	17 44 32.	- 74 00			NEBULA
IC 4661	17 44 32.	- 74 01			NONSTELLAR OBJECT
MCG+06-39-019	17 44 33.	+ 35 35 30.	90	13.5	GALAXY
PK006+03.1	17 44 33.5	- 21 46 23.	6		PLANETARY NEBULA
MCG+03-45-021	17 44 36.	+ 20 48	96	14.5	GALAXY
ZWG 171.004	17 44 36.	+ 29 03		15.5	GALAXY
ZWG 171.005	17 44 36.	+ 30 43		14.4	GALAXY
UGC 10977	17 44 36.	+ 30 43	66	14.4	GALAXY S
UGC 10976	17 44 36.	+ 30 43	54	14.4	GALAXY SB
KARA.72 524A	17 44 36.	+ 30 43	54	14.4	PART OF DOUBLE GALAXY
LB 01058	17 44 36.	+ 44 20 54.		17.6	FAINT BLUE STAR
MCG+09-29-028	17 44 36.	+ 55 34	30	16.	GALAXY
PK006+03.2	17 44 37.4	- 22 05 19.	48		PLANETARY NEBULA
MCG+05-42-002	17 44 39.	+ 30 44	54	14.4	GALAXY
ZWG 112.038	17 44 42.	+ 20 47		14.5	GALAXY
UGC 10978	17 44 42.	+ 20 47	108	14.5	GALAXY E
KARA.72 524B	17 44 42.	+ 30 43	54		PART OF DOUBLE GALAXY
MCG+05-40-003	17 44 42.	+ 30 43 30.	48	14.4	GALAXY
MCG+11-21-026	17 44 42.	+ 66 32	36	17.	GALAXY
PK359-00.1	17 44 43.9	- 29 58 39.		13.6	PLANETARY NEBULA
RNGC 6442	17 44 44.	+ 20 47		14.5	GALAXY
1ZW 195	17 44 48.	+ 47 12			COMPACT GALAXY

OBJECT NAME	RIGHT ASCEN.	DECLINATION	DIAM.	MAGN.	TYPE OF OBJECT
ZWG 322.010	17 44 48.	+ 66 34		15.7	GALAXY
ZWG 321.040	17 44 48.	+ 66 34		15.7	GALAXY
MCG+03-45-022	17 44 51.	+ 20 54	60	15.1	GALAXY
ARC 2283	17 44 51.	+ 69 41		17.4	RICH CLUSTER OF GALAXIES
LB 03902	17 44 52.	+ 05 51 46.		18.6	FAINT BLUE STAR
ZC 1744.9+3836	17 44 54.	+ 38 36	1410		CLUSTER OF GALAXIES
ZWG 278.025	17 44 54.	+ 55 47		15.3	GALAXY
MCG+09-29-029	17 44 54.	+ 55 47 30.	36	16.	GALAXY
MCG+11-21-027	17 44 54.	+ 66 36	15	17.	GALAXY
ISS 0893	17 44 54.	- 23 29	192		STELLAR RING
MRSL 000-00/2	17 44 54.	- 29 06	840		HII REGION
SCHO 0606	17 44 54.	- 33 35 48.	620		ISOLATED DARK CLOUD
RNGC 6459	17 44 56.	+ 55 47		16.0	GALAXY
KHAV 402	17 45	- 04 25	4020		DARK NEBULA
LBN 0002	17 45	- 29 10	720		BRIGHT NEBULA
LB 09921	17 45	- 84 17		14.8	FAINT BLUE STAR
ZWG 112.039	17 45 00.	+ 20 53		15.1	GALAXY
UGC 10979	17 45 00.	+ 20 53	60	15.1	GALAXY Sc
MCG+07-36-034	17 45 00.	+ 41 40	24	15.	GALAXY
ZWG 226.045	17 45 00.	+ 41 42		15.7	GALAXY
1ZW 196	17 45 00.	+ 47 26			COMPACT GALAXY
LDN 0428	17 45 00.	- 04 30	2460		DARK NEBULA
LDN 0382	17 45 00.	- 09 00	1860		DARK NEBULA
LDN 0034	17 45 00.	- 28 00	20280		DARK NEBULA
TER 05	17 45 00.08	- 24 45 52.1	70		STAR CLUSTER
SG 3.110	17 45 01.	- 22 31	36000		DIFFUSE EMISSION NEBULA
LB 01059	17 45 03.	+ 43 40 06.		17.2	FAINT BLUE STAR
SCHO 0607	17 45 04.	- 23 12 24.	230		ISOLATED DARK CLOUD
PK001-00.1	17 45 05.7	- 27 59 49.7		10.38	PLANETARY NEBULA
PK005+02.1	17 45 05.9	- 22 45 50.			PLANETARY NEBULA
MCG+10-25-086	17 45 06.	+ 57 16	18	16.	GALAXY
ZWG 300.064	17 45 06.	+ 57 18		15.3	GALAXY
UGC 10980	17 45 06.	+ 57 18	60	15.3	GALAXY COMPACT
MCG+11-21-028	17 45 06.	+ 66 34	39	17.	GALAXY
MCG+11-21-029	17 45 06.	+ 66 34 30.	45	17.	GALAXY
BIGO 542	17 45 07.	+ 55 48			NEBULA
IC 4666	17 45 08.	+ 55 48			NONSTELLAR OBJECT
LB 03903	17 45 10.	+ 05 50 48.		14.8	FAINT BLUE STAR
LB 03904	17 45 12.	+ 06 04 18.		18.7	FAINT BLUE STAR
ZWG 083.021	17 45 12.	+ 09 34		15.3	GALAXY
UGC 10981	17 45 12.	+ 09 34	60	15.3	GALAXY Sb-c
KARA.73B 0830	17 45 12.	+ 09 34	66	15.3	ISOLATED GALAXY S
UGC 10982	17 45 12.	+ 59 22	72	16.0	GALAXY Sb
ZWG 300.065	17 45 12.	+ 60 55		15.3	GALAXY
MCG+10-25-087	17 45 12.	+ 60 55	30	15.	GALAXY
RNGC 6464	17 45 13.	+ 60 55		15.5	GALAXY
SG 3.111	17 45 17.	- 21 01	3600		DIFFUSE EMISSION NEBULA
RNGC 6450	17 45 18.	+ 18 36			NON-EXISTENT OBJECT
MCG+08-32-019	17 45 18.	+ 45 40	72	16.	GALAXY
SCHO 0608	17 45 18.	- 36 49 00.	350		ISOLATED DARK CLOUD
RNGC 6439	17 45 21.	- 16 28		14.0	PLANETARY NEBULA
LB 03905	17 45 22.	+ 05 42 30.		19.1	FAINT BLUE STAR
HN 0048	17 45 23.	- 16 30			NEBULA
LB 03906	17 45 24.	+ 06 03 24.		18.4	FAINT BLUE STAR
MCG+09-29-030	17 45 24.	+ 51 17	72	14.	GALAXY
BIGO 543	17 45 24.	+ 55 54			NEBULA
IC 4667	17 45 24.	+ 55 54			NONSTELLAR OBJECT
MCG+10-25-088	17 45 24.	+ 59 20	66	16.	GALAXY
ZWG 322.011	17 45 24.	+ 64 47		15.7	GALAXY
ZWG 321.041	17 45 24.	+ 64 47		15.7	GALAXY
PK011+05.1	17 45 26.4	- 16 27 46.2	6	13.8	PLANETARY NEBULA
LB 01060	17 45 29.	+ 79 25 24.		17.2	FAINT BLUE STAR
ZWG 055.017	17 45 30.	+ 06 58		15.6	GALAXY
ZWG 083.022	17 45 30.	+ 14 42		15.2	GALAXY
ZWG 112.040	17 45 30.	+ 18 09		15.7	GALAXY
ZWG 112.041	17 45 30.	+ 18 17		15.6	GALAXY
MCG+03-45-023	17 45 30.	+ 18 17	60	15.6	GALAXY
MCG+08-32-020	17 45 30.	+ 45 00	42	16.	GALAXY
ZWG 253.034	17 45 30.	+ 45 42		15.6	GALAXY
UGC 10983	17 45 30.	+ 45 42	72	15.6	GALAXY SB:a-b
ZWG 278.026	17 45 30.	+ 51 20		15.1	GALAXY
UGC 10984	17 45 30.	+ 51 20	84	15.1	GALAXY SB:a-b
LDN 0159	17 45 30.	- 23 40	1080		DARK NEBULA
SCHO 0609	17 45 30.	- 33 20 00.	360		ISOLATED DARK CLOUD
SCHO 0610	17 45 32.	- 17 40 24.	420		ISOLATED DARK CLOUD
PK004+01.1	17 45 32.5	- 24 15 39.	6		PLANETARY NEBULA
B 282	17 45 34.	- 23 27	1080		DARK OBJECT
LB 01061	17 45 35.	+ 44 56 12.		16.0	FAINT BLUE STAR
OCL 0075	17 45 36.	+ 01 19	2700	6.4	OPEN STAR CLUSTER
MCG+02-45-002	17 45 36.	+ 14 42	36	15.2	GALAXY
MCG+03-45-024	17 45 36.	+ 17 41	36	15.3	GALAXY
ZC 1745.6+6703	17 45 36.	+ 67 03	10750		CLUSTER OF GALAXIES
MCG+12-17-008	17 45 36.	+ 70 48	33	16.	GALAXY
PK006+02.1	17 45 36.3	- 22 15 53.		14.2	PLANETARY NEBULA
SCHO 0611	17 45 38.	- 23 27 30.	480		ISOLATED DARK CLOUD
PK355-03.2	17 45 38.6	- 34 20 57.			PLANETARY NEBULA
ZWG 112.042	17 45 42.	+ 17 41		15.3	GALAXY
ZWG 253.035	17 45 42.	+ 45 01		15.3	GALAXY
RNGC 6437	17 45 43.	- 35 26			NON-EXISTENT OBJECT
MCG+08-32-021	17 45 45.	+ 45 42	21	15.	GALAXY
ZWG 083.023	17 45 48.	+ 14 45		13.9	GALAXY
UGC 10985	17 45 48.	+ 14 45	90	13.9	GALAXY
MCG+02-45-003	17 45 48.	+ 14 45	72	13.9	GALAXY
ZWG 112.043	17 45 48.	+ 20 51		15.3	GALAXY
ZWG 226.046	17 45 48.	+ 40 22		15.5	GALAXY
ZC 1745.8+4513	17 45 48.	+ 45 13	1550		CLUSTER OF GALAXIES
MCG+10-25-089	17 45 48.	+ 58 39	30	16.	GALAXY
ZWG 300.066	17 45 48.	+ 58 40		15.5	GALAXY
OCL 0002	17 45 48.	- 29 12	600	11.	OPEN STAR CLUSTER
RNGC 6452	17 45 50.	+ 20 51		15.5	GALAXY
ZWG 253.036	17 45 54.	+ 45 43		15.3	GALAXY
VHE 92	17 45 54.	- 31 24	18		REFLECTION NEBULA
SCHO 0612	17 45 54.	- 35 27 24.	740		ISOLATED DARK CLOUD
REIN 2.264	17 45 54.21	- 20 20 42.5			NEBULA
PK355-03.3	17 45 54.6	- 34 21 59.	5		PLANETARY NEBULA
RNGC 6440	17 45 55.	- 20 21		12.0	GLOBULAR CLUSTER
SCHO 0613	17 45 56.	- 36 10 00.	300		ISOLATED DARK CLOUD
MCG+03-45-025	17 45 57.	+ 17 39	30	14.7	GALAXY
MCG+03-45-026	17 45 57.	+ 17 41	15	15.5	GALAXY
PK004+02.1	17 45 57.5	- 23 42 00.	5		PLANETARY NEBULA
SG 3.112	17 45 58.	- 20 21			DIFFUSE EMISSION NEBULA
KHAV 404	17 46	- 03 13	6190		DARK NEBULA
KHAV 403	17 46	- 10 25	5540		DARK NEBULA
KHAV 405	17 46	- 20 37			DARK NEBULA
LBN 1120	17 46	- 31 20	1260		BRIGHT NEBULA
ZWG 112.044	17 46 00.	+ 17 07		15.4	GALAXY
MCG+03-45-027	17 46 00.	+ 17 07	36	15.4	GALAXY
ZWG 112.045	17 46 00.	+ 17 38		14.7	GALAXY

OBJECT NAME	RIGHT ASCEN.	DECLINATION	DIAM.	MAGN.	TYPE OF OBJECT
ZWG 278.027	17 46 00.	+ 54 54		15.1	GALAXY
UGC 10986	17 46 00.	+ 57 07	72	17.	GALAXY Sc
MCG+10-25-090	17 46 00.	+ 59 12	24	16.	GALAXY
MCG+10-25-091	17 46 00.	+ 59 14 30.	24	14.	GALAXY
LDN 0270	17 46 00.	- 18 20	2700		DARK NEBULA
LDN 0264	17 46 00.	- 18 50	1500		DARK NEBULA
GCL 077	17 46 00.	- 20 21	348	12.05	GLOBULAR STAR CLUSTER
MRSL 358-01/1	17 46 00.	- 31 15	1800		HII REGION
SCHO 0614	17 46 00.	- 33 22 00.	300		ISOLATED DARK CLOUD
RNGC 6474	17 46 05.	+ 57 20			NON-EXISTENT OBJECT
ZWG 112.046	17 46 06.	+ 17 58		15.1	GALAXY
UGC 10987	17 46 06.	+ 17 58	78	15.1	GALAXY TRP SYS
MCG+03-45-028	17 46 06.	+ 17 58	24	15.1	GALAXY
MCG+09-29-031	17 46 06.	+ 54 53	42	15.	GALAXY
MCG+10-25-092	17 46 06.	+ 57 20	48	15.	GALAXY
ZWG 300.067	17 46 06.	+ 59 16		14.4	GALAXY
UGC 10988	17 46 06.	+ 59 16	78	14.4	GALAXY E
BIGO 544	17 46 08.	+ 57 25			NEBULA
IC 4668	17 46 08.	+ 57 25			NONSTELLAR OBJECT
RNGC 6473	17 46 11.	+ 57 20		15.0	GALAXY
ZWG 083.024	17 46 12.	+ 12 32		15.3	GALAXY
KARA.73B 0831	17 46 12.	+ 12 32	36	15.3	ISOLATED GALAXY S
ZWG 112.047	17 46 12.	+ 16 12		15.6	GALAXY
ZC 1746.2+5429	17 46 12.	+ 54 29	810		CLUSTER OF GALAXIES
ZWG 300.068	17 46 12.	+ 57 20		15.0	GALAXY
UGC 10989	17 46 12.	+ 57 20	60	15.0	GALAXY S
OCL 0005	17 46 12.	- 28 43	540	9.3	OPEN STAR CLUSTER
OCL 1023	17 46 12.	- 34 48	720	11.	OPEN STAR CLUSTER
LV.56 I4662A	17 46 12.	- 64 58	72		S GALAXY
REIN 2.265	17 46 16.34	- 19 59 30.9			NEBULA
REIN 2.266	17 46 17.19	- 19 59 45.3			NEBULA
PK008+03.1	17 46 17.29	- 19 59 42.7	206	13.2	PLANETARY NEBULA
COU 007	17 46 18.	- 31 20	1500		EMISSION NEBULA
RNGC 6445	17 46 19.	- 20 00		13.0	PLANETARY NEBULA
PK001-00.2	17 46 22.5	- 27 45 26.4		15.65	PLANETARY NEBULA
ZWG 199.020	17 46 24.	+ 34 06		15.5	GALAXY
UGC 10990	17 46 24.	+ 34 06	108	15.5	GALAXY Sa-b
ZWG 278.028	17 46 24.	+ 51 28		15.7	GALAXY
PK353-04.1	17 46 24.1	- 37 00 36.	10		PLANETARY NEBULA
ZWG 199.021	17 46 30.	+ 34 01		15.6	GALAXY
MCG+06-39-020	17 46 30.	+ 34 07	102	14.	GALAXY
UGC 10991	17 46 30.	+ 67 21	96	17.	GALAXY DWRF IR
7ZW 741	17 46 30.	+ 67 23			COMPACT GALAXY
ACK 356-03.4	17 46 30.	- 33 31			PLANETARY NEBULA
ISS 1035	17 46 30.	- 33 44	83		STELLAR RING
PK355-03.1	17 46 34.	- 34 01	5		PLANETARY NEBULA
ZWG 112.048	17 46 36.	+ 17 59		15.5	GALAXY
ZWG 141.011	17 46 36.	+ 24 40		15.6	GALAXY
ZWG 199.022	17 46 36.	+ 34 05		15.5	GALAXY
8ZW 1746+55.2	17 46 36.	+ 55 13		15.6	COMPACT GALAXY
ZWG 300.069	17 46 36.	+ 61 27		15.1	GALAXY
UGC 10992	17 46 36.	+ 61 27	84	15.1	GALAXY SBb
MCG+11-22-001	17 46 36.	+ 67 22	45	17.	GALAXY
SCHO 0615	17 46 36.	- 36 06 36.	330		ISOLATED DARK CLOUD
SCHO 0616	17 46 37.	- 34 44 24.	520		ISOLATED DARK CLOUD
MAI 100	17 46 39.	+ 67 31	47		DWARF SPHEROIDAL GALAXY
IC 4669	17 46 41.1	+ 61 26 59.			GALAXY SB(s)
ZWG 055.018	17 46 42.	+ 08 07		15.5	GALAXY
MCG+06-39-021	17 46 42.	+ 34 05	42	15.	GALAXY
UGC 10993	17 46 42.	+ 55 12	60	16.0	GALAXY SB?
ZWG 300.070	17 46 42.	+ 61 19		15.3	GALAXY
SCHO 0617	17 46 46.	- 32 46 12.	560		ISOLATED DARK CLOUD
ZWG 083.025	17 46 48.	+ 13 20		15.5	GALAXY
ZWG 278.029	17 46 48.	+ 51 49		15.6	GALAXY
GCL 078	17 46 48.	- 37 02	180	8.93	GLOBULAR STAR CLUSTER
VHA 248	17 46 48.	- 37 02	150		GLOBULAR STAR CLUSTER
RNGC 6441	17 46 48.	- 37 02		9.0	GLOBULAR CLUSTER
SCHO 0618	17 46 50.	- 34 12 42.	610		ISOLATED DARK CLOUD
MCG+03-45-030	17 46 54.	+ 16 07 30.	48	15.7	GALAXY
ZWG 112.049	17 46 54.	+ 16 08		15.7	GALAXY
MCG+03-45-029	17 46 54.	+ 20 51	78	14.9	GALAXY
ZC 1746.9+4935	17 46 54.	+ 49 35	1410		CLUSTER OF GALAXIES
ZWG 300.071	17 46 54.	+ 59 26		15.6	GALAXY
MCG+10-25-093	17 46 54.	+ 61 18 30.	18	16.	GALAXY
SCHO 0619	17 46 55.	- 36 34 24.	400		ISOLATED DARK CLOUD
RNGC 6444	17 46 56.	- 34 52			NON-EXISTENT OBJECT
KHAV 407	17 47	- 21 49	5300		DARK NEBULA
LBN 1121	17 47	- 31 20	2400		BRIGHT NEBULA
B 283	17 47	- 33 52	5400		DARK OBJECT
KHAV 406	17 47	- 37 25	3450		DARK NEBULA
ZWG 112.050	17 47 00.	+ 17 59		15.5	GALAXY
ZWG 112.051	17 47 00.	+ 20 49		14.9	GALAXY
UGC 10994	17 47 00.	+ 20 49	66	14.9	GALAXY S0
KARA.72 525A	17 47 00.	+ 20 49	72	14.9	PART OF DOUBLE GALAXY
ZWG 226.047	17 47 00.	+ 39 40		15.4	GALAXY
ZWG 278.030	17 47 00.	+ 51 26		15.0	GALAXY
ZWG 278.031	17 47 00.	+ 52 15		15.5	GALAXY
MCG+10-25-094	17 47 00.	+ 59 56	30	16.	GALAXY
LDN 0367	17 47 00.	- 11 00	3600		DARK NEBULA
LDN 0326	17 47 00.	- 14 25	660		DARK NEBULA
LDN 0305	17 47 00.	- 16 15	8280		DARK NEBULA
LDN 0254	17 47 00.	- 19 40	1920		DARK NEBULA
LDN 1801	17 47 00.	- 29 30	6720		DARK NEBULA
RNGC 6466	17 47 01.	+ 51 26		15.0	GALAXY
RNGC 6458	17 47 02.	+ 20 49		15.0	GALAXY
PK359-01.1	17 47 04.8	- 30 34 06.	12	13.5	PLANETARY NEBULA
ZWG 322.012	17 47 06.	+ 64 02		14.8	GALAXY
ZWG 321.042	17 47 06.	+ 64 02		14.8	GALAXY
UGC 10995	17 47 06.	+ 64 02	66	14.8	GALAXY S
ZWG 112.052	17 47 12.	+ 18 35		14.9	GALAXY
MCG+03-45-032	17 47 12.	+ 18 35	24	14.9	GALAXY
KARA.73B 0832	17 47 12.	+ 18 35	24	14.9	ISOLATED GALAXY S0
MCG+03-45-031	17 47 12.	+ 20 49	120	14.4	GALAXY
MRSL 000-00/1	17 47 12.	- 28 53	300		HII REGION
PK000-01.1	17 47 12.8	- 29 24 29.	4	16.	PLANETARY NEBULA
B 284	17 47 16.	- 14 22	2100		DARK OBJECT
PK351-06.1	17 47 16.2	- 39 16 37.	10		PLANETARY NEBULA
RNGC 6479	17 47 17.	+ 54 10		14.5	GALAXY
ZWG 112.053	17 47 18.	+ 16 36		15.6	GALAXY
ZWG 112.054	17 47 18.	+ 16 42		15.3	GALAXY
MCG+03-45-034	17 47 18.	+ 16 42	30	15.3	GALAXY
MCG+03-45-033	17 47 18.	+ 16 42	18	15.3	GALAXY
MCG+09-29-032	17 47 18.	+ 51 09	114	13.	GALAXY
ZWG 278.032	17 47 18.	+ 54 10		14.5	GALAXY
UGC 10996	17 47 18.	+ 54 10	66	14.5	GALAXY Sc
ZWG 300.072	17 47 18.	+ 61 41		15.7	GALAXY
MCG+10-25-095	17 47 18.	+ 61 41 30.	24	17.	GALAXY
7ZW 742	17 47 18.	+ 68 38			COMPACT GALAXY

OBJECT NAME	RIGHT ASCEN.	DECLINATION	DIAM.	MAGN.	TYPE OF OBJECT
ZWG 322.013	17 47 18.	+ 68 38		15.5	GALAXY
ZWG 321.043	17 47 18.	+ 68 38		15.5	GALAXY
SCHO 0620	17 47 19.	- 19 58 18.	610		ISOLATED DARK CLOUD
IC 1270	17 47 20.	+ 62 14 23.			SINGLE STAR
PK353-05.1	17 47 20.4	- 37 23 05.	9		PLANETARY NEBULA
MCG+11-22-002	17 47 21.	+ 64 03	57	15.	GALAXY
KARA.72 525B	17 47 24.	+ 20 46	114	14.4	PART OF DOUBLE GALAXY
ZWG 083.026	17 47 24.	+ 13 23		15.4	GALAXY
ZWG 112.055	17 47 24.	+ 20 46		14.4	GALAXY
UGC 10997	17 47 24.	+ 20 46	114	14.4	GALAXY Sc
7ZW 743	17 47 24.	+ 64 28			COMPACT GALAXY
VHA 249	17 47 24.	- 31 17	30		OPEN STAR CLUSTER
RNGC 6460	17 47 26.	+ 20 46		14.5	GALAXY
ZWG 278.033	17 47 30.	+ 51 12		14.1	GALAXY
UGC 10998	17 47 30.	+ 51 12	108	14.1	GALAXY Sc
PK008+03.2	17 47 30.	- 19 52		13.7	PLANETARY NEBULA
OCL 1035	17 47 30.	- 30 12	1020	8.4	OPEN STAR CLUSTER
VHA 250	17 47 30.	- 30 12	360		OPEN STAR CLUSTER
RNGC 6451	17 47 30.	- 30 12		8.5	OPEN CLUSTER
RNGC 6478	17 47 31.	+ 51 10		14.0	GALAXY
PK009+04.1	17 47 32.	- 19 02		13.5	PLANETARY NEBULA
TER 06	17 47 32.12	- 31 15 44.0	30		STAR CLUSTER
SCHO 0621	17 47 33.	- 38 07 18.	600		ISOLATED DARK CLOUD
MCG+10-25-096	17 47 36.	+ 58 30	18	16.	GALAXY
ZWG 300.073	17 47 36.	+ 58 31		15.2	GALAXY
LDN 0330	17 47 36.	- 14 16	420		DARK NEBULA
SCHO 0622	17 47 40.	- 18 38 12.	480		ISOLATED DARK CLOUD
ZWG 055.019	17 47 42.	+ 07 18		15.6	GALAXY
ZWG 112.056	17 47 42.	+ 17 56		15.7	GALAXY
ZWG 141.012	17 47 42.	+ 25 48		14.8	GALAXY
UGC 10999	17 47 42.	+ 25 48	60	14.8	GALAXY
ZWG 199.023	17 47 42.	+ 36 09		14.0	GALAXY
UGC 11000	17 47 42.	+ 36 09	48	14.0	GALAXY
MCG+06-39-022	17 47 42.	+ 36 09	36	14.	GALAXY
ASS 01	17 47 42.	- 29 26			OB ASSOCIATION SGR OB5
MIN.48 04	17 47 44.	- 34 57			DIFFUSE NEBULA
PATH 1.802	17 47 46.	+ 44 56	8		NEBULA
SCHO 0623	17 47 47.	- 29 21 48.	590		ISOLATED DARK CLOUD
MCG+04-42-006	17 47 48.	+ 25 48 30.	36	14.8	GALAXY
MCG+10-25-097	17 47 48.	+ 58 23 30.	72	15.	GALAXY
SCHO 0624	17 47 51.	- 29 46 48.	530		ISOLATED DARK CLOUD
PK355-04.1	17 47 51.9	- 34 54 40.	11		PLANETARY NEBULA
SCHO 0625	17 47 53.	- 18 03 12.	570		ISOLATED DARK CLOUD
ZWG 083.027	17 47 54.	+ 14 18		14.4	GALAXY
UGC 11001	17 47 54.	+ 14 18	90	14.4	GALAXY Sc-IRR
MCG+02-45-004	17 47 54.	+ 14 18	72	14.4	GALAXY
MCG+09-29-033	17 47 54.	+ 55 06	30	17.	GALAXY
ZWG 300.074	17 47 54.	+ 58 26		15.4	GALAXY
UGC 11002	17 47 54.	+ 58 26	72	15.4	GALAXY Sb
HOLM 771A	17 47 59.	+ 14 18	60	13.9	PART OF MULTIPLE GALAXY
HOLM 771B	17 47 59.	+ 14 20	36	14.7	PART OF MULTIPLE GALAXY
PK006+02.2	17 47 59.7	- 22 18 48.	7		PLANETARY NEBULA
LBN 0004	17 48	- 28 55	180		BRIGHT NEBULA
KHAV 408	17 48	- 29 13	2730		DARK NEBULA
ZWG 141.013	17 48 00.	+ 21 50		15.7	GALAXY
1ZW 197	17 48 00.	+ 47 18			COMPACT GALAXY
MCG+09-29-034	17 48 00.	+ 55 10	48	16.	GALAXY
MCG+11-22-003	17 48 00.	+ 65 50	39	17.	GALAXY
ZC 1748.0+7125	17 48 00.	+ 71 25	1280		CLUSTER OF GALAXIES
7ZW 744	17 48 00.	+ 83 09			COMPACT GALAXY
ZWG 367.024	17 48 00.	+ 83 09		15.3	GALAXY
KARA.73B 0833	17 48 00.	+ 83 09	18	15.3	ISOLATED GALAXY E
LDN 0312	17 48 00.	- 16 00	6540		DARK NEBULA
LDN 0161	17 48 00.	- 24 00	1380		DARK NEBULA
GCL 079	17 48 00.	- 34 37	216	11.4	GLOBULAR STAR CLUSTER
VHA 251	17 48 00.	- 34 37	90		GLOBULAR STAR CLUSTER
RNGC 6453	17 48 02.	- 34 37		11.5	GLOBULAR CLUSTER
SCHO 0626	17 48 03.	- 18 53 12.	400		ISOLATED DARK CLOUD
PK359-01.2	17 48 05.9	- 30 23 09.	5	16.0	PLANETARY NEBULA
ZWG 055.020	17 48 06.	+ 05 40		15.7	GALAXY
7ZW 745	17 48 06.	+ 67 58			COMPACT GALAXY
ISS 0982	17 48 06.	- 32 30	145		STELLAR RING
SCHO 0627	17 48 07.	- 19 39 24.	460		ISOLATED DARK CLOUD
SCHO 0628	17 48 09.	- 34 43 00.	420		ISOLATED DARK CLOUD
ZWG 083.028	17 48 12.	+ 14 24		14.7	GALAXY
MCG+02-45-005	17 48 12.	+ 14 24	36	14.7	GALAXY
ZWG 141.014	17 48 12.	+ 22 58		15.5	GALAXY
7ZW 746	17 48 12.	+ 69 05			COMPACT GALAXY
VHE 93A	17 48 12.	- 31 15	84		REFLECTION NEBULA
ZWG 055.021	17 48 18.	+ 06 03		15.6	GALAXY
LDN 0349	17 48 18.	- 13 00	420		DARK NEBULA
VHE 93B	17 48 18.	- 31 17			REFLECTION NEBULA
ZWG 083.029	17 48 24.	+ 14 50		15.1	GALAXY
UGC 11003	17 48 24.	+ 14 50	66	15.1	GALAXY Sc
MCG+02-45-006	17 48 24.	+ 14 50	60	15.1	GALAXY
ZWG 112.057	17 48 24.	+ 17 13		15.1	GALAXY
MCG+03-45-036	17 48 24.	+ 17 13	36	15.1	GALAXY
MCG+03-45-035	17 48 24.	+ 17 34	150	14.5	GALAXY
ZWG 199.024	17 48 24.	+ 35 32		15.4	GALAXY
KARA.73B 0834	17 48 24.	+ 35 32	60	15.4	ISOLATED GALAXY S
MCG+08-32-022	17 48 24.	+ 45 00	36	16.	GALAXY
ZC 1748.4+5949	17 48 24.	+ 59 49	2490		CLUSTER OF GALAXIES
PK009+04.2	17 48 24.	- 18 46		14.4	PLANETARY NEBULA
ISS 0983	17 48 24.	- 30 05	152		STELLAR RING
IC 1268	17 48 26.	+ 17 13 19.			NONSTELLAR OBJECT
SCHO 0629	17 48 28.	- 36 10 36.	550		ISOLATED DARK CLOUD
BIGO 545	17 48 29.	+ 17 34			NEBULA
HOLM 772B	17 48 30.	+ 17 29	12	14.5	PART OF MULTIPLE GALAXY
ZWG 112.058	17 48 30.	+ 17 33		14.5	GALAXY
RNGC 6468	17 48 30.	+ 17 33			NON-EXISTENT OBJECT
RNGC 6467	17 48 30.	+ 17 33		14.5	GALAXY
UGC 11004	17 48 30.	+ 17 33	138	14.5	GALAXY S
ZWG 199.025	17 48 30.	+ 36 22		15.6	GALAXY
MCG+10-25-098	17 48 30.	+ 62 15	30	15.	GALAXY
PK358-02.1	17 48 30.	- 31 35 18.	5		PLANETARY NEBULA
PK357-03.1	17 48 30.	- 32 54 02.	10		PLANETARY NEBULA
HOLM 772A	17 48 31.	+ 17 29	12	14.0	PART OF MULTIPLE GALAXY
PK356-03.2	17 48 32.2	- 33 46 51.			PLANETARY NEBULA
SCHO 0630	17 48 33.	- 17 55 54.	500		ISOLATED DARK CLOUD
SCHO 0631	17 48 34.	- 29 59 48.	400		ISOLATED DARK CLOUD
ZWG 141.015	17 48 36.	+ 23 09		15.3	GALAXY
PATH 1.803	17 48 36.	+ 45 00	22		NEBULA
ZWG 278.034	17 48 36.	+ 55 31		15.3	GALAXY
ZWG 300.075	17 48 36.	+ 61 41		15.2	GALAXY
7ZW 747	17 48 36.	+ 61 42			COMPACT GALAXY
7ZW 748	17 48 36.	+ 66 42			COMPACT GALAXY
RNGC 6455	17 48 37.	- 35 23			NON-EXISTENT OBJECT
SCHO 0632	17 48 38.	- 27 58 12.	490		ISOLATED DARK CLOUD
PK001-00.3	17 48 39.0	- 27 46 57.4		16.07	PLANETARY NEBULA
SCHO 0633	17 48 41.	- 18 35 24.	420		ISOLATED DARK CLOUD
ZWG 112.059	17 48 42.	+ 17 09		15.3	GALAXY
ZWG 141.016	17 48 42.	+ 25 27		15.7	GALAXY
MCG+09-29-035	17 48 42.	+ 53 35	21	16.	GALAXY
MCG+09-29-036	17 48 42.	+ 55 30	36	15.	GALAXY
7ZW 749	17 48 42.	+ 66 13			COMPACT GALAXY
B 285	17 48 43.	- 12 51	900		DARK OBJECT
PATH 1.804	17 48 45.	+ 45 24	14		NEBULA
ZWG 112.060	17 48 48.	+ 20 26		15.7	GALAXY
MRK 507	17 48 48.	+ 68 43	10	16.	GALAXY WITH UV CONTINUUM
KW 47	17 48 48.	+ 68 43	9		SEYFERT GALAXY
ZC 1748.8+6928	17 48 48.	+ 69 28	810		CLUSTER OF GALAXIES
ISS 0894	17 48 48.	- 25 44	141		STELLAR RING
RNGC 6488	17 48 53.	+ 62 14		14.5	GALAXY
PK359-01.3	17 48 53.5	- 30 04 33.	6	16.	PLANETARY NEBULA
ZWG 083.030	17 48 54.	+ 14 07		15.6	GALAXY
ZWG 112.061	17 48 54.	+ 16 40		15.7	GALAXY
ZWG 171.006	17 48 54.	+ 29 38		15.7	GALAXY
ZWG 171.007	17 48 54.	+ 29 43		15.6	GALAXY
MCG+10-25-099	17 48 54.	+ 60 05	15	16.	GALAXY
ZWG 300.076	17 48 54.	+ 62 14		14.6	GALAXY
ZWG 322.014	17 48 54.	+ 63 46		15.5	GALAXY
UGC 11005	17 48 54.	+ 63 46	72	15.5	GALAXY SBb
LDN 0360	17 48 54.	- 12 15	360		DARK NEBULA
KHAV 410	17 49	- 16 49	6650		DARK NEBULA
KHAV 409	17 49	- 33 49	4470		DARK NEBULA
ZWG 112.062	17 49 00.	+ 15 50		15.4	GALAXY
KARA.73B 0835	17 49 00.	+ 15 50	36	15.4	ISOLATED GALAXY S
ZWG 171.008	17 49 00.	+ 31 53		15.7	GALAXY
ZWG 227.001	17 49 00.	+ 39 58		15.6	GALAXY
KARA.73B 0836	17 49 00.	+ 39 58	18	15.6	ISOLATED GALAXY F
1ZW 198	17 49 00.	+ 57 00			COMPACT GALAXY
LDN 0355	17 49 00.	- 12 40	1080		DARK NEBULA
OCL 1038	17 49 00.	- 30 04	540		OPEN STAR CLUSTER
MCG+03-45-037	17 49 06.	+ 18 46	60	15.2	GALAXY
ZWG 112.063	17 49 06.	+ 20 28		15.7	GALAXY
MCG+09-29-037	17 49 06.	+ 56 41	48	17.	GALAXY
MCG+10-25-100	17 49 06.	+ 59 37	24	16.	GALAXY
ZWG 300.077	17 49 06.	+ 59 38		15.7	GALAXY
MCG+10-25-101	17 49 06.	+ 60 22	24	17.	GALAXY
ZWG 300.078	17 49 06.	+ 60 23		15.4	GALAXY
MCG+11-22-004	17 49 06.	+ 63 47	36	16.	GALAXY
PK010+04.1	17 49 10.	- 17 35 24.	7		PLANETARY NEBULA
ZWG 112.064	17 49 12.	+ 18 46		15.2	GALAXY
ZWG 278.035	17 49 12.	+ 56 41		15.2	GALAXY
LDN 0361	17 49 12.	- 12 10	420		DARK NEBULA
ISS 0984	17 49 12.	- 29 30	110		STELLAR RING
PATH 1.805	17 49 14.	+ 44 48	8		NEBULA
PK000-01.4	17 49 14.1	- 29 45 19.4		14.93	PLANETARY NEBULA
MCG+10-25-102	17 49 15.	+ 61 36 30.	60	17.	GALAXY
SCHO 0634	17 49 15.	- 16 48 06.	260		ISOLATED DARK CLOUD
PK357-03.2	17 49 17.7	- 32 45 12.	6		PLANETARY NEBULA
MCG+09-29-038	17 49 18.	+ 51 01	72	16.	GALAXY
1ZW 199	17 49 18.	+ 56 41			COMPACT GALAXY
MCG+09-29-040	17 49 18.	+ 56 41	15	16.	GALAXY
MCG+09-29-039	17 49 18.	+ 56 41	30	16.	GALAXY
KEEN 6493A	17 49 18.	+ 61 30	96		Sb GALAXY
MCG+10-25-103	17 49 18.	+ 61 32 30.	72	15.	GALAXY
OCL 0006	17 49 18.	- 28 42	318	12.	OPEN STAR CLUSTER
SCHO 0635	17 49 20.	- 34 13 00.	620		ISOLATED DARK CLOUD
RNGC 6493	17 49 21.	+ 61 30		15.5	GALAXY
PK006+02.3	17 49 22.1	- 21 50 33.		14.7	PLANETARY NEBULA
RNGC 6489	17 49 23.	+ 60 06		15.5	GALAXY
SCHO 0636	17 49 23.	- 33 58 06.	730		ISOLATED DARK CLOUD
MCG+08-32-023	17 49 24.	+ 47 11	24	16.	GALAXY
ZWG 278.036	17 49 24.	+ 51 04		15.6	GALAXY
UGC 11006	17 49 24.	+ 51 04	72	15.6	GALAXY S
ZWG 300.079	17 49 24.	+ 60 06		15.3	GALAXY
UGC 11007	17 49 24.	+ 61 37	60	16.5	GALAXY SB
ZWG 322.015	17 49 24.	+ 67 21		15.7	GALAXY
ZWG 321.044	17 49 24.	+ 67 21		15.7	GALAXY
PK000-01.2	17 49 24.8	- 29 05 57.8		13.85	PLANETARY NEBULA
SCHO 0637	17 49 25.	- 32 52 00.	240		ISOLATED DARK CLOUD
PATH 1.806	17 49 28.	+ 45 00	11		NEBULA
ZWG 253.037	17 49 30.	+ 47 12		15.2	GALAXY
ZWG 300.080	17 49 30.	+ 61 32		14.5	GALAXY
UGC 11008	17 49 30.	+ 61 32	78	14.5	GALAXY Sa-b
PK359-02.2	17 49 32.4	- 30 48 57.	9	15.5	PLANETARY NEBULA
RNGC 6491	17 49 33.	+ 61 32		14.5	GALAXY
RNGC 6493A	17 49 33.	+ 61 33			GALAXY
MCG+10-25-104	17 49 36.	+ 59 38	30	15.	GALAXY
ZWG 300.081	17 49 36.	+ 59 39		15.1	GALAXY
MCG+10-25-105	17 49 36.	+ 61 34	66	15.	GALAXY
OCL 0004	17 49 36.	- 29 33	312	12.	OPEN STAR CLUSTER
VHA 252	17 49 36.	- 29 33	300		OPEN STAR CLUSTER
SCHO 0638	17 49 36.	- 36 25 42.	610		ISOLATED DARK CLOUD
RNGC 6482	17 49 39.	+ 23 05		12.5	GALAXY
PK355-04.2	17 49 39.1	- 34 37 43.	5		PLANETARY NEBULA
RNGC 6484	17 49 40.	+ 24 30		13.5	GALAXY
SCHO 0639	17 49 40.	- 32 10 54.	790		ISOLATED DARK CLOUD
PK006+02.5	17 49 40.2	- 22 21 18.	25		PLANETARY NEBULA
ZWG 141.017	17 49 42.	+ 23 05		12.8	GALAXY
UGC 11009	17 49 42.	+ 23 05	132	12.8	GALAXY E
ZWG 141.018	17 49 42.	+ 23 48		15.7	GALAXY
MCG+04-42-007	17 49 42.	+ 24 29	108	13.5	GALAXY
ZWG 141.019	17 49 42.	+ 24 30		13.5	GALAXY
UGC 11010	17 49 42.	+ 24 30	120	13.5	GALAXY Sb
MCG+09-29-041	17 49 42.	+ 51 00	30	16.	GALAXY
MCG+10-25-106	17 49 42.	+ 57 18	48	16.	GALAXY
ZWG 300.082	17 49 42.	+ 57 19		15.5	GALAXY
ZWG 300.083	17 49 42.	+ 59 49		15.6	GALAXY
PK007+02.1	17 49 42.	- 21 14		15.	PLANETARY NEBULA
OCL 0007	17 49 42.	- 28 27	204	12.	OPEN STAR CLUSTER
B 286	17 49 42.	- 35 36	900		DARK OBJECT
PK358-02.2	17 49 42.3	- 31 18 42.2		12.65	PLANETARY NEBULA
SCHO 0640	17 49 43.	- 34 53 18.	580		ISOLATED DARK CLOUD
SCHO 0641	17 49 45.	- 35 35 18.	520		ISOLATED DARK CLOUD
ZWG 055.022	17 49 48.	+ 05 27		15.5	GALAXY
MCG+04-42-008	17 49 48.	+ 23 06	48	12.8	GALAXY
ZWG 278.037	17 49 48.	+ 51 03		15.5	GALAXY
MCG+10-25-107	17 49 48.	+ 60 01	48	17.	GALAXY
ZWG 300.084	17 49 48.	+ 61 34		15.5	GALAXY
UGC 11011	17 49 48.	+ 61 34	72	15.5	GALAXY SB:c
ISS 0985	17 49 48.	- 29 28	114		STELLAR RING
RNGC 6465	17 49 51.	- 25 23			NON-EXISTENT OBJECT
RNGC 6469	17 49 53.	- 22 20		8.0	OPEN CLUSTER
SCHO 0642	17 49 53.	- 32 27 00.	290		ISOLATED DARK CLOUD

OBJECT NAME	RIGHT ASCEN.	DECLINATION	DIAM.	MAGN.	TYPE OF OBJECT
ZWG 112.065	17 49 54.	+ 18 41		15.5	GALAXY
ZWG 340.019	17 49 54.	+ 70 10		10.9	GALAXY
UGC 11012	17 49 54.	+ 70 10	480	10.9	GALAXY Sc
KARA.73B 0837	17 49 54.	+ 70 10	504	10.9	ISOLATED GALAXY S
OCL 0021	17 49 54.	- 22 20	2520	8.4	OPEN STAR CLUSTER
ISS 0986	17 49 54.	- 29 05	113		STELLAR RING
SCHO 0643	17 49 55.	- 32 38 00.	250		ISOLATED DARK CLOUD
SCHO 0644	17 49 56.	- 32 45 00.	250		ISOLATED DARK CLOUD
REIN 2.267	17 49 57.77	+ 70 09 24.5			NEBULA
RNGC 6485	17 49 58.	+ 31 28		14.0	GALAXY
RNGC 6503	17 49 58.	+ 70 09		11.5	GALAXY
IC 1269	17 49 59.	+ 21 32 32.			NONSTELLAR OBJECT
KHAV 415	17 50	- 05 43	5450		DARK NEBULA
KHAV 416	17 50	- 07 13	6010		DARK NEBULA
KHAV 417	17 50	- 10 31	5840		DARK NEBULA
KHAV 414	17 50	- 13 37	4860		DARK NEBULA
KHAV 413	17 50	- 14 55	7870		DARK NEBULA
MCG+04-42-009	17 50 00.	+ 21 34	96	14.5	GALAXY
ZWG 141.020	17 50 00.	+ 21 35		14.5	GALAXY
UGC 11013	17 50 00.	+ 21 35	102	14.5	GALAXY Sb/Sc
KARA.73B 0838	17 50 00.	+ 21 35	108	14.5	ISOLATED GALAXY S
ZWG 171.009	17 50 00.	+ 31 28		14.2	GALAXY
UGC 11014	17 50 00.	+ 31 28	84	14.2	GALAXY Sb
MCG+05-42-004	17 50 00.	+ 31 29	96	14.2	GALAXY
UGC 11015	17 50 00.	+ 37 22	60	16.5	GALAXY Sc
MCG+06-39-024	17 50 00.	+ 37 24	57	16.	GALAXY
MCG+06-39-023	17 50 00.	+ 37 25	36	15.	GALAXY
MCG+11-22-005	17 50 00.	+ 68 25	18	17.	GALAXY
MCG+12-17-009	17 50 00.	+ 70 10	336	10.7	GALAXY
LDN 0358	17 50 00.	- 12 30	3480		DARK NEBULA
LDN 0279	17 50 00.	- 18 30	2760		DARK NEBULA
COU 008	17 50 00.	- 20 40	15000		EMISSION NEBULA
LDN 0188	17 50 00.	- 23 10	2160		DARK NEBULA
LDN 0140	17 50 00.	- 25 00	4080		DARK NEBULA
LDN 1788	17 50 00.	- 30 40	1140		DARK NEBULA
COU 009	17 50 00.	- 35 00	18000		EMISSION NEBULA
PK357-02.3	17 50 00.2	- 32 40 04.	10		PLANETARY NEBULA
PK359-02.1	17 50 00.7	- 30 17 25.		16.	PLANETARY NEBULA
PK356-03.3	17 50 02.1	- 33 55 21.	5		PLANETARY NEBULA
SG 3.113	17 50 05.	- 25 11	4500		DIFFUSE EMISSION NEBULA
MCG+04-42-010	17 50 06.	+ 24 31	36	15.1	GALAXY
MCG+04-42-011	17 50 06.	+ 24 31 30.	24	15.1	GALAXY
ZWG 141.021	17 50 06.	+ 24 32		15.1	GALAXY
ZWG 171.010	17 50 06.	+ 28 55		15.7	GALAXY
UGC 11016	17 50 06.	+ 28 55	84	15.7	GALAXY Sc
ZWG 322.016	17 50 06.	+ 68 25		15.2	GALAXY
ZWG 321.045	17 50 06.	+ 68 25		15.2	GALAXY
OCL 0008	17 50 06.	- 27 21	180		OPEN STAR CLUSTER
VHA 253	17 50 06.	- 27 21	180		OPEN STAR CLUSTER
SCHO 0645	17 50 06.	- 34 45 00.	290		ISOLATED DARK CLOUD
MCG+11-22-006	17 50 09.	+ 68 26	27	16.	GALAXY
SCHO 0646	17 50 09.	- 31 05 48.	560		ISOLATED DARK CLOUD
ZWG 171.011	17 50 12.	+ 29 52		15.5	GALAXY
UGC 11017	17 50 12.	+ 29 52	108	15.5	GALAXY SBc-IRR
ZWG 300.085	17 50 12.	+ 57 02		15.7	GALAXY
MRSL 004+00/1	17 50 12.	- 24 56	4800		HII REGION
FATE 1.807	17 50 13.	+ 44 36			NEBULA
SCHO 0647	17 50 17.	- 34 56 00.	430		ISOLATED DARK CLOUD
UGC 11018	17 50 18.	+ 22 53	60	17.	GALAXY
MCG+05-42-005	17 50 18.	+ 29 53	90	15.5	GALAXY
ZWG 322.017	17 50 18.	+ 62 53		15.5	GALAXY
ZWG 300.086	17 50 18.	+ 62 53		15.5	GALAXY
UGC 11019	17 50 18.	+ 62 53	60	15.5	GALAXY Sa-b
MCG+10-25-108	17 50 18.	+ 62 54	60	16.	GALAXY
PK357-03.4	17 50 20.8	- 32 58 12.	5		PLANETARY NEBULA
RNGC 6481	17 50 22.	+ 04 10			NON-EXISTENT OBJECT
RNGC 6498	17 50 22.	+ 59 29			NON-EXISTENT OBJECT
MCG+10-25-109	17 50 24.	+ 59 30	84	14.	GALAXY
ZWG 340.020	17 50 24.	+ 72 14		15.2	GALAXY
SCHO 0648	17 50 27.	- 34 39 18.	400		ISOLATED DARK CLOUD
ZWG 112.066	17 50 30.	+ 20 56		15.6	GALAXY
SCHO 0649	17 50 32.	- 31 22 54.	670		ISOLATED DARK CLOUD
PK000-01.5	17 50 33.7	- 29 43 13.	7	15.5	PLANETARY NEBULA
RNGC 6497	17 50 34.	+ 59 30		14.5	GALAXY
ZWG 141.022	17 50 36.	+ 23 45		15.7	GALAXY
ZWG 171.012	17 50 36.	+ 29 50		15.0	GALAXY
1ZW 200	17 50 36.	+ 49 02			COMPACT GALAXY
ZWG 253.038	17 50 36.	+ 49 32		15.2	GALAXY
MCG+08-32-024	17 50 36.	+ 49 32	36	16.	GALAXY
1ZW 201	17 50 36.	+ 54 00			COMPACT GALAXY
ZWG 300.087	17 50 36.	+ 59 30		14.3	GALAXY
UGC 11020	17 50 36.	+ 59 30	96	14.3	GALAXY SBb
ZWG 300.088	17 50 36.	+ 62 50		15.6	GALAXY
OCL 1028	17 50 36.	- 34 48	7920	5.4	OPEN STAR CLUSTER
VHA 254	17 50 36.	- 34 48	900		OPEN STAR CLUSTER
PK006+02.4	17 50 36.4	- 21 58 07.	4		PLANETARY NEBULA
RNGC 6476	17 50 37.	- 29 08			NON-EXISTENT OBJECT
PK001-01.1	17 50 37.2	- 28 26 45.	4	16.	PLANETARY NEBULA
RNGC 6488	17 50 38.	+ 29 50		15.0	GALAXY
RNGC 6475	17 50 38.	- 34 48		3.5	OPEN CLUSTER
PK000-01.3	17 50 38.0	- 28 58 43.		16.	PLANETARY NEBULA
ZWG 141.023	17 50 42.	+ 23 01		15.6	GALAXY
ZWG 171.013	17 50 42.	+ 29 05		15.7	GALAXY
UGC 11021	17 50 42.	+ 29 05	72	15.7	GALAXY Sb
MCG+05-42-006	17 50 42.	+ 29 50	15	15.0	GALAXY
ZWG 171.014	17 50 42.	+ 29 51		14.0	GALAXY
UGC 11022	17 50 42.	+ 29 51	114	14.0	GALAXY E
ZWG 340.021	17 50 42.	+ 72 02		14.0	GALAXY
UGC 11023	17 50 42.	+ 72 02	84	14.0	GALAXY (E)
MCG+12-17-010	17 50 42.	+ 72 02	30	14.	GALAXY
ZC 1750.7+7321	17 50 42.	+ 73 21	1610		CLUSTER OF GALAXIES
RNGC 6487	17 50 44.	+ 29 51		14.0	GALAXY
RNGC 6508	17 50 44.	+ 72 02		14.0	GALAXY
MCG+05-42-007	17 50 45.	+ 29 49	12	15.0	GALAXY
SCHO 0650	17 50 47.	- 35 13 00.	420		ISOLATED DARK CLOUD
ZWG 141.024	17 50 48.	+ 23 14		15.0	GALAXY
UGC 11024	17 50 48.	+ 23 14	102	15.0	GALAXY S?
MCG+05-42-008	17 50 49.	+ 29 51 30.	30	14.0	GALAXY
VDB.66N 112	17 50 50.	- 05 35	72		REFLECTION NEBULA
MCG+04-42-012	17 50 51.	+ 23 13 30.	72	15.0	GALAXY
ZWG 171.015	17 50 54.	+ 27 41		14.6	GALAXY
UGC 11025	17 50 54.	+ 27 41	66	14.6	GALAXY S0
ZWG 199.026	17 50 54.	+ 37 45		14.8	GALAXY
MCG+09-29-042	17 50 54.	+ 52 27	48	15.	GALAXY
ZWG 340.022	17 50 54.	+ 71 50		15.1	GALAXY
SCHO 0651	17 50 54.	- 30 38 12.	750		ISOLATED DARK CLOUD
MCG+11-22-007	17 50 57.	+ 65 33	18	16.	GALAXY
KHAV 411	17 51	- 26 49	8470		DARK NEBULA

OBJECT NAME	RIGHT ASCEN.	DECLINATION	DIAM.	MAGN.	TYPE OF OBJECT
KHAV 412	17 51	- 32 13	2760		DARK NEBULA
ZWG 141.025	17 51 00.	+ 22 55		15.7	GALAXY
MCG+05-42-009	17 51 00.	+ 27 40 30.	60	14.6	GALAXY
ZWG 171.016	17 51 00.	+ 29 38		15.5	GALAXY
ZWG 278.038	17 51 00.	+ 52 28		15.4	GALAXY
LDN 0392	17 51 00.	- 08 30	360		DARK NEBULA
LDN 0345	17 51 00.	- 13 40	1980		DARK NEBULA
LDN 0325	17 51 00.	- 15 00	3840		DARK NEBULA
LDN 1753	17 51 00.	- 33 00	4560		DARK NEBULA
SCHO 0652	17 51 00.	- 33 18 12.	740		ISOLATED DARK CLOUD
SCHO 0653	17 51 02.	- 35 08 48.	560		ISOLATED DARK CLOUD
HUB B22	17 51 03.	- 23 57			DIFFUSE NEBULA
ZWG 171.017	17 51 06.	+ 27 31		15.7	GALAXY
MCG+10-25-110	17 51 06.	+ 59 39	30	17.	GALAXY
ZWG 322.018	17 51 06.	+ 65 33		15.4	GALAXY
UGC 11026	17 51 06.	+ 65 33	66	15.4	GALAXY
ISS 0987	17 51 06.	- 32 35	272		STELLAR RING
B 287	17 51 06.	- 35 11	1800		DARK OBJECT
RNGC 6505	17 51 09.	+ 65 33		15.5	GALAXY
PK036+17.1	17 51 12.	+ 10 38	83	14.7	PLANETARY NEBULA
ZWG 141.026	17 51 12.	+ 24 35		14.9	GALAXY
UGC 11027	17 51 12.	+ 24 35	102	14.9	GALAXY IRR
MCG+05-42-010	17 51 12.	+ 27 31	36	15.7	GALAXY
ZWG 171.018	17 51 12.	+ 31 53		15.7	GALAXY
ZWG 253.039	17 51 12.	+ 44 59		15.5	GALAXY
ZWG 227.002	17 51 12.	+ 44 59		15.5	GALAXY
1ZW 202	17 51 12.	+ 46 00			COMPACT GALAXY
FATH 1.808	17 51 13.	+ 44 59			NEBULA
PK000-01.6	17 51 13.6	- 29 35 38.	25	13.5	PLANETARY NEBULA
PK356-04.1	17 51 13.6	- 34 21 50.	5	13.9	PLANETARY NEBULA
MCG+04-42-013	17 51 15.	+ 24 35 30.	72	14.9	GALAXY
SCHO 0654	17 51 18.	- 36 27 00.	480		ISOLATED DARK CLOUD
ARC 2284	17 51 19.	+ 54 18		16.9	RICH CLUSTER OF GALAXIES
ZWG 171.019	17 51 24.	+ 32 03		15.6	GALAXY
MCG+07-37-001	17 51 24.	+ 41 05	48	15.5	GALAXY
ZC 1751.4+5414	17 51 24.	+ 54 14	1340		CLUSTER OF GALAXIES
ZWG 300.089	17 51 24.	+ 62 28		15.7	GALAXY
RNGC 6480	17 51 24.	- 30 26			NON-EXISTENT OBJECT
PK001-01.2	17 51 25.3	- 28 12 13.		16.	PLANETARY NEBULA
ZC 1751.5+5501	17 51 30.	+ 55 01	940		CLUSTER OF GALAXIES
ZWG 300.090	17 51 30.	+ 60 37		15.6	GALAXY
UGC 11028	17 51 30.	+ 60 37	108	15.6	GALAXY Sb
LDN 1795	17 51 30.	- 30 30	1140		DARK NEBULA
SCHO 0655	17 51 30.	- 32 42 00.	390		ISOLATED DARK CLOUD
FATE 1.809	17 51 33.	+ 44 59	16		NEBULA
ARC 2290	17 51 35.	+ 73 21		17.1	RICH CLUSTER OF GALAXIES
ZWG 083.031	17 51 36.	+ 14 03		15.3	GALAXY
ZC 1751.6+1415	17 51 36.	+ 14 15	2290		CLUSTER OF GALAXIES
1ZW 203	17 51 36.	+ 47 45			COMPACT GALAXY
MCG+10-25-111	17 51 36.	+ 60 36	60	15.	GALAXY
LDN 0421	17 51 36.	- 05 48	540		DARK NEBULA
PK014+06.1	17 51 37.	- 12 48	30		PLANETARY NEBULA
ZWG 084.001	17 51 42.	+ 14 52		15.0	GALAXY
ZWG 083.032	17 51 42.	+ 14 52		15.0	GALAXY
MCG+02-45-007	17 51 42.	+ 14 52	36	15.0	GALAXY
MCG+04-42-014	17 51 45.	+ 24 30	66	15.0	GALAXY
SCHO 0656	17 51 46.	- 35 30	250		ISOLATED DARK CLOUD
ZWG 141.027	17 51 48.	+ 21 25		15.7	GALAXY
ZWG 141.028	17 51 48.	+ 24 28		15.0	GALAXY
UGC 11029	17 51 48.	+ 24 28	84	15.0	GALAXY SBc
ISS 0895	17 51 48.	- 22 21	62		STELLAR RING
ISS 0988	17 51 48.	- 30 51	79		STELLAR RING
SCHO 0657	17 51 49.	- 35 28 48.	500		ISOLATED DARK CLOUD
SCHO 0658	17 51 51.	- 34 57 54.	450		ISOLATED DARK CLOUD
PK359-02.4	17 51 51.5	- 31 11 50.	5		PLANETARY NEBULA
KHAV 423	17 52	- 18 43	6270		DARK NEBULA
LBN 0014	17 52	- 24 50	3000		BRIGHT NEBULA
KHAV 418	17 52	- 28 13	3250		DARK NEBULA
KHAV 419	17 52	- 35 07	3540		DARK NEBULA
KHAV 420	17 52	- 36 13	3170		DARK NEBULA
ZWG 141.029	17 52 00.	+ 21 29		15.6	GALAXY
ZC 1752.0+5158	17 52 00.	+ 51 58	3230		CLUSTER OF GALAXIES
MCG+10-25-112	17 52 00.	+ 58 04	12	16.	GALAXY
LDN 0218	17 52 00.	- 21 58	600		DARK NEBULA
COU 010	17 52 00.	- 25 30	3000		EMISSION NEBULA
LDN 1758	17 52 00.	- 32 40	360		DARK NEBULA
FATH 1.810	17 52 02.	+ 45 20	16		NEBULA
LB 01062	17 52 04.	+ 43 28 24.		17.7	FAINT BLUE STAR
ZWG 028.001	17 52 06.	+ 02 53		14.9	GALAXY
UGC 11030	17 52 06.	+ 02 53	108	14.9	GALAXY Sc
MCG+00-46-001	17 52 06.	+ 02 53	96	14.9	GALAXY
ZWG 171.020	17 52 06.	+ 30 42		15.6	GALAXY
UGC 11031	17 52 06.	+ 30 42	60	15.6	GALAXY IRR
ZC 1752.1+4914	17 52 06.	+ 49 14	810		CLUSTER OF GALAXIES
PK007+01.1	17 52 06.5	- 21 44 11.	7	13.6	PLANETARY NEBULA
SCHO 0659	17 52 07.	- 28 18 00.	660		ISOLATED DARK CLOUD
SCHO 0660	17 52 08.	- 17 46 12.	400		ISOLATED DARK CLOUD
IC 4670	17 52 09.	- 21 47 41.	276		DIFFUSE NEBULA
PK000-02.2	17 52 09.0	- 29 57 13.9		13.35	PLANETARY NEBULA
FATH 1.811	17 52 11.	+ 45 26	16		NEBULA
ZWG 113.001	17 52 12.	+ 15 37		15.7	GALAXY
ZWG 112.067	17 52 12.	+ 15 37		15.7	GALAXY
UGC 11032	17 52 12.	+ 51 29	66	17.	GALAXY
SCHO 0661	17 52 13.	- 32 37 06.	740		ISOLATED DARK CLOUD
MCG+05-42-011	17 52 15.	+ 30 43	48	15.6	GALAXY
ARC 2286	17 52 15.	+ 52 06		16.9	RICH CLUSTER OF GALAXIES
SCHO 0662	17 52 15.	- 18 03 30.	380		ISOLATED DARK CLOUD
ZWG 113.002	17 52 18.	+ 18 23		14.7	GALAXY
ZWG 112.068	17 52 18.	+ 18 23		14.7	GALAXY
RNGC 6490	17 52 18.	+ 18 23		14.5	GALAXY
UGC 11033	17 52 18.	+ 18 23	60	14.7	GALAXY E-S
MCG+03-45-038	17 52 18.	+ 18 23	60	14.7	GALAXY
ZWG 113.003	17 52 18.	+ 18 27		15.7	GALAXY
ZWG 112.069	17 52 18.	+ 18 27		15.7	GALAXY
ZWG 171.021	17 52 18.	+ 29 04		15.2	GALAXY
ZWG 300.091	17 52 18.	+ 58 05		15.7	GALAXY
ZWG 340.023	17 52 18.	+ 74 48		15.0	GALAXY
IC 4671	17 52 22.	- 10 16			NONSTELLAR OBJECT
HW 0449	17 52 22.	- 10 17			NEBULA
SCHO 0663	17 52 22.	- 35 53 42.	630		ISOLATED DARK CLOUD
PK359-02.3	17 52 22.9	- 30 33 07.		17.59	PLANETARY NEBULA
LB 01063	17 52 23.	+ 42 41 48.		17.1	FAINT BLUE STAR
SCHO 0664	17 52 23.	- 32 40 00.	690		ISOLATED DARK CLOUD
MCG+05-42-012	17 52 24.	+ 29 03 30.	39	15.2	GALAXY
PK053+24.1	17 52 24.7	+ 28 00 29.	5		PLANETARY NEBULA
ARC 2288	17 52 27.	+ 59 43		16.9	RICH CLUSTER OF GALAXIES
SCHO 0665	17 52 27.	- 36 18 30.	770		ISOLATED DARK CLOUD
ZC 1752.5+5627	17 52 30.	+ 56 27	810		CLUSTER OF GALAXIES

OBJECT NAME	RIGHT ASCEN.	DECLINATION	DIAM.	MAGN.	TYPE OF OBJECT
ZWG 113.004	17 52 36.	+ 18 20		13.8	GALAXY
ZWG 112.070	17 52 36.	+ 18 20		13.8	GALAXY
RNGC 6495	17 52 36.	+ 18 20		14.0	GALAXY
UGC 11034	17 52 36.	+ 18 20	96	13.8	GALAXY E
ZC 1752.6+1842	17 52 36.	+ 18 42	4030		CLUSTER OF GALAXIES
ZWG 171.022	17 52 36.	+ 32 53		14.3	GALAXY
ZCG 1752+32	17 52 36.	+ 32 53		14.3	COMPACT GALAXY
UGC 11035	17 52 36.	+ 32 53	102	14.3	GALAXY PECULR
MCG+05-42-013	17 52 36.	+ 32 55	96	14.3	GALAXY
MCG+03-45-039	17 52 39.	+ 18 20 30.	102	13.8	GALAXY
ZWG 171.023	17 52 42.	+ 27 31		15.6	GALAXY
UGC 11036	17 52 42.	+ 58 23	108	16.5	GALAXY
MRSL 005+00/1	17 52 42.	- 24 21	3900		HII REGION
SCHO 0666	17 52 42.	- 35 06 54.	660		ISOLATED DARK CLOUD
SCHO 0667	17 52 44.	- 18 15 12.	380		ISOLATED DARK CLOUD
UGC 11037	17 52 48.	+ 18 17	78	16.0	GALAXY S-IRR
MCG+05-42-014	17 52 48.	+ 27 31	54	15.6	GALAXY
ABC 2285	17 52 48.	+ 42 51		17.0	RICH CLUSTER OF GALAXIES
MCG+07-37-002	17 52 48.	+ 44 33	39	16.	GALAXY
ZWG 340.024	17 52 48.	+ 72 09		15.1	GALAXY
UGC 11038	17 52 48.	+ 72 09	60	15.1	GALAXY Sc
SCHO 0668	17 52 49.	- 28 36 00.	200		ISOLATED DARK CLOUD
PK000-02.1	17 52 50.7	- 29 10 53.9		15.69	PLANETARY NEBULA
ZWG 113.005	17 52 54.	+ 18 33		15.7	GALAXY
UGC 11039	17 52 54.	+ 18 33	60	15.7	GALAXY DWRF IR
MCG+10-25-113	17 52 54.	+ 58 20	90	15.	GALAXY
ISS 0673	17 52 54.	- 15 04	107		STELLAR RING
MRSL 009+03/1	17 52 54.	- 19 09	4800		HII REGION
PK001-01.3	17 52 54.2	- 28 13 26.			PLANETARY NEBULA
SCHO 0669	17 52 55.	- 17 39 30.	390		ISOLATED DARK CLOUD
PK358-03.2	17 52 57.2	- 32 37 10.	11		PLANETARY NEBULA
PK348-09.1	17 52 59.	- 43 04 04.	5		PLANETARY NEBULA
KHAV 425	17 53	- 04 01	6270		DARK NEBULA
KHAV 424	17 53	- 08 31	6460		DARK NEBULA
KHAV 422	17 53	- 29 01	1350		DARK NEBULA
KHAV 421	17 53	- 30 37	4670		DARK NEBULA
MCG+03-46-001	17 53 00.	+ 18 32	54	15.7	GALAXY
ZWG 171.024	17 53 00.	+ 28 31		15.4	GALAXY
UGC 11040	17 53 00.	+ 28 31	90	15.4	GALAXY S
ZWG 171.025	17 53 00.	+ 28 58		15.7	GALAXY
MCG+06-39-025	17 53 00.	+ 34 46	48	14.5	GALAXY
ZWG 199.027	17 53 00.	+ 34 47		13.9	GALAXY
UGC 11041	17 53 00.	+ 34 47	78	13.9	GALAXY Sa-b
ZC 1753.0+3456	17 53 00.	+ 34 56	1340		CLUSTER OF GALAXIES
MCG+11-22-008	17 53 00.	+ 63 56	24	18.	GALAXY
ZC 1753.0+8648	17 53 00.	+ 86 48	2350		CLUSTER OF GALAXIES
LDN 0400	17 53 00.	- 08 15	1200		DARK NEBULA
LDN 0338	17 53 00.	- 14 30	720		DARK NEBULA
LDN 0285	17 53 00.	- 18 20	1860		DARK NEBULA
PK009+02.1	17 53 00.	- 19 28		14.8	PLANETARY NEBULA
COU 012	17 53 00.	- 24 40	2400		EMISSION NEBULA
COU 011	17 53 00.	- 25 00	7200		EMISSION NEBULA
SCHO 0670	17 53 04.	- 29 02 00.	960		ISOLATED DARK CLOUD
ZWG 056.001	17 53 06.	+ 06 12		15.5	GALAXY
RNGC 6499	17 53 06.	+ 18 23			NON-EXISTENT OBJECT
OCL 1025	17 53 06.	- 35 19	1980	8.9	OPEN STAR CLUSTER
ABC 2289	17 53 10.	+ 58 06		17.5	RICH CLUSTER OF GALAXIES
SCHO 0671	17 53 10.	- 29 01 30.	1050		ISOLATED DARK CLOUD
PK359-03.1	17 53 11.7	- 31 03 56.	5		PLANETARY NEBULA
PK000-02.3	17 53 11.9	- 29 37 49.4		15.33	PLANETARY NEBULA
ZWG 141.030	17 53 12.	+ 26 22		14.9	GALAXY
UGC 11042	17 53 12.	+ 26 22	72	14.9	GALAXY Sa?
MCG+11-22-009	17 53 12.	+ 64 10	54	16.	GALAXY
ZWG 322.019	17 53 12.	+ 64 11		15.4	GALAXY
SCHO 0672	17 53 14.	- 18 19 18.	380		ISOLATED DARK CLOUD
HUB E21	17 53 14.	- 23 05			DIFFUSE NEBULA
ZWG 141.031	17 53 18.	+ 25 26		15.5	GALAXY
UGC 11043	17 53 18.	+ 25 26	108	15.5	GALAXY TRP SYS
MCG+04-42-015	17 53 18.	+ 26 23 30.	66	14.9	GALAXY
ZWG 171.026	17 53 18.	+ 28 41		15.6	GALAXY
LDN 0035	17 53 18.	- 29 04	540		DARK NEBULA
ZWG 113.006	17 53 24.	+ 18 56		15.0	GALAXY
UGC 11044	17 53 24.	+ 18 56	84	15.0	GALAXY TRP SYS
ZWG 171.027	17 53 24.	+ 31 51		15.1	GALAXY
UGC 11045	17 53 24.	+ 31 51	66	15.1	GALAXY Sa
MCG+08-32-025	17 53 24.	+ 48 10	42	16.	GALAXY
PK011+04.1	17 53 26.	- 16 28 39.	8		PLANETARY NEBULA
B 289	17 53 26.	- 29 01	2100		DARK OBJECT
MCG+05-42-015	17 53 27.	+ 31 53	72	15.1	GALAXY
SCHO 0673	17 53 27.	- 36 52 24.	350		ISOLATED DARK CLOUD
MCG+03-46-002	17 53 30.	+ 18 55	66	15.0	GALAXY
ZWG 171.028	17 53 30.	+ 28 50		14.6	GALAXY
UGC 11046	17 53 30.	+ 28 50	78	14.6	GALAXY SBa
ISS 0750	17 53 30.	- 20 58	139		STELLAR RING
SCHO 0674	17 53 30.	- 37 09 12.	440		ISOLATED DARK CLOUD
MCG+05-42-016	17 53 36.	+ 28 50	72	14.6	GALAXY
ZC 1753.6+3759	17 53 36.	+ 37 59	3360		CLUSTER OF GALAXIES
MCG+11-22-010	17 53 36.	+ 64 08	45	17.	GALAXY
7ZW 750	17 53 36.	+ 65 48			COMPACT GALAXY
ZWG 141.032	17 53 42.	+ 26 36		15.4	GALAXY
ZWG 171.029	17 53 42.	+ 30 01		15.7	GALAXY
UGC 11047	17 53 42.	+ 30 01	66	15.7	GALAXY S
LDN 0339	17 53 42.	- 14 35	480		DARK NEBULA
B 288	17 53 42.	- 37 05	120		DARK OBJECT
SCHO 0675	17 53 45.	- 36 05 30.	470		ISOLATED DARK CLOUD
ZWG 113.007	17 53 48.	+ 16 12		15.7	GALAXY
ZWG 113.008	17 53 48.	+ 18 21		13.4	GALAXY
RNGC 6500	17 53 48.	+ 18 21		13.5	GALAXY
UGC 11048	17 53 48.	+ 18 21	150	13.4	GALAXY Sa
KARA.72 526A	17 53 48.	+ 18 21	108	13.4	PART OF DOUBLE GALAXY
ISS 0989	17 53 48.	- 31 32	196		STELLAR RING
MCG+03-46-003	17 53 51.	+ 18 20	120	13.4	GALAXY
SCHO 0676	17 53 52.	- 29 40 48.	280		ISOLATED DARK CLOUD
SCHO 0677	17 53 52.	- 29 54 54.	340		ISOLATED DARK CLOUD
ZWG 056.002	17 53 54.	+ 05 44		15.7	GALAXY
ZWG 084.002	17 53 54.	+ 13 12		15.6	GALAXY
MCG+03-46-004	17 53 54.	+ 18 22	84	13.4	GALAXY
ZWG 113.009	17 53 54.	+ 18 23		13.4	GALAXY
RNGC 6501	17 53 54.	+ 18 23		13.5	GALAXY
UGC 11049	17 53 54.	+ 18 23	108	13.4	GALAXY S0/Sa
KARA.72 526B	17 53 54.	+ 18 23	84	13.4	PART OF DOUBLE GALAXY
MCG+06-39-026	17 53 54.	+ 34 34 30.	78	15.	GALAXY
ZWG 199.028	17 53 54.	+ 34 35		15.7	GALAXY
UGC 11050	17 53 54.	+ 34 35	96	15.7	GALAXY Sb-c
OCL 0030	17 53 54.	- 19 01	2820	7.0	OPEN STAR CLUSTER
KHAV 430	17 54	- 17 36	1220		DARK NEBULA
KHAV 426	17 54	- 22 37	4140		DARK NEBULA
KHAV 427	17 54	- 27 43	2570		DARK NEBULA

OBJECT NAME	RIGHT ASCEN.	DECLINATION	DIAM.	MAGN.	TYPE OF OBJECT
KHAV 428	17 54	- 29 43	1770		DARK NEBULA
ZC 1754.0+3707	17 54 00.	+ 37 07	2220		CLUSTER OF GALAXIES
MCG+09-29-043	17 54 00.	+ 56 31	54	16.	GALAXY
MCG+10-25-114	17 54 00.	+ 60 49	60	14.	GALAXY
ZWG 300.092	17 54 00.	+ 60 50		14.3	GALAXY
UGC 11051	17 54 00.	+ 60 50	60	14.3	GALAXY SB
COU 013	17 54 00.	- 20 45	6000		EMISSION NEBULA
LDN 0150	17 54 00.	- 25 00	14940		DARK NEBULA
LDN 0040	17 54 00.	- 29 04	660		DARK NEBULA
PK356-04.2	17 54 00.1	- 34 09 30.	13		PLANETARY NEBULA
RNGC 6510	17 54 01.	+ 60 50		14.5	GALAXY
RNGC 6494	17 54 02.	- 19 01		6.0	OPEN CLUSTER
SCHO 0678	17 54 03.	- 29 25 00.	500		ISOLATED DARK CLOUD
ZWG 113.010	17 54 06.	+ 18 49		15.1	GALAXY
UGC 11052	17 54 06.	+ 18 49	66	15.1	GALAXY S
MCG+09-29-044	17 54 06.	+ 56 31	18	16.	GALAXY
ISS 0990	17 54 06.	- 29 21	78		STELLAR RING
PK357-04.1	17 54 07.1	- 33 35 24.	6	14.3	PLANETARY NEBULA
SCHO 0679	17 54 08.	- 29 21 36.	590		ISOLATED DARK CLOUD
MCG+03-46-005	17 54 09.	+ 18 49	66	15.1	GALAXY
SCHO 0680	17 54 10.	- 29 33 06.	250		ISOLATED DARK CLOUD
RNGC 6504	17 54 11.	+ 33 12		13.5	GALAXY
PK010+03.1	17 54 11.	- 18 06	25	13.3	PLANETARY NEBULA
SCHO 0681	17 54 11.	- 31 30 00.	280		ISOLATED DARK CLOUD
ZWG 199.029	17 54 12.	+ 33 12		13.4	GALAXY
UGC 11053	17 54 12.	+ 33 12	132	13.4	GALAXY S
ZC 1754.2+6120	17 54 12.	+ 61 20	1140		CLUSTER OF GALAXIES
UGC 11054	17 54 12.	+ 71 34	66	16.5	GALAXY
MCG+12-17-011	17 54 12.	+ 71 34	45	17.	GALAXY
MCG+06-39-027	17 54 15.	+ 33 11	138	13.	GALAXY
PK007+01.2	17 54 15.6	- 21 41 09.	5		PLANETARY NEBULA
ZWG 141.033	17 54 18.	+ 26 22		15.5	GALAXY
SCHO 0682	17 54 19.	- 17 53 00.	340		ISOLATED DARK CLOUD
SCHO 0683	17 54 20.	- 30 09 42.	450		ISOLATED DARK CLOUD
REIN 1.124	17 54 21.09	+ 62 38 46.4			NEBULA
SCHO 0684	17 54 22.	- 18 31 36.	370		ISOLATED DARK CLOUD
REIN 1.125	17 54 23.67	+ 62 38 09.1			NEBULA
REIN 1.126	17 54 23.84	+ 62 42 43.4			NEBULA
ZWG 300.093	17 54 24.	+ 62 39		14.8	GALAXY
RNGC 6512	17 54 24.	+ 62 39		15.0	GALAXY
RNGC 6511	17 54 25.	+ 60 49			NON-EXISTENT OBJECT
REIN 1.127	17 54 25.68	+ 62 38 42.9			NEBULA
REIN 1.128	17 54 25.74	+ 62 39 04.3			NEBULA
SCHO 0685	17 54 26.	- 29 04 00.	350		ISOLATED DARK CLOUD
MCG+10-25-115	17 54 27.	+ 62 39	18	15.	GALAXY
SCHO 0636	17 54 28.	- 31 22 00.	240		ISOLATED DARK CLOUD
MCG+10-25-116	17 54 30.	+ 62 42 30.	15	17.	GALAXY
ZC 1754.5+6807	17 54 30.	+ 68 07	4500		CLUSTER OF GALAXIES
ISS 0751	17 54 30.	- 21 03	77		STELLAR RING
SCHO 0687	17 54 30.	- 29 01 42.	350		ISOLATED DARK CLOUD
SCHO 0688	17 54 33.	- 29 51 12.	240		ISOLATED DARK CLOUD
ZWG 322.020	17 54 36.	+ 64 25		15.7	GALAXY
LDN 0302	17 54 36.	- 17 40	780		DARK NEBULA
PK000-02.5	17 54 36.	- 30 03			PLANETARY NEBULA
LDN 0010	17 54 36.	- 30 06	180		DARK NEBULA
ZWG 084.003	17 54 42.	+ 12 15		14.4	GALAXY
UGC 11055	17 54 42.	+ 12 15	60	14.4	GALAXY SB
KARA.72 527A	17 54 42.	+ 12 15	66	14.4	PART OF DOUBLE GALAXY
MCG+02-46-001	17 54 42.	+ 12 15	48	14.4	GALAXY
MCG+08-33-001	17 54 42.	+ 49 00	30	16.	GALAXY
ZC 1754.7+5654	17 54 42.	+ 56 54	1080		CLUSTER OF GALAXIES
B 084a	17 54 42.	- 17 40	960		DARK OBJECT
RNGC 6483	17 54 42.	- 63 40			GALAXY
SCHO 0689	17 54 45.	- 31 24 00.	240		ISOLATED DARK CLOUD
SCHO 0690	17 54 46.	- 28 29 06.	420		ISOLATED DARK CLOUD
REIN 1.129	17 54 47.71	+ 62 36 40.4			NEBULA
ZWG 141.034	17 54 48.	+ 24 01		15.7	GALAXY
ZWG 171.030	17 54 48.	+ 28 05		15.6	GALAXY
UGC 11056	17 54 48.	+ 28 05	66	15.6	GALAXY S0-a
ZWG 254.001	17 54 48.	+ 49 02		15.5	GALAXY
ZWG 253.040	17 54 48.	+ 49 02		15.5	GALAXY
KARA.73B 0839	17 54 48.	+ 49 02	24	15.5	ISOLATED GALAXY S
ZWG 301.001	17 54 48.	+ 62 40		15.7	GALAXY
ZWG 300.094	17 54 48.	+ 62 40		15.7	GALAXY
RNGC 6516	17 54 48.	+ 62 40		15.5	GALAXY
7ZW 751	17 54 48.	+ 68 03			COMPACT GALAXY
LDN 0292	17 54 48.	- 18 00	900		DARK NEBULA
SCHO 0691	17 54 49.	- 17 45 54.	1390		ISOLATED DARK CLOUD
REIN 1.130	17 54 49.08	+ 62 36 26.5			NEBULA
MCG+05-42-017	17 54 51.	+ 31 19	39	15.5	GALAXY
REIN 1.132	17 54 52.37	+ 62 40 32.1			NEBULA
REIN 1.131	17 54 52.44	+ 62 37 55.8			NEBULA
ZWG 084.004	17 54 54.	+ 12 11		13.8	GALAXY
UGC 11057	17 54 54.	+ 12 11	120	13.8	GALAXY Sc
KARA.72 527B	17 54 54.	+ 12 11	120	13.8	PART OF DOUBLE GALAXY
MCG+02-46-002	17 54 54.	+ 12 11	120	13.8	GALAXY
MCG+10-25-117	17 54 54.	+ 61 37	18	17.	GALAXY
ZC 1754.9+6230	17 54 54.	+ 62 30	5710		CLUSTER OF GALAXIES
MCG+11-22-011	17 54 54.	+ 65 55	39	16.	GALAXY
PK358-03.1	17 54 55.7	- 31 42 42.	4		PLANETARY NEBULA
MCG+10-25-118	17 54 57.	+ 62 40	30	16.	GALAXY
PK357-04.3	17 54 57.1	- 33 47 31.	25		PLANETARY NEBULA
PK000-02.4	17 54 57.8	- 29 44 06.	10		PLANETARY NEBULA
REIN 1.133	17 54 58.58	+ 62 39 50.3			NEBULA
PK359-03.2	17 54 58.6	- 31 07 53.	7		PLANETARY NEBULA
KHAV 432	17 55	- 17 24	1860		DARK NEBULA
KHAV 429	17 55	- 31 54	2230		DARK NEBULA
LB 09922	17 55	- 83 02		14.7	FAINT BLUE STAR
ZWG 171.031	17 55 00.	+ 31 48		15.5	GALAXY
ZWG 171.032	17 55 00.	+ 32 38		14.4	GALAXY
UGC 11058	17 55 00.	+ 32 38	96	14.4	GALAXY SBb
KARA.73B 0840	17 55 00.	+ 32 38	96	14.4	ISOLATED GALAXY S
ZWG 227.003	17 55 00.	+ 40 15		15.5	GALAXY
ZWG 227.004	17 55 00.	+ 41 42		15.7	GALAXY
LDN 0442	17 55 00.	- 04 30	480		DARK NEBULA
LDN 0337	17 55 00.	- 14 51	600		DARK NEBULA
COU 014	17 55 00.	- 16 30	15000		EMISSION NEBULA
LDN 0293	17 55 00.	- 18 00	5760		DARK NEBULA
LDN 0288	17 55 00.	- 18 30	900		DARK NEBULA
ISS 0752	17 55 00.	- 20 35	110		STELLAR RING
LDN 0215	17 55 00.	- 22 30	12060		DARK NEBULA
LDN 0048	17 55 00.	- 29 00	10020		DARK NEBULA
LDN 0012	17 55 00.	- 30 07	180		DARK NEBULA
LDN 1783	17 55 00.	- 31 30	1140		DARK NEBULA
LDN 1766	17 55 00.	- 32 20	1200		DARK NEBULA
REIN 1.134	17 55 02.80	+ 62 39 08.6			NEBULA
MCG+05-42-018	17 55 03.	+ 32 40	84	14.4	GALAXY
ISS 0753	17 55 06.	- 20 27	118		STELLAR RING

OBJECT NAME	RIGHT ASCEN.	DECLINATION	DIAM.	MAGN.	TYPE OF OBJECT
ISS 0896	17 55 06.	- 21 03	64		STELLAR RING
PK000-02.6	17 55 07.1	- 30 00 27.	5		PLANETARY NEBULA
ZWG 084.005	17 55 12.	+ 11 45		15.7	GALAXY
UGC 11059	17 55 12.	+ 11 45	66	15.7	GALAXY S
ZWG 084.006	17 55 12.	+ 14 12		15.4	GALAXY
ZWG 171.033	17 55 12.	+ 27 58		14.9	GALAXY
UGC 11060	17 55 12.	+ 27 58	90	14.9	GALAXY Sa
PK002-02.1	17 55 12.2	- 28 14 38.	10		PLANETARY NEBULA
PK357-04.2	17 55 14.7	- 33 28 23.	6		PLANETARY NEBULA
MCG+05-42-019	17 55 15.	+ 27 58	72	14.9	GALAXY
PK001-02.1	17 55 17.9	- 28 33 28.			PLANETARY NEBULA
ZWG 056.003	17 55 18.	+ 04 37		15.6	GALAXY
7ZW 752	17 55 18.	+ 68 39			COMPACT GALAXY
REIN 1.135	17 55 19.74	+ 62 37 00.0			NEBULA
PK002-01.1	17 55 22.5	- 27 36 52.	6		PLANETARY NEBULA
RNGC 6538	17 55 23.	+ 73 25		14.0	GALAXY
REIN 1.136	17 55 23.81	+ 62 37 01.7			NEBULA
ZWG 084.007	17 55 24.	+ 13 33		15.6	GALAXY
ZWG 301.002	17 55 24.	+ 62 37		14.3	GALAXY
ZWG 300.095	17 55 24.	+ 62 37		14.3	GALAXY
RNGC 6521	17 55 24.	+ 62 37		14.5	GALAXY
UGC 11061	17 55 24.	+ 62 37	96	14.3	GALAXY E
MCG+10-25-119	17 55 24.	+ 62 37	24	13.	GALAXY
MCG+11-22-012	17 55 24.	+ 65 16	39	16.	GALAXY
ZWG 322.021	17 55 24.	+ 66 33		15.4	GALAXY
ZWG 340.025	17 55 24.	+ 73 25		14.1	GALAXY
UGC 11062	17 55 24.	+ 73 25	66	14.1	GALAXY S
MCG+12-17-012	17 55 24.	+ 73 26	60	15.	GALAXY
OCL 0049	17 55 24.	- 11 41	672	12.	OPEN STAR CLUSTER
GCL 080	17 55 24.	- 44 15	762	10.3	GLOBULAR STAR CLUSTER
SCHO 0692	17 55 29.	- 28 05 00.	250		ISOLATED DARK CLOUD
RNGC 6496	17 55 29.	- 44 13		10.5	GLOBULAR CLUSTER
REIN 1.137	17 55 29.80	+ 62 37 18.0			NEBULA
ZWG 141.035	17 55 30.	+ 23 02		15.5	GALAXY
UGC 11063	17 55 30.	+ 23 02	66	15.5	GALAXY Sc
MCG+04-42-016	17 55 30.	+ 23 02	60	15.5	GALAXY
REIN 1.138	17 55 33.39	+ 62 36 59.5			NEBULA
REIN 1.139	17 55 33.42	+ 62 38 13.4			NEBULA
ZWG 028.002	17 55 36.	+ 01 20		15.7	GALAXY
MCG+00-46-002	17 55 36.	+ 01 20	30	15.7	GALAXY
LB 01064	17 55 36.	+ 44 27 18.		17.6	FAINT BLUE STAR
7ZW 753	17 55 36.	+ 65 47			COMPACT GALAXY
REIN 1.140	17 55 37.99	+ 62 37 17.7			NEBULA
SCHO 0693	17 55 39.	- 27 52 00.	320		ISOLATED DARK CLOUD
REIN 1.141	17 55 40.83	+ 62 36 57.2			NEBULA
ZWG 056.004	17 55 42.	+ 03 44		15.4	GALAXY
ZWG 141.036	17 55 42.	+ 23 53		15.4	GALAXY
MCG+05-42-020	17 55 42.	+ 27 50	108	14.5	GALAXY
ZWG 171.034	17 55 42.	+ 27 51		14.5	GALAXY
UGC 11064	17 55 42.	+ 27 51	120	14.5	GALAXY Sc
ZWG 227.005	17 55 42.	+ 43 45		15.7	GALAXY
ZWG 254.002	17 55 42.	+ 47 45		14.9	GALAXY
UGC 11065	17 55 42.	+ 47 45	60	14.9	GALAXY S
ZWG 300.096	17 55 42.	+ 58 42		15.4	GALAXY
ZWG 301.003	17 55 42.	+ 62 21		15.7	GALAXY
ZWG 300.097	17 55 42.	+ 62 21		15.7	GALAXY
ZWG 340.026	17 55 42.	+ 71 33		15.5	GALAXY
UGC 11066	17 55 42.	+ 71 33	96	15.5	GALAXY
ABC 2291	17 55 44.	+ 51 09		17.6	RICH CLUSTER OF GALAXIES
SCHO 0694	17 55 46.	- 33 19 24.	790		ISOLATED DARK CLOUD
PK358-04.1	17 55 46.3	- 32 21 33.	10		PLANETARY NEBULA
RNGC 6534	17 55 47.	+ 64 19		15.5	GALAXY
SCHO 0696	17 55 47.	- 28 08 00.	260		ISOLATED DARK CLOUD
SCHO 0695	17 55 47.	- 33 28 00.	350		ISOLATED DARK CLOUD
MCG+08-33-002	17 55 48.	+ 47 45	66	15.	GALAXY
MCG+10-25-120	17 55 48.	+ 62 20	24	16.	GALAXY
MCG+11-22-013	17 55 48.	+ 64 16	45	16.	GALAXY
ZWG 322.022	17 55 48.	+ 64 19		15.5	GALAXY
7ZW 754	17 55 48.	+ 65 21			COMPACT GALAXY
MCG+12-17-013	17 55 48.	+ 71 32 30.	66	16.	GALAXY
ISS 0897	17 55 48.	- 26 29	91		STELLAR RING
SCHO 0697	17 55 50.	- 29 36 00.	460		ISOLATED DARK CLOUD
ZWG 141.037	17 55 54.	+ 21 17		15.0	GALAXY
ZWG 171.035	17 55 54.	+ 31 49		15.6	GALAXY
ZWG 278.039	17 55 54.	+ 54 24		15.3	GALAXY
ZWG 278.040	17 55 54.	+ 55 49		15.5	GALAXY
ZC 1755.9+5556	17 55 54.	+ 55 56	2290		CLUSTER OF GALAXIES
B 290	17 55 56.	- 37 09	1500		DARK OBJECT
KHAV 433	17 56	- 15 06	4140		DARK NEBULA
KHAV 434	17 56	- 17 54	1000		DARK NEBULA
KHAV 435	17 56	- 20 36			DARK NEBULA
KHAV 431	17 56	- 21 36	4760		DARK NEBULA
ZWG 084.008	17 56 00.	+ 09 42		15.0	GALAXY
UGC 11067	17 56 00.	+ 09 42	60	15.0	GALAXY S
MCG+02-46-003	17 56 00.	+ 09 42	60	15.0	GALAXY
ZWG 084.009	17 56 00.	+ 14 58		15.5	GALAXY
MCG+02-46-004	17 56 00.	+ 14 58	48	15.5	GALAXY
ZWG 113.011	17 56 00.	+ 15 16		15.1	GALAXY
MCG+03-46-006	17 56 00.	+ 15 16	36	15.1	GALAXY
MCG+04-42-017	17 56 00.	+ 21 16	24	15.0	GALAXY
ZWG 171.036	17 56 00.	+ 28 15		15.5	GALAXY
UGC 11068	17 56 00.	+ 28 15	96	15.5	GALAXY S
MCG+05-42-021	17 56 00.	+ 31 51	42	15.6	GALAXY
ZC 1756.0+5114	17 56 00.	+ 51 14	2020		CLUSTER OF GALAXIES
UGC 11069	17 56 00.	+ 64 39	78	16.5	GALAXY S
ZC 1756.0+7117	17 56 00.	+ 71 17	4570		CLUSTER OF GALAXIES
LDN 0460	17 56 00.	- 03 45	300		DARK NEBULA
LDN 0432	17 56 00.	- 05 45	360		DARK NEBULA
LDN 0363	17 56 00.	- 13 00	6000		DARK NEBULA
LDN 0340	17 56 00.	- 14 36	720		DARK NEBULA
LDN 0298	17 56 00.	- 18 00	1320		DARK NEBULA
LDN 0133	17 56 00.	- 26 00	1380		DARK NEBULA
SCHO 0698	17 56 00.	- 29 32 18.	410		ISOLATED DARK CLOUD
SCHO 0699	17 56 02.	- 33 58 00.	310		ISOLATED DARK CLOUD
SCHO 0700	17 56 04.	- 34 07 00.	250		ISOLATED DARK CLOUD
ZWG 171.037	17 56 06.	+ 27 16		15.4	GALAXY
UGC 11070	17 56 06.	+ 27 16	66	15.4	GALAXY SBb
ZWG 254.003	17 56 06.	+ 45 16		15.7	GALAXY
ZWG 301.004	17 56 06.	+ 62 37		15.5	GALAXY
ZWG 300.098	17 56 06.	+ 62 37		15.5	GALAXY
MCG+10-25-121	17 56 06.	+ 62 38	48	15.	GALAXY
PK013+04.1	17 56 07.	- 15 32 06.	5		PLANETARY NEBULA
SCHO 0701	17 56 07.	- 33 53 18.	360		ISOLATED DARK CLOUD
MCG+05-42-022	17 56 09.	+ 28 15 30.	78	15.5	GALAXY
PK002-02.2	17 56 09.7	- 28 13 38.	10		PLANETARY NEBULA
ABC 2292	17 56 10.	+ 53 50		17.1	RICH CLUSTER OF GALAXIES
MCG+08-33-003	17 56 12.	+ 50 43	24	14.	GALAXY
ZWG 254.004	17 56 12.	+ 50 44		14.3	GALAXY

OBJECT NAME	RIGHT ASCEN.	DECLINATION	DIAM.	MAGN.	TYPE OF OBJECT
RNGC 6515	17 56 12.	+ 50 44		14.5	GALAXY
UGC 11071	17 56 12.	+ 50 44	102	14.3	GALAXY E
ISS 0754	17 56 12.	- 20 08	83		STELLAR RING
OCL 0015	17 56 12.	- 24 42	132	13.	OPEN STAR CLUSTER
ISS 0898	17 56 12.	- 25 28	107		STELLAR RING
LB 01065	17 56 13.	+ 63 24 12.		17.5	FAINT BLUE STAR
SCHO 0702	17 56 17.	- 16 43 00.	380		ISOLATED DARK CLOUD
ZWG 254.005	17 56 18.	+ 45 08		15.4	GALAXY
ZWG 322.023	17 56 18.	+ 65 13		15.6	GALAXY
MCG+11-22-014	17 56 18.	+ 65 13	15	16.	GALAXY
ISS 0991	17 56 18.	- 30 34	169		STELLAR RING
MCG+05-42-023	17 56 24.	+ 27 17	51	15.4	GALAXY
ZWG 171.038	17 56 24.	+ 27 50		15.7	GALAXY
UGC 11072	17 56 24.	+ 43 23	66	16.0	GALAXY Sa-b
ZWG 301.005	17 56 24.	+ 62 07		15.6	GALAXY
ZWG 300.099	17 56 24.	+ 62 07		15.6	GALAXY
ISS 0755	17 56 24.	- 20 07	114		STELLAR RING
OCL 0014	17 56 24.	- 24 47	150	14.	OPEN STAR CLUSTER
B 291	17 56 25.	- 33 54	300		DARK OBJECT
SCHO 0703	17 56 26.	- 32 28 00.	440		ISOLATED DARK CLOUD
ABC 2296	17 56 28.	+ 77 42		15.9	RICH CLUSTER OF GALAXIES
ZC 1756.5+2904	17 56 30.	+ 29 04	13640		CLUSTER OF GALAXIES
MCG+07-37-003	17 56 30.	+ 43 23	63	15.5	GALAXY
MCG+10-25-122	17 56 30.	+ 62 07	48	16.	GALAXY
ZWG 301.006	17 56 30.	+ 62 35		15.7	GALAXY
ZWG 300.100	17 56 30.	+ 62 35		15.7	GALAXY
7ZW 755	17 56 30.	+ 62 36			COMPACT GALAXY
LDN 0341	17 56 30.	- 14 40	420		DARK NEBULA
ZWG 084.010	17 56 36.	+ 10 33		14.8	GALAXY
UGC 11073	17 56 36.	+ 10 33	90	14.8	GALAXY Sb-c
MCG+02-46-005	17 56 36.	+ 10 33	60	14.8	GALAXY
MCG+07-37-004	17 56 36.	+ 43 24 30.	21	15.	GALAXY
ZWG 227.006	17 56 36.	+ 43 25		15.6	GALAXY
7ZW 756	17 56 36.	+ 65 13			COMPACT GALAXY
ZWG 322.024	17 56 36.	+ 65 16		15.6	GALAXY
MCG+11-22-015	17 56 36.	+ 65 16	39	16.	GALAXY
ISS 0756	17 56 36.	- 16 23	100		STELLAR RING
ISS 0757	17 56 36.	- 18 03	58		STELLAR RING
SCHO 0704	17 56 39.	- 29 32 18.	270		ISOLATED DARK CLOUD
SCHO 0705	17 56 40.	- 33 29 42.	360		ISOLATED DARK CLOUD
LB 01066	17 56 41.	+ 43 13 48.		17.6	FAINT BLUE STAR
PK359-04.1	17 56 41.4	- 31 54 27.	7		PLANETARY NEBULA
ZWG 056.005	17 56 42.	+ 07 09		15.2	GALAXY
UGC 11074	17 56 42.	+ 07 09	162	15.2	GALAXY
MCG+01-46-001	17 56 42.	+ 07 09	120	15.2	GALAXY
ZWG 171.039	17 56 42.	+ 28 20		15.7	GALAXY
MCG+08-33-004	17 56 42.	+ 50 52	36	16.	GALAXY
ZWG 278.041	17 56 42.	+ 54 27		15.5	GALAXY
OCL 0032	17 56 42.	- 17 24	450	9.6	OPEN STAR CLUSTER
ISS 0758	17 56 42.	- 18 14	206		STELLAR RING
OCL 0009	17 56 42.	- 28 11	600	10.1	OPEN STAR CLUSTER
SCHO 0706	17 56 42.	- 33 25 00.	320		ISOLATED DARK CLOUD
PK352-07.1	17 56 44.4	- 38 49 45.	25	11.4	PLANETARY NEBULA
RNGC 6507	17 56 45.	- 17 23		9.5	OPEN CLUSTER
RNGC 6506	17 56 46.	- 24 39			NON-EXISTENT OBJECT
ZWG 084.011	17 56 48.	+ 10 55		15.7	GALAXY
7ZW 757	17 56 48.	+ 68 03			COMPACT GALAXY
ISS 0759	17 56 48.	- 20 14	130		STELLAR RING
OCL 0013	17 56 48.	- 25 11	180	13.	OPEN STAR CLUSTER
LB 01067	17 56 51.	+ 45 16 00.		17.3	FAINT BLUE STAR
OCL 0016	17 56 54.	- 24 41	240	13.	OPEN STAR CLUSTER
PK356-05.1	17 56 58.8	- 34 27 35.	11		PLANETARY NEBULA
RNGC 6509	17 56 59.	+ 06 17		13.5	GALAXY
KHAV 438	17 57	- 18 00	1220		DARK NEBULA
ZWG 056.006	17 57 00.	+ 06 17		13.4	GALAXY
UGC 11075	17 57 00.	+ 06 17	102	13.4	GALAXY Sc
MCG+01-46-002	17 57 00.	+ 06 17	96	13.4	GALAXY
MCG+06-39-028	17 57 00.	+ 33 59	48	14.	GALAXY
ZWG 199.030	17 57 00.	+ 34 00		14.8	GALAXY
UGC 11076	17 57 00.	+ 34 00	78	14.8	GALAXY Sb-c
MCG+11-22-016	17 57 00.	+ 64 57	72	15.	GALAXY
LDN 0439	17 57 00.	- 05 00	1680		DARK NEBULA
LDN 0320	17 57 00.	- 16 30	4800		DARK NEBULA
LDN 0318	17 57 00.	- 16 50	360		DARK NEBULA
ISS 0760	17 57 00.	- 20 43	103		STELLAR RING
LDN 1760	17 57 00.	- 33 00	900		DARK NEBULA
ACK 340-14.1	17 57 00.	- 52 44			PLANETARY NEBULA
SCHO 0707	17 57 02.	- 35 20 00.	270		ISOLATED DARK CLOUD
SCHO 0708	17 57 02.	- 35 32 00.	270		ISOLATED DARK CLOUD
SCHO 0709	17 57 03.	- 29 24 00.	330		ISOLATED DARK CLOUD
ZWG 227.007	17 57 06.	+ 41 56		15.6	GALAXY
ZWG 322.025	17 57 06.	+ 64 56		14.3	GALAXY
UGC 11077	17 57 06.	+ 64 56	84	14.3	GALAXY SBb
RNGC 6536	17 57 07.	+ 64 56		14.5	GALAXY
MCG+07-37-005	17 57 12.	+ 41 56	12	16.	GALAXY
MCG+07-37-006	17 57 12.	+ 42 24	42	16.	GALAXY
PK011+02.1	17 57 13.	- 17 41	5	13.2	PLANETARY NEBULA
MCG+06-39-029	17 57 15.	+ 36 58	39	15.	GALAXY
LB 01068	17 57 15.	+ 46 49 54.		15.6	FAINT BLUE STAR
B 292	17 57 16.	- 33 21	3600		DARK OBJECT
ZWG 340.027	17 57 18.	+ 70 18		15.0	GALAXY
SCHO 0710	17 57 21.	- 17 18 36.	400		ISOLATED DARK CLOUD
SCHO 0711	17 57 22.	- 33 45 00.	570		ISOLATED DARK CLOUD
ZWG 113.012	17 57 24.	+ 17 33		15.4	GALAXY
MCG+07-37-007	17 57 24.	+ 42 24 30.	24	16.	GALAXY
7ZW 758	17 57 24.	+ 69 19			COMPACT GALAXY
ISS 0761	17 57 24.	- 20 46	257		STELLAR RING
SG 3.114	17 57 26.	- 23 15	2700		DIFFUSE EMISSION NEBULA
LB 01069	17 57 27.	+ 47 04 54.		17.0	FAINT BLUE STAR
SCHO 0712	17 57 27.	- 31 42 00.	250		ISOLATED DARK CLOUD
RNGC 6513	17 57 29.	+ 24 54		14.5	GALAXY
MCG+03-46-007	17 57 30.	+ 17 32	30	15.4	GALAXY
ZWG 141.038	17 57 30.	+ 24 54		14.7	GALAXY
UGC 11078	17 57 30.	+ 24 54	84	14.7	GALAXY SB:0
ZWG 171.040	17 57 30.	+ 27 48		15.6	GALAXY
ISS 0762	17 57 30.	- 16 03	113		STELLAR RING
MIL 57	17 57 30.	- 23 25	1800		SUPERNOVA REMNANT
SCHO 0713	17 57 32.	- 32 55 00.	300		ISOLATED DARK CLOUD
MCG+04-42-018	17 57 33.	+ 24 55	27	14.7	GALAXY
MCG+07-37-008	17 57 33.	+ 42 24	27	16.	GALAXY
SCHO 0714	17 57 34.	- 31 24 00.	310		ISOLATED DARK CLOUD
MCG+07-37-009	17 57 36.	+ 42 54	48	16.	GALAXY
ISS 0763	17 57 36.	- 16 02	82		STELLAR RING
VMT 19	17 57 36.	- 23 25	1800		SUPERNOVA REMNANT
URA 10	17 57 38.	- 20 17	192		STELLAR RING
HN 0450	17 57 38.	- 62 50			NEBULA
IC 4672	17 57 38.	- 62 50			NONSTELLAR OBJECT
RNGC 6492	17 57 39.	- 66 25			GALAXY

OBJECT NAME	RIGHT ASCEN.	DECLINATION	DIAM.	MAGN.	TYPE OF OBJECT
LB 01070	17 57 41.	+ 46 00 30.		18.3	FAINT BLUE STAR
SCHO 0715	17 57 41.	- 31 34 00.	240		ISOLATED DARK CLOUD
ZWG 171.041	17 57 42.	+ 28 52		15.1	GALAXY
MCG+05-42-024	17 57 42.	+ 28 52	24	15.1	GALAXY
SCHO 0716	17 57 42.	- 33 04 00.	370		ISOLATED DARK CLOUD
RNGC 6518	17 57 44.	+ 28 52		15.0	GALAXY
LB 01071	17 57 45.	+ 44 19 30.		17.2	FAINT BLUE STAR
SCHO 0717	17 57 45.	- 32 52 18.	490		ISOLATED DARK CLOUD
MCG+05-42-025	17 57 48.	+ 27 17	18	16.	GALAXY
ZC 1757.8+3842	17 57 48.	+ 38 42	4100		CLUSTER OF GALAXIES
MCG+08-33-005	17 57 48.	+ 45 54	42	14.	GALAXY
ZWG 254.006	17 57 48.	+ 45 55		14.0	GALAXY
RNGC 6524	17 57 48.	+ 45 55		14.0	GALAXY
UGC 11079	17 57 48.	+ 45 55	102	14.0	GALAXY E-SO
KARA.73B 0841	17 57 48.	+ 45 55	78	14.0	ISOLATED GALAXY S
B 293	17 57 52.	- 35 21	1080		DARK OBJECT
ZWG 141.039	17 57 54.	+ 26 20		15.7	GALAXY
UGC 11080	17 57 54.	+ 26 20	60	15.7	GALAXY S
KARA.72 528A	17 57 54.	+ 26 20	54	15.7	PART OF DOUBLE GALAXY
ZC 1757.9+4238	17 57 54.	+ 42 38	7460		CLUSTER OF GALAXIES
ZWG 227.008	17 57 54.	+ 44 17		15.4	GALAXY
MRSL 006-00/1	17 57 54.	- 23 20	1500		HII REGION
PK001-03.1	17 57 56.2	- 29 21 46.	5		PLANETARY NEBULA
PK003-01.1	17 57 59.	- 26 21	25		PLANETARY NEBULA
KHAV 439	17 58	- 19 00	8570		DARK NEBULA
KHAV 437	17 58	- 29 24	2370		DARK NEBULA
KHAV 436	17 58	- 33 18	5590		DARK NEBULA
ZWG 141.040	17 58 00.	+ 26 21		14.4	GALAXY
UGC 11082	17 58 00.	+ 26 21	48	14.4	GALAXY SO
KARA.72 528B	17 58 00.	+ 26 21	54	14.4	PART OF DOUBLE GALAXY
MCG+04-42-019	17 58 00.	+ 26 22	45	15.7	GALAXY
ZWG 199.031	17 58 00.	+ 36 40		15.3	GALAXY
MCG+07-37-010	17 58 00.	+ 42 30	42	15.5	GALAXY
ZWG 227.009	17 58 00.	+ 42 31		15.7	GALAXY
UGC 11083	17 58 00.	+ 57 15	60	17.	GALAXY S-IRR
UGC 11081	17 58 00.	+ 81 08	60	17.	GALAXY DWARF
LDN 0468	17 58 00.	- 03 30	480		DARK NEBULA
LDN 0289	17 58 00.	- 18 50	660		DARK NEBULA
LDN 0237	17 58 00.	- 22 10	2940		DARK NEBULA
LDN 0078	17 58 00.	- 28 11	120		DARK NEBULA
PK002-02.3	17 58 01.	- 27 38 42.	7		PLANETARY NEBULA
MCG+07-37-011	17 58 03.	+ 43 23	36	15.5	GALAXY
PK357-05.1	17 58 04.7	- 33 17 43.	13		PLANETARY NEBULA
MCG+06-39-030	17 58 06.	+ 36 40	42	15.	GALAXY
ISS 0764	17 58 06.	- 17 41	146		STELLAR RING
OCL 0020	17 58 06.	- 23 33	600	12.	OPEN STAR CLUSTER
MCG+04-42-020	17 58 09.	+ 26 23	42	14.4	GALAXY
MCG+10-25-123	17 58 09.	+ 62 44	18	17.	GALAXY
LB 01072	17 58 10.	+ 46 25 30.		15.9	FAINT BLUE STAR
SCHO 0718	17 58 11.	- 26 57 00.	540		ISOLATED DARK CLOUD
UGC 11084	17 58 12.	+ 57 08	66	16.5	GALAXY Sc
RNGC 6532	17 58 15.	+ 56 13		15.0	GALAXY
IC 4677	17 58 17.	+ 66 38 11.			NONSTELLAR OBJECT
SCHO 0719	17 58 17.	- 31 32 00.	270		ISOLATED DARK CLOUD
SCHO 0720	17 58 17.	- 32 36 00.	400		ISOLATED DARK CLOUD
ZWG 278.042	17 58 18.	+ 56 13		15.0	GALAXY
UGC 11085	17 58 18.	+ 56 13	126	15.0	GALAXY Sc
MCG+09-29-045	17 58 18.	+ 56 13	96	14.	GALAXY
ZWG 301.007	17 58 18.	+ 62 43		15.7	GALAXY
ZWG 300.101	17 58 18.	+ 62 43		15.7	GALAXY
VV 121	17 58 18.	+ 66 38	48	15.	INTERACTING GALAXY
MCG+11-22-017	17 58 18.	+ 66 38	60	15.	GALAXY
ISS 0674	17 58 18.	- 14 13	95		STELLAR RING
ISS 0992	17 58 18.	- 31 35	104		STELLAR RING
SCHO 0721	17 58 18.	- 33 08 00.	280		ISOLATED DARK CLOUD
SCHO 0722	17 58 18.	- 35 30 00.	500		ISOLATED DARK CLOUD
PK001-03.2	17 58 18.6	- 29 19 31.			PLANETARY NEBULA
SCHO 0723	17 58 19.	- 28 34 42.	180		ISOLATED DARK CLOUD
B 294	17 58 20.	- 28 36	180		DARK OBJECT
SCHO 0724	17 58 21.	- 32 34 30.	550		ISOLATED DARK CLOUD
MIN.46 07	17 58 23.	- 33 15			DIFFUSE NEBULA
ZWG 171.042	17 58 24.	+ 28 47		15.0	GALAXY
UGC 11086	17 58 24.	+ 28 47	84	15.0	GALAXY S
HOLM 773B	17 58 24.	+ 28 48	18	15.3	PART OF MULTIPLE GALAXY
ZWG 199.032	17 58 24.	+ 34 30		15.6	GALAXY
ZC 1758.4+5649	17 58 24.	+ 56 49	1810		CLUSTER OF GALAXIES
ZC 1758.4+5813	17 58 24.	+ 58 13	1610		CLUSTER OF GALAXIES
MCG+10-25-124	17 58 24.	+ 62 45	24	17.	GALAXY
PK358-05.1	17 58 25.3	- 33 15 26.	10		PLANETARY NEBULA
HOLM 773A	17 58 27.	+ 28 47	36	14.2	PART OF MULTIPLE GALAXY
MCG+05-42-026	17 58 27.	+ 28 48	66	15.0	GALAXY
SCHO 0725	17 58 29.	- 17 16 18.	330		ISOLATED DARK CLOUD
MCG+06-39-031	17 58 30.	+ 34 37	36	14.5	GALAXY
ZWG 199.033	17 58 30.	+ 34 38		14.4	GALAXY
UGC 11087	17 58 30.	+ 34 38	39	14.4	GALAXY S
ZWG 227.010	17 58 30.	+ 44 51		15.3	GALAXY
UGC 11088	17 58 30.	+ 44 51	66	15.3	GALAXY SBb
MCG+07-37-012	17 58 30.	+ 44 52	66	14.5	GALAXY
MCG+10-25-125	17 58 30.	+ 58 30	60	15.	GALAXY
7ZW 759	17 58 30.	+ 66 38			COMPACT GALAXY
ZWG 340.028	17 58 30.	+ 69 18		15.4	GALAXY
SCHO 0726	17 58 32.	- 16 51 18.	350		ISOLATED DARK CLOUD
PK002-02.4	17 58 32.7	- 28 25 46.	10		PLANETARY NEBULA
MCG+07-37-013	17 58 33.	+ 44 32	66	15.5	GALAXY
PK096+29.1	17 58 34.24	+ 66 38 05.3	300	8.8	PLANETARY NEBULA
OCL 0084	17 58 36.	+ 02 54	15000	3.2	OPEN STAR CLUSTER
ZWG 141.041	17 58 36.	+ 23 03		15.6	GALAXY
KARA.73B 0842	17 58 36.	+ 23 03	48	15.6	ISOLATED GALAXY S
MCG+06-39-032	17 58 36.	+ 34 37	9	17.	GALAXY
MCG+07-37-014	17 58 36.	+ 44 30	48	15.	GALAXY
ZWG 278.043	17 58 36.	+ 52 23		15.2	GALAXY
ZWG 301.008	17 58 36.	+ 58 33		15.2	GALAXY
ZWG 300.102	17 58 36.	+ 58 33		15.2	GALAXY
UGC 11089	17 58 36.	+ 58 33	60	15.2	GALAXY S
KARA.73B 0843	17 58 36.	+ 58 33	54	15.2	ISOLATED GALAXY S
RNGC 6543	17 58 36.	+ 66 38		9.0	PLANETARY NEBULA
ZC 1758.6+7111	17 58 36.	+ 71 11	1340		CLUSTER OF GALAXIES
SCHO 0727	17 58 37.	- 29 22 00.	340		ISOLATED DARK CLOUD
SCHO 0728	17 58 39.	- 33 28 00.	260		ISOLATED DARK CLOUD
ZWG 171.043	17 58 42.	+ 28 43		14.8	GALAXY
UGC 11090	17 58 42.	+ 28 43	90	14.8	GALAXY Sc
ZC 1758.7+6920	17 58 42.	+ 69 20	2890		CLUSTER OF GALAXIES
ZWG 340.029	17 58 42.	+ 69 38		15.6	GALAXY
TER 09	17 58 42.	- 26 52 00.	35		STAR CLUSTER
PK356-05.2	17 58 43.3	- 34 27 49.	11		PLANETARY NEBULA
MIL 56	17 58 44.	- 24 54	1080		SUPERNOVA REMNANT
SCHO 0729	17 58 45.	- 35 15 00.	290		ISOLATED DARK CLOUD
MCG+05-42-027	17 58 48.	+ 28 44	72	14.8	GALAXY
ZWG 340.030	17 58 48.	+ 70 40		15.2	GALAXY
SCHO 0730	17 58 52.	- 17 00 36.	390		ISOLATED DARK CLOUD
ZWG 141.042	17 58 54.	+ 22 34		15.5	GALAXY
MCG+06-39-033	17 58 54.	+ 34 59 30.	54	14.5	GALAXY
ZWG 199.034	17 58 54.	+ 35 00		15.7	GALAXY
UGC 11091	17 58 54.	+ 35 00	66	15.7	GALAXY Sc
KARA.73B 0844	17 58 54.	+ 35 00	66	15.7	ISOLATED GALAXY S
RNGC 6502	17 58 54.	- 65 25			GALAXY
SCHO 0731	17 58 56.	- 27 26 00.	260		ISOLATED DARK CLOUD
KHAV 443	17 59	+ 00 12	8110		DARK NEBULA
KHAV 442	17 59	- 13 24	7960		DARK NEBULA
LBN 0027	17 59	- 23 00	1200		BRIGHT NEBULA
KHAV 441	17 59	- 31 00	3250		DARK NEBULA
KHAV 440	17 59	- 34 00	3070		DARK NEBULA
LB 09923	17 59	- 81 14		14.6	FAINT BLUE STAR
MCG+04-42-021	17 59 00.	+ 22 33	48	15.5	GALAXY
ZWG 141.043	17 59 00.	+ 24 44		15.2	GALAXY
ZWG 171.044	17 59 00.	+ 29 00		15.6	GALAXY
LDN 0319	17 59 00.	- 16 50	840		DARK NEBULA
LDN 0286	17 59 00.	- 19 10	1320		DARK NEBULA
LDN 0166	17 59 00.	- 25 10	360		DARK NEBULA
LDN 0135	17 59 00.	- 26 20	360		DARK NEBULA
LDN 0049	17 59 00.	- 29 30	4140		DARK NEBULA
LDN 1762	17 59 00.	- 33 10	600		DARK NEBULA
ZWG 028.003	17 59 06.	+ 01 55		15.4	GALAXY
ZWG 113.013	17 59 06.	+ 15 20		15.6	GALAXY
MCG+03-46-008	17 59 06.	+ 15 20	12	15.6	GALAXY
ZWG 254.007	17 59 06.	+ 46 05		15.7	GALAXY
ZWG 301.009	17 59 06.	+ 61 21		14.0	GALAXY
ZWG 300.103	17 59 06.	+ 61 21		14.0	GALAXY
UGC 11092	17 59 06.	+ 61 21	90	14.0	GALAXY SO-a
MCG+10-25-126	17 59 06.	+ 61 22	72	14.	GALAXY
GCL 081	17 59 06.	- 08 57	462	12.90	GLOBULAR STAR CLUSTER
REIN 2.268	17 59 06.37	- 08 57 34.2			NEBULA
PK000-03.1	17 59 06.5	- 30 14 29.	8		PLANETARY NEBULA
RNGC 6542	17 59 08.	+ 61 21		14.0	GALAXY
RNGC 6517	17 59 08.	- 08 57		13.0	GLOBULAR CLUSTER
SG 3.115	17 59 08.	- 08 57	90		DIFFUSE EMISSION NEBULA
PK355-06.1	17 59 08.8	- 36 39 14.	5		PLANETARY NEBULA
PK356-06.1	17 59 11.1	- 35 13 18.	11		PLANETARY NEBULA
ZWG 171.045	17 59 12.	+ 28 53		15.3	GALAXY
MCG+08-33-006	17 59 12.	+ 46 04	30	16.	GALAXY
SCHO 0732	17 59 13.	- 33 32 00.	270		ISOLATED DARK CLOUD
MCG+08-33-007	17 59 15.	+ 48 24	36	16.	GALAXY
RNGC 6514	17 59 17.	- 23 02		5.0	CLUSTER WITH NEBULOSITY
ZWG 254.008	17 59 18.	+ 48 13		15.2	GALAXY
ZWG 301.010	17 59 18.	+ 62 46		15.6	GALAXY
ZWG 300.104	17 59 18.	+ 62 46		15.6	GALAXY
MCG+10-25-127	17 59 18.	+ 62 47	24	16.	GALAXY
ISS 0765	17 59 18.	- 17 43	137		STELLAR RING
OCL 0023	17 59 18.	- 23 02	2160	6.41	OPEN STAR CLUSTER
WRAY 19.54	17 59 21.6	- 23 01 59.		10.0	STAR-NEBULA ASSOCIATION
B 085	17 59 23.	- 23 01			DARK OBJECT
LB 01073	17 59 25.	+ 79 49 18.		10.5	FAINT BLUE STAR
ZWG 056.007	17 59 30.	+ 06 58		14.6	GALAXY
UGC 11093	17 59 30.	+ 06 58	204	14.6	GALAXY Sc
MCG+01-46-003	17 59 30.	+ 06 58	180	14.6	GALAXY
ZWG 141.044	17 59 30.	+ 26 15		14.8	GALAXY
MCG+04-42-022	17 59 30.	+ 26 15	21	14.8	GALAXY
MCG+08-33-008	17 59 30.	+ 48 13	18	16.	GALAXY
MRSL 007-00/2	17 59 30.	- 23 00	1200		HII REGION
PK359-04.3	17 59 30.6	- 32 09 32.	16		PLANETARY NEBULA
SG 3.117	17 59 32.	- 23 02	1200		DIFFUSE EMISSION NEBULA
SG 3.116	17 59 35.	- 25 20	2400		DIFFUSE EMISSION NEBULA
ZWG 113.014	17 59 36.	+ 19 44		14.9	GALAXY
UGC 11094	17 59 36.	+ 19 44	120	14.9	GALAXY Sa
MCG+03-46-009	17 59 36.	+ 19 44	78	14.9	GALAXY
ZWG 141.045	17 59 36.	+ 22 35		15.7	GALAXY
UGC 11095	17 59 36.	+ 22 35	72	15.7	GALAXY SO
ZWG 227.011	17 59 36.	+ 39 42		15.4	GALAXY
ZWG 227.012	17 59 36.	+ 42 21		15.6	GALAXY
7ZW 760	17 59 36.	+ 69 13			COMPACT GALAXY
LDN 0093	17 59 36.	- 27 50	300		DARK NEBULA
RNGC 6527	17 59 37.	+ 19 44		15.0	GALAXY
SCHO 0733	17 59 37.	- 27 56 00.	280		ISOLATED DARK CLOUD
CED 151	17 59 38.	- 23 02	1740		DIFFUSE GALACTIC NEBULA
ZC 1759.7+5504	17 59 42.	+ 55 04	1550		CLUSTER OF GALAXIES
7ZW 761	17 59 42.	+ 70 33			COMPACT GALAXY
ISS 0766	17 59 42.	- 19 45	95		STELLAR RING
RNGC 6525	17 59 44.	+ 11 03			NON-EXISTENT OBJECT
MCG+08-33-009	17 59 45.	+ 48 34	24	16.	GALAXY
ZWG 084.012	17 59 48.	+ 12 51		15.5	GALAXY
MCG+07-37-016	17 59 48.	+ 39 41	42	15.	GALAXY
ZWG 227.013	17 59 48.	+ 42 17		15.7	GALAXY
MCG+07-37-015	17 59 48.	+ 42 18	21	16.	GALAXY
ISS 0675	17 59 48.	- 13 56	100		STELLAR RING
B 086	17 59 52.	- 27 52	300		DARK OBJECT
ZWG 199.035	17 59 54.	+ 33 23		15.7	GALAXY
ZC 1759.9+6027	17 59 54.	+ 60 27	870		CLUSTER OF GALAXIES
ISS 0767	17 59 54.	- 16 18	106		STELLAR RING
ISS 0768	17 59 54.	- 16 25	106		STELLAR RING
ISS 0769	17 59 54.	- 16 52	101		STELLAR RING
ISS 0770	17 59 54.	- 17 02	89		STELLAR RING
SCHO 0734	17 59 56.	- 35 07 00.	300		ISOLATED DARK CLOUD
KHAV 444	18 00	- 04 48	8020		DARK NEBULA
ZWG 171.046	18 00 00.	+ 29 06		15.4	GALAXY
LDN 0444	18 00 00.	- 05 00	4080		DARK NEBULA
LDN 0397	18 00 00.	- 09 20	2160		DARK NEBULA
LDN 0384	18 00 00.	- 10 45	10860		DARK NEBULA
LDN 0383	18 00 00.	- 11 00	3480		DARK NEBULA
ISS 0771	18 00 00.	- 16 19	127		STELLAR RING
LDN 0262	18 00 00.	- 21 00	360		DARK NEBULA
LDN 0180	18 00 00.	- 24 45	6720		DARK NEBULA
LDN 0171	18 00 00.	- 25 10	2280		DARK NEBULA
LDN 0119	18 00 00.	- 27 00	2340		DARK NEBULA
LDN 0003	18 00 00.	- 31 00	9600		DARK NEBULA
SCHO 0735	18 00 02.	- 27 27 00.	240		ISOLATED DARK CLOUD
RNGC 6523	18 00 04.	- 24 23		5.0	CLUSTER WITH NEBULOSITY
PK003-02.2	18 00 04.2	- 26 58 38.	13		PLANETARY NEBULA
ZC 1800.1+3512	18 00 06.	+ 35 12	740		CLUSTER OF GALAXIES
MCG+09-29-046	18 00 06.	+ 53 54	36	17.	GALAXY
ZWG 322.026	18 00 06.	+ 66 36		14.6	GALAXY
RNGC 6552	18 00 06.	+ 66 36		14.5	GALAXY
UGC 11096	18 00 06.	+ 66 36	60	14.6	GALAXY SB
MCG+11-22-018	18 00 06.	+ 66 37	57	14.	GALAXY
OCL 0018	18 00 06.	- 24 23	5400	5.20	OPEN STAR CLUSTER
SCHO 0736	18 00 07.	- 17 01 00.	770		ISOLATED DARK CLOUD
PK358-05.2	18 00 07.2	- 32 42 30.	10		PLANETARY NEBULA

OBJECT NAME	RIGHT ASCEN.	DECLINATION	DIAM.	MAGN.	TYPE OF OBJECT
IC 4673	18 00 10.0	- 27 06 24.			PLANETARY NEBULA
PK003-02.3	18 00 10.4	- 27 06 30.	18		PLANETARY NEBULA
ZWG 227.014	18 00 12.	+ 44 34		15.7	GALAXY
ISS 0676	18 00 12.	- 13 25	174		STELLAR RING
ISS 0772	18 00 12.	- 19 38	113		STELLAR RING
OCL 0010	18 00 12.	- 27 54	600	9.0	OPEN STAR CLUSTER
VHA 255	18 00 12.	- 27 54	180		OPEN STAR CLUSTER
RNGC 6519	18 00 12.	- 29 48			NON-EXISTENT OBJECT
LB 01074	18 00 13.	+ 44 56 30.		17.5	FAINT BLUE STAR
ARC 2295	18 00 17.	+ 69 13		16.2	RICH CLUSTER OF GALAXIES
ZWG 141.046	18 00 18.	+ 26 02		14.6	GALAXY
UGC 11097	18 00 18.	+ 26 02	84	14.6	GALAXY
MCG+11-22-019	18 00 18.	+ 65 30	39	17.	GALAXY
RNGC 6520	18 00 19.	- 27 54		7.5	OPEN CLUSTER
MCG+04-42-023	18 00 24.	+ 26 04 30.	60	14.6	GALAXY
ZWG 254.009	18 00 24.	+ 45 18		15.3	GALAXY
MCG+08-33-010	18 00 24.	+ 45 19	24	16.	GALAXY
GCL 082	18 00 24.	- 30 02	246	10.40	GLOBULAR STAR CLUSTER
VHA 256	18 00 24.	- 30 02	150		GLOBULAR STAR CLUSTER
RNGC 6522	18 00 24.	- 30 02		10.5	GLOBULAR CLUSTER
IC 4675	18 00 27.	- 09 15			NONSTELLAR OBJECT
IC 4676	18 00 28.	+ 11 48 53.			NONSTELLAR OBJECT
ZWG 084.013	18 00 30.	+ 11 49		15.6	GALAXY
ZWG 227.015	18 00 30.	+ 44 28		15.6	GALAXY
MCG+07-37-017	18 00 30.	+ 44 28 30.	30	15.5	GALAXY
MCG+11-22-020	18 00 30.	+ 68 36	39	17.	GALAXY
TER 10	18 00 30.	- 26 04 30.	20		STAR CLUSTER
ARC 2293	18 00 31.	+ 57 29		16.2	RICH CLUSTER OF GALAXIES
PK003-02.1	18 00 31.9	- 26 43 44.	5		PLANETARY NEBULA
SCHO 0737	18 00 32.	- 32 37 00.	620		ISOLATED DARK CLOUD
ZWG 113.015	18 00 36.	+ 19 56		15.7	GALAXY
ZWG 254.010	18 00 36.	+ 45 17		15.1	GALAXY
MCG+08-33-011	18 00 36.	+ 45 17 30.	12	16.	GALAXY
ISS 0773	18 00 36.	- 20 51	220		STELLAR RING
PK358-05.3	18 00 36.7	- 32 41 50.	10	14.5	PLANETARY NEBULA
PK359-04.2	18 00 38.1	- 31 17 55.			PLANETARY NEBULA
CED 152A	18 00 40.	- 24 23	5400		DIFFUSE GALACTIC NEBULA
SG 3.118	18 00 40.	- 24 23	5640		DIFFUSE EMISSION NEBULA
PK357-06.1	18 00 45.6	- 34 28 50.	7		PLANETARY NEBULA
URA 09	18 00 46.	- 21 03	144		STELLAR RING
ZWG 171.047	18 00 48.	+ 29 18		15.6	GALAXY
UGC 11098	18 00 48.	+ 29 18	84	15.6	GALAXY S
ISS 0774	18 00 48.	- 17 05	99		STELLAR RING
MRSL 005-01/1	18 00 48.	- 24 20	5700		HII REGION
MCG+05-42-028	18 00 51.	+ 29 19	60	15.6	GALAXY
B 295	18 00 51.	- 31 10	3000		DARK OBJECT
ZWG 171.048	18 00 54.	+ 29 18		15.2	GALAXY
LB 01075	18 00 54.	+ 44 38 42.		16.1	FAINT BLUE STAR
ZC 1800.9+5401	18 00 54.	+ 54 01	6650		CLUSTER OF GALAXIES
ZWG 322.027	18 00 54.	+ 63 01		15.7	GALAXY
MCG+11-22-021	18 00 54.	+ 65 13	12	17.	GALAXY
ZWG 028.004	18 00 54.	- 00 18		15.6	GALAXY
B 087	18 00 54.	- 32 30			DARK OBJECT
SCHO 0738	18 00 57.	- 35 01 00.	240		ISOLATED DARK CLOUD
B 297	18 01	- 18 45	5400		DARK OBJECT
LBN 0025	18 01	- 24 20	2220		BRIGHT NEBULA
MCG+05-42-029	18 01 00.	+ 29 19 30.	30	15.2	GALAXY
ZWG 171.049	18 01 00.	+ 29 22		15.5	GALAXY
MCG+11-22-025	18 01 00.	+ 63 00	30	16.	GALAXY
MCG+11-22-022	18 01 00.	+ 65 16	24	16.	GALAXY
MCG+11-22-024	18 01 00.	+ 65 18	42	17.	GALAXY
MCG+11-22-023	18 01 00.	+ 65 21	18	17.	GALAXY
LDN 0441	18 01 00.	- 05 20	6480		DARK NEBULA
ISS 0775	18 01 00.	- 17 07	214		STELLAR RING
ISS 0776	18 01 00.	- 17 43	68		STELLAR RING
B 296	18 01 00.	- 24 32	360		DARK OBJECT
LDN 0147	18 01 00.	- 26 10	840		DARK NEBULA
LDN 1798	18 01 00.	- 31 30	540		DARK NEBULA
LDN 1771	18 01 00.	- 32 40	1320		DARK NEBULA
ZWG 200.001	18 01 06.	+ 34 48		15.7	GALAXY
ZWG 199.036	18 01 06.	+ 34 48		15.7	GALAXY
ZWG 227.016	18 01 06.	+ 42 47		15.0	GALAXY
ZWG 301.011	18 01 06.	+ 61 25		15.4	GALAXY
ZWG 300.105	18 01 06.	+ 61 25		15.4	GALAXY
PK356-06.2	18 01 08.7	- 34 58 12.	13		PLANETARY NEBULA
MCG+06-40-001	18 01 09.	+ 34 48	18	16.	GALAXY
MCG+07-37-018	18 01 09.	+ 42 47	36	15.5	GALAXY
ZC 1801.2+5136	18 01 12.	+ 51 36	1410		CLUSTER OF GALAXIES
7ZW 762	18 01 12.	+ 61 33			COMPACT GALAXY
REIN 2.269	18 01 14.06	- 00 17 40.5			NEBULA
REIN 2.270	18 01 16.64	- 00 17 59.5			NEBULA
ZWG 322.028	18 01 18.	+ 67 25		15.0	GALAXY
UGC 11099	18 01 18.	+ 67 25	78	15.0	GALAXY SB?c
MCG+11-22-026	18 01 18.	+ 67 26	66	14.	GALAXY
GCL 083	18 01 18.	- 00 18	108	11.90	GLOBULAR STAR CLUSTER
ISS 0777	18 01 18.	- 17 05	298		STELLAR RING
PK002-03.2	18 01 18.0	- 28 38 12.	6		PLANETARY NEBULA
RNGC 6535	18 01 19.	- 00 18		12.0	GLOBULAR CLUSTER
PK002-03.1	18 01 19.8	- 28 12 49.	11		PLANETARY NEBULA
WRAY 19.55	18 01 21.7	- 24 23 20.		10.0	STAR-NEBULA ASSOCIATION
ZWG 171.050	18 01 24.	+ 29 05		15.3	GALAXY
UGC 11100	18 01 24.	+ 29 05	60	15.3	GALAXY S
ZC 1801.4+3739	18 01 24.	+ 37 39	810		CLUSTER OF GALAXIES
ZC 1801.4+5339	18 01 24.	+ 53 39	740		CLUSTER OF GALAXIES
ZWG 322.029	18 01 24.	+ 65 21		15.6	GALAXY
MCG+05-42-030	18 01 27.	+ 29 06	48	15.3	GALAXY
LB 01076	18 01 27.	+ 43 21 48.		16.8	FAINT BLUE STAR
B 088	18 01 31.	- 24 07	162		DARK OBJECT
SCHO 0739	18 01 32.	- 30 49 00.	310		ISOLATED DARK CLOUD
PK354-07.1	18 01 32.8	- 37 33 21.	13		PLANETARY NEBULA
SG 3.119	18 01 33.	- 23 24	3000		DIFFUSE EMISSION NEBULA
RNGC 6531	18 01 35.	- 22 30		7.0	OPEN CLUSTER
ZWG 171.051	18 01 36.	+ 29 05		15.3	GALAXY
UGC 11101	18 01 36.	+ 45 15	78	16.0	GALAXY Sc
MCG+08-33-012	18 01 36.	+ 45 15	66	16.	GALAXY
ZC 1801.6+6144	18 01 36.	+ 61 44	670		CLUSTER OF GALAXIES
OCL 0026	18 01 36.	- 22 30	1500	7.4	OPEN STAR CLUSTER
GCL 084	18 01 36.	- 30 04	180	11.4	GLOBULAR STAR CLUSTER
VHA 257	18 01 36.	- 30 04	90		GLOBULAR STAR CLUSTER
RNGC 6528	18 01 36.	- 30 04		11.0	GLOBULAR CLUSTER
SCHO 0740	18 01 37.	- 30 12 00.	310		ISOLATED DARK CLOUD
PK358-05.4	18 01 39.7	- 32 54 14.	12		PLANETARY NEBULA
RNGC 6530	18 01 40.	- 24 20		7.5	CLUSTER WITH NEBULOSITY
ZC 1801.7+6136	18 01 42.	+ 61 36	2490		CLUSTER OF GALAXIES
MCG+10-26-001	18 01 42.	+ 62 44	60	17.	GALAXY
MRSL 006-01/2	18 01 42.	- 23 40	1800		HII REGION
OCL 0019	18 01 42.	- 24 20	1680	7.6	OPEN STAR CLUSTER
BC PKS1801+01	18 01 43.	+ 01 01 18.		19.	QUASI-STELLAR OBJECT
SHB 345	18 01 44.	+ 01 01 18.		19.	QUASI-STELLAR OBJECT
SCHO 0741	18 01 46.	- 28 30 00.	420		ISOLATED DARK CLOUD
RNGC 6526	18 01 47.	- 23 40			DIFFUSE NEBULA
ZWG 056.008	18 01 48.	+ 07 02		15.7	GALAXY
ZWG 084.014	18 01 48.	+ 09 20		15.6	GALAXY
UGC 11102	18 01 48.	+ 52 08	78	17.	GALAXY Sb-c
MCG+11-22-027	18 01 48.	+ 65 54	45	16.	GALAXY
PK000-04.1	18 01 48.8	- 30 58 32.	8		PLANETARY NEBULA
SCHO 0742	18 01 52.	- 30 54 00.	300		ISOLATED DARK CLOUD
SG 3.120	18 01 55.	- 07 36	150		DIFFUSE EMISSION NEBULA
B 089	18 01 56.	- 24 22	30		DARK OBJECT
B 298	18 01 59.	- 30 06	240		DARK OBJECT
KHAV 447	18 02	- 17 30	7600		DARK NEBULA
LBN 0026	18 02	- 24 20	3900		BRIGHT NEBULA
KHAV 445	18 02	- 30 36	7090		DARK NEBULA
ZWG 056.009	18 02 00.	+ 07 17		15.0	GALAXY
MCG+01-46-004	18 02 00.	+ 07 17	48	15.0	GALAXY
ZWG 171.052	18 02 00.	+ 29 30		15.7	GALAXY
UGC 11103	18 02 00.	+ 29 30	60	15.7	GALAXY Sb-c
LB 01077	18 02 00.	+ 43 14 48.		17.0	FAINT BLUE STAR
MCG+08-33-013	18 02 00.	+ 45 42	48	16.	GALAXY
LDN 0281	18 02 00.	- 20 00	4560		DARK NEBULA
LDN 0232	18 02 00.	- 22 50	1140		DARK NEBULA
LDN 0041	18 02 00.	- 30 00	360		DARK NEBULA
LDN 0026	18 02 00.	- 30 30	720		DARK NEBULA
RNGC 6533	18 02 04.	- 24 53			NON-EXISTENT OBJECT
MCG+10-26-002	18 02 06.	+ 59 09	54	16.	GALAXY
GCL 085	18 02 06.	- 07 35	828	12.39	GLOBULAR STAR CLUSTER
RNGC 6529	18 02 07.	- 36 18			NON-EXISTENT OBJECT
REIN 2.271	18 02 07.43	- 07 35 18.7			NEBULA
RNGC 6539	18 02 09.	- 07 35		12.5	GLOBULAR CLUSTER
PK010+00.1	18 02 15.48	- 19 50 51.9	147	12.5	PLANETARY NEBULA
RN 0045	18 02 16.	- 19 51			NEBULA
PK002-03.3	18 02 16.0	- 28 22 22.	11		PLANETARY NEBULA
MCG+07-37-019	18 02 18.	+ 43 15	54	15.5	GALAXY
RNGC 6537	18 02 19.	- 19 51		12.5	PLANETARY NEBULA
LB 01078	18 02 20.	+ 45 05 36.		17.8	FAINT BLUE STAR
SCHO 0743	18 02 20.	- 30 50 00.	290		ISOLATED DARK CLOUD
ZWG 113.016	18 02 24.	+ 20 05		15.3	GALAXY
UGC 11104	18 02 24.	+ 20 05	84	15.3	GALAXY SB
ZWG 141.047	18 02 24.	+ 21 38		15.7	GALAXY
UGC 11105	18 02 24.	+ 21 38	150	15.7	GALAXY
MCG+10-26-003	18 02 24.	+ 60 07	36	16.	GALAXY
ZWG 340.031	18 02 24.	+ 69 43		15.2	GALAXY
UGC 11106	18 02 24.	+ 69 43	72	15.2	GALAXY SBb
IC 1271	18 02 26.	- 24 27 15.			NONSTELLAR OBJECT
MCG+03-46-010	18 02 27.	+ 20 06	54	15.3	GALAXY
MCG+04-42-024	18 02 27.	+ 21 38	120	15.7	GALAXY
MCG+12-17-014	18 02 30.	+ 69 41	51	15.	GALAXY
LDN 0368	18 02 30.	- 13 10	5040		DARK NEBULA
SCHO 0744	18 02 32.	- 26 47	260		ISOLATED DARK CLOUD
SCHO 0745	18 02 33.	- 30 58 00.	420		ISOLATED DARK CLOUD
ZWG 084.015	18 02 36.	+ 09 28		15.4	GALAXY
ZWG 278.044	18 02 36.	+ 54 42		15.6	GALAXY
ZWG 340.032	18 02 36.	+ 69 11		15.7	GALAXY
ISS 0770	18 02 36.	- 16 39	145		STELLAR RING
SCHO 0746	18 02 44.	- 27 18	240		ISOLATED DARK CLOUD
ZC 1802.8+1301	18 02 48.	+ 13 01	2690		CLUSTER OF GALAXIES
ZWG 113.017	18 02 48.	+ 17 16		15.6	GALAXY
UGC 11107	18 02 48.	+ 17 16	66	15.6	GALAXY SBb
UGC 11108	18 02 48.	+ 20 02	60	16.5	GALAXY Sc
UGC 11109	18 02 48.	+ 46 45	66	17.	GALAXY DWARF
MCG+09-29-047	18 02 48.	+ 54 41	42	16.	GALAXY
LB 01079	18 02 48.	+ 63 47 54.		17.4	FAINT BLUE STAR
PK004-02.1	18 02 50.4	- 26 30 01.	25		PLANETARY NEBULA
IC 1272	18 02 51.	+ 25 05			OPEN CLUSTER
PK002-03.4	18 02 51.5	- 28 17 24.	10		PLANETARY NEBULA
MCG+03-46-011	18 02 54.	+ 17 16	42	15.6	GALAXY
MCG+09-29-048	18 02 54.	+ 55 44	48	16.	GALAXY
PK002-03.5	18 02 55.6	- 28 40 53.	3		PLANETARY NEBULA
MCG+10-26-004	18 02 57.	+ 60 40	36	16.	GALAXY
KHAV 449	18 03	- 11 18	8680		DARK NEBULA
LBN 0058	18 03	- 14 10	1500		BRIGHT NEBULA
KHAV 446	18 03	- 28 36	2840		DARK NEBULA
IC 1273	18 03 00.	+ 25 07			OPEN CLUSTER
ZWG 227.017	18 03 00.	+ 41 30		15.6	GALAXY
MCG+09-29-049	18 03 00.	+ 53 00	30	17.	GALAXY
ZWG 278.045	18 03 00.	+ 53 01		15.7	GALAXY
ZC 1803.0+5747	18 03 00.	+ 57 47	4500		CLUSTER OF GALAXIES
ISS 0677	18 03 00.	- 12 24	143		STELLAR RING
COU 016	18 03 00.	- 14 15	1500		EMISSION NEBULA
SCHO 0747	18 03 00.	- 16 46	550		ISOLATED DARK CLOUD
MRSL 012+02/1	18 03 00.	- 17 00	2400		HII REGION
COU 015	18 03 00.	- 23 50	3000		EMISSION NEBULA
LDN 0167	18 03 00.	- 25 40	1500		DARK NEBULA
LDN 0151	18 03 00.	- 26 10	2520		DARK NEBULA
RNGC 6547	18 03 05.	+ 25 13		14.5	GALAXY
SCHO 0748	18 03 05.	- 27 26 00.	170		ISOLATED DARK CLOUD
ZWG 142.001	18 03 06.	+ 25 13		14.3	GALAXY
ZWG 141.048	18 03 06.	+ 25 13		14.3	GALAXY
UGC 11110	18 03 06.	+ 25 13	90	14.3	GALAXY S0-a
MCG+07-37-020	18 03 06.	+ 41 30	36	15.	GALAXY
MCG+08-33-014	18 03 06.	+ 47 36	30	16.	GALAXY
MCG+04-43-001	18 03 12.	+ 25 13	60	14.3	GALAXY
MCG+10-26-005	18 03 12.	+ 60 29	30	16.	GALAXY
B 299	18 03 12.	- 27 18	180		DARK OBJECT
OCL 0011	18 03 12.	- 27 49	60	14.6	OPEN STAR CLUSTER
VHA 258	18 03 12.	- 27 49	90		OPEN STAR CLUSTER
RNGC 6540	18 03 14.	- 27 49		14.5	OPEN CLUSTER
MCG+04-43-002	18 03 18.	+ 23 06	30	17.	GALAXY
UGC 11111	18 03 18.	+ 23 07	72	17.	GALAXY DWRF SP
ZWG 142.002	18 03 18.	+ 23 08		15.7	GALAXY
ZWG 200.002	18 03 18.	+ 34 42		15.5	GALAXY
MCG+07-37-021	18 03 18.	+ 43 57	24	15.	GALAXY
MCG+08-33-015	18 03 18.	+ 46 15	60	17.	GALAXY
ZWG 254.011	18 03 18.	+ 46 16		15.3	GALAXY
UGC 11112	18 03 18.	+ 46 16	72	15.3	GALAXY Sb/SBb
MCG+10-26-006	18 03 18.	+ 61 54	36	15.	GALAXY
MRSL 015+03/1	18 03 18.	- 14 10	1500		HII REGION
LDN 0307	18 03 18.	- 18 27	120		DARK NEBULA
LDN 0120	18 03 18.	- 27 25	360		DARK NEBULA
SCHO 0749	18 03 18.	- 32 44	1030		ISOLATED DARK CLOUD
ZWG 142.003	18 03 24.	+ 23 16		15.7	GALAXY
UGC 11113	18 03 24.	+ 23 16	102	15.7	GALAXY Sc/SBc
MCG+06-40-002	18 03 24.	+ 34 42 30.	48	15.	GALAXY
MCG+08-33-016	18 03 24.	+ 46 17	66	16.	GALAXY
ZWG 254.012	18 03 24.	+ 50 25		15.7	GALAXY
ZC 1803.4+5716	18 03 24.	+ 57 16	1210		CLUSTER OF GALAXIES

OBJECT NAME	RIGHT ASCEN.	DECLINATION	DIAM.	MAGN.	TYPE OF OBJECT
7ZW 763	18 03 24.	+ 58 00			COMPACT GALAXY
ZWG 301.012	18 03 24.	+ 60 28		15.4	GALAXY
ISS 0779	18 03 24.	- 15 59	160		STELLAR RING
ISS 0780	18 03 24.	- 16 16	523		STELLAR RING
LDN 0116	18 03 24.	- 27 30	180		DARK NEBULA
LDN 1786	18 03 24.	- 32 23	420		DARK NEBULA
SG 3.121	18 03 26.	- 14 05	1140		DIFFUSE EMISSION NEBULA
PK342-14.1	18 03 26.	- 51 02	36	11.9	PLANETARY NEBULA
ZWG 113.018	18 03 30.	+ 15 53		15.5	GALAXY
MCG+04-43-003	18 03 30.	+ 23 17	132	15.7	GALAXY
ZWG 254.013	18 03 30.	+ 47 45		15.6	GALAXY
MCG+08-33-017	18 03 30.	+ 47 45	15	16.	GALAXY
MCG+08-33-018	18 03 30.	+ 50 25	48	16.	GALAXY
ZWG 301.013	18 03 30.	+ 56 57		15.6	GALAXY
MCG+09-29-050	18 03 30.	+ 56 57	12	16.	GALAXY
MRSL 008-00/1	18 03 30.	- 21 40	5400		HII REGION
OCL 0012	18 03 30.	- 27 28	90	13.4	OPEN STAR CLUSTER
HN 0451	18 03 30.	- 62 25			NEBULA
IC 4674	18 03 30.	- 62 25			NONSTELLAR OBJECT
SCHO 0750	18 03 34.	- 30 36	330		ISOLATED DARK CLOUD
PK004-03.1	18 03 34.7	- 26 55 23.	6		PLANETARY NEBULA
ZWG 113.019	18 03 36.	+ 18 32		14.8	GALAXY
RNGC 6550	18 03 36.	+ 18 32			NON-EXISTENT OBJECT
RNGC 6549	18 03 36.	+ 18 32		15.0	GALAXY
UGC 11114	18 03 36.	+ 18 32	84	14.8	GALAXY S
KARA.72 529A	18 03 36.	+ 18 32	72	14.8	PART OF DOUBLE GALAXY
ZWG 254.014	18 03 36.	+ 45 27		15.5	GALAXY
LDN 0029	18 03 36.	- 30 38	660		DARK NEBULA
SCHO 0751	18 03 37.	- 29 28	460		ISOLATED DARK CLOUD
MCG+03-46-012	18 03 39.	+ 18 32 30.	84	14.8	GALAXY
SCHO 0752	18 03 39.	- 26 29	330		ISOLATED DARK CLOUD
SCHO 0753	18 03 40.	- 24 37	320		ISOLATED DARK CLOUD
B 300	18 03 45.	- 32 40			DARK OBJECT
ZWG 113.020	18 03 48.	+ 18 35		13.1	GALAXY
RNGC 6548	18 03 48.	+ 18 35		13.0	GALAXY
UGC 11115	18 03 48.	+ 18 35	168	13.1	GALAXY SBO
KARA.72 529B	18 03 48.	+ 18 35	138	13.1	PART OF DOUBLE GALAXY
MCG+03-46-013	18 03 48.	+ 18 35 30.	138	13.1	GALAXY
MCG+10-26-007	18 03 48.	+ 60 14	72	16.	GALAXY
ZWG 322.030	18 03 48.	+ 67 02		15.3	GALAXY
UGC 11116	18 03 48.	+ 67 02	72	15.3	GALAXY Sb
LDN 0308	18 03 48.	- 18 26	120		DARK NEBULA
PK000-04.2	18 03 52.9	- 30 34 41.	11		PLANETARY NEBULA
ZWG 200.003	18 03 54.	+ 34 45		15.1	GALAXY
ZWG 254.015	18 03 54.	+ 46 52		14.2	GALAXY
UGC 11117	18 03 54.	+ 46 52	72	14.2	GALAXY S
MCG+08-33-019	18 03 54.	+ 46 52	66	14.	GALAXY
LDN 0068	18 03 54.	- 29 22	300		DARK NEBULA
RNGC 6560	18 03 55.	+ 46 52		14.0	GALAXY
LB 01080	18 03 56.	+ 62 15 00.		16.2	FAINT BLUE STAR
SCHO 0754	18 03 56.	- 27 55	210		ISOLATED DARK CLOUD
PK002-04.1	18 03 56.1	- 29 13 20.	5		PLANETARY NEBULA
SCHO 0755	18 03 59.	- 24 31 00.	300		ISOLATED DARK CLOUD
KHAV 451	18 04	- 16 30	7990		DARK NEBULA
LBN 0050	18 04	- 17 00	660		BRIGHT NEBULA
KHAV 448	18 04	- 29 24	4080		DARK NEBULA
ZWG 142.004	18 04 00.	+ 22 39		15.7	GALAXY
ZWG 142.005	18 04 00.	+ 23 28		15.1	GALAXY
MCG+04-43-004	18 04 00.	+ 23 28	48	15.1	GALAXY
ZWG 200.004	18 04 00.	+ 34 00		15.3	GALAXY
UGC 11118	18 04 00.	+ 34 00	66	15.3	GALAXY Sb
MCG+06-40-003	18 04 00.	+ 34 45 30.	36	15.	GALAXY
UGC 11119	18 04 00.	+ 60 13	78	16.5	GALAXY Sc
ZWG 340.033	18 04 00.	+ 70 14		15.3	GALAXY
LDN 0538	18 04 00.	- 02 50	1440		DARK NEBULA
LDN 0486	18 04 00.	- 03 00	4980		DARK NEBULA
LDN 0462	18 04 00.	- 04 40	780		DARK NEBULA
LDN 0310	18 04 00.	- 18 22	240		DARK NEBULA
COU 017	18 04 00.	- 21 40	7200		EMISSION NEBULA
SG 3.122	18 04 00.	- 21 50	10200		DIFFUSE EMISSION NEBULA
LDN 0002	18 04 00.	- 31 20	4500		DARK NEBULA
PK001-04.1	18 04 02.7	- 29 41 49.	5		PLANETARY NEBULA
MCG+06-40-004	18 04 03.	+ 34 01	60	15.	GALAXY
RNGC 6562	18 04 03.	+ 56 15		14.5	GALAXY
ZWG 279.001	18 04 06.	+ 56 15		14.7	GALAXY
ZWG 278.046	18 04 06.	+ 56 15		14.7	GALAXY
MCG+09-29-051	18 04 06.	+ 56 15	36	15.	GALAXY
ZWG 340.034	18 04 06.	+ 69 51		15.5	GALAXY
SCHO 0903	18 04 09.	+ 09 38 06.	750		ISOLATED DARK CLOUD
RNGC 6546	18 04 11.	- 23 19		8.5	OPEN CLUSTER
OCL 0024	18 04 12.	- 24 19	2580	8.6	OPEN STAR CLUSTER
LDN 0097	18 04 12.	- 28 11	420		DARK NEBULA
SCHO 0756	18 04 15.	- 26 13	270		ISOLATED DARK CLOUD
RNGC 6544	18 04 16.	- 25 01			GLOBULAR CLUSTER
LB 01081	18 04 18.	+ 46 32 30.		14.3	FAINT BLUE STAR
ISS 0993	18 04 18.	- 29 52	158		STELLAR RING
PK356-07.1	18 04 18.	- 36 06 12.	10		PLANETARY NEBULA
RNGC 6541	18 04 23.	- 43 44		8.0	GLOBULAR CLUSTER
ZWG 142.006	18 04 24.	+ 23 42		15.6	GALAXY
MCG+09-29-052	18 04 24.	+ 54 00	36	15.	GALAXY
7ZW 764	18 04 24.	+ 66 55			COMPACT GALAXY
OCL 0017	18 04 24.	- 25 00	180	10.3	OPEN STAR CLUSTER
GCL 087	18 04 24.	- 25 01	504		GLOBULAR STAR CLUSTER
GCL 086	18 04 24.	- 43 44	1392	7.9	GLOBULAR STAR CLUSTER
ARC 2297	18 04 26.	+ 42 22		17.0	RICH CLUSTER OF GALAXIES
MCG+12-17-015	18 04 30.	+ 71 14 30.	27	17.	GALAXY
LDN 0123	18 04 30.	- 27 25	180		DARK NEBULA
LDN 0535	18 04 30.	- 32 56	660		DARK NEBULA
SCHO 0757	18 04 31.	- 27 48	240		ISOLATED DARK CLOUD
SC 1800-5204.6	18 04 31.	- 52 04 25.	60		NEBULA
SCHO 0758	18 04 32.	- 24 24	290		ISOLATED DARK CLOUD
ISS 0678	18 04 36.	- 15 05	210		STELLAR RING
LB 01082	18 04 39.	+ 43 30 30.		16.5	FAINT BLUE STAR
PK015+03.1	18 04 41.	- 13 29 12.	10		PLANETARY NEBULA
PK001-04.2	18 04 41.9	- 29 45 01.	3		PLANETARY NEBULA
MCG+10-26-008	18 04 42.	+ 60 41	48	16.	GALAXY
ISS 0781	18 04 42.	- 20 46	277		STELLAR RING
PK359-04.4	18 04 46.	- 31 40			PLANETARY NEBULA
LB 01083	18 04 47.	+ 79 52 48.		16.0	FAINT BLUE STAR
MCG+03-46-014	18 04 48.	+ 20 30	90	15.6	GALAXY
ZWG 301.014	18 04 48.	+ 60 42		15.6	GALAXY
PK005-02.1	18 04 48.5	- 25 24 30.	10		PLANETARY NEBULA
ZWG 113.021	18 04 54.	+ 20 29		15.6	GALAXY
UGC 11120	18 04 54.	+ 20 29	84	15.6	GALAXY IRR
LDN 0502	18 04 54.	- 01 52	240		DARK NEBULA
ASS 02	18 04 54.	- 21 28	36000		OB ASSOCIATION SGR OB1
SCHO 0759	18 04 54.	- 26 22	310		ISOLATED DARK CLOUD
IC 4678	18 04 55.	- 23 53			NONSTELLAR OBJECT
SCHO 0760	18 04 56.	- 24 13	300		ISOLATED DARK CLOUD
CED 152B	18 04 57.	- 23 53	270		DIFFUSE GALACTIC NEBULA
PK002-03.6	18 04 57.	- 28 24			PLANETARY NEBULA
KHAV 452	18 05	- 02 48	11500		DARK NEBULA
LBN 0038	18 05	- 22 00	2700		BRIGHT NEBULA
LBN 0031	18 05	- 23 27	180		BRIGHT NEBULA
CED 153	18 05	- 26 15			DIFFUSE GALACTIC NEBULA
KHAV 450	18 05	- 27 48	2840		DARK NEBULA
7ZW 765	18 05 00.	+ 59 45			COMPACT GALAXY
MCG+10-26-009	18 05 00.	+ 60 30	48	16.	GALAXY
LDN 0347	18 05 00.	- 15 30	1740		DARK NEBULA
LDN 0303	18 05 00.	- 19 00	17220		DARK NEBULA
ISS 0782	18 05 00.	- 21 03	167		STELLAR RING
LDN 0127	18 05 00.	- 27 19	300		DARK NEBULA
SCHO 0761	18 05 00.	- 28 10	240		ISOLATED DARK CLOUD
LDN 0096	18 05 00.	- 28 20	660		DARK NEBULA
LDN 0075	18 05 00.	- 29 10	1020		DARK NEBULA
LDN 1789	18 05 00.	- 32 30	3540		DARK NEBULA
SCHO 0762	18 05 01.	- 27 23	240		ISOLATED DARK CLOUD
LB 01084	18 05 03.	+ 45 44 42.		16.9	FAINT BLUE STAR
SCHO 0763	18 05 04.	- 24 01	300		ISOLATED DARK CLOUD
SCHO 0764	18 05 04.	- 25 33	230		ISOLATED DARK CLOUD
MCG+08-33-020	18 05 06.	+ 46 31	15	16.	GALAXY
ZWG 254.016	18 05 06.	+ 46 32		15.7	GALAXY
SG 3.123	18 05 06.	- 25 50	8400		DIFFUSE EMISSION NEBULA
SCHO 0765	18 05 06.	- 27 40	210		ISOLATED DARK CLOUD
SCHO 0766	18 05 10.	- 28 00	240		ISOLATED DARK CLOUD
MCG+10-26-010	18 05 12.	+ 59 40	15	16.	GALAXY
DG 147	18 05 12.	- 23 27	180		REFLECTION NEBULA
MCG+10-26-011	18 05 15.	+ 59 40	18	16.	GALAXY
IC 4681	18 05 15.	- 23 25			NONSTELLAR OBJECT
CED 154A	18 05 15.	- 23 27			DIFFUSE GALACTIC NEBULA
PK000-05.1	18 05 15.6	- 31 37 06.	15		PLANETARY NEBULA
SCHO 0767	18 05 16.	- 26 47	250		ISOLATED DARK CLOUD
MCG+10-26-012	18 05 18.	+ 59 37	30	16.	GALAXY
7ZW 766	18 05 18.	+ 66 54			COMPACT GALAXY
PK008-01.1	18 05 24.9	- 22 17 23.	6	11.	PLANETARY NEBULA
HOLM 774A	18 05 30.	+ 17 35	66	12.4	PART OF MULTIPLE GALAXY
ZWG 322.031	18 05 30.	+ 68 19		15.6	GALAXY
ISS 0679	18 05 30.	- 14 57	209		STELLAR RING
VDB.66N 113	18 05 30.	- 21 26	1020		REFLECTION NEBULA
LDN 0224	18 05 30.	- 23 30	480		DARK NEBULA
HOLM 774B	18 05 31.	+ 17 36	30	13.9	PART OF MULTIPLE GALAXY
VDB.66N 114	18 05 34.	- 18 20	72		REFLECTION NEBULA
ZWG 113.022	18 05 36.	+ 17 35		12.7	GALAXY
UGC 11121	18 05 36.	+ 17 35	126	12.7	GALAXY Sc
RNGC 6555	18 05 36.	+ 17 36		12.5	GALAXY
MCG+03-46-015	18 05 36.	+ 17 36	126	12.7	GALAXY
IC 4683	18 05 36.	- 26 15			NONSTELLAR OBJECT
ARC 2298	18 05 37.	+ 50 14		17.4	RICH CLUSTER OF GALAXIES
ZWG 084.016	18 05 42.	+ 09 42		15.7	GALAXY
UGC 11122	18 05 42.	+ 09 42	60	15.7	GALAXY Sc
ZWG 172.001	18 05 42.	+ 28 28		15.4	GALAXY
UGC 11123	18 05 42.	+ 28 28	66	15.4	GALAXY
MCG+05-43-001	18 05 42.	+ 28 29	54	15.4	GALAXY
ZWG 172.002	18 05 42.	+ 28 57		15.6	GALAXY
ZWG 200.005	18 05 42.	+ 35 33		13.9	GALAXY
UGC 11124	18 05 42.	+ 35 33	156	13.9	GALAXY SBc
MCG+06-40-005	18 05 42.	+ 35 34	150	13.	GALAXY
ISS 0680	18 05 42.	- 14 32	121		STELLAR RING
SCHO 0768	18 05 43.	- 27 52	260		ISOLATED DARK CLOUD
MCG+10-26-013	18 05 45.	+ 62 06	18	16.	GALAXY
IC 4688	18 05 46.	+ 11 41 41.			NONSTELLAR OBJECT
ZWG 084.017	18 05 48.	+ 11 12		15.7	GALAXY
ZWG 084.018	18 05 48.	+ 11 41		14.7	GALAXY
UGC 11125	18 05 48.	+ 11 41	96	14.7	GALAXY Sc
MCG+02-46-006	18 05 48.	+ 11 41	84	14.7	GALAXY
ZC 1805.8+5014	18 05 48.	+ 50 14	1140		CLUSTER OF GALAXIES
MCG+09-30-001	18 05 48.	+ 52 13	42	16.	GALAXY
ZWG 279.002	18 05 48.	+ 52 16		15.5	GALAXY
KARA.73B 0845	18 05 48.	+ 52 16	30	15.5	ISOLATED GALAXY E
MCG+10-26-014	18 05 48.	+ 62 05	30	16.	GALAXY
MRSL 011+00/1	18 05 48.	- 18 17	180		HII REGION
ISS 0783	18 05 48.	- 21 02	199		STELLAR RING
RNGC 6551	18 05 48.	- 29 33			NON-EXISTENT OBJECT
LB 01085	18 05 49.	+ 46 01 48.		15.5	FAINT BLUE STAR
RNGC 6566	18 05 50.	+ 52 16		15.5	GALAXY
SCHO 0769	18 05 51.	- 30 09	260		ISOLATED DARK CLOUD
SCHO 0770	18 05 53.	- 21 37	330		ISOLATED DARK CLOUD
B 301	18 05 54.	- 18 43	2700		DARK OBJECT
CED 155	18 05 55.	+ 45 50			DIFFUSE GALACTIC NEBULA
SG 3.124	18 05 57.	- 23 49	2700		DIFFUSE EMISSION NEBULA
LBN 0048	18 06	- 18 10	180		BRIGHT NEBULA
KHAV 456	18 06	- 20 24	8480		DARK NEBULA
LBN 0040	18 06	- 21 30	480		BRIGHT NEBULA
LBN 0034	18 06	- 23 27	120		BRIGHT NEBULA
ZWG 142.007	18 06 00.	+ 21 38		15.6	GALAXY
MCG+05-43-002	18 06 00.	+ 30 35	36	16.	GALAXY
ZWG 200.006	18 06 00.	+ 35 52		14.8	GALAXY
MCG+10-26-015	18 06 00.	+ 61 35 30.	12	16.	GALAXY
ZC 1806.0+8351	18 06 00.	+ 83 51	940		CLUSTER OF GALAXIES
MCG+15-01-018	18 06 00.	+ 88 19 30.	27	16.	GALAXY
LDN 0375	18 06 00.	- 13 00	1380		DARK NEBULA
ISS 0784	18 06 00.	- 17 50	134		STELLAR RING
DG 148	18 06 00.	- 23 27	180		REFLECTION NEBULA
LDN 0230	18 06 00.	- 23 28	180		DARK NEBULA
LDN 0214	18 06 00.	- 23 59	120		DARK NEBULA
LDN 0213	18 06 00.	- 24 01	120		DARK NEBULA
ISS 0899	18 06 00.	- 24 53	90		STELLAR RING
LDN 0157	18 06 00.	- 26 25	360		DARK NEBULA
CED 154P	18 06 03.	- 23 27			DIFFUSE GALACTIC NEBULA
IC 4684	18 06 04.	- 23 26			REFLECTION NEBULA
VDB.66N 115	18 06 04.	- 23 27	108		REFLECTION NEBULA
MCG+06-40-006	18 06 06.	+ 35 53	36	15.	GALAXY
7ZW 767	18 06 06.	+ 61 35			COMPACT GALAXY
LDN 0210	18 06 06.	- 24 08	120		DARK NEBULA
PK005-03.1	18 06 07.5	- 26 03 02.			PLANETARY NEBULA
URA 15	18 06 08.	- 18 11	348		STELLAR RING
RNGC 6554	18 06 08.	- 18 26			NON-EXISTENT OBJECT
B 302	18 06 10.	- 23 59	30		DARK OBJECT
IC 4685	18 06 14.	- 23 59 51.			EMISSION NEBULA
CED 154C	18 06 15.	- 24 01			DIFFUSE GALACTIC NEBULA
SCHO 0771	18 06 15.	- 27 43	230		ISOLATED DARK CLOUD
ZC 1806.3+5939	18 06 18.	+ 59 39	2150		CLUSTER OF GALAXIES
ISS 0785	18 06 18.	- 17 25	260		STELLAR RING
ISS 0786	18 06 18.	- 19 25	164		STELLAR RING
LDN 0274	18 06 18.	- 20 58	540		DARK NEBULA
MRSL 006-02/1	18 06 18.	- 24 01	2400		HII REGION

OBJECT NAME	RIGHT ASCEN.	DECLINATION	DIAM.	MAGN.	TYPE OF OBJECT
GCL 088	18 06 18.	- 25 56	492	10.20	GLOBULAR STAR CLUSTER
SCHO 0772	18 06 19.	- 20 45	380		ISOLATED DARK CLOUD
SCHO 0773	18 06 20.	- 26 56	280		ISOLATED DARK CLOUD
RNGC 6553	18 06 21.	- 25 56		10.0	GLOBULAR CLUSTER
ZWG 084.019	18 06 24.	+ 11 49		15.5	GALAXY
UGC 11126	18 06 24.	+ 25 43	84	17.	GALAXY
IC 4691	18 06 25.	+ 11 49			NONSTELLAR OBJECT
B 303	18 06 25.	- 24 01	60		DARK OBJECT
SCHO 0774	18 06 25.	- 30 22	340		ISOLATED DARK CLOUD
PK006-02.1	18 06 26.1	- 24 13 03.	43		PLANETARY NEBULA
PK356-07.2	18 06 27.4	- 35 44 49.	13		PLANETARY NEBULA
IC 1274	18 06 29.	- 23 45			NONSTELLAR OBJECT
ZWG 084.020	18 06 30.	+ 09 56		15.7	GALAXY
MCG+02-46-007	18 06 30.	+ 09 56	36	15.7	GALAXY
OCL 0022	18 06 30.	- 24 00	2250	6.8	OPEN STAR CLUSTER
LDN 0211	18 06 30.	- 24 10	360		DARK NEBULA
ISS 0900	18 06 30.	- 24 32	79		STELLAR RING
SCHO 0775	18 06 31.	- 30 10	260		ISOLATED DARK CLOUD
ZWG 200.007	18 06 36.	+ 34 58		15.7	GALAXY
KARA.73B 0846	18 06 36.	+ 34 58	60	15.7	ISOLATED GALAXY S
MCG+06-40-007	18 06 36.	+ 34 58 30.	48	15.	GALAXY
ISS 0787	18 06 36.	- 18 11	152		STELLAR RING
ISS 0994	18 06 36.	- 29 12	169		STELLAR RING
PK358-06.1	18 06 36.	- 33 19 56.	10	12.	PLANETARY NEBULA
URA 12	18 06 38.	- 16 15	804		STELLAR RING
ARC 2300	18 06 39.	+ 76 40		17.1	RICH CLUSTER OF GALAXIES
SG 3.125	18 06 39.	- 24 00	540		DIFFUSE EMISSION NEBULA
MRSL 007-02/2	18 06 42.	- 23 40	480		HII REGION
IC 1274A	18 06 45.	- 23 41	324		REFLECTION NEBULA
CED 154D	18 06 45.	- 23 41	540		DIFFUSE GALACTIC NEBULA
CED 154E	18 06 45.	- 23 51	570		DIFFUSE GALACTIC NEBULA
ZWG 172.003	18 06 48.	+ 28 02		14.8	GALAXY
UGC 11127	18 06 48.	+ 28 02	66	14.8	GALAXY DBL SYS
MCG+05-43-003	18 06 48.	+ 28 02	36	14.8	GALAXY
ZWG 254.017	18 06 48.	+ 47 09		15.7	GALAXY
OCL 0003	18 06 48.	- 31 47	120	11.6	OPEN STAR CLUSTER
VHA 259	18 06 48.	- 31 47	60		GLOBULAR STAR CLUSTER
SCHO 0776	18 06 48.	- 32 32	250		ISOLATED DARK CLOUD
ACK 345-13.1	18 06 48.	- 48 26			PLANETARY NEBULA
LB 01086	18 06 49.	+ 45 41 54.		17.0	FAINT BLUE STAR
RNGC 6557	18 06 50.	- 26 36			UNVERIFIED SOUTHERN OBJECT
RNGC 6556	18 06 50.	- 27 31			NON-EXISTENT OBJECT
CED 154F	18 06 52.	- 24 08	480		DIFFUSE GALACTIC NEBULA
RNGC 6564	18 06 54.	+ 17 24			NON-EXISTENT OBJECT
OCL 0119	18 06 54.	+ 31 31	2040		OPEN STAR CLUSTER
ZC 1806.9+5853	18 06 54.	+ 58 53	2550		CLUSTER OF GALAXIES
ISS 0788	18 06 54.	- 15 01	85		STELLAR RING
SCHO 0777	18 06 55.	- 18 48	360		ISOLATED DARK CLOUD
IC 1275	18 06 55.	- 23 50			EMISSION NEBULA
ARC 2303	18 06 56.	+ 82 54		16.7	RICH CLUSTER OF GALAXIES
IC 4693	18 06 58.	+ 17 20			NONSTELLAR OBJECT
RNGC 6558	18 06 58.	- 31 47			GLOBULAR CLUSTER
SCHO 0778	18 06 59.	- 23 45	310		ISOLATED DARK CLOUD
LBN 0029	18 07	- 24 00	2700		BRIGHT NEBULA
LBN 0028	18 07	- 24 00	720		BRIGHT NEBULA
KHAV 453	18 07	- 24 18	12510		DARK NEBULA
KHAV 455	18 07	- 26 54	3170		DARK NEBULA
KHAV 454	18 07	- 28 24	1720		DARK NEBULA
ZWG 172.004	18 07 00.	+ 28 15		15.7	GALAXY
ZWG 254.018	18 07 00.	+ 50 01		15.6	GALAXY
MCG+12-17-016	18 07 00.	+ 69 30	39	16.	GALAXY
ZC 1807.0+8254	18 07 00.	+ 82 54	1680		CLUSTER OF GALAXIES
LDN 0593	18 07 00.	- 00 40	2760		DARK NEBULA
LDN 0508	18 07 00.	- 01 40	720		DARK NEBULA
LDN 0461	18 07 00.	- 05 10	1380		DARK NEBULA
LDN 0348	18 07 00.	- 15 40	660		DARK NEBULA
LDN 0249	18 07 00.	- 22 30	1740		DARK NEBULA
LDN 0227	18 07 00.	- 23 40	900		DARK NEBULA
COU 018	18 07 00.	- 24 20	3000		EMISSION NEBULA
LDN 0200	18 07 00.	- 24 50	4140		DARK NEBULA
LDN 0196	18 07 00.	- 25 00	180		DARK NEBULA
LDN 0176	18 07 00.	- 25 50	2220		DARK NEBULA
LDN 0143	18 07 00.	- 27 00	2820		DARK NEBULA
LDN 0108	18 07 00.	- 28 20	180		DARK NEBULA
GCL 089	18 07 00.	- 31 47			GLOBULAR STAR CLUSTER
SG 3.126	18 07 03.	- 23 42	420		DIFFUSE EMISSION NEBULA
B 091	18 07 04.	- 23 42	300		DARK OBJECT
RNGC 6559	18 07 04.	- 24 08			DIFFUSE NEBULA
B 090	18 07 04.	- 28 17	180		DARK OBJECT
ZWG 113.023	18 07 06.	+ 18 17		15.2	GALAXY
UGC 11128	18 07 06.	+ 18 17	66	15.2	GALAXY Sa-b
MCG+03-46-016	18 07 06.	+ 18 18	60	15.2	GALAXY
MRSL 007-02/1	18 07 06.	- 23 49	480		HII REGION
HN 0452	18 07 08.	- 56 17			NEBULA
IC 4679	18 07 08.	- 56 17			NONSTELLAR OBJECT
LB 01087	18 07 11.	+ 46 05 18.		18.4	FAINT BLUE STAR
SCHO 0779	18 07 11.	- 32 41	430		ISOLATED DARK CLOUD
ZWG 113.024	18 07 12.	+ 19 06		15.1	GALAXY
ZWG 172.005	18 07 12.	+ 30 18		14.9	GALAXY
UGC 11129	18 07 12.	+ 30 18	78	14.9	GALAXY Sa
ZC 1807.2+5633	18 07 12.	+ 56 33	5780		CLUSTER OF GALAXIES
ZWG 340.035	18 07 12.	+ 69 49		14.4	GALAXY
UGC 11130	18 07 12.	+ 69 49	11	14.4	GALAXY EX CMPT
SCHO 0780	18 07 13.	- 28 26	510		ISOLATED DARK CLOUD
LB 01088	18 07 15.	+ 80 39 18.		10.2	FAINT BLUE STAR
SG 3.127	18 07 16.	- 24 08	960		DIFFUSE EMISSION NEBULA
7ZW 768	18 07 18.	+ 69 50			COMPACT GALAXY
PK003-04.2	18 07 19.5	- 28 08 20.	10		PLANETARY NEBULA
MCG+05-43-004	18 07 21.	+ 30 20	72	14.9	GALAXY
ZWG 084.021	18 07 24.	+ 11 02		15.7	GALAXY
ZWG 254.019	18 07 24.	+ 46 49		15.3	GALAXY
MCG+08-33-021	18 07 24.	+ 46 50	42	16.	GALAXY
MCG+10-26-016	18 07 24.	+ 60 38 30.	30	16.	GALAXY
MRSL 013+01/1	18 07 24.	- 16 50	180		HII REGION
URA 14	18 07 26.	- 16 21	138		STELLAR RING
SG 3.128	18 07 27.	- 23 47	360		DIFFUSE EMISSION NEBULA
RNGC 6545	18 07 29.	- 63 46			NON-EXISTENT OBJECT
MCG+08-33-022	18 07 30.	+ 48 50	30	16.	GALAXY
URA 02	18 07 30.	- 18 09	156		STELLAR RING
IC 4690	18 07 30.	- 19 49 52.			NONSTELLAR OBJECT
PK003-04.1	18 07 34.6	- 27 58 31.	13	12.	PLANETARY NEBULA
ZC 1807.6+4400	18 07 36.	+ 44 00	1080		CLUSTER OF GALAXIES
ARC 2299	18 07 38.	+ 43 57		17.3	RICH CLUSTER OF GALAXIES
RNGC 6561	18 07 39.	- 16 48			NON-EXISTENT OBJECT
PK018+04.1	18 07 40.	- 10 29 48.	12		PLANETARY NEBULA
ZWG 028.005	18 07 46.	+ 01 34		15.7	GALAXY
UGC 11131	18 07 48.	+ 01 34	84	15.7	GALAXY
MCG+00-46-003	18 07 48.	+ 01 34	90	15.7	GALAXY
ZWG 200.008	18 07 48.	+ 38 46		15.6	GALAXY
UGC 11132	18 07 48.	+ 38 46	132	15.6	GALAXY Sb
KARA.73B 0847	18 07 48.	+ 38 46	132	15.6	ISOLATED GALAXY S
MCG+09-30-002	18 07 48.	+ 53 23	42	17.	GALAXY
UGC 11133	18 07 48.	+ 79 33	72	16.0	GALAXY S
URA 08	18 07 50.	- 16 16	126		STELLAR RING
PK003-04.6	18 07 51.6	- 28 33 21.	12	12.	PLANETARY NEBULA
MCG+06-40-008	18 07 54.	+ 38 48 30.	114	14.5	GALAXY
ZWG 322.032	18 07 54.	+ 63 41		15.5	GALAXY
ISS 0901	18 07 54.	- 26 51	110		STELLAR RING
PK002-04.2	18 07 54.3	- 28 59 40.	14		PLANETARY NEBULA
VDB.66N 116	18 07 56.	- 17 44	36		REFLECTION NEBULA
SCHO 0781	18 07 56.	- 18 39	300		ISOLATED DARK CLOUD
LBN 0053	18 08	- 16 40	1800		BRIGHT NEBULA
LBN 0033	18 08	- 23 45	720		BRIGHT NEBULA
ZWG 340.036	18 08 00.	+ 70 15		15.6	GALAXY
LDN 0513	18 08 00.	- 01 34	1440		DARK NEBULA
GCL 090	18 08 00.	- 07 14	360		GLOBULAR STAR CLUSTER
SG 3.129	18 08 00.	- 07 14	180		DIFFUSE EMISSION NEBULA
LDN 0408	18 08 00.	- 09 30	2760		DARK NEBULA
LDN 0333	18 08 00.	- 16 50	2820		DARK NEBULA
LDN 0271	18 08 00.	- 21 20	480		DARK NEBULA
COU 019	18 08 00.	- 21 25	3600		EMISSION NEBULA
LDN 0257	18 08 00.	- 22 25	1080		DARK NEBULA
LDN 0221	18 08 00.	- 24 00	5160		DARK NEBULA
LDN 0212	18 08 00.	- 24 20	1200		DARK NEBULA
UGC 11134	18 08 06.	+ 05 43	84	17.	GALAXY S
MCG+12-17-017	18 08 06.	+ 69 41	39	17.	GALAXY
ASS 05	18 08 06.	- 16 50			OB ASSOCIATION SGR OB6
SCHO 0808	18 08 06.	- 18 14	620		ISOLATED DARK CLOUD
ISS 0995	18 08 06.	- 27 05	131		STELLAR RING
SCHO 0782	18 08 08.	- 30 25	250		ISOLATED DARK CLOUD
ISS 0996	18 08 12.	- 27 04	68		STELLAR RING
MCG+11-22-028	18 08 18.	+ 63 17 30.	30	17.	GALAXY
HUB E23	18 08 18.	- 14 45			DIFFUSE NEBULA
IC 1277	18 08 20.	+ 30 59			OPEN CLUSTER
PK003-04.3	18 08 20.3	- 27 46 59.	7		PLANETARY NEBULA
ZWG 142.008	18 08 24.	+ 21 13		15.5	GALAXY
MCG+04-43-005	18 08 24.	+ 21 13	42	15.5	GALAXY
ZWG 356.007	18 08 24.	+ 75 04		15.6	GALAXY
IC 1276	18 08 25.	- 07 14 37.			GLOBULAR CLUSTER
PK003-04.7	18 08 25.3	- 28 23 20.	12	11.	PLANETARY NEBULA
ZWG 172.006	18 08 30.	+ 29 59		15.6	GALAXY
MCG+08-33-023	18 08 30.	+ 49 55	42	16.	GALAXY
LDN 0241	18 08 30.	- 23 20	540		DARK NEBULA
IC 4680	18 08 30.	- 64 29			NONSTELLAR OBJECT
HN 0453	18 08 30.	- 64 30			NEBULA
MIL 58	18 08 32.	- 19 26 54.	222		SUPERNOVA REMNANT
IC 1278	18 08 35.	+ 31 08			NONSTELLAR OBJECT
ZWG 113.025	18 08 36.	+ 18 25		15.7	GALAXY
MCG+03-46-017	18 08 36.	+ 18 25	42	15.7	GALAXY
ZWG 142.009	18 08 36.	+ 25 06		15.2	GALAXY
ZWG 172.007	18 08 36.	+ 28 55		15.6	GALAXY
ZWG 172.008	18 08 35.	+ 30 59		15.0	GALAXY
UGC 11135	18 08 36.	+ 30 59	102	15.0	GALAXY Sc
KARA.72 530A	18 08 36.	+ 30 59	72	15.0	PART OF DOUBLE GALAXY
MRSL 010-00/1	18 08 36.	- 20 20	1500		HII REGION
MCG+05-43-005	18 08 39.	+ 31 01	90	15.0	GALAXY
ZWG 142.010	18 08 42.	+ 21 14		15.4	GALAXY
ZWG 142.011	18 08 42.	+ 25 11		15.7	GALAXY
UGC 11136	18 08 42.	+ 25 11	72	15.7	GALAXY Sb
RNGC 6565	18 08 43.	- 28 11		13.0	PLANETARY NEBULA
PK003-04.5	18 08 43.0	- 28 11 27.	10	13.2	PLANETARY NEBULA
PK358-07.1	18 06 43.9	- 33 52 50.	60	13.8	PLANETARY NEBULA
RNGC 6571	18 08 44.	+ 21 14		15.5	GALAXY
HN 0042	18 08 45.	- 28 11			NEBULA
RNGC 6563	18 08 45.	- 33 53		14.0	PLANETARY NEBULA
RNGC 6570	18 08 46.	+ 14 05		13.0	GALAXY
ZWG 084.022	18 08 48.	+ 14 05		13.2	GALAXY
UGC 11137	18 08 48.	+ 14 05	108	13.2	GALAXY Sc-IRR
MCG+02-46-008	18 08 48.	+ 14 05	108	13.2	GALAXY
MCG+04-43-006	18 08 48.	+ 21 12	18	15.4	GALAXY
LDN 0426	18 08 48.	- 07 58	120		DARK NEBULA
SCHO 0783	18 08 51.	- 18 33	270		ISOLATED DARK CLOUD
PK011-00.1	18 08 52.	- 18 47 08.	7		PLANETARY NEBULA
ZC 1808.9+2531	18 08 54.	+ 25 31	8470		CLUSTER OF GALAXIES
MCG+08-33-024	18 08 54.	+ 47 51	30	17.	GALAXY
RNGC 6575	18 08 57.	+ 31 05		14.5	GALAXY
KHAV 457	18 09	- 13 05	10330		DARK NEBULA
LBN 0049	18 09	- 17 45	720		BRIGHT NEBULA
ZWG 172.009	18 09 00.	+ 31 05		14.4	GALAXY
UGC 11138	18 09 00.	+ 31 05	108	14.4	GALAXY E
KARA.72 530B	18 09 00.	+ 31 05	108	14.4	PART OF DOUBLE GALAXY
ZWG 301.015	18 09 00.	+ 58 22		15.6	GALAXY
MCG+10-26-017	18 09 00.	+ 58 22	54	17.	GALAXY
7ZW 769	18 09 00.	+ 62 23			COMPACT GALAXY
ZWG 370.012	18 09 00.	+ 87 40		15.4	GALAXY
ZWG 370.007	18 09 00.	+ 87 40		15.4	GALAXY
LDN 0430	18 09 00.	- 07 38	180		DARK NEBULA
LDN 0422	18 09 00.	- 08 10	300		DARK NEBULA
LDN 0275	18 09 00.	- 21 12	120		DARK NEBULA
LDN 0243	18 09 00.	- 23 20	2940		DARK NEBULA
LDN 0187	18 09 00.	- 25 45	3360		DARK NEBULA
SCHO 0784	18 09 02.	- 25 06	290		ISOLATED DARK CLOUD
MAI 101	18 09 04.	+ 80 47	33		DWARF SPHEROIDAL GALAXY
MCG+05-43-006	18 09 06.	+ 31 08	33	14.4	GALAXY
MCG+08-33-025	18 09 06.	+ 48 14	42	16.	GALAXY
ZWG 254.020	18 09 06.	+ 48 15		15.7	GALAXY
7ZW 770	18 09 06.	+ 58 22			COMPACT GALAXY
MRSL 012+00/1	18 09 06.	- 17 46	900		HII REGION
LDN 0242	18 09 06.	- 23 25	360		DARK NEBULA
RNGC 6598	18 09 10.	+ 69 04		14.0	GALAXY
ZWG 254.021	18 09 12.	+ 49 54		15.4	GALAXY
ZWG 340.037	18 09 12.	+ 69 04		14.2	GALAXY
UGC 11139	18 09 12.	+ 69 04	132	14.2	GALAXY PECULR
ISS 0997	18 09 12.	- 28 02	250		STELLAR RING
PK003-04.4	18 09 14.7	- 27 53 01.	9		PLANETARY NEBULA
MCG+08-33-026	18 09 15.	+ 49 53	30	16.	GALAXY
PK005-03.2	18 09 15.2	- 26 33 51.	17		PLANETARY NEBULA
PK004-04.1	18 09 16.2	- 27 29 42.	4		PLANETARY NEBULA
7ZW 771	18 09 18.	+ 61 25			COMPACT GALAXY
ZWG 301.016	18 09 18.	+ 61 25		15.5	GALAXY
MCG+10-26-018	18 09 18.	+ 61 25	18	15.	GALAXY
MCG+12-17-018	18 09 18.	+ 69 03 30.	66	14.	GALAXY
ACK 351-10.1	18 09 18.	- 41 31			PLANETARY NEBULA
HN 1153	18 09 19.	- 57 44		14.	NEBULA
IC 4687	18 09 19.	- 57 44			NONSTELLAR OBJECT
IC 4686	18 09 19.	- 57 45			NONSTELLAR OBJECT

OBJECT NAME	RIGHT ASCEN.	DECLINATION	DIAM.	MAGN.	TYPE OF OBJECT
RNGC 6592	18 09 20.	+ 61 25		15.5	GALAXY
HN 1152	18 09 20.	- 57 45		14.	NEBULA
IC 1279	18 09 21.	+ 36 01 13.			NONSTELLAR OBJECT
RNGC 6438	18 09 21.	- 85 26			GALAXY
PK018+03.1	18 09 23.	- 10 43 48.			PLANETARY NEBULA
ZWG 227.018	18 09 24.	+ 39 10		15.4	GALAXY
UGC 11140	18 09 24.	+ 39 10	150	15.4	GALAXY Sc
ZWG 279.003	18 09 24.	+ 54 13		14.9	GALAXY
7ZW 772	18 09 24.	+ 61 33			COMPACT GALAXY
PK003-04.8	18 09 24.9	- 27 59 01.	10		PLANETARY NEBULA
LB 01089	18 09 25.	+ 45 07 12.		16.0	FAINT BLUE STAR
IC 4689	18 09 25.	- 57 46			NONSTELLAR OBJECT
HN 1154	18 09 26.	- 57 46		14.	NEBULA
SCHO 0785	18 09 28.	- 32 26	330		ISOLATED DARK CLOUD
PK006-03.1	18 09 29.3	- 24 50 48.	25		PLANETARY NEBULA
ZWG 084.023	18 09 30.	+ 12 04		15.4	GALAXY
UGC 11141	18 09 30.	+ 12 04	120	15.4	GALAXY
MCG+02-46-009	18 09 30.	+ 12 04	120	15.4	GALAXY
ZWG 142.012	18 09 30.	+ 21 34		15.6	GALAXY
ZWG 142.013	18 09 30.	+ 25 38		15.4	GALAXY
UGC 11142	18 09 30.	+ 25 38	102	15.4	GALAXY Sc
MCG+04-43-007	18 09 30.	+ 25 38	90	15.4	GALAXY
ZWG 200.009	18 09 30.	+ 35 59		14.5	GALAXY
UGC 11143	18 09 30.	+ 35 59	150	14.5	GALAXY Sb
ZWG 227.019	18 09 30.	+ 44 45		15.3	GALAXY
MCG+10-26-019	18 09 30.	+ 61 08	54	15.	GALAXY
URA 03	18 09 30.	- 17 14	84		STELLAR RING
SCHO 0786	18 09 32.	- 32 18	300		ISOLATED DARK CLOUD
RNGC 6574	18 09 34.	+ 14 58		13.5	GALAXY
SCHO 0787	18 09 35.	- 18 28	290		ISOLATED DARK CLOUD
ZWG 113.026	18 09 36.	+ 14 57		12.5	GALAXY
ZWG 084.024	18 09 36.	+ 14 57		12.5	GALAXY
UGC 11144	18 09 36.	+ 14 57	84	12.5	GALAXY S
MCG+02-46-010	18 09 36.	+ 14 57	72	12.5	GALAXY
ZWG 142.014	18 09 36.	+ 21 25		15.5	GALAXY
MCG+06-40-009	18 09 36.	+ 36 01	162	14.	GALAXY
ZWG 301.017	18 09 36.	+ 61 07		15.4	GALAXY
TER 11	18 09 36.	- 22 46 00.	50		STAR CLUSTER
RNGC 6576	18 09 38.	+ 21 25		15.5	GALAXY
RNGC 6594	18 09 39.	+ 61 07		15.5	GALAXY
PK003-04.9	18 09 38.2	- 28 20 49.	9	12.	PLANETARY NEBULA
MCG+07-37-022	18 09 39.	+ 39 09	138	14.	GALAXY
PK034+11.1	18 09 40.57	+ 06 50 25.9	16	9.6	PLANETARY NEBULA
RNGC 6572	18 09 41.	+ 06 50		9.5	PLANETARY NEBULA
ZWG 084.025	18 09 42.	+ 09 25		15.7	GALAXY
ZWG 113.027	18 09 42.	+ 15 08		15.5	GALAXY
ZWG 142.015	18 09 42.	+ 21 30		15.7	GALAXY
ZWG 254.022	18 09 42.	+ 49 30		15.5	GALAXY
UGC 11145	18 09 42.	+ 49 30	72	15.5	GALAXY
KARA.72 531A	18 09 42.	+ 49 53	48	14.8	PART OF DOUBLE GALAXY
MCG+08-33-027	18 09 42.	+ 49 59	9	17.	GALAXY
HUB E24	18 09 45.	- 17 14			DIFFUSE NEBULA
SCHO 0788	18 09 46.	- 25 16	300		ISOLATED DARK CLOUD
RNGC 6582	18 09 47.	+ 49 53		15.0	GALAXY
ZWG 142.016	18 09 48.	+ 21 33		15.4	GALAXY
MCG+04-43-008	18 09 48.	+ 21 33	15	15.4	GALAXY
ZWG 200.010	18 09 48.	+ 35 58		15.5	GALAXY
MCG+08-33-028	18 09 48.	+ 49 30	18	16.	GALAXY
ZWG 254.023	18 09 48.	+ 49 53		14.8	GALAXY
UGC 11146	18 09 48.	+ 49 53	108	14.8	GALAXY DBL SYS
KARA.72 531B	18 09 48.	+ 49 53	60		PART OF DOUBLE GALAXY
ZWG 279.004	18 09 48.	+ 53 20		15.6	GALAXY
UGC 11147	18 09 48.	+ 53 20	66	15.6	GALAXY Sb
RNGC 6568	18 09 48.	- 21 36		8.5	OPEN CLUSTER
OCL 0028	18 09 46.	- 21 37	1740	8.8	OPEN STAR CLUSTER
URA 01	18 09 50.	- 17 24	516		STELLAR RING
IC 1281	18 09 51.	+ 36 00 42.			MAY NOT EXIST
MCG+08-33-029	18 09 51.	+ 49 53	24	16.	GALAXY
RNGC 6438A	18 09 51.	- 85 27			GALAXY
ZWG 142.017	18 09 54.	+ 21 28		14.7	GALAXY
UGC 11148	18 09 54.	+ 21 28	90	14.7	GALAXY E
MCG+04-43-009	18 09 54.	+ 21 28	24	14.7	GALAXY
UGC 11149	18 09 54.	+ 49 51	78	16.0	GALAXY
MCG+08-33-030	18 09 54.	+ 49 53	36	16.	GALAXY
MCG+09-30-003	18 09 54.	+ 54 22	42	16.	GALAXY
SCHO 0789	18 09 54.	- 21 44	290		ISOLATED DARK CLOUD
SCHO 0790	18 09 55.	- 25 01	270		ISOLATED DARK CLOUD
RNGC 6577	18 09 56.	+ 21 28		14.5	GALAXY
KHAV 459	18 10	- 02 35	10800		DARK NEBULA
LBN 0061	18 10	- 14 30	1020		BRIGHT NEBULA
KHAV 460	18 10	- 18 35	2230		DARK NEBULA
B 304	18 10	- 18 44	5400		DARK OBJECT
KHAV 458	18 10	- 31 53	6050		DARK NEBULA
ZWG 142.018	18 10 00.	+ 25 35		15.5	GALAXY
UGC 11150	18 10 00.	+ 25 35	102	15.5	GALAXY S?
ZC 1810.0+3538	18 10 00.	+ 35 39	1480		CLUSTER OF GALAXIES
MCG+09-30-004	18 10 00.	+ 53 18	66	15.	GALAXY
MCG+15-01-019	18 10 00.	+ 87 40	30	16.	GALAXY
LDN 0519	18 10 00.	- 01 10	1500		DARK NEBULA
LDN 0493	18 10 00.	- 03 20	1140		DARK NEBULA
LDN 0476	18 10 00.	- 04 30	1260		DARK NEBULA
LDN 0419	18 10 00.	- 08 40	1320		DARK NEBULA
LDN 0386	18 10 00.	- 12 00	17520		DARK NEBULA
LDN 0282	18 10 00.	- 21 00	900		DARK NEBULA
COU 020	18 10 00.	- 22 15	4800		EMISSION NEBULA
LDN 0208	18 10 00.	- 24 40	720		DARK NEBULA
LDN 0197	18 10 00.	- 25 21	480		DARK NEBULA
LDN 0194	18 10 00.	- 25 30	2280		DARK NEBULA
LDN 0124	18 10 00.	- 28 00	8400		DARK NEBULA
PK359-06.1	18 10 02.0	- 32 20 34.	25		PLANETARY NEBULA
ZWG 142.019	18 10 06.	+ 21 53		14.9	GALAXY
ZC 1810.1+5223	18 10 06.	+ 52 23	1210		CLUSTER OF GALAXIES
SCHO 0791	18 10 07.	- 32 01	280		ISOLATED DARK CLOUD
SCHO 0792	18 10 08.	- 25 24	270		ISOLATED DARK CLOUD
PK006-03.3	18 10 10.5	- 25 30 56.	5		PLANETARY NEBULA
ZWG 142.020	18 10 12.	+ 21 35		15.7	GALAXY
ZWG 142.021	18 10 12.	+ 25 39		15.4	GALAXY
MCG+04-43-010	18 10 12.	+ 25 39	24	15.4	GALAXY
ZWG 172.010	18 10 12.	+ 29 08		14.9	GALAXY
UGC 11151	18 10 12.	+ 29 08	102	14.9	GALAXY S
MCG+05-43-007	18 10 12.	+ 29 09	96	14.9	GALAXY
ZC 1810.2+4949	18 10 12.	+ 49 49	7260		CLUSTER OF GALAXIES
ZWG 322.033	18 10 12.	+ 64 02		15.1	GALAXY
KARA.73B 0848	18 10 12.	+ 64 02	42	15.1	ISOLATED GALAXY S
SCHO 0793	18 10 12.	- 31 38	610		ISOLATED DARK CLOUD
SCHO 0794	18 10 16.	- 22 34	250		ISOLATED DARK CLOUD
IC 1280	18 10 17.	+ 25 39			NONSTELLAR OBJECT
ZWG 113.028	18 10 18.	+ 18 35		15.0	GALAXY
UGC 11152	18 10 18.	+ 18 35	132	15.0	GALAXY
MCG+03-46-018	18 10 18.	+ 18 36	132	15.0	GALAXY
ZC 1810.3+5659	18 10 18.	+ 56 59	1340		CLUSTER OF GALAXIES
URA 07	18 10 19.	- 16 13	42		STELLAR RING
RNGC 6569	18 10 22.	- 31 50		10.5	GLOBULAR CLUSTER
HOLM 775B	18 10 23.	+ 21 25	18	14.1	PART OF MULTIPLE GALAXY
IC 4692	18 10 23.	- 58 42			NONSTELLAR OBJECT
ZWG 142.022	18 10 24.	+ 21 25		14.5	GALAXY
UGC 11153	18 10 24.	+ 21 25	90	14.5	GALAXY TRP SYS
ZWG 142.023	18 10 24.	+ 25 25		15.1	GALAXY
UGC 11154	18 10 24.	+ 42 38	66	16.0	GALAXY Sa-b
ZC 1810.4+6824	18 10 24.	+ 68 24	1140		CLUSTER OF GALAXIES
ISS 0789	18 10 24.	- 19 24	90		STELLAR RING
GCL 091	18 10 24.	- 31 50	396	10.63	GLOBULAR STAR CLUSTER
VHA 260	18 10 24.	- 31 50	90		GLOBULAR STAR CLUSTER
HN 0455	18 10 24.	- 58 42			NEBULA
HOLM 775A	18 10 25.	+ 21 25	30	13.9	PART OF MULTIPLE GALAXY
IC 4697	18 10 25.	+ 25 24 38.			NONSTELLAR OBJECT
RNGC 6579	18 10 26.	+ 21 24		14.0	GALAXY
MCG+04-43-011	18 10 27.	+ 21 24	24	14.5	GALAXY
ZWG 056.010	18 10 30.	+ 06 42		15.4	GALAXY
ZWG 113.029	18 10 30.	+ 18 38		15.4	GALAXY
MCG+04-43-012	18 10 30.	+ 21 24 30.	60	14.5	GALAXY
ZWG 142.024	18 10 30.	+ 25 25		15.5	GALAXY
UGC 11155	18 10 30.	+ 25 25	96	15.5	GALAXY Sc?
ZWG 142.025	18 10 30.	+ 25 31		14.7	GALAXY
UGC 11156	18 10 30.	+ 25 31	78	14.7	GALAXY E?
ZWG 172.011	18 10 30.	+ 30 11		15.6	GALAXY
UGC 11157	18 10 30.	+ 30 11	90	15.6	GALAXY
MCG+07-37-023	18 10 30.	+ 42 29	60	15.5	GALAXY
LDN 0198	18 10 30.	- 25 23	360		DARK NEBULA
IC 4682	18 10 30.	- 71 36			NONSTELLAR OBJECT
HN 0454	18 10 31.	- 71 36			NEBULA
RNGC 6580	18 10 32.	+ 21 24		14.0	GALAXY
MCG+04-43-013	18 10 33.	+ 25 24	48	15.1	GALAXY
MCG+05-43-008	18 10 33.	+ 30 11 30.	96	15.6	GALAXY
PK005-04.1	18 10 34.1	- 26 09 32.	8		PLANETARY NEBULA
RNGC 6581	18 10 35.	+ 25 25		15.5	GALAXY
MCG+04-43-014	18 10 36.	+ 25 25	72	15.5	GALAXY
ZWG 340.038	18 10 36.	+ 69 17		15.7	GALAXY
SCHO 0795	18 10 38.	- 22 21	260		ISOLATED DARK CLOUD
SCHO 0796	18 10 41.	- 25 24	250		ISOLATED DARK CLOUD
ZWG 113.030	18 10 42.	+ 20 40		15.1	GALAXY
UGC 11158	18 10 42.	+ 20 40	60	15.1	GALAXY S
MCG+03-46-019	18 10 42.	+ 20 40	54	15.1	GALAXY
ZWG 142.026	18 10 42.	+ 25 24		15.5	GALAXY
MCG+04-43-015	18 10 42.	+ 25 30	18	14.7	GALAXY
ZWG 227.020	18 10 42.	+ 39 37		13.6	GALAXY
UGC 11159	18 10 42.	+ 39 37	120	13.6	GALAXY S
ZWG 279.005	18 10 42.	+ 56 47		15.7	GALAXY
UGC 11160	18 10 42.	+ 56 47	66	15.7	GALAXY Sb
ZWG 301.018	18 10 42.	+ 61 10		15.7	GALAXY
ISS 0790	18 10 42.	- 17 02	152		STELLAR RING
RNGC 6597	18 10 44.	+ 61 10		15.5	GALAXY
MCG+09-30-005	18 10 45.	+ 56 49	42	16.	GALAXY
HN 0049	18 10 45.	- 19 05			NEBULA
RNGC 6585	18 10 47.	+ 39 37		13.5	GALAXY
ZWG 142.027	18 10 49.	+ 25 29		15.7	GALAXY
MCG+10-26-020	18 10 48.	+ 61 10	18	16.	GALAXY
LDN 0469	18 10 48.	- 05 07	180		DARK NEBULA
RNGC 6573	18 10 48.	- 22 10			NON-EXISTENT OBJECT
PK011-00.2	18 10 48.1	- 19 05 26.7	11	11.7	PLANETARY NEBULA
URA 05	18 10 50.	- 16 22	1338		STELLAR RING
RNGC 6567	18 10 50.	- 19 05		11.5	PLANETARY NEBULA
SCHO 0797	18 10 50.	- 24 05	310		ISOLATED DARK CLOUD
MCG+07-37-024	18 10 51.	+ 39 37	108	13.5	GALAXY
MCG+05-43-009	18 10 54.	+ 28 11 30.	60	15.	GALAXY
ZC 1810.9+5830	18 10 54.	+ 58 30	870		CLUSTER OF GALAXIES
SCHO 0798	18 10 56.	- 23 50	540		ISOLATED DARK CLOUD
LDN 0431	18 11 00.	- 07 50	1200		DARK NEBULA
ISS 0998	18 11 00.	- 28 27	125		STELLAR RING
VHA 261	18 11 00.	- 28 40	90		OPEN STAR CLUSTER
SCHO 0799	18 11 03.	- 25 10	280		ISOLATED DARK CLOUD
HN 0456	18 11 03.	- 58 13			NEBULA
IC 4694	18 11 04.	- 58 13			NONSTELLAR OBJECT
ZWG 172.012	18 11 06.	+ 28 25		15.7	GALAXY
UGC 11161	18 11 06.	+ 28 25	60	15.7	GALAXY S
ZWG 172.013	18 11 06.	+ 29 41		15.3	GALAXY
UGC 11162	18 11 06.	+ 29 41	96	15.3	GALAXY S
ZC 1811.1+5011	18 11 06.	+ 50 11	1280		CLUSTER OF GALAXIES
SCHO 0800	18 11 08.	- 25 38	280		ISOLATED DARK CLOUD
MCG+05-43-010	18 11 09.	+ 29 41 30.	66	15.3	GALAXY
ZWG 200.011	18 11 18.	+ 38 53		15.7	GALAXY
MCG+09-30-006	18 11 18.	+ 52 57	48	16.	GALAXY
MCG+10-26-021	18 11 18.	+ 60 05	30	16.	GALAXY
ZWG 301.019	18 11 18.	+ 61 26		15.6	GALAXY
MCG+10-26-022	18 11 18.	+ 61 27	18	16.	GALAXY
LDN 0203	18 11 18.	- 25 11	240		DARK NEBULA
RNGC 6601	18 11 20.	+ 61 26		15.5	GALAXY
SCHO 0801	18 11 20.	- 31 56	240		ISOLATED DARK CLOUD
SCHO 0802	18 11 22.	- 18 19	260		ISOLATED DARK CLOUD
B 305	18 11 23.	- 31 49	780		DARK OBJECT
ZWG 084.026	18 11 24.	+ 14 52		15.6	GALAXY
ZWG 200.012	18 11 24.	+ 33 48		14.7	GALAXY
UGC 11163	18 11 24.	+ 33 48	72	14.7	GALAXY B
KARA.72 532A	18 11 24.	+ 33 48	66	14.7	PART OF DOUBLE GALAXY
ZC 1811.4+6941	18 11 24.	+ 69 41	5980		CLUSTER OF GALAXIES
ZWG 340.039	18 11 24.	+ 74 55		15.6	GALAXY
PK002-05.1	18 11 24.	- 29 50 35.	10	11.5	PLANETARY NEBULA
PK006-03.2	18 11 24.3	- 24 44 34.	5		PLANETARY NEBULA
LB 01090	18 11 25.	+ 62 35 12.		17.7	FAINT BLUE STAR
SCHO 0803	18 11 26.	- 25 29	280		ISOLATED DARK CLOUD
URA 04	18 11 29.	- 17 23	366		STELLAR RING
ZWG 142.028	18 11 30.	+ 21 05		14.6	GALAXY
UGC 11164	18 11 30.	+ 21 05	60	14.6	GALAXY S
MCG+06-40-010	18 11 30.	+ 33 49 30.	66	14.5	GALAXY
UGC 11165	18 11 30.	+ 68 42	96	16.0	GALAXY Sc
LDN 0472	18 11 30.	- 05 00	720		DARK NEBULA
LDN 0438	18 11 30.	- 07 10	360		DARK NEBULA
RNGC 6586	18 11 32.	+ 21 05		14.5	GALAXY
MCG+04-43-016	18 11 33.	+ 21 03 30.	42	14.6	GALAXY
VDB.66N 117	18 11 34.	- 17 25	132		REFLECTION NEBULA
PK359-07.1	18 11 34.1	- 32 37 52.	10		PLANETARY NEBULA
ZWG 113.031	18 11 36.	+ 18 49		14.3	GALAXY
UGC 11166	18 11 36.	+ 18 49	120	14.3	GALAXY S0
MCG+03-46-020	18 11 36.	+ 18 49	36	14.3	GALAXY
ZWG 200.013	18 11 36.	+ 33 43		15.7	GALAXY
UGC 11167	18 11 36.	+ 33 43	60	15.7	GALAXY S

OBJECT NAME	RIGHT ASCEN.	DECLINATION	DIAM.	MAGN.	TYPE OF OBJECT
KARA.72 532B	18 11 36.	+ 33 43	54	15.7	PART OF DOUBLE GALAXY
SCHO 0804	18 11 38.	- 18 18	1360		ISOLATED DARK CLOUD
PK024+05.1	18 11 39.	- 05 00 18.	48		PLANETARY NEBULA
SCHO 0805	18 11 40.	- 23 47	300		ISOLATED DARK CLOUD
ZWG 084.027	18 11 42.	+ 13 15		14.2	GALAXY
UGC 11168	18 11 42.	+ 13 15	84	14.2	GALAXY Sc-IRR
MCG+02-46-011	18 11 42.	+ 13 15	72	14.2	GALAXY
ZC 1811.7+4100	18 11 42.	+ 41 00	940		CLUSTER OF GALAXIES
MCG+09-30-007	18 11 42.	+ 52 52 30.	36	17.	GALAXY
ZWG 301.020	18 11 42.	+ 61 19		15.6	GALAXY
MCG+10-26-023	18 11 42.	+ 61 19 30.	36	15.	GALAXY
UGC 11169	18 11 42.	+ 64 59	66	16.0	GALAXY Sa-b
ASS 03	18 11 42.	- 20 28			OB ASSOCIATION SGR OB7
RNGC 6587	18 11 43.	+ 18 49		14.5	GALAXY
RNGC 6607	18 11 44.	+ 61 19		15.0	GALAXY
ISS 0681	18 11 48.	- 10 28	79		STELLAR RING
RNGC 6591	18 11 50.	+ 21 02			GALAXY
SCHO 0806	18 11 50.	- 22 07	260		ISOLATED DARK CLOUD
HN 0047	18 11 52.	- 20 18			NEBULA
PK002-06.1	18 11 53.8	- 30 16 32.	5		PLANETARY NEBULA
ZWG 084.028	18 11 54.	+ 10 05		15.5	GALAXY
KARA.72 533A	18 11 54.	+ 10 05	42	15.5	PART OF DOUBLE GALAXY
ZWG 142.029	18 11 54.	+ 21 30		15.6	GALAXY
UGC 11170	18 11 54.	+ 21 30	84	15.6	GALAXY
MCG+04-43-017	18 11 54.	+ 21 30	72	15.6	GALAXY
IC 1282	18 11 56.	+ 21 05			MAY NOT EXIST
MCG+10-26-024	18 11 57.	+ 61 17 30.	48	16.	GALAXY
LBN 0102	18 12	+ 07 03	360		BRIGHT NEBULA
KHAV 462	18 12	- 21 17	3750		DARK NEBULA
LB 09924	18 12	- 86 30		14.0	FAINT BLUE STAR
ZWG 084.029	18 12 00.	+ 09 59		15.5	GALAXY
KARA.72 533B	18 12 00.	+ 09 59	48	15.5	PART OF DOUBLE GALAXY
ZWG 113.032	18 12 00.	+ 15 29		14.8	GALAXY
MCG+03-46-021	18 12 00.	+ 15 29	60	14.8	GALAXY
ZWG 142.030	18 12 00.	+ 22 16		15.3	GALAXY
MCG+04-43-018	18 12 00.	+ 22 16	21	15.3	GALAXY
ZWG 254.024	18 12 00.	+ 46 07		15.4	GALAXY
7ZW 773	18 12 00.	+ 61 19			COMPACT GALAXY
ZWG 301.021	18 12 00.	+ 61 19		15.3	GALAXY
MCG+10-26-025	18 12 00.	+ 61 19 30.	30	15.	GALAXY
LDN 0510	18 12 00.	- 02 15	900		DARK NEBULA
LDN 0436	18 12 00.	- 07 25	300		DARK NEBULA
LDN 0424	18 12 00.	- 08 30	1440		DARK NEBULA
LDN 0418	18 12 00.	- 09 10	1620		DARK NEBULA
LDN 0256	18 12 00.	- 23 00	2460		DARK NEBULA
LDN 0199	18 12 00.	- 25 28	120		DARK NEBULA
RNGC 6609	18 12 02.	+ 61 19		15.5	GALAXY
RNGC 6608	18 12 02.	+ 61 19		15.5	GALAXY
RNGC 6593	18 12 03.	+ 22 16		15.5	GALAXY
MCG+10-26-026	18 12 03.	+ 57 09	30	17.	GALAXY
SCHO 0807	18 12 04.	- 23 53	260		ISOLATED DARK CLOUD
7ZW 774	18 12 06.	+ 60 51			COMPACT GALAXY
7ZW 775	18 12 06.	+ 73 02			COMPACT GALAXY
PK001-06.1	18 12 07.	- 30 33	10	14.0	PLANETARY NEBULA
MRSL 035+11/1	18 12 12.	+ 07 03	300		HII REGION
ZWG 172.014	18 12 12.	+ 30 40		14.8	GALAXY
UGC 11171	18 12 12.	+ 30 40	90	14.8	GALAXY S
ZWG 172.015	18 12 12.	+ 31 36		15.7	GALAXY
ZWG 301.022	18 12 12.	+ 57 10		15.7	GALAXY
7ZW 776	18 12 12.	+ 69 55			COMPACT GALAXY
UGC 11172	18 12 12.	+ 69 55	12	14.0	GALAXY VY CMPT
CED 156	18 12 17.	- 20 45			DIFFUSE GALACTIC NEBULA
ZC 1812.3+2237	18 12 18.	+ 22 37	12430		CLUSTER OF GALAXIES
MCG+05-43-011	18 12 18.	+ 30 40	72	14.8	GALAXY
ISS 0682	18 12 18.	- 14 56	78		STELLAR RING
OCL 0033	18 12 18.	- 19 01	120	13.	OPEN STAR CLUSTER
MCG+10-26-027	18 12 21.	+ 57 05 30.	39	17.	GALAXY
SCHO 0809	18 12 22.	- 31 57	280		ISOLATED DARK CLOUD
ZC 1812.4+5254	18 12 24.	+ 52 54	870		CLUSTER OF GALAXIES
7ZW 777	18 12 24.	+ 60 30			COMPACT GALAXY
ZWG 322.034	18 12 24.	+ 63 53		15.5	GALAXY
LDN 0507	18 12 24.	- 02 38	480		DARK NEBULA
ISS 0683	18 12 24.	- 15 05	169		STELLAR RING
ASS 04	18 12 24.	- 19 04			OB ASSOCIATION SGR OB4
ISS 0902	18 12 24.	- 23 32	80		STELLAR RING
LDN 0490	18 12 30.	- 03 50	180		DARK NEBULA
URA 11	18 12 30.	- 16 04	432		STELLAR RING
PK009-02.1	18 12 30.	- 21 36 10.	10		PLANETARY NEBULA
SCHO 0810	18 12 30.	- 32 06	240		ISOLATED DARK CLOUD
SC 1808-5256.7	18 12 30.	- 52 55 56.	18		NEBULA
PK019+03.1	18 12 31.	- 10 11 12.	4		PLANETARY NEBULA
ZC 1812.6+6225	18 12 36.	+ 62 25	1810		CLUSTER OF GALAXIES
LDN 0323	18 12 36.	- 18 12	480		DARK NEBULA
ISS 0999	18 12 36.	- 30 17	260		STELLAR RING
B 092	18 12 39.	- 18 15	900		DARK OBJECT
B 306	18 12 39.	- 25 44	240		DARK OBJECT
ISS 1000	18 12 42.	- 29 11	90		STELLAR RING
PK004-05.2	18 12 45.	- 27 55 29.	5	11.	PLANETARY NEBULA
URA 06	18 12 47.	- 15 53	402		STELLAR RING
ZWG 254.025	18 12 48.	+ 49 02		15.4	GALAXY
UGC 11173	18 12 48.	+ 49 02	72	15.4	GALAXY S
ZWG 340.040	18 12 48.	+ 70 50		15.7	GALAXY
ISS 0684	18 12 48.	- 13 27	251		STELLAR RING
OCL 0027	18 12 48.	- 22 09	228	12.4	OPEN STAR CLUSTER
RNGC 6583	18 12 48.	- 22 09		12.0	OPEN CLUSTER
LDN 0195	18 12 48.	- 25 45	180		DARK NEBULA
MCG+08-33-031	18 12 54.	+ 49 01	60	15.	GALAXY
MCG+11-22-029	18 12 54.	+ 63 30	36	17.	GALAXY
MRSL 012-00/1	18 12 54.	- 18 15	5400		HII REGION
PK001-06.2	18 12 58.6	- 30 53 13.	5	11.8	PLANETARY NEBULA
SCHO 0811	18 12 59.	- 18 13	260		ISOLATED DARK CLOUD
SCHO 0812	18 12 59.	- 18 22	280		ISOLATED DARK CLOUD
LBN 0056	18 13	- 16 40	180		BRIGHT NEBULA
LBN 0055	18 13	- 16 40	3000		BRIGHT NEBULA
CED 158	18 13	- 16 44			DIFFUSE GALACTIC NEBULA
KHAV 463	18 13	- 17 23	6150		DARK NEBULA
LBN 0051	18 13	- 18 00	9900		BRIGHT NEBULA
KHAV 464	18 13	- 18 11	700		DARK NEBULA
LBN 0042	18 13	- 20 20	480		BRIGHT NEBULA
KHAV 461	18 13	- 28 53	2570		DARK NEBULA
ZWG 301.023	18 13 00.	+ 59 49		15.5	GALAXY
ZWG 322.035	18 13 00.	+ 63 30		15.7	GALAXY
7ZW 778	18 13 00.	+ 68 19			COMPACT GALAXY
ZWG 340.041	18 13 00.	+ 74 10		15.5	GALAXY
LDN 0492	18 13 00.	- 03 50	360		DARK NEBULA
LDN 0447	18 13 00.	- 06 40	3000		DARK NEBULA
COU 022	18 13 00.	- 15 40	12000		EMISSION NEBULA
LDN 0357	18 13 00.	- 15 45	1560		DARK NEBULA
COU 021	18 13 00.	- 18 30	9000		EMISSION NEBULA
COU 023	18 13 00.	- 20 20	6000		EMISSION NEBULA
LDN 0165	18 13 00.	- 27 00	900		DARK NEBULA
HN 0457	18 13 00.	- 58 56			NEBULA
IC 4695	18 13 00.	- 58 56			NONSTELLAR OBJECT
PK004-05.1	18 13 03.3	- 27 16 01.	8		PLANETARY NEBULA
ZWG 172.016	18 13 06.	+ 29 45		15.4	GALAXY
ZWG 227.021	18 13 06.	+ 43 15		14.4	GALAXY
UGC 11174	18 13 06.	+ 43 15	60	14.4	GALAXY S
MCG+07-37-025	18 13 06.	+ 43 16 30.	48	14.	GALAXY
MCG+10-26-028	18 13 06.	+ 59 49	30	16.	GALAXY
ISS 0791	18 13 06.	- 20 48	139		STELLAR RING
PK002-06.2	18 13 06.8	- 30 08 40.	5		PLANETARY NEBULA
RNGC 6606	18 13 09.	+ 43 15		14.5	GALAXY
PK004-04.2	18 13 09.5	- 27 05 37.	4		PLANETARY NEBULA
MCG+05-43-012	18 13 12.	+ 29 46	21	15.4	GALAXY
ZWG 301.024	18 13 12.	+ 61 08		15.5	GALAXY
MCG+11-22-032	18 13 12.	+ 63 03	42	16.	GALAXY
ZWG 322.036	18 13 12.	+ 68 20		13.6	GALAXY
UGC 11175	18 13 12.	+ 68 20	120	13.6	GALAXY DBL SYS
KARA.72 534B	18 13 12.	+ 68 20	48		PART OF DOUBLE GALAXY
KARA.72 534A	18 13 12.	+ 68 21	60	13.6	PART OF DOUBLE GALAXY
MCG+11-22-031	18 13 12.	+ 68 21	30	16.	GALAXY
MCG+11-22-030	18 13 12.	+ 68 21	114	14.	GALAXY
RNGC 6622	18 13 13.	+ 68 20		13.5	GALAXY
RNGC 6621	18 13 13.	+ 68 21		13.5	GALAXY
IC 4708	18 13 14.	+ 61 08 34.			NONSTELLAR OBJECT
PK010-01.1	18 13 17.	- 20 28 03.	9	13.1	PLANETARY NEBULA
ZC 1813.3+3452	18 13 18.	+ 34 52	2080		CLUSTER OF GALAXIES
DG 149	18 13 13.	- 19 48	240		REFLECTION NEBULA
RNGC 6578	18 13 19.	- 20 28			PLANETARY NEBULA
CED 157A	18 13 22.	- 19 48	300		DIFFUSE GALACTIC NEBULA
SCHO 0813	18 13 27.	- 18 03	380		ISOLATED DARK CLOUD
ZC 1813.5+4113	18 13 30.	+ 41 13	1340		CLUSTER OF GALAXIES
ZWG 301.025	18 13 30.	+ 61 18		15.6	GALAXY
UGC 11176	18 13 30.	+ 61 18	84	15.6	GALAXY Sc
MCG+10-26-029	18 13 30.	+ 61 18	60	15.	GALAXY
VV 247B	18 13 30.	+ 68 18	30	15.	INTERACTING GALAXY
VV 247A	18 13 30.	+ 68 18	78	14.	INTERACTING GALAXY
ARP 081	18 13 30.	+ 68 13			PECULIAR GALAXY
MRSL 013-00/1	18 13 30.	- 17 25	900		HII REGION
OCL 0035	18 13 30.	- 18 14	780	10.9	OPEN STAR CLUSTER
MRSL 011-01/1	18 13 30.	- 20 19	6600		HII REGION
RNGC 6617	18 13 32.	+ 61 18		15.5	GALAXY
RNGC 6599	18 13 35.	+ 24 54		13.5	GALAXY
ZWG 056.011	18 13 36.	+ 06 44		15.5	GALAXY
UGC 11177	18 13 36.	+ 06 44	90	15.5	GALAXY SBb
ZWG 084.030	18 13 36.	+ 13 45		14.7	GALAXY
MCG+02-46-012	18 13 36.	+ 13 45	36	14.7	GALAXY
ZWG 142.031	18 13 36.	+ 24 54		13.7	GALAXY
UGC 11178	18 13 36.	+ 24 54	78	13.7	GALAXY S0
MRSL 014-00/1	18 13 36.	- 16 42	3600		HII REGION
RNGC 6600	18 13 41.	+ 25 00			NON-EXISTENT OBJECT
MCG+04-43-019	18 13 42.	+ 24 52	30	13.7	GALAXY
ZWG 142.032	18 13 42.	+ 25 20		15.6	GALAXY
UGC 11179	18 13 42.	+ 25 20	66	15.6	GALAXY Sc
ZWG 254.026	18 13 42.	+ 49 24		15.6	GALAXY
7ZW 779	18 13 42.	+ 58 53			COMPACT GALAXY
MCG+11-22-033	18 13 42.	+ 64 45	30	17.	GALAXY
MRSL 024+05/1	18 13 42.	- 04 58	3000		HII REGION
MCG+04-43-020	18 13 48.	+ 25 20	63	15.6	GALAXY
ZWG 172.017	18 13 48.	+ 30 17		15.6	GALAXY
ZCG 1813+58	18 13 48.	+ 58 54		17.8	COMPACT GALAXY
MRSL 012-01/1	18 13 48.	- 18 40	180		HII REGION
VDB.66N 118	18 13 50.	- 19 46	312		REFLECTION NEBULA
PK337-18.1	18 13 52.	- 57 12 57.	10		PLANETARY NEBULA
MCG+07-37-026	18 13 54.	+ 41 33	36	16.	GALAXY
ZWG 254.027	18 13 54.	+ 47 31		15.6	GALAXY
UGC 11180	18 13 54.	+ 47 31	78	15.6	GALAXY S
MCG+09-30-008	18 13 54.	+ 56 29	66	16.	GALAXY
SG 3.130	18 13 54.	- 16 44	2700		DIFFUSE EMISSION NEBULA
VV 304B	18 13 54.	- 60 35	120		INTERACTING GALAXY
VV 304A	18 13 54.	- 60 35	120	12.7	INTERACTING GALAXY
SCHO 0814	18 13 58.	- 18 02	290		ISOLATED DARK CLOUD
B 093	18 13 58.	- 18 05	900		DARK OBJECT
LBN 0059	18 14	- 15 30	240		BRIGHT NEBULA
LBN 0047	18 14	- 19 30	900		BRIGHT NEBULA
LBN 0046	18 14	- 19 40	180		BRIGHT NEBULA
LBN 0043	18 14	- 19 45	180		BRIGHT NEBULA
SIV 01	18 14	- 21 31	9000		FAINT H EMISSION REGION
ZWG 172.018	18 14 00.	+ 28 42		15.2	GALAXY
UGC 11181	18 14 00.	+ 43 18	96	16.0	GALAXY MLT SYS
MCG+07-37-027	18 14 00.	+ 43 19	9	17.	GALAXY
MCG+08-33-032	18 14 00.	+ 47 30	72	15.	GALAXY
UGC 11182	18 14 00.	+ 56 28	60	16.5	GALAXY Sb-c
MCG+11-22-034	18 14 00.	+ 64 51	24	17.	GALAXY
LDN 0500	18 14 00.	- 03 20	240		DARK NEBULA
LDN 0488	18 14 00.	- 04 15	240		DARK NEBULA
LDN 0429	18 14 00.	- 08 20	1020		DARK NEBULA
COU 025	18 14 00.	- 16 40	2400		EMISSION NEBULA
LDN 0331	18 14 00.	- 17 50	240		DARK NEBULA
LDN 0328	18 14 00.	- 18 01	120		DARK NEBULA
LDN 0327	18 14 00.	- 18 05	420		DARK NEBULA
OCL 0031	18 14 00.	- 19 54	750	7.4	OPEN STAR CLUSTER
VDB.66N 119	18 14 00.	- 19 54	336		REFLECTION NEBULA
DG 150	18 14 00.	- 19 54	240		REFLECTION NEBULA
COU 024	18 14 00.	- 21 40	4800		EMISSION NEBULA
LDN 0265	18 14 00.	- 22 30	2280		DARK NEBULA
COU 026	18 14 00.	- 22 40	10800		EMISSION NEBULA
RNGC 6589	18 14 01.	- 19 40			DIFFUSE NEBULA
RNGC 6590	18 14 01.	- 19 45			DIFFUSE NEBULA
RNGC 6595	18 14 01.	- 19 54		7.0	CLUSTER WITH NEBULOSITY
PK030+08.1	18 14 01.0	+ 01 52 04.			PLANETARY NEBULA
MCG+05-43-013	18 14 03.	+ 28 41	54	15.2	GALAXY
MCG+07-37-028	18 14 03.	+ 43 18 30.	36	17.	GALAXY
CED 157B	18 14 04.	- 19 54	210		DIFFUSE GALACTIC NEBULA
IC 1285	18 14 06.	+ 25 05			OPEN CLUSTER
IC 4700	18 14 08.	- 19 53 30.			SAME AS NGC 6595
ZWG 084.031	18 14 12.	+ 14 49		15.7	GALAXY
LDN 0487	18 14 12.	- 04 20	240		DARK NEBULA
OCL 0047	18 14 12.	- 14 59		6.0	OPEN STAR CLUSTER
ISS 0792	18 14 12.	- 20 57	51		STELLAR RING
PK007-03.1	18 14 12.5	- 24 00 03.	13		PLANETARY NEBULA
IC 4701	18 14 13.	- 16 43			HII REGION
MCG+07-37-029	18 14 15.	+ 44 02	36	14.	GALAXY
MCG+10-26-030	18 14 18.	+ 60 41	30	16.	GALAXY
ISS 1001	18 14 18.	- 27 19	150		STELLAR RING
IC 1283	18 14 19.	- 19 45 28.			NONSTELLAR OBJECT

OBJECT NAME	RIGHT ASCEN.	DECLINATION	DIAM.	MAGN.	TYPE OF OBJECT
CED 157C	18 14 22.	- 19 45	240		DIFFUSE GALACTIC NEBULA
PK000-07.1	18 14 22.	- 31 57 58.	25		PLANETARY NEBULA
ZWG 142.033	18 14 24.	+ 22 05		15.6	GALAXY
1ZW 204	18 14 24.	+ 36 03			COMPACT GALAXY
ZWG 200.014	18 14 24.	+ 36 03		15.6	GALAXY
UGC 11183	18 14 24.	+ 68 00	84	16.5	GALAXY
DG 151	18 14 24.	- 19 43	1080		REFLECTION NEBULA
SC 1810-5443.8	18 14 24.	- 54 42 54.	42		NEBULA
RNGC 6612	18 14 25.	+ 36 03		15.5	GALAXY
VDB.66N 120	18 14 27.	- 17 00	72		REFLECTION NEBULA
ZWG 142.034	18 14 30.	+ 22 05		15.6	GALAXY
ZWG 142.035	18 14 30.	+ 25 01		14.6	GALAXY
UGC 11184	18 14 30.	+ 25 01	72	14.6	GALAXY S
ZWG 172.019	18 14 30.	+ 29 53		15.5	GALAXY
MCG+06-40-011	18 14 30.	+ 36 04 30.	36	15.	GALAXY
UGC 11185	18 14 30.	+ 42 37	102	15.9	GALAXY DBL SYS
1ZW 205	18 14 30.	+ 42 38			COMPACT GALAXY
ZWG 279.006	18 14 30.	+ 51 46		15.3	GALAXY
MCG+10-26-031	18 14 30.	+ 61 17 30.	18	16.	GALAXY
ISS 0793	18 14 30.	- 17 58	81		STELLAR RING
SCHO 0815	18 14 30.	- 18 00	260		ISOLATED DARK CLOUD
EM5	18 14 30.	- 29 08		13.00	PLANETARY NEBULA
PK003-06.1	18 14 30.5	- 29 09 31.	9	13.00	PLANETARY NEBULA
PK004-05.3	18 14 32.4	- 28 18 29.	12		PLANETARY NEBULA
RNGC 6596	18 14 33.	- 16 41			OPEN CLUSTER
RNGC 6602	18 14 35.	+ 25 00		14.5	GALAXY
MCG+04-43-021	18 14 36.	+ 25 00	36	14.6	GALAXY
MCG+07-37-030	18 14 36.	+ 42 39	18	15.5	GALAXY
MCG+07-37-031	18 14 36.	+ 42 39 30.	48	15.5	GALAXY
OCL 0041	18 14 36.	- 16 41			OPEN STAR CLUSTER
COU 027	18 14 36.	- 19 42	1200		EMISSION NEBULA
GCL 092	18 14 36.	- 52 14	582	9.4	GLOBULAR STAR CLUSTER
RNGC 6584	18 14 36.	- 52 14		9.5	GLOBULAR CLUSTER
CED 157D	18 14 40.	- 19 41	990		DIFFUSE GALACTIC NEBULA
SG 3.131	18 14 40.	- 19 41	1020		DIFFUSE EMISSION NEBULA
MCG+11-22-035	18 14 42.	+ 68 00	39	17.	GALAXY
OCL 0055	18 14 42.	- 13 22	1080	12.2	OPEN STAR CLUSTER
IC 1284	18 14 43.	- 19 41 07.	600		DIFFUSE NEBULA
MCG+09-30-009	18 14 48.	+ 51 44	48	15.	GALAXY
ZC 1814.8+5814	18 14 48.	+ 58 14	3760		CLUSTER OF GALAXIES
MCG+12-17-019	18 14 48.	+ 69 38	9	16.	GALAXY
MRSL 011-01/2	18 14 48.	- 19 41	1200		HII REGION
HN 0098	18 14 49.	- 46 01			NEBULA
PK348-13.1	18 14 49.	- 46 01	10	11.9	PLANETARY NEBULA
IC 4699	18 14 49.	- 46 01	5	11.9	PLANETARY NEBULA
MCG+11-22-036	18 14 54.	+ 67 57	39	17.	GALAXY
RNGC 6610	18 14 58.	+ 14 58			NON-EXISTENT OBJECT
LBN 0072	18 15	- 12 00	2700		BRIGHT NEBULA
LBN 0057	18 15	- 16 00	7200		BRIGHT NEBULA
KHAV 465	18 15	- 22 23	4220		DARK NEBULA
KHAV 466	18 15	- 31 29	1910		DARK NEBULA
ZWG 084.032	18 15 00.	+ 12 56		15.5	GALAXY
ZWG 254.028	18 15 00.	+ 47 49		14.6	GALAXY
UGC 11186	18 15 00.	+ 47 49	60	14.6	GALAXY S
MCG+08-33-033	18 15 00.	+ 47 49	54	14.	GALAXY
LDN 0501	18 15 00.	- 03 20	600		DARK NEBULA
LDN 0483	18 15 00.	- 04 50	780		DARK NEBULA
SG 3.132	18 15 00.	- 11 47	360		DIFFUSE EMISSION NEBULA
LDN 0315	18 15 00.	- 19 30	9900		DARK NEBULA
LDN 0272	18 15 00.	- 22 10	1020		DARK NEBULA
SCHO 0816	18 15 04.	- 17 58	250		ISOLATED DARK CLOUD
MRSL 018+02/1	18 15 06.	- 11 45	240		HII REGION
MIN.46 08	18 15 06.	- 11 46			DIFFUSE NEBULA
PK006-04.1	18 15 07.7	- 25 39 22.	7		PLANETARY NEBULA
ARC 2301	18 15 11.	+ 69 38		15.8	RICH CLUSTER OF GALAXIES
ZWG 301.026	18 15 12.	+ 61 40		15.5	GALAXY
KARA.73B 0849	18 15 12.	+ 61 40	54	15.5	ISOLATED GALAXY S
UGC 11187	18 15 12.	+ 62 15	66	16.0	GALAXY S
MRSL 018+01/1	18 15 12.	- 11 58	4200		HII REGION
SCHO 0817	18 15 12.	- 14 24	400		ISOLATED DARK CLOUD
PK038+12.1	18 15 13.	+ 10 07 37.	7	12.9	PLANETARY NEBULA
MCG+10-26-032	18 15 15.	+ 61 40 30.	60	16.	GALAXY
MCG+10-26-033	18 15 15.	+ 62 16 30.	66	16.	GALAXY
ZWG 113.033	18 15 18.	+ 18 53		15.6	GALAXY
UGC 11188	18 15 18.	+ 18 53	66	15.6	GALAXY Sc
ZWG 114.001	18 15 18.	+ 19 25		15.5	GALAXY
ZWG 113.034	18 15 18.	+ 19 25		15.5	GALAXY
MCG+03-46-022	18 15 18.	+ 19 26	54	15.5	GALAXY
UGC 11189	18 15 18.	+ 38 46	72	16.0	GALAXY Sb-c
UGC 11190	18 15 18.	+ 40 47	66	16.0	GALAXY Sc
ZWG 279.007	18 15 18.	+ 55 33		14.8	GALAXY
UGC 11191	18 15 18.	+ 55 33	90	14.8	GALAXY Sa-b
OCL 0056	18 15 18.	- 12 15	2400	9.	OPEN STAR CLUSTER
RNGC 6604	18 15 18.	- 12 15		8.0	OPEN CLUSTER
HN 0458	18 15 19.	- 64 46			NEBULA
SG 3.133	18 15 20.	- 13 39	480		DIFFUSE EMISSION NEBULA
IC 4696	18 15 20.	- 64 47			NONSTELLAR OBJECT
PK008-03.1	18 15 21.3	- 23 26 12.	8		PLANETARY NEBULA
IC 1286	18 15 22.	+ 55 34 30.			NONSTELLAR OBJECT
PK000-07.2	18 15 22.2	- 31 56 00.			PLANETARY NEBULA
ZWG 172.020	18 15 24.	+ 32 31		15.5	GALAXY
ZC 1815.4+3346	18 15 24.	+ 33 46	2490		CLUSTER OF GALAXIES
MCG+07-37-032	18 15 24.	+ 40 48	72	15.	GALAXY
MCG+09-30-010	18 15 24.	+ 55 34	78	14.	GALAXY
7ZW 780	18 15 24.	+ 58 17			COMPACT GALAXY
MCG+11-22-037	18 15 24.	+ 67 03	39	17.	GALAXY
7ZW 781	18 15 24.	+ 76 20			COMPACT GALAXY
ZWG 356.008	18 15 24.	+ 76 20		15.5	GALAXY
MRSL 015+00/1	18 15 24.	- 15 37	300		HII REGION
RNGC 6605	18 15 28.	- 14 59			NON-EXISTENT OBJECT
PK004-05.5	18 15 28.7	- 28 09 20.	25		PLANETARY NEBULA
COU 028	18 15 30.	- 12 00	4200		EMISSION NEBULA
OCL 0036	18 15 30.	- 18 26	660	11.8	OPEN STAR CLUSTER
SCHO 0818	18 15 31.	- 15 48	300		ISOLATED DARK CLOUD
RNGC 6603	18 15 32.	- 18 26		11.5	OPEN CLUSTER
SCHO 0819	18 15 33.	- 22 10	620		ISOLATED DARK CLOUD
PK004-05.4	18 15 33.	- 28 08	12		PLANETARY NEBULA
ZWG 084.033	18 15 36.	+ 12 45		15.6	GALAXY
MCG+04-43-022	18 15 36.	+ 22 12	84	15.2	GALAXY
ZWG 142.036	18 15 36.	+ 22 13		15.2	GALAXY
UGC 11192	18 15 36.	+ 22 13	90	15.2	GALAXY Sa-b
RNGC 6616	18 15 39.	+ 22 13		15.0	GALAXY
SCHO 0820	18 15 41.	- 17 50	250		ISOLATED DARK CLOUD
ZWG 227.022	18 15 42.	+ 40 01		15.2	GALAXY
MCG+12-17-020	18 15 42.	+ 70 59 30.	66	15.	GALAXY
ZWG 340.042	18 15 42.	+ 71 00		15.1	GALAXY
UGC 11193	18 15 42.	+ 71 00	96	15.1	GALAXY DWRF IR
B 307	18 15 43.	- 17 59	360		DARK OBJECT
URA 13	18 15 46.	- 17 25	354		STELLAR RING
ZWG 254.029	18 15 48.	+ 48 55		15.7	GALAXY
MCG+08-33-034	18 15 48.	+ 48 55	30	15.	GALAXY
ZWG 322.037	18 15 48.	+ 67 25		15.6	GALAXY
ASS 09	18 15 48.	- 11 59	18000		OB ASSOCIATION SER OB2
SG 3.135	18 15 48.	- 12 01	2700		DIFFUSE EMISSION NEBULA
SG 3.134	18 15 48.	- 12 09	7800		DIFFUSE EMISSION NEBULA
ZWG 114.002	18 15 54.	+ 18 05		15.2	GALAXY
ZWG 113.035	18 15 54.	+ 18 05		15.2	GALAXY
MCG+03-46-023	18 15 54.	+ 18 05	54	15.2	GALAXY
ZWG 142.037	18 15 54.	+ 26 44		14.4	GALAXY
UGC 11194	18 15 54.	+ 26 44	102	14.4	GALAXY SBb-c
MCG+07-37-033	18 15 54.	+ 40 01	48	15.	GALAXY
MCG+07-37-034	18 15 54.	+ 40 59	39	15.	GALAXY
ISS 0903	18 15 54.	- 22 22	125		STELLAR RING
SCHO 0821	18 15 58.	- 26 29	280		ISOLATED DARK CLOUD
RNGC 6611	18 15 59.	- 13 48		6.5	CLUSTER WITH NEBULOSITY
LBN 0071	18 16	- 12 30	3300		BRIGHT NEBULA
LBN 0067	18 16	- 13 50	4320		BRIGHT NEBULA
KHAV 467	18 16	- 23 35	2110		DARK NEBULA
KHAV 468	18 16	- 26 29	1990		DARK NEBULA
LDN 0569	18 16 00.	+ 01 30	22560		DARK NEBULA
MCG+04-43-023	18 16 00.	+ 26 43	96	14.4	GALAXY
ZWG 172.021	18 16 00.	+ 28 35		15.6	GALAXY
ZWG 172.022	18 16 00.	+ 30 37		14.6	GALAXY
UGC 11195	18 16 00.	+ 30 37	72	14.6	GALAXY S+COMP
KARA.72 535B	18 16 00.	+ 30 37	30		PART OF DOUBLE GALAXY
KARA.72 535A	18 16 00.	+ 30 37	36	14.6	PART OF DOUBLE GALAXY
ZWG 172.023	18 16 00.	+ 31 58		15.4	GALAXY
7ZW 782	18 16 00.	+ 72 22			COMPACT GALAXY
LDN 0504	18 16 00.	- 03 10	180		DARK NEBULA
LDN 0476	18 16 00.	- 05 25	240		DARK NEBULA
LDN 0470	18 16 00.	- 05 45	300		DARK NEBULA
OCL 0054	18 16 00.	- 13 48	2280	6.7	OPEN STAR CLUSTER
COU 029	18 16 00.	- 13 49	5400		EMISSION NEBULA
MRSL 016+00/1	18 16 00.	- 13 54	5400		HII REGION
ISS 0794	18 16 00.	- 19 42	155		STELLAR RING
CED 159	18 16 02.	- 13 48	2100		DIFFUSE GALACTIC NEBULA
SG 3.136	18 16 02.	- 13 48	4680		DIFFUSE EMISSION NEBULA
MCG+05-43-014	18 16 03.	+ 30 38	66	14.6	GALAXY
IC 4703	18 16 06.	- 13 48			OPEN CLUSTER
ISS 0795	18 16 06.	- 19 00	139		STELLAR RING
B 308	18 16 07.	- 22 15	420		DARK OBJECT
ZWG 084.034	18 16 12.	+ 13 15		14.8	GALAXY
UGC 11196	18 16 12.	+ 13 15	72	14.8	GALAXY SB0
MCG+02-46-013	18 16 12.	+ 13 15	48	14.8	GALAXY
ZWG 142.038	18 16 12.	+ 21 16		14.8	GALAXY
UGC 11197	18 16 12.	+ 21 16	66	14.8	GALAXY S0
LDN 0335	18 16 12.	- 17 50	240		DARK NEBULA
RNGC 6615	18 16 15.	+ 13 15		15.0	GALAXY
IC 4698	18 16 16.	- 63 23			NONSTELLAR OBJECT
HN 0459	18 16 17.	- 63 23			NEBULA
MCG+04-43-024	18 16 18.	+ 21 15	18	14.8	GALAXY
ZWG 227.023	18 16 18.	+ 41 20		15.6	GALAXY
LDN 0467	18 16 18.	- 06 00	300		DARK NEBULA
PK005-05.1	18 16 18.0	- 26 36 37.	9		PLANETARY NEBULA
MCG+07-37-035	18 16 21.	+ 41 20	24	16.	GALAXY
ZWG 227.024	18 16 24.	+ 39 48		15.0	GALAXY
SCHO 0822	18 16 24.	- 15 12	210		ISOLATED DARK CLOUD
MRSL 013-01/1	18 16 24.	- 17 35	34200		HII REGION
SCHO 0823	18 16 27.	- 16 10	500		ISOLATED DARK CLOUD
ZWG 227.025	18 16 30.	+ 39 46		15.5	GALAXY
MCG+07-37-036	18 16 30.	+ 39 48	42	14.5	GALAXY
LDN 0482	18 16 30.	- 05 08	480		DARK NEBULA
LDN 0479	18 16 30.	- 05 10	180		DARK NEBULA
LDN 0477	18 16 30.	- 05 25	300		DARK NEBULA
LDN 0475	18 16 30.	- 05 30	660		DARK NEBULA
ISS 0796	18 16 30.	- 19 28	351		STELLAR RING
RNGC 6588	18 16 35.	- 63 50			NON-EXISTENT OBJECT
LB 01091	18 16 36.	+ 65 12 48.		17.5	FAINT BLUE STAR
ISS 0797	18 16 36.	- 19 08	79		STELLAR RING
ZWG 114.003	18 16 42.	+ 16 14		15.1	GALAXY
UGC 11198	18 16 42.	+ 16 14	78	15.1	GALAXY Sa-b
MCG+03-46-024	18 16 42.	+ 16 14	60	15.1	GALAXY
ZWG 200.015	18 16 42.	+ 36 17		14.6	GALAXY
UGC 11199	18 16 42.	+ 36 17	84	14.6	GALAXY Sa
RNGC 6619	18 16 46.	+ 23 38		14.5	GALAXY
IC 4706	18 16 46.	- 16 02 15.			NONSTELLAR OBJECT
CED 160A	18 16 47.	- 16 02	390		DIFFUSE GALACTIC NEBULA
SG 3.137	18 16 47.	- 16 02	300		DIFFUSE EMISSION NEBULA
ZWG 142.039	18 16 48.	+ 23 38		14.3	GALAXY
UGC 11200	18 16 48.	+ 23 38	90	14.3	GALAXY E?
ZWG 172.024	18 16 48.	+ 27 17		15.7	GALAXY
MCG+06-40-012	18 16 48.	+ 36 18	90	14.5	GALAXY
SG 3.138	18 16 52.	- 15	7200		DIFFUSE EMISSION NEBULA
MCG+04-43-025	18 16 54.	+ 23 37	36	14.3	GALAXY
ZWG 254.030	18 16 54.	+ 47 41		15.7	GALAXY
SCHO 0824	18 16 58.	- 15 12	270		ISOLATED DARK CLOUD
KHAV 476	18 17	+ 00 55	11570		DARK NEBULA
KHAV 473	18 17	- 04 41	8540		DARK NEBULA
LBN 0070	18 17	- 13 00	9000		BRIGHT NEBULA
LBN 0068	18 17	- 13 55	1920		BRIGHT NEBULA
LBN 0052	18 17	- 18 00	1020		BRIGHT NEBULA
KHAV 474	18 17	- 20 41	7370		DARK NEBULA
KHAV 469	18 17	- 24 29	2230		DARK NEBULA
KHAV 470	18 17	- 27 23	3090		DARK NEBULA
KHAV 472	18 17	- 31 17	1000		DARK NEBULA
KHAV 471	18 17	- 32 05	1720		DARK NEBULA
7ZW 783	18 17 00.	+ 59 42			COMPACT GALAXY
LDN 0505	18 17 00.	- 03 15	300		DARK NEBULA
LDN 0471	18 17 00.	- 05 50	960		DARK NEBULA
LDN 0466	18 17 00.	- 06 10	420		DARK NEBULA
LDN 0440	18 17 00.	- 07 30	480		DARK NEBULA
COU 030	18 17 00.	- 12 20	9000		EMISSION NEBULA
MRSL 015-00/2	18 17 00.	- 15 58	1500		HII REGION
OCL 0040	18 17 00.	- 17 09	1320	8.5	OPEN STAR CLUSTER
LDN 0336	18 17 00.	- 17 55	840		DARK NEBULA
LDN 0322	18 17 00.	- 19 00	4560		DARK NEBULA
LDN 0291	18 17 00.	- 21 00	2100		DARK NEBULA
LDN 0201	18 17 00.	- 26 00	7500		DARK NEBULA
RNGC 6613	18 17 03.	- 17 09		8.0	OPEN CLUSTER
SG 3.139	18 17 05.	- 16 03	3600		DIFFUSE EMISSION NEBULA
PK007-04.1	18 17 05.1	- 24 16 27.	9		PLANETARY NEBULA
UGC 11201	18 17 06.	+ 54 28	84	16.0	GALAXY Sa-b
SCHO 0825	18 17 06.	- 11 44	380		ISOLATED DARK CLOUD
ZWG 172.025	18 17 12.	+ 28 28		15.7	GALAXY
VDB.66N 121	18 17 15.	- 17 01	36		REFLECTION NEBULA
IC 4707	18 17 16.	- 16 02 17.			NONSTELLAR OBJECT

OBJECT NAME	RIGHT ASCEN.	DECLINATION	DIAM.	MAGN.	TYPE OF OBJECT
CED 160B	18 17 17.	- 16 02	390		DIFFUSE GALACTIC NEBULA
SG 3.140	18 17 17.	- 16 03	300		DIFFUSE EMISSION NEBULA
MCG+09-30-011	18 17 18.	+ 54 29	72	14.	GALAXY
ZC 1817.3+6544	18 17 18.	+ 65 44	1480		CLUSTER OF GALAXIES
MCG+11-22-038	18 17 18.	+ 66 17	24	17.	GALAXY
ZWG 254.031	18 17 24.	+ 50 15		14.4	GALAXY
UGC 11202	18 17 24.	+ 50 15	66	14.4	GALAXY S0
ZC 1817.4+6037	18 17 24.	+ 60 37	2080		CLUSTER OF GALAXIES
LB 01092	18 17 27.	+ 79 56 00.		16.8	FAINT BLUE STAR
RNGC 6623	18 17 28.	+ 23 41		14.5	GALAXY
ZWG 142.040	18 17 30.	+ 23 41		14.4	GALAXY
UGC 11203	18 17 30.	+ 23 41	90	14.4	GALAXY E
MCG+08-33-035	18 17 30.	+ 50 15	60	15.	GALAXY
LDN 0437	18 17 30.	- 08 00	1620		DARK NEBULA
SCHO 0826	18 17 31.	- 26 40	300		ISOLATED DARK CLOUD
MCG+07-37-037	18 17 36.	+ 42 48	66	15.	GALAXY
ZWG 254.032	18 17 36.	+ 47 44		15.6	GALAXY
UGC 11204	18 17 36.	+ 47 44	66	15.6	GALAXY Sc
SCHO 0827	18 17 37.	- 11 23	850		ISOLATED DARK CLOUD
MCG+04-43-027	18 17 42.	+ 23 39 30.	9	14.4	GALAXY
MCG+04-43-026	18 17 42.	+ 23 40	60	14.4	GALAXY
ZWG 228.001	18 17 42.	+ 42 46		15.5	GALAXY
ZWG 227.026	18 17 42.	+ 42 46		15.5	GALAXY
UGC 11205	18 17 42.	+ 42 46	72	15.5	GALAXY Sc
MCG+08-33-036	18 17 42.	+ 47 43	48	15.	GALAXY
MCG+02-47-001	18 17 45.	+ 14 40	48	16.	GALAXY
MCG+04-43-028	18 17 45.	+ 23 40 30.	30	14.4	GALAXY
ZWG 228.002	18 17 48.	+ 39 56		15.7	GALAXY
ZWG 227.027	18 17 48.	+ 39 56		15.7	GALAXY
ZWG 228.003	18 17 48.	+ 40 57		15.4	GALAXY
ZWG 227.028	18 17 48.	+ 40 57		15.4	GALAXY
ISS 0574	18 17 48.	- 09 04	112		STELLAR RING
SCHO 0828	18 17 51.	- 26 40	240		ISOLATED DARK CLOUD
RNGC 6618	18 17 52.	- 16 12		7.0	CLUSTER WITH NEBULOSITY
PK005-05.2	18 17 52.	- 26 50	10		PLANETARY NEBULA
CED 161	18 17 53.	- 16 12	2760		DIFFUSE GALACTIC NEBULA
SG 3.141	18 17 53.	- 16 12	3240		DIFFUSE EMISSION NEBULA
MCG+07-37-038	18 17 54.	+ 40 57 30.	36	15.	GALAXY
7ZW 784	18 17 54.	+ 58 17			COMPACT GALAXY
ISS 0508	18 17 54.	- 02 39	76		STELLAR RING
ASS 06	18 17 54.	- 14 36	18000	8.4	OB ASSOCIATION SER OB1
OCL 0044	18 17 54.	- 16 12	3600	8.	OPEN STAR CLUSTER
KHAV 478	18 18	- 07 53	7960		DARK NEBULA
LBN 0062	18 18	- 14 50	1200		BRIGHT NEBULA
KHAV 479	18 18	- 15 11	8310		DARK NEBULA
LBN 0060	18 18	- 16 00	2100		BRIGHT NEBULA
KHAV 477	18 18	- 24 53	2230		DARK NEBULA
KHAV 475	18 18	- 28 23	3880		DARK NEBULA
LDN 0706	18 18 00.	+ 21 44	360		DARK NEBULA
ZC 1818.0+3306	18 18 00.	+ 33 06	740		CLUSTER OF GALAXIES
MCG+07-38-001	18 18 00.	+ 39 56	45	17.	GALAXY
MRSL 016-00/1	18 18 00.	- 15 09	4800		HII REGION
LDN 0332	18 18 00.	- 18 10	480		DARK NEBULA
LDN 0314	18 18 00.	- 20 00	7680		DARK NEBULA
SCHO 0829	18 18 08.	+ 05 50 12.	390		ISOLATED DARK CLOUD
ZC 1818.2+5401	18 18 12.	+ 54 01	540		CLUSTER OF GALAXIES
ZC 1818.2+6233	18 18 12.	+ 62 33	2080		CLUSTER OF GALAXIES
SCHO 0830	18 18 14.	- 26 39	190		ISOLATED DARK CLOUD
SCHO 0831	18 18 15.	- 13 43 24.	260		ISOLATED DARK CLOUD
ZWG 254.033	18 18 18.	+ 48 32		14.5	GALAXY
UGC 11206	18 18 18.	+ 48 32	54	14.5	GALAXY S
EM7	18 18 18.	- 31 33		13.09	EMISSION OBJECT
MCG+08-33-037	18 18 24.	+ 48 31	60	15.	GALAXY
MCG+11-22-039	18 18 24.	+ 63 40	45	16.	GALAXY
ISS 0685	18 18 24.	- 12 25	101		STELLAR RING
MRSL 015-00/1	18 18 24.	- 16 15	1800		HII REGION
OCL 0001	18 18 30.	- 33 16	210	11.	OPEN STAR CLUSTER
ISS 0798	18 18 36.	- 19 18	163		STELLAR RING
HN 0460	18 18 38.	- 59 16			NEBULA
IC 4702	18 18 38.	- 59 16			NONSTELLAR OBJECT
SCHO 0832	18 18 39.	- 26 40	240		ISOLATED DARK CLOUD
PK024+03.1	18 18 43.	- 06 03 26.	6		PLANETARY NEBULA
ZWG 172.026	18 18 48.	+ 30 46		15.3	GALAXY
ZWG 200.016	18 18 48.	+ 38 09		15.6	GALAXY
UGC 11207	18 18 48.	+ 47 54	60	16.5	GALAXY S
MCG+11-22-040	18 18 48.	+ 68 20	72	15.	GALAXY
ISS 0686	18 18 48.	- 14 32	201		STELLAR RING
ZC 1818.9+5705	18 18 54.	+ 57 05	940		CLUSTER OF GALAXIES
ZWG 322.038	18 18 54.	+ 68 19		15.1	GALAXY
UGC 11208	18 18 54.	+ 68 19	66	15.1	GALAXY Sc
PK008-04.1	18 18 57.5	- 24 12 09.	3		PLANETARY NEBULA
KHAV 482	18 19	- 12 29	5110		DARK NEBULA
KHAV 481	18 19	- 18 53	5110		DARK NEBULA
KHAV 480	18 19	- 27 41	2840		DARK NEBULA
LB 09925	18 19	- 84 40		14.1	FAINT BLUE STAR
ZWG 200.017	18 19 00.	+ 38 07		15.2	GALAXY
MCG+06-40-013	18 19 00.	+ 38 10	18	15.	GALAXY
MCG+07-38-002	18 19 00.	+ 40 54	30	16.	GALAXY
MCG+08-33-038	18 19 00.	+ 47 52	60	17.	GALAXY
LDN 0520	18 19 00.	- 02 20	1260		DARK NEBULA
LDN 0451	18 19 00.	- 07 10	600		DARK NEBULA
ISS 0687	18 19 00.	- 09 38	257		STELLAR RING
LDN 0394	18 19 00.	- 12 15	4920		DARK NEBULA
ISS 0688	18 19 00.	- 15 01	136		STELLAR RING
LDN 0342	18 19 00.	- 17 40	3600		DARK NEBULA
ZWG 200.018	18 19 06.	+ 38 06		15.5	GALAXY
ZWG 254.034	18 19 06.	+ 47 33		15.6	GALAXY
ARC 2302	18 19 06.	+ 57 08		17.4	RICH CLUSTER OF GALAXIES
MCG+06-40-015	18 19 09.	+ 38 07	36	15.5	GALAXY
MCG+06-40-014	18 19 09.	+ 38 07	12	16.	GALAXY
MCG+06-40-016	18 19 15.	+ 38 07	33	15.5	GALAXY
MCG+08-33-039	18 19 15.	+ 47 58	48	16.	GALAXY
MCG+11-22-043	18 19 18.	+ 63 39	24	16.	GALAXY
MCG+11-22-042	18 19 18.	+ 66 16	18	17.	GALAXY
MCG+11-22-041	18 19 18.	+ 66 16	18	17.	GALAXY
SCHO 0833	18 19 21.	- 14 48	420		ISOLATED DARK CLOUD
PK002-07.1	18 19 21.2	- 30 45 01.	25		PLANETARY NEBULA
ZC 1819.4+3325	18 19 24.	+ 33 25	3630		CLUSTER OF GALAXIES
ZWG 322.039	18 19 24.	+ 63 40		15.0	GALAXY
7ZW 785	18 19 24.	+ 66 17			COMPACT GALAXY
MCG+11-22-045	18 19 24.	+ 66 17	18	17.	GALAXY
MCG+11-22-044	18 19 24.	+ 66 17	18	17.	GALAXY
SG 3.142	18 19 27.	- 14 35	420		DIFFUSE EMISSION NEBULA
PK008-04.2	18 19 28.5	- 24 11 00.	5		PLANETARY NEBULA
UGC 11209	18 19 30.	+ 26 56	60	16.5	GALAXY S
MCG+15-01-020	18 19 30.	+ 87 44	12	17.	GALAXY
ISS 0904	18 19 30.	- 21 14	77		STELLAR RING
SCHO 0834	18 19 34.	+ 04 58 54.	510		ISOLATED DARK CLOUD
ZC 1819.6+6901	18 19 36.	+ 69 01	1610		CLUSTER OF GALAXIES
MRSL 016-00/2	18 19 36.	- 14 37	600		HII REGION
ZWG 322.040	18 19 42.	+ 63 40		15.7	GALAXY
HN 0043	18 19 43.	- 26 52			NEBULA
YM 11	18 19 45.	- 16 36	108		SYMMETRIC GALACTIC NEBULA
PK005-06.1	18 19 46.8	- 26 50 51.	5	15.0	PLANETARY NEBULA
UGC 11210	18 19 48.	+ 21 08	72	17.	GALAXY
ISS 0689	18 19 48.	- 13 30	269		STELLAR RING
RNGC 6620	18 19 50.	- 26 51		15.0	PLANETARY NEBULA
ZC 1819.9+5717	18 19 54.	+ 57 17	670		CLUSTER OF GALAXIES
ARC 2304	18 19 54.	+ 68 55		17.0	RICH CLUSTER OF GALAXIES
LBN 0064	18 20	- 14 40	780		BRIGHT NEBULA
KHAV 483	18 20	- 25 41	4700		DARK NEBULA
ZWG 279.008	18 20 00.	+ 55 31		15.6	GALAXY
LDN 0537	18 20 00.	- 01 00	2940		DARK NEBULA
LDN 0405	18 20 00.	- 11 30	13140		DARK NEBULA
LDN 0380	18 20 00.	- 14 00	2040		DARK NEBULA
LDN 0376	18 20 00.	- 14 40	3420		DARK NEBULA
LDN 0359	18 20 00.	- 16 30	9060		DARK NEBULA
SCHO 0835	18 20 01.	- 24 03	300		ISOLATED DARK CLOUD
B 309	18 20 05.	- 24 03	300		DARK OBJECT
7ZW 786	18 20 06.	+ 57 59			COMPACT GALAXY
ISS 0509	18 20 06.	- 02 09	63		STELLAR RING
LDN 0252	18 20 06.	- 24 05	240		DARK NEBULA
HN 0461	18 20 07.	- 56 14			NEBULA
IC 4709	18 20 07.	- 56 14			NONSTELLAR OBJECT
RNGC 6628	18 20 10.	+ 23 27		15.0	GALAXY
PK012-02.1	18 20 10.9	- 19 18 41.	10		PLANETARY NEBULA
SCHO 0836	18 20 11.	- 15 35 30.	370		ISOLATED DARK CLOUD
ZWG 142.041	18 20 12.	+ 23 27		14.8	GALAXY
UGC 11211	18 20 12.	+ 23 27	114	14.8	GALAXY S0?
ZWG 172.027	18 20 12.	+ 27 35		15.5	GALAXY
LDN 0306	18 20 12.	- 20 40	600		DARK NEBULA
SCHO 0837	18 20 15.	+ 07 25 12.	500		ISOLATED DARK CLOUD
MCG+04-43-029	18 20 18.	+ 23 27	90	14.8	GALAXY
ZWG 200.019	18 20 18.	+ 36 35		15.4	GALAXY
ZWG 254.035	18 20 18.	+ 48 22		15.3	GALAXY
RNGC 6625	18 20 18.	- 12 04			NON-EXISTENT OBJECT
LDN 0250	18 20 18.	- 24 10	180		DARK NEBULA
ISS 0905	18 20 18.	- 26 06	82		STELLAR RING
SCHO 0838	18 20 19.	- 24 33	370		ISOLATED DARK CLOUD
RNGC 6627	18 20 23.	+ 15 40		14.5	GALAXY
ZWG 114.004	18 20 24.	+ 15 40		14.5	GALAXY
UGC 11212	18 20 24.	+ 15 40	90	14.5	GALAXY SBb
MCG+03-47-001	18 20 24.	+ 15 40	66	14.5	GALAXY
MCG+06-40-017	18 20 24.	+ 36 35	42	15.	GALAXY
OCL 0058	18 20 24.	- 12 05		9.0	OPEN STAR CLUSTER
SCHO 0839	18 20 24.	- 25 37	210		ISOLATED DARK CLOUD
RNGC 6614	18 20 25.	- 63 16			GALAXY
PK010-03.1	18 20 28.	- 21 55	10		PLANETARY NEBULA
ZWG 279.009	18 20 30.	+ 52 20		15.4	GALAXY
UGC 11213	18 20 30.	+ 52 20	78	15.4	GALAXY SBa-b
ISS 0690	18 20 30.	- 13 20	265		STELLAR RING
ISS 0799	18 20 30.	- 19 34	141		STELLAR RING
LDN 0268	18 20 30.	- 23 10	240		DARK NEBULA
LDN 0267	18 20 30.	- 23 10	240		DARK NEBULA
GCL 093	18 20 30.	- 30 23	252	9.53	GLOBULAR STAR CLUSTER
VHA 262	18 20 30.	- 30 23	120		GLOBULAR STAR CLUSTER
RNGC 6624	18 20 30.	- 30 23		9.5	GLOBULAR CLUSTER
SCHO 0840	18 20 32.	+ 05 23 30.	300		ISOLATED DARK CLOUD
UGC 11214	18 20 36.	+ 12 24	90	15.	GALAXY Sc
MCG+02-47-002	18 20 36.	+ 12 24	72	15.	GALAXY
MCG+09-30-012	18 20 42.	+ 52 24	72	15.	GALAXY
ISS 0575	18 20 42.	- 07 39	320		STELLAR RING
UGC 11215	18 20 48.	+ 60 51	78	16.0	GALAXY Sc
7ZW 787	18 20 48.	+ 60 53			COMPACT GALAXY
7ZW 788	18 20 48.	+ 63 22			COMPACT GALAXY
ISS 0691	18 20 48.	- 12 57	105		STELLAR RING
ISS 0800	18 20 48.	- 16 55	164		STELLAR RING
PK032+07.1	18 20 52.1	+ 03 34 52.			PLANETARY NEBULA
UGC 11216	18 20 54.	+ 21 13	60	16.5	GALAXY SB
ZWG 254.036	18 20 54.	+ 48 05		14.6	GALAXY
UGC 11217	18 20 54.	+ 48 05	60	14.6	GALAXY Sb
MCG+10-26-034	18 20 54.	+ 60 52	42	15.	GALAXY
SCHO 0841	18 20 55.	+ 05 32 54.	360		ISOLATED DARK CLOUD
KHAV 486	18 21	- 13 29	5170		DARK NEBULA
KHAV 484	18 21	- 16 35	5360		DARK NEBULA
LDN 0585	18 21 00.	+ 02 00	4620		DARK NEBULA
ZWG 172.028	18 21 00.	+ 30 38		15.4	GALAXY
MCG+08-33-040	18 21 00.	+ 48 05	48	15.	GALAXY
LDN 0533	18 21 00.	- 01 20	600		DARK NEBULA
ISS 0576	18 21 00.	- 07 39	277		STELLAR RING
SCHO 0842	18 21 00.	- 10 05 30.	370		ISOLATED DARK CLOUD
LDN 0351	18 21 00.	- 17 15	660		DARK NEBULA
LDN 0247	18 21 00.	- 24 35	720		DARK NEBULA
ZWG 172.029	18 21 06.	+ 29 53		14.9	GALAXY
MCG+05-43-015	18 21 06.	+ 30 39 30.	48	15.4	GALAXY
ZWG 172.030	18 21 06.	+ 30 43		15.7	GALAXY
MCG+05-43-016	18 21 09.	+ 29 53	48	14.9	GALAXY
HN 0021	18 21 09.	+ 74 33			NEBULA
ZWG 340.043	18 21 12.	+ 74 32		11.8	GALAXY
UGC 11218	18 21 12.	+ 74 32	240	11.8	GALAXY Sc
KARA.73B 0850	18 21 12.	+ 74 32	204	11.8	ISOLATED GALAXY S
REIN 2.273	18 21 13.23	+ 74 32 36.0			NEBULA
RNGC 6643	18 21 14.	+ 74 33		12.5	GALAXY
PK007-06.1	18 21 14.4	- 25 43 55.	9		PLANETARY NEBULA
ZWG 172.031	18 21 18.	+ 30 14		15.3	GALAXY
UGC 11219	18 21 18.	+ 30 14	96	15.3	GALAXY DBL SYS
MCG+05-43-017	18 21 18.	+ 30 14	48	15.3	GALAXY
PK019+00.1	18 21 21.	- 11 08 21.	6		PLANETARY NEBULA
MIL 59	18 21 21.	- 12 22 12.	840		SUPERNOVA REMNANT
ZWG 254.037	18 21 24.	+ 47 35		15.2	GALAXY
MCG+08-33-041	18 21 24.	+ 47 35	42	15.	GALAXY
LDN 0539	18 21 24.	- 01 06	600		DARK NEBULA
RNGC 6626	18 21 28.	- 24 54		8.5	GLOBULAR CLUSTER
ZWG 228.004	18 21 30.	+ 43 38		15.7	GALAXY
MCG+12-17-021	18 21 30.	+ 74 31	204	11.6	GALAXY
LDN 0414	18 21 30.	- 10 52	360		DARK NEBULA
GCL 094	18 21 30.	- 25 54	900	8.48	GLOBULAR STAR CLUSTER
LB 01093	18 21 34.	+ 64 20 30.		15.9	FAINT BLUE STAR
ZWG 279.010	18 21 36.	+ 51 19		15.3	GALAXY
ZC 1821.6+6501	18 21 36.	+ 65 01	1750		CLUSTER OF GALAXIES
ISS 0801	18 21 36.	- 16 55	78		STELLAR RING
PK094+27.1	18 21 36.9	+ 64 20 18.	130		PLANETARY NEBULA
ZWG 142.042	18 21 42.	+ 21 01		15.5	GALAXY
MCG+04-43-030	18 21 42.	+ 21 01	42	15.5	GALAXY
SCHO 0843	18 21 42.	- 24 43	310		ISOLATED DARK CLOUD
UGC 11220	18 21 48.	+ 40 55	84	17.	GALAXY DWRF IP

OBJECT NAME	RIGHT ASCEN.	DECLINATION	DIAM.	MAGN.	TYPE OF OBJECT
ISS 0692	18 21 48.	- 12 05	118		STELLAR RING
OCL 0034	18 21 48.	- 19 43	480	9.0	OPEN STAR CLUSTER
SCHO 0844	18 21 48.	- 24 58	230		ISOLATED DARK CLOUD
7ZW 790	18 21 54.	+ 66 35			COMPACT GALAXY
MCG+11-22-046	18 21 54.	+ 66 35	120	14.	GALAXY
ISS 0577	18 21 54.	- 07 08	174		STELLAR RING
HN 0462	18 21 54.	- 71 38			NEBULA
IC 4704	18 21 54.	- 71 38			NONSTELLAR OBJECT
VDB.66N 122	18 21 56.	- 13 39	72		REFLECTION NEBULA
LBN 0093	18 22	+ 00 50	360		BRIGHT NEBULA
KHAV 487	18 22	- 07 58	11100		DARK NEBULA
CED 162	18 22	- 18 27			DIFFUSE GALACTIC NEBULA
KHAV 485	18 22	- 23 59	4700		DARK NEBULA
LB 09926	18 22	- 81 01		14.4	FAINT BLUE STAR
MCG+11-22-047	18 22 00.	+ 66 34	15	16.	GALAXY
KARA.72 536A	18 22 00.	+ 66 35	96	14.2	PART OF DOUBLE GALAXY
ZWG 322.041	18 22 00.	+ 66 36		14.2	GALAXY
RNGC 6636	18 22 00.	+ 66 36		14.0	GALAXY
UGC 11221	18 22 00.	+ 66 36	126	14.2	GALAXY DBL SYS
KARA.72 536B	18 22 00.	+ 66 36	30		PART OF DOUBLE GALAXY
MCG+11-22-048	18 22 00.	+ 68 06	78	16.	GALAXY
LDN 0453	18 22 00.	- 07 30	6300		DARK NEBULA
LDN 0415	18 22 00.	- 10 40	1500		DARK NEBULA
MRSL 015-01/1	18 22 00.	- 16 36	900		HII REGION
LDN 0354	18 22 00.	- 17 17	660		DARK NEBULA
LDN 0112	18 22 00.	- 21 50	720		DARK NEBULA
LDN 0259	18 22 00.	- 24 11	600		DARK NEBULA
PK009-04.1	18 22 03.6	- 22 36 35.	6		PLANETARY NEBULA
ZWG 322.042	18 22 06.	+ 68 05		14.9	GALAXY
UGC 11222	18 22 06.	+ 68 05	90	14.9	GALAXY SBc
HN 0463	18 22 07.	- 71 43			NEBULA
IC 4705	18 22 07.	- 71 43			NONSTELLAR OBJECT
SCHO 0845	18 22 10.	- 25 04	270		ISOLATED DARK CLOUD
ZC 1822.2+3322	18 22 12.	+ 33 22	1880		CLUSTER OF GALAXIES
PK032+07.2	18 22 13.6	+ 02 27 48.			PLANETARY NEBULA
SCHO 0846	18 22 16.	- 11 49	1110		ISOLATED DARK CLOUD
ZWG 114.005	18 22 18.	+ 20 53		15.2	GALAXY
UGC 11223	18 22 18.	+ 36 26	60	16.0	GALAXY Sb-c
ZWG 228.005	18 22 18.	+ 40 47		15.6	GALAXY
ZC 1822.3+5806	18 22 18.	+ 58 06	1340		CLUSTER OF GALAXIES
MAI 102	18 22 18.	+ 65 43	121		DWARF SPHEROIDAL GALAXY
ISS 0578	18 22 18.	- 06 48	145		STELLAR RING
EM6	18 22 18.	- 28 37		12.13	EMISSION OBJECT
SCHO 0847	18 22 20.	- 24 45	270		ISOLATED DARK CLOUD
SCHO 0848	18 22 23.	- 10 37	820		ISOLATED DARK CLOUD
UGC 11224	18 22 24.	+ 36 46	66	17.	GALAXY DWARF
MCG+10-26-035	18 22 24.	+ 62 55	36	16.	GALAXY
OCL 0064	18 22 24.	- 10 05	150	13.	OPEN STAR CLUSTER
MRSL 018-00/1	18 22 24.	- 13 16	900		HII REGION
PK028+05.1	18 22 25.1	- 01 32 34.			PLANETARY NEBULA
LDN 0417	18 22 30.	- 10 38	180		DARK NEBULA
LDN 0416	18 22 30.	- 10 40	240		DARK NEBULA
OCL 0053	18 22 30.	- 14 41	720		OPEN STAR CLUSTER
SG 3.143	18 22 31.	- 13 13	900		DIFFUSE EMISSION NEBULA
YM 15	18 22 32.	+ 00 49	360		SYMMETRIC GALACTIC NEBULA
PK013-03.1	18 22 33.3	- 19 07 48.	5		PLANETARY NEBULA
MRSL 030+06/1	18 22 36.	+ 00 49	480		HII REGION
ZWG 254.038	18 22 36.	+ 48 33		15.5	GALAXY
ISS 0802	18 22 36.	- 14 44	223		STELLAR RING
ISS 0803	18 22 36.	- 18 32	152		STELLAR RING
PK009-05.1	18 22 40.05	- 23 13 57.5	16	11.9	PLANETARY NEBULA
REIN 2.272	18 22 40.35	- 23 13 56.6			NEBULA
RNGC 6629	18 22 41.	- 23 14		12.0	PLANETARY NEBULA
ZWG 142.043	18 22 42.	+ 23 02		14.9	GALAXY
7ZW 791	18 22 42.	+ 59 48			COMPACT GALAXY
ZWG 301.027	18 22 42.	+ 60 43		15.4	GALAXY
UGC 11225	18 22 42.	+ 60 43	78	15.4	GALAXY Sc-IRR
ZC 1822.7+7116	18 22 42.	+ 71 16	3700		CLUSTER OF GALAXIES
ASS 08	18 22 42.	- 14 16			OB ASSOCIATION SCT OB3
MRSL 016-01/1	18 22 42.	- 14 45	2100		HII REGION
ISS 0906	18 22 42.	- 21 59	116		STELLAR RING
MCG+04-43-031	18 22 45.	+ 23 02	48	14.9	GALAXY
MCG+10-26-036	18 22 48.	+ 60 44	72	15.	GALAXY
ISS 0579	18 22 48.	- 07 00	205		STELLAR RING
B 095	18 22 48.	- 11 47	1800		DARK OBJECT
ZWG 254.039	18 22 54.	+ 50 23		15.5	GALAXY
SCHO 0849	18 22 58.	- 10 19	800		ISOLATED DARK CLOUD
LBN 0066	18 23	- 14 50	480		BRIGHT NEBULA
KHAV 489	18 23	- 21 34	3010		DARK NEBULA
ZWG 172.032	18 23 00.	+ 27 30		13.2	GALAXY
RNGC 6632	18 23 00.	+ 27 30		13.0	GALAXY
UGC 11226	18 23 00.	+ 27 30	180	13.2	GALAXY Sb
MCG+05-43-018	18 23 00.	+ 27 30	180	13.2	GALAXY
ZWG 172.033	18 23 00.	+ 31 23		15.5	GALAXY
LDN 0511	18 23 00.	- 03 40	3660		DARK NEBULA
LDN 0481	18 23 00.	- 06 00	10920		DARK NEBULA
COU 031	18 23 00.	- 09 50	6000		EMISSION NEBULA
LDN 0420	18 23 00.	- 10 20	360		DARK NEBULA
LDN 0406	18 23 00.	- 11 50	3480		DARK NEBULA
LDN 0236	18 23 00.	- 25 20	660		DARK NEBULA
LDN 0205	18 23 00.	- 26 25	600		DARK NEBULA
ZWG 114.006	18 23 06.	+ 20 53		15.4	GALAXY
ZC 1823.1+6110	18 23 06.	+ 61 10	2490		CLUSTER OF GALAXIES
UGC 11227	18 23 06.	+ 62 14	60	16.0	GALAXY S0-a
B 094	18 23 07.	- 10 42	900		DARK OBJECT
HN 0464	18 23 08.	- 64 58			NEBULA
IC 4711	18 23 09.	- 64 59			NONSTELLAR OBJECT
MCG+07-38-004	18 23 12.	+ 40 51 30.	6	17.	GALAXY
MCG+07-38-005	18 23 12.	+ 40 52	18	17.	GALAXY
MCG+07-38-003	18 23 12.	+ 41 27	54	16.	GALAXY
ZWG 228.006	18 23 12.	+ 41 28		14.5	GALAXY
UGC 11228	18 23 12.	+ 41 28	60	14.5	GALAXY SB0
7ZW 792	18 23 12.	+ 62 15			COMPACT GALAXY
MCG+10-26-037	18 23 12.	+ 62 15	48	16.	GALAXY
MIN.46 09	18 23 14.	- 18 10			DIFFUSE NEBULA
SG 3.144	18 23 15.	- 09 29	3000		DIFFUSE EMISSION NEBULA
IC 4715	18 23 17.	- 18 27			NONSTELLAR OBJECT
ISS 0907	18 23 18.	- 26 38	59		STELLAR RING
ISS 0580	18 23 24.	- 07 04	160		STELLAR RING
HN 0465	18 23 29.	- 67 00			NEBULA
ZWG 142.044	18 23 30.	+ 22 18		14.8	GALAXY
UGC 11229	18 23 30.	+ 22 18	60	14.8	GALAXY S0-a
MCG+04-43-032	18 23 30.	+ 22 18	24	14.8	GALAXY
UGC 11230	18 23 30.	+ 65 18	156	16.0	GALAXY Sc
ISS 0581	18 23 30.	- 03 39	64		STELLAR RING
IC 4710	18 23 30.	- 67 01	192	12.0	GALAXY SBd
ARC 2305	18 23 33.	+ 71 20		17.0	RICH CLUSTER OF GALAXIES
UGC 11231	18 23 36.	+ 29 22	66	17.	GALAXY Sc

OBJECT NAME	RIGHT ASCEN.	DECLINATION	DIAM.	MAGN.	TYPE OF OBJECT
ZWG 172.034	18 23 36.	+ 32 36		15.7	GALAXY
MCG+05-43-019	18 23 36.	+ 32 36	42	15.7	GALAXY
B 096	18 23 39.	- 10 20			DARK OBJECT
SCHO 0850	18 23 40.	+ 07 53 42.	410		ISOLATED DARK CLOUD
ZWG 172.035	18 23 42.	+ 31 58		15.0	GALAXY
MCG+05-43-020	18 23 42.	+ 31 58	30	15.0	GALAXY
MCG+11-22-049	18 23 42.	+ 65 16	180	16.	GALAXY
MCG+09-30-013	18 23 45.	+ 52 50	36	16.	GALAXY
UGC 11232	18 23 48.	+ 30 43	66	16.5	GALAXY Sc
ZWG 172.036	18 23 48.	+ 32 08		15.1	GALAXY
UGC 11233	18 23 48.	+ 32 08	72	15.1	GALAXY S0
MCG+05-43-021	18 23 48.	+ 32 08	18	15.1	GALAXY
LDN 0413	18 23 48.	- 11 51	360		DARK NEBULA
MCG+11-22-050	18 23 54.	+ 64 04	39	16.	GALAXY
KHAV 492	18 24	- 03 46	17800		DARK NEBULA
KHAV 490	18 24	- 24 40			DARK NEBULA
KHAV 488	18 24	- 30 22	5720		DARK NEBULA
SER 127.01	18 24	- 67 01	150		3 GLXIES, INCLUDES IC4710
ISS 0510	18 24 00.	+ 03 40	249		STELLAR RING
ZWG 142.045	18 24 00.	+ 23 17		15.1	GALAXY
LDN 0570	18 24 00.	- 00 30	1020		DARK NEBULA
LDN 0526	18 24 00.	- 02 20	3840		DARK NEBULA
LDN 0427	18 24 00.	- 09 50	240		DARK NEBULA
LDN 0399	18 24 00.	- 12 30	4140		DARK NEBULA
ISS 0804	18 24 00.	- 19 14	104		STELLAR RING
PK027+04.1	18 24 03.1	- 02 44 47.			PLANETARY NEBULA
MCG+04-43-033	18 24 06.	+ 23 17 30.	48	15.1	GALAXY
MCG+12-17-022	18 24 06.	+ 71 34	84	14.	GALAXY
SCHO 0851	18 24 09.	+ 06 18 24.	280		ISOLATED DARK CLOUD
ZWG 172.037	18 24 19.	+ 32 33		15.4	GALAXY
SCHO 0852	18 24 23.	- 15 14 00.	360		ISOLATED DARK CLOUD
ZWG 142.046	18 24 24.	+ 22 23		15.5	GALAXY
UGC 11234	18 24 24.	+ 22 23	66	15.5	GALAXY
ZWG 200.020	18 24 24.	+ 35 23		15.6	GALAXY
UGC 11235	18 24 24.	+ 46 08	90	16.0	GALAXY SBc-IRR
ZWG 254.040	18 24 24.	+ 46 51		15.6	GALAXY
OCL 0059	18 24 24.	- 12 04	960	11.9	OPEN STAR CLUSTER
RNGC 6631	18 24 24.	- 12 05		11.5	OPEN CLUSTER
SCHO 0853	18 24 30.	+ 09 39 06.	450		ISOLATED DARK CLOUD
MCG+08-33-042	18 24 30.	+ 45 06	90	16.	GALAXY
ZC 1824.5+5423	18 24 30.	+ 54 23	1080		CLUSTER OF GALAXIES
SCHO 0854	18 24 34.	+ 06 26 06.	210		ISOLATED DARK CLOUD
PK031+05.1	18 24 37.1	+ 01 12 37.	10		PLANETARY NEBULA
SCHO 0855	18 24 39.	+ 09 16 42.	430		ISOLATED DARK CLOUD
ISS 0805	18 24 42.	- 19 22	87		STELLAR RING
HN 0467	18 24 42.	- 67 15			NEBULA
IC 4713	18 24 43.	- 67 15			NONSTELLAR OBJECT
MCG+10-26-038	18 24 48.	+ 60 27 30.	15	16.	GALAXY
ZWG 301.028	18 24 48.	+ 60 28		15.7	GALAXY
ISS 0806	18 24 49.	- 17 07	86		STELLAR RING
PK007-06.2	18 24 53.2	- 26 08 43.	7	13.	PLANETARY NEBULA
1ZW 206	18 24 54.	+ 34 18			COMPACT GALAXY
MRSL 026+03/1	18 24 54.	- 03 53	240		HII REGION
ISS 0807	18 24 54.	- 18 03	152		STELLAR RING
KHAV 495	18 25	+ 14 56			DARK NEBULA
LBN 0085	18 25	- 03 50	180		BRIGHT NEBULA
KHAV 494	18 25	- 09 58	5500		DARK NEBULA
KHAV 499	18 25	- 13 40	3810		DARK NEBULA
KHAV 497	18 25	- 17 58	4490		DARK NEBULA
KHAV 493	18 25	- 20 16			DARK NEBULA
KHAV 491	18 25	- 28 16	5930		DARK NEBULA
MCG+03-47-002	18 25 00.	+ 19 54	27	15.3	GALAXY
LDN 0491	18 25 00.	- 05 30	2040		DARK NEBULA
LDN 0423	18 25 00.	- 10 15	420		DARK NEBULA
LDN 0411	18 25 00.	- 11 30	360		DARK NEBULA
LDN 0350	18 25 00.	- 17 50	1500		DARK NEBULA
LDN 0217	18 25 00.	- 26 10	360		DARK NEBULA
SCHO 0856	18 25 03.	+ 06 46 06.	410		ISOLATED DARK CLOUD
PK016-07.1	18 25 04.	- 15 34 51.	58		PLANETARY NEBULA
RNGC 6633	18 25 05.	+ 06 32		5.5	OPEN CLUSTER
RNGC 6651	18 25 05.	+ 71 34		15.0	GALAXY
ZWG 114.007	18 25 06.	+ 19 55		15.3	GALAXY
ZWG 228.007	18 25 06.	+ 42 39		15.1	GALAXY
ZWG 279.011	18 25 06.	+ 51 35		15.7	GALAXY
ZWG 340.044	18 25 06.	+ 71 35		13.7	GALAXY
UGC 11236	18 25 06.	+ 71 35	108	13.7	GALAXY Sb-c
HN 0466	18 25 07.	- 71 43			NEBULA
IC 4712	18 25 08.	- 71 43			NONSTELLAR OBJECT
ZWG 142.047	18 25 12.	+ 24 50		15.4	GALAXY
UGC 11237	18 25 12.	+ 24 50	84	15.4	GALAXY Sc
MCG+12-17-023	18 25 12.	+ 73 08	150	12.5	GALAXY
7ZW 793	18 25 12.	+ 73 09			COMPACT GALAXY
ZWG 340.045	18 25 12.	+ 73 09		12.7	GALAXY
UGC 11238	18 25 12.	+ 73 09	174	12.7	GALAXY SB0
KARA.73B 0851	18 25 12.	+ 73 09	150	12.7	ISOLATED GALAXY S
ISS 0693	18 25 12.	- 14 44	255		STELLAR RING
RNGC 6654	18 25 15.	+ 73 09		12.5	GALAXY
RNGC 6635	18 25 16.	+ 14 47		14.5	GALAXY
OCL 0090	18 25 18.	+ 06 32	2040	6.35	OPEN STAR CLUSTER
UGC 11239	18 25 18.	+ 14 44	90	14.5	GALAXY S0
MCG+02-47-003	18 25 18.	+ 14 47	48	14.	GALAXY
ISS 0694	18 25 18.	- 14 40	134		STELLAR RING
RNGC 6648	18 25 24.	+ 64 57			NON-EXISTENT OBJECT
UGC 11240	18 25 30.	+ 33 54	60	17.	GALAXY Sc
7ZW 794	18 25 30.	+ 67 57			COMPACT GALAXY
LDN 0575	18 25 30.	- 00 40	1080		DARK NEBULA
ISS 0695	18 25 30.	- 14 33	175		STELLAR RING
PK043+11.1	18 25 31.6	+ 14 27 11.			PLANETARY NEBULA
RNGC 6650	18 25 35.	+ 67 58		15.0	GALAXY
ZWG 279.012	18 25 36.	+ 51 07		15.2	GALAXY
UGC 11241	18 25 36.	+ 51 07	72	15.2	GALAXY Sb
7ZW 795	18 25 36.	+ 67 07			COMPACT GALAXY
ZWG 322.043	18 25 36.	+ 67 58		14.8	GALAXY
ISS 0582	18 25 36.	- 08 50	123		STELLAR RING
ZWG 114.008	18 25 42.	+ 16 06		15.7	GALAXY
UGC 11242	18 25 42.	+ 16 06	72	15.7	GALAXY Sc
ZWG 340.046	18 25 42.	+ 70 30		15.6	GALAXY
HN 0468	18 25 44.	- 66 41			NEBULA
IC 4714	18 25 44.	- 66 41			NONSTELLAR OBJECT
PK008-06.1	18 25 46.5	- 24 34 01.	7		PLANETARY NEBULA
MCG+03-47-003	18 25 48.	+ 16 05	66	15.7	GALAXY
UGC 11243	18 25 48.	+ 22 41	66	17.	GALAXY Sc
MCG+08-33-043	18 25 48.	+ 46 53	84	15.	GALAXY
MCG+09-30-014	18 25 48.	+ 51 05	54	14.	GALAXY
7ZW 796	18 25 48.	+ 67 23			COMPACT GALAXY
ISS 0696	18 25 48.	- 14 41	209		STELLAR RING
SCHO 0857	18 25 52.	- 09 59	3460		ISOLATED DARK CLOUD
ZWG 254.041	18 25 54.	+ 46 53		15.3	GALAXY

OBJECT NAME	RIGHT ASCEN.	DECLINATION	DIAM.	MAGN.	TYPE OF OBJECT
UGC 11244	18 25 54.	+ 46 53	102	15.3	GALAXY Sc-IRR
SCHO 0858	18 25 56.	- 16 40	370		ISOLATED DARK CLOUD
PK002-09.1	18 25 57.2	- 31 22 00.	5	11.89	PLANETARY NEBULA
KHAV 500	18 26	- 22 52	3090		DARK NEBULA
KHAV 498	18 26	- 25 34	2370		DARK NEBULA
KHAV 496	18 26	- 26 16	2660		DARK NEBULA
LDN 0592	18 26 00.	+ 02 00	420		DARK NEBULA
ZWG 228.006	18 26 00.	+ 41 44		15.4	GALAXY
7ZW 797	18 26 00.	+ 61 23			COMPACT GALAXY
LDN 0573	18 26 00.	- 00 25	600		DARK NEBULA
LDN 0489	18 26 00.	- 05 40	1980		DARK NEBULA
LDN 0454	18 26 00.	- 08 00	1500		DARK NEBULA
LDN 0343	18 26 00.	- 18 26	4020		DARK NEBULA
LDN 0251	18 26 00.	- 24 50	1740		DARK NEBULA
MCG+07-38-006	18 26 06.	+ 41 42	33	15.	GALAXY
ZWG 279.013	18 26 06.	+ 56 02		15.4	GALAXY
UGC 11245	18 26 06.	+ 56 02	66	15.4	GALAXY Sa-b
KARA.73B 0852	18 26 06.	+ 56 02	60	15.4	ISOLATED GALAXY S
MCG+09-30-015	18 26 06.	+ 56 03	66	15.	GALAXY
7ZW 798	18 26 06.	+ 57 55			COMPACT GALAXY
7ZW 799	18 26 06.	+ 60 25			COMPACT GALAXY
PK011-05.1	18 26 10.9	- 21 48 55.	5		PLANETARY NEBULA
ZWG 142.048	18 26 12.	+ 22 42		15.1	GALAXY
UGC 11246	18 26 12.	+ 22 42	78	15.1	GALAXY Sa-b
MCG+04-43-034	18 26 12.	+ 22 42	60	15.1	GALAXY
ZWG 172.038	18 26 12.	+ 32 38		15.6	GALAXY
RM1	18 26 12.	- 31 33		11.89	PLANETARY NEBULA
MCG-07-38-001	18 26 12.	- 42 49	60	15.	GALAXY
ZC 1826.3+3210	18 26 18.	+ 32 10	3430		CLUSTER OF GALAXIES
ZWG 200.021	18 26 18.	+ 34 16		14.2	GALAXY
RNGC 6640	18 26 18.	+ 34 16		14.0	GALAXY
UGC 11247	18 26 18.	+ 34 16	72	14.2	GALAXY Sc
B 097	18 26 20.	- 09 57	3600		DARK OBJECT
MCG+06-40-018	18 26 21.	+ 34 15	66	14.	GALAXY
IC 4733	18 26 21.	+ 64 56			NONSTELLAR OBJECT
ZC 1826.4+3410	18 26 24.	+ 34 10	7260		CLUSTER OF GALAXIES
ZWG 254.042	18 26 24.	+ 49 41		15.7	GALAXY
SCHO 0859	18 26 30.	+ 04 57 12.	460		ISOLATED DARK CLOUD
7ZW 800	18 26 30.	+ 72 07			COMPACT GALAXY
LDN 0503	18 26 30.	- 04 40	360		DARK NEBULA
LDN 0269	18 26 30.	- 23 50	780		DARK NEBULA
SCHO 0860	18 26 35.	- 16 46	260		ISOLATED DARK CLOUD
ISS 0479	18 26 36.	+ 04 53	369		STELLAR RING
UGC 11248	18 26 36.	+ 48 10	72	16.0	GALAXY Sc
ARC 2306	18 26 36.	+ 74 42		17.0	RICH CLUSTER OF GALAXIES
RNGC 6634	18 26 39.	- 33 27			NON-EXISTENT OBJECT
MCG+03-47-004	18 26 42.	+ 19 45	24	17.	GALAXY
UGC 11249	18 26 42.	+ 34 07	60	16.5	GALAXY S
7ZW 801	18 26 42.	+ 60 09			COMPACT GALAXY
RNGC 6641	18 26 45.	+ 22 52		14.5	GALAXY
MCG+04-43-035	18 26 48.	+ 22 51	42	14.3	GALAXY
ZWG 142.049	18 26 48.	+ 22 52		14.3	GALAXY
UGC 11250	18 26 48.	+ 22 52	66	14.3	GALAXY PECULR?
ZC 1826.8+5745	18 26 48.	+ 57 45	3700		CLUSTER OF GALAXIES
ISS 0583	18 26 48.	- 06 25	333		STELLAR RING
ZWG 279.014	18 26 54.	+ 53 05		15.5	GALAXY
ZC 1826.9+7621	18 26 54.	+ 76 21	1950		CLUSTER OF GALAXIES
KHAV 502	18 27	+ 12 56	3540		DARK NEBULA
KHAV 503	18 27	- 21 10	2230		DARK NEBULA
KHAV 501	18 27	- 24 16	5010		DARK NEBULA
ZWG 200.022	18 27 00.	+ 38 00		15.7	GALAXY
UGC 11251	18 27 00.	+ 38 00	90	15.7	GALAXY PECULR
KARA.73B 0853	18 27 00.	+ 38 00	78	15.7	ISOLATED GALAXY S
ZWG 301.029	18 27 00.	+ 62 05		15.3	GALAXY
LDN 0404	18 27 00.	- 12 30	420		DARK NEBULA
LDN 0402	18 27 00.	- 12 45	300		DARK NEBULA
LDN 0374	18 27 00.	- 15 50	2340		DARK NEBULA
LDN 0362	18 27 00.	- 17 10	2760		DARK NEBULA
SCHO 0861	18 27 02.	+ 04 41 06.	650		ISOLATED DARK CLOUD
PK013-04.1	18 27 02.0	- 19 42 42.	7		PLANETARY NEBULA
MCG+10-26-039	18 27 06.	+ 62 05	42	15.	GALAXY
ZC 1827.1+7435	18 27 06.	+ 74 35	2020		CLUSTER OF GALAXIES
MRSL 016-02/1	18 27 06.	- 15 24	600		HII REGION
MCG+05-40-019	18 27 12.	+ 38 01	60	15.	GALAXY
ISS 0697	18 27 12.	- 14 08	100		STELLAR RING
MRSL 013-04/1	18 27 12.	- 19 11	5400		HII REGION
MCG+08-33-044	18 27 15.	+ 48 16	60	18.	GALAXY
PK015-03.1	18 27 16.4	- 16 47 22.	67	17.4	PLANETARY NEBULA
RNGC 6639	18 27 17.	- 13 14			OPEN CLUSTER
B 310	18 27 17.	- 18 37	120		DARK OBJECT
ZWG 200.023	18 27 18.	+ 33 50		15.6	GALAXY
MCG+08-33-045	18 27 18.	+ 48 12	60	15.	GALAXY
ISS 0584	18 27 18.	- 07 23	249		STELLAR RING
OCL 0057	18 27 18.	- 13 14			OPEN STAR CLUSTER
SN 1953C	18 27 23.	+ 48 14		19.0	SUPERNOVA
PK018-01.1	18 27 23.	- 13 56 05.	10		PLANETARY NEBULA
MCG+00-47-001	18 27 24.	+ 03 00	48	13.	GALAXY
ZWG 255.001	18 27 24.	+ 48 13		14.9	GALAXY
ZWG 254.043	18 27 24.	+ 48 13		14.9	GALAXY
UGC 11252	18 27 24.	+ 48 13	78	14.9	GALAXY Sc
ZWG 142.050	18 27 30.	+ 23 01		15.7	GALAXY
UGC 11253	18 27 30.	+ 23 01	60	15.7	GALAXY Sc
MCG+12-17-024	18 27 30.	+ 72 43	27	16.	GALAXY
PK020-00.1	18 27 30.	- 11 39	330	14.3	PLANETARY NEBULA
LDN 0356	18 27 30.	- 17 45	420		DARK NEBULA
B 311	18 27 32.	- 17 42	360		DARK OBJECT
SCHO 0862	18 27 33.	- 17 39	720		ISOLATED DARK CLOUD
ZWG 142.051	18 27 36.	+ 22 58		15.2	GALAXY
ZWG 255.002	18 27 36.	+ 50 20		15.0	GALAXY
ZWG 254.044	18 27 36.	+ 50 20		15.0	GALAXY
MCG+08-33-046	18 27 36.	+ 50 20	18	16.	GALAXY
ZWG 340.047	18 27 36.	+ 72 45		15.7	GALAXY
ISS 0585	18 27 36.	- 06 50	106		STELLAR RING
ISS 0808	18 27 36.	- 16 27	101		STELLAR RING
ACK 013-04.2	18 27 36.	- 19 16			PLANETARY NEBULA
IC 1283	18 27 38.	+ 39 42 26.			NONSTELLAR OBJECT
MCG+04-43-036	18 27 42.	+ 22 58	36	15.2	GALAXY
ZWG 172.039	18 27 42.	+ 30 25		15.7	GALAXY
UGC 11254	18 27 42.	+ 30 25	102	15.7	GALAXY Sc
MCG+09-30-016	18 27 42.	+ 51 35	78	15.	GALAXY
ZWG 279.015	18 27 42.	+ 51 37		15.6	GALAXY
UGC 11255	18 27 42.	+ 51 37	96	15.6	GALAXY Sc
LDN 0370	18 27 42.	- 16 15	360		DARK NEBULA
LDN 0372	18 27 42.	- 16 36	120		DARK NEBULA
MCG+07-38-007	18 27 48.	+ 39 40	54	15.	GALAXY
ZWG 228.009	18 27 48.	+ 39 41		14.3	GALAXY
UGC 11256	18 27 48.	+ 39 41	78	14.3	GALAXY SBa
ISS 0586	18 27 48.	- 06 50	163		STELLAR RING
MRSL 022+00/3	18 27 48.	- 08 39	120		HII REGION
RNGC 6630	18 27 49.	- 63 19			NON-EXISTENT OBJECT
REIN 2.274	18 27 50.70	- 25 32 00.1			NEBULA
RNGC 6638	18 27 51.	- 25 32		10.0	GLOBULAR CLUSTER
SCHO 0863	18 27 52.	- 23 46	300		ISOLATED DARK CLOUD
DG 152	18 27 54.	+ 01 12	300		REFLECTION NEBULA
UGC 11257	18 27 54.	+ 53 53	72	16.0	GALAXY Sb
MCG+09-30-017	18 27 54.	+ 53 53	48	15.	GALAXY
7ZW 802	18 27 54.	+ 57 44			COMPACT GALAXY
7ZW 803	18 27 54.	+ 57 47			COMPACT GALAXY
MCG+12-17-025	18 27 54.	+ 72 48	33	16.	GALAXY
GCL 095	18 27 54.	- 25 32	258	10.24	GLOBULAR STAR CLUSTER
VDB.66N 123	18 27 57.	+ 01 09	204		REFLECTION NEBULA
RNGC 6646	18 27 59.	+ 39 50		13.5	GALAXY
LBN 0096	18 28	+ 01 00	360		BRIGHT NEBULA
LBN 0078	18 28	- 08 00	120		BRIGHT NEBULA
LBN 0077	18 28	- 09 40	540		BRIGHT NEBULA
KHAV 507	18 28	- 15 40	4960		DARK NEBULA
LBN 0063	18 28	- 16 00	4200		BRIGHT NEBULA
KHAV 504	18 28	- 22 16	3330		DARK NEBULA
LB 09927	18 28	- 81 42		14.2	FAINT BLUE STAR
LDN 0571	18 28 00.	+ 00 00	1500		DARK NEBULA
LDN 0572	18 28 00.	+ 00 00	15180		DARK NEBULA
LDN 0600	18 28 00.	+ 02 50	660		DARK NEBULA
ZWG 228.010	18 28 00.	+ 39 50		13.7	GALAXY
UGC 11258	18 28 00.	+ 39 50	90	13.7	GALAXY Sa
ZWG 340.048	18 28 00.	+ 72 48		15.6	GALAXY
LDN 0446	18 28 00.	- 08 40	1500		DARK NEBULA
LDN 0379	18 28 00.	- 15 10	1680		DARK NEBULA
LDN 0309	18 28 00.	- 21 30	480		DARK NEBULA
LDN 0266	18 28 00.	- 24 10	2580		DARK NEBULA
LDN 0245	18 28 00.	- 25 30	2640		DARK NEBULA
LDN 0181	18 28 00.	- 28 07	360		DARK NEBULA
RNGC 6637	18 28 04.	- 32 23		9.0	GLOBULAR CLUSTER
MCG+07-38-008	18 28 06.	+ 39 49	72	14.	GALAXY
GCL 096	18 28 06.	- 32 23	408	8.94	GLOBULAR STAR CLUSTER
7ZW 804	18 28 12.	+ 58 05			COMPACT GALAXY
MRSL 024+01/1	18 28 12.	- 06 27	2700		HII REGION
ISS 0587	18 28 12.	- 06 45	139		STELLAR RING
ISS 0588	18 28 12.	- 09 07	167		STELLAR RING
IC 1289	18 28 13.	+ 39 57 30.			NONSTELLAR OBJECT
SHB 346	18 28 13.4	+ 48 42 39.		16.8	QUASI-STELLAR OBJECT
BC 3CR380	18 28 13.51	+ 48 42 40.0		16.81	QUASI-STELLAR OBJECT
RNGC 6642	18 28 23.	- 23 30		10.5	GLOBULAR CLUSTER
MCG+07-38-009	18 28 24.	+ 39 55	36	15.	GALAXY
ZWG 228.011	18 28 24.	+ 39 56		15.3	GALAXY
MRSL 022+00/1	18 28 24.	- 09 45	420		HII REGION
VDB.66N 124	18 28 27.	- 10 52	2160		REFLECTION NEBULA
HN 0469	18 28 28.	- 57 00			NEBULA
IC 4716	18 28 28.	- 57 00			NONSTELLAR OBJECT
PK032+05.1	18 28 29.3	+ 02 23 20.	26		PLANETARY NEBULA
DG 153	18 28 30.	+ 01 23	900		REFLECTION NEBULA
UGC 11259	18 28 30.	+ 33 52	66	16.5	GALAXY Sc
MCG+06-40-020	18 28 30.	+ 33 52	42	17.	GALAXY
MRSL 022+00/2	18 28 30.	- 09 09	420		HII REGION
OCL 0060	18 28 30.	- 12 21	420	12.	OPEN STAR CLUSTER
YM 14	18 28 33.	- 02 07	360		SYMMETRIC GALACTIC NEBULA
ZWG 201.001	18 28 36.	+ 34 03		15.5	GALAXY
ZWG 200.024	18 28 36.	+ 34 03		15.5	GALAXY
UGC 11260	18 28 36.	+ 34 03	78	15.5	GALAXY SBa
MCG+06-40-021	18 28 36.	+ 34 03	42	15.5	GALAXY
ZWG 201.002	18 28 36.	+ 34 18		15.3	GALAXY
ZWG 200.025	18 28 36.	+ 34 18		15.3	GALAXY
LDN 0578	18 28 36.	- 00 32	360		DARK NEBULA
DG 154	18 28 36.	- 10 50	1620		REFLECTION NEBULA
OCL 0045	18 28 36.	- 17 23		9.0	OPEN STAR CLUSTER
MCG+10-26-040	18 28 39.	+ 60 40	54	16.	GALAXY
IC 1287	18 28 39.	- 10 49 47.	3000		DIFFUSE NEBULA
CED 163	18 28 40.	- 10 50	2640		DIFFUSE GALACTIC NEBULA
OCL 0070	18 28 42.	- 06 40	1080		OPEN STAR CLUSTER
MRSL 023+00/1	18 28 42.	- 08 30	480		HII REGION
OCL 0033	18 28 42.	- 19 17	3780	6.5	OPEN STAR CLUSTER
SCHO 0864	18 28 44.	+ 08 03 36.	300		ISOLATED DARK CLOUD
SCHO 0865	18 28 46.	- 08 53	670		ISOLATED DARK CLOUD
OCL 0029	18 28 48.	- 23 30	150	10.5	OPEN STAR CLUSTER
GCL 097	18 28 49.	- 23 30	48	10.3	GLOBULAR STAR CLUSTER
PK008-07.1	18 28 49.	- 24 48 37.	25		PLANETARY NEBULA
IC 4720	18 28 50.	- 58 26	180		GALAXY SBc
REIN 2.275	18 28 51.72	- 23 30 45.4			NEBULA
MCG+10-26-041	18 28 54.	+ 57 58 30.	36	16.	GALAXY
LDN 0576	18 28 54.	- 00 25	240		DARK NEBULA
HN 0470	18 28 56.	- 58 00			NEBULA
IC 4717	18 28 56.	- 58 00			NONSTELLAR OBJECT
HN 0471	18 28 57.	- 56 46			NEBULA
IC 4719	18 28 57.	- 56 46			NONSTELLAR OBJECT
KHAV 510	18 29	+ 00 56	19360		DARK NEBULA
LBN 0097	18 29	+ 01 11	420		BRIGHT NEBULA
LBN 0098	18 29	+ 01 20	1260		BRIGHT NEBULA
KHAV 513	18 29	+ 12 26	5270		DARK NEBULA
LBN 0090	18 29	- 02 10	840		BRIGHT NEBULA
KHAV 512	18 29	- 08 10	3470		DARK NEBULA
LBN 0079	18 29	- 08 30	360		BRIGHT NEBULA
LBN 0076	18 29	- 10 50	1800		BRIGHT NEBULA
LBN 0075	18 29	- 10 50	900		BRIGHT NEBULA
KHAV 511	18 29	- 14 04	700		DARK NEBULA
B 312	18 29	- 15 37	6300		DARK OBJECT
KHAV 509	18 29	- 18 40	4860		DARK NEBULA
IC 4725	18 29	- 19 10	2100		OPEN CLUSTER
KHAV 506	18 29	- 26 10	1990		DARK NEBULA
KHAV 505	18 29	- 29 10	2470		DARK NEBULA
ZWG 114.009	18 29 00.	+ 19 10		15.4	GALAXY
MCG+03-47-005	18 29 00.	+ 19 10	42	15.4	GALAXY
MCG+04-44-001	18 29 00.	+ 22 21	54	15.5	GALAXY
ZWG 143.001	18 29 00.	+ 22 23		15.5	GALAXY
ZWG 142.052	18 29 00.	+ 22 23		15.5	GALAXY
UGC 11261	18 29 00.	+ 22 23	60	15.5	GALAXY
ZWG 173.001	18 29 00.	+ 29 51		15.5	GALAXY
ZWG 172.040	18 29 00.	+ 29 51		15.5	GALAXY
ZWG 201.003	18 29 00.	+ 34 18		15.7	GALAXY
ZWG 200.026	18 29 00.	+ 34 18		15.7	GALAXY
ZWG 228.012	18 29 00.	+ 42 40		15.5	GALAXY
UGC 11262	18 29 00.	+ 42 40	102	15.5	GALAXY Sc
MCG+07-38-010	18 29 00.	+ 42 40	78	15.5	GALAXY
7ZW 805	18 29 00.	+ 57 28			COMPACT GALAXY
MRSL 029+03/1	18 29 00.	- 01 57	1500		HII REGION
LDN 0463	18 29 00.	- 08 00	4020		DARK NEBULA
SCHO 0866	18 29 00.	- 11 51	390		ISOLATED DARK CLOUD
LDN 0412	18 29 00.	- 12 00	540		DARK NEBULA

OBJECT NAME	RIGHT ASCEN.	DECLINATION	DIAM.	MAGN.	TYPE OF OBJECT
LDN 0294	18 29 00.	- 22 20	180		DARK NEBULA
SN 1934A	18 29 00.	- 56 46		13.6	SUPERNOVA
SCHO 0867	18 29 08.	- 09 11	780		ISOLATED DARK CLOUD
HN 0472	18 29 10.	- 58 26			NEBULA
ZWG 114.010	18 29 12.	+ 16 05		15.4	GALAXY
PK055+16.1	18 29 12.	+ 26 53	90	15.6	PLANETARY NEBULA
LDN 0396	18 29 12.	- 13 15	240		DARK NEBULA
ISS 0809	18 29 12.	- 16 27	203		STELLAR RING
PK034+06.1	18 29 16.8	+ 04 02 59.	9		PLANETARY NEBULA
HN 0473	18 29 17.	- 60 09			NEBULA
IC 4718	18 29 17.	- 60 09			NONSTELLAR OBJECT
UGC 11263	18 29 18.	+ 30 57	60	16.0	GALAXY Sc
ZWG 255.003	18 29 18.	+ 48 16		15.3	GALAXY
ZC 1829.3+6912	18 29 18.	+ 69 12	1140		CLUSTER OF GALAXIES
OCL 0061	18 29 18.	- 12 17	240	13.	OPEN STAR CLUSTER
OCL 0050	18 29 18.	- 16 05	420	14.	OPEN STAR CLUSTER
MCG-07-38-002	18 29 18.	- 41 35	120	13.	GALAXY
PK005-08.1	18 29 20.7	- 28 45 36.	19		PLANETARY NEBULA
MCG+08-34-001	18 29 24.	+ 48 15	54	15.	GALAXY
MASL 020+01/1	18 29 24.	- 11 48	300		HII REGION
AGU 52	18 29 24.	- 41 32 00.		12.5	PAIR OF GALAXIES
RNGC 6644	18 29 28.	- 25 10		12.0	PLANETARY NEBULA
PK008-07.2	18 29 29.8	- 25 09 59.	3	12.2	PLANETARY NEBULA
ZWG 201.004	18 29 30.	+ 34 22		15.4	GALAXY
ZWG 200.027	18 29 30.	+ 34 22		15.4	GALAXY
7ZW 806	18 29 30.	+ 60 06			COMPACT GALAXY
LDN 0296	18 29 30.	- 22 20	180		DARK NEBULA
MCG-07-38-003	18 29 30.	- 41 33	90	14.	GALAXY
MCG+11-22-051	18 29 36.	+ 64 57	12	17.	GALAXY
RNGC 6645	18 29 39.	- 16 56		8.5	OPEN CLUSTER
ISS 0698	18 29 42.	- 14 35	154		STELLAR RING
OCL 0048	18 29 42.	- 16 56	1740	9.0	OPEN STAR CLUSTER
MCG+06-41-001	18 29 45.	+ 33 53	36	15.	GALAXY
MCG+11-22-052	18 29 45.	+ 64 57 30.	12	17.	GALAXY
ZWG 173.002	18 29 48.	+ 31 02		15.2	GALAXY
ZWG 172.041	18 29 48.	+ 31 02		15.2	GALAXY
UGC 11264	18 29 48.	+ 31 02	72	15.2	GALAXY Sb-c
ZWG 201.005	18 29 48.	+ 33 53		13.9	GALAXY
UGC 11265	18 29 48.	+ 33 53	54	13.9	GALAXY COMPACT
ISS 0589	18 29 48.	- 04 17	121		STELLAR RING
ISS 0590	18 29 48.	- 04 18	229		STELLAR RING
OCL 0062	18 29 48.	- 12 10	336	14.	OPEN STAR CLUSTER
LDN 0295	18 29 48.	- 22 23	180		DARK NEBULA
SG 3.145	18 29 49.	- 13 18	36000		DIFFUSE EMISSION NEBULA
PK037+07.1	18 29 49.8	+ 07 11 54.			PLANETARY NEBULA
MCG+05-44-001	18 29 51.	+ 31 02 30.	72	15.2	GALAXY
UGC 11266	18 29 54.	+ 67 51	60	16.5	GALAXY Sc
ISS 0591	18 29 54.	- 06 14	167		STELLAR RING
B 101	18 29 56.	- 08 51	780		DARK OBJECT
PK021-00.1	18 29 56.	- 10 08 08.	32		PLANETARY NEBULA
B 100	18 29 57.	- 09 11	960		DARK OBJECT
KHAV 514	18 30	+ 04 08	11040		DARK NEBULA
LBN 0074	18 30	- 11 48	480		BRIGHT NEBULA
LBN 0073	18 30	- 11 48	1200		BRIGHT NEBULA
KHAV 508	18 30	- 27 34	3750		DARK NEBULA
MCG+10-26-042	18 30 00.	+ 61 37 30.	30	16.	GALAXY
UGC 11267	18 30 00.	+ 87 49	108	16.0	GALAXY Sc
LDN 0559	18 30 00.	- 01 00	16740		DARK NEBULA
LDN 0450	18 30 00.	- 08 40	840		DARK NEBULA
LDN 0443	18 30 00.	- 09 10	840		DARK NEBULA
LDN 0435	18 30 00.	- 10 00	11760		DARK NEBULA
LDN 0398	18 30 00.	- 13 20	420		DARK NEBULA
LDN 0389	18 30 00.	- 14 10	1380		DARK NEBULA
LDN 0388	18 30 00.	- 14 15	9840		DARK NEBULA
LDN 0353	18 30 00.	- 18 20	7380		DARK NEBULA
SC 1826-4637.3	18 30 00.	- 46 35 15.	18		NEBULA
HN 0474	18 30 04.	- 58 32			NEBULA
ISS 0810	18 30 06.	- 15 02	143		STELLAR RING
ISS 0811	18 30 06.	- 19 08	127		STELLAR RING
SER 128.01	18 30 06.	- 58 32	480	13.	INTERACTING GALAXIES
IC 4721	18 30 06.	- 58 32	252	12.0	GALAXY SB(s)
MCG+05-44-002	18 30 09.	+ 31 00	24	14.9	GALAXY
B 098	18 30 10.	- 26 06	180		DARK OBJECT
ZWG 173.003	18 30 12.	+ 31 00		14.9	GALAXY
1ZW 207	18 30 12.	+ 55 14			COMPACT GALAXY
OCL 0073	18 30 12.	- 06 04	1080		OPEN STAR CLUSTER
HN 0475	18 30 13.	- 57 49			NEBULA
IC 4722	18 30 13.	- 57 49			NONSTELLAR OBJECT
PK018-02.1	18 30 14.	- 13 46 41.			PLANETARY NEBULA
MIL 60	18 30 16.	- 10 13 00.	960		SUPERNOVA REMNANT
B 099	18 30 18.	- 21 31	660		DARK OBJECT
LDN 0239	18 30 18.	- 26 03	540		DARK NEBULA
RNGC 6647	18 30 21.	- 17 22			NON-EXISTENT OBJECT
ZWG 201.006	18 30 24.	+ 37 34		14.7	GALAXY
UGC 11268	18 30 24.	+ 37 34	84	14.7	GALAXY Sb-c
PK014-04.1	18 30 25.2	- 18 18 56.	6		PLANETARY NEBULA
MCG+06-41-002	18 30 27.	+ 37 35	66	14.5	GALAXY
7ZW 807	18 30 30.	+ 59 20			COMPACT GALAXY
PK021-00.2	18 30 30.	- 10 17 39.	8		PLANETARY NEBULA
LDN 0313	18 30 30.	- 21 35	540		DARK NEBULA
ZWG 173.004	18 30 36.	+ 31 27		15.7	GALAXY
ZWG 279.016	18 30 36.	+ 53 32		15.7	GALAXY
OCL 0063	18 30 36.	- 11 28	300	12.	OPEN STAR CLUSTER
MIL 61	18 30 38.	- 08 48	3600		SUPERNOVA REMNANT
ZWG 201.007	18 30 42.	+ 34 01		15.5	GALAXY
OCL 0066	18 30 42.	- 10 26	1320	10.6	OPEN STAR CLUSTER
PK021-01.1	18 30 42.	- 11 09 46.	13		PLANETARY NEBULA
RNGC 6649	18 30 43.	- 10 26		10.5	OPEN CLUSTER
PK030+04.1	18 30 44.0	+ 00 09 30.			PLANETARY NEBULA
MIN.47 10	18 30 45.	- 05 01			DIFFUSE NEBULA
RNGC 6667	18 30 47.	+ 67 57		13.5	GALAXY
ACK 038+07.1	18 30 48.	+ 08 16			PLANETARY NEBULA
7ZW 808	18 30 48.	+ 62 31			COMPACT GALAXY
ZWG 322.044	18 30 48.	+ 67 56		13.7	GALAXY
UGC 11269	18 30 48.	+ 67 56	192	13.7	GALAXY PECULR
MCG+11-22-053	18 30 48.	+ 67 57	114	14.	GALAXY
MCG+13-13-001	18 30 48.	+ 78 43	51	15.	GALAXY
ISS 0592	18 30 48.	- 05 22	178		STELLAR RING
ISS 0593	18 30 48.	- 08 42	277		STELLAR RING
RNGC 6668	18 30 50.	+ 67 07			NON-EXISTENT OBJECT
PK026+01.1	18 30 51.	- 04 59	120		PLANETARY NEBULA
SCHO 0868	18 30 52.	- 13 14	270		ISOLATED DARK CLOUD
BC 4C28.45	18 30 52.8	+ 28 33 00.		17.	QUASI-STELLAR OBJECT
SHB 347	18 30 52.8	+ 28 33 00.		17.	QUASI-STELLAR OBJECT
IC 4732	18 30 53.3	- 22 40 57.	2	13.3	PLANETARY NEBULA
ZWG 173.005	18 30 54.	+ 32 00		15.6	GALAXY
7ZW 809	18 30 54.	+ 58 33			COMPACT GALAXY
ZC 1830.9+6444	18 30 54.	+ 64 44	2150		CLUSTER OF GALAXIES

OBJECT NAME	RIGHT ASCEN.	DECLINATION	DIAM.	MAGN.	TYPE OF OBJECT
ISS 0812	18 30 54.	- 20 27	177		STELLAR RING
HN 0099	18 30 55.	- 22 41			NEBULA
LBN 0084	18 31	- 05 00	120		BRIGHT NEBULA
KHAV 517	18 31	- 10 34	6920		DARK NEBULA
KHAV 515	18 31	- 13 22	1990		DARK NEBULA
KHAV 518	18 31	- 14 22	3940		DARK NEBULA
LBN 0069	18 31	- 15 30	1200		BRIGHT NEBULA
KHAV 516	18 31	- 19 40	5110		DARK NEBULA
LB 09928	18 31	- 82 58		13.3	FAINT BLUE STAR
LDN 0583	18 31 00.	+ 00 35	1140		DARK NEBULA
ZWG 356.009	18 31 00.	+ 78 44		15.2	GALAXY
UGC 11270	18 31 00.	+ 78 44	96	15.2	GALAXY Sa-b
ISS 0594	18 31 00.	- 09 02	170		STELLAR RING
ISS 0813	18 31 00.	- 16 01	273		STELLAR RING
PK010-06.1	18 31 01.9	- 22 39 03.	3	13.3	PLANETARY NEBULA
IC 4723	18 31 03.	- 63 24	14	15.4	PLANETARY NEBULA
HN 0476	18 31 04.	- 63 25			NEBULA
ISS 0595	18 31 06.	- 05 51	130		STELLAR RING
PK017-02.1	18 31 07.	- 14 54 48.	7		PLANETARY NEBULA
SCHO 0869	18 31 07.	- 17 25	260		ISOLATED DARK CLOUD
SCHO 0870	18 31 10.	- 13 13	300		ISOLATED DARK CLOUD
RNGC 6657	18 31 11.	+ 34 00		14.0	GALAXY
ZC 1831.2+3154	18 31 12.	+ 31 54	4440		CLUSTER OF GALAXIES
ZWG 201.008	18 31 12.	+ 34 00		14.2	GALAXY
UGC 11271	18 31 12.	+ 34 00	60	14.2	GALAXY SB
MCG+06-41-003	18 31 15.	+ 34 01 30.	42	14.5	GALAXY
SCHO 0871	18 31 16.	- 17 34	260		ISOLATED DARK CLOUD
ISS 0480	18 31 18.	+ 08 03	190		STELLAR RING
MCG+09-30-018	18 31 18.	+ 52 29	33	16.	GALAXY
ISS 0814	18 31 18.	- 20 59	118		STELLAR RING
SCHO 0872	18 31 19.	- 13 29	310		ISOLATED DARK CLOUD
PK036+06.1	18 31 22.7	+ 05 50 49.			PLANETARY NEBULA
SCHO 0873	18 31 24.	+ 08 11 30.	440		ISOLATED DARK CLOUD
MCG+09-30-019	18 31 24.	+ 54 56	54	16.	GALAXY
ZC 1831.4+7310	18 31 24.	+ 73 10	1550		CLUSTER OF GALAXIES
PK019-02.1	18 31 25.	- 13 14 46.			PLANETARY NEBULA
UGC 11272	18 31 30.	+ 54 57	60	16.5	GALAXY Sb-c
7ZW 810	18 31 30.	+ 59 27			COMPACT GALAXY
ISS 0596	18 31 30.	- 04 52	143		STELLAR RING
SCHO 0874	18 31 34.	- 13 36	270		ISOLATED DARK CLOUD
1ZW 208	18 31 36.	+ 54 29			COMPACT GALAXY
PK028+02.1	18 31 37.1	- 02 29 59.			PLANETARY NEBULA
ZWG 114.011	18 31 42.	+ 19 03		15.7	GALAXY
UGC 11273	18 31 42.	+ 19 03	60	15.7	GALAXY S
ZWG 301.030	18 31 42.	+ 59 50		15.0	GALAXY
MCG+10-26-043	18 31 42.	+ 59 50	24	15.	GALAXY
ZWG 341.001	18 31 42.	+ 74 05		15.4	GALAXY
ZWG 340.049	18 31 42.	+ 74 05		15.4	GALAXY
RNGC 6658	18 31 45.	+ 22 51		15.0	GALAXY
SCHO 0875	18 31 45.	- 13 14	300		ISOLATED DARK CLOUD
ALL 006-08.1	18 31 47.	- 27 08 48.	14	12.	PLANETARY NEBULA
MCG+04-44-002	18 31 48.	+ 22 50	78	14.1	GALAXY
ZWG 143.002	18 31 48.	+ 22 51		15.0	GALAXY
UGC 11274	18 31 48.	+ 22 51	90	15.0	GALAXY S0-a
ZWG 173.006	18 31 48.	+ 31 05		15.7	GALAXY
UGC 11275	18 31 48.	+ 32 06	102	16.5	GALAXY Sc
ZWG 228.013	18 31 48.	+ 39 19		15.7	GALAXY
7ZW 811	18 31 48.	+ 59 25			COMPACT GALAXY
ISS 0699	18 31 48.	- 11 24	127		STELLAR RING
ACK 006-08.1	18 31 48.	- 27 09			PLANETARY NEBULA
PK010-06.2	18 31 50.3	- 22 45 44.	8		PLANETARY NEBULA
RNGC 6659	18 31 52.	+ 23 32			NON-EXISTENT OBJECT
RNGC 6663	18 31 53.	+ 40 01		15.0	GALAXY
ZWG 114.012	18 31 54.	+ 20 16		15.7	GALAXY
ZWG 228.014	18 31 54.	+ 40 01		14.8	GALAXY
UGC 11276	18 31 54.	+ 40 01	66	14.8	GALAXY Sc
LBN 0095	18 32	+ 00 00	660		BRIGHT NEBULA
LBN 0081	18 32	- 06 55	1620		BRIGHT NEBULA
MCG+07-38-011	18 32 00.	+ 40 00	60	14.5	GALAXY
UGC 11277	18 32 00.	+ 52 49	66	16.0	GALAXY Sb
MCG+09-30-020	18 32 00.	+ 52 53	66	16.	GALAXY
LDN 0536	18 32 00.	- 02 35	2040		DARK NEBULA
LDN 0407	18 32 00.	- 13 00	2340		DARK NEBULA
ISS 0700	18 32 00.	- 13 04	119		STELLAR RING
LDN 0385	18 32 00.	- 14 53	420		DARK NEBULA
LDN 0364	18 32 00.	- 17 40	1260		DARK NEBULA
LDN 0304	18 32 00.	- 22 10	540		DARK NEBULA
MCG+03-47-006	18 32 06.	+ 18 10	30	17.5	GALAXY
ZWG 201.009	18 32 06.	+ 35 32		15.3	GALAXY
ZWG 201.010	18 32 06.	+ 38 01		15.7	GALAXY
ZWG 201.011	18 32 06.	+ 38 34		15.1	GALAXY
UGC 11278	18 32 06.	+ 38 34	60	15.1	GALAXY S0
ISS 0815	18 32 06.	- 15 37	230		STELLAR RING
RNGC 6662	18 32 10.	+ 32 01		15.0	GALAXY
RNGC 6655	18 32 10.	- 06 01			NON-EXISTENT OBJECT
ZC 1832.2+3348	18 32 12.	+ 33 48	1140		CLUSTER OF GALAXIES
UGC 11279	18 32 12.	+ 37 09	60	17.	GALAXY DWARF
ZWG 228.015	18 32 12.	+ 42 46		14.9	GALAXY
MCG+08-34-002	18 32 12.	+ 48 12	60	15.	GALAXY
OCL 0071	18 32 12.	- 06 53	1080		OPEN STAR CLUSTER
IC 4726	18 32 13.	- 62 53			NONSTELLAR OBJECT
HN 0478	18 32 13.	- 62 54			NEBULA
RNGC 6672	18 32 14.	+ 42 46		15.0	GALAXY
MCG+05-44-003	18 32 15.	+ 32 01	90	14.9	GALAXY
ZWG 114.013	18 32 18.	+ 18 56		15.2	GALAXY
ZWG 173.007	18 32 18.	+ 32 01		14.9	GALAXY
UGC 11280	18 32 18.	+ 32 01	108	14.9	GALAXY SB?a-b
ZWG 173.008	18 32 18.	+ 32 08		15.6	GALAXY
ZWG 255.004	18 32 18.	+ 48 11		15.3	GALAXY
MCG+08-34-003	18 32 18.	+ 49 16	18	16.	GALAXY
ISS 0816	18 32 18.	- 15 08	199		STELLAR RING
ZWG 173.009	18 32 24.	+ 32 09		15.4	GALAXY
ZWG 201.012	18 32 24.	+ 38 33		14.9	GALAXY
UGC 11281	18 32 24.	+ 38 33	96	14.9	GALAXY Sb
ZWG 255.005	18 32 24.	+ 48 15		15.6	GALAXY
LDN 0301	18 32 24.	- 22 32	540		DARK NEBULA
BC 4C31.51	18 32 25.7	+ 31 34 01.		18.2	QUASI-STELLAR OBJECT
SHB 348	18 32 26.	+ 31 34 01.		18.	QUASI-STELLAR OBJECT
PK035+05.1	18 32 26.1	+ 05 02 30.			PLANETARY NEBULA
RNGC 6660	18 32 27.	+ 22 42			NON-EXISTENT OBJECT
RNGC 6661	18 32 27.	+ 22 52		14.0	GALAXY
MCG+06-41-004	18 32 27.	+ 38 34	90	14.5	GALAXY
RNGC 6652	18 32 28.	- 33 02		10.0	GLOBULAR CLUSTER
MCG+04-44-003	18 32 30.	+ 22 51 30.	108	14.1	GALAXY
ZWG 143.003	18 32 30.	+ 22 52		14.1	GALAXY
UGC 11282	18 32 30.	+ 22 52	126	14.1	GALAXY S0
MCG+15-01-021	18 32 30.	+ 67 48	78	16.	GALAXY
ISS 0817	18 32 30.	- 15 21	100		STELLAR RING

OBJECT NAME	RIGHT ASCEN.	DECLINATION	DIAM.	MAGN.	TYPE OF OBJECT
SCHO 0876	18 32 30.	- 18 38	300		ISOLATED DARK CLOUD
GCL 098	18 32 30.	- 33 02	252	9.86	GLOBULAR STAR CLUSTER
RNGC 6665	18 32 33.	+ 30 40		14.5	GALAXY
MCG+05-44-004	18 32 33.	+ 30 40	60	14.6	GALAXY
IC 1291	18 32 35.	+ 49 14 07.			NONSTELLAR OBJECT
ZWG 173.010	18 32 36.	+ 30 41		14.6	GALAXY
ZWG 255.006	18 32 36.	+ 49 14		14.2	GALAXY
UGC 11283	18 32 36.	+ 49 14	138	14.2	GALAXY
MCG+08-34-004	18 32 36.	+ 49 14	102	14.	GALAXY
ISS 0818	18 32 36.	- 15 36	286		STELLAR RING
IC 4762	18 32 38.	+ 67 48			OPEN CLUSTER
ZWG 173.011	18 32 48.	+ 32 29		15.5	GALAXY
MCG+08-34-005	18 32 48.	+ 49 20	42	16.	GALAXY
ZC 1832.8+6827	18 32 48.	+ 68 27	610		CLUSTER OF GALAXIES
PK030+03.1	18 32 48.	- 00 16	17	20.9	PLANETARY NEBULA
ASS 12	18 32 48.	- 09 10			OB ASSOCIATION SCT OB2
RNGC 6670	18 32 52.	+ 59 51		15.5	GALAXY
RNGC 6666	18 32 53.	+ 33 33			NON-EXISTENT OBJECT
PK015-04.1	18 32 53.5	- 17 38 38.	7		PLANETARY NEBULA
7ZW 812	18 32 54.	+ 59 51			COMPACT GALAXY
ZWG 301.031	18 32 54.	+ 59 51		15.3	GALAXY
UGC 11284	18 32 54.	+ 59 51	60	15.3	GALAXY TRP SYS
ISS 0597	18 32 54.	- 04 28	314		STELLAR RING
SCHO 0877	18 32 56.	- 18 38	290		ISOLATED DARK CLOUD
KHAV 521	18 33	- 08 46	3170		DARK NEBULA
KHAV 523	18 33	- 09 58	4860		DARK NEBULA
KHAV 520	18 33	- 13 22	700		DARK NEBULA
KHAV 519	18 33	- 23 28	2930		DARK NEBULA
MCG+10-26-044	18 33 00.	+ 59 50	60	15.	GALAXY
MCG+11-22-054	18 33 00.	+ 66 54	90	15.	GALAXY
ZC 1833.0+7952	18 33 00.	+ 79 52	1810		CLUSTER OF GALAXIES
LDN 0390	18 33 00.	- 14 30	1500		DARK NEBULA
LDN 0284	18 33 00.	- 23 30	1800		DARK NEBULA
LDN 0263	18 33 00.	- 25 00	6300		DARK NEBULA
LDN 0248	18 33 00.	- 25 50	900		DARK NEBULA
HN 0477	18 33 01.	- 70 10			NEBULA
IC 4724	18 33 02.	- 70 09			NONSTELLAR OBJECT
B 313	18 33 04.	- 15 44	900		DARK OBJECT
ZWG 143.004	18 33 06.	+ 22 28		15.3	GALAXY
UGC 11285	18 33 06.	+ 22 28	96	15.3	GALAXY Sc-IRR
ZWG 173.012	18 33 06.	+ 27 44		15.5	GALAXY
ZWG 173.013	18 33 06.	+ 30 35		15.2	GALAXY
ZC 1833.1+6840	18 33 06.	+ 68 40	1140		CLUSTER OF GALAXIES
7ZW 813	18 33 06.	+ 68 56			COMPACT GALAXY
ISS 0598	18 33 06.	- 08 32	118		STELLAR RING
IC 4727	18 33 06.	- 62 44			NONSTELLAR OBJECT
HN 0479	18 33 06.	- 62 45			NEBULA
MCG+04-44-004	18 33 09.	+ 22 26	84	15.3	GALAXY
MCG+05-44-005	18 33 09.	+ 30 35	48	15.2	GALAXY
HN 0480	18 33 11.	- 62 35			NEBULA
ZWG 173.014	18 33 12.	+ 32 39		15.5	GALAXY
ZWG 173.015	18 33 12.	+ 32 57		15.4	GALAXY
ZWG 322.045	18 33 12.	+ 66 55		15.3	GALAXY
UGC 11286	18 33 12.	+ 66 55	96	15.3	GALAXY Sb-c
ISS 0599	18 33 12.	- 04 45	91		STELLAR RING
ISS 0600	18 33 12.	- 08 31	61		STELLAR RING
ISS 0601	18 33 12.	- 08 32	100		STELLAR RING
IC 4728	18 33 12.	- 62 34			NONSTELLAR OBJECT
RNGC 6676	18 33 13.	+ 66 55		15.5	GALAXY
PK016-04.1	18 33 14.4	- 17 02 29.	17		PLANETARY NEBULA
RNGC 6656	18 33 16.	- 23 58		6.5	GLOBULAR CLUSTER
ISS 0602	18 33 18.	- 03 32	164		STELLAR RING
RNGC 6678	18 33 22.	+ 67 49			NON-EXISTENT OBJECT
ZWG 255.007	18 33 24.	+ 47 00		15.5	GALAXY
UGC 11287	18 33 24.	+ 47 00	108	15.5	GALAXY S
KARA.73B 0854	18 33 24.	+ 47 00	102	15.5	ISOLATED GALAXY S
MCG+11-22-055	18 33 24.	+ 67 05	24	15.	GALAXY
GCL 099	18 33 24.	- 23 58	1572	6.48	GLOBULAR STAR CLUSTER
MCG+11-22-056	18 33 27.	+ 67 05	15	16.	GALAXY
ABC 2307	18 33 28.	+ 61 09		17.8	RICH CLUSTER OF GALAXIES
MCG+08-34-006	18 33 30.	+ 48 25	30	16.	GALAXY
ZWG 301.032	18 33 30.	+ 59 49		15.2	GALAXY
MCG+11-22-057	18 33 30.	+ 67 03	45	14.	GALAXY
LDN 0588	18 33 30.	- 00 30	720		DARK NEBULA
IC 4763	18 33 31.	+ 67 05			NONSTELLAR OBJECT
SCHO 0878	18 33 32.	- 13 28	300		ISOLATED DARK CLOUD
PK011-06.1	18 33 33.6	- 21 51 35.	25		PLANETARY NEBULA
ZWG 201.013	18 33 36.	+ 38 24		15.4	GALAXY
MCG+08-34-007	18 33 36.	+ 50 05	30	16.	GALAXY
MCG+08-34-008	18 33 36.	+ 50 07 30.	24	16.	GALAXY
7ZW 814	18 33 36.	+ 67 06			COMPACT GALAXY
ZWG 323.001	18 33 36.	+ 67 06		13.6	GALAXY
ZWG 322.046	18 33 36.	+ 67 06		13.6	GALAXY
UGC 11288	18 33 36.	+ 67 06	36	13.6	GALAXY DBL SYS
ACK 014-05.1	18 33 36.	- 19 22			PLANETARY NEBULA
RNGC 6679	18 33 37.	+ 67 06		13.5	GALAXY
RNGC 6677	18 33 37.	+ 67 06		13.5	GALAXY
MCG+06-41-005	18 33 39.	+ 38 26	15	15.	GALAXY
ZWG 143.005	18 33 42.	+ 22 25		15.1	GALAXY
UGC 11289	18 33 42.	+ 22 25	102	15.1	GALAXY S
MCG+04-44-005	18 33 42.	+ 22 25	78	15.1	GALAXY
LB 04023	18 33 42.	+ 35 37		18.0	FAINT BLUE STAR
MCG+09-34-009	18 33 42.	+ 47 00	60	14.	GALAXY
ZC 1833.7+6106	18 33 42.	+ 61 06	1880		CLUSTER OF GALAXIES
7ZW 815	18 33 42.	+ 66 34			COMPACT GALAXY
ZWG 323.002	18 33 42.	+ 67 04		13.9	GALAXY
ZWG 322.047	18 33 42.	+ 67 04		13.9	GALAXY
UGC 11290	18 33 42.	+ 67 04	54	13.9	GALAXY S
ISS 0511	18 33 42.	- 00 42	503		STELLAR RING
LDN 0541	18 33 42.	- 02 30	640		DARK NEBULA
LB 04025	18 33 48.	+ 35 26		19.0	FAINT BLUE STAR
LB 04024	18 33 48.	+ 35 28		18.8	FAINT BLUE STAR
ZWG 228.016	18 33 48.	+ 44 44		15.5	GALAXY
MCG+08-34-010	18 33 48.	+ 48 25	36	16.	GALAXY
ISS 0603	18 33 48.	- 08 19	482		STELLAR RING
ISS 0819	18 33 48.	- 18 18	61		STELLAR RING
ISS 0820	18 33 48.	- 18 53	173		STELLAR RING
ISS 0512	18 33 54.	+ 03 01	114		STELLAR RING
ZWG 114.014	18 33 54.	+ 20 44		15.7	GALAXY
LB 04027	18 33 54.	+ 35 04		18.1	FAINT BLUE STAR
LB 04026	18 33 54.	+ 35 28		16.1	FAINT BLUE STAR
ISS 0821	18 33 54.	- 17 32	182		STELLAR RING
HN 0481	18 33 56.	- 62 59			NEBULA
IC 4731	18 33 56.	- 62 59			NONSTELLAR OBJECT
HN 0482	18 33 58.	- 63 24			NEBULA
IC 4730	18 33 58.	- 63 24			NONSTELLAR OBJECT
LBN 0082	18 34	- 06 40	900		BRIGHT NEBULA
KHAV 528	18 34	- 06 52	3400		DARK NEBULA
LBN 0080	18 34	- 07 40	1320		BRIGHT NEBULA
KHAV 527	18 34	- 12 16	5270		DARK NEBULA
KHAV 526	18 34	- 19 10	2370		DARK NEBULA
KHAV 522	18 34	- 24 40	3010		DARK NEBULA
LB 09929	18 34	- 80 39		14.6	FAINT BLUE STAR
LB 09930	18 34	- 80 55		13.4	FAINT BLUE STAR
ISS 0481	18 34 00.	+ 03 06	115		STELLAR RING
LB 04029	18 34 00.	+ 34 29		19.8	FAINT BLUE STAR
LB 04028	18 34 00.	+ 34 42		17.0	FAINT BLUE STAR
ZC 1834.0+6241	18 34 00.	+ 62 41	2150		CLUSTER OF GALAXIES
7ZW 816	18 34 00.	+ 80 07			COMPACT GALAXY
ZC 1834.0+8259	18 34 00.	+ 82 59	3090		CLUSTER OF GALAXIES
LDN 0591	18 34 00.	- 00 50	2280		DARK NEBULA
LDN 0538	18 34 00.	- 02 50	1440		DARK NEBULA
MRSL 025+00/1	18 34 00.	- 06 45	1200		HII REGION
LDN 0485	18 34 00.	- 07 00	3300		DARK NEBULA
OCL 0068	18 34 00.	- 08 16	2700	9.3	OPEN STAR CLUSTER
LDN 0448	18 34 00.	- 09 20	660		DARK NEBULA
LDN 0403	18 34 00.	- 13 30	420		DARK NEBULA
LDN 0378	18 34 00.	- 16 00	1200		DARK NEBULA
LDN 0240	18 34 00.	- 26 25	600		DARK NEBULA
RNGC 6664	18 34 02.	- 08 16		9.0	OPEN CLUSTER
ZWG 173.016	18 34 06.	+ 30 47		15.4	GALAXY
UGC 11291	18 34 06.	+ 30 47	126	15.4	GALAXY
HN 0484	18 34 06.	- 57 32			NEBULA
IC 4734	18 34 06.	- 57 32			NONSTELLAR OBJECT
MCG+05-44-006	18 34 09.	+ 30 47	132	15.4	GALAXY
ZWG 114.015	18 34 12.	+ 20 50		15.7	GALAXY
LB 04030	18 34 12.	+ 35 09		14.3	FAINT BLUE STAR
ZWG 201.014	18 34 12.	+ 38 05		14.6	GALAXY
ZWG 279.017	18 34 12.	+ 52 41		14.3	GALAXY
UGC 11292	18 34 12.	+ 52 41	60	14.3	GALAXY COMPACT
ABC 2308	18 34 12.	+ 71 00		16.4	RICH CLUSTER OF GALAXIES
SG 3.146	18 34 12.	- 06 47	720		DIFFUSE EMISSION NEBULA
ACK 014-05.3	18 34 12.	- 19 05			PLANETARY NEBULA
UGC 11293	18 34 18.	+ 10 23	66	15.	GALAXY
MCG+02-47-004	18 34 18.	+ 10 23	48	15.	GALAXY
LB 04033	18 34 18.	+ 35 04		17.0	FAINT BLUE STAR
LB 04032	18 34 18.	+ 35 18		17.4	FAINT BLUE STAR
LB 04031	18 34 18.	+ 35 25		17.9	FAINT BLUE STAR
MCG+06-41-006	18 34 18.	+ 38 06 30.	30	15.	GALAXY
MCG+09-30-021	18 34 18.	+ 52 40	42	15.	GALAXY
7ZW 817	18 34 18.	+ 71 32			COMPACT GALAXY
ISS 0604	18 34 18.	- 04 42	368		STELLAR RING
LDN 0494	18 34 18.	- 06 30	420		DARK NEBULA
MRSL 024-00/1	18 34 18.	- 07 39	1200		HII REGION
ISS 0701	18 34 18.	- 14 44	356		STELLAR RING
HN 0485	18 34 19.	- 57 56			NEBULA
IC 4736	18 34 19.	- 57 56			NONSTELLAR OBJECT
YM 13	18 34 20.	- 07 44	720		SYMMETRIC GALACTIC NEBULA
B 314	18 34 20.	- 09 45	2100		DARK OBJECT
ZWG 114.016	18 34 24.	+ 19 35		15.7	GALAXY
LB 04036	18 34 24.	+ 34 24		19.0	FAINT BLUE STAR
LB 04035	18 34 24.	+ 34 36		19.6	FAINT BLUE STAR
LB 04034	18 34 24.	+ 35 27		18.5	FAINT BLUE STAR
ZWG 201.015	18 34 24.	+ 38 06		15.7	GALAXY
ACK 014-05.2	18 34 24.	- 19 10			PLANETARY NEBULA
ZWG 114.017	18 34 30.	+ 19 41		14.8	GALAXY
MCG+03-47-008	18 34 30.	+ 19 41	72	14.8	GALAXY
ZWG 114.018	18 34 30.	+ 19 53		15.0	GALAXY
UGC 11294	18 34 30.	+ 19 53	84	15.0	GALAXY S?
MCG+03-47-007	18 34 30.	+ 19 53	72	15.0	GALAXY
LB 04039	18 34 30.	+ 34 32		20.3	FAINT BLUE STAR
LB 04038	18 34 30.	+ 34 36		19.6	FAINT BLUE STAR
LB 04037	18 34 30.	+ 35 08		18.8	FAINT BLUE STAR
UGC 11295	18 34 30.	+ 75 20	102	16.5	GALAXY
SHAH 168	18 34 30.	+ 83 04	198		GROUP OF COMPACT GALAXIES
ISS 0702	18 34 30.	- 09 21	86		STELLAR RING
LDN 0381	18 34 30.	- 15 50	240		DARK NEBULA
LB 04045	18 34 36.	+ 34 53		20.4	FAINT BLUE STAR
LB 04044	18 34 36.	+ 34 54		17.6	FAINT BLUE STAR
LB 04043	18 34 36.	+ 35 09		18.6	FAINT BLUE STAR
LB 04042	18 34 36.	+ 35 10		18.7	FAINT BLUE STAR
LB 04041	18 34 36.	+ 35 10		16.8	FAINT BLUE STAR
LB 04040	18 34 36.	+ 35 21		19.2	FAINT BLUE STAR
7ZW 818	18 34 36.	+ 57 52			COMPACT GALAXY
ZC 1834.6+6925	18 34 36.	+ 69 25	2550		CLUSTER OF GALAXIES
SCHO 0879	18 34 38.	- 13 46	370		ISOLATED DARK CLOUD
ZWG 114.019	18 34 42.	+ 19 48		15.7	GALAXY
LB 04047	18 34 42.	+ 34 21		18.6	FAINT BLUE STAR
LB 04046	18 34 42.	+ 35 27		18.3	FAINT BLUE STAR
MCG+08-34-011	18 34 42.	+ 50 18	54	16.	GALAXY
ZC 1834.7+7053	18 34 42.	+ 70 53	2290		CLUSTER OF GALAXIES
HN 0483	18 34 43.	- 67 28			NEBULA
IC 4729	18 34 43.	- 67 28			NONSTELLAR OBJECT
LB 04049	18 34 48.	+ 34 32		18.0	FAINT BLUE STAR
LB 04048	18 34 48.	+ 34 53		14.8	FAINT BLUE STAR
MCG+09-30-022	18 34 48.	+ 51 22	30	16.	GALAXY
ZWG 279.018	18 34 48.	+ 51 25		15.7	GALAXY
B 102	18 34 52.	- 13 47	480		DARK OBJECT
PK016-04.2	18 34 52.2	- 17 08 25.	10		PLANETARY NEBULA
LB 04050	18 34 54.	+ 35 19		19.1	FAINT BLUE STAR
MCG+09-30-023	18 34 54.	+ 51 23	42	15.	GALAXY
7ZW 819	18 34 54.	+ 58 45			COMPACT GALAXY
7ZW 820	18 34 54.	+ 62 41			COMPACT GALAXY
PK028+01.1	18 34 59.6	- 03 08 33.	8		PLANETARY NEBULA
KHAV 529	18 35	- 15 16	3810		DARK NEBULA
KHAV 525	18 35	- 22 10	3750		DARK NEBULA
KHAV 524	18 35	- 25 52	6270		DARK NEBULA
LDN 0660	18 35 00.	+ 14 30	6120		DARK NEBULA
ZWG 173.017	18 35 00.	+ 30 34		15.1	GALAXY
UGC 11296	18 35 00.	+ 30 34	60	15.1	GALAXY S
MCG+05-44-007	18 35 00.	+ 30 34	60	15.1	GALAXY
LB 04051	18 35 00.	+ 35 19		19.0	FAINT BLUE STAR
7ZW 821	18 35 00.	+ 62 49			COMPACT GALAXY
LDN 0564	18 35 00.	- 01 15	1800		DARK NEBULA
LDN 0558	18 35 00.	- 01 40	2640		DARK NEBULA
LDN 0515	18 35 00.	- 05 00	11760		DARK NEBULA
LDN 0473	18 35 00.	- 08 00	1920		DARK NEBULA
LDN 0455	18 35 00.	- 09 10	540		DARK NEBULA
LDN 0445	18 35 00.	- 09 40	1140		DARK NEBULA
HN 0486	18 35 02.	- 63 00			NEBULA
IC 4735	18 35 02.	- 63 00			NONSTELLAR OBJECT
ZWG 114.020	18 35 06.	+ 19 56		15.6	GALAXY
UGC 11297	18 35 06.	+ 19 56	84	15.6	GALAXY Sb-c
LB 04054	18 35 06.	+ 34 40		18.4	FAINT BLUE STAR
LB 04053	18 35 06.	+ 35 09		18.4	FAINT BLUE STAR
LB 04052	18 35 06.	+ 35 21		18.7	FAINT BLUE STAR

OBJECT NAME	RIGHT ASCEN.	DECLINATION	DIAM.	MAGN.	TYPE OF OBJECT
IC 4737	18 35 11.	- 62 39			NONSTELLAR OBJECT
ZWG 114.021	18 35 12.	+ 18 48		15.4	GALAXY
ZWG 173.018	18 35 12.	+ 30 35		15.7	GALAXY
ZWG 173.019	18 35 12.	+ 32 58		15.6	GALAXY
LB 04059	18 35 12.	+ 34 35		18.9	FAINT BLUE STAR
LB 04058	18 35 12.	+ 34 59		19.0	FAINT BLUE STAR
LB 04057	18 35 12.	+ 35 02		18.5	FAINT BLUE STAR
LB 04056	18 35 12.	+ 35 03		19.0	FAINT BLUE STAR
LB 04055	18 35 12.	+ 35 34		18.8	FAINT BLUE STAR
ZWG 279.019	18 35 12.	+ 51 26		14.9	GALAXY
UGC 11298	18 35 12.	+ 51 26	84	14.9	GALAXY E
ISS 0605	18 35 12.	- 08 59	106		STELLAR RING
ASS 07	18 35 12.	- 16 26			OB ASSOCIATION
HN 0487	18 35 12.	- 62 39			NEBULA
LB 04062	18 35 18.	+ 34 42		18.7	FAINT BLUE STAR
LB 04061	18 35 18.	+ 34 43		18.0	FAINT BLUE STAR
LB 04060	18 35 18.	+ 35 09		19.0	FAINT BLUE STAR
MCG+09-30-024	18 35 18.	+ 51 03	42	16.	GALAXY
MCG+09-30-025	18 35 18.	+ 51 23	60	14.	GALAXY
ZWG 279.020	18 35 18.	+ 53 47		15.3	GALAXY
MCG+09-30-026	18 35 18.	+ 53 47	54	16.	GALAXY
ZWG 301.033	18 35 18.	+ 59 50		15.7	GALAXY
7ZW 822	18 35 18.	+ 59 51			COMPACT GALAXY
ISS 0606	18 35 18.	- 03 03	72		STELLAR RING
LDN 0401	18 35 18.	- 13 50	300		DARK NEBULA
RNGC 6690	18 35 22.	+ 70 29		12.5	GALAXY
RNGC 6689	18 35 22.	+ 70 29			NON-EXISTENT OBJECT
ZWG 114.022	18 35 24.	+ 19 58		15.5	GALAXY
MCG+04-44-006	18 35 24.	+ 26 20 30.	48	13.8	GALAXY
ZWG 143.006	18 35 24.	+ 26 22		13.8	GALAXY
RNGC 6671	18 35 24.	+ 26 22		14.0	GALAXY
UGC 11299	18 35 24.	+ 26 22	114	13.8	GALAXY S
LB 04065	18 35 24.	+ 35 09		18.6	FAINT BLUE STAR
LB 04064	18 35 24.	+ 35 11		18.9	FAINT BLUE STAR
LB 04063	18 35 24.	+ 35 31		19.1	FAINT BLUE STAR
MCG+07-38-012	18 35 24.	+ 42 24	30	15.5	GALAXY
ZC 1835.4+5955	18 35 24.	+ 59 55	4770		CLUSTER OF GALAXIES
MCG+12-17-026	18 35 24.	+ 70 28	234	13.	GALAXY
ZWG 340.050	18 35 24.	+ 70 29		12.6	GALAXY
UGC 11300	18 35 24.	+ 70 29	276	12.6	GALAXY Sc
KARA.73B 0855	18 35 24.	+ 70 29	270	12.6	ISOLATED GALAXY S
ISS 0513	18 35 24.	- 02 57	76		STELLAR RING
IC 1290	18 35 25.	- 24 09 26.			NONSTELLAR OBJECT
ZWG 114.023	18 35 30.	+ 20 09		15.6	GALAXY
ZWG 143.007	18 35 30.	+ 22 31		15.6	GALAXY
ZWG 228.017	18 35 30.	+ 42 25		15.2	GALAXY
ZWG 255.008	18 35 30.	+ 48 59		15.7	GALAXY
ISS 0607	18 35 30.	- 03 02	78		STELLAR RING
MCG+08-34-012	18 35 36.	+ 49 00	36	16.	GALAXY
ISS 0608	13 35 36.	- 08 30	289		STELLAR RING
EM4	18 35 36.	- 29 57		12.99	EMISSION OBJECT
MCG+10-26-045	18 35 39.	+ 58 10	30	16.	GALAXY
ZWG 114.024	18 35 42.	+ 17 29		15.5	GALAXY
UGC 11301	18 35 42.	+ 17 29	144	15.5	GALAXY S
MCG+03-47-009	18 35 42.	+ 17 29	15	17.	GALAXY
UGC 11302	18 35 42.	+ 22 03	60	16.0	GALAXY SBc-IRR
MCG+06-41-007	18 35 42.	+ 33 13 30.	54	15.	GALAXY
ZWG 201.016	18 35 42.	+ 33 14		14.8	GALAXY
UGC 11303	18 35 42.	+ 33 14	78	14.8	GALAXY Sb
MCG+09-30-027	18 35 42.	+ 51 22	24	17.	GALAXY
MCG+09-30-028	18 35 42.	+ 53 42	36	16.	GALAXY
ISS 0609	18 35 42.	- 03 02	79		STELLAR RING
HN 0488	18 35 44.	- 61 57			NEBULA
IC 4738	18 35 44.	- 61 57			NONSTELLAR OBJECT
RNGC 6675	18 35 47.	+ 40 02		13.5	GALAXY
LB 04068	18 35 48.	+ 34 22		16.8	FAINT BLUE STAR
LB 04067	18 35 48.	+ 35 06		17.7	FAINT BLUE STAR
LB 04066	18 35 48.	+ 35 16		20.5	FAINT BLUE STAR
ZWG 201.017	18 35 48.	+ 36 48		14.6	GALAXY
UGC 11304	18 35 48.	+ 36 48	96	14.6	GALAXY Sa
MCG+06-41-008	18 35 48.	+ 36 49	66	14.5	GALAXY
ZWG 228.018	18 35 48.	+ 39 20		15.5	GALAXY
MCG+07-38-013	18 35 48.	+ 40 00	90	13.	GALAXY
ZWG 228.019	18 35 48.	+ 40 02		13.3	GALAXY
UGC 11305	18 35 49.	+ 40 02	120	13.3	GALAXY Sb-c
UGC 11306	18 35 48.	+ 53 43	66	16.0	GALAXY Sa-b
RNGC 6669	18 35 51.	+ 22 02			GALAXY
ZWG 173.020	18 35 54.	+ 27 45		15.2	GALAXY
UGC 11307	18 35 54.	+ 27 45	90	15.2	GALAXY SBc
LB 04070	18 35 54.	+ 34 53		17.2	FAINT BLUE STAR
LB 04069	18 35 54.	+ 35 33		18.8	FAINT BLUE STAR
IC 4756	18 36	+ 05 24			OPEN CLUSTER
KHAV 532	18 36	- 06 09	3400		DARK NEBULA
KHAV 533	18 36	- 11 21	3470		DARK NEBULA
KHAV 531	18 36	- 17 09	6870		DARK NEBULA
KHAV 530	18 36	- 27 27	3750		DARK NEBULA
MCG+05-44-008	18 36 00.	+ 27 43	72	15.2	GALAXY
LB 04072	18 36 00.	+ 34 26		16.2	FAINT BLUE STAR
LB 04071	18 36 00.	+ 35 28		18.0	FAINT BLUE STAR
ZWG 201.018	18 36 00.	+ 37 18		15.0	GALAXY
MCG+09-30-030	18 36 00.	+ 54 13	30	17.	GALAXY
MCG+09-30-029	18 36 00.	+ 54 13	42	16.	GALAXY
LDN 0557	18 36 00.	- 01 50	1680		DARK NEBULA
LDN 0525	18 36 00.	- 04 00	5940		DARK NEBULA
LDN 0395	18 36 00.	- 14 10	2580		DARK NEBULA
LB 04076	18 36 06.	+ 34 43		18.7	FAINT BLUE STAR
LB 04075	18 36 06.	+ 34 52		18.6	FAINT BLUE STAR
LB 04074	18 36 06.	+ 35 02		18.9	FAINT BLUE STAR
LB 04073	18 36 06.	+ 35 06		18.1	FAINT BLUE STAR
ZWG 201.019	18 36 06.	+ 37 24		15.0	GALAXY
ZWG 341.002	18 36 06.	+ 74 00		15.3	GALAXY
ZWG 340.051	18 36 06.	+ 74 00		15.3	GALAXY
HN 0489	18 36 08.	- 61 57			NEBULA
IC 4739	18 36 08.	- 61 57			NONSTELLAR OBJECT
ZWG 114.025	18 36 12.	+ 17 09		15.2	GALAXY
MCG+03-47-010	18 36 12.	+ 17 09	48	15.2	GALAXY
LB 04077	18 36 12.	+ 35 10		16.2	FAINT BLUE STAR
LDN 0495	18 36 12.	- 06 42	180		DARK NEBULA
PK004-11.1	18 36 12.9	- 30 43 21.	8	13.12	PLANETARY NEBULA
LB 04083	18 36 18.	+ 34 38		18.9	FAINT BLUE STAR
LB 04082	18 36 18.	+ 35 07		18.2	FAINT BLUE STAR
LB 04081	18 36 18.	+ 35 17		19.1	FAINT BLUE STAR
LB 04080	18 36 18.	+ 35 21		18.6	FAINT BLUE STAR
LB 04079	18 36 18.	+ 35 27		18.6	FAINT BLUE STAR
LB 04078	18 36 18.	+ 35 29		18.9	FAINT BLUE STAR
ISS 0610	18 36 18.	- 05 25	146		STELLAR RING
SCHO 0680	18 36 18.	- 06 15	2650		ISOLATED DARK CLOUD
LB 04085	18 36 24.	+ 35 15		17.6	FAINT BLUE STAR

OBJECT NAME	RIGHT ASCEN.	DECLINATION	DIAM.	MAGN.	TYPE OF OBJECT
LB 04084	18 36 24.	+ 35 29		18.8	FAINT BLUE STAR
ZWG 201.020	18 36 24.	+ 37 10		15.5	GALAXY
ZC 1836.4+6430	18 36 24.	+ 64 30	1410		CLUSTER OF GALAXIES
EM2	18 36 24.	- 30 45		13.12	PLANETARY NEBULA
RNGC 6674	18 36 29.	+ 25 20		13.5	GALAXY
LDN 0586	18 36 30.	+ 00 00	720		DARK NEBULA
OCL 0094	18 36 30.	+ 05 24	5400	5.6	OPEN STAR CLUSTER
LDN 0630	18 36 30.	+ 06 00	2220		DARK NEBULA
MCG+04-44-007	18 36 30.	+ 25 18	240	13.7	GALAXY
ZWG 143.008	18 36 30.	+ 25 20		13.7	GALAXY
UGC 11308	18 36 30.	+ 25 20	270	13.7	GALAXY SBb
LB 04087	18 36 30.	+ 35 18		19.3	FAINT BLUE STAR
LB 04086	18 36 30.	+ 35 22		17.3	FAINT BLUE STAR
LDN 0497	18 36 30.	- 06 40	660		DARK NEBULA
RNGC 6687	18 36 33.	+ 59 35		15.0	GALAXY
LB 04092	18 36 36.	+ 35 02		19.0	FAINT BLUE STAR
LB 04091	18 36 36.	+ 35 03		18.7	FAINT BLUE STAR
LB 04090	18 36 36.	+ 35 18		17.1	FAINT BLUE STAR
LB 04089	18 36 36.	+ 35 31		19.8	FAINT BLUE STAR
LB 04088	18 36 36.	+ 35 35		20.0	FAINT BLUE STAR
ZWG 301.034	18 36 36.	+ 59 35		14.9	GALAXY
UGC 11309	18 36 36.	+ 59 35	108	14.9	GALAXY Sc
LB 04094	18 36 42.	+ 34 53		18.5	FAINT BLUE STAR
LB 04093	18 36 42.	+ 34 59		19.2	FAINT BLUE STAR
MCG+10-26-046	18 36 42.	+ 59 35	60	14.	GALAXY
7ZW 823	18 36 42.	+ 63 42			COMPACT GALAXY
ISS 0611	18 36 42.	- 05 52	233		STELLAR RING
HN 0491	18 36 43.	- 61 49			NEBULA
IC 4743	18 36 43.	- 61 49			NONSTELLAR OBJECT
PK027+00.1	18 36 43.1	- 04 22 36.	7		PLANETARY NEBULA
B 103	18 36 44.	- 06 43	240		DARK OBJECT
ZC 1836.8+3306	18 36 48.	+ 33 06	2490		CLUSTER OF GALAXIES
LB 04097	18 36 48.	+ 34 27		17.4	FAINT BLUE STAR
LB 04096	18 36 48.	+ 35 12		20.0	FAINT BLUE STAR
LB 04095	18 36 48.	+ 35 25		19.0	FAINT BLUE STAR
HN 0490	18 36 49.	- 64 00			NEBULA
IC 4741	18 36 49.	- 64 00			NONSTELLAR OBJECT
ZWG 143.009	18 36 54.	+ 23 15		15.4	GALAXY
LB 04098	18 36 54.	+ 34 42		17.9	FAINT BLUE STAR
SCHO 0881	18 36 54.	- 05 12 42.	340		ISOLATED DARK CLOUD
KHAV 535	18 37	- 20 39	3540		DARK NEBULA
KHAV 534	18 37	- 21 39	4140		DARK NEBULA
LDN 0587	18 37 00.	+ 00 00	1200		DARK NEBULA
LDN 0648	18 37 00.	+ 12 40	420		DARK NEBULA
UGC 11310	18 37 00.	+ 22 10	60	16.0	GALAXY S
LB 04103	18 37 00.	+ 34 38		20.1	FAINT BLUE STAR
LB 04102	18 37 00.	+ 34 45		16.6	FAINT BLUE STAR
LB 04101	18 37 00.	+ 34 46		18.5	FAINT BLUE STAR
LB 04100	18 37 00.	+ 35 19		18.5	FAINT BLUE STAR
LB 04099	18 37 00.	+ 35 22		18.5	FAINT BLUE STAR
ZWG 201.021	18 37 00.	+ 36 32		15.5	GALAXY
ZWG 201.022	18 37 00.	+ 37 00		15.4	GALAXY
UGC 11311	18 37 00.	+ 37 00	108	15.4	GALAXY S
MCG+06-41-009	18 37 00.	+ 37 01	48	15.5	GALAXY
7ZW 824	18 37 00.	+ 63 41			COMPACT GALAXY
ZC 1837.0+8319	18 37 00.	+ 83 19	540		CLUSTER OF GALAXIES
LDN 0563	18 37 00.	- 01 35	6420		DARK NEBULA
LDN 0484	18 37 00.	- 07 30	5520		DARK NEBULA
LDN 0334	18 37 00.	- 20 30	1080		DARK NEBULA
LDN 0253	18 37 00.	- 26 00	3360		DARK NEBULA
IC 4742	18 37 00.	- 63 55			NONSTELLAR OBJECT
HN 0492	18 37 01.	- 63 54			NEBULA
HN 0493	18 37 03.	- 63 16			NEBULA
IC 4744	18 37 03.	- 63 17			NONSTELLAR OBJECT
LB 04105	18 37 06.	+ 35 18		15.0	FAINT BLUE STAR
LB 04104	18 37 06.	+ 35 26		18.8	FAINT BLUE STAR
ZC 1837.1+3633	18 37 06.	+ 36 33	8400		CLUSTER OF GALAXIES
ZWG 201.023	18 37 06.	+ 37 54		15.1	GALAXY
UGC 11312	18 37 06.	+ 37 54	72	15.1	GALAXY S0
OCL 0069	18 37 06.	- 08 32	840	11.6	OPEN STAR CLUSTER
LB 04111	18 37 12.	+ 34 49		20.3	FAINT BLUE STAR
LB 04110	18 37 12.	+ 34 54		19.1	FAINT BLUE STAR
LB 04109	18 37 12.	+ 35 01		18.0	FAINT BLUE STAR
LB 04108	18 37 12.	+ 35 07		19.7	FAINT BLUE STAR
LB 04107	18 37 12.	+ 35 25		19.2	FAINT BLUE STAR
LB 04106	18 37 12.	+ 35 28		17.7	FAINT BLUE STAR
MCG+06-41-010	18 37 12.	+ 37 55	42	15.	GALAXY
ISS 0514	18 37 12.	- 02 27	113		STELLAR RING
ISS 0612	18 37 12.	- 06 12	54		STELLAR RING
MCG-07-38-004	18 37 12.	- 41 39	72	14.	GALAXY
ISS 0482	18 37 18.	+ 08 03	92		STELLAR RING
LB 04114	18 37 18.	+ 34 39		18.7	FAINT BLUE STAR
LB 04113	18 37 18.	+ 35 15		17.9	FAINT BLUE STAR
LB 04112	18 37 18.	+ 35 35		17.8	FAINT BLUE STAR
LDN 0498	18 37 13.	- 06 42	240		DARK NEBULA
LB 04119	18 37 24.	+ 34 45		19.8	FAINT BLUE STAR
LB 04118	18 37 24.	+ 34 51		19.0	FAINT BLUE STAR
LB 04117	18 37 24.	+ 34 56		18.5	FAINT BLUE STAR
LB 04116	18 37 24.	+ 35 18		19.2	FAINT BLUE STAR
LB 04115	18 37 24.	+ 35 25		19.4	FAINT BLUE STAR
ZC 1837.4+6333	18 37 24.	+ 63 33	3560		CLUSTER OF GALAXIES
LDN 0650	18 37 30.	+ 13 00	900		DARK NEBULA
LDN 0665	18 37 30.	+ 14 50	1080		DARK NEBULA
LB 04120	18 37 30.	+ 34 56		18.0	FAINT BLUE STAR
ISS 0515	18 37 30.	- 02 34	78		STELLAR RING
ISS 0613	18 37 30.	- 06 06	60		STELLAR RING
LDN 0329	18 37 30.	- 20 55	900		DARK NEBULA
PK022-02.1	18 37 34.	- 10 42 37.	9		PLANETARY NEBULA
ZWG 143.010	18 37 36.	+ 22 16		15.4	GALAXY
LB 04122	18 37 36.	+ 34 50		19.1	FAINT BLUE STAR
LB 04121	18 37 36.	+ 35 05		18.4	FAINT BLUE STAR
7ZW 825	18 37 36.	+ 60 15			COMPACT GALAXY
ISS 0614	18 37 36.	- 06 20	154		STELLAR RING
ISS 0615	18 37 36.	- 08 39	134		STELLAR RING
MCG-07-38-005	18 37 36.	- 41 52	60	14.	GALAXY
HN 0495	18 37 37.	- 64 59			NEBULA
IC 4745	18 37 38.	- 65 00			NONSTELLAR OBJECT
RNGC 6680	18 37 39.	+ 22 16		15.5	GALAXY
PK023-01.1	18 37 40.3	- 08 46 36.			PLANETARY NEBULA
LB 04123	18 37 42.	+ 34 42		18.9	FAINT BLUE STAR
ZWG 201.024	18 37 42.	+ 38 52		15.5	GALAXY
UGC 11313	18 37 42.	+ 38 52	138	15.5	GALAXY Sb
MCG+06-41-011	13 37 42.	+ 38 55 30.	102	14.5	GALAXY
ZC 1837.7+7724	18 37 42.	+ 77 24	1210		CLUSTER OF GALAXIES
HN 0494	18 37 45.	- 68 24			NEBULA
IC 4740	18 37 45.	- 68 25			NONSTELLAR OBJECT
LB 04126	18 37 48.	+ 34 49		19.3	FAINT BLUE STAR
LB 04125	18 37 48.	+ 34 51		19.2	FAINT BLUE STAR

OBJECT NAME	RIGHT ASCEN.	DECLINATION	DIAM.	MAGN.	TYPE OF OBJECT
LB 04124	18 37 48.	+ 34 56		16.7	FAINT BLUE STAR
OCL 0080	18 37 48.	- 04 09	720		OPEN STAR CLUSTER
ISS 0616	18 37 48.	- 05 54	94		STELLAR RING
ISS 0617	18 37 48.	- 06 06	121		STELLAR RING
PK044+08.1	18 37 49.8	+ 14 08 59.			PLANETARY NEBULA
HN 0496	18 37 50.	- 64 07			NEBULA
IC 4748	18 37 50.	- 64 08			NONSTELLAR OBJECT
YM 09	18 37 51.	- 20 03	168		SYMMETRIC GALACTIC NEBULA
ZWG 143.011	18 37 54.	+ 23 50		15.6	GALAXY
MCG+05-44-009	18 37 54.	+ 32 46	54	14.9	GALAXY
ZWG 173.021	18 37 54.	+ 32 49		15.7	GALAXY
LB 04129	18 37 54.	+ 34 30		19.3	FAINT BLUE STAR
LB 04128	18 37 54.	+ 35 14		18.8	FAINT BLUE STAR
LB 04127	18 37 54.	+ 35 16		18.2	FAINT BLUE STAR
ZWG 279.021	18 37 54.	+ 55 22		15.6	GALAXY
MRSL 031+02/1	18 37 54.	- 00 48	2100		HII REGION
SCHO 0882	18 37 54.	- 07 25	280		ISOLATED DARK CLOUD
MRSL 017-05/1	18 37 54.	- 16 36	2100		HII REGION
LBN 0065	18 38	- 16 50	1500		BRIGHT NEBULA
MCG+04-44-008	18 38 00.	+ 21 24 30.	60	15.7	GALAXY
ZWG 143.012	18 38 00.	+ 21 27		15.7	GALAXY
UGC 11314	18 38 00.	+ 21 27	60	15.7	GALAXY IRR
ZWG 173.022	18 38 00.	+ 32 45		14.9	GALAXY
LB 04132	18 38 00.	+ 34 23		17.9	FAINT BLUE STAR
LB 04131	18 38 00.	+ 34 53		17.1	FAINT BLUE STAR
LB 04130	18 38 00.	+ 35 24		20.6	FAINT BLUE STAR
SHAH 169	18 38 00.	+ 79 05	78		GROUP OF COMPACT GALAXIES
ISS 0618	18 38 00.	- 04 20	771		STELLAR RING
OCL 0079	18 38 00.	- 04 27	420		OPEN STAR CLUSTER
LDN 0499	18 38 00.	- 06 30	900		DARK NEBULA
SC 1833-5338.1	18 38 00.	- 53 35 29.			NEBULA
LB 04135	18 38 06.	+ 34 32		19.0	FAINT BLUE STAR
LB 04134	18 38 06.	+ 34 52		19.7	FAINT BLUE STAR
LB 04133	18 38 06.	+ 35 07		19.0	FAINT BLUE STAR
7ZW 826	18 38 06.	+ 60 30			COMPACT GALAXY
ISS 0703	18 38 06.	- 09 40	235		STELLAR RING
SCHO 0883	18 38 08.	- 06 42 30.	570		ISOLATED DARK CLOUD
HN 0497	18 38 09.	- 63 15			NEBULA
IC 4749	18 38 09.	- 63 16			NONSTELLAR OBJECT
RNGC 6685	18 38 11.	+ 40 00		15.0	GALAXY
SCHO 0884	18 38 12.	+ 08 21 36.	860		ISOLATED DARK CLOUD
ZWG 143.013	18 38 12.	+ 24 09		14.2	GALAXY
UGC 11315	18 38 12.	+ 24 09	138	14.2	GALAXY Sb
LB 04138	18 38 12.	+ 35 28		20.6	FAINT BLUE STAR
LB 04137	18 38 12.	+ 35 30		17.9	FAINT BLUE STAR
LB 04136	18 38 12.	+ 35 35		18.8	FAINT BLUE STAR
ZWG 228.020	18 38 12.	+ 40 00		15.0	GALAXY
YM 10	18 38 13.	- 19 53	156		SYMMETRIC GALACTIC NEBULA
SCHO 0885	18 38 14.	- 05 59 18.	690		ISOLATED DARK CLOUD
MCG+04-44-009	18 38 15.	+ 24 07	42	14.2	GALAXY
MCG+04-44-010	18 38 18.	+ 23 43 30.	114	15.6	GALAXY
UGC 11316	18 38 18.	+ 24 37	60	16.5	GALAXY S
LB 04142	18 38 18.	+ 34 36		16.8	FAINT BLUE STAR
LB 04141	18 38 18.	+ 35 12		18.7	FAINT BLUE STAR
LB 04140	18 38 18.	+ 35 33		17.6	FAINT BLUE STAR
LB 04139	18 38 18.	+ 35 34		15.7	FAINT BLUE STAR
ZWG 201.025	18 38 18.	+ 37 46		15.6	GALAXY
ZWG 228.021	18 38 18.	+ 39 57		14.7	GALAXY
UGC 11317	18 38 18.	+ 39 57	66	14.7	GALAXY E-SO
MCG+07-38-014	18 38 18.	+ 39 58	24	17.	GALAXY
ZWG 279.022	18 38 18.	+ 55 35		14.1	GALAXY
UGC 11318	18 38 18.	+ 55 35	96	14.1	GALAXY SBb
MCG+09-30-031	18 38 18.	+ 55 35	90	13.	GALAXY
SC 1834-5626.6	18 38 18.	- 56 23 58.			NEBULA
IC 4772	18 38 19.	+ 39 58 49.			NONSTELLAR OBJECT
RNGC 6691	18 38 19.	+ 55 35		14.0	GALAXY
HN 0498	18 38 19.	- 63 00			NEBULA
IC 4750	18 38 19.	- 63 01			NONSTELLAR OBJECT
MCG+06-41-012	18 38 21.	+ 37 47 30.	30	15.	GALAXY
PK017-04.1	18 38 22.9	- 15 36 40.	17		PLANETARY NEBULA
PK023-01.2	18 38 23.2	- 08 58 51.			PLANETARY NEBULA
ZWG 143.014	18 38 24.	+ 23 45		15.6	GALAXY
LB 04144	18 38 24.	+ 35 05		18.8	FAINT BLUE STAR
LB 04143	18 38 24.	+ 35 14		16.3	FAINT BLUE STAR
MCG+07-38-015	18 38 24.	+ 39 56	18	16.	GALAXY
UGC 11319	18 38 24.	+ 40 05	102	16.0	GALAXY S
MCG+07-38-016	18 38 24.	+ 40 06	72	16.	GALAXY
ISS 0704	18 38 24.	- 09 32	95		STELLAR RING
SCHO 0886	18 38 24.	- 09 36	350		ISOLATED DARK CLOUD
ISS 0705	18 38 24.	- 14 46	118		STELLAR RING
RNGC 6653	18 38 24.	- 73 18			UNVERIFIED SOUTHERN OBJECT
RNGC 6686	18 38 29.	+ 40 06		15.0	GALAXY
LDN 0633	18 38 30.	+ 07 00	1080		DARK NEBULA
LDN 0666	18 38 30.	+ 15 00	1440		DARK NEBULA
ZWG 173.023	18 38 30.	+ 28 34		15.5	GALAXY
MCG+07-38-017	18 38 30.	+ 40 05	48	16.	GALAXY
ZWG 228.022	18 38 30.	+ 40 06		14.9	GALAXY
MIL 62	18 38 30.	- 05 01	1920		SUPERNOVA REMNANT
GCL 100	18 38 30.	- 19 52	96		GLOBULAR STAR CLUSTER
HN 0501	18 38 33.	- 62 09			NEBULA
IC 4751	18 38 33.	- 62 10			NONSTELLAR OBJECT
YM 12	18 38 35.	- 17 55	108		SYMMETRIC GALACTIC NEBULA
LB 04147	18 38 36.	+ 34 49		20.1	FAINT BLUE STAR
LB 04146	18 38 36.	+ 34 51		18.9	FAINT BLUE STAR
LB 04145	18 38 36.	+ 34 54		19.0	FAINT BLUE STAR
ZWG 356.010	18 38 36.	+ 77 03		15.7	GALAXY
SCHO 0887	18 38 40.	- 07 21	1060		ISOLATED DARK CLOUD
SCHO 0888	18 38 41.	- 06 19 42.	350		ISOLATED DARK CLOUD
ZWG 143.015	18 38 42.	+ 23 38		15.3	GALAXY
UGC 11320	18 38 42.	+ 23 38	114	15.3	GALAXY Sb-c
LB 04152	18 38 42.	+ 34 27		20.2	FAINT BLUE STAR
LB 04151	18 38 42.	+ 34 55		19.2	FAINT BLUE STAR
LB 04150	18 38 42.	+ 35 17		18.5	FAINT BLUE STAR
LB 04149	18 38 42.	+ 35 24		18.3	FAINT BLUE STAR
LB 04148	18 38 42.	+ 35 25		18.9	FAINT BLUE STAR
MCG+06-41-013	18 38 42.	+ 36 05 30.	60	15.	GALAXY
UGC 11321	18 38 42.	+ 44 13	84	17.	GALAXY DWARF
LB 04153	18 38 48.	+ 35 34		20.3	FAINT BLUE STAR
ZWG 201.026	18 38 48.	+ 36 04		15.1	GALAXY
UGC 11322	18 38 48.	+ 36 04	66	15.1	GALAXY Sb-c
ISS 0706	18 38 48.	- 12 52	152		STELLAR RING
MCG+04-44-011	18 38 51.	+ 23 00 30.	90	15.5	GALAXY
HN 0503	18 38 51.	- 62 09			NEBULA
IC 4753	18 38 51.	- 62 09			NONSTELLAR OBJECT
ZWG 143.016	18 38 54.	+ 23 02		15.5	GALAXY
UGC 11323	18 38 54.	+ 23 02	96	15.5	GALAXY SBb-c
LB 04154	18 38 54.	+ 34 41		16.2	FAINT BLUE STAR
MCG+06-41-014	18 38 54.	+ 36 07	138	14.	GALAXY
ZWG 201.027	18 38 54.	+ 36 14		13.9	GALAXY
UGC 11324	18 38 54.	+ 36 14	108	13.9	GALAXY SO
MCG+06-41-015	18 38 54.	+ 36 15	90	13.	GALAXY
LDN 0555	18 38 54.	- 02 20	360		DARK NEBULA
ACK 013-07.2	18 38 54.	- 20 35			PLANETARY NEBULA
RNGC 6688	18 38 55.	+ 36 14		14.0	GALAXY
HN 0502	18 38 56.	- 64 08			NEBULA
IC 4752	18 38 56.	- 64 08			NONSTELLAR OBJECT
RNGC 6682	18 38 58.	- 04 48			NON-EXISTENT OBJECT
SCHO 0889	18 38 58.	- 20 00	400		ISOLATED DARK CLOUD
KHAV 536	18 39	- 24 03	2110		DARK NEBULA
LDN 0634	18 39 00.	+ 07 30	960		DARK NEBULA
LDN 0637	18 39 00.	+ 08 10	660		DARK NEBULA
LDN 0659	18 39 00.	+ 14 00	4860		DARK NEBULA
LB 04156	18 39 00.	+ 35 00		18.9	FAINT BLUE STAR
LB 04155	18 39 00.	+ 35 16		19.0	FAINT BLUE STAR
ZWG 201.028	18 39 00.	+ 36 06		14.6	GALAXY
UGC 11325	18 39 00.	+ 36 06	138	14.6	GALAXY Sb
ZC 1839.0+8436	18 39 00.	+ 84 36	2550		CLUSTER OF GALAXIES
LDN 0560	18 39 00.	- 02 00	840		DARK NEBULA
ISS 0619	18 39 00.	- 07 10	259		STELLAR RING
LDN 0409	18 39 00.	- 13 30	4860		DARK NEBULA
LDN 0344	18 39 00.	- 20 00	480		DARK NEBULA
PK056+14.1	18 39 06.	+ 26 52	30		PLANETARY NEBULA
LB 04159	18 39 06.	+ 34 24		18.1	FAINT BLUE STAR
LB 04158	18 39 06.	+ 34 33		20.5	FAINT BLUE STAR
LB 04157	18 39 06.	+ 34 48		18.4	FAINT BLUE STAR
ISS 0620	18 39 06.	- 07 12	136		STELLAR RING
IC 4768	18 39 10.	- 05 34			OPEN CLUSTER
UGC 11326	18 39 12.	+ 08 05	120	15.	GALAXY
MCG+01-47-001	18 39 12.	+ 08 05	108	15.	GALAXY
LB 04164	18 39 12.	+ 35 01		15.8	FAINT BLUE STAR
LB 04163	18 39 12.	+ 35 02		18.7	FAINT BLUE STAR
LB 04162	18 39 12.	+ 35 10		19.0	FAINT BLUE STAR
LB 04161	18 39 12.	+ 35 12		18.9	FAINT BLUE STAR
LB 04160	18 39 12.	+ 35 26		20.0	FAINT BLUE STAR
ZWG 201.029	18 39 12.	+ 36 19		15.1	GALAXY
UGC 11327	18 39 12.	+ 36 19	60	15.1	GALAXY S
UGC 11328	18 39 12.	+ 37 57	78	16.0	GALAXY Sc
B 316	18 39 14.	- 02 11	360		DARK OBJECT
HN 0504	18 39 14.	- 62 02			NEBULA
IC 4754	18 39 14.	- 62 02			NONSTELLAR OBJECT
RNGC 6696	18 39 15.	+ 59 17		16.0	GALAXY
MCG+10-26-047	18 39 15.	+ 59 17	42	16.	GALAXY
LB 04170	18 39 18.	+ 34 39		16.9	FAINT BLUE STAR
LB 04169	18 39 18.	+ 34 41		18.1	FAINT BLUE STAR
LB 04168	18 39 18.	+ 34 42		17.4	FAINT BLUE STAR
LB 04167	18 39 18.	+ 35 26		20.3	FAINT BLUE STAR
LB 04166	18 39 18.	+ 35 26		18.1	FAINT BLUE STAR
LB 04165	18 39 18.	+ 35 27		20.1	FAINT BLUE STAR
MCG+06-41-016	18 39 18.	+ 36 21	60	15.	GALAXY
MCG+08-34-013	18 39 18.	+ 45 12	90	15.	GALAXY
MCG+09-30-032	18 39 18.	+ 52 55	36	16.	GALAXY
OCL 0078	18 39 18.	- 04 38	240		OPEN STAR CLUSTER
B 315	18 39 21.	- 20 05	300		DARK OBJECT
LB 04177	18 39 24.	+ 34 27		18.5	FAINT BLUE STAR
LB 04176	18 39 24.	+ 34 54		20.6	FAINT BLUE STAR
LB 04175	18 39 24.	+ 35 02		17.9	FAINT BLUE STAR
LB 04174	18 39 24.	+ 35 10		18.2	FAINT BLUE STAR
LB 04173	18 39 24.	+ 35 24		19.8	FAINT BLUE STAR
LB 04172	18 39 24.	+ 35 25		18.5	FAINT BLUE STAR
LB 04171	18 39 24.	+ 35 32		17.7	FAINT BLUE STAR
MCG+09-30-033	18 39 24.	+ 52 51	36	16.	GALAXY
PK025-00.1	18 39 26.5	- 06 43 53.	5		PLANETARY NEBULA
SCHO 0890	18 39 27.	- 07 31 00.	570		ISOLATED DARK CLOUD
RNGC 6683	18 39 28.	- 06 20		9.5	OPEN CLUSTER
LB 04180	18 39 30.	+ 34 29		20.6	FAINT BLUE STAR
LB 04179	18 39 30.	+ 34 36		19.3	FAINT BLUE STAR
ZWG 201.030	18 39 30.	+ 35 00		15.0	GALAXY
MCG+06-41-017	18 39 30.	+ 35 00	36	15.	GALAXY
LB 04178	18 39 30.	+ 35 19		19.0	FAINT BLUE STAR
ZWG 255.009	18 39 30.	+ 45 13		15.5	GALAXY
UGC 11329	18 39 30.	+ 45 13	114	15.5	GALAXY Sb-c
MCG+12-17-027	18 39 30.	+ 73 37	51	17.	GALAXY
OCL 0074	18 39 30.	- 06 20	660	9.57	OPEN STAR CLUSTER
PK012-07.1	18 39 35.	- 21 20 32.	10	12.	PLANETARY NEBULA
ZWG 173.024	18 39 36.	+ 22 18		15.7	GALAXY
LB 04186	18 39 36.	+ 34 48		19.1	FAINT BLUE STAR
LB 04185	18 39 36.	+ 34 51		16.6	FAINT BLUE STAR
LB 04184	18 39 36.	+ 34 52		18.9	FAINT BLUE STAR
LB 04183	18 39 36.	+ 35 07		18.5	FAINT BLUE STAR
LB 04182	18 39 36.	+ 35 23		19.9	FAINT BLUE STAR
LB 04181	18 39 36.	+ 35 29		20.0	FAINT BLUE STAR
ZC 1839.6+6530	18 39 36.	+ 65 30	870		CLUSTER OF GALAXIES
ISS 0621	18 39 36.	- 06 26	184		STELLAR RING
LB 04188	18 39 42.	+ 34 29		16.2	FAINT BLUE STAR
LB 04187	18 39 42.	+ 35 28		18.7	FAINT BLUE STAR
ZWG 201.031	18 39 42.	+ 36 10		15.6	GALAXY
MCG-07-38-006	18 39 42.	- 41 29	36	15.	GALAXY
HN 0506	18 39 46.	- 57 13			NEBULA
IC 4757	18 39 46.	- 57 13			NONSTELLAR OBJECT
ZWG 201.032	18 39 48.	+ 33 44		15.4	GALAXY
LB 04191	18 39 48.	+ 34 25		18.6	FAINT BLUE STAR
LB 04190	18 39 48.	+ 35 00		19.3	FAINT BLUE STAR
LB 04189	18 39 48.	+ 35 18		18.9	FAINT BLUE STAR
ISS 0622	18 39 48.	- 06 45	100		STELLAR RING
ISS 0707	18 39 48.	- 10 08	97		STELLAR RING
HN 0499	18 39 48.	- 72 43			NEBULA
IC 4746	18 39 48.	- 72 44			NONSTELLAR OBJECT
RNGC 6693	18 39 50.	+ 36 52			NON-EXISTENT OBJECT
HN 0500	18 39 53.	- 72 40			NEBULA
IC 4747	18 39 53.	- 72 41			NONSTELLAR OBJECT
ZWG 114.026	18 39 54.	+ 20 05		15.5	GALAXY
ZWG 201.033	18 39 54.	+ 34 47		14.3	GALAXY
RNGC 6692	18 39 54.	+ 34 47		14.5	GALAXY
UGC 11330	18 39 54.	+ 34 47	66	14.3	GALAXY
MCG+06-41-018	18 39 54.	+ 34 48	30	14.5	GALAXY
ISS 0623	18 39 54.	- 07 35	140		STELLAR RING
HN 1155	18 39 55.	- 52 54			NEBULA
IC 4761	18 39 55.	- 52 54			NONSTELLAR OBJECT
RNGC 6681	18 39 58.	- 32 21		9.0	GLOBULAR CLUSTER
SCHO 0891	18 39 59.	- 09 39	380		ISOLATED DARK CLOUD
KHAV 539	18 40	- 10 33	2470		DARK NEBULA
KHAV 538	18 40	- 18 09	3250		DARK NEBULA
KHAV 537	18 40	- 24 39	2370		DARK NEBULA
LDN 0653	18 40 00.	+ 13 00	3240		DARK NEBULA
ZWG 341.003	18 40 00.	+ 73 34		15.4	GALAXY
ZWG 340.052	18 40 00.	+ 73 34		15.4	GALAXY

OBJECT NAME	RIGHT ASCEN.	DECLINATION	DIAM.	MAGN.	TYPE OF OBJECT
UGC 11331	18 40 00.	+ 73 34	90	15.4	GALAXY DWRF SP
LDN 0459	18 40 00.	- 09 40	7020		DARK NEBULA
LDN 0434	18 40 00.	- 11 30	4260		DARK NEBULA
LDN 0433	18 40 00.	- 11 30	5280		DARK NEBULA
LDN 0410	18 40 00.	- 13 30	9060		DARK NEBULA
LDN 0366	18 40 00.	- 18 30	1200		DARK NEBULA
LDN 0365	18 40 00.	- 18 30	1200		DARK NEBULA
LDN 0346	18 40 00.	- 20 00	480		DARK NEBULA
GCL 101	18 40 00.	- 32 20	306	8.95	GLOBULAR STAR CLUSTER
MCG+12-17-028	18 40 06.	+ 73 34	84	16.	GALAXY
PK022-03.1	18 40 10.	- 11 09 53.	7		PLANETARY NEBULA
HN 0505	18 40 11.	- 63 44			NEBULA
ZWG 143.017	18 40 12.	+ 24 50		15.5	GALAXY
PK029+00.1	18 40 12.	- 03 16	43	18.7	PLANETARY NEBULA
IC 4755	18 40 12.	- 63 44			NONSTELLAR OBJECT
OCL 0083	18 40 18.	- 04 11	1380	10.3	OPEN STAR CLUSTER
ZWG 341.004	18 40 24.	+ 73 32		13.6	GALAXY
ZWG 340.053	18 40 24.	+ 73 32		13.6	GALAXY
UGC 11332	18 40 24.	+ 73 32	150	13.6	GALAXY
RNGC 6673	18 40 29.	- 62 21			GALAXY
PK031+01.1	18 40 29.3	- 00 19 37.			PLANETARY NEBULA
ZWG 173.025	18 40 30.	+ 32 19		15.4	GALAXY
UGC 11333	18 40 30.	+ 32 19	84	15.4	GALAXY S
ZC 1840.5+6126	18 40 30.	+ 61 26	4100		CLUSTER OF GALAXIES
7ZW 827	18 40 30.	+ 71 00			COMPACT GALAXY
UGC 11334	18 40 30.	+ 73 48	126	16.0	GALAXY PECULR?
RNGC 6654A	18 40 34.	+ 73 32			GALAXY
PK023-02.1	18 40 35.9	- 09 07 52.	5		PLANETARY NEBULA
LB 04204	18 40 36.	+ 37 13		15.2	FAINT BLUE STAR
UGC 11335	18 40 42.	+ 50 11	66	16.0	GALAXY Sb
7ZW 828	18 40 42.	+ 61 30			COMPACT GALAXY
MCG+12-17-029	18 40 42.	+ 73 32 30.	168	14.	GALAXY
IC 1293	18 40 46.	+ 56 15 49.			NONSTELLAR OBJECT
ZWG 201.034	18 40 48.	+ 34 25		15.5	GALAXY
UGC 11336	18 40 48.	+ 34 25	66	15.5	GALAXY Sc
MCG+06-41-019	18 40 48.	+ 34 25	48	15.	GALAXY
MCG+08-34-014	18 40 48.	+ 46 02	60	16.	GALAXY
MCG+08-34-015	18 40 48.	+ 50 12	72	16.	GALAXY
PK019-04.1	18 40 48.	- 13 47 51.			PLANETARY NEBULA
UGC 11337	18 40 54.	+ 18 40	96	15.	GALAXY SBa
ZWG 201.035	18 40 54.	+ 35 34		14.5	GALAXY
UGC 11338	18 40 54.	+ 35 34	30	14.5	GALAXY PECULR
7ZW 829	18 40 54.	+ 68 18			COMPACT GALAXY
ISS 0624	18 40 54.	- 08 39	133		STELLAR RING
HN 0507	18 40 56.	- 63 08			NEBULA
IC 4759	18 40 56.	- 63 08			NONSTELLAR OBJECT
KEEN 6654A	18 41	+ 73 32	156	13.2	Sc GALAXY
LBN 0099	18 41	- 00 20	720		BRIGHT NEBULA
KHAV 541	18 41	- 12 09	6190		DARK NEBULA
KHAV 540	18 41	- 19 39	1410		DARK NEBULA
MCG+03-48-001	18 41 00.	+ 18 41	84	15.	GALAXY
ZWG 255.010	18 41 00.	+ 46 04		15.7	GALAXY
KARA.73B 0856	18 41 00.	+ 46 04	48	15.7	ISOLATED GALAXY S
ZWG 370.013	18 41 00.	+ 86 52		15.4	GALAXY
ZWG 368.001	18 41 00.	+ 86 52		15.4	GALAXY
UGC 11339	18 41 00.	+ 86 52	72	15.4	GALAXY E-SO
KARA.73B 0857	18 41 00.	+ 86 52	42	15.4	ISOLATED GALAXY S
PK004-11.2	18 41 01.	- 30 22 18.			PLANETARY NEBULA
HN 0508	18 41 01.	- 63 00			NEBULA
IC 4760	18 41 01.	- 63 00			NONSTELLAR OBJECT
RNGC 6695	18 41 05.	+ 40 20		14.5	GALAXY
ZWG 143.018	18 41 06.	+ 21 55		15.7	GALAXY
MCG+07-38-018	18 41 06.	+ 40 19	66	14.	GALAXY
ZWG 228.023	18 41 06.	+ 40 20		14.3	GALAXY
UGC 11340	18 41 06.	+ 40 20	66	14.3	GALAXY SBb
KARA.73B 0858	18 41 06.	+ 40 20	66	14.3	ISOLATED GALAXY S
SCHO 0892	18 41 08.	- 07 03 30.	410		ISOLATED DARK CLOUD
SCHO 0893	18 41 10.	- 09 17	330		ISOLATED DARK CLOUD
ZWG 228.024	18 41 12.	+ 39 30		15.5	GALAXY
UGC 11341	18 41 12.	+ 39 30	90	16.5	GALAXY Sc
MCG+07-38-019	18 41 12.	+ 39 30	66	16.	GALAXY
PK014-07.1	18 41 12.	- 19 57	10		PLANETARY NEBULA
ISS 0625	18 41 18.	- 07 17	108		STELLAR RING
MCG+06-41-020	18 41 24.	+ 37 22	39	15.	GALAXY
ZWG 356.011	18 41 24.	+ 77 48		15.7	GALAXY
HN 0509	18 41 24.	- 65 48			NEBULA
IC 4758	18 41 24.	- 65 48			NONSTELLAR OBJECT
HN 0074	18 41 32.	- 27 52			NEBULA
ZWG 228.025	18 41 36.	+ 43 54		15.3	GALAXY
ZWG 280.001	18 41 36.	+ 56 32		15.7	GALAXY
ZWG 279.023	18 41 36.	+ 56 32		15.7	GALAXY
MCG+12-17-030	18 41 36.	+ 70 27 30.	24	16.	GALAXY
ZWG 341.005	18 41 36.	+ 70 30		15.5	GALAXY
ZWG 340.054	18 41 36.	+ 70 30		15.5	GALAXY
UGC 11342	18 41 36.	+ 70 30	108	15.5	GALAXY
KARA.73B 0859	18 41 36.	+ 70 30	24	15.5	ISOLATED GALAXY S
PK009-09.1	18 41 36.	- 25 23	7		PLANETARY NEBULA
IC 1292	18 41 37.	- 27 52			NONSTELLAR OBJECT
ZWG 255.011	18 41 42.	+ 50 37		15.6	GALAXY
7ZW 830	18 41 42.	+ 61 28			COMPACT GALAXY
LDN 0316	18 41 42.	- 22 40	1320		DARK NEBULA
PK038+04.1	18 41 46.2	+ 06 43 59.			PLANETARY NEBULA
ZWG 279.024	18 41 48.	+ 52 59		15.5	GALAXY
UGC 11343	18 41 48.	+ 52 59	72	15.5	GALAXY S
MRSL 031+01/1	18 41 48.	- 00 21	1200		HII REGION
ISS 0626	18 41 48.	- 07 23	107		STELLAR RING
MCG+09-30-034	18 41 54.	+ 52 58	48	15.	GALAXY
SC 1837-5238.1	18 41 54.	- 52 35 12.			NEBULA
SC 1837-5355.7	18 41 55.	- 53 52 49.			NEBULA
MCG-07-38-007	18 41 57.	- 39 15 30.	60	14.5	GALAXY
ISS 0483	18 42 00.	+ 03 18	118		STELLAR RING
MCG+10-26-048	18 42 00.	+ 62 26	42	16.	GALAXY
ZC 1842.0+8104	18 42 00.	+ 81 04	2760		CLUSTER OF GALAXIES
ISS 0516	18 42 00.	- 02 52	124		STELLAR RING
LDN 0480	18 42 00.	- 08 30	2640		DARK NEBULA
ZWG 143.019	18 42 06.	+ 24 05		14.8	GALAXY
UGC 11344	18 42 06.	+ 24 05	132	14.8	GALAXY Sa-b
MCG+09-30-035	18 42 06.	+ 56 09	60	16.	GALAXY
UGC 11345	18 42 06.	+ 58 51	72	16.0	GALAXY Sb-c
MCG+10-26-049	18 42 06.	+ 58 52	78	17.	GALAXY
7ZW 831	18 42 06.	+ 61 22			COMPACT GALAXY
ZWG 323.003	18 42 06.	+ 63 58		15.7	GALAXY
SCHO 0894	18 42 06.	- 09 20 06.	270		ISOLATED DARK CLOUD
MCG+04-44-012	18 42 09.	+ 24 04	36	14.8	GALAXY
MCG+04-44-013	18 42 12.	+ 25 10	72	15.6	GALAXY
ZWG 143.020	18 42 12.	+ 25 12		15.6	GALAXY
UGC 11346	18 42 12.	+ 25 12	66	15.6	GALAXY SBb
MCG+07-38-020	18 42 12.	+ 42 35	42	16.	GALAXY
UGC 11347	18 42 12.	+ 56 08	60	16.5	GALAXY Sc
7ZW 832	18 42 12.	+ 61 25			COMPACT GALAXY
ISS 0708	18 42 12.	- 10 55	503		STELLAR RING
MCG+08-34-016	18 42 15.	+ 45 00	42	16.	GALAXY
SCHO 0895	18 42 15.	- 07 06 42.	260		ISOLATED DARK CLOUD
HN 0510	18 42 16.	- 63 32			NEBULA
IC 4764	18 42 16.	- 63 32			NONSTELLAR OBJECT
ZWG 228.026	18 42 18.	+ 42 35		15.5	GALAXY
ISS 0517	18 42 18.	- 02 32	210		STELLAR RING
PK037+04.1	18 42 19.3	+ 06 03 56.			PLANETARY NEBULA
ZWG 302.001	18 42 24.	+ 62 25		15.5	GALAXY
ZWG 301.035	18 42 24.	+ 62 25		15.5	GALAXY
SC 1838-5215.6	18 42 26.	- 52 12 40.	24		NEBULA
HN 0511	18 42 27.	- 63 23			NEBULA
IC 4765	18 42 29.	- 63 23			NONSTELLAR OBJECT
MCG+10-26-050	18 42 30.	+ 60 36	90	13.	GALAXY
LDN 0517	18 42 30.	- 05 40	2040		DARK NEBULA
OCL 0067	18 42 30.	- 09 27	900	10.1	OPEN STAR CLUSTER
SBR 129.01	18 42 30.	- 63 25	2280	14.	CLUSTER OF GALAXIES
PK013-07.1	18 42 34.	- 20 35	35		PLANETARY NEBULA
IC 4776	18 42 34.1	- 33 23 52.	8	12.6	PLANETARY NEBULA
HN 0077	18 42 35.	- 33 24			NEBULA
PK002-13.1	18 42 35.	- 33 24	8	12.5	PLANETARY NEBULA
ZWG 302.002	18 42 36.	+ 60 37		12.9	GALAXY
ZWG 301.036	18 42 36.	+ 60 37		12.9	GALAXY
RNGC 6701	18 42 36.	+ 60 37		13.0	GALAXY
UGC 11348	18 42 36.	+ 60 37	96	12.9	GALAXY SBa
OCL 0086	18 42 36.	- 01 16	600	15.	OPEN STAR CLUSTER
SCHO 0896	18 42 38.	- 05 58 06.	230		ISOLATED DARK CLOUD
RNGC 6694	18 42 38.	- 09 27		9.5	OPEN CLUSTER
HN 0512	18 42 45.	- 63 20			NEBULA
IC 4766	18 42 45.	- 63 20			NONSTELLAR OBJECT
PK026-01.2	18 42 45.8	- 07 00 10.	3		PLANETARY NEBULA
7ZW 833	18 42 48.	+ 64 44			COMPACT GALAXY
SCHO 0897	18 42 49.	- 05 42	270		ISOLATED DARK CLOUD
IC 4767	18 42 51.	- 63 28			NONSTELLAR OBJECT
HN 0513	18 42 52.	- 63 28			NEBULA
PK034+02.1	18 42 53.0	+ 01 58 15.			PLANETARY NEBULA
MRSL 030+00/1	18 42 54.	- 02 03	480		HII REGION
ISS 0627	18 42 54.	- 04 27	469		STELLAR RING
ISS 0628	18 42 54.	- 04 29	157		STELLAR RING
PK026-01.1	18 42 54.1	- 06 21 46.			PLANETARY NEBULA
B 317	18 42 56.	- 14 16	1800		DARK OBJECT
HN 0514	18 42 56.	- 63 12			NEBULA
IC 4769	18 42 56.	- 63 12			NONSTELLAR OBJECT
LBN 0092	18 43	- 02 00	600		BRIGHT NEBULA
KHAV 545	18 43	- 07 09	4490		DARK NEBULA
KHAV 544	18 43	- 08 09	3940		DARK NEBULA
KHAV 543	18 43	- 14 57	4860		DARK NEBULA
KHAV 542	18 43	- 21 21	4320		DARK NEBULA
LDN 0529	18 43 00.	- 04 30	2040		DARK NEBULA
LDN 0524	18 43 00.	- 05 00	3120		DARK NEBULA
ISS 0709	18 43 00.	- 11 11	125		STELLAR RING
PK019-05.1	18 43 04.	- 14 30 52.			PLANETARY NEBULA
SCHO 0898	18 43 09.	- 05 53 12.	190		ISOLATED DARK CLOUD
RNGC 6697	18 43 11.	+ 25 27		14.5	GALAXY
SCHO 0899	18 43 11.	- 08 27 36.	300		ISOLATED DARK CLOUD
SCHO 0900	18 43 12.	+ 09 56 06.	400		ISOLATED DARK CLOUD
ISS 0437	18 43 12.	+ 10 03	199		STELLAR RING
MCG+04-44-014	18 43 12.	+ 25 25 30.	30	14.5	GALAXY
ZWG 143.021	18 43 12.	+ 25 27		14.5	GALAXY
UGC 11349	18 43 12.	+ 25 27	72	14.5	GALAXY F
LDN 0369	18 43 12.	- 18 15	480		DARK NEBULA
HN 1855	18 43 16.6	- 60 24 20.	24		NEBULA
ISS 0518	18 43 18.	- 02 23	205		STELLAR RING
HN 0515	18 43 21.	- 63 26			NEBULA
IC 4770	18 43 21.	- 63 26			NONSTELLAR OBJECT
ZWG 228.027	18 43 24.	+ 42 40		15.3	GALAXY
PK011-09.1	18 43 32.9	- 23 30 05.	5		PLANETARY NEBULA
HN 0516	18 43 33.	- 63 18			NEBULA
IC 4771	18 43 33.	- 63 18			NONSTELLAR OBJECT
ZWG 228.028	18 43 36.	+ 39 03		15.6	GALAXY
7ZW 834	18 43 42.	+ 62 00			COMPACT GALAXY
PK025-02.1	18 43 42.5	- 07 17 51.	22		PLANETARY NEBULA
LDN 0603	18 43 48.	+ 00 50	240		DARK NEBULA
MCG+07-38-021	18 43 48.	+ 40 39	36	16.	GALAXY
LDN 0527	18 43 48.	- 04 50	240		DARK NEBULA
MCG-07-38-008	18 43 48.	- 41 47	90	14.5	GALAXY
MIL 63	18 43 49.	- 03 02 06.	132		SUPERNOVA REMNANT
SCHO 0901	18 43 49.	- 05 46 12.	240		ISOLATED DARK CLOUD
SCHO 0902	18 43 50.	- 08 00 30.	330		ISOLATED DARK CLOUD
PK024-02.1	18 43 51.0	- 08 31 18.	6		PLANETARY NEBULA
ZWG 228.029	18 43 54.	+ 40 39		15.4	GALAXY
IC 4774	18 43 54.	- 57 59			NONSTELLAR OBJECT
HN 0517	18 43 55.	- 57 59			NEBULA
MCG+09-31-001	18 43 57.	+ 51 46	36	16.	GALAXY
LBN 0091	18 44	- 03 47	360		BRIGHT NEBULA
KHAV 547	18 44	- 09 15	2230		DARK NEBULA
KHAV 546	18 44	- 13 21	4600		DARK NEBULA
LB 09992	18 44	- 87 53		13.8	FAINT BLUE STAR
ZC 1844.0+6613	18 44 00.	+ 66 13	7060		CLUSTER OF GALAXIES
LDN 0297	18 44 00.	- 24 00	1680		DARK NEBULA
ZWG 143.022	18 44 06.	+ 22 34		15.7	GALAXY
UGC 11350	18 44 06.	+ 22 34	90	15.7	GALAXY Sc
ZWG 280.002	18 44 06.	+ 52 56		15.6	GALAXY
KARA.73B 0860	18 44 06.	+ 52 56	36	15.6	ISOLATED GALAXY S
MCG+09-31-002	18 44 06.	+ 52 57 30.	42	16.	GALAXY
7ZW 835	18 44 06.	+ 66 45			COMPACT GALAXY
IC 4777	18 44 07.	- 53 12			NONSTELLAR OBJECT
HN 1156	18 44 08.	- 53 12			NEBULA
RNGC 6684	18 44 08.	- 65 15		11.5	GLOBULAR CLUSTER
MCG+04-44-015	18 44 09.	+ 22 31	72	15.7	GALAXY
RNGC 6700	18 44 10.	+ 32 13		14.0	GALAXY
SCHO 0904	18 44 10.	- 08 23 12.	390		ISOLATED DARK CLOUD
LDN 0598	18 44 12.	+ 00 40	120		DARK NEBULA
LDN 0601	18 44 12.	+ 00 45	120		DARK NEBULA
LDN 0605	18 44 12.	+ 00 50	180		DARK NEBULA
ISS 0519	18 44 12.	+ 01 08	278		STELLAR RING
ZWG 173.026	18 44 12.	+ 32 13		14.2	GALAXY
UGC 11351	18 44 12.	+ 32 13	96	14.2	GALAXY SBc
MCG+05-44-010	18 44 12.	+ 32 13	84	14.2	GALAXY
HN 0518	18 44 16.	- 57 14			NEBULA
IC 4775	18 44 16.	- 57 14			NONSTELLAR OBJECT
MCG+09-31-003	18 44 18.	+ 54 59	54	16.	GALAXY
MRSL 029-00/1	18 44 18.	- 03 48	420		HII REGION
SCHO 0905	18 44 20.	- 04 55 48.	300		ISOLATED DARK CLOUD
MCG+08-34-017	18 44 24.	+ 49 43	18	17.	GALAXY
ISS 0629	18 44 24.	- 07 03	245		STELLAR RING

OBJECT NAME	RIGHT ASCEN.	DECLINATION	DIAM.	MAGN.	TYPE OF OBJECT
MCG+08-34-018	18 44 27.	+ 49 42 30.	24	17.	GALAXY
LDN 0532	18 44 30.	- 04 25	480		DARK NEBULA
LDN 0528	18 44 30.	- 04 50	2040		DARK NEBULA
ARC 2309	18 44 32.	+ 77 39		15.8	RICH CLUSTER OF GALAXIES
SCHO 0906	18 44 32.	- 06 13 18.	260		ISOLATED DARK CLOUD
ISS 0438	18 44 36.	+ 11 20	176		STELLAR RING
7ZW 836	18 44 36.	+ 60 43			COMPACT GALAXY
ISS 0710	18 44 36.	- 09 28	80		STELLAR RING
B 104	18 44 40.	- 04 35			DARK OBJECT
ZWG 228.030	18 44 42.	+ 40 32		15.0	GALAXY
MCG+07-38-022	18 44 42.	+ 40 32	30	16.	GALAXY
SCHO 0907	18 44 44.	- 04 23 06.	360		ISOLATED DARK CLOUD
SCHO 0908	18 44 45.	- 04 34 12.	280		ISOLATED DARK CLOUD
ZWG 255.012	18 44 48.	+ 48 47		15.4	GALAXY
UGC 11352	18 44 48.	+ 48 47	66	15.4	GALAXY SBb-c
KARA.73B 0861	18 44 48.	+ 48 47	60	15.4	ISOLATED GALAXY S
MCG+10-27-001	18 44 48.	+ 59 55	48	15.	GALAXY
ISS 0711	18 44 48.	- 08 54	76		STELLAR RING
PK026-02.1	18 44 50.6	- 06 57 24.	8		PLANETARY NEBULA
ZWG 302.003	18 44 54.	+ 59 56		15.4	GALAXY
ZWG 301.037	18 44 54.	+ 59 56		15.4	GALAXY
SCHO 0909	18 44 58.	- 09 07	790		ISOLATED DARK CLOUD
KHAV 548	18 45	- 23 39	10400		DARK NEBULA
LB 09931	18 45	- 85 21		13.5	FAINT BLUE STAR
LDN 0616	18 45 00.	+ 02 00	6240		DARK NEBULA
LDN 0621	18 45 00.	+ 03 10	27240		DARK NEBULA
7ZW 837	18 45 00.	+ 68 17			COMPACT GALAXY
LDN 0506	18 45 00.	- 03 55	120		DARK NEBULA
LDN 0496	18 45 00.	- 07 50	4500		DARK NEBULA
B 105	18 45 03.	- 06 58	30		DARK OBJECT
PK024-03.1	18 45 04.4	- 09 12 30.	15		PLANETARY NEBULA
ISS 0439	18 45 06.	+ 09 19	204		STELLAR RING
ZWG 143.023	18 45 06.	+ 25 44		15.4	GALAXY
SCHO 0910	18 45 06.	- 06 13 00.	280		ISOLATED DARK CLOUD
PK009-10.1	18 45 06.	- 25 32 18.	6		PLANETARY NEBULA
RNGC 6698	18 45 09.	- 25 59			NON-EXISTENT OBJECT
ZWG 280.003	18 45 12.	+ 55 02		15.6	GALAXY
MCG+09-31-004	18 45 15.	+ 55 03	24	16.	GALAXY
MCG+08-34-019	18 45 24.	+ 45 38	42	14.0	GALAXY
MCG+09-31-005	18 45 24.	+ 54 14	30	16.	GALAXY
HN 0520	18 45 24.	- 61 46			NEBULA
IC 4778	18 45 24.	- 61 46			NONSTELLAR OBJECT
RNGC 6702	18 45 29.	+ 45 39		14.0	GALAXY
7ZW 838	18 45 30.	+ 79 43			COMPACT GALAXY
LDN 0607	18 45 30.	- 00 45	1500		DARK NEBULA
OCL 0077	18 45 30.	- 05 54	540		OPEN STAR CLUSTER
LDN 0596	18 45 36.	+ 00 10	960		DARK NEBULA
ZWG 143.024	18 45 36.	+ 23 17		15.0	GALAXY
UGC 11353	18 45 36.	+ 23 17	78	15.0	GALAXY S0-a
ZC 1845.6+3606	18 45 36.	+ 36 06	2820		CLUSTER OF GALAXIES
ZWG 255.013	18 45 36.	+ 45 40		13.8	GALAXY
UGC 11354	18 45 36.	+ 45 40	126	13.8	GALAXY E
HN 0522	18 45 36.	- 59 18			NEBULA
IC 4780	18 45 36.	- 59 18			NONSTELLAR OBJECT
MCG+04-44-016	18 45 39.	+ 23 16	72	15.0	GALAXY
HN 0519	18 45 46.	- 69 59			NEBULA
IC 4773	18 45 46.	- 69 59			NONSTELLAR OBJECT
PK034+01.1	18 45 47.2	+ 01 39 43.			PLANETARY NEBULA
MCG+04-44-017	18 45 48.	+ 22 51 30.	60	15.4	GALAXY
ZWG 143.025	18 45 48.	+ 22 53		15.4	GALAXY
UGC 11355	18 45 48.	+ 22 53	102	15.4	GALAXY S
ZWG 201.036	18 45 48.	+ 33 16		15.6	GALAXY
MCG+08-34-021	18 45 48.	+ 45 20	60	16.	GALAXY
MCG+08-34-020	18 45 48.	+ 45 28	120	12.5	GALAXY
HN 0521	18 45 49.	- 63 04			NEBULA
IC 4779	18 45 49.	- 63 04			NONSTELLAR OBJECT
RNGC 6703	18 45 52.	+ 45 30		12.5	GALAXY
RNGC 6714	18 45 53.	+ 66 40			NON-EXISTENT OBJECT
ZWG 228.031	18 45 54.	+ 42 56		15.6	GALAXY
ZWG 255.014	18 45 54.	+ 45 30		12.4	GALAXY
UGC 11356	18 45 54.	+ 45 30	150	12.4	GALAXY S0
PK026-02.2	18 45 54.2	- 06 44 34.			PLANETARY NEBULA
SCHO 0911	18 45 55.	- 06 19 36.	210		ISOLATED DARK CLOUD
SCHO 0912	18 45 56.	- 06 17	250		ISOLATED DARK CLOUD
SCHO 0913	18 45 57.	- 09 10 48.	760		ISOLATED DARK CLOUD
KHAV 550	18 46	- 14 45	4860		DARK NEBULA
KHAV 549	18 46	- 20 21	5500		DARK NEBULA
LB 09932	18 46	- 80 43		14.0	FAINT BLUE STAR
LB 09933	18 46	- 83 29		14.0	FAINT BLUE STAR
LB 09934	18 46	- 86 18		11.5	FAINT BLUE STAR
ISS 0712	18 46 00.	- 14 03	228		STELLAR RING
FATH 2.163	18 46 03.	+ 45 43	11		NEBULA
KARA.72 537B	18 46 06.	+ 50 20	48	15.3	PART OF DOUBLE GALAXY
ZWG 255.015	18 46 06.	+ 50 21		15.6	GALAXY
KARA.72 537A	18 46 06.	+ 50 21	48	15.6	PART OF DOUBLE GALAXY
PK027-02.1	18 46 06.9	- 05 59 19.	7	14.	PLANETARY NEBULA
SC 1842-5121.9	18 46 09.	- 51 18 42.			NEBULA
B 106	18 46 10.	- 05 08	120		DARK OBJECT
PK042+05.1	18 46 11.2	+ 10 32 27.			PLANETARY NEBULA
OCL 0136	18 46 12.	+ 36 48	6600		OPEN STAR CLUSTER
MCG+08-34-023	18 46 12.	+ 50 19	15	16.	GALAXY
ZWG 255.016	18 46 12.	+ 50 20		15.3	GALAXY
MCG+08-34-022	18 46 12.	+ 50 20	42	16.	GALAXY
MCG+10-27-002	18 46 12.	+ 62 29	36	16.	GALAXY
ZWG 341.006	18 46 12.	+ 72 08		15.7	GALAXY
VVI 85	18 46 12.	+ 80 00	115	15.30	SEYFERT GALAXY
KW 70	18 46 12.	+ 80 00 00.	12		SEYFERT GALAXY
ACK 015-08.1	18 46 12.	- 19 38			PLANETARY NEBULA
ZWG 143.026	18 46 18.	+ 23 06		15.6	GALAXY
MCG+09-31-006	18 46 18.	+ 52 54	60	16.	GALAXY
CED 164	18 46 21.	+ 00 31			DIFFUSE GALACTIC NEBULA
MCG+08-34-024	18 46 24.	+ 45 22	72	15.	GALAXY
FATH 2.164	18 46 24.	+ 45 34	19		NEBULA
ISS 0520	18 46 24.	- 01 17	400		STELLAR RING
UGC 11357	18 46 30.	+ 45 22	96	16.5	GALAXY
ISS 0521	18 46 30.	- 00 48	88		STELLAR RING
LDN 0523	18 46 30.	- 05 30	960		DARK NEBULA
LDN 0311	18 46 30.	- 23 30	1920		DARK NEBULA
GCL 102	18 46 30.	- 65 12			GLOBULAR STAR CLUSTER
SC 1842-5046.5	18 46 33.	- 50 43 16.	30		NEBULA
SC 1842-5134.9	18 46 35.	- 51 31 40.	18		NEBULA
MCG+01-48-001	18 46 36.	+ 05 23	72	11.	GALAXY
IC 4782	18 46 45.	- 55 32			NONSTELLAR OBJECT
HN 0524	18 46 45.	- 55 33			NEBULA
MIL 64	18 46 47.	- 00 58 42.	210		SUPERNOVA REMNANT
ZWG 173.027	18 46 48.	+ 31 00		15.6	GALAXY
UGC 11358	18 46 48.	+ 31 00	66	15.6	GALAXY S?
BRSL 030-00/1	18 46 48.	- 02 25	600		HII REGION

OBJECT NAME	RIGHT ASCEN.	DECLINATION	DIAM.	MAGN.	TYPE OF OBJECT
ISS 0713	18 46 48.	- 09 32	76		STELLAR RING
IC 4791	18 46 50.	+ 19 16 21.			NONSTELLAR OBJECT
B 107	18 46 51.	- 05 04	300		DARK OBJECT
B 108	18 46 53.	- 06 22	180		DARK OBJECT
ISS 0630	18 46 54.	- 05 36	149		STELLAR RING
B 109	18 46 54.	- 07 37	42		DARK OBJECT
SCHO 0914	18 46 59.	- 06 22 36.	200		ISOLATED DARK CLOUD
IC 4781	18 46 59.	- 62 50			NONSTELLAR OBJECT
LBN 0094	18 47	- 02 23	900		BRIGHT NEBULA
KHAV 552	18 47	- 02 57	7960		DARK NEBULA
KHAV 551	18 47	- 04 51	8270		DARK NEBULA
B 111	18 47	- 05 01	7200		DARK OBJECT
LB 09935	18 47	- 80 47		14.3	FAINT BLUE STAR
UGC 11359	18 47 00.	+ 81 21	84	16.0	GALAXY
LDN 0610	18 47 00.	- 00 55	1440		DARK NEBULA
LDN 0548	18 47 00.	- 03 45	4020		DARK NEBULA
LDN 0530	18 47 00.	- 04 50	1380		DARK NEBULA
HN 0523	18 47 00.	- 62 51			NEBULA
B 318	18 47 02.	- 06 27			DARK OBJECT
PK026-02.3	18 47 02.8	- 07 05 06.	6		PLANETARY NEBULA
RNGC 6684A	18 47 03.	- 64 55			GALAXY
SCHO 0915	18 47 04.	+ 10 11 42.	300		ISOLATED DARK CLOUD
ZWG 341.007	18 47 06.	+ 70 18		15.1	GALAXY
ISS 0631	18 47 06.	- 08 31	345		STELLAR RING
DV.56 N6684A	18 47 05.	- 64 55			GALAXY
SC 1843-5052.7	18 47 09.	- 50 49 26.			NEBULA
SCHO 0916	18 47 10.	+ 10 04 36.	370		ISOLATED DARK CLOUD
IC 4783	18 47 10.	- 58 51			NONSTELLAR OBJECT
HN 0525	18 47 10.	- 58 52			NEBULA
OCL 0095	18 47 12.	+ 04 52	840		OPEN STAR CLUSTER
ZWG 201.037	18 47 12.	+ 34 24		15.3	GALAXY
ZC 1847.2+7711	18 47 12.	+ 77 11	9810		CLUSTER OF GALAXIES
SCHO 0917	18 47 14.	+ 10 12 00.	210		ISOLATED DARK CLOUD
MCG+06-41-021	18 47 15.	+ 34 24	30	15.	GALAXY
UGC 11360	18 47 18.	+ 18 39	66	17.	GALAXY IRR
LDN 0321	18 47 18.	- 22 42	600		DARK NEBULA
SC 1843-5240.3	18 47 18.	- 52 37 01.	36		NEBULA
SCHO 0918	18 47 21.	+ 10 04 48.	330		ISOLATED DARK CLOUD
SCHO 0919	18 47 23.	+ 10 26 30.	380		ISOLATED DARK CLOUD
PK012-09.1	18 47 25.0	- 22 37 56.	4		PLANETARY NEBULA
B 110	18 47 29.	- 04 51	660		DARK OBJECT
SER 129.03	18 47 30.	- 64 54	6000		LOW SURF. BRGHTNSS GALAXY
MCG+08-34-025	18 47 36.	+ 47 35	72	14.	GALAXY
ZWG 255.017	18 47 36.	+ 47 36		14.1	GALAXY
UGC 11361	18 47 36.	+ 47 36	66	14.1	GALAXY SB:b-c
KARA.73B 0862	18 47 36.	+ 47 36	66	14.1	ISOLATED GALAXY S
ISS 0522	18 47 36.	- 00 55	108		STELLAR RING
OCL 0052	18 47 36.	- 18 09	2100	10.	OPEN STAR CLUSTER
RNGC 6711	18 47 37.	+ 47 36		14.0	GALAXY
PK051+09.1	18 47 38.3	+ 20 46 56.	3	12.2	PLANETARY NEBULA
SC 1843-5159.5	18 47 40.	- 51 56 12.	12		NEBULA
ASS 10	18 47 42.	- 05 57			OB ASSOCIATION
RNGC 6699	18 47 46.	- 57 23		12.5	GALAXY
MCG+04-44-018	18 47 48.	+ 23 10	50	16.	GALAXY
UGC 11362	18 47 48.	+ 23 12	60	16.5	GALAXY
ZWG 229.001	18 47 48.	+ 40 04		15.7	GALAXY
ZWG 228.032	18 47 48.	+ 40 04		15.7	GALAXY
SCHO 0920	18 47 49.	- 06 35	520		ISOLATED DARK CLOUD
ACK 020-05.1	18 47 54.	- 13 35			PLANETARY NEBULA
KHAV 553	18 48	- 06 39	1570		DARK NEBULA
KHAV 555	18 48	- 07 09	2760		DARK NEBULA
KHAV 554	18 48	- 10 39	5540		DARK NEBULA
LDN 0638	18 48 00.	+ 07 15	18120		DARK NEBULA
ZWG 341.008	18 48 00.	+ 70 41		14.9	GALAXY
UGC 11363	18 48 00.	+ 70 41	72	14.9	GALAXY S0
LDN 0566	18 48 00.	- 02 50	900		DARK NEBULA
LDN 0547	18 48 00.	- 03 55	6120		DARK NEBULA
ISS 0632	18 48 00.	- 05 07	187		STELLAR RING
LDN 0300	18 48 00.	- 24 20	2460		DARK NEBULA
ZCG 1848+77	18 48 06.	+ 77 47		18.0	COMPACT GALAXY
ISS 0633	18 48 06.	- 08 51	104		STELLAR RING
HN 0526	18 48 09.	- 63 29			NEBULA
RNGC 6704	18 48 10.	- 05 16		9.5	OPEN CLUSTER
IC 4784	18 48 10.	- 63 28			NONSTELLAR OBJECT
LDN 0597	18 48 12.	+ 00 01	540		DARK NEBULA
PK064+15.1	18 48 12.	+ 35 11	20		PLANETARY NEBULA
ZWG 229.002	18 48 12.	+ 41 37		15.3	GALAXY
ZWG 228.033	18 48 12.	+ 41 37		15.3	GALAXY
7ZW 839	18 48 12.	+ 77 44			COMPACT GALAXY
OCL 0082	18 48 12.	- 05 16	600	9.8	OPEN STAR CLUSTER
IC 1294	18 48 13.	+ 40 11 11.			NONSTELLAR OBJECT
MCG+07-39-001	18 48 18.	+ 41 38	48	15.5	GALAXY
ISS 0634	18 48 18.	- 04 55	229		STELLAR RING
SCHO 0921	18 48 21.	+ 10 14 30.	350		ISOLATED DARK CLOUD
RNGC 6705	18 48 22.	- 06 20		7.0	OPEN CLUSTER
ZWG 280.004	18 48 24.	+ 51 50		15.7	GALAXY
MCG+09-31-007	18 48 24.	+ 51 50	42	15.	GALAXY
OCL 0076	18 48 24.	- 06 20	2400	6.9	OPEN STAR CLUSTER
SCHO 0922	18 48 24.	- 09 58 12.	280		ISOLATED DARK CLOUD
B 112	18 48 27.	- 06 44	1080		DARK OBJECT
MCG+04-44-019	18 48 30.	+ 26 46 30.	90	14.6	GALAXY
ZC 1848.5+7025	18 48 30.	+ 70 25	3430		CLUSTER OF GALAXIES
ISS 0635	18 48 30.	- 08 54	230		STELLAR RING
ARC 2310	18 48 31.	+ 73 16		17.0	RICH CLUSTER OF GALAXIES
IC 4786	18 48 32.	- 56 45			NONSTELLAR OBJECT
HN 0528	18 48 32.	- 56 46			NEBULA
HN 1856	18 48 33.1	- 47 04 13.	12		NEBULA
ZWG 143.027	18 48 36.	+ 26 46		14.6	GALAXY
UGC 11364	18 48 36.	+ 26 46	108	14.6	GALAXY Sa?
RNGC 6710	18 48 36.	+ 26 47		14.5	GALAXY
IC 4785	18 48 36.	- 59 18			NONSTELLAR OBJECT
HN 0527	18 48 36.	- 59 19			NEBULA
ISS 0440	18 48 42.	+ 10 17	131		STELLAR RING
LDN 0516	18 48 42.	- 06 40	660		DARK NEBULA
PK021-05.1	18 48 42.	- 13 14 14.	4		PLANETARY NEBULA
SC 1844-5039.8	18 48 45.	- 50 36 25.			NEBULA
B 113	18 48 46.	- 04 23	960		DARK OBJECT
MIL 66	18 48 50.	- 00 05	1860		SUPERNOVA REMNANT
RNGC 6713	18 48 53.	+ 33 53		14.0	GALAXY
SCHO 0926	18 48 54.	+ 11 13 00.	320		ISOLATED DARK CLOUD
ZWG 201.038	18 48 54.	+ 33 53		14.2	GALAXY
UGC 11365	18 48 54.	+ 33 53	30	14.2	GALAXY PECULR
SCHO 0925	18 48 54.	- 04 22 30.	1410		ISOLATED DARK CLOUD
SCHO 0924	18 48 54.	- 06 42 30.	450		ISOLATED DARK CLOUD
SCHO 0923	18 48 54.	- 06 55 54.	330		ISOLATED DARK CLOUD
SC 1845-5047.6	18 48 56.	- 50 44 12.			NEBULA
SCHO 0927	18 48 57.	- 05 50	1020		ISOLATED DARK CLOUD
LDN 0599	18 49 00.	+ 00 05	600		DARK NEBULA

OBJECT NAME	RIGHT ASCEN.	DECLINATION	DIAM.	MAGN.	TYPE OF OBJECT
LDN 0604	18 49 00.	+ 00 10	1560		DARK NEBULA
ISS 0441	18 49 00.	+ 12 32	227		STELLAR RING
UGC 11367	18 49 00.	+ 35 18	102	16.0	GALAXY SB+2CMP
UGC 11366	18 49 00.	+ 81 40	72	16.5	GALAXY Sb-c
LDN 0602	18 49 00.	- 00 08	180		DARK NEBULA
LDN 0606	18 49 00.	- 00 15	180		DARK NEBULA
LDN 0608	18 49 00.	- 00 30	240		DARK NEBULA
LDN 0609	18 49 00.	- 00 35	180		DARK NEBULA
LDN 0522	18 49 00.	- 05 50	1680		DARK NEBULA
LDN 0474	18 49 00.	- 09 45	5340		DARK NEBULA
LDN 0449	18 49 00.	- 11 10	900		DARK NEBULA
OCL 0100	18 49 06.	+ 10 17	1320	8.2	OPEN STAR CLUSTER
RNGC 6709	18 49 07.	+ 10 17		7.5	OPEN CLUSTER
MCG+08-34-026	18 49 12.	+ 46 50	54	16.	GALAXY
LB 04196	18 49 18.	+ 36 36		19.4	FAINT BLUE STAR
LB 04195	18 49 18.	+ 36 43		18.6	FAINT BLUE STAR
LB 04194	18 49 18.	+ 36 45		19.7	FAINT BLUE STAR
LB 04193	18 49 18.	+ 36 52		16.7	FAINT BLUE STAR
LB 04192	18 49 18.	+ 37 04		19.0	FAINT BLUE STAR
PK041+04.1	18 49 19.2	+ 09 51 13.			PLANETARY NEBULA
B 319	18 49 23.	- 01 20	420		DARK OBJECT
ZWG 143.028	18 49 24.	+ 26 25		15.1	GALAXY
UGC 11368	18 49 24.	+ 26 25	96	15.1	GALAXY S0-a
LB 04199	18 49 24.	+ 36 22		18.8	FAINT BLUE STAR
LB 04198	18 49 24.	+ 36 25		18.7	FAINT BLUE STAR
LB 04197	18 49 24.	+ 37 07		16.6	FAINT BLUE STAR
ISS 0636	18 49 24.	- 05 12	118		STELLAR RING
ISS 0637	18 49 24.	- 06 47	142		STELLAR RING
MCG-07-38-009	18 49 24.	- 42 37	60	14.	GALAXY
MCG-07-38-009	18 49 24.	- 42 37	60	14.	GALAXY
MCG+04-44-020	18 49 27.	+ 26 25	72	15.1	GALAXY
MCG+04-44-021	18 49 30.	+ 23 33	72	15.2	GALAXY
ZWG 143.029	18 49 30.	+ 23 34		15.2	GALAXY
UGC 11369	18 49 30.	+ 23 34	66	15.2	GALAXY SBa
LB 04202	18 49 30.	+ 36 45		18.3	FAINT BLUE STAR
LB 04201	18 49 30.	+ 36 52		20.2	FAINT BLUE STAR
LB 04200	18 49 30.	+ 36 55		19.1	FAINT BLUE STAR
MCG+07-39-002	18 49 30.	+ 39 52 30.	30	16.	GALAXY
ISS 0714	18 49 30.	- 14 10	155		STELLAR RING
HN 1857	18 49 31.8	- 44 47 33.			NEBULA
SCHO 0928	18 49 32.	- 07 03	620		ISOLATED DARK CLOUD
HUB C29	18 49 32.	- 36 51			DIFFUSE NEBULA
HUB C28	18 49 32.	- 36 51			DIFFUSE NEBULA
SCHO 0929	18 49 34.	+ 11 14 48.	250		ISOLATED DARK CLOUD
ISS 0442	18 49 36.	+ 09 12	131		STELLAR RING
MCG+04-44-022	18 49 36.	+ 26 29 30.	60	15.7	GALAXY
LB 04206	18 49 36.	+ 36 57		19.5	FAINT BLUE STAR
LB 04205	18 49 36.	+ 37 06		18.3	FAINT BLUE STAR
LB 04203	18 49 36.	+ 37 30		18.0	FAINT BLUE STAR
OCL 0039	18 49 36.	- 21 11	180	12.	OPEN STAR CLUSTER
ZWG 143.030	18 49 42.	+ 26 29		15.7	GALAXY
UGC 11370	18 49 42.	+ 26 29	90	15.7	GALAXY
LB 04208	18 49 42.	+ 37 29		18.7	FAINT BLUE STAR
LB 04207	18 49 42.	+ 37 29		15.5	FAINT BLUE STAR
7ZW 840	18 49 42.	+ 77 24			COMPACT GALAXY
SCHO 0930	18 49 43.	- 06 46	300		ISOLATED DARK CLOUD
SCHO 0931	18 49 47.	- 05 43 42.	270		ISOLATED DARK CLOUD
ZWG 341.009	18 49 48.	+ 73 18		14.6	GALAXY
ZCG 1849+73	18 49 48.	+ 73 18		14.6	COMPACT GALAXY
ISS 0638	18 49 48.	- 03 45	492		STELLAR RING
LDN 0521	18 49 48.	- 05 58	480		DARK NEBULA
SCHO 0932	18 49 48.	- 06 39	340		ISOLATED DARK CLOUD
ISS 0639	18 49 48.	- 06 44	209		STELLAR RING
MCG+04-44-023	18 49 51.	+ 26 25	78	15.5	GALAXY
SCHO 0933	18 49 52.	- 07 24	240		ISOLATED DARK CLOUD
HN 1858	18 49 53.1	- 46 07 56.	18		NEBULA
ZWG 143.031	18 49 54.	+ 26 25		15.5	GALAXY
UGC 11371	18 49 54.	+ 26 25	114	15.5	GALAXY S
LB 04210	18 49 54.	+ 36 41		18.6	FAINT BLUE STAR
LB 04209	18 49 54.	+ 36 52		19.5	FAINT BLUE STAR
ZWG 280.005	18 49 54.	+ 55 42		15.7	GALAXY
IC 4788	18 49 57.	- 63 30			NONSTELLAR OBJECT
HN 0529	18 49 57.	- 63 31			NEBULA
KHAV 556	18 50	- 07 14	4320		DARK NEBULA
LB 04214	18 50 00.	+ 36 34		19.0	FAINT BLUE STAR
LB 04213	18 50 00.	+ 36 43		16.2	FAINT BLUE STAR
LB 04212	18 50 00.	+ 36 46		19.4	FAINT BLUE STAR
LB 04211	18 50 00.	+ 36 48		19.6	FAINT BLUE STAR
ZC 1850.0+6426	18 50 00.	+ 64 26	4910		CLUSTER OF GALAXIES
LDN 0589	18 50 00.	- 01 30	5220		DARK NEBULA
LDN 0582	18 50 00.	- 02 00	2160		DARK NEBULA
LDN 0534	18 50 00.	- 05 00	7200		DARK NEBULA
LDN 0465	18 50 00.	- 10 40	720		DARK NEBULA
LDN 0464	18 50 00.	- 10 40	720		DARK NEBULA
LDN 0457	18 50 00.	- 11 00	720		DARK NEBULA
LDN 0456	18 50 00.	- 11 00	720		DARK NEBULA
SCHO 0934	18 50 03.	+ 11 32 42.	300		ISOLATED DARK CLOUD
MIL 67	18 50 04.	+ 00 35 54.	438		SUPERNOVA REMNANT
LB 04216	18 50 06.	+ 36 31		18.4	FAINT BLUE STAR
LB 04215	18 50 06.	+ 36 32		18.7	FAINT BLUE STAR
B 320	18 50 07.	- 05 55	900		DARK OBJECT
LB 04220	18 50 12.	+ 36 43		20.4	FAINT BLUE STAR
LB 04219	18 50 12.	+ 36 57		19.7	FAINT BLUE STAR
LB 04218	18 50 12.	+ 36 59		19.5	FAINT BLUE STAR
LB 04217	18 50 12.	+ 37 02		20.0	FAINT BLUE STAR
7ZW 841	18 50 12.	+ 64 11			COMPACT GALAXY
ABC 2311	18 50 12.	+ 70 20		16.0	RICH CLUSTER OF GALAXIES
LDN 0514	18 50 12.	- 07 00	240		DARK NEBULA
ZWG 173.028	18 50 18.	+ 28 43		15.6	GALAXY
LB 04223	18 50 18.	+ 36 23		18.8	FAINT BLUE STAR
LB 04222	18 50 18.	+ 36 47		17.4	FAINT BLUE STAR
LB 04221	18 50 18.	+ 36 56		18.9	FAINT BLUE STAR
SCHO 0935	18 50 18.	- 07 24	190		ISOLATED DARK CLOUD
OCL 0072	18 50 18.	- 08 46	540	9.2	OPEN STAR CLUSTER
GCL 103	18 50 18.	- 08 47	738	9.98	GLOBULAR STAR CLUSTER
RNGC 6712	18 50 20.	- 08 47		10.0	GLOBULAR CLUSTER
SCHO 0936	18 50 22.	- 05 57 06.	520		ISOLATED DARK CLOUD
LB 04227	18 50 24.	+ 36 49		19.2	FAINT BLUE STAR
LB 04226	18 50 24.	+ 36 57		19.8	FAINT BLUE STAR
LB 04225	18 50 24.	+ 37 04		17.2	FAINT BLUE STAR
LB 04224	18 50 24.	+ 37 08		10.	FAINT BLUE STAR
7ZW 842	18 50 24.	+ 64 14			COMPACT GALAXY
7ZW 843	18 50 24.	+ 73 16			COMPACT GALAXY
SCHO 0937	18 50 29.	- 07 07	240		ISOLATED DARK CLOUD
LB 04228	18 50 30.	+ 37 26		17.2	FAINT BLUE STAR
7ZW 844	18 50 30.	+ 64 28			COMPACT GALAXY
LDN 0518	18 50 30.	- 06 40	180		DARK NEBULA
B 114	18 50 30.	- 07 01	360		DARK OBJECT

OBJECT NAME	RIGHT ASCEN.	DECLINATION	DIAM.	MAGN.	TYPE OF OBJECT
SCHO 0938	18 50 30.	- 07 04 42.	340		ISOLATED DARK CLOUD
LDN 0512	18 50 30.	- 07 10	1020		DARK NEBULA
OCL 0042	18 50 30.	- 20 27	1350	6.8	OPEN STAR CLUSTER
SCHO 0939	18 50 31.	- 06 44	260		ISOLATED DARK CLOUD
LB 04231	18 50 36.	+ 36 21		19.3	FAINT BLUE STAR
LB 04230	18 50 36.	+ 36 37		19.4	FAINT BLUE STAR
LB 04229	18 50 36.	+ 36 52		18.3	FAINT BLUE STAR
ISS 0523	18 50 36.	- 01 19	139		STELLAR RING
B 115	18 50 38.	- 06 44			DARK OBJECT
SCHO 0940	18 50 38.	- 07 14 36.	290		ISOLATED DARK CLOUD
SCHO 0941	18 50 39.	+ 11 16 00.	250		ISOLATED DARK CLOUD
PK044+05.1	18 50 41.9	+ 12 12 17.			PLANETARY NEBULA
ZWG 201.039	18 50 42.	+ 33 46		15.0	GALAXY
UGC 11372	18 50 42.	+ 33 46	102	15.0	GALAXY S
LB 04234	18 50 42.	+ 36 57		20.0	FAINT BLUE STAR
LB 04233	18 50 42.	+ 37 00		17.0	FAINT BLUE STAR
LB 04232	18 50 42.	+ 37 28		18.7	FAINT BLUE STAR
ZWG 302.004	18 50 42.	+ 58 02		14.8	GALAXY
UGC 11373	18 50 42.	+ 58 02	90	14.8	GALAXY Sb
7ZW 845	18 50 42.	+ 63 33			COMPACT GALAXY
SCHO 0942	18 50 44.	- 07 09	250		ISOLATED DARK CLOUD
SCHO 0943	18 50 45.	- 06 52 12.	320		ISOLATED DARK CLOUD
IC 4787	18 50 45.	- 68 45			NONSTELLAR OBJECT
HN 0530	18 50 46.	- 68 45			NEBULA
LB 04237	18 50 48.	+ 36 31		11.	FAINT BLUE STAR
LB 04236	18 50 48.	+ 36 35		17.2	FAINT BLUE STAR
LB 04235	18 50 48.	+ 37 24		18.5	FAINT BLUE STAR
MCG+10-27-003	18 50 48.	+ 58 00	48	15.	GALAXY
ISS 0640	18 50 48.	- 06 19	118		STELLAR RING
PK027-03.1	18 50 48.	- 06 33	44	16.7	PLANETARY NEBULA
SCHO 0944	18 50 49.	- 06 43 48.	280		ISOLATED DARK CLOUD
B 116	18 50 51.	- 07 15	1200		DARK OBJECT
LB 04244	18 50 54.	+ 36 27		19.4	FAINT BLUE STAR
LB 04243	18 50 54.	+ 36 37		20.1	FAINT BLUE STAR
LB 04242	18 50 54.	+ 36 43		17.3	FAINT BLUE STAR
LB 04241	18 50 54.	+ 36 52		16.6	FAINT BLUE STAR
LB 04240	18 50 54.	+ 36 56		18.3	FAINT BLUE STAR
LB 04239	18 50 54.	+ 36 58		19.2	FAINT BLUE STAR
LB 04238	18 50 54.	+ 37 04		18.5	FAINT BLUE STAR
SCHO 0945	18 50 55.	- 07 30 18.	260		ISOLATED DARK CLOUD
KHAV 557	18 51	- 01 14	1990		DARK NEBULA
KHAV 558	18 51	- 05 08	2930		DARK NEBULA
LB 04250	18 51 00.	+ 36 21		18.5	FAINT BLUE STAR
LB 04249	18 51 00.	+ 36 22		19.3	FAINT BLUE STAR
LB 04248	18 51 00.	+ 36 25		20.0	FAINT BLUE STAR
LB 04247	18 51 00.	+ 36 58		20.5	FAINT BLUE STAR
LB 04246	18 51 00.	+ 37 20		18.4	FAINT BLUE STAR
LB 04245	18 51 00.	+ 37 24		18.8	FAINT BLUE STAR
ZWG 280.006	18 51 00.	+ 55 40		15.2	GALAXY
MCG+09-31-008	18 51 00.	+ 55 40	30	16.	GALAXY
LDN 0509	18 51 00.	- 07 22	180		DARK NEBULA
B 117	18 51 01.	- 07 29	60		DARK OBJECT
SCHO 0946	18 51 02.	+ 11 32 30.	330		ISOLATED DARK CLOUD
B 117a	18 51 03.	- 04 55	420		DARK OBJECT
HN 0531	18 51 03.	- 68 38			NEBULA
IC 4789	18 51 03.	- 68 38			NONSTELLAR OBJECT
LB 04254	18 51 06.	+ 36 28		16.9	FAINT BLUE STAR
LB 04253	18 51 06.	+ 36 54		18.8	FAINT BLUE STAR
LB 04252	18 51 06.	+ 37 03		14.0	FAINT BLUE STAR
LB 04251	18 51 06.	+ 37 20		18.7	FAINT BLUE STAR
SC 1847-5427.4	18 51 07.	- 54 23 51.			NEBULA
HN 1859	18 51 09.5	- 57 47 39.	24		NEBULA
LB 04260	18 51 12.	+ 36 29		17.0	FAINT BLUE STAR
LB 04259	18 51 12.	+ 36 31		18.5	FAINT BLUE STAR
LB 04258	18 51 12.	+ 36 36		17.5	FAINT BLUE STAR
LB 04257	18 51 12.	+ 36 39		17.8	FAINT BLUE STAR
LB 04256	18 51 12.	+ 36 49		16.9	FAINT BLUE STAR
LB 04255	18 51 12.	+ 37 06		15.0	FAINT BLUE STAR
ISS 0641	18 51 12.	- 06 33	85		STELLAR RING
B 118	18 51 13.	- 07 31	120		DARK OBJECT
B 321	18 51 15.	- 11 22	900		DARK OBJECT
RNGC 6707	18 51 17.	- 53 53			GALAXY
LB 04265	18 51 18.	+ 36 45		19.2	FAINT BLUE STAR
LB 04264	18 51 18.	+ 37 03		18.6	FAINT BLUE STAR
LB 04263	18 51 18.	+ 37 12		20.4	FAINT BLUE STAR
LB 04262	18 51 18.	+ 37 12		18.8	FAINT BLUE STAR
LB 04261	18 51 18.	+ 37 16		16.0	FAINT BLUE STAR
ZC 1851.3+6032	18 51 18.	+ 60 32	3560		CLUSTER OF GALAXIES
ISS 0642	18 51 18.	- 06 34	105		STELLAR RING
ISS 0643	18 51 18.	- 06 36	200		STELLAR RING
PK027-03.2	18 51 20.8	- 06 30 07.		13.4	PLANETARY NEBULA
MCG+06-41-022	18 51 21.	+ 33 01	66	15.	GALAXY
ZWG 201.040	18 51 24.	+ 33 00		15.4	GALAXY
UGC 11374	18 51 24.	+ 33 00	78	15.4	GALAXY SBb
LB 04267	18 51 24.	+ 36 50		18.8	FAINT BLUE STAR
LB 04266	18 51 24.	+ 37 11		18.5	FAINT BLUE STAR
ZWG 341.010	18 51 24.	+ 73 17		14.8	GALAXY
7ZW 846	18 51 24.	+ 73 18			COMPACT GALAXY
ISS 0524	18 51 24.	- 01 00	164		STELLAR RING
ISS 0525	18 51 24.	- 02 19	67		STELLAR RING
ISS 0644	18 51 24.	- 06 36	109		STELLAR RING
SCHO 0947	18 51 24.	- 07 15 30.	340		ISOLATED DARK CLOUD
IC 1296	18 51 27.	+ 32 59 59.			NONSTELLAR OBJECT
LB 04273	18 51 30.	+ 36 31		19.9	FAINT BLUE STAR
LB 04272	18 51 30.	+ 36 43		20.0	FAINT BLUE STAR
LB 04271	18 51 30.	+ 37 00		20.2	FAINT BLUE STAR
LB 04270	18 51 30.	+ 37 02		16.6	FAINT BLUE STAR
LB 04269	18 51 30.	+ 37 08		20.3	FAINT BLUE STAR
LB 04268	18 51 30.	+ 37 16		16.8	FAINT BLUE STAR
SCHO 0948	18 51 30.	- 07 30 24.	230		ISOLATED DARK CLOUD
HN 0533	18 51 30.	- 56 28			NEBULA
PK029-02.1	18 51 30.9	- 04 42 41.		16.	PLANETARY NEBULA
IC 4792	18 51 31.	- 56 29			NONSTELLAR OBJECT
PK024-05.1	18 51 32.5	- 10 09 02.	22		PLANETARY NEBULA
SCHO 0949	18 51 34.	- 06 18	300		ISOLATED DARK CLOUD
RNGC 6708	18 51 35.	- 53 47			GALAXY
ZWG 173.029	18 51 36.	+ 30 36		15.5	GALAXY
LB 04277	18 51 36.	+ 36 31		13.8	FAINT BLUE STAR
LB 04276	18 51 36.	+ 36 41		17.9	FAINT BLUE STAR
LB 04275	18 51 36.	+ 36 50		20.2	FAINT BLUE STAR
LB 04274	18 51 36.	+ 37 02		20.4	FAINT BLUE STAR
OCL 0046	18 51 36.	- 19 57	600	8.7	OPEN STAR CLUSTER
PK025-04.1	18 51 36.1	- 08 51 22.			PLANETARY NEBULA
SCHO 0950	18 51 37.	- 07 19 30.	200		ISOLATED DARK CLOUD
RNGC 6716	18 51 37.	- 19 57		7.0	OPEN CLUSTER
RNGC 6720	18 51 40.	+ 32 58		9.5	PLANETARY NEBULA
LB 04279	18 51 42.	+ 36 28		19.0	FAINT BLUE STAR
LB 04278	18 51 42.	+ 37 24		18.5	FAINT BLUE STAR

OBJECT NAME	RIGHT ASCEN.	DECLINATION	DIAM.	MAGN.	TYPE OF OBJECT
HN 0532	18 51 42.	- 64 59			NEBULA
IC 4790	18 51 42.	- 65 00			NONSTELLAR OBJECT
PK058+12.1	18 51 42.6	+ 28 28 27.			PLANETARY NEBULA
PK063+13.1	18 51 43.49	+ 32 57 54.8	200	9.7	PLANETARY NEBULA
LB 04282	18 51 48.	+ 36 22		19.0	FAINT BLUE STAR
OCL 0137	18 51 48.	+ 36 51	1200		OPEN STAR CLUSTER
LB 04281	18 51 48.	+ 36 56		18.2	FAINT BLUE STAR
LB 04280	18 51 48.	+ 37 08		20.3	FAINT BLUE STAR
ISS 0526	18 51 48.	- 01 47	139		STELLAR RING
SCHO 0951	18 51 50.	- 04 42 06.	330		ISOLATED DARK CLOUD
LB 04285	18 51 54.	+ 36 19		17.0	FAINT BLUE STAR
LB 04284	18 51 54.	+ 36 45		18.8	FAINT BLUE STAR
LB 04283	18 51 54.	+ 37 30		15.3	FAINT BLUE STAR
OCL 0087	18 51 54.	- 01 19	240	15.	OPEN STAR CLUSTER
SCHO 0952	18 51 54.	- 06 26	300		ISOLATED DARK CLOUD
IC 1295	18 51 54.	- 08 51	180	15.0	GLOBULAR CLUSTER
GCL 104	18 51 54.	- 30 32	330	8.74	GLOBULAR STAR CLUSTER
SCHO 0954	18 51 55.	- 07 15 06.	210		ISOLATED DARK CLOUD
SCHO 0953	18 51 56.	- 07 21 48.	230		ISOLATED DARK CLOUD
B 119	18 51 57.	- 04 37			DARK OBJECT
PK003-14.1	18 51 57.	- 32 19	4	10.9	PLANETARY NEBULA
KHAV 559	18 52	+ 04 28	29740		DARK NEBULA
LBN 0170	18 52	+ 05 59	240		BRIGHT NEBULA
KHAV 560	18 52	- 08 14	3010		DARK NEBULA
LB 09993	18 52	- 88 51		13.5	FAINT BLUE STAR
ISS 0484	18 52 00.	+ 08 25	271		STELLAR RING
ZWG 201.041	18 52 00.	+ 34 56		15.2	GALAXY
LB 04287	18 52 00.	+ 36 30		18.9	FAINT BLUE STAR
LB 04286	18 52 00.	+ 37 28		13.8	FAINT BLUE STAR
B 119a	18 52 00.	- 05 15	1800		DARK OBJECT
RNGC 6715	18 52 00.	- 30 32		8.5	GLOBULAR CLUSTER
HN 1860	18 52 00.3	- 46 50 41.	6		NEBULA
SCHO 0955	18 52 04.	- 04 38 30.	210		ISOLATED DARK CLOUD
MAI 103	18 52 05.	+ 76 53	40		DWARF SPHEROIDAL GALAXY
ISS 0443	18 52 06.	+ 13 57	78		STELLAR RING
LB 04290	18 52 06.	+ 36 49		19.1	FAINT BLUE STAR
LB 04289	18 52 06.	+ 37 07		19.0	FAINT BLUE STAR
LB 04288	18 52 06.	+ 37 21		18.7	FAINT BLUE STAR
ISS 0645	18 52 06.	- 04 12	104		STELLAR RING
SCHO 0956	18 52 06.	- 05 14 54.	910		ISOLATED DARK CLOUD
OCL 0037	18 52 06.	- 22 46	240	12.1	OPEN STAR CLUSTER
IC 4802	18 52 06.	- 22 46			NONSTELLAR OBJECT
RNGC 6717	18 52 06.	- 22 47			GLOBULAR CLUSTER
PK025-04.2	18 52 07.62	- 08 47 56.4	140	11.6	PLANETARY NEBULA
SCHO 0957	18 52 08.	- 06 22	280		ISOLATED DARK CLOUD
RNGC 6706	18 52 09.	- 63 14			UNVERIFIED SOUTHERN OBJECT
ISS 0445	18 52 12.	+ 09 32	221		STELLAR RING
ISS 0444	18 52 12.	+ 11 42	101		STELLAR RING
ZWG 143.032	18 52 12.	+ 24 35		15.6	GALAXY
UGC 11375	18 52 12.	+ 24 35	78	15.6	GALAXY Sc
LB 04299	18 52 12.	+ 36 27		17.8	FAINT BLUE STAR
LB 04298	18 52 12.	+ 36 32		18.8	FAINT BLUE STAR
LB 04297	18 52 12.	+ 36 41		19.0	FAINT BLUE STAR
LB 04296	18 52 12.	+ 36 47		19.6	FAINT BLUE STAR
LB 04295	18 52 12.	+ 37 08		18.8	FAINT BLUE STAR
LB 04294	18 52 12.	+ 37 11		17.4	FAINT BLUE STAR
LB 04293	18 52 12.	+ 37 12		18.1	FAINT BLUE STAR
LB 04292	18 52 12.	+ 37 16		20.0	FAINT BLUE STAR
LB 04291	18 52 12.	+ 37 30		20.6	FAINT BLUE STAR
7ZW 847	18 52 12.	+ 64 23			COMPACT GALAXY
ZC 1852.2+6818	18 52 12.	+ 68 18	2620		CLUSTER OF GALAXIES
ISS 0527	18 52 12.	- 00 52	228		STELLAR RING
GCL 105	18 52 12.	- 22 46	156	12.1	GLOBULAR STAR CLUSTER
SC 1846-5004.8	18 52 13.	- 50 01 10.	18		NEBULA
PK038+02.1	18 52 15.	+ 05 58	350		PLANETARY NEBULA
B 120	18 52 15.	- 04 40			DARK OBJECT
HN 1157	18 52 16.	- 54 16		14.	NEBULA
IC 4796	18 52 16.	- 54 17			NONSTELLAR OBJECT
MCG+04-44-024	18 52 18.	+ 24 34	90	15.6	GALAXY
LB 04303	18 52 18.	+ 37 23		19.8	FAINT BLUE STAR
LB 04302	18 52 18.	+ 37 24		19.2	FAINT BLUE STAR
LB 04301	18 52 18.	+ 37 25		18.5	FAINT BLUE STAR
LB 04300	18 52 18.	+ 37 29		19.2	FAINT BLUE STAR
ZWG 302.005	18 52 18.	+ 57 15		15.1	GALAXY
MCG+10-27-004	18 52 18.	+ 57 15	42	16.	GALAXY
ISS 0646	18 52 18.	- 03 37	100		STELLAR RING
ISS 0647	18 52 18.	- 05 02	195		STELLAR RING
ISS 0648	18 52 18.	- 05 07	108		STELLAR RING
IC 4797	18 52 18.	- 54 22	108	11.8	GALAXY E5
YM 16	18 52 19.	+ 05 57	300		SYMMETRIC GALACTIC NEBULA
HN 0534	18 52 22.	- 61 26			NEBULA
IC 4793	18 52 22.	- 61 27			NONSTELLAR OBJECT
HN 1158	18 52 23.	- 54 21		14.	NEBULA
LB 04306	18 52 24.	+ 36 43		17.8	FAINT BLUE STAR
LB 04305	18 52 24.	+ 37 09		17.6	FAINT BLUE STAR
LB 04304	18 52 24.	+ 37 26		19.6	FAINT BLUE STAR
ISS 0528	18 52 24.	- 01 16	142		STELLAR RING
ISS 0649	18 52 24.	- 08 07	85		STELLAR RING
SER 130.03	18 52 24.	- 54 15	3300	13.	LOOSE GROUP OF 5 GALAXIES
SCHO 0958	18 52 26.	- 04 39 18.	230		ISOLATED DARK CLOUD
LB 04308	18 52 30.	+ 36 21		20.3	FAINT BLUE STAR
LB 04307	18 52 30.	+ 36 56		17.9	FAINT BLUE STAR
ZWG 280.007	18 52 30.	+ 55 18		15.5	GALAXY
ACK 013-10.1	18 52 30.	- 21 54			PLANETARY NEBULA
HN 0535	18 52 31.	- 62 08			NEBULA
IC 4794	18 52 32.	- 62 09			NONSTELLAR OBJECT
ISS 0446	18 52 36.	+ 10 38	90		STELLAR RING
LB 04312	18 52 36.	+ 36 27		19.0	FAINT BLUE STAR
LB 04311	18 52 36.	+ 36 36		20.2	FAINT BLUE STAR
LB 04310	18 52 36.	+ 37 19		18.0	FAINT BLUE STAR
ZWG 201.042	18 52 36.	+ 37 20		15.2	GALAXY
LB 04309	18 52 36.	+ 37 26		19.2	FAINT BLUE STAR
ZC 1852.6+7820	18 52 36.	+ 78 20	470		CLUSTER OF GALAXIES
SCHO 0959	18 52 38.	+ 11 02 12.	280		ISOLATED DARK CLOUD
HN 0536	18 52 41.	- 61 39			NEBULA
IC 4795	18 52 41.	- 61 40			NONSTELLAR OBJECT
ISS 0373	18 52 42.	+ 18 27	724		STELLAR RING
ZWG 201.043	18 52 42.	+ 33 45		15.1	GALAXY
LB 04313	18 52 42.	+ 36 25		17.3	FAINT BLUE STAR
ZWG 255.018	18 52 42.	+ 45 32		15.7	GALAXY
KARA.73B 0863	18 52 42.	+ 45 32	36	15.7	ISOLATED GALAXY S
MCG+09-31-009	18 52 42.	+ 51 26	42	16.	GALAXY
HN 1861	18 52 45.7	- 43 12 49.	36		NEBULA
B 121	18 52 47.	- 04 40			DARK OBJECT
ISS 0529	18 52 48.	- 00 52	116		STELLAR RING
SCHO 0960	18 52 48.	- 04 21 12.	550		ISOLATED DARK CLOUD
MCG-07-39-001	18 52 48.	- 41 04	48	13.5	GALAXY
SC 1848-5440.4	18 52 52.	- 54 36 44.	30		NEBULA

OBJECT NAME	RIGHT ASCEN.	DECLINATION	DIAM.	MAGN.	TYPE OF OBJECT
LB 04315	18 52 54.	+ 36 31		19.8	FAINT BLUE STAR
LB 04314	18 52 54.	+ 37 07		17.7	FAINT BLUE STAR
ZWG 255.019	18 52 54.	+ 48 51		14.3	GALAXY
UGC 11376	18 52 54.	+ 48 51	42	14.3	GALAXY S
ISS 0650	18 52 54.	- 05 57	115		STELLAR RING
ISS 0651	18 52 54.	- 06 12	143		STELLAR RING
ARC 2313	18 52 58.	+ 78 21		17.4	RICH CLUSTER OF GALAXIES
KHAV 564	18 53	- 09 32	2110		DARK NEBULA
KHAV 561	18 53	- 11 44	6460		DARK NEBULA
KHAV 563	18 53	- 14 56	5540		DARK NEBULA
KHAV 562	18 53	- 18 50	1860		DARK NEBULA
LB 04317	18 53 00.	+ 36 26		16.0	FAINT BLUE STAR
LB 04316	18 53 00.	+ 36 34		16.8	FAINT BLUE STAR
ZWG 229.003	18 53 00.	+ 39 33		15.5	GALAXY
MCG+08-34-027	18 53 00.	+ 48 52	42	16.	GALAXY
ZC 1853.0+7226	18 53 00.	+ 72 26	5650		CLUSTER OF GALAXIES
LDN 0579	18 53 00.	- 02 30	5280		DARK NEBULA
LDN 0565	18 53 00.	- 03 35	4740		DARK NEBULA
LB 04318	18 53 06.	+ 36 34		18.2	FAINT BLUE STAR
LDN 0611	18 53 06.	- 00 20	480		DARK NEBULA
ISS 0530	18 53 06.	- 00 35	115		STELLAR RING
ISS 0652	18 53 06.	- 07 57	115		STELLAR RING
B 322	18 53 08.	- 04 31	120		DARK OBJECT
LB 04320	18 53 12.	+ 36 27		18.6	FAINT BLUE STAR
LB 04319	18 53 12.	+ 37 28		18.6	FAINT BLUE STAR
ZWG 229.004	18 53 12.	+ 41 54		15.3	GALAXY
ISS 0653	18 53 12.	- 08 57	67		STELLAR RING
SCHO 0961	18 53 13.	- 09 55 30.	300		ISOLATED DARK CLOUD
LB 04324	18 53 18.	+ 36 43		18.8	FAINT BLUE STAR
LB 04323	18 53 18.	+ 37 11		18.5	FAINT BLUE STAR
LB 04322	18 53 18.	+ 37 21		17.9	FAINT BLUE STAR
LB 04321	18 53 18.	+ 37 29		19.2	FAINT BLUE STAR
ISS 0531	18 53 18.	- 00 12	51		STELLAR RING
SCHO 0962	18 53 21.	+ 11 50 54.	290		ISOLATED DARK CLOUD
LB 04325	18 53 24.	+ 37 11		20.2	FAINT BLUE STAR
MCG+07-39-003	18 53 24.	+ 39 27	36	15.	GALAXY
ZWG 229.005	18 53 24.	+ 39 28		15.3	GALAXY
ZWG 341.011	18 53 24.	+ 73 07		15.6	GALAXY
UGC 11377	18 53 24.	+ 73 07	96	15.6	GALAXY Sc
LDN 0612	18 53 24.	- 00 30	180		DARK NEBULA
HN 1862	18 53 24.6	- 42 46 10.	12		NEBULA
LB 04326	18 53 30.	+ 36 47		19.1	FAINT BLUE STAR
ZC 1853.5+3744	18 53 30.	+ 37 44	5380		CLUSTER OF GALAXIES
ZWG 323.004	18 53 30.	+ 66 10		15.3	GALAXY
KARA.73B 0864	18 53 30.	+ 66 10	54	15.3	ISOLATED GALAXY S
ISS 0447	18 53 36.	+ 11 16	64		STELLAR RING
LB 04332	18 53 36.	+ 36 22		18.4	FAINT BLUE STAR
LB 04331	18 53 36.	+ 36 32		18.6	FAINT BLUE STAR
LB 04330	18 53 36.	+ 36 38		18.8	FAINT BLUE STAR
LB 04329	18 53 36.	+ 37 09		18.6	FAINT BLUE STAR
LB 04328	18 53 36.	+ 37 19		18.8	FAINT BLUE STAR
LB 04327	18 53 36.	+ 37 20		14.0	FAINT BLUE STAR
ZWG 229.006	18 53 36.	+ 39 54		15.7	GALAXY
ISS 0654	18 53 36.	- 07 54	80		STELLAR RING
ISS 0715	18 53 36.	- 10 27	208		STELLAR RING
SN 1971R	18 53 36.	- 62 10		13.	SUPERNOVA
ZWG 323.005	18 53 42.	+ 68 24		15.4	GALAXY
7ZW 848	18 53 42.	+ 68 25			COMPACT GALAXY
ISS 0716	18 53 42.	- 10 49	115		STELLAR RING
HN 0537	18 53 43.	- 62 10			NEBULA
SCHO 0963	18 53 44.	+ 15 28 48.	490		ISOLATED DARK CLOUD
IC 4798	18 53 44.	- 62 11			NONSTELLAR OBJECT
MIL 68	18 53 45.	+ 01 13 00.	2100		SUPERNOVA REMNANT
LB 04334	18 53 48.	+ 36 20		19.1	FAINT BLUE STAR
LB 04333	18 53 48.	+ 37 21		18.6	FAINT BLUE STAR
ARC 2312	18 53 49.	+ 68 18		15.8	RICH CLUSTER OF GALAXIES
SCHO 0964	18 53 49.	- 09 27 36.	260		ISOLATED DARK CLOUD
PK039+02.1	18 53 52.5	+ 07 03 30.	19		PLANETARY NEBULA
LB 04338	18 53 54.	+ 36 26		18.6	FAINT BLUE STAR
LB 04337	18 53 54.	+ 36 29		19.5	FAINT BLUE STAR
LB 04336	18 53 54.	+ 36 48		16.3	FAINT BLUE STAR
LB 04335	18 53 54.	+ 37 21		17.7	FAINT BLUE STAR
7ZW 849	18 53 54.	+ 68 19			COMPACT GALAXY
ISS 0655	18 53 54.	- 09 01	139		STELLAR RING
HN 1863	18 53 55.5	- 47 36 50.	48		NEBULA
LBN 0110	18 54	+ 07 45	360		BRIGHT NEBULA
MIL 78	18 54	+ 15 45	3600		SUPERNOVA REMNANT
MRSL 040+02/1	18 54 00.	+ 07 45	420		HII REGION
ZC 1854.0+3552	18 54 00.	+ 35 52	2820		CLUSTER OF GALAXIES
LB 04339	18 54 00.	+ 36 39		20.1	FAINT BLUE STAR
ISS 0532	18 54 00.	- 00 03	64		STELLAR RING
LDN 0584	18 54 00.	- 02 20	1080		DARK NEBULA
LDN 0458	18 54 00.	- 11 30	4560		DARK NEBULA
LDN 0452	18 54 00.	- 11 40	4560		DARK NEBULA
HN 0538	18 54 01.	- 63 12			NEBULA
IC 4800	18 54 01.	- 63 13			NONSTELLAR OBJECT
SCHO 0965	18 54 02.	+ 14 53 42.	570		ISOLATED DARK CLOUD
HN 0539	18 54 05.	- 63 59			NEBULA
LB 04341	18 54 06.	+ 36 49		16.8	FAINT BLUE STAR
LB 04340	18 54 06.	+ 37 12		18.7	FAINT BLUE STAR
LDN 0545	18 54 06.	- 04 50	300		DARK NEBULA
IC 4799	18 54 06.	- 64 00			NONSTELLAR OBJECT
SC 1850-5343.0	18 54 07.	- 53 39 14.	42		NEBULA
B 122	18 54 10.	- 04 49	240		DARK OBJECT
PK043+03.1	18 54 11.	+ 10 48 24.	4		PLANETARY NEBULA
LDN 0614	18 54 12.	- 00 25	720		DARK NEBULA
LDN 0613	18 54 12.	- 00 25	300		DARK NEBULA
SER 130.01	18 54 12.	- 53 00	70		LOW SURF. BRGHTNSS GALAXY
MCG+07-39-004	18 54 15.	+ 39 53	36	15.5	GALAXY
SCHO 0966	18 54 17.	+ 12 35 00.	250		ISOLATED DARK CLOUD
LB 04347	18 54 18.	+ 36 38		14.3	FAINT BLUE STAR
LB 04346	18 54 18.	+ 36 55		18.5	FAINT BLUE STAR
LB 04345	18 54 18.	+ 36 58		18.1	FAINT BLUE STAR
LB 04344	18 54 18.	+ 37 01		16.8	FAINT BLUE STAR
LB 04343	18 54 18.	+ 37 04		19.2	FAINT BLUE STAR
LB 04342	18 54 18.	+ 37 30		20.1	FAINT BLUE STAR
ZWG 229.007	18 54 18.	+ 39 53		15.5	GALAXY
ISS 0656	18 54 18.	- 07 27	86		STELLAR RING
SC 1850-4908.7	18 54 21.	- 49 04 55.	18		NEBULA
SCHO 0967	18 54 22.	- 04 47 42.	380		ISOLATED DARK CLOUD
SCHO 0968	18 54 23.	- 09 26 24.	420		ISOLATED DARK CLOUD
ISS 0374	18 54 24.	+ 16 26	128		STELLAR RING
LB 04348	18 54 24.	+ 36 54		17.2	FAINT BLUE STAR
ZWG 280.008	18 54 24.	+ 52 11		15.7	GALAXY
ISS 0657	18 54 24.	- 07 33	192		STELLAR RING
MCG+09-31-010	18 54 27.	+ 52 10	30	17.	GALAXY
LB 04351	18 54 30.	+ 36 27		19.2	FAINT BLUE STAR
LB 04350	18 54 30.	+ 37 05		16.0	FAINT BLUE STAR

OBJECT NAME	RIGHT ASCEN.	DECLINATION	DIAM.	MAGN.	TYPE OF OBJECT
LB 04349	18 54 30.	+ 37 24		18.8	FAINT BLUE STAR
ZWG 280.009	18 54 30.	+ 51 11		15.7	GALAXY
ZWG 280.010	18 54 30.	+ 52 05		15.5	GALAXY
UGC 11378	18 54 30.	+ 52 05	66	15.5	GALAXY PECULR
ARC 2314	18 54 32.	+ 78 57		17.2	RICH CLUSTER OF GALAXIES
SCHO 0969	18 54 33.	+ 12 32 48.	240		ISOLATED DARK CLOUD
ZWG 143.033	18 54 36.	+ 25 10		15.3	GALAXY
UGC 11379	18 54 36.	+ 25 10	60	15.3	GALAXY Sc
MCG+04-44-025	18 54 36.	+ 25 10	36	15.3	GALAXY
LB 04353	18 54 36.	+ 36 20		17.6	FAINT BLUE STAR
LB 04352	18 54 36.	+ 37 00		18.2	FAINT BLUE STAR
PK028-04.1	18 54 36.8	- 06 03 53.	11	15.	PLANETARY NEBULA
RNGC 6724	18 54 37.	+ 10 18			NON-EXISTENT OBJECT
HN 0540	18 54 40.	- 64 44			NEBULA
IC 4801	18 54 40.	- 64 44			NONSTELLAR OBJECT
SCHO 0970	18 54 42.	+ 12 27 42.	230		ISOLATED DARK CLOUD
LB 04357	18 54 42.	+ 36 24		17.4	FAINT BLUE STAR
LB 04356	18 54 42.	+ 36 44		19.6	FAINT BLUE STAR
LB 04355	18 54 42.	+ 36 50		18.3	FAINT BLUE STAR
LB 04354	18 54 42.	+ 36 59		18.6	FAINT BLUE STAR
ISS 0533	18 54 42.	- 01 24	208		STELLAR RING
ISS 0658	18 54 42.	- 05 35	166		STELLAR RING
SCHO 0971	18 54 44.	+ 12 06 00.	970		ISOLATED DARK CLOUD
LB 04359	18 54 48.	+ 36 28		18.6	FAINT BLUE STAR
LB 04358	18 54 48.	+ 36 59		18.9	FAINT BLUE STAR
LB 04361	18 54 54.	+ 36 41		18.3	FAINT BLUE STAR
LB 04360	18 54 54.	+ 37 19		18.1	FAINT BLUE STAR
LDN 0546	18 54 54.	- 04 50	300		DARK NEBULA
ISS 0659	18 54 54.	- 06 57	164		STELLAR RING
B 323	18 54 56.	- 03 29	1020		DARK OBJECT
SCHO 0972	18 54 58.	- 03 27 42.	540		ISOLATED DARK CLOUD
KHAV 568	18 55	+ 17 52	2660		DARK NEBULA
KHAV 567	18 55	- 00 38	1860		DARK NEBULA
KHAV 566	18 55	- 13 20	9850		DARK NEBULA
KHAV 565	18 55	- 17 14	8150		DARK NEBULA
LDN 0617	18 55 00.	+ 01 00	3480		DARK NEBULA
ISS 0448	18 55 00.	+ 10 14	219		STELLAR RING
LDN 0662	18 55 00.	+ 12 20	2580		DARK NEBULA
ZC 1855.0+8426	18 55 00.	+ 84 26	400		CLUSTER OF GALAXIES
LDN 0561	18 55 00.	- 04 00	5040		DARK NEBULA
B 123	18 55 01.	- 04 47	1080		DARK OBJECT
B 124	18 55 03.	- 04 25	180		DARK OBJECT
SCHO 0973	18 55 04.	- 03 48 36.	240		ISOLATED DARK CLOUD
LB 04362	18 55 06.	+ 36 25		18.4	FAINT BLUE STAR
ZWG 201.044	18 55 06.	+ 36 33		13.8	GALAXY
UGC 11380	18 55 06.	+ 36 33	96	13.8	GALAXY Sa-b
MCG+06-41-023	18 55 06.	+ 36 23	90	14.	GALAXY
PK028-03.1	18 55 09.7	- 05 31 44.	10	16.	PLANETARY NEBULA
ZWG 280.011	18 55 12.	+ 52 18		14.4	GALAXY
UGC 11381	18 55 12.	+ 52 18	54	14.4	GALAXY E?
MCG+09-31-011	18 55 12.	+ 52 18	60	15.	GALAXY
UGC 11382	18 55 12.	+ 74 36	72	16.5	GALAXY IRR
LDN 0551	18 55 12.	- 04 30	180		DARK NEBULA
RNGC 6732	18 55 13.	+ 52 18		14.5	GALAXY
MIL 69	18 55 18.	+ 02 06	600		SUPERNOVA REMNANT
MCG+08-34-028	18 55 18.	+ 50 56 30.	42	16.	GALAXY
ZWG 255.020	18 55 18.	+ 50 57		15.3	GALAXY
ISS 0660	18 55 18.	- 05 32	160		STELLAR RING
ISS 0449	18 55 24.	+ 12 41	155		STELLAR RING
UGC 11383	18 55 24.	+ 59 03	66	16.5	GALAXY Sc
RNGC 6731	18 55 25.	+ 43 00			NON-EXISTENT OBJECT
SC 1851-5429.5	18 55 29.	- 54 25 38.	36		NEBULA
SCHO 0974	18 55 30.	+ 10 17 00.	780		ISOLATED DARK CLOUD
ZC 1855.5+7953	18 55 30.	+ 79 53	2150		CLUSTER OF GALAXIES
ZWG 229.008	18 55 36.	+ 42 08		15.6	GALAXY
ISS 0534	18 55 36.	- 00 38	124		STELLAR RING
ISS 0661	18 55 36.	- 08 50	109		STELLAR RING
PK035-00.1	18 55 38.4	+ 01 32 54.	40		PLANETARY NEBULA
SCHO 0975	18 55 42.	- 00 49 42.	660		ISOLATED DARK CLOUD
B 125	18 55 43.	- 04 27	540		DARK OBJECT
SCHO 0976	18 55 44.	- 04 31 12.	520		ISOLATED DARK CLOUD
ZC 1855.8+7859	18 55 48.	+ 78 59	2080		CLUSTER OF GALAXIES
MCG+09-31-012	18 55 51.	+ 51 04 30.	60	15.	GALAXY
PK032-02.1	18 55 51.	- 01 07 55.			PLANETARY NEBULA
ZWG 202.001	18 55 54.	+ 37 56		14.9	GALAXY
ZWG 201.045	18 55 54.	+ 37 56		14.9	GALAXY
MCG+06-41-024	18 55 54.	+ 37 57	15	15.	GALAXY
KHAV 572	18 56	+ 10 58	6500		DARK NEBULA
KHAV 573	18 56	+ 12 16	4020		DARK NEBULA
KHAV 570	18 56	- 03 38	5270		DARK NEBULA
KHAV 571	18 56	- 05 08			DARK NEBULA
KHAV 569	18 56	- 07 14	3810		DARK NEBULA
LB 09936	18 56	- 80 45		13.9	FAINT BLUE STAR
LB 09994	18 56	- 88 13		12.9	FAINT BLUE STAR
LDN 0628	18 56 00.	+ 03 22	5700		DARK NEBULA
ZWG 280.012	18 56 00.	+ 51 05		15.6	GALAXY
ISS 0662	18 56 00.	- 09 02	131		STELLAR RING
MCG-05-45-001	18 56 00.	- 30 09	48	15.5	GALAXY
HN 0541	18 56 01.	- 62 08			NEBULA
IC 4803	18 56 01.	- 62 08			NONSTELLAR OBJECT
ISS 0450	18 56 06.	+ 12 28	106		STELLAR RING
AGU 55	18 56 06.	- 41 48 18.	48	12.5	PECULIAR GALAXY
MCG-07-39-002	18 56 06.	- 41 51	48	15.	GALAXY
PK028-04.2	18 56 10.	- 05 27		15.	PLANETARY NEBULA
ISS 0451	18 56 12.	+ 09 47	118		STELLAR RING
MCG+10-27-005	18 56 12.	+ 57 20	24	17.	GALAXY
7ZW 850	18 56 12.	+ 72 07			COMPACT GALAXY
GCL 106	18 56 12.	- 36 42	702	7.75	GLOBULAR STAR CLUSTER
RNGC 6723	18 56 13.	- 36 42		7.5	GLOBULAR CLUSTER
RNGC 6747	18 56 15.	+ 72 43		15.0	GALAXY
ISS 0375	18 56 18.	+ 20 48	449		STELLAR RING
ZWG 341.012	18 56 18.	+ 72 43		15.0	GALAXY
B 126	18 56 23.	- 04 36	480		DARK OBJECT
RNGC 6718	18 56 24.	- 66 10			UNVERIFIED SOUTHERN OBJECT
RNGC 6721	18 56 28.	- 57 51		13.0	GALAXY
HN 0542	18 56 29.	- 61 53			NEBULA
IC 4804	18 56 29.	- 61 53			NONSTELLAR OBJECT
LDN 0556	18 56 30.	- 04 35	1440		DARK NEBULA
ZWG 229.009	18 56 36.	+ 43 52		15.1	GALAXY
ZWG 255.021	18 56 42.	+ 46 12		15.7	GALAXY
ZWG 302.006	18 56 42.	+ 60 22		15.7	GALAXY
ZWG 341.013	18 56 42.	+ 69 50		15.4	GALAXY
ISS 0535	18 56 42.	- 01 55	130		STELLAR RING
B 324	18 56 46.	- 03 04			DARK OBJECT
MRSL 040+01/1	18 56 48.	+ 07 03	600		HII REGION
ZWG 202.002	18 56 48.	+ 36 20		15.7	GALAXY
ZWG 201.046	18 56 48.	+ 36 20		15.7	GALAXY
ZC 1856.8+6616	18 56 48.	+ 66 16	1280		CLUSTER OF GALAXIES
LDN 0568	18 56 48.	- 03 50	360		DARK NEBULA
SCHO 0977	18 56 48.	- 04 32 12.	680		ISOLATED DARK CLOUD
SC 1853-4914.0	18 56 51.	- 49 10 02.	18		NEBULA
PK052+07.1	18 56 54.2	+ 20 22 52.			PLANETARY NEBULA
SC 1852-5406.4	18 56 59.	- 54 02 26.	36		NEBULA
KHAV 575	18 57	+ 15 40	3400		DARK NEBULA
KHAV 574	18 57	+ 18 40	5670		DARK NEBULA
KHAV 576	18 57	- 16 32	2110		DARK NEBULA
LDN 0712	18 57 00.	+ 18 30	3720		DARK NEBULA
PK033-02.1	18 57 02.84	- 00 31 36.1	9	11.7	PLANETARY NEBULA
ZWG 229.010	18 57 06.	+ 43 24		15.4	GALAXY
MCG+08-34-029	18 57 06.	+ 46 33	18	16.	GALAXY
SC 1853-4720.3	18 57 06.	- 47 16 19.	42		NEBULA
HN 0544	18 57 10.	- 57 36			NEBULA
IC 4806	18 57 10.	- 57 36	144		GALAXY S0
SCHO 0978	18 57 11.	- 01 05 00.	800		ISOLATED DARK CLOUD
ISS 0376	18 57 12.	+ 15 53	82		STELLAR RING
UGC 11384	18 57 12.	+ 45 42	60	16.5	GALAXY Sb-c
ZWG 255.022	18 57 12.	+ 46 34		15.6	GALAXY
RNGC 6728	18 57 14.	- 09 01			NON-EXISTENT OBJECT
B 325	18 57 16.	- 04 08	900		DARK OBJECT
HN 0543	18 57 18.	- 63 07			NEBULA
IC 4805	18 57 18.	- 63 07			NONSTELLAR OBJECT
HN 0545	18 57 21.	- 45 23			NEBULA
IC 4808	18 57 21.	- 45 23			NONSTELLAR OBJECT
UGC 11385	18 57 24.	+ 19 22	72	16.0	GALAXY Sc-IRR
MCG+03-48-002	18 57 24.	+ 19 22	78	15.	GALAXY
ISS 0633	18 57 24.	- 08 13	200		STELLAR RING
ISS 0664	18 57 30.	- 05 42	176		STELLAR RING
SCHO 0979	18 57 34.	+ 10 34 00.	950		ISOLATED DARK CLOUD
SCHO 0981	18 57 38.	+ 12 41 12.	620		ISOLATED DARK CLOUD
SCHO 0980	18 57 38.	+ 13 29 42.	240		ISOLATED DARK CLOUD
HN 0547	18 57 41.	- 37 08			NEBULA
CED 165A	18 57 41.	- 37 08			DIFFUSE GALACTIC NEBULA
RNGC 6725	18 57 41.	- 54 00			NON-EXISTENT OBJECT
MIL 70	18 57 42.	+ 04 04	1800		SUPERNOVA REMNANT
ZWG 302.007	18 57 42.	+ 58 33		15.6	GALAXY
KARA.73B 0865	18 57 42.	+ 58 33	18	15.6	ISOLATED GALAXY E
IC 4812	18 57 42.	- 37 07 54.			NONSTELLAR OBJECT
RNGC 6719	18 57 45.	- 68 40			UNVERIFIED SOUTHERN OBJECT
SCHO 0982	18 57 47.	- 07 03 12.	300		ISOLATED DARK CLOUD
HN 1864	18 57 47.4	- 43 47 16.	12		NEBULA
7ZW 851	18 57 48.	+ 65 08			COMPACT GALAXY
ISS 0665	18 57 48.	- 08 01	103		STELLAR RING
SC 1853-5337.5	18 57 51.	- 53 33 28.	12		NEBULA
ZWG 341.014	18 57 54.	+ 72 09		15.6	GALAXY
SC 1853-5400.2	18 57 54.	- 53 56 10.	78		NEBULA
PK022-03.1	18 57 58.5	- 02 16 13.	6		PLANETARY NEBULA
KHAV 580	18 58	- 00 20	3090		DARK NEBULA
KHAV 579	18 58	- 01 20	2230		DARK NEBULA
KHAV 578	18 58	- 09 38	5010		DARK NEBULA
KHAV 577	18 58	- 15 14	4540		DARK NEBULA
SER 131.01	18 58	- 57 01	50	14.5	PEC. GALAXY
LB 09937	18 58	- 80 37		14.4	FAINT BLUE STAR
ZWG 229.011	18 58 00.	+ 41 48		15.7	GALAXY
PK078+18.1	18 58 00.	+ 48 24	36	15.0	PLANETARY NEBULA
7ZW 852	18 58 00.	+ 65 27			COMPACT GALAXY
7ZW 853	18 58 00.	+ 66 30			COMPACT GALAXY
ZC 1858.0+6801	18 58 00.	+ 68 01	1280		CLUSTER OF GALAXIES
LDN 0594	18 58 00.	- 01 50	240		DARK NEBULA
SCHO 0983	18 58 02.	+ 13 27 30.	280		ISOLATED DARK CLOUD
RNGC 6742	18 58 02.	+ 48 24			PLANETARY NEBULA
HN 0546	18 58 02.	- 57 00			NEBULA
IC 4807	18 58 02.	- 57 00			NONSTELLAR OBJECT
HN 1866	18 58 02.6	- 43 50 27.	30		NEBULA
PK017-10.1	18 58 06.	- 18 17	70	15.4	PLANETARY NEBULA
UGC 11386	18 58 12.	+ 73 13	78	16.0	GALAXY
RNGC 6735	18 58 13.	- 00 32			NON-EXISTENT OBJECT
CED 165B	18 58 16.	- 36 58			DIFFUSE GALACTIC NEBULA
UGC 11387	18 58 18.	+ 42 13	60	17.	GALAXY Sc
ISS 0666	18 58 18.	- 05 35	305		STELLAR RING
SCHO 0984	18 58 21.	- 00 47 00.	350		ISOLATED DARK CLOUD
RNGC 6727	18 58 25.	- 36 58			DIFFUSE NEBULA
RNGC 6726	18 58 25.	- 36 58			DIFFUSE NEBULA
RNGC 6729	18 58 25.	- 37 02			DIFFUSE NEBULA
SCHO 0985	18 58 28.	+ 17 44 24.	540		ISOLATED DARK CLOUD
MCG+07-39-007	18 58 30.	+ 40 50	24	17.	GALAXY
MCG+07-39-006	18 58 30.	+ 40 50	12	15.5	GALAXY
MCG+07-39-005	18 58 30.	+ 40 50 30.	24	15.5	GALAXY
CED 165C	18 58 35.	- 37 01			DIFFUSE GALACTIC NEBULA
ZWG 229.012	18 58 36.	+ 43 22		15.3	GALAXY
MCG+07-39-008	18 58 36.	+ 43 22	42	15.	GALAXY
HN 1865	18 58 40.9	- 60 29 38.	30		NEBULA
MCG+09-31-013	18 58 42.	+ 55 19	36	16.	GALAXY
MCG+08-34-030	18 58 45.	+ 48 47	30	16.	GALAXY
RNGC 6722	18 58 46.	- 64 58			UNVERIFIED SOUTHERN OBJECT
HN 0548	18 58 47.	- 56 14			NEBULA
IC 4810	18 58 47.	- 56 14	240		GALAXY
UGC 11388	18 58 48.	+ 28 42	60	15.5	GALAXY S
RNGC 6740	18 58 49.	+ 28 41		15.0	GALAXY
MCG+05-45-001	18 58 51.	+ 28 41	54	15.	GALAXY
B 127	18 58 52.	- 05 31	270		DARK OBJECT
LDN 0544	18 58 54.	- 05 30	300		DARK NEBULA
ARC 2316	18 58 55.	+ 79 59		17.4	RICH CLUSTER OF GALAXIES
LBN 0103	18 59	+ 02 05	60		BRIGHT NEBULA
ASS 11	18 59 00.	+ 03 32			OB ASSOCIATION AQL OB1
MCG+08-34-031	18 59 00.	+ 47 27	60	16.	GALAXY
ZWG 323.006	18 59 00.	+ 64 33		15.7	GALAXY
ZC 1859.0+6957	18 59 00.	+ 69 57	3020		CLUSTER OF GALAXIES
OCL 0088	18 59 00.	- 00 35	480	15.	OPEN STAR CLUSTER
LDN 0542	18 59 00.	- 05 10	1080		DARK NEBULA
SCHO 0986	18 59 00.	- 05 31 18.	270		ISOLATED DARK CLOUD
B 128	18 59 01.	- 04 39	600		DARK OBJECT
HN 0081	18 59 01.	- 13 14			NEBULA
IC 4816	18 59 01.	- 13 14			NONSTELLAR OBJECT
PK032-02.2	18 59 01.3	- 01 23 27.			PLANETARY NEBULA
RNGC 6738	18 59 02.	+ 11 32		8.0	OPEN CLUSTER
OCL 0101	18 59 06.	+ 11 32	930	8.5	OPEN STAR CLUSTER
ZWG 255.023	18 59 06.	+ 47 36		15.7	GALAXY
SCHO 0987	18 59 09.	- 00 53 24.	340		ISOLATED DARK CLOUD
ISS 0452	18 59 12.	+ 12 00	101		STELLAR RING
SC 1855-5155.2	18 59 13.	- 51 51 04.	18		NEBULA
HN 1867	18 59 13.9	- 61 11 59.	12		NEBULA
RNGC 6737	18 59 14.	- 18 36			NON-EXISTENT OBJECT
B 130	18 59 16.	- 05 39	420		DARK OBJECT
LB 00337	18 59 17.	- 25 52 48.		16.0	FAINT BLUE STAR
SC 1855-5336.7	18 59 19.	- 53 32 34.			NEBULA
MIN.46 10	18 59 25.	+ 02 05			DIFFUSE NEBULA

OBJECT NAME	RIGHT ASCEN.	DECLINATION	DIAM.	MAGN.	TYPE OF OBJECT
RNGC 6743	18 59 25.	+ 29 12			NON-EXISTENT OBJECT
B 129	18 59 25.	- 05 23	300		DARK OBJECT
HN 0549	18 59 25.	- 62 15			NEBULA
IC 4809	18 59 25.	- 62 15			NONSTELLAR OBJECT
SCHO 0988	18 59 26.	- 04 33 06.	770		ISOLATED DARK CLOUD
SCHO 0989	18 59 26.	- 05 37 36.	260		ISOLATED DARK CLOUD
PK036-01.1	18 59 28.9	+ 02 04 51.	278		PLANETARY NEBULA
ZWG 255.024	18 59 30.	+ 50 48		15.1	GALAXY
KARA.73B 0866	18 59 30.	+ 50 48	54	15.1	ISOLATED GALAXY S
LDN 0549	18 59 30.	- 05 20	300		DARK NEBULA
SCHO 0990	18 59 32.	+ 17 34 36.	380		ISOLATED DARK CLOUD
PK032-03.2	18 59 34.3	- 01 53 07.			PLANETARY NEBULA
SC 1855-5319.0	18 59 36.	- 53 14 51.	30		NEBULA
B 131	18 59 37.	- 04 27	240		DARK OBJECT
MCG+08-34-032	18 59 42.	+ 50 49	60	16.	GALAXY
SCHO 0991	18 59 45.	- 05 23 18.	440		ISOLATED DARK CLOUD
MCG+10-27-006	18 59 51.	+ 59 05	60	14.	GALAXY
ZWG 302.008	18 59 54.	+ 59 06		13.7	GALAXY
UGC 11389	18 59 54.	+ 59 06	60	13.7	GALAXY S
RNGC 6750	18 59 55.	+ 59 06		13.5	GALAXY
RNGC 6745	18 59 59.	+ 40 41		13.5	GALAXY
KHAV 581	19 00	+ 03 46	2110		DARK NEBULA
B 326	19 00	- 00 28	1500		DARK OBJECT
LDN 0697	19 00 00.	+ 16 00	3000		DARK NEBULA
LDN 0708	19 00 00.	+ 17 30	540		DARK NEBULA
LDN 0711	19 00 00.	+ 18 00	9900		DARK NEBULA
LDN 0710	19 00 00.	+ 18 00	840		DARK NEBULA
ZWG 229.013	19 00 00.	+ 40 41		13.3	GALAXY
UGC 11391	19 00 00.	+ 40 41	96	13.3	GALAXY TRP SYS
UGC 11390	19 00 00.	+ 84 30	84	16.5	GALAXY Sc
MCG+14-09-001	19 00 00.	+ 84 31	78	16.	GALAXY
LDN 0562	19 00 00.	- 04 35	540		DARK NEBULA
RNGC 6741	19 00 01.	- 00 31		11.5	PLANETARY NEBULA
HW 0050	19 00 04.	- 00 31			NEBULA
SCHO 0992	19 00 11.	+ 18 05 48.	970		ISOLATED DARK CLOUD
SCHO 0993	19 00 13.	+ 17 17 30.	670		ISOLATED DARK CLOUD
SC 1856-5155.2	19 00 13.	- 51 51 00.	18		NEBULA
7ZW 854	19 00 18.	+ 65 30			COMPACT GALAXY
PK047+04.1	19 00 23.1	+ 14 24 26.	7		PLANETARY NEBULA
UGC 11392	19 00 30.	+ 34 43	60	17.	GALAXY IRR
LDN 0553	19 00 30.	- 05 10	540		DARK NEBULA
SC 1856-5407.9	19 00 30.	- 54 03 41.	30		NEBULA
SCHO 0994	19 00 31.	- 07 02 24.	380		ISOLATED DARK CLOUD
SCHO 0995	19 00 34.	- 03 56 36.	970		ISOLATED DARK CLOUD
HN 1869	19 00 38.8	- 46 19 28.			NEBULA
HN 0552	19 00 39.	- 58 49			NEBULA
IC 4814	19 00 39.	- 58 49			NONSTELLAR OBJECT
ISS 0453	19 00 42.	+ 09 23	168		STELLAR RING
ISS 0377	19 00 42.	+ 15 08	230		STELLAR RING
MCG+05-45-002	19 00 42.	+ 27 12	72	15.	GALAXY
UGC 11393	19 00 42.	+ 27 14	78	15.5	GALAXY Sc
SHAH 170	19 00 42.	+ 79 37	264		GROUP OF COMPACT GALAXIES
HN 0550	19 00 44.	- 67 12			NEBULA
IC 4811	19 00 44.	- 67 12			NONSTELLAR OBJECT
HN 0551	19 00 45.	- 66 36			NEBULA
IC 4813	19 00 45.	- 66 36			NONSTELLAR OBJECT
HN 1868	19 00 47.9	- 59 01 58.			NEBULA
SER 129.04	19 00 48.	- 66 35	100		LOW SURF. BRGHTNSS GALAXY
HN 1870	19 00 49.2	- 46 19 57.	30		NEBULA
MRSL 046+03/1	19 00 54.	+ 14 03	720		HII REGION
MCG+09-31-014	19 00 54.	+ 51 03	30	16.	GALAXY
LB 03114	19 00 54.	- 62 37		11.5	FAINT BLUE STAR
SCHO 0996	19 00 56.	- 03 35 36.	340		ISOLATED DARK CLOUD
KHAV 583	19 01	- 07 44	3940		DARK NEBULA
KHAV 582	19 01	- 08 08	4380		DARK NEBULA
LDN 0719	19 01 00.	+ 19 40	1200		DARK NEBULA
7ZW 855	19 01 00.	+ 77 02			COMPACT GALAXY
SC 1857-4759.6	19 01 02.	- 47 55 20.	42		NEBULA
ISS 0536	19 01 06.	- 01 30	133		STELLAR RING
SCHO 0997	19 01 10.	- 01 46 30.	720		ISOLATED DARK CLOUD
PK042+01.1	19 01 10.1	+ 08 39 59.			PLANETARY NEBULA
MRSL 057+09/1	19 01 12.	+ 25 45	360		HII REGION
ABC 2315	19 01 13.	+ 69 53		16.3	RICH CLUSTER OF GALAXIES
MRSL 036-01/1	19 01 18.	+ 02 14	1500		HII REGION
ISS 0378	19 01 24.	+ 16 23	58		STELLAR RING
SCHO 0998	19 01 25.	+ 17 23 48.	320		ISOLATED DARK CLOUD
PK051+06.1	19 01 26.2	+ 19 16 53.	45		PLANETARY NEBULA
UGC 11394	19 01 30.	+ 27 32	120	16.0	GALAXY Sc
UGC 11395	19 01 30.	+ 71 42	60	16.5	GALAXY Sc
LDN 0567	19 01 30.	- 04 33	600		DARK NEBULA
RNGC 6733	19 01 30.	- 62 16			UNVERIFIED SOUTHERN OBJECT
MIL 71	19 01 31.	+ 05 22 30.	276		SUPERNOVA REMNANT
SCHO 0999	19 01 32.	- 04 32 18.	710		ISOLATED DARK CLOUD
PK048+04.1	19 01 32.1	+ 16 21 49.			PLANETARY NEBULA
SCHO 1000	19 01 34.	+ 13 04 48.	700		ISOLATED DARK CLOUD
ISS 0485	19 01 36.	+ 04 28	163		STELLAR RING
ISS 0379	19 01 36.	+ 18 57	169		STELLAR RING
MCG+05-45-003	19 01 36.	+ 27 31	126	15.	GALAXY
RNGC 6748	19 01 38.	+ 21 32			NON-EXISTENT OBJECT
LDN 0552	19 01 42.	- 05 20	720		DARK NEBULA
B 327	19 01 47.	- 05 13	1800		DARK OBJECT
B 132	19 01 49.	- 04 31			DARK OBJECT
SC 1857-5415.4	19 01 49.	- 54 11 06.	24		NEBULA
PK043+01.1	19 01 49.0	+ 10 06 16.			PLANETARY NEBULA
ISS 0537	19 01 54.	+ 01 32	119		STELLAR RING
MCG+08-34-033	19 01 54.	+ 45 32	42	16.	GALAXY
IC 4818	19 01 54.	- 55 10			NONSTELLAR OBJECT
HN 0554	19 01 55.	- 55 11			NEBULA
PK003-17.1	19 01 58.	- 33 15	5	13.4	PLANETARY NEBULA
IC 4817	19 01 58.	- 56 13			NONSTELLAR OBJECT
HN 0555	19 01 58.	- 56 14			NEBULA
LBN 0104	19 02	+ 02 20	1620		BRIGHT NEBULA
KHAV 585	19 02	+ 21 16	2230		DARK NEBULA
KHAV 584	19 02	- 06 26	4020		DARK NEBULA
ISS 0538	19 02 00.	+ 00 47	76		STELLAR RING
UGC 11396	19 02 00.	+ 24 17	60	17.	GALAXY
ZWG 202.003	19 02 00.	+ 33 45		14.0	GALAXY
UGC 11397	19 02 00.	+ 33 45	72	14.0	GALAXY SBa
ZWG 256.001	19 02 00.	+ 45 33		15.0	GALAXY
ZWG 255.025	19 02 00.	+ 45 33		15.0	GALAXY
UGC 11398	19 02 00.	+ 54 00	60	16.5	GALAXY Sc
MCG+10-27-007	19 02 00.	+ 62 22	24	17.	GALAXY
MCG+09-31-015	19 02 03.	+ 54 01	48	16.	GALAXY
SCHO 1001	19 02 04.	+ 16 19 48.	300		ISOLATED DARK CLOUD
PK035-02.1	19 02 07.4	+ 01 17 48.			PLANETARY NEBULA
IC 4815	19 02 10.	- 61 45			NONSTELLAR OBJECT
HN 0553	19 02 10.	- 61 46			NEBULA
SCHO 1002	19 02 11.	+ 17 14 24.	750		ISOLATED DARK CLOUD

OBJECT NAME	RIGHT ASCEN.	DECLINATION	DIAM.	MAGN.	TYPE OF OBJECT
B 328	19 02 11.	- 04 19	240		DARK OBJECT
ISS 0486	19 02 12.	+ 03 05	61		STELLAR RING
ZWG 256.002	19 02 12.	+ 50 42		15.4	GALAXY
ZWG 255.026	19 02 12.	+ 50 42		15.4	GALAXY
MCG-07-39-003	19 02 12.	- 42 28	42	15.	GALAXY
SER 132.02	19 02 12.	- 65 33	300	16.5	COMPACT GROUP OF 3 GLXIES
RNGC 6730	19 02 14.	- 68 59			UNVERIFIED SOUTHERN OBJECT
ISS 0539	19 02 18.	+ 00 24	82		STELLAR RING
PK050+05.1	19 02 18.	+ 17 53	40	16.5	PLANETARY NEBULA
MCG-07-39-004	19 02 18.	- 43 49	30	14.	GALAXY
HN 1872	19 02 20.2	- 43 48 08.	12		NEBULA
PK034-02.1	19 02 20.8	+ 00 15 55.			PLANETARY NEBULA
RNGC 6734	19 02 21.	- 65 32			UNVERIFIED SOUTHERN OBJECT
ZWG 229.014	19 02 24.	+ 40 41		15.6	GALAXY
HN 1873	19 02 26.8	- 42 26 38.	18		NEBULA
GCL 107	19 02 30.	+ 01 42			GLOBULAR STAR CLUSTER
MCG-07-39-005	19 02 30.	- 42 28	72	13.	GALAXY
RNGC 6736	19 02 33.	- 65 30			UNVERIFIED SOUTHERN OBJECT
PK048+04.2	19 02 35.8	+ 15 43 03.			PLANETARY NEBULA
OCL 0091	19 02 36.	+ 01 48	360	18.	OPEN STAR CLUSTER
ISS 0540	19 02 36.	+ 02 20	345		STELLAR RING
ZWG 323.007	19 02 36.	+ 63 46		15.3	GALAXY
ZCG 1902+63	19 02 36.	+ 63 46		15.3	COMPACT GALAXY
RNGC 6749	19 02 38.	+ 01 48			GLOBULAR CLUSTER
HN 1871	19 02 39.9	- 60 27 39.	18		NEBULA
ISS 0541	19 02 42.	+ 02 02	125		STELLAR RING
7ZW 856	19 02 42.	+ 63 47			COMPACT GALAXY
ZWG 323.008	19 02 42.	+ 63 47		15.0	GALAXY
IC 4819	19 02 42.	- 59 32			NONSTELLAR OBJECT
HN 0556	19 02 42.	- 59 33			NEBULA
ISS 0542	19 02 48.	+ 00 15	233		STELLAR RING
HN 1874	19 02 53.5	- 46 39 54.	36		NEBULA
MCG+09-31-016	19 02 54.	+ 51 04	60	16.	GALAXY
ISS 0543	19 02 54.	- 00 23	85		STELLAR RING
ISS 0544	19 02 54.	- 02 22	95		STELLAR RING
SC 1859-4647.2	19 02 54.	- 46 42 48.	42		NEBULA
KHAV 586	19 03	- 12 14	4910		DARK NEBULA
LDN 0624	19 03 00.	+ 01 30	2820		DARK NEBULA
ISS 0545	19 03 00.	+ 01 39	123		STELLAR RING
LDN 0667	19 03 00.	+ 12 00	6540		DARK NEBULA
LDN 0715	19 03 00.	+ 19 00	1320		DARK NEBULA
LDN 0577	19 03 00.	- 04 00	3540		DARK NEBULA
ZWG 202.004	19 03 06.	+ 34 22		14.6	GALAXY
UGC 11399	19 03 06.	+ 34 22	96	14.7	GALAXY S
RNGC 6751	19 03 10.	- 06 04		12.0	PLANETARY NEBULA
ISS 0547	19 03 12.	+ 00 32	81		STELLAR RING
ISS 0546	19 03 12.	+ 03 16	158		STELLAR RING
UGC 11400	19 03 12.	+ 78 58	66	16.5	GALAXY IRR
ISS 0548	19 03 12.	- 00 57	93		STELLAR RING
RNGC 6739	19 03 14.	- 61 26			UNVERIFIED SOUTHERN OBJECT
PK029-05.1	19 03 14.98	- 06 04 10.0	21	12.2	PLANETARY NEBULA
ZWG 256.003	19 03 18.	+ 49 34		15.1	GALAXY
ZWG 255.027	19 03 18.	+ 49 34		15.1	GALAXY
LDN 0540	19 03 24.	- 06 30	1320		DARK NEBULA
LDN 0531	19 03 24.	- 06 55	240		DARK NEBULA
SCHO 1003	19 03 26.	+ 15 43 36.	480		ISOLATED DARK CLOUD
ISS 0549	19 03 30.	+ 03 05	315		STELLAR RING
ISS 0550	19 03 30.	- 01 04	97		STELLAR RING
B 133	19 03 31.	- 06 58			DARK OBJECT
MCG+09-31-017	19 03 33.	+ 52 23	60	16.	GALAXY
ISS 0487	19 03 36.	+ 02 58	58		STELLAR RING
SC 1859-5207.5	19 03 36.	- 52 03 04.	54		NEBULA
SCHO 1005	19 03 38.	+ 17 09 36.	530		ISOLATED DARK CLOUD
SCHO 1004	19 03 38.	- 06 54 12.	480		ISOLATED DARK CLOUD
ISS 0552	19 03 42.	+ 00 10	183		STELLAR RING
ISS 0551	19 03 42.	+ 01 52	245		STELLAR RING
SCHO 1006	19 03 42.	+ 16 12 24.	470		ISOLATED DARK CLOUD
MCG+09-31-018	19 03 51.	+ 55 09	36	16.	GALAXY
ZWG 229.015	19 03 54.	+ 42 23		15.4	GALAXY
SCHO 1007	19 03 56.	+ 03 36 42.	420		ISOLATED DARK CLOUD
KHAV 587	19 04	- 03 55	4020		DARK NEBULA
LDN 0625	19 04 00.	+ 01 30	180		DARK NEBULA
LDN 0626	19 04 00.	+ 01 35	180		DARK NEBULA
LDN 0636	19 04 00.	+ 04 35	1860		DARK NEBULA
7ZW 857	19 04 00.	+ 67 45			COMPACT GALAXY
SHAH 171	19 04 00.	+ 79 30	144		GROUP OF COMPACT GALAXIES
LDN 0590	19 04 00.	- 03 05	12840		DARK NEBULA
LDN 0543	19 04 00.	- 06 15	300		DARK NEBULA
HN 1875	19 04 01.5	- 61 59 03.	18		NEBULA
ISS 0488	19 04 06.	+ 03 07	149		STELLAR RING
ZWG 256.004	19 04 06.	+ 50 17		15.7	GALAXY
ZWG 280.013	19 04 06.	+ 55 39		14.0	GALAXY
RNGC 6757	19 04 06.	+ 55 39		14.0	GALAXY
UGC 11401	19 04 06.	+ 55 39	114	14.0	GALAXY SB0/SBa
MCG+09-31-019	19 04 06.	+ 55 39	78	13.	GALAXY
ISS 0667	19 04 06.	- 05 32	173		STELLAR RING
ISS 0553	19 04 12.	+ 01 51	266		STELLAR RING
MCG+08-35-001	19 04 12.	+ 50 17	42	16.	GALAXY
ZWG 341.015	19 04 12.	+ 72 58		15.2	GALAXY
UGC 11402	19 04 12.	+ 72 58	66	15.2	GALAXY S
B 134	19 04 12.	- 06 19	360		DARK OBJECT
ZWG 341.016	19 04 18.	+ 73 43		15.3	GALAXY
SCHO 1008	19 04 18.	- 06 19 00.	240		ISOLATED DARK CLOUD
ISS 0668	19 04 18.	- 07 06	229		STELLAR RING
PK040-00.1	19 04 19.3	+ 06 19 08.	35	16.9	PLANETARY NEBULA
SCHO 1009	19 04 22.	+ 17 05 42.	390		ISOLATED DARK CLOUD
7ZW 858	19 04 24.	+ 65 24			COMPACT GALAXY
SCHO 1010	19 04 28.	+ 11 32 24.	730		ISOLATED DARK CLOUD
B 329	19 04 29.	+ 03 07	360		DARK OBJECT
MCG+07-39-009	19 04 30.	+ 40 47 30.	69	15.	GALAXY
ZWG 229.016	19 04 30.	+ 40 49		15.7	GALAXY
UGC 11403	19 04 30.	+ 40 49	84	15.7	GALAXY Sb?
LDN 0574	19 04 30.	- 04 30	960		DARK NEBULA
HN 0557	19 04 31.	- 63 33			NEBULA
MIL 65	19 04 32.	- 03 06 30.	3600		SUPERNOVA REMNANT
IC 4820	19 04 32.	- 63 32			NONSTELLAR OBJECT
ISS 0554	19 04 42.	+ 00 29	128		STELLAR RING
MCG+07-39-010	19 04 45.	+ 42 22	30	17.	GALAXY
RNGC 6754A	19 04 45.	- 51 08			GALAXY
ISS 0557	19 04 48.	+ 03 09	287		STELLAR RING
SCHO 1011	19 04 48.	+ 16 24 06.	490		ISOLATED DARK CLOUD
LDN 0581	19 04 48.	- 04 00	1080		DARK NEBULA
DV.56 N6754A	19 04 48.	- 51 08			GALAXY
SCHO 1012	19 04 51.	- 00 11 18.	480		ISOLATED DARK CLOUD
SCHO 1013	19 04 51.	- 00 48 12.	720		ISOLATED DARK CLOUD
B 135	19 04 54.	- 04 00	780		DARK OBJECT
SCHO 1014	19 04 58.	+ 04 02 30.	420		ISOLATED DARK CLOUD
KHAV 588	19 05	+ 18 29	2660		DARK NEBULA

OBJECT NAME	RIGHT ASCEN.	DECLINATION	DIAM.	MAGN.	TYPE OF OBJECT
KHAV 590	19 05	+ 20 05	3610		DARK NEBULA
KHAV 589	19 05	+ 22 23	2570		DARK NEBULA
UGC 11404	19 05 00.	+ 28 56	132	14.0	GALAXY Sb
LDN 0550	19 05 00.	- 05 50	7140		DARK NEBULA
RNGC 6763	19 05 01.	+ 63 52			NON-EXISTENT OBJECT
RNGC 6744	19 05 02.	- 63 56		10.0	GALAXY
MIL 73	19 05 05.	+ 07 03 36.	600		SUPERNOVA REMNANT
ISS 0555	19 05 06.	+ 00 39	139		STELLAR RING
OCL 0097	19 05 06.	+ 04 13	420		OPEN STAR CLUSTER
MCG+05-45-004	19 05 06.	+ 28 55	150	13.5	GALAXY
CED 166	19 05 11.	+ 04 16	60		DIFFUSE GALACTIC NEBULA
ISS 0558	19 05 12.	+ 03 14	278		STELLAR RING
SCHO 1015	19 05 12.	+ 03 30 18.	370		ISOLATED DARK CLOUD
ZWG 323.009	19 05 12.	+ 63 52		14.2	GALAXY
UGC 11405	19 05 12.	+ 64 08	96	14.2	GALAXY S0-a
ZC 1905.2+7535	19 05 12.	+ 75 35	2350		CLUSTER OF GALAXIES
MCG-05-45-002	19 05 12.	- 32 19	24	15.5	GALAXY
RNGC 6762	19 05 13.	+ 63 52		14.0	GALAXY
RNGC 6755	19 05 15.	+ 04 08		9.0	OPEN CLUSTER
SCHO 1016	19 05 16.	+ 17 14 42.	700		ISOLATED DARK CLOUD
OCL 0096	19 05 18.	+ 04 09	2160	9.44	OPEN STAR CLUSTER
SC 1901-4809.1	19 05 23.	- 48 04 32.	30		NEBULA
ISS 0556	19 05 24.	+ 00 15	418		STELLAR RING
LDN 0554	19 05 24.	- 05 45	660		DARK NEBULA
HN 0558	19 05 24.	- 55 05			NEBULA
IC 4821	19 05 24.	- 55 05			NONSTELLAR OBJECT
HN 1876	19 05 27.6	- 44 47 49.	78		NEBULA
SCHO 1017	19 05 30.	+ 16 31 48.	410		ISOLATED DARK CLOUD
MCG+07-39-011	19 05 30.	+ 41 39	48	15.	GALAXY
ZWG 229.017	19 05 30.	+ 41 41		15.6	GALAXY
MCG-07-39-006	19 05 30.	- 44 48	78	15.	GALAXY
LB 00338	19 05 34.	- 26 42 00.		16.3	FAINT BLUE STAR
ISS 0454	19 05 36.	+ 13 11	101		STELLAR RING
ZWG 256.005	19 05 36.	+ 46 06		15.3	GALAXY
MCG+08-35-002	19 05 36.	+ 50 16	54	15.	GALAXY
SCHO 1018	19 05 37.	- 00 04 00.	390		ISOLATED DARK CLOUD
SCHO 1019	19 05 39.	+ 04 03 42.	490		ISOLATED DARK CLOUD
RNGC 6759	19 05 40.	+ 50 15		15.0	GALAXY
SC 1901-5112.0	19 05 40.	- 51 07 25.	42		NEBULA
ZWG 229.018	19 05 42.	+ 39 51		15.5	GALAXY
ZWG 229.019	19 05 42.	+ 40 55		15.5	GALAXY
ZWG 256.006	19 05 42.	+ 50 15		15.2	GALAXY
RNGC 6746	19 05 43.	- 62 03			UNVERIFIED SOUTHERN OBJECT
ISS 0489	19 05 54.	+ 03 18	99		STELLAR RING
ISS 0380	19 05 54.	+ 15 10	113		STELLAR RING
HN 1877	19 05 56.4	- 45 35 59.	18		NEBULA
KHAV 591	19 06	- 06 07	4540		DARK NEBULA
LDN 0734	19 06 00.	+ 22 00	3300		DARK NEBULA
LDN 0580	19 06 00.	- 04 10	780		DARK NEBULA
SCHO 1020	19 06 06.	+ 17 28 00.	620		ISOLATED DARK CLOUD
SCHO 1021	19 06 07.	+ 14 37 00.	380		ISOLATED DARK CLOUD
RNGC 6756	19 06 10.	+ 04 35		10.5	OPEN CLUSTER
B 136	19 06 11.	- 04 05	480		DARK OBJECT
OCL 0099	19 06 12.	+ 04 36	660	10.7	OPEN STAR CLUSTER
SER 129.02	19 06 12.	- 64 17	100		LOW SURF. BRGHTNSS GALAXY
MRSL 036-01/2	19 06 24.	+ 05 32	180		HII REGION
SCHO 1022	19 06 24.	+ 15 25 36.	380		ISOLATED DARK CLOUD
MCG+07-39-012	19 06 24.	+ 41 20	48	15.	GALAXY
ZWG 229.020	19 06 24.	+ 41 21		15.4	GALAXY
GCL 108	19 06 24.	- 60 04	2514	7.2	GLOBULAR STAR CLUSTER
RNGC 6752	19 06 24.	- 60 04		7.0	GLOBULAR CLUSTER
SC 1902-4815.5	19 06 26.	- 48 10 51.	6		NEBULA
HN 1878	19 06 28.8	- 44 45 15.	24		NEBULA
LB 00339	19 06 31.	- 22 56 00.		15.9	FAINT BLUE STAR
ISS 0490	19 06 36.	+ 03 38	175		STELLAR RING
PK055+06.1	19 06 36.	+ 22 54	73	17.1	PLANETARY NEBULA
7ZW 859	19 06 36.	+ 76 44			COMPACT GALAXY
MCG+07-39-013	19 06 45.	+ 42 59	120	14.	GALAXY
ZWG 229.021	19 06 48.	+ 43 00		14.7	GALAXY
UGC 11406	19 06 48.	+ 43 00	126	14.7	GALAXY SB
PK033-04.1	19 06 49.4	- 01 13 59.			PLANETARY NEBULA
MCG-03-49-001	19 06 51.	- 18 00	36	15.	GALAXY
MCG+07-39-014	19 06 54.	+ 41 32	45	15.	GALAXY
ZWG 229.022	19 06 54.	+ 41 34		15.2	GALAXY
AGU 56	19 06 54.	- 44 15 00.	72	12.5	PECULIAR GALAXY
RNGC 6764	19 06 58.	+ 50 51		13.0	GALAXY
LBN 0109	19 07	+ 05 30	120		BRIGHT NEBULA
KHAV 592	19 07	+ 16 47	3810		DARK NEBULA
LDN 0635	19 07 00.	+ 04 03	120		DARK NEBULA
LDN 0728	19 07 00.	+ 21 00	4020		DARK NEBULA
MCG+08-35-003	19 07 00.	+ 50 02	126	14.	GALAXY
ZWG 256.007	19 07 00.	+ 50 50		13.2	GALAXY
UGC 11407	19 07 00.	+ 50 50	138	13.2	GALAXY SBb
MCG+09-31-020	19 07 00.	+ 52 26 30.	36	15.	GALAXY
SCHO 1023	19 07 02.	+ 17 45 42.	520		ISOLATED DARK CLOUD
HN 1879	19 07 05.7	- 44 15 36.	60		NEBULA
MCG-07-39-007	19 07 06.	- 44 16	78	15.	GALAXY
PK045+01.1	19 07 06.3	+ 11 55 54.			PLANETARY NEBULA
PK033-05.1	19 07 09.	- 02 26	54	15.4	PLANETARY NEBULA
ZWG 229.023	19 07 12.	+ 43 16		15.2	GALAXY
MCG+09-31-021	19 07 12.	+ 52 08	60	15.	GALAXY
ZWG 280.014	19 07 12.	+ 52 28		15.4	GALAXY
SER 130.02	19 07 12.	- 52 59	50	17.	PEC. GALAXY
RNGC 6753	19 07 12.	- 57 08		12.0	GALAXY
HN 1880	19 07 16.2	- 44 53 30.	36		NEBULA
ZWG 280.015	19 07 18.	+ 52 10		15.5	GALAXY
PK044+01.1	19 07 21.8	+ 11 00 23.	3		PLANETARY NEBULA
ZWG 280.016	19 07 24.	+ 53 17		15.2	GALAXY
KARA.73B 0867	19 07 24.	+ 53 17	60	15.2	ISOLATED GALAXY S
MRSL 031-05/1	19 07 24.	- 03 57	2400		HII REGION
MCG+09-31-022	19 07 27.	+ 53 16	54	16.	GALAXY
SCHO 1024	19 07 27.	- 01 59 42.	1310		ISOLATED DARK CLOUD
IC 4823	19 07 27.	- 64 05			NONSTELLAR OBJECT
RNGC 6754	19 07 28.	- 50 43		13.5	GALAXY
HN 0559	19 07 28.	- 64 04			NEBULA
SC 1903-5302.0	19 07 29.	- 52 57 18.	36		NEBULA
SCHO 1025	19 07 41.	+ 14 35 24.	480		ISOLATED DARK CLOUD
ISS 0455	19 07 42.	+ 12 44	100		STELLAR RING
ARC 2318	19 07 46.	+ 78 05		17.0	RICH CLUSTER OF GALAXIES
MCG+07-39-015	19 07 48.	+ 42 58	24	15.5	GALAXY
ZWG 229.024	19 07 48.	+ 42 59		14.8	GALAXY
SCHO 1027	19 07 54.	+ 00 16 18.	430		ISOLATED DARK CLOUD
SCHO 1026	19 07 54.	+ 01 04 18.	740		ISOLATED DARK CLOUD
ISS 0381	19 07 54.	+ 18 54	110		STELLAR RING
SCHO 1028	19 07 56.	+ 15 55 12.	290		ISOLATED DARK CLOUD
KHAV 594	19 08	+ 17 53	4020		DARK NEBULA
KHAV 593	19 08	+ 23 53	1570		DARK NEBULA
LB 09995	19 08	- 87 10		13.4	FAINT BLUE STAR
LDN 0693	19 08 00.	+ 14 25	16740		DARK NEBULA
ISS 0333	19 08 00.	+ 20 57	168		STELLAR RING
LDN 0759	19 08 00.	+ 24 00	2400		DARK NEBULA
MCG+05-45-005	19 08 00.	+ 28 51	54	15.	GALAXY
SCHO 1029	19 08 04.	+ 11 49 48.	260		ISOLATED DARK CLOUD
PK037-02.1	19 08 05.4	+ 02 44 33.			PLANETARY NEBULA
HN 0561	19 08 08.	- 57 17			NEBULA
IC 4826	19 08 08.	- 57 18			NONSTELLAR OBJECT
SC 1904-4732.9	19 08 11.	- 47 28 08.	72		NEBULA
ISS 0382	19 08 12.	+ 18 42	168		STELLAR RING
ISS 0334	19 08 12.	+ 24 14	86		STELLAR RING
MCG+07-39-016	19 08 12.	+ 41 28	78	14.5	GALAXY
ZWG 229.025	19 08 12.	+ 41 30		14.9	GALAXY
UGC 11408	19 08 12.	+ 41 30	84	14.9	GALAXY Sa-b
HN 1882	19 08 12.1	- 45 16 02.	18		NEBULA
MCG+09-31-023	19 08 18.	+ 52 48	48	15.	GALAXY
ZWG 280.017	19 08 18.	+ 52 50		15.5	GALAXY
7ZW 860	19 08 18.	+ 78 00			COMPACT GALAXY
IC 4829	19 08 22.	- 56 38			NONSTELLAR OBJECT
HN 0563	19 08 23.	- 56 37			NEBULA
MIL 72	19 08 24.	+ 05 04	3000		SUPERNOVA REMNANT
ZWG 229.026	19 08 24.	+ 39 56		15.4	GALAXY
LB 00340	19 08 26.	- 25 39 30.		15.0	FAINT BLUE STAR
HN 1881	19 08 26.5	- 60 28 14.	24		NEBULA
RNGC 6766	19 08 28.	+ 46 11			NON-EXISTENT OBJECT
ZWG 229.027	19 08 30.	+ 42 24		15.7	GALAXY
ZC 1908.5+7805	19 08 30.	+ 78 05	1010		CLUSTER OF GALAXIES
ARC 2317	19 08 31.	+ 68 59		17.6	RICH CLUSTER OF GALAXIES
HN 0562	19 08 35.	- 62 10			NEBULA
IC 4824	19 08 35.	- 62 11			NONSTELLAR OBJECT
GCL 109	19 08 36.	+ 00 57			GLOBULAR STAR CLUSTER
ISS 0383	19 08 36.	+ 17 09	163		STELLAR RING
ZWG 229.028	19 08 36.	+ 42 19		15.7	GALAXY
RNGC 6760	19 08 38.	+ 00 57		11.0	GLOBULAR CLUSTER
MIL 75	19 08 39.	+ 09 00 30.	324		SUPERNOVA REMNANT
SC 1904-5402.1	19 08 41.	- 53 57 19.	60		NEBULA
ISS 0384	19 08 42.	+ 15 48	142		STELLAR RING
ZC 1908.7+6854	19 08 42.	+ 68 54	1080		CLUSTER OF GALAXIES
SCHO 1030	19 08 46.	- 00 26 48.	690		ISOLATED DARK CLOUD
HN 0564	19 08 47.	- 60 56			NEBULA
IC 4827	19 08 47.	- 60 57			NONSTELLAR OBJECT
ZC 1908.8+6742	19 08 48.	+ 67 42	4700		CLUSTER OF GALAXIES
MCG-05-45-003	19 08 48.	- 32 14	96	14.5	GALAXY
UGC 11409	19 08 54.	+ 37 33	72	16.5	GALAXY S
ZC 1908.9+6416	19 08 54.	+ 64 16	1280		CLUSTER OF GALAXIES
ZWG 341.017	19 08 54.	+ 73 00		15.7	GALAXY
UGC 11410	19 08 54.	+ 73 00	72	15.7	GALAXY SB:a
IC 4822	19 08 57.	- 72 31			NONSTELLAR OBJECT
SCHO 1031	19 08 58.	+ 11 56 24.	290		ISOLATED DARK CLOUD
HN 0560	19 08 58.	- 72 31			NEBULA
KHAV 595	19 09	+ 15 17	4490		DARK NEBULA
LBN 0127	19 09	+ 16 45	60		BRIGHT NEBULA
ISS 0385	19 09 00.	+ 15 03	115		STELLAR RING
ZWG 229.029	19 09 00.	+ 44 13		15.6	GALAXY
ZWG 256.008	19 09 00.	+ 49 05		15.6	GALAXY
MCG+08-35-004	19 09 00.	+ 49 05	42	16.	GALAXY
7ZW 861	19 09 00.	+ 65 54			COMPACT GALAXY
ZWG 341.018	19 09 00.	+ 70 12		14.4	GALAXY
UGC 11411	19 09 00.	+ 70 12	54	14.4	GALAXY PECULR
LDN 0622	19 09 00.	- 00 25	2160		DARK NEBULA
MCG+07-39-017	19 09 03.	+ 44 13	48	16.	GALAXY
HN 0565	19 09 05.	- 62 09			NEBULA
IC 4828	19 09 05.	- 62 10			NONSTELLAR OBJECT
OCL 0103	19 09 06.	+ 12 59	240	14.	OPEN STAR CLUSTER
UGC 11411A	19 09 06.	+ 60 03	132	16.0	GALAXY Sc
HN 1883	19 09 08.4	- 46 05 46.	36		NEBULA
SC 1905-4610.8	19 09 10.	- 46 05 58.	72		NEBULA
PK062+09.1	19 09 10.88	+ 30 27 53.8	38		PLANETARY NEBULA
OCL 0092	19 09 12.	+ 00 58	420	10.7	OPEN STAR CLUSTER
ISS 0386	19 09 12.	+ 15 32	108		STELLAR RING
SER 132.01	19 09 12.	- 62 22	1800	14.	LOOSE GROUP OF 5 GALAXIES
RNGC 6765	19 09 14.	+ 30 27			PLANETARY NEBULA
MCG+09-31-024	19 09 15.	+ 55 05 30.	12	16.	GALAXY
PK050+03.1	19 09 16.5	+ 16 46 32.	99		PLANETARY NEBULA
ISS 0335	19 09 18.	+ 22 13	98		STELLAR RING
SC 1906-2912.3	19 09 20.	- 29 07 26.	30		NEBULA
MCG+10-27-008	19 09 24.	+ 60 02	120	15.	GALAXY
MCG-05-45-004	19 09 24.	- 29 07	30	15.	GALAXY
BON 3	19 09 26.	+ 52 08		17.	VARIABLE GALAXY
SC 1905-4755.8	19 09 27.	- 47 50 57.	30		NEBULA
HN 0566	19 09 28.	- 59 22			NEBULA
IC 4830	19 09 28.	- 59 23			NONSTELLAR OBJECT
MCG+09-31-025	19 09 33.	+ 52 02 30.	60	14.	GALAXY
ZWG 280.018	19 09 36.	+ 52 05		14.6	GALAXY
UGC 11412	19 09 36.	+ 52 05	72	14.6	GALAXY Sb-c
KARA.73B 0868	19 09 36.	+ 52 05	60	14.6	ISOLATED GALAXY S
SC 1905-4951.5	19 09 37.	- 49 46 38.	18		NEBULA
IC 4831	19 09 37.	- 62 22			NONSTELLAR OBJECT
SC 1906-2334.4	19 09 38.	- 23 29 30.	18		NEBULA
PK037-03.1	19 09 38.1	+ 02 32 56.			PLANETARY NEBULA
RNGC 6758	19 09 49.	- 56 24		13.0	GALAXY
PK048+02.1	19 09 49.3	+ 15 03 59.	6		PLANETARY NEBULA
HN 1885	19 09 49.6	- 47 08 49.	18		NEBULA
RNGC 6767	19 09 50.	+ 37 37			NON-EXISTENT OBJECT
SC 1906-4713.9	19 09 51.	- 47 09 01.	36		NEBULA
HN 0569	19 09 53.	- 56 42			NEBULA
IC 4832	19 09 53.	- 56 42			NONSTELLAR OBJECT
KHAV 596	19 10	- 04 49			DARK NEBULA
LDN 0631	19 10 00.	+ 02 30	540		DARK NEBULA
LDN 0639	19 10 00.	+ 05 00	4740		DARK NEBULA
LDN 0649	19 10 00.	+ 08 35	19440		DARK NEBULA
LDN 0713	19 10 00.	+ 17 00	8100		DARK NEBULA
LDN 0714	19 10 00.	+ 17 50	360		DARK NEBULA
LDN 0718	19 10 00.	+ 18 30	2940		DARK NEBULA
LDN 0724	19 10 00.	+ 20 00	9900		DARK NEBULA
LDN 0777	19 10 00.	+ 25 00	1920		DARK NEBULA
SCHO 1032	19 10 04.	+ 14 15 12.	290		ISOLATED DARK CLOUD
HN 0568	19 10 06.	- 62 21			NEBULA
HN 1884	19 10 07.5	- 58 20 43.	18		NEBULA
SCHO 1033	19 10 10.	+ 13 31 00.	620		ISOLATED DARK CLOUD
SCHO 1034	19 10 10.	- 00 51 00.	760		ISOLATED DARK CLOUD
PK032-06.1	19 10 26.	- 03 37	24		PLANETARY NEBULA
SC 1906-4634.6	19 10 33.	- 46 29 40.	48		NEBULA
PK037-03.2	19 10 36.	+ 02 48	216	15.5	PLANETARY NEBULA
7ZW 862	19 10 36.	+ 63 45			COMPACT GALAXY
ISS 0559	19 10 42.	+ 01 03	244		STELLAR RING
ZWG 357.001	19 10 42.	+ 76 37		15.6	GALAXY
ZWG 356.012	19 10 42.	+ 76 37		15.6	GALAXY

OBJECT NAME	RIGHT ASCEN.	DECLINATION	DIAM.	MAGN.	TYPE OF OBJECT
ZWG 256.009	19 10 48.	+ 46 10		15.3	GALAXY
PK049+02.1	19 10 49.5	+ 15 41 32.	37		PLANETARY NEBULA
PK038-03.1	19 10 52.6	+ 03 19 53.			PLANETARY NEBULA
HN 1887	19 10 53.8	- 46 41 03.	18		NEBULA
ISS 0456	19 10 54.	+ 13 22	208		STELLAR RING
ISS 0336	19 10 54.	+ 21 46	354		STELLAR RING
SC 1907-4646.0	19 10 54.	- 46 41 03.	78		NEBULA
SC 1907-4707.7	19 10 55.	- 47 02 45.	72		NEBULA
KHAV 597	19 11	+ 17 29	1410		DARK NEBULA
LDN 0679	19 11 00.	+ 12 40	3900		DARK NEBULA
HN 0570	19 11 00.	- 62 25			NEBULA
IC 4833	19 11 00.	- 62 25			NONSTELLAR OBJECT
PK037-03.3	19 11 02.3	+ 02 13 03.			PLANETARY NEBULA
HN 0571	19 11 05.	- 58 19			NEBULA
PK039-02.1	19 11 06.	+ 04 32 55.	10		PLANETARY NEBULA
IC 4835	19 11 06.	- 58 19			NONSTELLAR OBJECT
PK042-01.1	19 11 06.1	+ 07 22 10.			PLANETARY NEBULA
ACK 005-18.1	19 11 12.	- 32 40			PLANETARY NEBULA
DV.56 I4837A	19 11 12.	- 54 13	240		S GALAXY
HN 1886	19 11 13.9	- 54 12 44.	60		NEBULA
SC 1907-5418.4	19 11 14.	- 54 13 26.	180		NEBULA
SC 1907-5050.1	19 11 15.	- 50 45 08.			NEBULA
HN 0573	19 11 16.	- 54 45			NEBULA
MCG+09-31-026	19 11 18.	+ 53 04	60	15.	GALAXY
ZWG 280.019	19 11 18.	+ 53 06		15.6	GALAXY
UGC 11413	19 11 18.	+ 53 06	78	15.6	GALAXY Sc
IC 4837	19 11 18.	- 54 46	204	12.2	GALAXY SBc
HN 0567	19 11 19.	- 72 50			NEBULA
IC 4825	19 11 20.	- 72 50			NONSTELLAR OBJECT
PK048+01.1	19 11 21.8	+ 14 54 08.	4		PLANETARY NEBULA
RNGC 6761	19 11 22.	- 50 46			GALAXY
PK038-03.2	19 11 24.	+ 03 32 33.			PLANETARY NEBULA
SCHO 1035	19 11 24.	+ 17 24 42.	1150		ISOLATED DARK CLOUD
HN 0575	19 11 28.	- 54 43			NEBULA
IC 4839	19 11 28.	- 54 43	150		GALAXY SB(s)
LDN 0627	19 11 30.	+ 00 45	7020		DARK NEBULA
7ZW 863	19 11 30.	+ 72 53			COMPACT GALAXY
7ZW 864	19 11 30.	+ 73 20			COMPACT GALAXY
HN 0576	19 11 39.	- 56 18			NEBULA
IC 4840	19 11 40.	- 56 18			NONSTELLAR OBJECT
ZWG 256.010	19 11 42.	+ 45 14		15.4	GALAXY
MCG-05-45-005	19 11 42.	- 31 04	30	15.	GALAXY
HN 0574	19 11 44.	- 60 17			NEBULA
IC 4836	19 11 44.	- 60 17	60		GALAXY SB(s)
HN 0572	19 11 45.	- 64 06			NEBULA
IC 4834	19 11 45.	- 64 06			NONSTELLAR OBJECT
SCHO 1036	19 11 46.	+ 16 24 00.	470		ISOLATED DARK CLOUD
PK038-03.3	19 11 48.	+ 03 30	14		PLANETARY NEBULA
7ZW 865	19 11 48.	+ 68 09			COMPACT GALAXY
LB 03115	19 11 48.	- 63 59		15.2	FAINT BLUE STAR
PK051+03.1	19 11 50.4	+ 17 26 20.	10	11.	PLANETARY NEBULA
RNGC 6786	19 11 53.	+ 73 18		13.5	GALAXY
ZWG 341.019	19 11 54.	+ 73 18		13.7	GALAXY
ZCG 1911+73	19 11 54.	+ 73 18		13.7	COMPACT GALAXY
UGC 11414	19 11 54.	+ 73 18	72	13.7	GALAXY SB
KARA.72 538A	19 11 54.	+ 73 18	60	13.7	PART OF DOUBLE GALAXY
SC 1908-4803.7	19 11 54.	- 47 58 41.	78		NEBULA
SCHO 1037	19 11 57.	+ 25 31 24.	1070		ISOLATED DARK CLOUD
MIL 76	19 11 59.	+ 11 04 18.	234		SUPERNOVA REMNANT
REIN 2.276	19 11 59.41	- 02 47 44.6			NEBULA
KHAV 598	19 12	+ 25 29	1720		DARK NEBULA
IC 4895	19 12	- 15 00 24.			SAME AS NGC 6822
LDN 0646	19 12 00.	+ 08 00	6420		DARK NEBULA
LDN 0709	19 12 00.	+ 16 20	180		DARK NEBULA
LDN 0736	19 12 00.	+ 21 30	3300		DARK NEBULA
LDN 0780	19 12 00.	+ 25 20	1440		DARK NEBULA
MCG+08-35-005	19 12 00.	+ 50 48	60	16.	GALAXY
7ZW 866	19 12 00.	+ 66 03			COMPACT GALAXY
ZC 1912.0+6945	19 12 00.	+ 69 45	3970		CLUSTER OF GALAXIES
ZWG 341.020	19 12 00.	+ 73 19		15.1	GALAXY
ZCG 1912+73	19 12 00.	+ 73 19		15.1	COMPACT GALAXY
UGC 11415	19 12 00.	+ 73 19	78	15.1	GALAXY S
KARA.72 538B	19 12 00.	+ 73 19	54	15.1	PART OF DOUBLE GALAXY
RNGC 6772	19 12 00.	- 02 47		14.0	PLANETARY NEBULA
PK033-06.1	19 12 00.18	- 02 47 35.0	88	14.2	PLANETARY NEBULA
PK035-05.1	19 12 05.6	+ 00 08 22.			PLANETARY NEBULA
ISS 0337	19 12 06.	+ 22 59	161		STELLAR RING
PK045-00.1	19 12 08.9	+ 10 45 34.			PLANETARY NEBULA
ZWG 256.011	19 12 12.	+ 45 44		15.0	GALAXY
ZWG 256.012	19 12 12.	+ 48 58		15.6	GALAXY
UGC 11416	19 12 12.	+ 50 46	66	16.0	GALAXY Sc
7ZW 867	19 12 12.	+ 65 18			COMPACT GALAXY
HN 0577	19 12 14.	- 61 42			NEBULA
IC 4838	19 12 15.	- 61 42			NONSTELLAR OBJECT
MCG+08-35-006	19 12 18.	+ 45 14	24	16.	GALAXY
MCG+09-31-027	19 12 18.	+ 56 39 30.	36	16.	GALAXY
ISS 0339	19 12 24.	+ 22 32	114		STELLAR RING
ISS 0338	19 12 24.	+ 24 58	350		STELLAR RING
UGC 11417	19 12 24.	+ 29 54	66	16.0	GALAXY Sc
MCG+07-39-018	19 12 27.	+ 40 02	54	16.	GALAXY
ZWG 229.030	19 12 30.	+ 40 05		15.7	GALAXY
UGC 11418	19 12 30.	+ 40 05	66	15.7	GALAXY Sb-c
PK061+08.1	19 12 30.9	+ 28 35 33.	16		PLANETARY NEBULA
RNGC 6773	19 12 34.	+ 04 47			NON-EXISTENT OBJECT
ISS 0491	19 12 36.	+ 05 55	139		STELLAR RING
ZWG 341.021	19 12 36.	+ 70 17		15.7	GALAXY
ISS 0492	19 12 48.	+ 05 57	186		STELLAR RING
PK037-04.1	19 12 50.5	+ 02 27 48.			PLANETARY NEBULA
ISS 0560	19 12 54.	+ 02 57	190		STELLAR RING
MCG-07-39-008	19 12 54.	- 40 13 30.	60	15.	GALAXY
SC 1909-4652.1	19 12 55.	- 46 47 00.	54		NEBULA
SC 1908-5352.7	19 12 56.	- 53 47 37.	132		NEBULA
HN 1889	19 12 57.6	- 53 47 31.	72		NEBULA
B 138	19 13	+ 00 08	10800		DARK OBJECT
KHAV 602	19 13	+ 15 05	1410		DARK NEBULA
KHAV 601	19 13	+ 16 41	3250		DARK NEBULA
KHAV 600	19 13	+ 20 35	6690		DARK NEBULA
KHAV 599	19 13	- 00 55	1990		DARK NEBULA
ISS 0493	19 13 00.	+ 03 27	188		STELLAR RING
LDN 0749	19 13 00.	+ 22 20	2160		DARK NEBULA
LDN 0618	19 13 00.	- 01 20	540		DARK NEBULA
MCG-07-39-009	19 13 00.	- 40 19	30	14.5	GALAXY
PK027-09.1	19 13 00.5	- 09 04 58.0		12.7	PLANETARY NEBULA
MCG+09-31-028	19 13 03.	+ 53 38 30.	60	16.	GALAXY
MCG-07-39-010	19 13 03.	- 40 18	72	13.5	GALAXY
RNGC 6768	19 13 04.	- 40 17			GALAXY
SCHO 1038	19 13 08.	+ 00 01 42.	2710		ISOLATED DARK CLOUD
SCHO 1039	19 13 11.	+ 13 30 18.	570		ISOLATED DARK CLOUD
OCL 0102	19 13 12.	+ 11 08	360	15.	OPEN STAR CLUSTER
ZWG 280.020	19 13 12.	+ 53 40		15.6	GALAXY
UGC 11419	19 13 12.	+ 53 40	60	15.6	GALAXY Sa-b
ZWG 323.010	19 13 12.	+ 63 32		15.7	GALAXY
KARA.73B 0869	19 13 12.	+ 63 32	24	15.7	ISOLATED GALAXY S
PK048+01.2	19 13 12.6	+ 13 58 33.			PLANETARY NEBULA
HN 1888	19 13 13.0	- 59 31 06.	42		NEBULA
B 137	19 13 25.	- 01 25			DARK OBJECT
ISS 0387	19 13 30.	+ 15 57	140		STELLAR RING
SC 1909-4652.1	19 13 34.	- 46 46 57.	72		NEBULA
UGC 11420	19 13 36.	+ 42 49	90	15.1	GALAXY Sb
ISS 0388	19 13 42.	+ 16 09	125		STELLAR RING
HN 0100	19 13 44.	- 09 09			NEBULA
IC 4846	19 13 44.9	- 09 08 06.	2	12.7	PLANETARY NEBULA
RNGC 6774	19 13 46.	- 16 22			OPEN CLUSTER
OCL 0065	19 13 48.	- 16 22	2880	9.	OPEN STAR CLUSTER
HN 1890	19 13 51.3	- 60 56 52.	24		NEBULA
RNGC 6769	19 13 53.	- 60 35		12.5	GALAXY
ZWG 229.031	19 13 54.	+ 42 49		15.1	GALAXY
PK358-21.1	19 13 55.3	- 39 40 50.	8		PLANETARY NEBULA
HN 0073	19 13 56.	- 39 42			NEBULA
PK040-03.1	19 13 59.7	+ 05 07 58.			PLANETARY NEBULA
KHAV 605	19 14	+ 10 59	28100		DARK NEBULA
KHAV 604	19 14	+ 14 41	1570		DARK NEBULA
KHAV 603	19 14	+ 22 53	2110		DARK NEBULA
LDN 0641	19 14 00.	+ 05 00	2460		DARK NEBULA
LDN 0717	19 14 00.	+ 17 55	1500		DARK NEBULA
ISS 0340	19 14 00.	+ 25 20	62		STELLAR RING
MCG+07-39-019	19 14 00.	+ 42 49	72	15.	GALAXY
ZWG 229.032	19 14 00.	+ 44 01		14.6	GALAXY
UGC 11421	19 14 00.	+ 44 01	66	14.6	GALAXY (SB0)?
ZWG 280.021	19 14 00.	+ 51 27		15.4	GALAXY
MCG+13-14-001	19 14 00.	+ 77 45	18	17.	GALAXY
IC 1297	19 14 00.	- 39 42			GLOBULAR CLUSTER
MCG+07-39-020	19 14 03.	+ 44 01	60	14.	GALAXY
LB 03116	19 14 06.	- 64 41		11.7	FAINT BLUE STAR
RNGC 6770	19 14 11.	- 60 36			GALAXY
ZWG 229.033	19 14 12.	+ 43 21		15.7	GALAXY
UGC 11422	19 14 12.	+ 43 21	72	15.7	GALAXY Sc
MCG+07-39-021	19 14 12.	+ 43 21	66	15.	GALAXY
ZWG 280.022	19 14 12.	+ 51 19		15.7	GALAXY
RNGC 6775	19 14 13.	- 01 01			NON-EXISTENT OBJECT
SC 1910-4547.6	19 14 14.	- 45 42 25.	18		NEBULA
HN 1891	19 14 20.9	- 53 34 07.	60		NEBULA
RNGC 6771	19 14 23.	- 60 41			GALAXY
ISS 0341	19 14 24.	+ 22 29	81		STELLAR RING
ISS 0561	19 14 24.	+ 00 12	308		STELLAR RING
ISS 1002	19 14 24.	- 26 47	293		STELLAR RING
TER 07	19 14 24.49	- 34 44 52.8	80		STAR CLUSTER
SCHO 1040	19 14 26.	+ 15 35 12.	500		ISOLATED DARK CLOUD
MCG+09-31-029	19 14 30.	+ 52 19 30.	54	16.	GALAXY
GCL 110	19 14 36.	+ 30 05	606	9.55	GLOBULAR STAR CLUSTER
RNGC 6779	19 14 38.	+ 30 05		9.5	GLOBULAR CLUSTER
ZWG 357.002	19 14 42.	+ 77 50		15.4	GALAXY
ZWG 356.013	19 14 42.	+ 77 50		15.4	GALAXY
UGC 11423	19 14 42.	+ 77 50	66	15.4	GALAXY PECULR
MCG+07-39-022	19 14 48.	+ 42 35	36	17.	GALAXY
MCG+13-14-002	19 14 48.	+ 77 50 30.	51	15.	GALAXY
HN 1895	19 14 48.1	- 30 17 39.	30		NEBULA
HN 0581	19 14 50.	- 56 07			NEBULA
IC 4844	19 14 51.	- 56 07			NONSTELLAR OBJECT
ISS 0389	19 14 54.	+ 15 16	169		STELLAR RING
HN 0580	19 14 57.	- 59 24			NEBULA
IC 4843	19 14 57.	- 59 24			NONSTELLAR OBJECT
HN 0579	19 14 57.	- 60 44			NEBULA
IC 4842	19 14 58.	- 60 44	72		GALAXY E5
HN 0578	19 14 59.	- 72 19			NEBULA
KHAV 607	19 15	+ 00 53	4540		DARK NEBULA
KHAV 606	19 15	- 01 31	700		DARK NEBULA
LB 09938	19 15	- 85 15		14.2	FAINT BLUE STAR
LDN 0629	19 15 00.	+ 01 00	2520		DARK NEBULA
ISS 0390	19 15 00.	+ 19 22	151		STELLAR RING
OCL 0110	19 15 00.	+ 19 28	360	16.	OPEN STAR CLUSTER
PK058+06.1	19 15 00.	+ 25 32	40	17.5	PLANETARY NEBULA
ZWG 341.022	19 15 00.	+ 70 06		15.6	GALAXY
IC 4841	19 15 00.	- 72 19			NONSTELLAR OBJECT
ISS 0342	19 15 06.	+ 26 42	268		STELLAR RING
HN 1893	19 15 09.9	- 53 30 15.	24		NEBULA
RNGC 6783	19 15 10.	+ 45 55		15.5	GALAXY
ZWG 229.034	19 15 12.	+ 44 42		15.5	GALAXY
ZWG 256.013	19 15 12.	+ 45 55		15.4	GALAXY
PK038-04.1	19 15 17.4	+ 02 43 59.			PLANETARY NEBULA
MCG+08-35-007	19 15 18.	+ 45 55	18	16.	GALAXY
HN 1894	19 15 23.9	- 51 30 08.	18		NEBULA
ZWG 202.005	19 15 24.	+ 33 20		15.6	GALAXY
MCG+10-27-009	19 15 24.	+ 60 20	72	15.	GALAXY
B 139	19 15 25.	- 01 30			DARK OBJECT
ZWG 302.009	19 15 30.	+ 60 19		14.9	GALAXY
UGC 11424	19 15 30.	+ 60 19	72	14.9	GALAXY Sb/SBb
LDN 0619	19 15 30.	- 01 33	480		DARK NEBULA
SCHO 1041	19 15 31.	+ 08 20 24.	590		ISOLATED DARK CLOUD
SC 1911-4756.6	19 15 31.	- 47 51 20.	48		NEBULA
RNGC 6787	19 15 33.	+ 60 19		15.0	GALAXY
7ZW 868	19 15 36.	+ 69 40			COMPACT GALAXY
HN 1892	19 15 36.7	- 60 54 45.	12		NEBULA
ISS 0391	19 15 42.	+ 16 06	190		STELLAR RING
PK037-05.1	19 15 48.	+ 01 42	44	18.8	PLANETARY NEBULA
MIL 77	19 15 48.	+ 12 06	720		SUPERNOVA REMNANT
RNGC 6778	19 15 48.	- 01 40		13.5	PLANETARY NEBULA
PK034-06.1	19 15 49.39	- 01 41 17.4	25	13.3	PLANETARY NEBULA
PK026-11.1	19 15 53.1	- 11 11 42.		13.3	PLANETARY NEBULA
IC 1298	19 15 56.	- 01 43			OPEN CLUSTER
HN 0582	19 15 56.	- 60 29			NEBULA
IC 4845	19 15 56.	- 60 29			NONSTELLAR OBJECT
RNGC 6781	19 15 59.	+ 06 27		12.5	PLANETARY NEBULA
KHAV 610	19 16	+ 07 29			DARK NEBULA
KHAV 609	19 16	+ 07 35	2660		DARK NEBULA
KHAV 608	19 16	+ 26 29	1950		DARK NEBULA
LDN 0652	19 16 00.	+ 08 20	1860		DARK NEBULA
GCL 111	19 16 00.	+ 18 28	186		GLOBULAR STAR CLUSTER
LDN 0723	19 16 00.	+ 19 08	360		DARK NEBULA
LDN 0750	19 16 00.	+ 22 00	11460		DARK NEBULA
ISS 0343	19 16 00.	+ 24 26	282		STELLAR RING
REIN 2.277	19 16 00.74	+ 06 26 22.2			NEBULA
REIN 2.278	19 16 01.19	+ 06 26 38.8			NEBULA
PK041-02.1	19 16 01.42	+ 06 26 37.2	180	12.5	PLANETARY NEBULA
REIN 2.279	19 16 05.13	+ 06 27 23.7			NEBULA
HN 1896	19 16 06.0	- 51 48 59.	18		NEBULA

OBJECT NAME	RIGHT ASCEN.	DECLINATION	DIAM.	MAGN.	TYPE OF OBJECT
PK049+00.1	19 16 17.8	+ 14 54 24.			PLANETARY NEBULA
ISS 0392	19 16 18.	+ 15 12	105		STELLAR RING
MCG+07-39-023	19 16 18.	+ 42 22	30	16.	GALAXY
ZWG 256.014	19 16 18.	+ 46 44		15.5	GALAXY
MCG+08-35-008	19 16 18.	+ 46 44	42	16.	GALAXY
MCG+11-23-001	19 16 18.	+ 63 52	57	14.	GALAXY
ZWG 323.011	19 16 18.	+ 63 54		13.7	GALAXY
RNGC 6789	19 16 18.	+ 63 54		13.5	GALAXY
UGC 11425	19 16 18.	+ 63 54	108	13.7	GALAXY IRR
ZWG 202.006	19 16 24.	+ 36 18		15.3	GALAXY
ZWG 256.015	19 16 24.	+ 50 34		15.1	GALAXY
7ZW 869	19 16 24.	+ 67 17			COMPACT GALAXY
PK053+03.1	19 16 30.	+ 19 28	94	17.2	PLANETARY NEBULA
LDN 0763	19 16 30.	+ 23 20	360		DARK NEBULA
ZWG 202.007	19 16 30.	+ 34 45		14.4	GALAXY
UGC 11426	19 16 30.	+ 34 45	54	14.4	GALAXY S
PK025-11.1	19 16 30.	- 12 22	88	15.7	PLANETARY NEBULA
MCG+09-31-030	19 16 36.	+ 54 25	42	17.	GALAXY
7ZW 871	19 16 42.	+ 64 57			COMPACT GALAXY
SCHO 1042	19 16 44.	+ 23 18 54.	580		ISOLATED DARK CLOUD
ZWG 229.035	19 16 48.	+ 41 21		15.7	GALAXY
ZC 1916.8+4855	19 16 48.	+ 48 55	49530		CLUSTER OF GALAXIES
ZWG 341.023	19 16 48.	+ 72 40		14.7	GALAXY
UGC 11427	19 16 48.	+ 72 40	66	14.7	GALAXY S
PK052+02.1	19 16 50.6	+ 18 56 48.			PLANETARY NEBULA
OCL 0107	19 16 54.	+ 15 37	240	15.	OPEN STAR CLUSTER
ISS 0344	19 16 54.	+ 21 26	127		STELLAR RING
MCG+12-18-001	19 16 54.	+ 72 40	51	16.	GALAXY
MCG+09-31-031	19 16 57.	+ 52 05	42	16.	GALAXY
KHAV 612	19 17	+ 01 47	2470		DARK NEBULA
KHAV 613	19 17	+ 05 05	13060		DARK NEBULA
KHAV 614	19 17	+ 23 17	1720		DARK NEBULA
KHAV 611	19 17	- 01 55	4380		DARK NEBULA
LDN 0642	19 17 00.	+ 05 00	3000		DARK NEBULA
LDN 0644	19 17 00.	+ 06 30	6180		DARK NEBULA
LDN 0645	19 17 00.	+ 07 15	9060		DARK NEBULA
LDN 0651	19 17 00.	+ 08 00	600		DARK NEBULA
LDN 0764	19 17 00.	+ 23 20	1140		DARK NEBULA
ZC 1917.0+8013	19 17 00.	+ 80 13	4570		CLUSTER OF GALAXIES
LDN 0615	19 17 00.	- 02 10	4680		DARK NEBULA
ARC 2320	19 17 02.	+ 70 55		16.9	RICH CLUSTER OF GALAXIES
ZWG 341.024	19 17 06.	+ 74 22		15.7	GALAXY
ZWG 357.003	19 17 06.	+ 76 00		15.7	GALAXY
SCHO 1043	19 17 07.	+ 07 29 12.	1520		ISOLATED DARK CLOUD
B 330	19 17 08.	+ 07 28	1800		DARK OBJECT
7ZW 872	19 17 12.	+ 65 57			COMPACT GALAXY
ZWG 341.025	19 17 12.	+ 70 26		15.5	GALAXY
LDN 0744	19 17 18.	+ 21 20	660		DARK NEBULA
B 140	19 17 21.	+ 05 08	3600		DARK OBJECT
LB 00341	19 17 22.	+ 58 41 42.		15.8	FAINT BLUE STAR
ZWG 230.001	19 17 24.	+ 44 08		15.7	GALAXY
ZWG 229.036	19 17 24.	+ 44 08		15.7	GALAXY
ZWG 302.010	19 17 24.	+ 59 51		15.7	GALAXY
HN 1899	19 17 24.0	- 53 45 30.	18		NEBULA
PK056+04.1	19 17 29.3	+ 22 29 03.			PLANETARY NEBULA
LDN 0647	19 17 30.	+ 07 25	1800		DARK NEBULA
LDN 0732	19 17 30.	+ 20 05	540		DARK NEBULA
ISS 0345	19 17 30.	+ 23 45	127		STELLAR RING
ZC 1917.5+7050	19 17 30.	+ 70 50	2420		CLUSTER OF GALAXIES
HN 1897	19 17 33.7	- 59 52 00.	18		NEBULA
ZWG 230.002	19 17 36.	+ 44 04		15.4	GALAXY
ZWG 229.037	19 17 36.	+ 44 04		15.4	GALAXY
HN 1898	19 17 37.0	- 60 08 12.	18		NEBULA
SCHO 1044	19 17 39.	+ 23 23 24.	660		ISOLATED DARK CLOUD
B 141	19 17 40.	+ 01 48	1200		DARK OBJECT
LB 00342	19 17 41.	+ 59 54 18.		14.3	FAINT BLUE STAR
PK077+14.1	19 17 42.	+ 46 09	201	14.4	PLANETARY NEBULA
ZC 1917.7+6340	19 17 42.	+ 63 40	4700		CLUSTER OF GALAXIES
HN 1900	19 17 42.3	- 52 56 47.	18		NEBULA
LDN 0762	19 17 48.	+ 23 05	360		DARK NEBULA
ZWG 202.008	19 17 48.	+ 37 18		15.7	GALAXY
LB 03117	19 17 48.	- 64 01		13.0	FAINT BLUE STAR
IC 4850	19 17 52.	- 00 13			NONSTELLAR OBJECT
HN 0082	19 17 52.	- 00 14			NEBULA
LDN 0655	19 17 54.	+ 08 27	540		DARK NEBULA
KHAV 615	19 18	+ 20 48	9060		DARK NEBULA
LDN 0632	19 18 00.	+ 01 30	3540		DARK NEBULA
LDN 0730	19 18 00.	+ 19 50	600		DARK NEBULA
LDN 0620	19 18 00.	- 01 15	540		DARK NEBULA
RNGC 6785	19 18 01.	- 01 11			NON-EXISTENT OBJECT
MCG+09-31-032	19 18 06.	+ 52 13	60	15.	GALAXY
7ZW 873	19 18 06.	+ 68 43			COMPACT GALAXY
HH 32A	19 18 07.9	+ 10 56 21.			HERBIG-HARO OBJECT
HH 32B	19 18 08.4	+ 10 56 17.			HERBIG-HARO OBJECT
ISS 0393	19 18 12.	+ 14 59	115		STELLAR RING
ZWG 280.023	19 18 12.	+ 52 15		15.3	GALAXY
ISS 0394	19 18 18.	+ 15 09	115		STELLAR RING
7ZW 874	19 18 18.	+ 67 07			COMPACT GALAXY
ZWG 230.003	19 18 24.	+ 44 48		15.1	GALAXY
ZWG 229.038	19 18 24.	+ 44 48		15.1	GALAXY
7ZW 875	19 18 24.	+ 67 47			COMPACT GALAXY
7ZW 876	19 18 24.	+ 68 44			COMPACT GALAXY
ISS 0494	19 18 30.	+ 07 16	64		STELLAR RING
LDN 0658	19 18 30.	+ 09 00	480		DARK NEBULA
LDN 0664	19 18 30.	+ 09 40	1680		DARK NEBULA
LDN 0673	19 18 30.	+ 11 10	1800		DARK NEBULA
ISS 0346	19 18 30.	+ 23 25	89		STELLAR RING
MCG+05-45-006	19 18 30.	+ 30 42 30.	48	14.5	GALAXY
UGC 11428	19 18 30.	+ 30 44	72	14.5	GALAXY Sc
MCG+07-40-001	19 18 30.	+ 44 50	18	16.	GALAXY
ZWG 256.016	19 18 30.	+ 45 21		15.7	GALAXY
HN 0583	19 18 35.	- 65 37			NEBULA
PK043-03.1	19 18 35.3	+ 07 31 12.	27		PLANETARY NEBULA
LDN 0771	19 18 36.	+ 23 26	240		DARK NEBULA
IC 4847	19 18 36.	- 65 36			NONSTELLAR OBJECT
HN 1903	19 18 38.2	- 54 40 43.	30		NEBULA
RNGC 6780	19 18 39.	- 55 53		13.0	GALAXY
IC 4848	19 18 41.	- 56 52			NONSTELLAR OBJECT
HN 0584	19 18 41.	- 56 53			NEBULA
MCG-03-49-002	19 18 42.	- 17 14 30.	36	15.5	GALAXY
HN 1904	19 18 42.7	- 51 36 43.	18		NEBULA
HN 1901	19 18 47.8	- 60 34 31.	18		NEBULA
LDN 0656	19 18 54.	+ 08 22	120		DARK NEBULA
LDN 0751	19 18 54.	+ 21 39	180		DARK NEBULA
HN 1905	19 18 55.4	- 51 05 54.			NEBULA
HN 1908	19 18 55.6	- 29 42 34.	24		NEBULA
HN 1907	19 18 56.5	- 30 28 46.	18		NEBULA
PK048-00.1	19 18 56.8	+ 14 00 32.			PLANETARY NEBULA

OBJECT NAME	RIGHT ASCEN.	DECLINATION	DIAM.	MAGN.	TYPE OF OBJECT
HN 1902	19 18 59.5	- 60 34 13.	24		NEBULA
ISS 0495	19 19 00.	+ 07 12	206		STELLAR RING
LDN 0657	19 19 00.	+ 08 48	1080		DARK NEBULA
LDN 0683	19 19 00.	+ 11 50	1260		DARK NEBULA
LDN 0738	19 19 00.	+ 20 50	960		DARK NEBULA
LDN 0741	19 19 00.	+ 21 00	10740		DARK NEBULA
OCL 0142	19 19 00.	+ 37 45	960	16.	OPEN STAR CLUSTER
ZWG 230.004	19 19 00.	+ 43 57		15.6	GALAXY
ZWG 229.039	19 19 00.	+ 43 57		15.6	GALAXY
ZWG 368.002	19 19 00.	+ 83 49		14.8	GALAXY
KARA.73B 0870	19 19 00.	+ 83 49	36	14.8	ISOLATED GALAXY E
MCG-05-45-006	19 19 00.	- 30 29	9	15.5	GALAXY
ACK 006-19.1	19 19 00.	- 31 36			PLANETARY NEBULA
RNGC 6791	19 19 02.	+ 37 45			OPEN CLUSTER
ARC 2319	19 19 09.	+ 43 53		15.4	RICH CLUSTER OF GALAXIES
LDN 0676	19 19 12.	+ 11 25	300		DARK NEBULA
7ZW 877	19 19 12.	+ 68 33			COMPACT GALAXY
ISS 0347	19 19 18.	+ 23 42	166		STELLAR RING
PK032-08.1	19 19 20.	- 04 18			PLANETARY NEBULA
HN 1906	19 19 21.3	- 51 50 52.	30		NEBULA
LDN 0661	19 19 24.	+ 09 00	360		DARK NEBULA
ZWG 230.005	19 19 24.	+ 43 02		13.4	GALAXY
UGC 11429	19 19 24.	+ 43 02	144	13.4	GALAXY SBb
MCG+07-40-002	19 19 24.	+ 43 02	126	13.	GALAXY
MCG+09-31-033	19 19 24.	+ 54 06	36	16.	GALAXY
ZWG 280.024	19 19 24.	+ 54 07		15.3	GALAXY
KARA.73B 0871	19 19 24.	+ 54 07	36	15.3	ISOLATED GALAXY S
RNGC 6792	19 19 25.	+ 43 02		13.5	GALAXY
HN 1911	19 19 27.5	- 29 09 20.	48		NEBULA
HN 1912	19 19 28.4	- 29 40 14.	18		NEBULA
LDN 0684	19 19 30.	+ 12 20	1740		DARK NEBULA
RNGC 6782	19 19 31.	- 60 02		13.0	GALAXY
MCG+07-40-003	19 19 36.	+ 43 13	66	14.	GALAXY
ZWG 230.006	19 19 36.	+ 43 14		14.5	GALAXY
ZWG 230.007	19 19 36.	+ 43 52		15.3	GALAXY
UGC 11430	19 19 36.	+ 46 14	66	14.4	GALAXY Sc
LB 03118	19 19 36.	- 64 36		14.6	FAINT BLUE STAR
MCG+07-40-004	19 19 39.	+ 43 51	12	16.5	GALAXY
ISS 0348	19 19 42.	+ 22 02	193		STELLAR RING
7ZW 878	19 19 42.	+ 63 13			COMPACT GALAXY
SC 1917+1728.2	19 19 45.	+ 17 33 49.	6		NEBULA
HN 1910	19 19 45.9	- 52 41 14.	18		NEBULA
ZWG 230.008	19 19 48.	+ 43 55		15.7	GALAXY
UGC 11431	19 19 48.	+ 73 42	78	16.0	GALAXY Sa-b
7ZW 879	19 19 48.	+ 76 49			COMPACT GALAXY
HN 1909	19 19 48.6	- 54 39 14.	18		NEBULA
PK036-06.1	19 19 51.4	+ 00 06 53.			PLANETARY NEBULA
MIL 74	19 20	+ 06 00	9900		SUPERNOVA REMNANT
MIL 79	19 20	+ 14 00	1440		SUPERNOVA REMNANT
KHAV 616	19 20	+ 17 48	10160		DARK NEBULA
KHAV 617	19 20	+ 27 30	1000		DARK NEBULA
LDN 0640	19 20 00.	+ 04 00	5160		DARK NEBULA
LDN 0677	19 20 00.	+ 11 28	300		DARK NEBULA
LDN 0692	19 20 00.	+ 12 50	900		DARK NEBULA
LDN 0735	19 20 00.	+ 20 20	540		DARK NEBULA
LDN 0746	19 20 00.	+ 21 12	720		DARK NEBULA
ZWG 230.009	19 20 00.	+ 43 55		15.4	GALAXY
SC 1916-4806.5	19 20 02.	- 48 00 55.	48		NEBULA
MCG+07-40-005	19 20 03.	+ 43 54	54	16.	GALAXY
PK045-01.1	19 20 04.8	+ 10 35 36.			PLANETARY NEBULA
SCHO 1046	19 20 06.	+ 21 55 18.	760		ISOLATED DARK CLOUD
ZWG 230.010	19 20 12.	+ 43 20		15.7	GALAXY
RNGC 6776A	19 20 22.	- 63 48			GALAXY
LDN 0773	19 20 24.	+ 23 20	240		DARK NEBULA
DV.56 N6776A	19 20 24.	- 63 48	78		S GALAXY
PK037-06.1	19 20 24.93	+ 01 24 56.7	10	11.4	PLANETARY NEBULA
HN 0046	19 20 26.	+ 01 25			NEBULA
IC 1299	19 20 27.	+ 20 39			OPEN CLUSTER
HN 1915	19 20 33.6	- 29 02 21.	12		NEBULA
HN 1913	19 20 34.1	- 52 03 29.	60		NEBULA
RNGC 6776	19 20 39.	- 63 59		13.0	GALAXY
ZWG 230.011	19 20 42.	+ 44 10		15.4	GALAXY
MCG+10-27-010	19 20 42.	+ 61 03	102	13.	GALAXY
RNGC 6790	19 20 44.	+ 01 25		11.5	PLANETARY NEBULA
RNGC 6796	19 20 46.	+ 61 03		13.0	GALAXY
ISS 0497	19 20 48.	+ 08 10	229		STELLAR RING
ISS 0496	19 20 48.	+ 08 12	146		STELLAR RING
LDN 0774	19 20 48.	+ 23 20	180		DARK NEBULA
ZWG 302.011	19 20 48.	+ 61 03		13.5	GALAXY
UGC 11432	19 20 48.	+ 64 03	126	13.5	GALAXY S
HN 0585	19 20 49.	- 63 00			NEBULA
IC 4849	19 20 49.	- 63 00			NONSTELLAR OBJECT
LDN 0686	19 21 00.	+ 12 20	360		DARK NEBULA
LDN 0716	19 21 00.	+ 17 00	15180		DARK NEBULA
OCL 0115	19 21 00.	+ 22 05			OPEN STAR CLUSTER
ZC 1921.0+7546	19 21 00.	+ 75 46	4440		CLUSTER OF GALAXIES
PK037-06.2	19 21 01.5	+ 00 32 11.			PLANETARY NEBULA
RNGC 6793	19 21 02.	+ 22 05			OPEN CLUSTER
HN 1919	19 21 13.5	- 32 19 13.	24		NEBULA
HN 1914	19 21 13.9	- 53 41 50.	18		NEBULA
PK055+02.1	19 21 15.2	+ 21 02 11.	10		PLANETARY NEBULA
ZWG 323.012	19 21 18.	+ 63 10		15.6	GALAXY
KARA.73B 0872	19 21 18.	+ 63 10	36	15.6	ISOLATED GALAXY S
HN 0586	19 21 19.	- 57 45			NEBULA
IC 4851	19 21 19.	- 57 45			NONSTELLAR OBJECT
FATH 1.812	19 21 23.	+ 60 53	19		NEBULA
HN 1916	19 21 26.5	- 53 42 32.			NEBULA
LDN 0775	19 21 30.	+ 23 20	600		DARK NEBULA
MCG+10-27-011	19 21 30.	+ 58 09	30	15.	GALAXY
ZWG 302.012	19 21 30.	+ 58 10		15.2	GALAXY
HN 1917	19 21 32.3	- 53 24 43.	12		NEBULA
LDN 0643	19 21 36.	+ 04 44	480		DARK NEBULA
LDN 0675	19 21 36.	+ 11 00	180		DARK NEBULA
PK055+02.2	19 21 37.1	+ 21 00 46.	13		PLANETARY NEBULA
RNGC 6784	19 21 40.	- 65 43			UNVERIFIED SOUTHERN OBJECT
ISS 0457	19 21 42.	+ 10 00	119		STELLAR RING
RNGC 6788	19 21 46.	- 55 04			UNVERIFIED SOUTHERN OBJECT
HN 1918	19 21 53.6	- 53 55 24.			NEBULA
ZC 1921.9+3901	19 21 54.	+ 39 01	1750		CLUSTER OF GALAXIES
RNGC 6777	19 21 57.	- 71 36			GALAXY
PK059+04.1	19 21 58.5	+ 25 12 54.	11		PLANETARY NEBULA
PK045-02.1	19 21 59.	+ 09 47 59.		12.40	PLANETARY NEBULA
LDN 0722	19 22 00.	+ 18 20	600		DARK NEBULA
LDN 0739	19 22 00.	+ 20 30	2460		DARK NEBULA
LDN 0769	19 22 00.	+ 23 00	480		DARK NEBULA
ZWG 202.009	19 22 00.	+ 34 42		15.6	GALAXY
UGC 11433	19 22 00.	+ 34 42	60	15.6	GALAXY S
7ZW 880	19 22 00.	+ 63 04			COMPACT GALAXY

OBJECT NAME	RIGHT ASCEN.	DECLINATION	DIAM.	MAGN.	TYPE OF OBJECT
IC 4852	19 22 00.	- 60 26			NONSTELLAR OBJECT
HN 0587	19 22 01.	- 60 26			NEBULA
ZWG 230.012	19 22 06.	+ 39 28		15.1	GALAXY
HN 1921	19 22 09.1	- 28 59 51.	24		NEBULA
OCL 0106	19 22 12.	+ 13 36	300		OPEN STAR CLUSTER
ISS 0349	19 22 18.	+ 23 43	193		STELLAR RING
MRSL 055+02/1	19 22 24.	+ 20 42	120		HII REGION
ZWG 230.013	19 22 24.	+ 42 14		15.2	GALAXY
HN 1920	19 22 25.9	- 52 49 57.	18		NEBULA
SC 1918-5109.7	19 22 29.	- 51 03 57.	30		NEBULA
ISS 0395	19 22 42.	+ 19 35	277		STELLAR RING
DG 155	19 22 42.	+ 22 41	300		REFLECTION NEBULA
SCHO 1047	19 22 42.	+ 23 01 06.	1810		ISOLATED DARK CLOUD
ZWG 230.014	19 22 42.	+ 43 35		15.5	GALAXY
ZWG 281.001	19 22 48.	+ 53 32		14.5	GALAXY
UGC 11434	19 22 48.	+ 53 32	96	14.5	GALAXY S0
MCG+09-32-001	19 22 48.	+ 55 43	120	14.	GALAXY
ZWG 281.002	19 22 48.	+ 55 53		15.1	GALAXY
ZWG 280.025	19 22 48.	+ 55 53		15.1	GALAXY
UGC 11435	19 22 48.	+ 55 53	120	15.1	GALAXY Sc
RNGC 6798	19 22 49.	+ 53 32		14.5	GALAXY
IC 1300	19 22 52.	+ 52 33 00.			MAY NOT EXIST
KHAV 618	19 23	+ 00 48	6350		DARK NEBULA
B 331	19 23	+ 07 29	3600		DARK OBJECT
LBN 0133	19 23	+ 22 40	240		BRIGHT NEBULA
LDN 0748	19 23 00.	+ 21 00	3060		DARK NEBULA
LDN 0767	19 23 00.	+ 22 40	1800		DARK NEBULA
LDN 0772	19 23 00.	+ 23 00	1020		DARK NEBULA
MCG+09-32-002	19 23 00.	+ 53 31	42	14.	GALAXY
HN 0588	19 23 02.	- 59 24			NEBULA
IC 4854	19 23 03.	- 59 24			NONSTELLAR OBJECT
HN 1923	19 23 04.0	- 53 01 31.	12		NEBULA
HN 0589	19 23 08.	- 59 24			NEBULA
IC 4855	19 23 08.	- 59 24			NONSTELLAR OBJECT
OCL 0113	19 23 12.	+ 20 05	5400	5.1	OPEN STAR CLUSTER
HN 1924	19 23 18.7	- 52 20 06.			NEBULA
HN 1925	19 23 21.3	- 52 53 36.	12		NEBULA
ISS 0458	19 23 24.	+ 10 07	88		STELLAR RING
HN 1928	19 23 24.7	- 31 26 52.	18		NEBULA
HN 0590	19 23 27.	- 55 00			NEBULA
IC 4856	19 23 27.	- 55 00			NONSTELLAR OBJECT
RNGC 6795	19 23 33.	+ 03 25			NON-EXISTENT OBJECT
HN 1926	19 23 33.1	- 53 33 11.			NEBULA
ISS 0498	19 23 36.	+ 05 54	201		STELLAR RING
ISS 0459	19 23 36.	+ 10 05	105		STELLAR RING
MCG+10-27-012	19 23 36.	+ 59 36	54	15.	GALAXY
ZWG 341.026	19 23 36.	+ 70 16		15.5	GALAXY
UGC 11436	19 23 36.	+ 70 16	90	15.5	GALAXY SB
HN 1922	19 23 39.7	- 62 11 54.	36		NEBULA
VDB.66N 125	19 23 40.	+ 15 23	72		REFLECTION NEBULA
ZWG 302.013	19 23 42.	+ 59 36		15.3	GALAXY
7ZW 881	19 23 42.	+ 67 05			COMPACT GALAXY
LDN 0698	19 23 48.	+ 13 00	420		DARK NEBULA
ISS 0396	19 23 48.	+ 20 53	107		STELLAR RING
MCG+13-14-003	19 23 48.	+ 80 00	33	17.	GALAXY
ISS 0350	19 23 54.	+ 21 32	124		STELLAR RING
HN 1931	19 23 57.0	- 30 00 26.	12		NEBULA
LBN 0134	19 24	+ 22 40	300		BRIGHT NEBULA
LBN 0465	19 24	+ 70 10	3000		BRIGHT NEBULA
OCL 0105	19 24 00.	+ 11 30	480		OPEN STAR CLUSTER
VDB.66N 126	19 24 00.	+ 22 37	480		REFLECTION NEBULA
7ZW 882	19 24 00.	+ 78 57			COMPACT GALAXY
CED 167	19 24 02.	+ 22 39	390		DIFFUSE GALACTIC NEBULA
HN 1929	19 24 16.2	- 53 57 50.			NEBULA
HN 0592	19 24 17.	- 58 52			NEBULA
IC 4857	19 24 17.	- 58 52			NONSTELLAR OBJECT
PK031-10.1	19 24 22.	- 06 41 10.	6		PLANETARY NEBULA
HN 0593	19 24 23.	- 58 51			NEBULA
IC 4858	19 24 23.	- 58 51			NONSTELLAR OBJECT
ZC 1924.4+4821	19 24 24.	+ 48 21	1680		CLUSTER OF GALAXIES
HN 1936	19 24 27.8	- 31 14 30.	12		NEBULA
PK055+02.3	19 24 28.0	+ 21 03 23.	5		PLANETARY NEBULA
HN 1927	19 24 29.0	- 61 25 02.	36		NEBULA
LDN 0704	19 24 30.	+ 13 40	1260		DARK NEBULA
ISS 0397	19 24 30.	+ 19 46	123		STELLAR RING
RNGC 6794	19 24 30.	- 38 59			GALAXY
SC 1920-5127.7	19 24 31.	- 51 21 49.	12		NEBULA
HN 1930	19 24 32.3	- 53 29 55.			NEBULA
ISS 0499	19 24 36.	+ 07 24	169		STELLAR RING
SCHO 1048	19 24 36.	+ 08 35 42.	390		ISOLATED DARK CLOUD
MCG-07-40-001	19 24 39.	- 39 03	72	14.5	GALAXY
SC 1920-5134.8	19 24 40.	- 51 28 54.			NEBULA
DG 157	19 24 42.	+ 20 41	480		REFLECTION NEBULA
ZC 1924.7+6500	19 24 42.	+ 65 00	1280		CLUSTER OF GALAXIES
HN 1932	19 24 46.4	- 52 58 24.	12		NEBULA
HN 1939	19 24 47.2	- 28 30 34.	18		NEBULA
SER 131.02	19 24 48.	- 58 48	120		INTERACTING GALAXIES
LB 03119	19 24 48.	- 62 18		12.6	FAINT BLUE STAR
HN 1933	19 24 50.4	- 53 11 12.	18		NEBULA
HN 1940	19 24 51.5	- 29 45 10.	36		NEBULA
MCG+09-32-003	19 24 54.	+ 51 02	18	16.	GALAXY
B 332	19 25	+ 08 39			DARK OBJECT
LBN 0130	19 25	+ 20 41	360		BRIGHT NEBULA
KHAV 619	19 25	+ 24 00	1000		DARK NEBULA
DG 158	19 25 00.	+ 20 09	240		REFLECTION NEBULA
LDN 0755	19 25 00.	+ 21 20	1620		DARK NEBULA
DG 156	19 25 00.	+ 22 39	300		REFLECTION NEBULA
LDN 0778	19 25 00.	+ 23 35	840		DARK NEBULA
SCHO 1049	19 25 00.	+ 23 56 24.	1090		ISOLATED DARK CLOUD
LDN 0783	19 25 00.	+ 24 00	900		DARK NEBULA
MCG+07-40-006	19 25 00.	+ 44 34	42	16.	GALAXY
IC 4863	19 25 00.	- 36 18 10.			NONSTELLAR OBJECT
IC 4861	19 25 00.	- 57 41			NONSTELLAR OBJECT
HN 1941	19 25 00.6	- 29 36 57.	6		NEBULA
HN 0594	19 25 01.	- 57 40			NEBULA
HN 1934	19 25 01.3	- 50 30 17.			NEBULA
HN 1942	19 25 01.8	- 29 48 21.	36		NEBULA
RNGC 6800	19 25 04.	+ 25 02			OPEN CLUSTER
OCL 0123	19 25 06.	+ 25 02		10.	OPEN STAR CLUSTER
IC 4867	19 25 08.	+ 50 02			NONSTELLAR OBJECT
HN 1945	19 25 08.4	- 29 37 45.	12		NEBULA
PATH 1.813	19 25 09.	+ 60 43	8		NEBULA
HN 1935	19 25 09.8	- 52 38 04.	42		NEBULA
HN 1944	19 25 11.3	- 32 13 57.	36		NEBULA
HN 1946	19 25 11.5	- 31 35 21.	18		NEBULA
ISS 0351	19 25 12.	+ 26 27	218		STELLAR RING
MCG+08-35-009	19 25 12.	+ 49 40	66	15.	GALAXY
ZWG 256.017	19 25 12.	+ 50 01		14.3	GALAXY
UGC 11437	19 25 12.	+ 50 01	60	14.3	GALAXY S0-a
KARA.72 539A	19 25 12.	+ 50 01	72	14.3	PART OF DOUBLE GALAXY
MCG+08-35-010	19 25 12.	+ 50 01	60	15.	GALAXY
MCG+08-35-011	19 25 12.	+ 50 02	54	16.	GALAXY
HN 1937	19 25 14.2	- 52 32 10.	18		NEBULA
HN 1938	19 25 15.6	- 52 23 46.	18		NEBULA
ZWG 256.018	19 25 18.	+ 49 39		14.7	GALAXY
UGC 11438	19 25 18.	+ 49 39	78	14.7	GALAXY Sc
KARA.72 539B	19 25 18.	+ 50 02	42		PART OF DOUBLE GALAXY
MCG-05-46-001	19 25 18.	- 29 38	12	15.5	GALAXY
HN 0591	19 25 20.	- 71 11			NEBULA
IC 4853	19 25 20.	- 71 11			NONSTELLAR OBJECT
PK046-03.1	19 25 22.3	+ 10 18 12.			PLANETARY NEBULA
IC 1301	19 25 23.	+ 49 11 12.			NONSTELLAR OBJECT
SC 1921-5051.3	19 25 28.	- 50 45 21.	18		NEBULA
LDN 0756	19 25 30.	+ 21 20	300		DARK NEBULA
MCG+10-27-013	19 25 30.	+ 61 41	48	17.	GALAXY
ZC 1925.5+6546	19 25 30.	+ 65 46	2690		CLUSTER OF GALAXIES
SC 1921-4647.8	19 25 30.	- 46 41 50.	54		NEBULA
PK056+02.1	19 25 34.6	+ 21 23 55.			PLANETARY NEBULA
HN 1948	19 25 35.6	- 29 42 31.	6		NEBULA
ZWG 230.015	19 25 36.	+ 43 46		14.9	GALAXY
UGC 11439	19 25 36.	+ 43 46	60	14.9	GALAXY S
MCG+07-40-007	19 25 36.	+ 43 47	48	16.	GALAXY
GCL 112	19 25 36.	- 30 27			GLOBULAR STAR CLUSTER
HOLM 776C	19 25 40.	- 14 48	18	14.7	PART OF MULTIPLE GALAXY
SC 1921-5102.5	19 25 40.	- 50 56 32.			NEBULA
HOLM 776A	19 25 44.	- 14 49	18	13.5	PART OF MULTIPLE GALAXY
HOLM 776B	19 25 44.	- 14 50	12	13.5	PART OF MULTIPLE GALAXY
UGC 11440	19 25 48.	+ 65 57	60	17.	GALAXY Sc
HN 1943	19 25 50.7	- 52 57 31.	12		NEBULA
HN 0595	19 25 51.	- 66 25			NEBULA
IC 4859	19 25 51.	- 66 26			NONSTELLAR OBJECT
RNGC 6797	19 25 52.	- 25 46			NON-EXISTENT OBJECT
MRSL 064+06/1	19 25 54.	+ 31 22	1500		HII REGION
7ZW 883	19 25 54.	+ 62 39			COMPACT GALAXY
PK048-02.1	19 25 54.1	+ 12 13 28.			PLANETARY NEBULA
SCHO 1050	19 25 56.	+ 08 49 24.	490		ISOLATED DARK CLOUD
HN 1947	19 25 56.7	- 52 08 43.	12		NEBULA
B 333	19 26	+ 10 34	3600		DARK OBJECT
LBN 0148	19 26	+ 31 20	1560		BRIGHT NEBULA
LDN 0674	19 26 00.	+ 10 20	2700		DARK NEBULA
LDN 0725	19 26 00.	+ 18 04	240		DARK NEBULA
LDN 0740	19 26 00.	+ 20 00	7860		DARK NEBULA
LDN 0768	19 26 00.	+ 22 30	1320		DARK NEBULA
LDN 0781	19 26 00.	+ 23 40	840		DARK NEBULA
LDN 0782	19 26 00.	+ 23 50	360		DARK NEBULA
ZWG 230.016	19 26 00.	+ 42 17		15.7	GALAXY
UGC 11441	19 26 00.	+ 80 55	72	16.0	GALAXY Sa-b
HN 1951	19 26 08.3	- 32 17 35.	36		NEBULA
HN 1953	19 26 13.5	- 30 15 52.	24		NEBULA
MCG+09-32-004	19 26 15.	+ 54 41	60	16.	GALAXY
UGC 11442	19 26 18.	+ 54 43	60	16.0	GALAXY Sa-b
HN 0596	19 26 22.	- 67 28			NEBULA
IC 4860	19 26 23.	- 67 28			NONSTELLAR OBJECT
OCL 0109	19 26 24.	+ 17 18	360	16.	OPEN STAR CLUSTER
LDN 0682	19 26 30.	+ 10 50	900		DARK NEBULA
ZWG 281.003	19 26 30.	+ 54 16		14.8	GALAXY
UGC 11443	19 26 30.	+ 54 16	90	14.8	GALAXY Sc
MCG+09-32-005	19 26 30.	+ 54 16	78	14.	GALAXY
LB 03120	19 26 30.	- 62 32		13.2	FAINT BLUE STAR
RNGC 6801	19 26 32.	+ 54 16		15.0	GALAXY
HN 0597	19 26 34.	- 67 26			NEBULA
IC 4862	19 26 35.	- 67 26			NONSTELLAR OBJECT
SC 1922-5128.5	19 26 39.	- 51 22 28.			NEBULA
HN 1949	19 26 41.5	- 52 27 28.	12		NEBULA
OCL 0108	19 26 42.	+ 14 46	60		OPEN STAR CLUSTER
ISS 0352	19 26 42.	+ 21 54	300		STELLAR RING
HN 1950	19 26 43.1	- 52 23 40.	18		NEBULA
SC 1924+1806.3	19 26 44.	+ 18 12 24.	54		NEBULA
MRSL 066+07/1	19 26 48.	+ 32 35	1500		HII REGION
ZWG 357.004	19 26 48.	+ 77 10		15.5	GALAXY
KARA.73B 0873	19 26 48.	+ 77 10	42	15.5	ISOLATED GALAXY S
ISS 0398	19 26 54.	+ 19 09	80		STELLAR RING
ZWG 281.004	19 26 54.	+ 52 47		15.3	GALAXY
UGC 11444	19 26 54.	+ 52 47	60	15.3	GALAXY Sc
7ZW 884	19 26 54.	+ 61 50			COMPACT GALAXY
ZWG 323.013	19 26 54.	+ 65 12		15.4	GALAXY
KARA.72 540A	19 26 54.	+ 65 12	36	15.4	PART OF DOUBLE GALAXY
KHAV 620	19 27	+ 09 18	6650		DARK NEBULA
KHAV 621	19 27	+ 23 36	1570		DARK NEBULA
KHAV 622	19 27	+ 26 18	4600		DARK NEBULA
LBN 0150	19 27	+ 32 35	900		BRIGHT NEBULA
LDN 0654	19 27 00.	+ 07 00	8640		DARK NEBULA
LDN 0691	19 27 00.	+ 11 50	960		DARK NEBULA
LDN 0696	19 27 00.	+ 12 30	540		DARK NEBULA
LDN 0699	19 27 00.	+ 12 40	540		DARK NEBULA
LDN 0784	19 27 00.	+ 24 00	2520		DARK NEBULA
ZWG 323.014	19 27 00.	+ 65 13		15.6	GALAXY
KARA.72 540B	19 27 00.	+ 65 13	48	15.6	PART OF DOUBLE GALAXY
HN 1952	19 27 00.7	- 53 39 03.	36		NEBULA
ZC 1927.1+7024	19 27 06.	+ 70 24	1410		CLUSTER OF GALAXIES
IC 4865	19 27 13.	- 46 48			NONSTELLAR OBJECT
HN 1958	19 27 14.5	- 31 58 42.	36		NEBULA
HN 1954	19 27 17.9	- 52 34 19.	12		NEBULA
DG 159	19 27 24.	+ 18 20	240		REFLECTION NEBULA
PK320-28.1	19 27 29.	- 74 39	10		PLANETARY NEBULA
LDN 0685	19 27 30.	+ 11 20	780		DARK NEBULA
ISS 0400	19 27 30.	+ 19 08	113		STELLAR RING
ISS 0399	19 27 30.	+ 19 17	82		STELLAR RING
UGC 11445	19 27 30.	+ 56 35	66	16.0	GALAXY S
ZCG 1927+66	19 27 30.	+ 66 16		18.3	COMPACT GALAXY
7ZW 885	19 27 30.	+ 66 24			COMPACT GALAXY
ZC 1927.6+6726	19 27 36.	+ 67 26	4170		CLUSTER OF GALAXIES
HN 1956	19 27 38.8	- 52 30 06.	18		NEBULA
HN 1955	19 27 39.5	- 53 39 30.	60		NEBULA
MCG+09-32-006	19 27 42.	+ 56 35	60	15.	GALAXY
SC 1923-5127.3	19 27 44.	- 51 21 12.			NEBULA
HN 1957	19 27 48.0	- 52 29 59.			NEBULA
SC 1923-5132.9	19 27 49.	- 51 26 47.	18		NEBULA
ZWG 230.017	19 27 54.	+ 41 13		14.2	GALAXY
UGC 11446	19 27 54.	+ 41 13	84	14.2	GALAXY SB
UGC 11447	19 27 54.	+ 50 18	60	17.	GALAXY Sc
SCHO 1051	19 27 55.	+ 24 17 00.	350		ISOLATED DARK CLOUD
MCG+07-40-008	19 27 57.	+ 41 11	72	14.5	GALAXY
PK047-03.1	19 27 57.1	+ 11 17 19.			PLANETARY NEBULA
PK050-01.1	19 27 59.1	+ 14 41 04.			PLANETARY NEBULA
LBN 0128	19 28	+ 18 09	600		BRIGHT NEBULA

OBJECT NAME	RIGHT ASCEN.	DECLINATION	DIAM.	MAGN.	TYPE OF OBJECT
LBN 0129	19 28	+ 18 10	420		BRIGHT NEBULA
KHAV 623	19 28	+ 24 00	7600		DARK NEBULA
KHAV 624	19 28	+ 25 06	2930		DARK NEBULA
LDN 0760	19 28 00.	+ 21 45	360		DARK NEBULA
LDN 0779	19 28 00.	+ 23 15	1500		DARK NEBULA
HN 1960	19 28 00.0	- 49 44 22.	18		NEBULA
HN 1959	19 28 01.4	- 53 40 35.	12		NEBULA
MRSL 053+00/1	19 28 06.	+ 18 10	540		HII REGION
RNGC 6799	19 28 09.	- 56 01			UNVERIFIED SOUTHERN OBJECT
CED 168	19 28 12.	+ 18 05	390		DIFFUSE GALACTIC NEBULA
SG 2.147	19 28 13.	+ 18 09	360		DIFFUSE EMISSION NEBULA
SCHO 1052	19 28 17.	+ 09 17 42.	450		ISOLATED DARK CLOUD
7ZW 886	19 28 18.	+ 65 24			COMPACT GALAXY
7ZW 887	19 28 18.	+ 67 42			COMPACT GALAXY
HN 1961	19 28 21.2	- 53 41 33.	12		NEBULA
OCL 0114	19 28 24.	+ 20 10	1080	12.5	OPEN STAR CLUSTER
ZWG 230.018	19 28 24.	+ 43 22		15.6	GALAXY
RNGC 6802	19 28 25.	+ 20 10		12.0	OPEN CLUSTER
HN 1964	19 28 30.9	- 29 47 37.	18		NEBULA
PK057+02.1	19 28 31.0	+ 22 57 56.			PLANETARY NEBULA
LB 00343	19 28 32.	- 08 54 42.		15.2	FAINT BLUE STAR
HN 1962	19 28 34.8	- 51 17 56.	6		NEBULA
MCG-03-49-003	19 28 36.	- 18 50	30	17.	GALAXY
HN 1965	19 28 40.8	- 31 14 00.	18		NEBULA
SCHO 1053	19 28 41.	+ 07 38 54.	280		ISOLATED DARK CLOUD
ISS 0460	19 28 42.	+ 15 08	185		STELLAR RING
MCG+06-43-001	19 28 42.	+ 35 40 30.	60	15.	GALAXY
MCG+08-35-012	19 28 42.	+ 47 24	42	16.	GALAXY
MCG+13-14-004	19 28 42.	+ 79 50	36	16.	GALAXY
UGC 11448	19 28 48.	+ 35 40	72	16.0	GALAXY Sc
ZWG 357.005	19 28 48.	+ 79 50		15.6	GALAXY
PK004-22.1	19 28 50.	- 34 20 17.	10		PLANETARY NEBULA
PK046-04.1	19 28 53.6	+ 09 56 58.8	6	12.4	PLANETARY NEBULA
UGC 11449	19 28 54.	+ 64 50	60	16.5	GALAXY Sc
7ZW 888	19 28 54.	+ 65 43			COMPACT GALAXY
7ZW 889	19 28 54.	+ 66 44			COMPACT GALAXY
RNGC 6803	19 28 55.	+ 09 57		11.5	PLANETARY NEBULA
HN 1963	19 28 55.2	- 53 33 43.	18		NEBULA
SCHO 1054	19 28 56.	+ 24 15 24.	410		ISOLATED DARK CLOUD
HN 0052	19 28 59.	+ 09 58			NEBULA
KHAV 625	19 29	+ 21 30	11040		DARK NEBULA
KHAV 626	19 29	+ 25 18	2760		DARK NEBULA
KHAV 627	19 29	+ 29 30	1000		DARK NEBULA
LDN 0721	19 29 00.	+ 16 35	17700		DARK NEBULA
LDN 0727	19 29 00.	+ 18 10	540		DARK NEBULA
IC 1302	19 29 00.	+ 35 39			GALAXY Sb
MCG+06-43-002	19 29 00.	+ 35 41	48	14.	GALAXY
ZWG 230.019	19 29 00.	+ 43 19		15.4	GALAXY
MCG+07-40-009	19 29 00.	+ 43 19	30	16.	GALAXY
ZCG 1929+83	19 29 00.	+ 83 54		18.2	COMPACT GALAXY
MCG+15-01-022	19 29 00.	+ 88 30	30	17.	GALAXY
HN 1966	19 29 04.3	- 28 02 11.	18		NEBULA
ZWG 230.020	19 29 06.	+ 41 13		14.6	GALAXY
BC PKS1929-457	19 29 08.7	- 45 43 08.		18.5	QUASI-STELLAR OBJECT
HN 1969	19 29 11.5	- 29 29 34.	12		NEBULA
PK045-04.1	19 29 11.64	+ 09 07 03.1	65	13.3	PLANETARY NEBULA
PK045-04.1	19 29 11.64	+ 09 07 03.1	65	13.3	PLANETARY NEBULA
RNGC 6804	19 29 12.	+ 09 07		13.5	PLANETARY NEBULA
HN 1968	19 29 12.6	- 30 22 52.	36		NEBULA
PK009-21.1	19 29 15.	- 29 31	34		PLANETARY NEBULA
LDN 0377	19 29 18.	- 22 30	180		DARK NEBULA
HN 1970	19 29 18.4	- 30 02 46.	30		NEBULA
7ZW 890	19 29 24.	+ 67 35			COMPACT GALAXY
MCG-05-46-002	19 29 24.	- 29 29	60	15.	GALAXY
HN 1971	19 29 28.8	- 31 27 33.	18		NEBULA
LDN 0786	19 29 30.	+ 24 20	780		DARK NEBULA
MCG+06-43-003	19 29 30.	+ 36 57	24	15.5	GALAXY
ZWG 256.019	19 29 30.	+ 46 36		15.1	GALAXY
UGC 11450	19 29 30.	+ 46 36	60	15.1	GALAXY SB:a-b
MCG+08-35-013	19 29 30.	+ 46 37 30.	60	16.	GALAXY
IC 1303	19 29 33.	+ 35 45 03.	90	15.51	GALAXY Sc
MCG+06-43-004	19 29 36.	+ 35 47	72	14.	GALAXY
UGC 11451	19 29 36.	+ 42 06	90	16.5	GALAXY Sc
ZC 1929.6+5852	19 29 36.	+ 58 52	2620		CLUSTER OF GALAXIES
UGC 11452	19 29 42.	+ 35 46	84	15.0	GALAXY Sc
LB 00344	19 29 47.	- 06 37 00.		15.7	FAINT BLUE STAR
ZWG 230.021	19 29 48.	+ 41 48		15.6	GALAXY
ZWG 341.027	19 29 48.	+ 72 01		15.7	GALAXY
KARA.72 541A	19 29 48.	+ 72 01	42	15.7	PART OF DOUBLE GALAXY
SCHO 1055	19 29 52.	+ 09 03 36.	430		ISOLATED DARK CLOUD
ISS 0461	19 29 54.	+ 11 17	60		STELLAR RING
KARA.72 542B	19 29 54.	+ 53 59	48		PART OF DOUBLE GALAXY
ZWG 281.005	19 29 54.	+ 54 00		14.4	GALAXY
UGC 11453	19 29 54.	+ 54 00	102	14.4	GALAXY Sb
KARA.72 542A	19 29 54.	+ 54 00	102	14.4	PART OF DOUBLE GALAXY
IC 4868	19 29 56.	- 46 00 06.			NONSTELLAR OBJECT
HN 1967	19 29 57.1	- 53 20 03.			NEBULA
KHAV 628	19 30	+ 16 12	12140		DARK NEBULA
LBN 0481	19 30	+ 41 30	2640		BRIGHT NEBULA
LBN 0225	19 30	+ 44 30	7800		BRIGHT NEBULA
LDN 0770	19 30 00.	+ 22 00	2340		DARK NEBULA
MCG+09-32-008	19 30 00.	+ 53 58	18	16.	GALAXY
MCG+09-32-007	19 30 00.	+ 53 59	120	14.	GALAXY
ZWG 303.001	19 30 00.	+ 61 41		15.6	GALAXY
ZWG 302.014	19 30 00.	+ 61 41		15.6	GALAXY
MCG+10-28-001	19 30 00.	+ 61 42	24	17.	GALAXY
7ZW 891	19 30 00.	+ 67 40			COMPACT GALAXY
MCG+12-18-002	19 30 00.	+ 72 02	33	16.	GALAXY
SCHO 1056	19 30 02.	+ 07 43 48.	240		ISOLATED DARK CLOUD
MCG+08-35-014	19 30 06.	+ 49 48	54	16.	GALAXY
7ZW 892	19 30 06.	+ 62 42			COMPACT GALAXY
HN 0598	19 30 09.	- 61 16			NEBULA
IC 4866	19 30 09.	- 61 16			NONSTELLAR OBJECT
UGC 11454	19 30 12.	+ 49 45	60	17.	GALAXY
PK044-05.1	19 30 13.7	+ 07 21 25.			PLANETARY NEBULA
HN 1976	19 30 17.2	- 29 59 42.	12		NEBULA
HN 1972	19 30 25.4	- 53 45 07.	18		NEBULA
SCHO 1057	19 30 28.	+ 07 42 00.	180		ISOLATED DARK CLOUD
ZWG 281.006	19 30 30.	+ 56 15		15.7	GALAXY
ZWG 341.028	19 30 30.	+ 72 00		15.3	GALAXY
UGC 11455	19 30 30.	+ 72 00	168	15.3	GALAXY Sc
KARA.72 541B	19 30 30.	+ 72 00	138	15.3	PART OF DOUBLE GALAXY
HN 1974	19 30 35.8	- 53 07 30.	18		NEBULA
ISS 0462	19 30 42.	+ 09 17	114		STELLAR RING
MCG+12-18-003	19 30 42.	+ 72 01	150	15.	GALAXY
MCG+07-40-010	19 30 48.	+ 41 47	63	15.	GALAXY
ZWG 230.022	19 30 48.	+ 41 48		14.8	GALAXY
MIN.46 11	19 30 51.	+ 26 48			DIFFUSE NEBULA
ISS 0463	19 30 54.	+ 10 08	182		STELLAR RING
MCG+06-43-005	19 30 54.	+ 37 48	72	15.	GALAXY
7ZW 893	19 30 54.	+ 63 36			COMPACT GALAXY
PK061+03.1	19 30 54.9	+ 26 46 13.	16		PLANETARY NEBULA
KHAV 629	19 31	+ 12 06			DARK NEBULA
KHAV 630	19 31	+ 24 24	4140		DARK NEBULA
PK047-04.1	19 31 00.	+ 10 31	163	14.8	PLANETARY NEBULA
LDN 0742	19 31 00.	+ 19 30	840		DARK NEBULA
UGC 11456	19 31 00.	+ 37 46	84	15.5	GALAXY S
7ZW 894	19 31 00.	+ 83 56			COMPACT GALAXY
PK057+01.1	19 31 01.2	+ 22 52 02.			PLANETARY NEBULA
HN 1973	19 31 02.4	- 61 20 41.	12		NEBULA
ISS 0464	19 31 06.	+ 10 32	173		STELLAR RING
HN 1977	19 31 07.1	- 53 02 46.			NEBULA
HN 1978	19 31 09.3	- 52 57 10.	30		NEBULA
HN 1979	19 31 10.9	- 52 58 34.	36		NEBULA
HN 1159	19 31 14.	- 46 15	18		NEBULA
IC 4873	19 31 14.	- 46 15			NONSTELLAR OBJECT
HN 1980	19 31 16.1	- 53 09 45.	12		NEBULA
ZC 1931.3+6424	19 31 18.	+ 64 24	1080		CLUSTER OF GALAXIES
SCHO 1058	19 31 22.	+ 07 37 12.	310		ISOLATED DARK CLOUD
HN 1975	19 31 23.7	- 59 54 04.	18		NEBULA
SCHO 1059	19 31 24.	+ 07 45 30.	230		ISOLATED DARK CLOUD
ISS 0465	19 31 24.	+ 11 16	53		STELLAR RING
SCHO 1060	19 31 26.	+ 09 05 54.	370		ISOLATED DARK CLOUD
HN 0600	19 31 29.	- 57 38			NEBULA
IC 4871	19 31 29.	- 57 38			NONSTELLAR OBJECT
LDN 0783	19 31 30.	+ 24 40	300		DARK NEBULA
7ZW 895	19 31 30.	+ 63 24			COMPACT GALAXY
MCG-03-50-001	19 31 30.	- 15 39	30	15.	GALAXY
HN 0602	19 31 35.	- 57 38			NEBULA
IC 4872	19 31 35.	- 57 38			NONSTELLAR OBJECT
MCG+10-28-002	19 31 39.	+ 59 34	30	16.	GALAXY
PK059+02.1	19 31 40.9	+ 24 25 54.			PLANETARY NEBULA
7ZW 896	19 31 42.	+ 63 22			COMPACT GALAXY
HN 0601	19 31 45.	- 61 19			NEBULA
IC 4869	19 31 45.	- 61 19			NONSTELLAR OBJECT
LBN 0464	19 32	+ 69 50	2520		BRIGHT NEBULA
LDN 0785	19 32 00.	+ 24 00	300		DARK NEBULA
ZC 1932.0+7916	19 32 00.	+ 79 16	1550		CLUSTER OF GALAXIES
RNGC 6807	19 32 04.	+ 05 34		14.0	PLANETARY NEBULA
HN 0051	19 32 04.	+ 05 35			NEBULA
MCG-07-40-002	19 32 06.	- 40 10	48	16.5	GALAXY
PK042-06.1	19 32 06.1	+ 05 34 24.1		13.8	PLANETARY NEBULA
HN 1981	19 32 08.4	- 53 06 06.	6		NEBULA
ISS 0353	19 32 12.	+ 22 54	114		STELLAR RING
ABC 2321	19 32 13.	+ 73 19		17.6	RICH CLUSTER OF GALAXIES
HN 1982	19 32 16.0	- 52 32 35.	12		NEBULA
ISS 0354	19 32 18.	+ 24 21	161		STELLAR RING
HN 1983	19 32 27.3	- 52 50 17.	12		NEBULA
LDN 0701	19 32 36.	+ 12 10	480		DARK NEBULA
HN 0603	19 32 40.	- 65 56			NEBULA
IC 4870	19 32 40.	- 65 56			NONSTELLAR OBJECT
HN 1160	19 32 41.	- 47 23	18		NEBULA
IC 4874	19 32 41.	- 47 23			NONSTELLAR OBJECT
ZWG 256.020	19 32 42.	+ 49 08		15.4	GALAXY
B 334	19 32 46.	+ 12 13	180		DARK OBJECT
SCHO 1061	19 32 46.	+ 12 15 36.	420		ISOLATED DARK CLOUD
PK064+05.1	19 32 47.53	+ 30 24 20.6	13	9.6	PLANETARY NEBULA
ZWG 256.021	19 32 48.	+ 49 50		15.3	GALAXY
MCG+08-35-015	19 32 48.	+ 49 52 30.	15	17.	GALAXY
ZWG 303.002	19 32 48.	+ 61 58		15.5	GALAXY
HN 1984	19 32 56.9	- 53 37 27.	6		NEBULA
KHAV 632	19 33	+ 12 07			DARK NEBULA
KHAV 631	19 33	+ 12 13	1000		DARK NEBULA
KHAV 633	19 33	+ 21 49	14220		DARK NEBULA
KHAV 634	19 33	+ 30 31	1000		DARK NEBULA
LDN 0776	19 33 00.	+ 22 00	2940		DARK NEBULA
LDN 0790	19 33 00.	+ 25 35	240		DARK NEBULA
MCG+10-28-003	19 33 00.	+ 61 59	30	15.	GALAXY
ZWG 368.003	19 33 00.	+ 83 57		15.5	GALAXY
MCG+15-01-023	19 33 00.	+ 88 37	36	16.	GALAXY
HN 1985	19 33 00.4	- 53 24 44.	12		NEBULA
PK053-01.1	19 33 03.4	+ 17 06 22.	4		PLANETARY NEBULA
HN 0599	19 33 06.	- 77 41			NEBULA
IC 4864	19 33 06.	- 77 41			NONSTELLAR OBJECT
HN 1988	19 33 06.0	- 50 31 56.	12		NEBULA
HN 1986	19 33 07.1	- 53 00 38.	12		NEBULA
HN 1989	19 33 09.0	- 50 32 43.	6		NEBULA
HN 1987	19 33 11.5	- 53 00 50.	6		NEBULA
ZC 1933.2+6236	19 33 12.	+ 62 36	3290		CLUSTER OF GALAXIES
ZWG 341.029	19 33 12.	+ 72 21		15.2	GALAXY
ZWG 357.006	19 33 12.	+ 77 16		15.4	GALAXY
ISS 0293	19 33 18.	+ 27 28	166		STELLAR RING
UGC 11457	19 33 18.	+ 52 59	78	17.	GALAXY
RNGC 6805	19 33 25.	- 38 28			GALAXY
HN 1990	19 33 29.6	- 53 02 18.	6		NEBULA
RNGC 6806	19 33 33.	- 42 25			GALAXY
MRSL 064+04/1	19 33 36.	+ 29 30	7200		HII REGION
AGU 58	19 33 36.	- 42 24 42.	90	12.5	PECULIAR GALAXY
MCG-07-40-003	19 33 39.	- 42 26	66	14.5	GALAXY
HN 1161	19 33 41.	- 52 10	18		NEBULA
IC 4875	19 33 41.	- 52 10			NONSTELLAR OBJECT
OCL 0127	19 33 42.	+ 25 06	3600		OPEN STAR CLUSTER
SCHO 1062	19 33 46.	+ 12 24 18.	250		ISOLATED DARK CLOUD
IC 1304	19 33 47.	+ 40 56			NONSTELLAR OBJECT
SC 1929-5217.8	19 33 47.	- 52 11 17.	18		NEBULA
ISS 0401	19 33 48.	+ 15 37	130		STELLAR RING
HN 1991	19 33 48.8	- 51 16 17.	18		NEBULA
HN 1162	19 33 49.	- 52 57	30		NEBULA
IC 4876	19 33 49.	- 52 57			NONSTELLAR OBJECT
PK059+02.2	19 33 49.0	+ 24 48 10.			PLANETARY NEBULA
BC PKS1933-400	19 33 51.0	- 40 04 48.		18.	QUASI-STELLAR OBJECT
SHB 349	19 33 51.0	- 40 04 48.		18.	QUASI-STELLAR OBJECT
SCHO 1063	19 33 52.	+ 07 42 06.	240		ISOLATED DARK CLOUD
SCHO 1064	19 33 54.	+ 07 31 54.	240		ISOLATED DARK CLOUD
ZWG 230.023	19 33 54.	+ 41 47		15.6	GALAXY
HN 1163	19 33 59.	- 52 05	18		NEBULA
IC 4877	19 33 59.	- 52 05			NONSTELLAR OBJECT
KHAV 636	19 34	+ 07 25			DARK NEBULA
KHAV 635	19 34	+ 09 49	6050		DARK NEBULA
LB 09996	19 34	- 87 15		13.2	FAINT BLUE STAR
ISS 0466	19 34 00.	+ 12 58	112		STELLAR RING
LDN 0761	19 34 00.	+ 21 00	1380		DARK NEBULA
LDN 0816	19 34 00.	+ 30 25	600		DARK NEBULA
ZC 1934.0+6700	19 34 00.	+ 67 00	1410		CLUSTER OF GALAXIES
HN 1996	19 34 03.1	- 31 34 03.			NEBULA
SC 1930-5212.9	19 34 04.	- 52 06 22.	30		NEBULA

OBJECT NAME	RIGHT ASCEN.	DECLINATION	DIAM.	MAGN.	TYPE OF OBJECT
LDN 0702	19 34 06.	+ 12 10	240		DARK NEBULA
ISS 0402	19 34 06.	+ 19 30	146		STELLAR RING
HN 1992	19 34 06.6	- 49 52 52.	12		NEBULA
MIN.46 12	19 34 11.	+ 29 26			DIFFUSE NEBULA
ISS 0403	19 34 12.	+ 17 19	173		STELLAR RING
7ZW 897	19 34 12.	+ 68 49			COMPACT GALAXY
PK058+01.1	19 34 14.7	+ 23 33 05.			PLANETARY NEBULA
PK055-00.1	19 34 16.0	+ 19 35 38.	5		PLANETARY NEBULA
SCHO 1065	19 34 19.	+ 12 30 24.	270		ISOLATED DARK CLOUD
HN 1993	19 34 20.8	- 52 56 03.	12		NEBULA
B 336	19 34 21.	+ 12 14	120		DARK OBJECT
ISS 0405	19 34 24.	+ 17 45	173		STELLAR RING
ISS 0404	19 34 24.	+ 19 43	119		STELLAR RING
SCHO 1066	19 34 27.	+ 12 13 12.	260		ISOLATED DARK CLOUD
HN 1994	19 34 28.6	- 54 15 57.			NEBULA
B 335	19 34 29.	+ 07 30	360		DARK OBJECT
HN 1995	19 34 29.4	- 53 15 02.	12		NEBULA
LDN 0663	19 34 30.	+ 07 27	240		DARK NEBULA
LDN 0705	19 34 30.	+ 12 20	1140		DARK NEBULA
MCG-03-50-002	19 34 36.	- 17 34 30.	84	15.5	GALAXY
ISS 0406	19 34 36.	- 17 38	239		STELLAR RING
HN 0604	19 34 37.	- 58 19			NEBULA
IC 4878	19 34 37.	- 58 19			NONSTELLAR OBJECT
SCHO 1067	19 34 39.	+ 07 29 30.	450		ISOLATED DARK CLOUD
B 337	19 34 41.	+ 12 17	1020		DARK OBJECT
ZWG 371.001	19 34 42.	+ 02 37		15.6	GALAXY
ISS 0407	19 34 42.	+ 18 28	90		STELLAR RING
ISS 0294	19 34 42.	+ 29 27	82		STELLAR RING
ZC 1934.7+5902	19 34 42.	+ 59 02	610		CLUSTER OF GALAXIES
ZWG 341.030	19 34 42.	+ 74 00		15.6	GALAXY
PK050-03.1	19 34 45.8	+ 13 34 29.			PLANETARY NEBULA
SCHO 1068	19 34 50.	+ 12 21 42.	380		ISOLATED DARK CLOUD
MCG-07-40-004	19 34 51.	- 39 02	30	15.5	GALAXY
HN 1997	19 34 52.1	- 53 16 55.	6		NEBULA
KHAV 638	19 35	+ 01 13	15440		DARK NEBULA
KHAV 639	19 35	+ 04 37	7470		DARK NEBULA
KHAV 637	19 35	+ 25 25	2660		DARK NEBULA
LBN 0147	19 35	+ 29 30	3660		BRIGHT NEBULA
LDN 0671	19 35 00.	+ 08 20	1260		DARK NEBULA
LDN 0720	19 35 00.	+ 15 20	10740		DARK NEBULA
LDN 0792	19 35 00.	+ 25 35	1500		DARK NEBULA
LDN 0795	19 35 00.	+ 26 30	2100		DARK NEBULA
LDN 0794	19 35 00.	+ 26 30	2100		DARK NEBULA
HN 1998	19 35 04.4	- 53 14 48.	18		NEBULA
ZWG 397.001	19 35 06.	+ 06 15		15.7	GALAXY
ZC 1935.1+6729	19 35 06.	+ 67 29	1140		CLUSTER OF GALAXIES
UGC 11458	19 35 06.	+ 69 54	60	17.	GALAXY DWRF IP
LB 00345	19 35 08.	- 08 00 12.		15.1	FAINT BLUE STAR
HN 2010	19 35 09.1	- 30 39 10.	12		NEBULA
ISS 0467	19 35 12.	+ 12 25	113		STELLAR RING
LDN 0821	19 35 12.	+ 32 20	300		DARK NEBULA
HN 1999	19 35 16.9	- 53 04 59.	42		NEBULA
ISS 0408	19 35 18.	+ 15 08	127		STELLAR RING
7ZW 898	19 35 18.	+ 58 10			COMPACT GALAXY
MCG-05-46-003	19 35 18.	- 30 39	24	16.	GALAXY
HN 2000	19 35 21.8	- 53 11 11.	18		NEBULA
HN 2002	19 35 22.4	- 50 20 23.	12		NEBULA
ZC 1935.4+5755	19 35 24.	+ 57 55	2890		CLUSTER OF GALAXIES
ISS 0295	19 35 30.	+ 27 35	215		STELLAR RING
HN 2001	19 35 30.3	- 53 33 10.	12		NEBULA
HN 2005	19 35 35.3	- 51 49 22.	18		NEBULA
ISS 0410	19 35 36.	+ 15 28	203		STELLAR RING
ISS 0409	19 35 36.	+ 15 37	190		STELLAR RING
HN 2004	19 35 36.5	- 53 33 34.	12		NEBULA
HN 2003	19 35 39.8	- 54 55 46.	42		NEBULA
MCG+07-40-011	19 35 39.	+ 40 34	120	14.5	GALAXY
ISS 0411	19 35 42.	+ 16 50	223		STELLAR RING
ZWG 230.024	19 35 42.	+ 40 36		14.4	GALAXY
UGC 11459	19 35 42.	+ 40 36	144	14.4	GALAXY Sc
HN 2009	19 35 45.1	- 50 38 39.	18		NEBULA
IC 4879	19 35 47.	- 52 29			NONSTELLAR OBJECT
OCL 0098	19 35 48.	+ 00 13	90	7.0	OPEN STAR CLUSTER
HN 1164	19 35 48.	- 52 29	24		NEBULA
HN 2007	19 35 48.9	- 53 18 57.	6		NEBULA
HN 2006	19 35 49.1	- 54 28 57.			NEBULA
HN 2008	19 35 50.3	- 52 55 45.	12		NEBULA
ISS 0355	19 35 54.	+ 25 33	159		STELLAR RING
HN 2011	19 35 59.8	- 52 36 32.	12		NEBULA
KHAV 640	19 36	+ 08 13			DARK NEBULA
LDN 0731	19 36 00.	+ 17 30	2100		DARK NEBULA
PK060+01.1	19 36 03.2	+ 25 08 51.	25		PLANETARY NEBULA
MCG+07-40-012	19 36 06.	+ 40 53	30	15.	GALAXY
SC 1932-5138.8	19 36 11.	- 51 32 07.	18		NEBULA
LDN 0687	19 36 12.	+ 10 18	300		DARK NEBULA
ISS 0412	19 36 12.	+ 17 56	247		STELLAR RING
LDN 0743	19 36 12.	+ 18 50	480		DARK NEBULA
ZWG 230.025	19 36 12.	+ 40 53		14.0	GALAXY
UGC 11460	19 36 12.	+ 40 53	84	14.0	GALAXY E-SO
HN 2012	19 36 19.5	- 54 12 07.	18		NEBULA
MIL 80	19 36 20.	+ 17 13	1200		SUPERNOVA REMNANT
HN 0605	19 36 20.	- 55 18			NEBULA
IC 4882	19 36 21.	- 55 18			NONSTELLAR OBJECT
HN 0606	19 36 23.	- 55 58			NEBULA
IC 4881	19 36 23.	- 55 58			NONSTELLAR OBJECT
LDN 0688	19 36 24.	+ 10 22	480		DARK NEBULA
ISS 0413	19 36 24.	+ 15 36	146		STELLAR RING
MCG+10-28-004	19 36 24.	+ 59 36	42	16.	GALAXY
ZWG 303.003	19 36 24.	+ 59 37		15.7	GALAXY
7ZW 899	19 36 24.	+ 62 23			COMPACT GALAXY
HN 0607	19 36 24.	- 56 31			NEBULA
IC 4880	19 36 24.	- 56 31			NONSTELLAR OBJECT
ISS 0468	19 36 30.	+ 11 48	194		STELLAR RING
VMT 20	19 36 30.	+ 17 08	360		SUPERNOVA REMNANT
HN 2013	19 36 38.0	- 51 07 30.	12		NEBULA
RNGC 6811	19 36 39.	+ 46 27		9.0	OPEN CLUSTER
HN 2016	19 36 39.4	- 28 43 34.	18		NEBULA
ZWG 397.002	19 36 42.	+ 08 42		15.5	GALAXY
UGC 11461	19 36 42.	+ 08 42	78	15.5	GALAXY Sc
LDN 0752	19 36 42.	+ 19 25	420		DARK NEBULA
OCL 0185	19 36 42.	+ 46 27	1260	9.9	OPEN STAR CLUSTER
RNGC 6817	19 36 47.	+ 62 16		15.5	GALAXY
ZWG 303.004	19 36 48.	+ 62 16		15.6	GALAXY
UGC 11462	19 36 48.	+ 71 14	66	16.0	GALAXY S
ZC 1936.8+7648	19 36 48.	+ 76 48	610		CLUSTER OF GALAXIES
ZWG 371.002	19 36 48.	- 01 41		15.7	GALAXY
PK052-02.2	19 36 53.5	+ 15 49 52.	12	12.6	PLANETARY NEBULA
ZWG 357.007	19 36 54.	+ 75 54		15.7	GALAXY
GCL 113	19 36 54.	- 31 03	1266	7.05	GLOBULAR STAR CLUSTER
RNGC 6809	19 36 55.	- 31 03		7.0	GLOBULAR CLUSTER
PK052-02.1	19 36 59.8	+ 16 13 55.			PLANETARY NEBULA
KHAV 641	19 37	+ 17 25	2570		DARK NEBULA
KHAV 642	19 37	+ 26 37	1570		DARK NEBULA
KHAV 643	19 37	+ 27 37	3610		DARK NEBULA
ZWG 397.003	19 37 00.	+ 06 14		15.7	GALAXY
7ZW 900	19 37 00.	+ 59 13			COMPACT GALAXY
MCG+10-28-005	19 37 00.	+ 62 16	36	16.	GALAXY
ZC 1937.0+6605	19 37 00.	+ 66 05	5170		CLUSTER OF GALAXIES
ZWG 341.031	19 37 00.	+ 70 26		15.7	GALAXY
HN 2014	19 37 00.4	- 52 07 58.	6		NEBULA
IC 1305	19 37 04.	+ 20 06			NONSTELLAR OBJECT
CED 169	19 37 05.	+ 20 06	120		DIFFUSE GALACTIC NEBULA
HN 2015	19 37 08.8	- 53 17 58.	18		NEBULA
ZWG 371.003	19 37 12.	- 00 37		15.7	GALAXY
B 142	19 37 19.	+ 10 24	2400		DARK OBJECT
SCHO 1069	19 37 22.	+ 08 06 48.	400		ISOLATED DARK CLOUD
PK056-00.1	19 37 24.5	+ 20 12 07.	3		PLANETARY NEBULA
SCHO 1070	19 37 25.	+ 10 24 12.	870		ISOLATED DARK CLOUD
MCG+10-28-006	19 37 27.	+ 62 16	36	17.	GALAXY
LDN 0793	19 37 30.	+ 25 20	1020		DARK NEBULA
PK019-19.1	19 37 33.	- 20 34	140		PLANETARY NEBULA
PK061+02.1	19 37 35.	+ 26 23 39.	10		PLANETARY NEBULA
ZWG 303.005	19 37 36.	+ 61 29		15.5	GALAXY
UGC 11463	19 37 36.	+ 61 29	60	15.5	GALAXY Sb
SCHO 1071	19 37 42.	+ 17 19 30.	370		ISOLATED DARK CLOUD
MCG+10-28-007	19 37 42.	+ 61 29	60	16.	GALAXY
ZC 1937.7+6445	19 37 42.	+ 64 45	2960		CLUSTER OF GALAXIES
7ZW 901	19 37 54.	+ 60 29			COMPACT GALAXY
HN 0608	19 37 57.	- 55 39			NEBULA
IC 4883	19 37 57.	- 55 39			NONSTELLAR OBJECT
KHAV 644	19 38	+ 10 19			DARK NEBULA
KHAV 645	19 38	+ 10 55	3400		DARK NEBULA
LBN 0469	19 38	+ 70 50	2400		BRIGHT NEBULA
LDN 0689	19 38 00.	+ 10 10	960		DARK NEBULA
LDN 0737	19 38 00.	+ 18 10	1080		DARK NEBULA
LDN 0811	19 38 00.	+ 28 45	2040		DARK NEBULA
MCG+10-28-008	19 38 00.	+ 62 33	30	17.	GALAXY
SCHO 1072	19 38 05.	+ 17 28 00.	440		ISOLATED DARK CLOUD
LDN 0729	19 38 06.	+ 17 10	600		DARK NEBULA
ZWG 257.001	19 38 06.	+ 50 46		15.5	GALAXY
HN 2017	19 38 07.1	- 52 21 00.	12		NEBULA
PK055-01.1	19 38 12.3	+ 18 42 14.	3		PLANETARY NEBULA
RNGC 6813	19 38 17.	+ 27 11			DIFFUSE NEBULA
LDN 0690	19 38 18.	+ 10 13	720		DARK NEBULA
LDN 0694	19 38 18.	+ 10 50	1320		DARK NEBULA
SG 2.148	19 38 20.	+ 27 11	90		DIFFUSE EMISSION NEBULA
HN 2018	19 38 20.9	- 52 15 47.	18		NEBULA
SCHO 1073	19 38 28.	+ 07 47 36.	660		ISOLATED DARK CLOUD
RNGC 6808	19 38 28.	- 70 47		13.5	GALAXY
SCHO 1074	19 38 29.	+ 18 45 36.	300		ISOLATED DARK CLOUD
HN 0609	19 38 30.	- 58 14			NEBULA
IC 4884	19 38 30.	- 58 14			NONSTELLAR OBJECT
TER 08	19 38 30.26	- 34 07 05.1	200		STAR CLUSTER
MIN.46 13	19 38 32.	+ 27 12			DIFFUSE NEBULA
SCHO 1075	19 38 36.	+ 10 07 48.	420		ISOLATED DARK CLOUD
ZWG 257.002	19 38 36.	+ 50 50		15.1	GALAXY
MCG+09-32-009	19 38 36.	+ 54 33	60	17.	GALAXY
7ZW 902	19 38 36.	+ 57 53			COMPACT GALAXY
PK055-01.2	19 38 41.0	+ 18 37 51.			PLANETARY NEBULA
7ZW 903	19 38 48.	+ 60 52			COMPACT GALAXY
SC 1935-5102.2	19 38 49.	- 50 55 21.			NEBULA
HN 2024	19 38 50.3	- 29 10 19.	24		NEBULA
PK051-03.1	19 38 51.6	+ 14 49 50.	5		PLANETARY NEBULA
SCHO 1076	19 38 52.	+ 10 53 06.	1130		ISOLATED DARK CLOUD
RNGC 6815	19 38 52.	+ 26 43			OPEN CLUSTER
LDN 0700	19 38 54.	+ 11 08	240		DARK NEBULA
OCL 0131	19 38 54.	+ 26 43			OPEN STAR CLUSTER
ZWG 303.006	19 38 54.	+ 59 15		15.6	GALAXY
MCG-05-46-004	19 38 54.	- 29 10	24	16.	GALAXY
HN 2019	19 38 55.7	- 52 49 45.	18		NEBULA
SCHO 1077	19 38 58.	+ 18 51 00.	290		ISOLATED DARK CLOUD
KHAV 646	19 39	+ 27 19	5830		DARK NEBULA
KHAV 649	19 39	+ 29 31	1000		DARK NEBULA
KHAV 648	19 39	+ 32 01	1410		DARK NEBULA
SCHO 1078	19 39 00.	+ 11 08 48.	300		ISOLATED DARK CLOUD
LDN 0747	19 39 00.	+ 18 45	480		DARK NEBULA
LDN 0817	19 39 00.	+ 30 00	9060		DARK NEBULA
ZC 1939.0+7946	19 39 00.	+ 79 46	1340		CLUSTER OF GALAXIES
B 143	19 39 03.	+ 10 53	1800		DARK OBJECT
HN 2020	19 39 04.3	- 54 01 14.	12		NEBULA
ISS 0296	19 39 06.	+ 31 15	212		STELLAR RING
ZWG 397.004	19 39 12.	+ 03 32		15.5	GALAXY
ISS 0414	19 39 18.	+ 17 46	133		STELLAR RING
PK054-02.1	19 39 19.6	+ 17 38 09.	10		PLANETARY NEBULA
HN 2021	19 39 20.7	- 54 41 37.	24		NEBULA
RNGC 6810	19 39 23.	- 58 47		12.5	GALAXY
ISS 0415	19 39 24.	+ 17 34	236		STELLAR RING
LB 03121	19 39 24.	- 61 49		12.4	FAINT BLUE STAR
SCHO 1079	19 39 27.	+ 07 34 42.	520		ISOLATED DARK CLOUD
HN 1165	19 39 27.	- 51 55	18		NEBULA
IC 4886	19 39 27.	- 51 55			NONSTELLAR OBJECT
ISS 0416	19 39 30.	+ 15 40	92		STELLAR RING
ZWG 230.026	19 39 30.	+ 43 18		15.2	GALAXY
ZWG 257.003	19 39 30.	+ 50 47		15.3	GALAXY
HN 2022	19 39 30.8	- 52 31 54.			NEBULA
HN 2023	19 39 34.5	- 52 13 48.	12		NEBULA
ISS 0417	19 39 35.	+ 15 58	164		STELLAR RING
OCL 0155	19 39 36.	+ 40 04	840	10.2	OPEN STAR CLUSTER
MCG+07-40-013	19 39 36.	+ 43 16	36	15.	GALAXY
MCG+08-36-001	19 39 36.	+ 50 47	60	16.	GALAXY
7ZW 904	19 39 36.	+ 61 02			COMPACT GALAXY
MCG-03-50-003	19 39 36.	- 15 59	120	14.	GALAXY
RNGC 6819	19 39 39.	+ 40 04		10.0	OPEN CLUSTER
7ZW 905	19 39 42.	+ 61 44			COMPACT GALAXY
ZWG 357.008	19 39 42.	+ 79 43		15.7	GALAXY
PK051-04.1	19 39 44.6	+ 13 43 32.			PLANETARY NEBULA
HN 2027	19 39 46.6	- 28 37 04.	24		NEBULA
ISS 0297	19 39 48.	+ 30 27	104		STELLAR RING
MCG-02-50-001	19 39 48.	- 10 25	240	12.	GALAXY
MCG-05-46-005	19 39 48.	- 28 36	36	16.	GALAXY
HN 0610	19 39 53.	- 60 46			NEBULA
IC 4885	19 39 53.	- 60 46			NONSTELLAR OBJECT
PK053-03.1	19 39 54.	+ 16 58	40	17.1	PLANETARY NEBULA
ZWG 257.004	19 39 54.	+ 49 48		15.4	GALAXY
VVI 86	19 39 54.	- 10 25	240	11.50	SEYFERT GALAXY
REIN 2.280	19 39 54.13	- 10 26 35.6			NEBULA
REIN 2.281	19 39 55.77	- 10 26 31.6			NEBULA

OBJECT NAME	RIGHT ASCEN.	DECLINATIOF	DIAM.	MAGN.	TYPE OF OBJECT
IC 1306	19 39 56.	+ 37 32			MAY NOT EXIST
RNGC 6814	19 39 56.	- 10 25		12.5	GALAXY
REIN 2.282	19 39 56.99	- 10 27 49.8			NEBULA
SCHO 1080	19 39 57.	+ 11 18 00.	320		ISOLATED DARK CLOUD
KHAV 647	19 40	+ 07 19			DARK NEBULA
KHAV 654	19 40	+ 32 01	1220		DARK NEBULA
LDN 0757	19 40 00.	+ 19 27	720		DARK NEBULA
ZWG 281.007	19 40 00.	+ 51 43		15.2	GALAXY
UGC 11464	19 40 00.	+ 51 43	90	15.2	GALAXY S
PK052-04.1	19 40 01.3	+ 15 01 57.	9		PLANETARY NEBULA
PK060+00.1	19 40 01.4	+ 24 23 06.			PLANETARY NEBULA
ISS 0418	19 40 06.	+ 17 51	86		STELLAR RING
MCG+09-32-010	19 40 06.	+ 51 40	84	15.	GALAXY
SCHO 1081	19 40 12.	+ 11 01 48.	340		ISOLATED DARK CLOUD
ISS 0356	19 40 18.	+ 22 18	119		STELLAR RING
LDN 0803	19 40 18.	+ 27 10	540		DARK NEBULA
ISS 0298	19 40 18.	+ 31 18	88		STELLAR RING
ZWG 257.005	19 40 18.	+ 50 11		15.5	GALAXY
CED 170	19 40 20.	+ 22 58	30		DIFFUSE GALACTIC NEBULA
HN 2025	19 40 21.6	- 54 02 39.			NEBULA
OCL 0118	19 40 24.	+ 21 04	300		OPEN STAR CLUSTER
SCHO 1082	19 40 24.	+ 39 50 06.	280		ISOLATED DARK CLOUD
ZWG 257.006	19 40 24.	+ 50 32		14.4	GALAXY
UGC 11465	19 40 24.	+ 50 32	90	14.4	GALAXY
SG 3.149	19 40 26.	+ 22 58	30		DIFFUSE EMISSION NEBULA
ISS 0469	19 40 30.	+ 14 55	318		STELLAR RING
OCL 0122	19 40 30.	+ 22 58	42	14.9	OPEN STAR CLUSTER
LDN 0799	19 40 30.	+ 26 50	600		DARK NEBULA
ISS 0299	19 40 30.	+ 26 59	238		STELLAR RING
ZWG 257.007	19 40 30.	+ 50 30		14.9	GALAXY
MCG+08-36-002	19 40 30.	+ 50 31	30	15.	GALAXY
RNGC 6820	19 40 32.	+ 22 58		15.0	CLUSTER WITH NEBULOSITY
MCG+08-36-003	19 40 36.	+ 50 28	42	16.	GALAXY
MCG-01-50-001	19 40 36.	- 07 04	66	14.	GALAXY
B 338	19 40 37.	+ 07 20	480		DARK OBJECT
SCHO 1083	19 40 38.	+ 07 25 18.	480		ISOLATED DARK CLOUD
CED 171	19 40 44.	+ 27 23			DIFFUSE GALACTIC NEBULA
IC 1307	19 40 45.	+ 27 23			MAY NOT EXIST
HN 2029	19 40 45.0	- 51 46 43.	18		NEBULA
HN 2030	19 40 46.2	- 51 43 25.	24		NEBULA
HN 2026	19 40 47.7	- 58 09 26.			NEBULA
LDN 0804	19 40 48.	+ 27 08	540		DARK NEBULA
HN 2028	19 40 49.4	- 57 11 49.			NEBULA
RNGC 6816	19 40 51.	- 28 31		16.0	GALAXY
HN 1166	19 40 53.	- 54 35	12		NEBULA
IC 4888	19 40 53.	- 54 35			NONSTELLAR OBJECT
ZWG 371.004	19 40 54.	- 01 17		15.1	GALAXY
MCG+00-50-001	19 40 54.	- 01 18	42	13.5	GALAXY
HN 2035	19 40 55.4	- 28 31 11.	12		NEBULA
B 339	19 41	+ 08 11	3600		DARK OBJECT
KHAV 653	19 41	+ 08 25			DARK NEBULA
KHAV 652	19 41	+ 18 55	3680		DARK NEBULA
LBN 0135	19 41	+ 23 10	2100		BRIGHT NEBULA
KHAV 651	19 41	+ 27 01	2990		DARK NEBULA
KHAV 650	19 41	+ 28 55	1570		DARK NEBULA
KHAV 656	19 41	+ 32 19	1860		DARK NEBULA
LBN 0204	19 41	+ 41 40	780		BRIGHT NEBULA
LB 09939	19 41	- 86 50		13.5	FAINT BLUE STAR
LDN 0668	19 41 00.	+ 07 15	780		DARK NEBULA
BRSL 059-00/1	19 41 00.	+ 23 10	2400		HII REGION
OCL 0124	19 41 00.	+ 23 11	2400	9.8	OPEN STAR CLUSTER
ASS 14	19 41 00.	+ 24 27			OB ASSOCIATION VUL OB4
COU 032	19 41 00.	+ 25 12	1800		EMISSION NEBULA
LDN 0813	19 41 00.	+ 29 00	1920		DARK NEBULA
MCG-05-46-006	19 41 00.	- 28 30	30	15.	GALAXY
RNGC 6823	19 41 02.	+ 23 11		10.0	CLUSTER WITH NEBULOSITY
SCHO 1084	19 41 02.	+ 23 54 18.	1340		ISOLATED DARK CLOUD
HN 2031	19 41 02.9	- 56 31 30.	30		NEBULA
HN 2032	19 41 03.0	- 51 52 36.	18		NEBULA
RNGC 6818	19 41 06.	- 14 16		10.0	PLANETARY NEBULA
HN 2033	19 41 06.7	- 51 54 06.	12		NEBULA
SG 3.150	19 41 08.	+ 23 10	2400		DIFFUSE EMISSION NEBULA
PK025-17.1	19 41 08.68	- 14 16 26.2	25	10.0	PLANETARY NEBULA
HN 2034	19 41 09.6	- 51 53 48.	18		NEBULA
HN 1167	19 41 11.	- 54 29		10.	NEBULA
RNGC 6825	19 41 15.	+ 63 57		15.5	GALAXY
HN 0611	19 41 16.	- 54 19			NEBULA
IC 4891	19 41 17.	- 54 19			NONSTELLAR OBJECT
7ZW 906	19 41 18.	+ 63 57			COMPACT GALAXY
ZWG 324.001	19 41 18.	+ 63 57		15.3	GALAXY
KW 57	19 41 18.	- 10 21	136		SEYFERT GALAXY
IC 4889	19 41 18.	- 54 29	96	12.0	GALAXY E5
RNGC 6812	19 41 23.	- 55 28			UNVERIFIED SOUTHERN OBJECT
ZWG 257.008	19 41 24.	+ 45 11		13.5	GALAXY
UGC 11466	19 41 24.	+ 45 11	132	13.5	GALAXY PECULR
ZWG 357.009	19 41 24.	+ 78 15		15.5	GALAXY
MCG-07-40-005	19 41 24.	- 39 26	36	15.5	GALAXY
HN 0612	19 41 24.	- 56 39			NEBULA
IC 4890	19 41 24.	- 56 39			NONSTELLAR OBJECT
SCHO 1085	19 41 27.	+ 08 40 12.	400		ISOLATED DARK CLOUD
SCHO 1086	19 41 34.	+ 16 54 24.	280		ISOLATED DARK CLOUD
SCHO 1087	19 41 36.	+ 08 27 18.	420		ISOLATED DARK CLOUD
LDN 0807	19 41 36.	+ 27 15	600		DARK NEBULA
ZWG 230.027	19 41 36.	+ 41 55		15.5	GALAXY
UGC 11467	19 41 36.	+ 41 55	78	15.5	GALAXY IRR
MCG+07-40-014	19 41 36.	+ 41 55	72	15.	GALAXY
MCG+13-14-005	19 41 36.	+ 76 45	6	17.	GALAXY
RNGC 6821	19 41 40.	- 06 57		14.0	GALAXY
ISS 0419	19 41 42.	+ 19 39	87		STELLAR RING
MCG+07-40-015	19 41 42.	+ 41 31	42	15.5	GALAXY
ZWG 230.028	19 41 42.	+ 41 32		15.3	GALAXY
MCG+07-40-016	19 41 42.	+ 41 48	48	15.	GALAXY
ZWG 230.029	19 41 42.	+ 41 49		15.3	GALAXY
UGC 11468	19 41 42.	+ 41 49	60	15.3	GALAXY S
7ZW 907	19 41 42.	+ 71 52			COMPACT GALAXY
MCG-05-46-007	19 41 42.	- 28 28	24	15.5	GALAXY
HN 2037	19 41 43.2	- 50 27 33.	12		NEBULA
MCG-01-50-002	19 41 45.	- 06 57	60	14.	GALAXY
HN 2036	19 41 47.8	- 54 36 33.	12		NEBULA
MCG-02-50-002	19 41 48.	- 14 49	36	15.	GALAXY
MCG-02-50-003	19 41 51.	- 14 49 30.	60	15.	GALAXY
ASS 13	19 41 54.	+ 24 06			OB ASSOCIATION VUL OB1
7ZW 908	19 41 54.	+ 71 49			COMPACT GALAXY
MCG+13-14-006	19 41 54.	+ 76 41	36	17.	GALAXY
SCHO 1088	19 41 56.	+ 18 32 36.	340		ISOLATED DARK CLOUD
PK059-00.1	19 41 57.4	+ 23 19 35.	5		PLANETARY NEBULA
KHAV 655	19 42	+ 17 37	2230		DARK NEBULA
KHAV 658	19 42	+ 32 19	1720		DARK NEBULA

OBJECT NAME	RIGHT ASCEN.	DECLINATION	DIAM.	MAGN.	TYPE OF OBJECT
LBN 0165	19 42	+ 36 30	6720		BRIGHT NEBULA
LBN 0176	19 42	+ 37 30	1080		BRIGHT NEBULA
VDB.66G 209	19 42	- 14 57	840		DWARF GALAXY
LDN 0678	19 42 00.	+ 08 20	900		DARK NEBULA
LDN 0681	19 42 00.	+ 08 40	1020		DARK NEBULA
SCHO 1089	19 42 00.	+ 09 02 48.	240		ISOLATED DARK CLOUD
LDN 0695	19 42 00.	+ 10 20	2400		DARK NEBULA
LDN 0733	19 42 00.	+ 17 00	2520		DARK NEBULA
LDN 0765	19 42 00.	+ 20 10	780		DARK NEBULA
LDN 0796	19 42 00.	+ 26 15	480		DARK NEBULA
LDN 0800	19 42 00.	+ 26 40	720		DARK NEBULA
RNGC 6822	19 42 05.	- 14 53		10.0	GALAXY
MCG-02-50-004	19 42 06.	- 13 03 30.	48	16.	GALAXY
HN 2039	19 42 06.8	- 50 25 02.	18		NEBULA
SCHO 1090	19 42 08.	+ 18 40 00.	330		ISOLATED DARK CLOUD
MCG-02-50-005	19 42 09.	- 13 09 30.	48	16.	GALAXY
ISS 0300	19 42 12.	+ 27 27	82		STELLAR RING
ZWG 257.009	19 42 12.	+ 50 38		15.7	GALAXY
MCG-02-50-006	19 42 12.	- 14 55	600	8.	GALAXY
SG 3.151	19 42 13.	+ 23 33	600		DIFFUSE EMISSION NEBULA
IC 1308	19 42 16.1	- 14 51 32.			NONSTELLAR OBJECT
LDN 0700	19 42 19.	+ 08 31	300		DARK NEBULA
LDN 0806	19 42 18.	+ 27 00	300		DARK NEBULA
ZWG 257.010	19 42 18.	+ 50 26		14.8	GALAXY
MCG+09-32-011	19 42 18.	+ 53 02	60	16.	GALAXY
ZWG 303.007	19 42 18.	+ 57 30		15.7	GALAXY
HN 2038	19 42 18.5	- 55 50 55.			NEBULA
PK043+37.1	19 42 23.67	+ 23 53 29.1	48	9.7	PLANETARY NEBULA
ISS 0301	19 42 24.	+ 27 34	209		STELLAR RING
SCHO 1091	19 42 25.	+ 18 55 48.	330		ISOLATED DARK CLOUD
SCHO 1092	19 42 28.	+ 16 46 24.	360		ISOLATED DARK CLOUD
HN 2040	19 42 28.7	- 54 13 55.	18		NEBULA
UGC 11469	19 42 30.	+ 47 53	72	16.0	GALAXY SBa
HN 2041	19 42 31.2	- 52 15 18.	12		NEBULA
RNGC 6824	19 42 34.	+ 55 59		13.0	GALAXY
ZWG 281.008	19 42 36.	+ 55 59		13.1	GALAXY
UGC 11470	19 42 36.	+ 55 59	156	13.1	GALAXY Sa-b
MCG+09-32-012	19 42 36.	+ 56 00	96	12.9	GALAXY
GCL 114	19 42 35.	- 08 09	120		GLOBULAR STAR CLUSTER
HN 2042	19 42 40.9	- 52 13 24.	42		NEBULA
ISS 0302	19 42 42.	+ 28 47	398		STELLAR RING
MCG+07-40-017	19 42 42.	+ 44 02	36	16.	GALAXY
MCG+09-32-013	19 42 42.	+ 56 17	30	16.	GALAXY
HN 2043	19 42 45.9	- 53 04 17.	24		NEBULA
OCL 0112	19 42 48.	+ 17 24			OPEN STAR CLUSTER
7ZW 909	19 42 48.	+ 65 04			COMPACT GALAXY
SCHO 1093	19 42 49.	+ 23 17 06.	770		ISOLATED DARK CLOUD
ISS 0303	19 42 54.	+ 28 01	128		STELLAR RING
HN 2044	19 42 54.9	- 52 36 47.	12		NEBULA
KHAV 657	19 43	+ 27 43	1220		DARK NEBULA
KHAV 659	19 43	+ 30 49			DARK NEBULA
LBN 0101	19 43	- 06 00	8100		BRIGHT NEBULA
LBN 0100	19 43	- 07 30	1560		BRIGHT NEBULA
LBN 0089	19 43	- 12 20	120		BRIGHT NEBULA
LBN 0087	19 43	- 12 50	540		BRIGHT NEBULA
LBN 0083	19 43	- 15 00	540		BRIGHT NEBULA
LDN 0808	19 43 00.	+ 27 10	480		DARK NEBULA
LDN 0810	19 43 00.	+ 27 50	840		DARK NEBULA
LDN 0812	19 43 00.	+ 28 25	1500		DARK NEBULA
LDN 0820	19 43 00.	+ 30 50	2460		DARK NEBULA
LDN 0595	19 43 00.	- 07 30	900		DARK NEBULA
SCHO 1094	19 43 04.	+ 32 41 00.	300		ISOLATED DARK CLOUD
SC 1939-5205.4	19 43 05.	- 51 58 16.			NEBULA
ZWG 397.005	19 43 06.	+ 05 28		15.3	GALAXY
ZWG 341.032	19 43 06.	+ 71 17		15.6	GALAXY
KARA.73B 0874	19 43 06.	+ 71 17	30	15.6	ISOLATED GALAXY S
MCG-07-40-006	19 43 06.	- 40 35	24	15.	GALAXY
IC 4887	19 43 06.	- 69 42			NONSTELLAR OBJECT
HN 0613	19 43 06.	- 69 43			NEBULA
SCHO 1095	19 43 07.	+ 30 34 30.	300		ISOLATED DARK CLOUD
HN 1168	19 43 09.	- 51 58		15.	NEBULA
IC 4894	19 43 09.	- 51 58			NONSTELLAR OBJECT
SC 1939-5205.6	19 43 09.	- 51 58 28.	36		NEBULA
PK044-09.1	19 43 09.3	+ 05 27 19.	40	15.3	PLANETARY NEBULA
PK057-01.1	19 43 11.7	+ 21 12 46.	25		PLANETARY NEBULA
UGC 11471	19 43 12.	+ 47 53	78	16.0	GALAXY Sa-b
7ZW 910	19 43 12.	+ 67 12			COMPACT GALAXY
ISS 0420	19 43 18.	+ 17 12	122		STELLAR RING
OCL 0126	19 43 18.	+ 23 48			OPEN STAR CLUSTER
MCG+08-36-005	19 43 18.	+ 47 51	78	16.	GALAXY
MCG+08-36-004	19 43 18.	+ 49 22	30	16.	GALAXY
MCG+09-32-014	19 43 18.	+ 52 43	30	16.	GALAXY
ZWG 281.009	19 43 18.	+ 52 45		15.5	GALAXY
7ZW 911	19 43 18.	+ 60 48			COMPACT GALAXY
UGC 11472	19 43 18.	+ 72 56	78	16.0	GALAXY SBb
MCG-07-40-007	19 43 18.	- 40 30	60	15.	GALAXY
7ZW 912	19 43 24.	+ 63 37			COMPACT GALAXY
ZCG 1943+67	19 43 24.	+ 67 12		18.2	COMPACT GALAXY
ZC 1943.4+7007	19 43 24.	+ 70 07	1410		CLUSTER OF GALAXIES
PK083+12.1	19 43 27.16	+ 50 24 10.8	128	9.8	PLANETARY NEBULA
RNGC 6826	19 43 32.	+ 50 24		9.0	PLANETARY NEBULA
HN 2046	19 43 32.1	- 51 28 08.	18		NEBULA
HN 2045	19 43 33.1	- 52 13 32.	12		NEBULA
PK017-21.1	19 43 36.	- 23 16	136	15.2	PLANETARY NEBULA
MCG-04-46-001	19 43 36.	- 23 18	132	14.	GALAXY
HN 2048	19 43 41.7	- 51 23 37.	18		NEBULA
PK064+02.1	19 43 42.2	+ 28 30 51.			PLANETARY NEBULA
HN 2049	19 43 45.6	- 51 00 49.	18		NEBULA
SCHO 1096	19 43 46.	+ 07 37 18.	490		ISOLATED DARK CLOUD
SCHO 1097	19 43 46.	+ 32 42 30.	290		ISOLATED DARK CLOUD
HN 2050	19 43 46.6	- 48 02 31.	18		NEBULA
SCHO 1098	19 43 48.	+ 16 52 12.	310		ISOLATED DARK CLOUD
MCG+09-32-015	19 43 48.	+ 56 47	42	16.	GALAXY
MCG+07-40-018	19 43 51.	+ 43 00	96	14.5	GALAXY
HN 2047	19 43 53.7	- 57 06 07.	12		NEBULA
BRSL 061+00/1	19 43 54.	+ 25 13	1500		HII REGION
ZWG 230.030	19 43 54.	+ 43 01		14.6	GALAXY
UGC 11473	19 43 54.	+ 43 01	132	14.6	GALAXY Sc
SCHO 1099	19 43 56.	+ 30 41 00.	430		ISOLATED DARK CLOUD
KHAV 660	19 44	+ 07 01			DARK NEBULA
LBN 0136	19 44	+ 24 35	720		BRIGHT NEBULA
LBN 0137	19 44	+ 25 03	120		BRIGHT NEBULA
LBN 0138	19 44	+ 25 04	120		BRIGHT NEBULA
LBN 0139	19 44	+ 25 13	720		BRIGHT NEBULA
LBN 0145	19 44	+ 28 10	1920		BRIGHT NEBULA
KHAV 661	19 44	+ 31 49	2230		DARK NEBULA
LBN 0173	19 44	+ 37 10	1140		BRIGHT NEBULA
LB 03122	19 44	- 83 28		14.0	FAINT BLUE STAR

OBJECT NAME	RIGHT ASCEN.	DECLINATION	DIAM.	MAGN.	TYPE OF OBJECT
LDN 0726	19 44 00.	+ 15 50	660		DARK NEBULA
SCHO 1100	19 44 02.	+ 30 53 12.	260		ISOLATED DARK CLOUD
HN 0614	19 44 05.	- 70 22			NEBULA
ISS 0262	19 44 06.	+ 33 05	85		STELLAR RING
IC 4892	19 44 06.	- 70 21			NONSTELLAR OBJECT
PK060-00.1	19 44 08.7	+ 24 03 43.	7		PLANETARY NEBULA
PK054-03.1	19 44 10.5	+ 16 53 43.			PLANETARY NEBULA
SCHO 1101	19 44 15.	+ 18 52 48.	300		ISOLATED DARK CLOUD
HN 2051	19 44 15.9	- 51 46 41.	12		NEBULA
MRSL 060-00/1	19 44 18.	+ 24 30	600		HII REGION
SG 3.152	19 44 18.	+ 24 39	3000		DIFFUSE EMISSION NEBULA
SCHO 1102	19 44 29.	+ 07 07 06.	900		ISOLATED DARK CLOUD
SG 2.028	19 44 29.	+ 25 13	900		DIFFUSE EMISSION NEBULA
LDN 0758	19 44 30.	+ 18 50	660		DARK NEBULA
ISS 0304	19 44 30.	+ 29 34	132		STELLAR RING
IC 4898	19 44 34.	- 33 26 48.			NONSTELLAR OBJECT
HN 2054	19 44 34.3	- 46 26 52.	12		NEBULA
MRSL 064+01/1	19 44 36.	+ 28 07	3000		HII REGION
7ZW 913	19 44 36.	+ 61 19			COMPACT GALAXY
HN 2052	19 44 41.1	- 55 14 22.			NEBULA
ISS 0421	19 44 42.	+ 19 15	148		STELLAR RING
MCG+12-18-004	19 44 42.	+ 72 55	33	17.	GALAXY
HN 0616	19 44 50.	- 59 06			NEBULA
IC 4896	19 44 50.	- 59 06			NONSTELLAR OBJECT
SCHO 1103	19 44 52.	+ 31 47 54.	1130		ISOLATED DARK CLOUD
HN 0615	19 44 53.	- 72 39			NEBULA
UGC 11474	19 44 54.	+ 48 55	60	16.5	GALAXY Sa-b
IC 4893	19 44 54.	- 72 38			NONSTELLAR OBJECT
SCHO 1104	19 44 58.	+ 32 39 12.	360		ISOLATED DARK CLOUD
HN 2055	19 44 58.7	- 51 56 02.	12		NEBULA
KHAV 663	19 45	+ 30 19	1720		DARK NEBULA
LDN 0672	19 45 00.	+ 07 00	3360		DARK NEBULA
LDN 0754	19 45 00.	+ 18 35	360		DARK NEBULA
LDN 0798	19 45 00.	+ 26 00	6420		DARK NEBULA
LDN 0815	19 45 00.	+ 29 00	6180		DARK NEBULA
ISS 0305	19 45 00.	+ 31 38	122		STELLAR RING
LDN 0822	19 45 00.	+ 32 00	6300		DARK NEBULA
MCG+11-24-001	19 45 00.	+ 67 46	57	15.	GALAXY
HN 2053	19 45 03.4	- 58 09 51.			NEBULA
SCHO 1105	19 45 06.	+ 18 16 18.	260		ISOLATED DARK CLOUD
SCHO 1106	19 45 07.	+ 18 49 54.	260		ISOLATED DARK CLOUD
LDN 0703	19 45 12.	+ 10 43	540		DARK NEBULA
HN 2058	19 45 15.8	- 52 21 25.	18		NEBULA
HN 2056	19 45 16.0	- 54 38 50.	18		NEBULA
VDB.66N 127	19 45 17.	+ 18 26	3960		REFLECTION NEBULA
SG 2.029	19 45 20.	+ 28 04	1200		DIFFUSE EMISSION NEBULA
HN 2057	19 45 20.2	- 54 19 25.	60	9.	NEBULA
HN 2061	19 45 23.0	- 50 27 25.	18		NEBULA
URA 48	19 45 24.	+ 40 00	270		STELLAR RING
7ZW 914	19 45 24.	+ 66 23			COMPACT GALAXY
PATH 1.814	19 45 27.	+ 45 00	22		NEBULA
HN 2059	19 45 29.1	- 54 28 55.	12		NEBULA
COU 033	19 45 30.	+ 28 05	1800		EMISSION NEBULA
ZWG 324.002	19 45 30.	+ 64 01		15.2	GALAXY
MCG-03-50-004	19 45 30.	- 18 14	150	14.5	GALAXY
HN 2060	19 45 30.1	- 54 18 31.			NEBULA
HN 1169	19 45 33.	- 51 59		16.	NEBULA
IC 4897	19 45 33.	- 51 59			NONSTELLAR OBJECT
HN 2062	19 45 35.9	- 53 13 42.	18		NEBULA
ASS 15	19 45 36.	+ 28 42			OB ASSOCIATION VUL OB2
ISS 0306	19 45 36.	+ 28 50	127		STELLAR RING
ZC 1945.6+6509	19 45 36.	+ 65 09	1550		CLUSTER OF GALAXIES
HN 2063	19 45 39.0	- 50 37 54.	18		NEBULA
MRSL 040-12/1	19 45 42.	+ 01 01	480		HII REGION
ISS 0307	19 45 42.	+ 27 28	266		STELLAR RING
SCHO 1107	19 45 43.	+ 23 21 00.	730		ISOLATED DARK CLOUD
HN 2066	19 45 46.3	- 48 19 29.	12		NEBULA
HN 2067	19 45 46.4	- 48 13 35.	12		NEBULA
SCHO 1108	19 45 48.	+ 30 48 00.	330		ISOLATED DARK CLOUD
MCG+10-28-009	19 45 48.	+ 59 34	72	15.	GALAXY
ZWG 303.008	19 45 48.	+ 59 35		15.4	GALAXY
UGC 11475	19 45 48.	+ 59 35	96	15.4	GALAXY SBc
ZWG 324.003	19 45 48.	+ 67 51		15.6	GALAXY
UGC 11476	19 45 43.	+ 67 51	96	15.6	GALAXY S
HN 2064	19 45 50.3	- 52 07 41.	12		NEBULA
HN 2065	19 45 53.7	- 52 32 29.	12		NEBULA
MCG+09-32-016	19 45 54.	+ 56 48	66	15.	GALAXY
HN 2068	19 45 59.8	- 51 56 34.	12		NEBULA
LBN 0131	19 46	+ 18 15	540		BRIGHT NEBULA
KHAV 662	19 46	+ 27 43	2760		DARK NEBULA
LDN 0745	19 46 00.	+ 17 40	2820		DARK NEBULA
LDN 0753	19 46 00.	+ 18 10	240		DARK NEBULA
LDN 0819	19 46 00.	+ 30 15	1380		DARK NEBULA
ZWG 281.010	19 46 00.	+ 56 47		15.4	GALAXY
UGC 11477	19 46 00.	+ 56 47	84	15.4	GALAXY Sc
MCG-02-50-007	19 46 00.	- 10 42	42	15.	GALAXY
ISS 0357	19 46 06.	+ 21 04	62		STELLAR RING
HN 2069	19 46 07.0	- 54 22 40.	12		NEBULA
PK059-01.1	19 46 16.7	+ 22 01 07.	8		PLANETARY NEBULA
RNGC 6829	19 46 17.	+ 59 47		15.0	GALAXY
ZWG 397.006	19 46 18.	+ 03 51		15.7	GALAXY
LDN 0707	19 46 18.	+ 11 20	600		DARK NEBULA
7ZW 915	19 46 18.	+ 59 47			COMPACT GALAXY
ZWG 303.009	19 46 18.	+ 59 47		15.0	GALAXY
UGC 11478	19 46 18.	+ 59 47	102	15.0	GALAXY Sb
SCHO 1109	19 46 20.	+ 07 31 54.	330		ISOLATED DARK CLOUD
MCG+10-28-010	19 46 21.	+ 59 46	102	14.	GALAXY
HN 2070	19 46 21.7	- 50 36 27.	48		NEBULA
B 340	19 46 23.	+ 11 17	900		DARK OBJECT
ZWG 397.007	19 46 30.	+ 03 22		15.5	GALAXY
ZWG 303.010	19 46 30.	+ 62 24		15.3	GALAXY
MCG-04-46-002	19 46 30.	- 26 35	60	15.	GALAXY
MCG-07-40-008	19 46 30.	- 40 02	18	15.	GALAXY
HN 1170	19 46 31.	- 51 28		15.	NEBULA
IC 4900	19 46 31.	- 51 28			NONSTELLAR OBJECT
SCHO 1110	19 46 32.	+ 11 14 30.	390		ISOLATED DARK CLOUD
UGC 11479	19 46 36.	+ 78 14	66	16.0	GALAXY Sc
MCG-02-50-008	19 46 36.	- 11 03	48	14.	GALAXY
PK057-02.1	19 46 41.3	+ 19 59 38.			PLANETARY NEBULA
OCL 0120	19 46 42.	+ 21 04	240	15.	OPEN STAR CLUSTER
RNGC 6827	19 46 43.	+ 21 05			OPEN CLUSTER
SG 3.153	19 46 44.	+ 18 16	300		DIFFUSE EMISSION NEBULA
ZWG 397.008	19 46 48.	+ 03 23		15.7	GALAXY
SCHO 1111	19 46 48.	+ 11 26 36.	300		ISOLATED DARK CLOUD
MRSL 055-03/1	19 46 48.	+ 18 16	900		HII REGION
ZWG 257.011	19 46 48.	+ 46 18		15.5	GALAXY
UGC 11480	19 46 48.	+ 46 18	72	15.5	GALAXY Sc
HN 2071	19 46 48.1	- 51 32 13.	12		NEBULA
HN 2072	19 46 50.3	- 51 32 25.	12		NEBULA
ZWG 397.009	19 46 54.	+ 04 02		15.1	GALAXY
UGC 11481	19 46 54.	+ 04 02	66	15.1	GALAXY S
ZWG 397.010	19 46 54.	+ 07 02		15.7	GALAXY
UGC 11482	19 46 54.	+ 07 02	120	15.7	GALAXY IRR
LBN 0126	19 47	+ 11 20	840		BRIGHT NEBULA
LBN 0144	19 47	+ 26 41	300		BRIGHT NEBULA
MCG+01-50-001	19 47 00.	+ 07 01	60	15.	GALAXY
LDN 0766	19 47 00.	+ 19 30	4140		DARK NEBULA
LDN 0826	19 47 00.	+ 33 25	900		DARK NEBULA
ZWG 257.012	19 47 00.	+ 49 12		15.6	GALAXY
HN 2073	19 47 01.1	- 50 36 06.	12		NEBULA
IC 4899	19 47 03.	- 70 44			NONSTELLAR OBJECT
SG 2.030	19 47 04.	+ 26 43	360		DIFFUSE EMISSION NEBULA
RNGC 6831	19 47 04.	+ 59 46		14.5	GALAXY
PK044-10.1	19 47 05.6	+ 05 11 08.			PLANETARY NEBULA
ZWG 303.011	19 47 06.	+ 59 46		14.7	GALAXY
UGC 11483	19 47 06.	+ 59 46	96	14.7	GALAXY S0
HN 2076	19 47 11.4	- 47 29 59.	36		NEBULA
MRSL 063+00/1	19 47 12.	+ 26 44	360		HII REGION
UGC 11484	19 47 12.	+ 54 05	66	16.5	GALAXY Sc-IRR
MCG+10-28-011	19 47 12.	+ 59 45 30.	54	14.	GALAXY
RNGC 6832	19 47 15.	+ 59 17			NON-EXISTENT OBJECT
HN 2074	19 47 16.5	- 49 58 47.	18		NEBULA
HN 2075	19 47 16.9	- 49 58 47.			NEBULA
SCHO 1112	19 47 21.	+ 07 41 30.	340		ISOLATED DARK CLOUD
MCG-04-46-003	19 47 21.	- 22 42 30.	60	15.	GALAXY
MCG+09-32-017	19 47 24.	+ 54 04	54	16.	GALAXY
LDN 0789	19 47 30.	+ 23 30	480		DARK NEBULA
LDN 0791	19 47 30.	+ 23 50	1200		DARK NEBULA
ZWG 257.013	19 47 30.	+ 49 49		15.7	GALAXY
ZWG 303.012	19 47 30.	+ 58 06		15.3	GALAXY
MCG-07-40-009	19 47 30.	- 40 22	36	15.	GALAXY
ZWG 257.014	19 47 36.	+ 50 11		14.5	GALAXY
UGC 11485	19 47 36.	+ 50 11	84	14.5	GALAXY SBa
MCG+10-28-012	19 47 36.	+ 58 05	42	16.	GALAXY
MCG-04-46-004	19 47 36.	- 24 38	36	16.	GALAXY
MCG-05-46-009	19 47 36.	- 28 22	24	16.5	GALAXY
MCG-05-46-008	19 47 36.	- 30 58	9	15.5	GALAXY
HN 2077	19 47 43.1	- 50 32 10.	12		NEBULA
HN 2078	19 47 53.1	- 44 58 14.		12.	NEBULA
MCG-07-40-010	19 47 54.	- 45 00	36	14.5	GALAXY
RNGC 6828	19 47 59.	+ 07 47			NON-EXISTENT OBJECT
SCHO 1113	19 47 59.	+ 33 07 30.	320		ISOLATED DARK CLOUD
KHAV 664	19 48	+ 22 08	5930		DARK NEBULA
KHAV 665	19 48	+ 29 14	3810		DARK NEBULA
KHAV 666	19 48	+ 30 20	2230		DARK NEBULA
LBN 0201	19 48	+ 40 40	3000		BRIGHT NEBULA
LBN 0221	19 48	+ 42 20	1860		BRIGHT NEBULA
MRSL 062+00/1	19 48 00.	+ 26 21	300		HII REGION
HN 0054	19 48 02.	+ 48 50			NEBULA
PK069+03.1	19 48 06.0	+ 33 38 16.	71		PLANETARY NEBULA
SCHO 1114	19 48 07.	+ 22 38 48.	510		ISOLATED DARK CLOUD
B 341	19 48 08.	+ 34 09	1800		DARK OBJECT
HN 2079	19 48 12.9	- 52 09 26.	12		NEBULA
HN 2080	19 48 16.1	- 52 50 08.	12		NEBULA
RNGC 6833	19 48 17.	+ 48 50		14.0	PLANETARY NEBULA
SCHO 1115	19 48 18.	+ 25 15 12.	790		ISOLATED DARK CLOUD
PK082+11.1	19 48 20.9	+ 48 50 01.		13.8	PLANETARY NEBULA
SCHO 1116	19 48 22.	+ 22 54 00.	400		ISOLATED DARK CLOUD
PK064+00.1	19 48 23.9	+ 28 03 48.			PLANETARY NEBULA
PK062-00.1	19 48 24.	+ 25 46 48.	11		PLANETARY NEBULA
HN 2081	19 48 27.0	- 52 07 31.	6		NEBULA
OCL 0130	19 48 30.	+ 25 02	540		OPEN STAR CLUSTER
ZWG 257.015	19 48 30.	+ 50 35		15.5	GALAXY
UGC 11486	19 48 30.	+ 50 35	60	15.5	GALAXY S
HN 2082	19 48 34.5	- 49 32 42.		14.	NEBULA
SCHO 1117	19 48 36.	+ 33 24 12.	370		ISOLATED DARK CLOUD
7ZW 916	19 48 36.	+ 65 33			COMPACT GALAXY
HN 2083	19 48 39.5	- 50 36 24.	24		NEBULA
SCHO 1118	19 48 42.	+ 22 32 36.	330		ISOLATED DARK CLOUD
ISS 0308	19 48 42.	+ 32 57	240		STELLAR RING
MCG-05-47-001	19 48 42.	- 31 01 30.	18	15.5	GALAXY
ZWG 324.004	19 48 48.	+ 63 23		15.4	GALAXY
UGC 11487	19 48 48.	+ 63 23	84	15.4	GALAXY SBb
ZWG 371.005	19 48 48.	- 02 23		15.4	GALAXY
MCG-05-47-003	19 48 48.	- 30 58	24	15.	GALAXY
MCG-05-47-002	19 48 48.	- 31 02	24	15.5	GALAXY
OCL 0125	19 48 54.	+ 22 56	1260	9.0	OPEN STAR CLUSTER
7ZW 917	19 48 54.	+ 66 05			COMPACT GALAXY
RNGC 6830	19 48 56.	+ 22 57		9.0	OPEN CLUSTER
SCHO 1119	19 48 58.	+ 22 21 24.	420		ISOLATED DARK CLOUD
HN 2085	19 48 58.1	- 53 14 53.	18		NEBULA
LBN 0186	19 49	+ 38 35	900		BRIGHT NEBULA
ZWG 397.011	19 49 00.	+ 03 26		15.6	GALAXY
UGC 11488	19 49 00.	+ 03 26	78	15.6	GALAXY Sb-c
LDN 0828	19 49 00.	+ 33 25	540		DARK NEBULA
MCG-05-47-004	19 49 00.	- 31 00	36	15.	GALAXY
HN 2086	19 49 00.2	- 49 12 04.	12		NEBULA
SCHO 1120	19 49 01.	+ 10 07 42.	430		ISOLATED DARK CLOUD
HN 0617	19 49 01.	- 70 44			NEBULA
HN 2084	19 49 01.4	- 54 24 59.	12		NEBULA
PK066+02.1	19 49 02.4	+ 30 54 48.			PLANETARY NEBULA
ISS 0309	19 49 06.	+ 28 28	182		STELLAR RING
SCHO 1121	19 49 08.	+ 22 56 18.	250		ISOLATED DARK CLOUD
PK057-03.1	19 49 08.5	+ 19 49 54.			PLANETARY NEBULA
ZWG 257.016	19 49 12.	+ 46 42		15.4	GALAXY
1ZW 209	19 49 12.	+ 53 10			COMPACT GALAXY
7ZW 918	19 49 12.	+ 71 26			COMPACT GALAXY
HN 2087	19 49 13.9	- 52 09 58.	12		NEBULA
HN 2089	19 49 22.2	- 52 15 45.	12		NEBULA
SCHO 1122	19 49 26.	+ 40 24 24.	300		ISOLATED DARK CLOUD
HN 2088	19 49 26.8	- 55 49 51.			NEBULA
ZWG 257.017	19 49 30.	+ 47 56		14.9	GALAXY
HN 2092	19 49 34.5	- 47 43 50.	18		NEBULA
MCG+01-50-002	19 49 36.	+ 04 38	60	14.	GALAXY
ZWG 397.012	19 49 36.	+ 04 40		15.5	GALAXY
UGC 11489	19 49 36.	+ 04 40	90	15.5	GALAXY Sc
HN 2090	19 49 36.6	- 53 50 14.	18		NEBULA
HN 2091	19 49 38.0	- 52 06 14.	18		NEBULA
ISS 0358	19 49 42.	+ 22 00	108		STELLAR RING
HN 2093	19 49 47.7	- 52 34 14.	6		NEBULA
MCG+11-24-002	19 49 48.	+ 63 25 30.	45	16.	GALAXY
ASS 16	19 49 54.	+ 30 57			OB ASSOCIATION CYG OB5
7ZW 919	19 49 54.	+ 59 39			COMPACT GALAXY
ZWG 303.013	19 49 54.	+ 59 40		15.7	GALAXY
SCHO 1123	19 49 56.	+ 30 14 42.	630		ISOLATED DARK CLOUD
HN 2094	19 49 56.0	- 45 48 30.	12		NEBULA

OBJECT NAME	RIGHT ASCEN.	DECLINATION	DIAM.	MAGN.	TYPE OF OBJECT
PK068+03.1	19 49 57.2	+ 32 51 33.	10		PLANETARY NEBULA
LBN 0183	19 50	+ 38 10	780		BRIGHT NEBULA
LBN 0572	19 50	+ 83 30	6120		BRIGHT NEBULA
LDN 0825	19 50 00.	+ 33 00	5760		DARK NEBULA
LDN 0830	19 50 00.	+ 33 39	480		DARK NEBULA
PK063+00.1	19 50 05.8	+ 27 10 47.			PLANETARY NEBULA
SCHO 1124	19 50 06.	+ 40 27 54.	310		ISOLATED DARK CLOUD
ZC 1950.1+6207	19 50 06.	+ 62 07	340		CLUSTER OF GALAXIES
OCL 0134	19 50 12.	+ 29 17	840	10.3	OPEN STAR CLUSTER
RNGC 6834	19 50 12.	+ 29 17		10.0	OPEN CLUSTER
HN 0618	19 50 12.	- 58 50			NEBULA
IC 4901	19 50 12.	- 58 51			NONSTELLAR OBJECT
HN 2096	19 50 21.3	- 46 02 41.	48		NEBULA
HN 0619	19 50 22.	- 56 30			NEBULA
IC 4902	19 50 22.	- 56 31			NONSTELLAR OBJECT
ISS 0310	19 50 24.	+ 28 25	100		STELLAR RING
ZWG 397.013	19 50 30.	+ 05 08		15.7	GALAXY
ZWG 257.018	19 50 36.	+ 48 00		15.0	GALAXY
SCHO 1125	19 50 41.	+ 36 31 54.	350		ISOLATED DARK CLOUD
ISS 0422	19 50 42.	+ 21 05	64		STELLAR RING
HN 2095	19 50 47.1	- 58 51 28.			NEBULA
UGC 11490	19 50 48.	+ 57 52	72	17.	GALAXY DWARF
ZWG 324.005	19 50 48.	+ 63 15		15.7	GALAXY
HN 2099	19 50 48.1	- 47 10 15.	12		NEBULA
HN 2098	19 50 48.2	- 51 02 16.	12		NEBULA
BC PKS1950-613	19 50 52.0	- 61 23 11.		18.	QUASI-STELLAR OBJECT
HN 2097	19 50 52.7	- 55 14 22.			NEBULA
ZWG 397.014	19 50 54.	+ 08 13		15.6	GALAXY
OCL 0116	19 50 54.	+ 18 12	600	9.6	OPEN STAR CLUSTER
ISS 0263	19 50 54.	+ 33 25	78		STELLAR RING
PK060-02.1	19 50 57.1	+ 23 05 52.			PLANETARY NEBULA
KHAV 667	19 51	+ 39 20	2660		DARK NEBULA
LBN 0196	19 51	+ 39 50	840		BRIGHT NEBULA
LDN 0787	19 51 00.	+ 22 00	19860		DARK NEBULA
ISS 0359	19 51 00.	+ 26 24	509		STELLAR RING
RNGC 6837	19 51 07.	+ 11 34			NON-EXISTENT OBJECT
SCHO 1126	19 51 12.	+ 39 09 36.	270		ISOLATED DARK CLOUD
ZC 1951.2+6436	19 51 12.	+ 64 36	2350		CLUSTER OF GALAXIES
SC 1948-2458.5	19 51 13.	- 24 50 48.	6		NEBULA
HN 2102	19 51 14.0	- 50 30 14.	18		NEBULA
LDN 0829	19 51 18.	+ 33 19	420		DARK NEBULA
ZWG 257.019	19 51 18.	+ 48 34		14.8	GALAXY
MCG+08-36-006	19 51 18.	+ 48 54	48	16.	GALAXY
ZWG 303.014	19 51 18.	+ 60 48		15.6	GALAXY
HN 2100	19 51 23.9	- 57 38 26.	42		NEBULA
MCG+10-28-013	19 51 24.	+ 58 50	66	16.	GALAXY
UGC 11491	19 51 24.	+ 58 51	72	16.0	GALAXY Sc
MCG+10-28-014	19 51 24.	+ 60 47	66	16.	GALAXY
HN 2103	19 51 27.8	- 51 34 01.	12		NEBULA
RNGC 6838	19 51 29.	+ 18 39		8.5	GLOBULAR CLUSTER
OCL 0117	19 51 30.	+ 18 39	600	9.2	OPEN STAR CLUSTER
GCL 115	19 51 30.	+ 18 39	612	8.3	GLOBULAR STAR CLUSTER
LDN 0835	19 51 30.	+ 34 33	480		DARK NEBULA
LDN 0836	19 51 30.	+ 34 35	420		DARK NEBULA
ZC 1951.5+6148	19 51 30.	+ 61 48	9410		CLUSTER OF GALAXIES
SCHO 1127	19 51 31.	+ 39 13 36.	250		ISOLATED DARK CLOUD
HN 2101	19 51 32.9	- 57 06 37.	18		NEBULA
SC 1948-2459.6	19 51 33.	- 24 51 53.	48		NEBULA
HN 2104	19 51 35.3	- 52 34 37.	6		NEBULA
ZWG 303.015	19 51 36.	+ 57 20		14.2	GALAXY
UGC 11492	19 51 36.	+ 57 20	126	14.2	GALAXY SBb/Sc
MCG+10-28-015	19 51 42.	+ 57 19	72	14.	GALAXY
MCG-02-50-009	19 51 45.	- 12 40	138	13.	GALAXY
SN 1962J	19 51 45.	- 12 42		13.6	SUPERNOVA
SCHO 1128	19 51 47.	+ 24 21 18.	740		ISOLATED DARK CLOUD
HN 0620	19 51 47.	- 61 21			NEBULA
IC 4905	19 51 47.	- 61 21			NONSTELLAR OBJECT
SCHO 1129	19 51 48.	+ 36 29 18.	200		ISOLATED DARK CLOUD
MCG+11-24-003	19 51 48.	+ 63 17	39	17.	GALAXY
RNGC 6835	19 51 49.	- 12 42		13.5	GALAXY
HN 2107	19 51 50.1	- 50 00 53.	24		NEBULA
MCG-02-50-010	19 51 51.	- 12 47	66	13.	GALAXY
MRSL 077+07/2	19 51 54.	+ 42 25	1200		HII REGION
RNGC 6836	19 51 54.	- 12 49		13.0	GALAXY
KHAV 668	19 52	+ 30 50	3610		DARK NEBULA
LBN 0200	19 52	+ 40 10	2400		BRIGHT NEBULA
LBN 0228	19 52	+ 42 30	1500		BRIGHT NEBULA
LDN 0831	19 52 00.	+ 33 35	420		DARK NEBULA
LDN 0834	19 52 00.	+ 34 16	480		DARK NEBULA
LDN 0623	19 52 00.	- 05 00	7200		DARK NEBULA
HN 2105	19 52 03.0	- 58 06 59.			NEBULA
PK069+02.1	19 52 05.4	+ 33 14 20.			PLANETARY NEBULA
SCHO 1130	19 52 06.	+ 35 16 48.	370		ISOLATED DARK CLOUD
HN 2106	19 52 08.4	- 57 02 17.			NEBULA
PK062-01.1	19 52 08.6	+ 24 50 02.			PLANETARY NEBULA
HN 2110	19 52 09.2	- 50 10 16.	18		NEBULA
MRSL 075+06/1	19 52 12.	+ 40 07	2466		HII REGION
MCG-04-47-001	19 52 12.	- 23 24	36	15.	GALAXY
HN 2109	19 52 12.5	- 55 25 40.	18		NEBULA
HN 2108	19 52 13.8	- 56 58 05.	18		NEBULA
RNGC 6839	19 52 17.	+ 17 46			NON-EXISTENT OBJECT
7ZW 920	19 52 18.	+ 62 55			COMPACT GALAXY
HN 1171	19 52 21.	- 52 35		16.	NEBULA
IC 4907	19 52 21.	- 52 35			NONSTELLAR OBJECT
LDN 0840	19 52 24.	+ 35 00	360		DARK NEBULA
MCG-04-47-002	19 52 24.	- 23 23 30.	72	15.	GALAXY
ZWG 397.015	19 52 30.	+ 05 45		15.2	GALAXY
UGC 11493	19 52 30.	+ 05 45	138	15.2	GALAXY Sc
MCG+15-01-024	19 52 30.	+ 87 09 30.	36	16.	GALAXY
HN 0623	19 52 31.	- 60 35			NEBULA
IC 4906	19 52 31.	- 60 35			NONSTELLAR OBJECT
HN 2111	19 52 35.7	- 53 15 15.	12		NEBULA
MCG+01-50-003	19 52 36.	+ 05 44	84	14.	GALAXY
ZWG 257.020	19 52 36.	+ 49 48		15.3	GALAXY
UGC 11494	19 52 36.	+ 49 48	114	15.3	GALAXY DBL SYS
SCHO 1131	19 52 48.	+ 36 37 30.	150		ISOLATED DARK CLOUD
MCG+08-36-007	19 52 48.	+ 49 48	60	15.	GALAXY
SC 1949-2456.4	19 52 48.	- 24 48 36.	12		NEBULA
HN 0621	19 52 52.	- 70 35			NEBULA
IC 4903	19 52 53.	- 70 35			NONSTELLAR OBJECT
MRSL 077+07/1	19 52 54.	+ 42 28	936		HII REGION
MCG-05-47-005	19 52 54.	- 27 36	24	15.5	GALAXY
RNGC 6840	19 52 55.	+ 11 58			NON-EXISTENT OBJECT
HN 0624	19 52 55.	- 55 55			NEBULA
IC 4908	19 52 56.	- 55 55			NONSTELLAR OBJECT
MIN.46 14	19 52 58.	+ 27 05			DIFFUSE NEBULA
SG 3.154	19 52 58.	+ 27 05			DIFFUSE EMISSION NEBULA
LBN 0146	19 53	+ 27 09	60		BRIGHT NEBULA

OBJECT NAME	RIGHT ASCEN.	DECLINATION	DIAM.	MAGN.	TYPE OF OBJECT
LBN 0149	19 53	+ 29 10	60		BRIGHT NEBULA
KHAV 669	19 53	+ 33 56	3880		DARK NEBULA
MRSL 064-00/1	19 53 00.	+ 27 05	60		HII REGION
RNGC 6842	19 53 00.	+ 29 09		13.5	PLANETARY NEBULA
LDN 0832	19 53 00.	+ 33 38	600		DARK NEBULA
LDN 0833	19 53 00.	+ 33 40	4920		DARK NEBULA
UGC 11496	19 53 00.	+ 67 33	138	17.	GALAXY DWRF SP
UGC 11495	19 53 00.	+ 87 09	78	16.0	GALAXY Sc
PK065+00.1	19 53 02.9	+ 29 09 17.	57	13.6	PLANETARY NEBULA
ZWG 372.001	19 53 06.	+ 02 02		14.9	GALAXY
UGC 11497	19 53 06.	+ 02 02	72	14.9	GALAXY SBb
2ZW 078	19 53 06.	+ 09 24			COMPACT GALAXY
HN 2112	19 53 08.9	- 55 32 19.	24		NEBULA
HN 1172	19 53 09.	- 50 10		15.	NEBULA
IC 4909	19 53 09.	- 50 10			NONSTELLAR OBJECT
SCHO 1132	19 53 10.	+ 36 39 12.	210		ISOLATED DARK CLOUD
HN 0622	19 53 14.	- 70 19			NEBULA
IC 4904	19 53 14.	- 70 19			NONSTELLAR OBJECT
MCG-05-47-006	19 53 18.	- 32 20	30	15.5	GALAXY
SCHO 1133	19 53 24.	+ 21 05 42.	320		ISOLATED DARK CLOUD
HN 2113	19 53 24.3	- 55 16 12.	42		NEBULA
IC 4913	19 53 29.	- 37 27 46.			NONSTELLAR OBJECT
ZWG 398.001	19 53 30.	+ 05 52		15.3	GALAXY
MCG-05-47-007	19 53 30.	- 32 15	48	16.	GALAXY
SCHO 1134	19 53 35.	+ 36 06 42.	260		ISOLATED DARK CLOUD
HN 2114	19 53 35.3	- 55 42 47.			NEBULA
MCG-05-47-008	19 53 36.	- 31 38	36	15.	GALAXY
HN 2116	19 53 40.5	- 48 26 34.	12		NEBULA
MCG-04-47-003	19 53 42.	- 23 59	24	16.	GALAXY
RNGC 6843	19 53 43.	+ 12 01			NON-EXISTENT OBJECT
SCHO 1135	19 53 45.	+ 35 36 24.	260		ISOLATED DARK CLOUD
HN 0625	19 53 47.	- 57 00			NEBULA
ZWG 398.002	19 53 48.	+ 09 16		15.1	GALAXY
IC 4910	19 53 49.	- 57 00			NONSTELLAR OBJECT
SCHO 1136	19 53 51.	+ 36 38 06.	240		ISOLATED DARK CLOUD
HN 2117	19 53 54.1	- 52 39 58.	12		NEBULA
HN 1173	19 53 56.	- 52 06		16.	NEBULA
IC 4911	19 53 56.	- 52 06			NONSTELLAR OBJECT
SCHO 1137	19 53 57.	+ 25 25 12.	770		ISOLATED DARK CLOUD
LBN 0151	19 54	+ 30 00	600		BRIGHT NEBULA
LBN 0202	19 54	+ 40 00	1620		BRIGHT NEBULA
7ZW 921	19 54 00.	+ 62 14			COMPACT GALAXY
HN 2122	19 54 01.1	- 48 37 09.	18		NEBULA
SCHO 1138	19 54 02.	+ 38 46 36.	400		ISOLATED DARK CLOUD
SCHO 1045	19 54 04.	+ 22 03 42.	590		ISOLATED DARK CLOUD
MRSL 066+00/1	19 54 06.	+ 30 08	600		HII REGION
HN 2115	19 54 06.0	- 57 28 27.	36		NEBULA
HN 2121	19 54 07.2	- 51 39 39.	12		NEBULA
MCG-05-47-009	19 54 12.	- 32 40	60	15.	GALAXY
HN 2118	19 54 13.5	- 55 48 57.	60		NEBULA
HN 2119	19 54 15.6	- 55 34 51.			NEBULA
ZWG 398.003	19 54 18.	+ 07 42		15.1	GALAXY
HN 2120	19 54 18.8	- 55 54 56.			NEBULA
HN 1174	19 54 21.	- 50 15	12		NEBULA
IC 4914	19 54 21.	- 50 15			NONSTELLAR OBJECT
SHB 350	19 54 22.8	+ 51 23 46.		19.5	QUASI-STELLAR OBJECT
OCL 0139	19 54 24.	+ 30 13	42	14.2	OPEN STAR CLUSTER
SCHO 1139	19 54 26.	+ 38 24 12.	300		ISOLATED DARK CLOUD
HN 2123	19 54 27.7	- 55 43 08.			NEBULA
SCHO 1140	19 54 29.	+ 34 11 24.	470		ISOLATED DARK CLOUD
RNGC 6846	19 54 30.	+ 20 13		14.0	OPEN CLUSTER
MCG+07-41-001	19 54 30.	+ 40 16	18	16.	GALAXY
HN 2130	19 54 30.0	- 45 27 37.	48		NEBULA
HN 2127	19 54 33.5	- 51 02 49.	12		NEBULA
ISS 0211	19 54 36.	+ 42 19	238		STELLAR RING
PK019-23.1	19 54 36.	- 21 45	295	14.9	PLANETARY NEBULA
MCG-05-47-010	19 54 36.	- 30 54	54	15.5	GALAXY
HN 2124	19 54 36.4	- 57 08 01.	18		NEBULA
PK068+01.1	19 54 37.2	+ 32 14 09.			PLANETARY NEBULA
HN 2128	19 54 37.8	- 51 51 13.	12		NEBULA
HN 1175	19 54 39.	- 52 47		15.	NEBULA
IC 4915	19 54 39.	- 52 47			NONSTELLAR OBJECT
BC PKS1954-388	19 54 39.0	- 38 53 13.		17.5	QUASI-STELLAR OBJECT
MCG+01-51-001	19 54 42.	+ 05 45	180	13.	GALAXY
MCG-05-47-011	19 54 42.	- 31 59	24	14.	GALAXY
MCG-05-47-012	19 54 42.	- 32 32	30	15.5	GALAXY
MCG-05-47-013	19 54 42.	- 32 38	60	15.	GALAXY
HN 2133	19 54 42.0	- 46 14 48.	60		NEBULA
RNGC 6841	19 54 43.	- 31 59		14.0	GALAXY
HN 2125	19 54 43.9	- 57 57 55.			NEBULA
HN 1176	19 54 45.	- 50 24	18		NEBULA
HN 2129	19 54 45.9	- 52 43 12.			NEBULA
IC 4916	19 54 46.	- 50 24			NONSTELLAR OBJECT
HN 2126	19 54 46.5	- 55 43 25.	12		NEBULA
ZWG 398.004	19 54 48.	+ 04 27		15.5	GALAXY
ZWG 398.005	19 54 48.	+ 05 45		14.6	GALAXY
UGC 11498	19 54 48.	+ 05 45	204	14.6	GALAXY SB:b
ZC 1954.8+7824	19 54 48.	+ 78 24	11020		CLUSTER OF GALAXIES
MCG-05-47-014	19 54 48.	- 32 52	36	15.	GALAXY
HN 2131	19 54 53.0	- 52 29 24.			NEBULA
MRSL 027-20/1	19 54 54.	- 14 15	3300		HII REGION
MCG-07-41-001	19 54 54.	- 43 11	12	15.5	GALAXY
HN 2134	19 54 55.5	- 48 42 47.	18		NEBULA
MCG+00-51-001	19 54 57.	- 00 22	48	16.	GALAXY
MCG+00-51-002	19 54 57.	- 01 03 30.	48	15.	GALAXY
HN 2135	19 54 58.3	- 48 13 05.	12		NEBULA
KHAV 670	19 55	+ 31 56	6150		DARK NEBULA
LBN 0156	19 55	+ 33 05	720		BRIGHT NEBULA
LBN 0086	19 55	- 14 30	3300		BRIGHT NEBULA
RNGC 6847	19 55 00.	+ 29 12			NON-EXISTENT OBJECT
COU 034	19 55 00.	+ 36 30	9000		EMISSION NEBULA
ZWG 372.002	19 55 00.	- 00 22		15.6	GALAXY
MCG-05-47-015	19 55 00.	- 32 03	48	15.	GALAXY
HN 1177	19 55 02.	- 52 25		16.	NEBULA
IC 4917	19 55 02.	- 52 25			NONSTELLAR OBJECT
SCHO 1141	19 55 04.	+ 21 18 48.	440		ISOLATED DARK CLOUD
HN 2132	19 55 05.9	- 56 03 47.	24		NEBULA
MCG+07-41-002	19 55 06.	+ 40 15	15	16.	GALAXY
ZC 1955.1+6510	19 55 06.	+ 65 10	1550		CLUSTER OF GALAXIES
ZWG 372.003	19 55 06.	- 01 03		15.5	GALAXY
HN 2137	19 55 10.3	- 51 24 11.	6		NEBULA
ISS 0311	19 55 12.	+ 29 53	231		STELLAR RING
MCG-05-47-016	19 55 12.	- 31 57	72	15.	GALAXY
MCG-07-41-002	19 55 12.	- 40 58	66	15.	GALAXY
ACK 075+05.1	19 55 18.	+ 39 41			PLANETARY NEBULA
HN 2142	19 55 19.4	- 44 17 52.	18		NEBULA
HN 2140	19 55 21.8	- 48 44 40.	18		NEBULA
SG 2.031	19 55 23.	+ 38 30	2400		DIFFUSE EMISSION NEBULA

OBJECT NAME	RIGHT ASCEN.	DECLINATION	DIAM.	MAGN.	TYPE OF OBJECT
ZWG 257.021	19 55 24.	+ 49 45		15.7	GALAXY
UGC 11499	19 55 24.	+ 49 45	72	15.7	GALAXY Sc-IRR
HN 1178	19 55 26.	- 52 25		16.	NEBULA
IC 4918	19 55 26.	- 52 25			NONSTELLAR OBJECT
HN 2136	19 55 27.1	- 57 07 58.	12		NEBULA
HN 2138	19 55 28.7	- 57 07 22.	18		NEBULA
HN 2141	19 55 29.6	- 50 36 27.	12		NEBULA
ZWG 257.022	19 55 30.	+ 47 09		15.1	GALAXY
MCG+08-36-008	19 55 30.	+ 49 44	42	16.	GALAXY
ZWG 257.023	19 55 30.	+ 50 02		15.2	GALAXY
MRSL 074+05/1	19 55 36.	+ 38 35	1536		HII REGION
ZWG 257.024	19 55 36.	+ 50 03		15.5	GALAXY
UGC 11500	19 55 36.	+ 50 03	60	15.5	GALAXY Sc
SCHO 1142	19 55 38.	+ 32 26 24.	310		ISOLATED DARK CLOUD
SCHO 1143	19 55 41.	+ 35 42 48.	380		ISOLATED DARK CLOUD
ZWG 423.001	19 55 42.	+ 12 20		15.5	GALAXY
MCG+08-36-009	19 55 42.	+ 50 02	36	16.	GALAXY
MCG-04-47-004	19 55 42.	- 20 57	24	15.5	GALAXY
MCG+08-36-010	19 55 45.	+ 47 08	48	15.	GALAXY
MCG+08-36-011	19 55 45.	+ 50 02 30.	60	15.	GALAXY
HN 2139	19 55 45.0	- 58 06 45.			NEBULA
HN 2144	19 55 47.3	- 50 57 14.	48		NEBULA
SCHO 1144	19 55 48.	+ 40 08 42.	410		ISOLATED DARK CLOUD
HN 2146	19 55 51.0	- 49 18 44.	12		NEBULA
PK043-13.1	19 55 54.	+ 02 55	74	16.0	PLANETARY NEBULA
ZWG 257.025	19 55 54.	+ 50 14		15.7	GALAXY
HN 2143	19 55 57.8	- 56 23 20.	12		NEBULA
HN 2151	19 55 58.2	- 46 06 07.	12		NEBULA
HN 2149	19 55 59.6	- 49 18 55.	42		NEBULA
KHAV 671	19 56	+ 34 56	6350		DARK NEBULA
B 144	19 56	+ 35 12	21600		DARK OBJECT
LDN 0854	19 56 00.	+ 36 15	1320		DARK NEBULA
ZWG 257.026	19 56 00.	+ 50 17		15.2	GALAXY
ZC 1956.0+5746	19 56 00.	+ 57 46	2290		CLUSTER OF GALAXIES
SHAH 172	19 56 00.	+ 89 28	72		GROUP OF COMPACT GALAXIES
IC 4922	19 56 05.	- 40 30			NONSTELLAR OBJECT
HN 0626	19 56 05.	- 55 31			NEBULA
ZWG 372.004	19 56 06.	+ 02 28		14.5	GALAXY
UGC 11501	19 56 06.	+ 02 28	90	14.5	GALAXY Sb
ISS 0212	19 56 06.	+ 39 04	175		STELLAR RING
HN 0627	19 56 06.	- 40 30			NEBULA
IC 4919	19 56 06.	- 55 31			NONSTELLAR OBJECT
HN 2147	19 56 06.1	- 53 01 55.	30		NEBULA
HN 2145	19 56 06.7	- 56 03 07.	30		NEBULA
HN 2150	19 56 06.8	- 52 05 49.	6		NEBULA
MCG+00-51-003	19 56 09.	+ 02 29	66	14.5	GALAXY
HN 2152	19 56 11.3	- 47 15 18.	18		NEBULA
ISS 0312	19 56 12.	+ 31 28	378		STELLAR RING
HN 2154	19 56 14.5	- 47 09 42.	36		NEBULA
IC 4920	19 56 16.	- 53 31			NONSTELLAR OBJECT
HN 2153	19 56 16.7	- 48 11 18.			NEBULA
HN 1179	19 56 17.	- 53 31			NEBULA
HN 2156	19 56 18.0	- 46 23 54.	18		NEBULA
HN 2155	19 56 19.4	- 48 45 30.	18		NEBULA
HN 2148	19 56 21.3	- 57 08 19.	18		NEBULA
ISS 0313	19 56 24.	+ 29 58	300		STELLAR RING
MCG+09-32-018	19 56 24.	+ 52 44	60	15.	GALAXY
HN 2159	19 56 25.4	- 46 29 41.	30		NEBULA
HN 0629	19 56 26.	- 41 41			NEBULA
IC 4924	19 56 26.	- 41 41			NONSTELLAR OBJECT
OCL 0121	19 56 30.	+ 20 21			OPEN STAR CLUSTER
ZWG 281.011	19 56 30.	+ 52 44		15.3	GALAXY
UGC 11502	19 56 30.	+ 52 44	60	15.3	GALAXY COMPACT
HN 2161	19 56 33.9	- 46 29 35.	48		NEBULA
SCHO 1145	19 56 34.	+ 36 11 30.	180		ISOLATED DARK CLOUD
CED 172	19 56 36.	+ 29 48			DIFFUSE GALACTIC NEBULA
LDN 0855	19 56 42.	+ 36 15	360		DARK NEBULA
ISS 0213	19 56 42.	+ 40 03	136		STELLAR RING
MCG-05-47-017	19 56 42.	- 31 12	84	16.	GALAXY
HN 2164	19 56 43.3	- 48 19 22.	18		NEBULA
HN 2157	19 56 45.0	- 55 34 53.	30	12.	NEBULA
SG 3.155	19 56 46.	+ 35 42	3600		DIFFUSE EMISSION NEBULA
HN 2160	19 56 47.4	- 52 50 47.	18		NEBULA
MRSL 068+01/1	19 56 48.	+ 31 17	900		HII REGION
ZWG 372.005	19 56 48.	- 00 31		15.5	GALAXY
MCG-07-41-003	19 56 48.	- 40 53	60	16.5	GALAXY
HN 2163	19 56 51.0	- 53 04 10.			NEBULA
IC 4926	19 56 53.	- 38 42 52.			NONSTELLAR OBJECT
ZWG 372.006	19 56 54.	+ 00 57		14.8	GALAXY
ISS 0214	19 56 54.	+ 39 08	164		STELLAR RING
MCG-05-47-018	19 56 54.	- 32 16	72	15.	GALAXY
HN 2158	19 56 58.2	- 58 50 04.	30		NEBULA
HN 2162	19 56 58.7	- 55 46 16.			NEBULA
LBN 0154	19 57	+ 31 10	600		BRIGHT NEBULA
LBN 0178	19 57	+ 36 00	4200		BRIGHT NEBULA
LBN 0207	19 57	+ 40 10	1500		BRIGHT NEBULA
LBN 0237	19 57	+ 42 35	2160		BRIGHT NEBULA
ISS 0215	19 57 00.	+ 39 35	134		STELLAR RING
ISS 0264	19 57 06.	+ 36 03	95		STELLAR RING
SER 135.03	19 57 06.	- 47 14	228	14.	3 INTERACTING GALAXIES
IC 4923	19 57 08.	- 52 46			NONSTELLAR OBJECT
MCG+00-51-004	19 57 09.	- 00 05	36	16.	GALAXY
HN 1180	19 57 09.	- 52 46		14.	NEBULA
MRSL 077+06/1	19 57 12.	+ 42 32	1998		HII REGION
ZWG 257.027	19 57 12.	+ 49 54		14.5	GALAXY
UGC 11503	19 57 12.	+ 49 54	42	14.5	GALAXY
HN 2169	19 57 17.9	- 47 13 50.	12		NEBULA
MCG-05-47-019	19 57 18.	- 32 05	30	15.5	GALAXY
MCG-07-41-004	19 57 18.	- 39 48	72	14.5	GALAXY
RNGC 6845	19 57 18.	- 47 13			GALAXY
PK068+01.2	19 57 20.0	+ 31 47 03.	25	10.5	PLANETARY NEBULA
HN 1181	19 57 21.	- 53 00	30		NEBULA
IC 4925	19 57 21.	- 53 00			NONSTELLAR OBJECT
HN 2170	19 57 21.6	- 47 12 44.	60		NEBULA
MCG+08-36-012	19 57 24.	+ 49 53	60	16.	GALAXY
7ZW 922	19 57 24.	+ 61 39			COMPACT GALAXY
RNGC 6853	19 57 25.	+ 22 35		7.5	PLANETARY NEBULA
PK060-03.1	19 57 26.	+ 22 35 07.	840	7.6	PLANETARY NEBULA
KLEM 30	19 57 26.	- 47 12	240	13.	CMPT GROUP OF 4 GALAXIES
HN 2172	19 57 29.3	- 47 11 43.	24		NEBULA
HN 2167	19 57 29.9	- 53 08 44.	18		NEBULA
ZWG 423.002	19 57 30.	+ 14 38		14.9	GALAXY
SCHO 1146	19 57 30.	+ 37 23 18.	350		ISOLATED DARK CLOUD
IC 4931	19 57 31.	- 38 42 42.			NONSTELLAR OBJECT
HN 2173	19 57 31.1	- 46 22 07.			NEBULA
HN 2166	19 57 32.7	- 54 25 02.	18		NEBULA
HN 2168	19 57 33.8	- 54 25 44.			NEBULA
HN 2171	19 57 35.4	- 50 40 31.	24		NEBULA
LDN 0802	19 57 36.	+ 24 50	240		DARK NEBULA
ZWG 257.028	19 57 36.	+ 50 40		15.2	GALAXY
KARA.72 543B	19 57 36.	+ 50 40	48	15.2	PART OF DOUBLE GALAXY
ZWG 257.029	19 57 36.	+ 50 41		15.4	GALAXY
KARA.72 543A	19 57 36.	+ 50 41	24	15.4	PART OF DOUBLE GALAXY
HN 2174	19 57 40.3	- 46 17 07.		13.	NEBULA
HN 2165	19 57 40.9	- 58 49 20.	18		NEBULA
BMS 1.27	19 57 42.	+ 40 35			GALAXY
MCG+07-41-003	19 57 42.	+ 40 35	6	18.	GALAXY
7ZW 923	19 57 42.	+ 61 05			COMPACT GALAXY
SCHO 1147	19 57 44.	+ 32 37 24.	400		ISOLATED DARK CLOUD
VV 072	19 57 45.	+ 40 36	30	17.9	INTERACTING GALAXY
SCHO 1148	19 57 51.	+ 37 49 12.	210		ISOLATED DARK CLOUD
ISS 0265	19 57 54.	+ 33 23	381		STELLAR RING
OCL 0144	19 57 54.	+ 34 28	240	16.	OPEN STAR CLUSTER
HN 1182	19 57 54.	- 54 04	30		NEBULA
IC 4927	19 57 55.	- 54 04			NONSTELLAR OBJECT
HN 2175	19 57 59.8	- 48 22 42.	12		NEBULA
KHAV 672	19 58	+ 24 44			DARK NEBULA
LBN 0158	19 58	+ 33 04	660		BRIGHT NEBULA
LBN 0168	19 58	+ 35 10	1020		BRIGHT NEBULA
ZWG 398.006	19 58 00.	+ 06 33		15.5	GALAXY
PK060-04.1	19 58 00.	+ 21 35	46	16.6	PLANETARY NEBULA
LDN 0839	19 58 00.	+ 34 00	4800		DARK NEBULA
COU 035	19 58 00.	+ 35 08	1200		EMISSION NEBULA
LDN 0869	19 58 00.	+ 39 20	720		DARK NEBULA
HN 2178	19 58 01.9	- 46 43 29.	12		NEBULA
CED 174	19 58 02.	+ 36 54			DIFFUSE GALACTIC NEBULA
HN 2180	19 58 03.7	- 46 24 29.			NEBULA
CED 173	19 58 05.	+ 35 08	930		DIFFUSE GALACTIC NEBULA
MRSL 071+02/1	19 58 06.	+ 35 09	1200		HII REGION
LDN 0867	19 58 06.	+ 38 35	420		DARK NEBULA
ZWG 257.030	19 58 06.	+ 50 20		15.5	GALAXY
PK042-14.1	19 58 07.1	+ 01 35 23.	28		PLANETARY NEBULA
RNGC 6852	19 58 08.	+ 01 35			PLANETARY NEBULA
SCHO 1149	19 58 08.	+ 37 36 12.	260		ISOLATED DARK CLOUD
RNGC 6856	19 58 08.	+ 56 00			NON-EXISTENT OBJECT
RNGC 6844	19 58 10.	- 65 23			UNVERIFIED SOUTHERN OBJECT
MIN.46 15	19 58 11.	+ 35 08			DIFFUSE NEBULA
SG 2.032	19 58 11.	+ 35 11	1080		DIFFUSE EMISSION NEBULA
MCG-03-51-001	19 58 12.	- 17 12 30.	36	15.5	GALAXY
HN 2177	19 58 17.3	- 53 11 05.	18		NEBULA
ZWG 257.031	19 58 18.	+ 49 50		15.0	GALAXY
MCG+08-36-013	19 58 19.	+ 50 20	36	16.	GALAXY
HN 2184	19 58 19.5	- 47 24 16.	18		NEBULA
SCHO 1150	19 58 20.	+ 27 52 30.	180		ISOLATED DARK CLOUD
HN 2185	19 58 21.9	- 46 17 28.			NEBULA
HN 2179	19 58 22.1	- 53 56 05.	12		NEBULA
SCHO 1151	19 58 23.	+ 22 27 00.	350		ISOLATED DARK CLOUD
HN 0630	19 58 23.	- 67 59			NEBULA
IC 4921	19 58 23.	- 67 59			NONSTELLAR OBJECT
ZWG 257.032	19 58 24.	+ 50 21		15.6	GALAXY
SCHO 1152	19 58 25.	+ 32 57 18.	180		ISOLATED DARK CLOUD
HN 2176	19 58 25.5	- 55 59 47.	18		NEBULA
HN 1183	19 58 27.	- 52 59		15.	NEBULA
IC 4932	19 58 27.	- 52 59			NONSTELLAR OBJECT
ISS 0360	19 58 30.	+ 23 15	118		STELLAR RING
MCG+08-36-015	19 58 30.	+ 49 50	36	16.	GALAXY
SCHO 1153	19 58 31.	+ 37 38 54.	240		ISOLATED DARK CLOUD
HN 2181	19 58 31.8	- 55 27 52.	18		NEBULA
HN 0631	19 58 32.	- 54 27			NEBULA
IC 4930	19 58 32.	- 54 27			NONSTELLAR OBJECT
LDN 0861	19 58 36.	+ 36 38	240		DARK NEBULA
ZWG 342.001	19 58 36.	+ 69 27		15.7	GALAXY
HN 2187	19 58 36.8	- 48 16 33.	12		NEBULA
HN 2182	19 58 37.2	- 55 57 52.			NEBULA
HN 2188	19 58 38.1	- 48 12 51.	18		NEBULA
HN 2183	19 58 38.4	- 56 04 04.			NEBULA
HN 2189	19 58 41.5	- 45 35 21.	18		NEBULA
RNGC 6848	19 58 42.	- 56 13			UNVERIFIED SOUTHERN OBJECT
HN 2186	19 58 42.3	- 51 53 21.	18		NEBULA
SCHO 1154	19 58 48.	+ 24 32 42.	410		ISOLATED DARK CLOUD
MRSL 051-09/1	19 58 54.	+ 11 39	600		HII REGION
HN 2191	19 58 54.9	- 47 18 14.	18		NEBULA
MIN.46 16	19 58 55.	+ 33 21			DIFFUSE NEBULA
SCHO 1155	19 58 57.	+ 32 58 00.	240		ISOLATED DARK CLOUD
KAZ 2	19 59	+ 33 08	115	12.86	PLANETARY NEBULA
LBN 0160	19 59	+ 33 20	180		BRIGHT NEBULA
LBN 0171	19 59	+ 35 20	900		BRIGHT NEBULA
KHAV 674	19 59	+ 39 08	5800		DARK NEBULA
LBN 0199	19 59	+ 39 20	1320		BRIGHT NEBULA
KHAV 673	19 59	- 02 16	15210		DARK NEBULA
ACK 069+01.1	19 59 00.	+ 33 21			PLANETARY NEBULA
COU 036	19 59 00.	+ 42 40	10800		EMISSION NEBULA
MCG-05-47-020	19 59 00.	- 27 43	36	14.5	GALAXY
SG 3.156	19 59 02.	+ 33 19	60		DIFFUSE EMISSION NEBULA
HN 2192	19 59 11.9	- 53 07 37.	18		NEBULA
SCHO 1156	19 59 15.	+ 23 37 30.	480		ISOLATED DARK CLOUD
HN 2193	19 59 16.2	- 52 21 37.			NEBULA
OCL 0135	19 59 18.	+ 28 29	240	17.	OPEN STAR CLUSTER
SG 3.157	19 59 20.	+ 33 01	4800		DIFFUSE EMISSION NEBULA
HN 2197	19 59 21.0	- 47 39 24.			NEBULA
HN 2190	19 59 21.3	- 58 19 25.	18		NEBULA
HN 2194	19 59 24.6	- 54 44 49.	18		NEBULA
HN 2196	19 59 25.7	- 52 00 48.	24		NEBULA
IC 4933	19 59 27.	- 55 06			NONSTELLAR OBJECT
HN 0632	19 59 27.	- 55 07			NEBULA
HN 2195	19 59 27.7	- 52 59 18.	12		NEBULA
LDN 0805	19 59 30.	+ 24 37	240		DARK NEBULA
ISS 0216	19 59 30.	+ 39 44	145		STELLAR RING
MCG-03-51-002	19 59 30.	- 20 24	48	15.	GALAXY
SG 2.033	19 59 35.	+ 35 21	3600		DIFFUSE EMISSION NEBULA
HN 2198	19 59 36.3	- 52 09 12.	12		NEBULA
RNGC 6850	19 59 38.	- 54 59			GALAXY
UGC 11504	19 59 42.	+ 07 39	66	17.	GALAXY DWARF
ZWG 257.033	19 59 42.	+ 50 53		15.4	GALAXY
MCG-04-47-005	19 59 42.	- 23 25 30.	30	14.5	GALAXY
MIN.46 17	19 59 43.	+ 33 28			DIFFUSE NEBULA
SCHO 1157	19 59 44.	+ 24 51 00.	270		ISOLATED DARK CLOUD
MIN.46 18	19 59 50.	+ 33 23			DIFFUSE NEBULA
SG 3.158	19 59 50.	+ 33 23	60		DIFFUSE EMISSION NEBULA
PK070+01.1	19 59 50.2	+ 33 24 19.			PLANETARY NEBULA
HN 0628	19 59 52.	- 77 30			NEBULA
IC 4912	19 59 52.	- 77 30			NONSTELLAR OBJECT
REIN 2.284	19 59 52.10	+ 33 23 08.0			NEBULA
REIN 2.283	19 59 52.10	+ 33 23 14.2			NEBULA
PK070+01.2	19 59 52.18	+ 33 23 11.3	59	11.4	PLANETARY NEBULA
HN 2200	19 59 52.5	- 50 13 34.			NEBULA

OBJECT NAME	RIGHT ASCEN.	DECLINATION	DIAM.	MAGN.	TYPE OF OBJECT
RNGC 6851	19 59 53.	- 48 25		13.0	GALAXY
ZWG 372.007	19 59 54.	+ 01 30		15.2	GALAXY
UGC 11505	19 59 54.	+ 01 30	108	15.2	GALAXY S
MRSL 076+05/2	19 59 54.	+ 40 10	1440		HII REGION
SER 138.08	19 59 54.	- 55 09	480		CMPCT,SO,AND 3 SAB GLXIES
SER 138.08	19 59 54.	- 55 09	480		TRIO OF SAB GALAXIES
RNGC 6857	19 59 56.	+ 33 23			PLANETARY NEBULA
HN 2203	19 59 56.0	- 48 50 10.	36		NEBULA
MCG+00-51-005	19 59 57.	+ 01 30	78	14.5	GALAXY
HN 2199	19 59 57.2	- 56 08 23.	42		NEBULA
LBN 0159	20 00	+ 33 00	3300		BRIGHT NEBULA
LBN 0161	20 00	+ 33 20	180		BRIGHT NEBULA
KHAV 675	20 00	+ 37 38	1410		DARK NEBULA
LBN 0232	20 00	+ 41 50	3900		BRIGHT NEBULA
LBN 0244	20 00	+ 43 00	15300		BRIGHT NEBULA
LBN 0297	20 00	+ 46 00	2100		BRIGHT NEBULA
LBN 0108	20 00	- 02 00	6000		BRIGHT NEBULA
LBN 0054	20 00	- 28 00	9000		BRIGHT NEBULA
LBN 0041	20 00	- 31 30	9000		BRIGHT NEBULA
COU 037	20 00 00.	+ 33 10	6000		EMISSION NEBULA
LDN 0847	20 00 00.	+ 35 00	7020		DARK NEBULA
LDN 0857	20 00 00.	+ 36 00	3600		DARK NEBULA
ISS 0217	20 00 00.	+ 39 07	146		STELLAR RING
LDN 0872	20 00 00.	+ 40 20	1140		DARK NEBULA
ZC 2000.0+6449	20 00 00.	+ 64 49	3900		CLUSTER OF GALAXIES
KARA.73 48	20 00 02.	- 42 54	27		DWARF GALAXY
HN 2202	20 00 03.5	- 51 53 40.	18		NEBULA
ARC 2323	20 00 08.	+ 80 03		17.3	RICH CLUSTER OF GALAXIES
MCG+00-51-006	20 00 09.	+ 01 43 30.	15	15.	GALAXY
BC PKS2000-329	20 00 09.3	- 32 59 40.		18.	QUASI-STELLAR OBJECT
HN 2201	20 00 10.7	- 55 22 46.	18		NEBULA
IC 1309	20 00 11.	- 17 22 20.			NONSTELLAR OBJECT
ZWG 372.008	20 00 12.	+ 01 44		15.2	GALAXY
ISS 0266	20 00 18.	+ 33 33	1408		STELLAR RING
ZWG 324.006	20 00 18.	+ 66 05		12.8	GALAXY
RNGC 6869	20 00 18.	+ 66 05		13.0	GALAXY
UGC 11506	20 00 18.	+ 66 05	96	12.8	GALAXY S0
HN 2205	20 00 19.5	- 49 27 51.	12		NEBULA
PK056-06.1	20 00 20.7	+ 17 28 26.			PLANETARY NEBULA
ISS 0361	20 00 24.	+ 21 08	164		STELLAR RING
ZWG 257.034	20 00 24.	+ 49 00		15.6	GALAXY
UGC 11507	20 00 24.	+ 49 00	60	15.6	GALAXY Sc
HN 2207	20 00 27.6	- 48 59 14.	36		NEBULA
MCG+01-51-002	20 00 30.	+ 07 00	18	16.	GALAXY
MCG-05-47-021	20 00 30.	- 30 02	24	15.5	GALAXY
HN 0634	20 00 30.	- 57 44			NEBULA
IC 4935	20 00 31.	- 57 43			NONSTELLAR OBJECT
HN 2206	20 00 35.7	- 52 45 50.	18		NEBULA
MCG+08-36-015	20 00 36.	+ 49 00	54	16.	GALAXY
ZWG 398.007	20 00 42.	+ 06 57		15.3	GALAXY
UGC 11508	20 00 42.	+ 53 39	66	16.0	GALAXY Sc
ZWG 282.001	20 00 42.	+ 56 52		15.2	GALAXY
ZWG 281.012	20 00 42.	+ 56 52		15.2	GALAXY
UGC 11509	20 00 42.	+ 56 52	60	15.2	GALAXY Sb-c
MCG+09-33-001	20 00 42.	+ 56 53	60	14.	GALAXY
HN 2204	20 00 42.3	- 57 49 20.	12		NEBULA
RNGC 6858	20 00 43.	+ 11 08			NON-EXISTENT OBJECT
OCL 0172	20 00 48.	+ 41 58	840		OPEN STAR CLUSTER
MCG-04-47-006	20 00 48.	- 26 36	42	15.	GALAXY
SCHO 1158	20 00 50.	+ 37 33 12.	890		ISOLATED DARK CLOUD
HN 2211	20 00 51.4	- 47 37 01.	18		NEBULA
MCG+09-33-002	20 00 54.	+ 53 40	54		GALAXY
MCG+11-24-004	20 00 54.	+ 66 09	78	14.	GALAXY
HN 2209	20 00 54.8	- 50 27 01.			NEBULA
HN 2208	20 00 54.9	- 56 30 55.	24		NEBULA
SCHO 1159	20 00 55.	+ 34 21 54.	490		ISOLATED DARK CLOUD
HN 2212	20 00 56.8	- 46 53 48.			NEBULA
LBN 0184	20 01	+ 36 50	600		BRIGHT NEBULA
LBN 0242	20 01	+ 42 40	720		BRIGHT NEBULA
LBN 0271	20 01	+ 44 40	2880		BRIGHT NEBULA
LB 09997	20 01	- 87 12		13.6	FAINT BLUE STAR
B 145	20 01 00.	+ 37 33	2700		DARK OBJECT
7ZW 924	20 01 00.	+ 85 57			COMPACT GALAXY
ARC 2322	20 01 01.	+ 73 04		17.6	RICH CLUSTER OF GALAXIES
ISS 0314	20 01 06.	+ 28 15	116		STELLAR RING
MRSL 073+03/1	20 01 06.	+ 36 57	401		HII REGION
IC 4929	20 01 07.	- 71 49			NONSTELLAR OBJECT
HN 0633	20 01 07.	- 71 50			NEBULA
HN 2210	20 01 10.2	- 55 44 48.	24		NEBULA
ZWG 398.008	20 01 12.	+ 06 51		15.4	GALAXY
ZWG 398.009	20 01 12.	+ 07 01		15.6	GALAXY
MCG+01-51-003	20 01 12.	+ 07 01	24	16.	GALAXY
OCL 0154	20 01 12.	+ 37 33	480		OPEN STAR CLUSTER
MCG-07-41-005	20 01 12.	- 39 14	18	15.	GALAXY
HN 2214	20 01 13.9	- 46 57 41.	24		NEBULA
HN 0006	20 01 15.	+ 00 18			NEBULA
PK064-02.1	20 01 17.9	+ 26 52 28.			PLANETARY NEBULA
LDN 0801	20 01 18.	+ 24 10	780		DARK NEBULA
MCG-04-47-007	20 01 18.	- 26 30	36	15.5	GALAXY
RNGC 6859	20 01 19.	+ 00 18			NON-EXISTENT OBJECT
IC 4937	20 01 19.	- 56 23			NONSTELLAR OBJECT
HN 0638	20 01 19.	- 56 24			NEBULA
RNGC 6861A	20 01 23.	- 48 09			GALAXY
ZWG 257.035	20 01 24.	+ 49 11		14.6	GALAXY
DV.56 N6861A	20 01 24.	- 48 09	108		S GALAXY
LB 03635	20 01 30.	+ 22 50 24.		19.2	FAINT BLUE STAR
UHA 49	20 01 30.	+ 40 15	60		STELLAR RING
MRSL 078+06/1	20 01 30.	+ 42 37	1398		HII REGION
MRSL 078+06/2	20 01 30.	+ 42 52	1998		HII REGION
MCG+08-36-016	20 01 30.	+ 49 11	24	16.	GALAXY
MCG-03-51-003	20 01 30.	- 19 14	84	15.	GALAXY
IC 4936	20 01 33.	- 61 34			NONSTELLAR OBJECT
HN 0637	20 01 33.	- 61 35			NEBULA
HN 2213	20 01 35.3	- 56 17 11.	18		NEBULA
ZWG 282.002	20 01 36.	+ 53 44		15.6	GALAXY
UGC 11510	20 01 36.	+ 53 44	126	15.6	GALAXY Sc
B 146	20 01 37.	+ 35 53			DARK OBJECT
SG 2.034	20 01 39.	+ 43 38	18000		DIFFUSE EMISSION NEBULA
MCG+09-33-003	20 01 42.	+ 53 45	120	14.	GALAXY
SG 2.035	20 01 44.	+ 40 38	7200		DIFFUSE EMISSION NEBULA
RNGC 6854	20 01 45.	- 54 32		13.0	GALAXY
UGC 11511	20 01 48.	+ 07 16	108	17.	GALAXY Sc
MRSL 077+05/1	20 01 48.	+ 41 30	1536		HII REGION
HN 2215	20 01 51.7	- 56 24 57.	30		NEBULA
MRSL 074+03/2	20 01 54.	+ 38 10	533		HII REGION
ISS 0218	20 01 54.	+ 39 39	387		STELLAR RING
HN 2218	20 01 54.0	- 48 39 57.	18		NEBULA
IC 4938	20 01 57.	- 60 20			NONSTELLAR OBJECT
HN 2217	20 01 57.3	- 53 51 03.	12		NEBULA
HN 0639	20 01 58.	- 60 21			NEBULA
LBN 0155	20 02	+ 32 04	240		BRIGHT NEBULA
LBN 0164	20 02	+ 34 00	1440		BRIGHT NEBULA
LBN 0174	20 02	+ 35 00	3900		BRIGHT NEBULA
LBN 0180	20 02	+ 36 10	2700		BRIGHT NEBULA
KHAV 676	20 02	+ 36 32	3540		DARK NEBULA
LBN 0190	20 02	+ 37 50	2400		BRIGHT NEBULA
LBN 0213	20 02	+ 40 00	720		BRIGHT NEBULA
LBN 0260	20 02	+ 43 50	2220		BRIGHT NEBULA
LBN 0262	20 02	+ 44 00	3300		BRIGHT NEBULA
LBN 0294	20 02	+ 45 40	2220		BRIGHT NEBULA
LDN 0809	20 02 00.	+ 25 00	6000		DARK NEBULA
LDN 0814	20 02 00.	+ 26 25	720		DARK NEBULA
COU 038	20 02 00.	+ 33 40	7200		EMISSION NEBULA
LDN 0887	20 02 00.	+ 42 20	540		DARK NEBULA
LDN 0890	20 02 00.	+ 42 45	900		DARK NEBULA
LB 00172	20 02 00.	- 14 16		18.0	FAINT BLUE STAR
HN 2221	20 02 00.6	- 48 17 44.	18		NEBULA
HN 2219	20 02 02.9	- 51 01 08.	12		NEBULA
HN 2220	20 02 03.0	- 51 00 26.	12		NEBULA
HN 2216	20 02 03.8	- 57 00 33.	6		NEBULA
IC 4934	20 02 04.	- 69 36			NONSTELLAR OBJECT
HN 0636	20 02 04.	- 69 37			NEBULA
RNGC 6866	20 02 05.	+ 43 51		9.0	OPEN CLUSTER
ISS 0315	20 02 06.	+ 33 11	132		STELLAR RING
OCL 0183	20 02 06.	+ 43 51	1440	9.5	OPEN STAR CLUSTER
HN 2224	20 02 11.6	- 46 44 55.	12		NEBULA
ZWG 423.003	20 02 12.	+ 12 35		14.4	GALAXY
UGC 11512	20 02 12.	+ 12 35	66	14.4	GALAXY SB?
MCG-07-41-006	20 02 12.	- 42 52	60	16.	GALAXY
DV.56 N6861B	20 02 12.	- 48 38	78		GALAXY
HN 2225	20 02 12.6	- 46 54 25.	18		NEBULA
HN 0641	20 02 13.	- 44 51			NEBULA
IC 4940	20 02 13.	- 44 51			NONSTELLAR OBJECT
RNGC 6851B	20 02 17.	- 48 08			GALAXY
RNGC 6851A	20 02 17.	- 48 08			GALAXY
ISS 0316	20 02 18.	+ 33 09	226		STELLAR RING
MRSL 074+03/1	20 02 18.	+ 37 42	1296		HII REGION
MRSL 077+05/2	20 02 18.	+ 41 47	1536		HII REGION
7ZW 925	20 02 18.	+ 60 10			COMPACT GALAXY
DV.56 N6851B	20 02 18.	- 48 08			S GALAXY
DV.56 N6851A	20 02 18.	- 48 08			Sd GALAXY
HN 2222	20 02 20.6	- 54 19 19.	24		NEBULA
HN 2223	20 02 22.7	- 53 27 01.	18		NEBULA
ZWG 423.004	20 02 24.	+ 13 58		15.7	GALAXY
UGC 11513	20 02 24.	+ 13 58	78	15.7	GALAXY SBc
LDN 0865	20 02 24.	+ 37 38	1080		DARK NEBULA
ZC 2002.4+6822	20 02 24.	+ 68 22	940		CLUSTER OF GALAXIES
HN 2227	20 02 28.7	- 48 37 06.	18		NEBULA
RNGC 6863	20 02 29.	- 03 42			NON-EXISTENT OBJECT
SG 3.159	20 02 30.	+ 34 46	6000		DIFFUSE EMISSION NEBULA
MRSL 076+05/1	20 02 30.	+ 40 48	726		HII REGION
ACK 084+09.1	20 02 30.	+ 49 11			PLANETARY NEBULA
IC 4946	20 02 30.	- 44 10 56.			NONSTELLAR OBJECT
MIN.48 05	20 02 31.	+ 33 34			DIFFUSE NEBULA
PK075+04.1	20 02 34.	+ 39 26	30		PLANETARY NEBULA
RNGC 6861B	20 02 35.	- 48 38			GALAXY
MRSL 078+05/3	20 02 36.	+ 42 24	1200		HII REGION
HN 2229	20 02 36.5	- 46 36 00.	24		NEBULA
VDB.66N 128	20 02 38.	+ 32 05	480		REFLECTION NEBULA
IC 4943	20 02 39.	- 48 33 56.			NONSTELLAR OBJECT
HN 2226	20 02 39.2	- 53 41 06.	18		NEBULA
DG 160	20 02 42.	+ 29 06	300		REFLECTION NEBULA
LDN 0849	20 02 42.	+ 34 58	420		DARK NEBULA
MCG-04-47-008	20 02 42.	- 25 16	36	15.	GALAXY
RNGC 6849	20 02 43.	- 40 21			GALAXY
CED 175C	20 02 44.	+ 29 06	30		DIFFUSE GALACTIC NEBULA
CED 175B	20 02 44.	+ 29 06	60		DIFFUSE GALACTIC NEBULA
CED 175A	20 02 44.	+ 29 06	60		DIFFUSE GALACTIC NEBULA
HN 2231	20 02 44.4	- 48 13 29.	18		NEBULA
IC 4954	20 02 45.	+ 29 06 26.			NONSTELLAR OBJECT
MCG-03-51-004	20 02 45.	- 20 15	42	15.	GALAXY
PK068-00.1	20 02 45.1	+ 31 18 51.	59		PLANETARY NEBULA
RNGC 6855	20 02 48.	- 56 32			GALAXY
HN 2233	20 02 50.6	- 48 28 17.	18		NEBULA
IC 4955	20 02 51.	+ 29 02 51.			NONSTELLAR OBJECT
PK063-03.1	20 02 52.0	+ 25 18 04.			PLANETARY NEBULA
HN 2228	20 02 53.6	- 56 55 06.	12		NEBULA
OCL 0138	20 02 54.	+ 29 04			OPEN STAR CLUSTER
OCL 0143	20 02 54.	+ 33 43	240	16.	OPEN STAR CLUSTER
SG 2.036	20 02 54.	+ 34 36	7200		DIFFUSE EMISSION NEBULA
MCG-07-41-007	20 02 54.	- 40 22	78	13.5	GALAXY
HN 0640	20 02 54.	- 60 53			NEBULA
IC 4939	20 02 54.	- 60 53			NONSTELLAR OBJECT
IC 4942	20 02 55.	- 52 46			NONSTELLAR OBJECT
HN 2235	20 02 55.4	- 46 41 41.	18		NEBULA
HN 1184	20 02 56.	- 52 46	12		NEBULA
HN 2232	20 02 56.6	- 51 52 29.	60		NEBULA
MCG-02-51-001	20 02 57.	- 10 42	60	16.5	GALAXY
RNGC 6861C	20 02 58.	- 48 50			GALAXY
IC 4948	20 02 59.	- 43 46			NONSTELLAR OBJECT
HN 1185	20 02 59.	- 53 48	18		NEBULA
IC 4941	20 02 59.	- 53 48			NONSTELLAR OBJECT
HN 2230	20 02 59.6	- 54 26 53.			NEBULA
LBN 0153	20 03	+ 29 02	120		BRIGHT NEBULA
LBN 0177	20 03	+ 35 02	720		BRIGHT NEBULA
LBN 0194	20 03	+ 38 20	180		BRIGHT NEBULA
LBN 0217	20 03	+ 40 10	600		BRIGHT NEBULA
LBN 0238	20 03	+ 42 10	600		BRIGHT NEBULA
LDN 0860	20 03 00.	+ 35 55	120		DARK NEBULA
COU 039	20 03 00.	+ 38 00	3000		EMISSION NEBULA
MRSL 076+04/3	20 03 00.	+ 40 10	666		HII REGION
MCG-07-41-008	20 03 00.	- 42 55	48	15.	GALAXY
DV.56 N6861C	20 03 00.	- 48 50	90		GALAXY
HN 2237	20 03 00.2	- 46 45 04.		13.	NEBULA
SG 3.160	20 03 01.	+ 33 59	4800		DIFFUSE EMISSION NEBULA
HN 2236	20 03 01.9	- 48 47 28.	42		NEBULA
LDN 0845	20 03 06.	+ 34 13	1800		DARK NEBULA
REIN 2.285	20 03 08.08	- 22 03 56.6			NEBULA
HN 2234	20 03 09.0	- 53 29 40.	12		NEBULA
HOLM 777B	20 03 12.	- 10 35	36	14.6	PART OF MULTIPLE GALAXY
GCL 116	20 03 12.	- 22 04	294	9.5	GLOBULAR STAR CLUSTER
HN 1186	20 03 13.	- 54 36		14.	NEBULA
IC 4944	20 03 13.	- 54 36			NONSTELLAR OBJECT
HN 2238	20 03 13.5	- 46 27 46.	18		NEBULA
HOLM 777A	20 03 14.	- 10 36	36	14.2	PART OF MULTIPLE GALAXY
RNGC 6864	20 03 14.	- 22 04		9.5	GLOBULAR CLUSTER

OBJECT NAME	RIGHT ASCEN.	DECLINATION	DIAM.	MAGN.	TYPE OF OBJECT
RNGC 6865	20 03 15.	- 09 10			GALAXY
ZWG 398.010	20 03 18.	+ 06 36		15.5	GALAXY
ISS 0267	20 03 18.	+ 33 19	114		STELLAR RING
MRSL 078+05/2	20 03 18.	+ 42 05	1001		HII REGION
MCG-02-51-002	20 03 18.	- 10 33	30	15.5	GALAXY
MCG-02-51-003	20 03 18.	- 10 34	42	15.	GALAXY
LDN 0848	20 03 24.	+ 34 43	420		DARK NEBULA
IC 4928	20 03 25.	- 77 26			NONSTELLAR OBJECT
HN 0635	20 03 25.	- 77 27			NEBULA
LDN 0797	20 03 30.	+ 23 20	420		DARK NEBULA
ISS 0219	20 03 30.	+ 39 23	390		STELLAR RING
UGC 11514	20 03 30.	+ 55 32	84	16.5	GALAXY
HN 2239	20 03 30.4	- 46 54 32.			NEBULA
SC 2000-2128.3	20 03 31.	- 21 19 49.	72		NEBULA
HN 1187	20 03 33.	- 53 18		15.	NEBULA
IC 4947	20 03 33.	- 53 18			NONSTELLAR OBJECT
ZWG 398.011	20 03 36.	+ 08 01		15.7	GALAXY
IC 4949	20 03 38.	- 48 27 20.			NONSTELLAR OBJECT
RNGC 6861	20 03 41.	- 48 31		12.5	GALAXY
MRSL 073+02/2	20 03 42.	+ 37 05	798		HII REGION
MCG-07-41-009	20 03 48.	- 42 58	48	15.	GALAXY
HN 2240	20 03 49.4	- 53 02 38.	12		NEBULA
HN 2241	20 03 52.7	- 53 41 56.	12		NEBULA
RNGC 6871	20 03 58.	+ 35 38		6.0	OPEN CLUSTER
HN 2243	20 03 59.8	- 46 41 19.	12		NEBULA
LBN 0170	20 04	+ 34 30	3300		BRIGHT NEBULA
LBN 0192	20 04	+ 38 00	360		BRIGHT NEBULA
LBN 0193	20 04	+ 38 20	240		BRIGHT NEBULA
LBN 0251	20 04	+ 43 00	660		BRIGHT NEBULA
LDN 0852	20 04 00.	+ 35 00	660		DARK NEBULA
OCL 0148	20 04 00.	+ 35 38	2250	6.0	OPEN STAR CLUSTER
SCHO 1160	20 04 00.	+ 37 43 30.	300		ISOLATED DARK CLOUD
LDN 0871	20 04 00.	+ 39 40	1020		DARK NEBULA
OCL 0166	20 04 00.	+ 40 23	1200		OPEN STAR CLUSTER
OCL 0170	20 04 00.	+ 41 03	1080		OPEN STAR CLUSTER
LDN 0884	20 04 00.	+ 41 20	4500		DARK NEBULA
MCG+00-51-007	20 04 00.	- 01 55	36	15.	GALAXY
MCG-05-47-022	20 04 00.	- 30 34	72	14.5	GALAXY
ISS 0268	20 04 06.	+ 34 53	287		STELLAR RING
HN 2242	20 04 11.4	- 52 30 18.	18		NEBULA
MRSL 078+05/1	20 04 12.	+ 41 55	863		HII REGION
ZWG 303.016	20 04 12.	+ 62 38		14.8	GALAXY
UGC 11515	20 04 12.	+ 62 38	96	14.8	GALAXY Sc
MCG+10-28-016	20 04 12.	+ 62 38	72	14.	GALAXY
ZWG 372.009	20 04 12.	- 01 52		15.4	GALAXY
ZWG 398.012	20 04 18.	+ 06 58		15.7	GALAXY
HN 2245	20 04 20.8	- 48 06 29.	18		NEBULA
ZWG 398.013	20 04 24.	+ 05 32		15.7	GALAXY
HN 0642	20 04 24.	- 56 18			NEBULA
IC 4950	20 04 24.	- 56 19			NONSTELLAR OBJECT
HN 2244	20 04 24.0	- 52 30 42.	12		NEBULA
RNGC 6860	20 04 27.	- 61 14			UNVERIFIED SOUTHERN OBJECT
HN 2246	20 04 29.7	- 48 59 17.	12		NEBULA
MRSL 077+04/5	20 04 30.	+ 41 04	1931		HII REGION
OCL 0918	20 04 30.	- 79 28	4260	5.7	OPEN STAR CLUSTER
PK063-03.2	20 04 33.8	+ 24 51 38.			PLANETARY NEBULA
SG 2.037	20 04 35.	+ 35 36	5400		DIFFUSE EMISSION NEBULA
SCHO 1161	20 04 36.	+ 33 46 54.	430		ISOLATED DARK CLOUD
MCG-05-47-023	20 04 36.	- 29 58	54	14.	GALAXY
HN 0644	20 04 40.	- 55 35			NEBULA
IC 4952	20 04 40.	- 55 36			NONSTELLAR OBJECT
RNGC 6861D	20 04 41.	- 48 23			GALAXY
2ZW 079	20 04 42.	+ 05 35			COMPACT GALAXY
MRSL 073+02/1	20 04 42.	+ 37 00	401		HII REGION
MRSL 074+02/8	20 04 42.	+ 37 11	533		HII REGION
DV.56 N6861D	20 04 42.	- 48 23	144		SA0 GALAXY
HN 2249	20 04 42.4	- 48 21 16.	72		NEBULA
HN 2248	20 04 43.0	- 49 51 58.	12		NEBULA
ZWG 372.010	20 04 48.	+ 02 59		15.3	GALAXY
MRSL 076+04/2	20 04 48.	+ 40 26	533		HII REGION
MCG-04-47-009	20 04 48.	- 21 15	72	13.5	GALAXY
HN 2247	20 04 48.5	- 53 23 16.	18		NEBULA
SCHO 1162	20 04 53.	+ 23 08 48.	540		ISOLATED DARK CLOUD
ZWG 398.014	20 04 54.	+ 06 19		15.3	GALAXY
RNGC 6862	20 04 54.	- 56 31			UNVERIFIED SOUTHERN OBJECT
HN 2263	20 04 55.3	- 49 22 03.	60		NEBULA
B 147	20 04 57.	+ 35 14	660		DARK OBJECT
PK069+00.1	20 04 58.3	+ 32 07 52.			PLANETARY NEBULA
KHAV 677	20 05	+ 31 57	9540		DARK NEBULA
LBN 0162	20 05	+ 33 30	10800		BRIGHT NEBULA
LBN 0166	20 05	+ 34 00	1800		BRIGHT NEBULA
KHAV 678	20 05	+ 34 57	2230		DARK NEBULA
BA 06	20 05	+ 38 12	870		STELLAR GROUP
LBN 0205	20 05	+ 39 00	13500		BRIGHT NEBULA
LBN 0216	20 05	+ 39 50	1920		BRIGHT NEBULA
LBN 0227	20 05	+ 41 00	3300		BRIGHT NEBULA
SER 135.05	20 05	- 48 35	5760	13.	CLSTR OF ABOUT 20 GALXIES
COU 041	20 05 00.	+ 34 10	6000		EMISSION NEBULA
ISS 0269	20 05 00.	+ 34 24	291		STELLAR RING
LDN 0853	20 05 00.	+ 35 05	180		DARK NEBULA
LDN 0862	20 05 00.	+ 36 50	3600		DARK NEBULA
COU 040	20 05 00.	+ 37 00	3600		EMISSION NEBULA
MCG-05-47-024	20 05 00.	- 29 29	66	15.	GALAXY
SCHO 1163	20 05 04.	+ 37 39 30.	280		ISOLATED DARK CLOUD
HN 0645	20 05 04.	- 62 00			NEBULA
IC 4951	20 05 04.	- 62 01			NONSTELLAR OBJECT
HN 2253	20 05 04.0	- 48 21 39.	12		NEBULA
MRSL 077+04/4	20 05 06.	+ 40 55	1668		HII REGION
ISS 0220	20 05 06.	+ 41 13	82		STELLAR RING
HN 2250	20 05 10.1	- 54 32 09.			NEBULA
PK107+21.1	20 05 12.	+ 74 18	240		PLANETARY NEBULA
MCG+00-51-008	20 05 12.	- 00 25 30.	60	15.	GALAXY
HN 2251	20 05 13.7	- 54 31 39.	18		NEBULA
HN 2252	20 05 15.6	- 53 28 14.	18		NEBULA
PK079+06.1	20 05 17.0	+ 44 05 37.	4		PLANETARY NEBULA
LDN 0866	20 05 18.	+ 37 38	1680		DARK NEBULA
MCG+07-41-004	20 05 18.	+ 44 23	15	17.5	GALAXY
ZWG 372.011	20 05 18.	- 00 23		15.5	GALAXY
UGC 11516	20 05 18.	- 00 23	60	15.5	GALAXY
SCHO 1164	20 05 19.	+ 38 26 06.	290		ISOLATED DARK CLOUD
HN 2255	20 05 20.8	- 50 17 02.			NEBULA
HN 2257	20 05 21.7	- 49 10 08.	18		NEBULA
MCG+08-36-017	20 05 24.	+ 49 37	60	17.	GALAXY
MCG-03-51-005	20 05 24.	- 17 50	60	15.5	GALAXY
HN 2258	20 05 25.1	- 50 17 50.			NEBULA
HN 2259	20 05 30.8	- 48 15 55.	24		NEBULA
HN 2260	20 05 31.8	- 46 31 49.	18		NEBULA
OCL 0149	20 05 36.	+ 35 32	900		OPEN STAR CLUSTER

OBJECT NAME	RIGHT ASCEN.	DECLINATION	DIAM.	MAGN.	TYPE OF OBJECT
MCG-04-47-010	20 05 36.	- 25 35 30.	78	14.5	GALAXY
HN 0646	20 05 38.	- 62 56			NEBULA
HN 2256	20 05 38.1	- 55 51 19.	24		NEBULA
IC 4953	20 05 39.	- 62 57			NONSTELLAR OBJECT
HN 2261	20 05 40.5	- 46 29 48.	18		NEBULA
SCHO 1165	20 05 41.	+ 37 58 18.	380		ISOLATED DARK CLOUD
HN 2254	20 05 41.9	- 59 09 31.	18		NEBULA
SER 136.11	20 05 42.	- 46 28	300	15.	GALAXY WITH 2 SATELLITES
BC 3C407	20 05 46.	- 04 27 18.		18.	QUASI-STELLAR OBJECT
SHB 351	20 05 46.0	- 04 27 18.		18.	QUASI-STELLAR OBJECT
MRSL 074+02/6	20 05 48.	+ 37 05	863		HII REGION
MRSL 076+04/1	20 05 48.	+ 39 50	2663		HII REGION
IC 4956	20 05 48.	- 45 47 10.			NONSTELLAR OBJECT
HN 2262	20 05 48.0	- 48 25 54.	24		NEBULA
HN 0643	20 05 53.	- 71 09			NEBULA
IC 4945	20 05 53.	- 71 10			NONSTELLAR OBJECT
LDN 0863	20 05 54.	+ 36 58	240		DARK NEBULA
HN 2264	20 05 56.2	- 49 06 35.	36		NEBULA
LBN 0179	20 06	+ 35 15	1020		BRIGHT NEBULA
LBN 0187	20 06	+ 36 40	3000		BRIGHT NEBULA
LBN 0189	20 06	+ 37 00	2400		BRIGHT NEBULA
LBN 0214	20 06	+ 39 35	180		BRIGHT NEBULA
LBN 0222	20 06	+ 40 20	8100		BRIGHT NEBULA
LBN 0264	20 06	+ 43 40	2220		BRIGHT NEBULA
ZC 2006.0+0304	20 06 00.	+ 03 04	5780		CLUSTER OF GALAXIES
MCG+01-51-004	20 06 00.	+ 05 49	60	16.	GALAXY
RNGC 6874	20 06 00.	+ 38 06			NON-EXISTENT OBJECT
LDN 0875	20 06 00.	+ 39 50	240		DARK NEBULA
COU 042	20 06 00.	+ 39 55	4800		EMISSION NEBULA
ISS 0221	20 06 00.	+ 41 34	197		STELLAR RING
MRSL 079+06/1	20 06 00.	+ 43 27	2333		HII REGION
HN 2265	20 06 00.0	- 49 28 47.	30		NEBULA
HN 2268	20 06 01.2	- 46 10 47.	24		NEBULA
ZWG 398.015	20 06 06.	+ 05 49		15.6	GALAXY
UGC 11517	20 06 06.	+ 05 49	66	15.6	GALAXY S-IRR
ZWG 423.005	20 06 06.	+ 14 52		15.5	GALAXY
UGC 11518	20 06 06.	+ 14 52	60	15.5	GALAXY S
RNGC 6873	20 06 06.	+ 20 58			NON-EXISTENT OBJECT
ZC 2006.1+7407	20 06 06.	+ 74 07	3020		CLUSTER OF GALAXIES
HN 2269	20 06 06.2	- 48 19 53.	24		NEBULA
HN 2270	20 06 09.4	- 47 43 59.			NEBULA
SER 138.06	20 06 12.	- 52 46	150	18.	FAINT CLSTRS OF GALAXIES
HN 2266	20 06 15.2	- 54 48 11.	18		NEBULA
HN 2271	20 06 15.9	- 46 14 22.			NEBULA
RNGC 6868	20 06 17.	- 48 31		12.5	GALAXY
LDN 0856	20 06 18.	+ 35 12	480		DARK NEBULA
OCL 0153	20 06 18.	+ 36 24	360		OPEN STAR CLUSTER
ZC 2006.3+6056	20 06 18.	+ 60 56	2150		CLUSTER OF GALAXIES
MCG-07-41-010	20 06 18.	- 44 19	60	15.	GALAXY
MCG-04-47-011	20 06 21.	- 21 33	42	15.	GALAXY
HN 2267	20 06 22.8	- 56 20 58.	12		NEBULA
MRSL 074+02/7	20 06 24.	+ 37 20	600		HII REGION
ISS 0270	20 06 24.	+ 38 04	461		STELLAR RING
DV.56 N6876A	20 06 24.	- 71 09			GALAXY
RNGC 6876A	20 06 26.	- 71 09			GALAXY
RNGC 6870	20 06 29.	- 48 26			GALAXY
ZWG 398.016	20 06 30.	+ 07 38		15.6	GALAXY
UGC 11519	20 06 30.	+ 07 38	60	15.6	GALAXY Sc
ZC 2006.5+7920	20 06 30.	+ 79 20	870		CLUSTER OF GALAXIES
RNGC 6867	20 06 33.	- 54 55			GALAXY
ZWG 398.017	20 06 36.	+ 08 31		15.4	GALAXY
HN 2272	20 06 37.3	- 53 38 57.	18		NEBULA
ISS 0318	20 06 42.	+ 30 04	113		STELLAR RING
ISS 0317	20 06 42.	+ 30 38	272		STELLAR RING
HN 1188	20 06 46.	- 55 50		14.	NEBULA
IC 4957	20 06 46.	- 55 51			NONSTELLAR OBJECT
UGC 11520	20 06 48.	+ 73 31	66	16.5	GALAXY Sc
HN 2273	20 06 51.0	- 56 48 57.			NEBULA
MRSL 074+02/3	20 06 54.	+ 37 15	468		HII REGION
ZWG 257.036	20 06 54.	+ 50 32		15.3	GALAXY
ZWG 303.017	20 06 54.	+ 59 29		15.7	GALAXY
KHAV 679	20 07	+ 35 33	4410		DARK NEBULA
KHAV 680	20 07	+ 40 21	3010		DARK NEBULA
LBN 0233	20 07	+ 41 10	5700		BRIGHT NEBULA
LDN 0823	20 07 00.	+ 30 00	4860		DARK NEBULA
COU 043	20 07 00.	+ 35 40	7200		EMISSION NEBULA
LDN 0878	20 07 00.	+ 40 31	180		DARK NEBULA
ACK 078+05.1	20 07 00.	+ 42 21			PLANETARY NEBULA
LDN 0903	20 07 00.	+ 44 55	240		DARK NEBULA
HN 2274	20 07 00.3	- 56 54 44.	18		NEBULA
HN 2278	20 07 00.5	- 46 41 43.	18		NEBULA
HN 2279	20 07 00.6	- 46 37 55.			NEBULA
HN 2280	20 07 03.7	- 47 14 31.			NEBULA
CED 176A	20 07 05.	+ 39 39			DIFFUSE GALACTIC NEBULA
ZWG 372.012	20 07 06.	+ 03 03		15.7	GALAXY
SG 2.161	20 07 07.	+ 34 11	2100		DIFFUSE EMISSION NEBULA
HN 0647	20 07 08.	- 53 14			NEBULA
IC 4959	20 07 08.	- 53 15			NONSTELLAR OBJECT
HN 2275	20 07 11.4	- 56 03 49.	18		NEBULA
MRSL 074+02/4	20 07 12.	+ 37 28	798		HII REGION
MRSL 077+04/3	20 07 12.	+ 41 02	1133		HII REGION
MRSL 080+06/1	20 07 12.	+ 44 50	4463		HII REGION
MCG+12-19-001	20 07 12.	+ 71 48	9	17.	GALAXY
HN 2276	20 07 15.2	- 55 46 25.	18		NEBULA
RNGC 6861E	20 07 17.	- 48 47			GALAXY
OCL 0150	20 07 18.	+ 35 20	780	16.	OPEN STAR CLUSTER
DV.56 N6861E	20 07 18.	- 48 48	69		S GALAXY
HN 2277	20 07 18.2	- 54 09 37.	12		NEBULA
HN 2283	20 07 18.6	- 47 20 42.	18		NEBULA
PK079+05.1	20 07 22.2	+ 43 34 54.	45		PLANETARY NEBULA
HN 2284	20 07 24.2	- 48 50 12.	36		NEBULA
HN 2285	20 07 24.4	- 46 21 48.			NEBULA
HN 2286	20 07 26.1	- 46 40 54.	18		NEBULA
HN 2282	20 07 27.2	- 52 17 18.	12		NEBULA
LDN 0876	20 07 30.	+ 39 45	660		DARK NEBULA
LDN 0885	20 07 30.	+ 41 00	180		DARK NEBULA
HN 1189	20 07 32.	- 53 17	60		NEBULA
IC 4961	20 07 32.	- 53 18			NONSTELLAR OBJECT
CED 177	20 07 33.	+ 36 42			DIFFUSE GALACTIC NEBULA
SG 2.038	20 07 33.	+ 36 42	7200		DIFFUSE EMISSION NEBULA
SCHO 1166	20 07 34.	+ 39 31 42.	400		ISOLATED DARK CLOUD
RNGC 6861F	20 07 35.	- 48 25			GALAXY
HN 2287	20 07 35.4	- 48 25 23.	60		NEBULA
MRSL 075+03/1	20 07 36.	+ 38 45	2201		HII REGION
7ZW 926	20 07 36.	+ 62 24			COMPACT GALAXY
HN 2281	20 07 41.7	- 59 23 42.			NEBULA
DV.56 N6861F	20 07 42.	- 48 26	93		S GALAXY
SG 2.039	20 07 45.	+ 40 42	10800		DIFFUSE EMISSION NEBULA

OBJECT NAME	RIGHT ASCEN.	DECLINATION	DIAM.	MAGN.	TYPE OF OBJECT
B 342	20 07 46.	+ 41 03	240		DARK OBJECT
HN 2289	20 07 46.1	- 47 43 53.	18		NEBULA
HN 2294	20 07 59.8	- 45 44 34.			NEBULA
LBN 0163	20 08	+ 33 10	900		BRIGHT NEBULA
LBN 0172	20 08	+ 34 10	1740		BRIGHT NEBULA
LBN 0182	20 08	+ 35 40	3300		BRIGHT NEBULA
LBN 0193	20 08	+ 37 40	3600		BRIGHT NEBULA
LBN 0208	20 08	+ 38 50	4200		BRIGHT NEBULA
COU 044	20 08 00.	+ 40 30	12000		EMISSION NEBULA
MRSL 077+04/1	20 08 00.	+ 40 38	1068		HII REGION
LDN 0905	20 08 00.	+ 45 10	540		DARK NEBULA
HN 2291	20 08 04.4	- 51 32 10.	24		NEBULA
HN 2290	20 08 04.5	- 53 38 28.	18		NEBULA
OCL 0145	20 08 06.	+ 33 37			OPEN STAR CLUSTER
CED 178	20 08 06.	+ 34 49			DIFFUSE GALACTIC NEBULA
MRSL 074+02/5	20 08 06.	+ 37 37	1596		HII REGION
ISS 0222	20 08 06.	+ 40 30	77		STELLAR RING
HN 2292	20 08 07.2	- 51 53 34.	18		NEBULA
SCHO 1167	20 08 08.	+ 36 09 54.	240		ISOLATED DARK CLOUD
HN 0648	20 08 08.	- 55 23			NEBULA
IC 4963	20 08 08.	- 55 23			NONSTELLAR OBJECT
HN 2288	20 08 08.1	- 56 40 34.	12		NEBULA
IC 1310	20 08 09.	+ 34 49			NONSTELLAR OBJECT
PK057-08.1	20 08 09.95	+ 16 46 25.5	8	12.1	PLANETARY NEBULA
RNGC 6879	20 08 10.	+ 16 46		12.0	PLANETARY NEBULA
HN 0055	20 08 11.	+ 16 47			NEBULA
HN 2293	20 08 11.3	- 53 24 21.	12		NEBULA
OCL 0174	20 08 12.	+ 41 13	600		OPEN STAR CLUSTER
ZWG 372.013	20 08 18.	+ 01 59		14.9	GALAXY
UGC 11521	20 08 18.	+ 01 59	66	14.9	GALAXY S0
MRSL 074+02/1	20 08 18.	+ 37 00	731		HII REGION
SG 2.040	20 08 18.	+ 38 49	2400		DIFFUSE EMISSION NEBULA
MCG-05-47-025	20 08 18.	- 29 10	42	15.5	GALAXY
DV.56 N6875A	20 08 18.	- 46 19	150		S GALAXY
RNGC 6875A	20 08 20.	- 46 18			GALAXY
HN 1190	20 08 20.	- 56 59		15.	NEBULA
IC 4965	20 08 20.	- 56 59			NONSTELLAR OBJECT
LB 03607	20 08 22.	+ 26 32 48.		18.9	FAINT BLUE STAR
HN 2297	20 08 23.1	- 46 17 32.	78		NEBULA
ISS 0223	20 08 24.	+ 41 12	127		STELLAR RING
HN 2298	20 08 26.1	- 47 10 26.	30		NEBULA
HN 0650	20 08 27.	- 53 45			NEBULA
IC 4966	20 08 27.	- 53 45			NONSTELLAR OBJECT
HN 2296	20 08 28.9	- 53 10 20.	18		NEBULA
OCL 0147	20 08 30.	+ 34 47	240	14.	OPEN STAR CLUSTER
HN 2295	20 08 31.1	- 54 30 32.		10.	NEBULA
SCHO 1168	20 08 32.	+ 34 39 18.	490		ISOLATED DARK CLOUD
OCL 0173	20 08 36.	+ 41 04	780	13.1	OPEN STAR CLUSTER
SER 136.03	20 08 36.	- 44 17	40	15.5	ELLIPT GLXY AND SATELLITE
SCHO 1169	20 08 45.	+ 35 54 36.	240		ISOLATED DARK CLOUD
HN 0053	20 08 46.	+ 46 19			NEBULA
SCHO 1170	20 08 47.	+ 35 04 00.	280		ISOLATED DARK CLOUD
ISS 0271	20 08 48.	+ 36 02	280		STELLAR RING
MCG-04-47-012	20 08 48.	- 20 57	78	15.	GALAXY
MCG-07-41-011	20 08 48.	- 44 18	48	15.	GALAXY
RNGC 6884	20 08 49.	+ 46 19		12.5	PLANETARY NEBULA
VDB.66N 129	20 08 49.	- 00 58	5760		REFLECTION NEBULA
PK082+07.1	20 08 49.1	+ 46 18 44.	8	12.6	PLANETARY NEBULA
7ZW 927	20 08 54.	+ 62 24			COMPACT GALAXY
HN 2299	20 08 54.7	- 51 41 55.	12		NEBULA
SCHO 1171	20 08 58.	+ 40 01 06.	380		ISOLATED DARK CLOUD
IC 4977	20 08 58.	- 21 47			NONSTELLAR OBJECT
HN 1191	20 08 58.	- 54 04	12		NEBULA
IC 4969	20 08 58.	- 54 04			NONSTELLAR OBJECT
RNGC 6881	20 08 59.	+ 37 16		14.5	PLANETARY NEBULA
KHAV 681	20 09	+ 38 51	8820		DARK NEBULA
LBN 0243	20 09	+ 41 50	1500		BRIGHT NEBULA
LBN 0116	20 09	- 00 30	6600		BRIGHT NEBULA
LDN 0841	20 09 00.	+ 33 00	22200		DARK NEBULA
URA 50	20 09 00.	+ 41 02	222		STELLAR RING
ISS 0224	20 09 00.	+ 41 07	214		STELLAR RING
ZC 2009.0+8351	20 09 00.	+ 83 51	2220		CLUSTER OF GALAXIES
PK074+02.1	20 09 01.30	+ 37 15 47.4	5	14.3	PLANETARY NEBULA
HN 0044	20 09 03.	+ 37 16			NEBULA
HN 2300	20 09 06.4	- 51 43 18.	18		NEBULA
HN 2302	20 09 06.7	- 45 54 41.			NEBULA
IC 1311	20 09 08.	+ 41 02			NONSTELLAR OBJECT
LDN 0858	20 09 12.	+ 34 48	1800		DARK NEBULA
URA 51	20 09 12.	+ 41 01	102		STELLAR RING
MCG+01-51-005	20 09 15.	+ 05 36	60	14.	GALAXY
HN 2301	20 09 15.2	- 54 16 06.	18		NEBULA
MRSL 074+02/2	20 09 18.	+ 37 15	731		HII REGION
MRSL 077+04/2	20 09 18.	+ 41 11	401		HII REGION
RNGC 6883	20 09 21.	+ 35 42		8.0	OPEN CLUSTER
MCG-07-41-012	20 09 21.	- 40 18	78	15.	GALAXY
MCG+01-51-006	20 09 24.	+ 05 22	48	15.	GALAXY
ZWG 398.018	20 09 24.	+ 05 37		14.8	GALAXY
UGC 11522	20 09 24.	+ 05 37	66	14.8	GALAXY Sb-c
OCL 0152	20 09 24.	+ 35 42	1920	8.3	OPEN STAR CLUSTER
ZWG 398.019	20 09 30.	+ 05 22		15.4	GALAXY
UGC 11523	20 09 30.	+ 05 22	66	15.4	GALAXY Sc
MCG+01-51-007	20 09 30.	+ 05 36	72	14.	GALAXY
ZWG 398.020	20 09 30.	+ 05 41		15.7	GALAXY
RNGC 6882	20 09 33.	+ 26 24		5.5	OPEN CLUSTER
HN 2303	20 09 33.2	- 55 59 41.	18		NEBULA
HN 2304	20 09 35.4	- 52 38 52.	12		NEBULA
ZWG 398.021	20 09 36.	+ 05 37		14.9	GALAXY
UGC 11524	20 09 36.	+ 05 37	66	14.9	GALAXY Sc
SCHO 1172	20 09 36.	+ 25 56 00.	480		ISOLATED DARK CLOUD
OCL 0133	20 09 36.	+ 26 24	1080	5.8	OPEN STAR CLUSTER
MRSL 078+04/1	20 09 36.	+ 41 52	1266		HII REGION
RNGC 6875	20 09 38.	- 46 19		13.0	GALAXY
SG 2.041	20 09 43.	+ 41 49	1080		DIFFUSE EMISSION NEBULA
PK060-07.1	20 09 43.7	+ 20 10 42.	39		PLANETARY NEBULA
ZWG 372.014	20 09 48.	+ 01 48		15.4	GALAXY
UGC 11525	20 09 48.	+ 01 48	102	15.4	GALAXY Sa-b
OCL 0151	20 09 48.	+ 35 29	240	12.	OPEN STAR CLUSTER
MRSL 081+06/2	20 09 49.	+ 45 20	600		HII REGION
MCG+00-51-009	20 09 51.	+ 01 47 30.	90	14.5	GALAXY
RNGC 6885	20 09 51.	+ 26 20		9.0	OPEN CLUSTER
OCL 0132	20 09 54.	+ 26 20	1800	9.1	OPEN STAR CLUSTER
ISS 0272	20 09 54.	+ 35 20	268		STELLAR RING
LB 03608	20 09 55.	+ 26 11 00.		19.8	FAINT BLUE STAR
RNGC 6878A	20 09 57.	- 44 59			GALAXY
HN 0649	20 09 57.	- 72 51			NEBULA
IC 4958	20 09 57.	- 72 51			NONSTELLAR OBJECT
KHAV 682	20 10	+ 42 45	7500		DARK NEBULA
LBN 0313	20 10	+ 45 40	1860		BRIGHT NEBULA

OBJECT NAME	RIGHT ASCEN.	DECLINATION	DIAM.	MAGN.	TYPE OF OBJECT
COU 046	20 10 00.	+ 33 40	4200		EMISSION NEBULA
LDN 0850	20 10 00.	+ 34 00	3600		DARK NEBULA
COU 045	20 10 00.	+ 41 00	4800		EMISSION NEBULA
7ZW 929	20 10 00.	+ 84 08			COMPACT GALAXY
MCG+15-01-025	20 10 00.	+ 88 27	24	16.	GALAXY
DV.56 N6878A	20 10 00.	- 44 59	108		SAc GALAXY
HN 0651	20 10 04.	- 70 41			NEBULA
IC 4960	20 10 04.	- 70 41			NONSTELLAR OBJECT
SG 3.162	20 10 05.	+ 36 01	3600		DIFFUSE EMISSION NEBULA
HN 0653	20 10 05.	- 64 56			NEBULA
IC 4968	20 10 05.	- 64 56			NONSTELLAR OBJECT
OCL 0146	20 10 06.	+ 34 12	240	15.	OPEN STAR CLUSTER
MCG+12-19-002	20 10 06.	+ 72 08	33	17.	GALAXY
MCG+12-19-002	20 10 06.	+ 72 08	33	17.	GALAXY
HN 2306	20 10 07.0	- 44 57 56.	90		NEBULA
PK067-00.1	20 10 11.5	+ 30 24 07.			PLANETARY NEBULA
MCG-07-41-013	20 10 12.	- 44 59	90	15.	GALAXY
IC 4975	20 10 12.	- 52 53			NONSTELLAR OBJECT
SG 1.13	20 10 13.	+ 38 12	780		DIFFUSE EMISSION NEBULA
HN 1192	20 10 13.	- 52 53		15.	NEBULA
RNGC 6878	20 10 15.	- 44 41		13.5	GALAXY
HN 2305	20 10 16.7	- 57 39 02.	18		NEBULA
ISS 0319	20 10 24.	+ 29 04	214		STELLAR RING
DG 161	20 10 24.	- 02 32	960		REFLECTION NEBULA
MCG-07-41-014	20 10 24.	- 40 13	60	15.	GALAXY
HN 2307	20 10 27.3	- 50 41 01.	18		NEBULA
RNGC 6886	20 10 29.	+ 19 50		12.0	PLANETARY NEBULA
PK060-07.2	20 10 29.46	+ 19 50 18.6	9	12.2	PLANETARY NEBULA
ZWG 372.015	20 10 30.	+ 01 05		15.6	GALAXY
MRSL 075+02/1	20 10 30.	+ 38 12	1001		HII REGION
ACK 077+03.1	20 10 30.	+ 40 36			PLANETARY NEBULA
MRSL 077+03/3	20 10 30.	+ 40 55	1536		HII REGION
MCG-07-41-015	20 10 30.	- 44 42	66	14.4	GALAXY
HN 0656	20 10 30.	- 58 31			NEBULA
IC 4973	20 10 31.	- 58 31			NONSTELLAR OBJECT
MRSL 076+03/1	20 10 36.	+ 39 51	1001		HII REGION
CED 179	20 10 37.	+ 38 15	1080		DIFFUSE GALACTIC NEBULA
SG 2.042	20 10 37.	+ 38 15	1080		DIFFUSE EMISSION NEBULA
HN 0652	20 10 38.	- 71 09			NEBULA
IC 4962	20 10 40.	- 71 09			NONSTELLAR OBJECT
HN 1193	20 10 41.	- 54 35		12.	NEBULA
IC 4978	20 10 41.	- 54 35			NONSTELLAR OBJECT
HN 2310	20 10 43.1	- 44 46 41.			NEBULA
PATH 1.815	20 10 44.	- 14 38	14		NEBULA
HN 1194	20 10 45.	- 53 38	30		NEBULA
IC 4979	20 10 45.	- 53 38			NONSTELLAR OBJECT
CED 176B	20 10 46.	+ 40 07			DIFFUSE GALACTIC NEBULA
PK072+00.1	20 10 52.0	+ 34 11 29.			PLANETARY NEBULA
HN 2309	20 10 52.9	- 51 20 29.	12		NEBULA
MRSL 081+06/3	20 10 54.	+ 45 31	1668		HII REGION
HN 2308	20 10 56.0	- 54 19 41.	12		NEBULA
RNGC 6888	20 10 59.	+ 38 10			DIFFUSE NEBULA
LBN 0203	20 11	+ 38 10	900		BRIGHT NEBULA
LBN 0223	20 11	+ 39 50	720		BRIGHT NEBULA
LBN 0113	20 11	- 01 00	2700		BRIGHT NEBULA
LBN 0111	20 11	- 01 30	900		BRIGHT NEBULA
MCG+01-51-008	20 11 00.	+ 07 07 30.	60	15.	GALAXY
COU 047	20 11 00.	+ 34 45	2400		EMISSION NEBULA
LDN 0893	20 11 00.	+ 42 05	720		DARK NEBULA
1ZW 210	20 11 00.	+ 52 50			COMPACT GALAXY
LB 03609	20 11 01.	+ 26 24 48.		17.6	FAINT BLUE STAR
HN 0659	20 11 02.	- 62 01			NEBULA
IC 4974	20 11 02.	- 62 01			NONSTELLAR OBJECT
SCHO 1173	20 11 04.	+ 40 09 42.	520		ISOLATED DARK CLOUD
ZWG 398.022	20 11 06.	+ 07 07		15.3	GALAXY
UGC 11526	20 11 06.	+ 07 07	108	15.3	GALAXY Sa
ASS 19	20 11 06.	+ 40 49			OB ASSOCIATION CYG OB8
MCG+00-51-010	20 11 09.	- 01 21	96	13.	GALAXY
HN 0655	20 11 10.	- 70 43			NEBULA
IC 4967	20 11 10.	- 70 43			NONSTELLAR OBJECT
PATH 1.816	20 11 13.	- 14 54	11		NEBULA
HN 0660	20 11 14.	- 62 02			NEBULA
IC 4976	20 11 14.	- 62 02			NONSTELLAR OBJECT
PATH 1.817	20 11 15.	- 14 54	11		NEBULA
UGC 11528	20 11 18.	+ 55 25	66	16.5	GALAXY S
ZWG 372.016	20 11 18.	- 01 18		14.4	GALAXY
UGC 11527	20 11 18.	- 01 18	108	14.4	GALAXY S
SG 3.163	20 11 20.	+ 33 32	7200		DIFFUSE EMISSION NEBULA
HN 2311	20 11 20.3	- 54 32 34.	18		NEBULA
LB 03610	20 11 21.	+ 26 30 54.		18.2	FAINT BLUE STAR
HUB B25	20 11 23.	+ 37 50			DIFFUSE NEBULA
ZWG 372.017	20 11 24.	+ 01 07		15.7	GALAXY
HN 0662	20 11 28.	- 58 03			NEBULA
IC 4980	20 11 28.	- 58 03			NONSTELLAR OBJECT
HN 2312	20 11 33.7	- 52 24 51.	18		NEBULA
B 343	20 11 40.	+ 40 07	780		DARK OBJECT
HN 0654	20 11 40.	- 74 02			NEBULA
IC 4964	20 11 40.	- 74 02			NONSTELLAR OBJECT
RNGC 6872	20 11 41.	- 70 55			GALAXY
ZWG 372.018	20 11 42.	+ 03 04		15.2	GALAXY
ZWG 398.023	20 11 42.	+ 06 13		15.7	GALAXY
HN 0657	20 11 42.	- 70 55			NEBULA
VV 297A	20 11 42.	- 70 56			INTERACTING GALAXY
IC 4970	20 11 43.	- 70 55			NONSTELLAR OBJECT
HN 0658	20 11 46.	- 70 47			NEBULA
IC 4971	20 11 46.	- 70 47			NONSTELLAR OBJECT
SCHO 1174	20 11 48.	+ 35 23 18.	490		ISOLATED DARK CLOUD
ISS 0273	20 11 48.	+ 36 07	101		STELLAR RING
MRSL 083+07/2	20 11 43.	+ 47 25	666		HII REGION
ZWG 304.001	20 11 48.	+ 61 32		15.4	GALAXY
ZWG 303.018	20 11 48.	+ 61 32		15.4	GALAXY
MCG+10-28-017	20 11 48.	+ 61 32	42	15.	GALAXY
VV 297B	20 11 48.	- 70 55			INTERACTING GALAXY
MCG+11-24-005	20 11 54.	+ 66 04	78	16.	GALAXY
SER 137.03	20 11 54.	- 70 55	390	12.	INTERACTING GALAXIES
PK068-02.1	20 11 55.7	+ 29 24 48.	10		PLANETARY NEBULA
LBN 0175	20 12	+ 33 40	2040		BRIGHT NEBULA
LBN 0181	20 12	+ 34 50	540		BRIGHT NEBULA
LBN 0239	20 12	+ 41 10	2400		BRIGHT NEBULA
LBN 0331	20 12	+ 47 30	3660		BRIGHT NEBULA
KHAV 683	20 12	+ 48 15	4260		DARK NEBULA
SER 136.05	20 12	- 44 44	3600	15.	LOOSE GROUP OF 20 GALXIES
ISS 0362	20 12 00.	+ 23 11	174		STELLAR RING
COU 048	20 12 00.	+ 39 55	3000		EMISSION NEBULA
LDN 0880	20 12 00.	+ 40 10	480		DARK NEBULA
MRSL 078+03/2	20 12 00.	+ 41 02	1331		HII REGION
LDN 0891	20 12 00.	+ 41 50	420		DARK NEBULA
COU 050	20 12 00.	+ 43 30	1800		EMISSION NEBULA

OBJECT NAME	RIGHT ASCEN.	DECLINATION	DIAM.	MAGN.	TYPE OF OBJECT
COU 049	20 12 00.	+ 45 10	9000		EMISSION NEBULA
MCG-03-51-006	20 12 00.	- 18 11	48	15.5	GALAXY
FATH 1.818	20 12 01.	- 14 57	5		NEBULA
HN 2313	20 12 03.8	- 51 10 49.	18		NEBULA
HN 2314	20 12 05.5	- 51 09 49.	30		NEBULA
MCG-03-51-007	20 12 06.	- 20 10	36	15.	GALAXY
MCG-07-41-016	20 12 06.	- 40 10	60	15.	GALAXY
ZWG 398.024	20 12 12.	+ 06 22		15.6	GALAXY
MCG+01-51-009	20 12 12.	+ 06 27 30.	30	16.	GALAXY
MRSL 076+02/4	20 12 12.	+ 38 39	1133		HII REGION
OCL 0140	20 12 18.	+ 28 49	240	18.	OPEN STAR CLUSTER
MRSL 077+03/1	20 12 18.	+ 39 57	1536		HII REGION
MRSL 077+03/2	20 12 18.	+ 40 36	1133		HII REGION
MRSL 080+05/2	20 12 18.	+ 43 57	2268		HII REGION
MRSL 081+05/2	20 12 18.	+ 44 36	5598		HII REGION
SG 3.164	20 12 19.	+ 34 47	1800		DIFFUSE EMISSION NEBULA
HN 1195	20 12 22.	- 52 14	42		NEBULA
IC 4983	20 12 22.	- 52 14			NONSTELLAR OBJECT
HN 2316	20 12 23.1	- 46 18 17.			NEBULA
ISS 0225	20 12 24.	+ 44 52	184		STELLAR RING
ZC 2012.4+6008	20 12 24.	+ 60 08	1550		CLUSTER OF GALAXIES
HN 0661	20 12 25.	- 71 04			NEBULA
IC 4972	20 12 25.	- 71 04			NONSTELLAR OBJECT
LDN 0877	20 12 30.	+ 39 45	300		DARK NEBULA
MCG-07-41-017	20 12 30.	- 44 40	42	15.	GALAXY
HN 1196	20 12 30.	- 52 52	24		NEBULA
IC 4984	20 12 30.	- 52 52			NONSTELLAR OBJECT
HN 2315	20 12 34.2	- 53 03 23.	18		NEBULA
SCHO 1175	20 12 35.	+ 24 16 12.	300		ISOLATED DARK CLOUD
MRSL 079+04/1	20 12 36.	+ 42 38	666		HII REGION
ZC 2012.6-0232	20 12 36.	- 02 32	2290		CLUSTER OF GALAXIES
MCG-07-41-018	20 12 42.	- 44 35	48	15.5	GALAXY
HN 2318	20 12 42.0	- 46 45 10.	30		NEBULA
HN 2317	20 12 42.3	- 48 15 22.	18		NEBULA
PK054-12.1	20 12 47.65	+ 12 33 02.1	74	11.4	PLANETARY NEBULA
MRSL 083+07/1	20 12 48.	+ 47 27	1463		HII REGION
SER 136.06	20 12 48.	- 44 25	50	15.	TWO INTERACTING GALAXIES
RNGC 6891	20 12 49.	+ 12 33		11.5	PLANETARY NEBULA
MRSL 078+03/4	20 12 54.	+ 41 38	1001		HII REGION
UGC 11529	20 12 54.	+ 66 08	108	16.5	GALAXY Sb-c
MCG-02-51-004	20 12 54.	- 13 47	72	14.5	GALAXY
MCG-07-41-019	20 12 54.	- 44 28	36	14.	GALAXY
KHAV 684	20 13	+ 44 45	3680		DARK NEBULA
SER 135.02	20 13	- 46 45	198	14.5	PEC GLXY AND 2 SATELLITES
HN 2319	20 13 02.5	- 53 11 21.	12		NEBULA
HN 2321	20 13 03.2	- 47 07 15.			NEBULA
RNGC 6876	20 13 04.	- 71 01		13.0	GALAXY
HN 2320	20 13 04.3	- 53 43 45.	18		NEBULA
MRSL 081+06/1	20 13 06.	+ 45 35	3000		HII REGION
UGC 11530	20 13 06.	+ 73 32	66	17.	GALAXY DWRF SP
MCG-07-41-020	20 13 06.	- 44 52	36	16.	GALAXY
SER 135.01	20 13 06.	- 47 03	25	15.	RING GALAXY
MCG-01-51-001	20 13 12.	- 03 03	108	15.	GALAXY
SER 137.02	20 13 12.	- 71 01	2400	12.	LOOSE GROUP OF 15 GALXIES
HN 2325	20 13 16.1	- 46 41 20.	12		NEBULA
HN 2326	20 13 17.1	- 47 01 56.	18		NEBULA
SER 136.10	20 13 18.	- 46 38	126	13.	PEC. GLXY AND SATELLITE
HN 2322	20 13 18.4	- 52 49 32.	12		NEBULA
HN 0664	20 13 19.	- 55 11			NEBULA
IC 4986	20 13 19.	- 55 11			NONSTELLAR OBJECT
HN 2327	20 13 19.2	- 47 29 08.	18		NEBULA
HN 2323	20 13 22.1	- 52 18 44.	24		NEBULA
MIN.47 11	20 13 23.	+ 39 42			DIFFUSE NEBULA
RNGC 6877	20 13 23.	- 71 00			GALAXY
2ZW 080	20 13 24.	+ 06 56			COMPACT GALAXY
MCG-07-41-021	20 13 24.	- 44 51	48	16.	GALAXY
RNGC 6887	20 13 24.	- 52 56		12.5	GALAXY
FATH 1.819	20 13 26.	- 14 31	41		NEBULA
MCG-02-51-005	20 13 27.	- 14 32	48	15.	GALAXY
HN 2329	20 13 27.1	- 47 55 31.	48		NEBULA
HN 2324	20 13 29.2	- 54 06 08.	6		NEBULA
COU 051	20 13 30.	+ 37 40	1200		EMISSION NEBULA
MRSL 076+02/3	20 13 30.	+ 38 56	2933		HII REGION
HN 2328	20 13 34.1	- 53 26 19.	12		NEBULA
HN 1197	20 13 35.	- 52 27	12		NEBULA
IC 4987	20 13 35.	- 52 27			NONSTELLAR OBJECT
ACK 077+03.2	20 13 36.	+ 40 26			PLANETARY NEBULA
MCG-02-51-006	20 13 36.	- 11 53 30.	36	16.	GALAXY
SCHO 1176	20 13 39.	+ 28 08 36.	550		ISOLATED DARK CLOUD
MCG-07-41-022	20 13 42.	- 43 45	12	15.	GALAXY
HN 2330	20 13 47.3	- 53 18 55.			NEBULA
ISS 0320	20 13 48.	+ 31 39	262		STELLAR RING
OCL 0156	20 13 48.	+ 36 38	900		OPEN STAR CLUSTER
SER 136.08	20 13 48.	- 45 55	75	14.	PECULIAR GALAXY
SER 135.04	20 13 48.	- 48 10	50	16.	LOW SURF. BRGHTNSS GALAXY
SG 1.14	20 13 49.	+ 38 39			DIFFUSE EMISSION NEBULA
SG 2.043	20 13 49.	+ 38 44	2700		DIFFUSE EMISSION NEBULA
MRSL 080+04/3	20 13 54.	+ 43 14	666		HII REGION
SG 3.165	20 13 57.	+ 47 33	7200		DIFFUSE EMISSION NEBULA
LBN 0206	20 14	+ 38 00	2400		BRIGHT NEBULA
LBN 0215	20 14	+ 38 40	1800		BRIGHT NEBULA
LBN 0236	20 14	+ 40 30	4500		BRIGHT NEBULA
LBN 0447	20 14	+ 64 10	360		BRIGHT NEBULA
LDN 0824	20 14 00.	+ 29 15	1860		DARK NEBULA
COU 052	20 14 00.	+ 41 38	3600		EMISSION NEBULA
LDN 0892	20 14 00.	+ 41 40	900		DARK NEBULA
LDN 0894	20 14 00.	+ 41 45	240		DARK NEBULA
LDN 0895	20 14 00.	+ 42 30	9600		DARK NEBULA
UGC 11531	20 14 00.	+ 58 18	60	17.	GALAXY
SCHO 1177	20 14 04.	+ 27 21 42.	400		ISOLATED DARK CLOUD
MIL 83	20 14 04.	+ 37 03 48.	558		SUPERNOVA REMNANT
MCG+00-51-011	20 14 06.	+ 00 26	90	15.	GALAXY
MRSL 080+05/1	20 14 06.	+ 44 03	2268		HII REGION
MRSL 082+06/1	20 14 06.	+ 45 39	3210		HII REGION
SCHO 1178	20 14 10.	+ 24 32 18.	280		ISOLATED DARK CLOUD
ZWG 398.025	20 14 12.	+ 07 14		15.1	GALAXY
UGC 11532	20 14 12.	+ 07 14	60	15.1	GALAXY (S0)
CED 176C	20 14 14.	+ 41 39	1470		DIFFUSE GALACTIC NEBULA
MCG-02-51-007	20 14 15.	- 12 16	36	17.	GALAXY
HOLM 778B	20 14 16.	- 13 12	18	13.1	PART OF MULTIPLE GALAXY
HOLM 778A	20 14 17.	- 13 13	36	12.2	PART OF MULTIPLE GALAXY
RNGC 6880	20 14 17.	- 71 01			GALAXY
MRSL 075+01/1	20 14 18.	+ 37 52	1398		HII REGION
MRSL 076+02/1	20 14 18.	+ 38 35	1200		HII REGION
MRSL 078+03/3	20 14 18.	+ 41 50	666		HII REGION
MRSL 080+04/4	20 14 18.	+ 43 31	666		HII REGION
HN 2331	20 14 18.3	- 53 01 35.	12		NEBULA
MCG-02-51-008	20 14 21.	- 12 15	30	14.5	GALAXY
IC 4991	20 14 22.	- 41 44 18.			NONSTELLAR OBJECT
PK069-02.1	20 14 22.73	+ 30 24 32.3	150	14.4	PLANETARY NEBULA
UGC 11533	20 14 24.	+ 00 27	108	17.	GALAXY DWRF IR
ISS 0322	20 14 24.	+ 30 03	194		STELLAR RING
RNGC 6894	20 14 24.	+ 30 25		14.5	PLANETARY NEBULA
MRSL 082+05/3	20 14 24.	+ 45 31	2663		HII REGION
B 344	20 14 25.	+ 40 04	420		DARK OBJECT
SG 2.044	20 14 27.	+ 37 45	900		DIFFUSE EMISSION NEBULA
MCG-02-51-009	20 14 27.	- 12 01 30.	42	16.	GALAXY
SG 2.045	20 14 29.	+ 43 36	1500		DIFFUSE EMISSION NEBULA
HN 0663	20 14 29.	- 71 00			NEBULA
IC 4981	20 14 30.	- 71 00			NONSTELLAR OBJECT
RNGC 6892	20 14 34.	+ 17 52			NON-EXISTENT OBJECT
IC 1312	20 14 34.	+ 17 52			SAME AS NGC 6892
PK074+01.1	20 14 34.	+ 36 56 48.	10		PLANETARY NEBULA
ISS 0323	20 14 36.	+ 28 26	239		STELLAR RING
OCL 0158	20 14 36.	+ 37 29	840	7.2	OPEN STAR CLUSTER
IC 4996	20 14 36.	+ 37 29	360		OPEN CLUSTER
OCL 0159	20 14 36.	+ 37 43	780		OPEN STAR CLUSTER
MIL 88	20 14 37.	+ 41 45 06.	2280		SUPERNOVA REMNANT
MRSL 080+04/5	20 14 42.	+ 43 31	1398		HII REGION
MCG+01-51-010	20 14 45.	+ 05 05	36	15.5	GALAXY
RNGC 6890	20 14 46.	- 44 58		13.5	GALAXY
UGC 11534	20 14 48.	+ 05 08	72	16.0	GALAXY Sc-IRR
ISS 0321	20 14 48.	+ 33 14	159		STELLAR RING
MRSL 077+02/3	20 14 48.	+ 40 25	2801		HII REGION
SER 138.04	20 14 48.	- 52 37	40	17.	LOW SURF. BRGHTNSS GALAXY
HN 2332	20 14 49.1	- 54 09 33.	18		NEBULA
ZWG 398.026	20 14 54.	+ 05 05		15.5	GALAXY
UGC 11535	20 14 54.	+ 05 05	66	15.5	GALAXY SB:a-b
MRSL 080+04/6	20 14 54.	+ 43 37	666		HII REGION
MCG+00-51-012	20 14 54.	- 00 12	54	14.	GALAXY
MCG-07-41-023	20 14 54.	- 44 58	72	12.7	GALAXY
RNGC 6895	20 14 58.	+ 50 04			NON-EXISTENT OBJECT
LBN 0252	20 15	+ 41 40	2100		BRIGHT NEBULA
LBN 0310	20 15	+ 45 00	14400		BRIGHT NEBULA
LBN 0459	20 15	+ 67 10	8400		BRIGHT NEBULA
LDN 0818	20 15 00.	+ 26 00	12840		DARK NEBULA
COU 053	20 15 00.	+ 43 35	2400		EMISSION NEBULA
7ZW 929	20 15 00.	+ 79 15			COMPACT GALAXY
ZWG 357.010	20 15 00.	+ 79 15		15.3	GALAXY
UGC 11536	20 15 00.	+ 79 15	66	15.3	GALAXY SBa
KARA.73B 0875	20 15 00.	+ 79 15	54	15.3	ISOLATED GALAXY S
ZWG 372.019	20 15 00.	- 00 08		15.1	GALAXY
SER 136.07	20 15 00.	- 45 42	162	17.	RING GLXY AND ELLIPT GLXY
HN 2333	20 15 00.4	- 52 50 44.			NEBULA
SG 2.046	20 15 05.	+ 40 12	5400		DIFFUSE EMISSION NEBULA
RNGC 6889	20 15 05.	- 54 07			UNVERIFIED SOUTHERN OBJECT
SG 3.166	20 15 06.	+ 43 00	900		DIFFUSE EMISSION NEBULA
MCG-07-41-024	20 15 06.	- 41 13	90	12.5	GALAXY
HN 0665	20 15 06.	- 71 10			NEBULA
IC 4982	20 15 07.	- 71 10			NONSTELLAR OBJECT
ZC 2015.2+5915	20 15 12.	+ 59 15	2350		CLUSTER OF GALAXIES
MCG-07-41-025	20 15 12.	- 39 30	60	14.5	GALAXY
PK065-05.1	20 15 13.	+ 25 12 56.	42		PLANETARY NEBULA
HN 2334	20 15 14.7	- 53 03 43.	12		NEBULA
ISS 0324	20 15 18.	+ 28 32	172		STELLAR RING
7ZW 930	20 15 18.	+ 78 47			COMPACT GALAXY
MCG-07-41-026	20 15 18.	- 41 13	78	15.	GALAXY
HN 2335	20 15 19.5	- 53 34 13.	18		NEBULA
HN 2336	20 15 19.7	- 51 01 07.	6		NEBULA
SG 2.047	20 15 20.	+ 41 47	2160		DIFFUSE EMISSION NEBULA
IC 1315	20 15 21.	+ 30 32			NONSTELLAR OBJECT
LB 00346	20 15 22.	- 12 53 00.		15.0	FAINT BLUE STAR
ACK 076+01.1	20 15 24.	+ 38 41			PLANETARY NEBULA
HN 1198	20 15 24.	- 58 43	48		NEBULA
IC 4989	20 15 24.	- 58 43			NONSTELLAR OBJECT
MIN.48 06	20 15 25.	+ 38 57			DIFFUSE NEBULA
YC 2015-39	20 15 28.	- 39 32 06.			UNUSUAL SOUTHERN GALAXY
MRSL 078+03/1	20 15 30.	+ 41 42	2736		HII REGION
MRSL 080+04/2	20 15 30.	+ 43 02	1001		HII REGION
MRSL 080+04/7	20 15 30.	+ 43 58	336		HII REGION
MRSL 083+06/1	20 15 30.	+ 47 39	1068		HII REGION
HN 0666	20 15 30.	- 71 09			NEBULA
IC 4985	20 15 30.	- 71 09			NONSTELLAR OBJECT
IC 1314	20 15 32.	+ 25 02			NONSTELLAR OBJECT
SG 2.048	20 15 35.	+ 36 33	360		DIFFUSE EMISSION NEBULA
URA 47	20 15 36.	+ 39 26	66		STELLAR RING
MCG-07-41-027	20 15 36.	- 41 26	30	15.5	GALAXY
VDB.66N 130	20 15 44.	+ 39 12	36		REFLECTION NEBULA
HN 2337	20 15 44.0	- 53 39 42.	12		NEBULA
MRSL 074+00/1	20 15 48.	+ 36 37	336		HII REGION
MCG-03-51-008	20 15 48.	- 17 07	60	14.5	GALAXY
MIL 82	20 15 49.	+ 36 36 06.	282		SUPERNOVA REMNANT
IC 1313	20 15 53.	- 17 06 23.			NONSTELLAR OBJECT
ZWG 372.020	20 15 54.	+ 02 09		15.7	GALAXY
OCL 0157	20 15 54.	+ 36 31	420		OPEN STAR CLUSTER
ASS 18	20 15 54.	+ 37 29	46800		OB ASSOCIATION CYG OB1
HN 1299	20 15 55.	- 53 37		15.	NEBULA
IC 4994	20 15 56.	- 53 37			NONSTELLAR OBJECT
MCG+00-51-013	20 15 57.	- 00 21	144	13.	GALAXY
RNGC 6896	20 15 59.	+ 30 29			NON-EXISTENT OBJECT
LBN 0195	20 16	+ 36 40	420		BRIGHT NEBULA
LBN 0209	20 16	+ 37 50	420		BRIGHT NEBULA
LBN 0301	20 16	+ 44 30	900		BRIGHT NEBULA
LDN 0827	20 16 00.	+ 29 33	900		DARK NEBULA
LDN 0881	20 16 00.	+ 39 40	240		DARK NEBULA
MCG-04-48-001	20 16 00.	- 22 20	42	14.5	GALAXY
ISS 0274	20 16 06.	+ 37 50	677		STELLAR RING
MRSL 081+05/1	20 16 06.	+ 44 25	930		HII REGION
ZWG 372.021	20 16 06.	- 00 18		14.8	GALAXY
UGC 11537	20 16 06.	- 00 18	138	14.8	GALAXY Sc
MCG-07-41-028	20 16 06.	- 41 30	72	14.	GALAXY
MCG-07-41-029	20 16 12.	- 40 54	66	13.5	GALAXY
HN 1200	20 16 17.	- 52 47		14.	NEBULA
MRSL 076+01/3	20 16 18.	+ 38 50	533		HII REGION
MRSL 076+02/2	20 16 18.	+ 39 06	1331		HII REGION
MRSL 077+02/2	20 16 18.	+ 39 33	1463		HII REGION
OCL 0177	20 16 18.	+ 40 34	270	5.4	OPEN STAR CLUSTER
MRSL 082+05/1	20 16 18.	+ 45 15	533		HII REGION
IC 4995	20 16 18.	- 52 47			NONSTELLAR OBJECT
OCL 0162	20 16 24.	+ 37 41	720		OPEN STAR CLUSTER
MCG-05-48-001	20 16 24.	- 31 46	36	15.5	GALAXY
MCG-07-41-030	20 16 24.	- 39 27	78	15.	GALAXY
HOLM 779B	20 16 26.	+ 11 22	36	14.7	PART OF MULTIPLE GALAXY
HOLM 779A	20 16 35.	+ 11 23	60	14.5	PART OF MULTIPLE GALAXY
ZC 2016.6+1457	20 16 36.	+ 14 57	1280		CLUSTER OF GALAXIES
MRSL 081+04/5	20 16 36.	+ 44 07	336		HII REGION

OBJECT NAME	RIGHT ASCEN.	DECLINATION	DIAM.	MAGN.	TYPE OF OBJECT
MCG-03-51-009	20 16 36.	- 18 19 30.	42	15.5	GALAXY
HN 2339	20 16 40.6	- 51 56 50.	24		NEBULA
ZWG 447.001	20 16 42.	+ 17 42		15.3	GALAXY
SER 138.02	20 16 42.	- 51 56	60	16.5	LOW SURF. BRGHTNSS GALAXY
HN 2338	20 16 44.0	- 53 39 14.	12		NEBULA
HN 0668	20 16 46.	- 67 04			NEBULA
IC 4990	20 16 46.	- 67 04			NONSTELLAR OBJECT
HN 0667	20 16 46.	- 69 33			NEBULA
IC 4988	20 16 46.	- 69 33			NONSTELLAR OBJECT
HN 2340	20 16 48.5	- 52 19 56.	12		NEBULA
MRSL 078+02/4	20 16 54.	+ 40 40	1200		HII REGION
MRSL 082+05/4	20 16 54.	+ 46 03	1866		HII REGION
KHAV 686	20 17	+ 36 57	15160		DARK NEBULA
KHAV 685	20 17	+ 40 15	6230		DARK NEBULA
LBN 0259	20 17	+ 42 00	2400		BRIGHT NEBULA
LBN 0238	20 17	+ 43 00	1500		BRIGHT NEBULA
LBN 0325	20 17	+ 46 10	1020		BRIGHT NEBULA
OCL 0160	20 17 00.	+ 37 33	480	15.	OPEN STAR CLUSTER
LDN 0886	20 17 00.	+ 39 50	4140		DARK NEBULA
COU 055	20 17 00.	+ 40 25	2100		EMISSION NEBULA
LDN 0888	20 17 00.	+ 40 30	8520		DARK NEBULA
COU 054	20 17 00.	+ 46 00	2400		EMISSION NEBULA
SG 3.167	20 17 00.	+ 46 03	7200		DIFFUSE EMISSION NEBULA
MRSL 082+05/2	20 17 06.	+ 45 38	1133		HII REGION
HN 2341	20 17 07.5	- 52 50 18.	12		NEBULA
RNGC 6893	20 17 12.	- 48 25		12.5	GALAXY
SER 138.05	20 17 12.	- 52 58	3600	15.	CLOUD OF 100 FAINT GLXIES
MIL 90	20 17 15.	+ 45 24 36.	4800		SUPERNOVA REMNANT
HN 0669	20 17 16.	- 67 10			NEBULA
IC 4993	20 17 16.	- 67 10			NONSTELLAR OBJECT
HN 2342	20 17 19.2	- 52 59 42.	12		NEBULA
OCL 0163	20 17 24.	+ 37 35	660		OPEN STAR CLUSTER
UGC 11538	20 17 24.	+ 54 15	60	17.	GALAXY
SG 3.168	20 17 27.	+ 37 47	2400		DIFFUSE EMISSION NEBULA
MRSL 076+01/1	20 17 30.	+ 37 58	533		HII REGION
MRSL 077+01/6	20 17 30.	+ 39 19	1931		HII REGION
LDN 0882	20 17 30.	+ 39 30	180		DARK NEBULA
MRSL 083+05/5	20 17 30.	+ 46 20	1463		HII REGION
HN 2343	20 17 30.7	- 53 55 17.			NEBULA
HN 2344	20 17 31.9	- 52 37 59.	12		NEBULA
PK066-05.1	20 17 32.2	+ 26 50 44.	5		PLANETARY NEBULA
MCG+10-29-001	20 17 36.	+ 62 32	18	15.	GALAXY
ZWG 373.001	20 17 42.	+ 02 33		15.6	GALAXY
KARA.73B 0876	20 17 42.	+ 02 33	24	15.6	ISOLATED GALAXY IR
ZWG 399.001	20 17 42.	+ 07 37		15.6	GALAXY
ZWG 424.001	20 17 42.	+ 15 14		15.3	GALAXY
MRSL 078+02/3	20 17 42.	+ 40 40	533		HII REGION
MRSL 080+04/1	20 17 42.	+ 43 10	731		HII REGION
MRSL 081+04/4	20 17 42.	+ 44 12	858		HII REGION
MCG+10-29-002	20 17 42.	+ 59 19	36	16.	GALAXY
ZWG 304.002	20 17 42.	+ 62 31		15.2	GALAXY
UGC 11539	20 17 42.	+ 62 31	72	15.2	GALAXY (E)
7ZW 931	20 17 42.	+ 62 32			COMPACT GALAXY
MCG+01-52-001	20 17 45.	+ 07 39	36	16.	GALAXY
SG 2.049	20 17 46.	+ 40 37	1800		DIFFUSE EMISSION NEBULA
MRSL 084+06/2	20 17 48.	+ 47 35	1200		HII REGION
MRSL 084+06/3	20 17 48.	+ 48 00	2663		HII REGION
MCG-04-48-002	20 17 48.	- 24 19	72	14.	GALAXY
SG 2.050	20 17 49.	+ 39 04	3000		DIFFUSE EMISSION NEBULA
IC 4997	20 17 51.0	+ 16 34 27.	2	11.4	PLANETARY NEBULA
HN 0078	20 17 53.	+ 16 34			NEBULA
PK058-10.1	20 17 53.	+ 16 35	2	11.4	PLANETARY NEBULA
OCL 0164	20 17 54.	+ 37 58	660		OPEN STAR CLUSTER
LBN 0212	20 18	+ 37 50	1500		BRIGHT NEBULA
LBN 0241	20 18	+ 40 30	1800		BRIGHT NEBULA
LBN 0240	20 18	+ 40 30	5400		BRIGHT NEBULA
LBN 0298	20 18	+ 44 10	300		BRIGHT NEBULA
LBN 0303	20 18	+ 44 20	1500		BRIGHT NEBULA
LBN 0326	20 18	+ 46 20	2700		BRIGHT NEBULA
LBN 0444	20 18	+ 63 40	600		BRIGHT NEBULA
LDN 0842	20 18 00.	+ 31 50	480		DARK NEBULA
COU 056	20 18 00.	+ 39 00	3600		EMISSION NEBULA
LDN 0883	20 18 00.	+ 39 30	180		DARK NEBULA
ISS 0227	20 18 00.	+ 39 49	338		STELLAR RING
LDN 0909	20 18 00.	+ 45 00	7920		DARK NEBULA
LDN 0908	20 18 00.	+ 45 00	2400		DARK NEBULA
LDN 0910	20 18 00.	+ 45 10	420		DARK NEBULA
ZWG 370.014	20 18 00.	+ 88 28		15.7	GALAXY
ZWG 370.008	20 18 00.	+ 88 28		15.7	GALAXY
ZWG 373.003	20 18 00.	- 00 50		14.9	GALAXY
ZWG 373.002	20 18 00.	- 00 53		14.8	GALAXY
PK076+01.1	20 18 06.	+ 38 15	25	20.1	PLANETARY NEBULA
IC 4992	20 18 10.	- 71 43			NONSTELLAR OBJECT
HN 0670	20 18 10.	- 71 44			NEBULA
MCG-02-52-001	20 18 12.	- 12 26	48	15.	GALAXY
HN 2346	20 18 12.4	- 49 51 38.			NEBULA
RNGC 6897	20 18 13.	- 12 26		15.0	GALAXY
MIL 85	20 18 16.	+ 40 36 30.	798		SUPERNOVA REMNANT
HN 2345	20 18 17.3	- 51 57 26.	18		NEBULA
MRSL 076+01/2	20 18 18.	+ 38 55	863		HII REGION
MRSL 081+04/6	20 18 18.	+ 44 47	1200		HII REGION
MRSL 077+01/5	20 18 24.	+ 39 09	1200		HII REGION
MRSL 078+02/2	20 18 24.	+ 40 26	1200		HII REGION
MCG-02-52-002	20 18 24.	- 12 31	60	14.5	GALAXY
MCG-05-48-002	20 18 24.	- 27 29	24	15.	GALAXY
RNGC 6898	20 18 25.	- 12 31		14.0	GALAXY
HN 2349	20 18 27.5	- 51 52 13.	30		NEBULA
HN 2347	20 18 29.7	- 52 56 01.	18		NEBULA
HN 2348	20 18 31.3	- 53 20 31.	12		NEBULA
HN 2350	20 18 31.5	- 53 08 13.	18		NEBULA
MCG-04-48-003	20 18 33.	- 22 18 30.	54	15.	GALAXY
OCL 0167	20 18 36.	+ 38 32	480	13.	OPEN STAR CLUSTER
MCG+11-24-006	20 18 36.	+ 66 32	51	15.	GALAXY
MCG-05-48-003	20 18 36.	- 31 27	48	15.5	GALAXY
MCG+02-52-001	20 18 42.	+ 12 30	36	16.	GALAXY
LDN 0843	20 18 42.	+ 31 46	300		DARK NEBULA
OCL 0171	20 18 42.	+ 39 13	420		OPEN STAR CLUSTER
HN 2351	20 18 44.9	- 53 25 37.	42		NEBULA
SG 3.170	20 18 46.	+ 41 11	1440		DIFFUSE EMISSION NEBULA
MRSL 075+00/2	20 18 48.	+ 36 54	798		HII REGION
MRSL 077+02/1	20 18 48.	+ 39 48	1331		HII REGION
MRSL 079+03/3	20 18 48.	+ 42 21	600		HII REGION
7ZW 932	20 18 48.	+ 79 33			COMPACT GALAXY
HN 2352	20 18 51.7	- 50 55 00.	18		NEBULA
SG 3.169	20 18 52.	+ 36 51	1200		DIFFUSE EMISSION NEBULA
MCG-02-52-003	20 18 57.	- 10 38	96	14.5	GALAXY
LBN 0248	20 19	+ 41 00	1800		BRIGHT NEBULA
LBN 0314	20 19	+ 44 50	1200		BRIGHT NEBULA
LDN 0879	20 19 00.	+ 39 10	240		DARK NEBULA
OCL 0182	20 19 00.	+ 41 13	600		OPEN STAR CLUSTER
MRSL 081+04/2	20 19 00.	+ 43 55	1800		HII REGION
RNGC 6900	20 19 00.	- 02 44		14.0	GALAXY
MCG+00-52-001	20 19 00.	- 02 44	42	14.	GALAXY
AGU 62	20 19 00.	- 39 25 00.	90	12.5	PECULIAR GALAXY
MCG-03-52-001	20 19 03.	- 18 33	24	15.	GALAXY
KARA.73 49	20 19 03.	- 44 30	27		DWARF GALAXY
PK071-02.1	20 19 04.8	+ 32 19 47.	5		PLANETARY NEBULA
ZWG 373.004	20 19 06.	+ 03 27		15.7	GALAXY
MCG-05-48-004	20 19 06.	- 30 02	24	16.	GALAXY
MCG-07-41-031	20 19 06.	- 42 04	72	14.	GALAXY
SER 136.09	20 19 06.	- 46 55	90		LOW SURF. BRGHTNSS GALAXY
MCG+00-52-002	20 19 09.	- 00 55 30.	42	16.	GALAXY
RNGC 6911	20 19 11.	+ 66 35		15.5	GALAXY
ISS 0326	20 19 12.	+ 27 28	86		STELLAR RING
ISS 0325	20 19 12.	+ 29 48	85		STELLAR RING
MRSL 076+00/1	20 19 12.	+ 37 41	336		HII REGION
MRSL 079+03/1	20 19 12.	+ 41 50	401		HII REGION
MRSL 079+03/2	20 19 12.	+ 41 57	468		HII REGION
MRSL 080+-3/5	20 19 12.	+ 42 35	600		HII REGION
MRSL 081+04/3	20 19 12.	+ 44 15	1133		HII REGION
ZWG 325.001	20 19 12.	+ 66 35		15.7	GALAXY
ZWG 324.007	20 19 12.	+ 66 35		15.7	GALAXY
UGC 11540	20 19 12.	+ 66 35	132	15.7	GALAXY SBb
MCG-01-52-001	20 19 12.	- 02 43	36	15.	GALAXY
HN 2354	20 19 16.6	- 49 40 04.			NEBULA
MRSL 077+01/3	20 19 18.	+ 38 54	533		HII REGION
MRSL 078+02/5	20 19 18.	+ 41 13	1463		HII REGION
MCG-03-52-002	20 19 18.	- 16 56	120	15.	GALAXY
B 345	20 19 24.	+ 46 24	900		DARK OBJECT
MCG-05-48-005	20 19 24.	- 30 00	24	15.5	GALAXY
DV.56 N6902A	20 19 24.	- 44 27	78		GALAXY
HN 2353	20 19 24.0	- 54 05 22.	18		NEBULA
SER 136.04	20 19 30.	- 44 25	108	15.5	PEC. GLXY AND COMPANION
MRSL 075+00/1	20 19 36.	+ 36 38	533		HII REGION
MRSL 080+03/6	20 19 36.	+ 42 46	666		HII REGION
ISS 0226	20 19 36.	+ 45 19	133		STELLAR RING
MCG-05-48-006	20 19 36.	- 31 44	30	15.5	GALAXY
RNGC 6904	20 19 38.	+ 25 35			NON-EXISTENT OBJECT
HN 2355	20 19 38.5	- 52 14 27.	18		NEBULA
RNGC 6902A	20 19 40.	- 44 27			GALAXY
SCHO 1179	20 19 42.	+ 28 57 18.	410		ISOLATED DARK CLOUD
MCG-05-48-007	20 19 42.	- 31 42	36	15.	GALAXY
MCG-07-41-032	20 19 42.	- 44 27	48	15.	GALAXY
HN 2356	20 19 45.2	- 50 42 27.	6		NEBULA
RNGC 6902B	20 19 47.	- 44 03			GALAXY
UGC 11541	20 19 48.	+ 00 08	114	16.5	GALAXY SBc
MCG+01-52-002	20 19 48.	+ 06 17	66	14.5	GALAXY
OCL 0161	20 19 48.	+ 37 12	720	13.	OPEN STAR CLUSTER
LDN 0901	20 19 48.	+ 43 10	300		DARK NEBULA
MCG-07-41-033	20 19 48.	- 44 02	36	15.	GALAXY
MCG+00-52-003	20 19 51.	+ 00 08 30.	78	16.	GALAXY
HN 2357	20 19 51.3	- 51 18 20.	6		NEBULA
RNGC 6901	20 19 52.	+ 06 17		15.0	GALAXY
IC 5000	20 19 53.9	+ 06 16 10.	120		GALAXY SB(r)
ZWG 399.002	20 19 54.	+ 06 16		14.9	GALAXY
UGC 11542	20 19 54.	+ 06 16	72	14.9	GALAXY SBb
OCL 0175	20 19 54.	+ 39 47	1200		OPEN STAR CLUSTER
PK069-03.1	20 19 56.0	+ 29 49 46.	19		PLANETARY NEBULA
HN 2358	20 19 56.7	- 50 56 26.	12		NEBULA
IC 1316	20 19 59.	+ 06 22			SAME AS IC 5000
LBN 0234	20 20	+ 39 30	5100		BRIGHT NEBULA
LBN 0336	20 20	+ 47 00	2400		BRIGHT NEBULA
LBN 0345	20 20	+ 47 30	6000		BRIGHT NEBULA
LB 09998	20 20	- 88 30		14.4	FAINT BLUE STAR
ISS 0327	20 20 00.	+ 27 37	130		STELLAR RING
MRSL 077+01/4	20 20 00.	+ 39 22	936		HII REGION
SG 2.051	20 20 00.	+ 39 39	7200		DIFFUSE EMISSION NEBULA
LDN 0902	20 20 00.	+ 43 10	120		DARK NEBULA
MRSL 080+03/8	20 20 00.	+ 43 10	731		HII REGION
COU 057	20 20 00.	+ 44 20	18000		EMISSION NEBULA
OCL 0195	20 20 00.	+ 47 55	180	15.	OPEN STAR CLUSTER
2ZW 081	20 20 06.	+ 10 03			COMPACT GALAXY
MRSL 083+05/2	20 20 06.	+ 46 04	666		HII REGION
PK061-09.1	20 20 09.27	+ 19 56 34.8	100	11.9	PLANETARY NEBULA
RNGC 6905	20 20 11.	+ 19 57		12.0	PLANETARY NEBULA
HN 2359	20 20 11.4	- 49 48 49.	18		NEBULA
ZWG 399.003	20 20 12.	+ 03 46		15.7	GALAXY
ZWG 399.004	20 20 12.	+ 09 25		15.4	GALAXY
UGC 11543	20 20 12.	+ 09 25	66	15.4	GALAXY Sb
ZWG 424.002	20 20 12.	+ 09 59		15.0	GALAXY
MCG-05-48-008	20 20 12.	- 31 41	24	15.5	GALAXY
MCG+02-52-002	20 20 18.	+ 09 26	48	16.	GALAXY
MRSL 077+01/1	20 20 18.	+ 38 56	863		HII REGION
MRSL 083+05/3	20 20 18.	+ 46 22	1398		HII REGION
MRSL 083+05/4	20 20 18.	+ 47 00	1001		HII REGION
ZWG 282.003	20 20 18.	+ 56 05		15.7	GALAXY
MCG-05-48-009	20 20 18.	- 27 52	30	14.	GALAXY
HN 2360	20 20 20.0	- 52 11 07.	18		NEBULA
UGC 11544	20 20 24.	+ 07 58	60	16.5	GALAXY Sc
UGC 11545	20 20 30.	+ 52 15	78	16.5	GALAXY S
MCG+10-29-003	20 20 30.	+ 59 24 30.	36	15.	GALAXY
ZWG 304.003	20 20 30.	+ 59 35		15.7	GALAXY
MCG-07-42-001	20 20 30.	- 44 09	90	12.3	GALAXY
BC PKS2020-370	20 20 31.99	- 37 05 02.8		17.46	QUASI-STELLAR OBJECT
SHB 352	20 20 32.0	- 37 05 02.		17.5	QUASI-STELLAR OBJECT
HN 2361	20 20 32.2	- 50 32 18.			NEBULA
HBLW 481	20 20 33.	- 44 09 29.			NEBULA
KLEM 31	20 20 35.	- 37 06	480	14.	GROUP OF 6 GALAXIES
ZWG 399.005	20 20 36.	+ 09 22		15.7	GALAXY
MCG-05-48-010	20 20 36.	- 28 26	90	13.	GALAXY
SA 5	20 20 36.	- 44 10		12.3	GALAXY
VMT 21	20 20 38.	+ 40 03 24.	180		SUPERNOVA REMNANT
HN 2362	20 20 41.6	- 51 38 23.	24		NEBULA
2ZW 082	20 20 42.	+ 00 30			COMPACT GALAXY
ZWG 373.005	20 20 42.	+ 00 30		14.5	GALAXY
UGC 11546	20 20 42.	+ 00 30	48	14.5	GALAXY COMPACT
MCG+00-52-004	20 20 42.	+ 00 30	24	14.5	GALAXY
ISS 0328	20 20 42.	+ 29 14	146		STELLAR RING
MRSL 076+00/2	20 20 42.	+ 38 30	1800		HII REGION
IC 4999	20 20 42.	- 26 10 16.			NONSTELLAR OBJECT
IC 1317	20 20 42.3	+ 00 30 12.			GALAXY SB0
HN 2363	20 20 42.5	- 49 50 11.		10.	NEBULA
IC 4998	20 20 43.	- 38 24 24.			NONSTELLAR OBJECT
MIL 84	20 20 44.	+ 40 02 18.	1800		SUPERNOVA REMNANT
RNGC 6903	20 20 46.	- 19 29		13.0	GALAXY
RNGC 6899	20 20 46.	- 50 35			UNVERIFIED SOUTHERN OBJECT

OBJECT NAME	RIGHT ASCEN.	DECLINATION	DIAM.	MAGN.	TYPE OF OBJECT
SCHO 1180	20 20 48.	+ 34 12 30.	570		ISOLATED DARK CLOUD
ASS 17	20 20 48.	+ 35 41			OB ASSOCIATION CYG OB3
MRSL 078+01/7	20 20 48.	+ 40 03	336		HII REGION
MRSL 080+03/9	20 20 48.	+ 43 20	798		HII REGION
MCG-04-48-004	20 20 48.	- 26 11	102	13.5	GALAXY
SER 136.02	20 20 48.	- 43 39	402	13.	HIGH SURF. BRGHTNSS GLXY
HN 2365	20 20 50.2	- 50 33 05.			NEBULA
MCG-03-52-003	20 20 51.	- 19 29	30	13.5	GALAXY
HN 2364	20 20 52.9	- 53 00 17.			NEBULA
MRSL 079+02/4	20 20 54.	+ 41 30	1266		HII REGION
MRSL 082+04/5	20 20 54.	+ 45 03	600		HII REGION
LBN 0300	20 21	+ 43 50	2100		BRIGHT NEBULA
LBN 0334	20 21	+ 46 50	420		BRIGHT NEBULA
LBN 0370	20 21	+ 49 10	3600		BRIGHT NEBULA
MRSL 080+03/1	20 21 00.	+ 42 15	1068		HII REGION
MRSL 080+03/4	20 21 00.	+ 42 42	1001		HII REGION
MRSL 080+03/7	20 21 00.	+ 43 02	1001		HII REGION
UGC 11547	20 21 00.	+ 60 35	60	16.0	GALAXY SB
MCG-07-42-002	20 21 00.	- 43 49	72	12.4	GALAXY
SG 2.052	20 21 02.	+ 38 24	1440		DIFFUSE EMISSION NEBULA
RNGC 6906	20 21 04.	+ 06 17		13.5	GALAXY
ZWG 399.006	20 21 06.	+ 06 17		13.7	GALAXY
UGC 11548	20 21 06.	+ 06 17	96	13.7	GALAXY Sb
ZWG 447.002	20 21 06.	+ 15 47		15.7	GALAXY
MCG-07-42-003	20 21 06.	- 40 32	42	15.	GALAXY
RNGC 6902	20 21 11.	- 43 50		12.5	GALAXY
MCG+01-52-003	20 21 12.	+ 06 17	96	12.	GALAXY
MRSL 078+01/5	20 21 12.	+ 39 55	1001		HII REGION
MCG-04-48-005	20 21 12.	- 23 38 30.	48	15.	GALAXY
IC 5006	20 21 17.	+ 06 18 33.			NONSTELLAR OBJECT
OCL 0181	20 21 18.	+ 40 37	2340	7.7	OPEN STAR CLUSTER
MCG-03-52-004	20 21 18.	- 20 20	72	15.5	GALAXY
RNGC 6910	20 21 19.	+ 40 37		7.5	OPEN CLUSTER
SG 2.053	20 21 23.	+ 40 39	1800		DIFFUSE EMISSION NEBULA
MRSL 078+02/1	20 21 24.	+ 40 38	401		HII REGION
MCG-07-42-004	20 21 24.	- 41 41	120	15.	GALAXY
ARC 2324	20 21 25.	- 20 29		16.8	RICH CLUSTER OF GALAXIES
ASS 20	20 21 30.	+ 39 46			OB ASSOCIATION CYG OB9
MRSL 078+01/8	20 21 30.	+ 40 18	533		HII REGION
OCL 0184	20 21 30.	+ 41 32	360	10.2	OPEN STAR CLUSTER
MRSL 082+04/2	20 21 30.	+ 44 52	533		HII REGION
MCG-02-52-004	20 21 30.	- 11 14	72	14.5	GALAXY
DV.56 N6902B	20 21 36.	- 44 02	84		S GALAXY
HN 2366	20 21 36.9	- 53 30 56.	6		NEBULA
ZWG 447.003	20 21 42.	+ 15 35		15.0	GALAXY
MRSL 082+04/4	20 21 42.	+ 45 02	401		HII REGION
HN 2367	20 21 44.8	- 51 41 44.	24		NEBULA
MRSL 081+03/3	20 21 48.	+ 43 48	1331		HII REGION
ZWG 399.007	20 21 54.	+ 07 00		15.3	GALAXY
UGC 11549	20 21 54.	+ 07 00	66	15.3	GALAXY Sb-c
LDN 0837	20 21 54.	+ 30 37	360		DARK NEBULA
MRSL 080+03/2	20 21 54.	+ 42 37	863		HII REGION
ISS 0228	20 21 54.	+ 43 57	210		STELLAR RING
SG 2.171	20 21 54.	+ 49 19	7200		DIFFUSE EMISSION NEBULA
MCG-02-52-005	20 21 54.	- 12 09 30.	72	16.	GALAXY
SG 2.054	20 21 56.	+ 34 50	7200		DIFFUSE EMISSION NEBULA
MCG+01-52-004	20 21 57.	+ 07 02	60	15.	GALAXY
LBN 0185	20 22	+ 34 10	3600		BRIGHT NEBULA
LBN 0270	20 22	+ 42 00	8100		BRIGHT NEBULA
ZWG 424.003	20 22 00.	+ 12 25		15.7	GALAXY
MCG+02-52-003	20 22 00.	+ 12 25	60	15.5	GALAXY
LDN 0838	20 22 00.	+ 30 37	360		DARK NEBULA
ISS 0275	20 22 00.	+ 35 52	259		STELLAR RING
MRSL 077+01/2	20 22 00.	+ 39 37	2268		HII REGION
LDN 0899	20 22 00.	+ 42 30	540		DARK NEBULA
MCG-04-48-006	20 22 00.	- 25 00	180	12.	GALAXY
RNGC 6913	20 22 05.	+ 38 22		9.0	OPEN CLUSTER
UGC 11550	20 22 06.	+ 12 16	84	16.5	GALAXY Sc
OCL 0168	20 22 06.	+ 38 22	1920	9.0	OPEN STAR CLUSTER
IC 5003	20 22 06.	- 30 02 14.			NONSTELLAR OBJECT
RNGC 6907	20 22 07.	- 24 58		12.0	GALAXY
MCG+01-52-005	20 22 09.	+ 06 45	96	14.	GALAXY
IC 5005	20 22 09.	- 25 59 19.			NONSTELLAR OBJECT
ZWG 399.008	20 22 12.	+ 06 45		15.1	GALAXY
UGC 11551	20 22 12.	+ 06 45	102	15.1	GALAXY Sb-c
ZWG 424.004	20 22 12.	+ 12 17		15.4	GALAXY
UGC 11552	20 22 12.	+ 12 17	126	15.4	GALAXY Sa-b
MCG+02-52-004	20 22 12.	+ 12 17	108	15.	GALAXY
MRSL 081+04/1	20 22 12.	+ 44 30	936		HII REGION
RNGC 6908	20 22 13.	- 24 58			NON-EXISTENT OBJECT
MCG+06-45-001	20 22 15.	+ 35 42	48	15.5	GALAXY
IC 5004	20 22 16.	- 31 02 01.			NONSTELLAR OBJECT
MCG+00-52-006	20 22 18.	+ 00 43	24	16.	GALAXY
UGC 11553	20 22 18.	+ 01 16	90	16.0	GALAXY Sc
MCG+00-52-005	20 22 18.	+ 01 16	66	15.5	GALAXY
HN 2368	20 22 20.5	- 51 44 47.	12		NEBULA
RNGC 6916	20 22 21.	+ 58 12		15.5	GALAXY
MCG-04-48-007	20 22 21.	- 26 00	90	13.5	GALAXY
SG 2.055	20 22 23.	+ 40 43	7200		DIFFUSE EMISSION NEBULA
DG 162	20 22 24.	+ 42 07	180		REFLECTION NEBULA
MCG+10-29-004	20 22 24.	+ 58 10	96	14.	GALAXY
ZWG 304.004	20 22 24.	+ 58 12		15.3	GALAXY
UGC 11554	20 22 24.	+ 58 12	120	15.3	GALAXY SBb-c
MCG-05-48-011	20 22 24.	- 22 44	78	15.	GALAXY
HN 2369	20 22 26.2	- 50 42 59.	12		NEBULA
VDB.66N 131	20 22 28.	+ 42 05	336		REFLECTION NEBULA
MRSL 078+01/9	20 22 30.	+ 40 32	1068		HII REGION
IC 5007	20 22 30.	- 29 52 11.			NONSTELLAR OBJECT
CED 176D	20 22 33.	+ 42 08	360		DIFFUSE GALACTIC NEBULA
HN 1201	20 22 34.	- 54 55		15.	NEBULA
IC 5001	20 22 34.	- 54 55			NONSTELLAR OBJECT
HN 2370	20 22 34.3	- 52 33 59.		10.	NEBULA
LDN 0913	20 22 36.	+ 45 10	420		DARK NEBULA
MRSL 084+06/1	20 22 36.	+ 47 58	936		HII REGION
MCG-03-52-005	20 22 36.	- 16 01	54	15.	GALAXY
HN 2371	20 22 37.9	- 51 50 46.			NEBULA
HN 2372	20 22 39.5	- 52 22 46.	30		NEBULA
MCG+01-52-006	20 22 42.	+ 05 06	120	14.	GALAXY
OCL 0189	20 22 42.	+ 42 06	360		OPEN STAR CLUSTER
MRSL 083+05/1	20 22 42.	+ 46 27	936		HII REGION
HN 1202	20 22 46.	- 54 57		13.	NEBULA
IC 5002	20 22 46.	- 54 57			NONSTELLAR OBJECT
ZWG 399.009	20 22 48.	+ 05 06		14.7	GALAXY
UGC 11555	20 22 48.	+ 05 06	114	14.7	GALAXY Sb/Sc
MCG+02-52-005	20 22 48.	+ 12 13	36	17.	GALAXY
UGC 11556	20 22 48.	+ 12 14	60	17.	GALAXY
MRSL 079+02/2	20 22 48.	+ 41 18	468		HII REGION
MCG-07-42-005	20 22 48.	- 41 06	60	14.	GALAXY

OBJECT NAME	RIGHT ASCEN.	DECLINATION	DIAM.	MAGN.	TYPE OF OBJECT
DG 163	20 22 54.	+ 42 12	300		REFLECTION NEBULA
MRSL 081+03/4	20 22 54.	+ 44 02	533		HII REGION
MRSL 082+04/3	20 22 54.	+ 45 08	1068		HII REGION
MCG-03-52-006	20 22 54.	- 18 42	30	15.	GALAXY
LBN 0122	20 23	+ 00 41	1620		BRIGHT NEBULA
LBN 0132	20 23	+ 14 00	4500		BRIGHT NEBULA
LBN 0211	20 23	+ 37 00	9000		BRIGHT NEBULA
LBN 0224	20 23	+ 38 20	6300		BRIGHT NEBULA
LBN 0268	20 23	+ 41 42	240		BRIGHT NEBULA
LBN 0274	20 23	+ 42 09	180		BRIGHT NEBULA
LBN 0292	20 23	+ 43 00	3600		BRIGHT NEBULA
LBN 0343	20 23	+ 47 00	3600		BRIGHT NEBULA
LBN 0364	20 23	+ 48 30	4200		BRIGHT NEBULA
LBN 0377	20 23	+ 49 20	3420		BRIGHT NEBULA
LBN 0115	20 23	- 01 40	1320		BRIGHT NEBULA
LDN 0889	20 23 00.	+ 40 00	3600		DARK NEBULA
LDN 0897	20 23 00.	+ 42 00	1320		DARK NEBULA
LDN 0900	20 23 00.	+ 42 35	480		DARK NEBULA
MRSL 080+03/3	20 23 00.	+ 43 10	1398		HII REGION
OCL 0193	20 23 00.	+ 45 49	300	15.	OPEN STAR CLUSTER
ZWG 304.005	20 23 00.	+ 60 02		14.5	GALAXY
UGC 11557	20 23 00.	+ 60 02	144	14.5	GALAXY
7ZW 933	20 23 00.	+ 62 38			COMPACT GALAXY
RNGC 6914	20 23 02.	+ 42 09			DIFFUSE NEBULA
CED 176E	20 23 02.	+ 42 13	240		DIFFUSE GALACTIC NEBULA
MCG+02-52-006	20 23 06.	+ 10 24	54	14.5	GALAXY
ZWG 424.005	20 23 06.	+ 10 25		14.9	GALAXY
ISS 0329	20 23 06.	+ 27 36	77		STELLAR RING
ACK 073-02.1	20 23 06.	+ 33 25			PLANETARY NEBULA
MRSL 078+01/3	20 23 06.	+ 40 05	2598		HII REGION
MRSL 078+01/6	20 23 06.	+ 40 17	1266		HII REGION
VDB.66N 132	20 23 06.	+ 42 15	336		REFLECTION NEBULA
MCG+10-29-005	20 23 06.	+ 60 02	120	13.	GALAXY
HN 2373	20 23 07.0	- 52 39 45.	12		NEBULA
MCG-03-52-007	20 23 09.	- 18 41	36	14.5	GALAXY
IC 1319	20 23 10.	- 18 40 11.			NONSTELLAR OBJECT
MRSL 082+04/1	20 23 12.	+ 45 05	533		HII REGION
SG 2.056	20 23 13.	+ 43 10	21600		DIFFUSE EMISSION NEBULA
ISS 0330	20 23 18.	+ 28 44	199		STELLAR RING
MRSL 080+02/3	20 23 18.	+ 42 31	1536		HII REGION
MCG-05-48-012	20 23 18.	- 32 40	72	15.5	GALAXY
MCG-04-48-008	20 23 24.	- 23 18 30.	18	15.5	GALAXY
HN 2374	20 23 24.3	- 50 40 20.	6		NEBULA
PK081+03.1	20 23 28.5	+ 43 42 45.			PLANETARY NEBULA
UGC 11558	20 23 30.	- 01 31	60	16.0	GALAXY
ZWG 373.006	20 23 36.	+ 01 00		14.5	GALAXY
UGC 11559	20 23 36.	+ 01 00	78	14.5	GALAXY S0
KARA.72 544A	20 23 36.	+ 01 00	66	14.5	PART OF DOUBLE GALAXY
MCG+00-52-007	20 23 36.	+ 01 00	24	15.	GALAXY
MCG+00-52-008	20 23 42.	+ 00 56	36	15.5	GALAXY
ZWG 373.007	20 23 42.	+ 00 57		15.7	GALAXY
KARA.72 544B	20 23 42.	+ 00 57	48	15.7	PART OF DOUBLE GALAXY
HN 2375	20 23 45.5	- 52 32 06.			NEBULA
MRSL 080+02/1	20 23 48.	+ 42 01	863		HII REGION
MRSL 081+03/2	20 23 48.	+ 43 41	1001		HII REGION
MCG-01-52-002	20 23 48.	- 05 15	36	15.	GALAXY
MCG-05-48-013	20 23 48.	- 30 43	24	15.5	GALAXY
HN 2376	20 23 49.8	- 51 49 18.	60		NEBULA
HN 2377	20 23 53.2	- 51 58 18.	48		NEBULA
ZWG 373.008	20 23 54.	+ 02 45		14.5	GALAXY
UGC 11560	20 23 54.	+ 02 45	60	14.5	GALAXY S?
MCG+00-52-009	20 23 54.	+ 02 45	42	14.5	GALAXY
MRSL 078+01/2	20 23 54.	+ 39 58	533		HII REGION
OCL 0188	20 23 54.	+ 41 46	480		OPEN STAR CLUSTER
IC 1320	20 23 56.	+ 02 44 38.			NONSTELLAR OBJECT
RNGC 6912	20 23 58.	- 18 48		14.0	GALAXY
KHAV 687	20 24	+ 39 52	2840		DARK NEBULA
LBN 0249	20 24	+ 40 20	2400		BRIGHT NEBULA
LBN 0253	20 24	+ 40 50	1020		BRIGHT NEBULA
LBN 0273	20 24	+ 42 00	960		BRIGHT NEBULA
LBN 0280	20 24	+ 42 13	180		BRIGHT NEBULA
LBN 0281	20 24	+ 42 19	180		BRIGHT NEBULA
LB 03123	20 24	- 81 08		13.2	FAINT BLUE STAR
LB 09940	20 24	- 85 10		12.0	FAINT BLUE STAR
MCG+00-52-010	20 24 00.	+ 02 29	60	15.	GALAXY
SG 2.057	20 24 00.	+ 40 09	3000		DIFFUSE EMISSION NEBULA
COU 058	20 24 00.	+ 40 15	3000		EMISSION NEBULA
MCG-03-52-008	20 24 00.	- 18 48	72	14.	GALAXY
HN 2379	20 24 02.7	- 52 32 53.	60		NEBULA
IC 5011	20 24 04.	- 36 11 14.			NONSTELLAR OBJECT
HN 2378	20 24 04.2	- 54 47 54.			NEBULA
ZWG 373.009	20 24 06.	+ 02 29		14.9	GALAXY
UGC 11561	20 24 06.	+ 02 29	78	14.9	GALAXY S0?
ZWG 424.006	20 24 06.	+ 14 51		15.7	GALAXY
MCG-01-52-003	20 24 06.	- 05 15	66	15.	GALAXY
RNGC 6909	20 24 08.	- 47 12		13.0	GALAXY
ZWG 373.010	20 24 12.	+ 02 32		14.7	GALAXY
UGC 11562	20 24 12.	+ 02 32	84	14.7	GALAXY Sb
MCG+00-52-011	20 24 12.	+ 02 32	66	13.5	GALAXY
MRSL 078+01/1	20 24 12.	+ 39 52	533		HII REGION
HN 2380	20 24 12.3	- 52 47 11.	12		NEBULA
ZWG 424.007	20 24 18.	+ 10 32		15.1	GALAXY
ZWG 447.004	20 24 18.	+ 15 33		15.7	GALAXY
MRSL 079+02/3	20 24 18.	+ 41 45	1733		HII REGION
MCG+02-52-007	20 24 24.	+ 10 31	36	16.	GALAXY
ZWG 373.011	20 24 24.	- 02 25		15.2	GALAXY
KARA.73B 0877	20 24 24.	- 02 25	24	15.2	ISOLATED GALAXY E
2ZW 083	20 24 24.	- 02 26			COMPACT GALAXY
SER 136.01	20 24 24.	- 43	228	16.	INTERACTING GALAXIES
HN 2381	20 24 24.6	- 51 51 22.	30		NEBULA
SG 2.058	20 24 25.	+ 39 52			DIFFUSE EMISSION NEBULA
OCL 0180	20 24 30.	+ 39 57	960		OPEN STAR CLUSTER
HN 2382	20 24 31.9	- 52 50 40.	36		NEBULA
MIN.47 12	20 24 33.	+ 45 20			DIFFUSE NEBULA
MRSL 079+02/1	20 24 36.	+ 41 29	468		HII REGION
MRSL 080+02/2	20 24 36.	+ 42 20	1463		HII REGION
SER 138.03	20 24 36.	- 52 50	150	15.	INTERACTING GALAXIES
HN 2383	20 24 38.6	- 52 35 27.			NEBULA
HN 2384	20 24 39.9	- 52 00 33.	6		NEBULA
MCG+00-52-012	20 24 42.	+ 01 18	42	16.	GALAXY
ISS 0276	20 24 42.	+ 33 27	344		STELLAR RING
MRSL 078+01/4	20 24 42.	+ 40 25	2598		HII REGION
OCL 0186	20 24 42.	+ 41 17	480		OPEN STAR CLUSTER
MCG-03-52-009	20 24 42.	- 19 00	54	15.	GALAXY
DVDV 5	20 24 42.	- 33 12	42		GALAXY
SER 137.04	20 24 48.	- 72 01	100	16.	LOW SURF. BRGHTNSS GALAXY
MRSL 081+02/2	20 24 54.	+ 43 01	936		HII REGION
RNGC 6917	20 24 59.	+ 07 56		14.5	GALAXY

OBJECT NAME	RIGHT ASCEN.	DECLINATION	DIAM.	MAGN.	TYPE OF OBJECT
LBN 0123	20 25	+ 01 00	900		BRIGHT NEBULA
LBN 0141	20 25	+ 19 50	6600		BRIGHT NEBULA
LBN 0261	20 25	+ 41 05	1200		BRIGHT NEBULA
LBN 0279	20 25	+ 42 00	2700		BRIGHT NEBULA
KHAV 688	20 25	+ 45 58	1720		DARK NEBULA
LBN 0119	20 25	- 01 00	4500		BRIGHT NEBULA
ZWG 399.010	20 25 00.	+ 07 56		14.3	GALAXY
UGC 11563	20 25 00.	+ 07 56	96	14.3	GALAXY S
MCG+01-52-007	20 25 00.	+ 07 57	72	13.	GALAXY
COU 059	20 25 00.	+ 42 00	9000		EMISSION NEBULA
B 346	20 25 04.	+ 43 35	600		DARK OBJECT
RNGC 6915	20 25 06.	- 03 13			GALAXY
MCG-05-48-014	20 25 06.	- 28 15	48	15.5	GALAXY
2ZW 084	20 25 12.	+ 10 35			COMPACT GALAXY
ZWG 424.009	20 25 12.	+ 10 35		15.1	GALAXY
UGC 11564	20 25 12.	+ 10 35	96	15.1	GALAXY DBL SYS
ZWG 424.008	20 25 12.	+ 11 54		15.7	GALAXY
MCG-03-52-010	20 25 12.	- 18 57	48	15.	GALAXY
IC 5013	20 25 14.	- 36 12 30.			NONSTELLAR OBJECT
SG 2.059	20 25 15.	+ 38 38	1500		DIFFUSE EMISSION NEBULA
MCG+02-52-008	20 25 18.	+ 10 35	90	15.	GALAXY
IC 1321	20 25 19.	- 18 27 10.			NONSTELLAR OBJECT
MCG-03-52-011	20 25 21.	- 18 28	36	15.	GALAXY
MIL 87	20 25 25.	+ 39 56 06.	900		SUPERNOVA REMNANT
LB 01094	20 25 27.	- 15 12 36.		16.7	FAINT BLUE STAR
HN 2386	20 25 27.6	- 50 50 12.	18		NEBULA
IC 5015	20 25 28.	- 31 52 04.			NONSTELLAR OBJECT
HN 2385	20 25 29.6	- 53 30 00.			NEBULA
MCG+01-52-008	20 25 30.	+ 04 47	72	14.5	GALAXY
ZWG 424.010	20 25 30.	+ 13 21		15.4	GALAXY
MRSL 078+00/1	20 25 30.	+ 39 32	2333		HII REGION
MRSL 081+03/1	20 25 30.	+ 43 20	1668		HII REGION
MCG+00-52-013	20 25 33.	+ 00 52	30	16.	GALAXY
MIN.46 19	20 25 35.	+ 37 15			DIFFUSE NEBULA
SG 3.172	20 25 35.	+ 37 15	30		DIFFUSE EMISSION NEBULA
MCG+00-52-014	20 25 36.	+ 00 19	42	15.	GALAXY
ZWG 373.012	20 25 36.	+ 00 52		15.7	GALAXY
KARA.73B 0878	20 25 36.	+ 00 52	42	15.7	ISOLATED GALAXY S
ZWG 399.011	20 25 36.	+ 04 48		15.0	GALAXY
UGC 11565	20 25 36.	+ 04 48	78	15.0	GALAXY IRR
MRSL 076-00/1	20 25 36.	+ 37 14	180		HII REGION
MCG+00-52-015	20 25 39.	+ 00 08	30	15.	GALAXY
HN 1203	20 25 39.	- 56 54	60		NEBULA
IC 5012	20 25 40.	- 56 55			NONSTELLAR OBJECT
ZWG 373.013	20 25 42.	+ 00 07		14.2	GALAXY
UGC 11566	20 25 42.	+ 00 07	48	14.2	GALAXY
ZWG 373.014	20 25 42.	+ 00 18		15.5	GALAXY
UGC 11567	20 25 42.	+ 00 18	60	15.5	GALAXY
KARA.72 545A	20 25 42.	+ 00 19	48	15.5	PART OF DOUBLE GALAXY
ZWG 373.015	20 25 42.	+ 00 20		15.6	GALAXY
KARA.72 545B	20 25 42.	+ 00 20	30	15.6	PART OF DOUBLE GALAXY
MRSL 081+02/1	20 25 42.	+ 42 57	1331		HII REGION
MRSL 082+03/5	20 25 42.	+ 44 31	600		HII REGION
MCG-03-52-012	20 25 42.	- 19 10	30	15.	GALAXY
HN 2387	20 25 43.5	- 52 16 17.	6		NEBULA
HN 2388	20 25 47.3	- 52 48 41.	12		NEBULA
ZWG 424.011	20 25 48.	+ 14 07		15.3	GALAXY
ISS 0277	20 25 48.	+ 36 23	184		STELLAR RING
ZWG 424.012	20 25 54.	+ 10 35		15.0	GALAXY
UGC 11568	20 25 54.	+ 10 35	138	15.0	GALAXY Sc
7ZW 934	20 25 54.	+ 60 20			COMPACT GALAXY
LB 01095	20 25 55.	- 13 00 18.		16.3	FAINT BLUE STAR
PK064-09.1	20 25 57.4	+ 22 41 19.			PLANETARY NEBULA
LBN 0210	20 26	+ 36 30	3000		BRIGHT NEBULA
LBN 0245	20 26	+ 39 50	1920		BRIGHT NEBULA
LBN 0305	20 26	+ 43 20	720		BRIGHT NEBULA
MCG+02-52-010	20 26 00.	+ 10 32	24	18.	GALAXY
MCG+02-52-009	20 26 00.	+ 10 35	120	14.	GALAXY
ZWG 424.013	20 26 00.	+ 14 10		15.7	GALAXY
COU 060	20 26 00.	+ 39 50	3600		EMISSION NEBULA
IC 1318	20 26 00.	+ 39 50			HII REGION
SCHO 1181	20 26 01.	+ 30 58 36.	660		ISOLATED DARK CLOUD
MCG+01-52-009	20 26 06.	+ 04 10	36	15.	GALAXY
MRSL 079+01/5	20 26 06.	+ 41 22	263		HII REGION
LB 01096	20 26 08.	- 15 54 00.		17.0	FAINT BLUE STAR
HN 2389	20 26 13.4	- 51 53 04.	18		NEBULA
ZWG 399.012	20 26 18.	+ 04 12		15.2	GALAXY
ZC 2026.3+7639	20 26 18.	+ 76 39	1680		CLUSTER OF GALAXIES
MCG-04-48-009	20 26 18.	- 22 52	18	15.	GALAXY
RNGC 6921	20 26 20.	+ 25 33		14.5	GALAXY
HN 2390	20 26 21.9	- 50 30 39.	6		NEBULA
UGC 11569	20 26 24.	+ 10 28	66	16.0	GALAXY Sc
UGC 11570	20 26 24.	+ 25 33	102	15.0	GALAXY S
MCG+04-48-001	20 26 24.	+ 25 33	39	15.	GALAXY
SER 138.01	20 26 24.	- 51 52	40	14.5	HIGH SURF. BRGHTNSS GALXY
HN 0671	20 26 24.	- 66 05			NEBULA
IC 5010	20 26 24.	- 66 06			NONSTELLAR OBJECT
MCG+04-48-002	20 26 30.	+ 25 33	24	18.	GALAXY
MRSL 083+04/1	20 26 30.	+ 45 17	666		HII REGION
ZC 2026.5+7425	20 26 30.	+ 74 25	5170		CLUSTER OF GALAXIES
MCG+02-52-011	20 26 36.	+ 10 30	120	14.5	GALAXY
MCG-01-52-004	20 26 36.	- 08 12	48	15.	GALAXY
MCG-04-48-010	20 26 36.	- 22 17	48	15.5	GALAXY
MCG-07-42-006	20 26 36.	- 39 10	36	15.	GALAXY
CED 176F	20 26 37.	+ 39 47	600		DIFFUSE GALACTIC NEBULA
B 347	20 26 39.	+ 39 45	60		DARK OBJECT
SCHO 1182	20 26 40.	+ 31 51 42.	290		ISOLATED DARK CLOUD
ZWG 424.014	20 26 42.	+ 10 31		15.4	GALAXY
UGC 11571	20 26 42.	+ 10 31	126	15.4	GALAXY SBc
ZWG 447.005	20 26 42.	+ 18 13		15.6	GALAXY
MRSL 079+01/4	20 26 42.	+ 41 10	863		HII REGION
HN 2391	20 26 42.2	- 50 37 14.	18		NEBULA
MCG-02-52-006	20 26 45.	- 12 41 30.	60	14.	GALAXY
SCHO 1183	20 26 46.	+ 28 18 06.	380		ISOLATED DARK CLOUD
MCG+02-52-012	20 26 48.	+ 10 34	36	15.	GALAXY
ZWG 424.015	20 26 48.	+ 10 35		14.9	GALAXY
UGC 11572	20 26 48.	+ 10 35	78	14.9	GALAXY E
MRSL 078+00/2	20 26 48.	+ 39 49	2598		HII REGION
MCG+00-52-016	20 26 48.	- 00 21	48	15.	GALAXY
MCG-05-48-015	20 26 48.	- 32 25	30	15.5	GALAXY
SG 2.060	20 26 50.	+ 38 58	1200		DIFFUSE EMISSION NEBULA
MCG-02-52-007	20 26 51.	- 10 17 30.	36	15.	GALAXY
MRSL 080+01/4	20 26 54.	+ 41 43	468		HII REGION
ZWG 373.016	20 26 54.	- 00 20		14.8	GALAXY
LBN 0198	20 27	+ 35 30	5700		BRIGHT NEBULA
OCL 0178	20 27 00.	+ 39 13			OPEN STAR CLUSTER
ACK 078-00.1	20 27 00.	+ 40 15			PLANETARY NEBULA
SG 2.061	20 27 01.	+ 39 39	2700		DIFFUSE EMISSION NEBULA

OBJECT NAME	RIGHT ASCEN.	DECLINATION	DIAM.	MAGN.	TYPE OF OBJECT
SCHO 1184	20 27 03.	+ 29 53 36.	490		ISOLATED DARK CLOUD
UGC 11573	20 27 06.	+ 14 10	66	16.0	GALAXY Sc
HN 2392	20 27 09.3	- 50 30 24.	12		NEBULA
ARC 2325	20 27 10.	- 25 05		16.8	RICH CLUSTER OF GALAXIES
ZWG 399.013	20 27 12.	+ 03 33		15.1	GALAXY
MRSL 082+03/4	20 27 12.	+ 44 34	863		HII REGION
MCG+00-52-017	20 27 15.	- 02 12 30.	42	14.5	GALAXY
IC 5018	20 27 16.	- 38 23 00.			NONSTELLAR OBJECT
IC 5009	20 27 17.	- 72 20			NONSTELLAR OBJECT
ISS 0229	20 27 18.	+ 44 39	771		STELLAR RING
MRSL 079+01/1	20 27 18.	+ 40 26	600		HII REGION
MRSL 080+01/2	20 27 18.	+ 41 25	401		HII REGION
ZWG 373.018	20 27 18.	- 02 11		15.7	GALAXY
ZWG 373.017	20 27 18.	- 02 21		14.4	GALAXY
RNGC 6922	20 27 18.	- 02 21		14.5	GALAXY
UGC 11574	20 27 18.	- 02 21	72	14.4	GALAXY Sc
MCG-02-52-009	20 27 18.	- 08 53	60	15.5	GALAXY
MCG-02-52-008	20 27 18.	- 08 56	9	16.	GALAXY
MCG-07-42-008	20 27 18.	- 44 33 30.	36	15.	GALAXY
MCG-07-42-007	20 27 18.	- 44 35	36	15.	GALAXY
HN 0672	20 27 19.	- 72 20			NEBULA
IC 1322	20 27 20.	- 15 23 33.			NONSTELLAR OBJECT
MCG+00-52-018	20 27 21.	- 02 22 30.	66	14.	GALAXY
HN 2393	20 27 22.1	- 52 46 48.	36		NEBULA
MCG-03-52-013	20 27 24.	- 19 17	36	15.	GALAXY
HN 0673	20 27 24.	- 72 52			NEBULA
IC 5008	20 27 24.	- 72 52			NONSTELLAR OBJECT
ACK 078+00.1	20 27 30.	+ 40 05			PLANETARY NEBULA
ZWG 373.019	20 27 30.	- 01 01		15.7	GALAXY
MCG+00-52-019	20 27 30.	- 01 02	24	15.	GALAXY
MCG+00-52-020	20 27 33.	+ 02 54	66	14.	GALAXY
MCG-07-42-009	20 27 33.	- 39 26	60	15.	GALAXY
ZWG 373.020	20 27 36.	+ 02 54		14.8	GALAXY
UGC 11575	20 27 36.	+ 02 54	96	14.8	GALAXY Sc-IRR
KARA.73B 0879	20 27 36.	+ 02 54	96	14.8	ISOLATED GALAXY S
MRSL 080+01/3	20 27 36.	+ 41 30	131		HII REGION
MCG-01-52-005	20 27 36.	- 08 05	72	15.	GALAXY
MCG-03-52-014	20 27 36.	- 16 54 30.	60	15.5	GALAXY
HN 2394	20 27 38.4	- 53 43 11.	12		NEBULA
IC 5019	20 27 39.	- 36 29 15.			NONSTELLAR OBJECT
IC 1323	20 27 40.	- 15 20 54.			NONSTELLAR OBJECT
MCG-05-48-016	20 27 42.	- 30 52	48	15.5	GALAXY
MCG-07-42-010	20 27 42.	- 44 26	15	16.	GALAXY
ZWG 373.021	20 27 48.	- 01 10		15.2	GALAXY
UGC 11576	20 27 48.	- 01 10	96	15.2	GALAXY
MCG+00-52-021	20 27 48.	- 01 11	18	15.	GALAXY
MCG-01-52-006	20 27 48.	- 03 56	24	15.5	GALAXY
MCG-04-48-012	20 27 48.	- 25 33	36	15.5	GALAXY
MCG-04-48-011	20 27 48.	- 25 33	60	15.5	GALAXY
IC 5020	20 27 48.	- 33 42	42	13.1	GALAXY E
SCHO 1185	20 27 51.	+ 30 00 54.	700		ISOLATED DARK CLOUD
MCG+00-52-022	20 27 51.	- 01 00	48	15.5	GALAXY
MCG-03-52-015	20 27 51.	- 19 19	42	15.	GALAXY
SG 2.062	20 27 53.	+ 40 57	3000		DIFFUSE EMISSION NEBULA
MCG+00-52-023	20 27 54.	+ 01 42	42	15.	GALAXY
ZWG 424.016	20 27 54.	+ 13 29		15.7	GALAXY
OCL 0190	20 27 54.	+ 41 33	720		OPEN STAR CLUSTER
ZWG 373.022	20 27 54.	- 01 00		15.6	GALAXY
MCG-01-52-007	20 27 54.	- 04 05	45	15.5	GALAXY
LBN 0219	20 28	+ 36 55	1500		BRIGHT NEBULA
LBN 0257	20 28	+ 40 30	5400		BRIGHT NEBULA
LBN 0267	20 28	+ 41 00	2400		BRIGHT NEBULA
LBN 0293	20 28	+ 42 30	2700		BRIGHT NEBULA
LBN 0352	20 28	+ 47 00	3420		BRIGHT NEBULA
LBN 0379	20 28	+ 48 50	660		BRIGHT NEBULA
ZWG 373.024	20 28 00.	+ 01 42		14.9	GALAXY
MCG+01-52-010	20 28 00.	+ 04 34	36	16.	GALAXY
ISS 0331	20 28 00.	+ 28 37	119		STELLAR RING
MRSL 079+01/3	20 28 00.	+ 41 02	1200		HII REGION
ZWG 373.023	20 28 00.	- 01 10		15.5	GALAXY
MCG+00-52-024	20 28 00.	- 01 11 30.	12	15.5	GALAXY
RNGC 6918	20 28 02.	- 47 38			GALAXY
MCG+00-52-025	20 28 03.	+ 01 13	84	14.	GALAXY
HN 2395	20 28 05.4	- 53 56 33.	18		NEBULA
ZWG 373.025	20 28 06.	+ 01 13		14.6	GALAXY
UGC 11577	20 28 06.	+ 01 13	102	14.6	GALAXY S
LB 01097	20 28 06.	- 15 12 06.		16.8	FAINT BLUE STAR
MCG-07-42-011	20 28 06.	- 44 23	78	15.	GALAXY
RNGC 6919	20 28 11.	- 44 23			GALAXY
HN 1204	20 28 11.	- 57 45		14.5	NEBULA
IC 5017	20 28 11.	- 57 45			NONSTELLAR OBJECT
ZWG 399.014	20 28 12.	+ 04 35		15.6	GALAXY
MRSL 082+03/1	20 28 12.	+ 44 10	1733		HII REGION
ZWG 373.026	20 28 12.	- 02 30		15.5	GALAXY
HN 2396	20 28 16.8	- 53 48 09.	30		NEBULA
HN 2397	20 28 17.8	- 53 54 44.	18		NEBULA
MCG+01-52-011	20 28 18.	+ 04 35	24	16.	GALAXY
UGC 11578	20 28 18.	+ 09 00	102	16.5	GALAXY
ZWG 447.006	20 28 18.	+ 20 02		15.6	GALAXY
UGC 11580	20 28 24.	+ 07 55	66	16.0	GALAXY Sc
LDN 0851	20 28 24.	+ 31 30	540		DARK NEBULA
MRSL 082+03/3	20 28 24.	+ 44 27	666		HII REGION
UGC 11579	20 28 24.	- 00 49	96	16.0	GALAXY Sc
MCG+00-52-026	20 28 24.	- 00 49	78	14.5	GALAXY
MCG-05-48-017	20 28 24.	- 31 03	120	12.5	GALAXY
SER 138.07	20 28 24.	- 53 55	1020	14.	LOOSE GROUP OF 8 GALAXIES
HN 2398	20 28 24.2	- 51 03 20.	18		NEBULA
ZWG 399.015	20 28 30.	+ 04 36		15.5	GALAXY
ZWG 373.027	20 28 30.	- 02 05		15.4	GALAXY
UGC 11581	20 28 30.	- 02 05	78	15.4	GALAXY SB0/SBa
MCG+00-52-027	20 28 30.	- 02 07	66	15.	GALAXY
LDN 0844	20 28 32.	+ 30 30	540		DARK NEBULA
MIL 89	20 28 32.	+ 41 04 18.	1740		SUPERNOVA REMNANT
RNGC 6923	20 28 33.	- 31 01		12.0	GALAXY
ZWG 447.007	20 28 36.	+ 20 07		15.6	GALAXY
UGC 11582	20 28 36.	+ 20 07	126	15.6	GALAXY Sc-IRR
MCG-03-52-016	20 28 36.	- 19 31	48	15.	GALAXY
MCG-05-48-018	20 28 36.	- 31 00	60	15.	GALAXY
MCG+03-52-001	20 28 39.	+ 20 08	108	14.5	GALAXY
MCG+00-52-028	20 28 42.	+ 00 05	60	15.	GALAXY
ZWG 373.028	20 28 42.	+ 00 07		15.7	GALAXY
KARA.73B 0880	20 28 42.	+ 00 07	96	15.7	ISOLATED GALAXY S
ISS 0332	20 28 42.	- 28 19	124		STELLAR RING
MCG+00-52-029	20 28 45.	+ 00 36	36	16.	GALAXY
VDB.66N 133	20 28 45.	+ 36 46	684		REFLECTION NEBULA
MRSL 079+01/2	20 28 48.	+ 40 48	1733		HII REGION
VDB.66N 134	20 28 48.	+ 48 47	1320		REFLECTION NEBULA
MCG-04-48-013	20 28 48.	- 20 56 30.	30	16.	GALAXY

OBJECT NAME	RIGHT ASCEN.	DECLINATION	DIAM.	MAGN.	TYPE OF OBJECT
ZWG 399.016	20 28 54.	+ 04 27		15.0	GALAXY
ZWG 373.029	20 28 54.	- 02 07		15.5	GALAXY
MCG+00-52-030	20 28 54.	- 02 09	15	16.	GALAXY
PK038-25.1	20 28 54.	- 07 16	47	14.3	PLANETARY NEBULA
MCG-02-52-010	20 28 54.	- 12 27	66	15.	GALAXY
LBN 0125	20 29	+ 01 10	1020		BRIGHT NEBULA
LBN 0218	20 29	+ 36 45	600		BRIGHT NEBULA
LBN 0315	20 29	+ 44 00	4500		BRIGHT NEBULA
LBN 0114	20 29	- 02 40	3420		BRIGHT NEBULA
ZWG 399.017	20 29 00.	+ 06 57		15.6	GALAXY
KARA.72 546A	20 29 00.	+ 06 57	42	15.6	PART OF DOUBLE GALAXY
ZWG 424.017	20 29 00.	+ 14 27		15.7	GALAXY
ZWG 399.018	20 29 06.	+ 06 59		15.5	GALAXY
KARA.72 546B	20 29 06.	+ 06 59	48	15.5	PART OF DOUBLE GALAXY
CED 180	20 29 06.	+ 36 46			DIFFUSE GALACTIC NEBULA
SG 2.063	20 29 06.	+ 47 06	3000		DIFFUSE EMISSION NEBULA
UGC 11583	20 29 12.	+ 60 17	150	17.	GALAXY
LB 01098	20 29 14.	- 13 41 42.		16.6	FAINT BLUE STAR
MCG+00-52-031	20 29 18.	+ 01 21 30.	90	15.5	GALAXY
ZWG 373.030	20 29 18.	+ 01 22		15.7	GALAXY
UGC 11584	20 29 18.	+ 01 22	108	15.7	GALAXY Sc
MRSL 082+03/2	20 29 18.	+ 44 27	666		HII REGION
HN 2399	20 29 19.5	- 51 29 59.	12		NEBULA
ZWG 424.018	20 29 24.	+ 09 53		15.7	GALAXY
LDN 0859	20 29 24.	+ 32 10	480		DARK NEBULA
SG 2.064	20 29 24.	+ 43 59	3600		DIFFUSE EMISSION NEBULA
MCG-02-52-011	20 29 24.	- 08 54	66	15.	GALAXY
MCG-02-52-012	20 29 24.	- 09 14	84	14.5	GALAXY
MCG-03-52-017	20 29 27.	- 19 51 30.	30	15.5	GALAXY
LDN 0939	20 29 30.	+ 48 00	18120		DARK NEBULA
IC 1324	20 29 30.	- 09 14 25.			NONSTELLAR OBJECT
MCG-03-52-018	20 29 30.	- 19 54	15	15.	GALAXY
MIL 86	20 29 31.	+ 39 13 06.	570		SUPERNOVA REMNANT
LB 00347	20 29 33.	- 10 29 06.		15.3	FAINT BLUE STAR
MCG+01-52-012	20 29 36.	+ 04 40	72	16.	GALAXY
ZWG 447.008	20 29 36.	+ 18 15		15.7	GALAXY
HN 2400	20 29 38.8	- 52 09 04.	18		NEBULA
LB 01099	20 29 41.	- 14 29 24.		16.1	FAINT BLUE STAR
ZWG 399.019	20 29 42.	+ 04 41		15.6	GALAXY
UGC 11586	20 29 42.	+ 04 41	66	15.6	GALAXY Sa
MRSL 071-05/1	20 29 42.	+ 30 26	2400		HII REGION
ZWG 373.031	20 29 42.	- 02 24		14.8	GALAXY
HOLM 780B	20 29 42.	- 02 24	18	15.0	PART OF MULTIPLE GALAXY
UGC 11585	20 29 42.	- 02 24	120	14.8	GALAXY Sb
HOLM 780A	20 29 42.	- 02 25	42	14.3	PART OF MULTIPLE GALAXY
MCG+00-52-032	20 29 42.	- 02 25	96	14.	GALAXY
HN 1205	20 29 43.	- 54 41	12	14.	NEBULA
HN 0674	20 29 43.	- 73 38			NEBULA
IC 5014	20 29 43.	- 73 38			NONSTELLAR OBJECT
IC 5021	20 29 44.	- 54 41			NONSTELLAR OBJECT
HN 2401	20 29 44.9	- 50 47 33.	18		NEBULA
MRSL 082+02/3	20 29 48.	+ 43 58	1668		HII REGION
MCG+02-52-014	20 29 54.	+ 11 11	96	14.	GALAXY
ZWG 424.019	20 29 54.	+ 11 12		14.7	GALAXY
UGC 11587	20 29 54.	+ 11 12	108	14.7	GALAXY S0
MCG+02-52-013	20 29 54.	+ 11 13	36	17.	GALAXY
MCG-05-48-019	20 29 54.	- 31 31	24	15.5	GALAXY
HN 2402	20 29 54.6	- 49 44 45.	24		NEBULA
ABC 2326	20 29 58.	+ 69 42		17.4	RICH CLUSTER OF GALAXIES
LBN 0226	20 30	+ 37 20	5100		BRIGHT NEBULA
LBN 0288	20 30	+ 41 50	1920		BRIGHT NEBULA
LBN 0308	20 30	+ 43 00	1560		BRIGHT NEBULA
LBN 0317	20 30	+ 44 00	1260		BRIGHT NEBULA
LBN 0382	20 30	+ 49 00	10800		BRIGHT NEBULA
LBN 0386	20 30	+ 49 20	1800		BRIGHT NEBULA
LBN 0124	20 30	- 00 23	1020		BRIGHT NEBULA
LB 09999	20 30	- 88 06		12.5	FAINT BLUE STAR
COU 061	20 30 00.	+ 44 15	3600		EMISSION NEBULA
ZC 2030.0+6937	20 30 00.	+ 69 37	3760		CLUSTER OF GALAXIES
SCHO 1186	20 30 01.	+ 28 36 18.	360		ISOLATED DARK CLOUD
ZWG 373.032	20 30 06.	+ 00 27		15.7	GALAXY
ZWG 447.009	20 30 06.	+ 20 28		15.7	GALAXY
MCG-05-48-020	20 30 06.	- 31 29	18	16.	GALAXY
MCG-02-52-013	20 30 09.	- 12 03	30	15.	GALAXY
HMS 1.28	20 30 12.	+ 09 42			E7 GALAXY
MCG+02-52-015	20 30 12.	+ 09 42	15	17.	GALAXY
RNGC 6927A	20 30 12.	+ 09 43			GALAXY
MCG+02-52-016	20 30 12.	+ 09 43	24	16.	GALAXY
ZWG 424.020	20 30 12.	+ 09 45		15.6	GALAXY
RNGC 6927	20 30 12.	+ 09 45		15.5	GALAXY
HN 0675	20 30 12.	- 73 05			NEBULA
IC 5016	20 30 12.	- 73 05			NONSTELLAR OBJECT
RNGC 6924	20 30 13.	- 25 40		14.0	GALAXY
HN 2404	20 30 14.9	- 51 21 01.	36		NEBULA
MCG-04-48-014	20 30 15.	- 25 40	36	14.	GALAXY
HN 2403	20 30 13.2	- 53 09 07.	90		NEBULA
RNGC 6928	20 30 24.	+ 09 45		13.5	GALAXY
MCG+02-52-017	20 30 24.	+ 09 45	120	13.5	GALAXY
ZZW 935	20 30 24.	+ 60 05			COMPACT GALAXY
OCL 0217	20 30 24.	+ 60 28	1920	10.2	OPEN STAR CLUSTER
RNGC 6939	20 30 24.	+ 60 28		10.0	OPEN CLUSTER
HOLM 781A	20 30 24.	- 02 12	78	12.2	PART OF MULTIPLE GALAXY
HN 2405	20 30 25.6	- 52 18 31.	18		NEBULA
IC 1325	20 30 26.	+ 09 42 20.			SAME AS NGC 6927
MCG-02-52-014	20 30 27.	- 11 27	72	14.5	GALAXY
LB 00348	20 30 29.	- 10 19 12.		14.8	FAINT BLUE STAR
ZWG 424.021	20 30 30.	+ 09 45		13.7	GALAXY
UGC 11589	20 30 30.	+ 09 45	132	13.7	GALAXY SBa-b
MRSL 082+02/5	20 30 30.	+ 44 19	600		HII REGION
ZWG 373.033	20 30 30.	- 02 11		13.6	GALAXY
RNGC 6926	20 30 30.	- 02 11		13.5	GALAXY
UGC 11588	20 30 30.	- 02 11	126	13.6	GALAXY IRR
MCG-04-48-015	20 30 30.	- 24 48	60	15.	GALAXY
SER 141.01	20 30 30.	- 69 40	80	15.	INTERACTING GALAXIES
SG 2.066	20 30 35.	+ 41 28	300		DIFFUSE EMISSION NEBULA
ZWG 424.022	20 30 36.	+ 09 41		14.3	GALAXY
UGC 11590	20 30 36.	+ 09 41	78	14.3	GALAXY S+COMP
MCG+02-52-018	20 30 36.	+ 09 41	72	15.	GALAXY
RNGC 6930	20 30 36.	+ 09 42		14.5	GALAXY
IC 1326	20 30 36.	+ 09 43 15.			SAME AS NGC 6928
ASS 21	20 30 36.	+ 41 07	1800		OB ASSOCIATION CYG OB2
MCG+00-52-033	20 30 36.	- 02 12	108	13.	GALAXY
MCG-03-52-019	20 30 36.	- 19 45	60	15.5	GALAXY
MCG-05-48-021	20 30 36.	- 31 14	36	15.5	GALAXY
SG 2.065	20 30 37.	+ 40 01	600		DIFFUSE EMISSION NEBULA
HOLM 781B	20 30 40.	- 02 12	18	13.8	PART OF MULTIPLE GALAXY
ZWG 373.034	20 30 42.	+ 01 35		15.3	GALAXY
UGC 11591	20 30 42.	+ 01 35	72	15.3	GALAXY S

OBJECT NAME	RIGHT ASCEN.	DECLINATION	DIAM.	MAGN.	TYPE OF OBJECT
ZZW 085	20 30 42.	+ 09 04			COMPACT GALAXY
MCG-02-52-015	20 30 42.	- 10 52 30.	72	15.	GALAXY
MCG+00-52-034	20 30 48.	+ 01 35	54	15.	GALAXY
MRSL 080+01/1	20 30 48.	+ 41 30	798		HII REGION
MRSL 082+02/2	20 30 48.	+ 43 54	1068		HII REGION
MRSL 084+04/2	20 30 48.	+ 46 54	3336		HII REGION
PK085+04.1	20 30 48.	+ 47 11	168	15.2	PLANETARY NEBULA
ZWG 373.035	20 30 48.	- 02 12		14.9	GALAXY
RNGC 6929	20 30 48.	- 02 12		15.0	GALAXY
MCG+00-52-035	20 30 48.	- 02 12 30.	24	15.	GALAXY
MCG-02-52-016	20 30 48.	- 11 33 30.	60	14.5	GALAXY
RNGC 6931	20 30 50.	- 11 33		14.0	GALAXY
ZWG 399.020	20 30 54.	+ 05 52		15.7	GALAXY
LBN 0169	20 31	+ 30 30	2520		BRIGHT NEBULA
LBN 0258	20 31	+ 40 05	1200		BRIGHT NEBULA
LBN 0286	20 31	+ 41 30	420		BRIGHT NEBULA
LBN 0120	20 31	- 01 30	1920		BRIGHT NEBULA
ZWG 373.036	20 31 00.	+ 02 28		15.6	GALAXY
ISS 0278	20 31 00.	+ 33 59	88		STELLAR RING
LDN 0920	20 31 00.	+ 45 10	120		DARK NEBULA
LDN 0919	20 31 00.	+ 45 42	120		DARK NEBULA
MCG-07-42-012	20 31 00.	- 43 34	48	15.5	GALAXY
ZWG 447.010	20 31 06.	+ 15 53		15.3	GALAXY
ZWG 447.011	20 31 06.	+ 18 55		15.4	GALAXY
MRSL 079+00/1	20 31 06.	+ 39 48	198		HII REGION
MCG-05-48-022	20 31 06.	- 32 11	210	12.	GALAXY
SG 2.067	20 31 07.	+ 47 11	120		DIFFUSE EMISSION NEBULA
RNGC 6925	20 31 09.	- 32 09		12.0	GALAXY
RNGC 6933	20 31 11.	+ 07 13			NON-EXISTENT OBJECT
ABC 2327	20 31 11.	+ 82 27		16.8	RICH CLUSTER OF GALAXIES
MRSL 082+02/4	20 31 12.	+ 44 12	863		HII REGION
HN 2406	20 31 12.4	- 52 47 40.	12		NEBULA
SCHO 1187	20 31 14.	+ 28 01 18.	440		ISOLATED DARK CLOUD
MCG-04-48-016	20 31 21.	- 26 43	36	16.	GALAXY
MCG-04-48-017	20 31 24.	- 26 44	15	15.5	GALAXY
LB 00349	20 31 27.	- 11 18 30.		14.2	FAINT BLUE STAR
MCG-04-48-018	20 31 27.	- 26 44	24	16.5	GALAXY
LDN 0921	20 31 30.	+ 45 36	180		DARK NEBULA
UGC 11592	20 31 30.	+ 54 27	66	16.0	GALAXY S
YM 17	20 31 32.	+ 41 33	720		SYMMETRIC GALACTIC NEBULA
RNGC 6934	20 31 41.	+ 07 14		10.0	GLOBULAR CLUSTER
HN 2407	20 31 41.3	- 50 02 08.	24		NEBULA
MRSL 084+04/1	20 31 42.	+ 46 48	666		HII REGION
MCG-07-42-013	20 31 42.	- 41 52	60	13.5	GALAXY
GCL 117	20 31 48.	+ 07 14	372	10.01	GLOBULAR STAR CLUSTER
ZWG 424.023	20 31 48.	+ 10 14		15.7	GALAXY
SG 2.069	20 31 50.	+ 39 49	240		DIFFUSE EMISSION NEBULA
MRSL 083+03/2	20 31 54.	+ 45 32	336		HII REGION
MCG-04-48-019	20 31 54.	- 23 19	48	15.	GALAXY
LBN 0290	20 32	+ 41 20	2820		BRIGHT NEBULA
B 348	20 32	+ 41 55	3600		DARK OBJECT
LBN 0337	20 32	+ 45 30	780		BRIGHT NEBULA
LBN 0362	20 32	+ 47 00	1920		BRIGHT NEBULA
KHAV 690	20 32	+ 63 22	1990		DARK NEBULA
KHAV 689	20 32	+ 65 16	1220		DARK NEBULA
MRSL 082+02/1	20 32 00.	+ 43 51	798		HII REGION
LDN 0925	20 32 00.	+ 45 10	180		DARK NEBULA
LDN 1089	20 32 00.	+ 63 20	1980		DARK NEBULA
LDN 1094	20 32 00.	+ 63 50	180		DARK NEBULA
LDN 1122	20 32 00.	+ 65 10	780		DARK NEBULA
MCG-04-48-020	20 32 00.	- 25 32	96	14.5	GALAXY
MCG-07-42-014	20 32 00.	- 43 37	36	15.	GALAXY
ZWG 399.021	20 32 06.	+ 08 58		15.3	GALAXY
ISS 0279	20 32 06.	+ 33 28	150		STELLAR RING
SG 2.068	20 32 10.	+ 45 29	540		DIFFUSE EMISSION NEBULA
HN 2408	20 32 11.1	- 54 28 19.	12		NEBULA
LB 00350	20 32 12.	+ 01 00 30.		15.5	FAINT BLUE STAR
ZWG 399.022	20 32 12.	+ 07 48		14.7	GALAXY
UGC 11593	20 32 12.	+ 07 48	72	14.7	GALAXY DBL SYS
KARA.72 547A	20 32 12.	+ 07 49	36	14.7	PART OF DOUBLE GALAXY
MCG+01-52-013	20 32 12.	+ 08 59	36	14.5	GALAXY
ZWG 424.024	20 32 12.	+ 12 46		15.6	GALAXY
UGC 11594	20 32 12.	+ 12 46	72	15.6	GALAXY S
MRSL 084+03/4	20 32 12.	+ 46 32	1001		HII REGION
MCG+01-52-014	20 32 15.	+ 07 49	12	15.5	GALAXY
CED 181	20 32 16.	+ 45 29	540		DIFFUSE GALACTIC NEBULA
KARA.72 547B	20 32 18.	+ 07 48	54		PART OF DOUBLE GALAXY
MCG+01-52-015	20 32 18.	+ 07 49	60	15.	GALAXY
ZZW 086	20 32 18.	+ 09 47			COMPACT GALAXY
MCG+02-52-019	20 32 18.	+ 12 46	60	15.	GALAXY
MRSL 083+03/1	20 32 18.	+ 45 28	863		HII REGION
COU 062	20 32 18.	+ 45 29	600		EMISSION NEBULA
HN 2411	20 32 21.3	- 50 21 48.			NEBULA
HN 2410	20 32 23.7	- 51 34 12.	12		NEBULA
HN 2412	20 32 23.8	- 49 46 36.	18		NEBULA
MRSL 083+02/1	20 32 24.	+ 44 35	798		HII REGION
RNGC 6940	20 32 27.	+ 28 08		6.5	OPEN CLUSTER
RNGC 6940	20 32 27.	+ 28 08		6.5	OPEN CLUSTER
HN 2409	20 32 28.7	- 53 36 12.	18		NEBULA
ZWG 373.037	20 32 30.	+ 01 46		15.6	GALAXY
UGC 11595	20 32 30.	+ 01 46	126	15.6	GALAXY Sb-c
MCG+00-52-036	20 32 30.	+ 01 47	102	15.	GALAXY
OCL 0141	20 32 30.	+ 28 08	1950	8.4	OPEN STAR CLUSTER
COU 063	20 32 30.	+ 46 40	3600		EMISSION NEBULA
LB 00351	20 32 30.	- 00 53 12.		16.2	FAINT BLUE STAR
MCG-02-52-017	20 32 30.	- 09 48	72	15.	GALAXY
RNGC 6938	20 32 36.	+ 22 05			NON-EXISTENT OBJECT
MCG+00-52-037	20 32 36.	- 02 52	42	16.	GALAXY
MRSL 084+03/1	20 32 42.	+ 46 22	936		HII REGION
MCG-01-52-008	20 32 42.	- 06 26	120	14.5	GALAXY
MCG-02-52-018	20 32 45.	- 09 37	60	15.	GALAXY
SG 2.173	20 32 53.	+ 44 49	1800		DIFFUSE EMISSION NEBULA
MRSL 085+04/1	20 32 54.	+ 46 58	798		HII REGION
LB 00352	20 32 54.	- 11 34 30.		15.7	FAINT BLUE STAR
MCG-04-48-021	20 32 54.	- 25 29	24	15.	GALAXY
RNGC 6936	20 32 55.	- 25 29		15.0	GALAXY
LBN 0330	20 33	+ 44 50	1800		BRIGHT NEBULA
LBN 0358	20 33	+ 46 40	3000		BRIGHT NEBULA
LBN 0357	20 33	+ 46 40	1500		BRIGHT NEBULA
LBN 0381	20 33	+ 48 30	3900		BRIGHT NEBULA
LB 10000	20 33	- 89 43		12.2	FAINT BLUE STAR
ZZW 087	20 33 00.	+ 16 45			COMPACT GALAXY
LDN 0924	20 33 00.	+ 45 00	900		DARK NEBULA
LDN 0923	20 33 00.	+ 45 00	900		DARK NEBULA
MRSL 084+03/5	20 33 00.	+ 46 40	336		HII REGION
MCG-01-52-009	20 33 00.	- 02 55	120	15.	GALAXY
MCG+00-52-038	20 33 00.	- 02 57	78	14.	GALAXY
MCG-04-48-022	20 33 00.	- 22 02	42	15.	GALAXY

OBJECT NAME	RIGHT ASCEN.	DECLINATION	DIAM.	MAGN.	TYPE OF OBJECT
ZWG 373.038	20 33 06.	- 00 09		15.0	GALAXY
KARA.73B 0881	20 33 06.	- 00 09	36	15.0	ISOLATED GALAXY S
MCG-04-48-023	20 33 06.	- 22 04	24	15.	GALAXY
IC 1327	20 33 07.	- 00 10 37.			NONSTELLAR OBJECT
SG 2.070	20 33 08.	+ 46 45	2880		DIFFUSE EMISSION NEBULA
MCG-04-48-024	20 33 09.	- 24 22 30.	72	15.	GALAXY
MCG+00-52-039	20 33 15.	+ 01 10 30.	60	16.	GALAXY
MCG+00-52-040	20 33 15.	- 02 50	48	15.	GALAXY
UGC 11596	20 33 18.	+ 01 13	60	16.0	GALAXY S
YC 2033-34	20 33 18.	- 34 03 54.			UNUSUAL SOUTHERN GALAXY
HN 2413	20 33 21.3	- 51 20 15.	36		NEBULA
HN 0677	20 33 29.	- 67 22			NEBULA
MRSL 084+03/2	20 33 30.	+ 46 33	600		HII REGION
IC 5023	20 33 30.	- 67 22			NONSTELLAR OBJECT
ZWG 373.039	20 33 36.	+ 01 10		15.7	GALAXY
ZWG 373.040	20 33 36.	+ 01 33		15.0	GALAXY
MCG+10-29-006	20 33 36.	+ 59 58	540	9.6	GALAXY
SC 2031+0123.6	20 33 38.	+ 01 33 54.	24		NEBULA
LB 00353	20 33 38.	+ 01 54 06.		14.4	FAINT BLUE STAR
SN 1939C	20 33 41.	+ 59 59		13.0	SUPERNOVA
RNGC 6941	20 33 41.	- 04 48		13.5	GALAXY
LDN 0926	20 33 42.	+ 45 09	180		DARK NEBULA
OCL 0196	20 33 42.	+ 46 38	360	14.	OPEN STAR CLUSTER
MRSL 084+03/3	20 33 42.	+ 46 43	438		HII REGION
TON-S 0001	20 33 42.	- 27 30		16.0	BLUE STAR
PK063-12.1	20 33 44.	+ 20 00 18.	10		PLANETARY NEBULA
MCG-01-52-010	20 33 45.	- 04 48	120	13.5	GALAXY
RNGC 6946	20 33 46.	+ 59 59		10.5	GALAXY
SN 1917A	20 33 47.	+ 59 57		14.6	SUPERNOVA
ZWG 399.023	20 33 48.	+ 08 03		15.7	GALAXY
MRSL 085+04/2	20 33 48.	+ 47 33	468		HII REGION
SG 2.071	20 33 48.	+ 47 44	3600		DIFFUSE EMISSION NEBULA
ZWG 304.006	20 33 48.	+ 59 59		10.5	GALAXY
UGC 11597	20 33 48.	+ 59 59	840	10.5	GALAXY Sc
MCG-05-48-023	20 33 48.	- 29 52	60	15.	GALAXY
SN 1968D	20 33 50.	+ 59 59		13.5	SUPERNOVA
ARP 029	20 33 53.	+ 59 58			PECULIAR GALAXY
MRSL 080+00/1	20 33 54.	+ 40 40	401		HII REGION
SN 1948B	20 33 55.	+ 60 00		14.9	SUPERNOVA
KHAV 691	20 34	+ 39 10	17470		DARK NEBULA
LBN 0272	20 34	+ 40 30	1500		BRIGHT NEBULA
LBN 0276	20 34	+ 40 40	360		BRIGHT NEBULA
LBN 0368	20 34	+ 47 30	360		BRIGHT NEBULA
LBN 0367	20 34	+ 47 30	6300		BRIGHT NEBULA
LBN 0372	20 34	+ 47 40	960		BRIGHT NEBULA
LBN 0121	20 34	- 01 00	2160		BRIGHT NEBULA
SER 137.01	20 34	- 69 21	80		LOW SURF. BRGHTNSS GALAXY
LP 09941	20 34	- 80 49		13.0	FAINT BLUE STAR
LB 03124	20 34	- 83 36		13.4	FAINT BLUE STAR
ZWG 447.012	20 34 00.	+ 19 01		15.3	GALAXY
LDN 0929	20 34 00.	+ 45 10	180		DARK NEBULA
LDN 1100	20 34 00.	+ 63 50	300		DARK NEBULA
LDN 1152	20 34 00.	+ 67 50	600		DARK NEBULA
LB 00354	20 34 01.	- 01 49 06.		14.2	FAINT BLUE STAR
MCG+02-52-021	20 34 06.	+ 09 16	60	16.	GALAXY
ZWG 424.025	20 34 06.	+ 13 15		15.5	GALAXY
UGC 11598	20 34 06.	+ 13 15	60	15.5	GALAXY Sc
MCG+02-52-020	20 34 06.	+ 13 16	60	15.5	GALAXY
PK088+06.1	20 34 06.	+ 51 23	39		PLANETARY NEBULA
MCG+02-52-022	20 34 12.	+ 11 19	132	13.5	GALAXY
ZWG 424.026	20 34 12.	+ 11 20		14.0	GALAXY
UGC 11599	20 34 12.	+ 11 20	150	14.2	GALAXY Sb
MRSL 085+04/3	20 34 12.	+ 47 44	2466		HII REGION
ZWG 447.013	20 34 18.	+ 16 57		15.7	GALAXY
ZWG 325.002	20 34 18.	+ 64 38		15.0	GALAXY
UGC 11600	20 34 18.	+ 64 38	102	15.0	GALAXY S
RNGC 6949	20 34 20.	+ 64 38		15.0	GALAXY
ZWG 373.041	20 34 24.	+ 02 38		15.6	GALAXY
MRSL 085+03/1	20 34 24.	+ 46 51	798		HII REGION
MCG-01-52-011	20 34 24.	- 06 12	21	15.5	GALAXY
SNO 27	20 34 25.	- 22 13 09.	360	19.	CLUSTER OF 15 GALAXIES
ZWG 373.042	20 34 30.	- 02 07		15.2	GALAXY
MCG-05-48-024	20 34 30.	- 27 46	60	14.5	GALAXY
SG 2.072	20 34 31.	+ 40 27	1080		DIFFUSE EMISSION NEBULA
MCG-05-48-025	20 34 33.	- 27 45	36	14.5	GALAXY
MRSL 080+00/1	20 34 36.	+ 40 36	401		HII REGION
MCG+11-25-001	20 34 36.	+ 64 37	60	14.	GALAXY
LB 00355	20 34 39.	- 01 17 54.		16.1	FAINT BLUE STAR
MCG-03-52-020	20 34 39.	- 16 43	42	15.	GALAXY
RNGC 6935	20 34 40.	- 52 17		13.0	GALAXY
YC 2034-52	20 34 40.	- 52 17 12.			UNUSUAL SOUTHERN GALAXY
LB 00356	20 34 41.	- 13 13 00.		14.7	FAINT BLUE STAR
ZWG 424.027	20 34 42.	+ 12 28		15.1	GALAXY
ISS 0280	20 34 42.	+ 33 09	72		STELLAR RING
VDB.66N 135	20 34 45.	+ 32 14	168		REFLECTION NEBULA
ZZW 088	20 34 48.	+ 13 03			COMPACT GALAXY
ZWG 424.028	20 34 48.	+ 14 20		15.7	GALAXY
HN 0676	20 34 59.	- 76 38			NEBULA
IC 5022	20 34 59.	- 76 38			NONSTELLAR OBJECT
LBN 0229	20 35	+ 37 00	1500		BRIGHT NEBULA
LBN 0275	20 35	+ 40 30	480		BRIGHT NEBULA
LBN 0351	20 35	+ 46 00	9000		BRIGHT NEBULA
LBN 0376	20 35	+ 47 50	1560		BRIGHT NEBULA
LBN 0380	20 35	+ 48 00	6300		BRIGHT NEBULA
LDN 0870	20 35 00.	+ 35 00	20280		DARK NEBULA
LDN 0906	20 35 00.	+ 42 00	16080		DARK NEBULA
COU 065	20 35 00.	+ 45 15	7200		EMISSION NEBULA
COU 064	20 35 00.	+ 48 00	12000		EMISSION NEBULA
IC 5024	20 35 03.	- 71 17			NONSTELLAR OBJECT
RNGC 6937	20 35 04.	- 52 20			GALAXY
HN 0678	20 35 04.	- 71 17			NEBULA
LB 01100	20 35 05.	- 13 13 54.		16.5	FAINT BLUE STAR
ZWG 447.014	20 35 06.	+ 20 01		15.7	GALAXY
SG 3.174	20 35 11.	+ 48 27	2880		DIFFUSE EMISSION NEBULA
LDN 0930	20 35 12.	+ 45 10	180		DARK NEBULA
MCG-01-52-012	20 35 12.	- 06 18	24	15.5	GALAXY
MCG-05-48-026	20 35 12.	- 32 08	48	15.	GALAXY
ZWG 373.043	20 35 24.	- 02 12		15.7	GALAXY
UGC 11601	20 35 24.	- 02 12	60	15.7	GALAXY S
KARA.73B 0882	20 35 24.	- 02 12	66	15.7	ISOLATED GALAXY S
MCG-04-48-025	20 35 24.	- 25 19 30.	24	15.	GALAXY
MCG+00-52-041	20 35 27.	- 02 13 30.	54	15.5	GALAXY
ZWG 424.029	20 35 30.	+ 10 27		15.1	GALAXY
UGC 11602	20 35 30.	+ 10 27	90	15.1	GALAXY S0
MCG+02-52-023	20 35 30.	+ 10 27	60	15.	GALAXY
UGC 11603	20 35 30.	+ 63 31	84	17.	GALAXY S
MCG-07-42-015	20 35 36.	- 41 18	42	16.5	GALAXY
MCG-04-48-026	20 35 39.	- 25 20	18	15.	GALAXY
RNGC 6944A	20 35 40.	+ 06 44		15.0	GALAXY
MCG+01-52-016	20 35 42.	+ 06 42 30.	48	14.	GALAXY
ZWG 399.024	20 35 42.	+ 06 44		15.0	GALAXY
MCG-01-52-013	20 35 42.	- 04 19	24	15.5	GALAXY
ZWG 424.030	20 35 48.	+ 10 33		15.5	GALAXY
MCG+02-52-024	20 35 48.	+ 10 35	36	16.5	GALAXY
ZWG 424.031	20 35 48.	+ 10 36		15.7	GALAXY
ZWG 447.015	20 35 48.	+ 19 56		15.6	GALAXY
MCG-01-52-014	20 35 48.	- 04 19	72	15.	GALAXY
MCG-07-42-016	20 35 48.	- 44 48	60	15.	GALAXY
RNGC 6944	20 35 52.	+ 06 49		14.5	GALAXY
MCG+01-52-017	20 35 54.	+ 06 49	21	14.	GALAXY
ZWG 399.025	20 35 54.	+ 06 50		14.6	GALAXY
KHAV 692	20 36	+ 31 52	13450		DARK NEBULA
KHAV 694	20 36	+ 42 59	14030		DARK NEBULA
LBN 0339	20 36	+ 45 10	1620		BRIGHT NEBULA
LBN 0375	20 36	+ 47 40	780		BRIGHT NEBULA
KHAV 693	20 36	+ 56 52	6050		DARK NEBULA
LDN 1033	20 36 00.	+ 57 00	3420		DARK NEBULA
MCG+00-52-042	20 36 06.	+ 01 32 30.	48	16.	GALAXY
ZWG 373.044	20 36 06.	+ 01 33		15.4	GALAXY
KARA.73B 0883	20 36 06.	+ 01 33	60	15.4	ISOLATED GALAXY S
ZWG 424.032	20 36 06.	+ 10 22		15.7	GALAXY
ISS 0281	20 36 06.	+ 33 53	434		STELLAR RING
ISS 0282	20 36 06.	+ 34 02	307		STELLAR RING
ISS 0220	20 36 06.	+ 45 09	168		STELLAR RING
RNGC 6945	20 36 17.	- 05 11		13.5	GALAXY
ZWG 424.033	20 36 18.	+ 09 33		15.1	GALAXY
ZWG 424.034	20 36 18.	+ 12 00		14.9	GALAXY
MCG+02-52-025	20 36 18.	+ 12 00	48	14.5	GALAXY
MCG-01-52-015	20 36 18.	- 05 11	60	13.5	GALAXY
LB 01101	20 36 18.	- 14 34 00.		15.7	FAINT BLUE STAR
MCG-03-52-021	20 36 18.	- 20 25 30.	66	15.	GALAXY
MCG-01-52-016	20 36 24.	- 05 51	150	14.	GALAXY
MCG-05-48-027	20 36 24.	- 28 42	60	14.	GALAXY
VDB.66N 136	20 36 28.	+ 41 54	516		REFLECTION NEBULA
VVI 87	20 36 30.	+ 65 56	186	12.85	SEYFERT GALAXY
RNGC 6920	20 36 33.	- 80 10			UNVERIFIED SOUTHERN OBJECT
RNGC 6951	20 36 35.	+ 65 56		12.0	GALAXY
ISS 0155	20 36 36.	+ 48 33	125		STELLAR RING
ZWG 325.003	20 36 36.	+ 65 55		12.3	GALAXY
UGC 11604	20 36 36.	+ 65 55	240	12.3	GALAXY Sb/SBc
RNGC 6932	20 36 39.	- 73 48			UNVERIFIED SOUTHERN OBJECT
ZWG 373.045	20 36 48.	+ 01 55		15.2	GALAXY
MCG+11-25-002	20 36 48.	+ 65 56	192	11.8	GALAXY
MCG+00-52-043	20 36 54.	+ 01 50	42	15.	GALAXY
ZWG 373.046	20 36 54.	+ 01 51		15.2	GALAXY
ZWG 399.026	20 36 54.	+ 08 44		15.2	GALAXY
ZWG 424.035	20 36 54.	+ 10 37		15.2	GALAXY
UGC 11605	20 36 54.	+ 10 37	72	15.2	GALAXY E
LBN 0311	20 37	+ 42 10	1800		BRIGHT NEBULA
SC 2034+0140.9	20 37 00.	+ 01 51 24.	48		NEBULA
LDN 0922	20 37 00.	+ 44 25	240		DARK NEBULA
LDN 1036	20 37 00.	+ 57 00	1200		DARK NEBULA
LDN 1041	20 37 00.	+ 57 20	600		DARK NEBULA
RNGC 6952	20 37 00.	+ 66 16			NON-EXISTENT OBJECT
TON-S 0002	20 37 00.	- 27 28		15.5	BLUE STAR
RNGC 6942	20 37 01.	- 54 30		13.0	GALAXY
SG 3.175	20 37 11.	+ 41 58	1200		DIFFUSE EMISSION NEBULA
MCG+00-52-045	20 37 12.	+ 00 52 30.	60	15.5	GALAXY
ZWG 373.047	20 37 12.	+ 00 54		15.3	GALAXY
UGC 11606	20 37 12.	+ 00 54	60	15.3	GALAXY Sc
MCG+00-52-044	20 37 12.	+ 01 51	36	15.	GALAXY
ZWG 373.048	20 37 12.	+ 01 52		14.8	GALAXY
UGC 11607	20 37 12.	+ 01 52	60	14.8	GALAXY S?
ZWG 424.036	20 37 12.	+ 10 37		15.7	GALAXY
4ZW 064	20 37 12.	+ 27 04			COMPACT GALAXY
UGC 11609	20 37 12.	+ 27 04	33	10.5	GALAXY PAIR
UGC 11608	20 37 12.	+ 27 04	33	10.5	GALAXY PAIR
MCG-07-42-017	20 37 12.	- 40 16	30	15.5	GALAXY
IC 5029	20 37 14.	- 30 01 08.			MAY NOT EXIST
SC 2034+0141.6	20 37 15.	+ 01 52 07.	48		NEBULA
RNGC 6953	20 37 16.	+ 65 35			NON-EXISTENT OBJECT
ZWG 424.037	20 37 18.	+ 13 47		15.7	GALAXY
MRSL 081+00/1	20 37 19.	+ 42 03	1601		HII REGION
IC 5027	20 37 20.	- 55 39			NONSTELLAR OBJECT
HN 1206	20 37 20.	- 55 40	12	15.	NEBULA
ZC 2037.4+7800	20 37 24.	+ 78 00	5170		CLUSTER OF GALAXIES
ISS 0231	20 37 30.	+ 45 08	194		STELLAR RING
IC 5030	20 37 34.	- 30 02 06.			NONSTELLAR OBJECT
ZWG 399.027	20 37 36.	+ 07 05		14.9	GALAXY
UGC 11610	20 37 36.	+ 07 05	78	14.9	GALAXY Sb-0
ZWG 373.049	20 37 36.	- 01 47		15.5	GALAXY
MCG-04-48-027	20 37 36.	- 26 05	36	15.	GALAXY
SCHO 1188	20 37 39.	+ 47 28 30.	120		ISOLATED DARK CLOUD
SHAP 173	20 37 48.	+ 83 20	66		GROUP OF COMPACT GALAXIES
MCG-05-48-028	20 37 48.	- 32 42	60	14.	GALAXY
RNGC 6947	20 37 51.	- 32 42		14.0	GALAXY
LP 09942	20 38	- 80 48		12.5	FAINT BLUE STAR
MCG-05-48-029	20 38 00.	- 30 40	72	15.	GALAXY
ZWG 424.038	20 38 06.	+ 14 05		15.4	GALAXY
UGC 11611	20 38 06.	+ 14 05	78	15.4	GALAXY SBc
MCG-03-52-022	20 38 06.	- 20 40	30	15.	GALAXY
MCG-04-48-028	20 38 06.	- 22 25	42	15.	GALAXY
SCHO 1189	20 38 11.	+ 46 39 18.	400		ISOLATED DARK CLOUD
MCG+02-52-026	20 38 12.	+ 14 07	72	15.	GALAXY
ISS 0156	20 38 12.	+ 48 47	184		STELLAR RING
MCG+00-52-046	20 38 18.	+ 00 28	66	15.5	GALAXY
ZWG 373.050	20 38 18.	+ 00 29		15.7	GALAXY
UGC 11612	20 38 18.	+ 00 29	66	15.7	GALAXY Sb-c
KARA.73B 0884	20 38 18.	+ 00 29	102	15.7	ISOLATED GALAXY S
MCG+02-52-027	20 38 18.	+ 10 36	36	15.5	GALAXY
ZWG 424.039	20 38 18.	+ 10 37		15.5	GALAXY
LB 00201	20 38 21.	+ 00 19 48.		16.6	FAINT BLUE STAR
ISS 0232	20 38 24.	+ 44 50	177		STELLAR RING
LDN 0917	20 38 30.	+ 43 58	120		DARK NEBULA
TON-S 0003	20 38 30.	- 28 17		15.3	BLUE STAR
SCHO 1190	20 38 35.	+ 47 33 48.	170		ISOLATED DARK CLOUD
ISS 0233	20 38 36.	+ 45 14	204		STELLAR RING
SER 143.08	20 38 36.	- 50 23	20	15.5	PAIR OF PEC. GALAXIES
IC 5025	20 38 36.	- 77 09			NONSTELLAR OBJECT
HN 0679	20 38 36.	- 77 10			NEBULA
HN 2414	20 38 36.8	- 59 26 57.	18		NEBULA
LB 00202	20 38 45.	+ 00 06 00.		15.3	FAINT BLUE STAR
RNGC 6950	20 38 51.	+ 16 28			NON-EXISTENT OBJECT
LB 00357	20 38 51.	- 00 18 18.		15.4	FAINT BLUE STAR
UGC 11613	20 38 54.	+ 63 27	78	17.	GALAXY Sc
HN 0680	20 38 54.	- 65 49			NEBULA

OBJECT NAME	RIGHT ASCEN.	DECLINATION	DIAM.	MAGN.	TYPE OF OBJECT
IC 5028	20 38 54.	- 65 49			NONSTELLAR OBJECT
LBN 0143	20 39	+ 18 00	4200		BRIGHT NEBULA
LDN 0873	20 39 00.	+ 35 00	4020		DARK NEBULA
LDN 0911	20 39 00.	+ 42 45	420		DARK NEBULA
LDN 1157	20 39 00.	+ 67 50	240		DARK NEBULA
MCG-02-52-019	20 39 00.	- 14 02	48	15.	GALAXY
IC 1328	20 39 04.	- 19 48 50.			NONSTELLAR OBJECT
ZWG 424.040	20 39 06.	+ 12 53		15.6	GALAXY
UGC 11614	20 39 06.	+ 12 53	66	15.6	GALAXY Sb
MCG-03-52-023	20 39 06.	- 19 48 30.	48	15.	GALAXY
MCG+02-52-028	20 39 12.	+ 12 53	66	15.	GALAXY
SCHO 1191	20 39 12.	+ 37 34 30.	750		ISOLATED DARK CLOUD
LDN 0918	20 39 12.	+ 43 55	240		DARK NEBULA
MRSL 086+03/1	20 39 12.	+ 47 18	533		HII REGION
MCG-04-48-029	20 39 12.	- 24 24 30.	48	15.	GALAXY
LB C0203	20 39 13.	+ 00 34 54.		16.9	FAINT BLUE STAR
ZWG 399.028	20 39 18.	+ 05 25		15.0	GALAXY
ZWG 447.016	20 39 18.	+ 19 01		15.7	GALAXY
UGC 11615	20 39 18.	+ 19 01	84	15.7	GALAXY IRR
LDN 0928	20 39 18.	+ 44 25	180		DARK NEBULA
PK075-04.1	20 39 21.	+ 34 33 26.	10		PLANETARY NEBULA
ZWG 424.041	20 39 24.	+ 11 20		15.7	GALAXY
MCG+03-52-002	20 39 24.	+ 19 01	84	15.	GALAXY
MCG-04-48-030	20 39 30.	- 24 21	72	15.	GALAXY
ISS 0117	20 39 36.	+ 53 27	182		STELLAR RING
MRK 508	20 39 42.	- 05 48	22	16.	GALAXY WITH UV CONTINUUM
ZWG 424.042	20 39 48.	+ 11 23		15.5	GALAXY
MCG+02-52-029	20 39 48.	+ 11 23	36	16.	GALAXY
OCL 0165	20 39 48.	+ 35 22	3000	8.	OPEN STAR CLUSTER
MCG-01-52-017	20 39 48.	- 07 32	66	14.	GALAXY
RNGC 6943	20 39 49.	- 68 55		12.5	GALAXY
RNGC 6948	20 39 51.	- 53 32			UNVERIFIED SOUTHRN OBJECT
ZCG 2039+07	20 39 54.	+ 07 24		17.4	COMPACT GALAXY
2ZW 089	20 39 54.	+ 11 23			COMPACT GALAXY
HN 1207	20 39 54.	- 57 12		14.5	NEBULA
IC 5034	20 39 54.	- 57 12			NONSTELLAR OBJECT
HN 1208	20 39 55.	- 57 30	12	15.	NEBULA
IC 5033	20 39 55.	- 57 30			NONSTELLAR OBJECT
LBN 0247	20 40	+ 38 00	3600		BRIGHT NEBULA
LBN 0361	20 40	+ 45 50	2700		BRIGHT NEBULA
LBN 0579	20 40	+ 84 50	3600		BRIGHT NEBULA
LDN 0846	20 40 00.	+ 29 00	7020		DARK NEBULA
LDN 0864	20 40 00.	+ 32 00	9060		DARK NEBULA
LDN 0874	20 40 00.	+ 35 00	4020		DARK NEBULA
LDN 0896	20 40 00.	+ 39 30	660		DARK NEBULA
LDN 0898	20 40 00.	+ 39 40	1020		DARK NEBULA
LDN 1044	20 40 00.	+ 57 10	300		DARK NEBULA
LDN 1148	20 40 00.	+ 67 10	480		DARK NEBULA
LDN 1147	20 40 00.	+ 67 10	1440		DARK NEBULA
HUB E26	20 40 02.	+ 29 36			DIFFUSE NEBULA
SG 3.176	20 40 05.	+ 38 11	5400		DIFFUSE EMISSION NEBULA
IC 5039	20 40 11.3	- 30 02 03.	72	13.1	GALAXY E1
ZWG 399.029	20 40 12.	+ 06 22		15.7	GALAXY
2ZW 090	20 40 12.	+ 07 22			COMPACT GALAXY
MCG-05-49-001	20 40 12.	- 30 02	132	13.	GALAXY
UGC 11616	20 40 18.	+ 63 19	66	17.	GALAXY Sc
ACK 078-02.1	20 40 24.	+ 37 30			PLANETARY NEBULA
MCG+09-34-001	20 40 24.	+ 55 31	48		GALAXY
HN 1209	20 40 24.	- 57 19	12	15.	NEBULA
IC 5035	20 40 24.	- 57 19			NONSTELLAR OBJECT
IC 5041	20 40 31.	- 29 53 06.			NONSTELLAR OBJECT
SNO 28	20 40 33.	- 26 34 32.	720	17.	LINEAR GRP OF 7 GALAXIES
ISS 0283	20 40 36.	+ 35 18	92		STELLAR RING
ISS 0234	20 40 36.	+ 43 19	173		STELLAR RING
MCG-05-49-002	20 40 36.	- 29 52	120	13.	GALAXY
SER 141.02	20 40 36.	- 67 45	1020	17.	RICH CLUSTER OF GALAXIES
HN 0682	20 40 40.	- 67 43			NEBULA
IC 5031	20 40 40.	- 67 43			MAY NOT EXIST
LP 00358	20 40 42.	+ 01 02 24.		15.8	FAINT BLUE STAR
SG 3.177	20 40 46.	+ 42 51	1800		DIFFUSE EMISSION NEBULA
HN 0683	20 40 46.	- 67 43			NEBULA
IC 5032	20 40 46.	- 67 43			NONSTELLAR OBJECT
MRSL 077-03/1	20 40 48.	+ 36 10	300		HII REGION
MCG-01-52-018	20 40 48.	- 04 41	15	15.	GALAXY
MCG-02-52-020	20 40 48.	- 10 16	24	15.5	GALAXY
IC 5036	20 40 49.	- 57 48			NONSTELLAR OBJECT
SC 2038-0029.2	20 40 50.	- 00 18 29.	12		NEBULA
HN 1210	20 40 50.	- 57 48	60		NEBULA
ISS 0157	20 40 54.	+ 46 53	293		STELLAR RING
LBN 0231	20 41	+ 36 10	480		BRIGHT NEBULA
LBN 0235	20 41	+ 36 30	9900		BRIGHT NEBULA
LBN 0265	20 41	+ 39 00	2400		BRIGHT NEBULA
LBN 0277	20 41	+ 39 40	1800		BRIGHT NEBULA
LBN 0322	20 41	+ 42 50	360		BRIGHT NEBULA
LBN 0323	20 41	+ 42 55	900		BRIGHT NEBULA
ZWG 448.001	20 41 00.	+ 21 27		15.7	GALAXY
ZWG 447.017	20 41 00.	+ 21 27		15.7	GALAXY
LDN 1039	20 41 00.	+ 56 40	1140		DARK NEBULA
LDN 1049	20 41 00.	+ 57 20	240		DARK NEBULA
ZWG 448.002	20 41 06.	+ 16 48		15.3	GALAXY
ZWG 447.018	20 41 06.	+ 16 48		15.3	GALAXY
ZWG 374.001	20 41 06.	- 01 59		15.7	GALAXY
ACK 084+02.1	20 41 12.	+ 45 46			PLANETARY NEBULA
HN 2415	20 41 13.7	- 46 09 24.	90		NEBULA
MCG-01-53-002	20 41 18.	- 03 23	36	14.5	GALAXY
MCG-01-53-001	20 41 18.	- 03 23 30.	30	16.	GALAXY
MCG-01-53-003	20 41 18.	- 03 24	12	18.	GALAXY
IC 1329	20 41 22.	+ 15 24 21.			NONSTELLAR OBJECT
UGC 11617	20 41 24.	+ 14 08	60	16.0	GALAXY Sc
ZWG 374.003	20 41 24.	- 00 33		15.7	GALAXY
KARA.73B 0885	20 41 24.	- 00 33	48	15.7	ISOLATED GALAXY S
ZWG 374.002	20 41 24.	- 02 04		15.5	GALAXY
SC 2038-0115.2	20 41 29.	- 01 04 27.	12		NEBULA
ZC 2041.5+0702	20 41 30.	+ 07 02	1210		CLUSTER OF GALAXIES
MCG+02-53-001	20 41 30.	+ 12 20	132	12.	GALAXY
LDN 1051	20 41 30.	+ 57 20	360		DARK NEBULA
MRK 509	20 41 30.	- 10 54	10	13.	GALAXY WITH UV CONTINUUM
VVI 88	20 41 30.	- 10 54		13.	SEYFERT GAL1XY
RNGC 6954	20 41 33.	+ 03 02		14.0	GALAXY
MCG+00-53-001	20 41 36.	+ 03 01	48	14.	GALAXY
ZWG 374.004	20 41 36.	+ 03 02		14.2	GALAXY
UGC 11618	20 41 36.	+ 03 02	60	14.2	GALAXY S
KARA.73B 0886	20 41 36.	+ 03 02	60	14.2	ISOLATED GALAXY E
ZWG 425.001	20 41 36.	+ 12 20		13.5	GALAXY
UGC 11619	20 41 36.	+ 12 20	102	13.5	GALAXY SBb
OCL 0176	20 41 36.	+ 36 52	300	14.	OPEN STAR CLUSTER
RNGC 6956	20 41 37.	+ 12 20		13.5	GALAXY
HN 1211	20 41 40.	- 58 37	60		NEBULA
IC 5037	20 41 40.	- 58 37			NONSTELLAR OBJECT
MCG+02-53-002	20 41 42.	+ 12 15	36	14.5	GALAXY
IC 5046	20 41 42.	- 30 05 47.			NONSTELLAR OBJECT
HN 0681	20 41 43.	- 78 15			NEBULA
IC 5026	20 41 43.	- 78 15			NONSTELLAR OBJECT
ZWG 374.005	20 41 48.	+ 02 25		14.9	GALAXY
UGC 11621	20 41 48.	+ 02 25	90	14.9	GALAXY Sb
ZWG 425.002	20 41 48.	+ 12 14		14.5	GALAXY
UGC 11620	20 41 48.	+ 12 14	42	14.5	GALAXY
IC 5047	20 41 48.	- 29 55 46.			NONSTELLAR OBJECT
RNGC 6955	20 41 50.	+ 02 25		15.0	GALAXY
SG 3.178	20 41 52.	+ 39 06	4200		DIFFUSE EMISSION NEBULA
SG 3.179	20 41 53.	+ 38 30	600		DIFFUSE EMISSION NEBULA
HN 2416	20 41 55.2	- 45 47 45.			NEBULA
LBN 0324	20 42	+ 42 50	4800		BRIGHT NEBULA
KHAV 695	20 42	+ 67 29	2230		DARK NEBULA
LBN 0468	20 42	+ 67 40	3300		BRIGHT NEBULA
MCG+02-53-003	20 42 00.	+ 12 19	60	14.5	GALAXY
ZWG 425.003	20 42 06.	+ 12 19		15.2	GALAXY
UGC 11623	20 42 06.	+ 12 19	66	14.5	GALAXY SBa
MRSL 098+12/1	20 42 06.	+ 63 03	180		HII REGION
ZWG 374.006	20 42 06.	- 01 54		15.2	GALAXY
UGC 11622	20 42 06.	- 01 54	66	15.2	GALAXY SB
SCHO 1192	20 42 07.	+ 46 24 48.	330		ISOLATED DARK CLOUD
ZWG 374.007	20 42 18.	+ 02 24		15.3	GALAXY
RNGC 6957	20 42 20.	+ 02 24		15.5	GALAXY
MCG-02-53-001	20 42 24.	- 11 17	72	14.5	GALAXY
ZWG 374.008	20 42 30.	+ 01 20		15.7	GALAXY
KARA.73B 0887	20 42 30.	+ 01 20	60	15.7	ISOLATED GALAXY S
COU 066	20 42 30.	+ 38 50	3600		EMISSION NEBULA
IC 5038	20 42 30.	- 65 12			NONSTELLAR OBJECT
HN 0684	20 42 31.	- 65 12			NEBULA
IC 5050	20 42 37.	- 05 48 45.			NONSTELLAR OBJECT
ISS 0235	20 42 42.	+ 39 17	137		STELLAR RING
MCG-01-53-004	20 42 42.	- 05 42	60	14.5	GALAXY
MCG-01-53-005	20 42 42.	- 05 50	60	14.	GALAXY
SCHO 1193	20 42 50.	+ 46 23 12.	240		ISOLATED DARK CLOUD
HN 1212	20 42 53.	- 57 10	30		NEBULA
IC 5043	20 42 53.	- 57 10			NONSTELLAR OBJECT
ZC 2042.9+0531	20 42 54.	+ 05 31	5170		CLUSTER OF GALAXIES
ZWG 400.001	20 42 54.	+ 05 53		15.3	GALAXY
LBN 0263	20 43	+ 38 40	4800		BRIGHT NEBULA
LBN 0302	20 43	+ 40 50	1800		BRIGHT NEBULA
KHAV 696	20 43	+ 48 22	3540		DARK NEBULA
2ZW 091	20 43 00.	+ 12 46			COMPACT GALAXY
ZWG 425.004	20 43 00.	+ 12 47		15.6	GALAXY
ISS 0284	20 43 00.	+ 36 37	1139		STELLAR RING
COU 067	20 43 00.	+ 40 40	2100		EMISSION NEBULA
LDN 1155	20 43 00.	+ 67 30	300		DARK NEBULA
ZC 2043.1+0204	20 43 06.	+ 02 04	1080		CLUSTER OF GALAXIES
MCG-01-53-006	20 43 06.	- 05 08	72	15.	GALAXY
MCG-01-53-007	20 43 06.	- 05 48	36	15.	GALAXY
SCHO 1194	20 43 08.	+ 30 35 00.	250		ISOLATED DARK CLOUD
SG 3.180	20 43 10.	+ 30 13	480		DIFFUSE EMISSION NEBULA
ZWG 448.003	20 43 12.	+ 19 26		15.6	GALAXY
OCL 0169	20 43 12.	+ 35 19	540	11.	OPEN STAR CLUSTER
SCHO 1195	20 43 13.	+ 46 36 12.	300		ISOLATED DARK CLOUD
HELW 225	20 43 16.	- 00 00 03.			NEBULA
HELW 226	20 43 17.	- 00 00 03.			NEBULA
ZWG 374.009	20 43 18.	+ 00 00		14.9	GALAXY
ZWG 470.001	20 43 18.	+ 23 03		15.7	GALAXY
HN 2417	20 43 22.5	- 44 48 58.	18		NEBULA
MCG+04-49-001	20 43 24.	+ 22 32	24	15.5	GALAXY
ZWG 470.002	20 43 24.	+ 22 33		15.5	GALAXY
LDN 0907	20 43 24.	+ 40 50	840		DARK NEBULA
ACK 084+01.1	20 43 24.	+ 44 28			PLANETARY NEBULA
IC 5042	20 43 24.	- 65 16			NONSTELLAR OBJECT
HN 0685	20 43 25.	- 65 16			NEBULA
IC 1330	20 43 29.	- 14 12 28.			NONSTELLAR OBJECT
HN 2418	20 43 29.6	- 44 23 40.	24		NEBULA
ISS 0158	20 43 30.	+ 49 12	2681		STELLAR RING
MCG-02-53-002	20 43 30.	- 14 13	48	15.	GALAXY
RNGC 6960	20 43 34.	+ 30 32			DIFFUSE NEBULA
CED 182A	20 43 34.	+ 30 33	4200		DIFFUSE GALACTIC NEBULA
SG 3.181	20 43 34.	+ 30 33	4200		DIFFUSE EMISSION NEBULA
ZWG 491.001	20 43 36.	+ 28 00		15.5	GALAXY
LDN 0915	20 43 36.	+ 42 55	300		DARK NEBULA
IC 5049	20 43 39.	- 38 39 48.			NONSTELLAR OBJECT
MCG+00-53-002	20 43 42.	- 00 23 30.	36	14.5	GALAXY
SC 2041-0034.4	20 43 43.	- 00 23 32.	12		NEBULA
UGC 11624	20 43 48.	+ 06 30	66	17.	GALAXY DWARF
ACK 088+04.1	20 43 48.	+ 50 12			PLANETARY NEBULA
ZWG 374.010	20 43 48.	- 00 24		15.2	GALAXY
MCG-01-53-008	20 43 48.	- 02 59	60	15.	GALAXY
MCG-04-49-001	20 43 48.	- 23 50	42	14.5	GALAXY
ZWG 448.004	20 43 54.	+ 18 45		15.0	GALAXY
UGC 11625	20 43 54.	+ 28 22	72	16.5	GALAXY S
HELW 482	20 43 56.	- 38 16 20.			NEBULA
LBN 0112	20 44	- 05 50	1440		BRIGHT NEBULA
VDB.66G 210	20 44	- 13 04	100		DWARF GALAXY
UGC 11626	20 44 00.	+ 00 03	60	16.0	GALAXY Sc-IRR
MCG+03-53-001	20 44 00.	+ 18 47	24	15.	GALAXY
LDN 1158	20 44 00.	+ 67 30	1320		DARK NEBULA
HELW 227	20 44 01.	- 00 00 37.			NEBULA
ZWG 374.011	20 44 06.	+ 00 09		15.6	GALAXY
HELW 228	20 44 06.	+ 00 09 24.			NEBULA
HELW 483	20 44 11.	- 38 35 37.			NEBULA
ISS 0159	20 44 12.	+ 45 32	251		STELLAR RING
ZWG 374.012	20 44 12.	- 01 32		15.3	GALAXY
SCHO 1196	20 44 13.	+ 31 12 18.	340		ISOLATED DARK CLOUD
SC 2041-0144.0	20 44 15.	- 01 33 06.	12		NEBULA
MCG-02-53-003	20 44 18.	- 13 03	120	17.	GALAXY
SER 141.03	20 44 18.	- 69 16		14.	HIGH SURF BRGHTNSS GALAXY
UGC 11627	20 44 24.	+ 05 27	78	17.	GALAXY
ZWG 425.005	20 44 24.	+ 11 06		15.5	GALAXY
ZWG 374.013	20 44 30.	+ 00 15		14.6	GALAXY
MCG-04-49-002	20 44 30.	- 22 44	30	15.5	GALAXY
RNGC 6959	20 44 31.	+ 00 15		14.5	GALAXY
HELW 229	20 44 31.	+ 00 15 19.			NEBULA
HELW 229	20 44 31.	+ 00 15 19.			NEBULA
ZWG 374.014	20 44 36.	+ 00 11		14.8	GALAXY
SCHO 1197	20 44 36.	+ 37 24 00.	400		ISOLATED DARK CLOUD
RNGC 6961	20 44 37.	+ 00 11		15.0	GALAXY
RNGC 6965	20 44 37.	+ 00 15		14.0	GALAXY
BIGO 546	20 44 41.	+ 00 08			NEBULA
IC 5057	20 44 41.	+ 00 08			SINGLE STAR
ISS 0118	20 44 42.	+ 53 30	195		STELLAR RING
ISS 0119	20 44 42.	+ 51 34	226		STELLAR RING

OBJECT NAME	RIGHT ASCEN.	DECLINATION	DIAM.	MAGN.	TYPE OF OBJECT
MCG-03-53-001	20 44 42.	- 15 43	72	15.	GALAXY
MCG-04-49-003	20 44 42.	- 22 42	24	16.	GALAXY
TON-S 0004	20 44 42.	- 29 27		15.0	BLUE STAR
SCHO 1198	20 44 43.	+ 31 15 24.	380		ISOLATED DARK CLOUD
HELW 484	20 44 44.	- 38 18 35.			NEBULA
MCG+00-53-003	20 44 45.	+ 00 10	162	12.8	GALAXY
ZWG 374.015	20 44 48.	+ 00 08		13.5	GALAXY
UGC 11628	20 44 48.	+ 00 08	180	13.5	GALAXY Sa
KARA.72 548A	20 44 48.	+ 00 08	168	13.5	PART OF DOUBLE GALAXY
ZWG 374.016	20 44 48.	+ 00 18		15.2	GALAXY
IC 5058	20 44 48.	+ 00 18			SINGLE STAR
MCG+00-53-004	20 44 48.	+ 00 20	30	15.2	GALAXY
ZWG 425.006	20 44 48.	+ 10 54		15.7	GALAXY
RNGC 6964	20 44 49.	+ 00 07		14.0	GALAXY
RNGC 6962	20 44 49.	+ 00 08		13.5	GALAXY
RNGC 6966	20 44 49.	+ 00 10			NON-EXISTENT OBJECT
BIGO 547	20 44 49.	+ 00 11			NEBULA
RNGC 6963	20 44 49.	+ 00 18		15.0	GALAXY
MCG+00-53-005	20 44 51.	+ 00 08	36	14.2	GALAXY
MCG-03-53-002	20 44 51.	- 20 06	60	15.5	GALAXY
SC 2042-0110.4	20 44 52.	- 00 59 28.	12		NEBULA
ZWG 374.017	20 44 54.	+ 00 07		14.2	GALAXY
UGC 11629	20 44 54.	+ 00 07	96	14.2	GALAXY E
KARA.72 548B	20 44 54.	+ 00 07	90	14.2	PART OF DOUBLE GALAXY
4ZW 065	20 44 54.	+ 25 15			COMPACT GALAXY
ISS 0285	20 44 54.	+ 35 50	1139		STELLAR RING
ZC 2044.9-0315	20 44 54.	- 03 15	20360		CLUSTER OF GALAXIES
MCG-05-49-003	20 44 54.	- 26 52	48	15.	GALAXY
MCG+00-53-006	20 44 57.	+ 00 15	42	14.	GALAXY
KHAV 697	20 45	+ 38 29	6270		DARK NEBULA
LBN 0309	20 45	+ 40 50	4500		BRIGHT NEBULA
LBN 0341	20 45	+ 44 00	3300		BRIGHT NEBULA
MIL 91	20 45	+ 50 30	7200		SUPERNOVA REMNANT
ZWG 374.018	20 45 00.	+ 00 13		14.3	GALAXY
UGC 11630	20 45 00.	+ 00 13	60	14.3	GALAXY S?
LDN 0868	20 45 00.	+ 32 00	3240		DARK NEBULA
COU 069	20 45 00.	+ 40 25	6000		EMISSION NEBULA
COU 068	20 45 00.	+ 42 10	6000		EMISSION NEBULA
COU 070	20 45 00.	+ 43 30	2400		EMISSION NEBULA
LDN 1038	20 45 00.	+ 56 10	3300		DARK NEBULA
MCG-02-53-004	20 45 00.	- 10 08	48	15.5	GALAXY
RNGC 6967	20 45 01.	+ 00 13		14.5	GALAXY
SG 3.182	20 45 03.	+ 31 18	600		DIFFUSE EMISSION NEBULA
BIGO 548	20 45 05.	+ 00 09			NEBULA
IC 5061	20 45 05.	+ 00 09			THREE STARS
ZWG 425.007	20 45 06.	+ 12 54		14.7	GALAXY
IC 1331	20 45 06.	- 10 10 36.			NONSTELLAR OBJECT
HELW 485	20 45 07.	- 38 13 46.			NEBULA
ZWG 448.005	20 45 12.	+ 16 13		15.7	GALAXY
ISS 0236	20 45 12.	+ 39 53	90		STELLAR RING
MCG-02-53-005	20 45 12.	- 10 10 30.	108	14.5	GALAXY
HN 2419	20 45 15.3	- 43 28 10.	18		NEBULA
MCG-03-53-003	20 45 18.	- 20 03 30.	9	15.5	GALAXY
TON-S 0005	20 45 18.	- 28 30		15.8	BLUE STAR
HELW 230	20 45 23.	+ 00 02 46.			NEBULA
ZWG 448.006	20 45 24.	+ 20 36		15.3	GALAXY
4ZW 066	20 45 24.	+ 25 17			COMPACT GALAXY
ARC 2328	20 45 24.	- 18 00		16.4	RICH CLUSTER OF GALAXIES
MCG-03-53-004	20 45 24.	- 20 02	36	16.	GALAXY
MCG-03-53-005	20 45 24.	- 20 03	60	15.	GALAXY
RNGC 6958	20 45 24.	- 38 12		12.5	GALAXY
MCG+00-53-007	20 45 27.	- 00 21	66	14.	GALAXY
SC 2042-0032.1	20 45 29.	- 00 21 08.	72		NEBULA
IC 5062	20 45 29.	- 08 33			NONSTELLAR OBJECT
IC 5044	20 45 29.	- 72 09			NONSTELLAR OBJECT
COU 070A	20 45 30.	+ 42 50	2100		EMISSION NEBULA
ZWG 374.019	20 45 30.	- 00 22		15.0	GALAXY
UGC 11631	20 45 30.	- 00 22	66	15.0	GALAXY IRR
HN 0687	20 45 30.	- 72 08			NEBULA
HN 2420	20 45 33.5	- 47 28 51.	60		NEBULA
SG 3.183	20 45 34.	+ 30 27	360		DIFFUSE EMISSION NEBULA
IC 5045	20 45 35.	- 72 10			NONSTELLAR OBJECT
2ZW 092	20 45 36.	- 00 06			COMPACT GALAXY
HN 0688	20 45 36.	- 72 09			NEBULA
B 349	20 45 38.	+ 43 46	360		DARK OBJECT
ZWG 425.008	20 45 42.	+ 13 33		15.4	GALAXY
KARA.73B 0888	20 45 42.	+ 13 33	48	15.4	ISOLATED GALAXY S
MCG+12-19-003	20 45 42.	+ 69 05	45	16.	GALAXY
MCG-01-53-009	20 45 42.	- 04 03	60	13.	GALAXY
MCG-01-53-010	20 45 42.	- 04 03 30.	48	15.	GALAXY
TON-S 0006	20 45 42.	- 28 19		15.2	BLUE STAR
MCG-01-53-011	20 45 45.	- 06 00	72	15.	GALAXY
RNGC 6968	20 45 45.	- 08 32		14.0	GALAXY
HN 0691	20 45 45.	- 39 22			NEBULA
IC 5056	20 45 45.	- 39 22			NONSTELLAR OBJECT
SG 3.184	20 45 46.	+ 30 45	900		DIFFUSE EMISSION NEBULA
ZWG 448.007	20 45 48.	+ 16 11		15.6	GALAXY
UGC 11632	20 45 48.	+ 69 06	66	16.0	GALAXY SBa-b
MCG+13-15-001	20 45 48.	+ 79 58	132	13.	GALAXY
MCG-02-53-006	20 45 51.	- 08 32	60	14.	GALAXY
HELW 231	20 45 52.	+ 00 11 47.			NEBULA
HN 2421	20 45 52.1	- 44 23 32.	18		NEBULA
SG 2.073	20 45 53.	+ 42 46	1800		DIFFUSE EMISSION NEBULA
MCG+01-53-001	20 45 54.	+ 07 32	60	14.5	GALAXY
SG 2.074	20 45 54.	+ 42 26	2100		DIFFUSE EMISSION NEBULA
SG 3.185	20 45 58.	+ 30 19	420		DIFFUSE EMISSION NEBULA
RNGC 6969	20 45 59.	+ 07 33		15.0	GALAXY
LBN 0312	20 46	+ 40 50	2400		BRIGHT NEBULA
LBN 0320	20 46	+ 42 00	3600		BRIGHT NEBULA
LBN 0335	20 46	+ 43 30	3600		BRIGHT NEBULA
LBN 0338	20 46	+ 43 40	1620		BRIGHT NEBULA
SER 143.03	20 46	- 47 40	5120	13.	CLOUD OF GALAXIES
ZWG 400.002	20 46 00.	+ 07 33		15.0	GALAXY
UGC 11633	20 46 00.	+ 07 33	72	15.0	GALAXY Sa
KARA.73B 0889	20 46 00.	+ 07 33	78	15.0	ISOLATED GALAXY S
ZWG 448.008	20 46 00.	+ 16 32		14.7	GALAXY
UGC 11634	20 46 00.	+ 16 32	90	14.7	GALAXY DBL SYS
LDN 0931	20 46 00.	+ 43 40	420		DARK NEBULA
ZWG 358.001	20 46 00.	+ 79 58		14.2	GALAXY
ZWG 357.011	20 46 00.	+ 79 58		14.2	GALAXY
UGC 11635	20 46 00.	+ 79 58	210	14.2	GALAXY Sb
KARA.73B 0890	20 46 00.	+ 79 58	180	14.2	ISOLATED GALAXY S
TON-S 0008	20 46 00.	- 25 25		14.9	BLUE STAR
TON-S 0007	20 46 00.	- 27 28		15.3	BLUE STAR
CED 183A	20 46 03.	+ 44 11			DIFFUSE GALACTIC NEBULA
IC 5067	20 46 05.	+ 44 11			NONSTELLAR OBJECT
ZC 2046.1+2242	20 46 06.	+ 22 42	8530		CLUSTER OF GALAXIES
HN 0686	20 46 12.	- 76 52			NEBULA
IC 5040	20 46 12.	- 76 52			NONSTELLAR OBJECT
MCG-02-53-007	20 46 15.	- 08 37	36	15.	GALAXY
SCHO 1199	20 46 16.	+ 31 32 06.	190		ISOLATED DARK CLOUD
SG 3.186	20 46 21.	+ 31 46	1200		DIFFUSE EMISSION NEBULA
HN 2422	20 46 21.1	- 47 00 42.	24		NEBULA
SCHO 1200	20 46 29.	+ 31 03 48.	400		ISOLATED DARK CLOUD
HN 2423	20 46 29.2	- 43 35 42.	12		NEBULA
ZWG 448.009	20 46 30.	+ 16 31		15.6	GALAXY
MCG+04-49-002	20 46 30.	+ 25 05	33	16.5	GALAXY
UGC 11636	20 46 30.	+ 64 17	90	16.0	GALAXY SBc
SER 143.06	20 46 30.	- 50 18	100	14.	LOW SURF. BRGHTNSS GALAXY
SG 2.075	20 46 31.	+ 41 31	2100		DIFFUSE EMISSION NEBULA
MCG+04-49-003	20 46 33.	+ 25 05	12	17.	GALAXY
SG 3.187	20 46 34.	+ 30 56	1200		DIFFUSE EMISSION NEBULA
B 148	20 46 34.	+ 59 27	240		DARK OBJECT
HN 0689	20 46 34.	- 71 59			NEBULA
IC 5048	20 46 34.	- 71 59			NONSTELLAR OBJECT
SG 3.188	20 46 39.	+ 31 22	2400		DIFFUSE EMISSION NEBULA
HELW 232	20 46 41.	+ 00 08 02.			NEBULA
SER 141.04	20 46 42.	- 68 05	180	17.	LOOSE GROUP OF GALAXIES
SC 2044-0252.3	20 46 44.	- 02 41 16.	12		NEBULA
SG 3.189	20 46 47.	+ 29 40	3600		DIFFUSE EMISSION NEBULA
HN 2424	20 46 47.7	- 47 39 29.	18		NEBULA
MCG+01-53-002	20 46 48.	+ 05 47	60	14.	GALAXY
ZC 2046.8+2506	20 46 48.	+ 25 06	4230		CLUSTER OF GALAXIES
FEIG 113	20 46 48.	+ 30 05		12.0	FAINT BLUE STAR
SCHO 1201	20 46 49.	+ 31 31 00.	200		ISOLATED DARK CLOUD
SG 3.192	20 46 50.	+ 32 03	420		DIFFUSE EMISSION NEBULA
SG 3.194	20 46 51.	+ 40 16	6000		DIFFUSE EMISSION NEBULA
SG 3.191	20 46 52.	+ 30 50	720		DIFFUSE EMISSION NEBULA
SG 3.190	20 46 53.	+ 30 08	480		DIFFUSE EMISSION NEBULA
2ZW 093	20 46 54.	+ 08 16			COMPACT GALAXY
ZWG 425.009	20 46 54.	+ 11 07		15.2	GALAXY
HN 2425	20 46 54.1	- 43 41 53.	12		NEBULA
RNGC 6971	20 46 58.	+ 05 48		15.0	GALAXY
LBN 0329	20 47	+ 42 50	2400		BRIGHT NEBULA
LB 09943	20 47	- 86 49		14.4	FAINT BLUE STAR
ZWG 400.003	20 47 00.	+ 05 48		14.8	GALAXY
UGC 11637	20 47 00.	+ 05 48	72	14.8	GALAXY Sb
ZWG 448.010	20 47 00.	+ 16 53		15.6	GALAXY
COU 072	20 47 00.	+ 31 25	2400		EMISSION NEBULA
COU 071	20 47 00.	+ 40 20	3000		EMISSION NEBULA
MRSL 084+00/1	20 47 00.	+ 43 38			HII REGION
MCG-04-49-004	20 47 00.	- 25 53	36	15.	GALAXY
SG 3.193	20 47 04.	+ 30 35	3600		DIFFUSE EMISSION NEBULA
SC 2044-0252.2	20 47 05.	- 02 41 09.	18		NEBULA
MCG+03-53-002	20 47 06.	+ 16 55	36	16.5	GALAXY
SG 3.196	20 47 08.	+ 48 02	3000		DIFFUSE EMISSION NEBULA
ISS 0286	20 47 12.	+ 33 51	155		STELLAR RING
IC 5065	20 47 13.	- 30 00 44.			NONSTELLAR OBJECT
YC 2047-33	20 47 15.	- 33 00 12.			UNUSUAL SOUTHERN NEBULA
HN 0690	20 47 15.	- 71 58			NEBULA
IC 5051	20 47 15.	- 71 56			NONSTELLAR OBJECT
MCG+01-53-003	20 47 18.	+ 06 00	18	15.	GALAXY
ZWG 448.011	20 47 18.	+ 16 40		14.3	GALAXY
UGC 11638	20 47 18.	+ 16 40	126	14.3	GALAXY SBb
MCG+03-53-003	20 47 21.	+ 16 41	120	13.	GALAXY
ZWG 400.004	20 47 24.	+ 06 02		14.6	GALAXY
UGC 11639	20 47 24.	+ 06 02	72	14.6	GALAXY COMPACT
MCG+01-53-004	20 47 24.	+ 03 27	30	16.	GALAXY
ZC 2047.4+2050	20 47 24.	+ 20 50	1010		CLUSTER OF GALAXIES
B 350	20 47 24.	+ 45 42	180		DARK OBJECT
MCG-01-53-012	20 47 24.	- 07 13	180	14.	GALAXY
MCG-02-53-008	20 47 24.	- 10 00	60	14.	GALAXY
MCG-03-53-006	20 47 24.	- 19 38	150	15.	GALAXY
HN 1213	20 47 24.	- 57 52	18		NEBULA
IC 5059	20 47 24.	- 57 52			NONSTELLAR OBJECT
HN 0692	20 47 24.	- 69 24			NEBULA
SG 3.195	20 47 29.	+ 30 08	1200		DIFFUSE EMISSION NEBULA
ZWG 425.010	20 47 30.	+ 09 46		15.5	GALAXY
IC 5052	20 47 30.	- 69 25	360	12.3	GALAXY Sc
SER 141.05	20 47 30.	- 69 54	20	14.	INTERACTING GALAXIES
RNGC 6972	20 47 35.	+ 09 43		14.5	GALAXY
MCG+02-53-004	20 47 36.	+ 09 42	60	14.	GALAXY
ZWG 425.011	20 47 36.	+ 09 43		14.3	GALAXY
UGC 11640	20 47 36.	+ 09 43	72	14.3	GALAXY S0-a
ZWG 425.012	20 47 36.	+ 12 51		15.6	GALAXY
FEIG 114	20 47 36.	+ 29 54		12.5	FAINT BLUE STAR
SG 3.197	20 47 39.	+ 31 56	360		DIFFUSE EMISSION NEBULA
HN 1214	20 47 40.	- 47 15		13.	NEBULA
MCG+02-53-005	20 47 42.	+ 13 20	60	15.	GALAXY
PK059-18.1	20 47 42.	+ 13 22	161	14.6	PLANETARY NEBULA
SC 2045-0129.3	20 47 42.	- 01 18 13.	12		NEBULA
B 149	20 47 52.	+ 59 21	120		DARK OBJECT
ZWG 425.013	20 47 54.	+ 13 10		15.6	GALAXY
TON-S 0009	20 47 54.	- 26 18		15.3	BLUE STAR
SCHO 1202	20 47 56.	+ 37 43 48.	370		ISOLATED DARK CLOUD
LBN 0230	20 48	+ 35 00	1500		BRIGHT NEBULA
KHAV 698	20 48	+ 59 53	1860		DARK NEBULA
LB 10001	20 48	- 87 57		15.0	FAINT BLUE STAR
ZWG 400.005	20 48 00.	+ 07 05		15.6	GALAXY
COU 073	20 48 00.	+ 42 21	1800		EMISSION NEBULA
COU 074	20 48 00.	+ 43 00	10800		EMISSION NEBULA
ISS 0160	20 48 00.	+ 45 18	277		STELLAR RING
ZWG 425.014	20 48 06.	+ 09 38		15.4	GALAXY
CED 183B	20 48 06.	+ 42 21	2580		DIFFUSE GALACTIC NEBULA
IC 5068	20 48 07.	+ 42 21			NONSTELLAR OBJECT
LDN 1076	20 48 12.	+ 59 40	360		DARK NEBULA
VVI 89	20 48 12.	- 57 16	80	13.16	SEYFERT GALAXY
IC 5063	20 48 12.	- 57 16	78	12.6	GALAXY SAO
HN 0693	20 48 12.	- 68 38			NEBULA
IC 5055	20 48 13.	- 68 38			NONSTELLAR OBJECT
MIN.47 13	20 48 15.	+ 44 11			DIFFUSE NEBULA
ZWG 448.012	20 48 18.	+ 16 25		15.7	GALAXY
ZWG 374.020	20 48 18.	- 00 32		15.4	GALAXY
KARA.73B 0891	20 48 18.	- 00 32	30	15.4	ISOLATED GALAXY E
SG 3.198	20 48 21.	+ 31 52	1200		DIFFUSE EMISSION NEBULA
SG 2.076	20 48 24.	+ 42 21	1080		DIFFUSE EMISSION NEBULA
2ZW 094	20 48 24.	- 00 30			COMPACT GALAXY
SG 3.199	20 48 29.	+ 29 59	3000		DIFFUSE EMISSION NEBULA
HN 2426	20 48 29.1	- 44 04 47.	48		NEBULA
ZWG 448.013	20 48 30.	+ 21 18		15.6	GALAXY
RNGC 6970	20 48 33.	- 48 59		13.5	GALAXY
HN 0694	20 48 33.	- 71 20			NEBULA
IC 5053	20 48 33.	- 71 20			NONSTELLAR OBJECT
ZWG 470.003	20 48 36.	+ 21 41		15.7	GALAXY
UGC 11641	20 48 36.	+ 21 41	60	15.7	GALAXY SB
UGC 11642	20 48 36.	+ 70 03	7[illegible]	16.0	GALAXY S

OBJECT NAME	RIGHT ASCEN.	DECLINATION	DIAM.	MAGN.	TYPE OF OBJECT
MCG-03-53-007	20 48 36.	- 16 06	60	14.5	GALAXY
HN 0695	20 48 37.	- 71 13			NEBULA
IC 5054	20 48 38.	- 71 13			NONSTELLAR OBJECT
SG 3.200	20 48 39.	+ 31 22	720		DIFFUSE EMISSION NEBULA
HN 1215	20 48 40.	- 57 25		14.	NEBULA
IC 5064	20 48 40.	- 57 25			NONSTELLAR OBJECT
ZWG 448.014	20 48 42.	+ 16 23		15.6	GALAXY
MCG+12-19-004	20 48 42.	+ 70 01	66	16.	GALAXY
MCG+12-19-004	20 48 42.	+ 70 01	66	16.	GALAXY
MCG-05-49-004	20 48 42.	- 30 02	18	15.	GALAXY
SG 2.077	20 48 43.	+ 42 06	1080		DIFFUSE EMISSION NEBULA
RNGC 6974	20 48 46.	+ 31 40			DIFFUSE NEBULA
SG 3.201	20 48 47.	+ 34 51	2400		DIFFUSE EMISSION NEBULA
7ZW 936	20 48 48.	+ 63 44			COMPACT GALAXY
MCG-01-53-013	20 48 48.	- 07 30	36	14.5	GALAXY
MCG-05-49-005	20 48 48.	- 31 05	54	14.5	GALAXY
MCG-04-49-005	20 48 54.	- 26 28	24	15.5	GALAXY
HELW 486	20 48 55.	- 06 02 57.			NEBULA
SG 3.202	20 48 57.	+ 31 42	900		DIFFUSE EMISSION NEBULA
RNGC 6979	20 48 59.	+ 31 57			DIFFUSE NEBULA
MIL 81	20 49	+ 30 30	12000		SUPERNOVA REMNANT
LBN 0246	20 49	+ 36 20	2220		BRIGHT NEBULA
LBN 0328	20 49	+ 42 20	2100		BRIGHT NEBULA
LBN 0344	20 49	+ 43 30	1500		BRIGHT NEBULA
LBN 0350	20 49	+ 44 00	3300		BRIGHT NEBULA
LBN 0353	20 49	+ 44 10	1020		BRIGHT NEBULA
LBN 0422	20 49	+ 56 35	840		BRIGHT NEBULA
OCL 0179	20 49 00.	+ 35 46	780		OPEN STAR CLUSTER
DG 164	20 49 00.	+ 56 37	240		REFLECTION NEBULA
MCG-02-53-009	20 49 00.	- 09 29	72	15.	GALAXY
MCG-02-53-010	20 49 00.	- 12 23 30.	54	15.	GALAXY
MCG-04-49-006	20 49 00.	- 25 24	24	16.	GALAXY
CED 184	20 49 01.	+ 56 37	600		DIFFUSE GALACTIC NEBULA
IC 1332	20 49 04.	- 13 53 55.			NONSTELLAR OBJECT
ZWG 448.015	20 49 06.	+ 18 47		15.7	GALAXY
ZCG 2049+18	20 49 06.	+ 18 47		14.8	COMPACT GALAXY
KARA.72 549A	20 49 06.	+ 18 47	36	15.7	PART OF DOUBLE GALAXY
ACK 078-05.1	20 49 06.	+ 35 24			PLANETARY NEBULA
SCHO 1203	20 49 06.	+ 46 22 18.	210		ISOLATED DARK CLOUD
MCG-02-53-011	20 49 06.	- 13 54	48	15.	GALAXY
CED 183C	20 49 10.	+ 44 11	5100		DIFFUSE GALACTIC NEBULA
IC 5070	20 49 10.	+ 44 11			EMISSION NEBULA
ZWG 448.016	20 49 12.	+ 18 46		14.8	GALAXY
UGC 11643	20 49 12.	+ 18 46	66	14.8	GALAXY SBb
KARA.72 549B	20 49 12.	+ 18 46	66	14.8	PART OF DOUBLE GALAXY
MCG+03-53-004	20 49 12.	+ 18 49 30.	24	16.	GALAXY
TON-S 0010	20 49 12.	- 29 18		14.8	BLUE STAR
SG 3.203	20 49 17.	+ 30 01	480		DIFFUSE EMISSION NEBULA
ZWG 400.006	20 49 18.	+ 06 13		15.7	GALAXY
MCG+03-53-005	20 49 18.	+ 18 48	54	15.	GALAXY
ISS 0237	20 49 18.	+ 45 03	219		STELLAR RING
SG 3.204	20 49 19.	+ 28 38	2400		DIFFUSE EMISSION NEBULA
SG 3.205	20 49 19.	+ 45 41	6480		DIFFUSE EMISSION NEBULA
MCG-03-53-008	20 49 24.	- 16 28 30.	66	15.	GALAXY
IC 1333	20 49 28.	- 16 26 05.			NONSTELLAR OBJECT
RNGC 6973	20 49 29.	- 06 05			NON-EXISTENT OBJECT
IC 1334	20 49 29.	- 16 28 05.			NONSTELLAR OBJECT
VMT 22	20 49 30.	+ 30 45	12600		SUPERNOVA REMNANT
B 150	20 49 32.	+ 60 07	3600		DARK OBJECT
ZWG 425.015	20 49 36.	+ 13 13		15.3	GALAXY
HN 0696	20 49 36.	- 71 49			NEBULA
IC 5060	20 49 37.	- 71 49			NONSTELLAR OBJECT
RNGC 6975	20 49 41.	- 06 00		15.0	GALAXY
2ZW 095	20 49 42.	+ 09 10			COMPACT GALAXY
SCHO 1204	20 49 42.	+ 47 25 24.	260		ISOLATED DARK CLOUD
MCG-01-53-014	20 49 42.	- 06 00	60	15.	GALAXY
SER 141.06	20 49 42.	- 69 13	30		LOW SURF. BRGHTNSS GALAXY
SG 1.16	20 49 46.	+ 43 46	4800		DIFFUSE EMISSION NEBULA
RNGC 6976	20 49 47.	- 05 58		14.5	GALAXY
SG 3.206	20 49 48.	+ 29 05	5400		DIFFUSE EMISSION NEBULA
ISS 0287	20 49 48.	+ 37 36	217		STELLAR RING
MCG-01-53-015	20 49 48.	- 05 58	72	14.5	GALAXY
MCG+00-53-008	20 49 51.	- 00 06	48	14.5	GALAXY
RNGC 6977	20 49 53.	- 05 56		14.0	GALAXY
SCHO 1205	20 49 54.	+ 47 13 12.	300		ISOLATED DARK CLOUD
ZWG 374.021	20 49 54.	- 00 06		15.0	GALAXY
KARA.73B 0892	20 49 54.	- 00 06	54	15.0	ISOLATED GALAXY S
MCG-01-53-016	20 49 54.	- 05 56	66	14.	GALAXY
HN 2427	20 49 56.7	- 44 30 25.	18		NEBULA
SG 3.207	20 49 58.	+ 31 13	600		DIFFUSE EMISSION NEBULA
RNGC 6978	20 49 59.	- 05 54		14.0	GALAXY
LBN 0191	20 50	+ 31 00	11100		BRIGHT NEBULA
LBN 0255	20 50	+ 37 00	2400		BRIGHT NEBULA
LBN 0289	20 50	+ 38 50	2520		BRIGHT NEBULA
LBN 0327	20 50	+ 42 00	3300		BRIGHT NEBULA
LBN 0388	20 50	+ 47 00	10800		BRIGHT NEBULA
LB 09944	20 50	- 81 25		14.5	FAINT BLUE STAR
LDN 1082	20 50 00.	+ 30 00	1320		DARK NEBULA
LDN 0914	20 50 00.	+ 41 30	9240		DARK NEBULA
COU 075	20 50 00.	+ 44 00	5100		EMISSION NEBULA
LDN 0936	20 50 00.	+ 45 00	3840		DARK NEBULA
COU 076	20 50 00.	+ 46 40	10800		EMISSION NEBULA
MCG-01-53-017	20 50 00.	- 05 54	78	14.	GALAXY
RNGC 6980	20 50 11.	- 06 01			NON-EXISTENT OBJECT
SG 3.208	20 50 16.	+ 30 46	420		DIFFUSE EMISSION NEBULA
ASS 23	20 50 18.	+ 45 48			OB ASSOCIATION CYG OB6
IC 1335	20 50 18.	- 16 31 06.			NONSTELLAR OBJECT
HN 2428	20 50 18.6	- 44 59 41.	12		NEBULA
SG 3.209	20 50 23.	+ 39 18	5400		DIFFUSE EMISSION NEBULA
ZWG 400.007	20 50 30.	+ 06 55		15.7	GALAXY
MCG+01-53-005	20 50 30.	+ 06 57	72	14.	GALAXY
OCL 0187	20 50 30.	+ 37 44			OPEN STAR CLUSTER
ZWG 400.008	20 50 36.	+ 06 57		15.1	GALAXY
UGC 11644	20 50 36.	+ 06 57	72	15.1	GALAXY SBb
ZWG 425.016	20 50 42.	+ 13 24		15.7	GALAXY
REIK 2.286	20 50 42.61	- 12 43 34.6			NEBULA
RNGC 6981	20 50 44.	- 12 44		10.0	GLOBULAR CLUSTER
HN 2429	20 50 44.9	- 45 59 46.	18		NEBULA
B 351	20 50 47.	+ 47 12	1500		DARK OBJECT
GCL 118	20 50 48.	- 12 44	384	10.24	GLOBULAR STAR CLUSTER
SG 3.211	20 50 51.	+ 40 56	1680		DIFFUSE EMISSION NEBULA
ZWG 374.022	20 50 54.	+ 00 27		15.5	GALAXY
UGC 11645	20 50 54.	+ 00 27	60	15.5	GALAXY Sb-c
ZWG 400.009	20 50 54.	+ 06 03		15.7	GALAXY
LBN 0319	20 51	+ 41 00	1500		BRIGHT NEBULA
LBN 0359	20 51	+ 44 10	2100		BRIGHT NEBULA
KHAV 699	20 51	+ 55 17	6110		DARK NEBULA
LBN 0483	20 51	+ 68 00	900		BRIGHT NEBULA

OBJECT NAME	RIGHT ASCEN.	DECLINATION	DIAM.	MAGN.	TYPE OF OBJECT
SG 3.210	20 51 00.	+ 29 29	480		DIFFUSE EMISSION NEBULA
LDN 0904	20 51 00.	+ 39 00	11460		DARK NEBULA
LDN 0933	20 51 00.	+ 44 00	1440		DARK NEBULA
HN 2430	20 51 00.0	- 44 15 03.	18		NEBULA
HN 2431	20 51 01.3	- 43 15 15.	12		NEBULA
MCG-07-43-001	20 51 03.	- 43 17	24	15.	GALAXY
MCG-04-49-007	20 51 06.	- 25 40	72	14.5	GALAXY
MCG-07-43-002	20 51 06.	- 44 17	60	13.	GALAXY
HUB E27	20 51 10.	+ 30 10			DIFFUSE NEBULA
ISS 0161	20 51 18.	+ 47 28	168		STELLAR RING
MCG-07-43-003	20 51 18.	- 41 03	60	15.	GALAXY
ZWG 400.010	20 51 24.	+ 07 38		15.4	GALAXY
ISS 0162	20 51 24.	+ 47 35	239		STELLAR RING
SCHO 1206	20 51 26.	+ 47 18 48.	280		ISOLATED DARK CLOUD
ZWG 374.023	20 51 36.	+ 00 35		15.2	GALAXY
UGC 11646	20 51 36.	+ 00 35	60	15.2	GALAXY S
HN 2432	20 51 39.9	- 43 45 13.	12		NEBULA
ZWG 425.017	20 51 48.	+ 10 22		15.3	GALAXY
ACK 092+05.1	20 51 48.	+ 53 34			PLANETARY NEBULA
HN 0697	20 51 50.	- 73 21			NEBULA
IC 5066	20 51 50.	- 73 21			NONSTELLAR OBJECT
ZWG 448.017	20 51 54.	+ 17 35		15.6	GALAXY
UGC 11647	20 51 54.	+ 17 35	90	15.6	GALAXY Sc
HN 2433	20 51 58.1	- 45 30 30.	12		NEBULA
LBN 0295	20 52	+ 38 58	660		BRIGHT NEBULA
LBN 0118	20 52	- 05 00	3120		BRIGHT NEBULA
LDN 0940	20 52 00.	+ 45 24	1560		DARK NEBULA
OCL 0200	20 52 00.	+ 45 51	1200		OPEN STAR CLUSTER
ZWG 374.025	20 52 00.	- 00 08		15.3	GALAXY
ZWG 374.024	20 52 00.	- 00 18		15.6	GALAXY
MCG-05-49-006	20 52 00.	- 31 17	48	15.	GALAXY
HN 2434	20 52 04.8	- 43 52 18.	12		NEBULA
MCG-02-53-012	20 52 06.	- 08 49	54	15.	GALAXY
HN 2435	20 52 10.2	- 43 47 53.	18		NEBULA
HN 2436	20 52 10.5	- 43 39 11.	18		NEBULA
HN 2437	20 52 11.1	- 43 35 05.	18		NEBULA
HN 2438	20 52 11.3	- 43 46 59.			NEBULA
ISS 0239	20 52 12.	+ 39 14	96		STELLAR RING
ZWG 374.026	20 52 12.	- 00 14		15.5	GALAXY
AGU 63	20 52 12.	- 41 00 54.	48	12.5	COMPACT GALAXY
HN 2439	20 52 12.7	- 43 31 17.	12		NEBULA
7ZW 937	20 52 13.	+ 66 13			COMPACT GALAXY
MCG-03-53-009	20 52 15.	- 18 15	36	15.	GALAXY
IC 1336	20 52 16.	- 18 13 49.			NONSTELLAR OBJECT
MCG+01-53-006	20 52 18.	+ 05 47	30	16.	GALAXY
UGC 11648	20 52 18.	+ 66 58	102	17.	GALAXY IRR
HN 2440	20 52 18.0	- 43 49 47.	12		NEBULA
RNGC 6989	20 52 20.	+ 45 05			NON-EXISTENT OBJECT
SCHO 1207	20 52 22.	+ 46 01 24.	390		ISOLATED DARK CLOUD
ISS 0240	20 52 24.	+ 39 45	118		STELLAR RING
SCHO 1208	20 52 30.	+ 47 07 54.	240		ISOLATED DARK CLOUD
TON-S 0011	20 52 30.	- 26 33		14.8	BLUE STAR
HN 2441	20 52 39.4	- 43 31 28.	12		NEBULA
MCG+04-49-004	20 52 42.	+ 23 48 30.	27	16.	GALAXY
MCG-02-53-013	20 52 42.	- 13 38	72	15.	GALAXY
MCG-03-53-010	20 52 42.	- 19 31 30.	36	17.	GALAXY
HN 2442	20 52 45.9	- 43 56 27.	12		NEBULA
ZWG 374.027	20 52 48.	+ 00 21		15.2	GALAXY
ZWG 470.004	20 52 48.	+ 23 50		15.6	GALAXY
MCG-04-49-008	20 52 48.	- 22 14	36	16.	GALAXY
ZWG 374.029	20 52 54.	+ 02 09		15.1	GALAXY
MCG+00-53-009	20 52 54.	- 01 24	66	13.	GALAXY
ZWG 374.028	20 52 54.	- 01 25		14.5	GALAXY
UGC 11649	20 52 54.	- 01 25	90	14.5	GALAXY SBa
LBN 0166	20 53	+ 26 30	3000		BRIGHT NEBULA
LBN 0250	20 53	+ 36 00	2100		BRIGHT NEBULA
LBN 0283	20 53	+ 38 00	1320		BRIGHT NEBULA
KHAV 701	20 53	+ 41 23	8830		DARK NEBULA
LBN 0332	20 53	+ 42 20	2400		BRIGHT NEBULA
LBN 0342	20 53	+ 42 50	2700		BRIGHT NEBULA
KHAV 700	20 53	+ 43 41	7600		DARK NEBULA
LBN 0371	20 53	+ 45 00	4200		BRIGHT NEBULA
LB 03125	20 53	- 83 23		13.6	FAINT BLUE STAR
MCG+01-53-007	20 53 00.	+ 08 54	30	16.	GALAXY
LDN 0992	20 53 00.	+ 51 50	1140		DARK NEBULA
LDN 1037	20 53 00.	+ 55 15	3120		DARK NEBULA
LDN 1171	20 53 00.	+ 68 08	180		DARK NEBULA
HN 2444	20 53 02.7	- 45 33 09.	12		NEBULA
ABC 2329	20 53 03.	- 10 12		17.5	RICH CLUSTER OF GALAXIES
ZWG 400.011	20 53 06.	+ 07 17		15.5	GALAXY
ZWG 374.030	20 53 06.	- 00 49		15.5	GALAXY
KARA.73B 0893	20 53 06.	- 00 49	36	15.5	ISOLATED GALAXY S
MCG+01-53-008	20 53 12.	+ 08 55	48	16.	GALAXY
ZWG 425.018	20 53 12.	+ 15 19		15.2	GALAXY
ZWG 425.019	20 53 12.	+ 15 20		15.6	GALAXY
RNGC 6985	20 53 14.	- 11 17			NON-EXISTENT OBJECT
SCHO 1209	20 53 19.	+ 47 12 48.	240		ISOLATED DARK CLOUD
SG 3.212	20 53 22.	+ 36 01	10800		DIFFUSE EMISSION NEBULA
RNGC 6986	20 53 23.	- 18 47		15.0	GALAXY
ZWG 425.020	20 53 24.	+ 10 20		15.0	GALAXY
RNGC 6988	20 53 24.	+ 10 20		15.0	GALAXY
SG 2.078	20 53 25.	+ 42 19	2160		DIFFUSE EMISSION NEBULA
RNGC 6983	20 53 26.	- 44 10			GALAXY
HUB E28	20 53 28.	+ 44 19			DIFFUSE NEBULA
COU 077	20 53 30.	+ 43 05	1800		EMISSION NEBULA
MCG-03-53-011	20 53 30.	- 18 47	24	15.	GALAXY
MCG-07-43-004	20 53 30.	- 44 12	36	15.	GALAXY
HOLM 782B	20 53 38.	- 13 58	24	14.6	PART OF MULTIPLE GALAXY
HOLM 782A	20 53 40.	- 13 59	36	14.5	PART OF MULTIPLE GALAXY
MCG-02-53-014	20 53 42.	- 14 30	24	15.5	GALAXY
HN 2443	20 53 43.0	- 61 37 38.	18		NEBULA
MCG-01-53-018	20 53 45.	- 04 05	54	13.5	GALAXY
RNGC 6982	20 53 48.	- 52 03			GALAXY
MCG-03-53-012	20 53 57.	- 16 47	36	14.5	GALAXY
LBN 0349	20 54	+ 43 00	1920		BRIGHT NEBULA
LBN 0348	20 54	+ 43 00	1800		BRIGHT NEBULA
LBN 0394	20 54	+ 47 12	420		BRIGHT NEBULA
ZWG 448.018	20 54 00.	+ 17 07		15.6	GALAXY
ACK 086+00.1	20 54 00.	+ 46 22			PLANETARY NEBULA
ZWG 325.004	20 54 00.	+ 64 58		15.4	GALAXY
MCG-01-53-019	20 54 00.	- 04 00	42	14.	GALAXY
ZC 2054.1+0002	20 54 06.	+ 00 02	870		CLUSTER OF GALAXIES
MCG+00-53-010	20 54 06.	+ 00 30 30.	42	15.	GALAXY
ZWG 425.021	20 54 06.	+ 11 51		15.7	GALAXY
UGC 11650	20 54 06.	+ 11 51	84	15.7	GALAXY Sc
MCG+11-25-003	20 54 06.	+ 64 58	45	15.	GALAXY
ZWG 374.031	20 54 06.	- 01 01		15.7	GALAXY
IC 1337	20 54 06.	- 16 46 14.			NONSTELLAR OBJECT

OBJECT NAME	RIGHT ASCEN.	DECLINATION	DIAM.	MAGN.	TYPE OF OBJECT
MCG-05-49-007	20 54 06.	- 28 09	84	15.	GALAXY
IC 1340	20 54 09.	+ 30 52 08.			NONSTELLAR OBJECT
VDB.66N 137	20 54 09.	+ 47 12	816		REFLECTION NEBULA
IC 1338	20 54 10.	- 16 41 20.			NONSTELLAR OBJECT
CED 185	20 54 11.	+ 47 13	540		DIFFUSE GALACTIC NEBULA
ZWG 448.019	20 54 12.	+ 17 02		15.7	GALAXY
DG 165	20 54 12.	+ 47 13	720		REFLECTION NEBULA
ZC 2054.2-0148	20 54 12.	- 01 48	1080		CLUSTER OF GALAXIES
IC 5076	20 54 13.	+ 47 13 35.			NONSTELLAR OBJECT
HN 2446	20 54 14.3	- 46 47 41.	72		NEBULA
RNGC 6992	20 54 16.	+ 31 30			DIFFUSE NEBULA
CED 182B	20 54 16.	+ 31 30	4380		DIFFUSE GALACTIC NEBULA
SG 3.213	20 54 16.	+ 31 30	3600		DIFFUSE EMISSION NEBULA
SER 143.07	20 54 18.	- 50 52	420	17.	LOOSE GROUP OF 4 GALAXIES
RNGC 6984	20 54 18.	- 52 04		13.5	GALAXY
ZWG 374.032	20 54 24.	+ 02 05		15.7	GALAXY
KARA.72 550A	20 54 24.	+ 02 05	60	15.7	PART OF DOUBLE GALAXY
MCG+00-53-011	20 54 27.	+ 02 07	36	15.	GALAXY
HN 2448	20 54 29.9	- 43 32 34.	24		NEBULA
ZWG 400.012	20 54 30.	+ 06 37		15.1	GALAXY
KARA.73B 0894	20 54 30.	+ 06 37	30	15.1	ISOLATED GALAXY E
ISS 0241	20 54 30.	+ 44 56	119		STELLAR RING
ISS 0163	20 54 30.	+ 46 38	218		STELLAR RING
HN 2449	20 54 30.7	- 43 33 52.	120		NEBULA
HN 2445	20 54 33.9	- 60 58 35.	6		NEBULA
HOLM 783A	20 54 34.	- 00 25	48	14.4	PART OF MULTIPLE GALAXY
RNGC 6967	20 54 34.	- 48 49			GALAXY
ZWG 374.033	20 54 36.	+ 02 05		15.7	GALAXY
KARA.72 550B	20 54 36.	+ 02 05	42	15.7	PART OF DOUBLE GALAXY
MCG-07-43-006	20 54 36.	- 43 34	48	14.	GALAXY
ASU 64	20 54 36.	- 43 34 00.		12.5	PAIR OF GALAXIES
MCG-07-43-005	20 54 36.	- 43 35	150	14.	GALAXY
HOLM 783B	20 54 40.	- 00 27	48	14.5	PART OF MULTIPLE GALAXY
OCL 0197	20 54 42.	+ 44 26	900	10.1	OPEN STAR CLUSTER
RNGC 6996	20 54 43.	+ 44 26			OPEN CLUSTER
MCG+02-53-006	20 54 45.	+ 13 19	54	15.	GALAXY
ZWG 425.022	20 54 48.	+ 13 20		15.5	GALAXY
MCG-05-49-008	20 54 48.	- 32 30	48	15.	GALAXY
RNGC 6997	20 54 49.	+ 44 27			NON-EXISTENT OBJECT
ZWG 400.013	20 54 54.	+ 07 16		15.4	GALAXY
SCHO 1210	20 54 54.	+ 45 41 36.	890		ISOLATED DARK CLOUD
ABC 2330	20 54 54.	- 22 14		17.4	RICH CLUSTER OF GALAXIES
SS 67	20 54 55.	+ 52 22			DIFFUSE GALACTIC NEBULA
RNGC 6991	20 54 57.	+ 47 13			OPEN CLUSTER
RNGC 6995	20 54 58.	+ 31 01			DIFFUSE NEBULA
HN 0698	20 54 59.	- 72 00			NEBULA
IC 5069	20 54 59.	- 72 00			NONSTELLAR OBJECT
LBN 0378	20 55	+ 45 10	3600		BRIGHT NEBULA
KHAV 702	20 55	+ 45 54	1410		DARK NEBULA
LBN 0475	20 55	+ 67 10	6000		BRIGHT NEBULA
ZWG 448.020	20 55 00.	+ 16 55		15.2	GALAXY
ZZW 096	20 55 00.	+ 16 56			COMPACT GALAXY
MCG+04-49-005	20 55 00.	+ 25 45	168	14.5	GALAXY
LDN 0935	20 55 00.	+ 43 40	6060		DARK NEBULA
ISS 0242	20 55 00.	+ 40 05	89		STELLAR RING
LDN 0941	20 55 00.	+ 45 38	1260		DARK NEBULA
OCL 0202	20 55 00.	+ 47 13			OPEN STAR CLUSTER
OCL 0207	20 55 00.	+ 50 50	720	16.	OPEN STAR CLUSTER
LDN 1002	20 55 00.	+ 52 30	5700		DARK NEBULA
MCG-03-53-013	20 55 00.	- 18 09	60	14.	GALAXY
MCG-04-49-009	20 55 00.	- 24 48	90	16.	GALAXY
HN 2447	20 55 02.4	- 61 04 03.	18		NEBULA
HN 2450	20 55 03.2	- 54 22 51.	18		NEBULA
SG 1.15	20 55 04.	+ 31 17			DIFFUSE EMISSION NEBULA
CED 182C	20 55 05.	+ 31 02			DIFFUSE GALACTIC NEBULA
SG 3.214	20 55 05.	+ 31 02	1200		DIFFUSE EMISSION NEBULA
ZWG 470.005	20 55 06.	+ 22 59		15.6	GALAXY
ZWG 470.006	20 55 06.	+ 25 46		15.4	GALAXY
UGC 11651	20 55 06.	+ 25 46	240	15.4	GALAXY
IC 1339	20 55 07.	- 18 08 20.			NONSTELLAR OBJECT
HN 2453	20 55 09.0	- 42 50 38.	24		NEBULA
HN 2452	20 55 11.4	- 52 59 50.	42		NEBULA
ZWG 425.023	20 55 12.	+ 11 42		15.7	GALAXY
ZWG 425.024	20 55 12.	+ 14 02		15.3	GALAXY
UGC 11652	20 55 12.	+ 14 02	60	15.3	GALAXY SB
ZWG 448.021	20 55 12.	+ 18 37		15.7	GALAXY
UGC 11653	20 55 12.	+ 18 37	90	15.7	GALAXY
ZC 2055.2+2035	20 55 12.	+ 20 35	540		CLUSTER OF GALAXIES
PK095+07.1	20 55 12.	+ 57 15	80	17.4	PLANETARY NEBULA
MCG-05-49-009	20 55 12.	- 32 16	36	15.	GALAXY
SER 143.05	20 55 12.	- 49 29			COMPACT GROUP OF GALAXIES
SER 141.07	20 55 12.	- 66 59	150	17.	LOOSE GROUP OF GALAXIES
SCHO 1211	20 55 14.	+ 46 52 36.	320		ISOLATED DARK CLOUD
MCG+03-53-006	20 55 15.	+ 18 38 30.	60	16.	GALAXY
ISS 0164	20 55 18.	+ 49 48	277		STELLAR RING
MCG-05-49-010	20 55 18.	- 31 41	15	16.	GALAXY
HN 2454	20 55 18.4	- 43 28 13.	12		NEBULA
HN 2455	20 55 19.1	- 43 47 19.	24		NEBULA
ZWG 425.025	20 55 24.	+ 13 47		15.4	GALAXY
MCG+02-53-007	20 55 24.	+ 13 47	36	15.	GALAXY
HN 2451	20 55 24.4	- 60 51 38.	12		NEBULA
B 352	20 55 26.	+ 45 42	1320		DARK OBJECT
SCHO 1212	20 55 27.	+ 45 18 54.	410		ISOLATED DARK CLOUD
VDB.66N 138	20 55 27.	+ 48 04	444		REFLECTION NEBULA
ABC 2331	20 55 29.	- 07 57		16.3	RICH CLUSTER OF GALAXIES
MCG-03-53-014	20 55 30.	- 20 13	66	14.5	GALAXY
B 353	20 55 37.	+ 45 17	720		DARK OBJECT
HN 2456	20 55 37.8	- 52 11 49.	12		NEBULA
HN 2457	20 55 38.2	- 45 30 42.	12		NEBULA
UGC 11654	20 55 42.	+ 04 17	60	16.5	GALAXY PECULR
MCG+02-53-009	20 55 42.	+ 10 50	60	14.5	GALAXY
ZWG 425.026	20 55 42.	+ 10 52		15.1	GALAXY
MCG+02-53-008	20 55 42.	+ 13 13	36	15.	GALAXY
ZWG 470.007	20 55 42.	+ 24 54		15.6	GALAXY
UGC 11655	20 55 42.	+ 25 27	78	16.5	GALAXY
MCG-07-43-007	20 55 42.	- 39 41	72	15.	GALAXY
SCHO 1213	20 55 46.	+ 44 05 00.	190		ISOLATED DARK CLOUD
ZWG 400.014	20 55 48.	+ 08 10		15.7	GALAXY
ZWG 400.015	20 55 48.	+ 08 14		15.6	GALAXY
ZWG 425.027	20 55 48.	+ 13 19		15.5	GALAXY
ISS 0243	20 55 48.	+ 39 27	151		STELLAR RING
HN 2458	20 55 54.9	- 52 51 36.	30		NEBULA
SCHO 1214	20 55 55.	+ 43 36 06.	1930		ISOLATED DARK CLOUD
HN 2459	20 55 58.7	- 45 30 47.	18		NEBULA
LBN 0291	20 56	+ 37 55	540		BRIGHT NEBULA
LBN 0354	20 56	+ 43 10	2400		BRIGHT NEBULA
LBN 0403	20 56	+ 48 03	240		BRIGHT NEBULA
B 354	20 56	+ 57 57	3600		DARK OBJECT
KHAV 707	20 56	+ 77 18	4860		DARK NEBULA
ZWG 448.022	20 56 00.	+ 17 38		15.6	GALAXY
ISS 0244	20 56 00.	+ 39 29	187		STELLAR RING
LDN 0916	20 56 00.	+ 41 00	4860		DARK NEBULA
LDN 0945	20 56 00.	+ 46 20	1620		DARK NEBULA
LDN 0955	20 56 00.	+ 47 40	1140		DARK NEBULA
DG 166	20 56 00.	+ 48 03	180		REFLECTION NEBULA
LDN 0996	20 56 00.	+ 51 45	1140		DARK NEBULA
LDN 1168	20 56 00.	+ 67 25	180		DARK NEBULA
MCG+14-09-002	20 56 00.	+ 86 42	54	18.	GALAXY
MCG-05-49-011	20 56 00.	- 32 55	18	15.	GALAXY
ZWG 374.034	20 56 06.	+ 00 43		15.4	GALAXY
MCG+00-53-012	20 56 06.	+ 00 43 30.	12	15.5	GALAXY
ZC 2056.1+0131	20 56 06.	+ 01 31	1080		CLUSTER OF GALAXIES
TON-S 0012	20 56 06.	- 27 31		15.8	BLUE STAR
MCG-02-53-015	20 56 09.	- 14 14	54	15.	GALAXY
HN 2460	20 56 10.2	- 42 57 59.	12		NEBULA
ZWG 470.008	20 56 12.	+ 23 48		15.7	GALAXY
SCHO 1215	20 56 12.	+ 44 07 48.	170		ISOLATED DARK CLOUD
HN 0699	20 56 12.	- 72 51			NEBULA
IC 5071	20 56 12.	- 72 51			NONSTELLAR OBJECT
HN 2461	20 56 13.0	- 42 58 28.	24		NEBULA
RNGC 6994	20 56 14.	- 12 49		9.0	OPEN CLUSTER
RNGC 6990	20 56 14.	- 55 46			UNVERIFIED SOUTHERN OBJECT
MCG+04-49-006	20 56 15.	+ 23 46	30	15.5	GALAXY
ZC 2056.3+0554	20 56 18.	+ 05 54	940		CLUSTER OF GALAXIES
ZC 2056.3+3107	20 56 18.	+ 31 07	2350		CLUSTER OF GALAXIES
OCL 0069	20 56 18.	- 12 50	168	9.7	OPEN STAR CLUSTER
SCHO 1216	20 56 26.	+ 44 56 48.	200		ISOLATED DARK CLOUD
RNGC 6993	20 56 32.	- 25 53			NON-EXISTENT OBJECT
MCG+02-53-010	20 56 36.	+ 10 56	36	15.	GALAXY
ZWG 425.028	20 56 42.	+ 10 57		15.7	GALAXY
ZWG 425.029	20 56 42.	+ 11 06		15.7	GALAXY
UGC 11656	20 56 42.	+ 11 06	60	15.7	GALAXY Sc
ZWG 448.023	20 56 42.	+ 17 29		15.2	GALAXY
ZWG 448.024	20 56 42.	+ 18 09		15.7	GALAXY
ABC 2332	20 56 45.	- 17 08		17.5	RICH CLUSTER OF GALAXIES
HN 1216	20 56 46.	- 63 19		14.	NEBULA
IC 5074	20 56 46.	- 63 19			NONSTELLAR OBJECT
SG 3.215	20 56 47.	+ 44	11880		DIFFUSE EMISSION NEBULA
SG 1.17	20 56 47.	+ 44 12	7200		DIFFUSE EMISSION NEBULA
ZC 2056.8+0058	20 56 48.	+ 00 58	2550		CLUSTER OF GALAXIES
HN 2462	20 56 48.0	- 45 39 39.	18		NEBULA
SCHO 1217	20 56 50.	+ 44 11 12.	250		ISOLATED DARK CLOUD
IC 5072	20 56 50.	- 73 10			NONSTELLAR OBJECT
HN 0700	20 56 51.	- 73 10			NEBULA
HN 2463	20 56 52.3	- 43 13 38.	24		NEBULA
MCG-03-53-015	20 56 54.	- 16 50	48	15.	GALAXY
CED 183C	20 56 59.	+ 44 08	7200		DIFFUSE GALACTIC NEBULA
LBN 0384	20 57	+ 45 30	6600		BRIGHT NEBULA
KHAV 703	20 57	+ 46 42	2110		DARK NEBULA
KHAV 704	20 57	+ 57 54	3330		DARK NEBULA
ZWG 374.035	20 57 00.	+ 01 59		15.1	GALAXY
MCG+02-53-011	20 57 00.	+ 13 43	48	15.	GALAXY
MNSL 084-01/1	20 57 00.	+ 44 03	14400		HII REGION
LDN 1023	20 57 00.	+ 54 10	2100		DARK NEBULA
LDN 1056	20 57 00.	+ 55 50	720		DARK NEBULA
LDN 1061	20 57 00.	+ 57 10	3180		DARK NEBULA
LDN 1071	20 57 00.	+ 58 00	1200		DARK NEBULA
MCG-03-53-016	20 57 00.	- 15 35	36	15.5	GALAXY
MCG+00-53-013	20 57 06.	+ 01 59 30.	60	14.5	GALAXY
ZWG 425.030	20 57 06.	+ 13 45		15.7	GALAXY
SCHO 1218	20 57 07.	+ 46 53 30.	290		ISOLATED DARK CLOUD
ZC 2057.2-0108	20 57 12.	- 01 08	610		CLUSTER OF GALAXIES
KARA.72 551B	20 57 12.	- 02 03	48		PART OF DOUBLE GALAXY
MCG+00-53-014	20 57 12.	- 02 03 30.	30	15.	GALAXY
ZWG 374.036	20 57 12.	- 02 04		14.4	GALAXY
UGC 11658	20 57 12.	- 02 04	156	14.4	GALAXY DBL SYS
UGC 11657	20 57 12.	- 02 04	156	14.4	GALAXY DBL SYS
KARA.72 551A	20 57 12.	- 02 04	48	14.4	PART OF DOUBLE GALAXY
MCG+00-53-015	20 57 12.	- 02 04 30.	60	14.	GALAXY
ZZW 097	20 57 12.	- 02 05			COMPACT GALAXY
HN 2464	20 57 12.3	- 42 37 13.	18		NEBULA
LDN 0958	20 57 18.	+ 47 55	420		DARK NEBULA
LDN 0990	20 57 18.	+ 51 05	900		DARK NEBULA
MCG-07-43-008	20 57 18.	- 40 07	30	15.	GALAXY
HN 2465	20 57 18.8	- 44 50 31.	12		NEBULA
LDN 0985	20 57 30.	+ 50 23	720		DARK NEBULA
MCG-03-53-017	20 57 30.	- 14 40	30	15.	GALAXY
IC 1341	20 57 31.	- 14 10 48.			NONSTELLAR OBJECT
MCG+01-53-009	20 57 36.	+ 09 22	72	15.	GALAXY
ZWG 400.016	20 57 36.	+ 09 23		15.2	GALAXY
UGC 11659	20 57 36.	+ 09 23	96	15.2	GALAXY S
SCHO 1219	20 57 36.	+ 45 00 06.	200		ISOLATED DARK CLOUD
MCG-02-53-016	20 57 36.	- 14 10	30	15.5	GALAXY
MCG-07-43-009	20 57 36.	- 41 35		16.	GALAXY
IC 1342	20 57 41.	- 14 41 41.			NONSTELLAR OBJECT
ZWG 425.031	20 57 42.	+ 09 44		15.5	GALAXY
MCG-07-43-010	20 57 42.	- 38 48	30	15.	GALAXY
ZWG 400.017	20 57 48.	+ 09 22		15.0	GALAXY
ZWG 448.025	20 57 48.	+ 18 39		15.6	GALAXY
SCHO 1220	20 57 48.	+ 44 53 54.	200		ISOLATED DARK CLOUD
B 355	20 57 49.	+ 42 59	300		DARK OBJECT
HN 2475	20 57 50.4	+ 27 24 27.	90		NEBULA
ZWG 400.018	20 57 54.	+ 09 15		15.6	GALAXY
MCG-02-53-017	20 57 54.	- 13 25	54	15.	GALAXY
HELW 061	20 57 55.	- 17 18			NEBULA
HN 2466	20 57 55.4	- 52 12 35.	60		NEBULA
SCHO 1221	20 57 56.	+ 45 07 06.	330		ISOLATED DARK CLOUD
SCHO 1222	20 57 59.	+ 46 55 06.	240		ISOLATED DARK CLOUD
LBN 0256	20 58	+ 35 40	420		BRIGHT NEBULA
LBN 0266	20 58	+ 36 15	420		BRIGHT NEBULA
LBN 0305	20 58	+ 38 30	4260		BRIGHT NEBULA
LBN 0356	20 58	+ 43 00	2400		BRIGHT NEBULA
LBN 0387	20 58	+ 45 40	3000		BRIGHT NEBULA
KHAV 705	20 58	+ 47 54	6580		DARK NEBULA
KHAV 706	20 59	+ 55 12	2230		DARK NEBULA
COU 078	20 58 00.	+ 44 30	12000		EMISSION NEBULA
LDN 0981	20 58 00.	+ 50 00	720		DARK NEBULA
LDN 1003	20 58 00.	+ 52 10	1140		DARK NEBULA
LDN 1053	20 58 00.	+ 55 30	720		DARK NEBULA
HN 2468	20 58 00.8	- 46 06 29.	24		NEBULA
ABC 2333	20 58 01.	- 19 26		16.8	RICH CLUSTER OF GALAXIES
HN 2469	20 58 01.1	- 43 37 59.	12		NEBULA
MCG+03-53-007	20 58 03.	+ 16 29 30.	54	15.	GALAXY
UGC 11660	20 58 06.	+ 13 19	84	16.0	GALAXY S
ZWG 448.026	20 58 06.	+ 16 39		14.8	GALAXY
UGC 11661	20 58 06.	+ 16 39	72	14.8	GALAXY Sb-c

OBJECT NAME	RIGHT ASCEN.	DECLINATION	DIAM.	MAGN.	TYPE OF OBJECT
HN 2467	20 58 07.3	- 52 41 59.	12		NEBULA
HN 2471	20 58 08.5	- 46 13 58.			NEBULA
SCHO 1223	20 58 11.	+ 44 58 30.	200		ISOLATED DARK CLOUD
ISS 0165	20 58 12.	+ 46 23	130		STELLAR RING
SG 3.216	20 58 14.	+ 38 27	7200		DIFFUSE EMISSION NEBULA
B 356	20 58 15.	+ 46 29	5400		DARK OBJECT
HN 0701	20 58 17.	- 72 53			NEBULA
IC 5073	20 58 17.	- 72 53			NONSTELLAR OBJECT
SCHO 1224	20 58 18.	+ 43 38 12.	180		ISOLATED DARK CLOUD
MCG-03-53-018	20 58 18.	- 18 13 30.	36	15.	GALAXY
ZWG 425.032	20 58 24.	+ 13 07		15.5	GALAXY
ZWG 448.027	20 58 24.	+ 17 36		13.8	GALAXY
UGC 11662	20 58 24.	+ 17 36	66	13.8	GALAXY Sb-?
MCG+03-53-008	20 58 24.	+ 17 36	72	14.	GALAXY
IC 1343	20 58 24.	- 15 35 43.			NONSTELLAR OBJECT
MCG-07-43-011	20 58 24.	- 38 43	9	15.5	GALAXY
HELW 062	20 58 25.	- 17 19			NEBULA
RNGC 7003	20 58 27.	+ 17 36		14.0	GALAXY
B 357	20 58 27.	+ 55 23	1800		DARK OBJECT
HN 2472	20 58 28.7	- 44 31 33.	18		NEBULA
SCHO 1225	20 58 29.	+ 46 21 54.	690		ISOLATED DARK CLOUD
ZWG 425.033	20 58 30.	+ 10 07		15.5	GALAXY
KARA.73B 0895	20 58 30.	+ 10 07	18	15.5	ISOLATED GALAXY S
HMS 1.29	20 58 30.	+ 16 07			E1 GALAXY
ZWG 448.028	20 58 30.	+ 16 07		15.3	GALAXY
MCG+03-53-009	20 58 30.	+ 16 07	15	15.	GALAXY
LDN 0966	20 58 30.	+ 48 50	1680		DARK NEBULA
MCG-02-53-018	20 58 30.	- 13 34	36	15.	GALAXY
IC 1344	20 58 30.	- 13 34 25.			NONSTELLAR OBJECT
MCG+00-53-016	20 58 36.	- 00 22 30.	78	13.	GALAXY
ZWG 374.037	20 58 36.	- 00 23		14.0	GALAXY
UGC 11663	20 58 36.	- 00 23	78	14.0	GALAXY Sb
TON-S 0013	20 58 36.	- 27 32		13.7	BLUE STAR
RNGC 6998	20 58 36.	- 28 13			GALAXY
SCHO 1226	20 58 37.	+ 43 32 30.	120		ISOLATED DARK CLOUD
RNGC 7001	20 58 37.	- 00 23		14.0	GALAXY
IC 1345	20 58 38.	- 13 35 30.			NONSTELLAR OBJECT
SCHO 1227	20 58 40.	+ 44 31 30.	310		ISOLATED DARK CLOUD
HN 2470	20 58 41.4	- 60 18 10.	24		NEBULA
ZWG 374.038	20 58 42.	+ 02 08		15.6	GALAXY
ZWG 400.019	20 58 42.	+ 07 05		15.6	GALAXY
HMS 1.30	20 58 48.	+ 15 56			Sa GALAXY
MCG+03-53-010	20 58 48.	+ 15 56	24	16.5	GALAXY
ZWG 470.009	20 58 48.	+ 22 14		15.7	GALAXY
LDN 0950	20 58 48.	+ 46 25	1320		DARK NEBULA
MCG-03-53-019	20 58 48.	- 14 44	54	15.	GALAXY
MCG+03-53-011	20 58 51.	+ 15 58	36	16.5	GALAXY
SCHO 1228	20 58 53.	+ 43 39 54.	370		ISOLATED DARK CLOUD
ZWG 448.029	20 58 54.	+ 15 54		15.4	GALAXY
ZWG 448.030	20 58 54.	+ 15 57		15.7	GALAXY
ISS 0166	20 58 54.	+ 47 23	235		STELLAR RING
OCL 0233	20 58 54.	+ 67 58	240	13.8	OPEN STAR CLUSTER
MCG-02-53-019	20 58 54.	- 14 09 30.	36	15.	GALAXY
BC PKS2058-17	20 58 54.8	- 18 00 00.		18.	QUASI-STELLAR OBJECT
IC 1346	20 58 56.	- 14 03 40.			NONSTELLAR OBJECT
HN 2473	20 58 57.8	- 52 13 14.	18		NEBULA
IC 1347	20 58 59.	- 13 30 34.			NONSTELLAR OBJECT
IC 1348	20 58 59.	- 13 23 10.			NONSTELLAR OBJECT
HN 2474	20 58 59.8	- 44 38 08.	12		NEBULA
LBN 0289	20 59	+ 37 00	840		BRIGHT NEBULA
LBN 0290	20 59	+ 37 20	300		BRIGHT NEBULA
LBN 0218	20 59	+ 39 40	1800		BRIGHT NEBULA
LBN 0552	20 59	+ 78 25	1320		BRIGHT NEBULA
ZWG 374.039	20 59 00.	+ 01 25		15.7	GALAXY
KARA.73B 0896	20 59 00.	+ 01 25	36	15.7	ISOLATED GALAXY E
LDN 0957	20 59 00.	+ 47 35	2040		DARK NEBULA
LDN 1013	20 59 00.	+ 53 10	840		DARK NEBULA
LDN 1015	20 59 00.	+ 53 20	840		DARK NEBULA
UGC 11664	20 59 00.	+ 83 53	60	16.0	GALAXY DBL SYS
MCG-02-53-020	20 59 00.	- 13 30	48	15.	GALAXY
RNGC 6999	20 59 01.	- 28 03			GALAXY
SCHO 1229	20 59 02.	+ 44 51 24.	190		ISOLATED DARK CLOUD
REIN 2.287	20 59 02.88	+ 54 20 58.8			NEBULA
PK093+05.1	20 59 03.0	+ 54 20 59.			PLANETARY NEBULA
RNGC 7008	20 59 04.	+ 54 21		13.5	PLANETARY NEBULA
CED 186	20 59 04.	+ 54 22	900		DIFFUSE GALACTIC NEBULA
IC 1349	20 59 04.	- 13 27 27.			NONSTELLAR OBJECT
REIN 2.288	20 59 04.99	+ 54 20 48.5			NEBULA
PK093+05.2	20 59 05.1	+ 54 20 41.	98	13.3	PLANETARY NEBULA
GCL 119	20 59 06.	+ 16 00	180	11.45	GLOBULAR STAR CLUSTER
IC 1351	20 59 06.	- 13 23 39.			NONSTELLAR OBJECT
IC 1350	20 59 06.	- 14 02 39.			NONSTELLAR OBJECT
REIN 2.289	20 59 06.12	+ 54 21 15.0			NEBULA
HN 2476	20 59 06.3	- 43 41 07.	18		NEBULA
REIN 2.290	20 59 07.85	+ 54 21 00.8			NEBULA
RNGC 7006	20 59 08.	+ 16 00		11.5	GLOBULAR CLUSTER
SHB 353	20 59 08.8	+ 03 29 49.		18.	QUASI-STELLAR OBJECT
IC 1353	20 59 10.	- 13 28 15.			NONSTELLAR OBJECT
IC 1352	20 59 10.	- 13 34 45.			NONSTELLAR OBJECT
SCHO 1230	20 59 12.	+ 44 47 36.	140		ISOLATED DARK CLOUD
LDN 0971	20 59 12.	+ 49 42	360		DARK NEBULA
LDN 0987	20 59 12.	+ 50 29	360		DARK NEBULA
IC 1355	20 59 12.	- 13 22 03.			NONSTELLAR OBJECT
IC 1354	20 59 12.	- 13 57 15.			NONSTELLAR OBJECT
MCG-02-53-021	20 59 12.	- 14 03	36	15.	GALAXY
S2R 141.08	20 59 12.	- 68 46	1200	16.	LOOSE CLUSTER OF GALAXIES
RNGC 7005	20 59 14.	- 13 05			NON-EXISTENT OBJECT
SC 2056-1845.9	20 59 17.	- 18 34 12.	12		NEBULA
SCHO 1231	20 59 20.	+ 45 06 24.	260		ISOLATED DARK CLOUD
HN 2477	20 59 23.2	- 45 48 13.	12		NEBULA
ZWG 400.020	20 59 24.	+ 07 47		15.7	GALAXY
HN 2479	20 59 28.2	- 45 34 18.	6		NEBULA
LDN 0978	20 59 30.	+ 49 30	1800		DARK NEBULA
ZWG 374.040	20 59 30.	- 02 22		15.7	GALAXY
MCG-01-53-020	20 59 30.	- 06 29	78	13.	GALAXY
MCG-05-49-012	20 59 30.	- 28 22	48	15.	GALAXY
HN 0702	20 59 32.	- 72 03			NEBULA
IC 5075	20 59 33.	- 72 03			NONSTELLAR OBJECT
HN 2481	20 59 34.9	- 44 02 30.			NEBULA
HMS 1.31	20 59 36.	+ 15 56			SO GALAXY
MCG+03-53-012	20 59 36.	+ 15 56	6	17.5	GALAXY
MCG-03-53-021	20 59 36.	- 17 00	240	13.	GALAXY
MCG-03-53-020	20 59 36.	- 19 40	48	15.5	GALAXY
HN 2482	20 59 36.1	- 44 02 24.			NEBULA
MCG-01-53-021	20 59 42.	- 05 18	36	15.	GALAXY
UGC 11665	20 59 48.	+ 07 15	66	16.0	GALAXY S
MCG-02-53-022	20 59 48.	- 14 00	84	15.	GALAXY
HN 0703	20 59 48.	- 17 01			NEBULA
IC 5078	20 59 48.	- 17 01			NONSTELLAR OBJECT
MCG-05-49-013	20 59 48.	- 31 12	14	15.5	GALAXY
ZWG 425.034	20 59 54.	+ 10 47		15.4	GALAXY
ZWG 425.035	20 59 54.	+ 14 57		15.1	GALAXY
UGC 11666	20 59 54.	+ 14 57	66	15.1	GALAXY Sc
OCL 0235	20 59 54.	+ 67 58	1080	7.1	OPEN STAR CLUSTER
RNGC 7023	20 59 54.	+ 67 58		7.0	CLUSTER WITH NEBULOSITY
HELW 063	20 59 55.	- 17 28			NEBULA
MCG-03-53-022	20 59 57.	- 16 00	24	15.	GALAXY
HN 2478	20 59 59.2	- 59 46 30.	12		NEBULA
LBN 0269	21 00	+ 36 00	4500		BRIGHT NEBULA
KHAV 709	21 00	+ 37 54	6310		DARK NEBULA
LBN 0373	21 00	+ 44 00	4500		BRIGHT NEBULA
KHAV 708	21 00	+ 45 48	3470		DARK NEBULA
ZC 2100.0+1623	21 00 00.	+ 16 23	8270		CLUSTER OF GALAXIES
RNGC 7000	21 00 00.	+ 44 00			DIFFUSE NEBULA
LDN 0954	21 00 00.	+ 47 00	2700		DARK NEBULA
LDN 0962	21 00 00.	+ 48 10	3300		DARK NEBULA
LDN 0984	21 00 00.	+ 50 00	960		DARK NEBULA
LDN 1004	21 00 00.	+ 52 10	1680		DARK NEBULA
LDN 1026	21 00 00.	+ 53 50	840		DARK NEBULA
LDN 1029	21 00 00.	+ 54 00	12840		DARK NEBULA
LDN 1228	21 00 00.	+ 77 20	1140		DARK NEBULA
HN 2480	21 00 03.0	- 59 47 29.	18		NEBULA
IC 1356	21 00 05.	- 16 00 16.			NONSTELLAR OBJECT
HN 2484	21 00 05.9	- 52 10 41.	18		NEBULA
ZWG 400.021	21 00 06.	+ 08 07		15.5	GALAXY
MCG-05-49-014	21 00 06.	- 28 32	24	15.	GALAXY
RNGC 7011	21 00 09.	+ 47 07			NON-EXISTENT OBJECT
RNGC 7002	21 00 10.	- 49 13			GALAXY
ZWG 448.031	21 00 12.	+ 19 01		15.6	GALAXY
MCG-03-53-023	21 00 12.	- 14 42	36	15.5	GALAXY
MCG-03-53-024	21 00 12.	- 16 59	12	15.5	GALAXY
HELW 064	21 00 13.	- 17 22			NEBULA
MCG-03-53-025	21 00 15.	- 14 42	36	15.	GALAXY
HN 2483	21 00 15.7	- 57 39 28.	12		NEBULA
IC 5080	21 00 16.	+ 19 01 36.			NONSTELLAR OBJECT
HN 2485	21 00 17.8	- 43 56 28.	48		NEBULA
MCG-02-53-023	21 00 18.	- 14 28	60	15.	GALAXY
HELW 065	21 00 18.	- 17 00			NEBULA
HELW 066	21 00 18.	- 17 13			NEBULA
MCG-07-43-012	21 00 18.	- 39 40	78	15.	GALAXY
MCG-07-43-013	21 00 18.	- 43 58	60	16.5	GALAXY
LB 01102	21 00 22.	+ 00 05 00.		17.4	FAINT BLUE STAR
UGC 11667	21 00 24.	+ 17 51	66	17.	GALAXY Sc
ZC 2100.4+2233	21 00 24.	+ 22 33	2690		CLUSTER OF GALAXIES
4ZW 067	21 00 24.	+ 36 30			COMPACT GALAXY
UGC 11668	21 00 24.	+ 36 30	27	15.5	GALAXY PAIR
HELW 067	21 00 24.	- 17 13			NEBULA
RNGC 7004	21 00 28.	- 49 19			GALAXY
ZWG 425.036	21 00 30.	+ 15 12		14.9	GALAXY
ZWG 448.032	21 00 30.	+ 19 00		15.7	GALAXY
ISS 0245	21 00 30.	+ 44 12	118		STELLAR RING
HELW 068	21 00 31.	- 17 27			NEBULA
ZWG 425.037	21 00 36.	+ 14 57		15.7	GALAXY
ISS 0167	21 00 36.	+ 45 51	105		STELLAR RING
MCG-05-49-015	21 00 36.	- 28 03	48	15.	GALAXY
UGC 11669	21 00 42.	+ 12 58	66	16.0	GALAXY S
ZWG 448.033	21 00 42.	+ 15 51		14.9	GALAXY
ZWG 448.034	21 00 42.	+ 20 17		15.7	GALAXY
ISS 0246	21 00 42.	+ 39 48	157		STELLAR RING
LDN 0991	21 00 42.	+ 50 41	480		DARK NEBULA
SCHO 1232	21 00 43.	+ 45 49 42.	420		ISOLATED DARK CLOUD
IC 5081	21 00 44.	+ 19 00 14.			NONSTELLAR OBJECT
ISS 0168	21 00 48.	+ 45 44	91		STELLAR RING
LDN 0982	21 00 48.	+ 49 43	540		DARK NEBULA
HN 2487	21 00 50.2	- 43 44 02.	18		NEBULA
HN 2488	21 00 52.3	- 43 47 26.	36		NEBULA
HN 2489	21 00 52.6	- 43 53 26.	18		NEBULA
MCG-07-43-014	21 00 54.	- 43 49	42	15.	GALAXY
HN 2490	21 00 54.2	- 43 32 02.	12		NEBULA
SG 3.217	21 00 57.	+ 37 50	3600		DIFFUSE EMISSION NEBULA
HN 2486	21 00 57.2	- 55 00 32.	12		NEBULA
LBN 0296	21 01	+ 37 30	900		BRIGHT NEBULA
LBN 0385	21 01	+ 45 00	2700		BRIGHT NEBULA
LBN 0408	21 01	+ 50 00	120		BRIGHT NEBULA
KHAV 710	21 01	+ 55 42	4020		DARK NEBULA
KHAV 712	21 01	+ 67 30	3250		DARK NEBULA
LBN 0487	21 01	+ 68 00	540		BRIGHT NEBULA
LBN 0550	21 01	+ 78 00	5100		BRIGHT NEBULA
LBN 0555	21 01	+ 79 00	5520		BRIGHT NEBULA
LDN 1058	21 01 00.	+ 55 50	3780		DARK NEBULA
LDN 1170	21 01 00.	+ 67 25	1860		DARK NEBULA
SCHO 1233	21 01 04.	+ 44 40 00.	260		ISOLATED DARK CLOUD
ASS 24	21 01 06.	+ 49 31			OB ASSOCIATION CYG OB7
VDB.66N 139	21 01 10.	+ 68 00	1020		REFLECTION NEBULA
ISS 0247	21 01 12.	+ 43 53	164		STELLAR RING
MCG-07-43-015	21 01 12.	- 40 06	42	13.5	GALAXY
AGU 65	21 01 12.	- 40 06 06.	24	13.5	D GALAXY
CED 187	21 01 14.	+ 67 58	1080		DIFFUSE GALACTIC NEBULA
SG 3.218	21 01 14.	+ 67 58	600		DIFFUSE EMISSION NEBULA
ARC 2334	21 01 17.	- 25 28		17.4	RICH CLUSTER OF GALAXIES
ZWG 448.035	21 01 18.	+ 15 52		15.7	GALAXY
OCL 0194	21 01 18.	+ 40 16	360	17.	OPEN STAR CLUSTER
DG 167	21 01 18.	+ 67 58	1200		REFLECTION NEBULA
HN 2491	21 01 18.6	- 52 01 13.	12		NEBULA
HN 2493	21 01 19.7	- 43 52 30.	18		NEBULA
LB 01103	21 01 21.	+ 01 16 30.		16.7	FAINT BLUE STAR
ZWG 425.038	21 01 24.	+ 11 34		15.5	GALAXY
ZWG 491.002	21 01 24.	+ 29 42		12.9	GALAXY
UGC 11670	21 01 24.	+ 29 42	312	12.9	GALAXY Sa
ISS 0248	21 01 24.	+ 42 11	239		STELLAR RING
OCL 0201	21 01 24.	+ 44 23	1200	8.9	OPEN STAR CLUSTER
HN 2492	21 01 26.9	- 52 08 48.	48		NEBULA
IC 5083	21 01 27.	+ 11 33 48.			NONSTELLAR OBJECT
RNGC 7013	21 01 27.	+ 29 42		13.0	GALAXY
MCG+05-49-001	21 01 27.	+ 29 42	252	13.	GALAXY
HN 2505	21 01 27.0	+ 29 41 56.	60		NEBULA
REIN 2.291	21 01 27.50	- 11 33 46.8			NEBULA
PK037-34.1	21 01 27.53	- 11 33 46.6	51	8.9	PLANETARY NEBULA
FATH 2.165	21 01 28.	+ 29 40	68		NEBULA
HN 2494	21 01 29.2	- 47 18 54.			NEBULA
DG 168	21 01 30.	+ 59 21	60		REFLECTION NEBULA
MCG-04-49-010	21 01 30.	- 21 57 30.	60	14.	GALAXY
TON-S 0014	21 01 30.	- 27 25		15.0	BLUE STAR
SCHO 1234	21 01 31.	+ 45 42 24.	350		ISOLATED DARK CLOUD
RNGC 7009	21 01 32.	- 11 34		8.5	PLANETARY NEBULA
SS 62	21 01 33.	+ 59 21			DIFFUSE GALACTIC NEBULA

OBJECT NAME	RIGHT ASCEN.	DECLINATION	DIAM.	MAGN.	TYPE OF OBJECT
ZWG 425.039	21 01 36.	+ 12 56		15.6	GALAXY
DG 169	21 01 36.	+ 50 01	60		REFLECTION NEBULA
HN 2495	21 01 45.0	- 46 16 29.	12		NEBULA
HN 2497	21 01 46.9	- 43 37 11.	18		NEBULA
IC 5079	21 01 47.	- 56 27			OPEN CLUSTER
MCG-07-43-016	21 01 48.	- 43 39	36	15.	GALAXY
MCG-02-53-024	21 01 51.	- 12 32	72	14.5	GALAXY
HN 2498	21 01 52.1	- 44 30 41.	12		NEBULA
MCG+01-53-010	21 01 54.	+ 07 46	48	15.5	GALAXY
RNGC 7007	21 01 55.	- 52 45		13.0	GALAXY
HN 2499	21 01 55.1	- 44 26 41.	12		NEBULA
RNGC 7010	21 01 56.	- 12 32		14.0	GALAXY
RIGO 549	21 01 56.	- 12 33			NEBULA
IC 5082	21 01 56.	- 12 33			SAME AS NGC 7010
LBN 0299	21 02	+ 37 40	1800		BRIGHT NEBULA
LBN 0363	21 02	+ 43 00	5100		BRIGHT NEBULA
KHAV 711	21 02	+ 46 36	1720		DARK NEBULA
LBN 0409	21 02	+ 49 55	120		BRIGHT NEBULA
LBN 0413	21 02	+ 50 03	120		BRIGHT NEBULA
LBN 0441	21 02	+ 59 21	60		BRIGHT NEBULA
ZWG 400.022	21 02 00.	+ 07 46		15.3	GALAXY
ZWG 448.036	21 02 00.	+ 21 12		15.7	GALAXY
LDN 0927	21 02 00.	+ 40 54	600		DARK NEBULA
DG 170	21 02 00.	+ 50 04	60		REFLECTION NEBULA
LDN 1032	21 02 00.	+ 53 55	840		DARK NEBULA
ISS 0120	21 02 00.	+ 56 33	317		STELLAR RING
LDN 1172	21 02 00.	+ 67 30	360		DARK NEBULA
LDN 1174	21 02 00.	+ 68 00	2160		DARK NEBULA
MCG-04-49-011	21 02 00.	- 21 55	30	15.	GALAXY
HOLM 784A	21 02 02.	+ 15 53	18	14.8	PART OF MULTIPLE GALAXY
HN 2496	21 02 02.7	- 52 34 29.	18		NEBULA
HOLM 784B	21 02 04.	+ 15 53	18	15.1	PART OF MULTIPLE GALAXY
HN 2502	21 02 04.5	- 43 36 28.	24		NEBULA
ZWG 400.023	21 02 06.	+ 09 26		14.4	GALAXY
UGC 11671	21 02 06.	+ 09 26	72	14.4	GALAXY DBL SYS
VV 102B	21 02 06.	+ 15 52	9	16.	INTERACTING GALAXY
VV 102A	21 02 06.	+ 15 52	9	16.	INTERACTING GALAXY
VV 102	21 02 06.	+ 15 52	84		INTERACTING GALAXY
MCG+03-53-013	21 02 06.	+ 15 52 30.	90	14.5	GALAXY
ZWG 448.037	21 02 06.	+ 15 53		15.1	GALAXY
UGC 11672	21 02 06.	+ 15 53	90	15.1	GALAXY DBL SYS
MRSL 090+02/1	21 02 06.	+ 49 41	60		HII REGION
HN 2501	21 02 07.3	- 47 14 28.			NEBULA
HN 2503	21 02 10.5	- 44 18 40.	12		NEBULA
HN 2500	21 02 11.0	- 51 54 28.	42		NEBULA
2ZW 098	21 02 12.	- 00 24			COMPACT GALAXY
ZWG 374.041	21 02 12.	- 00 25		15.6	GALAXY
UGC 11673	21 02 12.	- 00 25	60	15.6	GALAXY DBL SYS
LB 01104	21 02 14.	+ 00 38 06.		16.6	FAINT BLUE STAR
ZWG 374.042	21 02 18.	+ 00 15		15.5	GALAXY
HN 2504	21 02 22.7	- 42 53 51.	18		NEBULA
MCG-07-43-017	21 02 24.	- 43 00	30	15.	GALAXY
SCHO 1235	21 02 29.	+ 46 33 42.	270		ISOLATED DARK CLOUD
LDN 0998	21 02 30.	+ 51 00	1680		DARK NEBULA
ZC 2102.6+1351	21 02 36.	+ 13 51	1010		CLUSTER OF GALAXIES
HN 2524	21 02 38.6	+ 29 18 18.	18		NEBULA
ZWG 400.024	21 02 42.	+ 05 24		15.7	GALAXY
ZWG 400.025	21 02 42.	+ 05 42		15.7	GALAXY
DG 171	21 02 42.	+ 49 57	240		REFLECTION NEBULA
MCG-01-53-022	21 02 42.	- 08 03	42	14.	GALAXY
SG 3.219	21 02 44.	+ 43 05	6000		DIFFUSE EMISSION NEBULA
HN 2506	21 02 44.1	- 45 38 02.			NEBULA
MCG-04-49-012	21 02 45.	- 26 12 30.	36	16.	GALAXY
MCG-01-53-023	21 02 46.	- 07 57 30.	90	14.5	GALAXY
ISS 0121	21 02 54.	+ 55 24	669		STELLAR RING
SER 143.01	21 02 54.	- 47 24	1020	15.	CLOUD OF GALAXIES
KHAV 714	21 03	+ 43 12	1000		DARK NEBULA
KHAV 713	21 03	+ 44 12	5930		DARK NEBULA
KHAV 715	21 03	+ 52 24	18630		DARK NEBULA
KHAV 716	21 03	+ 64 24	8970		DARK NEBULA
ZWG 448.038	21 03 00.	+ 19 32		15.7	GALAXY
LDN 0988	21 03 00.	+ 50 00	1860		DARK NEBULA
LDN 1018	21 03 00.	+ 53 05	480		DARK NEBULA
LDN 1167	21 03 00.	+ 66 50	960		DARK NEBULA
SCHO 1236	21 03 02.	+ 47 41 54.	310		ISOLATED DARK CLOUD
MCG+01-53-011	21 03 06.	+ 07 30	36	15.5	GALAXY
MCG+02-53-012	21 03 06.	+ 11 12	120	12.	GALAXY
MCG-02-53-026	21 03 06.	- 14 31	12	15.5	GALAXY
MCG-07-43-018	21 03 06.	- 43 44	15	14.5	GALAXY
HN 2508	21 03 06.0	- 43 42 37.	12		NEBULA
HN 2509	21 03 07.0	- 43 36 01.	24		NEBULA
HN 2510	21 03 09.4	- 44 59 25.	36		NEBULA
ABC 2335	21 03 10.	- 22 00		17.4	RICH CLUSTER OF GALAXIES
ZWG 400.026	21 03 12.	+ 07 30		14.9	GALAXY
ZWG 400.027	21 03 12.	+ 09 01		15.2	GALAXY
MCG-02-53-025	21 03 12.	- 11 44 30.	36	15.	GALAXY
IC 1357	21 03 13.	- 10 55 10.			NONSTELLAR OBJECT
HN 2507	21 03 14.0	- 52 57 31.	18		NEBULA
LB 01105	21 03 15.	+ 02 05 00.		16.4	FAINT BLUE STAR
ZWG 425.040	21 03 18.	+ 11 13		13.2	GALAXY
RNGC 7015	21 03 18.	+ 11 13		13.0	GALAXY
UGC 11674	21 03 18.	+ 11 13	108	13.2	GALAXY Sb
MCG+03-53-014	21 03 18.	+ 19 32	30	15.	GALAXY
SG 3.220	21 03 18.	+ 44 15	2400		DIFFUSE EMISSION NEBULA
RNGC 7012	21 03 20.	- 45 01			UNVERIFIED SOUTHRN OBJECT
ZWG 400.028	21 03 24.	+ 07 27		15.0	GALAXY
UGC 11675	21 03 24.	+ 07 27	60	15.0	GALAXY Sa
MCG+01-53-012	21 03 24.	+ 07 27	48	15.5	GALAXY
ZWG 448.039	21 03 24.	+ 18 16		14.8	GALAXY
UGC 11676	21 03 24.	+ 18 16	60	14.8	GALAXY S?
MCG-07-43-019	21 03 24.	- 42 47	72	14.	GALAXY
HN 2512	21 03 24.4	- 47 22 54.			NEBULA
HN 2513	21 03 24.6	- 42 45 24.	18		NEBULA
RNGC 7019	21 03 27.	- 24 34			GALAXY
MCG-07-43-020	21 03 27.	- 40 43	42	15.5	GALAXY
MCG+03-53-015	21 03 30.	+ 18 16	27	15.	GALAXY
HN 2511	21 03 30.9	- 54 06 12.			NEBULA
HN 2516	21 03 31.7	- 45 01 30.	6		NEBULA
HN 2517	21 03 32.0	- 45 06 12.	30		NEBULA
HN 2518	21 03 33.0	- 43 12 12.	18		NEBULA
MRSL 090+01/1	21 03 36.	+ 49 27	60		HII REGION
MCG-03-53-027	21 03 36.	- 15 37	48	15.5	GALAXY
UGC 11677	21 03 42.	+ 15 45	84	16.0	GALAXY
SHAH 174	21 03 42.	+ 85 33	216		GROUP OF COMPACT GALAXIES
IC 1358	21 03 43.	- 16 24 32.			NONSTELLAR OBJECT
HN 2520	21 03 43.8	- 45 05 11.	18		NEBULA
HN 2522	21 03 44.7	- 45 24 35.	18		NEBULA
HUB C30	21 03 47.	+ 41 13			DIFFUSE NEBULA

OBJECT NAME	RIGHT ASCEN.	DECLINATION	DIAM.	MAGN.	TYPE OF OBJECT
YC 2103-47	21 03 47.	- 47 45 30.			UNUSUAL SOUTHERN GALAXY
HN 6704	21 03 47.	- 73 51			NEBULA
IC 5077	21 03 47.	- 73 51			NONSTELLAR OBJECT
ZWG 374.043	21 03 48.	- 00 58		15.7	GALAXY
ZC 2103.8-0113	21 03 48.	- 01 13	3230		CLUSTER OF GALAXIES
HN 2515	21 03 48.2	- 53 57 59.	18		NEBULA
HN 2523	21 03 49.3	- 44 47 23.	12		NEBULA
B 358	21 03 50.	+ 43 05	1200		DARK OBJECT
SCHO 1237	21 03 50.	+ 43 05 54.	510		ISOLATED DARK CLOUD
HN 2514	21 03 57.9	- 58 30 23.	18		NEBULA
HN 2521	21 03 58.1	- 52 32 11.	12		NEBULA
LBN 0365	21 04	+ 43 00	4200		BRIGHT NEBULA
LBN 0411	21 04	+ 49 43	120		BRIGHT NEBULA
SER 143.02	21 04	- 47	60	14.	INTERACTING GALAXIES
ZC 2104.0+1313	21 04 00.	+ 13 13	340		CLUSTER OF GALAXIES
LDN 0932	21 04 00.	+ 41 40	240		DARK NEBULA
LDN 1028	21 04 00.	+ 53 20	900		DARK NEBULA
LDN 1034	21 04 00.	+ 53 50	840		DARK NEBULA
LDN 1050	21 04 00.	+ 54 40	1200		DARK NEBULA
LDN 1173	21 04 00.	+ 67 30	300		DARK NEBULA
LB 01106	21 04 05.	+ 03 13 24.		16.2	FAINT BLUE STAR
HN 2525	21 04 09.2	- 52 15 10.	18		NEBULA
RNGC 7024	21 04 10.	+ 41 18			NON-EXISTENT OBJECT
MCG+11-25-004	21 04 12.	+ 66 35	27	15.	GALAXY
HN 2519	21 04 14.2	- 59 06 46.	12		NEBULA
SCHO 1238	21 04 18.	+ 44 51 36.	370		ISOLATED DARK CLOUD
2ZW 099	21 04 18.	- 01 03			COMPACT GALAXY
ZWG 325.005	21 04 24.	+ 66 35		15.5	GALAXY
UGC 11678	21 04 24.	+ 66 35	96	15.5	GALAXY
MCG-04-49-013	21 04 24.	- 25 42	24	15.	GALAXY
RNGC 7016	21 04 26.	- 25 42		15.0	GALAXY
ABC 2336	21 04 27.	- 21 22		17.4	RICH CLUSTER OF GALAXIES
ZWG 400.029	21 04 30.	+ 03 38		15.7	GALAXY
MCG-04-49-014	21 04 30.	- 25 42	30	15.	GALAXY
RNGC 7014	21 04 30.	- 47 24		13.0	GALAXY
RNGC 7017	21 04 32.	- 25 42		15.0	GALAXY
IC 5086	21 04 32.	- 30 14 34.			NONSTELLAR OBJECT
HN 2526	21 04 32.4	- 57 42 21.	24		NEBULA
SCHO 1239	21 04 35.	+ 47 31 48.	410		ISOLATED DARK CLOUD
PK089+00.1	21 04 35.5	+ 47 39 03.	29	12.7	PLANETARY NEBULA
MCG-04-49-015	21 04 36.	- 25 40	48	14.5	GALAXY
HN 2529	21 04 37.4	- 43 27 38.	12		NEBULA
RNGC 7018	21 04 38.	- 25 40		14.0	GALAXY
RNGC 7026	21 04 39.	+ 47 39		12.5	PLANETARY NEBULA
ZWG 425.041	21 04 42.	+ 10 48		15.7	GALAXY
ZWG 449.001	21 04 42.	+ 17 49		15.7	GALAXY
ZWG 448.040	21 04 42.	+ 17 49		15.7	GALAXY
MCG+03-53-016	21 04 42.	+ 17 49	36	16.5	GALAXY
ISS 0249	21 04 42.	+ 44 22	125		STELLAR RING
LB 01107	21 04 44.	- 16 38 00.		16.0	FAINT BLUE STAR
MCG-04-49-016	21 04 48.	- 25 26	24	16.	GALAXY
HN 2528	21 04 51.4	- 52 22 20.	12		NEBULA
LDN 1068	21 04 54.	+ 56 55	660		DARK NEBULA
HN 2527	21 04 56.1	- 59 01 44.	12		NEBULA
HN 2530	21 04 56.7	- 42 46 07.	18		NEBULA
HN 2532	21 04 59.9	- 44 07 01.	24		NEBULA
ZC 2105.0+0703	21 05 00.	+ 07 03	1010		CLUSTER OF GALAXIES
LDN 1000	21 05 00.	+ 50 40	2040		DARK NEBULA
LDN 1011	21 05 00.	+ 52 10	1080		DARK NEBULA
LDN 1027	21 05 00.	+ 53 10	600		DARK NEBULA
UGC 11679	21 05 00.	+ 86 43	72	17.	GALAXY Sc
SG 3.221	21 05 01.	+ 44 00	3300		DIFFUSE EMISSION NEBULA
IC 5084	21 05 01.	- 63 30			NONSTELLAR OBJECT
HN 0705	21 05 02.	- 63 30			NEBULA
2ZW 101	21 05 06.	+ 03 41			COMPACT GALAXY
2ZW 100	21 05 06.	+ 17 45			COMPACT GALAXY
ZWG 449.002	21 05 06.	+ 17 51		15.0	GALAXY
ZWG 448.041	21 05 06.	+ 17 51		15.0	GALAXY
ZC 2105.1+1910	21 05 06.	+ 19 10	3230		CLUSTER OF GALAXIES
MCG-07-43-021	21 05 06.	- 44 08	36	17.	GALAXY
SER 141.09	21 05 06.	- 69 19	3000		LOW SURF. BRGHTNSS GALAXY
HUB C31	21 05 09.	+ 68 21			DIFFUSE NEBULA
PK084-03.1	21 05 09.40	+ 42 02 03.1	18	10.4	PLANETARY NEBULA
RNGC 7027	21 05 10.	+ 42 02		10.5	PLANETARY NEBULA
HN 2533	21 05 11.7	- 43 41 12.	18		NEBULA
2ZW 102	21 05 12.	+ 03 40			COMPACT GALAXY
ZWG 401.001	21 05 12.	+ 03 40		14.5	GALAXY
UGC 11680	21 05 12.	+ 03 40	126	14.5	GALAXY Sc+COMP
KARA.72 552A	21 05 12.	+ 03 40	60	14.5	PART OF DOUBLE GALAXY
ZWG 426.001	21 05 12.	+ 12 07		15.4	GALAXY
ZWG 425.042	21 05 12.	+ 12 07		15.4	GALAXY
LDN 0934	21 05 12.	+ 41 55	420		DARK NEBULA
MCG-07-43-022	21 05 12.	- 43 43	12	15.	GALAXY
HN 2531	21 05 14.6	- 52 16 43.	12		NEBULA
KARA.72 552B	21 05 18.	+ 03 40	36		PART OF DOUBLE GALAXY
ZWG 426.002	21 05 18.	+ 14 58		15.5	GALAXY
ZWG 425.043	21 05 18.	+ 14 58		15.5	GALAXY
MCG-02-54-001	21 05 18.	- 13 07 30.	48	15.	GALAXY
TON-S 0015	21 05 18.	- 28 37		15.2	BLUE STAR
MCG-05-50-001	21 05 18.	- 29 53	36	15.5	GALAXY
HN 2535	21 05 21.8	- 43 41 00.	18		NEBULA
MCG+03-54-001	21 05 24.	+ 16 08	96	14.	GALAXY
B 359	21 05 24.	+ 56 58	1200		DARK OBJECT
MCG-07-43-025	21 05 24.	- 40 06	24	15.	GALAXY
MCG-07-43-023	21 05 24.	- 43 43	48	14.5	GALAXY
MCG-07-43-024	21 05 24.	- 43 45	18	15.5	GALAXY
HN 2536	21 05 24.4	- 43 42 48.	18		NEBULA
ZWG 449.003	21 05 30.	+ 16 06		14.1	GALAXY
UGC 11681	21 05 30.	+ 16 08	136	14.1	GALAXY Sa
KARA.73B 0897	21 05 30.	+ 16 08	108	14.1	ISOLATED GALAXY S
MCG-07-43-026	21 05 30.	- 40 07	36	15.	GALAXY
HN 2537	21 05 31.4	- 44 29 24.	12		NEBULA
RNGC 7025	21 05 32.	+ 16 08		14.0	GALAXY
HN 2534	21 05 35.6	- 54 48 36.	12		NEBULA
MCG-05-50-002	21 05 36.	- 30 00	48	14.	GALAXY
HN 2540	21 05 38.6	- 43 41 35.	30		NEBULA
HN 2541	21 05 39.5	- 43 53 23.	18		NEBULA
HN 2542	21 05 40.3	- 43 50 11.	18		NEBULA
RNGC 7031	21 05 41.	+ 50 38		11.0	OPEN CLUSTER
ZWG 426.003	21 05 42.	+ 11 50		15.7	GALAXY
OCL 0210	21 05 42.	+ 50 38	840	11.4	OPEN STAR CLUSTER
MCG-07-43-027	21 05 42.	- 43 54	42	15.	GALAXY
HN 2539	21 05 42.4	- 47 04 35.			NEBULA
ACK 084-04.1	21 05 48.	+ 40 46			PLANETARY NEBULA
ISS 0169	21 05 48.	+ 45 55	240		STELLAR RING
MCG-01-54-001	21 05 48.	- 06 01	60	15.5	GALAXY
HN 2543	21 05 52.7	- 44 45 22.	6		NEBULA
HN 2538	21 05 53.9	- 53 07 59.	12		NEBULA

OBJECT NAME	RIGHT ASCEN.	DECLINATION	DIAM.	MAGN.	TYPE OF OBJECT
RNGC 7028	21 05 57.	+ 18 17			NON-EXISTENT OBJECT
SCHO 1240	21 05 58.	+ 45 40 00.	260		ISOLATED DARK CLOUD
LBN 0410	21 06	+ 49 25	120		BRIGHT NEBULA
LDN 0912	21 06 00.	+ 38 30	3600		DARK NEBULA
COU 079	21 06 00.	+ 46 50	7200		EMISSION NEBULA
LDN 0999	21 06 00.	+ 50 20	10860		DARK NEBULA
LDN 1063	21 06 00.	+ 56 10	600		DARK NEBULA
LDN 1065	21 06 00.	+ 56 20	600		DARK NEBULA
LB 00359	21 06 00.	- 15 35 24.		13.4	FAINT BLUE STAR
MCG+03-54-003	21 06 03.	+ 17 37	48	14.5	GALAXY
MCG+03-54-002	21 06 03.	+ 18 00	24	15.	GALAXY
RNGC 7022	21 06 04.	- 49 30			GALAXY
ZC 2106.1+0654	21 06 06.	+ 06 54	1010		CLUSTER OF GALAXIES
ZWG 449.004	21 06 06.	+ 17 36		14.6	GALAXY
UGC 11682	21 06 06.	+ 17 36	102	14.6	GALAXY S?
ZWG 449.005	21 06 06.	+ 18 00		14.9	GALAXY
KARA.72 553A	21 06 06.	+ 18 00	72	14.9	PART OF DOUBLE GALAXY
HN 2544	21 06 08.9	- 51 12 34.	12		NEBULA
MCG+03-54-004	21 06 09.	+ 17 59	54	15.	GALAXY
ZWG 449.006	21 06 12.	+ 17 59		15.3	GALAXY
UGC 11683	21 06 12.	+ 17 59	78	15.3	GALAXY Sc
KARA.72 553B	21 06 12.	+ 17 59	66	15.3	PART OF DOUBLE GALAXY
MCG-05-50-003	21 06 12.	- 27 07	36	14.5	GALAXY
HN 2545	21 06 14.8	- 50 30 58.	12		NEBULA
HN 2547	21 06 15.0	- 42 44 33.	30		NEBULA
SCHO 1241	21 06 17.	+ 45 34 24.	200		ISOLATED DARK CLOUD
UGC 11684	21 06 18.	+ 12 15	66	14.8	GALAXY S
2ZW 103	21 06 18.	+ 12 16			COMPACT GALAXY
ZWG 426.004	21 06 18.	+ 12 16		14.8	GALAXY
IC 1359	21 06 18.	+ 12 16 27.			NONSTELLAR OBJECT
HN 2546	21 06 18.8	- 52 16 33.	12		NEBULA
HN 2550	21 06 22.8	- 45 11 33.	12		NEBULA
MCG+02-54-001	21 06 24.	+ 12 17 30.	48	14.5	GALAXY
HN 2552	21 06 26.5	- 44 04 39.	12		NEBULA
B 360	21 06 27.	+ 56 18			DARK OBJECT
ZWG 401.002	21 06 30.	+ 05 12		15.7	GALAXY
HN 2553	21 06 31.6	- 45 43 51.	48		NEBULA
IC 5088	21 06 33.	- 23 04 59.			NONSTELLAR OBJECT
HN 2549	21 06 33.4	- 51 22 15.	18		NEBULA
MCG+00-54-001	21 06 36.	- 02 01	66	14.5	GALAXY
MCG-04-50-001	21 06 36.	- 23 07	18	14.5	GALAXY
HN 2548	21 06 39.0	- 54 23 09.	12		NEBULA
SCHO 1242	21 06 40.	+ 44 34 42.	1270		ISOLATED DARK CLOUD
HN 2551	21 06 40.3	- 54 09 15.	12		NEBULA
ZWG 375.001	21 06 42.	- 02 00		15.0	GALAXY
UGC 11685	21 06 42.	- 02 00	66	15.0	GALAXY S
SCHO 1243	21 06 43.	+ 45 30 54.	240		ISOLATED DARK CLOUD
HN 2554	21 06 43.7	- 44 46 32.	18		NEBULA
SG 3.222	21 06 45.	+ 42 35	1800		DIFFUSE EMISSION NEBULA
LDN 0960	21 06 48.	+ 46 58	180		DARK NEBULA
B 151	21 06 48.	+ 56 07	840		DARK OBJECT
ZWG 449.007	21 06 54.	+ 15 53		15.5	GALAXY
ZWG 375.002	21 06 54.	- 02 03		15.4	GALAXY
MRK 510	21 06 54.	- 02 03	20	16.	GALAXY WITH UV CONTINUUM
LBN 0307	21 07	+ 37 00	9000		BRIGHT NEBULA
LBN 0369	21 07	+ 42 40	1500		BRIGHT NEBULA
LBN 0392	21 07	+ 45 00	1740		BRIGHT NEBULA
LB 10002	21 07	- 87 20		14.0	FAINT BLUE STAR
ZWG 426.005	21 07 00.	+ 11 34		15.7	GALAXY
OCL 0191	21 07 00.	+ 37 24	1080		OPEN STAR CLUSTER
LDN 0938	21 07 00.	+ 42 30	2280		DARK NEBULA
COU 080	21 07 00.	+ 43 00	9000		EMISSION NEBULA
LDN 1069	21 07 00.	+ 56 40	540		DARK NEBULA
UGC 11686	21 07 00.	+ 65 32	78	17.	GALAXY SBc
RNGC 7021	21 07 03.	- 63 44			NON-EXISTENT OBJECT
LDN 0961	21 07 06.	+ 47 02	300		DARK NEBULA
MCG-01-54-002	21 07 06.	- 03 42	36	16.5	GALAXY
ZWG 426.006	21 07 12.	+ 14 55		15.2	GALAXY
KARA.72 554A	21 07 12.	+ 14 55	54	15.2	PART OF DOUBLE GALAXY
LB 01108	21 07 12.	- 19 06 06.		17.2	FAINT BLUE STAR
HN 2557	21 07 12.8	- 44 25 24.	18		NEBULA
RNGC 7033	21 07 13.	+ 14 55		15.0	GALAXY
HN 2555	21 07 16.7	- 53 37 25.	30		NEBULA
ZWG 426.007	21 07 18.	+ 14 57		15.3	GALAXY
UGC 11687	21 07 18.	+ 14 57	60	15.3	GALAXY E
KARA.72 554B	21 07 18.	+ 14 57	66	15.3	PART OF DOUBLE GALAXY
MCG+02-54-002	21 07 18.	+ 14 57	36	15.	GALAXY
RNGC 7034	21 07 19.	+ 14 57		15.5	GALAXY
RNGC 7020	21 07 20.	- 64 15		12.5	GALAXY
HN 2556	21 07 20.0	- 51 40 48.	24		NEBULA
MCG+02-54-003	21 07 24.	+ 14 58	48	15.	GALAXY
MCG-04-50-002	21 07 24.	- 24 28 30.	66	15.	GALAXY
MCG-05-50-004	21 07 24.	- 29 45	84	14.5	GALAXY
HN 2559	21 07 28.7	- 46 17 48.	18		NEBULA
LB 01109	21 07 34.	- 14 38 18.		16.0	FAINT BLUE STAR
HN 2558	21 07 34.5	- 50 07 54.	18		NEBULA
MRK 511	21 07 36.	- 01 48	7	15.	GALAXY WITH UV CONTINUUM
MCG-01-54-003	21 07 36.	- 03 53	120	14.5	GALAXY
HN 2560	21 07 36.6	- 46 15 53.	18		NEBULA
HN 2561	21 07 36.9	- 44 31 17.			NEBULA
HN 2563	21 07 38.3	- 42 50 17.	48		NEBULA
HN 2564	21 07 40.5	- 43 54 41.	24		NEBULA
HN 2562	21 07 41.3	- 46 22 35.	18		NEBULA
ZWG 426.008	21 07 42.	+ 11 51		15.6	GALAXY
LB 01110	21 07 42.	- 14 46 24.		17.9	FAINT BLUE STAR
MCG-07-43-029	21 07 42.	- 42 52		16.	GALAXY
MCG-07-43-028	21 07 42.	- 42 52	18	16.	GALAXY
LDN 0964	21 07 48.	+ 47 10	300		DARK NEBULA
RNGC 7036	21 07 49.	+ 15 15			NON-EXISTENT OBJECT
HN 2565	21 07 51.4	- 47 01 05.	36		NEBULA
LB 01111	21 07 56.	+ 02 25 42.		15.9	FAINT BLUE STAR
HN 2567	21 07 57.1	- 44 19 22.	18		NEBULA
SCHO 1244	21 07 58.	+ 47 07 54.	380		ISOLATED DARK CLOUD
LBN 0316	21 08	+ 38 00	5400		BRIGHT NEBULA
KHAV 717	21 08	+ 41 18	8620		DARK NEBULA
KHAV 718	21 08	+ 42 42			DARK NEBULA
2ZW 104	21 08 00.	+ 18 34			COMPACT GALAXY
LDN 0937	21 08 00.	+ 42 15	11580		DARK NEBULA
LDN 1022	21 08 00.	+ 52 40	2460		DARK NEBULA
LDN 1067	21 08 00.	+ 56 30	420		DARK NEBULA
ZWG 449.008	21 08 06.	+ 18 35		15.5	GALAXY
SCHO 1245	21 08 06.	+ 47 29 00.	260		ISOLATED DARK CLOUD
HN 2566	21 08 06.7	- 50 56 34.	12		NEBULA
LB 01112	21 08 09.	- 14 05 48.		17.0	FAINT BLUE STAR
IC 5085	21 08 12.	- 74 18			NONSTELLAR OBJECT
HE 0706	21 08 13.	- 74 19			NEBULA
HN 2570	21 08 15.5	- 45 27 15.	18		NEBULA
ZC 2108.3+0732	21 08 18.	+ 07 32	3560		CLUSTER OF GALAXIES

OBJECT NAME	RIGHT ASCEN.	DECLINATION	DIAM.	MAGN.	TYPE OF OBJECT
ISS 0170	21 08 18.	+ 48 51	95		STELLAR RING
ZWG 375.003	21 08 18.	- 02 11		15.1	GALAXY
LB 00204	21 08 20.	+ 15 17 42.		15.5	FAINT BLUE STAR
IC 1360	21 08 20.3	+ 04 51 59.			GALAXY Sa
IC 5089	21 08 21.	- 04 03 17.			NONSTELLAR OBJECT
HN 2572	21 08 22.6	- 44 11 27.	18		NEBULA
RNGC 7030	21 08 23.	- 20 42			GALAXY
RNGC 7029	21 08 23.	- 49 30		12.5	GALAXY
ZWG 401.003	21 08 24.	+ 04 52		15.6	GALAXY
MCG-04-50-003	21 08 24.	- 20 42 30.	24	15.	GALAXY
HN 2569	21 08 26.3	- 51 54 15.	12		NEBULA
HN 2573	21 08 26.5	- 46 14 27.	18		NEBULA
RNGC 7035	21 08 27.	- 23 25		15.0	GALAXY
HN 2568	21 08 27.2	- 57 28 51.	42		NEBULA
HN 2571	21 08 29.8	- 52 57 15.	12		NEBULA
LDN 0967	21 08 30.	+ 47 22	1560		DARK NEBULA
MCG+00-54-002	21 08 30.	- 02 14	15	15.	GALAXY
MCG-04-50-005	21 08 30.	- 23 24	72	15.5	GALAXY
MCG-04-50-004	21 08 30.	- 23 25 30.	18	15.5	GALAXY
ZWG 426.009	21 08 36.	+ 15 00		15.7	GALAXY
ZWG 375.004	21 08 36.	- 02 15		15.2	GALAXY
HN 2574	21 08 36.6	- 46 06 50.	12		NEBULA
RNGC 7037	21 08 40.	+ 33 31			NON-EXISTENT OBJECT
ZWG 375.005	21 08 42.	+ 03 18		15.3	GALAXY
MCG+01-54-001	21 08 42.	+ 04 11	48	15.	GALAXY
ZWG 401.004	21 08 42.	+ 04 12		15.7	GALAXY
UGC 11688	21 08 42.	+ 04 12	78	15.7	GALAXY Sb
MCG-04-50-006	21 08 45.	- 23 25	9	15.5	GALAXY
2ZW 105	21 08 48.	+ 06 34			COMPACT GALAXY
ZWG 325.006	21 08 48.	+ 65 57		15.1	GALAXY
UGC 11689	21 08 48.	+ 65 57	84	15.1	GALAXY SBb
MCG+11-25-005	21 08 48.	+ 65 59	78	15.	GALAXY
MCG-01-54-004	21 08 48.	- 04 31	36	15.5	GALAXY
MCG-04-50-008	21 08 48.	- 23 23	12	15.	GALAXY
MCG-04-50-007	21 08 48.	- 23 24	24	15.	GALAXY
IC 1363	21 08 49.	+ 46 39			MAY NOT EXIST
CED 188	21 08 52.	+ 46 39			DIFFUSE GALACTIC NEBULA
HN 2576	21 08 53.2	- 44 42 55.	18		NEBULA
ZWG 426.010	21 08 54.	+ 11 14		15.7	GALAXY
UGC 11690	21 08 54.	+ 22 43	78	16.0	GALAXY S0-a
4ZW 068	21 08 54.	+ 28 42			COMPACT GALAXY
MCG+00-54-003	21 08 54.	- 02 14	54	14.5	GALAXY
IC 5090	21 08 56.	- 02 14 20.			NONSTELLAR OBJECT
HN 2575	21 08 57.8	- 53 45 32.	12		NEBULA
HN 2578	21 08 57.9	- 46 27 25.	18		NEBULA
IC 1361	21 08 59.4	+ 04 05 57.			GALAXY SB(s)0
LBN 0340	21 09	+ 40 05	420		BRIGHT NEBULA
KHAV 719	21 09	+ 46 00	1410		DARK NEBULA
MCG+01-54-002	21 09 00.	+ 04 50	30	15.	GALAXY
ZWG 401.005	21 09 00.	+ 04 51		15.4	GALAXY
UGC 11692	21 09 00.	+ 04 51	66	15.4	GALAXY S
ZWG 375.006	21 09 00.	- 02 14		14.5	GALAXY
UGC 11691	21 09 00.	- 02 14	66	14.5	GALAXY S
MCG-02-54-002	21 09 00.	- 14 18	60	15.	GALAXY
SCHO 1246	21 09 01.	+ 44 15 48.	640		ISOLATED DARK CLOUD
ZWG 426.011	21 09 06.	+ 11 07		15.7	GALAXY
ZWG 426.012	21 09 12.	+ 11 17		15.6	GALAXY
ZWG 426.013	21 09 12.	+ 13 06		15.6	GALAXY
UGC 11693	21 09 12.	+ 37 42	90	15.5	GALAXY Sa-b
MRK 512	21 09 12.	- 01 35	12	15.	GALAXY WITH UV CONTINUUM
HN 2581	21 09 12.0	- 45 05 24.	36		NEBULA
HE 0707	21 09 14.	- 73 59			NEBULA
IC 5087	21 09 15.	- 73 59			NONSTELLAR OBJECT
ZWG 375.007	21 09 18.	+ 02 08		15.7	GALAXY
KARA.73B 0898	21 09 18.	+ 02 08	18	15.7	ISOLATED GALAXY E
ZWG 449.009	21 09 18.	+ 20 07		15.6	GALAXY
MCG-07-43-030	21 09 18.	- 39 33	18	15.	GALAXY
IC 1362	21 09 21.	+ 02 07 24.			NONSTELLAR OBJECT
SCHO 1248	21 09 23.	+ 45 42 48.	460		ISOLATED DARK CLOUD
SCHO 1247	21 09 23.	+ 46 16 42.	200		ISOLATED DARK CLOUD
HN 2577	21 09 23.3	- 58 38 25.	18		NEBULA
ZWG 426.014	21 09 24.	+ 11 03		15.2	GALAXY
UGC 11694	21 09 24.	+ 11 03	108	15.2	GALAXY
OCL 0203	21 09 24.	+ 45 27	1500	7.1	OPEN STAR CLUSTER
MCG-03-54-001	21 09 24.	- 20 02	24	15.	GALAXY
HN 2579	21 09 25.7	- 58 38 06.	18		NEBULA
SCHO 1249	21 09 27.	+ 46 04 18.	280		ISOLATED DARK CLOUD
HN 2583	21 09 29.1	- 44 49 48.	18		NEBULA
MCG+02-54-004	21 09 30.	+ 11 04	96	13.	GALAXY
ISS 0171	21 09 30.	+ 45 13	157		STELLAR RING
HN 2580	21 09 34.5	- 58 55 12.	12		NEBULA
UGC 11696	21 09 36.	+ 11 09	60	16.0	GALAXY Sc-IRR
ZWG 426.015	21 09 36.	+ 11 17		15.5	GALAXY
ISS 0250	21 09 36.	+ 42 22	203		STELLAR RING
ZWG 375.008	21 09 36.	- 01 40		14.7	GALAXY
UGC 11695	21 09 36.	- 01 40	84	14.7	GALAXY Sb
MCG+00-54-004	21 09 36.	- 01 40	66	14.	GALAXY
MCG+02-54-005	21 09 39.	+ 11 17 30.	60	15.	GALAXY
ZWG 426.016	21 09 42.	+ 11 26		15.4	GALAXY
UGC 11697	21 09 42.	+ 11 26	60	15.4	GALAXY SB
ZWG 426.017	21 09 42.	+ 12 47		15.4	GALAXY
HN 2582	21 09 42.3	- 58 20 54.	18		NEBULA
LB 00205	21 09 44.	+ 15 27 42.		17.1	FAINT BLUE STAR
MCG+02-54-007	21 09 48.	+ 11 27 30.	60	14.	GALAXY
ZWG 426.018	21 09 48.	+ 12 48		14.6	GALAXY
UGC 11698	21 09 48.	+ 12 48	60	14.6	GALAXY S+COMP
MCG+02-54-006	21 09 48.	+ 12 48	48	15.	GALAXY
ISS 0251	21 09 48.	+ 39 52	107		STELLAR RING
LDN 0956	21 09 48.	+ 45 46	180		DARK NEBULA
ZWG 375.009	21 09 48.	- 01 32		15.2	GALAXY
ZWG 426.019	21 09 54.	+ 12 24		15.6	GALAXY
UGC 11699	21 09 54.	+ 12 24	60	15.6	GALAXY SB:c
MCG+02-54-009	21 09 57.	+ 12 50	48	14.	GALAXY
MCG+02-54-008	21 09 57.	+ 12 50	36	13.	GALAXY
LBN 0321	21 10	+ 38 00	2820		BRIGHT NEBULA
KHAV 721	21 10	+ 39 36	6190		DARK NEBULA
KHAV 720	21 10	+ 47 24	2230		DARK NEBULA
LBN 0445	21 10	+ 59 05	1920		BRIGHT NEBULA
LBN 0449	21 10	+ 59 30	7800		BRIGHT NEBULA
ZWG 426.020	21 10 00.	+ 11 11		14.9	GALAXY
UGC 11700	21 10 00.	+ 11 11	66	14.9	GALAXY SBb
MCG+02-54-011	21 10 00.	+ 11 12	54	14.	GALAXY
MCG+02-54-010	21 10 00.	+ 12 25	60	14.5	GALAXY
COU 081	21 10 00.	+ 38 00	10800		EMISSION NEBULA
LDN 0959	21 10 00.	+ 46 20	1140		DARK NEBULA
LDN 0975	21 10 00.	+ 47 40	1080		DARK NEBULA
OCL 0206	21 10 00.	+ 48 16	360	16.	OPEN STAR CLUSTER
LDN 1060	21 10 00.	+ 55 05	660		DARK NEBULA

OBJECT NAME	RIGHT ASCEN.	DECLINATION	DIAM.	MAGN.	TYPE OF OBJECT
MCG-04-50-009	21 10 00.	- 20 46 30.	36	15.	GALAXY
HN 2584	21 10 01.2	- 54 12 52.	12		NEBULA
ZC 2110.2+1238	21 10 12.	+ 12 38	12570		CLUSTER OF GALAXIES
ZWG 426.021	21 10 12.	+ 13 46		14.9	GALAXY
LB C0360	21 10 15.	- 16 02 12.		15.0	FAINT BLUE STAR
HN 2585	21 10 16.8	- 46 36 33.			NEBULA
ZWG 401.006	21 10 18.	+ 08 34		15.5	GALAXY
ZC 2110.3+1203	21 10 18.	+ 12 03	2220		CLUSTER OF GALAXIES
2ZW 106	21 10 18.	+ 13 47			COMPACT GALAXY
SG 3.223	21 10 22.	+ 37 58	9000		DIFFUSE EMISSION NEBULA
ZWG 375.010	21 10 24.	+ 01 27		15.6	GALAXY
MCG+01-54-003	21 10 24.	+ 08 36	48	16.	GALAXY
MCG+02-54-012	21 10 24.	+ 13 47	48	14.5	GALAXY
RNGC 7039	21 10 24.	+ 45 27		7.0	OPEN CLUSTER
OCL 0204	21 10 24.	+ 47 32	540	12.3	OPEN STAR CLUSTER
MCG-02-54-003	21 10 24.	- 11 02	48	15.	GALAXY
IC 1369	21 10 26.	+ 47 32 17.	504		OPEN CLUSTER
CED 189	21 10 26.	+ 47 33	120		DIFFUSE GALACTIC NEBULA
HN 2586	21 10 29.3	- 44 48 39.	18		NEBULA
LB 00206	21 10 30.	+ 15 19 30.		16.6	FAINT BLUE STAR
MRSL 098+08/1	21 10 30.	+ 59 45	8400		HII REGION
CED 190	21 10 34.	+ 59 47	6000		DIFFUSE GALACTIC NEBULA
SG 2.079	21 10 34.	+ 59 47	8280		DIFFUSE EMISSION NEBULA
SG 1.18	21 10 34.	+ 59 47	8280		DIFFUSE EMISSION NEBULA
ZWG 449.010	21 10 36.	+ 19 10		15.7	GALAXY
KARA.73B 0899	21 10 36.	+ 19 10	24	15.7	ISOLATED GALAXY S
MCG+00-54-005	21 10 36.	- 00 34	36	15.	GALAXY
BC PKS2110-16	21 10 36.9	- 16 01 30.		19.	QUASI-STELLAR OBJECT
SS 63	21 10 37.	+ 51 35			DIFFUSE GALACTIC NEBULA
ZWG 375.011	21 10 42.	- 00 34		15.2	GALAXY
LB 01113	21 10 45.	+ 00 06 00.		16.2	FAINT BLUE STAR
RNGC 7040	21 10 47.	+ 08 39		15.0	GALAXY
MCG+00-54-006	21 10 48.	+ 02 35	18	15.	GALAXY
ZWG 401.007	21 10 48.	+ 06 32		15.4	GALAXY
ZWG 401.008	21 10 48.	+ 08 39		14.9	GALAXY
UGC 11701	21 10 48.	+ 08 39	66	14.9	GALAXY S
MCG+01-54-004	21 10 48.	+ 08 41	48	15.	GALAXY
ZWG 375.012	21 10 48.	- 00 22		15.4	GALAXY
IC 1364	21 10 53.	+ 02 33 27.			NONSTELLAR OBJECT
ZC 2110.9+0203	21 10 54.	+ 02 03	6520		CLUSTER OF GALAXIES
2ZW 107	21 10 54.	+ 02 34			COMPACT GALAXY
ZWG 375.013	21 10 54.	+ 02 34		14.7	GALAXY
RNGC 7032	21 10 59.	- 68 30			UNVERIFIED SOUTHERN OBJECT
KHAV 722	21 11	+ 43 30	1720		DARK NEBULA
KHAV 723	21 11	+ 61 18	1860		DARK NEBULA
LDN 0970	21 11 00.	+ 47 10	1380		DARK NEBULA
ZWG 375.014	21 11 00.	- 01 35		15.0	GALAXY
B 361	21 11 01.	+ 47 13	1800		DARK OBJECT
RNGC 7044	21 11 04.	+ 42 17		11.5	OPEN CLUSTER
ASS 22	21 11 06.	+ 37 40			OB ASSOCIATION CYG OB4
OCL 0198	21 11 06.	+ 42 17	660	12.0	OPEN STAR CLUSTER
SCHO 1250	21 11 08.	+ 47 17 18.	1550		ISOLATED DARK CLOUD
ZWG 426.022	21 11 12.	+ 13 39		15.6	GALAXY
LB 01114	21 11 16.	- 16 19 54.		17.6	FAINT BLUE STAR
ZWG 401.009	21 11 18.	+ 04 38		15.7	GALAXY
MCG-04-50-010	21 11 18.	- 26 17 30.	24	15.5	GALAXY
LB 01115	21 11 22.	- 20 21 24.		17.1	FAINT BLUE STAR
ZWG 375.015	21 11 24.	+ 02 21		15.1	GALAXY
IC 1365	21 11 24.	+ 02 21 06.			NONSTELLAR OBJECT
2ZW 108	21 11 24.	+ 02 22			COMPACT GALAXY
MCG+00-54-007	21 11 24.	+ 02 23	12	15.	GALAXY
ZWG 426.023	21 11 24.	+ 13 21		13.0	GALAXY
RNGC 7042	21 11 24.	+ 13 21		13.0	GALAXY
UGC 11702	21 11 24.	+ 13 21	126	13.0	GALAXY Sb
KARA.72 555A	21 11 24.	+ 13 21	120	13.0	PART OF DOUBLE GALAXY
MCG+02-54-013	21 11 24.	+ 13 22	120	12.	GALAXY
ZC 2111.4-0239	21 11 24.	- 02 39	1080		CLUSTER OF GALAXIES
ZWG 375.016	21 11 30.	+ 00 20		14.9	GALAXY
LB 01116	21 11 33.	- 14 57 54.		17.2	FAINT BLUE STAR
ZWG 375.017	21 11 36.	+ 01 33		15.7	GALAXY
MCG-05-50-005	21 11 36.	- 30 17	36	15.	GALAXY
IC 1366	21 11 37.	+ 01 34 07.			NONSTELLAR OBJECT
IC 1367	21 11 40.	+ 02 47 13.			NONSTELLAR OBJECT
ZWG 375.018	21 11 42.	+ 01 58		14.3	GALAXY
UGC 11703	21 11 42.	+ 01 58	78	14.3	GALAXY S
ZWG 375.019	21 11 42.	+ 02 47		15.3	GALAXY
ZWG 426.024	21 11 42.	+ 13 24		14.9	GALAXY
UGC 11704	21 11 42.	+ 13 24	66	14.9	GALAXY SB:a
KARA.72 555B	21 11 42.	+ 13 24	72	14.9	PART OF DOUBLE GALAXY
IC 1368	21 11 43.	+ 01 57 13.			NONSTELLAR OBJECT
RNGC 7043	21 11 43.	+ 13 24		15.0	GALAXY
RNGC 7038	21 11 43.	- 47 26		12.5	GALAXY
MCG+00-54-008	21 11 45.	+ 01 59	48	14.	GALAXY
MCG+02-54-014	21 11 45.	+ 13 25	60	14.5	GALAXY
MCG-01-54-005	21 11 48.	- 08 25	42	18.	GALAXY
SCHO 1251	21 11 51.	+ 46 12 24.	290		ISOLATED DARK CLOUD
ISS 0252	21 11 54.	+ 39 32	121		STELLAR RING
MCG-05-50-006	21 11 54.	- 31 49	54	15.	GALAXY
LBN 0383	21 12	+ 43 00	900		BRIGHT NEBULA
KHAV 724	21 12	+ 46 36	1860		DARK NEBULA
LBN 0117	21 12	- 08 00	14400		BRIGHT NEBULA
UGC 11705	21 12 00.	+ 12 27	60	16.5	GALAXY
ZWG 426.025	21 12 00.	+ 15 02		15.7	GALAXY
UGC 11706	21 12 00.	+ 15 02	102	15.7	GALAXY Sb
2ZW 109	21 12 00.	+ 21 20			COMPACT GALAXY
COU 082	21 12 00.	+ 45 00	7200		EMISSION NEBULA
LDN 0963	21 12 00.	+ 46 20	1260		DARK NEBULA
LDN 1005	21 12 00.	+ 50 30	10740		DARK NEBULA
LDN 1062	21 12 00.	+ 55 20	180		DARK NEBULA
LDN 1064	21 12 00.	+ 55 25	120		DARK NEBULA
LDN 1119	21 12 00.	+ 61 30	360		DARK NEBULA
SCHO 1252	21 12 03.	+ 46 25 48.	450		ISOLATED DARK CLOUD
IC 5092	21 12 03.	- 64 41	150		GALAXY SB(rs)
HN 1217	21 12 04.	- 64 41	90		NEBULA
MCG+02-54-015	21 12 06.	+ 15 02 30.	108	15.	GALAXY
ZWG 449.011	21 12 06.	+ 21 21		15.5	GALAXY
KARA.73B 0900	21 12 06.	+ 21 21	30	15.5	ISOLATED GALAXY S
HN 2588	21 12 06.9	- 45 11 10.	18		NEBULA
HN 2587	21 12 08.7	- 59 25 52.	12		NEBULA
HN 2589	21 12 11.9	- 46 59 16.			NEBULA
MCG-01-54-006	21 12 12.	- 07 40	48	16.	GALAXY
2ZW 110	21 12 18.	+ 10 39			COMPACT GALAXY
ZWG 449.012	21 12 18.	+ 18 47		14.9	GALAXY
ZWG 471.001	21 12 18.	+ 26 32		15.6	GALAXY
UGC 11707	21 12 18.	+ 26 32	216	15.6	GALAXY
MCG+04-50-001	21 12 18.	+ 26 32	210	13.	GALAXY
MCG-04-50-011	21 12 18.	- 22 31	42	15.	GALAXY
PK089-00.1	21 12 19.	+ 47 32	208		PLANETARY NEBULA

OBJECT NAME	RIGHT ASCEN.	DECLINATION	DIAM.	MAGN.	TYPE OF OBJECT
RNGC 7045	21 12 21.	+ 04 18			NON-EXISTENT OBJECT
MCG+03-54-005	21 12 21.	+ 18 49	24	15.	GALAXY
LB 01117	21 12 22.	- 16 13 18.		17.5	FAINT BLUE STAR
ZWG 375.020	21 12 24.	+ 02 37		14.2	GALAXY
UGC 11708	21 12 24.	+ 02 37	120	14.2	GALAXY SB
MCG+00-54-009	21 12 24.	+ 02 38	114	13.5	GALAXY
ZWG 426.026	21 12 24.	+ 10 38		15.4	GALAXY
KARA.73B 0901	21 12 24.	+ 10 38	42	15.4	ISOLATED GALAXY IR
RNGC 7048	21 12 24.	+ 46 04		11.5	PLANETARY NEBULA
ISS 0172	21 12 24.	+ 48 19	109		STELLAR RING
MCG-04-50-012	21 12 24.	- 21 22	42	15.	GALAXY
RNGC 7046	21 12 26.	+ 02 38		14.0	GALAXY
BIGO 550	21 12 27.	+ 04 15			NEBULA
PK089-01.1	21 12 27.75	+ 46 04 50.4	78	11.3	PLANETARY NEBULA
IC 5097	21 12 28.	+ 04 15			THREE STARS
IC 5098	21 12 30.	+ 04 17			TWO STARS
MCG-07-43-031	21 12 30.	- 42 39	48	16.	GALAXY
SCHO 1253	21 12 33.	+ 47 46 36.	360		ISOLATED DARK CLOUD
MCG+04-50-002	21 12 39.	+ 25 39	48	16.	GALAXY
LB 01118	21 12 40.	- 14 29 42.		16.8	FAINT BLUE STAR
IC 1370	21 12 42.	+ 01 58 30.			NONSTELLAR OBJECT
2ZW 111	21 12 42.	+ 01 59			COMPACT GALAXY
ZWG 375.021	21 12 42.	+ 01 59		15.1	GALAXY
2ZW 112	21 12 42.	+ 06 55			COMPACT GALAXY
ZWG 401.010	21 12 42.	+ 06 55		15.6	GALAXY
2ZW 113	21 12 42.	+ 11 28			COMPACT GALAXY
2ZW 114	21 12 42.	+ 12 27			COMPACT GALAXY
ZWG 426.027	21 12 42.	+ 12 27		15.0	GALAXY
PHL 2153	21 12 42.	- 19 11		18.6	BLUE STELLAR OBJECT
UGC 11709	21 12 48.	+ 25 40	96	16.5	GALAXY DWARF
DV.56 N7038A	21 12 48.	- 47 49			GALAXY
RNGC 7038A	21 12 49.	- 47 49			GALAXY
LB 01119	21 12 50.	- 13 15 54.		17.7	FAINT BLUE STAR
LB 01120	21 12 53.	- 15 22 00.		17.4	FAINT BLUE STAR
ZWG 401.011	21 12 54.	+ 07 35		15.6	GALAXY
ZWG 449.013	21 12 54.	+ 16 58		15.7	GALAXY
SCHO 1254	21 12 57.	+ 46 23 54.	230		ISOLATED DARK CLOUD
LB C0361	21 12 57.	- 15 35 30.		14.2	FAINT BLUE STAR
LBN 0390	21 13	+ 43 40	900		BRIGHT NEBULA
LBN 0389	21 13	+ 43 40	600		BRIGHT NEBULA
KHAV 725	21 13	+ 48 06	1410		DARK NEBULA
RNGC 7041	21 13 00.	- 48 35		12.5	GALAXY
HN 0708	21 13 02.	- 70 52			NEBULA
IC 5091	21 13 03.	- 70 52			NONSTELLAR OBJECT
RNGC 7050	21 13 05.	+ 36 00			NON-EXISTENT OBJECT
LB C0207	21 13 05.	- 14 38 30.		15.2	FAINT BLUE STAR
HN 2590	21 13 05.0	- 59 43 08.	12		NEBULA
MCG-08-43-032	21 13 06.	- 39 12	60	15.	GALAXY
MCG+04-50-003	21 13 09.	+ 24 08	72	16.5	GALAXY
SCHO 1255	21 13 09.	+ 46 41 00.	300		ISOLATED DARK CLOUD
UGC 11710	21 13 12.	+ 24 09	90	16.5	GALAXY
SCHO 1256	21 13 12.	+ 46 18 18.	260		ISOLATED DARK CLOUD
ZWG 426.028	21 13 18.	+ 15 27		15.6	GALAXY
B 152	21 13 18.	+ 61 32	900		DARK OBJECT
MCG-05-50-007	21 13 18.	- 30 54	36	15.	GALAXY
FATH 1.820	21 13 19.	- 14 35			NEBULA
HN 2591	21 13 20.5	- 59 32 31.	12		NEBULA
UGC 11711	21 13 24.	+ 17 10	66	16.0	GALAXY Sc
HN 1218	21 13 28.	- 60 09	60		NEBULA
IC 5095	21 13 28.	- 60 09			NONSTELLAR OBJECT
ZWG 401.012	21 13 30.	+ 09 21		15.5	GALAXY
KARA.73B 0902	21 13 30.	+ 09 21	42	15.5	ISOLATED GALAXY S
LDN 1225	21 13 30.	+ 61 30	360		DARK NEBULA
MCG-07-43-033	21 13 30.	- 42 29	60	14.	GALAXY
ZWG 375.022	21 13 36.	+ 00 48		15.2	GALAXY
AGU 66	21 13 36.	- 42 28 48.		12.5	PAIR OF GALAXIES
HN 2592	21 13 41.2	- 42 28 11.	18		NEBULA
ISS 0173	21 13 42.	+ 47 14	92		STELLAR RING
MCG-07-43-034	21 13 42.	- 42 29	90	13.	GALAXY
HN 0709	21 13 42.	- 66 39			NEBULA
LB 01121	21 13 43.	+ 19 13 54.		16.4	FAINT BLUE STAR
IC 5094	21 13 43.	- 66 39			NONSTELLAR OBJECT
LB 00208	21 13 53.	- 14 28 36.		15.5	FAINT BLUE STAR
ZWG 375.024	21 13 54.	+ 00 52		15.7	GALAXY
ISS 0174	21 13 54.	+ 46 54	372		STELLAR RING
MCG+00-54-010	21 13 54.	- 01 01	72	13.5	GALAXY
ZWG 375.023	21 13 54.	- 01 02		14.3	GALAXY
UGC 11712	21 13 54.	- 01 02	78	14.3	GALAXY S
RNGC 7047	21 13 55.	- 01 02		14.5	GALAXY
LB 01122	21 13 56.	- 17 26 42.		15.1	FAINT BLUE STAR
LB 01123	21 13 57.	+ 01 21 48.		16.6	FAINT BLUE STAR
LBN 0360	21 14	+ 40 20	1260		BRIGHT NEBULA
LBN 0393	21 14	+ 43 50	1020		BRIGHT NEBULA
LBN 0453	21 14	+ 60 00	3360		BRIGHT NEBULA
ZWG 426.029	21 14 00.	+ 10 03		15.5	GALAXY
KARA.73B 0903	21 14 00.	+ 10 03	36	15.5	ISOLATED GALAXY S0
ISS 0253	21 14 00.	+ 45 19	151		STELLAR RING
7ZW 938	21 14 00.	+ 87 42			COMPACT GALAXY
ZWG 370.015	21 14 00.	+ 87 42		15.6	GALAXY
ZWG 370.009	21 14 00.	+ 87 42		15.6	GALAXY
LB 01124	21 14 00.	- 13 19 48.		17.6	FAINT BLUE STAR
PHL 2173	21 14 00.	- 19 59		17.3	BLUE STELLAR OBJECT
MCG-07-43-035	21 14 00.	- 42 03	60	15.5	GALAXY
HN 2593	21 14 00.3	- 52 21 11.	12		NEBULA
LB 01125	21 14 01.	- 15 48 12.		16.0	FAINT BLUE STAR
SG 2.080	21 14 03.	+ 43 41	2880		DIFFUSE EMISSION NEBULA
ZWG 426.030	21 14 06.	+ 11 16		15.3	GALAXY
ZC 2114.1+1735	21 14 06.	+ 17 35	870		CLUSTER OF GALAXIES
MCG-05-50-008	21 14 06.	- 28 33	72	15.	GALAXY
HN 0710	21 14 07.	- 70 51			NEBULA
IC 5093	21 14 07.	- 70 51			NONSTELLAR OBJECT
MCG+02-54-016	21 14 12.	+ 11 17	60	13.	GALAXY
MCG-07-43-036	21 14 12.	- 42 33	15	15.	GALAXY
HN 0711	21 14 17.	- 63 58			NEBULA
ZWG 426.031	21 14 18.	+ 14 50		15.3	GALAXY
IC 5096	21 14 18.	- 63 58			NONSTELLAR OBJECT
LB 01126	21 14 23.	- 14 11 54.		17.3	FAINT BLUE STAR
ZWG 426.032	21 14 24.	+ 14 49		15.3	GALAXY
MCG-01-54-007	21 14 24.	- 06 58	66	15.5	GALAXY
DV.56 N7041B	21 14 24.	- 48 37	18		GALAXY
DV.56 N7041A	21 14 24.	- 48 37	54		GALAXY
SG 3.224	21 14 26.	+ 40 14	1800		DIFFUSE EMISSION NEBULA
HN 2594	21 14 27.3	- 50 21 57.	18		NEBULA
ZWG 426.033	21 14 30.	+ 13 02		15.1	GALAXY
2ZW 116	21 14 30.	+ 13 03			COMPACT GALAXY
MCG+02-54-018	21 14 30.	+ 14 51	18	15.	GALAXY
MCG+02-54-017	21 14 30.	+ 14 51 30.	36	14.5	GALAXY
ZWG 449.014	21 14 30.	+ 20 06		15.4	GALAXY

OBJECT NAME	RIGHT ASCEN.	DECLINATION	DIAM.	MAGN.	TYPE OF OBJECT
LDN 0942	21 14 30.	+ 43 00	300		DARK NEBULA
2ZW 115	21 14 30.	- 01 32			COMPACT GALAXY
MCG-01-54-008	21 14 30.	- 06 57	24	15.	GALAXY
RNGC 7041B	21 14 30.	- 48 37			GALAXY
ARC 2337	21 14 34.	- 22 33		17.4	RICH CLUSTER OF GALAXIES
HN 2595	21 14 34.3	- 51 56 09.	30		NEBULA
PK072-17.1	21 14 36.	+ 23 57	871	12.2	PLANETARY NEBULA
MCG+04-50-004	21 14 36.	+ 23 59 30.	24	15.	GALAXY
ZC 2114.6+2801	21 14 36.	+ 28 01	3020		CLUSTER OF GALAXIES
MCG-01-54-009	21 14 36.	- 06 52 30.	36	17.	GALAXY
MCG-04-50-013	21 14 36.	- 23 03	72	15.	GALAXY
RNGC 7041A	21 14 36.	- 48 37			GALAXY
LB 01127	21 14 39.	- 18 31 30.		18.2	FAINT BLUE STAR
ZWG 471.002	21 14 42.	+ 24 00		15.0	GALAXY
TON-S 0016	21 14 42.	- 28 42		15.3	BLUE STAR
LB 00362	21 14 43.	- 14 06 36.		14.4	FAINT BLUE STAR
HN 2597	21 14 46.7	- 46 30 44.			NEBULA
LDN 0943	21 14 48.	+ 43 11	360		DARK NEBULA
MCG-04-50-014	21 14 48.	- 23 00	36	15.	GALAXY
ZC 2114.9+0720	21 14 54.	+ 07 20	1880		CLUSTER OF GALAXIES
ZWG 449.015	21 14 54.	+ 18 12		15.7	GALAXY
ZC 2114.9+2010	21 14 54.	+ 20 10	1010		CLUSTER OF GALAXIES
MCG-04-50-015	21 14 57.	- 23 04	54	15.	GALAXY
KHAV 726	21 15	+ 43 07	2760		DARK NEBULA
LBN 0391	21 15	+ 43 30	6300		BRIGHT NEBULA
LBN 0495	21 15	+ 68 15	600		BRIGHT NEBULA
LBN 0545	21 15	+ 76 50	300		BRIGHT NEBULA
ZWG 426.034	21 15 00.	+ 15 20		15.3	GALAXY
UGC 11713	21 15 00.	+ 15 20	96	15.3	GALAXY SBc
LDN 1072	21 15 00.	+ 56 00	4980		DARK NEBULA
MCG-04-50-016	21 15 00.	- 23 16 30.	24	15.5	GALAXY
ZWG 426.035	21 15 06.	+ 15 15		15.5	GALAXY
OCL 0212	21 15 06.	+ 51 34	360	14.	OPEN STAR CLUSTER
MCG-04-50-017	21 15 06.	- 23 04	90	14.5	GALAXY
LB 01128	21 15 08.	- 16 23 54.		16.1	FAINT BLUE STAR
HN 2596	21 15 08.6	- 57 50 56.	72		NEBULA
SHB 354	21 15 11.1	- 30 31 50.		16.5	QUASI-STELLAR OBJECT
BC PKS2115-30	21 15 11.17	- 30 31 49.5		16.47	QUASI-STELLAR OBJECT
ZWG 375.025	21 15 12.	+ 02 09		15.7	GALAXY
UGC 11714	21 15 12.	+ 02 09	60	15.7	GALAXY Sc
ZC 2115.2+0951	21 15 12.	+ 09 51	1680		CLUSTER OF GALAXIES
2ZW 117	21 15 12.	+ 15 15			COMPACT GALAXY
LB 01129	21 15 12.	- 13 54 06.		15.8	FAINT BLUE STAR
MCG+00-54-011	21 15 15.	+ 02 11	66	15.	GALAXY
HN 2598	21 15 15.3	- 50 00 07.			NEBULA
ZWG 471.003	21 15 18.	+ 24 09		15.7	GALAXY
VDB.66N 141	21 15 21.	+ 67 59	504		REFLECTION NEBULA
ZWG 426.036	21 15 24.	+ 13 22		15.6	GALAXY
UGC 11715	21 15 24.	+ 13 22	60	15.6	GALAXY Sc
KARA.73B 0904	21 15 24.	+ 13 22	60	15.6	ISOLATED GALAXY S
2ZW 118	21 15 24.	+ 19 13			COMPACT GALAXY
MCG+04-50-005	21 15 24.	+ 25 44 30.	48	15.	GALAXY
DG 172	21 15 24.	+ 68 15	180		REFLECTION NEBULA
SCHO 1257	21 15 26.	+ 43 03 48.	650		ISOLATED DARK CLOUD
HN 2599	21 15 28.3	- 54 52 06.	12		NEBULA
UGC 11716	21 15 30.	+ 25 45	90	16.5	GALAXY
MCG-05-50-009	21 15 30.	- 27 33	36	15.	GALAXY
ISS 0175	21 15 36.	+ 45 21	206		STELLAR RING
RNGC 7049	21 15 36.	- 48 47		12.0	GALAXY
OCL 0199	21 15 48.	+ 41 41	480	16.	OPEN STAR CLUSTER
LDN 0944	21 15 48.	+ 43 05	180		DARK NEBULA
MRSL 105+13/1	21 15 48.	+ 68 03	300		HII REGION
SG 3.225	21 15 51.	+ 68 03	30		DIFFUSE EMISSION NEBULA
DG 173	21 15 54.	+ 58 23	660		REFLECTION NEBULA
CED 191	21 15 59.	+ 58 23	930		DIFFUSE GALACTIC NEBULA
LBN 0446	21 16	+ 58 23	660		BRIGHT NEBULA
COU 083	21 16 00.	+ 59 45	5400		EMISSION NEBULA
SCHO 1258	21 16 02.	+ 43 13 48.	400		ISOLATED DARK CLOUD
ZWG 426.037	21 16 06.	+ 11 51		15.7	GALAXY
2ZW 119	21 16 05.	+ 13 37			COMPACT GALAXY
VDB.66N 140	21 16 06.	+ 58 23	1200		REFLECTION NEBULA
ZWG 449.016	21 16 12.	+ 19 30		15.1	GALAXY
UGC 11717	21 16 12.	+ 19 30	84	15.1	GALAXY S
2ZW 120	21 16 12.	+ 22 28			COMPACT GALAXY
MRK 513	21 16 18.	+ 02 03	13	16.	GALAXY WITH UV CONTINUUM
ZC 2116.3+1830	21 16 18.	+ 18 30	1480		CLUSTER OF GALAXIES
MCG+03-54-006	21 16 18.	+ 19 31	66	15.	GALAXY
ZWG 471.004	21 16 18.	+ 22 32		14.8	GALAXY
ZWG 471.005	21 16 18.	+ 26 14		14.0	GALAXY
RNGC 7052	21 16 18.	+ 26 14		14.0	GALAXY
UGC 11718	21 16 18.	+ 26 14	180	14.0	GALAXY E
MCG+04-50-006	21 16 18.	+ 26 14	42	13.5	GALAXY
ZC 2116.4+0020	21 16 24.	+ 00 20	740		CLUSTER OF GALAXIES
ZWG 401.013	21 16 24.	+ 05 38		15.5	GALAXY
KARA.72 556A	21 16 24.	+ 05 38	60	15.5	PART OF DOUBLE GALAXY
MCG+01-54-005	21 16 24.	+ 05 38	42	15.	GALAXY
ZWG 449.017	21 16 24.	+ 16 26		15.7	GALAXY
KARA.73B 0905	21 16 24.	+ 16 26	24	15.7	ISOLATED GALAXY S
2ZW 121	21 16 24.	+ 19 45			COMPACT GALAXY
ZWG 426.038	21 16 30.	+ 15 27		15.7	GALAXY
UGC 11719	21 16 30.	+ 15 27	90	15.7	GALAXY Sa-b
ZWG 375.026	21 16 30.	- 01 10		15.5	GALAXY
HN 2600	21 16 33.6	- 51 16 09.	24		NEBULA
MRSL 087-03/1	21 16 36.	+ 43 44	9600		HII REGION
MCG-05-50-010	21 16 36.	- 26 34	60	15.	GALAXY
SG 2.081	21 16 40.	+ 43 45	9000		DIFFUSE EMISSION NEBULA
HN 2602	21 16 40.3	- 50 49 09.	12		NEBULA
HN 2601	21 16 41.6	- 52 41 57.	12		NEBULA
MCG+02-54-019	21 16 42.	+ 15 28	84	15.	GALAXY
MCG+04-50-007	21 16 42.	+ 25 12 30.	54	14.5	GALAXY
ZWG 471.006	21 16 42.	+ 25 13		15.0	GALAXY
ISS 0254	21 16 42.	+ 44 12	221		STELLAR RING
MCG+01-54-006	21 16 45.	+ 05 49	60	14.	GALAXY
MCG-01-54-010	21 16 45.	- 04 46	18	15.5	GALAXY
ZWG 401.014	21 16 48.	+ 05 48		14.8	GALAXY
UGC 11720	21 16 48.	+ 05 48	66	14.8	GALAXY SBc
KARA.72 556B	21 16 48.	+ 05 48	60	14.8	PART OF DOUBLE GALAXY
MCG-04-50-018	21 16 54.	- 24 30	30	15.	GALAXY
MCG-07-43-037	21 16 54.	- 40 00	60	14.	GALAXY
SCHO 1259	21 16 55.	+ 43 18 18.	300		ISOLATED DARK CLOUD
LBN 0255	21 17	+ 32 00	3600		BRIGHT NEBULA
KHAV 727	21 17	+ 47 13	8660		DARK NEBULA
KHAV 728	21 17	+ 52 37	2370		DARK NEBULA
ZWG 426.039	21 17 00.	+ 13 51		15.3	GALAXY
UGC 11721	21 17 00.	+ 13 51	90	15.3	GALAXY IRR
HN 2605	21 17 04.6	- 50 53 32.			NEBULA
LDN 0946	21 17 06.	+ 43 15	120		DARK NEBULA
MCG-01-54-011	21 17 06.	- 07 46	72	15.	GALAXY

OBJECT NAME	RIGHT ASCEN.	DECLINATION	DIAM.	MAGN.	TYPE OF OBJECT
RNGC 7051	21 17 10.	- 08 59		14.0	GALAXY
MCG+02-54-020	21 17 12.	+ 13 50	96	14.5	GALAXY
MCG-02-54-004	21 17 12.	- 08 59	66	14.	GALAXY
HN 0712	21 17 14.	- 71 12			NEBULA
IC 5099	21 17 15.	- 71 13			NONSTELLAR OBJECT
HN 2603	21 17 15.5	- 60 30 44.	12		NEBULA
MCG-07-43-038	21 17 18.	- 43 40	9	15.5	GALAXY
AGU 67	21 17 18.	- 43 40 06.		12.5	2 INTERACTING GALAXIES
PK089-02.1	21 17 19.	+ 46 06 05.	7		PLANETARY NEBULA
MCG-07-43-039	21 17 21.	- 43 40	12	16.	GALAXY
HN 2604	21 17 21.1	- 60 19 31.	18		NEBULA
ZWG 471.007	21 17 24.	+ 21 45		15.3	GALAXY
UGC 11722	21 17 24.	+ 21 45	102	15.3	GALAXY S
MCG+04-50-008	21 17 24.	+ 21 45	96	14.5	GALAXY
UGC 11723	21 17 24.	- 01 53	108	14.9	GALAXY Sb
LDN 0949	21 17 30.	+ 43 19	480		DARK NEBULA
LDN 0974	21 17 30.	+ 46 25	1740		DARK NEBULA
MCG-01-54-012	21 17 30.	- 04 07 30.	48	15.	GALAXY
MCG-02-54-005	21 17 36.	- 13 23	54	15.	GALAXY
IC 1371	21 17 38.	- 05 05 16.			NONSTELLAR OBJECT
HN 0713	21 17 38.	- 66 09			NEBULA
IC 5100	21 17 38.	- 66 09			NONSTELLAR OBJECT
IC 1372	21 17 39.	- 05 49 22.			NONSTELLAR OBJECT
ZWG 426.040	21 17 42.	+ 11 42		15.7	GALAXY
ZC 2117.7+3150	21 17 42.	+ 31 50	2550		CLUSTER OF GALAXIES
ZWG 375.027	21 17 42.	- 01 53		14.9	GALAXY
KARA.73B 0906	21 17 42.	- 01 53	126	14.9	ISOLATED GALAXY S
MCG+00-54-012	21 17 42.	- 01 54	114	14.	GALAXY
MCG-01-54-013	21 17 42.	- 05 06	48	15.	GALAXY
HN 2606	21 17 45.9	- 51 56 30.	18		NEBULA
2ZW 122	21 17 48.	+ 18 04			COMPACT GALAXY
MCG-05-50-011	21 17 48.	- 32 37	48	15.	GALAXY
HN 0714	21 17 49.	- 66 03			NEBULA
IC 5101	21 17 50.	- 66 03			NONSTELLAR OBJECT
HN 2607	21 17 53.5	- 53 06 17.	12		NEBULA
2ZW 123	21 17 54.	+ 02 33			COMPACT GALAXY
ZWG 426.041	21 17 54.	+ 10 06		15.6	GALAXY
KARA.73B 0907	21 17 54.	+ 10 06	42	15.6	ISOLATED GALAXY S
DG 174	21 17 54.	+ 64 47	1040		REFLECTION NEBULA
PHL 5037	21 17 54.	- 10 29		17.3	BLUE STELLAR OBJECT
MCG-05-50-012	21 17 57.	- 30 47	12	15.	GALAXY
RNGC 7055	21 17 59.	+ 57 23			NON-EXISTENT OBJECT
LBN 0401	21 18	+ 44 00	2700		BRIGHT NEBULA
MRK 514	21 18 00.	+ 02 33	10	15.5	GALAXY WITH UV CONTINUUM
LDN 0965	21 18 00.	+ 45 40	2640		DARK NEBULA
LDN 1177	21 18 00.	+ 68 03	240		DARK NEBULA
MCG-05-50-013	21 18 00.	- 30 45	36	15.	GALAXY
ARC 2338	21 18 04.	- 26 20		17.0	RICH CLUSTER OF GALAXIES
IC 1373	21 18 05.	+ 00 52 41.			NONSTELLAR OBJECT
ZWG 375.028	21 18 06.	+ 00 50		15.3	GALAXY
UGC 11724	21 18 06.	+ 00 50	102	15.3	GALAXY Sc
MCG+00-54-014	21 18 06.	+ 00 50	90	14.	GALAXY
ZWG 375.029	21 18 06.	+ 00 52		15.4	GALAXY
MCG+00-54-013	21 18 06.	+ 00 52 30.	18	15.	GALAXY
4ZW 069	21 18 06.	+ 31 53			COMPACT GALAXY
LDN 0947	21 18 06.	+ 43 06	360		DARK NEBULA
LDN 1016	21 18 06.	+ 50 48	180		DARK NEBULA
MCG-05-50-014	21 18 06.	- 30 47	12	15.5	GALAXY
SG 2.082	21 18 09.	+ 44 14	3000		DIFFUSE EMISSION NEBULA
PHL 5043	21 18 12.	- 09 50		17.6	BLUE STELLAR OBJECT
ARC 2339	21 18 19.	- 21 40		17.0	RICH CLUSTER OF GALAXIES
ZWG 375.030	21 18 24.	+ 03 27		15.0	GALAXY
ZWG 426.042	21 18 24.	+ 15 09		15.7	GALAXY
IC 1375	21 18 29.	+ 03 46 43.			NONSTELLAR OBJECT
ARC 2340	21 18 29.	- 13 00		17.1	RICH CLUSTER OF GALAXIES
ZWG 375.031	21 18 30.	+ 01 30		15.7	GALAXY
KARA.73B 0908	21 18 30.	+ 01 30	18	15.7	ISOLATED GALAXY S
ZWG 401.015	21 18 30.	+ 03 47		15.1	GALAXY
ZWG 401.016	21 18 30.	+ 08 58		15.4	GALAXY
UGC 11725	21 18 30.	+ 08 58	66	15.4	GALAXY Sc-IRR
ZWG 512.001	21 18 30.	+ 37 22		15.3	GALAXY
UGC 11726	21 18 30.	+ 37 22	84	15.3	GALAXY Sb-c
MCG+06-47-001	21 18 30.	+ 37 24	66	15.	GALAXY
MCG-04-50-019	21 18 30.	- 24 36	54	15.	GALAXY
HN 2608	21 18 30.8	- 59 54 34.	24		NEBULA
IC 1374	21 18 31.	+ 01 29 49.			NONSTELLAR OBJECT
HN 2609	21 18 35.5	- 52 02 39.	30		NEBULA
MCG+01-54-007	21 18 36.	+ 08 59	72	14.5	GALAXY
LDN 0952	21 18 36.	+ 43 15	120		DARK NEBULA
URA 65	21 18 36.	+ 54 06	264		STELLAR RING
HN 2610	21 18 39.9	- 53 41 51.	12		NEBULA
LB 01130	21 18 40.	- 17 09 30.		16.0	FAINT BLUE STAR
ZC 2118.7+0508	21 18 42.	+ 05 08	940		CLUSTER OF GALAXIES
MRSL 079-12/1	21 18 42.	+ 32 15	3000		HII REGION
RNGC 7054	21 18 43.	+ 38 58			NON-EXISTENT OBJECT
ARC 2341	21 18 46.	- 23 24		17.0	RICH CLUSTER OF GALAXIES
2ZW 124	21 18 48.	+ 22 53			COMPACT GALAXY
MRSL 083-08/1	21 18 48.	+ 37 53	900		HII REGION
LDN 0948	21 18 48.	+ 43 02	120		DARK NEBULA
HN 2611	21 18 49.2	- 53 11 45.	12		NEBULA
MCG+04-50-009	21 18 51.	+ 22 52 30.	72	14.	GALAXY
MCG+04-50-010	21 18 51.	+ 26 03 30.	72	14.5	GALAXY
RNGC 7053	21 18 52.	+ 22 52		14.5	GALAXY
ZWG 471.008	21 18 54.	+ 22 52		14.3	GALAXY
UGC 11727	21 18 54.	+ 22 52	96	14.3	GALAXY COMPACT
ZWG 471.009	21 18 54.	+ 26 03		15.6	GALAXY
UGC 11728	21 18 54.	+ 26 03	90	15.6	GALAXY Sc
MCG-07-43-040	21 18 54.	- 41 04	36	15.	GALAXY
SCHO 1260	21 18 55.	+ 50 52 18.	330		ISOLATED DARK CLOUD
LBN 0333	21 19	+ 37 50	720		BRIGHT NEBULA
LBN 0347	21 19	+ 38 35	540		BRIGHT NEBULA
LBN 0400	21 19	+ 43 30	3120		BRIGHT NEBULA
LBN 0472	21 19	+ 64 40	1320		BRIGHT NEBULA
UGC 11729	21 19 00.	+ 24 49	72	16.5	GALAXY SB:c
LDN 1162	21 19 00.	+ 64 50	240		DARK NEBULA
PK093+01.1	21 19 05.5	+ 51 40 41.	12		PLANETARY NEBULA
UGC 11730	21 19 06.	+ 29 15	60	17.	GALAXY DWARF
ARC 2342	21 19 11.	- 12 53		17.5	RICH CLUSTER OF GALAXIES
ZWG 426.043	21 19 12.	+ 11 05		15.6	GALAXY
KARA.73B 0909	21 19 12.	+ 11 05	54	15.6	ISOLATED GALAXY S
2ZW 125	21 19 12.	+ 16 38			COMPACT GALAXY
ZWG 449.018	21 19 12.	+ 21 01		14.4	GALAXY
UGC 11731	21 19 12.	+ 21 01	108	14.4	GALAXY SB?a-b
KARA.73B 0910	21 19 12.	+ 21 01	102	14.4	ISOLATED GALAXY S
MCG+03-54-007	21 19 12.	+ 21 02	84	14.	GALAXY
UGC 11732	21 19 12.	+ 29 15	72	16.5	GALAXY S
MRSL 084-07/1	21 19 12.	+ 38 30	540		HII REGION
LDN 0951	21 19 12.	+ 43 36	240		DARK NEBULA

OBJECT NAME	RIGHT ASCEN.	DECLINATION	DIAM.	MAGN.	TYPE OF OBJECT
IC 5104	21 19 13.	+ 20 59 06.			NONSTELLAR OBJECT
HN 2612	21 19 13.3	- 50 39 31.	30		NEBULA
LDN 1019	21 19 18.	+ 50 44	180		DARK NEBULA
LB 01131	21 19 20.	- 19 50 54.		17.0	FAINT BLUE STAR
RNGC 7058	21 19 21.	+ 50 36			NON-EXISTENT OBJECT
ZWG 426.044	21 19 24.	+ 10 26		15.3	GALAXY
MCG+02-54-021	21 19 24.	+ 10 26	48	14.	GALAXY
2ZW 126	21 19 24.	+ 12 55			COMPACT GALAXY
HN 2614	21 19 27.5	- 49 58 37.	18		NEBULA
UGC 11733	21 19 30.	+ 06 56	72	16.0	GALAXY SB?b
2ZW 127	21 19 30.	+ 08 19			COMPACT GALAXY
ZWG 401.017	21 19 30.	+ 08 21		15.6	GALAXY
LDN 0953	21 19 30.	+ 43 10	120		DARK NEBULA
LDN 0972	21 19 30.	+ 45 55	900		DARK NEBULA
TON-S 0017	21 19 30.	- 26 45		15.0	BLUE STAR
DV.56 NEW6A	21 19 30.	- 46 13	93		S GALAXY
B 153	21 19 34.	+ 56 14	300		DARK OBJECT
HN 2613	21 19 34.5	- 53 53 31.	18		NEBULA
ISS 0176	21 19 36.	+ 50 17	248		STELLAR RING
URA 66	21 19 36.	+ 51 17	312		STELLAR RING
PHL 5071	21 19 36.	- 12 35		18.0	BLUE STELLAR OBJECT
ZWG 401.018	21 19 42.	+ 06 53		15.6	GALAXY
LB 01132	21 19 44.	- 15 26 24.		16.1	FAINT BLUE STAR
HN 2616	21 19 46.6	- 52 52 12.	12		NEBULA
SG 2.083	21 19 47.	+ 43 32	3600		DIFFUSE EMISSION NEBULA
ZWG 449.019	21 19 48.	+ 18 27		13.8	GALAXY
UGC 11734	21 19 48.	+ 18 27	60	13.8	GALAXY SBb
KARA.73B 0911	21 19 48.	+ 18 27	60	13.8	ISOLATED GALAXY S
MCG+03-54-008	21 19 48.	+ 18 28	54	13.5	GALAXY
ARC 2343	21 19 48.	- 05 37		17.1	RICH CLUSTER OF GALAXIES
RNGC 7056	21 19 50.	+ 18 27		14.0	GALAXY
HN 2615	21 19 52.2	- 59 50 36.	18		NEBULA
B 154	21 19 53.	+ 56 24	480		DARK OBJECT
LDN 1021	21 19 54.	+ 50 49	120		DARK NEBULA
PHL 5079	21 19 54.	- 12 08		18.3	BLUE STELLAR OBJECT
KHAV 729	21 20	+ 50 49	1000		DARK NEBULA
ZWG 426.045	21 20 00.	+ 12 48		15.6	GALAXY
LDN 0989	21 20 00.	+ 47 20	4800		DARK NEBULA
LDN 1080	21 20 00.	+ 56 20	120		DARK NEBULA
MCG-02-54-006	21 20 00.	- 11 40	36	15.	GALAXY
SA 6	21 20 00.	- 46 00	240	12.9	GALAXY
HN 2618	21 20 04.6	- 51 45 11.	18		NEBULA
ZWG 426.046	21 20 06.	+ 15 04		14.9	GALAXY
UGC 11735	21 20 06.	+ 15 04	60	14.9	GALAXY E?
ZC 2120.1+2256	21 20 06.	+ 22 56	2080		CLUSTER OF GALAXIES
HN 2620	21 20 09.6	- 53 09 53.	18		NEBULA
MRSL 093+01/1	21 20 12.	+ 51 58	30		HII REGION
MCG-04-50-020	21 20 12.	- 22 44	36	15.5	GALAXY
HN 2617	21 20 13.5	- 60 37 35.	18		NEBULA
LB 01133	21 20 15.	- 16 56 00.		15.6	FAINT BLUE STAR
SCHO 1261	21 20 16.	+ 50 47 18.	350		ISOLATED DARK CLOUD
ZWG 426.047	21 20 18.	+ 14 39		15.3	GALAXY
MCG+02-54-022	21 20 18.	+ 15 04	18	15.	GALAXY
ACK 083-08.1	21 20 18.	+ 37 54			PLANETARY NEBULA
ZWG 375.032	21 20 24.	+ 00 49		15.7	GALAXY
UGC 11736	21 20 24.	+ 03 18	228		GALAXY CHAIN
UGC 11737	21 20 24.	+ 29 12	90	16.5	GALAXY S
HN 2621	21 20 25.0	- 53 58 46.	18		NEBULA
BC 3CR432	21 20 25.64	+ 16 51 46.0		17.96	QUASI-STELLAR OBJECT
SHB 355	21 20 25.7	+ 16 51 46.		18.0	QUASI-STELLAR OBJECT
MCG+01-54-008	21 20 30.	+ 08 21	24	16.	GALAXY
DVDV 6	21 20 30.	- 49 33	72		GALAXY
HN 2619	21 20 31.8	- 62 01 22.	18		NEBULA
ZWG 401.019	21 20 36.	+ 06 54		15.7	GALAXY
ZWG 426.048	21 20 42.	+ 15 10		15.2	GALAXY
LDN 0968	21 20 42.	+ 45 20	480		DARK NEBULA
MCG-02-54-007	21 20 42.	- 11 20	36	15.	GALAXY
MCG+00-54-015	21 20 45.	+ 03 20	24	15.	GALAXY
SHB 356	21 20 45.	+ 09 56		19.	QUASI-STELLAR OBJECT
BC PKS2120+09	21 20 46.5	+ 09 54 58.		18.	QUASI-STELLAR OBJECT
ZWG 375.033	21 20 48.	+ 01 58		14.9	GALAXY
MCG+02-54-023	21 20 48.	+ 15 10	48	15.5	GALAXY
MRK 515	21 20 48.	- 07 57	13	16.	GALAXY WITH UV CONTINUUM
LBN 0346	21 21	+ 38 10	3900		BRIGHT NEBULA
LBN 0401	21 21	+ 43 10	4200		BRIGHT NEBULA
KHAV 730	21 21	+ 58 01			DARK NEBULA
ZWG 426.049	21 21 00.	+ 09 54		15.5	GALAXY
KARA.73B 0912	21 21 00.	+ 09 54	48	15.5	ISOLATED GALAXY S
ZWG 449.020	21 21 00.	+ 17 02		15.2	GALAXY
MCG+14-10-001	21 21 00.	+ 83 36	57	15.	GALAXY
ZWG 369.001	21 21 00.	+ 83 37		15.2	GALAXY
ZWG 368.004	21 21 00.	+ 83 37		15.2	GALAXY
UGC 11738	21 21 00.	+ 83 37	66	15.2	GALAXY Sb-c
KARA.73B 0913	21 21 00.	+ 83 37	54	15.2	ISOLATED GALAXY S
YC 2121-49	21 21 03.	- 49 24 42.			UNUSUAL SOUTHERN GALAXY
MCG-01-54-014	21 21 06.	- 05 39	12	16.	GALAXY
BC CX036	21 21 11.	+ 05 42		17.5	QUASI-STELLAR OBJECT
IC 5105	21 21 11.	- 40 50 31.	78	13.0	GALAXY E4
ZWG 449.021	21 21 12.	+ 17 22		15.0	GALAXY
MCG+03-54-009	21 21 12.	+ 17 22	30	15.	GALAXY
SCHO 1262	21 21 12.	+ 46 35 12.	400		ISOLATED DARK CLOUD
SHB 357	21 21 14.7	+ 05 22 27.		17.5	QUASI-STELLAR OBJECT
MCG-07-44-001	21 21 18.	- 40 45	78	13.0	GALAXY
RNGC 7062	21 21 23.	+ 46 10		11.5	OPEN CLUSTER
OCL 0205	21 21 24.	+ 46 10	420	11.7	OPEN STAR CLUSTER
MCG-07-44-002	21 21 24.	- 42 48	12	15.5	GALAXY
HN 0715	21 21 25.	- 73 32			NEBULA
IC 5102	21 21 25.	- 73 32			NONSTELLAR OBJECT
ZWG 449.022	21 21 30.	+ 17 21		15.0	GALAXY
MCG+03-54-010	21 21 30.	+ 17 21 30.	42	15.	GALAXY
ZWG 449.023	21 21 30.	+ 17 43		15.6	GALAXY
2ZW 128	21 21 30.	+ 19 02			COMPACT GALAXY
IC 1378	21 21 30.	+ 55 14			NONSTELLAR OBJECT
CED 192	21 21 32.	+ 55 14			DIFFUSE GALACTIC NEBULA
SCHO 1263	21 21 33.	+ 47 03 06.	340		ISOLATED DARK CLOUD
PHL 5769	21 21 36.	- 09 22		17.2	BLUE STELLAR OBJECT
MCG-07-44-003	21 21 36.	- 42 49	15	15.5	GALAXY
ZWG 426.050	21 21 42.	+ 10 29		15.7	GALAXY
ZWG 512.002	21 21 42.	+ 39 23		15.7	GALAXY
SCHO 1264	21 21 42.	+ 44 43 42.	350		ISOLATED DARK CLOUD
MCG+03-54-011	21 21 48.	+ 15 45 30.	42	15.	GALAXY
ZWG 449.024	21 21 48.	+ 15 46		15.5	GALAXY
KARA.73B 0914	21 21 48.	+ 15 46	48	15.5	ISOLATED GALAXY S
MCG-07-44-004	21 21 48.	- 42 40	60	13.5	GALAXY
RNGC 7057	21 21 48.	- 42 41			GALAXY
HN 2622	21 21 50.9	- 50 28 12.	18		NEBULA
ZC 2121.9+3233	21 21 54.	+ 32 33	5310		CLUSTER OF GALAXIES
SG 2.084	21 21 54.	+ 43 04	5400		DIFFUSE EMISSION NEBULA
ISS 0177	21 21 54.	+ 46 50	127		STELLAR RING
HN 2623	21 21 54.6	- 52 01 42.	12		NEBULA
LBN 0404	21 22	+ 44 30	4500		BRIGHT NEBULA
KHAV 731	21 22	+ 50 01	700		DARK NEBULA
ZC 2122.0+0107	21 22 00.	+ 01 07	1280		CLUSTER OF GALAXIES
LB 01134	21 22 01.	- 15 11 18.		15.9	FAINT BLUE STAR
IC 1376	21 22 03.	- 05 57 36.			NONSTELLAR OBJECT
HN 2624	21 22 11.3	- 51 35 23.	12		NEBULA
ISS 0178	21 22 12.	+ 48 14	132		STELLAR RING
TON-S 0018	21 22 12.	- 26 28		14.7	BLUE STAR
B 362	21 22 15.	+ 49 59	900		DARK OBJECT
SCHO 1265	21 22 16.	+ 49 59 42.	590		ISOLATED DARK CLOUD
PHL 0001	21 22 18.	+ 02 27		16.3	BLUE STELLAR OBJECT
UGC 11739	21 22 18.	+ 20 28	60	16.5	GALAXY
LDN 0976	21 22 18.	+ 45 48			DARK NEBULA
LDN 1014	21 22 18.	+ 49 47	240		DARK NEBULA
MCG-04-50-021	21 22 21.	- 25 10	18	15.5	GALAXY
LB 01135	21 22 22.	- 14 34 24.		16.0	FAINT BLUE STAR
RNGC 7063	21 22 23.	+ 36 17		9.0	OPEN CLUSTER
SCHO 1266	21 22 23.	+ 48 28 42.	430		ISOLATED DARK CLOUD
OCL 0192	21 22 24.	+ 36 17	1020	9.0	OPEN STAR CLUSTER
OCL 0208	21 22 24.	+ 47 48	420	13.0	OPEN STAR CLUSTER
LB 01136	21 22 24.	- 15 42 36.		15.9	FAINT BLUE STAR
MCG-07-44-005	21 22 24.	- 40 29	90	13.	GALAXY
RNGC 7067	21 22 25.	+ 47 48		13.0	OPEN CLUSTER
ZWG 401.020	21 22 30.	+ 03 38		15.3	GALAXY
ZWG 426.051	21 22 36.	+ 10 38		15.4	GALAXY
MCG+02-54-024	21 22 36.	+ 12 12 30.	42	15.	GALAXY
ZWG 426.052	21 22 36.	+ 12 13		15.6	GALAXY
2ZW 129	21 22 36.	+ 20 54			COMPACT GALAXY
PHL 0002	21 22 36.	- 00 14		16.8	BLUE STELLAR OBJECT
DV.56 I5105A	21 22 36.	- 40 29	132		Sc GALAXY
PHL 4500	21 22 42.	+ 01 57		17.9	BLUE STELLAR OBJECT
PHL 4499	21 22 42.	- 00 22		17.5	BLUE STELLAR OBJECT
RNGC 7060	21 22 42.	- 42 37			GALAXY
MCG-07-44-006	21 22 42.	- 42 37	48	13.5	GALAXY
ARC 2344	21 22 48.	- 21 01		17.6	RICH CLUSTER OF GALAXIES
MCG-07-44-007	21 22 48.	- 40 13	42	14.5	GALAXY
SCHO 1267	21 22 52.	+ 47 12 30.	290		ISOLATED DARK CLOUD
ZWG 401.021	21 22 54.	+ 04 06		14.8	GALAXY
MCG+01-54-009	21 22 54.	+ 04 06	36	14.	GALAXY
KARA.73B 0915	21 22 54.	+ 04 06	36	14.8	ISOLATED GALAXY S
ZC 2122.9+1605	21 22 54.	+ 16 05	270		CLUSTER OF GALAXIES
ZWG 449.025	21 22 54.	+ 17 32		15.7	GALAXY
PHL 4501	21 22 54.	- 02 48		17.6	BLUE STELLAR OBJECT
DV.56 I5105B	21 22 54.	- 41 03	78		S GALAXY
MCG-07-44-008	21 22 54.	- 41 03	78	14.5	GALAXY
IC 1377	21 22 57.	+ 04 06 30.			NONSTELLAR OBJECT
KHAV 732	21 23	+ 48 55	1410		DARK NEBULA
KHAV 733	21 23	+ 62 49	2930		DARK NEBULA
LDN 0980	21 23 00.	+ 46 00			DARK NEBULA
PHL 0003	21 23 00.	- 01 20		17.6	BLUE STELLAR OBJECT
PHL 1570	21 23 00.	- 02 06		18.5	BLUE STELLAR OBJECT
MCG-04-50-022	21 23 00.	- 23 00	60	14.5	GALAXY
SCHO 1268	21 23 06.	+ 44 11 42.	460		ISOLATED DARK CLOUD
MCG-04-50-023	21 23 06.	- 22 58 30.	54	15.	GALAXY
B 363	21 23 08.	+ 48 43	2400		DARK OBJECT
HELW 069	21 23 10.	- 15 39			NEBULA
PHL 4506	21 23 12.	+ 00 28		17.8	BLUE STELLAR OBJECT
PHL 4505	21 23 12.	+ 00 33		17.2	BLUE STELLAR OBJECT
PHL 4504	21 23 12.	+ 00 33		17.2	BLUE STELLAR OBJECT
ZC 2123.2+0041	21 23 12.	+ 00 41	1550		CLUSTER OF GALAXIES
PHL 4503	21 23 12.	+ 01 20		17.7	BLUE STELLAR OBJECT
PHL 1571	21 23 12.	+ 01 20		18.2	BLUE STELLAR OBJECT
ISS 0179	21 23 12.	+ 48 08	134		STELLAR RING
PHL 1572	21 23 12.	- 00 05		18.0	BLUE STELLAR OBJECT
PHL 4502	21 23 12.	- 00 32		17.0	BLUE STELLAR OBJECT
MCG-02-54-008	21 23 15.	- 09 15	30	15.5	GALAXY
ZC 2123.3+1557	21 23 18.	+ 15 57	610		CLUSTER OF GALAXIES
ISS 0180	21 23 18.	+ 47 18	127		STELLAR RING
MCG-01-54-015	21 23 18.	- 04 10	42	16.	GALAXY
MCG-07-44-009	21 23 18.	- 40 06	9	15.5	GALAXY
HN 2625	21 23 18.6	- 51 20 56.			NEBULA
PHL 1573	21 23 24.	- 01 30		18.2	BLUE STELLAR OBJECT
MCG-01-54-016	21 23 24.	- 04 01	72	15.	GALAXY
IC 1379	21 23 27.	+ 02 52 44.			NONSTELLAR OBJECT
SCHO 1269	21 23 27.	+ 48 38 18.	340		ISOLATED DARK CLOUD
ZWG 375.034	21 23 30.	+ 02 53		15.7	GALAXY
LDN 1007	21 23 30.	+ 48 40	1560		DARK NEBULA
HN 2627	21 23 30.0	- 50 31 07.			NEBULA
PHL 4508	21 23 36.	+ 02 02		17.2	BLUE STELLAR OBJECT
ZWG 401.022	21 23 36.	+ 05 14		15.6	GALAXY
ZWG 401.023	21 23 36.	+ 06 30		15.6	GALAXY
MCG+01-54-010	21 23 36.	+ 06 31	24	15.	GALAXY
ISS 0181	21 23 36.	+ 46 00	172		STELLAR RING
LDN 1017	21 23 36.	+ 49 57	660		DARK NEBULA
RNGC 7059	21 23 37.	- 60 15		13.0	GALAXY
OCL 0220	21 23 42.	+ 57 17	240	15.	OPEN STAR CLUSTER
LB 00363	21 23 43.	- 18 00 54.		15.6	FAINT BLUE STAR
HN 2628	21 23 44.6	- 52 38 07.	12		NEBULA
HN 2626	21 23 45.7	- 61 02 13.	12		NEBULA
PHL 1574	21 23 48.	+ 00 14		16.6	BLUE STELLAR OBJECT
PHL 0004	21 23 48.	+ 00 47		13.9	BLUE STELLAR OBJECT
PHL 4509	21 23 48.	+ 01 29		15.0	BLUE STELLAR OBJECT
ZWG 401.024	21 23 48.	+ 04 53		15.7	GALAXY
ZWG 426.053	21 23 48.	+ 09 35		14.7	GALAXY
UGC 11740	21 23 48.	+ 09 35	96	14.7	GALAXY S
ZWG 426.054	21 23 48.	+ 13 57		15.2	GALAXY
RNGC 7066	21 23 48.	+ 13 57		15.0	GALAXY
UGC 11741	21 23 48.	+ 13 57	66	15.2	GALAXY S
2ZW 130	21 23 48.	+ 13 58			COMPACT GALAXY
ZWG 512.003	21 23 48.	+ 37 31		15.7	GALAXY
MCG+06-47-002	21 23 49.	+ 37 33	36	16.5	GALAXY
PHL 4510	21 23 48.	- 03 08		17.8	BLUE STELLAR OBJECT
MCG-04-50-024	21 23 48.	- 22 04	15	15.	GALAXY
HN 2630	21 23 50.2	- 51 33 24.	24		NEBULA
HELW 070	21 23 51.	- 15 03			NEBULA
UGC 11742	21 23 54.	+ 01 50	72	16.5	GALAXY S
MCG+02-54-026	21 23 54.	+ 09 33	84	14.5	GALAXY
MCG+02-54-025	21 23 54.	+ 13 57	48	14.5	GALAXY
ZWG 492.001	21 23 54.	+ 30 16		15.5	GALAXY
UGC 11743	21 23 54.	+ 30 16	72	15.5	GALAXY DISTRBD
HN 2631	21 23 55.7	- 50 21 18.	18		NEBULA
SCHO 1270	21 23 58.	+ 45 08 00.	670		ISOLATED DARK CLOUD
LBN 0417	21 24	+ 48 20	1500		BRIGHT NEBULA
KHAV 735	21 24	+ 49 31	1860		DARK NEBULA
KHAV 734	21 24	+ 50 37	4260		DARK NEBULA
PHL 4512	21 24 00.	+ 01 03		17.4	BLUE STELLAR OBJECT

OBJECT NAME	RIGHT ASCEN.	DECLINATION	DIAM.	MAGN.	TYPE OF OBJECT
PHL 4511	21 24 00.	+ 02 16		16.5	BLUE STELLAR OBJECT
MCG+05-50-001	21 24 00.	+ 30 16 30.	60	15.	GALAXY
MCG-01-54-017	21 24 00.	- 07 12	60	15.	GALAXY
RNGC 7065	21 24 05.	- 07 15			GALAXY
MCG+02-54-027	21 24 06.	+ 11 57 30.	36	14.5	GALAXY
PHL 4513	21 24 06.	- 00 18		17.7	BLUE STELLAR OBJECT
RNGC 7061	21 24 07.	- 49 17			UNVERIFIED SOUTHRN OBJECT
HN 0717	21 24 08.	- 71 03			NEBULA
IC 5106	21 24 08.	- 71 03			NONSTELLAR OBJECT
MCG-04-50-025	21 24 09.	- 23 13 30.	48	15.	GALAXY
HN 0718	21 24 09.	- 65 57			NEBULA
IC 5107	21 24 09.	- 65 57			NONSTELLAR OBJECT
HN 2629	21 24 11.4	- 61 05 18.	12		NEBULA
PHL 4515	21 24 12.	+ 01 14		16.9	BLUE STELLAR OBJECT
PHL 0005	21 24 12.	+ 01 54		17.5	BLUE STELLAR OBJECT
PHL 4516	21 24 12.	+ 02 18		18.9	BLUE STELLAR OBJECT
ZWG 426.055	21 24 12.	+ 11 59		14.8	GALAXY
RNGC 7068	21 24 12.	+ 11 59		15.0	GALAXY
LDN 0994	21 24 12.	+ 47 26	360		DARK NEBULA
PHL 4514	21 24 12.	- 01 19		18.4	BLUE STELLAR OBJECT
RNGC 7065A	21 24 17.	- 07 17		15.0	GALAXY
PHL 4517	21 24 13.	+ 02 06		17.2	BLUE STELLAR OBJECT
MCG-02-54-009	21 24 18.	- 09 44 30.	48	15.	GALAXY
HN 0716	21 24 18.	- 74 18			NEBULA
IC 5103	21 24 18.	- 74 18			NONSTELLAR OBJECT
SCHO 1271	21 24 20.	+ 48 50 30.	530		ISOLATED DARK CLOUD
ARC 2345	21 24 22.	- 12 22		16.9	RICH CLUSTER OF GALAXIES
PHL 0006	21 24 24.	- 02 30		16.1	BLUE STELLAR OBJECT
MCG-01-54-018	21 24 24.	- 07 15	72	15.	GALAXY
MCG-05-50-015	21 24 24.	- 30 16	48	15.	GALAXY
HN 2632	21 24 26.0	- 51 32 41.	24		NEBULA
IC 1382	21 24 35.	+ 18 26 09.			NONSTELLAR OBJECT
ZWG 375.035	21 24 36.	+ 02 30		15.5	GALAXY
RNGC 7076	21 24 36.	+ 62 34			DIFFUSE NEBULA
PHL 0007	21 24 36.	- 00 01		13.9	BLUE STELLAR OBJECT
IC 1380	21 24 39.	+ 02 29 14.			NONSTELLAR OBJECT
ZWG 426.056	21 24 42.	+ 15 03		15.5	GALAXY
RNGC 7071	21 24 42.	+ 47 43			NON-EXISTENT OBJECT
KLEM 32	21 24 44.	- 39 44	900	15.	CLUSTER OF 10 GALAXIES
SCHO 1272	21 24 46.	+ 61 43 48.	500		ISOLATED DARK CLOUD
PHL 1575	21 24 48.	+ 00 34		17.3	BLUE STELLAR OBJECT
PHL 4518	21 24 48.	+ 00 38		17.9	BLUE STELLAR OBJECT
LDN 0993	21 24 48.	+ 47 12	120		DARK NEBULA
LDN 0997	21 24 48.	+ 47 28	120		DARK NEBULA
MCG-04-50-027	21 24 48.	- 23 54	48	15.	GALAXY
MCG-04-50-026	21 24 48.	- 26 13	48	15.5	GALAXY
ZC 2124.9+1008	21 24 54.	+ 10 08	1410		CLUSTER OF GALAXIES
4ZW 070	21 24 54.	+ 37 18			COMPACT GALAXY
UGC 11744	21 24 54.	+ 37 28	60	16.5	GALAXY Sc
UGC 11745	21 24 54.	+ 37 43	90	16.0	GALAXY Sb-c
LDN 0995	21 24 54.	+ 47 18	120		DARK NEBULA
PHL 1576	21 25 00.	+ 00 22		18.5	BLUE STELLAR OBJECT
PHL 4519	21 25 00.	+ 00 44		17.2	BLUE STELLAR OBJECT
LDN 1008	21 25 00.	+ 48 50	420		DARK NEBULA
LDN 1108	21 25 00.	+ 59 20	840		DARK NEBULA
ZWG 358.002	21 25 00.	+ 79 41		15.5	GALAXY
UGC 11746	21 25 00.	+ 82 06	66	16.0	GALAXY Sb
ZWG 375.036	21 25 00.	- 01 23		15.6	GALAXY
MCG+00-54-016	21 25 00.	- 01 24	30	15.	GALAXY
IC 1381	21 25 00.	- 01 24 27.			NONSTELLAR OBJECT
HN 2633	21 25 02.0	- 53 54 15.	12		NEBULA
HELW 071	21 25 03.	- 15 02			NEBULA
ZWG 375.038	21 25 06.	+ 03 02		15.5	GALAXY
KARA.73B 0916	21 25 06.	+ 03 02	54	15.5	ISOLATED GALAXY S
ZWG 375.037	21 25 06.	- 01 18		15.7	GALAXY
IC 1383	21 25 06.	- 01 19 20.			NONSTELLAR OBJECT
HN 2634	21 25 06.1	- 53 51 45.	12		NEBULA
ARC 2346	21 25 11.	- 13 15		16.5	RICH CLUSTER OF GALAXIES
PK101+08.1	21 25 12.	+ 62 40	67	17.0	PLANETARY NEBULA
MCG+00-54-017	21 25 15.	- 01 35	36	14.	GALAXY
IC 1384	21 25 16.	- 01 34 13.			NONSTELLAR OBJECT
PHL 0009	21 25 18.	+ 00 17		17.7	BLUE STELLAR OBJECT
PHL 0008	21 25 18.	+ 02 18		17.3	BLUE STELLAR OBJECT
MCG+00-54-018	21 25 18.	+ 03 02 30.	48	15.5	GALAXY
ZC 2125.3+1543	21 25 18.	+ 15 43	810		CLUSTER OF GALAXIES
ZWG 375.039	21 25 18.	- 01 34		15.4	GALAXY
HN 2635	21 25 19.2	- 53 08 02.	12		NEBULA
HN 2636	21 25 22.3	- 51 20 20.	12		NEBULA
PHL 1577	21 25 24.	+ 00 06		18.4	BLUE STELLAR OBJECT
ZC 2125.4+0411	21 25 24.	+ 04 11	3490		CLUSTER OF GALAXIES
PHL 4520	21 25 24.	- 02 22		17.4	BLUE STELLAR OBJECT
MCG-04-50-028	21 25 24.	- 22 00	36	15.	GALAXY
RNGC 7064	21 25 28.	- 53 00		13.0	GALAXY
PHL 4522	21 25 30.	+ 01 10		17.8	BLUE STELLAR OBJECT
PHL 4521	21 25 30.	+ 02 20		17.8	BLUE STELLAR OBJECT
LDN 1001	21 25 30.	+ 47 28	120		DARK NEBULA
ZWG 375.040	21 25 30.	- 01 52		14.8	GALAXY
UGC 11747	21 25 30.	- 01 52	78	14.8	GALAXY E-S0
MCG+00-54-019	21 25 30.	- 01 52	24	14.	GALAXY
RNGC 7069	21 25 31.	- 01 52		15.0	GALAXY
YC 2125-52	21 25 34.	- 52 59 12.			UNUSUAL SOUTHERN GALAXY
MCG+00-54-020	21 25 36.	+ 00 05	18	16.	GALAXY
PHL 4523	21 25 36.	+ 00 13		17.8	BLUE STELLAR OBJECT
ZWG 375.041	21 25 36.	+ 02 15		15.1	GALAXY
KARA.73B 0917	21 25 36.	+ 02 15	42	15.1	ISOLATED GALAXY S
ZWG 401.025	21 25 36.	+ 08 23		15.5	GALAXY
ZCG 2125+08.5	21 25 36.	+ 08 31		17.0	COMPACT GALAXY
ZWG 426.057	21 25 36.	+ 12 32		15.6	GALAXY
IC 5111	21 25 40.	+ 02 15 43.			NONSTELLAR OBJECT
MCG+00-54-021	21 25 42.	+ 02 16	48	15.	GALAXY
ZCG 2125+08.6	21 25 42.	+ 08 32		17.0	COMPACT GALAXY
ZC 2125.7+2526	21 25 42.	+ 25 26	2620		CLUSTER OF GALAXIES
2ZW 131	21 25 48.	+ 08 33			COMPACT GALAXY
UGC 11748	21 25 48.	+ 45 16	84	17.	GALAXY S
UPA 67	21 25 48.	+ 57 18	84		STELLAR RING
SG 3.226	21 25 51.	+ 63 51	3000		DIFFUSE EMISSION NEBULA
HN 2637	21 25 51.2	- 51 32 07.	12		NEBULA
MCG-07-44-010	21 25 54.	- 41 59	42	15.	GALAXY
PK098+04.1	21 25 57.9	+ 57 26 03.			PLANETARY NEBULA
PHL 4524	21 26 00.	+ 00 50		18.5	BLUE STELLAR OBJECT
PHL 4525	21 26 00.	- 00 49		18.7	BLUE STELLAR OBJECT
PHL 0010	21 26 00.	- 01 48		15.4	BLUE STELLAR OBJECT
PHL 4526	21 26 06.	+ 00 56		18.3	BLUE STELLAR OBJECT
ZWG 401.026	21 26 06.	+ 03 56		15.6	GALAXY
HN 2638	21 26 11.6	- 52 44 18.	12		NEBULA
PHL 4527	21 26 12.	+ 00 27		18.0	BLUE STELLAR OBJECT
PHL 4528	21 26 12.	+ 02 12		16.8	BLUE STELLAR OBJECT
PHL 4529	21 26 12.	- 01 06		18.5	BLUE STELLAR OBJECT
SER 145.03	21 26 12.	- 43 21	456	13.	COMPACT GROUP OF GALAXIES
MCG-03-54-002	21 26 15.	- 19 53	24	15.	GALAXY
IC 1385	21 26 17.	- 01 17 20.			NONSTELLAR OBJECT
ZWG 375.043	21 26 18.	+ 01 02		15.2	GALAXY
ZC 2126.3+3149	21 26 18.	+ 31 49	1080		CLUSTER OF GALAXIES
ZWG 375.042	21 26 18.	- 01 17		15.4	GALAXY
MCG+00-54-022	21 26 18.	- 01 17 30.	24	15.	GALAXY
HELW 487	21 26 21.	- 43 27 47.			NEBULA
HN 2639	21 26 22.2	- 50 55 17.	12		NEBULA
PHL 4530	21 26 24.	+ 01 59		18.6	BLUE STELLAR OBJECT
2ZW 132	21 26 24.	+ 11 58			COMPACT GALAXY
ZWG 449.026	21 26 24.	+ 20 17		14.6	GALAXY
UGC 11749	21 26 24.	+ 20 17	90	14.6	GALAXY SBb
ZWG 512.004	21 26 24.	+ 35 05		15.6	GALAXY
UPA 68	21 26 24.	+ 52 51	270		STELLAR RING
PHL 0011	21 26 24.	- 01 28		16.0	BLUE STELLAR OBJECT
HN 2644	21 26 29.1	- 15 55 46.	24		NEBULA
MCG+03-54-012	21 26 30.	+ 20 17 30.	72	14.5	GALAXY
MCG+00-54-023	21 26 30.	- 00 41	30	15.5	GALAXY
MCG-03-54-003	21 26 30.	- 19 54	24	15.	GALAXY
HN 2640	21 26 31.5	- 51 07 41.	12		NEBULA
PHL 4531	21 26 36.	+ 00 11		17.3	BLUE STELLAR OBJECT
PHL 0012	21 26 36.	+ 00 33		16.1	BLUE STELLAR OBJECT
UGC 11750	21 26 36.	+ 02 35	66	16.0	GALAXY S
MCG+00-54-024	21 26 36.	+ 02 35	66	15.	GALAXY
ZWG 426.058	21 26 36.	+ 09 46		15.7	GALAXY
KARA.72 557A	21 26 36.	+ 11 09	42	14.7	PART OF DOUBLE GALAXY
ZWG 426.059	21 26 36.	+ 11 10		14.7	GALAXY
UGC 11751	21 26 36.	+ 11 10	60	14.7	GALAXY SB
KARA.72 557B	21 26 36.	+ 11 10	42		PART OF DOUBLE GALAXY
ZC 2126.6+1857	21 26 36.	+ 18 57	670		CLUSTER OF GALAXIES
UGC 11752	21 26 36.	+ 24 48	66	16.5	GALAXY Sb-c
PHL 1578	21 26 36.	- 00 05		18.7	BLUE STELLAR OBJECT
ZWG 375.044	21 26 36.	- 00 30		15.3	GALAXY
MCG-04-50-029	21 26 36.	- 23 21	36	15.	GALAXY
MCG-07-44-012	21 26 36.	- 39 53 30.	18	16.	GALAXY
MCG-07-44-011	21 26 36.	- 39 53 30.	12	16.5	GALAXY
ZWG 492.002	21 26 42.	+ 31 36		15.6	GALAXY
UGC 11753	21 26 42.	+ 31 36	84	15.6	GALAXY SBb
MCG+05-50-002	21 26 42.	+ 31 38	36	15.	GALAXY
MCG+00-54-025	21 26 42.	- 00 30	30	15.	GALAXY
MCG-02-54-010	21 26 42.	- 11 42 30.	48	14.	GALAXY
MCG-04-50-030	21 26 42.	- 21 25	36	14.5	GALAXY
ARC 2347	21 26 43.	- 22 26		16.4	RICH CLUSTER OF GALAXIES
RNGC 7073	21 26 45.	- 11 42		14.0	GALAXY
HN 2641	21 26 45.3	- 50 47 16.	12		NEBULA
LB 00364	21 26 46.	- 21 09 06.		15.3	FAINT BLUE STAR
IC 1386	21 26 47.	- 21 24 37.			NONSTELLAR OBJECT
ZC 2126.8+0709	21 26 48.	+ 07 09	1210		CLUSTER OF GALAXIES
ZC 2126.8-0028	21 26 48.	- 00 28	1880		CLUSTER OF GALAXIES
PHL 4533	21 26 48.	- 02 02		18.2	BLUE STELLAR OBJECT
PHL 4532	21 26 48.	- 02 29		15.9	BLUE STELLAR OBJECT
HN 2642	21 26 48.6	- 50 38 46.	12		NEBULA
MCG-07-44-013	21 26 51.	- 39 47	9	15.5	GALAXY
HN 1219	21 26 53.	- 60 13			NEBULA
PHL 4534	21 26 54.	+ 01 37		18.3	BLUE STELLAR OBJECT
PHL 0013	21 26 54.	- 01 57		17.9	BLUE STELLAR OBJECT
IC 5110	21 26 56.	- 60 13			NONSTELLAR OBJECT
IC 1387	21 26 57.	- 01 33 05.			NONSTELLAR OBJECT
HN 2643	21 26 57.6	- 52 54 16.	12		NEBULA
LBN 0436	21 27	+ 54 20	60		BRIGHT NEBULA
LBN 0476	21 27	+ 64 10	3000		BRIGHT NEBULA
PHL 4535	21 27 00.	+ 00 00		16.0	BLUE STELLAR OBJECT
IC 5112	21 27 00.	+ 06 34			OPEN CLUSTER
LDN 1086	21 27 00.	+ 57 20	1200		DARK NEBULA
ZWG 375.045	21 27 00.	- 01 34		15.0	GALAXY
MCG+00-54-026	21 27 00.	- 01 34	24	15.	GALAXY
ARC 2348	21 27 02.	- 11 16		17.1	RICH CLUSTER OF GALAXIES
RNGC 7074	21 27 04.	+ 06 28			GALAXY
MRSL 096+02/1	21 27 06.	+ 54 24	120		HII REGION
MIL 92	21 27 11.	+ 50 30	3240		SUPERNOVA REMNANT
HELW 488	21 27 11.	- 43 25 15.			NEBULA
2ZW 133	21 27 12.	+ 06 27			COMPACT GALAXY
ZWG 401.027	21 27 12.	+ 06 27		15.0	GALAXY
BIGO 551	21 27 12.	+ 06 36			NEBULA
IC 5113	21 27 12.	+ 06 36			OPEN CLUSTER
ZC 2127.2+1710	21 27 12.	+ 17 10	1080		CLUSTER OF GALAXIES
MCG+04-50-011	21 27 12.	+ 27 09	96	15.	GALAXY
MCG-07-44-014	21 27 12.	- 40 40		15.	GALAXY
MCG-07-44-015	21 27 12.	- 40 59		15.	GALAXY
HN 2645	21 27 14.8	- 53 22 51.	24		NEBULA
ZWG 471.010	21 27 18.	+ 27 06		15.6	GALAXY
UGC 11754	21 27 18.	+ 27 06	126	15.6	GALAXY Sc
ZWG 375.046	21 27 18.	- 00 50		15.0	GALAXY
MCG+00-54-027	21 27 18.	- 00 51	21	15.	GALAXY
PHL 4536	21 27 18.	- 01 08		17.1	BLUE STELLAR OBJECT
MCG-07-44-016	21 27 18.	- 43 18	90	12.6	GALAXY
RNGC 7070	21 27 18.	- 43 20		12.5	GALAXY
IC 1388	21 27 20.	- 00 53 09.			NONSTELLAR OBJECT
MCG-07-44-017	21 27 21.	- 43 26	30	17.	GALAXY
ZWG 375.047	21 27 24.	+ 02 12		14.3	GALAXY
UGC 11755	21 27 24.	+ 02 12	48	14.3	GALAXY E?
MCG+00-54-028	21 27 24.	+ 02 13	24	14.5	GALAXY
PHL 1579	21 27 24.	- 01 15		18.1	BLUE STELLAR OBJECT
PHL 4537	21 27 24.	- 03 17		13.9	BLUE STELLAR OBJECT
RNGC 7072	21 27 24.	- 43 22			GALAXY
AGU 69	21 27 24.	- 43 22 00.	30	12.5	PECULIAR GALAXY
RNGC 7072A	21 27 24.	- 43 26			GALAXY
RNGC 7077	21 27 26.	+ 02 12		14.5	GALAXY
HN 2646	21 27 26.4	- 54 09 51.			NEBULA
HN 2647	21 27 29.4	- 51 38 32.	12		NEBULA
PK050-36.1	21 27 30.	- 03 01	13	15.6	PLANETARY NEBULA
MCG-07-44-018	21 27 30.	- 43 22	78	15.	GALAXY
HN 2653	21 27 32.9	- 16 15 19.	18		NEBULA
REIN 2.292	21 27 33.32	+ 11 56 52.7			NEBULA
HN 2654	21 27 33.7	- 15 54 37.	24		NEBULA
PK065-27.1	21 27 34.45	+ 11 57 14.5	1	13.39	PLANETARY NEBULA
RNGC 7082	21 27 35.	+ 46 52			OPEN CLUSTER
GCL 120	21 27 36.	+ 11 57	738	7.33	GLOBULAR STAR CLUSTER
RNGC 7078	21 27 36.	+ 11 57		7.5	GLOBULAR CLUSTER
OCL 0209	21 27 36.	+ 46 52	1500		OPEN STAR CLUSTER
MCG-01-54-019	21 27 36.	- 08 18	66	15.5	GALAXY
PHL 0014	21 27 36.	- 20 46		18.0	BLUE STELLAR OBJECT
HN 2648	21 27 41.8	- 50 38 14.			NEBULA
MCG+04-50-012	21 27 42.	+ 26 32	108	13.5	GALAXY
PHL 4539	21 27 42.	- 01 25		18.1	BLUE STELLAR OBJECT
PHL 4538	21 27 42.	- 02 44		16.5	BLUE STELLAR OBJECT
PHL 1580	21 27 42.	- 19 35		12.2	BLUE STELLAR OBJECT

OBJECT NAME	RIGHT ASCEN.	DECLINATION	DIAM.	MAGN.	TYPE OF OBJECT
HN 2655	21 27 45.2	- 16 20 49.	18		NEBULA
RNGC 7080	21 27 47.	+ 26 30		14.0	GALAXY
PHL 4540	21 27 48.	+ 00 48		16.4	BLUE STELLAR OBJECT
ZWG 471.011	21 27 48.	+ 26 30		14.1	GALAXY
UGC 11756	21 27 48.	+ 26 30	108	14.1	GALAXY SBb
PHL 0015	21 27 48.	- 03 14		16.8	BLUE STELLAR OBJECT
MRSL 103+09/1	21 27 54.	+ 64 05	4800		HII REGION
PHL 0016	21 27 54.	- 08 04		17.2	BLUE STELLAR OBJECT
HN 2649	21 27 55.4	- 53 20 13.	12		NEBULA
LBN 0416	21 28	+ 47 30	2700		BRIGHT NEBULA
PHL 4541	21 28 00.	+ 00 34		15.6	BLUE STELLAR OBJECT
PHL 1581	21 28 00.	+ 02 10		18.9	BLUE STELLAR OBJECT
ISS 0078	21 28 00.	+ 58 08	700		STELLAR RING
PHL 1582	21 28 00.	- 01 56		18.4	BLUE STELLAR OBJECT
PHL 0017	21 28 00.	- 03 16		13.9	BLUE STELLAR OBJECT
HN 2651	21 28 01.0	- 51 53 01.	30		NEBULA
HN 2650	21 28 01.4	- 52 54 25.	12		NEBULA
HN 0719	21 28 02.	- 72 53			NEBULA
IC 5108	21 28 02.	- 72 53			NONSTELLAR OBJECT
HN 2652	21 28 04.7	- 50 57 55.	12		NEBULA
IC 5115	21 28 05.	+ 11 33 26.			NONSTELLAR OBJECT
2ZW 134	21 28 06.	+ 19 40			COMPACT GALAXY
PHL 0018	21 28 12.	+ 01 17		18.9	BLUE STELLAR OBJECT
ZWG 401.028	21 28 12.	+ 03 58		15.7	GALAXY
ZWG 512.005	21 28 12.	+ 35 28		15.6	GALAXY
PHL 0019	21 28 12.	- 05 18		16.6	BLUE STELLAR OBJECT
PHL 1583	21 28 12.	- 06 34		16.4	BLUE STELLAR OBJECT
MCG-07-44-019	21 28 12.	- 38 50	42	15.	GALAXY
HELW 489	21 28 12.	- 43 28 54.			NEBULA
PHL 1584	21 28 18.	- 00 11		18.7	BLUE STELLAR OBJECT
HN 2659	21 28 20.6	- 15 47 41.	18		NEBULA
PK096+02.1	21 28 23.8	+ 54 14 17.	6		PLANETARY NEBULA
PHL 4543	21 28 24.	+ 00 41		17.9	BLUE STELLAR OBJECT
ZWG 529.001	21 28 24.	+ 41 38		15.7	GALAXY
UGC 11757	21 28 24.	+ 41 38	96	15.7	GALAXY SBa-b
ZWG 375.048	21 28 24.	- 00 12		15.2	GALAXY
MCG+00-54-029	21 28 24.	- 00 13	48	14.5	GALAXY
PHL 0020	21 28 24.	- 00 55		18.0	BLUE STELLAR OBJECT
PHL 4542	21 28 24.	- 00 59		16.6	BLUE STELLAR OBJECT
MCG-01-54-020	21 28 24.	- 04 12	42	15.	GALAXY
PHL 4544	21 28 24.	- 07 46		18.1	BLUE STELLAR OBJECT
RNGC 7075	21 28 27.	- 38 50			GALAXY
ZWG 426.060	21 28 30.	+ 11 33		15.7	GALAXY
ZWG 426.061	21 28 30.	+ 13 46		15.5	GALAXY
UGC 11758	21 28 30.	+ 13 46	96	15.5	GALAXY S
PHL 1585	21 28 30.	- 06 24		18.9	BLUE STELLAR OBJECT
TON-S 0019	21 28 30.	- 27 00		13.7	BLUE STAR
MCG-05-50-016	21 28 30.	- 29 27	36	15.5	GALAXY
MCG-07-44-020	21 28 30.	- 38 50	30	14.	GALAXY
HN 2656	21 28 30.4	- 50 39 00.	12		NEBULA
MCG+02-54-028	21 28 36.	+ 13 44	84	14.5	GALAXY
PHL 1588	21 28 36.	- 00 49		18.2	BLUE STELLAR OBJECT
PHL 4545	21 28 36.	- 01 04		18.5	BLUE STELLAR OBJECT
PHL 4546	21 28 36.	- 01 20		18.1	BLUE STELLAR OBJECT
PHL 4547	21 28 36.	- 03 45		18.0	BLUE STELLAR OBJECT
PHL 1586	21 28 36.	- 09 30		18.0	BLUE STELLAR OBJECT
PHL 1587	21 28 36.	- 16 08		16.6	BLUE STELLAR OBJECT
PHL 0021	21 28 36.	- 18 05		18.0	BLUE STELLAR OBJECT
RNGC 7070A	21 28 36.	- 43 01			GALAXY
MCG-07-44-021	21 28 36.	- 43 04	72	13.	GALAXY
SER 145.02	21 28 36.	- 43 07	100	13.	SO GLXY WITH ABSORP. LANE
HELW 490	21 28 39.	- 43 04 29.			NEBULA
PHL 1589	21 28 42.	- 06 44		13.9	BLUE STELLAR OBJECT
PHL 4548	21 28 42.	- 07 45		18.4	BLUE STELLAR OBJECT
PHL 1590	21 28 42.	- 15 06		15.3	BLUE STELLAR OBJECT
PHL 0022	21 28 48.	+ 01 24		17.8	BLUE STELLAR OBJECT
PHL 4550	21 28 48.	+ 01 27		17.9	BLUE STELLAR OBJECT
ZWG 375.049	21 28 48.	+ 02 16		13.7	GALAXY
UGC 11759	21 28 48.	+ 02 16	66	13.7	GALAXY S?
MCG+00-54-030	21 28 48.	+ 02 17 30.	78	14.	GALAXY
OCL 0214	21 28 48.	+ 51 22	780	11.7	OPEN STAR CLUSTER
ISS 0122	21 28 48.	+ 57 21	233		STELLAR RING
PHL 4549	21 28 48.	- 02 22		16.9	BLUE STELLAR OBJECT
PHL 1591	21 28 48.	- 08 38		17.8	BLUE STELLAR OBJECT
PHL 4552	21 28 48.	- 09 52		18.4	BLUE STELLAR OBJECT
PHL 1592	21 28 48.	- 13 04		18.0	BLUE STELLAR OBJECT
PHL 4551	21 28 48.	- 17 16		17.9	BLUE STELLAR OBJECT
RNGC 7081	21 28 50.	+ 02 16		13.5	GALAXY
HN 2657	21 28 50.6	- 53 13 29.	18		NEBULA
RNGC 7086	21 28 51.	+ 51 22		11.5	OPEN CLUSTER
SCHO 1273	21 28 51.	+ 63 00 54.	980		ISOLATED DARK CLOUD
BC PKS2128-12	21 28 52.5	- 12 20 19.		15.98	QUASI-STELLAR OBJECT
SHB 358	21 28 52.5	- 12 20 19.		16.0	QUASI-STELLAR OBJECT
ARC 2349	21 28 53.	+ 03 44		17.1	RICH CLUSTER OF GALAXIES
BC A02128+08	21 28 54.	+ 08 59 11.			QUASI-STELLAR OBJECT
ZWG 402.001	21 28 54.	+ 09 12		15.6	GALAXY
ZWG 401.029	21 28 54.	+ 09 12		15.6	GALAXY
PHL 0023	21 28 54.	- 00 52		17.9	BLUE STELLAR OBJECT
PHL 1593	21 28 54.	- 02 50		17.8	BLUE STELLAR OBJECT
PHL 1594	21 28 54.	- 09 30		18.4	BLUE STELLAR OBJECT
PHL 4553	21 28 54.	- 20 35		16.1	BLUE STELLAR OBJECT
SHB 359	21 28 54.9	+ 08 59 42.		18.	QUASI-STELLAR OBJECT
IC 5109	21 28 56.	- 74 20			NONSTELLAR OBJECT
HN 0720	21 28 57.	- 74 20			NEBULA
HN 2663	21 28 57.8	- 16 18 03.	24		NEBULA
HN 2658	21 28 57.8	- 52 02 16.			NEBULA
KHAV 736	21 29	+ 44 55	2760		DARK NEBULA
PHL 4558	21 29 00.	+ 01 05		18.4	BLUE STELLAR OBJECT
ZWG 450.001	21 29 00.	+ 15 35		15.7	GALAXY
ZWG 449.027	21 29 00.	+ 15 35		15.7	GALAXY
KARA.73B 0918	21 29 00.	+ 15 35	24	15.7	ISOLATED GALAXY S
PHL 4554	21 29 00.	- 00 17		17.6	BLUE STELLAR OBJECT
PHL 4556	21 29 00.	- 03 01		17.0	BLUE STELLAR OBJECT
PHL 4555	21 29 00.	- 04 33		18.7	BLUE STELLAR OBJECT
PHL 1595	21 29 00.	- 05 32		18.3	BLUE STELLAR OBJECT
PHL 4557	21 29 00.	- 16 40		17.0	BLUE STELLAR OBJECT
PHL 4559	21 29 00.	- 19 57		18.2	BLUE STELLAR OBJECT
HN 2660	21 29 04.1	- 51 47 34.			NEBULA
HOLM 785B	21 29 05.	+ 02 14	24	15.0	PART OF MULTIPLE GALAXY
MCG+00-55-001	21 29 06.	+ 00 08	36	15.	GALAXY
PHL 0024	21 29 06.	+ 00 30		17.1	BLUE STELLAR OBJECT
ACK 093-00.1	21 29 06.	+ 49 47			PLANETARY NEBULA
PHL 1597	21 29 06.	- 01 03		18.1	BLUE STELLAR OBJECT
PHL 1596	21 29 06.	- 05 24		18.4	BLUE STELLAR OBJECT
HOLM 785A	21 29 08.	+ 02 14	36	14.2	PART OF MULTIPLE GALAXY
IC 5114	21 29 10.	- 36 55 50.			NONSTELLAR OBJECT
ZWG 376.001	21 29 12.	+ 00 08		15.3	GALAXY
MCG+00-55-002	21 29 12.	+ 02 14	60	14.	GALAXY

OBJECT NAME	RIGHT ASCEN.	DECLINATION	DIAM.	MAGN.	TYPE OF OBJECT
ZWG 376.002	21 29 12.	+ 02 15		14.6	GALAXY
UGC 11760	21 29 12.	+ 02 15	78	14.6	GALAXY S
ZWG 427.001	21 29 12.	+ 11 37		15.5	GALAXY
ZWG 426.062	21 29 12.	+ 11 37		15.5	GALAXY
PHL 4560	21 29 12.	- 02 29		17.8	BLUE STELLAR OBJECT
PHL 4562	21 29 12.	- 02 48		17.8	BLUE STELLAR OBJECT
PHL 1599	21 29 12.	- 08 04		18.5	BLUE STELLAR OBJECT
PHL 1598	21 29 12.	- 12 19		15.5	BLUE STELLAR OBJECT
PHL 0025	21 29 12.	- 17 32		11.8	BLUE STELLAR OBJECT
PHL 4561	21 29 12.	- 18 52		18.3	BLUE STELLAR OBJECT
HN 2661	21 29 14.8	- 51 48 04.	18		NEBULA
RNGC 7079	21 29 17.	- 44 18		12.5	GALAXY
ZWG 376.003	21 29 18.	+ 03 22		15.0	GALAXY
MCG+00-55-003	21 29 18.	+ 03 22	24	15.5	GALAXY
ZWG 427.002	21 29 18.	+ 12 15		15.7	GALAXY
ZWG 426.063	21 29 18.	+ 12 15		15.7	GALAXY
ZC 2129.3+3036	21 29 18.	+ 30 36	810		CLUSTER OF GALAXIES
PHL 1600	21 29 18.	- 02 51		18.0	BLUE STELLAR OBJECT
SCHO 1274	21 29 19.	+ 47 47 12.	280		ISOLATED DARK CLOUD
IC 1389	21 29 20.	- 18 14 36.			NONSTELLAR OBJECT
2ZW 135	21 29 24.	- 02 46			COMPACT GALAXY
RNGC 7064	21 29 25.	+ 17 11			NON-EXISTENT OBJECT
ARC 2350	21 29 25.	- 06 07		17.1	RICH CLUSTER OF GALAXIES
HN 2666	21 29 29.3	- 15 56 44.	18		NEBULA
LDN 1006	21 29 30.	+ 47 35	780		DARK NEBULA
MCG-03-55-001	21 29 30.	- 18 14	36	15.	GALAXY
ZC 2129.6+0352	21 29 36.	+ 03 52	3630		CLUSTER OF GALAXIES
4ZW 071	21 29 36.	+ 34 18			COMPACT GALAXY
UGC 11761	21 29 36.	+ 34 18	15	13.8	GALAXY COMPACT
PHL 4564	21 29 36.	- 00 11		16.0	BLUE STELLAR OBJECT
PHL 1601	21 29 36.	- 00 15		18.5	BLUE STELLAR OBJECT
PHL 0026	21 29 36.	- 02 37		18.6	BLUE STELLAR OBJECT
PHL 4563	21 29 36.	- 05 06		18.1	BLUE STELLAR OBJECT
MCG-01-55-001	21 29 36.	- 05 58	54	15.	GALAXY
PHL 0027	21 29 36.	- 11 00		17.3	BLUE STELLAR OBJECT
DV.56 N7070A	21 29 36.	- 43 01	72		PEC GALAXY
HN 2662	21 29 37.6	- 52 49 33.	18		NEBULA
LB 03907	21 29 39.	+ 47 58 18.		17.4	FAINT BLUE STAR
PHL 0028	21 29 42.	+ 00 02		14.8	BLUE STELLAR OBJECT
MCG+00-55-004	21 29 42.	- 02 07	36	15.	GALAXY
PHL 4565	21 29 42.	- 02 14		17.8	BLUE STELLAR OBJECT
ZWG 376.004	21 29 42.	- 02 23		15.3	GALAXY
PHL 4566	21 29 42.	- 05 34		18.3	BLUE STELLAR OBJECT
PHL 4567	21 29 42.	- 05 48		18.0	BLUE STELLAR OBJECT
MCG-07-44-022	21 29 42.	- 44 17	108	12.3	GALAXY
HN 2664	21 29 43.6	- 51 31 50.	12		NEBULA
PHL 1602	21 29 48.	+ 00 19		18.1	BLUE STELLAR OBJECT
4ZW 072	21 29 48.	+ 29 55			COMPACT GALAXY
UGC 11762	21 29 48.	+ 29 55	9	13.0	GALAXY EX CMPT
IC 1390	21 29 48.	- 02 04 57.			NONSTELLAR OBJECT
PHL 4568	21 29 48.	- 02 10		16.5	BLUE STELLAR OBJECT
RNGC 7085	21 29 51.	+ 06 22		15.0	GALAXY
HN 2670	21 29 52.1	- 15 46 37.	18		NEBULA
MCG+01-55-001	21 29 54.	+ 06 21	54	14.5	GALAXY
ZWG 402.002	21 29 54.	+ 06 22		15.2	GALAXY
KARA.73B 0919	21 29 54.	+ 06 22	66	15.2	ISOLATED GALAXY S
ZC 2129.9+2832	21 29 54.	+ 28 32	1080		CLUSTER OF GALAXIES
ISS 0182	21 29 54.	+ 47 57	609		STELLAR RING
PHL 0029	21 29 54.	- 01 02		17.2	BLUE STELLAR OBJECT
ZWG 376.005	21 29 54.	- 02 05		15.1	GALAXY
PHL 4569	21 29 54.	- 04 04		17.9	BLUE STELLAR OBJECT
PHL 4570	21 29 54.	- 04 12		17.9	BLUE STELLAR OBJECT
HN 2672	21 29 58.2	- 15 49 19.	42		NEBULA
HN 2673	21 29 59.9	- 16 01 55.	12		NEBULA
PHL 1604	21 30 00.	+ 01 46		18.9	BLUE STELLAR OBJECT
2ZW 136	21 30 00.	+ 09 56			COMPACT GALAXY
KW 64	21 30 00.	+ 09 56	30		SEYFERT GALAXY
VVI 90	21 30 00.	+ 09 56	25	14.64	SEYFERT GALAXY
UGC 11763	21 30 00.	+ 09 56	30	14.3	GALAXY VY CMPT
LDN 1075	21 30 00.	+ 54 10	1500		DARK NEBULA
LDN 1096	21 30 00.	+ 57 50	120		DARK NEBULA
LDN 1146	21 30 00.	+ 62 30	1680		DARK NEBULA
LDN 1145	21 30 00.	+ 62 30	1680		DARK NEBULA
LDN 1176	21 30 00.	+ 66 30	2700		DARK NEBULA
PHL 1605	21 30 00.	- 01 10		18.2	BLUE STELLAR OBJECT
PHL 4571	21 30 00.	- 01 30		17.0	BLUE STELLAR OBJECT
PHL 1603	21 30 00.	- 07 42		17.5	BLUE STELLAR OBJECT
PHL 0030	21 30 00.	- 16 50		13.7	BLUE STELLAR OBJECT
HN 2674	21 30 03.8	- 14 44 30.	30		NEBULA
SCHO 1275	21 30 04.	+ 44 57 18.	230		ISOLATED DARK CLOUD
PHL 1606	21 30 05.	+ 01 14		17.5	BLUE STELLAR OBJECT
ZWG 376.006	21 30 06.	+ 02 19		15.2	GALAXY
ZWG 376.007	21 30 06.	+ 02 29		14.8	GALAXY
2ZW 137	21 30 06.	+ 10 44			COMPACT GALAXY
LB 03908	21 30 06.	+ 48 21 48.		20.2	FAINT BLUE STAR
MCG+00-55-005	21 30 06.	- 00 22	48	15.	GALAXY
PHL 4573	21 30 06.	- 03 01		17.4	BLUE STELLAR OBJECT
PHL 4572	21 30 06.	- 06 25		18.0	BLUE STELLAR OBJECT
PHL 4574	21 30 06.	- 09 40		17.7	BLUE STELLAR OBJECT
TON-S 0020	21 30 06.	- 29 53		14.5	BLUE STAR
PK095+00.1	21 30 09.0	+ 52 20 37.			PLANETARY NEBULA
SCHO 1276	21 30 10.	+ 45 00 36.	210		ISOLATED DARK CLOUD
HN 2665	21 30 10.3	- 53 31 25.	12		NEBULA
SCHO 1278	21 30 12.	+ 45 05 00.	180		ISOLATED DARK CLOUD
ISS 0183	21 30 12.	+ 45 26	308		STELLAR RING
SCHO 1277	21 30 12.	+ 54 34 18.	410		ISOLATED DARK CLOUD
PHL 1608	21 30 12.	- 00 12		18.2	BLUE STELLAR OBJECT
ZWG 376.008	21 30 12.	- 00 20		15.3	GALAXY
PHL 1607	21 30 12.	- 00 32		15.0	BLUE STELLAR OBJECT
PHL 4576	21 30 12.	- 00 56		16.2	BLUE STELLAR OBJECT
PHL 4575	21 30 12.	- 01 08		17.9	BLUE STELLAR OBJECT
PHL 4577	21 30 12.	- 04 32		18.0	BLUE STELLAR OBJECT
PHL 0031	21 30 12.	- 05 54		16.3	BLUE STELLAR OBJECT
PHL 0032	21 30 12.	- 06 01		18.3	BLUE STELLAR OBJECT
B 155	21 30 15.	+ 44 45	780		DARK OBJECT
HN 2667	21 30 16.0	- 53 57 49.			NEBULA
LB 03909	21 30 17.	+ 48 33 24.		18.8	FAINT BLUE STAR
PHL 4578	21 30 18.	- 04 02		18.5	BLUE STELLAR OBJECT
PHL 4579	21 30 18.	- 08 20		17.5	BLUE STELLAR OBJECT
SCHO 1279	21 30 19.	+ 49 21 36.	240		ISOLATED DARK CLOUD
LB 03910	21 30 20.	+ 48 31 12.		19.9	FAINT BLUE STAR
HN 2669	21 30 21.6	- 52 07 07.	12		NEBULA
SCHO 1280	21 30 22.	+ 49 35 36.	240		ISOLATED DARK CLOUD
HN 2668	21 30 23.1	- 53 30 01.	12		NEBULA
OCL 0211	21 30 24.	+ 48 13	4800	7.5	OPEN STAR CLUSTER
RNGC 7092	21 30 24.	+ 48 13		5.5	OPEN CLUSTER
PHL 1609	21 30 24.	- 01 23		15.8	BLUE STELLAR OBJECT
PHL 1610	21 30 24.	- 03 44		18.5	BLUE STELLAR OBJECT

OBJECT NAME	RIGHT ASCEN.	DECLINATION	DIAM.	MAGN.	TYPE OF OBJECT
PHL 0033	21 30 24.	- 04 55		16.4	BLUE STELLAR OBJECT
PHL 1611	21 30 24.	- 05 27		18.2	BLUE STELLAR OBJECT
PHL 1612	21 30 24.	- 15 58		17.7	BLUE STELLAR OBJECT
SCHO 1281	21 30 28.	+ 44 45 48.	790		ISOLATED DARK CLOUD
SCHO 1282	21 30 28.	+ 53 36 18.	410		ISOLATED DARK CLOUD
LB 00365	21 30 29.	- 20 51 18.		14.1	FAINT BLUE STAR
LDN 0983	21 30 30.	+ 44 50	1020		DARK NEBULA
LDN 0968	21 30 30.	+ 45 00	1140		DARK NEBULA
ISS 0184	21 30 30.	+ 50 21	318		STELLAR RING
ZWG 376.009	21 30 30.	- 01 43		15.7	GALAXY
PHL 4581	21 30 30.	- 07 36		18.1	BLUE STELLAR OBJECT
PHL 4580	21 30 30.	- 12 15		17.0	BLUE STELLAR OBJECT
PK097+03.1	21 30 35.8	+ 55 39 33.	80	16.4	PLANETARY NEBULA
PHL 4582	21 30 36.	+ 01 12		16.7	BLUE STELLAR OBJECT
ZC 2130.6+0545	21 30 36.	+ 05 45	1340		CLUSTER OF GALAXIES
MCG+02-55-001	21 30 36.	+ 11 09	36	16.	GALAXY
HN 0104	21 30 36.	+ 44 23			NEBULA
PHL 4583	21 30 36.	- 05 36		17.9	BLUE STELLAR OBJECT
MCG-05-50-017	21 30 36.	- 29 43	36	15.	GALAXY
PK089-05.1	21 30 36.8	+ 44 22 29.		13.3	PLANETARY NEBULA
IC 5117	21 30 36.8	+ 44 22 29.	2	13.3	PLANETARY NEBULA
SCHO 1283	21 30 38.	+ 49 43 18.	210		ISOLATED DARK CLOUD
HN 2671	21 30 38.1	- 53 47 36.	12		NEBULA
SCHO 1284	21 30 40.	+ 47 40 48.	410		ISOLATED DARK CLOUD
PHL 0034	21 30 42.	+ 01 54		17.4	BLUE STELLAR OBJECT
UGC 11764	21 30 42.	+ 07 47	60	18.	GALAXY DWRF SP
ZC 2130.7+1034	21 30 42.	+ 10 34	810		CLUSTER OF GALAXIES
MCG-05-50-018	21 30 42.	- 32 32	24	16.	GALAXY
LB 03911	21 30 46.	+ 48 36 00.		18.2	FAINT BLUE STAR
CED 193	21 30 46.	- 00 57	4500		DIFFUSE GALACTIC NEBULA
PHL 1613	21 30 48.	+ 01 46		18.1	BLUE STELLAR OBJECT
2ZW 138	21 30 48.	+ 11 24			COMPACT GALAXY
PHL 4584	21 30 48.	- 02 54		16.6	BLUE STELLAR OBJECT
PHL 4585	21 30 48.	- 07 18		17.7	BLUE STELLAR OBJECT
PHL 1614	21 30 48.	- 07 32		18.6	BLUE STELLAR OBJECT
LB 03912	21 30 49.	+ 48 00 36.		18.4	FAINT BLUE STAR
RNGC 7088	21 30 49.	- 00 37			NON-EXISTENT OBJECT
REIN 2.293	21 30 52.39	- 01 02 41.7			NEBULA
PHL 4586	21 30 54.	+ 01 42		17.2	BLUE STELLAR OBJECT
PHL 4587	21 30 54.	+ 01 43		17.5	BLUE STELLAR OBJECT
GCL 121	21 30 54.	- 01 03	738	7.3	GLOBULAR STAR CLUSTER
PHL 0035	21 30 54.	- 04 46		13.9	BLUE STELLAR OBJECT
PHL 1615	21 30 54.	- 09 53		7.8	BLUE STELLAR OBJECT
PHL 0036	21 30 54.	- 20 32		16.8	BLUE STELLAR OBJECT
HN 2675	21 30 54.1	- 52 52 59.	24		NEBULA
HN 2676	21 30 54.5	- 50 14 53.	24		NEBULA
RNGC 7089	21 30 55.	- 01 03		7.5	GLOBULAR CLUSTER
BIN.48 07	21 30 58.	+ 55 44			DIFFUSE NEBULA
HN 2677	21 30 59.8	- 52 41 11.	12		NEBULA
LBN 0366	21 31	+ 38 00	5400		BRIGHT NEBULA
KHAV 737	21 31	+ 47 43	5170		DARK NEBULA
LBN 0443	21 31	+ 55 40	60		BRIGHT NEBULA
KHAV 738	21 31	+ 57 13			DARK NEBULA
KHAV 739	21 31	+ 62 55	3470		DARK NEBULA
LDN 1079	21 31 00.	+ 54 30	420		DARK NEBULA
LDN 1102	21 31 00.	+ 57 50	360		DARK NEBULA
PHL 1616	21 31 00.	- 01 12		17.9	BLUE STELLAR OBJECT
PHL 1618	21 31 00.	- 03 14		18.2	BLUE STELLAR OBJECT
PHL 1617	21 31 00.	- 05 15		17.9	BLUE STELLAR OBJECT
RNGC 7091	21 31 04.	- 36 52			GALAXY
SCHO 1285	21 31 05.	+ 54 51 54.	400		ISOLATED DARK CLOUD
2ZW 139	21 31 06.	+ 00 52			COMPACT GALAXY
ZWG 402.003	21 31 06.	+ 08 33		15.3	GALAXY
UGC 11765	21 31 06.	+ 08 33	72	15.3	GALAXY Sb
PHL 4588	21 31 06.	- 08 38		17.9	BLUE STELLAR OBJECT
PHL 4589	21 31 06.	- 19 00		18.0	BLUE STELLAR OBJECT
PK086-08.1	21 31 07.4	+ 39 24 40.	5	12.7	PLANETARY NEBULA
MCG+01-55-002	21 31 09.	+ 08 33	48	15.	GALAXY
ISS 0079	21 31 12.	+ 59 57	208		STELLAR RING
PHL 4591	21 31 12.	- 01 31		18.4	BLUE STELLAR OBJECT
PHL 4592	21 31 12.	- 04 41		18.2	BLUE STELLAR OBJECT
PHL 4590	21 31 12.	- 07 14		17.7	BLUE STELLAR OBJECT
MCG-07-44-023	21 31 12.	- 41 03	48	15.	GALAXY
MCG-07-44-024	21 31 12.	- 44 32		17.	GALAXY
HN 2678	21 31 17.8	- 49 55 58.	18		NEBULA
ZWG 402.004	21 31 18.	+ 07 12		15.6	GALAXY
KARA.73B 0920	21 31 18.	+ 07 12	30	15.6	ISOLATED GALAXY E
ZC 2131.3+0930	21 31 18.	+ 09 30	1280		CLUSTER OF GALAXIES
PHL 4593	21 31 18.	- 01 18		17.9	BLUE STELLAR OBJECT
PHL 1619	21 31 18.	- 07 25		18.6	BLUE STELLAR OBJECT
PHL 1620	21 31 18.	- 09 08		18.2	BLUE STELLAR OBJECT
PHL 4594	21 31 18.	- 19 06		18.8	BLUE STELLAR OBJECT
SER 145.01	21 31 18.	- 41 05			LOOSE GROUP OF 6 GALAXIES
BV 02	21 31 20.	- 19 06 42.		18.5	FAINT BLUE VARIABLE
ZC 2131.4+0224	21 31 24.	+ 02 24	1410		CLUSTER OF GALAXIES
ZC 2131.4+0954	21 31 24.	+ 09 54	940		CLUSTER OF GALAXIES
PHL 0037	21 31 24.	- 00 59		18.1	BLUE STELLAR OBJECT
PHL 4597	21 31 24.	- 01 42		17.7	BLUE STELLAR OBJECT
PHL 1621	21 31 24.	- 06 16		17.7	BLUE STELLAR OBJECT
PHL 0038	21 31 24.	- 16 27		17.8	BLUE STELLAR OBJECT
PHL 4595	21 31 24.	- 18 31		16.6	BLUE STELLAR OBJECT
PHL 4596	21 31 24.	- 18 50		18.6	BLUE STELLAR OBJECT
MCG-07-44-025	21 31 24.	- 41 02 30.	60	14.	GALAXY
RNGC 7087	21 31 26.	- 41 03			GALAXY
HN 2679	21 31 29.4	- 53 48 28.	12		NEBULA
ZC 2131.5+1014	21 31 30.	+ 10 14	3760		CLUSTER OF GALAXIES
PHL 1623	21 31 30.	- 01 18		18.2	BLUE STELLAR OBJECT
PHL 1622	21 31 30.	- 07 44		18.4	BLUE STELLAR OBJECT
PHL 4598	21 31 30.	- 18 50		18.8	BLUE STELLAR OBJECT
PHL 1624	21 31 36.	- 01 17		18.2	BLUE STELLAR OBJECT
PHL 4599	21 31 36.	- 01 23		17.7	BLUE STELLAR OBJECT
PHL 0039	21 31 36.	- 04 38		17.9	BLUE STELLAR OBJECT
PHL 1625	21 31 36.	- 15 04		18.2	BLUE STELLAR OBJECT
PHL 4600	21 31 36.	- 15 43		18.4	BLUE STELLAR OBJECT
PHL 4601	21 31 36.	- 16 40		18.2	BLUE STELLAR OBJECT
PHL 4602	21 31 36.	- 17 28		18.4	BLUE STELLAR OBJECT
HN 2680	21 31 37.8	- 52 09 21.	12		NEBULA
LB 03913	21 31 38.	+ 48 36 30.		20.0	FAINT BLUE STAR
MCG+01-55-003	21 31 39.	+ 08 35	48	16.	GALAXY
IC 5119	21 31 40.	+ 21 34 31.			NONSTELLAR OBJECT
SCHO 1286	21 31 41.	+ 51 31 12.	190		ISOLATED DARK CLOUD
ARC 2351	21 31 41.	- 13 38		17.1	RICH CLUSTER OF GALAXIES
ZWG 402.005	21 31 42.	+ 08 33		15.2	GALAXY
ZWG 472.001	21 31 42.	+ 21 36		15.3	GALAXY
UGC 11766	21 31 42.	+ 21 36	60	15.3	GALAXY Sa
PHL 0040	21 31 42.	- 01 10		18.6	BLUE STELLAR OBJECT
MCG+01-55-004	21 31 45.	+ 08 34	30	15.	GALAXY
ARC 2352	21 31 45.	- 16 04		17.5	RICH CLUSTER OF GALAXIES

OBJECT NAME	RIGHT ASCEN.	DECLINATION	DIAM.	MAGN.	TYPE OF OBJECT
SCHO 1287	21 31 47.	+ 49 39 48.	230		ISOLATED DARK CLOUD
ZWG 402.006	21 31 48.	+ 08 26		15.6	GALAXY
2ZW 140	21 31 48.	+ 08 27			COMPACT GALAXY
UGC 11767	21 31 48.	+ 41 36	72	17.	GALAXY
PHL 1626	21 31 48.	- 00 05		18.9	BLUE STELLAR OBJECT
PHL 4604	21 31 48.	- 02 45		18.0	BLUE STELLAR OBJECT
PHL 4603	21 31 48.	- 22 45		15.7	BLUE STELLAR OBJECT
ARC 2353	21 31 49.	- 01 50		16.8	RICH CLUSTER OF GALAXIES
RNGC 7083	21 31 50.	- 64 07		12.0	GALAXY
LB 03914	21 31 52.	+ 48 04 54.		16.4	FAINT BLUE STAR
ZWG 427.003	21 31 54.	+ 10 59		15.7	GALAXY
PHL 4605	21 31 54.	- 02 40		16.9	BLUE STELLAR OBJECT
PHL 4606	21 31 54.	- 13 53		17.2	BLUE STELLAR OBJECT
MCG-05-51-001	21 31 54.	- 30 00	18	15.	GALAXY
HN 2681	21 31 54.9	- 53 55 20.	12		NEBULA
KHAV 740	21 32	+ 54 07	4200		DARK NEBULA
B 364	21 32	+ 54 20	4500		DARK OBJECT
LDN 0969	21 32 00.	+ 43 22	540		DARK NEBULA
LDN 1009	21 32 00.	+ 47 40	960		DARK NEBULA
LDN 1083	21 32 00.	+ 55 45	3540		DARK NEBULA
LDN 1085	21 32 00.	+ 56 32	360		DARK NEBULA
SCHO 1288	21 32 00.	+ 57 17 54.	370		ISOLATED DARK CLOUD
LDN 1093	21 32 00.	+ 57 25	720		DARK NEBULA
PHL 4607	21 32 00.	- 00 32		17.4	BLUE STELLAR OBJECT
PHL 4608	21 32 00.	- 01 04		16.6	BLUE STELLAR OBJECT
PHL 1627	21 32 00.	- 01 28		16.7	BLUE STELLAR OBJECT
PHL 1628	21 32 00.	- 02 34		18.9	BLUE STELLAR OBJECT
PHL 0041	21 32 00.	- 06 07		17.8	BLUE STELLAR OBJECT
SCHO 1289	21 32 01.	+ 47 55 24.	400		ISOLATED DARK CLOUD
HN 2682	21 32 01.4	- 53 57 32.	18		NEBULA
HELW 072	21 32 02.	- 14 31			NEBULA
B 157	21 32 04.	+ 54 27	300		DARK OBJECT
PHL 4610	21 32 05.	+ 00 34		18.7	BLUE STELLAR OBJECT
MCG+04-51-001	21 32 06.	+ 26 09	66	15.5	GALAXY
B 156	21 32 06.	+ 45 22	480		DARK OBJECT
PHL 1629	21 32 06.	- 03 21		17.8	BLUE STELLAR OBJECT
PHL 4611	21 32 06.	- 06 30		18.5	BLUE STELLAR OBJECT
PHL 4609	21 32 06.	- 12 30		17.5	BLUE STELLAR OBJECT
LB 03915	21 32 08.	+ 48 23 48.		15.5	FAINT BLUE STAR
HN 2684	21 32 10.1	- 50 09 32.	12		NEBULA
HN 2683	21 32 11.1	- 52 29 32.	12		NEBULA
PHL 4612	21 32 12.	+ 01 48		18.0	BLUE STELLAR OBJECT
ZWG 427.004	21 32 12.	+ 11 12		15.6	GALAXY
UGC 11768	21 32 12.	+ 18 44	66	16.0	GALAXY Sb-c
ZWG 472.002	21 32 12.	+ 26 08		15.5	GALAXY
UGC 11769	21 32 12.	+ 26 08	96	15.5	GALAXY Sb
ZWG 493.001	21 32 12.	+ 33 02		15.7	GALAXY
ZWG 492.003	21 32 12.	+ 33 02		15.7	GALAXY
PHL 4615	21 32 12.	- 02 54		17.8	BLUE STELLAR OBJECT
PHL 4614	21 32 12.	- 02 54		17.5	BLUE STELLAR OBJECT
PHL 0042	21 32 12.	- 10 28		16.5	BLUE STELLAR OBJECT
PHL 4616	21 32 12.	- 14 37		18.8	BLUE STELLAR OBJECT
PHL 4613	21 32 12.	- 17 56		18.0	BLUE STELLAR OBJECT
MCG-03-55-002	21 32 12.	- 20 29	36	15.	GALAXY
HN 2685	21 32 12.5	- 51 32 08.	12		NEBULA
SCHO 1290	21 32 13.	+ 51 28 12.	200		ISOLATED DARK CLOUD
LDN 1012	21 32 18.	+ 47 53	540		DARK NEBULA
PHL 4617	21 32 18.	- 02 26		18.1	BLUE STELLAR OBJECT
PHL 4618	21 32 18.	- 03 40		18.0	BLUE STELLAR OBJECT
SCHO 1292	21 32 19.	+ 56 58 24.	440		ISOLATED DARK CLOUD
SCHO 1291	21 32 19.	+ 62 46 24.	1130		ISOLATED DARK CLOUD
HELW 073	21 32 20.	- 14 32			NEBULA
SCHO 1293	21 32 22.	+ 54 30 06.	290		ISOLATED DARK CLOUD
HN 2686	21 32 22.0	- 51 38 49.	36		NEBULA
MCG+00-55-006	21 32 24.	+ 01 12 30.	60	15.	GALAXY
MCG+00-55-007	21 32 24.	- 00 46	12	16.	GALAXY
PHL 4621	21 32 24.	- 04 50		17.9	BLUE STELLAR OBJECT
PHL 0043	21 32 24.	- 05 40		18.5	BLUE STELLAR OBJECT
PHL 4620	21 32 24.	- 12 52		18.2	BLUE STELLAR OBJECT
PHL 4619	21 32 24.	- 20 42		16.7	BLUE STELLAR OBJECT
TON-S 0021	21 32 24.	- 25 48		16.0	BLUE STAR
HN 2688	21 32 26.6	- 15 31 54.	42		NEBULA
IC 1391	21 32 28.	- 00 43 32.			NONSTELLAR OBJECT
ZWG 376.011	21 32 30.	+ 01 14		15.6	GALAXY
UGC 11770	21 32 30.	+ 01 14	66	15.6	GALAXY S
ZWG 427.005	21 32 30.	+ 14 14		15.6	GALAXY
KARA.73B 0921	21 32 30.	+ 14 14	18	15.6	ISOLATED GALAXY E
PHL 4622	21 32 30.	- 00 22		16.7	BLUE STELLAR OBJECT
ZWG 376.010	21 32 30.	- 00 43		15.7	GALAXY
PHL 0044	21 32 30.	- 13 46		13.0	BLUE STELLAR OBJECT
MCG-05-51-002	21 32 30.	- 26 55	48	15.	GALAXY
HELW 074	21 32 32.	- 14 41			NEBULA
HN 2689	21 32 32.0	- 17 59 18.	18		NEBULA
IC 5116	21 32 34.	- 71 12			NONSTELLAR OBJECT
HN 0721	21 32 34.	- 71 13			NEBULA
SCHO 1294	21 32 35.	+ 53 39 42.	440		ISOLATED DARK CLOUD
ZWG 376.012	21 32 36.	+ 00 20		15.7	GALAXY
PHL 4625	21 32 36.	+ 02 04		18.5	BLUE STELLAR OBJECT
ZC 2132.6+3317	21 32 36.	+ 33 17	1410		CLUSTER OF GALAXIES
PHL 4623	21 32 36.	- 03 18		16.9	BLUE STELLAR OBJECT
PHL 4626	21 32 36.	- 03 26		17.7	BLUE STELLAR OBJECT
PHL 0045	21 32 36.	- 24 57		17.8	BLUE STELLAR OBJECT
PHL 4624	21 32 36.	- 26 21		17.9	BLUE STELLAR OBJECT
SCHO 1295	21 32 40.	+ 56 31 36.	260		ISOLATED DARK CLOUD
PHL 4627	21 32 42.	+ 00 24		16.9	BLUE STELLAR OBJECT
ZWG 450.002	21 32 42.	+ 20 22		15.4	GALAXY
PHL 1631	21 32 42.	- 01 06		18.1	BLUE STELLAR OBJECT
PHL 1630	21 32 42.	- 02 59		17.8	BLUE STELLAR OBJECT
PHL 4628	21 32 42.	- 03 20		17.7	BLUE STELLAR OBJECT
PHL 0046	21 32 42.	- 16 45		15.5	BLUE STELLAR OBJECT
PHL 4629	21 32 42.	- 16 56		18.5	BLUE STELLAR OBJECT
SCHO 1296	21 32 44.	+ 50 13 18.	380		ISOLATED DARK CLOUD
PHL 0047	21 32 48.	- 00 45		18.9	BLUE STELLAR OBJECT
ZC 2132.8-0117	21 32 48.	- 01 17	1210		CLUSTER OF GALAXIES
PHL 4630	21 32 48.	- 01 46		17.0	BLUE STELLAR OBJECT
PHL 0048	21 32 48.	- 07 12		13.4	BLUE STELLAR OBJECT
PHL 4631	21 32 48.	- 15 07		17.0	BLUE STELLAR OBJECT
MCG-03-55-003	21 32 48.	- 19 15	42	15.	GALAXY
ARC 2355	21 32 51.	+ 01 10		17.7	RICH CLUSTER OF GALAXIES
RNGC 7090	21 32 51.	- 54 47		11.5	GALAXY
PHL 4632	21 32 54.	- 03 53		17.0	BLUE STELLAR OBJECT
PHL 0049	21 32 54.	- 05 39		17.7	BLUE STELLAR OBJECT
PHL 0050	21 32 54.	- 07 37		17.3	BLUE STELLAR OBJECT
RNGC 7093	21 32 58.	+ 45 47			NON-EXISTENT OBJECT
LBN 0355	21 33	+ 36 20	3900		BRIGHT NEBULA
KHAV 741	21 33	+ 56 25	700		DARK NEBULA
LBN 0451	21 33	+ 57 15	480		BRIGHT NEBULA
LDN 1054	21 33 00.	+ 50 30	9900		DARK NEBULA

OBJECT NAME	RIGHT ASCEN.	DECLINATION	DIAM.	MAGN.	TYPE OF OBJECT
LDN 1074	21 33 00.	+ 53 40	660		DARK NEBULA
LDN 1098	21 33 00.	+ 57 25	600		DARK NEBULA
PHL 1633	21 33 00.	- 00 59		17.3	BLUE STELLAR OBJECT
PHL 0051	21 33 00.	- 01 07		18.6	BLUE STELLAR OBJECT
PHL 4635	21 33 00.	- 02 50		17.8	BLUE STELLAR OBJECT
PHL 4634	21 33 00.	- 03 00		17.9	BLUE STELLAR OBJECT
PHL 1632	21 33 00.	- 04 36		18.1	BLUE STELLAR OBJECT
PHL 1634	21 33 00.	- 05 31		18.9	BLUE STELLAR OBJECT
PHL 4633	21 33 00.	- 11 02		16.7	BLUE STELLAR OBJECT
PHL 0052	21 33 00.	- 16 18		15.3	BLUE STELLAR OBJECT
HN 2687	21 33 01.7	- 53 49 42.			NEBULA
HELW 075	21 33 02.	- 14 21			NEBULA
PHL 0053	21 33 06.	- 01 08		18.6	BLUE STELLAR OBJECT
MCG-07-44-026	21 33 06.	- 38 46	42	14.5	GALAXY
ARC 2354	21 33 07.	- 15 09		17.1	RICH CLUSTER OF GALAXIES
ARC 2356	21 33 10.	- 00 07		17.1	RICH CLUSTER OF GALAXIES
PHL 1635	21 33 12.	+ 02 23		17.9	BLUE STELLAR OBJECT
UGC 11771	21 33 12.	+ 23 15	90	16.0	GALAXY Sc
ZC 2133.2-0006	21 33 12.	- 00 06	2020		CLUSTER OF GALAXIES
PHL 4638	21 33 12.	- 01 02		17.8	BLUE STELLAR OBJECT
PHL 4637	21 33 12.	- 01 08		17.3	BLUE STELLAR OBJECT
PHL 1636	21 33 12.	- 09 54		18.0	BLUE STELLAR OBJECT
PHL 4639	21 33 12.	- 18 13		18.4	BLUE STELLAR OBJECT
PHL 4636	21 33 12.	- 22 56		9.0	BLUE STELLAR OBJECT
PHL 0054	21 33 12.	- 23 08		18.5	BLUE STELLAR OBJECT
HN 2690	21 33 14.2	- 53 43 17.	24		NEBULA
HN 2691	21 33 14.7	- 53 39 29.	12		NEBULA
PHL 1637	21 33 18.	+ 01 38		16.5	BLUE STELLAR OBJECT
PHL 4640	21 33 18.	- 08 34		18.6	BLUE STELLAR OBJECT
B 365	21 33 20.	+ 56 30	1320		DARK OBJECT
PK081-14.1	21 33 24.	+ 31 28	113	16.0	PLANETARY NEBULA
MCG+06-47-003	21 33 24.	+ 35 10	33	14.	GALAXY
PHL 4641	21 33 24.	- 05 34		17.8	BLUE STELLAR OBJECT
PHL 1638	21 33 24.	- 06 08		18.3	BLUE STELLAR OBJECT
PHL 4643	21 33 24.	- 07 32		18.0	BLUE STELLAR OBJECT
PHL 4642	21 33 24.	- 09 28		16.6	BLUE STELLAR OBJECT
SCHO 1297	21 33 25.	+ 56 46 30.	490		ISOLATED DARK CLOUD
SCHO 1298	21 33 26.	+ 57 00 12.	240		ISOLATED DARK CLOUD
HELW 076	21 33 26.	- 14 38			NEBULA
IC 1392	21 33 29.	+ 35 10 29.			NONSTELLAR OBJECT
ZWG 512.006	21 33 30.	+ 35 11		13.0	GALAXY
UGC 11772	21 33 30.	+ 35 11	102	13.0	GALAXY E-SO
LDN 1048	21 33 30.	+ 50 42	1020		DARK NEBULA
PHL 4644	21 33 30.	- 21 40		16.5	BLUE STELLAR OBJECT
PHL 4648	21 33 36.	+ 01 15		18.4	BLUE STELLAR OBJECT
PHL 0055	21 33 36.	+ 01 26		17.8	BLUE STELLAR OBJECT
ZWG 402.007	21 33 36.	+ 07 43		14.9	GALAXY
MCG+01-55-005	21 33 36.	+ 07 43	30	15.	GALAXY
PHL 4645	21 33 36.	- 00 30		16.6	BLUE STELLAR OBJECT
PHL 4646	21 33 36.	- 00 51		17.1	BLUE STELLAR OBJECT
PHL 4647	21 33 36.	- 02 28		17.5	BLUE STELLAR OBJECT
PHL 0056	21 33 36.	- 02 38		15.3	BLUE STELLAR OBJECT
PHL 0057	21 33 36.	- 05 26		18.2	BLUE STELLAR OBJECT
PHL 1639	21 33 36.	- 06 20		17.7	BLUE STELLAR OBJECT
PHL 4649	21 33 36.	- 06 42		18.3	BLUE STELLAR OBJECT
MCG-05-51-003	21 33 36.	- 27 59	18	15.	GALAXY
MCG+06-47-004	21 33 39.	+ 35 07	24	14.5	GALAXY
PK096+01.1	21 33 40.9	+ 53 33 46.			PLANETARY NEBULA
ZWG 402.008	21 33 42.	+ 09 26		15.3	GALAXY
UGC 11773	21 33 42.	+ 09 26	60	15.3	GALAXY Sa
PHL 4650	21 33 42.	- 02 04		16.0	BLUE STELLAR OBJECT
PHL 4651	21 33 42.	- 02 44		18.4	BLUE STELLAR OBJECT
PHL 1640	21 33 42.	- 24 54		17.7	BLUE STELLAR OBJECT
ARC 2359	21 33 45.	+ 14 13		17.6	RICH CLUSTER OF GALAXIES
PHL 4653	21 33 48.	+ 00 52		17.9	BLUE STELLAR OBJECT
PHL 4652	21 33 48.	+ 01 22		17.9	BLUE STELLAR OBJECT
MCG+02-55-002	21 33 48.	+ 09 26	72	15.	GALAXY
ZC 2133.8+1414	21 33 48.	+ 14 14	1410		CLUSTER OF GALAXIES
UGC 11774	21 33 48.	+ 16 50	60	16.0	GALAXY
ZWG 512.007	21 33 48.	+ 35 08		14.5	GALAXY
UGC 11775	21 33 48.	+ 35 08	72	14.5	GALAXY SO
PHL 0058	21 33 48.	- 00 46		18.0	BLUE STELLAR OBJECT
PHL 1641	21 33 48.	- 03 20		10.9	BLUE STELLAR OBJECT
PHL 4655	21 33 48.	- 03 28		18.1	BLUE STELLAR OBJECT
PHL 1642	21 33 48.	- 04 28		18.5	BLUE STELLAR OBJECT
PHL 4654	21 33 48.	- 05 35		17.5	BLUE STELLAR OBJECT
ARC 2357	21 33 50.	- 23 29		17.0	RICH CLUSTER OF GALAXIES
SCHO 1299	21 33 52.	+ 50 01 06.	300		ISOLATED DARK CLOUD
ZWG 427.006	21 33 54.	+ 12 01		15.6	GALAXY
UGC 11776	21 33 54.	+ 12 01	66	15.6	GALAXY Sc
MCG+03-55-001	21 33 54.	+ 16 50	42	16.	GALAXY
UGC 11777	21 33 54.	+ 27 53	84	17.	GALAXY DWRF SP
PHL 4656	21 33 54.	- 05 21		18.2	BLUE STELLAR OBJECT
PHL 4657	21 33 54.	- 08 26		18.6	BLUE STELLAR OBJECT
PHL 4658	21 33 54.	- 09 30		18.0	BLUE STELLAR OBJECT
SCHO 1301	21 33 57.	+ 50 09 24.	230		ISOLATED DARK CLOUD
SCHO 1300	21 33 57.	+ 50 15 54.	210		ISOLATED DARK CLOUD
SCHO 1302	21 33 59.	+ 48 19 06.	300		ISOLATED DARK CLOUD
LBN 0374	21 34	+ 37 50	3720		BRIGHT NEBULA
LBN 0407	21 34	+ 44 00	3960		BRIGHT NEBULA
LBN 0421	21 34	+ 50 00	1260		BRIGHT NEBULA
LBN 0452	21 34	+ 57 10	480		BRIGHT NEBULA
KHAV 742	21 34	+ 57 13	1000		DARK NEBULA
PHL 0059	21 34 00.	+ 01 54		18.0	BLUE STELLAR OBJECT
UGC 11778	21 34 00.	+ 05 31	66	16.5	GALAXY S
MCG+02-55-003	21 34 00.	+ 12 00	72	15.	GALAXY
2ZW 141	21 34 00.	+ 12 37			COMPACT GALAXY
ACK 094-00.1	21 34 00.	+ 50 41			PLANETARY NEBULA
LDN 1087	21 34 00.	+ 56 20	420		DARK NEBULA
LDN 1090	21 34 00.	+ 56 30	660		DARK NEBULA
PHL 4660	21 34 00.	- 03 22		17.9	BLUE STELLAR OBJECT
PHL 4659	21 34 00.	- 09 48		17.8	BLUE STELLAR OBJECT
PHL 4661	21 34 00.	- 11 38		17.9	BLUE STELLAR OBJECT
PHL 1643	21 34 00.	- 11 42		18.7	BLUE STELLAR OBJECT
PHL 0060	21 34 00.	- 12 32		18.8	BLUE STELLAR OBJECT
MCG-04-51-001	21 34 00.	- 23 02	60	15.	GALAXY
BC PKS2134+004	21 34 04.	+ 00 28 12.		17.	QUASI-STELLAR OBJECT
SHB 360	21 34 04.	+ 00 28 12.		18.	QUASI-STELLAR OBJECT
PHL 0061	21 34 06.	+ 00 28		17.4	BLUE STELLAR OBJECT
UGC 11779	21 34 06.	+ 70 34	72	16.5	GALAXY Sc-IRR
PHL 4662	21 34 06.	- 02 26		18.3	BLUE STELLAR OBJECT
PHL 1645	21 34 06.	- 02 45		18.5	BLUE STELLAR OBJECT
PHL 1644	21 34 06.	- 05 50		19.4	BLUE STELLAR OBJECT
PHL 4663	21 34 06.	- 06 46		18.4	BLUE STELLAR OBJECT
PHL 0062	21 34 06.	- 21 30		18.2	BLUE STELLAR OBJECT
SCHO 1303	21 34 07.	+ 56 26 24.	540		ISOLATED DARK CLOUD
HN 2692	21 34 08.1	- 52 10 09.	12		NEBULA
2ZW 142	21 34 12.	+ 01 02			COMPACT GALAXY

OBJECT NAME	RIGHT ASCEN.	DECLINATION	DIAM.	MAGN.	TYPE OF OBJECT
ZWG 376.013	21 34 12.	+ 01 02		15.6	GALAXY
PHL 0063	21 34 12.	+ 01 31		17.1	BLUE STELLAR OBJECT
PHL 4664	21 34 12.	- 03 06		16.5	BLUE STELLAR OBJECT
PHL 1646	21 34 12.	- 06 28		16.5	BLUE STELLAR OBJECT
PHL 1647	21 34 12.	- 10 41		18.9	BLUE STELLAR OBJECT
ZC 2134.3+1007	21 34 18.	+ 10 07	1080		CLUSTER OF GALAXIES
PHL 4665	21 34 18.	- 02 18		17.5	BLUE STELLAR OBJECT
PHL 1650	21 34 18.	- 02 47		17.9	BLUE STELLAR OBJECT
PHL 1649	21 34 18.	- 06 31		14.0	BLUE STELLAR OBJECT
PHL 1648	21 34 18.	- 19 41		4.6	BLUE STELLAR OBJECT
YC 2134-56	21 34 19.	- 56 27 12.			UNUSUAL SOUTHERN GALAXY
ARC 2358	21 34 20.	- 16 07		17.5	RICH CLUSTER OF GALAXIES
HN 2693	21 34 22.6	- 51 27 44.	12		NEBULA
MCG+00-55-008	21 34 24.	+ 00 12 30.	36	15.5	GALAXY
PHL 1651	21 34 24.	+ 01 34		18.0	BLUE STELLAR OBJECT
PHL 4666	21 34 24.	- 00 48		18.8	BLUE STELLAR OBJECT
PHL 1653	21 34 24.	- 09 10		17.8	BLUE STELLAR OBJECT
PHL 4667	21 34 24.	- 14 46		18.3	BLUE STELLAR OBJECT
PHL 1652	21 34 24.	- 15 48		18.3	BLUE STELLAR OBJECT
ARC 2360	21 34 25.	- 15 18		17.7	RICH CLUSTER OF GALAXIES
PK066-28.1	21 34 27.9	+ 12 33 49.	99		PLANETARY NEBULA
ZWG 376.015	21 34 30.	+ 00 13		15.6	GALAXY
UGC 11780	21 34 30.	+ 00 13	60	15.6	GALAXY S
RNGC 7094	21 34 30.	+ 12 34			PLANETARY NEBULA
MCG+06-47-005	21 34 30.	+ 35 23	66	14.5	GALAXY
ZWG 376.014	21 34 30.	- 00 48		15.4	GALAXY
MCG-05-51-004	21 34 30.	- 29 43	18	15.5	GALAXY
MCG+00-55-009	21 34 36.	+ 00 13	12	15.5	GALAXY
ZWG 376.016	21 34 36.	+ 00 14		15.2	GALAXY
PHL 4676	21 34 36.	+ 00 32		18.1	BLUE STELLAR OBJECT
PHL 4675	21 34 36.	+ 01 01		18.4	BLUE STELLAR OBJECT
PHL 4674	21 34 36.	+ 02 12		18.7	BLUE STELLAR OBJECT
PHL 4669	21 34 36.	+ 02 24		18.0	BLUE STELLAR OBJECT
ZWG 512.008	21 34 36.	+ 35 28		13.7	GALAXY
UGC 11781	21 34 36.	+ 35 23	84	13.7	GALAXY SO
ZC 2134.6+4253	21 34 36.	+ 42 53	10890		CLUSTER OF GALAXIES
ISS 0123	21 34 36.	+ 55 14	124		STELLAR RING
SCHO 1304	21 34 36.	+ 57 36 48.	680		ISOLATED DARK CLOUD
PHL 4677	21 34 36.	- 04 44		18.2	BLUE STELLAR OBJECT
PHL 4671	21 34 36.	- 07 59		17.5	BLUE STELLAR OBJECT
PHL 0064	21 34 36.	- 13 40		16.6	BLUE STELLAR OBJECT
PHL 4672	21 34 36.	- 14 56		17.8	BLUE STELLAR OBJECT
PHL 4673	21 34 36.	- 15 16		17.8	BLUE STELLAR OBJECT
PHL 4670	21 34 36.	- 15 46		18.3	BLUE STELLAR OBJECT
PHL 4678	21 34 36.	- 16 34		18.1	BLUE STELLAR OBJECT
PHL 4668	21 34 36.	- 17 20		17.2	BLUE STELLAR OBJECT
HN 2695	21 34 37.5	- 52 01 55.	12		NEBULA
BB 6.09	21 34 41.65	- 14 47 36.5			GALAXY NEAR QSO PKS2135
HN 2694	21 34 41.8	- 54 22 07.	12		NEBULA
HN 2700	21 34 47.0	- 15 58 42.	30		NEBULA
ZWG 376.017	21 34 48.	+ 00 11		15.5	GALAXY
PHL 1654	21 34 48.	+ 01 48		17.1	BLUE STELLAR OBJECT
ISS 0124	21 34 48.	+ 53 00	149		STELLAR RING
PHL 4679	21 34 48.	- 02 34		17.0	BLUE STELLAR OBJECT
PHL 4681	21 34 48.	- 04 53		18.2	BLUE STELLAR OBJECT
PHL 4684	21 34 49.	- 06 26		17.0	BLUE STELLAR OBJECT
PHL 4683	21 34 48.	- 06 48		16.8	BLUE STELLAR OBJECT
PHL 4680	21 34 48.	- 08 55		17.9	BLUE STELLAR OBJECT
PHL 1655	21 34 48.	- 15 04		17.9	BLUE STELLAR OBJECT
PHL 4682	21 34 48.	- 19 27		7.1	BLUE STELLAR OBJECT
MCG-04-51-002	21 34 48.	- 25 10 30.	42	15.5	GALAXY
BB 6.12	21 34 50.7	- 14 56 10.5			GALAXY NEAR QSO PKS2135
HN 2696	21 34 50.9	- 52 02 19.	18		NEBULA
HN 1220	21 34 52.	- 64 35			NEBULA
IC 5120	21 34 52.	- 64 35			NONSTELLAR OBJECT
PHL 4685	21 34 54.	+ 01 30		17.9	BLUE STELLAR OBJECT
PHL 1656	21 34 54.	- 10 49		18.6	BLUE STELLAR OBJECT
HN 2701	21 34 55.2	- 16 17 12.	18		NEBULA
SCHO 1305	21 34 58.	+ 53 48 42.	400		ISOLATED DARK CLOUD
BB 6.10	21 34 59.0	- 14 57 28.			GALAXY NEAR QSO PKS2135
LBN 0152	21 35	+ 12 30	6120		BRIGHT NEBULA
KHAV 743	21 35	+ 57 31	1000		DARK NEBULA
ZWG 376.018	21 35 00.	+ 00 23		15.6	GALAXY
PHL 4686	21 35 00.	+ 00 28		15.9	BLUE STELLAR OBJECT
ZWG 450.003	21 35 00.	+ 19 23		15.7	GALAXY
MRSL 087-08/1	21 35 00.	+ 39 59	28800		HII REGION
LDN 1078	21 35 00.	+ 53 50	540		DARK NEBULA
LDN 1081	21 35 00.	+ 54 10	3060		DARK NEBULA
LDN 1099	21 35 00.	+ 57 10	360		DARK NEBULA
LDN 1116	21 35 00.	+ 58 20	720		DARK NEBULA
LDN 1199	21 35 00.	+ 68 20	1920		DARK NEBULA
ZC 2135.0+8302	21 35 00.	+ 83 02	470		CLUSTER OF GALAXIES
PHL 4688	21 35 00.	- 07 18		17.7	BLUE STELLAR OBJECT
PHL 1658	21 35 00.	- 11 25		18.3	BLUE STELLAR OBJECT
PHL 1657	21 35 00.	- 14 46		16.1	BLUE STELLAR OBJECT
PHL 4687	21 35 00.	- 24 58		18.2	BLUE STELLAR OBJECT
MCG-07-44-027	21 35 00.	- 44 09	60	17.	GALAXY
DV.56 N7096A	21 35 00.	- 64 35			GALAXY
RNGC 7095	21 35 01.	- 42 50			GALAXY
SHB 361	21 35 01.1	- 14 46 27.		15.5	QUASI-STELLAR OBJECT
HN 2697	21 35 01.2	- 52 02 06.	12		NEBULA
BC PKS2135-14	21 35 01.21	- 14 46 27.3		15.53	QUASI-STELLAR OBJECT
RNGC 7096A	21 35 02.	- 64 35			GALAXY
BB 6.04	21 35 02.25	- 14 46 36.			GALAXY NEAR QSO PKS2135
BB 6.11	21 35 02.4	- 14 55 56.			GALAXY NEAR QSO PKS2135
BB 6.05	21 35 03.8	- 14 43 57.			GALAXY NEAR QSO PKS2135
BB 6.06	21 35 03.9	- 14 40 30.			GALAXY NEAR QSO PKS2135
ZWG 427.007	21 35 06.	+ 14 22		15.1	GALAXY
4ZW 073	21 35 06.	+ 27 25			COMPACT GALAXY
PHL 4689	21 35 06.	- 04 14		18.5	BLUE STELLAR OBJECT
PHL 1660	21 35 06.	- 09 15		18.8	BLUE STELLAR OBJECT
PHL 1659	21 35 06.	- 13 17		18.7	BLUE STELLAR OBJECT
BB 6.03	21 35 06.1	- 14 47 39.			GALAXY NEAR QSO PKS2135
BB 6.02	21 35 06.7	- 14 48 15.5			GALAXY NEAR QSO PKS2135
BB 6.07	21 35 10.4	- 14 39 05.5			GALAXY NEAR QSO PKS2135
PHL 0065	21 35 12.	+ 01 56		17.0	BLUE STELLAR OBJECT
PHL 1662	21 35 12.	- 00 31		18.2	BLUE STELLAR OBJECT
PHL 1663	21 35 12.	- 01 30		16.4	BLUE STELLAR OBJECT
PHL 0066	21 35 12.	- 01 55		17.8	BLUE STELLAR OBJECT
PHL 0067	21 35 12.	- 02 43		18.2	BLUE STELLAR OBJECT
PHL 0068	21 35 12.	- 05 06		18.1	BLUE STELLAR OBJECT
PHL 4692	21 35 12.	- 06 02		18.2	BLUE STELLAR OBJECT
PHL 4693	21 35 12.	- 06 13		18.0	BLUE STELLAR OBJECT
PHL 4691	21 35 12.	- 07 11		17.8	BLUE STELLAR OBJECT
PHL 4690	21 35 12.	- 07 49		16.8	BLUE STELLAR OBJECT
PHL 1661	21 35 12.	- 13 20		17.9	BLUE STELLAR OBJECT
PHL 0069	21 35 12.	- 15 08		18.3	BLUE STELLAR OBJECT
PK093-02.1	21 35 12.6	+ 48 42 37.	56		PLANETARY NEBULA

OBJECT NAME	RIGHT ASCEN.	DECLINATION	DIAM.	MAGN.	TYPE OF OBJECT
HN 2698	21 35 16.9	- 53 48 18.	6		NEBULA
B 158	21 35 17.	+ 43 11	180		DARK OBJECT
BB 6.01	21 35 17.05	- 14 47 24.5			GALAXY NEAR QSO PKS2135
ZC 2135.3+0948	21 35 18.	+ 09 48	1410		CLUSTER OF GALAXIES
PHL 1664	21 35 18.	- 16 06		16.6	BLUE STELLAR OBJECT
HN 2699	21 35 19.9	- 54 18 41.	12		NEBULA
PHL 4695	21 35 24.	- 03 32		16.0	BLUE STELLAR OBJECT
PHL 4694	21 35 24.	- 11 41		17.3	BLUE STELLAR OBJECT
PHL 1665	21 35 24.	- 20 14		16.4	BLUE STELLAR OBJECT
PHL 4696	21 35 24.	- 20 20		17.6	BLUE STELLAR OBJECT
PHL 0070	21 35 24.	- 22 49		17.8	BLUE STELLAR OBJECT
BB 6.08	21 35 28.05	- 14 41 22.			GALAXY NEAR QSO PKS2135
LDN 0973	21 35 30.	+ 43 00	660		DARK NEBULA
VDB.66N 142	21 35 33.	+ 57 15	72		REFLECTION NEBULA
PHL 4697	21 35 36.	+ 01 36		18.3	BLUE STELLAR OBJECT
ZWG 427.008	21 35 36.	+ 14 20		15.6	GALAXY
LDN 1105	21 35 36.	+ 57 20	360		DARK NEBULA
PHL 4698	21 35 36.	- 05 50		18.7	BLUE STELLAR OBJECT
PHL 1698	21 35 36.	- 06 44		18.4	BLUE STELLAR OBJECT
PHL 0071	21 35 36.	- 07 55		17.1	BLUE STELLAR OBJECT
PHL 0072	21 35 36.	- 09 54		18.6	BLUE STELLAR OBJECT
PHL 0073	21 35 36.	- 11 03		17.7	BLUE STELLAR OBJECT
PHL 1666	21 35 36.	- 12 16		18.2	BLUE STELLAR OBJECT
PHL 0074	21 35 36.	- 20 08		18.4	BLUE STELLAR OBJECT
PHL 0075	21 35 36.	- 22 32		17.9	BLUE STELLAR OBJECT
SCHO 1306	21 35 40.	+ 61 50 30.	410		ISOLATED DARK CLOUD
HN 2704	21 35 40.6	- 49 50 16.	18		NEBULA
ZWG 402.009	21 35 42.	+ 06 31		15.7	GALAXY
ZWG 402.010	21 35 42.	+ 08 45		15.2	GALAXY
UGC 11782	21 35 42.	+ 08 45	150	15.3	GALAXY
PHL 4699	21 35 42.	- 02 52		18.0	BLUE STELLAR OBJECT
PHL 0076	21 35 42.	- 05 06		18.6	BLUE STELLAR OBJECT
HN 2702	21 35 42.4	- 50 56 52.			NEBULA
HN 2716	21 35 43.8	- 14 23 03.	24		NEBULA
HN 2705	21 35 44.5	- 51 54 46.			NEBULA
HN 0722	21 35 46.	- 71 37			NEBULA
IC 5[illegible]	21 35 46.	- 71 37			NONSTELLAR OBJECT
HN 27[illegible]	21 35 47.3	- 51 17 40.	12		NEBULA
HN 27[illegible]	21 35 47.9	- 53 52 10.	18		NEBULA
PHL 0[illegible]	21 35 48.	+ 00 59		17.4	BLUE STELLAR OBJECT
MCG+01[illegible]-006	21 35 48.	+ 08 46	120	14.	GALAXY
UGC 11[illegible]	21 35 48.	+ 30 58	78	16.0	GALAXY Sa-b
DG 175	21 35 48.	+ 67 57	540		REFLECTION NEBULA
PHL 470[illegible]	21 35 48.	- 01 32		16.5	BLUE STELLAR OBJECT
PHL 166[illegible]	21 35 48.	- 03 06		17.9	BLUE STELLAR OBJECT
PHL 0078	21 35 48.	- 03 06		18.8	BLUE STELLAR OBJECT
PHL 4702	21 35 48.	- 06 21		17.0	BLUE STELLAR OBJECT
PHL 0079	21 35 48.	- 13 24		14.2	BLUE STELLAR OBJECT
PHL 4701	21 35 48.	- 22 44		18.2	BLUE STELLAR OBJECT
HN 2719	21 35 48.5	- 14 23 51.	24		NEBULA
HN 2718	21 35 50.2	- 17 21 15.	36		NEBULA
HN 2706	21 35 50.6	- 53 32 58.	12		NEBULA
MCG-03-55-004	21 35 51.	- 17 22	36	15.	GALAXY
ZWG 402.011	21 35 54.	+ 08 35		15.7	GALAXY
PHL 0080	21 35 54.	- 03 12		17.8	BLUE STELLAR OBJECT
MCG-05-51-005	21 35 54.	- 28 15	36	15.	GALAXY
HN 2708	21 35 55.5	- 51 10 40.	12		NEBULA
CED 194	21 35 57.	+ 67 57	480		DIFFUSE GALACTIC NEBULA
LBN 0287	21 36	+ 29 50	2400		BRIGHT NEBULA
LBN 0395	21 36	+ 39 50	120		BRIGHT NEBULA
LBN 0396	21 36	+ 39 55	120		BRIGHT NEBULA
LBN 0399	21 36	+ 40 08	120		BRIGHT NEBULA
KHAV 744	21 36	+ 43 20	4960		DARK NEBULA
LBN 0423	21 36	+ 50 08	660		BRIGHT NEBULA
KHAV 745	21 36	+ 55 56	1860		DARK NEBULA
LBN 0504	21 36	+ 67 55	120		BRIGHT NEBULA
PHL 4703	21 36 00.	+ 01 57		18.0	BLUE STELLAR OBJECT
LDN 1077	21 36 00.	+ 53 30	180		DARK NEBULA
LDN 1092	21 36 00.	+ 56 45	480		DARK NEBULA
PHL 1669	21 36 00.	- 03 00		18.9	BLUE STELLAR OBJECT
PHL 4704	21 36 00.	- 16 17		18.2	BLUE STELLAR OBJECT
PHL 1668	21 36 00.	- 23 07		17.5	BLUE STELLAR OBJECT
HN 2709	21 36 02.0	- 50 56 40.			NEBULA
SCHO 1307	21 36 04.	+ 56 02 06.	1560		ISOLATED DARK CLOUD
PHL 0081	21 36 06.	- 00 48		17.1	BLUE STELLAR OBJECT
PHL 1670	21 36 06.	- 04 49		16.6	BLUE STELLAR OBJECT
PHL 4705	21 36 06.	- 07 16		18.7	BLUE STELLAR OBJECT
PHL 1671	21 36 06.	- 07 24		13.2	BLUE STELLAR OBJECT
MCG-04-51-003	21 36 06.	- 22 26 30.	15	15.5	GALAXY
HN 2710	21 36 10.4	- 51 37 21.	12		NEBULA
PHL 4708	21 36 12.	+ 00 07		18.5	BLUE STELLAR OBJECT
PHL 4712	21 36 12.	+ 00 10		18.4	BLUE STELLAR OBJECT
LB 00366	21 36 12.	+ 00 54 36.		15.5	FAINT BLUE STAR
PHL 4706	21 36 12.	+ 01 31		17.9	BLUE STELLAR OBJECT
PHL 1672	21 36 12.	+ 01 43		18.0	BLUE STELLAR OBJECT
PHL 4710	21 36 12.	- 04 00		16.6	BLUE STELLAR OBJECT
PHL 4711	21 36 12.	- 05 06		17.4	BLUE STELLAR OBJECT
PHL 4709	21 36 12.	- 07 19		18.0	BLUE STELLAR OBJECT
PHL 4707	21 36 12.	- 07 19		17.6	BLUE STELLAR OBJECT
PHL 0082	21 36 12.	- 09 04		17.9	BLUE STELLAR OBJECT
PHL 0083	21 36 12.	- 09 57		17.5	BLUE STELLAR OBJECT
HN 2723	21 36 13.3	- 16 01 56.			NEBULA
HN 2711	21 36 13.4	- 50 54 45.	18		NEBULA
VDB.66N 143	21 36 14.	+ 67 58	576		REFLECTION NEBULA
MCG-04-51-004	21 36 15.	- 23 00 30.	42	14.5	GALAXY
HN 2712	21 36 15.2	- 51 00 03.			NEBULA
HN 2713	21 36 17.3	- 52 11 09.	6		NEBULA
PHL 4713	21 36 18.	- 01 20		16.2	BLUE STELLAR OBJECT
PHL 4714	21 36 18.	- 01 35		18.1	BLUE STELLAR OBJECT
MCG-05-51-006	21 36 18.	- 30 06	18	15.5	GALAXY
HN 2715	21 36 22.7	- 52 00 39.	12		NEBULA
B 160	21 36 23.	+ 56 00	1860		DARK OBJECT
HN 2714	21 36 23.2	- 53 49 03.	12		NEBULA
PHL 4715	21 36 24.	+ 00 23		16.8	BLUE STELLAR OBJECT
ZC 2136.4+1210	21 36 24.	+ 12 10	1280		CLUSTER OF GALAXIES
ZC 2136.4+1536	21 36 24.	+ 15 36	1010		CLUSTER OF GALAXIES
ZWG 493.002	21 36 24.	+ 31 52		15.6	GALAXY
PHL 4716	21 36 24.	- 02 38		16.8	BLUE STELLAR OBJECT
PHL 4717	21 36 24.	- 08 17		16.6	BLUE STELLAR OBJECT
PHL 1674	21 36 24.	- 10 17		17.9	BLUE STELLAR OBJECT
PHL 1673	21 36 24.	- 12 52		18.5	BLUE STELLAR OBJECT
ARC 2361	21 36 24.	- 14 33		16.7	RICH CLUSTER OF GALAXIES
MCG-05-51-007	21 36 24.	- 30 08	18	15.5	GALAXY
B 159	21 36 25.	+ 43 00	1500		DARK OBJECT
HN 2726	21 36 25.2	- 15 38 38.	18		NEBULA
HN 2717	21 36 25.7	- 52 10 03.	12		NEBULA
ZWG 493.003	21 36 30.	+ 29 37		15.5	GALAXY
TON-S 0022	21 36 30.	- 32 57		15.9	BLUE STAR

OBJECT NAME	RIGHT ASCEN.	DECLINATION	DIAM.	MAGN.	TYPE OF OBJECT
HN 2720	21 36 30.6	- 50 10 26.	12		NEBULA
ZWG 472.003	21 36 36.	+ 24 35		15.7	GALAXY
ISS 0185	21 36 36.	+ 45 28	259		STELLAR RING
BRSL 094-01/1	21 36 36.	+ 50 07	4200		HII REGION
PHL 4721	21 36 36.	- 03 06		18.0	BLUE STELLAR OBJECT
PHL 4720	21 36 36.	- 04 11		18.0	BLUE STELLAR OBJECT
PHL 0085	21 36 36.	- 10 42		16.6	BLUE STELLAR OBJECT
PHL 0084	21 36 36.	- 12 12		17.1	BLUE STELLAR OBJECT
PHL 4718	21 36 36.	- 14 29		17.9	BLUE STELLAR OBJECT
PHL 4719	21 36 36.	- 21 51		16.5	BLUE STELLAR OBJECT
HN 2721	21 36 37.1	- 52 02 02.	12		NEBULA
BC OX161	21 36 37.37	+ 14 10 00.4		18.5	QUASI-STELLAR OBJECT
SHB 362	21 36 37.4	+ 14 10 00.		18.5	QUASI-STELLAR OBJECT
SCHO 1308	21 36 38.	+ 42 55 54.	670		ISOLATED DARK CLOUD
ZWG 376.019	21 36 42.	- 00 43		15.4	GALAXY
HN 2722	21 36 43.2	- 51 47 38.	30		NEBULA
PHL 4722	21 36 48.	+ 01 53		18.1	BLUE STELLAR OBJECT
ZWG 427.009	21 36 48.	+ 10 06		15.7	GALAXY
ZC 2136.8+1738	21 36 48.	+ 17 38	870		CLUSTER OF GALAXIES
UGC 11784	21 36 48.	+ 38 17	60	17.	GALAXY
PHL 0086	21 36 48.	- 02 42		12.9	BLUE STELLAR OBJECT
PHL 0087	21 36 48.	- 03 15		18.0	BLUE STELLAR OBJECT
PHL 4723	21 36 48.	- 06 15		17.7	BLUE STELLAR OBJECT
PHL 0088	21 36 48.	- 11 08		17.1	BLUE STELLAR OBJECT
PHL 1675	21 36 48.	- 19 02		15.9	BLUE STELLAR OBJECT
MCG-07-44-028	21 36 48.	- 43 06	36	15.	GALAXY
HN 2729	21 36 48.7	- 16 22 19.	24		NEBULA
SG 2.085	21 36 53.	+ 50 09	720		DIFFUSE EMISSION NEBULA
ZWG 376.020	21 36 54.	+ 02 36		15.4	GALAXY
UGC 11785	21 36 54.	+ 02 36	84	15.4	GALAXY Sc
KARA.73B 0922	21 36 54.	+ 02 36	132	15.4	ISOLATED GALAXY S
MCG+02-55-004	21 36 54.	+ 10 06	48	15.5	GALAXY
ACK 091-04.1	21 36 54.	+ 45 47			PLANETARY NEBULA
PHL 0089	21 36 54.	- 10 01		18.2	BLUE STELLAR OBJECT
PHL 1676	21 36 54.	- 18 54		14.1	BLUE STELLAR OBJECT
HN 2731	21 36 54.2	- 17 02 06.			NEBULA
IC 5122	21 36 57.	- 22 37 54.			NONSTELLAR OBJECT
RNGC 7101	21 36 58.	+ 08 40			GALAXY
SCHO 1309	21 36 59.	+ 46 55 30.	600		ISOLATED DARK CLOUD
LBN 0292	21 37	+ 29 20	2700		BRIGHT NEBULA
LBN 0397	21 37	+ 39 48	120		BRIGHT NEBULA
LBN 0398	21 37	+ 39 50	120		BRIGHT NEBULA
LDN 0977	21 37 00.	+ 43 00	720		DARK NEBULA
LDN 1010	21 37 00.	+ 46 50	1680		DARK NEBULA
LDN 1038	21 37 00.	+ 56 00	540		DARK NEBULA
COU 034	21 37 00.	+ 57 00	14400		EMISSION NEBULA
IC 1396	21 37 00.	+ 57 00	7800		OPEN CLUSTER
LDN 1112	21 37 00.	+ 57 50	420		DARK NEBULA
LDN 1117	21 37 00.	+ 58 05	240		DARK NEBULA
PHL 4724	21 37 00.	- 21 32		17.3	BLUE STELLAR OBJECT
HN 2724	21 37 01.4	- 53 47 01.	12		NEBULA
HN 2725	21 37 01.9	- 51 59 37.	12		NEBULA
SCHO 1310	21 37 02.	+ 58 04 48.	440		ISOLATED DARK CLOUD
RNGC 7100	21 37 04.	+ 08 39		14.5	GALAXY
RNGC 7103	21 37 05.	- 22 42		15.0	GALAXY
HN 2732	21 37 05.4	- 14 53 30.	18		NEBULA
ZWG 402.012	21 37 06.	+ 08 39		14.7	GALAXY
PHL 0090	21 37 06.	- 00 24		17.4	BLUE STELLAR OBJECT
PHL 4725	21 37 06.	- 07 15		18.2	BLUE STELLAR OBJECT
PHL 0091	21 37 06.	- 13 18		18.4	BLUE STELLAR OBJECT
IC 5124	21 37 06.	- 22 39 11.			NONSTELLAR OBJECT
MCG-04-51-006	21 37 06.	- 22 42	24	15.	GALAXY
MCG-04-51-005	21 37 06.	- 25 30	48	15.	GALAXY
MCG-07-44-029	21 37 06.	- 42 46	72	12.6	GALAXY
RNGC 7097	21 37 07.	- 42 46		13.0	GALAXY
RNGC 7102	21 37 09.	+ 06 04		14.5	GALAXY
HN 2733	21 37 11.4	- 15 06 42.	24		NEBULA
PHL 4727	21 37 12.	+ 01 10		15.9	BLUE STELLAR OBJECT
ZWG 402.013	21 37 12.	+ 06 04		14.4	GALAXY
UGC 11786	21 37 12.	+ 06 04	102	14.4	GALAXY SBb
MCG+01-55-007	21 37 12.	+ 08 40	36	15.	GALAXY
PHL 4726	21 37 12.	- 05 00		17.9	BLUE STELLAR OBJECT
PHL 1679	21 37 12.	- 06 18		17.3	BLUE STELLAR OBJECT
PHL 4729	21 37 12.	- 07 11		16.0	BLUE STELLAR OBJECT
PHL 4730	21 37 12.	- 07 38		18.8	BLUE STELLAR OBJECT
PHL 0092	21 37 12.	- 07 56		16.6	BLUE STELLAR OBJECT
PHL 1678	21 37 12.	- 10 45		18.7	BLUE STELLAR OBJECT
PHL 1677	21 37 12.	- 13 00		16.8	BLUE STELLAR OBJECT
PHL 4728	21 37 12.	- 14 31		17.0	BLUE STELLAR OBJECT
MCG-04-51-007	21 37 12.	- 22 56	24	15.5	GALAXY
HN 2727	21 37 14.9	- 50 53 36.	12		NEBULA
MCG+01-55-008	21 37 15.	+ 06 03	96	13.	GALAXY
HN 2728	21 37 15.2	- 50 01 48.	18		NEBULA
IC 5127	21 37 16.	+ 06 00			SAME AS NGC 7102
SCHO 1311	21 37 17.	+ 47 30 30.	370		ISOLATED DARK CLOUD
RNGC 7104	21 37 17.	- 22 39		15.0	GALAXY
ZWG 427.010	21 37 18.	+ 10 06		15.4	GALAXY
2ZW 143	21 37 18.	+ 10 07			COMPACT GALAXY
PHL 1680	21 37 18.	- 01 52		18.0	BLUE STELLAR OBJECT
PHL 4732	21 37 18.	- 03 28		17.0	BLUE STELLAR OBJECT
PHL 4731	21 37 18.	- 04 36		17.9	BLUE STELLAR OBJECT
PHL 0093	21 37 18.	- 08 04		16.0	BLUE STELLAR OBJECT
PHL 1681	21 37 18.	- 17 48		18.6	BLUE STELLAR OBJECT
PHL 4733	21 37 18.	- 18 16		18.6	BLUE STELLAR OBJECT
MCG-04-51-008	21 37 18.	- 22 39 30.	24	15.5	GALAXY
SCHO 1312	21 37 19.	+ 57 49 36.	260		ISOLATED DARK CLOUD
RNGC 7096	21 37 21.	- 64 08		13.0	GALAXY
IC 1393	21 37 23.	- 22 38 28.			NONSTELLAR OBJECT
2ZW 144	21 37 24.	+ 05 08			COMPACT GALAXY
MCG+02-55-005	21 37 24.	+ 10 06	36	15.	GALAXY
PHL 1683	21 37 24.	- 01 20		17.8	BLUE STELLAR OBJECT
PHL 4734	21 37 24.	- 07 50		17.5	BLUE STELLAR OBJECT
PHL 4735	21 37 24.	- 08 04		18.0	BLUE STELLAR OBJECT
PHL 1682	21 37 24.	- 09 44		17.9	BLUE STELLAR OBJECT
RNGC 7097A	21 37 25.	- 42 42			GALAXY
CED 195A	21 37 27.	+ 57 14	10800		DIFFUSE GALACTIC NEBULA
HN 1221	21 37 27.	- 64 38		13.	NEBULA
IC 5121	21 37 27.	- 64 38			NONSTELLAR OBJECT
IC 5126	21 37 28.	- 05 35 45.			NONSTELLAR OBJECT
RNGC 7099	21 37 29.	- 23 25		8.5	GLOBULAR CLUSTER
ZWG 376.021	21 37 30.	+ 00 59		15.0	GALAXY
PHL 4736	21 37 30.	+ 01 16		17.7	BLUE STELLAR OBJECT
LDN 0979	21 37 30.	+ 43 08	180		DARK NEBULA
OCL 0222	21 37 30.	+ 57 16	6900	5.1	OPEN STAR CLUSTER
BRSL 099+03/1	21 37 30.	+ 57 16	10200		HII REGION
PHL 1684	21 37 30.	- 04 12		18.9	BLUE STELLAR OBJECT
MCG-04-51-009	21 37 30.	- 22 38 30.	24	15.5	GALAXY
MCG-07-44-030	21 37 30.	- 42 43	12	15.5	GALAXY

OBJECT NAME	RIGHT ASCEN.	DECLINATION	DIAM.	MAGN.	TYPE OF OBJECT
HN 2730	21 37 30.5	- 51 55 06.			NEBULA
REIN 2.294	21 37 31.54	- 23 24 30.4			NEBULA
HN 2738	21 37 32.9	- 14 27 53.	24		NEBULA
SG 1.19	21 37 33.	+ 57 14	10800		DIFFUSE EMISSION NEBULA
PK098+02.1	21 37 34.8	+ 55 32 29.	8		PLANETARY NEBULA
PHL 4737	21 37 36.	+ 02 10		17.8	BLUE STELLAR OBJECT
2ZW 145	21 37 36.	+ 12 33			COMPACT GALAXY
ZWG 472.004	21 37 36.	+ 24 25		15.7	GALAXY
PHL 0094	21 37 36.	- 04 56		18.2	BLUE STELLAR OBJECT
PHL 1685	21 37 36.	- 07 01		18.3	BLUE STELLAR OBJECT
PHL 1686	21 37 36.	- 09 14		18.6	BLUE STELLAR OBJECT
PHL 0095	21 37 36.	- 13 22		18.2	BLUE STELLAR OBJECT
PHL 1687	21 37 36.	- 15 10		18.5	BLUE STELLAR OBJECT
GCL 122	21 37 36.	- 23 25	534	8.58	GLOBULAR STAR CLUSTER
MCG-05-51-008	21 37 36.	- 26 47	150	14.	GALAXY
ZC 2137.7+0209	21 37 42.	+ 02 09	870		CLUSTER OF GALAXIES
PHL 4738	21 37 42.	- 00 40		16.8	BLUE STELLAR OBJECT
IC 1394	21 37 43.	+ 14 24 42.			NONSTELLAR OBJECT
SCHO 1313	21 37 46.	+ 49 56 18.	250		ISOLATED DARK CLOUD
ZWG 427.011	21 37 48.	+ 14 24		15.3	GALAXY
MCG+02-55-006	21 37 48.	+ 14 24	48	15.	GALAXY
ZWG 472.005	21 37 48.	+ 24 49		15.7	GALAXY
UGC 11787	21 37 48.	+ 24 49	60	15.7	GALAXY Sa
MCG+04-51-002	21 37 48.	+ 24 49	30	15.5	GALAXY
SCHO 1314	21 37 49.	+ 58 24 00.	830		ISOLATED DARK CLOUD
PHL 4740	21 37 48.	- 06 06		18.2	BLUE STELLAR OBJECT
PHL 4739	21 37 48.	- 17 50		3.9	BLUE STELLAR OBJECT
RNGC 7105	21 37 52.	- 10 34			NON-EXISTENT OBJECT
HN 2734	21 37 52.9	- 51 59 17.	18		NEBULA
ZWG 427.012	21 37 54.	+ 12 07		15.7	GALAXY
MCG+02-55-007	21 37 54.	+ 12 07	42	16.	GALAXY
KARA.73B 0923	21 37 54.	+ 12 07	168	15.7	ISOLATED GALAXY S
SCHO 1315	21 37 54.	+ 51 23 30.	200		ISOLATED DARK CLOUD
PHL 1688	21 37 54.	- 02 47		18.0	BLUE STELLAR OBJECT
PHL 1689	21 37 54.	- 11 50		18.0	BLUE STELLAR OBJECT
HN 2735	21 37 55.7	- 51 58 29.			NEBULA
HN 2737	21 37 59.5	- 51 07 28.	12		NEBULA
LBN 0455	21 38	+ 57 30	8400		BRIGHT NEBULA
LBN 0456	21 38	+ 58 05	360		BRIGHT NEBULA
KHAV 746	21 38	+ 59 32	3610		DARK NEBULA
LDN 1043	21 38 00.	+ 49 00	4320		DARK NEBULA
LDN 1101	21 38 00.	+ 56 45	1200		DARK NEBULA
LDN 1111	21 38 00.	+ 57 40	180		DARK NEBULA
LDN 1110	21 38 00.	+ 57 40	300		DARK NEBULA
LDN 1123	21 38 00.	+ 58 15	180		DARK NEBULA
LDN 1126	21 38 00.	+ 59 20	300		DARK NEBULA
LDN 1135	21 38 00.	+ 60 25	900		DARK NEBULA
PHL 0096	21 38 00.	- 05 56		17.3	BLUE STELLAR OBJECT
PHL 4743	21 38 00.	- 12 53		18.1	BLUE STELLAR OBJECT
ARC 2362	21 38 00.	- 14 30		16.9	RICH CLUSTER OF GALAXIES
PHL 4741	21 38 00.	- 18 25		16.4	BLUE STELLAR OBJECT
PHL 4742	21 38 00.	- 23 19		16.3	BLUE STELLAR OBJECT
PHL 1690	21 38 00.	- 26 49		15.7	BLUE STELLAR OBJECT
TON-S 0023	21 38 00.	- 26 48		14.6	BLUE STAR
HN 2736	21 38 05.3	- 54 55 10.	18		NEBULA
ZWG 402.014	21 38 06.	+ 06 17		15.7	GALAXY
ZWG 472.006	21 38 06.	+ 24 57		15.7	GALAXY
UGC 11788	21 38 06.	+ 24 57	108	15.7	GALAXY S+COMP
PHL 0097	21 38 06.	- 06 32		17.5	BLUE STELLAR OBJECT
PHL 1691	21 38 06.	- 11 58		18.3	BLUE STELLAR OBJECT
SCHO 1318	21 38 09.	+ 57 44 24.	280		ISOLATED DARK CLOUD
HN 2742	21 38 11.9	- 16 04 21.	18		NEBULA
ZC 2138.2+0700	21 38 12.	+ 07 00	3490		CLUSTER OF GALAXIES
MCG+04-51-003	21 38 12.	+ 24 56	36	15.5	GALAXY
PHL 1692	21 38 12.	- 01 16		18.7	BLUE STELLAR OBJECT
PHL 1693	21 38 12.	- 04 34		18.0	BLUE STELLAR OBJECT
PHL 1694	21 38 12.	- 09 26		18.9	BLUE STELLAR OBJECT
PHL 4744	21 38 12.	- 09 30		15.6	BLUE STELLAR OBJECT
PHL 0098	21 38 12.	- 18 47		16.9	BLUE STELLAR OBJECT
PHL 0099	21 38 12.	- 21 51		16.6	BLUE STELLAR OBJECT
HN 2741	21 38 12.4	- 16 56 57.	18		NEBULA
HN 2743	21 38 12.5	- 14 15 09.	30		NEBULA
ZC 2138.3+0331	21 38 18.	+ 03 31	1010		CLUSTER OF GALAXIES
ZC 2138.3+1711	21 38 18.	+ 17 11	870		CLUSTER OF GALAXIES
PHL 4745	21 38 18.	- 02 58		17.7	BLUE STELLAR OBJECT
PHL 4746	21 38 18.	- 15 30		18.1	BLUE STELLAR OBJECT
PHL 4747	21 38 18.	- 17 01		18.4	BLUE STELLAR OBJECT
PHL 1695	21 38 18.	- 23 41		17.9	BLUE STELLAR OBJECT
HN 2744	21 38 18.8	- 16 55 39.	18		NEBULA
SCHO 1316	21 38 19.	+ 50 41 12.	150		ISOLATED DARK CLOUD
ARC 2363	21 38 21.	- 08 33		17.1	RICH CLUSTER OF GALAXIES
MCG-07-44-031	21 38 21.	- 39 59	72	15.5	GALAXY
HN 2739	21 38 22.1	- 52 57 40.	18		NEBULA
HN 2746	21 38 22.3	- 14 34 45.	18		NEBULA
HN 2747	21 38 23.8	- 15 01 39.	24		NEBULA
PHL 4754	21 38 24.	+ 00 12		17.0	BLUE STELLAR OBJECT
PHL 4750	21 38 24.	+ 00 56		16.7	BLUE STELLAR OBJECT
MCG+00-55-010	21 38 24.	+ 01 05	60	15.	GALAXY
PHL 4753	21 38 24.	+ 01 32		16.6	BLUE STELLAR OBJECT
ZC 2138.4+0809	21 38 24.	+ 08 09	870		CLUSTER OF GALAXIES
ZC 2138.4+2140	21 38 24.	+ 21 40	1550		CLUSTER OF GALAXIES
SCHO 1317	21 38 24.	+ 49 37 48.	360		ISOLATED DARK CLOUD
PHL 4755	21 38 24.	- 05 40		17.8	BLUE STELLAR OBJECT
PHL 0100	21 38 24.	- 07 02		15.4	BLUE STELLAR OBJECT
PHL 4751	21 38 24.	- 11 15		17.8	BLUE STELLAR OBJECT
PHL 4748	21 38 24.	- 13 28		12.0	BLUE STELLAR OBJECT
PHL 4749	21 38 24.	- 14 44		15.4	BLUE STELLAR OBJECT
PHL 4752	21 38 24.	- 15 43		18.2	BLUE STELLAR OBJECT
TON-S 0024	21 38 24.	- 34 03		14.0	BLUE STAR
HN 2745	21 38 24.6	- 16 06 45.	24		NEBULA
IC 5125	21 38 25.	- 53 01			NONSTELLAR OBJECT
SCHO 1319	21 38 26.	+ 50 35 24.	200		ISOLATED DARK CLOUD
HN 0723	21 38 26.	- 53 00			NEBULA
ZWG 376.022	21 38 30.	+ 01 06		14.9	GALAXY
UGC 11789	21 38 30.	+ 01 06	60	14.9	GALAXY Sb
LDN 1131	21 38 30.	+ 59 20	840		DARK NEBULA
PHL 1697	21 38 36.	- 00 40		17.8	BLUE STELLAR OBJECT
PHL 1696	21 38 36.	- 00 40		16.7	BLUE STELLAR OBJECT
PHL 4756	21 38 36.	- 01 51		16.6	BLUE STELLAR OBJECT
PHL 4779	21 38 36.	- 12 51		16.9	BLUE STELLAR OBJECT
PHL 4757	21 38 36.	- 13 07		17.9	BLUE STELLAR OBJECT
PHL 4758	21 38 36.	- 13 33		17.8	BLUE STELLAR OBJECT
MCG-05-51-009	21 38 36.	- 26 51	48	15.	GALAXY
HN 2749	21 38 39.9	- 15 31 38.	24		NEBULA
HN 2750	21 38 40.9	- 14 18 44.	30		NEBULA
PHL 4759	21 38 42.	+ 00 28		16.6	BLUE STELLAR OBJECT
PHL 1699	21 38 42.	+ 01 03		14.0	BLUE STELLAR OBJECT
PHL 1700	21 38 42.	- 02 09		17.9	BLUE STELLAR OBJECT

OBJECT NAME	RIGHT ASCEN.	DECLINATION	DIAM.	MAGN.	TYPE OF OBJECT
PHL 1701	21 38 42.	- 10 40		18.6	BLUE STELLAR OBJECT
HN 2740	21 38 42.5	- 51 59 51.	12		NEBULA
ZC 2136.8+1008	21 38 48.	+ 10 08	3630		CLUSTER OF GALAXIES
PHL 1702	21 38 48.	- 20 34		18.4	BLUE STELLAR OBJECT
PHL 0101	21 38 48.	- 21 57		14.4	BLUE STELLAR OBJECT
PHL 4760	21 38 48.	- 23 36		16.0	BLUE STELLAR OBJECT
B 161	21 38 49.	+ 57 35			DARK OBJECT
BC PKS2138-377	21 38 49.24	- 37 42 54.4		17.	QUASI-STELLAR OBJECT
SCHO 1320	21 38 52.	+ 47 25 36.	300		ISOLATED DARK CLOUD
VDB.66N 144	21 38 53.	+ 54 36	204		REFLECTION NEBULA
B 366	21 38 53.	+ 59 20	600		DARK OBJECT
MCG+00-55-011	21 38 54.	+ 00 38	84	14.	GALAXY
ZC 2138.9+1941	21 38 54.	+ 19 41	1080		CLUSTER OF GALAXIES
PHL 0102	21 38 54.	- 02 50		17.4	BLUE STELLAR OBJECT
PHL 4761	21 38 54.	- 04 32		17.1	BLUE STELLAR OBJECT
RNGC 7109	21 38 54.	- 34 41			GALAXY
MCG-07-44-032	21 38 54.	- 39 03 30.	42	15.	GALAXY
SCHO 1322	21 38 57.	+ 50 41 06.	180		ISOLATED DARK CLOUD
KHAV 747	21 39	+ 44 08	1220		DARK NEBULA
KHAV 750	21 39	+ 47 02	5270		DARK NEBULA
KHAV 749	21 39	+ 49 26	2930		DARK NEBULA
LBN 0426	21 39	+ 50 10	3600		BRIGHT NEBULA
KHAV 748	21 39	+ 57 38	1220		DARK NEBULA
ZWG 376.023	21 39 00.	+ 00 40		15.3	GALAXY
UGC 11790	21 39 00.	+ 00 40	96	15.3	GALAXY Sc
KARA.73B 0924	21 39 00.	+ 00 40	72	15.3	ISOLATED GALAXY S
PHL 4762	21 39 00.	+ 01 48		16.1	BLUE STELLAR OBJECT
ZC 2139.0+0245	21 39 00.	+ 02 45	1140		CLUSTER OF GALAXIES
ISS 0186	21 39 00.	+ 49 32	194		STELLAR RING
LDN 1059	21 39 00.	+ 50 30	960		DARK NEBULA
LDN 1121	21 39 00.	+ 58 03	360		DARK NEBULA
LDN 1140	21 39 00.	+ 60 40	1980		DARK NEBULA
PHL 4763	21 39 00.	- 01 33		16.1	BLUE STELLAR OBJECT
PHL 4764	21 39 00.	- 06 56		18.2	BLUE STELLAR OBJECT
MCG-02-55-001	21 39 00.	- 10 52	60	15.	GALAXY
MCG-04-51-010	21 39 00.	- 21 00	36	15.5	GALAXY
HN 2755	21 39 02.9	- 17 40 49.	30		NEBULA
SCHO 1321	21 39 03.	+ 50 21 42.	350		ISOLATED DARK CLOUD
SCHO 1323	21 39 04.	+ 57 35 36.	240		ISOLATED DARK CLOUD
ZWG 427.013	21 39 06.	+ 10 13		15.7	GALAXY
ZWG 450.004	21 39 06.	+ 16 47		15.7	GALAXY
PHL 0103	21 39 06.	- 03 16		15.3	BLUE STELLAR OBJECT
MCG-03-55-005	21 39 06.	- 17 40	48	16.	GALAXY
IC 1395	21 39 09.	+ 03 52 42.			NONSTELLAR OBJECT
HN 2748	21 39 10.3	- 51 27 31.	12		NEBULA
RNGC 7111	21 39 11.	- 06 57		15.0	GALAXY
PHL 4766	21 39 12.	+ 00 54		17.9	BLUE STELLAR OBJECT
PHL 1703	21 39 12.	+ 01 26		17.0	BLUE STELLAR OBJECT
ZWG 402.015	21 39 12.	+ 03 53		15.5	GALAXY
PHL 4767	21 39 12.	- 01 15		18.1	BLUE STELLAR OBJECT
PHL 4765	21 39 12.	- 02 48		16.6	BLUE STELLAR OBJECT
PHL 0104	21 39 12.	- 03 22		15.9	BLUE STELLAR OBJECT
MCG-01-55-002	21 39 12.	- 06 57	30	15.	GALAXY
PHL 0105	21 39 12.	- 12 45		17.8	BLUE STELLAR OBJECT
PHL 4768	21 39 12.	- 16 54		18.1	BLUE STELLAR OBJECT
PHL 1704	21 39 12.	- 24 15		17.9	BLUE STELLAR OBJECT
RNGC 7110	21 39 12.	- 34 24			GALAXY
RNGC 7107	21 39 12.	- 45 02		13.0	GALAXY
RNGC 7106	21 39 12.	- 52 56			UNVERIFIED SOUTHERN OBJECT
ARC 2364	21 39 14.	- 20 32		17.4	RICH CLUSTER OF GALAXIES
YC 2139-45	21 39 14.	- 45 01 18.			UNUSUAL SOUTHERN GALAXY
HN 2759	21 39 15.9	- 17 09 06.	18		NEBULA
URA 69	21 39 18.	+ 52 37	222		STELLAR RING
PHL 4769	21 39 18.	- 02 35		16.9	BLUE STELLAR OBJECT
PHL 1705	21 39 18.	- 12 46		18.1	BLUE STELLAR OBJECT
HN 2762	21 39 18.5	- 14 40 54.			NEBULA
SCHO 1324	21 39 21.	+ 50 38 18.	240		ISOLATED DARK CLOUD
HN 2751	21 39 22.1	- 52 30 37.	18		NEBULA
ZC 2139.4+0914	21 39 24.	+ 09 14	1210		CLUSTER OF GALAXIES
LDN 1127	21 39 24.	+ 58 20	240		DARK NEBULA
PHL 4774	21 39 24.	- 02 24		17.0	BLUE STELLAR OBJECT
PHL 4775	21 39 24.	- 02 36		17.0	BLUE STELLAR OBJECT
PHL 0106	21 39 24.	- 07 14		18.6	BLUE STELLAR OBJECT
PHL 4776	21 39 24.	- 07 48		17.9	BLUE STELLAR OBJECT
PHL 4770	21 39 24.	- 09 50		17.6	BLUE STELLAR OBJECT
PHL 4771	21 39 24.	- 14 29		18.4	BLUE STELLAR OBJECT
PHL 4777	21 39 24.	- 16 18		17.8	BLUE STELLAR OBJECT
PHL 4773	21 39 24.	- 18 54		16.7	BLUE STELLAR OBJECT
PHL 4772	21 39 24.	- 19 24		18.7	BLUE STELLAR OBJECT
MCG-07-44-033	21 39 24.	- 38 46 30.	9	15.5	GALAXY
HN 2752	21 39 25.6	- 51 30 37.	12		NEBULA
HN 2763	21 39 29.5	- 17 10 00.	36		NEBULA
UGC 11791	21 39 30.	+ 25 11	60	16.5	GALAXY Sc
ISS 0125	21 39 30.	+ 53 18	287		STELLAR RING
MCG-01-55-003	21 39 30.	- 05 12	45	15.	GALAXY
B 162	21 39 33.	+ 56 05	780		DARK OBJECT
IC 5128	21 39 35.	- 39 13 25.			NONSTELLAR OBJECT
PHL 4780	21 39 36.	- 00 20		17.3	BLUE STELLAR OBJECT
PHL 4781	21 39 36.	- 10 52		17.1	BLUE STELLAR OBJECT
PHL 4778	21 39 36.	- 15 32		18.3	BLUE STELLAR OBJECT
MCG-03-55-006	21 39 36.	- 17 11	36	15.5	GALAXY
PHL 4782	21 39 36.	- 17 34		18.2	BLUE STELLAR OBJECT
MCG-05-51-010	21 39 36.	- 29 38	66	15.	GALAXY
HN 2753	21 39 36.3	- 52 36 36.	12		NEBULA
HN 2754	21 39 38.2	- 52 54 36.	24		NEBULA
MCG+04-51-004	21 39 39.	+ 22 29	72	14.5	GALAXY
SCHO 1325	21 39 39.	+ 56 19 00.	190		ISOLATED DARK CLOUD
ZWG 376.024	21 39 42.	+ 02 26		15.5	GALAXY
ZWG 402.016	21 39 42.	+ 05 23		15.6	GALAXY
UGC 11792	21 39 42.	+ 05 23	96	15.6	GALAXY Sc
ZWG 472.007	21 39 42.	+ 22 29		15.0	GALAXY
UGC 11793	21 39 42.	+ 22 29	84	15.0	GALAXY SBb
4ZW 074	21 39 42.	+ 25 04			COMPACT GALAXY
MRSL 081-17/1	21 39 42.	+ 29 53	5400		HII REGION
ZWG 493.004	21 39 42.	+ 31 38		15.7	GALAXY
RNGC 7114	21 39 43.	+ 42 36			NON-EXISTENT OBJECT
HN 2756	21 39 43.3	- 51 45 24.			NEBULA
HN 2769	21 39 43.4	- 17 12 53.	18		NEBULA
HN 2770	21 39 44.3	- 16 20 23.	30		NEBULA
HN 2757	21 39 44.3	- 52 14 06.	12		NEBULA
RNGC 7108	21 39 47.	- 07 01			GALAXY
SCHO 1326	21 39 48.	+ 56 25 24.	230		ISOLATED DARK CLOUD
PHL 1706	21 39 48.	- 02 26		17.8	BLUE STELLAR OBJECT
PHL 0107	21 39 48.	- 04 50		18.4	BLUE STELLAR OBJECT
PHL 1707	21 39 48.	- 20 14		17.7	BLUE STELLAR OBJECT
HN 2758	21 39 50.0	- 52 15 18.	12		NEBULA
RNGC 7112	21 39 53.	+ 12 16		15.5	GALAXY
ZWG 427.014	21 39 54.	+ 12 16		15.5	GALAXY

OBJECT NAME	RIGHT ASCEN.	DECLINATION	DIAM.	MAGN.	TYPE OF OBJECT
UGC 11794	21 39 54.	+ 12 16	72	15.5	GALAXY Sa-b
KARA.72 558A	21 39 54.	+ 12 16	66	15.5	PART OF DOUBLE GALAXY
ZWG 427.015	21 39 54.	+ 14 35		15.4	GALAXY
PHL 1708	21 39 54.	- 06 17		17.4	BLUE STELLAR OBJECT
HN 2760	21 39 56.4	- 52 43 36.	12		NEBULA
SCHO 1327	21 39 58.	+ 56 10 30.	430		ISOLATED DARK CLOUD
HN 2761	21 39 58.3	- 52 11 23.	12		NEBULA
RNGC 7113	21 39 59.	+ 12 21		15.0	GALAXY
LBN 0140	21 40	+ 06 00	6900		BRIGHT NEBULA
LBN 0405	21 40	+ 42 00	10800		BRIGHT NEBULA
LBN 0414	21 40	+ 44 12	720		BRIGHT NEBULA
KHAV 751	21 40	+ 56 32	700		DARK NEBULA
LBN 0515	21 40	+ 70 20	1620		BRIGHT NEBULA
LBN 0553	21 40	+ 76 43	300		BRIGHT NEBULA
PHL 0108	21 40 00.	+ 01 13		16.1	BLUE STELLAR OBJECT
MCG+02-55-010	21 40 00.	+ 12 16	72	14.5	GALAXY
MCG+02-55-009	21 40 00.	+ 12 20	21	14.	GALAXY
ZWG 427.016	21 40 00.	+ 12 21		15.2	GALAXY
KARA.72 558B	21 40 00.	+ 12 21	72	15.2	PART OF DOUBLE GALAXY
ZWG 427.017	21 40 00.	+ 13 40		15.4	GALAXY
MCG+02-55-008	21 40 00.	+ 13 40	60	15.	GALAXY
ISS 0187	21 40 00.	+ 47 22	391		STELLAR RING
LDN 1084	21 40 00.	+ 54 40	1260		DARK NEBULA
LDN 1124	21 40 00.	+ 58 00	480		DARK NEBULA
LDN 1128	21 40 00.	+ 58 20	480		DARK NEBULA
LDN 1134	21 40 00.	+ 60 00	6540		DARK NEBULA
ZC 2140.0+8255	21 40 00.	+ 82 55	1340		CLUSTER OF GALAXIES
PHL 4784	21 40 00.	- 01 40		16.9	BLUE STELLAR OBJECT
PHL 0109	21 40 00.	- 04 52		17.5	BLUE STELLAR OBJECT
PHL 4785	21 40 00.	- 05 14		18.1	BLUE STELLAR OBJECT
PHL 4786	21 40 00.	- 07 31		18.5	BLUE STELLAR OBJECT
PHL 1709	21 40 00.	- 09 30		17.9	BLUE STELLAR OBJECT
PHL 4783	21 40 00.	- 13 42		15.5	BLUE STELLAR OBJECT
PHL 4787	21 40 00.	- 25 34		18.3	BLUE STELLAR OBJECT
HN 2774	21 40 04.9	- 14 14 34.	30		NEBULA
LDN 1103	21 40 06.	+ 56 30	180		DARK NEBULA
PHL 1710	21 40 06.	- 06 30		18.4	BLUE STELLAR OBJECT
MCG-01-55-004	21 40 06.	- 07 05	30	16.	GALAXY
PHL 4788	21 40 06.	- 11 54		17.8	BLUE STELLAR OBJECT
MCG-07-44-034	21 40 06.	- 38 51 30.	36	15.5	GALAXY
MCG-07-44-035	21 40 06.	- 39 12	36	14.5	GALAXY
RNGC 7129	21 40 08.	+ 65 52		11.5	CLUSTER WITH NEBULOSITY
HN 2765	21 40 11.7	- 51 35 53.	12		NEBULA
PHL 4789	21 40 12.	+ 01 58		16.1	BLUE STELLAR OBJECT
UGC 11795	21 40 12.	+ 23 24	72	16.0	GALAXY
OCL 0240	21 40 12.	+ 65 52	168	11.5	OPEN STAR CLUSTER
PHL 1711	21 40 12.	- 00 37		18.3	BLUE STELLAR OBJECT
PHL 4791	21 40 12.	- 01 52		18.4	BLUE STELLAR OBJECT
PHL 1712	21 40 12.	- 09 39		18.2	BLUE STELLAR OBJECT
PHL 4790	21 40 12.	- 09 41		16.6	BLUE STELLAR OBJECT
PHL 4793	21 40 12.	- 10 49		18.3	BLUE STELLAR OBJECT
PHL 4792	21 40 12.	- 15 30		16.6	BLUE STELLAR OBJECT
ARC 2365	21 40 12.	- 18 55		17.0	RICH CLUSTER OF GALAXIES
MCG-07-44-036	21 40 12.	- 39 25	24	15.	GALAXY
HN 2764	21 40 12.8	- 52 37 35.	12		NEBULA
ARC 2366	21 40 13.	- 07 06		15.9	RICH CLUSTER OF GALAXIES
HOLM 786B	21 40 14.	+ 41 08	12	13.5	PART OF MULTIPLE GALAXY
ARC 2367	21 40 15.	- 08 19		17.1	RICH CLUSTER OF GALAXIES
HN 2767	21 40 15.5	- 51 25 11.	12		NEBULA
HN 2768	21 40 15.7	- 50 24 29.			NEBULA
HN 2776	21 40 16.5	- 17 46 58.	36		NEBULA
PHL 4796	21 40 18.	+ 01 38		18.3	BLUE STELLAR OBJECT
HOLM 786A	21 40 18.	+ 41 08	30	12.3	PART OF MULTIPLE GALAXY
SCHO 1328	21 40 18.	+ 50 00 12.	240		ISOLATED DARK CLOUD
PHL 4795	21 40 18.	- 02 10		16.5	BLUE STELLAR OBJECT
PHL 4794	21 40 18.	- 07 27		15.5	BLUE STELLAR OBJECT
TON-S 0025	21 40 18.	- 27 32		15.7	BLUE STAR
MCG-07-44-037	21 40 18.	- 40 21 30.	60	15.	GALAXY
HN 2766	21 40 19.5	- 53 41 59.	12		NEBULA
HN 0724	21 40 21.	- 72 39			NEBULA
IC 5123	21 40 21.	- 72 40			NONSTELLAR OBJECT
PHL 4798	21 40 24.	+ 00 22		16.0	BLUE STELLAR OBJECT
PHL 4804	21 40 24.	+ 00 29		18.6	BLUE STELLAR OBJECT
PHL 1713	21 40 24.	+ 02 21		18.1	BLUE STELLAR OBJECT
ZC 2140.4+2311	21 40 24.	+ 23 11	1480		CLUSTER OF GALAXIES
ZWG 493.005	21 40 24.	+ 28 43		14.2	GALAXY
RNGC 7116	21 40 24.	+ 28 43		14.0	GALAXY
UGC 11796	21 40 24.	+ 28 43	72	14.2	GALAXY S
MCG+05-51-001	21 40 24.	+ 28 43	60	14.	GALAXY
BRSL 091-06/1	21 40 24.	+ 44 19	780		HII REGION
LDN 1025	21 40 24.	+ 47 18	420		DARK NEBULA
LDN 1024	21 40 24.	+ 47 18	420		DARK NEBULA
PHL 4797	21 40 24.	- 02 04		16.3	BLUE STELLAR OBJECT
PHL 4805	21 40 24.	- 03 50		18.0	BLUE STELLAR OBJECT
PHL 1714	21 40 24.	- 04 00		17.2	BLUE STELLAR OBJECT
PHL 4801	21 40 24.	- 05 24		17.9	BLUE STELLAR OBJECT
PHL 4802	21 40 24.	- 11 51		17.4	BLUE STELLAR OBJECT
PHL 4799	21 40 24.	- 12 13		15.6	BLUE STELLAR OBJECT
PHL 0110	21 40 24.	- 12 19		12.7	BLUE STELLAR OBJECT
PHL 4800	21 40 24.	- 12 57		15.8	BLUE STELLAR OBJECT
PHL 0111	21 40 24.	- 14 49		18.5	BLUE STELLAR OBJECT
PHL 4803	21 40 24.	- 23 57		17.4	BLUE STELLAR OBJECT
HN 2771	21 40 24.8	- 50 31 58.	12		NEBULA
LDN 1104	21 40 30.	+ 56 30	300		DARK NEBULA
HN 2772	21 40 32.5	- 50 31 34.			NEBULA
B 163	21 40 34.	+ 56 28			DARK OBJECT
PHL 1717	21 40 36.	+ 00 34		18.2	BLUE STELLAR OBJECT
PHL 0113	21 40 36.	+ 00 37		18.9	BLUE STELLAR OBJECT
PHL 4808	21 40 36.	+ 01 10		17.9	BLUE STELLAR OBJECT
PHL 0112	21 40 36.	+ 02 10		18.4	BLUE STELLAR OBJECT
PHL 4306	21 40 36.	- 01 16		16.7	BLUE STELLAR OBJECT
PHL 4807	21 40 36.	- 01 42		17.1	BLUE STELLAR OBJECT
PHL 4809	21 40 36.	- 03 18		18.1	BLUE STELLAR OBJECT
PHL 1715	21 40 36.	- 06 16		18.2	BLUE STELLAR OBJECT
PHL 1716	21 40 36.	- 07 14		14.0	BLUE STELLAR OBJECT
HN 2773	21 40 39.2	- 49 46 40.			NEBULA
PHL 4812	21 40 42.	+ 01 34		18.9	BLUE STELLAR OBJECT
PHL 1718	21 40 42.	- 04 02		18.4	BLUE STELLAR OBJECT
PHL 4810	21 40 42.	- 17 45		16.6	BLUE STELLAR OBJECT
PHL 0114	21 40 42.	- 20 40		18.0	BLUE STELLAR OBJECT
PHL 4811	21 40 42.	- 21 53		17.8	BLUE STELLAR OBJECT
RNGC 7115	21 40 46.	- 25 35		14.0	GALAXY
HN 2775	21 40 46.2	- 52 27 09.			NEBULA
ZC 2140.8+3240	21 40 48.	+ 32 40	3900		CLUSTER OF GALAXIES
PHL 0115	21 40 48.	- 06 09		15.1	BLUE STELLAR OBJECT
PHL 1719	21 40 48.	- 17 15		18.5	BLUE STELLAR OBJECT
MCG-04-51-011	21 40 48.	- 25 35	96	14.	GALAXY
TON-S 0026	21 40 48.	- 32 34		14.1	BLUE STAR
HN 2781	21 40 51.4	- 16 48 44.	24		NEBULA
SCHO 1329	21 40 52.	+ 51 13 30.	680		ISOLATED DARK CLOUD
CED 195B	21 40 52.	+ 57 31	9000		DIFFUSE GALACTIC NEBULA
LB 04009	21 40 54.	+ 46 35 42.		18.3	FAINT BLUE STAR
MCG+12-20-001	21 40 54.	+ 73 45	15	16.	GALAXY
PHL 4813	21 40 54.	- 14 12		17.7	BLUE STELLAR OBJECT
PHL 4814	21 40 54.	- 15 46		17.7	BLUE STELLAR OBJECT
MCG-07-44-038	21 40 54.	- 43 26	30	16.	GALAXY
HN 2778	21 40 54.1	- 43 26 21.	24		NEBULA
SCHO 1330	21 40 56.	+ 49 52 06.	300		ISOLATED DARK CLOUD
HN 2777	21 40 59.1	- 50 38 45.	12		NEBULA
LBN 0497	21 41	+ 65 50	120		BRIGHT NEBULA
MCG+00-55-012	21 41 00.	+ 02 23	66	15.	GALAXY
2ZW 146	21 41 00.	+ 13 52			COMPACT GALAXY
ZWG 529.002	21 41 00.	+ 44 09		15.7	GALAXY
LDN 1020	21 41 00.	+ 47 00	1440		DARK NEBULA
LDN 1095	21 41 00.	+ 56 05	420		DARK NEBULA
LDN 1106	21 41 00.	+ 56 40	360		DARK NEBULA
LDN 1181	21 41 00.	+ 65 55	360		DARK NEBULA
LDN 1183	21 41 00.	+ 66 00	1200		DARK NEBULA
PHL 0116	21 41 00.	- 01 02		18.4	BLUE STELLAR OBJECT
PHL 4815	21 41 00.	- 22 41		18.0	BLUE STELLAR OBJECT
ZWG 376.025	21 41 06.	+ 02 24		15.2	GALAXY
KARA.73B 0925	21 41 06.	+ 02 24	60	15.2	ISOLATED GALAXY S
PHL 4817	21 41 06.	- 02 30		17.9	BLUE STELLAR OBJECT
PHL 1720	21 41 06.	- 05 58		18.8	BLUE STELLAR OBJECT
PHL 4816	21 41 06.	- 20 04		18.3	BLUE STELLAR OBJECT
PHL 0117	21 41 06.	- 23 43		13.6	BLUE STELLAR OBJECT
HN 2779	21 41 09.0	- 51 16 02.	12		NEBULA
IC 5132	21 41 12.	+ 65 55			NONSTELLAR OBJECT
MCG-01-55-005	21 41 12.	- 08 23	48	16.	GALAXY
PHL 4818	21 41 12.	- 26 08		18.0	BLUE STELLAR OBJECT
SHB 363	21 41 13.8	+ 17 30 02.		16.5	QUASI-STELLAR OBJECT
HN 2783	21 41 13.9	- 15 48 43.	42		NEBULA
LB 04010	21 41 14.	+ 46 44 36.		15.0	FAINT BLUE STAR
IC 5133	21 41 17.	+ 65 56			NONSTELLAR OBJECT
HN 2785	21 41 17.8	- 15 45 07.	24		NEBULA
MCG+07-44-001	21 41 18.	+ 43 21	36	16.	GALAXY
ZWG 376.027	21 41 18.	- 00 33		15.3	GALAXY
ZWG 376.026	21 41 18.	- 01 12		15.5	GALAXY
PHL 4819	21 41 18.	- 01 36		15.6	BLUE STELLAR OBJECT
PHL 4820	21 41 18.	- 10 17		17.5	BLUE STELLAR OBJECT
HN 2780	21 41 18.2	- 52 02 32.	12		NEBULA
ARC 2368	21 41 19.	- 20 12		17.5	RICH CLUSTER OF GALAXIES
HN 2786	21 41 19.3	- 15 52 07.	36		NEBULA
ARC 2387	21 41 20.	+ 82 53		17.0	RICH CLUSTER OF GALAXIES
LB 04011	21 41 21.	+ 46 25 18.		18.0	FAINT BLUE STAR
MCG+07-44-002	21 41 24.	+ 43 19	72	15.	GALAXY
PHL 1721	21 41 24.	- 03 07		17.9	BLUE STELLAR OBJECT
PHL 4821	21 41 24.	- 17 49		18.4	BLUE STELLAR OBJECT
RNGC 7133	21 41 25.	+ 65 49			DIFFUSE NEBULA
IC 1397	21 41 25.	- 05 06 56.			NONSTELLAR OBJECT
PK095-02.1	21 41 29.3	+ 50 11 29.			PLANETARY NEBULA
ZWG 529.003	21 41 30.	+ 43 19		15.2	GALAXY
UGC 11798	21 41 30.	+ 43 19	120	15.2	GALAXY S
UGC 11797	21 41 30.	+ 43 19	96	15.2	GALAXY
MCG+07-44-003	21 41 30.	+ 43 19	27	15.	GALAXY
UGC 11799	21 41 30.	+ 43 27	102	16.0	GALAXY SB
LDN 1046	21 41 30.	+ 48 30	1200		DARK NEBULA
PHL 4824	21 41 30.	- 01 48		18.3	BLUE STELLAR OBJECT
PHL 4822	21 41 30.	- 11 34		17.0	BLUE STELLAR OBJECT
PHL 4823	21 41 30.	- 26 25		17.8	BLUE STELLAR OBJECT
MCG-05-51-011	21 41 30.	- 30 10	78	16.	GALAXY
HN 2782	21 41 30.2	- 50 25 25.	18		NEBULA
MCG+07-44-004	21 41 33.	+ 43 27 30.	54	17.	GALAXY
ZC 2141.6+0454	21 41 36.	+ 04 54	740		CLUSTER OF GALAXIES
ZC 2141.6+0750	21 41 36.	+ 07 50	2820		CLUSTER OF GALAXIES
ZWG 493.006	21 41 36.	+ 30 51		15.7	GALAXY
ZWG 529.004	21 41 36.	+ 43 18		15.3	GALAXY
UGC 11801	21 41 36.	+ 43 18	60	15.3	GALAXY S
MCG+07-44-005	21 41 36.	+ 43 19	8	18.5	GALAXY
ZWG 544.001	21 41 36.	+ 46 01		15.7	GALAXY
UGC 11800	21 41 36.	+ 46 01	90	15.7	GALAXY S
PHL 4827	21 41 36.	- 04 38		18.3	BLUE STELLAR OBJECT
PHL 4328	21 41 36.	- 05 48		18.2	BLUE STELLAR OBJECT
PHL 1722	21 41 36.	- 08 12		17.8	BLUE STELLAR OBJECT
PHL 4829	21 41 36.	- 13 03		18.1	BLUE STELLAR OBJECT
PHL 0118	21 41 36.	- 16 49		18.0	BLUE STELLAR OBJECT
PHL 0119	21 41 36.	- 19 00		16.5	BLUE STELLAR OBJECT
PHL 0120	21 41 36.	- 22 00		14.3	BLUE STELLAR OBJECT
PHL 4825	21 41 36.	- 22 32		18.4	BLUE STELLAR OBJECT
PHL 4826	21 41 36.	- 25 55		18.1	BLUE STELLAR OBJECT
MCG-07-44-039	21 41 36.	- 39 25	30	15.	GALAXY
LB 04012	21 41 39.	+ 46 23 48.		19.0	FAINT BLUE STAR
VDB.66N 146	21 41 41.	+ 65 53	540		REFLECTION NEBULA
REIN 2.295	21 41 41.54	+ 65 52 48.8			NEBULA
PHL 4831	21 41 42.	- 03 34		17.7	BLUE STELLAR OBJECT
MCG-03-55-007	21 41 42.	- 20 18	36	15.	GALAXY
PHL 4830	21 41 42.	- 20 39		16.4	BLUE STELLAR OBJECT
REIN 2.296	21 41 42.36	+ 65 51 58.5			NEBULA
REIN 2.297	21 41 45.98	+ 65 52 37.6			NEBULA
RNGC 7120	21 41 47.	- 06 47		15.0	GALAXY
ARC 2369	21 41 47.	- 18 34		17.0	RICH CLUSTER OF GALAXIES
PHL 0121	21 41 48.	+ 01 21		17.4	BLUE STELLAR OBJECT
PHL 1723	21 41 48.	- 00 37		18.4	BLUE STELLAR OBJECT
PHL 4832	21 41 48.	- 01 35		16.0	BLUE STELLAR OBJECT
MCG-01-55-006	21 41 48.	- 06 47	42	15.	GALAXY
IC 5134	21 41 49.	+ 65 53			NONSTELLAR OBJECT
REIN 2.298	21 41 49.15	+ 65 52 07.7			NEBULA
REIN 2.299	21 41 49.88	+ 65 52 23.9			NEBULA
HN 2784	21 41 50.1	- 49 55 37.	18		NEBULA
LB 04013	21 41 51.	+ 46 33 30.		18.1	FAINT BLUE STAR
CED 196	21 41 51.	+ 65 53	450		DIFFUSE GALACTIC NEBULA
MCG-03-55-008	21 41 51.	- 20 19	30	15.	GALAXY
LB 04014	21 41 52.	+ 46 28 30.		17.8	FAINT BLUE STAR
REIN 2.300A	21 41 52.50	+ 65 53 22.3			NEBULA
REIN 2.300B	21 41 52.72	+ 65 53 22.1			NEBULA
SCHO 1331	21 41 53.	+ 59 04 48.	650		ISOLATED DARK CLOUD
ZC 2141.9+2010	21 41 54.	+ 20 10	1810		CLUSTER OF GALAXIES
ISS 0188	21 41 54.	+ 45 33	347		STELLAR RING
DG 176	21 41 54.	+ 65 53	480		REFLECTION NEBULA
PHL 0122	21 41 54.	- 05 28		18.4	BLUE STELLAR OBJECT
PHL 0123	21 41 54.	- 09 02		18.0	BLUE STELLAR OBJECT
PHL 4833	21 41 54.	- 12 35		18.5	BLUE STELLAR OBJECT
PHL 0124	21 41 54.	- 16 05		17.6	BLUE STELLAR OBJECT
VDB.66N 145	21 41 55.	+ 48 41	516		REFLECTION NEBULA
SCHO 1332	21 41 56.	+ 60 31 30.	570		ISOLATED DARK CLOUD
RNGC 7098	21 41 56.	- 75 21			GALAXY

OBJECT NAME	RIGHT ASCEN.	DECLINATION	DIAM.	MAGN.	TYPE OF OBJECT
REIN 2.301	21 41 58.07	+ 65 53 07.3			NEBULA
LBN 0306	21 42	+ 29 40	1500		BRIGHT NEBULA
KHAV 752	21 42	+ 53 32	3540		DARK NEBULA
PHL 1725	21 42 00.	+ 01 43		18.2	BLUE STELLAR OBJECT
PHL 4834	21 42 00.	+ 01 49		16.6	BLUE STELLAR OBJECT
MCG+08-39-001	21 42 00.	+ 46 23	42	15.	GALAXY
ZWG 544.002	21 42 00.	+ 46 24		15.1	GALAXY
UGC 11802	21 42 00.	+ 46 24	60	15.1	GALAXY SB:c
KARA.72 559A	21 42 00.	+ 46 24	72	15.1	PART OF DOUBLE GALAXY
LDN 1057	21 42 00.	+ 49 20	900		DARK NEBULA
COU 085	21 42 00.	+ 54 20	6000		EMISSION NEBULA
MCG-01-55-007	21 42 00.	- 07 18	15	15.5	GALAXY
PHL 4836	21 42 00.	- 09 37		17.9	BLUE STELLAR OBJECT
PHL 1724	21 42 00.	- 09 37		18.9	BLUE STELLAR OBJECT
PHL 4835	21 42 00.	- 11 56		18.2	BLUE STELLAR OBJECT
PHL 1726	21 42 00.	- 13 42		18.7	BLUE STELLAR OBJECT
HN 2787	21 42 03.2	- 51 59 18.			NEBULA
ZWG 427.018	21 42 06.	+ 14 40		15.4	GALAXY
UGC 11803	21 42 06.	+ 14 40	60	15.4	GALAXY Sb-c
4ZW 075	21 42 06.	+ 37 18			COMPACT GALAXY
PHL 4837	21 42 06.	- 03 58		17.8	BLUE STELLAR OBJECT
ARC 2370	21 42 06.	- 19 42		17.1	RICH CLUSTER OF GALAXIES
TON-S 0027	21 42 06.	- 28 54		15.9	BLUE STAR
HN 2789	21 42 08.0	- 51 50 18.			NEBULA
ARC 2373	21 42 09.	+ 00 51		17.5	RICH CLUSTER OF GALAXIES
RNGC 7127	21 42 10.	+ 54 23			OPEN CLUSTER
HN 2788	21 42 10.7	- 53 42 18.	12		NEBULA
PHL 0125	21 42 12.	+ 00 46		18.8	BLUE STELLAR OBJECT
ZC 2142.2+0059	21 42 12.	+ 00 59	1480		CLUSTER OF GALAXIES
PHL 1728	21 42 12.	+ 02 12		15.5	BLUE STELLAR OBJECT
MCG+02-55-011	21 42 12.	+ 14 40	48	15.	GALAXY
OCL 0219	21 42 12.	+ 54 23	168		OPEN STAR CLUSTER
MCG+12-20-002	21 42 12.	+ 74 12	15	17.	GALAXY
MCG+12-20-002	21 42 12.	+ 74 12	15	17.	GALAXY
PHL 4838	21 42 12.	- 00 04		16.6	BLUE STELLAR OBJECT
PHL 4839	21 42 12.	- 02 50		17.2	BLUE STELLAR OBJECT
PHL 1727	21 42 12.	- 05 12		18.2	BLUE STELLAR OBJECT
PHL 4840	21 42 12.	- 21 24		17.7	BLUE STELLAR OBJECT
MCG-04-51-012	21 42 12.	- 25 15	54	14.5	GALAXY
TON-S 0028	21 42 12.	- 31 20		14.8	BLUE STAR
MCG+00-55-013	21 42 18.	+ 01 17	12	16.	GALAXY
ZWG 376.028	21 42 18.	+ 01 18		15.3	GALAXY
SN 1968S	21 42 18.	+ 02 51		16.5	SUPERNOVA
MCG+08-39-002	21 42 18.	+ 46 00	120		GALAXY
ZWG 544.003	21 42 18.	+ 46 02		15.1	GALAXY
UGC 11805	21 42 18.	+ 46 02	96	15.1	GALAXY
UGC 11804	21 42 18.	+ 46 02	90	15.1	GALAXY
ZWG 544.004	21 42 18.	+ 46 24		15.0	GALAXY
UGC 11806	21 42 18.	+ 46 24	102	15.0	GALAXY SBb
KARA.72 559B	21 42 18.	+ 46 24	60	15.0	PART OF DOUBLE GALAXY
SCHO 1333	21 42 18.	+ 50 09 48.	240		ISOLATED DARK CLOUD
OCL 0218	21 42 18.	+ 53 29	540	11.4	OPEN STAR CLUSTER
PHL 1729	21 42 18.	- 02 12		18.2	BLUE STELLAR OBJECT
PHL 0126	21 42 18.	- 03 12		18.1	BLUE STELLAR OBJECT
RNGC 7121	21 42 18.	- 03 51		14.0	GALAXY
MCG-01-55-008	21 42 18.	- 03 51	60	14.	GALAXY
PHL 0127	21 42 18.	- 04 23		17.7	BLUE STELLAR OBJECT
PHL 0128	21 42 18.	- 05 06		17.0	BLUE STELLAR OBJECT
PHL 4841	21 42 18.	- 05 37		17.9	BLUE STELLAR OBJECT
PHL 4843	21 42 18.	- 08 13		18.6	BLUE STELLAR OBJECT
PHL 4842	21 42 18.	- 24 46		17.7	BLUE STELLAR OBJECT
ARC 2371	21 42 19.	- 24 26		17.1	RICH CLUSTER OF GALAXIES
RNGC 7128	21 42 21.	+ 53 29		11.5	OPEN CLUSTER
HN 2790	21 42 21.9	- 51 49 41.	12		NEBULA
PHL 4844	21 42 24.	+ 00 47		17.8	BLUE STELLAR OBJECT
ISS 0189	21 42 24.	+ 45 32	178		STELLAR RING
MCG+08-39-003	21 42 24.	+ 46 23	96	15.	GALAXY
PHL 1731	21 42 24.	- 08 46		18.3	BLUE STELLAR OBJECT
PHL 1730	21 42 24.	- 09 14		17.6	BLUE STELLAR OBJECT
PHL 4845	21 42 24.	- 25 33		18.4	BLUE STELLAR OBJECT
HN 2796	21 42 25.0	- 16 45 28.	18		NEBULA
CED 197	21 42 26.	+ 52 43			DIFFUSE GALACTIC NEBULA
IC 1400	21 42 26.	+ 52 43			NONSTELLAR OBJECT
RNGC 7117	21 42 28.	- 48 38			GALAXY
SC 2138-5637.5	21 42 28.	- 56 23 47.			NEBULA
ZC 2142.5+0855	21 42 30.	+ 08 55	1610		CLUSTER OF GALAXIES
OCL 0216	21 42 30.	+ 50 59			OPEN STAR CLUSTER
PHL 1732	21 42 30.	- 06 52		17.2	BLUE STELLAR OBJECT
PHL 1733	21 42 30.	- 16 20		17.6	BLUE STELLAR OBJECT
ARC 2372	21 42 31.	- 20 12		16.5	RICH CLUSTER OF GALAXIES
SC 2139-5735.1	21 42 33.	- 57 21 23.			NEBULA
PHL 4847	21 42 36.	+ 00 35		18.0	BLUE STELLAR OBJECT
PHL 0129	21 42 36.	+ 02 06		16.8	BLUE STELLAR OBJECT
ZWG 376.029	21 42 36.	+ 02 53		15.2	GALAXY
ZWG 493.007	21 42 36.	+ 32 40		15.7	GALAXY
PHL 4848	21 42 36.	- 07 02		18.0	BLUE STELLAR OBJECT
PHL 0130	21 42 36.	- 09 56		18.7	BLUE STELLAR OBJECT
PHL 1734	21 42 36.	- 22 41		18.5	BLUE STELLAR OBJECT
PHL 4846	21 42 36.	- 26 12		17.9	BLUE STELLAR OBJECT
MCG-05-51-012	21 42 36.	- 29 21	36	15.	GALAXY
SCHO 1334	21 42 37.	+ 58 13 12.	790		ISOLATED DARK CLOUD
HN 2791	21 42 37.2	- 53 50 05.	24		NEBULA
ARC 2374	21 42 38.	- 07 40		17.1	RICH CLUSTER OF GALAXIES
SC 2138-6353.7	21 42 41.	- 63 39 59.			NEBULA
ZC 2142.7+0202	21 42 42.	+ 02 02	870		CLUSTER OF GALAXIES
ZC 2142.7+0243	21 42 42.	+ 02 43	1210		CLUSTER OF GALAXIES
ZWG 376.030	21 42 42.	- 02 25		15.5	GALAXY
KARA.73B 0926	21 42 42.	- 02 25	24	15.5	ISOLATED GALAXY S
PHL 0131	21 42 42.	- 16 57		16.0	BLUE STELLAR OBJECT
PHL 4849	21 42 42.	- 23 04		17.6	BLUE STELLAR OBJECT
TON-S 0029	21 42 42.	- 33 59		14.0	BLUE STAR
SC 2139-5851.6	21 42 44.	- 58 37 53.			NEBULA
HN 2793	21 42 44.4	- 52 29 52.	24		NEBULA
HN 2794	21 42 45.0	- 52 01 52.	12		NEBULA
HN 2795	21 42 45.4	- 50 28 46.	18		NEBULA
HN 2792	21 42 46.1	- 54 00 04.	12		NEBULA
HN 2803	21 42 46.7	- 18 01 04.	18		NEBULA
SC 2139-5904.4	21 42 47.	- 58 50 41.	24		NEBULA
ZWG 427.019	21 42 48.	+ 09 52		15.0	GALAXY
KARA.73B 0927	21 42 48.	+ 09 52	36	15.0	ISOLATED GALAXY S
ZWG 427.020	21 42 48.	+ 15 21		15.7	GALAXY
LDN 1035	21 42 48.	+ 47 30	1320		DARK NEBULA
ISS 0190	21 42 48.	+ 49 06	257		STELLAR RING
PHL 1735	21 42 48.	- 10 43		18.4	BLUE STELLAR OBJECT
TON-S 0030	21 42 48.	- 28 11		14.8	BLUE STAR
B 367	21 42 49.	+ 56 57	300		DARK OBJECT
RNGC 7118	21 42 52.	- 48 35			GALAXY
SC 2139-5744.7	21 42 52.	- 57 30 58.			NEBULA
BC M2142+110	21 42 52.1	+ 11 01 42.		19.	QUASI-STELLAR OBJECT
SHB 364	21 42 52.1	+ 11 01 42.		19.	QUASI-STELLAR OBJECT
ZC 2142.9-0109	21 42 54.	- 01 09	1480		CLUSTER OF GALAXIES
PHL 4851	21 42 54.	- 05 54		18.2	BLUE STELLAR OBJECT
PHL 4850	21 42 54.	- 09 14		17.8	BLUE STELLAR OBJECT
LB 04015	21 42 55.	+ 46 48 00.		17.7	FAINT BLUE STAR
SC 2139-5911.4	21 42 57.	- 58 57 40.	18		NEBULA
KHAV 753	21 43	+ 51 26	4200		DARK NEBULA
LBN 0457	21 43	+ 57 50	3900		BRIGHT NEBULA
PHL 4853	21 43 00.	+ 01 02		18.9	BLUE STELLAR OBJECT
ZC 2143.0+0326	21 43 00.	+ 03 26	2290		CLUSTER OF GALAXIES
ZC 2143.0+2118	21 43 00.	+ 21 18	1880		CLUSTER OF GALAXIES
LDN 1113	21 43 00.	+ 56 58	180		DARK NEBULA
LDN 1130	21 43 00.	+ 58 05	420		DARK NEBULA
PHL 0132	21 43 00.	- 01 42		15.8	BLUE STELLAR OBJECT
MCG-01-55-009	21 43 00.	- 07 19	36	16.	GALAXY
PHL 4854	21 43 00.	- 10 06		18.1	BLUE STELLAR OBJECT
PHL 1736	21 43 00.	- 21 28		17.6	BLUE STELLAR OBJECT
PHL 4852	21 43 00.	- 24 00		17.8	BLUE STELLAR OBJECT
YC 2143-46	21 43 02.	- 46 44 54.			UNUSUAL SOUTHERN GALAXY
RNGC 7122	21 43 04.	- 09 03			NON-EXISTENT OBJECT
IC 1402	21 43 05.	+ 53 02			NONSTELLAR OBJECT
RNGC 7119B	21 43 05.	- 46 45			GALAXY
RNGC 7119A	21 43 05.	- 46 45			GALAXY
SC 2139-6346.3	21 43 05.	- 63 32 34.	9		NEBULA
PHL 1737	21 43 06.	- 01 30		17.7	BLUE STELLAR OBJECT
PHL 4855	21 43 06.	- 13 08		18.7	BLUE STELLAR OBJECT
SC 2139-5146.9	21 43 06.	- 51 33 10.			NEBULA
SC 2139-5815.0	21 43 06.	- 58 01 16.			NEBULA
LB 04016	21 43 09.	+ 46 35 42.		18.4	FAINT BLUE STAR
ARC 2376	21 43 11.	- 09 41		17.1	RICH CLUSTER OF GALAXIES
HN 2798	21 43 11.3	- 51 58 45.	12		NEBULA
PHL 0133	21 43 12.	- 01 32		18.1	BLUE STELLAR OBJECT
PHL 4856	21 43 12.	- 04 59		18.3	BLUE STELLAR OBJECT
PHL 0134	21 43 12.	- 05 35		17.9	BLUE STELLAR OBJECT
PHL 1738	21 43 12.	- 07 32		17.9	BLUE STELLAR OBJECT
PHL 0135	21 43 12.	- 12 23		18.7	BLUE STELLAR OBJECT
PHL 4857	21 43 12.	- 15 21		18.2	BLUE STELLAR OBJECT
ARC 2375	21 43 12.	- 19 23		17.1	RICH CLUSTER OF GALAXIES
SC 2140-3940.0	21 43 12.	- 39 26 15.			NEBULA
HN 2797	21 43 12.4	- 53 55 15.	36		NEBULA
LB 04017	21 43 13.	+ 46 36 12.		18.6	FAINT BLUE STAR
SC 2139-6244.2	21 43 13.	- 62 30 28.	12		NEBULA
HN 2809	21 43 13.4	- 15 24 38.	24		NEBULA
HN 2799	21 43 14.5	- 52 38 51.	12		NEBULA
HN 2800	21 43 15.3	- 51 12 51.	12		NEBULA
ARC 2377	21 43 17.	- 10 17		16.9	RICH CLUSTER OF GALAXIES
ZWG 427.021	21 43 18.	+ 15 04		15.7	GALAXY
PHL 4858	21 43 18.	- 17 23		17.1	BLUE STELLAR OBJECT
HN 2804	21 43 22.5	- 51 32 45.	6		NEBULA
HN 2801	21 43 22.9	- 53 50 15.	12		NEBULA
ZWG 402.017	21 43 24.	+ 09 15		15.4	GALAXY
ZC 2143.4+1350	21 43 24.	+ 13 50	1280		CLUSTER OF GALAXIES
ISS 0191	21 43 24.	+ 49 09	125		STELLAR RING
PHL 4859	21 43 24.	- 01 08		16.0	BLUE STELLAR OBJECT
PHL 1740	21 43 24.	- 03 42		18.8	BLUE STELLAR OBJECT
PHL 4864	21 43 24.	- 04 21		18.2	BLUE STELLAR OBJECT
PHL 4862	21 43 24.	- 06 34		17.9	BLUE STELLAR OBJECT
PHL 4863	21 43 24.	- 08 00		17.8	BLUE STELLAR OBJECT
PHL 4865	21 43 24.	- 08 20		18.6	BLUE STELLAR OBJECT
PHL 4866	21 43 24.	- 09 12		18.6	BLUE STELLAR OBJECT
PHL 1739	21 43 24.	- 10 08		15.5	BLUE STELLAR OBJECT
PHL 4860	21 43 24.	- 12 12		17.5	BLUE STELLAR OBJECT
PHL 4861	21 43 24.	- 16 15		17.7	BLUE STELLAR OBJECT
HN 2802	21 43 24.2	- 54 11 03.	18		NEBULA
IC 1398	21 43 25.	+ 09 14 37.			NONSTELLAR OBJECT
LB 04018	21 43 26.	+ 46 26 48.		18.8	FAINT BLUE STAR
HN 2805	21 43 26.1	- 51 45 45.	12		NEBULA
MCG+01-55-009	21 43 27.	+ 09 14 30.	48	15.	GALAXY
SC 2139-5741.5	21 43 27.	- 57 27 45.			NEBULA
LB 04019	21 43 28.	+ 46 28 12.		17.5	FAINT BLUE STAR
MCG+00-55-014	21 43 30.	+ 00 48	36	16.	GALAXY
MCG+01-55-010	21 43 30.	+ 07 37	36	15.	GALAXY
ZWG 402.018	21 43 30.	+ 07 38		15.3	GALAXY
ZWG 427.022	21 43 30.	+ 11 26		15.7	GALAXY
KARA.73B 0928	21 43 30.	+ 11 26	30	15.7	ISOLATED GALAXY S
ZC 2143.5+2014	21 43 30.	+ 20 14	1080		CLUSTER OF GALAXIES
PHL 4867	21 43 30.	- 07 20		17.9	BLUE STELLAR OBJECT
TON-S 0031	21 43 30.	- 28 22		15.8	BLUE STAR
MCG-05-51-013	21 43 30.	- 30 18	24	15.	GALAXY
SC 2140-3942.4	21 43 32.	- 39 28 38.	12		NEBULA
LB 04020	21 43 33.	+ 46 47 42.		11.0	FAINT BLUE STAR
PHL 0136	21 43 36.	+ 00 36		17.7	BLUE STELLAR OBJECT
ZWG 402.019	21 43 36.	+ 04 10		15.6	GALAXY
PHL 4868	21 43 36.	- 01 02		18.8	BLUE STELLAR OBJECT
PHL 4870	21 43 36.	- 04 14		16.9	BLUE STELLAR OBJECT
PHL 0137	21 43 36.	- 04 34		17.4	BLUE STELLAR OBJECT
PHL 4869	21 43 36.	- 06 04		18.4	BLUE STELLAR OBJECT
TON-S 0032	21 43 36.	- 27 54		16.1	BLUE STAR
TON-S 0033	21 43 36.	- 27 55		16.3	BLUE STAR
MCG-05-51-014	21 43 36.	- 29 48	30	15.5	GALAXY
SCHO 1335	21 43 39.	+ 54 42 18.	350		ISOLATED DARK CLOUD
IC 1399	21 43 40.	+ 04 10 08.			NONSTELLAR OBJECT
HN 2810	21 43 40.8	- 14 55 31.	30		NEBULA
SC 2139-6329.0	21 43 41.	- 63 15 15.			NEBULA
HN 2807	21 43 41.9	- 51 46 44.	12		NEBULA
HN 2806	21 43 41.9	- 52 19 32.	12		NEBULA
ZWG 402.020	21 43 42.	+ 09 20		15.5	GALAXY
2ZW 147	21 43 42.	+ 16 24			COMPACT GALAXY
ISS 0126	21 43 42.	+ 55 08	79		STELLAR RING
PHL 1741	21 43 42.	- 03 19		16.2	BLUE STELLAR OBJECT
PHL 4871	21 43 42.	- 03 51		18.6	BLUE STELLAR OBJECT
PHL 0138	21 43 42.	- 13 40		17.6	BLUE STELLAR OBJECT
PHL 1742	21 43 42.	- 15 38		17.6	BLUE STELLAR OBJECT
GCL 123	21 43 42.	- 21 28	126		GLOBULAR STAR CLUSTER
SC 2140-5747.1	21 43 42.	- 57 33 20.			NEBULA
SC 2140-5750.4	21 43 43.	- 57 36 38.			NEBULA
MCG+01-55-011	21 43 45.	+ 09 19	24	15.5	GALAXY
SCHO 1336	21 43 46.	+ 49 37 18.	310		ISOLATED DARK CLOUD
HN 2808	21 43 46.4	- 53 50 56.	18		NEBULA
SC 2140-3900.6	21 43 47.	- 38 46 49.	18		NEBULA
UGC 11807	21 43 48.	+ 06 52	60	13.	GALAXY DWRF IR
ZWG 402.021	21 43 48.	+ 09 27		15.6	GALAXY
UGC 11808	21 43 48.	+ 41 02	138	16.5	GALAXY S
PHL 4872	21 43 48.	- 03 12		17.9	BLUE STELLAR OBJECT
PHL 4877	21 43 48.	- 04 44		18.6	BLUE STELLAR OBJECT
PHL 4875	21 43 48.	- 05 18		17.9	BLUE STELLAR OBJECT
PHL 4874	21 43 48.	- 06 40		15.9	BLUE STELLAR OBJECT

OBJECT NAME	RIGHT ASCEN.	DECLINATION	DIAM.	MAGN.	TYPE OF OBJECT
PHL 4873	21 43 48.	- 07 18		17.3	BLUE STELLAR OBJECT
PHL 4876	21 43 48.	- 09 16		17.2	BLUE STELLAR OBJECT
PHL 1743	21 43 48.	- 10 14		18.3	BLUE STELLAR OBJECT
MCG-04-51-013	21 43 43.	- 21 28	420		GALAXY
HN 2811	21 43 48.6	- 14 53 13.	24		NEBULA
SC 2140-5711.9	21 43 52.	- 56 58 08.			NEBULA
HN 0725	21 43 52.	- 65 37			NEBULA
IC 5129	21 43 52.	- 65 37			NONSTELLAR OBJECT
PHL 4878	21 43 54.	- 10 56		18.5	BLUE STELLAR OBJECT
PHL 1744	21 43 54.	- 12 57		17.6	BLUE STELLAR OBJECT
SC 2140-3940.4	21 43 54.	- 39 26 37.			NEBULA
SC 2140-6358.2	21 43 54.	- 63 44 26.			NEBULA
SC 2140-3924.8	21 43 56.	- 39 11 01.			NEBULA
SC 2140-5734.3	21 43 56.	- 57 20 32.			NEBULA
SC 2140-6100.8	21 43 56.	- 60 47 02.			NEBULA
SC 2140-5519.8	21 43 57.	- 55 06 02.			NEBULA
SC 2140-5916.8	21 43 59.	- 59 03 02.			NEBULA
KHAV 754	21 44	+ 49 08			DARK NEBULA
LB 04021	21 44 00.	+ 46 29 24.		19.0	FAINT BLUE STAR
LDN 1031	21 44 00.	+ 47 00	540		DARK NEBULA
LDN 1136	21 44 00.	+ 59 45	420		DARK NEBULA
PHL 4379	21 44 00.	- 00 54		18.4	BLUE STELLAR OBJECT
PHL 4880	21 44 00.	- 01 53		18.0	BLUE STELLAR OBJECT
PHL 4881	21 44 00.	- 02 04		18.0	BLUE STELLAR OBJECT
KARA.73 50	21 44 05.	- 37 10	27		DWARF GALAXY
PHL 4883	21 44 06.	+ 02 14		17.4	BLUE STELLAR OBJECT
ZWG 402.022	21 44 06.	+ 08 04		15.4	GALAXY
UGC 11809	21 44 06.	+ 08 04	60	15.4	GALAXY Sb
KARA.73B 0929	21 44 06.	+ 08 04	66	15.4	ISOLATED GALAXY S
PHL 4884	21 44 06.	- 02 32		17.8	BLUE STELLAR OBJECT
PHL 4882	21 44 06.	- 20 47		12.5	BLUE STELLAR OBJECT
SC 2140-5746.2	21 44 06.	- 57 32 25.			NEBULA
SC 2141+1822.1	21 44 11.	+ 18 35 55.	6		NEBULA
PHL 0139	21 44 12.	+ 00 28		14.1	BLUE STELLAR OBJECT
PHL 4885	21 44 12.	+ 01 36		14.9	BLUE STELLAR OBJECT
MCG+01-55-012	21 44 12.	+ 08 05	48	14.5	GALAXY
PHL 1746	21 44 12.	- 00 10		18.8	BLUE STELLAR OBJECT
PHL 4886	21 44 12.	- 02 36		18.5	BLUE STELLAR OBJECT
PHL 1745	21 44 12.	- 20 22		12.0	BLUE STELLAR OBJECT
PHL 1747	21 44 12.	- 21 36		18.0	BLUE STELLAR OBJECT
SC 2140-5922.3	21 44 17.	- 59 08 31.			NEBULA
BC PKS2144-17	21 44 17.62	- 17 54 05.6		19.	QUASI-STELLAR OBJECT
SWB 365	21 44 17.7	- 17 54 05.		19.5	QUASI-STELLAR OBJECT
PHL 4887	21 44 18.	+ 01 36		14.9	BLUE STELLAR OBJECT
2ZW 148	21 44 18.	+ 12 55			COMPACT GALAXY
PHL 0140	21 44 18.	- 02 33		18.6	BLUE STELLAR OBJECT
PHL 4888	21 44 18.	- 02 44		17.9	BLUE STELLAR OBJECT
PHL 1749	21 44 18.	- 15 29		17.9	BLUE STELLAR OBJECT
PHL 1748	21 44 18.	- 23 24		18.3	BLUE STELLAR OBJECT
BV 06	21 44 20.	- 02 32 12.		18.9	FAINT BLUE VARIABLE
SC 2140-5856.2	21 44 21.	- 58 42 25.			NEBULA
SCHO 1337	21 44 23.	+ 59 14 54.	300		ISOLATED DARK CLOUD
MCG+00-55-015	21 44 24.	+ 01 27 30.	108	13.5	GALAXY
ZWG 376.031	21 44 24.	+ 01 28		14.7	GALAXY
UGC 11810	21 44 24.	+ 01 28	126	14.7	GALAXY Sb
KARA.73B 0930	21 44 24.	+ 01 28	120	14.7	ISOLATED GALAXY S
LDN 1030	21 44 24.	+ 46 52	660		DARK NEBULA
PHL 0141	21 44 24.	- 04 42		17.3	BLUE STELLAR OBJECT
PHL 4889	21 44 24.	- 08 52		15.8	BLUE STELLAR OBJECT
PHL 0142	21 44 24.	- 10 00		17.8	BLUE STELLAR OBJECT
IC 1401	21 44 25.	+ 01 27 59.			NONSTELLAR OBJECT
IC 5131	21 44 26.	- 35 06 56.	78		GALAXY SB0
SC 2140-6221.1	21 44 26.	- 62 07 19.	9		NEBULA
MCG+06-47-006	21 44 27.	+ 38 15	36	15.5	GALAXY
SC 2140-5747.3	21 44 27.	- 57 33 30.			NEBULA
PHL 1750	21 44 30.	- 06 56		16.6	BLUE STELLAR OBJECT
PHL 0143	21 44 30.	- 09 26		18.3	BLUE STELLAR OBJECT
PHL 1751	21 44 30.	- 12 16		17.2	BLUE STELLAR OBJECT
PHL 1752	21 44 30.	- 12 44		18.7	BLUE STELLAR OBJECT
PHL 4890	21 44 30.	- 15 40		17.9	BLUE STELLAR OBJECT
PHL 0144	21 44 30.	- 19 11		17.4	BLUE STELLAR OBJECT
ABC 2378	21 44 31.	- 20 14		16.5	RICH CLUSTER OF GALAXIES
SCHO 1338	21 44 32.	+ 51 40 36.	640		ISOLATED DARK CLOUD
HN 2814	21 44 34.1	- 42 21 18.	36		NEBULA
HN 2813	21 44 34.9	- 49 59 06.	18		NEBULA
ZWG 427.023	21 44 36.	+ 09 53		15.7	GALAXY
ZWG 512.009	21 44 36.	+ 38 16		15.2	GALAXY
MCG-05-51-015	21 44 36.	- 29 58	18	15.	GALAXY
PK104+07.1	21 44 37.	+ 63 35 23.	86	13.	PLANETARY NEBULA
RNGC 7129	21 44 38.	+ 63 25		13.0	PLANETARY NEBULA
MCG+06-47-007	21 44 39.	+ 38 16	27	15.	GALAXY
HN 2812	21 44 40.7	- 54 13 30.	12		NEBULA
ZC 2144.7+3333	21 44 42.	+ 33 33	2690		CLUSTER OF GALAXIES
LDN 1040	21 44 42.	+ 47 30	480		DARK NEBULA
SCHO 1339	21 44 42.	+ 51 03 24.	2000		ISOLATED DARK CLOUD
OCL 0241	21 44 42.	+ 65 34	1020	10.4	OPEN STAR CLUSTER
RNGC 7142	21 44 42.	+ 65 34		10.0	OPEN CLUSTER
PHL 1753	21 44 42.	- 06 35		17.3	BLUE STELLAR OBJECT
PHL 4892	21 44 42.	- 07 12		17.9	BLUE STELLAR OBJECT
PHL 4891	21 44 42.	- 09 48		17.3	BLUE STELLAR OBJECT
LB 04022	21 44 43.	+ 46 42 48.		18.6	FAINT BLUE STAR
B 164	21 44 44.	+ 50 52	1200		DARK OBJECT
HN 2815	21 44 44.1	- 14 57 05.	24		NEBULA
RNGC 7124	21 44 45.	- 50 48		13.0	GALAXY
HN 2816	21 44 45.1	- 14 56 29.	18		NEBULA
RNGC 7132	21 44 46.	+ 10 00		14.5	GALAXY
SCHO 1340	21 44 47.	+ 61 42 30.	430		ISOLATED DARK CLOUD
PHL 4894	21 44 48.	+ 00 56		18.6	BLUE STELLAR OBJECT
PHL 1754	21 44 48.	+ 01 48		17.8	BLUE STELLAR OBJECT
ZWG 427.024	21 44 48.	+ 10 00		14.6	GALAXY
2ZW 149	21 44 48.	+ 13 23			COMPACT GALAXY
MCG+02-55-012	21 44 48.	+ 13 20	30	16.	GALAXY
ZC 2144.8+2404	21 44 48.	+ 24 04	1280		CLUSTER OF GALAXIES
PHL 4895	21 44 48.	- 04 52		18.2	BLUE STELLAR OBJECT
PHL 0145	21 44 48.	- 08 00		14.1	BLUE STELLAR OBJECT
PHL 4893	21 44 48.	- 12 42		16.6	BLUE STELLAR OBJECT
PHL 4896	21 44 48.	- 14 02		18.0	BLUE STELLAR OBJECT
SC 2141-5547.7	21 44 48.	- 55 33 53.			NEBULA
SC 2141-5526.6	21 44 49.	- 55 12 47.	30		NEBULA
KLEM 33	21 44 50.	- 46 14	1200	15.	CLUSTER OF 20 GALAXIES
ABC 2379	21 44 52.	+ 00 31		17.1	RICH CLUSTER OF GALAXIES
MCG+02-55-013	21 44 54.	+ 10 00	60	14.	GALAXY
MCG-01-55-010	21 44 54.	- 06 42	36	15.	GALAXY
PHL 4900	21 44 54.	- 07 18		18.2	BLUE STELLAR OBJECT
PHL 4697	21 44 54.	- 10 56		16.2	BLUE STELLAR OBJECT
PHL 4899	21 44 54.	- 11 10		17.7	BLUE STELLAR OBJECT
MCG-02-55-002	21 44 54.	- 13 25	21	14.5	GALAXY
PHL 4898	21 44 54.	- 17 32		7.2	BLUE STELLAR OBJECT

OBJECT NAME	RIGHT ASCEN.	DECLINATION	DIAM.	MAGN.	TYPE OF OBJECT
SC 2141-5731.0	21 44 54.	- 57 17 11.			NEBULA
SC 2141-6100.8	21 44 55.	- 60 46 59.			NEBULA
SC 2141-5526.0	21 44 56.	- 55 12 11.	18		NEBULA
RNGC 7131	21 44 57.	- 13 25		14.0	GALAXY
SC 2141-5916.8	21 44 59.	- 59 02 59.			NEBULA
BC M2145+100B	21 45	+ 10 00		19.	QUASI-STELLAR OBJECT
LBN 0220	21 45	+ 22 10	1020		BRIGHT NEBULA
KHAV 755	21 45	+ 57 44	1570		DARK NEBULA
KHAV 756	21 45	+ 59 26			DARK NEBULA
PHL 1755	21 45 00.	+ 01 48		18.2	BLUE STELLAR OBJECT
UGC 11811	21 45 00.	+ 18 31	60	16.0	GALAXY Sc-IRR
UGC 11812	21 45 00.	+ 31 53	60	17.	GALAXY DWARF
LDN 1070	21 45 00.	+ 50 50	900		DARK NEBULA
LDN 1097	21 45 00.	+ 55 30	180		DARK NEBULA
LDN 1115	21 45 00.	+ 56 45	180		DARK NEBULA
LDN 1118	21 45 00.	+ 56 59	120		DARK NEBULA
LDN 1129	21 45 00.	+ 57 40	780		DARK NEBULA
LDN 1132	21 45 00.	+ 58 30	180		DARK NEBULA
PHL 1756	21 45 00.	- 05 11		18.3	BLUE STELLAR OBJECT
PHL 0146	21 45 00.	- 07 03		17.9	BLUE STELLAR OBJECT
PHL 4901	21 45 00.	- 08 40		17.6	BLUE STELLAR OBJECT
PHL 0147	21 45 00.	- 21 30		17.5	BLUE STELLAR OBJECT
SC 2141-5731.0	21 45 00.	- 57 17 11.			NEBULA
SC 2141-5920.6	21 45 00.	- 59 06 47.			NEBULA
SC 2141-5932.8	21 45 02.	- 59 18 59.			NEBULA
MCG+03-55-002	21 45 03.	+ 18 31	54	16.	GALAXY
PHL 0148	21 45 06.	- 01 27		15.6	BLUE STELLAR OBJECT
MCG-01-55-011	21 45 06.	- 04 25	60	14.5	GALAXY
PHL 1757	21 45 06.	- 24 00		18.4	BLUE STELLAR OBJECT
TON-S 0034	21 45 06.	- 30 33		14.0	BLUE STAR
HN 2817	21 45 06.5	- 17 01 40.	18		NEBULA
SC 2141-5622.3	21 45 07.	- 56 08 29.	12		NEBULA
SCHO 1341	21 45 10.	+ 60 46 54.	330		ISOLATED DARK CLOUD
SC 2141-5642.5	21 45 10.	- 56 28 41.	72		NEBULA
SCHO 1342	21 45 11.	+ 49 41 12.	330		ISOLATED DARK CLOUD
PHL 1767	21 45 12.	+ 00 24		17.4	BLUE STELLAR OBJECT
UGC 11813	21 45 12.	+ 21 56	60	17.	GALAXY IRR
4ZW 076	21 45 12.	+ 33 33			COMPACT GALAXY
PHL 4902	21 45 12.	- 03 36		16.0	BLUE STELLAR OBJECT
PHL 4903	21 45 12.	- 05 24		16.6	BLUE STELLAR OBJECT
PHL 4905	21 45 12.	- 06 08		18.0	BLUE STELLAR OBJECT
PHL 0149	21 45 12.	- 12 50		14.2	BLUE STELLAR OBJECT
PHL 4904	21 45 12.	- 26 08		18.0	BLUE STELLAR OBJECT
TON-S 0035	21 45 12.	- 32 18		15.2	BLUE STAR
HN 2318	21 45 12.2	- 15 01 04.	30		NEBULA
PHL 0150	21 45 18.	+ 01 34		18.4	BLUE STELLAR OBJECT
PHL 4906	21 45 18.	- 12 28		17.4	BLUE STELLAR OBJECT
RNGC 7130	21 45 18.	- 35 11			GALAXY
SCHO 1343	21 45 19.	+ 57 31 36.	620		ISOLATED DARK CLOUD
SCHO 1344	21 45 20.	+ 65 52 48.	410		ISOLATED DARK CLOUD
SC 2141-6438.4	21 45 22.	- 64 24 35.	12		NEBULA
PHL 0151	21 45 24.	- 00 59		18.2	BLUE STELLAR OBJECT
PHL 1758	21 45 24.	- 17 14		18.0	BLUE STELLAR OBJECT
SC 2141-5933.2	21 45 27.	- 59 19 22.	18		NEBULA
SC 2141-6235.9	21 45 28.	- 62 22 04.	12		NEBULA
ZC 2145.5+1332	21 45 30.	+ 13 32	1080		CLUSTER OF GALAXIES
MCG-01-55-012	21 45 30.	- 02 58	48	15.5	GALAXY
PHL 0152	21 45 30.	- 04 58		18.9	BLUE STELLAR OBJECT
TON-S 0036	21 45 30.	- 34 05		14.6	BLUE STAR
SC 2141-5930.0	21 45 34.	- 59 16 10.			NEBULA
RNGC 7125	21 45 34.	- 60 56		12.5	GALAXY
SC 2141-6203.6	21 45 34.	- 61 49 46.	6		NEBULA
SC 2142-5624.1	21 45 35.	- 56 10 16.	36		NEBULA
SWB 366	21 45 35.9	+ 06 43 43.		16.5	QUASI-STELLAR OBJECT
PHL 0153	21 45 36.	- 03 59		17.1	BLUE STELLAR OBJECT
MCG-02-55-003	21 45 36.	- 13 29	48	15.5	GALAXY
PHL 0154	21 45 36.	- 20 18		18.6	BLUE STELLAR OBJECT
PHL 4907	21 45 36.	- 25 46		17.5	BLUE STELLAR OBJECT
MCG-05-51-016	21 45 36.	- 32 27	60	14.	GALAXY
TON-S 0037	21 45 36.	- 34 24		14.2	BLUE STAR
BC PKS2145+06	21 45 36.07	+ 06 43 41.3		16.47	QUASI-STELLAR OBJECT
SC 2142-5629.2	21 45 37.	- 56 15 21.	18		NEBULA
RNGC 7126	21 45 40.	- 60 50		13.0	GALAXY
SCHO 1345	21 45 41.	+ 51 05 24.	1050		ISOLATED DARK CLOUD
PHL 4908	21 45 42.	- 13 44		16.1	BLUE STELLAR OBJECT
PHL 4909	21 45 42.	- 15 44		16.1	BLUE STELLAR OBJECT
SC 2142-5608.7	21 45 44.	- 55 54 51.	18		NEBULA
SC 2142-5935.8	21 45 44.	- 59 21 57.			NEBULA
MCG+00-55-016	21 45 45.	- 01 56	36	14.	GALAXY
SC 2142-6117.0	21 45 45.	- 61 03 09.	18		NEBULA
SC 2142-5707.8	21 45 47.	- 56 53 57.			NEBULA
PHL 0155	21 45 48.	+ 00 25		15.2	BLUE STELLAR OBJECT
MCG+00-55-017	21 45 48.	+ 01 37	12	15.5	GALAXY
ZWG 376.032	21 45 48.	- 01 54		14.3	GALAXY
UGC 11814	21 45 48.	- 01 54	54	14.3	GALAXY
PHL 4911	21 45 48.	- 05 05		17.9	BLUE STELLAR OBJECT
PHL 0156	21 45 48.	- 05 37		18.2	BLUE STELLAR OBJECT
PHL 4910	21 45 49.	- 18 51		16.6	BLUE STELLAR OBJECT
PHL 1759	21 45 48.	- 18 58		18.0	BLUE STELLAR OBJECT
SCHO 1246	21 45 52.	+ 58 05 54.	290		ISOLATED DARK CLOUD
IC 5136	21 45 52.	- 33 53 08.			NONSTELLAR OBJECT
SC 2142-5549.3	21 45 52.	- 55 35 27.			NEBULA
REIN 2.302	21 45 52.13	+ 21 55 49.3			NEBULA
HN 0726	21 45 53.	- 74 14			NEBULA
ZC 2145.9+0017	21 45 54.	+ 00 17	870		CLUSTER OF GALAXIES
ZWG 376.033	21 45 54.	+ 01 38		15.0	GALAXY
PHL 4912	21 45 54.	+ 01 58		18.5	BLUE STELLAR OBJECT
ZWG 472.008	21 45 54.	+ 21 56		13.3	GALAXY
UGC 11815	21 45 54.	+ 21 56	96	13.3	GALAXY S
MCG+04-51-005	21 45 54.	+ 21 56	60	13.5	GALAXY
4ZW 077	21 45 54.	+ 26 13			COMPACT GALAXY
PHL 1760	21 45 54.	- 06 01		17.7	BLUE STELLAR OBJECT
SC 2142-5947.4	21 45 54.	- 59 33 33.	30		NEBULA
IC 5130	21 45 54.	- 74 14			NONSTELLAR OBJECT
REIN 2.303	21 45 54.14	+ 21 55 36.8			NEBULA
RNGC 7137	21 45 57.	+ 21 56		13.5	GALAXY
KHAV 757	21 46	+ 47 20	3010		DARK NEBULA
KHAV 758	21 46	+ 60 08			DARK NEBULA
PHL 1761	21 46 00.	+ 01 48		17.9	BLUE STELLAR OBJECT
LDN 1052	21 46 00.	+ 47 54	540		DARK NEBULA
LDN 1066	21 46 00.	+ 50 30	1800		DARK NEBULA
LDN 1120	21 46 00.	+ 56 55	120		DARK NEBULA
PHL 0157	21 46 00.	- 03 42		17.8	BLUE STELLAR OBJECT
PHL 1763	21 46 00.	- 06 11		18.9	BLUE STELLAR OBJECT
PHL 4913	21 46 00.	- 10 03		18.4	BLUE STELLAR OBJECT
PHL 1762	21 46 00.	- 16 42		17.9	BLUE STELLAR OBJECT
SC 2142-6150.6	21 46 05.	- 61 36 45.	6		NEBULA
ZC 2146.1+2611	21 46 06.	+ 26 11	2820		CLUSTER OF GALAXIES

OBJECT NAME	RIGHT ASCEN.	DECLINATION	DIAM.	MAGN.	TYPE OF OBJECT
PHL 1764	21 46 06.	- 00 04		18.1	BLUE STELLAR OBJECT
PHL 1765	21 46 06.	- 04 15		17.1	BLUE STELLAR OBJECT
PHL 0158	21 46 06.	- 22 06		16.1	BLUE STELLAR OBJECT
SC 2143-3936.7	21 46 07.	- 39 22 50.			NEBULA
SC 2142-5810.2	21 46 08.	- 57 56 20.			NEBULA
RNGC 7134	21 46 09.	- 13 13			NON-EXISTENT OBJECT
SC 2142-5709.0	21 46 10.	- 56 55 08.			NEBULA
SC 2142-5545.9	21 46 11.	- 55 32 02.			NEBULA
PHL 0159	21 46 12.	+ 01 43		10.8	BLUE STELLAR OBJECT
ZWG 493.008	21 46 12.	+ 29 10		15.0	GALAXY
LDN 1047	21 46 12.	+ 47 38	240		DARK NEBULA
ISS 0127	21 46 12.	+ 54 25	239		STELLAR RING
ISS 0080	21 46 12.	+ 58 39	143		STELLAR RING
PHL 0160	21 46 12.	- 08 58		17.5	BLUE STELLAR OBJECT
PHL 4914	21 46 12.	- 11 03		18.0	BLUE STELLAR OBJECT
PHL 1766	21 46 12.	- 18 06		18.6	BLUE STELLAR OBJECT
PHL 4915	21 46 12.	- 20 58		17.8	BLUE STELLAR OBJECT
PHL 1768	21 46 12.	- 22 08		16.1	BLUE STELLAR OBJECT
SC 2142-6021.4	21 46 12.	- 60 07 32.	18		NEBULA
SC 2142-5609.4	21 46 13.	- 55 55 32.			NEBULA
HN 2819	21 46 13.7	- 54 24 14.	12		NEBULA
HN 2820	21 46 14.4	- 53 55 50.	12		NEBULA
RNGC 7136	21 46 15.	- 12 01			NON-EXISTENT OBJECT
SC 2143-3907.1	21 46 15.	- 38 53 13.	60		NEBULA
SC 2142-5800.7	21 46 15.	- 57 46 50.			NEBULA
SC 2142-5710.9	21 46 16.	- 56 57 02.			NEBULA
SC 2142-5947.7	21 46 16.	- 59 33 50.			NEBULA
PHL 4917	21 46 18.	+ 00 20		16.6	BLUE STELLAR OBJECT
SCHO 1347	21 46 18.	+ 47 16 06.	1690		ISOLATED DARK CLOUD
PHL 0161	21 46 18.	- 15 16		18.3	BLUE STELLAR OBJECT
PHL 4916	21 46 18.	- 16 04		16.7	BLUE STELLAR OBJECT
SC 2142-6157.0	21 46 23.	- 61 43 08.	6		NEBULA
MCG+00-55-018	21 46 24.	+ 00 12	72	14.5	GALAXY
ASS 27	21 46 24.	+ 60 50	27600		OB ASSOCIATION CEP OB2
PHL 4918	21 46 24.	- 04 26		16.1	BLUE STELLAR OBJECT
MCG-01-55-013	21 46 24.	- 07 32	48	15.	GALAXY
PHL 4919	21 46 24.	- 11 34		17.4	BLUE STELLAR OBJECT
PHL 1769	21 46 24.	- 11 35		18.2	BLUE STELLAR OBJECT
SC 2142-5556.3	21 46 25.	- 55 42 25.			NEBULA
SC 2142-6157.1	21 46 25.	- 61 43 14.	6		NEBULA
HN 2821	21 46 26.9	- 52 54 13.	12		NEBULA
ZWG 376.034	21 46 30.	+ 00 13		14.8	GALAXY
UGC 11816	21 46 30.	+ 00 13	96	14.8	GALAXY SBb-c
KARA.73B 0931	21 46 30.	+ 00 13	78	14.8	ISOLATED GALAXY S
LDN 1045	21 46 30.	+ 47 25	540		DARK NEBULA
LDN 1073	21 46 30.	+ 51 10	1620		DARK NEBULA
PHL 4920	21 46 30.	- 06 05		18.2	BLUE STELLAR OBJECT
PHL 4921	21 46 30.	- 13 02		16.5	BLUE STELLAR OBJECT
SCHO 1348	21 46 32.	+ 52 32 30.	410		ISOLATED DARK CLOUD
SC 2143-5543.9	21 46 32.	- 55 30 01.			NEBULA
SC 2142-6018.0	21 46 33.	- 60 04 07.	18		NEBULA
RNGC 7123	21 46 34.	- 70 34			UNVERIFIED SOUTHERN OBJECT
RNGC 7138	21 46 35.	+ 12 16		15.5	GALAXY
HN 2823	21 46 35.7	- 52 56 19.			NEBULA
ZC 2146.6+0155	21 46 36.	+ 01 55	1080		CLUSTER OF GALAXIES
ZWG 427.025	21 46 36.	+ 12 16		15.4	GALAXY
UGC 11817	21 46 36.	+ 12 16	72	15.4	GALAXY SBa
MCG+02-55-014	21 46 36.	+ 12 16	72	14.5	GALAXY
KARA.73B 0932	21 46 36.	+ 12 16	72	15.4	ISOLATED GALAXY S
PHL 0162	21 46 36.	- 04 05		17.3	BLUE STELLAR OBJECT
PHL 4922	21 46 36.	- 24 00		17.8	BLUE STELLAR OBJECT
HN 2822	21 46 36.8	- 53 59 49.	18		NEBULA
SC 2142-6239.1	21 46 37.	- 62 25 13.	6		NEBULA
SC 2143-3930.9	21 46 38.	- 39 17 00.	48		NEBULA
SC 2143-5542.6	21 46 39.	- 55 28 43.			NEBULA
RNGC 7143	21 46 42.	+ 29 43			NON-EXISTENT OBJECT
PHL 4924	21 46 42.	- 04 45		18.3	BLUE STELLAR OBJECT
PHL 1770	21 46 42.	- 05 51		18.7	BLUE STELLAR OBJECT
PHL 4923	21 46 42.	- 16 48		16.6	BLUE STELLAR OBJECT
PHL 1771	21 46 42.	- 22 56		18.2	BLUE STELLAR OBJECT
SC 2143-5613.2	21 46 44.	- 55 59 19.			NEBULA
SC 2142-6236.9	21 46 44.	- 62 23 01.			NEBULA
SC 2142-6318.2	21 46 44.	- 63 04 19.			NEBULA
SC 2143-6238.2	21 46 45.	- 62 24 19.	6		NEBULA
SCHO 1349	21 46 46.	+ 62 22 54.	330		ISOLATED DARK CLOUD
SHB 367	21 46 46.1	- 13 18 24.		20.	QUASI-STELLAR OBJECT
BC PKS2146-13	21 46 46.37	- 13 18 26.7		20.	QUASI-STELLAR OBJECT
SC 2143-3815.0	21 46 47.	- 38 01 06.	9		NEBULA
SC 2143-6133.1	21 46 47.	- 61 19 13.	30		NEBULA
ZWG 343.001	21 46 48.	+ 72 15		15.7	GALAXY
UGC 11818	21 46 48.	+ 72 15	120	15.7	GALAXY S
PHL 1772	21 46 48.	- 04 25		16.2	BLUE STELLAR OBJECT
PHL 0163	21 46 48.	- 07 44		17.4	BLUE STELLAR OBJECT
PHL 4925	21 46 48.	- 11 46		16.5	BLUE STELLAR OBJECT
PHL 4926	21 46 48.	- 20 25		16.7	BLUE STELLAR OBJECT
PHL 4927	21 46 48.	- 22 50		18.0	BLUE STELLAR OBJECT
RNGC 7135	21 46 48.	- 35 07		13.0	GALAXY
SCHO 1350	21 46 49.	+ 52 29 24.	800		ISOLATED DARK CLOUD
SC 2143-5544.6	21 46 51.	- 55 30 42.			NEBULA
SC 2143-5949.7	21 46 52.	- 59 35 49.			NEBULA
PHL 1773	21 46 54.	- 01 40		17.1	BLUE STELLAR OBJECT
PHL 4928	21 46 54.	- 05 08		17.9	BLUE STELLAR OBJECT
PHL 0164	21 46 54.	- 19 15		17.8	BLUE STELLAR OBJECT
SCHO 1351	21 46 55.	+ 52 51 30.	230		ISOLATED DARK CLOUD
SC 2143-6144.5	21 46 55.	- 61 30 37.			NEBULA
SC 2143-6202.2	21 46 57.	- 61 48 19.	18		NEBULA
ZC 2147.0+0807	21 47 00.	+ 08 07	2350		CLUSTER OF GALAXIES
ZC 2147.0+2148	21 47 00.	+ 21 48	3970		CLUSTER OF GALAXIES
UGC 11819	21 47 00.	+ 41 43	96	17.	GALAXY DWARF
ZC 2147.0+8155	21 47 00.	+ 81 55	1340		CLUSTER OF GALAXIES
PHL 1774	21 47 00.	- 03 04		18.9	BLUE STELLAR OBJECT
PHL 0165	21 47 00.	- 13 04		17.8	BLUE STELLAR OBJECT
PHL 4930	21 47 00.	- 21 26		18.4	BLUE STELLAR OBJECT
PHL 4929	21 47 00.	- 22 22		16.8	BLUE STELLAR OBJECT
SC 2143-6134.7	21 47 02.	- 61 20 48.	18		NEBULA
SC 2143-6316.8	21 47 04.	- 63 02 54.	6		NEBULA
UGC 11820	21 47 06.	+ 13 59	90	17.	GALAXY DWRF SP
MCG+02-55-015	21 47 06.	+ 14 00	120	16.	GALAXY
ZWG 450.005	21 47 06.	+ 15 33		15.3	GALAXY
UGC 11821	21 47 06.	+ 15 33	114	15.3	GALAXY Sa
PHL 1775	21 47 06.	- 03 13		18.7	BLUE STELLAR OBJECT
PHL 4931	21 47 06.	- 03 57		16.4	BLUE STELLAR OBJECT
SC 2143-5617.3	21 47 07.	- 56 03 24.			NEBULA
SC 2143-6213.8	21 47 08.	- 61 59 54.	60		NEBULA
IC 5135	21 47 09.	- 35 02 37.	84	13.1	GALAXY IRR
SC 2143-6238.2	21 47 11.	- 62 24 18.			NEBULA
SC 2143-6238.4	21 47 11.	- 62 24 30.			NEBULA
HN 2824	21 47 11.8	- 52 28 47.	12		NEBULA
SCHO 1352	21 47 12.	+ 52 08 42.	420		ISOLATED DARK CLOUD
ZC 2147.2-0229	21 47 12.	- 02 29	1280		CLUSTER OF GALAXIES
PHL 4933	21 47 12.	- 05 18		16.4	BLUE STELLAR OBJECT
PHL 4932	21 47 12.	- 07 50		18.0	BLUE STELLAR OBJECT
PHL 0166	21 47 12.	- 09 50		17.0	BLUE STELLAR OBJECT
PHL 4934	21 47 12.	- 11 44		18.0	BLUE STELLAR OBJECT
PHL 1776	21 47 12.	- 19 13		18.0	BLUE STELLAR OBJECT
SC 2143-6239.8	21 47 12.	- 62 25 54.	6		NEBULA
HN 2825	21 47 12.9	- 46 49 53.	12		NEBULA
HUB C52	21 47 14.	+ 65 21			DIFFUSE NEBULA
MCG+03-55-003	21 47 15.	+ 15 23	114	15.	GALAXY
LDN 1042	21 47 18.	+ 47 10	480		DARK NEBULA
ISS 0192	21 47 18.	+ 48 50	660		STELLAR RING
PHL 4935	21 47 18.	- 02 36		16.2	BLUE STELLAR OBJECT
PHL 4936	21 47 18.	- 04 55		16.9	BLUE STELLAR OBJECT
PHL 0167	21 47 18.	- 14 06		16.0	BLUE STELLAR OBJECT
PHL 1777	21 47 18.	- 23 08		18.3	BLUE STELLAR OBJECT
TON-S 0038	21 47 18.	- 31 04		13.8	BLUE STAR
SC 2143-5542.8	21 47 19.	- 55 28 53.			NEBULA
SC 2143-5538.2	21 47 20.	- 55 24 17.			NEBULA
SCHO 1353	21 47 21.	+ 51 40 00.	690		ISOLATED DARK CLOUD
ZC 2147.4+0915	21 47 24.	+ 09 15	1480		CLUSTER OF GALAXIES
PHL 1778	21 47 24.	- 03 50		18.2	BLUE STELLAR OBJECT
PHL 1779	21 47 24.	- 05 35		18.4	BLUE STELLAR OBJECT
PHL 0168	21 47 24.	- 16 17		17.9	BLUE STELLAR OBJECT
PHL 4937	21 47 24.	- 20 18		16.4	BLUE STELLAR OBJECT
SC 2143-5548.5	21 47 25.	- 55 34 35.			NEBULA
B 165	21 47 26.	+ 59 59	1080		DARK OBJECT
SCHO 1354	21 47 27.	+ 51 36 42.	890		ISOLATED DARK CLOUD
SC 2144-5546.6	21 47 28.	- 55 32 41.			NEBULA
PHL 0169	21 47 30.	- 00 04		17.6	BLUE STELLAR OBJECT
PHL 1780	21 47 30.	- 05 01		15.8	BLUE STELLAR OBJECT
PHL 0170	21 47 30.	- 07 08		17.3	BLUE STELLAR OBJECT
MCG-05-51-017	21 47 30.	- 31 16	36	14.	GALAXY
SC 2144-5617.1	21 47 30.	- 56 03 11.			NEBULA
SC 2143-6139.7	21 47 30.	- 61 25 47.	18		NEBULA
SC 2144-5655.3	21 47 31.	- 56 41 23.			NEBULA
SC 2144-4129.8	21 47 32.	- 41 15 52.	18		NEBULA
SC 2143-6140.2	21 47 33.	- 61 26 17.	6		NEBULA
HN 2826	21 47 33.8	- 54 46 59.	18		NEBULA
SC 2144-5501.1	21 47 34.	- 54 47 11.	30		NEBULA
ZWG 427.026	21 47 36.	+ 13 51		15.7	GALAXY
ZWG 472.009	21 47 36.	+ 22 35		15.7	GALAXY
UGC 11822	21 47 36.	+ 40 26	66	16.5	GALAXY SB0
PHL 4938	21 47 36.	- 05 32		17.5	BLUE STELLAR OBJECT
PHL 0171	21 47 36.	- 06 36		17.1	BLUE STELLAR OBJECT
PHL 4939	21 47 36.	- 07 00		18.0	BLUE STELLAR OBJECT
SC 2143-6137.6	21 47 36.	- 61 23 41.	36		NEBULA
SC 2143-6405.0	21 47 39.	- 63 51 05.			NEBULA
HN 2827	21 47 39.1	- 54 33 52.	12		NEBULA
ZWG 493.009	21 47 42.	+ 21 23		15.7	GALAXY
PHL 4940	21 47 42.	- 25 22		18.3	BLUE STELLAR OBJECT
PHL 0172	21 47 42.	- 25 52		17.6	BLUE STELLAR OBJECT
MCG-05-51-018	21 47 42.	- 29 27	42	15.	GALAXY
SC 2144-3752.4	21 47 42.	- 37 38 28.	12		NEBULA
SC 2144-5548.0	21 47 43.	- 55 34 04.			NEBULA
SC 2144-5754.8	21 47 44.	- 57 40 52.			NEBULA
HN 0727	21 47 44.	- 65 49			NEBULA
SC 2144-6126.6	21 47 45.	- 61 12 41.	12		NEBULA
SC 2144-6127.0	21 47 45.	- 61 13 05.	12		NEBULA
SC 2144-6241.5	21 47 45.	- 62 27 35.	24		NEBULA
IC 5137	21 47 45.	- 65 49			NONSTELLAR OBJECT
ARC 2380	21 47 46.	- 04 58		17.7	RICH CLUSTER OF GALAXIES
IC 5139	21 47 46.	- 31 12 53.			NONSTELLAR OBJECT
ZWG 450.006	21 47 48.	+ 15 38		15.7	GALAXY
4ZW 078	21 47 48.	+ 34 43			COMPACT GALAXY
UGC 11823	21 47 48.	+ 34 43	18	14.0	GALAXY EX CMPT
ZC 2147.8-0101	21 47 48.	- 01 01	870		CLUSTER OF GALAXIES
PHL 4941	21 47 48.	- 03 02		15.9	BLUE STELLAR OBJECT
PHL 4942	21 47 48.	- 16 23		18.0	BLUE STELLAR OBJECT
PHL 0173	21 47 48.	- 18 02		16.7	BLUE STELLAR OBJECT
PHL 1731	21 47 48.	- 20 06		17.8	BLUE STELLAR OBJECT
HN 2828	21 47 48.5	- 54 00 16.	12		NEBULA
SC 2144-6027.6	21 47 51.	- 60 13 40.			NEBULA
IC 1403	21 47 52.	- 02 56 58.			NONSTELLAR OBJECT
SC 2144-6231.2	21 47 53.	- 62 17 16.	4		NEBULA
SCHO 1355	21 47 55.	+ 50 40 48.	660		ISOLATED DARK CLOUD
HN 2829	21 47 55.7	- 46 30 40.	12		NEBULA
LBN 0157	21 48	+ 13 00	3300		BRIGHT NEBULA
UGC 11824	21 48 00.	+ 30 37	60	16.0	GALAXY SB?c
LDN 1114	21 48 00.	+ 56 10	480		DARK NEBULA
7ZW 939	21 48 00.	+ 84 02			COMPACT GALAXY
PHL 4945	21 48 00.	- 06 09		17.3	BLUE STELLAR OBJECT
PHL 4946	21 48 00.	- 18 56		14.8	BLUE STELLAR OBJECT
PHL 0174	21 48 00.	- 19 56		14.1	BLUE STELLAR OBJECT
PHL 4943	21 48 00.	- 21 02		16.6	BLUE STELLAR OBJECT
PHL 4944	21 48 00.	- 21 46		16.6	BLUE STELLAR OBJECT
SC 2144-5535.9	21 48 02.	- 55 21 58.			NEBULA
KARA.73 51	21 48 04.	- 43 22	67		DWARF GALAXY
SC 2144-6241.2	21 48 04.	- 62 27 16.			NEBULA
MCG+00-55-019	21 48 06.	+ 00 39	48	15.	GALAXY
ZWG 376.035	21 48 06.	+ 00 41		15.4	GALAXY
UGC 11825	21 48 06.	+ 00 41	66	15.4	GALAXY Sa?
SC 2144-5522.4	21 48 06.	- 55 08 27.	24		NEBULA
SC 2144-5548.1	21 48 06.	- 55 34 09.			NEBULA
SC 2144-6205.3	21 48 06.	- 61 51 22.	4		NEBULA
HN 2830	21 48 08.7	- 52 52 15.	18		NEBULA
SC 2144-6405.0	21 48 11.	- 63 51 04.	12		NEBULA
2ZW 150	21 48 12.	+ 02 35			COMPACT GALAXY
ZWG 427.027	21 48 12.	+ 13 02		15.4	GALAXY
PHL 4949	21 48 12.	- 02 44		15.2	BLUE STELLAR OBJECT
PHL 4952	21 48 12.	- 04 32		19.5	BLUE STELLAR OBJECT
PHL 1782	21 48 12.	- 04 54		16.7	BLUE STELLAR OBJECT
PHL 4948	21 48 12.	- 06 07		17.8	BLUE STELLAR OBJECT
PHL 1784	21 48 12.	- 06 16		18.0	BLUE STELLAR OBJECT
PHL 4947	21 48 12.	- 07 41		18.2	BLUE STELLAR OBJECT
PHL 4951	21 48 12.	- 09 34		17.2	BLUE STELLAR OBJECT
PHL 4950	21 48 12.	- 13 00		16.8	BLUE STELLAR OBJECT
PHL 1783	21 48 12.	- 15 50		18.0	BLUE STELLAR OBJECT
SC 2145-3903.0	21 48 13.	- 38 49 03.	30		NEBULA
SC 2144-6029.3	21 48 13.	- 60 15 21.			NEBULA
SC 2144-6219.0	21 48 16.	- 62 05 03.	6		NEBULA
IC 1404	21 48 17.	- 09 30 20.			NONSTELLAR OBJECT
MCG+00-55-020	21 48 18.	+ 01 46	48	15.	GALAXY
ZWG 376.037	21 48 18.	+ 01 47		15.0	GALAXY
UGC 11826	21 48 18.	+ 01 47	60	15.0	GALAXY SBa-b
IC 1405	21 48 18.	+ 01 47 29.			NONSTELLAR OBJECT
ZC 2148.3+0202	21 48 18.	+ 02 02	870		CLUSTER OF GALAXIES

OBJECT NAME	RIGHT ASCEN.	DECLINATION	DIAM.	MAGN.	TYPE OF OBJECT
MCG+02-55-016	21 48 18.	+ 13 03	48	15.5	GALAXY
ZWG 472.010	21 48 18.	+ 22 37		15.2	GALAXY
UGC 11827	21 48 18.	+ 22 37	60	15.2	GALAXY PECULR?
ZWG 376.036	21 48 18.	- 01 04		15.5	GALAXY
PHL 4953	21 48 18.	- 04 31		16.0	BLUE STELLAR OBJECT
PHL 0175	21 48 18.	- 15 30		16.0	BLUE STELLAR OBJECT
PHL 0176	21 48 18.	- 25 04		17.1	BLUE STELLAR OBJECT
SER 149.05	21 48 18.	- 57 43	480	18.	CLUSTER OF GALAXIES
ZWG 427.028	21 48 24.	+ 13 51		15.5	GALAXY
PHL 4955	21 48 24.	- 03 28		18.0	BLUE STELLAR OBJECT
PHL 4954	21 48 24.	- 04 41		17.9	BLUE STELLAR OBJECT
PHL 4956	21 48 24.	- 08 50		18.0	BLUE STELLAR OBJECT
PHL 0177	21 48 24.	- 10 01		17.8	BLUE STELLAR OBJECT
PHL 0178	21 48 24.	- 21 20		12.7	BLUE STELLAR OBJECT
HN 2831	21 48 25.0	- 52 30 15.			NEBULA
ARC 2381	21 48 26.	+ 02 03		17.1	RICH CLUSTER OF GALAXIES
ZWG 376.038	21 48 30.	+ 01 45		15.5	GALAXY
ZWG 376.039	21 48 30.	+ 02 32		15.4	GALAXY
UGC 11828	21 48 30.	+ 02 32	60	15.4	GALAXY Sa-b
ZWG 402.023	21 48 30.	+ 06 00		15.5	GALAXY
KARA.73B 0933	21 48 30.	+ 06 00	36	15.5	ISOLATED GALAXY S
PHL 4958	21 48 30.	- 04 05		17.9	BLUE STELLAR OBJECT
PHL 4957	21 48 30.	- 04 06		17.9	BLUE STELLAR OBJECT
SC 2144-6348.9	21 48 32.	- 63 34 57.	6		NEBULA
IC 1406	21 48 33.	+ 01 45 30.			NONSTELLAR OBJECT
HN 0001	21 48 33.	+ 49 31			NEBULA
SC 2145-5603.5	21 48 33.	- 55 49 32.			NEBULA
SC 2145-5757.4	21 48 33.	- 57 43 26.			NEBULA
RNGC 7150	21 48 34.	+ 49 31			NON-EXISTENT OBJECT
SC 2145-5759.0	21 48 35.	- 57 45 02.			NEBULA
ZWG 376.040	21 48 36.	+ 02 28		15.3	GALAXY
MCG+01-55-013	21 48 36.	+ 06 01	24	14.5	GALAXY
ZWG 427.029	21 48 36.	+ 14 17		15.6	GALAXY
UGC 11829	21 48 36.	+ 44 28	90	18.	GALAXY DWARF
PHL 0179	21 48 36.	- 02 58		16.5	BLUE STELLAR OBJECT
PHL 4959	21 48 36.	- 07 39		17.4	BLUE STELLAR OBJECT
MCG-02-55-004	21 48 36.	- 12 30	36	15.	GALAXY
PHL 4960	21 48 36.	- 12 44		17.4	BLUE STELLAR OBJECT
TON-S 0039	21 48 36.	- 25 54		16.2	BLUE STAR
YC 2148-43	21 48 38.	- 43 21 36.			UNUSUAL SOUTHERN GALAXY
MCG+00-55-021	21 48 42.	+ 02 16	18	15.5	GALAXY
ZWG 376.043	21 48 42.	+ 02 17		15.4	GALAXY
5ZW 378	21 48 42.	+ 41 45			COMPACT GALAXY
UPA 70	21 48 42.	+ 54 17	264		STELLAR RING
ZWG 376.042	21 48 42.	- 01 11		15.1	GALAXY
MCG+00-55-022	21 48 42.	- 01 13	36	16.	GALAXY
ZWG 376.041	21 48 42.	- 01 53		15.1	GALAXY
MCG+00-55-023	21 48 42.	- 01 55	15	15.5	GALAXY
RNGC 7141	21 48 47.	- 55 48			GALAXY
ZC 2148.8+1334	21 48 48.	+ 13 34	940		CLUSTER OF GALAXIES
ZWG 472.011	21 48 48.	+ 25 37		15.0	GALAXY
UGC 11830	21 48 48.	+ 25 37	84	15.0	GALAXY Sb
ISS 0081	21 48 48.	+ 58 35	215		STELLAR RING
PHL 1785	21 48 48.	- 07 38		18.5	BLUE STELLAR OBJECT
PHL 4961	21 48 48.	- 09 04		17.1	BLUE STELLAR OBJECT
MCG+04-51-006	21 48 51.	+ 25 38	78	14.5	GALAXY
RNGC 7140	21 48 52.	- 56 47			NON-EXISTENT OBJECT
MCG+01-55-014	21 48 54.	+ 05 11 30.	24	15.5	GALAXY
UGC 11831	21 48 54.	+ 44 19	66	16.0	GALAXY E
ISS 0193	21 48 54.	+ 49 02	222		STELLAR RING
PHL 1786	21 48 54.	- 20 42		16.9	BLUE STELLAR OBJECT
KHAV 759	21 49	+ 59 50	1570		DARK NEBULA
LDN 1144	21 49 00.	+ 59 53	660		DARK NEBULA
PHL 0180	21 49 00.	- 05 18		15.9	BLUE STELLAR OBJECT
PHL 1787	21 49 00.	- 10 15		16.6	BLUE STELLAR OBJECT
PHL 4962	21 49 00.	- 12 53		18.7	BLUE STELLAR OBJECT
PHL 0181	21 49 00.	- 13 34		18.3	BLUE STELLAR OBJECT
PHL 0182	21 49 00.	- 15 16		17.9	BLUE STELLAR OBJECT
HN 2832	21 49 00.9	- 52 38 55.	24		NEBULA
MCG+07-45-001	21 49 03.	+ 44 17	36	17.	GALAXY
KARA.73 52	21 49 03.	- 33 14	27		DWARF GALAXY
BC PKS2149-20	21 49 03.7	- 20 00 14.			QUASI-STELLAR OBJECT
ZC 2149.1+2036	21 49 06.	+ 20 36	2490		CLUSTER OF GALAXIES
ZWG 472.012	21 49 06.	+ 22 36		15.7	GALAXY
ISS 0194	21 49 06.	+ 48 36	309		STELLAR RING
PHL 4963	21 49 06.	- 19 17		15.8	BLUE STELLAR OBJECT
SC 2145-5553.0	21 49 07.	- 55 39 01.			NEBULA
ZC 2149.2+0236	21 49 12.	+ 02 36	540		CLUSTER OF GALAXIES
ZWG 376.044	21 49 12.	+ 02 47		15.5	GALAXY
ZWG 450.007	21 49 12.	+ 16 07		15.7	GALAXY
ZWG 472.013	21 49 12.	+ 22 25		15.6	GALAXY
MCG+04-51-007	21 49 12.	+ 22 36	36	15.5	GALAXY
PHL 4964	21 49 12.	- 10 18		18.1	BLUE STELLAR OBJECT
RNGC 7146	21 49 14.	+ 02 47		15.5	GALAXY
MCG+00-55-024	21 49 15.	+ 02 45	36	15.	GALAXY
SC 2145-6300.8	21 49 15.	- 62 46 49.	9		NEBULA
MCG+02-55-017	21 49 18.	+ 11 20	72	15.5	GALAXY
UGC 11832	21 49 18.	+ 11 21	72	16.0	GALAXY Sb-c
MCG+04-51-008	21 49 18.	+ 22 25	36	15.5	GALAXY
UGC 11833	21 49 18.	+ 33 53	60	17.	GALAXY Sc
ARC 2382	21 49 18.	- 15 53		16.0	RICH CLUSTER OF GALAXIES
MCG-05-51-019	21 49 18.	- 27 16	60	15.5	GALAXY
SC 2145-5750.2	21 49 20.	- 57 36 13.			NEBULA
SC 2145-6258.7	21 49 20.	- 62 44 43.	6		NEBULA
B 368	21 49 21.	+ 58 45	840		DARK OBJECT
HN 0728	21 49 22.	- 69 11			NEBULA
IC 5138	21 49 23.	- 69 11			NONSTELLAR OBJECT
2ZW 151	21 49 24.	+ 02 46			COMPACT GALAXY
MCG+00-55-025	21 49 24.	+ 02 49	36	15.	GALAXY
ZWG 376.045	21 49 24.	+ 02 50		14.8	GALAXY
ZWG 472.014	21 49 24.	+ 25 01		14.5	GALAXY
UGC 11834	21 49 24.	+ 25 01	66	14.5	GALAXY
PHL 4966	21 49 24.	- 04 22		16.6	BLUE STELLAR OBJECT
PHL 4965	21 49 24.	- 05 50		15.7	BLUE STELLAR OBJECT
PHL 0183	21 49 24.	- 07 00		17.9	BLUE STELLAR OBJECT
PHL 0184	21 49 24.	- 22 16		17.9	BLUE STELLAR OBJECT
RNGC 7147	21 49 26.	+ 02 50		15.0	GALAXY
ARC 2383	21 49 26.	- 21 26		16.9	RICH CLUSTER OF GALAXIES
MCG+04-51-009	21 49 27.	+ 25 01 30.	54	14.5	GALAXY
SC 2146-3845.4	21 49 27.	- 38 31 24.			NEBULA
ARC 2384	21 49 29.	- 19 47		15.9	RICH CLUSTER OF GALAXIES
RNGC 7144	21 49 29.	- 48 29		12.0	GALAXY
SC 2145-6127.6	21 49 29.	- 61 13 36.	30		NEBULA
SC 2145-6232.7	21 49 29.	- 62 18 42.	6		NEBULA
ZWG 376.046	21 49 30.	+ 02 49		15.4	GALAXY
ARC 2386	21 49 32.	+ 24 55		17.6	RICH CLUSTER OF GALAXIES
SC 2146-5621.6	21 49 32.	- 56 07 36.			NEBULA
B 166	21 49 33.	+ 59 51	300		DARK OBJECT
SC 2145-6219.9	21 49 34.	- 62 05 54.	12		NEBULA
ZWG 376.047	21 49 36.	+ 03 04		14.6	GALAXY
UGC 11835	21 49 36.	+ 03 04	90	14.6	GALAXY E
ZC 2149.6+0319	21 49 36.	+ 03 19	6550		CLUSTER OF GALAXIES
ZWG 402.024	21 49 36.	+ 08 10		15.3	GALAXY
ZWG 427.030	21 49 36.	+ 12 26		15.7	GALAXY
PHL 1788	21 49 36.	- 06 28		16.6	BLUE STELLAR OBJECT
PHL 4967	21 49 36.	- 08 04		18.6	BLUE STELLAR OBJECT
SC 2146-5532.1	21 49 36.	- 55 18 06.			NEBULA
HN 1222	21 49 36.	- 59 42		15.	NEBULA
IC 5141	21 49 37.	- 59 42			NONSTELLAR OBJECT
RNGC 7149	21 49 38.	+ 03 04		14.5	GALAXY
SC 2146-5626.7	21 49 41.	- 56 12 42.			NEBULA
SC 2146-5628.3	21 49 41.	- 56 14 18.			NEBULA
MCG+00-55-026	21 49 42.	+ 03 02 30.	21	15.	GALAXY
SC 2146-5824.1	21 49 45.	- 58 10 06.			NEBULA
SC 2146-6046.1	21 49 46.	- 60 32 06.			NEBULA
SC 2146-6222.2	21 49 46.	- 62 08 12.	24		NEBULA
2ZW 152	21 49 48.	+ 03 12			COMPACT GALAXY
ZWG 376.048	21 49 48.	+ 03 12		15.1	GALAXY
2ZW 153	21 49 48.	+ 30 12			COMPACT GALAXY
ISS 0195	21 49 48.	+ 47 34	143		STELLAR RING
PHL 4969	21 49 48.	- 06 56		18.4	BLUE STELLAR OBJECT
MCG-02-55-005	21 49 48.	- 10 53 30.	60	15.	GALAXY
PHL 1789	21 49 48.	- 14 53		18.0	BLUE STELLAR OBJECT
PHL 4968	21 49 48.	- 17 56		14.1	BLUE STELLAR OBJECT
PHL 0185	21 49 48.	- 18 17		17.8	BLUE STELLAR OBJECT
SC 2146-5814.9	21 49 48.	- 58 00 53.	18		NEBULA
SC 2146-5909.7	21 49 49.	- 58 55 41.	18		NEBULA
RNGC 7148	21 49 50.	+ 03 12			GALAXY
SC 2146-5600.0	21 49 50.	- 55 45 59.	24		NEBULA
IC 1407	21 49 51.	+ 03 10 47.			NONSTELLAR OBJECT
SC 2146-3807.1	21 49 52.	- 37 53 05.	9		NEBULA
2ZW 154	21 49 54.	+ 06 54			COMPACT GALAXY
ZWG 427.031	21 49 54.	+ 12 25		15.7	GALAXY
4ZW 079	21 49 54.	+ 23 11			COMPACT GALAXY
UGC 11836	21 49 54.	+ 46 25	96	17.	GALAXY S-IRR
HN 2833	21 49 54.4	- 52 07 35.	12		NEBULA
SC 2146-5754.6	21 49 57.	- 57 40 35.			NEBULA
SC 2146-5855.3	21 49 59.	- 58 41 17.			NEBULA
MCG+02-55-019	21 50 00.	+ 12 10	24	16.	GALAXY
MCG+02-55-018	21 50 00.	+ 12 25	36	15.5	GALAXY
ZC 2150.0+1332	21 50 00.	+ 13 32	1880		CLUSTER OF GALAXIES
LDN 1109	21 50 00.	+ 55 35	1560		DARK NEBULA
LDN 1138	21 50 00.	+ 58 50	540		DARK NEBULA
LDN 1137	21 50 00.	+ 58 50	540		DARK NEBULA
LDN 1241	21 50 00.	+ 76 30	4740		DARK NEBULA
LDN 1289	21 50 00.	+ 88 10	420		DARK NEBULA
PHL 0186	21 50 00.	- 02 39		16.0	BLUE STELLAR OBJECT
PHL 1790	21 50 00.	- 06 03		17.1	BLUE STELLAR OBJECT
RNGC 7145	21 50 05.	- 48 07		12.5	GALAXY
HN 2834	21 50 05.7	- 52 06 05.	12		NEBULA
MCG+02-55-020	21 50 06.	+ 12 11	36	15.5	GALAXY
ZC 2150.1+2430	21 50 06.	+ 24 30	3490		CLUSTER OF GALAXIES
UGC 11837	21 50 06.	+ 35 52	84	16.0	GALAXY Sc
PHL 0187	21 50 06.	- 05 01		18.5	BLUE STELLAR OBJECT
PHL 4971	21 50 06.	- 05 26		17.9	BLUE STELLAR OBJECT
PHL 4970	21 50 06.	- 06 34		14.7	BLUE STELLAR OBJECT
SC 2146-5552.4	21 50 06.	- 55 38 23.	18		NEBULA
SCHO 1356	21 50 08.	+ 59 51 12.	640		ISOLATED DARK CLOUD
MCG+06-48-001	21 50 09.	+ 35 51	66	15.5	GALAXY
ZC 2150.2+0820	21 50 12.	+ 08 20	3630		CLUSTER OF GALAXIES
ZWG 427.032	21 50 12.	+ 12 17		15.6	GALAXY
MCG+02-55-021	21 50 12.	+ 12 18	36	15.5	GALAXY
PHL 4972	21 50 12.	- 06 10		18.0	BLUE STELLAR OBJECT
PHL 4973	21 50 12.	- 08 43		18.0	BLUE STELLAR OBJECT
MCG-02-55-006	21 50 12.	- 10 45 30.	84	15.	GALAXY
HN 0729	21 50 14.	- 67 34			NEBULA
IC 5140	21 50 14.	- 67 34			NONSTELLAR OBJECT
SCHO 1357	21 50 16.	+ 47 14 42.	950		ISOLATED DARK CLOUD
SC 2147-4154.1	21 50 16.	- 41 40 04.	6		NEBULA
ARC 2385	21 50 17.	- 23 46		17.1	RICH CLUSTER OF GALAXIES
ZWG 493.010	21 50 18.	+ 28 04		15.4	GALAXY
UGC 11838	21 50 18.	+ 28 04	126	15.4	GALAXY Sc
PHL 1791	21 50 18.	- 10 20		17.1	BLUE STELLAR OBJECT
HN 2835	21 50 20.6	- 17 07 51.	18		NEBULA
MCG+05-51-002	21 50 24.	+ 28 02	120	15.	GALAXY
PHL 1794	21 50 24.	- 02 34		18.8	BLUE STELLAR OBJECT
PHL 0188	21 50 24.	- 03 08		18.4	BLUE STELLAR OBJECT
PHL 4975	21 50 24.	- 03 22		18.8	BLUE STELLAR OBJECT
PHL 1792	21 50 24.	- 04 39		15.9	BLUE STELLAR OBJECT
PHL 4976	21 50 24.	- 05 16		18.4	BLUE STELLAR OBJECT
PHL 0189	21 50 24.	- 07 36		17.1	BLUE STELLAR OBJECT
PHL 4977	21 50 24.	- 08 10		18.8	BLUE STELLAR OBJECT
PHL 1793	21 50 24.	- 14 46		17.8	BLUE STELLAR OBJECT
PHL 4974	21 50 24.	- 20 55		17.6	BLUE STELLAR OBJECT
SER 149.06	21 50 24.	- 57 50	258		LOW SURF. BRGHTNSS GALAXY
SCHO 1358	21 50 26.	+ 51 18 54.	370		ISOLATED DARK CLOUD
B 167	21 50 26.	+ 59 50	300		DARK OBJECT
IC 1408	21 50 27.	- 13 34 59.			NONSTELLAR OBJECT
MCG+00-55-027	21 50 30.	+ 03 27	36	15.5	GALAXY
ZWG 376.049	21 50 30.	+ 03 28		15.6	GALAXY
ZWG 402.025	21 50 30.	+ 04 03		15.6	GALAXY
ZWG 427.033	21 50 30.	+ 09 58		15.7	GALAXY
ZWG 427.034	21 50 30.	+ 15 20		15.6	GALAXY
MCG+02-55-022	21 50 30.	+ 15 20	72	15.	GALAXY
SCHO 1359	21 50 30.	+ 66 48 42.	690		ISOLATED DARK CLOUD
PHL 4978	21 50 30.	- 07 34		17.8	BLUE STELLAR OBJECT
PHL 0190	21 50 30.	- 07 38		17.3	BLUE STELLAR OBJECT
MCG-02-55-007	21 50 30.	- 13 35	36	15.	GALAXY
SER 149.10	21 50 30.	- 58 06	480	18.5	FAINT CLUSTER OF GALAXIES
SCHO 1360	21 50 32.	+ 47 03 30.	290		ISOLATED DARK CLOUD
MCG+01-55-015	21 50 33.	+ 04 02 30.	48	15.	GALAXY
SC 2147-4156.5	21 50 34.	- 41 42 27.	9		NEBULA
ZWG 513.001	21 50 36.	+ 36 13		15.7	GALAXY
UGC 11839	21 50 36.	+ 36 13	60	15.7	GALAXY Sb-c
PHL 1795	21 50 36.	- 04 15		18.1	BLUE STELLAR OBJECT
PHL 4979	21 50 36.	- 06 42		17.9	BLUE STELLAR OBJECT
SC 2147-5820.4	21 50 39.	- 58 06 21.	18		NEBULA
SC 2147-5601.9	21 50 40.	- 55 47 51.	6		NEBULA
SC 2147-5614.2	21 50 41.	- 56 00 09.			NEBULA
SC 2147-5907.2	21 50 41.	- 58 53 09.			NEBULA
UGC 11840	21 50 42.	+ 04 00	60	17.	GALAXY S
ZC 2150.7+0710	21 50 42.	+ 07 10	1280		CLUSTER OF GALAXIES
4ZW 080	21 50 42.	+ 38 43			COMPACT GALAXY
ZWG 376.050	21 50 42.	- 02 27		15.5	GALAXY
KARA.73B 0934	21 50 42.	- 02 27	42	15.5	ISOLATED GALAXY S
IC 1409	21 50 42.	- 07 44 04.			NONSTELLAR OBJECT

OBJECT NAME	RIGHT ASCEN.	DECLINATION	DIAM.	MAGN.	TYPE OF OBJECT
PHL 4980	21 50 42.	- 08 40		12.7	BLUE STELLAR OBJECT
PHL 1796	21 50 42.	- 12 32		16.5	BLUE STELLAR OBJECT
VDB.66N 147	21 50 44.	+ 47 03	108		REFLECTION NEBULA
SC 2147-5819.2	21 50 45.	- 58 05 09.			NEBULA
SC 2147-5819.1	21 50 46.	- 58 05 03.	30		NEBULA
ZWG 376.051	21 50 48.	+ 00 08		15.7	GALAXY
ZC 2150.8+1040	21 50 48.	+ 10 40	2150		CLUSTER OF GALAXIES
ZWG 427.035	21 50 48.	+ 15 22		15.5	GALAXY
MCG+04-51-010	21 50 48.	+ 22 12 30.	15	15.5	GALAXY
UGC 11841	21 50 48.	+ 38 42	180	16.5	GALAXY
ISS 0238	21 50 48.	+ 45 06	128		STELLAR RING
ZC 2150.8-0002	21 50 48.	- 00 02	610		CLUSTER OF GALAXIES
PHL 0191	21 50 48.	- 04 02		18.1	BLUE STELLAR OBJECT
PHL 0192	21 50 48.	- 05 30		18.7	BLUE STELLAR OBJECT
PHL 4982	21 50 48.	- 07 25		17.9	BLUE STELLAR OBJECT
PHL 4981	21 50 48.	- 14 20		17.7	BLUE STELLAR OBJECT
SC 2147-5543.9	21 50 49.	- 55 29 51.	12		NEBULA
SC 2147-5904.3	21 50 49.	- 58 50 15.			NEBULA
ZWG 427.036	21 50 54.	+ 15 17		15.5	GALAXY
ZWG 427.037	21 50 54.	+ 15 18		15.3	GALAXY
MCG+02-55-024	21 50 54.	+ 15 19	24	15.	GALAXY
MCG+02-55-023	21 50 54.	+ 15 22	60	15.	GALAXY
PHL 1797	21 50 54.	- 10 32		6.5	BLUE STELLAR OBJECT
PHL 4983	21 50 54.	- 12 06		16.6	BLUE STELLAR OBJECT
PHL 4984	21 50 54.	- 21 31		16.4	BLUE STELLAR OBJECT
SC 2147-5908.7	21 50 54.	- 58 54 39.			NEBULA
SC 2147-5626.2	21 50 56.	- 56 12 09.			NEBULA
SC 2147-3931.6	21 50 57.	- 39 17 32.	6		NEBULA
ZC 2151.0+1325	21 51 00.	+ 13 25	810		CLUSTER OF GALAXIES
2ZW 155	21 51 00.	+ 15 18			COMPACT GALAXY
MCG+02-55-025	21 51 00.	+ 15 18	24	15.	GALAXY
ZWG 472.015	21 51 00.	+ 25 16		15.5	GALAXY
UGC 11842	21 51 00.	+ 25 16	60	15.5	GALAXY Sa-b
PHL 1798	21 51 00.	- 02 58		16.8	BLUE STELLAR OBJECT
PHL 4985	21 51 00.	- 04 34		17.5	BLUE STELLAR OBJECT
PHL 4986	21 51 00.	- 05 44		17.5	BLUE STELLAR OBJECT
PHL 4987	21 51 00.	- 09 24		17.8	BLUE STELLAR OBJECT
RNGC 7152	21 51 03.	- 29 30		14.0	GALAXY
SC 2147-6334.9	21 51 04.	- 63 20 51.	18		NEBULA
SC 2148-4032.2	21 51 05.	- 40 18 08.	30		NEBULA
HN 2836	21 51 05.6	- 53 42 26.	18		NEBULA
3ZW 089	21 51 06.	+ 03 23			COMPACT GALAXY
MCG+04-51-011	21 51 06.	+ 25 17	48	15.	GALAXY
SCHO 1361	21 51 06.	+ 51 33 06.	380		ISOLATED DARK CLOUD
MCG-05-51-020	21 51 06.	- 29 34	60	14.5	GALAXY
ARC 2388	21 51 07.	+ 08 01		16.5	RICH CLUSTER OF GALAXIES
ZWG 427.038	21 51 12.	+ 09 48		15.5	GALAXY
PHL 1799	21 51 12.	- 02 40		17.8	BLUE STELLAR OBJECT
PHL 1800	21 51 12.	- 03 22		17.6	BLUE STELLAR OBJECT
PHL 4988	21 51 12.	- 04 00		18.0	BLUE STELLAR OBJECT
PHL 0193	21 51 12.	- 07 10		17.2	BLUE STELLAR OBJECT
PHL 1801	21 51 12.	- 18 45		18.4	BLUE STELLAR OBJECT
SC 2147-5642.9	21 51 12.	- 56 28 50.	18		NEBULA
ARC 2390	21 51 13.	+ 17 27		17.6	RICH CLUSTER OF GALAXIES
MCG+02-55-026	21 51 18.	+ 09 48	36	15.	GALAXY
ZC 2151.3+1727	21 51 18.	+ 17 27	1210		CLUSTER OF GALAXIES
LDN 1055	21 51 18.	+ 46 58	1800		DARK NEBULA
PHL 1802	21 51 18.	- 03 31		18.1	BLUE STELLAR OBJECT
PHL 4992	21 51 18.	- 06 03		18.9	BLUE STELLAR OBJECT
PHL 4989	21 51 18.	- 09 55		17.8	BLUE STELLAR OBJECT
PHL 0194	21 51 18.	- 12 34		18.6	BLUE STELLAR OBJECT
PHL 4990	21 51 18.	- 14 30		17.8	BLUE STELLAR OBJECT
PHL 0195	21 51 18.	- 15 14		16.9	BLUE STELLAR OBJECT
PHL 4991	21 51 18.	- 16 23		16.7	BLUE STELLAR OBJECT
SC 2148-3909.8	21 51 19.	- 38 55 43.			NEBULA
2ZW 156	21 51 24.	+ 00 39			COMPACT GALAXY
ZWG 402.026	21 51 24.	+ 06 30		15.7	GALAXY
ZC 2151.4+1702	21 51 24.	+ 17 02	1010		CLUSTER OF GALAXIES
ZWG 493.011	21 51 24.	+ 32 33		15.7	GALAXY
ISS 0082	21 51 24.	+ 58 36	314		STELLAR RING
PHL 4993	21 51 24.	- 04 02		17.6	BLUE STELLAR OBJECT
PHL 1803	21 51 24.	- 05 27		17.8	BLUE STELLAR OBJECT
MCG-03-55-010	21 51 24.	- 16 24	15	16.	GALAXY
MCG-03-55-009	21 51 24.	- 16 24	12	16.	GALAXY
B 168	21 51 25.	+ 47 02			DARK OBJECT
ARC 2389	21 51 27.	- 04 12		17.4	RICH CLUSTER OF GALAXIES
SC 2147-5735.9	21 51 28.	- 57 21 49.			NEBULA
ZC 2151.5+0156	21 51 30.	+ 01 56	940		CLUSTER OF GALAXIES
OCL 0213	21 51 30.	+ 47 02	570	8.3	OPEN STAR CLUSTER
IC 5146	21 51 30.	+ 47 02	540		OPEN CLUSTER
PHL 4995	21 51 30.	- 10 42		17.6	BLUE STELLAR OBJECT
PHL 4994	21 51 30.	- 11 10		16.6	BLUE STELLAR OBJECT
PHL 0196	21 51 30.	- 24 12		15.4	BLUE STELLAR OBJECT
MCG-05-51-021	21 51 30.	- 32 15	60	15.5	GALAXY
SER 149.02	21 51 30.	- 55 38	3000	14.5	CLSTRS OF GLXIES: 50 MEMB
HN 0730	21 51 30.	- 65 45			NEBULA
CED 198	21 51 31.	+ 47 02	720		DIFFUSE GALACTIC NEBULA
SG 3.227	21 51 31.	+ 47 02	600		DIFFUSE EMISSION NEBULA
IC 5142	21 51 31.	- 65 45			NONSTELLAR OBJECT
IC 5144	21 51 33.	+ 14 51 49.			NONSTELLAR OBJECT
RNGC 7153	21 51 33.	- 29 16		14.0	GALAXY
SC 2148-5807.5	21 51 33.	- 57 53 25.			NEBULA
SC 2147-6227.9	21 51 34.	- 62 13 49.			NEBULA
HOLM 787B	21 51 35.	- 16 40	12	14.4	PART OF MULTIPLE GALAXY
MRSL 094-05/1	21 51 36.	+ 47 02	540		HII REGION
ZC 2151.6-0008	21 51 36.	- 00 08	610		CLUSTER OF GALAXIES
PHL 1804	21 51 36.	- 05 01		18.2	BLUE STELLAR OBJECT
PHL 0197	21 51 36.	- 06 27		13.3	BLUE STELLAR OBJECT
PHL 0198	21 51 36.	- 15 30		18.3	BLUE STELLAR OBJECT
PHL 0199	21 51 36.	- 22 50		18.0	BLUE STELLAR OBJECT
MCG-05-51-023	21 51 36.	- 28 37	24	15.	GALAXY
MCG-05-51-022	21 51 36.	- 29 20	120	14.5	GALAXY
SC 2148-5559.1	21 51 37.	- 55 45 01.			NEBULA
MIN.47 14	21 51 38.	+ 46 59			DIFFUSE NEBULA
HOLM 787A	21 51 38.	- 16 41	54	13.6	PART OF MULTIPLE GALAXY
SC 2148-5615.2	21 51 40.	- 56 01 07.	12		NEBULA
SCHO 1362	21 51 41.	+ 51 40 48.	310		ISOLATED DARK CLOUD
MCG+00-55-028	21 51 42.	+ 00 06	42	15.	GALAXY
ZWG 376.052	21 51 42.	+ 00 07		15.3	GALAXY
ZWG 427.039	21 51 42.	+ 14 47		15.4	GALAXY
ZC 2151.7+2236	21 51 42.	+ 22 36	2690		CLUSTER OF GALAXIES
PHL 4996	21 51 42.	- 06 09		19.0	BLUE STELLAR OBJECT
SC 2147-6452.7	21 51 43.	- 64 38 37.	60		NEBULA
ARC 2392	21 51 46.	+ 00 24		17.7	RICH CLUSTER OF GALAXIES
RNGC 7151	21 51 46.	- 50 54			GALAXY
4ZW 081	21 51 48.	+ 28 47			COMPACT GALAXY
PHL 1806	21 51 48.	- 08 15		18.2	BLUE STELLAR OBJECT
PHL 4998	21 51 48.	- 09 48		18.2	BLUE STELLAR OBJECT

OBJECT NAME	RIGHT ASCEN.	DECLINATION	DIAM.	MAGN.	TYPE OF OBJECT
PHL 0200	21 51 48.	- 10 00		16.0	BLUE STELLAR OBJECT
PHL 4997	21 51 48.	- 11 28		17.7	BLUE STELLAR OBJECT
PHL 1805	21 51 48.	- 20 16		17.5	BLUE STELLAR OBJECT
HN 2837	21 51 49.5	- 53 51 48.	18		NEBULA
IC 5145	21 51 50.	+ 14 59 08.			NONSTELLAR OBJECT
HN 2839	21 51 52.8	- 45 38 36.	18		NEBULA
ARC 2391	21 51 53.	- 15 29		17.2	RICH CLUSTER OF GALAXIES
SC 2148-5602.9	21 51 53.	- 55 48 48.			NEBULA
SC 2148-5639.1	21 51 53.	- 56 25 00.			NEBULA
ZWG 427.040	21 51 54.	+ 14 47		15.7	GALAXY
MCG+02-55-027	21 51 54.	+ 14 48	36	15.5	GALAXY
ISS 0196	21 51 54.	+ 45 40	134		STELLAR RING
ZC 2151.9-0244	21 51 54.	- 02 44	1340		CLUSTER OF GALAXIES
PHL 0201	21 51 54.	- 10 23		17.9	BLUE STELLAR OBJECT
SCHO 1363	21 51 59.	+ 58 26 00.	300		ISOLATED DARK CLOUD
LBN 0424	21 52	+ 47 00	600		BRIGHT NEBULA
KHAV 760	21 52	+ 56 26	9930		DARK NEBULA
MCG+00-55-029	21 52 00.	+ 02 41	96	13.	GALAXY
ZWG 376.053	21 52 00.	+ 02 42		13.5	GALAXY
UGC 11843	21 52 00.	+ 02 42	102	13.5	GALAXY Sc
KARA.73B 0935	21 52 00.	+ 02 42	102	13.5	ISOLATED GALAXY S
ZWG 427.041	21 52 00.	+ 14 54		14.7	GALAXY
UGC 11844	21 52 00.	+ 14 54	108	14.7	GALAXY Sa-b
MCG+02-55-028	21 52 00.	+ 14 55	96	14.	GALAXY
PHL 1807	21 52 00.	- 03 22		18.2	BLUE STELLAR OBJECT
PHL 0203	21 52 00.	- 04 32		18.7	BLUE STELLAR OBJECT
PHL 0202	21 52 00.	- 04 34		15.9	BLUE STELLAR OBJECT
PHL 0204	21 52 00.	- 10 01		16.0	BLUE STELLAR OBJECT
PHL 1808	21 52 00.	- 15 47		18.4	BLUE STELLAR OBJECT
PHL 4999	21 52 00.	- 24 46		17.8	BLUE STELLAR OBJECT
PHL 5000	21 52 00.	- 26 48		17.0	BLUE STELLAR OBJECT
SC 2148-4013.0	21 52 00.	- 39 58 54.	12		NEBULA
SCHO 1364	21 52 01.	+ 51 27 48.	210		ISOLATED DARK CLOUD
RNGC 7156	21 52 02.	+ 02 42		13.5	GALAXY
HN 2838	21 52 02.6	- 53 55 00.	18		NEBULA
SC 2148-5613.9	21 52 03.	- 55 59 48.			NEBULA
SC 2148-5820.9	21 52 05.	- 58 06 48.			NEBULA
ZC 2152.1+1150	21 52 06.	+ 11 50	1080		CLUSTER OF GALAXIES
UGC 11845	21 52 06.	+ 14 47	66	16.0	GALAXY Sb-c
MCG+02-55-029	21 52 06.	+ 14 47	72	16.	GALAXY
PHL 0205	21 52 06.	- 04 22		18.8	BLUE STELLAR OBJECT
PHL 0206	21 52 06.	- 12 26		18.8	BLUE STELLAR OBJECT
PHL 5001	21 52 06.	- 26 14		15.9	BLUE STELLAR OBJECT
SC 2149-3809.5	21 52 09.	- 37 55 23.	12		NEBULA
SCHO 1365	21 52 11.	+ 60 40 42.	780		ISOLATED DARK CLOUD
ZWG 402.027	21 52 12.	+ 06 10		15.2	GALAXY
UGC 11846	21 52 12.	+ 06 10	60	15.2	GALAXY Sb-c
MCG+02-55-030	21 52 12.	+ 12 37	36	15.5	GALAXY
PHL 5002	21 52 12.	- 07 22		17.5	BLUE STELLAR OBJECT
PHL 1809	21 52 12.	- 22 22		18.3	BLUE STELLAR OBJECT
PHL 0207	21 52 12.	- 22 24		18.0	BLUE STELLAR OBJECT
PHL 1810	21 52 12.	- 23 56		18.4	BLUE STELLAR OBJECT
PHL 5003	21 52 12.	- 26 19		17.9	BLUE STELLAR OBJECT
SER 149.03	21 52 12.	- 56 22	150	15.	INTERACTING GALAXIES
ARC 2393	21 52 14.	- 03 35		17.7	RICH CLUSTER OF GALAXIES
MCG+01-55-016	21 52 15.	+ 06 10	48	14.5	GALAXY
RNGC 7160	21 52 17.	+ 62 22		6.5	OPEN CLUSTER
SC 2148-5520.4	21 52 17.	- 55 06 17.	90		NEBULA
SC 2148-5616.3	21 52 17.	- 56 02 11.	12		NEBULA
SC 2148-5634.7	21 52 17.	- 56 20 35.			NEBULA
OCL 0236	21 52 18.	+ 62 22	960	6.6	OPEN STAR CLUSTER
PHL 5005	21 52 18.	- 05 00		17.9	BLUE STELLAR OBJECT
PHL 0208	21 52 18.	- 06 28		16.6	BLUE STELLAR OBJECT
PHL 5004	21 52 18.	- 07 17		18.5	BLUE STELLAR OBJECT
PHL 1811	21 52 18.	- 09 36		14.2	BLUE STELLAR OBJECT
PHL 1812	21 52 18.	- 20 13		17.8	BLUE STELLAR OBJECT
HUP C33	21 52 20.	+ 46 23			DIFFUSE NEBULA
SCHO 1366	21 52 20.	+ 61 43 18.	480		ISOLATED DARK CLOUD
HN 2840	21 52 20.9	- 43 29 17.	18		NEBULA
SC 2148-6038.5	21 52 23.	- 60 24 23.	12		NEBULA
ISS 0197	21 52 24.	+ 49 28	151		STELLAR RING
ISS 0128	21 52 24.	+ 54 33	298		STELLAR RING
PHL 5006	21 52 24.	- 06 56		18.0	BLUE STELLAR OBJECT
PHL 0209	21 52 24.	- 17 32		16.7	BLUE STELLAR OBJECT
RNGC 7154	21 52 25.	- 35 03			GALAXY
HN 2841	21 52 25.8	- 44 06 11.	12		NEBULA
SC 2149-5610.2	21 52 28.	- 55 56 05.			NEBULA
ZC 2152.5+0834	21 52 30.	+ 08 34	1950		CLUSTER OF GALAXIES
ISS 0129	21 52 30.	+ 51 22	208		STELLAR RING
PHL 0210	21 52 30.	- 04 47		18.2	BLUE STELLAR OBJECT
ZC 2152.6+0405	21 52 36.	+ 04 05	1550		CLUSTER OF GALAXIES
ZC 2152.6+2118	21 52 36.	+ 21 18	2760		CLUSTER OF GALAXIES
MCG+04-51-012	21 52 36.	+ 21 33	18	15.5	GALAXY
PHL 0211	21 52 36.	- 10 38		13.6	BLUE STELLAR OBJECT
SC 2149-5614.0	21 52 41.	- 55 59 53.	12		NEBULA
HN 2842	21 52 41.1	- 43 27 28.	30		NEBULA
SN 1951A	21 52 43.	- 04 32		16.6	SUPERNOVA
MCG-07-45-001	21 52 45.	- 43 28	66	15.	GALAXY
ARC 2395	21 52 48.	+ 08 32		17.1	RICH CLUSTER OF GALAXIES
ZWG 493.012	21 52 48.	+ 30 21		15.4	GALAXY
PHL 5010	21 52 48.	- 06 08		18.1	BLUE STELLAR OBJECT
PHL 5009	21 52 48.	- 07 46		15.6	BLUE STELLAR OBJECT
PHL 5008	21 52 48.	- 08 54		17.9	BLUE STELLAR OBJECT
PHL 5007	21 52 48.	- 11 24		16.4	BLUE STELLAR OBJECT
MCG-02-55-008	21 52 48.	- 12 15	36	14.5	GALAXY
SC 2149-5627.5	21 52 48.	- 56 13 22.			NEBULA
SC 2149-5634.7	21 52 51.	- 56 20 34.	9		NEBULA
SC 2149-5946.0	21 52 51.	- 59 31 52.	36		NEBULA
ARC 2394	21 52 52.	- 19 28		17.1	RICH CLUSTER OF GALAXIES
RNGC 7155	21 52 53.	- 49 46		13.0	GALAXY
PHL 5011	21 52 54.	- 03 17		17.7	BLUE STELLAR OBJECT
PHL 0212	21 52 54.	- 15 39		18.0	BLUE STELLAR OBJECT
SC 2149-4014.5	21 52 54.	- 40 00 22.	12		NEBULA
SC 2149-5634.6	21 52 54.	- 56 20 28.			NEBULA
SC 2149-6037.4	21 52 56.	- 60 23 16.	12		NEBULA
LBN 0569	21 53	+ 80 10	3900		BRIGHT NEBULA
ZWG 403.001	21 53 00.	+ 06 22		15.5	GALAXY
ZWG 402.028	21 53 00.	+ 06 22		15.5	GALAXY
UGC 11847	21 53 00.	+ 30 16	60	17.	GALAXY DWARF
SCHO 1367	21 53 00.	+ 54 12 30.	310		ISOLATED DARK CLOUD
PHL 1814	21 53 00.	- 07 56		17.8	BLUE STELLAR OBJECT
PHL 1816	21 53 00.	- 15 13		18.2	BLUE STELLAR OBJECT
PHL 1815	21 53 00.	- 15 40		18.0	BLUE STELLAR OBJECT
PHL 0213	21 53 00.	- 18 14		17.2	BLUE STELLAR OBJECT
PHL 5012	21 53 00.	- 20 55		18.3	BLUE STELLAR OBJECT
PHL 1813	21 53 00.	- 21 46		18.5	BLUE STELLAR OBJECT
HN 2843	21 53 00.2	- 43 32 09.	18		NEBULA
IC 5143	21 53 01.	- 49 17 44.			NONSTELLAR OBJECT

OBJECT NAME	RIGHT ASCEN.	DECLINATION	DIAM.	MAGN.	TYPE OF OBJECT
ZC 2153.1+0701	21 53 06.	+ 07 01	2690		CLUSTER OF GALAXIES
ZC 2153.1+1347	21 53 06.	+ 13 47	810		CLUSTER OF GALAXIES
PHL 1817	21 53 06.	- 24 54		18.4	BLUE STELLAR OBJECT
ZC 2153.2+0921	21 53 12.	+ 09 21	810		CLUSTER OF GALAXIES
UGC 11848	21 53 12.	+ 10 14	66	16.0	GALAXY
ZC 2153.2+1216	21 53 12.	+ 12 16	940		CLUSTER OF GALAXIES
PHL 5014	21 53 12.	- 06 52		17.9	BLUE STELLAR OBJECT
PHL 5013	21 53 12.	- 09 52		16.3	BLUE STELLAR OBJECT
SC 2149-5601.8	21 53 12.	- 55 47 39.			NEBULA
ARC 2396	21 53 13.	+ 12 16		17.5	RICH CLUSTER OF GALAXIES
HN 2844	21 53 13.9	- 52 16 09.	18		NEBULA
MCG+01-56-001	21 53 15.	+ 05 33	42	15.	GALAXY
ZWG 403.002	21 53 18.	+ 05 34		15.5	GALAXY
ZWG 472.016	21 53 18.	+ 24 39		14.8	GALAXY
UGC 11849	21 53 18.	+ 24 39	102	14.8	GALAXY SBc
ZC 2153.3+2701	21 53 18.	+ 27 01	870		CLUSTER OF GALAXIES
PHL 5015	21 53 18.	- 18 29		18.5	BLUE STELLAR OBJECT
PHL 5016	21 53 18.	- 22 38		18.3	BLUE STELLAR OBJECT
SC 2149-6105.4	21 53 18.	- 60 51 15.	6		NEBULA
SC 2149-6234.4	21 53 19.	- 62 20 15.	18		NEBULA
MCG+00-56-001	21 53 21.	- 01 45	36	15.	GALAXY
SC 2149-5452.9	21 53 21.	- 54 38 45.			NEBULA
SC 2149-5636.6	21 53 21.	- 56 22 27.			NEBULA
HN 2845	21 53 23.1	- 54 06 21.	24		NEBULA
ZC 2153.4+0806	21 53 24.	+ 08 06	400		CLUSTER OF GALAXIES
ZWG 428.001	21 53 24.	+ 14 31		15.7	GALAXY
MCG+04-51-013	21 53 24.	+ 24 39	90	14.5	GALAXY
PHL 0214	21 53 24.	- 04 30		17.5	BLUE STELLAR OBJECT
PHL 0215	21 53 24.	- 06 06		15.8	BLUE STELLAR OBJECT
HN 2846	21 53 25.6	- 45 14 08.	30		NEBULA
SCHO 1368	21 53 26.	+ 66 09 42.	300		ISOLATED DARK CLOUD
IC 1410	21 53 26.	- 03 07 46.			NONSTELLAR OBJECT
IC 1411	21 53 27.	- 01 45 10.			NONSTELLAR OBJECT
SC 2150-5608.5	21 53 29.	- 55 54 21.			NEBULA
MCG+01-56-002	21 53 30.	+ 05 39	60	15.	GALAXY
ZWG 403.003	21 53 30.	+ 05 40		14.7	GALAXY
UGC 11851	21 53 30.	+ 05 40	60	14.7	GALAXY Sa
ZWG 377.001	21 53 30.	- 01 45		14.5	GALAXY
UGC 11850	21 53 30.	- 01 45	54	14.5	GALAXY E
SC 2150-5541.2	21 53 33.	- 55 27 02.			NEBULA
ISS 0198	21 53 36.	+ 46 53	211		STELLAR RING
SC 2150-4027.8	21 53 39.	- 40 13 38.	12		NEBULA
SC 2149-6203.6	21 53 38.	- 61 49 27.	12		NEBULA
ARC 2397	21 53 39.	+ 01 07		17.9	RICH CLUSTER OF GALAXIES
MCG+00-56-002	21 53 39.	+ 01 57	42	15.	GALAXY
ARC 2398	21 53 39.	+ 06 18		17.1	RICH CLUSTER OF GALAXIES
MCG+00-56-003	21 53 39.	- 01 25	60	15.	GALAXY
HN 2847	21 53 39.6	- 45 04 14.	12		NEBULA
ZWG 428.002	21 53 42.	+ 14 30		15.5	GALAXY
ZWG 493.013	21 53 42.	+ 27 40		15.0	GALAXY
UGC 11852	21 53 42.	+ 27 40	96	15.0	GALAXY SBa?
ZWG 377.002	21 53 42.	- 01 53		15.5	GALAXY
PHL 1818	21 53 42.	- 22 02		17.9	BLUE STELLAR OBJECT
SC 2150-5811.4	21 53 45.	- 57 57 14.			NEBULA
SC 2150-5554.2	21 53 47.	- 55 40 02.			NEBULA
BC PKS2153-204	21 53 47.1	- 20 26 50.		17.5	QUASI-STELLAR OBJECT
ZC 2153.8+0109	21 53 48.	+ 01 09	740		CLUSTER OF GALAXIES
ZWG 377.004	21 53 48.	+ 01 56		14.7	GALAXY
ZWG 377.005	21 53 48.	+ 02 07		15.7	GALAXY
ZWG 377.003	21 53 48.	- 01 25		15.4	GALAXY
UGC 11853	21 53 48.	- 01 25	66	15.4	GALAXY Sc
SHB 368	21 53 48.	- 20 26 30.		17.5	QUASI-STELLAR OBJECT
PHL 1819	21 53 48.	- 20 28		17.0	BLUE STELLAR OBJECT
MCG-05-51-024	21 53 48.	- 28 41	36	15.	GALAXY
ZWG 377.006	21 53 54.	+ 01 07		15.6	GALAXY
ZWG 403.004	21 53 54.	+ 07 07		15.4	GALAXY
MRK 516	21 53 54.	+ 07 07	20	16.	GALAXY WITH UV CONTINUUM
MCG-04-51-014	21 53 54.	- 21 27	36	15.	GALAXY
SC 2150-5942.0	21 53 55.	- 59 27 50.			NEBULA
LB 01137	21 53 56.	+ 13 06 24.		17.8	FAINT BLUE STAR
RNGC 7159	21 53 59.	+ 13 20		15.0	GALAXY
RNGC 7157	21 53 59.	- 25 37		14.0	GALAXY
ZWG 428.003	21 54 00.	+ 13 20		15.2	GALAXY
LDN 1139	21 54 00.	+ 58 20	480		DARK NEBULA
LDN 1141	21 54 00.	+ 58 30	360		DARK NEBULA
SC 2150-5854.4	21 54 00.	- 58 40 14.			NEBULA
SC 2150-5634.9	21 54 01.	- 56 20 43.			NEBULA
RNGC 7164	21 54 02.	+ 01 08			GALAXY
SC 2150-5655.4	21 54 03.	- 56 41 13.			NEBULA
SC 2150-5943.6	21 54 04.	- 59 29 25.	24		NEBULA
HN 2848	21 54 04.8	- 51 27 19.	24		NEBULA
SC 2150-5937.2	21 54 05.	- 59 23 01.			NEBULA
UGC 11854	21 54 06.	+ 15 43	66	16.0	GALAXY S
PHL 0216	21 54 06.	- 22 00		18.0	BLUE STELLAR OBJECT
MCG-04-51-015	21 54 06.	- 25 36	66	14.5	GALAXY
SC 2151-3843.5	21 54 07.	- 38 29 19.	36		NEBULA
MCG+05-51-003	21 54 09.	+ 30 32 30.	60	15.	GALAXY
FELW 044	21 54 10.	- 34 49			NEBULA
ZWG 377.007	21 54 12.	+ 02 22		15.3	GALAXY
ZWG 493.014	21 54 12.	+ 30 33		15.7	GALAXY
UGC 11855	21 54 12.	+ 30 33	90	15.7	GALAXY Sb-c
MCG-05-51-025	21 54 12.	- 28 53	18	15.	GALAXY
SCHO 1369	21 54 13.	+ 54 21 12.	260		ISOLATED DARK CLOUD
SC 2150-5919.5	21 54 16.	- 59 05 19.	12		NEBULA
UGC 11856	21 54 18.	+ 06 41	60	16.0	GALAXY S
MCG+01-56-003	21 54 18.	+ 06 41	42	15.	GALAXY
PHL 5017	21 54 18.	- 21 37		18.5	BLUE STELLAR OBJECT
SC 2150-5631.7	21 54 18.	- 56 17 31.	30		NEBULA
LB 01138	21 54 21.	- 00 18 30.		15.7	FAINT BLUE STAR
RNGC 7159	21 54 22.	- 11 54			GALAXY
MCG+02-56-001	21 54 24.	+ 12 05	60	15.	GALAXY
7ZW 940	21 54 24.	+ 74 15			COMPACT GALAXY
ZWG 343.002	21 54 24.	+ 74 15		15.5	GALAXY
RNGC 7161	21 54 26.	+ 02 43			NON-EXISTENT OBJECT
ZWG 403.005	21 54 30.	+ 06 25		14.8	GALAXY
MRK 517	21 54 30.	+ 06 25	18	16.	GALAXY WITH UV CONTINUUM
ZC 2154.5+1045	21 54 30.	+ 10 45	1340		CLUSTER OF GALAXIES
ZWG 428.004	21 54 30.	+ 12 06		15.6	GALAXY
UGC 11857	21 54 30.	+ 12 06	66	15.6	GALAXY S
MCG-05-51-026	21 54 30.	- 29 00	30	15.	GALAXY
HN 2849	21 54 31.1	- 44 25 30.	12		NEBULA
PHL 5018	21 54 36.	- 20 04		18.4	BLUE STELLAR OBJECT
MCG-05-51-027	21 54 36.	- 29 41	24	15.5	GALAXY
LB 01139	21 54 37.	+ 13 08 18.		17.5	FAINT BLUE STAR
SC 2151-4053.7	21 54 37.	- 40 39 30.	60		NEBULA
SC 2151-5614.8	21 54 37.	- 56 00 36.			NEBULA
SC 2151-5614.8	21 54 37.	- 56 00 36.			NEBULA
SC 2151-3838.8	21 54 41.	- 38 24 35.	60		NEBULA
SC 2151-3837.8	21 54 45.	- 38 23 35.	48		NEBULA
SC 2151-3837.8	21 54 45.	- 38 23 35.	48		NEBULA
SHB 369	21 54 45.1	+ 10 00 06.		19.	QUASI-STELLAR OBJECT
SC 2151-5614.1	21 54 47.	- 55 59 54.	12		NEBULA
4ZW 083	21 54 48.	+ 38 11			COMPACT GALAXY
OCL 0239	21 54 48.	+ 63 42	240	16.	OPEN STAR CLUSTER
PHL 1820	21 54 48.	- 14 15		18.5	BLUE STELLAR OBJECT
PHL 5019	21 54 48.	- 24 16		18.0	BLUE STELLAR OBJECT
HN 2850	21 54 49.7	- 52 19 17.	24		NEBULA
SC 2151-6052.2	21 54 51.	- 60 38 00.	6		NEBULA
SC 2151-5903.2	21 54 52.	- 58 49 00.	18		NEBULA
ZWG 493.015	21 54 54.	+ 30 30		15.0	GALAXY
SCHO 1370	21 54 54.	+ 58 47 30.	300		ISOLATED DARK CLOUD
SC 2151-3953.3	21 54 54.	- 39 39 05.	6		NEBULA
SC 2151-6053.1	21 54 54.	- 60 38 54.	6		NEBULA
ARC 2399	21 54 56.	- 08 02		15.6	RICH CLUSTER OF GALAXIES
SC 2151-5614.0	21 54 59.	- 55 59 47.	30		NEBULA
LBN 0498	21 55	+ 64 20	5400		BRIGHT NEBULA
COU 086	21 55 00.	+ 52 30	5400		EMISSION NEBULA
LDN 1107	21 55 00.	+ 54 15	13440		DARK NEBULA
LDN 1142	21 55 00.	+ 58 50	180		DARK NEBULA
COU 087	21 55 00.	+ 59 30	12000		EMISSION NEBULA
MCG-05-51-028	21 55 00.	- 29 04	24	14.5	GALAXY
SC 2151-6107.6	21 55 01.	- 60 53 23.	6		NEBULA
SC 2151-5909.0	21 55 05.	- 58 54 47.	30		NEBULA
ZWG 493.016	21 55 06.	+ 30 29		15.7	GALAXY
SCHO 1371	21 55 06.	+ 61 43 12.	660		ISOLATED DARK CLOUD
ARC 2400	21 55 06.	- 11 37		16.5	RICH CLUSTER OF GALAXIES
ZWG 451.001	21 55 12.	+ 16 24		15.5	GALAXY
SC 2151-5939.6	21 55 12.	- 59 25 23.	18		NEBULA
SCHO 1372	21 55 14.	+ 54 24 48.	290		ISOLATED DARK CLOUD
SCHO 1373	21 55 16.	+ 48 23 36.	300		ISOLATED DARK CLOUD
ZC 2155.3+1402	21 55 18.	+ 14 02	1210		CLUSTER OF GALAXIES
MCG-05-51-029	21 55 18.	- 27 39	36	14.	GALAXY
MCG-05-51-030	21 55 18.	- 30 36	24	15.5	GALAXY
SC 2151-5927.5	21 55 18.	- 59 13 17.			NEBULA
SC 2151-5614.1	21 55 21.	- 55 59 52.			NEBULA
ZC 2155.4+0053	21 55 24.	+ 00 53	1610		CLUSTER OF GALAXIES
ZWG 530.001	21 55 24.	+ 41 00		15.1	GALAXY
UGC 11858	21 55 24.	+ 41 00	108	15.1	GALAXY S0-a
SHAH 081	21 55 24.	- 01 59	78	17.8	GROUP OF COMPACT GALAXIES
MCG-04-51-016	21 55 24.	- 25 08	24	15.	GALAXY
SC 2151-6223.4	21 55 25.	- 62 09 10.	48		NEBULA
SC 2151-6111.4	21 55 26.	- 60 57 10.			NEBULA
SC 2152-3919.1	21 55 27.	- 39 04 52.	48		NEBULA
SC 2151-6307.8	21 55 27.	- 62 53 34.	18		NEBULA
ZWG 403.006	21 55 30.	+ 06 12		15.7	GALAXY
MCG+07-45-002	21 55 30.	+ 41 01	48	15.5	GALAXY
MCG-05-51-031	21 55 30.	- 32 31	48	15.5	GALAXY
SC 2151-6300.4	21 55 31.	- 62 46 10.			NEBULA
MCG+00-56-004	21 55 33.	+ 00 46 30.	180	15.	GALAXY
SC 2151-6118.3	21 55 34.	- 61 04 04.	6		NEBULA
IC 1412	21 55 35.	- 17 25 07.			NONSTELLAR OBJECT
SC 2152-5922.5	21 55 35.	- 59 08 16.			NEBULA
ZWG 377.008	21 55 36.	+ 00 46		15.6	GALAXY
UGC 11859	21 55 36.	+ 00 46	150	15.6	GALAXY Sb
KARA.73B 0936	21 55 36.	+ 00 46	216	15.6	ISOLATED GALAXY S
ZC 2155.6+1109	21 55 36.	+ 11 09	1340		CLUSTER OF GALAXIES
MCG-03-56-001	21 55 36.	- 17 25	42	15.	GALAXY
PHL 5020	21 55 36.	- 20 06		17.8	BLUE STELLAR OBJECT
LB 01140	21 55 40.	+ 15 59 18.		16.5	FAINT BLUE STAR
IC 5147	21 55 40.	- 65 41			NONSTELLAR OBJECT
EN 0731	21 55 40.	- 65 42			NEBULA
SC 2152-5825.8	21 55 41.	- 58 11 34.			NEBULA
ZWG 403.007	21 55 42.	+ 08 08		15.7	GALAXY
UGC 11860	21 55 42.	+ 24 01	96	16.0	GALAXY
MCG+04-51-014	21 55 42.	+ 24 01 30.	90	15.5	GALAXY
ZWG 343.003	21 55 42.	+ 73 01		15.2	GALAXY
UGC 11861	21 55 42.	+ 73 01	228	15.2	GALAXY
MCG-05-51-032	21 55 42.	- 28 43	60	14.5	GALAXY
SCHO 1374	21 55 46.	+ 47 28 18.	300		ISOLATED DARK CLOUD
2ZW 157	21 55 48.	+ 08 11			COMPACT GALAXY
UGC 11862	21 55 48.	+ 38 42	108	16.5	GALAXY
MPSL 105+07/1	21 55 48.	+ 64 27	5400		HII REGION
PHL 0217	21 55 48.	- 20 46		17.7	BLUE STELLAR OBJECT
IC 1414	21 55 50.	+ 08 10			NONSTELLAR OBJECT
IC 1413	21 55 51.	- 03 20 36.			NONSTELLAR OBJECT
PK097-02.1	21 55 51.2	+ 51 27 20.	5		PLANETARY NEBULA
IC 5149	21 55 52.	- 27 37 13.			NONSTELLAR OBJECT
SC 2152-5639.6	21 55 53.	- 56 25 21.			NEBULA
ZWG 403.008	21 55 54.	+ 08 11		14.9	GALAXY
MCG+06-48-002	21 55 54.	+ 36 43 30.	24	15.5	GALAXY
ZWG 530.002	21 55 54.	+ 42 03		15.7	GALAXY
UGC 11864	21 55 54.	+ 42 03	150	15.7	GALAXY
MCG+07-45-003	21 55 54.	+ 42 04	120	16.	GALAXY
ZWG 377.009	21 55 54.	- 00 58		15.7	GALAXY
UGC 11863	21 55 54.	- 00 58	60	15.7	GALAXY DWPF SP
KARA.73B 0937	21 55 54.	- 00 58	66	15.7	ISOLATED GALAXY S
SC 2152-3957.8	21 55 55.	- 39 43 33.	12		NEBULA
SC 2152-5233.6	21 55 56.	- 52 19 21.	18		NEBULA
ARC 2402	21 55 58.	- 10 01		16.5	RICH CLUSTER OF GALAXIES
KHAV 761	21 56	+ 45 32			DARK NEBULA
ZWG 472.017	21 56 00.	+ 26 57		15.7	GALAXY
LDN 1143	21 56 00.	+ 58 45	540		DARK NEBULA
LDN 1149	21 56 00.	+ 58 53	480		DARK NEBULA
PHL 1821	21 56 00.	- 24 40		15.9	BLUE STELLAR OBJECT
IC 5148	21 56 02.	- 39 38 31.			NONSTELLAR OBJECT
ARC 2401	21 56 05.	- 20 21		16.5	RICH CLUSTER OF GALAXIES
SC 2152-5629.4	21 56 05.	- 56 15 09.			NEBULA
MCG+01-56-004	21 56 06.	+ 07 11	30	14.5	GALAXY
MCG+02-56-002	21 56 06.	+ 13 53	72	15.	GALAXY
PHL 1822	21 56 06.	- 24 16		18.1	BLUE STELLAR OBJECT
MCG-05-51-033	21 56 06.	- 27 40	60	15.	GALAXY
SC 2152-5920.8	21 56 06.	- 59 06 33.			NEBULA
SC 2152-5514.0	21 56 08.	- 54 59 45.	66		NEBULA
IC 1415	21 56 12.	+ 01 07			NONSTELLAR OBJECT
ZWG 403.009	21 56 12.	+ 07 12		14.9	GALAXY
MCG+01-56-005	21 56 12.	+ 08 12 30.	42	14.	GALAXY
ZWG 403.010	21 56 12.	+ 08 13		15.5	GALAXY
ZC 2156.2+0834	21 56 12.	+ 08 34	940		CLUSTER OF GALAXIES
ZWG 428.005	21 56 12.	+ 11 48		14.3	GALAXY
MRK 518	21 56 12.	+ 11 48	20	14.5	GALAXY WITH UV CONTINUUM
UGC 11865	21 56 12.	+ 11 48	30	14.3	GALAXY
ZWG 428.006	21 56 12.	+ 13 53		15.6	GALAXY
UGC 11866	21 56 12.	+ 13 53	90	15.6	GALAXY PECULR
KARA.73B 0938	21 56 12.	+ 13 53	90	15.6	ISOLATED GALAXY SPEC
HN 2851	21 56 12.5	- 44 06 26.	78		NEBULA
SCHO 1375	21 56 13.	+ 54 23 54.	260		ISOLATED DARK CLOUD

OBJECT NAME	RIGHT ASCEN.	DECLINATION	DIAM.	MAGN.	TYPE OF OBJECT
SCHO 1376	21 56 13.	+ 58 41 48.	550		ISOLATED DARK CLOUD
IC 5150	21 56 13.	- 39 39 24.			NONSTELLAR OBJECT
ARC 2403	21 56 14.	- 18 27		17.1	RICH CLUSTER OF GALAXIES
MCG+00-56-005	21 56 15.	- 02 14 30.	66	15.	GALAXY
MCG-07-45-002	21 56 15.	- 44 06	120	14.	GALAXY
IC 1416	21 56 17.	+ 01 13			NONSTELLAR OBJECT
ZWG 377.010	21 56 18.	- 02 14		15.4	GALAXY
UGC 11867	21 56 18.	- 02 14	66	15.4	GALAXY Sb-c
MCG-05-51-034	21 56 18.	- 28 57	84	14.5	GALAXY
RNGC 7163	21 56 20.	- 32 08		14.0	GALAXY
SC 2153-4128.5	21 56 22.	- 41 14 14.	30		NEBULA
IC 5151	21 56 23.	+ 03 31 03.			NONSTELLAR OBJECT
ZWG 403.011	21 56 24.	+ 03 31		15.5	GALAXY
LB 01141	21 56 25.	+ 00 24 42.		17.4	FAINT BLUE STAR
B 170	21 56 26.	+ 58 43	1560		DARK OBJECT
SC 2152-6308.4	21 56 27.	- 62 54 08.			NEBULA
SC 2152-6055.8	21 56 28.	- 60 41 32.	36		NEBULA
SC 2152-6105.8	21 56 29.	- 60 51 32.	12		NEBULA
ZC 2156.5+1740	21 56 30.	+ 17 40	940		CLUSTER OF GALAXIES
MCG-05-51-035	21 56 30.	- 32 10	72	14.	GALAXY
SC 2153-5728.5	21 56 30.	- 57 14 14.			NEBULA
ARC 2406	21 56 33.	+ 11 03		17.7	RICH CLUSTER OF GALAXIES
MCG+04-51-015	21 56 33.	+ 25 15 30.	27	15.5	GALAXY
MCG-04-51-018	21 56 33.	- 23 30	18	16.	GALAXY
MCG-04-51-017	21 56 33.	- 23 30	36	15.5	GALAXY
PK002-52.1	21 56 33.	- 39 37 25.	120	13.	PLANETARY NEBULA
ZC 2156.6+0902	21 56 36.	+ 09 02	1210		CLUSTER OF GALAXIES
PHL 5021	21 56 36.	- 24 12		17.0	BLUE STELLAR OBJECT
MCG-07-45-003	21 56 36.	- 43 33	90	13.1	GALAXY
RNGC 7162	21 56 39.	- 43 33		13.0	GALAXY
ARC 2404	21 56 40.	- 14 40		17.8	RICH CLUSTER OF GALAXIES
SC 2153-5755.7	21 56 40.	- 57 41 25.			NEBULA
2ZW 158	21 56 42.	+ 17 56			COMPACT GALAXY
UGC 11868	21 56 42.	+ 17 56	156	15.1	GALAXY
ZC 2156.7+3202	21 56 42.	+ 32 02	740		CLUSTER OF GALAXIES
MCG-03-56-002	21 56 42.	- 16 45	48	14.5	GALAXY
TON-S 0040	21 56 42.	- 26 34		15.5	BLUE STAR
SC 2153-6102.7	21 56 42.	- 60 48 25.	18		NEBULA
SC 2153-5755.7	21 56 43.	- 57 41 25.	24		NEBULA
RNGC 7165	21 56 44.	- 16 45		14.0	GALAXY
MCG+03-56-001	21 56 45.	+ 17 55 30.	120	14.5	GALAXY
SC 2153-5510.8	21 56 46.	- 54 56 31.			NEBULA
ZWG 428.007	21 56 48.	+ 13 25		15.7	GALAXY
MCG-02-56-001	21 56 48.	- 12 29	36	15.	GALAXY
ARC 2405	21 56 50.	- 18 05		17.1	RICH CLUSTER OF GALAXIES
ZWG 428.008	21 56 54.	+ 10 14		15.5	GALAXY
MRK 519	21 56 54.	+ 10 14	12	15.5	GALAXY WITH UV CONTINUUM
ZCG 2156+17	21 56 54.	+ 17 59		18.0	COMPACT GALAXY
4ZW 082	21 56 54.	+ 25 47			COMPACT GALAXY
SCHO 1377	21 56 58.	+ 61 48 48.	330		ISOLATED DARK CLOUD
KHAV 762	21 57	+ 58 20	2660		DARK NEBULA
ZC 2157.0+0744	21 57 00.	+ 07 44	1340		CLUSTER OF GALAXIES
MCG+01-56-006	21 57 00.	+ 09 10	42	15.	GALAXY
ZC 2157.0+3126	21 57 00.	+ 31 26	2020		CLUSTER OF GALAXIES
SC 2153-5639.4	21 57 00.	- 56 25 07.			NEBULA
RNGC 7175	21 57 01.	+ 54 35			NON-EXISTENT OBJECT
ZWG 403.012	21 57 06.	+ 09 10		15.2	GALAXY
SC 2153-6247.8	21 57 09.	- 62 33 31.			NEBULA
SC 2153-5613.5	21 57 11.	- 55 59 12.	18		NEBULA
ZWG 473.001	21 57 12.	+ 23 28		15.5	GALAXY
KARA.73B 0939	21 57 12.	+ 23 28	48	15.5	ISOLATED GALAXY S
MCG-05-52-001	21 57 12.	- 31 46	24	15.5	GALAXY
MCG+02-56-003	21 57 15.	+ 10 05	24	15.	GALAXY
MCG+04-52-001	21 57 15.	+ 23 28	45	15.	GALAXY
B 169	21 57 15.	+ 58 31	3600		DARK OBJECT
ZWG 428.009	21 57 18.	+ 10 06		15.6	GALAXY
5ZW 379	21 57 18.	+ 44 05			COMPACT GALAXY
SCHO 1378	21 57 18.	+ 49 10 42.	200		ISOLATED DARK CLOUD
ISS 0130	21 57 18.	+ 51 17	151		STELLAR RING
MCG-02-56-002	21 57 18.	- 14 02	24	15.5	GALAXY
SC 2153-5535.3	21 57 22.	- 55 21 00.			NEBULA
SC 2153-5808.4	21 57 22.	- 57 54 06.			NEBULA
MCG+02-56-004	21 57 24.	+ 13 20	48	15.	GALAXY
SC 2154-5207.5	21 57 28.	- 51 53 11.			NEBULA
SC 2154-4025.1	21 57 29.	- 40 10 47.			NEBULA
SC 2153-6139.4	21 57 29.	- 61 25 06.	12		NEBULA
ZWG 428.010	21 57 30.	+ 13 21		15.6	GALAXY
MCG-07-45-004	21 57 30.	- 43 38	96	11.6	GALAXY
HN 2852	21 57 30.6	- 43 22 59.	36		NEBULA
LB 01142	21 57 32.	+ 14 03 18.		16.4	FAINT BLUE STAR
RNGC 7166	21 57 33.	- 43 39		13.0	GALAXY
HN 2853	21 57 34.8	- 44 43 47.	12		NEBULA
MCG-02-56-003	21 57 36.	- 13 22	54	14.5	GALAXY
MCG-05-52-002	21 57 36.	- 27 12	48	15.5	GALAXY
MCG-07-45-005	21 57 36.	- 43 23	78	16.	GALAXY
SCHO 1379	21 57 38.	+ 49 01 48.	200		ISOLATED DARK CLOUD
MCG-02-56-004	21 57 39.	- 14 04	36	15.	GALAXY
RNGC 7162A	21 57 39.	- 43 23			GALAXY
IC 1417	21 57 39.8	- 13 23 15.			GALAXY S
RNGC 7167	21 57 41.	- 24 52		13.0	GALAXY
SC 2154-5813.2	21 57 41.	- 57 58 53.			NEBULA
UGC 11869	21 57 42.	+ 44 08	60	16.5	GALAXY S
SCHO 1380	21 57 42.	+ 49 09 42.	250		ISOLATED DARK CLOUD
MCG-04-52-001	21 57 42.	- 24 52 30.	84	13.5	GALAXY
DV.56 N7162A	21 57 42.	- 43 23			GALAXY
BC M2157+103	21 57 42.1	+ 10 16 10.		19.	QUASI-STELLAR OBJECT
SHB 370	21 57 42.1	+ 10 16 10.		19.	QUASI-STELLAR OBJECT
SC 2154-6301.6	21 57 43.	- 62 47 17.	30		NEBULA
2ZW 159	21 57 48.	+ 18 16			COMPACT GALAXY
MCG-05-52-003	21 57 48.	- 28 30	36	15.	GALAXY
SC 2154-5118.5	21 57 52.	- 51 04 10.			NEBULA
MCG+01-56-007	21 57 54.	+ 06 12	60	15.	GALAXY
SC 2154-5533.3	21 57 55.	- 55 18 58.			NEBULA
SC 2154-5549.7	21 57 56.	- 55 35 22.			NEBULA
SC 2154-5837.6	21 57 58.	- 58 23 16.	60		NEBULA
IC 5153	21 57 59.	+ 17 37			MAY NOT EXIST
ZWG 403.013	21 58 00.	+ 06 12		15.6	GALAXY
UGC 11870	21 58 00.	+ 06 12	66	15.6	GALAXY S
ISS 0131	21 58 00.	+ 53 37	280		STELLAR RING
LDN 1151	21 58 00.	+ 58 50	660		DARK NEBULA
MCG-03-56-003	21 58 00.	- 15 08	36	16.	GALAXY
SC 2154-4104.9	21 58 00.	- 40 50 34.	6		NEBULA
LB 01143	21 58 01.	+ 16 01 36.		16.8	FAINT BLUE STAR
SCHO 1381	21 58 02.	+ 58 23 42.	1940		ISOLATED DARK CLOUD
LB 01144	21 58 03.	+ 16 44 24.		16.3	FAINT BLUE STAR
SCHO 1382	21 58 04.	+ 49 27 48.	310		ISOLATED DARK CLOUD
SCHO 1383	21 58 05.	+ 52 37 18.	290		ISOLATED DARK CLOUD
ZC 2158.1+2437	21 58 06.	+ 24 37	5980		CLUSTER OF GALAXIES
MCG-05-52-004	21 58 06.	- 28 40	48	15.	GALAXY
SC 2155-3818.3	21 58 06.	- 38 03 58.	30		NEBULA
ZWG 428.011	21 58 12.	+ 10 19		14.7	GALAXY
MRK 520	21 58 12.	+ 10 19	15	15.	GALAXY WITH UV CONTINUUM
UGC 11871	21 58 12.	+ 10 19	60	14.7	GALAXY DBL SYS
KARA.73B 0940	21 58 12.	+ 10 19	18	14.7	ISOLATED GALAXY S
ZC 2158.2+1101	21 58 12.	+ 11 01	1550		CLUSTER OF GALAXIES
ZWG 428.012	21 58 12.	+ 11 08		15.6	GALAXY
URA 81	21 58 12.	+ 52 06	648		STELLAR RING
LB 00367	21 58 13.	- 00 03 12.		15.4	FAINT BLUE STAR
MCG+02-56-005	21 58 15.	+ 10 18	60	16.	GALAXY
RNGC 7171	21 58 15.	- 13 31		12.5	GALAXY
MCG-04-52-002	21 58 15.	- 25 31	54	15.	GALAXY
SC 2154-6248.1	21 58 17.	- 62 33 46.			NEBULA
SN 1960L	21 58 18.	+ 17 29		16.0	SUPERNOVA
MCG+03-56-003	21 58 18.	+ 17 29	210	12.5	GALAXY
ZWG 451.002	21 58 18.	+ 17 30		12.2	GALAXY
UGC 11872	21 58 18.	+ 17 30	198	12.2	GALAXY Sb
MCG+03-56-002	21 58 18.	+ 19 25	66	15.	GALAXY
ISS 0199	21 58 18.	+ 50 43	132		STELLAR RING
ZWG 377.011	21 58 18.	- 02 04		15.3	GALAXY
MCG-02-56-005	21 58 18.	- 13 30	132	12.	GALAXY
RNGC 7177	21 58 19.	+ 17 30		12.5	GALAXY
UGC 11873	21 58 24.	+ 19 26	66	16.0	GALAXY S
HELW 305	21 58 25.	- 31 45 57.			NEBULA
HN 2854	21 58 28.5	- 42 41 45.	24		NEBULA
ZWG 403.014	21 58 30.	+ 08 23		15.7	GALAXY
URA 71	21 58 30.	+ 53 49	300		STELLAR RING
MCG-05-52-005	21 58 30.	- 28 46	48	15.	GALAXY
MCG-05-52-006	21 58 30.	- 31 51	48	15.	GALAXY
SC 2155-4134.9	21 58 30.	- 41 20 33.			NEBULA
MCG-07-45-006	21 58 30.	- 42 42	36	15.5	GALAXY
SC 2155-5707.0	21 58 33.	- 56 52 39.			NEBULA
SC 2155-5722.0	21 58 33.	- 57 07 39.	18		NEBULA
ARC 2409	21 58 34.	+ 20 43		16.8	RICH CLUSTER OF GALAXIES
SC 2154-6246.2	21 58 35.	- 62 31 51.			NEBULA
HN 2355	21 58 35.6	- 42 40 27.	12		NEBULA
URA 33	21 58 36.	+ 51 58	258		STELLAR RING
TON-S 0041	21 58 36.	- 25 39		15.6	BLUE STAR
MCG-07-45-007	21 58 36.	- 42 41	30	16.	GALAXY
ARC 2407	21 58 38.	+ 07 09		17.1	RICH CLUSTER OF GALAXIES
SC 2155-5653.4	21 58 38.	- 56 39 03.	18		NEBULA
SC 2154-6310.3	21 58 38.	- 62 55 57.			NEBULA
SC 2155-3754.5	21 58 41.	- 37 40 08.	36		NEBULA
SC 2155-5925.2	21 58 41.	- 59 10 51.	12		NEBULA
ZC 2158.7+2051	21 58 42.	+ 20 51	1610		CLUSTER OF GALAXIES
ISS 0200	21 58 42.	+ 48 26	146		STELLAR RING
SC 2155-4014.2	21 58 42.	- 39 59 50.			NEBULA
SC 2155-5922.9	21 58 43.	- 59 08 33.			NEBULA
ARC 2408	21 58 45.	+ 05 56		17.1	RICH CLUSTER OF GALAXIES
SC 2155-6041.0	21 58 46.	- 60 26 39.	6		NEBULA
SC 2155-6056.9	21 58 46.	- 60 42 33.	6		NEBULA
RNGC 7170	21 58 48.	- 05 42			GALAXY
MCG-04-52-003	21 58 48.	- 22 20	72	14.5	GALAXY
DV.56 I5152A	21 58 48.	- 52 02	63		E1 GALAXY
SC 2155-3654.1	21 58 50.	- 36 39 44.	9		NEBULA
SCHO 1384	21 58 52.	+ 52 31 12.	350		ISOLATED DARK CLOUD
RNGC 7168	21 58 52.	- 52 00		13.0	GALAXY
SC 2155-5231.5	21 58 53.	- 52 17 08.			NEBULA
URA 82	21 58 54.	+ 52 21	420		STELLAR RING
ZWG 377.012	21 58 54.	- 02 21		15.4	GALAXY
MCG-04-52-004	21 58 54.	- 21 45	72	15.	GALAXY
SC 2155-6039.7	21 58 54.	- 60 25 20.	6		NEBULA
RNGC 7169	21 58 55.	- 47 56			UNVERIFIED SOUTHRN OBJECT
SC 2155-5549.7	21 58 56.	- 55 35 20.			NEBULA
SC 2155-5638.6	21 58 56.	- 56 24 14.	90		NEBULA
SCHO 1385	21 58 59.	+ 52 43 24.	290		ISOLATED DARK CLOUD
ZWG 545.001	21 59 00.	+ 51 20		15.6	GALAXY
UGC 11874	21 59 00.	+ 51 20	78	15.6	GALAXY S
LDN 1091	21 59 00.	+ 52 40	1380		DARK NEBULA
LDN 1153	21 59 00.	+ 58 45	1140		DARK NEBULA
MCG-04-52-005	21 59 00.	- 23 58	54	15.	GALAXY
RNGC 7172	21 59 03.	- 32 07		12.0	GALAXY
MCG+09-36-001	21 59 06.	+ 51 21	48	16.	GALAXY
ZWG 377.013	21 59 06.	- 02 20		15.6	GALAXY
MCG-04-52-006	21 59 06.	- 22 44	42	14.5	GALAXY
RNGC 7186	21 59 07.	+ 34 52			NON-EXISTENT OBJECT
SC 2155-5933.8	21 59 07.	- 59 19 26.	15		NEBULA
RNGC 7174	21 59 09.	- 32 15		12.0	GALAXY
KLEM 34	21 59 11.	- 32 12	600	12.	BRIGHT GRP OF 4 GALAXIES
MCG+02-56-006	21 59 12.	+ 09 44	54	15.	GALAXY
ZWG 428.013	21 59 12.	+ 09 45		14.8	GALAXY
ZWG 377.014	21 59 12.	- 02 12		15.4	GALAXY
MCG+00-56-006	21 59 12.	- 02 28	42	15.	GALAXY
SC 2156-4134.3	21 59 12.	- 41 19 55.	60		NEBULA
MCG-07-45-008	21 59 12.	- 41 21	60	15.	GALAXY
RNGC 7181	21 59 13.	- 02 12		15.5	GALAXY
MCG+02-56-007	21 59 15.	+ 11 37	36	15.	GALAXY
MCG-05-52-007	21 59 15.	- 32 10	120	12.	GALAXY
ZWG 428.014	21 59 18.	+ 11 37		15.6	GALAXY
URA 72	21 59 18.	+ 55 06	600		STELLAR RING
ZWG 377.015	21 59 18.	- 02 26		15.6	GALAXY
RNGC 7182	21 59 18.	- 02 26		15.5	GALAXY
MCG-02-56-006	21 59 18.	- 10 30	30	15.5	GALAXY
MCG-05-52-009	21 59 18.	- 31 31	60	15.5	GALAXY
MCG-05-52-008	21 59 18.	- 32 16	30	13.	GALAXY
SC 2155-6132.7	21 59 19.	- 61 18 20.			NEBULA
MCG-02-56-007	21 59 21.	- 12 25	120	14.	GALAXY
HELW 306	21 59 21.	- 31 59 49.			NEBULA
RNGC 7173	21 59 21.	- 32 16		13.0	GALAXY
MCG-05-52-010	21 59 21.	- 32 17	48	12.	GALAXY
ARC 2410	21 59 22.	- 10 09		16.0	RICH CLUSTER OF GALAXIES
ZWG 403.015	21 59 24.	+ 04 47		15.7	GALAXY
MCG+01-56-008	21 59 24.	+ 04 47	15	15.5	GALAXY
UGC 11875	21 59 24.	+ 09 58	60	16.0	GALAXY IRR
MCG-05-52-012	21 59 24.	- 30 46	30	15.5	GALAXY
MCG-05-52-011	21 59 24.	- 32 17	24	13.	GALAXY
SC 2155-5922.3	21 59 24.	- 59 07 55.	18		NEBULA
SCHO 1386	21 59 27.	+ 49 42 00.	310		ISOLATED DARK CLOUD
RNGC 7176	21 59 27.	- 32 17		13.0	GALAXY
IC 1418	21 59 28.	+ 04 07 51.			NONSTELLAR OBJECT
BN 0732	21 59 29.	- 51 32			NEBULA
MCG+01-56-009	21 59 30.	+ 04 07	48	14.	GALAXY
ZWG 403.016	21 59 30.	+ 04 08		15.3	GALAXY
MCG-02-56-008	21 59 30.	- 09 33	36	15.5	GALAXY
RNGC 7180	21 59 31.	- 20 48		14.0	GALAXY
RNGC 7178	21 59 31.	- 36 03			GALAXY
SC 2155-6240.5	21 59 33.	- 62 26 07.	24		NEBULA

OBJECT NAME	RIGHT ASCEN.	DECLINATION	DIAM.	MAGN.	TYPE OF OBJECT
IC 5155	21 59 34.	+ 00 16			MAY NOT EXIST
SC 2156-5750.9	21 59 34.	- 57 36 31.	12		NEBULA
SC 2156-4133.7	21 59 35.	- 41 19 18.	18		NEBULA
ZWG 428.015	21 59 36.	+ 14 57		15.5	GALAXY
ZC 2159.6-0115	21 59 36.	- 01 15	1140		CLUSTER OF GALAXIES
MCG-03-56-004	21 59 36.	- 19 10	132	14.	GALAXY
IC 5152	21 59 36.	- 51 32	276	11.5	GALAXY IRR
SC 2156-6106.9	21 59 36.	- 60 52 31.			NEBULA
SC 2155-6309.2	21 59 36.	- 62 54 49.	12		NEBULA
RNGC 7183	21 59 37.	- 19 10		14.0	GALAXY
MCG-04-52-007	21 59 39.	- 21 15 30.	66	15.	GALAXY
B 171	21 59 40.	+ 58 38	1140		DARK OBJECT
SC 2156-5711.5	21 59 40.	- 56 57 07.			NEBULA
SC 2156-5933.5	21 59 41.	- 59 19 07.	6		NEBULA
SC 2156-6058.1	21 59 41.	- 60 43 43.			NEBULA
MCG-04-52-008	21 59 42.	- 20 47	72	14.	GALAXY
SC 2156-6104.9	21 59 44.	- 60 50 31.	9		NEBULA
ZC 2159.8+0921	21 59 48.	+ 09 21	1410		CLUSTER OF GALAXIES
ZC 2159.8+2253	21 59 48.	+ 22 53	670		CLUSTER OF GALAXIES
MCG-05-52-015	21 59 48.	- 28 23	15	15.	GALAXY
MCG-05-52-013	21 59 48.	- 29 17	72	15.	GALAXY
MCG-05-52-014	21 59 48.	- 32 18	48	15.	GALAXY
SER 143.04	21 59 48.	- 48 34	50	17.	LOW SURF. BRGHTNSS GALAXY
SC 2156-5055.2	21 59 48.	- 50 40 48.	48		NEBULA
HELW 307	21 59 49.	- 32 13 54.			NEBULA
LB 01145	21 59 50.	+ 14 32 42.		17.2	FAINT BLUE STAR
IC 1420	21 59 50.	+ 19 30 29.			NONSTELLAR OBJECT
REIN 2.304	21 59 53.30	- 21 03 19.0			NEBULA
UGC 11876	21 59 54.	+ 02 36	60	16.0	GALAXY Sc
ZWG 428.016	21 59 54.	+ 11 37		15.7	GALAXY
UGC 11877	21 59 54.	+ 11 37	60	15.7	GALAXY Sc
MCG-04-52-009	21 59 54.	- 21 03	330	12.	GALAXY
KLEM 35	21 59 54.	- 22 43	1080	16.	CLUSTER OF 20 GALAXIES
SC 2156-5415.3	21 59 54.	- 54 00 54.	12		NEBULA
SC 2156-5627.7	21 59 54.	- 56 13 18.	36		NEBULA
SC 2156-6126.2	21 59 54.	- 61 11 48.			NEBULA
RNGC 7184	21 59 55.	- 21 04		12.0	GALAXY
HN 2856	21 59 55.3	- 44 24 24.	12		NEBULA
ARC 2411	21 59 56.	- 08 47		17.5	RICH CLUSTER OF GALAXIES
MCG+03-56-004	21 59 57.	+ 18 04	66	15.	GALAXY
KHAV 763	22 00	+ 52 38	3680		DARK NEBULA
LBN 0535	22 00	+ 72 40	5520		BRIGHT NEBULA
KHAV 764	22 00	+ 77 02	7640		DARK NEBULA
UGC 11878	22 00 00.	+ 18 00	84	14.8	GALAXY SB?
2ZW 160	22 00 00.	+ 18 04			COMPACT GALAXY
ZWG 451.003	22 00 00.	+ 18 05		14.8	GALAXY
KARA.72 560B	22 00 00.	+ 18 05	24		PART OF DOUBLE GALAXY
KARA.72 560A	22 00 00.	+ 18 05	36	14.8	PART OF DOUBLE GALAXY
ZC 2200.0+3117	22 00 00.	+ 31 17	200		CLUSTER OF GALAXIES
COU 088	22 00 00.	+ 57 20	12000		EMISSION NEBULA
LDN 1292	22 00 00.	+ 88 20	2760		DARK NEBULA
TON-S 0042	22 00 00.	- 28 57		15.4	BLUE STAR
RNGC 7187	22 00 02.	- 33 02			GALAXY
SC 2157-3811.9	22 00 02.	- 37 57 29.	30		NEBULA
IC 5159	22 00 09.	+ 00 06			SINGLE STAR
MCG+03-56-005	22 00 09.	+ 19 30	84	13.5	GALAXY
YC 2200-37	22 00 09.	- 37 41 06.			UNUSUAL SOUTHERN NEBULA
ZWG 428.017	22 00 12.	+ 15 01		15.3	GALAXY
UGC 11879	22 00 12.	+ 15 01	60	15.3	GALAXY S
ZWG 451.004	22 00 12.	+ 19 30		14.5	GALAXY
UGC 11880	22 00 12.	+ 19 30	90	14.5	GALAXY SB
KARA.72 561B	22 00 12.	+ 19 30	48		PART OF DOUBLE GALAXY
KARA.72 561A	22 00 12.	+ 19 30	42	14.5	PART OF DOUBLE GALAXY
ISS 0083	22 00 12.	+ 60 02	133		STELLAR RING
SER 150.04	22 00 12.	- 64 21	30	17.5	LOW SURF. BRGHTNSS GALAXY
SC 2156-6242.8	22 00 13.	- 62 28 24.	30		NEBULA
MCG-04-52-010	22 00 15.	- 22 44	108	15.	GALAXY
SC 2157-3832.9	22 00 16.	- 38 18 29.	9		NEBULA
MCG-04-52-011	22 00 18.	- 20 42 30.	24	14.	GALAXY
SER 150.09	22 00 18.	- 66 23	50	14.	HIGH SURF BRGHTNSS GALAXY
IC 1419	22 00 19.	- 10 09 36.			NONSTELLAR OBJECT
RNGC 7185	22 00 19.	- 20 42		14.0	GALAXY
ZWG 403.017	22 00 24.	+ 09 20		15.5	GALAXY
SCHO 1387	22 00 24.	+ 65 49 48.	330		ISOLATED DARK CLOUD
MCG-02-56-009	22 00 24.	- 10 12	15	15.5	GALAXY
MCG-05-52-016	22 00 24.	- 26 41	30	15.5	GALAXY
IC 5156	22 00 24.	- 34 02	78	13.2	GALAXY E7
IC 1421	22 00 25.	- 10 13 05.			NONSTELLAR OBJECT
IC 1422	22 00 27.	+ 02 21 49.			NONSTELLAR OBJECT
IC 5157	22 00 28.	- 35 11 36.			NONSTELLAR OBJECT
SC 2156-6330.3	22 00 29.	- 63 15 53.	18		NEBULA
ZWG 377.016	22 00 30.	+ 02 22		15.7	GALAXY
UGC 11881	22 00 30.	+ 17 40	66	16.0	GALAXY
ZWG 494.001	22 00 30.	+ 31 27		15.7	GALAXY
KARA.73B 0941	22 00 30.	+ 31 27	48	15.7	ISOLATED GALAXY S
MCG-05-52-017	22 00 30.	- 29 04	60	15.	GALAXY
SC 2157-5222.0	22 00 31.	- 52 07 35.			NEBULA
SC 2157-5540.6	22 00 31.	- 55 26 11.			NEBULA
SC 2157-3804.9	22 00 32.	- 37 50 28.	66		NEBULA
SC 2157-5825.8	22 00 32.	- 58 11 23.			NEBULA
SC 2156-6214.3	22 00 33.	- 61 59 53.	6		NEBULA
ARP 325	22 00 35.	- 21 18			PECULIAR GALAXY
ZC 2200.6+2925	22 00 36.	+ 29 25	3230		CLUSTER OF GALAXIES
URA 73	22 00 36.	+ 52 32	72		STELLAR RING
IC 5160	22 00 39.	+ 10 40 26.			NONSTELLAR OBJECT
BC VRO42.22.01	22 00 39.42	+ 42 02 08.0		15.	QUASI-STELLAR OBJECT
RNGC 7190	22 00 40.	+ 10 58		15.0	GALAXY
ZWG 377.017	22 00 42.	+ 00 20		14.4	GALAXY
UGC 11882	22 00 42.	+ 00 20	54	14.4	GALAXY SBb
MCG+00-56-007	22 00 42.	+ 00 20	48	14.	GALAXY
ZWG 403.018	22 00 42.	+ 04 03		14.8	GALAXY
UGC 11883	22 00 42.	+ 04 03	66	14.8	GALAXY SBb
IC 1423	22 00 42.	+ 04 03 20.			NONSTELLAR OBJECT
ZWG 428.018	22 00 42.	+ 10 41		15.6	GALAXY
UGC 11884	22 00 42.	+ 10 41	66	15.0	GALAXY SB0
ZWG 428.019	22 00 42.	+ 10 58		15.0	GALAXY
UGC 11885	22 00 42.	+ 10 58	60	15.0	GALAXY S0-a
ZWG 428.020	22 00 42.	+ 12 43		15.7	GALAXY
ZC 2200.7+3752	22 00 42.	+ 37 52	3490		CLUSTER OF GALAXIES
RNGC 7189	22 00 43.	+ 00 20		14.5	GALAXY
IC 1424	22 00 43.	+ 10 57			NONSTELLAR OBJECT
RNGC 7188	22 00 43.	- 20 34		14.0	GALAXY
MCG+01-56-010	22 00 45.	+ 04 02 30.	54	14.	GALAXY
MCG-04-52-012	22 00 45.	- 20 34	66	14.	GALAXY
IC 5154	22 00 45.	- 66 21			NONSTELLAR OBJECT
HN 0733	22 00 45.	- 66 22			NEBULA
UGC 11886	22 00 48.	+ 44 28	66	16.0	GALAXY Sb/SBb
MCG-02-56-010	22 00 48.	- 10 57	42	15.	GALAXY

OBJECT NAME	RIGHT ASCEN.	DECLINATION	DIAM.	MAGN.	TYPE OF OBJECT
MCG-05-52-018	22 00 48.	- 28 05	72	15.	GALAXY
MCG-05-52-019	22 00 48.	- 28 14	60	15.5	GALAXY
RNGC 7197	22 00 51.	+ 40 49		14.5	GALAXY
MCG+07-45-004	22 00 51.	+ 44 27	54	16.	GALAXY
IC 1425	22 00 53.	+ 02 21 33.			NONSTELLAR OBJECT
HELW 309	22 00 53.	- 32 11 27.			NEBULA
HELW 308	22 00 53.	- 32 31 27.			NEBULA
ZWG 377.018	22 00 54.	+ 02 22		15.0	GALAXY
ZWG 428.021	22 00 54.	+ 12 24		15.7	GALAXY
ZWG 530.003	22 00 54.	+ 40 49		14.5	GALAXY
UGC 11887	22 00 54.	+ 40 49	126	14.5	GALAXY Sa
MCG+07-45-005	22 00 54.	+ 40 49	60	14.5	GALAXY
MCG-05-52-020	22 00 54.	- 32 36	96	14.	GALAXY
SC 2157-6227.7	22 00 54.	- 62 13 16.			NEBULA
IC 1427	22 00 55.	+ 14 52 22.			NONSTELLAR OBJECT
RNGC 7195	22 00 59.	+ 12 25		15.5	GALAXY
LB 09945	22 01	- 83 50		14.1	FAINT BLUE STAR
ZC 2201.0+0825	22 01 00.	+ 08 25	1610		CLUSTER OF GALAXIES
MCG+02-56-008	22 01 00.	+ 12 23	48	14.5	GALAXY
ZWG 428.022	22 01 00.	+ 12 25		15.6	GALAXY
MCG+02-56-009	22 01 00.	+ 12 25	24	15.	GALAXY
ZWG 428.023	22 01 00.	+ 13 51		15.7	GALAXY
KARA.73B 0942	22 01 00.	+ 13 51	24	15.7	ISOLATED GALAXY
LDN 1133	22 01 00.	+ 56 00	1980		DARK NEBULA
MCG-05-52-021	22 01 00.	- 32 16	54	15.	GALAXY
SHB 371	22 01 01.1	+ 31 31 10.		16.5	QUASI-STELLAR OBJECT
SC 2157-5327.7	22 01 02.	- 53 13 15.			NEBULA
SC 2156-7603.3	22 01 04.	- 75 48 53.	18		NEBULA
RNGC 7194	22 01 05.	+ 12 23		14.5	GALAXY
SC 2157-6334.1	22 01 05.	- 63 19 40.			NEBULA
SC 2157-6344.9	22 01 05.	- 63 30 28.	12		NEBULA
ZC 2201.1+0808	22 01 06.	+ 08 08	5650		CLUSTER OF GALAXIES
ZWG 428.024	22 01 06.	+ 12 23		14.5	GALAXY
UGC 11888	22 01 06.	+ 12 23	78	14.5	GALAXY E
ZC 2201.1+2339	22 01 06.	+ 23 39	200		CLUSTER OF GALAXIES
MCG-03-56-005	22 01 06.	- 20 07 30.	60	15.	GALAXY
RNGC 7179	22 01 09.	- 64 18			GALAXY
RNGC 7193	22 01 10.	+ 10 35			NON-EXISTENT OBJECT
SC 2157-5833.2	22 01 11.	- 58 18 45.			NEBULA
ZWG 428.025	22 01 12.	+ 14 52		15.2	GALAXY
UGC 11889	22 01 12.	+ 14 52	78	15.2	GALAXY E
MCG+02-56-010	22 01 12.	+ 14 52	36	15.	GALAXY
IC 1426	22 01 14.	- 10 09 08.			NONSTELLAR OBJECT
SC 2157-5800.8	22 01 16.	- 57 46 21.			NEBULA
ISS 0084	22 01 18.	+ 59 32	80		STELLAR RING
MCG-01-56-001	22 01 18.	- 02 38	48	16.	GALAXY
SER 149.11	22 01 18.	- 58 32	300	15.	LOOSE GROUP OF 4 GALAXIES
ZWG 513.002	22 01 24.	+ 38 18		14.8	GALAXY
UGC 11890	22 01 24.	+ 38 18	60	14.8	GALAXY Sa-b
ARC 2412	22 01 24.	- 21 42		15.9	RICH CLUSTER OF GALAXIES
SER 149.07	22 01 24.	- 57 42	1020	15.	LOOSE GROUP OF 6 GALAXIES
SC 2157-5835.8	22 01 24.	- 58 21 21.			NEBULA
SC 2157-5846.7	22 01 24.	- 58 32 15.	48		NEBULA
SC 2158-5319.8	22 01 25.	- 53 05 21.			NEBULA
SC 2157-5806.4	22 01 26.	- 57 51 57.			NEBULA
MCG+06-48-003	22 01 27.	+ 38 20	66	15.	GALAXY
SC 2158-4144.7	22 01 29.	- 41 30 14.	12		NEBULA
SC 2157-6059.9	22 01 29.	- 60 45 27.	12		NEBULA
ZC 2201.5+1906	22 01 30.	+ 19 06	3020		CLUSTER OF GALAXIES
5ZW 380	22 01 30.	+ 43 30			COMPACT GALAXY
ZWG 530.004	22 01 30.	+ 43 30		15.5	GALAXY
UGC 11891	22 01 30.	+ 43 30	240	15.5	GALAXY DWRF IR
MCG+07-45-006	22 01 30.	+ 43 30	270		GALAXY
MCG-02-56-011	22 01 30.	- 11 36	48	15.	GALAXY
SC 2158-5756.5	22 01 31.	- 57 42 03.	54		NEBULA
SC 2157-6256.5	22 01 34.	- 62 42 03.	6		NEBULA
2ZW 161	22 01 36.	+ 05 57			COMPACT GALAXY
SC 2158-4130.4	22 01 39.	- 41 15 56.	24		NEBULA
HN 2857	22 01 39.1	- 43 45 56.	12		NEBULA
ARC 2413	22 01 40.	+ 11 04		17.4	RICH CLUSTER OF GALAXIES
SC 2158-5846.5	22 01 40.	- 58 32 02.	48		NEBULA
4ZW 084	22 01 42.	+ 24 16			COMPACT GALAXY
ZWG 513.003	22 01 42.	+ 35 45		15.1	GALAXY
UGC 11892	22 01 42.	+ 35 45	66	15.1	GALAXY E-S0
MCG+06-48-004	22 01 42.	+ 35 45	36	15.	GALAXY
ZWG 377.019	22 01 42.	- 00 16		15.3	GALAXY
KARA.73B 0943	22 01 42.	- 00 16	36	15.3	ISOLATED GALAXY S
BC PKS2201+04	22 01 43.	+ 04 25 54.		16.	QUASI-STELLAR OBJECT
SC 2158-6109.7	22 01 44.	- 60 55 14.	78		NEBULA
SC 2158-5758.5	22 01 45.	- 57 44 02.	18		NEBULA
ZWG 403.019	22 01 48.	+ 04 25		15.2	GALAXY
ZC 2201.8+1101	22 01 48.	+ 11 01	1210		CLUSTER OF GALAXIES
MCG-05-52-022	22 01 48.	- 26 41	60	15.	GALAXY
SCHO 1388	22 01 52.	+ 62 02 42.	410		ISOLATED DARK CLOUD
ZWG 377.020	22 01 54.	+ 00 28		15.3	GALAXY
ZWG 377.021	22 01 54.	+ 02 23		15.6	GALAXY
ZWG 428.026	22 01 54.	+ 12 43		15.5	GALAXY
ZWG 473.002	22 01 54.	+ 26 07		15.6	GALAXY
KARA.73B 0944	22 01 54.	+ 26 07	24	15.6	ISOLATED GALAXY S0
MCG+06-48-005	22 01 54.	+ 35 41	108	15.	GALAXY
IC 1428	22 01 55.	+ 02 23 43.			NONSTELLAR OBJECT
ZWG 428.027	22 02 00.	+ 12 13		15.3	GALAXY
UGC 11893	22 02 00.	+ 35 42	138	16.0	GALAXY Sc
MCG+07-45-007	22 02 00.	+ 41 34	66	16.	GALAXY
ZWG 530.005	22 02 00.	+ 41 35		15.7	GALAXY
UGC 11894	22 02 00.	+ 41 35	84	15.7	GALAXY Sa
MCG-02-56-012	22 02 00.	- 11 09	48	15.	GALAXY
SC 2158-6059.5	22 02 01.	- 60 45 02.	12		NEBULA
SC 2158-6337.2	22 02 02.	- 63 22 44.			NEBULA
URA 80	22 02 06.	+ 52 43	660		STELLAR RING
SC 2158-6119.2	22 02 06.	- 61 04 43.	6		NEBULA
SC 2156-7910.9	22 02 08.	- 78 56 27.	18		NEBULA
SC 2158-5214.0	22 02 10.	- 51 59 31.			NEBULA
ZWG 403.020	22 02 12.	+ 09 00		15.6	GALAXY
MCG+01-56-011	22 02 12.	+ 09 01	24	15.	GALAXY
ZWG 530.006	22 02 12.	+ 39 30		14.8	GALAXY
ZWG 513.004	22 02 12.	+ 39 30		14.8	GALAXY
UGC 11895	22 02 12.	+ 39 30	84	14.8	GALAXY Sb
ARC 2414	22 02 13.	+ 08 34		17.1	RICH CLUSTER OF GALAXIES
MCG+07-45-008	22 02 15.	+ 39 30	66	14.5	GALAXY
ZWG 377.022	22 02 18.	+ 02 16		15.6	GALAXY
TON-S 0043	22 02 18.	- 28 45		15.2	BLUE STAR
SC 2156-7911.8	22 02 19.	- 78 57 21.	18		NEBULA
SCHO 1389	22 02 20.	+ 67 00 54.	520		ISOLATED DARK CLOUD
SC 2159-5212.0	22 02 21.	- 51 57 31.			NEBULA
MCG+03-56-006	22 02 24.	+ 15 32	42	15.	GALAXY
ZWG 451.005	22 02 24.	+ 15 33		15.5	GALAXY
UGC 11896	22 02 24.	+ 15 33	60	15.5	GALAXY Sb-c

OBJECT NAME	RIGHT ASCEN.	DECLINATION	DIAM.	MAGN.	TYPE OF OBJECT
ZWG 530.007	22 02 24.	+ 41 10		14.3	GALAXY
UGC 11897	22 02 24.	+ 41 10	90	14.3	GALAXY SB:b-c
MCG+07-45-009	22 02 24.	+ 41 10	78	14.5	GALAXY
HELW 310	22 02 27.	- 32 14 12.			NEBULA
2ZW 162	22 02 30.	+ 03 57			COMPACT GALAXY
ZWG 530.008	22 02 30.	+ 42 05		15.7	GALAXY
UGC 11898	22 02 30.	+ 42 05	60	15.7	GALAXY SBa-b
MCG+04-52-002	22 02 33.	+ 23 00	36	15.5	GALAXY
SC 2159-5826.5	22 02 33.	- 58 12 00.			NEBULA
UGC 11899	22 02 36.	+ 44 44	60	16.0	GALAXY S
MCG+07-45-010	22 02 36.	+ 44 44	30	17.	GALAXY
MCG+00-56-008	22 02 36.	- 00 54	21	14.5	GALAXY
RNGC 7196	22 02 36.	- 50 22		12.5	GALAXY
IC 5158	22 02 37.	- 67 45			NONSTELLAR OBJECT
HN 0734	22 02 37.	- 67 46			NEBULA
SC 2159-5415.1	22 02 40.	- 54 00 36.			NEBULA
ZC 2202.7+1628	22 02 42.	+ 16 28	6050		CLUSTER OF GALAXIES
MCG+07-45-011	22 02 42.	+ 42 05	36	15.5	GALAXY
MCG+07-45-012	22 02 42.	+ 44 43	48	17.	GALAXY
URA 74	22 02 42.	+ 52 28	408		STELLAR RING
ZWG 377.023	22 02 42.	- 00 53		14.8	GALAXY
RNGC 7198	22 02 43.	- 00 53		15.0	GALAXY
SC 2159-5758.5	22 02 44.	- 57 44 00.	18		NEBULA
SC 2159-6344.5	22 02 45.	- 63 30 00.	36		NEBULA
ARC 2415	22 02 47.	- 05 50		15.9	RICH CLUSTER OF GALAXIES
MCG+01-56-012	22 02 48.	+ 04 56	54	15.	GALAXY
ZWG 473.003	22 02 48.	+ 26 22		15.0	GALAXY
KARA.73B 0945	22 02 48.	+ 26 22	42	15.0	ISOLATED GALAXY S
UGC 11900	22 02 48.	+ 43 50	60	17.	GALAXY
MCG-04-52-013	22 02 48.	- 22 31	48	15.5	GALAXY
SC 2159-5239.7	22 02 48.	- 52 25 12.			NEBULA
MCG+04-52-003	22 02 51.	+ 26 21 30.	48	15.	GALAXY
ZWG 403.021	22 02 54.	+ 04 56		15.4	GALAXY
UGC 11901	22 02 54.	+ 04 56	60	15.4	GALAXY S
LBN 0420	22 03	+ 42 30	8400		BRIGHT NEBULA
LBN 0482	22 03	+ 59 30	900		BRIGHT NEBULA
LDN 1160	22 03 00.	+ 58 45	540		DARK NEBULA
UGC 11902	22 03 06.	+ 16 29	66	16.0	GALAXY S
UGC 11903	22 03 06.	+ 34 45	60	16.0	GALAXY SB
ISS 0085	22 03 06.	+ 59 04	238		STELLAR RING
SCHO 1390	22 03 08.	+ 58 40 36.	840		ISOLATED DARK CLOUD
SC 2200-4114.8	22 03 08.	- 41 00 17.	54		NEBULA
RNGC 7192	22 03 09.	- 64 33		13.0	GALAXY
SC 2157-7912.8	22 03 09.	- 78 58 19.	18		NEBULA
IC 5161	22 03 10.	+ 09 18 54.			NONSTELLAR OBJECT
ARC 2416	22 03 10.	- 25 27		17.5	RICH CLUSTER OF GALAXIES
ZC 2203.2+0122	22 03 12.	+ 01 22	1210		CLUSTER OF GALAXIES
ZWG 403.022	22 03 12.	+ 09 23		15.5	GALAXY
ZWG 451.006	22 03 12.	+ 16 33		14.8	GALAXY
RNGC 7206	22 03 12.	+ 16 33		15.0	GALAXY
UGC 11904	22 03 12.	+ 16 33	108	14.8	GALAXY (S0)
MCG+06-48-006	22 03 12.	+ 34 44	36	17.	GALAXY
OCL 0215	22 03 12.	+ 46 15	2400	8.4	OPEN STAR CLUSTER
RNGC 7209	22 03 12.	+ 46 15		8.0	OPEN CLUSTER
SC 2159-5542.5	22 03 15.	- 55 27 59.	12		NEBULA
RNGC 7191	22 03 15.	- 64 53			GALAXY
ZC 2203.3+1251	22 03 18.	+ 12 51	1410		CLUSTER OF GALAXIES
RNGC 7207	22 03 18.	+ 16 31			GALAXY
ZWG 451.007	22 03 18.	+ 16 32		15.6	GALAXY
MCG+03-56-007	22 03 18.	+ 16 32	60	14.	GALAXY
SER 148.02	22 03 18.	- 50 05	78	18.5	FAINT CLSTR OF GALAXIES
HN 2858	22 03 20.9	- 44 04 58.	12		NEBULA
SC 2200-3924.5	22 03 22.	- 39 09 58.	12		NEBULA
HELW 233	22 03 23.	- 28 11 34.			NEBULA
ZC 2203.4+2324	22 03 24.	+ 23 24	1280		CLUSTER OF GALAXIES
HELW 234	22 03 24.	- 28 12 10.			NEBULA
MCG-05-52-023	22 03 24.	- 28 14 30.	60	15.5	GALAXY
MCG-05-52-024	22 03 24.	- 28 15	48	15.	GALAXY
MCG-07-45-009	22 03 24.	- 44 05	9	15.5	GALAXY
SHB 372	22 03 25.8	- 18 50 16.		19.	QUASI-STELLAR OBJECT
SC 2200-5827.8	22 03 28.	- 59 13 16.			NEBULA
IC 5163	22 03 29.	+ 26 50			NONSTELLAR OBJECT
SCHO 1392	22 03 29.	+ 47 52 36.	360		ISOLATED DARK CLOUD
2ZW 163	22 03 30.	+ 20 23			COMPACT GALAXY
ZWG 451.008	22 03 30.	+ 20 24		15.1	GALAXY
UGC 11905	22 03 30.	+ 20 24	120	15.1	GALAXY PECULR
MCG+03-56-008	22 03 30.	+ 20 24	48	14.5	GALAXY
KARA.73B 0946	22 03 30.	+ 20 24	48	15.1	ISOLATED GALAXY SPEC
UGC 11906	22 03 30.	+ 44 38	60	17.	GALAXY Sb-c
SCHO 1391	22 03 30.	+ 61 43 42.	440		ISOLATED DARK CLOUD
LB 00368	22 03 30.	- 00 03 12.		14.8	FAINT BLUE STAR
SC 2200-5601.8	22 03 31.	- 55 47 16.	60		NEBULA
RNGC 7211	22 03 35.	- 08 21			GALAXY
SC 2200-5925.0	22 03 35.	- 59 10 28.			NEBULA
ZC 2203.6+0530	22 03 36.	+ 05 30	1280		CLUSTER OF GALAXIES
ZWG 428.028	22 03 36.	+ 14 03		15.5	GALAXY
4ZW 085	22 03 36.	+ 23 02			COMPACT GALAXY
MCG-03-56-006	22 03 36.	- 15 38 30.	72	15.5	GALAXY
SG 1.20	22 03 38.	+ 62 03	33600		DIFFUSE EMISSION NEBULA
SC 2200-5534.6	22 03 38.	- 55 20 04.	54		NEBULA
VV 167	22 03 39.	- 21 18	72		INTERACTING GALAXY
KLEM 36	22 03 39.	- 31 23	960	13.	LINEAR GRP OF 4 GALAXIES
RNGC 7201	22 03 39.	- 31 34		14.0	GALAXY
IC 5164	22 03 41.	+ 26 48			NONSTELLAR OBJECT
KLEM 37	22 03 41.	- 28 11	420	16.	LINEAR GRP OF 4 GALAXIES
HN 2859	22 03 41.2	- 46 25 34.	18		NEBULA
MCG-04-52-014	22 03 42.	- 21 19	84	14.5	GALAXY
MCG-05-52-025	22 03 42.	- 28 17	48	15.	GALAXY
MCG-05-52-026	22 03 42.	- 31 34	72	14.	GALAXY
HELW 235	22 03 43.	- 28 11 09.			NEBULA
RNGC 7203	22 03 45.	- 31 27		13.0	GALAXY
SC 2200-5842.4	22 03 47.	- 58 27 52.			NEBULA
ZWG 377.024	22 03 48.	+ 00 52		15.7	GALAXY
ZWG 377.025	22 03 48.	+ 02 06		15.0	GALAXY
UGC 11907	22 03 48.	+ 02 06	66	15.0	GALAXY S0-a
ZC 2203.8-0046	22 03 48.	- 00 46	1210		CLUSTER OF GALAXIES
MCG-05-52-027	22 03 48.	- 31 27	78	13.5	GALAXY
IC 5166	22 03 50.	+ 26 49			NONSTELLAR OBJECT
SCHO 1393	22 03 55.	+ 67 15 06.	420		ISOLATED DARK CLOUD
RNGC 7204	22 03 57.	- 31 22		14.0	GALAXY
MCG-05-52-028	22 03 57.	- 31 22	42	14.	GALAXY
RNGC 7202	22 03 57.	- 31 26			NON-EXISTENT OBJECT
SC 2200-5118.8	22 03 57.	- 51 04 15.			NEBULA
SC 2200-5340.8	22 03 58.	- 53 26 15.			NEBULA
LBN 0528	22 04	+ 70 40	4200		BRIGHT NEBULA
LDN 1166	22 04 00.	+ 59 20	120		DARK NEBULA
ZWG 377.026	22 04 00.	- 00 10		15.6	GALAXY
MCG-05-52-029	22 04 00.	- 31 22	48	14.	GALAXY

OBJECT NAME	RIGHT ASCEN.	DECLINATION	DIAM.	MAGN.	TYPE OF OBJECT
SC 2200-6058.2	22 04 00.	- 60 43 39.			NEBULA
RNGC 7210	22 04 03.	+ 26 52			NON-EXISTENT OBJECT
RNGC 7200	22 04 06.	- 50 15			GALAXY
RNGC 7205A	22 04 06.	- 57 41			GALAXY
SC 2200-6220.6	22 04 07.	- 62 06 03.	6		NEBULA
SC 2200-5747.6	22 04 09.	- 57 33 03.			NEBULA
ZWG 403.023	22 04 12.	+ 09 06		15.4	GALAXY
ZWG 428.029	22 04 12.	+ 11 24		15.0	GALAXY
ZC 2204.2+1204	22 04 12.	+ 12 04	810		CLUSTER OF GALAXIES
MCG+03-56-010	22 04 12.	+ 15 48	90	16.	GALAXY
UGC 11908	22 04 12.	+ 15 49	72	16.0	GALAXY
ZWG 451.009	22 04 12.	+ 17 13		15.4	GALAXY
MCG+03-56-009	22 04 12.	+ 17 13	36	15.	GALAXY
MCG-07-47-019	22 04 12.	- 43 00	72	16.	GALAXY
SC 2200-5757.1	22 04 12.	- 57 42 33.	48		NEBULA
HELW 236	22 04 14.	- 28 06 56.			NEBULA
HELW 237	22 04 17.	- 31 41 08.			NEBULA
ZWG 428.030	22 04 18.	+ 14 50		15.7	GALAXY
ZWG 545.002	22 04 18.	+ 47 00		13.3	GALAXY
UGC 11909	22 04 18.	+ 47 00	192	13.3	GALAXY S
MCG+08-40-001	22 04 18.	+ 47 00	162	14.	GALAXY
SC 2200-6130.5	22 04 23.	- 61 15 56.	12		NEBULA
MCG-05-52-030	22 04 24.	- 30 07	42	15.	GALAXY
IC 1429	22 04 25.	+ 09 51			NONSTELLAR OBJECT
MCG+02-56-011	22 04 30.	+ 10 00	60	14.5	GALAXY
ZWG 428.031	22 04 30.	+ 15 29		15.7	GALAXY
ZC 2204.5+2507	22 04 30.	+ 25 07	5310		CLUSTER OF GALAXIES
MCG-07-45-010	22 04 30.	- 43 51	18	16.	GALAXY
ARC 2417	22 04 33.	- 24 40		17.1	RICH CLUSTER OF GALAXIES
RNGC 7212	22 04 34.	+ 10 00		15.0	GALAXY
SCHO 1394	22 04 35.	+ 50 49 00.	310		ISOLATED DARK CLOUD
SC 2201-5847.7	22 04 35.	- 58 33 08.	12		NEBULA
MCG+01-56-013	22 04 36.	+ 06 14	24	15.	GALAXY
ZWG 428.032	22 04 36.	+ 10 00		15.1	GALAXY
UGC 11910	22 04 36.	+ 10 00	96	15.1	GALAXY S
MCG-02-56-013	22 04 36.	- 12 27	30	15.	GALAXY
SCHO 1395	22 04 37.	+ 65 45 00.	310		ISOLATED DARK CLOUD
ZC 2204.7+0518	22 04 42.	+ 05 18	810		CLUSTER OF GALAXIES
ZWG 403.024	22 04 42.	+ 06 14		15.5	GALAXY
ISS 0201	22 04 42.	+ 48 16	299		STELLAR RING
ZWG 377.027	22 04 42.	- 00 50		15.6	GALAXY
MCG-04-52-015	22 04 45.	- 21 05	60	15.	GALAXY
SC 2201-6327.6	22 04 45.	- 63 13 02.	6		NEBULA
HN 2860	22 04 47.6	- 43 31 19.	24		NEBULA
ZWG 428.033	22 04 48.	+ 14 10		15.7	GALAXY
MCG-02-56-014	22 04 48.	- 10 57	36	15.	GALAXY
SC 2201-5735.8	22 04 49.	- 57 21 13.	42		NEBULA
RNGC 7199	22 04 51.	- 64 57			GALAXY
IC 5167	22 04 53.	- 08 23			NONSTELLAR OBJECT
HN 0735	22 04 53.	- 52 57			NEBULA
IC 5162	22 04 53.	- 52 58			NONSTELLAR OBJECT
ZWG 403.025	22 04 54.	+ 06 47		15.4	GALAXY
SC 2201-6125.2	22 04 55.	- 61 10 37.	6		NEBULA
KHAV 766	22 05	+ 59 03	1410		DARK NEBULA
ZC 2205.0+0139	22 05 00.	+ 01 39	1010		CLUSTER OF GALAXIES
ZC 2205.0+0740	22 05 00.	+ 07 40	1680		CLUSTER OF GALAXIES
ZC 2205.0+1230	22 05 00.	+ 12 30	1410		CLUSTER OF GALAXIES
LDN 1159	22 05 00.	+ 58 20	6780		DARK NEBULA
LDN 1164	22 05 00.	+ 58 55	540		DARK NEBULA
MCG-02-56-015	22 05 00.	- 13 45	36	15.	GALAXY
SC 2201-5240.7	22 05 03.	- 52 26 07.	12		NEBULA
MCG+01-56-014	22 05 06.	+ 08 41	60	14.	GALAXY
ZWG 530.009	22 05 06.	+ 44 03		15.7	GALAXY
UGC 11911	22 05 06.	+ 44 03	102	15.7	GALAXY COMPACT
MCG+07-45-013	22 05 06.	+ 44 03	39	15.5	GALAXY
MCG-02-56-016	22 05 06.	- 10 59	54	15.5	GALAXY
RNGC 7205	22 05 06.	- 57 40		11.5	GALAXY
SER 149.08	22 05 06.	- 57 50	60	17.	LOW SURF. BRGHTNSS GALAXY
SC 2201-5805.8	22 05 07.	- 57 51 13.			NEBULA
SC 2201-5745.8	22 05 11.	- 57 31 13.			NEBULA
ZWG 403.026	22 05 12.	+ 08 40		15.3	GALAXY
MCG+06-48-007	22 05 12.	+ 38 31	66	16.	GALAXY
SC 2202-3841.8	22 05 16.	- 38 27 12.	30		NEBULA
MCG+00-56-009	22 05 18.	+ 00 07	36	15.	GALAXY
ZWG 377.028	22 05 18.	+ 01 07		15.7	GALAXY
ZC 2205.3+0246	22 05 18.	+ 02 46	1210		CLUSTER OF GALAXIES
ZC 2205.3+0529	22 05 18.	+ 05 29	1010		CLUSTER OF GALAXIES
UGC 11912	22 05 18.	+ 38 30	108	16.0	GALAXY S
UGC 11913	22 05 18.	+ 40 48	60	17.	GALAXY
MCG-05-52-031	22 05 18.	- 32 46	72	14.5	GALAXY
ZWG 377.029	22 05 24.	+ 00 07		15.4	GALAXY
ZWG 377.030	22 05 24.	+ 01 08		15.7	GALAXY
ZWG 451.010	22 05 24.	+ 18 17		15.7	GALAXY
MCG+03-56-011	22 05 24.	+ 18 17	36	16.	GALAXY
SC 2202-5246.4	22 05 25.	- 52 31 48.	18		NEBULA
ZWG 403.027	22 05 30.	+ 08 59		15.7	GALAXY
ZWG 428.034	22 05 30.	+ 13 10		15.6	GALAXY
2ZW 164	22 05 30.	+ 13 44			COMPACT GALAXY
ZWG 428.035	22 05 30.	+ 13 44		15.7	GALAXY
VDB.66N 148	22 05 30.	+ 55 56	204		REFLECTION NEBULA
LDN 1165	22 05 30.	+ 58 50	540		DARK NEBULA
LDN 1169	22 05 30.	+ 59 30	240		DARK NEBULA
MCG-04-52-016	22 05 30.	- 25 18	36	15.	GALAXY
SC 2202-5716.8	22 05 32.	- 57 02 12.			NEBULA
SC 2202-5747.6	22 05 32.	- 57 33 00.			NEBULA
SC 2202-5312.5	22 05 33.	- 52 57 54.	6		NEBULA
RNGC 7208	22 05 34.	- 29 20		14.0	GALAXY
RNGC 7217	22 05 35.	+ 31 07		11.5	GALAXY
ZWG 494.002	22 05 36.	+ 31 07		11.0	GALAXY
UGC 11914	22 05 36.	+ 31 07	240	11.0	GALAXY Sb
MCG+05-52-001	22 05 36.	+ 31 07	192	11.	GALAXY
KARA.73B 0947	22 05 36.	+ 31 07	204	11.0	ISOLATED GALAXY S
MCG-05-52-032	22 05 36.	- 29 20	36	14.5	GALAXY
SC 2202-5325.2	22 05 36.	- 53 10 36.			NEBULA
B 174	22 05 39.	+ 58 50	1140		DARK OBJECT
MCG+01-56-015	22 05 42.	+ 04 26	72	14.5	GALAXY
ZWG 403.028	22 05 42.	+ 04 27		14.7	GALAXY
UGC 11915	22 05 42.	+ 04 27	84	14.7	GALAXY S
SN 1953I	22 05 42.	+ 04 27		18.5	SUPERNOVA
MCG-03-56-007	22 05 42.	- 19 19	24	16.	GALAXY
SC 2202-5822.1	22 05 47.	- 58 07 29.			NEBULA
LB 01146	22 05 49.	+ 00 08 54.		15.4	FAINT BLUE STAR
B 173	22 05 49.	+ 59 26	240		DARK OBJECT
IC 1430	22 05 49.	- 13 49 02.			NONSTELLAR OBJECT
SC 2202-6356.7	22 05 49.	- 63 42 06.	42		NEBULA
SCHO 1396	22 05 52.	+ 59 25 24.	380		ISOLATED DARK CLOUD
ZC 2205.9+0920	22 05 54.	+ 09 20	1080		CLUSTER OF GALAXIES
ZWG 451.011	22 05 54.	+ 18 13		15.6	GALAXY

OBJECT NAME	RIGHT ASCEN.	DECLINATION	DIAM.	MAGN.	TYPE OF OBJECT
UGC 11916	22 05 54.	+ 18 13	60	15.6	GALAXY
MCG-05-52-033	22 05 54.	- 28 07	66	15.	GALAXY
IC 5168	22 05 55.0	- 28 06 08.			GALAXY
SC 2202-6147.8	22 05 56.	- 61 33 11.	12		NEBULA
MCG+03-56-012	22 05 57.	+ 18 13	66	15.	GALAXY
IC 1431	22 05 59.	- 13 44 56.			NONSTELLAR OBJECT
KHAV 765	22 06	+ 45 03	13510		DARK NEBULA
LBN 0474	22 06	+ 58 10	1620		BRIGHT NEBULA
ZWG 377.031	22 06 00.	+ 00 16		15.1	GALAXY
2ZW 165	22 06 00.	+ 16 33			COMPACT GALAXY
2ZW 166	22 06 00.	+ 18 12			COMPACT GALAXY
UGC 11917	22 06 00.	+ 45 21	60	17.	GALAXY S
UGC 11918	22 06 00.	+ 74 51	66	17.	GALAXY S
RNGC 7215	22 06 01.	+ 00 16		15.0	GALAXY
MCG+07-45-014	22 06 03.	+ 40 56	84	15.	GALAXY
ZWG 530.010	22 06 06.	+ 40 56		14.8	GALAXY
UGC 11919	22 06 06.	+ 40 56	108	14.8	GALAXY Sb/SBc
ZWG 530.011	22 06 06.	+ 41 02		15.6	GALAXY
MCG-05-52-033A	22 06 06.	- 27 30	36	14.5	GALAXY
HELW 238	22 06 06.	- 28 01 28.			NEBULA
SC 2202-6158.3	22 06 07.	- 61 43 41.	12		NEBULA
SC 2202-6159.6	22 06 07.	- 61 44 59.	12		NEBULA
ARC 2418	22 06 11.	- 26 26		17.5	RICH CLUSTER OF GALAXIES
ZC 2206.2+1400	22 06 12.	+ 14 00	1280		CLUSTER OF GALAXIES
OCL 0255	22 06 12.	+ 71 45	7800		OPEN STAR CLUSTER
2ZW 167	22 06 12.	- 00 55			COMPACT GALAXY
MCG-02-56-017	22 06 12.	- 12 06	36	15.	GALAXY
KLEM 38	22 06 12.	- 34 50	600	15.	CLUSTER OF 10 GALAXIES
SC 2202-5656.2	22 06 12.	- 56 41 34.			NEBULA
RNGC 7213	22 06 14.	- 47 25		12.5	GALAXY
SC 2203-5159.3	22 06 15.	- 51 44 40.			NEBULA
RNGC 7214	22 06 17.	- 28 03		12.0	GALAXY
HN 2861	22 06 17.0	- 47 20 40.			NEBULA
MCG-02-56-018	22 06 18.	- 11 11	42	15.5	GALAXY
MCG-05-52-035	22 06 18.	- 28 04 30.	18	15.	GALAXY
MCG-05-52-034	22 06 18.	- 28 05	90	12.	GALAXY
MCG-05-52-038	22 06 24.	- 27 26	96	14.5	GALAXY
MCG-05-52-037	22 06 24.	- 27 41	48	15.	GALAXY
HELW 239	22 06 24.	- 27 48 39.			NEBULA
MCG-05-52-039	22 06 24.	- 28 00	60	15.	GALAXY
MCG-05-52-036	22 06 24.	- 28 03	48	14.5	GALAXY
LB 00369	22 06 25.	+ 00 46 30.		13.9	FAINT BLUE STAR
HELW 240	22 06 25.	- 27 38 57.			NEBULA
HELW 241	22 06 25.	- 28 01 33.			NEBULA
HN 0736	22 06 25.	- 64 49			NEBULA
IC 5165	22 06 25.	- 64 50			NONSTELLAR OBJECT
SC 2203-5654.4	22 06 26.	- 56 39 46.			NEBULA
HELW 242	22 06 27.	- 27 58 27.			NEBULA
SC 2202-6504.0	22 06 27.	- 64 49 22.	60		NEBULA
ZWG 473.004	22 06 30.	+ 24 26		15.7	GALAXY
ZWG 513.005	22 06 30.	+ 37 50		15.5	GALAXY
MCG+08-40-002	22 06 30.	+ 48 11	138	14.	GALAXY
ZWG 545.003	22 06 30.	+ 48 12		13.0	GALAXY
UGC 11920	22 06 30.	+ 48 12	168	13.0	GALAXY S0/SBa
SCHO 1397	22 06 30.	+ 59 17 00.	400		ISOLATED DARK CLOUD
MCG-04-52-017	22 06 30.	- 20 43	48	15.	GALAXY
SER 148.03	22 06 30.	- 50 03		15.	PEC. GLXY AND COMPANION
SC 2203-5746.1	22 06 30.	- 57 31 28.			NEBULA
MCG+00-56-010	22 06 33.	+ 01 45 30.	48	14.5	GALAXY
MCG+06-48-008	22 06 33.	+ 37 51	24	16.	GALAXY
SC 2202-6358.2	22 06 33.	- 63 43 34.	18		NEBULA
KLEM 39	22 06 34.	- 27 20	900	15.	CLUSTER OF 10 GALAXIES
KLEM 40	22 06 34.	- 28 02	720	13.	LINEAR GRP OF 5 GALAXIES
SC 2203-5656.6	22 06 34.	- 56 41 58.	18		NEBULA
SC 2203-5623.8	22 06 35.	- 56 09 10.			NEBULA
ZWG 377.032	22 06 36.	+ 01 46		15.6	GALAXY
ZWG 451.012	22 06 36.	+ 17 34		15.7	GALAXY
KARA.72 562A	22 06 36.	+ 17 34	36	15.7	PART OF DOUBLE GALAXY
ZWG 451.013	22 06 36.	+ 17 35		15.5	GALAXY
KARA.72 562B	22 06 36.	+ 17 35	36	15.5	PART OF DOUBLE GALAXY
ZWG 451.014	22 06 36.	+ 18 44		15.5	GALAXY
SC 2203-5258.8	22 06 36.	- 52 44 09.			NEBULA
SC 2203-5315.8	22 06 37.	- 53 01 09.	12		NEBULA
LB 00370	22 06 39.	- 17 46 12.		14.6	FAINT BLUE STAR
MCG+02-56-013	22 06 42.	+ 13 16	36	15.5	GALAXY
MCG+02-56-012	22 06 42.	+ 14 07	96	14.	GALAXY
MCG-05-52-040	22 06 42.	- 27 32	48	15.	GALAXY
ZWG 428.036	22 06 48.	+ 13 16		15.5	GALAXY
ZWG 428.037	22 06 48.	+ 14 07		14.7	GALAXY
UGC 11921	22 06 48.	+ 14 07	114	14.7	GALAXY IRR
KARA.73B 0949	22 06 48.	+ 14 07	102	14.7	ISOLATED GALAXY S
ZWG 451.015	22 06 48.	+ 20 08		15.7	GALAXY
MCG+03-56-013	22 06 48.	+ 20 09	48	15.	GALAXY
MCG+07-45-015	22 06 49.	+ 40 06	48	17.	GALAXY
UGC 11922	22 06 48.	+ 40 07	66	16.0	GALAXY Sc
UGC 11923	22 06 48.	+ 44 18	78	16.0	GALAXY S
MCG+07-45-016	22 06 48.	+ 44 18	45	17.	GALAXY
ZWG 377.033	22 06 48.	- 01 53		15.4	GALAXY
KARA.73B 0948	22 06 48.	- 01 53	24	15.4	ISOLATED GALAXY S
MCG-02-56-019	22 06 48.	- 13 51	42	15.5	GALAXY
MCG+03-56-014	22 06 51.	+ 21 17	96	14.	GALAXY
ZWG 451.016	22 06 54.	+ 21 16		14.8	GALAXY
UGC 11924	22 06 54.	+ 21 16	102	14.8	GALAXY Sc
KARA.73B 0950	22 06 54.	+ 21 16	102	14.8	ISOLATED GALAXY S
SC 2203-6255.4	22 06 56.	- 62 40 45.	42		NEBULA
ARC 2419	22 06 57.	+ 17 34		16.9	RICH CLUSTER OF GALAXIES
SC 2203-5730.0	22 06 58.	- 57 15 21.			NEBULA
VDB.66G 211	22 07	- 19 07	70		DWARF GALAXY
UGC 11925	22 07 00.	+ 41 10	60	16.5	GALAXY SBc
MCG-01-56-002	22 07 00.	- 06 23	90	13.	GALAXY
HELW 243	22 07 01.	- 27 46 44.			NEBULA
SC 2203-6213.1	22 07 02.	- 61 58 27.	12		NEBULA
SCHO 1398	22 07 04.	+ 53 34 42.	240		ISOLATED DARK CLOUD
ZC 2207.1+0059	22 07 06.	+ 00 59	3630		CLUSTER OF GALAXIES
ZC 2207.1+0635	22 07 06.	+ 06 35	1410		CLUSTER OF GALAXIES
MCG+01-56-016	22 07 06.	+ 06 55	30	15.	GALAXY
ZC 2207.1+1733	22 07 06.	+ 17 33	1080		CLUSTER OF GALAXIES
MCG+03-56-015	22 07 06.	+ 18 27	72	17.	GALAXY
SC 2203-5132.2	22 07 06.	- 51 17 32.			NEBULA
SC 2203-5330.7	22 07 06.	- 53 16 02.			NEBULA
HN 0737	22 07 09.	- 36 19			NEBULA
IC 5169	22 07 10.	- 36 20			NONSTELLAR OBJECT
MCG+02-56-014	22 07 12.	+ 13 23	48	15.5	GALAXY
2ZW 168	22 07 12.	+ 17 24			COMPACT GALAXY
ZWG 451.017	22 07 12.	+ 17 25		15.3	GALAXY
UGC 11926	22 07 12.	+ 18 27	84	16.0	GALAXY
SC 2203-5710.0	22 07 14.	- 56 55 20.			NEBULA
MCG+01-56-017	22 07 18.	+ 09 20	24	15.	GALAXY

OBJECT NAME	RIGHT ASCEN.	DECLINATION	DIAM.	MAGN.	TYPE OF OBJECT
ZWG 428.038	22 07 18.	+ 13 23		15.7	GALAXY
ZWG 513.006	22 07 18.	+ 38 58		15.6	GALAXY
MCG+07-45-017	22 07 18.	+ 40 44	60	15.	GALAXY
ZWG 530.012	22 07 18.	+ 40 45		15.4	GALAXY
UGC 11927	22 07 18.	+ 40 45	108	15.4	GALAXY Sb
HELW 244	22 07 21.	- 28 26 19.			NEBULA
ZWG 451.018	22 07 24.	+ 16 38		15.7	GALAXY
UGC 11928	22 07 24.	+ 16 38	60	15.7	GALAXY SB
MCG+03-56-016	22 07 24.	+ 16 38	48	15.5	GALAXY
KARA.73B 0951	22 07 24.	+ 16 38	54	15.7	ISOLATED GALAXY S
ZWG 451.019	22 07 24.	+ 18 15		15.4	GALAXY
ZC 2207.4+2033	22 07 24.	+ 20 33	1080		CLUSTER OF GALAXIES
ZWG 513.007	22 07 24.	+ 39 02		14.4	GALAXY
UGC 11929	22 07 24.	+ 39 02	66	14.4	GALAXY S0
MCG-03-56-008	22 07 24.	- 16 54	132	12.5	GALAXY
SC 2204-5249.3	22 07 24.	- 52 34 38.			NEBULA
IC 5172	22 07 26.	+ 12 32 29.			NONSTELLAR OBJECT
RNGC 7218	22 07 26.	- 16 54		13.0	GALAXY
SC 2204-5307.0	22 07 26.	- 52 52 20.			NEBULA
SC 2204-5621.6	22 07 28.	- 56 06 56.			NEBULA
ZWG 428.039	22 07 30.	+ 12 34		15.6	GALAXY
ZWG 451.020	22 07 30.	+ 16 15		15.6	GALAXY
KARA.73B 0952	22 07 30.	+ 16 15	42	15.6	ISOLATED GALAXY S
ZWG 513.008	22 07 30.	+ 37 57		14.9	GALAXY
MCG+06-48-009	22 07 30.	+ 39 03 30.	60	14.5	GALAXY
LDN 1178	22 07 30.	+ 62 05	240		DARK NEBULA
MCG-03-56-009	22 07 30.	- 19 07	96	16.	GALAXY
SC 2204-5212.8	22 07 31.	- 51 58 08.			NEBULA
IC 1432	22 07 32.	+ 03 26 29.			NONSTELLAR OBJECT
SC 2204-5127.5	22 07 32.	- 51 12 49.	6		NEBULA
MCG+06-48-010	22 07 33.	+ 37 57	42	15.	GALAXY
HN 2862	22 07 34.3	- 43 29 55.	60		NEBULA
MCG+03-56-017	22 07 36.	+ 16 14	42	15.	GALAXY
MCG-05-52-041	22 07 36.	- 29 15	42	15.5	GALAXY
IC 5170	22 07 38.	- 47 25 21.			NONSTELLAR OBJECT
UGC 11930	22 07 42.	+ 20 43	60	16.0	GALAXY Sa-b
ARC 2420	22 07 42.	- 12 26		16.8	RICH CLUSTER OF GALAXIES
SC 2204-6409.6	22 07 44.	- 63 54 55.	18		NEBULA
SCHO 1399	22 07 46.	+ 53 14 30.	260		ISOLATED DARK CLOUD
ZC 2207.8+0528	22 07 48.	+ 05 28	2080		CLUSTER OF GALAXIES
ZWG 513.009	22 07 48.	+ 38 18		15.7	GALAXY
ZC 2207.8+4114	22 07 48.	+ 41 14	9880		CLUSTER OF GALAXIES
ISS 0086	22 07 48.	+ 62 17	178		STELLAR RING
MCG-04-52-018	22 07 48.	- 22 55	90	14.5	GALAXY
MCG-05-52-042	22 07 48.	- 28 10	48	14.	GALAXY
SC 2204-3808.0	22 07 50.	- 37 53 19.	18		NEBULA
HN 2863	22 07 50.7	- 46 19 31.	90		NEBULA
MCG+06-48-011	22 07 51.	+ 38 19 30.	42	17.	GALAXY
MCG+00-56-011	22 07 51.	- 00 17	48	15.	GALAXY
HELW 245	22 07 51.	- 28 09 00.			NEBULA
SC 2204-5330.4	22 07 51.	- 53 15 43.			NEBULA
SC 2204-5224.2	22 07 53.	- 52 09 31.	12		NEBULA
ZC 2207.9+1216	22 07 54.	+ 12 16	1410		CLUSTER OF GALAXIES
ZWG 473.005	22 07 54.	+ 25 12		15.7	GALAXY
KARA.73B 0953	22 07 54.	+ 25 12	24	15.7	ISOLATED GALAXY E
ZWG 377.034	22 07 54.	- 00 16		15.6	GALAXY
SER 143.04	22 07 54.	- 52 02	1800	18.5	FAINT CLSTRS OF GALAXIES
SN 1937B	22 07 56.	- 22 57		15.3	SUPERNOVA
SC 2204-5406.4	22 07 56.	- 53 51 43.			NEBULA
HOLM 788B	22 07 57.	+ 40 47	30	14.6	PART OF MULTIPLE GALAXY
ARC 2421	22 07 58.	- 10 56		16.8	RICH CLUSTER OF GALAXIES
LBN 0488	22 08	+ 60 00	1140		BRIGHT NEBULA
LBN 0494	22 08	+ 61 00	4500		BRIGHT NEBULA
LBN 0541	22 08	+ 72 38	240		BRIGHT NEBULA
LBN 0546	22 08	+ 73 08	420		BRIGHT NEBULA
ZC 2208.0+1533	22 08 00.	+ 15 33	1480		CLUSTER OF GALAXIES
ZWG 513.010	22 08 00.	+ 36 24		15.5	GALAXY
ZWG 530.013	22 08 00.	+ 40 46		13.5	GALAXY
HOLM 788A	22 08 00.	+ 40 46	60	12.9	PART OF MULTIPLE GALAXY
UGC 11931	22 08 00.	+ 40 46	120	13.5	GALAXY SBc
MCG+07-45-018	22 08 00.	+ 40 46	90	13.	GALAXY
UGC 11932	22 08 00.	+ 44 19	60	18.	GALAXY DWARF
ISS 0132	22 08 00.	+ 55 42	188		STELLAR RING
COU 089	22 08 00.	+ 61 00	2400		EMISSION NEBULA
MCG-02-56-020	22 08 00.	- 10 43	9	15.	GALAXY
SCHO 1400	22 08 02.	+ 53 14 48.	300		ISOLATED DARK CLOUD
RNGC 7223	22 08 03.	+ 40 46		13.5	GALAXY
ARC 2422	22 08 04.	+ 06 07		17.7	RICH CLUSTER OF GALAXIES
ZC 2208.1+0606	22 08 06.	+ 06 06	1340		CLUSTER OF GALAXIES
MCG+07-45-019	22 08 06.	+ 41 19	42	17.	GALAXY
UGC 11933	22 08 06.	+ 41 20	66	16.0	GALAXY S
SCHO 1401	22 08 06.	+ 52 24 06.	190		ISOLATED DARK CLOUD
SC 2204-5304.1	22 08 09.	- 52 49 24.			NEBULA
ARC 2423	22 08 10.	+ 05 32		17.7	RICH CLUSTER OF GALAXIES
LB 01147	22 08 11.	+ 01 26 30.		16.2	FAINT BLUE STAR
MCG-04-52-019	22 08 12.	- 25 19	72	15.	GALAXY
SER 151.03	22 08 12.	- 46 54	20	17.	PEC GLXY WITH COMPANION
SC 2204-5153.6	22 08 12.	- 51 38 54.			NEBULA
SER 150.01	22 08 12.	- 62 59	70	16.	PAIR OF GALAXIES
DG 177	22 08 18.	+ 72 50	300		REFLECTION NEBULA
IC 5171	22 08 19.	- 46 21 18.			NONSTELLAR OBJECT
SC 2204-7325.7	22 08 20.	- 73 11 01.	18		NEBULA
MCG+00-56-012	22 08 21.	+ 01 51	66	14.	GALAXY
SNO 29	22 08 22.	- 27 58 58.		17.	GROUP OF 4 GALAXIES
RNGC 7221	22 08 22.	- 30 52		12.0	GALAXY
ZWG 377.035	22 08 24.	+ 01 51		15.1	GALAXY
UGC 11934	22 08 24.	+ 01 51	66	15.1	GALAXY SBa-b
ZWG 530.014	22 08 24.	+ 40 39		15.1	GALAXY
URA 75	22 08 24.	+ 54 15	258		STELLAR RING
MCG-05-52-043	22 08 24.	- 30 52	96	12.5	GALAXY
RNGC 7222	22 08 26.	+ 01 51		15.0	GALAXY
MCG+07-45-020	22 08 30.	+ 41 46	15	15.	GALAXY
ZWG 530.015	22 08 30.	+ 41 47		15.0	GALAXY
UGC 11935	22 08 30.	+ 41 47	60	15.0	GALAXY (S0)
SER 151.05	22 08 30.	- 46 54	480	15.5	LOOSE GROUP OF 6 GALAXIES
VDB.66N 149	22 08 33.	+ 72 39	312		REFLECTION NEBULA
VDB.66N 150	22 08 33.	+ 73 09	684		REFLECTION NEBULA
SCHO 1402	22 08 35.	+ 52 27 06.	200		ISOLATED DARK CLOUD
ZWG 377.036	22 08 36.	+ 01 52		15.6	GALAXY
OCL 0223	22 08 36.	+ 52 35	660	10.0	OPEN STAR CLUSTER
SER 151.01	22 08 36.	- 45 50	1020	14.	5 LOW SURF BRGHTNSS GLXYS
SCHO 1403	22 08 37.	+ 50 50 48.	290		ISOLATED DARK CLOUD
RNGC 7216	22 08 39.	- 68 54			UNVERIFIED SOUTHERN OBJECT
UGC 11936	22 08 42.	+ 41 01	60	16.5	GALAXY Sc
OCL 0226	22 08 42.	+ 55 10	240	13.3	OPEN STAR CLUSTER
RNGC 7226	22 08 42.	+ 55 10		13.5	OPEN CLUSTER
ISS 0087	22 08 42.	+ 61 03	350		STELLAR RING
DG 178	22 08 42.	+ 73 20	540		REFLECTION NEBULA

OBJECT NAME	RIGHT ASCEN.	DECLINATION	DIAM.	MAGN.	TYPE OF OBJECT
SC 2205-5357.6	22 08 42.	- 53 42 53.	42		NEBULA
RNGC 7220	22 08 43.	- 23 12		15.0	GALAXY
IC 1434	22 08 44.	+ 52 34			OPEN CLUSTER
MCG-04-52-020	22 08 45.	- 23 12	30	15.	GALAXY
SC 2205-6312.8	22 08 45.	- 62 58 05.	9		NEBULA
ZWG 428.040	22 08 48.	+ 10 10		15.7	GALAXY
ZWG 494.003	22 08 48.	+ 29 22		15.7	GALAXY
KARA.72 563A	22 08 48.	+ 29 22	48	15.7	PART OF DOUBLE GALAXY
MCG+05-52-002	22 08 48.	+ 29 22	42	15.5	GALAXY
ZWG 494.004	22 08 48.	+ 29 24		15.7	GALAXY
KARA.72 563B	22 08 48.	+ 29 24	36	15.7	PART OF DOUBLE GALAXY
MCG-05-52-044	22 08 48.	- 29 05	24	15.	GALAXY
HELW 246	22 08 50.	- 28 13 16.			NEBULA
MCG+05-52-003	22 08 51.	+ 29 24	30	15.5	GALAXY
SC 2205-5207.2	22 08 52.	- 51 52 29.			NEBULA
ZWG 377.037	22 08 54.	+ 01 41		15.4	GALAXY
ZWG 530.016	22 08 54.	+ 41 03		15.3	GALAXY
UGC 11937	22 08 54.	+ 41 03	60	15.3	GALAXY Sa
SC 2205-5856.6	22 08 59.	- 58 41 53.			NEBULA
ZWG 513.011	22 09 00.	+ 38 40		14.8	GALAXY
UGC 11938	22 09 00.	+ 38 40	66	14.8	GALAXY (E)
MCG-05-52-045	22 09 00.	- 28 14	36	15.	GALAXY
HELW 247	22 09 02.	- 28 13 28.			NEBULA
MCG+06-48-013	22 09 03.	+ 38 39 30.	42	17.	GALAXY
IC 5180	22 09 03.	+ 38 40			NONSTELLAR OBJECT
MCG+06-48-012	22 09 03.	+ 38 41 30.	18	15.	GALAXY
SC 2205-5249.9	22 09 03.	- 52 35 10.			NEBULA
SCHO 1404	22 09 05.	+ 52 45 30.	310		ISOLATED DARK CLOUD
ZWG 428.041	22 09 06.	+ 10 38		15.5	GALAXY
MCG+02-56-015	22 09 06.	+ 11 32 30.	96	15.	GALAXY
ZWG 428.042	22 09 06.	+ 11 33		15.2	GALAXY
UGC 11939	22 09 06.	+ 11 33	96	15.2	GALAXY SBb
MCG-07-45-011	22 09 06.	- 42 02	48	15.	GALAXY
IC 5177	22 09 09.	+ 11 34 41.			NONSTELLAR OBJECT
SC 2205-5145.7	22 09 10.	- 51 30 58.			NEBULA
ZWG 377.038	22 09 12.	+ 02 10		15.3	GALAXY
2ZW 169	22 09 12.	+ 04 15			COMPACT GALAXY
2ZW 170	22 09 12.	+ 10 38			COMPACT GALAXY
ZWG 473.006	22 09 12.	+ 25 36		14.7	GALAXY
UGC 11940	22 09 12.	+ 25 36	96	14.7	GALAXY E
MCG+06-48-014	22 09 12.	+ 38 39	30	16.	GALAXY
MCG-04-52-021	22 09 12.	- 24 44	60	15.	GALAXY
SCHO 1405	22 09 14.	+ 53 22 06.	320		ISOLATED DARK CLOUD
RNGC 7224	22 09 15.	+ 25 36		14.5	GALAXY
MCG+00-56-013	22 09 15.	- 00 09 30.	48	15.	GALAXY
MCG+04-52-004	22 09 18.	+ 25 37	27	14.5	GALAXY
UGC 11941	22 09 18.	+ 29 37	66	16.5	GALAXY Sb-c
ZWG 513.012	22 09 18.	+ 38 28		15.0	GALAXY
UGC 11942	22 09 18.	+ 38 28	78	15.0	GALAXY S0
ZWG 377.039	22 09 18.	- 00 08		15.0	GALAXY
RNGC 7227	22 09 19.	+ 38 28		15.0	GALAXY
SC 2205-6216.7	22 09 19.	- 62 01 58.	12		NEBULA
HN 2864	22 09 22.3	- 47 28 10.	60		NEBULA
MCG+06-48-015	22 09 24.	+ 38 28 30.	18	15.	GALAXY
MCG+00-56-014	22 09 24.	- 00 31	36	15.	GALAXY
MCG-05-52-046	22 09 24.	- 28 06	15	15.5	GALAXY
MCG-05-52-047	22 09 24.	- 28 08	48	15.5	GALAXY
IC 1433	22 09 28.	- 13 00 49.			NONSTELLAR OBJECT
2ZW 171	22 09 30.	+ 18 26			COMPACT GALAXY
UGC 11943	22 09 30.	+ 25 53	72	16.0	GALAXY Sa
MCG-02-56-021	22 09 30.	- 13 01	24	15.	GALAXY
SC 2205-6214.9	22 09 30.	- 62 00 10.	9		NEBULA
SC 2205-6233.6	22 09 30.	- 62 18 52.	120		NEBULA
SC 2204-7809.8	22 09 31.	- 77 55 05.	12		NEBULA
SC 2206-5823.4	22 09 32.	- 58 08 39.			NEBULA
SC 2206-6121.9	22 09 32.	- 61 07 10.			NEBULA
SHB 373	22 09 32.3	+ 08 04 26.		18.5	QUASI-STELLAR OBJECT
MCG+03-56-018	22 09 33.	+ 17 39	150	16.	GALAXY
SC 2206-5136.4	22 09 33.	- 51 21 39.			NEBULA
UGC 11944	22 09 36.	+ 17 40	138	16.0	GALAXY DWRF IP
ZC 2209.6+1753	22 09 36.	+ 17 53	1410		CLUSTER OF GALAXIES
ZWG 473.007	22 09 36.	+ 24 50		15.7	GALAXY
MCG+04-52-005	22 09 36.	+ 25 53	48	15.	GALAXY
ZWG 513.013	22 09 36.	+ 38 27		15.0	GALAXY
UGC 11945	22 09 36.	+ 38 27	102	15.0	GALAXY SBa
MCG+08-40-003	22 09 36.	+ 46 02 30.	66	15.	GALAXY
ZWG 545.004	22 09 36.	+ 46 04		14.7	GALAXY
UGC 11946	22 09 36.	+ 46 04	66	14.7	GALAXY Sc
URA 76	22 09 36.	+ 56 37	354		STELLAR RING
RNGC 7228	22 09 37.	+ 38 27		15.0	GALAXY
SC 2206-5150.9	22 09 39.	- 51 36 09.			NEBULA
IC 5174	22 09 40.	- 38 25			NONSTELLAR OBJECT
RNGC 7219	22 09 40.	- 65 05			GALAXY
SCHO 1406	22 09 41.	+ 52 28 30.	250		ISOLATED DARK CLOUD
HN 0738	22 09 41.	- 38 24			NEBULA
ZC 2209.7+2123	22 09 42.	+ 21 23	740		CLUSTER OF GALAXIES
MCG+06-48-016	22 09 42.	+ 38 27 30.	72	15.	GALAXY
MCG-03-56-010	22 09 42.	- 17 57	30	15.	GALAXY
MCG+02-56-016	22 09 45.	+ 11 15	72	15.	GALAXY
SC 2206-5344.3	22 09 45.	- 53 29 33.	9		NEBULA
IC 5175	22 09 46.	- 38 23			NONSTELLAR OBJECT
IC 5178	22 09 47.	- 23 12 18.			NONSTELLAR OBJECT
HN 0739	22 09 47.	- 38 22			NEBULA
SC 2206-3839.9	22 09 47.	- 38 25 09.	42		NEBULA
ZWG 428.043	22 09 48.	+ 11 15		15.1	GALAXY
UGC 11947	22 09 48.	+ 11 15	72	15.1	GALAXY SBb
MRSL 103+02/1	22 09 48.	+ 59 10	9600		HII REGION
CED 199	22 09 48.	+ 59 11			DIFFUSE GALACTIC NEBULA
SG 1.21	22 09 48.	+ 59 11	36000		DIFFUSE EMISSION NEBULA
MCG-04-52-022	22 09 48.	- 23 12	42	14.	GALAXY
TON-S 0044	22 09 48.	- 27 58		15.0	BLUE STAR
SC 2206-3837.1	22 09 51.	- 38 22 20.	18		NEBULA
SC 2206-5207.2	22 09 52.	- 51 52 27.			NEBULA
ZWG 428.044	22 09 54.	+ 13 47		15.2	GALAXY
UGC 11948	22 09 54.	+ 13 47	72	15.2	GALAXY SBb
MCG+02-56-017	22 09 54.	+ 13 47	60	15.	GALAXY
ZWG 451.021	22 09 54.	+ 16 25		15.7	GALAXY
SC 2206-5157.6	22 09 58.	- 51 42 50.			NEBULA
LBN 0425	22 10	+ 43 00	16200		BRIGHT NEBULA
ZC 2210.0+0445	22 10 00.	+ 04 45	1210		CLUSTER OF GALAXIES
MCG+04-52-006	22 10 00.	+ 26 57 30.	42	15.	GALAXY
ZC 2210.0+3745	22 10 00.	+ 37 45	10480		CLUSTER OF GALAXIES
COU 090	22 10 00.	+ 62 30	10800		EMISSION NEBULA
LDN 1243	22 10 00.	+ 75 05	1140		DARK NEBULA
ZWG 473.008	22 10 06.	+ 26 57		15.5	GALAXY
KARA.73B 0954	22 10 06.	+ 26 57	48	15.5	ISOLATED GALAXY S
ZWG 451.022	22 10 12.	+ 17 47		15.6	GALAXY
ZWG 513.014	22 10 12.	+ 39 02		14.8	GALAXY

OBJECT NAME	RIGHT ASCEN.	DECLINATION	DIAM.	MAGN.	TYPE OF OBJECT
UGC 11949	22 10 12.	+ 39 02	72	14.8	GALAXY Sa-b
ZC 2210.2-0304	22 10 12.	- 03 04	4640		CLUSTER OF GALAXIES
MCG-02-56-022	22 10 12.	- 13 50	42	15.	GALAXY
SC 2206-5428.1	22 10 12.	- 54 13 20.			NEBULA
SC 2207-5238.3	22 10 14.	- 52 23 32.	12		NEBULA
HN 2865	22 10 15.8	- 46 15 50.	78		NEBULA
MCG+02-56-018	22 10 18.	+ 11 18	24	15.5	GALAXY
ZWG 513.015	22 10 18.	+ 38 25		14.3	GALAXY
UGC 11950	22 10 18.	+ 38 25	102	14.3	GALAXY (E)
SCHO 1407	22 10 19.	+ 49 07 30.	320		ISOLATED DARK CLOUD
RNGC 7234	22 10 19.	+ 56 44			NON-EXISTENT OBJECT
MCG+06-48-017	22 10 21.	+ 39 03 30.	66	15.	GALAXY
RNGC 7231	22 10 23.	+ 45 05		14.0	GALAXY
IC 5179	22 10 23.	- 37 07 40.			NONSTELLAR OBJECT
MCG+02-56-019	22 10 24.	+ 11 09	18	17.	GALAXY
MCG+02-56-020	22 10 24.	+ 11 10 30.	24	16.	GALAXY
ZWG 428.045	22 10 24.	+ 11 19		15.7	GALAXY
MCG+06-48-018	22 10 24.	+ 38 27	33	14.5	GALAXY
ZWG 530.017	22 10 24.	+ 45 05		14.0	GALAXY
UGC 11951	22 10 24.	+ 45 05	126	14.0	GALAXY SBa
MCG-03-56-011	22 10 24.	- 16 53	30	15.5	GALAXY
MCG-04-52-024	22 10 24.	- 22 41	48	14.	GALAXY
RNGC 7225	22 10 24.	- 26 23		13.0	GALAXY
MCG-04-52-023	22 10 24.	- 26 23	96	13.5	GALAXY
ARC 2424	22 10 27.	+ 12 49		17.4	RICH CLUSTER OF GALAXIES
ZWG 428.046	22 10 30.	+ 11 11		15.5	GALAXY
LDN 1219	22 10 30.	+ 70 45	180		DARK NEBULA
MCG-05-52-048	22 10 30.	- 27 29	60	15.	GALAXY
RNGC 7232A	22 10 33.	- 46 01			GALAXY
SC 2207-5348.5	22 10 35.	- 53 33 43.	9		NEBULA
TON-S 0045	22 10 36.	- 27 24		14.8	BLUE STAR
MCG-05-52-049	22 10 36.	- 27 50	78	14.	GALAXY
DV.56 N7232A	22 10 36.	- 46 02			S GALAXY
HN 2866	22 10 36.2	- 46 08 25.	60		NEBULA
SCHO 1408	22 10 37.	+ 53 45 36.	190		ISOLATED DARK CLOUD
IC 1435	22 10 41.	- 22 20 26.			NONSTELLAR OBJECT
MCG-04-52-025	22 10 42.	- 22 20 30.	54	14.	GALAXY
SCHO 1409	22 10 44.	+ 53 42 00.	140		ISOLATED DARK CLOUD
SC 2207-5731.3	22 10 47.	- 57 16 31.	54		NEBULA
ZC 2210.8+1251	22 10 48.	+ 12 51	1480		CLUSTER OF GALAXIES
OCL 0229	22 10 48.	+ 57 02	720	9.4	OPEN STAR CLUSTER
MCG-02-56-023	22 10 48.	- 09 32	36	15.5	GALAXY
SCHO 1410	22 10 49.	+ 53 42 00.	290		ISOLATED DARK CLOUD
RNGC 7235	22 10 49.	+ 57 02		9.0	OPEN CLUSTER
HN 0740	22 10 52.	- 69 36			NEBULA
IC 5173	22 10 52.	- 69 36			NONSTELLAR OBJECT
MCG-04-52-026	22 10 54.	- 21 59	72	15.5	GALAXY
MCG-07-45-012	22 10 54.	- 42 50	36	14.5	GALAXY
SC 2207-5739.9	22 10 54.	- 57 25 07.	6		NEBULA
HN 2868	22 10 56.1	- 42 50 24.	18		NEBULA
HN 2867	22 10 59.0	- 46 03 30.	18		NEBULA
LBN 0479	22 11	+ 57 40	1920		BRIGHT NEBULA
LBN 0485	22 11	+ 58 46	420		BRIGHT NEBULA
LB 09946	22 11	- 84 15		13.6	FAINT BLUE STAR
ZC 2211.0+0538	22 11 00.	+ 05 38	1210		CLUSTER OF GALAXIES
SC 2207-5437.0	22 11 01.	- 54 22 12.	6		NEBULA
ARC 2425	22 11 04.	+ 05 42		17.7	RICH CLUSTER OF GALAXIES
ZWG 428.047	22 11 06.	+ 13 52		15.6	GALAXY
ZWG 513.016	22 11 06.	+ 38 35		15.4	GALAXY
SC 2207-5705.0	22 11 06.	- 56 50 12.			NEBULA
RNGC 7229	22 11 11.	- 29 40		12.0	GALAXY
ZWG 428.048	22 11 12.	+ 13 58		15.7	GALAXY
UGC 11952	22 11 12.	+ 13 58	102	15.7	GALAXY SB?c
MCG-02-56-024	22 11 12.	- 10 26	12	15.	GALAXY
MCG-05-52-050	22 11 12.	- 27 13	150	14.	GALAXY
MCG-05-52-051	22 11 12.	- 29 40	72	12.5	GALAXY
SER 150.07	22 11 12.	- 65 51	780	15.	LOOSE GROUP OF GALAXIES
IC 1436	22 11 14.	- 10 26 36.			NONSTELLAR OBJECT
SC 2208-5159.5	22 11 14.	- 51 44 42.			NEBULA
HN 0741	22 11 17.	- 67 05			NEBULA
MCG+00-56-015	22 11 18.	+ 00 45	30	15.	GALAXY
ZWG 428.049	22 11 18.	+ 13 33		15.6	GALAXY
ZWG 428.050	22 11 18.	+ 13 45		15.7	GALAXY
IC 5181	22 11 18.	- 46 09	168	12.6	GALAXY SA0
IC 5176	22 11 18.	- 67 05			NONSTELLAR OBJECT
CED 200	22 11 20.	+ 55 04			DIFFUSE GALACTIC NEBULA
KLEM 41	22 11 26.	- 36 56	240	16.	COMPACT GRP OF 5 GALAXIES
SC 2208-5720.0	22 11 29.	- 57 05 11.	12		NEBULA
ZWG 428.051	22 11 30.	+ 13 30		15.4	GALAXY
UGC 11953	22 11 30.	+ 13 30	60	15.4	GALAXY Sa-b
ZC 2211.5+2051	22 11 30.	+ 20 51	1280		CLUSTER OF GALAXIES
UGC 11954	22 11 30.	+ 40 32	60	18.	GALAXY DWARF
LDN 1182	22 11 30.	+ 61 40	240		DARK NEBULA
MCG-03-56-012	22 11 30.	- 17 18 30.	36	14.5	GALAXY
SC 2208-4109.4	22 11 31.	- 40 54 35.	30		NEBULA
RNGC 7230	22 11 32.	- 17 18		14.0	GALAXY
MCG+02-56-021	22 11 36.	+ 13 14	36	15.	GALAXY
ZWG 513.017	22 11 36.	+ 36 47		15.0	GALAXY
ZWG 513.018	22 11 36.	+ 38 59		14.7	GALAXY
UGC 11955	22 11 36.	+ 38 59	72	14.7	GALAXY Sc
MCG-02-56-025	22 11 36.	- 10 54	42	15.5	GALAXY
MCG-05-52-052	22 11 36.	- 30 17	60	14.5	GALAXY
MCG+06-48-019	22 11 39.	+ 39 00	60	15.	GALAXY
ZC 2211.7+0951	22 11 42.	+ 09 51	740		CLUSTER OF GALAXIES
MCG+02-56-022	22 11 42.	+ 13 01	36	15.	GALAXY
ZWG 428.052	22 11 42.	+ 13 14		15.4	GALAXY
UGC 11956	22 11 42.	+ 28 40	72	16.5	GALAXY
MCG+06-48-020	22 11 42.	+ 36 46	30	15.	GALAXY
SCHO 1411	22 11 42.	+ 52 46 12.	270		ISOLATED DARK CLOUD
MCG-05-52-054	22 11 42.	- 27 44	66	13.	GALAXY
MCG-05-52-053	22 11 42.	- 27 44	30	15.	GALAXY
MCG-05-52-055	22 11 42.	- 27 56	48	15.	GALAXY
MCG-05-52-056	22 11 42.	- 29 27	24	14.	GALAXY
SC 2208-3918.0	22 11 44.	- 39 03 11.	18		NEBULA
ARC 2426	22 11 45.	- 10 38		16.8	RICH CLUSTER OF GALAXIES
SC 2208-5628.5	22 11 47.	- 56 13 41.			NEBULA
ZWG 377.040	22 11 48.	+ 00 46		15.5	GALAXY
ZWG 428.053	22 11 48.	+ 13 02		15.5	GALAXY
ASS 26	22 11 48.	+ 53 46	14400		OB ASSOCIATION CEP-LACOB1
SC 2208-5155.3	22 11 49.	- 51 40 29.			NEBULA
SC 2208-5207.8	22 11 50.	- 51 52 59.			NEBULA
SC 2208-5419.6	22 11 52.	- 54 04 47.			NEBULA
ZWG 428.054	22 11 54.	+ 13 42		15.5	GALAXY
ZWG 428.055	22 11 54.	+ 15 08		15.5	GALAXY
SC 2208-5512.9	22 11 55.	- 54 58 05.			NEBULA
SCHO 1412	22 11 57.	+ 52 46 00.	270		ISOLATED DARK CLOUD
SC 2208-5512.7	22 11 58.	- 54 57 52.			NEBULA
KHAV 767	22 12	+ 53 27	5010		DARK NEBULA

OBJECT NAME	RIGHT ASCEN.	DECLINATION	DIAM.	MAGN.	TYPE OF OBJECT
B 175	22 12	+ 69 41	3600		DARK OBJECT
LBN 0531	22 12	+ 70 01	360		BRIGHT NEBULA
ZWG 403.029	22 12 00.	+ 04 47		15.5	GALAXY
UGC 11957	22 12 00.	+ 04 47	66	15.5	GALAXY Sb-c
MCG+01-56-018	22 12 00.	+ 04 47	60	15.	GALAXY
ZC 2212.0+1326	22 12 00.	+ 13 26	5310		CLUSTER OF GALAXIES
ZWG 428.056	22 12 00.	+ 15 06		15.6	GALAXY
URA 84	22 12 00.	+ 55 45	1200		STELLAR RING
LDN 1175	22 12 00.	+ 60 30	480		DARK NEBULA
LDN 1186	22 12 00.	+ 61 53	240		DARK NEBULA
LDN 1217	22 12 00.	+ 70 30	1740		DARK NEBULA
LB 00371	22 12 00.	- 15 55 48.		14.8	FAINT BLUE STAR
SC 2208-5706.1	22 12 00.	- 56 51 16.			NEBULA
VDB.66N 151	22 12 01.	+ 39 27	648		REFLECTION NEBULA
ZWG 428.057	22 12 06.	+ 13 32		15.7	GALAXY
MCG-05-52-057	22 12 06.	- 28 13	60	14.	GALAXY
DG 179	22 12 12.	+ 69 59	660		REFLECTION NEBULA
SC 2208-6326.5	22 12 12.	- 63 11 40.	6		NEBULA
RNGC 7237	22 12 17.	+ 13 36		15.0	GALAXY
RNGC 7236	22 12 17.	+ 13 36		15.0	GALAXY
CED 201	22 12 17.	+ 70 00	690		DIFFUSE GALACTIC NEBULA
ZWG 377.041	22 12 18.	+ 00 02		15.7	GALAXY
2ZW 172	22 12 18.	+ 13 25			COMPACT GALAXY
ZWG 428.058	22 12 18.	+ 13 35		14.3	GALAXY
UGC 11958	22 12 18.	+ 13 35	138	14.3	GALAXY TRP SYS
KARA.72 564B	22 12 18.	+ 13 35	54		PART OF DOUBLE GALAXY
KARA.72 564A	22 12 18.	+ 13 35	54	14.3	PART OF DOUBLE GALAXY
MCG+02-56-024	22 12 18.	+ 13 36	30	15.	GALAXY
MCG+02-56-023	22 12 18.	+ 13 36	30	15.	GALAXY
ZC 2212.3+2813	22 12 18.	+ 28 13	610		CLUSTER OF GALAXIES
UGC 11959	22 12 18.	+ 48 11	60	17.	GALAXY
RNGC 7239	22 12 18.	- 05 19			GALAXY
ARP 169	22 12 19.	+ 13 36			PECULIAR GALAXY
VDB.66N 152	22 12 20.	+ 70 03	816		REFLECTION NEBULA
MCG+05-52-004	22 12 21.	+ 32 43	42	15.	GALAXY
SC 2209-5200.0	22 12 23.	- 51 45 10.			NEBULA
ZWG 428.059	22 12 24.	+ 13 28		15.7	GALAXY
ZWG 494.005	22 12 24.	+ 32 25		15.7	GALAXY
ZWG 494.006	22 12 24.	+ 32 43		15.6	GALAXY
URA 85	22 12 24.	+ 55 59	240		STELLAR RING
MCG-02-56-026	22 12 24.	- 11 46	60	14.	GALAXY
MCG-04-52-027	22 12 24.	- 21 13	54	14.5	GALAXY
TON-S 0046	22 12 24.	- 27 57		15.4	BLUE STAR
SC 2209-5638.2	22 12 24.	- 56 23 22.	12		NEBULA
IC 5183	22 12 26.	- 36 05 26.			NONSTELLAR OBJECT
ZWG 428.060	22 12 30.	+ 13 30		15.7	GALAXY
UGC 11960	22 12 30.	+ 41 12	60	16.5	GALAXY SB:c
LDN 1191	22 12 30.	+ 62 10	240		DARK NEBULA
SER 151.02	22 12 30.	- 46 05			INTERACTING GALAXIES
SC 2209-5255.5	22 12 30.	- 52 40 39.			NEBULA
HN 0742	22 12 30.	- 65 42			NEBULA
IC 5182	22 12 30.	- 65 42			NONSTELLAR OBJECT
IC 5184	22 12 32.	- 37 07 32.			NONSTELLAR OBJECT
RNGC 7232	22 12 34.	- 46 05		13.0	GALAXY
SC 2208-6330.8	22 12 34.	- 63 15 58.	6		NEBULA
ZWG 428.061	22 12 36.	+ 15 09		15.3	GALAXY
MCG+02-56-025	22 12 36.	+ 15 10	48	15.	GALAXY
ZWG 451.023	22 12 36.	+ 16 06		15.6	GALAXY
2ZW 173	22 12 36.	+ 21 12			COMPACT GALAXY
ZWG 530.018	22 12 36.	+ 41 56		15.7	GALAXY
UGC 11961	22 12 36.	+ 41 56	96	15.7	GALAXY Sb
SC 2209-5336.0	22 12 37.	- 53 21 09.			NEBULA
MCG+07-45-021	22 12 39.	+ 41 56	108	16.	GALAXY
RNGC 7233	22 12 40.	- 46 05			GALAXY
ZWG 494.007	22 12 42.	+ 31 31		15.5	GALAXY
KARA.73B 0955	22 12 42.	+ 31 31	36	15.5	ISOLATED GALAXY S
UGC 11962	22 12 42.	+ 41 45	60	16.5	GALAXY S
KARA.68 236	22 12 42.	- 10 44	40		DWARF GALAXY
ABC 2427	22 12 42.	- 24 07		17.7	RICH CLUSTER OF GALAXIES
SER 150.05	22 12 42.	- 64 18	100	16.	CMPCT GROUP OF 4 GALAXIES
SC 2209-5518.1	22 12 45.	- 55 03 15.	6		NEBULA
SC 2209-5037.9	22 12 46.	- 50 23 03.	12		NEBULA
ZWG 513.019	22 12 48.	+ 35 38		15.6	GALAXY
ZWG 513.020	22 12 48.	+ 37 03		15.0	GALAXY
UGC 11963	22 12 48.	+ 37 03	66	15.0	GALAXY S0-a
MCG+06-48-021	22 12 48.	+ 37 03	60	15.5	GALAXY
SC 2209-5204.0	22 12 48.	- 51 49 09.			NEBULA
SCHO 1413	22 12 49.	+ 53 50 48.	580		ISOLATED DARK CLOUD
SC 2209-5314.0	22 12 51.	- 52 59 09.			NEBULA
IC 5191	22 12 51.4	+ 37 03 04.	45		GALAXY SA0
RNGC 7232B	22 12 52.	- 45 55			GALAXY
ZC 2212.9+2537	22 12 54.	+ 25 37	1010		CLUSTER OF GALAXIES
4ZW 086	22 12 54.	+ 38 04			COMPACT GALAXY
MCG-05-52-058	22 12 54.	- 27 47	48	15.	GALAXY
DV.56 N7232B	22 12 54.	- 45 56			GALAXY
RNGC 7238	22 12 55.	+ 22 16			NON-EXISTENT OBJECT
SC 2209-5812.0	22 12 59.	- 57 57 08.			NEBULA
LBN 0496	22 13	+ 61 10	240		BRIGHT NEBULA
LBN 0500	22 13	+ 62 50	360		BRIGHT NEBULA
ZC 2213.0+1127	22 13 00.	+ 11 27	1210		CLUSTER OF GALAXIES
ZWG 428.062	22 13 00.	+ 15 20		15.3	GALAXY
MCG+06-48-022	22 13 00.	+ 37 01 30.	24	17.	GALAXY
LDN 1193	22 13 00.	+ 62 10	180		DARK NEBULA
MCG-04-52-028	22 13 00.	- 26 13	60	15.5	GALAXY
MCG-07-45-013	22 13 00.	- 41 19	36	16.	GALAXY
SC 2209-5144.0	22 13 00.	- 51 29 08.	30		NEBULA
MCG+03-56-019	22 13 03.	+ 18 58	108	14.5	GALAXY
MCG+06-48-023	22 13 03.	+ 37 03	30	15.5	GALAXY
IC 5192	22 13 03.0	+ 37 01 26.	12		GALAXY E2
UGC 11964	22 13 06.	+ 18 58	126	16.0	GALAXY Sc
ZWG 514.001	22 13 06.	+ 37 03		15.3	GALAXY
ZWG 513.021	22 13 06.	+ 37 03		15.3	GALAXY
URA 77	22 13 06.	+ 56 38	600		STELLAR RING
IC 1441	22 13 08.3	+ 37 03 08.	30		GALAXY SBa
MCG+06-48-024	22 13 09.	+ 37 02	24	15.5	GALAXY
ZWG 377.042	22 13 12.	+ 01 48		14.9	GALAXY
UGC 11965	22 13 12.	+ 01 48	66	14.9	GALAXY S0-a
MCG+00-56-016	22 13 12.	+ 01 48	60	14.	GALAXY
KARA.73B 0956	22 13 12.	+ 01 48	60	14.9	ISOLATED GALAXY S
IC 1437	22 13 12.	+ 01 48 56.			NONSTELLAR OBJECT
HMS 1.32	22 13 12.	+ 37 02			S0 GALAXY
ZWG 514.002	22 13 12.	+ 37 02		15.6	GALAXY
ZWG 513.022	22 13 12.	+ 37 02		15.6	GALAXY
RNGC 7240	22 13 12.	+ 37 02		15.5	GALAXY
4ZW 087	22 13 12.	+ 37 05			COMPACT GALAXY
MCG-05-52-059	22 13 12.	- 26 57	72	14.5	GALAXY
ZWG 403.030	22 13 18.	+ 08 24		15.4	GALAXY
UGC 11966	22 13 18.	+ 08 24	60	15.4	GALAXY IRR
MCG+01-56-019	22 13 18.	+ 08 25	48	15.	GALAXY
4ZW 088	22 13 18.	+ 30 58			COMPACT GALAXY
ZWG 494.008	22 13 18.	+ 33 23		15.5	GALAXY
UGC 11967	22 13 18.	+ 33 23	72	15.5	GALAXY Sc-IRR
MCG+05-52-005	22 13 18.	+ 33 23 30.	60	15.	GALAXY
OCL 0221	22 13 18.	+ 49 38	9000	7.4	OPEN STAR CLUSTER
MCG-05-52-060	22 13 18.	- 30 40	54	14.5	GALAXY
RNGC 7243	22 13 19.	+ 49 38		6.5	OPEN CLUSTER
RNGC 7245	22 13 22.	+ 54 05		11.5	OPEN CLUSTER
HOLM 789A	22 13 23.	+ 37 03	30	13.7	PART OF MULTIPLE GALAXY
2ZW 174	22 13 24.	+ 18 58			COMPACT GALAXY
ZWG 451.024	22 13 24.	+ 18 59		13.8	GALAXY
RNGC 7241	22 13 24.	+ 18 59		14.0	GALAXY
UGC 11968	22 13 24.	+ 18 59	210	13.8	GALAXY S-IRR
MCG+03-56-020	22 13 24.	+ 18 59	198	13.	GALAXY
ZWG 514.003	22 13 24.	+ 37 03		14.6	GALAXY
ZWG 513.023	22 13 24.	+ 37 03		14.6	GALAXY
UGC 11969	22 13 24.	+ 37 03	162	14.6	GALAXY E
4ZW 089	22 13 24.	+ 37 46			COMPACT GALAXY
5ZW 381	22 13 24.	+ 39 43			COMPACT GALAXY
OCL 0225	22 13 24.	+ 54 05	360		OPEN STAR CLUSTER
MCG-05-52-061	22 13 24.	- 27 40	60	14.	GALAXY
IC 5186	22 13 24.	- 37 05	144	12.5	GALAXY SA(rs)
HOLM 789B	22 13 25.	+ 37 03	18	14.9	PART OF MULTIPLE GALAXY
MCG+06-48-025	22 13 27.	+ 37 03 30.	18	13.5	GALAXY
SC 2210-5208.5	22 13 27.	- 51 53 37.			NEBULA
SC 2210-5231.1	22 13 29.	- 52 16 13.			NEBULA
ZC 2213.5+0804	22 13 30.	+ 08 04	3090		CLUSTER OF GALAXIES
ZWG 428.063	22 13 30.	+ 13 53		14.6	GALAXY
4ZW 090	22 13 30.	+ 37 03			COMPACT GALAXY
BIGO 553	22 13 30.	+ 37 03			NEBULA
RNGC 7242	22 13 30.	+ 37 03		14.5	GALAXY
IC 5195	22 13 30.5	+ 37 03 15.			GALAXY E0
SC 2210-1309.9	22 13 32.	- 12 55 01.	24		NEBULA
IC 5193	22 13 32.5	+ 36 59 39.			GALAXY E0
MCG+06-48-026	22 13 33.	+ 37 00 30.	13	16.	GALAXY
SCHO 1414	22 13 33.	+ 57 51 42.	590		ISOLATED DARK CLOUD
ZC 2213.6+0926	22 13 36.	+ 09 26	3630		CLUSTER OF GALAXIES
4ZW 091	22 13 36.	+ 24 56			COMPACT GALAXY
OCL 0227	22 13 36.	+ 54 09	150	18.	OPEN STAR CLUSTER
BIGO 552	22 13 37.	- 05 15			NEBULA
IC 5189	22 13 37.	- 05 15			NONSTELLAR OBJECT
ABC 2428	22 13 38.	- 09 37		17.2	RICH CLUSTER OF GALAXIES
MCG-05-52-062	22 13 39.	- 27 05	36	14.5	GALAXY
4ZW 092	22 13 42.	+ 27 14			COMPACT GALAXY
IC 1438	22 13 44.	- 21 40 33.			NONSTELLAR OBJECT
MCG-04-52-029	22 13 45.	- 21 40	132	13.	GALAXY
ZC 2213.8+1652	22 13 48.	+ 16 52	1140		CLUSTER OF GALAXIES
4ZW 093	22 13 48.	+ 22 41			COMPACT GALAXY
MCG+05-52-006	22 13 48.	+ 27 39	54	15.	GALAXY
DG 180	22 13 48.	+ 61 10	180		REFLECTION NEBULA
MCG-03-56-013	22 13 48.	- 16 15	36	15.5	GALAXY
SC 2210-6203.1	22 13 50.	- 61 48 13.	12		NEBULA
SS 64	22 13 51.	+ 61 11			DIFFUSE GALACTIC NEBULA
IC 1440	22 13 51.	- 16 15 45.			NONSTELLAR OBJECT
ZWG 451.025	22 13 54.	+ 16 13		14.9	GALAXY
MRK 303	22 13 54.	+ 16 13	30	15.	GALAXY WITH UV CONTINUUM
RNGC 7244	22 13 54.	+ 16 13		15.0	GALAXY
4ZW 094	22 13 54.	+ 25 36			COMPACT GALAXY
ZWG 494.009	22 13 54.	+ 27 40		15.7	GALAXY
KARA.73B 0957	22 13 54.	+ 27 40	66	15.7	ISOLATED GALAXY S
ZC 2213.9+3011	22 13 54.	+ 30 11	870		CLUSTER OF GALAXIES
MCG-04-52-030	22 13 54.	- 21 43 30.	72	15.	GALAXY
SC 2210-6304.7	22 13 54.	- 62 49 49.			NEBULA
IC 1439	22 13 55.	- 21 44 09.			NONSTELLAR OBJECT
SC 2210-6159.8	22 13 58.	- 61 44 55.	6		NEBULA
VDB.66G 212	22 14	- 21 27	100		DWARF GALAXY
SER 150.06	22 14	- 65 27	199	18.5	FAINT CLUSTER OF GALAXIES
ZWG 377.043	22 14 00.	+ 03 09		15.3	GALAXY
UGC 11970	22 14 00.	+ 03 09	60	15.3	GALAXY SBb
MCG+01-56-020	22 14 00.	+ 05 21	30	16.	GALAXY
MCG+03-56-021	22 14 00.	+ 16 13	36	14.5	GALAXY
UGC 11971	22 14 00.	+ 18 34	66	16.0	GALAXY Sb-c
MCG+06-49-001	22 14 00.	+ 37 15		15.	GALAXY
LDN 1150	22 14 00.	+ 55 45	300		DARK NEBULA
LDN 1235	22 14 00.	+ 73 10	780		DARK NEBULA
SER 151.04	22 14 00.	- 48 42	980	15.	GROUP OF 15 GALAXIES
SCHO 1415	22 14 02.	+ 55 47 24.	330		ISOLATED DARK CLOUD
MCG+00-56-017	22 14 03.	+ 03 09	36	15.	GALAXY
SG 3.228	22 14 03.	+ 58 22	3600		DIFFUSE EMISSION NEBULA
B 369	22 14 04.	+ 55 46	300		DARK OBJECT
SC 2210-5119.1	22 14 05.	- 51 04 12.	6		NEBULA
ZC 2214.1+1030	22 14 06.	+ 10 30	1550		CLUSTER OF GALAXIES
ZC 2214.1+1445	22 14 06.	+ 14 45	1010		CLUSTER OF GALAXIES
MCG-04-52-031	22 14 06.	- 21 30	120	14.	GALAXY
HN 0743	22 14 06.	- 66 06			NEBULA
IC 5185	22 14 06.	- 66 06			NONSTELLAR OBJECT
ZWG 514.004	22 14 12.	+ 36 56		15.6	GALAXY
ZWG 513.024	22 14 12.	+ 36 56		15.6	GALAXY
5ZW 382	22 14 12.	+ 50 25			COMPACT GALAXY
MCG-01-56-003	22 14 12.	- 05 01	36	15.	GALAXY
MCG-02-56-027	22 14 12.	- 09 16	42	15.	GALAXY
HELW 496	22 14 13.	- 37 12 00.			NEBULA
SC 2210-5926.0	22 14 15.	- 59 11 06.	12		NEBULA
ZWG 514.005	22 14 18.	+ 36 13		15.6	GALAXY
ZWG 513.025	22 14 18.	+ 36 13		15.6	GALAXY
SCHO 1416	22 14 18.	+ 55 43 18.	550		ISOLATED DARK CLOUD
PK103+00.1	22 14 18.6	+ 57 13 43.	87		PLANETARY NEBULA
SC 2211-5224.5	22 14 21.	- 52 09 36.			NEBULA
URA 78	22 14 24.	+ 52 45	360		STELLAR RING
SER 148.01	22 14 24.	- 48 35	1500	15.	CLUSTER OF 20 GALAXIES
BIGO 554	22 14 25.	- 16 12			NEBULA
IC 5194	22 14 26.	- 16 12	10		GALAXY
MCG+02-56-026	22 14 30.	+ 13 52	30	15.	GALAXY
4ZW 095	22 14 30.	+ 37 14			COMPACT GALAXY
MCG-01-56-004	22 14 30.	- 05 15	24	15.5	GALAXY
SCHO 1417	22 14 32.	+ 56 00 42.	270		ISOLATED DARK CLOUD
SC 2211-5300.9	22 14 32.	- 52 45 59.			NEBULA
CED 202	22 14 35.	+ 53 48			DIFFUSE GALACTIC NEBULA
ZWG 428.064	22 14 36.	+ 13 07		15.6	GALAXY
MCG+02-56-027	22 14 36.	+ 13 07	12	15.5	GALAXY
OCL 0224	22 14 36.	+ 53 48		12.	OPEN STAR CLUSTER
IC 1442	22 14 36.	+ 53 48	210		OPEN CLUSTER
DG 181	22 14 36.	+ 60 34	120		REFLECTION NEBULA
ABC 2429	22 14 37.	+ 08 49		17.7	RICH CLUSTER OF GALAXIES
SC 2211-6405.9	22 14 39.	- 63 50 59.	6		NEBULA
SS 65	22 14 41.	+ 60 35			DIFFUSE GALACTIC NEBULA
2ZW 175	22 14 42.	+ 13 59			COMPACT GALAXY

OBJECT NAME	RIGHT ASCEN.	DECLINATION	DIAM.	MAGN.	TYPE OF OBJECT
ZWG 428.065	22 14 42.	+ 13 59		14.6	GALAXY
MRK 304	22 14 42.	+ 13 59	9	15.5	GALAXY WITH UV CONTINUUM
KW 29	22 14 42.	+ 13 59	11		SEYFERT GALAXY
VVI 91	22 14 42.	+ 13 59	9	14.67	SEYFERT GAL1XY
ZWG 428.066	22 14 42.	+ 14 15		15.5	GALAXY
ZWG 530.019	22 14 42.	+ 40 15		13.6	GALAXY
UGC 11972	22 14 42.	+ 40 15	108	13.6	GALAXY S0
MCG+07-45-022	22 14 42.	+ 40 15	24	14.5	GALAXY
ZWG 530.020	22 14 42.	+ 41 15		13.5	GALAXY
UGC 11973	22 14 42.	+ 41 15	198	13.5	GALAXY Sb/SBc
MCG+07-45-023	22 14 42.	+ 41 15	192	14.	GALAXY
MCG-05-52-063	22 14 42.	- 27 37	24	15.	GALAXY
RNGC 7248	22 14 44.	+ 40 15		13.5	GALAXY
SC 2211-5319.9	22 14 47.	- 53 04 59.			NEBULA
MCG+01-56-021	22 14 48.	+ 08 58	42	15.	GALAXY
ZWG 473.009	22 14 48.	+ 24 57		15.7	GALAXY
SC 2211-5215.5	22 14 48.	- 52 00 35.			NEBULA
FATH 1.821	22 14 50.	+ 15 58			NEBULA
RNGC 7246	22 14 51.	- 15 47		15.0	GALAXY
SC 2211-5302.3	22 14 53.	- 52 47 23.			NEBULA
ZWG 403.031	22 14 54.	+ 08 57		15.2	GALAXY
ZWG 428.067	22 14 54.	+ 14 58		15.7	GALAXY
ZC 2214.9+1717	22 14 54.	+ 17 17	1550		CLUSTER OF GALAXIES
MCG+04-52-007	22 14 54.	+ 24 58	48	15.	GALAXY
MCG-03-56-014	22 14 54.	- 15 47 30.	48	15.	GALAXY
HN 1223	22 14 56.	- 59 52	12	14.5	NEBULA
IC 5187	22 14 56.	- 59 52			NONSTELLAR OBJECT
SC 2211-5530.2	22 14 57.	- 55 15 16.	18		NEBULA
LBN 0486	22 15	+ 58 34	840		BRIGHT NEBULA
LBN 0489	22 15	+ 58 55	300		BRIGHT NEBULA
LBN 0490	22 15	+ 59 00	10200		BRIGHT NEBULA
ZWG 428.068	22 15 00.	+ 09 33		15.6	GALAXY
ZWG 428.069	22 15 00.	+ 14 52		15.7	GALAXY
FATH 1.822	22 15 00.	+ 14 58	35		NEBULA
ZWG 494.010	22 15 00.	+ 33 16		14.7	GALAXY
UGC 11974	22 15 00.	+ 33 16	72	14.7	GALAXY SBb
MCG+05-52-007	22 15 00.	+ 33 16	54	14.5	GALAXY
ZWG 530.021	22 15 00.	+ 40 51		15.0	GALAXY
LDN 1154	22 15 00.	+ 55 58	300		DARK NEBULA
COU 091	22 15 00.	+ 58 30	6000		EMISSION NEBULA
LDN 1138	22 15 00.	+ 61 30	2520		DARK NEBULA
MCG-04-52-032	22 15 00.	- 23 59	84	13.5	GALAXY
SC 2211-5416.1	22 15 00.	- 54 01 10.			NEBULA
RNGC 7247	22 15 01.	- 23 59		13.0	GALAXY
IC 5198	22 15 02.	- 15 55			NONSTELLAR OBJECT
BIGO 555	22 15 02.	- 15 57			NEBULA
HN 0744	22 15 02.	- 59 54			NEBULA
IC 5188	22 15 02.	- 59 54			NONSTELLAR OBJECT
MCG+06-49-002	22 15 06.	+ 35 19	84	14.	GALAXY
ZWG 514.006	22 15 06.	+ 35 20		15.2	GALAXY
UGC 11975	22 15 06.	+ 35 20	96	15.2	GALAXY S0/Sa
5ZW 383	22 15 06.	+ 48 23			COMPACT GALAXY
DG 182	22 15 06.	+ 60 35	120		REFLECTION NEBULA
CED 203	22 15 08.	+ 55 56	1800		DIFFUSE GALACTIC NEBULA
SS 66	22 15 11.	+ 60 36			DIFFUSE GALACTIC NEBULA
ZWG 514.007	22 15 12.	+ 37 33		15.7	GALAXY
ZWG 513.026	22 15 12.	+ 37 33		15.7	GALAXY
SC 2211-5540.7	22 15 14.	- 55 25 46.	6		NEBULA
SCHO 1418	22 15 16.	+ 56 58 54.	250		ISOLATED DARK CLOUD
MCG+05-52-008	22 15 18.	+ 28 01 30.	54	15.	GALAXY
UGC 11976	22 15 18.	+ 28 43	66	16.5	GALAXY S
UGC 11977	22 15 18.	+ 33 01	72	16.5	GALAXY
ZC 2215.4+0656	22 15 24.	+ 06 56	2490		CLUSTER OF GALAXIES
ZWG 473.010	22 15 24.	+ 27 22		15.4	GALAXY
ZWG 494.011	22 15 24.	+ 28 02		15.7	GALAXY
UGC 11978	22 15 24.	+ 28 02	78	15.7	GALAXY Sc
KARA.73B 0958	22 15 24.	+ 28 02	54	15.7	ISOLATED GALAXY S
SC 2212-5301.2	22 15 29.	- 52 46 15.			NEBULA
ABC 2431	22 15 31.	+ 08 41		17.7	RICH CLUSTER OF GALAXIES
HN 1224	22 15 32.	- 60 08		15.	NEBULA
IC 5190	22 15 32.	- 60 08			NONSTELLAR OBJECT
ABC 2432	22 15 33.	+ 07 12		17.7	RICH CLUSTER OF GALAXIES
ZC 2215.6+1238	22 15 36.	+ 12 38	2550		CLUSTER OF GALAXIES
ABC 2430	22 15 38.	- 09 33		17.6	RICH CLUSTER OF GALAXIES
SC 2212-5214.0	22 15 40.	- 51 59 03.	18		NEBULA
UBA 79	22 15 42.	+ 55 02	528		STELLAR RING
ZC 2215.8+0446	22 15 48.	+ 04 46	1610		CLUSTER OF GALAXIES
MCG+02-56-028	22 15 48.	+ 13 21	36	15.5	GALAXY
ISS 0133	22 15 48.	+ 55 00	436		STELLAR RING
PHL 0218	22 15 48.	- 20 30		15.9	BLUE STELLAR OBJECT
SC 2212-5225.3	22 15 49.	- 52 10 21.			NEBULA
SC 2212-5724.2	22 15 50.	- 57 09 15.			NEBULA
HELW 497	22 15 53.	- 37 03 02.			NEBULA
ZWG 428.070	22 15 54.	+ 13 22		15.6	GALAXY
LBN 0491	22 16	+ 59 00	540		BRIGHT NEBULA
ZWG 403.032	22 16 00.	+ 04 02		15.3	GALAXY
ZC 2216.0+1343	22 16 00.	+ 13 43	1880		CLUSTER OF GALAXIES
MCG+08-40-004	22 16 00.	+ 45 26	84	16.	GALAXY
UGC 11979	22 16 00.	+ 45 28	120	17.	GALAXY
SCHO 1419	22 16 00.	+ 55 50 06.	250		ISOLATED DARK CLOUD
ZWG 377.044	22 16 00.	- 01 49		15.7	GALAXY
PHL 0219	22 16 00.	- 19 46		17.5	BLUE STELLAR OBJECT
ABC 2433	22 16 02.	+ 13 46		17.1	RICH CLUSTER OF GALAXIES
FATH 1.823	22 16 04.	+ 15 36	8		NEBULA
ZWG 530.022	22 16 06.	+ 40 19		13.1	GALAXY
UGC 11980	22 16 06.	+ 40 19	78	13.1	GALAXY S-IRR
MCG+07-45-024	22 16 06.	+ 40 19	66	14.	GALAXY
RNGC 7250	22 16 08.	+ 40 19		13.0	GALAXY
PHL 0220	22 16 12.	+ 03 05		17.0	BLUE STELLAR OBJECT
MCG+05-52-009	22 16 12.	+ 28 59 30.	78	14.5	GALAXY
ZWG 494.012	22 16 12.	+ 29 00		15.4	GALAXY
UGC 11981	22 16 12.	+ 29 00	96	15.4	GALAXY Sc
MCG+08-40-005	22 16 12.	+ 47 30	24	16.	GALAXY
MCG-01-56-005	22 16 12.	- 03 46	48	14.	GALAXY
PHL 5022	22 16 12.	- 19 09		17.6	BLUE STELLAR OBJECT
SC 2212-5430.4	22 16 14.	- 54 15 26.			NEBULA
SG 3.229	22 16 15.	+ 59 01	720		DIFFUSE EMISSION NEBULA
BC PKS2216-03	22 16 16.3	- 03 50 43.		16.38	QUASI-STELLAR OBJECT
SHB 374	22 16 16.3	- 03 50 43.		16.4	QUASI-STELLAR OBJECT
ZWG 377.045	22 16 18.	- 01 18		15.7	GALAXY
UGC 11982	22 16 18.	- 01 18	78	15.7	GALAXY Sc
KARA.73B 0959	22 16 18.	- 01 18	102	15.7	ISOLATED GALAXY S
PHL 5023	22 16 18.	- 13 18		17.8	BLUE STELLAR OBJECT
MCG-04-52-033	22 16 18.	- 21 11	54	14.	GALAXY
IC 1443	22 16 18.	- 21 11 24.			NONSTELLAR OBJECT
SNO 30	22 16 18.	- 28 36 40.	420	16.	GROUP OF 4 GALAXIES
HN 1225	22 16 20.	- 60 24		15.	NEBULA
IC 5197	22 16 21.	- 60 24			NONSTELLAR OBJECT

OBJECT NAME	RIGHT ASCEN.	DECLINATION	DIAM.	MAGN.	TYPE OF OBJECT
PHL 1823	22 16 24.	+ 03 24		17.2	BLUE STELLAR OBJECT
MCG-04-52-034	22 16 24.	- 24 26	15	15.5	GALAXY
MCG-05-52-064	22 16 24.	- 28 42	54	14.5	GALAXY
SC 2212-6242.2	22 16 24.	- 62 27 14.	12		NEBULA
SC 2213-5459.6	22 16 29.	- 54 44 37.	12		NEBULA
PHL 5024	22 16 30.	+ 00 36		18.2	BLUE STELLAR OBJECT
ZWG 428.071	22 16 30.	+ 12 55		15.7	GALAXY
KARA.72 565B	22 16 30.	+ 12 55	48	15.7	PART OF DOUBLE GALAXY
ZWG 428.072	22 16 30.	+ 12 56		15.7	GALAXY
KARA.72 565A	22 16 30.	+ 12 56	30	15.7	PART OF DOUBLE GALAXY
MCG+04-52-008	22 16 30.	+ 24 28	27	15.5	GALAXY
ABC 2435	22 16 31.	+ 08 52		17.7	RICH CLUSTER OF GALAXIES
SC 2213-5142.4	22 16 31.	- 51 27 25.			NEBULA
IC 5199	22 16 32.	- 37 46			NONSTELLAR OBJECT
SG 3.230	22 16 33.	+ 58 52	360		DIFFUSE EMISSION NEBULA
HN 0746	22 16 33.	- 37 46			NEBULA
HN 0745	22 16 33.	- 65 39			NEBULA
IC 5196	22 16 33.	- 65 39			NONSTELLAR OBJECT
SC 2213-6235.2	22 16 35.	- 62 20 13.	12		NEBULA
PHL 5025	22 16 36.	+ 03 18		17.5	BLUE STELLAR OBJECT
2ZW 176	22 16 36.	+ 06 04			COMPACT GALAXY
8ZW 2216+06.1	22 16 36.	+ 06 05		16.7	COMPACT GALAXY
PHL 0221	22 16 36.	- 19 26		18.2	BLUE STELLAR OBJECT
ABC 2434	22 16 37.	- 14 29		17.0	RICH CLUSTER OF GALAXIES
ZWG 473.011	22 16 42.	+ 24 20		15.5	GALAXY
KARA.73B 0960	22 16 42.	+ 24 20	54	15.5	ISOLATED GALAXY S
PHL 1824	22 16 42.	- 02 15		18.5	BLUE STELLAR OBJECT
SC 2213-5428.9	22 16 42.	- 54 13 55.			NEBULA
RNGC 7249	22 16 46.	- 55 23			GALAXY
ZWG 377.046	22 16 48.	+ 02 11		15.7	GALAXY
PHL 5026	22 16 48.	+ 02 32		16.6	BLUE STELLAR OBJECT
ZWG 403.033	22 16 48.	+ 08 46		15.6	GALAXY
2ZW 177	22 16 48.	+ 11 53			COMPACT GALAXY
ZWG 494.013	22 16 48.	+ 33 05		15.6	GALAXY
UGC 11983	22 16 48.	+ 33 05	66	15.6	GALAXY Sb-c
PHL 5027	22 16 48.	- 07 30		17.6	BLUE STELLAR OBJECT
PHL 0222	22 16 48.	- 07 56		17.3	BLUE STELLAR OBJECT
PHL 5028	22 16 48.	- 11 32		17.2	BLUE STELLAR OBJECT
MCG-07-45-014	22 16 48.	- 40 50	30	15.	GALAXY
CED 204	22 16 49.	+ 62 58			DIFFUSE GALACTIC NEBULA
SG 2.086	22 16 49.	+ 62 58	3600		DIFFUSE EMISSION NEBULA
SC 2213-5459.0	22 16 50.	- 54 44 01.			NEBULA
MRSL 102-00/1	22 16 54.	+ 55 53	5400		HII REGION
PHL 0223	22 16 54.	- 07 30		18.2	BLUE STELLAR OBJECT
SC 2213-5220.9	22 16 55.	- 52 05 55.			NEBULA
LBN 0470	22 17	+ 55 50	1500		BRIGHT NEBULA
LBN 0505	22 17	+ 62 50	1800		BRIGHT NEBULA
KHAV 768	22 17	+ 63 21	6050		DARK NEBULA
PHL 5030	22 17 00.	- 09 30		17.5	BLUE STELLAR OBJECT
PHL 5029	22 17 00.	- 09 30		18.0	BLUE STELLAR OBJECT
PHL 1825	22 17 00.	- 13 00		18.4	BLUE STELLAR OBJECT
MCG-05-52-065	22 17 00.	- 26 36	36	14.5	GALAXY
SC 2213-5553.2	22 17 01.	- 55 38 12.			NEBULA
RNGC 7253B	22 17 03.	+ 29 08		14.5	GALAXY
HOLM 790A	22 17 05.	+ 29 08	120	13.2	PART OF MULTIPLE GALAXY
VV 242B	22 17 06.	+ 29 08	96	13.5	INTERACTING GALAXY
VV 242A	22 17 06.	+ 29 08	90	13.2	INTERACTING GALAXY
VV 242	22 17 06.	+ 29 08	120		INTERACTING GALAXY
KARA.72 566A	22 17 06.	+ 29 09	102	14.4	PART OF DOUBLE GALAXY
MCG+05-52-010	22 17 06.	+ 29 09	96	13.	GALAXY
5ZW 364	22 17 06.	+ 44 59			COMPACT GALAXY
PHL 5031	22 17 06.	- 06 58		16.5	BLUE STELLAR OBJECT
PHL 5032	22 17 06.	- 11 40		18.4	BLUE STELLAR OBJECT
SC 2213-6232.2	22 17 06.	- 62 17 13.	12		NEBULA
HOLM 790B	22 17 08.	+ 29 07	96	13.5	PART OF MULTIPLE GALAXY
RNGC 7253A	22 17 09.	+ 29 08		14.5	GALAXY
MCG+05-52-011	22 17 09.	+ 29 08	90	13.5	GALAXY
ARP 278	22 17 11.	+ 29 03			PECULIAR GALAXY
ZC 2217.2+0108	22 17 12.	+ 01 08	740		CLUSTER OF GALAXIES
PHL 5033	22 17 12.	+ 03 28		17.7	BLUE STELLAR OBJECT
UGC 11984	22 17 12.	+ 28 08	108	14.4	GALAXY DBL SYS
ZWG 494.014	22 17 12.	+ 29 08		14.4	GALAXY
UGC 11985	22 17 12.	+ 29 08	102	14.4	GALAXY DBL SYS
KARA.72 566B	22 17 12.	+ 29 08	96		PART OF DOUBLE GALAXY
4ZW 096	22 17 12.	+ 35 45			COMPACT GALAXY
SC 2214-5537.6	22 17 17.	- 55 22 36.	12		NEBULA
ZWG 429.001	22 17 18.	+ 14 44		15.7	GALAXY
ZWG 428.073	22 17 18.	+ 14 44		15.7	GALAXY
UGC 11986	22 17 18.	+ 41 00	78	16.5	GALAXY
MCG-01-57-001	22 17 18.	- 07 55	54	15.	GALAXY
PHL 0224	22 17 19.	- 08 53		15.3	BLUE STELLAR OBJECT
SC 2214-5524.6	22 17 19.	- 55 09 36.			NEBULA
MIN.47 15	22 17 21.	+ 55 51			DIFFUSE NEBULA
SC 2214-5535.4	22 17 23.	- 55 20 24.	12		NEBULA
PHL 1826	22 17 24.	+ 02 26		17.8	BLUE STELLAR OBJECT
ZC 2217.4+3030	22 17 24.	+ 30 30	2350		CLUSTER OF GALAXIES
ZWG 514.008	22 17 24.	+ 35 16		15.6	GALAXY
MCG+06-49-003	22 17 24.	+ 35 16	48	16.	GALAXY
SER 149.01	22 17 24.	- 55 20	1980	15.	CLUSTER OF 50 GALAXIES
MCG+01-57-001	22 17 30.	+ 08 45	36	15.	GALAXY
MRSL 106+05/1	22 17 30.	+ 63 02	1800		HII REGION
PHL 1827	22 17 30.	- 08 04		5.3	BLUE STELLAR OBJECT
PHL 5035	22 17 30.	- 08 27		19.0	BLUE STELLAR OBJECT
PHL 5034	22 17 30.	- 09 23		18.4	BLUE STELLAR OBJECT
MCG-03-57-001	22 17 33.	- 16 12 30.	84	14.	GALAXY
ZWG 404.001	22 17 36.	+ 08 45		15.4	GALAXY
UGC 11987	22 17 36.	+ 08 45	96	15.4	GALAXY SB?
MCG-01-57-002	22 17 36.	- 03 42	48	16.	GALAXY
PHL 1828	22 17 36.	- 08 41		17.5	BLUE STELLAR OBJECT
MCG-04-52-035	22 17 36.	- 22 18 30.	24	15.5	GALAXY
MCG-05-52-066	22 17 36.	- 28 39	48	15.5	GALAXY
ZWG 404.002	22 17 42.	+ 07 37		15.5	GALAXY
KARA.73B 0961	22 17 42.	+ 07 37	36	15.5	ISOLATED GALAXY S
UGC 11988	22 17 42.	+ 32 55	60	17.	GALAXY
ISS 0134	22 17 42.	+ 55 23	342		STELLAR RING
PHL 0225	22 17 42.	- 09 48		17.8	BLUE STELLAR OBJECT
PHL 0226	22 17 42.	- 19 32		17.8	BLUE STELLAR OBJECT
SC 2214-5336.7	22 17 44.	- 53 21 41.			NEBULA
SC 2215-1156.2	22 17 45.	- 11 41 10.	6		NEBULA
RNGC 7251	22 17 45.	- 16 02		13.0	GALAXY
SC 2214-5548.0	22 17 46.	- 55 32 59.			NEBULA
PHL 0227	22 17 48.	+ 01 44		12.8	BLUE STELLAR OBJECT
ZWG 452.001	22 17 48.	+ 18 36		15.7	GALAXY
PHL 5036	22 17 48.	- 10 10		17.4	BLUE STELLAR OBJECT
PHL 1829	22 17 48.	- 12 15		17.9	BLUE STELLAR OBJECT
MCG-03-57-002	22 17 48.	- 16 02	120	13.	GALAXY
SG 1.23	22 17 51.	+ 55 55	1440		DIFFUSE EMISSION NEBULA
PHL 5038	22 17 54.	- 00 58		18.0	BLUE STELLAR OBJECT

OBJECT NAME	RIGHT ASCEN.	DECLINATION	DIAM.	MAGN.	TYPE OF OBJECT
PHL 0228	22 17 54.	- 02 18		16.7	BLUE STELLAR OBJECT
ARC 2436	22 17 55.	- 03 03		16.9	RICH CLUSTER OF GALAXIES
SC 2214-5351.1	22 17 57.	- 53 36 05.			NEBULA
ARP 226	22 17 59.	- 24 56			PECULIAR GALAXY
LBN 0473	22 18	+ 55 40	4500		BRIGHT NEBULA
KHAV 769	22 18	+ 62 21	3010		DARK NEBULA
ZWG 429.002	22 18 00.	+ 15 17		15.7	GALAXY
ZWG 452.002	22 18 00.	+ 16 48		15.7	GALAXY
ZWG 452.003	22 18 00.	+ 18 41		15.2	GALAXY
LDN 1156	22 18 00.	+ 55 30	840		DARK NEBULA
LDN 1161	22 18 00.	+ 55 53	300		DARK NEBULA
PHL 1830	22 18 00.	- 16 10		18.0	BLUE STELLAR OBJECT
PHL 0229	22 18 00.	- 19 08		17.9	BLUE STELLAR OBJECT
MCG-04-52-036	22 18 00.	- 24 55 30.	96	13.	GALAXY
SC 2214-5431.5	22 18 00.	- 54 16 29.	12		NEBULA
IC 5204	22 18 01.	- 14 39 06.			NONSTELLAR OBJECT
RNGC 7252	22 18 01.	- 24 56		13.0	GALAXY
PHL 1831	22 18 06.	+ 00 18		8.4	BLUE STELLAR OBJECT
PHL 5039	22 18 06.	- 10 04		17.4	BLUE STELLAR OBJECT
PHL 1832	22 18 06.	- 10 38		18.2	BLUE STELLAR OBJECT
PHL 1833	22 18 06.	- 15 28		18.5	BLUE STELLAR OBJECT
PHL 5040	22 18 06.	- 19 10		18.9	BLUE STELLAR OBJECT
8ZW 2218+04.2	22 18 12.	+ 04 14		16.5	COMPACT GALAXY
UGC 11989	22 18 12.	+ 32 42	60	17.	GALAXY DWRF SP
PHL 5045	22 18 12.	- 08 53		18.0	BLUE STELLAR OBJECT
PHL 5044	22 18 12.	- 10 06		17.8	BLUE STELLAR OBJECT
PHL 5042	22 18 12.	- 12 24		18.1	BLUE STELLAR OBJECT
PHL 5046	22 18 12.	- 13 46		18.1	BLUE STELLAR OBJECT
PHL 5047	22 18 12.	- 14 20		18.7	BLUE STELLAR OBJECT
PHL 5041	22 18 12.	- 14 52		16.4	BLUE STELLAR OBJECT
PHL 1834	22 18 12.	- 15 26		17.9	BLUE STELLAR OBJECT
SC 2215-5225.7	22 18 17.	- 52 10 40.			NEBULA
ZWG 514.009	22 18 18.	+ 35 44		15.6	GALAXY
UGC 11990	22 18 18.	+ 35 44	66	15.6	GALAXY (E)
MCG+08-40-006	22 18 18.	+ 47 26	96	15.	GALAXY
ZWG 545.005	22 18 18.	+ 47 27		14.4	GALAXY
UGC 11991	22 18 18.	+ 47 27	126	14.4	GALAXY Sc
PHL 5048	22 18 18.	- 13 33		17.9	BLUE STELLAR OBJECT
PHL 1835	22 18 18.	- 16 34		18.6	BLUE STELLAR OBJECT
IC 5201	22 18 18.	- 46 19	480	12.8	GALAXY Sc
ARC 2437	22 18 21.	+ 12 51		17.5	RICH CLUSTER OF GALAXIES
PHL 5049	22 18 24.	+ 03 12		17.8	BLUE STELLAR OBJECT
ZC 2218.4+0511	22 18 24.	+ 05 11	2350		CLUSTER OF GALAXIES
UGC 11992	22 18 24.	+ 13 59	66	16.0	GALAXY
PHL 1837	22 18 24.	- 02 33		18.7	BLUE STELLAR OBJECT
PHL 5051	22 18 24.	- 08 58		18.1	BLUE STELLAR OBJECT
PHL 5050	22 18 24.	- 08 58		18.2	BLUE STELLAR OBJECT
PHL 0230	22 18 24.	- 11 27		17.2	BLUE STELLAR OBJECT
PHL 1836	22 18 24.	- 11 36		18.7	BLUE STELLAR OBJECT
PHL 5052	22 18 24.	- 15 44		18.4	BLUE STELLAR OBJECT
AGU 73	22 18 24.	- 42 10 30.	42	17.	IRREGULAR DWARF GALAXY
SCHO 1420	22 18 26.	+ 51 33 36.	280		ISOLATED DARK CLOUD
PK104+00.1	22 18 28.5	+ 57 59 08.		17.	PLANETARY NEBULA
UGC 11993	22 18 30.	+ 34 58	72	16.0	GALAXY Sc
COU 092	22 18 30.	+ 56 00	3000		EMISSION NEBULA
PHL 1838	22 18 30.	- 09 21		18.0	BLUE STELLAR OBJECT
PHL 1839	22 18 30.	- 10 16		17.8	BLUE STELLAR OBJECT
PHL 0231	22 18 30.	- 20 27		16.6	BLUE STELLAR OBJECT
8ZW 2218+05.1	22 18 36.	+ 05 06		16.8	COMPACT GALAXY
ZC 2218.6+1246	22 18 36.	+ 12 46	1480		CLUSTER OF GALAXIES
ZWG 494.015	22 18 36.	+ 33 03		15.0	GALAXY
UGC 11994	22 18 36.	+ 33 03	150	15.0	GALAXY Sb-c
MCG+05-52-012	22 18 36.	+ 33 03 30.	120	14.5	GALAXY
OCL 0237	22 18 36.	+ 57 50	600	9.8	OPEN STAR CLUSTER
RNGC 7261	22 18 36.	+ 57 50		10.0	OPEN CLUSTER
PHL 1841	22 18 36.	- 10 42		18.5	BLUE STELLAR OBJECT
PHL 1840	22 18 36.	- 12 45		17.6	BLUE STELLAR OBJECT
PHL 5054	22 18 36.	- 14 26		18.1	BLUE STELLAR OBJECT
PHL 5053	22 18 36.	- 19 40		17.7	BLUE STELLAR OBJECT
TON-S 0047	22 18 36.	- 27 06		14.9	BLUE STAR
SC 2215-5150.2	22 18 37.	- 51 35 09.			NEBULA
SC 2216-1104.8	22 18 39.	- 10 49 45.	6		NEBULA
HN 0747	22 18 39.	- 66 00			NEBULA
IC 5200	22 18 39.	- 66 00			NONSTELLAR OBJECT
SC 2216-1101.6	22 18 41.	- 10 46 33.	6		NEBULA
PK103+00.2	22 18 41.8	+ 57 21 11.	16		PLANETARY NEBULA
PHL 5056	22 18 42.	+ 00 39		17.8	BLUE STELLAR OBJECT
PHL 5055	22 18 42.	- 13 58		17.8	BLUE STELLAR OBJECT
PHL 0232	22 18 48.	+ 02 00		13.6	BLUE STELLAR OBJECT
ZWG 404.003	22 18 48.	+ 09 23		15.7	GALAXY
PHL 0233	22 18 48.	- 00 34		18.4	BLUE STELLAR OBJECT
ZWG 514.010	22 18 54.	+ 36 20		14.8	GALAXY
UGC 11995	22 18 54.	+ 36 20	96	14.8	GALAXY Sa
SC 2215-6147.1	22 18 57.	- 61 32 03.	18		NEBULA
LBN 0492	22 19	+ 58 25	1080		BRIGHT NEBULA
LBN 0493	22 19	+ 58 30	180		BRIGHT NEBULA
LBN 0538	22 19	+ 70 45	420		BRIGHT NEBULA
PHL 0234	22 19 00.	+ 01 05		17.0	BLUE STELLAR OBJECT
LDN 1163	22 19 00.	+ 55 55	300		DARK NEBULA
LDN 1184	22 19 00.	+ 60 30	1380		DARK NEBULA
PHL 5058	22 19 00.	- 02 58		16.2	BLUE STELLAR OBJECT
PHL 1843	22 19 00.	- 03 37		18.4	BLUE STELLAR OBJECT
PHL 1844	22 19 00.	- 12 44		18.4	BLUE STELLAR OBJECT
PHL 1845	22 19 00.	- 12 48		18.9	BLUE STELLAR OBJECT
PHL 1842	22 19 00.	- 12 48		17.0	BLUE STELLAR OBJECT
PHL 5057	22 19 00.	- 13 15		16.6	BLUE STELLAR OBJECT
PHL 0235	22 19 00.	- 13 44		18.3	BLUE STELLAR OBJECT
SC 2215-5301.1	22 19 00.	- 52 46 03.			NEBULA
SC 2215-5111.8	22 19 04.	- 50 56 44.	6		NEBULA
UGC 11996	22 19 06.	+ 12 14	60	16.5	GALAXY S
ZWG 514.011	22 19 06.	+ 34 20		15.5	GALAXY
PHL 5059	22 19 06.	- 17 46		18.2	BLUE STELLAR OBJECT
PHL 1848	22 19 12.	+ 00 05		15.3	BLUE STELLAR OBJECT
PHL 5060	22 19 12.	+ 01 44		17.8	BLUE STELLAR OBJECT
ZWG 404.004	22 19 12.	+ 06 40		15.1	GALAXY
MCG+04-52-009	22 19 12.	+ 25 01 30.	12	15.5	GALAXY
MCG+08-40-007	22 19 12.	+ 45 12 30.	24	16.	GALAXY
PHL 5062	22 19 12.	- 13 04		18.0	BLUE STELLAR OBJECT
PHL 1846	22 19 12.	- 14 14		17.8	BLUE STELLAR OBJECT
PHL 5061	22 19 12.	- 14 34		17.7	BLUE STELLAR OBJECT
PHL 1847	22 19 12.	- 17 50		18.2	BLUE STELLAR OBJECT
MCG-04-52-037	22 19 12.	- 25 49	12	15.	GALAXY
MCG-04-52-038	22 19 12.	- 25 50	24	15.	GALAXY
MCG-07-45-015	22 19 12.	- 40 21	48	14.5	GALAXY
SC 2216-5309.3	22 19 12.	- 52 54 14.			NEBULA
SC 2216-5319.3	22 19 12.	- 53 04 14.			NEBULA
HN 1226	22 19 12.	- 60 02		15.	NEBULA
RNGC 7254	22 19 13.	- 21 59			NON-EXISTENT OBJECT
IC 5203	22 19 13.	- 60 02			NONSTELLAR OBJECT
ARC 2438	22 19 14.	- 15 50		17.2	RICH CLUSTER OF GALAXIES
SC 2216-5258.1	22 19 14.	- 52 43 02.			NEBULA
RNGC 7255	22 19 15.	- 15 49			NON-EXISTENT OBJECT
SC 2216-5447.5	22 19 17.	- 54 32 26.	24		NEBULA
ZWG 531.001	22 19 18.	+ 45 15		15.5	GALAXY
ZWG 530.023	22 19 18.	+ 45 15		15.5	GALAXY
UGC 11997	22 19 18.	+ 45 15	60	15.5	GALAXY (S0)
PHL 1849	22 19 18.	- 08 37		13.8	BLUE STELLAR OBJECT
PHL 1850	22 19 18.	- 10 42		18.2	BLUE STELLAR OBJECT
PHL 5064	22 19 18.	- 11 32		17.0	BLUE STELLAR OBJECT
PHL 5063	22 19 18.	- 17 26		16.9	BLUE STELLAR OBJECT
IC 5202	22 19 20.	- 66 03			NONSTELLAR OBJECT
HN 0748	22 19 21.	- 66 03			NEBULA
MIN.47 16	22 19 23.	+ 58 36			DIFFUSE NEBULA
PHL 0236	22 19 24.	+ 03 18		17.8	BLUE STELLAR OBJECT
ZWG 473.012	22 19 24.	+ 25 03		15.7	GALAXY
PHL 0237	22 19 24.	- 00 48		18.0	BLUE STELLAR OBJECT
PHL 1851	22 19 24.	- 02 10		16.5	BLUE STELLAR OBJECT
PHL 0238	22 19 24.	- 09 42		17.8	BLUE STELLAR OBJECT
PHL 5065	22 19 24.	- 10 24		17.9	BLUE STELLAR OBJECT
PHL 1852	22 19 24.	- 16 30		18.0	BLUE STELLAR OBJECT
PHL 5066	22 19 24.	- 16 56		17.9	BLUE STELLAR OBJECT
MCG-03-57-003	22 19 24.	- 19 51	48	14.5	GALAXY
MCG-04-52-039	22 19 24.	- 23 52	12	16.	GALAXY
AGU 74	22 19 24.	- 42 16 18.	24	17.	DWARF GALAXY
SC 2216-5217.8	22 19 24.	- 52 02 44.			NEBULA
HN 1227	22 19 24.	- 60 03		16.	NEBULA
IC 5205	22 19 25.	- 60 03			NONSTELLAR OBJECT
MCG+04-52-010	22 19 30.	+ 25 06	48	15.	GALAXY
4ZW 097	22 19 30.	+ 36 06			COMPACT GALAXY
MCG+06-49-004	22 19 30.	+ 36 06	36	15.	GALAXY
SCHO 1421	22 19 30.	+ 51 39 00.	310		ISOLATED DARK CLOUD
PHL 1853	22 19 30.	- 09 08		18.0	BLUE STELLAR OBJECT
PHL 1854	22 19 30.	- 09 34		18.5	BLUE STELLAR OBJECT
PHL 0239	22 19 30.	- 12 36		17.7	BLUE STELLAR OBJECT
SC 2216-5105.6	22 19 34.	- 50 50 31.			NEBULA
RNGC 7263	22 19 35.	+ 36 06		15.5	GALAXY
SG 1.22	22 19 35.	+ 58 33	720		DIFFUSE EMISSION NEBULA
SC 2216-5207.4	22 19 35.	- 51 52 19.			NEBULA
PHL 5067	22 19 36.	+ 02 32		18.0	BLUE STELLAR OBJECT
ZWG 514.012	22 19 36.	+ 36 06		15.7	GALAXY
RNGC 7257	22 19 36.	- 04 17			NON-EXISTENT OBJECT
PHL 5070	22 19 36.	- 11 12		18.0	BLUE STELLAR OBJECT
PHL 1856	22 19 36.	- 12 32		18.7	BLUE STELLAR OBJECT
PHL 5069	22 19 36.	- 12 40		17.7	BLUE STELLAR OBJECT
PHL 5072	22 19 36.	- 12 52		18.5	BLUE STELLAR OBJECT
PHL 5068	22 19 36.	- 13 02		16.4	BLUE STELLAR OBJECT
PHL 1855	22 19 36.	- 13 20		18.2	BLUE STELLAR OBJECT
SC 2216-5325.0	22 19 36.	- 53 09 55.	24		NEBULA
SC 2216-5714.9	22 19 39.	- 56 59 49.	9		NEBULA
SC 2216-5643.7	22 19 41.	- 56 28 37.			NEBULA
3ZW 090	22 19 42.	+ 14 28			COMPACT GALAXY
PHL 1857	22 19 42.	- 09 56		17.2	BLUE STELLAR OBJECT
SC 2216-5616.1	22 19 42.	- 56 01 01.	6		NEBULA
SER 150.02	22 19 42.	- 63 25	1080	16.	CLUSTER OF FAINT GALAXIES
SC 2216-5527.1	22 19 43.	- 55 12 01.			NEBULA
SC 2216-5728.4	22 19 43.	- 57 13 19.			NEBULA
PHL 5075	22 19 48.	+ 01 03		18.1	BLUE STELLAR OBJECT
PHL 5073	22 19 48.	+ 02 40		17.5	BLUE STELLAR OBJECT
PHL 5074	22 19 48.	+ 02 52		17.8	BLUE STELLAR OBJECT
UGC 11998	22 19 48.	+ 04 53	66	16.0	GALAXY SBc
ZWG 514.013	22 19 48.	+ 35 25		15.5	GALAXY
UGC 11999	22 19 48.	+ 42 23	60	17.	GALAXY Sc
PHL 1858	22 19 48.	- 12 52		18.2	BLUE STELLAR OBJECT
PHL 1860	22 19 48.	- 13 13		18.3	BLUE STELLAR OBJECT
PHL 1859	22 19 48.	- 15 57		16.9	BLUE STELLAR OBJECT
IC 5210	22 19 48.	- 19 07 18.			NONSTELLAR OBJECT
MCG-04-52-041	22 19 48.	- 23 47	15	15.5	GALAXY
MCG-04-52-040	22 19 48.	- 23 47 30.	12	15.5	GALAXY
MCG-05-52-067	22 19 48.	- 27 38	54	15.	GALAXY
SC 2216-5729.7	22 19 50.	- 57 14 37.	6		NEBULA
IC 1444	22 19 53.	+ 04 53 13.			NONSTELLAR OBJECT
PHL 5078	22 19 54.	+ 01 10		17.8	BLUE STELLAR OBJECT
PHL 1861	22 19 54.	+ 02 30		18.6	BLUE STELLAR OBJECT
ZWG 378.001	22 19 54.	+ 02 36		15.7	GALAXY
UGC 12000	22 19 54.	+ 02 36	60	15.7	GALAXY S
ZWG 404.005	22 19 54.	+ 04 52		15.4	GALAXY
PHL 5076	22 19 54.	- 10 24		15.8	BLUE STELLAR OBJECT
PHL 5077	22 19 54.	- 17 16		16.3	BLUE STELLAR OBJECT
MCG-03-57-004	22 19 54.	- 19 07	48	14.5	GALAXY
MCG-04-52-042	22 19 54.	- 22 00	48	14.	GALAXY
RNGC 7256	22 19 56.	- 22 00		14.0	GALAXY
MCG+06-49-005	22 19 57.	+ 36 08	138	14.	GALAXY
ARC 2439	22 19 58.	+ 00 19		17.5	RICH CLUSTER OF GALAXIES
SC 2217+1709.6	22 19 58.	+ 17 24 42.	12		NEBULA
RNGC 7264	22 19 59.	+ 36 08		14.5	GALAXY
SG 2.087	22 19 59.	+ 58 33	360		DIFFUSE EMISSION NEBULA
SC 2216-5137.2	22 19 59.	- 51 22 07.			NEBULA
SC 2216-5203.7	22 19 59.	- 51 48 37.			NEBULA
LBN 0432	22 20	+ 42 00	9000		BRIGHT NEBULA
LBN 0558	22 20	+ 75 00	4920		BRIGHT NEBULA
LBN 0567	22 20	+ 77 55	1260		BRIGHT NEBULA
PHL 0240	22 20 00.	+ 00 36		16.5	BLUE STELLAR OBJECT
ZC 2220.0+0113	22 20 00.	+ 01 13	870		CLUSTER OF GALAXIES
ZWG 378.002	22 20 00.	+ 02 32		15.4	GALAXY
ZWG 514.014	22 20 00.	+ 36 08		14.7	GALAXY
UGC 12001	22 20 00.	+ 36 08	132	14.7	GALAXY Sb
ZWG 531.002	22 20 00.	+ 41 05		15.7	GALAXY
UGC 12002	22 20 00.	+ 41 05	60	15.7	GALAXY E
MCG+07-46-001	22 20 00.	+ 41 05	48	15.	GALAXY
COU 092A	22 20 00.	+ 55 00	6000		EMISSION NEBULA
LDN 1247	22 20 00.	+ 75 00	1620		DARK NEBULA
PHL 5082	22 20 00.	- 05 36		18.5	BLUE STELLAR OBJECT
PHL 5080	22 20 00.	- 06 23		17.9	BLUE STELLAR OBJECT
PHL 0242	22 20 00.	- 08 43		17.8	BLUE STELLAR OBJECT
PHL 0241	22 20 00.	- 08 43		16.9	BLUE STELLAR OBJECT
PHL 5081	22 20 00.	- 08 48		17.8	BLUE STELLAR OBJECT
PHL 1862	22 20 00.	- 12 14		17.7	BLUE STELLAR OBJECT
PHL 1863	22 20 00.	- 12 40		18.2	BLUE STELLAR OBJECT
PHL 5083	22 20 00.	- 12 52		17.9	BLUE STELLAR OBJECT
IC 5211	22 20 00.	- 19 07 59.			NONSTELLAR OBJECT
SC 2216-5952.8	22 20 00.	- 59 37 43.	12		NEBULA
SC 2216-5838.2	22 20 01.	- 58 23 07.	6		NEBULA
SC 2216-5714.7	22 20 03.	- 56 59 37.	9		NEBULA
SC 2216-5540.8	22 20 05.	- 55 25 43.	12		NEBULA
ZC 2220.1+0020	22 20 06.	+ 00 20	1410		CLUSTER OF GALAXIES
PHL 5084	22 20 06.	+ 03 38		16.9	BLUE STELLAR OBJECT

OBJECT NAME	RIGHT ASCEN.	DECLINATION	DIAM.	MAGN.	TYPE OF OBJECT
UGC 12003	22 20 06.	+ 73 24	60	17.	GALAXY DWARF?
RNGC 7260	22 20 06.	- 04 23		14.0	GALAXY
MCG-01-57-003	22 20 06.	- 04 23	90	14.	GALAXY
MCG-03-57-005	22 20 06.	- 19 08	42	15.	GALAXY
MCG-04-52-043	22 20 06.	- 20 36	60	15.	GALAXY
HN 1228	22 20 07.	- 60 50		16.	NEBULA
IC 5207	22 20 07.	- 60 50			NONSTELLAR OBJECT
HN 0750	22 20 08.	- 38 14			NEBULA
IC 5209	22 20 08.	- 38 14			NONSTELLAR OBJECT
SC 2216-6338.8	22 20 08.	- 63 23 43.	30		NEBULA
MCG+04-52-011	22 20 09.	+ 25 04	24	15.5	GALAXY
SN 1970H	22 20 11.	+ 35 46		17.	SUPERNOVA
RNGC 7265	22 20 11.	+ 35 58		13.5	GALAXY
ZWG 514.015	22 20 12.	+ 35 58		13.7	GALAXY
UGC 12004	22 20 12.	+ 35 58	156	13.7	GALAXY S0
PHL 0243	22 20 12.	- 06 33		16.3	BLUE STELLAR OBJECT
PHL 5085	22 20 12.	- 13 26		16.7	BLUE STELLAR OBJECT
PHL 0244	22 20 12.	- 18 34		13.8	BLUE STELLAR OBJECT
RNGC 7258	22 20 12.	- 28 38		14.0	GALAXY
MCG-05-52-068	22 20 12.	- 28 38	84	14.	GALAXY
SC 2216-5704.1	22 20 14.	- 56 49 00.	18		NEBULA
MCG+06-49-006	22 20 15.	+ 35 57	36	13.	GALAXY
SC 2217-5102.6	22 20 15.	- 50 47 30.			NEBULA
SC 2217-5524.8	22 20 15.	- 55 09 42.	18		NEBULA
RNGC 7259	22 20 17.	- 29 14		12.0	GALAXY
SC 2216-5719.9	22 20 17.	- 57 04 48.	12		NEBULA
ZWG 494.016	22 20 18.	+ 28 13		15.6	GALAXY
KARA.73B 0962	22 20 18.	+ 28 13	48	15.6	ISOLATED GALAXY S
UGC 12005	22 20 18.	+ 35 46	60	16.5	GALAXY
ZWG 514.016	22 20 18.	+ 36 51		14.5	GALAXY
UGC 12006	22 20 18.	+ 36 51	90	14.5	GALAXY SB:0-a
MCG+06-49-007	22 20 18.	+ 36 52	72	14.	GALAXY
PHL 1864	22 20 18.	- 07 26		14.8	BLUE STELLAR OBJECT
PHL 5086	22 20 18.	- 08 38		18.0	BLUE STELLAR OBJECT
MCG-05-52-069	22 20 18.	- 29 14	30	12.5	GALAXY
SC 2217-5833.2	22 20 22.	- 58 18 06.	18		NEBULA
SC 2217-5138.0	22 20 23.	- 51 22 54.			NEBULA
SC 2217-5656.4	22 20 23.	- 56 41 18.	12		NEBULA
PHL 1865	22 20 24.	+ 01 08		17.7	BLUE STELLAR OBJECT
ZC 2220.4+1139	22 20 24.	+ 11 39	2080		CLUSTER OF GALAXIES
ZWG 494.017	22 20 24.	+ 29 31		15.2	GALAXY
4ZW 098	22 20 24.	+ 35 57			COMPACT GALAXY
MRSL 104+01/1	22 20 24.	+ 58 29	900		HII REGION
PHL 5089	22 20 24.	- 06 41		17.5	BLUE STELLAR OBJECT
PHL 5087	22 20 24.	- 08 30		15.9	BLUE STELLAR OBJECT
PHL 5090	22 20 24.	- 10 52		17.7	BLUE STELLAR OBJECT
PHL 5088	22 20 24.	- 10 58		18.4	BLUE STELLAR OBJECT
PHL 1866	22 20 24.	- 12 48		18.7	BLUE STELLAR OBJECT
IC 5206	22 20 29.	- 67 06			NONSTELLAR OBJECT
ZC 2220.5+0602	22 20 30.	+ 06 02	1550		CLUSTER OF GALAXIES
8ZW 2220+08.1	22 20 30.	+ 08 04		16.5	COMPACT GALAXY
HOLM 791A	22 20 30.	+ 29 31	36	14.8	PART OF MULTIPLE GALAXY
MCG+05-52-013	22 20 30.	+ 29 31	42	15.	GALAXY
MCG+06-49-009	22 20 30.	+ 35 56 30.	90	14.	GALAXY
ZWG 514.017	22 20 30.	+ 35 57		15.3	GALAXY
UGC 12007	22 20 30.	+ 35 57	84	15.3	GALAXY
UGC 12008	22 20 30.	+ 37 27	60	16.0	GALAXY Sc
ZWG 514.018	22 20 30.	+ 37 44		13.8	GALAXY
UGC 12009	22 20 30.	+ 37 44	90	13.8	GALAXY
MCG+06-49-008	22 20 30.	+ 37 44	36	14.5	GALAXY
MCG-03-57-006	22 20 30.	- 15 48 30.	60	15.	GALAXY
MCG-05-52-070	22 20 30.	- 29 16	144	13.5	GALAXY
HN 0749	22 20 30.	- 67 06			NEBULA
HOLM 791B	22 20 31.	+ 29 32	18	15.6	PART OF MULTIPLE GALAXY
SCHO 1422	22 20 31.	+ 55 47 42.	240		ISOLATED DARK CLOUD
SC 2217-6003.5	22 20 31.	- 59 48 24.	18		NEBULA
HN 0752	22 20 32.	- 38 18			NEBULA
IC 5212	22 20 32.	- 38 18			NONSTELLAR OBJECT
SC 2217-5738.7	22 20 32.	- 57 23 36.	6		NEBULA
SC 2217-5801.7	22 20 32.	- 57 46 36.	12		NEBULA
RNGC 7262	22 20 34.	- 32 36			GALAXY
SC 2217-5814.3	22 20 34.	- 57 59 12.	12		NEBULA
PHL 1867	22 20 36.	- 09 10		18.6	BLUE STELLAR OBJECT
PHL 1868	22 20 36.	- 12 20		18.0	BLUE STELLAR OBJECT
PHL 0245	22 20 36.	- 16 42		17.8	BLUE STELLAR OBJECT
MCG-04-52-044	22 20 36.	- 22 47 30.	54	15.	GALAXY
MCG-07-46-001	22 20 36.	- 42 32	30	17.	GALAXY
SC 2217-5728.7	22 20 36.	- 57 13 36.			NEBULA
SC 2217-5320.4	22 20 37.	- 53 05 18.			NEBULA
SC 2217-5521.4	22 20 39.	- 55 06 18.	18		NEBULA
MCG+01-57-002	22 20 42.	+ 05 15	60	15.	GALAXY
ZWG 514.019	22 20 42.	+ 35 30		15.7	GALAXY
SG 2.088	22 20 42.	+ 58 25	900		DIFFUSE EMISSION NEBULA
PHL 0246	22 20 42.	- 00 10		16.9	BLUE STELLAR OBJECT
SER 149.09	22 20 42.	- 57 45	50	16.	INTERACTING GALAXIES
SC 2217-5710.6	22 20 44.	- 56 55 29.	12		NEBULA
ZWG 404.006	22 20 48.	+ 04 14		15.6	GALAXY
ZWG 404.007	22 20 48.	+ 05 17		15.4	GALAXY
UGC 12010	22 20 48.	+ 05 17	66	15.4	GALAXY Sa-b
ZWG 494.018	22 20 48.	+ 30 40		14.0	GALAXY
ZCG 2220+30	22 20 48.	+ 30 40		14.0	COMPACT GALAXY
UGC 12011	22 20 48.	+ 30 40	36	14.0	GALAXY PAIR
KARA.72 567B	22 20 48.	+ 30 40	24		PART OF DOUBLE GALAXY
KARA.72 567A	22 20 48.	+ 30 40	24	14.0	PART OF DOUBLE GALAXY
ZWG 494.019	22 20 48.	+ 31 54		15.6	GALAXY
MCG+06-49-010	22 20 48.	+ 35 34	60	15.	GALAXY
ZWG 514.020	22 20 48.	+ 35 35		15.3	GALAXY
UGC 12012	22 20 48.	+ 35 35	60	15.3	GALAXY Sb
ZWG 514.021	22 20 48.	+ 35 48		15.3	GALAXY
UGC 12013	22 20 48.	+ 35 48	66	15.3	GALAXY S
MCG+08-40-008	22 20 48.	+ 48 21	30	18.	GALAXY
OCL 0231	22 20 48.	+ 55 36	240	13.	OPEN STAR CLUSTER
PHL 5091	22 20 48.	- 10 00		18.4	BLUE STELLAR OBJECT
PHL 1869	22 20 48.	- 10 44		18.8	BLUE STELLAR OBJECT
MCG-05-52-071	22 20 48.	- 26 53	72	14.5	GALAXY
ZC 2220.9+1411	22 20 54.	+ 14 11	3490		CLUSTER OF GALAXIES
UGC 12014	22 20 54.	+ 48 23	66	17.	GALAXY
PHL 1870	22 20 54.	- 12 54		18.9	BLUE STELLAR OBJECT
PHL 1871	22 20 54.	- 13 46		18.8	BLUE STELLAR OBJECT
PHL 5092	22 20 54.	- 15 43		18.6	BLUE STELLAR OBJECT
IC 5214	22 20 54.	- 27 43 26.			NONSTELLAR OBJECT
SC 2217-5651.9	22 20 54.	- 56 36 47.	12		NEBULA
SC 2217-5532.5	22 20 57.	- 55 17 23.			NEBULA
SC 2217-5144.2	22 20 59.	- 51 29 05.			NEBULA
LBN 0508	22 21	+ 62 27	60		BRIGHT NEBULA
PHL 0247	22 21 00.	+ 00 44		17.7	BLUE STELLAR OBJECT
PHL 5096	22 21 00.	+ 01 56		18.1	BLUE STELLAR OBJECT
PHL 5093	22 21 00.	+ 02 40		9.5	BLUE STELLAR OBJECT

OBJECT NAME	RIGHT ASCEN.	DECLINATION	DIAM.	MAGN.	TYPE OF OBJECT
ZWG 452.004	22 21 00.	+ 18 29		15.5	GALAXY
ZWG 452.005	22 21 00.	+ 19 35		15.7	GALAXY
UGC 12015	22 21 00.	+ 19 35	78	15.7	GALAXY SBc
KARA.73B 0963	22 21 00.	+ 19 35	90	15.7	ISOLATED GALAXY S
MCG+03-57-001	22 21 00.	+ 19 37	78	15.	GALAXY
PHL 0248	22 21 00.	- 00 56		17.1	BLUE STELLAR OBJECT
PHL 5094	22 21 00.	- 01 29		17.8	BLUE STELLAR OBJECT
PHL 5097	22 21 00.	- 09 44		18.5	BLUE STELLAR OBJECT
PHL 5098	22 21 00.	- 10 50		18.7	BLUE STELLAR OBJECT
PHL 5095	22 21 00.	- 11 08		17.5	BLUE STELLAR OBJECT
PHL 1872	22 21 00.	- 19 36		18.4	BLUE STELLAR OBJECT
HN 0751	22 21 00.	- 65 28			NEBULA
IC 5208	22 21 00.	- 65 28			NONSTELLAR OBJECT
SC 2217-5632.1	22 21 01.	- 56 16 59.	18		NEBULA
VDB.66N 153	22 21 03.	+ 62 28	168		REFLECTION NEBULA
SC 2217-5545.3	22 21 04.	- 55 30 11.	12		NEBULA
ZC 2221.1+1722	22 21 06.	+ 17 22	1140		CLUSTER OF GALAXIES
ZWG 514.022	22 21 06.	+ 34 58		15.7	GALAXY
MCG+07-46-002	22 21 06.	+ 39 47	36	15.	GALAXY
ZWG 531.003	22 21 06.	+ 39 48		14.0	GALAXY
UGC 12016	22 21 06.	+ 39 48	72	14.0	GALAXY COMPACT
PHL 5099	22 21 06.	- 11 39		16.6	BLUE STELLAR OBJECT
SC 2218-1301.3	22 21 11.	- 12 46 10.	6		NEBULA
SC 2217-5839.7	22 21 11.	- 58 24 35.	6		NEBULA
SC 2217-5839.7	22 21 11.	- 58 24 35.	6		NEBULA
PHL 5100	22 21 12.	+ 02 46		18.2	BLUE STELLAR OBJECT
ZWG 429.003	22 21 12.	+ 11 36		15.2	GALAXY
UGC 12017	22 21 12.	+ 11 36	72	15.2	GALAXY SBb
MCG+02-57-001	22 21 12.	+ 11 37	60	14.	GALAXY
MCG-01-57-004	22 21 12.	- 03 40	36	14.5	GALAXY
PHL 1873	22 21 12.	- 13 41		18.2	BLUE STELLAR OBJECT
SC 2217-5638.5	22 21 14.	- 56 23 22.	6		NEBULA
SC 2217-5701.1	22 21 15.	- 56 45 58.	36		NEBULA
SC 2217-5730.8	22 21 17.	- 57 15 40.	42		NEBULA
MCG+02-57-002	22 21 18.	+ 11 32 30.	24	16.	GALAXY
ZWG 494.020	22 21 18.	+ 30 36		14.9	GALAXY
UGC 12018	22 21 18.	+ 30 36	108	14.9	GALAXY SB?b
MCG+05-52-014	22 21 18.	+ 30 36	108	14.5	GALAXY
ZWG 531.004	22 21 18.	+ 40 55		15.0	GALAXY
DG 183	22 21 18.	+ 62 27	60		REFLECTION NEBULA
ARC 2440	22 21 18.	- 01 51		16.0	RICH CLUSTER OF GALAXIES
MCG-01-57-005	22 21 18.	- 05 49	36	15.	GALAXY
PHL 1874	22 21 18.	- 10 20		17.0	BLUE STELLAR OBJECT
SER 150.03	22 21 18.	- 64 32	300	19.	CLUSTER OF FAINT GALAXIES
SC 2217-5927.3	22 21 19.	- 59 12 10.	36		NEBULA
RNGC 7270	22 21 22.	+ 32 08		15.0	GALAXY
SS 67	22 21 22.	+ 62 28			DIFFUSE GALACTIC NEBULA
SC 2218-1138.7	22 21 22.	- 11 23 34.	12		NEBULA
SC 2218-1302.2	22 21 23.	- 12 47 04.	12		NEBULA
SC 2218-5841.1	22 21 23.	- 58 25 58.	12		NEBULA
PHL 5101	22 21 24.	+ 02 11		17.8	BLUE STELLAR OBJECT
ZWG 494.021	22 21 24.	+ 32 08		15.0	GALAXY
UGC 12019	22 21 24.	+ 32 08	66	15.0	GALAXY S?
MCG-02-57-001	22 21 24.	- 11 24	12	15.5	GALAXY
PHL 5102	22 21 24.	- 15 16		18.0	BLUE STELLAR OBJECT
PHL 0249	22 21 24.	- 18 50		18.7	BLUE STELLAR OBJECT
PHL 1875	22 21 24.	- 19 33		18.0	BLUE STELLAR OBJECT
SER 149.04	22 21 24.	- 56 44	480	18.	RICH CLUSTER OF GALAXIES
SC 2218-5902.3	22 21 27.	- 58 47 10.	24		NEBULA
RNGC 7267	22 21 28.	- 33 57			GALAXY
ZC 2221.5+0149	22 21 30.	+ 01 49	1080		CLUSTER OF GALAXIES
PHL 1876	22 21 30.	+ 02 08		17.8	BLUE STELLAR OBJECT
ZWG 404.008	22 21 30.	+ 03 31		15.7	GALAXY
MCG+01-57-003	22 21 30.	+ 05 43	60	15.	GALAXY
ZWG 404.009	22 21 30.	+ 08 38		15.5	GALAXY
ZWG 429.004	22 21 30.	+ 14 58		15.6	GALAXY
MCG+05-52-015	22 21 30.	+ 32 09	48	14.5	GALAXY
MCG+06-49-011	22 21 30.	+ 35 08	90	15.	GALAXY
SCHO 1423	22 21 30.	+ 56 55 42.	300		ISOLATED DARK CLOUD
RNGC 7266	22 21 30.	- 04 20		15.0	GALAXY
MCG-01-57-006	22 21 30.	- 04 20	30	15.	GALAXY
SC 2218-5102.3	22 21 31.	- 50 47 10.			NEBULA
SC 2218-5658.5	22 21 31.	- 56 43 22.	6		NEBULA
SC 2218-5842.0	22 21 33.	- 58 26 52.	12		NEBULA
SC 2218-5855.9	22 21 33.	- 58 40 46.	12		NEBULA
RNGC 7271	22 21 34.	+ 32 06		15.5	GALAXY
PHL 5104	22 21 36.	+ 01 18		17.6	BLUE STELLAR OBJECT
ZWG 378.003	22 21 36.	+ 02 28		15.7	GALAXY
MCG+00-57-001	22 21 36.	+ 02 29	48	15.	GALAXY
MCG+01-57-004	22 21 36.	+ 05 05	36	15.	GALAXY
ZWG 404.010	22 21 36.	+ 08 26		15.7	GALAXY
ZWG 494.022	22 21 36.	+ 32 06		15.7	GALAXY
MCG+05-52-016	22 21 36.	+ 32 07	30	15.5	GALAXY
ZWG 514.023	22 21 36.	+ 35 09		15.6	GALAXY
UGC 12020	22 21 36.	+ 35 09	84	15.6	GALAXY SBa
PHL 5227	22 21 36.	- 14 50		6.5	BLUE STELLAR OBJECT
PHL 5103	22 21 36.	- 16 30		16.5	BLUE STELLAR OBJECT
PHL 5105	22 21 36.	- 18 56		17.9	BLUE STELLAR OBJECT
SC 2218-5057.1	22 21 36.	- 50 41 58.			NEBULA
HN 1229	22 21 36.	- 60 44		16.	NEBULA
IC 5213	22 21 36.	- 60 44			NONSTELLAR OBJECT
SC 2218-5638.6	22 21 39.	- 56 23 28.	6		NEBULA
SC 2218-5558.8	22 21 41.	- 55 43 40.	12		NEBULA
SC 2218-6243.1	22 21 41.	- 62 27 58.	6		NEBULA
ZC 2221.7+0005	22 21 42.	+ 00 05	810		CLUSTER OF GALAXIES
PHL 0251	22 21 42.	+ 00 55		17.8	BLUE STELLAR OBJECT
PHL 0250	22 21 42.	+ 03 14		17.1	BLUE STELLAR OBJECT
ZWG 404.011	22 21 42.	+ 05 44		15.3	GALAXY
UGC 12021	22 21 42.	+ 05 44	96	15.3	GALAXY Sb
ZWG 494.023	22 21 42.	+ 33 11		15.1	GALAXY
UGC 12022	22 21 42.	+ 33 11	96	15.1	GALAXY SB?
MCG+07-46-003	22 21 42.	+ 41 18	42	16.	GALAXY
PHL 5108	22 21 42.	- 00 41		18.2	BLUE STELLAR OBJECT
MCG-01-57-007	22 21 42.	- 03 44	96	14.5	GALAXY
PHL 5106	22 21 42.	- 09 36		17.8	BLUE STELLAR OBJECT
PHL 5107	22 21 42.	- 11 04		17.7	BLUE STELLAR OBJECT
PHL 1877	22 21 42.	- 14 40		18.5	BLUE STELLAR OBJECT
PHL 5109	22 21 48.	+ 01 48		18.4	BLUE STELLAR OBJECT
ZWG 404.012	22 21 48.	+ 05 06		15.6	GALAXY
UGC 12023	22 21 48.	+ 05 06	66	15.6	GALAXY SBb
ZWG 494.024	22 21 48.	+ 28 52		15.7	GALAXY
KARA.73B 0964	22 21 48.	+ 28 52	36	15.7	ISOLATED GALAXY S0
MCG+05-52-018	22 21 48.	+ 32 10 30.	48	16.5	GALAXY
MCG+05-52-017	22 21 48.	+ 33 12	84	14.5	GALAXY
UGC 12024	22 21 48.	+ 41 14	60	16.0	GALAXY S0
MCG+07-46-004	22 21 48.	+ 41 14	48	16.	GALAXY
PHL 5110	22 21 48.	- 10 14		18.0	BLUE STELLAR OBJECT
RNGC 7275	22 21 52.	+ 32 11		15.0	GALAXY

OBJECT NAME	RIGHT ASCEN.	DECLINATION	DIAM.	MAGN.	TYPE OF OBJECT
SC 2218-5218.1	22 21 52.	- 52 02 57.	18		NEBULA
RNGC 7273	22 21 53.	+ 35 57		15.0	GALAXY
ZWG 452.006	22 21 54.	+ 16 54		15.3	GALAXY
ZWG 473.013	22 21 54.	+ 21 58		15.6	GALAXY
ZWG 494.025	22 21 54.	+ 32 11		15.0	GALAXY
UGC 12025	22 21 54.	+ 32 11	60	15.0	GALAXY Sa
MCG+06-49-013	22 21 54.	+ 35 52	24	13.5	GALAXY
ZWG 514.024	22 21 54.	+ 35 57		14.8	GALAXY
MCG+06-49-012	22 21 54.	+ 35 57	24	14.5	GALAXY
HN 0102	22 21 54.	+ 50 43			NEBULA
PHL 5111	22 21 54.	- 02 22		17.2	BLUE STELLAR OBJECT
PK100-05.1	22 21 55.6	+ 50 42 52.	8	12.6	PLANETARY NEBULA
IC 5217	22 21 56.	+ 50 42 45.	3		PLANETARY NEBULA
SC 2218-6210.6	22 21 56.	- 61 55 27.	6		NEBULA
SC 2218-6004.4	22 21 57.	- 59 49 15.	24		NEBULA
RNGC 7276	22 21 59.	+ 35 50		15.0	GALAXY
RNGC 7274	22 21 59.	+ 35 53		14.0	GALAXY
PHL 5115	22 22 00.	+ 03 30		18.2	BLUE STELLAR OBJECT
ZC 2222.0+0950	22 22 00.	+ 09 50	1550		CLUSTER OF GALAXIES
2ZW 178	22 22 00.	+ 17 50			COMPACT GALAXY
ZWG 452.007	22 22 00.	+ 17 50		15.3	GALAXY
MCG+03-57-002	22 22 00.	+ 17 50	36	15.	GALAXY
ZWG 514.025	22 22 00.	+ 35 50		15.0	GALAXY
MCG+06-49-014	22 22 00.	+ 35 50	18	15.	GALAXY
ZWG 514.026	22 22 00.	+ 35 53		14.2	GALAXY
UGC 12026	22 22 00.	+ 35 53	102	14.2	GALAXY E
UGC 12027	22 22 00.	+ 41 00	60	17.	GALAXY
LDN 1201	22 22 00.	+ 63 15	360		DARK NEBULA
ZC 2222.0-0235	22 22 00.	- 02 35	1410		CLUSTER OF GALAXIES
PHL 5114	22 22 00.	- 04 51		17.9	BLUE STELLAR OBJECT
MCG-01-57-008	22 22 00.	- 05 01	60	15.	GALAXY
PHL 1878	22 22 00.	- 09 10		17.8	BLUE STELLAR OBJECT
PHL 5112	22 22 00.	- 09 30		16.8	BLUE STELLAR OBJECT
PHL 5113	22 22 00.	- 11 06		16.6	BLUE STELLAR OBJECT
PHL 5116	22 22 00.	- 15 41		18.5	BLUE STELLAR OBJECT
IC 5216	22 22 02.	- 18 20 28.			NONSTELLAR OBJECT
SC 2218-5735.8	22 22 04.	- 57 20 39.	12		NEBULA
SC 2218-5736.0	22 22 04.	- 57 20 51.	12		NEBULA
RNGC 7272	22 22 05.	+ 16 20		15.0	GALAXY
ZWG 452.008	22 22 06.	+ 16 20		15.0	GALAXY
UGC 12028	22 22 06.	+ 16 20	66	15.0	GALAXY SBa
MCG+05-52-019	22 22 06.	+ 32 11 30.	48	15.	GALAXY
PHL 5117	22 22 06.	- 05 23		16.9	BLUE STELLAR OBJECT
MCG-02-57-002	22 22 06.	- 13 58	36	15.	GALAXY
SC 2218-5640.1	22 22 07.	- 56 24 57.	12		NEBULA
SC 2219-1410.7	22 22 08.	- 13 55 32.	18		NEBULA
MCG+03-57-003	22 22 12.	+ 16 20	48	14.	GALAXY
MCG+03-57-004	22 22 12.	+ 17 50	48	17.	GALAXY
PHL 5118	22 22 12.	- 08 26		18.0	BLUE STELLAR OBJECT
MCG-02-57-003	22 22 12.	- 13 56 30.	36	15.	GALAXY
PHL 1879	22 22 12.	- 19 36		18.0	BLUE STELLAR OBJECT
MCG-05-52-072	22 22 12.	- 31 37	24	15.5	GALAXY
PHL 0252	22 22 12.	- 31 54		13.7	BLUE STELLAR OBJECT
SC 2218-5704.7	22 22 16.	- 56 49 33.	6		NEBULA
SC 2219-5713.1	22 22 17.	- 56 57 57.			NEBULA
UGC 12029	22 22 18.	+ 22 42	96	17.	GALAXY DWARF
ZC 2222.3-0205	22 22 18.	- 02 05	1340		CLUSTER OF GALAXIES
PHL 1880	22 22 18.	- 07 16		17.4	BLUE STELLAR OBJECT
PHL 1881	22 22 18.	- 10 30		17.8	BLUE STELLAR OBJECT
PHL 0253	22 22 18.	- 18 58		17.4	BLUE STELLAR OBJECT
PHL 5119	22 22 18.	- 28 22		16.8	BLUE STELLAR OBJECT
ZWG 404.013	22 22 24.	+ 09 15		15.1	GALAXY
PHL 5120	22 22 24.	- 02 59		14.1	BLUE STELLAR OBJECT
PHL 5121	22 22 24.	- 09 20		18.0	BLUE STELLAR OBJECT
SC 2219-1201.4	22 22 24.	- 11 46 14.	12		NEBULA
PHL 5122	22 22 24.	- 15 37		18.5	BLUE STELLAR OBJECT
SC 2218-6257.1	22 22 25.	- 62 41 56.	12		NEBULA
SC 2219-5559.5	22 22 26.	- 55 44 20.	18		NEBULA
8ZW 2222+07.5	22 22 30.	+ 07 29		17.7	COMPACT GALAXY
MCG+07-46-005	22 22 30.	+ 44 38	48	17.	GALAXY
UGC 12030	22 22 30.	+ 44 39	60	16.5	GALAXY Sc
PHL 1882	22 22 30.	- 09 37		18.0	BLUE STELLAR OBJECT
SC 2219-5939.9	22 22 30.	- 59 24 44.			NEBULA
SC 2219-6347.0	22 22 31.	- 63 31 50.	12		NEBULA
RNGC 7268	22 22 35.	- 31 40		16.0	GALAXY
4ZW 099	22 22 36.	+ 38 34			COMPACT GALAXY
DG 184	22 22 36.	+ 61 00	60		REFLECTION NEBULA
PHL 1884	22 22 36.	- 01 03		17.8	BLUE STELLAR OBJECT
PHL 0254	22 22 36.	- 02 50		16.7	BLUE STELLAR OBJECT
PHL 1883	22 22 36.	- 11 27		18.4	BLUE STELLAR OBJECT
PHL 5124	22 22 36.	- 19 14		18.1	BLUE STELLAR OBJECT
PHL 5123	22 22 36.	- 23 22		17.7	BLUE STELLAR OBJECT
MCG-05-52-073	22 22 36.	- 31 40	42	16.	GALAXY
SS 63	22 22 38.	+ 61 01			DIFFUSE GALACTIC NEBULA
PHL 0255	22 22 42.	+ 01 07		4.5	BLUE STELLAR OBJECT
PHL 1885	22 22 42.	+ 01 16		16.6	BLUE STELLAR OBJECT
ZWG 404.014	22 22 42.	+ 08 15		15.5	GALAXY
8ZW 2222+08.3	22 22 42.	+ 08 15		15.5	COMPACT GALAXY
ZWG 514.027	22 22 42.	+ 38 35		15.7	GALAXY
ASS 28	22 22 42.	+ 54 59	12600		OB ASSOCIATION CEP OB1
PHL 1886	22 22 42.	- 08 46		16.2	BLUE STELLAR OBJECT
PHL 5125	22 22 42.	- 11 06		18.2	BLUE STELLAR OBJECT
TON-S 0048	22 22 42.	- 27 52		14.8	BLUE STAR
SC 2219-5623.6	22 22 44.	- 56 08 26.			NEBULA
SC 2219-5558.9	22 22 47.	- 55 43 44.	30		NEBULA
SC 2219-6246.9	22 22 47.	- 62 31 44.	12		NEBULA
ZWG 404.015	22 22 48.	+ 08 09		15.7	GALAXY
PHL 5127	22 22 48.	- 09 42		18.6	BLUE STELLAR OBJECT
PHL 1887	22 22 48.	- 13 21		18.2	BLUE STELLAR OBJECT
PHL 0256	22 22 48.	- 16 54		18.4	BLUE STELLAR OBJECT
MCG-03-57-007	22 22 48.	- 17 30	72	14.	GALAXY
PHL 1888	22 22 48.	- 17 41		18.1	BLUE STELLAR OBJECT
PHL 5126	22 22 48.	- 20 31		17.7	BLUE STELLAR OBJECT
MCG-04-52-046	22 22 48.	- 24 29	54	14.5	GALAXY
MCG-04-52-045	22 22 48.	- 25 52 30.	36	15.	GALAXY
IC 1445	22 22 49.	- 17 30 14.			NONSTELLAR OBJECT
SC 2219-5625.4	22 22 51.	- 56 10 13.	6		NEBULA
SC 2219-5706.9	22 22 52.	- 56 51 43.	12		NEBULA
RNGC 7281	22 22 53.	+ 57 35			OPEN CLUSTER
OCL 0238	22 22 54.	+ 57 35			OPEN STAR CLUSTER
PHL 0257	22 22 54.	- 11 39		16.4	BLUE STELLAR OBJECT
PHL 0258	22 22 54.	- 16 00		16.9	BLUE STELLAR OBJECT
SC 2220-1027.1	22 22 58.	- 10 11 55.	30		NEBULA
SC 2219-5059.7	22 22 59.	- 50 44 31.			NEBULA
ZC 2223.0+0431	22 23 00.	+ 04 31	3430		CLUSTER OF GALAXIES
ZC 2223.0-0232	22 23 00.	- 02 32	8740		CLUSTER OF GALAXIES
PHL 0259	22 23 00.	- 02 38		17.8	BLUE STELLAR OBJECT
PHL 5128	22 23 00.	- 05 07		17.7	BLUE STELLAR OBJECT
PHL 5130	22 23 00.	- 09 49		17.2	BLUE STELLAR OBJECT
MCG-02-57-004	22 23 00.	- 10 12	108	15.	GALAXY
PHL 0260	22 23 00.	- 10 42		18.9	BLUE STELLAR OBJECT
PHL 1890	22 23 00.	- 11 22		18.6	BLUE STELLAR OBJECT
PHL 1889	22 23 00.	- 11 28		18.9	BLUE STELLAR OBJECT
PHL 5129	22 23 00.	- 15 24		16.5	BLUE STELLAR OBJECT
LB 01501	22 23 00.	- 49 25		14.7	FAINT BLUE STAR
SC 2219-6430.8	22 23 00.	- 64 15 37.	18		NEBULA
SG 2.089	22 23 03.	+ 63 19	1200		DIFFUSE EMISSION NEBULA
SC 2219-5326.2	22 23 03.	- 53 11 01.	60		NEBULA
BV 07	22 23 04.	- 11 29 00.		19.1	FAINT BLUE VARIABLE
KLEM 42	22 23 04.	- 31 25	1200	14.	GROUP OF 6 GALAXIES
SC 2219-5949.8	22 23 04.	- 59 34 37.	30		NEBULA
ZC 2223.1+0049	22 23 06.	+ 00 49	470		CLUSTER OF GALAXIES
UGC 12031	22 23 06.	+ 39 15	72	16.0	GALAXY Sb-c
PHL 5203	22 23 06.	- 02 44		18.1	BLUE STELLAR OBJECT
MCG-05-53-002	22 23 06.	- 31 30	9	16.	GALAXY
MCG-05-53-001	22 23 06.	- 31 30	15	14.5	GALAXY
ARC 2441	22 23 07.	- 03 32		17.2	RICH CLUSTER OF GALAXIES
SC 2220-1340.2	22 23 08.	- 13 25 00.	42		NEBULA
RNGC 7269	22 23 10.	- 13 25		14.0	GALAXY
ARC 2442	22 23 11.	- 06 50		17.2	RICH CLUSTER OF GALAXIES
BC 3C446	22 23 11.05	- 05 12 17.0		18.39	QUASI-STELLAR OBJECT
SWB 375	22 23 11.1	- 05 12 17.		18.4	QUASI-STELLAR OBJECT
PHL 5133	22 23 12.	+ 01 58		17.8	BLUE STELLAR OBJECT
ZWG 378.004	22 23 12.	+ 02 58		15.6	GALAXY
ZWG 429.005	22 23 12.	+ 11 24		15.6	GALAXY
ZC 2223.2+1207	22 23 12.	+ 12 07	940		CLUSTER OF GALAXIES
ZWG 531.005	22 23 12.	+ 42 27		15.3	GALAXY
UGC 12032	22 23 12.	+ 42 27	132	15.3	GALAXY S
MCG+07-46-006	22 23 12.	+ 42 27	114	15.	GALAXY
PHL 5134	22 23 12.	- 01 19		14.4	BLUE STELLAR OBJECT
PHL 5132	22 23 12.	- 04 23		16.2	BLUE STELLAR OBJECT
PHL 5135	22 23 12.	- 08 48		18.7	BLUE STELLAR OBJECT
PHL 0261	22 23 12.	- 10 14		16.6	BLUE STELLAR OBJECT
MCG-02-57-005	22 23 12.	- 13 25	48	14.5	GALAXY
PHL 1891	22 23 12.	- 14 18		18.2	BLUE STELLAR OBJECT
PHL 1892	22 23 12.	- 18 36		18.5	BLUE STELLAR OBJECT
PHL 5131	22 23 12.	- 29 36		14.4	BLUE STELLAR OBJECT
MCG-04-53-001	22 23 15.	- 25 01	48	15.	GALAXY
ZWG 404.016	22 23 18.	+ 07 36		15.5	GALAXY
ZWG 514.028	22 23 18.	+ 35 56		15.6	GALAXY
PHL 1893	22 23 18.	- 11 05		18.8	BLUE STELLAR OBJECT
PHL 5136	22 23 18.	- 26 54		17.3	BLUE STELLAR OBJECT
SC 2220-5238.4	22 23 18.	- 52 23 13.			NEBULA
MCG+06-49-015	22 23 21.	+ 39 02	42	15.5	GALAXY
SC 2220-5709.0	22 23 21.	- 56 53 49.	6		NEBULA
SC 2219-6432.8	22 23 23.	- 64 17 37.	12		NEBULA
PHL 1894	22 23 24.	+ 02 57		18.0	BLUE STELLAR OBJECT
PHL 5137	22 23 24.	- 04 40		17.0	BLUE STELLAR OBJECT
PHL 5138	22 23 24.	- 11 14		17.8	BLUE STELLAR OBJECT
PHL 0262	22 23 24.	- 12 44		17.4	BLUE STELLAR OBJECT
PHL 0263	22 23 24.	- 15 16		18.2	BLUE STELLAR OBJECT
PHL 5139	22 23 24.	- 17 45		18.4	BLUE STELLAR OBJECT
PHL 1895	22 23 24.	- 18 19		18.5	BLUE STELLAR OBJECT
HN 0753	22 23 25.	- 66 12			NEBULA
IC 5215	22 23 25.	- 66 12			NONSTELLAR OBJECT
ZWG 378.005	22 23 30.	+ 00 57		15.7	GALAXY
ZWG 404.017	22 23 30.	+ 08 11		15.6	GALAXY
ZWG 452.009	22 23 30.	+ 17 24		15.6	GALAXY
KARA.73B 0965	22 23 30.	+ 17 24	36	15.6	ISOLATED GALAXY S
MCG-05-53-003	22 23 30.	- 31 10	36	15.5	GALAXY
SC 2220-5636.8	22 23 32.	- 56 21 36.	6		NEBULA
SG 3.231	22 23 35.	+ 59 36	1500		DIFFUSE EMISSION NEBULA
RNGC 7277	22 23 35.	- 31 26		14.0	GALAXY
ZC 2223.6+1344	22 23 36.	+ 13 44	470		CLUSTER OF GALAXIES
4ZW 100	22 23 36.	+ 29 51			COMPACT GALAXY
UGC 12033	22 23 36.	+ 46 49	60	17.	GALAXY
PHL 0264	22 23 36.	- 09 00		17.2	BLUE STELLAR OBJECT
PHL 0265	22 23 36.	- 12 00		18.7	BLUE STELLAR OBJECT
PHL 0266	22 23 36.	- 25 30		17.8	BLUE STELLAR OBJECT
PHL 5141	22 23 36.	- 27 00		17.0	BLUE STELLAR OBJECT
PHL 5140	22 23 36.	- 29 18		16.7	BLUE STELLAR OBJECT
MCG-05-53-004	22 23 36.	- 31 26	60	14.	GALAXY
PHL 0267	22 23 36.	- 32 16		17.6	BLUE STELLAR OBJECT
SC 2220-1229.8	22 23 37.	- 12 14 36.	6		NEBULA
SC 2220-5108.7	22 23 39.	- 50 53 30.			NEBULA
ZWG 378.006	22 23 42.	+ 00 25		15.6	GALAXY
ARC 2443	22 23 42.	+ 17 06		16.5	RICH CLUSTER OF GALAXIES
ISS 0088	22 23 42.	+ 59 26	292		STELLAR RING
MCG-01-57-009	22 23 42.	- 04 35	42	15.	GALAXY
PHL 1896	22 23 42.	- 10 01		18.9	BLUE STELLAR OBJECT
MCG+07-46-007	22 23 45.	+ 40 03	144	14.	GALAXY
SC 2220-5351.1	22 23 47.	- 53 35 54.			NEBULA
PHL 5142	22 23 48.	+ 00 57		18.0	BLUE STELLAR OBJECT
ZWG 514.029	22 23 48.	+ 35 42		15.7	GALAXY
ZWG 531.006	22 23 48.	+ 40 03		15.3	GALAXY
UGC 12034	22 23 48.	+ 40 03	156	15.3	GALAXY SBb
PHL 0268	22 23 48.	- 06 43		17.8	BLUE STELLAR OBJECT
PHL 5143	22 23 48.	- 08 33		17.2	BLUE STELLAR OBJECT
MCG-07-46-002	22 23 48.	- 41 50	30	15.5	GALAXY
RNGC 7282	22 23 49.	+ 40 03		15.5	GALAXY
SC 2220-5147.4	22 23 52.	- 51 32 12.			NEBULA
ZWG 378.007	22 23 54.	+ 03 13		15.7	GALAXY
ZWG 452.010	22 23 54.	+ 19 19		15.6	GALAXY
BRSL 107+05/1	22 23 54.	+ 64 03	5400		HII REGION
MCG-01-57-010	22 23 54.	- 06 39	72	16.	GALAXY
PHL 1897	22 23 54.	- 08 34		16.0	BLUE STELLAR OBJECT
PHL 0269	22 23 54.	- 08 36		19.0	BLUE STELLAR OBJECT
PHL 5145	22 23 54.	- 17 36		18.3	BLUE STELLAR OBJECT
PHL 1898	22 23 54.	- 18 04		18.5	BLUE STELLAR OBJECT
MCG-04-53-002	22 23 54.	- 25 09	72	14.	GALAXY
PHL 5144	22 23 54.	- 27 39		18.0	BLUE STELLAR OBJECT
SC 2220-5600.3	22 23 58.	- 55 45 05.	30		NEBULA
RNGC 7280	22 23 59.	+ 15 54		13.5	GALAXY
KHAV 770	22 24	+ 60 45	4860		DARK NEBULA
LB 09947	22 24	- 81 11		13.7	FAINT BLUE STAR
ZC 2224.0+1236	22 24 00.	+ 12 36	1610		CLUSTER OF GALAXIES
MCG+03-57-005	22 24 00.	+ 15 54	120	13.	GALAXY
MCG+07-46-008	22 24 00.	+ 42 33	39	17.	GALAXY
PHL 1899	22 24 00.	- 01 12		18.0	BLUE STELLAR OBJECT
PHL 1900	22 24 00.	- 02 12		18.0	BLUE STELLAR OBJECT
PHL 1901	22 24 00.	- 04 49		18.5	BLUE STELLAR OBJECT
PHL 5146	22 24 00.	- 09 09		18.2	BLUE STELLAR OBJECT
PHL 5147	22 24 06.	+ 02 38		14.4	BLUE STELLAR OBJECT
ZWG 452.011	22 24 06.	+ 15 54		13.6	GALAXY
UGC 12035	22 24 06.	+ 15 54	120	13.6	GALAXY S0/Sa
KARA.72 568A	22 24 06.	+ 15 54	108	13.6	PART OF DOUBLE GALAXY

OBJECT NAME	RIGHT ASCEN.	DECLINATION	DIAM.	MAGN.	TYPE OF OBJECT
MCG-01-57-011	22 24 06.	- 03 57	48	16.	GALAXY
PHL 5148	22 24 06.	- 10 27		17.5	BLUE STELLAR OBJECT
SC 2220-5722.8	22 24 06.	- 57 07 35.	6		NEBULA
SC 2221-5109.1	22 24 09.	- 50 53 53.			NEBULA
SC 2220-5552.2	22 24 10.	- 55 36 59.	6		NEBULA
ZC 2224.2+0109	22 24 12.	+ 01 09	1210		CLUSTER OF GALAXIES
PHL 0270	22 24 12.	+ 03 22		17.8	BLUE STELLAR OBJECT
ZC 2224.2+1651	22 24 12.	+ 16 51	2960		CLUSTER OF GALAXIES
ZWG 474.001	22 24 12.	+ 24 52		15.0	GALAXY
UGC 12036	22 24 12.	+ 24 52	66	15.0	GALAXY E-S0
PHL 0271	22 24 12.	- 13 09		18.4	BLUE STELLAR OBJECT
PHL 5152	22 24 12.	- 14 04		17.2	BLUE STELLAR OBJECT
PHL 5149	22 24 12.	- 19 50		17.9	BLUE STELLAR OBJECT
PHL 0272	22 24 12.	- 21 36		18.0	BLUE STELLAR OBJECT
PHL 5151	22 24 12.	- 27 42		18.2	BLUE STELLAR OBJECT
PHL 5150	22 24 12.	- 30 18		17.0	BLUE STELLAR OBJECT
RNGC 7279	22 24 16.	- 35 24			GALAXY
SC 2221-5604.2	22 24 16.	- 55 48 59.	6		NEBULA
SC 2220-6136.5	22 24 17.	- 61 21 17.	6		NEBULA
MCG+03-57-006	22 24 18.	+ 15 56	60	15.5	GALAXY
MCG+04-53-001	22 24 18.	+ 24 50	30	15.	GALAXY
ZWG 514.030	22 24 18.	+ 36 29		15.5	GALAXY
UGC 12037	22 24 18.	+ 36 29	90	15.5	GALAXY SBb
MCG+07-46-009	22 24 18.	+ 39 53	45	15.5	GALAXY
ZWG 531.007	22 24 18.	+ 39 54		15.6	GALAXY
PHL 0273	22 24 18.	- 01 28		17.7	BLUE STELLAR OBJECT
PHL 1902	22 24 18.	- 05 09		18.3	BLUE STELLAR OBJECT
PHL 5153	22 24 18.	- 11 22		17.5	BLUE STELLAR OBJECT
PHL 5156	22 24 18.	- 15 18		17.9	BLUE STELLAR OBJECT
PHL 5154	22 24 18.	- 15 36		18.7	BLUE STELLAR OBJECT
PHL 5155	22 24 18.	- 27 44		18.0	BLUE STELLAR OBJECT
MCG-05-53-005	22 24 18.	- 31 11	36	16.	GALAXY
MIN.47 17	22 24 19.	+ 54 34			DIFFUSE NEBULA
SG 3.232	22 24 19.	+ 54 34	30		DIFFUSE EMISSION NEBULA
PK102-02.1	22 24 21.4	+ 54 34 20.	130	15.8	PLANETARY NEBULA
ZWG 452.012	22 24 24.	+ 15 55		15.7	GALAXY
KARA.72 568B	22 24 24.	+ 15 55	60	15.7	PART OF DOUBLE GALAXY
ZC 2224.4+2221	22 24 24.	+ 22 21	540		CLUSTER OF GALAXIES
ZWG 514.031	22 24 24.	+ 34 48		15.5	GALAXY
MCG+06-49-017	22 24 24.	+ 34 48	36	15.5	GALAXY
MCG+06-49-016	22 24 24.	+ 36 28	72	15.	GALAXY
PHL 5158	22 24 24.	- 09 50		17.0	BLUE STELLAR OBJECT
PHL 1903	22 24 24.	- 12 44		18.5	BLUE STELLAR OBJECT
PHL 1904	22 24 24.	- 27 00		17.8	BLUE STELLAR OBJECT
PHL 5159	22 24 24.	- 27 47		18.3	BLUE STELLAR OBJECT
PHL 5157	22 24 24.	- 28 44		17.3	BLUE STELLAR OBJECT
ARC 2445	22 24 27.	+ 25 35		17.8	RICH CLUSTER OF GALAXIES
PHL 5160	22 24 30.	+ 01 26		17.5	BLUE STELLAR OBJECT
ZWG 452.013	22 24 30.	+ 19 17		15.1	GALAXY
UGC 12038	22 24 30.	+ 19 17	78	15.1	GALAXY Sb
MCG+03-57-007	22 24 30.	+ 19 18	72	14.5	GALAXY
MCG+06-49-018	22 24 30.	+ 35 15 30.	90	14.	GALAXY
ZWG 514.032	22 24 30.	+ 35 17		14.1	GALAXY
UGC 12039	22 24 30.	+ 35 17	102	14.1	GALAXY SBb
PHL 5161	22 24 30.	- 11 09		17.8	BLUE STELLAR OBJECT
TON-S 0049	22 24 30.	- 28 43		15.8	BLUE STAR
SC 2221-5525.8	22 24 30.	- 55 10 34.	6		NEBULA
SG 3.233	22 24 31.	+ 64 18	3000		DIFFUSE EMISSION NEBULA
SC 2221-1310.5	22 24 33.	- 12 55 16.	30		NEBULA
SC 2221-5550.4	22 24 34.	- 55 35 10.	6		NEBULA
PHL 5165	22 24 36.	+ 00 10		17.7	BLUE STELLAR OBJECT
ZC 2224.6+0238	22 24 36.	+ 02 38	870		CLUSTER OF GALAXIES
ZC 2224.6+2532	22 24 36.	+ 25 32	1080		CLUSTER OF GALAXIES
PHL 0274	22 24 36.	- 09 45		16.6	BLUE STELLAR OBJECT
PHL 1905	22 24 36.	- 10 00		18.5	BLUE STELLAR OBJECT
PHL 5162	22 24 36.	- 11 01		15.8	BLUE STELLAR OBJECT
PHL 0275	22 24 36.	- 20 53		18.0	BLUE STELLAR OBJECT
PHL 5164	22 24 36.	- 25 42		17.8	BLUE STELLAR OBJECT
PHL 5163	22 24 36.	- 29 25		16.0	BLUE STELLAR OBJECT
LB 01502	22 24 36.	- 45 55		12.2	FAINT BLUE STAR
SC 2221-5514.8	22 24 37.	- 54 59 34.	12		NEBULA
SC 2221-5056.4	22 24 39.	- 50 41 10.	60		NEBULA
SC 2221-5753.4	22 24 40.	- 57 38 10.	6		NEBULA
MCG+07-46-010	22 24 42.	+ 39 31	63	16.	GALAXY
MCG-01-57-012	22 24 42.	- 07 29	54	16.	GALAXY
PHL 5166	22 24 42.	- 08 48		18.1	BLUE STELLAR OBJECT
MCG-02-57-006	22 24 42.	- 12 54	60	15.	GALAXY
PHL 5167	22 24 42.	- 15 27		18.0	BLUE STELLAR OBJECT
PHL 0276	22 24 42.	- 20 47		18.5	BLUE STELLAR OBJECT
TON-S 0050	22 24 42.	- 26 39		14.8	BLUE STAR
ARC 2444	22 24 46.	- 24 05		17.9	RICH CLUSTER OF GALAXIES
SC 2221-5839.8	22 24 46.	- 58 24 34.	6		NEBULA
IC 5218	22 24 46.	- 60 39			NONSTELLAR OBJECT
HN 1230	22 24 47.	- 60 39	30		NEBULA
PHL 5168	22 24 48.	+ 00 16		17.2	BLUE STELLAR OBJECT
ZWG 514.033	22 24 48.	+ 36 06		14.6	GALAXY
UGC 12040	22 24 48.	+ 36 06	84	14.6	GALAXY SB0
MCG+06-49-019	22 24 48.	+ 36 06 30.	42	14.5	GALAXY
UGC 12041	22 24 48.	+ 39 32	72	16.0	GALAXY S
ZWG 378.008	22 24 48.	- 01 36		15.6	GALAXY
MCG-05-53-006	22 24 48.	- 31 40	18	16.	GALAXY
PHL 5169	22 24 48.	- 32 02		17.0	BLUE STELLAR OBJECT
HN 1231	22 24 51.	- 59 59	42		NEBULA
IC 5220	22 24 52.	- 59 59			NONSTELLAR OBJECT
8ZW 2224+05.0	22 24 54.	+ 05 03		17.5	COMPACT GALAXY
ZWG 495.001	22 24 54.	+ 31 15		15.4	GALAXY
PHL 5174	22 24 54.	- 02 30		18.1	BLUE STELLAR OBJECT
PHL 5170	22 24 54.	- 02 36		17.9	BLUE STELLAR OBJECT
PHL 5171	22 24 54.	- 10 34		17.8	BLUE STELLAR OBJECT
PHL 0277	22 24 54.	- 17 28		17.8	BLUE STELLAR OBJECT
PHL 5173	22 24 54.	- 26 40		16.1	BLUE STELLAR OBJECT
PHL 5172	22 24 54.	- 31 46		18.0	BLUE STELLAR OBJECT
KHAV 771	22 25	+ 59 27	12390		DARK NEBULA
5ZW 385	22 25 00.	+ 51 15			COMPACT GALAXY
LDN 1180	22 25 00.	+ 59 00	10740		DARK NEBULA
COU 093	22 25 00.	+ 59 40	3600		EMISSION NEBULA
LDN 1196	22 25 00.	+ 61 00	660		DARK NEBULA
LDN 1195	22 25 00.	+ 61 00	480		DARK NEBULA
LDN 1202	22 25 00.	+ 62 50	240		DARK NEBULA
LDN 1204	22 25 00.	+ 63 00	6420		DARK NEBULA
COU 094	22 25 00.	+ 64 00	7200		EMISSION NEBULA
PHL 5176	22 25 00.	- 19 38		17.9	BLUE STELLAR OBJECT
PHL 5175	22 25 00.	- 26 36		17.5	BLUE STELLAR OBJECT
SC 2221-5344.6	22 25 00.	- 53 29 22.			NEBULA
RNGC 7278	22 25 02.	- 60 26			UNVERIFIED SOUTHERN OBJECT
MCG+05-53-001	22 25 06.	+ 31 16	60	15.	GALAXY
PHL 1906	22 25 06.	- 25 08		18.2	BLUE STELLAR OBJECT
MCG-05-53-007	22 25 06.	- 30 19	48	15.	GALAXY
PHL 1908	22 25 12.	+ 00 13		17.7	BLUE STELLAR OBJECT
PHL 5179	22 25 12.	- 02 16		16.2	BLUE STELLAR OBJECT
PHL 5182	22 25 12.	- 11 21		18.7	BLUE STELLAR OBJECT
PHL 1907	22 25 12.	- 11 22		18.4	BLUE STELLAR OBJECT
PHL 1909	22 25 12.	- 11 40		18.3	BLUE STELLAR OBJECT
PHL 5177	22 25 12.	- 12 30		16.0	BLUE STELLAR OBJECT
PHL 5180	22 25 12.	- 19 24		17.9	BLUE STELLAR OBJECT
PHL 5181	22 25 12.	- 23 00		17.3	BLUE STELLAR OBJECT
PHL 5178	22 25 12.	- 30 00		17.5	BLUE STELLAR OBJECT
HN 0754	22 25 12.	- 66 08			NEBULA
IC 5219	22 25 12.	- 66 08			NONSTELLAR OBJECT
MCG-02-57-007	22 25 15.	- 09 59	60	15.	GALAXY
SC 2222-5622.0	22 25 16.	- 56 06 45.	9		NEBULA
SC 2221-6346.4	22 25 17.	- 63 31 09.	18		NEBULA
PHL 1910	22 25 18.	+ 02 44		18.2	BLUE STELLAR OBJECT
PHL 5183	22 25 18.	- 10 46		17.6	BLUE STELLAR OBJECT
IC 5221	22 25 23.	- 66 09			NONSTELLAR OBJECT
PHL 1912	22 25 24.	- 09 03		18.4	BLUE STELLAR OBJECT
PHL 5185	22 25 24.	- 10 11		18.0	BLUE STELLAR OBJECT
PHL 5186	22 25 24.	- 12 19		18.6	BLUE STELLAR OBJECT
PHL 1911	22 25 24.	- 13 48		18.7	BLUE STELLAR OBJECT
PHL 1913	22 25 24.	- 15 10		18.0	BLUE STELLAR OBJECT
PHL 5184	22 25 24.	- 17 25		17.5	BLUE STELLAR OBJECT
SC 2222-6127.8	22 25 24.	- 61 12 33.	12		NEBULA
HN 0755	22 25 24.	- 66 09			NEBULA
SC 2222-5633.2	22 25 27.	- 56 17 57.	12		NEBULA
ZC 2225.5+1515	22 25 30.	+ 15 15	1280		CLUSTER OF GALAXIES
4ZW 101	22 25 30.	+ 28 39			COMPACT GALAXY
UGC 12042	22 25 30.	+ 37 12	66	16.5	GALAXY
LDN 1179	22 25 30.	+ 58 47	180		DARK NEBULA
PHL 1914	22 25 30.	- 06 36		17.4	BLUE STELLAR OBJECT
PHL 5187	22 25 30.	- 26 04		16.9	BLUE STELLAR OBJECT
MCG-05-53-008	22 25 30.	- 30 11	30	15.	GALAXY
SC 2222-5543.9	22 25 30.	- 55 28 39.	9		NEBULA
SC 2222-5525.6	22 25 32.	- 55 10 21.	18		NEBULA
ZWG 495.002	22 25 36.	+ 28 51		13.4	GALAXY
UGC 12043	22 25 36.	+ 28 51	114	13.4	GALAXY S0-a
ZWG 514.034	22 25 36.	+ 38 20		15.5	GALAXY
UGC 12044	22 25 36.	+ 38 20	108	15.5	GALAXY Sa-b
MCG+06-49-020	22 25 36.	+ 38 20	120	15.	GALAXY
PHL 5190	22 25 36.	- 02 48		18.3	BLUE STELLAR OBJECT
PHL 5189	22 25 36.	- 09 14		16.6	BLUE STELLAR OBJECT
PHL 0278	22 25 36.	- 11 59		18.8	BLUE STELLAR OBJECT
PHL 1915	22 25 36.	- 13 43		17.9	BLUE STELLAR OBJECT
PHL 5188	22 25 36.	- 19 58		15.7	BLUE STELLAR OBJECT
MCG-04-53-003	22 25 36.	- 22 42 30.	24	15.	GALAXY
SC 2222-6123.3	22 25 36.	- 61 08 03.	18		NEBULA
RNGC 7287	22 25 38.	- 22 41		15.0	GALAXY
RNGC 7286	22 25 39.	+ 28 51		13.5	GALAXY
MCG+05-53-002	22 25 42.	+ 28 51	90	14.5	GALAXY
ISS 0135	22 25 42.	+ 52 02	187		STELLAR RING
RNGC 7288	22 25 42.	- 03 07		14.5	GALAXY
MCG-01-57-013	22 25 42.	- 03 07	120	14.5	GALAXY
PHL 0279	22 25 42.	- 05 20		17.9	BLUE STELLAR OBJECT
PHL 1916	22 25 42.	- 08 15		17.9	BLUE STELLAR OBJECT
PHL 5191	22 25 42.	- 15 00		18.3	BLUE STELLAR OBJECT
PHL 1917	22 25 48.	+ 03 29		18.2	BLUE STELLAR OBJECT
8ZW 2225+05.6	22 25 48.	+ 05 35		19.5	COMPACT GALAXY
4ZW 102	22 25 48.	+ 22 11			COMPACT GALAXY
PHL 1918	22 25 48.	- 06 44		18.5	BLUE STELLAR OBJECT
PHL 5192	22 25 48.	- 08 00		11.5	BLUE STELLAR OBJECT
PHL 5195	22 25 48.	- 11 46		17.6	BLUE STELLAR OBJECT
PHL 5196	22 25 48.	- 14 34		18.2	BLUE STELLAR OBJECT
PHL 5194	22 25 48.	- 15 53		18.2	BLUE STELLAR OBJECT
PHL 5197	22 25 48.	- 17 48		18.9	BLUE STELLAR OBJECT
MCG-04-53-004	22 25 48.	- 25 07	24	13.5	GALAXY
PHL 5193	22 25 48.	- 28 10		17.5	BLUE STELLAR OBJECT
RNGC 7284	22 25 49.	- 25 07		13.0	GALAXY
SHB 376	22 25 50.6	- 05 20 36.		18.0	QUASI-STELLAR OBJECT
VV 074B	22 25 51.	- 25 07	60	14.	INTERACTING GALAXY
VV 074A	22 25 51.	- 25 07	150	14.	INTERACTING GALAXY
MCG-04-53-005	22 25 51.	- 25 07	120	13.5	GALAXY
ZWG 514.035	22 25 54.	+ 35 05		15.6	GALAXY
PHL 5200	22 25 54.	- 05 34		18.2	BLUE STELLAR OBJECT
PHL 5198	22 25 54.	- 10 32		18.5	BLUE STELLAR OBJECT
PHL 5199	22 25 54.	- 11 44		17.7	BLUE STELLAR OBJECT
ARP 093	22 25 54.	- 25 06			PECULIAR GALAXY
BC PHL5200	22 25 54.02	- 05 34 16.6		17.70	QUASI-STELLAR OBJECT
RNGC 7285	22 25 55.	- 25 07		13.0	GALAXY
RNGC 7290	22 25 59.	+ 16 54		14.0	GALAXY
LBN 0142	22 26	+ 20 20	6900		BRIGHT NEBULA
PHL 0280	22 26 00.	+ 03 22		18.4	BLUE STELLAR OBJECT
MCG+03-57-008	22 26 00.	+ 16 33	90	14.5	GALAXY
ZWG 452.014	22 26 00.	+ 16 53		13.8	GALAXY
UGC 12045	22 26 00.	+ 16 53	102	13.8	GALAXY Sb
MCG+03-57-009	22 26 00.	+ 16 53	102	13.	GALAXY
MCG+03-57-010	22 26 00.	+ 19 17	42	15.	GALAXY
UGC 12046	22 26 00.	+ 29 27	72	16.0	GALAXY Sb
LDN 1203	22 26 00.	+ 62 45	480		DARK NEBULA
PHL 1919	22 26 00.	- 06 28		17.3	BLUE STELLAR OBJECT
PHL 0281	22 26 00.	- 10 05		17.2	BLUE STELLAR OBJECT
PHL 5202	22 26 00.	- 16 14		17.7	BLUE STELLAR OBJECT
PHL 5201	22 26 00.	- 27 51		17.8	BLUE STELLAR OBJECT
KARA.73 53	22 26 02.	- 35 08	67		DWARF GALAXY
SC 2222-5527.4	22 26 03.	- 55 12 08.	9		NEBULA
SN 1964H	22 26 04.	+ 30 03		14.8	SUPERNOVA
RNGC 7291	22 26 05.	+ 16 31		15.0	GALAXY
ZC 2226.1+0718	22 26 06.	+ 07 18	1410		CLUSTER OF GALAXIES
ZWG 429.006	22 26 06.	+ 11 47		15.4	GALAXY
KARA.73B 0966	22 26 06.	+ 11 47	24	15.4	ISOLATED GALAXY S
ZWG 452.015	22 26 06.	+ 16 31		14.8	GALAXY
UGC 12047	22 26 06.	+ 16 31	78	14.8	GALAXY S0
ZWG 452.016	22 26 06.	+ 19 15		15.4	GALAXY
ZWG 495.003	22 26 06.	+ 30 03		13.1	GALAXY
UGC 12048	22 26 06.	+ 30 03	138	13.1	GALAXY IRR
KARA.73B 0967	22 26 06.	+ 30 03	132	13.1	ISOLATED GALAXY IR
PHL 5205	22 26 06.	- 07 51		17.1	BLUE STELLAR OBJECT
PHL 5204	22 26 06.	- 12 06		18.2	BLUE STELLAR OBJECT
RNGC 7292	22 26 09.	+ 30 03		13.0	GALAXY
MCG-03-57-008	22 26 09.	- 15 20	96	14.	GALAXY
RNGC 7283	22 26 11.	+ 17 14		15.0	GALAXY
PHL 0282	22 26 12.	+ 01 50		13.4	BLUE STELLAR OBJECT
MCG+03-57-011	22 26 12.	+ 16 41	42	15.	GALAXY
ZWG 452.017	22 26 12.	+ 17 13		15.5	GALAXY
MCG+03-57-012	22 26 12.	+ 17 14	45	15.	GALAXY
ZWG 514.036	22 26 12.	+ 33 45		14.9	GALAXY
OCL 0228	22 26 12.	+ 52 02	360	10.0	OPEN STAR CLUSTER
RNGC 7296	22 26 12.	+ 52 02		9.5	OPEN CLUSTER

OBJECT NAME	RIGHT ASCEN.	DECLINATION	DIAM.	MAGN.	TYPE OF OBJECT
PHL 1922	22 26 12.	- 02 47		18.5	BLUE STELLAR OBJECT
PHL 5206	22 26 12.	- 05 26		15.9	BLUE STELLAR OBJECT
PHL 1920	22 26 12.	- 12 25		18.9	BLUE STELLAR OBJECT
PHL 1923	22 26 12.	- 13 30		18.3	BLUE STELLAR OBJECT
PHL 1921	22 26 12.	- 21 18		18.7	BLUE STELLAR OBJECT
PHL 5207	22 26 12.	- 31 22		16.4	BLUE STELLAR OBJECT
ZWG 378.009	22 26 18.	+ 01 38		15.3	GALAXY
MCG+00-57-002	22 26 18.	+ 01 39	42	14.5	GALAXY
ZWG 452.018	22 26 18.	+ 16 40		15.5	GALAXY
ZWG 474.002	22 26 18.	+ 22 27		15.7	GALAXY
PHL 1924	22 26 18.	- 14 16		17.2	BLUE STELLAR OBJECT
PHL 1925	22 26 18.	- 15 04		18.2	BLUE STELLAR OBJECT
PHL 5208	22 26 18.	- 15 27		17.9	BLUE STELLAR OBJECT
SC 2223-5720.4	22 26 18.	- 57 05 07.	12		NEBULA
MCG+05-53-003	22 26 21.	+ 30 04	108	13.5	GALAXY
RNGC 7289	22 26 22.	- 35 43			GALAXY
SC 2223-5710.4	22 26 22.	- 56 55 07.	6		NEBULA
HN 0756	22 26 22.	- 65 54			NEBULA
IC 5222	22 26 22.	- 65 54			NONSTELLAR OBJECT
SC 2223-5724.3	22 26 23.	- 57 09 01.	9		NEBULA
PHL 0283	22 26 24.	+ 00 48		18.0	BLUE STELLAR OBJECT
PHL 1927	22 26 24.	+ 01 10		17.9	BLUE STELLAR OBJECT
ZC 2225.4+1404	22 26 24.	+ 14 04	1810		CLUSTER OF GALAXIES
UGC 12049	22 26 24.	+ 16 45	66	17.	GALAXY DWRF IR
RNGC 7295	22 26 24.	+ 52 34			NON-EXISTENT OBJECT
PHL 1928	22 26 24.	- 00 12		18.5	BLUE STELLAR OBJECT
PHL 5211	22 26 24.	- 05 32		18.3	BLUE STELLAR OBJECT
PHL 5209	22 26 24.	- 08 38		17.9	BLUE STELLAR OBJECT
PHL 5210	22 26 24.	- 10 02		17.8	BLUE STELLAR OBJECT
PHL 5212	22 26 24.	- 15 42		18.4	BLUE STELLAR OBJECT
MCG-03-57-009	22 26 24.	- 19 25 30.	24	16.	GALAXY
PHL 1926	22 26 24.	- 28 03		18.4	BLUE STELLAR OBJECT
ZWG 452.019	22 26 30.	+ 19 48		15.6	GALAXY
KARA.73B 0968	22 26 30.	+ 19 48	24	15.6	ISOLATED GALAXY S
OCL 0242	22 26 30.	+ 58 52	360	15.	OPEN STAR CLUSTER
PHL 1929	22 26 30.	- 18 40		17.8	BLUE STELLAR OBJECT
SC 2223-5711.6	22 26 30.	- 56 56 19.	12		NEBULA
SC 2223-5719.4	22 26 31.	- 57 04 07.	12		NEBULA
IC 1446	22 26 32.	- 01 27 25.			NONSTELLAR OBJECT
SC 2223-5717.2	22 26 32.	- 57 01 55.			NEBULA
SC 2223-6131.6	22 26 32.	- 61 16 19.	12		NEBULA
SC 2223-5717.6	22 26 34.	- 57 02 19.	12		NEBULA
ZWG 452.020	22 26 36.	+ 18 51		15.0	GALAXY
UGC 12050	22 26 36.	+ 18 51	60	15.0	GALAXY S
MCG+03-57-013	22 26 36.	+ 18 52	48	14.5	GALAXY
ZC 2226.6+2844	22 26 36.	+ 28 44	2890		CLUSTER OF GALAXIES
ZWG 514.037	22 26 36.	+ 38 50		15.0	GALAXY
MCG+06-49-021	22 26 36.	+ 38 50	36	14.5	GALAXY
PHL 1930	22 26 36.	- 00 59		16.0	BLUE STELLAR OBJECT
PHL 0284	22 26 36.	- 09 04		18.9	BLUE STELLAR OBJECT
PHL 1931	22 26 36.	- 16 32		18.4	BLUE STELLAR OBJECT
4ZW 103	22 26 42.	+ 28 40			COMPACT GALAXY
MPSL 106+03/1	22 26 42.	+ 61 23	300		HII REGION
DG 185	22 26 42.	+ 62 45	60		REFLECTION NEBULA
PHL 5213	22 26 42.	- 26 52		18.4	BLUE STELLAR OBJECT
SC 2223-6120.3	22 26 44.	- 61 05 01.	12		NEBULA
MCG+01-57-005	22 26 48.	+ 07 27	78	14.	GALAXY
PHL 0285	22 26 48.	- 05 30		18.0	BLUE STELLAR OBJECT
PHL 1933	22 26 48.	- 10 06		16.9	BLUE STELLAR OBJECT
PHL 1932	22 26 48.	- 12 35		18.4	BLUE STELLAR OBJECT
PHL 0286	22 26 48.	- 16 40		18.7	BLUE STELLAR OBJECT
PHL 0287	22 26 48.	- 21 06		13.0	BLUE STELLAR OBJECT
SG 3.234	22 26 51.	+ 61 23			DIFFUSE EMISSION NEBULA
MCG+06-49-022	22 26 54.	+ 38 17	66	16.	GALAXY
PHL 5214	22 26 54.	- 04 04		18.1	BLUE STELLAR OBJECT
PHL 0288	22 26 54.	- 13 24		17.4	BLUE STELLAR OBJECT
MCG-03-57-010	22 26 54.	- 20 06	36	15.	GALAXY
PK036-57.1	22 26 55.8	- 21 03 22.	1150		PLANETARY NEBULA
RNGC 7293	22 26 56.	- 21 03			PLANETARY NEBULA
SC 2223-5613.6	22 26 59.	- 55 58 18.	12		NEBULA
LBN 0509	22 27	+ 61 20	300		BRIGHT NEBULA
LBN 0510	22 27	+ 61 25	180		BRIGHT NEBULA
LBN 0532	22 27	+ 68 50	1200		BRIGHT NEBULA
PHL 0290	22 27 00.	+ 01 28		18.0	BLUE STELLAR OBJECT
PHL 5216	22 27 00.	+ 02 42		17.9	BLUE STELLAR OBJECT
ZWG 514.038	22 27 00.	+ 35 58		15.7	GALAXY
ZWG 514.039	22 27 00.	+ 37 30		15.1	GALAXY
UGC 12051	22 27 00.	+ 37 30	60	15.1	GALAXY Sb/SBc
ZWG 514.040	22 27 00.	+ 38 16		15.1	GALAXY
UGC 12053	22 27 00.	+ 38 16	72	15.1	GALAXY Sb
UGC 12052	22 27 00.	+ 38 16	60	15.1	GALAXY Sc
MCG+06-49-023	22 27 00.	+ 38 17	60	16.	GALAXY
SS 69	22 27 00.	+ 62 46			DIFFUSE GALACTIC NEBULA
LDN 1206	22 27 00.	+ 64 10	1140		DARK NEBULA
LDN 1221	22 27 00.	+ 68 45	540		DARK NEBULA
ZC 2227.0+8225	22 27 00.	+ 82 25	1680		CLUSTER OF GALAXIES
PHL 0289	22 27 00.	- 01 04		16.5	BLUE STELLAR OBJECT
PHL 5217	22 27 00.	- 03 06		17.9	BLUE STELLAR OBJECT
PHL 5215	22 27 00.	- 03 42		16.1	BLUE STELLAR OBJECT
PHL 5218	22 27 00.	- 11 07		18.6	BLUE STELLAR OBJECT
MCG-03-57-011	22 27 00.	- 20 01 30.	12	15.5	GALAXY
BC PKS2227-08	22 27 02.	- 08 48 18.		18.	QUASI-STELLAR OBJECT
MCG+01-57-006	22 27 06.	+ 05 25	48	15.5	GALAXY
ZWG 404.018	22 27 06.	+ 07 27		15.0	GALAXY
UGC 12054	22 27 06.	+ 07 27	102	15.0	GALAXY S
4ZW 104	22 27 06.	+ 35 22			COMPACT GALAXY
PHL 0291	22 27 06.	- 07 34		17.1	BLUE STELLAR OBJECT
PHL 1934	22 27 06.	- 12 58		18.5	BLUE STELLAR OBJECT
SCHO 1424	22 27 08.	+ 51 02 00.	360		ISOLATED DARK CLOUD
ARC 2446	22 27 09.	- 05 28		17.2	RICH CLUSTER OF GALAXIES
SC 2224-4721.8	22 27 11.	- 47 06 30.			NEBULA
PHL 1935	22 27 12.	+ 03 27		16.9	BLUE STELLAR OBJECT
PHL 5219	22 27 12.	- 00 25		16.6	BLUE STELLAR OBJECT
PHL 0292	22 27 12.	- 06 38		15.5	BLUE STELLAR OBJECT
MCG-04-53-006	22 27 12.	- 24 53 30.	60	15.	GALAXY
PHL 5220	22 27 12.	- 30 30		17.6	BLUE STELLAR OBJECT
IC 5223	22 27 14.	+ 07 44 19.			NONSTELLAR OBJECT
SC 2223-6130.4	22 27 14.	- 61 15 06.	12		NEBULA
SC 2223-6144.4	22 27 14.	- 61 29 06.	6		NEBULA
CED 205	22 27 15.	+ 58 09			DIFFUSE GALACTIC NEBULA
PHL 1936	22 27 18.	+ 00 16		17.8	BLUE STELLAR OBJECT
ZWG 404.019	22 27 18.	+ 07 43		15.6	GALAXY
UGC 12055	22 27 18.	+ 07 43	66	15.6	GALAXY TRP SYS
SC 2224-1216.6	22 27 18.	- 12 01 17.	12		NEBULA
MCG-05-53-009	22 27 18.	- 30 48	48	15.5	GALAXY
PHL 5222	22 27 24.	- 02 40		17.7	BLUE STELLAR OBJECT
PHL 5224	22 27 24.	- 03 18		13.4	BLUE STELLAR OBJECT
MCG-01-57-014	22 27 24.	- 05 24	78	13.	GALAXY
IC 1447	22 27 24.	- 05 24 05.			NONSTELLAR OBJECT
PHL 5223	22 27 24.	- 07 15		17.9	BLUE STELLAR OBJECT
PHL 5221	22 27 24.	- 07 26		17.5	BLUE STELLAR OBJECT
MCG-02-57-008	22 27 24.	- 08 32 30.	48	14.5	GALAXY
PHL 5225	22 27 24.	- 08 50		17.5	BLUE STELLAR OBJECT
SC 2224-6127.4	22 27 25.	- 61 12 05.	18		NEBULA
5ZW 386	22 27 30.	+ 50 36			COMPACT GALAXY
OCL 0232	22 27 30.	+ 55 09	120	13.	OPEN STAR CLUSTER
LDN 1185	22 27 30.	+ 58 50	300		DARK NEBULA
MCG-03-57-012	22 27 30.	- 14 33 30.	30	14.5	GALAXY
MCG-05-53-010	22 27 30.	- 27 02	24	15.	GALAXY
PHL 5226	22 27 30.	- 28 30		18.4	BLUE STELLAR OBJECT
PHL 1937	22 27 36.	+ 01 09		18.4	BLUE STELLAR OBJECT
PHL 5230	22 27 36.	+ 03 04		18.1	BLUE STELLAR OBJECT
PHL 5229	22 27 36.	+ 03 30		18.0	BLUE STELLAR OBJECT
PHL 5228	22 27 36.	+ 03 35		17.9	BLUE STELLAR OBJECT
MCG+06-49-024	22 27 36.	+ 36 27	96	15.	GALAXY
ZC 2227.6-0018	22 27 36.	- 00 18	1010		CLUSTER OF GALAXIES
PHL 5231	22 27 36.	- 10 03		18.2	BLUE STELLAR OBJECT
PHL 5232	22 27 36.	- 11 38		18.6	BLUE STELLAR OBJECT
SC 2224-4534.6	22 27 37.	- 45 19 17.			NEBULA
SC 2224-1101.4	22 27 38.	- 10 46 04.	30		NEBULA
SC 2224-4635.6	22 27 40.	- 46 20 17.			NEBULA
ZWG 514.041	22 27 42.	+ 36 01		15.7	GALAXY
ZWG 514.042	22 27 42.	+ 36 28		15.5	GALAXY
UGC 12056	22 27 42.	+ 36 28	96	15.5	GALAXY Sa
PHL 1938	22 27 42.	- 10 14		18.3	BLUE STELLAR OBJECT
SC 2224-4338.4	22 27 43.	- 43 23 05.			NEBULA
MCG-02-57-009	22 27 45.	- 10 46	42	15.	GALAXY
RNGC 7301	22 27 45.	- 17 50		14.0	GALAXY
SC 2224-4502.0	22 27 47.	- 44 46 40.			NEBULA
PHL 1940	22 27 48.	+ 01 44		18.0	BLUE STELLAR OBJECT
PHL 5234	22 27 48.	- 08 37		18.2	BLUE STELLAR OBJECT
PHL 1939	22 27 48.	- 09 00		17.6	BLUE STELLAR OBJECT
PHL 5233	22 27 48.	- 10 59		17.5	BLUE STELLAR OBJECT
MCG-03-57-014	22 27 48.	- 15 16	18	16.	GALAXY
MCG-03-57-013	22 27 48.	- 15 16 30.	18	16.	GALAXY
SC 2224-4333.7	22 27 51.	- 43 18 22.			NEBULA
SC 2224-4429.2	22 27 52.	- 44 13 52.			NEBULA
SC 2224-5457.0	22 27 53.	- 54 41 40.	12		NEBULA
ZC 2227.9+0354	22 27 54.	+ 03 54	1280		CLUSTER OF GALAXIES
ZC 2227.9+2046	22 27 54.	+ 20 46	1080		CLUSTER OF GALAXIES
PHL 5235	22 27 54.	- 02 00		17.9	BLUE STELLAR OBJECT
MCG-03-57-015	22 27 54.	- 17 50	48	14.5	GALAXY
SC 2224-5717.1	22 27 57.	- 57 01 46.	12		NEBULA
SC 2224-4534.8	22 27 58.	- 45 19 28.			NEBULA
LBN 0429	22 28	+ 38 50	3600		BRIGHT NEBULA
KHAV 772	22 28	+ 62 33	7680		DARK NEBULA
PHL 5236	22 28 00.	+ 01 44		18.1	BLUE STELLAR OBJECT
ZWG 495.004	22 28 00.	+ 31 26		15.6	GALAXY
ZWG 514.043	22 28 00.	+ 36 48		15.6	GALAXY
ZWG 514.044	22 28 00.	+ 38 55		15.7	GALAXY
LDN 1207	22 28 00.	+ 64 10	840		DARK NEBULA
LDN 1209	22 28 00.	+ 64 30	1320		DARK NEBULA
PHL 0293	22 28 00.	- 00 23		16.7	BLUE STELLAR OBJECT
PHL 5237	22 28 00.	- 09 40		18.1	BLUE STELLAR OBJECT
PHL 1941	22 28 00.	- 27 01		18.1	BLUE STELLAR OBJECT
MCG-05-53-011	22 28 00.	- 27 12	60	16.	GALAXY
PHL 0294	22 28 00.	- 32 46		16.8	BLUE STELLAR OBJECT
ARC 2447	22 28 01.	+ 03 49		17.7	RICH CLUSTER OF GALAXIES
SC 2224-5557.4	22 28 02.	- 55 42 04.			NEBULA
SC 2224-5621.1	22 28 02.	- 56 05 46.	12		NEBULA
RNGC 7297	22 28 03.	- 38 05			GALAXY
SC 2225-4642.1	22 28 03.	- 46 26 46.			NEBULA
4ZW 105	22 28 06.	+ 21 53			COMPACT GALAXY
ZWG 474.003	22 28 06.	+ 21 53		15.7	GALAXY
MCG+07-46-011	22 28 06.	+ 42 31	48	15.5	GALAXY
ZWG 531.008	22 28 06.	+ 42 32		15.2	GALAXY
UGC 12057	22 28 06.	+ 42 32	72	15.2	GALAXY (S0)
PHL 5238	22 28 06.	- 11 18		17.4	BLUE STELLAR OBJECT
SC 2224-5617.7	22 28 08.	- 56 02 22.			NEBULA
MCG+06-49-025	22 28 09.	+ 33 33	72	15.	GALAXY
RNGC 7298	22 28 10.	- 14 27		14.0	GALAXY
PHL 1942	22 28 12.	+ 01 26		16.7	BLUE STELLAR OBJECT
PHL 0295	22 28 12.	+ 02 54		17.0	BLUE STELLAR OBJECT
MCG+03-57-014	22 28 12.	+ 16 50	42	16.	GALAXY
UGC 12058	22 28 12.	+ 50 50	60	16.5	GALAXY S
MRSL 121+25/1	22 28 12.	+ 87 31	25200		HII REGION
PHL 5243	22 28 12.	- 09 42		18.1	BLUE STELLAR OBJECT
PHL 5242	22 28 12.	- 10 00		17.7	BLUE STELLAR OBJECT
PHL 5240	22 28 12.	- 11 16		17.5	BLUE STELLAR OBJECT
PHL 5239	22 28 12.	- 16 47		15.5	BLUE STELLAR OBJECT
PHL 5241	22 28 12.	- 31 46		18.0	BLUE STELLAR OBJECT
SC 2225-5603.8	22 28 13.	- 55 48 28.	30		NEBULA
HELW 045	22 28 14.	- 21 33			NEBULA
MCG-02-57-010	22 28 15.	- 14 27	72	14.	GALAXY
SC 2225-5616.7	22 28 15.	- 56 01 22.	6		NEBULA
RNGC 7300	22 28 16.	- 14 17		13.5	GALAXY
SC 2225-5713.2	22 28 17.	- 56 57 52.	12		NEBULA
ZWG 474.004	22 28 18.	+ 22 17		15.0	GALAXY
UGC 12059	22 28 18.	+ 22 17	72	15.0	GALAXY SBb
ZWG 514.045	22 28 18.	+ 33 34		15.5	GALAXY
UGC 12060	22 28 18.	+ 33 34	120	15.5	GALAXY IRR
ZWG 514.046	22 28 18.	+ 37 38		15.4	GALAXY
UGC 12061	22 28 18.	+ 37 38	72	15.4	GALAXY Sa-b
PHL 5244	22 28 18.	- 14 16		17.8	BLUE STELLAR OBJECT
SC 2225-4454.2	22 28 20.	- 44 38 52.			NEBULA
MCG+06-49-026	22 28 21.	+ 37 37	78	14.5	GALAXY
SC 2225-1431.2	22 28 21.	- 14 15 51.	72		NEBULA
MCG-04-53-007	22 28 21.	- 25 37	30	15.5	GALAXY
SC 2225-5620.2	22 28 22.	- 56 04 52.	18		NEBULA
SC 2225-4638.8	22 28 23.	- 46 23 27.			NEBULA
ZWG 514.047	22 28 24.	+ 39 22		15.3	GALAXY
PHL 5246	22 28 24.	- 09 03		18.1	BLUE STELLAR OBJECT
PHL 5245	22 28 24.	- 12 54		18.6	BLUE STELLAR OBJECT
MCG-02-57-011	22 28 24.	- 14 16	108	13.	GALAXY
SC 2225-4303.6	22 28 24.	- 42 48 15.			NEBULA
SC 2225-4532.1	22 28 24.	- 45 16 45.			NEBULA
HELW 046	22 28 26.	- 21 26			NEBULA
SC 2225-4235.9	22 28 29.	- 42 20 33.			NEBULA
SC 2225-5618.0	22 28 29.	- 56 02 39.	6		NEBULA
2ZW 180	22 28 30.	+ 00 01			COMPACT GALAXY
MCG+04-53-002	22 28 30.	+ 22 18	36	15.	GALAXY
MCG+06-49-027	22 28 30.	+ 39 03	24	15.5	GALAXY
PHL 0296	22 28 30.	- 08 24		18.0	BLUE STELLAR OBJECT
MCG-03-57-016	22 28 30.	- 17 35 30.	24	15.	GALAXY
PHL 1943	22 28 30.	- 31 55		18.1	BLUE STELLAR OBJECT
SC 2225-4620.0	22 28 30.	- 46 04 39.			NEBULA

OBJECT NAME	RIGHT ASCEN.	DECLINATION	DIAM.	MAGN.	TYPE OF OBJECT
SC 2225-4647.7	22 28 30.	- 46 32 21.	12		NEBULA
SC 2225-5617.1	22 28 30.	- 56 01 45.	6		NEBULA
IC 5224	22 28 31.	- 46 14 38.			NONSTELLAR OBJECT
SC 2225-5617.9	22 28 31.	- 56 02 33.	6		NEBULA
SC 2225-5919.2	22 28 32.	- 59 03 51.	30		NEBULA
MCG-02-57-012	22 28 33.	- 11 58	30	15.	GALAXY
RNGC 7299	22 28 33.	- 38 05			GALAXY
ZWG 514.048	22 28 36.	+ 36 18		15.4	GALAXY
MCG-05-53-012	22 28 36.	- 28 47	48	16.	GALAXY
SC 2225-4333.8	22 28 41.	- 43 18 27.			NEBULA
UGC 12062	22 28 42.	+ 39 13	66	16.0	GALAXY S0-a
PHL 0297	22 28 42.	- 00 28		17.3	BLUE STELLAR OBJECT
MCG-05-53-013	22 28 42.	- 28 40	72	15.	GALAXY
MCG-03-57-017	22 28 45.	- 19 13	84	14.5	GALAXY
SC 2225-4621.7	22 28 45.	- 46 06 21.			NEBULA
SC 2225-4245.6	22 28 46.	- 42 30 15.			NEBULA
IC 5225	22 28 47.	- 25 37 19.			NONSTELLAR OBJECT
8ZW 2228+04.6	22 28 48.	+ 04 36		18.2	COMPACT GALAXY
ZC 2228.8+2938	22 28 48.	+ 29 38	1340		CLUSTER OF GALAXIES
MCG+06-49-028	22 28 48.	+ 35 06	90	15.	GALAXY
PHL 1944	22 28 48.	- 00 18		18.2	BLUE STELLAR OBJECT
PHL 5247	22 28 48.	- 08 49		18.9	BLUE STELLAR OBJECT
PHL 1945	22 28 48.	- 12 43		18.6	BLUE STELLAR OBJECT
SC 2225-5906.6	22 28 49.	- 58 51 15.			NEBULA
SC 2225-4340.2	22 28 53.	- 43 24 51.			NEBULA
ZWG 514.049	22 28 54.	+ 35 08		15.6	GALAXY
UGC 12063	22 28 54.	+ 35 08	90	15.6	GALAXY Sa-b
ZC 2228.9-0116	22 28 54.	- 01 16	3020		CLUSTER OF GALAXIES
PHL 1946	22 28 54.	- 14 37		18.0	BLUE STELLAR OBJECT
PHL 0298	22 28 54.	- 14 38		17.8	BLUE STELLAR OBJECT
IC 5226	22 28 56.	- 24 55 00.			NONSTELLAR OBJECT
SC 2225-4439.4	22 28 58.	- 44 24 02.			NEBULA
LBN 0431	22 29	+ 38 50	1800		BRIGHT NEBULA
PHL 5250	22 29 00.	+ 02 04		17.5	BLUE STELLAR OBJECT
ISS 0136	22 29 00.	+ 57 25	334		STELLAR RING
LDN 1190	22 29 00.	+ 58 53	240		DARK NEBULA
LDN 1208	22 29 00.	+ 64 10	540		DARK NEBULA
PHL 5249	22 29 00.	- 00 16		17.9	BLUE STELLAR OBJECT
PHL 0299	22 29 00.	- 01 46		16.2	BLUE STELLAR OBJECT
PHL 5248	22 29 00.	- 02 45		14.7	BLUE STELLAR OBJECT
PHL 5251	22 29 00.	- 09 50		17.9	BLUE STELLAR OBJECT
PHL 1947	22 29 00.	- 11 02		18.1	BLUE STELLAR OBJECT
PHL 1949	22 29 00.	- 11 22		18.7	BLUE STELLAR OBJECT
PHL 1948	22 29 00.	- 14 18		18.1	BLUE STELLAR OBJECT
SC 2226-4624.6	22 29 03.	- 46 09 14.			NEBULA
SC 2226-4555.6	22 29 05.	- 45 40 14.	12		NEBULA
ZC 2229.1+1402	22 29 06.	+ 14 02	1410		CLUSTER OF GALAXIES
BIGO 556	22 29 06.	+ 30 41			NEBULA
ZWG 514.050	22 29 06.	+ 39 06		14.6	GALAXY
UGC 12064	22 29 06.	+ 39 06	72	14.6	GALAXY E-S0
DG 186	22 29 06.	+ 65 12	780		REFLECTION NEBULA
ARC 2448	22 29 06.	- 08 43		16.0	RICH CLUSTER OF GALAXIES
SC 2226-4336.1	22 29 06.	- 43 20 44.			NEBULA
SC 2226-4336.1	22 29 07.	- 43 20 44.			NEBULA
SC 2225-6359.8	22 29 08.	- 63 44 27.			NEBULA
RNGC 7304	22 29 09.	+ 30 42			NON-EXISTENT OBJECT
SC 2226-1055.4	22 29 09.	- 10 40 02.	12		NEBULA
ZWG 495.005	22 29 12.	+ 30 42		13.7	GALAXY
UGC 12065	22 29 12.	+ 30 42	102	13.7	GALAXY S
ISS 0089	22 29 12.	+ 57 42	644		STELLAR RING
PHL 5252	22 29 12.	- 00 41		14.0	BLUE STELLAR OBJECT
PHL 5257	22 29 12.	- 01 06		18.0	BLUE STELLAR OBJECT
PHL 5253	22 29 12.	- 01 09		17.8	BLUE STELLAR OBJECT
PHL 5258	22 29 12.	- 05 38		18.1	BLUE STELLAR OBJECT
PHL 1951	22 29 12.	- 07 00		16.1	BLUE STELLAR OBJECT
PHL 5255	22 29 12.	- 09 12		17.2	BLUE STELLAR OBJECT
PHL 0300	22 29 12.	- 10 25		18.2	BLUE STELLAR OBJECT
PHL 0301	22 29 12.	- 11 10		17.2	BLUE STELLAR OBJECT
PHL 5254	22 29 12.	- 11 14		17.9	BLUE STELLAR OBJECT
PHL 1950	22 29 12.	- 12 44		18.8	BLUE STELLAR OBJECT
PHL 5256	22 29 12.	- 15 18		17.0	BLUE STELLAR OBJECT
SC 2225-5847.6	22 29 14.	- 58 32 14.	6		NEBULA
RNGC 7303	22 29 15.	+ 30 42		13.5	GALAXY
MCG+06-49-029	22 29 15.	+ 39 07	18	15.	GALAXY
SC 2226-4618.0	22 29 15.	- 46 02 38.			NEBULA
SC 2226-5301.8	22 29 16.	- 52 46 26.	6		NEBULA
MRK 305	22 29 18.	+ 19 25	8	17.	GALAXY WITH UV CONTINUUM
MCG+03-57-015	22 29 18.	+ 19 27	60	14.	GALAXY
MCG+03-57-016	22 29 18.	+ 20 22	54	15.	GALAXY
ZWG 452.021	22 29 18.	+ 20 27		15.4	GALAXY
ZWG 514.051	22 29 18.	+ 34 29		15.4	GALAXY
PHL 0302	22 29 18.	- 01 01		18.3	BLUE STELLAR OBJECT
PHL 1952	22 29 18.	- 09 59		18.3	BLUE STELLAR OBJECT
PHL 1953	22 29 18.	- 14 50		18.1	BLUE STELLAR OBJECT
MCG-03-57-018	22 29 18.	- 17 25	78	15.	GALAXY
MCG-04-53-008	22 29 18.	- 25 38	36	15.5	GALAXY
SC 2225-6310.8	22 29 18.	- 62 55 26.			NEBULA
SC 2226-4556.0	22 29 19.	- 45 40 38.			NEBULA
SC 2226-4448.4	22 29 20.	- 44 33 02.			NEBULA
MCG-03-57-019	22 29 21.	- 15 41 30.	36	15.	GALAXY
SC 2226-1054.1	22 29 22.	- 10 38 43.	12		NEBULA
SC 2225-6339.1	22 29 23.	- 63 23 44.	12		NEBULA
ZWG 452.022	22 29 24.	+ 19 25		14.6	GALAXY
MRK 306	22 29 24.	+ 19 25	24	15.	GALAXY WITH UV CONTINUUM
UGC 12066	22 29 24.	+ 19 25	60	14.6	GALAXY DBL SYS
ZWG 452.023	22 29 24.	+ 20 22		15.3	GALAXY
UGC 12067	22 29 24.	+ 20 22	60	15.3	GALAXY Sa
MCG+04-53-003	22 29 24.	+ 24 30	36	16.	GALAXY
MCG+05-53-004	22 29 24.	+ 30 43 30.	90	14.	GALAXY
PHL 0303	22 29 24.	- 08 21		18.9	BLUE STELLAR OBJECT
PHL 5262	22 29 24.	- 12 42		17.7	BLUE STELLAR OBJECT
PHL 5259	22 29 24.	- 16 48		16.2	BLUE STELLAR OBJECT
PHL 5261	22 29 24.	- 18 22		18.5	BLUE STELLAR OBJECT
PHL 5263	22 29 24.	- 19 36		18.9	BLUE STELLAR OBJECT
MCG-04-53-009	22 29 24.	- 25 39	30	14.5	GALAXY
PHL 5260	22 29 24.	- 30 46		17.7	BLUE STELLAR OBJECT
SC 2226-4503.6	22 29 24.	- 44 48 14.			NEBULA
SC 2226-4556.2	22 29 24.	- 45 40 50.			NEBULA
RNGC 7294	22 29 25.	- 25 39		14.0	GALAXY
VDB.66N 154	22 29 26.	+ 65 09	612		REFLECTION NEBULA
SC 2226-5959.4	22 29 26.	- 59 44 02.	12		NEBULA
2ZW 179	22 29 30.	+ 03 35			COMPACT GALAXY
ZC 2229.5+0335	22 29 30.	+ 03 35	270		CLUSTER OF GALAXIES
MRSL 108+06/1	22 29 30.	+ 64 51	2400		HII REGION
MCG+00-57-003	22 29 30.	- 00 14	66	14.	GALAXY
LB 00372	22 29 33.	- 26 10 36.		15.2	FAINT BLUE STAR
SC 2226-4600.2	22 29 34.	- 45 44 49.			NEBULA
IC 5228	22 29 35.	- 14 22 46.			SAME AS NGC 7302
PHL 1956	22 29 36.	+ 01 44		18.0	BLUE STELLAR OBJECT
PHL 1954	22 29 36.	+ 01 53		17.8	BLUE STELLAR OBJECT
ZWG 514.052	22 29 36.	+ 34 11		15.7	GALAXY
ZWG 378.010	22 29 36.	- 00 13		15.6	GALAXY
UGC 12068	22 29 36.	- 00 13	66	15.6	GALAXY Sb
PHL 1955	22 29 36.	- 10 18		18.7	BLUE STELLAR OBJECT
PHL 5264	22 29 36.	- 15 48		17.0	BLUE STELLAR OBJECT
TON-S 0051	22 29 36.	- 29 33		16.0	BLUE STAR
PK100-08.1	22 29 36.4	+ 47 32 46.			PLANETARY NEBULA
SC 2226-4318.0	22 29 38.	- 43 02 37.			NEBULA
SC 2226-4359.1	22 29 39.	- 43 43 43.			NEBULA
RNGC 7302	22 29 40.	- 14 23		13.5	GALAXY
SVEN 442	22 29 40.	- 25 56	30	15.3	GALAXY
SC 2226-4503.4	22 29 40.	- 44 48 01.	18		NEBULA
SC 2226-4537.3	22 29 40.	- 45 21 55.			NEBULA
HELW 248	22 29 41.	- 25 55 43.			NEBULA
MCG+06-49-030	22 29 42.	+ 39 00	24	16.	GALAXY
ISS 0090	22 29 42.	+ 57 27	331		STELLAR RING
PHL 5265	22 29 42.	- 26 54		15.3	BLUE STELLAR OBJECT
TON-S 0052	22 29 42.	- 26 54		13.5	BLUE STAR
PHL 5266	22 29 42.	- 28 16		17.7	BLUE STELLAR OBJECT
PHL 0304	22 29 42.	- 29 32		17.8	BLUE STELLAR OBJECT
SC 2226-4334.2	22 29 43.	- 43 18 49.			NEBULA
SG 3.235	22 29 45.	+ 64 49	3000		DIFFUSE EMISSION NEBULA
MCG-04-53-010	22 29 45.	- 25 56	84	13.5	GALAXY
SC 2226-4446.6	22 29 45.	- 44 31 13.			NEBULA
RNGC 7305	22 29 46.	+ 11 27		15.0	GALAXY
MCG+01-57-007	22 29 48.	+ 07 57	24	15.5	GALAXY
ZWG 429.007	22 29 48.	+ 11 27		15.1	GALAXY
MCG+02-57-003	22 29 48.	+ 11 27 30.	15	15.	GALAXY
PHL 1957	22 29 48.	- 04 06		12.2	BLUE STELLAR OBJECT
PHL 5270	22 29 48.	- 10 25		17.0	BLUE STELLAR OBJECT
MCG-02-57-013	22 29 48.	- 14 23	84	13.	GALAXY
PHL 5269	22 29 48.	- 17 28		17.1	BLUE STELLAR OBJECT
PHL 5267	22 29 48.	- 20 36		12.7	BLUE STELLAR OBJECT
PHL 5268	22 29 48.	- 27 54		17.9	BLUE STELLAR OBJECT
LB 00373	22 29 51.	- 26 21 42.		14.9	FAINT BLUE STAR
SC 2226-4739.9	22 29 51.	- 47 24 31.	12		NEBULA
SC 2226-4547.2	22 29 52.	- 45 31 49.			NEBULA
ZWG 474.005	22 29 54.	+ 23 37		15.7	GALAXY
PHL 5271	22 29 54.	- 11 28		17.8	BLUE STELLAR OBJECT
PHL 0305	22 29 54.	- 12 02		18.8	BLUE STELLAR OBJECT
ARC 2449	22 29 57.	+ 14 38		17.1	RICH CLUSTER OF GALAXIES
SC 2226-4546.2	22 29 59.	- 45 30 49.			NEBULA
LBN 0428	22 30	+ 38 00	4200		BRIGHT NEBULA
LBN 0437	22 30	+ 40 20	2820		BRIGHT NEBULA
LBN 0520	22 30	+ 64 50	1200		BRIGHT NEBULA
LBN 0523	22 30	+ 65 12	720		BRIGHT NEBULA
DV.56 N2573A	22 30	- 89 26	132		SP GALAXY
PHL 1958	22 30 00.	+ 02 42		17.0	BLUE STELLAR OBJECT
ZWG 404.020	22 30 00.	+ 07 57		15.2	GALAXY
LDN 1248	22 30 00.	+ 62 00	8340		DARK NEBULA
LDN 1213	22 30 00.	+ 65 10	300		DARK NEBULA
LDN 1242	22 30 00.	+ 73 00	3600		DARK NEBULA
ZWG 359.001	22 30 00.	+ 76 15		15.7	GALAXY
UGC 12069	22 30 00.	+ 76 15	120	15.7	GALAXY
KARA.73B 0969	22 30 00.	+ 76 15	78	15.7	ISOLATED GALAXY S
UGC 12070	22 30 00.	+ 77 52	66	16.5	GALAXY IRR
MCG-02-57-014	22 30 00.	- 09 47	30	15.	GALAXY
PHL 5272	22 30 00.	- 11 50		18.3	BLUE STELLAR OBJECT
PHL 1959	22 30 00.	- 12 18		17.8	BLUE STELLAR OBJECT
SC 2227-4558.2	22 30 00.	- 45 42 48.			NEBULA
SC 2227-4558.2	22 30 00.	- 45 42 49.			NEBULA
SC 2227-4406.0	22 30 03.	- 43 50 37.			NEBULA
HELW 077	22 30 04.	- 14 22			NEBULA
ZWG 474.006	22 30 06.	+ 23 01		15.6	GALAXY
ZWG 495.006	22 30 06.	+ 30 35		15.1	GALAXY
UGC 12071	22 30 06.	+ 30 35	78	15.1	GALAXY SBb
PHL 1960	22 30 06.	- 30 48		18.4	BLUE STELLAR OBJECT
PHL 1961	22 30 06.	- 32 50		17.8	BLUE STELLAR OBJECT
SHB 377	22 30 07.7	+ 11 28 22.		17.3	QUASI-STELLAR OBJECT
RC CTA102	22 30 07.78	+ 11 28 22.8		17.32	QUASI-STELLAR OBJECT
SC 2227-4553.8	22 30 09.	- 45 38 24.			NEBULA
SC 2227-4559.0	22 30 09.	- 45 43 36.			NEBULA
SC 2227-4441.8	22 30 10.	- 44 26 24.			NEBULA
SC 2227-4255.2	22 30 11.	- 42 39 48.			NEBULA
PHL 5274	22 30 12.	+ 02 24		18.3	BLUE STELLAR OBJECT
PHL 1963	22 30 12.	- 04 37		17.9	BLUE STELLAR OBJECT
PHL 5275	22 30 12.	- 11 38		18.5	BLUE STELLAR OBJECT
PHL 1962	22 30 12.	- 14 05		18.5	BLUE STELLAR OBJECT
MCG-02-57-015	22 30 12.	- 14 22	48	15.	GALAXY
PHL 0306	22 30 12.	- 27 31		17.8	BLUE STELLAR OBJECT
PHL 5273	22 30 12.	- 28 04		17.1	BLUE STELLAR OBJECT
SC 2226-6038.7	22 30 12.	- 60 23 19.	18		NEBULA
SC 2227-5659.9	22 30 13.	- 56 44 30.	12		NEBULA
HELW 078	22 30 16.	- 14 14			NEBULA
SC 2227-5717.6	22 30 16.	- 57 02 12.	42		NEBULA
PHL 5276	22 30 16.	+ 02 34		17.7	BLUE STELLAR OBJECT
UGC 12072	22 30 18.	+ 23 40	60	16.5	GALAXY S
MCG+05-53-005	22 30 18.	+ 30 35 30.	72	15.	GALAXY
ZWG 514.053	22 30 18.	+ 38 58		14.8	GALAXY
UGC 12073	22 30 18.	+ 38 58	132	14.8	GALAXY SBb
PHL 0307	22 30 18.	- 08 56		17.9	BLUE STELLAR OBJECT
PHL 0308	22 30 18.	- 11 17		19.2	BLUE STELLAR OBJECT
PHL 0309	22 30 18.	- 12 37		17.9	BLUE STELLAR OBJECT
MCG-04-53-011	22 30 21.	- 25 01 30.	36	16.	GALAXY
SC 2227-4645.6	22 30 21.	- 46 30 12.			NEBULA
SC 2227-4313.7	22 30 23.	- 42 58 18.			NEBULA
SC 2227-5602.7	22 30 23.	- 55 47 18.	6		NEBULA
PHL 0310	22 30 24.	+ 01 26		18.4	BLUE STELLAR OBJECT
PHL 1964	22 30 24.	+ 01 41		18.0	BLUE STELLAR OBJECT
MCG+01-57-008	22 30 24.	+ 07 50	36	14.	GALAXY
ZC 2230.4+1433	22 30 24.	+ 14 33	1480		CLUSTER OF GALAXIES
MCG+06-49-031	22 30 24.	+ 36 57	48	15.5	GALAXY
MCG+06-49-032	22 30 24.	+ 38 58	120	14.5	GALAXY
PHL 0311	22 30 24.	- 06 58		17.5	BLUE STELLAR OBJECT
ARC 2450	22 30 24.	- 09 19		18.0	RICH CLUSTER OF GALAXIES
PHL 5278	22 30 24.	- 10 04		18.0	BLUE STELLAR OBJECT
PHL 5279	22 30 24.	- 14 45		18.5	BLUE STELLAR OBJECT
PHL 5277	22 30 24.	- 19 40		8.8	BLUE STELLAR OBJECT
MCG-03-57-020	22 30 24.	- 20 22 30.	72	15.	GALAXY
SC 2227-4739.2	22 30 24.	- 47 23 48.	6		NEBULA
SC 2227-4625.3	22 30 27.	- 46 19 54.			NEBULA
SVEN 443	22 30 28.	- 26 09	24	15.9	GALAXY
MCG-03-57-021	22 30 30.	- 17 20 30.	48	15.	GALAXY
TON-S 0053	22 30 30.	- 26 20		14.2	BLUE STAR
SC 2227-4342.5	22 30 34.	- 43 27 06.			NEBULA
2ZW 181	22 30 36.	+ 07 50			COMPACT GALAXY

OBJECT NAME	RIGHT ASCEN.	DECLINATION	DIAM.	MAGN.	TYPE OF OBJECT
ZWG 404.021	22 30 36.	+ 07 50		14.2	GALAXY
UGC 12074	22 30 36.	+ 07 50	36	14.2	GALAXY COMPACT
ZWG 514.054	22 30 36.	+ 38 58		15.1	GALAXY
UGC 12075	22 30 36.	+ 38 58	96	15.1	GALAXY SBc
MCG+06-49-033	22 30 36.	+ 38 58	78	14.5	GALAXY
PHL 5282	22 30 36.	- 00 55		18.2	BLUE STELLAR OBJECT
PHL 5280	22 30 36.	- 02 41		17.7	BLUE STELLAR OBJECT
PHL 1965	22 30 36.	- 09 42		18.8	BLUE STELLAR OBJECT
PHL 5281	22 30 36.	- 13 08		17.9	BLUE STELLAR OBJECT
HELW 491	22 30 36.	- 41 13 54.			NEBULA
SVEN 444	22 30 41.	- 26 30 05.	24	15.7	GALAXY
IC 5227	22 30 41.	- 64 56			NONSTELLAR OBJECT
HN 0757	22 30 41.	- 64 57			NEBULA
ZC 2230.7+0307	22 30 42.	+ 03 07	1210		CLUSTER OF GALAXIES
ZWG 514.055	22 30 42.	+ 38 58		15.7	GALAXY
MCG+06-49-034	22 30 42.	+ 38 58	36	16.	GALAXY
ISS 0137	22 30 42.	+ 57 02	131		STELLAR RING
PHL 1966	22 30 42.	- 08 22		18.3	BLUE STELLAR OBJECT
PHL 0312	22 30 42.	- 10 38		16.7	BLUE STELLAR OBJECT
PHL 0313	22 30 42.	- 11 44		17.9	BLUE STELLAR OBJECT
MCG-05-53-014	22 30 42.	- 27 30	72	13.5	GALAXY
SC 2227-4630.9	22 30 42.	- 46 15 29.			NEBULA
SC 2227-6113.0	22 30 42.	- 60 57 36.	6		NEBULA
ARC 2451	22 30 43.	+ 03 09		17.7	RICH CLUSTER OF GALAXIES
RNGC 7306	22 30 43.	- 27 30		13.0	GALAXY
HELW 249	22 30 44.	- 26 30 05.			NEBULA
PHL 0314	22 30 48.	+ 02 45		18.4	BLUE STELLAR OBJECT
MCG+07-46-012	22 30 48.	+ 43 15	48	17.	GALAXY
MRSL 105+00/1	22 30 48.	+ 58 13	60		HII REGION
PHL 5284	22 30 48.	- 00 10		18.1	BLUE STELLAR OBJECT
PHL 1967	22 30 48.	- 06 17		18.3	BLUE STELLAR OBJECT
PHL 5283	22 30 48.	- 10 10		17.5	BLUE STELLAR OBJECT
PHL 1968	22 30 48.	- 13 26		18.1	BLUE STELLAR OBJECT
PHL 5285	22 30 48.	- 17 56		18.0	BLUE STELLAR OBJECT
SC 2228-1241.2	22 30 51.	- 12 25 47.	18		NEBULA
SC 2227-4441.8	22 30 51.	- 44 26 23.			NEBULA
SC 2227-4725.6	22 30 52.	- 47 10 11.	12		NEBULA
SC 2227-4423.3	22 30 53.	- 44 07 53.			NEBULA
SC 2226-7324.7	22 30 53.	- 73 09 18.	12		NEBULA
ZWG 514.056	22 30 54.	+ 38 54		15.5	GALAXY
UGC 12076	22 30 54.	+ 43 17	72	16.0	GALAXY Sa-b
PHL 0315	22 30 54.	- 12 31		16.1	BLUE STELLAR OBJECT
MCG-04-53-012	22 30 54.	- 25 02	18	15.5	GALAXY
MCG-07-46-003	22 30 54.	- 41 13	180	13.1	GALAXY
SC 2227-5641.5	22 30 54.	- 56 26 05.			NEBULA
ARC 2453	22 30 56.	+ 16 20		17.1	RICH CLUSTER OF GALAXIES
RNGC 7307	22 30 56.	- 41 12		13.0	GALAXY
VDB.66G 213	22 31	+ 32 34	100		DWARF GALAXY
KHAV 773	22 31	+ 59 27	5400		DARK NEBULA
KHAV 774	22 31	+ 64 03	8510		DARK NEBULA
ZWG 378.011	22 31 00.	+ 01 35		15.6	GALAXY
ZC 2231.0+1616	22 31 00.	+ 16 16	2550		CLUSTER OF GALAXIES
MCG+06-49-035	22 31 00.	+ 38 54	30	16.	GALAXY
LDN 1194	22 31 00.	+ 58 50	360		DARK NEBULA
PHL 5286	22 31 00.	- 04 25		17.6	BLUE STELLAR OBJECT
PHL 5287	22 31 00.	- 08 06		18.4	BLUE STELLAR OBJECT
MCG-07-46-004	22 31 00.	- 39 39	30	15.5	GALAXY
SC 2228-4656.4	22 31 02.	- 46 40 59.	6		NEBULA
SC 2228-4534.3	22 31 03.	- 45 18 53.			NEBULA
SC 2228-4654.1	22 31 03.	- 46 38 41.			NEBULA
SC 2228-4415.2	22 31 05.	- 43 59 47.			NEBULA
UGC 12077	22 31 06.	+ 38 42	60	16.0	GALAXY S
ARC 2452	22 31 06.	- 09 04		18.0	RICH CLUSTER OF GALAXIES
SC 2227-5704.7	22 31 06.	- 56 49 17.	12		NEBULA
SC 2228-4626.0	22 31 08.	- 46 10 35.			NEBULA
SC 2228-4653.2	22 31 08.	- 46 37 47.			NEBULA
HELW 250	22 31 09.	- 26 41 05.			NEBULA
SVEN 445	22 31 11.	- 26 41 04.	18	16.5	GALAXY
UGC 12078	22 31 12.	+ 30 35	90	16.0	GALAXY S
ZC 2231.2+3732	22 31 12.	+ 37 32	21370		CLUSTER OF GALAXIES
MESL 095-16/1	22 31 12.	+ 38 19	9600		HII REGION
PHL 1969	22 31 12.	- 05 40		18.5	BLUE STELLAR OBJECT
PHL 1970	22 31 12.	- 16 54		18.6	BLUE STELLAR OBJECT
PHL 1972	22 31 18.	+ 00 15		18.2	BLUE STELLAR OBJECT
UGC 12079	22 31 18.	+ 38 43	60	16.5	GALAXY DWARF
PHL 5288	22 31 18.	- 02 33		17.9	BLUE STELLAR OBJECT
PHL 5289	22 31 18.	- 16 52		17.1	BLUE STELLAR OBJECT
PHL 0316	22 31 18.	- 26 49		15.3	BLUE STELLAR OBJECT
TON-S 0054	22 31 18.	- 26 50		13.7	BLUE STAR
MCG-05-53-015	22 31 18.	- 27 30	18	15.	GALAXY
PHL 1971	22 31 18.	- 32 24		7.1	BLUE STELLAR OBJECT
SC 2228-4309.1	22 31 18.	- 42 53 40.			NEBULA
HELW 079	22 31 22.	- 14 35			NEBULA
SC 2228-4451.8	22 31 22.	- 44 36 22.			NEBULA
SC 2228-4247.4	22 31 23.	- 42 31 58.			NEBULA
ZWG 452.024	22 31 24.	+ 20 23		15.4	GALAXY
PHL 1975	22 31 24.	- 00 49		18.2	BLUE STELLAR OBJECT
PHL 1973	22 31 24.	- 01 11		17.7	BLUE STELLAR OBJECT
PHL 1974	22 31 24.	- 02 15		17.8	BLUE STELLAR OBJECT
PHL 5291	22 31 24.	- 12 25		9.3	BLUE STELLAR OBJECT
PHL 5292	22 31 24.	- 12 44		17.8	BLUE STELLAR OBJECT
MCG-03-57-022	22 31 24.	- 14 34	60	14.5	GALAXY
PHL 5290	22 31 24.	- 18 50		17.8	BLUE STELLAR OBJECT
MCG-05-53-016	22 31 24.	- 32 42	48	15.	GALAXY
SC 2228-4512.6	22 31 25.	- 44 57 10.	12		NEBULA
SC 2228-4612.8	22 31 25.	- 45 57 22.			NEBULA
MCG+01-57-009	22 31 27.	+ 05 18	84	13.	GALAXY
HN 1232	22 31 27.	- 61 39		15.	NEBULA
SC 2228-4615.6	22 31 28.	- 46 00 10.			NEBULA
IC 5229	22 31 28.	- 61 38			NONSTELLAR OBJECT
ZC 2231.5+0052	22 31 30.	+ 00 52	7730		CLUSTER OF GALAXIES
SC 2228-4552.5	22 31 30.	- 45 37 04.			NEBULA
SC 2228-4615.6	22 31 31.	- 46 00 10.			NEBULA
SC 2228-5639.9	22 31 31.	- 56 24 28.	6		NEBULA
HELW 492	22 31 32.	- 41 23 28.			NEBULA
SC 2228-6248.4	22 31 33.	- 62 32 58.			NEBULA
RNGC 7309	22 31 35.	- 10 27		13.0	GALAXY
PHL 5295	22 31 36.	+ 02 54		17.3	BLUE STELLAR OBJECT
ZWG 404.022	22 31 36.	+ 05 19		13.4	GALAXY
UGC 12080	22 31 36.	+ 05 19	108	13.4	GALAXY Sa-b
ZWG 429.008	22 31 36.	+ 09 54		15.5	GALAXY
UGC 12081	22 31 36.	+ 09 54	78	15.5	GALAXY Sa-b
KARA.73B 0970	22 31 36.	+ 09 54	126	15.5	ISOLATED GALAXY S
ZWG 452.025	22 31 36.	+ 20 01		15.5	GALAXY
ZWG 474.007	22 31 36.	+ 23 05		14.8	GALAXY
MCG+04-53-004	22 31 36.	+ 23 05	24	15.	GALAXY
PHL 5293	22 31 36.	- 26 30		14.1	BLUE STELLAR OBJECT
TON-S 0055	22 31 36.	- 26 30		13.5	BLUE STAR
PHL 5294	22 31 36.	- 28 42		17.5	BLUE STELLAR OBJECT
PHL 1976	22 31 36.	- 29 34		16.6	BLUE STELLAR OBJECT
TON-S 0056	22 31 36.	- 29 36		14.7	BLUE STAR
SC 2228-4613.0	22 31 36.	- 45 57 34.			NEBULA
LB 01503	22 31 36.	- 46 58		13.9	FAINT BLUE STAR
IC 5231	22 31 37.	+ 23 04 45.			NONSTELLAR OBJECT
SC 2228-4304.8	22 31 37.	- 42 49 22.			NEBULA
RNGC 7311	22 31 38.	+ 05 19		13.5	GALAXY
SC 2228-4702.4	22 31 38.	- 46 46 58.	12		NEBULA
ARC 2454	22 31 41.	+ 05 32		17.1	RICH CLUSTER OF GALAXIES
SVEN 446	22 31 41.	- 26 41 04.	18	16.2	GALAXY
SC 2228-4547.8	22 31 41.	- 45 32 22.			NEBULA
PHL 1978	22 31 42.	+ 01 26		18.2	BLUE STELLAR OBJECT
PHL 5298	22 31 42.	+ 02 22		17.8	BLUE STELLAR OBJECT
ZWG 378.012	22 31 42.	+ 03 20		15.7	GALAXY
PHL 1977	22 31 42.	- 00 37		16.6	BLUE STELLAR OBJECT
PHL 5297	22 31 42.	- 08 02		16.8	BLUE STELLAR OBJECT
PHL 5299	22 31 42.	- 11 23		18.5	BLUE STELLAR OBJECT
PHL 5300	22 31 42.	- 18 24		18.1	BLUE STELLAR OBJECT
PHL 5296	22 31 42.	- 28 19		15.9	BLUE STELLAR OBJECT
SC 2228-4626.4	22 31 44.	- 46 10 58.			NEBULA
MCG-02-57-016	22 31 45.	- 10 37 30.	120	13.	GALAXY
MCG-04-53-013	22 31 45.	- 24 56	30	15.5	GALAXY
SC 2228-5854.3	22 31 46.	- 58 38 52.	12		NEBULA
SC 2228-5955.3	22 31 46.	- 59 39 52.	12		NEBULA
PHL 1979	22 31 48.	+ 00 19		13.0	BLUE STELLAR OBJECT
PHL 1980	22 31 48.	- 06 08		18.3	BLUE STELLAR OBJECT
PHL 5301	22 31 48.	- 07 51		13.8	BLUE STELLAR OBJECT
PHL 5303	22 31 48.	- 10 44		18.4	BLUE STELLAR OBJECT
PHL 5302	22 31 48.	- 11 56		17.5	BLUE STELLAR OBJECT
PHL 0317	22 31 48.	- 16 44		18.3	BLUE STELLAR OBJECT
MCG-04-53-014	22 31 48.	- 22 58	90	14.	GALAXY
PHL 0318	22 31 48.	- 25 11		17.6	BLUE STELLAR OBJECT
RNGC 7308	22 31 52.	- 13 12		14.0	GALAXY
SC 2228-4522.0	22 31 52.	- 45 06 33.			NEBULA
ZC 2231.9+0530	22 31 54.	+ 05 30	1550		CLUSTER OF GALAXIES
MCG+01-57-010	22 31 54.	+ 05 33	72	14.	GALAXY
ZC 2231.9+0552	22 31 54.	+ 05 52	400		CLUSTER OF GALAXIES
ZWG 474.008	22 31 54.	+ 23 21		15.7	GALAXY
ZWG 495.007	22 31 54.	+ 32 37		15.6	GALAXY
UGC 12082	22 31 54.	+ 32 37	210	15.6	GALAXY DWRF SP
KARA.73B 0971	22 31 54.	+ 32 37	144	15.6	ISOLATED GALAXY S
DG 187	22 31 54.	+ 40 29	1320		REFLECTION NEBULA
IC 1448	22 31 54.	- 13 11 33.			NONSTELLAR OBJECT
MCG-02-57-017	22 31 54.	- 13 12	60	14.	GALAXY
MCG-04-53-015	22 31 54.	- 22 45	48	14.	GALAXY
PHL 1981	22 31 54.	- 24 25		17.7	BLUE STELLAR OBJECT
PHL 5304	22 31 54.	- 29 52		16.8	BLUE STELLAR OBJECT
LB 01504	22 31 54.	- 48 38		13.2	FAINT BLUE STAR
SC 2228-5659.7	22 31 55.	- 56 44 16.	18		NEBULA
RNGC 7310	22 31 56.	- 22 45		14.0	GALAXY
SC 2228-4347.6	22 31 56.	- 43 32 09.			NEBULA
SC 2228-4347.4	22 31 57.	- 43 31 57.			NEBULA
KEEL 707	22 31 57.2	+ 34 38 20.			NEBULA
SC 2229-4320.4	22 31 58.	- 43 04 57.			NEBULA
SC 2229-4332.8	22 31 58.	- 43 17 21.			NEBULA
B 370	22 32	+ 56 23			DARK OBJECT
LBN 0526	22 32	+ 66 30	2100		BRIGHT NEBULA
PHL 0319	22 32 00.	+ 00 06		18.0	BLUE STELLAR OBJECT
PHL 5307	22 32 00.	+ 02 28		17.8	BLUE STELLAR OBJECT
LDN 1192	22 32 00.	+ 58 20	300		DARK NEBULA
LDN 1214	22 32 00.	+ 65 30	2040		DARK NEBULA
MCG+00-57-004	22 32 00.	- 00 07	42	15.5	GALAXY
PHL 5305	22 32 00.	- 00 11		17.8	BLUE STELLAR OBJECT
PHL 1982	22 32 00.	- 00 16		18.5	BLUE STELLAR OBJECT
PHL 1983	22 32 00.	- 00 54		18.7	BLUE STELLAR OBJECT
PHL 0320	22 32 00.	- 06 04		17.5	BLUE STELLAR OBJECT
PHL 5306	22 32 00.	- 16 54		16.8	BLUE STELLAR OBJECT
HELW 251	22 32 02.	- 25 56 15.			NEBULA
SC 2229-4326.6	22 32 02.	- 43 11 09.			NEBULA
SVEN 447	22 32 04.	- 25 56 03.	12	15.3	GALAXY
SVEN 448	22 32 04.	- 25 58 03.	12	16.1	GALAXY
SC 2229-4341.8	22 32 04.	- 43 26 21.			NEBULA
ZWG 404.023	22 32 06.	+ 05 33		14.5	GALAXY
UGC 12083	22 32 06.	+ 05 33	102	14.5	GALAXY SBb
ZWG 474.009	22 32 06.	+ 24 46		15.3	GALAXY
UGC 12084	22 32 06.	+ 24 46	96	15.3	GALAXY Sb/SBb
MCG+04-53-005	22 32 06.	+ 24 46	66	15.	GALAXY
ZWG 514.057	22 32 06.	+ 34 53		15.6	GALAXY
UGC 12085	22 32 06.	+ 34 53	60	15.6	GALAXY Sb
MCG+07-46-013	22 32 06.	+ 41 02	51	14.5	GALAXY
ZWG 531.009	22 32 06.	+ 41 03		14.8	GALAXY
UGC 12086	22 32 06.	+ 41 03	60	14.8	GALAXY S0-a
PHL 5308	22 32 06.	- 12 52		17.5	BLUE STELLAR OBJECT
PHL 5310	22 32 06.	- 18 14		17.9	BLUE STELLAR OBJECT
PHL 5309	22 32 06.	- 20 20		17.8	BLUE STELLAR OBJECT
MCG-04-53-016	22 32 06.	- 25 56 30.	78	14.	GALAXY
HELW 252	22 32 06.	- 25 58 15.			NEBULA
SC 2229-4341.6	22 32 06.	- 43 26 09.			NEBULA
SC 2229-4658.6	22 32 07.	- 46 43 09.	6		NEBULA
RNGC 7312	22 32 08.	+ 05 33		14.5	GALAXY
MCG+05-53-006	22 32 09.	+ 32 36 30.	150	15.	GALAXY
ARC 2455	22 32 10.	- 13 58		17.2	RICH CLUSTER OF GALAXIES
SC 2229-4441.4	22 32 10.	- 44 25 57.			NEBULA
UGC 12087	22 32 12.	+ 41 17	60	16.5	GALAXY S
ZWG 531.010	22 32 12.	+ 43 32		15.2	GALAXY
UGC 12088	22 32 12.	+ 43 32	66	15.2	GALAXY Sa
MCG+08-41-001	22 32 12.	+ 50 08	66	14.	GALAXY
ZWG 546.001	22 32 12.	+ 50 09		15.7	GALAXY
UGC 12089	22 32 12.	+ 50 09	66	15.7	GALAXY Sb-c
PHL 5312	22 32 12.	- 03 01		18.0	BLUE STELLAR OBJECT
TON-S 0057	22 32 12.	- 25 31		14.8	BLUE STAR
MCG-05-53-017	22 32 12.	- 27 03	18	16.	GALAXY
PHL 5311	22 32 12.	- 28 10		17.3	BLUE STELLAR OBJECT
PHL 1984	22 32 12.	- 31 56		18.0	BLUE STELLAR OBJECT
SC 2229-4422.8	22 32 12.	- 44 07 21.			NEBULA
SC 2229-4658.5	22 32 14.	- 46 43 03.			NEBULA
MCG+01-57-011	22 32 15.	+ 06 02	60	15.5	GALAXY
MCG-02-57-018	22 32 15.	- 13 11	72	14.	GALAXY
HN 0758	22 32 15.	- 61 48			NEBULA
IC 5230	22 32 16.	- 61 47			NONSTELLAR OBJECT
MCG+03-57-017	22 32 18.	+ 15 41	42	15.	GALAXY
ZWG 452.026	22 32 18.	+ 15 42		15.7	GALAXY
UGC 12090	22 32 18.	+ 15 42	60	15.7	GALAXY IRR
KARA.73B 0972	22 32 18.	+ 15 42	54	15.7	ISOLATED GALAXY S
MCG+03-57-018	22 32 18.	+ 18 23	84	14.5	GALAXY
MCG+07-46-014	22 32 18.	+ 43 32	54	14.5	GALAXY
MCG+08-41-002	22 32 18.	+ 50 10	18	14.	GALAXY

OBJECT NAME	RIGHT ASCEN.	DECLINATION	DIAM.	MAGN.	TYPE OF OBJECT
ZWG 546.002	22 32 18.	+ 50 11		15.7	GALAXY
UGC 12091	22 32 18.	+ 50 11	120	15.7	GALAXY (E)
PHL 1985	22 32 18.	- 01 32		16.8	BLUE STELLAR OBJECT
PHL 1986	22 32 18.	- 05 20		16.5	BLUE STELLAR OBJECT
PHL 5313	22 32 18.	- 18 44		18.8	BLUE STELLAR OBJECT
SC 2229-4601.7	22 32 22.	- 45 46 15.	12		NEBULA
SC 2229-5745.5	22 32 22.	- 57 30 03.	12		NEBULA
SC 2229-4303.6	22 32 23.	- 42 48 09.			NEBULA
SC 2229-4522.7	22 32 23.	- 45 07 15.			NEBULA
SC 2229-4533.1	22 32 23.	- 45 17 39.			NEBULA
SC 2229-5726.8	22 32 23.	- 57 11 21.	6		NEBULA
ZWG 404.024	22 32 24.	+ 06 03		15.7	GALAXY
UGC 12092	22 32 24.	+ 06 03	60	15.7	GALAXY
ZWG 452.027	22 32 24.	+ 18 23		15.4	GALAXY
UGC 12093	22 32 24.	+ 18 23	90	15.4	GALAXY Sc
UGC 12094	22 32 24.	+ 22 17	60	16.5	GALAXY
PHL 1987	22 32 24.	- 00 06		18.1	BLUE STELLAR OBJECT
PHL 1988	22 32 24.	- 02 40		18.4	BLUE STELLAR OBJECT
PHL 5314	22 32 24.	- 04 43		14.7	BLUE STELLAR OBJECT
MCG-01-57-015	22 32 24.	- 04 58	84	14.5	GALAXY
PHL 0321	22 32 24.	- 10 27		18.3	BLUE STELLAR OBJECT
PHL 5315	22 32 24.	- 12 46		18.7	BLUE STELLAR OBJECT
ARC 2456	22 32 24.	- 15 34		17.2	RICH CLUSTER OF GALAXIES
SC 2229-4438.2	22 32 26.	- 44 22 45.			NEBULA
SC 2229-5630.1	22 32 28.	- 56 14 39.	12		NEBULA
MCG+04-53-006	22 32 30.	+ 22 18	30	16.	GALAXY
IC 1449	22 32 30.	- 09 02 07.			NONSTELLAR OBJECT
MCG-03-57-023	22 32 30.	- 15 08 30.	48	15.	GALAXY
SC 2229-4626.8	22 32 31.	- 46 11 20.			NEBULA
SC 2229-4636.0	22 32 32.	- 46 20 32.			NEBULA
SC 2229-5547.3	22 32 32.	- 55 31 51.	18		NEBULA
SC 2229-5600.3	22 32 33.	- 55 44 51.	24		NEBULA
SC 2229-4630.1	22 32 34.	- 46 14 38.			NEBULA
ZWG 452.028	22 32 36.	+ 20 32		15.6	GALAXY
ZWG 546.003	22 32 36.	+ 49 55		15.6	GALAXY
UGC 12096	22 32 36.	+ 49 55	66	15.6	GALAXY S?
UGC 12095	22 32 36.	+ 49 55	60	15.6	GALAXY IRR
PHL 5317	22 32 36.	- 09 12		18.3	BLUE STELLAR OBJECT
PHL 1989	22 32 36.	- 12 06		18.4	BLUE STELLAR OBJECT
PHL 1992	22 32 36.	- 14 28		18.3	BLUE STELLAR OBJECT
PHL 1990	22 32 36.	- 16 54		18.1	BLUE STELLAR OBJECT
PHL 1991	22 32 36.	- 17 16		18.1	BLUE STELLAR OBJECT
PHL 5318	22 32 36.	- 18 25		18.1	BLUE STELLAR OBJECT
PHL 5316	22 32 36.	- 19 28		16.3	BLUE STELLAR OBJECT
SC 2229-4623.6	22 32 36.	- 46 08 08.			NEBULA
SC 2229-4627.1	22 32 36.	- 46 11 38.			NEBULA
SC 2229-4346.2	22 32 38.	- 43 30 44.			NEBULA
SC 2230-1320.4	22 32 40.	- 13 04 56.	36		NEBULA
SC 2229-4345.0	22 32 40.	- 43 29 32.			NEBULA
SC 2229-4648.3	22 32 40.	- 46 32 50.			NEBULA
SC 2229-5124.5	22 32 41.	- 51 09 02.	30		NEBULA
ZWG 404.025	22 32 42.	+ 07 55		15.5	GALAXY
KARA.73B 0973	22 32 42.	+ 07 55	24	15.5	ISOLATED GALAXY IR
MCG+06-49-036	22 32 42.	+ 36 56	36	15.	GALAXY
PK102-05.1	22 32 42.	+ 52 11	161	15.2	PLANETARY NEBULA
PHL 5319	22 32 42.	- 11 49		18.4	BLUE STELLAR OBJECT
MCG-04-53-017	22 32 42.	- 25 20	30	15.5	GALAXY
TON-S 0058	22 32 42.	- 26 57		15.2	BLUE STAR
SC 2229-4538.2	22 32 42.	- 45 22 44.			NEBULA
SC 2229-4628.0	22 32 42.	- 46 12 32.			NEBULA
SC 2229-4635.9	22 32 44.	- 46 20 26.			NEBULA
MCG-02-57-019	22 32 45.	- 13 05	30	15.5	GALAXY
SC 2229-4539.2	22 32 45.	- 45 23 44.			NEBULA
SVEN 449	22 32 46.	- 26 22	18	15.3	GALAXY
ZWG 514.058	22 32 48.	+ 36 57		15.3	GALAXY
PHL 1993	22 32 48.	- 00 23		4.1	BLUE STELLAR OBJECT
PHL 5322	22 32 48.	- 00 58		16.3	BLUE STELLAR OBJECT
PHL 0322	22 32 48.	- 12 46		18.0	BLUE STELLAR OBJECT
PHL 5321	22 32 48.	- 16 16		18.6	BLUE STELLAR OBJECT
PHL 5320	22 32 48.	- 28 17		16.9	BLUE STELLAR OBJECT
TON-S 0059	22 32 48.	- 28 19		14.9	BLUE STAR
SC 2229-4637.4	22 32 48.	- 46 21 56.			NEBULA
RNGC 7313	22 32 49.	- 26 23			GALAXY
SC 2229-4645.8	22 32 51.	- 46 30 20.			NEBULA
PHL 0323	22 32 54.	+ 02 37		18.0	BLUE STELLAR OBJECT
PHL 0324	22 32 54.	- 07 31		18.1	BLUE STELLAR OBJECT
PHL 5324	22 32 54.	- 10 39		18.4	BLUE STELLAR OBJECT
PHL 1995	22 32 54.	- 13 51		18.9	BLUE STELLAR OBJECT
MCG-03-57-024	22 32 54.	- 15 27	30	15.	GALAXY
PHL 1994	22 32 54.	- 29 46		15.9	BLUE STELLAR OBJECT
PHL 5323	22 32 54.	- 31 01		15.9	BLUE STELLAR OBJECT
KEEL 708	22 32 54.8	+ 34 14 05.			NEBULA
SC 2229-4526.0	22 32 55.	- 45 10 32.			NEBULA
SC 2229-4547.2	22 32 55.	- 45 31 44.			NEBULA
SC 2229-4646.4	22 32 57.	- 46 30 56.			NEBULA
SVEN 450	22 32 58.	- 26 19	180	12.0	GALAXY
VDB.66G 214	22 33	- 03 11	130		DWARF GALAXY
ZWG 429.009	22 33 00.	+ 13 08		15.6	GALAXY
ZWG 452.029	22 33 00.	+ 19 25		15.2	GALAXY
PHL 5325	22 33 00.	- 05 55		15.8	BLUE STELLAR OBJECT
PHL 5326	22 33 00.	- 09 47		18.8	BLUE STELLAR OBJECT
ARP 014	22 33 00.	- 26 18			PECULIAR GALAXY
MCG-04-53-018	22 33 00.	- 26 19	288	11.5	GALAXY
RNGC 7314	22 33 01.	- 26 18		12.0	GALAXY
SC 2230-4645.2	22 33 04.	- 46 29 44.			NEBULA
SC 2230-4645.1	22 33 05.	- 46 29 37.			NEBULA
SC 2229-5209.1	22 33 05.	- 51 53 38.	18		NEBULA
ZC 2233.1+1810	22 33 06.	+ 18 10	1550		CLUSTER OF GALAXIES
MCG+03-57-019	22 33 06.	+ 19 25	36	15.	GALAXY
MRSL 105-00/1	22 33 06.	+ 57 57	600		HII REGION
PHL 0325	22 33 06.	- 08 26		18.5	BLUE STELLAR OBJECT
PHL 5327	22 33 06.	- 12 20		17.9	BLUE STELLAR OBJECT
MCG-05-53-018	22 33 06.	- 31 56	24	16.	GALAXY
SC 2230-4413.0	22 33 08.	- 43 57 31.			NEBULA
SC 2230-4535.0	22 33 08.	- 45 19 31.			NEBULA
MCG+06-49-037	22 33 09.	+ 34 31	24	14.	GALAXY
SC 2230-4644.0	22 33 09.	- 46 28 31.			NEBULA
RNGC 7315	22 33 10.	+ 34 34		14.0	GALAXY
SC 2230-4341.0	22 33 11.	- 43 25 31.			NEBULA
KEEL 709	22 33 11.5	+ 34 15 50.			NEBULA
PHL 0326	22 33 12.	+ 01 18		18.5	BLUE STELLAR OBJECT
ARC 2458	22 33 12.	+ 18 16		16.9	RICH CLUSTER OF GALAXIES
ZWG 514.059	22 33 12.	+ 34 34		13.8	GALAXY
UGC 12097	22 33 12.	+ 34 34	102	13.8	GALAXY S0
PHL 5328	22 33 12.	- 12 02		17.8	BLUE STELLAR OBJECT
PHL 1996	22 33 12.	- 17 27		18.9	BLUE STELLAR OBJECT
TON-S 0060	22 33 12.	- 31 01		15.2	BLUE STAR
SC 2230-4416.4	22 33 13.	- 44 00 55.			NEBULA
ARC 2457	22 33 15.	+ 01 13		16.0	RICH CLUSTER OF GALAXIES
ZC 2233.3+1911	22 33 18.	+ 19 11	8600		CLUSTER OF GALAXIES
SG 2.090	22 33 18.	+ 58 11	1200		DIFFUSE EMISSION NEBULA
PHL 5329	22 33 18.	- 15 52		17.7	BLUE STELLAR OBJECT
PHL 0327	22 33 18.	- 19 42		18.1	BLUE STELLAR OBJECT
SC 2230-4613.6	22 33 18.	- 45 58 07.			NEBULA
KEEL 710	22 33 20.0	+ 34 00 28.			NEBULA
MCG+01-57-012	22 33 21.	+ 03 45 30.	36	15.5	GALAXY
HELW 253	22 33 21.	- 26 34 31.			NEBULA
SVEN 451	22 33 22.	- 26 35	30	15.2	GALAXY
SC 2230-6155.6	22 33 22.	- 61 40 07.			NEBULA
ZC 2233.4+2759	22 33 24.	+ 27 59	940		CLUSTER OF GALAXIES
PHL 5331	22 33 24.	- 00 54		17.3	BLUE STELLAR OBJECT
PHL 5332	22 33 24.	- 09 16		18.2	BLUE STELLAR OBJECT
PHL 5330	22 33 24.	- 12 08		17.8	BLUE STELLAR OBJECT
PHL 0328	22 33 24.	- 16 39		6.5	BLUE STELLAR OBJECT
SC 2230-4338.6	22 33 24.	- 43 23 07.			NEBULA
SC 2230-6154.7	22 33 24.	- 61 39 13.			NEBULA
SVEN 452	22 33 28.	- 26 33 01.	12	16.7	GALAXY
SC 2230-4319.2	22 33 29.	- 43 03 43.			NEBULA
2ZW 182	22 33 30.	+ 18 50			COMPACT GALAXY
ZWG 452.030	22 33 30.	+ 20 04		13.7	GALAXY
MRK 307	22 33 30.	+ 20 04	36	15.5	GALAXY WITH UV CONTINUUM
RNGC 7316	22 33 30.	+ 20 04		13.5	GALAXY
MCG+03-57-020	22 33 30.	+ 20 04	66	13.5	GALAXY
KARA.73B 0974	22 33 30.	+ 20 04	66	13.7	ISOLATED GALAXY S
UGC 12098	22 33 30.	+ 20 07	66	13.7	GALAXY S
4ZW 106	22 33 30.	+ 28 14			COMPACT GALAXY
MCG+06-49-038	22 33 30.	+ 33 39	24	15.	GALAXY
PHL 0329	22 33 30.	- 00 08		13.7	BLUE STELLAR OBJECT
PHL 0330	22 33 30.	- 01 58		15.7	BLUE STELLAR OBJECT
PHL 5333	22 33 30.	- 15 55		17.1	BLUE STELLAR OBJECT
HELW 254	22 33 30.	- 26 33 07.			NEBULA
HELW 254	22 33 30.	- 26 33 07.			NEBULA
MCG-05-53-019	22 33 30.	- 26 34	72	15.	GALAXY
SC 2230-4444.6	22 33 30.	- 44 29 07.			NEBULA
LB 01505	22 33 30.	- 46 34		14.1	FAINT BLUE STAR
SC 2230-4344.6	22 33 31.	- 43 29 07.			NEBULA
SC 2230-5836.7	22 33 31.	- 58 21 13.			NEBULA
RNGC 7317	22 33 33.	+ 33 41		15.5	GALAXY
SC 2230-4350.3	22 33 34.	- 43 34 49.			NEBULA
SC 2230-4417.6	22 33 34.	- 44 02 07.			NEBULA
SC 2230-4633.9	22 33 34.	- 46 18 25.			NEBULA
KEEL 711	22 33 34.3	+ 34 24 05.			NEBULA
HOLM 792D	22 33 35.	+ 33 41	12	14.8	PART OF MULTIPLE GALAXY
ARP 319	22 33 35.	+ 33 41			PECULIAR GALAXY
MCG+06-49-040	22 33 36.	+ 33 40	90	14.5	GALAXY
MCG+06-49-039	22 33 36.	+ 33 40 30.	24	14.5	GALAXY
ZWG 514.060	22 33 36.	+ 33 41		15.3	GALAXY
VV 288D	22 33 36.	+ 33 41	36	15.3	INTERACTING GALAXY
MCG+08-41-004	22 33 36.	+ 49 53	72	16.	GALAXY
MCG+08-41-003	22 33 36.	+ 49 53	48	16.	GALAXY
PHL 5334	22 33 36.	- 08 33		15.4	BLUE STELLAR OBJECT
PHL 5335	22 33 36.	- 11 06		18.3	BLUE STELLAR OBJECT
MCG-02-57-020	22 33 36.	- 13 07	24	15.5	GALAXY
PHL 0331	22 33 36.	- 17 52		18.5	BLUE STELLAR OBJECT
PHL 0332	22 33 35.	- 20 08		15.9	BLUE STELLAR OBJECT
PHL 0333	22 33 36.	- 23 32		14.1	BLUE STELLAR OBJECT
PHL 1997	22 33 36.	- 23 54		17.9	BLUE STELLAR OBJECT
PHL 0334	22 33 36.	- 31 58		12.5	BLUE STELLAR OBJECT
MCG-05-53-020	22 33 36.	- 32 02	36	15.	GALAXY
SC 2230-6110.2	22 33 36.	- 60 54 43.	6		NEBULA
LB 00374	22 33 37.	- 23 32 12.		14.5	FAINT BLUE STAR
SC 2230-4636.8	22 33 37.	- 46 21 19.	18		NEBULA
SC 2230-5839.1	22 33 38.	- 58 23 37.	12		NEBULA
MCG+06-49-042	22 33 39.	+ 33 39 30.	114	13.	GALAXY
MCG+06-49-041	22 33 39.	+ 33 41 30.	78	15.	GALAXY
RNGC 7318B	22 33 39.	+ 33 42		14.0	GALAXY
RNGC 7318A	22 33 39.	+ 33 42		15.0	GALAXY
SC 2230-4535.5	22 33 39.	- 45 20 01.			NEBULA
ZWG 429.010	22 33 42.	+ 11 58		15.7	GALAXY
ZC 2233.7+2945	22 33 42.	+ 29 45	1210		CLUSTER OF GALAXIES
HOLM 792C	22 33 42.	+ 33 42	12	14.8	PART OF MULTIPLE GALAXY
VV 288E	22 33 42.	+ 33 42	102	14.9	INTERACTING GALAXY
VV 288C	22 33 42.	+ 33 42	36	14.8	INTERACTING GALAXY
VV 288E	22 33 42.	+ 33 42	162	13.7	INTERACTING GALAXY
ZWG 514.062	22 33 42.	+ 33 43		14.4	GALAXY
ZWG 514.061	22 33 42.	+ 33 43		14.9	GALAXY
UGC 12100	22 33 42.	+ 33 43	102	14.4	GALAXY E+SB(s)
UGC 12099	22 33 42.	+ 33 43	102	14.9	GALAXY E+SB(s)
5ZW 387	22 33 42.	+ 40 20			COMPACT GALAXY
PHL 0335	22 33 42.	- 01 10		18.1	BLUE STELLAR OBJECT
PHL 5336	22 33 42.	- 10 04		18.8	BLUE STELLAR OBJECT
MCG-02-57-021	22 33 42.	- 13 22	72	15.	GALAXY
MCG-04-53-019	22 33 42.	- 24 36 30.	24	15.	GALAXY
RNGC 7320	22 33 45.	+ 33 41		14.0	GALAXY
RNGC 7319	22 33 45.	+ 33 43		15.0	GALAXY
SC 2230-4333.2	22 33 45.	- 43 17 42.			NEBULA
ARP 003	22 33 46.	- 03 12			PECULIAR GALAXY
SC 2230-6004.1	22 33 46.	- 59 48 37.	30		NEBULA
HOLM 792A	22 33 47.	+ 33 41	120	13.6	PART OF MULTIPLE GALAXY
HOLM 792B	22 33 47.	+ 33 43	30	14.6	PART OF MULTIPLE GALAXY
PHL 1998	22 33 48.	+ 01 46		17.6	BLUE STELLAR OBJECT
ZWG 514.063	22 33 48.	+ 33 41		13.8	GALAXY
UGC 12101	22 33 48.	+ 33 41	114	13.8	GALAXY Sc
ZWG 514.064	22 33 48.	+ 33 43		14.8	GALAXY
UGC 12102	22 33 48.	+ 33 43	96	14.8	GALAXY SB
VV 288A	22 33 48.	+ 33 43	114	13.6	INTERACTING GALAXY
PHL 5337	22 33 48.	- 06 02		14.1	BLUE STELLAR OBJECT
PHL 5340	22 23 48.	- 07 00		17.9	BLUE STELLAR OBJECT
PHL 0336	22 33 48.	- 12 54		18.7	BLUE STELLAR OBJECT
PHL 5338	22 33 48.	- 15 40		17.9	BLUE STELLAR OBJECT
MCG-04-53-020	22 33 48.	- 24 36	48	15.	GALAXY
PHL 5339	22 33 48.	- 30 50		18.0	BLUE STELLAR OBJECT
TON-S 0061	22 33 48.	- 31 57		13.3	BLUE STAR
SN 1971P	22 33 49.	+ 33 43		16.8	SUPERNOVA
SC 2230-4653.2	22 33 49.	- 46 37 42.			NEBULA
HELW 255	22 33 50.	- 26 31 12.			NEBULA
SC 2230-4704.5	22 33 50.	- 46 49 00.			NEBULA
SVEN 453	22 33 52.	- 26 31 00.	30	15.8	GALAXY
SC 2230-4706.2	22 33 52.	- 46 50 42.			NEBULA
PHL 5341	22 33 54.	+ 03 35		18.2	BLUE STELLAR OBJECT
ZWG 429.011	22 33 54.	+ 13 45		15.7	GALAXY
MCG+06-49-043	22 33 54.	+ 33 42	24	17.	GALAXY
MCG+06-49-044	22 33 54.	+ 34 15	42	15.	GALAXY
PHL 1999	22 33 54.	- 01 00		17.6	BLUE STELLAR OBJECT
PHL 0337	22 33 54.	- 06 26		16.8	BLUE STELLAR OBJECT
MCG-02-57-022	22 33 54.	- 12 50	72	14.5	GALAXY

OBJECT NAME	RIGHT ASCEN.	DECLINATION	DIAM.	MAGN.	TYPE OF OBJECT
ARC 2459	22 33 54.	- 15 56		16.0	RICH CLUSTER OF GALAXIES
SC 2230-4658.0	22 33 54.	- 46 42 30.			NEBULA
SC 2230-4704.9	22 33 56.	- 46 49 24.			NEBULA
RNGC 7320C	22 33 57.	+ 33 44			GALAXY
SC 2230-4553.2	22 33 57.	- 45 37 42.			NEBULA
SC 2231-4253.9	22 33 59.	- 42 38 24.			NEBULA
KHAV 775	22 34	+ 56 46	1720		DARK NEBULA
LBN 0539	22 34	+ 68 50	120		BRIGHT NEBULA
ZC 2234.0+0443	22 34 00.	+ 04 43	1410		CLUSTER OF GALAXIES
ZWG 452.031	22 34 00.	+ 21 22		14.0	GALAXY
RNGC 7321	22 34 00.	+ 21 22		14.0	GALAXY
UGC 12103	22 34 00.	+ 21 22	102	14.0	GALAXY SBb
MCG+03-57-021	22 34 00.	+ 21 22	96	13.5	GALAXY
MCG+05-53-007	22 34 00.	+ 32 16 30.	48	15.5	GALAXY
ZWG 514.065	22 34 00.	+ 36 03		15.6	GALAXY
PHL 5342	22 34 00.	- 02 28		17.6	BLUE STELLAR OBJECT
MCG-01-57-016	22 34 00.	- 03 09	120	13.5	GALAXY
PHL 5343	22 34 00.	- 05 40		17.6	BLUE STELLAR OBJECT
PHL 5344	22 34 00.	- 09 30		18.4	BLUE STELLAR OBJECT
PHL 0338	22 34 00.	- 16 58		18.6	BLUE STELLAR OBJECT
MCG-05-53-021	22 34 00.	- 26 31	36	15.	GALAXY
KEEL 712	22 34 00.7	+ 34 17 03.			NEBULA
HOLM 793B	22 34 03.	+ 21 22	24	15.2	PART OF MULTIPLE GALAXY
SC 2231-4705.6	22 34 04.	- 46 50 06.			NEBULA
SC 2230-6149.0	22 34 04.	- 61 33 30.	12		NEBULA
SC 2230-7315.6	22 34 04.	- 73 00 07.	12		NEBULA
SC 2231-4637.6	22 34 05.	- 46 22 06.			NEBULA
SC 2231-4639.6	22 34 05.	- 46 24 06.			NEBULA
SC 2231-4721.0	22 34 05.	- 47 05 30.			NEBULA
UGC 12104	22 34 06.	+ 02 09	72	16.0	GALAXY S
HOLM 793A	22 34 06.	+ 21 23	54	13.3	PART OF MULTIPLE GALAXY
ZWG 514.066	22 34 06.	+ 34 18		15.7	GALAXY
UGC 12105	22 34 06.	+ 75 23	72	14.7	GALAXY S0
SC 2231-4448.1	22 34 07.	- 44 32 36.			NEBULA
SC 2231-1303.9	22 34 08.	- 12 48 23.	18		NEBULA
RNGC 7326	22 34 09.	+ 34 17			GALAXY
SC 2230-6147.2	22 34 09.	- 61 31 42.	12		NEBULA
HN 0759	22 34 09.	- 69 08			NEBULA
SC 2231-4714.7	22 34 10.	- 46 59 12.	12		NEBULA
IC 5232	22 34 10.	- 69 07			NONSTELLAR OBJECT
IC 5233	22 34 11.	+ 25 29 29.			NONSTELLAR OBJECT
SC 2231-4313.2	22 34 11.	- 42 57 42.			NEBULA
MCG+03-57-022	22 34 12.	+ 19 16	42	15.	GALAXY
MCG+03-57-023	22 34 12.	+ 19 23	36	15.5	GALAXY
ZWG 452.032	22 34 12.	+ 19 24		15.5	GALAXY
ZWG 474.010	22 34 12.	+ 25 30		14.8	GALAXY
UGC 12106	22 34 12.	+ 25 30	66	14.8	GALAXY S
MCG+04-53-007	22 34 12.	+ 25 30	54	14.5	GALAXY
KARA.73B 0975	22 34 12.	+ 25 30	60	14.8	ISOLATED GALAXY S
PHL 0339	22 34 12.	- 08 38		18.6	BLUE STELLAR OBJECT
PHL 5345	22 34 12.	- 09 47		17.9	BLUE STELLAR OBJECT
PHL 2000	22 34 12.	- 10 15		18.6	BLUE STELLAR OBJECT
MCG-02-57-023	22 34 12.	- 12 49	54	14.5	GALAXY
SC 2231-4444.2	22 34 12.	- 44 28 42.			NEBULA
SC 2231-4450.9	22 34 12.	- 44 35 24.			NEBULA
SC 2230-6339.9	22 34 12.	- 63 24 24.	18		NEBULA
SC 2231-4448.6	22 34 14.	- 44 33 06.			NEBULA
RNGC 7320A	22 34 15.	+ 33 31			GALAXY
RNGC 7325	22 34 15.	+ 34 14			GALAXY
SC 2231-1305.1	22 34 15.	- 12 49 35.	18		NEBULA
MCG-03-57-025	22 34 15.	- 20 04 30.	54	14.5	GALAXY
MCG-04-53-021	22 34 15.	- 22 29	72	14.5	GALAXY
SC 2231-4421.2	22 34 16.	- 44 05 42.			NEBULA
SC 2231-4607.8	22 34 16.	- 45 52 18.			NEBULA
SC 2231-4624.0	22 34 16.	- 46 08 30.	12		NEBULA
SC 2231-4712.0	22 34 16.	- 46 56 30.	18		NEBULA
SC 2230-6422.9	22 34 16.	- 64 07 24.	48		NEBULA
ZC 2234.3+0726	22 34 18.	+ 07 26	740		CLUSTER OF GALAXIES
ZWG 452.033	22 34 18.	+ 19 07		15.3	GALAXY
UGC 12107	22 34 18.	+ 19 07	84	15.3	GALAXY S0
MCG+03-57-024	22 34 18.	+ 19 07	66	15.	GALAXY
4ZW 107	22 34 18.	+ 39 16			COMPACT GALAXY
MCG-02-57-024	22 34 18.	- 12 50	36	15.	GALAXY
PHL 0340	22 34 18.	- 24 10		18.7	BLUE STELLAR OBJECT
MCG-07-46-005	22 34 18.	- 43 58	24	15.5	GALAXY
KEEL 713	22 34 18.6	+ 33 57 46.			NEBULA
MCG-04-53-022	22 34 21.	- 25 30	120	14.5	GALAXY
SC 2231-4607.6	22 34 22.	- 45 52 05.			NEBULA
RNGC 7323	22 34 23.	+ 18 53		14.0	GALAXY
PHL 0341	22 34 24.	+ 03 24		16.5	BLUE STELLAR OBJECT
ZWG 429.012	22 34 24.	+ 14 07		15.7	GALAXY
MCG+02-57-004	22 34 24.	+ 14 09	24	16.	GALAXY
MCG+03-57-025	22 34 24.	+ 18 52	90	13.5	GALAXY
ZWG 452.034	22 34 24.	+ 18 53		14.0	GALAXY
UGC 12108	22 34 24.	+ 18 53	78	14.0	GALAXY Sb
KARA.72 569A	22 34 24.	+ 18 53	78	14.0	PART OF DOUBLE GALAXY
ZWG 452.035	22 34 24.	+ 19 12		15.5	GALAXY
BIGO 557	22 34 24.	+ 34 11			NEBULA
UGC 12109	22 34 24.	+ 38 35	60	16.0	GALAXY
PHL 0342	22 34 24.	- 05 26		18.3	BLUE STELLAR OBJECT
PHL 0343	22 34 24.	- 15 16		18.8	BLUE STELLAR OBJECT
MCG-04-53-023	22 34 24.	- 24 56 30.	12	16.	GALAXY
PHL 5347	22 34 24.	- 28 36		18.2	BLUE STELLAR OBJECT
PHL 5346	22 34 24.	- 29 52		16.1	BLUE STELLAR OBJECT
SC 2231-4448.4	22 34 24.	- 44 32 53.			NEBULA
SC 2231-4615.4	22 34 24.	- 45 59 53.			NEBULA
HOLM 794A	22 34 25.	+ 18 53	42	13.3	PART OF MULTIPLE GALAXY
SC 2231-5549.4	22 34 25.	- 55 33 53.	12		NEBULA
SC 2231-5121.1	22 34 27.	- 51 05 35.	6		NEBULA
SC 2231-4654.0	22 34 28.	- 46 38 29.			NEBULA
SC 2231-6105.9	22 34 29.	- 60 50 23.	18		NEBULA
ZWG 429.013	22 34 30.	+ 14 09		15.4	GALAXY
UGC 12110	22 34 30.	+ 14 09	72	15.4	GALAXY S
MCG+02-57-006	22 34 30.	+ 14 09	18	16.	GALAXY
MCG+02-57-005	22 34 30.	+ 14 10	54	15.	GALAXY
MCG-04-53-024	22 34 30.	- 24 57	36	15.	GALAXY
SC 2231-4448.4	22 34 30.	- 44 32 53.			NEBULA
HOLM 795F	22 34 31.	+ 34 06	12	14.8	PART OF MULTIPLE GALAXY
SC 2231-4256.0	22 34 32.	- 42 40 29.			NEBULA
SC 2231-4626.4	22 34 32.	- 46 10 53.	12		NEBULA
HOLM 794B	22 34 33.	+ 18 53	12	14.5	PART OF MULTIPLE GALAXY
RNGC 7327	22 34 33.	+ 34 13			NON-EXISTENT OBJECT
HOLM 795H	22 34 34.	+ 34 09	12	15.3	PART OF MULTIPLE GALAXY
SC 2231-4450.6	22 34 34.	- 44 35 05.			NEBULA
RNGC 7324	22 34 35.	+ 18 53		15.0	GALAXY
SC 2231-4421.4	22 34 35.	- 44 05 53.			NEBULA
ZC 2234.6+1242	22 34 36.	+ 12 42	3630		CLUSTER OF GALAXIES
ZWG 429.014	22 34 36.	+ 13 57		15.6	GALAXY

OBJECT NAME	RIGHT ASCEN.	DECLINATION	DIAM.	MAGN.	TYPE OF OBJECT
ZWG 452.036	22 34 36.	+ 18 53		15.1	GALAXY
KARA.72 569B	22 34 36.	+ 18 53	48	15.1	PART OF DOUBLE GALAXY
MCG+03-57-026	22 34 36.	+ 18 53	36	15.	GALAXY
ZWG 474.011	22 34 36.	+ 22 40		15.7	GALAXY
5ZW 388	22 34 36.	+ 47 10			COMPACT GALAXY
PHL 2001	22 34 36.	- 06 50		18.5	BLUE STELLAR OBJECT
MCG-02-57-025	22 34 36.	- 10 09	36	15.5	GALAXY
PHL 5348	22 34 36.	- 18 49		18.7	BLUE STELLAR OBJECT
HOLM 795G	22 34 37.	+ 34 11	36	14.9	PART OF MULTIPLE GALAXY
SC 2231-4332.3	22 34 37.	- 43 16 47.			NEBULA
SC 2231-4528.3	22 34 37.	- 45 12 47.			NEBULA
SC 2231-4640.2	22 34 38.	- 46 24 41.	12		NEBULA
MCG+06-49-045	22 34 39.	+ 34 06	600	10.	GALAXY
RNGC 7330	22 34 41.	+ 38 17		13.5	GALAXY
SN 1954T	22 34 42.	+ 03 28		19.0	SUPERNOVA
ZWG 514.067	22 34 42.	+ 38 17		13.6	GALAXY
UGC 12111	22 34 42.	+ 38 17	108	13.6	GALAXY E
MCG+06-49-046	22 34 42.	+ 38 17	24	14.	GALAXY
MCG+00-57-005	22 34 42.	- 01 19	48	16.5	GALAXY
PHL 5349	22 34 42.	- 11 56		18.2	BLUE STELLAR OBJECT
MCG-04-53-025	22 34 42.	- 24 30	48	15.	GALAXY
TON-S 0062	22 34 42.	- 31 32		15.2	BLUE STAR
SC 2231-4423.0	22 34 42.	- 44 07 29.			NEBULA
SC 2231-4445.3	22 34 42.	- 44 29 47.			NEBULA
SC 2231-4503.4	22 34 42.	- 44 47 53.			NEBULA
SC 2231-4650.2	22 34 44.	- 46 34 41.			NEBULA
RNGC 7331	22 34 45.	+ 34 10		10.5	GALAXY
SN 1959D	22 34 46.	+ 34 10		13.4	SUPERNOVA
KEEL 714	22 34 46.6	+ 34 34 45.			NEBULA
REIN 2.305	22 34 46.78	+ 34 09 20.5			NEBULA
HOLM 795A	22 34 47.	+ 34 09	450	10.8	PART OF MULTIPLE GALAXY
UGC 12112	22 34 48.	+ 11 42	60	16.0	GALAXY Sc
ZWG 514.068	22 34 48.	+ 34 10		10.4	GALAXY
UGC 12113	22 34 48.	+ 34 10	684	10.4	GALAXY Sb
PHL 0344	22 34 48.	- 09 34		18.3	BLUE STELLAR OBJECT
PHL 0345	22 34 48.	- 19 56		16.6	BLUE STELLAR OBJECT
LB 01506	22 34 48.	- 47 39		14.4	FAINT BLUE STAR
REIN 2.306	22 34 49.59	+ 34 08 21.8			NEBULA
KEEL 715	22 34 50.1	+ 34 21 37.			NEBULA
RNGC 7322	22 34 52.	- 37 29			GALAXY
HOLM 795I	22 34 54.	+ 34 10	6	15.3	PART OF MULTIPLE GALAXY
MCG+06-49-047	22 34 54.	+ 34 10	24	14.	GALAXY
UGC 12114	22 34 54.	- 01 17	60	16.5	GALAXY
PHL 5350	22 34 54.	- 13 45		18.3	BLUE STELLAR OBJECT
PHL 0346	22 34 54.	- 18 56		11.7	BLUE STELLAR OBJECT
SC 2231-6127.3	22 34 55.	- 61 11 47.	6		NEBULA
RNGC 7335	22 34 57.	+ 34 11		14.5	GALAXY
RNGC 7333	22 34 57.	+ 34 11			NON-EXISTENT OBJECT
RNGC 7336	22 34 57.	+ 34 13		15.0	GALAXY
SC 2232-4446.8	22 34 58.	- 44 31 16.			NEBULA
HOLM 796A	22 34 59.	+ 23 32	126	13.1	PART OF MULTIPLE GALAXY
LBN 0433	22 35	+ 37 50	4500		BRIGHT NEBULA
PHL 2003	22 35 00.	+ 03 04		17.9	BLUE STELLAR OBJECT
MCG+02-57-007	22 35 00.	+ 10 17	120	14.	GALAXY
ZC 2235.0+1731	22 35 00.	+ 17 31	2350		CLUSTER OF GALAXIES
MCG+03-57-027	22 35 00.	+ 18 32	60	15.	GALAXY
ZWG 474.012	22 35 00.	+ 23 32		12.0	GALAXY
RNGC 7332	22 35 00.	+ 23 32		12.5	GALAXY
UGC 12115	22 35 00.	+ 23 32	216	12.0	GALAXY S0
KARA.72 570A	22 35 00.	+ 23 32	204	12.0	PART OF DOUBLE GALAXY
MCG+04-53-008	22 35 00.	+ 23 32	210	12.5	GALAXY
MCG+06-49-050	22 35 00.	+ 34 05 30.	54	15.	GALAXY
MCG+06-49-049	22 35 00.	+ 34 11 30.	24	15.	GALAXY
ZWG 514.069	22 35 00.	+ 34 12		14.7	GALAXY
UGC 12116	22 35 00.	+ 34 12	84	14.7	GALAXY S0-a
MCG+06-49-048	22 35 00.	+ 38 22	48	15.5	GALAXY
ZWG 514.070	22 35 00.	+ 38 23		15.6	GALAXY
UGC 12117	22 35 00.	+ 49 56	60	16.0	GALAXY
LDN 1198	22 35 00.	+ 59 10	900		DARK NEBULA
LDN 1251	22 35 00.	+ 75 00	1740		DARK NEBULA
PHL 5351	22 35 00.	- 09 33		17.9	BLUE STELLAR OBJECT
PHL 0347	22 35 00.	- 13 34		18.4	BLUE STELLAR OBJECT
PHL 0348	22 35 00.	- 14 40		17.4	BLUE STELLAR OBJECT
PHL 2002	22 35 00.	- 25 45		16.2	BLUE STELLAR OBJECT
HOLM 795C	22 35 02.	+ 34 11	60	14.5	PART OF MULTIPLE GALAXY
RNGC 7328	22 35 03.	+ 10 16		14.5	GALAXY
KEEL 716	22 35 03.3	+ 34 06 52.			NEBULA
HOLM 795J	22 35 04.	+ 34 13	12	15.4	PART OF MULTIPLE GALAXY
SN 1973O	22 35 05.	+ 34 07		19.0	SUPERNOVA
PHL 2004	22 35 06.	+ 00 00		17.4	BLUE STELLAR OBJECT
ZWG 429.015	22 35 06.	+ 10 16		14.3	GALAXY
UGC 12118	22 35 06.	+ 10 16	126	14.3	GALAXY Sa-b
KARA.73B 0976	22 35 06.	+ 10 16	132	14.3	ISOLATED GALAXY S
UGC 12119	22 35 06.	+ 18 32	60	16.0	GALAXY S
ZWG 514.071	22 35 06.	+ 34 07		15.7	GALAXY
UGC 12120	22 35 06.	+ 34 07	72	15.7	GALAXY SBa
SC 2232-4459.8	22 35 06.	- 44 44 16.			NEBULA
KEEL 717	22 35 07.2	+ 34 29 26.			NEBULA
SC 2232-4718.6	22 35 08.	- 47 03 04.			NEBULA
RNGC 2573B	22 35 08.	- 89 26			GALAXY
RNGC 2573A	22 35 08.	- 89 26			GALAXY
RNGC 7320B	22 35 09.	+ 33 39			GALAXY
HOLM 795B	22 35 09.	+ 34 06	12	14.4	PART OF MULTIPLE GALAXY
RNGC 7337	22 35 09.	+ 34 07		15.5	GALAXY
RNGC 7338	22 35 09.	+ 34 10			NON-EXISTENT OBJECT
SC 2232-4313.2	22 35 11.	- 42 57 40.			NEBULA
PHL 5353	22 35 12.	+ 00 51		17.5	BLUE STELLAR OBJECT
ZC 2235.2+0637	22 35 12.	+ 06 37	740		CLUSTER OF GALAXIES
ZWG 452.037	22 35 12.	+ 19 45		15.3	GALAXY
MCG+03-57-028	22 35 12.	+ 19 46	42	15.	GALAXY
ZWG 514.072	22 35 12.	+ 33 40		15.7	GALAXY
ZWG 514.073	22 35 12.	+ 35 17		15.3	GALAXY
ZC 2235.2-0248	22 35 12.	- 02 48	2550		CLUSTER OF GALAXIES
PHL 5355	22 35 12.	- 12 31		18.0	BLUE STELLAR OBJECT
PHL 5352	22 35 12.	- 15 34		17.0	BLUE STELLAR OBJECT
PHL 5354	22 35 12.	- 20 02		17.9	BLUE STELLAR OBJECT
PHL 2005	22 35 12.	- 20 29		17.9	BLUE STELLAR OBJECT
PHL 0349	22 35 12.	- 26 47		16.9	BLUE STELLAR OBJECT
SC 2232-4221.2	22 35 12.	- 42 05 40.			NEBULA
SC 2232-4523.3	22 35 12.	- 45 07 46.			NEBULA
ARC 2460	22 35 13.	+ 17 40		16.9	RICH CLUSTER OF GALAXIES
HOLM 795D	22 35 14.	+ 34 09	12	14.5	PART OF MULTIPLE GALAXY
SC 2232-4329.3	22 35 14.	- 43 13 46.			NEBULA
SC 2232-4455.9	22 35 16.	- 44 40 22.			NEBULA
SC 2232-4659.0	22 35 16.	- 46 43 28.			NEBULA
SC 2232-4537.4	22 35 17.	- 45 21 52.			NEBULA
SC 2232-4634.8	22 35 17.	- 46 19 16.	12		NEBULA
SC 2232-4645.8	22 35 17.	- 46 30 16.			NEBULA

OBJECT NAME	RIGHT ASCEN.	DECLINATION	DIAM.	MAGN.	TYPE OF OBJECT
MCG+06-49-051	22 35 18.	+ 34 33 30.	42	15.	GALAXY
ZWG 514.074	22 35 18.	+ 34 36		14.9	GALAXY
UGC 12121	22 35 18.	+ 34 36	66	14.9	GALAXY S0
PHL 2006	22 35 18.	- 00 16		16.0	BLUE STELLAR OBJECT
PHL 5357	22 35 18.	- 11 14		18.1	BLUE STELLAR OBJECT
PHL 2007	22 35 18.	- 27 52		18.2	BLUE STELLAR OBJECT
PHL 5356	22 35 18.	- 27 59		17.2	BLUE STELLAR OBJECT
KEEL 718	22 35 18.2	+ 34 35 11.			NEBULA
RNGC 7340	22 35 21.	+ 34 10		15.0	GALAXY
HOLM 796B	22 35 22.	+ 23 31	120	13.6	PART OF MULTIPLE GALAXY
SC 2232-4546.4	22 35 23.	- 45 30 52.			NEBULA
PHL 5359	22 35 24.	+ 00 54		16.6	BLUE STELLAR OBJECT
PHL 5358	22 35 24.	+ 02 00		15.0	BLUE STELLAR OBJECT
UGC 12122	22 35 24.	+ 23 22	174	13.1	GALAXY Sb-c
ZWG 474.013	22 35 24.	+ 23 31		13.1	GALAXY
KARA.72 570B	22 35 24.	+ 23 31	180	13.1	PART OF DOUBLE GALAXY
MCG+04-53-009	22 35 24.	+ 23 31 30.	168	13.	GALAXY
RNGC 7339	22 35 24.	+ 23 32		13.0	GALAXY
MCG+06-49-052	22 35 24.	+ 34 07	24	14.5	GALAXY
ZWG 514.075	22 35 24.	+ 34 10		14.9	GALAXY
PHL 5361	22 35 24.	- 11 06		18.3	BLUE STELLAR OBJECT
PHL 5360	22 35 24.	- 13 03		17.1	BLUE STELLAR OBJECT
PHL 2008	22 35 24.	- 13 38		18.3	BLUE STELLAR OBJECT
PHL 0350	22 35 24.	- 26 14		14.1	BLUE STELLAR OBJECT
SC 2232-4455.0	22 35 24.	- 44 39 28.			NEBULA
SC 2232-4601.2	22 35 24.	- 45 45 40.			NEBULA
HOLM 795E	22 35 27.	+ 34 09	12	14.8	PART OF MULTIPLE GALAXY
SC 2232-4607.5	22 35 27.	- 45 51 58.			NEBULA
SC 2232-5743.9	22 35 27.	- 57 28 22.	30		NEBULA
8ZW 2235+04.1	22 35 30.	+ 04 07		17.9	COMPACT GALAXY
ZWG 474.014	22 35 30.	+ 24 56		15.6	GALAXY
UGC 12123	22 35 30.	+ 24 56	66	15.6	GALAXY SB?c
ZWG 474.015	22 35 30.	+ 25 05		15.0	GALAXY
UGC 12124	22 35 30.	+ 25 05	60	15.0	GALAXY S0?
MCG+07-46-015	22 35 30.	+ 40 01 30.	102	16.	GALAXY
UGC 12125	22 35 30.	+ 40 04	90	16.0	GALAXY Sa
LB 00375	22 35 30.	- 26 13 18.		15.0	FAINT BLUE STAR
SC 2232-4606.9	22 35 30.	- 45 51 22.			NEBULA
PHL 5363	22 35 36.	- 01 02		17.0	BLUE STELLAR OBJECT
PHL 2010	22 35 35.	- 06 00		17.2	BLUE STELLAR OBJECT
MCG-01-57-017	22 35 36.	- 07 18	48	14.5	GALAXY
PHL 0351	22 35 36.	- 11 55		17.9	BLUE STELLAR OBJECT
PHL 0352	22 35 36.	- 14 32		18.1	BLUE STELLAR OBJECT
PHL 5362	22 35 36.	- 23 28		17.0	BLUE STELLAR OBJECT
PHL 2009	22 35 36.	- 23 28		17.9	BLUE STELLAR OBJECT
MCG-05-53-022	22 35 36.	- 28 30	30	15.	GALAXY
SC 2232-4548.6	22 35 37.	- 45 33 03.			NEBULA
SC 2232-4711.4	22 35 38.	- 46 55 51.			NEBULA
SC 2232-5554.8	22 35 38.	- 55 39 15.	18		NEBULA
MCG+01-57-013	22 35 39.	+ 08 20	24	15.5	GALAXY
SC 2232-4616.0	22 35 39.	- 46 00 27.			NEBULA
SC 2232-4712.2	22 35 40.	- 46 56 39.			NEBULA
IC 1450	22 35 41.	+ 34 17			SINGLE STAR
SC 2232-4537.4	22 35 41.	- 45 21 51.			NEBULA
ZWG 474.016	22 35 42.	+ 24 42		15.6	GALAXY
PHL 2011	22 35 42.	- 09 54		17.9	BLUE STELLAR OBJECT
PHL 5364	22 35 42.	- 13 55		18.1	BLUE STELLAR OBJECT
SC 2232-4422.2	22 35 44.	- 44 06 39.			NEBULA
SC 2232-4617.6	22 35 44.	- 46 02 03.			NEBULA
MCG-02-57-026	22 35 45.	- 13 24	54	15.	GALAXY
PHL 0353	22 35 48.	+ 03 02		17.5	BLUE STELLAR OBJECT
ZWG 404.026	22 35 48.	+ 08 22		15.6	GALAXY
ZWG 452.038	22 35 48.	+ 18 20		15.4	GALAXY
MCG+06-49-054	22 35 48.	+ 35 12 30.	72	14.	GALAXY
MCG+06-49-053	22 35 48.	+ 39 20	60	16.5	GALAXY
PHL 0354	22 35 48.	- 01 13		18.0	BLUE STELLAR OBJECT
PHL 2012	22 35 48.	- 09 12		18.9	BLUE STELLAR OBJECT
PHL 2013	22 35 48.	- 09 25		18.7	BLUE STELLAR OBJECT
PHL 0355	22 35 48.	- 10 05		17.4	BLUE STELLAR OBJECT
PHL 2014	22 35 48.	- 11 11		18.5	BLUE STELLAR OBJECT
PHL 5365	22 35 48.	- 28 43		15.5	BLUE STELLAR OBJECT
PHL 0356	22 35 48.	- 29 00		17.0	BLUE STELLAR OBJECT
SC 2232-4609.5	22 35 48.	- 45 53 57.	12		NEBULA
SC 2232-4357.1	22 35 51.	- 43 41 33.			NEBULA
RNGC 7342	22 35 52.	+ 35 15		15.5	GALAXY
ZC 2235.9+0329	22 35 54.	+ 03 29	1340		CLUSTER OF GALAXIES
MCG+03-57-029	22 35 54.	+ 18 19	54	15.	GALAXY
ZWG 514.076	22 35 54.	+ 35 15		15.3	GALAXY
UGC 12126	22 35 54.	+ 35 15	90	15.3	GALAXY SBa
PHL 5367	22 35 54.	- 00 26		18.2	BLUE STELLAR OBJECT
PHL 5366	22 35 54.	- 12 14		18.2	BLUE STELLAR OBJECT
MCG-04-53-026	22 35 54.	- 26 07	132	14.	GALAXY
PHL 0357	22 35 54.	- 30 17		15.1	BLUE STELLAR OBJECT
KEEL 719	22 35 54.7	+ 34 36 35.			NEBULA
SC 2232-4606.3	22 35 55.	- 45 50 45.			NEBULA
HELW 256	22 35 58.	- 26 06 45.			NEBULA
SVEN 454	22 35 58.	- 26 06 57.	24	15.1	GALAXY
RNGC 7334	22 35 58.	- 37 28			NON-EXISTENT OBJECT
SC 2232-5554.7	22 35 59.	- 55 39 09.	12		NEBULA
LBN 0442	22 36	+ 40 50	4200		BRIGHT NEBULA
KHAV 776	22 36	+ 74 52	3090		DARK NEBULA
VDB.66G 215	22 36	- 05 00	70		DWARF GALAXY
PHL 5370	22 36 00.	+ 02 38		17.9	BLUE STELLAR OBJECT
PHL 5369	22 36 00.	+ 03 26		17.7	BLUE STELLAR OBJECT
MCG+06-49-055	22 36 00.	+ 35 40	42	16.	GALAXY
ZWG 514.077	22 36 00.	+ 35 42		15.7	GALAXY
LDN 1187	22 36 00.	+ 57 00	1500		DARK NEBULA
LDN 1197	22 36 00.	+ 58 40	360		DARK NEBULA
PHL 2015	22 36 00.	- 11 30		18.0	BLUE STELLAR OBJECT
PHL 5368	22 36 00.	- 17 45		12.7	BLUE STELLAR OBJECT
MCG+06-49-057	22 36 03.	+ 35 05	7	16.	GALAXY
MCG+06-49-056	22 36 03.	+ 35 05	42	15.	GALAXY
SC 2233-4748.2	22 36 03.	- 47 32 39.			NEBULA
ARC 2461	22 36 04.	- 21 21		17.1	RICH CLUSTER OF GALAXIES
MCG+06-49-058	22 36 06.	+ 35 02 30.	24	14.	GALAXY
ZWG 514.078	22 36 06.	+ 35 07		15.4	GALAXY
PHL 5371	22 36 06.	- 07 20		17.7	BLUE STELLAR OBJECT
KEEL 720	22 36 07.9	+ 33 58 50.			NEBULA
SC 2233-4239.6	22 36 08.	- 42 24 02.			NEBULA
SC 2233-4601.1	22 36 09.	- 45 45 33.	24		NEBULA
SC 2233-4349.9	22 36 11.	- 43 34 20.			NEBULA
PHL 2016	22 36 12.	+ 03 05		17.7	BLUE STELLAR OBJECT
PHL 2019	22 36 12.	+ 03 15		17.9	BLUE STELLAR OBJECT
ZC 2236.2+0509	22 36 12.	+ 05 09	1550		CLUSTER OF GALAXIES
ZC 2236.2+1844	22 36 12.	+ 18 44	1080		CLUSTER OF GALAXIES
MCG+06-49-059	22 36 12.	+ 33 47	60	14.	GALAXY
ZWG 514.079	22 36 12.	+ 33 59		15.0	GALAXY
MCG+06-49-060	22 36 12.	+ 35 03 30.	24	15.5	GALAXY
ZWG 514.080	22 36 12.	+ 35 05		15.0	GALAXY
UGC 12127	22 36 12.	+ 35 05	90	15.0	GALAXY E
MCG+06-49-061	22 36 12.	+ 35 06	48	15.5	GALAXY
ZWG 514.081	22 36 12.	+ 35 08		15.7	GALAXY
PHL 0358	22 36 12.	- 04 23		13.9	BLUE STELLAR OBJECT
PHL 2017	22 36 12.	- 11 46		18.9	BLUE STELLAR OBJECT
PHL 0359	22 36 12.	- 18 15		15.3	BLUE STELLAR OBJECT
PHL 2018	22 36 12.	- 24 09		11.7	BLUE STELLAR OBJECT
PHL 5372	22 36 12.	- 27 54		17.7	BLUE STELLAR OBJECT
SC 2233-4542.0	22 36 12.	- 45 26 26.			NEBULA
ARC 2463	22 36 13.	+ 03 28		17.7	RICH CLUSTER OF GALAXIES
RNGC 7343	22 36 15.	+ 33 49		14.5	GALAXY
KEEL 721	22 36 15.3	+ 34 19 43.			NEBULA
UGC 12128	22 36 18.	+ 28 23	66	17.	GALAXY S?
ZWG 514.082	22 36 18.	+ 33 48		14.3	GALAXY
UGC 12129	22 36 18.	+ 33 48	60	14.3	GALAXY SBb
MCG+06-49-062	22 36 18.	+ 35 02 30.	18	16.	GALAXY
MCG-01-57-018	22 36 18.	- 06 07	132	13.5	GALAXY
PHL 0360	22 36 18.	- 15 28		18.3	BLUE STELLAR OBJECT
KEEL 722	22 36 18.0	+ 34 04 40.			NEBULA
SC 2233-4244.0	22 36 19.	- 42 28 26.			NEBULA
RNGC 7345	22 36 22.	+ 35 17		15.0	GALAXY
SC 2233-4234.4	22 36 22.	- 42 18 50.			NEBULA
SC 2233-4545.6	22 36 22.	- 45 30 02.			NEBULA
PHL 5374	22 36 24.	+ 00 22		18.1	BLUE STELLAR OBJECT
PHL 5373	22 36 24.	+ 01 15		17.0	BLUE STELLAR OBJECT
MCG+06-49-064	22 36 24.	+ 35 15 30.	72	14.5	GALAXY
ZWG 514.083	22 36 24.	+ 35 17		15.1	GALAXY
UGC 12130	22 36 24.	+ 35 17	72	15.1	GALAXY Sa
MCG+06-49-063	22 36 24.	+ 35 35	48	16.	GALAXY
ZWG 514.084	22 36 24.	+ 35 37		15.6	GALAXY
ZWG 514.085	22 36 24.	+ 37 20		15.2	GALAXY
UGC 12131	22 36 24.	+ 37 20	66	15.2	GALAXY Sb
ZWG 531.011	22 36 24.	+ 39 30		15.7	GALAXY
PHL 5375	22 36 24.	- 07 28		18.0	BLUE STELLAR OBJECT
PHL 2020	22 36 24.	- 09 00		18.9	BLUE STELLAR OBJECT
MCG-02-57-027	22 36 24.	- 12 52	24	15.5	GALAXY
MCG-04-53-027	22 36 24.	- 22 57	132	13.	GALAXY
TON-S 0063	22 36 24.	- 30 19		14.8	BLUE STAR
ARC 2462	22 36 25.	- 17 37		16.2	RICH CLUSTER OF GALAXIES
RNGC 7341	22 36 26.	- 22 57		13.0	GALAXY
SC 2233-4720.7	22 36 26.	- 47 05 08.			NEBULA
MCG-03-57-026	22 36 27.	- 19 56	54	15.	GALAXY
SC 2233-4611.0	22 36 27.	- 45 55 26.			NEBULA
KEEL 723	22 36 28.2	+ 34 02 34.			NEBULA
ZWG 514.086	22 36 30.	+ 34 03		15.4	GALAXY
UGC 12132	22 36 30.	+ 34 03	60	15.4	GALAXY SB?b
MCG+06-49-065	22 36 30.	+ 37 19	54	14.5	GALAXY
SC 2233-4234.3	22 36 30.	- 42 18 44.			NEBULA
SC 2233-4344.0	22 36 30.	- 43 28 26.			NEBULA
SC 2233-4615.1	22 36 31.	- 45 59 32.			NEBULA
MCG-04-53-028	22 36 33.	- 26 07	36	15.5	GALAXY
MCG+06-49-066	22 36 36.	+ 35 09	27	15.5	GALAXY
MCG+00-57-006	22 36 36.	- 01 59	30	15.	GALAXY
PHL 0361	22 36 36.	- 02 02		18.1	BLUE STELLAR OBJECT
PHL 2021	22 36 36.	- 02 26		17.2	BLUE STELLAR OBJECT
MCG-01-57-019	22 36 36.	- 05 02	72	16.	GALAXY
PHL 2022	22 36 36.	- 10 00		18.4	BLUE STELLAR OBJECT
PHL 5377	22 36 36.	- 12 56		18.2	BLUE STELLAR OBJECT
MCG-05-53-023	22 36 36.	- 26 45	12	15.	GALAXY
PHL 5376	22 36 36.	- 28 54		17.9	BLUE STELLAR OBJECT
SC 2233-4234.4	22 36 36.	- 42 18 50.			NEBULA
ARC 2464	22 36 38.	- 04 13		17.8	RICH CLUSTER OF GALAXIES
SC 2233-4434.2	22 36 39.	- 44 18 28.			NEBULA
SC 2233-4554.7	22 36 39.	- 45 39 08.			NEBULA
PHL 5380	22 36 42.	+ 01 16		18.5	BLUE STELLAR OBJECT
PHL 2023	22 36 42.	+ 02 36		15.7	BLUE STELLAR OBJECT
ZWG 452.039	22 36 42.	+ 18 23		15.4	GALAXY
ZC 2236.7+2442	22 36 42.	+ 24 42	6520		CLUSTER OF GALAXIES
ZWG 514.087	22 36 42.	+ 35 11		15.4	GALAXY
ZWG 378.013	22 36 42.	- 01 58		15.7	GALAXY
KARA.73E 0977	22 36 42.	- 01 58	42	15.7	ISOLATED GALAXY S
PHL 5381	22 36 42.	- 14 36		18.2	BLUE STELLAR OBJECT
PHL 5378	22 36 42.	- 18 56		17.1	BLUE STELLAR OBJECT
PHL 5379	22 36 42.	- 28 32		17.2	BLUE STELLAR OBJECT
SC 2233-4531.2	22 36 42.	- 45 15 38.			NEBULA
SC 2233-4745.1	22 36 42.	- 47 29 32.			NEBULA
SC 2233-4530.4	22 36 44.	- 45 14 50.			NEBULA
SC 2232-7145.7	22 36 44.	- 71 30 08.	18		NEBULA
MCG+01-57-014	22 36 45.	+ 08 21	120	15.	GALAXY
IC 5224	22 36 46.	- 66 04			NONSTELLAR OBJECT
EN 0760	22 36 47.	- 66 04			NEBULA
ZWG 452.040	22 36 48.	+ 18 47		15.6	GALAXY
PHL 2024	22 36 48.	- 00 20		18.0	BLUE STELLAR OBJECT
PHL 0362	22 36 48.	- 06 56		16.6	BLUE STELLAR OBJECT
MCG-05-53-024	22 36 48.	- 26 39	30	16.	GALAXY
SC 2233-4555.4	22 36 50.	- 45 39 49.	12		NEBULA
LB 00376	22 36 53.	- 22 20 24.		14.8	FAINT BLUE STAR
ZWG 429.016	22 36 54.	+ 13 36		15.4	GALAXY
PHL 5382	22 36 54.	- 25 12		12.3	BLUE STELLAR OBJECT
MCG-07-46-006	22 36 54.	- 44 06 30.	36	15.	GALAXY
SC 2233-4423.0	22 36 54.	- 44 07 25.	18		NEBULA
SC 2233-4453.6	22 36 55.	- 44 38 01.			NEBULA
SC 2233-4607.8	22 36 55.	- 45 52 13.	24		NEBULA
SC 2233-4555.2	22 36 56.	- 45 39 37.			NEBULA
ARC 2465	22 36 57.	- 06 00		17.8	RICH CLUSTER OF GALAXIES
SC 2233-5334.2	22 36 59.	- 53 18 37.			NEBULA
LBN 0501	22 37	+ 58 10	1200		BRIGHT NEBULA
PHL 5383	22 37 00.	+ 00 36		15.2	BLUE STELLAR OBJECT
PHL 2026	22 37 00.	+ 03 09		18.2	BLUE STELLAR OBJECT
ZWG 404.027	22 37 00.	+ 08 22		15.7	GALAXY
UGC 12133	22 37 00.	+ 08 22	108	15.7	GALAXY Sc
4ZW 108	22 37 00.	+ 24 13			COMPACT GALAXY
ZWG 495.008	22 37 00.	+ 32 42		15.7	GALAXY
MCG+06-49-067	22 37 00.	+ 35 40 30.	30	15.	GALAXY
LDN 1189	22 37 00.	+ 57 00	2760		DARK NEBULA
PHL 2025	22 37 00.	- 01 00		17.9	BLUE STELLAR OBJECT
PHL 2027	22 37 00.	- 06 26		18.1	BLUE STELLAR OBJECT
PHL 2028	22 37 00.	- 11 44		18.8	BLUE STELLAR OBJECT
PHL 2029	22 37 00.	- 13 20		18.2	BLUE STELLAR OBJECT
SC 2233-5614.5	22 37 00.	- 55 53 55.	36		NEBULA
SC 2234-4623.0	22 37 01.	- 46 07 25.			NEBULA
SC 2234-4709.4	22 37 01.	- 46 53 49.			NEBULA
SG 1.24	22 37 02.	+ 38 48	18000		DIFFUSE EMISSION NEBULA
RNGC 7346	22 37 03.	+ 10 49		15.5	GALAXY
RNGC 7353	22 37 04.	+ 11 31		15.0	GALAXY
RNGC 7329	22 37 04.	- 66 44		12.5	GALAXY
ZWG 429.017	22 37 06.	+ 10 49		15.6	GALAXY

OBJECT NAME	RIGHT ASCEN.	DECLINATION	DIAM.	MAGN.	TYPE OF OBJECT
ZWG 429.018	22 37 06.	+ 11 31		14.9	GALAXY
UGC 12134	22 37 06.	+ 11 31	108	14.9	GALAXY Sb-c
ZWG 514.088	22 37 06.	+ 35 42		15.7	GALAXY
UGC 12135	22 37 06.	+ 72 15	84	16.0	GALAXY Sc
RNGC 7344	22 37 06.	- 04 26		14.0	GALAXY
MCG-01-57-020	22 37 06.	- 04 26	78	14.	GALAXY
PHL 0363	22 37 06.	- 05 10		16.9	BLUE STELLAR OBJECT
SC 2233-7143.0	22 37 07.	- 71 27 26.	12		NEBULA
IC 5237	22 37 08.	- 30 16 00.			NONSTELLAR OBJECT
SC 2234-4536.0	22 37 08.	- 45 20 25.			NEBULA
SC 2234-4528.4	22 37 09.	- 45 12 49.			NEBULA
SC 2234-4534.2	22 37 10.	- 45 18 37.			NEBULA
SC 2234-4624.0	22 37 10.	- 46 08 25.			NEBULA
PHL 5385	22 37 12.	+ 00 04		18.5	BLUE STELLAR OBJECT
MCG+02-57-008	22 37 12.	+ 11 31	108	14.	GALAXY
PHL 2030	22 37 12.	- 07 33		17.2	BLUE STELLAR OBJECT
PHL 5384	22 37 12.	- 08 28		16.9	BLUE STELLAR OBJECT
SC 2234-4634.3	22 37 15.	- 46 18 43.			NEBULA
SC 2234-4600.8	22 37 16.	- 45 45 13.			NEBULA
SC 2234-4629.8	22 37 16.	- 46 14 13.			NEBULA
SC 2234-4613.0	22 37 17.	- 45 57 25.			NEBULA
PHL 5386	22 37 18.	- 28 25		17.3	BLUE STELLAR OBJECT
PHL 2031	22 37 18.	- 32 46		17.3	BLUE STELLAR OBJECT
SC 2234-4601.5	22 37 19.	- 45 45 55.			NEBULA
SC 2234-4615.0	22 37 21.	- 45 59 25.			NEBULA
SC 2234-4635.0	22 37 21.	- 46 19 25.			NEBULA
SN 1960K	22 37 23.	+ 34 07		19.0	SUPERNOVA
SC 2234-5554.5	22 37 23.	- 55 38 55.	12		NEBULA
ZWG 514.089	22 37 24.	+ 34 07		15.2	GALAXY
PHL 0364	22 37 24.	- 01 08		14.8	BLUE STELLAR OBJECT
PHL 5387	22 37 24.	- 03 20		16.3	BLUE STELLAR OBJECT
PHL 0365	22 37 24.	- 18 30		17.8	BLUE STELLAR OBJECT
PHL 5388	22 37 24.	- 25 58		18.3	BLUE STELLAR OBJECT
SC 2234-4415.6	22 37 25.	- 44 00 00.			NEBULA
SC 2234-4635.0	22 37 26.	- 46 19 24.			NEBULA
RNGC 7347	22 37 27.	+ 10 46		14.5	GALAXY
SC 2234-4619.3	22 37 29.	- 46 03 42.			NEBULA
MCG+01-57-015	22 37 30.	+ 08 18	48	15.5	GALAXY
ZWG 429.019	22 37 30.	+ 10 46		14.7	GALAXY
UGC 12136	22 37 30.	+ 10 46	102	14.7	GALAXY S
MCG+02-57-009	22 37 30.	+ 10 47	84	14.	GALAXY
MCG+06-49-068	22 37 30.	+ 34 06 30.	36	15.	GALAXY
ZWG 546.004	22 37 30.	+ 48 49		15.6	GALAXY
PHL 0366	22 37 30.	- 00 36		17.5	BLUE STELLAR OBJECT
ZC 2237.5-0100	22 37 30.	- 01 00	6720		CLUSTER OF GALAXIES
PHL 5389	22 37 30.	- 02 26		18.1	BLUE STELLAR OBJECT
SC 2234-4612.2	22 37 32.	- 45 56 36.			NEBULA
MCG+01-57-016	22 37 33.	+ 07 47	48	15.	GALAXY
SC 2234-4502.2	22 37 34.	- 44 46 36.			NEBULA
SC 2234-4503.5	22 37 34.	- 44 47 54.			NEBULA
ARC 2467	22 37 35.	+ 05 49		17.1	RICH CLUSTER OF GALAXIES
PHL 0367	22 37 36.	+ 01 52		15.8	BLUE STELLAR OBJECT
MCG+06-49-069	22 37 36.	+ 37 56 30.	108	13.	GALAXY
ZWG 514.090	22 37 36.	+ 37 57		14.0	GALAXY
UGC 12137	22 37 36.	+ 37 57	114	14.0	GALAXY SBb/Sc
OCL 0243	22 37 36.	+ 58 45	360	11.	OPEN STAR CLUSTER
2ZW 183	22 37 36.	- 02 41			COMPACT GALAXY
PHL 0368	22 37 36.	- 25 06		17.1	BLUE STELLAR OBJECT
SC 2234-4607.5	22 37 39.	- 45 51 54.			NEBULA
SC 2234-4551.1	22 37 40.	- 45 35 30.			NEBULA
SC 2234-4600.6	22 37 40.	- 45 45 00.			NEBULA
ZWG 404.028	22 37 42.	+ 07 47		14.3	GALAXY
UGC 12138	22 37 42.	+ 07 47	54	14.3	GALAXY SBa
ZWG 404.029	22 37 42.	+ 08 18		15.5	GALAXY
UGC 12139	22 37 42.	+ 08 18	66	15.5	GALAXY SB
ZWG 514.091	22 37 42.	+ 35 22		15.7	GALAXY
MCG-01-57-021	22 37 42.	- 02 40	120	14.	GALAXY
PHL 5390	22 37 42.	- 13 27		18.6	BLUE STELLAR OBJECT
PHL 2032	22 37 42.	- 14 59		18.4	BLUE STELLAR OBJECT
PHL 5391	22 37 42.	- 31 17		18.0	BLUE STELLAR OBJECT
RNGC 7352	22 37 43.	+ 57 08			NON-EXISTENT OBJECT
ARC 2466	22 37 46.	- 21 10		17.7	RICH CLUSTER OF GALAXIES
ZC 2237.8+1649	22 37 48.	+ 16 49	2350		CLUSTER OF GALAXIES
ZC 2237.8+1700	22 37 48.	+ 17 00	870		CLUSTER OF GALAXIES
PHL 5392	22 37 48.	- 01 20		17.9	BLUE STELLAR OBJECT
PHL 2033	22 37 48.	- 08 30		17.9	BLUE STELLAR OBJECT
PHL 0369	22 37 48.	- 14 57		15.4	BLUE STELLAR OBJECT
IC 5239	22 37 52.	- 38 18 10.			NONSTELLAR OBJECT
SC 2234-6135.3	22 37 53.	- 61 19 42.	6		NEBULA
FATH 1.824	22 37 54.	+ 00 01			NEBULA
ZWG 378.014	22 37 54.	+ 02 56		15.7	GALAXY
ZWG 378.015	22 37 54.	+ 03 05		15.2	GALAXY
OCL 0245	22 37 54.	+ 59 37	180		OPEN STAR CLUSTER
PHL 0370	22 37 54.	- 01 24		17.7	BLUE STELLAR OBJECT
PHL 5393	22 37 54.	- 01 58		17.2	BLUE STELLAR OBJECT
PHL 2034	22 37 54.	- 13 45		18.0	BLUE STELLAR OBJECT
SC 2234-4555.5	22 37 55.	- 45 39 54.			NEBULA
SC 2235-4340.4	22 37 57.	- 43 24 48.			NEBULA
ARC 2468	22 37 58.	+ 07 58		17.7	RICH CLUSTER OF GALAXIES
LBN 0435	22 38	+ 37 20	960		BRIGHT NEBULA
LBN 0448	22 38	+ 41 20	1560		BRIGHT NEBULA
LBN 0502	22 38	+ 57 58	480		BRIGHT NEBULA
ZC 2238.0+0540	22 38 00.	+ 05 40	3290		CLUSTER OF GALAXIES
MCG+02-57-010	22 38 00.	+ 11 39	60	13.5	GALAXY
ARC 2469	22 38 00.	+ 12 01		17.1	RICH CLUSTER OF GALAXIES
UGC 12140	22 38 00.	+ 15 41	66	15.3	GALAXY S
ZWG 452.041	22 38 00.	+ 15 45		15.3	GALAXY
KARA.73B 0978	22 38 00.	+ 15 45	66	15.3	ISOLATED GALAXY S
ZWG 495.009	22 38 00.	+ 31 49		15.7	GALAXY
ZWG 514.092	22 38 00.	+ 33 10		15.7	GALAXY
ZWG 359.002	22 38 00.	+ 80 25		15.6	GALAXY
ZWG 358.003	22 38 00.	+ 80 25		15.6	GALAXY
UGC 12141	22 38 00.	+ 80 25	84	15.6	GALAXY Sc
KARA.73B 0979	22 38 00.	+ 80 25	72	15.6	ISOLATED GALAXY S
PHL 0371	22 38 00.	- 09 02		18.6	BLUE STELLAR OBJECT
PHL 2035	22 38 00.	- 27 19		4.2	BLUE STELLAR OBJECT
TON-S 0064	22 38 00.	- 31 06		15.2	BLUE STAR
SC 2235-4552.6	22 38 00.	- 45 37 00.			NEBULA
SC 2235-4610.5	22 38 00.	- 45 54 54.	18		NEBULA
HN 0761	22 38 00.	- 66 50			NEBULA
IC 5235	22 38 00.	- 66 50			NONSTELLAR OBJECT
SC 2235-4624.2	22 38 01.	- 46 08 36.	12		NEBULA
SC 2235-4553.9	22 38 02.	- 45 38 18.			NEBULA
SC 2235-4617.0	22 38 02.	- 46 01 24.			NEBULA
KARA.73 54	22 38 03.	- 42 19	34		DWARF GALAXY
RNGC 7348	22 38 04.	+ 11 39		15.0	GALAXY
SC 2235-4621.4	22 38 04.	- 46 05 47.	12		NEBULA
ZWG 404.030	22 38 06.	+ 03 32		15.7	GALAXY

OBJECT NAME	RIGHT ASCEN.	DECLINATION	DIAM.	MAGN.	TYPE OF OBJECT
ZWG 429.020	22 38 06.	+ 11 29		14.8	GALAXY
UGC 12142	22 38 06.	+ 11 39	78	14.5	GALAXY Sc
ZC 2238.1+1158	22 38 06.	+ 11 58	1550		CLUSTER OF GALAXIES
ZWG 495.010	22 38 06.	+ 31 35		15.2	GALAXY
UGC 12143	22 38 06.	+ 31 35	96	15.2	GALAXY Sa-b
MCG+06-49-070	22 38 06.	+ 38 09	30	15.	GALAXY
PHL 5394	22 38 06.	- 28 22		16.8	BLUE STELLAR OBJECT
SC 2235-4346.9	22 38 06.	- 43 31 17.			NEBULA
SC 2235-4717.2	22 38 06.	- 47 01 35.	6		NEBULA
HN 0762	22 38 06.	- 66 52			NEBULA
IC 5236	22 38 06.	- 66 52			NONSTELLAR OBJECT
ARC 2470	22 38 08.	+ 17 00		17.1	RICH CLUSTER OF GALAXIES
SC 2235-4737.2	22 38 09.	- 47 21 35.			NEBULA
SC 2235-4455.3	22 38 11.	- 44 39 41.			NEBULA
ZWG 495.011	22 38 12.	+ 33 12		15.4	GALAXY
UGC 12144	22 38 12.	+ 33 12	72	15.4	GALAXY SBb
ISS 0138	22 38 12.	+ 54 16	211		STELLAR RING
ISS 0091	22 38 12.	+ 57 48	570		STELLAR RING
PHL 5395	22 38 12.	- 12 30		17.8	BLUE STELLAR OBJECT
PHL 5396	22 38 12.	- 14 41		17.5	BLUE STELLAR OBJECT
PHL 5397	22 38 12.	- 15 38		18.3	BLUE STELLAR OBJECT
TON-S 0065	22 38 12.	- 28 23		15.3	BLUE STAR
SC 2235-4558.2	22 38 14.	- 45 42 35.			NEBULA
HN 1233	22 38 17.	- 61 01	12		NEBULA
IC 5238	22 38 17.	- 61 01			NONSTELLAR OBJECT
ZWG 474.017	22 38 18.	+ 25 22		15.6	GALAXY
UGC 12145	22 38 13.	+ 33 19	66	16.0	GALAXY S
PHL 5398	22 38 18.	- 16 47		18.5	BLUE STELLAR OBJECT
MCG-05-53-025	22 38 19.	- 27 26	24	15.	GALAXY
SC 2235-4528.4	22 38 20.	- 45 12 47.			NEBULA
SC 2235-4530.9	22 38 20.	- 45 15 17.			NEBULA
SC 2234-7255.5	22 38 20.	- 72 39 54.	12		NEBULA
MCG+05-53-008	22 38 21.	+ 31 35	54	15.5	GALAXY
SC 2235-4654.1	22 38 23.	- 46 38 29.	6		NEBULA
ZWG 474.018	22 38 24.	+ 24 37		15.6	GALAXY
PHL 0372	22 38 24.	- 04 34		17.2	BLUE STELLAR OBJECT
PHL 2036	22 38 24.	- 05 24		18.0	BLUE STELLAR OBJECT
PHL 5401	22 38 24.	- 10 34		18.3	BLUE STELLAR OBJECT
PHL 5402	22 38 24.	- 12 34		18.5	BLUE STELLAR OBJECT
PHL 5399	22 38 24.	- 14 04		17.4	BLUE STELLAR OBJECT
MCG-05-53-026	22 38 24.	- 27 27	24	15.	GALAXY
PHL 5400	22 38 24.	- 29 02		17.5	BLUE STELLAR OBJECT
SC 2235-4455.0	22 38 24.	- 44 39 23.			NEBULA
SC 2235-4624.0	22 38 24.	- 46 08 23.			NEBULA
SC 2235-4627.8	22 38 24.	- 46 12 11.			NEBULA
SC 2235-4647.2	22 38 25.	- 46 31 35.			NEBULA
SC 2235-4648.6	22 38 25.	- 46 32 59.			NEBULA
SC 2235-6315.7	22 38 27.	- 63 00 05.	12		NEBULA
PK107+02.1	22 38 27.90	+ 61 01 29.1	32	12.9	PLANETARY NEBULA
RNGC 7354	22 38 28.	+ 61 01		13.0	PLANETARY NEBULA
SC 2235-5807.7	22 38 29.	- 57 52 05.	60		NEBULA
ZWG 474.019	22 38 30.	+ 22 13		15.7	GALAXY
UGC 12146	22 38 30.	+ 22 13	66	15.7	GALAXY Sb-c
SC 2235-4350.9	22 38 30.	- 43 35 17.			NEBULA
SC 2235-4626.4	22 38 33.	- 46 10 47.			NEBULA
RNGC 7350	22 38 34.	+ 11 42			NON-EXISTENT OBJECT
SC 2235-4607.2	22 38 34.	- 45 51 35.			NEBULA
2ZW 184	22 38 36.	+ 01 30			COMPACT GALAXY
PHL 5404	22 38 36.	- 09 58		18.8	BLUE STELLAR OBJECT
MCG-04-53-029	22 38 36.	- 22 04	60	15.	GALAXY
PHL 5403	22 38 36.	- 26 34		17.2	BLUE STELLAR OBJECT
SC 2235-4751.9	22 38 36.	- 47 36 17.			NEBULA
SC 2235-4256.5	22 38 39.	- 42 40 53.			NEBULA
SC 2235-4629.0	22 38 39.	- 46 13 23.			NEBULA
SC 2234-7311.8	22 38 40.	- 72 56 11.	18		NEBULA
SC 2235-4305.0	22 38 41.	- 42 49 23.			NEBULA
ZWG 429.021	22 38 42.	+ 13 04		15.7	GALAXY
KARA.73B 0980	22 38 42.	+ 13 04	42	15.7	ISOLATED GALAXY S
ZWG 452.042	22 38 42.	+ 18 32		15.4	GALAXY
UGC 12147	22 38 42.	+ 18 32	66	15.4	GALAXY S
MCG+03-57-030	22 38 42.	+ 18 32	36	15.	GALAXY
MCG+06-49-071	22 38 42.	+ 33 57	48	15.	GALAXY
PHL 2038	22 38 42.	- 02 25		13.3	BLUE STELLAR OBJECT
RNGC 7351	22 38 42.	- 04 43		13.0	GALAXY
MCG-01-57-022	22 38 42.	- 04 43	60	13.	GALAXY
PHL 2037	22 38 42.	- 13 23		18.1	BLUE STELLAR OBJECT
PHL 0373	22 38 42.	- 17 30		17.8	BLUE STELLAR OBJECT
SC 2235-4627.0	22 38 45.	- 46 11 22.			NEBULA
SC 2235-4626.2	22 38 46.	- 46 10 34.			NEBULA
SC 2235-6138.9	22 38 46.	- 61 23 17.			NEBULA
SC 2235-4454.4	22 38 47.	- 44 38 46.			NEBULA
SC 2235-4619.9	22 38 47.	- 46 04 16.			NEBULA
ZWG 474.020	22 38 48.	+ 23 08		14.7	GALAXY
ZCG 2238+23	22 38 48.	+ 23 08		14.7	COMPACT GALAXY
UGC 12148	22 38 48.	+ 23 08	60	14.7	GALAXY COMPACT
KARA.72 571A	22 38 48.	+ 23 08	48	14.7	PART OF DOUBLE GALAXY
MCG+04-53-010	22 38 48.	+ 23 09	36	15.	GALAXY
IC 5242	22 38 48.	+ 23 08 18.			NONSTELLAR OBJECT
ZWG 495.012	22 38 48.	+ 31 55		14.5	GALAXY
UGC 12149	22 38 48.	+ 31 55	66	14.5	GALAXY SBa
PHL 2041	22 38 48.	- 07 26		17.3	BLUE STELLAR OBJECT
PHL 2040	22 38 48.	- 07 28		18.0	BLUE STELLAR OBJECT
PHL 2039	22 38 48.	- 14 37		18.3	BLUE STELLAR OBJECT
PHL 5405	22 38 48.	- 15 50		18.3	BLUE STELLAR OBJECT
PHL 5406	22 38 48.	- 18 12		18.3	BLUE STELLAR OBJECT
PHL 2042	22 38 48.	- 24 28		18.2	BLUE STELLAR OBJECT
SC 2235-4556.0	22 38 48.	- 45 40 22.			NEBULA
SC 2235-4629.8	22 38 48.	- 46 14 10.			NEBULA
SC 2235-5333.8	22 38 48.	- 53 18 10.	18		NEBULA
SC 2235-6343.1	22 38 48.	- 63 27 29.	18		NEBULA
LB 00377	22 38 49.	- 25 11 00.		14.7	FAINT BLUE STAR
SC 2235-4640.8	22 38 49.	- 46 25 10.			NEBULA
SC 2235-4258.3	22 38 50.	- 42 42 40.			NEBULA
SC 2235-4615.2	22 38 51.	- 45 59 34.			NEBULA
SC 2235-4612.2	22 38 52.	- 45 56 34.			NEBULA
SC 2235-4528.2	22 38 53.	- 45 12 34.			NEBULA
ZWG 429.022	22 38 54.	+ 09 29		15.1	GALAXY
ZWG 404.031	22 38 54.	+ 09 29		15.1	GALAXY
KARA.73B 0981	22 38 54.	+ 09 29	18	15.1	ISOLATED GALAXY E
2ZW 185	22 38 54.	+ 23 05			COMPACT GALAXY
MCG+04-53-011	22 38 54.	+ 23 06	30	15.	GALAXY
4ZW 109	22 38 54.	+ 23 43			COMPACT GALAXY
ZWG 514.093	22 38 54.	+ 34 00		15.0	GALAXY
UGC 12150	22 38 54.	+ 34 00	66	15.0	GALAXY SB0-a
MCG+06-49-072	22 38 54.	+ 34 20	48	15.	GALAXY
ASS 25	22 38 54.	+ 38 49	54000		OB ASSOCIATION LAC OB1
PHL 5407	22 38 54.	- 06 50		14.3	BLUE STELLAR OBJECT
PHL 5408	22 38 54.	- 08 58		18.4	BLUE STELLAR OBJECT

OBJECT NAME	RIGHT ASCEN.	DECLINATION	DIAM.	MAGN.	TYPE OF OBJECT
PHL 0374	22 38 54.	- 14 54		16.6	BLUE STELLAR OBJECT
IC 5243	22 38 59.	+ 23 06 25.			NONSTELLAR OBJECT
LBN 0438	22 39	+ 37 30	1020		BRIGHT NEBULA
UGC 12151	22 39 00.	+ 00 08	180	16.0	GALAXY DWARF
MCG+00-57-007	22 39 00.	+ 00 08	72	15.5	GALAXY
ZWG 378.016	22 39 00.	+ 02 22		14.9	GALAXY
UGC 12152	22 39 00.	+ 02 22	60	14.9	GALAXY S?
PHL 2044	22 39 00.	+ 02 54		17.2	BLUE STELLAR OBJECT
PHL 2043	22 39 00.	+ 02 58		17.8	BLUE STELLAR OBJECT
ZC 2239.0+0710	22 39 00.	+ 07 10	4500		CLUSTER OF GALAXIES
ZWG 474.021	22 39 00.	+ 23 07		14.3	GALAXY
UGC 12153	22 39 00.	+ 23 07	42	14.3	GALAXY PAIR
KARA.72 571B	22 39 00.	+ 23 07	42	14.3	PART OF DOUBLE GALAXY
MCG+05-53-009	22 39 00.	+ 31 55	42	15.	GALAXY
ZWG 495.013	22 39 00.	+ 33 22		15.3	GALAXY
ZWG 514.094	22 39 00.	+ 34 22		14.8	GALAXY
ISS 0092	22 39 00.	+ 62 48	328		STELLAR RING
UGC 12154	22 39 00.	+ 71 34	96	16.0	GALAXY Sb
MAI 104	22 39 00.	+ 88 17	47		DWARF SPHEROIDAL GALAXY
PHL 5409	22 39 00.	- 09 44		17.3	BLUE STELLAR OBJECT
PHL 2045	22 39 00.	- 15 47		18.2	BLUE STELLAR OBJECT
SC 2236-4307.0	22 39 00.	- 42 51 22.			NEBULA
IC 5240	22 39 00.	- 45 04	138	12.0	GALAXY SB(r)
IC 5241	22 39 05.	+ 02 22 49.			NONSTELLAR OBJECT
MCG+00-57-008	22 39 06.	+ 02 21	42	14.	GALAXY
ZWG 404.032	22 39 06.	+ 07 48		15.7	GALAXY
MCG+06-49-073	22 39 06.	+ 34 58 30.	60	15.	GALAXY
TON-S 0066	22 39 06.	- 31 10		15.1	BLUE STAR
SC 2236-4621.2	22 39 07.	- 46 05 34.			NEBULA
SC 2236-4710.3	22 39 09.	- 46 54 40.	12		NEBULA
SC 2235-7159.0	22 39 11.	- 71 43 22.	12		NEBULA
PHL 0376	22 39 12.	+ 00 06		18.6	BLUE STELLAR OBJECT
PHL 0375	22 39 12.	+ 00 57		11.9	BLUE STELLAR OBJECT
ZWG 404.033	22 39 12.	+ 08 29		15.6	GALAXY
ZWG 514.095	22 39 12.	+ 35 01		15.4	GALAXY
UGC 12155	22 39 12.	+ 35 01	66	15.4	GALAXY SBb
ZWG 514.096	22 39 12.	+ 39 02		14.8	GALAXY
UGC 12156	22 39 12.	+ 39 02	72	14.8	GALAXY SBc
MCG+06-49-074	22 39 12.	+ 39 02	48	14.5	GALAXY
PHL 2046	22 39 12.	- 06 07		18.1	BLUE STELLAR OBJECT
PHL 5410	22 39 12.	- 10 36		18.6	BLUE STELLAR OBJECT
SC 2236-4538.8	22 39 15.	- 45 23 10.			NEBULA
ABC 2471	22 39 17.	+ 07 01		17.7	RICH CLUSTER OF GALAXIES
SC 2236-4622.9	22 39 17.	- 46 07 16.			NEBULA
PHL 5411	22 39 18.	- 12 28		18.2	BLUE STELLAR OBJECT
PHL 2047	22 39 18.	- 17 11		18.9	BLUE STELLAR OBJECT
SC 2236-5333.0	22 39 21.	- 53 17 22.	18		NEBULA
SC 2236-4612.5	22 39 22.	- 45 56 51.			NEBULA
SC 2236-4635.2	22 39 22.	- 46 19 33.			NEBULA
SC 2236-4355.2	22 39 23.	- 43 39 33.			NEBULA
ZC 2239.4+1715	22 39 24.	+ 17 15	940		CLUSTER OF GALAXIES
ZCG 2239+34.4	22 39 24.	+ 34 43		19.8	COMPACT GALAXY
PHL 2048	22 39 24.	- 14 55		18.2	BLUE STELLAR OBJECT
SC 2236-4309.4	22 39 25.	- 42 53 45.			NEBULA
ABC 2472	22 39 26.	+ 17 17		17.9	RICH CLUSTER OF GALAXIES
SC 2236-4316.4	22 39 26.	- 43 00 45.			NEBULA
MCG+06-49-075	22 39 27.	+ 34 38	114	15.5	GALAXY
MCG+06-49-076	22 39 27.	+ 35 49	36	14.5	GALAXY
SC 2236-4608.8	22 39 28.	- 45 53 09.	12		NEBULA
SC 2236-4656.7	22 39 28.	- 46 41 03.	12		NEBULA
ZC 2239.5+0700	22 39 30.	+ 07 00	1080		CLUSTER OF GALAXIES
ZWG 404.034	22 39 30.	+ 07 58		15.6	GALAXY
ZWG 452.043	22 39 30.	+ 20 00		14.7	GALAXY
MRK 308	22 39 30.	+ 20 00	18	15.5	GALAXY WITH UV CONTINUUM
MCG+03-57-031	22 39 30.	+ 20 00	27	15.	GALAXY
ZWG 514.097	22 39 30.	+ 35 51		15.2	GALAXY
4ZW 110	22 39 30.	+ 35 52			COMPACT GALAXY
MCG+06-49-077	22 39 30.	+ 37 26	36	15.	GALAXY
ZWG 514.098	22 39 30.	+ 37 27		15.6	GALAXY
MCG-01-57-023	22 39 30.	- 03 02	60	14.	GALAXY
SC 2236-4543.3	22 39 30.	- 45 27 39.			NEBULA
RNGC 7361	22 39 31.	- 30 19		13.0	GALAXY
SC 2236-4614.7	22 39 32.	- 45 59 03.			NEBULA
SC 2236-4628.0	22 39 32.	- 46 12 21.			NEBULA
SC 2236-4626.5	22 39 33.	- 46 10 51.			NEBULA
ZWG 452.044	22 39 36.	+ 19 23		15.5	GALAXY
ZWG 514.099	22 39 36.	+ 34 40		15.5	GALAXY
ZCG 2239+34.1	22 39 36.	+ 34 40		15.5	COMPACT GALAXY
UGC 12157	22 39 36.	+ 34 40	96	15.5	GALAXY S
MCG+07-46-016	22 39 36.	+ 42 41	42	16.	GALAXY
PHL 5413	22 39 36.	- 00 40		16.9	BLUE STELLAR OBJECT
PHL 5412	22 39 36.	- 02 24		14.5	BLUE STELLAR OBJECT
PHL 5415	22 39 36.	- 16 51		18.5	BLUE STELLAR OBJECT
PHL 5414	22 39 36.	- 27 15		17.5	BLUE STELLAR OBJECT
MCG-05-53-027	22 39 36.	- 30 20	240	12.	GALAXY
SC 2236-4411.8	22 39 38.	- 43 56 09.			NEBULA
SC 2236-4358.6	22 39 40.	- 43 42 57.			NEBULA
ZWG 452.045	22 39 42.	+ 19 44		15.3	GALAXY
UGC 12158	22 39 42.	+ 19 44	78	15.3	GALAXY Sb
ZWG 495.014	22 39 42.	+ 30 27		15.0	GALAXY
UGC 12159	22 39 42.	+ 30 27	66	15.0	GALAXY Sb-c
ZCG 2239+34.2	22 39 42.	+ 34 40		16.3	COMPACT GALAXY
UGC 12160	22 39 42.	+ 74 53	138	15.5	GALAXY Sc
PHL 2049	22 39 42.	- 16 52		18.1	BLUE STELLAR OBJECT
RNGC 7356	22 39 44.	+ 30 27		15.0	GALAXY
SC 2236-4548.8	22 39 44.	- 45 33 09.			NEBULA
MCG+03-57-032	22 39 45.	+ 19 44	72	14.	GALAXY
MCG-07-46-007	22 39 45.	- 43 10	36	14.5	GALAXY
SC 2236-4410.1	22 39 45.	- 43 54 27.			NEBULA
SC 2236-4604.3	22 39 45.	- 45 48 39.			NEBULA
SC 2236-4645.5	22 39 46.	- 46 29 51.			NEBULA
ZC 2239.8+1222	22 39 48.	+ 12 22	4170		CLUSTER OF GALAXIES
MCG+05-53-010	22 39 48.	+ 30 26	72	14.5	GALAXY
PHL 5417	22 39 48.	- 08 46		18.5	BLUE STELLAR OBJECT
PHL 5416	22 39 48.	- 09 28		17.9	BLUE STELLAR OBJECT
PHL 5418	22 39 48.	- 11 04		17.1	BLUE STELLAR OBJECT
PHL 0377	22 39 48.	- 13 34		18.3	BLUE STELLAR OBJECT
PHL 2050	22 39 48.	- 17 02		18.5	BLUE STELLAR OBJECT
PHL 0378	22 39 48.	- 19 58		16.0	BLUE STELLAR OBJECT
MCG-05-53-028	22 39 48.	- 28 51	12	15.	GALAXY
SC 2236-4434.0	22 39 49.	- 44 18 21.			NEBULA
ABC 2473	22 39 51.	- 13 48		17.6	RICH CLUSTER OF GALAXIES
HBLW 493	22 39 51.	- 39 38 27.			NEBULA
ZWG 474.022	22 39 54.	+ 23 51		15.7	GALAXY
ZWG 495.015	22 39 54.	+ 32 57		15.3	GALAXY
UGC 12161	22 39 54.	+ 32 57	60	15.3	GALAXY S
ZCG 2239+34.3	22 39 54.	+ 34 42		16.5	COMPACT GALAXY
ZWG 514.100	22 39 54.	+ 35 40		15.5	GALAXY

OBJECT NAME	RIGHT ASCEN.	DECLINATION	DIAM.	MAGN.	TYPE OF OBJECT
MCG-02-57-028	22 39 54.	- 11 45	36	15.	GALAXY
PHL 2051	22 39 54.	- 30 50		18.2	BLUE STELLAR OBJECT
SC 2236-4323.9	22 39 54.	- 43 08 15.			NEBULA
FATH 1.825	22 39 55.	+ 00 56	14		NEBULA
SC 2236-4610.8	22 39 55.	- 45 55 09.			NEBULA
SC 2236-4612.3	22 39 56.	- 45 56 39.			NEBULA
SC 2237-4239.2	22 39 57.	- 42 23 33.			NEBULA
SC 2237-4653.9	22 39 58.	- 46 38 15.	12		NEBULA
SC 2237-4234.2	22 39 59.	- 42 18 32.			NEBULA
LBN 0450	22 40	+ 42 00	11400		BRIGHT NEBULA
FATH 1.826	22 40 00.	+ 00 48	5		NEBULA
PHL 5421	22 40 00.	+ 01 53		16.0	BLUE STELLAR OBJECT
MCG+04-53-012	22 40 00.	+ 23 51	15	16.	GALAXY
ZWG 495.016	22 40 00.	+ 29 55		15.1	GALAXY
UGC 12162	22 40 00.	+ 29 55	108	15.1	GALAXY Sb
ZCG 2240+34	22 40 00.	+ 34 39		17.8	COMPACT GALAXY
COU 095	22 40 00.	+ 57 00	10800		EMISSION NEBULA
PHL 0379	22 40 00.	- 10 40		18.6	BLUE STELLAR OBJECT
PHL 5419	22 40 00.	- 27 38		16.1	BLUE STELLAR OBJECT
PHL 5420	22 40 00.	- 28 21		17.7	BLUE STELLAR OBJECT
SC 2237-4535.2	22 40 01.	- 45 19 32.			NEBULA
RNGC 7357	22 40 02.	+ 29 55		15.0	GALAXY
SC 2237-4541.8	22 40 03.	- 45 26 08.			NEBULA
SC 2237-4628.8	22 40 03.	- 46 13 08.	12		NEBULA
ABC 2475	22 40 04.	+ 07 13		17.9	RICH CLUSTER OF GALAXIES
SC 2237-4614.6	22 40 05.	- 45 59 56.			NEBULA
5ZW 389	22 40 06.	+ 40 35			COMPACT GALAXY
PHL 2052	22 40 06.	- 06 19		18.3	BLUE STELLAR OBJECT
MCG-04-53-030	22 40 06.	- 21 26	72	14.	GALAXY
SC 2237-4633.2	22 40 10.	- 46 17 32.			NEBULA
MCG+05-53-011	22 40 12.	+ 29 54	66	14.5	GALAXY
PHL 2054	22 40 12.	- 03 00		17.9	BLUE STELLAR OBJECT
FEIG 106	22 40 12.	- 04 27		14.0	FAINT BLUE STAR
PHL 0380	22 40 12.	- 04 28		15.1	BLUE STELLAR OBJECT
PHL 2055	22 40 12.	- 06 16		17.4	BLUE STELLAR OBJECT
PHL 0381	22 40 12.	- 13 33		16.6	BLUE STELLAR OBJECT
PHL 5422	22 40 12.	- 13 49		17.7	BLUE STELLAR OBJECT
PHL 2053	22 40 12.	- 15 43		18.2	BLUE STELLAR OBJECT
SC 2237-4622.3	22 40 12.	- 46 06 38.			NEBULA
MCG-02-57-029	22 40 15.	- 10 56	36	15.	GALAXY
SC 2237-4709.5	22 40 15.	- 46 53 50.	12		NEBULA
ZC 2240.3+0716	22 40 18.	+ 07 16	940		CLUSTER OF GALAXIES
ZWG 495.017	22 40 18.	+ 29 15		15.6	GALAXY
ZWG 495.018	22 40 18.	+ 29 28		14.4	GALAXY
UGC 12163	22 40 18.	+ 29 28	48	14.4	GALAXY SB
5ZW 390	22 40 18.	+ 40 28			COMPACT GALAXY
ZWG 378.017	22 40 18.	- 02 20		15.6	GALAXY
MCG+00-57-009	22 40 18.	- 02 23	30	15.	GALAXY
PHL 0382	22 40 18.	- 15 06		11.8	BLUE STELLAR OBJECT
PHL 0383	22 40 18.	- 22 58		16.1	BLUE STELLAR OBJECT
AGU 75	22 40 18.	- 40 08 48.	54	14.5	IRREGULAR Sd DWARF GALAXY
SC 2237-4431.9	22 40 18.	- 44 16 14.			NEBULA
SC 2237-4629.3	22 40 18.	- 46 13 38.			NEBULA
ABC 2474	22 40 21.	- 20 28		17.3	RICH CLUSTER OF GALAXIES
MCG-04-53-031	22 40 21.	- 24 36	36	15.	GALAXY
HBLW 494	22 40 21.	- 39 44 20.			NEBULA
PHL 0384	22 40 24.	+ 01 36		13.8	BLUE STELLAR OBJECT
5ZW 391	22 40 24.	+ 39 27			COMPACT GALAXY
PHL 2056	22 40 24.	- 02 06		17.7	BLUE STELLAR OBJECT
PHL 2057	22 40 24.	- 02 57		8.3	BLUE STELLAR OBJECT
PHL 2058	22 40 24.	- 04 50		17.8	BLUE STELLAR OBJECT
PHL 2059	22 40 24.	- 11 02		18.9	BLUE STELLAR OBJECT
PHL 5423	22 40 24.	- 29 18		17.8	BLUE STELLAR OBJECT
MCG-07-46-008	22 40 24.	- 40 08	72	16.	GALAXY
SC 2237-4616.6	22 40 24.	- 46 00 56.	12		NEBULA
SC 2237-4655.5	22 40 26.	- 46 39 50.	11		NEBULA
MCG+05-53-012	22 40 30.	+ 29 27	36	15.	GALAXY
ZWG 495.019	22 40 30.	+ 30 15		15.6	GALAXY
UGC 12164	22 40 30.	+ 30 15	96	15.6	GALAXY Sb
ZWG 495.020	22 40 30.	+ 32 44		15.7	GALAXY
UGC 12165	22 40 30.	+ 32 44	66	15.7	GALAXY Sb-c
MCG-01-57-024	22 40 30.	- 04 03	72	15.	GALAXY
PHL 5424	22 40 30.	- 17 36		17.8	BLUE STELLAR OBJECT
SC 2237-4642.3	22 40 30.	- 46 26 38.			NEBULA
LB 01507	22 40 30.	- 47 16		13.3	FAINT BLUE STAR
SC 2237-4231.5	22 40 31.	- 42 15 50.			NEBULA
SC 2237-4617.5	22 40 32.	- 46 01 50.			NEBULA
SC 2237-4617.0	22 40 33.	- 46 01 20.			NEBULA
RNGC 7355	22 40 34.	- 38 08			NON-EXISTENT OBJECT
ABC 2476	22 40 35.	+ 13 32		17.5	RICH CLUSTER OF GALAXIES
SC 2237-4549.0	22 40 35.	- 45 33 20.			NEBULA
ZWG 378.018	22 40 36.	+ 03 12		15.7	GALAXY
UGC 12166	22 40 36.	+ 04 47	60	17.	GALAXY DWRF IR
ZC 2240.6+1332	22 40 36.	+ 13 32	1140		CLUSTER OF GALAXIES
MCG+05-53-013	22 40 36.	+ 30 14	72	15.	GALAXY
ZWG 514.101	22 40 36.	+ 35 51		15.7	GALAXY
PHL 0385	22 40 36.	- 00 20		15.6	BLUE STELLAR OBJECT
PHL 0386	22 40 36.	- 01 44		16.8	BLUE STELLAR OBJECT
PHL 0387	22 40 36.	- 07 00		16.6	BLUE STELLAR OBJECT
PHL 2060	22 40 36.	- 07 14		6.3	BLUE STELLAR OBJECT
PHL 0388	22 40 36.	- 09 02		18.7	BLUE STELLAR OBJECT
PHL 5426	22 40 36.	- 10 30		18.4	BLUE STELLAR OBJECT
PHL 0389	22 40 36.	- 14 39		18.1	BLUE STELLAR OBJECT
PHL 2061	22 40 36.	- 14 42		17.9	BLUE STELLAR OBJECT
PHL 5425	22 40 36.	- 18 33		18.3	BLUE STELLAR OBJECT
PHL 2062	22 40 36.	- 18 42		18.3	BLUE STELLAR OBJECT
MCG-05-53-029	22 40 36.	- 30 23	30	15.5	GALAXY
SC 2237-4551.1	22 40 36.	- 45 35 26.	42		NEBULA
SC 2237-4618.9	22 40 37.	- 46 03 14.			NEBULA
FATH 1.827	22 40 38.	+ 00 10			NEBULA
SC 2237-4425.2	22 40 40.	- 44 09 31.	54		NEBULA
FATH 1.828	22 40 41.	+ 00 10			NEBULA
FATH 1.829	22 40 42.	+ 00 09			NEBULA
PHL 0390	22 40 42.	- 26 39		16.0	BLUE STELLAR OBJECT
TON-S 0067	22 40 42.	- 26 39		14.7	BLUE STAR
PHL 5427	22 40 42.	- 28 30		17.9	BLUE STELLAR OBJECT
SC 2237-4617.2	22 40 44.	- 46 01 31.			NEBULA
FATH 1.830	22 40 46.	+ 00 12	8		NEBULA
SC 2237-4543.0	22 40 46.	- 45 27 19.			NEBULA
FATH 1.831	22 40 47.	+ 00 13	8		NEBULA
ZWG 474.023	22 40 48.	+ 22 30		15.7	GALAXY
4ZW 111	22 40 48.	+ 31 24			COMPACT GALAXY
PHL 0391	22 40 48.	- 07 58		14.8	BLUE STELLAR OBJECT
PHL 5428	22 40 48.	- 10 48		18.1	BLUE STELLAR OBJECT
TON-S 0068	22 40 49.	- 34 05		15.2	BLUE STAR
SC 2237-7201.8	22 40 49.	- 71 46 08.	12		NEBULA
MCG+06-49-078	22 40 51.	+ 33 41	48	14.	GALAXY
ZWG 405.001	22 40 54.	+ 08 25		15.5	GALAXY

OBJECT NAME	RIGHT ASCEN.	DECLINATION	DIAM.	MAGN.	TYPE OF OBJECT
ZWG 404.035	22 40 54.	+ 08 25		15.5	GALAXY
SC 2237-4430.8	22 40 54.	- 44 15 07.			NEBULA
SC 2237-4638.8	22 40 56.	- 46 23 07.			NEBULA
HN 0763	22 40 58.	- 64 17			NEBULA
IC 5244	22 40 58.	- 64 17			NONSTELLAR OBJECT
BV 08	22 40 59.	- 00 28 24.		17.9	FAINT BLUE VARIABLE
SC 2238-4556.1	22 40 59.	- 45 40 25.			NEBULA
MCG+01-58-001	22 41 00.	+ 03 52 30.	48	15.	GALAXY
ZWG 405.002	22 41 00.	+ 03 53		14.5	GALAXY
ZWG 404.036	22 41 00.	+ 03 53		14.5	GALAXY
UGC 12167	22 41 00.	+ 03 53	42	14.5	GALAXY
ZWG 514.102	22 41 00.	+ 33 45		14.6	GALAXY
PHL 0392	22 41 00.	- 00 29		15.5	BLUE STELLAR OBJECT
PHL 5429	22 41 00.	- 26 42		17.6	BLUE STELLAR OBJECT
RNGC 7360	22 41 02.	+ 03 53		14.5	GALAXY
RNGC 7363	22 41 03.	+ 33 44		14.5	GALAXY
SC 2238-4318.0	22 41 05.	- 43 02 19.			NEBULA
UGC 12168	22 41 06.	+ 05 52	60	16.0	GALAXY S
ZWG 474.024	22 41 06.	+ 23 40		15.7	GALAXY
ZC 2241.1+3703	22 41 06.	+ 37 03	1140		CLUSTER OF GALAXIES
OCL 0234	22 41 06.	+ 52 09	420	15.	OPEN STAR CLUSTER
PHL 0393	22 41 06.	- 21 59		17.3	BLUE STELLAR OBJECT
SC 2238-4609.6	22 41 08.	- 45 53 55.			NEBULA
SC 2238-4608.8	22 41 09.	- 45 53 07.			NEBULA
SC 2238-4607.8	22 41 11.	- 45 52 07.			NEBULA
ZWG 474.025	22 41 12.	+ 23 41		15.7	GALAXY
UGC 12169	22 41 12.	+ 23 41	66	15.7	GALAXY Sb
UGC 12170	22 41 12.	+ 78 42	66	16.0	GALAXY Sc
PHL 5430	22 41 12.	- 00 38		13.4	BLUE STELLAR OBJECT
PHL 5431	22 41 12.	- 08 36		15.3	BLUE STELLAR OBJECT
PHL 5432	22 41 12.	- 09 31		17.8	BLUE STELLAR OBJECT
PHL 0394	22 41 12.	- 25 17		16.6	BLUE STELLAR OBJECT
TON-S 0069	22 41 12.	- 33 31		15.8	BLUE STAR
SC 2238-4540.9	22 41 13.	- 45 25 13.			NEBULA
SC 2238-4612.7	22 41 13.	- 45 57 01.	12		NEBULA
SC 2238-4427.9	22 41 14.	- 44 12 13.			NEBULA
SC 2238-4542.5	22 41 15.	- 45 26 49.			NEBULA
ZWG 405.003	22 41 18.	+ 08 27		14.9	GALAXY
UGC 12171	22 41 18.	+ 08 27	90	14.9	GALAXY (E)
MCG+04-53-013	22 41 18.	+ 23 42	48	15.	GALAXY
PHL 0395	22 41 18.	- 02 02		17.7	BLUE STELLAR OBJECT
PHL 2063	22 41 18.	- 11 04		17.3	BLUE STELLAR OBJECT
ABC 2477	22 41 18.	- 17 23		16.9	RICH CLUSTER OF GALAXIES
PHL 5433	22 41 18.	- 30 40		16.4	BLUE STELLAR OBJECT
SC 2238-4539.4	22 41 19.	- 45 23 43.			NEBULA
RNGC 7362	22 41 21.	+ 08 27		15.0	GALAXY
SC 2238-4622.3	22 41 22.	- 46 06 36.			NEBULA
HELW 495	22 41 23.	- 39 46 00.			NEBULA
SC 2238-4538.0	22 41 23.	- 45 22 18.			NEBULA
MCG+01-58-002	22 41 24.	+ 08 27	48	15.	GALAXY
ZC 2241.4+2104	22 41 24.	+ 21 04	270		CLUSTER OF GALAXIES
PHL 5434	22 41 24.	- 14 23		16.6	BLUE STELLAR OBJECT
PHL 2064	22 41 24.	- 15 44		16.6	BLUE STELLAR OBJECT
MCG-05-53-030	22 41 24.	- 27 30	48	15.5	GALAXY
PHL 5435	22 41 24.	- 30 32		18.0	BLUE STELLAR OBJECT
SC 2238-4603.6	22 41 25.	- 45 47 54.			NEBULA
SC 2238-4609.4	22 41 26.	- 45 53 42.			NEBULA
SC 2238-4603.0	22 41 27.	- 45 47 18.			NEBULA
SC 2238-4550.8	22 41 29.	- 45 35 06.			NEBULA
ZWG 379.001	22 41 30.	+ 01 26		15.7	GALAXY
PHL 5436	22 41 30.	+ 01 29		18.5	BLUE STELLAR OBJECT
ZWG 405.004	22 41 30.	+ 07 07		15.5	GALAXY
UGC 12172	22 41 30.	+ 07 07	72	15.5	GALAXY Sc
KARA.73B 0982	22 41 30.	+ 07 07	66	15.5	ISOLATED GALAXY S
MCG+01-58-003	22 41 30.	+ 07 08	60	15.	GALAXY
4ZW 112	22 41 30.	+ 23 34			COMPACT GALAXY
SC 2238-4605.2	22 41 33.	- 45 49 30.	6		NEBULA
SC 2238-4358.1	22 41 34.	- 43 42 24.			NEBULA
SC 2238-4545.5	22 41 35.	- 45 29 48.			NEBULA
SC 2238-4621.2	22 41 35.	- 46 05 30.			NEBULA
ZWG 515.001	22 41 36.	+ 38 06		13.7	GALAXY
ZWG 514.103	22 41 36.	+ 38 06		13.7	GALAXY
UGC 12173	22 41 36.	+ 38 06	132	13.7	GALAXY Sc
KARA.73B 0983	22 41 36.	+ 38 06	126	13.7	ISOLATED GALAXY S
MCG+07-46-017	22 41 36.	+ 42 30	36	16.	GALAXY
TON-S 0070	22 41 36.	- 30 40		15.3	BLUE STAR
MCG+06-49-079	22 41 39.	+ 38 08	126	13.	GALAXY
SC 2238-4302.8	22 41 39.	- 42 47 06.			NEBULA
SC 2238-4623.8	22 41 41.	- 46 08 06.			NEBULA
MCG+01-58-004	22 41 42.	+ 06 10	180	13.5	GALAXY
ZC 2241.7+0657	22 41 42.	+ 06 57	1080		CLUSTER OF GALAXIES
ZWG 405.005	22 41 42.	+ 09 10		15.5	GALAXY
MCG+02-58-003	22 41 42.	+ 09 42	36	15.	GALAXY
ZWG 430.001	22 41 42.	+ 09 43		15.3	GALAXY
ZWG 430.002	22 41 42.	+ 09 45		15.6	GALAXY
MCG+02-58-002	22 41 42.	+ 09 45	48	15.	GALAXY
ZWG 430.003	22 41 42.	+ 09 48		15.4	GALAXY
MCG+02-58-001	22 41 42.	+ 09 48	12	15.	GALAXY
ZWG 453.001	22 41 42.	+ 15 36		15.7	GALAXY
HN 0764	22 41 42.	- 65 36			NEBULA
SC 2238-4610.0	22 41 43.	- 45 54 18.			NEBULA
IC 5245	22 41 43.	- 65 36			NONSTELLAR OBJECT
SC 2238-4632.6	22 41 44.	- 46 16 54.			NEBULA
ZC 2241.8+0047	22 41 48.	+ 00 47	6250		CLUSTER OF GALAXIES
ZWG 405.006	22 41 48.	+ 04 06		15.3	GALAXY
ZWG 405.007	22 41 48.	+ 05 07		15.6	GALAXY
MCG+01-58-005	22 41 48.	+ 09 10	36	15.5	GALAXY
MCG+00-58-001	22 41 48.	- 00 26	36	13.	GALAXY
PHL 5437	22 41 48.	- 20 10		12.5	BLUE STELLAR OBJECT
SC 2238-7201.8	22 41 48.	- 71 46 06.	12		NEBULA
SC 2238-4553.6	22 41 49.	- 45 37 54.			NEBULA
RNGC 7366	22 41 51.	+ 10 30		15.5	GALAXY
MCG+06-49-080	22 41 51.	+ 34 04 30.	48	14.5	GALAXY
SC 2238-4433.3	22 41 51.	- 44 17 36.			NEBULA
SC 2238-4612.7	22 41 51.	- 45 57 00.			NEBULA
MCG+02-58-004	22 41 54.	+ 10 30	18	15.5	GALAXY
ZWG 453.002	22 41 54.	+ 19 39		15.5	GALAXY
ZWG 474.026	22 41 54.	+ 24 51		15.6	GALAXY
4ZW 113	22 41 54.	+ 34 05			COMPACT GALAXY
ZWG 515.002	22 41 54.	+ 34 05		14.8	GALAXY
ZWG 514.104	22 41 54.	+ 34 05		14.8	GALAXY
ISS 0093	22 41 54.	+ 57 40	231		STELLAR RING
ZWG 379.002	22 41 54.	- 00 23		13.8	GALAXY
UGC 12174	22 41 54.	- 00 23	108	13.8	GALAXY Sa
ABC 2478	22 41 54.	- 17 58		17.3	RICH CLUSTER OF GALAXIES
MCG-04-53-033	22 41 54.	- 22 11	48	15.	GALAXY
MCG-04-53-032	22 41 54.	- 23 17	66	15.	GALAXY
PHL 0396	22 41 54.	- 32 34		16.4	BLUE STELLAR OBJECT

OBJECT NAME	RIGHT ASCEN.	DECLINATION	DIAM.	MAGN.	TYPE OF OBJECT
RNGC 7364	22 41 55.	- 00 23		14.0	GALAXY
RNGC 7369	22 41 57.	+ 34 05		15.0	GALAXY
SC 2239-4312.8	22 41 57.	- 42 57 06.			NEBULA
SC 2239-4557.4	22 41 57.	- 45 41 42.			NEBULA
SC 2239-4601.0	22 41 57.	- 45 45 18.			NEBULA
SC 2239-4400.6	22 41 59.	- 43 44 53.			NEBULA
SC 2239-4616.0	22 41 59.	- 46 00 18.			NEBULA
LBN 0440	22 42	+ 37 00	600		BRIGHT NEBULA
MCG+00-58-002	22 42 00.	+ 03 25	84	14.5	GALAXY
MCG+04-53-014	22 42 00.	+ 24 52	18	15.	GALAXY
PHL 2065	22 42 00.	- 24 19		17.5	BLUE STELLAR OBJECT
PHL 2066	22 42 00.	- 25 22		18.3	BLUE STELLAR OBJECT
PHL 5438	22 42 00.	- 27 00		16.9	BLUE STELLAR OBJECT
PHL 0397	22 42 00.	- 28 16		15.7	BLUE STELLAR OBJECT
SC 2239-4534.0	22 42 01.	- 45 18 17.			NEBULA
SC 2239-4535.0	22 42 01.	- 45 19 17.			NEBULA
PK117+18.1	22 42 02.	+ 80 10	34	14.8	PLANETARY NEBULA
IC 1454	22 42 02.	+ 80 10			NONSTELLAR OBJECT
SC 2239-4424.3	22 42 03.	- 44 08 35.			NEBULA
SC 2239-4428.2	22 42 03.	- 44 12 29.			NEBULA
ZWG 379.003	22 42 05.	+ 03 25		14.9	GALAXY
UGC 12175	22 42 06.	+ 03 25	96	14.9	GALAXY Sa-b
KARA.73B 0984	22 42 06.	+ 03 25	150	14.9	ISOLATED GALAXY S
MCG+08-41-005	22 42 06.	+ 48 30	78	16.	GALAXY
MCG-04-53-034	22 42 06.	- 23 58	120	13.5	GALAXY
TON-S 0071	22 42 06.	- 28 20		14.6	BLUE STAR
RNGC 7367	22 42 08.	+ 03 25		15.0	GALAXY
IC 5248	22 42 08.	- 00 37			NONSTELLAR OBJECT
RNGC 7359	22 42 08.	- 23 58		13.0	GALAXY
UGC 12176	22 42 12.	+ 48 30	72	16.0	GALAXY Sc
PHL 5439	22 42 12.	- 15 35		15.2	BLUE STELLAR OBJECT
PHL 2067	22 42 12.	- 18 40		17.9	BLUE STELLAR OBJECT
MCG-04-53-035	22 42 12.	- 24 08	48	15.5	GALAXY
PHL 5440	22 42 12.	- 27 28		15.4	BLUE STELLAR OBJECT
TON-S 0072	22 42 12.	- 32 35		15.7	BLUE STAR
SC 2239-4403.6	22 42 12.	- 43 47 53.			NEBULA
SC 2239-4546.4	22 42 15.	- 45 30 41.			NEBULA
SC 2239-4545.0	22 42 16.	- 45 29 17.			NEBULA
LB 01508	22 42 18.	- 48 24		12.7	FAINT BLUE STAR
SC 2239-4404.0	22 42 19.	- 43 48 17.			NEBULA
RNGC 7358	22 42 19.	- 65 23			GALAXY
RNGC 7365	22 42 21.	- 20 15		14.0	GALAXY
SC 2239-4406.0	22 42 21.	- 43 50 17.			NEBULA
SC 2239-4609.4	22 42 23.	- 45 53 41.			NEBULA
ZC 2242.4+1040	22 42 24.	+ 10 40	400		CLUSTER OF GALAXIES
MCG+03-58-001	22 42 24.	+ 18 52	36	15.	GALAXY
ZWG 495.021	22 42 24.	+ 33 12		13.5	GALAXY
UGC 12177	22 42 24.	+ 33 12	66	13.5	GALAXY S
PHL 5441	22 42 24.	- 26 30		17.1	BLUE STELLAR OBJECT
PHL 5442	22 42 24.	- 29 24		17.3	BLUE STELLAR OBJECT
MCG-03-58-001	22 42 27.	- 20 15	36	14.	GALAXY
SC 2239-4546.4	22 42 27.	- 45 30 41.			NEBULA
SC 2238-7150.8	22 42 27.	- 71 35 05.	12		NEBULA
ZWG 453.003	22 42 30.	+ 18 51		15.6	GALAXY
MCG-07-46-009	22 42 30.	- 44 40		15.	GALAXY
SC 2239-4533.1	22 42 30.	- 45 17 23.			NEBULA
SC 2239-4553.0	22 42 31.	- 45 37 17.			NEBULA
SC 2239-4638.6	22 42 33.	- 46 22 53.			NEBULA
SC 2239-4418.8	22 42 34.	- 44 03 05.			NEBULA
SC 2239-4551.6	22 42 34.	- 45 35 53.			NEBULA
ZWG 405.008	22 42 35.	+ 06 10		14.2	GALAXY
UGC 12178	22 42 36.	+ 06 10	198	14.2	GALAXY
KARA.73B 0985	22 42 36.	+ 06 10	168	14.2	ISOLATED GALAXY S
IC 5251	22 42 36.	+ 10 53			SINGLE STAR
MCG-04-53-036	22 42 36.	- 23 01	132	13.	GALAXY
BIGO 558	22 42 37.	+ 10 54			NEBULA
SC 2239-4630.0	22 42 37.	- 46 14 17.			NEBULA
SC 2239-4631.7	22 42 38.	- 46 15 59.			NEBULA
ABC 2479	22 42 39.	+ 16 57		16.5	RICH CLUSTER OF GALAXIES
RNGC 7349	22 42 39.	- 23 01		13.0	GALAXY
RNGC 7368	22 42 40.	- 39 36			GALAXY
SC 2239-4630.8	22 42 40.	- 46 15 04.			NEBULA
SC 2239-4308.3	22 42 41.	- 42 52 34.			NEBULA
SC 2239-4533.6	22 42 41.	- 45 17 52.			NEBULA
SC 2239-4615.2	22 42 41.	- 45 59 28.			NEBULA
MCG+05-53-014	22 42 42.	+ 33 11 30.	60	14.5	GALAXY
MCG+06-50-001	22 42 42.	+ 33 43	54	14.	GALAXY
MCG-07-46-010	22 42 42.	- 39 37	72	12.5	GALAXY
SC 2239-4536.2	22 42 42.	- 45 20 28.			NEBULA
SC 2239-4621.8	22 42 43.	- 46 06 04.			NEBULA
SC 2239-4631.2	22 42 45.	- 46 15 28.			NEBULA
SC 2239-4535.3	22 42 47.	- 45 19 34.			NEBULA
4ZW 114	22 42 48.	+ 23 12			COMPACT GALAXY
ZWG 515.003	22 42 48.	+ 33 44		14.5	GALAXY
UGC 12179	22 42 48.	+ 33 44	78	14.5	GALAXY S0
4ZW 115	22 42 48.	+ 33 56			COMPACT GALAXY
PHL 5443	22 42 48.	- 30 44		17.6	BLUE STELLAR OBJECT
SC 2239-4308.8	22 42 49.	- 42 53 04.			NEBULA
SC 2239-4614.5	22 42 51.	- 45 58 46.			NEBULA
ZC 2242.9+1647	22 42 54.	+ 16 47	2820		CLUSTER OF GALAXIES
MRSL 107+00/1	22 42 54.	+ 59 38	240		HII REGION
SC 2240-4625.2	22 42 57.	- 46 09 28.			NEBULA
SC 2240-4426.8	22 42 58.	- 44 11 04.			NEBULA
LBN 0506	22 43	+ 57 40	4260		BRIGHT NEBULA
LBN 0512	22 43	+ 59 34	480		BRIGHT NEBULA
KHAV 777	22 43	+ 60 52	11760		DARK NEBULA
ZWG 474.027	22 43 00.	+ 21 33		15.5	GALAXY
UGC 12180	22 43 00.	+ 21 33	60	15.5	GALAXY S0-a
ZWG 515.004	22 43 00.	+ 33 47		15.6	GALAXY
LDN 1210	22 43 00.	+ 61 50	420		DARK NEBULA
ZC 2243.0+8713	22 43 00.	+ 87 13	1810		CLUSTER OF GALAXIES
SC 2240-4401.8	22 43 00.	- 43 46 04.			NEBULA
SC 2240-4522.4	22 43 00.	- 45 06 40.			NEBULA
SC 2240-4309.0	22 43 01.	- 42 53 16.			NEBULA
RNGC 7370	22 43 03.	+ 10 47			GALAXY
SC 2240-4425.9	22 43 03.	- 44 10 10.			NEBULA
SC 2240-4617.7	22 43 03.	- 46 01 58.			NEBULA
IC 5253	22 43 05.	+ 21 32 24.			NONSTELLAR OBJECT
MCG-04-53-037	22 43 09.	- 21 07	48	15.	GALAXY
SC 2240-4406.8	22 43 11.	- 43 51 04.			NEBULA
PHL 5444	22 43 12.	- 27 00		17.3	BLUE STELLAR OBJECT
PHL 5445	22 43 12.	- 29 28		17.8	BLUE STELLAR OBJECT
LB C1148	22 43 14.	+ 22 57 54.		17.5	FAINT BLUE STAR
SC 2240-4455.6	22 43 14.	- 44 39 52.			NEBULA
RNGC 7372	22 43 15.	+ 10 52		14.5	GALAXY
SC 2240-4425.9	22 43 15.	- 44 10 10.			NEBULA
HN 0765	22 43 16.	- 65 09			NEBULA
IC 5246	22 43 16.	- 65 10			NONSTELLAR OBJECT

OBJECT NAME	RIGHT ASCEN.	DECLINATION	DIAM.	MAGN.	TYPE OF OBJECT
SC 2240-4538.6	22 43 17.	- 45 22 52.			NEBULA
MCG+02-58-005	22 43 18.	+ 10 51	54	14.	GALAXY
ZWG 430.004	22 43 18.	+ 10 52		14.6	GALAXY
ZWG 453.004	22 43 18.	+ 19 30		15.6	GALAXY
ZWG 495.022	22 43 18.	+ 30 46		15.7	GALAXY
MCG-03-58-002	22 43 18.	- 19 41	102	15.	GALAXY
PHL 2068	22 43 18.	- 24 40		18.7	BLUE STELLAR OBJECT
MCG-05-53-031	22 43 18.	- 28 18	48	15.	GALAXY
TON-S 0073	22 43 18.	- 31 36		15.9	BLUE STAR
RNGC 7371	22 43 23.	- 11 16		13.0	GALAXY
SC 2240-4550.0	22 43 23.	- 45 34 15.			NEBULA
MCG-02-58-001	22 43 24.	- 11 16	72	12.5	GALAXY
HOLM 797B	22 43 24.	- 11 17	30	15.0	PART OF MULTIPLE GALAXY
ARC 2480	22 43 24.	- 17 57		16.9	RICH CLUSTER OF GALAXIES
SC 2240-4524.0	22 43 24.	- 45 08 15.			NEBULA
SC 2240-4426.7	22 43 25.	- 44 10 57.			NEBULA
HOLM 797A	22 43 26.	- 11 16	72	13.8	PART OF MULTIPLE GALAXY
IC 1452	22 43 27.	+ 16 35			NONSTELLAR OBJECT
SC 2240-4415.6	22 43 27.	- 43 59 51.			NEBULA
HOLM 798B	22 43 28.	+ 10 38	18	14.6	PART OF MULTIPLE GALAXY
IC 5255	22 43 28.	+ 35 57 49.			NONSTELLAR OBJECT
SC 2240-4322.6	22 43 28.	- 43 06 51.			NEBULA
IC 1451	22 43 29.	- 10 38 00.			NONSTELLAR OBJECT
SC 2240-4600.0	22 43 29.	- 45 44 15.			NEBULA
ZWG 430.005	22 43 30.	+ 10 35		15.6	GALAXY
MCG+02-58-007	22 43 30.	+ 10 35	48	14.	GALAXY
KARA.72 572A	22 43 30.	+ 10 36	30	15.6	PART OF DOUBLE GALAXY
MCG+02-58-006	22 43 30.	+ 10 36	18	16.	GALAXY
HOLM 798A	22 43 30.	+ 10 37	42	14.2	PART OF MULTIPLE GALAXY
SC 2240-4638.3	22 43 31.	- 46 22 33.	12		NEBULA
RNGC 7374A	22 43 33.	+ 10 35		15.0	GALAXY
IC 5254	22 43 33.	+ 20 51 25.			NONSTELLAR OBJECT
SC 2240-4359.7	22 43 33.	- 43 43 57.			NEBULA
SC 2240-4313.8	22 43 34.	- 42 58 03.			NEBULA
HN 0766	22 43 35.	- 65 32			NEBULA
IC 5247	22 43 35.	- 65 33			NONSTELLAR OBJECT
ZWG 430.006	22 43 36.	+ 10 35		14.9	GALAXY
KARA.72 572B	22 43 36.	+ 10 35	48	14.9	PART OF DOUBLE GALAXY
MCG-02-58-002	22 43 36.	- 10 38	24	15.	GALAXY
SC 2240-4706.2	22 43 36.	- 46 50 27.			NEBULA
SC 2240-4323.8	22 43 37.	- 43 08 03.			NEBULA
SC 2240-4413.7	22 43 37.	- 43 57 57.			NEBULA
SC 2240-5314.8	22 43 37.	- 52 59 03.	12		NEBULA
MCG+02-58-008	22 43 42.	+ 10 17	24	15.5	GALAXY
ZWG 453.005	22 43 42.	+ 20 51		15.6	GALAXY
MCG+06-50-002	22 43 42.	+ 33 42	24	15.	GALAXY
ISS 0139	22 43 42.	+ 54 55	155		STELLAR RING
SC 2240-4415.2	22 43 45.	- 43 59 27.			NEBULA
SC 2240-4416.4	22 43 45.	- 44 00 39.			NEBULA
SC 2240-4420.2	22 43 45.	- 44 04 27.			NEBULA
ZWG 430.007	22 43 48.	+ 09 52		15.7	GALAXY
KARA.73B 0986	22 43 48.	+ 09 52	48	15.7	ISOLATED GALAXY S
ZWG 453.006	22 43 48.	+ 15 36		15.7	GALAXY
MCG-02-58-003	22 43 48.	- 14 27	108	15.	GALAXY
PHL 5446	22 43 48.	- 24 58		18.4	BLUE STELLAR OBJECT
SC 2240-4632.9	22 43 52.	- 46 17 09.			NEBULA
HN 0767	22 43 52.	- 65 05			NEBULA
IC 5249	22 43 52.	- 65 06			NONSTELLAR OBJECT
SC 2240-5847.7	22 43 53.	- 58 31 57.	30		NEBULA
ZWG 379.004	22 43 54.	+ 02 57		14.8	GALAXY
ZC 2243.9+0408	22 43 54.	+ 04 08	200		CLUSTER OF GALAXIES
MCG+03-58-002	22 43 54.	+ 15 27	36	15.	GALAXY
ZWG 515.005	22 43 54.	+ 37 48		14.5	GALAXY
UGC 12181	22 43 54.	+ 37 48	66	14.5	GALAXY Sc/SBc
MCG+06-50-003	22 43 54.	+ 37 48 30.	48	14.	GALAXY
SC 2240-6127.6	22 43 54.	- 61 11 51.			NEBULA
RNGC 7373	22 43 56.	+ 02 57		15.0	GALAXY
SC 2241-4402.6	22 43 58.	- 43 46 51.			NEBULA
HN 0768	22 43 58.	- 65 20			NEBULA
IC 5250	22 43 58.	- 65 21			NONSTELLAR OBJECT
ZWG 430.008	22 44 00.	+ 11 40		15.6	GALAXY
ZWG 495.023	22 44 00.	+ 28 23		15.6	GALAXY
KARA.73B 0987	22 44 00.	+ 28 23	42	15.6	ISOLATED GALAXY S
LDN 1200	22 44 00.	+ 58 20	540		DARK NEBULA
LDN 1205	22 44 00.	+ 60 10	1200		DARK NEBULA
UGC 12182	22 44 00.	+ 72 53	66	15.5	GALAXY S
SC 2240-7246.8	22 44 00.	- 72 31 03.	48		NEBULA
SC 2241-4657.6	22 44 01.	- 46 41 51.	6		NEBULA
ARC 2481	22 44 03.	- 21 55		17.5	RICH CLUSTER OF GALAXIES
SC 2241-4357.0	22 44 03.	- 43 41 14.			NEBULA
SC 2241-4419.7	22 44 03.	- 44 03 56.			NEBULA
SC 2241-4657.2	22 44 03.	- 46 41 26.			NEBULA
SC 2240-7234.7	22 44 04.	- 72 18 57.	18		NEBULA
SC 2241-4658.1	22 44 05.	- 46 42 20.			NEBULA
SC 2241-4701.8	22 44 05.	- 46 46 02.			NEBULA
ZC 2244.1+0839	22 44 06.	+ 08 39	1210		CLUSTER OF GALAXIES
MCG+03-58-003	22 44 06.	+ 20 50	42	15.	GALAXY
SC 2241-4631.6	22 44 08.	- 46 15 50.			NEBULA
RNGC 7374B	22 44 09.	+ 10 55		15.5	GALAXY
SC 2241-4419.4	22 44 09.	- 44 03 38.			NEBULA
SC 2241-4624.0	22 44 09.	- 46 08 14.			NEBULA
RNGC 7375	22 44 11.	+ 20 49		15.0	GALAXY
ZWG 430.009	22 44 12.	+ 10 55		15.5	GALAXY
MCG+02-58-009	22 44 12.	+ 10 55	36	15.	GALAXY
ZWG 453.007	22 44 12.	+ 20 49		14.9	GALAXY
MCG+12-21-001	22 44 12.	+ 72 53	39	15.	GALAXY
MCG-02-58-004	22 44 12.	- 13 43	36	15.	GALAXY
SC 2241-4626.7	22 44 14.	- 46 10 56.			NEBULA
SC 2240-7249.9	22 44 15.	- 72 34 09.	18		NEBULA
IC 1453	22 44 16.	- 13 42 34.			NONSTELLAR OBJECT
SC 2241-4622.2	22 44 17.	- 46 06 26.			NEBULA
SC 2241-4630.0	22 44 17.	- 46 14 14.			NEBULA
SC 2241-4548.4	22 44 18.	- 45 32 38.			NEBULA
SC 2241-4624.2	22 44 19.	- 46 08 26.			NEBULA
SC 2241-4707.2	22 44 22.	- 46 51 26.			NEBULA
SC 2241-4416.2	22 44 23.	- 44 00 26.			NEBULA
SC 2241-4602.2	22 44 23.	- 45 46 26.			NEBULA
ZC 2244.4+3558	22 44 24.	+ 35 58	1210		CLUSTER OF GALAXIES
SC 2241-4424.6	22 44 29.	- 44 08 50.			NEBULA
SC 2241-4549.6	22 44 29.	- 45 33 50.			NEBULA
ZWG 474.028	22 44 30.	+ 21 38		15.4	GALAXY
ZWG 379.005	22 44 30.	- 00 18		15.5	GALAXY
UGC 12183	22 44 30.	- 00 18	60	15.5	GALAXY S
MCG+00-58-003	22 44 30.	- 00 20	60	14.	GALAXY
MCG-05-53-032	22 44 30.	- 29 45	36	15.	GALAXY
ARC 2482	22 44 31.	- 03 17		17.8	RICH CLUSTER OF GALAXIES
SC 2241-4639.1	22 44 33.	- 46 23 20.			NEBULA
SC 2241-4547.9	22 44 34.	- 45 32 08.			NEBULA

OBJECT NAME	RIGHT ASCEN.	DECLINATION	DIAM.	MAGN.	TYPE OF OBJECT
SC 2241-4419.8	22 44 35.	- 44 04 02.			NEBULA
ZWG 430.010	22 44 36.	+ 11 20		15.7	GALAXY
UGC 12184	22 44 36.	+ 11 20	60	15.7	GALAXY Sc
MCG+02-58-010	22 44 36.	+ 11 20	60	15.	GALAXY
PHL 5447	22 44 36.	- 29 52		18.4	BLUE STELLAR OBJECT
SC 2241-4320.0	22 44 36.	- 43 04 14.			NEBULA
SC 2241-6155.7	22 44 36.	- 61 39 56.	12		NEBULA
SC 2241-4412.5	22 44 38.	- 43 56 44.			NEBULA
SC 2241-4607.9	22 44 39.	- 45 52 08.			NEBULA
SC 2241-4639.9	22 44 41.	- 46 24 08.			NEBULA
ZC 2244.7+0446	22 44 42.	+ 04 46	810		CLUSTER OF GALAXIES
COU 096	22 44 42.	+ 57 50	1800		EMISSION NEBULA
SC 2241-4305.3	22 44 42.	- 42 49 31.			NEBULA
SC 2241-4313.6	22 44 42.	- 42 57 49.			NEBULA
SC 2241-4603.1	22 44 42.	- 45 47 20.			NEBULA
PN 0769	22 44 44.	- 69 10			NEBULA
IC 5252	22 44 44.	- 69 11			NONSTELLAR OBJECT
SC 2241-4644.8	22 44 45.	- 46 29 01.			NEBULA
SC 2241-6155.7	22 44 45.	- 61 39 56.	12		NEBULA
SC 2241-4252.5	22 44 46.	- 42 36 43.			NEBULA
ZWG 379.006	22 44 43.	+ 03 24		15.6	GALAXY
ZWG 495.024	22 44 48.	+ 31 16		15.7	GALAXY
ARC 2483	22 44 49.	+ 04 44		17.9	RICH CLUSTER OF GALAXIES
RNGC 7376	22 44 50.	+ 03 24		15.5	GALAXY
SC 2241-4616.8	22 44 50.	- 46 01 01.			NEBULA
SC 2241-4456.4	22 44 53.	- 44 40 37.			NEBULA
PHL 5448	22 44 54.	- 28 06		17.7	BLUE STELLAR OBJECT
SC 2242-4417.0	22 44 55.	- 44 01 13.			NEBULA
SC 2241-4554.2	22 44 55.	- 45 38 25.			NEBULA
SC 2242-4401.2	22 44 56.	- 43 45 25.			NEBULA
SC 2242-4509.8	22 44 57.	- 44 54 01.			NEBULA
SC 2242-4457.4	22 44 58.	- 44 41 37.			NEBULA
LBN 0454	22 45	+ 41 30	9000		BRIGHT NEBULA
KHAV 778	22 45	+ 52 28	4430		DARK NEBULA
LBN 0511	22 45	+ 57 45	1320		BRIGHT NEBULA
LBN 0522	22 45	+ 62 00	18900		BRIGHT NEBULA
ZWG 405.009	22 45 00.	+ 05 13		15.7	GALAXY
MCG+03-58-004	22 45 00.	+ 21 00	30	16.	GALAXY
ZWG 495.025	22 45 00.	+ 31 07		14.8	GALAXY
UGC 12185	22 45 00.	+ 31 07	108	14.8	GALAXY SBb
CED 206	22 45 00.	+ 57 37	1080		DIFFUSE GALACTIC NEBULA
SG 2.091	22 45 00.	+ 57 37	1800		DIFFUSE EMISSION NEBULA
OCL 0244	22 45 00.	+ 57 50	1800	8.8	OPEN STAR CLUSTER
RNGC 7380	22 45 00.	+ 57 50		9.0	CLUSTER WITH NEBULOSITY
LDN 1211	22 45 00.	+ 61 55	420		DARK NEBULA
PHL 5449	22 45 00.	- 28 34		18.5	BLUE STELLAR OBJECT
SC 2242-4644.0	22 45 02.	- 46 28 13.			NEBULA
MCG-03-58-003	22 45 03.	- 17 37	60	14.5	GALAXY
RNGC 7377	22 45 03.	- 22 35		12.5	GALAXY
SC 2242-4505.1	22 45 05.	- 44 49 19.			NEBULA
MCG+01-58-006	22 45 06.	+ 05 12	24	15.	GALAXY
ZWG 531.012	22 45 06.	+ 40 36		15.1	GALAXY
UGC 12186	22 45 06.	+ 49 52	60	16.0	GALAXY
RNGC 7379	22 45 10.	+ 39 58		14.5	GALAXY
SC 2242-4248.0	22 45 11.	- 42 32 13.			NEBULA
SC 2241-6230.4	22 45 11.	- 62 14 37.			NEBULA
ZWG 453.008	22 45 12.	+ 21 22		15.6	GALAXY
ZWG 531.013	22 45 12.	+ 39 58		14.4	GALAXY
UGC 12187	22 45 12.	+ 39 58	84	14.4	GALAXY SBa
57W 392	22 45 12.	+ 40 58			COMPACT GALAXY
MCG-03-58-004	22 45 12.	- 17 21	72	15.	GALAXY
MCG-04-53-038	22 45 12.	- 22 36	48	12.5	GALAXY
SC 2242-4418.7	22 45 12.	- 44 02 55.			NEBULA
SC 2242-4333.2	22 45 13.	- 43 17 25.			NEBULA
MCG+07-46-018	22 45 15.	+ 39 57	66	14.	GALAXY
SC 2242-5853.2	22 45 16.	- 58 37 25.	60		NEBULA
RNGC 7378	22 45 17.	- 12 05		13.0	GALAXY
SC 2242-4509.9	22 45 17.	- 44 54 07.			NEBULA
ZWG 453.009	22 45 18.	+ 18 30		15.7	GALAXY
MCG+05-53-015	22 45 18.	+ 31 06	72	14.5	GALAXY
ZWG 531.014	22 45 18.	+ 39 37		14.4	GALAXY
UGC 12188	22 45 18.	+ 39 37	54	14.4	GALAXY S+COMP
MCG-02-58-005	22 45 18.	- 12 05	78	13.5	GALAXY
SC 2242-4551.2	22 45 18.	- 45 35 25.			NEBULA
SC 2241-7309.3	22 45 18.	- 72 53 31.	12		NEBULA
SC 2242-4635.9	22 45 19.	- 46 20 07.	18		NEBULA
MCG+07-46-019	22 45 21.	+ 39 35	45	14.5	GALAXY
SC 2242-4352.3	22 45 23.	- 43 36 31.			NEBULA
PHL 5450	22 45 24.	- 32 45		13.9	BLUE STELLAR OBJECT
SC 2242-5946.9	22 45 24.	- 59 31 07.	12		NEBULA
SC 2242-4522.5	22 45 26.	- 45 06 42.			NEBULA
MCG+07-46-020	22 45 30.	+ 39 50	36	16.	GALAXY
ZWG 531.015	22 45 30.	+ 39 51		15.7	GALAXY
SC 2242-4337.8	22 45 30.	- 43 22 00.			NEBULA
SC 2242-4340.8	22 45 32.	- 43 25 00.			NEBULA
SC 2242-6300.1	22 45 32.	- 62 44 19.	36		NEBULA
SC 2242-4650.5	22 45 33.	- 46 34 42.			NEBULA
SC 2242-5855.7	22 45 34.	- 58 39 54.	12		NEBULA
UGC 12189	22 45 36.	+ 03 39	60	16.0	GALAXY SB
MCG+01-58-007	22 45 36.	+ 03 39	48	15.5	GALAXY
MCG+01-58-007	22 45 36.	+ 03 39	48	15.5	GALAXY
ZC 2245.6+3516	22 45 36.	+ 35 16	1010		CLUSTER OF GALAXIES
MRSL 107-01/2	22 45 36.	+ 57 48	1800		HII REGION
TON-S 0074	22 45 36.	- 32 46		14.2	BLUE STAR
LB 01509	22 45 36.	- 45 42		14.1	FAINT BLUE STAR
SC 2242-4344.8	22 45 38.	- 43 29 00.			NEBULA
SC 2242-4549.2	22 45 38.	- 45 33 24.			NEBULA
SC 2242-4549.6	22 45 38.	- 45 33 48.			NEBULA
ARC 2484	22 45 39.	+ 01 09		18.0	RICH CLUSTER OF GALAXIES
SC 2242-4543.9	22 45 39.	- 45 28 06.			NEBULA
ZWG 495.026	22 45 42.	+ 28 01		15.6	GALAXY
UGC 12190	22 45 42.	+ 28 01	132	15.6	GALAXY Sc
KARA.73B 0988	22 45 42.	+ 28 01	132	15.6	ISOLATED GALAXY S
MCG-07-46-011	22 45 42.	- 39 56	120	14.5	GALAXY
SC 2242-4347.5	22 45 42.	- 43 31 42.			NEBULA
SC 2242-4348.4	22 45 42.	- 43 32 36.			NEBULA
SC 2242-6218.5	22 45 44.	- 62 02 42.	12		NEBULA
HELW 498	22 45 45.	- 39 54 54.			NEBULA
SC 2242-4624.6	22 45 46.	- 46 08 48.			NEBULA
ZWG 495.027	22 45 48.	+ 31 42		15.6	GALAXY
PHL 5451	22 45 48.	- 23 48		16.9	BLUE STELLAR OBJECT
PHL 5452	22 45 48.	- 31 24		17.7	BLUE STELLAR OBJECT
SC 2242-4540.1	22 45 48.	- 45 24 18.			NEBULA
SC 2242-4456.6	22 45 49.	- 44 40 48.			NEBULA
SC 2242-5825.3	22 45 50.	- 58 09 30.	24		NEBULA
SC 2242-4640.3	22 45 51.	- 46 24 30.			NEBULA
ARC 2485	22 45 52.	- 16 23		17.2	RICH CLUSTER OF GALAXIES
SC 2242-4548.1	22 45 53.	- 45 32 18.			NEBULA

OBJECT NAME	RIGHT ASCEN.	DECLINATION	DIAM.	MAGN.	TYPE OF OBJECT
SC 2242-4619.6	22 45 53.	- 46 03 48.			NEBULA
ZWG 405.010	22 45 54.	+ 09 11		15.6	GALAXY
KARA.73B 0989	22 45 54.	+ 09 11	24	15.6	ISOLATED GALAXY S
ZWG 474.029	22 45 54.	+ 24 33		15.5	GALAXY
ZWG 495.028	22 45 54.	+ 30 35		15.6	GALAXY
PHL 5453	22 45 54.	- 20 38		18.4	BLUE STELLAR OBJECT
LB 01510	22 45 54.	- 46 19		13.5	FAINT BLUE STAR
SC 2243-4425.0	22 45 57.	- 44 09 12.			NEBULA
SC 2242-6218.4	22 45 57.	- 62 02 36.	12		NEBULA
SC 2243-4249.2	22 45 58.	- 42 33 24.			NEBULA
SC 2243-4614.2	22 45 58.	- 45 58 24.			NEBULA
KHAV 779	22 46	+ 57 10	2230		DARK NEBULA
LB 09948	22 46	- 85 08		13.4	FAINT BLUE STAR
ZWG 453.010	22 46 00.	+ 15 39		15.4	GALAXY
LDN 1212	22 46 00.	+ 61 57	480		DARK NEBULA
PHL 5454	22 46 00.	- 22 42		16.3	BLUE STELLAR OBJECT
PHL 5455	22 46 00.	- 27 24		16.9	BLUE STELLAR OBJECT
SC 2242-6223.7	22 46 01.	- 62 07 54.	18		NEBULA
SC 2243-4538.4	22 46 02.	- 45 22 36.			NEBULA
SC 2243-4557.4	22 46 02.	- 45 41 36.			NEBULA
SC 2243-4415.5	22 46 03.	- 43 59 42.			NEBULA
SC 2243-4652.4	22 46 03.	- 46 36 36.	12		NEBULA
SC 2243-4303.9	22 46 04.	- 42 48 06.			NEBULA
SC 2243-4528.5	22 46 04.	- 45 12 42.			NEBULA
SC 2243-4520.6	22 46 05.	- 45 04 48.			NEBULA
SC 2243-4609.4	22 46 06.	- 45 53 36.	12		NEBULA
SC 2243-4429.8	22 46 07.	- 44 13 59.			NEBULA
SC 2242-7218.6	22 46 07.	- 72 02 48.			NEBULA
SC 2243-4338.0	22 46 08.	- 43 22 11.			NEBULA
SC 2243-4653.0	22 46 08.	- 46 37 11.			NEBULA
SC 2243-4524.2	22 46 09.	- 45 08 23.			NEBULA
ZC 2246.2-0049	22 46 12.	- 00 49	1140		CLUSTER OF GALAXIES
MCG-05-53-033	22 46 12.	- 27 52	54	15.5	GALAXY
SC 2243-4607.8	22 46 12.	- 45 51 59.			NEBULA
SC 2243-4414.4	22 46 13.	- 43 58 35.			NEBULA
SC 2243-4459.2	22 46 13.	- 44 43 23.			NEBULA
SC 2243-4302.0	22 46 14.	- 42 46 11.			NEBULA
SC 2243-4652.3	22 46 15.	- 46 36 29.	12		NEBULA
ABC 2486	22 46 16.	+ 16 54		17.1	RICH CLUSTER OF GALAXIES
ZWG 474.030	22 46 18.	+ 27 21		14.9	GALAXY
UGC 12191	22 46 18.	+ 27 21	90	14.9	GALAXY S
KARA.72 573A	22 46 18.	+ 27 21	78	14.9	PART OF DOUBLE GALAXY
MCG+04-53-015	22 46 18.	+ 27 23	72	15.	GALAXY
UGC 12192	22 46 18.	+ 36 57	60	17.	GALAXY
HN 0770	22 46 18.	- 68 57			NEBULA
SC 2243-4249.9	22 46 19.	- 42 34 05.			NEBULA
SC 2243-4451.3	22 46 19.	- 44 35 29.			NEBULA
IC 5256	22 46 19.	- 68 58			NONSTELLAR OBJECT
SC 2243-4652.0	22 46 20.	- 46 36 11.			NEBULA
ZWG 453.011	22 46 24.	+ 18 18		15.3	GALAXY
ZWG 474.031	22 46 24.	+ 27 19		14.1	GALAXY
UGC 12193	22 46 24.	+ 27 19	78	14.1	GALAXY SBc
KARA.72 573B	22 46 24.	+ 27 19	90	14.1	PART OF DOUBLE GALAXY
MCG+04-53-016	22 46 24.	+ 27 21	66	14.5	GALAXY
ZWG 495.029	22 46 24.	+ 27 42		15.7	GALAXY
UGC 12194	22 46 24.	+ 27 42	66	15.7	GALAXY S
ZWG 495.030	22 46 24.	+ 32 58		15.7	GALAXY
UGC 12195	22 46 24.	+ 32 58	66	15.7	GALAXY SBa-b
MCG-03-58-005	22 46 24.	- 19 59	36	15.5	GALAXY
SC 2243-4647.1	22 46 26.	- 46 31 17.			NEBULA
MCG-03-58-006	22 46 27.	- 16 51	18	15.	GALAXY
SC 2243-4654.4	22 46 28.	- 46 38 35.	12		NEBULA
ZWG 405.011	22 46 30.	+ 06 57		15.2	GALAXY
UGC 12196	22 46 30.	+ 06 57	66	15.2	GALAXY Sa
MCG+01-58-009	22 46 30.	+ 06 58	48	15.	GALAXY
MCG+01-58-008	22 46 30.	+ 06 59	24	17.	GALAXY
MCG+02-58-011	22 46 30.	+ 10 54	48	15.5	GALAXY
MCG+03-58-005	22 46 30.	+ 18 19	36	15.	GALAXY
UGC 12197	22 46 30.	+ 24 35	72	17.	GALAXY Sb-c
ZWG 474.032	22 46 30.	+ 27 18		15.7	GALAXY
SC 2243-4638.6	22 46 31.	- 46 22 47.			NEBULA
SC 2243-4656.5	22 46 31.	- 46 40 41.	12		NEBULA
SC 2243-4559.5	22 46 32.	- 45 43 41.			NEBULA
ZC 2246.6+1645	22 46 36.	+ 16 45	2350		CLUSTER OF GALAXIES
ZC 2246.6+4731	22 46 36.	+ 47 31	1810		CLUSTER OF GALAXIES
ACK 107-00.1	22 46 36.	+ 58 13			PLANETARY NEBULA
ABC 2489	22 46 38.	- 05 42		18.0	RICH CLUSTER OF GALAXIES
SC 2243-4633.2	22 46 39.	- 46 17 23.	12		NEBULA
SC 2243-4527.1	22 46 41.	- 45 11 17.			NEBULA
SC 2243-4528.3	22 46 41.	- 45 12 29.			NEBULA
SC 2243-4604.2	22 46 41.	- 45 48 23.			NEBULA
MCG+02-58-012	22 46 42.	+ 11 17	36	15.5	GALAXY
ZWG 495.031	22 46 42.	+ 31 20		15.5	GALAXY
ABC 2490	22 46 43.	- 04 04		17.2	RICH CLUSTER OF GALAXIES
SC 2243-4513.6	22 46 43.	- 44 57 47.			NEBULA
ABC 2487	22 46 44.	- 21 14		17.5	RICH CLUSTER OF GALAXIES
ABC 2488	22 46 46.	- 23 50		17.5	RICH CLUSTER OF GALAXIES
SC 2243-4659.8	22 46 46.	- 46 43 59.			NEBULA
SC 2243-4704.0	22 46 47.	- 46 48 11.	12		NEBULA
SC 2243-4730.0	22 46 47.	- 47 14 11.	12		NEBULA
ZC 2246.8+0155	22 46 48.	+ 01 55	4170		CLUSTER OF GALAXIES
ZWG 495.032	22 46 48.	+ 27 36		15.7	GALAXY
UGC 12198	22 46 48.	+ 27 36	60	15.7	GALAXY S
MCG+07-46-021	22 46 48.	+ 39 42	90	14.5	GALAXY
ZWG 531.016	22 46 48.	+ 39 44		14.5	GALAXY
UGC 12199	22 46 48.	+ 39 44	90	14.5	GALAXY SBb
PHL 0398	22 46 48.	- 27 23		16.4	BLUE STELLAR OBJECT
LB 01511	22 46 48.	- 45 45		13.3	FAINT BLUE STAR
ZC 2246.9+1415	22 46 54.	+ 14 15	470		CLUSTER OF GALAXIES
ZWG 453.012	22 46 54.	+ 19 01		14.8	GALAXY
UGC 12200	22 46 54.	+ 19 01	60	14.8	GALAXY Sb
MCG+03-58-006	22 46 54.	+ 19 02	48	14.5	GALAXY
ZWG 515.006	22 46 54.	+ 34 44		15.0	GALAXY
UGC 12201	22 46 54.	+ 34 44	96	15.0	GALAXY Sa-b
MCG+06-50-004	22 46 54.	+ 34 44	90	15.	GALAXY
MCG-03-58-007	22 46 54.	- 19 34	30	15.	GALAXY
PHL 5456	22 46 54.	- 23 42		18.0	BLUE STELLAR OBJECT
PHL 2069	22 46 54.	- 26 50		18.3	BLUE STELLAR OBJECT
SC 2244-4417.6	22 46 55.	- 44 01 46.			NEBULA
SC 2243-4730.7	22 46 55.	- 47 14 52.	12		NEBULA
SC 2244-4605.2	22 46 56.	- 45 49 22.			NEBULA
SC 2244-4633.0	22 46 58.	- 46 17 10.			NEBULA
SC 2244-4633.2	22 46 58.	- 46 17 22.			NEBULA
SC 2243-6147.3	22 46 59.	- 61 31 28.	12		NEBULA
MCG+02-58-013	22 47 00.	+ 11 01	72	14.	GALAXY
ZWG 430.011	22 47 00.	+ 11 03		14.8	GALAXY
UGC 12202	22 47 00.	+ 11 03	72	14.8	GALAXY Sa
ZC 2247.0+1833	22 47 00.	+ 18 33	1550		CLUSTER OF GALAXIES

OBJECT NAME	RIGHT ASCEN.	DECLINATION	DIAM.	MAGN.	TYPE OF OBJECT
ZC 2247.0+3000	22 47 00.	+ 30 00	2490		CLUSTER OF GALAXIES
4ZW 116	22 47 00.	+ 30 01			COMPACT GALAXY
TON-S 0075	22 47 00.	- 27 24		14.9	BLUE STAR
ABC 2491	22 47 02.	+ 18 37		17.5	RICH CLUSTER OF GALAXIES
RNGC 7383	22 47 03.	+ 11 17		15.0	GALAXY
SC 2244-4704.2	22 47 03.	- 46 48 22.			NEBULA
SC 2244-4559.2	22 47 04.	- 45 43 22.			NEBULA
REIN 7.195	22 47 05.87	+ 11 17 28.3			NEBULA
ZWG 430.012	22 47 06.	+ 11 17		15.1	GALAXY
MCG+02-58-014	22 47 06.	+ 11 17	24	15.	GALAXY
ZWG 430.013	22 47 06.	+ 11 44		15.7	GALAXY
ZWG 453.013	22 47 06.	+ 17 10		15.7	GALAXY
MCG+03-58-007	22 47 06.	+ 17 10	42	15.	GALAXY
UGC 12203	22 47 06.	+ 32 33	60	16.5	GALAXY S
UGC 12204	22 47 06.	+ 39 58	96	16.0	GALAXY Sb-c
SC 2244-4420.5	22 47 06.	- 44 04 40.			NEBULA
SC 2244-4648.6	22 47 08.	- 46 32 46.			NEBULA
SC 2244-4619.7	22 47 09.	- 46 03 52.			NEBULA
REIN 7.196	22 47 10.22	+ 11 14 19.2			NEBULA
MCG+00-58-004	22 47 12.	+ 02 52	48	15.	GALAXY
ZWG 430.014	22 47 12.	+ 14 49		15.4	GALAXY
UGC 12205	22 47 12.	+ 14 49	54	15.4	GALAXY IRR
KARA.73B 0990	22 47 12.	+ 14 49	66	15.4	ISOLATED GALAXY S
MCG+02-58-015	22 47 12.	+ 14 50	48	15.	GALAXY
ZC 2247.2+2650	22 47 12.	+ 26 50	1550		CLUSTER OF GALAXIES
ZWG 495.033	22 47 12.	+ 32 06		15.5	GALAXY
ZWG 495.034	22 47 12.	+ 33 06		15.5	GALAXY
UGC 12206	22 47 12.	+ 33 06	180	15.5	GALAXY SBa
REIN 7.197	22 47 12.66	+ 11 13 22.3			NEBULA
SC 2244-4636.6	22 47 13.	- 46 20 46.			NEBULA
SC 2244-4328.6	22 47 14.	- 43 12 46.			NEBULA
RNGC 7384	22 47 15.	+ 11 17			GALAXY
SC 2244-5714.2	22 47 15.	- 56 58 22.	12		NEBULA
BC PKS2247+13	22 47 15.5	+ 13 15 24.		18.	QUASI-STELLAR OBJECT
LB 01149	22 47 16.	+ 25 04 12.		17.1	FAINT BLUE STAR
ZC 2247.3+1107	22 47 18.	+ 11 07	8200		CLUSTER OF GALAXIES
4ZW 117	22 47 18.	+ 32 05			COMPACT GALAXY
ACK 111+06.1	22 47 18.	+ 66 46			PLANETARY NEBULA
SC 2244-4525.8	22 47 18.	- 45 09 58.			NEBULA
REIN 7.198	22 47 18.46	+ 11 19 15.1			NEBULA
LB 01150	22 47 19.	+ 26 01 24.		17.5	FAINT BLUE STAR
RNGC 7385	22 47 21.	+ 11 21		14.5	GALAXY
RNGC 7381	22 47 21.	- 19 59		15.0	GALAXY
SC 2244-4451.4	22 47 21.	- 44 35 34.			NEBULA
SC 2244-4640.4	22 47 22.	- 46 24 34.			NEBULA
SC 2244-4653.4	22 47 22.	- 46 37 34.			NEBULA
SC 2244-4523.6	22 47 23.	- 45 07 46.			NEBULA
ZWG 430.015	22 47 24.	+ 11 20		14.4	GALAXY
UGC 12207	22 47 24.	+ 11 20	114	14.4	GALAXY E
MCG+02-58-017	22 47 24.	+ 11 20	27	13.	GALAXY
MCG+02-58-016	22 47 24.	+ 11 47	36	15.	GALAXY
MCG+05-53-017	22 47 24.	+ 32 07 30.	30	15.	GALAXY
ZC 2247.4+3235	22 47 24.	+ 32 35	1010		CLUSTER OF GALAXIES
MCG+05-53-016	22 47 24.	+ 33 05 30.	60	15.5	GALAXY
MCG+07-46-022	22 47 24.	+ 43 43	30	15.	GALAXY
ZWG 532.001	22 47 24.	+ 43 44		15.6	GALAXY
ZWG 531.017	22 47 24.	+ 43 44		15.6	GALAXY
MCG-03-58-007A	22 47 24.	- 20 01	36	14.	GALAXY
PHL 2070	22 47 24.	- 26 45		16.0	BLUE STELLAR OBJECT
PHL 5457	22 47 24.	- 28 22		16.7	BLUE STELLAR OBJECT
SC 2244-4318.8	22 47 24.	- 43 02 58.			NEBULA
SC 2244-4515.5	22 47 24.	- 44 59 40.			NEBULA
REIN 7.199	22 47 24.25	+ 11 20 18.5			NEBULA
REIN 7.200	22 47 24.79	+ 11 20 35.7			NEBULA
RNGC 7386	22 47 27.	+ 11 26		14.5	GALAXY
HOLM 799A	22 47 29.	- 04 40	30	14.9	PART OF MULTIPLE GALAXY
RNGC 7382	22 47 29.	- 37 07			GALAXY
MCG+02-58-018	22 47 30.	+ 11 25	24	13.	GALAXY
MRSL 108-00/1	22 47 30.	+ 59 40	120		HII REGION
MCG-01-58-001	22 47 30.	- 04 04	36	15.	GALAXY
HOLM 799B	22 47 30.	- 04 39	18	15.1	PART OF MULTIPLE GALAXY
TON-S 0076	22 47 30.	- 26 46		14.8	BLUE STAR
SC 2244-4323.7	22 47 30.	- 43 07 52.			NEBULA
SC 2244-4501.1	22 47 30.	- 44 45 16.			NEBULA
SC 2244-4659.9	22 47 30.	- 46 44 04.			NEBULA
REIN 7.201	22 47 32.41	+ 11 26 01.4			NEBULA
SC 2244-4420.6	22 47 34.	- 44 04 45.			NEBULA
SC 2244-4555.3	22 47 34.	- 45 39 27.			NEBULA
SC 2244-4541.9	22 47 35.	- 45 26 03.			NEBULA
ZC 2247.6+0815	22 47 36.	+ 08 15	400		CLUSTER OF GALAXIES
ZWG 430.016	22 47 36.	+ 11 25		14.6	GALAXY
UGC 12209	22 47 36.	+ 11 25	126	14.6	GALAXY E-SO
ZWG 430.017	22 47 36.	+ 11 47		15.4	GALAXY
ZWG 379.007	22 47 36.	- 01 17		15.2	GALAXY
UGC 12208	22 47 36.	- 01 17	66	15.2	GALAXY SBc
MCG+00-58-005	22 47 36.	- 01 20	66	14.	GALAXY
SC 2244-4543.8	22 47 39.	- 45 27 57.			NEBULA
SC 2244-4544.3	22 47 39.	- 45 28 27.			NEBULA
REIN 7.202	22 47 39.13	+ 11 23 01.3			NEBULA
ZC 2247.7+2503	22 47 42.	+ 25 03	1480		CLUSTER OF GALAXIES
ZC 2247.7+3208	22 47 42.	+ 32 08	1080		CLUSTER OF GALAXIES
SC 2244-4509.6	22 47 43.	- 44 53 45.			NEBULA
SC 2244-4648.6	22 47 44.	- 46 32 45.			NEBULA
RNGC 7389	22 47 45.	+ 11 19		15.0	GALAXY
RNGC 7387	22 47 45.	+ 11 22		15.5	GALAXY
RNGC 7388	22 47 45.	+ 11 28			GALAXY
SC 2244-4559.8	22 47 45.	- 45 43 57.			NEBULA
ABC 2494	22 47 46.	+ 16 15		17.3	RICH CLUSTER OF GALAXIES
SG 2.092	22 47 46.	+ 59 35	18		DIFFUSE EMISSION NEBULA
REIN 7.203	22 47 46.14	+ 11 18 02.3			NEBULA
REIN 7.204	22 47 47.86	+ 11 22 18.2			NEBULA
MCG+02-58-021	22 47 48.	+ 10 37	36	16.	GALAXY
MCG+02-58-020	22 47 48.	+ 11 15	48	15.5	GALAXY
MCG+02-58-019	22 47 48.	+ 11 17	60	15.	GALAXY
ZWG 430.018	22 47 48.	+ 11 19		15.2	GALAXY
MCG+02-58-023	22 47 48.	+ 11 21	15	17.	GALAXY
ZWG 430.019	22 47 48.	+ 11 22		15.3	GALAXY
MCG+02-58-022	22 47 48.	+ 11 22	36	15.	GALAXY
ZWG 495.035	22 47 48.	+ 31 07		15.4	GALAXY
UGC 12210	22 47 48.	+ 31 07	78	15.4	GALAXY SBa
FEIG 107	22 47 48.	- 13 34		12.0	FAINT BLUE STAR
PHL 2071	22 47 48.	- 25 06		18.0	BLUE STELLAR OBJECT
PHL 5458	22 47 48.	- 28 52		17.5	BLUE STELLAR OBJECT
ABC 2492	22 47 49.	- 19 32		16.5	RICH CLUSTER OF GALAXIES
REIN 7.205	22 47 49.83	+ 11 15 57.6			NEBULA
SC 2244-4557.2	22 47 50.	- 45 41 21.			NEBULA
SC 2244-4644.2	22 47 50.	- 46 28 21.			NEBULA
RNGC 7390	22 47 51.	+ 11 16		15.5	GALAXY

OBJECT NAME	RIGHT ASCEN.	DECLINATION	DIAM.	MAGN.	TYPE OF OBJECT
REIN 7.206	22 47 51.32	+ 11 26 43.6			NEBULA
SC 2244-4359.2	22 47 52.	- 43 43 21.			NEBULA
SC 2244-4649.2	22 47 52.	- 46 33 21.			NEBULA
ZWG 430.020	22 47 54.	+ 11 16		15.7	GALAXY
PHL 5459	22 47 54.	- 27 48		14.3	BLUE STELLAR OBJECT
SC 2245-4359.6	22 47 54.	- 43 43 45.			NEBULA
SC 2244-4648.8	22 47 54.	- 46 32 57.			NEBULA
LB 01151	22 47 55.	+ 25 10 18.		18.2	FAINT BLUE STAR
SC 2245-4500.8	22 47 56.	- 44 44 57.			NEBULA
ARC 2495	22 47 57.	+ 10 38		16.5	RICH CLUSTER OF GALAXIES
SC 2245-4655.0	22 47 57.	- 46 39 09.			NEBULA
BC 4C14.82	22 47 57.1	+ 14 04 06.		17.	QUASI-STELLAR OBJECT
SHB 378	22 47 57.4	+ 14 03 07.		17.	QUASI-STELLAR OBJECT
SC 2245-4700.6	22 47 58.	- 46 44 45.	12		NEBULA
SC 2245-4358.0	22 47 59.	- 43 42 09.			NEBULA
LBN 0513	22 48	+ 58 40	2400		BRIGHT NEBULA
LBN 0514	22 48	+ 59 38	120		BRIGHT NEBULA
MCG+05-53-018	22 48 00.	+ 31 06 30.	48	15.	GALAXY
COU 097	22 48 00.	+ 61 00	1800		EMISSION NEBULA
MCG+00-58-006	22 48 00.	- 01 49	27	13.	GALAXY
ARC 2493	22 48 00.	- 26 19		17.1	RICH CLUSTER OF GALAXIES
PHL 5460	22 48 00.	- 30 21		18.5	BLUE STELLAR OBJECT
SC 2245-4550.4	22 48 01.	- 45 34 33.			NEBULA
REIN 7.207	22 48 01.62	+ 11 17 40.4			NEBULA
SC 2245-4313.6	22 48 03.	- 42 57 45.			NEBULA
SC 2245-4522.2	22 48 04.	- 45 06 21.			NEBULA
SC 2245-4632.0	22 48 04.	- 46 16 09.			NEBULA
SC 2245-4700.4	22 48 04.	- 46 44 33.			NEBULA
UGC 12212	22 48 06.	+ 28 52	150	16.0	GALAXY DWRF SP
ZWG 379.008	22 48 06.	- 01 47		13.7	GALAXY
UGC 12211	22 48 06.	- 01 47	120	13.7	GALAXY E
RNGC 7391	22 48 07.	- 01 47		13.5	GALAXY
SC 2245-4524.8	22 48 07.	- 45 08 57.			NEBULA
SC 2245-4550.6	22 48 07.	- 45 34 45.			NEBULA
SC 2245-4654.0	22 48 11.	- 46 38 09.			NEBULA
ZC 2248.2+0756	22 48 12.	+ 07 56	3290		CLUSTER OF GALAXIES
MCG+02-58-024	22 48 12.	+ 11 08	24	15.5	GALAXY
MCG+05-53-019	22 48 12.	+ 28 52	60	16.	GALAXY
MRSL 107-01/1	22 48 12.	+ 57 27	240		HII REGION
PHL 5461	22 48 12.	- 28 06		18.3	BLUE STELLAR OBJECT
SC 2245-4550.0	22 48 13.	- 45 34 09.			NEBULA
SC 2245-4700.3	22 48 14.	- 46 44 27.	6		NEBULA
SC 2245-4734.0	22 48 14.	- 47 18 09.			NEBULA
ARC 2496	22 48 16.	- 16 41		17.2	RICH CLUSTER OF GALAXIES
ZWG 430.021	22 48 18.	+ 12 17		15.5	GALAXY
MCG+02-58-025	22 48 18.	+ 12 17	48	15.	GALAXY
ZC 2248.3-0203	22 48 18.	- 02 03	1080		CLUSTER OF GALAXIES
SC 2245-4549.7	22 48 19.	- 45 33 50.			NEBULA
SC 2245-4504.6	22 48 20.	- 44 48 44.			NEBULA
SC 2245-4640.5	22 48 20.	- 46 24 38.			NEBULA
MCG-04-53-039	22 48 21.	- 20 32	90	14.	GALAXY
SC 2245-4446.0	22 48 23.	- 44 30 08.			NEBULA
MCG+02-58-027	22 48 24.	+ 11 47	24	16.5	GALAXY
MCG+02-58-026	22 48 24.	+ 11 47	36	15.	GALAXY
RNGC 7394	22 48 26.	+ 51 54			NON-EXISTENT OBJECT
UGC 12213	22 48 30.	+ 07 02	126	16.0	GALAXY
MCG+02-58-028	22 48 30.	+ 10 57	48	15.	GALAXY
ZWG 430.022	22 48 30.	+ 10 58		15.4	GALAXY
ZWG 430.023	22 48 30.	+ 11 48		15.4	GALAXY
SC 2245-4610.8	22 48 30.	- 45 54 56.			NEBULA
ARC 2497	22 48 31.	- 20 12		17.4	RICH CLUSTER OF GALAXIES
SC 2245-4555.1	22 48 33.	- 45 39 14.			NEBULA
SC 2245-4615.2	22 48 33.	- 45 59 20.			NEBULA
SC 2245-4731.0	22 48 34.	- 47 15 08.			NEBULA
MCG+01-58-010	22 48 35.	+ 07 03	120	15.	GALAXY
ZWG 495.036	22 48 36.	+ 31 06		14.1	GALAXY
UGC 12214	22 48 36.	+ 31 06	84	14.1	GALAXY COMPACT
KARA.72 574A	22 48 36.	+ 31 06	60	14.1	PART OF DOUBLE GALAXY
4ZW 118	22 48 36.	+ 34 35			COMPACT GALAXY
ZWG 515.007	22 48 36.	+ 34 35		15.5	GALAXY
UGC 12215	22 48 36.	+ 34 35	96	15.5	GALAXY COMPACT
MCG+06-50-005	22 48 36.	+ 34 36	48	15.	GALAXY
SC 2245-4449.2	22 48 41.	- 44 33 20.			NEBULA
MCG+03-58-008	22 48 42.	+ 18 36	42	15.	GALAXY
ZWG 495.037	22 48 42.	+ 31 06		15.7	GALAXY
KARA.72 574B	22 48 42.	+ 31 06	30	15.7	PART OF DOUBLE GALAXY
SC 2245-4705.3	22 48 43.	- 46 49 26.	12		NEBULA
MCG+05-53-020	22 48 48.	+ 31 06	48	14.5	GALAXY
ZWG 515.008	22 48 48.	+ 36 49		15.3	GALAXY
UGC 12216	22 48 48.	+ 36 49	72	15.3	GALAXY S0?
MCG+06-50-006	22 48 48.	+ 36 50	24	14.5	GALAXY
SC 2245-4600.6	22 48 48.	- 45 44 44.			NEBULA
SC 2245-4706.9	22 48 50.	- 46 51 02.			NEBULA
MCG+05-53-021	22 48 51.	+ 31 06	18	15.	GALAXY
RNGC 7395	22 48 51.	+ 36 49		15.5	GALAXY
SC 2245-4557.4	22 48 51.	- 45 41 32.			NEBULA
SC 2245-4601.4	22 48 52.	- 45 45 32.			NEBULA
ACK 108+00.1	22 48 54.	+ 59 14			PLANETARY NEBULA
SC 2246-4442.2	22 48 54.	- 44 26 20.			NEBULA
SC 2245-4705.7	22 48 54.	- 46 49 50.			NEBULA
SC 2245-4707.5	22 48 54.	- 46 51 38.			NEBULA
SC 2245-4750.2	22 48 54.	- 47 34 20.	18		NEBULA
HN 0771	22 48 55.	- 67 41			NEBULA
IC 5257	22 48 55.	- 67 41			NONSTELLAR OBJECT
SC 2246-4330.0	22 48 57.	- 43 14 08.			NEBULA
SC 2246-4457.0	22 48 57.	- 44 41 08.			NEBULA
SC 2246-4524.2	22 48 57.	- 45 08 20.			NEBULA
ARP 015	22 48 58.	- 05 49			PECULIAR GALAXY
SC 2246-4555.0	22 48 59.	- 45 29 08.			NEBULA
LBN 0521	22 49	+ 61 00	2220		BRIGHT NEBULA
ZWG 474.033	22 49 00.	+ 22 49		14.1	GALAXY
UGC 12217	22 49 00.	+ 22 49	96	14.1	GALAXY E-S0
MCG+05-53-022	22 49 00.	+ 31 07	21	16.	GALAXY
ZWG 496.001	22 49 00.	+ 32 05		14.9	GALAXY
ZWG 495.038	22 49 00.	+ 32 05		14.9	GALAXY
UGC 12218	22 49 00.	+ 32 05	60	14.9	GALAXY IRR?
RNGC 7393	22 49 00.	- 05 49		14.0	GALAXY
SC 2246-4336.2	22 49 00.	- 43 20 19.			NEBULA
SC 2246-4322.6	22 49 01.	- 43 06 43.			NEBULA
SC 2246-4448.1	22 49 02.	- 44 32 13.			NEBULA
SC 2246-4721.9	22 49 02.	- 47 06 01.	12		NEBULA
MCG-04-53-040	22 49 03.	- 20 52	96	12.5	GALAXY
SC 2246-4707.3	22 49 04.	- 46 51 25.			NEBULA
SC 2246-4713.0	22 49 04.	- 46 57 07.			NEBULA
ZWG 379.009	22 49 06.	+ 01 13		15.7	GALAXY
MCG+04-53-017	22 49 06.	+ 22 46 30.	24	14.5	GALAXY
ZC 2249.1+4211	22 49 06.	+ 42 11	2960		CLUSTER OF GALAXIES
VV 068	22 49 06.	- 05 50	60	11.5	INTERACTING GALAXY
MCG-01-58-002	22 49 06.	- 05 50	108	14.	GALAXY
IC 5258	22 49 07.	+ 22 45 39.			NONSTELLAR OBJECT
SC 2246-4529.1	22 49 07.	- 45 13 13.			NEBULA
BC 3CR454	22 49 07.86	+ 18 32 46.6		18.40	QUASI-STELLAR OBJECT
SHB 379	22 49 07.9	+ 18 32 46.		18.4	QUASI-STELLAR OBJECT
RNGC 7392	22 49 09.	- 20 53		13.0	GALAXY
SC 2246-4652.8	22 49 10.	- 46 36 55.	12		NEBULA
SC 2246-4612.0	22 49 11.	- 45 56 07.			NEBULA
SC 2246-4703.8	22 49 11.	- 46 47 55.			NEBULA
MCG+05-53-023	22 49 12.	+ 32 06	48	15.	GALAXY
4ZW 119	22 49 12.	+ 33 21			COMPACT GALAXY
PHL 2072	22 49 12.	- 20 52		16.6	BLUE STELLAR OBJECT
SC 2246-4700.4	22 49 12.	- 46 44 31.			NEBULA
SC 2246-4322.2	22 49 14.	- 43 06 19.			NEBULA
SC 2246-4448.4	22 49 14.	- 44 32 31.			NEBULA
SC 2246-4554.4	22 49 17.	- 45 38 31.	12		NEBULA
SC 2246-4720.8	22 49 17.	- 47 04 55.	18		NEBULA
ZC 2249.3+1403	22 49 18.	+ 14 03	810		CLUSTER OF GALAXIES
ZWG 453.014	22 49 18.	+ 17 00		15.6	GALAXY
UGC 12219	22 49 18.	+ 17 00	60	15.6	GALAXY DBL SYS
4ZW 120	22 49 18.	+ 22 54			COMPACT GALAXY
SC 2246-4316.7	22 49 21.	- 43 00 49.			NEBULA
MCG+02-58-029	22 49 24.	+ 15 24	36	15.	GALAXY
MRSL 109+01/1	22 49 24.	+ 60 55	3600		HII REGION
PHL 0399	22 49 24.	- 27 57		17.5	BLUE STELLAR OBJECT
MCG-05-54-001	22 49 24.	- 28 03	48	15.	GALAXY
SC 2246-4606.0	22 49 25.	- 45 50 07.			NEBULA
SC 2246-4401.5	22 49 26.	- 43 45 37.			NEBULA
LB 01152	22 49 27.	+ 23 28 18.		18.0	FAINT BLUE STAR
SC 2246-4649.1	22 49 28.	- 46 33 13.			NEBULA
ARC 2498	22 49 30.	+ 14 04		17.9	RICH CLUSTER OF GALAXIES
PK104-06.1	22 49 30.	+ 51 34 46.			PLANETARY NEBULA
TON-S 0077	22 49 30.	- 27 58		15.8	BLUE STAR
MCG-05-54-002	22 49 30.	- 28 52	42	15.	GALAXY
SC 2246-4550.0	22 49 33.	- 45 34 07.			NEBULA
SC 2246-4500.6	22 49 34.	- 44 44 43.			NEBULA
MCG-03-58-008	22 49 36.	- 15 10	21	14.5	GALAXY
SC 2246-4714.0	22 49 36.	- 46 58 07.			NEBULA
SC 2246-4355.8	22 49 37.	- 43 39 55.			NEBULA
SC 2246-4648.8	22 49 38.	- 46 32 55.			NEBULA
SG 2.093	22 49 39.	+ 60 55	1200		DIFFUSE EMISSION NEBULA
ZC 2249.7+1601	22 49 42.	+ 16 01	1880		CLUSTER OF GALAXIES
SC 2246-4612.8	22 49 44.	- 45 56 55.			NEBULA
SC 2246-4712.6	22 49 46.	- 46 56 42.			NEBULA
MCG+02-58-030	22 49 48.	+ 11 17	36	15.5	GALAXY
MCG-07-46-012	22 49 48.	- 40 37	9	15.5	GALAXY
MCG+00-58-007	22 49 51.	+ 00 50	114	13.	GALAXY
SC 2246-4557.3	22 49 52.	- 45 41 24.	12		NEBULA
LB 01153	22 49 53.	+ 22 12 00.		16.4	FAINT BLUE STAR
ZWG 379.010	22 49 54.	+ 00 50		14.1	GALAXY
UGC 12220	22 49 54.	+ 00 50	120	14.1	GALAXY Sa
ZC 2249.9+1116	22 49 54.	+ 11 16	1140		CLUSTER OF GALAXIES
RNGC 7396	22 49 55.	+ 00 50		14.0	GALAXY
SC 2247-4701.6	22 49 56.	- 46 45 42.			NEBULA
ZWG 430.024	22 50 00.	+ 11 23		15.2	GALAXY
UGC 12222	22 50 00.	+ 11 23	72	15.2	GALAXY Sb
MCG+02-58-031	22 50 00.	+ 11 23	72	14.	GALAXY
UGC 12223	22 50 00.	+ 49 11	72	16.0	GALAXY S
LDN 1215	22 50 00.	+ 61 50	540		DARK NEBULA
LDN 1216	22 50 00.	+ 62 00	1500		DARK NEBULA
ZWG 369.002	22 50 00.	+ 82 38		14.9	GALAXY
ZWG 360.001	22 50 00.	+ 82 38		14.9	GALAXY
UGC 12221	22 50 00.	+ 82 38	150	14.9	GALAXY Sc
KARA.73B 0991	22 50 00.	+ 82 38	132	14.9	ISOLATED GALAXY S
SC 2247-4440.2	22 50 02.	- 44 24 18.			NEBULA
SC 2246-6113.6	22 50 02.	- 60 57 42.	6		NEBULA
SC 2247-4328.6	22 50 04.	- 43 12 42.			NEBULA
ZWG 405.012	22 50 06.	+ 05 49		15.3	GALAXY
UGC 12224	22 50 06.	+ 05 49	126	15.3	GALAXY Sc
KARA.73B 0992	22 50 06.	+ 05 49	120	15.3	ISOLATED GALAXY S
ZWG 475.001	22 50 06.	+ 24 27		15.3	GALAXY
MRK 309	22 50 06.	+ 24 27	8	16.	GALAXY WITH UV CONTINUUM
KARA.73B 0993	22 50 06.	+ 24 27	36	15.3	ISOLATED GALAXY S
OCL 0247	22 50 06.	+ 58 01	240	12.	OPEN STAR CLUSTER
RNGC 7399	22 50 06.	- 09 32		14.0	GALAXY
MCG-02-58-006	22 50 06.	- 09 32	48	14.5	GALAXY
MCG-05-54-003	22 50 06.	- 29 19	36	15.	GALAXY
SC 2247-4554.4	22 50 07.	- 45 38 30.	12		NEBULA
SC 2247-4439.5	22 50 09.	- 44 23 36.			NEBULA
SC 2246-6124.4	22 50 09.	- 61 08 30.			NEBULA
SC 2247-4425.0	22 50 11.	- 44 09 06.			NEBULA
MCG+01-58-011	22 50 12.	+ 05 50	132	14.	GALAXY
4ZW 121	22 50 12.	+ 24 28			COMPACT GALAXY
ZWG 496.002	22 50 12.	+ 30 35		15.7	GALAXY
ZWG 495.039	22 50 12.	+ 30 35		15.7	GALAXY
SC 2247-4244.0	22 50 16.	- 42 28 06.			NEBULA
SC 2247-4421.0	22 50 16.	- 44 05 06.			NEBULA
SC 2247-4336.1	22 50 17.	- 43 20 12.			NEBULA
SC 2247-4344.4	22 50 17.	- 43 28 30.			NEBULA
ZWG 379.011	22 50 18.	+ 00 52		15.3	GALAXY
MCG+00-58-008	22 50 18.	+ 00 53	36	14.	GALAXY
ZWG 379.012	22 50 18.	+ 00 57		14.9	GALAXY
UGC 12225	22 50 18.	+ 00 57	72	14.9	GALAXY Sa
ZWG 496.003	22 50 18.	+ 29 03		15.7	GALAXY
ZWG 495.040	22 50 18.	+ 29 03		15.7	GALAXY
KARA.73B 0994	22 50 18.	+ 29 03	102	15.7	ISOLATED GALAXY S
ZC 2250.3+2919	22 50 18.	+ 29 19	1140		CLUSTER OF GALAXIES
RNGC 7397	22 50 19.	+ 00 52		15.5	GALAXY
RNGC 7398	22 50 19.	+ 00 57		15.0	GALAXY
MCG+00-58-009	22 50 21.	+ 00 57	66	14.	GALAXY
RNGC 7405	22 50 21.	+ 12 19			GALAXY
SC 2247-4345.9	22 50 21.	- 43 30 00.			NEBULA
LB 01154	22 50 22.	+ 24 19 12.		17.8	FAINT BLUE STAR
SC 2247-4656.9	22 50 23.	- 46 41 00.			NEBULA
MCG+00-58-010	22 50 24.	+ 00 54	42	14.	GALAXY
MCG+03-58-009	22 50 24.	+ 15 32 30.	42	15.	GALAXY
ZWG 453.015	22 50 24.	+ 15 33		15.3	GALAXY
UGC 12226	22 50 24.	+ 15 33	66	15.3	GALAXY SBc
ARC 2499	22 50 24.	- 26 15		17.1	RICH CLUSTER OF GALAXIES
PHL 5462	22 50 24.	- 27 44		16.8	BLUE STELLAR OBJECT
SC 2247-4527.6	22 50 24.	- 45 11 42.			NEBULA
RNGC 7402	22 50 25.	+ 00 54		14.0	GALAXY
SC 2247-4702.0	22 50 25.	- 46 46 06.			NEBULA
IC 5259	22 50 26.	+ 36 26 43.			NONSTELLAR OBJECT
SC 2247-4356.4	22 50 29.	- 43 40 29.			NEBULA
ZWG 379.013	22 50 30.	+ 00 53		15.7	GALAXY
ZWG 430.025	22 50 30.	+ 11 33		15.6	GALAXY
UGC 12227	22 50 30.	+ 34 13	66	16.5	GALAXY Sc

OBJECT NAME	RIGHT ASCEN.	DECLINATION	DIAM.	MAGN.	TYPE OF OBJECT
SC 2247-6135.6	22 50 30.	- 61 19 42.			NEBULA
RNGC 7401	22 50 31.	+ 00 54			GALAXY
RNGC 7403	22 50 31.	+ 01 13			NON-EXISTENT OBJECT
HN 0022	22 50 33.	+ 01 13			NEBULA
UGC 12228	22 50 36.	+ 50 17	60	17.	GALAXY
MCG-02-58-007	22 50 36.	- 10 46	30	15.	GALAXY
SC 2247-4507.0	22 50 36.	- 44 51 05.			NEBULA
SC 2247-4701.6	22 50 37.	- 46 45 41.			NEBULA
SC 2247-4439.8	22 50 38.	- 44 23 53.			NEBULA
SC 2247-4417.0	22 50 39.	- 44 01 05.			NEBULA
SC 2247-4704.2	22 50 41.	- 46 48 17.			NEBULA
ZWG 453.016	22 50 42.	+ 16 29		15.5	GALAXY
SC 2247-4355.0	22 50 43.	- 43 39 05.			NEBULA
ZWG 453.017	22 50 48.	+ 20 33		15.7	GALAXY
ZWG 496.004	22 50 48.	+ 32 30		15.6	GALAXY
ZWG 495.041	22 50 48.	+ 32 30		15.6	GALAXY
UGC 12229	22 50 48.	+ 32 30	66	15.6	GALAXY S
KARA.68 237	22 50 48.	- 20 32	40		DWARF GALAXY
SC 2247-4504.8	22 50 49.	- 44 48 53.			NEBULA
SC 2247-4416.6	22 50 50.	- 44 00 41.			NEBULA
SC 2248-4353.1	22 50 53.	- 43 37 11.			NEBULA
ZWG 496.005	22 50 54.	+ 31 52		14.2	GALAXY
ZWG 495.042	22 50 54.	+ 31 52		14.2	GALAXY
UGC 12230	22 50 54.	+ 31 52	126	14.2	GALAXY Sc
MCG+05-54-001	22 50 54.	+ 32 31	60	15.5	GALAXY
ZWG 496.006	22 50 54.	+ 33 13		15.7	GALAXY
ZWG 495.043	22 50 54.	+ 33 13		15.7	GALAXY
LB 01512	22 50 54.	- 46 19		14.5	FAINT BLUE STAR
RNGC 7407	22 50 55.	+ 31 52		14.0	GALAXY
RNGC 7400	22 50 57.	- 45 37			UNVERIFIED SOUTHERN OBJECT
SC 2247-6324.5	22 50 59.	- 63 08 35.	42		NEBULA
LBN 0461	22 51	+ 42 50	2220		BRIGHT NEBULA
LBN 0524	22 51	+ 61 55	180		BRIGHT NEBULA
LDN 1236	22 51 00.	+ 68 40	840		DARK NEBULA
MCG+14-01-001	22 51 00.	+ 82 39	144	14.	GALAXY
MCG-03-58-009	22 51 00.	- 17 47	54	15.	GALAXY
SC 2248-4340.7	22 51 00.	- 43 24 47.			NEBULA
SC 2248-4522.0	22 51 00.	- 45 06 05.			NEBULA
SC 2248-4557.7	22 51 02.	- 45 41 47.			NEBULA
SC 2248-4421.8	22 51 03.	- 44 05 53.			NEBULA
SC 2248-4424.2	22 51 03.	- 44 08 17.			NEBULA
ABC 2500	22 51 05.	- 25 45		16.9	RICH CLUSTER OF GALAXIES
4ZW 122	22 51 06.	+ 31 23			COMPACT GALAXY
ZWG 496.007	22 51 06.	+ 31 23		14.7	GALAXY
MCG+05-54-002	22 51 06.	+ 31 53 30.	114	14.5	GALAXY
ZWG 515.009	22 51 06.	+ 33 45		15.7	GALAXY
KARA.73B 0995	22 51 06.	+ 33 45	54	15.7	ISOLATED GALAXY S
ZWG 515.010	22 51 06.	+ 36 54		15.0	GALAXY
MCG+06-50-007	22 51 06.	+ 36 56	42	14.5	GALAXY
ZC 2251.1-0119	22 51 06.	- 01 19	2020		CLUSTER OF GALAXIES
SC 2248-4408.8	22 51 06.	- 43 52 53.			NEBULA
SC 2248-4524.3	22 51 07.	- 45 08 23.			NEBULA
SC 2248-4346.4	22 51 08.	- 43 30 29.			NEBULA
SC 2248-4506.0	22 51 09.	- 44 50 05.			NEBULA
VDB.66N 155	22 51 11.	+ 61 54	372		REFLECTION NEBULA
ZC 2251.2+0331	22 51 12.	+ 03 31	1080		CLUSTER OF GALAXIES
ZWG 496.008	22 51 12.	+ 30 08		15.7	GALAXY
KARA.73B 0996	22 51 12.	+ 30 08	24	15.7	ISOLATED GALAXY S0
ZWG 496.009	22 51 12.	+ 31 21		15.5	GALAXY
UGC 12231	22 51 12.	+ 31 21	90	15.5	GALAXY SBc
IC 1455	22 51 13.	+ 01 06 26.			NONSTELLAR OBJECT
SC 2248-4517.4	22 51 14.	- 45 01 28.			NEBULA
MCG+00-58-011	22 51 15.	+ 01 08	48	14.	GALAXY
SC 2248-4536.8	22 51 16.	- 45 20 52.			NEBULA
SC 2248-4536.9	22 51 16.	- 45 20 58.			NEBULA
SC 2248-6124.5	22 51 16.	- 61 08 35.			NEBULA
RNGC 7409	22 51 17.	+ 19 56		15.5	GALAXY
ZWG 379.014	22 51 18.	+ 01 07		14.9	GALAXY
UGC 12232	22 51 18.	+ 01 07	60	14.9	GALAXY SBa
ZC 2251.3+1252	22 51 18.	+ 12 52	1210		CLUSTER OF GALAXIES
ZWG 453.018	22 51 18.	+ 19 56		15.6	GALAXY
ZWG 475.002	22 51 18.	+ 25 35		15.4	GALAXY
UGC 12233	22 51 18.	+ 25 35	66	15.4	GALAXY SBb
MCG+05-54-003	22 51 18.	+ 31 24	24	15.5	GALAXY
ZWG 496.010	22 51 18.	+ 33 26		13.3	GALAXY
ZWG 495.044	22 51 18.	+ 33 26		13.3	GALAXY
UGC 12234	22 51 18.	+ 33 26	78	13.3	GALAXY Sb-c
MCG+05-54-004	22 51 21.	+ 31 21 30.	72	15.	GALAXY
SC 2248-4500.5	22 51 22.	- 44 44 34.			NEBULA
ZC 2251.4+2634	22 51 24.	+ 26 34	2150		CLUSTER OF GALAXIES
ZWG 496.011	22 51 24.	+ 31 30		15.7	GALAXY
RNGC 7406	22 51 24.	- 06 52		15.0	GALAXY
MCG-01-58-003	22 51 24.	- 06 52	45	15.	GALAXY
MCG-03-58-010	22 51 24.	- 15 34	30	16.	GALAXY
SC 2248-4419.2	22 51 24.	- 44 03 16.			NEBULA
SC 2248-4419.2	22 51 25.	- 44 03 16.			NEBULA
SC 2248-4431.0	22 51 25.	- 44 15 04.			NEBULA
ARP 110	22 51 26.	- 15 28			PECULIAR GALAXY
MCG+04-54-001	22 51 27.	+ 25 33	36	15.	GALAXY
RNGC 7404	22 51 29.	- 39 35			GALAXY
SHB 380	22 51 29.6	+ 15 52 53.		16.1	QUASI-STELLAR OBJECT
BC 3CR454.3	22 51 29.61	+ 15 52 53.6		16.10	QUASI-STELLAR OBJECT
ABC 2501	22 51 30.	+ 15 12		17.9	RICH CLUSTER OF GALAXIES
MCG+05-54-005	22 51 30.	+ 33 27	66	14.5	GALAXY
MCG-03-58-011	22 51 30.	- 15 32	36	15.	GALAXY
MCG-07-47-001	22 51 30.	- 39 35	72	13.5	GALAXY
SC 2248-4419.0	22 51 31.	- 44 03 04.			NEBULA
SS 70	22 51 33.	+ 61 52			DIFFUSE GALACTIC NEBULA
SC 2248-4527.4	22 51 33.	- 45 11 28.			NEBULA
SC 2248-4531.4	22 51 35.	- 45 15 28.			NEBULA
ZWG 405.013	22 51 36.	+ 08 15		15.4	GALAXY
DG 188	22 51 36.	+ 61 52	300		REFLECTION NEBULA
MCG-02-58-008	22 51 36.	- 11 27	36	15.	GALAXY
SC 2248-4545.6	22 51 38.	- 45 29 40.			NEBULA
BC PKS2251+11	22 51 40.55	+ 11 20 38.7		15.80	QUASI-STELLAR OBJECT
SHB 381	22 51 40.6	+ 11 20 39.		15.8	QUASI-STELLAR OBJECT
SC 2248-4247.6	22 51 41.	- 42 31 40.			NEBULA
SC 2248-4420.5	22 51 41.	- 44 04 34.			NEBULA
SC 2248-4420.8	22 51 41.	- 44 04 52.			NEBULA
ZWG 453.019	22 51 42.	+ 19 56		15.7	GALAXY
ZWG 496.012	22 51 42.	+ 32 06		15.1	GALAXY
UGC 12235	22 51 42.	+ 32 06	72	15.1	GALAXY S0
UGC 12236	22 51 42.	+ 35 58	96	16.5	GALAXY
MCG+06-50-008	22 51 42.	+ 36 00	72	14.5	GALAXY
MCG-02-58-009	22 51 42.	- 10 26	24	15.5	GALAXY
SC 2248-4521.8	22 51 42.	- 45 05 52.			NEBULA
SC 2248-4731.0	22 51 46.	- 47 15 04.	24		NEBULA
ZWG 430.026	22 51 48.	+ 11 30		15.7	GALAXY
UGC 12237	22 51 48.	+ 11 30	90	15.7	GALAXY Sb-c
MCG+02-58-032	22 51 48.	+ 11 30	84	14.5	GALAXY
MCG+05-54-006	22 51 48.	+ 32 08	48	15.5	GALAXY
MCG-04-54-001	22 51 48.	- 20 38	72	14.	GALAXY
IC 5260	22 51 49.	- 37 37 33.			MAY NOT EXIST
IC 5261	22 51 51.	- 20 39 21.			NONSTELLAR OBJECT
ZWG 496.013	22 51 54.	+ 31 59		14.9	GALAXY
UGC 12238	22 51 54.	+ 31 59	72	14.9	GALAXY S0
UGC 12239	22 51 54.	+ 48 53	60	16.5	GALAXY
SC 2249-4419.4	22 51 54.	- 44 03 28.			NEBULA
SC 2249-4700.6	22 51 55.	- 46 44 40.			NEBULA
LB 01155	22 51 57.	+ 25 56 18.		17.9	FAINT BLUE STAR
SC 2249-4346.4	22 51 57.	- 43 30 28.			NEBULA
SC 2249-4425.8	22 51 57.	- 44 09 52.			NEBULA
SC 2249-4625.0	22 51 57.	- 46 09 04.			NEBULA
LBN 0458	22 52	+ 41 20	3300		BRIGHT NEBULA
ZC 2252.0+1756	22 52 00.	+ 17 56	1550		CLUSTER OF GALAXIES
UGC 12240	22 52 00.	+ 20 55	72	16.0	GALAXY SB?a-b
MCG+05-54-007	22 52 00.	+ 32 00	54	15.5	GALAXY
4ZW 123	22 52 00.	+ 32 13			COMPACT GALAXY
SC 2249-4625.0	22 52 02.	- 46 09 03.			NEBULA
SC 2249-4351.2	22 52 03.	- 43 35 15.			NEBULA
RNGC 7411	22 52 05.	+ 19 58		15.0	GALAXY
RNGC 7410	22 52 05.	- 39 56		12.0	GALAXY
SC 2248-6320.1	22 52 05.	- 63 04 10.	18		NEBULA
MCG+02-58-033	22 52 06.	+ 15 29	36	15.	GALAXY
ZWG 453.020	22 52 06.	+ 19 58		14.8	GALAXY
UGC 12241	22 52 06.	+ 19 58	66	14.8	GALAXY (E)
MCG+03-58-010	22 52 06.	+ 19 59	42	15.	GALAXY
ZWG 496.014	22 52 06.	+ 32 10		14.5	GALAXY
UGC 12242	22 52 06.	+ 32 10	84	14.5	GALAXY (E)
MCG+05-54-008	22 52 06.	+ 32 10	24	15.5	GALAXY
SC 2249-4508.0	22 52 06.	- 44 52 03.			NEBULA
SC 2249-4345.4	22 52 07.	- 43 29 27.			NEBULA
SC 2249-4330.8	22 52 11.	- 43 14 51.			NEBULA
ZWG 430.027	22 52 12.	+ 11 27		15.6	GALAXY
UGC 12243	22 52 12.	+ 11 27	60	15.6	GALAXY Sa-b
MCG+02-58-034	22 52 12.	+ 11 27	60	15.	GALAXY
ZWG 453.021	22 52 12.	+ 15 29		15.3	GALAXY
ZWG 430.028	22 52 12.	+ 15 29		15.3	GALAXY
MCG+05-54-009	22 52 12.	+ 32 12 30.	60	14.5	GALAXY
MCG-05-54-004	22 52 12.	- 27 07	36	15.5	GALAXY
MCG-07-47-002	22 52 12.	- 39 57	240	11.5	GALAXY
SC 2249-4510.2	22 52 16.	- 44 54 15.			NEBULA
ZWG 496.015	22 52 18.	+ 32 03		15.5	GALAXY
MCG+05-54-010	22 52 18.	+ 32 32	24	16.	GALAXY
OCL 0250	22 52 18.	+ 60 34	660	13.1	OPEN STAR CLUSTER
RNGC 7419	22 52 18.	+ 60 34		13.0	OPEN CLUSTER
VV 017D	22 52 18.	- 16 24	9	16.	INTERACTING GALAXY
VV 017C	22 52 18.	- 16 24	6	17.	INTERACTING GALAXY
VV 017B	22 52 18.	- 16 24	9	16.	INTERACTING GALAXY
VV 017A	22 52 18.	- 16 24	15	15.	INTERACTING GALAXY
VV 017	22 52 18.	- 16 24	84		INTERACTING GALAXY
MCG-03-58-012	22 52 18.	- 16 26 30.	78	14.5	GALAXY
MCG-07-47-003	22 52 18.	- 38 52	96	16.	GALAXY
SC 2249-4510.2	22 52 19.	- 44 54 15.			NEBULA
SC 2249-4610.6	22 52 21.	- 45 54 39.			NEBULA
ABC 2502	22 52 22.	- 16 50		16.6	RICH CLUSTER OF GALAXIES
SC 2249-4349.6	22 52 22.	- 43 33 39.			NEBULA
RNGC 7415	22 52 23.	+ 19 59		15.0	GALAXY
SC 2249-4341.6	22 52 23.	- 43 25 39.			NEBULA
ZWG 453.022	22 52 24.	+ 19 07		15.5	GALAXY
ZWG 453.023	22 52 24.	+ 19 59		15.0	GALAXY
UGC 12244	22 52 24.	+ 19 59	66	15.0	GALAXY Sa-b
MCG+03-58-012	22 52 24.	+ 20 00	48	15.5	GALAXY
MCG+03-58-011	22 52 24.	+ 20 00	27	15.5	GALAXY
ZWG 475.003	22 52 24.	+ 21 33		15.6	GALAXY
UGC 12245	22 52 24.	+ 21 33	60	15.6	GALAXY Sc
KARA.73B 0997	22 52 24.	+ 21 33	54	15.6	ISOLATED GALAXY S
UGC 12246	22 52 24.	+ 37 58	60	16.5	GALAXY
UGC 12247	22 52 24.	+ 74 57	72	16.0	GALAXY Sc
SC 2249-4510.0	22 52 24.	- 44 54 03.			NEBULA
ABC 2503	22 52 25.	+ 29 12		16.4	RICH CLUSTER OF GALAXIES
SC 2249-4336.0	22 52 26.	- 43 20 03.			NEBULA
IC 5262	22 52 27.	- 34 06 02.			NONSTELLAR OBJECT
SC 2249-4531.5	22 52 29.	- 45 15 33.			NEBULA
MCG+02-58-035	22 52 30.	+ 12 57	15	15.	GALAXY
MCG+04-54-002	22 52 30.	+ 21 31 30.	51	15.5	GALAXY
MCG+05-54-012	22 52 30.	+ 31 58	39	15.5	GALAXY
MCG+05-54-013	22 52 30.	+ 32 04	30	15.5	GALAXY
MCG+05-54-011	22 52 30.	+ 32 10	24	16.	GALAXY
ZWG 496.016	22 52 30.	+ 32 32		15.6	GALAXY
UGC 12248	22 52 30.	+ 36 16	60	16.5	GALAXY PECULR
MCG+06-50-009	22 52 30.	+ 37 58	66	16.	GALAXY
RNGC 7413	22 52 33.	+ 12 56		15.0	GALAXY
MCG+05-54-014	22 52 33.	+ 32 05 30.	33	15.5	GALAXY
BC 3C455	22 52 34.5	+ 12 57 34.		19.	QUASI-STELLAR OBJECT
SHB 382	22 52 34.5	+ 12 57 34.		19.7	QUASI-STELLAR OBJECT
ZWG 430.029	22 52 36.	+ 12 56		15.2	GALAXY
ZC 2252.6+1646	22 52 36.	+ 16 46	1080		CLUSTER OF GALAXIES
ZC 2252.6+3135	22 52 36.	+ 31 35	7590		CLUSTER OF GALAXIES
ZWG 496.017	22 52 36.	+ 31 47		15.7	GALAXY
SC 2249-4401.3	22 52 38.	- 43 45 21.			NEBULA
IC 1456	22 52 41.	- 12 59 25.			NONSTELLAR OBJECT
SC 2249-4402.2	22 52 41.	- 43 46 15.			NEBULA
ZWG 453.024	22 52 42.	+ 18 48		15.6	GALAXY
ZWG 496.018	22 52 42.	+ 31 32		15.5	GALAXY
MCG+05-54-015	22 52 42.	+ 32 33	42	15.	GALAXY
ACK 106-04.1	22 52 42.	+ 54 40			PLANETARY NEBULA
RNGC 7408	22 52 42.	- 63 58			GALAXY
SC 2249-4616.8	22 52 44.	- 46 00 51.			NEBULA
MCG+00-58-012	22 52 45.	+ 02 00	48	14.	GALAXY
SC 2249-4348.8	22 52 46.	- 43 32 50.			NEBULA
LB 01156	22 52 47.	+ 22 52 48.		17.7	FAINT BLUE STAR
ZWG 379.015	22 52 48.	+ 01 59		15.4	GALAXY
ZWG 496.019	22 52 48.	+ 31 03		15.2	GALAXY
MCG+05-54-016	22 52 48.	+ 31 03 30.	48	15.	GALAXY
ZWG 496.020	22 52 48.	+ 31 55		15.3	GALAXY
IC 1457	22 52 48.	- 05 49			SINGLE STAR
SC 2249-6112.9	22 52 48.	- 60 56 57.	12		NEBULA
RNGC 7414	22 52 51.	+ 12 59			GALAXY
MCG+05-54-017	22 52 51.	+ 31 56 30.	18	15.	GALAXY
SC 2249-4622.4	22 52 53.	- 46 06 26.			NEBULA
ZWG 515.011	22 52 54.	+ 36 24		14.8	GALAXY
OCL 0248	22 52 54.	+ 58 54	180	11.	OPEN STAR CLUSTER
MCG-04-54-002	22 52 54.	- 24 14 30.	42	14.	GALAXY
SC 2250-4247.2	22 52 54.	- 42 31 14.			NEBULA
MCG-07-47-004	22 52 54.	- 42 55	150	12.0	GALAXY

OBJECT NAME	RIGHT ASCEN.	DECLINATION	DIAM.	MAGN.	TYPE OF OBJECT
SC 2250-4354.6	22 52 55.	- 43 38 38.			NEBULA
SC 2250-4359.2	22 52 55.	- 43 43 14.			NEBULA
SC 2250-4403.2	22 52 56.	- 43 47 14.			NEBULA
RNGC 7412	22 52 58.	- 42 55		12.0	GALAXY
LBN 0462	22 53	+ 42 20	1500		BRIGHT NEBULA
KHAV 780	22 53	+ 62 16			DARK NEBULA
MCG+02-58-036	22 53 00.	+ 12 31	96	14.	GALAXY
ZWG 453.025	22 53 00.	+ 19 01		15.1	GALAXY
LB 01157	22 53 00.	+ 22 44 36.		17.6	FAINT BLUE STAR
UGC 12249	22 53 00.	+ 28 04	60	16.0	GALAXY Sc
MCG+06-50-010	22 53 00.	+ 36 25	33	14.5	GALAXY
ACK 107-02.1	22 53 00.	+ 56 26			PLANETARY NEBULA
DV.56 I5267A	22 53 00.	- 43 41	186		Sb GALAXY
LB 01513	22 53 00.	- 46 07		15.5	FAINT BLUE STAR
SC 2250-4249.4	22 53 02.	- 42 33 26.			NEBULA
SC 2250-4523.0	22 53 02.	- 45 07 02.			NEBULA
SC 2250-4517.9	22 53 05.	- 45 01 56.			NEBULA
ZWG 430.030	22 53 06.	+ 12 32		14.4	GALAXY
UGC 12250	22 53 06.	+ 12 32	96	14.4	GALAXY SBb
ZWG 453.026	22 53 06.	+ 18 53		15.5	GALAXY
4ZW 124	22 53 06.	+ 24 29			COMPACT GALAXY
ZWG 496.021	22 53 06.	+ 31 27		15.7	GALAXY
ZWG 496.022	22 53 06.	+ 31 36		15.5	GALAXY
RNGC 7416	22 53 06.	- 05 46		13.0	GALAXY
TON-S 0078	22 53 06.	- 28 03		13.3	BLUE STAR
DV.56 I5269A	22 53 06.	- 36 39	84		SBm GALAXY
MCG-07-47-005	22 53 06.	- 43 42	72	13.5	GALAXY
SC 2250-4358.4	22 53 06.	- 43 42 26.	102		NEBULA
SC 2249-6119.5	22 53 07.	- 61 03 32.			NEBULA
SVEN 455	22 53 09.	- 40 56	36	15.2	GALAXY
SC 2250-4459.6	22 53 09.	- 44 43 38.			NEBULA
SC 2250-4500.6	22 53 09.	- 44 44 38.			NEBULA
RNGC 7423	22 53 10.	+ 56 52			NON-EXISTENT OBJECT
SC 2250-4234.2	22 53 11.	- 42 18 14.			NEBULA
SC 2250-4336.8	22 53 11.	- 43 20 50.			NEBULA
MCG+03-58-013	22 53 12.	+ 19 02	36	15.	GALAXY
ZWG 453.027	22 53 12.	+ 19 05		15.7	GALAXY
ZWG 496.023	22 53 12.	+ 29 32		14.9	GALAXY
KARA.73B 0998	22 53 12.	+ 29 32	42	14.9	ISOLATED GALAXY S
MCG+05-54-018	22 53 12.	+ 29 33	48	14.5	GALAXY
OCL 0246	22 53 12.	+ 56 52	360	15.	OPEN STAR CLUSTER
MCG-01-58-004	22 53 12.	- 05 46	180	13.	GALAXY
RNGC 7420	22 53 13.	+ 29 32		15.0	GALAXY
MCG+08-41-006	22 53 15.	+ 46 10	18	18.	GALAXY
MCG-01-58-005	22 53 18.	- 06 42	36	15.	GALAXY
ZWG 405.014	22 53 24.	+ 06 07		15.4	GALAXY
UGC 12251	22 53 24.	+ 06 07	66	15.4	GALAXY S0?
KARA.73B 0999	22 53 24.	+ 06 07	72	15.4	ISOLATED GALAXY S
UGC 12252	22 53 24.	+ 31 30	96	16.0	GALAXY Sc
MCG+08-41-007	22 53 24.	+ 46 11	30	18.	GALAXY
MRSL 108-01/1	22 53 24.	+ 58 12	120		HII REGION
LB 01514	22 53 24.	- 49 16		12.7	FAINT BLUE STAR
MCG+01-58-012	22 53 30.	+ 06 07	30	15.	GALAXY
MCG+02-58-037	22 53 30.	+ 12 29	108	15.5	GALAXY
UGC 12253	22 53 30.	+ 12 31	102	16.0	GALAXY Sb
MCG+06-50-011	22 53 30.	+ 39 02	42	15.5	GALAXY
ARC 2504	22 53 32.	- 15 12		17.2	RICH CLUSTER OF GALAXIES
ZWG 405.015	22 53 36.	+ 03 40		14.3	GALAXY
UGC 12254	22 53 36.	+ 03 40	54	14.3	GALAXY S
MCG+02-58-038	22 53 36.	+ 12 31	36	16.	GALAXY
ZWG 496.024	22 53 36.	+ 31 24		14.9	GALAXY
MCG+05-54-019	22 53 36.	+ 31 25	48	15.	GALAXY
SC 2250-4609.0	22 53 36.	- 45 53 01.			NEBULA
RNGC 7422	22 53 38.	+ 03 40		14.5	GALAXY
SC 2250-4340.6	22 53 38.	- 43 24 37.			NEBULA
SC 2250-4625.0	22 53 38.	- 46 09 01.			NEBULA
SC 2250-4647.7	22 53 39.	- 46 31 43.			NEBULA
MCG+01-58-013	22 53 42.	+ 03 38	48	14.	GALAXY
ZWG 405.016	22 53 42.	+ 05 07		15.5	GALAXY
UGC 12255	22 53 42.	+ 05 07	72	15.5	GALAXY S
ZWG 430.031	22 53 42.	+ 12 49		15.7	GALAXY
ZWG 515.012	22 53 42.	+ 36 05		13.6	GALAXY
UGC 12256	22 53 42.	+ 36 05	108	13.6	GALAXY E
MCG+06-50-012	22 53 42.	+ 36 07	42	13.	GALAXY
MCG-02-58-010	22 53 42.	- 09 35	60	16.	GALAXY
MCG-03-58-013	22 53 42.	- 16 52	48	14.5	GALAXY
TON-S 0079	22 53 42.	- 27 26		13.7	BLUE STAR
SC 2250-4416.9	22 53 43.	- 44 00 55.			NEBULA
SC 2250-4712.5	22 53 43.	- 46 56 31.			NEBULA
RNGC 7426	22 53 44.	+ 36 05		13.5	GALAXY
MCG+03-58-014	22 53 45.	+ 19 07 30.	48	16.	GALAXY
MCG+01-58-014	22 53 48.	+ 05 07	18	15.	GALAXY
ZWG 453.028	22 53 48.	+ 18 51		15.5	GALAXY
ZWG 453.029	22 53 48.	+ 19 06		15.7	GALAXY
UGC 12257	22 53 48.	+ 19 06	66	15.7	GALAXY Sc
ZC 2253.8-0057	22 53 48.	- 00 57	2150		CLUSTER OF GALAXIES
RNGC 7418	22 53 48.	- 37 17		12.0	GALAXY
SC 2250-4649.3	22 53 48.	- 46 33 19.			NEBULA
SC 2250-6246.8	22 53 50.	- 62 30 49.			NEBULA
IC 5268	22 53 53.	+ 36 19 51.			NONSTELLAR OBJECT
RNGC 7429	22 53 53.	+ 59 43			OPEN CLUSTER
ARC 2505	22 53 53.	- 00 49		17.6	RICH CLUSTER OF GALAXIES
IC 5264	22 53 53.	- 36 52 46.			MAY NOT EXIST
MCG+05-54-020	22 53 54.	+ 28 11	48	15.	GALAXY
OCL 0249	22 53 54.	+ 59 43			OPEN STAR CLUSTER
DV.56 I5269B	22 53 54.	- 36 31	252		S GALAXY
DV.56 N7418A	22 53 54.	- 37 01			SB(s)c GALAXY
RNGC 7418A	22 53 54.	- 37 01			GALAXY
SC 2251-4230.1	22 53 55.	- 42 14 07.			NEBULA
SC 2251-4649.7	22 53 55.	- 46 33 43.			NEBULA
IC 5265	22 53 58.	- 36 47 46.			SAME AS IC 1459
RNGC 7412A	22 53 58.	- 43 04			GALAXY
LBN 0516	22 54	+ 58 15	120		BRIGHT NEBULA
KHAV 781	22 54	+ 61 16			DARK NEBULA
LBN 0527	22 54	+ 62 20	540		BRIGHT NEBULA
SER 150.08	22 54	- 66 22	300	18.	COMPACT GROUP OF GALAXIES
MCG+00-58-013	22 54 00.	+ 01 10	42	15.	GALAXY
ZC 2254.0+1309	22 54 00.	+ 13 09	1480		CLUSTER OF GALAXIES
ZWG 453.030	22 54 00.	+ 17 31		15.1	GALAXY
UGC 12258	22 54 00.	+ 17 31	96	15.1	GALAXY Sa
MCG+03-58-015	22 54 00.	+ 17 31	90	14.5	GALAXY
UGC 12259	22 54 00.	+ 18 27	96	16.0	GALAXY SB
4ZW 125	22 54 00.	+ 27 27			COMPACT GALAXY
ZWG 496.025	22 54 00.	+ 28 10		15.4	GALAXY
MCG-01-58-006	22 54 00.	- 07 06	30	15.5	GALAXY
TON-S 0080	22 54 00.	- 30 56		15.8	BLUE STAR
DV.56 N7412A	22 54 00.	- 43 04	210		S GALAXY
SC 2251-4317.4	22 54 01.	- 43 01 25.			NEBULA
MIN.46 20	22 54 03.	+ 58 13			DIFFUSE NEBULA
MCG-04-54-003	22 54 03.	- 25 14	72	14.5	GALAXY
SC 2251-4435.9	22 54 03.	- 44 19 55.			NEBULA
SC 2251-4528.6	22 54 03.	- 45 12 37.			NEBULA
IC 1458	22 54 05.	- 07 38 39.			NONSTELLAR OBJECT
SC 2251-4539.6	22 54 05.	- 45 23 37.			NEBULA
ZWG 515.013	22 54 06.	+ 35 58		15.6	GALAXY
MRSL 108-01/2	22 54 06.	+ 58 15	120		HII REGION
HELW 047	22 54 06.	- 36 49			NEBULA
RNGC 7421	22 54 06.	- 37 37		13.0	GALAXY
SC 2251-4417.8	22 54 06.	- 44 01 49.	54		NEBULA
DV.56 I5267B	22 54 06.	- 44 02	96		S GALAXY
ARC 2506	22 54 07.	+ 13 04		17.1	RICH CLUSTER OF GALAXIES
MCG-07-47-006	22 54 09.	- 44 02	72	13.5	GALAXY
SC 2251-4424.2	22 54 10.	- 44 08 13.			NEBULA
SC 2251-4315.0	22 54 11.	- 42 59 01.			NEBULA
ZWG 515.014	22 54 12.	+ 37 28		15.7	GALAXY
UGC 12260	22 54 12.	+ 37 28	102	15.7	GALAXY Sc
KARA.73B 1000	22 54 12.	+ 37 28	102	15.7	ISOLATED GALAXY S
MCG+06-50-013	22 54 12.	+ 37 29	78	16.	GALAXY
ZWG 532.002	22 54 12.	+ 40 27		15.7	GALAXY
MCG-01-58-007	22 54 12.	- 07 39	72	14.	GALAXY
ARC 2507	22 54 13.	+ 05 14		17.6	RICH CLUSTER OF GALAXIES
PK107-02.1	22 54 14.6	+ 56 53 19.	8	14.0	PLANETARY NEBULA
SC 2251-4320.6	22 54 16.	- 43 04 37.	222		NEBULA
SC 2251-4434.7	22 54 17.	- 44 18 43.			NEBULA
ZC 2254.3+0515	22 54 18.	+ 05 15	670		CLUSTER OF GALAXIES
ZWG 496.026	22 54 18.	+ 31 20		15.7	GALAXY
MRSL 108-01/3	22 54 18.	+ 58 16	60		HII REGION
MCG-02-58-011	22 54 18.	- 09 13	114	14.	GALAXY
HELW 499	22 54 18.	- 43 04 13.			NEBULA
SG 3.236	22 54 21.	+ 58 16	30		DIFFUSE EMISSION NEBULA
IC 5267	22 54 22.2	- 43 39 54.	300	11.0	GALAXY SA(s)
ZC 2254.4+2429	22 54 24.	+ 24 29	1550		CLUSTER OF GALAXIES
HELW 048	22 54 24.	- 36 44			NEBULA
MCG-07-47-007	22 54 24.	- 43 40	180	11.4	GALAXY
SVEN 456	22 54 27.	- 41 20	240	12.2	GALAXY
SC 2251-4316.8	22 54 27.	- 43 00 48.			NEBULA
SC 2251-4401.0	22 54 27.	- 43 45 00.			NEBULA
SC 2251-4605.9	22 54 28.	- 45 49 54.			NEBULA
RNGC 7424	22 54 29.	- 41 20		11.0	GALAXY
SC 2251-4419.0	22 54 29.	- 44 03 00.			NEBULA
ZWG 405.017	22 54 30.	+ 04 25		15.0	GALAXY
MCG-02-58-012	22 54 30.	- 09 12	12	17.	GALAXY
IC 1459	22 54 30.	- 36 41	60	11.3	GALAXY E3
MCG-07-47-008	22 54 30.	- 41 21	90	11.4	GALAXY
SC 2251-4424.2	22 54 30.	- 44 08 12.			NEBULA
IC 1460	22 54 31.6	+ 04 24 37.		15.2	GALAXY S0
SC 2251-4542.3	22 54 33.	- 45 26 18.			NEBULA
SC 2251-4505.0	22 54 35.	- 44 49 00.			NEBULA
RNGC 7417	22 54 35.	- 65 17			GALAXY
MCG+01-58-015	22 54 36.	+ 04 24	18	15.5	GALAXY
ZWG 405.018	22 54 36.	+ 08 14		15.4	GALAXY
MRK 521	22 54 36.	+ 08 14	12	16.	GALAXY WITH UV CONTINUUM
ZWG 430.032	22 54 36.	+ 12 54		15.6	GALAXY
MCG+02-58-039	22 54 36.	+ 12 55	30	15.	GALAXY
UGC 12261	22 54 36.	+ 71 11	60	16.0	GALAXY IRR
MCG-01-58-008	22 54 36.	- 03 24	24	15.	GALAXY
MCG-05-54-005	22 54 36.	- 31 44	30	15.	GALAXY
MCG-07-47-009	22 54 36.	- 40 19	36	16.	GALAXY
AGU 76	22 54 36.	- 40 19 00.		17.	IRREGULAR DWARF GALAXY
LB 01158	22 54 38.	+ 25 31 12.		15.0	FAINT BLUE STAR
RNGC 7427	22 54 39.	+ 08 14		15.5	GALAXY
SC 2251-4501.8	22 54 40.	- 44 45 48.			NEBULA
RNGC 7425	22 54 41.	- 11 13		15.0	GALAXY
MCG+01-58-016	22 54 42.	+ 08 15	36	15.5	GALAXY
ZWG 379.016	22 54 42.	- 01 18		13.8	GALAXY
UGC 12262	22 54 42.	- 01 18	150	13.8	GALAXY SBa
KARA.73B 1001	22 54 42.	- 01 18	168	13.8	ISOLATED GALAXY S
MCG-02-58-013	22 54 42.	- 11 13	42	15.	GALAXY
RNGC 7428	22 54 43.	- 01 18		14.0	GALAXY
HN 0772	22 54 43.	- 69 20			NEBULA
IC 5263	22 54 43.	- 69 20			NONSTELLAR OBJECT
SHB 383	22 54 44.	+ 02 27 18.		18.0	QUASI-STELLAR OBJECT
RC PKS2254+024	22 54 44.6	+ 02 27 12.		18.	QUASI-STELLAR OBJECT
MCG+00-58-014	22 54 45.	- 01 19	126	13.	GALAXY
MCG-03-58-014	22 54 45.	- 18 10 30.	48	14.5	GALAXY
SC 2251-4606.8	22 54 45.	- 45 50 48.			NEBULA
IC 5269	22 54 47.	- 36 11 36.	84	13.1	GALAXY SA0
MRSL 110+02/1	22 54 48.	+ 62 21	3600		HII REGION
ARC 2508	22 54 49.	+ 14 13		17.5	RICH CLUSTER OF GALAXIES
SG 2.094	22 54 52.	+ 62 21	3600		DIFFUSE EMISSION NEBULA
IC 5270	22 54 53.	- 36 21 38.			NONSTELLAR OBJECT
ZWG 405.019	22 54 54.	+ 08 32		15.1	GALAXY
SC 2252-4430.8	22 54 56.	- 44 14 48.			NEBULA
RNGC 7430	22 54 57.	+ 08 32		15.0	GALAXY
CED 207	22 54 57.	+ 62 28			DIFFUSE GALACTIC NEBULA
LBN 0418	22 55	+ 24 00	9600		BRIGHT NEBULA
LBN 0460	22 55	+ 41 00	1620		BRIGHT NEBULA
LBN 0529	22 55	+ 62 10	2400		BRIGHT NEBULA
ZWG 475.004	22 55 00.	+ 23 01		15.7	GALAXY
ZWG 475.005	22 55 00.	+ 25 51		15.5	GALAXY
ZC 2255.0+3531	22 55 00.	+ 35 31	1950		CLUSTER OF GALAXIES
COU 099	22 55 00.	+ 62 25	2700		EMISSION NEBULA
COU 098	22 55 00.	+ 62 40	6600		EMISSION NEBULA
LDN 1223	22 55 00.	+ 64 00	4080		DARK NEBULA
PK114+10.1	22 55 00.	+ 71 12	42		PLANETARY NEBULA
UGC 12263	22 55 00.	+ 72 25	180	17.	GALAXY DWARF?
SC 2252-4520.2	22 55 01.	- 45 04 12.			NEBULA
HN 0773	22 55 04.	- 65 23			NEBULA
IC 5266	22 55 04.	- 65 23			NONSTELLAR OBJECT
MCG+01-58-017	22 55 06.	+ 08 31	24	15.5	GALAXY
UGC 12264	22 55 06.	+ 12 39	60	17.	GALAXY Sc
ZWG 453.031	22 55 06.	+ 19 31		14.5	GALAXY
UGC 12265	22 55 06.	+ 19 31	48	14.5	GALAXY DBL SYS
SC 2252-4309.4	22 55 07.	- 42 53 23.			NEBULA
SC 2252-4602.8	22 55 07.	- 45 46 47.			NEBULA
ZC 2255.2+0309	22 55 12.	+ 03 09	1010		CLUSTER OF GALAXIES
ZC 2255.2+1414	22 55 12.	+ 14 14	340		CLUSTER OF GALAXIES
5ZW 393	22 55 12.	+ 45 25			COMPACT GALAXY
SC 2252-4442.0	22 55 12.	- 44 25 59.			NEBULA
SC 2252-4638.6	22 55 12.	- 46 22 35.	24		NEBULA
ARC 2509	22 55 13.	- 22 01		17.1	RICH CLUSTER OF GALAXIES
SC 2252-4401.6	22 55 13.	- 43 45 35.			NEBULA
SC 2252-4438.7	22 55 13.	- 44 22 41.			NEBULA
ARP 314	22 55 16.	- 04 03			PECULIAR GALAXY
SC 2252-4439.1	22 55 16.	- 44 23 05.			NEBULA
ZWG 453.032	22 55 18.	+ 16 30		15.6	GALAXY

OBJECT NAME	RIGHT ASCEN.	DECLINATION	DIAM.	MAGN.	TYPE OF OBJECT
MRK 310	22 55 18.	+ 16 30	20	15.5	GALAXY WITH UV CONTINUUM
2ZW 186	22 55 18.	+ 22 01			COMPACT GALAXY
IC 5271	22 55 18.	- 34 01	126	12.6	GALAXY
ZWG 405.020	22 55 24.	+ 05 48		14.9	GALAXY
UGC 12266	22 55 24.	+ 05 48	84	14.9	GALAXY SBa-b
ZWG 475.006	22 55 24.	+ 25 54		15.6	GALAXY
RNGC 7431	22 55 24.	+ 25 54		15.5	GALAXY
ZC 2255.4+3005	22 55 24.	+ 30 05	1080		CLUSTER OF GALAXIES
SC 2252-4322.4	22 55 24.	- 43 06 23.			NEBULA
RNGC 7438	22 55 26.	+ 54 05			NON-EXISTENT OBJECT
SC 2252-4631.3	22 55 29.	- 46 15 17.	18		NEBULA
MCG+02-58-040	22 55 30.	+ 12 51	72	14.	GALAXY
ZC 2255.5+2041	22 55 30.	+ 20 41	940		CLUSTER OF GALAXIES
ZWG 475.007	22 55 30.	+ 25 52		15.5	GALAXY
RNGC 7435	22 55 30.	+ 25 52		15.5	GALAXY
UGC 12267	22 55 30.	+ 25 52	90	15.5	GALAXY SBa
MCG+04-54-003	22 55 30.	+ 25 52	30	15.5	GALAXY
MCG-01-58-009	22 55 30.	- 04 02	54	14.	GALAXY
VV 295A	22 55 30.	- 04 02 30.	48	12.	INTERACTING GALAXY
SC 2252-4706.0	22 55 32.	- 46 49 59.	18		NEBULA
RNGC 7432	22 55 33.	+ 12 52		15.0	GALAXY
MCG+04-54-004	22 55 33.	+ 25 51	60	14.5	GALAXY
HOLM 800C	22 55 33.	+ 25 52	12	15.3	PART OF MULTIPLE GALAXY
VV 084	22 55 33.	+ 25 53	90	14.2	INTERACTING GALAXY
ZC 2255.6+0000	22 55 36.	+ 00 00	1140		CLUSTER OF GALAXIES
MCG+01-58-018	22 55 36.	+ 05 48	60	15.	GALAXY
ZWG 430.033	22 55 36.	+ 12 52		15.1	GALAXY
UGC 12268	22 55 36.	+ 12 52	96	15.1	GALAXY E
MCG+04-54-007	22 55 36.	+ 25 51	9	16.5	GALAXY
MCG+04-54-006	22 55 36.	+ 25 52	24	14.	GALAXY
MCG+04-54-005	22 55 36.	+ 25 52	36	15.	GALAXY
ZWG 475.008	22 55 36.	+ 25 53		14.0	GALAXY
RNGC 7436B	22 55 36.	+ 25 53		15.0	GALAXY
RNGC 7436A	22 55 36.	+ 25 53		14.0	GALAXY
RNGC 7433	22 55 36.	+ 25 53		14.0	GALAXY
UGC 12269	22 55 36.	+ 25 53	120	14.0	GALAXY E
MCG-01-58-010	22 55 36.	- 04 03	60	14.	GALAXY
MCG-01-58-011	22 55 36.	- 04 05	60	16.	GALAXY
SC 2252-4311.0	22 55 36.	- 42 54 59.			NEBULA
HOLM 800A	22 55 37.	+ 25 52	18	14.1	PART OF MULTIPLE GALAXY
RNGC 7437	22 55 39.	+ 14 02		14.5	GALAXY
VV 295C	22 55 39.	- 04 03	60		INTERACTING GALAXY
VV 295B	22 55 39.	- 04 03	60	12.	INTERACTING GALAXY
ARC 2510	22 55 40.	+ 00 10		17.8	RICH CLUSTER OF GALAXIES
HOLM 800B	22 55 40.	+ 25 53	18	14.2	PART OF MULTIPLE GALAXY
ZWG 430.034	22 55 42.	+ 14 02		14.4	GALAXY
UGC 12270	22 55 42.	+ 14 02	108	14.4	GALAXY Sc
MCG+02-58-041	22 55 42.	+ 14 02	108	13.	GALAXY
ZWG 453.033	22 55 42.	+ 15 37		15.6	GALAXY
ZC 2255.7-0022	22 55 42.	- 00 22	870		CLUSTER OF GALAXIES
MCG-05-54-006	22 55 42.	- 32 32	72	14.	GALAXY
SC 2252-4358.9	22 55 44.	- 43 42 53.			NEBULA
SC 2252-4406.6	22 55 44.	- 43 50 35.			NEBULA
SC 2252-4435.6	22 55 44.	- 44 19 35.			NEBULA
MCG+00-58-015	22 55 45.	+ 02 03	90	15.	GALAXY
MCG+05-54-021	22 55 45.	+ 28 58	30	15.5	GALAXY
MCG+00-58-016	22 55 45.	- 01 27	24	15.	GALAXY
SC 2252-4411.6	22 55 46.	- 43 55 35.			NEBULA
SC 2252-4714.3	22 55 46.	- 46 58 17.	60		NEBULA
ZWG 379.018	22 55 48.	+ 02 02		14.9	GALAXY
UGC 12271	22 55 48.	+ 02 02	96	14.9	GALAXY Sc
ZC 2255.8+1350	22 55 48.	+ 13 50	9880		CLUSTER OF GALAXIES
4ZW 126	22 55 48.	+ 24 57			COMPACT GALAXY
ZWG 475.009	22 55 48.	+ 24 57		15.2	GALAXY
ZWG 475.010	22 55 48.	+ 25 30		15.0	GALAXY
UGC 12272	22 55 48.	+ 25 30	72	15.0	GALAXY Sa
ZWG 496.027	22 55 48.	+ 28 58		15.2	GALAXY
RNGC 7439	22 55 48.	+ 28 58		15.0	GALAXY
UGC 12273	22 55 48.	+ 28 58	78	15.2	GALAXY SB0
ZWG 379.017	22 55 48.	- 01 27		15.7	GALAXY
RNGC 7434	22 55 49.	- 01 27		15.5	GALAXY
MCG+04-54-008	22 55 51.	+ 25 29	48	15.	GALAXY
SC 2253-4540.4	22 55 53.	- 45 24 23.			NEBULA
MCG+02-58-042	22 55 54.	+ 10 27	15	15.	GALAXY
MCG+04-54-009	22 55 54.	+ 24 56	33	15.	GALAXY
MCG+04-54-010	22 55 54.	+ 25 46	72	15.	GALAXY
ZWG 475.011	22 55 54.	+ 25 48		15.4	GALAXY
UGC 12274	22 55 54.	+ 25 48	78	15.4	GALAXY S
IC 5274	22 55 59.	+ 18 39 31.			NONSTELLAR OBJECT
LBN 0463	22 56	+ 41 20	2220		BRIGHT NEBULA
KHAV 792	22 56	+ 69 16	7300		DARK NEBULA
LBN 0188	22 56	- 00 10	1020		BRIGHT NEBULA
ZWG 430.035	22 56 00.	+ 10 28		15.3	GALAXY
ZWG 430.036	22 56 00.	+ 14 53		15.0	GALAXY
MRK 311	22 56 00.	+ 14 53	14	15.	GALAXY WITH UV CONTINUUM
ZWG 453.034	22 56 00.	+ 18 39		15.3	GALAXY
UGC 12275	22 56 00.	+ 18 39	56	15.3	GALAXY (S0)
ZWG 475.012	22 56 00.	+ 26 22		15.3	GALAXY
4ZW 127	22 56 00.	+ 35 15			COMPACT GALAXY
MCG-01-58-012	22 56 00.	- 05 12	54	16.	GALAXY
SC 2253-4446.4	22 56 01.	- 44 30 22.			NEBULA
ZWG 379.019	22 56 06.	+ 02 39		15.6	GALAXY
MCG+00-58-017	22 56 06.	+ 02 41	30	15.5	GALAXY
IC 1462	22 56 06.	+ 08 08			NONSTELLAR OBJECT
IC 1461	22 56 06.	+ 14 54 37.			NONSTELLAR OBJECT
MCG+07-47-001	22 56 06.	+ 40 08	30	15.	GALAXY
IC 5275	22 56 10.	+ 18 36 08.			NONSTELLAR OBJECT
IC 5276	22 56 11.	+ 18 33 38.			NONSTELLAR OBJECT
ZWG 453.035	22 56 12.	+ 18 33		15.2	GALAXY
ZWG 515.015	22 56 12.	+ 35 32		14.6	GALAXY
UGC 12276	22 56 12.	+ 35 32	102	14.6	GALAXY SBa
MCG+06-50-014	22 56 12.	+ 35 32	102	14.	GALAXY
ZWG 532.003	22 56 12.	+ 40 10		14.8	GALAXY
UGC 12277	22 56 12.	+ 50 33	60	16.5	GALAXY Sc
SC 2253-4555.4	22 56 13.	- 45 39 22.			NEBULA
RNGC 7440	22 56 14.	+ 35 32		14.5	GALAXY
ZWG 453.036	22 56 18.	+ 20 02		15.3	GALAXY
UGC 12278	22 56 18.	+ 20 02	60	15.3	GALAXY Sc
MCG+03-58-016	22 56 18.	+ 20 02	42	15.	GALAXY
UGC 12279	22 56 18.	+ 21 27	66	16.0	GALAXY Sc
ZWG 475.013	22 56 18.	+ 24 22		15.5	GALAXY
UGC 12280	22 56 18.	+ 24 22	60	15.5	GALAXY DBL SYS
SC 2253-4439.5	22 56 20.	- 44 23 28.			NEBULA
MCG+04-54-011	22 56 21.	+ 24 21	48	15.5	GALAXY
ARC 2511	22 56 21.	- 07 54		16.0	RICH CLUSTER OF GALAXIES
MCG-04-54-004	22 56 21.	- 25 48	60	14.	GALAXY
HW 0774	22 56 21.	- 65 27			NEBULA
IC 5272	22 56 21.	- 65 27			NONSTELLAR OBJECT
SC 2253-4536.4	22 56 23.	- 45 20 22.			NEBULA
SC 2253-4634.0	22 56 23.	- 46 17 58.			NEBULA
ZWG 430.037	22 56 24.	+ 14 25		15.7	GALAXY
MCG-05-54-007	22 56 24.	- 30 46	60	14.5	GALAXY
BC PKS2256+017	22 56 24.59	+ 01 47 35.6		19.	QUASI-STELLAR OBJECT
SHB 384	22 56 24.6	+ 01 47 35.		19.	QUASI-STELLAR OBJECT
SC 2253-6200.2	22 56 28.	- 61 44 10.	6		NEBULA
ZWG 430.038	22 56 30.	+ 13 32		15.6	GALAXY
ZC 2256.5+1933	22 56 30.	+ 19 33	12230		CLUSTER OF GALAXIES
ZWG 453.037	22 56 30.	+ 20 07		15.4	GALAXY
IC 5273	22 56 30.	- 38 02	150	12.0	GALAXY SB(rs)
SVEN 457	22 56 32.	- 41 02	60	14.1	GALAXY
SC 2253-4710.5	22 56 34.	- 46 54 28.	18		NEBULA
MCG+07-47-002	22 56 36.	+ 40 39	114	14.	GALAXY
MRSL 108-00/1	22 56 36.	+ 58 31	120		HII REGION
KLEM 43	22 56 36.	- 07 55	900	16.	CLUSTER OF 15 GALAXIES
SC 2253-4458.2	22 56 36.	- 44 42 10.			NEBULA
SG 3.237	22 56 40.	+ 58 32	30		DIFFUSE EMISSION NEBULA
ZWG 430.039	22 56 42.	+ 13 19		15.2	GALAXY
UGC 12281	22 56 42.	+ 13 19	210	15.2	GALAXY
MCG+02-58-043	22 56 42.	+ 13 19	204	14.	GALAXY
MCG+05-54-022	22 56 42.	+ 28 41	24	15.	GALAXY
ZWG 532.004	22 56 42.	+ 40 40		14.7	GALAXY
UGC 12282	22 56 42.	+ 40 40	126	14.7	GALAXY Sa
IC 1463	22 56 43.	- 10 48 09.			MAY NOT EXIST
MIN.46 21	22 56 46.	+ 58 30			DIFFUSE NEBULA
SC 2253-6206.1	22 56 46.	- 61 50 04.	60		NEBULA
MCG+02-58-044	22 56 48.	+ 12 26	60	15.	GALAXY
ZWG 475.014	22 56 48.	+ 23 51		15.7	GALAXY
UGC 12283	22 56 48.	+ 23 51	72	15.7	GALAXY SBa-b
ZC 2256.8+2445	22 56 48.	+ 24 45	10950		CLUSTER OF GALAXIES
ZWG 496.028	22 56 48.	+ 28 41		15.6	GALAXY
UGC 12284	22 56 48.	+ 28 41	78	15.6	GALAXY
SG 1.25	22 56 52.	+ 58 28	300		DIFFUSE EMISSION NEBULA
ZWG 430.040	22 56 54.	+ 12 26		15.6	GALAXY
UGC 12285	22 56 54.	+ 12 26	60	15.6	GALAXY S
ZWG 430.041	22 56 54.	+ 14 53		15.7	GALAXY
ZWG 430.042	22 56 54.	+ 15 17		14.2	GALAXY
UGC 12286	22 56 54.	+ 15 17	60	14.2	GALAXY Sc
MCG+02-58-045	22 56 54.	+ 15 17	60	14.	GALAXY
ISS 0140	22 56 54.	+ 56 15	389		STELLAR RING
RNGC 7441	22 56 54.	- 07 20		15.5	GALAXY
MCG-01-58-013	22 56 54.	- 07 20	72	15.5	GALAXY
SC 2254-4452.4	22 56 54.	- 44 36 21.			NEBULA
MCG+04-54-012	22 56 57.	+ 23 50	48	15.	GALAXY
RNGC 7442	22 56 58.	+ 15 17		14.0	GALAXY
SC 2254-4448.4	22 56 58.	- 44 32 21.			NEBULA
ARC 2513	22 56 59.	+ 25 59		17.9	RICH CLUSTER OF GALAXIES
LBN 0518	22 57	+ 58 29	300		BRIGHT NEBULA
LBN 0519	22 57	+ 58 31	120		BRIGHT NEBULA
MCG+02-58-046	22 57 00.	+ 13 48	48	15.	GALAXY
SC 2254-4536.5	22 57 00.	- 45 20 27.			NEBULA
SC 2254-4452.9	22 57 02.	- 44 36 51.			NEBULA
SC 2253-6354.2	22 57 02.	- 63 38 09.	18		NEBULA
ARC 2512	22 57 04.	+ 09 50		17.1	RICH CLUSTER OF GALAXIES
MCG+00-58-018	22 57 06.	+ 02 06	42	15.	GALAXY
ZWG 430.043	22 57 06.	+ 13 48		15.5	GALAXY
ZWG 453.038	22 57 06.	+ 17 42		15.1	GALAXY
MCG+03-58-017	22 57 06.	+ 19 53	54	17.	GALAXY
ZWG 515.016	22 57 06.	+ 38 50		15.6	GALAXY
UGC 12287	22 57 06.	+ 53 26	96	15.5	GALAXY Sc
MCG+09-37-001	22 57 06.	+ 53 26	78	15.	GALAXY
SC 2254-4409.2	22 57 06.	- 43 53 09.			NEBULA
LB 01515	22 57 06.	- 46 30		12.6	FAINT BLUE STAR
RNGC 7445	22 57 08.	+ 38 50		15.5	GALAXY
SC 2254-4404.0	22 57 09.	- 43 47 57.			NEBULA
SG 3.238	22 57 10.	+ 58 30	300		DIFFUSE EMISSION NEBULA
ZC 2257.2+0945	22 57 12.	+ 09 45	2080		CLUSTER OF GALAXIES
ZWG 453.039	22 57 12.	+ 17 45		15.7	GALAXY
ZWG 453.040	22 57 12.	+ 19 53		15.7	GALAXY
UGC 12288	22 57 12.	+ 19 53	66	15.7	GALAXY S
UGC 12289	22 57 12.	+ 23 50	84	16.0	GALAXY Sc
ZWG 475.015	22 57 12.	+ 24 35		15.7	GALAXY
UGC 12290	22 57 12.	+ 24 35	96	15.7	GALAXY Sb
ZWG 475.016	22 57 12.	+ 25 56		15.5	GALAXY
ZWG 515.017	22 57 12.	+ 38 48		15.7	GALAXY
MCG+06-50-015	22 57 12.	+ 38 50	15	15.	GALAXY
MRSL 108-01/4	22 57 12.	+ 58 28	300		HII REGION
SC 2254-4411.7	22 57 12.	- 43 55 39.			NEBULA
RNGC 7446	22 57 14.	+ 38 48		15.5	GALAXY
SC 2254-4606.6	22 57 15.	- 45 50 33.			NEBULA
SC 2254-4522.0	22 57 16.	- 45 05 57.			NEBULA
MCG+04-54-013	22 57 18.	+ 23 47 30.	72	15.	GALAXY
ZWG 475.017	22 57 18.	+ 26 02		15.6	GALAXY
UGC 12291	22 57 18.	+ 26 02	96	15.6	GALAXY Sc
ZWG 515.018	22 57 19.	+ 38 52		15.4	GALAXY
UGC 12292	22 57 18.	+ 38 52	66	15.4	GALAXY (E)
MCG+06-50-016	22 57 18.	+ 38 55	48	15.	GALAXY
ARC 2515	22 57 19.	+ 30 47		17.6	RICH CLUSTER OF GALAXIES
RNGC 7449	22 57 20.	+ 38 52		15.5	GALAXY
SC 2254-4412.2	22 57 20.	- 43 56 09.			NEBULA
SC 2254-6205.0	22 57 23.	- 61 48 57.	12		NEBULA
ZWG 475.018	22 57 24.	+ 25 45		15.5	GALAXY
UGC 12293	22 57 24.	+ 25 45	66	15.5	GALAXY Sa
ZC 2257.4+3047	22 57 24.	+ 30 47	740		CLUSTER OF GALAXIES
MCG+04-54-014	22 57 27.	+ 26 00 30.	30	15.5	GALAXY
ARC 2516	22 57 28.	+ 18 15		17.1	RICH CLUSTER OF GALAXIES
RNGC 7443	22 57 29.	- 13 04		14.0	GALAXY
RNGC 7444	22 57 29.	- 13 05		14.0	GALAXY
MCG+01-58-019	22 57 30.	+ 06 41	12	15.5	GALAXY
MCG+02-58-047	22 57 30.	+ 15 25	36	15.	GALAXY
ZWG 453.041	22 57 30.	+ 17 37		15.3	GALAXY
MCG+04-54-015	22 57 30.	+ 25 45	42	15.	GALAXY
MCG-02-58-014	22 57 30.	- 08 55	48	15.	GALAXY
SC 2254-6220.0	22 57 30.	- 62 03 57.	6		NEBULA
REIN 2.307	22 57 31.21	- 13 04 37.3			NEBULA
REIN 2.308	22 57 31.31	- 13 06 12.9			NEBULA
SC 2254-4426.2	22 57 32.	- 44 10 09.			NEBULA
REIN 2.309	22 57 33.72	- 13 07 01.7			NEBULA
RNGC 7448	22 57 34.	+ 15 43		12.5	GALAXY
SC 2254-4449.2	22 57 34.	- 44 33 08.			NEBULA
SC 2254-6220.6	22 57 34.	- 62 04 33.			NEBULA
ZWG 453.042	22 57 36.	+ 15 42		12.0	GALAXY
UGC 12294	22 57 36.	+ 15 42	162	12.0	GALAXY Sc
MCG+03-58-018	22 57 36.	+ 15 42 30.	138	12.5	GALAXY
MCG-02-58-015	22 57 36.	- 13 04 30.	72	14.5	GALAXY
MCG-02-58-016	22 57 36.	- 13 06	84	14.5	GALAXY
DV.56 I5269C	22 57 36.	- 35 38			SB(s) GALAXY

OBJECT NAME	RIGHT ASCEN.	DECLINATION	DIAM.	MAGN.	TYPE OF OBJECT
SC 2254-4445.0	22 57 36.	- 44 28 56.			NEBULA
SC 2254-4454.8	22 57 36.	- 44 38 44.			NEBULA
ARP 013	22 57 37.	+ 15 43			PECULIAR GALAXY
ARC 2514	22 57 38.	- 23 29		17.6	RICH CLUSTER OF GALAXIES
MCG-03-58-015	22 57 39.	- 16 45	30	15.	GALAXY
ARC 2517	22 57 40.	+ 10 22		17.5	RICH CLUSTER OF GALAXIES
SC 2254-4546.4	22 57 40.	- 45 30 20.	12		NEBULA
ZC 2257.7+1020	22 57 42.	+ 10 20	1080		CLUSTER OF GALAXIES
ZWG 475.019	22 57 42.	+ 25 50		15.6	GALAXY
MCG-01-58-014	22 57 42.	- 08 28	60	15.5	GALAXY
MCG-02-58-018	22 57 42.	- 12 54	36	15.5	GALAXY
MCG-02-58-017	22 57 42.	- 13 14	60	15.	GALAXY
SC 2254-4336.4	22 57 42.	- 43 20 20.			NEBULA
SC 2254-6222.8	22 57 44.	- 62 06 44.	6		NEBULA
SC 2254-4408.8	22 57 45.	- 43 52 44.			NEBULA
SC 2254-4414.0	22 57 45.	- 43 57 56.			NEBULA
SC 2254-6223.5	22 57 45.	- 62 07 26.	6		NEBULA
SC 2254-4629.8	22 57 46.	- 46 13 44.			NEBULA
RNGC 7447	22 57 47.	- 10 48			NON-EXISTENT OBJECT
ZWG 379.020	22 57 48.	+ 01 22		15.5	GALAXY
UGC 12295	22 57 48.	+ 01 22	78	15.5	GALAXY Sc
MCG+00-58-019	22 57 48.	+ 01 22	66	14.	GALAXY
ZC 2257.8+0125	22 57 48.	+ 01 25	1550		CLUSTER OF GALAXIES
ZC 2257.8+0902	22 57 48.	+ 09 02	1410		CLUSTER OF GALAXIES
MCG+02-58-048	22 57 48.	+ 13 19	48	15.	GALAXY
MRK 522	22 57 48.	+ 16 06	9	17.	GALAXY WITH UV CONTINUUM
ZWG 453.043	22 57 48.	+ 16 26		15.6	GALAXY
ZWG 453.044	22 57 48.	+ 18 51		15.5	GALAXY
UGC 12296	22 57 48.	+ 18 51	72	15.5	GALAXY Sc
4ZW 128	22 57 48.	+ 26 32			COMPACT GALAXY
ZWG 496.029	22 57 48.	+ 31 06		15.5	GALAXY
ZWG 496.030	22 57 48.	+ 31 08		15.0	GALAXY
UGC 12297	22 57 48.	+ 31 08	78	15.0	GALAXY S0?
MCG-03-58-017	22 57 48.	- 15 07	24	15.	GALAXY
MCG-03-58-016	22 57 48.	- 15 39	30	15.	GALAXY
SC 2254-4416.2	22 57 48.	- 44 00 08.			NEBULA
MCG+05-54-023	22 57 51.	+ 31 07 30.	39	15.5	GALAXY
MCG+05-54-024	22 57 51.	+ 31 09 30.	48	15.	GALAXY
SC 2254-6411.1	22 57 51.	- 63 55 02.	18		NEBULA
MCG+03-58-019	22 57 54.	+ 18 51	63	15.	GALAXY
ZWG 515.019	22 57 54.	+ 38 58		15.0	GALAXY
UGC 12298	22 57 54.	+ 38 58	90	15.0	GALAXY SBa-b
MCG+02-58-049	22 58 00.	+ 12 30	60	14.5	GALAXY
ZWG 430.044	22 58 00.	+ 12 45		15.5	GALAXY
MCG+06-50-017	22 58 00.	+ 39 00	72	15.	GALAXY
ISS 0094	22 58 00.	+ 62 55	169		STELLAR RING
IC 5278	22 58 02.	- 08 24			NONSTELLAR OBJECT
KARA.73 55	22 58 04.	- 37 12	27		DWARF GALAXY
SC 2255-4426.0	22 58 05.	- 44 09 56.			NEBULA
ZWG 405.021	22 58 06.	+ 07 01		15.3	GALAXY
MRK 523	22 58 06.	+ 07 01	20	15.5	GALAXY WITH UV CONTINUUM
ZWG 405.022	22 58 06.	+ 08 12		15.0	GALAXY
UGC 12299	22 58 06.	+ 08 12	66	15.0	GALAXY SBb-c
KARA.73B 1002	22 58 06.	+ 08 12	60	15.0	ISOLATED GALAXY S
ZWG 430.045	22 58 06.	+ 12 31		15.3	GALAXY
UGC 12300	22 58 06.	+ 12 31	78	15.3	GALAXY SB:a
ZC 2256.1+1339	22 58 06.	+ 13 39	1340		CLUSTER OF GALAXIES
MRK 312	22 58 06.	+ 16 05	20	16.	GALAXY WITH UV CONTINUUM
RNGC 7455	22 58 08.	+ 07 01		15.5	GALAXY
RNGC 7451	22 58 08.	+ 08 12		15.0	GALAXY
ARC 2518	22 58 08.	- 24 26		17.2	RICH CLUSTER OF GALAXIES
SC 2255-4542.4	22 58 10.	- 45 26 20.	180		NEBULA
SC 2255-6203.8	22 58 11.	- 61 47 44.	6		NEBULA
ZWG 379.021	22 58 12.	+ 02 14		15.7	GALAXY
UGC 12301	22 58 12.	+ 02 14	78	15.7	GALAXY DBL SYS
MCG+01-58-020	22 58 12.	+ 08 12	60	14.5	GALAXY
MCG+02-58-050	22 58 12.	+ 13 20	48	14.5	GALAXY
ZC 2258.2+1601	22 58 12.	+ 16 01	1280		CLUSTER OF GALAXIES
ARC 2520	22 58 14.	+ 13 45		17.7	RICH CLUSTER OF GALAXIES
SC 2255-4429.9	22 58 15.	- 44 13 50.			NEBULA
RNGC 7450	22 58 17.	- 13 12		13.0	GALAXY
SC 2255-4455.6	22 58 17.	- 44 39 32.			NEBULA
UGC 12302	22 58 18.	+ 06 34	138	16.0	GALAXY DBL SYS
ZWG 430.046	22 58 18.	+ 13 21		15.3	GALAXY
ZC 2258.3+1406	22 58 18.	+ 14 06	1080		CLUSTER OF GALAXIES
3ZW 091	22 58 18.	+ 16 31			COMPACT GALAXY
ZWG 475.020	22 58 18.	+ 25 25		15.6	GALAXY
ZWG 496.031	22 58 18.	+ 30 28		15.1	GALAXY
MCG-02-58-019	22 58 18.	- 12 12	96	13.5	GALAXY
SC 2255-4412.8	22 58 18.	- 43 56 44.			NEBULA
SC 2255-4401.9	22 58 19.	- 43 45 50.			NEBULA
SC 2255-4435.0	22 58 19.	- 44 18 56.			NEBULA
ARC 2519	22 58 20.	- 15 21		17.2	RICH CLUSTER OF GALAXIES
SC 2255-4428.5	22 58 20.	- 44 12 26.			NEBULA
SC 2255-4439.4	22 58 20.	- 44 23 20.			NEBULA
MCG+05-54-025	22 58 21.	+ 30 29	60	15.	GALAXY
SC 2255-4436.2	22 58 21.	- 44 20 08.			NEBULA
ZWG 475.021	22 58 24.	+ 26 28		14.8	GALAXY
UGC 12303	22 58 24.	+ 26 28	102	14.8	GALAXY Sb
ASS 30	22 58 24.	+ 63 47	43500		OB ASSOCIATION CEP OB3
MCG+04-54-016	22 58 27.	+ 26 27	96	14.	GALAXY
SC 2255-4458.4	22 58 27.	- 44 42 19.			NEBULA
SC 2255-6010.3	22 58 28.	- 59 54 14.	12		NEBULA
MCG+01-58-021	22 58 30.	+ 06 29	24	15.	GALAXY
ZCG 2258+33	22 58 30.	+ 33 04		17.4	COMPACT GALAXY
SC 2255-6010.4	22 58 30.	- 59 54 20.	12		NEBULA
RNGC 7452	22 58 32.	+ 06 29		15.0	GALAXY
SC 2255-4406.4	22 58 33.	- 43 50 19.			NEBULA
SC 2255-4434.3	22 58 33.	- 44 18 13.			NEBULA
RNGC 7454	22 58 34.	+ 16 07		13.5	GALAXY
SC 2255-4619.8	22 58 35.	- 46 03 43.			NEBULA
ZWG 405.023	22 58 36.	+ 05 23		15.4	GALAXY
UGC 12304	22 58 36.	+ 05 23	84	15.4	GALAXY S
KARA.73B 1003	22 58 36.	+ 05 23	102	15.4	ISOLATED GALAXY S
ZWG 430.047	22 58 36.	+ 10 46		15.6	GALAXY
ZWG 453.045	22 58 36.	+ 16 07		13.6	GALAXY
UGC 12305	22 58 36.	+ 16 07	114	13.6	GALAXY E
MCG+02-58-020	22 58 36.	+ 16 07 30.	42	13.5	GALAXY
ZWG 453.046	22 58 36.	+ 19 20		15.2	GALAXY
ZWG 496.032	22 58 36.	+ 29 53		12.3	GALAXY
RNGC 7457	22 58 36.	+ 29 53		12.0	GALAXY
UGC 12306	22 58 36.	+ 29 53	252	12.3	GALAXY S0
MCG+05-54-026	22 58 36.	+ 29 53 30.	90	12.	GALAXY
SC 2255-4412.5	22 58 39.	- 43 56 25.			NEBULA
SC 2255-4711.0	22 58 39.	- 46 54 55.	72		NEBULA
SC 2255-6201.8	22 58 41.	- 61 45 43.	42		NEBULA
ZWG 405.024	22 58 42.	+ 09 20		14.6	GALAXY
MRK 524	22 58 42.	+ 09 20	24	15.5	GALAXY WITH UV CONTINUUM
MCG+02-58-051	22 58 42.	+ 11 36	36	16.	GALAXY
MCG+02-58-053	22 58 42.	+ 12 27	84	16.	GALAXY
UGC 12307	22 58 42.	+ 12 28	96	16.0	GALAXY DWRF IP
MCG+02-58-052	22 58 42.	+ 14 04	156	14.	GALAXY
SC 2255-4547.5	22 58 42.	- 45 31 25.			NEBULA
SC 2255-4452.7	22 58 43.	- 44 36 37.			NEBULA
IC 5277	22 58 43.	- 65 28			NONSTELLAR OBJECT
SC 2255-4433.6	22 58 44.	- 44 17 31.			NEBULA
HW 0775	22 58 44.	- 65 28			NEBULA
SC 2255-6104.8	22 58 45.	- 60 48 43.	12		NEBULA
SC 2255-4432.1	22 58 46.	- 44 16 01.			NEBULA
ZC 2258.8+0327	22 58 48.	+ 03 27	940		CLUSTER OF GALAXIES
MCG+01-58-022	22 58 48.	+ 09 19	36	15.	GALAXY
ZWG 430.048	22 58 48.	+ 10 05		15.7	GALAXY
MCG+02-58-054	22 58 48.	+ 10 05	48	15.	GALAXY
ZWG 430.049	22 58 48.	+ 14 04		15.3	GALAXY
UGC 12308	22 58 48.	+ 14 04	150	15.3	GALAXY Sc
ZC 2258.8+2719	22 58 48.	+ 27 19	870		CLUSTER OF GALAXIES
RNGC 7453	22 58 48.	- 06 37			NON-EXISTENT OBJECT
SC 2255-4412.5	22 58 49.	- 43 56 25.			NEBULA
HOLM 801B	22 58 52.	+ 24 55	30	14.2	PART OF MULTIPLE GALAXY
ZWG 379.022	22 58 54.	+ 01 29		13.9	GALAXY
UGC 12309	22 58 54.	+ 01 29	90	13.9	GALAXY E
MCG+00-58-020	22 58 54.	+ 01 30	72	13.	GALAXY
MCG+02-58-055	22 58 54.	+ 11 48	30	15.5	GALAXY
MCG+05-54-027	22 58 54.	+ 28 08	18	16.	GALAXY
ZWG 496.033	22 58 54.	+ 30 16		15.6	GALAXY
RNGC 7458	22 58 55.	+ 01 29		14.0	GALAXY
SC 2255-6333.1	22 58 57.	- 63 17 01.			NEBULA
ZWG 430.050	22 59 00.	+ 14 20		15.7	GALAXY
ZWG 453.047	22 59 00.	+ 18 27		15.7	GALAXY
ZWG 496.034	22 59 00.	+ 28 08		15.0	GALAXY
UGC 12310	22 59 00.	+ 28 08	60	15.0	GALAXY COMPACT
ZWG 496.035	22 59 00.	+ 29 58		15.4	GALAXY
UGC 12311	22 59 00.	+ 29 58	96	15.4	GALAXY S?
COU 100	22 59 00.	+ 56 50	2400		EMISSION NEBULA
HOLM 801A	22 59 01.	+ 24 54	72	13.3	PART OF MULTIPLE GALAXY
SC 2256-4451.4	22 59 03.	- 44 35 19.			NEBULA
MCG+05-54-028	22 59 06.	+ 29 58 30.	84	15.	GALAXY
4ZW 129	22 59 06.	+ 32 04			COMPACT GALAXY
SC 2256-4308.1	22 59 06.	- 42 52 01.			NEBULA
RNGC 7459	22 59 08.	+ 06 28			NON-EXISTENT OBJECT
ZC 2259.2+0001	22 59 12.	+ 00 01	1810		CLUSTER OF GALAXIES
ZWG 379.023	22 59 12.	+ 02 00		14.2	GALAXY
UGC 12312	22 59 12.	+ 02 00	90	14.2	GALAXY Sb
MCG+00-58-021	22 59 12.	+ 02 01	60	13.	GALAXY
ZC 2259.2+0223	22 59 12.	+ 02 23	1140		CLUSTER OF GALAXIES
MCG+03-58-021	22 59 12.	+ 15 47	72	17.	GALAXY
UGC 12313	22 59 12.	+ 15 48	108	16.0	GALAXY DWARF
SC 2256-4327.0	22 59 12.	- 43 10 55.			NEBULA
LB 01516	22 59 12.	- 48 16		12.5	FAINT BLUE STAR
RNGC 7460	22 59 13.	+ 02 00		14.0	GALAXY
MCG+00-58-022	22 59 15.	+ 01 30	48	14.5	GALAXY
RNGC 7461	22 59 16.	+ 15 19		14.5	GALAXY
MCG+02-58-056	22 59 18.	+ 15 18	48	15.	GALAXY
ZWG 430.051	22 59 18.	+ 15 19		14.5	GALAXY
UGC 12314	22 59 18.	+ 15 19	54	14.5	GALAXY SB0
ZWG 453.049	22 59 18.	+ 15 42		14.5	GALAXY
UGC 12315	22 59 18.	+ 15 42	30	14.5	GALAXY
MCG+03-58-022	22 59 18.	+ 15 42 30.	180	13.	GALAXY
ZWG 453.048	22 59 18.	+ 15 43		13.5	GALAXY
UGC 12316	22 59 18.	+ 15 43	192	13.5	GALAXY S
RNGC 7456	22 59 19.	- 39 51		12.0	GALAXY
MCG-07-47-010	22 59 19.	- 41 27	48	16.	GALAXY
SC 2256-6305.9	22 59 20.	- 62 49 49.			NEBULA
MCG+03-58-023	22 59 21.	+ 15 42	18	14.5	GALAXY
RNGC 7464	22 59 22.	+ 15 42		14.5	GALAXY
RNGC 7463	22 59 22.	+ 15 43		13.5	GALAXY
HOLM 802A	22 59 23.	+ 15 43	54	13.2	PART OF MULTIPLE GALAXY
TON-S 0081	22 59 24.	- 26 48		14.7	BLUE STAR
TON-S 0082	22 59 24.	- 32 13		15.8	BLUE STAR
MCG-05-54-008	22 59 24.	- 32 17	30	15.	GALAXY
MCG-07-47-011	22 59 24.	- 39 51	150	12.5	GALAXY
HOLM 802C	22 59 25.	+ 15 43	18	13.8	PART OF MULTIPLE GALAXY
ARC 2522	22 59 26.	+ 13 48		17.7	RICH CLUSTER OF GALAXIES
RNGC 7465	22 59 29.	+ 15 42		13.5	GALAXY
MCG+03-58-024	22 59 30.	+ 15 41	60	13.	GALAXY
ZWG 453.050	22 59 30.	+ 15 42		13.3	GALAXY
MRK 313	22 59 30.	+ 15 42	25	14.	GALAXY WITH UV CONTINUUM
UGC 12317	22 59 30.	+ 15 42	72	13.3	GALAXY SB0
HOLM 802B	22 59 32.	+ 15 44	36	13.4	PART OF MULTIPLE GALAXY
SC 2256-4458.8	22 59 33.	- 44 42 42.			NEBULA
VDB.66N 156	22 59 35.	+ 42 03	11280		REFLECTION NEBULA
SC 2256-4403.4	22 59 35.	- 43 47 18.			NEBULA
ZC 2259.6+0746	22 59 36.	+ 07 46	10010		CLUSTER OF GALAXIES
UGC 12318	22 59 36.	+ 25 24	72	16.0	GALAXY SBb
ZWG 475.022	22 59 36.	+ 25 29		15.5	GALAXY
ZWG 475.023	22 59 36.	+ 26 47		14.4	GALAXY
RNGC 7466	22 59 36.	+ 26 47		14.5	GALAXY
UGC 12319	22 59 36.	+ 26 47	96	14.4	GALAXY Sb
ARC 2521	22 59 36.	- 22 15		15.9	RICH CLUSTER OF GALAXIES
MCG-07-47-012	22 59 36.	- 41 07	42	14.5	GALAXY
MCG+04-54-017	22 59 39.	+ 26 45 30.	90	14.5	GALAXY
SC 2256-4503.8	22 59 41.	- 44 47 42.			NEBULA
ZWG 430.052	22 59 42.	+ 10 12		15.7	GALAXY
ZC 2259.7+1350	22 59 42.	+ 13 50	1410		CLUSTER OF GALAXIES
UGC 12320	22 59 42.	+ 30 30	66	16.0	GALAXY Sc
IC 5279	22 59 45.	- 69 29			NONSTELLAR OBJECT
HW 0776	22 59 46.	- 69 29			NEBULA
ZWG 453.051	22 59 48.	+ 15 45		15.7	GALAXY
UGC 12321	22 59 48.	+ 15 45	60	15.7	GALAXY Sb-c
MCG-04-54-005	22 59 51.	- 24 35 30.	36	15.	GALAXY
RNGC 7467	22 59 52.	+ 15 17		15.5	GALAXY
MCG+01-58-023	22 59 54.	+ 06 39	24	16.	GALAXY
ZWG 430.053	22 59 54.	+ 15 17		15.6	GALAXY
MCG+02-58-057	22 59 54.	+ 15 17	12	15.5	GALAXY
VDB.66N 157	22 59 54.	+ 72 28	276		REFLECTION NEBULA
MCG-05-54-009	22 59 54.	- 31 21	48	15.	GALAXY
MCG-07-47-013	22 59 54.	- 41 07	210	12.7	GALAXY
SC 2257-4316.2	22 59 57.	- 43 00 06.			NEBULA
IC 5281	22 59 58.	+ 26 46			NONSTELLAR OBJECT
SC 2257-4717.5	22 59 58.	- 47 01 24.			NEBULA
BIGO 559	22 59 59.	+ 26 46			NEBULA
KHAV 783	23 00	+ 65 28	9600		DARK NEBULA
LBN 0606	23 00	+ 88 30	6300		BRIGHT NEBULA
ZWG 496.036	23 00 00.	+ 31 55		15.7	GALAXY
ASS 29	23 00 00.	+ 56 45			OB ASSOCIATION CEP OB5
LDN 1218	23 00 00.	+ 62 00	5100		DARK NEBULA

OBJECT NAME	RIGHT ASCEN.	DECLINATION	DIAM.	MAGN.	TYPE OF OBJECT
ZC 2300.0+7617	23 00 00.	+ 76 17	2350		CLUSTER OF GALAXIES
RNGC 7462	23 00 00.	- 41 06		13.0	GALAXY
SC 2257-4447.7	23 00 03.	- 44 31 36.			NEBULA
SC 2257-4714.6	23 00 03.	- 46 58 30.			NEBULA
MCG+02-58-058	23 00 06.	+ 13 03	36	15.	GALAXY
ZC 2300.1+1724	23 00 06.	+ 17 24	4370		CLUSTER OF GALAXIES
ZC 2300.2+1042	23 00 12.	+ 10 42	2290		CLUSTER OF GALAXIES
ZWG 430.054	23 00 12.	+ 13 03		15.3	GALAXY
UGC 12322	23 00 12.	+ 20 04	60	16.0	GALAXY S
ZWG 496.037	23 00 12.	+ 32 20		15.0	GALAXY
UGC 12323	23 00 12.	+ 32 20	72	15.0	GALAXY Sc
ZWG 515.020	23 00 12.	+ 38 26		15.5	GALAXY
MCG+06-50-018	23 00 12.	+ 38 28	24	15.	GALAXY
SC 2257-4449.0	23 00 14.	- 44 32 53.			NEBULA
SC 2257-4428.4	23 00 15.	- 44 12 17.			NEBULA
UGC 12324	23 00 18.	+ 08 19	78	16.0	GALAXY Sc
ZWG 475.024	23 00 18.	+ 21 37		15.2	GALAXY
UGC 12325	23 00 18.	+ 21 37	90	15.2	GALAXY Sc
ZWG 475.025	23 00 18.	+ 21 50		15.6	GALAXY
UGC 12326	23 00 18.	+ 21 50	72	15.6	GALAXY Sc
UGC 12327	23 00 18.	+ 25 46	90	16.5	GALAXY S
MCG+05-54-029	23 00 18.	+ 32 21	60	14.5	GALAXY
LB 01517	23 00 18.	- 47 27		13.6	FAINT BLUE STAR
IC 5282	23 00 20.	+ 21 36 23.			NONSTELLAR OBJECT
SC 2257-4347.6	23 00 22.	- 43 31 29.			NEBULA
MCG+04-54-019	23 00 24.	+ 21 34 30.	72	14.5	GALAXY
MCG+04-54-018	23 00 24.	+ 21 48	48	14.5	GALAXY
ZWG 475.026	23 00 24.	+ 26 34		15.3	GALAXY
UGC 12328	23 00 24.	+ 26 34	66	15.3	GALAXY S0-a
IC 1465	23 00 26.	+ 16 19			OPEN CLUSTER
RNGC 7468	23 00 28.	+ 16 20		14.0	GALAXY
ZC 2300.5+0407	23 00 30.	+ 04 07	1140		CLUSTER OF GALAXIES
ZWG 453.052	23 00 30.	+ 16 20		14.0	GALAXY
MRK 314	23 00 30.	+ 16 20	30	14.5	GALAXY WITH UV CONTINUUM
UGC 12329	23 00 30.	+ 16 20	54	14.0	GALAXY PECULR
ZWG 405.025	23 00 36.	+ 09 05		15.6	GALAXY
UGC 12330	23 00 36.	+ 09 05	60	15.6	GALAXY S?
ZWG 430.055	23 00 36.	+ 09 42		15.6	GALAXY
MCG+06-50-019	23 00 36.	+ 34 27	48	15.	GALAXY
ZWG 515.021	23 00 36.	+ 34 28		15.5	GALAXY
UGC 12331	23 00 36.	+ 34 28	66	15.5	GALAXY S
MCG-02-58-021	23 00 36.	- 09 15	36	15.5	GALAXY
MCG-02-58-020	23 00 36.	- 09 15 30.	30	16.5	GALAXY
RN 0777	23 00 36.	- 65 29			NEBULA
IC 5280	23 00 36.	- 65 29			NONSTELLAR OBJECT
IC 1464	23 00 37.	- 09 15 43.			NONSTELLAR OBJECT
SC 2257-4634.2	23 00 38.	- 46 18 05.	78		NEBULA
ARC 2524	23 00 41.	+ 17 29		16.5	RICH CLUSTER OF GALAXIES
FATH 1.832	23 00 41.	+ 30 24	5		NEBULA
ZWG 405.026	23 00 42.	+ 08 35		13.0	GALAXY
UGC 12332	23 00 42.	+ 08 35	108	13.0	GALAXY SBa
KARA.72 575A	23 00 42.	+ 08 35	96	13.0	PART OF DOUBLE GALAXY
VVI 92	23 00 42.	+ 08 36	85	13.62	SEYFERT GAL1XY
ARP 298	23 00 42.	+ 08 36			PECULIAR GALAXY
MCG+01-58-024	23 00 42.	+ 09 07	54	15.5	GALAXY
ZWG 453.053	23 00 42.	+ 19 45		15.7	GALAXY
ZWG 475.027	23 00 42.	+ 23 01		15.6	GALAXY
ZWG 475.028	23 00 42.	+ 23 29		15.7	GALAXY
ZC 2300.7+2810	23 00 42.	+ 28 10	1750		CLUSTER OF GALAXIES
5ZW 394	23 00 42.	+ 44 25			COMPACT GALAXY
SC 2257-4625.4	23 00 42.	- 46 09 17.			NEBULA
LB 01518	23 00 42.	- 47 00		14.2	FAINT BLUE STAR
SC 2257-6303.7	23 00 42.	- 62 47 35.	90		NEBULA
HOLM 803B	23 00 46.	+ 08 36	18	14.7	PART OF MULTIPLE GALAXY
IC 5283	23 00 47.0	+ 08 37 26.			GALAXY IRR
MCG+00-58-023	23 00 48.	+ 01 37 30.	42	15.	GALAXY
ZWG 405.027	23 00 48.	+ 08 36		15.2	GALAXY
KARA.72 575B	23 00 48.	+ 08 36	42	15.2	PART OF DOUBLE GALAXY
MCG+01-58-025	23 00 48.	+ 08 37	84	13.5	GALAXY
ZWG 453.054	23 00 48.	+ 19 42		15.5	GALAXY
UGC 12333	23 00 48.	+ 19 42	84	15.5	GALAXY SBb
MCG+04-54-020	23 00 48.	+ 23 00	12	15.	GALAXY
HOLM 303A	23 00 49.	+ 08 37	42	14.0	PART OF MULTIPLE GALAXY
SC 2257-4438.8	23 00 49.	- 44 22 41.			NEBULA
MCG+01-58-026	23 00 51.	+ 08 38	36	15.	GALAXY
SC 2258-4451.7	23 00 51.	- 44 35 35.			NEBULA
ZC 2300.9+1653	23 00 54.	+ 16 53	870		CLUSTER OF GALAXIES
ZWG 453.055	23 00 54.	+ 19 40		15.6	GALAXY
MCG+03-58-025	23 00 54.	+ 19 41	66	15.	GALAXY
ZWG 453.056	23 00 54.	+ 19 59		15.7	GALAXY
ZWG 475.029	23 00 54.	+ 23 08		15.7	GALAXY
MFSL 108-02/1	23 00 54.	+ 56 48	1200		HII REGION
SC 2258-4434.6	23 00 55.	- 44 18 29.			NEBULA
ARC 2523	23 00 57.	- 17 26		17.0	RICH CLUSTER OF GALAXIES
ARC 2525	23 00 59.	- 10 51		16.0	RICH CLUSTER OF GALAXIES
LBN 0517	23 01	+ 56 50	1560		BRIGHT NEBULA
MCG+00-58-024	23 01 00.	+ 01 35	36	15.	GALAXY
ZWG 405.028	23 01 00.	+ 09 01		14.9	GALAXY
ZWG 430.056	23 01 00.	+ 10 07		15.5	GALAXY
ZC 2301.0+3656	23 01 00.	+ 36 56	2760		CLUSTER OF GALAXIES
MCG+07-47-003	23 01 00.	+ 41 06	42	15.	GALAXY
MCG+09-01-002	23 01 00.	+ 51 29	12	16.	GALAXY
SC 2257-7317.9	23 01 00.	- 73 01 47.	18		NEBULA
SG 3.239	23 01 01.	+ 56 51	600		DIFFUSE EMISSION NEBULA
IC 1466	23 01 05.	- 03 02 30.			NONSTELLAR OBJECT
ZWG 379.024	23 01 06.	+ 00 47		15.7	GALAXY
ZWG 379.025	23 01 06.	+ 01 35		15.4	GALAXY
MCG+02-58-059	23 01 06.	+ 10 23	48	15.	GALAXY
ZC 2301.1+1505	23 01 06.	+ 15 05	1010		CLUSTER OF GALAXIES
ZWG 532.005	23 01 06.	+ 41 09		15.0	GALAXY
MCG+01-58-027	23 01 12.	+ 09 02	36	14.5	GALAXY
ZWG 430.057	23 01 12.	+ 10 23		15.7	GALAXY
MCG+09-01-003	23 01 12.	+ 51 27	27	16.	GALAXY
MCG-01-58-015	23 01 12.	- 07 30	66	14.5	GALAXY
SC 2258-6358.0	23 01 13.	- 63 41 52.	12		NEBULA
RNGC 7472	23 01 14.	+ 02 47			NON-EXISTENT OBJECT
RNGC 7471	23 01 16.	- 23 11			NON-EXISTENT OBJECT
MCG+06-50-020	23 01 18.	+ 34 42	48	14.	GALAXY
MCG-04-54-006	23 01 21.	- 24 06	30	17.	GALAXY
SC 2258-4442.2	23 01 21.	- 44 26 04.			NEBULA
ZWG 515.022	23 01 24.	+ 34 43		15.1	GALAXY
TON-S 0083	23 01 24.	- 30 00		15.7	BLUE STAR
ARC 2526	23 01 25.	- 24 18		17.4	RICH CLUSTER OF GALAXIES
RNGC 7470	23 01 27.	- 50 23			UNVERIFIED SOUTHERN OBJECT
SC 2258-4305.5	23 01 28.	- 42 49 22.			NEBULA
SC 2258-4432.6	23 01 29.	- 44 16 28.			NEBULA
MCG+00-58-025	23 01 30.	+ 01 39	15	14.	GALAXY
ZWG 453.057	23 01 30.	+ 19 55		15.5	GALAXY
MRK 315	23 01 30.	+ 22 21	15	15.	GALAXY WITH UV CONTINUUM
KW 20	23 01 30.	+ 22 21	19		SEYFERT GALAXY
ZWG 475.030	23 01 30.	+ 27 02		15.5	GALAXY
UGC 12334	23 01 30.	+ 27 02	78	15.5	GALAXY SB0
ZWG 496.038	23 01 30.	+ 29 54		14.8	GALAXY
RNGC 7473	23 01 30.	+ 29 54		15.0	GALAXY
UGC 12335	23 01 30.	+ 29 54	66	14.8	GALAXY SB0
ZC 2301.5+3040	23 01 30.	+ 30 40	1080		CLUSTER OF GALAXIES
MCG+05-54-030	23 01 33.	+ 29 54	48	14.5	GALAXY
RNGC 7474	23 01 34.	+ 19 46		15.0	GALAXY
SS 71	23 01 34.	+ 60 06			DIFFUSE GALACTIC NEBULA
SC 2258-4427.6	23 01 34.	- 44 11 28.			NEBULA
ZWG 379.026	23 01 36.	+ 01 38		15.2	GALAXY
UGC 12336	23 01 36.	+ 01 38	66	15.2	GALAXY E-S0
ZWG 453.058	23 01 36.	+ 19 46		15.2	GALAXY
MCG+03-58-026	23 01 36.	+ 19 49	30	14.5	GALAXY
2ZW 187	23 01 36.	+ 22 21			COMPACT GALAXY
MCG+04-54-021	23 01 36.	+ 27 01	18	15.5	GALAXY
FATH 2.166	23 01 36.	+ 29 53			NEBULA
DG 189	23 01 36.	+ 60 06	60		REFLECTION NEBULA
MCG-05-54-010	23 01 36.	- 28 40	24	15.5	GALAXY
RNGC 7475	23 01 40.	+ 19 47		15.0	GALAXY
MRK 525	23 01 42.	+ 01 28	7	17.	GALAXY WITH UV CONTINUUM
ZWG 453.059	23 01 42.	+ 19 47		15.1	GALAXY
UGC 12337	23 01 42.	+ 19 47	90	15.1	GALAXY DBL SYS
MCG+03-58-028	23 01 42.	+ 19 50	24	15.	GALAXY
MCG+03-58-027	23 01 42.	+ 19 50	60	14.5	GALAXY
RNGC 7469	23 01 44.	+ 08 36		13.5	GALAXY
SC 2258-4343.8	23 01 44.	- 43 27 40.			NEBULA
SC 2258-4407.1	23 01 46.	- 43 50 58.	30		NEBULA
ZC 2301.8+1325	23 01 48.	+ 13 25	3020		CLUSTER OF GALAXIES
UGC 12338	23 01 48.	+ 17 38	66	16.0	GALAXY Sc
ZWG 453.060	23 01 48.	+ 17 41		15.5	GALAXY
ZWG 475.031	23 01 48.	+ 22 16		15.5	GALAXY
ZWG 475.032	23 01 48.	+ 27 21		15.6	GALAXY
MCG-01-58-016	23 01 48.	- 05 08	72	15.	GALAXY
SC 2259-4415.5	23 01 51.	- 43 59 21.			NEBULA
ZWG 379.027	23 01 54.	+ 02 02		15.6	GALAXY
MCG+04-54-022	23 01 54.	+ 22 15 30.	48	15.	GALAXY
MCG-07-47-014	23 01 54.	- 43 52	36	16.	GALAXY
LBN 0477	23 02	+ 43 10	1800		BRIGHT NEBULA
5ZW 395	23 02 00.	+ 46 16			COMPACT GALAXY
LDN 1220	23 02 00.	+ 61 35	300		DARK NEBULA
LDN 1224	23 02 00.	+ 63 30	480		DARK NEBULA
ZWG 475.033	23 02 06.	+ 26 53		15.6	GALAXY
UGC 12340	23 02 06.	+ 26 53	66	15.6	GALAXY S
MCG+04-54-023	23 02 06.	+ 27 04 30.	42	15.5	GALAXY
3ZW 092	23 02 06.	- 01 45			COMPACT GALAXY
UGC 12339	23 02 06.	- 01 45	15	14.0	GALAXY COMPACT
LB 01519	23 02 06.	- 47 52		13.8	FAINT BLUE STAR
ZWG 405.029	23 02 12.	+ 07 31		15.7	GALAXY
ZC 2302.2+2156	23 02 12.	+ 21 56	1210		CLUSTER OF GALAXIES
ZC 2302.2+2309	23 02 12.	+ 23 09	1140		CLUSTER OF GALAXIES
MCG+04-54-024	23 02 12.	+ 26 52	36	15.5	GALAXY
ZC 2302.2+2729	23 02 12.	+ 27 29	1950		CLUSTER OF GALAXIES
ZWG 532.006	23 02 12.	+ 43 57		15.7	GALAXY
UGC 12341	23 02 12.	+ 43 57	102	15.7	GALAXY SBc
MCG-05-54-011	23 02 12.	- 30 42	42	15.5	GALAXY
RNGC 7477	23 02 14.	+ 02 51			NON-EXISTENT OBJECT
MCG+07-47-004	23 02 15.	+ 43 57	66	16.	GALAXY
IC 1467	23 02 16.	- 03 29 52.			NONSTELLAR OBJECT
MCG+01-58-028	23 02 18.	+ 07 31	42	15.	GALAXY
ZWG 453.061	23 02 18.	+ 16 24		15.0	GALAXY
UGC 12342	23 02 18.	+ 16 24	90	15.0	GALAXY PECULR
MCG-01-58-017	23 02 18.	- 03 29	42	15.	GALAXY
MCG-02-58-022	23 02 18.	- 09 00	30	15.5	GALAXY
MCG-05-54-012	23 02 18.	- 31 34	36	15.	GALAXY
RNGC 7478	23 02 20.	+ 02 19			GALAXY
SHB 385	23 02 20.0	- 71 19 23.		17.5	QUASI-STELLAR OBJECT
BC PKS2302-713	23 02 20.03	- 71 19 23.3		17.5	QUASI-STELLAR OBJECT
MCG-07-47-015	23 02 21.	- 43 22	60	14.	GALAXY
RNGC 7476	23 02 23.	- 43 23			GALAXY
ZWG 430.058	23 02 24.	+ 12 03		11.7	GALAXY
UGC 12343	23 02 24.	+ 12 03	264	11.7	GALAXY SBb
MCG+02-58-060	23 02 24.	+ 12 03	240	11.	GALAXY
KARA.73B 1004	23 02 24.	+ 12 03	246	11.7	ISOLATED GALAXY S
MCG+03-58-030	23 02 24.	+ 16 24	72	15.	GALAXY
MCG+03-58-029	23 02 24.	+ 18 27	120	15.	GALAXY
ZWG 453.062	23 02 24.	+ 19 16		15.2	GALAXY
MCG-01-58-018	23 02 24.	- 06 49	72	15.	GALAXY
RNGC 7479	23 02 27.	+ 12 03		11.5	GALAXY
SC 2259-6251.7	23 02 28.	- 62 35 33.			NEBULA
UGC 12344	23 02 30.	+ 18 27	138	16.0	GALAXY
ZWG 475.034	23 02 30.	+ 27 05		15.6	GALAXY
MCG+06-50-021	23 02 30.	+ 35 45	42	14.5	GALAXY
UGC 12345	23 02 30.	+ 44 17	84	17.	GALAXY SB
ARC 2527	23 02 32.	- 25 35		17.7	RICH CLUSTER OF GALAXIES
HOLM 804B	23 02 33.	+ 02 16	36	15.5	PART OF MULTIPLE GALAXY
IC 1468	23 02 34.	- 03 28 21.			NONSTELLAR OBJECT
ZWG 379.028	23 02 36.	+ 00 33		15.7	GALAXY
UGC 12346	23 02 36.	+ 00 33	84	15.7	GALAXY Sc
ZC 2302.6+1530	23 02 36.	+ 15 30	940		CLUSTER OF GALAXIES
ZWG 453.063	23 02 36.	+ 18 35		15.5	GALAXY
UGC 12347	23 02 36.	+ 18 35	60	15.5	GALAXY IRR
MCG+03-58-031	23 02 36.	+ 18 36	48	15.	GALAXY
ZC 2302.6+2044	23 02 36.	+ 20 44	1610		CLUSTER OF GALAXIES
ZWG 515.023	23 02 36.	+ 35 45		15.0	GALAXY
MCG-01-58-019	23 02 36.	- 03 27	36	15.5	GALAXY
SC 2259-4432.4	23 02 37.	- 44 16 15.			NEBULA
RNGC 7480B	23 02 38.	+ 02 21		15.5	GALAXY
MCG+00-58-026	23 02 42.	+ 00 34	84	15.	GALAXY
ZWG 379.030	23 02 42.	+ 02 16		15.1	GALAXY
UGC 12349	23 02 42.	+ 02 16	90	15.1	GALAXY Sa
4ZW 130	23 02 42.	+ 39 06			COMPACT GALAXY
ZWG 379.029	23 02 42.	- 00 05		15.5	GALAXY
UGC 12348	23 02 42.	- 00 05	72	15.5	GALAXY Sa
RNGC 7480A	23 02 44.	+ 02 16		15.0	GALAXY
HOLM 804A	23 02 45.	+ 02 17	72	14.8	PART OF MULTIPLE GALAXY
MCG+00-58-027	23 02 45.	+ 02 18	66	14.	GALAXY
MCG+03-58-032	23 02 45.	+ 16 36	132	15.5	GALAXY
MCG+00-58-028	23 02 45.	- 00 03	54	14.5	GALAXY
ZWG 430.059	23 02 48.	+ 13 53		15.5	GALAXY
MCG+02-58-061	23 02 48.	+ 13 54	48	15.	GALAXY
ZWG 453.064	23 02 48.	+ 16 35		15.1	GALAXY
UGC 12350	23 02 48.	+ 16 35	180	15.1	GALAXY
ZC 2302.8+3209	23 02 48.	+ 32 09	870		CLUSTER OF GALAXIES
MCG+07-47-005	23 02 48.	+ 41 39	42	16.	GALAXY
ZWG 532.007	23 02 48.	+ 41 41		15.4	GALAXY

OBJECT NAME	RIGHT ASCEN.	DECLINATION	DIAM.	MAGN.	TYPE OF OBJECT
MCG-01-58-020	23 02 48.	- 03 12	48	15.5	GALAXY
MCG-02-58-023	23 02 48.	- 10 20	30	15.	GALAXY
ZC 2302.9+1104	23 02 54.	+ 11 04	1210		CLUSTER OF GALAXIES
ZWG 453.065	23 02 54.	+ 20 53		15.1	GALAXY
ARC 2528	23 02 59.	- 21 40		16.8	RICH CLUSTER OF GALAXIES
LBN 0439	23 03	+ 28 20	1020		BRIGHT NEBULA
KW 58	23 03 00.	+ 08 44	89		SEYFERT GALAXY
ZWG 453.066	23 03 00.	+ 16 12		15.3	GALAXY
ZWG 453.067	23 03 00.	+ 18 55		15.7	GALAXY
LDN 1222	23 03 00.	+ 61 30	180		DARK NEBULA
CED 208	23 03 05.	+ 59 58	900		DIFFUSE GALACTIC NEBULA
IC 1470	23 03 05.0	+ 59 58 10.	60		HII REGION
ZWG 379.031	23 03 06.	+ 02 47		14.6	GALAXY
MCG+00-58-029	23 03 06.	+ 02 49	21	14.	GALAXY
ISS 0095	23 03 06.	+ 57 51	146		STELLAR RING
MRSL 110+00/1	23 03 06.	+ 59 59	120		HII REGION
RNGC 7482	23 03 08.	+ 02 47		14.5	GALAXY
SC 2259-7319.6	23 03 09.	- 73 03 26.	12		NEBULA
RNGC 7481	23 03 10.	- 20 13			NON-EXISTENT OBJECT
ZWG 405.030	23 03 12.	+ 07 45		15.7	GALAXY
ZWG 453.068	23 03 12.	+ 18 43		15.6	GALAXY
UGC 12351	23 03 12.	+ 18 43	96	15.6	GALAXY IRR
ZC 2303.2+1919	23 03 12.	+ 19 19	1480		CLUSTER OF GALAXIES
4ZW 131	23 03 12.	+ 24 21			COMPACT GALAXY
ZWG 475.035	23 03 12.	+ 27 23		14.7	GALAXY
UGC 12352	23 03 12.	+ 27 23	108	14.7	GALAXY S
MCG-05-54-013	23 03 12.	- 30 54	120	14.5	GALAXY
ZWG 379.032	23 03 18.	+ 03 16		14.3	GALAXY
UGC 12353	23 03 18.	+ 03 16	96	14.3	GALAXY SBa
MCG+00-58-030	23 03 18.	+ 03 19	90	14.	GALAXY
MCG+01-58-029	23 03 18.	+ 07 46	12	15.5	GALAXY
ZWG 453.069	23 03 18.	+ 16 37		15.5	GALAXY
MCG+03-58-033	23 03 18.	+ 18 42 30.	66	15.5	GALAXY
MCG+04-54-025	23 03 18.	+ 27 23	48	14.5	GALAXY
ISS 0141	23 03 18.	+ 55 23	359		STELLAR RING
RNGC 7483	23 03 20.	+ 03 16		14.5	GALAXY
ZWG 379.033	23 03 24.	+ 02 42		15.2	GALAXY
MCG+00-58-031	23 03 24.	+ 02 44	42	14.5	GALAXY
ZWG 430.060	23 03 24.	+ 14 05		15.3	GALAXY
UGC 12354	23 03 24.	+ 14 05	66	15.3	GALAXY Sc
MCG+02-58-062	23 03 24.	+ 14 06	60	14.5	GALAXY
UGC 12355	23 03 24.	+ 24 20	72	16.5	GALAXY Sc
ZWG 496.039	23 03 24.	+ 28 25		15.6	GALAXY
MCG+05-54-031	23 03 24.	+ 28 25	72	15.	GALAXY
MCG+05-54-032	23 03 24.	+ 30 45 30.	24	14.5	GALAXY
ZWG 496.040	23 03 24.	+ 30 50		14.6	GALAXY
UGC 12356	23 03 24.	+ 30 50	60	14.6	GALAXY COMPACT
KARA.72 576A	23 03 24.	+ 30 50	66	14.6	PART OF DOUBLE GALAXY
FATH 1.833	23 03 28.	+ 30 49	8		NEBULA
MCG+05-54-033	23 03 30.	+ 30 47	48	15.5	GALAXY
ZWG 496.041	23 03 30.	+ 30 48		15.7	GALAXY
UGC 12357	23 03 30.	+ 30 48	60	15.7	GALAXY
KARA.72 576B	23 03 30.	+ 30 48	60	15.7	PART OF DOUBLE GALAXY
UGC 12358	23 03 30.	+ 33 53	60	16.0	GALAXY Sb-c
ARC 2530	23 03 35.	+ 19 21		17.1	RICH CLUSTER OF GALAXIES
PHL 0400	23 03 36.	+ 01 43		15.8	BLUE STELLAR OBJECT
ZWG 405.031	23 03 36.	+ 07 42		15.6	GALAXY
ZWG 405.032	23 03 36.	+ 07 43		15.6	GALAXY
MCG+01-58-030	23 03 36.	+ 07 44	30	15.5	GALAXY
MCG+02-58-063	23 03 36.	+ 14 37	84	14.5	GALAXY
MCG+06-50-022	23 03 36.	+ 32 50	36	14.5	GALAXY
MCG+01-58-031	23 03 42.	+ 07 43	36	15.5	GALAXY
UGC 12359	23 03 42.	+ 14 36	84	16.5	GALAXY S
ZWG 515.024	23 03 42.	+ 33 50		14.2	GALAXY
UGC 12360	23 03 42.	+ 32 50	72	14.2	GALAXY S0
ARC 2529	23 03 42.	- 13 31		17.2	RICH CLUSTER OF GALAXIES
RNGC 7485	23 03 43.	+ 32 50		14.0	GALAXY
UGC 12361	23 03 48.	+ 11 00	60	16.0	GALAXY IRR
ZC 2303.8+2813	23 03 48.	+ 28 13	1280		CLUSTER OF GALAXIES
ZWG 496.042	23 03 48.	+ 31 37		15.3	GALAXY
UGC 12362	23 03 48.	+ 31 57	102	15.3	GALAXY SBb
MRSL 111+04/1	23 03 48.	+ 64 24	4800		HII REGION
TON-S 0084	23 03 48.	- 26 20		15.8	BLUE STAR
RNGC 7486	23 03 49.	+ 33 48			GALAXY
IC 1469	23 03 50.	- 13 48 18.			NONSTELLAR OBJECT
MCG+07-47-006	23 03 51.	+ 41 18	36	17.	GALAXY
PHL 0401	23 03 54.	+ 01 54		16.2	BLUE STELLAR OBJECT
ZWG 405.033	23 03 54.	+ 08 07		15.1	GALAXY
MCG+02-58-064	23 03 54.	+ 11 17	36	15.	GALAXY
MCG+05-54-034	23 03 54.	+ 31 37	96	14.5	GALAXY
ZWG 379.034	23 03 54.	- 00 07		14.6	GALAXY
UGC 12363	23 03 54.	- 00 07	66	14.6	GALAXY S
SC 2300-6350.5	23 03 58.	- 63 34 19.			NEBULA
PHL 5463	23 04 00.	+ 02 05		17.1	BLUE STELLAR OBJECT
ZC 2304.0+1016	23 04 00.	+ 10 16	1410		CLUSTER OF GALAXIES
MCG+00-58-032	23 04 00.	- 00 06	66	13.5	GALAXY
PHL 0402	23 04 06.	+ 01 18		15.9	BLUE STELLAR OBJECT
ZC 2304.1-0158	23 04 06.	- 01 58	340		CLUSTER OF GALAXIES
MCG-07-47-017	23 04 06.	- 43 10	9	15.5	GALAXY
MCG-07-47-016	23 04 06.	- 43 10	12	14.5	GALAXY
ARC 2532	23 04 11.	+ 28 15		17.5	RICH CLUSTER OF GALAXIES
GCL 124	23 04 12.	+ 12 28	240		GLOBULAR STAR CLUSTER
ZWG 430.061	23 04 12.	+ 12 28		15.6	GALAXY
ZWG 453.070	23 04 12.	+ 18 50		14.9	GALAXY
UGC 12364	23 04 12.	+ 18 50	66	14.9	GALAXY Sa
ZC 2304.2+2846	23 04 12.	+ 28 46	940		CLUSTER OF GALAXIES
MCG+00-58-033	23 04 12.	- 01 07	42	16.	GALAXY
MCG-07-47-018	23 04 12.	- 43 11	12	15.	GALAXY
MCG+03-58-034	23 04 15.	+ 18 51	60	14.5	GALAXY
SG 3.240	23 04 17.	+ 60 00	60		DIFFUSE EMISSION NEBULA
ARC 2531	23 04 17.	- 21 57		17.1	RICH CLUSTER OF GALAXIES
ZWG 430.062	23 04 18.	+ 14 45		15.6	GALAXY
IC 5284	23 04 18.0	+ 18 51 06.			GALAXY
RNGC 7484	23 04 19.	- 36 32			GALAXY
SC 2301-6201.7	23 04 20.	- 61 45 31.	48		NEBULA
ZWG 475.036	23 04 24.	+ 22 40		14.4	GALAXY
UGC 12365	23 04 24.	+ 22 40	102	14.4	GALAXY COMPACT
MCG+05-54-035	23 04 24.	+ 27 53 30.	24	14.5	GALAXY
PHL 0403	23 04 24.	- 01 40		18.3	BLUE STELLAR OBJECT
PHL 5464	23 04 24.	- 12 48		17.6	BLUE STELLAR OBJECT
PHL 2073	23 04 24.	- 13 52		18.6	BLUE STELLAR OBJECT
ARC 2535	23 04 25.	+ 40 24		16.9	RICH CLUSTER OF GALAXIES
MCG-02-58-024	23 04 27.	- 09 34	30	15.	GALAXY
RNGC 7487	23 04 29.	+ 27 55		15.0	GALAXY
PHL 0404	23 04 30.	+ 02 08		18.0	BLUE STELLAR OBJECT
UGC 12366	23 04 30.	+ 05 41	72	16.0	GALAXY S
MCG+02-58-065	23 04 30.	+ 09 41	60	15.	GALAXY
ZC 2304.5+1614	23 04 30.	+ 16 14	200		CLUSTER OF GALAXIES
UGC 12367	23 04 30.	+ 16 52	60	16.5	GALAXY Sc
2ZW 188	23 04 30.	+ 22 40			COMPACT GALAXY
ZWG 496.043	23 04 30.	+ 27 55		14.8	GALAXY
UGC 12368	23 04 30.	+ 27 55	138	14.8	GALAXY COMPACT
MCG-05-54-014	23 04 30.	- 28 53	48	15.	GALAXY
MCG+04-54-026	23 04 33.	+ 22 39	90	14.	GALAXY
IC 5285	23 04 33.	+ 22 40 07.			NONSTELLAR OBJECT
UGC 12369	23 04 36.	+ 05 43	60	17.	GALAXY DWARF
ZWG 405.034	23 04 36.	+ 08 28		15.3	GALAXY
ZWG 430.063	23 04 36.	+ 09 40		15.5	GALAXY
UGC 12370	23 04 36.	+ 09 40	90	15.5	GALAXY Sc
KARA.73B 1005	23 04 36.	+ 09 40	108	15.5	ISOLATED GALAXY S
ZWG 431.001	23 04 36.	+ 14 44		15.7	GALAXY
ZWG 430.064	23 04 36.	+ 14 44		15.7	GALAXY
UGC 12371	23 04 36.	+ 16 51	60	16.5	GALAXY Sc
ZWG 515.025	23 04 36.	+ 35 30		14.5	GALAXY
UGC 12372	23 04 36.	+ 35 30	48	14.5	GALAXY S
KARA.73B 1006	23 04 36.	+ 35 30	54	14.5	ISOLATED GALAXY SPEC
PHL 2074	23 04 36.	- 14 55		18.5	BLUE STELLAR OBJECT
PHL 5465	23 04 36.	- 18 46		18.1	BLUE STELLAR OBJECT
PHL 5466	23 04 36.	- 19 21		18.1	BLUE STELLAR OBJECT
ARC 2533	23 04 37.	- 15 29		17.2	RICH CLUSTER OF GALAXIES
MCG+01-58-032	23 04 42.	+ 08 28	48	15.	GALAXY
MCG+06-50-023	23 04 42.	+ 35 30	42	14.	GALAXY
MCG-01-58-021	23 04 42.	- 06 34	30	17.	GALAXY
PHL 5467	23 04 42.	- 08 12		18.1	BLUE STELLAR OBJECT
MCG+00-58-034	23 04 45.	+ 01 57	48	15.	GALAXY
SC 2301-6145.4	23 04 45.	- 61 29 12.			NEBULA
ZWG 379.035	23 04 48.	+ 01 55		15.6	GALAXY
PHL 2075	23 04 48.	+ 03 30		18.6	BLUE STELLAR OBJECT
UGC 12373	23 04 48.	+ 13 23	60	16.0	GALAXY
ZWG 431.002	23 04 48.	+ 15 05		15.7	GALAXY
ZWG 430.065	23 04 48.	+ 15 05		15.7	GALAXY
UGC 12374	23 04 48.	+ 15 05	72	15.7	GALAXY SBb
ZC 2304.8+2330	23 04 48.	+ 23 30	1210		CLUSTER OF GALAXIES
UGC 12375	23 04 48.	+ 24 59	72	16.0	GALAXY S
MCG+06-50-024	23 04 48.	+ 34 58	66	15.	GALAXY
MCG-01-58-022	23 04 48.	- 05 25	72	15.	GALAXY
PHL 5468	23 04 48.	- 16 38		18.5	BLUE STELLAR OBJECT
MCG-05-54-015	23 04 48.	- 27 36	24	15.	GALAXY
ARC 2534	23 04 53.	- 22 56		17.5	RICH CLUSTER OF GALAXIES
MCG+02-58-066	23 04 54.	+ 15 06	72	15.5	GALAXY
3ZW 093	23 04 54.	+ 15 36			COMPACT GALAXY
UGC 12376	23 04 54.	+ 15 36	12	14.0	GALAXY COMPACT
ZWG 475.037	23 04 54.	+ 22 33		15.6	GALAXY
ZC 2304.9+2657	23 04 54.	+ 26 57	2290		CLUSTER OF GALAXIES
MCG+06-50-025	23 04 54.	+ 36 05	24	15.	GALAXY
ZC 2304.9+4025	23 04 54.	+ 40 25	1410		CLUSTER OF GALAXIES
PHL 5469	23 04 54.	- 09 44		18.2	BLUE STELLAR OBJECT
RNGC 7489	23 04 58.	+ 22 44		14.5	GALAXY
ZC 2305.0+0114	23 05 00.	+ 01 14	400		CLUSTER OF GALAXIES
MCG+04-54-027	23 05 00.	+ 22 30 30.	45	15.	GALAXY
ZWG 475.038	23 05 00.	+ 22 44		14.3	GALAXY
UGC 12378	23 05 00.	+ 22 44	132	14.3	GALAXY Sc
ZWG 496.044	23 05 00.	+ 32 06		13.5	GALAXY
RNGC 7490	23 05 00.	+ 32 06		13.5	GALAXY
UGC 12379	23 05 00.	+ 32 06	180	13.5	GALAXY Sb
ZWG 515.026	23 05 00.	+ 36 05		14.8	GALAXY
MCG+07-47-007	23 05 00.	+ 41 54	42	16.	GALAXY
ZWG 532.008	23 05 00.	+ 41 55		15.6	GALAXY
5ZW 396	23 05 00.	+ 41 56			COMPACT GALAXY
MCG+07-47-008	23 05 00.	+ 43 20	72	14.5	GALAXY
UGC 12377	23 05 00.	+ 86 30	66	16.0	GALAXY
MCG-01-59-001	23 05 00.	- 05 10	18	15.	GALAXY
ARC 2536	23 05 05.	- 22 42		17.5	RICH CLUSTER OF GALAXIES
ZWG 431.003	23 05 06.	+ 11 16		15.7	GALAXY
ZWG 430.066	23 05 06.	+ 11 16		15.7	GALAXY
UGC 12380	23 05 06.	+ 11 16	66	15.7	GALAXY Sc
ZWG 475.039	23 05 06.	+ 27 24		15.5	GALAXY
MCG+05-54-036	23 05 06.	+ 32 07	150	13.	GALAXY
5ZW 097	23 05 06.	+ 43 20			COMPACT GALAXY
ZWG 532.009	23 05 06.	+ 43 20		15.5	GALAXY
UGC 12381	23 05 06.	+ 43 20	108	15.5	GALAXY SBc
MCG+04-54-028	23 05 09.	+ 22 42	120	14.	GALAXY
MCG+00-59-001	23 05 12.	+ 00 40	18	15.	GALAXY
PHL 5471	23 05 12.	+ 02 30		18.0	BLUE STELLAR OBJECT
PHL 2076	23 05 12.	- 02 56		18.5	BLUE STELLAR OBJECT
PHL 5472	23 05 12.	- 10 12		18.4	BLUE STELLAR OBJECT
PHL 5470	23 05 12.	- 11 18		13.4	BLUE STELLAR OBJECT
PHL 2077	23 05 12.	- 16 12		18.2	BLUE STELLAR OBJECT
ZWG 380.001	23 05 18.	+ 00 40		14.9	GALAXY
UGC 12382	23 05 18.	+ 04 53	72	16.0	GALAXY Sc
ZWG 406.001	23 05 18.	+ 08 13		15.6	GALAXY
ZWG 475.040	23 05 18.	+ 22 26		15.7	GALAXY
UGC 12383	23 05 18.	+ 22 26	90	15.7	GALAXY SBb/Sb
ISS 0096	23 05 18.	+ 57 20	237		STELLAR RING
ISS 0023	23 05 18.	+ 65 04	142		STELLAR RING
PHL 2078	23 05 18.	- 12 19		18.3	BLUE STELLAR OBJECT
RNGC 7488	23 05 19.	+ 00 40		15.0	GALAXY
UGC 12384	23 05 24.	+ 24 41	60	16.0	GALAXY SBc
ZWG 496.045	23 05 24.	+ 30 03		15.5	GALAXY
UGC 12385	23 05 24.	+ 30 03	60	15.5	GALAXY SBc
MCG+05-54-037	23 05 24.	+ 30 03	48	15.	GALAXY
ZC 2305.4-0153	23 05 24.	- 01 53	1080		CLUSTER OF GALAXIES
PHL 2080	23 05 24.	- 06 03		18.6	BLUE STELLAR OBJECT
PHL 2079	23 05 24.	- 08 56		18.9	BLUE STELLAR OBJECT
PHL 5473	23 05 24.	- 09 16		17.8	BLUE STELLAR OBJECT
PHL 5476	23 05 24.	- 11 42		18.2	BLUE STELLAR OBJECT
PHL 5474	23 05 24.	- 18 40		18.6	BLUE STELLAR OBJECT
PHL 5475	23 05 24.	- 19 44		18.8	BLUE STELLAR OBJECT
SC 2302-6453.6	23 05 29.	- 64 37 24.	42		NEBULA
PHL 0405	23 05 30.	- 01 54		17.3	BLUE STELLAR OBJECT
RNGC 7491	23 05 30.	- 06 14		14.5	GALAXY
MCG-01-59-002	23 05 30.	- 06 14	48	14.5	GALAXY
MCG-02-59-001	23 05 30.	- 10 48	78	15.5	GALAXY
PHL 2081	23 05 30.	- 18 16		18.8	BLUE STELLAR OBJECT
MCG+05-54-038	23 05 33.	+ 28 17	30	15.	GALAXY
FATH 1.834	23 05 34.	+ 29 54			NEBULA
UGC 12386	23 05 36.	+ 24 42	72	16.0	GALAXY Sc
PHL 5481	23 05 36.	- 05 08		17.9	BLUE STELLAR OBJECT
PHL 5478	23 05 36.	- 07 08		18.6	BLUE STELLAR OBJECT
PHL 5477	23 05 36.	- 09 26		17.9	BLUE STELLAR OBJECT
PHL 5479	23 05 36.	- 11 09		18.3	BLUE STELLAR OBJECT
ARC 2537	23 05 41.	- 02 27		18.0	RICH CLUSTER OF GALAXIES
RNGC 7492	23 05 41.	- 15 54		12.5	GLOBULAR CLUSTER
ZWG 496.046	23 05 42.	+ 28 18		15.7	GALAXY
ISS 0097	23 05 42.	+ 57 32	146		STELLAR RING
ACK 113+06.1	23 05 42.	+ 66 43			PLANETARY NEBULA

OBJECT NAME	RIGHT ASCEN.	DECLINATION	DIAM.	MAGN.	TYPE OF OBJECT
PHL 5482	23 05 42.	- 11 04		17.7	BLUE STELLAR OBJECT
MCG-02-59-002	23 05 42.	- 12 29 30.	48	15.	GALAXY
PHL 5483	23 05 42.	- 15 52		18.3	BLUE STELLAR OBJECT
GCL 125	23 05 42.	- 15 54	258	12.33	GLOBULAR STAR CLUSTER
PHL 5484	23 05 48.	+ 01 14		17.7	BLUE STELLAR OBJECT
ZWG 431.004	23 05 48.	+ 12 47		15.4	GALAXY
MCG+03-59-001	23 05 48.	+ 17 49	24	15.5	GALAXY
4ZW 132	23 05 48.	+ 27 32			COMPACT GALAXY
ZC 2305.6+3935	23 05 48.	+ 39 35	1080		CLUSTER OF GALAXIES
PHL 2082	23 05 48.	- 11 40		17.9	BLUE STELLAR OBJECT
PHL 5485	23 05 48.	- 15 49		18.3	BLUE STELLAR OBJECT
OCL 0111	23 05 48.	- 15 54	240	16.	OPEN STAR CLUSTER
PHL 5486	23 05 48.	- 15 56		18.3	BLUE STELLAR OBJECT
ZWG 406.002	23 05 54.	+ 04 23		15.4	GALAXY
MCG+02-59-001	23 05 54.	+ 12 32 30.	108	15.	GALAXY
ZC 2305.9-0229	23 05 54.	- 02 29	1080		CLUSTER OF GALAXIES
PHL 5487	23 05 54.	- 09 08		16.5	BLUE STELLAR OBJECT
PHL 0406	23 05 54.	- 11 56		15.8	BLUE STELLAR OBJECT
ARC 2538	23 05 57.	- 20 09		16.5	RICH CLUSTER OF GALAXIES
LBN 0431	23 06	+ 25 10	1740		BRIGHT NEBULA
LBN 0476	23 06	+ 41 30	1800		BRIGHT NEBULA
PHL 5488	23 06 00.	+ 03 00		18.5	BLUE STELLAR OBJECT
UGC 12388	23 06 00.	+ 12 33	108	16.0	GALAXY
MCG+02-59-002	23 06 00.	+ 12 56	48	15.	GALAXY
ZWG 431.005	23 06 00.	+ 12 57		15.7	GALAXY
ZWG 370.010	23 06 00.	+ 86 29		15.6	GALAXY
ZWG 369.003	23 06 00.	+ 86 29		15.6	GALAXY
ZWG 360.002	23 06 00.	+ 86 29		15.6	GALAXY
UGC 12387	23 06 00.	+ 86 29	72	15.6	GALAXY SBa
MCG+14-01-002	23 06 00.	+ 86 30	45	17.	GALAXY
PHL 2083	23 06 00.	- 05 30		17.7	BLUE STELLAR OBJECT
PHL 5489	23 06 00.	- 15 47		18.2	BLUE STELLAR OBJECT
PHL 5490	23 06 00.	- 15 53		18.4	BLUE STELLAR OBJECT
RNGC 7493	23 06 01.	+ 00 38			NON-EXISTENT OBJECT
ARC 2539	23 06 04.	- 21 45		16.9	RICH CLUSTER OF GALAXIES
SC 2303-6158.8	23 06 05.	- 61 42 35.	36		NEBULA
4ZW 133	23 06 06.	+ 38 47			COMPACT GALAXY
5ZW 398	23 06 06.	+ 46 28			COMPACT GALAXY
UGC 12389	23 06 06.	+ 46 28	42	13.8	GALAXY PR CMPT
MCG-02-59-003	23 06 06.	- 12 55	42	14.5	GALAXY
TON-S 0085	23 06 06.	- 23 48		14.3	BLUE STAR
IC 1471	23 06 07.	- 12 54 50.			NONSTELLAR OBJECT
PHL 0407	23 06 12.	+ 03 31		17.5	BLUE STELLAR OBJECT
OCL 0254	23 06 12.	+ 60 15	420	12.	OPEN STAR CLUSTER
PHL 2084	23 06 12.	- 02 12		18.5	BLUE STELLAR OBJECT
PHL 5491	23 06 12.	- 16 52		17.4	BLUE STELLAR OBJECT
PHL 2085	23 06 12.	- 17 02		18.6	BLUE STELLAR OBJECT
PHL 5492	23 06 12.	- 18 26		17.9	BLUE STELLAR OBJECT
TON-S 0086	23 06 12.	- 27 26		15.8	BLUE STAR
RNGC 7494	23 06 16.	- 24 38		15.0	GALAXY
MRSL 089-41/1	23 06 18.	+ 14 39	2400		HII REGION
UGC 12390	23 06 18.	+ 23 07	66	16.0	GALAXY S0
ZC 2306.3+2543	23 06 18.	+ 25 43	1480		CLUSTER OF GALAXIES
PHL 2086	23 06 18.	- 08 46		17.6	BLUE STELLAR OBJECT
PHL 5493	23 06 18.	- 13 52		18.3	BLUE STELLAR OBJECT
MCG-04-54-007	23 06 18.	- 24 33 30.	24	15.5	GALAXY
MCG-05-54-016	23 06 18.	- 31 08	84	14.	GALAXY
SC 2302-7233.9	23 06 22.	- 72 17 41.	42		NEBULA
PHL 5497	23 06 24.	+ 00 21		17.8	BLUE STELLAR OBJECT
PHL 5496	23 06 24.	+ 01 38		18.1	BLUE STELLAR OBJECT
PHL 0408	23 06 24.	+ 02 28		16.8	BLUE STELLAR OBJECT
MCG+02-59-003	23 06 24.	+ 11 46	120	13.	GALAXY
PHL 5495	23 06 24.	- 03 34		17.5	BLUE STELLAR OBJECT
PHL 5494	23 06 24.	- 05 51		16.9	BLUE STELLAR OBJECT
PHL 5498	23 06 24.	- 14 32		17.2	BLUE STELLAR OBJECT
SC 2303-6232.3	23 06 24.	- 62 16 05.	24		NEBULA
SC 2303-7232.3	23 06 26.	- 72 16 05.	24		NEBULA
SN 1973N	23 06 29.	+ 11 48		15.5	SUPERNOVA
ZWG 431.006	23 06 30.	+ 11 48		14.7	GALAXY
UGC 12391	23 06 30.	+ 11 48	120	14.7	GALAXY Sc
ZWG 454.001	23 06 30.	+ 19 47		15.7	GALAXY
KARA.73B 1007	23 06 30.	+ 19 47	24	15.7	ISOLATED GALAXY S
MCG+04-54-029	23 06 30.	+ 23 05	12	15.5	GALAXY
RNGC 7495	23 06 33.	+ 11 48		14.5	GALAXY
RNGC 7497	23 06 34.	+ 17 54		13.5	GALAXY
MCG+03-59-002	23 06 36.	+ 17 53	300	12.	GALAXY
MCG+03-59-003	23 06 36.	+ 18 16	30	17.	GALAXY
MCG+00-59-002	23 06 36.	- 01 14	24	16.	GALAXY
PHL 2087	23 06 36.	- 02 12		18.6	BLUE STELLAR OBJECT
PHL 5499	23 06 36.	- 18 52		18.2	BLUE STELLAR OBJECT
IC 1472	23 06 38.	+ 16 58			NONSTELLAR OBJECT
HN 0778	23 06 38.	- 68 32			NEBULA
IC 5266	23 06 38.	- 68 32			NONSTELLAR OBJECT
MCG+01-59-001	23 06 42.	+ 07 16	24	16.0	GALAXY
ZC 2306.7+1215	23 06 42.	+ 12 15	810		CLUSTER OF GALAXIES
ZWG 454.002	23 06 42.	+ 16 58		15.5	GALAXY
MCG+03-59-004	23 06 42.	+ 16 58	54	15.	GALAXY
ZWG 454.003	23 06 42.	+ 17 54		13.3	GALAXY
UGC 12342	23 06 42.	+ 17 54	270	13.3	GALAXY Sc
ZC 2306.7+3813	23 06 42.	+ 38 13	340		CLUSTER OF GALAXIES
PHL 5500	23 06 42.	- 00 20		17.0	BLUE STELLAR OBJECT
ZC 2306.7-0055	23 06 42.	- 00 55	1810		CLUSTER OF GALAXIES
ZWG 380.002	23 06 42.	- 01 12		15.7	GALAXY
PHL 0409	23 06 42.	- 06 32		18.0	BLUE STELLAR OBJECT
PHL 2088	23 06 42.	- 19 10		17.0	BLUE STELLAR OBJECT
MCG+00-59-003	23 06 45.	+ 00 29	54	14.5	GALAXY
IC 5287	23 06 46.	+ 00 29 54.			NONSTELLAR OBJECT
ARC 2540	23 06 47.	- 22 26		17.1	RICH CLUSTER OF GALAXIES
ZWG 380.003	23 06 48.	+ 00 29		15.2	GALAXY
UGC 12393	23 06 48.	+ 00 29	66	15.2	GALAXY SBb
KARA.73B 1008	23 06 48.	+ 00 29	66	15.2	ISOLATED GALAXY S
ZC 2306.8+1626	23 06 48.	+ 16 26	1480		CLUSTER OF GALAXIES
PHL 0410	23 06 48.	- 00 48		17.9	BLUE STELLAR OBJECT
PHL 5503	23 06 48.	- 12 53		18.1	BLUE STELLAR OBJECT
PHL 2089	23 06 48.	- 13 56		18.6	BLUE STELLAR OBJECT
PHL 5501	23 06 48.	- 14 12		16.9	BLUE STELLAR OBJECT
PHL 5502	23 06 48.	- 15 20		16.6	BLUE STELLAR OBJECT
PHL 2090	23 06 48.	- 16 28		18.2	BLUE STELLAR OBJECT
MCG-05-54-017	23 06 48.	- 29 04	24	15.	GALAXY
PHL 5505	23 06 54.	+ 01 10		18.6	BLUE STELLAR OBJECT
PHL 5504	23 06 54.	- 13 56		17.7	BLUE STELLAR OBJECT
TON-S 0087	23 06 54.	- 22 58		16.0	BLUE STAR
LB 01520	23 06 54.	- 48 02		14.8	FAINT BLUE STAR
LB 09949	23 07	- 82 02		14.2	FAINT BLUE STAR
ZC 2307.0+0044	23 07 00.	+ 00 44	1010		CLUSTER OF GALAXIES
MCG+05-54-039	23 07 00.	+ 29 12	54	15.	GALAXY
MCG+14-01-003	23 07 00.	+ 86 29	36	16.	GALAXY
RNGC 7496	23 07 00.	- 43 42		12.0	GALAXY
MCG-07-47-020	23 07 00.	- 43 42	144	12.1	GALAXY
SG 2.095	23 07 01.	+ 64 28	3000		DIFFUSE EMISSION NEBULA
ZWG 406.003	23 07 06.	+ 05 24		15.6	GALAXY
ZWG 496.047	23 07 06.	+ 29 13		15.7	GALAXY
ZWG 496.048	23 07 06.	+ 32 24		14.8	GALAXY
UGC 12394	23 07 06.	+ 32 24	60	14.8	GALAXY SBa-b
PHL 5506	23 07 06.	- 11 00		18.5	BLUE STELLAR OBJECT
PHL 5507	23 07 06.	- 17 42		18.2	BLUE STELLAR OBJECT
TON-S 0088	23 07 06.	- 32 28		15.7	BLUE STAR
PK108-05.1	23 07 07.4	+ 54 28 36.			PLANETARY NEBULA
RNGC 7498	23 07 10.	- 24 41		15.0	GALAXY
ZWG 431.007	23 07 12.	+ 15 24		15.7	GALAXY
UGC 12395	23 07 12.	+ 15 24	60	15.7	GALAXY Sc-IRR
MCG+05-54-040	23 07 12.	+ 32 24 30.	54	14.5	GALAXY
PHL 2091	23 07 12.	- 12 31		18.9	BLUE STELLAR OBJECT
MCG-04-54-008	23 07 15.	- 24 41 30.	48	15.	GALAXY
PHL 0411	23 07 18.	+ 00 56		16.9	BLUE STELLAR OBJECT
3ZW 094	23 07 18.	+ 08 14			COMPACT GALAXY
PHL 5508	23 07 18.	- 14 19		16.3	BLUE STELLAR OBJECT
MCG-04-54-009	23 07 18.	- 25 45	36	15.5	GALAXY
ARC 2541	23 07 23.	- 23 14		17.1	RICH CLUSTER OF GALAXIES
ZWG 406.004	23 07 24.	+ 06 31		15.2	GALAXY
MCG+01-59-002	23 07 24.	+ 07 14	36	15.	GALAXY
ZWG 406.005	23 07 24.	+ 07 15		15.3	GALAXY
PHL 2092	23 07 24.	- 14 52		18.6	BLUE STELLAR OBJECT
ARC 2543	23 07 24.	- 15 11		17.2	RICH CLUSTER OF GALAXIES
PHL 5509	23 07 24.	- 18 38		18.3	BLUE STELLAR OBJECT
ARC 2542	23 07 24.	- 24 42		17.1	RICH CLUSTER OF GALAXIES
SC 2304-7325.7	23 07 26.	- 73 09 28.	18		NEBULA
MCG-04-54-010	23 07 27.	- 25 46	15	15.	GALAXY
ZC 2307.5+1703	23 07 30.	+ 17 03	1550		CLUSTER OF GALAXIES
TON-S 0089	23 07 30.	- 33 00		14.8	BLUE STAR
MCG+03-59-005	23 07 33.	+ 18 10	78	17.	GALAXY
ZC 2307.6+0713	23 07 36.	+ 07 13	4100		CLUSTER OF GALAXIES
MCG+01-59-003	23 07 36.	+ 07 17 30.	36	15.5	GALAXY
MCG+07-47-009	23 07 36.	+ 42 16	42	15.	GALAXY
PHL 2093	23 07 36.	- 01 26		18.8	BLUE STELLAR OBJECT
PHL 5513	23 07 36.	- 08 36		18.4	BLUE STELLAR OBJECT
PHL 5511	23 07 36.	- 10 14		17.6	BLUE STELLAR OBJECT
PHL 5510	23 07 36.	- 12 22		16.0	BLUE STELLAR OBJECT
PHL 5514	23 07 36.	- 17 49		18.6	BLUE STELLAR OBJECT
PHL 5512	23 07 36.	- 18 06		17.1	BLUE STELLAR OBJECT
ARC 2545	23 07 37.	+ 05 08		17.6	RICH CLUSTER OF GALAXIES
SC 2304-6313.8	23 07 37.	- 62 57 33.			NEBULA
MCG+05-54-041	23 07 39.	+ 29 56	27	15.	GALAXY
SC 2304-6313.9	23 07 39.	- 62 57 39.	9		NEBULA
ARC 2544	23 07 40.	- 11 05		17.2	RICH CLUSTER OF GALAXIES
MCG+01-59-004	23 07 42.	+ 07 05	24	16.5	GALAXY
ZWG 496.049	23 07 42.	+ 29 58		14.6	GALAXY
ZWG 532.010	23 07 42.	+ 42 18		15.5	GALAXY
UGC 12396	23 07 42.	+ 42 18	72	15.5	GALAXY COMPACT
PHL 5516	23 07 42.	- 11 23		17.5	BLUE STELLAR OBJECT
PHL 5517	23 07 42.	- 11 59		17.7	BLUE STELLAR OBJECT
PHL 5515	23 07 42.	- 19 20		17.6	BLUE STELLAR OBJECT
MCG+05-54-042	23 07 45.	+ 30 18	30	16.	GALAXY
MCG+00-59-004	23 07 48.	+ 02 22	60	15.5	GALAXY
ZWG 406.006	23 07 48.	+ 07 05		15.7	GALAXY
ZWG 406.007	23 07 48.	+ 07 18		15.0	GALAXY
UGC 12397	23 07 48.	+ 07 18	96	15.0	GALAXY S0
ZC 2307.8+2327	23 07 48.	+ 23 27	1340		CLUSTER OF GALAXIES
ZWG 475.041	23 07 48.	+ 26 21		15.1	GALAXY
4ZW 134	23 07 48.	+ 29 39			COMPACT GALAXY
ZWG 496.050	23 07 48.	+ 29 39		14.8	GALAXY
ZWG 496.051	23 07 48.	+ 30 18		15.7	GALAXY
PHL 0412	23 07 48.	- 03 40		15.9	BLUE STELLAR OBJECT
MCG-02-59-004	23 07 48.	- 08 57 30.	108	14.5	GALAXY
MCG-02-59-005	23 07 48.	- 08 58	12	15.	GALAXY
PHL 5518	23 07 48.	- 19 56		17.9	BLUE STELLAR OBJECT
MCG-05-54-018	23 07 48.	- 30 00	96	14.	GALAXY
RNGC 7499	23 07 50.	+ 07 19		15.0	GALAXY
ZWG 380.004	23 07 54.	+ 02 22		15.7	GALAXY
UGC 12398	23 07 54.	+ 02 22	66	15.7	GALAXY S
KARA.68 238	23 07 54.	+ 06 53	34		DWARF GALAXY
MCG+01-59-006	23 07 54.	+ 07 07	24	16.0	GALAXY
MCG+01-59-005	23 07 54.	+ 07 18	60	14.5	GALAXY
ZC 2307.9+1111	23 07 54.	+ 11 11	1340		CLUSTER OF GALAXIES
LBN 0427	23 08	+ 23 20	840		BRIGHT NEBULA
KHAV 784	23 08	+ 46 22	1000		DARK NEBULA
ZWG 406.008	23 08 00.	+ 07 18		15.3	GALAXY
MCG+01-59-007	23 08 00.	+ 07 19	24	14.5	GALAXY
ZWG 406.009	23 08 00.	+ 07 52		15.5	GALAXY
MCG+02-59-004	23 08 00.	+ 10 44	18	14.	GALAXY
ZWG 431.008	23 08 00.	+ 10 45		14.9	GALAXY
UGC 12399	23 08 00.	+ 10 45	132	14.9	GALAXY S0
LDN 1226	23 08 00.	+ 62 00	360		DARK NEBULA
LDN 1227	23 08 00.	+ 62 03	360		DARK NEBULA
PHL 0413	23 08 00.	- 01 18		18.3	BLUE STELLAR OBJECT
PHL 2095	23 08 00.	- 02 54		16.7	BLUE STELLAR OBJECT
PHL 5519	23 08 00.	- 12 52		16.8	BLUE STELLAR OBJECT
PHL 2094	23 08 00.	- 20 21		18.1	BLUE STELLAR OBJECT
RNGC 7501	23 08 02.	+ 07 19		15.5	GALAXY
RNGC 7500	23 08 03.	+ 10 45		15.0	GALAXY
RNGC 7504	23 08 03.	+ 14 08			NON-EXISTENT OBJECT
ARC 2546	23 08 05.	- 22 56		17.1	RICH CLUSTER OF GALAXIES
ZWG 406.010	23 08 06.	+ 07 55		15.6	GALAXY
ZWG 406.011	23 08 06.	+ 08 55		15.6	GALAXY
ZWG 454.004	23 08 06.	+ 21 26		15.5	GALAXY
UGC 12400	23 08 06.	+ 21 26	66	15.5	GALAXY SBa
MCG+03-59-006	23 08 06.	+ 21 26	72	15.	GALAXY
UGC 12401	23 08 06.	+ 34 08	60	17.	GALAXY DWRF IR
PHL 2096	23 08 06.	- 05 47		18.1	BLUE STELLAR OBJECT
PHL 5520	23 08 06.	- 06 14		7.0	BLUE STELLAR OBJECT
ARC 2547	23 08 10.	- 21 24		16.9	RICH CLUSTER OF GALAXIES
RNGC 7502	23 08 10.	- 22 02			NON-EXISTENT OBJECT
PHL 2097	23 08 12.	+ 00 02		17.8	BLUE STELLAR OBJECT
ZC 2308.2+0104	23 08 12.	+ 01 04	1210		CLUSTER OF GALAXIES
ZWG 406.012	23 08 12.	+ 07 17		14.9	GALAXY
UGC 12402	23 08 12.	+ 48 21	60	17.	GALAXY
PHL 5524	23 08 12.	- 05 17		17.2	BLUE STELLAR OBJECT
PHL 5525	23 08 12.	- 14 39		17.1	BLUE STELLAR OBJECT
PHL 5522	23 08 12.	- 14 49		18.5	BLUE STELLAR OBJECT
PHL 5521	23 08 12.	- 15 00		17.8	BLUE STELLAR OBJECT
PHL 5523	23 08 12.	- 16 48		18.9	BLUE STELLAR OBJECT
SC 2305-6314.7	23 08 12.	- 62 58 27.			NEBULA
SC 2305-6314.8	23 08 12.	- 62 58 33.	9		NEBULA
RNGC 7503	23 08 14.	+ 07 18		15.0	GALAXY
IC 5289	23 08 14.	- 32 48 58.			NONSTELLAR OBJECT
MCG+01-59-008	23 08 15.	+ 07 17 30.	36	14.5	GALAXY

OBJECT NAME	RIGHT ASCEN.	DECLINATION	DIAM.	MAGN.	TYPE OF OBJECT
PHL 0414	23 08 18.	+ 02 37		18.8	BLUE STELLAR OBJECT
ZWG 406.013	23 08 18.	+ 08 52		14.8	GALAXY
KARA.72 577A	23 08 18.	+ 08 52	30	14.8	PART OF DOUBLE GALAXY
KARA.72 577B	23 08 18.	+ 08 53	30		PART OF DOUBLE GALAXY
5ZW 399	23 08 18.	+ 40 17			COMPACT GALAXY
PHL 0415	23 08 18.	- 13 22		17.1	BLUE STELLAR OBJECT
PHL 5526	23 08 18.	- 16 24		17.6	BLUE STELLAR OBJECT
MCG-05-54-019	23 08 18.	- 28 22	48	15.	GALAXY
PHL 2098	23 08 24.	+ 01 50		18.4	BLUE STELLAR OBJECT
PHL 5528	23 08 24.	- 08 46		17.5	BLUE STELLAR OBJECT
PHL 5527	23 08 24.	- 09 20		18.2	BLUE STELLAR OBJECT
PHL 5529	23 08 24.	- 15 08		17.9	BLUE STELLAR OBJECT
PHL 2099	23 08 24.	- 15 42		18.9	BLUE STELLAR OBJECT
PHL 2100	23 08 24.	- 20 40		18.0	BLUE STELLAR OBJECT
MCG-07-47-021	23 08 24.	- 38 33	42	15.	GALAXY
AGU 77	23 08 24.	- 38 33 18.		16.	IRREGULAR DWARF GALAXY
ZC 2308.5+3400	23 08 30.	+ 34 00	1080		CLUSTER OF GALAXIES
MCG-01-59-003	23 08 30.	- 04 46	30	15.	GALAXY
PHL 0416	23 08 30.	- 12 25		17.9	BLUE STELLAR OBJECT
PHL 2101	23 08 30.	- 13 14		18.4	BLUE STELLAR OBJECT
SC 2305-6401.5	23 08 30.	- 63 45 15.	36		NEBULA
HN 0779	23 08 31.	- 68 22			NEBULA
IC 5288	23 08 31.	- 68 22			NONSTELLAR OBJECT
ZWG 406.014	23 08 36.	+ 09 17		15.6	GALAXY
ZC 2308.6+0944	23 08 36.	+ 09 44	1410		CLUSTER OF GALAXIES
UGC 12403	23 08 36.	+ 12 30	66	16.0	GALAXY SB:c
ZWG 431.009	23 08 36.	+ 13 21		15.6	GALAXY
MCG+03-59-007	23 08 36.	+ 16 03	27	16.	GALAXY
MCG+05-54-043	23 08 36.	+ 29 21	90	14.	GALAXY
ZWG 380.005	23 08 36.	- 00 25		15.4	GALAXY
PHL 0417	23 08 36.	- 01 54		16.9	BLUE STELLAR OBJECT
PHL 2102	23 08 36.	- 07 06		18.4	BLUE STELLAR OBJECT
PHL 5480	23 08 36.	- 13 17		18.5	BLUE STELLAR OBJECT
PHL 2103	23 08 36.	- 19 02		18.8	BLUE STELLAR OBJECT
MCG-05-54-021	23 08 36.	- 32 45	24	16.5	GALAXY
MCG-05-54-020	23 08 36.	- 32 45	36	15.	GALAXY
SC 2305-7237.4	23 08 36.	- 72 21 09.	12		NEBULA
RNGC 7505	23 08 39.	+ 13 21		15.5	GALAXY
ARC 2548	23 08 39.	- 20 42		16.9	RICH CLUSTER OF GALAXIES
ARC 2549	23 08 41.	- 13 05		17.0	RICH CLUSTER OF GALAXIES
ZWG 496.052	23 08 42.	+ 29 22		14.2	GALAXY
UGC 12404	23 08 42.	+ 29 22	126	14.2	GALAXY S0
PHL 5530	23 08 42.	- 11 06		18.1	BLUE STELLAR OBJECT
IC 1473	23 08 47.	+ 29 21 15.			NONSTELLAR OBJECT
BC 4C9.72	23 08 47.2	+ 09 51 56.		15.	QUASI-STELLAR OBJECT
SHB 386	23 08 47.2	+ 09 51 56.		15.	QUASI-STELLAR OBJECT
ZWG 406.015	23 08 48.	+ 04 42		15.2	GALAXY
MCG+01-59-009	23 08 48.	+ 04 42	36	15.	GALAXY
UGC 12405	23 08 48.	+ 05 58	60	16.0	GALAXY S
ARC 2551	23 08 48.	+ 07 37		17.5	RICH CLUSTER OF GALAXIES
ZC 2308.8+1240	23 08 48.	+ 12 40	1080		CLUSTER OF GALAXIES
PHL 0418	23 08 48.	- 09 04		15.3	BLUE STELLAR OBJECT
PHL 2105	23 08 48.	- 12 07		18.2	BLUE STELLAR OBJECT
PHL 2104	23 08 48.	- 13 26		17.9	BLUE STELLAR OBJECT
MCG+03-59-008	23 08 51.	+ 17 36	30	16.5	GALAXY
ARC 2550	23 08 52.	- 22 01		16.9	RICH CLUSTER OF GALAXIES
SC 2305-6152.5	23 08 53.	- 61 36 14.			NEBULA
PHL 5531	23 08 54.	- 10 58		18.3	BLUE STELLAR OBJECT
PHL 2106	23 08 54.	- 16 43		18.7	BLUE STELLAR OBJECT
MCG-04-54-011	23 08 54.	- 24 37 30.	54	14.5	GALAXY
ARC 2552	23 08 56.	+ 03 20		18.0	RICH CLUSTER OF GALAXIES
SC 2305-6158.9	23 08 59.	- 61 42 38.	6		NEBULA
ZC 2309.0+0004	23 09 00.	+ 00 04	1280		CLUSTER OF GALAXIES
PHL 2107	23 09 00.	+ 02 48		17.1	BLUE STELLAR OBJECT
ZWG 406.016	23 09 00.	+ 07 27		15.6	GALAXY
ZWG 547.001	23 09 00.	+ 43 11		15.7	GALAXY
LDN 1240	23 09 00.	+ 66 10	1440		DARK NEBULA
ZC 2309.0+8317	23 09 00.	+ 83 17	1340		CLUSTER OF GALAXIES
PHL 5532	23 09 00.	- 08 30		16.1	BLUE STELLAR OBJECT
PHL 5533	23 09 00.	- 15 27		18.1	BLUE STELLAR OBJECT
MCG-03-59-001	23 09 00.	- 15 43	78	15.	GALAXY
PHL 2108	23 09 00.	- 19 31		18.7	BLUE STELLAR OBJECT
TON-S 0090	23 09 00.	- 27 28		13.7	BLUE STAR
PHL 0419	23 09 06.	+ 01 18		18.4	BLUE STELLAR OBJECT
MCG+00-59-005	23 09 06.	- 02 27	90	13.5	GALAXY
PHL 2110	23 09 06.	- 04 30		18.3	BLUE STELLAR OBJECT
PHL 5534	23 09 06.	- 08 24		18.4	BLUE STELLAR OBJECT
PHL 2109	23 09 06.	- 18 45		18.9	BLUE STELLAR OBJECT
SG 1.26	23 09 08.	+ 60	36000		DIFFUSE EMISSION NEBULA
MCG+03-59-009	23 09 09.	+ 18 30	36	15.	GALAXY
PHL 5535	23 09 12.	+ 02 40		18.2	BLUE STELLAR OBJECT
ZC 2309.2+0324	23 09 12.	+ 03 24	470		CLUSTER OF GALAXIES
ZWG 431.010	23 09 12.	+ 14 50		15.6	GALAXY
ZWG 454.005	23 09 12.	+ 18 30		15.7	GALAXY
PHL 0420	23 09 12.	- 01 58		16.6	BLUE STELLAR OBJECT
ZWG 380.006	23 09 12.	- 02 26		14.3	GALAXY
UGC 12406	23 09 12.	- 02 26	126	14.3	GALAXY S0
PHL 2111	23 09 12.	- 04 24		18.5	BLUE STELLAR OBJECT
RNGC 7506	23 09 13.	- 02 26		14.5	GALAXY
MCG+02-59-005	23 09 15.	+ 12 40	54	16.	GALAXY
ZWG 406.017	23 09 18.	+ 09 14		14.2	GALAXY
UGC 12407	23 09 18.	+ 09 14	72	14.2	GALAXY S
ZC 2309.3+1118	23 09 18.	+ 11 18	400		CLUSTER OF GALAXIES
ZWG 431.011	23 09 18.	+ 12 42		15.7	GALAXY
UGC 12408	23 09 18.	+ 12 42	66	15.7	GALAXY
PHL 5536	23 09 18.	- 11 42		16.3	BLUE STELLAR OBJECT
PHL 0421	23 09 18.	- 12 44		18.0	BLUE STELLAR OBJECT
PHL 2112	23 09 18.	- 14 31		18.3	BLUE STELLAR OBJECT
PHL 5537	23 09 18.	- 20 02		18.1	BLUE STELLAR OBJECT
RNGC 7508	23 09 21.	+ 12 42		15.5	GALAXY
SC 2306-6149.7	23 09 23.	- 61 33 26.	6		NEBULA
MCG+01-59-010	23 09 24.	+ 09 15	72	15.	GALAXY
UGC 12409	23 09 24.	+ 23 47	66	16.0	GALAXY Sa
OCL 0256	23 09 24.	+ 60 18	1020	9.6	OPEN STAR CLUSTER
PHL 2115	23 09 24.	- 05 19		15.0	BLUE STELLAR OBJECT
PHL 5538	23 09 24.	- 06 36		17.8	BLUE STELLAR OBJECT
PHL 2114	23 09 24.	- 07 58		18.5	BLUE STELLAR OBJECT
PHL 2116	23 09 24.	- 10 19		18.4	BLUE STELLAR OBJECT
PHL 5539	23 09 24.	- 11 12		18.7	BLUE STELLAR OBJECT
PHL 2113	23 09 24.	- 16 02		17.2	BLUE STELLAR OBJECT
PHL 2117	23 09 24.	- 17 51		18.9	BLUE STELLAR OBJECT
MCG-04-54-012	23 09 24.	- 23 50	18	15.5	GALAXY
RNGC 7510	23 09 26.	+ 60 18		9.5	OPEN CLUSTER
RNGC 7507	23 09 27.	- 28 49		12.5	GALAXY
ZC 2309.5+1527	23 09 30.	+ 15 27	1340		CLUSTER OF GALAXIES
MCG+05-54-044	23 09 30.	+ 30 44 30.	120	15.	GALAXY
UGC 12410	23 09 30.	+ 30 45	108	16.0	GALAXY Sc
MCG+00-59-006	23 09 30.	- 01 51	36	15.	GALAXY
PHL 2118	23 09 30.	- 09 44		18.5	BLUE STELLAR OBJECT
MCG-04-54-013	23 09 30.	- 23 53	24	15.5	GALAXY
MCG-05-54-022	23 09 30.	- 28 49	90	11.	GALAXY
ZWG 380.008	23 09 36.	+ 03 22		15.4	GALAXY
MCG+00-59-007	23 09 36.	+ 03 23	48	14.	GALAXY
ZWG 380.007	23 09 36.	- 01 50		15.6	GALAXY
PHL 2119	23 09 36.	- 12 32		18.4	BLUE STELLAR OBJECT
PHL 2120	23 09 36.	- 12 46		18.1	BLUE STELLAR OBJECT
DV.56 N7496A	23 09 36.	- 43 03			SB(s):c GALAXY
RNGC 7496A	23 09 36.	- 43 03			GALAXY
MCG-07-47-022	23 09 36.	- 44 08		16.5	GALAXY
PHL 5540	23 09 42.	- 08 44		17.7	BLUE STELLAR OBJECT
PHL 2121	23 09 42.	- 19 46		17.7	BLUE STELLAR OBJECT
ARC 2553	23 09 42.	- 25 13		17.3	RICH CLUSTER OF GALAXIES
SER 159.04	23 09 42.	- 44 05	168	14.	LOW SURF. BRGHTNSS GALAXY
SC 2306-6200.6	23 09 42.	- 61 44 19.	9		NEBULA
MCG+02-59-006	23 09 45.	+ 14 20	60	14.	GALAXY
ARC 2554	23 09 46.	- 21 45		16.9	RICH CLUSTER OF GALAXIES
PHL 5542	23 09 48.	+ 00 17		17.8	BLUE STELLAR OBJECT
MCG+02-59-007	23 09 48.	+ 12 27	60	14.5	GALAXY
MCG+05-54-045	23 09 48.	+ 28 26	60	15.	GALAXY
UGC 12411	23 09 48.	+ 48 32	138	16.5	GALAXY
PHL 5541	23 09 48.	- 01 16		15.8	BLUE STELLAR OBJECT
ZWG 380.009	23 09 48.	- 01 48		15.2	GALAXY
PHL 2122	23 09 48.	- 17 04		18.6	BLUE STELLAR OBJECT
RNGC 7509	23 09 51.	+ 14 21		15.0	GALAXY
RNGC 7512	23 09 53.	+ 30 52		14.0	GALAXY
ZWG 431.012	23 09 54.	+ 13 27		15.1	GALAXY
UGC 12412	23 09 54.	+ 13 27	66	15.1	GALAXY S?
ZWG 431.013	23 09 54.	+ 14 21		15.0	GALAXY
ZWG 496.053	23 09 54.	+ 28 27		15.7	GALAXY
UGC 12413	23 09 54.	+ 28 27	66	15.7	GALAXY SB0?
4ZW 135	23 09 54.	+ 30 51			COMPACT GALAXY
MCG+05-54-046	23 09 54.	+ 30 51	27	13.5	GALAXY
ZWG 496.054	23 09 54.	+ 30 52		14.1	GALAXY
UGC 12414	23 09 54.	+ 30 52	126	14.1	GALAXY P
MCG+06-50-026	23 09 54.	+ 34 36	48	13.5	GALAXY
PHL 2124	23 09 54.	- 08 57		18.9	BLUE STELLAR OBJECT
PHL 2123	23 09 54.	- 15 54		18.8	BLUE STELLAR OBJECT
RNGC 7511	23 09 57.	+ 13 27		15.0	GALAXY
SC 2306-6158.4	23 09 59.	- 61 42 07.	18		NEBULA
LBN 0419	23 10	+ 19 00	10800		BRIGHT NEBULA
KHAV 785	23 10	+ 50 10	2110		DARK NEBULA
LBN 0551	23 10	+ 64 20	3420		BRIGHT NEBULA
PHL 2125	23 10 00.	+ 00 38		17.1	BLUE STELLAR OBJECT
ZC 2310.0+0215	23 10 00.	+ 02 15	3160		CLUSTER OF GALAXIES
ZWG 496.055	23 10 00.	+ 30 52		15.7	GALAXY
UGC 12415	23 10 00.	+ 34 27	96	13.5	GALAXY S
ZWG 515.027	23 10 00.	+ 34 37		13.5	GALAXY
RNGC 7514	23 10 00.	+ 34 37		13.5	GALAXY
KARA.73B 1009	23 10 00.	+ 34 37	78	13.5	ISOLATED GALAXY S
4ZW 136	23 10 00.	+ 35 01			COMPACT GALAXY
LDN 1225	23 10 00.	+ 61 20	720		DARK NEBULA
COU 101	23 10 00.	+ 64 20	6000		EMISSION NEBULA
PHL 2129	23 10 00.	- 02 09		18.5	BLUE STELLAR OBJECT
PHL 2126	23 10 00.	- 06 28		18.1	BLUE STELLAR OBJECT
PHL 0422	23 10 00.	- 08 56		18.4	BLUE STELLAR OBJECT
PHL 2127	23 10 00.	- 09 14		18.3	BLUE STELLAR OBJECT
PHL 2128	23 10 00.	- 17 59		18.8	BLUE STELLAR OBJECT
SC 2306-6158.5	23 10 00.	- 61 42 13.			NEBULA
ARC 2555	23 10 04.	- 22 29		16.9	RICH CLUSTER OF GALAXIES
ZC 2310.1+1602	23 10 06.	+ 16 02	2420		CLUSTER OF GALAXIES
PHL 2130	23 10 06.	- 04 34		18.5	BLUE STELLAR OBJECT
PHL 5543	23 10 06.	- 18 38		18.2	BLUE STELLAR OBJECT
SC 2307-6123.1	23 10 10.	- 61 06 49.			NEBULA
ZWG 431.014	23 10 12.	+ 10 28		15.4	GALAXY
MRK 526	23 10 12.	+ 10 28	24	15.	GALAXY WITH UV CONTINUUM
UGC 12416	23 10 12.	+ 10 28	66	15.4	GALAXY PECULR
ZWG 496.056	23 10 12.	+ 33 26		15.3	GALAXY
ACK 112+03.1	23 10 12.	+ 64 23			PLANETARY NEBULA
PHL 2131	23 10 12.	- 03 56		18.5	BLUE STELLAR OBJECT
PHL 5544	23 10 12.	- 17 08		18.2	BLUE STELLAR OBJECT
MCG-04-54-014	23 10 12.	- 23 45 30.	60	14.	GALAXY
IC 5290	23 10 14.	- 23 44 30.			NONSTELLAR OBJECT
MCG+02-59-008	23 10 15.	+ 12 23	90	13.	GALAXY
ARC 2558	23 10 17.	+ 10 03		17.1	RICH CLUSTER OF GALAXIES
PHL 2132	23 10 18.	+ 00 01		18.5	BLUE STELLAR OBJECT
ZWG 406.018	23 10 18.	+ 05 31		14.9	GALAXY
UGC 12417	23 10 18.	+ 05 31	66	14.9	GALAXY Sc
ZWG 406.019	23 10 18.	+ 08 53		15.5	GALAXY
ZWG 431.015	23 10 18.	+ 12 25		14.0	GALAXY
UGC 12418	23 10 18.	+ 12 25	102	14.0	GALAXY S
3ZW 095	23 10 18.	+ 15 38			COMPACT GALAXY
UGC 12419	23 10 18.	+ 15 38	12	14.0	GALAXY COMPACT
ZWG 475.042	23 10 18.	+ 25 56		15.7	GALAXY
LDN 1229	23 10 18.	+ 65 48	240		DARK NEBULA
PHL 2133	23 10 18.	- 19 57		18.2	BLUE STELLAR OBJECT
IC 1474	23 10 19.	+ 05 32 00.			NONSTELLAR OBJECT
RNGC 7515	23 10 21.	+ 12 25		14.0	GALAXY
RNGC 7516	23 10 22.	+ 19 58		14.5	GALAXY
ARC 2556	23 10 22.	- 21 54		16.9	RICH CLUSTER OF GALAXIES
MCG+01-59-011	23 10 24.	+ 05 31	54	14.	GALAXY
ZWG 454.006	23 10 24.	+ 19 58		14.6	GALAXY
UGC 12420	23 10 24.	+ 19 58	66	14.6	GALAXY S0
MCG+03-59-010	23 10 24.	+ 19 58	54	14.	GALAXY
MCG+05-54-047	23 10 24.	+ 33 26	48	15.5	GALAXY
ZWG 359.003	23 10 24.	+ 77 19		15.4	GALAXY
ZWG 344.001	23 10 24.	+ 77 19		15.4	GALAXY
UGC 12421	23 10 24.	+ 77 19	84	15.4	GALAXY Sc
PHL 5545	23 10 24.	- 08 26		17.9	BLUE STELLAR OBJECT
PHL 2134	23 10 24.	- 08 57		18.5	BLUE STELLAR OBJECT
PHL 5546	23 10 24.	- 10 22		18.6	BLUE STELLAR OBJECT
PHL 2135	23 10 24.	- 11 00		18.7	BLUE STELLAR OBJECT
ARC 2557	23 10 25.	- 17 15		17.2	RICH CLUSTER OF GALAXIES
RNGC 7513	23 10 27.	- 28 38		12.0	GALAXY
ARC 2559	23 10 29.	- 13 58		17.0	RICH CLUSTER OF GALAXIES
ZC 2310.5+1002	23 10 30.	+ 10 02	1880		CLUSTER OF GALAXIES
PHL 2136	23 10 30.	- 09 43		18.4	BLUE STELLAR OBJECT
SC 2307-6501.7	23 10 33.	- 64 45 25.			NEBULA
ZWG 406.020	23 10 36.	+ 06 03		14.5	GALAXY
MRK 527	23 10 36.	+ 06 03	12	16.	GALAXY WITH UV CONTINUUM
UGC 12422	23 10 36.	+ 06 03	90	14.5	GALAXY Sa
ZWG 406.021	23 10 36.	+ 06 08		14.8	GALAXY
UGC 12423	23 10 36.	+ 06 08	216	14.8	GALAXY Sc
PHL 5547	23 10 36.	- 01 00		16.5	BLUE STELLAR OBJECT
MCG+00-59-008	23 10 36.	- 02 23	24	15.	GALAXY
PHL 5549	23 10 36.	- 13 18		18.3	BLUE STELLAR OBJECT
ARC 2560	23 10 36.	- 16 15		17.5	RICH CLUSTER OF GALAXIES

OBJECT NAME	RIGHT ASCEN.	DECLINATION	DIAM.	MAGN.	TYPE OF OBJECT
TON-S 0092	23 10 36.	- 25 27		16.1	BLUE STAR
MCG-05-54-023	23 10 36.	- 28 37	216	12.	GALAXY
TON-S 0091	23 10 36.	- 32 15		15.9	BLUE STAR
RNGC 7518	23 10 38.	+ 06 03		14.5	GALAXY
MCG+02-59-009	23 10 42.	+ 10 30	72	14.	GALAXY
ZWG 431.016	23 10 42.	+ 10 31		15.2	GALAXY
UGC 12424	23 10 42.	+ 10 31	84	15.2	GALAXY Sb
MCG+03-59-011	23 10 42.	+ 17 46	30	15.	GALAXY
ZWG 475.043	23 10 42.	+ 23 58		15.0	GALAXY
UGC 12425	23 10 42.	+ 23 58	102	15.0	GALAXY Sa-b
MCG+04-54-030	23 10 42.	+ 23 58	84	15.	GALAXY
ZWG 380.010	23 10 42.	- 02 23		15.5	GALAXY
PHL 0423	23 10 42.	- 09 53		18.1	BLUE STELLAR OBJECT
PHL 5550	23 10 42.	- 15 04		18.2	BLUE STELLAR OBJECT
PHL 2137	23 10 42.	- 16 03		18.6	BLUE STELLAR OBJECT
RNGC 7517	23 10 43.	- 02 23		15.5	GALAXY
SC 2307-6251.2	23 10 43.	- 62 34 54.	18		NEBULA
RNGC 7519	23 10 44.	+ 10 31		15.0	GALAXY
MCG+01-59-012	23 10 45.	+ 06 02 30.	108	14.	GALAXY
MCG+01-59-013	23 10 45.	+ 06 09	204	14.	GALAXY
PHL 0424	23 10 48.	+ 00 18		16.6	BLUE STELLAR OBJECT
ZC 2310.8+1106	23 10 48.	+ 11 06	810		CLUSTER OF GALAXIES
ZWG 431.017	23 10 48.	+ 14 38		15.7	GALAXY
ZWG 454.007	23 10 48.	+ 17 47		15.6	GALAXY
KARA.73B 1010	23 10 48.	+ 17 47	24	15.6	ISOLATED GALAXY S
ZWG 475.044	23 10 48.	+ 22 35		15.7	GALAXY
MCG+13-16-001	23 10 48.	+ 77 19	78	15.	GALAXY
PHL 5552	23 10 48.	- 12 11		18.3	BLUE STELLAR OBJECT
PHL 5551	23 10 48.	- 16 28		18.2	BLUE STELLAR OBJECT
TON-S 0093	23 10 48.	- 27 33		16.2	BLUE STAR
SC 2307-6209.0	23 10 49.	- 61 52 42.			NEBULA
ZC 2310.9+1436	23 10 54.	+ 14 36	1950		CLUSTER OF GALAXIES
MCG+05-54-048	23 10 54.	+ 28 40	63	15.	GALAXY
ZWG 516.001	23 10 54.	+ 36 12		15.5	GALAXY
ZWG 515.028	23 10 54.	+ 36 12		15.5	GALAXY
PHL 2138	23 10 54.	- 09 21		18.4	BLUE STELLAR OBJECT
PHL 5553	23 10 54.	- 13 45		17.9	BLUE STELLAR OBJECT
PHL 2139	23 10 54.	- 15 02		18.5	BLUE STELLAR OBJECT
LB 09950	23 11	- 80 07		12.4	FAINT BLUE STAR
UGC 12426	23 11 00.	+ 06 17	78	16.0	GALAXY Sc
ZWG 454.008	23 11 00.	+ 20 22		15.7	GALAXY
4ZW 137	23 11 00.	+ 22 52			COMPACT GALAXY
ZWG 496.057	23 11 00.	+ 28 41		15.4	GALAXY
UGC 12427	23 11 00.	+ 28 41	66	15.4	GALAXY Sb-0
MCG+00-59-009	23 11 00.	- 02 00	24	15.	GALAXY
PHL 2140	23 11 00.	- 11 56		18.2	BLUE STELLAR OBJECT
SER 159.03	23 11 00.	- 42 55	1500	17.	CLUSTER OF GALAXIES
KEEL 724	23 11 01.9	+ 04 16 56.			NEBULA
RNGC 7520	23 11 04.	- 24 04			NON-EXISTENT OBJECT
SC 2308-6306.6	23 11 05.	- 62 50 18.	9		NEBULA
SC 2308-6309.4	23 11 05.	- 62 53 06.			NEBULA
ZC 2311.1+0828	23 11 06.	+ 08 28	340		CLUSTER OF GALAXIES
ZWG 431.018	23 11 06.	+ 13 44		15.7	GALAXY
ZC 2311.1+3830	23 11 06.	+ 38 30	2690		CLUSTER OF GALAXIES
ZWG 380.011	23 11 06.	- 02 00		14.9	GALAXY
IC 5291	23 11 07.	+ 08 58 13.			NONSTELLAR OBJECT
RNGC 7521	23 11 07.	- 02 00		15.0	GALAXY
RNGC 7523	23 11 09.	+ 13 44		15.5	GALAXY
ARC 2561	23 11 09.	+ 14 29		17.5	RICH CLUSTER OF GALAXIES
RNGC 7527	23 11 10.	+ 24 38		14.5	GALAXY
ARC 2562	23 11 10.	+ 32 10		17.4	RICH CLUSTER OF GALAXIES
RNGC 7522	23 11 10.	- 23 10			NON-EXISTENT OBJECT
SHAH 032	23 11 12.	+ 01 30	54	16.8	GROUP OF COMPACT GALAXIES
ZWG 406.022	23 11 12.	+ 08 57		15.6	GALAXY
ZWG 431.019	23 11 12.	+ 13 45		15.2	GALAXY
MRK 316	23 11 12.	+ 13 45	30	16.	GALAXY WITH UV CONTINUUM
ZWG 475.045	23 11 12.	+ 24 38		14.7	GALAXY
UGC 12428	23 11 12.	+ 24 38	78	14.7	GALAXY E
4ZW 138	23 11 12.	+ 24 41			COMPACT GALAXY
ZC 2311.2+3211	23 11 12.	+ 32 11	470		CLUSTER OF GALAXIES
PHL 5555	23 11 12.	- 09 07		18.3	BLUE STELLAR OBJECT
PHL 0425	23 11 12.	- 13 56		18.0	BLUE STELLAR OBJECT
PHL 5554	23 11 12.	- 18 56		18.8	BLUE STELLAR OBJECT
TON-S 0094	23 11 12.	- 26 03		15.8	BLUE STAR
MCG-07-47-023	23 11 12.	- 43 00	30	16.	GALAXY
RNGC 7524	23 11 13.	- 02 01			GALAXY
IC 5292	23 11 15.	+ 13 25			NONSTELLAR OBJECT
RNGC 7525	23 11 15.	+ 13 45		15.0	GALAXY
MCG+04-54-031	23 11 15.	+ 24 37	18	14.5	GALAXY
MCG+00-59-010	23 11 15.	- 02 00	42	15.5	GALAXY
ZWG 431.020	23 11 18.	+ 13 25		15.5	GALAXY
ZWG 475.046	23 11 18.	+ 22 25		15.6	GALAXY
UGC 12429	23 11 18.	+ 22 25	90	15.6	GALAXY Sc
MRK 317	23 11 18.	+ 23 32	10	16.	GALAXY WITH UV CONTINUUM
ZWG 496.058	23 11 18.	+ 28 44		14.9	GALAXY
UGC 12430	23 11 18.	+ 28 44	150	14.9	GALAXY Sc
PHL 5557	23 11 18.	- 09 27		18.6	BLUE STELLAR OBJECT
RNGC 7526	23 11 18.	- 09 28			NON-EXISTENT OBJECT
MCG-02-59-006	23 11 18.	- 11 47	36	15.	GALAXY
PHL 5556	23 11 18.	- 14 22		15.5	BLUE STELLAR OBJECT
PHL 5558	23 11 18.	- 15 18		18.4	BLUE STELLAR OBJECT
PHL 5559	23 11 18.	- 16 21		18.4	BLUE STELLAR OBJECT
MCG-05-54-024	23 11 18.	- 29 53	24	15.	GALAXY
MCG+05-54-049	23 11 21.	+ 28 43 30.	120	14.5	GALAXY
IC 1475	23 11 21.	- 28 42			NONSTELLAR OBJECT
KEEL 725	23 11 22.1	+ 04 36 45.			NEBULA
MCG+04-54-032	23 11 24.	+ 22 25	78	15.	GALAXY
ZWG 475.047	23 11 24.	+ 25 10		15.5	GALAXY
PHL 5560	23 11 24.	- 10 04		18.4	BLUE STELLAR OBJECT
MIK.46 22	23 11 26.	+ 61 14			DIFFUSE NEBULA
CED 209	23 11 26.	+ 61 14	600		DIFFUSE GALACTIC NEBULA
MCG-07-47-024	23 11 27.	- 40 54	90	17.	GALAXY
ZWG 380.012	23 11 30.	+ 03 26		15.4	GALAXY
ZWG 406.023	23 11 30.	+ 08 15		15.6	GALAXY
ZWG 406.024	23 11 30.	+ 08 42		14.6	GALAXY
UGC 12431	23 11 30.	+ 08 42	66	14.6	GALAXY S
ZWG 406.025	23 11 30.	+ 09 11		15.6	GALAXY
UGC 12432	23 11 30.	+ 24 38	60	16.0	GALAXY Sc
MCG+04-54-034	23 11 30.	+ 25 10 30.	30	15.5	GALAXY
MCG+04-54-033	23 11 30.	+ 25 11	21	16.	GALAXY
UGC 12433	23 11 30.	+ 49 24	120	16.0	GALAXY SB0-a
MRSL 111+00/2	23 11 30.	+ 61 14	600		HII REGION
PHL 2141	23 11 30.	- 19 06		18.6	BLUE STELLAR OBJECT
RNGC 7529	23 11 32.	+ 08 42		14.5	GALAXY
SG 1.27	23 11 32.	+ 61 16			DIFFUSE EMISSION NEBULA
MCG+01-59-014	23 11 36.	+ 08 44	48	14.	GALAXY
ZWG 431.021	23 11 36.	+ 12 54		15.4	GALAXY
MRK 528	23 11 36.	+ 12 54	20	16.	GALAXY WITH UV CONTINUUM
UGC 12434	23 11 36.	+ 12 54	66	15.4	GALAXY Sa?
MCG+02-59-011	23 11 36.	+ 13 09	108	14.	GALAXY
MCG+02-59-010	23 11 36.	+ 13 18	96	14.	GALAXY
ZWG 496.059	23 11 36.	+ 31 22		15.7	GALAXY
UGC 12435	23 11 36.	+ 31 47	60	16.0	GALAXY SBa
PHL 5561	23 11 36.	- 13 26		18.8	BLUE STELLAR OBJECT
MCG-03-59-002	23 11 39.	- 19 15 30.	24	16.	GALAXY
REIN 2.310	23 11 41.46	+ 04 14 57.1			NEBULA
ZWG 406.026	23 11 42.	+ 05 05		15.6	GALAXY
UGC 12436	23 11 42.	+ 05 05	60	15.6	GALAXY DBL SYS
ZWG 431.022	23 11 42.	+ 13 10		14.8	GALAXY
UGC 12437	23 11 42.	+ 13 10	132	14.8	GALAXY SBb
ZWG 431.023	23 11 42.	+ 13 19		15.6	GALAXY
UGC 12438	23 11 42.	+ 13 19	102	15.6	GALAXY Sc
MCG+05-54-050	23 11 42.	+ 28 20	27	16.	GALAXY
UGC 12439	23 11 42.	+ 48 03	78	16.0	GALAXY SBb
MCG-01-59-004	23 11 42.	- 03 01	27	15.	GALAXY
MCG-02-59-007	23 11 42.	- 10 07	84	15.	GALAXY
PHL 5562	23 11 42.	- 11 23		17.8	BLUE STELLAR OBJECT
TON-S 0095	23 11 42.	- 26 08		16.1	BLUE STAR
RNGC 7530	23 11 43.	- 03 01		15.0	GALAXY
RNGC 7528	23 11 44.	+ 09 57			GALAXY
RNGC 7536	23 11 45.	+ 13 10		15.0	GALAXY
RNGC 7535	23 11 45.	+ 13 19		15.5	GALAXY
ARC 2563	23 11 47.	- 14 33		17.2	RICH CLUSTER OF GALAXIES
SC 2308-7226.2	23 11 47.	- 72 09 53.	12		NEBULA
ZWG 454.009	23 11 48.	+ 16 11		15.6	GALAXY
UGC 12440	23 11 48.	+ 16 11	60	15.6	GALAXY Sa-b
ZWG 496.060	23 11 48.	+ 30 17		15.6	GALAXY
ZWG 496.061	23 11 48.	+ 31 13		15.6	GALAXY
PHL 5564	23 11 48.	- 00 34		18.4	BLUE STELLAR OBJECT
PHL 2144	23 11 48.	- 00 56		17.9	BLUE STELLAR OBJECT
ZWG 380.013	23 11 48.	- 02 18		15.7	GALAXY
MRK 529	23 11 48.	- 02 58	12	14.	GALAXY WITH UV CONTINUUM
MCG-01-59-005	23 11 48.	- 02 58	72	14.5	GALAXY
PHL 2142	23 11 48.	- 06 49		16.2	BLUE STELLAR OBJECT
PHL 5565	23 11 48.	- 13 22		18.2	BLUE STELLAR OBJECT
PHL 5563	23 11 48.	- 15 22		18.0	BLUE STELLAR OBJECT
PHL 2143	23 11 48.	- 17 22		18.0	BLUE STELLAR OBJECT
RNGC 7533	23 11 49.	- 02 17		15.5	GALAXY
RNGC 7532	23 11 49.	- 02 58		14.5	GALAXY
MCG+05-54-051	23 11 51.	+ 30 15 30.	36	15.5	GALAXY
SC 2308-6442.1	23 11 52.	- 64 25 47.			NEBULA
PHL 2145	23 11 54.	+ 00 54		18.0	BLUE STELLAR OBJECT
ZWG 406.027	23 11 54.	+ 03 55		15.1	GALAXY
MCG+03-59-012	23 11 54.	+ 16 10	60	15.	GALAXY
4ZW 139	23 11 54.	+ 24 41			COMPACT GALAXY
ZC 2311.9+2724	23 11 54.	+ 27 24	1280		CLUSTER OF GALAXIES
ZWG 496.062	23 11 54.	+ 28 10		15.4	GALAXY
UGC 12441	23 11 54.	+ 30 52	60	16.0	GALAXY SB?b-c
MCG-01-59-006	23 11 54.	- 02 55	39	14.5	GALAXY
PHL 5567	23 11 54.	- 11 06		18.6	BLUE STELLAR OBJECT
RNGC 7534	23 11 55.	- 02 55		14.5	GALAXY
REIN 2.311	23 11 56.23	+ 03 55 11.3			NEBULA
RNGC 7539	23 11 58.	+ 23 24		13.5	GALAXY
SC 2308-6201.1	23 11 58.	- 61 44 47.	60		NEBULA
LBN 0533	23 12	+ 59 30	1080		BRIGHT NEBULA
LBN 0542	23 12	+ 61 13	420		BRIGHT NEBULA
LBN 0573	23 12	+ 73 40	1500		BRIGHT NEBULA
PHL 5569	23 12 00.	+ 02 05		18.2	BLUE STELLAR OBJECT
ZWG 406.028	23 12 00.	+ 04 13		13.8	GALAXY
UGC 12442	23 12 00.	+ 04 13	126	13.8	GALAXY Sb
KARA.72 578A	23 12 00.	+ 04 13	126	13.8	PART OF DOUBLE GALAXY
ZWG 406.029	23 12 00.	+ 06 16		15.7	GALAXY
ZWG 475.048	23 12 00.	+ 23 24		13.7	GALAXY
UGC 12443	23 12 00.	+ 23 24	90	13.7	GALAXY S0
MCG+04-54-035	23 12 00.	+ 23 24	66	14.5	GALAXY
MCG+05-54-052	23 12 00.	+ 28 01 30.	66	14.5	GALAXY
ZWG 496.063	23 12 00.	+ 31 17		14.5	GALAXY
UGC 12444	23 12 00.	+ 31 17	66	14.5	GALAXY
ZC 2312.0+3502	23 12 00.	+ 35 02	5310		CLUSTER OF GALAXIES
ZWG 547.002	23 12 00.	+ 49 32		15.7	GALAXY
UGC 12445	23 12 00.	+ 49 32	72	15.7	GALAXY S0
LDN 1229	23 12 00.	+ 61 43	240		DARK NEBULA
LDN 1230	23 12 00.	+ 61 45	180		DARK NEBULA
MCG+00-59-011	23 12 00.	- 00 03	30	14.	GALAXY
PHL 0426	23 12 00.	- 06 26		17.9	BLUE STELLAR OBJECT
PHL 2146	23 12 00.	- 20 25		18.2	BLUE STELLAR OBJECT
PHL 5568	23 12 00.	- 20 34		17.9	BLUE STELLAR OBJECT
MCG-07-47-025	23 12 00.	- 43 52	132	12.1	GALAXY
HOLM 805B	23 12 01.	+ 04 13	78	13.1	PART OF MULTIPLE GALAXY
REIN 2.312	23 12 01.49	+ 23 24 45.6			NEBULA
RNGC 7537	23 12 02.	+ 04 14		14.0	GALAXY
RNGC 7538	23 12 02.	+ 61 13			DIFFUSE NEBULA
LB 01159	23 12 03.	+ 12 16 42.		16.0	FAINT BLUE STAR
RNGC 7540	23 12 03.	+ 15 40		15.5	GALAXY
MCG+01-59-016	23 12 06.	+ 04 12	132	13.5	GALAXY
HOLM 805B	23 12 06.	+ 04 13	66	12.8	PART OF MULTIPLE GALAXY
MCG+01-59-015	23 12 06.	+ 06 17	42	15.	GALAXY
ZWG 454.010	23 12 06.	+ 15 40		15.7	GALAXY
ZWG 496.064	23 12 06.	+ 30 54		15.7	GALAXY
ZWG 380.014	23 12 06.	- 00 02		14.4	GALAXY
UGC 12446	23 12 06.	- 00 02	42	14.4	GALAXY
PHL 2147	23 12 06.	- 11 54		17.8	BLUE STELLAR OBJECT
RNGC 7531	23 12 07.	- 43 53		12.0	GALAXY
HOLM 805A	23 12 09.	+ 04 15	150	11.6	PART OF MULTIPLE GALAXY
RNGC 7551	23 12 09.	+ 15 39			GALAXY
SG 1.28	23 12 09.	+ 60 11	3000		DIFFUSE EMISSION NEBULA
LB 01160	23 12 10.	+ 10 59 42.		14.5	FAINT BLUE STAR
RNGC 7543	23 12 11.	+ 28 03		14.0	GALAXY
ZC 2312.2+0221	23 12 12.	+ 02 21	270		CLUSTER OF GALAXIES
PHL 0427	23 12 12.	+ 03 14		18.7	BLUE STELLAR OBJECT
ZWG 406.030	23 12 12.	+ 04 15		12.7	GALAXY
UGC 12447	23 12 12.	+ 04 15	204	12.7	GALAXY Sc
KARA.72 578B	23 12 12.	+ 04 15	186	12.7	PART OF DOUBLE GALAXY
ZWG 406.031	23 12 12.	+ 07 26		15.7	GALAXY
SN 1964K	23 12 12.	+ 07 26		18.0	SUPERNOVA
ZWG 431.024	23 12 12.	+ 10 08		15.4	GALAXY
ZWG 431.025	23 12 12.	+ 10 23		15.7	GALAXY
ZC 2312.2+1205	23 12 12.	+ 12 05	870		CLUSTER OF GALAXIES
UGC 12448	23 12 12.	+ 14 38	60	16.0	GALAXY Sc
UGC 12449	23 12 12.	+ 23 03	60	16.0	GALAXY S
ZWG 475.049	23 12 12.	+ 24 52		15.6	GALAXY
ZWG 496.065	23 12 12.	+ 28 03		14.1	GALAXY
UGC 12450	23 12 12.	+ 28 03	90	14.1	GALAXY S
PHL 0428	23 12 12.	- 00 10		17.9	BLUE STELLAR OBJECT
PHL 2148	23 12 12.	- 07 51		16.0	BLUE STELLAR OBJECT
PHL 5755	23 12 12.	- 08 28		16.0	BLUE STELLAR OBJECT

OBJECT NAME	RIGHT ASCEN.	DECLINATION	DIAM.	MAGN.	TYPE OF OBJECT
PHL 5570	23 12 12.	- 18 43		18.7	BLUE STELLAR OBJECT
PHL 2149	23 12 12.	- 19 08		18.9	BLUE STELLAR OBJECT
MCG-04-54-015	23 12 12.	- 21 16	120	15.	GALAXY
RNGC 7541	23 12 14.	+ 04 16		12.5	GALAXY
RNGC 7542	23 12 14.	+ 10 23		15.5	GALAXY
KEEL 726	23 12 14.6	+ 04 22 02.			NEBULA
MCG+01-59-017	23 12 15.	+ 04 14	192	12.	GALAXY
HOLM 805A	23 12 15.	+ 04 15	180	11.3	PART OF MULTIPLE GALAXY
MCG+01-59-018	23 12 15.	+ 05 08	96	14.5	GALAXY
ARC 2564	23 12 15.	+ 13 50		17.1	RICH CLUSTER OF GALAXIES
IC 5293	23 12 15.	+ 24 51 52.			NONSTELLAR OBJECT
ZC 2312.3+0457	23 12 18.	+ 04 57	340		CLUSTER OF GALAXIES
UGC 12451	23 12 18.	+ 05 09	96	16.0	GALAXY DWRF IR
MCG+01-59-019	23 12 18.	+ 07 27	42	15.5	GALAXY
ZWG 406.032	23 12 18.	+ 08 47		15.7	GALAXY
ZC 2312.3+1350	23 12 18.	+ 13 50	2760		CLUSTER OF GALAXIES
ZC 2312.3+2134	23 12 18.	+ 21 34	1080		CLUSTER OF GALAXIES
ZWG 496.066	23 12 18.	+ 28 17		15.5	GALAXY
ZWG 496.067	23 12 18.	+ 31 02		15.7	GALAXY
ZWG 380.015	23 12 18.	- 00 31		15.7	GALAXY
PHL 2150	23 12 18.	- 01 32		18.3	BLUE STELLAR OBJECT
PHL 5571	23 12 18.	- 15 25		18.4	BLUE STELLAR OBJECT
MCG+00-59-012	23 12 24.	+ 01 10	120	14.	GALAXY
ZWG 431.026	23 12 24.	+ 09 43		15.5	GALAXY
MCG+02-59-012	23 12 24.	+ 14 43	54	15.	GALAXY
ZC 2312.4+2629	23 12 24.	+ 26 29	1880		CLUSTER OF GALAXIES
ZWG 496.068	23 12 24.	+ 32 45		15.5	GALAXY
KARA.73B 1011	23 12 24.	+ 32 45	36	15.5	ISOLATED GALAXY S
ZC 2312.4+4012	23 12 24.	+ 40 12	1010		CLUSTER OF GALAXIES
PHL 5572	23 12 24.	- 10 21		14.9	BLUE STELLAR OBJECT
PHL 0429	23 12 24.	- 15 34		17.3	BLUE STELLAR OBJECT
PHL 5573	23 12 24.	- 18 36		17.8	BLUE STELLAR OBJECT
RNGC 7544	23 12 25.	- 02 28			GALAXY
SC 2309-1236.7	23 12 25.	- 12 20 23.	24		NEBULA
RNGC 7547	23 12 27.	+ 18 42		15.0	GALAXY
ZWG 380.016	23 12 30.	+ 01 10		15.7	GALAXY
UGC 12452	23 12 30.	+ 01 10	120	15.7	GALAXY Sc
ZWG 431.027	23 12 30.	+ 12 52		15.5	GALAXY
ZWG 454.011	23 12 30.	+ 18 42		14.9	GALAXY
UGC 12453	23 12 30.	+ 18 42	66	14.9	GALAXY SBa
ZC 2312.5-0229	23 12 30.	- 02 29	8600		CLUSTER OF GALAXIES
PHL 2151	23 12 30.	- 10 48		18.9	BLUE STELLAR OBJECT
MCG-04-54-016	23 12 33.	- 21 38	60	15.	GALAXY
KEEL 727	23 12 33.8	+ 04 06 02.			NEBULA
3ZW 096	23 12 36.	+ 09 24			COMPACT GALAXY
ZWG 406.033	23 12 36.	+ 09 24		15.0	GALAXY
UGC 12454	23 12 36.	+ 09 24	84	15.0	GALAXY S0
MCG+03-59-013	23 12 36.	+ 18 42	63	14.	GALAXY
ZWG 496.069	23 12 36.	+ 28 12		15.7	GALAXY
MRSL 111-00/2	23 12 36.	+ 59 55	5400		HII REGION
PHL 2152	23 12 36.	- 00 06		18.1	BLUE STELLAR OBJECT
MCG-01-59-008	23 12 36.	- 02 32	42	16.	GALAXY
MCG-01-59-007	23 12 36.	- 02 34	60	14.5	GALAXY
PHL 5575	23 12 36.	- 17 12		17.9	BLUE STELLAR OBJECT
RNGC 7546	23 12 37.	- 02 34		14.5	GALAXY
MCG+05-54-053	23 12 39.	+ 28 40	63	15.	GALAXY
RNGC 7548	23 12 40.	+ 25 00		14.5	GALAXY
MCG+01-59-020	23 12 42.	+ 09 25	60	14.5	GALAXY
SN 1972I	23 12 42.	+ 14 42		17.5	SUPERNOVA
ZWG 475.050	23 12 42.	+ 25 00		14.5	GALAXY
UGC 12455	23 12 42.	+ 25 00	60	14.5	GALAXY S0
ZWG 496.070	23 12 42.	+ 28 41		15.1	GALAXY
3ZW 097	23 12 42.	- 01 30			COMPACT GALAXY
ZWG 380.017	23 12 42.	- 01 31		15.4	GALAXY
MCG-07-47-026	23 12 42.	- 38 48	60	14.	GALAXY
RNGC 7545	23 12 44.	- 38 48			GALAXY
SC 2309-6333.9	23 12 44.	- 63 17 34.	12		NEBULA
RNGC 7550	23 12 45.	+ 18 41		14.0	GALAXY
RNGC 7549	23 12 45.	+ 18 46		14.0	GALAXY
MCG+04-54-036	23 12 45.	+ 25 00	54	14.5	GALAXY
ZWG 454.012	23 12 48.	+ 18 41		13.9	GALAXY
UGC 12456	23 12 48.	+ 18 41	96	13.9	GALAXY E-S0
ARP 099	23 12 48.	+ 18 41			PECULIAR GALAXY
MCG+03-59-015	23 12 48.	+ 18 41	63	13.	GALAXY
ZWG 454.013	23 12 48.	+ 18 46		14.1	GALAXY
UGC 12457	23 12 48.	+ 18 46	168	14.1	GALAXY SB
MCG+03-59-014	23 12 48.	+ 18 46	174	13.	GALAXY
UGC 12458	23 12 48.	+ 30 41	60	16.5	GALAXY Sb-c
PHL 5576	23 12 48.	- 14 54		16.1	BLUE STELLAR OBJECT
PHL 0430	23 12 48.	- 15 02		16.2	BLUE STELLAR OBJECT
IC 1476	23 12 50.	+ 30 16 17.			OPEN CLUSTER
RNGC 7553	23 12 51.	+ 18 45			GALAXY
ZWG 454.014	23 12 54.	+ 15 46		15.7	GALAXY
ZWG 475.051	23 12 54.	+ 23 31		15.7	GALAXY
PHL 5577	23 12 54.	- 03 26		17.2	BLUE STELLAR OBJECT
PHL 5578	23 12 54.	- 04 38		18.5	BLUE STELLAR OBJECT
PHL 2154	23 12 54.	- 07 44		17.2	BLUE STELLAR OBJECT
MCG-07-47-027	23 12 54.	- 44 04	9	16.	GALAXY
KEEL 728	23 12 57.0	+ 04 37 40.			NEBULA
IC 5295	23 12 59.	+ 24 50 47.			NONSTELLAR OBJECT
SC 2309-6423.4	23 12 59.	- 64 07 04.	12		NEBULA
LBN 0537	23 13	+ 59 30	180		BRIGHT NEBULA
LBN 0536	23 13	+ 59 30	3600		BRIGHT NEBULA
LB 09951	23 13	- 80 21		13.0	FAINT BLUE STAR
ZWG 406.034	23 13 00.	+ 04 37		15.1	GALAXY
ZWG 475.052	23 13 00.	+ 24 51		15.7	GALAXY
COU 102	23 13 00.	+ 59 40	4200		EMISSION NEBULA
MRSL 113-00/1	23 13 00.	+ 60 31	600		HII REGION
COU 103	23 13 00.	+ 61 30	5400		EMISSION NEBULA
PHL 5579	23 13 00.	- 01 24		16.0	BLUE STELLAR OBJECT
PHL 2155	23 13 00.	- 09 28		18.3	BLUE STELLAR OBJECT
PHL 2156	23 13 00.	- 12 07		18.3	BLUE STELLAR OBJECT
PHL 2157	23 13 00.	- 16 15		18.6	BLUE STELLAR OBJECT
AGU 78	23 13 00.	- 38 47 12.	48	12.5	PECULIAR GALAXY
KARA.73 56	23 13 00.	- 42 46	60		DWARF GALAXY
KEEL 729	23 13 00.6	+ 04 41 34.			NEBULA
RNGC 7555	23 13 03.	+ 12 19			NON-EXISTENT OBJECT
HOLM 806B	23 13 06.	+ 00 09	18	14.7	PART OF MULTIPLE GALAXY
3ZW 098	23 13 06.	+ 01 22			COMPACT GALAXY
ZWG 406.035	23 13 06.	+ 06 25		15.0	GALAXY
ZWG 406.036	23 13 06.	+ 07 01		15.6	GALAXY
SN 1972M	23 13 06.	+ 07 01		18.0	SUPERNOVA
ZC 2313.1+1201	23 13 06.	+ 12 01	270		CLUSTER OF GALAXIES
ZC 2313.1+1420	23 13 06.	+ 14 20	1080		CLUSTER OF GALAXIES
ZWG 454.015	23 13 06.	+ 18 47		15.7	GALAXY
OCL 0257	23 13 06.	+ 60 12	300		OPEN STAR CLUSTER
PHL 5580	23 13 06.	- 03 47		5.6	BLUE STELLAR OBJECT
PHL 5647	23 13 06.	- 10 59		18.5	BLUE STELLAR OBJECT
PHL 2158	23 13 06.	- 15 32		18.0	BLUE STELLAR OBJECT
RNGC 7554	23 13 07.	- 02 39			GALAXY
RNGC 7557	23 13 08.	+ 06 25		15.0	GALAXY
RNGC 7564	23 13 08.	+ 07 01		15.5	GALAXY
LB 01161	23 13 08.	- 06 24 48.		17.6	FAINT BLUE STAR
RNGC 7558	23 13 09.	+ 18 39		16.0	GALAXY
HOLM 806A	23 13 10.	+ 00 08	42	14.0	PART OF MULTIPLE GALAXY
ZWG 380.018	23 13 12.	+ 00 09		14.9	GALAXY
MCG+00-59-013	23 13 12.	+ 00 09	36	14.5	GALAXY
ZC 2313.2+0036	23 13 12.	+ 00 36	870		CLUSTER OF GALAXIES
ZWG 380.019	23 13 12.	+ 00 45		15.4	GALAXY
MCG+01-59-021	23 13 12.	+ 06 26	36	14.5	GALAXY
MCG+02-59-014	23 13 12.	+ 13 00	12	16.	GALAXY
MCG+02-59-013	23 13 12.	+ 13 00	23	14.	GALAXY
MCG+03-59-016	23 13 12.	+ 18 39	18	16.	GALAXY
ZWG 475.053	23 13 12.	+ 24 37		15.6	GALAXY
UGC 12459	23 13 12.	+ 24 37	66	15.6	GALAXY Sb-c
ZWG 475.054	23 13 12.	+ 24 50		15.5	GALAXY
UGC 12460	23 13 12.	+ 24 50	60	15.5	GALAXY SBb
3ZW 099	23 13 12.	- 02 39			COMPACT GALAXY
PHL 5581	23 13 12.	- 06 42		18.7	BLUE STELLAR OBJECT
PHL 0431	23 13 12.	- 30 55		17.1	BLUE STELLAR OBJECT
IC 5296	23 13 13.	+ 24 49 17.			NONSTELLAR OBJECT
MCG+00-59-014	23 13 15.	+ 00 45	30	15.5	GALAXY
ARC 2565	23 13 15.	- 21 24		16.9	RICH CLUSTER OF GALAXIES
SG 3.241	23 13 16.	+ 59 27	1200		DIFFUSE EMISSION NEBULA
IC 5294	23 13 16.	- 42 52 07.			MAY NOT EXIST
ZWG 406.037	23 13 18.	+ 04 51		15.6	GALAXY
UGC 12461	23 13 18.	+ 04 51	60	15.6	GALAXY S0?
ZWG 406.038	23 13 18.	+ 09 14		15.5	GALAXY
UGC 12462	23 13 18.	+ 09 14	72	15.5	GALAXY Sb
ZWG 431.028	23 13 18.	+ 13 01		14.7	GALAXY
UGC 12463	23 13 13.	+ 13 01	78	14.7	GALAXY E-S0
MCG+04-54-037	23 13 18.	+ 24 49	42	15.	GALAXY
MRSL 111+01/1	23 13 18.	+ 61 35	3300		HII REGION
PHL 2161	23 13 18.	- 00 44		18.5	BLUE STELLAR OBJECT
MCG-01-59-009	23 13 18.	- 02 36	42	14.	GALAXY
PHL 5582	23 13 18.	- 03 05		13.0	BLUE STELLAR OBJECT
PHL 2162	23 13 18.	- 04 54		18.7	BLUE STELLAR OBJECT
PHL 2159	23 13 18.	- 08 32		18.7	BLUE STELLAR OBJECT
PHL 2160	23 13 18.	- 18 24		17.9	BLUE STELLAR OBJECT
RNGC 7556	23 13 19.	- 02 36		14.0	GALAXY
ARC 2566	23 13 20.	- 20 40		16.9	RICH CLUSTER OF GALAXIES
LB 01162	23 13 21.	+ 11 17 06.		16.6	FAINT BLUE STAR
RNGC 7559B	23 13 21.	+ 13 01		16.0	GALAXY
RNGC 7559A	23 13 21.	+ 13 01		14.5	GALAXY
FATH 1.835	23 13 21.	+ 15 12	16		NEBULA
FATH 1.836	23 13 23.	+ 15 44	11		NEBULA
PHL 0432	23 13 24.	+ 02 08		18.4	BLUE STELLAR OBJECT
MCG+01-59-022	23 13 24.	+ 04 51	60	15.	GALAXY
MCG+01-59-025	23 13 24.	+ 06 21	84	15.	GALAXY
ZWG 406.039	23 13 24.	+ 06 24		13.0	GALAXY
UGC 12464	23 13 24.	+ 06 24	138	13.0	GALAXY E
MCG+01-59-024	23 13 24.	+ 06 24	43	13.	GALAXY
MCG+01-59-023	23 13 24.	+ 09 16	72	15.	GALAXY
ZWG 431.029	23 13 24.	+ 12 55		14.5	GALAXY
UGC 12465	23 13 24.	+ 12 55	126	14.5	GALAXY SBa
MCG+02-59-015	23 13 24.	+ 12 55	108	14.	GALAXY
MCG+03-59-017	23 13 24.	+ 15 36	45	16.	GALAXY
MCG+00-59-015	23 13 24.	- 02 08	60	14.	GALAXY
PHL 2164	23 13 24.	- 02 46		17.6	BLUE STELLAR OBJECT
LB 01163	23 13 24.	- 06 47 48.		17.3	FAINT BLUE STAR
PHL 2163	23 13 24.	- 09 02		18.2	BLUE STELLAR OBJECT
MCG-07-47-028	23 13 24.	- 42 51	150	11.5	GALAXY
RNGC 7560	23 13 26.	+ 04 14			NON-EXISTENT OBJECT
RNGC 7561	23 13 26.	+ 04 15			NON-EXISTENT OBJECT
RNGC 7562	23 13 26.	+ 06 25		13.0	GALAXY
RNGC 7563	23 13 27.	+ 12 55		14.5	GALAXY
IC 5297	23 13 29.	+ 24 45 30.			NONSTELLAR OBJECT
UGC 12467	23 13 30.	+ 06 22	90	16.0	GALAXY
ZWG 475.055	23 13 30.	+ 24 45		15.7	GALAXY
IC 5298	23 13 30.	+ 25 16 48.			NONSTELLAR OBJECT
ZWG 475.056	23 13 30.	+ 25 17		15.0	GALAXY
ZWG 475.057	23 13 30.	+ 27 10		15.4	GALAXY
KARA.73B 1012	23 13 30.	+ 27 10	36	15.4	ISOLATED GALAXY S
MCG+05-54-054	23 13 30.	+ 28 24 30.	18	16.	GALAXY
MRSL 111+00/1	23 13 30.	+ 60 52	420		HII REGION
ZWG 380.020	23 13 30.	- 02 07		14.9	GALAXY
UGC 12466	23 13 30.	- 02 07	60	14.9	GALAXY SBb
TON-S 0096	23 13 30.	- 24 09		14.2	BLUE STAR
RNGC 7552	23 13 31.	- 42 53		12.0	GALAXY
RNGC 7562A	23 13 32.	+ 06 23		15.0	GALAXY
MCG+04-54-038	23 13 36.	+ 25 17	36	15.5	GALAXY
MCG+04-54-039	23 13 36.	+ 27 11	30	15.	GALAXY
PHL 0433	23 13 36.	- 02 06		12.3	BLUE STELLAR OBJECT
FEIG 108	23 13 36.	- 02 07		13.2	FAINT BLUE STAR
PHL 0434	23 13 36.	- 16 58		18.2	BLUE STELLAR OBJECT
RNGC 7567	23 13 39.	+ 15 36		15.5	GALAXY
SG 1.29	23 13 40.	+ 59 46	2100		DIFFUSE EMISSION NEBULA
RNGC 7573	23 13 40.	- 22 26		14.0	GALAXY
FATH 2.167	23 13 41.	+ 15 34	54		NEBULA
UGC 12468	23 13 42.	+ 15 34	60	15.4	GALAXY
ZWG 454.016	23 13 42.	+ 15 36		15.4	GALAXY
PHL 5584	23 13 42.	- 02 00		16.8	BLUE STELLAR OBJECT
PHL 2165	23 13 42.	- 02 40		18.4	BLUE STELLAR OBJECT
PHL 2166	23 13 42.	- 13 28		18.8	BLUE STELLAR OBJECT
PHL 5583	23 13 42.	- 15 32		18.9	BLUE STELLAR OBJECT
MCG-04-54-017	23 13 42.	- 22 26	72	14.	GALAXY
RNGC 7565	23 13 43.	- 00 19			NON-EXISTENT OBJECT
ARC 2567	23 13 43.	- 06 26		18.0	RICH CLUSTER OF GALAXIES
KEEL 730	23 13 44.1	+ 04 21 40.			NEBULA
ZC 2313.8+1458	23 13 48.	+ 14 58	1140		CLUSTER OF GALAXIES
MCG+03-59-019	23 13 48.	+ 15 34	45	14.5	GALAXY
MCG+03-59-018	23 13 48.	+ 18 53	27	16.	GALAXY
PHL 5585	23 13 48.	- 08 46		17.7	BLUE STELLAR OBJECT
PHL 2167	23 13 48.	- 09 20		17.9	BLUE STELLAR OBJECT
PHL 2168	23 13 48.	- 15 08		17.7	BLUE STELLAR OBJECT
PHL 5586	23 13 48.	- 17 06		17.5	BLUE STELLAR OBJECT
PHL 2169	23 13 48.	- 24 09		15.0	BLUE STELLAR OBJECT
IC 5299	23 12 50.	+ 20 34 36.			NONSTELLAR OBJECT
MCG+05-54-055	23 13 51.	+ 28 11	48	15.	GALAXY
SG 1.30	23 13 51.	+ 60 54	300		DIFFUSE EMISSION NEBULA
RNGC 7568	23 13 52.	+ 24 13		14.5	GALAXY
MIN.46 23	23 13 52.	+ 59 46			DIFFUSE NEBULA
FATH 1.837	23 13 53.	+ 15 35			NEBULA
ZC 2313.9+0139	23 13 54.	+ 01 39	1210		CLUSTER OF GALAXIES
ZWG 431.030	23 13 54.	+ 15 29		15.7	GALAXY
ZWG 454.017	23 13 54.	+ 15 35		15.7	GALAXY

OBJECT NAME	RIGHT ASCEN.	DECLINATION	DIAM.	MAGN.	TYPE OF OBJECT
ZWG 475.058	23 13 54.	+ 24 13		14.5	GALAXY
UGC 12469	23 13 54.	+ 24 13	60	14.5	GALAXY S
ZWG 496.071	23 13 54.	+ 28 13		15.4	GALAXY
UGC 12470	23 13 54.	+ 28 13	66	15.4	GALAXY Sa
ZC 2313.9+3038	23 13 54.	+ 30 38	940		CLUSTER OF GALAXIES
MFSL 111-00/1	23 13 54.	+ 59 46	180		HII REGION
PHL 0435	23 13 54.	- 02 43		17.1	BLUE STELLAR OBJECT
PHL 2170	23 13 54.	- 17 59		18.6	BLUE STELLAR OBJECT
PHL 2171	23 13 54.	- 19 00		18.4	BLUE STELLAR OBJECT
PHL 2172	23 13 54.	- 19 29		18.1	BLUE STELLAR OBJECT
PHL 5587	23 13 54.	- 28 12		17.8	BLUE STELLAR OBJECT
FATH 1.838	23 13 55.	+ 15 28			NEBULA
MCG+03-59-020	23 13 57.	+ 15 35	24	15.5	GALAXY
LBN 0540	23 14	+ 60 00	2100		BRIGHT NEBULA
LBN 0543	23 14	+ 60 50	420		BRIGHT NEBULA
PHL 2174	23 14 00.	+ 01 35		18.2	BLUE STELLAR OBJECT
ZC 2314.0+1042	23 14 00.	+ 10 42	340		CLUSTER OF GALAXIES
LB 01164	23 14 00.	+ 11 07 54.		16.5	FAINT BLUE STAR
ZWG 431.031	23 14 00.	+ 11 47		15.6	GALAXY
MCG+02-59-016	23 14 00.	+ 15 19	48	16.	GALAXY
UGC 12471	23 14 00.	+ 18 17	60	16.0	GALAXY S
MCG+03-59-021	23 14 00.	+ 18 18	48	15.	GALAXY
MCG+06-51-001	23 14 00.	+ 36 00	48	16.	GALAXY
PHL 2175	23 14 00.	- 00 45		17.2	BLUE STELLAR OBJECT
PHL 2176	23 14 00.	- 03 40		17.9	BLUE STELLAR OBJECT
PHL 5588	23 14 00.	- 14 10		17.4	BLUE STELLAR OBJECT
PHL 2177	23 14 00.	- 14 24		18.5	BLUE STELLAR OBJECT
IC 5300	23 14 05.	+ 20 32 55.			NONSTELLAR OBJECT
SC 2311-6131.8	23 14 05.	- 61 15 27.	42		NEBULA
ZWG 454.018	23 14 06.	+ 20 33		15.7	GALAXY
ISS 0098	23 14 06.	+ 58 33	230		STELLAR RING
PHL 5589	23 14 06.	- 11 49		18.0	BLUE STELLAR OBJECT
PHL 2227	23 14 06.	- 20 50		17.8	BLUE STELLAR OBJECT
FATH 1.839	23 14 07.	+ 15 37	35		NEBULA
SG 2.096	23 14 09.	+ 61 30	3300		DIFFUSE EMISSION NEBULA
PHL 2179	23 14 12.	+ 01 00		13.9	BLUE STELLAR OBJECT
ZWG 406.040	23 14 12.	+ 04 43		15.7	GALAXY
3ZW 100	23 14 12.	+ 08 37			COMPACT GALAXY
ZWG 406.041	23 14 12.	+ 08 37		14.8	GALAXY
UGC 12472	23 14 12.	+ 08 37	60	14.8	GALAXY S0?
MCG+01-59-026	23 14 12.	+ 08 40	72	14.5	GALAXY
MCG+02-59-018	23 14 12.	+ 13 12	84	13.5	GALAXY
ZWG 431.032	23 14 12.	+ 13 13		14.3	GALAXY
UGC 12473	23 14 12.	+ 13 13	102	14.3	GALAXY SBa
MCG+02-59-017	23 14 12.	+ 15 18	15	16.	GALAXY
ZWG 454.019	23 14 12.	+ 15 37		15.4	GALAXY
MCG+03-59-022	23 14 12.	+ 15 37	33	15.	GALAXY
ZWG 454.020	23 14 12.	+ 18 08		15.7	GALAXY
PHL 5591	23 14 12.	- 01 00		18.4	BLUE STELLAR OBJECT
MCG-01-59-010	23 14 12.	- 02 34	60	14.5	GALAXY
PHL 5590	23 14 12.	- 23 23		17.9	BLUE STELLAR OBJECT
PHL 2178	23 14 12.	- 23 40		17.9	BLUE STELLAR OBJECT
RNGC 7566	23 14 13.	- 02 34		14.5	GALAXY
RNGC 7569	23 14 14.	+ 10 39			NON-EXISTENT OBJECT
RNGC 7570	23 14 15.	+ 13 13		14.5	GALAXY
RNGC 7571	23 14 15.	+ 18 43			NON-EXISTENT OBJECT
FATH 1.840	23 14 16.	+ 15 18			NEBULA
MCG+05-54-056	23 14 18.	+ 29 18	24	15.	GALAXY
ZWG 496.072	23 14 18.	+ 29 19		15.5	GALAXY
MCG+06-51-002	23 14 18.	+ 33 42	63	14.5	GALAXY
ZWG 516.002	23 14 18.	+ 33 43		14.1	GALAXY
UGC 12474	23 14 18.	+ 33 43	72	14.1	GALAXY Sa
KARA.73B 1013	23 14 18.	+ 33 43	78	14.1	ISOLATED GALAXY S
LDN 1233	23 14 18.	+ 62 04	120		DARK NEBULA
PHL 2181	23 14 18.	- 18 15		18.0	BLUE STELLAR OBJECT
PHL 2180	23 14 18.	- 18 46		18.2	BLUE STELLAR OBJECT
PHL 5592	23 14 18.	- 19 14		17.3	BLUE STELLAR OBJECT
RNGC 7572	23 14 21.	+ 18 13		15.0	GALAXY
ARC 2568	23 14 21.	- 22 29		16.9	RICH CLUSTER OF GALAXIES
RNGC 7574	23 14 22.	+ 23 44			NON-EXISTENT OBJECT
PHL 5596	23 14 24.	+ 00 00		18.3	BLUE STELLAR OBJECT
PHL 5595	23 14 24.	+ 02 45		17.4	BLUE STELLAR OBJECT
MCG+00-59-016	23 14 24.	+ 03 27	114	15.	GALAXY
ZWG 454.021	23 14 24.	+ 18 13		15.0	GALAXY
MCG+03-59-023	23 14 24.	+ 18 13	42	15.	GALAXY
ZC 2314.4+3147	23 14 24.	+ 31 47	1480		CLUSTER OF GALAXIES
PHL 0436	23 14 24.	- 09 44		15.4	BLUE STELLAR OBJECT
PHL 5597	23 14 24.	- 12 56		18.6	BLUE STELLAR OBJECT
PHL 5598	23 14 24.	- 13 01		18.6	BLUE STELLAR OBJECT
PHL 2189	23 14 24.	- 14 00		18.4	BLUE STELLAR OBJECT
PHL 5574	23 14 24.	- 16 05		18.4	BLUE STELLAR OBJECT
PHL 5599	23 14 24.	- 17 20		18.0	BLUE STELLAR OBJECT
PHL 5594	23 14 24.	- 17 36		16.5	BLUE STELLAR OBJECT
PHL 2184	23 14 24.	- 20 38		18.8	BLUE STELLAR OBJECT
PHL 2182	23 14 24.	- 20 47		18.4	BLUE STELLAR OBJECT
PHL 5593	23 14 24.	- 27 10		13.4	BLUE STELLAR OBJECT
PHL 2183	23 14 24.	- 28 53		18.1	BLUE STELLAR OBJECT
HUB E29	23 14 29.	+ 59 24			DIFFUSE NEBULA
ZWG 380.021	23 14 30.	+ 03 26		15.7	GALAXY
UGC 12475	23 14 30.	+ 03 26	96	15.7	GALAXY Sb
PHL 2185	23 14 30.	- 03 15		17.6	BLUE STELLAR OBJECT
PHL 2186	23 14 30.	- 13 58		18.6	BLUE STELLAR OBJECT
PHL 5600	23 14 30.	- 21 20		17.2	BLUE STELLAR OBJECT
PHL 2187	23 14 30.	- 27 38		18.4	BLUE STELLAR OBJECT
MCG-03-59-003	23 14 33.	- 18 02 30.	30	16.	GALAXY
PHL 5603	23 14 36.	+ 01 40		17.9	BLUE STELLAR OBJECT
PHL 5602	23 14 36.	+ 03 13		17.8	BLUE STELLAR OBJECT
ZWG 406.042	23 14 36.	+ 06 50		15.7	GALAXY
MCG+01-59-027	23 14 36.	+ 06 51	48	15.	GALAXY
MCG+05-54-057	23 14 36.	+ 30 02 30.	90	14.5	GALAXY
ZWG 496.073	23 14 36.	+ 30 04		14.1	GALAXY
UGC 12476	23 14 36.	+ 30 04	84	14.1	GALAXY S0-a
PHL 2190	23 14 36.	- 00 54		18.3	BLUE STELLAR OBJECT
RNGC 7596	23 14 35.	- 07 11			GALAXY
MCG-01-59-011	23 14 36.	- 07 12	36	15.	GALAXY
PHL 0437	23 14 36.	- 10 51		18.1	BLUE STELLAR OBJECT
PHL 5601	23 14 36.	- 14 36		16.9	BLUE STELLAR OBJECT
PHL 0438	23 14 36.	- 17 50		18.4	BLUE STELLAR OBJECT
PHL 5605	23 14 36.	- 20 26		18.5	BLUE STELLAR OBJECT
PHL 5606	23 14 36.	- 21 46		18.0	BLUE STELLAR OBJECT
PHL 2188	23 14 36.	- 26 00		16.0	BLUE STELLAR OBJECT
PHL 5604	23 14 36.	- 26 12		17.9	BLUE STELLAR OBJECT
IC 1477	23 14 33.	- 07 11 11.			NONSTELLAR OBJECT
FATH 1.841	23 14 41.	+ 15 34			NEBULA
ZWG 454.022	23 14 42.	+ 18 25		15.3	GALAXY
UGC 12477	23 14 42.	+ 18 25	108	15.3	GALAXY DBL SYS
MCG+03-59-025	23 14 42.	+ 18 26	48	15.	GALAXY
MCG+03-59-024	23 14 42.	+ 18 26	66	15.5	GALAXY

OBJECT NAME	RIGHT ASCEN.	DECLINATION	DIAM.	MAGN.	TYPE OF OBJECT
ZC 2314.7+2723	23 14 42.	+ 27 23	3630		CLUSTER OF GALAXIES
4ZW 140	23 14 42.	+ 28 42			COMPACT GALAXY
PHL 5607	23 14 42.	- 10 37		17.1	BLUE STELLAR OBJECT
PHL 2191	23 14 42.	- 11 39		18.9	BLUE STELLAR OBJECT
PHL 5608	23 14 42.	- 17 55		18.6	BLUE STELLAR OBJECT
PHL 0439	23 14 42.	- 18 12		18.1	BLUE STELLAR OBJECT
HOLM 807A	23 14 44.	+ 05 22	18	13.9	PART OF MULTIPLE GALAXY
RNGC 7575	23 14 44.	+ 06 23			NON-EXISTENT OBJECT
HOLM 807B	23 14 45.	+ 05 21	12	14.6	PART OF MULTIPLE GALAXY
RNGC 7578B	23 14 45.	+ 18 26		15.0	GALAXY
RNGC 7578A	23 14 45.	+ 18 26		15.0	GALAXY
MCG+01-59-029	23 14 48.	+ 05 22	24	15.	GALAXY
ZWG 406.043	23 14 48.	+ 05 23		15.4	GALAXY
KARA.72 579A	23 14 48.	+ 05 23	36	15.4	PART OF DOUBLE GALAXY
MCG+01-59-028	23 14 48.	+ 05 23	30	14.5	GALAXY
ZWG 406.044	23 14 48.	+ 05 24		15.0	GALAXY
KARA.72 579B	23 14 48.	+ 05 24	42	15.0	PART OF DOUBLE GALAXY
ZWG 431.033	23 14 48.	+ 12 04		15.7	GALAXY
3ZW 101	23 14 48.	+ 17 46			COMPACT GALAXY
ZWG 454.023	23 14 48.	+ 18 09		15.7	GALAXY
ARP 170	23 14 48.	+ 18 25			PECULIAR GALAXY
ZWG 454.024	23 14 48.	+ 18 26		15.0	GALAXY
UGC 12478	23 14 48.	+ 18 26	108	15.0	GALAXY DBL SYS
MCG+05-54-058	23 14 48.	+ 28 18 30.	36	15.	GALAXY
ZWG 496.074	23 14 48.	+ 28 20		15.6	GALAXY
ZC 2314.8+3325	23 14 48.	+ 33 25	1410		CLUSTER OF GALAXIES
MCG+00-59-017	23 14 48.	- 01 52	60	14.5	GALAXY
PHL 2193	23 14 48.	- 03 26		18.6	BLUE STELLAR OBJECT
MCG-01-59-012	23 14 48.	- 05 00	60	14.	GALAXY
PHL 0440	23 14 48.	- 05 14		18.6	BLUE STELLAR OBJECT
PHL 2194	23 14 48.	- 08 58		18.7	BLUE STELLAR OBJECT
PHL 0441	23 14 48.	- 11 40		18.3	BLUE STELLAR OBJECT
PHL 5609	23 14 48.	- 14 16		18.2	BLUE STELLAR OBJECT
PHL 2192	23 14 48.	- 16 26		6.5	BLUE STELLAR OBJECT
PHL 2195	23 14 48.	- 20 40		18.6	BLUE STELLAR OBJECT
REIN 4.299	23 14 48.02	- 05 00 05.3			NEBULA
RNGC 7577	23 14 50.	+ 07 06			GALAXY
MCG-02-59-008	23 14 51.	- 10 19	54	14.5	GALAXY
UGC 12480	23 14 54.	+ 07 22	60	17.	GALAXY DWRF IR
PBIG 109	23 14 54.	+ 07 35		13.6	FAINT BLUE STAR
MCG+03-59-027	23 14 54.	+ 17 46	42	15.5	GALAXY
ZWG 454.025	23 14 54.	+ 17 47		15.7	GALAXY
MCG+03-59-026	23 14 54.	+ 18 09	42	15.5	GALAXY
ZWG 380.022	23 14 54.	- 01 52		15.3	GALAXY
UGC 12479	23 14 54.	- 01 52	60	15.3	GALAXY Sa
PHL 5611	23 14 54.	- 06 00		18.1	BLUE STELLAR OBJECT
PHL 5610	23 14 54.	- 08 26		17.9	BLUE STELLAR OBJECT
RNGC 7576	23 14 55.	- 05 01		14.0	GALAXY
LBN 0547	23 15	+ 61 30	4200		BRIGHT NEBULA
MCG+01-59-030	23 15 00.	+ 07 22	60	16.	GALAXY
MCG+02-59-019	23 15 00.	+ 13 43	42	14.5	GALAXY
VV 181B	23 15 00.	+ 18 25	15	16.	INTERACTING GALAXY
VV 181A	23 15 00.	+ 18 25	18	16.	INTERACTING GALAXY
VV 181	23 15 00.	+ 18 25	72		INTERACTING GALAXY
MCG+03-59-028	23 15 00.	+ 18 55	30	16.5	GALAXY
ZWG 475.059	23 15 00.	+ 21 52		15.7	GALAXY
ZC 2315.0+2254	23 15 00.	+ 22 54	2080		CLUSTER OF GALAXIES
ZWG 496.075	23 15 00.	+ 28 47		15.7	GALAXY
LDN 1232	23 15 00.	+ 61 30	240		DARK NEBULA
LDN 1234	23 15 00.	+ 62 08	240		DARK NEBULA
PHL 0442	23 15 00.	- 00 32		16.6	BLUE STELLAR OBJECT
PHL 0443	23 15 00.	- 09 19		13.7	BLUE STELLAR OBJECT
PHL 5614	23 15 00.	- 10 57		18.4	BLUE STELLAR OBJECT
PHL 5612	23 15 00.	- 13 10		15.9	BLUE STELLAR OBJECT
PHL 2196	23 15 00.	- 14 58		18.6	BLUE STELLAR OBJECT
PHL 5613	23 15 00.	- 27 15		17.8	BLUE STELLAR OBJECT
REIN 4.300	23 15 03.19	- 04 56 04.4			NEBULA
SN 1970T	23 15 05.	+ 05 44		18.0	SUPERNOVA
PHL 5615	23 15 06.	+ 00 48		16.0	BLUE STELLAR OBJECT
ZWG 406.045	23 15 06.	+ 07 59		15.1	GALAXY
ZWG 406.046	23 15 06.	+ 09 09		15.4	GALAXY
MCG+01-59-031	23 15 06.	+ 09 11	12	15.	GALAXY
ZWG 431.034	23 15 06.	+ 13 44		14.8	GALAXY
MRK 318	23 15 06.	+ 13 44	35	16.	GALAXY WITH UV CONTINUUM
UGC 12481	23 15 06.	+ 13 44	48	14.8	GALAXY S?
MCG+05-54-059	23 15 06.	+ 28 44	18	14.5	GALAXY
ZWG 496.076	23 15 06.	+ 28 45		14.4	GALAXY
UGC 12482	23 15 06.	+ 28 45	96	14.4	GALAXY (E)
PHL 5616	23 15 06.	- 13 52		18.3	BLUE STELLAR OBJECT
RNGC 7586	23 15 08.	+ 07 59		15.0	GALAXY
RNGC 7579	23 15 08.	+ 09 09		15.5	GALAXY
MCG+01-59-032	23 15 09.	+ 09 10	24	16.	GALAXY
RNGC 7580	23 15 09.	+ 13 44		15.0	GALAXY
FATH 1.842	23 15 11.	+ 15 51			NEBULA
ZWG 454.026	23 15 12.	+ 15 52		15.7	GALAXY
ZWG 454.027	23 15 12.	+ 16 26		15.7	GALAXY
ZWG 454.028	23 15 12.	+ 18 08		15.7	GALAXY
PHL 5618	23 15 12.	- 14 54		18.4	BLUE STELLAR OBJECT
PHL 5617	23 15 12.	- 15 41		18.4	BLUE STELLAR OBJECT
PHL 5619	23 15 12.	- 19 22		18.2	BLUE STELLAR OBJECT
RNGC 7581	23 15 14.	+ 04 24			NON-EXISTENT OBJECT
MCG+03-59-029	23 15 15.	+ 16 25	30	15.5	GALAXY
ARC 2569	23 15 16.	- 13 09		16.6	RICH CLUSTER OF GALAXIES
ZWG 406.047	23 15 18.	+ 07 05		14.8	GALAXY
MCG+01-59-034	23 15 18.	+ 07 07	36	14.5	GALAXY
ZWG 406.048	23 15 18.	+ 07 08		15.2	GALAXY
MCG+01-59-033	23 15 18.	+ 07 10	18	15.	GALAXY
ZWG 454.029	23 15 18.	+ 16 26		15.6	GALAXY
MCG+05-54-060	23 15 18.	+ 28 44	15	15.5	GALAXY
ZWG 497.001	23 15 18.	+ 28 45		15.6	GALAXY
ZWG 496.077	23 15 18.	+ 28 45		15.6	GALAXY
MCG-01-59-013	23 15 18.	- 03 45	36	15.	GALAXY
PHL 2197	23 15 18.	- 09 20		18.6	BLUE STELLAR OBJECT
PHL 5620	23 15 18.	- 16 28		18.0	BLUE STELLAR OBJECT
PHL 2198	23 15 18.	- 24 24		18.4	BLUE STELLAR OBJECT
RNGC 7583	23 15 20.	+ 07 08		15.0	GALAXY
RNGC 7604	23 15 20.	+ 07 10		15.0	GALAXY
MCG+03-59-030	23 15 21.	+ 16 25 30.	24	15.5	GALAXY
RNGC 7588	23 15 21.	+ 18 29		15.5	GALAXY
IC 5303	23 15 22.	- 00 01			NONSTELLAR OBJECT
ARP 223	23 15 22.	- 04 56			PECULIAR GALAXY
PHL 2199	23 15 24.	+ 00 30		18.4	BLUE STELLAR OBJECT
PHL 2201	23 15 24.	+ 01 20		18.4	BLUE STELLAR OBJECT
PHL 5622	23 15 24.	+ 02 02		18.5	BLUE STELLAR OBJECT
ZWG 406.049	23 15 24.	+ 09 09		15.3	GALAXY
ZWG 406.050	23 15 24.	+ 09 10		15.7	GALAXY
MCG+01-59-035	23 15 24.	+ 09 11	18	15.	GALAXY
ZWG 406.051	23 15 24.	+ 09 23		15.6	GALAXY

OBJECT NAME	RIGHT ASCEN.	DECLINATION	DIAM.	MAGN.	TYPE OF OBJECT
KARA.72 580B	23 15 24.	+ 09 23	42	15.6	PART OF DOUBLE GALAXY
KARA.72 580A	23 15 24.	+ 09 24	78	14.9	PART OF DOUBLE GALAXY
MCG+02-59-020	23 15 24.	+ 11 03	60	14.	GALAXY
ZWG 431.035	23 15 24.	+ 11 05		14.6	GALAXY
UGC 12483	23 15 24.	+ 11 05	60	14.6	GALAXY S
ZWG 454.030	23 15 24.	+ 18 29		15.7	GALAXY
PHL 5623	23 15 24.	- 03 38		18.3	BLUE STELLAR OBJECT
PHL 2200	23 15 24.	- 09 28		4.4	BLUE STELLAR OBJECT
PHL 5621	23 15 24.	- 27 49		17.8	BLUE STELLAR OBJECT
RNGC 7585	23 15 25.	- 04 56		13.0	GALAXY
RNGC 7584	23 15 26.	+ 09 09		15.5	GALAXY
RNGC 7593	23 15 26.	+ 11 05		14.5	GALAXY
REIF 2.313	23 15 26.67	- 04 55 27.2			NEBULA
MCG+01-59-036	23 15 27.	+ 09 12 30.	18	15.5	GALAXY
MCG+01-59-037	23 15 27.	+ 09 25	72	13.5	GALAXY
MCG+03-59-031	23 15 27.	+ 18 28 30.	18	16.5	GALAXY
LB 01165	23 15 29.	- 05 10 24.		15.5	FAINT BLUE STAR
ZWG 406.052	23 15 30.	+ 09 24		14.9	GALAXY
UGC 12484	23 15 30.	+ 09 24	72	14.9	GALAXY Sa
MCG+02-59-021	23 15 30.	+ 10 00	24	15.	GALAXY
ZC 2315.5+2854	23 15 30.	+ 28 54	8270		CLUSTER OF GALAXIES
ZC 2315.5+3038	23 15 30.	+ 30 38	1080		CLUSTER OF GALAXIES
MCG-01-59-014	23 15 30.	- 03 56	84	15.	GALAXY
MCG-01-59-015	23 15 30.	- 04 55	72	12.	GALAXY
PHL 0444	23 15 30.	- 05 10		15.7	BLUE STELLAR OBJECT
PHL 5625	23 15 30.	- 10 53		17.8	BLUE STELLAR OBJECT
PHL 0445	23 15 30.	- 11 43		17.5	BLUE STELLAR OBJECT
PHL 0446	23 15 30.	- 17 52		18.3	BLUE STELLAR OBJECT
PHL 5624	23 15 30.	- 19 42		19.0	BLUE STELLAR OBJECT
PHL 2202	23 15 30.	- 26 35		18.3	BLUE STELLAR OBJECT
IC 5305	23 15 31.	+ 10 01 21.			NONSTELLAR OBJECT
RNGC 7587	23 15 32.	+ 09 25		15.0	GALAXY
HN 0781	23 15 32.	- 64 51			NEBULA
IC 5302	23 15 32.	- 64 51			NONSTELLAR OBJECT
ABC 2570	23 15 33.	+ 01 42		17.5	RICH CLUSTER OF GALAXIES
ZC 2315.6+0142	23 15 36.	+ 01 42	940		CLUSTER OF GALAXIES
ZWG 431.036	23 15 36.	+ 09 58		15.6	GALAXY
IC 5306	23 15 36.	+ 09 58 21.			NONSTELLAR OBJECT
ZWG 431.037	23 15 36.	+ 10 02		15.4	GALAXY
ZC 2315.6+3525	23 15 36.	+ 35 25	2290		CLUSTER OF GALAXIES
ZWG 380.023	23 15 36.	- 00 39		15.0	GALAXY
MCG+00-59-018	23 15 36.	- 00 40	21	15.	GALAXY
PHL 5626	23 15 36.	- 01 40		17.3	BLUE STELLAR OBJECT
PHL 2203	23 15 36.	- 06 54		18.6	BLUE STELLAR OBJECT
PHL 5629	23 15 36.	- 08 54		18.4	BLUE STELLAR OBJECT
IC 5304	23 15 36.	- 10 34 15.			NONSTELLAR OBJECT
PHL 0447	23 15 36.	- 11 30		18.6	BLUE STELLAR OBJECT
PHL 5628	23 15 36.	- 14 36		18.2	BLUE STELLAR OBJECT
PHL 2205	23 15 36.	- 18 29		18.0	BLUE STELLAR OBJECT
PHL 2206	23 15 36.	- 19 15		18.4	BLUE STELLAR OBJECT
PHL 2204	23 15 36.	- 20 50		18.0	BLUE STELLAR OBJECT
PHL 5627	23 15 36.	- 22 42		17.9	BLUE STELLAR OBJECT
PHL 2207	23 15 36.	- 28 02		18.0	BLUE STELLAR OBJECT
REIN 4.302	23 15 39.04	- 04 57 06.7			NEBULA
HN 0780	23 15 40.	- 69 51			NEBULA
IC 5301	23 15 40.	- 69 51			NONSTELLAR OBJECT
ZCG 2315+03	23 15 42.	+ 03 54		17.0	COMPACT GALAXY
MCG+02-59-022	23 15 42.	+ 09 57	36	16.	GALAXY
MCG+02-59-023	23 15 42.	+ 10 00	96	14.	GALAXY
ZWG 431.038	23 15 42.	+ 10 02		15.3	GALAXY
UGC 12485	23 15 42.	+ 10 02	102	15.3	GALAXY Sb
IC 1478	23 15 42.	+ 10 02			NONSTELLAR OBJECT
MCG+00-59-019	23 15 42.	- 00 01	48	14.	GALAXY
PHL 0448	23 15 42.	- 01 10		18.4	BLUE STELLAR OBJECT
PHL 2208	23 15 42.	- 04 48		18.1	BLUE STELLAR OBJECT
PHL 5630	23 15 42.	- 11 00		18.0	BLUE STELLAR OBJECT
PHL 5631	23 15 42.	- 15 33		18.5	BLUE STELLAR OBJECT
MCG-07-47-029	23 15 42.	- 42 38	198	11.6	GALAXY
RNGC 7594	23 15 44.	+ 09 57		16.0	GALAXY
IC 5307	23 15 47.	+ 09 57 33.			NONSTELLAR OBJECT
FATH 1.843	23 15 47.	+ 15 02	5		NEBULA
REIN 4.303	23 15 47.92	- 04 41 26.2			NEBULA
ZWG 380.024	23 15 48.	+ 00 00		15.4	GALAXY
ZWG 406.053	23 15 48.	+ 06 18		13.8	GALAXY
UGC 12486	23 15 48.	+ 06 18	114	13.8	GALAXY SBb
MCG+01-59-038	23 15 48.	+ 06 19	120	13.	GALAXY
ZWG 406.054	23 15 48.	+ 06 33		15.5	GALAXY
MCG+05-54-061	23 15 48.	+ 28 56	48	15.	GALAXY
ZWG 497.002	23 15 48.	+ 28 57		15.3	GALAXY
ZWG 496.078	23 15 48.	+ 28 57		15.3	GALAXY
ZC 2315.8+3727	23 15 48.	+ 37 27	2420		CLUSTER OF GALAXIES
MCG-01-59-016	23 15 48.	- 03 52	60	15.	GALAXY
PHL 2211	23 15 48.	- 06 57		18.5	BLUE STELLAR OBJECT
PHL 5635	23 15 48.	- 09 31		18.2	BLUE STELLAR OBJECT
PHL 0449	23 15 48.	- 09 56		18.3	BLUE STELLAR OBJECT
PHL 5632	23 15 48.	- 10 04		16.9	BLUE STELLAR OBJECT
PHL 5633	23 15 48.	- 10 34		16.2	BLUE STELLAR OBJECT
PHL 2212	23 15 48.	- 10 37		18.0	BLUE STELLAR OBJECT
PHL 5636	23 15 48.	- 16 28		18.1	BLUE STELLAR OBJECT
PHL 5634	23 15 48.	- 23 22		17.7	BLUE STELLAR OBJECT
PHL 2210	23 15 48.	- 23 48		17.5	BLUE STELLAR OBJECT
PHL 2209	23 15 48.	- 30 14		15.7	BLUE STELLAR OBJECT
RNGC 7589	23 15 49.	- 00 00		15.5	GALAXY
RNGC 7582	23 15 49.	- 42 38		12.0	GALAXY
RNGC 7591	23 15 50.	+ 06 18		14.0	GALAXY
REIN 4.304	23 15 50.99	- 04 45 36.5			NEBULA
LB 01166	23 15 52.	+ 10 30 48.		16.7	FAINT BLUE STAR
SG 1.31	23 15 53.	+ 60 16	900		DIFFUSE EMISSION NEBULA
ZWG 454.031	23 15 54.	+ 19 54		15.7	GALAXY
MCG-01-59-017	23 15 54.	- 04 41	60	14.	GALAXY
PHL 0450	23 15 54.	- 09 53		18.7	BLUE STELLAR OBJECT
PHL 5638	23 15 54.	- 13 55		18.0	BLUE STELLAR OBJECT
PHL 5639	23 15 54.	- 20 30		18.3	BLUE STELLAR OBJECT
PHL 5637	23 15 54.	- 30 45		18.0	BLUE STELLAR OBJECT
FATH 1.844	23 15 55.	+ 15 10	8		NEBULA
RNGC 7592	23 15 55.	- 04 42		14.0	GALAXY
RNGC 7595	23 15 56.	+ 09 38			GALAXY
ABC 2572	23 15 56.	+ 18 28		15.3	RICH CLUSTER OF GALAXIES
RNGC 7597	23 15 57.	+ 18 25		15.5	GALAXY
ABC 2571	23 15 59.	- 02 33		17.6	RICH CLUSTER OF GALAXIES
LBN 0544	23 16	+ 60 20	660		BRIGHT NEBULA
ZWG 406.055	23 16 00.	+ 05 23		15.6	GALAXY
VV 020B	23 16 00.	+ 09 13	12	16.5	INTERACTING GALAXY
VV 020A	23 16 00.	+ 09 13	18	16.	INTERACTING GALAXY
VV 020	23 16 00.	+ 09 13	42		INTERACTING GALAXY
ZWG 454.032	23 16 00.	+ 18 25		15.3	GALAXY
MCG+03-59-032	23 16 00.	+ 18 25	48	14.5	GALAXY
LDN 1231	23 16 00.	+ 61 00	420		DARK NEBULA
PHL 2213	23 16 00.	- 07 30		18.2	BLUE STELLAR OBJECT
PHL 5642	23 16 00.	- 08 40		18.3	BLUE STELLAR OBJECT
PHL 5640	23 16 00.	- 10 16		15.8	BLUE STELLAR OBJECT
MCG-02-59-009	23 16 00.	- 10 42	78	15.5	GALAXY
PHL 5644	23 16 00.	- 11 39		16.6	BLUE STELLAR OBJECT
PHL 5645	23 16 00.	- 20 57		18.4	BLUE STELLAR OBJECT
PHL 0451	23 16 00.	- 23 12		18.0	BLUE STELLAR OBJECT
PHL 5641	23 16 00.	- 25 02		17.8	BLUE STELLAR OBJECT
MCG-05-55-001	23 16 00.	- 27 24	42	14.5	GALAXY
PHL 5643	23 16 00.	- 30 47		18.2	BLUE STELLAR OBJECT
RNGC 7598	23 16 03.	+ 18 28		15.5	GALAXY
MCG+03-59-033	23 16 03.	+ 18 29	12	16.	GALAXY
FATH 1.845	23 16 06.	+ 15 01	19		NEBULA
ZWG 431.039	23 16 06.	+ 15 02		15.7	GALAXY
KARA.73B 1014	23 16 06.	+ 15 02	36	15.7	ISOLATED GALAXY S0
ZWG 454.033	23 16 06.	+ 18 28		15.6	GALAXY
MCG+07-47-010	23 16 06.	+ 42 57	30	16.	GALAXY
PHL 2214	23 16 06.	- 02 06		17.5	BLUE STELLAR OBJECT
MCG-02-59-010	23 16 06.	- 10 41	48	14.	GALAXY
PHL 5646	23 16 06.	- 26 54		17.7	BLUE STELLAR OBJECT
FATH 1.846	23 16 09.	+ 15 41	8		NEBULA
IC 1479	23 16 11.	- 10 40 08.			NONSTELLAR OBJECT
PHL 2215	23 16 12.	+ 01 18		18.7	BLUE STELLAR OBJECT
ZWG 406.056	23 16 12.	+ 08 57		14.7	GALAXY
UGC 12487	23 16 12.	+ 08 57	78	14.7	GALAXY Sb-c
MCG+03-59-034	23 16 12.	+ 18 25 30.	30	15.	GALAXY
UGC 12488	23 16 12.	+ 20 42	66	16.5	GALAXY
UGC 12489	23 16 12.	+ 22 37	78	16.0	GALAXY Sc
ZWG 476.001	23 16 12.	+ 24 57		14.0	GALAXY
ZWG 475.060	23 16 12.	+ 24 57		14.0	GALAXY
MRK 319	23 16 12.	+ 24 57	25	15.5	GALAXY WITH UV CONTINUUM
UGC 12490	23 16 12.	+ 24 57	60	14.0	GALAXY SBa
KARA.72 581A	23 16 12.	+ 24 57	78	14.0	PART OF DOUBLE GALAXY
ZWG 476.002	23 16 12.	+ 24 59		15.7	GALAXY
ZWG 475.061	23 16 12.	+ 24 59		15.7	GALAXY
KARA.72 581B	23 16 12.	+ 24 59	48	15.7	PART OF DOUBLE GALAXY
MCG+07-47-011	23 16 12.	+ 42 39	15	15.	GALAXY
PHL 5648	23 16 12.	- 12 08		17.3	BLUE STELLAR OBJECT
MCG-07-47-030	23 16 12.	- 42 30	138	12.2	GALAXY
SC 2313-6329.4	23 16 13.	- 63 13 01.	24		NEBULA
MCG+01-59-039	23 16 15.	+ 08 59	78	14.5	GALAXY
RNGC 7602	23 16 15.	+ 18 26		15.5	GALAXY
MCG+04-55-001	23 16 15.	+ 24 57	48	14.5	GALAXY
MCG-02-59-011	23 16 15.	- 10 32	60	14.	GALAXY
ARP 092	23 16 17.	- 00 02			PECULIAR GALAXY
ZWG 406.057	23 16 18.	+ 05 56		15.6	GALAXY
ZWG 406.058	23 16 18.	+ 06 46		14.9	GALAXY
MCG+03-59-035	23 16 18.	+ 16 20 30.	54	14.5	GALAXY
ZWG 454.034	23 16 18.	+ 18 26		15.5	GALAXY
ZWG 532.011	23 16 18.	+ 42 41		15.3	GALAXY
UGC 12491	23 16 18.	+ 42 41	72	15.3	GALAXY (E)
MCG+00-59-020	23 16 18.	- 01 20	60	14.	GALAXY
MCG-01-59-018	23 16 18.	- 03 15	48	15.	GALAXY
RNGC 7600	23 16 18.	- 07 52		13.0	GALAXY
MCG-01-59-019	23 16 18.	- 07 52	66	13.	GALAXY
PHL 5651	23 16 18.	- 09 26		17.7	BLUE STELLAR OBJECT
PHL 5652	23 16 18.	- 19 56		18.6	BLUE STELLAR OBJECT
MCG-04-55-001	23 16 18.	- 20 58	42	15.	GALAXY
PHL 5649	23 16 18.	- 27 57		18.0	BLUE STELLAR OBJECT
MCG-07-47-032	23 16 18.	- 42 22		15.	GALAXY
MCG-07-47-031	23 16 18.	- 42 23	30	15.	GALAXY
REIN 2.314	23 16 18.53	- 07 51 14.0			NEBULA
RNGC 7590	23 16 19.	- 42 31		12.5	GALAXY
RNGC 7605	23 16 20.	+ 07 09			NON-EXISTENT OBJECT
RNGC 7601	23 16 20.	+ 08 58		14.5	GALAXY
MCG+07-47-012	23 16 21.	+ 41 45	54	16.	GALAXY
LB 01167	23 16 23.	+ 12 19 18.		15.2	FAINT BLUE STAR
PHL 2216	23 16 24.	+ 03 02		18.4	BLUE STELLAR OBJECT
ZWG 406.059	23 16 24.	+ 06 35		15.0	GALAXY
UGC 12494	23 16 24.	+ 06 35	90	15.0	GALAXY Sc/SBc
MCG+01-59-040	23 16 24.	+ 06 36	72	14.	GALAXY
MCG+01-59-041	23 16 24.	+ 06 37	30	15.	GALAXY
ZWG 454.035	23 16 24.	+ 16 21		15.2	GALAXY
UGC 12495	23 16 24.	+ 16 21	84	15.2	GALAXY Sc-IRR
ZWG 380.026	23 16 24.	- 00 01		14.4	GALAXY
MRK 530	23 16 24.	- 00 01	10	14.5	GALAXY WITH UV CONTINUUM
UGC 12493	23 16 24.	- 00 01	96	14.4	GALAXY S
VVI 93	23 16 24.	- 00 02	95	14.4	SEYFERT GAL1XY
MCG+00-59-021	23 16 24.	- 00 02 30.	96	13.	GALAXY
PHL 5657	23 16 24.	- 00 44		18.6	BLUE STELLAR OBJECT
PHL 5658	23 16 24.	- 00 46		18.1	BLUE STELLAR OBJECT
ZWG 380.025	23 16 24.	- 01 20		14.9	GALAXY
UGC 12492	23 16 24.	- 01 20	66	14.9	GALAXY S0
ZC 2316.4-0238	23 16 24.	- 02 38	1610		CLUSTER OF GALAXIES
PHL 5653	23 16 24.	- 08 40		18.3	BLUE STELLAR OBJECT
MCG-02-59-012	23 16 24.	- 08 46	288	11.5	GALAXY
PHL 5654	23 16 24.	- 09 02		18.9	BLUE STELLAR OBJECT
PHL 2218	23 16 24.	- 09 02		18.2	BLUE STELLAR OBJECT
PHL 2217	23 16 24.	- 10 12		17.9	BLUE STELLAR OBJECT
PHL 5659	23 16 24.	- 10 36		18.7	BLUE STELLAR OBJECT
PHL 5655	23 16 24.	- 12 26		7.1	BLUE STELLAR OBJECT
PHL 5656	23 16 24.	- 14 28		17.5	BLUE STELLAR OBJECT
RNGC 7603	23 16 25.	- 00 01		14.5	GALAXY
RNGC 7607	23 16 26.	+ 11 05			NON-EXISTENT OBJECT
MCG+00-59-022	23 16 27.	+ 02 35	36	15.	GALAXY
SC 2313-6145.2	23 16 27.	- 61 28 49.	66		NEBULA
IC 1480	23 16 28.	+ 11 03			OPEN CLUSTER
SE 1965M	23 16 28.	- 08 45		16.0	SUPERNOVA
FATH 1.847	23 16 29.	+ 15 35	16		NEBULA
REIF 2.315	23 16 29.18	- 08 45 33.5			NEBULA
ZC 2316.5+0046	23 16 30.	+ 00 46	6920		CLUSTER OF GALAXIES
ZWG 380.027	23 16 30.	+ 02 35		15.5	GALAXY
ZC 2316.5+1057	23 16 30.	+ 10 57	940		CLUSTER OF GALAXIES
4ZW 141	23 16 30.	+ 25 47			COMPACT GALAXY
ZWG 532.012	23 16 30.	+ 41 48		14.8	GALAXY
UGC 12496	23 16 30.	+ 41 48	66	14.8	GALAXY Sa
PHL 5660	23 16 30.	- 08 35		17.8	BLUE STELLAR OBJECT
RNGC 7606	23 16 30.	- 08 46		11.5	GALAXY
PHL 2219	23 16 30.	- 21 01		17.7	BLUE STELLAR OBJECT
MCG-04-55-002	23 16 30.	- 23 00	60	14.	GALAXY
LB 01168	23 16 31.	+ 10 05 00.		17.0	FAINT BLUE STAR
LB 00209	23 16 34.	- 14 30 00.		16.6	FAINT BLUE STAR
PHL 0452	23 16 36.	+ 01 57		16.9	BLUE STELLAR OBJECT
ZWG 406.060	23 16 36.	+ 07 25		15.6	GALAXY
UGC 12497	23 16 36.	+ 07 25	66	15.6	GALAXY IRR
ZWG 406.061	23 16 36.	+ 07 50		15.0	GALAXY
UGC 12498	23 16 36.	+ 07 50	90	15.0	GALAXY Sb
MCG+02-59-024	23 16 36.	+ 10 30	60	15.5	GALAXY

OBJECT NAME	RIGHT ASCEN.	DECLINATION	DIAM.	MAGN.	TYPE OF OBJECT
UGC 12499	23 16 36.	+ 25 53	66	17.	GALAXY
PHL 5663	23 16 36.	- 02 32		14.0	BLUE STELLAR OBJECT
LB 01169	23 16 36.	- 06 21 36.		17.2	FAINT BLUE STAR
PHL 5662	23 16 36.	- 15 57		18.6	BLUE STELLAR OBJECT
PHL 5661	23 16 36.	- 19 08		17.7	BLUE STELLAR OBJECT
PHL 2220	23 16 36.	- 19 42		18.5	BLUE STELLAR OBJECT
MCG-04-55-003	23 16 36.	- 23 44	48	15.	GALAXY
PHL 5664	23 16 36.	- 25 16		17.6	BLUE STELLAR OBJECT
MCG-05-55-002	23 16 36.	- 28 17	36	15.	GALAXY
MCG-05-55-003	23 16 36.	- 29 13	30	15.	GALAXY
SER 159.02	23 16 36.	- 42 21	600	18.	FAINT CLUSTER OF GALAXIES
IC 5308	23 16 36.	- 42 33 26.			TWO STARS
REIN 4.305	23 16 36.10	- 07 46 58.2			NEBULA
IC 5309	23 16 38.	+ 07 42 53.			NONSTELLAR OBJECT
MCG-07-47-033	23 16 39.	- 42 32	210	12.2	GALAXY
REIN 4.307	23 16 39.30	- 07 46 24.1			NEBULA
REIN 4.308A	23 16 39.62	+ 07 49 42.6			NEBULA
REIN 4.308B	23 16 39.63	+ 07 49 42.6			NEBULA
REIN 4.309A	23 16 39.89	+ 07 50 04.2			NEBULA
REIN 4.309B	23 16 39.89	+ 07 50 05.1			NEBULA
FATH 1.848	23 16 41.	+ 15 26	8		NEBULA
ARC 2573	23 16 41.	- 02 44		17.6	RICH CLUSTER OF GALAXIES
MCG+01-59-042	23 16 42.	+ 07 50	60	14.	GALAXY
ZWG 406.062	23 16 42.	+ 08 04		15.2	GALAXY
UGC 12500	23 16 42.	+ 08 04	96	15.2	GALAXY S
ZWG 406.063	23 16 42.	+ 09 00		15.5	GALAXY
MCG+01-59-043	23 16 42.	+ 09 02 30.	36	15.	GALAXY
ZWG 431.040	23 16 42.	+ 10 32		15.7	GALAXY
UGC 12501	23 16 42.	+ 10 32	78	15.7	GALAXY S
ZWG 476.003	23 16 42.	+ 22 09		15.5	GALAXY
UGC 12502	23 16 42.	+ 22 09	66	15.5	GALAXY Sa
PHL 2221	23 16 42.	- 21 05		17.8	BLUE STELLAR OBJECT
MCG-04-55-004	23 16 42.	- 22 55	48	14.5	GALAXY
PHL 0453	23 16 42.	- 24 18		18.0	BLUE STELLAR OBJECT
REIN 4.310A	23 16 43.54	+ 08 04 35.2			NEBULA
REIN 4.310B	23 16 43.54	+ 08 04 35.5			NEBULA
RNGC 7608	23 16 44.	+ 08 04		15.0	GALAXY
RNGC 7599	23 16 44.	- 42 32		12.0	GALAXY
MCG+00-59-023	23 16 45.	+ 00 40	72	15.	GALAXY
ARC 2574	23 16 45.	+ 02 18		17.8	RICH CLUSTER OF GALAXIES
MCG+01-59-045	23 16 45.	+ 07 25	66	15.	GALAXY
MCG+01-59-044	23 16 45.	+ 08 06	90	14.	GALAXY
MCG+04-55-002	23 16 45.	+ 25 47	36	15.5	GALAXY
ZWG 380.028	23 16 48.	+ 00 40		15.7	GALAXY
UGC 12503	23 16 48.	+ 00 40	78	15.7	GALAXY S
ZC 2316.8+0224	23 16 48.	+ 02 24	1410		CLUSTER OF GALAXIES
ZWG 431.041	23 16 48.	+ 13 38		15.7	GALAXY
SN 1973T	23 16 48.	+ 13 38		18.0	SUPERNOVA
ZWG 454.036	23 16 48.	+ 18 45		15.5	GALAXY
ZWG 454.037	23 16 48.	+ 19 16		15.4	GALAXY
ZWG 476.004	23 16 48.	+ 25 47		14.8	GALAXY
ZWG 359.004	23 16 48.	+ 77 03		15.6	GALAXY
ZWG 344.002	23 16 48.	+ 77 03		15.6	GALAXY
UGC 12504	23 16 48.	+ 77 03	96	15.6	GALAXY Sc
PHL 5670	23 16 48.	- 01 22		18.5	BLUE STELLAR OBJECT
ZC 2316.8-0135	23 16 48.	- 01 35	870		CLUSTER OF GALAXIES
PHL 0454	23 16 48.	- 04 02		17.2	BLUE STELLAR OBJECT
PHL 5666	23 16 48.	- 04 50		15.8	BLUE STELLAR OBJECT
PHL 0455	23 16 48.	- 10 20		18.9	BLUE STELLAR OBJECT
PHL 2223	23 16 48.	- 12 35		17.9	BLUE STELLAR OBJECT
PHL 5671	23 16 48.	- 14 32		18.6	BLUE STELLAR OBJECT
PHL 2222	23 16 48.	- 20 28		17.8	BLUE STELLAR OBJECT
PHL 5667	23 16 48.	- 21 08		17.5	BLUE STELLAR OBJECT
PHL 5665	23 16 48.	- 23 04		17.1	BLUE STELLAR OBJECT
PHL 5668	23 16 48.	- 25 22		17.7	BLUE STELLAR OBJECT
PHL 5669	23 16 48.	- 27 03		17.8	BLUE STELLAR OBJECT
IC 1481	23 16 52.	+ 05 37 47.			NONSTELLAR OBJECT
ZWG 330.030	23 16 54.	+ 00 50		15.6	GALAXY
ZWG 406.064	23 16 54.	+ 05 38		14.5	GALAXY
UGC 12505	23 16 54.	+ 05 38	54	14.5	GALAXY PECULR?
ZWG 454.038	23 16 54.	+ 16 17		15.7	GALAXY
ZWG 476.005	23 16 54.	+ 23 33		15.7	GALAXY
ZWG 380.029	23 16 54.	- 02 05		15.3	GALAXY
PHL 2224	23 16 54.	- 02 31		18.0	BLUE STELLAR OBJECT
PHL 0457	23 16 54.	- 09 09		13.0	BLUE STELLAR OBJECT
PHL 0456	23 16 54.	- 12 46		18.6	BLUE STELLAR OBJECT
PHL 5672	23 16 54.	- 15 23		18.1	BLUE STELLAR OBJECT
PHL 0459	23 16 54.	- 17 22		14.2	BLUE STELLAR OBJECT
PHL 0458	23 16 54.	- 18 00		18.1	BLUE STELLAR OBJECT
RNGC 7616	23 16 56.	+ 09 53			GALAXY
SC 2313-6350.6	23 16 57.	- 63 34 13.	12		NEBULA
ARP 150	23 16 59.	+ 09 13			PECULIAR GALAXY
PHL 2225	23 17 00.	+ 00 12		16.9	BLUE STELLAR OBJECT
MCG+00-59-024	23 17 00.	+ 01 12 30.	48	14.	GALAXY
ZWG 406.065	23 17 00.	+ 09 13		15.3	GALAXY
MCG+01-59-048	23 17 00.	+ 09 14 30.	36	15.5	GALAXY
MCG+01-59-047	23 17 00.	+ 09 15	60	14.5	GALAXY
MCG+01-59-046	23 17 00.	+ 09 15	36	17.	GALAXY
ZWG 454.039	23 17 00.	+ 15 48		15.6	GALAXY
UGC 12506	23 17 00.	+ 15 48	168	15.6	GALAXY Sc
ZWG 532.013	23 17 00.	+ 43 42		14.2	GALAXY
UGC 12507	23 17 00.	+ 43 42	72	14.2	GALAXY Sa
MCG+13-01-001	23 17 00.	+ 77 02 30.	90	16.	GALAXY
ZWG 380.031	23 17 00.	- 02 07		15.3	GALAXY
PHL 5673	23 17 00.	- 17 42		18.0	BLUE STELLAR OBJECT
PHL 0460	23 17 00.	- 22 36		11.5	BLUE STELLAR OBJECT
PHL 0461	23 17 00.	- 28 36		17.9	BLUE STELLAR OBJECT
SN 1973M	23 17 01.	+ 09 13		19.0	SUPERNOVA
RNGC 7609	23 17 02.	+ 09 13		15.5	GALAXY
FATH 1.849	23 17 02.	+ 15 48	54		NEBULA
MCG+03-59-036	23 17 03.	+ 15 47	180	15.	GALAXY
REIN 4.311A	23 17 03.51	+ 08 20 12.9			NEBULA
REIN 4.311B	23 17 03.52	+ 08 20 13.4			NEBULA
ZWG 380.032	23 17 06.	+ 01 12		14.5	GALAXY
UGC 12508	23 17 06.	+ 01 12	54	14.5	GALAXY S
ZWG 406.066	23 17 06.	+ 07 46		14.0	GALAXY
UGC 12509	23 17 06.	+ 07 46	72	14.0	GALAXY S0
MCG+01-59-049	23 17 06.	+ 07 48	72	13.5	GALAXY
UGC 12510	23 17 06.	+ 07 59	78	16.0	GALAXY DWRF?EL
ZWG 406.067	23 17 06.	+ 09 12		15.7	GALAXY
ZWG 431.042	23 17 06.	+ 09 54		14.9	GALAXY
UGC 12511	23 17 06.	+ 09 54	162	14.9	GALAXY Sc
MCG+02-59-025	23 17 06.	+ 09 54	180	13.5	GALAXY
ZC 2317.1+1210	23 17 06.	+ 12 10	940		CLUSTER OF GALAXIES
ASS 31	23 17 06.	+ 60 53	7200		OB ASSOCIATION CAS OB2
PHL 5674	23 17 06.	- 03 30		18.5	BLUE STELLAR OBJECT
PHL 2226	23 17 06.	- 04 04		18.1	BLUE STELLAR OBJECT
PHL 0462	23 17 06.	- 21 38		17.0	BLUE STELLAR OBJECT

OBJECT NAME	RIGHT ASCEN.	DECLINATION	DIAM.	MAGN.	TYPE OF OBJECT
SC 2314-6230.0	23 17 06.	- 62 13 37.			NEBULA
RNGC 7611	23 17 08.	+ 07 47		14.0	GALAXY
RNGC 7610	23 17 08.	+ 09 54		15.0	GALAXY
ZWG 406.068	23 17 12.	+ 08 17		14.3	GALAXY
UGC 12512	23 17 12.	+ 08 17	96	14.3	GALAXY S0
PHL 2228	23 17 12.	- 20 14		17.0	BLUE STELLAR OBJECT
RNGC 7612	23 17 14.	+ 08 17		14.5	GALAXY
ARC 2575	23 17 14.	- 22 22		17.9	RICH CLUSTER OF GALAXIES
MCG+01-59-050	23 17 15.	+ 08 20	72	14.	GALAXY
UGC 12513	23 17 18.	+ 00 56	60	16.0	GALAXY S
ZWG 406.069	23 17 18.	+ 08 17		15.7	GALAXY
MCG+07-47-013	23 17 18.	+ 42 33	24	14.5	GALAXY
PHL 0463	23 17 18.	- 09 57		17.3	BLUE STELLAR OBJECT
PHL 2231	23 17 18.	- 15 24		17.7	BLUE STELLAR OBJECT
PHL 2230	23 17 18.	- 20 32		18.6	BLUE STELLAR OBJECT
PHL 2229	23 17 18.	- 22 30		17.8	BLUE STELLAR OBJECT
RNGC 7613	23 17 19.	+ 00 08			GALAXY
LB 01170	23 17 21.	+ 12 30 30.		16.9	FAINT BLUE STAR
ARC 2576	23 17 21.	- 22 48		17.5	RICH CLUSTER OF GALAXIES
SC 2314-7114.7	23 17 22.	- 70 58 18.	12		NEBULA
REIN 4.312A	23 17 22.93	+ 08 07 24.8			NEBULA
REIN 4.312B	23 17 22.94	+ 08 07 25.2			NEBULA
ZWG 406.070	23 17 24.	+ 08 07		15.3	GALAXY
MCG+01-59-051	23 17 24.	+ 08 09	48	15.	GALAXY
ZWG 476.006	23 17 24.	+ 25 44		15.6	GALAXY
UGC 12514	23 17 24.	+ 25 44	66	15.6	GALAXY Sc-IRR
MCG+04-55-003	23 17 24.	+ 25 59	78	14.5	GALAXY
ZWG 476.007	23 17 24.	+ 26 00		14.1	GALAXY
UGC 12515	23 17 24.	+ 26 00	84	14.1	GALAXY S0
ZWG 532.014	23 17 24.	+ 42 35		14.3	GALAXY
UGC 12516	23 17 24.	+ 42 35	90	14.3	GALAXY E
PHL 2232	23 17 24.	- 05 23		18.4	BLUE STELLAR OBJECT
PHL 0464	23 17 24.	- 05 26		12.3	BLUE STELLAR OBJECT
REIG 110	23 17 24.	- 05 26		11.5	FAINT BLUE STAR
PHL 2233	23 17 24.	- 16 18		18.5	BLUE STELLAR OBJECT
PHL 5675	23 17 24.	- 21 36		17.9	BLUE STELLAR OBJECT
RNGC 7618	23 17 25.	+ 42 35		14.5	GALAXY
RNGC 7614	23 17 25.	- 00 03			NON-EXISTENT OBJECT
RNGC 7615	23 17 26.	+ 08 08		15.5	GALAXY
ZWG 431.043	23 17 30.	+ 12 37		15.7	GALAXY
MCG+07-47-014	23 17 30.	+ 43 40	21	14.5	GALAXY
ZWG 532.015	23 17 30.	+ 43 41		14.5	GALAXY
UGC 12517	23 17 30.	+ 43 41	84	14.5	GALAXY E?
MCG+00-59-025	23 17 30.	- 00 37	42	15.	GALAXY
PHL 2234	23 17 30.	- 10 53		18.4	BLUE STELLAR OBJECT
PHL 5676	23 17 30.	- 13 48		18.3	BLUE STELLAR OBJECT
FATH 1.850	23 17 33.	+ 15 40	60		NEBULA
RNGC 7620	23 17 34.	+ 23 56		13.5	GALAXY
ZWG 406.071	23 17 36.	+ 07 39		15.7	GALAXY
UGC 12518	23 17 36.	+ 07 39	78	15.7	GALAXY Sb
ZWG 406.072	23 17 36.	+ 07 53		15.1	GALAXY
MCG+01-59-51A	23 17 36.	+ 07 53	42	15.0	GALAXY
MCG+03-59-037	23 17 36.	+ 15 40	78	14.5	GALAXY
ZWG 454.040	23 17 36.	+ 15 41		14.3	GALAXY
UGC 12519	23 17 36.	+ 15 41	90	14.3	GALAXY S
ZWG 454.041	23 17 36.	+ 15 59		15.7	GALAXY
ZWG 454.042	23 17 36.	+ 16 01		15.7	GALAXY
ZWG 476.008	23 17 36.	+ 23 56		13.5	GALAXY
MRK 321	23 17 36.	+ 23 56	24	14.5	GALAXY WITH UV CONTINUUM
UGC 12520	23 17 36.	+ 23 56	78	13.5	GALAXY Sc
ZWG 476.009	23 17 36.	+ 25 57		15.0	GALAXY
MRK 322	23 17 36.	+ 25 57	16	15.5	GALAXY WITH UV CONTINUUM
4ZW 142	23 17 36.	+ 25 58			COMPACT GALAXY
MRK 320	23 17 36.	+ 26 50	6	16.5	GALAXY WITH UV CONTINUUM
ZWG 497.003	23 17 36.	+ 32 40		15.6	GALAXY
5ZW 400	23 17 36.	+ 40 54			COMPACT GALAXY
ZC 2317.6+4129	23 17 36.	+ 41 29	870		CLUSTER OF GALAXIES
ZWG 380.033	23 17 36.	- 00 37		15.7	GALAXY
PHL 0465	23 17 36.	- 03 40		17.0	BLUE STELLAR OBJECT
PHL 2235	23 17 36.	- 07 18		17.9	BLUE STELLAR OBJECT
PHL 5680	23 17 36.	- 12 48		18.5	BLUE STELLAR OBJECT
PHL 5679	23 17 36.	- 20 58		17.9	BLUE STELLAR OBJECT
PHL 5681	23 17 36.	- 21 14		18.0	BLUE STELLAR OBJECT
PHL 5677	23 17 36.	- 27 39		16.1	BLUE STELLAR OBJECT
PHL 5678	23 17 36.	- 31 03		17.0	BLUE STELLAR OBJECT
RNGC 7617	23 17 38.	+ 07 54		15.0	GALAXY
SN 1970J	23 17 40.	+ 07 54		14.5	SUPERNOVA
REIN 4.313A	23 17 40.98	+ 07 39 28.3			NEBULA
REIN 4.313B	23 17 40.99	+ 07 39 29.1			NEBULA
UGC 12522	23 17 42.	+ 07 43	102	16.5	GALAXY DWRF SP
ZWG 406.073	23 17 42.	+ 07 55		12.7	GALAXY
UGC 12523	23 17 42.	+ 07 55	180	12.7	GALAXY E
ZWG 476.010	23 17 42.	+ 25 19		15.5	GALAXY
ZWG 476.011	23 17 42.	+ 25 44		15.5	GALAXY
4ZW 143	23 17 42.	+ 27 37			COMPACT GALAXY
MCG+05-55-001	23 17 42.	+ 32 38	48	15.5	GALAXY
MCG+07-47-015	23 17 42.	+ 42 40	51	17.	GALAXY
ZWG 380.034	23 17 42.	- 02 07		14.4	GALAXY
UGC 12521	23 17 42.	- 02 07	96	14.4	GALAXY Sc
MCG+00-59-026	23 17 42.	- 02 08	60	13.	GALAXY
PHL 5684	23 17 42.	- 09 11		18.2	BLUE STELLAR OBJECT
PHL 5682	23 17 42.	- 09 38		18.2	BLUE STELLAR OBJECT
PHL 0466	23 17 42.	- 18 26		14.2	BLUE STELLAR OBJECT
PHL 5683	23 17 42.	- 20 52		18.5	BLUE STELLAR OBJECT
MCG-07-47-034	23 17 42.	- 42 02	9	15.5	GALAXY
SC 2314-6224.0	23 17 42.	- 62 07 36.	42		NEBULA
REIN 2.316	23 17 42.64	+ 07 55 52.9			NEBULA
RNGC 7619	23 17 44.	+ 07 56		13.0	GALAXY
MCG+01-59-053	23 17 45.	+ 07 40	78	15.	GALAXY
MCG+01-59-052	23 17 45.	+ 07 57	48	12.	GALAXY
LB 01171	23 17 45.	+ 10 35 42.		16.9	FAINT BLUE STAR
MCG+00-59-027	23 17 48.	+ 01 17 30.	30	15.	GALAXY
MCG+01-59-054	23 17 48.	+ 07 44	42	16.	GALAXY
ZWG 406.074	23 17 48.	+ 08 04		15.6	GALAXY
MCG+02-59-026	23 17 48.	+ 11 47 30.	60	15.	GALAXY
ZWG 532.016	23 17 48.	+ 42 41		15.7	GALAXY
UGC 12524	23 17 48.	+ 42 41	84	15.7	GALAXY SBb
PHL 5686	23 17 43.	- 00 54		18.0	BLUE STELLAR OBJECT
MCG-01-59-020	23 17 48.	- 05 30	18	16.	GALAXY
PHL 2236	23 17 48.	- 09 11		18.4	BLUE STELLAR OBJECT
PHL 0467	23 17 48.	- 11 30		17.6	BLUE STELLAR OBJECT
PHL 5687	23 17 48.	- 11 50		18.0	BLUE STELLAR OBJECT
PHL 0468	23 17 48.	- 18 14		18.1	BLUE STELLAR OBJECT
PHL 5685	23 17 48.	- 29 54		17.5	BLUE STELLAR OBJECT
KW 59	23 17 49.	+ 00 07	55		SEYFERT GALAXY
RNGC 7621	23 17 50.	+ 08 04		15.5	GALAXY
RNGC 7624	23 17 52.	+ 27 03		13.5	GALAXY
SG 3.242	23 17 52.	+ 61 54	1800		DIFFUSE EMISSION NEBULA

OBJECT NAME	RIGHT ASCEN.	DECLINATION	DIAM.	MAGN.	TYPE OF OBJECT
REIN 4.314A	23 17 52.73	+ 08 05 28.0			NEBULA
REIN 4.314B	23 17 52.73	+ 08 05 28.3			NEBULA
ZWG 380.036	23 17 54.	+ 01 17		15.1	GALAXY
UGC 12525	23 17 54.	+ 01 17	60	15.1	GALAXY TRP SYS
ZWG 406.075	23 17 54.	+ 08 06		13.9	GALAXY
UGC 12526	23 17 54.	+ 08 06	102	13.9	GALAXY S0
MCG+01-59-055	23 17 54.	+ 08 07	36	15.5	GALAXY
VV 280B	23 17 54.	+ 16 57	18	15.	INTERACTING GALAXY
VV 280A	23 17 54.	+ 16 57	48	13.	INTERACTING GALAXY
VV 280	23 17 54.	+ 16 57	90	13.2	INTERACTING GALAXY
ZC 2317.9+2233	23 17 54.	+ 22 33	670		CLUSTER OF GALAXIES
ZWG 476.012	23 17 54.	+ 27 03		13.7	GALAXY
MRK 323	23 17 54.	+ 27 03	40	15.5	GALAXY WITH UV CONTINUUM
UGC 12527	23 17 54.	+ 27 03	66	13.7	GALAXY Sc
MCG+06-51-003	23 17 54.	+ 35 12 30.	72	15.	GALAXY
ZWG 516.003	23 17 54.	+ 35 14		15.3	GALAXY
UGC 12528	23 17 54.	+ 35 14	72	15.3	GALAXY Sb-c
ZWG 380.035	23 17 54.	- 01 16		15.1	GALAXY
MCG+00-59-028	23 17 54.	- 01 17 30.	15	15.	GALAXY
PHL 2237	23 17 54.	- 01 58		18.5	BLUE STELLAR OBJECT
PHL 5688	23 17 54.	- 18 34		18.0	BLUE STELLAR OBJECT
MCG+03-59-038	23 17 57.	+ 16 57	102	13.	GALAXY
MCG+03-59-039	23 17 57.	+ 18 44	30	16.5	GALAXY
MCG+04-55-004	23 17 57.	+ 27 03	48	14.5	GALAXY
LBN 0549	23 18	+ 60 55	1620		BRIGHT NEBULA
LBN 0548	23 18	+ 60 55	660		BRIGHT NEBULA
KHAV 786	23 18	+ 67 10	10690		DARK NEBULA
KHAV 788	23 18	+ 70 40	3330		DARK NEBULA
KHAV 787	23 18	+ 73 46	1570		DARK NEBULA
SER 159.01	23 18	- 42 02	180		LOW SURF. BRGHTNSS GALAXY
LB 09952	23 18	- 80 12		14.1	FAINT BLUE STAR
MCG+01-59-056	23 18 00.	+ 08 09	72	14.	GALAXY
ZC 2318.0+0812	23 18 00.	+ 08 12	1210		CLUSTER OF GALAXIES
3ZW 102	23 18 00.	+ 16 57			COMPACT GALAXY
ZWG 454.043	23 18 00.	+ 16 57		12.8	GALAXY
UGC 12529	23 18 00.	+ 16 57	90	12.8	GALAXY
ARP 212	23 18 00.	+ 16 57			PECULIAR GALAXY
ZC 2318.0+1910	23 18 00.	+ 19 10	12900		CLUSTER OF GALAXIES
ZC 2318.0+8243	23 18 00.	+ 82 43	2080		CLUSTER OF GALAXIES
PHL 2239	23 18 00.	- 07 30		17.8	BLUE STELLAR OBJECT
PHL 5689	23 18 00.	- 09 32		18.3	BLUE STELLAR OBJECT
PHL 0469	23 18 00.	- 11 18		16.6	BLUE STELLAR OBJECT
PHL 5690	23 18 00.	- 12 16		17.3	BLUE STELLAR OBJECT
PHL 2238	23 18 00.	- 12 38		16.6	BLUE STELLAR OBJECT
PHL 2240	23 18 00.	- 13 06		18.1	BLUE STELLAR OBJECT
PHL 5691	23 18 00.	- 21 20		18.2	BLUE STELLAR OBJECT
PHL 5692	23 18 00.	- 21 52		18.3	BLUE STELLAR OBJECT
KARA.73 57	23 18 01.	- 42 01	47		DWARF GALAXY
SC 2315-6133.8	23 18 01.	- 61 17 24.	12		NEBULA
RNGC 7623	23 18 02.	+ 08 07		14.0	GALAXY
RNGC 7625	23 18 03.	+ 16 57		13.0	GALAXY
IC 5311	23 18 05.	+ 17 00			SINGLE STAR
ZWG 454.044	23 18 06.	+ 15 56		15.4	GALAXY
4ZW 144	23 18 06.	+ 25 05			COMPACT GALAXY
ZWG 497.004	23 18 06.	+ 29 02		14.9	GALAXY
UGC 12530	23 18 06.	+ 29 02	90	14.9	GALAXY SBc
PHL 2242	23 18 06.	- 05 40		16.6	BLUE STELLAR OBJECT
PHL 5693	23 18 06.	- 22 49		17.6	BLUE STELLAR OBJECT
PHL 2241	23 18 06.	- 25 26		9.2	BLUE STELLAR OBJECT
FATH 1.851	23 18 09.	+ 15 55	27		NEBULA
MCG+05-55-002	23 18 09.	+ 29 01	84	14.5	GALAXY
IC 5310	23 18 09.	- 22 25 23.			NONSTELLAR OBJECT
ABC 2577	23 18 09.	- 23 15		17.5	RICH CLUSTER OF GALAXIES
REIN 2.317	23 18 10.67	+ 07 56 31.1			NEBULA
ZC 2318.2+0015	23 18 12.	+ 00 15	1280		CLUSTER OF GALAXIES
PHL 5694	23 18 12.	+ 01 11		16.8	BLUE STELLAR OBJECT
ZWG 406.076	23 18 12.	+ 07 56		12.8	GALAXY
UGC 12531	23 18 12.	+ 07 56	174	12.8	GALAXY E
ZC 2318.2+1138	23 18 12.	+ 11 38	940		CLUSTER OF GALAXIES
ABC 2578	23 18 12.	- 04 49		17.6	RICH CLUSTER OF GALAXIES
PHL 5697	23 18 12.	- 09 54		18.3	BLUE STELLAR OBJECT
PHL 5695	23 18 12.	- 18 40		17.9	BLUE STELLAR OBJECT
MCG-03-59-004	23 18 12.	- 20 26 30.	36	16.	GALAXY
PHL 5696	23 18 12.	- 28 47		18.0	BLUE STELLAR OBJECT
RNGC 7626	23 18 14.	+ 07 57		13.0	GALAXY
MCG+00-59-029	23 18 15.	+ 01 28	15	14.5	GALAXY
MCG+01-59-057	23 18 15.	+ 07 57 30.	48	12.	GALAXY
MCG+03-59-040	23 18 15.	+ 18 37 30.	42	15.5	GALAXY
IC 1482	23 18 16.	+ 01 27 37.			NONSTELLAR OBJECT
SG 1.32	23 18 17.	+ 60 59	180		DIFFUSE EMISSION NEBULA
ZWG 380.037	23 18 18.	+ 01 27		15.3	GALAXY
UGC 12532	23 18 18.	+ 02 14	66	17.	GALAXY
ZWG 454.045	23 18 18.	+ 18 38		15.5	GALAXY
ZWG 454.046	23 18 18.	+ 19 21		15.7	GALAXY
PHL 0470	23 18 18.	- 06 07		14.9	BLUE STELLAR OBJECT
PHL 5698	23 18 18.	- 09 15		16.9	BLUE STELLAR OBJECT
PHL 2243	23 18 18.	- 10 51		18.8	BLUE STELLAR OBJECT
PHL 0471	23 18 18.	- 12 57		16.6	BLUE STELLAR OBJECT
PHL 0472	23 18 18.	- 14 15		18.0	BLUE STELLAR OBJECT
PHL 5700	23 18 18.	- 21 21		18.2	BLUE STELLAR OBJECT
PHL 5699	23 18 18.	- 22 10		18.0	BLUE STELLAR OBJECT
PHL 5650	23 18 18.	- 28 10		18.3	BLUE STELLAR OBJECT
LB 01172	23 18 19.	- 02 45 48.		16.9	FAINT BLUE STAR
RNGC 7627	23 18 20.	+ 11 47			NON-EXISTENT OBJECT
LB 01173	23 18 21.	- 06 07 30.		15.2	FAINT BLUE STAR
RNGC 7628	23 18 22.	+ 25 37		14.0	GALAXY
ZWG 406.077	23 18 24.	+ 07 38		15.6	GALAXY
ZWG 476.013	23 18 24.	+ 23 32		15.6	GALAXY
UGC 12533	23 18 24.	+ 23 32	96	15.6	GALAXY Sb
ZWG 476.014	23 18 24.	+ 25 37		13.8	GALAXY
UGC 12534	23 18 24.	+ 25 37	78	13.8	GALAXY E
ZC 2318.4+3651	23 18 24.	+ 36 51	3020		CLUSTER OF GALAXIES
PHL 5702	23 18 24.	- 11 56		17.9	BLUE STELLAR OBJECT
PHL 5701	23 18 24.	- 22 54		17.7	BLUE STELLAR OBJECT
FATH 1.852	23 18 27.	+ 15 55	14		NEBULA
MCG+03-59-041	23 18 27.	+ 19 01 30.	15	15.	GALAXY
MCG+04-55-005	23 18 27.	+ 25 37	24	14.5	GALAXY
IC 5312	23 18 29.	+ 19 02 31.			NONSTELLAR OBJECT
REIN 4.315A	23 18 29.59	+ 07 38 24.7			NEBULA
REIN 4.315B	23 18 29.60	+ 07 38 25.3			NEBULA
REIN 4.316A	23 18 29.78	+ 07 54 11.3			NEBULA
REIN 4.316B	23 18 29.78	+ 07 54 11.5			NEBULA
MCG+00-59-030	23 18 30.	+ 02 37	48	15.	GALAXY
UGC 12535	23 18 30.	+ 07 54	66	16.0	GALAXY Sb-c
SN 1968K	23 18 30.	+ 14 57		15.4	SUPERNOVA
ZWG 454.047	23 18 30.	+ 19 03		15.0	GALAXY
UGC 12536	23 18 30.	+ 43 22	60	17.	GALAXY
COU 104	23 18 30.	+ 60 45	1500		EMISSION NEBULA
SG 1.33A	23 18 30.	+ 60 53			DIFFUSE EMISSION NEBULA
SG 1.33	23 18 30.	+ 60 54	180		DIFFUSE EMISSION NEBULA
RNGC 7635	23 18 30.	+ 60 55			DIFFUSE NEBULA
MRSL 112+00/1	23 18 30.	+ 60 55	2400		HII REGION
CED 210	23 18 30.	+ 60 55	930		DIFFUSE GALACTIC NEBULA
MCG-01-59-021	23 18 30.	- 05 10	60	14.5	GALAXY
ABC 2579	23 18 32.	- 21 51		17.1	RICH CLUSTER OF GALAXIES
REIN 4.317B	23 18 33.59	+ 07 49 39.6			NEBULA
REIN 4.317A	23 18 33.59	+ 07 49 39.6			NEBULA
ZWG 380.038	23 18 36.	+ 00 07		15.7	GALAXY
PHL 0474	23 18 36.	+ 00 48		18.4	BLUE STELLAR OBJECT
PHL 2244	23 18 36.	+ 02 28		17.5	BLUE STELLAR OBJECT
ZWG 380.039	23 18 36.	+ 02 36		15.6	GALAXY
ZWG 406.078	23 18 36.	+ 07 05		14.8	GALAXY
ZWG 406.079	23 18 36.	+ 07 49		15.6	GALAXY
MCG+01-59-058	23 18 36.	+ 07 50	48	15.	GALAXY
ZWG 406.080	23 18 36.	+ 07 54		15.6	GALAXY
ZWG 406.081	23 18 36.	+ 09 25		15.5	GALAXY
MCG+03-59-042	23 18 36.	+ 19 01	27	15.	GALAXY
ZC 2318.6+2224	23 18 36.	+ 22 24	670		CLUSTER OF GALAXIES
ZWG 476.015	23 18 36.	+ 26 08		15.7	GALAXY
SG 1.34	23 18 36.	+ 60 56	180		DIFFUSE EMISSION NEBULA
PHL 5703	23 18 36.	- 11 44		17.4	BLUE STELLAR OBJECT
RNGC 7622	23 18 38.	- 62 23			GALAXY
IC 5314	23 18 40.	+ 19 02 02.			NONSTELLAR OBJECT
PHL 5704	23 18 42.	+ 01 18		17.7	BLUE STELLAR OBJECT
MCG+01-59-059	23 18 42.	+ 07 05	48	14.5	GALAXY
ZWG 406.082	23 18 42.	+ 07 11		15.6	GALAXY
MCG+02-59-027	23 18 42.	+ 11 07	72	14.5	GALAXY
ZWG 454.048	23 18 42.	+ 19 02		15.6	GALAXY
4ZW 145	23 18 42.	+ 29 17			COMPACT GALAXY
ZWG 497.005	23 18 42.	+ 29 17		15.4	GALAXY
UGC 12537	23 18 42.	+ 29 17	78	15.4	GALAXY COMPACT
ZWG 497.006	23 18 42.	+ 33 08		14.7	GALAXY
UGC 12538	23 18 42.	+ 32 08	60	14.7	GALAXY PECULR
MCG-04-55-005	23 18 42.	- 21 29	36	15.	GALAXY
TON-S 0097	23 18 42.	- 22 36		15.1	BLUE STAR
MCG+00-59-031	23 18 45.	+ 01 08	18	15.	GALAXY
MCG+05-55-003	23 18 45.	+ 29 16	72	15.	GALAXY
LB 01174	23 18 45.	- 06 57 36.		15.6	FAINT BLUE STAR
REIN 2.318	23 18 45.54	+ 07 55 51.9			NEBULA
ZWG 380.040	23 18 48.	+ 01 07		15.2	GALAXY
ZWG 406.083	23 18 48.	+ 07 56		13.8	GALAXY
UGC 12539	23 18 48.	+ 07 56	108	13.8	GALAXY Sb
ZWG 431.044	23 18 48.	+ 11 08		15.5	GALAXY
UGC 12540	23 18 48.	+ 11 08	66	15.5	GALAXY S
ZWG 476.016	23 18 48.	+ 25 06		14.6	GALAXY
ZCG 2318+25	23 18 48.	+ 25 06		14.6	COMPACT GALAXY
UGC 12541	23 18 48.	+ 25 06	60	14.6	GALAXY COMPACT
MCG+04-55-006	23 18 48.	+ 25 06	36	15.	GALAXY
KARA.72P 1015	23 18 48.	+ 25 06	42	14.6	ISOLATED GALAXY E
MCG+05-55-004	23 18 48.	+ 33 08	60	15.	GALAXY
PHL 5706	23 18 48.	- 08 30		18.5	BLUE STELLAR OBJECT
PHL 5707	23 18 48.	- 15 08		18.7	BLUE STELLAR OBJECT
PHL 2245	23 18 48.	- 18 34		18.3	BLUE STELLAR OBJECT
PHL 0475	23 18 48.	- 22 37		16.5	BLUE STELLAR OBJECT
PHL 5705	23 18 48.	- 26 35		17.8	BLUE STELLAR OBJECT
PHL 5708	23 18 48.	- 29 09		18.3	BLUE STELLAR OBJECT
RNGC 7629	23 18 49.	+ 01 07		15.0	GALAXY
RNGC 7630	23 18 50.	+ 11 08		15.5	GALAXY
IC 5315	23 18 52.	+ 25 06 56.			NONSTELLAR OBJECT
MCG+01-59-060	23 18 54.	+ 07 57 30.	96	13.	GALAXY
MCG+01-59-061	23 18 54.	+ 08 17	30	15.5	GALAXY
ZWG 406.084	23 18 54.	+ 09 17		15.7	GALAXY
ZWG 454.049	23 18 54.	+ 16 15		15.7	GALAXY
ZWG 476.017	23 18 54.	+ 21 56		15.7	GALAXY
ZWG 380.041	23 18 54.	- 01 15		15.7	GALAXY
PHL 0476	23 18 54.	- 04 04		17.9	BLUE STELLAR OBJECT
PHL 2247	23 18 54.	- 18 59		18.4	BLUE STELLAR OBJECT
PHL 2246	23 18 54.	- 26 24		18.0	BLUE STELLAR OBJECT
RNGC 7631	23 18 56.	+ 07 57		14.0	GALAXY
MCG+00-59-032	23 19 00.	+ 01 11	48	15.	GALAXY
PHL 2248	23 19 00.	+ 02 34		17.9	BLUE STELLAR OBJECT
ZC 2319.0+1353	23 19 00.	+ 13 53	1010		CLUSTER OF GALAXIES
MCG+04-55-007	23 19 00.	+ 26 50	48	15.5	GALAXY
SG 1.35	23 19 00.	+ 60 56	1500		DIFFUSE EMISSION NEBULA
PHL 2249	23 19 00.	- 01 20		17.4	BLUE STELLAR OBJECT
PHL 5710	23 19 00.	- 11 59		16.7	BLUE STELLAR OBJECT
PHL 5711	23 19 00.	- 16 35		18.8	BLUE STELLAR OBJECT
PHL 5712	23 19 00.	- 18 00		18.8	BLUE STELLAR OBJECT
PHL 5709	23 19 00.	- 30 15		18.0	BLUE STELLAR OBJECT
ABC 2580	23 19 02.	- 23 31		17.1	RICH CLUSTER OF GALAXIES
IC 5313	23 19 03.	- 42 46 52.			MAY NOT EXIST
ZWG 380.042	23 19 06.	+ 01 10		15.5	GALAXY
ZC 2319.1+0234	23 19 06.	+ 02 34	1010		CLUSTER OF GALAXIES
SN 1972J	23 19 06.	+ 08 35		14.3	SUPERNOVA
ZWG 406.085	23 19 06.	+ 08 36		13.7	GALAXY
UGC 12542	23 19 06.	+ 08 36	78	13.7	GALAXY SBO
ZWG 406.086	23 19 06.	+ 08 42		15.5	GALAXY
ZC 2319.1+1448	23 19 06.	+ 14 48	940		CLUSTER OF GALAXIES
ZWG 476.018	23 19 06.	+ 26 51		15.4	GALAXY
UGC 12543	23 19 06.	+ 26 51	60	15.4	GALAXY Sc
PHL 0477	23 19 06.	- 09 45		17.7	BLUE STELLAR OBJECT
PHL 2250	23 19 06.	- 23 16		15.4	BLUE STELLAR OBJECT
MCG-05-55-004	23 19 06.	- 26 34	36	15.	GALAXY
RNGC 7634	23 19 08.	+ 08 36		13.5	GALAXY
ABC 2581	23 19 11.	- 17 15		17.6	RICH CLUSTER OF GALAXIES
ZWG 406.087	23 19 12.	+ 03 33		15.3	GALAXY
MCG+01-59-062	23 19 12.	+ 08 38	30	13.5	GALAXY
MCG+01-59-063	23 19 12.	+ 08 44	48	14.5	GALAXY
ZWG 406.088	23 19 12.	+ 08 47		15.5	GALAXY
UGC 12544	23 19 12.	+ 08 47	72	15.5	GALAXY
KARA.68 239	23 19 12.	+ 08 52	34		DWARF GALAXY
ZWG 406.089	23 19 12.	+ 09 03		15.6	GALAXY
ZWG 476.019	23 19 12.	+ 26 12		15.7	GALAXY
MCG+04-55-009	23 19 12.	+ 26 47 30.	66	15.	GALAXY
ZWG 476.020	23 19 12.	+ 26 48		15.3	GALAXY
UGC 12545	23 19 12.	+ 26 48	84	15.3	GALAXY SBc
MCG+04-55-008	23 19 12.	+ 26 48 30.	66	15.	GALAXY
ZWG 476.021	23 19 12.	+ 26 49		15.3	GALAXY
UGC 12546	23 19 12.	+ 26 49	84	15.3	GALAXY Sb-c
ZWG 497.007	23 19 12.	+ 33 13		15.0	GALAXY
PHL 2251	23 19 12.	- 04 10		17.9	BLUE STELLAR OBJECT
PHL 0473	23 19 12.	- 10 02		7.4	BLUE STELLAR OBJECT
PHL 2252	23 19 12.	- 15 52		18.5	BLUE STELLAR OBJECT
MCG-03-59-005	23 19 12.	- 17 00	30	15.	GALAXY
PHL 0479	23 19 12.	- 31 57		17.7	BLUE STELLAR OBJECT
LB 01175	23 19 13.	- 05 38 12.		16.6	FAINT BLUE STAR

OBJECT NAME	RIGHT ASCEN.	DECLINATION	DIAM.	MAGN.	TYPE OF OBJECT
ARC 2582	23 19 15.	+ 02 40		17.6	RICH CLUSTER OF GALAXIES
MCG+01-59-064	23 19 15.	+ 08 49	48	14.	GALAXY
ZWG 406.090	23 19 18.	+ 04 43		15.2	GALAXY
UGC 12547	23 19 18.	+ 04 43	72	15.2	GALAXY S
KARA.72 582A	23 19 18.	+ 04 43	72	15.2	PART OF DOUBLE GALAXY
MCG+01-59-065	23 19 18.	+ 09 05	36	14.5	GALAXY
ZWG 380.043	23 19 18.	- 00 57		15.0	GALAXY
MCG+00-59-033	23 19 18.	- 00 57 30.	36	14.	GALAXY
MCG-07-47-036	23 19 18.	- 41 50		16.	GALAXY
MCG-07-47-035	23 19 18.	- 42 45	78	13.5	GALAXY
HOLM 808A	23 19 21.	+ 04 44	48	13.7	PART OF MULTIPLE GALAXY
MCG+01-59-066	23 19 24.	+ 04 43	72	14.	GALAXY
ZWG 406.091	23 19 24.	+ 04 45		14.9	GALAXY
UGC 12548	23 19 24.	+ 04 45	66	14.9	GALAXY Sa?
KARA.72 582B	23 19 24.	+ 04 45	72	14.9	PART OF DOUBLE GALAXY
ZWG 406.092	23 19 24.	+ 09 01		15.7	GALAXY
UGC 12549	23 19 24.	+ 17 10	66	16.0	GALAXY IRR
UGC 12550	23 19 24.	+ 43 44	72	16.0	GALAXY SBO?
PHL 2254	23 19 24.	- 11 47		18.8	BLUE STELLAR OBJECT
PHL 5717	23 19 24.	- 20 13		18.5	BLUE STELLAR OBJECT
PHL 5715	23 19 24.	- 20 14		18.5	BLUE STELLAR OBJECT
PHL 2253	23 19 24.	- 21 02		17.2	BLUE STELLAR OBJECT
PHL 5713	23 19 24.	- 22 50		17.7	BLUE STELLAR OBJECT
PHL 5716	23 19 24.	- 27 08		18.0	BLUE STELLAR OBJECT
PHL 5714	23 19 24.	- 31 43		17.8	BLUE STELLAR OBJECT
HOLM 808B	23 19 26.	+ 04 46	48	13.8	PART OF MULTIPLE GALAXY
IC 5316	23 19 26.	+ 20 56 09.			NONSTELLAR OBJECT
RNGC 7632	23 19 26.	- 42 46			GALAXY
ZWG 380.044	23 19 30.	+ 01 12		15.6	GALAXY
MCG+00-59-034	23 19 30.	+ 01 13	30	15.	GALAXY
ZWG 380.045	23 19 30.	+ 01 25		15.4	GALAXY
MCG+01-59-067	23 19 30.	+ 04 45	60	14.	GALAXY
MCG+02-59-028	23 19 30.	+ 12 45	108	15.	GALAXY
MCG+07-48-001	23 19 30.	+ 43 42	42	16.5	GALAXY
ISS 0024	23 19 30.	+ 63 48	237		STELLAR RING
PHL 5718	23 19 30.	- 21 14		17.8	BLUE STELLAR OBJECT
PHL 2255	23 19 30.	- 22 54		16.0	BLUE STELLAR OBJECT
MCG-04-55-006	23 19 30.	- 23 47 30.	48	14.5	GALAXY
MCG+03-59-043	23 19 33.	+ 19 07	24	17.	GALAXY
ARC 2584	23 19 34.	+ 27 17		17.1	RICH CLUSTER OF GALAXIES
SC 2316-6015.8	23 19 35.	- 59 59 23.	18		NEBULA
ZCG 2319+04	23 19 36.	+ 04 43		15.9	COMPACT GALAXY
ZWG 406.093	23 19 36.	+ 09 00		15.7	GALAXY
UGC 12551	23 19 36.	+ 09 00	72	15.7	GALAXY SBb
ZWG 431.045	23 19 36.	+ 12 46		15.7	GALAXY
UGC 12552	23 19 36.	+ 12 46	108	15.7	GALAXY Sa-b
ZC 2319.6+2712	23 19 36.	+ 27 12	1810		CLUSTER OF GALAXIES
ZWG 476.022	23 19 36.	+ 27 15		15.4	GALAXY
MCG+07-48-002	23 19 36.	+ 40 33	600	12.	GALAXY
ZC 2319.6+4109	23 19 36.	+ 41 09	1340		CLUSTER OF GALAXIES
OCL 0253	23 19 36.	+ 55 30	720		OPEN STAR CLUSTER
OCL 0271	23 19 36.	+ 71 29	420	14.	OPEN STAR CLUSTER
PHL 5719	23 19 36.	- 09 06		17.8	BLUE STELLAR OBJECT
PHL 2256	23 19 36.	- 09 39		17.5	BLUE STELLAR OBJECT
PHL 0481	23 19 36.	- 09 56		18.9	BLUE STELLAR OBJECT
PHL 0480	23 19 36.	- 10 54		17.9	BLUE STELLAR OBJECT
PHL 5721	23 19 36.	- 11 49		18.0	BLUE STELLAR OBJECT
PHL 2257	23 19 36.	- 16 39		18.9	BLUE STELLAR OBJECT
PHL 5720	23 19 36.	- 22 20		17.9	BLUE STELLAR OBJECT
ARC 2583	23 19 37.	- 20 43		17.1	RICH CLUSTER OF GALAXIES
RNGC 7636	23 19 40.	- 29 34		15.0	GALAXY
MCG+01-59-068	23 19 42.	+ 09 00	54	15.5	GALAXY
UGC 12553	23 19 42.	+ 09 07	84	17.	GALAXY DWARF
RNGC 7640	23 19 42.	+ 40 34		11.5	GALAXY
ZWG 533.001	23 19 42.	+ 40 35		11.6	GALAXY
ZWG 532.017	23 19 42.	+ 40 25		11.6	GALAXY
UGC 12554	23 19 42.	+ 40 35	660	11.6	GALAXY SBc
ZC 2319.7-0038	23 19 42.	- 00 38	740		CLUSTER OF GALAXIES
PHL 5722	23 19 42.	- 13 24		18.6	BLUE STELLAR OBJECT
PHL 2258	23 19 42.	- 21 07		17.7	BLUE STELLAR OBJECT
MCG-04-55-007	23 19 42.	- 23 58	54	15.	GALAXY
MCG-05-55-005	23 19 42.	- 29 34	36	15.	GALAXY
MCG+01-59-069	23 19 45.	+ 09 07 30.	120	16.	GALAXY
ZWG 380.046	23 19 48.	+ 03 13		15.7	GALAXY
MCG+02-59-029	23 19 48.	+ 11 37	108	14.5	GALAXY
ZWG 454.050	23 19 48.	+ 18 27		15.4	GALAXY
PHL 5723	23 19 48.	- 12 31		16.0	BLUE STELLAR OBJECT
PHL 0482	23 19 48.	- 31 20		14.2	BLUE STELLAR OBJECT
LB 01521	23 19 48.	- 46 16		11.7	FAINT BLUE STAR
LB 01176	23 19 49.	+ 11 50 18.		15.5	FAINT BLUE STAR
ZC 2319.9+0054	23 19 54.	+ 00 54	1410		CLUSTER OF GALAXIES
KARA.68 240	23 19 54.	+ 04 30	34		DWARF GALAXY
MCG+02-59-030	23 19 54.	+ 11 02	36	15.	GALAXY
PHL 5724	23 19 54.	- 02 24		17.4	BLUE STELLAR OBJECT
MCG-02-59-013	23 19 54.	- 13 22	78	14.	GALAXY
SC 2316-6124.6	23 19 55.	- 61 08 10.	6		NEBULA
RNGC 7638	23 19 56.	+ 11 02		15.0	GALAXY
LBN 0467	23 20	+ 33 30	600		BRIGHT NEBULA
LBN 0480	23 20	+ 35 00	3300		BRIGHT NEBULA
LBN 0575	23 20	+ 73 50	1620		BRIGHT NEBULA
PHL 2259	23 20 00.	+ 00 54		18.3	BLUE STELLAR OBJECT
UGC 12555	23 20 00.	+ 04 51	66	16.0	GALAXY Sc
ZC 2320.0+0845	23 20 00.	+ 08 45	22710		CLUSTER OF GALAXIES
ZWG 431.046	23 20 00.	+ 11 04		15.5	GALAXY
ZWG 431.047	23 20 00.	+ 11 37		15.2	GALAXY
UGC 12556	23 20 00.	+ 11 37	102	15.2	GALAXY Sa
ZWG 431.048	23 20 00.	+ 12 59		15.6	GALAXY
MCG+03-59-045	23 20 00.	+ 19 10	18	16.5	GALAXY
MCG+03-59-044	23 20 00.	+ 19 10	30	16.5	GALAXY
MCG+05-55-005	23 20 00.	+ 28 54	114	15.	GALAXY
ZWG 497.008	23 20 00.	+ 28 55		15.3	GALAXY
UGC 12557	23 20 00.	+ 28 55	90	15.3	GALAXY S
ZWG 547.003	23 20 00.	+ 49 54		14.9	GALAXY
UGC 12558	23 20 00.	+ 49 54	66	14.9	GALAXY Sc
LDN 1250	23 20 00.	+ 67 00	6540		DARK NEBULA
PHL 0483	23 20 00.	- 08 46		18.5	BLUE STELLAR OBJECT
PHL 5725	23 20 00.	- 09 28		18.2	BLUE STELLAR OBJECT
PHL 0484	23 20 00.	- 10 01		18.2	BLUE STELLAR OBJECT
PHL 0485	23 20 00.	- 13 57		17.5	BLUE STELLAR OBJECT
IC 1483	23 20 01.	+ 11 03 16.			NONSTELLAR OBJECT
BC PKS2319+07	23 20 02.	+ 07 55 48.		17.5	QUASI-STELLAR OBJECT
RNGC 7641	23 20 02.	+ 11 37		15.0	GALAXY
LB 01177	23 20 02.	+ 11 48 30.		15.6	FAINT BLUE STAR
SH3 387	23 20 03.	+ 07 57		18.5	QUASI-STELLAR OBJECT
RNGC 7633	23 20 05.	- 67 56			UNVERIFIED SOUTHERN OBJECT
ZWG 406.094	23 20 06.	+ 03 57		15.6	GALAXY
ZWG 454.051	23 20 06.	+ 17 10		15.7	GALAXY
ZWG 454.052	23 20 06.	+ 19 11		15.5	GALAXY

OBJECT NAME	RIGHT ASCEN.	DECLINATION	DIAM.	MAGN.	TYPE OF OBJECT
ZWG 476.023	23 20 06.	+ 23 06		15.7	GALAXY
ZC 2320.1+4134	23 20 06.	+ 41 34	1880		CLUSTER OF GALAXIES
UGC 12559	23 20 06.	+ 43 44	72	17.	GALAXY
ZWG 533.002	23 20 06.	+ 45 20		15.3	GALAXY
ZWG 532.018	23 20 06.	+ 45 20		15.3	GALAXY
ZWG 380.047	23 20 06.	- 01 15		15.6	GALAXY
PHL 0486	23 20 06.	- 02 50		15.3	BLUE STELLAR OBJECT
PHL 5727	23 20 06.	- 17 56		17.2	BLUE STELLAR OBJECT
PHL 0487	23 20 06.	- 32 42		18.0	BLUE STELLAR OBJECT
LB 01522	23 20 06.	- 47 34		15.2	FAINT BLUE STAR
IC 1484	23 20 08.	+ 11 06 34.			NONSTELLAR OBJECT
PHL 5726	23 20 09.	- 30 54		17.8	BLUE STELLAR OBJECT
ARC 2585	23 20 10.	- 26 32		17.5	RICH CLUSTER OF GALAXIES
PHL 5729	23 20 12.	+ 01 14		17.8	BLUE STELLAR OBJECT
MCG+02-59-032	23 20 12.	+ 11 05	15	15.5	GALAXY
MCG+02-59-031	23 20 12.	+ 11 29	72	15.	GALAXY
ZWG 431.049	23 20 12.	+ 15 07		15.7	GALAXY
ZC 2320.2+4309	23 20 12.	+ 43 09	6720		CLUSTER OF GALAXIES
PHL 2260	23 20 12.	- 02 33		16.6	BLUE STELLAR OBJECT
PHL 5730	23 20 12.	- 20 16		18.3	BLUE STELLAR OBJECT
PHL 5728	23 20 12.	- 29 56		17.9	BLUE STELLAR OBJECT
RNGC 7639	23 20 14.	+ 11 05		15.5	GALAXY
MCG+02-59-033	23 20 15.	+ 11 42	96	14.	GALAXY
IC 1485	23 20 16.	+ 11 05 52.			NONSTELLAR OBJECT
SC 2317-6331.4	23 20 16.	- 63 14 58.			NEBULA
ZWG 380.048	23 20 18.	+ 01 10		14.5	GALAXY
UGC 12560	23 20 18.	+ 01 10	27	14.5	GALAXY S
MCG+00-59-035	23 20 18.	+ 01 11	24	15.	GALAXY
UGC 12561	23 20 18.	+ 08 41	102	16.0	GALAXY
ZWG 431.050	23 20 18.	+ 11 07		15.6	GALAXY
UGC 12562	23 20 18.	+ 11 29	78	16.0	GALAXY Sc-IRR
ZWG 431.051	23 20 18.	+ 11 43		14.8	GALAXY
UGC 12563	23 20 18.	+ 11 43	90	14.8	GALAXY S
ZWG 476.024	23 20 18.	+ 22 39		15.3	GALAXY
UGC 12564	23 20 18.	+ 22 39	60	15.3	GALAXY Sc
ZWG 476.025	23 20 18.	+ 22 56		15.4	GALAXY
UGC 12565	23 20 18.	+ 22 56	66	15.4	GALAXY S
MCG+04-55-010	23 20 18.	+ 22 56 30.	36	15.5	GALAXY
ZWG 476.026	23 20 18.	+ 26 00		15.7	GALAXY
ZWG 497.009	23 20 18.	+ 28 52		15.0	GALAXY
UGC 12566	23 20 18.	+ 28 52	108	15.0	GALAXY Sa-b
MCG+07-48-003	23 20 18.	+ 43 39 30.	60	17.	GALAXY
UGC 12567	23 20 18.	+ 43 42	78	16.0	GALAXY SB
PHL 5733	23 20 18.	- 18 14		16.6	BLUE STELLAR OBJECT
PHL 5732	23 20 18.	- 18 26		17.2	BLUE STELLAR OBJECT
PHL 5731	23 20 19.	- 30 36		15.8	BLUE STELLAR OBJECT
RNGC 7642	23 20 19.	+ 01 10		14.5	GALAXY
RNGC 7643	23 20 20.	+ 11 43		15.0	GALAXY
LB 01178	23 20 23.	- 04 12 24.		16.1	FAINT BLUE STAR
ZWG 380.049	23 20 24.	+ 01 07		15.2	GALAXY
MCG+00-59-036	23 20 24.	+ 01 08	36	15.5	GALAXY
PHL 0488	23 20 24.	+ 01 22		17.9	BLUE STELLAR OBJECT
ZWG 406.095	23 20 24.	+ 06 01		15.7	GALAXY
MCG+05-55-006	23 20 24.	+ 28 51 30.	15	15.	GALAXY
LDN 1237	23 20 24.	+ 62 20	240		DARK NEBULA
PHL 2262	23 20 24.	- 01 11		17.8	BLUE STELLAR OBJECT
PHL 2261	23 20 24.	- 08 45		18.4	BLUE STELLAR OBJECT
PHL 5737	23 20 24.	- 09 38		17.5	BLUE STELLAR OBJECT
PHL 5735	23 20 24.	- 16 12		17.9	BLUE STELLAR OBJECT
PHL 5739	23 20 24.	- 18 04		18.7	BLUE STELLAR OBJECT
PHL 2263	23 20 24.	- 22 44		18.2	BLUE STELLAR OBJECT
PHL 5734	23 20 24.	- 27 14		16.9	BLUE STELLAR OBJECT
MCG-05-55-006	23 20 24.	- 29 02	30	15.5	GALAXY
PHL 5738	23 20 24.	- 30 19		17.8	BLUE STELLAR OBJECT
PHL 2264	23 20 24.	- 30 30		18.2	BLUE STELLAR OBJECT
PHL 5736	23 20 24.	- 31 56		17.0	BLUE STELLAR OBJECT
SC 2317-6242.8	23 20 29.	- 62 26 22.			NEBULA
3ZW 103	23 20 30.	+ 06 01			COMPACT GALAXY
MCG+01-59-070	23 20 30.	+ 08 44	36	15.	GALAXY
MCG+03-59-046	23 20 30.	+ 17 27	24	15.5	GALAXY
ZWG 454.053	23 20 30.	+ 17 28		15.6	GALAXY
MCG+05-55-007	23 20 30.	+ 28 56 30.	30	15.5	GALAXY
PHL 5740	23 20 30.	- 01 32		18.7	BLUE STELLAR OBJECT
MCG+03-59-047	23 20 33.	+ 17 44	36	15.	GALAXY
SC 2317-6229.1	23 20 33.	- 62 12 40.			NEBULA
PHL 0489	23 20 36.	+ 00 55		18.5	BLUE STELLAR OBJECT
3ZW 104	23 20 36.	+ 12 46			COMPACT GALAXY
ZWG 454.054	23 20 36.	+ 17 45		15.7	GALAXY
4ZW 146	23 20 36.	+ 28 30			COMPACT GALAXY
UGC 12568	23 20 35.	+ 42 25	60	17.	GALAXY
PHL 2266	23 20 36.	- 08 32		18.2	BLUE STELLAR OBJECT
PHL 5741	23 20 36.	- 15 16		17.9	BLUE STELLAR OBJECT
PHL 5742	23 20 36.	- 17 09		17.6	BLUE STELLAR OBJECT
PHL 5743	23 20 36.	- 19 44		18.5	BLUE STELLAR OBJECT
PHL 5744	23 20 36.	- 20 04		18.3	BLUE STELLAR OBJECT
PHL 2265	23 20 36.	- 20 16		18.6	BLUE STELLAR OBJECT
PK107-12.1	23 20 36.0	+ 46 37 32.	5	13.9	PLANETARY NEBULA
LB 01179	23 20 38.	+ 13 10 42.		17.4	FAINT BLUE STAR
PHL 5745	23 20 42.	+ 00 04		18.0	BLUE STELLAR OBJECT
ZC 2320.7+0609	23 20 42.	+ 06 09	1750		CLUSTER OF GALAXIES
ZC 2320.7+1138	23 20 42.	+ 11 38	940		CLUSTER OF GALAXIES
UGC 12569	23 20 42.	+ 14 38	60	16.0	GALAXY S
ZWG 454.055	23 20 42.	+ 17 15		15.5	GALAXY
MCG+03-59-049	23 20 42.	+ 20 18	30	15.	GALAXY
MCG+03-59-048	23 20 42.	+ 20 19	54	15.	GALAXY
ZWG 497.010	23 20 42.	+ 32 15		14.5	GALAXY
UGC 12570	23 20 42.	+ 32 15	36	14.5	GALAXY PECULR
PHL 2268	23 20 42.	- 07 49		17.9	BLUE STELLAR OBJECT
PHL 5746	23 20 42.	- 11 16		18.3	BLUE STELLAR OBJECT
PHL 2267	23 20 42.	- 12 14		18.0	BLUE STELLAR OBJECT
PHL 2269	23 20 42.	- 27 26		17.9	BLUE STELLAR OBJECT
MCG+02-59-034	23 20 45.	+ 13 02	54	15.	GALAXY
PHL 0490	23 20 48.	+ 02 02		18.6	BLUE STELLAR OBJECT
PHL 2272	23 20 48.	+ 02 42		17.9	BLUE STELLAR OBJECT
ZWG 431.052	23 20 48.	+ 13 03		15.5	GALAXY
UGC 12571	23 20 48.	+ 13 30	120	15.5	GALAXY SB
ZWG 454.056	23 20 48.	+ 20 17		15.3	GALAXY
ZWG 454.057	23 20 49.	+ 20 19		15.5	GALAXY
MCG+03-59-050	23 20 48.	+ 20 19 30.	27	16.	GALAXY
ZWG 476.027	23 20 48.	+ 24 28		15.6	GALAXY
UGC 12572	23 20 48.	+ 24 38	60	15.6	GALAXY Sa-b
MCG+05-55-008	23 20 48.	+ 32 15	27	15.	GALAXY
ZWG 533.003	23 20 48.	+ 43 41		15.0	GALAXY
UGC 12573	23 20 48.	+ 43 41	90	15.0	GALAXY (E)
PHL 5747	23 20 48.	- 03 52		16.6	BLUE STELLAR OBJECT
PHL 5748	23 20 48.	- 09 27		16.6	BLUE STELLAR OBJECT
PHL 2270	23 20 48.	- 09 38		15.3	BLUE STELLAR OBJECT
PHL 2271	23 20 48.	- 11 35		16.6	BLUE STELLAR OBJECT

OBJECT NAME	RIGHT ASCEN.	DECLINATION	DIAM.	MAGN.	TYPE OF OBJECT
PHL 2274	23 20 48.	- 19 12		18.0	BLUE STELLAR OBJECT
PHL 2273	23 20 48.	- 21 16		17.0	BLUE STELLAR OBJECT
PHL 5749	23 20 48.	- 24 07		17.8	BLUE STELLAR OBJECT
ARC 2586	23 20 49.	- 20 43		17.1	RICH CLUSTER OF GALAXIES
MCG+00-59-037	23 20 54.	+ 02 47 30.	36	16.	GALAXY
MCG+01-59-071	23 20 54.	+ 04 41	30	15.5	GALAXY
ZWG 454.058	23 20 54.	+ 20 53		15.1	GALAXY
ZWG 497.011	23 20 54.	+ 29 09		15.5	GALAXY
MCG+07-48-004	23 20 54.	+ 43 40	48	15.	GALAXY
MCG-02-59-014	23 20 54.	- 08 35	54	16.	GALAXY
PHL 5750	23 20 54.	- 12 51		16.6	BLUE STELLAR OBJECT
PHL 2275	23 20 54.	- 13 38		18.5	BLUE STELLAR OBJECT
PHL 5751	23 20 54.	- 28 20		17.1	BLUE STELLAR OBJECT
PHL 0491	23 20 54.	- 29 33		17.7	BLUE STELLAR OBJECT
AGU 79	23 20 54.	- 42 41 30.		15.	IRREGULAR DWARF GALAXY
ARC 2587	23 20 56.	- 22 42		17.2	RICH CLUSTER OF GALAXIES
MCG+05-55-009	23 20 57.	+ 29 09	24	15.5	GALAXY
LBN 0499	23 21	+ 41 20	1200		BRIGHT NEBULA
ZC 2321.0+0006	23 21 00.	+ 00 06	2490		CLUSTER OF GALAXIES
ZWG 431.053	23 21 00.	+ 14 03		15.7	GALAXY
ZWG 454.059	23 21 00.	+ 19 19		15.4	GALAXY
MCG+03-59-051	23 21 00.	+ 19 19	24	15.	GALAXY
IC 5317	23 21 00.	+ 20 53 41.			NONSTELLAR OBJECT
ISS 0099	23 21 00.	+ 61 41	282		STELLAR RING
LDN 1259	23 21 00.	+ 74 00	720		DARK NEBULA
PHL 0492	23 21 00.	- 05 20		17.7	BLUE STELLAR OBJECT
PHL 2277	23 21 00.	- 10 40		18.7	BLUE STELLAR OBJECT
PHL 5753	23 21 00.	- 18 20		18.7	BLUE STELLAR OBJECT
MCG-03-59-006	23 21 00.	- 19 19	36	14.5	GALAXY
PHL 5752	23 21 00.	- 23 45		17.8	BLUE STELLAR OBJECT
PHL 2276	23 21 00.	- 25 40		18.4	BLUE STELLAR OBJECT
MCG-07-48-001	23 21 00.	- 42 41	60	16.	GALAXY
SC 2318-6243.4	23 21 00.	- 62 26 57.			NEBULA
RNGC 7645	23 21 04.	- 29 39		14.0	GALAXY
PHL 2279	23 21 06.	+ 00 36		18.7	BLUE STELLAR OBJECT
ZC 2321.1+0851	23 21 06.	+ 08 51	1340		CLUSTER OF GALAXIES
ZWG 454.060	23 21 06.	+ 19 17		15.5	GALAXY
UGC 12574	23 21 06.	+ 19 17	72	15.5	GALAXY SBb
VV 305C	23 21 06.	+ 19 18	60	15.	INTERACTING GALAXY
VV 305B	23 21 06.	+ 19 18	12	18.	INTERACTING GALAXY
VV 305A	23 21 06.	+ 19 18	102	15.	INTERACTING GALAXY
ZWG 454.061	23 21 06.	+ 20 20		15.7	GALAXY
ZWG 476.028	23 21 06.	+ 24 55		15.1	GALAXY
KARA.73B 1016	23 21 06.	+ 24 55	30	15.1	ISOLATED GALAXY S
ZWG 476.029	23 21 06.	+ 25 20		15.7	GALAXY
ZWG 497.012	23 21 06.	+ 28 55		14.8	GALAXY
4ZW 147	23 21 06.	+ 30 33			COMPACT GALAXY
PHL 0493	23 21 06.	- 01 03		18.3	BLUE STELLAR OBJECT
PHL 2278	23 21 06.	- 03 31		17.5	BLUE STELLAR OBJECT
MCG-05-55-007	23 21 06.	- 29 39	78	14.	GALAXY
YC 2321-36	23 21 07.	- 36 33 06.			UNUSUAL SOUTHERN NEBULA
MCG+03-59-052	23 21 09.	+ 19 17	66	15.	GALAXY
MCG+05-55-010	23 21 09.	+ 28 53 30.	45	15.	GALAXY
VMT 23	23 21 10.	+ 58 32 24.	7800		SUPERNOVA REMNANT
MIL 93	23 21 11.	+ 58 32 48.	240		SUPERNOVA REMNANT
ZC 2321.2+0145	23 21 12.	+ 01 45	3090		CLUSTER OF GALAXIES
ARC 2588	23 21 12.	+ 08 53		17.8	RICH CLUSTER OF GALAXIES
PHL 0494	23 21 12.	- 06 38		17.7	BLUE STELLAR OBJECT
PHL 5758	23 21 12.	- 07 56		17.7	BLUE STELLAR OBJECT
PHL 2281	23 21 12.	- 08 46		18.5	BLUE STELLAR OBJECT
PHL 5759	23 21 12.	- 10 36		17.9	BLUE STELLAR OBJECT
PHL 2282	23 21 12.	- 11 40		18.4	BLUE STELLAR OBJECT
PHL 5760	23 21 12.	- 12 32		18.2	BLUE STELLAR OBJECT
MCG-03-59-007	23 21 12.	- 16 44	54	15.	GALAXY
PHL 2280	23 21 12.	- 20 50		17.0	BLUE STELLAR OBJECT
PHL 5756	23 21 12.	- 22 06		16.9	BLUE STELLAR OBJECT
PHL 5754	23 21 12.	- 25 17		12.0	BLUE STELLAR OBJECT
PHL 5757	23 21 12.	- 26 48		16.3	BLUE STELLAR OBJECT
ZWG 406.096	23 21 18.	+ 09 23		13.5	GALAXY
MRK 531	23 21 18.	+ 09 23	30	14.5	GALAXY WITH UV CONTINUUM
UGC 12575	23 21 18.	+ 09 23	96	13.5	GALAXY S0
ZC 2321.3+3035	23 21 18.	+ 30 35	1410		CLUSTER OF GALAXIES
5ZW 401	23 21 18.	+ 44 35			COMPACT GALAXY
PHL 0495	23 21 18.	- 02 12		15.2	BLUE STELLAR OBJECT
PHL 0496	23 21 18.	- 18 16		17.9	BLUE STELLAR OBJECT
LB 01180	23 21 19.	- 05 19 18.		15.9	FAINT BLUE STAR
RNGC 7648	23 21 20.	+ 09 23		13.5	GALAXY
SC 2318-6038.4	23 21 21.	- 60 21 57.	12		NEBULA
IC 1486	23 21 22.	+ 09 23			NONSTELLAR OBJECT
LB 01181	23 21 22.	+ 14 14 36.		14.3	FAINT BLUE STAR
PHL 0497	23 21 24.	+ 01 24		14.0	BLUE STELLAR OBJECT
ZC 2321.4+0746	23 21 24.	+ 07 46	1280		CLUSTER OF GALAXIES
MCG+01-59-072	23 21 24.	+ 09 24	72	13.	GALAXY
MCG+03-59-053	23 21 24.	+ 16 21	45	15.	GALAXY
ZWG 454.062	23 21 24.	+ 16 22		15.7	GALAXY
5ZW 402	23 21 24.	+ 42 52			COMPACT GALAXY
PHL 5761	23 21 24.	- 10 21		18.0	BLUE STELLAR OBJECT
RNGC 7646	23 21 24.	- 12 08			GALAXY
MCG-02-59-015	23 21 24.	- 12 09	54	14.	GALAXY
PHL 5765	23 21 24.	- 13 07		18.0	BLUE STELLAR OBJECT
PHL 2283	23 21 24.	- 18 46		15.4	BLUE STELLAR OBJECT
PHL 5762	23 21 24.	- 19 05		18.0	BLUE STELLAR OBJECT
PHL 5763	23 21 24.	- 20 14		18.1	BLUE STELLAR OBJECT
PHL 5764	23 21 24.	- 27 35		7.9	BLUE STELLAR OBJECT
MCG+03-59-055	23 21 27.	+ 16 29	84	14.5	GALAXY
ARC 2589	23 21 27.	+ 16 33		15.3	RICH CLUSTER OF GALAXIES
MCG+03-59-054	23 21 27.	+ 20 25	72	16.	GALAXY
LB 01182	23 21 29.	- 06 41 42.		16.9	FAINT BLUE STAR
ZWG 454.063	23 21 30.	+ 16 31		15.2	GALAXY
UGC 12576	23 21 30.	+ 16 31	96	15.2	GALAXY E
UGC 12577	23 21 30.	+ 20 26	78	16.0	GALAXY Sb
IC 5318	23 21 31.	- 12 08 06.			NONSTELLAR OBJECT
RNGC 7647	23 21 33.	+ 16 31		15.0	GALAXY
PHL 2284	23 21 36.	+ 01 44		16.0	BLUE STELLAR OBJECT
PHL 0498	23 21 36.	+ 02 26		17.9	BLUE STELLAR OBJECT
PHL 2285	23 21 36.	+ 02 38		18.0	BLUE STELLAR OBJECT
ZC 2321.6+1530	23 21 36.	+ 15 30	1480		CLUSTER OF GALAXIES
PHL 0499	23 21 36.	- 09 42		18.1	BLUE STELLAR OBJECT
PHL 5768	23 21 36.	- 11 28		16.7	BLUE STELLAR OBJECT
PHL 5766	23 21 36.	- 12 54		13.0	BLUE STELLAR OBJECT
PHL 5767	23 21 36.	- 14 18		14.0	BLUE STELLAR OBJECT
PHL 5770	23 21 36.	- 15 03		18.3	BLUE STELLAR OBJECT
SC 2318-6255.0	23 21 39.	- 62 38 33.	360		NEBULA
3ZW 105	23 21 42.	+ 06 08			COMPACT GALAXY
PHL 2286	23 21 42.	- 12 39		17.9	BLUE STELLAR OBJECT
RNGC 7667	23 21 43.	- 00 23		14.0	GALAXY
ARC 2590	23 21 45.	+ 01 49		17.5	RICH CLUSTER OF GALAXIES
MCG+02-59-035	23 21 45.	+ 14 22	60	15.	GALAXY
MCG+00-59-038	23 21 45.	- 00 23	90	14.	GALAXY
ARC 2591	23 21 46.	+ 00 01		17.6	RICH CLUSTER OF GALAXIES
LB 01183	23 21 47.	- 06 12 30.		17.1	FAINT BLUE STAR
PHL 2289	23 21 48.	+ 00 02		18.5	BLUE STELLAR OBJECT
ZWG 380.051	23 21 48.	+ 01 09		15.6	GALAXY
ZC 2321.8+0150	23 21 48.	+ 01 50	810		CLUSTER OF GALAXIES
MCG+02-59-036	23 21 48.	+ 13 41	15	15.	GALAXY
ZWG 431.054	23 21 48.	+ 14 22		15.7	GALAXY
UGC 12579	23 21 48.	+ 14 22	96	15.7	GALAXY E
ZC 2321.8+1758	23 21 48.	+ 17 58	1410		CLUSTER OF GALAXIES
ZWG 380.050	23 21 48.	- 00 23		15.1	GALAXY
UGC 12578	23 21 48.	- 00 23	96	15.1	GALAXY S
PHL 0500	23 21 48.	- 06 12		17.0	BLUE STELLAR OBJECT
PHL 2288	23 21 48.	- 21 12		17.4	BLUE STELLAR OBJECT
PHL 2290	23 21 48.	- 21 56		18.2	BLUE STELLAR OBJECT
PHL 5771	23 21 48.	- 22 16		17.0	BLUE STELLAR OBJECT
PHL 5774	23 21 48.	- 22 26		18.2	BLUE STELLAR OBJECT
PHL 5773	23 21 48.	- 23 28		18.2	BLUE STELLAR OBJECT
PHL 5775	23 21 48.	- 26 44		18.0	BLUE STELLAR OBJECT
PHL 5772	23 21 48.	- 28 09		17.5	BLUE STELLAR OBJECT
PHL 2287	23 21 48.	- 29 11		16.4	BLUE STELLAR OBJECT
LB 01184	23 21 49.	- 04 13 42.		18.1	FAINT BLUE STAR
RNGC 7649	23 21 51.	+ 14 22		15.5	GALAXY
ARC 2592	23 21 51.	+ 17 52		16.5	RICH CLUSTER OF GALAXIES
LB 01185	23 21 52.	+ 13 47 42.		15.1	FAINT BLUE STAR
RNGC 7656	23 21 53.	- 19 21		14.0	GALAXY
ZWG 431.055	23 21 54.	+ 13 42		15.0	GALAXY
ZWG 476.030	23 21 54.	+ 25 07		15.0	GALAXY
ACK 112-00.1	23 21 54.	+ 60 41			PLANETARY NEBULA
PHL 5777	23 21 54.	- 08 40		17.9	BLUE STELLAR OBJECT
PHL 5776	23 21 54.	- 12 42		16.7	BLUE STELLAR OBJECT
MCG-03-59-008	23 21 54.	- 19 21	78	14.5	GALAXY
RNGC 7651	23 21 57.	+ 13 42		15.0	GALAXY
RNGC 7644	23 21 57.	+ 13 42			GALAXY
ARC 2593	23 21 58.	+ 14 22		15.1	RICH CLUSTER OF GALAXIES
RNGC 7654	23 21 59.	+ 61 19		8.0	OPEN CLUSTER
HMS 2322+1425	23 22	+ 14 25			CLUSTER OF GALAXIES
PHL 2292	23 22 00.	+ 01 34		18.2	BLUE STELLAR OBJECT
ZWG 406.097	23 22 00.	+ 08 20		15.6	GALAXY
UGC 12580	23 22 00.	+ 08 20	78	15.6	GALAXY Sa?
ZWG 406.098	23 22 00.	+ 08 59		15.5	GALAXY
UGC 12581	23 22 00.	+ 08 59	90	15.6	GALAXY SBa-b
MCG+02-59-037	23 22 00.	+ 14 24	12	16.	GALAXY
MCG+03-59-056	23 22 00.	+ 16 35	54	15.	GALAXY
ZWG 454.064	23 22 00.	+ 16 36		15.6	GALAXY
UGC 12582	23 22 00.	+ 16 36	66	15.6	GALAXY SB
ZWG 476.031	23 22 00.	+ 23 42		15.7	GALAXY
UGC 12583	23 22 00.	+ 23 42	78	15.7	GALAXY Sc
OCL 0260	23 22 00.	+ 61 19	21000	8.9	OPEN STAR CLUSTER
MCG+00-59-039	23 22 00.	- 02 17 30.	18	15.5	GALAXY
PHL 5778	23 22 00.	- 03 20		15.3	BLUE STELLAR OBJECT
PHL 2291	23 22 00.	- 04 10		17.1	BLUE STELLAR OBJECT
PHL 2293	23 22 00.	- 15 41		18.3	BLUE STELLAR OBJECT
PHL 0501	23 22 00.	- 17 30		18.2	BLUE STELLAR OBJECT
ARC 2594	23 22 01.	+ 07 48		17.8	RICH CLUSTER OF GALAXIES
LB 01186	23 22 02.	+ 09 51 54.		16.1	FAINT BLUE STAR
MCG+00-59-040	23 22 03.	- 02 17 30.	54	15.5	GALAXY
MCG+00-59-041	23 22 06.	+ 00 45	42	15.5	GALAXY
ZWG 406.099	23 22 06.	+ 08 08		15.5	GALAXY
UGC 12585	23 22 06.	+ 08 08	96	15.5	GALAXY
MCG+01-59-073	23 22 06.	+ 08 21	48	15.	GALAXY
MCG+01-59-074	23 22 06.	+ 09 00	60	14.	GALAXY
IC 1487	23 22 06.	+ 14 21 55.			NONSTELLAR OBJECT
ZWG 380.052	23 22 06.	- 02 16		15.6	GALAXY
UGC 12584	23 22 06.	- 02 16	90	15.6	GALAXY DBL SYS
PHL 5779	23 22 06.	- 09 56		18.3	BLUE STELLAR OBJECT
SC 2319-6020.4	23 22 06.	- 60 03 57.	48		NEBULA
RNGC 7637	23 22 11.	- 82 10			GALAXY
ZWG 380.053	23 22 12.	+ 00 45		15.7	GALAXY
MCG+01-59-075	23 22 12.	+ 08 10	96	15.	GALAXY
ZWG 431.056	23 22 12.	+ 14 21		15.7	GALAXY
ZWG 497.013	23 22 12.	+ 27 34		15.7	GALAXY
ZWG 533.004	23 22 12.	+ 43 57		15.5	GALAXY
PHL 0502	23 22 12.	- 01 08		18.2	BLUE STELLAR OBJECT
PHL 5782	23 22 12.	- 07 08		17.0	BLUE STELLAR OBJECT
PHL 2294	23 22 12.	- 10 02		18.8	BLUE STELLAR OBJECT
PHL 2296	23 22 12.	- 16 30		18.6	BLUE STELLAR OBJECT
PHL 5780	23 22 12.	- 22 12		17.0	BLUE STELLAR OBJECT
PHL 5783	23 22 12.	- 22 41		17.9	BLUE STELLAR OBJECT
PHL 5781	23 22 12.	- 25 00		17.5	BLUE STELLAR OBJECT
PHL 5784	23 22 12.	- 28 00		17.7	BLUE STELLAR OBJECT
PHL 2295	23 22 12.	- 28 31		18.0	BLUE STELLAR OBJECT
MCG+02-59-038	23 22 15.	+ 15 00	96	13.	GALAXY
MCG+05-55-011	23 22 15.	+ 27 33 30.	30	16.	GALAXY
MCG+07-48-005	23 22 15.	+ 41 04	96	13.5	GALAXY
MCG+01-59-076	23 22 18.	+ 09 14	24	15.5	GALAXY
ZWG 406.100	23 22 18.	+ 09 15		15.6	GALAXY
ZWG 431.057	23 22 18.	+ 14 43		15.7	GALAXY
ZWG 431.058	23 22 18.	+ 15 00		13.8	GALAXY
UGC 12586	23 22 18.	+ 15 00	102	13.8	GALAXY Sb
ZWG 476.032	23 22 18.	+ 26 22		15.6	GALAXY
UGC 12587	23 22 18.	+ 26 22	72	15.6	GALAXY Sb
KARA.73B 1017	23 22 18.	+ 26 22	54	15.6	ISOLATED GALAXY S
ZWG 533.005	23 22 18.	+ 41 04		14.0	GALAXY
UGC 12588	23 22 18.	+ 41 04	108	14.0	GALAXY
MCG+07-48-006	23 22 18.	+ 43 58	30	15.	GALAXY
PHL 2297	23 22 18.	- 09 54		18.3	BLUE STELLAR OBJECT
PHL 5785	23 22 18.	- 12 45		17.6	BLUE STELLAR OBJECT
PHL 5786	23 22 18.	- 17 10		18.5	BLUE STELLAR OBJECT
ARC 2595	23 22 18.	- 20 49		17.2	RICH CLUSTER OF GALAXIES
PHL 5787	23 22 18.	- 24 24		18.2	BLUE STELLAR OBJECT
PHL 0503	23 22 18.	- 24 35		18.2	BLUE STELLAR OBJECT
PHL 5788	23 22 18.	- 29 11		18.1	BLUE STELLAR OBJECT
MCG-05-55-008	23 22 18.	- 29 45	24	15.5	GALAXY
LB 01523	23 22 13.	- 45 55		13.6	FAINT BLUE STAR
LB 01187	23 22 19.	+ 11 16 30.		16.2	FAINT BLUE STAR
IC 5319	23 22 19.	+ 13 43 19.			NONSTELLAR OBJECT
IC 1788	23 22 20.	+ 15 03 01.			NONSTELLAR OBJECT
RNGC 7653	23 22 21.	+ 15 00		14.0	GALAXY
ZWG 406.101	23 22 24.	+ 09 12		15.4	GALAXY
ZWG 431.059	23 22 24.	+ 12 41		15.7	GALAXY
ZC 2322.4+1427	23 22 24.	+ 14 27	6990		CLUSTER OF GALAXIES
MCG+04-55-011	23 22 24.	+ 26 21	54	15.	GALAXY
MCG+00-59-042	23 22 24.	- 00 17	78	15.	GALAXY
PHL 2298	23 22 24.	- 03 30		18.5	BLUE STELLAR OBJECT
PHL 5790	23 22 24.	- 07 28		17.9	BLUE STELLAR OBJECT
PHL 5792	23 22 24.	- 09 28		18.3	BLUE STELLAR OBJECT
PHL 0504	23 22 24.	- 17 22		18.9	BLUE STELLAR OBJECT

OBJECT NAME	RIGHT ASCEN.	DECLINATION	DIAM.	MAGN.	TYPE OF OBJECT
PHL 5791	23 22 24.	- 18 40		17.4	BLUE STELLAR OBJECT
PHL 2299	23 22 24.	- 19 52		18.7	BLUE STELLAR OBJECT
PHL 5789	23 22 24.	- 21 04		16.8	BLUE STELLAR OBJECT
PHL 5793	23 22 24.	- 23 02		18.5	BLUE STELLAR OBJECT
MCG-05-55-009	23 22 24.	- 32 23	24	16.	GALAXY
LB 01188	23 22 25.	+ 11 52 18.		16.0	FAINT BLUE STAR
ABC 2596	23 22 26.	- 23 42		17.1	RICH CLUSTER OF GALAXIES
RNGC 7650	23 22 29.	- 58 04			GALAXY
ZWG 406.102	23 22 30.	+ 07 58		15.2	GALAXY
ZC 2322.5+1738	23 22 30.	+ 17 38	1010		CLUSTER OF GALAXIES
ZWG 497.014	23 22 30.	+ 29 31		15.6	GALAXY
ZWG 380.054	23 22 30.	- 00 16		15.6	GALAXY
UGC 12589	23 22 30.	- 00 16	84	15.6	GALAXY S-IRR
PHL 5794	23 22 30.	- 09 06		18.4	BLUE STELLAR OBJECT
ZC 2322.6+0744	23 22 36.	+ 07 44	810		CLUSTER OF GALAXIES
MCG+01-59-077	23 22 36.	+ 08 00	36	15.	GALAXY
MCG+02-59-039	23 22 36.	+ 12 01	36	15.	GALAXY
ZWG 431.060	23 22 36.	+ 12 02		15.5	GALAXY
MCG+05-55-012	23 22 36.	+ 29 30 30.	42	15.	GALAXY
PHL 5796	23 22 36.	- 11 22		17.3	BLUE STELLAR OBJECT
PHL 5795	23 22 36.	- 20 24		18.2	BLUE STELLAR OBJECT
LB 01524	23 22 36.	- 46 00		14.4	FAINT BLUE STAR
SEP 160.01	23 22 36.	- 58 06	720	15.5	LOOSE GROUP OF GALAXIES
SN 19730	23 22 42.	+ 06 04		18.5	SUPERNOVA
ZWG 431.061	23 22 42.	+ 14 55		15.4	GALAXY
UGC 12590	23 22 42.	+ 14 55	60	15.4	GALAXY S0-a
MCG+05-55-013	23 22 42.	+ 29 30 30.	42	16.	GALAXY
PHL 0505	23 22 42.	- 08 35		13.8	BLUE STELLAR OBJECT
PHL 2300	23 22 42.	- 09 02		18.0	BLUE STELLAR OBJECT
PHL 2301	23 22 42.	- 16 02		18.2	BLUE STELLAR OBJECT
PHL 5797	23 22 42.	- 16 26		18.9	BLUE STELLAR OBJECT
PHL 0506	23 22 42.	- 18 08		15.9	BLUE STELLAR OBJECT
BV C3	23 22 44.	- 08 35 00.		14.8	FAINT BLUE VARIABLE
ABC 2597	23 22 45.	- 12 24		16.6	RICH CLUSTER OF GALAXIES
RNGC 7652	23 22 47.	- 58 10			GALAXY
ZWG 406.103	23 22 48.	+ 02 31		15.5	GALAXY
ZWG 431.062	23 22 48.	+ 13 37		15.7	GALAXY
ZWG 476.033	23 22 48.	+ 24 57		15.6	GALAXY
PHL 0507	23 22 48.	- 00 13		18.3	BLUE STELLAR OBJECT
PHL 5800	23 22 48.	- 14 56		18.3	BLUE STELLAR OBJECT
PHL 5799	23 22 48.	- 20 47		17.9	BLUE STELLAR OBJECT
PHL 5798	23 22 48.	- 21 56		9.1	BLUE STELLAR OBJECT
LB 01525	23 22 48.	- 48 17		16.0	FAINT BLUE STAR
ZWG 431.063	23 22 54.	+ 09 36		15.5	GALAXY
MCG+05-55-015	23 22 54.	+ 28 11 30.	108	14.5	GALAXY
ZWG 497.015	23 22 54.	+ 28 13		14.0	GALAXY
UGC 12591	23 22 54.	+ 28 13	96	14.0	GALAXY S0-a
MCG+05-55-014	23 22 54.	+ 28 34	60	15.	GALAXY
ZWG 497.016	23 22 54.	+ 28 35		15.0	GALAXY
PHL 0508	23 22 54.	- 06 17		15.8	BLUE STELLAR OBJECT
PHL 0509	23 22 54.	- 15 01		18.5	BLUE STELLAR OBJECT
PHL 5801	23 22 54.	- 30 32		13.0	BLUE STELLAR OBJECT
LB 01189	23 22 57.	- 06 17 48.		15.8	FAINT BLUE STAR
PHL 5802	23 23 00.	+ 01 18		17.5	BLUE STELLAR OBJECT
PHL 0510	23 23 00.	+ 03 02		18.4	BLUE STELLAR OBJECT
UGC 12592	23 23 00.	+ 05 11	60	17.	GALAXY DWARF
ZWG 406.104	23 23 00.	+ 08 33		15.7	GALAXY
MCG+05-55-016	23 23 00.	+ 32 23 30.	72	15.	GALAXY
ZWG 497.017	23 23 00.	+ 32 24		15.6	GALAXY
UGC 12593	23 23 00.	+ 32 24	78	15.6	GALAXY Sc
LDN 1244	23 23 00.	+ 62 30	300		DARK NEBULA
LDN 1246	23 23 00.	+ 63 20	180		DARK NEBULA
PHL 0511	23 23 00.	- 03 18		16.1	BLUE STELLAR OBJECT
PHL 2302	23 23 00.	- 08 24		18.6	BLUE STELLAR OBJECT
PHL 5803	23 23 00.	- 14 42		17.2	BLUE STELLAR OBJECT
PHL 2303	23 23 00.	- 16 38		18.8	BLUE STELLAR OBJECT
PHL 2304	23 23 00.	- 20 40		18.4	BLUE STELLAR OBJECT
PHL 5804	23 23 00.	- 23 56		17.9	BLUE STELLAR OBJECT
ZC 2323.1+0617	23 23 06.	+ 06 17	940		CLUSTER OF GALAXIES
ZWG 406.105	23 23 06.	+ 06 28		15.6	GALAXY
KARA.73B 1018	23 23 06.	+ 06 28	36	15.6	ISOLATED GALAXY S
PHL 5806	23 23 06.	- 07 08		16.6	BLUE STELLAR OBJECT
PHL 5807	23 23 06.	- 10 31		18.3	BLUE STELLAR OBJECT
PHL 5808	23 23 06.	- 21 44		18.1	BLUE STELLAR OBJECT
TON-S 0098	23 23 06.	- 23 53		15.1	BLUE STAR
PHL 5805	23 23 06.	- 27 59		17.9	BLUE STELLAR OBJECT
ABC 2598	23 23 11.	+ 27 33		17.1	RICH CLUSTER OF GALAXIES
ZWG 406.106	23 23 12.	+ 04 47		15.7	GALAXY
ZWG 406.107	23 23 12.	+ 07 45		15.4	GALAXY
ZWG 454.065	23 23 12.	+ 17 15		15.7	GALAXY
ZWG 476.034	23 23 12.	+ 22 38		15.0	GALAXY
PHL 0512	23 23 12.	- 03 18		17.5	BLUE STELLAR OBJECT
PHL 2305	23 23 12.	- 08 26		17.2	BLUE STELLAR OBJECT
PHL 0525	23 23 12.	- 11 19		8.5	BLUE STELLAR OBJECT
PHL 0513	23 23 12.	- 13 59		17.1	BLUE STELLAR OBJECT
PHL 0514	23 23 12.	- 15 58		18.1	BLUE STELLAR OBJECT
PHL 2306	23 23 12.	- 17 11		18.6	BLUE STELLAR OBJECT
PHL 0515	23 23 12.	- 23 52		16.3	BLUE STELLAR OBJECT
PHL 5809	23 23 12.	- 26 00		17.8	BLUE STELLAR OBJECT
MCG+01-59-078	23 23 15.	+ 04 47 30.	30	15.	GALAXY
RNGC 7660	23 23 16.	+ 26 45		14.0	GALAXY
SC 2320-6313.4	23 23 16.	- 62 56 56.			NEBULA
MCG+01-59-079	23 23 18.	+ 07 46	36	15.	GALAXY
ZWG 476.035	23 23 18.	+ 26 45		13.9	GALAXY
UGC 12594	23 23 18.	+ 26 45	90	13.9	GALAXY E
PHL 2307	23 23 18.	- 05 26		18.6	BLUE STELLAR OBJECT
PHL 2308	23 23 18.	- 09 53		18.2	BLUE STELLAR OBJECT
PHL 5810	23 23 18.	- 20 00		17.5	BLUE STELLAR OBJECT
PHL 0516	23 23 18.	- 24 58		18.1	BLUE STELLAR OBJECT
MCG+04-55-012	23 23 21.	+ 26 44	72	14.5	GALAXY
PHL 4507	23 23 24.	+ 00 56		18.4	BLUE STELLAR OBJECT
UGC 12595	23 23 24.	+ 13 55	66	15.1	GALAXY S0-a
ZWG 431.064	23 23 24.	+ 13 56		15.1	GALAXY
MCG+02-59-040	23 23 24.	+ 13 56	36	15.	GALAXY
ZC 2323.4+1700	23 23 24.	+ 17 00	2420		CLUSTER OF GALAXIES
ZWG 454.066	23 23 24.	+ 17 13		15.6	GALAXY
ZC 2323.4+2734	23 23 24.	+ 27 34	1410		CLUSTER OF GALAXIES
PHL 5812	23 23 24.	- 00 52		16.6	BLUE STELLAR OBJECT
PHL 5811	23 23 24.	- 12 19		14.0	BLUE STELLAR OBJECT
PHL 5813	23 23 24.	- 14 32		16.8	BLUE STELLAR OBJECT
PHL 2309	23 23 24.	- 19 14		18.5	BLUE STELLAR OBJECT
RNGC 7659	23 23 26.	+ 13 56		15.0	GALAXY
PK106-17.1	23 23 29.59	+ 42 15 39.6	53	9.2	PLANETARY NEBULA
MCG+02-59-041	23 23 30.	+ 12 54	72	14.5	GALAXY
ZWG 431.065	23 23 30.	+ 12 55		15.4	GALAXY
UGC 12596	23 23 30.	+ 12 55	66	15.4	GALAXY S
RNGC 7662	23 23 30.	+ 42 16		9.0	PLANETARY NEBULA
OCL 0261	23 23 30.	+ 61 02	840		OPEN STAR CLUSTER
PHL 5814	23 23 30.	- 21 52		18.2	BLUE STELLAR OBJECT
PHL 2310	23 23 30.	- 25 14		16.5	BLUE STELLAR OBJECT
MCG-05-55-010	23 23 30.	- 32 07	30	16.	GALAXY
PHL 2311	23 23 36.	+ 01 15		18.0	BLUE STELLAR OBJECT
ZWG 476.036	23 23 36.	+ 22 31		15.5	GALAXY
LG 190	23 23 36.	+ 81 01	600		REFLECTION NEBULA
PHL 5815	23 23 36.	- 00 13		17.2	BLUE STELLAR OBJECT
SHAH 083	23 23 36.	- 02 01	114	16.5	GROUP OF COMPACT GALAXIES
PHL 2315	23 23 36.	- 03 16		17.9	BLUE STELLAR OBJECT
PHL 2312	23 23 36.	- 09 26		18.2	BLUE STELLAR OBJECT
PHL 2313	23 23 36.	- 15 20		18.6	BLUE STELLAR OBJECT
PHL 2314	23 23 36.	- 19 03		18.7	BLUE STELLAR OBJECT
PHL 5817	23 23 36.	- 20 22		17.7	BLUE STELLAR OBJECT
PHL 0517	23 23 36.	- 21 04		17.9	BLUE STELLAR OBJECT
PHL 5816	23 23 36.	- 25 30		18.0	BLUE STELLAR OBJECT
UGC 12597	23 23 42.	+ 21 29	60	16.5	GALAXY
ZC 2323.7+2715	23 23 42.	+ 27 15	470		CLUSTER OF GALAXIES
ZWG 497.018	23 23 42.	+ 31 43		15.7	GALAXY
ZWG 380.055	23 23 42.	- 02 15		15.6	GALAXY
MCG+00-59-043	23 23 42.	- 02 16	30	15.	GALAXY
PHL 0518	23 23 42.	- 08 28		17.2	BLUE STELLAR OBJECT
IC 5321	23 23 42.	- 18 13 52.			NONSTELLAR OBJECT
MCG-03-59-009	23 23 42.	- 18 15	60	14.	GALAXY
PHL 5818	23 23 42.	- 18 16		17.7	BLUE STELLAR OBJECT
MCG-05-55-011	23 23 42.	- 27 56	48	15.5	GALAXY
MCG-05-55-012	23 23 42.	- 32 40	78	15.	GALAXY
MCG-07-48-002	23 23 42.	- 39 29	15	15.	GALAXY
MCG-07-48-003	23 23 42.	- 39 30	36	15.	GALAXY
RNGC 7655	23 23 42.	- 68 18			NON-EXISTENT OBJECT
PHL 0519	23 23 48.	+ 02 42		17.6	BLUE STELLAR OBJECT
4ZW 148	23 23 48.	+ 32 40			COMPACT GALAXY
PHL 2316	23 23 48.	- 03 19		17.8	BLUE STELLAR OBJECT
PHL 5819	23 23 48.	- 04 32		13.6	BLUE STELLAR OBJECT
MCG-01-59-022	23 23 48.	- 05 14	18	15.5	GALAXY
PHL 0520	23 23 48.	- 11 08		18.8	BLUE STELLAR OBJECT
PHL 5821	23 23 48.	- 17 32		18.7	BLUE STELLAR OBJECT
PHL 5820	23 23 48.	- 20 35		17.7	BLUE STELLAR OBJECT
SS 72	23 23 49.	+ 62 40			DIFFUSE GALACTIC NEBULA
RNGC 7658	23 23 51.	- 39 30			GALAXY
RNGC 7657	23 23 53.	- 58 05			GALAXY
ZWG 380.056	23 23 54.	+ 01 40		15.7	GALAXY
MCG+00-59-044	23 23 54.	+ 02 03	36	15.5	GALAXY
PHL 2318	23 23 54.	- 05 30		18.7	BLUE STELLAR OBJECT
MCG-02-59-016	23 23 54.	- 12 48	24	14.5	GALAXY
PHL 2317	23 23 54.	- 22 37		17.4	BLUE STELLAR OBJECT
PHL 5822	23 23 54.	- 29 00		17.9	BLUE STELLAR OBJECT
PHL 5823	23 23 54.	- 30 24		17.9	BLUE STELLAR OBJECT
IC 1489	23 23 55.	- 12 47 21.			NONSTELLAR OBJECT
MCG+00-59-045	23 23 57.	+ 02 04	24	15.5	GALAXY
PK111-02.1	23 23 57.2	+ 57 54 24.	10	13.98	PLANETARY NEBULA
LBN 0471	23 24	+ 32 00	10800		BRIGHT NEBULA
ZWG 380.057	23 24 00.	+ 00 33		15.6	GALAXY
ZWG 380.058	23 24 00.	+ 02 02		15.5	GALAXY
KARA.72 583A	23 24 00.	+ 02 02	36	15.5	PART OF DOUBLE GALAXY
ZWG 380.059	23 24 00.	+ 02 03		15.6	GALAXY
KARA.72 583B	23 24 00.	+ 02 03	36	15.6	PART OF DOUBLE GALAXY
PHL 2320	23 24 00.	+ 03 10		18.3	BLUE STELLAR OBJECT
ZWG 431.066	23 24 00.	+ 12 55		15.7	GALAXY
PHL 0521	23 24 00.	- 09 09		18.3	BLUE STELLAR OBJECT
PHL 5824	23 24 00.	- 09 34		17.4	BLUE STELLAR OBJECT
MCG-02-59-017	23 24 00.	- 11 54	72	14.5	GALAXY
PHL 0522	23 24 00.	- 13 18		18.5	BLUE STELLAR OBJECT
PHL 5827	23 24 00.	- 18 00		19.0	BLUE STELLAR OBJECT
PHL 2319	23 24 00.	- 22 54		18.3	BLUE STELLAR OBJECT
PHL 2321	23 24 00.	- 23 32		18.2	BLUE STELLAR OBJECT
PHL 5826	23 24 00.	- 24 13		18.1	BLUE STELLAR OBJECT
PHL 0523	23 24 00.	- 29 11		14.2	BLUE STELLAR OBJECT
PHL 5828	23 24 00.	- 29 16		18.3	BLUE STELLAR OBJECT
PHL 5825	23 24 00.	- 30 22		17.1	BLUE STELLAR OBJECT
ABC 2599	23 24 02.	- 24 04		17.1	RICH CLUSTER OF GALAXIES
PHL 2322	23 24 06.	+ 02 12		18.0	BLUE STELLAR OBJECT
ZWG 380.060	23 24 06.	+ 03 28		15.3	GALAXY
ZWG 431.067	23 24 06.	+ 11 28		15.6	GALAXY
MRK 324	23 24 06.	+ 17 59	14	15.5	GALAXY WITH UV CONTINUUM
ZWG 476.037	23 24 06.	+ 25 22		15.7	GALAXY
ZC 2324.1+3840	23 24 06.	+ 38 40	670		CLUSTER OF GALAXIES
PHL 0524	23 24 06.	- 11 14		17.9	BLUE STELLAR OBJECT
RNGC 7663	23 24 07.	- 05 12			GALAXY
ABC 2600	23 24 07.	- 22 42		17.1	RICH CLUSTER OF GALAXIES
ABC 2601	23 24 08.	- 24 43		17.7	RICH CLUSTER OF GALAXIES
RNGC 7664	23 24 10.	+ 24 48		13.5	GALAXY
MCG+04-55-013	23 24 12.	+ 24 47	156	13.5	GALAXY
ZWG 476.038	23 24 12.	+ 24 48		13.3	GALAXY
UGC 12598	23 24 12.	+ 24 48	198	13.3	GALAXY Sc
KARA.73B 1019	23 24 12.	+ 24 48	204	13.3	ISOLATED GALAXY S
MCG+05-55-017	23 24 12.	+ 32 34 30.	54	15.5	GALAXY
ZWG 497.019	23 24 12.	+ 32 35		15.7	GALAXY
MCG-01-59-023	23 24 12.	- 05 14	42	15.5	GALAXY
PHL 2323	23 24 12.	- 09 56		18.7	BLUE STELLAR OBJECT
PHL 5829	23 24 12.	- 11 55		18.0	BLUE STELLAR OBJECT
PHL 2324	23 24 12.	- 17 26		18.6	BLUE STELLAR OBJECT
ABC 2602	23 24 14.	+ 21 02		17.7	RICH CLUSTER OF GALAXIES
RNGC 7661	23 24 14.	- 65 33			UNVERIFIED SOUTHERN OBJECT
LB 01190	23 24 17.	+ 12 13 54.		13.6	FAINT BLUE STAR
PHL 5831	23 24 18.	+ 01 30		18.0	BLUE STELLAR OBJECT
ZWG 406.108	23 24 18.	+ 06 26		15.7	GALAXY
ZWG 431.068	23 24 18.	+ 11 05		15.6	GALAXY
MRK 532	23 24 18.	+ 11 05	10	16.	GALAXY WITH UV CONTINUUM
ZWG 476.039	23 24 18.	+ 25 24		15.4	GALAXY
UGC 12599	23 24 18.	+ 25 24	66	15.4	GALAXY S0-a
ZWG 476.040	23 24 18.	+ 26 20		15.5	GALAXY
5ZW 403	23 24 18.	+ 47 02			COMPACT GALAXY
JSS 0100	23 24 18.	+ 59 31	290		STELLAR RING
PHL 5832	23 24 18.	- 06 26		17.7	BLUE STELLAR OBJECT
PHL 2325	23 24 18.	- 10 15		17.6	BLUE STELLAR OBJECT
MCG-02-59-018	23 24 18.	- 13 22	36	15.	GALAXY
PHL 0526	23 24 18.	- 13 36		17.9	BLUE STELLAR OBJECT
PHL 2326	23 24 18.	- 20 44		18.5	BLUE STELLAR OBJECT
PHL 5830	23 24 18.	- 31 04		16.3	BLUE STELLAR OBJECT
PHL 5833	23 24 24.	+ 00 59		5.0	BLUE STELLAR OBJECT
ZWG 406.109	23 24 24.	+ 08 30		15.7	GALAXY
MCG+02-59-042	23 24 24.	+ 15 27	36	15.	GALAXY
ZC 2324.4+2602	23 24 24.	+ 26 02	670		CLUSTER OF GALAXIES
MCG+05-55-018	23 24 24.	+ 32 01	24	16.	GALAXY
UGC 12600	23 24 24.	+ 48 55	78	17.	GALAXY
OCL 0264	23 24 24.	+ 63 28	240	16.	OPEN STAR CLUSTER
PHL 5834	23 24 24.	- 00 41		17.8	BLUE STELLAR OBJECT
PHL 2327	23 24 24.	- 11 17		18.4	BLUE STELLAR OBJECT

OBJECT NAME	RIGHT ASCEN.	DECLINATION	DIAM.	MAGN.	TYPE OF OBJECT
PHL 2328	23 24 24.	- 11 22		18.0	BLUE STELLAR OBJECT
PHL 0527	23 24 24.	- 25 14		18.4	BLUE STELLAR OBJECT
PHL 5835	23 24 24.	- 26 24		18.0	BLUE STELLAR OBJECT
IC 5320	23 24 25.	- 68 01			NONSTELLAR OBJECT
HN 1234	23 24 25.	- 68 02		14.	NEBULA
PHL 5836	23 24 30.	+ 00 30		16.9	BLUE STELLAR OBJECT
MCG+02-59-043	23 24 30.	+ 14 45	72	15.	GALAXY
PHL 0528	23 24 30.	- 14 52		17.7	BLUE STELLAR OBJECT
HN 1235	23 24 30.	- 68 02		14.	NEBULA
IC 5322	23 24 31.	- 68 01			NONSTELLAR OBJECT
MCG+03-59-057	23 24 33.	+ 17 32	36	15.5	GALAXY
UGC 12601	23 24 36.	+ 14 46	96	16.0	GALAXY
ZWG 454.067	23 24 36.	+ 17 32		15.7	GALAXY
ZWG 497.020	23 24 36.	+ 28 51		15.6	GALAXY
ZC 2324.6-0059	23 24 36.	- 00 59	1680		CLUSTER OF GALAXIES
PHL 5838	23 24 36.	- 03 56		18.2	BLUE STELLAR OBJECT
PHL 0529	23 24 36.	- 04 40		14.1	BLUE STELLAR OBJECT
RNGC 7665	23 24 36.	- 09 40		13.0	GALAXY
MCG-02-59-019	23 24 36.	- 09 40	36	13.5	GALAXY
PHL 5837	23 24 36.	- 23 28		16.9	BLUE STELLAR OBJECT
RNGC 7670	23 24 37.	- 00 27			NON-EXISTENT OBJECT
RNGC 7669	23 24 37.	- 00 27			NON-EXISTENT OBJECT
RNGC 7668	23 24 37.	- 00 27			NON-EXISTENT OBJECT
MCG+05-55-019	23 24 39.	+ 32 01	15	16.	GALAXY
ZWG 406.110	23 24 42.	+ 04 29		15.2	GALAXY
ZCG 2324+04	23 24 42.	+ 04 29		15.2	COMPACT GALAXY
ZC 2324.7+3145	23 24 42.	+ 31 45	9680		CLUSTER OF GALAXIES
PHL 2329	23 24 42.	- 03 48		18.5	BLUE STELLAR OBJECT
PHL 5839	23 24 42.	- 11 05		17.2	BLUE STELLAR OBJECT
MCG-04-55-008	23 24 42.	- 23 46	36	15.	GALAXY
SC 2321-6042.9	23 24 42.	- 60 26 25.	48		NEBULA
HN 0782	23 24 42.	- 68 07			NEBULA
IC 5323	23 24 43.	- 68 06			NONSTELLAR OBJECT
MCG+02-59-044	23 24 45.	+ 12 11	36	14.	GALAXY
LB 01191	23 24 47.	+ 12 48 24.		16.0	FAINT BLUE STAR
ZC 2324.8+0756	23 24 48.	+ 07 56	2080		CLUSTER OF GALAXIES
ZWG 431.069	23 24 48.	+ 12 12		14.3	GALAXY
UGC 12602	23 24 48.	+ 12 12	96	14.3	GALAXY SO
ZC 2324.8+2613	23 24 48.	+ 26 13	610		CLUSTER OF GALAXIES
5ZW 404	23 24 48.	+ 47 34			COMPACT GALAXY
PHL 5842	23 24 48.	- 03 32		18.1	BLUE STELLAR OBJECT
PHL 2330	23 24 48.	- 11 01		17.2	BLUE STELLAR OBJECT
PHL 2331	23 24 48.	- 15 28		18.9	BLUE STELLAR OBJECT
PHL 5841	23 24 48.	- 15 46		17.5	BLUE STELLAR OBJECT
PHL 5840	23 24 48.	- 27 39		18.3	BLUE STELLAR OBJECT
PHL 5843	23 24 48.	- 29 03		18.0	BLUE STELLAR OBJECT
RNGC 7666	23 24 49.	- 04 27			NON-EXISTENT OBJECT
RNGC 7671	23 24 50.	+ 12 12		14.5	GALAXY
MCG+06-51-004	23 24 51.	+ 36 57	45	14.5	GALAXY
MCG+02-59-045	23 24 54.	+ 12 07	48	14.5	GALAXY
ZWG 516.004	23 24 54.	+ 36 57		14.8	GALAXY
PHL 5845	23 24 54.	- 16 28		17.9	BLUE STELLAR OBJECT
PHL 5844	23 24 54.	- 28 00		17.5	BLUE STELLAR OBJECT
VDB.66G 216	23 25	+ 14 26	270		DWARF GALAXY
LB 09953	23 25	- 81 06		15.0	FAINT BLUE STAR
PHL 2332	23 25 00.	+ 03 08		17.2	BLUE STELLAR OBJECT
ZWG 431.070	23 25 00.	+ 12 07		14.8	GALAXY
UGC 12603	23 25 00.	+ 29 35	66	16.5	GALAXY S
ZWG 497.021	23 25 00.	+ 29 47		15.5	GALAXY
LDN 1262	23 25 00.	+ 74 00	1020		DARK NEBULA
LDN 1261	23 25 00.	+ 74 00	660		DARK NEBULA
PHL 0530	23 25 00.	- 22 38		18.4	BLUE STELLAR OBJECT
RNGC 7672	23 25 02.	+ 12 07		15.0	GALAXY
ZWG 476.041	23 25 06.	+ 22 29		15.6	GALAXY
UGC 12604	23 25 06.	+ 22 29	60	15.6	GALAXY SBa
UGC 12605	23 25 06.	+ 39 07	60	16.5	GALAXY SB:c
ZWG 533.006	23 25 06.	+ 43 50		15.6	GALAXY
UGC 12606	23 25 06.	+ 43 52	66	16.5	GALAXY S
PHL 2333	23 25 06.	- 18 28		18.1	BLUE STELLAR OBJECT
SC 2322-6306.6	23 25 07.	- 62 50 06.			NEBULA
RNGC 7673	23 25 09.	+ 23 19		12.5	GALAXY
SC 2322-6227.7	23 25 09.	- 62 11 12.	24		NEBULA
ZWG 431.071	23 25 12.	+ 14 16		15.7	GALAXY
SN 1973L	23 25 12.	+ 14 16		18.5	SUPERNOVA
ZWG 454.068	23 25 12.	+ 18 18		15.5	GALAXY
MCG+04-55-014	23 25 12.	+ 23 18 30.	66	14.	GALAXY
4ZW 149	23 25 12.	+ 23 19			COMPACT GALAXY
ZWG 476.042	23 25 12.	+ 23 19		12.7	GALAXY
MRK 325	23 25 12.	+ 23 19	25	13.5	GALAXY WITH UV CONTINUUM
UGC 12607	23 25 12.	+ 23 19	102	12.7	GALAXY COMPACT
KARA.72 584A	23 25 12.	+ 23 19	102	12.7	PART OF DOUBLE GALAXY
PHL 0531	23 25 12.	- 06 29		13.8	BLUE STELLAR OBJECT
PHL 2334	23 25 12.	- 16 19		18.6	BLUE STELLAR OBJECT
PHL 5847	23 25 12.	- 19 38		18.1	BLUE STELLAR OBJECT
PHL 0532	23 25 12.	- 21 48		17.8	BLUE STELLAR OBJECT
PHL 5848	23 25 12.	- 25 42		18.0	BLUE STELLAR OBJECT
MCG-05-55-013	23 25 12.	- 28 14	48	15.	GALAXY
PHL 5846	23 25 12.	- 32 43		17.9	BLUE STELLAR OBJECT
ARP 182	23 25 17.	+ 08 30			PECULIAR GALAXY
PHL 5849	23 25 18.	+ 02 16		16.8	BLUE STELLAR OBJECT
ZC 2325.3-0005	23 25 18.	- 00 05	870		CLUSTER OF GALAXIES
PHL 2335	23 25 18.	- 19 00		18.7	BLUE STELLAR OBJECT
PHL 5850	23 25 18.	- 30 44		16.9	BLUE STELLAR OBJECT
IC 5324	23 25 18.	- 68 06			NONSTELLAR OBJECT
HN 0783	23 25 18.	- 68 07			NEBULA
ARC 2603	23 25 20.	- 25 38		17.7	RICH CLUSTER OF GALAXIES
ZWG 406.111	23 25 24.	+ 07 33		15.7	GALAXY
ZWG 406.112	23 25 24.	+ 08 30		13.6	GALAXY
MRK 533	23 25 24.	+ 08 30	10	16.	GALAXY WITH UV CONTINUUM
UGC 12608	23 25 24.	+ 08 30	60	13.6	GALAXY SBb
VV 343B	23 25 24.	+ 08 30	18	15.	INTERACTING GALAXY
VV 343A	23 25 24.	+ 08 30	66	13.	INTERACTING GALAXY
MCG+01-59-080	23 25 24.	+ 08 30	60	14.	GALAXY
ZWG 406.113	23 25 24.	+ 08 38		15.7	GALAXY
ZC 2325.4+1930	23 25 24.	+ 19 30	4300		CLUSTER OF GALAXIES
PHL 5851	23 25 24.	- 08 39		16.2	BLUE STELLAR OBJECT
PHL 2337	23 25 24.	- 08 46		18.7	BLUE STELLAR OBJECT
PHL 2338	23 25 24.	- 10 38		18.5	BLUE STELLAR OBJECT
PHL 5852	23 25 24.	- 19 50		17.9	BLUE STELLAR OBJECT
PHL 2336	23 25 24.	- 26 14		17.9	BLUE STELLAR OBJECT
PHL 2339	23 25 24.	- 29 16		18.2	BLUE STELLAR OBJECT
RNGC 7674	23 25 26.	+ 08 30		13.5	GALAXY
SC 2322-6232.0	23 25 26.	- 62 15 30.	60		NEBULA
MCG+01-59-081	23 25 27.	+ 08 30	12	16.	GALAXY
BC 4C27.52	23 25 28.5	+ 26 59 26.		17.5	QUASI-STELLAR OBJECT
SHE 388	23 25 29.	+ 26 59 26.		18.	QUASI-STELLAR OBJECT
MCG+01-59-082	23 25 30.	+ 07 35	36	15.	GALAXY
ZWG 406.114	23 25 30.	+ 08 29		14.8	GALAXY

OBJECT NAME	RIGET ASCEN.	DECLINATION	DIAM.	MAGN.	TYPE OF OBJECT
ZWG 497.022	23 25 30.	+ 30 44		15.7	GALAXY
UGC 12609	23 25 30.	+ 30 44	78	15.7	GALAXY
ZWG 497.023	23 25 30.	+ 31 54		15.1	GALAXY
PHL 2340	23 25 30.	- 02 47		18.2	BLUE STELLAR OBJECT
PHL 2341	23 25 30.	- 10 40		18.6	BLUE STELLAR OBJECT
TON-S 0099	23 25 30.	- 23 11		14.8	BLUE STAR
PHL 5853	23 25 30.	- 24 58		18.0	BLUE STELLAR OBJECT
RNGC 7675	23 25 32.	+ 08 29		15.0	GALAXY
MCG+03-59-058	23 25 33.	+ 18 15	42	14.	GALAXY
RNGC 7677	23 25 33.	+ 23 15		14.0	GALAXY
MCG+05-55-020	23 25 33.	+ 30 44 30.	60	15.	GALAXY
MCG+06-51-005	23 25 33.	+ 34 34 30.	42	15.	GALAXY
MCG+01-59-083	23 25 36.	+ 08 30	30	15.	GALAXY
ZWG 454.069	23 25 36.	+ 18 15		14.9	GALAXY
ZC 2325.6+1830	23 25 36.	+ 18 30	1210		CLUSTER OF GALAXIES
ZWG 476.043	23 25 36.	+ 23 15		13.9	GALAXY
MRK 326	23 25 36.	+ 23 15	22	15.	GALAXY WITH UV CONTINUUM
UGC 12610	23 25 36.	+ 23 15	102	13.9	GALAXY SB?b
KARA.72 584B	23 25 36.	+ 23 15	102	13.9	PART OF DOUBLE GALAXY
MCG+04-55-015	23 25 36.	+ 23 15	96	13.5	GALAXY
4ZW 150	23 25 36.	+ 24 33			COMPACT GALAXY
MCG+05-55-021	23 25 36.	+ 31 53	48	15.	GALAXY
ZWG 516.005	23 25 36.	+ 34 35		15.7	GALAXY
PHL 2345	23 25 36.	- 08 32		18.1	BLUE STELLAR OBJECT
PHL 2344	23 25 36.	- 09 22		17.9	BLUE STELLAR OBJECT
PHL 5854	23 25 36.	- 14 13		17.1	BLUE STELLAR OBJECT
PHL 2343	23 25 36.	- 24 11		18.5	BLUE STELLAR OBJECT
PHL 2342	23 25 36.	- 24 44		17.6	BLUE STELLAR OBJECT
PHL 5855	23 25 36.	- 32 30		16.8	BLUE STELLAR OBJECT
LB 01192	23 25 40.	+ 10 43 36.		15.6	FAINT BLUE STAR
BC CTD141	23 25 41.3	+ 29 20 36.		17.30	QUASI-STELLAR OBJECT
SHB 389	23 25 41.3	+ 29 20 36.		17.3	QUASI-STELLAR OBJECT
PHL 2346	23 25 42.	+ 01 00		18.1	BLUE STELLAR OBJECT
ZWG 406.115	23 25 42.	+ 07 20		15.3	GALAXY
MCG+01-59-084	23 25 42.	+ 09 20	30	15.	GALAXY
ZWG 476.044	23 25 42.	+ 27 24		15.5	GALAXY
UGC 12611	23 25 42.	+ 27 24	72	15.5	GALAXY Sa
ZWG 497.024	23 25 42.	+ 32 12		15.5	GALAXY
ZWG 533.007	23 25 42.	+ 40 11		15.6	GALAXY
PHL 0533	23 25 42.	- 15 28		17.9	BLUE STELLAR OBJECT
PHL 5856	23 25 42.	- 17 03		18.1	BLUE STELLAR OBJECT
PHL 5857	23 25 42.	- 23 11		15.6	BLUE STELLAR OBJECT
MCG+04-55-016	23 25 45.	+ 27 24	72	16.5	GALAXY
PHL 5861	23 25 48.	+ 00 06		18.5	BLUE STELLAR OBJECT
PHL 5858	23 25 48.	+ 00 56		17.2	BLUE STELLAR OBJECT
ZWG 454.070	23 25 48.	+ 19 45		15.6	GALAXY
MCG+05-55-022	23 25 48.	+ 32 12	42	15.5	GALAXY
UGC 12612	23 25 48.	+ 35 02	60	17.	GALAXY Sc
5ZW 405	23 25 48.	+ 44 55			COMPACT GALAXY
PHL 5862	23 25 48.	- 16 04		18.1	BLUE STELLAR OBJECT
PHL 5859	23 25 48.	- 29 13		17.9	BLUE STELLAR OBJECT
PHL 5860	23 25 48.	- 31 06		17.0	BLUE STELLAR OBJECT
MCG+03-59-059	23 25 54.	+ 18 10 30.	27	15.	GALAXY
SN 1971C	23 25 54.	+ 18 11		16.8	SUPERNOVA
ZWG 454.071	23 25 54.	+ 19 35		14.9	GALAXY
MCG+04-55-017	23 25 54.	+ 22 07	120	12.5	GALAXY
MCG-01-59-024	23 25 54.	- 03 05	72	14.	GALAXY
PHL 0534	23 25 54.	- 23 21		18.0	BLUE STELLAR OBJECT
PHL 2347	23 25 54.	- 23 22		17.9	BLUE STELLAR OBJECT
PHL 5863	23 25 54.	- 24 00		17.3	BLUE STELLAR OBJECT
ARC 2604	23 25 55.	- 22 49		17.1	RICH CLUSTER OF GALAXIES
RNGC 7678	23 25 57.	+ 22 09		13.0	GALAXY
MCG+02-59-047	23 26 00.	+ 11 54	48	15.	GALAXY
ZWG 431.072	23 26 00.	+ 14 27		13.2	GALAXY
UGC 12613	23 26 00.	+ 14 27	300	15.5	GALAXY DWRF IR
MCG+02-59-046	23 26 00.	+ 14 29	300	12.5	GALAXY
ZWG 454.072	23 26 00.	+ 17 00		15.0	GALAXY
ZWG 476.045	23 26 00.	+ 22 08		12.7	GALAXY
UGC 12614	23 26 00.	+ 22 08	168	12.7	GALAXY Sc/SBc
UGC 12615	23 26 00.	+ 28 30	60	17.	GALAXY
MCG-01-59-025	23 26 00.	- 03 04	36	16.	GALAXY
PHL 0535	23 26 00.	- 22 26		16.6	BLUE STELLAR OBJECT
IC 5325	23 26 00.	- 41 36	144	12.0	GALAXY SA(s)
MCG-07-48-004	23 26 00.	- 41 37	150	12.5	GALAXY
RNGC 7681	23 26 03.	+ 17 00		15.0	GALAXY
RNGC 7680	23 26 04.	+ 32 08		13.5	GALAXY
RNGC 7676	23 26 05.	- 59 59			GALAXY
PHL 5864	23 26 06.	+ 00 02		16.9	BLUE STELLAR OBJECT
MCG+00-59-046	23 26 06.	+ 03 16	114	13.2	GALAXY
ZWG 406.116	23 26 06.	+ 09 16		15.6	GALAXY
ZC 2326.1+1329	23 26 06.	+ 13 29	940		CLUSTER OF GALAXIES
MCG+03-59-060	23 26 06.	+ 16 59	42	15.	GALAXY
ZWG 454.073	23 26 06.	+ 17 00		15.7	GALAXY
ARP 028	23 26 06.	+ 22 08			PECULIAR GALAXY
ZWG 497.025	23 26 06.	+ 32 08		13.5	GALAXY
UGC 12616	23 26 06.	+ 32 08	120	13.5	GALAXY E-SO
4ZW 151	23 26 06.	+ 32 09			COMPACT GALAXY
UGC 12617	23 26 06.	+ 35 10	60	16.5	GALAXY S
PHL 2348	23 26 06.	- 04 44		16.9	BLUE STELLAR OBJECT
PHL 5867	23 26 06.	- 13 02		18.3	BLUE STELLAR OBJECT
PHL 0536	23 26 06.	- 21 46		18.2	BLUE STELLAR OBJECT
PHL 5866	23 26 06.	- 27 08		18.2	BLUE STELLAR OBJECT
PHL 5865	23 26 06.	- 29 04		18.1	BLUE STELLAR OBJECT
MCG-05-55-014	23 26 06.	- 29 43	24	15.	GALAXY
LB 01193	23 26 09.	- 04 45 18.		16.8	FAINT BLUE STAR
ARP 216	23 26 11.	+ 03 15			PECULIAR GALAXY
ZWG 380.061	23 26 12.	+ 03 14		13.2	GALAXY
MRK 534	23 26 12.	+ 03 14	25	13.	GALAXY WITH UV CONTINUUM
UGC 12618	23 26 12.	+ 03 14	102	13.2	GALAXY SO
4ZW 152	23 26 12.	+ 14 25			COMPACT GALAXY
ZWG 476.046	23 26 12.	+ 21 57		15.4	GALAXY
UGC 12619	23 26 12.	+ 21 57	96	15.4	GALAXY
PHL 0537	23 26 12.	- 07 09		17.7	BLUE STELLAR OBJECT
MCG-02-59-020	23 26 12.	- 08 57	48	14.	GALAXY
PHL 2349	23 26 12.	- 09 45		17.8	BLUE STELLAR OBJECT
PHL 2351	23 26 12.	- 12 40		18.8	BLUE STELLAR OBJECT
PHL 2350	23 26 12.	- 16 00		17.9	BLUE STELLAR OBJECT
PHL 5869	23 26 12.	- 22 57		17.0	BLUE STELLAR OBJECT
PHL 5868	23 26 12.	- 26 35		18.2	BLUE STELLAR OBJECT
RNGC 7679	23 26 13.	+ 03 14		13.5	GALAXY
MCG+03-59-061	23 26 15.	+ 16 56	42	15.5	GALAXY
MCG+05-55-023	23 26 15.	+ 32 08 30.	90	14.5	GALAXY
VV 329A	23 26 18.	+ 03 14	84	13.1	INTERACTING GALAXY
MCG+04-55-018	23 26 18.	+ 21 56	90	15.	GALAXY
ZWG 497.026	23 26 18.	+ 29 27		15.0	GALAXY
ZC 2326.3+3519	23 26 18.	+ 35 19	2690		CLUSTER OF GALAXIES
ZC 2326.3-0054	23 26 18.	- 00 54	1010		CLUSTER OF GALAXIES
PHL 2352	23 26 18.	- 10 14		17.4	BLUE STELLAR OBJECT

OBJECT NAME	RIGHT ASCEN.	DECLINATION	DIAM.	MAGN.	TYPE OF OBJECT
PHL 0528	23 26 18.	- 30 03		13.1	BLUE STELLAR OBJECT
IC 1497	23 26 19.	+ 11 43			NONSTELLAR OBJECT
BV C4	23 26 22.	- 30 03 24.		14.5	FAINT BLUE VARIABLE
PHL 5871	23 26 24.	+ 01 27		18.3	BLUE STELLAR OBJECT
MCG+00-59-047	23 26 24.	+ 03 17	66	14.	GALAXY
ZWG 406.117	23 26 24.	+ 08 36		15.6	GALAXY
ZWG 406.118	23 26 24.	+ 09 24		15.5	GALAXY
MCG+02-59-049	23 26 24.	+ 10 15	48	15.	GALAXY
MCG+02-59-048	23 26 24.	+ 11 10	30	14.	GALAXY
ZWG 454.074	23 26 24.	+ 17 03		14.2	GALAXY
UGC 12620	23 26 24.	+ 17 03	108	14.2	GALAXY S0
ZC 2326.4+2625	23 26 24.	+ 26 25	470		CLUSTER OF GALAXIES
MCG+05-55-024	23 26 24.	+ 29 26 30.	36	15.5	GALAXY
5ZW 406	23 26 24.	+ 43 36			COMPACT GALAXY
UGC 12621	23 26 24.	+ 46 43	90	16.0	GALAXY SB
ISS 0101	23 26 24.	+ 60 27	151		STELLAR RING
PHL 2353	23 26 24.	- 08 54		18.2	BLUE STELLAR OBJECT
PHL 5872	23 26 24.	- 09 02		18.5	BLUE STELLAR OBJECT
PHL 2354	23 26 24.	- 15 18		18.8	BLUE STELLAR OBJECT
PHL 0539	23 26 24.	- 15 21		16.7	BLUE STELLAR OBJECT
PHL 5870	23 26 24.	- 23 58		16.9	BLUE STELLAR OBJECT
PHL 5873	23 26 24.	- 27 24		18.2	BLUE STELLAR OBJECT
ABC 2605	23 26 25.	- 23 39		17.1	RICH CLUSTER OF GALAXIES
MCG+03-59-063	23 26 27.	+ 17 01 30.	90	14.	GALAXY
MCG+03-59-062	23 26 27.	+ 21 18	66	15.	GALAXY
IC 1490	23 26 28.	- 04 24 30.			NONSTELLAR OBJECT
ZWG 380.062	23 26 30.	+ 03 15		14.3	GALAXY
UGC 12622	23 26 30.	+ 03 15	78	14.3	GALAXY SBa
MCG+01-59-085	23 26 30.	+ 08 39	36	15.	GALAXY
ZWG 431.073	23 26 30.	+ 10 14		15.7	GALAXY
ZWG 431.074	23 26 30.	+ 11 10		14.6	GALAXY
UGC 12623	23 26 30.	+ 11 10	126	14.6	GALAXY S0
KARA.73B 1020	23 26 30.	+ 11 10	120	14.6	ISOLATED GALAXY S
3ZW 106	23 26 30.	+ 24 53			COMPACT GALAXY
PHL 2355	23 26 30.	- 15 20		18.6	BLUE STELLAR OBJECT
RNGC 7682	23 26 31.	+ 03 15		14.5	GALAXY
RNGC 7683	23 26 32.	+ 11 10		14.5	GALAXY
BC PKS2326-477	23 26 33.6	- 47 46 52.		16.	QUASI-STELLAR OBJECT
SHB 390	23 26 33.6	- 47 46 52.		16.	QUASI-STELLAR OBJECT
PHL 5874	23 26 36.	+ 02 48		17.7	BLUE STELLAR OBJECT
VV 329B	23 26 36.	+ 03 15	54	13.	INTERACTING GALAXY
UGC 12624	23 26 36.	+ 21 17	72	16.0	GALAXY IRR
ZWG 476.047	23 26 36.	+ 26 18		15.5	GALAXY
ZWG 497.027	23 26 36.	+ 29 30		15.0	GALAXY
UGC 12625	23 26 36.	+ 29 30	96	15.0	GALAXY Sb
MCG+06-51-006	23 26 36.	+ 34 45 30.	48	15.	GALAXY
PHL 0540	23 26 36.	- 10 22		13.6	BLUE STELLAR OBJECT
PHL 5876	23 26 36.	- 17 27		18.3	BLUE STELLAR OBJECT
TON-S 0100	23 26 36.	- 23 01		15.3	BLUE STAR
PHL 5875	23 26 36.	- 24 08		18.0	BLUE STELLAR OBJECT
MCG+05-55-025	23 26 39.	+ 29 29	90	15.	GALAXY
PHL 0541	23 26 42.	+ 02 00		18.5	BLUE STELLAR OBJECT
ZC 2326.7+0508	23 26 42.	+ 05 08	870		CLUSTER OF GALAXIES
ZWG 476.048	23 26 42.	+ 26 06		15.7	GALAXY
UGC 12626	23 26 42.	+ 26 06	72	15.7	GALAXY S
UGC 12627	23 26 42.	+ 34 47	60	16.0	GALAXY S0-a
PHL 5877	23 26 42.	- 02 38		15.3	BLUE STELLAR OBJECT
TON-S 0101	23 26 42.	- 24 26		15.3	BLUE STAR
MCG+00-59-048	23 26 45.	+ 03 09	90	15.	GALAXY
IC 1491	23 26 47.	- 16 35 24.			NONSTELLAR OBJECT
ZWG 380.063	23 26 48.	+ 03 07		15.6	GALAXY
UGC 12628	23 26 48.	+ 03 07	90	15.6	GALAXY SBc
ZWG 406.119	23 26 48.	+ 09 29		15.3	GALAXY
ZWG 431.075	23 26 49.	+ 13 55		15.5	GALAXY
PHL 5879	23 26 48.	- 03 52		8.7	BLUE STELLAR OBJECT
PHL 5881	23 26 48.	- 07 04		18.7	BLUE STELLAR OBJECT
PHL 0542	23 26 48.	- 10 23		18.1	BLUE STELLAR OBJECT
PHL 5878	23 26 48.	- 12 22		18.0	BLUE STELLAR OBJECT
PHL 2356	23 26 48.	- 14 11		13.4	BLUE STELLAR OBJECT
MCG-02-59-010	23 26 48.	- 16 35	30	15.	GALAXY
PHL 2358	23 26 48.	- 17 46		18.2	BLUE STELLAR OBJECT
PHL 5880	23 26 48.	- 20 38		17.9	BLUE STELLAR OBJECT
PHL 0543	23 26 48.	- 23 00		16.3	BLUE STELLAR OBJECT
PHL 2357	23 26 48.	- 29 55		17.7	BLUE STELLAR OBJECT
PHL 5882	23 26 54.	- 08 48		12.0	BLUE STELLAR OBJECT
PHL 2359	23 26 54.	- 11 38		18.6	BLUE STELLAR OBJECT
MCG-05-55-015	23 26 54.	- 29 05	48	14.	GALAXY
IC 5326	23 26 56.	- 29 06 29.			NONSTELLAR OBJECT
LBN 0503	23 27	+ 39 45	1020		BRIGHT NEBULA
VDB.66G 217	23 27	+ 40 41	170		DWARF GALAXY
PHL 2361	23 27 00.	+ 01 14		18.2	BLUE STELLAR OBJECT
ZC 2327.0+1059	23 27 00.	+ 10 59	1410		CLUSTER OF GALAXIES
ZWG 454.075	23 27 00.	+ 19 06		15.7	GALAXY
PHL 5884	23 27 00.	- 09 50		18.4	BLUE STELLAR OBJECT
PHL 2360	23 27 00.	- 17 51		18.7	BLUE STELLAR OBJECT
ABC 2606	23 27 00.	- 21 30		17.1	RICH CLUSTER OF GALAXIES
PHL 5883	23 27 00.	- 31 53		17.5	BLUE STELLAR OBJECT
MCG+03-59-064	23 27 03.	+ 19 05 30.	48	15.5	GALAXY
SC 2324-6049.2	23 27 05.	- 60 32 41.			NEBULA
ZWG 476.049	23 27 06.	+ 23 40		15.6	GALAXY
KARA.73B 1021	23 27 06.	+ 23 40	36	15.6	ISOLATED GALAXY S
PHL 5886	23 27 06.	- 09 17		17.8	BLUE STELLAR OBJECT
PHL 5885	23 27 06.	- 09 17		10.1	BLUE STELLAR OBJECT
PHL 5890	23 27 06.	- 12 53		18.5	BLUE STELLAR OBJECT
PHL 5887	23 27 06.	- 21 14		17.5	BLUE STELLAR OBJECT
MCG-04-55-009	23 27 06.	- 21 33 30.	12	15.5	GALAXY
PHL 5888	23 27 06.	- 23 04		18.0	BLUE STELLAR OBJECT
PHL 5889	23 27 06.	- 24 44		17.9	BLUE STELLAR OBJECT
ABC 2607	23 27 12.	+ 11 01		17.6	RICH CLUSTER OF GALAXIES
PHL 0544	23 27 12.	- 00 06		18.3	BLUE STELLAR OBJECT
PHL 5892	23 27 12.	- 15 54		18.1	BLUE STELLAR OBJECT
PHL 0545	23 27 12.	- 21 50		16.3	BLUE STELLAR OBJECT
PHL 5891	23 27 12.	- 25 18		17.9	BLUE STELLAR OBJECT
PHL 2363	23 27 12.	- 25 31		18.5	BLUE STELLAR OBJECT
PHL 2362	23 27 12.	- 26 19		10.0	BLUE STELLAR OBJECT
ZWG 497.028	23 27 18.	+ 32 25		15.6	GALAXY
UGC 12629	23 27 18.	+ 32 25	72	15.6	GALAXY Sa
PHL 0546	23 27 18.	- 03 58		18.6	BLUE STELLAR OBJECT
PHL 2364	23 27 18.	- 06 16		18.0	BLUE STELLAR OBJECT
MCG-02-59-021	23 27 18.	- 12 46	36	14.5	GALAXY
PHL 2365	23 27 18.	- 13 27		18.6	BLUE STELLAR OBJECT
PHL 5893	23 27 18.	- 23 40		18.4	BLUE STELLAR OBJECT
PHL 2366	23 27 18.	- 25 34		18.2	BLUE STELLAR OBJECT
PHL 5894	23 27 18.	- 28 28		18.2	BLUE STELLAR OBJECT
SC 2324-6317.7	23 27 20.	- 63 01 11.	30		NEBULA
ZWG 406.120	23 27 24.	+ 08 40		15.4	GALAXY
UGC 12630	23 27 24.	+ 08 40	60	15.4	GALAXY Sa-b
MCG+01-59-086	23 27 24.	+ 08 42 30.	48	15.	GALAXY
ZWG 476.050	23 27 24.	+ 22 55		15.6	GALAXY
ZWG 476.051	23 27 24.	+ 24 41		15.6	GALAXY
ZWG 497.029	23 27 24.	+ 31 49		15.0	GALAXY
ZWG 497.030	23 27 24.	+ 32 23		15.1	GALAXY
PHL 5896	23 27 24.	- 07 00		16.2	BLUE STELLAR OBJECT
PHL 2367	23 27 24.	- 09 22		17.9	BLUE STELLAR OBJECT
PHL 5898	23 27 24.	- 09 29		18.0	BLUE STELLAR OBJECT
PHL 5895	23 27 24.	- 11 43		15.2	BLUE STELLAR OBJECT
PHL 5897	23 27 24.	- 13 21		16.9	BLUE STELLAR OBJECT
PHL 5899	23 27 24.	- 19 48		18.1	BLUE STELLAR OBJECT
MCG-05-55-016	23 27 24.	- 31 25	36	15.	GALAXY
MCG+05-55-026	23 27 27.	+ 32 54	24	18.	GALAXY
ZWG 476.052	23 27 30.	+ 25 43		15.5	GALAXY
ZWG 476.053	23 27 30.	+ 25 55		15.7	GALAXY
MCG+04-55-019	23 27 30.	+ 26 47	48	15.	GALAXY
ZWG 476.054	23 27 30.	+ 26 49		14.8	GALAXY
UGC 12631	23 27 30.	+ 26 49	66	14.8	GALAXY Sb
MCG+05-55-027	23 27 30.	+ 31 48	39	15.	GALAXY
MCG+05-55-029	23 27 30.	+ 32 23	48	15.5	GALAXY
MCG+05-55-028	23 27 30.	+ 32 25 30.	48	15.5	GALAXY
MCG+07-48-007	23 27 30.	+ 40 42	240	16.	GALAXY
PHL 5900	23 27 30.	- 11 04		18.1	BLUE STELLAR OBJECT
MCG-05-55-017	23 27 30.	- 31 24	24	15.	GALAXY
MCG+02-59-050	23 27 36.	+ 15 29	48	14.5	GALAXY
4ZW 153	23 27 36.	+ 25 15			COMPACT GALAXY
ZWG 476.055	23 27 36.	+ 25 15		15.0	GALAXY
MCG+07-48-008	23 27 36.	+ 39 56	48	16.	GALAXY
ZWG 533.008	23 27 36.	+ 40 43		15.7	GALAXY
UGC 12632	23 27 36.	+ 40 43	300	15.7	GALAXY DWRF SP
PHL 0547	23 27 36.	- 03 25		17.8	BLUE STELLAR OBJECT
MCG-02-59-022	23 27 36.	- 09 10	48	15.	GALAXY
PHL 5903	23 27 36.	- 10 14		18.1	BLUE STELLAR OBJECT
PHL 2368	23 27 36.	- 10 56		17.9	BLUE STELLAR OBJECT
PHL 2370	23 27 36.	- 14 35		18.8	BLUE STELLAR OBJECT
PHL 2369	23 27 36.	- 17 43		18.6	BLUE STELLAR OBJECT
PHL 5902	23 27 36.	- 18 04		17.1	BLUE STELLAR OBJECT
PHL 5901	23 27 36.	- 27 00		12.5	BLUE STELLAR OBJECT
ZWG 454.076	23 27 42.	+ 15 29		15.4	GALAXY
ZWG 431.076	23 27 42.	+ 15 29		15.4	GALAXY
UGC 12633	23 27 42.	+ 15 29	60	15.4	GALAXY SB
ZC 2327.7+1610	23 27 42.	+ 16 10	670		CLUSTER OF GALAXIES
ZC 2327.7+1702	23 27 42.	+ 17 02	1340		CLUSTER OF GALAXIES
ZWG 476.056	23 27 42.	+ 22 33		15.5	GALAXY
3ZW 107	23 27 42.	+ 25 15			COMPACT GALAXY
ZWG 533.009	23 27 42.	+ 39 57		15.7	GALAXY
UGC 12634	23 27 42.	+ 39 57	96	15.7	GALAXY Sc
MCG-01-59-026	23 27 42.	- 02 43	36	15.5	GALAXY
PHL 2371	23 27 42.	- 08 54		18.1	BLUE STELLAR OBJECT
PHL 5904	23 27 42.	- 14 25		17.9	BLUE STELLAR OBJECT
PHL 5905	23 27 42.	- 27 11		16.2	BLUE STELLAR OBJECT
PHL 5906	23 27 42.	- 27 43		18.3	BLUE STELLAR OBJECT
IC 5227	23 27 43.	- 13 51 28.			NONSTELLAR OBJECT
ZC 2327.8+0707	23 27 48.	+ 07 07	1210		CLUSTER OF GALAXIES
ZC 2327.8+2401	23 27 48.	+ 24 01	1810		CLUSTER OF GALAXIES
MCG+04-55-021	23 27 48.	+ 26 48 30.	18	18.	GALAXY
MCG+04-55-020	23 27 48.	+ 26 50 30.	42	15.	GALAXY
ZWG 476.057	23 27 48.	+ 26 51		15.7	GALAXY
UGC 12636	23 27 48.	+ 26 51	66	15.7	GALAXY S0
ZC 2327.8+2824	23 27 48.	+ 28 24	2080		CLUSTER OF GALAXIES
ZWG 497.031	23 27 48.	+ 32 37		15.5	GALAXY
ZC 2327.8+3323	23 27 48.	+ 33 23	1480		CLUSTER OF GALAXIES
OCL 0251	23 27 48.	+ 48 51	900	8.4	OPEN STAR CLUSTER
RNGC 7686	23 27 48.	+ 48 51		8.0	OPEN CLUSTER
MCG+00-59-049	23 27 48.	- 00 06	66	15.	GALAXY
ZWG 380.064	23 27 48.	- 00 07		15.6	GALAXY
UGC 12635	23 27 48.	- 00 07	66	15.6	GALAXY Sc
PHL 0548	23 27 48.	- 07 14		18.5	BLUE STELLAR OBJECT
PHL 0549	23 27 48.	- 09 55		18.1	BLUE STELLAR OBJECT
MCG-02-59-023	23 27 48.	- 12 22	48	15.	GALAXY
PHL 2372	23 27 48.	- 16 34		18.7	BLUE STELLAR OBJECT
PHL 5908	23 27 48.	- 23 29		18.2	BLUE STELLAR OBJECT
PHL 5907	23 27 48.	- 24 27		16.5	BLUE STELLAR OBJECT
MCG-04-55-010	23 27 48.	- 26 13	36	15.5	GALAXY
MCG-05-55-018	23 27 48.	- 29 04	48	14.5	GALAXY
MCG-07-48-005	23 27 48.	- 39 34	30	15.	GALAXY
ABC 2610	23 27 52.	+ 17 00		17.6	RICH CLUSTER OF GALAXIES
ZWG 406.121	23 27 54.	+ 03 37		15.0	GALAXY
UGC 12638	23 27 54.	+ 03 37	120	15.0	GALAXY Sc
ZC 2327.9+1702	23 27 54.	+ 17 02	400		CLUSTER OF GALAXIES
ZC 2327.9+2020	23 27 54.	+ 20 20	1750		CLUSTER OF GALAXIES
ZWG 497.032	23 27 54.	+ 29 57		14.9	GALAXY
UGC 12639	23 27 54.	+ 29 57	84	14.9	GALAXY SBb
MCG+00-59-050	23 27 54.	- 00 11	78	14.	GALAXY
ZWG 380.065	23 27 54.	- 00 12		14.8	GALAXY
UGC 12637	23 27 54.	- 00 12	102	14.8	GALAXY Sa
ABC 2608	23 27 54.	- 21 57		17.1	RICH CLUSTER OF GALAXIES
RNGC 7684	23 27 55.	- 00 12		15.0	GALAXY
ABC 2609	23 27 56.	- 26 23		17.4	RICH CLUSTER OF GALAXIES
ABC 2611	23 27 57.	+ 20 21		17.2	RICH CLUSTER OF GALAXIES
MCG+05-55-030	23 27 57.	+ 32 37	39	15.5	GALAXY
IC 1493	23 27 59.	+ 14 10 38.			NONSTELLAR OBJECT
LBN 0507	23 28	+ 40 30	2100		BRIGHT NEBULA
KHAV 789	23 28	+ 58 47	14700		DARK NEBULA
LB 09954	23 28	- 86 45		13.9	FAINT BLUE STAR
ZC 2328.0+1104	23 28 00.	+ 11 04	1210		CLUSTER OF GALAXIES
ZC 2328.0+1629	23 28 00.	+ 16 29	1010		CLUSTER OF GALAXIES
MCG+05-55-031	23 28 00.	+ 29 56	84	15.	GALAXY
7ZW 941	23 28 00.	+ 85 18			COMPACT GALAXY
PHL 2374	23 28 00.	- 02 32		17.4	BLUE STELLAR OBJECT
MCG-01-59-027	23 28 00.	- 02 44	108	15.	GALAXY
PHL 2375	23 28 00.	- 12 08		18.5	BLUE STELLAR OBJECT
PHL 5909	23 28 00.	- 17 02		17.8	BLUE STELLAR OBJECT
RNGC 7685	23 28 01.	+ 03 38		15.0	GALAXY
IC 1492	23 28 01.	- 03 18 40.			NONSTELLAR OBJECT
ZWG 406.122	23 28 06.	+ 03 41		15.2	GALAXY
5ZW 407	23 28 06.	+ 54 50			COMPACT GALAXY
MCG-01-59-028	23 28 06.	- 03 19	48	14.	GALAXY
PHL 5911	23 28 06.	- 13 28		18.2	BLUE STELLAR OBJECT
PHL 2376	23 28 06.	- 19 52		16.7	BLUE STELLAR OBJECT
PHL 5910	23 28 06.	- 31 30		17.8	BLUE STELLAR OBJECT
LB 01194	23 28 08.	+ 10 45 48.		14.8	FAINT BLUE STAR
IC 1496	23 28 10.	- 03 13 10.			NONSTELLAR OBJECT
IC 1494	23 28 11.	- 13 00 04.			NONSTELLAR OBJECT
MCG+01-59-087	23 28 12.	+ 03 37 30.	132	13.	GALAXY
PHL 2377	23 28 12.	- 03 48		18.5	BLUE STELLAR OBJECT
IC 1495	23 28 12.	- 13 45 40.			NONSTELLAR OBJECT
MCG-02-59-024	23 28 12.	- 13 47	60	13.5	GALAXY
PHL 5912	23 28 12.	- 16 03		18.9	BLUE STELLAR OBJECT

OBJECT NAME	RIGHT ASCEN.	DECLINATION	DIAM.	MAGN.	TYPE OF OBJECT
PHL 2378	23 28 12.	- 24 32		18.5	BLUE STELLAR OBJECT
PHL 0550	23 28 12.	- 24 43		18.4	BLUE STELLAR OBJECT
MCG-05-55-019	23 28 12.	- 27 55	36	15.	GALAXY
ARC 2612	23 28 17.	- 18 56		17.7	RICH CLUSTER OF GALAXIES
ZWG 380.066	23 28 18.	+ 03 16		14.9	GALAXY
MCG+00-59-051	23 28 18.	+ 03 17 30.	66	14.	GALAXY
MCG+03-59-065	23 28 18.	+ 19 59	72	15.	GALAXY
PHL 2380	23 28 18.	- 10 12		18.1	BLUE STELLAR OBJECT
PHL 2379	23 28 18.	- 13 31		17.9	BLUE STELLAR OBJECT
PHL 5914	23 28 18.	- 14 46		18.9	BLUE STELLAR OBJECT
PHL 5913	23 28 18.	- 14 54		17.9	BLUE STELLAR OBJECT
MCG-05-55-020	23 28 18.	- 27 46	36	15.	GALAXY
PHL 5915	23 28 18.	- 28 25		18.1	BLUE STELLAR OBJECT
RNGC 7687	23 28 19.	+ 03 16		15.0	GALAXY
ZWG 431.077	23 28 24.	+ 09 32		15.6	GALAXY
ZWG 431.078	23 28 24.	+ 15 13		15.6	GALAXY
UGC 12640	23 28 24.	+ 15 13	60	15.6	GALAXY
ZWG 455.001	23 28 24.	+ 19 58		15.5	GALAXY
ZWG 454.077	23 28 24.	+ 19 58		15.5	GALAXY
UGC 12641	23 28 24.	+ 19 58	78	15.5	GALAXY Sb-c
4ZW 154	23 28 24.	+ 29 41			COMPACT GALAXY
5ZW 408	23 28 24.	+ 44 01			COMPACT GALAXY
MCG-01-59-029	23 28 24.	- 03 12	90	14.	GALAXY
PHL 5917	23 28 24.	- 15 24		18.5	BLUE STELLAR OBJECT
PHL 2381	23 28 24.	- 21 30		16.6	BLUE STELLAR OBJECT
PHL 5916	23 28 24.	- 28 30		17.9	BLUE STELLAR OBJECT
MCG-05-55-021	23 28 24.	- 30 02	48	15.	GALAXY
SC 2325-6128.9	23 28 24.	- 61 12 22.	72		NEBULA
ZC 2328.5+0027	23 28 30.	+ 00 27	1080		CLUSTER OF GALAXIES
ZWG 455.002	23 28 30.	+ 17 13		15.7	GALAXY
ZWG 454.078	23 28 30.	+ 17 13		15.7	GALAXY
ZWG 455.003	23 28 30.	+ 21 11		15.7	GALAXY
ZWG 454.079	23 28 30.	+ 21 11		15.7	GALAXY
ZWG 380.067	23 28 30.	- 02 30		15.4	GALAXY
PHL 5919	23 28 30.	- 11 58		18.6	BLUE STELLAR OBJECT
PHL 5913	23 28 30.	- 13 18		17.7	BLUE STELLAR OBJECT
SC 2325-7303.7	23 28 32.	- 72 47 10.	18		NEBULA
LB 01195	23 28 36.	+ 11 23 06.		15.9	FAINT BLUE STAR
ZWG 455.004	23 28 36.	+ 21 08		15.4	GALAXY
ZWG 454.080	23 28 36.	+ 21 08		15.4	GALAXY
ZC 2328.6+4604	23 28 36.	+ 46 04	1480		CLUSTER OF GALAXIES
PHL 0551	23 28 36.	- 11 32		17.5	BLUE STELLAR OBJECT
PHL 2382	23 28 36.	- 16 19		18.5	BLUE STELLAR OBJECT
ARC 2613	23 28 38.	- 13 14		17.2	RICH CLUSTER OF GALAXIES
RNGC 7683	23 28 39.	+ 21 08		15.5	GALAXY
ZWG 406.123	23 28 42.	+ 08 29		15.7	GALAXY
UGC 12642	23 28 42.	+ 08 57	72	16.5	GALAXY
ZWG 455.005	23 28 42.	+ 17 12		15.7	GALAXY
ZWG 454.081	23 28 42.	+ 17 12		15.7	GALAXY
PHL 5920	23 28 42.	- 13 06		17.7	BLUE STELLAR OBJECT
PHL 5921	23 28 42.	- 17 07		18.2	BLUE STELLAR OBJECT
PHL 5922	23 28 42.	- 24 20		16.5	BLUE STELLAR OBJECT
PHL 2384	23 28 48.	+ 00 22		17.6	BLUE STELLAR OBJECT
ZWG 407.001	23 28 48.	+ 09 11		15.6	GALAXY
ZWG 406.124	23 28 48.	+ 09 11		15.6	GALAXY
ZWG 455.006	23 28 48.	+ 17 58		15.7	GALAXY
ZWG 454.082	23 28 48.	+ 17 58		15.7	GALAXY
KARA.73B 1022	23 28 48.	+ 17 58	24	15.7	ISOLATED GALAXY S
PHL 2385	23 28 48.	- 01 24		18.3	BLUE STELLAR OBJECT
PHL 0552	23 28 48.	- 07 37		17.0	BLUE STELLAR OBJECT
PHL 5923	23 28 48.	- 08 46		18.5	BLUE STELLAR OBJECT
PHL 2383	23 28 48.	- 12 42		17.0	BLUE STELLAR OBJECT
PHL 5924	23 28 48.	- 15 42		17.9	BLUE STELLAR OBJECT
LB 01196	23 28 54.	+ 11 28 18.		17.1	FAINT BLUE STAR
LBN 0466	23 29	+ 28 20	720		BRIGHT NEBULA
ZWG 407.002	23 29 00.	+ 08 20		15.6	GALAXY
ZWG 406.125	23 29 00.	+ 08 20		15.6	GALAXY
UGC 12643	23 29 00.	+ 24 56	84	16.5	GALAXY
MCG+05-55-032	23 29 00.	+ 28 53 30.	72	15.	GALAXY
ZWG 497.033	23 29 00.	+ 28 55		15.7	GALAXY
UGC 12644	23 29 00.	+ 28 55	78	15.7	GALAXY SBc
ZWG 497.034	23 29 00.	+ 32 12		15.1	GALAXY
UGC 12645	23 29 00.	+ 32 12	66	15.1	GALAXY S
PHL 2386	23 29 00.	- 00 29		18.1	BLUE STELLAR OBJECT
PHL 0553	23 29 00.	- 02 05		17.5	BLUE STELLAR OBJECT
ZC 2329.0-0224	23 29 00.	- 02 24	9270		CLUSTER OF GALAXIES
PHL 5925	23 29 00.	- 08 54		17.9	BLUE STELLAR OBJECT
PHL 0554	23 29 00.	- 08 54		17.6	BLUE STELLAR OBJECT
PHL 2387	23 29 00.	- 13 59		18.3	BLUE STELLAR OBJECT
PHL 2388	23 29 00.	- 19 50		18.0	BLUE STELLAR OBJECT
PHL 5926	23 29 00.	- 22 30		17.8	BLUE STELLAR OBJECT
PHL 5927	23 29 00.	- 26 15		18.0	BLUE STELLAR OBJECT
MCG-05-55-022	23 29 00.	- 31 25	48	16.	GALAXY
ZWG 476.058	23 29 06.	+ 25 40		14.4	GALAXY
UGC 12646	23 29 06.	+ 25 40	114	14.4	GALAXY SBb
KARA.73B 1023	23 29 06.	+ 25 40	120	14.4	ISOLATED GALAXY S
MCG+05-55-033	23 29 06.	+ 32 12	48	15.	GALAXY
ZC 2329.1+3416	23 29 06.	+ 34 16	2490		CLUSTER OF GALAXIES
PHL 5928	23 29 06.	- 28 12		17.9	BLUE STELLAR OBJECT
MCG+04-55-022	23 29 09.	+ 25 38 30.	102	14.5	GALAXY
KARA.68 241	23 29 12.	+ 03 38	34		DWARF GALAXY
ZWG 476.059	23 29 12.	+ 27 03		15.5	GALAXY
KARA.73B 1024	23 29 12.	+ 27 03	48	15.5	ISOLATED GALAXY S
PHL 5929	23 29 12.	- 26 34		17.4	BLUE STELLAR OBJECT
PHL 0555	23 29 12.	- 29 09		13.7	BLUE STELLAR OBJECT
TON-S 0102	23 29 12.	- 29 09		13.7	BLUE STAR
SCHO 1425	23 29 17.	+ 55 48 06.	190		ISOLATED DARK CLOUD
IC 1498	23 29 17.	- 05 16 27.			NONSTELLAR OBJECT
ZWG 476.060	23 29 18.	+ 24 27		15.7	GALAXY
UGC 12647	23 29 18.	+ 47 10	60	16.5	GALAXY Sa-b
MCG-01-60-001	23 29 18.	- 05 55	66	16.	GALAXY
PHL 5930	23 29 18.	- 16 53		18.1	BLUE STELLAR OBJECT
PHL 5931	23 29 18.	- 30 24		18.3	BLUE STELLAR OBJECT
BC PKS2329-384	23 29 18.9	- 38 28 22.		17.	QUASI-STELLAR OBJECT
SHB 391	23 29 18.9	- 38 28 22.		17.	QUASI-STELLAR OBJECT
IC 1499	23 29 22.	- 13 43 09.			NONSTELLAR OBJECT
SC 2326-6202.7	23 29 22.	- 61 46 10.			NEBULA
ZWG 455.007	23 29 24.	+ 19 24		15.5	GALAXY
ZWG 454.083	23 29 24.	+ 19 24		15.5	GALAXY
ZWG 533.010	23 29 24.	+ 44 33		15.7	GALAXY
UGC 12649	23 29 24.	+ 44 33	66	15.7	GALAXY COMPACT
ZWG 381.001	23 29 24.	- 02 26		15.1	GALAXY
MCG-01-60-002	23 29 24.	- 05 17	84	14.	GALAXY
PHL 0556	23 29 24.	- 12 25		18.5	BLUE STELLAR OBJECT
MCG-05-55-023	23 29 24.	- 27 58	60	15.	GALAXY
PHL 5933	23 29 24.	- 28 28		18.3	BLUE STELLAR OBJECT
PHL 5932	23 29 24.	- 30 18		17.6	BLUE STELLAR OBJECT
SC 2326-6416.4	23 29 28.	- 63 59 52.	24		NEBULA
ZWG 476.061	23 29 30.	+ 23 17		15.4	GALAXY
MCG+04-55-023	23 29 30.	+ 23 37	30	15.5	GALAXY
ZWG 476.062	23 29 30.	+ 23 38		15.3	GALAXY
ZWG 497.035	23 29 30.	+ 32 08		15.5	GALAXY
UGC 12650	23 29 30.	+ 32 08	84	15.5	GALAXY Sc
MCG+07-48-009	23 29 30.	+ 44 33	72	16.	GALAXY
MRK 535	23 29 36.	+ 25 33	12	17.	GALAXY WITH UV CONTINUUM
MCG+06-51-007	23 29 36.	+ 35 06	60	15.	GALAXY
5ZW 409	23 29 36.	+ 44 26			COMPACT GALAXY
PHL 5934	23 29 36.	- 26 06		18.3	BLUE STELLAR OBJECT
MCG-05-55-024	23 29 36.	- 28 01	48	14.5	GALAXY
MCG+05-55-034	23 29 39.	+ 32 08	90	15.	GALAXY
PK116+08.1	23 29 41.	+ 70 05 39.	58		PLANETARY NEBULA
ZWG 381.002	23 29 42.	+ 02 08		15.0	GALAXY
MRK 536	23 29 42.	+ 02 08	13	16.	GALAXY WITH UV CONTINUUM
KARA.73B 1025	23 29 42.	+ 02 08	18	15.0	ISOLATED GALAXY E
ZWG 455.008	23 29 42.	+ 16 23		15.6	GALAXY
KARA.73B 1026	23 29 42.	+ 16 23	36	15.6	ISOLATED GALAXY S
UGC 12651	23 29 42.	+ 35 07	66	16.5	GALAXY
UGC 12648	23 29 42.	- 02 26	60	15.1	GALAXY IRR
PHL 2389	23 29 42.	- 19 54		18.6	BLUE STELLAR OBJECT
ZWG 476.063	23 29 48.	+ 23 14		15.6	GALAXY
MCG+04-55-024	23 29 48.	+ 23 14	36	15.5	GALAXY
PHL 0557	23 29 48.	- 09 04		18.0	BLUE STELLAR OBJECT
MCG-04-55-011	23 29 48.	- 24 12	48	15.	GALAXY
PHL 5935	23 29 48.	- 27 16		18.3	BLUE STELLAR OBJECT
PHL 2390	23 29 48.	- 27 35		17.9	BLUE STELLAR OBJECT
ZWG 407.003	23 29 54.	+ 05 35		15.6	GALAXY
MCG+03-60-001	23 29 54.	+ 15 33	108	13.	GALAXY
MCG+04-55-025	23 29 54.	+ 23 39 30.	60	14.5	GALAXY
5ZW 410	23 29 54.	+ 44 42			COMPACT GALAXY
ZWG 533.011	23 29 54.	+ 44 42		15.5	GALAXY
UGC 12652	23 29 54.	+ 44 42	78	15.5	GALAXY COMPACT
MCG+07-48-010	23 29 54.	+ 44 42	24	16.	GALAXY
PHL 5936	23 29 54.	- 27 34		17.4	BLUE STELLAR OBJECT
PHL 2391	23 29 54.	- 31 11		17.9	BLUE STELLAR OBJECT
RNGC 7689	23 29 55.	- 54 22		12.5	GALAXY
MCG+01-60-001	23 30 00.	+ 06 34	54	15.	GALAXY
MCG+02-60-001	23 30 00.	+ 14 32	108	13.5	GALAXY
ZWG 432.001	23 30 00.	+ 14 34		15.3	GALAXY
UGC 12653	23 30 00.	+ 14 34	96	15.3	GALAXY S
ZWG 455.009	23 30 00.	+ 15 34		14.2	GALAXY
UGC 12654	23 30 00.	+ 15 34	144	14.2	GALAXY SBb/Sc
ZWG 476.064	23 30 00.	+ 23 39		14.0	GALAXY
UGC 12655	23 30 00.	+ 23 39	90	14.0	GALAXY S0
LDN 1238	23 30 00.	+ 59 30	15180		DARK NEBULA
LDN 1245	23 30 00.	+ 60 30	1020		DARK NEBULA
PHL 5939	23 30 00.	- 24 18		18.3	BLUE STELLAR OBJECT
PHL 5938	23 30 00.	- 27 19		17.8	BLUE STELLAR OBJECT
PHL 5937	23 30 00.	- 29 17		17.9	BLUE STELLAR OBJECT
RNGC 7691	23 30 02.	+ 15 34		14.0	GALAXY
ZWG 407.004	23 30 06.	+ 06 35		15.0	GALAXY
VV 314B	23 30 06.	+ 29 46	15	17.	INTERACTING GALAXY
VV 314A	23 30 06.	+ 29 46	72	15.	INTERACTING GALAXY
ISS 0102	23 30 06.	+ 61 47	226		STELLAR RING
ZWG 381.003	23 30 06.	- 02 01		15.5	GALAXY
UGC 12656	23 30 06.	- 02 01	60	15.5	GALAXY S
HN 0002	23 30 11.	- 05 52			NEBULA
SHAH 027	23 30 12.	+ 01 17	180	17.	GROUP OF COMPACT GALAXIES
3ZW 108	23 30 12.	+ 19 05			COMPACT GALAXY
ZC 2330.2+2239	23 30 12.	+ 22 39	5980		CLUSTER OF GALAXIES
ZWG 497.036	23 30 12.	+ 29 11		14.5	GALAXY
UGC 12657	23 30 12.	+ 29 11	126	14.5	GALAXY S0
UGC 12658	23 30 12.	+ 30 52	78	16.0	GALAXY SBb?
5ZW 411	23 30 12.	+ 47 09			COMPACT GALAXY
SG 2.097	23 30 12.	+ 60 33	300		DIFFUSE EMISSION NEBULA
MCG+00-60-001	23 30 12.	- 02 01	78	15.	GALAXY
MCG-01-60-003	23 30 12.	- 05 52	27	15.	GALAXY
PHL 5941	23 30 12.	- 17 57		17.9	BLUE STELLAR OBJECT
PHL 5942	23 30 12.	- 18 04		18.0	BLUE STELLAR OBJECT
PHL 5940	23 30 12.	- 23 01		17.5	BLUE STELLAR OBJECT
RNGC 7692	23 30 13.	- 05 52		15.0	GALAXY
RNGC 7690	23 30 14.	- 51 58		13.0	GALAXY
MCG+05-55-035	23 30 15.	+ 29 10	24	15.	GALAXY
PHL 2392	23 30 16.	- 16 05		18.3	BLUE STELLAR OBJECT
ARC 2614	23 30 17.	- 21 52		17.5	RICH CLUSTER OF GALAXIES
ZC 2330.3+0630	23 30 18.	+ 06 30	470		CLUSTER OF GALAXIES
ZC 2330.3+1700	23 30 18.	+ 17 00	1080		CLUSTER OF GALAXIES
5ZW 412	23 30 18.	+ 48 40			COMPACT GALAXY
ZWG 381.004	23 30 18.	- 02 03		15.6	GALAXY
UGC 12659	23 30 18.	- 02 03	78	15.6	GALAXY Sc
PHL 0558	23 30 18.	- 03 18		17.3	BLUE STELLAR OBJECT
PHL 5943	23 30 18.	- 22 54		16.9	BLUE STELLAR OBJECT
PK104-01.1	23 30 21.4	+ 55 54 55.	17		PLANETARY NEBULA
ZWG 381.005	23 30 24.	+ 00 56		15.5	GALAXY
MCG+00-60-002	23 30 24.	- 02 04	78	14.5	GALAXY
PHL 0559	23 30 24.	- 21 12		17.0	BLUE STELLAR OBJECT
PHL 5944	23 30 24.	- 22 10		16.0	BLUE STELLAR OBJECT
PHL 5945	23 30 24.	- 22 32		17.4	BLUE STELLAR OBJECT
ARC 2615	23 30 24.	- 23 51		17.7	RICH CLUSTER OF GALAXIES
PHL 5946	23 30 24.	- 25 36		17.9	BLUE STELLAR OBJECT
SC 2327-6107.5	23 30 24.	- 60 50 57.	48		NEBULA
SCHO 1426	23 30 28.	+ 55 59 48.	260		ISOLATED DARK CLOUD
OCL 0259	23 30 30.	+ 55 12	600	8.	OPEN STAR CLUSTER
PHL 0560	23 30 30.	- 17 45		16.5	BLUE STELLAR OBJECT
IC 5328	23 30 30.	- 45 19	144	12.2	GALAXY E5
ZC 2330.6+0128	23 30 36.	+ 01 28	1810		CLUSTER OF GALAXIES
3ZW 109	23 30 36.	+ 09 29			COMPACT GALAXY
ZWG 407.005	23 30 36.	+ 09 29		15.1	GALAXY
MCG+03-60-002	23 30 36.	+ 20 58	90	15.	GALAXY
PHL 5947	23 30 36.	- 26 53		18.4	BLUE STELLAR OBJECT
PHL 5948	23 30 36.	- 23 08		17.8	BLUE STELLAR OBJECT
IC 1500	23 30 38.	+ 04 16 59.			NONSTELLAR OBJECT
IC 5329	23 30 40.	+ 20 56 59.			NONSTELLAR OBJECT
ZWG 407.006	23 30 42.	+ 04 17		15.4	GALAXY
ZWG 455.010	23 30 42.	+ 15 55		15.4	GALAXY
ZWG 476.065	23 30 42.	+ 25 22		15.4	GALAXY
ZWG 476.066	23 30 42.	+ 26 36		15.5	GALAXY
ISS 0103	23 30 42.	+ 60 46	157		STELLAR RING
ZWG 381.006	23 30 42.	- 01 33		14.7	GALAXY
SCHO 1427	23 30 43.	+ 55 52 48.	280		ISOLATED DARK CLOUD
RNGC 7693	23 30 43.	- 01 33		14.5	GALAXY
RNGC 7695	23 30 43.	- 02 59			NON-EXISTENT OBJECT
ARC 2616	23 30 44.	+ 05 20		17.2	RICH CLUSTER OF GALAXIES
MCG+01-60-002	23 30 45.	+ 07 34	30	15.5	GALAXY
ZWG 407.007	23 30 48.	+ 07 36		15.7	GALAXY
ZC 2330.8+0903	23 30 48.	+ 09 03	1480		CLUSTER OF GALAXIES
3ZW 110	23 30 48.	+ 12 00			COMPACT GALAXY

OBJECT NAME	RIGHT ASCEN.	DECLINATION	DIAM.	MAGN.	TYPE OF OBJECT
MCG+03-60-003	23 30 48.	+ 20 52	60	14.5	GALAXY
ZWG 455.011	23 30 48.	+ 20 57		15.7	GALAXY
UGC 12660	23 30 48.	+ 20 57	108	15.7	GALAXY Sc
KARA.72 585A	23 30 48.	+ 20 57	108	15.7	PART OF DOUBLE GALAXY
MCG+00-60-003	23 30 48.	- 01 34	60	14.	GALAXY
MCG-01-60-004	23 30 48.	- 02 58	72	14.	GALAXY
PHL 5949	23 30 48.	- 26 58		18.0	BLUE STELLAR OBJECT
MCG-07-48-006	23 30 48.	- 38 53	36	16.	GALAXY
ARC 2617	23 30 49.	+ 09 12		17.0	RICH CLUSTER OF GALAXIES
RNGC 7694	23 30 49.	- 02 58		14.0	GALAXY
MCG+01-60-003	23 30 54.	+ 04 05	48	16.	GALAXY
ZC 2330.9+1545	23 30 54.	+ 15 45	1280		CLUSTER OF GALAXIES
MCG+04-55-026	23 30 54.	+ 25 21	36	15.5	GALAXY
ZWG 476.067	23 30 54.	+ 25 22		15.1	GALAXY
SG 2.098	23 30 54.	+ 60 31	360		DIFFUSE EMISSION NEBULA
MCG+00-60-004	23 30 54.	- 00 44	12	16.	GALAXY
IC 5330	23 30 55.	- 03 08			NONSTELLAR OBJECT
IC 5331	23 30 56.	+ 20 50 23.			NONSTELLAR OBJECT
LBN 0557	23 31	+ 60 30	840		BRIGHT NEBULA
ZWG 432.002	23 31 00.	+ 14 05		15.3	GALAXY
ZWG 455.012	23 31 00.	+ 19 10		15.6	GALAXY
ZWG 455.013	23 31 00.	+ 20 51		15.1	GALAXY
UGC 12662	23 31 00.	+ 20 51	66	15.1	GALAXY Sa-b
KARA.72 585B	23 31 00.	+ 20 51	66	15.1	PART OF DOUBLE GALAXY
MCG+04-55-027	23 31 00.	+ 23 43	60	15.	GALAXY
UGC 12663	23 31 00.	+ 23 44	66	16.0	GALAXY
ZWG 497.037	23 31 00.	+ 33 22		15.7	GALAXY
4ZW 155	23 31 00.	+ 33 23			COMPACT GALAXY
OCL 0262	23 31 00.	+ 58 14	300	12.	OPEN STAR CLUSTER
MCG+00-60-005	23 31 00.	- 01 26	18	16.	GALAXY
ZWG 381.007	23 31 00.	- 02 15		15.0	GALAXY
UGC 12661	23 31 00.	- 02 15	66	15.0	GALAXY Sa-b
PHL 5950	23 31 00.	- 21 02		11.0	BLUE STELLAR OBJECT
MCG+00-60-006	23 31 03.	- 02 15	30	16.	GALAXY
ZWG 407.008	23 31 06.	+ 04 12		15.4	GALAXY
ZWG 407.009	23 31 06.	+ 08 33		15.4	GALAXY
ZWG 432.003	23 31 06.	+ 15 17		15.7	GALAXY
UGC 12664	23 31 06.	+ 15 17	72	15.7	GALAXY Sc
MCG+00-60-007	23 31 06.	- 02 16	60	14.	GALAXY
PHL 5951	23 31 06.	- 24 04		14.1	BLUE STELLAR OBJECT
ZC 2331.2+2144	23 31 12.	+ 21 44	1810		CLUSTER OF GALAXIES
MCG+04-55-028	23 31 12.	+ 25 29 30.	42	15.	GALAXY
ZWG 476.068	23 31 12.	+ 25 30		14.8	GALAXY
ARP 046	23 31 12.	+ 29 45			PECULIAR GALAXY
MCG+05-55-036	23 31 12.	+ 29 45	66	14.5	GALAXY
ZWG 497.038	23 31 12.	+ 29 46		15.1	GALAXY
UGC 12665	23 31 12.	+ 29 46	78	15.1	GALAXY S
KARA.72 586A	23 31 12.	+ 29 46	72	15.1	PART OF DOUBLE GALAXY
HOLM 809B	23 31 12.	+ 29 47	30	14.6	PART OF MULTIPLE GALAXY
ZWG 497.039	23 31 12.	+ 32 06		14.5	GALAXY
UGC 12666	23 31 12.	+ 32 06	90	14.5	GALAXY Sc
OCL 0265	23 31 12.	+ 61 38	360		OPEN STAR CLUSTER
OCL 0267	23 31 12.	+ 63 56	240	16.	OPEN STAR CLUSTER
PHL 5952	23 31 12.	- 26 53		16.1	BLUE STELLAR OBJECT
SN 1953E	23 31 13.	+ 29 46		19.5	SUPERNOVA
SC 2328-6047.9	23 31 13.	- 60 31 21.			NEBULA
MCG+05-55-037	23 31 15.	+ 32 06 30.	84	14.	GALAXY
SC 2328-6003.3	23 31 15.	- 59 46 44.	18		NEBULA
ZWG 407.010	23 31 18.	+ 04 27		14.8	GALAXY
ZWG 497.040	23 31 18.	+ 29 48		13.5	GALAXY
UGC 12667	23 31 18.	+ 29 48	96	13.5	GALAXY Sc
MCG-01-60-005	23 31 18.	- 03 21	66	16.	GALAXY
PC OZ-252	23 31 18.	- 24 00 14.			QUASI-STELLAR OBJECT
SHB 392	23 31 18.	- 24 00 14.		17.	QUASI-STELLAR OBJECT
TON-S 0103	23 31 18.	- 29 08		13.8	BLUE STAR
SCHO 1428	23 31 19.	+ 56 12 36.	280		ISOLATED DARK CLOUD
RNGC 7696	23 31 20.	+ 04 37		15.0	GALAXY
ARC 2618	23 31 21.	+ 22 44		15.9	RICH CLUSTER OF GALAXIES
HOLM 809A	23 31 23.	+ 29 48	66	12.6	PART OF MULTIPLE GALAXY
MCG+01-60-004	23 31 24.	+ 04 34	30	14.5	GALAXY
ZC 2331.4+1915	23 31 24.	+ 19 15	2350		CLUSTER OF GALAXIES
MCG+05-55-038	23 31 24.	+ 29 46	78	13.	GALAXY
KARA.72 586B	23 31 24.	+ 29 47	78	13.5	PART OF DOUBLE GALAXY
ZWG 533.012	23 31 24.	+ 44 54		15.7	GALAXY
ZWG 381.008	23 31 24.	- 01 36		15.5	GALAXY
MCG-02-60-001	23 31 24.	- 12 03	30	15.	GALAXY
PHL 5954	23 31 24.	- 26 35		18.1	BLUE STELLAR OBJECT
PHL 5953	23 31 24.	- 29 08		17.9	BLUE STELLAR OBJECT
PHL 0561	23 31 24.	- 29 08		14.3	BLUE STELLAR OBJECT
PHL 0562	23 31 24.	- 32 23		14.3	BLUE STELLAR OBJECT
SC 2328-6122.2	23 31 24.	- 61 05 38.			NEBULA
VV 187	23 31 27.	+ 01 18 30.	30		INTERACTING GALAXY
ARC 2619	23 31 27.	+ 21 42		16.5	RICH CLUSTER OF GALAXIES
RNGC 7698	23 31 27.	+ 24 40		14.5	GALAXY
BV 05	23 31 29.	- 23 06 36.		17.5	FAINT BLUE VARIABLE
MCG+04-55-029	23 31 30.	+ 24 39	48	15.	GALAXY
ZWG 476.069	23 31 30.	+ 24 40		14.5	GALAXY
UGC 12668	23 31 30.	+ 24 40	66	14.5	GALAXY S0
ZWG 533.013	23 31 30.	+ 44 25		15.3	GALAXY
UGC 12669	23 31 30.	+ 44 25	72	15.3	GALAXY Sb-c
5ZW 413	23 31 30.	+ 47 25			COMPACT GALAXY
PHL 2393	23 31 30.	- 22 52		17.6	BLUE STELLAR OBJECT
SC 2328-7111.0	23 31 30.	- 70 54 26.	36		NEBULA
SC 2328-6345.3	23 31 34.	- 63 28 44.	9		NEBULA
ZWG 407.011	23 31 36.	+ 09 24		15.7	GALAXY
MCG+07-48-011	23 31 36.	+ 44 24	66	14.5	GALAXY
MCG+00-60-008	23 31 36.	- 01 36	42	15.5	GALAXY
PHL 0563	23 31 36.	- 23 06		17.8	BLUE STELLAR OBJECT
PHL 5955	23 31 36.	- 29 57		18.0	BLUE STELLAR OBJECT
PHL 2394	23 31 36.	- 30 02		18.0	BLUE STELLAR OBJECT
MCG-05-55-025	23 31 36.	- 32 21	24	15.5	GALAXY
LB 01526	23 31 36.	- 47 30		13.0	FAINT BLUE STAR
ARC 2620	23 31 38.	+ 06 38		17.8	RICH CLUSTER OF GALAXIES
SVEF 458	23 31 40.	- 36 22	402	13.0	GALAXY
MCG+00-60-009	23 31 42.	+ 01 19	48	16.	GALAXY
ZC 2331.7+0705	23 31 42.	+ 07 05	540		CLUSTER OF GALAXIES
3ZW 111	23 31 42.	+ 14 42			COMPACT GALAXY
PHL 5956	23 31 42.	- 27 16		17.4	BLUE STELLAR OBJECT
IC 5332	23 31 42.	- 36 22	522	11.9	GALAXY SA(s)
MCG+02-60-002	23 31 48.	+ 12 16	72	14.5	GALAXY
ZWG 432.004	23 31 48.	+ 12 17		15.5	GALAXY
UGC 12670	23 31 48.	+ 12 17	60	15.5	GALAXY Sa-b
ZC 2331.8+1931	23 31 48.	+ 19 31	810		CLUSTER OF GALAXIES
ZWG 455.014	23 31 48.	+ 19 43		15.4	GALAXY
UGC 12671	23 31 48.	+ 43 07	66	16.0	GALAXY
MCG-02-60-002	23 31 48.	- 11 56	48	14.5	GALAXY
PHL 5957	23 31 48.	- 30 38		18.1	BLUE STELLAR OBJECT
RNGC 7699	23 31 49.	- 03 10			GALAXY
MCG+06-51-008	23 31 51.	+ 34 19 30.	60	14.	GALAXY
ARC 2621	23 31 52.	+ 19 37		17.9	RICH CLUSTER OF GALAXIES
ZWG 407.012	23 31 54.	+ 04 43		15.3	GALAXY
MRK 537	23 31 54.	+ 04 43	18	15.5	GALAXY WITH UV CONTINUUM
ZC 2331.9+0517	23 31 54.	+ 05 17	2960		CLUSTER OF GALAXIES
ZWG 516.006	23 31 54.	+ 34 21		14.6	GALAXY
UGC 12672	23 31 54.	+ 34 21	66	14.6	GALAXY S
PHL 5958	23 31 54.	- 24 18		18.0	BLUE STELLAR OBJECT
PHL 5959	23 31 54.	- 25 56		18.2	BLUE STELLAR OBJECT
VDB.66G 218	23 32	+ 17 56	70		DWARF GALAXY
UGC 12673	23 32 00.	+ 14 54	60	16.0	GALAXY Sc
ZWG 455.015	23 32 00.	+ 16 33		15.7	GALAXY
UGC 12674	23 32 00.	+ 28 27	66	16.5	GALAXY
MCG-01-60-007	23 32 00.	- 03 07	48	14.	GALAXY
MCG-01-60-006	23 32 00.	- 03 13	108	14.5	GALAXY
PHL 2395	23 32 00.	- 22 04		17.5	BLUE STELLAR OBJECT
RNGC 7701	23 32 01.	- 03 07		14.0	GALAXY
RNGC 7700	23 32 01.	- 03 13		14.5	GALAXY
IC 5334	23 32 02.	- 04 48 42.			NONSTELLAR OBJECT
SHAH 084	23 32 06.	+ 07 43	78	16.5	GROUP OF COMPACT GALAXIES
ZC 2332.1+1737	23 32 06.	+ 17 37	1550		CLUSTER OF GALAXIES
ZWG 476.070	23 32 06.	+ 22 57		15.7	GALAXY
UGC 12675	23 32 06.	+ 22 57	66	15.7	GALAXY
MCG-01-60-008	23 32 06.	- 04 48	90	14.	GALAXY
PHL 5961	23 32 06.	- 26 26		18.3	BLUE STELLAR OBJECT
PHL 5960	23 32 06.	- 30 24		17.9	BLUE STELLAR OBJECT
IC 1501	23 32 07.	- 03 25 54.			NONSTELLAR OBJECT
RNGC 7708	23 32 08.	+ 72 39			NON-EXISTENT OBJECT
IC 5333	23 32 10.	- 65 41			NONSTELLAR OBJECT
HN 0784	23 32 11.	- 65 41			NEBULA
ZWG 476.071	23 32 12.	+ 26 26		15.4	GALAXY
4ZW 156	23 32 12.	+ 34 57			COMPACT GALAXY
5ZW 414	23 32 12.	+ 47 13			COMPACT GALAXY
PHL 5963	23 32 12.	- 03 13		18.0	BLUE STELLAR OBJECT
MCG-01-60-009	23 32 12.	- 03 25	60	14.5	GALAXY
PHL 2396	23 32 12.	- 03 41		17.7	BLUE STELLAR OBJECT
PHL 5962	23 32 12.	- 30 02		17.7	BLUE STELLAR OBJECT
MCG+03-60-004	23 32 15.	+ 15 47	96	14.	GALAXY
ZWG 455.016	23 32 18.	+ 15 48		14.8	GALAXY
UGC 12676	23 32 18.	+ 15 48	138	14.8	GALAXY S0
MCG+03-60-005	23 32 18.	+ 16 32	33	15.	GALAXY
ZWG 455.017	23 32 18.	+ 16 35		15.4	GALAXY
UGC 12677	23 32 18.	+ 16 35	60	15.4	GALAXY Sb-c
MCG+03-60-006	23 32 18.	+ 17 00	78	15.	GALAXY
MCG+03-60-008	23 32 18.	+ 17 47 30.	54	14.5	GALAXY
MCG+03-60-007	23 32 18.	+ 17 57	72	14.	GALAXY
ZWG 476.072	23 32 18.	+ 26 02		15.4	GALAXY
UGC 12678	23 32 18.	+ 26 02	72	15.4	GALAXY Sb
ZC 2332.3+2700	23 32 18.	+ 27 00	3360		CLUSTER OF GALAXIES
UGC 12679	23 32 18.	+ 47 13	66	16.0	GALAXY S0?
MCG+00-60-010	23 32 18.	- 01 43	42	15.5	GALAXY
RNGC 7703	23 32 20.	+ 15 48		15.0	GALAXY
RNGC 7707	23 32 23.	+ 44 02		15.0	GALAXY
ZWG 455.018	23 32 24.	+ 17 01		15.7	GALAXY
UGC 12680	23 32 24.	+ 17 01	78	15.7	GALAXY SB?b-c
ZWG 455.019	23 32 24.	+ 17 48		15.2	GALAXY
UGC 12681	23 32 24.	+ 17 48	78	15.2	GALAXY SB
ZWG 455.020	23 32 24.	+ 17 57		14.7	GALAXY
UGC 12682	23 32 24.	+ 17 57	96	14.7	GALAXY IRR
ZWG 533.014	23 32 24.	+ 44 02		14.8	GALAXY
UGC 12683	23 32 24.	+ 44 02	78	14.8	GALAXY E-S0
ZWG 381.009	23 32 24.	- 02 01		15.7	GALAXY
PHL 5965	23 32 24.	- 23 10		15.9	BLUE STELLAR OBJECT
PHL 5964	23 32 24.	- 31 38		17.7	BLUE STELLAR OBJECT
ARC 2622	23 32 26.	+ 27 09		15.9	RICH CLUSTER OF GALAXIES
ZC 2332.5+0030	23 32 30.	+ 00 30	4570		CLUSTER OF GALAXIES
MCG+01-60-005	23 32 30.	+ 04 36	18	14.	GALAXY
MCG+07-48-012	23 32 30.	+ 44 01	63	14.5	GALAXY
MCG+00-60-011	23 32 30.	- 01 29	27	15.	GALAXY
ARC 2623	23 32 32.	+ 05 20		17.2	RICH CLUSTER OF GALAXIES
ZWG 407.013	23 32 36.	+ 04 32		15.4	GALAXY
ZWG 407.014	23 32 36.	+ 04 38		14.8	GALAXY
UGC 12684	23 32 36.	+ 04 38	72	14.8	GALAXY E-S0
ZC 2332.6+2810	23 32 36.	+ 28 10	1140		CLUSTER OF GALAXIES
PHL 5967	23 32 36.	- 22 20		18.0	BLUE STELLAR OBJECT
PHL 5966	23 32 36.	- 26 42		15.8	BLUE STELLAR OBJECT
PHL 5968	23 32 36.	- 28 20		18.0	BLUE STELLAR OBJECT
RNGC 7705	23 32 37.	+ 04 32		15.5	GALAXY
RNGC 7704	23 32 37.	+ 04 38		15.0	GALAXY
MCG+01-60-006	23 32 42.	+ 04 40	60	14.	GALAXY
ZWG 407.015	23 32 42.	+ 04 43		14.6	GALAXY
UGC 12686	23 32 42.	+ 04 43	66	14.6	GALAXY S0
ZWG 455.021	23 32 42.	+ 21 21		15.2	GALAXY
UGC 12685	23 32 42.	- 00 13	66	16.0	GALAXY Sc
MCG-03-60-001	23 32 42.	- 17 42	72	14.5	GALAXY
RNGC 7706	23 32 43.	+ 04 43		14.5	GALAXY
RNGC 7702	23 32 43.	- 56 17		13.0	GALAXY
ZWG 407.016	23 32 48.	+ 09 24		15.6	GALAXY
MCG+02-60-003	23 32 48.	+ 12 38	96	14.	GALAXY
ZWG 432.005	23 32 48.	+ 12 40		14.9	GALAXY
UGC 12687	23 32 48.	+ 12 40	120	14.9	GALAXY SBb
ZC 2332.8+1823	23 32 48.	+ 18 23	1550		CLUSTER OF GALAXIES
ZC 2332.8+2027	23 32 48.	+ 20 27	4770		CLUSTER OF GALAXIES
PHL 5971	23 32 48.	- 04 02		18.2	BLUE STELLAR OBJECT
RNGC 7709	23 32 48.	- 16 58		13.0	GALAXY
MCG-03-60-002	23 32 48.	- 16 58	84	13.5	GALAXY
MCG-05-55-026	23 32 48.	- 26 55	36	15.	GALAXY
PHL 5969	23 32 48.	- 27 00		15.8	BLUE STELLAR OBJECT
PHL 5970	23 32 48.	- 27 51		16.1	BLUE STELLAR OBJECT
MCG+00-60-012	23 32 51.	- 00 13	72	15.	GALAXY
SC 2330-6427.6	23 32 51.	- 64 11 02.	12		NEBULA
MCG+01-60-007	23 32 54.	+ 07 02	102	13.	GALAXY
ZWG 407.017	23 32 54.	+ 07 03		14.4	GALAXY
UGC 12688	23 32 54.	+ 07 03	96	14.4	GALAXY PECULR
KARA.73B 1027	23 32 54.	+ 07 03	102	14.4	ISOLATED GALAXY IR
MCG+03-60-009	23 32 54.	+ 18 01	36	16.	GALAXY
ZC 2332.9-0221	23 32 54.	- 02 21	740		CLUSTER OF GALAXIES
PHL 5972	23 32 54.	- 20 57		17.0	BLUE STELLAR OBJECT
TON-S 0104	23 32 54.	- 24 32		14.6	BLUE STAR
HN 0785	23 32 54.	- 67 41			NEBULA
SC 2330-6410.3	23 32 55.	- 63 53 44.	60		NEBULA
IC 5335	23 32 55.	- 67 41			NONSTELLAR OBJECT
LBN 0484	23 33	+ 28 50	3000		BRIGHT NEBULA
MCG+00-60-013	23 33 00.	+ 00 47 30.	12	15.	GALAXY
MCG+01-60-008	23 33 00.	+ 04 55	96	14.5	GALAXY
ZWG 407.018	23 33 00.	+ 04 57		14.5	GALAXY
UGC 12689	23 33 00.	+ 04 57	114	14.5	GALAXY Sa-b
ZWG 432.006	23 33 00.	+ 10 59		15.7	GALAXY

OBJECT NAME	RIGHT ASCEN.	DECLINATION	DIAM.	MAGN.	TYPE OF OBJECT
ZWG 455.022	23 33 00.	+ 18 02		15.7	GALAXY
5ZW 415	23 33 00.	+ 45 30			COMPACT GALAXY
MCG+00-60-014	23 33 00.	- 01 25	36	14.	GALAXY
MCG-03-60-003	23 33 00.	- 19 40	42	15.	GALAXY
PHL 0564	23 33 00.	- 24 31		15.8	BLUE STELLAR OBJECT
MCG+02-60-004	23 33 03.	+ 15 01	120	13.	GALAXY
UGC 12690	23 33 06.	+ 00 57	120	17.	GALAXY DWARF
ZWG 432.007	23 33 06.	+ 15 02		14.0	GALAXY
UGC 12691	23 33 06.	+ 15 02	240	14.0	GALAXY S0
MCG+03-60-010	23 33 06.	+ 20 00	54	14.5	GALAXY
ARC 2624	23 33 08.	+ 05 20		18.0	RICH CLUSTER OF GALAXIES
RNGC 7711	23 33 08.	+ 15 02		14.0	GALAXY
SC 2330-6351.7	23 33 08.	- 63 35 07.	6		NEBULA
PK114+03.1	23 33 11.	+ 64 36			PLANETARY NEBULA
ZWG 432.008	23 33 12.	+ 09 53		15.6	GALAXY
ZWG 455.023	23 33 12.	+ 20 01		15.6	GALAXY
UGC 12692	23 33 12.	+ 20 01	72	15.6	GALAXY S?
ZWG 497.041	23 33 12.	+ 32 06		15.6	GALAXY
UGC 12693	23 33 12.	+ 32 06	126	15.6	GALAXY SB?c
ZC 2333.2+7815	23 33 12.	+ 78 15	2080		CLUSTER OF GALAXIES
MCG-01-60-010	23 33 12.	- 03 09	72	15.	GALAXY
PHL 2399	23 33 12.	- 21 00		18.3	BLUE STELLAR OBJECT
PHL 2398	23 33 12.	- 21 52		18.0	BLUE STELLAR OBJECT
PHL 2397	23 33 12.	- 22 14		17.9	BLUE STELLAR OBJECT
RNGC 7710	23 33 13.	- 03 09		15.0	GALAXY
MCG+00-60-015	23 33 18.	+ 00 56	66	16.	GALAXY
ZC 2333.3+0521	23 33 18.	+ 05 21	340		CLUSTER OF GALAXIES
MCG+03-60-011	23 33 18.	+ 16 05	30	15.	GALAXY
ZWG 455.024	23 33 18.	+ 16 06		15.3	GALAXY
MCG+04-55-030	23 33 18.	+ 23 19 30.	48	14.	GALAXY
ZWG 476.073	23 33 18.	+ 23 20		13.7	GALAXY
UGC 12694	23 33 18.	+ 23 20	54	13.7	GALAXY PECULR
KARA.73B 1028	23 33 18.	+ 23 20	48	13.7	ISOLATED GALAXY E
ZWG 476.074	23 33 18.	+ 24 18		15.7	GALAXY
ZWG 476.075	23 33 18.	+ 26 43		15.6	GALAXY
MCG+05-55-039	23 33 18.	+ 32 06	138	15.	GALAXY
5ZW 416	23 33 18.	+ 42 06			COMPACT GALAXY
SC 2330-6338.5	23 33 18.	- 63 21 55.			NEBULA
RNGC 7712	23 33 21.	+ 23 20		13.5	GALAXY
ZC 2333.4+0546	23 33 24.	+ 05 46	1340		CLUSTER OF GALAXIES
PK104-29.1	23 33 24.	+ 30 11	350	15.1	PLANETARY NEBULA
PHL 5973	23 33 24.	- 22 54		17.0	BLUE STELLAR OBJECT
PHL 5974	23 33 24.	- 30 06		17.1	BLUE STELLAR OBJECT
MCG-07-48-007	23 33 24.	- 40 49	12	15.	GALAXY
SC 2330-6129.9	23 33 26.	- 61 13 19.			NEBULA
MCG+04-55-031	23 33 27.	+ 22 45	54	15.	GALAXY
MCG-03-60-004	23 33 27.	- 19 45	30	15.	GALAXY
UGC 12695	23 33 30.	+ 12 35	72	17.	GALAXY DWRF SP
MCG+02-60-005	23 33 30.	+ 12 35	96	16.	GALAXY
ZWG 432.009	23 33 30.	+ 13 32		15.7	GALAXY
KARA.73B 1029	23 33 30.	+ 13 32	36	15.7	ISOLATED GALAXY E
ZWG 476.076	23 33 30.	+ 22 46		15.1	GALAXY
UGC 12696	23 33 30.	+ 22 46	60	15.1	GALAXY SBb/SBc
UGC 12697	23 33 30.	+ 35 40	60	17.	GALAXY DWARF
UGC 12698	23 33 30.	+ 48 36	60	16.5	GALAXY Sc
MCG-03-60-005	23 33 30.	- 19 58	30	15.	GALAXY
MCG-07-48-008	23 33 30.	- 40 03	42	15.	GALAXY
ZWG 407.019	23 33 36.	+ 09 09		15.6	GALAXY
MCG+07-48-013	23 33 36.	+ 43 51	30	16.5	GALAXY
ZWG 381.010	23 33 36.	- 02 10		15.4	GALAXY
SC 2330-6350.7	23 33 39.	- 63 34 07.	6		NEBULA
HOLM 810A	23 33 39.	+ 01 53	30	13.4	PART OF MULTIPLE GALAXY
ARP 284	23 33 41.	+ 01 53			PECULIAR GALAXY
ZWG 381.011	23 33 42.	+ 01 53		13.1	GALAXY
MRK 538	23 33 42.	+ 01 53	35	13.5	GALAXY WITH UV CONTINUUM
VVI 94	23 33 42.	+ 01 53	150	13.4	SEYFERT GALAXY
UGC 12699	23 33 42.	+ 01 53	138	13.1	GALAXY S
VV 051A	23 33 42.	+ 01 53	90	13.4	INTERACTING GALAXY
KARA.72 587A	23 33 42.	+ 01 53	96	13.1	PART OF DOUBLE GALAXY
ZWG 432.010	23 33 42.	+ 14 45		15.7	GALAXY
MCG-02-60-003	23 33 42.	- 11 44	36	14.5	GALAXY
MCG-07-48-009	23 33 42.	- 39 04	24	16.	GALAXY
RNGC 7714	23 33 43.	+ 01 53		13.0	GALAXY
MCG+00-60-016	23 33 45.	- 02 11	48	15.	GALAXY
ARC 2625	23 33 46.	+ 20 15		15.6	RICH CLUSTER OF GALAXIES
RNGC 7713	23 33 46.	- 38 13		12.0	GALAXY
HOLM 810B	23 33 47.	+ 01 53	90	14.3	PART OF MULTIPLE GALAXY
VV 051B	23 33 48.	+ 01 52	120	14.3	INTERACTING GALAXY
ZWG 381.012	23 33 48.	+ 01 53		14.9	GALAXY
UGC 12700	23 33 48.	+ 01 53	198	14.9	GALAXY S
KARA.72 587B	23 33 48.	+ 01 53	102	14.9	PART OF DOUBLE GALAXY
ZC 2333.8+0428	23 33 48.	+ 04 28	1140		CLUSTER OF GALAXIES
MCG+03-60-012	23 33 48.	+ 20 52 30.	42	15.	GALAXY
MCG+05-55-040	23 33 48.	+ 27 39	66	16.5	GALAXY
UGC 12701	23 33 48.	+ 27 41	60	16.5	GALAXY Sc
PHL 5975	23 33 48.	- 24 30		18.3	BLUE STELLAR OBJECT
PHL 2401	23 33 48.	- 24 48		18.5	BLUE STELLAR OBJECT
PHL 2400	23 33 48.	- 25 10		15.7	BLUE STELLAR OBJECT
RNGC 7715	23 33 49.	+ 01 53		15.0	GALAXY
IC 5336	23 33 50.	+ 20 48 26.			NONSTELLAR OBJECT
MCG+00-60-017	23 33 54.	+ 01 53	150	13.	GALAXY
ZWG 455.025	23 33 54.	+ 20 52		15.4	GALAXY
MCG+03-60-013	23 33 54.	+ 20 52 30.	42	15.	GALAXY
SN 19600	23 33 54.	+ 27 39		18.5	SUPERNOVA
MCG-01-60-011	23 33 54.	- 05 11	120	14.5	GALAXY
PHL 0565	23 33 54.	- 22 28		18.4	BLUE STELLAR OBJECT
IC 5337	23 33 55.	+ 20 51 38.			NONSTELLAR OBJECT
HELW 500	23 33 56.	- 05 10 55.			NEBULA
SC 2331-6332.1	23 33 56.	- 63 15 31.	48		NEBULA
ARC 2626	23 33 58.	+ 20 54		15.2	RICH CLUSTER OF GALAXIES
VDB.66G 219	23 34	+ 00 02	100		DWARF GALAXY
LP 09955	23 34	- 80 07		14.9	FAINT BLUE STAR
ZWG 381.013	23 34 00.	+ 00 02		12.9	GALAXY
UGC 12702	23 34 00.	+ 00 02	138	12.9	GALAXY SBb
MCG+00-60-018	23 34 00.	+ 01 53	156	14.	GALAXY
ZWG 455.026	23 34 00.	+ 20 52		15.5	GALAXY
UGC 12703	23 34 00.	+ 20 52	84	15.5	GALAXY E?
UGC 12704	23 34 00.	+ 50 00	66	16.0	GALAXY E?
MCG+08-43-001	23 34 00.	+ 50 00	18	16.	GALAXY
PHL 2403	23 34 00.	- 24 50		18.0	BLUE STELLAR OBJECT
PHL 2402	23 34 00.	- 26 29		17.5	BLUE STELLAR OBJECT
PHL 5976	23 34 00.	- 29 37		18.1	BLUE STELLAR OBJECT
RNGC 7716	23 34 01.	+ 00 01		13.0	GALAXY
IC 5338	23 34 01.	+ 20 51 15.			NONSTELLAR OBJECT
RNGC 7725	23 34 01.	- 05 12		14.5	GALAXY
MCG+00-60-019	23 34 06.	+ 00 02	132	12.9	GALAXY
MCG+02-60-006	23 34 06.	+ 13 52 30.	72	15.	GALAXY
ZWG 432.011	23 34 06.	+ 13 53		15.7	GALAXY

OBJECT NAME	RIGHT ASCEN.	DECLINATION	DIAM.	MAGN.	TYPE OF OBJECT
UGC 12705	23 34 06.	+ 13 53	90	15.7	GALAXY SBc
KARA.73B 1030	23 34 06.	+ 13 53	90	15.7	ISOLATED GALAXY S
MCG+03-60-014	23 34 06.	+ 17 13	84	14.	GALAXY
ZC 2334.1+1752	23 34 06.	+ 17 52	1340		CLUSTER OF GALAXIES
ZWG 359.005	23 34 06.	+ 75 23		14.7	GALAXY
ZWG 344.003	23 34 06.	+ 75 23		14.7	GALAXY
UGC 12706	23 34 06.	+ 75 23	78	14.7	GALAXY S0-a
ZC 2334.1-0146	23 34 06.	- 01 46	1010		CLUSTER OF GALAXIES
PHL 0566	23 34 06.	- 20 36		18.0	BLUE STELLAR OBJECT
MCG-05-55-027	23 34 06.	- 27 14	36	15.	GALAXY
MCG-05-55-028	23 34 06.	- 29 20	12	15.	GALAXY
ARC 2627	23 34 09.	+ 23 39		17.1	RICH CLUSTER OF GALAXIES
HELW 501	23 34 11.	- 05 19 01.			NEBULA
ZWG 455.027	23 34 12.	+ 17 15		15.0	GALAXY
UGC 12707	23 34 12.	+ 17 15	102	15.0	GALAXY SB
ZWG 455.028	23 34 12.	+ 20 50		15.6	GALAXY
ZC 2334.2+2337	23 34 12.	+ 23 37	2550		CLUSTER OF GALAXIES
MCG-01-60-012	23 34 12.	- 05 19	18	15.	GALAXY
PHL 5977	23 34 12.	- 20 46		18.0	BLUE STELLAR OBJECT
PHL 5978	23 34 12.	- 29 10		18.3	BLUE STELLAR OBJECT
IC 1502	23 34 14.	+ 75 22 39.			NONSTELLAR OBJECT
SCHO 1429	23 34 16.	+ 56 36 48.	230		ISOLATED DARK CLOUD
4ZW 157	23 34 18.	+ 23 46			COMPACT GALAXY
ZWG 476.077	23 34 18.	+ 26 13		15.5	GALAXY
UGC 12708	23 34 18.	+ 26 13	66	15.5	GALAXY
MCG+12-01-001	23 34 18.	+ 75 22	39	15.	GALAXY
ARC 2628	23 34 23.	- 24 26		17.7	RICH CLUSTER OF GALAXIES
ZWG 407.020	23 34 24.	+ 06 35		15.5	GALAXY
ZWG 455.029	23 34 24.	+ 21 10		15.6	GALAXY
MCG-01-60-013	23 34 24.	- 04 57	54	16.	GALAXY
PHL 5979	23 34 24.	- 23 13		17.7	BLUE STELLAR OBJECT
PHL 2404	23 34 24.	- 24 01		16.3	BLUE STELLAR OBJECT
HELW 502	23 34 25.	- 04 57 13.			NEBULA
RNGC 7713A	23 34 28.	- 37 53			GALAXY
MCG+01-60-009	23 34 30.	+ 04 37 30.	36	15.	GALAXY
ZWG 407.021	23 34 30.	+ 04 40		15.5	GALAXY
MRK 327	23 34 30.	+ 23 02	12	15.5	GALAXY WITH UV CONTINUUM
ZWG 476.078	23 34 30.	+ 23 03		15.7	GALAXY
MCG-04-55-012	23 34 30.	- 20 44	96	13.	GALAXY
PHL 2405	23 34 30.	- 26 09		8.6	BLUE STELLAR OBJECT
PHL 5980	23 34 30.	- 26 31		12.0	BLUE STELLAR OBJECT
DV.56 N7713A	23 34 30.	- 37 53	114		SAc GALAXY
MCG-02-60-004	23 34 33.	- 09 52	60	15.5	GALAXY
ZWG 407.022	23 34 36.	+ 08 57		15.7	GALAXY
ZWG 455.030	23 34 36.	+ 18 42		15.6	GALAXY
PHL 5981	23 34 36.	- 03 50		16.6	BLUE STELLAR OBJECT
MCG-01-60-014	23 34 36.	- 04 00	36	15.	GALAXY
MCG-03-60-006	23 34 36.	- 16 54	54	15.5	GALAXY
PHL 5982	23 34 36.	- 21 23		18.0	BLUE STELLAR OBJECT
PHL 2407	23 34 36.	- 24 24		18.0	BLUE STELLAR OBJECT
PHL 2406	23 34 36.	- 24 34		17.7	BLUE STELLAR OBJECT
5ZW 417	23 34 42.	+ 52 53			COMPACT GALAXY
ZWG 381.014	23 34 42.	- 01 37		15.6	GALAXY
PHL 5983	23 34 42.	- 29 42		17.7	BLUE STELLAR OBJECT
MCG+01-60-010	23 34 45.	+ 04 35	24	16.	GALAXY
MCG+05-55-041	23 34 45.	+ 31 31	36	15.	GALAXY
MCG+00-60-020	23 34 45.	- 01 38	42	15.	GALAXY
MCG+01-60-011	23 34 48.	+ 04 36	54	15.	GALAXY
ZWG 407.023	23 34 48.	+ 04 38		15.0	GALAXY
ZC 2334.8+1543	23 34 48.	+ 15 43	4030		CLUSTER OF GALAXIES
ZWG 476.079	23 34 48.	+ 26 36		15.6	GALAXY
MCG+07-48-014	23 34 48.	+ 43 52	30	15.	GALAXY
ZWG 533.015	23 34 48.	+ 43 53		15.2	GALAXY
5ZW 418	23 34 48.	+ 49 17			COMPACT GALAXY
ZWG 548.001	23 34 48.	+ 49 17		15.0	GALAXY
ZWG 547.004	23 34 48.	+ 49 17		15.0	GALAXY
OCL 0258	23 34 48.	+ 52 09	1200	8.	OPEN STAR CLUSTER
MCG+13-01-002	23 34 48.	+ 75 22 30.	66	15.	GALAXY
ZWG 381.015	23 34 48.	- 01 57		15.5	GALAXY
PHL 2408	23 34 48.	- 21 11		11.5	BLUE STELLAR OBJECT
PHL 0567	23 34 48.	- 22 54		15.3	BLUE STELLAR OBJECT
TON-S 0105	23 34 48.	- 22 54		14.5	BLUE STAR
PHL 5984	23 34 48.	- 28 10		17.9	BLUE STELLAR OBJECT
HOLM 811B	23 34 50.	+ 04 38	24	15.3	PART OF MULTIPLE GALAXY
HOLM 811A	23 34 51.	+ 04 39	36	14.5	PART OF MULTIPLE GALAXY
ZWG 381.016	23 34 54.	+ 00 08		15.7	GALAXY
UGC 12709	23 34 54.	+ 00 08	180	15.7	GALAXY DWRF SP
ZWG 407.024	23 34 54.	+ 08 52		15.7	GALAXY
ZWG 476.080	23 34 54.	+ 26 48		15.5	GALAXY
MCG+04-55-032	23 34 54.	+ 26 48	30	15.5	GALAXY
ASS 32	23 34 54.	+ 58 44			OB ASSOCIATION CAS OB9
ISS 0104	23 34 54.	+ 59 04	183		STELLAR RING
MCG+00-60-021	23 34 54.	- 01 58	48	15.	GALAXY
PHL 0568	23 34 54.	- 03 54		17.5	BLUE STELLAR OBJECT
MCG-03-60-007	23 34 57.	- 19 53 30.	24	15.	GALAXY
KHAV 790	23 35	+ 71 35	7400		DARK NEBULA
MCG+00-60-022	23 35 00.	+ 00 08	180	15.	GALAXY
ZWG 407.025	23 35 00.	+ 07 48		15.7	GALAXY
ARC 2630	23 35 00.	+ 15 34		15.2	RICH CLUSTER OF GALAXIES
MCG+03-60-015	23 35 00.	+ 17 42 30.	72	14.	GALAXY
ZWG 455.031	23 35 00.	+ 17 44		14.7	GALAXY
UGC 12710	23 35 00.	+ 17 44	90	14.7	GALAXY IRR
MCG-04-55-013	23 35 00.	- 23 46	30	15.5	GALAXY
PHL 2409	23 35 00.	- 23 54		17.0	BLUE STELLAR OBJECT
PHL 5985	23 35 00.	- 30 29		16.8	BLUE STELLAR OBJECT
HN 2869	23 35 03.3	- 47 46 48.			NEBULA
ARC 2631	23 35 04.	+ 00 02		18.0	RICH CLUSTER OF GALAXIES
SC 2332-6244.4	23 35 04.	- 62 27 48.	12		NEBULA
ARC 2629	23 35 05.	- 23 12		17.5	RICH CLUSTER OF GALAXIES
ZC 2335.1+0000	23 35 06.	+ 00 00	1280		CLUSTER OF GALAXIES
MCG+01-60-012	23 35 06.	+ 07 49	36	16.	GALAXY
SN 1966P	23 35 06.	+ 26 32		16.8	SUPERNOVA
ZC 2335.1+3423	23 35 06.	+ 34 23	4230		CLUSTER OF GALAXIES
5ZW 419	23 35 06.	+ 43 51			COMPACT GALAXY
RNGC 7717	23 35 06.	- 15 23		13.0	GALAXY
MCG-03-60-008	23 35 06.	- 15 23	60	13.5	GALAXY
PHL 5986	23 35 06.	- 24 16		17.7	BLUE STELLAR OBJECT
MCG-05-55-029	23 35 06.	- 26 30	42	15.	GALAXY
PHL 5987	23 35 06.	- 27 38		18.2	BLUE STELLAR OBJECT
MCG+07-48-015	23 35 09.	+ 43 51	42	17.	GALAXY
ZWG 407.026	23 35 12.	+ 04 46		15.7	GALAXY
ZC 2335.2+0512	23 35 12.	+ 05 12	1480		CLUSTER OF GALAXIES
ZWG 432.012	23 35 12.	+ 14 27		15.6	GALAXY
ZWG 455.032	23 35 12.	+ 18 37		15.4	GALAXY
ZWG 497.042	23 35 12.	+ 29 52		15.5	GALAXY
MRK 328	23 35 12.	+ 29 52	14	16.	GALAXY WITH UV CONTINUUM
ZCG 2335+29	23 35 12.	+ 29 52		15.5	COMPACT GALAXY
KARA.73B 1031	23 35 12.	+ 29 52	18	15.5	ISOLATED GALAXY E

OBJECT NAME	RIGHT ASCEN.	DECLINATION	DIAM.	MAGN.	TYPE OF OBJECT
PHL 5988	23 35 12.	- 03 31		17.6	BLUE STELLAR OBJECT
TON-S 0106	23 35 12.	- 31 21		14.2	BLUE STAR
PHL 0569	23 35 12.	- 31 22		15.3	BLUE STELLAR OBJECT
ARC 2632	23 35 13.	- 09 30		17.8	RICH CLUSTER OF GALAXIES
HN 1236	23 35 17.	- 68 43			NEBULA
ZWG 407.027	23 35 18.	+ 07 48		15.7	GALAXY
ZWG 432.013	23 35 18.	+ 11 39		15.6	GALAXY
ZWG 455.033	23 35 18.	+ 20 32		15.7	GALAXY
MCG+04-55-033	23 35 18.	+ 25 30	42	16.	GALAXY
ZWG 476.081	23 35 18.	+ 26 13		15.5	GALAXY
DG 191	23 35 18.	+ 48 22	4920		REFLECTION NEBULA
ZWG 381.017	23 35 18.	- 00 47		15.7	GALAXY
ZC 2335.3-0133	23 35 18.	- 01 33	270		CLUSTER OF GALAXIES
MCG-03-60-009	23 35 18.	- 19 51 30.	24	15.	GALAXY
PHL 5990	23 35 18.	- 22 39		18.3	BLUE STELLAR OBJECT
PHL 5989	23 35 18.	- 30 45		16.3	BLUE STELLAR OBJECT
IC 5339	23 35 18.	- 68 43			NONSTELLAR OBJECT
BC PKS2335-18	23 35 20.	- 18 08 48.			QUASI-STELLAR OBJECT
RNGC 7697	23 35 23.	- 65 50			GALAXY
ZC 2335.4+2431	23 35 24.	+ 24 31	540		CLUSTER OF GALAXIES
ZWG 497.043	23 35 24.	+ 31 43		14.1	GALAXY
UGC 12711	23 35 24.	+ 31 43	72	14.1	GALAXY
SN 1953P	23 35 24.	+ 31 43		19.0	SUPERNOVA
MCG+00-60-023	23 35 24.	- 00 47	48	15.	GALAXY
MCG-01-60-015	23 35 24.	- 03 45	48	15.	GALAXY
RNGC 7719	23 35 24.	- 23 14			GALAXY
PHL 5991	23 35 24.	- 28 22		18.2	BLUE STELLAR OBJECT
MCG+01-60-013	23 35 30.	+ 07 10	36	15.	GALAXY
ZC 2335.5+2449	23 35 30.	+ 24 49	21710		CLUSTER OF GALAXIES
ZWG 476.082	23 35 30.	+ 25 25		15.2	GALAXY
UGC 12712	23 35 30.	+ 25 25	84	15.2	GALAXY S
ZWG 476.083	23 35 30.	+ 26 10		15.4	GALAXY
RNGC 7718	23 35 33.	+ 25 25		15.0	GALAXY
MCG+04-55-034	23 35 33.	+ 25 26 30.	48	15.	GALAXY
MCG+05-55-042	23 35 33.	+ 31 42 30.	36	14.5	GALAXY
ZWG 407.028	23 35 36.	+ 07 13		15.6	GALAXY
ZWG 407.029	23 35 36.	+ 07 43		15.4	GALAXY
ZC 2335.6+1250	23 35 36.	+ 12 50	2150		CLUSTER OF GALAXIES
ZC 2335.6+1907	23 35 36.	+ 19 07	1080		CLUSTER OF GALAXIES
MCG+00-60-024	23 35 36.	- 02 19	60	16.	GALAXY
PHL 5992	23 35 36.	- 24 10		10.0	BLUE STELLAR OBJECT
PHL 0570	23 35 36.	- 29 40		17.0	BLUE STELLAR OBJECT
ARC 2633	23 35 37.	+ 12 56		17.6	RICH CLUSTER OF GALAXIES
VDB.66N 158	23 35 38.	+ 48 14	204		REFLECTION NEBULA
ZC 2335.7+0252	23 35 42.	+ 02 52	810		CLUSTER OF GALAXIES
MCG+01-60-014	23 35 42.	+ 07 31	36	15.	GALAXY
ZWG 407.030	23 35 42.	+ 07 32		15.3	GALAXY
ZWG 432.014	23 35 42.	+ 11 57		15.3	GALAXY
MCG+03-60-016	23 35 42.	+ 17 04	42	17.	GALAXY
ZWG 497.044	23 35 42.	+ 30 25		15.1	GALAXY
UGC 12713	23 35 42.	+ 30 25	72	15.1	GALAXY
ZWG 497.045	23 35 42.	+ 32 03		15.6	GALAXY
UGC 12714	23 35 42.	+ 32 03	78	15.6	GALAXY Sc
PHL 2410	23 35 42.	- 04 12		17.9	BLUE STELLAR OBJECT
MCG-04-55-014	23 35 42.	- 21 03	54	15.	GALAXY
SC 2332-6142.3	23 35 42.	- 61 25 42.			NEBULA
MCG+05-55-043	23 35 45.	+ 32 04	96	15.	GALAXY
ZC 2335.8+0541	23 35 48.	+ 05 41	1680		CLUSTER OF GALAXIES
MCG+01-60-015	23 35 48.	+ 08 56	36	14.5	GALAXY
ZWG 455.034	23 35 48.	+ 17 04		15.7	GALAXY
ZWG 476.084	23 35 48.	+ 26 32		15.7	GALAXY
ZWG 476.085	23 35 48.	+ 26 37		15.7	GALAXY
ZWG 476.086	23 35 48.	+ 26 53		15.5	GALAXY
MCG+05-55-044	23 35 48.	+ 30 25	48	14.5	GALAXY
PHL 2411	23 35 48.	- 25 40		17.0	BLUE STELLAR OBJECT
PHL 5994	23 35 48.	- 26 41		18.1	BLUE STELLAR OBJECT
PHL 5993	23 35 48.	- 26 45		17.9	BLUE STELLAR OBJECT
SCHO 1430	23 35 49.	+ 57 16 30.	260		ISOLATED DARK CLOUD
ARC 2634	23 35 51.	+ 26 46		13.8	RICH CLUSTER OF GALAXIES
SS 73	23 35 51.	+ 59 42			DIFFUSE GALACTIC NEBULA
IC 1503	23 35 53.	+ 04 31 46.			NONSTELLAR OBJECT
MCG+01-60-016	23 35 54.	+ 04 30	54	14.	GALAXY
ZWG 407.031	23 35 54.	+ 04 32		14.3	GALAXY
UGC 12715	23 35 54.	+ 04 32	60	14.3	GALAXY S-IRR
ZWG 476.087	23 35 54.	+ 26 43		15.5	GALAXY
MCG+04-55-035	23 35 54.	+ 26 43 30.	12	15.	GALAXY
ZC 2335.9+3318	23 35 54.	+ 33 18	1280		CLUSTER OF GALAXIES
4ZW 158	23 35 54.	+ 36 08			COMPACT GALAXY
PHL 2412	23 35 54.	- 24 14		18.0	BLUE STELLAR OBJECT
PHL 5995	23 35 54.	- 26 32		17.3	BLUE STELLAR OBJECT
LB 01527	23 35 54.	- 55 47		14.6	FAINT BLUE STAR
IC 5341	23 35 56.2	+ 26 42 34.			GALAXY E0
MCG+04-55-036	23 35 57.	+ 26 46	30	14.	GALAXY
IC 5340	23 35 59.	- 05 08			NONSTELLAR OBJECT
LBN 0554	23 36	+ 56 41	300		BRIGHT NEBULA
ZWG 476.088	23 36 00.	+ 26 19		15.6	GALAXY
MCG+04-55-038	23 36 00.	+ 26 20	42	16.	GALAXY
ZWG 476.089	23 36 00.	+ 26 33		15.4	GALAXY
ZWG 476.090	23 36 00.	+ 26 42		15.3	GALAXY
MCG+04-55-037	23 36 00.	+ 26 43	12	15.	GALAXY
ZWG 476.091	23 36 00.	+ 26 45		13.9	GALAXY
UGC 12716	23 36 00.	+ 26 45	168	13.9	GALAXY E+COMP
KARA.72 588B	23 36 00.	+ 26 45	36		PART OF DOUBLE GALAXY
KARA.72 588A	23 36 00.	+ 26 45	54	13.9	PART OF DOUBLE GALAXY
MRSL 113-01/1	23 36 00.	+ 59 42	180		HII REGION
PHL 5996	23 36 00.	- 22 20		18.3	BLUE STELLAR OBJECT
ARC 2635	23 36 02.	- 13 39		17.3	RICH CLUSTER OF GALAXIES
RNGC 7720	23 36 03.	+ 26 45		14.0	GALAXY
SCHO 1431	23 36 03.	+ 54 34 48.	310		ISOLATED DARK CLOUD
SG 3.243	23 36 03.	+ 59 42	90		DIFFUSE EMISSION NEBULA
ZC 2336.1+0321	23 36 06.	+ 03 21	1410		CLUSTER OF GALAXIES
MCG+01-60-017	23 36 06.	+ 05 09	60	14.5	GALAXY
ZWG 407.032	23 36 06.	+ 05 10		15.3	GALAXY
UGC 12717	23 36 06.	+ 05 10	66	15.3	GALAXY Sc
MCG+03-60-017	23 36 06.	+ 15 40	90	13.	GALAXY
ZC 2336.1+1630	23 36 06.	+ 16 30	1340		CLUSTER OF GALAXIES
ZWG 476.092	23 36 06.	+ 26 42		15.6	GALAXY
MCG+04-55-039	23 36 06.	+ 26 45	12	15.	GALAXY
5ZW 420	23 36 06.	+ 45 12			COMPACT GALAXY
PHL 5997	23 36 06.	- 22 00		18.2	BLUE STELLAR OBJECT
IC 5342	23 36 08.1	+ 26 44 05.			GALAXY E0
RNGC 7726	23 36 09.	+ 26 44		15.5	GALAXY
HOLM 812B	23 36 11.	- 06 48	90	15.0	PART OF MULTIPLE GALAXY
3ZW 112	23 36 12.	+ 08 47			COMPACT GALAXY
ZWG 407.033	23 36 12.	+ 08 47		15.2	GALAXY
ZWG 455.035	23 36 12.	+ 15 41		13.7	GALAXY
UGC 12718	23 36 12.	+ 15 41	132	13.7	GALAXY S0-a
ZC 2336.2+1555	23 36 12.	+ 15 55	1010		CLUSTER OF GALAXIES

OBJECT NAME	RIGHT ASCEN.	DECLINATION	DIAM.	MAGN.	TYPE OF OBJECT
ZWG 476.093	23 36 12.	+ 26 30		15.7	GALAXY
UGC 12719	23 36 12.	+ 26 30	66	15.7	GALAXY S0-a
ZWG 476.094	23 36 12.	+ 26 44		15.4	GALAXY
SN 1961N	23 36 12.	+ 26 44		16.8	SUPERNOVA
ZWG 476.095	23 36 12.	+ 26 56		15.5	GALAXY
PHL 2413	23 36 12.	- 04 08		17.6	BLUE STELLAR OBJECT
PHL 0571	23 36 12.	- 26 08		17.7	BLUE STELLAR OBJECT
TON-S 0107	23 36 12.	- 30 45		15.8	BLUE STAR
RNGC 7721	23 36 13.	- 06 48		12.5	GALAXY
ARC 2636	23 36 13.	- 09 50		17.8	RICH CLUSTER OF GALAXIES
RNGC 7722	23 36 14.	+ 15 41		13.5	GALAXY
HOLM 812A	23 36 17.	- 06 47	198	12.7	PART OF MULTIPLE GALAXY
ZWG 407.034	23 36 18.	+ 04 10		15.3	GALAXY
ZC 2336.3+2054	23 36 18.	+ 20 54	2960		CLUSTER OF GALAXIES
MRK 329	23 36 18.	+ 23 04	12	17.	GALAXY WITH UV CONTINUUM
ZWG 476.096	23 36 18.	+ 26 59		15.0	GALAXY
OCL 0263	23 36 18.	+ 56 22	360	18.	OPEN STAR CLUSTER
MCG-01-60-016	23 36 18.	- 06 02	96	15.	GALAXY
MCG-01-60-017	23 36 18.	- 06 48	156	12.	GALAXY
PHL 5998	23 36 18.	- 28 11		18.1	BLUE STELLAR OBJECT
MCG-02-60-005	23 36 21.	- 13 14	216	11.5	GALAXY
REIN 2.319	23 36 21.83	- 13 14 18.4			NEBULA
ZWG 381.018	23 36 24.	+ 01 31		15.3	GALAXY
MCG+00-60-025	23 36 24.	- 01 44	48	15.	GALAXY
RNGC 7723	23 36 24.	- 13 14		12.0	GALAXY
MCG-04-55-015	23 36 24.	- 21 05	48	15.	GALAXY
PHL 5999	23 36 24.	- 21 38		16.5	BLUE STELLAR OBJECT
LP 01528	23 36 24.	- 57 59		13.8	FAINT BLUE STAR
MCG+00-60-026	23 36 30.	+ 01 32	54	15.	GALAXY
MCG+01-60-018	23 36 30.	+ 07 31	36	15.	GALAXY
RNGC 7724	23 36 30.	- 12 30		13.0	GALAXY
MCG-02-60-006	23 36 30.	- 12 30	84	13.	GALAXY
MCG-04-55-016	23 36 30.	- 25 57 30.	48	14.5	GALAXY
MCG-05-55-030	23 36 30.	- 31 50	30	15.5	GALAXY
SC 2333-6045.7	23 36 30.	- 60 29 06.	12		NEBULA
ARC 2637	23 36 34.	+ 21 12		16.6	RICH CLUSTER OF GALAXIES
ZWG 381.019	23 36 36.	+ 00 11		15.4	GALAXY
ZC 2336.6+0225	23 36 36.	+ 02 25	1550		CLUSTER OF GALAXIES
MCG+01-60-019	23 36 36.	+ 05 11	72	15.	GALAXY
ZWG 497.046	23 36 36.	+ 31 49		14.8	GALAXY
PHL 2414	23 36 36.	- 23 14		18.3	BLUE STELLAR OBJECT
PHL 6000	23 36 36.	- 27 17		18.1	BLUE STELLAR OBJECT
MCG+04-55-040	23 36 39.	+ 26 51	90	15.	GALAXY
HELW 503	23 36 41.	- 04 48 36.			NEBULA
ZWG 407.035	23 36 42.	+ 04 08		15.7	GALAXY
ZWG 407.036	23 36 42.	+ 05 13		15.3	GALAXY
UGC 12720	23 36 42.	+ 05 13	66	15.3	GALAXY S
ZWG 455.036	23 36 42.	+ 20 24		15.6	GALAXY
ZWG 476.097	23 36 42.	+ 24 44		15.3	GALAXY
ZWG 476.098	23 36 42.	+ 26 50		15.0	GALAXY
UGC 12721	23 36 42.	+ 26 50	90	15.0	GALAXY SBb
MCG+05-55-045	23 36 42.	+ 31 48	21	15.	GALAXY
ZC 2336.7-0154	23 36 42.	- 01 54	1410		CLUSTER OF GALAXIES
MCG-01-60-018	23 36 42.	- 04 48	24	15.	GALAXY
IC 5344	23 36 42.	- 05 15			NONSTELLAR OBJECT
MCG-04-55-017	23 36 42.	- 21 47	36	17.5	GALAXY
PHL 2415	23 36 42.	- 22 42		17.9	BLUE STELLAR OBJECT
PHL 6001	23 36 42.	- 23 22		17.2	BLUE STELLAR OBJECT
PHL 6002	23 36 42.	- 27 02		15.0	BLUE STELLAR OBJECT
TON-S 0108	23 36 42.	- 27 02		14.1	BLUE STAR
PHL 6003	23 36 42.	- 27 42		15.3	BLUE STELLAR OBJECT
PK110-12.1	23 36 44.2	+ 47 55 52.	37		PLANETARY NEBULA
IC 5343	23 36 45.	- 22 46 31.			NONSTELLAR OBJECT
ZWG 407.037	23 36 48.	+ 08 57		15.3	GALAXY
UGC 12722	23 36 48.	+ 08 57	60	15.3	GALAXY S
ZWG 432.015	23 36 48.	+ 10 35		15.7	GALAXY
UGC 12723	23 36 48.	+ 10 35	60	15.7	GALAXY
ZWG 455.037	23 36 48.	+ 20 25		15.7	GALAXY
MCG-04-55-018	23 36 48.	- 21 33	36	15.5	GALAXY
MCG-04-55-019	23 36 48.	- 22 47 30.	36	14.	GALAXY
PHL 6004	23 36 48.	- 25 32		17.1	BLUE STELLAR OBJECT
PHL 6006	23 36 48.	- 25 46		18.3	BLUE STELLAR OBJECT
PHL 6005	23 36 48.	- 26 40		17.4	BLUE STELLAR OBJECT
PHL 2416	23 36 48.	- 28 46		18.0	BLUE STELLAR OBJECT
3ZW 113	23 36 54.	+ 08 35			COMPACT GALAXY
ZC 2336.9+3022	23 36 54.	+ 30 22	1140		CLUSTER OF GALAXIES
ZWG 381.020	23 36 54.	- 01 43		15.5	GALAXY
PHL 6007	23 36 54.	- 30 30		18.1	BLUE STELLAR OBJECT
IC 5345	23 36 55.	- 22 41 25.			NONSTELLAR OBJECT
LBN 0565	23 37	+ 61 39	480		BRIGHT NEBULA
KHAV 791	23 37	+ 63 29			DARK NEBULA
UGC 12724	23 37 00.	+ 28 33	60	17.	GALAXY DWARF
MCG+00-60-027	23 37 00.	- 01 43	48	14.	GALAXY
MCG-04-55-020	23 37 00.	- 22 42	30	14.5	GALAXY
SCHO 1432	23 37 03.	+ 61 18 24.	520		ISOLATED DARK CLOUD
ZWG 432.016	23 37 06.	+ 10 36		15.7	GALAXY
ZWG 476.099	23 37 06.	+ 21 39		15.1	GALAXY
UGC 12725	23 37 06.	+ 21 39	66	15.1	GALAXY S0
ZWG 476.100	23 37 06.	+ 24 49		15.7	GALAXY
PHL 0572	23 37 06.	- 21 45		17.9	BLUE STELLAR OBJECT
PHL 6009	23 37 06.	- 25 08		17.5	BLUE STELLAR OBJECT
PHL 6008	23 37 06.	- 27 23		16.3	BLUE STELLAR OBJECT
MCG-07-48-010	23 37 06.	- 38 43	12	15.	GALAXY
4ZW 159	23 37 12.	+ 32 16			COMPACT GALAXY
ZWG 381.021	23 37 12.	- 00 18		15.7	GALAXY
PHL 2417	23 37 12.	- 21 10		17.4	BLUE STELLAR OBJECT
PHL 2418	23 37 12.	- 26 22		17.5	BLUE STELLAR OBJECT
PHL 2419	23 37 12.	- 31 38		18.0	BLUE STELLAR OBJECT
MCG-07-48-011	23 37 12.	- 42 09	12	15.5	GALAXY
ARP 222	23 37 16.	- 12 34			PECULIAR GALAXY
ZC 2337.3+1919	23 37 18.	+ 19 19	3560		CLUSTER OF GALAXIES
ZWG 476.101	23 37 18.	+ 27 06		15.7	GALAXY
UGC 12726	23 37 18.	+ 31 07	78	16.0	GALAXY SA0-b
4ZW 160	23 37 18.	+ 34 09			COMPACT GALAXY
4ZW 161	23 37 18.	+ 38 54			COMPACT GALAXY
MCG-02-60-007	23 37 18.	- 12 05	60	14.5	GALAXY
MCG-02-60-008	23 37 18.	- 12 33 30.	180	11.5	GALAXY
RNGC 7727	23 37 18.	- 12 34		11.5	GALAXY
VV 067	23 37 18.	- 12 34	180	11.6	INTERACTING GALAXY
PHL 6010	23 37 18.	- 24 34		17.6	BLUE STELLAR OBJECT
REIN 2.320	23 37 18.70	- 12 34 14.0			NEBULA
ZWG 476.102	23 37 24.	+ 26 33		15.7	GALAXY
ZWG 476.103	23 37 24.	+ 26 52		14.3	GALAXY
UGC 12727	23 37 24.	+ 26 52	90	14.3	GALAXY E
MRSL 114+00/1	23 37 24.	+ 61 40	600		HII REGION
MCG-04-55-021	23 37 24.	- 23 04	30	15.	GALAXY
PHL 6011	23 37 24.	- 24 08		16.9	BLUE STELLAR OBJECT
PHL 2420	23 37 24.	- 24 20		17.8	BLUE STELLAR OBJECT

OBJECT NAME	RIGHT ASCEN.	DECLINATION	DIAM.	MAGN.	TYPE OF OBJECT
LB 01529	23 37 24.	- 55 48		12.4	FAINT BLUE STAR
RNGC 7728	23 37 27.	+ 26 52		14.5	GALAXY
SG 2.099	23 37 27.	+ 61 39	420		DIFFUSE EMISSION NEBULA
ZWG 455.038	23 37 30.	+ 21 26		15.2	GALAXY
MCG+04-55-041	23 37 30.	+ 26 52	24	15.	GALAXY
ZC 2337.5+3322	23 37 30.	+ 33 22	1010		CLUSTER OF GALAXIES
TON-S 0109	23 37 30.	- 31 11		15.8	BLUE STAR
PHL 2421	23 37 30.	- 32 02		17.5	BLUE STELLAR OBJECT
SC 2334-6158.3	23 37 31.	- 61 41 41.			NEBULA
SC 2334-6300.3	23 37 33.	- 62 43 41.			NEBULA
ZWG 548.002	23 37 36.	+ 46 33		15.7	GALAXY
PHL 2422	23 37 36.	- 28 15		17.9	BLUE STELLAR OBJECT
PHL 0573	23 37 36.	- 29 45		14.2	BLUE STELLAR OBJECT
ZC 2337.7+0821	23 37 42.	+ 08 21	1280		CLUSTER OF GALAXIES
ZWG 548.003	23 37 42.	+ 47 42		15.4	GALAXY
UGC 12728	23 37 42.	+ 47 42	60	15.4	GALAXY COMPACT
MCG+00-60-028	23 37 42.	- 00 54	48	15.	GALAXY
PHL 6012	23 37 42.	- 27 23		14.1	BLUE STELLAR OBJECT
TON-S 0110	23 37 42.	- 27 23		13.9	BLUE STAR
SER 162.03	23 37 42.	- 44 50	1200	15.	LOOSE GROUP OF 12 GALXIES
HN 2870	23 37 44.2	- 44 47 23.	36		NEBULA
ZWG 407.038	23 37 48.	+ 03 45		15.5	GALAXY
ZWG 407.039	23 37 48.	+ 05 59		15.5	GALAXY
ZC 2337.8+2140	23 37 48.	+ 21 40	1140		CLUSTER OF GALAXIES
ZWG 476.104	23 37 48.	+ 24 39		15.7	GALAXY
ZWG 476.105	23 37 48.	+ 27 17		15.5	GALAXY
MCG+08-43-003	23 37 48.	+ 47 41	42	17.	GALAXY
SC 2335-6301.0	23 37 51.	- 62 44 23.			NEBULA
ZWG 381.022	23 37 54.	+ 00 58		15.2	GALAXY
UGC 12729	23 37 54.	+ 00 58	78	15.2	GALAXY S0-a
KARA.73B 1032	23 37 54.	+ 00 58	96	15.2	ISOLATED GALAXY S
ZC 2337.9+1010	23 37 54.	+ 10 10	1550		CLUSTER OF GALAXIES
ZC 2337.9+2434	23 37 54.	+ 24 34	2890		CLUSTER OF GALAXIES
MCG+07-48-016	23 37 54.	+ 42 58	30	15.	GALAXY
ZWG 533.016	23 37 54.	+ 42 59		15.4	GALAXY
PHL 2373	23 37 54.	- 17 12		18.0	BLUE STELLAR OBJECT
PHL 6013	23 37 54.	- 27 17		18.0	BLUE STELLAR OBJECT
ARC 2638	23 37 55.	- 11 59		17.2	RICH CLUSTER OF GALAXIES
ARC 2639	23 37 56.	+ 10 15		17.7	RICH CLUSTER OF GALAXIES
SC 2335-6300.1	23 37 58.	- 62 43 29.	30		NEBULA
ARC 2640	23 37 59.	+ 19 24		16.7	RICH CLUSTER OF GALAXIES
LBN 0534	23 38	+ 48 30	3000		BRIGHT NEBULA
LB 09956	23 38	- 80 25		14.7	FAINT BLUE STAR
MCG+00-60-029	23 38 00.	+ 00 59	60	15.	GALAXY
ZWG 381.023	23 38 00.	+ 02 30		15.1	GALAXY
MCG+03-60-018	23 38 00.	+ 20 11	84	15.	GALAXY
ZC 2338.0+2036	23 38 00.	+ 20 36	740		CLUSTER OF GALAXIES
ZC 2338.0+2138	23 38 00.	+ 21 38	340		CLUSTER OF GALAXIES
MCG+05-55-046	23 38 00.	+ 28 54	108	14.5	GALAXY
ZWG 497.047	23 38 00.	+ 28 55		14.6	GALAXY
UGC 12730	23 38 00.	+ 28 55	114	14.6	GALAXY Sa
MCG+00-60-030	23 38 00.	- 00 10	42	15.	GALAXY
PHL 2423	23 38 00.	- 24 32		17.9	BLUE STELLAR OBJECT
SC 2335-6143.1	23 38 00.	- 61 26 29.			NEBULA
MCG+00-60-031	23 38 03.	+ 02 31	48	14.5	GALAXY
RNGC 7729	23 38 03.	+ 28 55		14.5	GALAXY
ZWG 455.039	23 38 06.	+ 20 10		15.7	GALAXY
UGC 12731	23 38 06.	+ 20 10	90	15.7	GALAXY Sb
ZWG 476.106	23 38 06.	+ 25 57		14.5	GALAXY
UGC 12732	23 38 06.	+ 25 57	180	14.5	GALAXY DWRF SP
KARA.72 589A	23 38 06.	+ 25 57	258	14.5	PART OF DOUBLE GALAXY
3ZW 114	23 38 06.	- 02 53			COMPACT GALAXY
MCG-04-55-022	23 38 06.	- 20 46	36	15.	GALAXY
PHL 6014	23 38 06.	- 23 14		17.9	BLUE STELLAR OBJECT
MCG+04-55-042	23 38 09.	+ 25 58	180	15.	GALAXY
SC 2335-6146.7	23 38 09.	- 61 30 05.			NEBULA
ZWG 407.040	23 38 12.	+ 04 20		14.9	GALAXY
ZWG 476.107	23 38 12.	+ 26 33		15.1	GALAXY
UGC 12733	23 38 12.	+ 26 33	66	15.1	GALAXY E
MCG+00-60-032	23 38 12.	- 00 50	42	14.5	GALAXY
MCG-05-55-031	23 38 12.	- 28 35	168	14.5	GALAXY
PHL 6015	23 38 12.	- 30 06		17.7	BLUE STELLAR OBJECT
SN 1969K	23 38 13.	+ 26 33		17.5	SUPERNOVA
MCG+04-55-043	23 38 15.	+ 26 33 30.	42	15.	GALAXY
MCG-04-55-023	23 38 15.	- 24 59	72	14.5	GALAXY
ARC 2641	23 38 16.	- 25 07		17.7	RICH CLUSTER OF GALAXIES
3ZW 115	23 38 18.	+ 09 12			COMPACT GALAXY
5ZW 421	23 38 18.	+ 47 25			COMPACT GALAXY
ZWG 359.006	23 38 18.	+ 77 58		15.5	GALAXY
ZWG 344.004	23 38 18.	+ 77 58		15.5	GALAXY
ZC 2338.3-0022	23 38 18.	- 00 22	3900		CLUSTER OF GALAXIES
PHL 6016	23 38 18.	- 26 30		18.0	BLUE STELLAR OBJECT
ARC 2642	23 38 19.	- 11 03		16.8	RICH CLUSTER OF GALAXIES
ZWG 476.108	23 38 24.	+ 25 20		15.5	GALAXY
ZWG 476.109	23 38 24.	+ 27 14		15.7	GALAXY
MCG-02-60-010	23 38 24.	- 13 38	72	14.	GALAXY
SER 160.02	23 38 24.	- 58 26	70		LOW SURF. BRGHTNSS GALAXY
MCG-02-60-011	23 38 27.	- 08 29 30.	30	15.	GALAXY
ARC 2643	23 38 29.	+ 20 11		17.7	RICH CLUSTER OF GALAXIES
FATH 1.853	23 38 30.	+ 00 45	11		NEBULA
FATH 1.853	23 38 30.	+ 00 45	11		NEBULA
ZC 2338.5+1610	23 38 30.	+ 16 10	2150		CLUSTER OF GALAXIES
ZC 2338.5+2014	23 38 30.	+ 20 14	1140		CLUSTER OF GALAXIES
ZWG 476.110	23 38 30.	+ 24 40		15.7	GALAXY
ZWG 476.111	23 38 30.	+ 24 53		15.4	GALAXY
MCG+04-55-044	23 38 33.	+ 24 53	48	15.5	GALAXY
IC 5346	23 38 34.	+ 24 40 25.			NONSTELLAR OBJECT
ARC 2644	23 38 34.	- 00 11		16.6	RICH CLUSTER OF GALAXIES
MCG-01-60-019	23 38 36.	- 06 54	48	15.5	GALAXY
ZWG 407.041	23 38 42.	+ 03 45		14.5	GALAXY
UGC 12734	23 38 42.	+ 03 45	114	14.5	GALAXY Sb
ZC 2338.7+1745	23 38 42.	+ 17 45	1340		CLUSTER OF GALAXIES
MCG+04-55-045	23 38 42.	+ 25 16 30.	45	15.	GALAXY
ZWG 476.112	23 38 42.	+ 25 17		14.7	GALAXY
ZWG 381.024	23 38 42.	- 01 18		15.2	GALAXY
MCG+01-60-020	23 38 45.	+ 03 43	96	13.	GALAXY
MCG+00-60-033	23 38 45.	- 01 19	48	14.	GALAXY
IC 1504	23 38 46.	+ 03 46 19.			NONSTELLAR OBJECT
ARC 2647	23 38 46.	+ 25 32		17.1	RICH CLUSTER OF GALAXIES
SC 2336-6405.2	23 38 47.	- 63 48 35.			NEBULA
ZC 2338.8+0225	23 38 48.	+ 02 25	400		CLUSTER OF GALAXIES
MCG+03-60-019	23 38 48.	+ 15 51	42	15.5	GALAXY
ZWG 455.040	23 38 48.	+ 15 52		15.3	GALAXY
UGC 12735	23 38 48.	+ 15 52	60	15.3	GALAXY SBa
UGC 12736	23 38 48.	+ 47 33	66	16.0	GALAXY SB:b
ARC 2645	23 38 48.	- 09 19		18.0	RICH CLUSTER OF GALAXIES
ARC 2646	23 38 48.	- 10 17		17.6	RICH CLUSTER OF GALAXIES
RNGC 7730	23 38 48.	- 20 30			NON-EXISTENT OBJECT
MCG-05-55-032	23 38 48.	- 29 34	18	15.	GALAXY
HOLM 813B	23 38 52.	+ 03 28	18	14.7	PART OF MULTIPLE GALAXY
ARP 295	23 38 52.	- 03 51			PECULIAR GALAXY
ZWG 407.042	23 38 54.	+ 03 55		15.5	GALAXY
ZC 2338.9+2532	23 38 54.	+ 25 32	870		CLUSTER OF GALAXIES
ZC 2338.9+2546	23 38 54.	+ 25 46	810		CLUSTER OF GALAXIES
5ZW 422	23 38 54.	+ 47 31			COMPACT GALAXY
MCG-02-60-009	23 38 54.	- 10 10	36	15.	GALAXY
PHL 6017	23 38 54.	- 27 57		17.9	BLUE STELLAR OBJECT
MCG-05-55-033	23 38 54.	- 29 30	9	15.5	GALAXY
HOLM 813A	23 38 57.	+ 03 27	84	13.9	PART OF MULTIPLE GALAXY
ARC 2649	23 38 58.	+ 24 25		16.9	RICH CLUSTER OF GALAXIES
ZWG 381.025	23 39 00.	+ 03 28		14.3	GALAXY
ZCG 2339+03.2	23 39 00.	+ 03 28		14.3	COMPACT GALAXY
UGC 12737	23 39 00.	+ 03 28	90	14.3	GALAXY SBa
KARA.72 590A	23 39 00.	+ 03 28	78	14.3	PART OF DOUBLE GALAXY
ZC 2339.0+0745	23 39 00.	+ 07 45	1950		CLUSTER OF GALAXIES
ZC 2339.0+1528	23 39 00.	+ 15 28	1610		CLUSTER OF GALAXIES
ZWG 476.113	23 39 00.	+ 24 36		15.7	GALAXY
ZWG 359.007	23 39 00.	+ 77 58		15.6	GALAXY
ZWG 344.005	23 39 00.	+ 77 58		15.6	GALAXY
ZC 2339.0+8209	23 39 00.	+ 82 09	2350		CLUSTER OF GALAXIES
PHL 6018	23 39 00.	- 24 11		17.7	BLUE STELLAR OBJECT
RNGC 7731	23 39 01.	+ 03 28		14.5	GALAXY
ARC 2648	23 39 01.	- 14 45		17.6	RICH CLUSTER OF GALAXIES
IC 1505	23 39 03.	- 03 50 23.			NONSTELLAR OBJECT
IC 5347	23 39 04.	+ 24 36 31.			NONSTELLAR OBJECT
ARC 2650	23 39 04.	+ 25 48		17.1	RICH CLUSTER OF GALAXIES
ZWG 381.026	23 39 06.	+ 03 27		14.5	GALAXY
ZCG 2339+03.1	23 39 06.	+ 03 27		14.5	COMPACT GALAXY
UGC 12738	23 39 06.	+ 03 27	114	14.5	GALAXY Sc-IRR
KARA.72 590B	23 39 06.	+ 03 27	102	14.5	PART OF DOUBLE GALAXY
MCG+00-60-035	23 39 06.	+ 03 29	114	13.9	GALAXY
MCG+00-60-034	23 39 06.	+ 03 29 30.	78	14.7	GALAXY
ZWG 407.043	23 39 06.	+ 05 54		15.6	GALAXY
MCG-01-60-020	23 39 06.	- 03 50	24	15.	GALAXY
PHL 2424	23 39 06.	- 23 26		18.2	BLUE STELLAR OBJECT
RNGC 7732	23 39 07.	+ 03 27		14.5	GALAXY
SC 2336-6344.9	23 39 09.	- 63 29 17.			NEBULA
ZC 2339.2+1425	23 39 12.	+ 14 25	4570		CLUSTER OF GALAXIES
ZWG 381.027	23 39 12.	- 01 36		15.0	GALAXY
VV 034B	23 39 12.	- 03 52	120	15.	INTERACTING GALAXY
MCG-01-60-021	23 39 12.	- 03 55	72	14.5	GALAXY
PHL 6021	23 39 12.	- 26 16		18.4	BLUE STELLAR OBJECT
PHL 6019	23 39 12.	- 27 01		17.6	BLUE STELLAR OBJECT
MCG-05-55-034	23 39 12.	- 28 16	48	15.	GALAXY
PHL 6020	23 39 12.	- 29 33		17.9	BLUE STELLAR OBJECT
UGC 12740	23 39 18.	+ 23 33	66	16.5	GALAXY Sc?
3ZW 116	23 39 18.	- 01 36			COMPACT GALAXY
ZWG 381.028	23 39 18.	- 01 37		14.2	GALAXY
UGC 12739	23 39 18.	- 01 37	54	14.2	GALAXY
HMS 1.33	23 39 18.	- 03 54			Sc GALAXY
MCG-02-60-012	23 39 18.	- 08 54 30.	84	14.5	GALAXY
HOLM 814B	23 39 19.	- 04 28	24	14.4	PART OF MULTIPLE GALAXY
MCG+07-48-017	23 39 21.	+ 44 42	108	15.	GALAXY
ARC 2651	23 39 23.	+ 20 48		16.9	RICH CLUSTER OF GALAXIES
ZWG 381.029	23 39 24.	+ 02 57		15.7	GALAXY
ZWG 498.001	23 39 24.	+ 30 19		14.4	GALAXY
ZWG 497.048	23 39 24.	+ 30 19		14.4	GALAXY
UGC 12741	23 39 24.	+ 30 19	60	14.4	GALAXY Sa
ZWG 533.017	23 39 24.	+ 44 42		15.5	GALAXY
UGC 12742	23 39 24.	+ 44 42	138	15.5	GALAXY SBc
PHL 6023	23 39 24.	- 22 00		17.3	BLUE STELLAR OBJECT
PHL 6022	23 39 24.	- 24 24		17.2	BLUE STELLAR OBJECT
MCG+05-55-047	23 39 27.	+ 30 17	72	15.	GALAXY
HOLM 814A	23 39 28.	- 04 29	66	13.7	PART OF MULTIPLE GALAXY
ZWG 407.044	23 39 30.	+ 04 03		15.6	GALAXY
ZWG 476.114	23 39 30.	+ 27 18		15.7	GALAXY
MCG+06-52-001	23 39 30.	+ 35 50	15	16.	GALAXY
5ZW 423	23 39 30.	+ 53 38			COMPACT GALAXY
VV 034A	23 39 30.	- 03 49	66	15.	INTERACTING GALAXY
HMS 1.34	23 39 30.	- 03 50			Sb GALAXY
MCG-01-60-022	23 39 30.	- 03 52	36	14.5	GALAXY
MCG-01-60-023	23 39 30.	- 04 20	60	14.5	GALAXY
ZC 2339.6+2132	23 39 36.	+ 21 32	940		CLUSTER OF GALAXIES
MCG+06-52-002	23 39 36.	+ 35 50	48	15.	GALAXY
UGC 12743	23 39 36.	+ 35 52	60	16.5	GALAXY SB
ISS 0105	23 39 36.	+ 59 19	479		STELLAR RING
PHL 6024	23 39 36.	- 30 20		17.0	BLUE STELLAR OBJECT
MCG-02-60-013	23 39 39.	- 12 20	36	15.5	GALAXY
ZWG 476.115	23 39 42.	+ 25 57		15.0	GALAXY
UGC 12744	23 39 42.	+ 25 57	78	15.0	GALAXY E
RNGC 7735	23 39 45.	+ 25 57		15.0	GALAXY
3ZW 117	23 39 48.	+ 06 39			COMPACT GALAXY
ZWG 407.045	23 39 48.	+ 06 39		15.2	GALAXY
ZC 2339.8+2015	23 39 48.	+ 20 15	5110		CLUSTER OF GALAXIES
MCG+04-55-046	23 39 48.	+ 25 57 30.	60	15.	GALAXY
ZWG 476.116	23 39 48.	+ 27 03		15.6	GALAXY
MRSL 114-00/1	23 39 48.	+ 60 42	600		HII REGION
RNGC 7736	23 39 48.	- 19 43		14.0	GALAXY
PHL 6025	23 39 48.	- 23 45		18.1	BLUE STELLAR OBJECT
PHL 6026	23 39 48.	- 28 58		18.2	BLUE STELLAR OBJECT
RNGC 7733	23 39 49.	- 66 15			UNVERIFIED SOUTHRN OBJECT
MCG-03-60-010	23 39 51.	- 19 43	24	14.	GALAXY
ZWG 359.008	23 39 54.	+ 77 54		15.7	GALAXY
ZWG 344.006	23 39 54.	+ 77 54		15.7	GALAXY
PHL 6028	23 39 54.	- 21 18		17.0	BLUE STELLAR OBJECT
PHL 6027	23 39 54.	- 21 21		14.2	BLUE STELLAR OBJECT
PHL 2425	23 39 54.	- 27 24		17.1	BLUE STELLAR OBJECT
RNGC 7734	23 39 55.	- 66 14			UNVERIFIED SOUTHRN OBJECT
HN 2871	23 39 56.8	- 45 10 46.	150		NEBULA
MCG+05-55-048	23 39 57.	+ 30 18	48	15.	GALAXY
SCHO 1433	23 39 58.	+ 56 45 12.	240		ISOLATED DARK CLOUD
LBN 0434	23 40	+ 09 00	3300		BRIGHT NEBULA
LBN 0566	23 40	+ 60 38	600		BRIGHT NEBULA
ISS 0025	23 40 00.	+ 63 55	263		STELLAR RING
PHL 0574	23 40 00.	- 24 08		18.2	BLUE STELLAR OBJECT
PHL 6030	23 40 00.	- 26 36		17.8	BLUE STELLAR OBJECT
PHL 0575	23 40 00.	- 28 07		15.3	BLUE STELLAR OBJECT
TON-S 0111	23 40 00.	- 28 07		14.9	BLUE STAR
PHL 0576	23 40 00.	- 31 05		17.3	BLUE STELLAR OBJECT
TON-S 0112	23 40 00.	- 31 05		15.8	BLUE STAR
PHL 6029	23 40 00.	- 31 28		9.8	BLUE STELLAR OBJECT
SCHO 1434	23 40 04.	+ 57 03 48.	210		ISOLATED DARK CLOUD
ZWG 476.117	23 40 06.	+ 26 49		15.0	GALAXY
MCG+04-55-047	23 40 06.	+ 26 49	30	15.5	GALAXY
PHL 2426	23 40 06.	- 21 58		17.1	BLUE STELLAR OBJECT
ZWG 381.030	23 40 12.	+ 01 41		15.6	GALAXY

OBJECT NAME	RIGHT ASCEN.	DECLINATION	DIAM.	MAGN.	TYPE OF OBJECT
KARA.73B 1033	23 40 12.	+ 01 41	54	15.6	ISOLATED GALAXY S
ZWG 455.041	23 40 12.	+ 18 26		15.7	GALAXY
ZWG 476.118	23 40 12.	+ 26 46		14.8	GALAXY
UGC 12745	23 40 12.	+ 26 46	66	14.8	GALAXY S0-a
ZWG 498.002	23 40 12.	+ 27 34		15.7	GALAXY
ZWG 497.049	23 40 12.	+ 27 34		15.7	GALAXY
MCG+00-60-036	23 40 15.	+ 01 42 30.	48	15.	GALAXY
RNGC 7737	23 40 15.	+ 26 47		15.0	GALAXY
MCG+04-55-048	23 40 15.	+ 26 47	48	15.	GALAXY
MCG+04-55-049	23 40 15.	+ 27 01	75	15.	GALAXY
ZWG 476.119	23 40 18.	+ 27 01		15.0	GALAXY
UGC 12746	23 40 18.	+ 27 01	84	15.0	GALAXY Sc
MCG-07-48-012	23 40 18.	- 43 34	6	16.	GALAXY
SHB 393	23 40 22.9	- 03 40 20.		17.	QUASI-STELLAR OBJECT
SC 2337-7307.1	23 40 23.	- 72 50 28.	18		NEBULA
ZWG 432.017	23 40 24.	+ 14 20		15.6	GALAXY
ZWG 476.120	23 40 24.	+ 27 12		15.0	GALAXY
MCG-01-60-024	23 40 24.	- 03 55	24	15.5	GALAXY
PHL 2427	23 40 24.	- 23 05		18.1	BLUE STELLAR OBJECT
MCG+03-60-020	23 40 30.	+ 19 08	48	14.5	GALAXY
ZWG 455.042	23 40 30.	+ 19 09		14.6	GALAXY
MRK 330	23 40 30.	+ 19 09	22	15.5	GALAXY WITH UV CONTINUUM
UGC 12747	23 40 30.	+ 19 09	66	14.6	GALAXY IRR
MCG-07-48-014	23 40 30.	- 43 11	9	17.5	GALAXY
MCG-07-48-013	23 40 30.	- 43 11	12	15.5	GALAXY
PHL 5548	23 40 36.	+ 01 52		18.6	BLUE STELLAR OBJECT
ZWG 455.043	23 40 36.	+ 18 23		14.7	GALAXY
4ZW 162	23 40 36.	+ 29 55			COMPACT GALAXY
PHL 2428	23 40 36.	- 23 39		18.3	BLUE STELLAR OBJECT
PHL 2429	23 40 36.	- 29 42		18.1	BLUE STELLAR OBJECT
HOLM 815B	23 40 38.	+ 18 23	18	14.8	PART OF MULTIPLE GALAXY
HOLM 815A	23 40 38.	+ 18 23	18	14.5	PART OF MULTIPLE GALAXY
ZWG 407.046	23 40 42.	+ 04 17		15.4	GALAXY
MCG+01-60-021	23 40 42.	+ 08 27	84	16.	GALAXY
ZWG 432.018	23 40 42.	+ 10 07		15.5	GALAXY
UGC 12748	23 40 42.	+ 10 07	72	15.5	GALAXY S
MCG+03-60-022	23 40 42.	+ 18 22	15	16.	GALAXY
MCG+03-60-021	23 40 42.	+ 18 22	30	15.	GALAXY
ZWG 476.121	23 40 42.	+ 22 44		15.7	GALAXY
ZWG 476.122	23 40 42.	+ 27 05		15.6	GALAXY
ARC 2652	23 40 42.	- 10 39		17.7	RICH CLUSTER OF GALAXIES
TON-S 0113	23 40 42.	- 21 09		14.8	BLUE STAR
RNGC 7740	23 40 45.	+ 27 05		15.5	GALAXY
MCG+08-43-004	23 40 45.	+ 49 42	72	15.	GALAXY
UGC 12749	23 40 48.	+ 08 26	108	16.0	GALAXY S
ZWG 498.003	23 40 48.	+ 28 13		15.7	GALAXY
ZWG 497.050	23 40 48.	+ 28 13		15.7	GALAXY
4ZW 163	23 40 48.	+ 28 14			COMPACT GALAXY
UGC 12750	23 40 48.	+ 49 42	108	15.0	GALAXY Sa
ZWG 548.004	23 40 48.	+ 49 43		15.0	GALAXY
PHL 2430	23 40 48.	- 21 08		16.3	BLUE STELLAR OBJECT
PHL 6031	23 40 48.	- 27 30		17.3	BLUE STELLAR OBJECT
ZWG 548.005	23 40 54.	+ 50 11		15.7	GALAXY
PHL 6032	23 40 54.	- 27 13		18.0	BLUE STELLAR OBJECT
PHL 6033	23 40 54.	- 29 01		17.1	BLUE STELLAR OBJECT
MCG-07-48-015	23 40 54.	- 41 35	42	16.	GALAXY
SC 2338-6209.1	23 40 58.	- 61 52 28.	18		NEBULA
UGC 12751	23 41 00.	+ 04 49	66	16.5	GALAXY
MCG+02-60-008	23 41 00.	+ 11 22	60	14.5	GALAXY
ZWG 432.019	23 41 00.	+ 11 23		15.5	GALAXY
UGC 12752	23 41 00.	+ 11 23	60	15.5	GALAXY
ZWG 432.020	23 41 00.	+ 12 55		15.1	GALAXY
UGC 12753	23 41 00.	+ 12 55	102	15.1	GALAXY Sb
MCG+02-60-007	23 41 00.	+ 12 56	120	13.	GALAXY
ZWG 476.123	23 41 00.	+ 27 02		14.9	GALAXY
ZC 2341.0+4058	23 41 00.	+ 40 59	2820		CLUSTER OF GALAXIES
MCG-07-48-016	23 41 00.	- 38 58		15.	GALAXY
RNGC 7739	23 41 01.	+ 00 12			NON-EXISTENT OBJECT
MCG+01-60-022	23 41 03.	+ 04 49	48	16.	GALAXY
ZC 2341.1+0000	23 41 06.	+ 00 00	1280		CLUSTER OF GALAXIES
ZC 2341.1+1135	23 41 06.	+ 11 35	1610		CLUSTER OF GALAXIES
ZC 2341.1+2544	23 41 06.	+ 25 44	1880		CLUSTER OF GALAXIES
5ZW 424	23 41 06.	+ 44 43			COMPACT GALAXY
ZC 2341.1+4532	23 41 06.	+ 45 32	1080		CLUSTER OF GALAXIES
MCG-03-60-011	23 41 06.	- 14 58	96	14.5	GALAXY
PHL 2431	23 41 06.	- 28 18		17.8	BLUE STELLAR OBJECT
CED 211	23 41 11.	- 15 23	90		DIFFUSE GALACTIC NEBULA
ZWG 476.124	23 41 12.	+ 26 25		15.7	GALAXY
PHL 2432	23 41 12.	- 23 15		14.8	BLUE STELLAR OBJECT
PATH 1.854	23 41 13.	+ 00 25	5		NEBULA
ARC 2653	23 41 13.	+ 13 57		17.7	RICH CLUSTER OF GALAXIES
ZC 2341.3+0423	23 41 18.	+ 04 23	740		CLUSTER OF GALAXIES
ZC 2341.3+2510	23 41 18.	+ 25 10	1680		CLUSTER OF GALAXIES
ZWG 476.125	23 41 18.	+ 25 48		11.8	GALAXY
UGC 12754	23 41 18.	+ 25 48	264	11.8	GALAXY SBc
KARA.72 589B	23 41 18.	+ 25 48	270	11.8	PART OF DOUBLE GALAXY
ZWG 498.004	23 41 18.	+ 28 04		15.0	GALAXY
ZWG 497.051	23 41 18.	+ 28 04		15.0	GALAXY
UGC 12755	23 41 18.	+ 28 04	96	15.0	GALAXY SB
MCG-01-60-025	23 41 18.	- 08 20	60	15.	GALAXY
MCG+04-55-050	23 41 21.	+ 25 48	228	12.	GALAXY
MCG+05-55-049	23 41 21.	+ 28 02	78	15.	GALAXY
REIN 2.321	23 41 23.07	+ 25 47 53.1			NEBULA
ZWG 381.032	23 41 24.	+ 02 28		15.5	GALAXY
MRK 539	23 41 24.	+ 02 28	20	16.	GALAXY WITH UV CONTINUUM
ZWG 407.047	23 41 24.	+ 08 03		15.7	GALAXY
MCG+01-60-023	23 41 24.	+ 08 04	36	15.	GALAXY
ZWG 432.021	23 41 24.	+ 11 15		15.7	GALAXY
UGC 12756	23 41 24.	+ 11 15	72	15.7	GALAXY SBb
MCG+05-55-051	23 41 24.	+ 28 48 30.	48	15.	GALAXY
MCG+05-55-050	23 41 24.	+ 28 50	36	15.	GALAXY
ZWG 498.005	23 41 24.	+ 28 52		15.7	GALAXY
ZWG 497.052	23 41 24.	+ 28 52		15.7	GALAXY
ZWG 533.018	23 41 24.	+ 45 09		15.6	GALAXY
ZWG 381.031	23 41 24.	- 00 03		15.6	GALAXY
MCG-05-56-001	23 41 24.	- 32 15	150	14.	GALAXY
TON-S 0114	23 41 24.	- 33 18		14.2	BLUE STAR
RNGC 7741	23 41 26.	+ 25 48		12.0	GALAXY
PATH 1.856	23 41 27.	+ 00 14	68		NEBULA
PATH 1.855	23 41 27.	+ 00 37	14		NEBULA
ZWG 381.033	23 41 30.	+ 00 14		14.4	GALAXY
UGC 12757	23 41 30.	+ 00 14	126	14.4	GALAXY SBb
MCG+00-60-037	23 41 30.	+ 02 30	48	15.	GALAXY
MCG+02-60-009	23 41 30.	+ 11 14	42	14.5	GALAXY
MCG-02-60-014	23 41 30.	- 09 20	54	15.5	GALAXY
RNGC 7738	23 41 31.	+ 00 14		14.5	GALAXY
MCG+05-55-052	23 41 33.	+ 28 48 30.	8	15.	GALAXY
SC 2338-6105.7	23 41 35.	- 60 49 04.	18		NEBULA
ZWG 381.034	23 41 36.	+ 01 11		15.6	GALAXY
PATH 1.857	23 41 36.	+ 01 12	8		NEBULA
ZC 2341.6+0443	23 41 36.	+ 04 43	1080		CLUSTER OF GALAXIES
ZC 2341.6+1812	23 41 36.	+ 18 12	1610		CLUSTER OF GALAXIES
MCG-01-60-026	23 41 36.	- 06 27	120	13.5	GALAXY
PATH 1.858	23 41 41.	+ 00 32	16		NEBULA
MCG+00-60-038	23 41 42.	+ 00 16	114	13.	GALAXY
ZWG 381.035	23 41 42.	+ 00 32		15.3	GALAXY
ZWG 381.036	23 41 42.	+ 02 00		15.2	GALAXY
UGC 12758	23 41 42.	+ 02 00	78	15.2	GALAXY Sb
MCG+01-60-024	23 41 42.	+ 08 54	9	16.	GALAXY
5ZW 425	23 41 42.	+ 42 38			COMPACT GALAXY
ARC 2654	23 41 42.	- 07 40		17.2	RICH CLUSTER OF GALAXIES
MCG-05-56-002	23 41 42.	- 31 56	36	16.	GALAXY
RNGC 7742	23 41 44.	+ 10 29		13.0	GALAXY
MCG+00-60-039	23 41 45.	+ 02 00	66	14.	GALAXY
MCG+01-60-025	23 41 45.	+ 08 53 30.	15	16.	GALAXY
ZWG 381.037	23 41 48.	+ 02 32		15.7	GALAXY
ZC 2341.8+0251	23 41 48.	+ 02 51	1480		CLUSTER OF GALAXIES
MCG+01-60-026	23 41 48.	+ 05 22 30.	66	15.5	GALAXY
ZWG 407.048	23 41 43.	+ 08 52		15.5	GALAXY
ZWG 432.022	23 41 48.	+ 09 39		12.9	GALAXY
UGC 12759	23 41 48.	+ 09 39	162	12.9	GALAXY SB0/SBa
ZWG 432.023	23 41 48.	+ 10 30		12.5	GALAXY
UGC 12760	23 41 48.	+ 10 30	120	12.5	GALAXY S0?
MCG+02-60-010	23 41 48.	+ 10 30	108	12.	GALAXY
ZC 2341.8+1556	23 41 48.	+ 15 56	1550		CLUSTER OF GALAXIES
RNGC 7743	23 41 50.	+ 09 39		12.5	GALAXY
ARC 2655	23 41 51.	- 22 09		17.1	RICH CLUSTER OF GALAXIES
ZWG 381.038	23 41 54.	+ 00 03		15.1	GALAXY
ZWG 407.049	23 41 54.	+ 05 24		15.4	GALAXY
KARA.73B 1034	23 41 54.	+ 05 24	60	15.4	ISOLATED GALAXY S
ZWG 407.050	23 41 54.	+ 09 00		15.6	GALAXY
MCG+02-60-011	23 41 54.	+ 09 40	150	10.	GALAXY
PATH 1.859	23 41 58.	+ 00 59	11		NEBULA
MCG+00-60-040	23 42 00.	+ 02 33	48	14.	GALAXY
UGC 12761	23 42 00.	+ 05 44	60	16.5	GALAXY Sc
MCG+01-60-027	23 42 00.	+ 05 45	24	15.	GALAXY
ZWG 407.051	23 42 00.	+ 08 34		15.7	GALAXY
ZWG 432.024	23 42 00.	+ 11 24		15.7	GALAXY
MCG+02-60-012	23 42 00.	+ 11 25	15	15.	GALAXY
ZWG 432.025	23 42 00.	+ 12 26		15.5	GALAXY
ZWG 432.026	23 42 00.	+ 13 15		15.7	GALAXY
MCG+03-60-023	23 42 00.	+ 17 28	48	15.5	GALAXY
ZWG 455.044	23 42 00.	+ 17 30		15.6	GALAXY
ZWG 477.001	23 42 00.	+ 21 40		15.6	GALAXY
ZWG 476.126	23 42 00.	+ 21 40		15.6	GALAXY
UGC 12762	23 42 00.	+ 21 40	78	15.6	GALAXY S
MCG+04-56-001	23 42 00.	+ 27 20	30	15.5	GALAXY
LDN 1249	23 42 00.	+ 58 40	840		DARK NEBULA
TON-S 0115	23 42 00.	- 30 33		14.8	BLUE STAR
IC 5348	23 42 04.	- 43 12 33.			NONSTELLAR OBJECT
MCG+01-60-028	23 42 06.	+ 04 32	36	15.5	GALAXY
ZWG 432.027	23 42 06.	+ 11 25		15.1	GALAXY
MCG+04-56-002	23 42 06.	+ 21 39	60	15.	GALAXY
MCG-04-56-002	23 42 06.	- 24 04	9	15.5	GALAXY
MCG-04-56-001	23 42 06.	- 24 04	12	15.5	GALAXY
MCG+04-56-003	23 42 09.	+ 21 43	36	15.	GALAXY
ZC 2342.2+0049	23 42 12.	+ 00 49	3090		CLUSTER OF GALAXIES
ZWG 407.052	23 42 12.	+ 04 28		15.4	GALAXY
MCG+02-60-013	23 42 12.	+ 11 25	60	14.5	GALAXY
UGC 12763	23 42 12.	+ 12 36	60	16.5	GALAXY
ZWG 477.002	23 42 12.	+ 21 44		15.6	GALAXY
UGC 12764	23 42 12.	+ 21 44	60	15.6	GALAXY Sc
PHL 2433	23 42 12.	- 30 11		18.0	BLUE STELLAR OBJECT
RNGC 7745	23 42 14.	+ 25 37			GALAXY
MCG+01-60-029	23 42 15.	+ 04 27	36	15.	GALAXY
IC 1506	23 42 15.	+ 04 28 09.			NONSTELLAR OBJECT
ARC 2656	23 42 17.	- 04 23		16.2	RICH CLUSTER OF GALAXIES
MCG+01-60-030	23 42 18.	+ 08 55	36	16.	GALAXY
MCG+02-60-014	23 42 18.	+ 12 35	72	15.	GALAXY
SN 1973S	23 42 18.	+ 25 27		18.5	SUPERNOVA
MCG+07-48-018	23 42 18.	+ 44 19	72	16.	GALAXY
ZWG 533.019	23 42 18.	+ 44 20		15.2	GALAXY
UGC 12765	23 42 18.	+ 44 20	78	15.2	GALAXY SB?b
MCG-01-60-027	23 42 18.	- 04 33	36	15.	GALAXY
ARC 2657	23 42 20.	+ 08 53		14.9	RICH CLUSTER OF GALAXIES
PATL 1.860	23 42 23.	+ 00 51	5		NEBULA
RNGC 7744	23 42 23.	- 43 12		13.0	GALAXY
ZWG 407.053	23 42 24.	+ 08 56		15.3	GALAXY
MCG+04-56-004	23 42 24.	+ 25 38 30.	36	15.5	GALAXY
ARC 2658	23 42 24.	- 12 35		17.4	RICH CLUSTER OF GALAXIES
MCG-07-48-017	23 42 24.	- 43 11	120	12.8	GALAXY
PATH 1.861	23 42 29.	+ 00 59	5		NEBULA
MCG+03-60-024	23 42 30.	+ 19 38	36	14.5	GALAXY
UGC 12766	23 42 30.	+ 25 15	66	16.0	GALAXY S
SCHO 1435	23 42 30.	+ 55 23 42.	240		ISOLATED DARK CLOUD
MCG-02-60-015	23 42 30.	- 13 11	24	15.	GALAXY
PATH 1.862	23 42 31.	+ 00 54	8		NEBULA
ARC 2659	23 42 31.	- 15 45		17.0	RICH CLUSTER OF GALAXIES
SC 2339-6025.3	23 42 33.	- 60 08 39.	72		NEBULA
PATH 1.863	23 42 35.	+ 00 59	16		NEBULA
ZWG 407.054	23 42 36.	+ 06 46		14.8	GALAXY
UGC 12767	23 42 36.	+ 06 46	126	14.8	GALAXY SBb
MCG+01-60-031	23 42 36.	+ 06 46	120	13.	GALAXY
ZC 2342.6+1338	23 42 36.	+ 13 38	870		CLUSTER OF GALAXIES
ZWG 455.045	23 42 36.	+ 19 37		15.2	GALAXY
KARA.73B 1035	23 42 36.	+ 19 37	36	15.2	ISOLATED GALAXY S
MCG+05-56-001	23 42 36.	+ 27 54	12	16.	GALAXY
ZWG 498.006	23 42 36.	+ 28 48		15.5	GALAXY
ZWG 497.053	23 42 36.	+ 28 48		15.5	GALAXY
ZWG 381.039	23 42 36.	- 02 18		15.6	GALAXY
MCG-07-48-019	23 42 36.	- 38 37	30	15.	GALAXY
MCG-07-48-018	23 42 36.	- 41 17	30	17.	GALAXY
CED 212	23 42 37.	+ 69 29	180		DIFFUSE GALACTIC NEBULA
RNGC 7748	23 42 38.	+ 69 29			NON-EXISTENT OBJECT
ARC 2660	23 42 39.	- 26 15		16.4	RICH CLUSTER OF GALAXIES
ZWG 498.007	23 42 42.	+ 27 56		15.3	GALAXY
HW 2872	23 42 44.1	- 45 11 57.	30		NEBULA
MCG+00-60-041	23 42 45.	- 00 32	60	15.	GALAXY
3ZW 118	23 42 48.	+ 26 46			COMPACT GALAXY
OCL 0266	23 42 48.	+ 59 02	240	15.	OPEN STAR CLUSTER
ZWG 381.040	23 42 48.	- 01 57		14.5	GALAXY
UGC 12768	23 42 48.	- 01 57	84	14.5	GALAXY S0
RNGC 7746	23 42 49.	- 01 57		14.5	GALAXY
SCHO 1436	23 42 50.	+ 61 49 24.	260		ISOLATED DARK CLOUD
MCG+01-60-032	23 42 54.	+ 04 18	48	15.5	GALAXY
ZC 2342.9+1920	23 42 54.	+ 19 20	3360		CLUSTER OF GALAXIES

OBJECT NAME	RIGHT ASCEN.	DECLINATION	DIAM.	MAGN.	TYPE OF OBJECT
UGC 12769	23 42 54.	- 01 32	90	16.0	GALAXY
MCG+00-60-042	23 42 54.	- 01 32 30.	66	15.	GALAXY
LB 01530	23 42 54.	- 56 32		15.1	FAINT BLUE STAR
MCG-04-56-003	23 42 57.	- 21 13	54	15.	GALAXY
IC 1507	23 42 59.	+ 01 24 22.			NONSTELLAR OBJECT
LB 10003	23 43	- 87 55		12.7	FAINT BLUE STAR
ZWG 381.041	23 43 00.	+ 01 25		14.8	GALAXY
UGC 12770	23 43 00.	+ 01 25	84	14.8	GALAXY S0-a
UGC 12771	23 43 00.	+ 16 58	66	16.5	GALAXY IRR
ZWG 477.003	23 43 00.	+ 27 06		14.5	GALAXY
UGC 12772	23 43 00.	+ 27 06	102	14.5	GALAXY SBb
ZC 2343.0+3921	23 43 00.	+ 39 21	1880		CLUSTER OF GALAXIES
5ZW 426	23 43 00.	+ 43 17			COMPACT GALAXY
LDN 1252	23 43 00.	+ 63 00	1680		DARK NEBULA
MCG+00-60-043	23 43 00.	- 01 57 30.	24	14.	GALAXY
RNGC 7747	23 43 02.	+ 27 06		14.5	GALAXY
MCG+00-60-044	23 43 06.	+ 01 26	66	14.	GALAXY
HN 2873	23 43 09.0	- 45 33 03.	24		NEBULA
MCG+02-60-015	23 43 12.	+ 11 18	48	15.5	GALAXY
ZC 2343.2+1713	23 43 12.	+ 17 13	1280		CLUSTER OF GALAXIES
MCG+04-56-005	23 43 12.	+ 27 06	84	14.5	GALAXY
MCG-01-60-028	23 43 12.	- 07 04	42	15.	GALAXY
ZC 2343.3+0421	23 43 18.	+ 04 21	1550		CLUSTER OF GALAXIES
ZWG 477.004	23 43 18.	+ 27 04		15.5	GALAXY
ZWG 477.005	23 43 18.	+ 27 20		15.3	GALAXY
MCG-01-60-029	23 43 18.	- 07 25	48	16.	GALAXY
RNGC 7749	23 43 18.	- 29 49		14.0	GALAXY
MCG-05-56-003	23 43 18.	- 29 49	60	14.5	GALAXY
HN 2874	23 43 18.2	- 44 32 45.	18		NEBULA
IC 1508	23 43 23.	+ 11 45 58.			NONSTELLAR OBJECT
ZC 2343.4+0845	23 43 24.	+ 08 45	7460		CLUSTER OF GALAXIES
ZWG 432.028	23 43 24.	+ 11 47		14.6	GALAXY
UGC 12773	23 43 24.	+ 11 47	108	14.6	GALAXY S-IRR
KARA.73B 1036	23 43 24.	+ 11 47	102	14.6	ISOLATED GALAXY S
SCHO 1437	23 43 24.	+ 55 15 36.	290		ISOLATED DARK CLOUD
PK114-04.1	23 43 24.	+ 56 47	100	15.2	PLANETARY NEBULA
MCG+04-56-006	23 43 27.	+ 27 19 30.	30	15.	GALAXY
MCG+02-60-016	23 43 30.	+ 11 47 30.	120	13.	GALAXY
ZWG 381.042	23 43 30.	- 02 10		14.7	GALAXY
UGC 12774	23 43 30.	- 02 10	60	14.7	GALAXY SBa
MCG-01-60-030	23 43 30.	- 07 26	48	15.	GALAXY
HN 2875	23 43 30.7	- 44 55 27.	12		NEBULA
ZWG 407.055	23 43 36.	+ 05 36		15.5	GALAXY
UGC 12775	23 43 36.	+ 05 36	60	15.5	GALAXY S
KARA.73B 1037	23 43 36.	+ 05 36	60	15.5	ISOLATED GALAXY S
MCG+05-56-002	23 43 36.	+ 33 06	132	14.5	GALAXY
OCL 0268	23 43 36.	+ 61 54	60		OPEN STAR CLUSTER
MCG-03-60-012	23 43 36.	- 19 38	48	15.	GALAXY
MCG-05-56-004	23 43 36.	- 28 24	48	15.	GALAXY
TON-S 0116	23 43 36.	- 29 44		15.0	BLUE STAR
HN 2876	23 43 37.9	- 44 25 39.	12		NEBULA
MCG+01-60-033	23 43 42.	+ 05 35	60	15.	GALAXY
ZC 2343.7+1215	23 43 42.	+ 12 15	810		CLUSTER OF GALAXIES
ZC 2343.7+2058	23 43 42.	+ 20 58	1810		CLUSTER OF GALAXIES
ZWG 498.008	23 43 42.	+ 33 06		14.2	GALAXY
UGC 12776	23 43 42.	+ 33 06	168	14.2	GALAXY SBb
KARA.73B 1038	23 43 42.	+ 33 06	156	14.2	ISOLATED GALAXY S
MCG-05-56-005	23 43 42.	- 28 17	36	15.	GALAXY
LB 01531	23 43 42.	- 56 58		14.0	FAINT BLUE STAR
MCG+00-60-045	23 43 45.	- 02 10	66	13.	GALAXY
IC 5349	23 43 46.	- 28 16 56.			NONSTELLAR OBJECT
FATH 1.864	23 43 49.	+ 00 21	5		NEBULA
ZC 2343.9+0029	23 43 54.	+ 00 29	810		CLUSTER OF GALAXIES
TON-S 0117	23 43 54.	- 30 03		15.2	BLUE STAR
LB 01532	23 43 54.	- 56 33		14.7	FAINT BLUE STAR
KHAV 792	23 44	+ 55 11	3470		DARK NEBULA
MCG+01-60-034	23 44 00.	+ 03 30	84	12.5	GALAXY
ZWG 432.029	23 44 00.	+ 14 37		15.6	GALAXY
ZC 2344.0+3601	23 44 00.	+ 36 01	870		CLUSTER OF GALAXIES
ZC 2344.0-0225	23 44 00.	- 02 25	1010		CLUSTER OF GALAXIES
MCG-07-48-020	23 44 03.	- 38 45	36	15.	GALAXY
SHB 394	23 44 03.4	+ 09 14 04.		16.0	QUASI-STELLAR OBJECT
BC PKS2344+09	23 44 03.71	+ 09 14 05.0		15.92	QUASI-STELLAR OBJECT
ZWG 407.056	23 44 06.	+ 03 30		13.8	GALAXY
ZWG 381.043	23 44 06.	+ 03 30		13.8	GALAXY
UGC 12777	23 44 06.	+ 03 30	102	13.8	GALAXY Sc
ZWG 517.001	23 44 06.	+ 39 07		15.7	GALAXY
ARC 2661	23 44 06.	- 10 42		17.8	RICH CLUSTER OF GALAXIES
MCG-05-56-006	23 44 06.	- 29 00	36	15.5	GALAXY
MCG-05-56-007	23 44 06.	- 29 22	24	15.	GALAXY
RNGC 7750	23 44 07.	+ 03 30		14.0	GALAXY
MCG+06-52-003	23 44 09.	+ 39 08	36	15.	GALAXY
MCG+03-60-025	23 44 12.	+ 15 45	27	15.	GALAXY
ZWG 455.046	23 44 12.	+ 15 46		15.4	GALAXY
4ZW 164	23 44 12.	+ 21 48			COMPACT GALAXY
SCHO 1438	23 44 12.	+ 62 15 06.	290		ISOLATED DARK CLOUD
SHAH 021	23 44 12.	- 02 02	156	16.	GROUP OF COMPACT GALAXIES
MCG+05-56-003	23 44 15.	+ 27 44	78	16.	GALAXY
PK112-06.1	23 44 18.	+ 54 28	54	17.6	PLANETARY NEBULA
MCG-01-60-031	23 44 18.	- 07 23	48	16.	GALAXY
MCG-07-48-021	23 44 18.	- 38 54	12	15.5	GALAXY
ZWG 381.044	23 44 24.	+ 01 29		15.7	GALAXY
ZWG 407.057	23 44 24.	+ 06 35		13.9	GALAXY
UGC 12778	23 44 24.	+ 06 35	96	13.9	GALAXY
MCG+01-60-035	23 44 24.	+ 06 35	36	14.	GALAXY
SHAH 024	23 44 24.	- 01 10	42	18.5	GROUP OF COMPACT GALAXIES
MCG-07-48-022	23 44 24.	- 43 36	12	15.5	GALAXY
RNGC 7751	23 44 25.	+ 06 35		14.0	GALAXY
HOLM 816B	23 44 25.	+ 29 11	24	13.6	PART OF MULTIPLE GALAXY
MCG+05-56-004	23 44 27.	+ 29 10	24	14.	GALAXY
SC 2341-6354.6	23 44 28.	- 63 37 57.	24		NEBULA
ZWG 381.046	23 44 30.	+ 02 59		15.4	GALAXY
ZWG 432.030	23 44 30.	+ 14 34		15.2	GALAXY
ZWG 498.009	23 44 30.	+ 29 10		14.3	GALAXY
UGC 12779	23 44 30.	+ 29 10	27	14.3	GALAXY
KARA.72 591A	23 44 30.	+ 29 10	54	14.3	PART OF DOUBLE GALAXY
VV 005B	23 44 30.	+ 29 13 30.	60	13.6	INTERACTING GALAXY
ZWG 381.045	23 44 30.	- 00 43		15.6	GALAXY
MRK 540	23 44 30.	- 00 43	12	16.5	GALAXY WITH UV CONTINUUM
MCG-05-56-008	23 44 30.	- 29 27	24	15.5	GALAXY
TON-S 0118	23 44 30.	- 30 18		14.6	BLUE STAR
HOLM 816A	23 44 31.	+ 29 12	180	12.6	PART OF MULTIPLE GALAXY
RNGC 7752	23 44 32.	+ 29 11		14.5	GALAXY
MCG+05-56-005	23 44 33.	+ 29 12 30.	150	12.5	GALAXY
MCG-03-60-013	23 44 33.	- 18 31	36	15.5	GALAXY
ZWG 432.031	23 44 36.	+ 15 20		15.5	GALAXY
ZC 2344.6+2118	23 44 36.	+ 21 18	870		CLUSTER OF GALAXIES
ZWG 498.010	23 44 36.	+ 29 12		13.2	GALAXY
UGC 12780	23 44 36.	+ 29 12	210	13.2	GALAXY Sb
VV 005A	23 44 36.	+ 29 12	180	12.6	INTERACTING GALAXY
KARA.72 591B	23 44 36.	+ 29 12	150	13.2	PART OF DOUBLE GALAXY
ARP 086	23 44 36.	+ 29 12			PECULIAR GALAXY
4ZW 165	23 44 36.	+ 29 13			COMPACT GALAXY
MCG-03-60-014	23 44 36.	- 15 34	78	15.	GALAXY
MCG-05-56-009	23 44 36.	- 28 15	15	15.	GALAXY
RNGC 7753	23 44 38.	+ 29 12		13.0	GALAXY
IC 5350	23 44 39.	- 28 14 01.			NONSTELLAR OBJECT
IC 1509	23 44 41.	- 15 35 19.			NONSTELLAR OBJECT
SC 2341-6419.7	23 44 41.	- 64 03 02.	30		NEBULA
SC 2341-6419.7	23 44 41.	- 64 03 02.	30		NEBULA
MCG+00-60-046	23 44 42.	+ 03 01	36	15.	GALAXY
MCG+02-60-018	23 44 42.	+ 14 35	48	15.	GALAXY
MCG+02-60-017	23 44 42.	+ 15 20	36	15.	GALAXY
MCG+05-56-006	23 44 42.	+ 32 30	48	15.	GALAXY
ZWG 498.011	23 44 42.	+ 32 31		14.4	GALAXY
UGC 12781	23 44 42.	+ 32 31	66	14.4	GALAXY SBc
KARA.73B 1039	23 44 42.	+ 32 31	60	14.4	ISOLATED GALAXY S
MRSL 120+18/1	23 44 42.	+ 80 40	600		HII REGION
ZWG 381.047	23 44 42.	- 00 45		15.4	GALAXY
MCG-03-60-015	23 44 42.	- 20 18	24	15.	GALAXY
IC 5355	23 44 43.	+ 32 30 29.			NONSTELLAR OBJECT
MCG+00-60-047	23 44 45.	- 00 53	24	15.	GALAXY
IC 5351	23 44 47.	- 02 35 01.			NONSTELLAR OBJECT
ZWG 381.048	23 44 48.	+ 01 38		14.7	GALAXY
5ZW 427	23 44 48.	+ 48 03			COMPACT GALAXY
MCG+08-43-002	23 44 48.	+ 49 16	24	15.	GALAXY
MCG+00-60-048	23 44 48.	- 00 45	48	14.	GALAXY
IC 5352	23 44 48.	- 02 33 07.			NONSTELLAR OBJECT
MCG-01-60-032	23 44 48.	- 02 34	17	15.	GALAXY
PPL 5566	23 44 48.	- 15 23		18.6	BLUE STELLAR OBJECT
SC 2341-7332.7	23 44 49.	- 73 16 02.	30		NEBULA
IC 5356	23 44 50.	- 02 37 37.			NONSTELLAR OBJECT
MCG+05-56-007	23 44 51.	+ 28 06	48	15.	GALAXY
IC 5357	23 44 51.	- 02 34 25.			NONSTELLAR OBJECT
IC 5353	23 44 53.	- 28 23 07.			NONSTELLAR OBJECT
IC 5354	23 44 53.	- 28 24 43.			NONSTELLAR OBJECT
ZWG 498.012	23 44 54.	+ 28 07		15.1	GALAXY
ZC 2344.9-0025	23 44 54.	- 00 25	340		CLUSTER OF GALAXIES
MCG-01-60-033	23 44 54.	- 02 33	24	14.5	GALAXY
MCG-01-60-034	23 44 54.	- 02 36	18	15.	GALAXY
SHAH 030	23 44 54.	- 02 38	270	14.	GROUP OF COMPACT GALAXIES
MCG-01-60-035	23 44 54.	- 07 28	54	16.	GALAXY
MCG-05-56-010	23 44 54.	- 28 24	18	14.	GALAXY
MCG-05-56-011	23 44 54.	- 28 25	36	15.	GALAXY
KLEM 44	23 44 55.	- 28 23	1800	14.	CLUSTER OF 30 GALAXIES
LBN 0598	23 45	+ 80 40	840		BRIGHT NEBULA
LB 09957	23 45	- 86 50		14.2	FAINT BLUE STAR
MCG+00-60-049	23 45 00.	+ 01 40	36	14.	GALAXY
ZWG 407.058	23 45 00.	+ 09 04		15.6	GALAXY
MCG+03-60-026	23 45 00.	+ 18 19	24	18.	GALAXY
ZWG 455.047	23 45 00.	+ 19 04		15.7	GALAXY
UGC 12782	23 45 00.	+ 50 43	66	16.0	GALAXY SB?
5ZW 428	23 45 00.	+ 54 45			COMPACT GALAXY
MCG-03-60-016	23 45 00.	- 18 50	60	15.	GALAXY
SC 2342-0310.6	23 45 01.	- 02 53 56.	36		NEBULA
UGC 12783	23 45 06.	+ 18 19	60	16.5	GALAXY DWARF
ZC 2345.1+2140	23 45 06.	+ 21 40	1210		CLUSTER OF GALAXIES
MCG-01-60-036	23 45 06.	- 02 34	60	15.5	GALAXY
IC 5359	23 45 06.	- 02 35 55.			NONSTELLAR OBJECT
MCG-01-60-039	23 45 06.	- 02 49	12	17.	GALAXY
MCG-01-60-038	23 45 06.	- 02 49	24	17.	GALAXY
MCG-01-60-037	23 45 06.	- 08 06	60	15.	GALAXY
IC 5358	23 45 09.	- 28 25 01.			NONSTELLAR OBJECT
ZWG 455.048	23 45 12.	+ 17 49		15.6	GALAXY
ZWG 477.006	23 45 12.	+ 27 09		15.3	GALAXY
3ZW 119	23 45 12.	+ 27 24			COMPACT GALAXY
PK112-10.1	23 45 12.	+ 51 07	147	14.4	PLANETARY NEBULA
MCG-05-56-012	23 45 12.	- 27 46	48	15.	GALAXY
IC 5360	23 45 17.	- 37 20 13.			NONSTELLAR OBJECT
ZWG 477.007	23 45 18.	+ 27 24		15.7	GALAXY
ZWG 548.006	23 45 18.	+ 49 11		15.7	GALAXY
MCG+08-43-005	23 45 18.	+ 49 11	48	16.	GALAXY
5ZW 429	23 45 18.	+ 51 00			COMPACT GALAXY
MCG-05-56-013	23 45 18.	- 28 26	72	14.	GALAXY
RNGC 7755	23 45 18.	- 30 48		12.0	GALAXY
SC 2342-6144.1	23 45 19.	- 61 27 26.			NEBULA
SC 2342-6315.9	23 45 19.	- 62 59 14.	12		NEBULA
MCG+03-60-027	23 45 24.	+ 16 12 30.	60	14.5	GALAXY
ZC 2345.4+1648	23 45 24.	+ 16 48	2490		CLUSTER OF GALAXIES
ZC 2345.4+1857	23 45 24.	+ 18 57	1550		CLUSTER OF GALAXIES
OCL 0279	23 45 24.	+ 68 21	210	17.	OPEN STAR CLUSTER
MCG-05-56-015	23 45 24.	- 28 36	36	15.	GALAXY
MCG-05-56-014	23 45 24.	- 30 49	216	11.	GALAXY
MCG+04-56-007	23 45 27.	+ 27 25	30	16.	GALAXY
MCG-03-60-017	23 45 27.	- 18 52 30.	36	14.5	GALAXY
BC PKS2345-16	23 45 27.56	- 16 47 52.5		18.	QUASI-STELLAR OBJECT
SHB 395	23 45 27.6	- 16 47 50.		18.	QUASI-STELLAR OBJECT
SC 2342-0258.4	23 45 29.	- 02 41 44.	30		NEBULA
SC 2342-0306.8	23 45 29.	- 02 50 08.	18		NEBULA
MCG+01-60-036	23 45 30.	+ 05 18	48	15.5	GALAXY
ZWG 455.049	23 45 30.	+ 17 12		15.2	GALAXY
UGC 12784	23 45 30.	+ 17 12	96	15.2	GALAXY SBb
ZWG 477.008	23 45 30.	+ 27 06		15.0	GALAXY
UGC 12785	23 45 30.	+ 27 06	72	15.0	GALAXY S0
UGC 12786	23 45 30.	+ 49 58	66	16.5	GALAXY S
MCG-07-48-023	23 45 30.	- 38 41 30.	30	15.	GALAXY
ISS 0106	23 45 36.	+ 61 23	308		STELLAR RING
MCG-01-60-040	23 45 36.	- 02 40	60	15.5	GALAXY
MCG+04-56-008	23 45 39.	+ 27 06	48	15.	GALAXY
SC 2343-0339.4	23 45 39.	- 03 22 44.	6		NEBULA
ZC 2345.7+0030	23 45 42.	+ 00 30	740		CLUSTER OF GALAXIES
ZWG 381.049	23 45 42.	+ 02 03		15.2	GALAXY
ZC 2345.7+0442	23 45 42.	+ 04 42	1140		CLUSTER OF GALAXIES
UGC 12787	23 45 42.	+ 20 58	60	16.0	GALAXY Sb-c
MCG-01-60-041	23 45 42.	- 02 41	24	15.5	GALAXY
MCG-02-60-016	23 45 42.	- 08 47	54	14.	GALAXY
MCG+00-60-050	23 45 48.	+ 02 05	42	15.	GALAXY
ZWG 477.009	23 45 48.	+ 23 49		15.7	GALAXY
KARA.73B 1040	23 45 48.	+ 23 49	42	15.7	ISOLATED GALAXY S
4ZW 166	23 45 48.	+ 32 14			COMPACT GALAXY
MCG+04-56-009	23 45 51.	+ 23 49 30.	36	15.	GALAXY
HOLM 817B	23 45 54.	+ 03 52	24	15.1	PART OF MULTIPLE GALAXY
ZWG 455.050	23 45 54.	+ 18 30		15.7	GALAXY
ZWG 498.013	23 45 54.	+ 27 40		15.6	GALAXY
ZWG 381.050	23 45 54.	- 01 47		15.6	GALAXY
MCG-05-56-016	23 45 54.	- 27 07	24	15.5	GALAXY

OBJECT NAME	RIGHT ASCEN.	DECLINATION	DIAM.	MAGN.	TYPE OF OBJECT
SHB 396	23 45 57.9	+ 06 11 42.		17.5	QUASI-STELLAR OBJECT
BC 4C06.76	23 45 58.53	+ 06 08 20.8		17.5	QUASI-STELLAR OBJECT
ARP 068	23 45 59.	+ 03 50			PECULIAR GALAXY
VDB.66G 220	23 46	+ 25 57	70		DWARF GALAXY
SHB 397	23 46	+ 38			QUASI-STELLAR OBJECT
ZWG 477.010	23 46 00.	+ 24 19		15.4	GALAXY
SC 2343-0254.2	23 46 01.	- 02 37 32.	6		NEBULA
ZWG 381.051	23 46 06.	+ 01 57		15.3	GALAXY
ZC 2346.1+0322	23 46 06.	+ 03 22	1210		CLUSTER OF GALAXIES
MCG+00-60-051	23 46 06.	- 01 47	42	15.	GALAXY
SC 2343-0255.5	23 46 06.	- 02 38 50.	30		NEBULA
HOLM 817A	23 46 09.	+ 03 54	120	13.2	PART OF MULTIPLE GALAXY
ZWG 407.059	23 46 12.	+ 03 55		13.9	GALAXY
UGC 12788	23 46 12.	+ 03 55	150	13.9	GALAXY Sc
ZWG 455.051	23 46 12.	+ 15 39		15.6	GALAXY
KARA.73B 1041	23 46 12.	+ 15 39	36	15.6	ISOLATED GALAXY S
MCG+03-60-028	23 46 12.	+ 17 30	30	15.5	GALAXY
UGC 12789	23 46 12.	+ 26 08	66	16.5	GALAXY Sb-c
ZC 2346.2+3440	23 46 12.	+ 34 40	1140		CLUSTER OF GALAXIES
UGC 12790	23 46 12.	+ 50 10	60	15.5	GALAXY S
MCG-02-60-017	23 46 12.	- 14 18 30.	24	15.5	GALAXY
RNGC 7757	23 46 13.	+ 03 54		12.0	GALAXY
RNGC 7756	23 46 13.	+ 03 55		14.0	GALAXY
MCG+01-60-037	23 46 15.	+ 03 52 30.	138	12.	GALAXY
3ZW 122	23 46 18.	+ 08 32			COMPACT GALAXY
ZWG 455.052	23 46 18.	+ 17 30		15.6	GALAXY
ZWG 477.011	23 46 18.	+ 22 25		15.7	GALAXY
ZWG 477.012	23 46 18.	+ 25 57		15.2	GALAXY
UGC 12791	23 46 18.	+ 25 57	102	15.2	GALAXY DWRF IR
MCG-03-60-019	23 46 18.	- 16 48	30	15.5	GALAXY
MCG-03-60-018	23 46 18.	- 16 48	48	14.5	GALAXY
RNGC 7758	23 46 18.	- 22 19			GALAXY
RNGC 7759	23 46 19.	- 16 48		14.0	GALAXY
MCG+01-60-038	23 46 24.	+ 05 51	30	15.	GALAXY
ZWG 407.060	23 46 24.	+ 05 52		15.4	GALAXY
MCG+04-56-010	23 46 24.	+ 22 26	36	15.5	GALAXY
MCG+04-56-011	23 46 24.	+ 25 56	90	15.	GALAXY
MCG-01-60-042	23 46 24.	- 07 20	48	16.	GALAXY
MCG-03-60-020	23 46 24.	- 19 57	42	14.5	GALAXY
SC 2343-0251.0	23 46 26.	- 02 34 20.	42		NEBULA
ZC 2346.5+1923	23 46 30.	+ 19 23	740		CLUSTER OF GALAXIES
ZWG 477.013	23 46 30.	+ 26 31		14.8	GALAXY
UGC 12792	23 46 30.	+ 26 31	90	14.8	GALAXY SBb
ZC 2346.5+2647	23 46 30.	+ 26 47	1750		CLUSTER OF GALAXIES
SHAH 175	23 46 30.	+ 86 15	132		GROUP OF COMPACT GALAXIES
MCG-01-60-043	23 46 30.	- 02 33	30	15.	GALAXY
MCG-01-60-044	23 46 30.	- 07 20	42	15.	GALAXY
SC 2343-6209.6	23 46 33.	- 61 52 56.	9		NEBULA
ZC 2346.6+0932	23 46 36.	+ 09 32	1080		CLUSTER OF GALAXIES
MCG+04-56-012	23 46 36.	+ 26 30 30.	78	14.5	GALAXY
MCG-03-60-021	23 46 36.	- 16 51 30.	24	15.5	GALAXY
RNGC 7754	23 46 37.	- 16 51		15.0	GALAXY
MCG+05-56-008	23 46 39.	+ 30 42 30.	42	15.	GALAXY
MCG+08-43-006	23 46 39.	+ 46 15 30.	60	14.	GALAXY
ZWG 432.032	23 46 42.	+ 11 32		15.6	GALAXY
UGC 12793	23 46 42.	+ 11 32	66	15.6	GALAXY Sc
3ZW 120	23 46 42.	+ 17 49			COMPACT GALAXY
ZWG 498.014	23 46 42.	+ 30 42		14.8	GALAXY
UGC 12794	23 46 42.	+ 30 42	90	14.8	GALAXY VY CMPT
KARA.73B 1042	23 46 42.	+ 30 42	42	14.8	ISOLATED GALAXY E
ZWG 548.007	23 46 42.	+ 46 17		14.9	GALAXY
UGC 12795	23 46 42.	+ 46 17	60	14.9	GALAXY S
MCG-05-56-018	23 46 42.	- 29 18	36	15.	GALAXY
MCG-05-56-017	23 46 42.	- 30 33	36	15.	GALAXY
RNGC 7760	23 46 44.	+ 30 42		15.0	GALAXY
MCG+02-60-019	23 46 45.	+ 11 32	48	14.5	GALAXY
SC 2344-0250.4	23 46 45.	- 02 33 44.	18		NEBULA
ZC 2346.8+0129	23 46 48.	+ 01 29	200		CLUSTER OF GALAXIES
ZC 2346.8+3511	23 46 48.	+ 35 11	1210		CLUSTER OF GALAXIES
MCG+08-43-007	23 46 48.	+ 47 38	90	14.	GALAXY
ZWG 548.008	23 46 48.	+ 47 39		15.2	GALAXY
UGC 12796	23 46 48.	+ 47 39	84	15.2	GALAXY SB
3ZW 121	23 46 54.	+ 10 20			COMPACT GALAXY
ZWG 498.015	23 46 54.	+ 32 53		15.7	GALAXY
KARA.73B 1043	23 46 54.	+ 32 53	24	15.7	ISOLATED GALAXY S
ARC 2662	23 46 55.	+ 16 09		17.5	RICH CLUSTER OF GALAXIES
5ZW 430	23 47 00.	+ 40 53			COMPACT GALAXY
VV 208B	23 47 00.	- 00 24	24	15.	INTERACTING GALAXY
VV 208A	23 47 00.	- 00 24	42	14.	INTERACTING GALAXY
MCG-07-48-024	23 47 00.	- 38 43	30	16.	GALAXY
ZC 2347.1+0212	23 47 06.	+ 02 12	1080		CLUSTER OF GALAXIES
UGC 12797	23 47 06.	+ 45 51	60	16.5	GALAXY SBc
MCG-07-48-025	23 47 06.	- 38 50	24	16.	GALAXY
ARC 2663	23 47 08.	- 25 00		17.5	RICH CLUSTER OF GALAXIES
MCG+05-56-009	23 47 09.	+ 29 38 30.	78	15.5	GALAXY
ZWG 407.061	23 47 12.	+ 07 55		15.5	GALAXY
ZC 2347.2+1447	23 47 12.	+ 14 47	2820		CLUSTER OF GALAXIES
ZWG 498.016	23 47 12.	+ 29 40		15.5	GALAXY
UGC 12798	23 47 12.	+ 29 40	72	15.5	GALAXY S
MCG+05-56-010	23 47 18.	+ 27 45 30.	30	16.	GALAXY
ZWG 498.017	23 47 18.	+ 29 46		15.3	GALAXY
MCG-02-60-018	23 47 18.	- 12 45	48	14.	GALAXY
MCG+05-56-011	23 47 21.	+ 29 45	30	15.5	GALAXY
SC 2344-6353.2	23 47 21.	- 63 36 32.	18		NEBULA
SC 2344-0357.2	23 47 22.	- 03 40 32.	12		NEBULA
RNGC 7762	23 47 23.	+ 67 45		10.0	OPEN CLUSTER
ZWG 381.052	23 47 24.	+ 02 25		15.7	GALAXY
UGC 12799	23 47 24.	+ 02 25	90	15.7	GALAXY Sc
MCG+00-60-052	23 47 24.	+ 02 27	78	14.	GALAXY
ZWG 407.062	23 47 24.	+ 06 10		15.7	GALAXY
MCG+05-56-012	23 47 24.	+ 28 23	24	16.	GALAXY
OCL 0270	23 47 24.	+ 62 26	150	10.	OPEN STAR CLUSTER
OCL 0280	23 47 24.	+ 67 45	1260	10.0	OPEN STAR CLUSTER
MCG-05-56-019	23 47 24.	- 28 13	72	15.	GALAXY
TON-S 0119	23 47 24.	- 33 04		14.6	BLUE STAR
SC 2344-6046.2	23 47 25.	- 60 29 32.	12		NEBULA
ZC 2347.5+0707	23 47 30.	+ 07 07	10750		CLUSTER OF GALAXIES
ZWG 517.002	23 47 30.	+ 37 33		15.6	GALAXY
MCG+08-43-008	23 47 30.	+ 46 00	54	15.	GALAXY
ZWG 548.009	23 47 30.	+ 46 01		15.7	GALAXY
TON-S 0120	23 47 30.	- 26 39		15.5	BLUE STAR
SC 2344-0343.6	23 47 33.	- 03 26 56.	12		NEBULA
ARC 2664	23 47 35.	- 03 54		17.7	RICH CLUSTER OF GALAXIES
ZWG 517.003	23 47 36.	+ 36 35		15.5	GALAXY
MCG-05-56-021	23 47 36.	- 29 17	24	15.5	GALAXY
MCG-05-56-020	23 47 36.	- 29 25	66	15.5	GALAXY
SC 2345-0433.0	23 47 38.	- 04 16 20.	12		NEBULA
ZC 2347.7+0332	23 47 42.	+ 03 32	1550		CLUSTER OF GALAXIES
MCG-03-60-022	23 47 42.	- 18 01	72	15.	GALAXY
RNGC 7763	23 47 43.	- 16 53			GALAXY
ZWG 432.033	23 47 48.	+ 10 29		15.4	GALAXY
UGC 12800	23 47 48.	+ 10 29	72	15.4	GALAXY SBa
MCG+02-60-020	23 47 48.	+ 10 30	84	13.5	GALAXY
ZWG 517.004	23 47 48.	+ 36 23		15.4	GALAXY
MCG+03-60-029	23 47 54.	+ 18 49	42	15.	GALAXY
ZWG 455.053	23 47 54.	+ 18 50		15.7	GALAXY
UGC 12801	23 47 54.	+ 25 53	60	16.0	GALAXY E-S0
TON-S 0121	23 47 54.	- 26 37		15.9	BLUE STAR
TON-S 0122	23 47 54.	- 31 46		15.8	BLUE STAR
IC 1510	23 47 58.	+ 01 48 01.			NONSTELLAR OBJECT
LBN 0562	23 48	+ 54 40	3900		BRIGHT NEBULA
ZWG 381.053	23 48 00.	+ 01 47		15.2	GALAXY
UGC 12802	23 48 00.	+ 14 13	66	16.5	GALAXY Sc
MCG+05-56-013	23 48 00.	+ 28 42	60	15.	GALAXY
ZWG 498.018	23 48 00.	+ 28 43		15.2	GALAXY
UGC 12803	23 48 00.	+ 28 43	90	15.2	GALAXY
MCG+00-60-053	23 48 06.	+ 01 50	36	15.	GALAXY
ZC 2348.1+2312	23 48 06.	+ 23 12	1480		CLUSTER OF GALAXIES
ZWG 477.014	23 48 06.	+ 24 16		15.6	GALAXY
UGC 12804	23 48 06.	+ 24 16	66	15.6	GALAXY E
MCG+04-56-013	23 48 06.	+ 24 16 30.	45	15.	GALAXY
SER 162.02	23 48 06.	- 41 01		12.5	HIGH SURF BRGHTNSS GALAXY
ARC 2665	23 48 09.	+ 05 51		15.8	RICH CLUSTER OF GALAXIES
HOLM 818B	23 48 09.	+ 26 49	42	14.7	PART OF MULTIPLE GALAXY
ZC 2348.2+0548	23 48 12.	+ 05 48	3490		CLUSTER OF GALAXIES
ZWG 432.034	23 48 12.	+ 15 23		15.7	GALAXY
ZWG 477.015	23 48 12.	+ 26 54		15.7	GALAXY
ZWG 477.016	23 48 12.	+ 27 01		14.9	GALAXY
ZWG 381.054	23 48 12.	- 02 28		15.4	GALAXY
MCG-05-56-022	23 48 12.	- 28 43	24	15.5	GALAXY
MCG-07-48-026	23 48 12.	- 39 09	30	16.	GALAXY
HOLM 818C	23 48 13.	+ 26 53	24	14.9	PART OF MULTIPLE GALAXY
SC 2345-0244.9	23 48 13.	- 02 28 13.	12		NEBULA
RNGC 7765	23 48 14.	+ 26 54		15.5	GALAXY
MCG-07-48-027	23 48 15.	- 41 00	78	12.8	GALAXY
HOLM 818D	23 48 17.	+ 26 51	18	15.0	PART OF MULTIPLE GALAXY
MCG+01-60-039	23 48 18.	+ 05 52 30.	12	15.	GALAXY
ZWG 477.017	23 48 18.	+ 26 49		14.2	GALAXY
UGC 12805	23 48 18.	+ 26 49	66	14.2	GALAXY S0-a
ZWG 477.018	23 48 18.	+ 26 51		15.7	GALAXY
RNGC 7767	23 48 20.	+ 26 49		14.0	GALAXY
RNGC 7766	23 48 20.	+ 26 51		15.5	GALAXY
HOLM 818A	23 48 20.	+ 26 52	18	14.5	PART OF MULTIPLE GALAXY
MCG+01-60-040	23 48 21.	+ 08 24	36	15.5	GALAXY
SN 1968Z	23 48 23.	+ 26 54		17.9	SUPERNOVA
ZWG 407.063	23 48 24.	+ 08 24		15.5	GALAXY
MCG+03-60-030	23 48 24.	+ 19 53	84	13.	GALAXY
ZWG 477.019	23 48 24.	+ 26 53		14.0	GALAXY
ARC 2666	23 48 24.	+ 26 53		13.8	RICH CLUSTER OF GALAXIES
UGC 12806	23 48 24.	+ 26 53	114	14.0	GALAXY E
ZWG 477.020	23 48 24.	+ 26 57		15.5	GALAXY
MCG+04-56-014	23 48 24.	+ 27 01	30	15.5	GALAXY
ZC 2348.4+2908	23 48 24.	+ 29 08	1950		CLUSTER OF GALAXIES
UGC 12807	23 48 24.	+ 35 30	60	17.	GALAXY
MCG+00-60-054	23 48 24.	- 02 27	12	15.	GALAXY
RNGC 7764	23 48 24.	- 41 01		13.0	GALAXY
SCHO 1439	23 48 25.	+ 58 13 00.	340		ISOLATED DARK CLOUD
RNGC 7768	23 48 26.	+ 26 53		14.0	GALAXY
MCG+04-56-015	23 48 27.	+ 26 53	36	14.9	GALAXY
HOLM 820C	23 48 28.	+ 19 52	48	12.5	PART OF MULTIPLE GALAXY
IC 1511	23 48 28.	+ 26 48			SAME AS NGC 7767
IC 1512	23 48 29.	+ 26 46			NONSTELLAR OBJECT
MCG+01-60-041	23 48 30.	+ 08 27	48	15.5	GALAXY
ZWG 432.035	23 48 30.	+ 09 46		15.6	GALAXY
ZWG 455.054	23 48 30.	+ 19 52		12.9	GALAXY
UGC 12808	23 48 30.	+ 19 52	150	12.9	GALAXY Sa-b
KARA.72 592A	23 48 30.	+ 19 52	168	12.9	PART OF DOUBLE GALAXY
MCG+03-60-031	23 48 30.	+ 20 00	24	15.	GALAXY
MCG+04-56-016	23 48 30.	+ 26 48	54	14.7	GALAXY
MCG+04-56-017	23 48 30.	+ 26 50	30	15.0	GALAXY
ZWG 549.001	23 48 30.	+ 46 27		15.3	GALAXY
ZWG 548.010	23 48 30.	+ 46 27		15.3	GALAXY
UGC 12809	23 48 30.	+ 46 27	108	15.3	GALAXY Sb
LB 01533	23 48 30.	- 59 21		14.4	FAINT BLUE STAR
RNGC 7769	23 48 32.	+ 19 52		13.0	GALAXY
SC 2345-0423.0	23 48 32.	- 04 06 19.	12		NEBULA
MCG+04-56-018	23 48 33.	+ 26 51 30.	18	14.5	GALAXY
MCG+08-01-001	23 48 33.	+ 46 26	72	15.	GALAXY
ZWG 381.055	23 48 36.	+ 00 46		14.4	GALAXY
UGC 12810	23 48 36.	+ 00 46	108	14.4	GALAXY Sb
ZWG 407.064	23 48 36.	+ 08 27		15.4	GALAXY
UGC 12811	23 48 36.	+ 08 27	78	15.4	GALAXY S0
MCG+03-60-033	23 48 36.	+ 20 19	24	16.5	GALAXY
MCG+03-60-032	23 48 36.	+ 20 19	48	15.	GALAXY
MCG+04-56-019	23 48 36.	+ 26 56 30.	39	15.5	GALAXY
MCG+07-48-019	23 48 36.	+ 43 15	27	16.	GALAXY
TON-S 0123	23 48 36.	- 31 48		15.1	BLUE STAR
HOLM 819B	23 48 39.	+ 43 16	12	15.3	PART OF MULTIPLE GALAXY
SCHO 1440	23 48 41.	+ 57 46 54.	300		ISOLATED DARK CLOUD
MCG+00-60-055	23 48 42.	+ 00 48	108	13.	GALAXY
MCG+03-60-034	23 48 42.	+ 19 50	36	14.5	GALAXY
MCG+03-60-035	23 48 42.	+ 19 51	138	13.	GALAXY
ZWG 455.055	23 48 42.	+ 19 57		15.7	GALAXY
ZWG 455.056	23 48 42.	+ 20 17		15.5	GALAXY
UGC 12812	23 48 42.	+ 20 17	60	15.5	GALAXY S
KARA.72 593A	23 48 42.	+ 20 17	48	15.5	PART OF DOUBLE GALAXY
ZWG 477.021	23 48 42.	+ 27 01		15.7	GALAXY
5ZW 431	23 48 42.	+ 39 46			COMPACT GALAXY
HOLM 819A	23 48 42.	+ 43 16	36	15.0	PART OF MULTIPLE GALAXY
SC 2346-0322.9	23 48 42.	- 03 06 13.	12		NEBULA
MCG-02-60-019	23 48 42.	- 13 02	54	15.5	GALAXY
LB 01534	23 48 42.	- 56 05		15.1	FAINT BLUE STAR
MCG+03-60-036	23 48 45.	+ 20 20	42	14.5	GALAXY
ZWG 455.057	23 48 48.	+ 19 49		14.5	GALAXY
HOLM 820B	23 48 48.	+ 19 49	18	13.8	PART OF MULTIPLE GALAXY
UGC 12813	23 48 48.	+ 19 49	60	14.5	GALAXY
ZWG 477.022	23 48 48.	+ 24 23		15.7	GALAXY
MCG+08-01-002	23 48 48.	+ 48 47	72	15.	GALAXY
ZWG 549.002	23 48 48.	+ 48 48		15.6	GALAXY
ZWG 548.011	23 48 48.	+ 48 48		15.6	GALAXY
UGC 12814	23 48 48.	+ 48 48	72	15.6	GALAXY Sc
RNGC 7761	23 48 49.	- 13 40		14.0	GALAXY
RNGC 7770	23 48 50.	+ 19 49		14.5	GALAXY
HOLM 820A	23 48 50.	+ 19 50	84	13.0	PART OF MULTIPLE GALAXY
SCHO 1441	23 48 51.	+ 59 19 54.	410		ISOLATED DARK CLOUD
MCG-02-60-020	23 48 51.	- 13 40	24	14.	GALAXY

OBJECT NAME	RIGHT ASCEN.	DECLINATION	DIAM.	MAGN.	TYPE OF OBJECT
ZWG 455.058	23 48 54.	+ 19 50		13.1	GALAXY
UGC 12815	23 48 54.	+ 19 50	168	13.1	GALAXY SBa
KARA.72 592B	23 48 54.	+ 19 50	144	13.1	PART OF DOUBLE GALAXY
ZWG 455.059	23 48 54.	+ 20 18		14.9	GALAXY
MRK 331	23 48 54.	+ 20 18	24	16.	GALAXY WITH UV CONTINUUM
KARA.72 593B	23 48 54.	+ 20 18	48	14.9	PART OF DOUBLE GALAXY
ZWG 477.023	23 48 54.	+ 26 57		15.7	GALAXY
IC 5361	23 48 54.	- 13 39 29.			NONSTELLAR OBJECT
SC 2346-0406.6	23 48 55.	- 03 49 55.			NEBULA
RNGC 7771	23 48 56.	+ 19 50		13.0	GALAXY
SCHO 1442	23 48 58.	+ 62 24 00.	550		ISOLATED DARK CLOUD
LB 09958	23 49	- 81 01		14.8	FAINT BLUE STAR
FEIG 111	23 49 00.	+ 08 25		11.2	FAINT BLUE STAR
ZWG 407.065	23 49 00.	+ 08 27		15.5	GALAXY
ZC 2349.0+1045	23 49 00.	+ 10 45	5650		CLUSTER OF GALAXIES
ZC 2349.0+1625	23 49 00.	+ 16 25	2490		CLUSTER OF GALAXIES
2ZW 189	23 49 00.	+ 18 47			COMPACT GALAXY
4ZW 167	23 49 00.	+ 25 16			COMPACT GALAXY
PHL 6034	23 49 00.	- 30 50		17.8	BLUE STELLAR OBJECT
TON-S 0124	23 49 00.	- 31 37		15.1	BLUE STAR
SC 2346-6408.1	23 49 00.	- 63 51 25.			NEBULA
IC 5362	23 49 01.	- 28 38 29.			NONSTELLAR OBJECT
ZWG 455.060	23 49 06.	+ 18 32		15.4	GALAXY
MCG-05-56-023	23 49 06.	- 28 39	21	14.5	GALAXY
ZWG 407.066	23 49 12.	+ 05 46		15.2	GALAXY
OCL 0230	23 49 12.	+ 15 59			OPEN STAR CLUSTER
ZWG 455.061	23 49 12.	+ 20 43		15.7	GALAXY
PHL 0577	23 49 12.	- 31 36		16.0	BLUE STELLAR OBJECT
RNGC 7772	23 49 14.	+ 15 59			OPEN CLUSTER
SCHO 1443	23 49 14.	+ 58 16 06.	240		ISOLATED DARK CLOUD
ARC 2667	23 49 14.	- 26 17		17.9	RICH CLUSTER OF GALAXIES
MCG+03-60-037	23 49 15.	+ 16 17 30.	30	16.	GALAXY
MCG-04-56-004	23 49 15.	- 26 06	36	16.	GALAXY
ZWG 381.056	23 49 18.	+ 02 48		15.0	GALAXY
UGC 12816	23 49 18.	+ 02 48	114	15.0	GALAXY Sc
ZWG 432.036	23 49 18.	+ 13 25		15.6	GALAXY
ZWG 498.019	23 49 18.	+ 29 00		15.7	GALAXY
UGC 12817	23 49 18.	+ 29 00	72	15.7	GALAXY Sb
TON-S 0125	23 49 13.	- 26 35		15.9	BLUE STAR
MCG+05-56-014	23 49 21.	+ 28 58 30.	66	16.	GALAXY
KW 71	23 49 23.	- 01 26 06.	12		SEYFERT GALAXY
MCG+00-60-056	23 49 24.	+ 02 50	90	14.	GALAXY
MCG+02-60-021	23 49 24.	+ 13 27	30	15.	GALAXY
ZC 2349.4+1348	23 49 24.	+ 13 48	1410		CLUSTER OF GALAXIES
ZWG 455.062	23 49 24.	+ 16 17		15.5	GALAXY
ZWG 455.063	23 49 24.	+ 21 06		15.6	GALAXY
ZC 2349.4+2158	23 49 24.	+ 21 58	470		CLUSTER OF GALAXIES
ZWG 477.024	23 49 24.	+ 27 04		15.7	GALAXY
ZWG 498.020	23 49 24.	+ 28 18		15.7	GALAXY
ZWG 517.005	23 49 24.	+ 33 32		15.6	GALAXY
MCG-05-56-025	23 49 24.	- 28 12	24	15.	GALAXY
MCG-05-56-024	23 49 24.	- 28 14	36	16.	GALAXY
IC 5363	23 49 35.	- 28 54 41.			NONSTELLAR OBJECT
ZWG 407.067	23 49 36.	+ 08 07		15.1	GALAXY
UGC 12818	23 49 36.	+ 08 07	78	15.1	GALAXY Sc
MCG+01-60-042	23 49 36.	+ 08 07 30.	84	14.	GALAXY
ZWG 432.037	23 49 36.	+ 11 12		14.6	GALAXY
UGC 12819	23 49 36.	+ 11 12	96	14.6	GALAXY E+COMP
KARA.72 594B	23 49 36.	+ 11 12	54		PART OF DOUBLE GALAXY
KARA.72 594A	23 49 36.	+ 11 12	66	14.6	PART OF DOUBLE GALAXY
ZWG 498.021	23 49 36.	+ 27 48		15.7	GALAXY
MCG+05-56-015	23 49 36.	+ 31 00	72	14.	GALAXY
MCG-04-56-005	23 49 36.	- 26 06 30.	30	16.	GALAXY
MCG-05-56-026	23 49 36.	- 29 22	18	15.	GALAXY
RNGC 7774	23 49 37.	+ 11 12		14.5	GALAXY
KEEL 731	23 49 37.8	+ 08 07 02.			SPIRAL NEBULA
ARC 2668	23 49 38.	+ 13 47		17.1	RICH CLUSTER OF GALAXIES
HOLM 821A	23 49 39.	+ 11 12	24	14.5	PART OF MULTIPLE GALAXY
SCHO 1444	23 49 40.	+ 57 46 06.	240		ISOLATED DARK CLOUD
HOLM 821B	23 49 41.	+ 11 12	18	14.8	PART OF MULTIPLE GALAXY
ZC 2349.7+0734	23 49 42.	+ 07 34	1080		CLUSTER OF GALAXIES
MCG+02-60-022	23 49 42.	+ 11 13	36	13.	GALAXY
ZC 2349.7+1653	23 49 42.	+ 16 53	7260		CLUSTER OF GALAXIES
ZWG 498.022	23 49 42.	+ 31 00		14.5	GALAXY
UGC 12820	23 49 42.	+ 31 00	72	14.5	GALAXY SBb
ZWG 498.023	23 49 42.	+ 31 11		15.6	GALAXY
ZC 2349.7+3259	23 49 42.	+ 32 59	3560		CLUSTER OF GALAXIES
MCG+08-01-003	23 49 42.	+ 47 06	72	16.	GALAXY
MCG-04-56-006	23 49 42.	- 25 39	78	14.	GALAXY
PHL 2434	23 49 42.	- 27 28		18.3	BLUE STELLAR OBJECT
PHL 6035	23 49 42.	- 28 02		18.5	BLUE STELLAR OBJECT
TON-S 0126	23 49 42.	- 28 20		15.0	BLUE STAR
PHL 0578	23 49 42.	- 28 21		16.0	BLUE STELLAR OBJECT
MCG-05-56-027	23 49 42.	- 29 44	42	15.	GALAXY
RNGC 7773	23 49 44.	+ 31 00		14.5	GALAXY
SG 3.244	23 49 44.	+ 67 47	4500		DIFFUSE EMISSION NEBULA
MCG+05-56-016	23 49 48.	+ 28 29	48	14.5	GALAXY
MCG+05-56-017	23 49 48.	+ 33 09	24	16.	GALAXY
ZCG 2349+39	23 49 48.	+ 39 38		17.4	COMPACT GALAXY
MCG-05-56-029	23 49 48.	- 29 18	24	15.5	GALAXY
MCG-05-56-028	23 49 48.	- 30 32	24	15.	GALAXY
SC 2347-6307.4	23 49 51.	- 62 50 43.	30		NEBULA
ZWG 498.024	23 49 54.	+ 28 30		13.9	GALAXY
UGC 12821	23 49 54.	+ 28 30	66	13.9	GALAXY Sc
5ZW 432	23 49 54.	+ 39 38			COMPACT GALAXY
MCG-05-56-030	23 49 54.	- 28 37	42	15.5	GALAXY
SC 2347-0326.0	23 49 55.	- 03 09 19.	18		NEBULA
RNGC 7775	23 49 56.	+ 28 30		14.0	GALAXY
MCG+08-01-004	23 49 57.	+ 47 05 30.	90	16.	GALAXY
LBN 0568	23 50	+ 60 10	300		BRIGHT NEBULA
LB 09959	23 50	- 80 03		14.4	FAINT BLUE STAR
ZWG 432.038	23 50 00.	+ 14 17		15.0	GALAXY
UGC 12822	23 50 00.	+ 14 17	66	15.0	GALAXY E-SO
MCG+02-60-023	23 50 00.	+ 14 18	72	14.5	GALAXY
ZCG 2350+39	23 50 00.	+ 39 38		18.0	COMPACT GALAXY
5ZW 433	23 50 00.	+ 47 06			COMPACT GALAXY
ZWG 549.003	23 50 00.	+ 47 07		14.6	GALAXY
ZWG 548.012	23 50 00.	+ 47 07		14.6	GALAXY
5ZW 434	23 50 00.	+ 50 03			COMPACT GALAXY
LDN 1264	23 50 00.	+ 68 00	1260		DARK NEBULA
PHL 0579	23 50 00.	- 29 56		18.5	BLUE STELLAR OBJECT
MCG-05-56-031	23 50 00.	- 30 28	54	15.	GALAXY
TON-S 0127	23 50 00.	- 32 02		15.8	BLUE STAR
PHL 2435	23 50 00.	- 32 03		16.6	BLUE STELLAR OBJECT
SC 2347-0335.0	23 50 04.	- 03 18 19.	6		NEBULA
ZC 2350.1+0052	23 50 06.	+ 00 52	1080		CLUSTER OF GALAXIES
ZC 2350.1+4131	23 50 06.	+ 41 31	1480		CLUSTER OF GALAXIES
PHL 2436	23 50 06.	- 21 35		17.7	BLUE STELLAR OBJECT

OBJECT NAME	RIGHT ASCEN.	DECLINATION	DIAM.	MAGN.	TYPE OF OBJECT
RNGC 7776	23 50 07.	- 13 40			NON-EXISTENT OBJECT
SC 2347-6111.9	23 50 08.	- 60 55 13.	48		NEBULA
ZWG 477.025	23 50 12.	+ 26 52		15.4	GALAXY
UGC 12823	23 50 12.	+ 26 52	60	15.4	GALAXY SB0
MCG-05-56-032	23 50 12.	- 28 51	12	15.5	GALAXY
MCG-06-01-001	23 50 12.	- 34 51	24	15.5	GALAXY
SC 2347-0530.1	23 50 14.	- 05 13 25.	6		NEBULA
MCG+08-01-005	23 50 15.	+ 46 30	48	15.	GALAXY
ARC 2669	23 50 16.	+ 02 55		17.8	RICH CLUSTER OF GALAXIES
HN 2877	23 50 16.2	- 44 25 55.	18		NEBULA
MCG+04-56-020	23 50 18.	+ 26 50	42	15.	GALAXY
3ZW 123	23 50 18.	+ 27 06			COMPACT GALAXY
ZWG 549.004	23 50 18.	+ 46 32		15.4	GALAXY
ZWG 548.013	23 50 18.	+ 46 32		15.4	GALAXY
5ZW 435	23 50 18.	+ 47 12			COMPACT GALAXY
SC 2347-0214.8	23 50 20.	- 01 58 07.	6		NEBULA
SC 2347-0256.8	23 50 23.	- 02 40 07.	6		NEBULA
4ZW 168	23 50 24.	+ 27 07			COMPACT GALAXY
SC 2347-0303.8	23 50 24.	- 02 47 07.	42		NEBULA
SC 2347-0317.1	23 50 24.	- 03 00 25.	18		NEBULA
PHL 0580	23 50 24.	- 24 51		15.4	BLUE STELLAR OBJECT
SC 2347-0256.4	23 50 25.	- 02 39 43.	6		NEBULA
SG 2.100	23 50 28.	+ 60 12	300		DIFFUSE EMISSION NEBULA
ZC 2350.5+0252	23 50 30.	+ 02 52	1410		CLUSTER OF GALAXIES
ZWG 455.064	23 50 30.	+ 15 47		15.7	GALAXY
UGC 12824	23 50 30.	+ 17 35	66	16.0	GALAXY S
ZC 2350.5+2513	23 50 30.	+ 25 13	1950		CLUSTER OF GALAXIES
OCL 0272	23 50 30.	+ 61 41	180	10.	OPEN STAR CLUSTER
MCG-01-60-045	23 50 30.	- 04 04	60	15.	GALAXY
SC 2347-0501.8	23 50 30.	- 04 45 07.	36		NEBULA
SER 162.01	23 50 30.	- 41 07			NGC 7764-A = AGUERO 81
SC 2347-0325.0	23 50 33.	- 03 08 19.	12		NEBULA
SC 2347-7358.1	23 50 34.	- 73 41 25.	36		NEBULA
SC 2348-0330.8	23 50 35.	- 03 14 07.	12		NEBULA
ZC 2350.6+0026	23 50 36.	+ 00 26	1140		CLUSTER OF GALAXIES
ZC 2350.6+0758	23 50 36.	+ 07 58	1810		CLUSTER OF GALAXIES
ZC 2350.6+0831	23 50 36.	+ 08 31	1210		CLUSTER OF GALAXIES
UGC 12825	23 50 36.	+ 10 38	66	16.5	GALAXY Sb-c
ZC 2350.6+1327	23 50 36.	+ 13 27	1080		CLUSTER OF GALAXIES
ZWG 498.025	23 50 36.	+ 29 18		15.7	GALAXY
UGC 12826	23 50 36.	+ 29 18	72	15.7	GALAXY
MRSL 115-01/2	23 50 36.	+ 60 12	420		HII REGION
MCG+05-56-018	23 50 39.	+ 28 00	60	14.5	GALAXY
MIN.47 18	23 50 40.	+ 60 12			DIFFUSE NEBULA
ZWG 381.057	23 50 42.	+ 03 26		14.9	GALAXY
ZWG 407.068	23 50 42.	+ 07 08		15.5	GALAXY
MCG+01-60-043	23 50 42.	+ 07 35	36	14.	GALAXY
ZWG 407.069	23 50 42.	+ 07 36		13.8	GALAXY
UGC 12827	23 50 42.	+ 07 36	66	13.8	GALAXY E
UGC 12828	23 50 42.	+ 19 07	72	16.0	GALAXY Sc
ZWG 477.026	23 50 42.	+ 24 40		15.6	GALAXY
ZWG 477.027	23 50 42.	+ 27 25		15.5	GALAXY
ZWG 498.026	23 50 42.	+ 28 01		14.5	GALAXY
UGC 12829	23 50 42.	+ 28 01	72	14.5	GALAXY S0
SC 2348-0155.3	23 50 42.	- 01 38 37.	6		NEBULA
SC 2348-0342.1	23 50 42.	- 03 25 25.	18		NEBULA
MCG-06-01-002	23 50 42.	- 34 19	12	15.	GALAXY
LB 01535	23 50 42.	- 57 33		14.7	FAINT BLUE STAR
RNGC 7778	23 50 43.	+ 07 36		14.0	GALAXY
RNGC 7777	23 50 44.	+ 28 01		14.5	GALAXY
MCG+01-60-044	23 50 45.	+ 07 52 30.	48	15.	GALAXY
SC 2348-7109.6	23 50 45.	- 70 52 55.	72		NEBULA
MCG+01-60-045	23 50 48.	+ 07 35	72	13.5	GALAXY
UGC 12830	23 50 48.	+ 10 41	60	16.5	GALAXY Sa-b
SC 2348-0358.9	23 50 48.	- 03 42 13.	6		NEBULA
PHL 6036	23 50 48.	- 24 02		18.6	BLUE STELLAR OBJECT
MCG-05-56-033	23 50 48.	- 29 50	36	15.	GALAXY
MCG-06-01-003	23 50 48.	- 34 19	12	15.5	GALAXY
DV.56 N7764A	23 50 48.	- 40 59			S GALAXY
RNGC 7764A	23 50 48.	- 40 59			GALAXY
AGU 83	23 50 48.	- 40 59 00.		12.5	PECULIAR GALAXY
YC 2350-41	23 50 48.	- 41 05 06.			UNUSUAL SOUTHERN GALAXY
HELW 504	23 50 48.	- 41 21 48.			NEBULA
LB 01536	23 50 48.	- 55 14		13.8	FAINT BLUE STAR
KEEL 732	23 50 48.9	+ 07 52 13.			SPIRAL NEBULA
SC 2348-0202.5	23 50 49.	- 01 45 49.			NEBULA
MCG-02-60-021	23 50 51.	- 12 06	42	14.5	GALAXY
ZWG 407.070	23 50 54.	+ 07 36		13.6	GALAXY
UGC 12831	23 50 54.	+ 07 36	96	13.6	GALAXY S0/Sa
MCG+01-60-046	23 50 54.	+ 07 50	48	14.5	GALAXY
ZWG 432.039	23 50 54.	+ 11 03		15.1	GALAXY
UGC 12832	23 50 54.	+ 11 03	60	15.1	GALAXY S
3ZW 124	23 50 54.	+ 15 50			COMPACT GALAXY
5ZW 436	23 50 54.	+ 46 31			COMPACT GALAXY
RNGC 7779	23 50 55.	+ 07 36		13.5	GALAXY
IC 1513	23 50 55.	+ 11 02 08.			NONSTELLAR OBJECT
KN 13.001	23 50 56.3	+ 11 02 27.			NEBULA
SC 2348-0117.0	23 50 58.	- 01 00 19.	12		NEBULA
SC 2348-0254.3	23 50 59.	- 02 37 37.	6		NEBULA
LBN 0571	23 51	+ 60 12	300		BRIGHT NEBULA
ZWG 407.071	23 51 00.	+ 07 50		14.8	GALAXY
UGC 12833	23 51 00.	+ 07 50	78	14.8	GALAXY Sa-b
MCG+02-60-024	23 51 00.	+ 11 03	60	14.	GALAXY
MCG+05-56-019	23 51 00.	+ 28 58	18	16.	GALAXY
LDN 1255	23 51 00.	+ 60 15	120		DARK NEBULA
MCG+14-01-004	23 51 00.	+ 85 46	54	17.	GALAXY
ZWG 381.058	23 51 00.	- 01 01		15.7	GALAXY
RNGC 7780	23 51 01.	+ 07 50		15.0	GALAXY
HN 2878	23 51 04.2	- 44 52 13.	12		NEBULA
SC 2348-0141.5	23 51 05.	- 01 24 49.	12		NEBULA
ZWG 498.027	23 51 06.	+ 28 59		15.3	GALAXY
MCG+05-56-020	23 51 09.	+ 28 58	36	15.5	GALAXY
ZWG 407.072	23 51 12.	+ 07 35		15.2	GALAXY
ZC 2351.2+1629	23 51 12.	+ 16 29	1340		CLUSTER OF GALAXIES
ZWG 477.028	23 51 12.	+ 27 08		15.3	GALAXY
ZWG 498.028	23 51 12.	+ 28 59		15.7	GALAXY
ZC 2351.2+3138	23 51 12.	+ 31 38	2690		CLUSTER OF GALAXIES
PHL 2437	23 51 12.	- 27 50		18.5	BLUE STELLAR OBJECT
PHL 6037	23 51 12.	- 27 54		16.5	BLUE STELLAR OBJECT
RNGC 7781	23 51 13.	+ 07 35		15.0	GALAXY
MCG+01-60-047	23 51 15.	+ 07 34	42	15.5	GALAXY
SC 2348-0132.3	23 51 16.	- 01 15 37.	36		NEBULA
MCG+01-60-048	23 51 18.	+ 07 41	120	13.	GALAXY
ZWG 407.073	23 51 18.	+ 07 42		13.2	GALAXY
UGC 12834	23 51 18.	+ 07 42	126	13.2	GALAXY Sb
ZWG 381.059	23 51 18.	- 01 15		15.2	GALAXY
SC 2348-7106.0	23 51 18.	- 70 49 19.	36		NEBULA
RNGC 7782	23 51 19.	+ 07 42		13.0	GALAXY

OBJECT NAME	RIGHT ASCEN.	DECLINATION	DIAM.	MAGN.	TYPE OF OBJECT
SG 3.245	23 51 21.	+ 67 00	1200		DIFFUSE EMISSION NEBULA
MCG+05-56-021	23 51 24.	+ 28 12	42	15.	GALAXY
ZWG 499.001	23 51 24.	+ 28 13		14.4	GALAXY
ZWG 498.029	23 51 24.	+ 28 13		14.4	GALAXY
UGC 12835	23 51 24.	+ 28 13	72	14.4	GALAXY (E)
UGC 12836	23 51 24.	+ 51 53	72	16.5	GALAXY Sc
MCG+00-60-057	23 51 24.	- 01 14	36	15.	GALAXY
PHL 6038	23 51 24.	- 16 28		17.8	BLUE STELLAR OBJECT
PHL 6040	23 51 24.	- 26 51		18.5	BLUE STELLAR OBJECT
PHL 6039	23 51 24.	- 28 42		18.1	BLUE STELLAR OBJECT
HN 2879	23 51 29.9	- 43 35 07.	18		NEBULA
MCG+08-01-006	23 51 30.	+ 47 12	60	14.	GALAXY
5ZW 437	23 51 30.	+ 47 13			COMPACT GALAXY
ZWG 549.005	23 51 30.	+ 47 13		15.5	GALAXY
ZWG 548.014	23 51 30.	+ 47 13		15.5	GALAXY
MRSL 115-01/1	23 51 30.	+ 60 06	300		HII REGION
PHL 2438	23 51 30.	- 21 24		18.5	BLUE STELLAR OBJECT
PHL 6041	23 51 30.	- 22 26		18.6	BLUE STELLAR OBJECT
PHL 6042	23 51 30.	- 23 29		18.4	BLUE STELLAR OBJECT
SC 2348-0132.8	23 51 31.	- 01 16 07.	12		NEBULA
SC 2348-0305.6	23 51 33.	- 02 48 55.	12		NEBULA
ARP 323	23 51 35.	+ 00 06			PECULIAR GALAXY
ABC 2670	23 51 35.	- 10 41		15.7	RICH CLUSTER OF GALAXIES
BC PKS2351-006	23 51 35.36	- 00 36 28.9		19.	QUASI-STELLAR OBJECT
SHB 398	23 51 35.4	- 00 36 28.		19.	QUASI-STELLAR OBJECT
ZWG 381.060	23 51 36.	+ 00 06		14.1	GALAXY
UGC 12837	23 51 36.	+ 00 06	96	14.1	GALAXY DBL SYS
KARA.72 595A	23 51 36.	+ 00 06	60	14.1	PART OF DOUBLE GALAXY
OCL 0273	23 51 36.	+ 61 29	300	9.0	OPEN STAR CLUSTER
PHL 0581	23 51 36.	- 16 20		18.3	BLUE STELLAR OBJECT
PHL 2439	23 51 36.	- 22 04		17.7	BLUE STELLAR OBJECT
PHL 6043	23 51 36.	- 24 56		18.6	BLUE STELLAR OBJECT
MCG-04-01-001	23 51 36.	- 25 41 30.	24	14.5	GALAXY
RNGC 7783B	23 51 37.	+ 00 08		15.0	GALAXY
RNGC 7783A	23 51 37.	+ 00 08		14.0	GALAXY
RNGC 7783C	23 51 37.	+ 00 09		16.0	GALAXY
SCHO 1445	23 51 37.	+ 55 53 48.	240		ISOLATED DARK CLOUD
SC 2349-0317.6	23 51 37.	- 03 00 55.	12		NEBULA
MCG-02-60-022	23 51 39.	- 13 52	48	15.	GALAXY
SCHO 1446	23 51 40.	+ 58 06 00.	530		ISOLATED DARK CLOUD
KARA.72 595B	23 51 42.	+ 00 06	42		PART OF DOUBLE GALAXY
MCG+00-60-059	23 51 42.	+ 00 08	18	15.	GALAXY
MCG+00-60-058	23 51 42.	+ 00 08	42	14.	GALAXY
MCG+00-60-060	23 51 42.	+ 00 09	12	16.	GALAXY
ZC 2351.7+3821	23 51 42.	+ 38 21	940		CLUSTER OF GALAXIES
SC 2349-0309.2	23 51 42.	- 02 52 31.	12		NEBULA
IC 1514	23 51 42.	- 13 52			NONSTELLAR OBJECT
ZWG 499.002	23 51 48.	+ 30 01		15.3	GALAXY
ZWG 498.030	23 51 48.	+ 30 01		15.3	GALAXY
MCG-03-60-023	23 51 48.	- 17 34 30.	24	15.	GALAXY
TON-S 0128	23 51 48.	- 23 59		15.2	BLUE STAR
MCG-05-01-001	23 51 48.	- 27 47	48	16.	GALAXY
SC 2349-0309.4	23 51 49.	- 02 52 43.	18		NEBULA
SC 2349-0430.8	23 51 53.	- 04 14 07.	12		NEBULA
PHL 2441	23 51 54.	+ 06 50		16.8	BLUE STELLAR OBJECT
ZC 2351.9+2434	23 51 54.	+ 24 34	610		CLUSTER OF GALAXIES
4ZW 169	23 51 54.	+ 26 20			COMPACT GALAXY
5ZW 438	23 51 54.	+ 40 53			COMPACT GALAXY
SC 2349-0229.4	23 51 54.	- 02 12 43.	36		NEBULA
ZWG 381.061	23 51 54.	- 02 13		14.9	GALAXY
UGC 12838	23 51 54.	- 02 13	60	14.9	GALAXY Sc
PHL 2440	23 51 54.	- 29 18		17.9	BLUE STELLAR OBJECT
SC 2349-0431.6	23 51 57.	- 04 14 55.	6		NEBULA
4ZW 170	23 52 00.	+ 21 32			COMPACT GALAXY
MCG+05-01-001	23 52 00.	+ 28 00 30.	45	15.	GALAXY
ZWG 499.003	23 52 00.	+ 28 02		14.7	GALAXY
ZWG 498.031	23 52 00.	+ 28 02		14.7	GALAXY
UGC 12839	23 52 00.	+ 28 05	90	14.7	GALAXY S0
MCG+05-01-002	23 52 00.	+ 28 34	72	15.	GALAXY
ZWG 499.004	23 52 00.	+ 28 36		14.3	GALAXY
ZWG 498.032	23 52 00.	+ 28 36		14.3	GALAXY
UGC 12840	23 52 00.	+ 28 36	72	14.3	GALAXY SB0
KARA.73B 1044	23 52 00.	+ 28 36	66	14.3	ISOLATED GALAXY S
ZC 2352.0-0016	23 52 00.	- 00 16	2150		CLUSTER OF GALAXIES
MCG-05-01-002	23 52 00.	- 27 50	12	15.5	GALAXY
ZWG 407.074	23 52 06.	+ 09 00		15.7	GALAXY
ZC 2352.1+1612	23 52 06.	+ 16 12	540		CLUSTER OF GALAXIES
ZC 2352.1+4718	23 52 06.	+ 47 18	11560		CLUSTER OF GALAXIES
MCG+00-60-061	23 52 09.	- 02 12	66	13.5	GALAXY
HN 2880	23 52 10.1	- 45 14 19.	12		NEBULA
PHL 6046	23 52 12.	+ 08 08		18.5	BLUE STELLAR OBJECT
ZWG 516.001	23 52 12.	+ 34 04		15.6	GALAXY
ZWG 517.006	23 52 12.	+ 34 04		15.6	GALAXY
ZC 2352.2+4016	23 52 12.	+ 40 16	3830		CLUSTER OF GALAXIES
PHL 6044	23 52 12.	- 13 00		16.7	BLUE STELLAR OBJECT
MCG-03-60-024	23 52 12.	- 18 02	48	15.5	GALAXY
PHL 6047	23 52 12.	- 24 25		18.3	BLUE STELLAR OBJECT
PHL 6048	23 52 12.	- 26 54		18.2	BLUE STELLAR OBJECT
PHL 6045	23 52 12.	- 29 56		18.6	BLUE STELLAR OBJECT
KEEL 733	23 52 12.6	+ 08 06 08.			NEBULA
SC 2349-0053.8	23 52 13.	- 00 37 07.	12		NEBULA
MCG-04-01-002	23 52 15.	- 25 59	72	16.	GALAXY
SCHO 1447	23 52 18.	+ 55 41 36.	240		ISOLATED DARK CLOUD
PHL 2442	23 52 18.	- 04 22		18.1	BLUE STELLAR OBJECT
PHL 0582	23 52 18.	- 13 18		18.3	BLUE STELLAR OBJECT
PHL 6049	23 52 18.	- 14 28		16.7	BLUE STELLAR OBJECT
KN 13.002	23 52 21.1	+ 13 01 26.			NEBULA
ZC 2352.4+0140	23 52 24.	+ 01 40	670		CLUSTER OF GALAXIES
ZC 2352.4+0508	23 52 24.	+ 05 08	870		CLUSTER OF GALAXIES
PHL 0583	23 52 24.	+ 06 16		17.1	BLUE STELLAR OBJECT
PHL 6050	23 52 24.	+ 07 54		18.0	BLUE STELLAR OBJECT
PHL 6053	23 52 24.	+ 09 13		18.0	BLUE STELLAR OBJECT
ZWG 499.005	23 52 24.	+ 29 56		15.6	GALAXY
ZWG 498.033	23 52 24.	+ 29 56		15.6	GALAXY
SC 2349-0151.0	23 52 24.	- 01 34 19.	6		NEBULA
PHL 6051	23 52 24.	- 06 21		18.2	BLUE STELLAR OBJECT
PHL 2443	23 52 24.	- 07 12		18.4	BLUE STELLAR OBJECT
PHL 6052	23 52 24.	- 11 42		18.1	BLUE STELLAR OBJECT
PHL 6054	23 52 24.	- 14 54		18.3	BLUE STELLAR OBJECT
MCG-06-01-005	23 52 24.	- 33 10	24	15.5	GALAXY
MCG-06-01-004	23 52 24.	- 34 52	36	15.5	GALAXY
SC 2349-0136.8	23 52 25.	- 01 20 07.	18		NEBULA
ABC 2671	23 52 27.	+ 05 09		16.8	RICH CLUSTER OF GALAXIES
HN 2881	23 52 28.3	- 44 18 07.	12		NEBULA
SC 2349-0418.2	23 52 29.	- 04 01 31.	12		NEBULA
SC 2349-0314.3	23 52 30.	- 02 57 37.	6		NEBULA
PHL 6055	23 52 30.	- 05 22		18.1	BLUE STELLAR OBJECT
MCG-04-01-003	23 52 33.	- 21 10	36	15.	GALAXY

OBJECT NAME	RIGHT ASCEN.	DECLINATION	DIAM.	MAGN.	TYPE OF OBJECT
PHL 2446	23 52 36.	+ 06 48		6.1	BLUE STELLAR OBJECT
PHL 2444	23 52 36.	+ 07 57		18.4	BLUE STELLAR OBJECT
MCG-01-01-001	23 52 36.	- 03 59	30	15.	GALAXY
PHL 2445	23 52 36.	- 04 14		18.1	BLUE STELLAR OBJECT
PHL 6056	23 52 36.	- 07 09		18.0	BLUE STELLAR OBJECT
PHL 0584	23 52 36.	- 12 50		17.6	BLUE STELLAR OBJECT
PHL 6057	23 52 36.	- 15 26		16.6	BLUE STELLAR OBJECT
PHL 2447	23 52 36.	- 21 44		18.8	BLUE STELLAR OBJECT
MCG-05-01-003	23 52 36.	- 31 52	36	15.	GALAXY
ABC 2672	23 52 37.	+ 26 11		17.1	RICH CLUSTER OF GALAXIES
SCHO 1448	23 52 38.	+ 58 41 48.	490		ISOLATED DARK CLOUD
MCG+06-01-001	23 52 39.	+ 35 28	36	15.	GALAXY
ABC 2673	23 52 40.	+ 01 40		18.0	RICH CLUSTER OF GALAXIES
SC 2350-0418.4	23 52 41.	- 04 01 43.	18		NEBULA
SC 2350-0452.1	23 52 41.	- 04 35 25.			NEBULA
MCG+02-60-025	23 52 42.	+ 14 07 30.	60	15.	GALAXY
ZWG 478.001	23 52 42.	+ 21 29		15.5	GALAXY
ZWG 477.029	23 52 42.	+ 21 29		15.5	GALAXY
ZC 2352.7+2607	23 52 42.	+ 26 07	1210		CLUSTER OF GALAXIES
ZWG 534.001	23 52 42.	+ 43 37		15.6	GALAXY
5ZW 439	23 52 42.	+ 43 39			COMPACT GALAXY
ZWG 381.062	23 52 42.	- 01 42		15.7	GALAXY
PHL 6059	23 52 42.	- 05 14		18.2	BLUE STELLAR OBJECT
PHL 2448	23 52 42.	- 05 58		18.2	BLUE STELLAR OBJECT
PHL 2450	23 52 42.	- 08 24		18.6	BLUE STELLAR OBJECT
PHL 6058	23 52 42.	- 12 24		17.6	BLUE STELLAR OBJECT
PHL 2449	23 52 42.	- 32 12		5.9	BLUE STELLAR OBJECT
SC 2350-0223.4	23 52 43.	- 02 16 43.	12		NEBULA
KN 13.003	23 52 43.1	+ 14 05 57.			NEBULA
RNGC 7784	23 52 44.	+ 21 29		15.5	GALAXY
MCG+01-60-049	23 52 45.	+ 05 40	48	13.	GALAXY
MCG+04-01-001	23 52 45.	+ 21 31	30	15.	GALAXY
MCG+05-01-003	23 52 45.	+ 30 07	48	15.5	GALAXY
MCG-01-01-002	23 52 45.	- 03 59	66	14.	GALAXY
MCG-04-01-004	23 52 45.	- 24 09	102	14.	GALAXY
ZWG 407.075	23 52 46.	+ 05 38		13.0	GALAXY
UGC 12841	23 52 46.	+ 05 38	138	13.0	GALAXY E
KARA.73B 1045	23 52 48.	+ 05 38	126	13.0	ISOLATED GALAXY E
PHL 0585	23 52 48.	+ 07 16		18.4	BLUE STELLAR OBJECT
PHL 6061	23 52 48.	+ 07 56		18.2	BLUE STELLAR OBJECT
ZWG 433.001	23 52 48.	+ 14 06		15.5	GALAXY
ZWG 432.040	23 52 48.	+ 14 06		15.5	GALAXY
ZWG 456.001	23 52 48.	+ 21 19		13.9	GALAXY
ZWG 455.065	23 52 48.	+ 21 19		13.9	GALAXY
UGC 12842	23 52 48.	+ 21 19	36	13.9	GALAXY
MCG+03-60-038	23 52 48.	+ 21 20	30	14.	GALAXY
ZWG 499.006	23 52 48.	+ 30 07		15.5	GALAXY
ZWG 498.034	23 52 48.	+ 30 07		15.5	GALAXY
KARA.72 596A	23 52 48.	+ 30 07	48	15.5	PART OF DOUBLE GALAXY
MCG+05-01-004	23 52 48.	+ 30 07 30.	30	15.5	GALAXY
ZWG 499.007	23 52 48.	+ 30 08		15.6	GALAXY
ZWG 498.035	23 52 48.	+ 30 08		15.6	GALAXY
KARA.72 596B	23 52 48.	+ 30 08	48	15.6	PART OF DOUBLE GALAXY
PHL 2451	23 52 48.	- 08 30		18.0	BLUE STELLAR OBJECT
PHL 6060	23 52 48.	- 09 40		15.8	BLUE STELLAR OBJECT
PHL 0586	23 52 48.	- 17 32		17.9	BLUE STELLAR OBJECT
PHL 2452	23 52 48.	- 21 20		18.3	BLUE STELLAR OBJECT
RNGC 7785	23 52 49.	+ 05 38		13.0	GALAXY
RNGC 7786	23 52 50.	+ 21 19		14.0	GALAXY
MCG-01-01-003	23 52 51.	- 06 30 30.	30	15.	GALAXY
SC 2350-0357.8	23 52 53.	- 03 41 06.	6		NEBULA
BC PKS2352-455	23 52 53.2	- 45 30 08.		17.5	QUASI-STELLAR OBJECT
SHB 399	23 52 53.2	- 45 30 08.		17.5	QUASI-STELLAR OBJECT
ZC 2352.9+0128	23 52 54.	+ 01 28	.240		CLUSTER OF GALAXIES
MCG+03-60-039	23 52 54.	+ 17 40	156	13.5	GALAXY
ZWG 499.008	23 52 54.	+ 29 59		15.7	GALAXY
ZWG 498.036	23 52 54.	+ 29 59		15.7	GALAXY
MCG+08-01-007	23 52 54.	+ 46 59	60	16.	GALAXY
ZWG 549.006	23 52 54.	+ 47 00		15.3	GALAXY
ZWG 548.015	23 52 54.	+ 47 00		15.3	GALAXY
TON-S 0129	23 52 54.	- 24 42		15.1	BLUE STAR
MCG-01-01-004	23 52 57.	- 03 59	54	15.	GALAXY
ABC 2674	23 52 58.	+ 01 29		17.9	RICH CLUSTER OF GALAXIES
KHAV 793	23 53	+ 58 17	2470		DARK NEBULA
PHL 2453	23 53 00.	+ 04 32		18.5	BLUE STELLAR OBJECT
ZC 2353.0+1535	23 53 00.	+ 15 35	1410		CLUSTER OF GALAXIES
ZWG 456.002	23 53 00.	+ 17 38		14.4	GALAXY
ZWG 455.066	23 53 00.	+ 17 38		14.4	GALAXY
UGC 12843	23 53 00.	+ 17 38	180	14.4	GALAXY
LDN 1256	23 53 00.	+ 59 45	1620		DARK NEBULA
PHL 6062	23 53 00.	- 03 42		18.4	BLUE STELLAR OBJECT
MCG-02-01-001	23 53 00.	- 09 55	30	15.5	GALAXY
PHL 0587	23 53 00.	- 11 43		16.6	BLUE STELLAR OBJECT
PHL 6063	23 53 00.	- 14 04		18.1	BLUE STELLAR OBJECT
SNO 31	23 53 00.	- 25 29 40.	420	16.	LINEAR GRP OF 3 GALAXIES
PHL 6064	23 53 00.	- 27 36		18.1	BLUE STELLAR OBJECT
ABC 2675	23 53 03.	+ 11 10		16.4	RICH CLUSTER OF GALAXIES
MCG+03-01-001	23 53 03.	+ 19 15	66	15.	GALAXY
ZWG 456.003	23 53 06.	+ 19 15		15.4	GALAXY
ZWG 455.067	23 53 06.	+ 19 15		15.4	GALAXY
UGC 12844	23 53 06.	+ 19 15	96	15.4	GALAXY Sc
KARA.73B 1046	23 53 06.	+ 19 15	66	15.4	ISOLATED GALAXY S
MCG+05-01-005	23 53 06.	+ 28 57	30	16.	GALAXY
ZWG 499.009	23 53 06.	+ 28 59		15.5	GALAXY
ZWG 498.037	23 53 06.	+ 28 59		15.5	GALAXY
MCG+05-01-006	23 53 06.	+ 31 37 30.	108	15.	GALAXY
PHL 2454	23 53 06.	- 17 06		17.4	BLUE STELLAR OBJECT
MCG-04-01-005	23 53 06.	- 21 08	36	15.5	GALAXY
ABC 2676	23 53 09.	+ 05 47		16.8	RICH CLUSTER OF GALAXIES
ZC 2353.2+0535	23 53 12.	+ 05 35	3630		CLUSTER OF GALAXIES
ZWG 499.010	23 53 12.	+ 29 56		15.2	GALAXY
ZWG 498.038	23 53 12.	+ 29 56		15.2	GALAXY
ZWG 499.011	23 53 12.	+ 31 38		15.4	GALAXY
ZWG 498.039	23 53 12.	+ 31 38		15.4	GALAXY
UGC 12845	23 53 12.	+ 31 38	144	15.4	GALAXY Sc
PHL 2455	23 53 12.	- 05 00		18.4	BLUE STELLAR OBJECT
PHL 6065	23 53 12.	- 05 24		17.7	BLUE STELLAR OBJECT
SC 2350-0158.8	23 53 13.	- 01 42 06.	6		NEBULA
MCG+03-01-002	23 53 15.	+ 18 10	33	15.	GALAXY
SC 2350-0122.2	23 53 17.	- 01 05 30.	12		NEBULA
MCG+00-01-001	23 53 18.	+ 00 19	15	14.	GALAXY
PHL 6067	23 53 18.	+ 04 32		18.3	BLUE STELLAR OBJECT
UGC 12846	23 53 18.	+ 18 10	120	17.	GALAXY DWRF SP
ZWG 478.002	23 53 18.	+ 25 14		15.0	GALAXY
ZWG 477.030	23 53 18.	+ 25 14		15.0	GALAXY
4ZW 171	23 53 18.	+ 26 16			COMPACT GALAXY
ZWG 478.003	23 53 18.	+ 26 35		15.6	GALAXY
ZWG 477.031	23 53 18.	+ 26 35		15.6	GALAXY

OBJECT NAME	RIGHT ASCEN.	DECLINATION	DIAM.	MAGN.	TYPE OF OBJECT
4ZW 172	23 53 18.	+ 36 57			COMPACT GALAXY
PHL 6066	23 53 18.	- 07 44		17.2	BLUE STELLAR OBJECT
MCG-02-01-002	23 53 18.	- 12 41	42	15.	GALAXY
MCG-02-01-003	23 53 18.	- 13 00	42	15.	GALAXY
MCG-05-01-004	23 53 18.	- 30 09	9	16.	GALAXY
ARC 2677	23 53 19.	+ 34 06		17.5	RICH CLUSTER OF GALAXIES
KN 13.004	23 53 20.2	+ 11 54 35.			NEBULA
BC 4C28.59	23 53 20.9	+ 28 19 25.		17.8	QUASI-STELLAR OBJECT
ARC 2678	23 53 21.	+ 11 24		16.9	RICH CLUSTER OF GALAXIES
SHB 400	23 53 21.	+ 28 19 25.		18.	QUASI-STELLAR OBJECT
SC 2350-0114.6	23 53 23.	- 00 57 54.	12		NEBULA
ZWG 382.001	23 53 24.	+ 00 17		15.7	GALAXY
UGC 12847	23 53 24.	+ 00 17	60	15.7	GALAXY VY CMPT
PHL 6072	23 53 24.	+ 05 01		18.3	BLUE STELLAR OBJECT
PHL 6068	23 53 24.	+ 08 18		16.1	BLUE STELLAR OBJECT
PHL 6071	23 53 24.	+ 08 33		16.8	BLUE STELLAR OBJECT
ZC 2353.4+1238	23 53 24.	+ 12 38	3090		CLUSTER OF GALAXIES
MCG+04-01-004	23 53 24.	+ 25 11 30.	24	16.	GALAXY
MCG+04-01-003	23 53 24.	+ 25 12 30.	36	16.	GALAXY
MCG+04-01-002	23 53 24.	+ 25 12 30.	21	16.	GALAXY
MCG+00-01-002	23 53 24.	- 00 27	36	15.5	GALAXY
MCG+00-01-003	23 53 24.	- 01 44	36	15.	GALAXY
PHL 6069	23 53 24.	- 06 22		18.4	BLUE STELLAR OBJECT
PHL 6073	23 53 24.	- 09 10		18.2	BLUE STELLAR OBJECT
PHL 2456	23 53 24.	- 15 24		18.7	BLUE STELLAR OBJECT
PHL 6070	23 53 24.	- 18 57		18.0	BLUE STELLAR OBJECT
MCG-04-01-006	23 53 24.	- 20 33	36	15.5	GALAXY
IC 1515	23 53 28.	- 01 15 57.			NONSTELLAR OBJECT
BC PKS2353-68	23 53 28.3	- 68 35 24.		17.	QUASI-STELLAR OBJECT
SHB 401	23 53 28.3	- 68 35 24.		17.	QUASI-STELLAR OBJECT
HN 2882	23 53 29.9	- 44 12 54.	12		NEBULA
SC 2350-0202.7	23 53 29.	- 01 46 00.	18		NEBULA
ZWG 408.001	23 53 30.	+ 07 15		15.5	GALAXY
MRK 541	23 53 30.	+ 07 15	10	15.5	GALAXY WITH UV CONTINUUM
IC 1516	23 53 30.	- 01 11 15.			NONSTELLAR OBJECT
MCG+00-01-004	23 53 30.	- 01 15	54	13.5	GALAXY
MCG+05-01-007	23 53 33.	+ 29 05 30.	48	15.5	GALAXY
MCG+08-01-008	23 53 33.	+ 49 12	72	15.	GALAXY
KN 13.005	23 53 33.0	+ 14 12 17.			NEBULA
HN 2883	23 53 33.7	- 47 53 48.	18		NEBULA
IC 5364	23 53 35.	- 29 21 09.			NONSTELLAR OBJECT
ZWG 382.003	23 53 36.	+ 00 17		15.7	GALAXY
UGC 12849	23 53 36.	+ 00 17	120	15.7	GALAXY S
MCG+00-01-005	23 53 36.	+ 00 18	36	13.5	GALAXY
ZC 2353.6+2224	23 53 36.	+ 22 24	1410		CLUSTER OF GALAXIES
ZWG 499.012	23 53 36.	+ 29 07		15.6	GALAXY
ZWG 498.040	23 53 36.	+ 29 07		15.6	GALAXY
UGC 12850	23 53 36.	+ 29 07	66	15.6	GALAXY S
ZC 2353.6+3410	23 53 36.	+ 34 10	1950		CLUSTER OF GALAXIES
MCG+08-01-009	23 53 36.	+ 45 27	48	16.	GALAXY
5ZW 440	23 53 36.	+ 49 13			COMPACT GALAXY
ZWG 549.007	23 53 36.	+ 49 13		15.5	GALAXY
ZWG 548.016	23 53 36.	+ 49 13		15.5	GALAXY
UGC 12851	23 53 36.	+ 49 13	72	15.5	GALAXY SBb
MCG+00-01-006	23 53 36.	- 01 11	90	13.5	GALAXY
ZWG 382.002	23 53 36.	- 01 16		14.8	GALAXY
UGC 12848	23 53 36.	- 01 16	72	14.8	GALAXY SBb
KARA.72 597A	23 53 36.	- 01 16	84	14.8	PART OF DOUBLE GALAXY
PHL 6074	23 53 36.	- 04 02		18.8	BLUE STELLAR OBJECT
PHL 6075	23 53 36.	- 17 10		18.2	BLUE STELLAR OBJECT
MCG-05-01-005	23 53 36.	- 27 44	60	15.	GALAXY
RNGC 7787	23 53 37.	+ 00 17		15.5	GALAXY
HN 2884	23 53 39.7	- 43 42 12.	60		NEBULA
SC 2351-0106.3	23 53 41.	- 00 49 36.	18		NEBULA
PHL 2458	23 53 42.	+ 04 02		18.4	BLUE STELLAR OBJECT
PHL 2457	23 53 42.	+ 07 46		16.6	BLUE STELLAR OBJECT
ZC 2353.7+2044	23 53 42.	+ 20 44	2150		CLUSTER OF GALAXIES
4ZW 173	23 53 42.	+ 24 43			COMPACT GALAXY
MCG+05-01-008	23 53 42.	+ 29 07	30	15.5	GALAXY
ZWG 499.013	23 53 42.	+ 29 09		15.4	GALAXY
ZWG 498.041	23 53 42.	+ 29 09		15.4	GALAXY
MCG+00-01-007	23 53 42.	- 00 47 30.	42	14.	GALAXY
ZWG 382.005	23 53 42.	- 00 48		15.4	GALAXY
ZWG 382.004	23 53 42.	- 01 12		14.3	GALAXY
UGC 12852	23 53 42.	- 01 12	78	14.3	GALAXY Sb-c
KARA.72 597B	23 53 42.	- 01 12	108	14.3	PART OF DOUBLE GALAXY
PHL 2459	23 53 42.	- 02 17		18.3	BLUE STELLAR OBJECT
PHL 6076	23 53 42.	- 02 26		17.2	BLUE STELLAR OBJECT
PHL 6077	23 53 42.	- 08 08		18.3	BLUE STELLAR OBJECT
PHL 2460	23 53 42.	- 22 10		18.5	BLUE STELLAR OBJECT
LB 01537	23 53 42.	- 59 18		14.0	FAINT BLUE STAR
IC 1517	23 53 43.	- 00 35 09.			NONSTELLAR OBJECT
MCG+05-01-009	23 53 45.	+ 29 06	30	17.	GALAXY
MCG+00-01-008	23 53 45.	- 00 33	36	14.	GALAXY
SCHO 1449	23 53 46.	+ 57 50 54.	400		ISOLATED DARK CLOUD
ZC 2353.8+1120	23 53 48.	+ 11 20	3760		CLUSTER OF GALAXIES
OCL 0278	23 53 48.	+ 64 16	180		OPEN STAR CLUSTER
ZWG 382.006	23 53 48.	- 00 34		14.8	GALAXY
PHL 2461	23 53 48.	- 10 22		18.6	BLUE STELLAR OBJECT
MCG-02-01-005	23 53 48.	- 11 22	96	15.	GALAXY
MCG-02-01-004	23 53 48.	- 11 23	18	15.	GALAXY
PHL 0588	23 53 48.	- 12 01		18.0	BLUE STELLAR OBJECT
PHL 6078	23 53 48.	- 18 48		18.0	BLUE STELLAR OBJECT
MCG-05-01-008	23 53 48.	- 29 18	36	14.5	GALAXY
MCG-05-01-007	23 53 48.	- 29 18	18	15.	GALAXY
MCG-05-01-009	23 53 48.	- 29 25	30	15.	GALAXY
MCG-05-01-006	23 53 48.	- 31 39	48	14.5	GALAXY
PHL 0589	23 53 54.	+ 03 24		18.3	BLUE STELLAR OBJECT
SC 2351-0124.6	23 53 54.	- 01 07 54.	12		NEBULA
PHL 6079	23 53 54.	- 12 09		16.5	BLUE STELLAR OBJECT
MCG-03-01-001	23 53 54.	- 20 24	36	15.	GALAXY
ARC 2679	23 53 54.	- 20 29		17.7	RICH CLUSTER OF GALAXIES
ARC 2680	23 53 54.	- 21 19		17.7	RICH CLUSTER OF GALAXIES
MCG-06-01-006	23 53 54.	- 34 52	18	16.	GALAXY
SC 2351-0339.6	23 53 56.	- 03 22 54.	18		NEBULA
LBN 0559	23 54	+ 49 20	720		BRIGHT NEBULA
KHAV 794	23 54	+ 58 11			DARK NEBULA
KHAV 795	23 54	+ 59 23	2110		DARK NEBULA
KHAV 796	23 54	+ 61 35	4490		DARK NEBULA
PHL 2463	23 54 00.	+ 04 22		18.7	BLUE STELLAR OBJECT
PHL 2464	23 54 00.	+ 05 21		18.7	BLUE STELLAR OBJECT
PHL 0590	23 54 00.	+ 08 16		18.1	BLUE STELLAR OBJECT
ZWG 478.004	23 54 00.	+ 26 55		15.6	GALAXY
ZWG 477.032	23 54 00.	+ 26 55		15.6	GALAXY
UGC 12853	23 54 00.	+ 45 08	72	16.0	GALAXY S
LDN 1253	23 54 00.	+ 58 15	360		DARK NEBULA
LDN 1254	23 54 00.	+ 58 30	1560		DARK NEBULA
ZC 2354.0-0143	23 54 00.	- 01 43	4500		CLUSTER OF GALAXIES
PHL 2462	23 54 00.	- 04 21		17.5	BLUE STELLAR OBJECT
PHL 6080	23 54 00.	- 18 06		18.5	BLUE STELLAR OBJECT
HOLM 822B	23 54 02.	- 02 57	48	14.8	PART OF MULTIPLE GALAXY
ZWG 433.002	23 54 06.	+ 13 30		14.3	GALAXY
UGC 12854	23 54 06.	+ 13 30	54	14.3	GALAXY S
MCG+02-01-001	23 54 06.	+ 13 30	48	15.	GALAXY
MCG+04-01-005	23 54 06.	+ 26 49	66	15.	GALAXY
ZWG 478.005	23 54 06.	+ 26 50		14.7	GALAXY
ZWG 477.033	23 54 06.	+ 26 50		14.7	GALAXY
UGC 12855	23 54 06.	+ 26 50	84	14.7	GALAXY S0/Sa
ZC 2354.1+4532	23 54 06.	+ 45 32	1010		CLUSTER OF GALAXIES
5ZW 441	23 54 06.	+ 46 33			COMPACT GALAXY
PHL 0591	23 54 06.	- 21 02		18.4	BLUE STELLAR OBJECT
PHL 0592	23 54 06.	- 30 22		16.9	BLUE STELLAR OBJECT
PK118+08.1	23 54 07.	+ 70 31 38.			PLANETARY NEBULA
SC 2351-6840.1	23 54 07.	- 68 23 24.	6		NEBULA
KN 13.006	23 54 08.8	+ 13 29 47.			NEBULA
HOLM 822A	23 54 10.	- 02 55	36	14.4	PART OF MULTIPLE GALAXY
KN 13.007	23 54 11.3	+ 13 37 55.			NEBULA
KN 13.008	23 54 11.8	+ 16 32 13.			NEBULA
PHL 2467	23 54 12.	+ 07 10		18.0	BLUE STELLAR OBJECT
PHL 2466	23 54 12.	+ 08 47		18.6	BLUE STELLAR OBJECT
ZWG 433.003	23 54 12.	+ 13 35		15.6	GALAXY
KARA.72 598A	23 54 12.	+ 16 32	42	14.5	PART OF DOUBLE GALAXY
MCG+03-01-004	23 54 12.	+ 16 32	66	15.	GALAXY
MCG+03-01-003	23 54 12.	+ 16 32	30	15.	GALAXY
ZWG 456.004	23 54 12.	+ 16 33		14.5	GALAXY
UGC 12856	23 54 12.	+ 16 33	156	14.5	GALAXY IRR
KARA.72 598B	23 54 12.	+ 16 33	84		PART OF DOUBLE GALAXY
ZWG 499.014	23 54 12.	+ 29 37		15.7	GALAXY
ZWG 498.042	23 54 12.	+ 29 37		15.7	GALAXY
OCL 0275	23 54 12.	+ 61 07	1200	9.5	OPEN STAR CLUSTER
SC 2351-0312.5	23 54 12.	- 02 55 48.	18		NEBULA
PHL 6081	23 54 12.	- 05 31		18.6	BLUE STELLAR OBJECT
PHL 6084	23 54 12.	- 06 48		18.5	BLUE STELLAR OBJECT
PHL 6082	23 54 12.	- 08 42		18.1	BLUE STELLAR OBJECT
PHL 2465	23 54 12.	- 10 51		18.2	BLUE STELLAR OBJECT
PHL 6083	23 54 12.	- 12 16		18.1	BLUE STELLAR OBJECT
PHL 2468	23 54 12.	- 22 09		18.5	BLUE STELLAR OBJECT
SC 2351-7256.5	23 54 12.	- 72 39 48.	30		NEBULA
KN 13.009	23 54 12.2	+ 13 35 36.			NEBULA
VV 255A	23 54 15.	+ 16 32	24	16.	INTERACTING GALAXY
VV 255B	23 54 15.	+ 16 33	36	15.	INTERACTING GALAXY
RNGC 7788	23 54 15.	+ 61 07		9.5	OPEN CLUSTER
HOLM 823A	23 54 16.	- 03 37	60	13.5	PART OF MULTIPLE GALAXY
HOLM 823B	23 54 16.	- 03 39	30	14.0	PART OF MULTIPLE GALAXY
ARP 262	23 54 17.	+ 16 32			PECULIAR GALAXY
SC 2351-0354.4	23 54 17.	- 03 37 42.	42		NEBULA
ZWG 382.007	23 54 18.	+ 01 05		14.7	GALAXY
UGC 12857	23 54 18.	+ 01 05	114	14.7	GALAXY S
KARA.73B 1047	23 54 18.	+ 01 05	120	14.7	ISOLATED GALAXY S
MCG+00-01-009	23 54 18.	+ 01 07	114	13.	GALAXY
ZWG 534.002	23 54 18.	+ 44 21		15.6	GALAXY
UGC 12858	23 54 18.	+ 44 21	66	15.6	GALAXY Sc
MCG+07-01-001	23 54 18.	+ 44 21	48	15.5	GALAXY
ISS 0107	23 54 18.	+ 63 20	271		STELLAR RING
MCG-01-01-005	23 54 18.	- 02 53	48	14.5	GALAXY
8ZW 2354-04.2	23 54 18.	- 04 13		17.1	COMPACT GALAXY
PHL 6085	23 54 18.	- 04 15		18.2	BLUE STELLAR OBJECT
PHL 2469	23 54 18.	- 09 00		18.3	BLUE STELLAR OBJECT
MCG-03-01-002	23 54 18.	- 16 46	18	14.5	GALAXY
MCG+01-01-001	23 54 21.	+ 05 12 30.	60	14.	GALAXY
MCG-01-01-006	23 54 21.	- 03 35	60	14.	GALAXY
MCG-01-01-007	23 54 21.	- 03 37	60	14.	GALAXY
SC 2351-0356.7	23 54 21.	- 03 40 00.	42		NEBULA
PHL 2471	23 54 24.	+ 05 10		17.3	BLUE STELLAR OBJECT
ZWG 408.002	23 54 24.	+ 05 14		15.5	GALAXY
UGC 12859	23 54 24.	+ 05 15	84	15.5	GALAXY Sb-c
PHL 6088	23 54 24.	+ 06 10		18.4	BLUE STELLAR OBJECT
PHL 6089	23 54 24.	+ 06 18		18.2	BLUE STELLAR OBJECT
PHL 6087	23 54 24.	+ 07 44		18.6	BLUE STELLAR OBJECT
ZWG 433.004	23 54 24.	+ 10 33		15.2	GALAXY
UGC 12860	23 54 24.	+ 10 33	96	15.2	GALAXY Sb
5ZW 442	23 54 24.	+ 47 41			COMPACT GALAXY
PHL 2470	23 54 24.	- 06 54		18.2	BLUE STELLAR OBJECT
PHL 6086	23 54 24.	- 27 20		16.2	BLUE STELLAR OBJECT
MCG-05-01-010	23 54 24.	- 32 27	36	15.	GALAXY
MCG-06-01-007	23 54 24.	- 34 56	42	15.	GALAXY
MCG-06-01-008	23 54 24.	- 35 01	42	14.5	GALAXY
KN 13.010	23 54 24.7	+ 10 32 49.			NEBULA
SC 2351-0429.4	23 54 25.	- 04 12 42.	6		NEBULA
SC 2351-0115.5	23 54 26.	- 00 58 48.			NEBULA
MCG+08-01-010	23 54 27.	+ 47 13 30.	72	15.	GALAXY
MCG-03-01-003	23 54 27.	- 17 02	48	15.	GALAXY
SC 2351-0400.8	23 54 29.	- 03 44 06.	30		NEBULA
MCG+02-01-002	23 54 30.	+ 10 32	108	14.5	GALAXY
ZWG 499.015	23 54 30.	+ 29 35		15.2	GALAXY
ZWG 498.043	23 54 30.	+ 29 35		15.2	GALAXY
UGC 12861	23 54 30.	+ 29 35	60	15.2	GALAXY Sa-b
ZWG 499.016	23 54 30.	+ 33 25		15.7	GALAXY
ZWG 498.044	23 54 30.	+ 33 25		15.7	GALAXY
ZWG 549.008	23 54 30.	+ 47 15		15.7	GALAXY
ZWG 548.017	23 54 30.	+ 47 15		15.7	GALAXY
UGC 12862	23 54 30.	+ 47 15	66	15.7	GALAXY SB
OCL 0269	23 54 30.	+ 56 27	3060	9.6	OPEN STAR CLUSTER
LDN 1260	23 54 30.	+ 64 40	240		DARK NEBULA
ZWG 382.008	23 54 30.	- 02 22		15.4	GALAXY
MRK 542	23 54 30.	- 02 22	11	15.5	GALAXY WITH UV CONTINUUM
PHL 0593	23 54 30.	- 03 06		14.4	BLUE STELLAR OBJECT
8ZW 2354-04.2	23 54 30.	- 04 13		16.5	COMPACT GALAXY
PHL 0594	23 54 30.	- 05 56		18.3	BLUE STELLAR OBJECT
PHL 2472	23 54 30.	- 26 30		17.8	BLUE STELLAR OBJECT
RNGC 7789	23 54 32.	+ 56 27		9.5	OPEN CLUSTER
KN 13.011	23 54 32.9	+ 12 11 21.			NEBULA
IC 1518	23 54 33.	+ 12 11 33.			NONSTELLAR OBJECT
MCG+05-01-010	23 54 33.	+ 29 34	48	15.5	GALAXY
IC 1519	23 54 35.	+ 12 11 03.			NONSTELLAR OBJECT
KN 13.012	23 54 35.0	+ 12 10 55.			NEBULA
PHL 6090	23 54 36.	+ 08 55		16.9	BLUE STELLAR OBJECT
ZWG 433.005	23 54 36.	+ 12 11		15.7	GALAXY
KARA.72 599B	23 54 36.	+ 12 11	36	15.7	PART OF DOUBLE GALAXY
ZWG 433.006	23 54 36.	+ 12 12		15.7	GALAXY
KARA.72 599A	23 54 36.	+ 12 12	36	15.7	PART OF DOUBLE GALAXY
ZWG 433.007	23 54 36.	+ 12 30		15.6	GALAXY
PHL 6091	23 54 36.	- 07 36		18.5	BLUE STELLAR OBJECT
PHL 2473	23 54 36.	- 08 32		18.0	BLUE STELLAR OBJECT
PHL 2474	23 54 36.	- 15 46		18.5	BLUE STELLAR OBJECT
KN 13.013	23 54 36.4	+ 12 30 46.			NEBULA

OBJECT NAME	RIGHT ASCEN.	DECLINATION	DIAM.	MAGN.	TYPE OF OBJECT
SCHO 1450	23 54 38.	+ 59 27 18.	500		ISOLATED DARK CLOUD
SC 2352-0142.6	23 54 38.	- 01 25 54.	12		NEBULA
PHL 2475	23 54 42.	+ 05 24		18.4	BLUE STELLAR OBJECT
PHL 6092	23 54 42.	+ 07 00		18.0	BLUE STELLAR OBJECT
ZC 2354.7+2436	23 54 42.	+ 24 36	1010		CLUSTER OF GALAXIES
ISS 0026	23 54 42.	+ 63 31	286		STELLAR RING
MCG-03-01-004	23 54 42.	- 17 34	48	15.	GALAXY
MCG-03-01-005	23 54 42.	- 18 54	18	15.5	GALAXY
SC 2352-7212.3	23 54 42.	- 71 55 36.	18		NEBULA
SHB 402	23 54 44.7	+ 14 29 26.		18.2	QUASI-STELLAR OBJECT
BC PKS2354+14	23 54 44.86	+ 14 29 27.1		18.18	QUASI-STELLAR OBJECT
MCG+01-01-002	23 54 45.	+ 08 13 30.	72	15.5	GALAXY
SC 2352-7156.9	23 54 46.	- 71 40 12.	12		NEBULA
SCHO 1451	23 54 47.	+ 58 38 54.	360		ISOLATED DARK CLOUD
PHL 6094	23 54 48.	+ 03 40		18.5	BLUE STELLAR OBJECT
UGC 12863	23 54 48.	+ 08 14	66	16.0	GALAXY Sb-c
MCG+05-01-011	23 54 48.	+ 30 42 30.	96	15.	GALAXY
ZWG 499.017	23 54 48.	+ 30 43		14.7	GALAXY
ZWG 498.045	23 54 48.	+ 30 43		14.7	GALAXY
UGC 12864	23 54 48.	+ 30 43	108	14.7	GALAXY SBb
KARA.73B 1048	23 54 48.	+ 30 43	102	14.7	ISOLATED GALAXY S
MCG+08-01-011	23 54 48.	+ 47 36 30.	15	16.	GALAXY
SC 2352-0212.9	23 54 48.	- 01 56 12.	12		NEBULA
PHL 2476	23 54 48.	- 03 59		18.5	BLUE STELLAR OBJECT
PHL 6093	23 54 48.	- 14 28		18.0	BLUE STELLAR OBJECT
ARC 2681	23 54 48.	- 24 37		17.9	RICH CLUSTER OF GALAXIES
SC 2352-0213.5	23 54 49.	- 01 56 48.	12		NEBULA
KN 13.014	23 54 49.0	+ 14 56 10.			NEBULA
PHL 2477	23 54 54.	+ 08 44		18.5	BLUE STELLAR OBJECT
ZWG 433.008	23 54 54.	+ 14 56		15.6	GALAXY
MCG+08-01-012	23 54 54.	+ 47 35	42	16.	GALAXY
5ZW 443	23 54 54.	+ 47 36			COMPACT GALAXY
ZC 2354.9+5212	23 54 54.	+ 52 12	11630		CLUSTER OF GALAXIES
SC 2352-0328.4	23 54 54.	- 03 11 42.	30		NEBULA
PHL 2478	23 54 54.	- 05 47		18.2	BLUE STELLAR OBJECT
ARC 2682	23 54 54.	- 20 50		17.5	RICH CLUSTER OF GALAXIES
PHL 2479	23 54 54.	- 21 08		18.4	BLUE STELLAR OBJECT
MCG-05-01-011	23 54 54.	- 29 20	108	14.	GALAXY
MCG-05-01-013	23 54 54.	- 30 44	18	15.	GALAXY
MCG-05-01-012	23 54 54.	- 30 44	30	16.5	GALAXY
SHB 403	23 54 57.1	- 11 42 23.		18.	QUASI-STELLAR OBJECT
LBN 0560	23 55	+ 49 30	6000		BRIGHT NEBULA
LBN 0563	23 55	+ 50 25	2100		BRIGHT NEBULA
PHL 0597	23 55 00.	+ 04 10		18.2	BLUE STELLAR OBJECT
PHL 0596	23 55 00.	+ 04 41		17.6	BLUE STELLAR OBJECT
ZC 2355.0+0516	23 55 00.	+ 05 16	870		CLUSTER OF GALAXIES
PHL 2480	23 55 00.	+ 06 34		18.6	BLUE STELLAR OBJECT
PHL 0595	23 55 00.	+ 08 54		17.9	BLUE STELLAR OBJECT
ZWG 456.005	23 55 00.	+ 18 02		14.8	GALAXY
ZWG 499.018	23 55 00.	+ 32 30		15.6	GALAXY
ZWG 498.046	23 55 00.	+ 32 30		15.6	GALAXY
ZWG 518.002	23 55 00.	+ 36 45		15.7	GALAXY
ZWG 517.007	23 55 00.	+ 36 45		15.7	GALAXY
KARA.73B 1049	23 55 00.	+ 36 45	36	15.7	ISOLATED GALAXY S
ZC 2355.0+3834	23 55 00.	+ 38 34	740		CLUSTER OF GALAXIES
MCG+08-01-013	23 55 00.	+ 47 09 30.	60	15.	GALAXY
ZWG 549.009	23 55 00.	+ 47 10		15.5	GALAXY
ZWG 548.018	23 55 00.	+ 47 10		15.5	GALAXY
LDN 1257	23 55 00.	+ 59 23	660		DARK NEBULA
LDN 1263	23 55 00.	+ 64 34	180		DARK NEBULA
LDN 1266	23 55 00.	+ 67 00	6420		DARK NEBULA
LDN 1274	23 55 00.	+ 70 40	720		DARK NEBULA
PHL 0598	23 55 00.	- 03 08		18.2	BLUE STELLAR OBJECT
PHL 0599	23 55 00.	- 15 09		15.6	BLUE STELLAR OBJECT
VV 332B	23 55 00.	- 22 17 30.	72	16.	INTERACTING GALAXY
MCG-04-01-007	23 55 00.	- 22 18	84	15.	GALAXY
ARC 2683	23 55 00.	- 25 50		16.9	RICH CLUSTER OF GALAXIES
IC 5365	23 55 00.	- 37 18 09.			NONSTELLAR OBJECT
HN 2885	23 55 00.1	- 46 26 06.	12		NEBULA
MCG-01-01-008	23 55 03.	- 03 08	48	14.5	GALAXY
VV 332A	23 55 03.	- 22 16	42	16.	INTERACTING GALAXY
MCG+00-01-010	23 55 06.	+ 00 22	54	15.	GALAXY
ZWG 456.006	23 55 06.	+ 17 55		15.3	GALAXY
MCG+05-01-012	23 55 06.	+ 32 29	36	16.	GALAXY
ZWG 499.019	23 55 06.	+ 32 30		15.7	GALAXY
ZWG 498.047	23 55 06.	+ 32 30		15.7	GALAXY
PHL 2481	23 55 06.	- 02 55		18.0	BLUE STELLAR OBJECT
PHL 6095	23 55 06.	- 04 27		18.2	BLUE STELLAR OBJECT
MCG-04-01-008	23 55 06.	- 22 16	48	15.	GALAXY
CED 213	23 55 07.	+ 52 31	1800		DIFFUSE GALACTIC NEBULA
SC 2352-0409.2	23 55 07.	- 03 52 30.	54		NEBULA
IC 5366	23 55 08.	+ 52 30 45.			MAY NOT EXIST
SC 2352-0425.5	23 55 09.	- 04 08 48.	18		NEBULA
UGC 12865	23 55 12.	+ 00 20	66	15.7	GALAXY Sa-b
ZWG 382.009	23 55 12.	+ 00 22		15.7	GALAXY
PHL 6098	23 55 12.	+ 07 07		18.1	BLUE STELLAR OBJECT
ZWG 408.003	23 55 12.	+ 09 12		15.7	GALAXY
PHL 6096	23 55 12.	- 08 20		17.8	BLUE STELLAR OBJECT
PHL 6097	23 55 12.	- 09 42		17.9	BLUE STELLAR OBJECT
VV 025B	23 55 12.	- 14 19	6	18.	INTERACTING GALAXY
VV 025A	23 55 12.	- 14 19	30	15.	INTERACTING GALAXY
MCG-01-01-009	23 55 15.	- 04 06 30.	24	15.5	GALAXY
SVEN 459	23 55 17.	- 32 52	360	10.4	GALAXY
PHL 2482	23 55 18.	+ 06 02		18.3	BLUE STELLAR OBJECT
PHL 2483	23 55 18.	+ 08 16		18.6	BLUE STELLAR OBJECT
MCG+03-01-005	23 55 18.	+ 16 15	12	17.	GALAXY
UGC 12866	23 55 18.	+ 22 04	60	16.0	GALAXY Sc
PHL 6100	23 55 18.	- 03 15		18.3	BLUE STELLAR OBJECT
PHL 6101	23 55 18.	- 08 26		18.5	BLUE STELLAR OBJECT
PHL 6099	23 55 18.	- 08 36		17.0	BLUE STELLAR OBJECT
PHL 2484	23 55 18.	- 16 04		18.5	BLUE STELLAR OBJECT
RNGC 7793	23 55 19.	- 32 51		10.0	GALAXY
SC 2352-0113.9	23 55 20.	- 00 57 12.	18		NEBULA
IC 1520	23 55 20.	- 14 19 03.			NONSTELLAR OBJECT
MCG+01-01-003	23 55 21.	+ 09 11 30.	12	15.5	GALAXY
MCG-03-01-006	23 55 21.	- 19 03	42	15.	GALAXY
ARP 050	23 55 22.	- 14 19			PECULIAR GALAXY
SVEN 460	23 55 23.	- 33 09	36	14.5	GALAXY
PHL 6102	23 55 24.	+ 05 48		17.8	BLUE STELLAR OBJECT
PHL 6104	23 55 24.	+ 06 45		18.5	BLUE STELLAR OBJECT
PHL 2485	23 55 24.	+ 07 32		18.7	BLUE STELLAR OBJECT
PHL 6103	23 55 24.	+ 09 21		18.4	BLUE STELLAR OBJECT
ZC 2355.4+3804	23 55 24.	+ 38 04	1280		CLUSTER OF GALAXIES
ZWG 534.003	23 55 24.	+ 43 55		15.6	GALAXY
MCG-02-01-006	23 55 24.	- 12 12	36	15.	GALAXY
MCG-02-01-007	23 55 24.	- 14 18 30.	30	14.5	GALAXY
PHL 6105	23 55 24.	- 17 27		17.9	BLUE STELLAR OBJECT
PHL 6107	23 55 24.	- 28 45		18.5	BLUE STELLAR OBJECT

OBJECT NAME	RIGHT ASCEN.	DECLINATION	DIAM.	MAGN.	TYPE OF OBJECT
PHL 6106	23 55 24.	- 30 10		18.2	BLUE STELLAR OBJECT
MCG-05-01-014	23 55 24.	- 30 10	30	15.	GALAXY
PHL 6108	23 55 24.	- 30 21		18.1	BLUE STELLAR OBJECT
MCG-05-01-015	23 55 24.	- 32 08	30	14.5	GALAXY
MCG-06-01-009	23 55 24.	- 32 51	360	9.7	GALAXY
RNGC 7791	23 55 25.	+ 10 29			NON-EXISTENT OBJECT
HELW 505	23 55 26.	- 33 09 25.			NEBULA
PHL 6109	23 55 30.	+ 09 10		18.2	BLUE STELLAR OBJECT
MCG+03-01-006	23 55 30.	+ 16 13	42	15.	GALAXY
ISS 0002	23 55 30.	+ 63 29	177		STELLAR RING
8ZW 2355-05.8	23 55 30.	- 05 47		18.0	COMPACT GALAXY
PHL 2486	23 55 30.	- 08 52		18.2	BLUE STELLAR OBJECT
MCG-03-01-007	23 55 30.	- 20 05	36	15.	GALAXY
MCG-05-01-016	23 55 30.	- 30 02	15	15.5	GALAXY
MCG-06-01-010	23 55 30.	- 34 34	30	14.5	GALAXY
KN 13.015	23 55 30.0	+ 16 13 27.			NEBULA
RNGC 7792	23 55 31.	+ 16 13		15.0	GALAXY
KN 13.016	23 55 31.5	+ 13 42 26.			NEBULA
ARC 2584	23 55 35.	- 10 48		17.6	RICH CLUSTER OF GALAXIES
ZC 2355.6+0351	23 55 36.	+ 03 51	2080		CLUSTER OF GALAXIES
PHL 6111	23 55 36.	+ 07 22		18.1	BLUE STELLAR OBJECT
PHL 6110	23 55 36.	+ 08 24		18.4	BLUE STELLAR OBJECT
MCG+03-01-007	23 55 36.	+ 16 15	18	17.	GALAXY
ZWG 456.007	23 55 36.	+ 16 44		14.9	GALAXY
PHL 6112	23 55 36.	- 04 39		18.1	BLUE STELLAR OBJECT
PHL 6113	23 55 36.	- 08 18		18.5	BLUE STELLAR OBJECT
PHL 0600	23 55 36.	- 10 36		15.3	BLUE STELLAR OBJECT
PHL 0601	23 55 36.	- 16 46		17.9	BLUE STELLAR OBJECT
PHL 6114	23 55 36.	- 17 56		18.6	BLUE STELLAR OBJECT
MCG-06-01-011	23 55 36.	- 33 10	36	15.5	GALAXY
SC 2353-0332.0	23 55 39.	- 03 15 18.	12		NEBULA
ZWG 499.020	23 55 42.	+ 28 11		15.7	GALAXY
ZWG 498.048	23 55 42.	+ 28 11		15.7	GALAXY
ZWG 499.021	23 55 42.	+ 29 06		15.4	GALAXY
ZWG 498.049	23 55 42.	+ 29 06		15.4	GALAXY
UGC 12867	23 55 42.	+ 29 06	72	15.4	GALAXY S0-a
ZWG 499.022	23 55 42.	+ 29 19		15.3	GALAXY
ZWG 498.050	23 55 42.	+ 29 19		15.3	GALAXY
MCG+08-01-014	23 55 42.	+ 47 34	18	16.	GALAXY
SCHO 1452	23 55 42.	+ 59 21 06.	280		ISOLATED DARK CLOUD
PHL 2487	23 55 42.	- 04 23		16.6	BLUE STELLAR OBJECT
PHL 6118	23 55 42.	- 04 53		18.5	BLUE STELLAR OBJECT
PHL 6117	23 55 42.	- 04 53		17.2	BLUE STELLAR OBJECT
PHL 6116	23 55 42.	- 18 21		16.8	BLUE STELLAR OBJECT
PHL 6115	23 55 42.	- 27 54		16.8	BLUE STELLAR OBJECT
MCG-05-01-017	23 55 42.	- 30 25	60	15.	GALAXY
TON-S 0130	23 55 42.	- 32 20		13.7	BLUE STAR
SC 2353-0240.5	23 55 44.	- 02 23 48.	6		NEBULA
MCG+05-01-013	23 55 45.	+ 29 04	48	15.	GALAXY
KN 13.017	23 55 45.2	+ 11 50 01.			NEBULA
LB 01197	23 55 47.	+ 31 23 24.		16.3	FAINT BLUE STAR
SVEN 461	23 55 47.	- 33 13	18	15.3	GALAXY
4ZW 174	23 55 48.	+ 27 41			COMPACT GALAXY
UGC 12868	23 55 48.	+ 43 47	66	16.5	GALAXY S
SC 2353-0120.8	23 55 48.	- 01 04 06.	30		NEBULA
PHL 6119	23 55 48.	- 05 33		18.3	BLUE STELLAR OBJECT
PHL 6120	23 55 48.	- 07 10		18.2	BLUE STELLAR OBJECT
PHL 0602	23 55 48.	- 10 51		18.0	BLUE STELLAR OBJECT
PHL 0603	23 55 48.	- 32 21		14.9	BLUE STELLAR OBJECT
LB 03126	23 55 48.	- 74 06		12.7	FAINT BLUE STAR
MCG+05-01-015	23 55 51.	+ 27 45	30	15.	GALAXY
MCG+05-01-014	23 55 51.	+ 28 18	30	15.	GALAXY
PHL 6121	23 55 54.	+ 05 46		18.6	BLUE STELLAR OBJECT
ZWG 499.023	23 55 54.	+ 27 45		15.5	GALAXY
ZWG 498.051	23 55 54.	+ 27 45		15.5	GALAXY
4ZW 175	23 55 54.	+ 27 46			COMPACT GALAXY
MCG+05-01-017	23 55 54.	+ 28 33	42	15.	GALAXY
MCG+05-01-016	23 55 54.	+ 31 57 30.	66	15.	GALAXY
ZWG 499.024	23 55 54.	+ 31 58		15.3	GALAXY
ZWG 498.052	23 55 54.	+ 31 58		15.3	GALAXY
UGC 12869	23 55 54.	+ 31 58	60	15.3	GALAXY SBc
ZWG 534.004	23 55 54.	+ 44 17		15.5	GALAXY
UGC 12870	23 55 54.	+ 44 17	60	15.5	GALAXY S
MCG+07-01-002	23 55 54.	+ 44 17	36	15.	GALAXY
5ZW 444	23 55 54.	+ 49 10			COMPACT GALAXY
OCL 0276	23 55 54.	+ 60 56	1050	7.2	OPEN STAR CLUSTER
ARC 2685	23 55 54.	- 24 41		17.9	RICH CLUSTER OF GALAXIES
RNGC 7790	23 55 56.	+ 60 56		7.0	OPEN CLUSTER
HN 0004	23 55 59.	- 32 51			NEBULA
LBN 0576	23 56	+ 62 00	960		BRIGHT NEBULA
PHL 2488	23 56 00.	+ 03 50		18.3	BLUE STELLAR OBJECT
ZWG 433.009	23 56 00.	+ 09 41		15.3	GALAXY
UGC 12871	23 56 00.	+ 09 41	66	15.3	GALAXY Sc
MCG+02-01-003	23 56 00.	+ 09 41	84	14.5	GALAXY
ZWG 433.010	23 56 00.	+ 10 26		13.8	GALAXY
UGC 12872	23 56 00.	+ 10 26	84	13.8	GALAXY S
MCG+02-01-004	23 56 00.	+ 10 27	96	13.	GALAXY
ZWG 478.006	23 56 00.	+ 25 56		15.6	GALAXY
ZWG 477.034	23 56 00.	+ 25 56		15.6	GALAXY
UGC 12873	23 56 00.	+ 25 56	90	15.6	GALAXY
MCG+04-01-006	23 56 00.	+ 25 56	72	15.	GALAXY
KARA.73B 1050	23 56 00.	+ 25 56	66	15.6	ISOLATED GALAXY S
MCG+08-01-015	23 56 00.	+ 47 39	72	15.	GALAXY
MCG+14-01-005	23 56 00.	+ 82 54	36	16.	GALAXY
MCG-03-01-008	23 56 00.	- 18 59	36	15.	GALAXY
PHL 6122	23 56 00.	- 22 16		18.2	BLUE STELLAR OBJECT
PHL 2489	23 56 00.	- 27 34		18.3	BLUE STELLAR OBJECT
KN 13.018	23 56 00.5	+ 10 27 01.			NEBULA
RNGC 7794	23 56 01.	+ 10 26		14.0	GALAXY
PHL 6123	23 56 06.	+ 03 56		17.0	BLUE STELLAR OBJECT
PHL 2490	23 56 06.	+ 07 10		18.3	BLUE STELLAR OBJECT
IC 5367	23 56 06.	+ 22 09 46.			NONSTELLAR OBJECT
ZWG 478.007	23 56 06.	+ 26 52		15.7	GALAXY
ZWG 477.035	23 56 06.	+ 26 52		15.7	GALAXY
UGC 12874	23 56 06.	+ 26 52	66	15.7	GALAXY Sc
8ZW 2356-05.3	23 56 06.	- 05 16		17.8	COMPACT GALAXY
PHL 6124	23 56 06.	- 07 01		18.0	BLUE STELLAR OBJECT
RNGC 7795	23 56 08.	+ 59 44			NON-EXISTENT OBJECT
KN 13.019	23 56 08.4	+ 15 34 09.			NEBULA
PHL 0604	23 56 12.	+ 05 16		18.2	BLUE STELLAR OBJECT
PHL 6127	23 56 12.	+ 07 46		18.1	BLUE STELLAR OBJECT
ZWG 408.004	23 56 12.	+ 08 07		15.6	GALAXY
UGC 12875	23 56 12.	+ 08 07	66	15.6	GALAXY S
MCG+03-01-008	23 56 12.	+ 20 54	27	17.	GALAXY
4ZW 176	23 56 12.	+ 26 39			COMPACT GALAXY
ZWG 499.025	23 56 12.	+ 27 49		15.7	GALAXY
ZWG 498.053	23 56 12.	+ 27 49		15.7	GALAXY
ZCG 2356+27	23 56 12.	+ 27 49		15.7	COMPACT GALAXY

OBJECT NAME	RIGHT ASCEN.	DECLINATION	DIAM.	MAGN.	TYPE OF OBJECT
ASS 33	23 56 12.	+ 60 05	9000		OB ASSOCIATION CAS OB5
SS 74	23 56 12.	+ 66 09			DIFFUSE GALACTIC NEBULA
PHL 2491	23 56 12.	- 06 40		18.7	BLUE STELLAR OBJECT
PHL 6126	23 56 12.	- 10 14		17.2	BLUE STELLAR OBJECT
PHL 0605	23 56 12.	- 12 48		18.4	BLUE STELLAR OBJECT
PHL 6125	23 56 12.	- 13 19		17.3	BLUE STELLAR OBJECT
PHL 2492	23 56 12.	- 17 50		18.4	BLUE STELLAR OBJECT
MCG-03-01-009	23 56 12.	- 17 50	36	14.5	GALAXY
MCG-05-01-018	23 56 12.	- 30 07	18	15.5	GALAXY
KN 13.020	23 56 12.1	+ 12 19 19.			NEBULA
MCG+01-01-004	23 56 15.	+ 08 08	60	15.	GALAXY
MCG+02-01-005	23 56 15.	+ 12 20	48	16.	GALAXY
MCG-03-01-010	23 56 15.	- 17 07	30	15.	GALAXY
KN 13.021	23 56 15.4	+ 12 04 15.			NEBULA
PHL 6129	23 56 18.	+ 07 42		18.0	BLUE STELLAR OBJECT
ZWG 433.011	23 56 18.	+ 12 05		15.7	GALAXY
UGC 12876	23 56 18.	+ 12 19	66	17.	GALAXY DWRF IR
MCG+07-01-003	23 56 18.	+ 41 17	30	17.	GALAXY
PHL 2494	23 56 18.	- 02 40		17.8	BLUE STELLAR OBJECT
PHL 2495	23 56 18.	- 02 58		17.9	BLUE STELLAR OBJECT
PHL 0606	23 56 18.	- 05 43		16.2	BLUE STELLAR OBJECT
SN 1954B	23 56 18.	- 05 53		17.8	SUPERNOVA
PHL 2493	23 56 18.	- 06 52		16.7	BLUE STELLAR OBJECT
PHL 2496	23 56 18.	- 11 15		18.6	BLUE STELLAR OBJECT
PHL 6130	23 56 18.	- 21 43		18.6	BLUE STELLAR OBJECT
PHL 6128	23 56 18.	- 30 58		17.8	BLUE STELLAR OBJECT
MCG+00-01-011	23 56 21.	+ 03 21 30.	54	13.	GALAXY
SC 2353-0407.0	23 56 22.	- 03 50 18.	6		NEBULA
ZWG 382.010	23 56 24.	+ 03 22		14.8	GALAXY
UGC 12877	23 56 24.	+ 03 22	72	14.8	GALAXY S
UGC 12878	23 56 24.	+ 41 17	66	16.5	GALAXY Sb/SBb
SC 2353-0258.2	23 56 24.	- 02 41 30.	54		NEBULA
8ZW 2356-04.2	23 56 24.	- 04 14		15.8	COMPACT GALAXY
PHL 0607	23 56 24.	- 06 27		17.5	BLUE STELLAR OBJECT
PHL 2497	23 56 24.	- 07 42		16.8	BLUE STELLAR OBJECT
RNGC 7797	23 56 25.	+ 03 22		15.0	GALAXY
IC 1521	23 56 28.	- 07 25 38.			NONSTELLAR OBJECT
IC 1522	23 56 29.	+ 01 25 46.			NONSTELLAR OBJECT
ZWG 382.011	23 56 30.	+ 01 26		15.3	GALAXY
ZWG 456.008	23 56 30.	+ 18 34		15.0	GALAXY
UGC 12879	23 56 30.	+ 18 34	60	15.0	GALAXY S
5ZW 445	23 56 30.	+ 48 05			COMPACT GALAXY
MCG+09-01-001	23 56 30.	+ 51 17	42	15.	GALAXY
LDN 1258	23 56 30.	+ 59 20	360		DARK NEBULA
8ZW 2356-04.3	23 56 30.	- 04 21		15.9	COMPACT GALAXY
PHL 2498	23 56 30.	- 04 58		18.6	BLUE STELLAR OBJECT
PHL 2499	23 56 30.	- 09 53		17.8	BLUE STELLAR OBJECT
PHL 6131	23 56 30.	- 14 02		17.5	BLUE STELLAR OBJECT
MCG-06-01-012	23 56 30.	- 35 21	30	15.	GALAXY
SER 164.02	23 56 30.	- 55 45		13.	CMPCT GROUP OF 3 GALAXIES
RNGC 7796	23 56 31.	- 55 44		13.0	GALAXY
IC 1523	23 56 33.	+ 06 35 28.			NONSTELLAR OBJECT
MCG-01-01-010	23 56 33.	- 02 31	84	15.	GALAXY
MCG+00-01-012	23 56 36.	+ 01 29	36	14.	GALAXY
PHL 2500	23 56 36.	+ 07 08		17.0	BLUE STELLAR OBJECT
PHL 2501	23 56 36.	+ 08 12		18.4	BLUE STELLAR OBJECT
ZC 2356.6+2219	23 56 36.	+ 22 19	1480		CLUSTER OF GALAXIES
ZWG 549.010	23 56 36.	+ 51 17		15.7	GALAXY
ZWG 548.019	23 56 36.	+ 51 17		15.7	GALAXY
UGC 12880	23 56 36.	+ 51 19	60	16.5	GALAXY Sc
5ZW 446	23 56 36.	+ 51 40			COMPACT GALAXY
PHL 0608	23 56 36.	- 05 11		18.2	BLUE STELLAR OBJECT
PHL 6132	23 56 36.	- 08 54		14.8	BLUE STELLAR OBJECT
PHL 2502	23 56 36.	- 14 58		18.8	BLUE STELLAR OBJECT
PHL 0609	23 56 36.	- 18 16		18.2	BLUE STELLAR OBJECT
MCG-04-01-009	23 56 36.	- 21 02 30.	9	15.5	GALAXY
MCG-04-01-010	23 56 36.	- 23 43 30.	30	16.	GALAXY
MCG-05-01-019	23 56 36.	- 28 53	36	15.	GALAXY
LB 01538	23 56 36.	- 50 10		12.4	FAINT BLUE STAR
MCG-04-01-011	23 56 39.	- 21 05	42	15.	GALAXY
MCG+01-01-005	23 56 42.	+ 04 28	60	14.	GALAXY
MCG+02-01-006	23 56 42.	+ 15 25	36	15.5	GALAXY
ZWG 478.008	23 56 42.	+ 25 39		15.5	GALAXY
ZWG 477.036	23 56 42.	+ 25 39		15.5	GALAXY
MCG+08-01-016	23 56 42.	+ 46 36 30.	120	12.	GALAXY
MCG-01-01-011	23 56 42.	- 04 22 30.	78	13.	GALAXY
MCG-01-01-012	23 56 42.	- 04 26	78	14.5	GALAXY
PHL 6133	23 56 42.	- 18 36		18.5	BLUE STELLAR OBJECT
PHL 0610	23 56 42.	- 26 53		12.3	BLUE STELLAR OBJECT
TON-S 0131	23 56 42.	- 26 53		13.3	BLUE STAR
IC 5368	23 56 43.	+ 06 35 22.			NONSTELLAR OBJECT
MCG+00-01-013	23 56 45.	+ 01 36	66	15.	GALAXY
VRT 24	23 56 45.	+ 62 10	1920		SUPERNOVA REMNANT
ARC 2686	23 56 47.	- 21 05		16.9	RICH CLUSTER OF GALAXIES
ZWG 382.012	23 56 48.	+ 01 36		15.6	GALAXY
ZWG 408.005	23 56 48.	+ 04 29		15.5	GALAXY
UGC 12881	23 56 48.	+ 04 29	108	15.5	GALAXY S-IRR
UGC 12882	23 56 48.	+ 31 00	60	16.0	GALAXY Sc
ZWG 549.011	23 56 48.	+ 46 37		13.3	GALAXY
ZWG 548.020	23 56 48.	+ 46 37		13.3	GALAXY
UGC 12883	23 56 48.	+ 46 37	114	13.3	GALAXY SBb
PHL 6135	23 56 48.	- 02 46		17.8	BLUE STELLAR OBJECT
PHL 6134	23 56 48.	- 03 21		14.0	BLUE STELLAR OBJECT
PHL 6136	23 56 48.	- 07 55		18.0	BLUE STELLAR OBJECT
PHL 2503	23 56 48.	- 19 00		17.1	BLUE STELLAR OBJECT
SC 2354-0404.4	23 56 50.	- 03 47 42.	12		NEBULA
SC 2354-0208.6	23 56 53.	- 01 51 54.	6		NEBULA
ZWG 382.013	23 56 54.	+ 01 18		15.5	GALAXY
PHL 6137	23 56 54.	+ 07 00		17.2	BLUE STELLAR OBJECT
MCG+03-01-009	23 56 54.	+ 17 56	42	15.	GALAXY
ZWG 456.009	23 56 54.	+ 20 29		12.7	GALAXY
UGC 12884	23 56 54.	+ 20 29	96	12.7	GALAXY S
MCG+03-01-010	23 56 54.	+ 20 30	66	13.	GALAXY
MCG-02-01-008	23 56 54.	- 10 11	24	16.	GALAXY
MCG-04-01-012	23 56 54.	- 24 42 30.	36	15.	GALAXY
PHL 6138	23 56 54.	- 26 16		17.5	BLUE STELLAR OBJECT
RNGC 7798	23 56 55.	+ 20 29		12.5	GALAXY
RNGC 7799	23 56 55.	+ 31 01			GALAXY
SC 2354-0208.2	23 56 56.	- 01 51 30.	12		NEBULA
IC 1524	23 56 57.	- 04 25 56.			NONSTELLAR OBJECT
SC 2354-0303.6	23 56 59.	- 02 46 54.	48		NEBULA
HOLM 824B	23 56 59.	- 11 43	30	14.4	PART OF MULTIPLE GALAXY
LBN 0561	23 57	+ 48 12	600		BRIGHT NEBULA
KHAV 797	23 57	+ 68 59	10280		DARK NEBULA
ZWG 433.012	23 57 00.	+ 14 32		13.4	GALAXY
UGC 12885	23 57 00.	+ 14 32	138	13.4	GALAXY IRR
MCG+02-01-007	23 57 00.	+ 14 32	144	12.5	GALAXY
ZWG 456.010	23 57 00.	+ 17 56		15.1	GALAXY

OBJECT NAME	RIGHT ASCEN.	DECLINATION	DIAM.	MAGN.	TYPE OF OBJECT
UGC 12886	23 57 00.	+ 17 56	96	15.1	GALAXY Sb-c
ZWG 456.011	23 57 00.	+ 21 08		15.3	GALAXY
KARA.73B 1051	23 57 00.	+ 21 08	24	15.3	ISOLATED GALAXY S
LDN 1268	23 57 00.	+ 67 10	1560		DARK NEBULA
ASS 34	23 57 00.	+ 67 18			OB ASSOCIATION CEP OB4
PHL 2504	23 57 00.	- 02 34		16.9	BLUE STELLAR OBJECT
PHL 2505	23 57 00.	- 03 53		18.2	BLUE STELLAR OBJECT
PHL 2506	23 57 00.	- 04 06		18.2	BLUE STELLAR OBJECT
PHL 6139	23 57 00.	- 05 40		18.0	BLUE STELLAR OBJECT
PHL 2507	23 57 00.	- 05 55		18.4	BLUE STELLAR OBJECT
PHL 6140	23 57 00.	- 06 22		18.3	BLUE STELLAR OBJECT
MCG-03-01-011	23 57 00.	- 15 28	36	15.	GALAXY
MCG-04-01-013	23 57 00.	- 21 55 30.	36	16.	GALAXY
LB 04887	23 57 00.	- 30 42		18.8	FAINT BLUE STAR
KN 13.022	23 57 00.4	+ 14 29 20.			NEBULA
RNGC 7800	23 57 01.	+ 14 32		13.5	GALAXY
KN 13.023	23 57 03.2	+ 14 31 35.			NEBULA
HOLM 824A	23 57 04.	- 11 44	150	13.8	PART OF MULTIPLE GALAXY
PHL 6141	23 57 06.	+ 04 15		17.4	BLUE STELLAR OBJECT
PHL 6142	23 57 06.	+ 06 10		18.3	BLUE STELLAR OBJECT
ZC 2357.1+3948	23 57 06.	+ 39 48	1210		CLUSTER OF GALAXIES
MCG-01-01-013	23 57 06.	- 02 44	78	15.5	GALAXY
MCG-01-01-014	23 57 06.	- 07 03	36	14.5	GALAXY
PHL 2508	23 57 06.	- 09 34		18.7	BLUE STELLAR OBJECT
LP 04868	23 57 06.	- 30 34		20.5	FAINT BLUE STAR
PHL 6144	23 57 12.	+ 07 02		18.3	BLUE STELLAR OBJECT
PHL 2510	23 57 12.	+ 07 10		18.6	BLUE STELLAR OBJECT
PHL 2509	23 57 12.	+ 08 02		18.6	BLUE STELLAR OBJECT
ZC 2357.2+1020	23 57 12.	+ 10 20	2150		CLUSTER OF GALAXIES
ZWG 456.012	23 57 12.	+ 21 21		15.7	GALAXY
ZWG 499.026	23 57 12.	+ 28 01		15.6	GALAXY
ZWG 498.054	23 57 12.	+ 28 01		15.6	GALAXY
ZWG 518.003	23 57 12.	+ 34 04		15.6	GALAXY
ZWG 517.008	23 57 12.	+ 34 04		15.6	GALAXY
UGC 12887	23 57 12.	+ 48 15	90	17.	GALAXY IRR
PHL 6143	23 57 12.	- 11 54		17.7	BLUE STELLAR OBJECT
PHL 0611	23 57 12.	- 16 26		18.2	BLUE STELLAR OBJECT
PHL 2511	23 57 12.	- 18 24		18.1	BLUE STELLAR OBJECT
PHL 6145	23 57 12.	- 20 25		18.2	BLUE STELLAR OBJECT
LB 04889	23 57 12.	- 29 54		17.9	FAINT BLUE STAR
LB 04890	23 57 12.	- 30 26		18.6	FAINT BLUE STAR
LB 04891	23 57 12.	- 30 31		17.5	FAINT BLUE STAR
MCG-05-01-020	23 57 12.	- 30 55	18	15.5	GALAXY
SER 164.03	23 57 12.	- 56 20		17.	LOW SURF. BRGHTNSS GALAXY
KN 13.024	23 57 13.3	+ 14 56 46.			NEBULA
SCHO 1453	23 57 15.	+ 57 18 18.	290		ISOLATED DARK CLOUD
IC 5369	23 57 16.	+ 32 24 28.			NONSTELLAR OBJECT
PHL 2512	23 57 18.	+ 06 56		16.7	BLUE STELLAR OBJECT
ZWG 499.027	23 57 18.	+ 32 25		15.3	GALAXY
ZWG 498.055	23 57 18.	+ 32 25		15.3	GALAXY
ZWG 518.004	23 57 18.	+ 35 09		15.7	GALAXY
ZWG 517.009	23 57 18.	+ 35 09		15.7	GALAXY
MCG+08-01-017	23 57 18.	+ 46 36	90	14.	GALAXY
ZWG 549.012	23 57 18.	+ 46 37		15.1	GALAXY
ZWG 548.021	23 57 18.	+ 46 37		15.1	GALAXY
UGC 12888	23 57 18.	+ 46 37	84	15.1	GALAXY Sc
PHL 6146	23 57 18.	- 10 52		18.6	BLUE STELLAR OBJECT
TON-S 0132	23 57 18.	- 22 21		14.8	BLUE STAR
PHL 0612	23 57 18.	- 22 22		14.7	BLUE STELLAR OBJECT
PHL 6147	23 57 18.	- 22 38		18.4	BLUE STELLAR OBJECT
MCG-05-01-021	23 57 18.	- 27 31	24	15.5	GALAXY
LB 04892	23 57 18.	- 30 48		19.1	FAINT BLUE STAR
SC 2354-0059.2	23 57 21.	- 00 42 30.	12		NEBULA
PHL 0613	23 57 24.	+ 08 11		18.3	BLUE STELLAR OBJECT
PHL 2515	23 57 24.	+ 08 35		18.3	BLUE STELLAR OBJECT
ZC 2357.4+1515	23 57 24.	+ 15 15	4030		CLUSTER OF GALAXIES
ZC 2357.4+2018	23 57 24.	+ 20 18	2220		CLUSTER OF GALAXIES
ZC 2357.4+3449	23 57 24.	+ 34 49	2150		CLUSTER OF GALAXIES
MCG+06-01-002	23 57 24.	+ 35 09	30	14.5	GALAXY
MCG+08-01-018	23 57 24.	+ 46 59 30.	120	12.	GALAXY
PHL 6150	23 57 24.	- 07 44		17.8	BLUE STELLAR OBJECT
PHL 6148	23 57 24.	- 08 48		18.4	BLUE STELLAR OBJECT
PHL 6151	23 57 24.	- 09 27		18.6	BLUE STELLAR OBJECT
PHL 6149	23 57 24.	- 13 01		18.1	BLUE STELLAR OBJECT
PHL 2513	23 57 24.	- 13 22		16.6	BLUE STELLAR OBJECT
PHL 0614	23 57 24.	- 16 52		17.8	BLUE STELLAR OBJECT
PHL 0615	23 57 24.	- 17 26		16.7	BLUE STELLAR OBJECT
PHL 2514	23 57 24.	- 18 34		17.9	BLUE STELLAR OBJECT
LB 04893	23 57 24.	- 30 45		18.9	FAINT BLUE STAR
IC 1525	23 57 28.	+ 46 35 52.			NONSTELLAR OBJECT
PHL 6152	23 57 30.	+ 05 07		16.6	BLUE STELLAR OBJECT
PHL 2517	23 57 30.	+ 06 44		18.7	BLUE STELLAR OBJECT
ZC 2357.5+0838	23 57 30.	+ 08 38	400		CLUSTER OF GALAXIES
PHL 0616	23 57 30.	+ 09 06		18.4	BLUE STELLAR OBJECT
ZWG 456.013	23 57 30.	+ 15 36		15.4	GALAXY
MCG+05-01-018	23 57 30.	+ 32 28 30.	42	15.	GALAXY
ZWG 549.013	23 57 30.	+ 47 00		14.0	GALAXY
ZWG 548.022	23 57 30.	+ 47 00		14.0	GALAXY
UGC 12889	23 57 30.	+ 47 00	138	14.0	GALAXY SBb
ISS 0108	23 57 30.	+ 61 12	334		STELLAR RING
MCG+00-01-014	23 57 30.	- 00 20	12	14.	GALAXY
ZWG 382.014	23 57 30.	- 00 21		15.6	GALAXY
KARA.72 600A	23 57 30.	- 00 21	24	15.6	PART OF DOUBLE GALAXY
PHL 2518	23 57 30.	- 03 00		18.5	BLUE STELLAR OBJECT
PHL 0617	23 57 30.	- 03 10		17.9	BLUE STELLAR OBJECT
MCG-03-01-012	23 57 30.	- 18 18	36	15.	GALAXY
PHL 2516	23 57 30.	- 18 34		17.9	BLUE STELLAR OBJECT
PHL 6153	23 57 30.	- 22 12		17.7	BLUE STELLAR OBJECT
LB 04894	23 57 30.	- 30 53		19.1	FAINT BLUE STAR
SG 3.246	23 57 31.	+ 66	9000		DIFFUSE EMISSION NEBULA
SN 1951F	23 57 31.	- 06 39		17.3	SUPERNOVA
ARC 2687	23 57 32.	+ 31 54		16.7	RICH CLUSTER OF GALAXIES
MCG+01-01-006	23 57 33.	+ 07 59 30.	36	15.	GALAXY
ARC 2688	23 57 33.	+ 15 34		17.5	RICH CLUSTER OF GALAXIES
SC 2354-0546.3	23 57 33.	- 05 29 36.	30		NEBULA
IC 5370	23 57 35.	+ 32 26 40.			NONSTELLAR OBJECT
ARC 2689	23 57 35.	- 16 01		18.0	RICH CLUSTER OF GALAXIES
PHL 6154	23 57 36.	+ 05 29		17.9	BLUE STELLAR OBJECT
PHL 2519	23 57 36.	+ 06 32		17.9	BLUE STELLAR OBJECT
UGC 12890	23 57 36.	+ 08 00	66	16.0	GALAXY E
ZWG 499.028	23 57 36.	+ 32 28		14.9	GALAXY
ZWG 498.056	23 57 36.	+ 32 28		14.9	GALAXY
ZWG 499.029	23 57 36.	+ 32 52		15.4	GALAXY
ZWG 498.057	23 57 36.	+ 32 52		15.4	GALAXY
MCG+08-01-019	23 57 36.	+ 46 40 30.	66	13.	GALAXY
OCL 0277	23 57 36.	+ 60 41	480	15.	OPEN STAR CLUSTER
ZWG 382.015	23 57 36.	- 00 19		15.5	GALAXY
KARA.72 600B	23 57 36.	- 00 19	24	15.5	PART OF DOUBLE GALAXY

OBJECT NAME	RIGHT ASCEN.	DECLINATION	DIAM.	MAGN.	TYPE OF OBJECT
MCG+00-01-015	23 57 36.	- 00 19	15	14.	GALAXY
MCG-01-01-015	23 57 36.	- 05 28	36	14.	GALAXY
PHL 2520	23 57 36.	- 06 25		18.3	BLUE STELLAR OBJECT
MCG-01-01-016	23 57 36.	- 06 37	90	14.	GALAXY
PHL 6155	23 57 36.	- 07 51		18.2	BLUE STELLAR OBJECT
PHL 0618	23 57 36.	- 26 45		18.3	BLUE STELLAR OBJECT
PHL 6156	23 57 36.	- 27 06		17.7	BLUE STELLAR OBJECT
LB 04895	23 57 36.	- 30 31		18.5	FAINT BLUE STAR
LB 04896	23 57 36.	- 30 45		19.2	FAINT BLUE STAR
PHL 6157	23 57 36.	- 31 56		18.3	BLUE STELLAR OBJECT
SC 2355-0543.3	23 57 38.	- 05 26 36.	54		NEBULA
KN 13.025	23 57 38.0	+ 13 48 41.			NEBULA
MCG-01-01-017	23 57 39.	- 05 25	48	14.	GALAXY
IC 5371	23 57 40.	+ 32 32 10.			NONSTELLAR OBJECT
ZWG 382.016	23 57 42.	+ 00 51		15.4	GALAXY
ZC 2357.7+2137	23 57 42.	+ 21 37	1280		CLUSTER OF GALAXIES
VV 186B	23 57 42.	+ 22 42	12	15.	INTERACTING GALAXY
VV 186A	23 57 42.	+ 22 42	12	16.	INTERACTING GALAXY
VV 186	23 57 42.	+ 22 42	60		INTERACTING GALAXY
IC 5372	23 57 42.	+ 32 29 52.			NONSTELLAR OBJECT
ZWG 499.030	23 57 42.	+ 32 34		15.7	GALAXY
ZWG 498.058	23 57 42.	+ 32 34		15.7	GALAXY
ZWG 549.014	23 57 42.	+ 46 41		14.8	GALAXY
ZWG 548.023	23 57 42.	+ 46 41		14.8	GALAXY
8ZW 2357-02.9	23 57 42.	- 02 55		17.1	COMPACT GALAXY
PHL 6158	23 57 42.	- 05 04		17.6	BLUE STELLAR OBJECT
PHL 2521	23 57 42.	- 18 06		18.3	BLUE STELLAR OBJECT
PHL 2522	23 57 42.	- 23 45		18.4	BLUE STELLAR OBJECT
LB 04897	23 57 42.	- 30 38		19.8	FAINT BLUE STAR
MCG+00-01-016	23 57 45.	+ 00 52	48	13.5	GALAXY
MCG+04-01-007	23 57 45.	+ 22 41 30.	72	15.	GALAXY
SC 2355-0310.4	23 57 46.	- 02 53 42.	12		NEBULA
ARC 2690	23 57 47.	- 25 25		17.2	RICH CLUSTER OF GALAXIES
PHL 6159	23 57 48.	+ 04 37		17.7	BLUE STELLAR OBJECT
MCG+01-01-007	23 57 48.	+ 07 33	36	15.	GALAXY
ZWG 478.009	23 57 48.	+ 22 43		15.1	GALAXY
ZWG 477.037	23 57 48.	+ 22 43		15.1	GALAXY
UGC 12891	23 57 48.	+ 22 43	66	15.1	GALAXY DBL SYS
4ZW 177	23 57 48.	+ 22 44			COMPACT GALAXY
MCG+05-01-019	23 57 48.	+ 32 30	42	15.	GALAXY
ZWG 518.005	23 57 48.	+ 34 20		15.5	GALAXY
ZWG 517.010	23 57 48.	+ 34 20		15.5	GALAXY
PHL 2524	23 57 48.	- 02 43		18.5	BLUE STELLAR OBJECT
MCG-01-01-018	23 57 48.	- 07 11	42	14.5	GALAXY
MCG-01-01-019	23 57 48.	- 08 09	54	16.	GALAXY
PHL 6160	23 57 48.	- 12 16		18.2	BLUE STELLAR OBJECT
PHL 2523	23 57 48.	- 12 58		17.0	BLUE STELLAR OBJECT
PHL 0619	23 57 48.	- 25 30		18.2	BLUE STELLAR OBJECT
PHL 0620	23 57 48.	- 28 08		17.9	BLUE STELLAR OBJECT
RNGC 7801	23 57 49.	+ 50 26			NON-EXISTENT OBJECT
SCHO 1454	23 57 50.	+ 56 25 00.	170		ISOLATED DARK CLOUD
SC 2355-0256.5	23 57 50.	- 02 39 48.	12		NEBULA
SC 2355-0101.6	23 57 51.	- 00 44 54.	6		NEBULA
LB 01198	23 57 52.	+ 32 11 12.		15.9	FAINT BLUE STAR
ARP 249	23 57 53.	+ 22 43			PECULIAR GALAXY
PHL 6162	23 57 54.	+ 05 42		18.1	BLUE STELLAR OBJECT
ZWG 408.006	23 57 54.	+ 07 34		15.7	GALAXY
UGC 12892	23 57 54.	+ 07 34	60	15.7	GALAXY SBa-b
ZWG 456.014	23 57 54.	+ 16 57		15.3	GALAXY
UGC 12893	23 57 54.	+ 16 57	120	15.3	GALAXY
IC 5373	23 57 54.	+ 32 29 16.			NONSTELLAR OBJECT
ZWG 499.031	23 57 54.	+ 32 31		15.1	GALAXY
ZWG 498.059	23 57 54.	+ 32 31		15.1	GALAXY
UGC 12894	23 57 54.	+ 39 14	60	17.	GALAXY DWRF IR
MCG+00-01-017	23 57 54.	- 02 02	42	14.	GALAXY
MCG-01-01-020	23 57 54.	- 02 51	60	15.	GALAXY
PHL 0621	23 57 54.	- 05 45		18.6	BLUE STELLAR OBJECT
8ZW 2357-06.0	23 57 54.	- 06 03		17.2	COMPACT GALAXY
PHL 6161	23 57 54.	- 10 14		15.4	BLUE STELLAR OBJECT
PHL 2525	23 57 54.	- 13 02		16.3	BLUE STELLAR OBJECT
KN 13.026	23 57 54.4	+ 11 18 09.			NEBULA
RNGC 7807	23 57 55.	- 19 07			GALAXY
MCG+04-01-009	23 57 57.	+ 26 01	12	16.	GALAXY
MCG+04-01-008	23 57 57.	+ 26 02	48	15.	GALAXY
SNO 32	23 57 58.	- 27 45 15.	600	17.	GROUP OF 7 GALAXIES
LBN 0580	23 58	+ 67 10	420		BRIGHT NEBULA
VDB.66G 221	23 58	- 15 43	740		DWARF GALAXY
PHL 2526	23 58 00.	+ 03 56		18.2	BLUE STELLAR OBJECT
PHL 6163	23 58 00.	+ 04 12		18.3	BLUE STELLAR OBJECT
MCG+03-01-011	23 58 00.	+ 16 57	90	14.5	GALAXY
UGC 12895	23 58 00.	+ 19 47	72	16.5	GALAXY
ZWG 478.010	23 58 00.	+ 26 03		14.7	GALAXY
ZWG 477.038	23 58 00.	+ 26 03		14.7	GALAXY
ZWG 499.032	23 58 00.	+ 28 07		14.9	GALAXY
ZWG 498.060	23 58 00.	+ 28 07		14.9	GALAXY
UGC 12897	23 58 00.	+ 28 07	72	14.9	GALAXY Sa-b
ZC 2358.0+3145	23 58 00.	+ 31 45	3090		CLUSTER OF GALAXIES
UGC 12896	23 58 00.	+ 36 03	60	14.7	GALAXY S
SCHO 1455	23 58 00.	+ 58 13 24.	360		ISOLATED DARK CLOUD
LDN 1267	23 58 00.	+ 66 30	540		DARK NEBULA
LDN 1269	23 58 00.	+ 66 53	600		DARK NEBULA
PHL 0622	23 58 00.	- 05 46		16.9	BLUE STELLAR OBJECT
PHL 2527	23 58 00.	- 13 28		18.2	BLUE STELLAR OBJECT
PHL 2528	23 58 00.	- 18 50		18.2	BLUE STELLAR OBJECT
SC 2355-0158.6	23 58 04.	- 01 41 54.			NEBULA
SC 2355-0219.8	23 58 04.	- 02 03 06.	24		NEBULA
MCG+05-01-020	23 58 06.	+ 28 06	66	14.5	GALAXY
ZWG 499.033	23 58 06.	+ 28 08		15.5	GALAXY
ZWG 498.061	23 58 06.	+ 28 08		15.5	GALAXY
UGC 12898	23 58 06.	+ 33 20	60	16.5	GALAXY Sc
ZWG 382.017	23 58 06.	- 02 03		15.3	GALAXY
PHL 2529	23 58 06.	- 08 14		18.4	BLUE STELLAR OBJECT
PHL 6164	23 58 06.	- 27 16		18.2	BLUE STELLAR OBJECT
LB 04898	23 58 06.	- 30 33		15.4	FAINT BLUE STAR
LB 04899	23 58 06.	- 30 59		19.8	FAINT BLUE STAR
KN 13.027	23 58 06.5	+ 14 16 18.			NEBULA
MCG+05-01-021	23 58 09.	+ 28 06 30.	42	15.	GALAXY
KN 13.028	23 58 11.9	+ 13 47 39.			NEBULA
MCG+05-01-022	23 58 12.	+ 28 07	60	15.	GALAXY
ZWG 499.034	23 58 12.	+ 28 08		14.4	GALAXY
ZWG 498.062	23 58 12.	+ 28 08		14.4	GALAXY
HOLM 825B	23 58 12.	+ 28 08	24	15.0	PART OF MULTIPLE GALAXY
UGC 12899	23 58 12.	+ 28 08	48	14.4	GALAXY COMPACT
PHL 0623	23 58 12.	- 02 45		17.9	BLUE STELLAR OBJECT
PHL 2531	23 58 12.	- 03 00		18.5	BLUE STELLAR OBJECT
SC 2355-0402.4	23 58 12.	- 03 45 42.	6		NEBULA
MCG-01-01-021	23 58 12.	- 05 51	30	14.5	GALAXY
PHL 2532	23 58 12.	- 05 58		18.0	BLUE STELLAR OBJECT
PHL 2530	23 58 12.	- 07 46		16.3	BLUE STELLAR OBJECT
PHL 2533	23 58 12.	- 15 10		18.0	BLUE STELLAR OBJECT
PHL 0624	23 58 12.	- 21 07		16.4	BLUE STELLAR OBJECT
LB 04900	23 58 12.	- 30 16		16.5	FAINT BLUE STAR
LB 04901	23 58 12.	- 30 27		17.9	FAINT BLUE STAR
LB 04902	23 58 12.	- 30 34		17.3	FAINT BLUE STAR
SC 2355-0254.5	23 58 14.	- 02 37 48.			NEBULA
KN 13.029	23 58 14.4	+ 13 59 41.			NEBULA
HOLM 825A	23 58 15.	+ 28 08	18	14.7	PART OF MULTIPLE GALAXY
KN 13.030	23 58 15.7	+ 11 11 30.			NEBULA
PHL 6165	23 58 18.	- 16 42		16.5	BLUE STELLAR OBJECT
PHL 2534	23 58 18.	- 23 33		17.9	BLUE STELLAR OBJECT
LB 04903	23 58 18.	- 29 50		18.2	FAINT BLUE STAR
LB 04904	23 58 18.	- 30 14		17.2	FAINT BLUE STAR
LB 04905	23 58 18.	- 30 27		17.6	FAINT BLUE STAR
LB 04906	23 58 18.	- 30 42		17.6	FAINT BLUE STAR
ZC 2358.4+0220	23 58 24.	+ 02 20	1610		CLUSTER OF GALAXIES
PHL 2535	23 58 24.	+ 07 08		18.6	BLUE STELLAR OBJECT
ZWG 456.015	23 58 24.	+ 20 04		15.7	GALAXY
UGC 12900	23 58 24.	+ 20 04	120	15.7	GALAXY Sc
MCG+03-01-012	23 58 24.	+ 20 05	102	14.5	GALAXY
MCG+05-01-023	23 58 24.	+ 28 37 30.	72	14.5	GALAXY
ZWG 499.035	23 58 24.	+ 28 39		14.8	GALAXY
ZWG 498.063	23 58 24.	+ 28 39		14.8	GALAXY
UGC 12901	23 58 24.	+ 28 39	108	14.8	GALAXY SBb
MCG+06-01-003	23 58 24.	+ 34 22	24	16.5	GALAXY
ISS 0029	23 58 24.	+ 60 35	451		STELLAR RING
PHL 2536	23 58 24.	- 04 44		18.5	BLUE STELLAR OBJECT
PHL 2537	23 58 24.	- 09 06		18.6	BLUE STELLAR OBJECT
PHL 6166	23 58 24.	- 11 34		18.1	BLUE STELLAR OBJECT
PHL 0625	23 58 24.	- 12 04		18.3	BLUE STELLAR OBJECT
PHL 6167	23 58 24.	- 12 18		17.9	BLUE STELLAR OBJECT
PHL 2538	23 58 24.	- 16 07		18.2	BLUE STELLAR OBJECT
MCG-03-01-013	23 58 24.	- 19 14	30	15.	GALAXY
MCG-06-01-013	23 58 24.	- 33 54	48	15.	GALAXY
MCG+01-01-008	23 58 27.	+ 05 57	60	14.	GALAXY
SC 2355-0452.8	23 58 27.	- 04 36 06.	18		NEBULA
SC 2355-0453.0	23 58 27.	- 04 36 18.	6		NEBULA
SNO 33	23 58 27.	- 25 29 45.	600	19.	CLUSTER OF 15 GALAXIES
HN 2886	23 58 28.7	- 43 36 24.	18		NEBULA
MCG+01-01-010	23 58 30.	+ 04 11 30.	36	14.5	GALAXY
IC 5374	23 58 30.	+ 04 12 52.			NONSTELLAR OBJECT
MCG+01-01-009	23 58 30.	+ 04 14 30.	54	14.5	GALAXY
IC 5375	23 58 30.	+ 04 15 16.			NONSTELLAR OBJECT
ZWG 408.007	23 58 30.	+ 05 57		14.7	GALAXY
UGC 12902	23 58 30.	+ 05 57	66	14.7	GALAXY S0
ZC 2358.5+1246	23 58 30.	+ 12 46	4440		CLUSTER OF GALAXIES
MCG+02-01-010	23 58 30.	+ 12 51	36	15.5	GALAXY
MCG+02-01-009	23 58 30.	+ 12 53	48	15.	GALAXY
MCG+02-01-008	23 58 30.	+ 12 54	18	19.	GALAXY
ZWG 518.006	23 58 30.	+ 34 22		15.7	GALAXY
ZWG 517.011	23 58 30.	+ 34 22		15.7	GALAXY
MCG+06-01-004	23 58 30.	+ 34 22 30.	36	14.5	GALAXY
LB 04907	23 58 30.	- 30 06		18.0	FAINT BLUE STAR
KN 13.031	23 58 30.0	+ 14 18 04.			NEBULA
KN 13.032	23 58 30.2	+ 15 21 04.			NEBULA
RNGC 7802	23 58 31.	+ 05 57		14.5	GALAXY
MCG+01-01-011	23 58 33.	+ 06 02 30.	60	15.	GALAXY
ZWG 408.008	23 58 36.	+ 04 13		15.4	GALAXY
KARA.72 601A	23 58 36.	+ 04 13	36	15.4	PART OF DOUBLE GALAXY
ZWG 408.009	23 58 36.	+ 04 15		15.3	GALAXY
KARA.72 601B	23 58 36.	+ 04 15	48	15.3	PART OF DOUBLE GALAXY
PHL 6169	23 58 36.	+ 04 56		17.8	BLUE STELLAR OBJECT
PHL 6168	23 58 36.	+ 07 02		17.0	BLUE STELLAR OBJECT
PHL 6170	23 58 36.	+ 07 59		18.3	BLUE STELLAR OBJECT
MCG+02-01-011	23 58 36.	+ 12 51	60	14.	GALAXY
PHL 0626	23 58 36.	- 07 57		13.6	BLUE STELLAR OBJECT
PHL 2539	23 58 36.	- 11 10		18.4	BLUE STELLAR OBJECT
PHL 2540	23 58 36.	- 19 24		18.4	BLUE STELLAR OBJECT
LB 04908	23 58 36.	- 30 54		20.2	FAINT BLUE STAR
MCG+06-01-005	23 58 39.	+ 34 24	66	14.5	GALAXY
MCG-05-01-022	23 58 39.	- 27 41	18	15.5	GALAXY
KN 13.033	23 58 39.5	+ 12 51 54.			NEBULA
HN 2887	23 58 39.7	- 44 17 18.	18		NEBULA
KN 13.034	23 58 41.4	+ 12 49 59.			NEBULA
PHL 6173	23 58 42.	+ 04 15		18.0	BLUE STELLAR OBJECT
ZWG 408.010	23 58 42.	+ 06 03		15.7	GALAXY
UGC 12903	23 58 42.	+ 06 03	66	15.7	GALAXY Sb-c
PHL 2542	23 58 42.	+ 08 03		18.6	BLUE STELLAR OBJECT
MCG+02-01-012	23 58 42.	+ 12 51	42	15.	GALAXY
ZWG 518.007	23 58 42.	+ 34 24		15.4	GALAXY
ZWG 517.012	23 58 42.	+ 34 24		15.4	GALAXY
UGC 12904	23 58 42.	+ 34 24	66	15.4	GALAXY SBa-b
MCG+06-01-006	23 58 42.	+ 36 42	21	15.	GALAXY
UGC 12905	23 58 42.	+ 80 23	72	16.5	GALAXY Sc
PHL 2541	23 58 42.	- 02 30		18.1	BLUE STELLAR OBJECT
PHL 6172	23 58 42.	- 03 28		16.5	BLUE STELLAR OBJECT
PHL 6171	23 58 42.	- 03 28		16.2	BLUE STELLAR OBJECT
PHL 6174	23 58 42.	- 06 22		18.9	BLUE STELLAR OBJECT
LB 04909	23 58 42.	- 30 43		16.6	FAINT BLUE STAR
MCG+02-01-013	23 58 45.	+ 12 50	24	18.	GALAXY
ARC 2691	23 58 46.	- 03 22		16.6	RICH CLUSTER OF GALAXIES
KN 13.035	23 58 46.3	+ 12 49 54.			NEBULA
KEEL 734	23 58 46.7	+ 16 12 16.			NEBULA
IC 5376	23 58 47.	+ 34 14 52.			NONSTELLAR OBJECT
ZC 2358.8+0207	23 58 48.	+ 02 07	270		CLUSTER OF GALAXIES
PHL 2543	23 58 48.	+ 06 16		17.0	BLUE STELLAR OBJECT
PHL 6175	23 58 48.	+ 06 22		17.7	BLUE STELLAR OBJECT
ZWG 433.013	23 58 48.	+ 12 50		13.8	GALAXY
UGC 12906	23 58 48.	+ 12 50	60	13.8	GALAXY S0-a
UGC 12907	23 58 48.	+ 17 55	66	17.	GALAXY DBL SYS
ZWG 499.036	23 58 48.	+ 31 09		14.3	GALAXY
ZWG 498.064	23 58 48.	+ 31 09		14.3	GALAXY
UGC 12908	23 58 48.	+ 31 09	66	14.3	GALAXY
KARA.72 602A	23 58 48.	+ 31 09	72	14.3	PART OF DOUBLE GALAXY
ZWG 518.008	23 58 48.	+ 34 15		14.7	GALAXY
ZWG 517.013	23 58 48.	+ 34 15		14.7	GALAXY
UGC 12909	23 58 48.	+ 34 15	126	14.7	GALAXY Sa-b
MCG+06-01-007	23 58 48.	+ 34 16	114	13.5	GALAXY
PHL 2544	23 58 48.	- 08 04		17.1	BLUE STELLAR OBJECT
PHL 6177	23 58 48.	- 11 36		16.0	BLUE STELLAR OBJECT
PHL 6178	23 58 48.	- 19 50		18.4	BLUE STELLAR OBJECT
PHL 2545	23 58 48.	- 30 11		18.5	BLUE STELLAR OBJECT
PHL 6176	23 58 48.	- 30 51		18.0	BLUE STELLAR OBJECT
PHL 6179	23 58 48.	- 31 51		18.4	BLUE STELLAR OBJECT
RNGC 7804	23 58 49.	+ 07 28			NON-EXISTENT OBJECT
RNGC 7803	23 58 49.	+ 12 50		14.0	GALAXY
RNGC 7805	23 58 49.	+ 31 09		14.5	GALAXY

OBJECT NAME	RIGHT ASCEN.	DECLINATION	DIAM.	MAGN.	TYPE OF OBJECT
KEEL 735	23 58 49.4	+ 16 00 18.			NEBULA
SG 3.247	23 58 50.	+ 65 56	1200		DIFFUSE EMISSION NEBULA
SG 3.248	23 58 50.	+ 68 12	1800		DIFFUSE EMISSION NEBULA
HOLM 826A	23 58 51.	+ 31 09	18	14.3	PART OF MULTIPLE GALAXY
MCG+05-01-024	23 58 51.	+ 31 09	66	15.	GALAXY
KN 13.036	23 58 52.5	+ 12 49 57.			NEBULA
ARP 112	23 58 53.	+ 31 10			PECULIAR GALAXY
HN 2888	23 58 53.5	- 44 11 42.	18		NEBULA
UGC 12910	23 58 54.	+ 05 05	66	17.	GALAXY
ZWG 433.014	23 58 54.	+ 12 50		15.3	GALAXY
VV 226B	23 58 54.	+ 31 09 36.	30	14.5	INTERACTING GALAXY
ZWG 499.037	23 58 54.	+ 31 10		14.4	GALAXY
ZWG 498.065	23 58 54.	+ 31 10		14.4	GALAXY
UGC 12911	23 58 54.	+ 31 10	78	14.4	GALAXY S
KARA.72 602B	23 58 54.	+ 31 10	66	14.4	PART OF DOUBLE GALAXY
MCG+05-01-026	23 58 54.	+ 31 10	30	17.	GALAXY
MCG+05-01-025	23 58 54.	+ 31 10	66	14.5	GALAXY
PHL 6180	23 58 54.	- 32 25		17.5	BLUE STELLAR OBJECT
RNGC 7806	23 58 55.	+ 31 10		14.5	GALAXY
HOLM 826B	23 58 55.	+ 31 10	30	14.5	PART OF MULTIPLE GALAXY
VV 226A	23 58 57.	+ 31 10 12.	78	14.3	INTERACTING GALAXY
KN 13.037	23 58 57.8	+ 11 03 59.			NEBULA
KN 13.038	23 58 59.8	+ 14 47 59.			NEBULA
LBN 0565	23 59	+ 48 20	2520		BRIGHT NEBULA
LBN 0577	23 59	+ 64 23	1200		BRIGHT NEBULA
LBN 0581	23 59	+ 66 55	900		BRIGHT NEBULA
LB 09960	23 59	- 80 10		14.6	FAINT BLUE STAR
PHL 2547	23 59 00.	+ 03 48		18.3	BLUE STELLAR OBJECT
PHL 2546	23 59 00.	+ 05 12		17.2	BLUE STELLAR OBJECT
PHL 6182	23 59 00.	+ 05 16		16.9	BLUE STELLAR OBJECT
PHL 6181	23 59 00.	+ 05 16		18.0	BLUE STELLAR OBJECT
ZWG 408.011	23 59 00.	+ 08 44		15.6	GALAXY
UGC 12912	23 59 00.	+ 08 44	60	15.6	GALAXY S
ZWG 433.015	23 59 00.	+ 11 04		15.1	GALAXY
IC 1526	23 59 00.	+ 11 04 10.			NONSTELLAR OBJECT
ZWG 433.016	23 59 00.	+ 14 48		15.7	GALAXY
MCG+02-01-014	23 59 00.	+ 14 49	48	15.	GALAXY
MCG+04-01-010	23 59 00.	+ 23 12 30.	108	14.	GALAXY
ZWG 478.011	23 59 00.	+ 26 39		15.6	GALAXY
ZWG 477.039	23 59 00.	+ 26 39		15.6	GALAXY
ZC 2359.0+2731	23 59 00.	+ 27 31	1280		CLUSTER OF GALAXIES
ZWG 499.038	23 59 00.	+ 33 18		15.0	GALAXY
ZWG 498.066	23 59 00.	+ 33 18		15.0	GALAXY
OCL 0281	23 59 00.	+ 64 22	300		OPEN STAR CLUSTER
LDN 1270	23 59 00.	+ 66 53	360		DARK NEBULA
PK118+08.2	23 59 00.	+ 70 26	70	16.7	PLANETARY NEBULA
PHL 6183	23 59 00.	- 03 00		18.2	BLUE STELLAR OBJECT
MCG-01-01-022	23 59 00.	- 04 35 30.	48	16.	GALAXY
PHL 0627	23 59 00.	- 05 36		13.9	BLUE STELLAR OBJECT
PHL 0628	23 59 00.	- 08 54		17.8	BLUE STELLAR OBJECT
PHL 2548	23 59 00.	- 26 38		18.3	BLUE STELLAR OBJECT
LB 04910	23 59 00.	- 29 59		20.5	FAINT BLUE STAR
LB 04911	23 59 00.	- 30 19		17.5	FAINT BLUE STAR
LB 01199	23 59 02.	+ 32 18 42.		16.9	FAINT BLUE STAR
MCG+00-01-018	23 59 03.	+ 03 14 30.	72	15.	GALAXY
KN 13.039	23 59 05.9	+ 13 28 26.			NEBULA
UGC 12913	23 59 06.	+ 03 14	84	16.0	GALAXY Sc
ZWG 478.012	23 59 06.	+ 23 13		13.2	GALAXY
ZWG 477.040	23 59 06.	+ 23 13		13.2	GALAXY
UGC 12914	23 59 06.	+ 23 13	162	13.2	GALAXY S
KARA.72 603A	23 59 06.	+ 23 13	126	13.2	PART OF DOUBLE GALAXY
ZWG 534.005	23 59 06.	+ 44 44		15.7	GALAXY
MRSL 117+02/1	23 59 06.	+ 64 21	1200		HII REGION
PHL 6184	23 59 06.	- 06 38		17.9	BLUE STELLAR OBJECT
MCG-03-01-014	23 59 06.	- 15 01	42	14.5	GALAXY
PHL 6185	23 59 06.	- 23 49		18.0	BLUE STELLAR OBJECT
PHL 0629	23 59 06.	- 28 44		16.6	BLUE STELLAR OBJECT
TON-S 0133	23 59 06.	- 28 44		15.8	BLUE STAR
LB 04912	23 59 06.	- 29 50		16.5	FAINT BLUE STAR
LB 04913	23 59 06.	- 30 32		17.0	FAINT BLUE STAR
AGU 84	23 59 06.	- 41 05 00.	36	14.5	PECULIAR GALAXY
SG 2.101	23 59 08.	+ 64 21	900		DIFFUSE EMISSION NEBULA
MCG+04-01-011	23 59 09.	+ 23 13 30.	90	14.5	GALAXY
SC 2356-0256.0	23 59 09.	- 02 39 18.	6		NEBULA
KEEL 736	23 59 10.4	+ 16 01 54.			NEBULA
KN 13.040	23 59 11.7	+ 12 49 13.			NEBULA
PHL 6186	23 59 12.	+ 05 20		18.3	BLUE STELLAR OBJECT
3ZW 125	23 59 12.	+ 23 13			COMPACT GALAXY
ZWG 478.013	23 59 12.	+ 23 14		13.9	GALAXY
ZWG 477.041	23 59 12.	+ 23 14		13.9	GALAXY
UGC 12915	23 59 12.	+ 23 14	96	13.9	GALAXY
KARA.72 603B	23 59 12.	+ 23 14	90	13.9	PART OF DOUBLE GALAXY
DG 192	23 59 12.	+ 64 17	960		REFLECTION NEBULA
PHL 2549	23 59 12.	- 10 20		18.7	BLUE STELLAR OBJECT
HMS 1.35	23 59 12.	- 15 43			IRR GALAXY
LB 04914	23 59 12.	- 29 44		18.3	FAINT BLUE STAR
LB 04915	23 59 12.	- 30 07		18.2	FAINT BLUE STAR
LB 04916	23 59 12.	- 30 35		17.2	FAINT BLUE STAR
MCG-06-01-014	23 59 12.	- 37 06	30	15.	GALAXY
SC 2356-0454.0	23 59 13.	- 04 37 18.	12		NEBULA
KEEL 737	23 59 13.6	+ 16 02 41.			NEBULA
MCG-01-01-023	23 59 15.	- 04 35 30.		15.5	GALAXY
SCHO 1456	23 59 16.	+ 61 54 24.	240		ISOLATED DARK CLOUD
PHL 6187	23 59 18.	- 03 01		18.2	BLUE STELLAR OBJECT
PHL 2552	23 59 18.	- 03 18		18.2	BLUE STELLAR OBJECT
PHL 2551	23 59 18.	- 03 18		5.1	BLUE STELLAR OBJECT
PHL 2550	23 59 18.	- 03 38		18.1	BLUE STELLAR OBJECT
PHL 6188	23 59 18.	- 04 46		16.4	BLUE STELLAR OBJECT
PHL 6189	23 59 18.	- 05 30		18.0	BLUE STELLAR OBJECT
PHL 6190	23 59 18.	- 13 06		18.6	BLUE STELLAR OBJECT
LB 01200	23 59 19.	+ 32 40 36.		16.4	FAINT BLUE STAR
UGC 12916	23 59 24.	+ 17 17	66	16.5	GALAXY
ZWG 456.016	23 59 24.	+ 21 21		15.4	GALAXY
ZC 2359.4+2506	23 59 24.	+ 25 06	1010		CLUSTER OF GALAXIES
ZWG 518.009	23 59 24.	+ 36 22		15.4	GALAXY
ZWG 517.014	23 59 24.	+ 36 22		15.4	GALAXY
UGC 12917	23 59 24.	+ 40 03	72	16.0	GALAXY SBb
ISS 0030	23 59 24.	+ 59 09	428		STELLAR RING
SC 2356-0350.0	23 59 24.	- 03 33 18.	30		NEBULA
PHL 6191	23 59 24.	- 04 34		18.2	BLUE STELLAR OBJECT
PHL 2554	23 59 24.	- 05 36		18.2	BLUE STELLAR OBJECT
PHL 2553	23 59 24.	- 07 20		18.0	BLUE STELLAR OBJECT
PHL 0630	23 59 24.	- 07 56		18.1	BLUE STELLAR OBJECT
PHL 6193	23 59 24.	- 10 48		18.0	BLUE STELLAR OBJECT
PHL 6192	23 59 24.	- 11 57		16.2	BLUE STELLAR OBJECT
MCG-05-01-023	23 59 24.	- 27 55	72	14.5	GALAXY
MCG-05-01-024	23 59 24.	- 28 17	72	15.	GALAXY
PHL 6194	23 59 24.	- 29 25		18.4	BLUE STELLAR OBJECT
LB 04917	23 59 24.	- 30 42		17.9	FAINT BLUE STAR
KEEL 738	23 59 26.3	+ 20 42 39.			NEBULA
MCG-03-01-015	23 59 27.	- 15 44	660	11.	GALAXY
ARC 2692	23 59 28.	+ 11 47		17.8	RICH CLUSTER OF GALAXIES
HN 2889	23 59 29.5	- 45 38 12.	18		NEBULA
ZC 2359.5+1147	23 59 30.	+ 11 47	2420		CLUSTER OF GALAXIES
ZC 2359.0+1147	23 59 30.	+ 11 47	2420		CLUSTER OF GALAXIES
ZWG 433.017	23 59 30.	+ 12 15		15.5	GALAXY
ZWG 456.017	23 59 30.	+ 16 19		15.6	GALAXY
UGC 12918	23 59 30.	+ 16 19	78	15.6	GALAXY IRR
MCG-02-01-009	23 59 30.	- 10 29	36	15.	GALAXY
PHL 2555	23 59 30.	- 11 06		18.3	BLUE STELLAR OBJECT
LB 04918	23 59 30.	- 30 07		18.2	FAINT BLUE STAR
LB 04919	23 59 30.	- 30 16		18.5	FAINT BLUE STAR
MCG-05-01-025	23 59 30.	- 30 55	12	16.	GALAXY
MCG-06-01-015	23 59 30.	- 33 44	78	15.	GALAXY
KN 13.041	23 59 30.0	+ 12 15 08.			NEBULA
KN 13.042	23 59 30.0	+ 15 07 31.			NEBULA
KN 13.043	23 59 31.6	+ 16 18 44.			NEBULA
IC 5377	23 59 33.	+ 16 17 40.			NONSTELLAR OBJECT
MCG+03-01-013	23 59 33.	+ 16 18	42	16.	GALAXY
ARC 2693	23 59 35.	- 19 50		17.5	RICH CLUSTER OF GALAXIES
3ZW 126	23 59 36.	+ 02 40			COMPACT GALAXY
ZWG 382.018	23 59 36.	+ 02 40		15.1	GALAXY
PHL 2556	23 59 36.	- 03 10		17.6	BLUE STELLAR OBJECT
PHL 6196	23 59 36.	- 03 42		17.9	BLUE STELLAR OBJECT
PHL 6195	23 59 36.	- 06 05		16.1	BLUE STELLAR OBJECT
PHL 0631	23 59 36.	- 06 23		16.6	BLUE STELLAR OBJECT
PHL 2557	23 59 36.	- 08 26		18.3	BLUE STELLAR OBJECT
PHL 2558	23 59 36.	- 17 12		18.0	BLUE STELLAR OBJECT
LB 04920	23 59 36.	- 29 58		13.4	FAINT BLUE STAR
KN 13.044	23 59 36.8	+ 14 53 24.			NEBULA
RNGC 7809	23 59 37.	+ 02 40		15.0	GALAXY
HN 2890	23 59 38.2	- 44 15 18.	12		NEBULA
KEEL 739	23 59 38.3	+ 16 05 51.			NEBULA
KN 13.045	23 59 38.6	+ 12 29 12.			NEBULA
MCG+00-01-019	23 59 39.	+ 02 40 30.	30	14.	GALAXY
MCG+02-01-015	23 59 45.	+ 12 41	72	14.5	GALAXY
KN 13.046	23 59 45.4	+ 12 41 24.			NEBULA
MCG+00-01-020	23 59 48.	+ 03 04	24	13.	GALAXY
MCG+01-01-012	23 59 48.	+ 03 46	36	14.5	GALAXY
IC 1527	23 59 48.	+ 03 50 10.			NONSTELLAR OBJECT
PHL 2563	23 59 48.	+ 07 16		18.1	BLUE STELLAR OBJECT
PHL 2561	23 59 48.	+ 07 16		18.2	BLUE STELLAR OBJECT
PHL 2560	23 59 48.	+ 09 20		18.6	BLUE STELLAR OBJECT
ZWG 433.018	23 59 48.	+ 12 41		14.3	GALAXY
UGC 12919	23 59 48.	+ 12 41	60	14.3	GALAXY S0
ZWG 478.014	23 59 48.	+ 26 56		15.5	GALAXY
ZWG 477.042	23 59 48.	+ 26 56		15.5	GALAXY
UGC 12920	23 59 48.	+ 26 56	78	15.5	GALAXY Sb-c
MCG+04-01-012	23 59 48.	+ 26 56	72	16.	GALAXY
ISS 0031	23 59 48.	+ 61 45	577		STELLAR RING
MIL 94	23 59 48.	+ 62 11	7800		SUPERNOVA REMNANT
PHL 6197	23 59 48.	- 03 20		18.5	BLUE STELLAR OBJECT
PHL 2559	23 59 48.	- 17 44		16.8	BLUE STELLAR OBJECT
PHL 2562	23 59 48.	- 29 59		4.9	BLUE STELLAR OBJECT
RNGC 7810	23 59 49.	+ 12 41		14.5	GALAXY
CED 214A	23 59 51.	+ 66 53	960		DIFFUSE GALACTIC NEBULA
3ZW 127	23 59 54.	+ 03 04			COMPACT GALAXY
ZWG 382.019	23 59 54.	+ 03 04		14.9	GALAXY
MRK 543	23 59 54.	+ 03 04	18	15.5	GALAXY WITH UV CONTINUUM
ZWG 408.012	23 59 54.	+ 03 48		15.4	GALAXY
ZWG 518.010	23 59 54.	+ 33 40		15.7	GALAXY
ZWG 517.015	23 59 54.	+ 33 40		15.7	GALAXY
MCG+06-01-008	23 59 54.	+ 33 40 30.	24	15.	GALAXY
ZWG 344.007	23 59 54.	+ 76 59		15.6	GALAXY
UGC 12921	23 59 54.	+ 76 59	102	15.6	GALAXY
8ZW 2359-07.9	23 59 54.	- 07 57		17.0	COMPACT GALAXY
PHL 0632	23 59 54.	- 08 00		18.4	BLUE STELLAR OBJECT
TON-S 0134	23 59 54.	- 22 28		15.6	BLUE STAR
RNGC 7811	23 59 55.	+ 03 04		15.0	GALAXY
SC 2357-7238.8	23 59 56.	- 72 22 06.	18		NEBULA
ARC 2694	23 59 58.	+ 08 10		17.0	RICH CLUSTER OF GALAXIES
VDB.66G 222	00 00	+ 15 02	100		DWARF GALAXY
LBN 0583	00 00	+ 67 00	6300		BRIGHT NEBULA
LBN 0582	00 00	+ 67 00	9900		BRIGHT NEBULA
LBN 0587	00 00	+ 68 20	3420		BRIGHT NEBULA
PHL 6201	00 00 00.	+ 04 00		18.2	BLUE STELLAR OBJECT
PHL 6200	00 00 00.	+ 05 44		18.4	BLUE STELLAR OBJECT
ZWG 456.018	00 00 00.	+ 16 22		14.9	GALAXY
UGC 00001	00 00 00.	+ 16 22	90	14.9	GALAXY DBL SYS
ZWG 456.019	00 00 00.	+ 21 19		15.7	GALAXY
MCG+03-01-014	00 00 00.	+ 21 20	27	17.	GALAXY
UGC 00002	00 00 00.	+ 44 39	66	17.	GALAXY
LDN 1272	00 00 00.	+ 67 00	11940		DARK NEBULA
OCL 0286	00 00 00.	+ 67 06	600	11.	OPEN STAR CLUSTER
LDN 1273	00 00 00.	+ 68 15	1800		DARK NEBULA
MCG+13-01-003	00 00 00.	+ 77 00	57	15.	GALAXY
SC 2357-0416.0	00 00 00.	- 03 59 18.	48		NEBULA
PHL 6199	00 00 00.	- 08 30		16.5	BLUE STELLAR OBJECT
PHL 2564	00 00 00.	- 18 38		17.8	BLUE STELLAR OBJECT
PHL 6198	00 00 00.	- 18 49		17.5	BLUE STELLAR OBJECT
PHL 6202	00 00 00.	- 28 54		18.6	BLUE STELLAR OBJECT
LB 04921	00 00 00.	- 29 51		18.5	FAINT BLUE STAR
LB 04922	00 00 00.	- 30 41		18.9	FAINT BLUE STAR
MCG-05-01-026	00 00 00.	- 30 55	30	15.	GALAXY
PHL 0633	00 00 00.	- 32 28		16.6	BLUE STELLAR OBJECT
RNGC 1638	00 00 01.	- 00 00		13.0	GALAXY
RNGC 1072	00 00 01.	- 00 00		14.5	GALAXY
KN 13.047	00 00 03.9	+ 16 22 01.			NEBULA
IC 5378	00 00 04.	+ 16 20 40.			NONSTELLAR OBJECT
SC 2357-0411.4	00 00 05.	- 03 54 42.	12		NEBULA
ZC 0000.1+0806	00 00 06.	+ 08 06	2420		CLUSTER OF GALAXIES
ZWG 408.013	00 00 06.	+ 08 27		15.7	GALAXY
MCG+01-01-013	00 00 06.	+ 08 27 30.	36	15.	GALAXY
ZWG 456.020	00 00 06.	+ 16 18		15.7	GALAXY
MCG+03-01-016	00 00 06.	+ 16 21	48	15.	GALAXY
MCG+03-01-015	00 00 06.	+ 16 21	30	14.5	GALAXY
VV 263B	00 00 06.	+ 16 22	48	16.	INTERACTING GALAXY
VV 263A	00 00 06.	+ 16 22	42	15.	INTERACTING GALAXY
VV 263	00 00 06.	+ 16 22	72		INTERACTING GALAXY
PHL 2565	00 00 06.	- 02 37		18.9	BLUE STELLAR OBJECT
MCG-01-01-024	00 00 06.	- 03 57 30.	60	14.	GALAXY
PHL 6203	00 00 06.	- 09 37		18.0	BLUE STELLAR OBJECT
PHL 6204	00 00 06.	- 12 18		18.0	BLUE STELLAR OBJECT
LB 04923	00 00 06.	- 30 50		18.0	FAINT BLUE STAR
KN 13.048	00 00 06.8	+ 16 19 26.			NEBULA
IC 5379	00 00 07.	+ 16 18 16.			NONSTELLAR OBJECT

OBJECT NAME	RIGHT ASCEN.	DECLINATION	DIAM.	MAGN.	TYPE OF OBJECT
ARC 2695	00 00 10.	+ 18 29		17.7	RICH CLUSTER OF GALAXIES
SC 2357-0409.9	00 00 10.	- 03 53 12.	60		NEBULA
HN 1237	00 00 10.	- 66 28			NEBULA
IC 5380	00 00 10.	- 66 28			NONSTELLAR OBJECT
SC 2357-6805.4	00 00 10.	- 67 48 42.	6		NEBULA
PHL 6207	00 00 12.	+ 04 11		18.3	BLUE STELLAR OBJECT
PHL 0634	00 00 12.	+ 07 22		17.8	BLUE STELLAR OBJECT
MCG+03-01-017	00 00 12.	+ 16 18	24	17.5	GALAXY
ZWG 456.021	00 00 12.	+ 18 37		14.8	GALAXY
UGC 00003	00 00 12.	+ 18 37	120	14.8	GALAXY SBa
SCHO 0001	00 00 12.	+ 61 30 30.	320		ISOLATED DARK CLOUD
PHL 2566	00 00 12.	- 03 18		18.8	BLUE STELLAR OBJECT
8ZW 0000-03.8	00 00 12.	- 03 50		15.5	COMPACT GALAXY
MCG-01-01-025	00 00 12.	- 03 52	36	16.	GALAXY
PHL 6205	00 00 12.	- 05 06		18.0	BLUE STELLAR OBJECT
PHL 6206	00 00 12.	- 05 12		17.8	BLUE STELLAR OBJECT
PHL 6208	00 00 12.	- 18 08		18.6	BLUE STELLAR OBJECT
LB 04924	00 00 12.	- 29 48		18.3	FAINT BLUE STAR
LB 04925	00 00 12.	- 30 54		17.6	FAINT BLUE STAR
MCG+03-01-018	00 00 15.	+ 18 38	66	14.5	GALAXY
SC 2357-0406.9	00 00 15.	- 03 50 12.	12		NEBULA
ZWG 382.020	00 00 18.	+ 03 13		15.6	GALAXY
ZWG 499.039	00 00 18.	+ 31 12		15.6	GALAXY
ZWG 498.067	00 00 18.	+ 31 12		15.6	GALAXY
MCG+05-01-027	00 00 18.	+ 31 12 30.	36	15.	GALAXY
MRSL 118+06/1	00 00 18.	+ 68 13	6600		HII REGION
PHL 0635	00 00 18.	- 08 14		18.0	BLUE STELLAR OBJECT
LB 04926	00 00 18.	- 29 46		18.4	FAINT BLUE STAR
LB 04927	00 00 18.	- 30 30		18.7	FAINT BLUE STAR
MCG-05-01-027	00 00 18.	- 32 30	12	16.	GALAXY
RNGC 7812	00 00 19.	- 34 31			GALAXY
KN 13.049	00 00 19.6	+ 13 08 31.			NEBULA
SCHO 0002	00 00 20.	+ 61 57 42.	250		ISOLATED DARK CLOUD
MCG-01-01-026	00 00 21.	- 03 50 30.	60	15.	GALAXY
ARP 130	00 00 23.	+ 16 22			PECULIAR GALAXY
MCG+01-01-014	00 00 24.	+ 03 54	36	14.5	GALAXY
PHL 2568	00 00 24.	+ 14 17		18.6	BLUE STELLAR OBJECT
ZC 0000.4+1931	00 00 24.	+ 19 31	4300		CLUSTER OF GALAXIES
PHL 6209	00 00 24.	- 12 16		18.1	BLUE STELLAR OBJECT
PHL 0636	00 00 24.	- 26 21		17.1	BLUE STELLAR OBJECT
PHL 2567	00 00 24.	- 28 10		18.5	BLUE STELLAR OBJECT
LB 04928	00 00 24.	- 30 08		18.7	FAINT BLUE STAR
LB 04929	00 00 24.	- 30 26		19.5	FAINT BLUE STAR
MCG-06-01-016	00 00 24.	- 34 31	36	13.5	GALAXY
HOLM 827B	00 00 25.	- 02 14	30	14.6	PART OF MULTIPLE GALAXY
ZWG 408.014	00 00 30.	+ 03 56		15.5	GALAXY
UGC 00004	00 00 30.	+ 03 56	78	15.5	GALAXY Sb-c
ZWG 456.022	00 00 30.	+ 18 36		15.5	GALAXY
MCG+00-01-021	00 00 30.	- 02 12	72	12.8	GALAXY
PHL 2569	00 00 30.	- 06 44		18.4	BLUE STELLAR OBJECT
MCG-02-01-010	00 00 30.	- 12 40	60	15.	GALAXY
PHL 6210	00 00 30.	- 29 12		18.4	BLUE STELLAR OBJECT
LB 04930	00 00 30.	- 29 51		18.5	FAINT BLUE STAR
SC 2357-0228.9	00 00 32.	- 02 12 12.	60		NEBULA
HOLM 827A	00 00 34.	- 02 11	66	12.8	PART OF MULTIPLE GALAXY
KN 13.050	00 00 34.9	+ 15 50 32.			NEBULA
MCG+00-01-022	00 00 36.	+ 03 20 30.	42	13.5	GALAXY
PHL 0637	00 00 36.	+ 07 00		18.2	BLUE STELLAR OBJECT
4ZW 001	00 00 36.	+ 21 42			COMPACT GALAXY
ZWG 478.015	00 00 36.	+ 21 42		14.4	GALAXY
ZWG 477.043	00 00 36.	+ 21 42		14.4	GALAXY
MRK 334	00 00 36.	+ 21 42	18	15.	GALAXY WITH UV CONTINUUM
UGC 00006	00 00 36.	+ 21 42	60	14.4	GALAXY PECULR
4ZW 002	00 00 36.	+ 24 54			COMPACT GALAXY
ZWG 478.016	00 00 36.	+ 24 54		15.5	GALAXY
ZWG 477.044	00 00 36.	+ 24 54		15.5	GALAXY
ZWG 499.040	00 00 36.	+ 30 45		15.0	GALAXY
ZWG 498.068	00 00 36.	+ 30 45		15.0	GALAXY
ZWG 382.021	00 00 36.	- 02 11		14.3	GALAXY
UGC 00005	00 00 36.	- 02 11	84	14.3	GALAXY SBb/Sc
KARA.73B 0001	00 00 36.	- 02 11	144	14.3	ISOLATED GALAXY S
PHL 6212	00 00 36.	- 03 09		17.0	BLUE STELLAR OBJECT
PHL 2570	00 00 36.	- 04 09		18.1	BLUE STELLAR OBJECT
PHL 6211	00 00 36.	- 12 58		15.8	BLUE STELLAR OBJECT
PHL 2571	00 00 36.	- 16 00		18.3	BLUE STELLAR OBJECT
PHL 2572	00 00 36.	- 28 08		18.7	BLUE STELLAR OBJECT
LB 01539	00 00 36.	- 48 03		14.2	FAINT BLUE STAR
KN 13.051	00 00 37.6	+ 15 41 12.			NEBULA
IC 5381	00 00 37.6	+ 15 41 14.			GALAXY
KEEL 740	00 00 37.6	+ 15 41 16.			NEBULA
MCG+03-01-019	00 00 39.	+ 15 41	60	15.	GALAXY
SC 2358-0404.3	00 00 40.	- 03 47 36.	12		NEBULA
KN 13.052	00 00 40.7	+ 15 51 52.			NEBULA
SC 2358-6555.4	00 00 41.	- 65 38 42.	66		NEBULA
ZWG 382.022	00 00 42.	+ 03 20		15.3	GALAXY
PHL 2574	00 00 42.	+ 05 42		18.4	BLUE STELLAR OBJECT
ZWG 456.023	00 00 42.	+ 15 42		14.9	GALAXY
UGC 00007	00 00 42.	+ 15 42	90	14.9	GALAXY PECULR
ZWG 456.024	00 00 42.	+ 15 52		12.0	GALAXY
UGC 00008	00 00 42.	+ 15 52	390	12.0	GALAXY Sa-b
MCG+03-01-020	00 00 42.	+ 15 52	300	11.7	GALAXY
MCG+04-01-013	00 00 42.	+ 21 40	60	14.5	GALAXY
MCG+05-01-028	00 00 42.	+ 29 31	30	15.	GALAXY
8ZW 0000-07.5	00 00 42.	- 07 28		17.5	COMPACT GALAXY
PHL 6214	00 00 42.	- 08 36		18.7	BLUE STELLAR OBJECT
MCG-02-01-011	00 00 42.	- 10 15	48	15.5	GALAXY
PHL 0638	00 00 42.	- 14 36		18.3	BLUE STELLAR OBJECT
PHL 2575	00 00 42.	- 20 24		18.5	BLUE STELLAR OBJECT
PHL 6213	00 00 42.	- 20 27		18.2	BLUE STELLAR OBJECT
TON-S 0135	00 00 42.	- 23 57		12.5	BLUE STAR
PHL 2573	00 00 42.	- 24 56		17.9	BLUE STELLAR OBJECT
LB 04931	00 00 42.	- 30 15		18.3	FAINT BLUE STAR
LB 04932	00 00 42.	- 30 46		17.5	FAINT BLUE STAR
MCG-06-01-017	00 00 42.	- 36 12	36	15.	GALAXY
MCG+01-01-015	00 00 45.	+ 08 20 30.	60	14.5	GALAXY
MCG-02-01-012	00 00 45.	- 11 02	48	14.5	GALAXY
ARC 2696	00 00 46.	+ 00 38		16.9	RICH CLUSTER OF GALAXIES
ARC 2697	00 00 46.	- 06 23		17.2	RICH CLUSTER OF GALAXIES
UGC 00009	00 00 48.	+ 04 21	78	16.5	GALAXY
ZC 0000.8+0452	00 00 48.	+ 04 52	8130		CLUSTER OF GALAXIES
MCG+01-01-016	00 00 48.	+ 05 25	36	14.	GALAXY
ZWG 408.015	00 00 48.	+ 08 20		15.4	GALAXY
UGC 00010	00 00 48.	+ 08 20	120	15.4	GALAXY Sc
MCG+04-01-014	00 00 48.	+ 21 48 30.	48	15.5	GALAXY
UGC 00011	00 00 48.	+ 21 50	66	16.0	GALAXY
ZWG 499.041	00 00 48.	+ 29 31		15.7	GALAXY
ZWG 498.069	00 00 48.	+ 29 31		15.7	GALAXY
UGC 00012	00 00 48.	+ 29 31	66	15.7	GALAXY Sc
KARA.73B 0002	00 00 48.	+ 29 31	42	15.7	ISOLATED GALAXY S
ZWG 499.042	00 00 48.	+ 30 30		15.7	GALAXY
ZWG 498.070	00 00 48.	+ 30 30		15.7	GALAXY
KARA.73B 0003	00 00 48.	+ 30 30	18	15.7	ISOLATED GALAXY S
ZCG 0000+34	00 00 48.	+ 34 29		17.6	COMPACT GALAXY
4ZW 003	00 00 48.	+ 34 37			COMPACT GALAXY
PHL 6215	00 00 48.	- 02 48		17.8	BLUE STELLAR OBJECT
PHL 6216	00 00 48.	- 04 20		18.1	BLUE STELLAR OBJECT
PHL 2576	00 00 48.	- 04 51		18.4	BLUE STELLAR OBJECT
PHL 2577	00 00 48.	- 05 17		18.3	BLUE STELLAR OBJECT
PHL 2578	00 00 48.	- 09 19		18.4	BLUE STELLAR OBJECT
PHL 2579	00 00 48.	- 09 41		13.3	BLUE STELLAR OBJECT
PHL 6217	00 00 48.	- 12 24		18.4	BLUE STELLAR OBJECT
PHL 6218	00 00 48.	- 14 29		18.0	BLUE STELLAR OBJECT
PHL 2580	00 00 48.	- 23 57		13.2	BLUE STELLAR OBJECT
PHL 6219	00 00 48.	- 25 26		18.2	BLUE STELLAR OBJECT
PHL 2581	00 00 48.	- 28 57		18.2	BLUE STELLAR OBJECT
RNGC 7815	00 00 49.	+ 20 25			NON-EXISTENT OBJECT
KEEL 741	00 00 51.7	+ 21 03 50.			NEBULA
ARC 2698	00 00 52.	+ 04 22		17.0	RICH CLUSTER OF GALAXIES
HN 1238	00 00 52.	- 65 28			NEBULA
IC 5382	00 00 52.	- 65 28			NONSTELLAR OBJECT
SCHO 0003	00 00 53.	+ 59 46 18.	270		ISOLATED DARK CLOUD
ZC 0000.9+0417	00 00 54.	+ 04 17	2890		CLUSTER OF GALAXIES
ZWG 408.016	00 00 54.	+ 05 25		15.2	GALAXY
PHL 6220	00 00 54.	+ 06 56		18.1	BLUE STELLAR OBJECT
ZWG 456.025	00 00 54.	+ 16 28		15.5	GALAXY
MCG+04-01-015	00 00 54.	+ 27 04	66	15.	GALAXY
ZWG 478.017	00 00 54.	+ 27 05		15.0	GALAXY
ZWG 477.045	00 00 54.	+ 27 05		15.0	GALAXY
UGC 00013	00 00 54.	+ 27 05	66	15.0	GALAXY SB0/a
OCL 0282	00 00 54.	+ 63 19	240	16.	OPEN STAR CLUSTER
PHL 2582	00 00 54.	- 04 02		17.8	BLUE STELLAR OBJECT
PHL 6223	00 00 54.	- 05 20		16.9	BLUE STELLAR OBJECT
PHL 6222	00 00 54.	- 09 36		13.9	BLUE STELLAR OBJECT
PHL 6221	00 00 54.	- 09 40		18.2	BLUE STELLAR OBJECT
MCG-02-01-013	00 00 54.	- 11 01	78	14.	GALAXY
PHL 0639	00 00 54.	- 17 00		14.8	BLUE STELLAR OBJECT
PHL 2583	00 00 54.	- 19 51		18.3	BLUE STELLAR OBJECT
LB 04933	00 00 54.	- 29 48		19.2	FAINT BLUE STAR
LB 04934	00 00 54.	- 30 25		18.4	FAINT BLUE STAR
RNGC 7808	00 00 55.	- 11 01		14.0	GALAXY
MCG+04-01-016	00 00 57.	+ 24 49	30	16.	GALAXY
SG 3.249	00 00 57.	+ 66 57	3000		DIFFUSE EMISSION NEBULA
LBN 0591	00 01	+ 65 20	120		BRIGHT NEBULA
LBN 0584	00 01	+ 66 40	2100		BRIGHT NEBULA
LBN 0586	00 01	+ 67 10	1320		BRIGHT NEBULA
LBN 0589	00 01	+ 68 20	720		BRIGHT NEBULA
PHL 2584	00 01 00.	+ 06 36		17.9	BLUE STELLAR OBJECT
MCG+02-01-016	00 01 00.	+ 10 20	36	15.	GALAXY
ZWG 433.019	00 01 00.	+ 11 13		15.6	GALAXY
ZWG 456.026	00 01 00.	+ 16 44		15.7	GALAXY
PHL 6225	00 01 00.	+ 17 55		18.2	BLUE STELLAR OBJECT
ZWG 478.018	00 01 00.	+ 21 49		15.7	GALAXY
ZWG 477.046	00 01 00.	+ 21 49		15.7	GALAXY
MCG+04-01-017	00 01 00.	+ 22 55	108	14.	GALAXY
ZWG 478.019	00 01 00.	+ 22 56		14.0	GALAXY
ZWG 477.047	00 01 00.	+ 22 56		14.0	GALAXY
UGC 00014	00 01 00.	+ 22 56	120	14.0	GALAXY Sc
MCG+06-01-009	00 01 00.	+ 37 04	30	15.	GALAXY
PHL 6226	00 01 00.	- 05 49		17.9	BLUE STELLAR OBJECT
MCG-01-01-027	00 01 00.	- 07 58	36	14.5	GALAXY
PHL 6224	00 01 00.	- 21 23		16.6	BLUE STELLAR OBJECT
LB 04935	00 01 00.	- 29 55		17.1	FAINT BLUE STAR
LB 04936	00 01 00.	- 30 40		18.6	FAINT BLUE STAR
RNGC 7822	00 01 01.	+ 68 20			DIFFUSE NEBULA
KN 13.053	00 01 02.2	+ 10 19 32.			NEBULA
KN 13.054	00 01 03.0	+ 13 57 46.			NEBULA
KN 13.055	00 01 05.0	+ 11 13 08.			NEBULA
ZC 0001.1+0030	00 01 06.	+ 00 30	3760		CLUSTER OF GALAXIES
MCG+01-01-017	00 01 06.	+ 03 59	60	15.	GALAXY
PHL 6230	00 01 06.	+ 04 50		18.4	BLUE STELLAR OBJECT
PHL 6229	00 01 06.	+ 05 01		18.3	BLUE STELLAR OBJECT
ZWG 408.017	00 01 06.	+ 08 21		15.6	GALAXY
MCG+02-01-017	00 01 06.	+ 14 57	108	16.	GALAXY
PHL 6227	00 01 06.	+ 18 38		18.4	BLUE STELLAR OBJECT
4ZW 004	00 01 06.	+ 24 20			COMPACT GALAXY
4ZW 005	00 01 06.	+ 25 41			COMPACT GALAXY
PHL 6231	00 01 06.	- 09 38		18.1	BLUE STELLAR OBJECT
PHL 2585	00 01 06.	- 12 08		18.3	BLUE STELLAR OBJECT
PHL 6228	00 01 06.	- 28 42		8.0	BLUE STELLAR OBJECT
LB 04937	00 01 06.	- 29 55		17.8	FAINT BLUE STAR
LB 01540	00 01 06.	- 57 49		14.2	FAINT BLUE STAR
KN 13.056	00 01 09.0	+ 14 56 13.			NEBULA
ARC 2699	00 01 10.	- 05 33		17.6	RICH CLUSTER OF GALAXIES
ZC 0001.2+0140	00 01 12.	+ 01 40	3430		CLUSTER OF GALAXIES
UGC 00015	00 01 12.	+ 04 00	90	16.0	GALAXY Sa-b
PHL 6233	00 01 12.	+ 04 40		15.2	BLUE STELLAR OBJECT
PHL 6235	00 01 12.	+ 07 04		17.1	BLUE STELLAR OBJECT
ZWG 408.018	00 01 12.	+ 07 11		14.0	GALAXY
UGC 00016	00 01 12.	+ 07 11	126	14.0	GALAXY Sb/Sc
MCG+01-01-018	00 01 12.	+ 07 11	84	13.5	GALAXY
PHL 2587	00 01 12.	+ 08 26		18.5	BLUE STELLAR OBJECT
ZWG 456.027	00 01 12.	+ 15 55		15.4	GALAXY
MCG+03-01-025	00 01 12.	+ 19 03	42	15.5	GALAXY
ZCG 0001+24.3	00 01 12.	+ 24 18		19.3	COMPACT GALAXY
PHL 2586	00 01 12.	- 05 54		17.1	BLUE STELLAR OBJECT
PHL 0640	00 01 12.	- 10 14		16.9	BLUE STELLAR OBJECT
PHL 6232	00 01 12.	- 17 37		4.6	BLUE STELLAR OBJECT
PHL 6234	00 01 12.	- 30 18		18.0	BLUE STELLAR OBJECT
LB 04938	00 01 12.	- 30 35		19.6	FAINT BLUE STAR
LB 04939	00 01 12.	- 30 51		19.5	FAINT BLUE STAR
SC 2358-6813.8	00 01 12.	- 67 57 06.	18		NEBULA
RNGC 7816	00 01 13.	+ 07 11		14.0	GALAXY
IC 5383	00 01 13.	+ 15 43 28.			NONSTELLAR OBJECT
ARC 2700	00 01 16.	+ 01 48		16.0	RICH CLUSTER OF GALAXIES
HN 2891	00 01 16.4	- 43 53 48.	24		NEBULA
UGC 00017	00 01 18.	+ 14 56	180	17.	GALAXY DWARF
ZC 0001.3+3230	00 01 18.	+ 32 30	610		CLUSTER OF GALAXIES
5ZW 001	00 01 18.	+ 51 28			COMPACT GALAXY
PHL 2590	00 01 18.	- 04 24		18.1	BLUE STELLAR OBJECT
PHL 2591	00 01 18.	- 05 00		18.3	BLUE STELLAR OBJECT
PHL 2588	00 01 18.	- 06 23		16.9	BLUE STELLAR OBJECT
PHL 0641	00 01 18.	- 14 39		17.2	BLUE STELLAR OBJECT
MCG-03-01-016	00 01 18.	- 14 58	36	14.5	GALAXY
PHL 2589	00 01 18.	- 22 58		17.8	BLUE STELLAR OBJECT
LB 04940	00 01 18.	- 30 08		18.2	FAINT BLUE STAR
KN 13.057	00 01 19.9	+ 12 29 15.			NEBULA

OBJECT NAME	RIGHT ASCEN.	DECLINATION	DIAM.	MAGN.	TYPE OF OBJECT
KN 13.058	00 01 22.5	+ 10 35 53.			NEBULA
PHL 0642	00 01 24.	+ 09 20		18.0	BLUE STELLAR OBJECT
ZWG 433.020	00 01 24.	+ 10 35		15.7	GALAXY
UGC 00018	00 01 24.	+ 10 35	66	15.7	GALAXY COMPACT
PHL 6238	00 01 24.	+ 17 58		18.0	BLUE STELLAR OBJECT
ZWG 456.028	00 01 24.	+ 20 28		12.7	GALAXY
UGC 00019	00 01 24.	+ 20 28	240	12.7	GALAXY Sb/Sc
KARA.73B 0004	00 01 24.	+ 20 28	216	12.7	ISOLATED GALAXY S
MCG+03-01-021	00 01 24.	+ 20 30	180	12.	GALAXY
MCG+13-01-004	00 01 24.	+ 80 01	39	16.	GALAXY
PHL 2593	00 01 24.	- 05 20		18.3	BLUE STELLAR OBJECT
PHL 6236	00 01 24.	- 06 12		18.5	BLUE STELLAR OBJECT
MCG-02-01-015	00 01 24.	- 11 27 30.	12	15.	GALAXY
MCG-02-01-014	00 01 24.	- 11 27 30.	60	14.5	GALAXY
PHL 6237	00 01 24.	- 17 42		18.9	BLUE STELLAR OBJECT
PHL 2592	00 01 24.	- 19 04		18.0	BLUE STELLAR OBJECT
PHL 2594	00 01 24.	- 22 39		18.5	BLUE STELLAR OBJECT
SMO 34	00 01 24.	- 27 32 03.	600	17.	GROUP OF 8 GALAXIES
LB 04941	00 01 24.	- 29 55		20.2	FAINT BLUE STAR
KN 13.059	00 01 24.9	+ 13 26 43.			NEBULA
RNGC 7817	00 01 25.	+ 20 28		12.5	GALAXY
HN 2892	00 01 25.6	- 46 28 30.	18		NEBULA
KN 13.060	00 01 26.3	+ 14 25 30.			NEBULA
KN 13.061	00 01 26.8	+ 15 44 30.			NEBULA
LIN.CL 001	00 01 27.	- 73 45 12.	246	11.7	STAR CLUSTER IN SMC
KEEL 742	00 01 27.0	+ 15 44 30.			NEBULA
MCG+01-01-019	00 01 30.	+ 07 05	60	14.	GALAXY
UGC 00020	00 01 30.	+ 80 01	132	16.0	GALAXY
MCG+00-01-023	00 01 30.	- 01 10 30.	42	14.5	GALAXY
PHL 6239	00 01 30.	- 13 25		16.7	BLUE STELLAR OBJECT
PHL 6240	00 01 30.	- 21 33		18.4	BLUE STELLAR OBJECT
TON-S 0136	00 01 30.	- 26 59		15.7	BLUE STAR
LB 04942	00 01 30.	- 29 54		16.7	FAINT BLUE STAR
LB 04943	00 01 30.	- 30 03		19.8	FAINT BLUE STAR
SC 2358-6820.4	00 01 30.	- 68 03 42.	12		NEBULA
KN 13.062	00 01 31.3	+ 10 01 09.			NEBULA
SCHO 0005	00 01 34.	+ 58 19 54.	250		ISOLATED DARK CLOUD
SCHO 0004	00 01 34.	+ 62 42 06.	300		ISOLATED DARK CLOUD
PHL 2597	00 01 36.	+ 01 55		18.6	BLUE STELLAR OBJECT
ZWG 408.019	00 01 36.	+ 07 05		15.1	GALAXY
UGC 00021	00 01 36.	+ 07 05	66	15.1	GALAXY Sc
MCG+02-01-018	00 01 36.	+ 10 00	72	15.	GALAXY
PHL 6241	00 01 36.	+ 16 12		17.0	BLUE STELLAR OBJECT
PHL 6242	00 01 36.	+ 17 36		18.7	BLUE STELLAR OBJECT
ZWG 382.023	00 01 36.	- 01 08		15.6	GALAXY
PHL 2595	00 01 36.	- 03 56		17.0	BLUE STELLAR OBJECT
PHL 2596	00 01 36.	- 11 02		18.5	BLUE STELLAR OBJECT
MCG-02-01-016	00 01 36.	- 12 15	48	15.	GALAXY
PHL 6244	00 01 36.	- 16 10		18.5	BLUE STELLAR OBJECT
PHL 6243	00 01 36.	- 16 12		18.3	BLUE STELLAR OBJECT
RNGC 7818	00 01 37.	+ 07 05		15.0	GALAXY
RNGC 7813	00 01 37.	- 12 15		15.0	GALAXY
IC 5384	00 01 37.	- 12 15 44.			NONSTELLAR OBJECT
KN 13.063	00 01 39.1	+ 10 30 54.			NEBULA
ARC 2701	00 01 40.	- 09 52		17.8	RICH CLUSTER OF GALAXIES
UGC 00022	00 01 42.	+ 10 02	78	16.0	GALAXY SO
ZWG 433.021	00 01 42.	+ 10 30		15.2	GALAXY
UGC 00023	00 01 42.	+ 10 30	78	15.2	GALAXY SBb
MCG+02-01-019	00 01 42.	+ 10 30	84	14.	GALAXY
PHL 0643	00 01 42.	+ 15 13		16.9	BLUE STELLAR OBJECT
MCG+04-01-018	00 01 42.	+ 22 18 30.	48	15.5	GALAXY
ZWG 478.020	00 01 42.	+ 22 19		15.4	GALAXY
ZWG 477.048	00 01 42.	+ 22 19		15.4	GALAXY
UGC 00024	00 01 42.	+ 22 19	72	15.4	GALAXY SBc
ZCG 0001+24.4	00 01 42.	+ 24 23		17.0	COMPACT GALAXY
ZC 0001.7+2746	00 01 42.	+ 27 46	1340		CLUSTER OF GALAXIES
ZWG 499.043	00 01 42.	+ 31 47		15.7	GALAXY
ZWG 498.071	00 01 42.	+ 31 47		15.7	GALAXY
SC 2359-0225.0	00 01 42.	- 02 08 18.	12		NEBULA
SC 2359-0228.4	00 01 42.	- 02 11 42.			NEBULA
MCG-02-01-017	00 01 42.	- 08 18	30	15.	GALAXY
PHL 6245	00 01 42.	- 08 34		11.4	BLUE STELLAR OBJECT
MCG-03-01-017	00 01 42.	- 14 47	54	15.	GALAXY
PHL 6246	00 01 42.	- 16 16		17.9	BLUE STELLAR OBJECT
PHL 6248	00 01 42.	- 29 19		18.2	BLUE STELLAR OBJECT
LB 04944	00 01 42.	- 29 39		19.6	FAINT BLUE STAR
OCL 0043	00 01 42.	- 30 13	5400	8.	OPEN STAR CLUSTER
PHL 6247	00 01 42.	- 30 25		7.0	BLUE STELLAR OBJECT
LB 04945	00 01 42.	- 30 30		19.5	FAINT BLUE STAR
MOHR 1	00 01 42.	- 73 38 48.		14.0	STAR CLUSTER IN SMC
UGC 00025	00 01 48.	+ 05 53	66	16.0	GALAXY Sc
MCG+01-01-020	00 01 48.	+ 05 53	60	15.	GALAXY
PHL 6253	00 01 48.	+ 17 12		18.6	BLUE STELLAR OBJECT
ZWG 499.044	00 01 48.	+ 31 12		14.3	GALAXY
ZWG 498.072	00 01 48.	+ 31 12		14.3	GALAXY
UGC 00026	00 01 48.	+ 31 12	120	14.3	GALAXY SBb
MCG+05-01-029	00 01 48.	+ 31 12	90	14.	GALAXY
MCG+06-01-010	00 01 48.	+ 37 43	36	16.	GALAXY
MCG+08-01-020	00 01 48.	+ 47 12	72	15.	GALAXY
OCL 0274	00 01 48.	+ 55 45	180	8.	OPEN STAR CLUSTER
PHL 6250	00 01 48.	- 02 28		17.4	BLUE STELLAR OBJECT
PHL 6249	00 01 48.	- 03 18		16.5	BLUE STELLAR OBJECT
PHL 6251	00 01 48.	- 03 50		18.7	BLUE STELLAR OBJECT
PHL 2598	00 01 48.	- 05 57		18.5	BLUE STELLAR OBJECT
PHL 0644	00 01 48.	- 09 45		17.3	BLUE STELLAR OBJECT
PHL 6252	00 01 48.	- 11 56		18.0	BLUE STELLAR OBJECT
LB 04946	00 01 48.	- 29 52		19.0	FAINT BLUE STAR
LB 04947	00 01 48.	- 30 14		18.2	FAINT BLUE STAR
LB 04949	00 01 48.	- 30 20		18.4	FAINT BLUE STAR
LB 04948	00 01 48.	- 30 20		18.2	FAINT BLUE STAR
LB 04950	00 01 48.	- 30 32		19.2	FAINT BLUE STAR
RNGC 7819	00 01 49.	+ 31 12		14.5	GALAXY
MCG+01-01-021	00 01 54.	+ 05 32 30.	72	14.	GALAXY
ZWG 408.020	00 01 54.	+ 05 34		15.3	GALAXY
UGC 00027	00 01 54.	+ 05 34	132	15.3	GALAXY Sc
PHL 6255	00 01 54.	+ 16 55		18.4	BLUE STELLAR OBJECT
ZC 0001.9+2351	00 01 54.	+ 23 51	740		CLUSTER OF GALAXIES
ZWG 499.045	00 01 54.	+ 27 42		15.7	GALAXY
ZWG 498.073	00 01 54.	+ 27 42		15.7	GALAXY
ZWG 549.015	00 01 54.	+ 47 13		15.1	GALAXY
ZWG 548.024	00 01 54.	+ 47 13		15.1	GALAXY
PHL 6256	00 01 54.	- 04 26		18.1	BLUE STELLAR OBJECT
PHL 6254	00 01 54.	- 13 20		15.9	BLUE STELLAR OBJECT
PHL 2599	00 01 54.	- 15 00		19.0	BLUE STELLAR OBJECT
LB 04951	00 01 54.	- 30 42		18.3	FAINT BLUE STAR
LB 04952	00 01 54.	- 30 55		18.8	FAINT BLUE STAR
KEEL 743	00 01 54.5	+ 15 42 14.			NEBULA
KEEL 744	00 01 56.9	+ 20 26 22.			NEBULA
MCG+01-01-022	00 01 57.	+ 04 54 30.	60	14.	GALAXY
KN 13.065	00 01 57.	+ 15 35			NEBULA
KN 13.064	00 01 57.3	+ 11 25 56.			NEBULA
KN 13.066	00 01 59.3	+ 11 25 07.			NEBULA
ZWG 408.021	00 02 00.	+ 04 55		13.9	GALAXY
UGC 00028	00 02 00.	+ 04 55	96	13.9	GALAXY S0-a
ZC 0002.0+1820	00 02 00.	+ 18 20	2490		CLUSTER OF GALAXIES
MCG+05-01-030	00 02 00.	+ 28 00 30.	21	15.	GALAXY
ZWG 499.046	00 02 00.	+ 28 02		15.2	GALAXY
ZWG 498.074	00 02 00.	+ 28 02		15.2	GALAXY
UGC 00029	00 02 00.	+ 28 02	78	15.2	GALAXY E
UGC 00030	00 02 00.	+ 33 17	72	16.0	GALAXY S
MRSL 118+04/1	00 02 00.	+ 66 53	10800		HII REGION
PHL 2600	00 02 00.	- 07 14		15.6	BLUE STELLAR OBJECT
MCG-02-01-018	00 02 00.	- 09 18	30	15.5	GALAXY
PHL 2601	00 02 00.	- 10 22		18.7	BLUE STELLAR OBJECT
PHL 6257	00 02 00.	- 16 24		18.2	BLUE STELLAR OBJECT
PHL 0645	00 02 00.	- 24 41		13.3	BLUE STELLAR OBJECT
TON-S 0137	00 02 00.	- 24 41		13.3	BLUE STAR
PHL 2602	00 02 00.	- 25 24		18.3	BLUE STELLAR OBJECT
PHL 2603	00 02 00.	- 26 51		18.3	BLUE STELLAR OBJECT
RNGC 7820	00 02 01.	+ 04 55		14.0	GALAXY
HN 2893	00 02 02.5	- 45 45 24.	12		NEBULA
MCG-01-01-028	00 02 03.	- 07 21 30.	138	13.	GALAXY
MCG-01-01-029	00 02 03.	- 08 22	30	16.	GALAXY
MCG-01-01-030	00 02 03.	- 08 23	24	13.5	GALAXY
CED 214B	00 02 04.	+ 66 53	1500		DIFFUSE GALACTIC NEBULA
SC 2359-0232.1	00 02 05.	- 02 15 24.	12		NEBULA
DG 001	00 02 06.	+ 66 53	3900		REFLECTION NEBULA
PHL 6258	00 02 06.	- 04 39		18.0	BLUE STELLAR OBJECT
PHL 2605	00 02 06.	- 06 23		18.8	BLUE STELLAR OBJECT
PHL 2604	00 02 06.	- 09 02		15.9	BLUE STELLAR OBJECT
PHL 2606	00 02 06.	- 21 23		18.3	BLUE STELLAR OBJECT
MCG-05-01-028	00 02 06.	- 30 46	48	14.5	GALAXY
HN 2894	00 02 06.8	- 45 54 24.	18		NEBULA
HN 2895	00 02 07.7	- 45 45 42.	12		NEBULA
SCHO 0006	00 02 08.	+ 58 50 48.	280		ISOLATED DARK CLOUD
CED 215	00 02 10.	+ 68 17	3600		DIFFUSE GALACTIC NEBULA
ZWG 478.021	00 02 12.	+ 26 33		14.6	GALAXY
ZWG 477.049	00 02 12.	+ 26 33		14.6	GALAXY
ZWG 499.047	00 02 12.	+ 32 00		15.6	GALAXY
ZWG 498.075	00 02 12.	+ 32 00		15.6	GALAXY
MCG+00-01-024	00 02 12.	- 01 48	24	14.	GALAXY
MCG+00-01-025	00 02 12.	- 01 51	36	14.	GALAXY
PHL 0646	00 02 12.	- 10 24		13.8	BLUE STELLAR OBJECT
PHL 2607	00 02 12.	- 14 19		13.9	BLUE STELLAR OBJECT
PHL 6259	00 02 12.	- 16 42		18.7	BLUE STELLAR OBJECT
RNGC 7823	00 02 14.	- 62 21			UNVERIFIED SOUTHERN OBJECT
MCG+01-01-023	00 02 15.	+ 05 25 30.	36	15.5	GALAXY
LB 00401	00 02 16.	+ 29 23 06.		17.4	FAINT BLUE STAR
ARC 2702	00 02 16.	+ 31 08		17.1	RICH CLUSTER OF GALAXIES
ZWG 408.022	00 02 18.	+ 05 27		15.7	GALAXY
PHL 6260	00 02 18.	+ 06 59		18.1	BLUE STELLAR OBJECT
PHL 0647	00 02 18.	+ 16 35		15.3	BLUE STELLAR OBJECT
ZWG 456.029	00 02 18.	+ 16 55		15.3	GALAXY
UGC 00031	00 02 18.	+ 16 55	84	15.3	GALAXY IRR
MCG+03-01-022	00 02 18.	+ 16 55	48	14.5	GALAXY
MCG+07-01-004	00 02 18.	+ 41 27	60	17.	GALAXY
ZWG 382.025	00 02 18.	- 01 46		15.0	GALAXY
MRK 544	00 02 18.	- 01 46	19	15.5	GALAXY WITH UV CONTINUUM
ZWG 382.024	00 02 18.	- 01 51		15.1	GALAXY
MCG-03-01-018	00 02 18.	- 16 19	78	14.	GALAXY
LB 04953	00 02 18.	- 29 48		18.8	FAINT BLUE STAR
LB 04954	00 02 18.	- 30 30		20.5	FAINT BLUE STAR
HN 2896	00 02 18.9	- 45 45 54.			NEBULA
HN 2897	00 02 19.2	- 45 45 54.	12		NEBULA
KN 13.067	00 02 20.3	+ 13 19 20.			NEBULA
SC 2359-0204.2	00 02 21.	- 01 47 30.	18		NEBULA
SC 2359-0425.8	00 02 23.	- 04 09 06.	36		NEBULA
MCG+01-01-024	00 02 24.	+ 04 49	48	14.5	GALAXY
PHL 2608	00 02 24.	+ 06 16		18.2	BLUE STELLAR OBJECT
PHL 6262	00 02 24.	+ 07 10		18.3	BLUE STELLAR OBJECT
ZC 0002.4+0744	00 02 24.	+ 07 44	9410		CLUSTER OF GALAXIES
PHL 2612	00 02 24.	+ 07 54		18.6	BLUE STELLAR OBJECT
PHL 6261	00 02 24.	+ 08 10		17.1	BLUE STELLAR OBJECT
PHL 6263	00 02 24.	+ 08 31		18.0	BLUE STELLAR OBJECT
UGC 00032	00 02 24.	+ 11 26	78	17.	GALAXY
ZWG 478.022	00 02 24.	+ 21 52		15.6	GALAXY
ZWG 477.050	00 02 24.	+ 21 52		15.6	GALAXY
PHL 2609	00 02 24.	- 02 42		18.6	BLUE STELLAR OBJECT
PHL 2610	00 02 24.	- 15 58		18.4	BLUE STELLAR OBJECT
PHL 2611	00 02 24.	- 29 58		17.3	BLUE STELLAR OBJECT
LB 04955	00 02 24.	- 30 07		17.5	FAINT BLUE STAR
MCG-06-01-018	00 02 24.	- 35 58	36	16.	GALAXY
KN 13.068	00 02 24.7	+ 11 25 22.			NEBULA
MCG+04-01-019	00 02 27.	+ 21 50	42	15.5	GALAXY
MCG-05-01-029	00 02 27.	- 28 00	24	14.5	GALAXY
MCG-05-01-030	00 02 27.	- 30 47	30	15.	GALAXY
SC 2359-0207.9	00 02 29.	- 01 51 12.	18		NEBULA
ZWG 408.023	00 02 30.	+ 04 51		15.4	GALAXY
UGC 00033	00 02 30.	+ 04 51	78	15.4	GALAXY SB0-a
MCG+01-01-025	00 02 30.	+ 06 37	72	14.	GALAXY
ZWG 478.023	00 02 30.	+ 21 55		15.7	GALAXY
ZWG 477.051	00 02 30.	+ 21 55		15.7	GALAXY
ZWG 382.026	00 02 30.	- 01 57		15.7	GALAXY
8ZW 0002-07.4	00 02 30.	- 07 25		15.7	COMPACT GALAXY
LB 04956	00 02 30.	- 29 40		17.7	FAINT BLUE STAR
LB 04957	00 02 30.	- 30 20		16.7	FAINT BLUE STAR
MCG-06-01-019	00 02 30.	- 35 58 30.	30	15.	GALAXY
IC 1528	00 02 31.	- 03 23 44.			NONSTELLAR OBJECT
SC 2359-0331.1	00 02 33.	- 03 14 24.			NEBULA
MIL 01	00 02 35.	+ 72 20	7500		SUPERNOVA REMNANT
SC 0000-0422.1	00 02 35.	- 04 05 24.	36		NEBULA
ZWG 408.024	00 02 36.	+ 04 56		15.5	GALAXY
MCG+01-01-026	00 02 36.	+ 06 28	42	14.5	GALAXY
ZWG 408.025	00 02 36.	+ 06 38		14.5	GALAXY
UGC 00034	00 02 36.	+ 06 38	120	14.5	GALAXY Sa/Sb
SCHO 0209	00 02 36.	+ 09 20 00.	510		ISOLATED DARK CLOUD
PHL 6266	00 02 36.	+ 09 29		18.3	BLUE STELLAR OBJECT
PHL 6268	00 02 36.	+ 17 44		19.0	BLUE STELLAR OBJECT
PHL 0648	00 02 36.	- 04 20		18.2	BLUE STELLAR OBJECT
PHL 2613	00 02 36.	- 06 58		18.3	BLUE STELLAR OBJECT
PHL 6265	00 02 36.	- 09 19		16.3	BLUE STELLAR OBJECT
PHL 6264	00 02 36.	- 13 12		14.1	BLUE STELLAR OBJECT
PHL 2614	00 02 36.	- 16 58		18.4	BLUE STELLAR OBJECT
PHL 6267	00 02 36.	- 20 05		18.9	BLUE STELLAR OBJECT
TON-S 0138	00 02 36.	- 26 49		15.6	BLUE STAR
PHL 0649	00 02 36.	- 26 50		15.8	BLUE STELLAR OBJECT

OBJECT NAME	RIGHT ASCEN.	DECLINATION	DIAM.	MAGN.	TYPE OF OBJECT
LB 04958	00 02 36.	- 29 56		18.6	FAINT BLUE STAR
LB 04959	00 02 36.	- 30 23		18.6	FAINT BLUE STAR
LB 01541	00 02 36.	- 57 13		14.0	FAINT BLUE STAR
LB 03127	00 02 36.	- 62 54		12.9	FAINT BLUE STAR
RNGC 7825	00 02 37.	+ 04 56		15.5	GALAXY
RNGC 7824	00 02 37.	+ 06 38		14.5	GALAXY
RNGC 7826	00 02 37.	- 21 00			NON-EXISTENT OBJECT
SCHO 0007	00 02 38.	+ 57 17 36.	230		ISOLATED DARK CLOUD
IC 1529	00 02 38.	- 11 47 08.			NONSTELLAR OBJECT
UGC 00035	00 02 42.	+ 05 59	102	17.	GALAXY DWRF SP
ZWG 408.026	00 02 42.	+ 06 30		14.7	GALAXY
UGC 00036	00 02 42.	+ 06 30	102	14.7	GALAXY Sa
PHL 2617	00 02 42.	+ 06 56		18.1	BLUE STELLAR OBJECT
ZWG 499.048	00 02 42.	+ 32 42		15.4	GALAXY
ZWG 498.076	00 02 42.	+ 32 42		15.4	GALAXY
PHL 2616	00 02 42.	- 03 11		18.6	BLUE STELLAR OBJECT
MCG-01-01-031	00 02 42.	- 04 04	48	15.	GALAXY
MCG-02-01-019	00 02 42.	- 11 45	90	14.	GALAXY
MCG-03-01-019	00 02 42.	- 16 45 30.	72	14.	GALAXY
PHL 2615	00 02 42.	- 17 04		17.9	BLUE STELLAR OBJECT
LB 04960	00 02 42.	- 30 16		17.5	FAINT BLUE STAR
RNGC 7814	00 02 43.	+ 15 52		12.0	GALAXY
RNGC 7821	00 02 43.	- 16 45		14.0	GALAXY
SCHO 0008	00 02 44.	+ 56 12 06.	260		ISOLATED DARK CLOUD
ARC 2703	00 02 46.	+ 15 50		17.1	RICH CLUSTER OF GALAXIES
SCHO 0009	00 02 47.	+ 58 56 12.	200		ISOLATED DARK CLOUD
PHL 0651	00 02 48.	+ 03 20		17.9	BLUE STELLAR OBJECT
MCG+01-01-028	00 02 48.	+ 04 53	48	15.	GALAXY
MCG+01-01-027	00 02 48.	+ 04 55 30.	48	14.	GALAXY
PHL 0650	00 02 48.	+ 05 06		16.3	BLUE STELLAR OBJECT
PHL 2619	00 02 48.	+ 14 28		18.3	BLUE STELLAR OBJECT
MCG+05-01-031	00 02 48.	+ 32 14 30.	42	15.	GALAXY
ZWG 499.049	00 02 48.	+ 32 15		15.0	GALAXY
ZWG 498.077	00 02 48.	+ 32 15		15.0	GALAXY
PHL 2618	00 02 48.	- 13 50		18.1	BLUE STELLAR OBJECT
PHL 0652	00 02 48.	- 15 19		17.9	BLUE STELLAR OBJECT
KN 13.069	00 02 50.1	+ 15 46 26.			NEBULA
SC 0000-6815.2	00 02 51.	- 67 58 29.	6		NEBULA
KN 13.070	00 02 53.1	+ 13 32 04.			NEBULA
ZWG 408.027	00 02 54.	+ 04 54		15.5	GALAXY
UGC 00037	00 02 54.	+ 04 54	78	15.5	GALAXY SBb
ZWG 408.028	00 02 54.	+ 04 57		14.6	GALAXY
UGC 00038	00 02 54.	+ 04 57	84	14.6	GALAXY SB0
ZWG 549.016	00 02 54.	+ 46 16		15.3	GALAXY
ZWG 548.025	00 02 54.	+ 46 16		15.3	GALAXY
PHL 6269	00 02 54.	- 05 10		18.0	BLUE STELLAR OBJECT
MCG-05-01-031	00 02 54.	- 28 22	24	16.	GALAXY
LB 04961	00 02 54.	- 30 18		17.6	FAINT BLUE STAR
MCG-06-01-020	00 02 54.	- 36 12	60	16.	GALAXY
RNGC 7827	00 02 55.	+ 04 57		14.5	GALAXY
ARC 2704	00 02 58.	- 12 09		17.7	RICH CLUSTER OF GALAXIES
SG 3.001	00 02 59.	+ 65 47	360		DIFFUSE EMISSION NEBULA
LBN 0588	00 03	+ 66 50	1320		BRIGHT NEBULA
PHL 0654	00 03 00.	+ 15 08		18.1	BLUE STELLAR OBJECT
PHL 6270	00 03 00.	+ 17 27		18.5	BLUE STELLAR OBJECT
ZC 0003.0+2008	00 03 00.	+ 20 08	1340		CLUSTER OF GALAXIES
UGC 00039	00 03 00.	+ 53 22	78	17.	GALAXY Sc
MCG+15-01-001	00 03 00.	+ 88 04	96	16.	GALAXY
PHL 2620	00 03 00.	- 06 00		16.8	BLUE STELLAR OBJECT
PHL 0653	00 03 00.	- 21 56		17.2	BLUE STELLAR OBJECT
LB 04962	00 03 00.	- 30 46		18.2	FAINT BLUE STAR
MCG-05-01-032	00 03 00.	- 30 52	30	16.	GALAXY
KEEL 001	00 03 01.2	+ 20 51 39.			NEBULA
SCHO 0010	00 03 02.	+ 57 37 00.	180		ISOLATED DARK CLOUD
MCG-02-01-020	00 03 04.	- 13 53	9	15.	GALAXY
ZC 0003.1+1114	00 03 06.	+ 11 14	2960		CLUSTER OF GALAXIES
PHL 2621	00 03 06.	- 04 02		18.2	BLUE STELLAR OBJECT
MCG-02-01-021	00 03 06.	- 13 52	60	16.5	GALAXY
LB 04963	00 03 06.	- 29 44		17.9	FAINT BLUE STAR
LB 04964	00 03 06.	- 29 54		18.5	FAINT BLUE STAR
MCG-06-01-021	00 03 06.	- 36 13	30	15.5	GALAXY
ARP 051	00 03 11.	- 13 43			PECULIAR GALAXY
PHL 0655	00 03 12.	+ 06 33		16.7	BLUE STELLAR OBJECT
ZC 0003.2+2410	00 03 12.	+ 24 10	740		CLUSTER OF GALAXIES
ZWG 478.024	00 03 12.	+ 27 10		15.2	GALAXY
ZWG 477.052	00 03 12.	+ 27 10		15.2	GALAXY
UGC 00040	00 03 12.	+ 27 10	72	15.2	GALAXY SB
PHL 2622	00 03 12.	- 05 21		18.6	BLUE STELLAR OBJECT
8ZW 0003-06.7	00 03 12.	- 06 44		15.1	COMPACT GALAXY
PHL 6271	00 03 12.	- 07 42		17.8	BLUE STELLAR OBJECT
MCG-01-01-032	00 03 12.	- 07 55	30	15.	GALAXY
MCG+04-01-020	00 03 15.	+ 22 13	36	15.5	GALAXY
MCG+04-01-021	00 03 15.	+ 27 10	66	14.5	GALAXY
PHL 2629	00 03 18.	- 04 47		17.9	BLUE STELLAR OBJECT
PHL 0656	00 03 18.	- 07 26		18.0	BLUE STELLAR OBJECT
PHL 0657	00 03 18.	- 10 19		18.2	BLUE STELLAR OBJECT
MCG-02-01-022	00 03 18.	- 14 16	60	15.	GALAXY
MCG-02-01-023	00 03 18.	- 14 17	60	15.	GALAXY
LB 04965	00 03 18.	- 29 43		19.0	FAINT BLUE STAR
LB 04966	00 03 18.	- 30 05		18.9	FAINT BLUE STAR
MCG+04-01-022	00 03 21.	+ 22 11	60	15.5	GALAXY
SC 0000-6828.4	00 03 21.	- 68 11 41.	6		NEBULA
ARC 2705	00 03 22.	+ 15 32		17.1	RICH CLUSTER OF GALAXIES
ARP 144	00 03 23.	- 13 41			PECULIAR GALAXY
PHL 6322	00 03 24.	+ 03 32		18.3	BLUE STELLAR OBJECT
ZC 0003.4+1035	00 03 24.	+ 10 35	940		CLUSTER OF GALAXIES
PHL 0658	00 03 24.	+ 15 51		16.2	BLUE STELLAR OBJECT
UGC 00041	00 03 24.	+ 22 12	72	16.0	GALAXY SBc
5ZW 002	00 03 24.	+ 47 07			COMPACT GALAXY
PHL 6272	00 03 24.	- 03 28		17.2	BLUE STELLAR OBJECT
PHL 2625	00 03 24.	- 06 39		18.4	BLUE STELLAR OBJECT
PHL 2623	00 03 24.	- 12 58		18.7	BLUE STELLAR OBJECT
PHL 6274	00 03 24.	- 25 28		18.2	BLUE STELLAR OBJECT
PHL 6273	00 03 24.	- 28 16		17.7	BLUE STELLAR OBJECT
PHL 2624	00 03 24.	- 30 39		16.6	BLUE STELLAR OBJECT
MCG-05-01-033	00 03 24.	- 31 24	36	15.	GALAXY
MCG-06-01-022	00 03 24.	- 36 23	9	15.	GALAXY
SER 164.01	00 03 24.	- 54 09	1500	16.	CLUSTER OF GALAXIES
SC 0000-0417.2	00 03 25.	- 04 00 30.	12		NEBULA
BC 4C15.01	00 03 25.01	+ 15 53 06C6		16.40	QUASI-STELLAR OBJECT
SHB 001	00 03 26.0	+ 15 53 00.		16.4	QUASI-STELLAR OBJECT
ARC 2706	00 03 28.	+ 10 52		17.2	RICH CLUSTER OF GALAXIES
KN 13.071	00 03 28.1	+ 14 08 14.			NEBULA
KN 13.072	00 03 28.2	+ 15 42 43.			NEBULA
UGC 00042	00 03 30.	+ 12 51	60	17.	GALAXY DISTRBD
UGC 00043	00 03 30.	+ 14 09	72	16.5	GALAXY
MCG+04-01-023	00 03 30.	+ 27 04	42	17.	GALAXY
PHL 0659	00 03 30.	- 02 26		17.8	BLUE STELLAR OBJECT

OBJECT NAME	RIGHT ASCEN.	DECLINATION	DIAM.	MAGN.	TYPE OF OBJECT
PHL 2626	00 03 30.	- 04 46		18.4	BLUE STELLAR OBJECT
LB 04967	00 03 30.	- 29 54		18.7	FAINT BLUE STAR
LB 04968	00 03 30.	- 30 09		18.5	FAINT BLUE STAR
LB 04969	00 03 30.	- 30 19		16.8	FAINT BLUE STAR
MCG-05-01-034	00 03 30.	- 30 55	30	15.	GALAXY
KN 13.073	00 03 35.3	+ 10 03 27.			NEBULA
PHL 2630	00 03 36.	+ 07 38		18.0	BLUE STELLAR OBJECT
MCG+01-01-029	00 03 36.	+ 08 37	60	14.5	GALAXY
PHL 2627	00 03 36.	+ 08 41		17.0	BLUE STELLAR OBJECT
ZC 0003.6+1331	00 03 36.	+ 13 31	1550		CLUSTER OF GALAXIES
UGC 00044	00 03 36.	+ 19 28	60	17.	GALAXY
ZC 0003.6+3250	00 03 36.	+ 32 50	4370		CLUSTER OF GALAXIES
ZWG 556.001	00 03 36.	+ 54 16		15.5	GALAXY
UGC 00045	00 03 36.	+ 54 16	60	15.5	GALAXY Sc-IRR
PHL 2628	00 03 36.	- 03 36		17.8	BLUE STELLAR OBJECT
PHL 0660	00 03 36.	- 05 01		18.1	BLUE STELLAR OBJECT
VV 272B	00 03 36.	- 13 41	18	16.	INTERACTING GALAXY
VV 272A	00 03 36.	- 13 41	48	14.	INTERACTING GALAXY
VV 272	00 03 36.	- 13 41	90		INTERACTING GALAXY
MCG-02-01-024	00 03 36.	- 13 42	30	15.	GALAXY
PHL 0661	00 03 36.	- 22 18		18.4	BLUE STELLAR OBJECT
LB 04970	00 03 36.	- 29 48		18.2	FAINT BLUE STAR
LB 04971	00 03 36.	- 30 24		18.6	FAINT BLUE STAR
RNGC 7830	00 03 37.	+ 08 06			NON-EXISTENT OBJECT
KN 13.074	00 03 38.6	+ 16 12 15.			NEBULA
PHL 2631	00 03 42.	+ 06 45		18.7	BLUE STELLAR OBJECT
ZWG 408.029	00 03 42.	+ 08 36		15.7	GALAXY
ZC 0003.7+2608	00 03 42.	+ 26 08	1080		CLUSTER OF GALAXIES
MCG-04-01-014	00 03 42.	- 23 09	24	15.	GALAXY
PHL 2632	00 03 42.	- 24 00		18.5	BLUE STELLAR OBJECT
PHL 0662	00 03 42.	- 29 29		18.1	BLUE STELLAR OBJECT
LB 04973	00 03 42.	- 30 24		18.4	FAINT BLUE STAR
LB 04972	00 03 42.	- 30 50		18.1	FAINT BLUE STAR
ARC 2707	00 03 46.	- 10 41		17.6	RICH CLUSTER OF GALAXIES
PHL 2633	00 03 48.	+ 03 24		18.6	BLUE STELLAR OBJECT
PHL 2636	00 03 48.	+ 12 20		17.5	BLUE STELLAR OBJECT
ZWG 456.030	00 03 48.	+ 17 09		14.5	GALAXY
UGC 00046	00 03 48.	+ 17 09	48	14.5	GALAXY DBL SYS
KARA.72 001A	00 03 48.	+ 17 09	36	14.5	PART OF DOUBLE GALAXY
MCG+03-01-024	00 03 48.	+ 17 09 30.	36	16.	GALAXY
KARA.72 001B	00 03 48.	+ 17 10	30		PART OF DOUBLE GALAXY
MCG+03-01-023	00 03 48.	+ 17 10	24	16.	GALAXY
PHL 2635	00 03 48.	+ 17 43		17.4	BLUE STELLAR OBJECT
MRK 335	00 03 48.	+ 19 55	15	14.	GALAXY WITH UV CONTINUUM
KW 31	00 03 48.	+ 19 55	16		SEYFERT GALAXY
VVI 01	00 03 48.	+ 19 55	15	14.18	SEYFERT GALAXY
PHL 6275	00 03 48.	- 06 05		17.6	BLUE STELLAR OBJECT
PHL 6278	00 03 48.	- 07 22		18.6	BLUE STELLAR OBJECT
PHL 2634	00 03 48.	- 07 22		18.2	BLUE STELLAR OBJECT
PHL 6276	00 03 48.	- 13 04		11.0	BLUE STELLAR OBJECT
MCG-02-01-025	00 03 48.	- 13 41 30.	48	14.	GALAXY
PHL 2637	00 03 48.	- 15 00		18.3	BLUE STELLAR OBJECT
PHL 2638	00 03 48.	- 15 46		18.3	BLUE STELLAR OBJECT
PHL 0663	00 03 48.	- 16 12		16.8	BLUE STELLAR OBJECT
PHL 2639	00 03 48.	- 18 48		18.4	BLUE STELLAR OBJECT
PHL 6277	00 03 48.	- 22 08		17.1	BLUE STELLAR OBJECT
TON-S 0139	00 03 48.	- 22 09		15.1	BLUE STAR
LB 04974	00 03 48.	- 29 35		17.4	FAINT BLUE STAR
LB 04975	00 03 48.	- 29 43		18.9	FAINT BLUE STAR
LB 04976	00 03 48.	- 30 35		17.4	FAINT BLUE STAR
AGU 01	00 03 48.	- 41 45 00.		12.5	2 INTERACTING GALAXIES
LB 01542	00 03 48.	- 48 03		14.3	FAINT BLUE STAR
SHB 002	00 03 48.7	- 00 21 06.		19.4	QUASI-STELLAR OBJECT
BC 3CR2	00 03 48.70	- 00 21 06.6		19.35	QUASI-STELLAR OBJECT
RNGC 7829	00 03 49.	- 13 41		14.0	GALAXY
RNGC 7828	00 03 49.	- 13 41		14.0	GALAXY
IC 5385	00 03 50.	- 00 21			NONSTELLAR OBJECT
PHL 6279	00 03 54.	- 06 14		18.1	BLUE STELLAR OBJECT
LB 04977	00 03 54.	- 30 23		19.2	FAINT BLUE STAR
LB 01543	00 03 54.	- 46 31		15.1	FAINT BLUE STAR
RNGC 7833	00 03 55.	+ 27 22			NON-EXISTENT OBJECT
IC 5386	00 03 55.	- 03 59 44.			SAME AS NGC 7832
MCG+08-01-021	00 03 57.	+ 47 35 30.	72	14.	GALAXY

Appendix A
The Luyten Search for Faint Blue Stars

The "Search for faint blue stars," by W. J. Luyten, consists of fifty papers of which the first six were published in the *Astronomical Journal* and the remainder in publications of the Observatory of the University of Minnesota. Thirty-four of the fifty papers contain positions and magnitudes for the 11,444 faint blue stars found in this search.

To obtain the information on the faint blue stars, Luyten blinked a large number of plate pairs and triplets from a number of different observatories.

Table 4 summarizes the contents of those papers used in the MOL, and table 5 summarizes the characteristics of the plates used by Luyten.

Figure 1 depicts the regions searched by Luyten. The fields were assumed to be circular unless otherwise stated. A total of 5,832 square degrees were searched at least once, and many areas were surveyed several times.

TABLE 4
PAPERS USED IN MOL

Paper Number	Luyten Blue (LB Number)	Number of Plate Pairs Blinked	Origin of the Plates	Original RA Precision	Original Dec. Precision	Notes
1	1–70	54	Steward Observatory	$0^{m}.1$	1′	
		4	University of Michigan			
2	71–175	2	University of Michigan	$0^{m}.1$	1′	Some redone in 39
		5	Harvard Southern Station			
3	176–210	45	Steward Observatory	1^{s}	0′.1	Some redone in 16
4	211–237	7	University of Michigan			Superceded by 10
		9	Steward Observatory			
5	238–251	4	University of Michigan			Superceded by 14
6	252–377	34	Dyer Observatory	1^{s}	0′.1	
7	378–400	10	Steward Observatory	1^{s}	0′.1	Some redone in 31 and 40
8	401–1200	55	University of Michigan	1^{s}	0′.1	Some redone in 14
9						Not a catalog of objects
10	1201–1464	N.S.*†	Tonantzintla Observatory	1^{s}	0′.1	
11	1465–1500	1†	Tonantzintla Observatory	1^{s}	0′.1	
		2	Palomar Schmidt			
12	1501–1800	50	Bruce Proper Motion Survey	$0^{m}.1$	1′	
13	1801–1900	1†	Tonantzintla Observatory	1^{s}	0′.1	Some redone in 31 and 40
14	1901–2696	30†	Tonantzintla Observatory	1^{s}	0′.1	
15						Not a catalog of objects
16	2697–3100	9†	Tonantzintla Observatory	1^{s}	0′.1	
17						Not a catalog of objects
18	3101–3500	48	Bruce Proper Motion Survey	$0^{m}.1$	1′	Some redone in 44
19	3501–3540	1	Palomar Survey	1^{s}	0′.1	
20	3541–3564	2	Palomar Survey	1^{s}	0′.1	
21						Not a catlog of objects

22						Not a catalog of objects
23	3565–3837	6	Palomar Survey	1^s	$0'\!.1$	Some redone in 32
24	3838–4022	5	Palomar Survey	1^s	$0'\!.1$	
25	4023–4362	1	Palomar Survey	$0^m\!.1$	$1'$	
26	4363–4886	1	Palomar Survey	$0^m\!.1$	$1'$	
27	4887–5100	1	Palomar Survey	$0^m\!.1$	$1'$	
28	5101–5312	1	Palomar Survey	$0^m\!.1$	$1'$	
29						Not a catalog of objects
30						Not a catalog of objects
31	5313–6308	1	Palomar Survey	$0^m\!.1$	$1'$	
32	6309–6383	6 plates	Mt. Wilson 200-inch	1^s	$0'\!.1$	
33						Not a catalog of objects
34						Not a catalog of objects
35						Not a catalog of objects
36						Not a catalog of objects
37						Not a catalog of objects
38	6384–7415	1	Palomar Survey	$0^m\!.1$	$1'$	
39	7416–8585	1	Palomar Survey	$0^m\!.1$	$1'$	
40	8586–9388	1 1†	Palomar Survey Palomar Schmidt	$0^m\!.1$	$1'$	
41						Not a catalog of objects
42						Not a catalog of objects
43	9389–9769	1†	Palomar Schmidt	$0^m\!.1$	$1'$	
44	9770–10003	N.S.*	Bruce Proper Motion Survey	$0^m\!.1$	$1'$	
45						Separately included with ZL designation
46	10004–10279	1†	Palomar Schmidt	$0^m\!.1$	$1'$	
47	10280–10747	1†	Palomar Schmidt	$0^m\!.1$	$1'$	
48	10748–11209	1†	Palomar Schmidt	$0^m\!.1$	$1'$	
49	11210–11444	1†	Palomar Schmidt	$0^m\!.1$	$1'$	

*N.S. indicates that number of plate pairs or triplets blinked was not stated.

†Indicates that only plate triplets were taken and reduced by Haro's three-image method.

FIGURE ONE

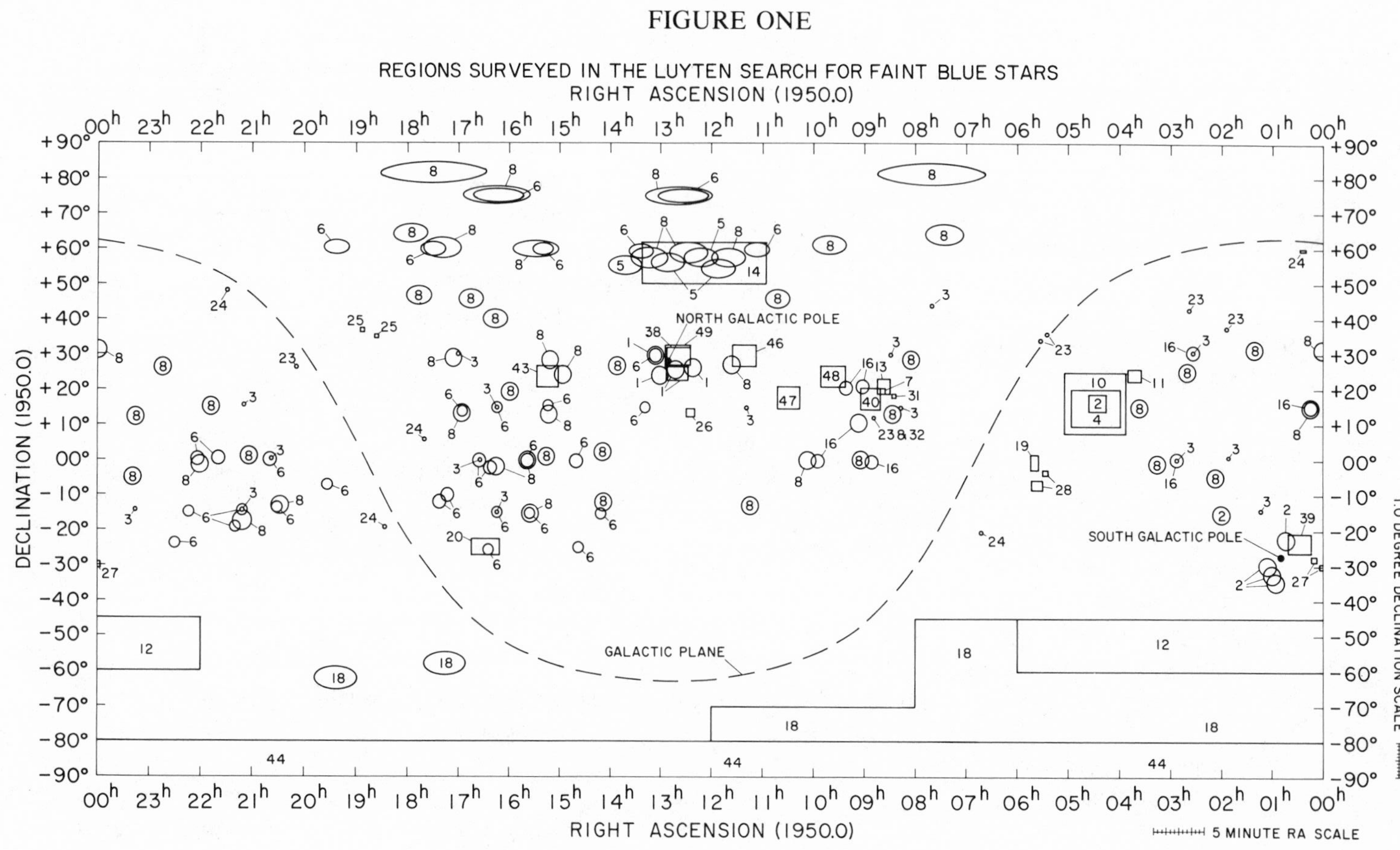
REGIONS SURVEYED IN THE LUYTEN SEARCH FOR FAINT BLUE STARS
RIGHT ASCENSION (1950.0)
DECLINATION (1950.0)
NORTH GALACTIC POLE
SOUTH GALACTIC POLE
GALACTIC PLANE
1.0 DEGREE DECLINATION SCALE
5 MINUTE RA SCALE

TABLE 5

CHARACTERISTICS OF PLATES USED BY LUYTEN

Plate Origin	Telescope	Plate Area (Square Degrees)
Steward Observatory	36-inch reflector	0.5
University of Michigan	24-inch Curtis Schmidt	20.0
Dyer Observatory	24-inch Baker Schmidt	12.0
Tonantizintla Observatory	66-cm. Schmidt	15.0
Bruce Proper Motion Survey	24-inch Bruce telescope	30.0
Palomar Survey	48-inch Schmidt	42.0
Palomar Schmidt	48-inch Schmidt	42.0
Harvard Southern Station	ADH telescope	18.0

Bibliography to Appendix A

P.O.U.M.—Publications of the Observatory, University of Minnesota
S.F.B.S.—Search for Faint Blue Stars
Luyten, W. J. 1953. *Astronomical Journal* 58:75.
Luyten, W. J. 1954. *Astronomical Journal* 59:224.
Luyten, W. J., and Carpenter, E. F. 1955. *Astronomical Journal* 60:429.
Luyten, W. J. 1956. *Astronomical Journal* 61:261.
Luyten, W. J., and Miller, F. D. 1956. *Astronomical Journal* 61:262.
Luyten, W. J., and Seyfert, C. K. 1956. *Astronomical Journal* 61:264.
Luyten, W. J. 1956. P.O.U.M., S.F.B.S. Paper 7.
Luyten, W. J., and Miller F. D. 1956. P.O.U.M., S.F.B.S. Paper 8.
Luyten, W. J. 1957. P.O.U.M., S.F.B.S. Paper 10.
Luyten, W. J. 1958. P.O.U.M., S.F.B.S. Paper 11.
Luyten, W. J., and Anderson, J. H. 1958. P.O.U.M., S.F.B.S. Paper 12.
Luyten, W. J. 1958. P.O.U.M., S.F.B.S. Paper 13.
Luyten, W. J. 1958. P.O.U.M., S.F.B.S. Paper 14.
Luyten, W. J. 1958. P.O.U.M., S.F.B.S. Paper 16.
Luyten, W. J., and Anderson, J. H. 1959. P.O.U.M., S.F.B.S. Paper 18.
Luyten, W. J., and Anderson, J. H. 1959. P.O.U.M., S.F.B.S. Paper 19.
Luyten, W. J. 1959. P.O.U.M., S.F.B.S. Paper 20.
Luyten, W. J. 1959. P.O.U.M., S.F.B.S. Paper 23.
Luyten, W. J. 1961. P.O.U.M., S.F.B.S. Paper 24.
Luyten, W. J. 1961. P.O.U.M., S.F.B.S. Paper 25.
Luyten, W. J. 1961. P.O.U.M., S.F.B.S. Paper 26.
Luyten, W. J. 1962. P.O.U.M., S.F.B.S. Paper 27.
Luyten, W. J. 1962. P.O.U.M., S.F.B.S. Paper 28.
Luyten, W. J. 1962. P.O.U.M., S.F.B.S. Paper 31.
Luyten, W. J. 1963. P.O.U.M., S.F.B.S. Paper 32.
Luyten, W. J. 1966. P.O.U.M., S.F.B.S. Paper 38.
Luyten, W. J. 1966. P.O.U.M., S.F.B.S. Paper 39.
Luyten, W. J., and Sandage, A. R. 1966. P.O.U.M., S.F.B.S. Paper 40.
Luyten, W. J., Anderson, J. H., and Sandage, A. R. 1967. P.O.U.M., S.F.B.S. Paper 43.
Luyten, W. J., and Anderson, J. H. 1967. P.O.U.M., S.F.B.S. Paper 44.
Luyten, W. J., Anderson, J. H., and Sandage, A. R. 1967. P.O.U.M., S.F.B.S. Paper 46.
Luyten, W. J., Anderson, J. H., and Sandage, A. R. 1967. P.O.U.M., S.F.B.S. Paper 47.
Luyten, W. J., Anderson, J. H., and Sandage, A. R. 1968. P.O.U.M., S.F.B.S. Paper 48.
Luyten, W. J., Anderson, J. H., and Sandage, A. R. 1968. P.O.U.M., S.F.B.S. Paper 49.

Appendix B

Palomar Sky Survey Fields for the *Morphological Catalog of Galaxies*

The galaxy numbers assigned by Vorontsov-Velyaminov et al. in The *Morphological Catalog of Galaxies* (MCG) are directly related to the Palomar Sky Survey field in which they lie. Table 6 (see inside back cover) lists the PSS field number for each possible combination of the first two pairs of digits of an MCG galaxy number.

'6	47	48	49	50	51	52	53	54	55	56	57	58	59	60
37	876	554												
79	269	815	778	838	873	405								
.75	771	332	757	803	1212	383	1174	1184	914	606				
185	289	1608	1103	290	286	204	817	1175	843	779				
29	808	287	793	190	372	276	831	1141	375	187	842	1161	320	1182
23	184	544	506	782	325	812	558	298	799	1137	875	800	313	318
164	166	264	202	171	805	315	836	552	860	795	1157	821	316	796
773	1084	193	323	167	297	794	305	575	1130	1146	364	905	834	431
156	300	1085	327	835	268	309	797	810	1126	312	1142	811	306	1147
'03	296	308	331	294	373	855	791	813	1123	833	1180	826	1152	788
'06	569	1111	157	163	790	1122	1150	1156	1163	1448	1171	1190	1228	334
160	1107	1112	1211	1108	194	1151	1172	1181	1167	1158				
S92	S2	S6	S10	S5	S8	S14	S12	S16	S15	S17				